PHYSICS

FOR SCIENTISTS AND ENGINEERS
WITH MODERN PHYSICS

SOME SI BASE UNITS

Physical Quantity	Name of Unit	Symbol
length	meter	m
mass	kilogram	kg
time	second	s
electric current	ampere	A
thermodynamic temperature	kelvin	K
amount of substance	mole	mol

SOME SI DERIVED UNITS

Physical Quantity	Name of Unit	Symbol	SI Unit
frequency	hertz	Hz	s^{-1}
energy	joule	J	$kg \cdot m^2/s^2$
force	newton	N	$kg \cdot m/s^2$
pressure	pascal	Pa	$kg/m \cdot s^2$
power	watt	W	$kg \cdot m^2/s^3$
electric charge	coulomb	C	$A \cdot s$
electric potential	volt	V	$kg \cdot m^2/A \cdot s^3$
electric resistance	ohm	Ω	$kg \cdot m^2/A^2 \cdot s^3$
capacitance	farad	F	$A^2 \cdot s^4/kg \cdot m^2$
inductance	henry	H	$kg \cdot m^2/A^2 \cdot s^2$
magnetic flux	weber	Wb	$kg \cdot m^2/A \cdot s^2$
magnetic flux density	tesla	T	$kg/A \cdot s^2$

SI UNITS OF SOME OTHER PHYSICAL QUANTITIES

Physical Quantity	SI Unit
speed	m/s
acceleration	m/s^2
angular speed	rad/s
angular acceleration	rad/s^2
torque	$kg \cdot m^2/s^2$, or $N \cdot m$
heat flow	J, or $kg \cdot m^2/s^2$, or $N \cdot m$
entropy	J/K, or $kg \cdot m^2/K \cdot s^2$, or $N \cdot m/K$
thermal conductivity	$W/m \cdot K$

SOME CONVERSIONS OF NON-SI UNITS TO SI UNITS

Energy:

1 electron-volt (eV) = 1.6022×10^{-19} J
1 erg = 10^{-7} J
1 British thermal unit (BTU) = 1055 J
1 calorie (cal) = 4.186 J
1 kilowatt-hour (kWh) = 3.6×10^6 J

Mass:

1 gram (g) = 10^{-3} kg
1 atomic mass unit (u) = 931.5 MeV/c^2
= 1.661×10^{-27} kg
1 MeV/c^2 = 1.783×10^{-30} kg

Force:

1 dyne = 10^{-5} N
1 pound (lb or #) = 4.448 N

Length:

1 centimeter (cm) = 10^{-2} m
1 kilometer (km) = 10^3 m
1 fermi = 10^{-15} m
1 Angstrom (Å) = 10^{-10} m
1 inch (in or ″) = 0.0254 m
1 foot (ft) = 0.3048 m
1 mile (mi) = 1609.3 m
1 astronomical unit (AU) = 1.496×10^{11} m
1 light-year (ly) = 9.46×10^{15} m
1 parsec (ps) = 3.09×10^{16} m

Angle:

1 degree (°) = 1.745×10^{-2} rad
1 min (′) = 2.909×10^{-4} rad
1 second (″) = 4.848×10^{-6} rad

Volume:

1 liter (L) = 10^{-3} m^3

Power:

1 kilowatt (kW) = 10^3 W
1 horsepower (hp) = 745.7 W

Pressure:

1 bar = 10^5 Pa
1 atmosphere (atm) = 1.013×10^5 Pa
1 pound per square inch (lb/in^2) = 6.895×10^3 Pa

Time:

1 year (yr) = 3.156×10^7 s
1 day (d) = 8.640×10^4 s
1 hour (h) = 3600 s
1 minute (min) = 60 s

Speed:

1 mile per hour (mi/h) = 0.447 m/s

Magnetic field:

1 gauss = 10^{-4} T

PHYSICS

FOR SCIENTISTS AND ENGINEERS
WITH MODERN PHYSICS

THIRD EDITION

Paul M. Fishbane
University of Virginia

Stephen G. Gasiorowicz
University of Minnesota

Stephen T. Thornton
University of Virginia

Upper Saddle River, New Jersey 07458

Library of Congress Cataloging-in-Publication Data
Fishbane, Paul M.
Physics for scientists and engineers / Paul M. Fishbane, Stephen G. Gasiorowicz, Stephen T. Thornton.—3rd ed.
p. cm.
Includes indexes.
ISBN 0-13-142094-1—ISBN 0-13-141883-1
ISBN 0-13-141881-5—ISBN 0-13-141882-3—ISBN 0-13-035299-3
1. Physics. I. Gasiorowicz, Stephen. II. Thornton, Stephen T. III. Title.

QC23.2.F58 2005
530—dc22

2003058209

Senior Editor: Erik Fahlgren
Editor in Chief, Science: John Challice
Development Editor: Catherine Flack
Executive Project Manager: Ann Heath
Editor in Chief, Development: Carol Trueheart
Executive Managing Editor: Kathleen Schiaparelli
Assistant Managing Editor: Beth Sweeten
Vice President ESM Production and Manufacturing: David W. Riccardi
Manufacturing Manager: Trudy Pisciotti
Manufacturing Buyer: Alan Fischer
Creative Director: Carole Anson
Art Director: John Christiana
Director of Creative Services: Paul Belfanti
Managing Editor of AV Management and Production: Patty Burns
AV Art Editor: Abigail Bass
Artwork: Imagineering
Media Editor: Paul Draper
Assistant Managing Editor, Science Media: Nicole Bush
Assistant Managing Editor, Science Supplements: Becca Richter
Executive Marketing Manager: Mark Pfaltzgraff
Associate Editor: Christian Botting
Editorial Assistant: Andrew Sobel
Interior Design: Dina Curro
Cover Design: Kiwi Design
Cover Photo: Antonio Mo/Getty Images, Inc. -Taxi
Photo Research: Yvonne Gerin
Photo Editor: Nancy Seise
Production Services/Composition: Preparé Inc.

Pearson Prentice Hall
Pearson Education, Inc.
Upper Saddle River, New Jersey 07458

Printed in the United States of America
10 9 8 7 6 5 4 3 2 1

ISBN 0-13-035299-3

Pearson Education Ltd., *London*
Pearson Education Australia Pty. *Limited, Sydney*
Pearson Education *Singapore Pte. Ltd.*
Pearson Education North Asia Ltd., *Hong Kong*
Pearson Education Canada, Ltd., *Toronto*
Pearson Educación de Mexico, *S.A. de C.V*
Pearson Education—*Japan, Tokyo*
Pearson Education *Malaysia, Pte. Ltd.*

Dedication

To our students, the most important element in the making of this book

Brief Contents

Contents

Note: Sections marked with a * can be considered optional.

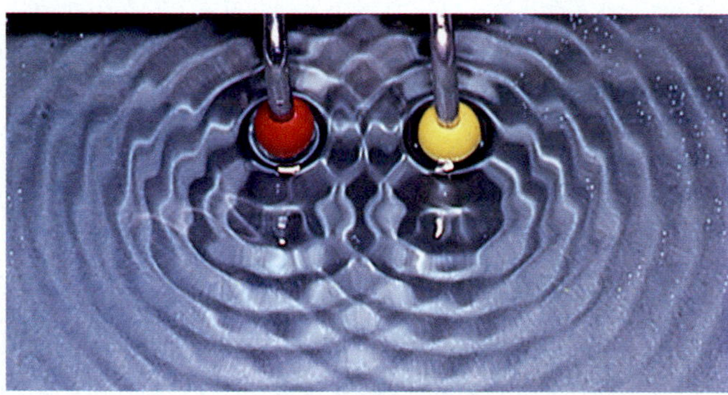

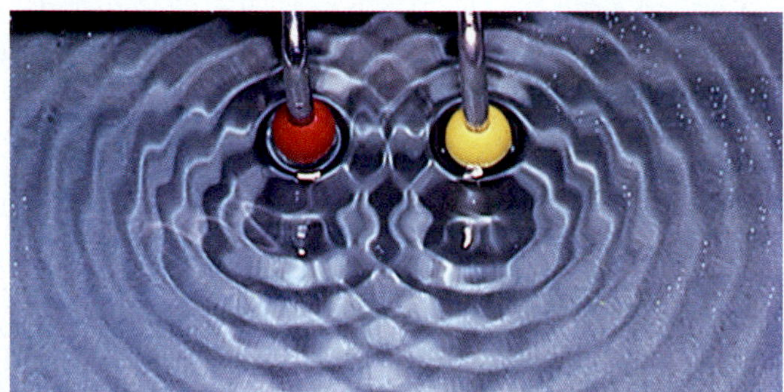

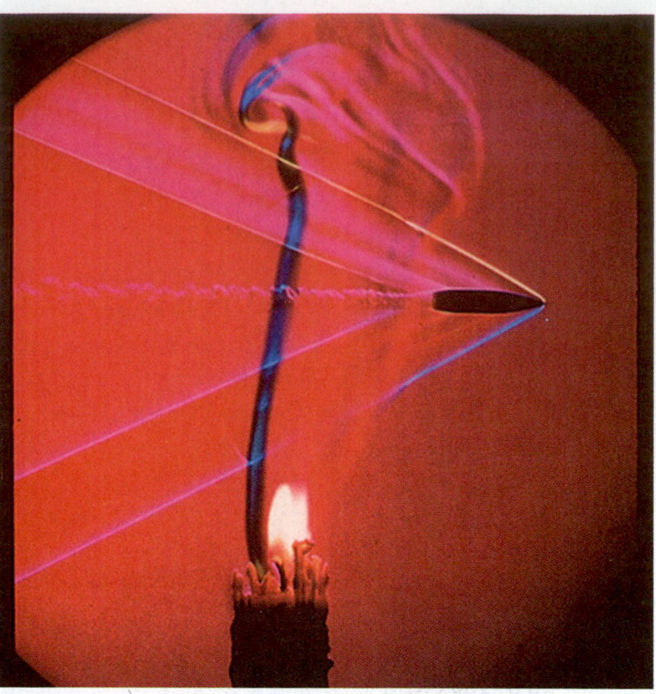

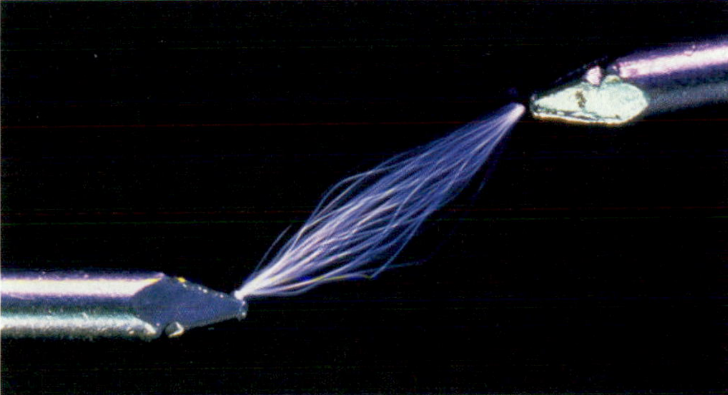

NASA
ERAST
AeroVironment Inc.
HELIOS

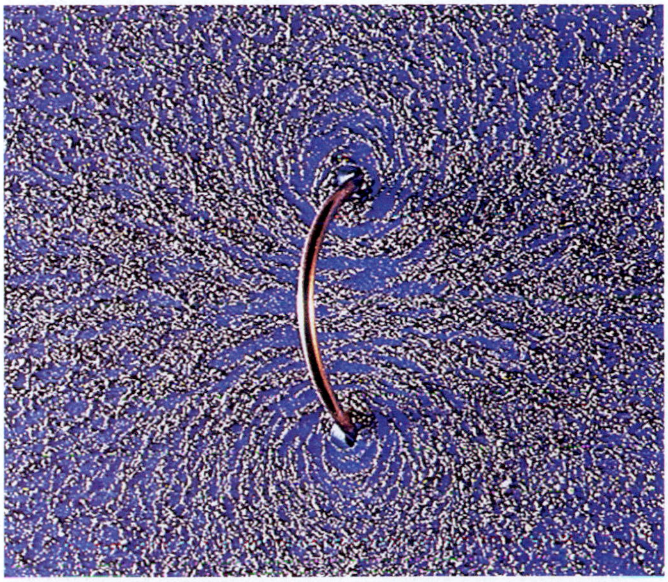

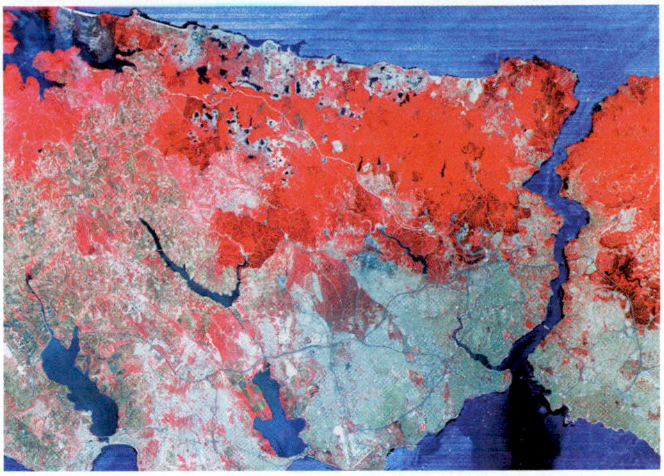

Applications in the Text

Note: This list includes applied topics that receive significant discussion in the chapter text, Think About This sections, Conceptual Examples and worked Examples. Many other applied topics are introduced briefly or appear in the end-of-chapter Understanding the Concepts Questions and Problems. Topics that appear in a Think About This section appear in blue.

Preface

This text is designed for a calculus-based physics course at the beginning university and college level. It is written with the expectation that students either have taken or are currently taking a beginning course in calculus. Students taking a physics course based on this book should leave with a solid conceptual understanding of the fundamental physical laws, how these laws can be applied to solve many problems, and how physics is relevant to the world around them.

"Understanding" encompasses our three overriding goals for this book:

Doing Things Correctly. First, we want a book that is fundamentally *right*. This is sometimes taken to mean "rigorous," but we want to emphasize that we do not feel "rigorous" is a synonym for "difficult." Rather, we associate "right" with showing all the evidence, with using the evidence correctly to support the point being made, and with making the point in a way that will allow the student to say, "I see where that comes from." We try to avoid the phrase "It can be shown that ..." and the attitude that "It is true because we say it is."

Conceptual Emphasis. Second, we want students to understand the material at as deep a conceptual level as possible. We are aware that there is a large gap between being able to get the right answer to a physics problem and comprehending the physical concepts that lie behind it. We want students to be able to answer the *why* as well as the *how*. The student who has a conceptual understanding can not only do problems for which he or she has models, but can also approach a new problem with confidence. Many of the changes to the third edition are designed to address this issue, including the addition of Conceptual Examples, a "What Do You Think" step at the end of most Examples, and a substantial increase in the number of end-of-chapter qualitative questions.

Modern Physics Integrated. Third, without sacrificing the essential aspects of classical physics, we have included modern notions throughout the book. Classical topics have lost none of their importance and must form the basis of any first course. However, what is traditionally called "modern" physics—the topics centered on relativity and quantum physics—began about a century ago. It hardly seems possible to ignore these topics in view of their importance for today's technology and, more critically, for understanding today's world. Many of the ideas of modern physics are not mathematically difficult. However, they can be nonintuitive, and we think it is important that students begin to develop intuition about this material as early as possible. Although much of this material appears in optional sections (marked with an asterisk), in many cases it is intertwined with the classical material. The uncertainty principle and its role in both classical and quantum physics, information on atomic structure and spectra, information on band structure or on blackbody radiation, and the nature and role of fundamental forces are a few of the topics that are included in this way. We conclude the text in what has become the traditional way, with chapters on modern physics. We think the preparation we laid down for this material in earlier chapters will make it more easily assimilated.

A few words about mathematics: The idea of getting it *right* applies to mathematical derivation as well as to qualitative aspects. Our approach is to introduce the mathematics that students need to know the first time they need to know it, in the context of the physics being presented. We try to make that material self-contained so that the student can understand the material without having to go elsewhere for mathematical help. In this way, the mathematics appears in progressive degrees of difficulty. We believe that this approach fosters better understanding and less reliance on formula memorization. We also feel that the ability to make quantitative estimates is one of the most important skills that a scientist or engineer can have. We have made the development of that ability an important part of our approach, both in the text and in the problems. Finally, as in real-world problems, we vary the number of significant figures in examples and in problems. We feel students should maintain an awareness of significant figures and not end up thinking all problems involve the same number of significant figures.

The Third Edition

With the help of reviewers and users of the second edition, we have made a thorough review of content and organization, with some material moved both within and among chapters to enhance the logical progression and structure of the material. We have rewritten much of the material with clarity in mind.

Organizational Changes.

Changes to the third edition include:

- Redistribution of the material in the 2nd-edition chapter "Properties of Solids"—for example, the material on heat conduction in solids now appears in the appropriate chapter on thermal physics, while material on stress and strain now appears in the chapter on statics.
- Redistribution of the material on waves between Chapters 14 and 15 to create a more logical division between single waves, and the superposition of several.
- Consolidation of some material that we feel does not affect the basic understanding of the subject—for example, both the "physical optics" and "magnetism in matter" chapters are more compact.

Conceptual Examples. We have added a new type of Example, designed to help the student think about the material in a way that emphasizes conceptual understanding of the content. These may have some modest algebraic content, such as a simple estimate or reasoning involving inequalities. There are two or more of these per chapter.

CONCEPTUAL EXAMPLE 17–4 We learn by experience that we can loosen a metal lid that is stuck on a glass jar by pouring hot water over the lid. Why does this work?

Answer As the lid's temperature rises, it expands. Movement occurs where the glass and metal are stuck and the lid releases (Fig. 17–11). In fact, you could dip the entire system (lid and glass container) in hot water, and the different expansions of the metal and the glass will lead to the same result. Note that the same difference in the thermal expansion is used in making a good seal: lids are placed on jars when the contents are hot.

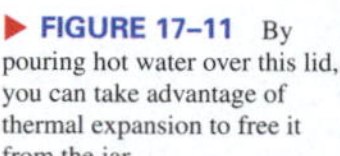

▶ **FIGURE 17–11** By pouring hot water over this lid, you can take advantage of thermal expansion to free it from the jar.

Think About This. The primary purpose of these sections is to pose and answer questions about a new idea or the application of the material discussed. When writing these sections, we ask the kinds of questions a good student might be asking on his or her own, and which the majority of students will find intriguing.

Worked Examples. We have introduced a new structure into the Worked Examples to serve as a model and to build problem-solving skills. The goal of this new structure is to emphasize visualization and the identification of the knowns, unknowns, and concepts to be used. We avoid using a stiff and uncompromising framework, but most examples are broken down into a series of steps:

- ***Setting It Up.*** In most cases this step begins with a sketch and, for mechanics problems, the preparation of a free-body diagram. The figures accompanying most Examples are in a student sketch style that the student can realistically be expected to emulate. To reinforce this first step, all the figures in the end-of-chapter Problems are also drawn in this style. Because a sketch represents a first step in problem solving, the sketches in the end-of-chapter problems sometimes provide a crucial hint to the student. This step also includes some reasoning on how to determine what is being asked when it may not be completely obvious.
- ***Strategy.*** Here the concepts used to solve the problem are outlined and applied to the situation; it is where we "talk it through." This is the heart of the solution, the part where most students will succeed or fail.
- ***Working It Out.*** The strategy outlined in the preceding part is carried through in a series of well-defined steps. This part should be straightforward if the previous steps have been done properly.
- ***What Do You Think?*** The final step in solving a problem is to confirm that the answer makes sense. This section of the solution reinforces the example by asking the student a thought-provoking conceptual question associated with the problem just solved. These questions should confirm that the student has understood the concepts or send them looking for checks to the answers. Answers to "What Do You Think?" questions are provided at the end of the book.

Vectors. These are now represented with an arrow over the letter rather than in boldface (for example, $\vec{F} = m\vec{a}$) to be more consistent with the way professors write them in lecture and the way students write them in homework and exams.

Questions. The end-of-chapter material includes qualitative questions under the heading "Understanding the Concepts." We have increased the number of these conceptual questions by nearly 50 percent.

THINK ABOUT THIS . . .

HOW DOES AN AIR BAG PROTECT YOU IN A CRASH?

Large forces imply large accelerations. A car accident or a fall from a great height may be deadly because of the rapid deceleration, the result of large forces that your body may not be equipped to withstand. For protection it is necessary to find a way to bring you to a stop by providing a smaller deceleration over a larger time. Air bags in automobiles work on this principle; when a collision stops a car very suddenly, a passenger would suffer a very sudden deceleration in a subsequent collision with the steering wheel or the windshield. This is mitigated by the very rapid release of an air bag, which is deep enough and "soft" enough to allow the passenger to slow down over a longer period of time. Firefighters similarly use large elastic safety nets to catch people who have to jump from burning buildings. When the deceleration is for fun, the same principle applies. Bungee cords are made of a very elastic material, and there are no bungee *chains*, which would have the unfortunate effect of stopping you "on a dime." Still another application is provided by airplane ejection seats, which in the past were powered by explosives beneath the seat. The rapid acceleration of these mechanisms often led to serious damage to the pilot. Today ejection seats are powered by small rockets that can supply a smaller acceleration over a longer period of time, rather than a large acceleration over a very short period of time and hence a safer ejection (Fig. 4–12). One other example comes to mind: You may have seen drawings in which Superman catches Lois Lane just before she hits the ground. That very action would imply a rapid deceleration that would be just as bad for Lois as hitting the ground. Superman would do better not to wait for the last instant and instead slow Lois down over a longer period of time.

▲ **FIGURE 4–12** Test ejection of a pilot from an AMX jet fighter. The jet plumes below the seat are due to the ejection rocket.

EXAMPLE 4–3 You need to deliver a box of bowling balls to a bowling alley. The balls will be placed in a box that is initially at rest but that you want to push into the bowling alley. The box itself has a mass that is very small compared to even one bowling ball. You start with one ball in the box, exert a given force of given strength upon the box for a time period Δt, and at the end of that time the box moves at a speed of 3.2 m/s. You then repeat the procedure with more bowling balls in the box; you exert the same amount of force on the box for the same period of time (Δt) and find the box to have a final speed of 0.4 m/s. How many balls are in the box now?

Setting It Up The two cases are shown in Figs. 4–11a and b at the particular time, after an interval Δt, when the speeds are v_1 and v_2, respectively. You know that an identical force of constant magnitude F acts on two different masses m_1 and m_2 for identical time periods Δt, where $m_2 = nm_1$. Here m_1 is the mass of one bowling ball and n is the number of balls in the box, which is the quantity we want to find. The resulting speeds after time Δt are v_1 and v_2, respectively, and are given.

Strategy In the two cases described the box containing the bowling ball(s) is subject to the same force. Moreover, we can ignore the mass of the box. Using Newton's second law, we can find the accelerations a_1 and a_2 during the period Δt when the force operates. These accelerations are

$$a_1 = \frac{F}{m_1} \quad \text{and} \quad a_2 = \frac{F}{m_2} = \frac{F}{nm_1}.$$

Although we do not know the numerical values of the two accelerations, we do know the speeds v_1 and v_2 after a fixed period of acceleration. Further, we learned in Chapter 2 that an object that starts at rest and undergoes a fixed acceleration $\vec{a}$ for a given period of time Δt has the velocity $\Delta\vec{v} = \vec{v} = \vec{a}\,\Delta t$. In our one-dimensional case, then, we have

$$v_1 = a_1\,\Delta t = \frac{F\,\Delta t}{m_1} \quad \text{and} \quad v_2 = a_2\,\Delta t = \frac{F\,\Delta t}{nm_1}.$$

We now have enough information to solve for the unknown, n.

Working It Out We can solve for the ratio F/m in terms of v_1 and Δt and substitute it into the equation for v_2, which we can then solve for n. Alternatively, we can simply take the ratio of the two speeds:

$$\frac{v_1}{v_2} = \frac{(F\,\Delta t/m_1)}{(F\,\Delta t/nm_1)} = n.$$

Numerical substitution gives $n = (3.2\ \text{m/s})/(0.4\ \text{m/s}) = 8$ bowling balls.

What Do You Think? Suppose the mass of the box is not negligible. What would be the effect?

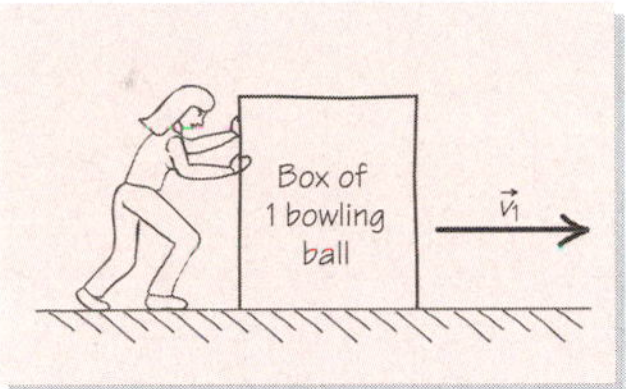

(a)

(b)

▲ **FIGURE 4–11** Delivering bowling balls to an alley. In (a) the box has one ball, and in (b) there is an unknown number of balls.

Problems. Approximately 40 percent of the end-of-chapter Problems are new or revised. We have separated the Problems into two categories. The first group of Problems is organized by section. The second group, called General Problems, resembles the situations that are met in real-life science and engineering. They also help to develop the student's appreciation of the links that exist throughout physics as well as how to approach problems for which the clues may be more obscure. All Problems are labeled I, II, or III. Level I Problems are the least difficult. These Problems develop student recognition of particular physics concepts and build confidence. Level II Problems typically have more than one step and require an increased understanding of the material; the General Problems carry this requirement a step further. Level III Problems are especially challenging, in some cases demanding significant synthesis of concepts in the text. The gradations in problem range and difficulty allow you to tailor the Problems you assign to the capabilities of your class and to the subjects that interest you the most.

Versions of the Text

The third edition is available in two hardback versions, Extended with Modern Physics (Chapters 1–45) and Standard (Chapters 1–40), as well as in three softcover volumes: Volume I (Chapters 1–20), Volume II (Chapters 21–38), and Volume III (Chapters 39–45).

Supplements

Instructor Supplements

Instructor's Resource Center on CD-ROM (0-13-039150-6) This CD-ROM set includes virtually every electronic asset you'll need in and out of the classroom. Though you can navigate freely through the CDs to find the resources you want, the included software allows you to browse and search through the catalog of assets. The CD-ROMs are organized by chapter and include all text illustrations and tables from *Physics for Scientists and Engineers, Third Edition*, in JPEG, Microsoft PowerPoint, and Adobe PDF formats. Instructors can preview, sequence, and play back images, as well as perform keyword searches, add lecture notes, and incorporate their own digital resources. The IRC/CDs also contain TestGen, a powerful dual-platform, fully networkable software program for creating tests ranging from short quizzes to long exams. Questions from the third edition *Test Item File*, including algorithmic versions, are supplied, and professors can use the Question Editor to modify existing questions or create new questions. The CD-ROMs also contain additional Powerpoint Presentations, electronic versions of the *Test Item File*, the *Instructor's Solutions Manual*, the *Instructor's Resource Manual*, and the end-of-chapter Understanding the Concepts questions and problems from *Physics for Scientists and Engineers, Third Edition.*

Instructor's Solutions Manual (Vol. I: 0-13-039157-3; Vol. II: 0-13-144741-6) Authored by *Jerry Shi (Pasadena City College)*, the ISM contains detailed, worked solutions to every problem from the text, as well as answers to the "Understanding the Concepts" questions.

Instructor's Resource Manual (0-13-141738-X) By *Prabha Ramakrishnan (North Carolina State University)*. This IRM contains lecture outlines, notes, demonstration suggestions, and other teaching resources.

Test Item File (0-13-039158-1) This test bank contains approximately 3000 multiple choice, short answer, and true/false questions, many conceptual in nature. All are referenced to the corresponding text section and ranked by level of difficulty.

Transparency Pack (0-13-039166-2) Includes approximately 400 full-color transparencies of images from the text.

PH GradeAssist Instructor's Quick Start Guide (0-13-141740-1) This guide will help adopting instructors register for and use PH GradeAssist, Prentice Hall's own online homework system. It also contains the access code necessary to create their accounts and access the course material.

Peer Instruction: A User's Manual (0-13-565441-6)
Eric Mazur (Harvard University)
A manual with ready-to-use resources for an innovative new approach to teaching introductory physics, developed by a well-known physicist and leader in physics education.

Just-in-Time Teaching: Blending Active Learning with Web Technology (0-13-085034-9)
Gregor Novak (Air Force Academy), Andrew Gavrin (IUPUI), Wolfgang Christian (Davidson College), and Evelyn Patterson (Air Force Academy)
In this resource book for educators, the four authors fully discuss just what *Just-in-Time Teaching* is. *Just-in-Time Teaching (JiTT)* is a teaching and learning methodology designed to engage students by using feedback from pre-class Web assignments to adjust classroom lessons. This allows students to receive rapid response to the specific questions and problems they are having instead of more generic lectures that may or may not address what students actually need help with. Many teachers have found that this process also encourages students to take more control of the learning process and become active and interested learners.

Physlets®: Teaching Physics with Interactive Curricular Material (0-13-029341-5)
Wolfgang Christian and Mario Belloni (Davidson College)
This manual/CD package shows physics instructors—both Web novices and Java-savvy programmers alike—how to author their own interactive curricular material using Physlets—Java applets written for physics pedagogy that can be embedded directly into HTML documents and that can interact with the user. It demonstrates the use of Physlets in conjunction with JavaScript to deliver a wide variety of Web-based interactive physics activities. It also provides examples of Physlets created for classroom demonstrations, traditional and Just-in-Time Teaching homework problems, pre- and post-laboratory exercises, and Interactive Engagement activities. More than just a technical how-to book, the manual gives instructors some ideas about the new possibilities that Physlets offer and is designed to make the transition to using Physlets quick and easy.

Student Supplements

Student Study Guide with Selected Solutions (Vol. I: 0-13-100070-5; Vol. II: 0-13-146500-7)
David Reid (Eastern Michigan University)
The print study guide provides the following for each chapter:

- Objectives
- Chapter Review with Examples and integrated quizzes
- Reference Tools & Resources (equation summaries, important tips, and tools)
- Practice Problems by *Carl Adler (East Carolina University)*
- Selected Solutions for several end-of-chapter problems

PH GradeAssist Student Quick Start Guide (0-13-141741-X) This guide contains information on how to register and use PH GradeAssist. It also contains the access code necessary for students to create their accounts and access the course.

Ranking Task Exercises in Physics: Student Edition (0-13-144851-X)
Thomas L. O'Kuma (Lee College), David P. Maloney (Indiana University–Purdue University at Fort Wayne), and Curtis J. Hieggelke (Joliet Junior College)
Ranking Task Exercises are a unique resource for physics instructors who are looking for tools to incorporate more conceptual analysis in their course. This supplement contains 218 Ranking Task Exercises that cover all classical physics topics. Ranking Tasks are an innovative type of conceptual exercise that ask students to make comparative judgments about a set of variations on a particular physical situation.

Physlet® Physics (0-13-101969-4)
Wolfgang Christian and Mario Belloni (Davidson College)
This text and CD-ROM package provides ready-to-run interactive Physlet-based curricular material for both teachers and students. Physlets, award-winning Java applets written by Christian and Belloni, have been widely adopted by the physics teaching community. This book provides the first class-tested collection of ready-to-run Physlet-based material that is easy to assign (like an end-of-chapter problem of a textbook) and easy to use (material is on a CD and in the book). Neither a Web server nor an Internet connection is required.

Interactive Physics Workbook, Second Edition (0-13-067108-8)
Cindy Schwarz (Vassar College), John Ertel (Naval Academy), MSC Software
This interactive workbook and hybrid CD-ROM package is designed to help students visualize and work with specific physics problems through simulations created with Interactive Physics files. Forty problems of varying degrees of difficulty require students to make predictions, change variables, run, and visualize motion on the computer. The workbook/study guide provides instructions, physics review, hints, and questions. The accompanying hybrid CD-ROM contains everything students need to run the simulations.

Tutorials in Introductory Physics and Homework Package (0-13-097069-7)
Lillian C. McDermott, Peter S. Shaffer, and the Physics Education Group (all of the University of Washington)
This landmark book presents a series of physics tutorials designed by a leading physics education research group. Emphasizing the development of concepts and scientific reasoning skills, the tutorials focus on the specific conceptual and reasoning difficulties that students tend to encounter. The tutorials cover a range of topics in Mechanics, Electricity and Magnetism, and Waves & Optics.

The Portable TA: A Physics Problem-Solving Guide, Second Edition (Vol. I: 0-13-231713-3; Vol. II: 0-13-231721-4)
Andrew Elby
This two-volume set contains a collection of problems with carefully detailed strategies and solutions that provide students with additional problem-solving techniques.

MCAT Physics Study Guide (0-13-627951-1)
Joseph Boone (California Polytechnic State University-San Luis Obispo)
Since most MCAT questions require more thought and reasoning than simply plugging numbers into an equation, this study guide is designed to refresh students' memories about the topics they've covered in class. Additional review, practice problems, and review questions are included.

Mathematics for Physics with Calculus (0-13-191336-0)
Biman Das (SUNY Potsdam)
Designed for concurrent self-study or remedial math work for students in introductory courses, this text is ideal for students who find themselves unable to keep pace because of a lack of familiarity with necessary mathematical tools. It not only shows them clearly how mathematics is directly applied to physics, but discusses math anxiety in general and how to overcome it.

Media Resources

Course Management Options

Course management systems offer you a robust architecture for communicating with your students and letting them communicate with each other, allowing you and your students to post course-related documents, managing your roster and gradebook, and providing on-line assessment. For schools with a local WebCT or Blackboard license, we offer a complete downloadable content cartridge that will give you a rapid start to your on-line course materials. Adapt and customize our materials to your needs, using the tools of these systems. For instructors without the benefit of a local course management system, we also offer OneKey, Pearson Education's own nationally hosted course management system, powered by Blackboard. OneKey combines the power of a full-featured course management system with a quick, easy-to-use interface.

The content cartridge for *Physics for Scientists and Engineers, Third Edition*, includes all the material from the Companion Website, selected resources from the Instructor's Resource Center on CD-ROMs, and all of the questions from the TestGenerator test bank, plus additional materials designed specifically to work in concert with innovative teaching methods. The latter include several activities for *Just-in-Time Teaching*, by *Gregor Novak* and *Andrew Gavrin*; and conceptual, quantitative, and MCAT practice problem sets for on-line assessment.

On-line Assessment Options

You may not need the full capabilities of a course management system (such as posting documents or managing a bulletin board) but may prefer to use an on-line assessment system. On-line assessment provides students instantaneous feedback and repeated practice, and it offers instructors relief from hours of grading and managing a gradebook. On-line assessment systems feature algorithmically-generated questions and quizzes, allow you to create and modify assignments and question pools, and provide tools for analyzing your gradebook entries. Prentice Hall has a powerful new entry in the on-line assessment space: PH GradeAssist. Ask your PH representative about getting a PHGA demonstration and an Instructor's Quick Start Guide. In addition, Prentice Hall partners with other systems, such as WebAssign, to provide an ample selection of book-specific questions to include in your assignments.

PH GradeAssist

This nationally hosted system includes assessment banks associated with Just-in-Time Teaching materials, and other conceptual and quantitative questions. In addition, most of the even-numbered end-of-chapter Problems and questions for *Physics for Scientists and Engineers, Third Edition*, have been converted for use in PHGA, and the majority of them have an algorithmically-generated variant. You select which questions to assign, and you may edit them or create new questions. You also control various important parameters, such as how much questions are worth and when a student can take a quiz.

WebAssign

WebAssign's nationally hosted homework delivery service harnesses the power of the Internet and puts it to work for you by collecting and grading homework. You can create assignments from a database of end-of-chapter questions and problems from the third edition of *Physics for Scientists and Engineers*, or write and customize your own. You have complete control over the homework your students receive, including due date, content, feedback, and question formats.

Companion Website (*http://physics.prenhall.com/fishbane*)

The Companion Website is a quick, interactive resource that allows students to check their understanding with practice quizzes and to explore the material of each chapter further by mining the World Wide Web.

- Reference Tools and Resources by *David Reid (Eastern Michigan University)*
- Practice Questions by *Carl Adler (East Carolina University)*
- Algorithmically generated Practice Problems by *Carl Adler*
- On-line Destinations (links to related sites) by *Carl Adler*
- Applications, with links to related Internet sites, by *Gregor Novak* and *Andrew Gavrin (Indiana University–Purdue University, Indianapolis)*

All quiz modules are scored by the computer; results can be automatically e-mailed to the student's professor or teaching assistant.

Acknowledgments

Three authors alone cannot accomplish a project of this magnitude. We are grateful to the many people who have contributed to making the third edition a better text. A special thanks goes to Professor Jerry Shi (Pasadena City College). In addition to writing the *Instructor's Solutions Manual*, he was an invaluable resource in checking and refining the problem sets and answers and coordinating the independent critique and feedback of the problems and answers from Jenny Quan (Pasadena City College) and David Curott (emeritus, University of North Alabama).

We would like to acknowledge and thank all the instructors who provided valuable feedback for the third edition:

H. Biritz
Georgia Institute of Technology

Tim Bolton
Kansas State University

David Boness
Seattle University

Brian Borovsky
Grinnell College

Chih-Yung Chien
Johns Hopkins University

Krishna M. Chowdary
Bucknell University

Stephane Coutu
Pennsylvania State University

J. William Dawicke
Milwaukee School of Engineering

Michael Dennin
University of California–Irvine

Kathryn Dimiduk
University of New Mexico

N. John DiNardo
Drexel University

J. Finkelstein
San Jose State University

Francis M. Gasparini
SUNY University at Buffalo

Michael Grady
SUNY College at Fredonia

Benjamin Grinstein
University of California–San Diego

John B. Gruber
San Jose State University

Paul Haines
Boston College

Randy Harris
University of California–Davis

Charles A. Hughes
University of Central Oklahoma

Kevin Kimberlin
Bradley University

Amitabh Lath
Rutgers University

Ronald S. MacTaylor
Salem State College

Daniel Marlow
Princeton University

William W. McNairy
Duke University

M. Howard Miles
Washington State University (emeritus)

John Milsom
University of Arizona

Gary A. Morris
Rice University

Elena Murino Di Ventra
Virginia Polytechnic Institute and State University

Michael G. Nichols
Creighton University

Michael B. Ottinger
Missouri Western State College

Kevin T. Pitts
University of Illinois

Andrea Raspini
SUNY College at Fredonia

Dennis Rioux
University of Wisconsin-Oshkosh

Joseph Rothberg
University of Washington

Andrew Scherbakov
Georgia Institute of Technology

Bartlett M. Sheinberg
Houston Community College

Marllin L. Simon
Auburn University

Ross L. Spencer
Brigham Young University

Mark W. Sprague
East Carolina University

Michael G. Strauss
University of Oklahoma

Laszlo Takacs
University of Maryland–Baltimore County

Michael J. Tammaro
University of Rhode Island

Anatoli Vankov
Rochester Institute of Technology

T.S. Venkataraman
Drexel University

Walter D. Wales
University of Pennsylvania

Jeffrey L. Wragg
College of Charleston

Scott A. Yost
Baylor University

R. K. P. Zia
Virginia Polytechnic Institute and State University

We would also like to acknowledge reviewers of the previous editions:

Maris Abolins
Michigan State University

V. K. Agarwal
Moorhead State University

Ricardo Alarcon
Arizona State University

Bradley Antanaitis
Lafayette College

Thomas Armstrong
University of Kansas

Philip S. Baringer
University of Kansas

John E. Bartelt
Vanderbilt University

William Bassichis
Texas A & M University

Benjamin F. Bayman
University of Minnesota

Carl Bender
Washington University

Hans-Uno Bengtsson
University of California, Los Angeles

Robert Bowden
Virginia Polytechnic Institute and State University

Bennet Brabson
Indiana University

Michael Browne
University of Idaho

Timothy Burns
Leeward Community College

Alice Chance
Western Connecticut State University

Edward Chang
University of Massachusetts

Robert Clark
Texas A & M University

Albert Claus
Loyola University

Robert Coakley
University of Southern Maine

Lucien Cremaldi
University of Mississippi

W. Lawrence Croft
Mississippi State University

Chris L. Davis
University of Louisville

Robin Davis
University of Kansas

Jack Denson
Mississippi State University

James Dicello
Clarkson University

P. E. Eastman
University of Waterloo

Robert J. Endorf
University of Cincinnati

Arnold Feldman
University of Hawaii at Manoa

A. L. Ford
Texas A & M University

Gabor Forgacs
Clarkson University

William Fickinger
Case Western Reserve University

Rex Gandy
Auburn University

Alexander B. Gardner
Howard University

Simon George
California State University, Long Beach

James Gerhart
University of Washington

Robert E. Gibbs
Eastern Washington University

Wallace L. Glab
Texas Tech University

James R. Goff
Pima Community College

Alan I. Goldman
Iowa State University

Phillip Gutierrez
University of Oklahoma

Frank Hagelberg
SUNY–Albany

Robert F. Harder
George Fox College

Bruce Harmon
Iowa State University

Warren W. Hein
South Dakota State University

Joseph Hemsky
Wright State University

Jerome Hosken
City College of San Francisco

Joey Houston
Michigan State University

Francis L. Howell
University of North Dakota

Alvin Jenkins
North Carolina State University

Karen Johnston
North Carolina State University

Evan W. Jones
Sierra College

Garth Jones
University of British Columbia

Leonard Kahn
University of Rhode Island

Alain E. Kaloyeros
SUNY–Albany

Charles Kaufman
University of Rhode Island

Robert J. Kearney
University of Idaho

Thomas Keil
Worcester Polytechnic University

Carl Kocher
Oregon State University

Arthur Z. Kovacs
Rochester Institute of Technology

Claude Laird
University of Kansas

Vance Gordon Lind
Utah State University

A. Eugene Livingston
University of Notre Dame

B. A. Logan
University of Ottawa

Karl Ludwig
Boston University

Robert Marande
Pennsylvania State University

David Markowitz
University of Connecticut

Erwin Marquit
University of Minnesota

Marvin L. Marshak
University of Minnesota

Charles R. McKenzie
Salisbury State University

Norman McNeal
Sauk Valley Community College

Forrest Meiere
Indiana University–Purdue University

Roy Middleton
University of Pennsylvania

Irvin A. Miller
Drexel University

George Miner
University of Dayton

Thomas Muller
University of California, Los Angeles

Richard Murphy
University of Missouri–Kansas City

Lorenzo Narducci
Drexel University

Peter Nemethy
New York University

David Ober
Ball State University

Gottlieb S. Oehrlein
SUNY–Albany

Jay Orear
Cornell University

Micheal J. O'Shea
Kansas State University

Dan Overcash
Auburn University

Patrick Papin
San Diego State University

Kwangjai Park
University of Oregon

Robert A. Pelcovits
Brown University

R. J. Peterson
University of Colorado

Frank Pinski
University of Cincinnati

Lawrence Pinsky
University of Houston

Stephen Pinsky
Ohio State University

Richard Plano
Rutgers University

Hans Plendl
Florida State University

Shafigur Rahman
Allegheny College

Don D. Reeder
University of Wisconsin–Madison

Peter Riley
University of Texas, Austin

John Lewis Robinson
University of British Columbia

L. David Roper
Virginia Polytechnic Institute and State University

Ernest Rost
University of Colorado

Richard Roth
Eastern Michigan University

Carl Rotter
West Virginia University

Mendel Sachs
SUNY–Buffalo

Francesca Sammarruca
University of Idaho

Charles Scherr
University of Texas

Eric Sheldon
University of Lowell

Charles Shirkey
Bowling Green State University

Robert Simpson
University of New Hampshire

James Smith
University of Illinois

J. C. Sprott
University of Wisconsin

Malcolm Steuer
University of Georgia

Thor Stromberg
New Mexico State University

William G. Sturrus
Youngstown State University

Richard E. Swanson
Sandhills Community College

Leo H. Takahashi
Pennsylvania State University

Smio Tani
Marquette University

Robert Tribble
Texas A & M University

Rod Varley
Hunter College

Gianfranco Vidali
Syracuse University

John Wahr
University of Colorado

William Walker
University of California, Santa Barbara

Fa-chung Wang
Prairie View A & M University

Gail S. Welsh
Salisbury State University

George Williams
University of Utah

Finally, we would like to thank the publishing team at Prentice Hall who have helped us to carry this project through. Our editor, Erik Fahlgren, who directed the project, has been a constant source of ideas, encouragement, and material help. He has been a most constructive listener and has played the devil's advocate very well when it was necessary. Catherine Flack, our development editor, reminded us that this book is meant for students first and that a failure to communicate at the appropriate level is the worst kind of failure. She too has been willing to listen, and her role as the surrogate student has been immensely helpful. Her help has been crucial to the project. Thanks to Christian Botting, who coordinated the supplements program, and Paul Draper, who managed the media, including the Web-based material. Each of them have been crucial. Special thanks to Andrew Sobel, who worked behind the scenes and handled a massive string of entangled cords like Horowitz playing Chopin. It would be difficult to see how this work could have been completed without his unerring attention to detail. Last but not least, the production of a book such as this one is an enormous task demanding the most careful attention. We want to thank John Christiana for creating a design that is appealing and economical. We also want to thank Beth Sweeten and Fran Daniele for the elaborate management job that has led this book to press. To all the individuals listed above, and to the many others at Prentice Hall who have worked to make this book a success, we extend our heartfelt thanks.

The cumulative and accelerating nature of science and technology make it more imperative than ever that our emerging scientists and engineers understand how few and how solid are the pillars of the enterprise. From this view, the distinctions between "science" and "engineering" and between "classical physics" and "modern physics" melt. We want this book to make evident the pillars of physics as well as the highly interconnected structure that has been erected on those pillars.

Paul Fishbane
pmf2r@virginia.edu

Stephen Gasiorowicz
gasior@umn.edu

Stephen Thornton
stt@virginia.edu

About the Authors

Paul M. Fishbane

Paul Fishbane has been teaching undergraduate courses at the University of Virginia, where he is Professor of Physics, for more than 30 years. In addition to this text, he is the co-author of a modern physics text also published by Prentice Hall. He received his doctoral degree from Princeton University in 1967 and has published over 100 papers in his field, theoretical high-energy physics. He has held visiting appointments at the State University of New York at Stony Brook, Los Alamos Scientific Lab, CERN Laboratory in Switzerland, Amsterdam's NIKHEF laboratory, France's Institut de Physique Nucleaire, l'Université Paris-Sud, and l'École Polytechnique. He has been active for many years at the Aspen Center for Physics, where current issues in physics are discussed with an international group of participants. Among his many other interests, we'll mention in alphabetical order antiques, biking, the kitchen, and music. The rest of his time is spent trying to keep up with his family, including his grandchildren, Ruby and Ivy.

Stephen G. Gasiorowicz

Stephen Gasiorowicz was born in Poland and received his Ph.D. in physics at the University of California, Los Angeles, in 1952. After spending eight years at the Lawrence Radiation Laboratory in Berkeley, California, he joined the faculty of the University of Minnesota, where he is now Professor Emeritus. His field of research was theoretical high-energy physics. As a visiting professor, he has spent extended periods of time at the Niels Bohr Institute, NORDITA in Copenhagen, the Max Planck Institute for Physics and Astrophysics in Munich, DESY in Hamburg, Fermilab in Batavia, and the Universities of Marseille and Tokyo. He has been a frequent visitor to and officer of the Aspen Center of Physics. He has written books on elementary particle physics and quantum physics, and co-authored a modern physics text also published by Prentice Hall. He enjoys biking, canoeing, and skiing, and he is constantly amazed watching his grandchildren Hannah, Becca, Kyle, and Eliza grow up.

Stephen T. Thornton

Stephen Thornton completed his doctoral research at Oak Ridge National Laboratory while completing his Ph.D. at the University of Tennessee. He joined the faculty at the University of Virginia in 1968 and became the first Director of its Institute of Nuclear and Particle Physics in 1984. He has held a Max Planck fellowship and two Fulbright fellowships and has performed nuclear physics experiments at accelerators throughout the United States and Europe. His recent interests include teaching and developing distance-learning courses in physics and physical science for K–12 teachers, and he helped establish the Center for Science Education at the University of Virginia. He has had over 120 research papers published and co-authored three textbooks. He has two grown sons and is married to former NASA astronaut Dr. Kathryn Thornton, with whom he has three daughters. His interests include keeping up with his children, snow skiing, scuba diving, and traveling.

Chapter 1

◀ A wide range of physical activity is implied in this set of tools. They exert forces and make things move; they allow one to measure and act with precision; they suggest control of a set of actions. The design and construction of these tools is itself a technological exploit. Some tools are designed with the aid of computers, which themselves are a formidable tool in the exploration of the world, and the materials from which they are constructed reveal a mastery of the properties of metals and plastics.

Tooling Up

Why physics? What is it that makes an understanding of the fundamentals of science so essential in today's world? Unlike our distant ancestors, we no longer rely on our innate senses but employ high technology to learn about the world around us. We no longer rely just on the strength of our own bodies to build or move but also on machines. We no longer gather food or building materials only from our local environment but also move raw materials around the globe on a daily basis. We strive to improve our lives by accessing information in ever greater amounts and at ever greater rates. Our society is heavily reliant on technology and all technology is rooted in the basic sciences. To gain an understanding of the workings of our society, to actively contribute to it as scientists or engineers, or to make use of it as consumers, we need to understand some of the basics of physics.

In this chapter we will gather a few of the essential ingredients, both mathematical and physical, that will help us to gain an understanding of the physical laws described in the rest of this book. We will discuss the different regimes of the physical world. We will describe the essentials of measurement and the meaning of accuracy in measurement. We will explain the role of the fundamental quantities of our physical world, namely space, time, and mass. We will describe the system of units that allows us to communicate the results of our observations in a universal way. Finally, in our discussion of vectors, we will introduce a significant descriptive tool for the real world.

1–1 A Little Background

To see forward, first look backward. The earliest steps toward the sciences arose out of the recognition of patterns of regularity. Many of these patterns—the sequence of day and night, the seasons, and the regular motion of the heavenly bodies—raised questions about their causes. But it was less the proposed answers to these questions than *the attention given to observation* that began to move us toward science. Ancient observations of the position and timing of objects and events in the sky were surprisingly accurate. The Babylonians could predict the motion of planets, and the description of the appearance of a brilliant "new star" in A.D. 1054—now known to be the Crab supernova—was recorded by Chinese astronomers in enough day-by-day detail to confirm today's nuclear physics calculations of the star's brightness.

Blind reverence for authority can cripple scientific progress. For example, based on the authority of the ancient Greek scientists, the idea that gravity makes heavier objects fall more quickly than lighter ones persisted from Greek times all the way to the Renaissance. An important component of the Renaissance was a reintroduction of the role of experiment in natural science. Among those who must be given credit for challenging authority and helping to reestablish experiment to its primary level are William Gilbert, for magnetism, and Galileo Galilei, for motion. Galileo had a more immediate impact on the rise of physics because of his influence on Isaac Newton, whose laws of motion of 1665 form the underpinning of much of the physics that we will study.

The give and take between experimental observation, the mathematical formulation of descriptive and predictive theories, and further experimental tests of predictions of these theories form the *scientific method.* Figure 1–1 evokes the operation of the scientific method. One does not start from scratch every time one encounters a new set of data; instead one builds on the body of known knowledge. That is why there is a balance in the figure; only in the most extreme circumstances is one forced to formulate hypotheses that break with what we already know, and even then any new hypothesis must be consistent with all that has gone before. We can cite quantum mechanics as a true scientific revolution; nevertheless it was a revolution that had to take into account the enormous success of Newtonian physics. In the scientific method, human imagination is subject to the checks and balances of experiment and of a long history of scientific development.

Progress in our understanding of the physical world has moved along two fronts. On one side, new concepts are built—sometimes in small steps, sometimes in large ones—on earlier concepts. These concepts summarize an ever-increasing body of experimental information and permit an extrapolation of our ideas into areas where patterns of regularity had not previously been seen or even suspected. We speak of opening up new fields. Thus, for example, our current deeper understanding of weather patterns is built on the existence of larger computers and on ever-improving techniques for modeling the behavior of fluids and gases. On the other side, progress is driven by an improvement of experimental techniques that has allowed scientists to probe nature in domains hitherto inaccessible to experimentation. For example, much modern science and technology can be traced to the development of pumps efficient enough to allow the creation of nearly

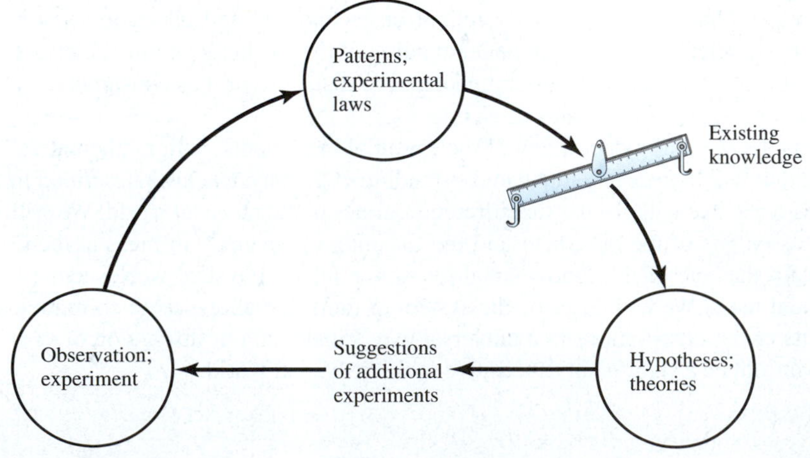

FIGURE 1–1 The scientific method is represented here as an interplay between observation and the formal structure of ideas and mathematical techniques that we know as a theory. Observation sometimes leads to new theory, always with the requirement that any new theory must remain consistent with the large body of existing knowledge, both experimental and theoretical, and theory suggests further experiments that can confirm or deny the acceptability of the theory.

perfect vacuums in sealed vessels. Among other things, these vacuums have allowed us to make the beams of particles that are used to probe nuclei and their constituents—pure science—as well as to use similar beams to help construct tiny integrated circuits.

Scales of the Observable World

Physics underwent a series of revolutionary developments in the period 1900 to 1925, and the field is sometimes divided into classical (pre-1900) and modern (post-1900) physics. Another way to think of physics is in terms of *scale* (Fig. 1–2). Newton's laws of motion were thought for a long time to be universally applicable. However, starting in the late nineteenth century observations began to reveal that there are scales at which these laws must be replaced by a different picture of motion. We need quantum mechanics to describe the behavior of matter at distances on the atomic scale and below; we need special relativity to describe motion at speeds on the scale of the speed of light; and we need general relativity when large masses (compared with, say, that of the Sun) are involved. The behavior of atoms and their constituents cannot be understood without quantum mechanics and special relativity, and the behavior of the universe as a whole cannot be understood without general relativity. Because matter is made up of atoms, and because some space-based technology demands that we take general relativity into account, the importance of understanding these laws of physics is obvious. Nevertheless, for objects that are large on an atomic scale or move slowly in comparison with the speed of light, these laws reduce to the simpler laws of motion set out by Newton. Thus for the description of most of the world, and for many—but by no means all—engineering applications, we can use Newton's laws without having to think about the fact that they are actually approximations to another, deeper set of laws.

The regime of nature that can be described in the context of Newton's laws includes the great subjects of classical physics—mechanics, waves, thermodynamics, and electricity and magnetism—and forms the content of most of this book. In this regime, the impact of laws other than Newton's laws reveal themselves only in the form of properties and constants that can be taken from experiment. Physical properties such as the ability to conduct electricity or to change the direction of light propagation can be calculated in terms of the underlying theory, quantum mechanics, only with the greatest difficulty, if at all. But this is unimportant as far as how they are used in classical physics, where these properties can be described by empirical quantities such as electrical conductivity or the index of refraction of light whose numerical values can be taken from experiment.

The realm of the very small (quantum physics) and the realm of the very fast (special relativity) form what we might call the frontier of physics, where we practice one of the permanent goals of science—to build our understanding. Where we can, we will consistently

(a)

(b)

(c)

▲ **FIGURE 1–2** (a) A scanning tunneling microscope image of a material surface, with individual atoms visible. This realm is the realm of quantum mechanics, a world very different from that in which we live. (b) An image from our daily lives, a regime with which we are all familiar, where Newton's laws can be used to understand physical phenomena. (c) We know from historical record that the Crab nebula is the result of a supernova, a great stellar explosion. At its heart is a very compact remnant—a neutron star—of the original star. Astronomical objects such as these serve as laboratories that allow us to test our understanding of many aspects of the natural world. A supernova brings in both very large and very small scales: it is explicable with an understanding of the gravitational and electromagnetic forces as well as of nuclear phenomena.

try to relate classical physics to the underlying frontier physics. We must do this because technology is pushing engineering practice into these frontier areas, and quantum physics is increasingly essential to modern engineering. It took very little time for the laser to move from the physics research laboratory into the CD player and grocery scanner, and the everyday medical technique of magnetic resonance imaging relies on a deep understanding of the behavior of nuclei in magnetic fields. In the near future few engineers will be able to function without some understanding of quantum mechanics and relativity.

Students sometimes feel that physics is somehow separate from engineering, but in fact it is only the goals of the two disciplines that differ. The aim of much of engineering is to use knowledge of the basic structure and functioning of materials to advance technology, and it is this knowledge that is the domain of physics. Both the advancement of technology and the attainment of the necessary basic knowledge are extremely important in the twenty-first century. That is why a knowledge of physics is so essential to future engineers.

There is one last point to make, and not the least. There are a lot of pages in this book, but there are only a few laws of physics. Beneath a lot of complexity, there is simplicity. The very same laws of physics that explain the structure of atomic nuclei explain the behavior of the Crab nebula. The behavior of an automobile engine is explained by the same laws of physics that describe the cooling of a container of hot water. Keep this in mind. Physics is not the accumulation of detail; it is the ability to understand the detail with a unifying overview that reveals the essential.

1–2 Fundamental Physical Quantities and Their Units

Scientific Notation

The range of numbers that appear in the physical world is truly enormous. For example, the mass of Earth is about 5,980,000,000,000,000,000,000,000 kilograms (kg), and the diameter of a proton is about 0.000000000000001 meter (m). This many zeros are inconvenient, and we employ a shorthand method of writing very large and very small numbers. By using powers of 10, Earth's mass is more easily written as 5.98×10^{24} kg, and the diameter of a proton is written as 10^{-15} m. In this notation, 10^3 represents 1000 and 10^{-4} means 0.0001. We shall use this standard *scientific notation* throughout this book.

A considerable advantage of scientific notation is that multiplication and division are easily performed by adding and subtracting exponents of 10. Thus the product $100 \times 100 = 10{,}000$ can be written as $10^2 \times 10^2 = 10^{2+2} = 10^4$. The awkward multiplication $0.00000055 \times 24{,}000$ can be done more easily as $(5.5 \times 10^{-7}) \times (2.4 \times 10^4) = (5.5 \times 2.4) \times 10^{-7+4} = 13 \times 10^{-3} = 1.3 \times 10^1 \times 10^{-3} = 1.3 \times 10^{-2}$. Where division is involved, we simply change the sign of an exponent and use the multiplication rules. For example,

$$\frac{7.5 \times 10^{-3}}{2.5 \times 10^{-4}} = \frac{7.5}{2.5} \times 10^{-3} \times 10^{+4} = 3.0 \times 10 = 30.$$

Fundamental Physical Quantities

The three quantities *distance*, *time*, and *mass* play a fundamental role in our exploration of the physical world (Fig. 1–3). As Tables 1–1, 1–2, and 1–3 show, these quantities cover an enormous range of values in our universe. These three quantities are already intuitively familiar, as is the idea that we measure them in a certain set of units. The purpose of units is to provide a common set of standards. The importance of standardization is evident. You may wear a size 7 shoe in the United States, but this unit is not of much use in Europe, where a different system is used; there, your shoe size would be 38.

Hundreds of years ago, people used what was readily available as standards for measurement. Length measurements such as the foot came into use in this manner. Over time, measurement systems have become both more precise and more universal. For an early example, French scientists established the forerunner of the International System of measurements in 1791. They defined the meter, the second, and the kilogram: The meter—roughly one yard—was defined as one ten-millionth (10^{-7}) of the distance along Earth's surface between the equator and the North Pole, the second as 1/86,400 of a mean solar day, and the kilogram as the mass of a certain volume of water. In 1889,

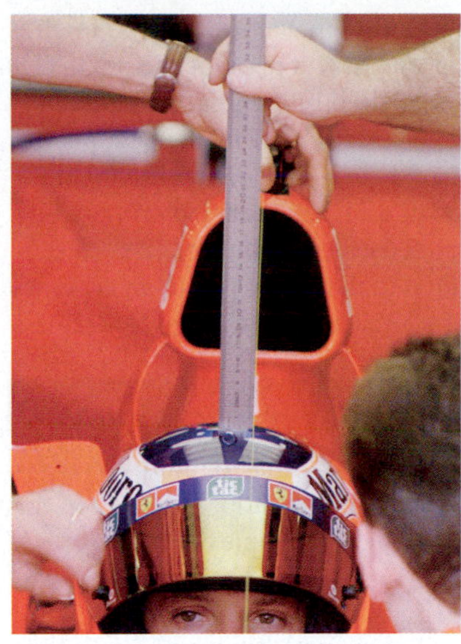

(a)

(b)

(c)

▲ **FIGURE 1–3** (a) Measuring the position of a racecar driver using a simple ruler. Length measurements have different instruments for different needs. (b) Clocks and watches that display time in a digital fashion are commonplace. (c) The entrants in the pumpkin contest demand an objective, repeatable way to determine mass.

an international organization called the General Conference on Weights and Measures was formed to meet periodically to refine these units of measure. In 1960, this organization decided to name a system of units based on the meter, second, and kilogram. These form the International System, with the abbreviation **SI** (for the French words **Système International**). This system is also known as the *metric*, or *mks, system* (after *m*eter, *k*ilogram, and *s*econd). As the tables suggest, we employ the SI throughout.

Length: The definition of the meter has changed several times. In 1889, one meter was defined as the distance between two finely engraved marks on a bar of platinum–iridium that was kept in a vault outside Paris. Even though several copies of this bar were distributed throughout the world, such a standard of length had many shortcomings. For instance, with progress in optical techniques, the scratches on the bar were seen to be fuzzy and imprecise. In 1960, the standard of length was changed to depend upon an atomic constant—the wavelength of a particular orange-red light emitted by an isotope of krypton $\left(^{86}\text{Kr}\right)$ gas. Because our ability (and need) to measure length has led us to require even greater accuracy, this standard also became insufficiently precise. Therefore, in 1983, the Seventeenth General Conference on Weights and Measures established a standard of length based on the speed of light in vacuum (denoted by the letter c). A **meter** (m) is now defined as the distance light travels in vacuum during 1/299,792,458 second. Some orders of magnitude for lengths are given in Table 1–1.

TABLE 1–1 • Orders of Magnitude for Length

Parameter	Length (m)	Parameter	Length (m)
Proton	10^{-15}	Earth–Moon distance	10^{9}
Hydrogen atom	10^{-10}	Earth–Sun distance	10^{11}
Flu virus	10^{-7}	Diameter of solar system	10^{13}
One bit on a DVD	10^{-6}	Distance to nearest star (Proxima Centauri)	10^{17}
Raindrop	10^{-3}	Diameter of our galaxy (Milky Way)	10^{21}
Height of person	10^{0}	Distance to nearest galaxy	10^{22}
One mile	10^{3}	Distance to edge of observable universe	10^{26}
Diameter of Earth	10^{7}		

Time: The second was originally defined as 1/86,400 of the mean solar day, which is the time interval, averaged over a year, from noon of one day to noon of the next. This definition is insufficient because Earth's rotation is both slightly irregular and gradually slowing down from year to year. Therefore, in 1967, a definition of the second was adopted that depends on an atomic standard. The **second** (s) is now defined as the duration of 9,192,631,770 periods of a particular vibration of a cesium atom isotope $\left({}^{133}\text{Cs}\right)$. Clocks based on this standard are, in effect, identical because all atoms of ${}^{133}\text{Cs}$ are indistinguishable and because frequency can be measured in the laboratory to an accuracy of about 4 parts in 10^{13}. Some orders of magnitude for time are given in Table 1–2.

TABLE 1–2 • Orders of Magnitude for Time

Parameter	Time (s)	Parameter	Time (s)
Time for light to cross proton	10^{-23}	Class lecture	10^{3}
Time for light to cross atom	10^{-19}	One Earth day	10^{5}
Period of visible light wave	10^{-15}	One Earth year	10^{7}
Period of vibration for standard cesium clock	10^{-10}	Age of Greek antiquities	10^{11}
Time required for one operation in a personal computer	10^{-9}	Age of first humanoids	10^{14}
Half-life of muon	10^{-6}	Age of Earth	10^{17}
Period of highest audible sound	10^{-4}	Age of universe	10^{18}
Period of human heartbeat	10^{0}		

Mass: The kilogram was originally defined as the mass of one liter of water under certain conditions of temperature and pressure. In 1901, the standard **kilogram** (kg) was defined as the mass of a particular cylinder of platinum–iridium alloy kept at the International Bureau of Weights and Measures in France. Duplicate copies of the cylinder made of this particularly stable alloy are kept in laboratories such as the National Institute of Standards and Technology in Maryland. Although the standards of time and length can be reproduced to precisions of 1 part in 10^{12}, the standard of mass can be reproduced only to perhaps 1 part in 10^{8} or 10^{9}. This standard of mass leaves much to be desired. We would like to find an atomic or natural standard for mass, but even though we know that all atoms of the same type have the same mass, nobody knows how to count atoms with the required accuracy. Some orders of magnitude for mass are given in Table 1–3.

TABLE 1–3 • Orders of Magnitude for Mass

Parameter	Mass (kg)	Parameter	Mass (kg)
Electron	10^{-30}	Battleship	10^{8}
Hydrogen atom	10^{-27}	Moon	10^{23}
Uranium atom	10^{-24}	Earth	10^{25}
Dust particle	10^{-13}	Sun	10^{30}
Raindrop	10^{-6}	Our galaxy (Milky Way)	10^{41}
Piece of paper	10^{-2}	Observable universe	10^{52}
Human	10^{2}		

THINK ABOUT THIS . . .

WHAT DOES THE GLOBAL POSITIONING SYSTEM MEASURE?

Many of you have used the Global Positioning System (or GPS for short) to keep track of where you are on Earth (Fig. 1–4). This system consists of a network of 24 satellites orbiting Earth at an altitude of 20,000 km. The satellites carry very accurate atomic clocks—accurate to about 4 parts in 10^{13}—and emit regular signals. Moreover, the satellite positions are closely tracked and known very accurately. The emitted signals arrive at the speed of light at a receiver that you carry, and because six of the satellites are in your line of sight at any point on Earth, your receiver can compare the times at which signals from different satellites arrive. In that way what is measured is in fact time. These time measurements can be translated to position as your apparatus recognizes, through the time differences for signal arrival, that you are closer to one satellite than to another and in this way triangulates your position. You can find your position to within a meter with the best receiver available, and with the type of receiver many hikers carry today, accuracy of a couple of hundred meters or less is commonplace.

The accuracy with which you learn your position depends on knowing accurately your distance from each satellite. This distance is determined by the travel time and the speed of light. But the speed of light is not so simple to determine, in part because the speed of light in air differs from its speed in vacuum. The accuracy of your position measurement is also limited by your receiver's capacity to measure the tiny difference in arrival times of signals from different satellites. It is interesting to know that some elements of Einstein's general theory of relativity, one of the most advanced and complex fields of research, are essential to the design of the GPS. The theory takes into account how Earth's mass affects the frequency of any ticking clock, and the correction is important. The utility of the GPS is a reply to anyone who tells you that the general theory of relativity is irrelevant to the real world.

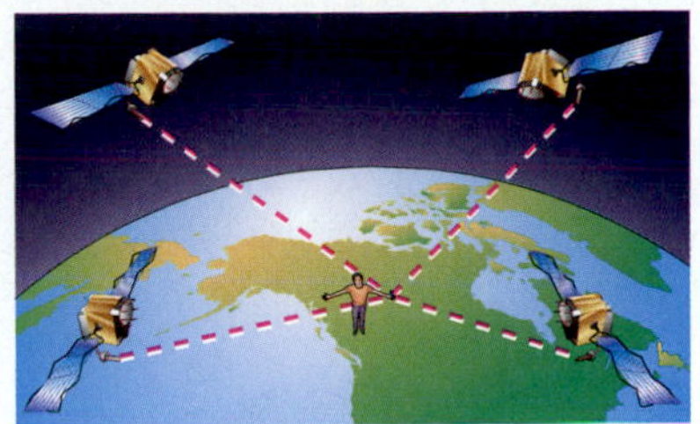

▲ **FIGURE 1–4** With the GPS, travelers—from airplane pilots to hikers—can locate their positions with remarkable accuracy.

Other Systems of Units

The SI is by far the most important and widely accepted system of units in the world today. Two other systems, however, are still in common use.

cgs: The *cgs system* is based on the centimeter, gram, and second and is a metric system derived directly from SI. Different systems may be used sometimes because they are easy to use; for example, density (mass per unit volume) is still normally quoted in grams per cubic centimeter (g/cm^3) because most densities are nearer unity in this system than in SI. For example, the density of water is 1 g/cm^3 in cgs but is 10^3 kg/m^3 in SI. The definition of the units of the cgs system is based on those of SI:

$$1\ \text{cm} \equiv 0.01\ \text{m} \quad \text{and} \quad 1\ \text{g} \equiv 0.001\ \text{kg},$$

where we have used the symbol ≡ to indicate a definition.

British Engineering System: The *British engineering system*, or British system, is based on units of the inch, pound, and second. This system is used only in the United States and in parts of the British Commonwealth, where it is in the process of being replaced by SI. Even in the United States, scientists seldom use the British system, but existing technology based on this system may require its use in engineering applications. The desirability of international trade suggests that even this limited use will eventually disappear.

The British system of units is now *defined* in terms of SI units. The unit of length, the *inch* (in.), is defined as 1 in. ≡ 2.54 cm (Fig. 1–5). The unit of mass in the British system is called a *slug*, but it is seldom used. A slug is equal to 14.5939 kg. (You are perhaps used to thinking of the unit of mass as the pound. In fact, the pound is not a unit of mass at all, but rather one of a force, here the particular force known as the *weight*. See Chapter 4 for a discussion of this issue.) Finally, the unit of time—the *second*—is the same in the British system as it is in SI.

(a)

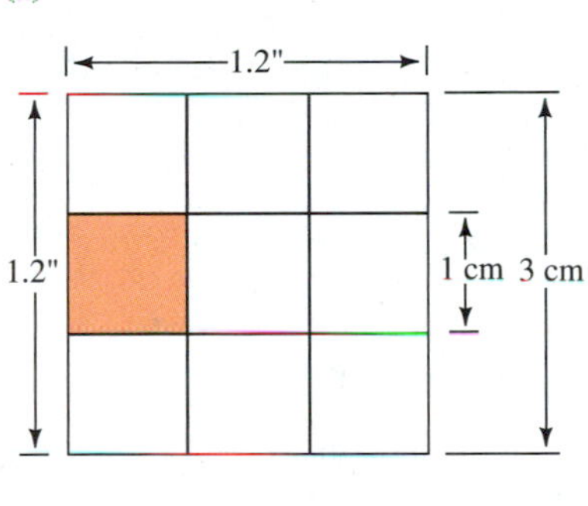

(b)

▲ **FIGURE 1–5** (a) One inch is exactly 2.54 cm. (b) A square 3.0 cm × 3.0 cm in size has an area 9.0 cm^2, or $(3.0/2.54)^2$ in^2 = 1.4 $in.^2$

Unit Prefixes

A useful set of prefixes in SI replace given powers of 10, as shown in Table 1–4. For example, the distance between New York City and Los Angeles is 4,483,000 m,

TABLE 1–4 • Unit Prefixes for Powers of 10

Prefix	Symbol	Multiple	Prefix	Symbol	Multiple	Prefix	Symbol	Multiple	Prefix	Symbol	Multiple
Exa†	E	10^{18}	Mega	M	10^6	Deci†	d	10^{-1}	Nano	n	10^{-9}
Peta†	P	10^{15}	Kilo	k	10^3	Centi	c	10^{-2}	Pico	p	10^{-12}
Tera	T	10^{12}	Hecto†	h	10^2	Milli	m	10^{-3}	Femto†	f	10^{-15}
Giga	G	10^9	Deka†	da	10^1	Micro	μ	10^{-6}	Atto†	a	10^{-18}

†Except for centi, the prefixes near 1 (10^0) are rarely used in the United States. Similarly, the very small and very large multiples (powers of ±15, ±18) are not normally used. You should become familiar with all others.

or 4.483×10^6 m. Neither of these quantities is convenient for everyday use. However, we can use the prefix *kilo-*, which stands for 10^3, to state this distance as 4483 kilometers (km). A finger's width, for example, would not normally be written as 0.015 m or as 1.5×10^{-2} m, but rather as 1.5 centimeters, or 1.5 cm. The standard of mass, the kilogram (kg), is often more convenient to use than the gram (g). In most countries of the world, produce prices are quoted per kilogram because that is a convenient mass unit. You will soon appreciate the simplicity of SI and obtain a feel for its units of length and mass.

Many special units, both within and outside SI, are present because of historical precedent or simply because they are easy to use. For example, the quantity 10^{-6} m is equivalent to a micrometer, or μm, in SI, but is also known as a micron. Another example is the angstrom, Å, a unit often used in atomic physics for the wavelength of light. One angstrom has the value 10^{-10} m, or 0.1 nm $[1 \text{ nanometer (nm)} = 10^{-9} \text{ m}]$. The range of visible light is about 3000 to 7000 Å, or 300 to 700 nm. Both the nanometer and the angstrom are used, although the angstrom is being used less and less frequently.

At the other extreme of the length scale, astronomers and astrophysicists find the meter to be too small for practical use, and they employ three other distance measures: (1) the *astronomical unit* (AU), which is the mean distance between Earth and the Sun, or 1.496×10^{11} m, is useful for distances within our solar system; (2) the *light-year* (ly), the distance that light travels in one year, roughly 0.95×10^{16} m, useful for dealing with interstellar distances; and (3) the *parsec* (pc), which is 3.0857×10^{16} m, equal to about 3.26 ly or 2.063×10^5 AU, also useful for interstellar distance measurements.

Units and Unit Conversions

The angstrom and the astronomical unit are examples of *derived units*, units defined in terms of the basic set (the meter, the kilogram, and the second). These units are simple changes in scale of the basic units. A more complicated type of derived unit involves combinations of the fundamental units. Sometimes these combinations are given names, sometimes not. Speed is measured in meters per second (m/s), but its unit is not given a separate name. When we study power, however, we will find it to be measured in kilogram-meters squared per second cubed $(\text{kg} \cdot \text{m}^2/\text{s}^3)$; this more complicated unit is given a name, the watt (W), and is a *derived* unit in SI.

There are still other units that had their own names from the start and were not obviously combinations of length, time, and mass when they first came into use. They measure quantities that only later were understood to be expressible in terms of length, time, and mass, so that the units could be expressed in SI in terms of the meter, second, and kilogram. These units are "adoptions" of the SI. One example is the unit of pressure, the pascal, (Pa).

We need to be aware of units when we work with equations involving physical quantities. If all the quantities in such an equation are expressed within a *single* system of units—for example, if all the units are SI units—then the units on both sides of such relations must match. This provides us with a very useful tool for checking these relations: If units on both sides of an equation cannot be made to match, there must be an error somewhere within the logic that led to the equation. Further, if a situation arises involving more than one unit for a particular quantity—such as both meters and inches for length—then we must find a simple way to *convert* between different systems of units so that only one unit appears. As we will now see, we can change units conveniently and systematically with the primary conversion equations that relate one set of units to another. Examples of such primary equations are the expressions

$$1 \text{ in} = 2.54 \text{ cm} \quad \text{and} \quad 100 \text{ cm} = 1 \text{ m}.$$

The first of these can be rewritten as

$$1 = \frac{2.54 \text{ cm}}{1 \text{ in}} \quad \text{or} \quad 1 = \frac{1 \text{ in}}{2.54 \text{ cm}}.$$

Any equation can be multiplied by the pure number 1 without change; by judicious choice of the factor 1, units can be canceled and replaced by others. For example,

$$15 \text{ in} = (15 \text{ in})(1) = (15 \cancel{\text{in}})\left(\frac{2.54 \text{ cm}}{1 \cancel{\text{in}}}\right).$$

The inch unit cancels, and

$$15 \text{ in} = (15)(2.54 \text{ cm}) = 38 \text{ cm}.$$

Some other examples of conversions of this type are

$$1 \text{ yd} = 1 \cancel{\text{yd}}\left(\frac{36 \cancel{\text{in}}}{1 \cancel{\text{yd}}}\right)\left(\frac{2.54 \cancel{\text{cm}}}{1 \cancel{\text{in}}}\right)\left(\frac{1 \text{ m}}{100 \cancel{\text{cm}}}\right) = 0.9144 \text{ m}$$

and

$$1 \text{ mi} = 1 \cancel{\text{mi}}\left(\frac{5280 \cancel{\text{ft}}}{1 \cancel{\text{mi}}}\right)\left(\frac{12 \cancel{\text{in}}}{1 \cancel{\text{ft}}}\right)\left(\frac{2.54 \cancel{\text{cm}}}{1 \cancel{\text{in}}}\right)\left(\frac{1 \cancel{\text{m}}}{100 \cancel{\text{cm}}}\right)\left(\frac{1 \text{ km}}{1000 \cancel{\text{m}}}\right) = 1.609 \text{ km}.$$

The technique described here is applicable to more complicated examples.

EXAMPLE 1–1 Given that the speed of light is 2.998×10^8 m/s, what is the distance in meters that light travels in 1 yr? [This distance is the *light-year* (ly).]

Strategy This is a problem involving the conversion of units. The fact that the speed of light c is 2.998×10^8 m/s tells us that light travels 2.998×10^8 m in 1 s. Thus we are interested in expressing c in meters per year to find the distance, in meters, that light travels in 1 yr.

Working It Out We want a conversion for time. We proceed by writing unity as a series of ratios; for example, we know 60 s = 1 min, so that

$$1 = \frac{60 \text{ s}}{1 \text{ min}}.$$

We want to go all the way from seconds to years. Thus we need

$$1 = \frac{60 \text{ s}}{1 \text{ min}}, \quad 1 = \frac{60 \text{ min}}{1 \text{ h}}, \quad 1 = \frac{24 \text{ h}}{1 \text{ d}}, \quad 1 = \frac{365.25 \text{ d}}{1 \text{ yr}}.$$

(For more precision, we use 365.25 days (d) as an average year instead of just 365 d.) The speed of light is then

$$c = (2.998 \times 10^8 \text{ m/s})\left(\frac{60 \text{ s}}{1 \text{ min}}\right)\left(\frac{60 \text{ min}}{1 \text{ h}}\right)\left(\frac{24 \text{ h}}{1 \text{ d}}\right)\left(\frac{365.25 \text{ d}}{1 \text{ yr}}\right)$$
$$= 9.461 \times 10^{15} \text{ m/yr}.$$

We have calculated that 1 ly = 9.461×10^{15} m.

What Do You Think? We went to a lot of trouble to find the distance light travels in a year. What is so special about the unit 1 yr, and if there is nothing special, why do we bother with such quantities? *Answers to* ***What Do You Think?*** *questions are given in the back of the book.*

1–3 Accuracy and Significant Figures

Uncertainty in Measurement

Physics rests on experiment, and experiment requires measurement. But measurements are, at best, only approximate, more or less so depending on the instrument doing the measurement. Although you might be happy to know the distance of the route you took between New York and Los Angeles to within the tenth of a mile possible on your car's odometer, you would use a tape measure rather than the car's odometer to measure the length of your driveway. An **uncertainty** is an indication of the accuracy of a measurement. The uncertainty depends on the accuracy and calibration of the instrument that is making the measurement and on how well the instrument can be read. We can best illustrate the meaning of uncertainty with an example. If the width of a page of paper is measured with a ruler to be 21.6 cm with an uncertainty of 1 mm (or 0.1 cm), which is about the best you could manage with a typical ruler, it would be correct to say that the width is 21.6 cm ± 0.1 cm, or 21.6 ± 0.1 cm. (The ± is read as "plus or minus.") Here, 21.6 cm is called the *central value* and 0.1 cm the *uncertainty* around that central value. In this case, the basis of the uncertainty lies in how well our eyes can read the ruler and on the precision with which the ruler was made. We often use the term *percentage uncertainty* as a measure of the ratio of the uncertainty of a quantity to its central value. The percentage uncertainty is found by multiplying this ratio by 100. The percentage uncertainty of our paper measurement is thus

$$(100\%)\left(\frac{0.1\ \cancel{\text{cm}}}{21.6\ \cancel{\text{cm}}}\right) = 0.5\%.$$

We also say that a quantity is known to one part in some total. For example, in saying that the paper width was 21.6 cm ± 0.1 cm, we could equally well say that the width is known to 1 part in 216.

We can find the area of the paper by measuring the length and multiplying by the width. Suppose that we measure the paper's length to be 27.9 ± 0.1 cm. The percentage uncertainty for the length is 0.4 percent. We find the area by multiplying 21.6 cm by 27.9 cm, which equals 603 cm^2. Because the measurements of width and length both contain uncertainties, there will also be an uncertainty in the paper's area. But what is this uncertainty? A correct way to proceed is to use a process called quadrature: If P_1 and P_2 are the percentage uncertainties of two quantities being multiplied (or divided), the net uncertainty is $\sqrt{P_1^2 + P_2^2}$. If P_1 and P_2 are about the same size, we can more simply get an *approximate* idea of the percentage uncertainty of the product by adding the percentage uncertainties. In the case of the paper's area, the percentage uncertainty is approximately $0.5\% + 0.4\% = 0.9\%$. This means an uncertainty of $(0.009)(603\ \text{cm}^2) = 5\ \text{cm}^2$, and so the area of the paper is $603 \pm 5\ \text{cm}^2$. If P_1 and P_2 are rather different from one another, then the quadrature process shows that it is the larger of the two that dominates in the product (or division) of the two. Thus, for example, if you used a ruler with a percentage uncertainty of 1 percent for measuring the length of your piece of paper but a ruler with a percentage uncertainty of 0.1 percent for the width measurement, the area percentage uncertainty will be approximately 1 percent.

Significant Figures

Physical quantities are never known with certainty unless they are merely definitions. The degree of uncertainty in a quantity is implied in the number of digits assigned to its numerical value. Thus, when we say that an object is 2.00 m long, we mean that it is between 1.995 and 2.005 m long. If we want to say that the length is somewhere between 1.9995 and 2.0005 m, we say that the length is 2.000 m. In the first case, three significant figures are used to describe the object's length; in the second case, the number of significant figures is 4. When we say that a sheet of paper has an area of 603 cm^2, we are using three significant figures, meaning that there is a good probability that the area lies between 602.5 and 603.5 cm^2.

Zeros that are used only to set a decimal point are not part of our count of significant figures. Thus 0.00035 has 2 significant figures, not 6. To take a more extreme example, we mentioned in Section 1–1 that the mass of Earth is 5,980,000,000,000,000,000,000,000 kg. Surely we do not know Earth's mass to 25 significant figures! Scientific notation provides a way to avoid this ambiguity. When we write the mass of Earth as 5.98×10^{24} kg, we indicate unambiguously that we know the mass to 3 significant figures; if we knew the mass to only 2 significant figures, we would write 6.0×10^{24} kg.

In many cases a quantity is made up of the product of several other quantities each known to a differing number of significant figures. Our discussion of percentage uncertainties above for P_1 and P_2 when they are very different from one another (one significant figure difference is a factor of 10 in the uncertainty) shows that the number of significant figures in the product is generally that of the least well-known factor. When sums are involved, we can use the same rule, but when there are differences and significant cancellations occur, one has to be more careful (see Example 1–4).

Keep in mind that definitions assign exact values. Thus the inch is *defined* as 2.54 cm, meaning that 1 in = 2.54000 . . . cm, or *exactly* 2.54 cm. In effect, there are an infinite number of significant figures in the number of centimeters in an inch.

EXAMPLE 1–2 How many centimeters are there in 1 mi?

Strategy The number of centimeters in an inch is given by definition, as is the number of inches in a foot and the number of feet in a mile. We can thus apply a string of unit conversions to connect a centimeter to a mile.

Working It Out There are 5280 ft in 1 mi by definition, there are 12 in in 1 ft by definition, and there are 2.54 cm in 1 in by definition. Thus the number of centimeters in a mile is

$$1\ \text{mi} \times \left(\frac{5280\ \text{ft}}{1\ \text{mi}}\right) \times \left(\frac{12\ \text{in}}{1\ \text{ft}}\right) \times \left(\frac{2.54\ \text{cm}}{1\ \text{in}}\right) = 160{,}934.4\ \text{cm}.$$

What Do You Think? There are 160,934.4 cm in a mile by definition. Are you obliged to keep all these figures when you have a unit conversion from miles to centimeters? *Answers to **What Do You Think?** questions are given in the back of the book.*

EXAMPLE 1–3 Suppose that you have an apparatus that can measure a length in millimeters to two significant figures and a time in seconds to five significant figures and that you measure speed by dividing a length by a time. A fellow student claims that one insect has been measured walking at 0.523 m/min and another at 0.516 m/min. Can the student really tell the difference between these two speeds?

Setting It Up In Figure 1–6 we indicate the tools necessary for this measurement. The question is properly interpreted as: Given the ability of your apparatus to measure speed, how many significant figures are there in a speed measurement and are the two numbers claimed different to within the number of significant figures you expect?

Strategy You are directly dividing a measured distance by the time measured to walk that distance: speed = distance/time. As distance is measured to two significant figures and time to five significant figures, the number of significant figures in the numerator is fewer than the number of significant figures in the denominator. It is the least accurate of the quantities in a calculation that dictates the number of significant figures in the answer, so there are two significant figures in the distance–time ratio.

Working It Out According to the reasoning above, the speed of 0.523 m/min should be rounded to 0.52 m/min and 0.516 m/min should similarly be rounded to 0.52 m/min, and there is no evidence for a difference in the speed of the two insects. To within the accuracy of the measurement, the insects are moving at the same speed.

What Do You Think? In working through the solution, we retained two significant figures when we converted millimeters to meters. Why is that allowed? *Answers to **What Do You Think?** questions are given in the back of the book.*

▲ **FIGURE 1–6** To measure the speed of a walking insect, both a ruler and a watch are necessary, each with its own precision.

It is tempting when using a hand calculator to keep many digits even when the problem calls for fewer. In fact, it is a good strategy to keep many digits through a calculation, because it is sometimes necessary to do this to get the right answer to a sufficient number of significant figures, particularly when subtractions are involved (see Example 1–4). At the end of a calculation, however, the result should be whittled down to only as many figures as are significant, and this number is generally dictated by the input parameter with the fewest significant figures.

EXAMPLE 1–4 An angle θ in radians (see Section 3–5) is known to one significant figure, $\theta = 0.005$ rad. Calculate the deviation from the small angle approximation

$$\frac{1}{\sin\theta} - \frac{1}{\theta}.$$

Working it out If we keep only one significant figure in each term, we would use $\sin(0.005) \cong 0.005$ and find

$$\frac{1}{\sin\theta} - \frac{1}{\theta} = \frac{1}{\sin(0.005)} - \frac{1}{0.005} = \frac{1}{0.005} - \frac{1}{0.005} = 0.$$

But if we keep more digits in our result for $\sin\theta$, $\sin(0.005) \cong 0.00499998$, then we find

$$\frac{1}{\sin\theta} - \frac{1}{\theta} \cong \frac{1}{0.00499998} - \frac{1}{0.005} \cong 0.0008.$$

This result has been rounded off to include one significant figure.

What Do You Think? When you try to use your hand calculator to check the result of the last example, you might find that it does not have enough digits available to do so. Does that mean you should toss your calculator in the wastebasket? *Answers to **What Do You Think?** questions are given in the back of the book.*

1–4 Dimensional Analysis

Dimensions

Three basic elements enter the description of any physical quantity: its spatial properties, how much matter is involved, and its temporal elements. All descriptions of matter, relationships, and events involve combinations of these three basic characteristics, and all measurements can be reduced ultimately to the measurement of three fundamental physical quantities: length, time, and mass. Any physical quantity, no matter how complex, can be expressed as an algebraic combination of these three basic quantities. Speed, for example, is a length per time.

Length, time, and mass therefore have significance far beyond that of providing the basis of a system of units. They specify the three **primary dimensions**. We use the abbreviations $[L]$, $[T]$, and $[M]$ for these primary dimensions (the square brackets here and below indicate that we are dealing with dimensions). The **dimension** of a physical quantity is the algebraic combination of $[L]$, $[T]$, and $[M]$ from which the quantity is formed. The speed v provides an example. The dimension of v is

$$[v] = [L/T], \quad \text{or} \quad [v] = [LT^{-1}].$$

Do not confuse the dimension of a quantity with the units in which it is measured. A speed may have units of meters per second, miles per hour, or, for that matter, light-years per century. All of these different choices of units are consistent with the dimension $[LT^{-1}]$.

Any physical quantity has dimensions that are algebraic combinations $[L^q T^r M^s]$ of the primary dimensions, where the superscripts q, r, and s refer to the order (or power) of the dimension. Thus, for example, an area has dimension $[L^2]$. If all of the exponents q, r, and s are zero, the combination will be dimensionless. The number π, which is the ratio of two quantities each of which have dimension of length (the circumference and the radius of a circle), is an example of a dimensionless quantity. The exponents q, r, and s can be positive integers, negative integers, or even fractional powers.

Matching Dimensions

Study of the dimensions of an equation—*dimensional analysis*—is an important exercise with several different roles in science and engineering. Any equation that relates physical quantities must have consistent dimensions; that is, *the dimensions on one side of an equation must be the same as those on the other side.* This provides a valuable check for any calculation. Dimensional analysis can also reveal how changes in one quantity in a physical situation leads to changes in dependent quantities. Finally, when there is reason to believe that only certain physical quantities can enter into a physical situation, dimensional analysis can provide us with powerful and confirming insights.

Let's look at some examples of dimensional analysis. In Chapter 7, we derive a relation between the height h of a dropped object and the speed of that object. This relation involves the *acceleration of gravity*, g, a quantity whose dimension is $[g] = [LT^{-2}]$. The relation reads

$$gh = \tfrac{1}{2}v^2.$$

Let's compare the dimensions on each side of this equation. The dimension of h is $[L]$, so the left-hand side has dimensions $[LT^{-2}][L] = [L^2T^{-2}]$. The right-hand side has dimensions of speed squared, $[LT^{-1}]^2 = [L^2T^{-2}]$. Thus the dimensions match. If, through error, we had written a relation $gh^2 = v^2/2$, then this check would have revealed the error. Note that dimensional analysis does not help us understand or calculate numerical factors, in this case $\frac{1}{2}$.

CONCEPTUAL EXAMPLE 1–5 A discussion among five friends results in five suggestions for formulas for the time t it will take you to walk home based on the distance d you need to walk and the average walking speed v that you can muster, namely

(a) $t = d/(2v)$, (b) $t = d/v^2$, (c) $t = v/d$,
(d) $t = d/v$, (e) $t = d \times v$.

Based on your knowledge that speed can be measured in miles per hour, which of the formulas above could be correct?

Answer We can use dimensional analysis here. Based on the fact that speed can have units of miles per hour, we deduce that the dimensions of speed are given by

$$[v] = [\text{miles}]/[\text{hours}] = [L/T].$$

The dimension of a correct formula for t must be $[T]$, and as the dimension of d is $[L]$, we can eliminate (b) $\{\text{dimensions } [L/(L/T)^2] = [T^2]\}$, (c) $\{\text{dimensions } [(L/T)/L] = [T^{-1}]\}$, and (e) $\{\text{dimensions } [L \times (L/T)] = [L^2/T]\}$. Both (a) and (d) are dimensionally correct and are possible formulas. Only (d) is in fact correct, but to show that you would have to go beyond dimensional analysis.

EXAMPLE 1–6 Newton's law of universal gravitation gives the magnitude of the force between two objects of mass m_1 and m_2 separated by a distance r as

$$F = G\frac{m_1 m_2}{r^2}.$$

The SI unit of the force is the newton, equivalent to 1 kg · m/s^2. Find the dimensions of the gravitational constant, G.

Strategy By solving the equation above for the quantity G, we can use the fact that the dimensions of both sides of the resulting equation match, and if we know the dimensions of each quantity in the expression for G, we can find the dimensions of the combination by algebra. From the information given on the unit of force, the dimensions of force must be $[MLT^{-2}]$. We now know the dimensions of every quantity in the expression for G.

Working It Out We have $G = Fr^2/m_1 m_2$. Writing the dimensions for both sides gives

$$[G] = \frac{[F][L^2]}{[M^2]} = \frac{[MLT^{-2}][L^2]}{[M^2]}$$
$$= [MLT^{-2}] \times [M^{-2}L^2] = [M^{-1}L^3T^{-2}].$$

Note that the individual dimensions can be consolidated inside the square brackets or left within their own brackets—whichever is easiest.

What Do You Think? Given the dimensions of G, are its units uniquely chosen? *Answers to **What Do You Think?** questions are given in the back of the book.*

To see how dimensional analysis can allow us to derive relations between physical quantities, consider the simple pendulum (Fig. 1–7). This system consists of a small bob of mass m on the end of a light string of length ℓ. A pendulum swings. When it is displaced away from the vertical direction, gravity pulls it back down. It overshoots the minimum, swings to the other side, swings all the way back to the starting point, then repeats its motion. One full cycle of this motion takes a time τ called the *period*. But how does the period depend on the mass of the pendulum bob?

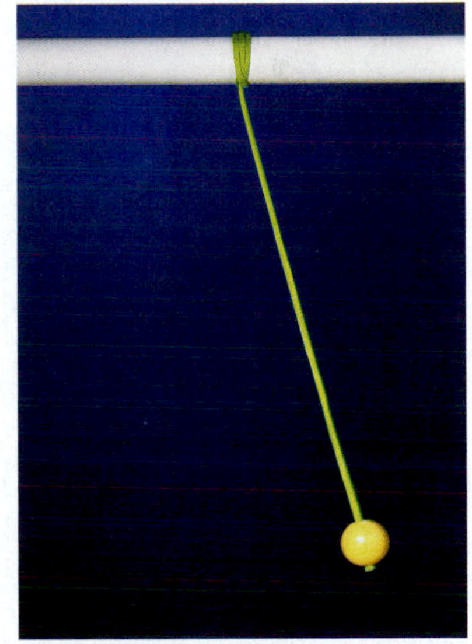

▲ **FIGURE 1–7** The simple pendulum consists of a light string from which a bob—a small dense object—is suspended.

To answer this, we must gather a list of physical parameters on which the period might depend. We would not expect air resistance to play a large role, because if you set your pendulum up within a vacuum, the motion will not be very different. That leaves us with a list of candidates consisting of the bob mass m with dimension $[M]$, the length ℓ of the string with dimension $[L]$, and the acceleration of gravity, g, discussed above. The latter quantity has dimension $[g] = [LT^{-2}]$. There are no other dimensional quantities on which the period of the pendulum should depend.

The dimension of the period is time $[T]$. We now look for an algebraic combination of m, ℓ, and g that has the dimension of τ. We want to find q, r, and s so that

$$[\tau] = [m^q][\ell^r][g^s],$$

or, in terms of the dimensions,

$$[T] = [M^q][L^r][L^sT^{-2s}].$$

There are no powers of $[M]$ on the left-hand side, so $q = 0$. From dimensional analysis alone, we have learned a remarkable fact: *The mass does not enter at all into the period.* In Problem 1–75 at the end of the chapter, we will continue this treatment and show that, in fact, the only combination of the parameters with the same dimension as that of the period is

$$\sqrt{\frac{\ell}{g}}.$$

The pendulum period τ must then be proportional to this quantity. We do not learn the dimensionless numerical coefficient by which the square root is multiplied to give an equality for τ rather than a proportionality. But we do learn the dependence of τ on ℓ and g, and we learn that the mass does not enter into the result. One measurement of the period of a pendulum with known length would determine the unknown numerical coefficient.

Warning: This example also illustrates the limits of dimensional analysis. If for example our system contained two masses, say m_1 and m_2, then the ratio m_1/m_2 is dimensionless and could appear in any way in any formula derived from dimensional analysis.

THINK ABOUT THIS . . .

WHAT GOVERNS THE PROPORTIONS OF THINGS?

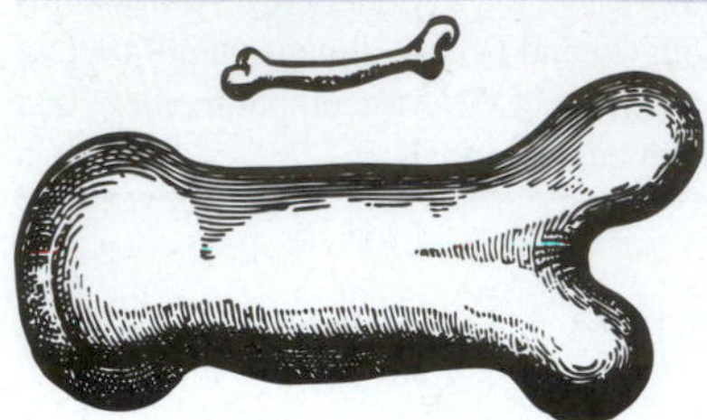

▲ **FIGURE 1–8** The scaling of bones. (From Galileo's *Dialogue Concerning Two New Sciences*, 1638.)

There is a basic scale within objects of all sizes, from a planet to a microbe: the atom, which is a building block from which our world is built up. Moreover, the strengths of materials depend on interactions between atoms, and as we do not know how to derive these interactions independently, we must take these interactions as given. In other words, we must take the strength of materials as given. Given this strength, Galileo, in a process closely related to dimensional analysis, studied the question of what sort of bones are required to support animals of different sizes (see his *Dialogue Concerning Two New Sciences,* 1638). He was able to understand that the bones necessary to support an elephant of several tons must be proportionally thicker than the bones of a human weighing around 150 lb (Fig. 1–8), and the basic morphology of an elephant—its proportions—accordingly differs from that of a human. This type of analysis, often referred to as a *scaling analysis*, shows that the type of insect exoskeleton that supports an ant 1 cm long would be completely inadequate for a hypothetical ant 1 m in length. More importantly, scaling analysis helps structural engineers determine what size of beam is necessary to support a large structure. It tells aeronautical engineers how to use small models to simulate the behavior of large aircraft and filmmakers how to film ripples in a bathtub yet convince us that they are mighty waves on the ocean. ■

1–5 Estimates: How a Little Reasoning Goes a Long Way

We may sometimes want to make a quick calculation. We may want to check a complicated numerical calculation to see if the answer is reasonable. We may want to make a cost estimate or an estimate of the amount of materials needed for a project. Or we may not have access to all the data needed, so that only an estimate, or a very rough approximation, is possible. In these cases, we perform an *order-of-magnitude* calculation, in which variables are rounded off to the nearest power of 10 or to some other easily handled number. The final result of a calculation with variables so dramatically rounded off is accurate only to within an order of magnitude, but such an estimate can often be extremely useful.

For example, the sight of a Christmas tree lot may make us think about how many natural trees are sold at Christmas time. There are roughly 100 million families in the United States, but perhaps only half of them—50 million—have a Christmas tree. It is reasonable to assume that about half of those families with a tree buy a natural tree. We thus arrive at an estimate of about 25 million natural trees sold.

▲ **FIGURE 1–9** Images such as this one allow us to make useful estimates—in this case, a set of questions revolving around the number of Christmas trees that are planted.

Now, let's say that we have a friend who owns a 200-acre Christmas tree farm (Fig. 1–9), and we wonder how many trees he can plant. We reasonably suppose that trees are planted about 6 ft apart, so that each tree takes up about 36 ft^2. Considering space occupied by roads, buildings, or other uses, we might change this estimate to one tree per 50 ft^2. But how big is 200 acres? Most of us do not remember the precise size of an acre, but we might hazard a guess that a typical suburban house lot is about a quarter-acre and that it is perhaps 100 ft across by 100 ft deep, or 10^4 ft^2. This indicates that an acre is about 40,000 ft^2. (In fact, 1 acre $=$ 43,560 ft^2, so our guess is not far off.) If each tree requires about 50 ft^2, then our friend can plant 40,000/50 $=$ 800 trees—let's take 1000 for simplicity—per acre, for a total of 200,000 trees. This is a reasonable order-of-magnitude calculation.

EXAMPLE 1–7 Estimate the average area available to each person in the United States and then to each person on Earth.

Setting It Up In a problem such as this it is sometimes just as important to know what is *not* given and must be found from other sources—prior knowledge, an appendix to this book, an encyclopedia, the Internet, and so forth. We shall need to know the number of people in the United States and the world and the surface area of each in order to make our estimate.

Strategy The average area per person means simply the total area divided by the number of persons. We'll need to calculate the ratio twice, once for the United States and once for the world.

Working It Out There are some 300 million people in the United States. The United States is roughly a rectangle about 3000 mi (5000 km) from east to west and about 2000 mi (3000 km) from north to south, giving an approximate total area of $15 \times 10^6\ km^2$. With $(1\ km)^2 = (10^3\ m)^2 = 10^6\ m^2$, this translates to $15 \times 10^{12}\ m^2$.

Dividing $15 \times 10^{12}\ m^2$ by 3×10^8 people, we obtain $5 \times 10^4\ m^2$ per person. This is a square roughly 200 m on a side.

For Earth, we might estimate a total of about 6 billion people. We should remember from geography that the circumference of Earth is 25,000 mi, or about 40,000 km. We divide 40,000 km by 6 (circumference $= 2\pi r$, and we will approximate 2π as 6) to obtain a radius of 7000 km. The total surface area of Earth is $4\pi r^2 = 12 \times 7 \times 7 \times 10^6\ km^2$, or roughly $5 \times 10^8\ km^2$. However, only about one-third of Earth's surface area is land—roughly $2 \times 10^8\ km^2$, or $2 \times 10^{14}\ m^2$. Dividing $2 \times 10^{14}\ m^2$ by 6×10^9 people, we estimate an area on Earth of about $3 \times 10^4\ m^2$ per person. The United States has roughly the same population density as the land masses of Earth as a whole.

What Do You Think? The estimate made in this example required you to find by one means or another the radius of Earth. Suppose your number for the radius were larger than the number used here. Would your estimate for the result of this example be larger or smaller? *Answers to **What Do You Think?** questions are given in the back of the book.*

EXAMPLE 1–8 To a good approximation, a human body consists mainly of water. One mole (mol) of water, which consists of about 6×10^{23} molecules, has a mass of 18 g. Assuming that the molecules of water in your body are closely packed together, make a rough estimate of the size of a molecule.

Strategy To start, if we know the volume V_1 of one water molecule, we can make an estimate of the linear size d of the molecule by taking the cube root. With the assumption that the molecules are "closely packed," we are approximating the volume of one molecule to be the average volume taken up by each molecule. And to estimate the volume taken up by one molecule under the conditions stated, we can estimate the volume of your body and divide by the estimated number N of molecules. This leaves us with the task of estimating the number of molecules N in your body.

The number N is determined by finding out how many moles n you contain. (A mole is about 6×10^{23} molecules. It is the quantity of a substance whose weight in grams equals the substance's molecular weight.) If you have a mass M in grams, the number of moles you contain is $n = M/(\text{mass of 1 mole})$, where for water the mass of 1 mol is 18 g. Then $N = (6 \times 10^{23})n$. With this information we can proceed to perform the calculation.

Working It Out For your mass, let's use 60 kg, which is equivalent to a weight of about 132 lb. Because $60\ kg = 60 \times 10^3\ g = 6.0 \times 10^4\ g$, the number of moles n is

$$n = \frac{\text{total mass}}{\text{mass of 1 mol}} = \frac{6.0 \times 10^4\ g}{18\ g} = 3.3 \times 10^3\ \text{mol}.$$

In turn, the total number of molecules N is

$$N = (6 \times 10^{23}\ \text{molecules/mol})(3.3 \times 10^3\ \text{mol}) = 2 \times 10^{27}\ \text{molecules}.$$

To find the volume taken up by one molecule, we divide the total volume V of your body by N. We might estimate that your body forms a solid with height 2 m, width 0.5 m, and depth 0.3 m, so that

$$V = (2\ m)(0.5\ m)(0.3\ m) = 0.3\ m^3.$$

Thus the volume V_1 taken up by each molecule is

$$V_1 = \frac{V}{N} = \frac{0.3\ m^3}{2 \times 10^{27}\ \text{molecules}} = 0.15 \times 10^{-27}\ m^3/\text{molecule},$$

and with the close-packing assumption this is the same as the volume of the molecule itself. Finally, if a water molecule is approximated by a little cube with sides of length d,

$$d = (V_1)^{1/3} = (0.15 \times 10^{-27}\ m^3)^{1/3} \cong 0.5 \times 10^{-9}\ m,$$

or about 5×10^{-10} m. Because 1 Å is 10^{-10} m, the size of a water molecule is estimated to be several angstroms. In fact, the separation between the hydrogen atoms in a water molecule is about 2 Å.

What Do You Think? Suppose that instead of being closely packed the molecules in your body are on the average separated by a distance 10 times greater than their typical radius. Is the typical radius of a molecule then (a) 10 times less than the answer above, (b) 1000 times less than the answer above, (c) 10 times greater than the answer above, or (d) 1000 times greater than the answer above? *Answers to **What Do You Think?** questions are given in the back of the book.*

1–6 Scalars and Vectors

The mathematical descriptions of physical systems in this book deal with two types of quantities. One type is an ordinary algebraic quantity called a **scalar**. A scalar has no direction associated with it. The statement that the mass of a ball is $\frac{1}{4}$ kg specifies all we need to know about its mass. The same is true of the time it takes the ball to travel a certain distance—the statement of a number for the time says it all. Other examples of scalar quantities include temperature, the energy of a moving body, and electric charge. Some scalar quantities, such as mass, are always positive, whereas others, such as electric charge, can be positive or negative.

There are physical quantities, however, that cannot be described by scalars; a direction is needed for a complete description of these quantities. Specifying the velocity of a ball requires specifying not only the speed (how fast it is going) but also the direction in which it is traveling. Quantities that must be described by both a magnitude (always positive) and a direction are called **vectors**. Vectors describe displacement, velocity, acceleration, force, electric fields, and numerous other quantities. They play an important role in physics, and this section summarizes some of their properties.

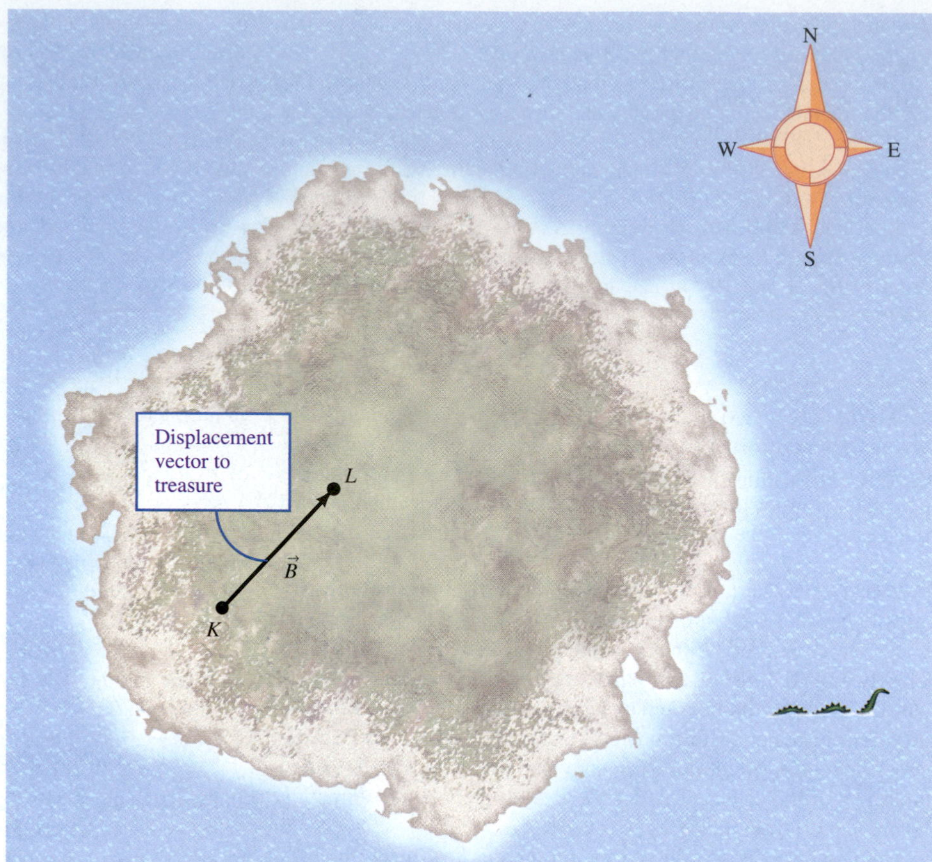

▶ **FIGURE 1–10** The displacement vector $\vec{B}$ from some point K to a second point L on the map for a treasure hunt.

The Displacement Vector

Displacement, which is the difference between two positions of an object, is an important vector quantity. We shall use displacement to describe many of the properties of vectors. To help understand displacement, imagine a treasure hunt in which we must proceed from some point K to a second point L that is 30 paces northeast of K. The displacement from K to L may be drawn as an arrow on a map (Fig. 1–10), and that arrow is the pictorial representation of the displacement vector from K to L.

We can give the vector a name—$\vec{B}$ in Fig. 1–10. The vector $\vec{B}$ has two attributes: a length, or **magnitude** (30 paces), and a direction (to the northeast). We refer to the point where the vector starts as the tail and the point where it ends as the tip, or head, of the vector. Although the length of the displacement vector and the direction in which it points are fixed, the position of the tail (and the tip) of the vector is not. We can shift a vector by moving it to another location in such a way that the vector retains its original direction and length. Thus vector $\vec{B}$ represents the displacement from *any* starting point to the point that is 30 paces away from and to the northeast of the starting point (Fig. 1–11).

In the text, we denote all vectors with an overarrow letter, such as $\vec{B}$. The magnitude of $\vec{B}$ is sometimes denoted by $|\vec{B}|$ but more usually by the unadorned symbol B.

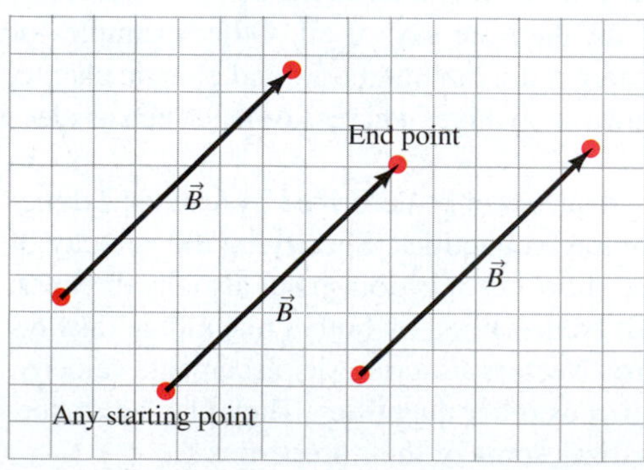

▶ **FIGURE 1–11** The vector $\vec{B}$ represents the displacement 30 paces to the northeast from *any* point. Vectors can be shifted about without changing their characteristics as long as their magnitude (length) and direction (orientation) are unchanged. Thus each of the vectors drawn is the same.

Addition and Subtraction of Vectors

Addition: The result of two successive displacements is also a displacement, which we call a *net displacement.* For example, on an organized day hike the path heads from Base to Coyote Point and then on to Sunset Rock (Fig. 1–12). The initial displacement from Base to Coyote Point is 3.5 km due east of Base. We call this displacement $\vec{A}$ in Figure 1–12. Our second displacement is specified by vector $\vec{B}$; it starts at Coyote Point and proceeds 3.0 km northeast to Sunset Rock. The net displacement takes us from Base to Sunset Rock, and we denote this by the vector $\vec{R}$. The vector $\vec{R}$ is the *sum of the two vectors* $\vec{A}$ and $\vec{B}$:

$$\vec{R} = \vec{A} + \vec{B} \tag{1–1}$$

The sum of the two vectors, which is known as the **resultant vector**, is formed as follows. Draw vector $\vec{A}$, then place the tail of vector $\vec{B}$ on the tip of vector $\vec{A}$. The line from the tail of $\vec{A}$ to the tip of $\vec{B}$ is vector $\vec{R}$. The addition of vectors is *commutative*, that is, the order of the vectors does not matter, so

$$\vec{A} + \vec{B} = \vec{B} + \vec{A}. \tag{1–2}$$

It is easy to see this in Figure 1–13 when we use the method of placing the tail of the second vector at the tip of the first. The figure shows the sum in both orders, and the result is the same. Note that the magnitude R is not the sum of the magnitudes A and B! A glance at Figure 1–12 shows you that Base and Sunset Rock are not separated by the sum of the lengths of the vectors $\vec{A}$ and $\vec{B}$; only if those vectors were aligned would that be the case.

Figure 1–14 shows the result of adding one more displacement vector to the series of displacements in Figure 1–12: The vector $\vec{C}$ takes us from Sunset Rock to Joe's Bar and Grill on the figure. Vector $\vec{C}$ has magnitude 1.0 km and points in a southerly direction. When $\vec{C}$ is added to $\vec{R} = \vec{A} + \vec{B}$, we obtain a new vector, $\vec{S}$, which represents the net displacement from Base to Joe's Bar and Grill:

$$\vec{S} = \vec{A} + \vec{B} + \vec{C}. \tag{1–3}$$

How To Draw

1. Draw first displacement

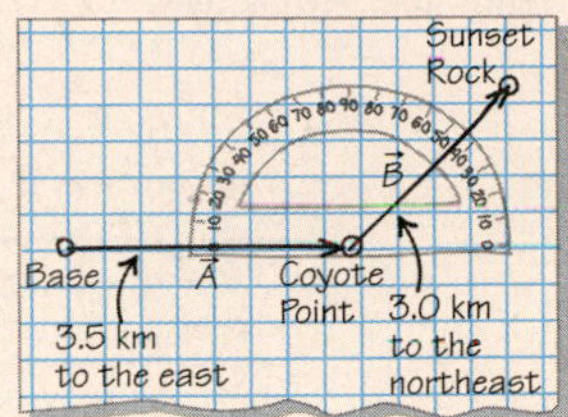

2. Draw second displacement

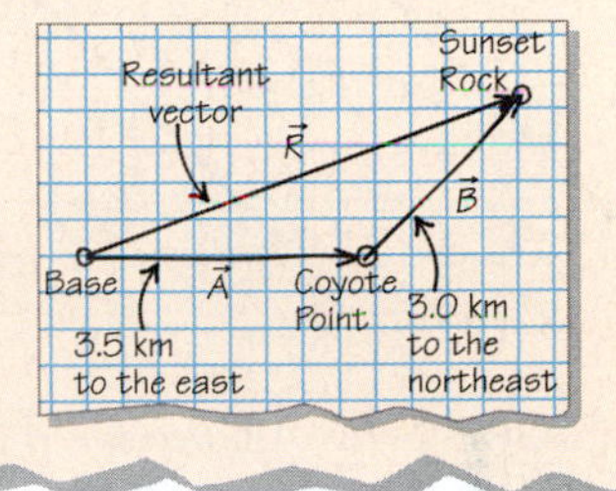

3. Draw resultant vector

▲ **FIGURE 1–12** A day hike involves first a walk from Base to Coyote Point, then a walk from Coyote Point to Sunset Rock. This sequence of two displacement vectors leads to a net displacement. Here, we use a series of steps to outline how to add vectors to arrive at a resultant vector.

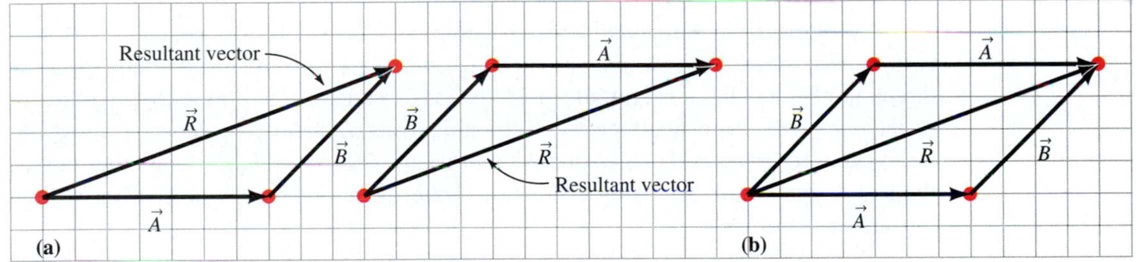

▲ **FIGURE 1–13** (a) We can find the sum $\vec{R}$ of $\vec{A}$ and $\vec{B}$ by placing $\vec{A}$ and $\vec{B}$ tail to head *in either order.* (b) By combining the two graphs in a parallelogram, the equality of the sum in either order is apparent.

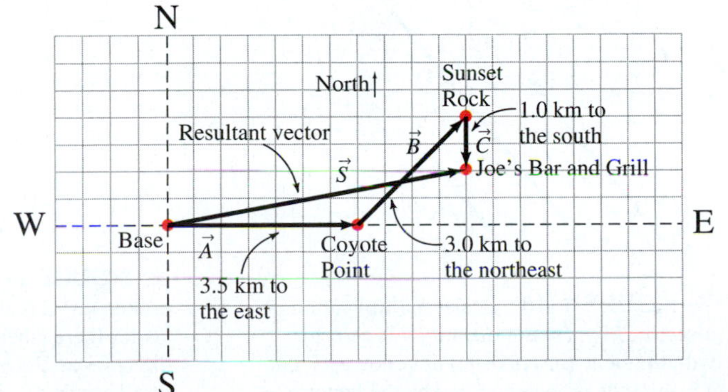

◀ **FIGURE 1–14** One more displacement vector, $\vec{C}$, is added to the displacements in Fig. 1–12. The net displacement is $\vec{S}$.

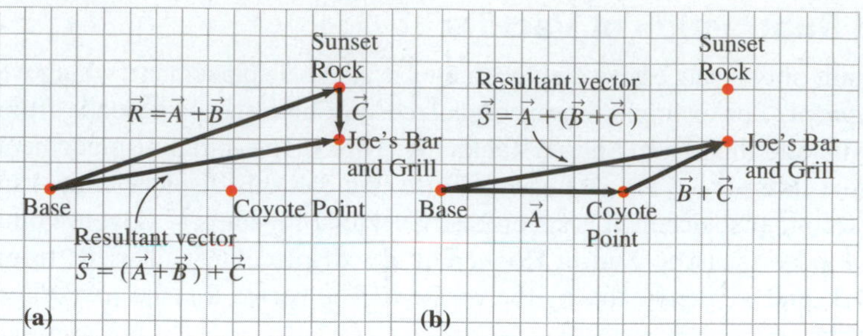

▶ **FIGURE 1–15** (a) The net displacement $\vec{S}$ is found by first adding $\vec{A}$ to $\vec{B}$, then adding the result to $\vec{C}$. (b) The net displacement $\vec{S}$ is found by first adding $\vec{B}$ to $\vec{C}$, then adding the result to $\vec{A}$. The net displacement is the same in both cases, which shows that vector addition is associative.

Figure 1–15 uses this vector to show that vector addition is *associative*. This term means that we can group the addition in any way we find convenient:

$$\vec{S} = (\vec{A} + \vec{B}) + \vec{C} = \vec{A} + (\vec{B} + \vec{C}). \tag{1–4}$$

Subtraction: The *null vector* $\vec{0}$ is a special vector with zero magnitude. It has the property that $\vec{A} + \vec{0} = \vec{A}$. With the help of this vector, we can define the negative, $-\vec{B}$, of a vector $\vec{B}$: When $-\vec{B}$ is added to the vector $\vec{B}$, the sum is the null vector:

$$\vec{0} = \vec{B} + (-\vec{B}). \tag{1–5}$$

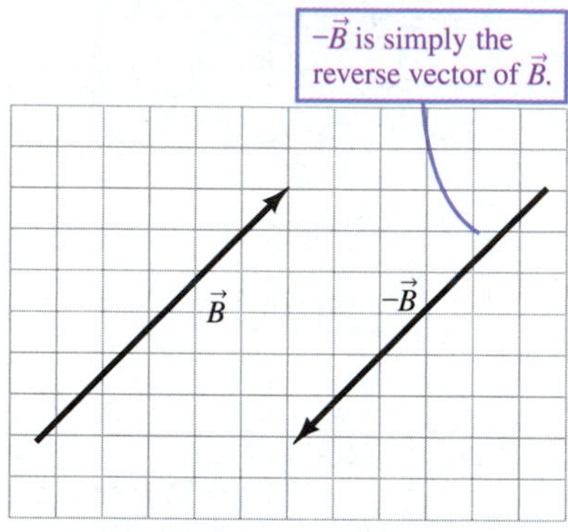

▲ **FIGURE 1–16** The vector $-\vec{B}$ is obtained from $\vec{B}$ by reversing the direction of $\vec{B}$ while leaving its magnitude unchanged.

This means, as shown in Figure 1–16, that $-\vec{B}$ is a vector that has the same magnitude as $\vec{B}$, but $-\vec{B}$ points in the opposite direction. The subtraction $\vec{A} - \vec{B}$ of two vectors is simply the addition of $\vec{A}$ and $-\vec{B}$:

$$\vec{T} = \vec{A} - \vec{B} = \vec{A} + (-\vec{B}). \tag{1–6}$$

This vector is shown in Figure 1–17.

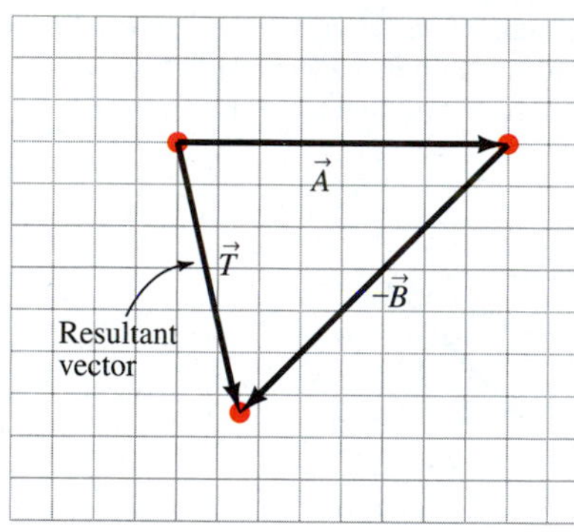

▲ **FIGURE 1–17** Vector difference $\vec{T} = \vec{A} - \vec{B}$.

Scalar Multiplication and Unit Vectors

When vectors are multiplied by scalars, the result is a vector. Twice the displacement $\vec{B}$ is just a displacement with twice the magnitude of $\vec{B}$, but $\vec{B}$ continues to point in the same direction. More generally, $b\vec{B}$ has a length b times that of the vector $\vec{B}$ (Fig. 1–18). The vector $4\vec{B}$ is formed by the *scalar multiplication* of 4 and $\vec{B}$.

This allows us to define *unit vectors* as follows (Fig. 1–19): Any vector $\vec{U}$ is written as $\vec{U} = U\hat{u}$, where U is the magnitude of $\vec{U}$ and $\hat{u}$ is a **unit vector** that points in the direction of $\vec{U}$ and has a magnitude of 1. We consistently denote unit vectors by the use of the caret over them. We include all the units (meter, meter per second, etc.) of our vector with the magnitude U, so that the unit vector $\hat{u}$ is *dimensionless*. In other words, the unit vector simply specifies a direction. Other notations for the unit vector will be introduced as needed.

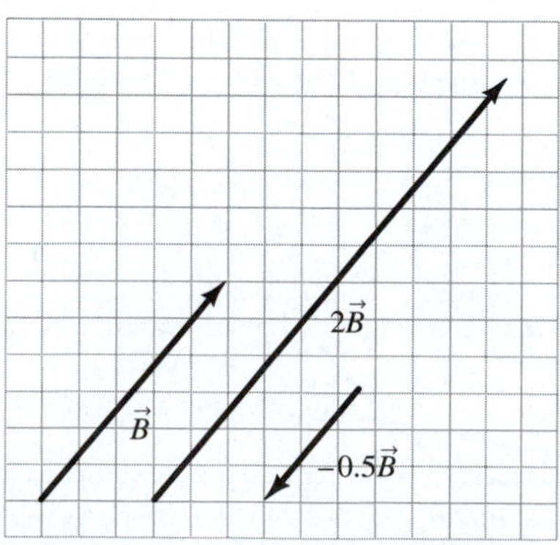

▲ **FIGURE 1–18** Scalar multiplication of a vector $\vec{B}$ by a scalar quantity b is a vector with the same (or opposite) direction as $\vec{B}$ but a magnitude or length scaled by the factor b.

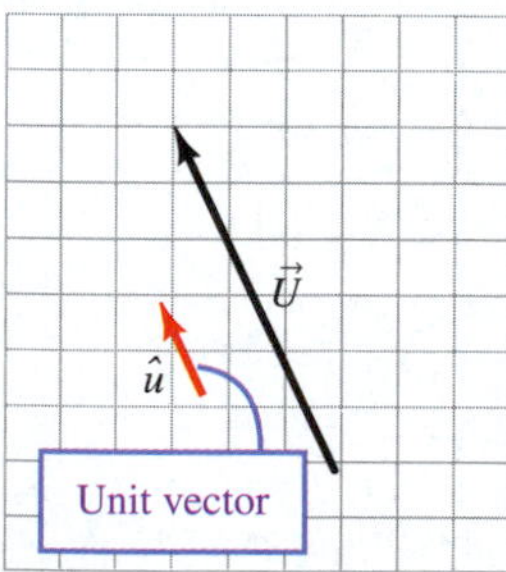

▲ **FIGURE 1–19** Any vector $\vec{U}$ can be broken into a scalar magnitude U times a unit vector (here labeled $\hat{u}$) that points in the same direction as $\vec{U}$ ($\vec{U} = U\hat{u}$). All units are placed into U so that $\hat{u}$ is dimensionless.

Components

It is convenient to use coordinate axes to describe the location of points. In two dimensions, such as the surface of a table, we shall often specify two perpendicular axes, the x-axis and the y-axis (Fig. 1–20a). With these axes, we can specify a point P by giving its coordinates (x_1, y_1), that is, how far it is along the x-direction and the y-direction from the origin O, rather than giving a direction and a distance. The point P may also be described with a *position vector* $\vec{D}$ that extends from the origin to P. We use the term **position vector** to denote the displacement *as measured from the origin of a particular coordinate frame*. The position vector differs from the displacements used above in that it is tied to a particular point.

Let us now use our axes as another way to think about a displacement. Using our knowledge of vector addition and scalar multiplication, we arrive at this displacement in two steps: We first make a displacement of magnitude x_1 along the x-axis, then we follow it with a displacement of magnitude y_1 along the y-axis (Fig. 1–20b). To write this in vectorial form, we employ a unit vector $\hat{i}$ that points in the x-direction and another unit vector $\hat{j}$ that points in the y-direction. (We shall use this notation throughout.) Then $x_1\hat{i}$ is a vector pointing in the x-direction whose magnitude is the absolute value of x_1, whereas $y_1\hat{j}$ is a vector pointing in the y-direction whose magnitude is the absolute value of y_1. Thus

$$\vec{D} = x_1\hat{i} + y_1\hat{j}. \tag{1–7}$$

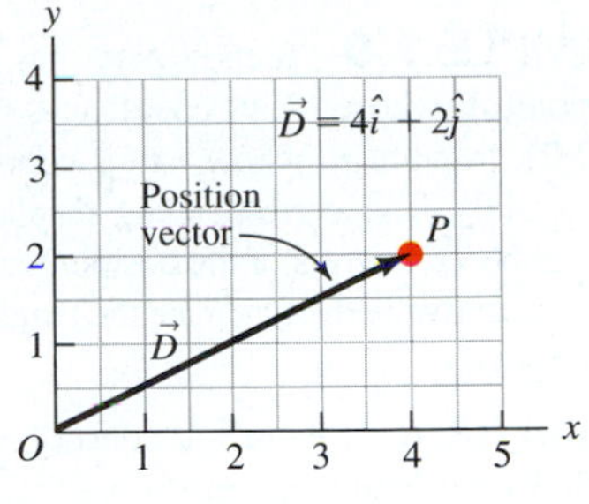

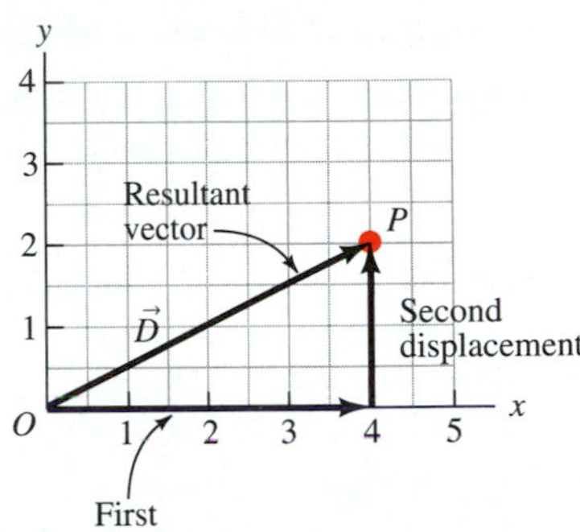

▲ **FIGURE 1–20** (a) A point P can be located in a two-dimensional coordinate system by specifying the two coordinates $x_1 = 4$ and $y_1 = 2$ along their x- and y-axes, respectively. The position of point P with respect to the origin, O, is denoted bythe vector $\vec{D}$. (b) The displacement $\vec{D}$ from O to P can be produced with two successive displacements taken in either order: a displacement in the x-direction and a displacement in the y-direction.

Figure 1–21 applies this reasoning to a more general vector $\vec{V}$. We draw the vector in the xy-plane, and as above

$$\vec{V} = V_x\hat{i} + V_y\hat{j}. \tag{1–8}$$

The **component vectors** of $\vec{V}$ are the vectors $V_x\hat{i}$ and $V_y\hat{j}$, while we refer to V_x and V_y as the **components** of the vector $\vec{V}$. The coordinate frame is specified in the figure, *as it always must be if we refer to particular components.* Figure 1–21 also shows another way to describe the vector $\vec{V}$. It follows from Pythagoras' theorem that the length of $\vec{V}$ is

$$V = \sqrt{V_x^2 + V_y^2}. \tag{1–9}$$

We see from the figure that $\vec{V}$ makes an angle θ with the $+x$-direction. The angle θ is given by

$$\tan\theta = \frac{V_y}{V_x} \quad \text{or} \quad \theta = \tan^{-1}\left(\frac{V_y}{V_x}\right). \tag{1–10}$$

Equivalently, the vector components are described in terms of V and θ by

$$V_x = V\cos\theta \quad \text{and} \quad V_y = V\sin\theta. \tag{1–11}$$

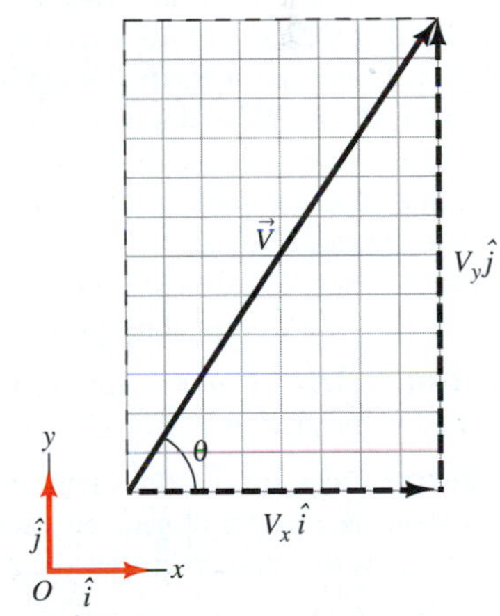

▲ **FIGURE 1–21** A vector $\vec{V}$ in two dimensions is described either by a magnitude V and an angle θ or by the components V_x and V_y.

We have therefore shown that:

> **In two dimensions, a vector may be described either with a magnitude V and an angle θ measured from the x-axis or with components V_x and V_y.**

We saw in our earlier discussion how to add vectors graphically, placing them in sequence tip to tail. But addition (and subtraction) are especially simple to handle in terms of components. The component vectors add independently, so the components do too. Thus, for example, if $\vec{V}$ and $\vec{W}$ are two vectors with components (V_x, V_y) and (W_x, W_y), then

$$\vec{V} + \vec{W} = (V_x + W_x)\hat{i} + (V_y + W_y)\hat{j}. \tag{1–12}$$

EXAMPLE 1–9 A chess piece has been moved on a chess board from point A, whose (x, y) coordinates (measured in centimeters) are $(2.0, 3.0)$, to point B, whose (x, y) coordinates, again in centimeters, are $(5.0, 4.0)$. The corresponding displacement vector is labeled $\vec{D}$. (a) Describe $\vec{D}$ in terms of the position vectors $\vec{A}$ and $\vec{B}$ of the points A and B, respectively. (b) Calculate the length of $\vec{D}$ and the angle it makes with the x-axis.

Setting It Up Figure 1–22 describes the situation, including the specified coordinate system as well as the two points A and B, the position vectors $\vec{A}$ and $\vec{B}$, and the displacement vector $\vec{D}$. When we say that we want to "describe" the vector $\vec{D}$, we mean that we want to specify it, and this can be done by specifying its components.

Strategy (a) A glance at our sketch tells us immediately that $\vec{A} + \vec{D} = \vec{B}$. We can now solve this equation for $\vec{D}$:

$$\vec{D} = \vec{B} - \vec{A}.$$

Since the coordinates of both $\vec{A}$ and $\vec{B}$ are specified, we merely need to carry out the required subtraction of coordinates. (b) Since we will have solved for the coordinates of $\vec{D}$ in part (a), this part is a simple application of Eqs. (1–9) and (1–10).

Working It Out (a) We have

$$\vec{A} = (2.0\text{ cm})\hat{i} + (3.0\text{ cm})\hat{j} \quad \text{and} \quad \vec{B} = (5.0\text{ cm})\hat{i} + (4.0\text{ cm})\hat{j}.$$

Then, from Eq. (1–12),

$$\begin{aligned}\vec{D} &= (5.0\text{ cm} - 2.0\text{ cm})\hat{i} + (4.0\text{ cm} - 3.0\text{ cm})\hat{j} \\ &= (3.0\text{ cm})\hat{i} + (1.0\text{ cm})\hat{j}.\end{aligned}$$

The minus signs appear because, in $\vec{B} - \vec{A}$, the components of $\vec{A}$ are *subtracted* from those of $\vec{B}$.

(b) We have, from Eq. (1–9),

$$D = \sqrt{D_x^2 + D_y^2} = \sqrt{(3.0\text{ cm})^2 + (1.0\text{ cm})^2} = 3.2\text{ cm}$$

and, from Eq. (1–10),

$$\tan\theta = \frac{D_y}{D_x} = \frac{1.0\ \cancel{\text{cm}}}{3.0\ \cancel{\text{cm}}} = 0.33;$$

that is, $\theta = 18°$.

What Do You Think? Suppose that you had changed the placement of the origin by moving it 2 cm in the negative x-direction. [Thus the (x, y) coordinates of point A (measured in centimeters) would be $(4.0, 3.0)$ whereas those of point B would be $(7.0, 4.0)$.] Without doing any calculations, what can you say about the displacement vector $\vec{D}$? *Answers to* ***What Do You Think?*** *questions are given in the back of the book.*

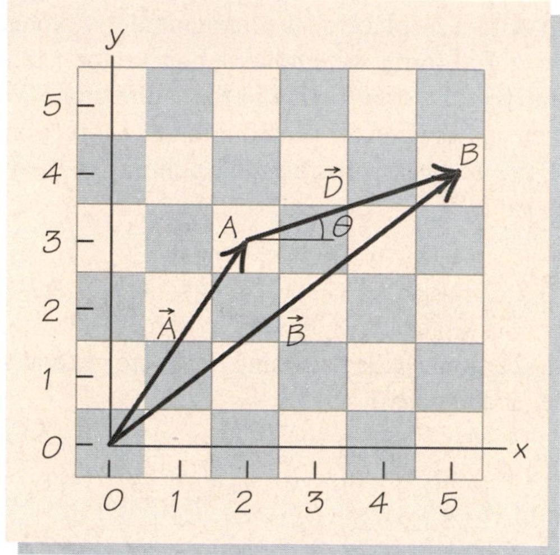

▲ **FIGURE 1–22** The two points A and B have position vectors $\vec{A}$ and $\vec{B}$, respectively. The vector $\vec{D}$ is the displacement from point A to point B.

EXAMPLE 1–10 Figure 1–23a specifies a set of points J, K, L, and M that refer to the day hike described earlier: J is Base, K is Coyote Point, L is Sunset Rock, and M is Joe's Bar and Grill. In the figure, distances are measured in tenths of kilometers. Express the vectors $\vec{A}$, $\vec{B}$, and $\vec{C}$ by describing the displacements from J to K, from K to L, and from L to M, respectively. Describe these displacements in component form relative to a set of axes in which the x-axis points to the east and the y-axis points north. Use your description to calculate the sum of the three vectors. Give the length and inclination with respect to the x-axis of the resultant vector.

Setting It Up Figure 1–23a includes the set of x- and y-axes specified in the problem.

Strategy The breakdown of the displacements as drawn into components is a matter of geometry and can be read off the sketch. After that, vector addition is carried out by adding components.

Working It Out The vector $\vec{A}$ lies along the x-direction; thus, it has no y-component and is given by

$$\vec{A} = (3.5\text{ km})\hat{i}.$$

The vector $\vec{B}$ points in the northeast direction; that is, it makes an angle $\phi = 45°$ with the x-axis. Thus $\sin\phi = \cos\phi = 1/\sqrt{2} \cong 0.71$. From Eqs. (1–11) we have $B_x = B\cos\phi = (3.0\text{ km})(0.71) = 2.1\text{ km}$ and B_y has the same value. Thus

$$\vec{B} = B_x\hat{i} + B_y\hat{j} = (2.1\text{ km})\hat{i} + (2.1\text{ km})\hat{j}.$$

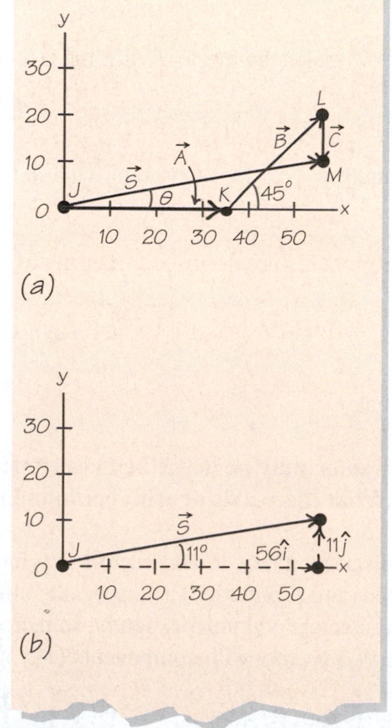

◀ **FIGURE 1–23** (a) A set of coordinate axes has been added to the map of Fig. 1–12. (b) The vector $\vec{S}$ is constructed by adding the components of $\vec{A}$, $\vec{B}$, and $\vec{C}$.

Finally, the vector $\vec{C}$ has a y-component only. Because it points to the $-y$-direction, it has a component that is -1.0 km:

$$\vec{C} = (-1.0\ \text{km})\hat{j}.$$

Now that we have the three vectors, we can find their sum $\vec{S}$ by adding the components:

$$\begin{aligned}\vec{S} = \vec{A} + \vec{B} + \vec{C} &= (3.5\ \text{km})\hat{i} + (2.1\ \text{km})\hat{i} + (2.1\ \text{km})\hat{j} \\ &\quad + (-1.0\ \text{km})\hat{j} \\ &= (3.5\ \text{km} + 2.1\ \text{km})\hat{i} + (2.1\ \text{km} - 1.0\ \text{km})\hat{j} \\ &= (5.6\ \text{km})\hat{i} + (1.1\ \text{km})\hat{j}.\end{aligned}$$

To find the length and angle of $\vec{S}$, given its components, we use Eqs. (1–9) and (1–10):

$$S = \sqrt{(5.6\ \text{km})^2 + (1.1\ \text{km})^2} = 5.7\ \text{km},$$

$$\tan\theta = \frac{1.1\ \cancel{\text{km}}}{5.6\ \cancel{\text{km}}} = 0.20,$$

or $\theta = 11°$. This is the angle the resultant vector makes with the x-axis.

What Do You Think? If you had chosen your x-axis tilted with respect to the one used in the example, you would have found a resultant vector whose inclination with respect to the x-axis would have differed from that found in the example. But we have argued that displacements are independent of coordinate systems. Is this a real conflict? *Answers to* ***What Do You Think?*** *questions are given in the back of the book.*

Vectors in Three-Dimensional Space

The vectors we have considered thus far are two dimensional. Vectors can also represent quantities in three dimensions. A vector in three dimensions can be specified with a *Cartesian* or *Euclidean* set of axes x, y, and z. (The two-dimensional x- and y-axes are also called Cartesian.) Figure 1–24 illustrates such a three-dimensional system, showing the conventional orientation of the three axes. The orientation of the axes is best described using a *right-hand rule*, as this figure shows. Start with the usual x- and y-axes (x to the east, say, and y to the north). To find the direction of the z-axis:

> **Point the fingers of your right hand along the x-axis. Now curl these fingers in the direction of the y-axis. Your thumb will point along the z-axis.**

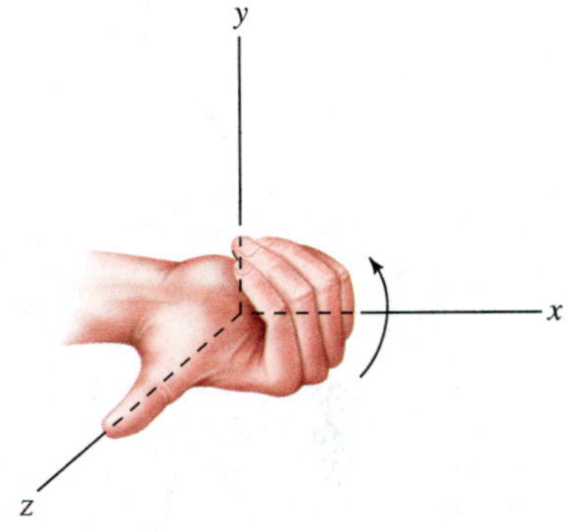

▲ **FIGURE 1–24** Cartesian coordinate system for three dimensions. A right-hand rule specifies the orientation of the three axes. The z-axis points out of the plane of the paper. The thumb points along the z-axis; the fingers curl from the x-axis toward the y-axis.

This right-hand rule has become a well-established convention and will appear in many places in this book. Figure 1–25 shows how a vector $\vec{V}$ is decomposed into three components along the three axes.

The three unit vectors for the three axes are denoted by $\hat{i}$, $\hat{j}$, and $\hat{k}$; the unit vector $\hat{k}$ points in the z-direction (Fig. 1–26). A point P in three-dimensional space is now assigned the coordinates (x, y, z) in a given frame, and its displacement vector from the origin—its position vector—is

$$\vec{D} = x\hat{i} + y\hat{j} + z\hat{k}. \tag{1–13}$$

Similarly, any vector $\vec{V}$ has the components (V_x, V_y, V_z) such that

$$\vec{V} = V_x\hat{i} + V_y\hat{j} + V_z\hat{k}. \tag{1–14}$$

Pythagoras' theorem again tells us the magnitude, or length, of $\vec{V}$:

$$V = \sqrt{V_x^2 + V_y^2 + V_z^2}. \tag{1–15}$$

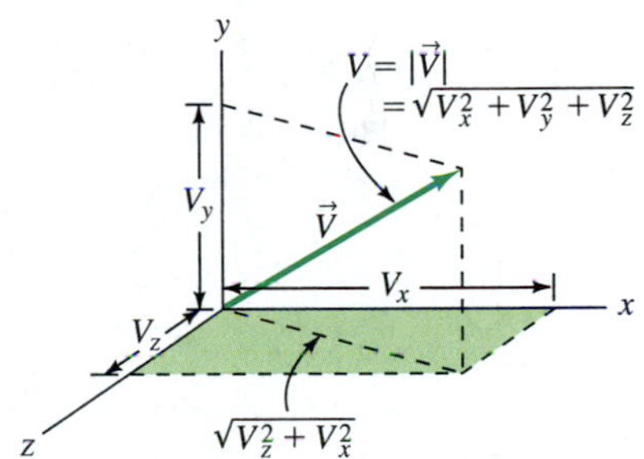

▲ **FIGURE 1–25** Vector $\vec{V}$ decomposed into three components in three-dimensional space.

Vector Equations

Vector equations are equations in which vector quantities appear on both sides of the equal sign. For example, we shall learn in Chapter 4 that a force, which is a vector quantity, causes an acceleration, which is also a vector quantity, proportional to the force. The equality force = a constant × acceleration is a vector equation. The equality in such an equation means either that the magnitudes and direction of the vector on each side are the same or, equivalently, that the respective components of the vector equation are equal. In two dimensions, a vector equation is really shorthand for two separate equations for the components; in three dimensions, a vector equation represents three separate equations.

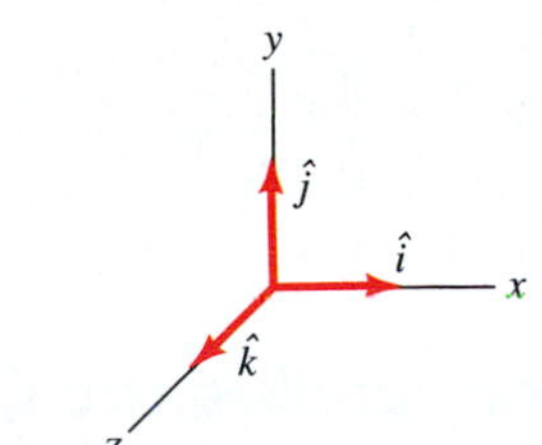

▲ **FIGURE 1–26** The unit vectors $\hat{i}$, $\hat{j}$, and $\hat{k}$ are aligned along the three Cartesian axes x, y, and z, respectively. Each has unit magnitude.

Summary

In this chapter we have given you a set of tools and methods that you can use to tackle the ideas about the physical world that are the real subject of this book. What is covered here is an indispensable minimum and we will add more tools as we go along. We have also touched on a number of ideas—that length, time, and mass are fundamental "dimensions"; the nature of the scientific method; and how useful estimates can be. You won't need to think about these consciously all the time, but we'll be using them throughout.

The range of quantities relevant to our understanding of the physical world is so large that it is useful to employ scientific notation. In this notation, any number can be represented by a decimal number from 1 to 10 multiplied by a power of 10. The quantities that appear in physics and engineering have units as well as sizes. The International System of units, or SI, provides reproducible and precise definitions of mass, length, and time. The SI units are the kilogram (kg), meter (m), and second (s). Some quantities are used so often that their units are given a special name (e.g., force is measured in newtons, N, in SI), but these units are derived units: They can always be expressed in terms of the three primary units (e.g., $1\ \text{N} = 1\ \text{kg} \cdot \text{m/s}^2$). Units that appear in equations can be manipulated algebraically, and after conversion to a single unit system, the units on both sides of any correct equation will match.

Mass, length, and time are the quantities with the three primary dimensions, abbreviated $[M]$, $[L]$, $[T]$. Dimensions should not be confused with units, which refer to a particular choice of unit system. Any physical quantity has dimensions that are rational combinations of the primary dimensions. Dimensions can be manipulated algebraically, and both sides of any correct equation will have the same dimensions. When we analyze the dimensions of an equation, we are performing a dimensional analysis. Dimensional analysis is useful for checking the answers to problems, for learning scaling laws, and for discovering relations between physical quantities.

Numbers that represent physical quantities can be measured only to a certain accuracy. An explicit way to indicate this accuracy is to write a physical quantity x as a central value $\pm$ an uncertainty. Calculations involving physical quantities are meaningful only to within the known accuracy of those quantities. When several numbers of different accuracies are involved in a calculation, the least accurate quantity primarily determines the accuracy of the result. A second way to indicate the known accuracy of a physical quantity is through the use of significant figures, that is, the number of digits in the value between 1 and 10, which are then multiplied by the power of 10 in scientific notation.

The ability to estimate is one that should be cultivated. An educated first guess is a valuable start to the solution of any problem. An order-of-magnitude calculation is such a guess. Similarly, when you arrive at an answer to a physical problem, it is always wise to ask yourself whether it makes sense.

Some, but not all, physical quantities include directional information. Displacement and velocity, for example, include directional information, and such quantities are represented by vectors, in contrast to others, such as mass or time, that are scalars. The vectors $\vec{A}$ and $\vec{B}$ are mathematical objects with both magnitude and direction. They obey the rule

$$\vec{A} + \vec{B} = \vec{B} + \vec{A}. \qquad (1\text{–}2)$$

Vectors can be expressed in graphical form; we draw them as arrows of length equal to their magnitude within a particular coordinate system. The simplest such system is the Cartesian system, with mutually perpendicular x-, y-, and z-axes for three-dimensional space. We can express any vector $\vec{V}$ in terms of the unit vectors—vectors of unit length—for a given coordinate system. Thus the unit vectors $\hat{i}$, $\hat{j}$, and $\hat{k}$ point along the x-, y-, and z-axes, respectively. Then

$$\vec{V} = V_x\hat{i} + V_y\hat{j} + V_z\hat{k}. \qquad (1\text{–}14)$$

The quantities V_x, V_y, and V_z are the components of $\vec{V}$. The magnitude of $\vec{V}$ is

$$V = \sqrt{V_x^2 + V_y^2 + V_z^2}. \qquad (1\text{–}15)$$

Vector equations are relations that equate different vectors. In such an equation, the components of the vectors are equal on both sides, so that, in three dimensions, a vector equation is an elegant and compact way to state three separate equations.

Understanding the Concepts

1. If space were somehow four dimensional instead of three dimensional, would the concept of displacement vectors still make sense? In what way could the idea of vector displacement be generalized?
2. How might the measurement of distance in centimeters, meters, and kilometers rather than in inches, feet, yards, and miles be more convenient? How might it be less convenient?
3. What is your height in centimeters and meters?

4. What is your mass (notice that we do not say *weight*) in kilograms?
5. Angles can be measured in units of degrees ($1° = 1/360$th of a full circle) or of radians ($1 \text{ rad} = 1/2\pi$ of a full circle). What are the *dimensions* of these units?
6. The human pulse and the swing of a pendulum are possible time units. Are they ideal ones?
7. How can length be defined in terms of the speeds of light and time?
8. Three vectors all have the same magnitude. Is it possible to add them together to obtain a null vector?
9. Three vectors have different magnitudes given by 4, 5, and 7 m. Is it possible for the three vectors to add to zero? Prove your result by drawing a figure.
10. It has been said that a recent measurement of the magnetization of an electron has been carried to an accuracy comparable to determining the distance between Los Angeles and New York to within the width of a human hair. What level of accuracy is that in parts per thousand, or million, or billion?
11. Is the vector sum of the two unit vectors $\hat{i} + \hat{j}$ also a unit vector? Under what conditions is it possible for two unit vectors to add to a resultant vector that also has unit magnitude?
12. Suppose we were in radio contact with some distant civilization. How would we communicate how large we are? What assumptions are you making in providing your suggestions?
13. A useful estimate is that there are $\pi \times 10^7$ s in a year. How accurate is this?
14. How does Pythagoras' theorem enter in the determination of the length of the vector $\vec{v} = a\hat{i} + b\hat{j}$?
15. The next time you weigh yourself, consider the uncertainty of the numbers you read. What would you estimate the percentage uncertainty in this measurement of your weight to be?
16. A small worm absorbs the oxygen it needs through its surface. Assume that the oxygen needed by an animal is proportional to its mass and estimate how much the absorption of oxygen per unit area would have to increase if the worm were to increase each of its dimensions by a factor of 10. You may find it interesting to know that the lungs of a human being have about 100 m^2 of surface available for the absorption of oxygen.
17. A mouse eats the equivalent of about one-quarter of its mass in food every day. You do not. Why?
18. The estimate made in Example 1–7 required you to find by one means or another the radius of Earth. Suppose your number for the radius were 10 percent larger than the number used here. How much would this change your estimate for the result of this example?
19. A 1-kg meteorite adds its mass to that of Earth when it comes crashing in from outer space. Is it necessary to revise all previous calculations involving the effect of Earth on other objects?
20. True or false: Two vectors that have exactly the same length will give a vector of zero length when one is subtracted from the other.
21. What is the minimum number of vectors needed to describe the position of an object in three-dimensional space? How many different sets of vectors can do this?
22. A river flows from west to east at 6 mph. A swimmer enters the river on the south bank and swims straight across at 3 mph relative to the water. Draw a vector diagram for the velocity of the swimmer relative to the bank. If the river were flowing twice as fast, what would change about the cross-river trip?

Problems

1–1 Scientific Notation

1. (I) Twenty thousand jelly beans are in a jar, and 15 percent of them are green. Express in scientific notation the number of green jelly beans.
2. (I) What is the product of 10^5 and 10^{-4}? The ratio $10^{-4}/10^5$?
3. (I) Your bank account contains \$356.00. How many cents is that, in scientific notation?
4. (II) Express the number e^{84} in standard scientific notation (powers of 10). The mathematical constant e has a value 2.718....
5. (II) Calculate the cube root of the number 10^{21} as well as the square of the resulting number.

1–2 Length, Time, and Mass

6. (I) The Empire State Building is 1472 ft high. Express this height in both meters and centimeters.
7. (I) What is your height in atomic diameters (see Table 1–1)?
8. (I) A 5-ft, 5-in-tall skier should use skis 5 cm longer than her height. How long should her skis be? Skis are made in 5-cm intervals (150 cm, 155 cm, etc.). What length skis should she buy if she rounds off to the nearest 5 cm?
9. (I) At one time, grapes sold in Italy for 1.25 euros per kilogram. If the conversion rate between dollars and euros was then 1 euro = \$0.94, what was the price of grapes in dollars per kilogram?
10. (II) Gasoline is heavily taxed in Europe, with a recent cost of 1.20 euros per liter. What was this price in dollars per gallon if the currency conversion is 1.04 euros = \$1 US? One gallon (gal) = 3.8 liters (L).
11. (II) The acceleration due to gravity, g, is 9.80 m/s^2 in SI. Convert this to the British system, where length is measured in feet rather than meters.
12. (II) The gravitational constant G is $6.67 \times 10^{-11} \text{ m}^3/\text{s}^2 \cdot \text{kg}$. What is G in units of $\text{cm}^3/\text{s}^2 \cdot \text{g}$?
13. (II) The radius of the Moon (assumed spherical) is 1.74×10^3 km, and its mass is 7.35×10^{22} kg. What is the density of the Moon in grams per cubic centimeter?
14. (II) A neutron star has a radius of 15 km and a mass of 1.4×10^{31} kg. What is the density of the neutron star in metric tons per cubic centimeter?
15. (II) Gasoline consumption in Europe is measured in liters (L) per 100 km. For example, a small Opel uses 7.0 L/100 km, while the gasoline consumption of a large Mercedes is 23 L/100 km. Convert these to miles per gallon.

1–3 Accuracy and Significant Figures

16. (I) A student wishes to make a measurement of the road distance from his dormitory to the physics building of his university. He uses his car's trip odometer, which measures distance only in units of a tenth of a mile. (a) He makes one trip and the odometer reads 0.3 mi. What can he say is the distance and, in particular, to how many significant figures? (b) One day he has nothing better to do with his time and, adding trips both to and from, he makes 100 trips; his odometer measures 27.2 mi. What can he now say is the distance? How is this result consistent with the result of part (a)? (c) A friend challenges the student on his measurement and, upon reflection, the student is not sure whether he made 99, 100, or 101 trips. How should he modify his statement of the distance?

17. (II) A well-known approximation to π is $\pi \cong 22/7$. What percentage error does this result have? How much better is the approximation $355/113$?

18. (II) The net force F on a mass m moving at speed v in a circular path of radius r has magnitude $F = mv^2/r$. The mass is measured to be 0.00535 kg, the radius is 0.3 m, and the speed is 1.1 m/s. Give the magnitude of the net force. Pay attention to the number of significant figures.

19. (II) A rectangular box is stated to have width 0.75 ± 0.02 m, length 0.5 ± 0.1 m, and height 0.582 ± 0.058 m. What is the volume, stated to the appropriate number of significant figures with an uncertainty?

20. (II) You wish to determine your density (mass per unit volume) by two measurements: by weighing yourself on a digital scale and by submerging yourself in a tank of water with vertical sides and noting the rise in water level on a scale marked in centimeters (Fig. 1–27). The surface area of water in the tank is 1.5 m^2. The digital scale gives the weight in pounds at 0.5-lb intervals, and you cannot make a reading of the water level to better than 0.5-cm accuracy. Suppose that the weight reading is 213 lb and the water level changes from 152 to 158.5 cm. What is your density? Express your result as a central value with a percentage error.

▲ **FIGURE 1–27** Problem 20.

21. (III) If you want to know the area of a circle to 10 percent accuracy, how accurately should you measure the diameter of the circle?

22. (III) Consider the infinite series $\sum_{n=0}^{\infty}(x^n/n!)$. [The symbol $n!$ (n factorial) means the product $1 \times 2 \times 3 \times \cdots \times n$. By definition, $0! = 1$.] If $x = 0.100000$, how many terms in the series suffice to give a result correct to six significant figures?

1–4 Dimensional Analysis

23. (I) The kinetic energy of a baseball is denoted by $mv^2/2 = p^2/2m$, where m is the baseball's mass and v is its speed. This relation can be used to define p, the baseball's momentum. Use dimensional analysis to find the dimensions of momentum.

24. (I) One of Einstein's most famous results is contained in the formula $E = mc^2$, where E is the energy content of the mass m and c is the speed of light. What are the dimensions of E?

25. (I) A length L that appears in atomic physics is given by the formula $L = h/m_e c$, where m_e is the mass of an electron, c is the speed of light, and h is a constant known as Planck's constant. What are the dimensions of h?

26. (II) What are the dimensions of h^2/m^3G, where h is a constant called Planck's constant, m is a mass, and G is the gravitational constant? The dimensions of the constants in this formula can be found in the list of physical constants given in Appendix II.

27. (II) A fellow student has proposed that an object accelerates under the influence of the gravity of the Sun according to the formula

$$a = \frac{MG}{r},$$

where a is the magnitude of the acceleration measured in feet per seconds squared, M is the mass of the Sun, the physical constant G is 6.67×10^{-11} m^3/s$^2 \cdot$kg, and r is the distance to the Sun's center. Can this formula be correct?

28. (II) You are told that the speed of sound in a metal depends only on the density $\rho([ML^{-3}])$ and on the bulk modulus of the metal, B, which has dimensions $[ML^{-1}T^{-2}]$. Express the sound speed in terms of ρ and B.

29. (III) A force F acting on a body of mass m a distance r from some origin has magnitude $F = Ame^{-\alpha r}/r^4$, where A and α are both constants. The constant $e = 2.718\cdots$. Given that the force has dimensions kilogram-meter per seconds squared, what are the dimensions of (a) the constant α and (b) the constant A?

30. (II) In quantum mechanics, the fundamental constant called Planck's constant, h, has dimensions of $[ML^2T^{-1}]$. Construct a quantity with the dimensions of length using h, a mass m, and c, the speed of light.

31. (II) It is known that the quantity Kq^2/hc is dimensionless (K is a numerical constant; h and c are as discussed in Problem 30; q is an unknown quantity). (a) What are the dimensions of q? (b) What are the dimensions of q^2/R, where R is a length?

1–5 How a Little Reasoning Goes a Long Way

32. (I) How many times does an average person's heart beat in a lifetime? Estimate the number of times an automobile tire rotates in a trip across the United States.

33. (II) A criminal posing as a tourist wants to smuggle \$25 million in gold across the U.S. border in his station wagon. Is he likely to make it? Use estimates and explain your reasoning.

34. (II) Estimate the volume of concrete used for the construction of the tunnel under the English Channel. The system consists of two railway tunnels and a service tunnel and is 30 km long.

35. (II) Estimate the area used for the storage of one bit on a 1.44-MB 3 1/2-in diameter floppy disk (1 byte = 8 bits).

36. (II) What is the weight of water in a full 5-gal can? About what volume of water can a typical person carry?

37. (II) Make separate estimates of the number of automobile mechanics in the United States based on (a) your total bill for automobile repairs compared with a reasonable average salary for a mechanic, (b) how many hours it takes to repair a car, and (c) the number of people you know who are automobile mechanics compared with the total number of people you know.

38. (II) The Sun is 93 million miles from Earth. What is the diameter of the Sun? [A dime held at arm's length will just about cover the surface of the Sun (Fig. 1–28).]

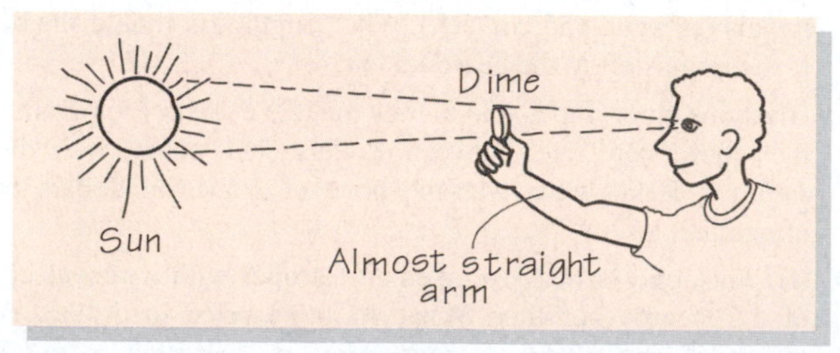

▲ **FIGURE 1–28** Problem 38.

39. (II) A typical cloud contains droplets of water with an average radius of 0.5×10^{-4} m. How many droplets are needed for a cloud that provides a rainfall of 0.5 cm in your city?

40. (II) Pine trees can only be purchased from the local state nursery in bundles of 100. You want to plant trees over 1 acre (0.405 hectares, or 4050 m^2) in 2-m intervals. Your plot of land is square. How many bundles do you order?

41. (II) Earth's radius is approximately 6400 km, and about two-thirds of the surface is water. Make a reasonable guess as to the average depth of the oceans and seas and calculate the volume of all the water. Express your answer in cubic centimeters. If 1 liter (L) of some medicine were poured into the oceans and stirred sufficiently to be *uniformly* distributed throughout all the oceans, what is the degree of dilution in powers of 10?

42. (II) A strong radioactive point source emits 10^8 gamma particles per second uniformly in all directions. What fraction of these particles will hit a circular detector of diameter 4 cm that is 2 m away from the source?

43. (II) There are some 200 million vehicles in the United States, and each one is driven about 15,000 mi/yr. Estimate the number of automobiles on the road at any one moment.

44. (II) Suppose that all the eighteen-wheeler trucks in the United States line up bumper to bumper on I-80. Can they form a continuous line from New York to San Francisco?

45. (II) Suppose that the circumference of Earth is a perfect circle of exactly 25,000 mi. Somebody prepares a wire that is supposed to go around the equator completely but makes it 2 m too long by mistake. If this 2-m-too-long wire were placed around the equator in a perfect circle with the ends of the wire just touching each other, by how much would the wire be off the ground?

46. (II) Make a rough estimate of the number of apples produced each year in a 10-acre orchard.

47. (II) The mass of one atom of hydrogen is 1.7×10^{-27} kg. Given that the mass of the Sun is 2.0×10^{30} kg, how many atoms of hydrogen would it contain if it consisted purely of hydrogen? Actually, it consists of 70 percent hydrogen and 30 percent helium by mass, and the mass of a helium atom is 6.6×10^{-27} kg. How many hydrogen atoms does the Sun have?

48. (II) Suppose that oil consists of molecules of CH_2. This means that 1 mol of oil, consisting of 6.02×10^{23} molecules, has a mass of 14 g. One milliliter (mL) of oil has a mass of 0.95 g. When this much oil is poured on water, it does not spread forever but spreads until it makes a circular film of area 1.5×10^7 cm^2. How does the fact that the oil does not spread forever support the idea of atoms and molecules? Assuming that the oil slick is one molecule thick (a *monomolecular layer*), with the molecules touching, and that the molecules are spheres, estimate the size of an oil molecule.

49. (II) By estimating the size of Manhattan Island and the average amount of living space that a person uses and by supposing that 2 million people live in Manhattan, make an estimate of the average building height there.

1–6 Scalars and Vectors

50. (I) Draw the vector $3\hat{i} + 4\hat{j}$ by first drawing the x-component vector, then the y-component vector, then adding them graphically. Multiply the vector by a factor of 2 and repeat the exercise.

51. (I) A girl runs around a circular lake. Devise a simple coordinate system to describe her position and direction of travel at any time assuming that she begins at the south end of the lake and runs clockwise at a speed of 3 m/s.

52. (I) What is the resultant vector when the vectors $\vec{A} = 6\hat{i} - 5\hat{j}$ and $\vec{B} = 8\hat{i} + 3\hat{j}$ are added together? When $\vec{B}$ is subtracted from $\vec{A}$?

53. (I) A drunken sailor stumbles four paces north, six paces northeast, two paces east, and five paces west. Describe the final location from the initial position by a single displacement vector.

54. (II) The access point to a septic system buried in a back yard is described by 50 ft northwest and 24 ft northeast from the corner of the house. Using a coordinate system with north in the y-direction and east in the x-direction, write out the two vectors with the magnitude in meters. Use unit vectors.

55. (II) A football player catches the kickoff on the 10-yd line and runs straight up the field for 15 yd, turns left for 15 yd, goes straight up the field for 10 yd, turns right for 20 yd, reverses his field (makes a 180° turn) for 5 yd, and then streaks straight up the field for a touchdown. Define a coordinate system and draw the entire path in vector form.

56. (II) Draw a vector $\vec{V}$ that points in the northwesterly direction, making an angle α with the northerly direction, as in Figure 1–29. If north is chosen as the y-direction and east as the x-direction, what is the x-component of $\vec{V}$?

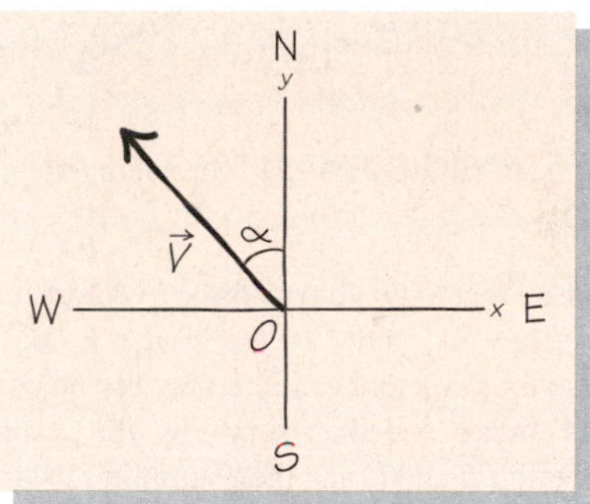

▲ **FIGURE 1–29** Problem 56.

57. (II) Suppose that in Problem 56 you choose north as the x-direction and west as the y-direction. What is the x-component of $\vec{V}$ in this case?

58. (II) Refer to the situation outlined in Problems 56 and 57. Choose the x-axis as the line that makes an angle of 45° with the northerly direction and is inclined to the east and the y-axis as the line that makes a 45° angle with the westerly direction and is inclined to the north. What is the x-component of $\vec{V}$ in this case?

59. (II) In computer-aided drafting programs, lines can be specified in either rectangular or polar coordinates (Fig. 1–30). In such programs, if the coordinates (x, y) are given, the cursor draws a line from its current position a distance x to the right and a distance y up. If polar coordinates (r, θ) are given, the line is drawn

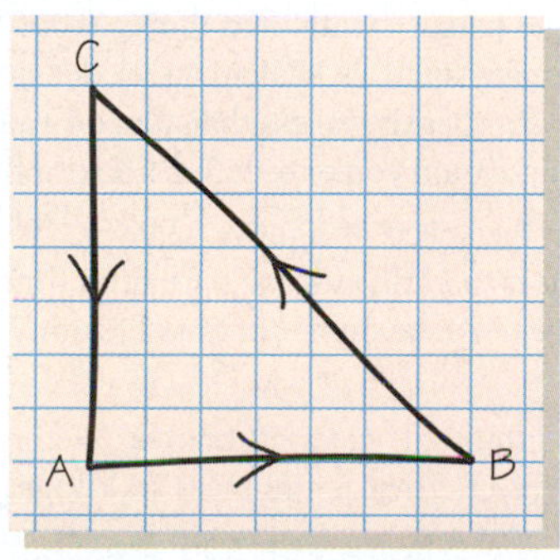

▲ **FIGURE 1–30** Problem 59.

from its current position through an angle θ in the counterclockwise direction from the positive x-axis through a distance r. Give the instructions both in terms of (x, y) and (r, θ) for the drafting of the triangle shown in the figure starting at point A.

60. (II) Consider the following vectors: $\vec{A} = -2\hat{\imath} - 3\hat{\jmath}$, $\vec{B} = \hat{\imath} + 2\hat{\jmath} + 3\hat{k}$, $\vec{C} = 3\hat{\jmath} + 3\hat{k}$, and $\vec{D} = -2\hat{\imath} - \hat{k}$. Find (a) $\vec{A} + \vec{B} + \vec{C} + \vec{D}$, (b) $\vec{A} - \vec{D}$, (c) $\vec{A} + \vec{D} - \vec{B}$, and (d) $|\vec{A} - \vec{C}|$.

61. (II) Vectors $\vec{A}, \vec{B}, \vec{C}$, and $\vec{D}$ are shown in Fig. 1–31. (a) Give the vectors in component form. (b) Determine the following quantities both algebraically and graphically: $2\vec{A} + \vec{C} - \vec{D}$, $\vec{B} + \vec{C}/2$, and $|\vec{D} - \vec{B}|$.

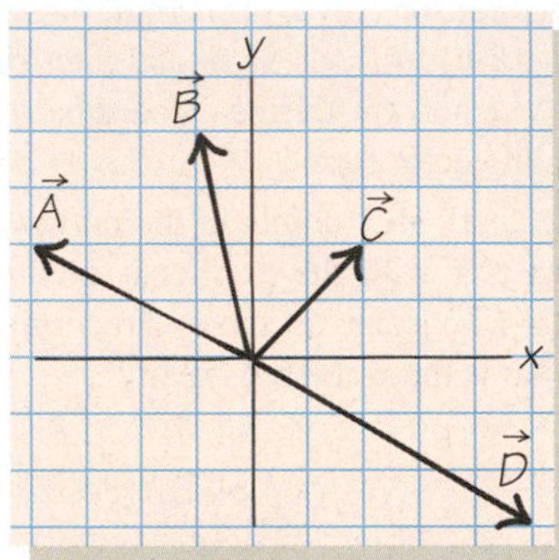

▲ **FIGURE 1–31** Problem 61.

62. (II) Suppose that you have three vectors, $\vec{A} = 3\hat{\imath} + 4\hat{\jmath}$, $\vec{B} = 2\hat{\imath} - 2\hat{\jmath} + 4\hat{k}$, and $\vec{C} = -\hat{\imath} + 5\hat{\jmath} - 3\hat{k}$. Show that the sum of these three vectors can alternatively be computed by first summing $\vec{A}$ and $\vec{B}$ and then summing the resultant with $\vec{C}$ or by first summing $\vec{B}$ and $\vec{C}$ and then summing the resultant with $\vec{A}$.

63. (II) In all of our sketches of vectors in a plane, we used *orthogonal* axes, the horizontal x-axis, and the vertical y-axis. Consider a pair of nonorthogonal axes, for example a horizontal x-axis and a y-axis that makes an angle of 60° with the horizontal. Show what the *components* of an arbitrary vector look like along these axes and express these components in terms of the length of the vector and the angle that it makes with the x- axis.

General Problems

64. (I) A human hair has a diameter of about 10^{-4} m. Given that all of the hair on a head can be gathered into a lock of diameter 4 cm, estimate the fraction of the human scalp area from which hairs spring and the number of hairs on the head.

65. (I) There are 100 cm in 1 m, 1000 kg in 1 metric ton, and 1000 g in 1 kg. (a) How many metric tons are in 1 g? (b) How many cubic meters are in 1 cm^3?

66. (I) Measure the height, width, and thickness of this book in centimeters. Estimate its mass. Calculate its density. An object will float in water if its density is less than that of water. Without doing the experiment, would you expect this book to float in water?

67. (I) Given that the speed of light is 2.998×10^8 m/s and the distance from Mars to Earth at some particular time is 1.5×10^8 mi, how long does it take light to travel from Mars to Earth at that time?

68. (I) Eighteen grams of water is known to contain 6.02×10^{23} molecules of H_2O. What is the mass in kilograms of one molecule of H_2O?

69. (II) The density of a human body is approximately 1 g/cm^3, which is also the density of water. Use the result of the calculation in Problem 68 to estimate the number of molecules of water that a typical human body would contain if it were made up entirely of water.

70. (II) A silver nucleus consists of 108 closely packed nucleons (protons and neutrons) and has a radius of 5×10^{-15} m. A neutron star is basically an overgrown nucleus, with neutrons only, closely packed in the same way as nucleons are in the silver nucleus. If the radius of a neutron star is 12 km, how many neutrons does it contain?

71. (II) Imagine that molecules of H_2O are stacked up in a cubic array, like a large number of cubical boxes, with a water molecule at the center of each cube. Let the side of each cube be L. Given that the density of water is 1 g/cm^3, estimate the distance L from the data given in Problem 68.

72. (II) The gasoline usage rate required to propel an automobile is very roughly proportional to the mass of the automobile. Assuming that the proportions and types of materials of an automobile do not change, calculate the percentage of gasoline savings that would be realized if cars were reduced by 20 percent in each of their three space dimensions.

73. (II) A typical star has a mass of about 2×10^{30} kg and there are about 10^{11} stars in a galaxy. What is the mass of this typical galaxy? Assume that stars are made primarily of hydrogen; the mass of a hydrogen atom is 1.67×10^{-27} kg. How many hydrogen atoms are there in a galaxy?

74. (II) Determine the thickness of a page of this book to an accuracy better than 10 percent. Explain your method and give your uncertainty.

75. (II) We have seen in the text that the period of a simple pendulum is independent of the mass of the pendulum bob. Further, we have seen that the dimensional relation between the period τ, the pendulum length ℓ, and the acceleration of gravity g takes the form

$$[\tau] = [\ell^r][g^s].$$

Use the fact that the dimension of τ is $[T]$, that of ℓ is $[L]$, and that of g is $[L/T^2]$ to show that

$$\tau \propto \sqrt{\frac{\ell}{g}}.$$

76. (II) You need to know the area of a square bedroom to 5 percent accuracy in order to purchase carpet. How accurately do you need to know the length of the side?

77. (II) The water supply of Pittsburgh is contaminated by an oil spill. Make some reasonable assumptions in order to estimate how many trucks per day are needed to bring in a minimum supply of water. What if each person were allowed to take a bath every three days?

78. (II) Assume that houses are set on quarter-acre lots and each house receives four pieces of mail a day. Estimate how far a postal carrier walks in one day and how much mail he or she carries. What is the mass of all this mail?

79. (II) According to Kepler's third law of planetary motion, the square of the period of a planet is proportional to the cube of its mean distance from the Sun. Given that Earth, whose period is 1 yr, is 1.5×10^8 km away from the Sun, calculate the distance from the Sun to Venus, whose period is 0.61 yr, and the period of Saturn, which is 14×10^8 km away from the Sun.

80. (II) When two parallel plates of area A separated by a distance y move with relative speed v with respect to each other in a fluid (Fig. 1–32), there is a frictional force (the *viscosity*) given by the formula

$$F = \eta \frac{vA}{y}.$$

What are the dimensions of the coefficient of viscosity η?

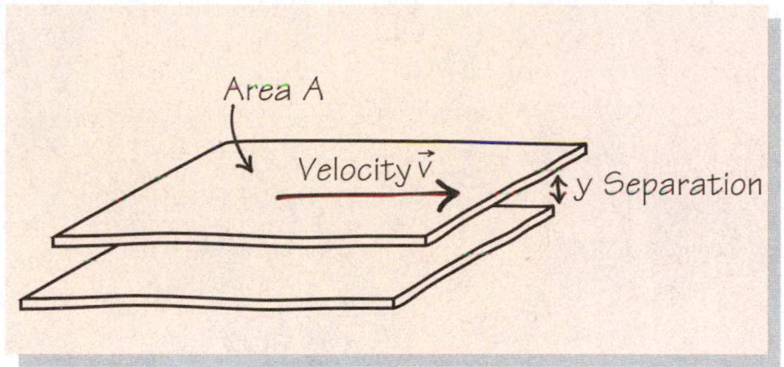

▲ **FIGURE 1–32** Problem 80.

81. (II) The mass of a vertical cylinder of atmosphere of cross section 1 cm^2 is approximately the same as the mass of a cylinder of the same area of water that is 30 ft high. Use this to estimate the number of "molecules of air" in the atmosphere given that, on average, one molecule of air is 1.6 times as massive as one molecule of water. (See Example 1–8.)

82. (II) Show that $(\vec{A} + \vec{B}) + \vec{C} = \vec{A} + (\vec{B} + \vec{C})$.

83. (II) A vector $\vec{u}$ in the xy-plane has x- and y-components $u \cos\theta$ and $u \sin\theta$, respectively, where θ is the angle that $\vec{u}$ makes with the $+x$-axis. A second vector, $\vec{v}$, also lies in the xy-plane, and it is perpendicular to $\vec{u}$. (a) Draw a figure to show that there are two possibilities for the vector $\vec{v}$: Its x- and y-components are either $-v\sin\theta$ and $v\cos\theta$ or $v\sin\theta$ and $-v\cos\theta$. (b) Check that $v_x u_x + v_y u_y = 0$ independent of θ.

84. (II) For objects that move in a circle about an origin O, it can be convenient to use the mutually perpendicular unit vectors $\hat{i}_r$ and $\hat{i}_t$, defined as in Figure 1–33. If a Cartesian coordinate system has its origin at O, with an x-axis chosen so that the angle between it and the line OP is θ, then (a) show that $\hat{i}_r = \hat{i}\cos\theta + \hat{j}\sin\theta$, (b) calculate the y-component of $\hat{i}_r$, and (c) express $\hat{i}_t$ as a combination of $\hat{i}$ and $\hat{j}$.

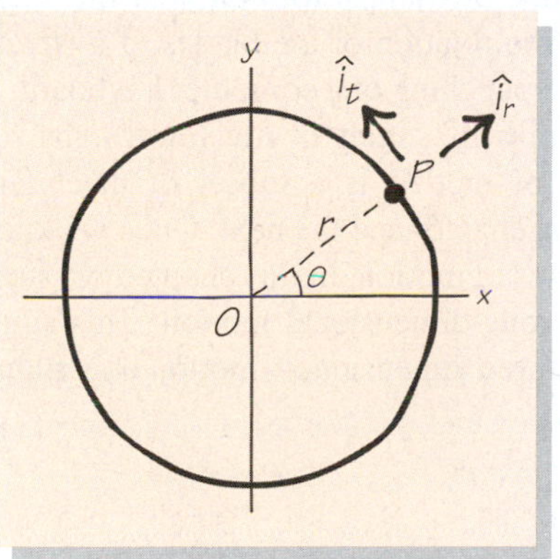

▲ **FIGURE 1–33** Problem 84.

85. (II) A vector $\vec{r}$ has length r and points in the direction shown in Figure 1–34. The angles θ and ϕ are drawn in the figure. (a) Show that the x-component of $\vec{r}$ is $r\sin\theta\cos\phi$. (b) Show that the z-component of $\vec{r}$ is $r\cos\theta$. (c) Find the y-component of $\vec{r}$ in terms of r, θ, and ϕ.

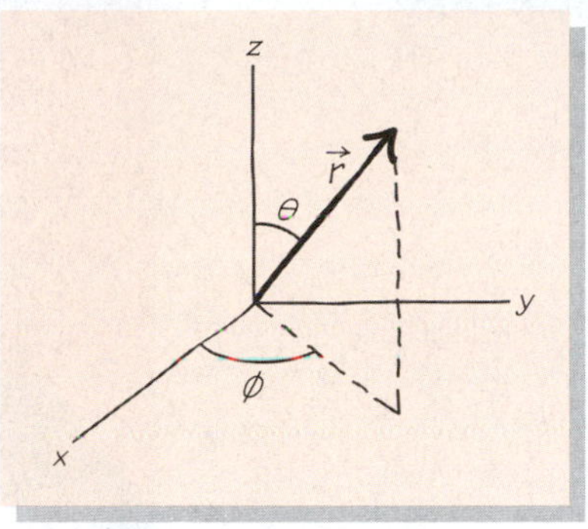

▲ **FIGURE 1–34** Problem 85.

86. (II) The number of molecules in 22.4 L of air at the surface of Earth is around 6.0×10^{23}. A liter is 10^{-3} m^3. Let us say that the volume of air you take in and then expel with each breath is 2.5 L. Let us also say that the air around Earth is a uniform blanket 8 km thick. (Actually, it drops as height increases, but our assumption is not far off.) Finally, Michelangelo Buonarotti lived for 91 yr and died sufficiently long ago that all the "molecules of air" he ever breathed are thoroughly dispersed throughout the atmosphere. Under all these conditions, how many molecules of air breathed by Michelangelo are in your lungs right now?

87. (III) A stretched wire has three physical attributes: the density λ, or mass per unit length; the total length ℓ; and the tension τ. The latter is related to how hard the wire is being pulled to keep it stretched and has dimensions of $[MLT^{-2}]$. Show by dimensional analysis that if the time t_0 of one back-and-forth vibration of the wire in a direction perpendicular to its length depends only on these three quantities, then t_0 has the form $t_0 = (\text{a constant})\ell\sqrt{\lambda/\tau}$.

Chapter 2

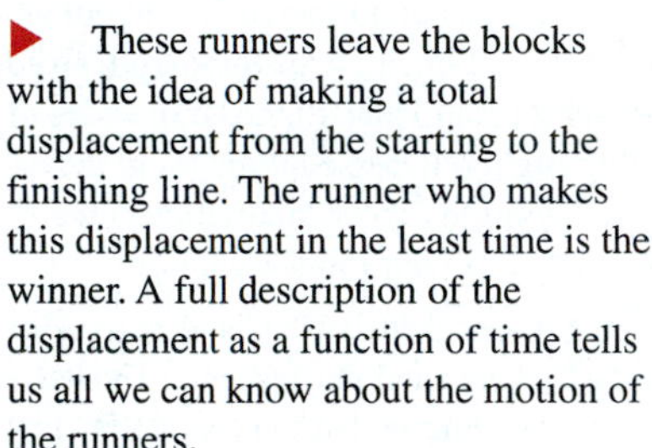

▶ These runners leave the blocks with the idea of making a total displacement from the starting to the finishing line. The runner who makes this displacement in the least time is the winner. A full description of the displacement as a function of time tells us all we can know about the motion of the runners.

Straight-Line Motion

Understanding motion is one of our key goals. Motion occurs over all the scales of our world, from the heavenly bodies to the motion of particles in atoms and nuclei. We need to understand the motion of airplanes and automobiles when we design airport runways and interstate off-ramps. Motion is at the heart of physical science, whether we are interested in the rotation of the blades of a steam turbine, the pulse of electricity that results from the punching of a computer keyboard, or the takeoff of a high jumper. That is why we begin with a study of **mechanics**, the science of motion and its causes. The *description* of motion is a subset of mechanics known as **kinematics**, which is the subject of this chapter and the next. Once we know how to describe motion, we will explore the *causes* of motion. In this chapter, we study the motion of an object in a straight line; that is, one-dimensional motion. This simpler situation will prepare us for motion in two and three dimensions—motion on a plane and motion in space—the topic of Chapter 3.

2–1 Displacement

Consider the motion of athletes running a 100-m dash. Some runners are able to attain a tremendous advantage right at the beginning of the race—they leap ahead of the others. Other runners have a late kick that allows them to take the lead at the end of the race (Fig. 2–1). To understand this motion, we want to break it down into variables that we can measure: distance and time. Suppose that we set up an electronic timing system along the path of the 100-m dash that records the times of the runners every 5 m. The timing system will give us the distance each runner travels as a function of time.

Let us first consider the distance traveled by the athletes. To do this, we construct a coordinate system with an x-axis using the variable x to indicate the distance traveled. As in our discussion of vectors in Chapter 1, the choice of origin of the x-axis is up to us, but it is convenient to take the origin at the starting line, which we therefore set at $x = 0$ m. We also choose the direction that the runners take to be the $+x$-direction. Figure 2–2a shows the data points that represent the distance (the x-value) traveled by one runner—actually, the distance traveled by a point on the runner's chest (we do this so that we do not have to worry about the complicated motion of the runner's entire body)—plotted against the time it takes the runner to reach that distance. We choose the starting time to be $t = 0$ s, again for convenience.

▲ **FIGURE 2–1** The motion of runners in the 100-m dash provides us with a test case to examine motion in detail and allows us to refine our ideas of motion.

Figure 2–2b shows a curve that interpolates the data points of Fig. 2–2a to reasonable accuracy. It also contains an extension (the dashed line) that shows the runner slowing down and coming to rest after a short time interval. The figure can be used to read the time taken to reach any distance x within the 100 m, even those values not given directly by the data. The curve, which we call $x(t)$, represents the distance from the starting point, x, as a function of time t. We use the notion of *displacement*, a concept we met in Chapter 1, to study this curve further. The displacement is the *change* in the position of an object. If we denote the first position at time t_1 by x_1 and the second position at time t_2 by x_2, then the displacement is the final position minus the initial position; this is defined mathematically by

$$\Delta x \equiv x_2 - x_1. \tag{2–1}$$

We use the capital Greek letter Δ (delta) to indicate a change (or difference) in a variable from one value to another. The time interval is, similarly,

$$\Delta t = t_2 - t_1. \tag{2–2}$$

There is an important distinction between Δx and Δt: The displacement Δx can have a sign, whereas the time interval Δt is normally positive. In our example of the runner, x increases steadily with time, and the possibility that Δx can have two signs does not play an important role. But if one of the runners were to set off in the opposite direction, his or her displacement in a given (positive) time interval would be negative—that very fact would tell us that the runner was moving in the $-x$-direction. If the runner who took off in the wrong direction realized there was a problem and turned around, then we would have a negative displacement during some time interval and a positive one during some other time interval.

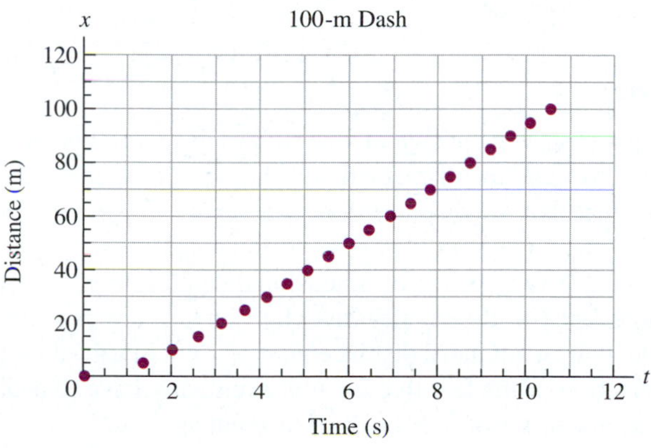

(a)

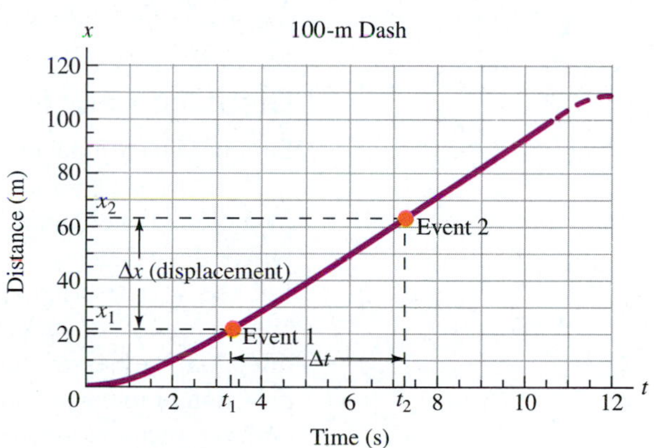

(b)

▲ **FIGURE 2–2** (a) Distance plotted against time at 5-m intervals for one athlete in the 100-m dash. (b) The curve is an interpolation of the data shown in part (a), with the dashed part an extrapolation. The runner is at position x_1 at time t_1 (event 1) and at position x_2 at time t_2 (event 2). In the time interval $\Delta t = t_2 - t_1$ between event 1 and event 2, the displacement of the runner is $\Delta x = x_2 - x_1$.

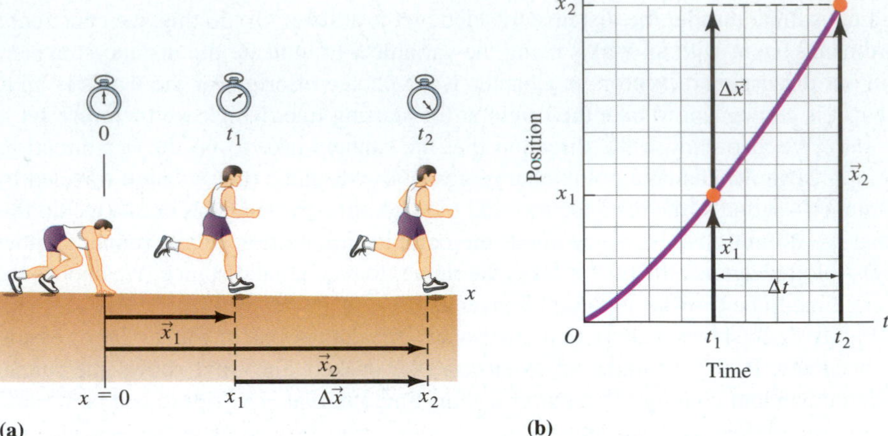

▶ **FIGURE 2–3** (a) The position vectors $\vec{x}_1$ and $\vec{x}_2$ run from the origin to the positions x_1 and x_2, respectively, of the runner. The displacement vector $\Delta\vec{x} = \vec{x}_2 - \vec{x}_1$ is also drawn. (b) The information contained in part (a) can be plotted as a curve of position versus time.

While the position depends on the choice of origin, the displacement in Eq. (2–1) does not. Suppose that the origin were placed 10 m back of the starting blocks. Then the position at time t_1 would be given by $x_1 + 10$ m, and the position at time t_2 would be given by $x_2 + 10$ m. The displacement would still be given by

$$\Delta x = (x_2 + 10\text{ m}) - (x_1 + 10\text{ m}) = x_2 - x_1. \tag{2–3}$$

This relationship is shown in Fig. 2–3.

Finally, let us mention the *net displacement* over some definite time interval. If a few minutes after the race is over the runners return to the starting blocks, then their net displacement is zero for that interval; we simply take the difference between the final and initial positions, and these positions are the same. Of course, there are many ways to have a net displacement of zero, or any other value, over a long time interval. The net displacement, or the displacement over a long time interval, contains less information about the details of the motion than data from a series of displacements over short time intervals.

EXAMPLE 2–1 Use Fig. 2–2 to find the runner's position for the dash at times 2 and 5 s. What is the displacement between these two times?

Setting It Up We label the positions x_1 at $t_1 = 2$ s and x_2 at $t_2 = 5$ s. These are given as data on the graph.

Strategy The displacement is simply $\Delta x = x_2 - x_1$. We can use the diagram in Fig. 2–2 to read off the distances traveled, x_1 and x_2.

Working It Out With the origin at the beginning of the race, the position at 2 s is $x_1 = 10$ m and that at 5 s is $x_2 = 40$ m. Then

$$\Delta x = x_2 - x_1 = (40\text{ m}) - (10\text{ m}) = 30\text{ m}.$$

This displacement is positive, indicating movement in the $+x$-direction.

Displacement Is a Vector

For one-dimensional motion, keeping track of the sign of x as well as its magnitude provides all possible information for both x and the displacement. The sign of x is the manifestation in one dimension of the fact that the displacement is a vector. For motion in two or three dimensions the displacement has both a magnitude and a *direction*, and the full vectorial description is important. A two- or three-dimensional displacement is not simply forward or backward along a line (positive or negative) but can take any possible direction. We can say, for example, that an ant had a displacement of 2 cm in a northeast direction or to the south. In anticipation of the fact that we will eventually have to deal with the vectorial nature of displacement, we define the displacement as

$$\Delta\vec{x} \equiv \vec{x}_2 - \vec{x}_1, \tag{2–4}$$

where $\vec{x}_2$ is the *position vector* at time t_2 (in the example of our runner, the vector from the starting line to the position of the runner at time t_2) and $\vec{x}_1$ is the position vector at t_1. In one dimension use of the full vector notation is unnecessary, although we could write the vector $\vec{x}_1$ as the product of the distance from the origin x_1 (including the sign) and the unit vector $\hat{i}$ in the positive x-direction, but we don't gain much by this. Instead, we can work with components like x_1, always recalling that the sign is important. This is how we proceed in this chapter; in later chapters, where more than one-dimensional motion occurs, we'll employ a more specific vector notation.

2–2 Speed and Velocity

Speed

The quantities *speed* and *velocity* both describe how fast the position of an object changes. The speed of an automobile or airplane is a familiar concept that we will refine here. The **average speed** for an object in motion is defined as the total distance traveled divided by the time taken to travel that distance:

$$\text{average speed} \equiv \frac{\text{total distance traveled}}{\text{time interval}}. \tag{2–5}$$

Speed is a scalar quantity and is always positive. As the time interval over which the object moves changes, the average speed may also change. For example, from the graph of Fig. 2–2, during the first 5.1 s of the 100-m dash, the runner has traveled 40 m. The average speed over this interval is

$$\text{for 0 to 5.1 s: average speed} = \frac{40\ \text{m}}{5.1\ \text{s}} = 7.8\ \text{m/s}.$$

For the last 5.4 s, however, the runner progresses 60 m, and the average speed is

$$\text{for 5.1 to 10.5 s: average speed} = \frac{60\ \text{m}}{5.4\ \text{s}} = 11\ \text{m/s}.$$

The runner completes the 100-m dash in 10.5 s. Over the entire 100-m dash, then, the runner's average speed is (100 m)/(10.5 s), or 9.5 m/s.

We can imagine taking the time interval to be a very small one. Then we would be calculating the *instantaneous speed*, or just speed for short. We'll learn more about this procedure in the discussion of velocity below. For example, if we find that (distance traveled in 1 ms), divided by (1 ms) at a clock time $t = 1.5$ s, we get the speed *at* $t = 1.5$ s. (Strictly speaking, we should take the time interval so small that there is no ambiguity about whether we were "at" or merely "around" $t = 1.5$ s.) This speed, which is what your car's speedometer measures, can vary from moment to moment.

EXAMPLE 2–2 A photographer is assigned the task of filming the 100-m dash of the runner whose data are given in Fig. 2–2. The photographer is stationary at a point 20 m in front of the starting blocks. When the runner reaches the 10-m mark, the photographer's cart starts traveling at 10 m/s in the same direction as the runner until the cart reaches the 100-m mark, at which point it stops. Superimpose on Fig. 2–2b the drawing of a continuous line representing the position of the photographer's cart from the beginning to the end of the race.

Setting It Up We must recognize here that, starting from the time the runner reaches the 10-m mark, the photographer's cart moves at a constant speed v, here 10 m/s.

Strategy Constant speed means that position changes *linearly* with time, and this "curve" is a straight line, one whose slope on the plot of x versus t is the value of the speed.

Working It Out We get the particular value of the slope from the value of the speed: 10 m/s. Thus the straight line must have a change of 10 units of length for every unit of time. The straight line that represents this motion is drawn in Fig. 2–4.

What Do You Think? Draw several straight lines through the origin on the position-versus-time graph to represent motions of different carts like the one the photographer used. Draw a line that shows you the order in which the carts arrive at a certain place; draw another line that shows you how far different carts have traveled in a fixed time. Which slope describes the fastest cart? ***Answers to What Do You Think? questions are given in the back of the book.***

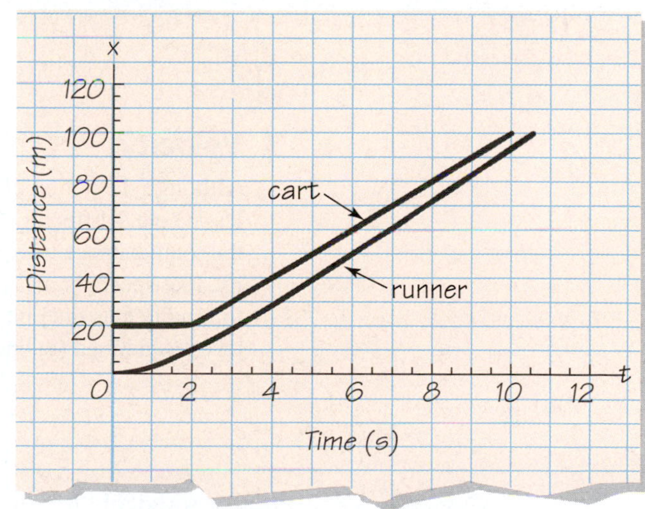

▲ **FIGURE 2–4** The motion of the photographer's cart is the straight line adjacent to the plot of the runner's motion.

Problem-Solving Techniques

Problem solving is important. The techniques we present here represent the basis for a plan of action on *any* problem. As we proceed, other techniques will fit into the general scheme.

1. Read the problem, then read it again. Failure to read the problem carefully is perhaps the source of more false starts and wrong answers than any other cause.
2. Draw a sketch or diagram of the problem to help you to visualize the situation presented by the problem. We illustrate this aspect of the technique in Fig. 2B1–1.
3. Write down the given and known quantities.
4. Make sure you understand which quantities are to be found.
5. There are generally only a few principles applicable to the solution of a problem. Think about which principles link the quantities to be determined to those that are known.
6. Use the principles that apply to the situation to guide you toward equation(s) that relate the quantities in the problem. Take care—at times, certain equations apply to a given situation and others do not. The rest is mathematics. Several equations may need to be manipulated together at times. Count the number of equations available to see if there are enough equations to determine the unknowns. There should be as many equations as there are unknowns.
7. When you solve for an unknown in terms of the known quantities, use symbols, not numbers. Wait until the end to replace symbols with numbers and units. It is important to include units; the answer may require them, and the proper cancellation of units provides a valuable accuracy check.
8. When you arrive at a number, think about it. Does it make sense? If you find that it takes 3 min to drive from New York City to Los Angeles, you have probably made a mistake!
9. Use *any* checks you can find for your result.

How To Draw Sketches

1. **Problem Statement:** A girl starts from rest at $x = -10$ m (time = 0), walks to $x = 20$ m at $t = 25$ s, turns around, and walks back to $x = -10$ m, where she stops at 45 s. Sketch a plot of position versus time.

2. **Thinking Process:** Let's first make a sketch of the motion along the x-axis and mark the start, turn around, and stop positions as 1, 2, and 3.

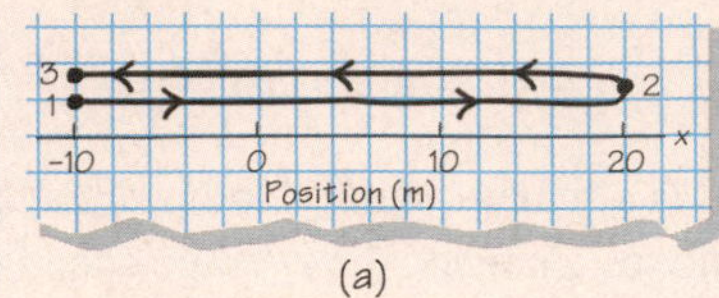

(a)

Note that the positions 1, 2, and 3 are "at rest." For these three positions, the speed is zero and their slope on the x-versus t-plot must be zero.

3. **Make a Sketch:**

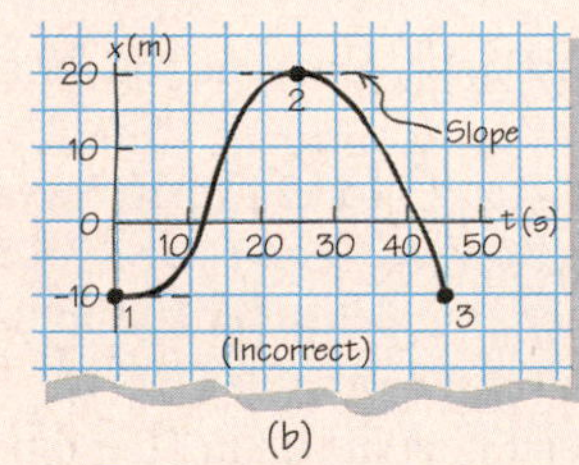

(b)

4. **Check Sketch:** The curve goes through all three known positions. However, the slope (remember the speed) at position 3 is not zero so we must fix that.

5. **Redraw Sketch:**

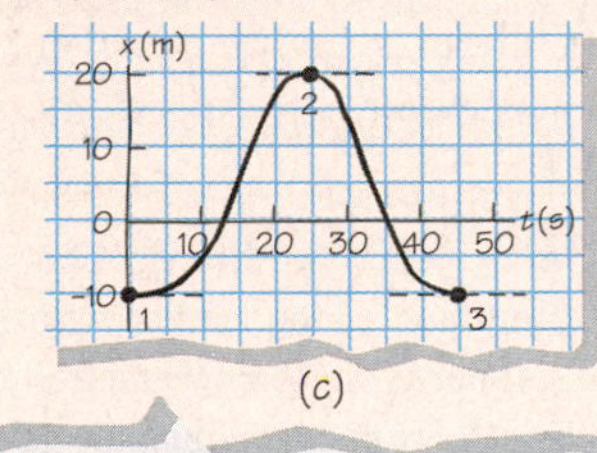

(c)

▶ **FIGURE 2B1–1** Some suggestions for illustrations.

Velocity

Although the notion of speed may be useful for sports, it is of limited utility in mechanics, where a different variable, velocity, plays an important role. Average speed refers to the total distance traveled; **velocity** refers to how fast the *displacement* changes. Velocity, like speed, is measured over a certain time interval. If a car has a displacement Δx in a particular time interval Δt, then the car's **average velocity**, v_{av}, over that time interval is defined (in one dimension) by

$$v_{av} \equiv \frac{\text{displacement}}{\text{time interval}} \tag{2–6}$$

$$= \frac{x_2 - x_1}{t_2 - t_1} \tag{2–7}$$

$$= \frac{\Delta x}{\Delta t}. \tag{2–8}$$

In one-dimensional motion, since the displacement Δx has a sign, *so does the velocity* v (or v_{av}). If v_{av} is positive, there is motion in the $+x$-direction, and if it is negative, there is motion in the $-x$-direction. The dimensions of velocity are $[LT^{-1}]$, with SI units of meters per second.

In our discussion of displacement, we pointed out that when we go from one-dimensional motion to two- or three-dimensional motion we will need to use the fact that the displacement is in fact a vector. In two- and three-dimensional motion, the vector nature of the velocity will likewise be crucial.

Warning: We caution the reader that it is *only* in this chapter—for one-dimensional motion—that we attach a sign to the symbol v to indicate whether the velocity is positive or negative. Everywhere else—for two- and three-dimensional motion—the symbol v will stand for the *magnitude* of the vector describing the velocity, and the magnitude is always positive. Similar remarks apply for displacement and acceleration.

Average Velocity: The average velocity $\Delta x/\Delta t$ provides us with only limited information of the details of the motion. Consider, for example, a car being driven along a straight road for 1 h (so that $\Delta t = 1$ h). If the displacement of the car in that time is $\Delta x = 30$ mi along the road, then the above expression will read

$$v_{av} = 30 \text{ mi/h}.$$

Since the displacement is in the positive direction, the average velocity is positive. Its magnitude only depends on the net displacement, so that any changes from the $+x$-direction to the $-x$-direction and any slowing down or speeding up are not reflected in that number. The car could have traveled at 60 mi/h to a point 45 mi down the road, then turned around and traveled 15 mi back to the finishing point at 60 mi/h. It could have traveled a steady 30 mi/h or it could have traveled the 30 mi at 120 mi/h, arriving in 15 min, and simply parked at the finish point until the hour was up. All that the definition in Eq. (2–8) has given us is the *average velocity* over the time interval in question. It is perfectly consistent to have a zero average velocity over a finite time interval even though a considerable distance may have been covered. On any round trip the net displacement is zero. Of course, an Olympic swimmer might resent being told that, whatever her efforts, her average velocity will be zero in a race that finishes at the same point as it starts. It is because of the limited information content of the average velocity that we turn to the concept of *instantaneous velocity.*

Instantaneous Velocity: The definition of the average velocity includes a time interval. We learn more about the details of the motion when smaller time intervals are used. For example, if we had asked about the average velocity during the 15-min intervals in the 1-h trip just discussed, we would already have a more detailed picture of the car's motion. Dividing the hour into sixty 1-min intervals and finding the average velocity during each minute would give us still more information.

It is possible to make the time interval Δt progressively smaller, finding the average velocity for each interval. Suppose that we look at the particular time interval from some time t to a time $t + \Delta t$. In this time interval, the displacement $\Delta x = x(t + \Delta t) - x(t)$ occurs. The average velocity during this interval is the ratio of Δx to Δt. Now, if Δt becomes very small, so does Δx, *but their ratio remains finite.* (For example, if a car travels at a constant velocity of 30 mi/h, you could learn that value by measuring the distance traveled over 1 h or one hundredth of a second—in each case the ratio displacement divided by the time interval is exactly the same.) We say that we are taking the *limit* as Δt approaches zero, symbolized by $\Delta t \to 0$. This limit refers to a particular time t and gives us the average velocity over a shorter and shorter time interval around that time. The **instantaneous velocity** at time t is the velocity of an object at that given instant of time, and it is defined as the limit of the average velocity as $\Delta t \to 0$:

$$v(t) = \lim_{\Delta t \to 0} \frac{x(t + \Delta t) - x(t)}{\Delta t} = \lim_{\Delta t \to 0} \frac{\Delta x}{\Delta t}. \tag{2–9}$$

When we use the term *velocity*, we mean instantaneous velocity, unless we state otherwise, and we refer to the instantaneous velocity without subscript in Eq. (2–9). The right-hand side of Eq. (2–9) is the *definition* of a derivative in calculus. Velocity is the derivative of displacement with respect to the time:

$$v(t) = \lim_{\Delta t \to 0} \frac{\Delta x}{\Delta t} = \frac{dx}{dt} \tag{2–10}$$

INSTANTANEOUS VELOCITY

The instantaneous speed v is the magnitude of the velocity:

$$v \equiv |\vec{v}|, \tag{2–11}$$

that is, the velocity stripped of any directional information. In one-dimensional motion, the speed is just the absolute value of the velocity. In more than one dimension, where the velocity is explicitly a vector, the speed is the length of that vector. We shall limit the use of the word *speed* in our discussion of mechanics to the definition Eq. (2–11).

Graphing the Motion: Figure 2–5 graphs the motion of an automobile as it moves along a straight road during a ride that begins at the time t_1 at point A and ends with the car parked at point Z at time t_2. The vertical axis shows the position, in kilometers, and the horizontal axis the time, in minutes. The curve allows us to study the average and instantaneous velocities of the car.

First of all we see that, before the start time t_1 and after the arrival time t_2, x does not change, so that $\Delta x = 0$ during these periods and therefore the velocities are zero. A flat line in the x-versus-t graph means zero derivative and hence zero velocity—the car is not moving. Next, we see that the total displacement A to Z is $+5.0$ km and the time of travel is 6.0 min, that is, 0.1 h. Thus the average velocity is $\Delta x/\Delta t = 5.0\ \text{km}/0.1\ \text{h} = 50\ \text{km/h}$. We can learn more detail about the motion by looking at shorter time intervals. For example, the displacement in the last minute of travel (from W to Z) is 1.15 km, and this allows us to calculate the average velocity in the last minute as

$$\frac{\Delta x}{\Delta t} = \frac{1.15\ \text{km}}{1.0\ \text{min}} = \frac{1.15\ \text{km}}{1.0\ \text{min}} \times \frac{60\ \text{min}}{1\ \text{h}} = 69\ \text{km/h}.$$

We note here an important point: The ratio $\Delta x/\Delta t$—the average velocity—is the *slope of the straight line that joins the initial and final points.*

As we shorten the time interval, we get closer to the instantaneous velocity. We can get close to the instantaneous velocity at 3.0 min (point C) by looking at the *short* time

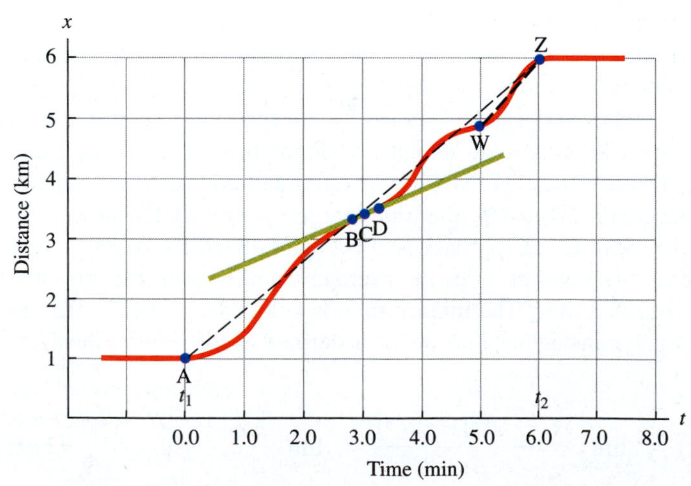

FIGURE 2–5 As two end points defining a time interval move closer together, the time interval moves toward zero, and the average velocity for that interval then becomes the slope of the curve of displacement versus time at that point. The slope of a line (in green) tangent to the curve (see points BCD) of displacement versus time gives the instantaneous velocity at a time corresponding to the tangent point. Here we see the line that determines the velocity at point C. At the end of the curve, the line tangent to the curve is flat (has zero slope), so that the instantaneous velocity there is zero. The automobile represented here has come to a stop.

interval of 24 s from 2 min 48 s (point *B*) to 3 min 12 s (point *D*). During that time the displacement is from 3.28 km to 3.48 km. Thus,

$$\frac{\Delta x}{\Delta t} = \frac{0.20\text{ km}}{24\text{ s}} = \frac{0.20\text{ km}}{24\text{ s}} \times \frac{60\text{ s}}{1\text{ min}} \times \frac{60\text{ min}}{1\text{ h}} = 30\text{ km/h}.$$

As the interval Δt becomes smaller, the line joining the two points and whose slope represents the average velocity between the two points becomes closer and closer to a tangent to the curve. In the limit of very tiny Δt, the slope of this line is the instantaneous velocity. As the slope of the tangent line changes from point to point on the curve, so does the instantaneous velocity change from moment to moment as you move along the curve. If you move a point on a transparent ruler along the curve so that the ruler's edge is always tangential to the curve, you can visualize the changes in instantaneous velocity during the trip.

All this is completely consistent with what we know from calculus: The derivative dx/dt at any given time t is the slope of the function $x(t)$ at that time, and the slope of the function $x(t)$ at time t is the slope of the tangent to the curve at that time. Thus *the slope of the tangent to our curve of x versus t at any time is the instantaneous velocity of the automobile at that time.*

Finally, let us write an explicit form for the position as a function of time when velocity has the constant value $v = v_0$. We know that for constant velocity the average velocity is the same over any interval, $v_{av} = v_0$. We could then, for example, solve for $x(t)$ in Eq. (2–7) written in the form $v_{av} = v_0 = [x(t) - x_0]/(t - 0)$. We find immediately

$$x(t) = v_0 t + x_0. \tag{2–12}$$

This should not be a surprising result. Starting at an initial point x_0, the position changes linearly with t, where v_0, the coefficient of t, is the slope of the x- versus t-curve, that is, the time derivative of x with respect to t.

EXAMPLE 2–3 We want to analyze a runner's motion around the 30.0-m mark in the 100-m dash using the data in Table 2–1. First calculate the average velocity of a runner over two time intervals $\Delta t = 3.91$ s and $\Delta t = 8.20$ s, with the time intervals centered around $x = 30.0$ m. Also calculate the average velocity over the smallest time interval available from the table near $x = 30.0$ m.

Setting It Up Calculation of an average velocity requires knowing displacements and the time intervals over which those displacements take place. This information is precisely what is given in Table 2–1.

TABLE 2–1 • Times for a 100-m Dash

Distance (m)	Time (s)	Distance (m)	Time (s)
0	0	55	6.37
5	1.36	60	6.83
10	2.01	65	7.28
15	2.57	70	7.74
20	3.09	75	8.20
25	3.60	80	8.65
30	4.09	85	9.11
35	4.55	90	9.57
40	5.01	95	10.04
45	5.47	100	10.50
50	5.92		

Strategy The average velocity can be calculated using Eqs. (2–6) through (2–8), with the sign of the average velocity determined by the sign of the displacement. This will require us to read off the displacements and their corresponding time intervals from the table.

Working It Out According to Table 2–1, the time points on either side of 30.0 m that give a time interval Δt of 3.91 s are $t_1 = 2.01$ s and $t_2 = 5.92$ s; for $\Delta t = 8.20$ s, the times are $t_1 = 0$ s and $t_2 = 8.20$ s. We obtain the distances that correspond to each of these time points from Table 2–1. The average velocities are then

$$\text{for } \Delta t = 3.91\text{ s: } v_{av} = \frac{x_2 - x_1}{t_2 - t_1} = \frac{50\text{ m} - 10\text{ m}}{5.92\text{ s} - 2.01\text{ s}} = 10\text{ m/s};$$

$$\text{for } \Delta t = 8.20\text{ s: } v_{av} = \frac{x_2 - x_1}{t_2 - t_1} = \frac{75\text{ m} - 0\text{ m}}{8.20\text{ s} - 0\text{ s}} = 9.2\text{ m/s}.$$

The smallest available time interval around 30.0 m in Table 2–1 is the 0.95-s interval from 25 m (3.60 s) to 35 m (4.55 s), and

$$\text{for } \Delta t = 0.95\text{ s: } v_{av} = \frac{x_2 - x_1}{t_2 - t_1} = \frac{35\text{ m} - 25\text{ m}}{4.55\text{ s} - 3.60\text{ s}} = 11\text{ m/s}.$$

What Do You Think? With the data in the table, we found the smallest time interval listed around a particular time (or place) and used it to calculate the average velocity for that interval. Could you use graphical techniques to find the velocity at a particular time, that is, the instantaneous velocity at that time? If so, how would you proceed? Is this more precise than using only the data in the table? *Answers to **What Do You Think?** questions are given in the back of the book.*

CONCEPTUAL EXAMPLE 2–4 Consider the one-dimensional motion described in parts (a) to (d) of Fig. 2–6. Match the graphs to the following descriptions: (1) a rock dropped off the roof of a building, (2) fooling around with a time machine, (3) a rock thrown vertically upward, and (4) a baton in a relay race being passed from one runner to the next.

Answer **1.** If x marks the height, the graph (b) describes a rock dropped off the roof of a building. You would start at a large value of x, which would decrease starting from the moment the rock is dropped, then stop, as the rock then stays at ground level.

2. You might suppose that a time machine (no, don't buy stock in the company selling them!) would take you back in time. You could start somewhere at some time, say, $t = t_0$, move, then somehow go back in time, move to a different position, and when the clock reached $t = t_0$ again, you would be in two different positions at the same time! This two-places-at-one-time property is visible in graph (d), as can be seen by following the vertical line that corresponds to a given time.
3. For vertical motion x again measures the height. Starting from when the rock is thrown, x increases, reaches a maximum at a later time, then decreases as the rock falls. That describes graph (a).
4. A straight line on a position–time plot represents constant velocity. Two runners will generally run at different speeds. Thus graph (c) represents a succession of two runners, the first slower than the second.

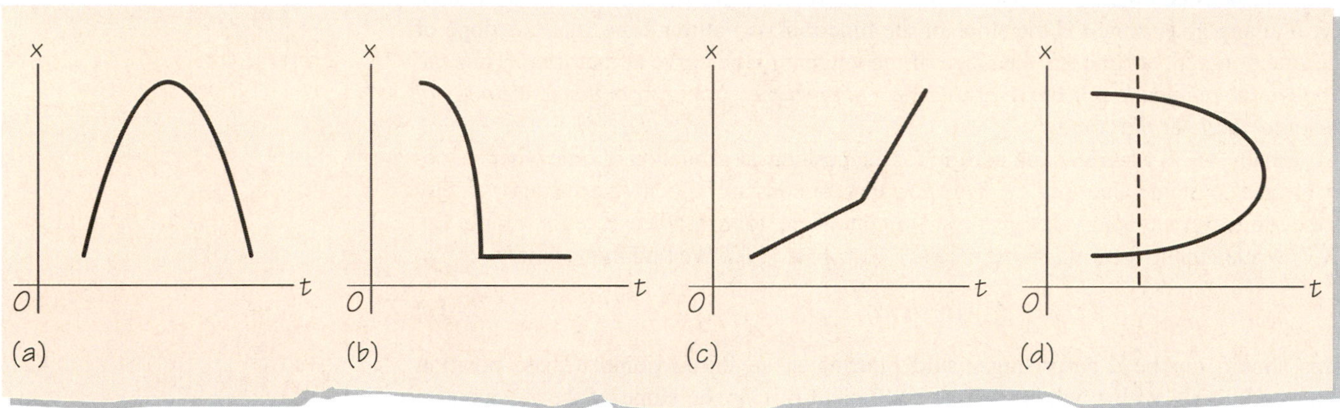

▲ **FIGURE 2–6** A series of curves on a graph of position versus time.

2–3 Acceleration

Just as velocity is defined as the rate of change of position, an object's acceleration is defined as the rate of change of its velocity with respect to time. Velocity can change because of a change in speed or because of a change in direction or both. An airplane taking off undergoes acceleration as its speed increases; an automobile going around a sharp curve undergoes acceleration because the direction of its motion changes. Since we are dealing with one-dimensional motion in this chapter, it is mainly changes in the magnitude of the velocity that count, along with what happens at "turn-around" places.

In our discussion of velocity, there is a particularly simple case: uniform, or constant, velocity, in which a position change is proportional to the time interval. There is also a particularly simple case for acceleration, in which the velocity change is proportional to the time interval; we refer to this case as uniform, or constant, acceleration.

Average Acceleration

The runner in the 100-m dash of Fig. 2–2 starts off with $v = 0$ m/s at $t = 0$ s. Two seconds later, he is moving with $v = 8$ m/s. His velocity has changed by $\Delta v = 8$ m/s over a time interval $\Delta t = 2$ s. We define the **average acceleration** over a time period, a_{av}, in terms of velocity v_1 at time t_1 and velocity v_2 at time t_2:

$$a_{av} \equiv \frac{v_2 - v_1}{t_2 - t_1} = \frac{\Delta v}{\Delta t}. \qquad (2\text{–}13)$$

In one-dimensional motion the velocity change may well be negative, even though both the initial and final velocities are positive—the change in velocity is the *difference* between two velocities. This would occur, for example, when the object is slowing down. More generally, the acceleration in one-dimensional motion can have a sign opposite to that of the velocity. The dimensions of acceleration are $[LT^{-2}]$, with SI units of meters per second squared.

EXAMPLE 2–5 A runner in the 100-m dash accelerates to 10 m/s at 4 s and maintains this velocity for the next 4 s. She then realizes that she is going to win and slows over the next 4.7 s to reach a velocity of 8 m/s at the end of the race. She has run the 100-m dash in 12.7 s. What is the runner's average acceleration over the time periods 0 to 4 s, 4 to 8 s, and 8 to 12.7 s?

Setting It Up We plot the velocity versus the time in Fig. 2–7 to match the description, assuming $t = 0$ at the start. The sections with linearly changing velocity correspond to the periods of uniform (constant) acceleration. All required velocities and times are given in the problem except for the additional magnitude $v = 0\ \text{m/s}$ at $t = 0$ s. The runner is moving to the right always, so that the velocities are always positive.

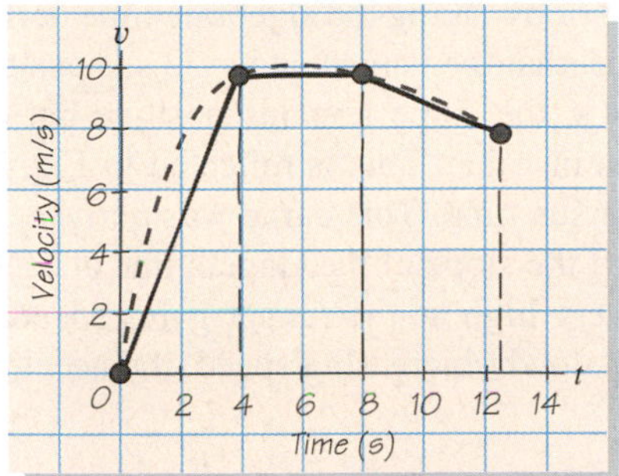

▲ **FIGURE 2–7** The dashed line could represent the actual velocity, which is not specified in the problem statement; the segmented straight lines are the means of finding the average acceleration.

Strategy Equation (2–13) is used to find the average acceleration in terms of the given times and velocities. We must be careful to keep track of signs.

Working It Out

$$\text{for 0 to 4 s: } a_{av} = \frac{v_2 - v_1}{t_2 - t_1} = \frac{10\ \text{m/s} - 0\ \text{m/s}}{4\ \text{s} - 0\ \text{s}} = 2.5\ \text{m/s}^2;$$

$$\text{for 4 to 8 s: } a_{av} = \frac{v_2 - v_1}{t_2 - t_1} = \frac{10\ \text{m/s} - 10\ \text{m/s}}{8\ \text{s} - 4\ \text{s}} = 0\ \text{m/s}^2;$$

$$\text{for 8 to 12.7 s: } a_{av} = \frac{v_2 - v_1}{t_2 - t_1} = \frac{8\ \text{m/s} - 10\ \text{m/s}}{12.7\ \text{s} - 8\ \text{s}} = -0.42\ \text{m/s}^2.$$

The greatest average acceleration occurs at the beginning of the race, when the runner is attempting to reach her greatest speed. Although she is running at her highest velocity during the middle part of the race, her average acceleration during this period is zero. During the time interval when she is slowing down at the end of the race, her average acceleration is negative.

What Do You Think? Consider the (one-dimensional) motion described in parts (a) to (d) of Fig. 2–8. Match the graphs to the following descriptions: (1) a somewhat idealized ping-pong match, (2) the baton being carried by one and then another relay runner, (3) a ball being thrown into the air, and (4) driving along a road with badly synchronized red lights. *Answers to **What Do You Think?** questions are given in the back of the book.*

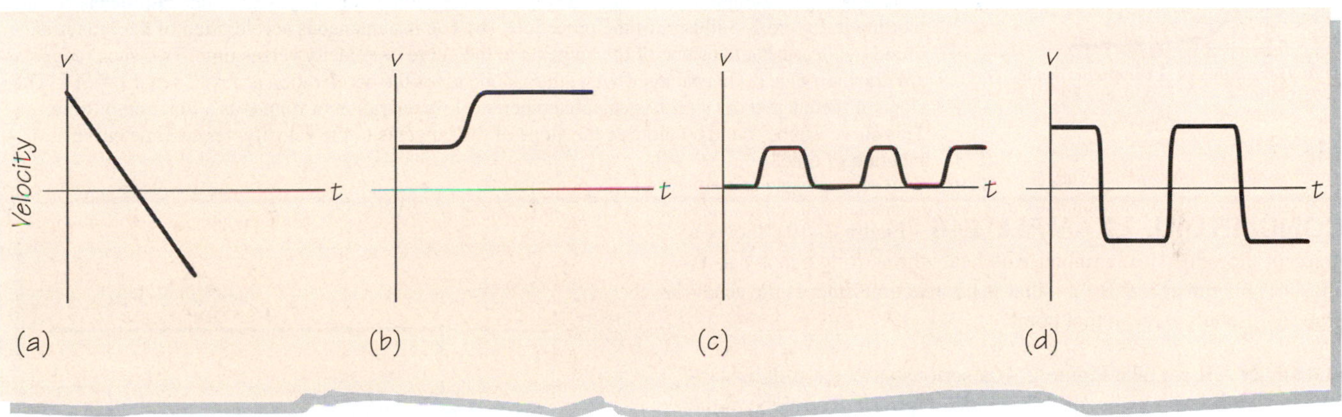

▲ **FIGURE 2–8** Series of curves on graph of velocity versus time.

In Example 2–5, the fact that the runner's average acceleration at the end of the race was negative means that the magnitude of the velocity is decreasing rather than increasing. You will sometimes see the term **deceleration** to describe situations in which the *magnitude* of the velocity decreases. For one-dimensional motion the sign of the acceleration determines its direction, just as it does for the velocity.

Instantaneous Acceleration

When we considered velocity, it was the instantaneous velocity that provided the most detailed and useful information about an object's motion, and for the same reasons we want to work with the **instantaneous acceleration**. In fact, when we use the term *acceleration*, we shall be referring to the instantaneous acceleration, unless otherwise

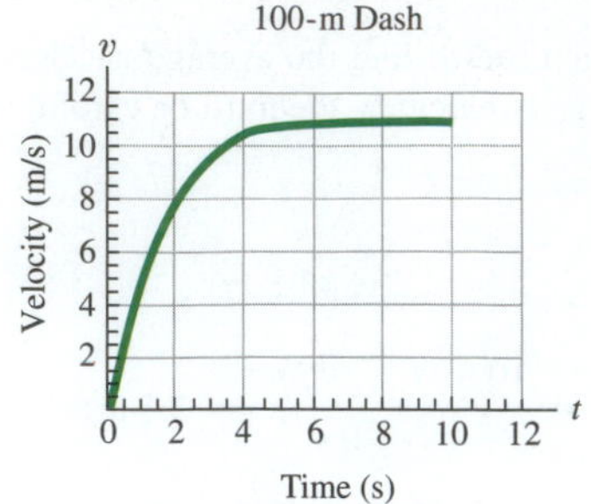

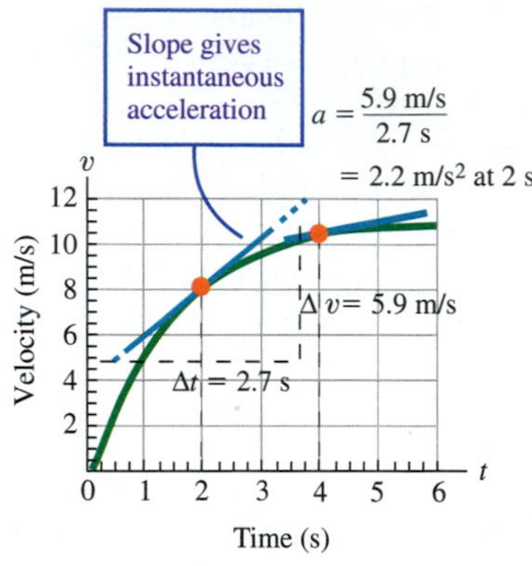

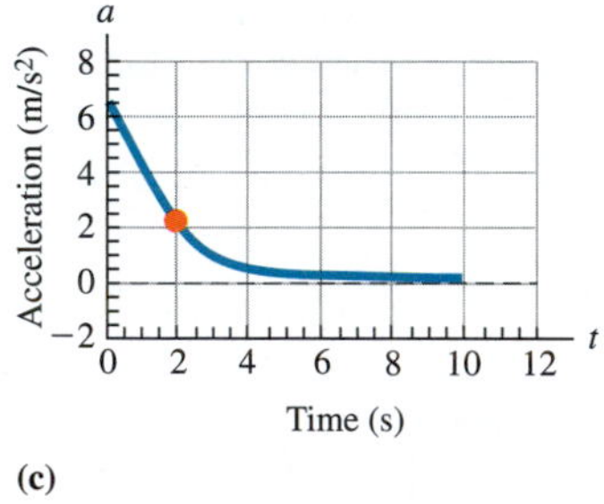

stated. We define the instantaneous acceleration as the limit of $\Delta v/\Delta t$ as the time interval Δt goes to zero. This is again a derivative:

$$a \equiv \lim_{\Delta t \to 0} \frac{\Delta v}{\Delta t} = \frac{dv}{dt}. \tag{2–14}$$

INSTANTANEOUS ACCELERATION

We have discovered that we can find the velocity of a runner from a plot of position versus time; analogously, we can determine the acceleration from a plot of velocity versus time. Consider again our 100-m-dash runner (Fig. 2–9a). We obtain the acceleration at any time t by finding the slope of the tangent to the curve of v as a function of t at the particular time t. In Fig. 2–9b, we show an enlarged view of the region around times $t = 2$ s and $t = 4$ s, with a drawing of the tangent to the curve of v versus t at $t = 2.0$ s. We can determine the acceleration of our runner either by finding the slopes of tangents to the velocity curve or by taking the algebraic time derivative of the function $v(t)$. You can see, for example, that the tangent at $t = 2$ s is positive (positive acceleration), and as it goes to $t = 4$ s, the slope remains positive but decreases in value. Thus the acceleration decreases in value. This is reflected in Fig. 2–9c, which plots the acceleration of the runner versus time. This curve was derived from the velocity curve of Fig. 2–9b by reading off the slope of the tangent line at each point. Notice that the acceleration is initially very high as the runner gains speed, but by 4 s, when the runner moves at only a very slowly increasing speed, the acceleration drops nearly to zero.

◀ **FIGURE 2–9** (a) Velocity of runner whose position is shown in Fig. 2–2. This curve can be determined, for example, by finding the slope at each point along the position-versus-time curve and plotting it. Figure 2–5 illustrates the procedure. (b) The instantaneous acceleration of the runner is found by measuring the slope of the tangents to the curve of velocity versus time. Two such tangents are drawn in blue in this enlarged view; their slope gives the acceleration at $t = 2$ s and $t = 4$ s. The slope of the tangent at $t = 2$ s is calculated here. (c) Acceleration of runner as a function of time. This curve can be found by plotting the slope of the tangents to the velocity-versus-time curve as a function of time.

CONCEPTUAL EXAMPLE 2–6

Figure 2–10a shows a graph of the velocity of a runner who hears the starting gun at $t = 0$. He starts his run at $t = 0.1$ s—that is his reaction time. Is the acceleration infinitely large at that point?

Answer If we take Figure 2–10a seriously, with a truly instantaneous jump at $t = 0.1$ s, then it is indeed infinite. The slope of the v-versus-t curve is perfectly vertical, and that means an infinite slope. This idealized situation corresponds to the runner making the transition from at rest to full speed instantaneously. Of course, if we were to look carefully at the velocity as a function of time near 0.1 s, it would look more like Fig. 2–10b, which shows a more gradual rise to full speed, with a slope that is never infinite. A good runner's acceleration can be large, but it is limited.

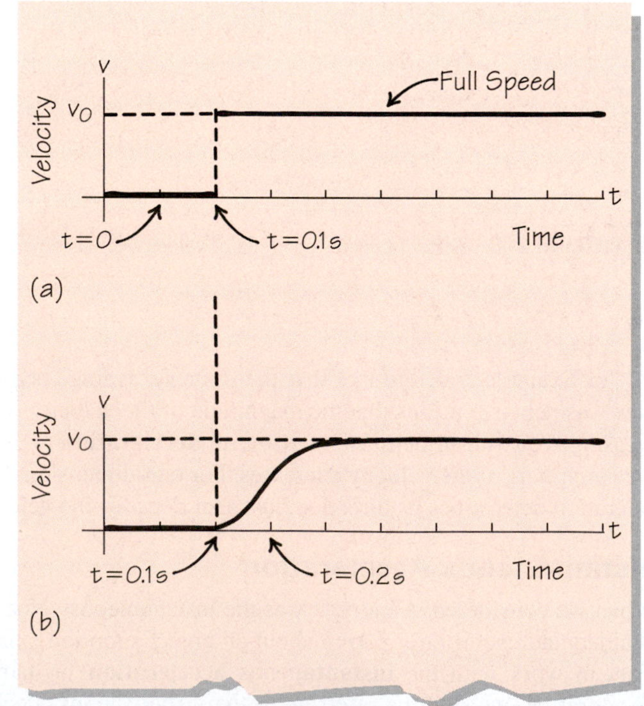

▶ **FIGURE 2–10** In (a) we see an idealized version of the acceleration that in reality looks more like that plotted in (b).

By using our discussion about how velocity depends on position and how acceleration depends on velocity, we can make a link here from acceleration to position. If we examine the derivatives presented in Eqs. (2–10) and (2–14), we find that

$$a = \frac{dv}{dt} = \frac{d}{dt}\left(\frac{dx}{dt}\right) = \frac{d^2x}{dt^2}. \tag{2–15}$$

The acceleration is the time derivative of the velocity; equivalently, the acceleration is *the second time derivative of the displacement.*

THINK ABOUT THIS . . .

WHAT IS AN ACCELEROMETER AND WHAT ARE ITS USES?

An accelerometer is a device that measures acceleration; it does so in its simplest form by using a spring to translate acceleration into a measurement of the position of a mass. Figure 2–11 shows the basic arrangement for a one-dimensional motion. As we shall see in Chapter 5, a spring has the characteristic that it will stretch (or compress if a is negative) by an amount $x_f = (\text{const})a$, where the constant is characteristic of the mass and the particular spring. The length x_f is measurable and can be used to read the acceleration.

Accelerometers have two basic uses. First, they can be used as navigation devices because the measurement of acceleration can be turned into a measurement of position, as we shall see below. Second, they can be used to test vibrations and shock, which are characterized by accelerations within the test system. This would include crash testing for passenger vehicles (cars and planes) as well as monitoring how buildings and other structures respond to events such as earthquakes or explosions. Air bags are triggered by accelerometers.

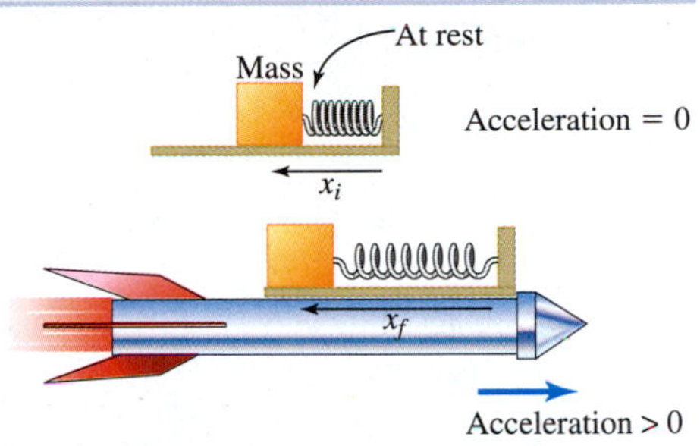

▲ **FIGURE 2–11** Schematic view of an accelerometer for one-dimensional motion. The amount by which the mass moves is a measure of the acceleration.

EXAMPLE 2–7 The position x of an experimental rocket moving along a long rail is measured to be $x(t) = (5\text{ m/s})t + (8\text{ m/s}^2)t^2 + (4\text{ m/s}^3)t^3 - (0.25\text{ m/s}^4)t^4$ over the first 10 s of its motion, where t is in seconds and x is in meters. Find the velocity and acceleration of the rocket over the first 10 s and display the results graphically.

Setting It Up A plot of the position x as a function of time is shown in Fig. 2–12a.

Strategy The velocity and acceleration of the rocket can be determined by taking the time derivatives in Eqs. (2–10) and (2–15), respectively. This is a straightforward exercise in calculus. In particular, we use the rule $(d/dt)At^n = nAt^{n-1}$.

Working It Out We have

$$v = \frac{dx}{dt} = (5\text{ m/s}) + (16\text{ m/s}^2)t + (12\text{ m/s}^3)t^2 - (1\text{ m/s}^4)t^3;$$

$$a = \frac{dv}{dt} = (16\text{ m/s}^2) + (24\text{ m/s}^3)t - (3\text{ m/s}^4)t^2.$$

Velocity and acceleration are measured in meters per second and meters per second squared, respectively, when time is measured in seconds. We plot these results over the time period 0 to 10 s in Figs. 2–12b and c, respectively. Note that although the position x is zero at time $t = 0$, neither the velocity nor the acceleration is zero at this time.

What Do You Think? The statement of this example specifies that $x(t)$ describes the motion over the first 10 s of motion. Suppose you wanted to use this formula to describe the motion for times beyond 10 s. What's to stop you from doing this? [*Hint*: In a real situation will the rocket keep accelerating? What are the consequences for its motion if it doesn't continue to accelerate?] *Answers to* ***What Do You Think?*** *questions are given in the back of the book.*

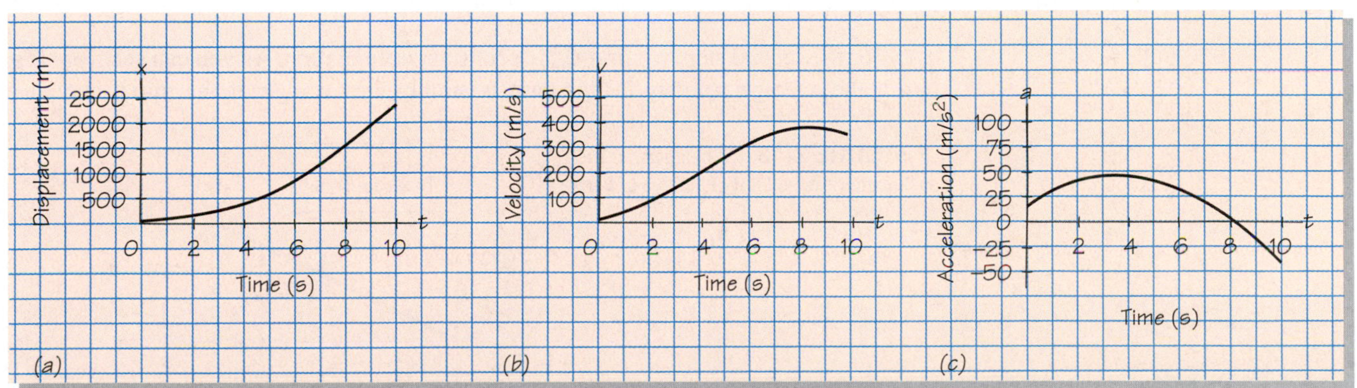

▲ **FIGURE 2–12** We have plotted the displacement, the instantaneous velocity, and the instantaneous acceleration as functions of time.

If we wished, we could continue and look at the rate of change of acceleration, da/dt, or even higher derivatives. But we don't need to do this because it is acceleration, not its changes, that plays a primary role in the laws of motion (see Chapter 4).

2–4 Motion with Constant Acceleration

The simplest example of acceleration is constant acceleration. It is a physically important case because the motion of freely falling objects near Earth's surface is motion with constant acceleration.

A constant acceleration implies that the value of the instantaneous acceleration equals the value of the average acceleration:

$$\text{for constant acceleration: } a_{av} = \frac{v(t_2) - v(t_1)}{t_2 - t_1} = a, \tag{2–16}$$

where a is the constant acceleration and t_2 and t_1 are any values of time that we take for a velocity measurement. For convenience, we'll give the last time t_2 the more general label t and choose the first time t_1 to be 0. We also relabel $v(t_1) = v(0) = v_0$, and with this relabeling, we can rearrange the above equation to read

$$v = at + v_0. \tag{2–17}$$

VELOCITY IN ONE DIMENSION UNDER CONSTANT ACCELERATION

The velocity's *linear* dependence on time is characteristic of constant acceleration (see also Section 2–6). The sign of the change in the velocity, whether it is increasing or decreasing, is the sign of a.

Let us now turn to the question of how the position changes with time under constant acceleration. We have already seen that when the velocity is constant the displacement changes linearly with time. Here, the velocity changes linearly with time; if velocity is increasing, then we would expect the displacement to increase even more rapidly with time. We analyze this in two steps: First we establish that the formula $x = v_0 t + x_0$ that applies for constant velocity [Eq. (2–12)] also applies for nonconstant velocity if v_0 is replaced by v_{av}, that is, the speed averaged over the time interval from 0 to t. Second, we find v_{av} for constant acceleration and insert it into the formula for the position.

For the first step, we write Eq. (2–7) as

$$v_{av} = \frac{x(t) - x(0)}{t - 0} = \frac{x(t) - x_0}{t}.$$

This equation is solved for x to yield

$$x = v_{av} t + x_0. \tag{2–18}$$

For the second step, we find v_{av} for constant acceleration. As you can see from Fig. 2–13, the average velocity in this case is simply the average of the initial and final

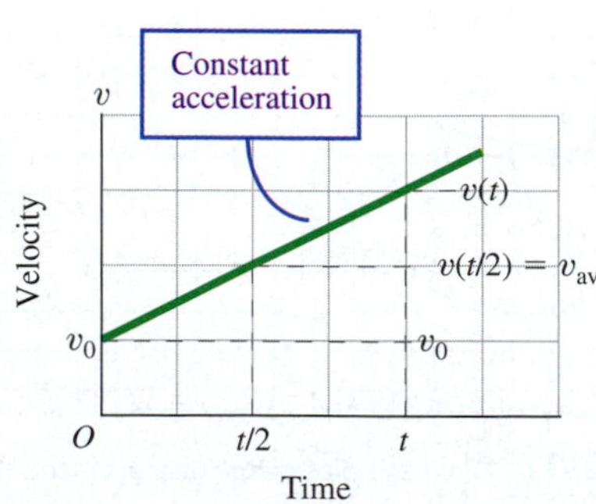

FIGURE 2–13 When acceleration is constant, the velocity changes linearly with time.

velocities over the total time period t. The initial velocity is v_0 and the final velocity at time t is $v = v(t)$, giving

$$v_{av} = \tfrac{1}{2}(v_0 + v). \qquad (2\text{–}19)$$

AVERAGE VELOCITY IN ONE DIMENSION UNDER CONSTANT ACCELERATION

For this case, v_{av} is just the velocity at time $t/2$. When we substitute this result for v_{av} into Eq. (2–18), we find that for constant acceleration

$$x = \tfrac{1}{2}(v_0 + v)t + x_0 = \tfrac{1}{2}vt + \tfrac{1}{2}v_0 t + x_0. \qquad (2\text{–}20)$$

We now substitute v from Eq. (2–17) into this equation and determine the position as a function of time:

$$x = \tfrac{1}{2}at^2 + v_0 t + x_0. \qquad (2\text{–}21)$$

POSITION IN ONE DIMENSION UNDER CONSTANT ACCELERATION

Part of this result, $v_0 t + x_0$, describes the position of the object if it moved with a constant velocity v_0 starting from the position x_0. The additional quadratic term in t is due to the constant acceleration a. As we had expected, the position varies more rapidly than linearly with time.

We should discuss one more useful relation. The labeled equations (2–17) and (2–21) above, which contain the variable time, allow us to deduce a relationship between displacement, speed, and acceleration that does not involve time. To do so, we first solve for the time t from Eq. (2–17):

$$\text{for constant acceleration: } t = \frac{v - v_0}{a}. \qquad (2\text{–}22)$$

If we substitute this expression for time into the first part of Eq. (2–20), we have

$$x = \left(\frac{v_0 + v}{2}\right)\left(\frac{v - v_0}{a}\right) + x_0$$

for constant acceleration, or

$$x - x_0 = \frac{v^2 - v_0^2}{2a}. \qquad (2\text{–}23)$$

This equation can be written as

$$v^2 = v_0^2 + 2a(x - x_0). \qquad (2\text{–}24)$$

SPEED AS A FUNCTION OF POSITION UNDER CONSTANT ACCELERATION

This result gives us the speed at any position x in terms of the constant acceleration a and the initial speed and position. Time does not enter into this result. Note the use of the word *speed* here, as only the magnitude of the velocity appears.

The labeled equations of this section are particularly useful, but you need not memorize them. You can deduce all the information in these equations from the simple statement that the acceleration is constant; alternatively, simply remember that the position changes quadratically with time and has a linear term in t as well [Eq. (2–21)]. The series of examples that follow illustrates some of the uses of these results.

EXAMPLE 2–8 An amateur bowler releases a ball with an initial velocity of 3.0 m/s; the ball slows down with a constant negative acceleration of $-0.20\ \text{m/s}^2$. How far does the ball roll before stopping, and how long does it take to stop?

Setting It Up Choose a coordinate system with $x = 0$ m at the point where the ball leaves the bowler's hand (Fig. 2–14a). The x-axis is along the direction of the ball's motion. In Fig. 2–14b, we have sketched the constant acceleration as well as the linearly decreasing velocity and the quadratically varying position that correspond to the acceleration. We start the clock at $t = 0$ s when the ball leaves the bowler's hand. The initial conditions are then $x_0 = 0$ m, $t_0 = 0$ s, and $v_0 = 3.0$ m/s. The constant acceleration a_0 is negative in this coordinate system.

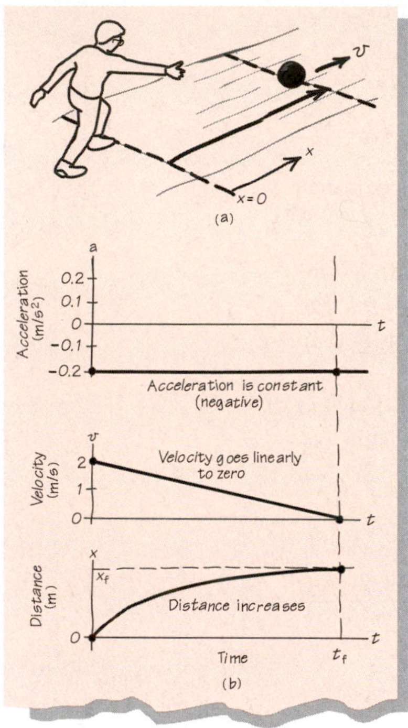

▲ **FIGURE 2–14** (a) The origin O marks the spot $(x = 0\ \text{m})$ where the ball leaves the bowler's hand. (b) Sketches of acceleration, velocity, and displacement as functions of time.

Strategy The crucial point is that the acceleration is constant here, so we can use our derived results. The first part of the question asks us to find the connection between speed and position, for which we can use Eq. (2–24) (an equation for which all is known except x, which is the displacement of the ball). Thus we can solve Eq. (2–24) for the x-value for which the final velocity v has become zero. For the second part, we can use the connection between velocity and time [Eq. (2–22)] to find the time when the velocity is zero.

Working It Out From Eq. (2–24), the solution for x is

$$x = x_0 + \frac{v^2 - v_0^2}{2a}.$$

All the quantities on the right-hand side are known, giving the numerical value

$$x = (0\ \text{m}) + \frac{(0\ \text{m/s})^2 - (3.0\ \text{m/s})^2}{2(-0.20\ \text{m/s}^2)} = 22.5\ \text{m}.$$

Use Eq. (2–22) to determine the time of motion,

$$t = \frac{(0\ \text{m/s}) - (3.0\ \text{m/s})}{-0.20\ \text{m/s}^2} = 15\ \text{s}.$$

Any checks you can find for your answer are helpful. In particular, here use Eq. (2–21) with $t = 15$ s to determine the displacement once more:

$$x = \frac{(-0.20\ \text{m/s}^2)(15\ \text{s})^2}{2} + (3.0\ \text{m/s})(15\ \text{s}) + (0\ \text{m}) = 22.5\ \text{m},$$

which confirms the earlier result. The equations for constant acceleration are consistent with one another.

What Do You Think? A bowling alley is usually polished very smoothly, as is a bowling ball. Why does a ball slow down after it leaves the bowler's hand? *Answers to* ***What Do You Think?*** *questions are given in the back of the book.*

EXAMPLE 2–9 A runner bursts out of the starting block 0.10 s after the gun signals the start of a race. She runs at constant acceleration for the next 1.9 s of the race. If she has gone 8.0 m after 2.0 s, what are her acceleration and velocity at this time?

Setting It Up We draw a coordinate system in Fig. 2–15, which also contains a sketch of the acceleration as a function of time. In this case the only thing we know about the acceleration is that it is constant and positive—we don't know its value a. We do, however, know that $x(t_f) = 8.0$ m, where t_f is the final time, 2.0 s. We also know the initial position x_0 and the initial velocity v_0.

Strategy The runner is not moving during the first 0.10 s of the race, and thus her acceleration is zero during this period of time. During the next 1.9 s, she has an acceleration that is not zero. We can use the results of this section *only if the acceleration is constant* during the entire time, so a correct strategy is to break up the motion into constant-acceleration pieces. In the first piece ($t = 0$ s to $t = 0.10$ s, nothing at all happens. For the second piece, consider a time

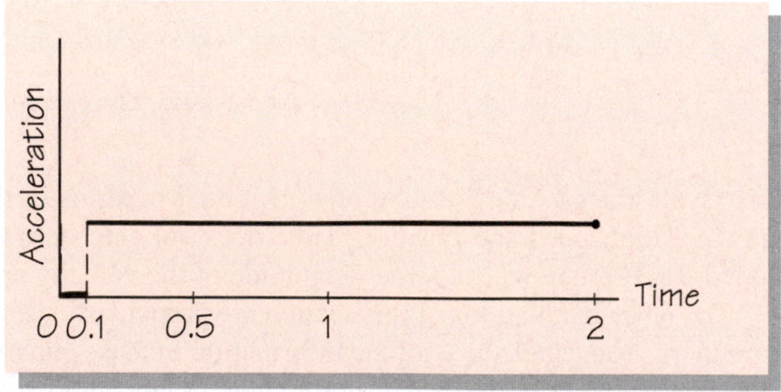

▲ **FIGURE 2–15** The acceleration of the runner is zero in the first one-tenth second and a positive constant thereafter.

$t' = t - 0.10$ s. (This is equivalent to starting your clock at 0.1 s rather than at 0 s.) For times t of 0.10 and 2.0 s, t' is 0 and 1.9 s, respectively. Over the time period $t' = 0$ s to $t' = 1.9$ s, we can use the results of this section, including that the initial time t_0' is 0. The initial conditions of the problem are therefore $t_0' = 0$ s, $x_0 = 0$ m, and $v_0 = 0$ m/s, and we want the acceleration and velocity at $t_f' = t_f - 0.1\text{ s} = 1.9$ s and $x = 8.0$ m. An examination of the labeled equations shows that there are two unknowns (v and a) in Eqs. (2–17) and (2–24). Equation (2–19) does not allow us to determine either v or a. Equation (2–21), however, allows us to determine the acceleration. Knowing the acceleration, we can then use Eq. (2–17) to determine the velocity at any time, in particular for the final time.

Working It Out We insert the known values (with the primed values of time) into Eq. (2–21) and solve for the acceleration:

$$8.0\text{ m} = \frac{a(1.9\text{ s})^2}{2} + (0\text{ m/s})(0\text{ s}) + (0\text{ m}) = 10\text{ m};$$

$$a = \frac{16\text{ m}}{3.6\text{ s}^2} = 4.4\text{ m/s}^2.$$

Now at $t' = 1.9$ s ($t = 2.0$ s) after the runner starts,

$$v = at' + v_0 = (4.4\text{ m/s}^2)(1.9\text{ s}) + 0\text{ m/s} = 8.4\text{ m/s}.$$

What happens if the initial time is not $t = 0$? There is nothing very special about the time $t = 0$ in motion under constant acceleration; in fact, Example 2–9 illustrated the difficulty with Eqs. (2–17) through (2–24) if the initial time must always be zero. That is why it is useful to allow t_0 to be an arbitrary time in the equations relating acceleration, velocity, and displacement. In this case, the important equations of this section are as follows:

$$v = a(t - t_0) + v_0, \qquad \text{(2–25a)}$$

FOR CONSTANT ACCELERATION a:

$$v_{av} = \tfrac{1}{2}(v_0 + v) \quad \text{(unchanged)}, \qquad \text{(2–25b)}$$

$$x = \tfrac{1}{2}a(t - t_0)^2 + v_0(t - t_0) + x_0, \qquad \text{(2–25c)}$$

$$v^2 = v_0^2 + 2a(x - x_0) \quad \text{(unchanged)}. \qquad \text{(2–25d)}$$

The only difference between the four labeled equations (2–17), (2–19), (2–21), and (2–24) and Eqs. (2–25a) to (2–25d) is that t is replaced by $t - t_0$ in the latter equations. Remember, Eqs. (2–25a) to (2–25d) are valid only when the acceleration is constant between the times t_0 and t. When $t_0 = 0$ in Eqs. (2–25a) to (2–25d), we obtain the four labeled equations. By letting $t_0 = 0.10$ s and $t = 2.0$ s, Example 2–9 can now be worked more easily, as the acceleration is constant during this time interval.

EXAMPLE 2–10 A T-38 training jet (Fig. 2–16a) has an acceleration of 3.6 m/s^2 that lasts 5.0 s during the initial phase of takeoff. The afterburner engines are then turned up to full power for an acceleration of 5.1 m/s^2. The speed needed for takeoff is 164 knots (1 m/s = 1.94 knots). Calculate the length of runway needed and the total time of takeoff.

Setting It Up We draw the acceleration as a function of time in Fig. 2–16b, labeling the two constant values of acceleration that enter, a_1 (3.6 m/s^2) and a_2 (5.1 m/s^2). We label the point where the acceleration changes, $t_1 = 5.0$ s.

Strategy There are two different constant accelerations in this example; therefore, we need to divide the problem into two parts. For the period from $t = 0$ to t_1, we have the values $x_0 = 0$ m and $v_0 = 0$ m/s at $t_0 = 0$ s. We then use Eqs. (2–25a) and (2–25d) to find the velocity and distance, respectively, at t_1. We then repeat the exercise for the second period of acceleration, with initial values of position and time coming from the final values of the first period. For this period we first find the time for which the velocity takes a final value $v_f = 164$ knots. We then use this value of time to find the distance traveled in the second period and add it to the distance traveled in the first period to obtain the runway length.

Working It Out For the first period, Eq. (2–25a) gives the velocity at 5.0 s:

$$\text{for } t = 5.0\text{ s: } v = (3.6\text{ m/s}^2)(5.0\text{ s} - 0\text{ s}) + (0\text{ m/s}) = 18\text{ m/s}.$$

Next we use Eq. (2–25c) to determine the distance the jet has traveled:

$$\text{for } t = 5.0\text{ s:}$$
$$x = \tfrac{1}{2}(3.6\text{ m/s}^2)(5.0\text{ s} - 0\text{ s})^2 + (0\text{ m/s})(5.0\text{ s} - 0\text{ s}) + (0\text{ m}) = 45\text{ m}.$$

We now move to the second phase of takeoff, where full power is applied. We have a new set of initial conditions beginning with the time $t_0 = 5.0$ s; namely, $x_0 = 45$ m and $v_0 = 18$ m/s. We want to find the time and distance corresponding to a final velocity of 164 knots. Let us first convert this value into SI units. The final velocity of 164 knots is

$$(164\text{ knots})\frac{1\text{ m/s}}{1.94\text{ knots}} = 84.4\text{ m/s}.$$

(continues on next page)

(a)

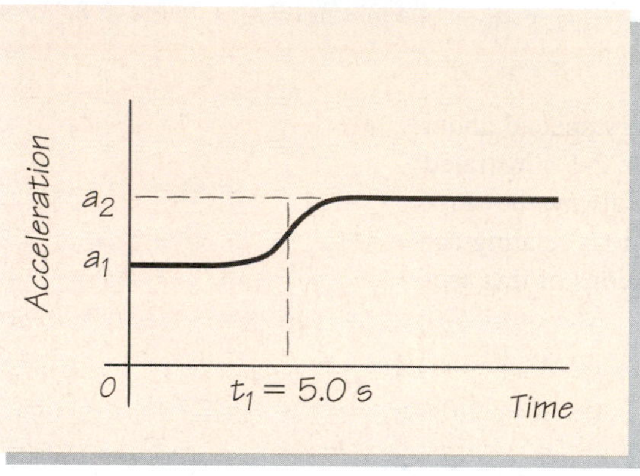

(b)

▲ **FIGURE 2–16** (a) A T-38 jet accelerating during takeoff. In part (b) we have made a graph of acceleration versus time.

We use Eq. (2–25a) to find the time t at takeoff; the acceleration is now 5.1 m/s^2. From Eq. (2–25a) we have

$$a(t - t_0) = v - v_0,$$

$$t = \frac{v - v_0}{a} + t_0 = \frac{84.4\text{ m/s} - 18\text{ m/s}}{5.1\text{ m/s}^2} + 5.0\text{ s} = 18\text{ s}.$$

Note that we waited until we had solved for the variable t before inserting the numerical values (with units) for v, v_0, a, and t_0. This technique also serves as a check as the cancellation of units gives the expected result, seconds in this case.

Equation (2–25c) can be used directly to determine the takeoff distance because all the variables for the second phase, except x, are now known:

$$x = \frac{(5.1\text{ m/s}^2)(18\text{ s} - 5.0\text{ s})^2}{2} + (18\text{ m/s})(18\text{ m/s} - 5.0\text{ s})$$
$$+ 45\text{ m} = 7.1 \times 10^2\text{ m}$$

Because we have included $x_0 = 45$ m from the first phase, 710 m (or 2330 ft) is the total amount of runway used.

What Do You Think? Figure 2–17 shows a sequence of constant accelerations undergone by a body in one-dimensional motion. Ignoring the sharp points of rapid change in the acceleration, sketch the velocity of a particle that starts at $t = 0$ with velocity $v = 0$. *Answers to **What Do You Think?** questions are given in the back of the book.*

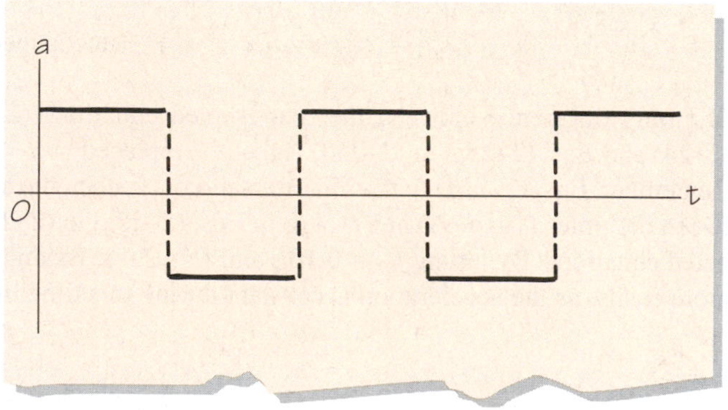

▶ **FIGURE 2–17**

2–5 Freely Falling Objects

In Section 2–4, we mentioned an important example of constant acceleration: gravity. The **acceleration due to gravity** is given the symbol $\vec{g}$, and its magnitude is approximately 9.80 m/s^2.[†] Ignoring the effects of air resistance, any object dropped in the vicinity of Earth's surface will move with constant acceleration $\vec{g}$. We call this motion **free fall**. The direction of $\vec{g}$ is down, toward Earth's center, a direction easily found using a plumb bob (a string with a mass at its end).

Galileo Galilei, who can be considered the first modern physicist, systematically investigated the motions of falling objects. Centuries earlier, Aristotle had suggested (incorrectly) that the speed of a falling object depends on the weight of the object and that this speed is proportional to the distance fallen. Galileo questioned this wisdom; his approach was to test by experiment, and he performed precise measurements whose results he described in mathematical language. He determined that the distance that objects fall

† There are variations of the order of 1 percent in the magnitude and direction of $\vec{g}$ over Earth's surface.

after starting from rest is proportional to the square of the time; equivalently, the speed of a falling object is proportional to the square root of the distance fallen. As Eqs. (2–21) and (2–24) show, this type of motion is characteristic of constant acceleration.

If we drop a hard rubber ball and a sheet of paper simultaneously from the same height, we observe that the paper floats down and the ball reaches the floor first (Figure 2–18a). The ball experiences a greater acceleration and a larger final velocity than does the sheet of paper. This is because air resistance affects the sheet of paper to a much greater degree than it affects the ball. If we wad the paper up and repeat the experiment (Fig. 2–18b), the effect of air resistance on the paper is decreased and the falling times for the paper and the ball in this experiment are more nearly equal. If the same experiment is done in a vacuum, the falling times are, as best as we can measure, the same for all objects. We shall assume for now that we are dealing with small, heavy objects ("particles") with negligible air resistance, although we must remember that in many situations air resistance can be an important effect. A falling particle near Earth's surface undergoes a constant acceleration that is the same for *all* particles, independent of their composition or their mass. We should emphasize that this fact applies to all free-fall motions, including motion that is initially *upward*. For such objects a constant acceleration downward will eventually produce a downward velocity—an initially upward velocity is only a question of the initial value. A tennis ball that is tossed up moves with the same constant acceleration downward as a dropped rock.

To study the effects of gravity quantitatively, it is easiest to set up a coordinate system with a direction perpendicular to Earth's surface. Let us align the y-axis with the vertical direction. We have two choices for this direction. If we choose the positive direction toward the center of Earth, then the acceleration of gravity is $+g$. If we choose the positive direction of the y-axis away from Earth's center, then the acceleration of gravity is $-g$. We could work problems by using either choice of axes, but here we choose the latter because it will facilitate our future discussion of motion in two and three dimensions. The vertical motion of any freely moving object for which air resistance can be ignored is then summarized by the four labeled equations of Section 2–4 for constant acceleration with $a = -g$. Thus for a freely falling body,

$$v = -gt + v_0, \qquad (2\text{–}26a)$$

$$v_{av} = \tfrac{1}{2}(v_0 + v) \quad \text{(unchanged)}, \qquad (2\text{–}26b)$$

$$y = \tfrac{1}{2}(-gt^2) + v_0 t + y_0, \qquad (2\text{–}26c)$$

$$v^2 = v_0^2 - 2g(y - y_0). \qquad (2\text{–}26d)$$

Similarly, g can be inserted into Eqs. (2–25a) to (2–25d). Remember that it is only the near constancy of the acceleration of gravity near Earth's surface that makes these equations applicable. Figure 2–19 represents the up-and-down motion of an object with the corresponding velocity and acceleration vectors.

(a)

(b)

▲ **FIGURE 2–18** Photos taken under a flashing strobe light allow us to see the position of a moving object at equal time intervals. (a) The rubber ball falls directly to the floor, whereas the sheet of paper floats down more slowly. (b) After the sheet of paper is wadded up and the experiment is repeated, the ball and the paper fall almost together.

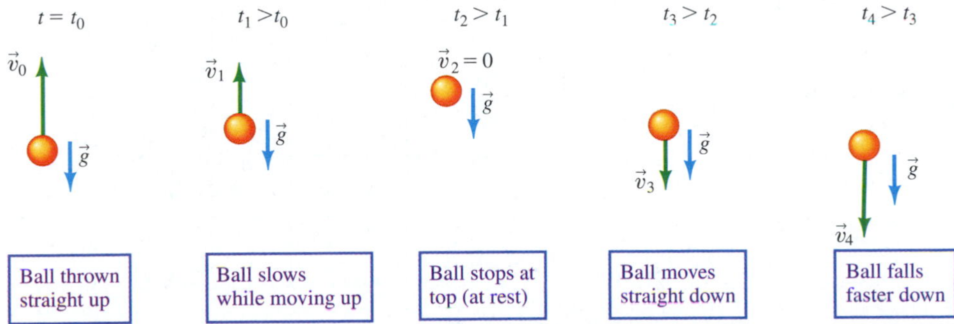

▲ **FIGURE 2–19** Details of the motion of a ball tossed in the air. It slows, stops, and falls back. Although position and velocity change with time, the acceleration is constant.

EXAMPLE 2–11 How much time elapses before a ball hits the ground after it has been dropped from rest from a height of 100 m (Fig. 2–20)? What is the ball's velocity just before it hits the ground?

Setting It Up In Fig. 2–20 we indicate that the ground is at a level $y = 0$ m, whereas $y = 100$ m is the position from which the ball is dropped. The given initial conditions at $t_0 = 0$ s are $y_0 = 100$ m and $v_0 = 0$ m/s.

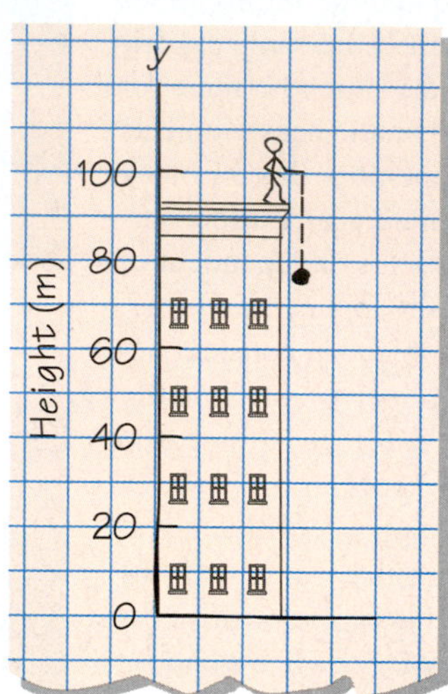

▶ **FIGURE 2–20** The ball is released at 100 m from the ground at $t = 0$ s.

Strategy To find the time of flight t knowing the displacement and the initial conditions, we use Eq. (2–26c). Knowing this time, we can then use Eq. (2–26a) to calculate the velocity.

Working It Out From Eq. (2–26c) we have

$$0 \text{ m} = \tfrac{1}{2}(-9.80 \text{ m/s}^2)t^2 + (0 \text{ m/s})t + 100 \text{ m};$$

$$t^2 = -\frac{200 \text{ m}}{-9.80 \text{ m/s}^2} = 20.4 \text{ s}^2, \qquad t = 4.52 \text{ s}.$$

Then Eq. (2–26c) gives the velocity:

$$v = -(9.80 \text{ m/s}^2)(4.52 \text{ s}) + 0 \text{ m/s} = -(44.3 \text{ m/s}).$$

Notice that the final velocity has a large magnitude and is *negative;* that is, it is in a downward direction. The negative sign appears because we chose the direction of positive displacement (along the y-axis) to be up.

What Do You Think? Rather than simply being dropped, the ball is tossed upward. Immediately after leaving the hand, the ball's acceleration is (a) upward, at least until reaching the top of its trajectory; (b) downward with magnitude g; or (c) zero, for a brief moment before gravity starts to act. *Answers to **What Do You Think?** questions are given in the back of the book.*

In Example 2–11 we calculated the velocity of an object falling freely from rest. Assume that an object starts at $y_0 = h$ with $v_0 = 0$ at $t_0 = 0$. What is the magnitude of the velocity when the object hits the ground $(y = 0)$? We use Eq. (2–26d) to determine

$$v^2 = 0 - 2g(0 - h) = 2gh;$$

$$\text{for constant acceleration:} \quad v = \sqrt{2gh}. \tag{2–27}$$

The positive sign of the square root is appropriate because it refers to the *magnitude* of the velocity. If the y-axis is up, the velocity $v = -\sqrt{2gh}$.

EXAMPLE 2–12 Calculate the time elapsed for the ball of Example 2–11 to drop from 100 to 75 m and from 75 to 50 m.

Setting It Up We are using the same data as in Example 2–11.

Strategy We calculate the times t_1 and t_2 at 75 and 50 m, respectively, given initial values $y_0 = 100$ m at $t_0 = 0$ s. We can use exactly the same procedure as was used in Example 2–11 to find the time t_1 to get to 75 m. We can also calculate directly the time elapsed $(t_2 - t_1)$ in falling from 75 to 50 m.

Working It Out Using Eq. (2–26c),

$$y = \frac{-gt^2}{2} + y_0, \qquad t^2 = \frac{2(y_0 - y)}{g};$$

fall from 100 to 75 m:

$$t_1^2 = \frac{2(100 \text{ m} - 75 \text{ m})}{9.80 \text{ m/s}^2} = 5.10 \text{ s}^2 \quad \text{or} \quad t_1 = 2.26 \text{ s};$$

fall from 100 to 50 m:

$$t_2^2 = \frac{2(100 \text{ m} - 50 \text{ m})}{9.80 \text{ m/s}^2} = 10.2 \text{ s}^2 \quad \text{or} \quad t_2 = 3.19 \text{ s}.$$

Thus the times of flight are

from 100 to 75 m: $t = 2.26 \text{ s} - 0 \text{ s} = 2.26$ s;

from 75 to 50 m: $t = 3.19 \text{ s} - 2.26 \text{ s} = 0.93$ s.

The ball's velocity increases as it falls, and it takes less time to drop a given distance as time progresses (Fig. 2–21).

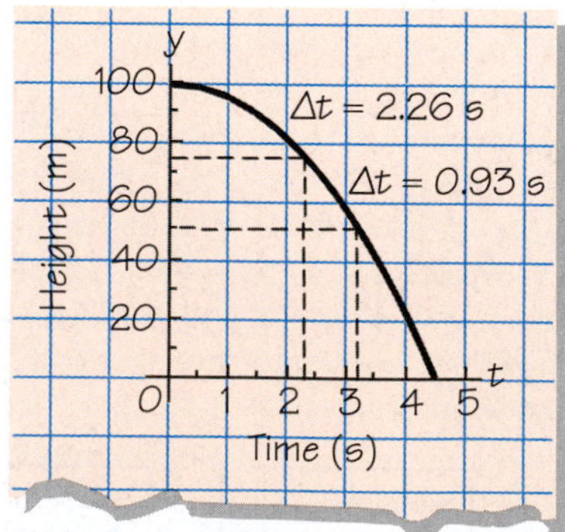

▲ **FIGURE 2–21** The time for a ball to drop a given distance decreases as the ball gains velocity.

EXAMPLE 2–13 Calculate the position of the ball at $t = 1$ s and $t = 2$ s under the conditions stated in Example 2–11.

Setting It Up We again use the data from Example 2–11.

Strategy The initial conditions are the same as before. In this case we can use Eq. (2–26c) to find y as a function of time and then at $t = 1$ s and $t = 2$ s.

Working It Out With the given initial conditions,

$$y = -\tfrac{1}{2}gt^2 + 100\text{ m};$$

$$\text{at } t = 1\text{ s: } y = \frac{(-9.80\text{ m/s}^2)(1\text{ s})^2}{2} + 100\text{ m} = 95.1\text{ m};$$

$$\text{at } t = 2\text{ s: } y = \frac{(-9.80\text{ m/s}^2)(2\text{ s})^2}{2} + 100\text{ m} = 80.4\text{ m}.$$

We show these distances in Fig. 2–22 for the 1-s equal time intervals. During the first time interval ($t = 0$ to $t = 1$ s), the ball travels 100 m − 95.1 m = 4.9 m,but during the second interval ($t = 1$ s to $t = 2$ s), the ball travels 95.1 m − 80.4 m = 14.7 m.

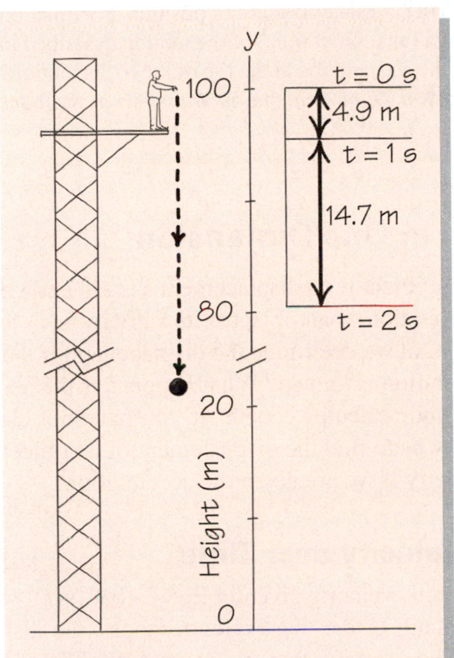

◀ **FIGURE 2–22** The distance traveled in a given time interval increases as the square of the total time elapsed for an object in free fall.

What Do You Think? Figure 2–23 illustrates the increasing distance a freely falling object travels in each time interval. What measurements could you make on this picture to verify that it corresponds to constant acceleration? *Answers to* ***What Do You Think?*** *questions are given in the back of the book.*

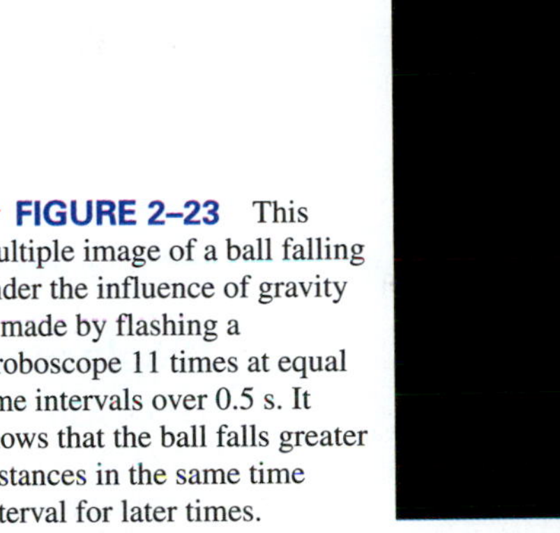

▶ **FIGURE 2–23** This multiple image of a ball falling under the influence of gravity is made by flashing a stroboscope 11 times at equal time intervals over 0.5 s. It shows that the ball falls greater distances in the same time interval for later times.

EXAMPLE 2–14 A ball is thrown straight up with a speed of 10.0 m/s from a third-floor window that is located 15.0 m above the ground. Calculate the maximum height of the ball, the ball's velocity when it hits the ground, and the total time it takes to reach the ground.

Setting It Up See Fig. 2–24, which also labels the y-axis, the initial ($t = 0$) height y_0, and the initial (upward) velocity v_0.

Strategy A ball thrown upward reaches its maximum height when its velocity is zero. At this point, the ball's direction of motion changes. We use this fact in Eq. (2–26a); we set $v = 0$ in that equation and solve for the time, call it t_{top}. Then we can find the maximum height above the ground, y_{max}, by substituting $t = t_{top}$ in Eq. (2–26c).

We can work the remainder of this example in two possible ways. We can either take a new initial condition to be at the maximum height of the ball with an initial velocity of zero or we can keep the original initial conditions. We'll work the example using the second strategy, so the initial conditions are as already given. We can use the connection between speed and distance, Eq. (2–26d), to find the final speed when the ball hits the ground ($y = 0$). It is not a problem that we have the speed and not the velocity because we know in this case that the ball is heading downward when it hits the ground. Finally, we can solve Eq. (2–26a) to find the flight time to the ground.

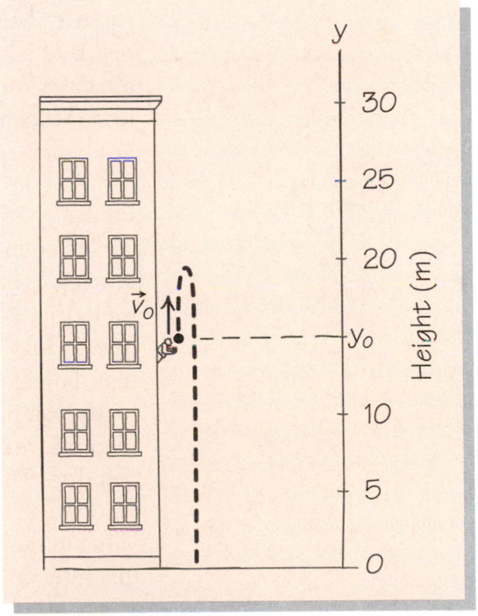

▲ **FIGURE 2–24** A ball is initially thrown up.

(continues on next page)

Working It Out To determine the time when the ball reaches its maximum height, use Eq. (2–26a):

$$v = 0\ \text{m/s} = gt_{top} + v_0 = -(9.80\ \text{m/s}^2)t_{top} + 10.0\ \text{m/s};$$

$$t_{top} = \frac{10.0\ \cancel{\text{m}}/\cancel{\text{s}}}{9.80\ \cancel{\text{m}}/\text{s}^{\cancel{2}}} = 1.02\ \text{s}.$$

The maximum height is found by using this value of t in Eq. (2–26c):

$$\begin{aligned} y_{max} &= \tfrac{1}{2}gt_{top}^2 + v_0 t_{top} + y_0 \\ &= \tfrac{1}{2}(-9.80\ \text{m/s}^2)(1.02\ \text{s})^2 + (10.0\ \text{m/s})(1.02\ \text{s}) \\ &\quad + 15.0\ \text{m} = 20.1\ \text{m}. \end{aligned}$$

To calculate the final velocity v, we use Eq. (2–26d) with $y = 0$:

$$\begin{aligned} v^2 &= v_0^2 + 2g(y - y_0) \\ &= (10.0\ \text{m/s})^2 + 2(-9.80\ \text{m/s}^2)(0\ \text{m} - 15.0\ \text{m}) \\ &= 100\ \text{m}^2/\text{s}^2 + 294\ \text{m}^2/\text{s}^2 = 394\ \text{m}^2/\text{s}^2; \end{aligned}$$

$$v = -19.9\ \text{m/s}.$$

In the last step we chose the negative root of v^2 based on our knowledge that the ball is moving downward. (We can choose either the positive or negative solution of the square root of v^2, whichever is physically sensible.)

Finally, from Eq. (2–26a) the flight time is as follows:

$$v = -19.9\ \text{m/s} = gt + v_0 = (-9.80\ \text{m/s}^2)t + 10.0\ \text{m/s};$$

$$t = \frac{19.9\ \cancel{\text{m}}/\cancel{\text{s}} + 10.0\ \cancel{\text{m}}/\cancel{\text{s}}}{9.80\ \cancel{\text{m}}/\text{s}^{\cancel{2}}} = 3.05\ \text{s}.$$

What Do You Think? Make sketches of position as a function of time and velocity as a function of time for the motion described in this example. What is the acceleration at the top of the ball's motion? *Answers to* ***What Do You Think?*** *questions are given in the back of the book.*

*2–6 Integration and Motion in One Dimension

We have learned how to find the velocity of an object if its displacement $x(t)$ is known $(v = dx/dt)$; we can also determine the acceleration of an object if its velocity $v(t)$ is known $(a = dv/dt)$. What about the inverse? Can we determine the displacement if the velocity is known or the velocity if the acceleration is known? What happens if the acceleration is not constant? You will learn in your calculus course (if you have not already) that the techniques of **integration** allow us to find the displacement of an object if we are given its velocity and to find its velocity if we are given the acceleration.

Displacement as an Integral of Velocity over Time

Let us start with the formula for the instantaneous velocity given in Eq. (2–10), $v(t) = dx(t)/dt$. Let us first consider $v(t)$ to be constant. Under these circumstances, the displacement is the time elapsed multiplied by the velocity, that is, $\Delta x = v\ \Delta t$. [This is just a rearrangement of the constant-velocity version of Eq. (2–10), namely $v = \Delta x/\Delta t$.] If we graph the relation $\Delta x = v\ \Delta t$ on a v–t plot, as in Fig. 2–25a, we see that the displacement Δx is *equal to the area under the curve of v as a function of time* (area = height × width). If the velocity changes from v_0 during the time interval $t = 0$ to $t = t_1$ to v_1 from the time interval $t = t_1$ to $t = T$ (Fig. 2–25b), as might be the case, for example, in a relay race in which two successive runners do not have the same velocity, then the total displacement is the sum of two displacements:

$$\Delta x = \Delta x_1 + \Delta x_2 = v_0 \times (t_1 - 0) + v_1 \times (T - t_1).$$

Figure 2–25b shows that this is again the sum of the areas under the plot of velocity versus time, that is, the total area under the v–t curve. The generalization of this process is straightforward. We may have a velocity that changes continuously with time (Fig. 2–25c). We then divide the total time interval from an initial time t_i to a final time t_f into many small equal time intervals Δt. These new time intervals are so short that the velocity cannot change very much over their duration; in other words, the velocity can be taken as a constant in these intervals. Let's concentrate on the tiny time interval from t_k to $t_k + \Delta t$ in Fig. 2–25c, where t_k is some intermediate time between t_i and t_f. The velocity $v(t_k)$ at the beginning of the time interval must be similar to the velocity $v(t_k + \Delta t)$ at the end; further, it should be nearly the same as the average value in the time interval. We therefore write this average value as just $v(t_k)$ itself. For this interval, then,

$$\Delta x_k = v(t_k)\ \Delta t.$$

This typical interval is shown as the darker area of the shaded curve in Fig. 2–25c.

Note: Sections marked with a * can be considered optional.

The total displacement $x_f - x_i$ over the total time interval t_i to t_f is then a sum over the small displacements $\Delta x_k = v_k\,\Delta t$. There must be N equal, tiny time intervals Δt to span the time interval from t_i to t_f:

$$N\,\Delta t = t_f - t_i;$$

$$x_f - x_i = \sum_{k=1}^{N} \Delta x_k = \sum_{k=1}^{N} v_k\,\Delta t. \qquad (2\text{–}28)$$

This equation approximates the full area under the curve in Fig. 2–25c; we have now seen that it also represents the displacement $x_f - x_i$ over the period from t_i to t_f. If we are concerned that our treatment of approximating the exact area by a sum over the areas of skinny rectangles isn't accurate enough because v changes significantly over the time interval Δt, we can simply make Δt smaller and smaller. In the limit that each time interval Δt goes to zero, we have an exact result:

$$x_f - x_i = \lim_{\Delta t \to 0} \sum_{k=1}^{N} v_k\,\Delta t. \qquad (2\text{–}29)$$

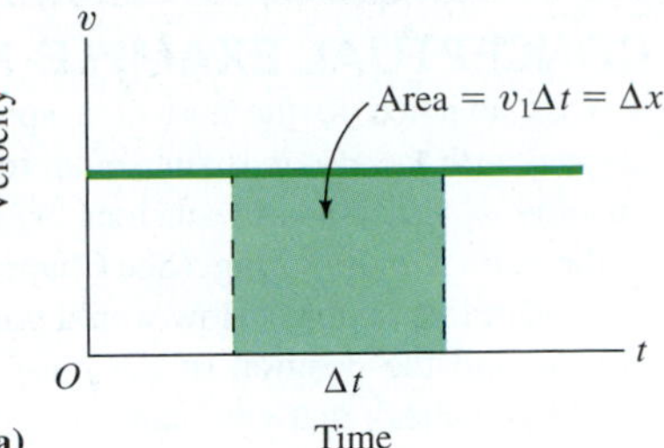

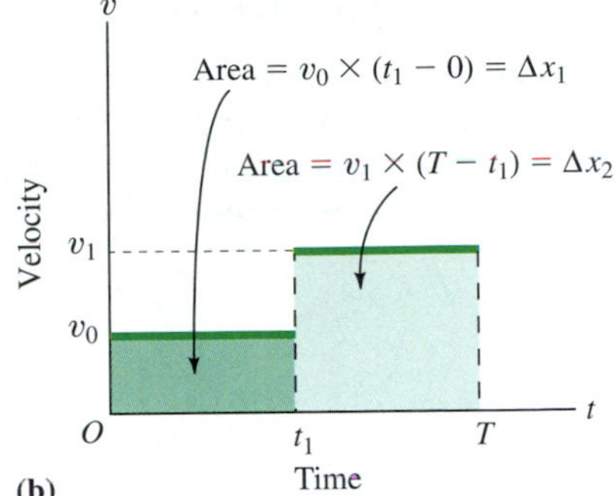

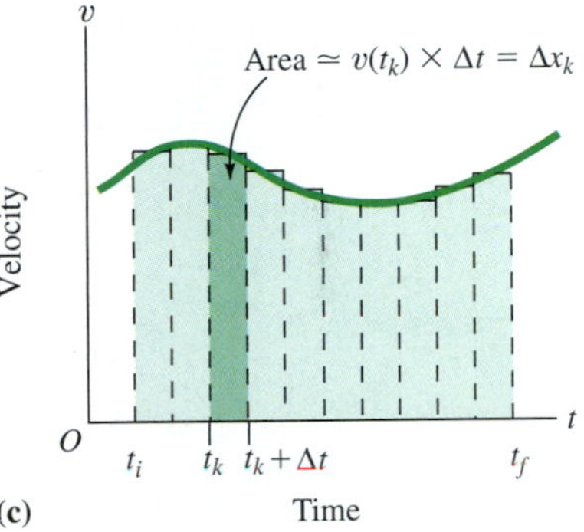

▲ **FIGURE 2–25** (a) The displacement over the dark area is $\Delta x_k = v\,\Delta t_k$, where v is the (constant) velocity in this interval. (b) If we have a sequence of two constant velocities, we can sum the two areas to find the displacement. (c) For varying velocity, we divide up the time into a series of intervals over which the velocity is (approximately) constant. When the areas of all the intervals are summed, we find the net displacement between initial time t_i and final time t_f, which is the shaded area under the curve of velocity versus time. This area is the integral

$$x_f - x_i = \int_{t_i}^{t_f} v\,dt.$$

Integration

The right-hand side of Eq. (2–29) is the definition of the integral of $v(t)$ over time:

$$x_f - x_i = \int_{t_i}^{t_f} v(t)\,dt. \qquad (2\text{–}30)$$

This expression is called a **definite integral** because the limits of integration—here t_f and t_i—are specified. The symbol $\int$ replaces the summation sign Σ when we sum over an infinite number of infinitesimal intervals of Δt. The area under the curve of velocity versus time between t_i and t_f in Fig. 2–25c is exactly the displacement $x_f - x_i$.

If the integration limits are not specified, we have

$$x = \int v(t)\,dt + C, \qquad (2\text{–}31)$$

where C is a constant of integration determined from the initial conditions discussed earlier (i.e., x_0 and v_0 at t_0). Equation (2–31) is called an **indefinite integral**. You will find a table of some indefinite integrals in Table IV–8 in the Appendix.

In Eqs. (2–30) and (2–31), we speak of taking the integral of $v(t)$ over the time. The function $v(t)$ in this case is called the *integrand.* The integral of a function between two limits is the area under the curve that represents a graph of the function versus its variable. For example, we have shown that the integral of $v(t)$ over t is the total area under the curve $v(t)$ versus t.

Velocity as an Integral of Acceleration over Time

The analysis of how velocity varies with time when the acceleration is constant shows that on a graph of acceleration versus time the change in velocity over a certain time interval is equal to the area under the a–t curve (Fig. 2–26). Just as displacement is the integral of the velocity over time, we find that the velocity is an integral of the acceleration over time:

$$v_f - v_i = \int_{t_i}^{t_f} a(t)\,dt, \qquad (2\text{–}32)$$

$$v = \int a(t)\,dt + C. \qquad (2\text{–}33)$$

We have again written both the definite and indefinite integral forms.

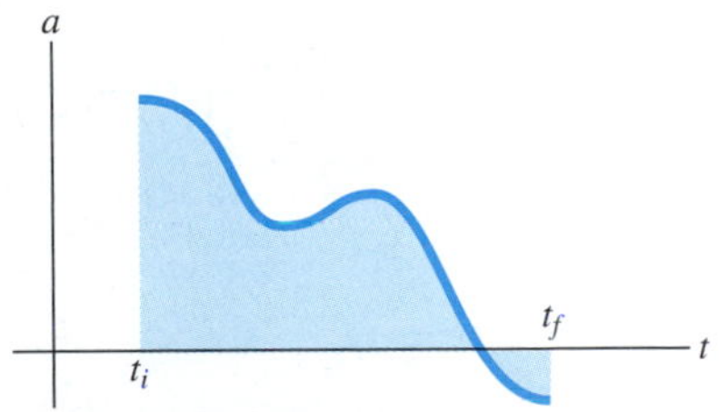

▲ **FIGURE 2–26** The change in velocity is found by integrating the acceleration over time.

CONCEPTUAL EXAMPLE 2–15 The acceleration a of a mass attached to the end of a spring and moving in a groove aligned with the x-axis is sinusoidal in time, $a(t) = A\sin(\omega t)$. The quantity ω, which has dimensions $[1/T]$, is a constant characteristic of the mass and the spring. (See Chapter 13 for a more complete discussion of this system.) How would you go about using this information to find the position of the mass as a function of time? How would you check that you made your calculation correctly? You can assume that you have a table of integrals and derivatives to help you with the mathematical details.

Answer We can put together the steps we have separately described in this section: The velocity is the integral over time of the acceleration, and the position is the integral over time of the velocity. Thus we want to carry out two integrals in succession to find the position of the mass as a function of time. Since the integral of a sine function is proportional to a cosine function (see Table IV–9 in the Appendix), that is the form of the velocity. And the integral of a cosine function is proportional to a sine function (see the same table), so the position of the mass—the integral of the velocity—is a sine function. Both the original acceleration and the position are sine functions. The mass moves back and forth with time, and its position is proportional to the original acceleration. As a good check, we could take a derivative with respect to time of the position to find the velocity, then a derivative of the velocity to find the acceleration (equivalent to taking two time derivatives of the position to find the acceleration).

More on Integration

Integration is the inverse of differentiation. We can see this for ourselves as follows: An integral of the form

$$I(T) \equiv \int_{t_0}^{T} f(t)\,dt$$

measures the area under the graph of $f(t)$ as a function of t extending from t_0 to T. (In fact, we have used this in finding, for example, a displacement for a time-varying velocity.) The derivative of our integral is defined by

$$\frac{dI(T)}{dT} = \lim_{\Delta T \to 0} \frac{I(T + \Delta T) - I(T)}{\Delta T}.$$

The first term in the numerator is the area under the curve of $f(t)$ extending from t_0 to $T + \Delta T$; the second is the area extending only to T. The numerator is the difference of these, and this difference is the area under the curve extending from T to $T + \Delta T$. As ΔT becomes smaller and smaller, the area is just the width, ΔT, multiplied by the height, which as $\Delta T \to 0$ is just the height, $f(T)$. Thus we clearly see from our graphical interpretation that $dI(T)/dT = f(T)$. Since T is arbitrary, the formula holds everywhere; in other words, we can just write it in terms of the time t:

$$\frac{dI(t)}{dt} = f(t).$$

The derivative of an integral with respect to the upper limit of the integration is the integrand.

The role of the lower limit on the integral is just to set a starting point. Suppose we were to shift the lower limit on the integral from t_0 to t_1. We choose $t_1 > t_0$ for convenience. We then have

$$(\text{area from } t_0 \text{ to } T) = (\text{area from } t_0 \text{ to } t_1) + (\text{area from } t_1 \text{ to } T),$$

or, equivalently,

$$\int_{t_0}^{T} f(t)\,dt = \int_{t_0}^{t_1} f(t)\,dt + \int_{t_1}^{T} f(t)\,dt.$$

As far as the dependence on T is concerned, the first integral on the right is just a number, independent of T. Thus when we differentiate our expression, this term plays no role—the derivative of a constant is zero. The significance of this is the following: If we know that some function of t may be written in the form $f(t) = dI(t)/dt$, it then follows that we can write $I(t)$ as an integral over $f(t)$. However, because of the independence of this result on the lower limit on the integral (a constant), we can only conclude that

$$I(T) = \int^{T} f(t)\,dt + C,$$

where C is an arbitrary constant. The integral does not tell us where the area calculation starts from (that is why the lower limit has been left off), and this ambiguity is represented by C. Once we decide on a lower limit on the integral, say t_0, then we can immediately fix C. Its value is $I(t_0)$, since the integral for which the upper and lower limits are equal is zero.

We can easily write some integrals based on some known derivatives and the inverse nature of integration and differentiation:

(i) $$\frac{dt^n}{dt} = nt^{n-1} \quad \text{leading to} \quad \int^T t^{n-1}\,dt = \frac{T^n}{n} + C; \tag{2–34}$$

(ii) $$\frac{d\sin t}{dt} = \cos t \quad \text{leading to} \quad \int^T (\cos t)\,dt = (\sin T) + C; \tag{2–35}$$

(iii) $$\frac{de^{at}}{dt} = ae^{at} \quad \text{leading to} \quad \int^T e^{at}\,dt = \frac{1}{a}e^{aT} + C. \tag{2–36}$$

Special Case of Constant Acceleration

The value of Eqs. (2–30) and (2–32) is that they can be used to find the position and the velocity of an object whether its acceleration is constant or not. We can demonstrate the consistency of these integration techniques by using them to derive once more the equations that describe the constant-acceleration situation. Suppose that the magnitude of the constant acceleration is a. Beginning with Eq. (2–32), with the initial conditions x_0 and v_0 for $t = t_0$, we have

$$v - v_0 = \int_{t_0}^{t} a\,dt.$$

The integral of a constant C_1 is, according to Eq. (2–35), $C_1 t$, so

$$\text{for constant acceleration: } v - v_0 = at\Big|_{t_0}^{t} = a(t - t_0),$$

where the vertical line indicates that the total function is to be evaluated at the two limits of integration (t_0 and t) and the results subtracted. Rearranging this result gives Eq. (2–25a),

$$\text{for constant acceleration: } v = v_0 + a(t - t_0).$$

We integrate this result once more to determine x:

for constant acceleration:

$$\begin{aligned} x - x_0 &= \int_{t_0}^{t} v(t)\,dt = \int_{t_0}^{t} (v_0 + at - at_0)\,dt \\ &= \left(v_0 t + \tfrac{1}{2}at^2 - at_0 t\right)\Big|_{t_0}^{t} \\ &= v_0(t - t_0) + \tfrac{1}{2}a(t^2 - t_0^2) - at_0(t - t_0) \\ &= v_0(t - t_0) + \tfrac{1}{2}at^2 - at_0 t + \tfrac{1}{2}at_0^2 = v_0(t - t_0) + \tfrac{1}{2}a(t - t_0)^2, \\ x &= \tfrac{1}{2}a(t - t_0)^2 + v_0(t - t_0) + x_0, \end{aligned}$$

which is Eq. (2–25c). Integration techniques allow us to obtain these results directly.

Summary

Kinematics is the mathematical description of motion. Quantities of particular physical significance are displacement, instantaneous velocity, and instantaneous acceleration. These are all vector quantities, expressible in one dimension as magnitudes with signs. The displacement Δx measures the change in an object's position; when an object moves from position x_1 to position x_2, its displacement is for one dimension,

$$\Delta x \equiv x_2 - x_1. \tag{2–1}$$

The instantaneous velocity at any given time is defined as the limit of the average velocity as $\Delta t \rightarrow 0$:

$$v(t) = \lim_{\Delta t \rightarrow 0} \frac{x(t + \Delta t) - x(t)}{\Delta t}. \tag{2–9}$$

It is thus the rate of change, or derivative, of the displacement at that time:

$$v(t) = \lim_{\Delta t \rightarrow 0} \frac{\Delta x}{\Delta t} = \frac{dx}{dt}. \tag{2–10}$$

The instantaneous acceleration is defined as the rate of change of velocity:

$$a \equiv \lim_{\Delta t \rightarrow 0} \frac{\Delta v}{\Delta t} = \frac{dv}{dt}. \tag{2–14}$$

We can also define the average speed and average velocity, but these concepts are used less frequently.

Useful relations between displacement, velocity, and acceleration that are valid for constant acceleration in one dimension are

for constant acceleration a:

$$v = a(t - t_0) + v_0, \tag{2–25a}$$

$$v_{av} = \tfrac{1}{2}(v_0 + v), \tag{2–19, 2–25b}$$

$$x = \tfrac{1}{2}a(t - t_0)^2 + v_0(t - t_0) + x_0, \tag{2–25c}$$

$$v^2 = v_0^2 + 2a(x - x_0). \tag{2–24, 2–25d}$$

Graphical techniques are useful in determining velocity. Graphically, velocity is the slope of a curve of displacement versus time. Similar analysis can determine acceleration from a curve of velocity versus time.

The acceleration due to gravity, which has magnitude g, is an important example of a constant acceleration. In this case, the acceleration points to Earth's center. (This result is accurate to the extent that there is no air resistance or friction in the motion.) A particle falling a distance h from rest has the final velocity

$$\text{for constant acceleration: } v = \sqrt{2gh}. \tag{2–27}$$

By using integration techniques, the displacement and velocity can be determined from the velocity and acceleration, respectively:

$$x_f - x_i = \int_{t_i}^{t_f} v(t)\, dt, \tag{2–30}$$

$$v_f - v_i = \int_{t_i}^{t_f} a(t)\, dt. \tag{2–32}$$

These last two expressions are valid even if the acceleration is not constant.

Understanding the Concepts

1. Why is it a good idea to increase the space between your car and the car in front of you when the speed of the cars increases?
2. A piece of chalk is thrown straight up; at some point it reaches a maximum height and begins to drop. What is the velocity at the maximum height? Can there be a nonzero acceleration at this point even though the velocity is zero?
3. In a series of thought experiments, an object is dropped from rest from a given height on a variety of planets. Each of these planets has a different acceleration due to gravity, g_x. Describe how the time of fall varies with g_x. How does the speed of the object at the end of the fall vary with g_x?
4. You are in the unfortunate position of being in an elevator with 20 bowling balls when the elevator cable breaks, causing both you and the elevator to fall under the acceleration of gravity. The emergency brake has not yet cut in. What is happening inside the elevator?

5. A falling object moves faster and faster the farther it falls. Does this mean that an object dropped from an arbitrarily large height will hit the ground moving with an arbitrarily high speed? If you answered yes, how do you explain the fact that meteors, which can come from very far away, don't hit the ground with a nearly infinite speed?
6. If an object that is restricted to moving along a straight line has a positive initial velocity and if the acceleration is always negative, can the velocity remain positive?
7. An astronaut in full gear can jump up 0.8 m on Earth. What arguments would you use to estimate the height of a jump on the moon, where the acceleration of gravity has magnitude 1.6 m/s^2?
8. What is the role of an air bag placed where a falling object is expected to land? How can an air bag prevent injury to someone who jumps from a height?
9. An object moving along the x-axis on a straight horizontal rail starts by moving rapidly to the right, slows, and comes to a stop, then starts moving more and more rapidly to the left. True or false: This description is consistent with motion with constant acceleration (even around the region where the object reverses the direction of its motion).
10. For the data shown in Table 2–1, will there be any difference between the average speed over some interval and the magnitude of the average velocity over the same interval? How would you answer the same question if the motion were not on a straight track?
11. What should the velocity of the runner in Fig. 2–2 be at $t = 0$ s? Do the data justify your conclusion? Explain.
12. "Zeno's paradox" comes to us from ancient Greece. It concerns the difficulty that a runner might have in catching a tortoise near the finish line of a race if the tortoise is ahead of the runner at one point, as follows. At some time, the tortoise is a distance L in front of the runner. After a time interval Δt, the runner is $L/2$ behind the tortoise. After a later time $\Delta t/2$, the runner is $L/4$ behind the tortoise. After a time $\Delta t/4$, the runner is $L/8$ behind the tortoise. The runner always appears to be behind! Where did the Greeks go wrong? By the way, the correct answer to this question was given only in Newton's time and lies behind the crucial concepts of calculus.
13. The velocity of an object moving in one dimension is measured at equal distance intervals. It is found that the magnitude of the velocity is proportional to the square root of the distance traveled. What can you say about the motion?
14. If the velocity of an object is positive, is its acceleration necessarily positive? Is there any connection between the sign of the velocity and the sign of the acceleration?
15. True or false: A freely falling body is moving with four times the speed when it has fallen twice as far.
16. A juggler tosses a beanbag straight up with initial speed v_0 under the influence of gravity, lets a second beanbag drop from rest, and tosses a third straight down with initial speed v_0. Compare the subsequent accelerations of the three beanbags.
17. In what order do the beanbags of the previous question hit the floor? What are the relative speeds of the three bags when they hit the floor?
18. A beanbag is tossed straight up. It rises, reaches a maximum height, then falls back down. What is the acceleration of the beanbag at its maximum height?
19. You measure the velocity of a bicycle moving in a straight line. How would you determine the rotations per second made by the wheel; in particular, what additional measurements are needed?
20. Given a stopwatch and a measuring rod, how would you determine the average acceleration experienced by someone jumping on a trampoline?
21. Describe some situations, other than free fall under the influence of gravity, in which an object could be undergoing constant acceleration.
22. You are given a measuring rod and a movie camera with a rather precisely known speed of the motion of the film. How would you use this to determine (to some degree of accuracy) the instantaneous velocity of a person jumping up and down on a trampoline? How would you use your apparatus to measure the instantaneous acceleration?
23. Consider a super-ball that drops from a certain height onto a rigid surface. It starts off with zero velocity at time $t = 0$. Sketch the velocity as a function of time from the time that the ball is dropped to just before it hits the floor at time $t = 10$ s. Assuming that the starting velocity on the rebound has the same magnitude (though opposite direction) as the ball when it reaches the ground, sketch the velocity as a function of time after the rebound. Do all this on the same graph. When will the ball have zero velocity again? What will its position be at that time?
24. If you consider the fact that in nature we can never have perfectly sharp angles in the velocity directions, round off the sketch in the question above at the point where the ball reverses direction. Describe in words what happens to the acceleration during the turn-around motion of the ball. We have restricted ourselves in sketching the motion as a function of time, with t along the horizontal axis and x along the vertical axis. Suppose we were to reverse these, so that x is along the horizontal axis and t along the vertical one. Sketch the following motions with your exchanged axes: (a) a car is at rest; (b) a car is moving slowly; (c) a car is moving rapidly. What is the main disadvantage of using this set of axes?

Problems

2–1 Displacement

1. (I) A grasshopper jumps along a groove aligned with the x-axis. Starting at the origin, the grasshopper's first jump has a displacement +32 cm, the second jump has a displacement −27 cm, the third a displacement −23 cm, and the fourth a displacement +39 cm. What is the net displacement? At what position is the grasshopper after all four jumps?
2. (I) Using the data in Table 2–1, draw position vectors to the runner for 40 and 80 m. Write and draw the displacement from 40 to 80 m.
3. (I) A gym teacher organizes a series of indoor races in the gym, which is 42 m in length. The students run from one end to the other and back again. After three round trips, what is the distance traveled by each student and what is the displacement vector? Draw a graph of the magnitude of the displacement vector as a function of time if it takes 7 s to run each 42-m leg. Assume that the speed is constant.

2–2 Speed and Velocity

4. (I) In 1991, Carl Lewis edged out Leroy Burrell in the World Championships in Tokyo to set a new world record in the 100-m dash. Their times at 10-m intervals are given here. Calculate the average velocity for Lewis's world record for 0 to 50 m, 50 to 100 m, and 0 to 100 m.

Distance (m)	Time (s) Lewis	Time (s) Burrell
10	1.88	1.83
20	2.96	2.89
30	3.88	3.79
40	4.77	4.68
50	5.61	5.55
60	6.46	6.41
70	7.30	7.28
80	8.13	8.12
90	9.00	9.01
100	9.86	9.88

5. (I) (a) Plot the path of an automobile that travels from a starting point to a point 15 km along a straight road at 75 km/h. It stops for 25 min, then continues on the same straight road for 40 km at 100 km/h. After a 5-min stop, it returns to its starting point at 60 km/h. Draw your position axis as horizontal and your time axis as vertical. (b) On the same plot, draw the path of an automobile that starts from the same spot 25 min after the first one and travels at 74 km/h in the original direction of the first automobile. Where and how often will the two cars meet?

6. (I) A car moving at 65 mi/h passes a pickup truck moving at 50 mi/h. The car goes on for 30 mi, then stops at a rest stop for 20 min. The car resumes its journey, again at 65 mi/h. Assuming the truck did not stop and maintained its speed, did it pass the car while it was stopped? If so, how long after the initial passing does it take for the car to catch the truck again? Solve this problem by graphical means.

7. (I) Redraw the same paths for the two vehicles of Problem 6 on a plot in which the horizontal axis is the time axis and the position, x, is along the vertical axis. Suppose that you took this new plot and simply relabeled the axes, so that the vertical direction represents time and the horizontal direction represents position. Interpret the paths of the two vehicles.

8. (I) An automobile travels north, covering a distance of 30 mi in 35 min, stops for 20 min, and then continues north for 20 mi, taking 25 min. Assume that the car moves uniformly during each segment of the trip. Calculate the average velocity of the total trip. Calculate the average velocity for the first half (by time) and the last half of the trip.

9. (II) An automobile driver travels north for 2 min at 30 mi/h, then stops at a red light for 30 s before proceeding again for 3 min at 45 mi/h. He then stops at a stop sign for 3 s, drives forward at 30 mi/h for 2 min, and finally stops for gas. (a) How far does the automobile travel? (b) What is the average velocity? Use units of miles and minutes.

10. (II) The position of a falling particle is given by $x = x_0 + v_0 t - \frac{1}{2} g t^2$. What is the velocity of the particle as a function of time? Calculate the average velocity during the time intervals t of 0 to 1 s, 1 to 2 s, and, more generally, t to $t + 1$ seconds.

11. (II) Use the velocity of a particle as a function of time, tabulated below, to calculate the position of the particle at each of the times; assume that at $t = 0$ s the particle was at the origin and at rest. [*Hint*: A graph is simplest.]

Time (s)	Velocity (m/s)
0.5	0.75
1.5	1.75
2.5	8.75
3.5	21.75
4.5	39.25
5.5	62.75
6.5	90.75
7.5	122.75

12. (II) The height of a bungee jumper above ground level is given as a function of time t by

$$y = (25\text{ m}) \cos[\pi t/(6\text{ s})] + (38\text{ m}).$$

(a) Sketch the function $y(t)$ from $t = 0$ s to $t = 5$ s. (b) Calculate the average velocity of the jumper between $t = 2$ s and $t = 3$ s and between $t = 3$ s and $t = 4$ s. (c) What is the instantaneous velocity of the jumper when he is closest to the ground?

13. (II) An old brain teaser reads as follows: Two trains leave different stations 80 km apart and travel toward each other on a straight track. One train has a speed of 80 km/h and the other has a speed of 160 km/h. A very fast insect leaves the slower train and heads toward the faster train at a speed of 240 km/h. Upon encountering the second train, it turns around and just as rapidly returns to the first train. It continues these maneuvers until it is squashed between the two trains when they collide. Graph what has happened and use your graph to determine how far the insect will have traveled. Can you think of a way to estimate your result or to calculate it rapidly?

14. (II) Traffic signals are placed along a straight road at positions $x = 0$ m, $x = 600$ m, and $x = 1200$ m (Fig. 2–27). The time intervals during which the signals are green are shown by the thick lines in the figure. (a) Draw the displacement-versus-time curves (fastest and slowest) for a car that passes through all the lights when the car moves with constant speed. (b) Draw a similar set of lines for a car traveling in the opposite direction. (c) Assuming that the lights are timed such that a car passes through all lights in the middle of the time interval, what is the speed for which the lights are timed? (d) What is the fastest constant speed of a car that makes it through all the signals, assuming it arrives at the first light at the optimal moment?

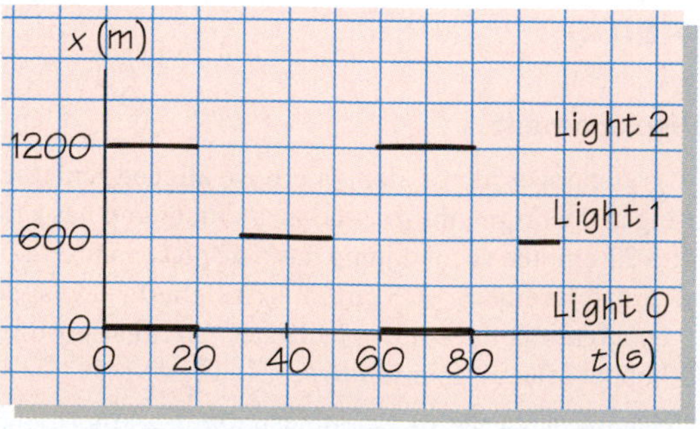

▲ **FIGURE 2–27** Problem 14.

15. (III) The distance an ant moves in a straight-line motion is given by $x = 0.010t^3 - 0.050t^2 + 1.5t$ centimeters, where t is in seconds. Calculate the velocities for t values of 1, 5, and 10 s. What is the average velocity for the first 10 s? Why is the formula unrealistic for long times?

16. (III) The displacement of a particle as a function of t is described by the equation $x = \sqrt{(2.0\ \text{m}^2/\text{s})(t + 1.0\ \text{s})}$, where t is measured in seconds. (a) Plot $x(t)$ between $t = 0$ s and $t = 5.0$ s. (b) Calculate the average velocity between $t = 1.0$ s and $t = 5.0$ s, between $t = 2.0$ s and $t = 4.0$ s, and between $t = 2.8$ s and 3.2 s. (c) Compare these results with the instantaneous velocity at $t = 3.0$ s.

2–3 Acceleration

17. (I) An automobile badly in need of repairs is able to accelerate at a constant value of $0.40\ \text{m/s}^2$. How long does it take the automobile to get to 35 mi/h?

18. (I) A bicyclist is pedaling at a constant speed of 10 m/s when she decides to slow down. She stops pedaling and sits up, and the combined effects of wind resistance and road friction cause a negative acceleration of $-0.3\ \text{m/s}^2$. If this acceleration does not change, how long would it take her to slow to 5 m/s?

19. (I) A car is said to go from rest to 60 mi/h in 9.0 s. Assuming that the acceleration is uniform, what is its value in units of g?

20. (II) Car A leaves a city and travels along a straight road for 1.5 min at 60 km/h. It then accelerates uniformly for 0.25 min until it reaches a speed of 80 km/h. It proceeds at that speed for 2.0 min, then decelerates uniformly for 0.50 min until it comes to rest. Car B leaves the same city along the same road and accelerates uniformly for 1.6 min until it reaches a speed of 120 km/h. It then decelerates uniformly until it comes to rest again after 1.6 min. (a) Plot the curve of the cars' motions on a graph in which the vertical axis is the speed v and the horizontal axis is the time t. (b) Plot the motions of the cars on a graph in which the vertical axis is the distance x from the city and the horizontal axis is time t. (c) How far will the two cars have traveled during the different stages?

21. (II) An automobile starting from rest at $t = 0$ s undergoes constant acceleration on a straight line (Fig. 2–28). It is observed to pass two marks separated by 64 m, the first at $t = 8$ s and the second at $t = 12$ s. What is the value of the acceleration?

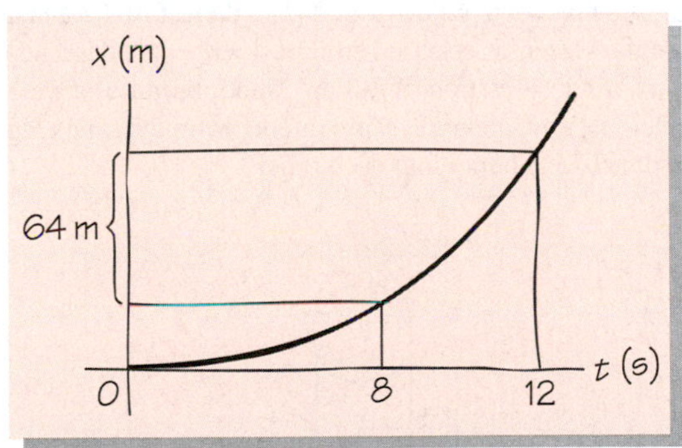

▲ **FIGURE 2–28** Problem 21.

22. (II) Inclined planes are convenient tools to study motion under a constant acceleration. The time of passage of a ball rolling on an inclined plane is measured by three light gates positioned 60 cm apart. The ball passes the light gates at 0.30, 1.15, and 1.70 s. Find the acceleration of the ball.

23. (II) Suppose the position of a particle is described by $x = A\sin(\omega t)$. Calculate the velocity and acceleration of the particle as a function of time.

24. (II) Consider the motion of the particle whose velocity is tabulated in Problem 11. Use these data to make a table of approximate values of the acceleration for $t = 1$ s, 2 s, . . . , 7 s.

25. (II) The position x of a block attached to a spring as a function of time is given by the formula $x = A\sin(\pi t/12)$, as shown in Fig. 2–29. Describe in words the motion of the block.

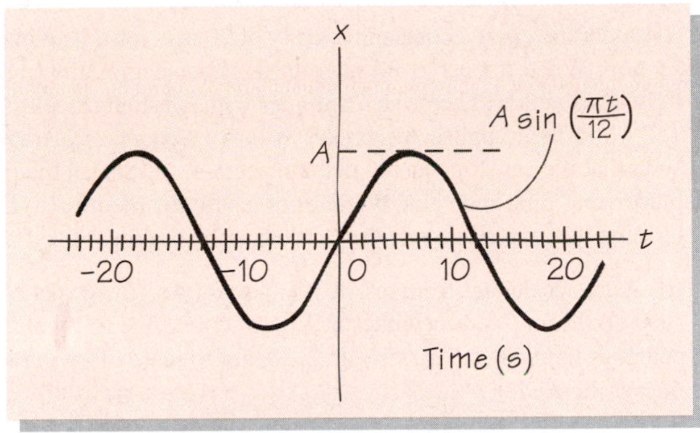

▲ **FIGURE 2–29** Problem 25.

26. (II) The position of a particle is given by $x = At^2 + Be^{\alpha t}$. The particle is initially ($t = 0$) at $x = -1.5$ cm with $v = 0.25$ cm/s. At 0.10 s, the velocity is observed to be 0.045 cm/s. What is the acceleration at 1.0 s?

27. (II) Consider an object whose acceleration is determined by its velocity, as in the equation $a = A - (v/t_0)$. Here t_0 and A are constants with the dimensions of time and acceleration, respectively. Assume that the object starts out at $t = 0$ s with an initial velocity v_0. Sketch the behavior of the acceleration and of the velocity. Describe the motion after a long time.

2–4 Motion with Constant Acceleration

28. (I) A drag racer reaches 128 mi/h in a $\frac{1}{4}$-mi race. Assuming a constant acceleration, what was the elapsed time?

29. (I) A car traveling 25 mi/h must reach a minimum of 50 mi/h within a 1000-ft access lane. What must the car's constant acceleration be?

30. (I) An airplane starting from rest reached its takeoff velocity of 212 mi/h over a runway of 6000 ft. How long did this take if the plane rolled with a constant acceleration?

31. (I) A ball rolling straight down a ramp undergoes a constant acceleration of $0.50\ \text{m/s}^2$. What is the average velocity over the period 1 to 2 s assuming the ball started from rest at $t = 0$ s?

32. (I) A rocket accelerates uniformly from rest to a speed of 4.2×10^3 mi/h in 125 s. Over what distance does the rocket accelerate?

33. (I) A soccer ball rolls with an initial velocity of 8.0 m/s in an easterly direction across a flat field. Friction slows the ball down at the rate of $0.50\ \text{m/s}^2$. (a) Express the ball's velocity as a function of time. (b) Where is the ball 5.0 s after it starts to roll?

34. (I) A bowling ball is rolled down the alley with an initial velocity of +10 m/s. There is a small amount of friction, and it produces an acceleration of $-0.2\ \text{m/s}^2$. (a) What is the velocity at 1 and 2 s? (b) What is the average velocity over the first 2 s?

35. (II) A lead weight falls from a height of 6 m onto a muddy surface. It comes to rest after penetrating 0.4 cm into the surface. What was the magnitude of the average acceleration during the impact? How long did it take to stop?

36. (II) The speed of a landing airplane is 80 m/s. After touching ground it rolls a distance of 400 m on the runway at a constant velocity. It then decelerates at $3.0\ \text{m/s}^2$ until it stops. (a) Sketch the displacement–time and velocity–time curves. (b) Calculate the distance traveled on the ground and the time interval between touch-down and full stop.

37. (II) A car travels at a constant velocity of 20 m/s toward an intersection. When the car is 80 m from the intersection, the traffic light turns yellow. The driver continues with constant velocity for 1.2 s and then applies the brakes with a constant acceleration such that the car stops just at the intersection. (a) Sketch the displacement–time and velocity–time curves for the motion. (b) Determine the acceleration of the car during the braking period.

38. (II) A car accelerates from rest at $3.0\ \text{m/s}^2$ for 4 s, travels at a constant speed for 7 s, accelerates at $1.0\ \text{m/s}^2$ for 15 s, and then decelerates to rest at $2.5\ \text{m/s}^2$ (Fig. 2–30, not to scale). How far has the car traveled?

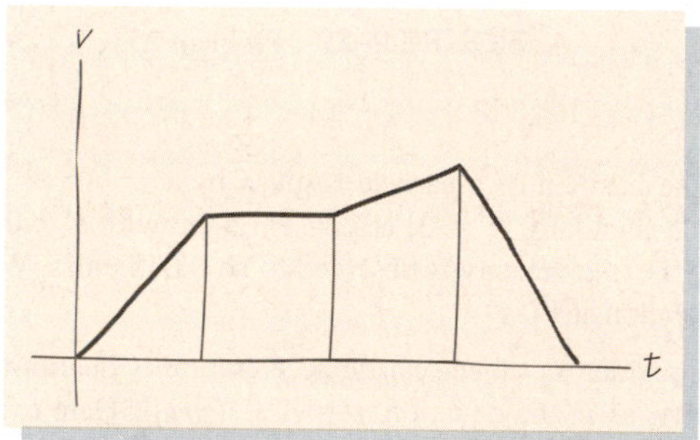

▲ **FIGURE 2–30** Problem 38.

39. (II) A child is in an open-cage elevator facing out on a hotel lobby. The elevator is descending at a constant speed of 1 m/s. The child lets a penny drop from his hand when the elevator is 20 m above the floor of the lobby. How much time does the penny spend in the air? Ignore air resistance.

40. (II) Suppose that a runner were capable of a constant acceleration of $2.8\ \text{m/s}^2$ for the entire length of a 100-m dash. (a) How long would it take the runner to run the first 10 m? (b) How long for the first 50 m? (c) For the second 50 m? (d) For the entire 100 m? (e) Compare to the times of Problem 4.

41. (II) In 1979, the Japanese tested a magnetically levitated train. The train is both suspended and propelled by magnetic forces. The train traveled on a straight 7000-m-long track starting from rest; it reached a peak speed of 144 m/s before it came to rest again. Both the acceleration and deceleration were constant and of the same magnitude. The entire length of the track was used. (a) What was the magnitude of the acceleration (and deceleration)? (b) How much time was spent on the trip from one end of the track to the other?

42. (II) Your bus is leaving the stop, accelerating at a constant rate of $0.6\ \text{m/s}^2$. You turn the corner to see the bus pulling out of the stop 30 m ahead of you. What is the minimum steady speed with which you must run to catch the bus? Olympic sprinters can run at 10 m/s.

43. (II) A speeder is traveling along a straight road at 75 mi/h. He passes a standing police car, which starts to chase him. The police car accelerates from 0 to 85 mi/h in 13 s and travels at 85 mi/h thereafter. (a) Sketch the positions of both cars on the same x-versus-t graph. (b) How far from its starting point does the police car overtake the speeder? (c) What is the elapsed time?

44. (II) A car is moving at 35 mi/h when the driver sees a light turn red. She hits the brake pedal when she is 90 ft from the light and the deceleration of the automobile has magnitude $3.0\ \text{m/s}^2$. Does the car stop before it arrives at the light? How far does it travel before stopping?

45. (II) In Problem 44, how long does it take to stop from the moment the brakes are applied?

46. (II) Two automobiles are geared quite differently and are to be used for a drag race over a distance of 400 m. Car A accelerates at a constant value of $5.0\ \text{m/s}^2$ for the first 200 m, then at a constant value of $2.5\ \text{m/s}^2$ for the remaining 200 m. Car B accelerates at a constant rate of $4.5\ \text{m/s}^2$ for the first 200 m but at $3.0\ \text{m/s}^2$ for the remaining distance. (a) Give the value of the speed of each automobile at the 200-m mark and the time it took each to get there. (b) What are the finishing times for the race and the values of the respective speeds at the end of the race?

47. (II) An electron in the picture tube of a TV set traveling in a straight line accelerates uniformly from speed 3×10^4 to 5×10^6 m/s along a length of 2 cm. (a) How much time does the electron spend in this 2-cm region? (b) What is the magnitude of the electron's acceleration?

48. (II) You have an old, heavy automobile that does not accelerate very rapidly but can maintain acceleration for a long period. Suppose that your car has a maximum acceleration that takes it from 0 to 50 mi/h in 18 s. What would the speed be if this average acceleration were maintained for 36 s? How far does your car go during the first 20 s, and how far would it travel in 36 s under the above conditions?

49. (II) An elevator accelerates from the ground with a uniform acceleration a. After 3 s, an object is dropped out of an opening in the floor of the elevator and that object hits the ground 3.5 s later. How large is the acceleration? How high was the elevator when the object was dropped?

50. (II) Two small objects A and B are suspended from the ends of a rope thrown over a pulley (Fig. 2–31). Object A is 1.2 m above B when the system is released from rest. Object A descends with a downward acceleration of $0.3\ \text{m/s}^2$ and, because of the rope, B accelerates upward at the same rate. How much time elapses before the objects bump into each other?

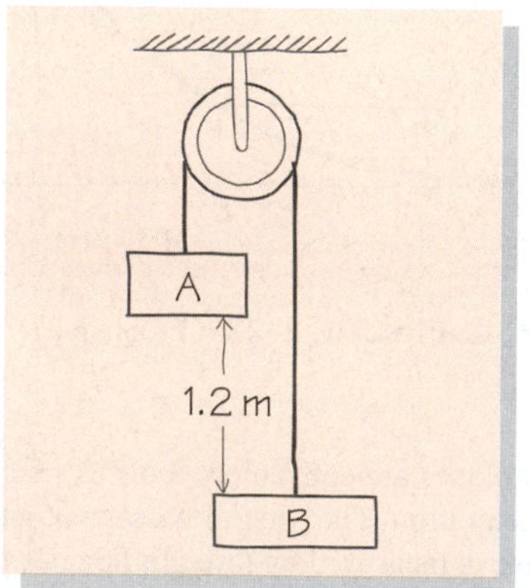

▲ **FIGURE 2–31** Problem 50.

51. (II) A car moving at 60 mi/h can be brought to rest in 4 s. Assuming that the deceleration is uniform, how far will the car travel between the time the brakes are applied and the time the car stops?

52. (II) A mountain climber is attached to a rope. She slips, and after she has fallen straight down 8 m, the rope starts to decelerate her. If the constant deceleration is $5g$ (five times the acceleration due to gravity), how much will the rope have to stretch? (In reality, the deceleration depends on the stretching of the rope.)

53. (II) A bullet traveling at 600 m/s penetrates a block of wood and comes to rest with a constant deceleration after traveling 20 cm (Fig. 2–32). What is the magnitude of the deceleration? How long does it take the bullet to stop?

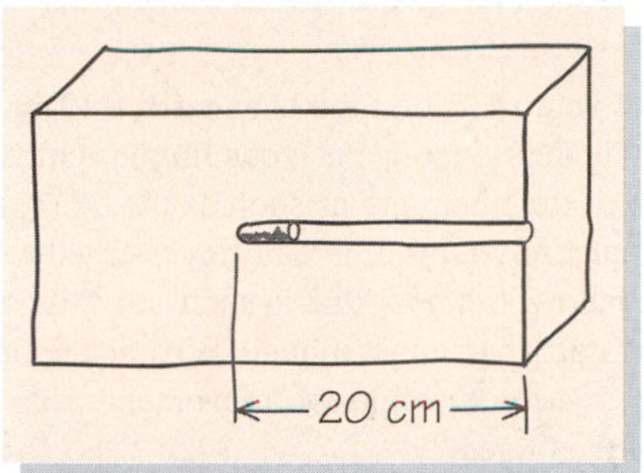

▲ **FIGURE 2–32** Problem 53.

2–5 Freely Falling Objects

54. (I) The tower of Pisa is 54.5 m tall. Assuming that Galileo dropped his object from rest from the top of the tower and that the effects of air resistance were negligible, how long would it have taken the object to fall?

55. (I) A story claims that someone who fell off New York City's Empire State Building (which has approximately 100 floors) was overheard to say "so far, so good" as he passed a third-floor window. Make some estimates to see if this is possible.

56. (I) A string is to have a series of lead sinkers tied to it. The first is tied at the bottom, and the second is tied 10 cm up from the bottom. The string can be held at its top and dropped from a height onto the top of a drum on which the first sinker already rests; each time a sinker hits the drum, a tap is heard. How far above the bottom sinker must the third, fourth, and fifth sinkers be tied so that the series of four taps is spaced by equal time intervals when the string is dropped?

57. (I) The acceleration due to gravity on the surface of the Moon is only about one-sixth the acceleration due to gravity on Earth's surface. In the celebrated experiment of the dropped feather performed by an astronaut on the Moon, how long did it take for the feather to drop 1 m to the surface if it started from rest?

58. (I) The acceleration due to gravity on the surface of Jupiter is 25.9 m/s^2. If it were possible to perform the experiment, how long would it take for an object that is initially at rest to fall a distance of 10 m on Jupiter's surface? Ignore any effects due to "air" resistance.

59. (II) A rock is thrown nearly straight upward from the edge of the top of a building at an initial speed of 22 m/s. Its trajectory will take it just past the edge, so that it lands on the ground. How much later must a second rock be dropped from rest at the same initial height of 10 m so that the two rocks hit the ground at the same time?

60. (II) An astronaut shipwrecked on a distant planet with unknown characteristics is on top of a cliff, which he wishes to descend. He does not know the acceleration due to gravity on the planet, and he has only a good watch with which to make measurements. He wants to learn the height of the cliff, and to do this, he makes two measurements (Fig. 2–33). First, he lets a rock fall from rest off the cliff edge; he finds that the rock takes 4.15 s to reach the distant ground. Second, he releases the rock from the same spot but tosses it upward so that it rises a height of what he estimates to be 2 m before it falls to the ground below. This time the rock takes 6.30 s to reach the ground. What is the height of the cliff?

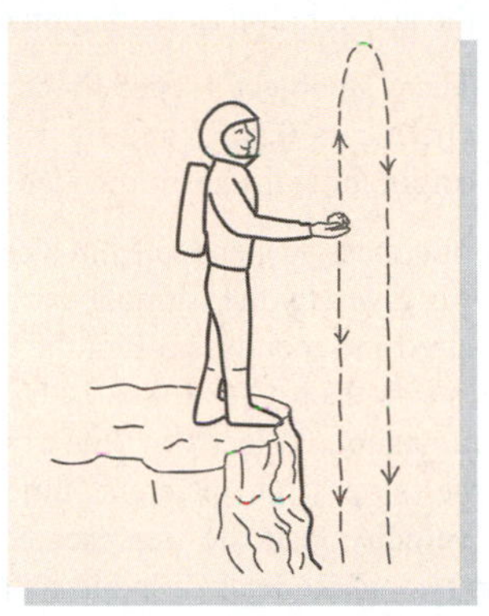

▲ **FIGURE 2–33** Problem 60.

61. (II) A ball is dropped from the roof of a 25-m-tall building. What is the velocity of the object when it touches the ground? Suppose that the ball is a perfect golf ball and it bounces such that the velocity as it leaves the ground has the same magnitude but the opposite direction as the velocity with which it reached the ground. How high will the ball bounce? Now suppose, instead, that the ball bounces back to a height of 20 m. What was the velocity with which it left the ground?

62. (II) A ball is thrown upward from the ground. It passes a window 10 m above the ground and is seen to descend past the window 2.2 s after it went by on its way up (Fig. 2–34). It reaches the ground 3.6 s after it was thrown. Use this information to calculate the acceleration due to gravity, g.

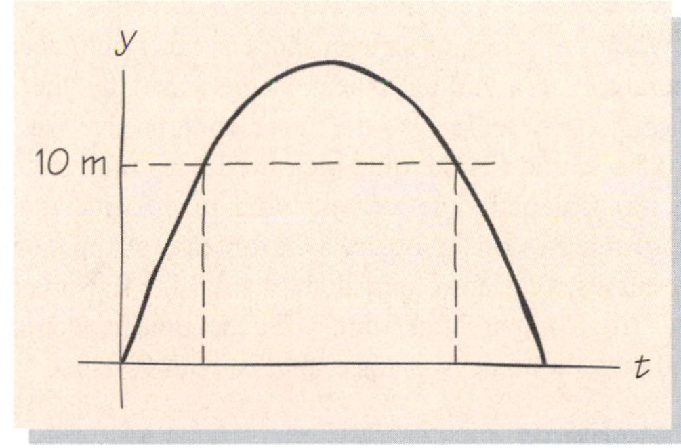

▲ **FIGURE 2–34** Problem 62.

*2–6 Integration and Motion in One Dimension

63. (II) An object moves with an acceleration that depends on time and has the form $a = -4\sqrt{t}$ meters per seconds squared, where t is measured in seconds. Its velocity at $t = 0$ s is 15 m/s. How far will the object travel before it comes to a stop?

64. (II) A horizontally moving rocket has a chemical burn rate that produces an acceleration given by $a = 10 - 2t$, where the units are in SI. The rocket starts from rest at $x = 0$ and $t = 0$ and the fuel burns out at $t = 5$ s. (a) What is the velocity of the rocket as a function of time? (b) What is the position of the rocket at 5 s?

65. (II) The velocity of an object moving in a viscous fluid is given by the expression

$$v(t) = (4.0 \text{ m/s}) + (8.0 \text{ m/s})e^{-0.5t},$$

where t is measured in seconds. Calculate the acceleration of the object as a function of time. Using this result and the expression for $v(t)$, write the acceleration as a function of the velocity.

66. (I) A machine causes an object's speed to increase exponentially, $v(t) = v_0 e^{at}$, where $a = 0.5 \text{ s}^{-1}$ and $v_0 = 1$ m/s. If the object starts from the origin, how far has it traveled after 2 s?

67. (II) The height of a mass suspended from a ceiling by a spring at different times t is given by the formula $y = (0.1 \text{ m}) \sin(3\pi t)$, where t is measured in seconds. (a) Plot the height as a function of time for times $t = 0$ s to $t = 1.0$ s. (b) Use your graph to determine the instantaneous velocity at time $t = 0.15$ s. (c) A measurement of the slopes at different times shows that the instantaneous velocity can be represented by the formula $\vec{v} = A[\cos(3\pi t)]\hat{j}$, where again t is measured in seconds. Use your measurement from part (b) to determine A. (Do not forget the units of A.) (d) Plot v as a function of t and use that graph to determine the instantaneous acceleration at time $t = 0.15$ s.

68. (II) The acceleration of gravity g is a constant only for a limited range of height differences. A better approximation, one that might hold over a larger range of height differences, is that g decreases linearly with height, $g = g_0 - hg'$, where h is the height measured from the ground surface and g' is a (small) constant of the appropriate dimensions. (a) Find the speed of a dropped object as a function of height assuming it was dropped starting from rest from a height h_0. (b) Find the speed of a dropped object as a function of time assuming it was dropped starting from rest from a height h_0.

69. (III) A powerful rocket moves for a short time with an acceleration that grows with time according to the formula $a = \alpha t^2$. If the rocket is to accelerate from rest in this way until it reaches a speed v_f, how long must the acceleration be maintained?

70. (III) When the effect of air resistance is taken into account, the acceleration of a falling object is described by the equation $a = ge^{-bt}$, where g is the acceleration due to gravity, $b = 0.5 \text{ s}^{-1}$, and t is the time measured from the moment of release. (a) Calculate the velocity and displacement (from the place of release) of the object as a function of time. (b) Sketch these curves. (c) How long does it take for the object to fall 50 m? (d) Compare this time with the time it would take to fall 50 m without air resistance, that is, with $b = 0$.

General Problems

71. (I) The simplest juggling act involves two objects, one of which is transferred from one hand to the other when the second object is tossed upward (Fig. 2–35). Perform an experiment that will tell you how fast you can transfer an object from one hand to the other (e.g., by transferring something back and forth 20 times while a friend times you). Estimate from this how high you would have to toss the second object to perform the juggling act. How high would you have to toss the objects if you wanted to juggle three of them?

▲ **FIGURE 2–35** Problem 71.

72. (I) A test of your reaction time is to catch a 12-in ruler held vertically by another person. Put your thumb and one finger near the bottom of the ruler, and as soon as the other person releases the ruler, squeeze your thumb and finger together to prevent the ruler from falling. Suppose that in such a test the ruler is grabbed after 5 in of it has passed your hand. What is the time interval between the visual detection of movement and the squeezing together of the fingers?

73. (I) A high jumper can jump 2 m on Earth. All other things being equal, how high could the same jumper jump on the surface of the Moon, where the acceleration of gravity is one-sixth that of Earth?

74. (I) You could probably jump off an 8-ft wall without hurting yourself (but do not try it!). Estimate what your deceleration would be when you hit the ground.

75. (II) A Moon rock is thrown upward with velocity 7 m/s. After 7 s, it has a downward velocity of 4 m/s. What is the acceleration due to gravity on the Moon? How high above the starting point did the rock go before it began to fall?

76. (II) Two long-separated friends, June and Bill, spot each other in an airport terminal from a distance of 20 m. They start to run toward each other (Fig. 2–36). Bill accelerates at a constant rate of 0.9 m/s^2 and June at a constant rate of 1.0 m/s^2. How far from June's initial position do they meet?

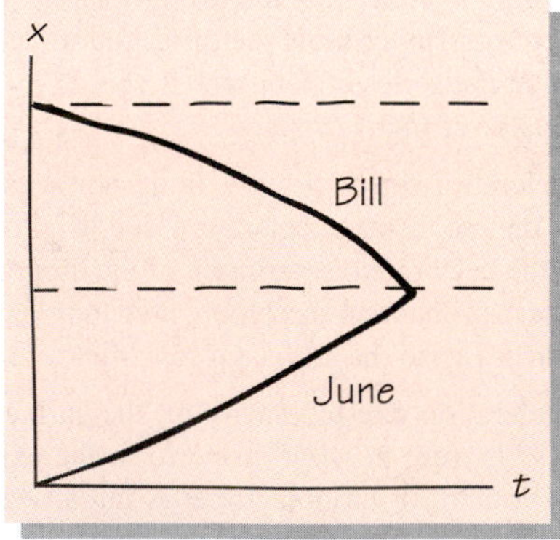

▲ **FIGURE 2–36** Problem 76.

77. (II) An object moves in one dimension (described by an x-value) with a constant value J of the *third* derivative of position with respect to time. Write an equation for the position as a function of time given an initial position x_0 and an initial velocity v_0 (the sign of v_0 indicates the initial direction of motion).

78. (II) The velocity of an object moving along a straight axis is given as a function of time by $v = (4\ \text{m/s}^4)t^3 - (1\ \text{m/s}^2)t$, where v is measured in meters per second and t is measured in seconds. How far did the object move in the period from 0.5 to 1.5 s? Find the average velocity in this period and compare it to the maximum and minimum values of the velocity in the same period. You may want to make a graph to help.

79. (II) A tennis ball is dropped from a height of 10 m. It falls onto an electrical switch and bounces back to a height of 9 m. The switch is connected to an electronic device that shows that the time of contact between the ball and the switch was 0.002 s. Calculate (a) the velocity with which the ball hit the switch, (b) the velocity with which the ball left the switch, and (c) the average acceleration during the time of contact with the switch.

80. (II) A water balloon is dropped from the top of a tower, 200 m off the ground. An alert archer at the base of the tower sees the balloon and shoots an arrow straight up toward the balloon 5 s after the balloon is dropped. The arrow's initial velocity is 40 m/s. Where does the arrow intercept the balloon?

81. (II) There are several known cases of paratroopers whose chutes did not open as they fell but who survived by falling into brush or snow or onto a steep hillside. It is possible to survive a fall when the deceleration on impact is some 500 m/s^2, equivalent to about 50g. What is the distance traveled within a snowbank while a paratrooper comes to a stop if the deceleration is constant with a magnitude of 50g? The speed with which a paratrooper would enter the snowbank is about 40 m/s. Over how much time does the paratrooper decelerate?

82. (III) One test for the effects of the acceleration of gravity is to tie a set of weights to a string, with the second lowest separated from the lowest by L_0, the third lowest from the second lowest by L_1, and so forth, and to then drop the string (Fig. 2–37). Supposing that free fall corresponds to motion with constant acceleration, how should the separations $L_1, L_2, \ldots, L_n$ (where n is the number of weights) be related to L_0 if the sounds made by the weights as they land form a steady beat? The lowest weight starts at the surface onto which the other weights fall.

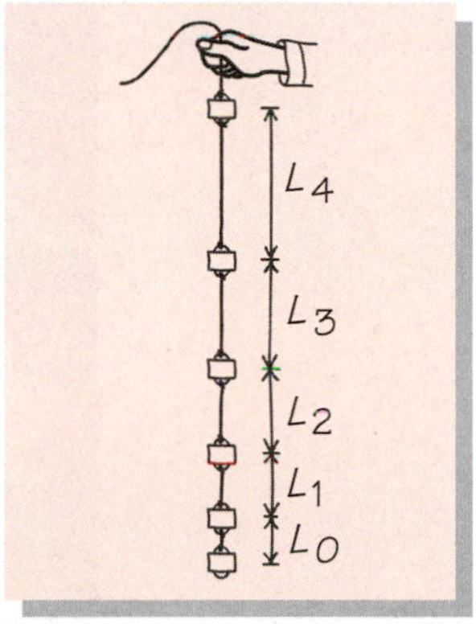

▲ **FIGURE 2–37** Problem 82.

Chapter 3

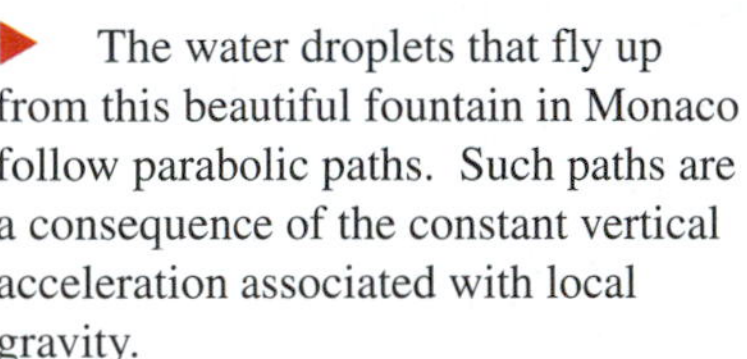

The water droplets that fly up from this beautiful fountain in Monaco follow parabolic paths. Such paths are a consequence of the constant vertical acceleration associated with local gravity.

Motion in Two and Three Dimensions

In Chapter 2 we considered the description of motion in one dimension—linear motion. We will now extend this description to motion in space. We observe such motion in the curved path of a thrown ball, in the swing of the pendulum of a grandfather clock, or in the orbits of the planets around the Sun. While for one-dimensional motion the directional aspect is encapsulated in signs, for motion in two or three dimensions we must use vectors to describe the directional aspect properly. With the help of the mathematical apparatus provided by vectors, we'll find that it is straightforward to describe the motion of objects in a plane or in space in a manner that builds on our earlier work with one-dimensional motion.

3–1 Position and Displacement

The motion of a planet orbiting the Sun traces out a path in space. Similarly, a rock thrown off a cliff follows a certain path, or **trajectory**, as does any pointlike object as it moves through space. For motion in a plane, think of a skater on a lake whose skates leave marks that specify the trajectory of the motion. Figure 3–1 depicts a particle, for example the skater, moving in a two-dimensional plane. We label the plane as the xy-plane and introduce a Cartesian coordinate system that contains an origin and x- and

(a)

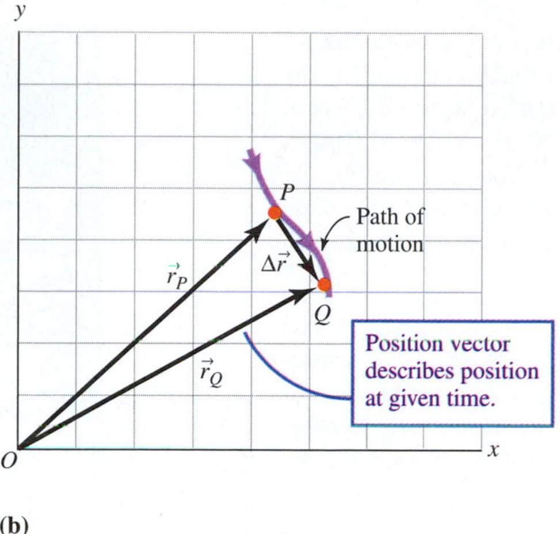

(b)

FIGURE 3–1 (a) The ice skater is gliding over the path indicated by the blue trajectory line. (b) Position vectors $\vec{r}_P$ and $\vec{r}_Q$ point from the origin to the positions P and Q at the two times t_1 and t_2, respectively, along the skater's path of motion. The displacement vector between these times is $\Delta\vec{r} \equiv \vec{r}_Q - \vec{r}_P$.

y-axes. The particle is at the position P at time t_1; this position is described by the position vector $\vec{r}_P$, which points from the origin to the point P. At a later time t_2 the particle is located at position Q and is described by the position vector $\vec{r}_Q$. The change in the particle's position between times t_1 and t_2—the final position minus the initial position—can be described by the displacement vector $\Delta\vec{r}$; this vector is defined by

$$\Delta\vec{r} \equiv \vec{r}_Q - \vec{r}_P. \tag{3–1}$$

The vector $\Delta\vec{r}$ points from the tip of vector $\vec{r}_P$ to the tip of vector $\vec{r}_Q$ and describes the direction of the displacement as well as its magnitude. Whereas the position vectors $\vec{r}_Q$ and $\vec{r}_P$ depend on the choice of origin, the displacement vector $\Delta\vec{r}$ is independent of the choice of origin. To see this clearly, let's imagine that there is a new origin O' such that the vector from O' to O is the fixed vector $\vec{b}$ (Fig. 3–2). The position vector of point P in the new coordinate system is $\vec{r}_P + \vec{b}$ and that of point Q in the new system is $\vec{r}_Q + \vec{b}$. If we calculate the displacement, which is the difference $(\vec{r}_Q + \vec{b}) - (\vec{r}_P + \vec{b})$, the vector $\vec{b}$ cancels. In other words, we have again arrived at the displacement vector defined in Eq. (3–1). This can also be seen in the graphical representation in Fig. 3–2. The displacement is independent of our choice of origin.

As a particle moves, the components of its position vector (with respect to the Cartesian coordinate axes) change with time:

$$\vec{r}(t) = x(t)\hat{i} + y(t)\hat{j}. \tag{3–2}$$

For three-dimensional motion, we would proceed exactly as in Section 1–6: We set up three axes, define three mutually perpendicular unit vectors $\hat{i}$, $\hat{j}$, and $\hat{k}$, and write a position vector in the form

$$\vec{r}(t) = x(t)\hat{i} + y(t)\hat{j} + z(t)\hat{k}. \tag{3–3}$$

The fact that there is more than one vector component to the motion is the only difference between one-dimensional motion and two- or three-dimensional motion.

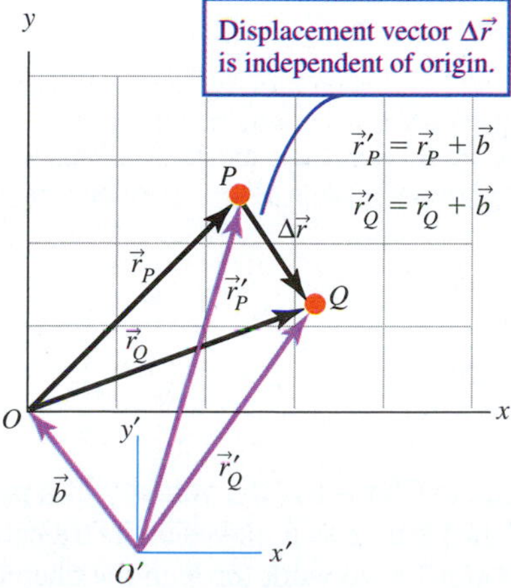

FIGURE 3–2 A displacement vector $\Delta\vec{r}$ is independent of the origin. Here O and O' are two origins, and although the initial and final position vectors to points P and Q do depend on the origins, the difference between these position vectors does not.

EXAMPLE 3–1 The position of a bumper car in an amusement park ride (Fig. 3–3a) is described as a function of time by the coordinates $x = c_1t^2 + c_2t + c_3$ and $y = d_1t^2 + d_2t + d_3$, where $c_1 = 0.20\ \text{m/s}^2$, $c_2 = 5.0\ \text{m/s}$, $c_3 = 0.50\ \text{m}$, $d_1 = -1.0\ \text{m/s}^2$, $d_2 = 10.0\ \text{m/s}$, and $d_3 = 2.0\ \text{m}$. Find the position vectors of the car at $t = 3.0$ s and $t = 6.0$ s and the displacement vector between these times. Plot the trajectory, that is, a curve of y versus x that traces the path of the car on the floor.

Setting It Up We plot the locations of the car along the x-axis and the y-axis as a function of time in Figs. 3–3b and 3–3c, respectively. In the first instance we are given the position as a function of time and are asked to find the position at a particular time; in other words, we are actually given what we must find! We'll label the desired displacement vector $\Delta\vec{r}$.

Strategy The first part is a straightforward numerical substitution. For the displacement, we calculate the difference between the position vectors (components x and y) at 6.0 and 3.0 s. This difference is the displacement vector. As for plotting the trajectories, the simplest way to proceed is to start with an xy-plane. We can then mark the x- and y-values at a given time as a point on this graph. A half-second later, say, there is another point that can be marked, and so forth. By connecting those consecutive points starting from the one at the earliest time, we mark out the trajectory.

Working It Out We insert the two values of time (3.0 and 6.0 s) into the equations for x and y:

for $t = 3.0$ s:

$$x(t) = (0.20\ \text{m/s}^2)(3.0\ \text{s})^2 + (5.0\ \text{m/s})(3.0\ \text{s}) + 0.50\ \text{m} = 17\ \text{m},$$
$$y(t) = (-1.0\ \text{m/s}^2)(3.0\ \text{s})^2 + (10.0\ \text{m/s})(3.0\ \text{s}) + 2.0\ \text{m} = 23\ \text{m};$$

for $t = 6.0$ s:

$$x(t) = (0.20\ \text{m/s}^2)(6.0\ \text{s})^2 + (5.0\ \text{m/s})(6.0\ \text{s}) + 0.50\ \text{m} = 38\ \text{m},$$
$$y(t) = (-1.0\ \text{m/s}^2)(6.0\ \text{s})^2 + (10.0\ \text{m/s})(6.0\ \text{s}) + 2.0\ \text{m} = 26\ \text{m}.$$

With these components, Eq. (3–2) gives us the position vectors of the car at the two times:

$$\text{for } t = 3.0\ \text{s}: \vec{r}(t) = (17\hat{i} + 23\hat{j})\ \text{m};$$
$$\text{for } t = 6.0\ \text{s}: \vec{r}(t) = (38\hat{i} + 26\hat{j})\ \text{m}.$$

Thus the displacement vector of the car between 3.0 and 6.0 s is [Eq. (3–1)]

$$\Delta\vec{r} = \vec{r}(t = 6.0\ \text{s}) - \vec{r}(t = 3.0\ \text{s})$$
$$= (38\hat{i} + 26\hat{j})\ \text{m} - (17\hat{i} + 23\hat{j})\ \text{m} = (21\hat{i} + 3\hat{j})\ \text{m}.$$

Finally, we plot y versus x, moment by moment, in Fig. 3–3d. This curve is the trajectory of the car.

What Do You Think? Why does the trajectory curve look so similar to the curve of y versus time? There are other vector descriptions of motion in a plane. Can you think of another such set? *Answers to* ***What Do You Think?*** *questions are given in the back of the book.*

(a)

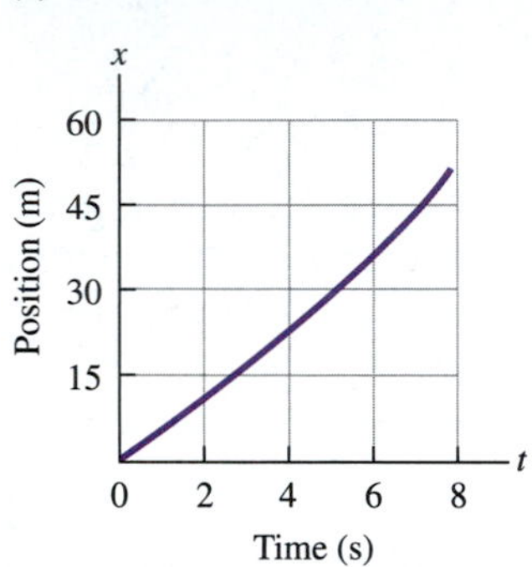

(b)

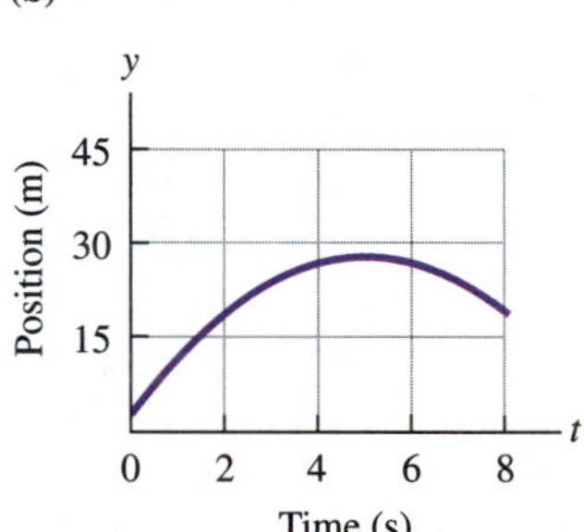

(c)

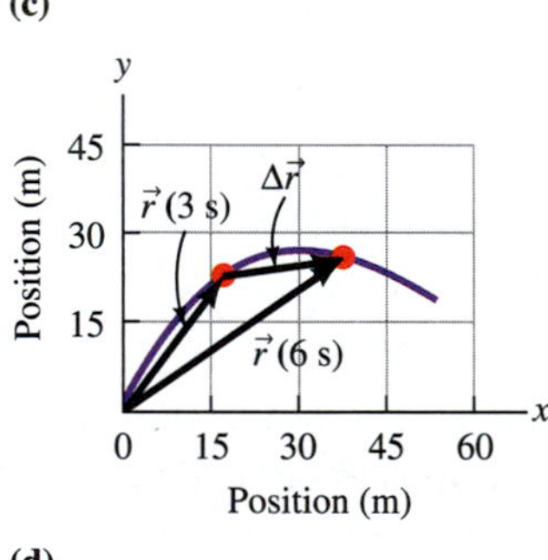

(d)

▲ **FIGURE 3–3** (a) An xy-coordinate system can be laid out on the floor. (b) Location along x-axis of car's motion from 0 to 8 s. (c) Car's location along y-axis. (d) By combining the results of part (b) and part (c), we can plot the trajectory, a graph of the y-position versus the x-position.

3–2 Velocity and Acceleration

Velocity

As for the one-dimensional motion described in Chapter 2, the velocity of a particle describes the rate of change of the position of the particle as it moves on its trajectory. Generally we will consider two-dimensional motion as we work through the chapter, as it is simpler than considering three-dimensional motion, but the approach applies perfectly well to three dimensions. Using Eq. (3–1) for the particle's displacement, the average velocity $\vec{v}_{av}$ over the finite time interval from t to $t + \Delta t$ is accordingly defined by

$$\vec{v}_{av} \equiv \frac{\vec{r}(t + \Delta t) - \vec{r}(t)}{\Delta t} = \frac{\Delta \vec{r}}{\Delta t}. \tag{3–4}$$

Equation (3–4) shows that the direction of $\vec{v}_{av}$ is the same as the direction of the displacement vector $\Delta \vec{r}$.

As the time Δt tends towards zero, the displacement over that interval becomes smaller and smaller, and as we'll describe in more detail later, the displacement vector $\Delta \vec{r}$ becomes tangent to the particle's trajectory at the location of the moving particle. Then, as in Eq. (2–11), the *instantaneous velocity* $\vec{v}(t)$ is obtained by letting Δt become infinitesimally small:

$$\vec{v}(t) \equiv \lim_{\Delta t \to 0} \frac{\vec{r}(t + \Delta t) - \vec{r}(t)}{\Delta t} = \frac{d\vec{r}}{dt}. \tag{3–5}$$

We have recognized that in the limit $\Delta t \to 0$ we arrive at the time derivative of the position vector. The instantaneous velocity can change from moment to moment. The direction of $\vec{v}$ at time t is tangent to the trajectory curve at that time (Fig. 3–4). Of course, we already know that its magnitude is by definition the particle's speed.

We can write the velocity vector in terms of components by using Eqs. (3–2) and (3–5):

$$\vec{v} = \frac{d}{dt}\vec{r}(t) = \frac{d}{dt}[x(t)\hat{i} + y(t)\hat{j}] \tag{3–6}$$

$$= \frac{dx}{dt}\hat{i} + \frac{dy}{dt}\hat{j}. \tag{3–7}$$

(The unit vectors $\hat{i}$ and $\hat{j}$ are constant in magnitude and direction, so their derivatives are zero.) We write Eq. (3–7) in the form

$$\vec{v} = v_x\hat{i} + v_y\hat{j} \tag{3–8}$$

$$= \vec{v}_x + \vec{v}_y, \tag{3–9}$$

where

$$v_x = \frac{dx}{dt}, \tag{3–10a}$$

$$v_y = \frac{dy}{dt}, \tag{3–10b}$$

and the component vectors are

$$\vec{v}_x = \frac{dx}{dt}\hat{i}, \tag{3–11a}$$

$$\vec{v}_y = \frac{dy}{dt}\hat{j}. \tag{3–11b}$$

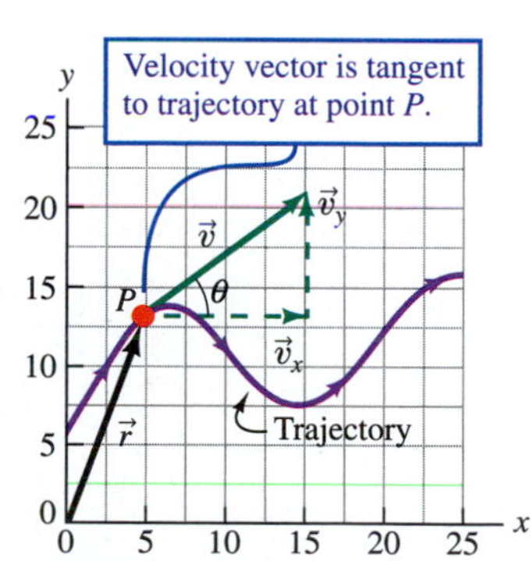

▲ **FIGURE 3–4** The velocity vector $\vec{v}$ at point P is tangent to the particle's trajectory at that point. The component vectors $\vec{v}_x$ and $\vec{v}_y$ of the velocity vector $\vec{v}$ at that point are also included.

The component vectors $\vec{v}_x$ and $\vec{v}_y$ of the velocity vector $\vec{v}$ are drawn in Fig. 3–4. The magnitude of the velocity $\vec{v}$ can be written in terms of the components of $\vec{v}$:

$$v = |\vec{v}| = \sqrt{v_x^2 + v_y^2}. \tag{3–12}$$

The angle θ that the velocity vector $\vec{v}$ makes with the x-axis is determined in terms of the components of the velocity by

$$\tan\theta = \frac{v_y}{v_x}. \tag{3–13}$$

EXAMPLE 3–2 Use the data presented in Example 3–1 to find the bumper car's average velocity over the period from 3.0 to 6.0 s and the car's instantaneous velocity at $t = 3.0$ s.

Setting It Up The data of Example 3–1 give the position of the car as a function of time.

Strategy The average velocity is given by Eq. (3–4), and this requires us to know the displacement for a given time interval. That information is available from Example 3–1. For the instantaneous velocity we use Eq. (3–7) and evaluate the derivatives of $x(t)$ and $y(t)$.

Working It Out Given the result of Example 3–1, that the displacement vector of the bumper car between $t = 3.0$ s and $t = 6.0$ s is $\Delta\vec{r} = (21\hat{i} + 3.0\hat{j})$ m, we have

$$\vec{v}_{\text{av}} = \frac{\Delta\vec{r}}{\Delta t} = \frac{(21\hat{i} + 3.0\hat{j})\text{ m}}{6.0\text{ s} - 3.0\text{ s}} = (7.0\hat{i} + 1.0\hat{j})\text{ m/s}.$$

As for the instantaneous velocity, we require

$$\frac{dx}{dt} = \frac{d}{dt}(c_1t^2 + c_2t + c_3) = 2c_1t + c_2,$$

$$\frac{dy}{dt} = \frac{d}{dt}(d_1t^2 + d_2t + d_3) = 2d_1t + d_2.$$

Substituting the numerical values of c_1, c_2, d_1, and d_2 at $t = 3.0$ s from Example 3–1, we have

$$\frac{dx}{dt} = 2(0.20\text{ m/s}^2)(3.0\text{ s}) + (5.0\text{ m/s}) = 6.2\text{ m/s},$$

$$\frac{dy}{dt} = 2(-1.0\text{ m/s}^2)(3.0\text{ s}) + (10.0\text{ m/s}) = 4.0\text{ m/s}.$$

Thus the velocity at $t = 3.0$ s is

$$\vec{v} = \frac{dx}{dt}\hat{i} + \frac{dy}{dt}\hat{j} = (6.2\text{ m/s})\hat{i} + (4.0\text{ m/s})\hat{j}.$$

This velocity vector is shown in Fig. 3–5.

What Do You Think? Could you have used the graph of the trajectory in Fig. 3–3d to read off the velocity of the car?

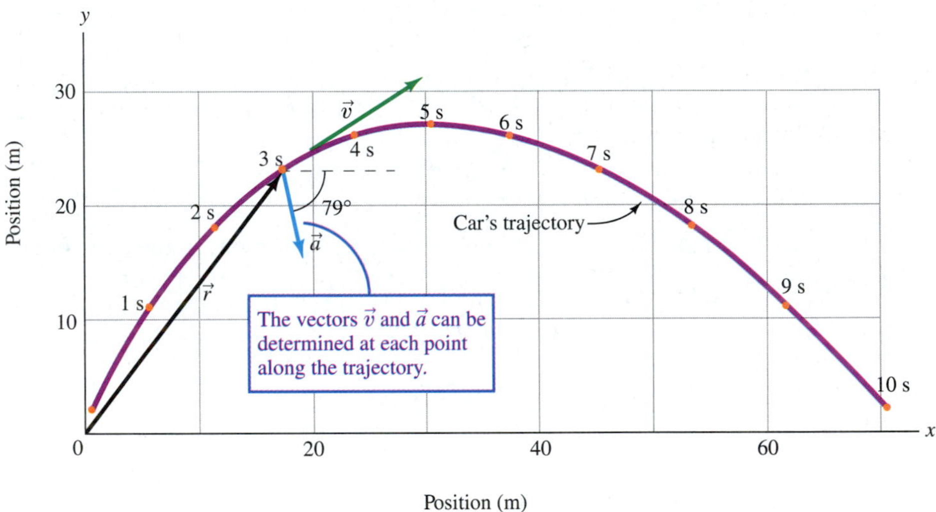

▶ **FIGURE 3–5** The trajectory of the car's path in Example 3–1 is plotted for times up to 10 s; also shown are the position vector $\vec{r}$, the velocity $\vec{v}$, and the acceleration $\vec{a}$ at $t = 3.0$ s.

Acceleration

Acceleration describes how rapidly velocity changes with time. This "change" could be in the magnitude (the speed) or the speed could remain the same while the *direction* of the velocity vector changes or both magnitude and direction may change. As for motion in one dimension, acceleration is found from velocity in the same way that velocity is found from displacement. For a finite time interval Δt, the average acceleration is defined as

$$\vec{a}_{\text{av}} \equiv \frac{\vec{v}(t + \Delta t) - \vec{v}(t)}{\Delta t} = \frac{\Delta\vec{v}}{\Delta t}. \quad (3\text{–}14)$$

The instantaneous acceleration at time t is the limit of the average acceleration as Δt approaches zero, which is a derivative:

$$\vec{a} \equiv \lim_{\Delta t \to 0} \frac{\vec{v}(t + \Delta t) - \vec{v}(t)}{\Delta t} = \frac{d\vec{v}}{dt}. \quad (3\text{–}15)$$

The instantaneous acceleration is in principle a function of time, meaning that its three components are generally functions of time. As for velocity, we can express acceleration in terms of its components; for two dimensions (again for economy) we have

$$\vec{a} = \frac{dv_x}{dt}\hat{i} + \frac{dv_y}{dt}\hat{j} \quad (3\text{–}16)$$

$$= a_x\hat{i} + a_y\hat{j}. \quad (3\text{–}17)$$

Here, the components of the acceleration vector are

$$a_x = \frac{dv_x}{dt} = \frac{d^2x}{dt^2}, \quad (3\text{–}18a)$$

$$a_y = \frac{dv_y}{dt} = \frac{d^2y}{dt^2}. \quad (3\text{–}18b)$$

EXAMPLE 3–3 Calculate the instantaneous acceleration, magnitude and direction, of the bumper car in Example 3–1 at $t = 1.0$ s and $t = 3.0$ s.

Setting It Up We will want to use the known velocity vector of the car calculated in Example 3–2 using data for the position vectors from Example 3–1.

Strategy The acceleration vector is the time derivative of the known velocity vector, Eqs. (3–18a) and (3–18b). The acceleration is a function of time, into which we will then substitute particular values of time.

Working It Out From Eqs. (3–18),

$$a_x = \frac{dv_x}{dt} = \frac{d}{dt}(2c_1t + c_2) = 2c_1$$

and

$$a_y = \frac{dv_y}{dt} = \frac{d}{dt}(2d_1t + d_2) = 2d_1.$$

Thus [Eq. (3–17)]

$$\vec{a} = 2c_1\hat{i} + 2d_1\hat{j}.$$

In this case, the car's acceleration is a constant—it is independent of time—and so is exactly the same for $t = 1.0$ s and for $t = 3.0$ s. Given the values of c_1 and d_1 from Example 3–1 (0.20 and $-1.0\ \text{m/s}^2$, respectively), the numerical value of the acceleration is

$$\vec{a} = (0.40\hat{i} - 2.0\hat{j})\ \text{m/s}^2.$$

The magnitude of the acceleration is

$$a = |\vec{a}| = \sqrt{a_x^2 + a_y^2} = \sqrt{(0.40)^2 + (-2.0)^2}\ \text{m/s}^2$$
$$= \sqrt{4.2}\ \text{m/s}^2 = 2.0\ \text{m/s}^2.$$

The acceleration vector makes an angle θ with the x-axis, which is shown in Fig. 3–5; the angle θ is derived from

$$\tan\theta = \frac{a_y}{a_x} = \frac{-2.0\ \text{m/s}^2}{0.40\ \text{m/s}^2} = -5.0;$$

so $\theta = -79°$; that is, the direction of the acceleration is at $-79°$ to the horizontal, almost directly toward the $-y$-direction.

What Do You Think? We started this example by taking the time derivative of the known velocity vector. Could we instead have started with the position vector as a function of time (given in Example 3–1)?

Representing Trajectories

Look again at the trajectory of the bumper car discussed in Examples 3–1, 3–2, and 3–3 (Fig. 3–5); this trajectory is a curve representing the car's position on the floor of the amusement park ride (its x-position versus its y-position). In Fig. 3–5, we show a position vector $\vec{r}$, a velocity vector $\vec{v}$, and an acceleration vector $\vec{a}$ at $t = 3$ s. Although the figure shows $\vec{a}$ at the point corresponding to $t = 3.0$ s, we in fact saw that the acceleration of the car is independent of time and hence would be drawn as the same vector everywhere along the curve.

We can create a graphical representation like that in Fig. 3–5 for any motion. As an object moves, its trajectory will be traced out by the tip of the position vector $\vec{r}$ as $\vec{r}$ changes with time. The velocity vector $\vec{v}$ at any time t is a vector of magnitude $|\vec{v}|$ that is *tangential* to the trajectory at time t. This is quite intuitive. However, it is not quite so obvious how to think about the acceleration.

Since the acceleration is to the velocity as the velocity is to the displacement, one thing we could do is to repeat the procedure of the preceding paragraph with velocity and acceleration. A plot of the tip of the vector $\vec{v}$ can be drawn; it is the curve of the points whose *horizontal coordinate* at time t is v_x and whose *vertical coordinate* at that time is v_y. We might call this the "velocity trajectory." The acceleration at time t is given by a vector whose magnitude is $|\vec{a}|$ and whose direction is tangential to the velocity trajectory at time t. *The acceleration vector $\vec{a}$ is tangent to the velocity trajectory but not to the trajectory itself.*

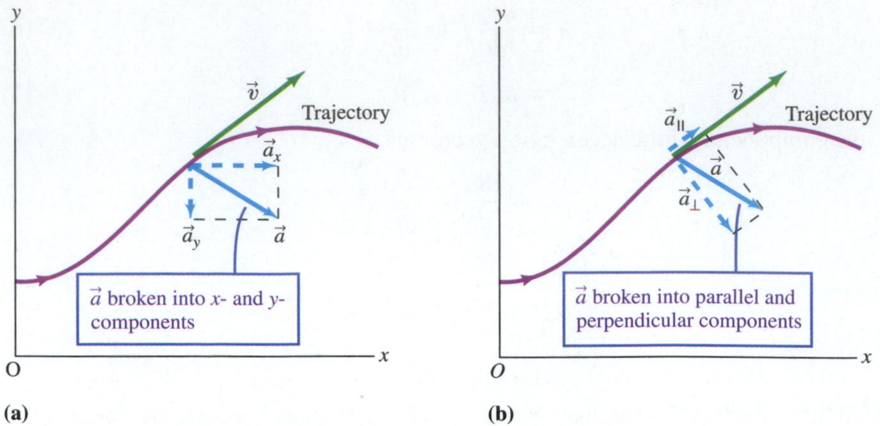

▲ **FIGURE 3–6** Velocity $\vec{v}$ and acceleration $\vec{a}$ of a particle following some trajectory. (a) The acceleration of the particle is separated into x- and y-components. (b) The acceleration of the particle is separated into components parallel and perpendicular to the path.

We can illustrate the consequences of these facts using Fig. 3–6, which shows the path of an object with $\vec{v}$ and $\vec{a}$ indicated at one time along the path. In Fig. 3–6a, the acceleration is separated into its a_x and a_y components. Alternatively, we can separate the acceleration $\vec{a}$ into components that are parallel (tangential) and perpendicular (normal) to the velocity vector (Fig. 3–6b). We label these components $a_\parallel$ and $a_\perp$, respectively. The component $a_\parallel$ of $\vec{a}$ that is parallel to $\vec{v}$ affects the magnitude but not the direction of $\vec{v}$. Similarly, the $a_\perp$ component changes the direction but not the magnitude of $\vec{v}$. It is useful to refer separately to the parallel and perpendicular components of an object's acceleration because they affect the velocity differently.

CONCEPTUAL EXAMPLE 3–4 The motion of bumper cars is extremely erratic: You are colliding with other cars or you are trying to use evasive techniques. Figure 3–7 shows the path of a bumper car. Consider the points *A, B, C, D*, and *E*. (a) At which point did a collision most likely take place? (b) At which point did evasive action most likely take place? (c) Can you determine where the magnitude of the velocity is the greatest? (d) Can you determine where the magnitude of the acceleration is the greatest?

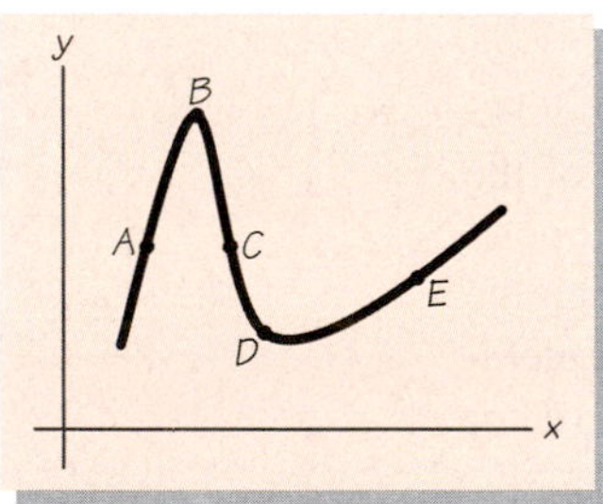

▲ **FIGURE 3–7** The motion of a bumper car can be quite erratic as it slams into other cars and takes evasive action to avoid collisions. Here we have the trajectory, or path, of such a car.

Answer (a) A collision most likely took place at point *B* because there is an abrupt change in direction. The driver could not change the direction so quickly without some outside effect.

(b) Evasive action probably took place at point *D* because there is a rapid but smooth change in direction as the driver turned quickly. The motion change is typically more abrupt at collisions.

(c, d) We can't tell where the magnitudes of the velocity and acceleration are the greatest from the trajectory alone because we do not know the times associated with points along the trajectory. We might guess that the acceleration was a maximum during the collision at point *B* because the velocity would change dramatically during the collision. On the other hand, if the bumper car were traveling very slowly at the time of a collision at point *B* then the collision might not be a very violent one and the acceleration would not necessarily be very large. You simply do not have enough information on a trajectory to tell. A plot such as Fig. 3–7 does not contain *all* the information about the motion.

3–3 Motion with Constant Acceleration

When an object moves with *constant acceleration*—meaning constant in both magnitude and direction—it can move only in a straight line (one dimension) or a plane (two dimensions). The plane of motion is formed by the initial velocity vector and the acceleration vector $\vec{a}$. The motion remains in this plane because, as Fig. 3–8 illustrates, the initial velocity vector has no component $v_{0\perp}$ perpendicular to the specified plane, and since the acceleration is in the plane, $v_\perp$ can never change and become nonzero. Motion near

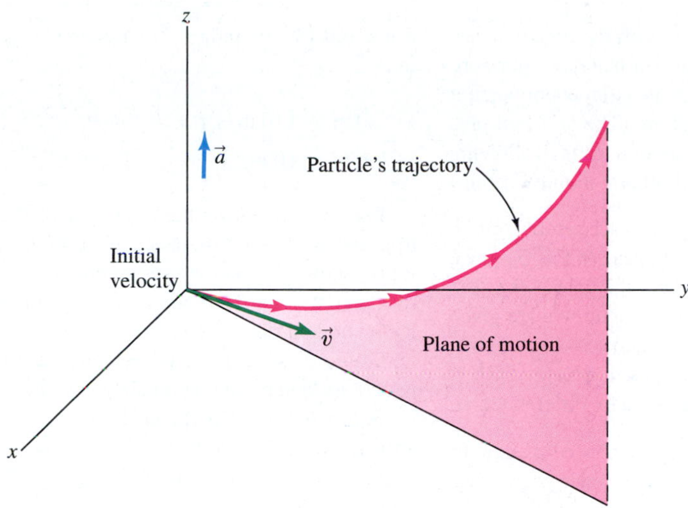

◀ **FIGURE 3–8** Projectile motion lies in a plane, the plane formed by the initial velocity vector and the acceleration vector.

Earth's surface, that is, motion under the sole influence of gravity, with air resistance neglected, provides an everyday example. If we throw a rock, it moves in a plane defined by the initial direction of the motion and the constant (vertical) acceleration of gravity.

At this point we'll simplify our notation by defining the plane of the motion as the xy-plane. We'll suppose for the moment that the initial velocity can have both x- and y-components, as can the (constant) acceleration. We can then use the results for one dimension from Chapter 2 to write *independently* the x- and y-components of position $\vec{r}$ and velocity $\vec{v}$ in terms of the constant-acceleration components. In other words, we can think of the x- and y-motions as separate from each other, governed only by their own separate constant accelerations. We use Eqs. (2–17) and (2–21) to find

$$x\text{-component of } \vec{r}: \quad x = x_0 + v_{0x}t + \tfrac{1}{2}a_x t^2, \tag{3–19}$$

$$x\text{-component of } \vec{v}: \quad v_x = v_{0x} + a_x t; \tag{3–20}$$

$$y\text{-component of } \vec{r}: \quad y = y_0 + v_{0y}t + \tfrac{1}{2}a_y t^2, \tag{3–21}$$

$$y\text{-component of } \vec{v}: \quad v_y = v_{0x} + a_x t. \tag{3–22}$$

Here, x_0 and y_0 are the components of $\vec{r} = \vec{r}_0$ at an initial time $t = 0$ and v_{0x} and v_{0y} are the components of $\vec{v} = \vec{v}_0$ at time $t = 0$. Together these quantities are the given **initial conditions**. In vectorial form, the initial conditions are

$$\vec{r}_0 = x_0\hat{i} + y_0\hat{j} \tag{3–23}$$

and

$$\vec{v}_0 = v_{0x}\hat{i} + v_{0y}\hat{j} \tag{3–24}$$

at $t = 0$.

Equations (3–19) through (3–22), which give position and velocity for motion with constant acceleration $\vec{a}$, can be written more compactly in vector form:

$$\vec{r} = \vec{r}_0 + \vec{v}_0 t + \tfrac{1}{2}\vec{a}t^2 \tag{3–25}$$

MOTION WITH CONSTANT ACCELERATION

$$\vec{v} = \vec{v}_0 + \vec{a}t \tag{3–26}$$

This form of the kinematic equations has the additional benefit that it does not refer to any particular set of axes. Remember that these important and useful results are valid *only* when $\vec{a}$ is constant. We can easily see the important features of these compact equations. In particular, you can see that for any direction for which the acceleration component is zero the position (or, equivalently, the displacement) component changes *linearly* with time, corresponding to a *constant*-velocity component. For any direction for which the acceleration component is not zero, the position component changes *quadratically* in time, corresponding to a *linearly* changing velocity component.

EXAMPLE 3–5 A wayward golf ball rolls off the edge of a vertical cliff overlooking the Pacific Ocean. The golf ball has a horizontal velocity component of 10 m/s and no vertical component when it leaves the cliff. Describe the subsequent motion. (The golf ball provides us with our first glimpse of projectile motion. In the following section we will look at this important type of motion in more detail.)

Setting It Up The displacement, velocity, and acceleration all lie in the same plane, which we assign to be the xy-plane. In Fig. 3–9a we include a coordinate system, placing the origin at the point where the ball leaves the cliff and with the y-direction pointing up. We are given initial values of velocity. A "description" of the motion consists of writing the position as a function of time. Given this, further quantities, such as velocity as a function of time, can be found by differentiation.

Strategy This is a case of motion under constant acceleration. In this case the ball's constant acceleration is that of gravity, and thus $\vec{a} = \vec{g}$. The vector $\vec{g}$ points toward Earth's center—vertically downward—and has magnitude 9.8 m/s^2. Because we have constant acceleration, we can use Eqs. (3–19) through (3–22) to describe the motion, for which we know the initial values (at $t = 0$, the moment when the ball rolls off the cliff). As emphasized above, we can say that because there is no component of acceleration in the x-direction, the horizontal velocity component is constant and will remain at its initial value.

Working It Out We start with initial values: The golf ball's initial position and velocity in our chosen coordinate system are $x_0 = 0\text{ m}$, $v_{0x} = 10\text{ m/s}$, $y_0 = 0\text{ m}$, and $v_{0y} = 0\text{ m/s}$. Next we specify that in our coordinate system the acceleration has components $a_x = 0\text{ m/s}^2$ and $a_y = -9.8\text{ m/s}^2$. We determine the velocity components as a function of time from Eqs. (3–20) and (3–22):

$$v_x = 10\text{ m/s}$$

and

$$v_y = 0\text{ m/s} + (-9.8\text{ m/s}^2)t = (-9.8\text{ m/s}^2)t. \qquad (3\text{–}27)$$

Equations (3–19) and (3–21) give the ball's position as a function of time:

$$\begin{aligned} x &= 0\text{ m} + (10\text{ m/s})t + \tfrac{1}{2}(0\text{ m/s}^2)t^2 = (10\text{ m/s})t, \\ y &= 0\text{ m} + (0\text{ m/s})t + \tfrac{1}{2}(-9.8\text{ m/s}^2)t^2 = (-4.9\text{ m/s}^2)t^2. \end{aligned} \qquad (3\text{–}28)$$

Figure 3–9a shows the trajectory of the golf ball. (We'll discuss trajectories under constant acceleration in more detail in Section 3–4.) It also shows the velocity vector and its components at 0.5-s intervals for the first 2 s of the motion. The horizontal component of the velocity stays constant, whereas the vertical component changes linearly with time. Further, the total velocity vector is a tangent to the ball's path of motion at each point along its trajectory.

Figure 3–9b shows the position vector $\vec{r}$, velocity $\vec{v}$, and acceleration $\vec{a}$ at $t = 1$ s and $t = 2$ s. Whereas $\vec{a}$ remains constant, $\vec{r}$ and $\vec{v}$ change with time. The three vectors $\vec{r}$, $\vec{v}$, and $\vec{a}$ do not generally point in the same direction at a given time during the golf ball's motion. The directions of $\vec{r}$ and $\vec{v}$ are specified by angles θ and θ', respectively, with respect to the x-axis. These angles are

$$\tan\theta = \frac{y}{x} = \frac{(-4.9\text{ m/s}^2)t^2}{(10\text{ m/s})t} = (-0.49\text{ s}^{-1})t$$

and

$$\tan\theta' = \frac{v_y}{v_x} = \frac{(-9.8\text{ m/s}^2)t}{(10\text{ m/s})} = (-0.98\text{ s}^{-1})t,$$

respectively. Both angles vary with time.

What Do You Think? According to Fig. 3–9, the ball appears to drop into the ocean about 25 m from the cliff. In this problem what determines how far from the base of the cliff the ball enters the water?

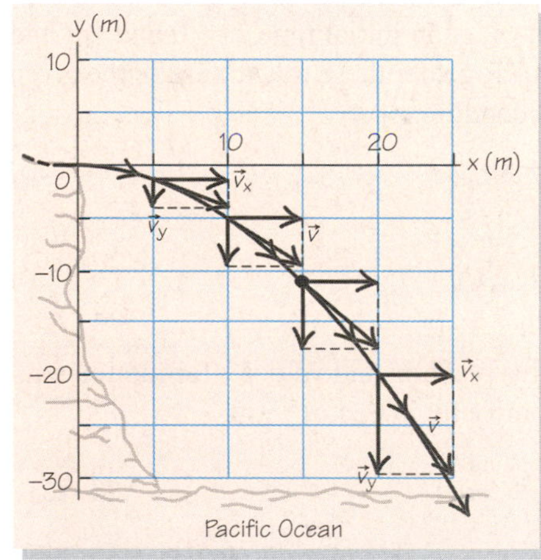

(a)

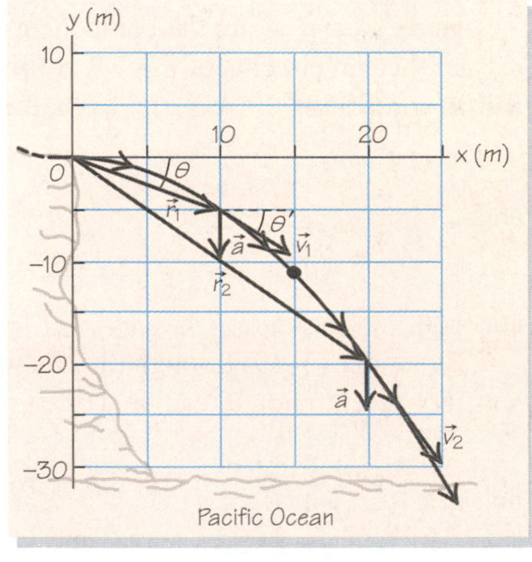

(b)

FIGURE 3–9 (a) The velocity vector $\vec{v}$ and components $\vec{v}_x$ and $\vec{v}_y$ of the golf ball are shown at 0.5-s intervals up to 2.0 s. (b) Position $\vec{r}$, velocity $\vec{v}$, and acceleration $\vec{a}$ of the golf ball for $t_1 = 1$ s and $t_2 = 2$ s.

3–4 Projectile Motion

A golf ball in motion is an example of a *projectile* that moves under the effect of gravity. In the absence of air resistance, what is the trajectory of a projectile? The motion is that of constant acceleration due to gravity, and this constant acceleration $\vec{g}$ has only a vertical component; we can use all the constant-acceleration results of the previous section to find the trajectory. The ball's motion is best described by separating it into horizontal and vertical components—as we have already emphasized, the horizontal motion is *independent* of the vertical motion—and then applying the kinematic equations for constant acceleration.

Usually it is easiest to place the origin at the starting point, assigning the y-direction vertically and the x-direction along the horizontal (Fig. 3–10), as we did in Example 3–5. The initial position of the ball is $x_0 = y_0 = 0$; the initial velocity at $t = 0$ is $\vec{v}_0$. The flight of the golf ball starts at an initial angle to the horizontal that we call the *elevation angle* θ_0. Then $\vec{v}_0$ has components

$$v_{0x} = v_0 \cos\theta_0 \quad \text{and} \quad v_{0y} = v_0 \sin\theta_0. \tag{3–29}$$

The components of the acceleration are the constants

$$a_x = 0 \quad \text{and} \quad a_y = -g. \tag{3–30}$$

Using Eqs. (3–19) through (3–22), the components of $\vec{r}$ and $\vec{v}$ (the position and velocity of the ball, respectively) are

$$x = 0 + (v_0 \cos\theta_0)t + \tfrac{1}{2}(0)t^2 = (v_0 \cos\theta_0)t, \tag{3–31}$$

$$y = 0 + (v_0 \sin\theta_0)t + \tfrac{1}{2}(-g)t^2 = (v_0 \sin\theta_0)t - \tfrac{1}{2}gt^2, \tag{3–32}$$

and

$$v_x = v_0 \cos\theta_0 + (0)t = v_0 \cos\theta_0, \tag{3–33}$$

$$v_y = v_0 \sin\theta_0 - gt. \tag{3–34}$$

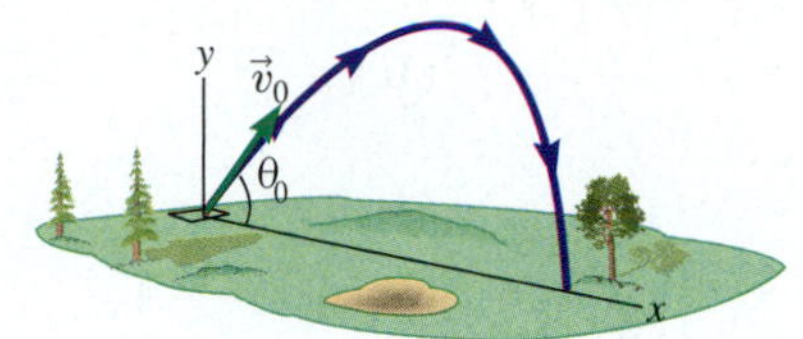

(a)

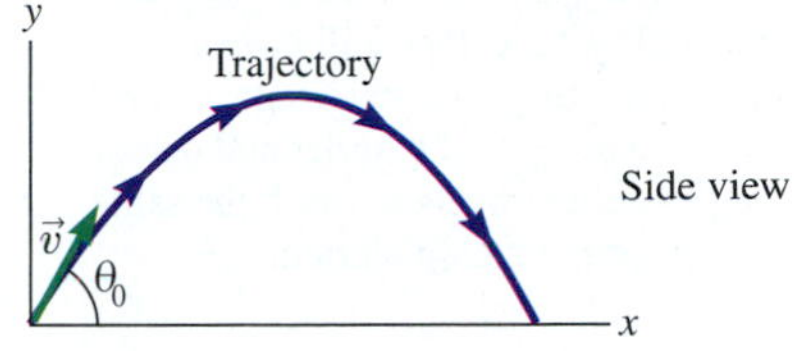

(b)

▲ **FIGURE 3–10** (a) A golf ball leaves a tee with an initial velocity of magnitude v_0 at an elevation angle θ_0. (b) The side view of the motion shows a parabolic trajectory.

The Trajectory

We can find the trajectory of the golf ball by plotting its height y versus its x-position. We know both x and y as functions of time, and we can eliminate the time dependence by using Eq. (3–31) to find the time t as a function of x. We then insert the result for t into Eq. (3–32) to find the trajectory, that is, the height y as a function of x, with the time dependence eliminated:

$$t = \frac{x}{v_0 \cos\theta_0}; \tag{3–35}$$

$$y = (v_0 \sin\theta_0)\frac{x}{v_0\cos\theta_0} - \frac{1}{2}g\left(\frac{x}{v_0\cos\theta_0}\right)^2$$

$$= (\tan\theta_0)x - \left(\frac{g}{2v_0^2\cos^2\theta_0}\right)x^2. \tag{3–36}$$

The coefficients of x and x^2 in Eq. (3–36) are both constants, so the trajectory has the form

$$y = C_1 x - C_2 x^2. \tag{3–37}$$

This is the equation of a parabola passing through the origin with its axis parallel to the y-axis. *The trajectory of all objects moving with constant acceleration is parabolic.* Parabolic motion is illustrated in the chapter-opening photograph and Fig. 3–10 as well as in Fig. 3–11, which shows the position of a ball at equal time intervals.

The trajectory and the time dependence of the components of displacement have some simple characteristics that can be useful in our study of projectile motion—range, flight time, and maximum height. These are easily extracted from the motion, and we discuss them further below.

▲ **FIGURE 3–11** Motion of a ball bouncing along the floor and moving under the influence of gravity. In the air, the ball moves with constant acceleration, which in this case is directed downward due to gravity. The velocity vector changes throughout the motion, although its horizontal component does not. The velocity's vertical component changes linearly with time. The resulting trajectory forms a series of parabolas.

Range: We define the **range** R of a projectile launched from the ground ($y = 0$) to be the horizontal distance that the projectile travels over level ground; that is, it lands at the same height from which it started. The quantity R is the value of x when the projectile has returned to the ground, that is, when y again equals zero. If we insert $y = 0$ into Eq. (3–37), we have

$$y = 0 = R(C_1 - C_2 R), \tag{3–38}$$

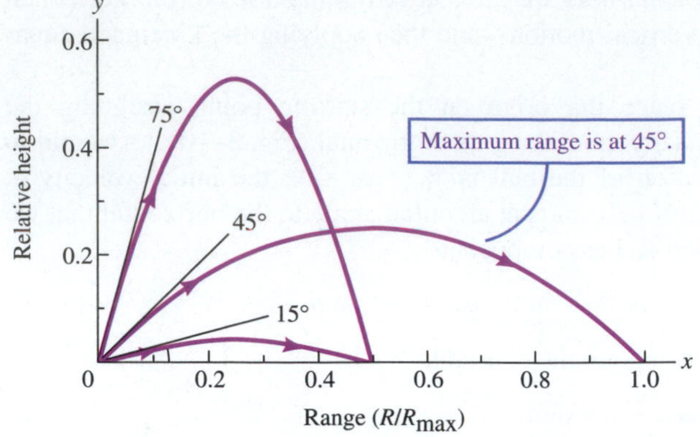

▶ **FIGURE 3–12** For a fixed initial speed and if air resistance is ignored, a projectile's trajectory will have a maximum range for an elevation angle of 45°. The range is the horizontal distance the projectile travels to reach the same height from which it started.

where we have set $x = R$. To find R, set the factor $C_1 - C_2R = 0$ in Eq. (3–38), or $R = C_1/C_2$. Inserting the values of C_1 and C_2 from Eq. (3–36) yields

$$R = \frac{C_1}{C_2} = \frac{\tan\theta_0(2v_0^2\cos^2\theta_0)}{g} = \frac{2v_0^2}{g}\left(\frac{\sin\theta_0}{\cancel{\cos\theta_0}}\right)\cos^{\cancel{2}}\theta_0 = \frac{v_0^2}{g}2\sin\theta_0\cos\theta_0.$$

From trigonometry, $\sin(2\theta_0) = 2\sin\theta_0\cos\theta_0$, and we find

$$R = \frac{v_0^2}{g}\sin 2\theta_0. \tag{3–39}$$

The range R depends on the initial speed v_0 and the elevation angle (the initial angle) of the projectile. As θ_0 increases progressively from 0° to 45° and then to 90°, the range $R\,[\propto \sin(2\theta_0)]$ starts out at zero, increases to a maximum at $\theta_0 = 45°$ [i.e., $\sin(2\theta_0) = 1$], then decreases back down to zero at $\theta_0 = 90°$. So, to throw or kick a ball over level ground as far as you can, send it upward at a 45° angle. For this case, which gives the maximum range, we have

$$R_{max} = \frac{v_0^2}{g}. \tag{3–40}$$

If the projectile is launched at an angle higher or lower than 45°, the range is shorter (Fig. 3–12). Note that according to Eq. (3–39) there are *two* initial angles for which a projectile has the same range for a given initial speed (Fig. 3–12). For example, in softball a pop fly at 75° and a line drive at 15° can both be caught by the shortstop (compare the two trajectories in Fig. 3–12).

Flight Time: Let T be the total flight time of a ball. Figure 3–13 shows that the ball reaches its maximum height exactly halfway through its trajectory, at time $t = T/2$. At this point, its motion is horizontal and the vertical component of velocity is zero. We can find $T/2$ by setting $v_y = 0$ in Eq. (3–34), $0 = v_0\sin\theta_0 - g(T/2)$. We solve for T to find that

$$T = \frac{2v_0}{g}\sin\theta_0. \tag{3–41}$$

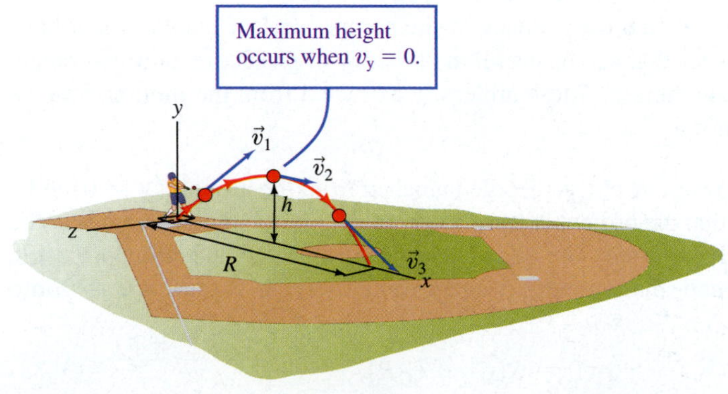

▶ **FIGURE 3–13** A projectile (a ball) moving under the force of gravity is at its maximum height when $v_y = 0$. At that moment, the ball is traveling horizontally. We have marked the velocity at this point as $\vec{v}_2$.

This time the simple factor $\sin\theta_0$ enters. Look again at the motion at 75° and 15° in Fig. 3–12 and you will understand that in softball the fly ball's flight time is greater than that of the line drive.

Maximum Height: The maximum height $y_{max} = h$ is reached at time $T/2$. From Eq. (3–32), we find the height at this time,

$$h = (v_0 \sin\theta_0)\frac{2v_0}{2g}\sin\theta_0 - \frac{1}{2}g\left(\frac{2v_0}{2g}\sin\theta_0\right)^2 = v_0^2\frac{\sin^2\theta_0}{g} - gv_0^2\frac{\sin^2\theta_0}{2g^2}$$
$$= v_0^2\frac{\sin^2\theta_0}{2g}. \qquad (3\text{–}42)$$

We use Eqs. (3–36), (3–39), (3–41), and (3–42) to determine a projectile's trajectory, range, flight time, and maximum height, respectively. The range and flight time refer to the special case where the ball returns to its original height. These equations need not be memorized; instead, it is important to understand how they were obtained. We apply these methods again in Examples 3–6 through 3–10.

EXAMPLE 3–6 To win a bet that he can drive a golf ball a horizontal distance of 250 m, an amateur golfer goes to a cliff overlooking the ocean. The cliff is 52 m above the ocean. The golfer strikes the golf ball so that the ball's initial speed is 48 m/s and the elevation angle (from the horizontal) is 36°. Does he win his bet? What is the horizontal distance actually covered by the ball?

Setting It Up Figure 3–14 shows the situation. We place the origin of our coordinate system at the tee where the ball's motion starts, letting y extend upward. We know the initial conditions ($t = 0$ when the ball is struck), which with our coordinate system are $x_0 = 0$ m, $y_0 = 0$ m, $v_0 = 48$ m/s, and $\theta_0 = 36°$. We want to determine the distance R' from the tee to the point at which the golf ball reaches the ocean ($y = -52$ m).

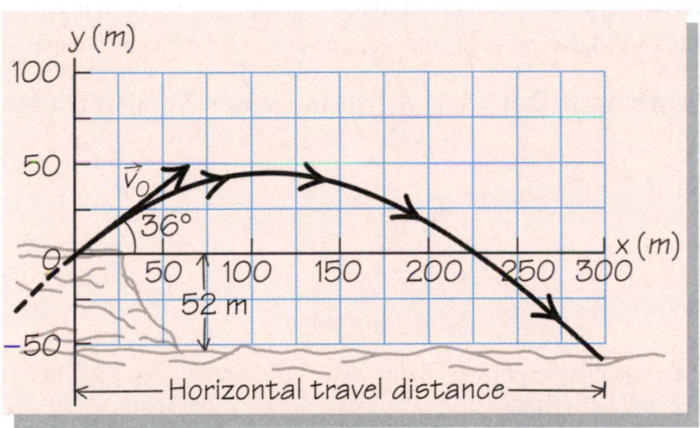

▲ **FIGURE 3–14** A golf ball is driven off a cliff into the ocean.

Strategy We use the trajectory equation to find the value of x at which the golf ball reaches the ocean surface. Note that we cannot use Eq. (3–39) to calculate the range because that result applies only to level ground; we don't want the horizontal distance when the ball returns to $y = 0$ m. However, we can still use Eq. (3–36) to find the value of x when $y = -52$ m.

Working It Out Equation (3–36) reads in our case

$$y = -52\text{ m} = (\tan\theta_0)R' - \left(\frac{g}{2v_0^2\cos^2\theta_0}\right)R'^2.$$

Rearranging this equation yields

$$R'^2 - \frac{2v_0^2\cos^2\theta_0\tan\theta_0}{g}R' + \frac{2yv_0^2\cos^2\theta_0}{g} = 0,$$
$$R'^2 + bR' + c = 0,$$

where

$$b = -\frac{2v_0^2\cos^2\theta_0\tan\theta_0}{g} \quad\text{and}\quad c = \frac{2yv_0^2\cos^2\theta_0}{g}.$$

Solving this quadratic equation to find R' gives

$$R' = \frac{-b \pm \sqrt{b^2 - 4c}}{2}.$$

Inserting the values of b and c, we obtain

$$R' = \frac{v_0^2\cos^2\theta_0\tan\theta_0}{g} \pm \frac{1}{2}\sqrt{\frac{4v_0^4\cos^4\theta_0\tan^2\theta_0}{g^2} - \frac{8yv_0^2\cos^2\theta_0}{g}}.$$

Now inserting $y = -52$ m and the initial values to determine R' yields

$$R' = \frac{(48\text{ m/s})^2\cos^2 36°\tan 36°}{9.8\text{ m/s}^2}$$
$$\pm\frac{1}{2}\sqrt{\frac{4(48\text{ m/s})^4\cos^4 36°\tan^2 36°}{(9.8\text{ m/s}^2)^2} - \frac{8(-52\text{ m})(48\text{ m/s})^2\cos^2 36°}{9.8\text{ m/s}^2}}$$
$$= 281\text{ m or } -57\text{ m}.$$

Now, did the golfer drive the ball a distance of 281 m or −57 m? The positive value must be correct. The golfer wins his bet.

What Do You Think? We stated the positive solution (281 m) must be the correct solution to the problem, but the negative solution (−57 m) also is a solution. What is the physical meaning of the negative solution?

EXAMPLE 3–7 What was the maximum height above the ocean of the golf ball in Example 3–6, and how long was the golf ball in flight?

Setting It Up We can again refer to Fig. 3–14. We denote the maximum height above the ground by h. In this case we want the maximum height *above the ocean*. We also want to find the total time T of the trip.

Strategy The maximum height of the golf ball occurs when the vertical component of the velocity is zero, and Eq. (3–42) will give h. The value we seek is $h + 52$ m. As for T, we can find it by using Eq. (3–31), together with the knowledge that the total horizontal distance traveled is 281 m.

Working It Out From Eq. (3–42),

$$h = \frac{v_0^2 \sin^2 \theta_0}{2g} = \frac{(48\text{ m/s})^2 \sin^2 36^\circ}{2(9.8\text{ m/s}^2)} = 41\text{ m}.$$

The answer is therefore 41 m + 52 m = 93 m.

From Eq. (3–31) with a horizontal distance traveled of 281 m,

$$281\text{ m} = (48\text{ m/s})(\cos 36^\circ)T;$$

$$T = \frac{281\text{ m}}{(48\text{ m/s})\cos 36^\circ} = 7.2\text{ s}.$$

What Do You Think? On level ground the horizontal distance covered by the golf ball depends on $\sin 2\theta_0$ [see Eq. (3–39)], where θ_0 is the initial elevation angle, and the range is a maximum for $\theta_0 = 45^\circ$. However, in this example we do not have level ground. Will the maximum horizontal travel distance still occur for $\theta_0 = 45^\circ$?

CONCEPTUAL EXAMPLE 3–8 A major league pitcher and you, the student, compete in throwing a baseball as far as possible in an initially horizontal direction. Assume that you each throw so that each ball leaves the hand at exactly the same height. Whose ball will go further and why?

Answer The time it takes the ball to hit the ground is determined by the height from which the ball starts. Since the ball leaves the hand horizontally, the initial vertical velocity component is the same for both of you, namely zero, and hence the time it takes to hit the ground is the same for both of you. But the initial horizontal velocity component of the ball is expected to be larger for the professional pitcher, so that in the same time it covers a larger distance. The independence of the two components of the motion is again key here. Figure 3–15 illustrates the equal fall time for two projectiles that fit this description; in this photograph one of the projectiles has an initial speed of zero.

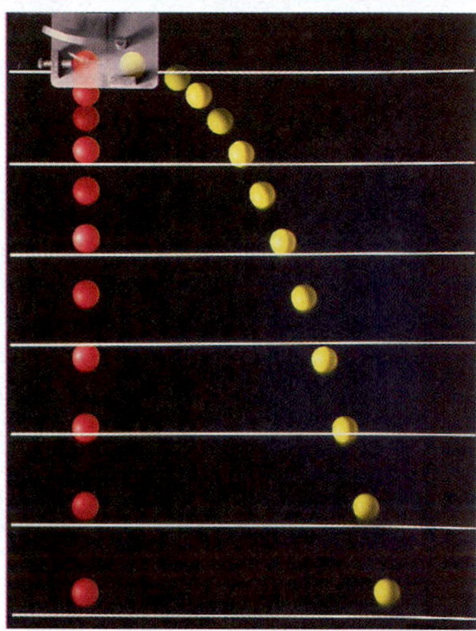

▶ **FIGURE 3–15** Two balls released simultaneously have two different trajectories, but in a given time each moves the same vertical distance. The difference in their motions is the magnitude of their (constant) x-components of velocity.

EXAMPLE 3–9 A group of engineering students constructs a slingshot device that lobs water balloons. The device is constructed so that the angle of the lob can be adjusted, and it has a launch speed (the balloon's initial speed) of 12 m/s. There is a target 14 m away at the same elevation. How should they adjust the initial angle so that they reach the target?

Setting It Up The slingshot setup is shown in Fig. 3–16. The students must find a value of launch angle θ_0 that will produce a given range R for a given initial speed v_0.

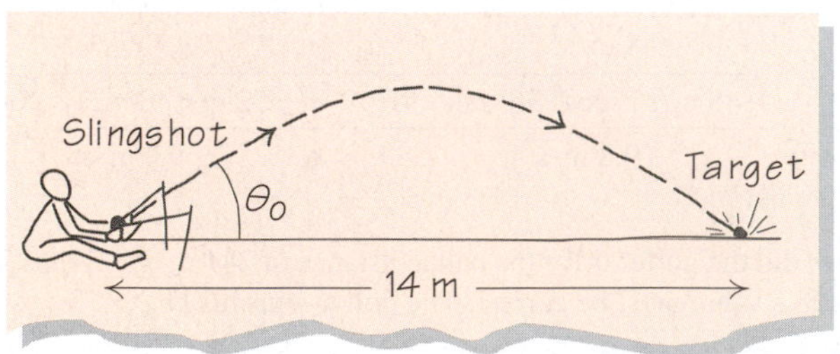

▲ **FIGURE 3–16** The students can orient their slingshot in two ways to get the same range for the same initial speed—just one is shown here.

Strategy In this case the range equation for horizontal ground, Eq. (3–39), can be used, and we can solve it for the launch angle.

Working It Out With $R = 14$ m and $v_0 = 12$ m/s, Eq. (3–39) gives

$$R = 14\text{ m} = \frac{(12\text{ m/s})^2 \sin 2\theta_0}{9.8\text{ m/s}^2},$$

or

$$\sin 2\theta_0 = 0.95.$$

This equation has *two* solutions, $2\theta_0 = 72^\circ$ and $2\theta_0 = 108^\circ$, or θ_0 is 36° and 54°. These are the two possible initial angles that result in a given range, as in Fig. 3–12. We have drawn one of these trajectories (Fig. 3–16). *A reminder*: There will always be two initial angles that generate the same range, except for maximum range, which is produced only by the limiting angle 45°.

What Do You Think? From the standpoint of surprise, which of the two solutions, 36° or 54°, might be best for the students to use if the target were human?

EXAMPLE 3–10 A boy would rather shoot coconuts down from a tree than climb the tree or wait for the coconuts to drop. The boy aims his slingshot directly at a coconut, but at the same moment that his rock leaves the slingshot, the coconut falls from the tree. Show that the rock will hit the coconut.

Setting It Up We establish the launch point of the rock as the origin of a suitable xy-coordinate system in Fig. 3–17. The coconut is at the point (x_0, y_0). We are asked if the two objects moving under the influence of gravity will be at the same spot at the same time. For one of the trajectories, the initial velocity is given by an angle that would take it to the coconut if there were no gravity.

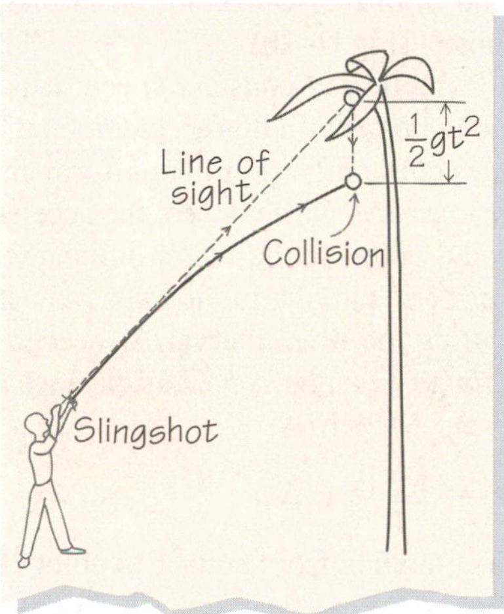

▲ **FIGURE 3–17** If the coconut falls at the same time the rock leaves the slingshot, both the coconut and rock fall the same distance.

Strategy We must compare a trajectory that includes both a horizontal and a vertical component (the rock) versus one that has only a vertical component (the coconut). For that reason it will be useful to think in terms of these components. It is useful to first consider what would happen if there were no gravity, then to see how the presence of gravity modifies the positions of both the rock and coconut.

Working It Out The rock has an initial velocity (v_{x0}, v_{y0}). If there were no gravity acting, the rock would follow a straight-line path that would place it at the point $(x_0 = v_{x0}t, y_0 = v_{y0}t)$ after a time t. This is the time necessary for the rock to reach the coconut (which is still at the tree since gravity has been ignored so far). Now let's include the effect of gravity. First, consider what happens to the coconut. During the time t that the rock travels toward the coconut, the coconut falls the distance $gt^2/2$ (Fig. 3–17). In other words, the height of the coconut after time t is [Eq. (3–21)]

$$y = y_0 - \tfrac{1}{2}gt^2.$$

Next, consider what happens to the rock when we include gravity. The rock's horizontal velocity component remains constant at v_{x0}. However, the vertical velocity component of the rock is changing under the effect of gravity and, after time t, Eq. (3–21) shows us that the rock's height is not $v_{y0}t$ but rather

$$y = v_{y0}t - \tfrac{1}{2}gt^2.$$

The rock is a height $gt^2/2$ below the height it would have if it followed a straight-line motion, which is precisely the distance the coconut falls (Fig. 3–17). Thus the rock will hit the coconut at the common point $\frac{1}{2}gt^2$ below the coconut's starting point. In effect, the parabolic path of the rock "tracks" the falling coconut.

What Do You Think? The real world is usually somewhat different than the idealized case discussed in textbooks. What are some reasons why the rock may not hit the coconut?

THINK ABOUT THIS . . .

IS IT POSSIBLE TO EXPERIENCE FREE FALL FOR LONG PERIODS?

Every jump puts you in free fall. Some of you may have done bungee jumping, where you can be in free fall for a couple of seconds until the cord starts to pull. What would it be like to be in free fall for longer periods? NASA has equipped a KC-135 airplane that allows training astronauts and others to experience longer periods of free fall. The plane is equipped to coast following a parabolic trajectory identical to that of a projectile. For the 25 s of the dive, the occupants are in free fall along with the airplane (see Fig. 3–18). Much of the film *Apollo 13*, which recounts the dramatic story of a mission to the Moon that barely made it back to Earth, was shot within the NASA plane. The best place to experience free fall is the International Space Station, which is orbiting around Earth in a free fall in which Earth's curvature allows the surface to "fall away" from the projectile's path as the projectile proceeds. The personnel inhabiting the station may be in free fall for months.

◀ **FIGURE 3–18** The interior of the airplane used by NASA for a free-fall environment during that part of the flight where the plane follows the same parabolic path taken by a projectile in free fall. In this photograph, astronauts in training are experiencing some of the same effects they will feel during a stay in the International Space Station.

▲ **FIGURE 3–19** This turntable is rotating with two orange capital letters "E". They are blurred due to the motion.

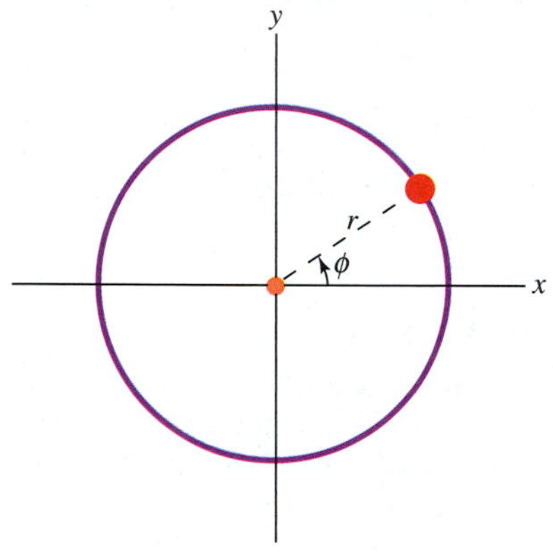

▲ **FIGURE 3–20** Radial variable r and angular variable ϕ of plane polar coordinates. The symbol ρ is sometimes used instead of r.

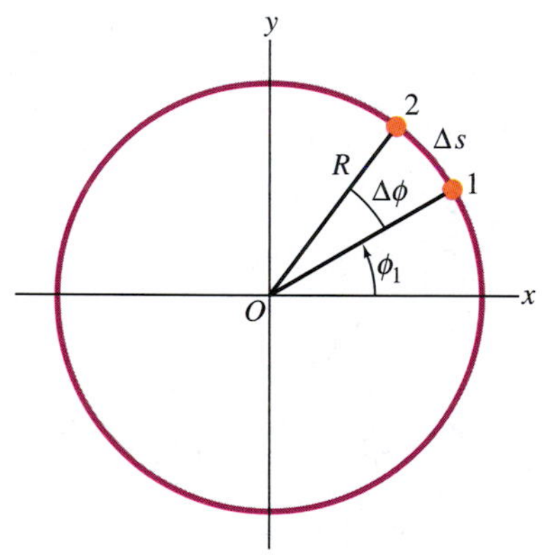

▲ **FIGURE 3–21** During the time interval Δt a particle in circular motion traverses an arc length Δs and an angle $\Delta\phi$.

3–5 Uniform Circular Motion

We are quite familiar with objects that move in circles at constant speeds, for example a carnival merry-go-round, the platter of a microwave oven, or a phonograph turntable. We are in fact being carried in a circular motion with Earth as it rotates each day. An object undergoes **uniform circular motion** when it travels at a constant speed along a circular path (see Fig. 3–19). Circular motion is a special case of motion on a plane. We can best describe uniform circular motion by specifying a radius r and an angular coordinate ϕ, as shown in Fig. 3–20. The origin of the coordinate system is at the center of the circle and the angle ϕ is measured from the $+x$-axis, increasing in the counterclockwise direction. We measure ϕ, which is dimensionless, in units of *radians* (rad) (an entire rotation around one circle is 2π radians or 360°). The coordinates r and ϕ are referred to as *plane polar coordinates*. For circular motion, the radius is fixed, so we can let the radius $r = R$, a constant, and in this case we only need one variable, the angle ϕ, to describe the position of an object (Fig. 3–20).

Although uniform circular motion is motion at constant speed, it is motion with *acceleration*, and this is a crucial point to remember. In this acceleration the *direction* of the velocity but not its magnitude changes constantly. (For nonuniform circular motion both the magnitude and the direction change.) As we shall see, the acceleration in uniform circular motion is a vector that always points to the center of the circular path.

Consider a point object moving in uniform circular motion at a radius R from point 1 to point 2, as shown in Fig. 3–21. During the time interval Δt needed to travel between these points, the object moves through an angle $\Delta\phi$. This distance along the circle is the *arc length* and we denote it by Δs. We have

$$\text{arc length} = \Delta s = R\,\Delta\phi. \tag{3–43}$$

The average speed of motion of the object moving from point 1 to point 2 is in turn

$$v = \frac{\Delta s}{\Delta t} = \frac{R\,\Delta\phi}{\Delta t} = R\frac{\Delta\phi}{\Delta t}. \tag{3–44}$$

If we take the limit of Δt, we arrive at the instantaneous speed

$$v = \frac{ds}{dt} = R\frac{d\phi}{dt}. \tag{3–45}$$

Because we are dealing with *uniform* circular motion here, the average speed and the instantaneous speed will be the same.

It is useful to define the **angular speed** ω of the object moving along the circle as the rate of change of the angle ϕ; in terms of the derivative this is

$$\omega \equiv \frac{d\phi}{dt}. \tag{3–46}$$

The speed v is simply related to ω,

$$v = \omega R. \tag{3–47}$$

In uniform circular motion, v is a constant, and hence, from Eq. (3–47), *ω is a constant in uniform circular motion*. This is simply understood by realizing that in uniform circular motion the angle is swept out at a steady rate. The angular speed ω is measured in radians per second (rad/s) in SI and has dimensions $[1/T]$.

The **period** T is the time an object takes to make one complete revolution, for example, the time it takes the friend you're watching on the carousel to come around again. The distance traveled in one revolution is $2\pi R$, and for uniform circular motion, the speed is a constant v. We can then determine the period T,

$$\text{distance traveled during one revolution} = 2\pi R = vT; \tag{3–48}$$

$$T = \frac{2\pi R}{v} = \frac{2\pi \not{R}}{\omega \not{R}} = \frac{2\pi}{\omega}. \tag{3–49}$$

The **frequency** f is the number of revolutions that the object makes per unit time; for example, Earth makes 1.156×10^{-5} revolutions per second (or 1 revolution every 86,400 s, i.e., a day). Frequency is the inverse of the period T:

$$f \equiv \frac{1}{T}. \tag{3–50}$$

This relation is easy to understand: If the period is 2 s, the frequency is $\frac{1}{2}$ revolution per second, or if the period is 0.1 s, the frequency is 10 revolutions per second. Using the last two equations, we see that the relation between the angular speed and frequency is

$$\omega = 2\pi f. \tag{3–51}$$

The SI unit of frequency f is the hertz (Hz), defined as one cycle (or revolution) per second (cps).

Acceleration in Uniform Circular Motion: Let's look in more detail at acceleration in uniform circular motion by considering how direction changes as an object moves in a circle at constant speed. We aim to show that the acceleration always points to the center of the circle and find its magnitude.

Figure 3–22a shows a particle located at point A, determined by angle ϕ_1 at time t_1. At a later time t_2 the particle is at point B and the particle's angular position is given by ϕ_2. The *direction* of the velocity vector is always tangential to the circle in the direction of motion and therefore changes continuously with time. Figure 3–22a includes the velocity vectors $\vec{v}_1$ at time t_1 and $\vec{v}_2$ at time t_2. These two vectors have the same magnitude but different directions. The *change* in the velocity vector over the time period Δt (where $\Delta t = t_2 - t_1$) is given by $\Delta\vec{v} = \vec{v}_2 - \vec{v}_1$. We have drawn this vector difference in Fig. 3–22b midway between points A and B. We see that $\Delta\vec{v}$, and hence $\vec{a}_{av} = \Delta\vec{v}/\Delta t$, points toward the center of the circle. To see this more clearly for $\Delta\vec{v}$, just mentally shift the position of $\Delta\vec{v}$ so that it acts at the midway point in Fig. 3–22b, which puts its direction toward $-y$ (down). If we make the time interval Δt smaller, as in Fig. 3–23a, then $\Delta\phi$ also gets smaller, and $\Delta\vec{v}$—and consequently $\vec{a}_{av}$—points ever more closely to the center of the circle (Fig. 3–23b). In the limit in which Δt goes to zero, the ratio $\Delta\vec{v}/\Delta t$ gives us the instantaneous acceleration. *The instantaneous acceleration points precisely to the center of the circle.* We accordingly call this a **centripetal acceleration**, from the Latin meaning "center seeking."

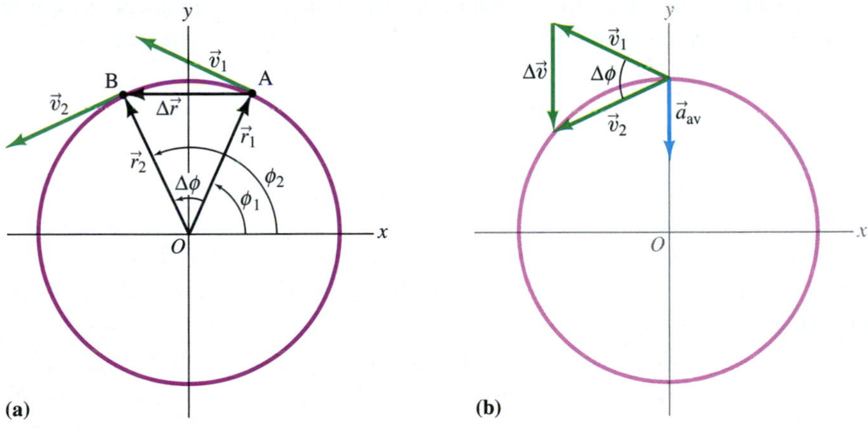

▲ **FIGURE 3–22** (a) During time Δt from t_1 to t_2 a particle in motion has changed its position vector from $\vec{r}_1$ to $\vec{r}_2$ and its velocity vector from $\vec{v}_1$ to $\vec{v}_2$. For uniform circular motion, the magnitude of $\vec{v}$ is constant; however, the direction is always perpendicular to $\vec{r}$ and is therefore changing continuously. (b) To form the vector difference $\Delta v = \vec{v}_2 - \vec{v}_1$, we translate the vector $\vec{v}_2$ so that its tail meets the tail of the vector $\vec{v}_1$. We do this at the midway point in the particle's path from time t_1 to time t_2. In part (a), the angle between $\vec{r}_1$ and $\vec{r}_2$ is $\Delta\phi$, the same angle as that between $\vec{v}_1$ and $\vec{v}_2$. We see from the figure that the average change in velocity, which is proportional to $\vec{a}_{av}$, points toward the center of the circle.

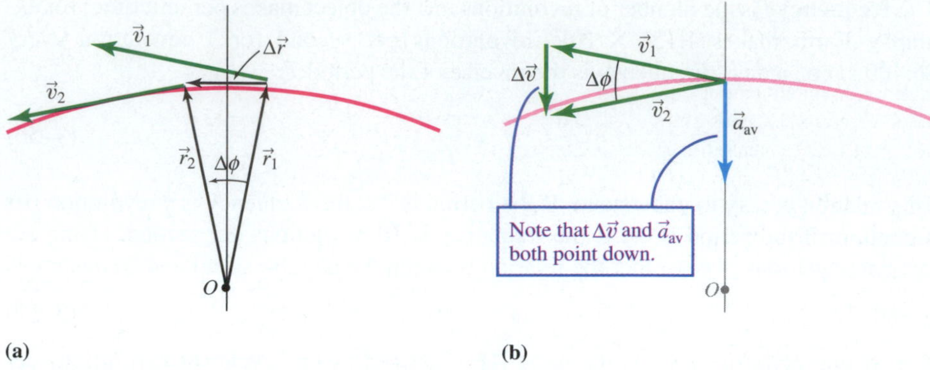

▶ **FIGURE 3–23** (a) The time difference Δt is smaller than in Figs. 3–21a and b; so too are $\Delta\phi$, $\Delta\vec{r}$, and $\Delta\vec{v}$. (b) The resulting $\vec{a}_{av}$ points more precisely to the center of the circle. In the limit that $\Delta t \rightarrow 0$, $\vec{a}_{av}$ becomes the instantaneous acceleration $\vec{a}(t)$ and points exactly to the center of the circle.

To obtain the magnitude of the acceleration, notice from Figs. 3–23a and b that the angle $\Delta\phi$ between $\vec{v}_1$ and $\vec{v}_2$ is the same as the angle between $\vec{r}_1$ and $\vec{r}_2$, so they form similar isosceles triangles (see also Fig. 3–22). Such triangles obey the relation

$$\frac{\Delta v}{v} = \frac{\Delta r}{r}. \tag{3–52}$$

(Note that here Δr is *not* a radius change; it is more closely related to the arc length covered in the time interval.) We write this in the form $\Delta v = (v/r)\,\Delta r$, and, after dividing by Δt, we obtain

$$\frac{\Delta v}{\Delta t} = \frac{v}{r}\frac{\Delta r}{\Delta t}.$$

By taking the limit as Δt goes to zero, we are led to the derivative form for the instantaneous acceleration:

$$a = \lim_{\Delta t \to 0} \frac{\Delta v}{\Delta t} = \frac{v}{r} \lim_{\Delta t \to 0} \frac{\Delta r}{\Delta t} = \frac{v}{r} v = \frac{v^2}{r}. \tag{3–53}$$

An alternative form for the acceleration is expressed in terms of the angular speed ω rather than v:

$$a = \frac{v^2}{r} = \frac{(\omega r)^2}{r} = r\omega^2. \tag{3–54}$$

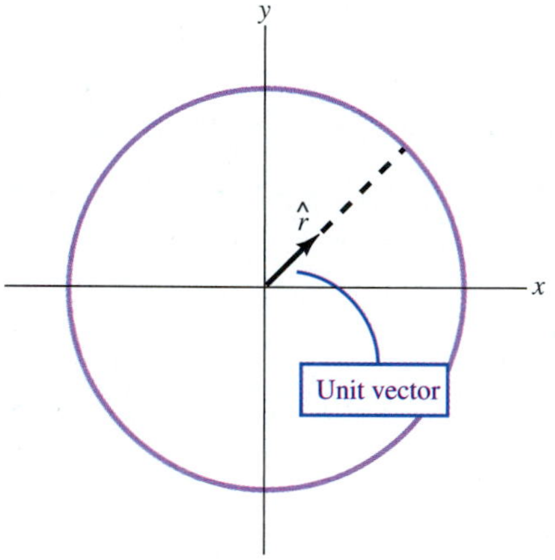

Equations (3–53) and (3–54) express the magnitude of the acceleration.

Since the instantaneous acceleration of an object undergoing uniform circular motion is centripetal, we use a unit vector in the radial direction to help us express this. We define this unit vector by

$$\hat{r} \equiv \frac{\vec{r}}{r}, \tag{3–55}$$

where the caret over the unit vector $\hat{r}$ distinguishes this vector from the position vector $\vec{r}$. In contrast to the Cartesian unit vectors $\hat{i}$ and $\hat{j}$, which are fixed in space, the unit vector in the radial direction varies with the angle ϕ. Notice that the unit vector $\hat{r}$ points outward from the origin (Fig. 3–24). In terms of this vector, the centripetal acceleration is

$$\vec{a} = -\frac{v^2}{r}\hat{r}. \tag{3–56}$$

THE ACCELERATION IN UNIFORM CIRCULAR MOTION

▲ **FIGURE 3–24** The unit vector $\hat{r}$ has length 1 and points away from the origin.

We emphasize the negative sign here: Since the unit vector $\hat{r}$ points away from the origin, the acceleration points *toward* the origin.

The relationship of $\vec{r}$, $\vec{v}$, and $\vec{a}$ is shown in Fig. 3–25 for various times during a particle's circular orbit. Note that $\vec{a}$ is always perpendicular to $\vec{v}$. All the vectors are constant in magnitude but *vary continuously in direction.*

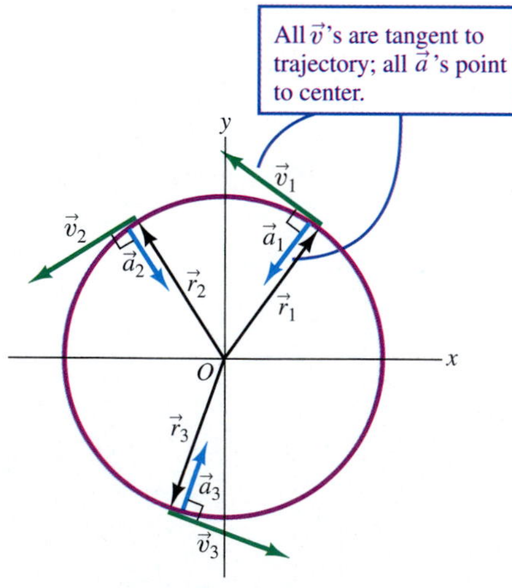

▲ **FIGURE 3–25** The instantaneous position $\vec{r}$, instantaneous velocity $\vec{v}$, and instantaneous acceleration $\vec{a}$ are shown for several positions around the path of a particle in uniform circular motion.

EXAMPLE 3–11 Automobiles include automatic protection systems such as side airbags and roll bars. One well-known German automobile manufacturer has a system that deploys when the "lateral acceleration" exceeds $3g$ in magnitude. Let's assume that this lateral (sideways) acceleration occurs when the car is rounding a curve. What is the relationship between the radius of the car's circular motion and its speed in this situation? How fast could the car go around a curve of radius 20 m, which might be a typical city street corner, before the automatic system deploys?

Setting It Up Figure 3–26 labels the unknown speed v and known radius R for the circular motion. These together imply a centripetal acceleration a (the "lateral acceleration") whose magnitude is given.

Strategy This is a straightforward use of the relation between a, R, and v. The relation is Eq. (3–54), and we can use it to solve for v.

Working It Out If $a = v^2/R = 3g$, then $v^2 = 3gR$. For $R = 20$ m, we have

$$v^2 = 3gR = 3(9.8\ \text{m/s}^2)(20\ \text{m}) = 590\ \text{m}^2/\text{s}^2,$$
$$v = 24\ \text{m/s} = 54\ \text{mph}.$$

Our experience tells us that this is indeed too high a speed with which to round a city street corner.

What Do You Think? Would it be a good idea for the manufacturer simply to set the speed as a criterion for deployment of their airbag?

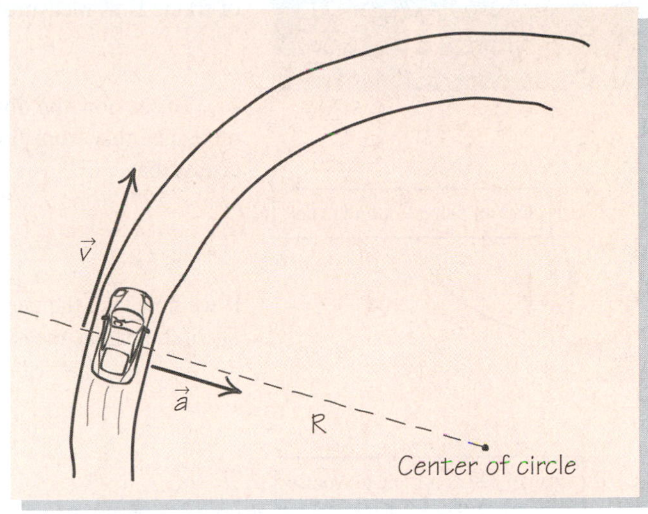

▲ **FIGURE 3–26** Car turning in circular motion.

EXAMPLE 3–12 A child's top spins uniformly at 4.0 Hz. What is the centripetal acceleration on the outside surface if the radius of the top is 3.0 cm?

Setting It Up Here we are given the rotation frequency f and the radius r of the circular motion and want the magnitude of the centripetal acceleration.

Strategy We can find ω immediately given f using $\omega = 2\pi f$. We can then use Eq. (3–54), which gives a in terms of ω and r.

Working It Out The angular speed ω is

$$\omega = (2\pi\ \text{rad})(4.0\ \text{Hz}) = 25\ \text{rad/s}.$$

The radius in question is 3.0 cm $= 3.0 \times 10^{-2}$ m, so the acceleration has magnitude

$$a = \omega^2 r = (25\ \text{rad/s})^2(3.0 \times 10^{-2}\ \text{m}) = 19\ \text{m/s}^2.$$

Remember that the unit *radian* is a dimensionless quantity. It is a measure of angle and we use it in the measurement of angular speed. However, the unit is dropped in this example during the calculation of the acceleration, as it would not be appropriate to write the answer for acceleration as 19 $\text{rad}^2 \cdot \text{m/s}^2$.

What Do You Think? Since the angular velocity is constant for all parts of the top, the expression $v = \omega r$ shows that the outermost parts of the top move the fastest. Which parts of the top have the largest acceleration?

3–6 Relative Motion

Consider, as in Fig. 3–27a, a boy in a bus moving with constant velocity. The boy is tossing a ball and catching it. He is observed by a woman standing on the sidewalk who sees the bus moving past her. The boy and the woman both observe the ball, and we will call them *observers* of the motion of the objects around them. Each observer sees the situation (or, more formally, measures the motion) from his or her **frame of reference**. Let us think about how the description of the ball's motion can vary according to the frame of reference of the observers and how these descriptions are related. This is an important exercise as some descriptions of physical systems are more easily performed in one frame of reference than in another.

Our two observers are said to have a **relative motion**. Let us place the origin of a frame of reference B on the boy and the origin of a second frame of reference W on the woman. From frame W, the bus, along with the boy in it, moves to the right with velocity $\vec{u}$. On the other hand, the boy, or any observer in frame B, observes the woman to be moving with velocity $-\vec{u}$ (the same speed but opposite direction).

(a)

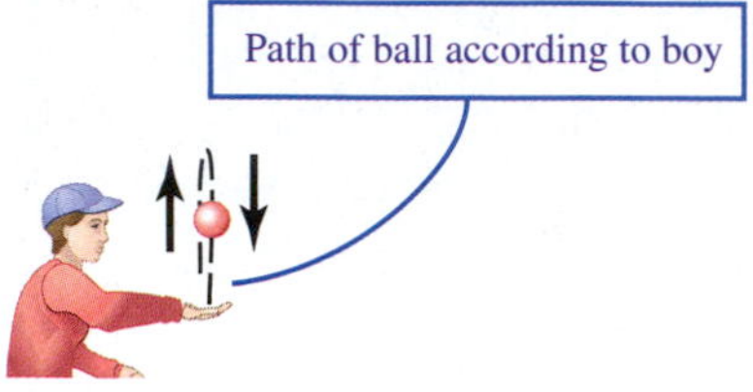

(b)

(c)

▲ **FIGURE 3–27** A boy and a woman are in relative motion. (a) The woman observes the bus and the boy on the bus tossing a ball to travel with constant velocity. (b) The boy observes the trajectory of his ball going directly up and down. (c) The woman sees the trajectory of the boy's ball as a parabola.

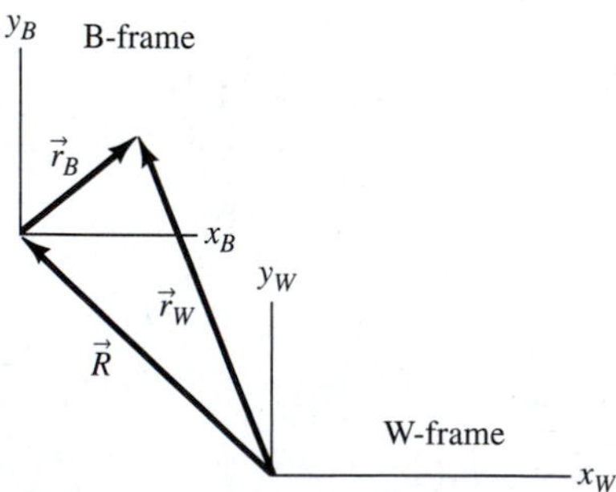

▲ **FIGURE 3–28** The position of an object (the ball) can be measured by an observer in frame W (whose origin is centered on the woman) or by an observer in frame B (whose origin is centered on the boy). The positions in the two frames can be related once the relative position of the two frames is known.

Let us now consider the motion of the ball according to our two observers (Figs. 3–27b and c). In Fig. 3–28 we set up position vectors in each system to denote the position of the ball. The position vector of the ball as measured by the boy is $\vec{r}_B$, and the position vector according to the woman is $\vec{r}_W$. In addition, we let position vector $\vec{R}$ describe the position of the boy according to the woman, that is, the position of the origin of frame B as measured from frame W. We have

$$\vec{r}_W = \vec{R} + \vec{r}_B. \tag{3–57}$$

To see how the observations of the ball compare, we need one more ingredient. We recognize that from frame W the position of the boy moves with speed u or, more precisely, that

$$\vec{u} = \frac{d\vec{R}}{dt}. \tag{3–58}$$

If we now take the derivatives of Eq. (3–57), we obtain the relation between the velocity $\vec{v}_W$ of the ball as measured by the woman and the velocity $\vec{v}_B$ that the boy measures:

$$\frac{d\vec{r}_W}{dt} = \frac{d\vec{R}}{dt} + \frac{d\vec{r}_B}{dt} = \vec{u} + \frac{d\vec{r}_B}{dt}, \tag{3–59}$$

or

$$\vec{v}_W = \vec{u} + \vec{v}_B. \tag{3–60}$$

The woman sees the ball moving at the velocity of the bus plus the velocity that the boy measures it to have.

While the two observers disagree on the velocity of the ball, we can show that they agree on the ball's *acceleration*. To do so, take the derivative of the velocities in Eq. (3–60) to obtain the accelerations:

$$\vec{a}_W = \frac{d\vec{v}_W}{dt} = \frac{d\vec{u}}{dt} + \frac{d\vec{v}_B}{dt} = \frac{d\vec{u}}{dt} + \vec{a}_B \tag{3–61}$$

or, since the speed of the bus is constant, $d\vec{u}/dt = 0$,

$$\vec{a}_W = \vec{a}_B. \tag{3–62}$$

So we find that if the relative speed of the two observers is a constant, then the two observers will agree as to the acceleration of an object, in this case the ball.

How will the two observers describe the ball's motion, assuming the boy tosses the ball vertically with respect to himself? The boy sees the ball go up and down, which is just the projectile motion described in Chapter 2 (or the full projectile motion of this chapter with no horizontal initial velocity). The woman, however, sees the ball in parabolic motion with an initial (and constant) horizontal velocity $\vec{u}$.

One more interesting case: If the bus accelerates, $\vec{u}$ is no longer constant. The boy will accelerate with the bus, but the ball in midair won't. As soon as the ball leaves the boy's hand, it only experiences the vertical acceleration of gravity, and its horizontal motion is determined at the instant that it left the boy's hand. While it is in the air, the boy's velocity will have changed, and when the ball comes down, the boy may no longer be underneath it. From the point of view of the woman, if she keeps her eyes on the ball, she will see exactly what she saw before because from her point of view the ball is only subject to a vertical acceleration and a certain initial horizontal velocity, that of the bus when the ball left the boy's hand.

In this situation $d\vec{u}/dt$ is no longer zero. Equations (3–57), (3–60), and (3–61) are still valid, but Eq. (3–62) becomes

$$\vec{a}_W = \vec{A} + \vec{a}_B, \tag{3–63}$$

where $\vec{A}$, the acceleration of the bus, is

$$\vec{A} = \frac{d\vec{u}}{dt} = \frac{d^2\vec{R}}{dt^2}.$$

EXAMPLE 3–13 A boat must cross a river that is 150 m wide. The river has a current of 3 km/h, and the boat can be rowed through the water with a uniform speed of 4 km/h with respect to the water. Set up two coordinate systems to describe the displacement of the boat: one fixed on the bank and the other fixed to a spot moving with the current of the river. Using these coordinate systems, express the position vector of the boat at time t; assume that the boat leaves the dock at an angle θ to the $-x$-axis with respect to a point moving with the water. Calculate θ such that the boat lands at a point exactly opposite the starting point. How long will the trip take?

Setting It Up The preparation of appropriate sketches is quite useful here. We draw two figures, one that includes the origin O and axes of a frame (the "dock frame") fixed on the dock on the river bank (Fig. 3–29a) and a second that includes the origin O' and axes of a frame (the "raft frame") fixed to a hypothetical spot in the water—a raft that drifts with the current (Fig. 3–29b). The river is assumed to move with a velocity $\vec{u}$ in the $+x$-direction in the dock frame.

Strategy The origin O' (the raft) is seen by the observer on the fixed dock to move with the velocity $\vec{u} = u\hat{i}$, where u is the speed of the current (3 km/h).

If the boat can move at speed $v' = 4$ km/h with respect to the water, then, as observed in the frame moving with the water (the raft frame), the boat moves with velocity

$$\vec{v}_b' = (-v' \cos\theta)\hat{i} + (v' \sin\theta)\hat{j},$$

where we do not distinguish between the unit vectors of the two frames. Here we have set the boat off at the angle specified in Fig. 3–29. The primes indicate that we are describing the boat in the raft frame because the speed of the raft is specified in this frame. As long as the rower does not look at the bank and fixes his vision on a spot in the water, he sees himself moving with speed v' at the angle θ through the water. An observer on the dock, however, sees the boat move with velocity

$$\vec{v}_b = \vec{u} + \vec{v}_b' = (u - v' \cos\theta)\hat{i} + (v' \sin\theta)\hat{j},$$

consistent with Eq. (3–57). At this point our problem is solved if angle θ is chosen such that the dockside observer sees no horizontal velocity (x-component of velocity); with this choice of θ, he will see the boat moving straight across the water to the point on the opposite shore, as in Fig. 3–29a.

Working It Out We choose θ so that the x-component of $\vec{v}_b$ vanishes,

$$\cos\theta = \frac{u}{v'} = \frac{3\ \text{km/h}}{4\ \text{km/h}} = 0.75;$$

$$\theta = 41°.$$

The dockside observer sees the boat move straight across the river with speed $v' \sin\theta$, and the trip takes a time

$$t = \frac{y}{v' \sin\theta} = \frac{(0.15\ \text{km})(60\ \text{min/h})}{(4\ \text{km/h})(\sin 41°)} \cong 3\ \text{min}.$$

What Do You Think? If there were no river current, would a trip straight across take (a) more time, (b) less time, or (c) the same time as worked out in the example?

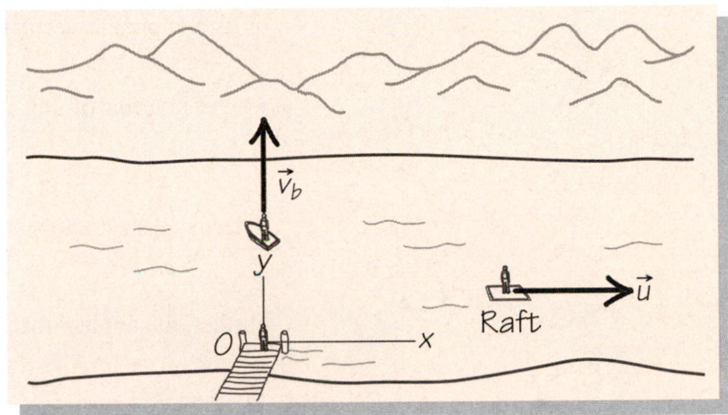

(a)

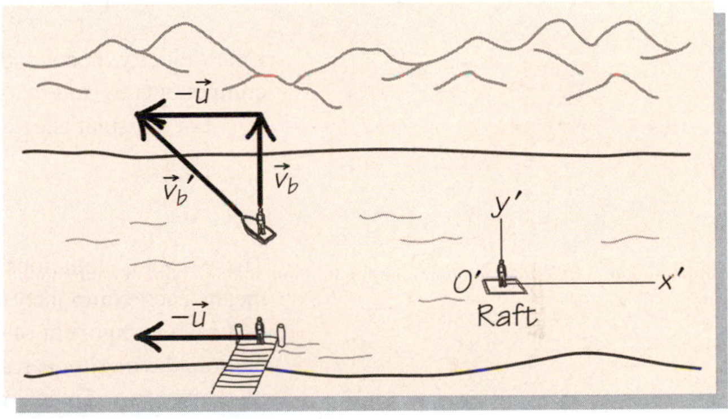

(b)

▲ **FIGURE 3–29** (a) As seen by an observer O standing on the dock, the boat moves straight across the river. (b) As seen by an observer O' moving with the water, the boat moves at an angle θ.

CONCEPTUAL EXAMPLE 3–14 A high-elevation wind called the jet stream flows on average from west to east. A U.S. coast-to-coast airplane flight takes about 5 h west to east but about 30 min longer east to west. Explain this in terms of the relative motion discussed in this section.

Answer Whatever the direction, the airplane flies at the same speed relative to the air. However, the air is not at rest with respect to Earth's surface. If we use the terminology of this section, the jet stream moves with velocity $\vec{u}$ with respect to an observer fixed on Earth with direction from west to east. The airplane has a velocity $\vec{v}_A$ with respect to the atmosphere, but it is positive in one direction (say, from Los Angeles to Washington) and negative for the opposite direction. The velocity $\vec{v}_B$ of the airplane with respect to an observer fixed on Earth is the sum $\vec{v}_A + \vec{u}$. When $\vec{v}_A$ and $\vec{u}$ have the same direction (west to east), their sum has a larger magnitude than when they point in opposite directions (east to west), so the speed is larger west to east than east to west. The jet stream can flow at over 100 mph in places, and it can significantly affect flight times.

What Do You Think? Suppose a plane encountered the west-to-east jet stream as it was flying north to south. Would this lengthen the flight time?

Summary

A particle moving in space follows a trajectory, or path. In three dimensions, the position vector of such a particle is

$$\vec{r} = x(t)\hat{i} + y(t)\hat{j} + z(t)\hat{k}. \tag{3–3}$$

The displacement vector $\Delta\vec{r}$ between times t and $t + \Delta t$ is the difference between the position vectors at these times. If P and Q label the particle's position at t and $t + \Delta t$, respectively,

$$\Delta\vec{r} \equiv \vec{r}_Q - \vec{r}_P. \tag{3–1}$$

The instantaneous velocity $\vec{v}$ is found from the displacement vector $\Delta\vec{r}$ over small time intervals:

$$\vec{v}(t) \equiv \lim_{\Delta t \to 0} \frac{\vec{r}(t + \Delta t) - \vec{r}(t)}{\Delta t} = \frac{d\vec{r}}{dt}. \tag{3–5}$$

In two dimensions, which we describe by x- and y-coordinates, the velocity is expressed in terms of its component vectors as

$$\vec{v} = \vec{v}_x + \vec{v}_y, \tag{3–9}$$

where, in terms of unit vectors,

$$\vec{v}_x = \frac{dx}{dt}\hat{i} \quad \text{and} \quad \vec{v}_y = \frac{dy}{dt}\hat{j}. \tag{3–11a,b}$$

In terms of its components, the magnitude of the velocity vector is

$$v = |\vec{v}| = \sqrt{v_x^2 + v_y^2}. \tag{3–12}$$

Further, the angle θ that $\vec{v}$ makes with the x-axis is given by

$$\tan\theta = \frac{v_y}{v_x}. \tag{3–13}$$

The instantaneous acceleration is a derivative of the velocity:

$$\vec{a} \equiv \lim_{\Delta t \to 0} \frac{\vec{v}(t + \Delta t) - \vec{v}(t)}{\Delta t} = \frac{d\vec{v}}{dt}. \tag{3–15}$$

Like velocity, acceleration can also be expressed in vector components. The acceleration has components a_x and a_y, which are derivatives of the x- and y-components of the velocity.

For constant acceleration, we have

$$\vec{r} = \vec{r}_0 + \vec{v}_0 t + \tfrac{1}{2}\vec{a}t^2, \tag{3–25}$$

$$\vec{v} = \vec{v}_0 + \vec{a}t \tag{3–26}$$

These vector equations represent a set of three component equations. Constant acceleration means each component of the acceleration is constant, and the kinematic equations above show that each component of the motion is independent of the other components. With constant acceleration, the motion is restricted to the plane formed by the initial velocity and the acceleration, so it is really two-dimensional motion.

In the absence of air resistance, a projectile moves under the influence of gravity in the vicinity of Earth's surface with a constant-acceleration vector $\vec{a} = \vec{g}$. The trajectory of such a projectile is a parabola. The range (the horizontal distance a projectile launched from the ground travels over level ground), the maximum height of the trajectory, and the projectile's flight time can all be calculated and follow directly from the equations for motion under constant acceleration.

In addition to motion under constant acceleration, we also looked at uniform circular motion. This motion is most simply described by means of plane polar coordinates, with r as the radial coordinate and ϕ as the angular coordinate. Angles are measured in radians (which are dimensionless), so that the arc length formed by the angle $\Delta\phi$ in a circle of radius R is

$$\text{arc length} = R\,\Delta\phi. \tag{3–43}$$

The angular speed of an object in circular motion is

$$\omega \equiv \frac{d\phi}{dt}. \tag{3–46}$$

The speed of a particle moving in circular motion with angular speed ω is

$$v = \omega R. \tag{3–47}$$

The period T is the time to make one complete revolution, and the frequency f is the inverse of T. The relation between angular speed and frequency is

$$\omega = 2\pi f. \tag{3–51}$$

The velocity of a particle in uniform circular motion is tangent to the circle. The acceleration of this particle has magnitude

$$a = \frac{v^2}{r} = r\omega^2. \qquad (3\text{–}53, 3\text{–}54)$$

The direction of the particle's acceleration is centripetal; that is, it points inward along a radius toward the center of the circle.

Finally, we looked at relative motion. Suppose that two observers, A and B, are in relative motion. These observers are said to be in different frames. Observer A measures the velocity of observer B to be $\vec{u}$, the relative velocity between A and B. Then, if the velocities of an object as measured by observers A and B are $\vec{v}_A$ and $\vec{v}_B$, respectively, these velocities are related by

$$\vec{v}_A = \vec{u} + \vec{v}_B. \qquad (3\text{–}60)$$

If $\vec{u}$ is a constant, then the two observers agree on the acceleration of the observed object.

Understanding the Concepts

1. A boy wants to knock down a coconut with a rock and a slingshot. He knows it is unlikely that the coconut will fall while he is shooting. Does he aim directly at the coconut or does he aim a little higher?
2. What factors determine how far an athlete travels in the long jump? Which factor do you think must be the most important?
3. Why is it important to have limits on wind speed for world-record track and field events?
4. Under what conditions is it possible to have a constant speed yet a nonzero acceleration?
5. You wish to row a boat to a point directly opposite your starting point on the bank of a swiftly moving river. The best speed you can make through the water—meaning with respect to the water—is 2 mi/h. The downstream current of the river is also 2 mi/h. Can you reach your desired goal in a finite time without doing some walking? Explain your answer.
6. There are two initial angles for which the range of a projectile is the same (except in the instance where the initial angle is 45°). What are the differences in the two trajectories that might lead a football quarterback to choose one of them over the other in throwing a pass? If you are familiar with football, you will know that the quarterback usually chooses one of these two trajectories.
7. Two cannons side by side have the same angle of inclination. One of them can shoot a shell with double the initial velocity of the other one. The ratio of the ranges turns out to be 4 : 1. Why is that?
8. A ball is thrown straight up in an elevator moving up with uniform velocity. The ball comes straight down. What is the value of its velocity relative to the ground at the top of its flight?
9. Your lecture demonstrations may include a car that, even while moving, shoots a ball vertically off a platform on the car. If the car is moving with constant speed on a horizontal surface, the ball ends up at the same position on the car that it left. What would happen if the car were decelerating as it released the ball?
10. The Earth–Sun distance is approximately 400 times the Earth–Moon distance. Make a rough estimate of the ratio of the centripetal acceleration of Earth around the Sun to that of the Moon around Earth.
11. Suppose that you measure the period for successive rotations of a phonograph turntable as being unchanging. Is this enough to assure you that the turntable has uniform circular motion?
12. With a given initial speed, there are two initial firing angles for which a projectile has the same range. How many initial velocities give the same maximum height? The same time of flight?
13. You are running in the rain. At what angle should you hold your umbrella for the most protection? Assume that no wind is blowing, so that the raindrops fall straight down. You may need to assume some value for the raindrops' speed as well as your own.
14. You are on the rotating platform of a merry-go-round and throw a ball straight up. Describe the motion of the ball according to a person standing on the ground beside the platform. If it is thrown up at the outside edge of the platform, where will it land?
15. Suppose you are standing on a moving merry-go-round and you throw a ball straight up in the air. Where will it land if you throw it (a) very high and (b) just above your head? Assume that you are standing about midway between the center and the edge of the rotating circular platform.
16. Earth is a rotating platform because it turns on an axis. If a projectile initially aimed at a particular point on the equator is shot from the North Pole, will it arrive at that point, assuming that it could travel such a great distance? Describe the motion of the projectile as seen by someone on Earth.
17. Suppose a projectile is fired from Washington, D.C., toward the equator. How does this situation differ from the one discussed in Question 16?
18. A motorcyclist rides on a large, rotating platform in an amusement park. The cyclist starts at the center of the rotating platform and travels in a radial direction with a constant speed. Is it correct to say that when he or she has traveled some distance from the platform center the acceleration due to the circular motion (which is directed in the inward radial direction) will stop the motorcycle? Analyze the motion.
19. The claim is made that the horizontal and vertical motions of a projectile are independent. Which of the following statements support this claim? (a) If a bullet is dropped from the hand while one is fired horizontally from the same height, they both hit the ground at the same time (ignoring air resistance). (b) A ball thrown horizontally will hit the ground earlier than one thrown at an upward angle with the same initial speed. (c) A thrown ball covers the maximum horizontal distance for a given initial speed when it is thrown at a 45° upward angle.
20. Any change in an object's velocity, whether the change is in the magnitude or the direction, means that there has been a (nonzero) acceleration. True or false?
21. Does the parabolic motion of projectiles apply to a missile traveling from a launch pad to a target area 5000 mi downrange?
22. You are playing a table tennis match, with the table placed centrally on a rotating platform. What sort of adjustments to your play do you think you will have to make due to the rotation?

Problems

3–1 Position and Displacement

1. (I) A car travels 21 km to the northeast, then 15 km to the east, before it travels 28 km to the north. Express the position vector from where the car starts to the point at which each turn occurs. What is the car's total displacement?

2. (I) A particle is located by the position described by the vector $\vec{r} = (c_1 - c_2 t)\hat{\imath} + (d_1 + d_2 t + d_3 t^2)\hat{\jmath}$, where $c_1 = 11$ m, $c_2 = 1.5$ m/s, $d_1 = -12$ m, $d_2 = -2.0$ m/s, and $d_3 = 0.85$ m/s^2. At what time(s) does the particle pass through the position $x = 0$ m? At what time(s), and where, does the particle cross the line $x = y$? Sketch the particle's trajectory from $t = -10$ s to $t = +10$ s.

3. (I) A gym teacher organizes a series of indoor races that follow along the walls of the gym; the race starts from corner A, continues to corner B, and so forth. The gym is a rectangle with the distance $AB = 25$ m and the distance $BC = 35$ m. Suppose that the origin of a coordinate system is at point A, leg AB is in the $+x$-direction, and leg BC is in the $+y$-direction. Express the position vector of a running student at each of the four corners.

4. (I) A runner races with a uniform speed of 27.0 km/h around a circular track of radius 172 m. Draw the track and the runner's displacement vector after 20 s, 40 s, 60 s, and 2 min assuming that at $t = 0$ s the runner is at the three o'clock position. Assume counterclockwise motion.

5. (I) A treasure map locates the site of a treasure by reference to two starting points, A and B. Point A is chosen to be the origin, and B is at the point $(2.5\text{ km})\hat{\imath}$. The instructions state that the treasure lies at the intersection of two lines. One line starts at A and passes through the point $(2.0\text{ km})\hat{\imath} + (4.0\text{ km})\hat{\jmath}$; the other line, starting at B, passes through the point $(6.0\text{ km})\hat{\imath} - (8.0\text{ km})\hat{\jmath}$. Sketch the instructions and find the location of the intersection graphically as well as algebraically. Express the vector that gives the displacement of the intersection point from point C, whose location relative to the coordinate system specified by A and B is given by $(1.2\text{ km})\hat{\imath} - (2.2\text{ km})\hat{\jmath}$.

6. (II) The position of a particle in a given coordinate system is $\vec{r}(t) = (-6 + 4t^2)\hat{\imath} + (-4 + 3t)\hat{\jmath}$, where the distances are in meters when t is in seconds. At what time will the particle cross the y-axis? At what time will it cross the x-axis? Can you find an equation that relates the y-coordinate to the x-coordinate and therefore gives the trajectory in the xy-plane? Where would the x- and y-axes have to be moved so that at $t = 0$ s the trajectory passes through the origin?

7. (II) Figure 3–30 shows the trajectory of a bicyclist traveling with a constant speed of 25.2 km/h. Express the position vector of the bicyclist as a function of time.

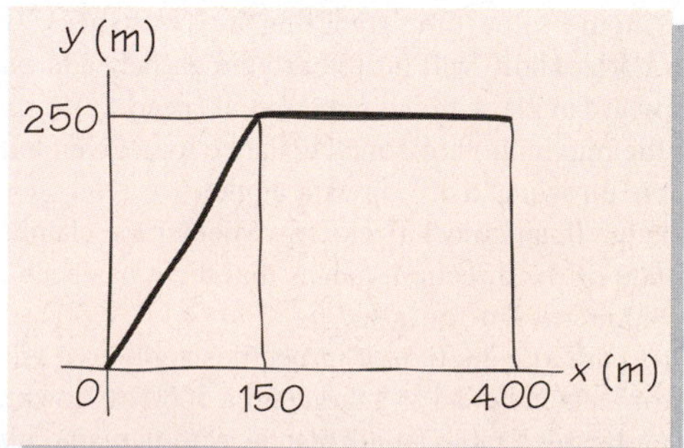

▲ **FIGURE 3–30** Problem 7.

8. (II) The x- and y-coordinates of the position of a land surveyor are shown on Fig. 3–31 as a function of time. (a) Plot the trajectory of the surveyor. (b) Calculate the area of the plot that she surveyed.

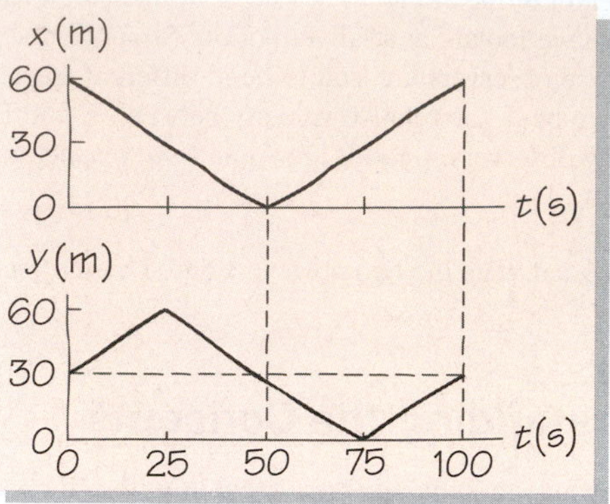

▲ **FIGURE 3–31** Problem 8.

9. (II) The position of a particle is given in a certain coordinate system by the vector $\vec{r}(t) = (4\text{ m})\cos(\pi t/T)\hat{\imath} - (4\text{ m})\sin(\pi t/T)\hat{\jmath}$. Find the displacement vector at times $t = T/3$, $t = T/2$, and $t = 2T$. In each case, compute the distance to the origin. What is the angle that the position vector makes with the $+x$-axis for arbitrary t? (The angles are in radians).

10. (III) In a "shoot the coconut" lecture demonstration, the position of the coconut is given by the vector $(h_0 - \frac{1}{2}gt^2)\hat{\jmath}$, whereas that of the projectile, aimed at the coconut at time $t = 0$, is given by $(-L + ut)\hat{\imath} - [(h_0 ut/L) - (\frac{1}{2}gt^2)]\hat{\jmath}$. Show that the two will always collide and find the time at which this takes place (Fig. 3–32). Express the displacement vector of the coconut relative to the projectile.

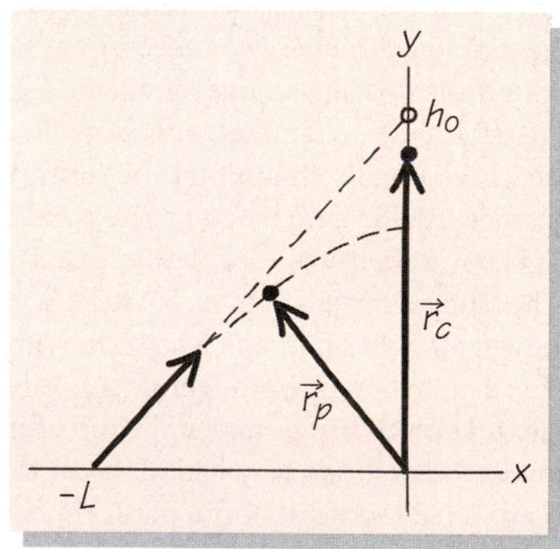

▲ **FIGURE 3–32** Problem 10.

3–2 Velocity and Acceleration

11. (I) A tennis ball leaves a racquet in some direction with an initial speed of 45 m/s. After 2.0 s, it has slowed down to 37 m/s. Assume that there is no bouncing on the ground involved. What is the magnitude of the average acceleration? What would you need to know to determine the direction of the acceleration?

12. (I) A particle is observed to move with the coordinates $x(t) = (1.5\text{ m/s})t + (-0.5\text{ m/s}^2)t^2$ and $y(t) = 6\text{ m} + (-3\text{ m/s})t + (1.5\text{ m/s}^2)t^2$. What are the particle's position, velocity, and acceleration? At what time(s) are the velocity's horizontal and vertical components equal?

13. (I) At a given moment, a fly moving through the air has a velocity vector that changes with time according to $v_x = 2.2\ \text{m/s}$, $v_y = (3.7\ \text{m/s}^2)t$, and $v_z = (-1.2\ \text{m/s}^3)t^2 + 3.3\ \text{m/s}$, where t is measured in seconds. What is the fly's acceleration?

14. (I) A particle moves in such a way that its coordinates are

$$x(t) = A\cos\omega t, \qquad y(t) = A\sin\omega t.$$

Calculate the x- and y-components of the velocity and the acceleration of the particle.

15. (II) A whale traveling southwest at 70 km/h is spotted 5.0 km to the northwest off the coast of Malibu. Photographers jump into a boat that can move at 30 km/h. With what velocity will the photographers intercept the whale assuming that their boat travels a straight-line path? What is the position vector of the whale from the original point on the coast when the photographers reach the whale?

16. (II) A lifeguard standing on a tower throws a buoy to a swimmer 5 m from the tower (Fig. 3–33). The lifeguard, positioned 3 m above the water, pulls in the rope at a speed of 1 m/s. How fast is the swimmer coming to the shore when he is (a) 4 m and (b) 3 m from the water's edge?

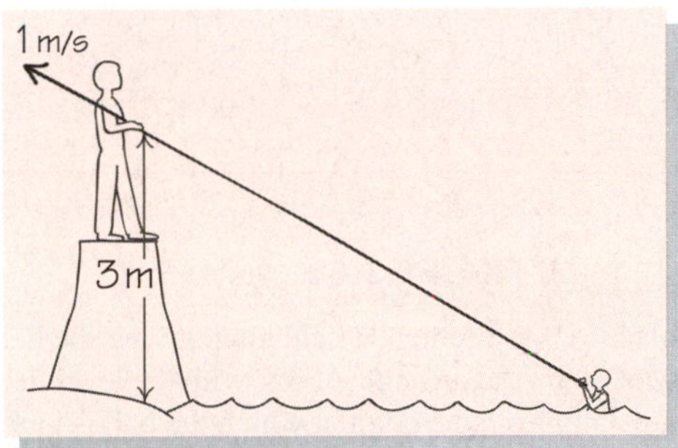

▲ **FIGURE 3–33** Problem 16.

17. (II) An engineering student holds her open compass perpendicular to the drafting board, touching the board with both tips of the compass (Fig. 3–34). She slowly closes the compass so that the tips move toward each other with a speed of $2v_0 = 0.060\ \text{m/s}$. Initially, the angle between the arms of the compass is $2\theta = 60°$. The arms of the compass are $L = 15\ \text{cm}$ long. (a) Calculate the velocity of the joint of the compass as a function of time. (b) Give the velocity of the joint at the moment that she starts to close the compass and at the moment that the two tips reach each other.

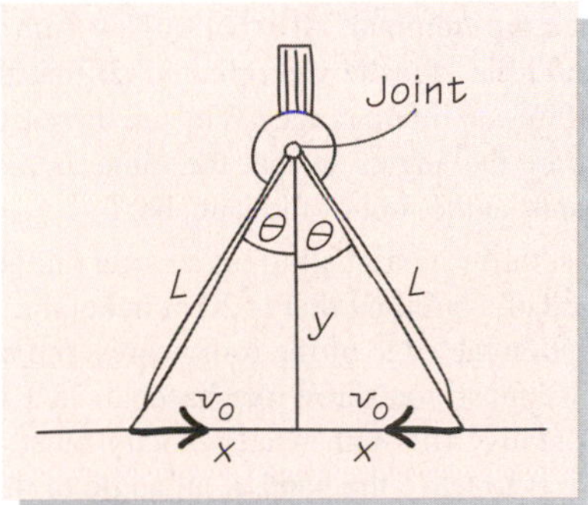

▲ **FIGURE 3–34** Problem 17.

18. (II) Calculate the velocity vectors for the coconut and projectile in Problem 10. Express the difference between the two velocity vectors (the relative velocity). What is the magnitude of the relative velocity? What is the angle that its direction makes with the x-axis?

19. (II) Calculate the velocity vector for the particle described by the position vector given in Problem 9. Calculate the angle that the velocity vector makes with the x-axis and show that the velocity vector is always perpendicular to the position vector.

20. (II) A particle moving with an initial velocity $\vec{v} = (30\ \text{m/s})\hat{j}$ undergoes an acceleration $\vec{a} = [3.5\ \text{m/s}^2 + (0.7\ \text{m/s}^5)t^3]\hat{i} + [2\ \text{m/s}^2 - (0.3\ \text{m/s}^4)t^2]\hat{j}$. What are the particle's position and velocity after 30 s assuming that it starts at the origin?

21. (II) The motion of a planet about a star is described by the vector $\vec{r}_1 = R\cos(2\pi ft)\hat{i} + R\sin(2\pi ft)\hat{j}$. The position vector of another planet about the same star is $\vec{r}_2 = 4R\cos(\pi ft/4)\hat{i} + 4R\sin(\pi ft/4)\hat{j}$. Show that both of these describe circular motion with constant speed. Calculate the position vector of the second planet relative to the first planet. Sketch the path by noting that the speed of one of the planets in its circular motion is eight times larger than the speed of the other. This is like the motion of a planet whose year is 8 Earth years.

22. (II) Calculate the acceleration vectors for the two planets in Problem 21. Express the vector that describes the acceleration of the second planet relative to the first one.

23. (II) A bag is dropped from a hot-air balloon. Its height is given by the formula $h = H - ut - (u/B)e^{-Bt}$. What are the dimensions of B? What is the initial velocity? What is the velocity as $t \to \infty$? Calculate the accelerations at $t = 0$ and at $t = \infty$.

24. (II) A car and a truck start from a common spot and travel in straight lines at respective speeds of 30 and 40 km/h. Exactly 1 h later they telephone each other and find that they are separated by exactly 50 km. At what relative directions did they travel?

3–3 Motion with Constant Acceleration

25. (I) A launching mechanism accelerates a baseball horizontally at $24\ \text{m/s}^2$ for 0.5 s. The baseball's initial velocity is zero. With what velocity will it leave the launcher? The baseball leaves the launcher at the same speed when the launcher is turned in the vertical direction. How high will the baseball go?

26. (I) An airplane is flying due south on a level course at a speed of 600 km/h. At an altitude of 7.50×10^3 m directly above a mountain top, the airplane meets severe turbulence and descends with a vertical acceleration of $4.00\ \text{m/s}^2$ for 15 s. (a) What is the total displacement of the plane in the 15 s? (b) What is the plane's velocity at the end of the 15 s? (c) What is the plane's position vector with respect to the mountain top at the end of the 15 s?

27. (II) A boy shoots a rock with an initial velocity of 21 m/s straight up from his slingshot. He quickly reloads and shoots another rock in the same way 3.0 s later. (a) At what time and (b) at what height do the rocks meet? (c) What is the velocity of each rock when they meet?

28. (II) A gymnast works out on a trampoline. At the instant that she leaves the trampoline, a point on her waist is 2.3 m above the floor and at the center of the trampoline. At that instant, the point has an upward velocity of 7.8 m/s and a horizontal velocity of 3.0 m/s. Write equations that describe the subsequent motion of that point and find its maximum height.

29. (II) A softball pitcher pitches a ball at a speed of 65 mph (roughly constant over the pitch distance). The distance from the pitcher's mound to home plate is 43 ft, but the pitcher releases the ball about 41 ft from home plate. (a) How long does it take for the ball to reach home plate? (b) A good pitcher can make the ball curve to the left about 6 in over the pitch distance. If this is due to a constant acceleration, what is its magnitude and in which direction does it act?

30. (II) A man in the crow's nest of a sailing ship moving through smooth seas at a steady 8 km/h accidentally lets a cannonball drop from his station, which is 6.5 m above the deck at the top of the main mast. (a) Assuming that he dropped the ball from a position immediately adjacent to the vertical mast, where does the ball land with respect to the mast? (b) How long does it take for the ball to fall to the deck? (c) In the time it takes the ball to fall, how far has the ball moved with respect to an observer fixed on the shore?

3–4 Projectile Motion

31. (I) A runner attempting a broad jump leaves the ground with a horizontal velocity of magnitude 9.0 m/s. Assuming the horizontal component of velocity is unaffected, what vertical component of velocity must the runner acquire to jump 9.5 m?

32. (I) An engineering student wants to throw a ball out of a third-story dormitory window (10 m off the ground) onto a target on the ground placed 8.0 m away from the building. (a) If the student throws the ball horizontally, with what velocity must it be thrown? (b) What must the velocity of the ball be if it is thrown up at an elevation angle of 29°? (c) What is the ball's time of flight in case (b)?

33. (I) A projectile is shot at an angle of 34° to the horizontal with an initial speed of 225 m/s. What is the speed at the maximum height of the trajectory?

34. (I) At what points in a projectile's trajectory above level ground is the magnitude of the velocity a maximum and a minimum? What are these velocities in terms of the initial speed v_0 and elevation angle θ_0?

35. (I) A projectile is shot at an angle of 25° to the horizontal over level ground. Assuming air resistance plays no role, what angle does the projectile make with the horizontal when it lands?

36. (I) Find the initial angle if the range of a projectile is twice its maximum height.

37. (II) (a) Show that the range R can be expressed in terms of the maximum height h and in particular that $R = 4h \cot \theta_0$. (b) Show that when the range is a maximum $h = R/4$.

38. (II) A grasshopper can jump a maximum horizontal distance of 65 cm. Assuming that it spends effectively no time on the ground, that it is tireless, and that it moves in a straight line through a succession of jumps, what is the distance that it can cover in an hour?

39. (II) At what angle must a projectile be fired so that its maximum height and its range are equal? What other angle will give the same range?

40. (II) A boy wants to knock down a coconut with a rock and his slingshot. He observes that the coconut is about 3.0 m above his slingshot and the tree is 4.0 m away along the ground. He knows from experience that the release speed of his rock is 20 m/s. How far above the coconut should he aim?

41. (II) Potato cannons are popular devices made out of PVC pipe that utilize hair spray as an explosive mechanism. One group built a cannon that shoots small potatoes over 100 m. Let's ignore friction in our analysis. (a) What will be the muzzle velocity for a range of 120 m? (b) What height is reached for this maximum horizontal distance? (c) The students want to use the device to get rid of pigeons roosting on tall buildings. How high can they reach?

42. (II) Galileo throws a rock from the top of the Leaning Tower of Pisa at an upward angle of 60° with speed v_0. The rock is in flight for 6.5 s and hits the ground 15 m from the base of the building. Ignore air resistance and ignore the fact that the tower tilts a bit. (a) What is the speed v_0? (b) How high off the ground is the top of the tower? (c) What is the speed of the rock just before it hits the ground?

43. (II) In the Battle of Hastings in A.D. 1066, during which the Normans of France defeated the Saxons in England, an important role was played by Norman archers who shot arrows over a wall of shields erected by the Saxons. If the Norman bows had a maximum range of 350 m and the arrows were shot at an elevation of 55°, how close were the Normans to the Saxons? Assume that the arrows reached their target.

44. (II) A punter kicks a football during a critical football game. The ball leaves his foot from ground level with a speed of 28 m/s at an angle of 50° to the horizontal. At the very top of its flight, the ball hits a wandering seagull. The ball and the seagull each stop dead and fall vertically from the point of collision (Fig. 3–35). In the following, ignore air resistance. (a) With what speed is the ball moving when it strikes the seagull? (b) How high was the unfortunate seagull when it met the ball? (c) What is the speed of the seagull when it hits the ground?

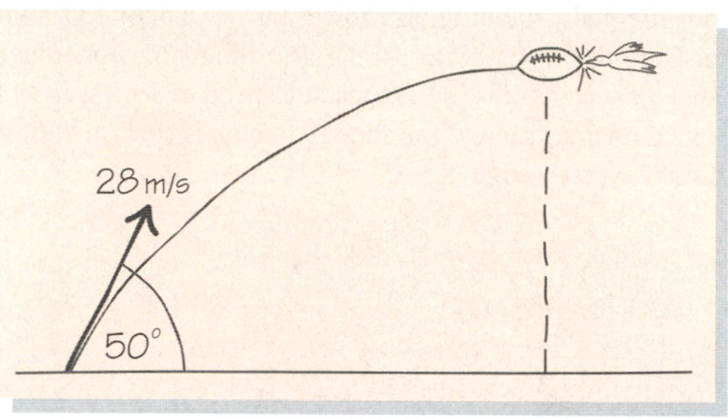

▲ **FIGURE 3–35** Problem 44.

45. (II) A place kicker attempts a field goal, giving the ball an initial velocity of 30 m/s at an angle of 32° with the level field. The uprights are 35 m from the point at which the ball is kicked and the horizontal bar is 4.0 m from the ground. (a) At what time after the kick will the ball pass the goal posts? (b) Is the kick successful, and by how many meters does the ball clear or pass beneath the bar?

46. (II) Astronaut Alan Shepard had the good fortune to play golf on the Moon. The acceleration of gravity on the surface of the Moon is only about one-sixth of that on the surface of Earth. Assuming that Shepard was not noticeably hampered by his space suit and that his best drive on Earth (unhampered by air resistance) was some 210 yd, how far could he drive a ball on the Moon? Derive a general answer to this question for planet X in terms of f, where g_x, the acceleration of gravity on that planet, is fg.

47. (II) A projectile is launched over flat ground and the effects of air resistance are minimal. At what angles with respect to the ground should the launcher be oriented so that the projectile's range is half its maximum range? Why are there two possible angles? What are the angles so that the range is zero, that is, the projectile lands at the foot of the launcher?

48. (II) You must throw a baseball to hit a target on the ground 50 m from the base of a building that is 20 m in height. You are standing at a point on the edge of the roof nearest the target. (a) With what velocity must you throw the baseball if it is to leave the hand horizontally? (b) With what velocity must you throw the baseball if it is to leave the hand at an angle of 45° up from the horizontal? (c) What is the horizontal component of the initial value of the velocity in case (b)?

3–5 Circular Motion

49. (I) The space shuttle is in a circular orbit 220 km above Earth's surface and completes an Earth revolution every 89 min. (a) What is the shuttle's speed? (b) Its acceleration?

50. (I) The Moon circles Earth at a distance of 3.84×10^5 km. The period is approximately 28 d. What is the magnitude of the moon's acceleration, in units of g, as the Moon orbits Earth?

51. (I) A runner in the 200-m dash must make part of the dash around a curve that forms the arc of a circle. This arc has a radius of curvature of 25 m. Assuming that she runs at a steady speed and completes the 200 m in 25.5 s, what is her centripetal acceleration while she is running the curve?

52. (I) A rock placed in a plastic bag is tied to a rope 1.2 m long. The rock is whirled in a horizontal circle. (a) What is the rock's centripetal acceleration if the period of motion is 1.8 s? (b) The plastic bag will break if the radial acceleration exceeds 56 m/s^2. With what speed must the rock be whirled if the plastic bag is to be broken?

53. (I) A passenger on the outer edge of a merry-go-round, 7.5 m from the central pivot, learns that when the merry-go-round is in steady motion his centripetal acceleration is 3.3 m/s^2. How long does it take to make one revolution?

54. (II) The shaft of the engine of a car rotates at 4000 rev/min. A flywheel 20 cm in diameter rotates with the shaft. Calculate the centripetal acceleration of a point on the rim of the flywheel and express it in units of g, the acceleration of gravity.

55. (II) Safety requires that the centripetal acceleration of cars traveling along highway curves may not exceed one-tenth of the acceleration of gravity. How small can the radius of curvature of a curve be for a road where the posted speed limit is 65 mph?

56. (II) Suppose that a point object is in uniform circular motion, moving steadily at a distance R from some central point. The time for one revolution is T. Use dimensional analysis to find the dependence of the centripetal acceleration on T and R. Compare this result to the acceleration derived from a detailed analysis of uniform circular motion performed in Section 3–6.

57. (II) A mass is tethered to a post and moves in a circular path of radius $r = 0.35$ m on an air table—friction free—at a constant speed $v = 18$ m/s. We employ the coordinate system shown in Fig. 3–36. (a) If at $t = 0$ s the mass is at $\theta = 0°$, what are the coordinates (x, y) of the mass at $t = 0.1$ s? (b) What is the acceleration vector of the mass at $t = 0$ s? (c) What is the acceleration vector of the mass when $\theta = 90°$?

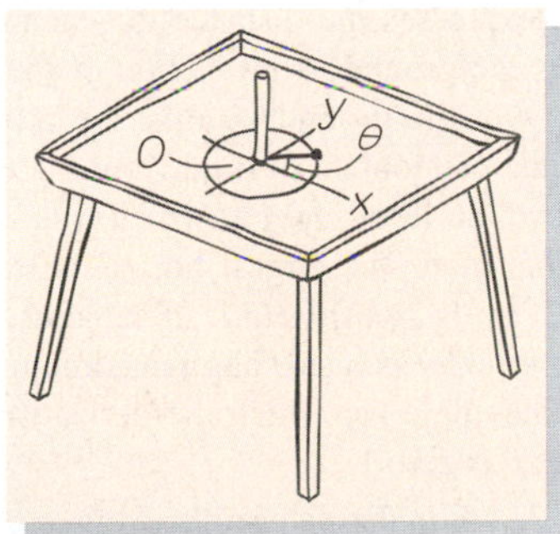

▲ **FIGURE 3–36** Problem 57.

58. (II) The space shuttle is moving in a circular orbit with a speed of 7.8 km/s and a period of 87 min. In order to return to Earth, the shuttle fires its retro engines opposite to its direction of motion. The engines provide a deceleration of 6 m/s^2 that is constant in magnitude and direction. What is the total acceleration of the shuttle?

59. (II) An electron in a research apparatus follows a circular path. On the electron's first circuit of the apparatus, its speed is v_0 and the radius of its circular path is R. Each time it makes one circuit, it passes a short region where it receives a "kick" and gains an additional speed of $v_0/100$. The electron follows a circular path such that the magnitude of its acceleration is always the same. What is the radius of the circular path after the electron has received 10 kicks?

60. (II) An automobile moves on a circular track of radius 1.00 km. It starts from rest from the point $(x, y) = (1.00 \text{ km}, 0 \text{ km})$ and moves counterclockwise with a steady *tangential* acceleration such that it returns to the starting point with a speed of 30.0 m/s after one lap. (The origin of the Cartesian coordinate system is at the center of the circular track.) What is the car's velocity (magnitude and direction) when it is one-eighth of the way around the track? Express the position and velocity at this point in terms of the unit vectors along the x- and y-axes.

3–6 Relative Motion

61. (I) A sailor wants to travel due east from Miami at a velocity of 15 km/h with respect to a coordinate system fixed on land. The sailor must contend with the Gulf Stream, which moves north at 5 km/h. With what velocity with respect to the water should the sailboat proceed under sail?

62. (I) During an uphill portion of a bicycle race, a cyclist reads a message on a board informing him that the leader is 30 s ahead and that the leader is traveling at 24 km/h. The cyclist's speedometer informs him that he is traveling at 21 km/h. (a) What is the speed of the leader with respect to the cyclist? (b) How far in front of the cyclist is the leader assuming that the speeds have not changed in the last half minute or so?

63. (I) Rain is falling steadily but there is no wind. You are in an automobile that moves at 80 km/h, and you see from the drops on a side window that the rain makes streaks at a 58° angle with respect to the vertical. What is the vertical velocity of the raindrops?

64. (II) A cyclist's top speed on a flat road is v. This is an "air speed" because the limiting speed for a cyclist is determined by the wind resistance. In other words, this is her top speed with respect to the air. She cycles a flat course straight north for a distance L, turns around, and cycles straight south for the same distance. In the following, ignore the time it takes her to turn around and assume that the cyclist can maintain her top speed with respect to the air. (a) Write a formula for the total course time t_0 in terms of L and v. (b) There is a north wind blowing at speed v_w. Write a new formula for the course time t_1 of the cyclist, including the effect of v_w. (c) Show that for $v_w \ll v$ the course time can be approximated by $t_1 = t_0 [1 + (v_w^2/v^2)]$. To show this result, you may want to use the approximation $(1 - x)^{-1} \cong 1 + x$ for $x \ll 1$. (d) Plot the time t_1 as a function of v_w and show that it is always greater than t_0. What happens at $v_w = v$ and why?

65. (II) An athlete can jump vertically a distance of 40 in starting from a standing position. (a) Find the speed with which the athlete left the ground. (b) Assuming that he can leave the ground with the same starting speed as in part (a), find the jump height when the athlete jumps at a 45° angle. What horizontal distance is covered in this jump? (Assume flat ground.)

66. (II) Consider Problem 24. What is the relative velocity of the car and the truck in this problem? More precisely, what is the velocity of the truck as seen by an observer stationed in the car?

67. (II) An airplane flies due south with respect to the ground at an air speed of 900 km/h for 2.0 h before turning and moving southwest with respect to the ground for 3.0 h. During the entire trip, a wind blows in the easterly direction at 120 km/h. (a) What is the plane's average speed with respect to the ground? (b) What is the plane's average velocity with respect to the ground? (c) What is the final position vector?

68. (II) An airplane is to fly due north from New Orleans to St. Louis, a distance of 673 mi. On that day and at the altitude of the flight, a wind blows from the west at a steady speed of 85 mi/h. The airplane can maintain an air speed of 320 mi/h. Ignore the periods of takeoff and landing. (a) In what direction must the airplane fly in order to arrive at St. Louis without changing direction? Draw a diagram and label this direction with an angle. Would this calculation change if the distance between the cities were twice as great? (b) What is the flying time for this flight? (c) Recalculate the flying time if the airplane heads due north until it reaches the latitude of St. Louis and then flies due west into the wind to reach the city.

69. (III) Earth has a radius of 6.4×10^6 m, and its orbit around the Sun has a radius of some 1.5×10^{11} m. Earth simultaneously rotates about its own axis and moves around the Sun (Fig. 3–37). Assume a circular orbit and that Earth's axis of rotation is perpendicular to its orbital plane. (a) What is the speed with respect to the Sun of the point on Earth's equator nearest the Sun? (b) Of the point on Earth's equator farthest from the Sun? (c) Of the two points on Earth's equator, midway between the points in parts (a) and (b)?

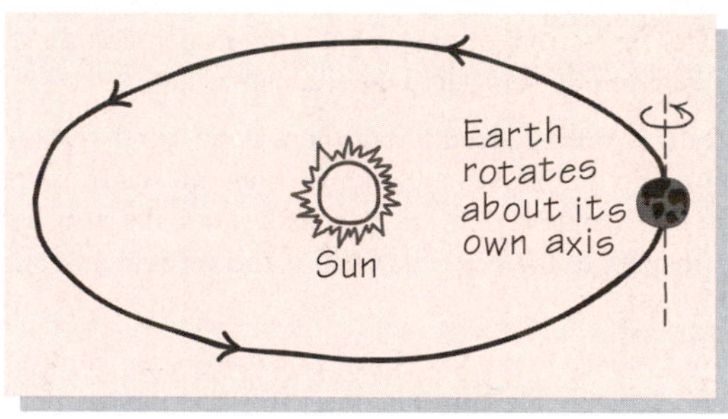

▲ **FIGURE 3–37** Problem 69.

General Problems

70. (I) Neutron stars are the very dense remnants of partially collapsed bright stars. One such (hypothetical) star has a radius of 40 km and rotates with a period of 26 s. What are the velocity and acceleration of a particle on the surface due to the rotation? Compare this acceleration with the acceleration of gravity on Earth.

71. (II) A golfer wants to land a golf ball on the green located 155 m away horizontally but 4.0 m higher. The golfer chooses an eight iron that he knows will result in the ball leaving the tee at an elevation angle of 65°. (a) With what velocity should the ball leave the tee? (b) What is the maximum height of the ball above the green?

72. (II) A softball player hits the ball when it is 0.9 m above home plate. The ball leaves the bat at an elevation angle of 30°. What initial speed must the ball have to clear a fence 1.8 m high located 65 m away in left field?

73. (II) A boat is required to traverse a river that is 150 m wide. The current in the river moves with a speed of 6 km/h. The boat can be rowed on still water with a speed of 10 km/h. Set up a convenient coordinate system in which to describe the various displacements. Using this coordinate system, write down the position vector of the boat at time t assuming that the boat moves with uniform speed and that it leaves one side with the velocity vector making an angle θ with the direction of the river. Calculate θ such that the boat lands at a point exactly opposite the starting point. How long will the trip take?

74. (II) A possible way to measure g is to send a mass traveling horizontally with a known speed off the edge of a vertical drop, such as a table. The table height can be measured, as can the distance from the base of the table where the mass strikes the floor. (a) If the speed of the mass just as it leaves the table is 2.50 m/s and the distance from the table base to the point at which the mass strikes the floor is 108 cm, how long was the mass in flight? (b) The table height is 86 cm. What is the value of g obtained from this data? (c) Use $g = 9.80\ \text{m/s}^2$ and neglect the air resistance to calculate the speed of the mass when it arrives at the floor.

75. (II) A sailor on top of a mast 26 m high drops a hammer. The ship is rolling with a maximum angle away from the vertical of no more than 15°. At the moment that the hammer is dropped, the mast is exactly vertical while the top of the mast is moving laterally at a speed of 3.6 m/s. Will the hammer fall into the sea or onto the deck given that the ship is 19 m wide?

76. (II) Someone standing on a down-tilting inclined plane making an angle of 30° with the horizontal throws a ball horizontally with a velocity of 10 m/s. How far down the plane will the ball strike the plane?

77. (III) A juggler is able to handle four balls simultaneously. He takes 0.3 s to cycle each ball through his hands, throw the ball, and be ready to catch the next ball. (a) With what velocity must he throw each ball vertically? (b) What is the position of the other three balls when he has just caught one of the balls? (c) How high must he throw the balls if he is to juggle five balls?

78. (III) A cannon can project a cannonball from its barrel with a certain muzzle speed v_0. Ignoring the effects of air resistance, what formula expresses the distance the cannonball travels before it reaches the ground as a function of the angle θ that the barrel makes with the ground? Unlike the projectile treated in Section 3–4, the cannonball is fired from the edge of a cliff of height h_0 above the level plain at which it is aimed. Show that the angle θ that gives the largest horizontal range is given by $\sin^2\theta = v_0^2/2(v_0^2 + 2gh_0)$. [*Hint*: In calculus, you learn that a function of a variable such as θ has a maximum (or minimum) at an angle θ_0, the angle for which the derivative of the function with respect to θ is zero.]

79. (III) A wheel 72 cm in diameter rolls along a road, with the center moving in a straight line at a uniform speed of 18 km/h. What are the position vector, the velocity vector, and the acceleration vector of a fixed point on the rim of the wheel relative to a fixed point on the straight line followed by the wheel on the road?

The idea that an apple falling from the tree might have something in common with the Moon rotating around the Earth originated with Isaac Newton, who showed that the same force governs the falling of an apple and the Moon's movement around the Earth. This observation, popularly linked with Newton watching an apple fall, is one of the truly significant discoveries in science.

Chapter 4

Newton's Laws

In the previous chapters, we explored quantities that *describe* motion, such as velocity and acceleration. This subject—kinematics—tells us such things as how, given an object's acceleration, we can find its velocity, or displacement. What it does not tell us is how it got that acceleration in the first place. What is it that makes an object fall with a constant acceleration? What governs the way a baseball's velocity changes when it connects with a baseball bat? The study of the *cause* of changes in motion is called **dynamics**, and we can only understand dynamics by understanding the concept of forces, "pushes" or "pulls" that act on an object and influence its motion.

The three laws that describe how forces govern motion were described by Isaac Newton in 1687. Now known as Newton's laws, they are based on careful and extensive observation. These laws provide an extraordinarily accurate description of the motions of all material objects—small or large, simple or complicated—on the everyday scale. It took physicists more than two centuries to discover that some motions cannot be treated using these laws. We must go beyond Newton's laws to describe motion within systems as small as the atom and for motion at speeds near the speed of light (3.0×10^8 m/s); we'll look at these questions more thoroughly in the latter chapters of this book. Newton's laws represent a tremendous achievement in their simplicity and breadth. In this book we will apply Newton's laws to the motion of the heavenly bodies, fluids, springs, projectiles, electric charges, and many other systems. In this chapter and the next we study the concept of forces, the basic three laws, and how these laws apply to some simple situations.

▶ **FIGURE 4–1** The woman exerts a force $\vec{F}$ on the wagon. Other forces must also act on the wagon if it travels with constant velocity as with a constant velocity the net force has to be zero.

4–1 Forces and Newton's First Law

For most of us, everyday experience implies that the "natural" state of motion of a body, when left alone, is the state of being at rest. A sled comes to rest at the bottom of a hill, a piece of furniture will not move unless it is pushed, a car will ultimately come to rest if the engine is turned off. In ancient times, Aristotle observed the motions around him and argued that a body would move only when subjected to a **force**; otherwise, in the absence of these forces, it would come to rest.

We have long since understood that Aristotle's ideas on this subject are wrong, but forces do play a central role in motion. In simplest terms, a force is something that acts to push or pull on an object. For example, Fig. 4–1 shows an adult pulling a child in a wagon. The adult exerts a force on the wagon by pulling it. The compression or extension of a spring represents another situation in which forces act (Fig. 4–2). If the spring is extended, it acts to pull in any object attached to its end; if the spring is compressed, it acts to push out the object at its end. You can on your own think of two or three forces you have experienced or observed today. Forces have two important characteristics (based on experiments that test these ideas):

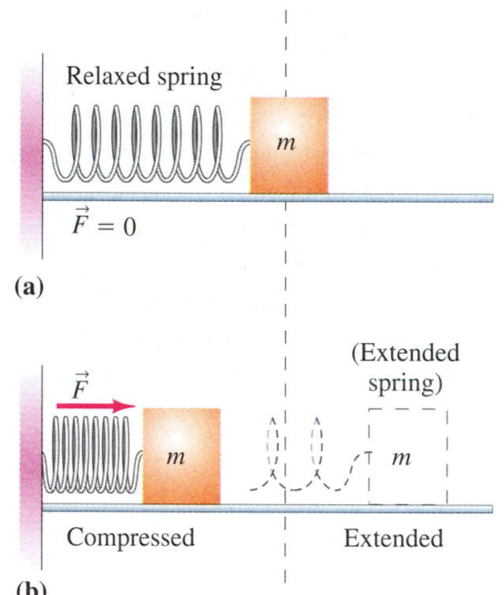

▲ **FIGURE 4–2** (a) A mass attached to the end of a relaxed spring is at rest. (b) The spring has been compressed to the left and pushes the mass to the right. What will happen next?

a. A force is a **vector**, so that it has a direction as well as a magnitude. In the case of the wagon, the adult's pulling force acts along the handle rod (as shown in Fig. 4–1), and the direction depends on the angle that the handle rod makes.

b. Forces are additive, which means that when several forces act together, the subsequent motion of the object acted on is the same as if a single force equal to the vector sum of the individual forces were acting. That single force is the **net force**.

Figure 4–3 shows a sled being pulled by two ropes. If the forces exerted by the ropes are denoted by $\vec{F}_1$ and $\vec{F}_2$, respectively, then the net force is the vector sum of these, so that

$$\vec{F}_{\text{net}} = \vec{F}_1 + \vec{F}_2. \tag{4–1}$$

More generally, for any number of forces acting, we have

$$\vec{F}_{\text{net}} = \sum_i \vec{F}_i. \tag{4–2}$$

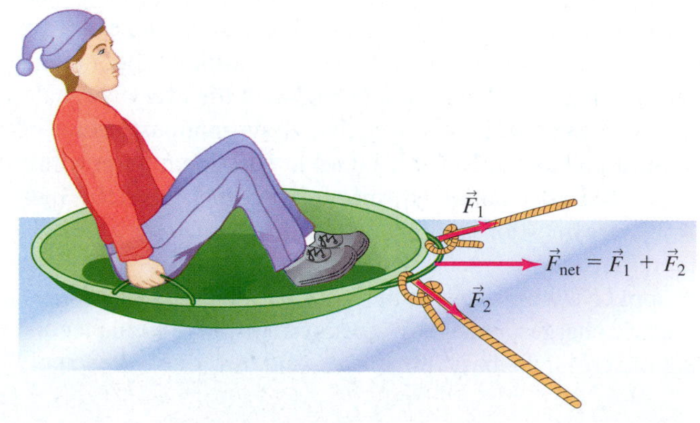

▶ **FIGURE 4–3** Ignoring gravity, two forces, $\vec{F}_1$ and $\vec{F}_2$, act on a sled, resulting in the net force $\vec{F}_{\text{net}}$.

EXAMPLE 4–1 Three small children each tug at the same teddy bear as in Fig. 4–4a. (All the forces are in the horizontal plane.) The three forces on the teddy bear are $\vec{F}_1 = (-5\text{ units})\hat{k}$, $\vec{F}_2 = (5\text{ units})\hat{\imath}$, and $\vec{F}_3 = (-5\text{ units})\hat{\imath} + (5\text{ units})\hat{k}$. What is the net force on the teddy bear? Ignore the force of gravity on it.

Strategy The unknown net force is a vectorial sum of three given individual forces. This is a straightforward application of vector addition; the net force on the teddy bear is the vector sum of the individual forces (Fig. 4–4b). This is most easily accomplished by adding the components separately.

Working It Out We have

$$\vec{F}_{\text{net}} = \vec{F}_1 + \vec{F}_2 + \vec{F}_3 = -5\hat{k} + 5\hat{\imath} + (-5\hat{\imath} + \hat{k})$$
$$= (-5 + 5)\hat{k} + (5 - 5)\hat{\imath} = 0.$$

The three forces add vectorially to zero.

We have not specified the units of force in this example. When the net force is zero, the units are irrelevant. We shall return to the question of dimensions and units in the next section.

What Do You Think? A large piece of machinery is being maneuvered into place by seven workers. Five workers pull on the machine with rods that have fixed direction (not all in the same plane) and are pulled with constant given forces, whereas two workers pull on ropes with a fixed pull strength (fixed magnitude), but the direction in which they act can be changed. Is it always possible to make the net force on the machine zero? *Answers to* ***What Do You Think?*** *questions are given in the back of the book.*

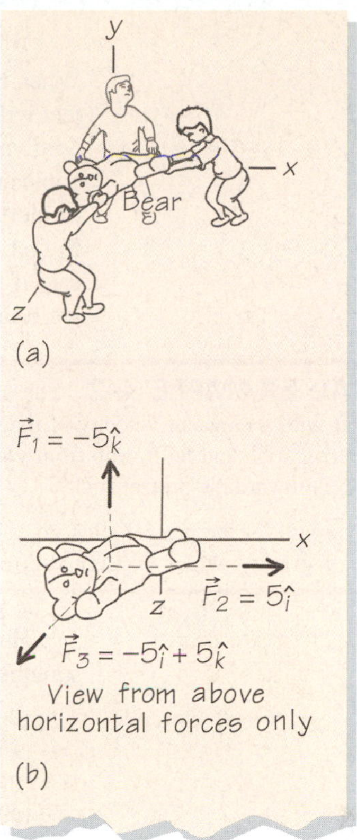

▲ **FIGURE 4–4** (a) Three children pull on a teddy bear in the xz-plane (horizontal). (b) Diagram for forces on the teddy bear. We have ignored the force of gravity.

The Aristotelian interpretation of experience was that to maintain motion a net force must act on a moving object. Galileo challenged this view in the first half of the seventeenth century. By carrying out experiments with objects sliding and rolling on a succession of ever smoother surfaces, so that friction became less and less important, he was led to the conclusion that moving objects possess an *inertia*, that is, they persist in their motion unless *something acts on them to make them change their motion*. Think of a hockey puck sliding across the ice, a situation in which friction is indeed small. The puck would go a long way if the surface were perfectly level.

Isaac Newton refined this within his **first law of motion**, also known as the **law of inertia**, which states:

> **When there is no net force acting on an object, that object maintains its motion with a constant velocity.**

We must also include the converse statement:

> **When an object moves with constant velocity, the net force acting on it must be zero.**

Constant velocity means that both the magnitude *and the direction* of the velocity do not change.

Recall the discussion of Section 3–6, where observers in different frames were moving at constant velocity with respect to each other. With this perspective there is another way to phrase the first law. In that situation, if one of these observers sees an object moving at constant velocity, then the other observers will all agree that the object moves at constant velocity, although they each will have a different value for that velocity. One of these observers, the one that is moving with the object, will say that the object is at rest. But that observer is not singled out in any significant way—zero velocity is a special case of constant velocity. From this perspective, the first law can be rephrased as:

When there is no net force on an object, it will be at rest in a reference frame that is one of a set of reference frames moving at constant velocity to one another.

When there is no net force on an object, there exists a reference frame in which the object will be at rest. It, and all other reference frames that move with a constant velocity relative to this one, forms the set of inertial reference frames. The name "inertial frame" comes from the fact that observers in inertial frames all agree that the first law, the law of inertia, holds. We see here that the first law is about the existence of inertial frames. In fact, the *converse* statement of the first law of motion, stated on p.89 above, is only true if the object has a constant velocity as seen in an inertial reference frame. (Later we'll talk about noninertial reference frames.)

CONCEPTUAL EXAMPLE 4–2 You are in an airplane flying at 30,000 ft with a constant velocity of magnitude 600 mph. You pick up a pretzel stick and let it drop from your hand. Are you traveling at 600 mph toward the pretzel stick?

Answer Once you let the pretzel stick go, the only force acting on it is that of gravity, which is aligned directly down. Both before and after it is dropped there is no net *horizontal* force acting on it, so the pretzel stick maintains its original horizontal motion of 600 mph relative to the ground and stays in the same horizontal position relative to you. Note the use of the qualifier "relative to the ground"—in fact, Newton's first law tells us that such qualifiers are necessary.

Having stated two versions of the law of inertia, let us now look in more detail at some issues raised by it. Mathematically we can express constant velocity by saying that

$$\frac{d\vec{v}}{dt} = 0,$$

or equivalently,

$$\vec{v} = \vec{v}_0, \tag{4–3}$$

where $\vec{v}_0$ is a fixed vector. It is important to realize that Eq. (4–3) is a vector equation, which means that each vectorial *component* of the velocity is a constant. An object at rest is just a special case, with $\vec{v}_0 = 0$, of Eq. (4–3).

Since acceleration is the rate of change of velocity, this also means that in the absence of a net force an object does not accelerate,

$$\vec{a} = 0. \tag{4–4}$$

Finally, we know from Chapter 3 that constant velocity means that an object's position vector $\vec{r}$ changes linearly with time; that is, $d\vec{r}/dt$ = a constant. In other words, all three position coordinates change linearly with time:

$$\vec{r} = \vec{r}_0 + \vec{v}_0 t. \tag{4–5}$$

At this point we might think about how this applies to the situation described in Example 4–1. What is the motion of the teddy bear as a result of the children tugging on it? We found that there are three forces on the toy, but the **net force** on it is zero. (Keep in mind that this means that the x-, y-, and z-components of the net force are all zero.) According to the first law, then, the bear's velocity is unchanging. If it is at rest when the three children first apply their forces, it remains at rest.

How does Newton's first law square with our experience? The answer to that question depends on how familiar you are with the action of friction. To understand the first law, you have to be able to visualize how friction acts and how it can be limited. Consider the horizontal motion depicted in Fig. 4–1. When we look at this figure, it may appear as though there is just one force acting upon the wagon, the pull of the woman's arm, and this would suggest the first law is wrong: The wagon is moving at constant speed, and you think that there is only a single force acting on it. However, if you consider the motion again, you may recognize the presence of friction (acting especially in the mechanical part of the axle), the force of gravity, and the force exerted by the ground that keeps the wagon from sinking. These must all add up to give a net force of zero. We can get a sense of a nearly friction-free world at a curling rink (Fig. 4–5), where the stone is subject to very little friction, and it is more apparent that the first law applies. We will look at the forces of friction and air resistance in more detail in Chapter 5.

▲ **FIGURE 4–5** In the game of curling, the broom melts a layer of ice and reduces the friction for the sliding rock. Even the reduced friction eventually brings the rock to rest.

A First Look at Some Common Forces and Their Properties

In our discussion of the first law, we have dealt with forces in a qualitative way. Now we'll look in a little more detail. Our qualitative notions about pushes and pulls arise to a large extent from *contact forces* that are associated with physical contact. Let's consider the wagon shown in Fig. 4–1. The child's mother pulls on a handle, and the handle pulls on the wagon to which it is attached. The force of friction, which we argued had to cancel the horizontal force acting on the wagon through the handle, is also associated with contact, either within the wheel system or in the contact of the wheel and the ground.

Some forces act without physical contact, and we say that they *act at a distance*. Consider the following application of the first law: A person leans over a railing and holds a short rope with a bowling ball attached to it (Fig. 4–6). There is no question that the person has to exert a force on the rope quite similar to the force she would have to exert when pulling a wagon. The rope in turn acts on the ball. The force that a taut rope can exert *as a pull* on whatever it is attached to is the force of **tension**. Yet the bowling ball is at rest, so it must be subject to a zero net force. There must be a force that just cancels the pull on the rope, and that is the force due to **gravity**, also known as the ball's **weight**. This force is due to the presence of Earth, and as the ball is not touching Earth, we must conclude that gravity acts at a distance. Now suppose the rope is cut (Fig. 4–6b). The ball no longer remains at rest. It falls with a changing velocity, as we know very well from our discussion of falling objects or projectile motion in Chapters 2 and 3. Thus, according to the first law, the net force on the ball is no longer zero. In fact, it is just the force of gravity, acting alone, without tension or any other force to cancel it and bring the net force to zero. If we watched the ball a little more, we would see it land and come to rest on the ground (Fig. 4–6c). The ball has arrived at another situation in which no net force acts on it. Yet gravity continues to act, so now there is a new force that cancels gravity: another contact force called the **normal force**. This force is associated with the material strength of the solids that make up the ground or any other solid surface. It is what keeps one solid object from penetrating the surface of another solid surface.

Is gravity the only force that acts at a distance? All the forces of electricity and magnetism also act at a distance, and these forces are the ones responsible for the properties of materials. This means that on a fundamental level these forces govern tension, friction, the normal force, and any other contact force. Is this a contradiction? Not at all: The distance involved in the contact forces we have discussed is of atomic size. The atoms of the mother's hand and those of the wagon handle interact—forces act between them—over imperceptibly small distances. Contact forces refer to forces that act over distances too short to be visible to the eye.

(a)

(b)

(c)

▲ **FIGURE 4–6** Bowling ball held suspended by means of a rope. (a) The tension in the rope is a force equal and opposite to the weight of the bowling ball yielding a zero net force. (b) After the rope is cut, the net force becomes the weight, and the bowling ball accelerates downward where (c) it comes to rest, and there is once more no net force acting on it.

Relative Motion and Reference Frames

In Section 3–6 we described the uniform (constant-velocity) motion of a bus. We can say now that no net force acts on the bus, and we can use this situation to look more deeply at the form of Newton's first law that describes the existence of inertial frames. We'll vary the situation of Section 3–6 a little. We continue to have an observer A standing at the curb, measuring the bus to move with constant velocity $\vec{v}_A$, but this time we'll place the second observer, we'll call her B, walking parallel to and at the same speed as the bus. Observer A measures observer B to have the same velocity $\vec{v}_1$ as the bus. Figure 4–7a shows the situation from the point of view of observer A, whereas Fig. 4–7b depicts observer B's rather different point of view. Observer B keeps up with the bus, so that she sees the bus at rest; that is, to observer B, the bus has velocity $\vec{v}_B = 0$. The coordinate system, or frame of reference, centered on observer B is the one referred to in the statement of Newton's first law in which an object with no net force acting on it is at rest.

Which of these observers sees the "true" situation? It is true that observer B sees observer A move with velocity $-\vec{v}_1$; observer B even sees the street moving with that same velocity. But so what? Our "commonsense" preference is for the point of view of observer A because his reference frame is more familiar. But if you have ever had the experience of sitting in a very slowly moving train leaving a station on a track adjacent to another train, you will recall a disorientation as to whether it is your own train or the other train that is moving—we really cannot distinguish who moves and who doesn't.

(a) According to person A

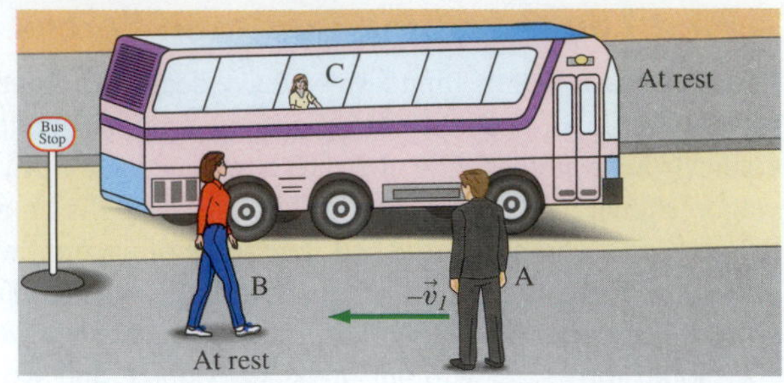

(b) According to Person B

▲ **FIGURE 4–7** Observer A stands watching a bus moving down a street, and observer B moves at the same speed as the moving bus but on the sidewalk. In (a) we see things from the point of view of observer A's frame; in (b) we see things from the point of view of observer B's (and C's) frame.

In other words, there is no "true" situation—A's point of view is no better or worse than that of B. As each of our observers measures the bus to have a constant velocity, each observer concludes that there is no net force acting on the bus, and Newton's law holds in the same way for each observer.

We have concluded that the first law provides no clue as to who is at rest. When observers A and B try to determine which one of them is at rest or exactly what his or her absolute velocity is by looking at the bus, they cannot do so. Another way to say this is that among a set of inertial reference frames there is no way, at least based on Newton's first law, to choose one frame as "special" or "preferred." While an object may be at rest only in a particular inertial frame—here the bus is at rest in the frame centered on observer B—it is not at rest in an absolute sense. We shall see in Section 4–2 that, *even in the presence of net forces*, there is no fundamental way to distinguish between different *inertial* frames—so there is no way to say if any particular inertial frame is at rest in an absolute sense.

4–2 Newton's Second Law of Motion

When the net force acting on an object is zero, the object maintains constant velocity. What happens to this object when a (nonzero) net force acts on it? Newton's second law answers this question both qualitatively and quantitatively. We can start by saying:

An object acted upon by a net force accelerates.

Further, the object will accelerate, or change its velocity, in the same direction as the net force.

Consider a sled on an ice surface slick enough to be friction free (Fig. 4–8). If the sled were at rest under these conditions and you could push it with a given force—say by bracing your back against a rock and pushing with your feet—the sled would start to move, that is, it would accelerate starting from zero velocity to some velocity that depends on the duration and strength of the push. In fact, the acceleration of the sled is proportional to the strength, or magnitude, of the push; if the sled is pushed twice as hard (and we'll show what "twice" means below), its acceleration will double.

(a)

(b)

▲ **FIGURE 4–8** You exert a fixed force on a sled. The resulting acceleration of the sled is smaller in (b) than in (a), because the sled has more mass. As a result, the empty sled aquires a much greater speed than does the full sled.

Which is easiest to push in this way, an empty sled or one loaded with kids (Figs. 4–8a and b)? Experience tells us that the specific response of an object to a given net force also depends on the mass of the object. Mass measures an object's resistance to a change in its motion, the property that we earlier called *inertia*, and to reinforce this point, the mass in this context is sometimes called the **inertial mass**. If the object is already moving, then the object's resistance to change can be its resistance either to speeding up or to slowing down; if it is at rest, we are dealing with its resistance to starting to move. The greater the mass of an object, the greater its resistance to a change in motion, that is, the greater its inertia. It is easier to set the empty sled in motion than the full one.

To state all this in a more quantitative fashion, if the mass of the sled in our example were doubled and then pushed with the same force you used the first time, the sled would move with half of the initial acceleration. For a given force the less the object's mass, the greater the acceleration (Fig. 4–8a); the greater the mass, the less the acceleration (Fig. 4–8b). As we'll see in more detail in the next subsection, we can even build a mass scale in this way. Experiments with a set of masses will then exhibit an *inverse* relation between mass and the rate of change of velocity (acceleration) for a given force.

This basic relation between the net force on an object, $\vec{F}_{net}$, the mass of the object, m, and the object's acceleration, $\vec{a}$, was discovered by Newton and has come to be known as **Newton's second law of motion**. It is stated quantitatively as $\vec{a} = \vec{F}_{net}/m$ or, in more standard form,

$$\vec{F}_{net} = m\vec{a}. \tag{4–6}$$

NEWTON'S SECOND LAW

Acceleration is the rate of change of velocity so $\vec{a} = d\vec{v}/dt$, and in terms of the velocity, Newton's second law becomes

$$\vec{F}_{net} = m\frac{d\vec{v}}{dt}. \tag{4–7}$$

It is important to remember that the force appearing here is the net force; that is, the vector sum of all the forces acting on the object. As with any vectorial expression, Newton's second law is equivalent to three scalar equations for the components:

$$F_{x,net} = ma_x = m\frac{dv_x}{dt}; \tag{4–8a}$$

$$F_{y,net} = ma_y = m\frac{dv_y}{dt}; \tag{4–8b}$$

$$F_{z,net} = ma_z = m\frac{dv_z}{dt}. \tag{4–8c}$$

We shall refer interchangeably to Newton's second law in its vector form, Eq. (4–6) or (4–7), or in its equivalent component form, Eqs. (4–8).

There are some very important properties of the inertial mass (or just mass for simplicity) to take into account. Mass is a scalar quantity; it has no direction associated with it. It is always positive. As a consequence, the acceleration of an object is in the same direction as the force exerted on it. Furthermore, masses are additive; that is, if an object contains two parts with masses m_1 and m_2, the total mass of that object is[†]

$$m = m_1 + m_2. \tag{4–9}$$

Is the Second Law Just a Definition of Force?

In Chapter 5 and beyond, we will explore many applications of Newton's second law. At this point, however, we must consider what the second law really means. Does Eq. (4–6) merely define the force? We can measure an object's acceleration with a meter stick and a watch. Do we use that measurement merely to define the quantity $\vec{F}/m$? The power of the second law lies in the fact that it goes beyond this mere definition. Here we shall explain how forces and masses are defined and how Newton's second law allows us to make predictions of the motion of objects.

We'll begin with the calibration of a set of masses. Suppose that we have a large supply of lumps of gold. We will label one of them A and call it our standard mass, $m_A = 1$ kg. We also have available a spring and a very smooth (frictionless) table (Fig. 4–9a). As you know from experience, a relaxed spring (one that is in an "equilibrium" position) neither pulls nor pushes, but a compressed spring exerts a force that will push away an object in contact with it. The force exerted by a given spring depends only on how much the spring is compressed. We compress the spring by 1 cm, place lump A against it, release the spring, and measure the instantaneous acceleration of the lump for the particular compression of the spring. (Actually a hand is needed to compress and then release the spring, although we have not included the hand in the figure.) Suppose that when

[†]When the object has a mass that changes with time (as in rocket motion), we need to proceed differently, but the argument will still remain valid.

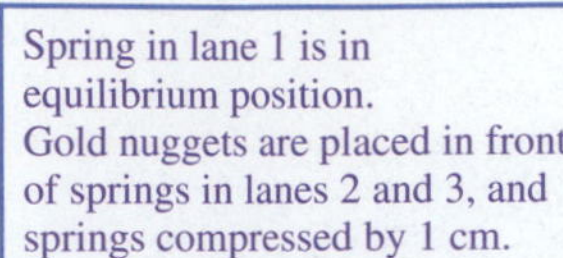

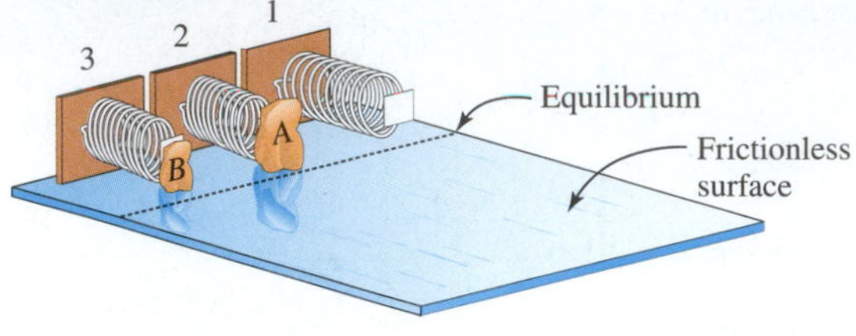

(a)

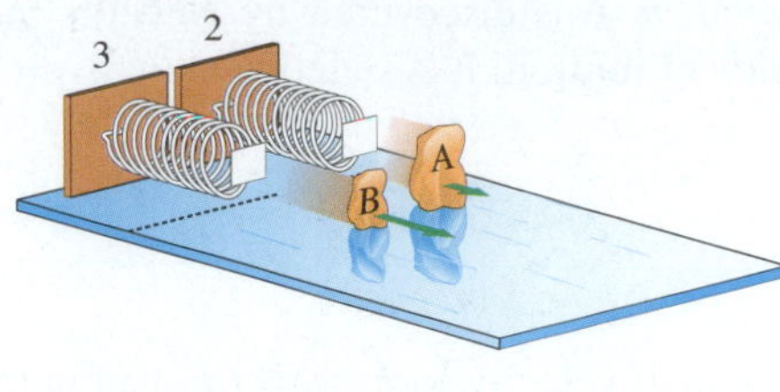

(b)

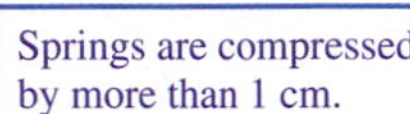

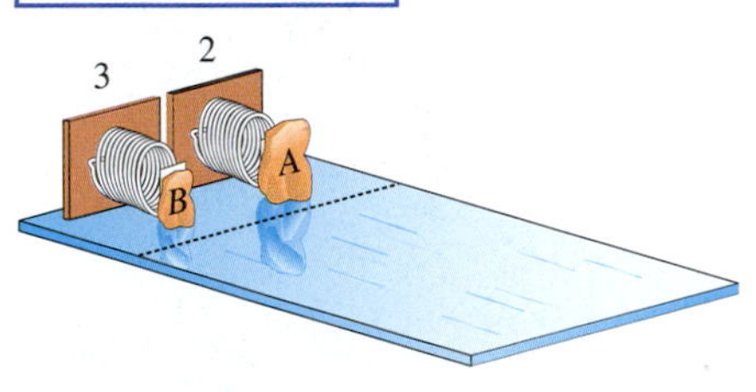

(c)

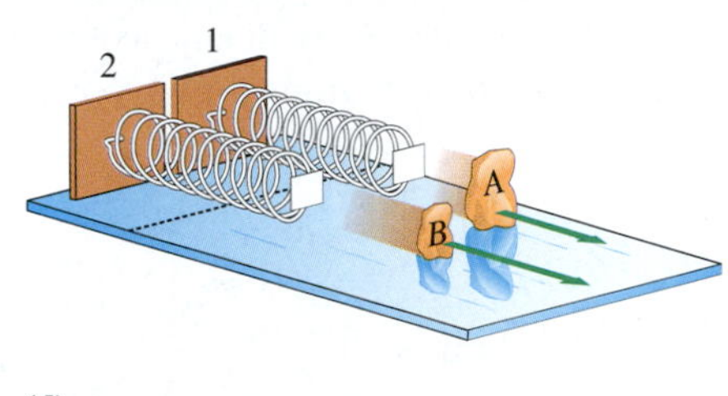

(d)

▲ **FIGURE 4–9** (a, b) A series of experiments with a spring and a set of masses allows us to construct a mass scale and to learn about the forces exerted by the spring. (c, d) Once we have established masses and know the force law, we can predict the motion of a mass.

the spring is compressed by 1 cm and then released, lump A has an acceleration of, say, $a_A = 2\text{ m/s}^2$, as in Fig. 4–9b. (Notice that this measurement must be made instantaneously, at the moment that the spring is released. Although this measurement is correlated with the measurement of the speed with which the mass leaves the spring, it is not the same measurement!) At this point, Newton's second law provides a precise definition of the force exerted by this spring when it is compressed by 1 cm. Because $F = m_A a_A$, the force is $F = (1\text{ kg})(2\text{ m/s}^2) = 2\text{ kg}\cdot\text{m/s}^2$. Every time the spring is compressed by 1 cm, it will exert this force.

We can now measure the masses of our other lumps of gold. Place a second lump, B, in front of the spring and compress it by the same 1 cm (Fig. 4–9a). When the spring is released (Fig. 4–9b), the initial instantaneous acceleration of lump B is measured to be, say, $a_B = 3\text{ m/s}^2$. By Newton's second law, the mass m_B is

$$m_B = \frac{F}{a_B} = \frac{m_A a_A}{a_B} = m_A \frac{2\,\cancel{\text{m}/\text{s}^2}}{3\,\cancel{\text{m}/\text{s}^2}} = \frac{2}{3} m_A = \frac{2}{3}\text{ kg}.$$

In this way, we can determine the mass of each of our lumps of gold—or of any object for that matter.

The force of the spring is always the same whenever it is compressed by 1 cm. Let's now compress the spring by a different amount, place lump A in front of it (Fig. 4–9c), then measure the acceleration of lump A when the spring is released (Fig. 4–9d). This time, we measure, say, $a'_A = 5\text{ m/s}^2$. We can again use Newton's second law to define the force that the spring exerts when it is compressed by the new amount; it is $F' = m_A a'_A = (1\text{ kg})(5\text{ m/s}^2) = 5\text{ kg}\cdot\text{m/s}^2$. We can continue this process to define the force exerted by the spring for different compressions. In the same way we can figure out the pulling force exerted by the spring when it is stretched. We have found a *force law* for the spring, which in this case is simply a catalog of how much force the particular spring exerts for any given compression or extension. A force law describes the force due to a particular source and how that force depends on variables such as position or time. Throughout this book, we shall study the force laws describing many phenomena, including friction, gravity, springs, and electric charges.

At this point, Newton's second law enables us to make predictions about the motion of an object for the first time. If we compress the spring to the position shown in

Fig. 4–9c and place lump B against it, we can predict lump B's acceleration. We have already determined that the force exerted by the spring in this second position has magnitude $F' = 5\ \text{kg}\cdot\text{m/s}^2$. When the spring is released, as in Fig. 4–9d, the second law *predicts* the instantaneous acceleration to be

$$a'_B = \frac{F'}{m_B} = \frac{5\ \cancel{\text{kg}}\cdot\text{m/s}^2}{2/3\ \cancel{\text{kg}}} = 7.5\ \text{m/s}^2.$$

Once we know how much force is exerted by the spring for a given compression or stretch, we can use the spring to learn about other forces. Suppose we have an unknown force X. Rather than studying the motions that force X causes, we act with it on a mass—any mass—which is also attached to our standard spring. The spring length is allowed to adjust so that the mass does not accelerate. According to the first law, the spring force is exactly canceling force X, and because we know the spring force, we now know force X. As an example, if we turn our spring to the vertical and suspend a block of known mass from it so that the block is motionless (Fig. 4–10), the force of gravity acting on the block must exactly balance the upward force of the stretched spring. By observing how much the spring is stretched, we have measured the force of gravity on the block. This is done in a way that involves no motion! We could now predict how the block would accelerate if there were no spring and only gravity acted on it.

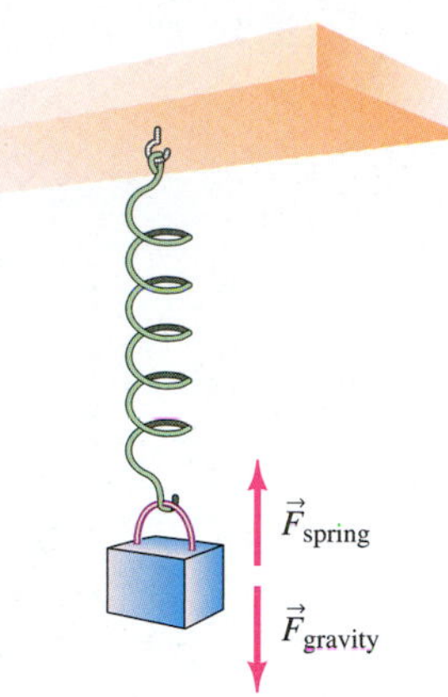

▲ **FIGURE 4–10** By using Newton's first law and a known force, such as that exerted by the stretched spring, we can measure other forces—in this case, the force of gravity acting on the (stationary) block.

EXAMPLE 4–3 You need to deliver a box of bowling balls to a bowling alley. The balls will be placed in a box that is initially at rest but that you want to push into the bowling alley. The box itself has a mass that is very small compared to even one bowling ball. You start with one ball in the box, exert a given force of given strength upon the box for a time period Δt, and at the end of that time the box moves at a speed of 3.2 m/s. You then repeat the procedure with more bowling balls in the box; you exert the same amount of force on the box for the same period of time (Δt) and find the box to have a final speed of 0.4 m/s. How many balls are in the box now?

Setting It Up The two cases are shown in Figs. 4–11a and b at the particular time, after an interval Δt, when the speeds are v_1 and v_2, respectively. You know that an identical force of constant magnitude F acts on two different masses m_1 and m_2 for identical time periods Δt, where $m_2 = nm_1$. Here m_1 is the mass of one bowling ball and n is the number of balls in the box, which is the quantity we want to find. The resulting speeds after time Δt are v_1 and v_2, respectively, and are given.

Strategy In the two cases described the box containing the bowling ball(s) is subject to the same force. Moreover, we can ignore the mass of the box. Using Newton's second law, we can find the accelerations a_1 and a_2 during the period Δt when the force operates. These accelerations are

$$a_1 = \frac{F}{m_1} \quad \text{and} \quad a_2 = \frac{F}{m_2} = \frac{F}{nm_1}.$$

Although we do not know the numerical values of the two accelerations, we do know the speeds v_1 and v_2 after a fixed period of acceleration. Further, we learned in Chapter 2 that an object that starts at rest and undergoes a fixed acceleration $\vec{a}$ for a given period of time Δt has the velocity $\Delta\vec{v} = \vec{v} = \vec{a}\,\Delta t$. In our one-dimensional case, then, we have

$$v_1 = a_1\,\Delta t = \frac{F\,\Delta t}{m_1} \quad \text{and} \quad v_2 = a_2\,\Delta t = \frac{F\,\Delta t}{nm_1}.$$

We now have enough information to solve for the unknown, n.

Working It Out We can solve for the ratio F/m in terms of v_1 and Δt and substitute it into the equation for v_2, which we can then solve for n. Alternatively, we can simply take the ratio of the two speeds:

$$\frac{v_1}{v_2} = \frac{(\cancel{F}\,\cancel{\Delta t}/\cancel{m_1})}{(\cancel{F}\,\cancel{\Delta t}/n\cancel{m_1})} = n.$$

Numerical substitution gives $n = (3.2\ \text{m/s})/(0.4\ \text{m/s}) = 8$ bowling balls.

What Do You Think? Suppose the mass of the box is not negligible. What would be the effect?

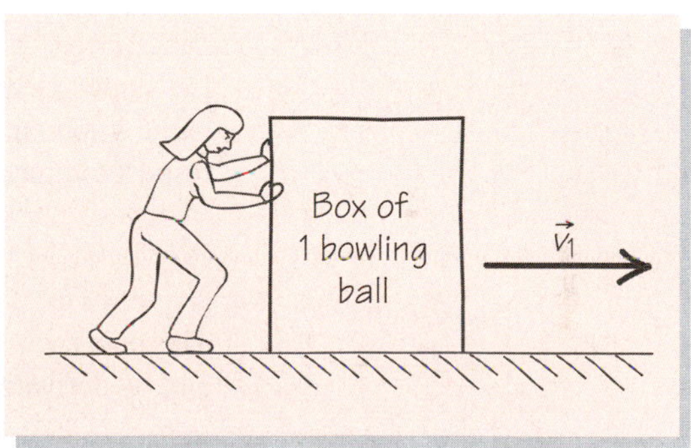

(a)

(b)

▲ **FIGURE 4–11** Delivering bowling balls to an alley. In (a) the box has one ball, and in (b) there is an unknown number of balls.

Units: We first discussed the dimensions and units of mass, length, and time in Chapter 1. Because acceleration has dimensions of $[LT^{-2}]$ and units of meters per seconds squared in SI, force has dimensions of $[M \cdot LT^{-2}]$ and, in SI, units of kilogram-meters per seconds squared, or **newtons** (N):

$$1\ \text{N} \equiv 1\ \text{kg} \cdot \text{m/s}^2. \tag{4–10}$$

In other words, a force of 1 N exerted upon an object with a mass of 1 kg will produce an acceleration of 1 m/s^2.

In the system of centimeters, grams, and seconds (the cgs system), the force 1 $\text{g} \cdot \text{cm/s}^2$ is called the *dyne*:

$$1\ \text{dyne} \equiv 1\ \text{g} \cdot \text{cm/s}^2 = 10^{-5}\ \text{N}. \tag{4–11}$$

Here, again, a force of 1 dyne acting on a mass of 1 g causes an acceleration of 1 cm/s^2.

Another force unit in everyday use is the pound (lb); 1 lb = 4.448 N. The pound is used in the British engineering system, in which mass is measured in slugs.

How Do Observers in Different Inertial Frames See Newton's Second Law?

Let's revisit the two streetside observers A and B of Section 4–1 (Fig. 4–7). They are in different inertial frames moving with fixed velocity with respect to one another. In Section 4–1, B was moving with the constant velocity of the bus, but here we are going to let the bus change its velocity, so we introduce a new label $\vec{u}$ for the fixed velocity of B with respect to A. These observers will generally measure the bus to have different velocities, $\vec{v}_A$ and $\vec{v}_B$, whether the bus is accelerating or not.

Now suppose there is a net force on the bus—the bus is accelerating. Do each of the observers see the same acceleration? If they do, then they will agree on the force acting. We have already seen in Eq. (3–60) how the two observers see the velocity of a bus: $\vec{v}_A = \vec{v}_B + \vec{u}$, or

$$\vec{v}_B = \vec{v}_A - \vec{u}.$$

Let's now see how observers A and B measure the *rate of change* in the bus's velocity—this is the important measurement for the second law. The rate of change is a derivative with respect to time, so we take the derivative with respect to time of both sides of the equation:

$$\frac{d\vec{v}_B}{dt} = \frac{d\vec{v}_A}{dt} - \frac{d\vec{u}}{dt}. \tag{4–12}$$

The observers are in inertial frames, meaning that their relative velocity $\vec{u}$ is constant and its derivative is zero. Therefore

$$\frac{d\vec{v}_B}{dt} = \frac{d\vec{v}_A}{dt}. \tag{4–13}$$

Our observers *agree* on the acceleration of the bus, a result we first encountered in Chapter 3. Assuming now that the mass m of the bus does not change with time,[†] we multiply both sides of Eq. (4–13) by m to find that

$$m\frac{d\vec{v}_B}{dt} = m\frac{d\vec{v}_A}{dt},$$

or, according to the second law,

$$\vec{F}_B = \vec{F}_A, \tag{4–14}$$

where $\vec{F}_B$ is the force on an accelerating object, here the bus, as measured by observer B and $\vec{F}_A$ is the force measured by observer A. We conclude that *observers in different inertial frames agree on the net force acting on an object*.

We saw in Section 4–1 that observers in different inertial frames could not use the first law as a way of deciding who was moving and who was standing still; here we see

[†] When the forces holding the system together are strong, special relativity implies a modification. In most of the situations we meet, the effect is very small.

that as they also agree on the net force acting, the second law cannot help us to answer this question either. **Observers in different inertial frames cannot by experiment tell which of them is moving and which of them is at rest.** The statement of the equivalence of all inertial frames is sometimes called the *relativity principle*. This result takes us a long way from Aristotle: not only is the state of being at rest not special, there is no fundamental way to distinguish that state from any state of steady motion. Once Einstein understood that the relativity principle applied to any physical situation, not just those involving the mechanical motion of objects under the influence of mechanical forces, he was led to the theory of relativity (see Chapter 39).

THINK ABOUT THIS . . .

HOW DOES AN AIR BAG PROTECT YOU IN A CRASH?

Large forces imply large accelerations. A car accident or a fall from a great height may be deadly because of the rapid deceleration, the result of large forces that your body may not be equipped to withstand. For protection it is necessary to find a way to bring you to a stop by providing a smaller deceleration over a larger time. Air bags in automobiles work on this principle; when a collision stops a car very suddenly, a passenger would suffer a very sudden deceleration in a subsequent collision with the steering wheel or the windshield. This is mitigated by the very rapid release of an air bag, which is deep enough and "soft" enough to allow the passenger to slow down over a longer period of time. Firefighters similarly use large elastic safety nets to catch people who have to jump from burning buildings. When the deceleration is for fun, the same principle applies. Bungee cords are made of a very elastic material, and there are no bungee *chains*, which would have the unfortunate effect of stopping you "on a dime." Still another application is provided by airplane ejection seats, which in the past were powered by explosives beneath the seat. The rapid acceleration of these mechanisms often led to serious damage to the pilot. Today ejection seats are powered by small rockets that can supply a smaller acceleration over a longer period of time, rather than a large acceleration over a very short period of time and hence a safer ejection (Fig. 4–12). One other example comes to mind: You may have seen drawings in which Superman catches Lois Lane just before she hits the ground. That very action would imply a rapid deceleration that would be just as bad for Lois as hitting the ground. Superman would do better not to wait for the last instant and instead slow Lois down over a longer period of time.

▲ **FIGURE 4–12** Test ejection of a pilot from an AMX jet fighter. The jet plumes below the seat are due to the ejection rocket.

4–3 Newton's Third Law of Motion

The first and second laws involve forces acting *on* objects, and such forces have a specific source. The push on the sled that we described earlier did not come from nowhere; it came from a person. Newton noticed, however, that forces always come in pairs. When you push on the sled, the sled pushes back on you. *When a force due to object B acts on object A, then a force due to object A also acts on object B.* It is obvious that when you push on a wall the wall pushes back on you—that is why you may have to dig in your feet and why you would move backward if you pushed on a wall while you were standing on ice. It is less obvious that when Earth tugs on an apple, causing it to fall (the force is gravity), the apple also tugs on Earth, causing Earth to accelerate toward the apple. Earth and the apple exert gravitational forces *on each other*. We can refine the notion we adopted when discussing the first two laws of motion—that forces act upon objects—to say that forces do not simply act *on* objects; rather, forces act *between* two objects or between an object and its surroundings. Objects are said to *interact* when forces act between them.

Newton extended this statement to a quantitative form in the **third law of motion**:

When a force due to object B acts on object A, then an equal and opposite force due to object A acts on object B.

According to Newton's third law, the force on Earth due to the apple is equal in magnitude but opposite in direction to the force on the apple due to Earth.

A mathematical statement of the third law is the following: Let the force on object A due to object B be $\vec{F}_{AB}$. Then an equal and opposite force $\vec{F}_{BA}$ acts on object B due to object A:

$$\vec{F}_{BA} = -\vec{F}_{AB} \tag{4–15}$$

NEWTON'S THIRD LAW

Notice that on the left we have a force acting on object A and on the right a force acting on object B. These forces act on different objects—this is not a pair of canceling forces acting on the same object! The third law is sometimes called *the law of equal action and reaction*. Do not let this particular phrasing mislead you into believing that the *accelerations* of the two objects are the same. Each object accelerates according to the second law, which means that the acceleration depends on the mass of the object. When the apple accelerates toward Earth, Earth accelerates towards the apple, but the Earth's acceleration is far far less than that of the apple's.

The third law is illustrated in Examples 4–4 and 4–5. In these examples, we imagine an outer space environment where we can think of astronauts and satellites as isolated from all forces except the forces they exert on one another. Such a situation is more difficult to arrange on Earth.

EXAMPLE 4–4 An astronaut and a satellite are in an environment where they can be considered to form an isolated system with no external forces acting on that system (Fig. 4–13a). The astronaut tugs on the satellite with a force of 10.0 N toward herself (to the right and down in the figure). What is the force on the astronaut?

Setting It Up The figure labels the the two objects involved as the astronaut A and the satellite S. We are given the force on the satellite due to the astronaut, $\vec{F}_{SA}$ = 10.0 N to the right and down.

Strategy This is an application of the third law: If we know the force on the satellite due to the astronaut, then the force on the astronaut due to the satellite, $\vec{F}_{AS}$, is equal in magnitude and oppositely directed.

Working It Out We have

$$\vec{F}_{AS} = -\vec{F}_{SA} = 10.0\text{ N to the left and up in the figure}$$

This force is drawn in Fig. 4–13b.

What Do You Think? An astronaut is at work repairing the Hubble Telescope in outer space. Having just finished a successful maneuver, she slaps the telescope-carrying space ship in a gesture of satisfaction. She is loosely tethered to the space ship. What will be the effect of the slap?

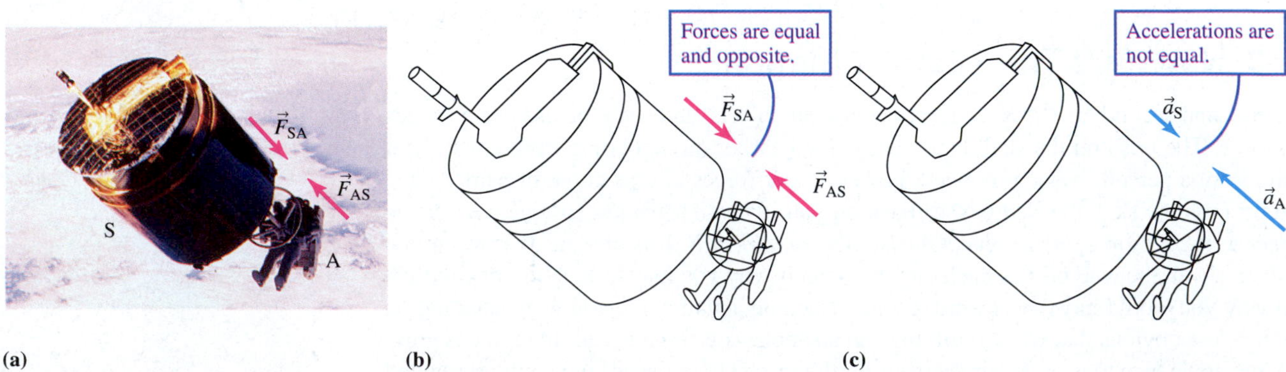

▲ **FIGURE 4–13** (a) An astronaut A is interacting with a satellite S. (b) According to Newton's third law, the force the astronaut exerts on the satellite is equal and opposite to the force the satellite exerts on the astronaut. (c) Even though the force on the astronaut has the same magnitude as the force on the satellite, the accelerations of astronaut and satellite are quite different as they have unequal masses.

EXAMPLE 4–5 Assume that the mass of the astronaut in Example 4–4 is 75.5 kg and that of the satellite is 755 kg. What is the acceleration of each?

Setting It Up We can refer to Figs. 4–13a and b. We label the known masses of astronaut and satellite by m_A and m_S, respectively. Here we want to know their accelerations given the forces on them and their masses.

Strategy In the previous example we assessed the forces acting on each of the two objects using the third law. In this example we must apply the second law twice to find the accelerations of the objects.

Working It Out The forces, and thus the accelerations, are all directed along the line between the astronaut and satellite, and we can therefore drop the vector notation. According to the second law, the satellite has an acceleration of magnitude

$$a_S = \frac{F_{SA}}{m_S} = \frac{10.0\ \text{N}}{755\ \text{kg}} = 0.0132\ \text{m/s}^2.$$

This acceleration of the satellite is directed to the right (and down). For the astronaut, the acceleration has magnitude

$$a_A = \frac{F_{AS}}{m_A} = \frac{10.0\ \text{N}}{75.5\ \text{kg}} = 0.132\ \text{m/s}^2.$$

This acceleration is directed to the left (and up). Thus, as Fig. 4–13c shows, the astronaut experiences an acceleration whose magnitude is 10 times larger than that of the satellite. The forces may be equal, but the result of those forces is very different. This difference follows from the difference in the masses of the satellite and the astronaut.

What Do You Think? Consider a sled containing a passenger on flat, slick ice. The sled is given a push by a person. Does the passenger exert a force on the sled when the push is given?

Examples 4–4 and 4–5 demonstrate why the force that the apple exerts on Earth is not directly observable. With a small apple of mass $m_a = 0.1$ kg, the force on the apple due to Earth ($\vec{F}_{aE}$) is approximately 1 N, directed downward. This leads to an acceleration of magnitude $F_{aE}/m_a = 10\ \text{m/s}^2$ when the apple falls from the tree. According to Newton's third law, the upward force $\vec{F}_{Ea}$ that the apple exerts on Earth also has magnitude 1 N. But because Earth's mass m_E is approximately 6×10^{24} kg, its upward acceleration has magnitude $F_{Ea}/m_E \cong 2 \times 10^{-25}\ \text{m/s}^2$, much too small to be observable. The evidence that led Newton to the third law involved the forces of the Earth–Moon system, where the law's effects are observable.

THINK ABOUT THIS . . .

HOW DOES A ROCKET WORK?

In Fig. 4–14, we see a very entertaining demonstration that works on the same principle as a rocket's propulsion system. When the valve of the fire extinguisher is opened, a force is exerted on a mass of CO_2 gas, causing it to be accelerated from the fire extinguisher. (The origin of this force is the pressure in the canister, but that is not what is important here; we only need *some* force to act on a mass and cause it to be expelled.) By Newton's third law, the expelled gas exerts an equal and opposite force on the extinguisher. If the extinguisher is held rigidly by the man on the wagon, the force due to the expelled gas acts on the entire system, which in the figure is accelerated to the left. This is the principle of the rocket: The rocket accelerates matter in one direction and by Newton's third law it is itself accelerated in the opposite direction.

▲ **FIGURE 4–14** Action and reaction go together thanks to the third law. In this case, the release of CO_2 from a fire extinguisher causes the initially stationary cart to be propelled in a direction opposite that of the released gas.

*4–4 Noninertial Frames

In the discussion of reference frames to this point, we focused on **inertial frames**, that is, frames moving with a uniform velocity. For such frames there is no net force on an observer that is at rest with respect to the coordinates that define the frame. We now consider accelerating or **noninertial frames of reference**. For example a set of coordinate axes centered on an airplane accelerating down a runway or on a car going around

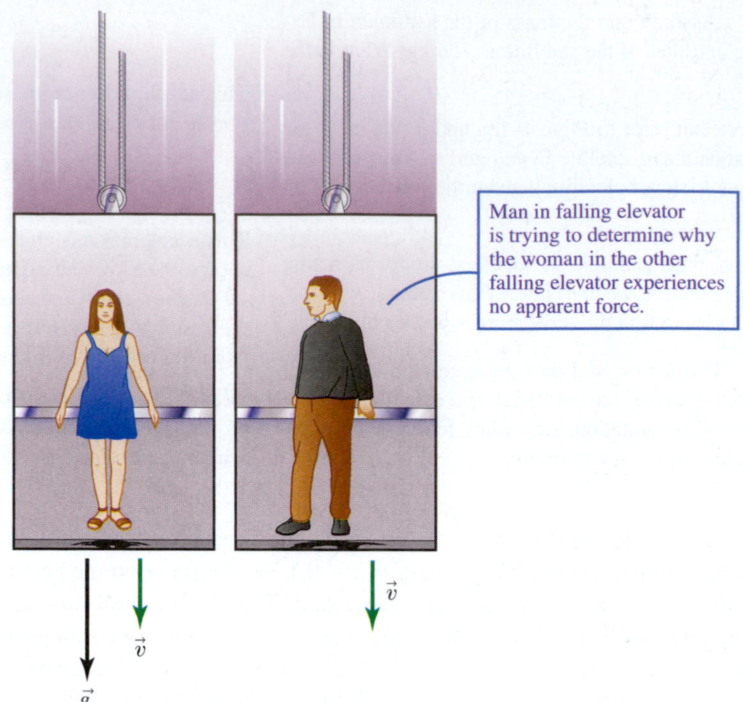

FIGURE 4–15 If the person in the right-hand elevator observing the person in the left-hand elevator does not know that each elevator is in free fall, then he will have to have recourse to a fictitious force to explain his observations.

a corner define noninertial frames. Consider another example: Two adjacent glass elevators start at rest, at the same height and time and drop in free fall. A person inside the first elevator moves under the influence of the force of gravity and falls with acceleration $\vec{g}$. An observer in the second elevator, who is also in free fall, looks at the first elevator (Fig. 4–15). When he does so, he has the choice of two explanations of his observations:

1. We are each in free fall, subject to the same gravitational force, and as a consequence we are both falling with acceleration $\vec{g}$. Since we started out together, with the same initial velocity, we are still together, at rest relative to each other. We are both in noninertial frames.
2. The person in the first elevator is at rest relative to me. She is however subject to a gravitational force and ought to be accelerating downward with acceleration $\vec{g}$. Since she is still at rest relative to me, it must be that there is some force that just counters gravity, so that she is *not* accelerating. I wonder what the source of that force is.

In the first interpretation, the observer in the second elevator knows somehow that he is in free fall and interprets the motion of the first elevator accordingly. In the second interpretation, the observer in the second elevator doesn't know he is accelerating under the influence of gravity and assumes that he is in an inertial frame—Newton's laws of motion are operative—and he must therefore invent a *fictitious force* (sometimes called a *pseudoforce* or a *noninertial force*) to make sense of his observations. The force is fictitious because the observer cannot find any source for it. More generally, it is necessary to invent such fictitious forces to make sense of observations made in in any noninertial frame of reference.

You may wonder if this is realistic. After all, wouldn't the observer "feel" that he is falling? But suppose the "observer" is just a camera or some other scientific instrument looking at the person in the first elevator and that in the film of that person you could see Earth below and thereby know that gravity was acting on him or her. Then you really couldn't decide from the film which interpretation was correct.

In the example above, the observer in the noninertial frame had to invent a fictitious force to "maintain" the first law. Similarly, fictitious forces may be needed to "maintain" the second law for an observer in a noninertial frame. Consider a wheeled food cart sitting in the aisle of an airplane taxiing along a runway with a constant velocity.

In a reference frame centered on the plane the cart is at rest, and in a reference frame fixed to Earth both the cart and the plane are moving with a constant velocity. In both cases the first law—the law of inertia—holds. But suppose now the airplane accelerates down the runway for takeoff. What happens to the wheeled cart? If you are inside the airplane, you will see the cart start to accelerate backward down the aisle, even though no visible identifiable force has acted upon it. You would have to invent a fictitious force to account for its motion. (From the reference frame fixed to Earth, this would pose no problem: The cart would be moving forward with the same constant velocity it had before the plane accelerated.)

To take a last example, consider an observer turning around on her heels (Fig. 4–16). She sees a parked automobile 3 m away, and as she turns—you can certainly try this yourself—the car appears to move in a circle about her: she sees the automobile as having a velocity that is constantly changing direction. According to the observer, it is accelerating (as in Chapter 3—all objects in circular motion are accelerating as the direction of their motion is always changing). If there is acceleration, then according to Newton's laws, there is a force acting. But what is the force that accelerates the automobile? The observer again has to invent a fictitious force to explain the phenomenon. This invented force would have to be a bizarre one: A second automobile 30 m away will have an acceleration 10 times as large as the car 3 m away ($a = \omega^2 r$, and the angular speed ω is set by the observer's rotation), so that the fictitious force that acts on that farthest automobile is, if we use the second law, 10 times as large! The conclusion, and the main point to be made here, is that Newton's laws do not apply to motions observed from noninertial frames of reference. A certain amount of care is necessary to determine whether all possible sources of force have been identified and whether an observer is in an inertial or noninertial frame. We shall normally deal with real forces in inertial frames and give ample warning when accelerating (noninertial) frames are involved.

▲ **FIGURE 4–16** A person pivots in the street. From her point of view the near car is accelerating in circular motion, and the far car is similarly accelerating with even greater magnitude.

CONCEPTUAL EXAMPLE 4–6 Consider a skateboarder who jumps a barrier while the skateboard moves under the barrier. We'll assume here that the boarder has practiced on a stationary board and always jumps straight up off the board. If the board with boarder moves with constant velocity along a flat surface, then by jumping straight up with respect to the board, both boarder and board will continue to move with their original horizontal component of motion, and after the barrier is passed, the boarder falls neatly back onto the board (Fig. 4–17). Consider now a variant: The same board has a little rocket engine triggered by the jump, and it accelerates as the boarder is jumping. What happens? Describe the process from the point of view of (a) the other boarders watching the stunt and (b) an observer *on* the board—for example, a small camera that sends a video to someone watching a monitor.

Answer The skateboarder will land behind the board.

(a) From the point of view of the audience, this is easy to understand. The skateboarder launches when the board has a certain horizontal speed. Because after leaving the board there are no further horizontal forces acting on the boarder, his horizontal velocity component remains fixed. But during the time the boarder moves in the air, the board speeds up, so the audience sees the boarder fall behind.

(b) The observer on the board will also see the skateboarder fall behind. In fact, he sees the skateboarder accelerating to the rear with a magnitude equal to the board's acceleration as the audience sees it. Being a believer in Newton's second law, he ascribes the boarder's acceleration as due to a force, but since he is in a noninertial frame, that force will have no apparent source. He might use that fact to decide he is indeed a noninertial observer.

▲ **FIGURE 4–17** As long as the skateboard is in uniform motion and the skateboarder jumps vertically off it, the boarder will end up right back on the board.

EXAMPLE 4–7 Professor A is standing on a train platform; her friend, professor B, is leaving the station in a train with acceleration α in the $+x$-direction (Fig. 4–18). Professor A considers herself to be at rest and states that there is no net force acting on her. What does professor B observe professor A's motion to be, and how might he interpret that motion?

Setting It Up Figure 4–18 includes the x-axis. We call the known velocity of the train as measured from the platform $\vec{u}$ (actually only its change is known, and that is all that matters here), and we call $\vec{v}_A$ the velocity of professor A as measured from a frame of reference at rest with respect to professor B, the motion we want to find.

Strategy We can first simplify matters by recognizing that the motion is in one dimension and drop the explicit vector notation. We also recognize that the velocity of the platform as seen by professor B is $-u$ and that since professor A is at rest with respect to the platform, this is also the velocity that professor B would ascribe to professor A. We still need to interpret this result, which we can do by translating it into the fact that professor B will see professor A accelerating.

Working It Out We have found that professor B measures professor A to have velocity

$$v_A = -u.$$

By taking a derivative of this, we find that professor B measures professor A to have acceleration

$$\frac{dv_A}{dt} = -\frac{du}{dt} = -\alpha.$$

Professor B sees professor A accelerating *backward* (minus sign) with an acceleration of magnitude α. He would then say that, according to the second law, there must be a force on professor A of $-m\alpha$ that is responsible for giving her this motion. He would not, however, be able to find an identifiable agent for this force and might in this way decide that he is in a noninertial frame.

What Do You Think? Suppose each professor drops a tennis ball on the floor and each observes their own ball bounce. What would each conclude?

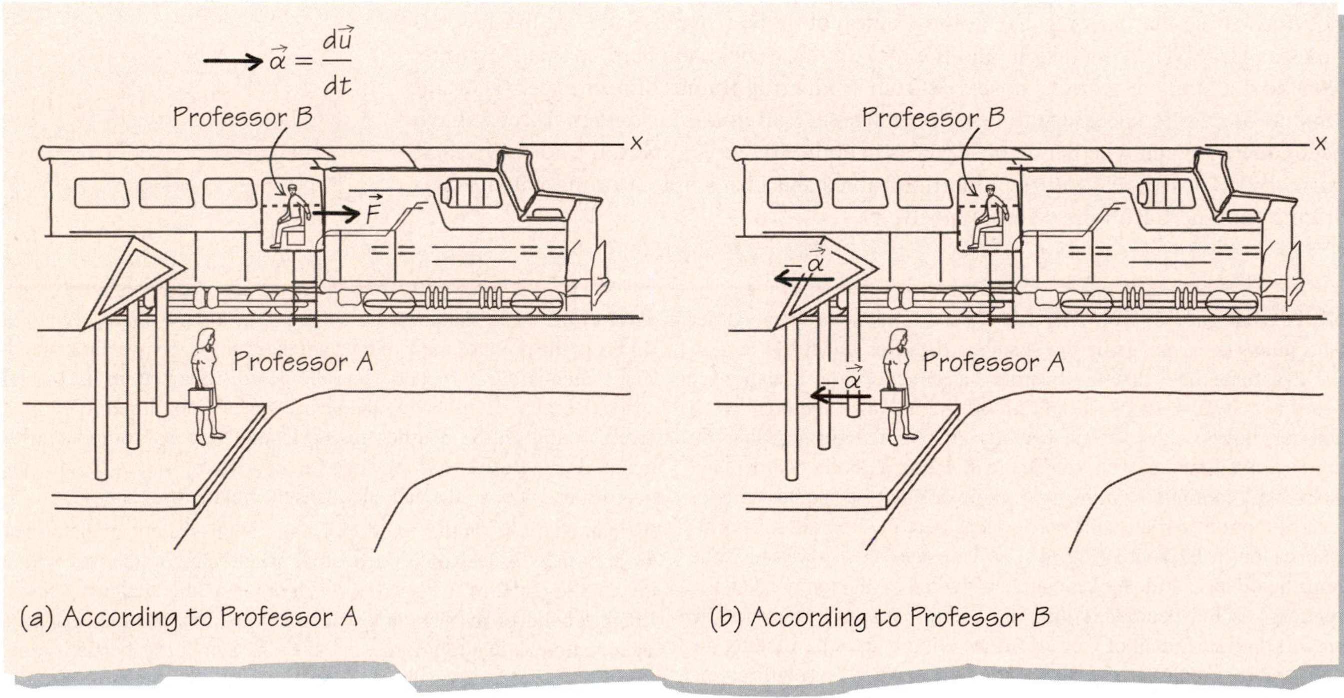

▲ **FIGURE 4–18** (a) Professor A's view. (b) Because professor B is in a noninertial frame, Newton's second law does not apply to his measurements of professor A's motion, and he sees professor A accelerating.

Suppose an observer sees an object (including perhaps himself) accelerating. How does the observer know if the object accelerates in response to a force or if instead the effect is associated with the observer being in a noninertial frame? The answer to this question is that many experiments have been performed with forces in many different situations, and as a consequence we know a good deal about the sources of force. We know, for example, that magnets and electric charges are associated with forces that act at a distance, and we know how to detect contact or friction forces of various kinds. Once we know the sources of forces, we can check to see if any such sources are present and then check whether they are acting in our given situation. If there are none, then the observed effect is due to measurements having been made in an accelerating frame.

THINK ABOUT THIS . . .

IS EARTH'S SURFACE AN INERTIAL FRAME?

The answer to this question is no. Earth rotates about the Sun, and it rotates about its own axis. In addition, there are minor wobbles associated with tidal effects and even the effects of other planets. These effects are detectable with precision equipment and are significant when motion on the planetary scale is concerned—Earth's rotation has a dominant effect on the weather. For most of the situations we encounter, however, the fact that a frame fixed to Earth's surface accelerates is a very minor effect, and we can treat a coordinate system fixed to Earth as an inertial frame. For example, at the surface Earth's rotation has a 1 percent effect on the value of g. We have to conclude that an inertial frame is an idealization that we can get quite close to in a laboratory but not actually achieve.

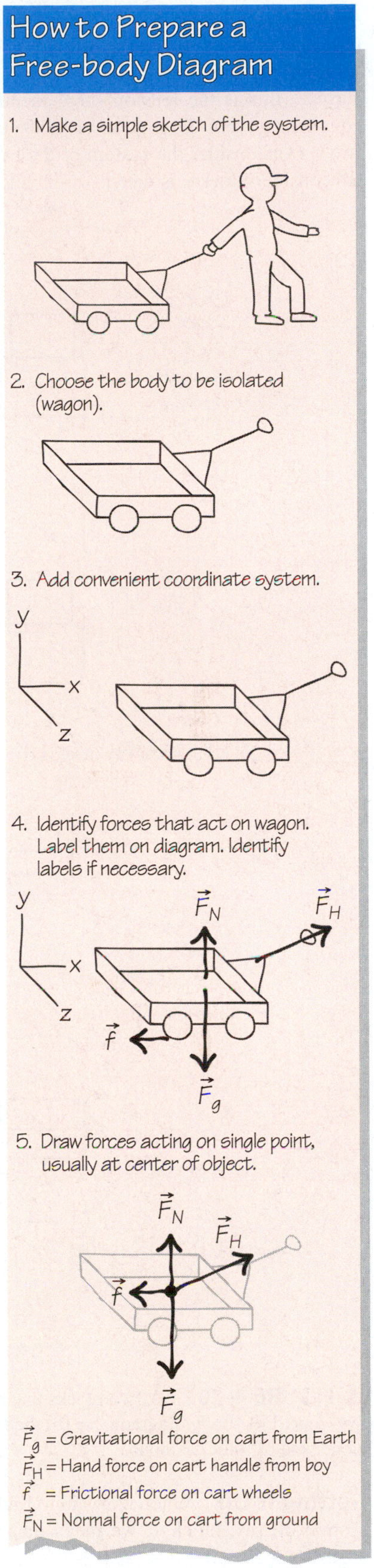

4–5 Using Newton's Laws: Identifying Forces and Free-Body Diagrams

We have seen that Newton's second law allows us to predict motion when we know the forces that are acting or, using the first law, allows us to adjust forces so that there will *not* be motion. Much of this text is devoted to exploring these issues, and in this section and the next we set out a framework for the study of motion in the presence of forces. In this section, we will discuss the first step in this approach, in which we identify all the forces that act on an object and find the net force. In the next section, we show how to use the second law to find the subsequent motion of the object.

Newton's second law, $\vec{F}_{\text{net}} = m\vec{a}$, relates the mass of an object, the net force acting on it, and its acceleration. To use the second law, *we must know exactly what object we are talking about.* This step may be less obvious than it first seems. For example, if we want to analyze the motion of a wagon being pulled by a child we must consider only the forces acting *on the wagon*, not the forces that act on the child. It is only the forces on the wagon and the mass of the wagon itself that will determine its motion according to Newton's second law. We must accordingly be able to isolate the wagon. Or we could isolate the system of wagon and child, find all the forces acting on that system, and determine the system's motion. After all, the wagon itself consists of several parts. So in this context, the word *system* refers to whatever is being isolated.

We can best isolate an object in a sketch that we shall refer to as a **free-body diagram** for the object, in this case the wagon. The free-body diagram starts with the isolated wagon and indicates each individual force, magnitude and direction, that acts on it. Figure 4–19 shows how to set up a free-body diagram. In addition to the forces, the free-body diagram should indicate a set of coordinate axes so that we can use this diagram to help us separate the vectors into their vector components. To avoid the effects of noninertial frames, place the axes in an inertial frame—usually attached to some fixed, stationary point—rather than attached to an accelerating point. In Fig. 4–19, we have placed the origin of the coordinate system at a spot on the ground adjacent to the wagon. It is often convenient to use a set of Cartesian axes: (x, y) for planar figures or (x, y, z) for three dimensions. Newton's second law breaks down into separate equations for the vector components along these mutually perpendicular axes. Any orientation of the axes is acceptable, but certain choices will be easier to use. For example, it is often convenient to orient the y-axis vertically in the study of falling objects so that the force of gravity has only a y-component.

While it is frequently convenient to draw the resulting acceleration of the object in question on the free-body diagram, you should take the utmost care *not to confuse the acceleration with a force*. The acceleration is the object's *response* to the net force acting on it; *it is not a force itself.* In figures in this book, we single out the acceleration vector in blue when we include it in the free-body diagram, whereas force vectors are always drawn in pink.

How to Prepare a Free-body Diagram

1. Make a simple sketch of the system.

2. Choose the body to be isolated (wagon).

3. Add convenient coordinate system.

4. Identify forces that act on wagon. Label them on diagram. Identify labels if necessary.

5. Draw forces acting on single point, usually at center of object.

$\vec{F}_g$ = Gravitational force on cart from Earth
$\vec{F}_H$ = Hand force on cart handle from boy
$\vec{f}$ = Frictional force on cart wheels
$\vec{F}_N$ = Normal force on cart from ground

▲ **FIGURE 4–19** How to draw a free-body diagram.

EXAMPLE 4–8 Block 1 is glued to the top of block 2 (Fig. 4–20a). The masses of the blocks are m_1 and m_2, respectively. A rope is attached to block 2, pulling it horizontally to the right with a force of constant magnitude T along a perfectly smooth horizontal surface (this is the tension force identified in Section 4–1). What equations govern the motion of block 1? Solve this problem in two ways: (a) Consider the system of the two blocks glued together and (b) consider block 1 isolated.

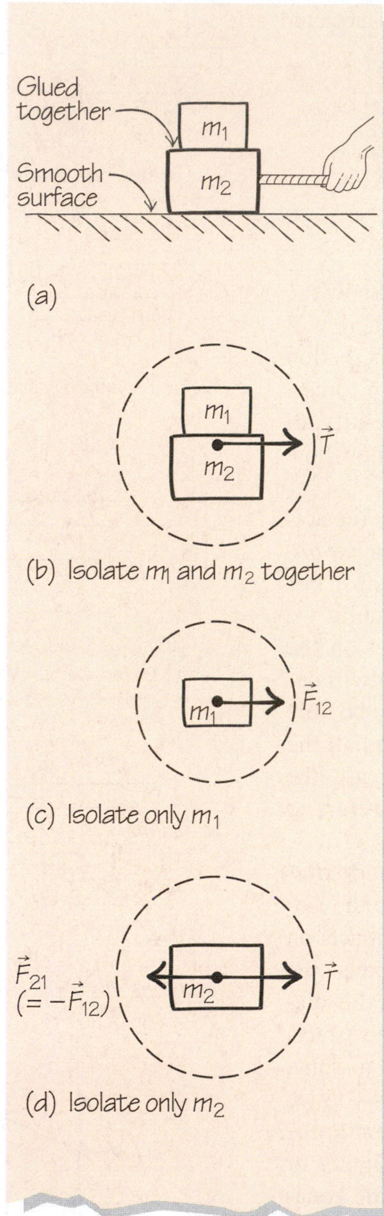

▲ **FIGURE 4–20** (a) Two blocks that are glued together are pulled by a rope. Free-body diagrams for (b) the two-block system, (c) block 1 alone, and (d) block 2 alone.

Setting It Up When we want to find the equation describing the motion of, say, block 1, we mean that we want to write Newton's second law for it.

Strategy This problem starts with the preparation of free-body diagrams. These diagrams involve first identifying the object or system, then identifying all the *external* forces on that object or system. Once this is done, we can immediately write the second law. Here we apply the method to two objects: (a) the system of the two blocks glued together and (b) block 1 alone. There is an additional piece of information for us: The acceleration of block 1 and of the system of two blocks will be the same, since they are glued together. That means that if in part (a) we find an expression for the acceleration of the system we have found an expression for the acceleration of block 1. We also note that whatever forces act in the vertical direction (e.g., gravity) they must cancel out entirely because there is no motion of either block 1 or the block 1–block 2 system in the vertical direction. Thus we can ignore all vertical forces.

For part (a), in which we consider the system of two blocks glued together, we find a direct result for the acceleration of the system and therefore the acceleration of block 1 alone. For part (b), in which we consider a free-body diagram for block 1 alone, we are going to discover as we work it through that we are going to have to consider a separate free-body diagram for block 2 as well. Method (b) turns out to be more complicated, but the results are the same.

Working It Out (a) Figure 4–20b is a free-body diagram for the block 1–block 2 system. The net force on the system is the force $\vec{T}$ due to the rope as all vertical forces cancel. The mass of the system is $m_1 + m_2$. Thus Newton's second law for the system is

$$\vec{T} = (m_1 + m_2)\vec{a}_{sys}.$$

This equation gives us $\vec{a}_{sys} = \vec{T}/(m_1 + m_2)$ and $\vec{a}_1 = \vec{a}_{sys}$.

(b) Figure 4–20c is a free-body diagram for block 1 alone. This time, the net (horizontal) force on block 1 is exclusively a contact force due to block 2. This contact force, which is at this point unknown, is written as $\vec{F}_{12}$, where the subscript specifies that we have a force *on* block 1 *due to* block 2. Thus we have

$$m_1\vec{a}_1 = \vec{F}_{12}.$$

This certainly does not appear to be equivalent to our result in part (a), at least superficially. To go further, we must also look at the free-body diagram for block 2 (Fig. 4–20d). From that diagram, we find

$$m_2\vec{a}_2 = \vec{T} + \vec{F}_{21}.$$

Here, $\vec{F}_{21}$ is the force on block 2 due to block 1. According to Newton's third law, however, $\vec{F}_{21} = -\vec{F}_{12}$; so, the expression for Newton's second law applied to block 2 is

$$m_2\vec{a}_2 = \vec{T} - \vec{F}_{12}.$$

We then solve for the unknown force $\vec{F}_{12}$, with the result $\vec{F}_{12} = \vec{T} - m_2\vec{a}_2$. We substitute this back into our equation for block 1:

$$m_1\vec{a}_1 = \vec{F}_{12} = \vec{T} - m_2\vec{a}_2.$$

Finally, the blocks move together and thus $\vec{a}_2 = \vec{a}_1 = \vec{a}_{sys}$, so that

$$m_1\vec{a}_1 = \vec{T} - m_2\vec{a}_1,$$

or $(m_1 + m_2)\vec{a}_{sys} = \vec{T}$. We can now recognize the equation we found in part (a) and discover that the two methods give the same answer.

What Do You Think? Did the presence of the glue have any important role to play in the solution to this problem? In other words, if there were no glue and no friction between the two blocks, would the motion of block 1 be different?

External and Internal Forces

Example 4–8 illustrates the important difference between *external* and *internal* forces. Internal forces are those that act within the system we are isolating. These forces may act to hold the system rigidly together or they may simply act between different parts of the system. If we consider the two blocks in Example 4–8 as a single, isolated system, the internal forces would be associated with the glue holding the two masses together. We can examine these internal forces, $\vec{F}_{12}$ and $\vec{F}_{21}$, in Fig. 4–21a. The fact that these forces are equal and opposite (Newton's third law) is, as we have seen in the example, an essential component of our analysis.

External forces, in contrast, are forces that act on the system from outside. In Example 4–8, the net force $\vec{T}$ is an external force acting on the two-block system (Fig. 4–21b). Once a system has been isolated, only the external forces acting on it influence its overall motion. As we shall see in more detail in Chapter 8, internal forces do not enter into the second law because, by Newton's third law, they cancel in pairs and so do not contribute to the net force. For example, we saw that the internal forces $\vec{F}_{12}$ and $\vec{F}_{21}$ of the two-block system canceled each other and did not enter into the solution of the problem. In the statement of the second law, *the net force is the net external force.* All of the objects we deal with are, in fact, complicated systems composed of many atoms held together by internal forces. But even though these internal forces act within the object, they cannot influence its overall acceleration.

If a given system is broken up into separate pieces, the internal forces for the original system may become external forces for the pieces. The motion of the smaller pieces is then governed by these external forces. In Example 4–8, the isolation of block 1 compelled us to treat $\vec{F}_{12}$ as an external force on block 1. In that same example, we made two different choices of the objects to be isolated; hence, we made different choices as to which forces to treat as external and which ones as internal. Each time, we found that only the external forces affected the motion of the blocks. We saw that a judicious choice of which object or system of objects to isolate can simplify the calculation of the object's motion considerably. Take a careful look at the free-body diagrams in Fig. 4–20 to learn how to isolate a given object and recognize the forces on it.

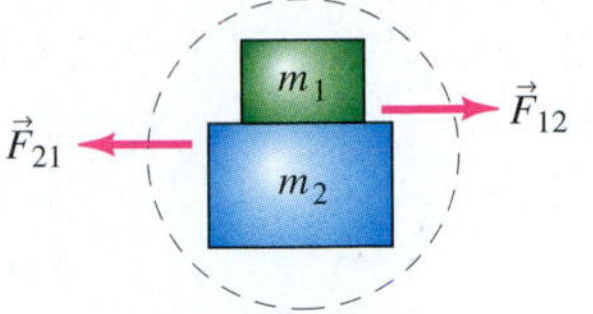

(a) Internal forces

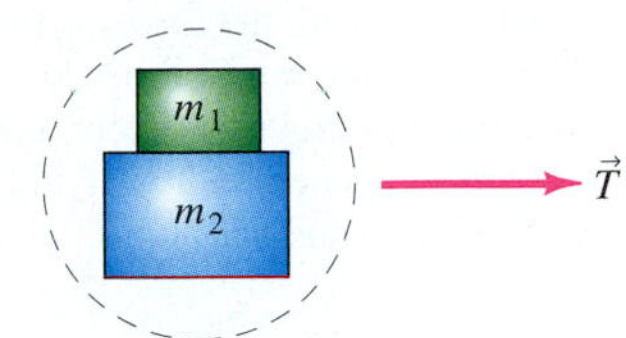

(b) Net external force

▲ **FIGURE 4–21** (a) Internal forces within system of two blocks studied in Example 4–8. (b) Net external force on that same system. Some vertical forces act (gravity and contact forces), but these do not contribute to the net external forces because they cancel.

CONCEPTUAL EXAMPLE 4–9 As a result of the tug, the astronaut and the satellite in Example 4–5 (Fig. 4–13) are accelerated toward each other. What happens when they collide, assuming that the astronaut sticks to the satellite?

Answer If we look at the total system, that is, the satellite and astronaut together, we see that there are no external forces acting on the system, and therefore the system will maintain its initial velocity. An observer who sees the system at rest at the beginning will see it at rest at the end. The tug is a force internal to the system that results in some rearrangement of the overall system but not in movement of the system as a whole. We will revisit this sort of question in more detail in Chapter 8.

A Second Look at Some Common Forces

In Section 4–1 we first mentioned some common forces that appear in many physical situations: gravity, normal forces, tension, and friction. We will now look again at these forces, this time within the context of free-body diagrams, both as a more complete illustration of the preparation of free-body diagrams and as a way to understand more about the forces themselves. We start with a familiar example: a sled of mass m moving down a snow-covered hill. What is the free-body diagram for the sled?

Three forces act on the sled, one of which acts at a distance. The force that acts at a distance is the force that pulls the sled vertically downward—the force of gravity $\vec{F}_g$, drawn downward in the diagram of this situation in Fig. 4–22a and in the free-body diagram of Fig. 4–22b. We draw this force vector with its tail at some point in the sled.

Figure 4–22 shows a second force, this time a contact force. It is the normal force $\vec{F}_N$, and it is exerted by the hill on the sled. As we stated earlier, this force must be present because in its absence the force of gravity would cause the sled to accelerate down *into* the surface of the hill. The direction of the normal force is *normal*, or perpendicular, to the surface of the hill because that is the direction with no component of motion of the sled or any object on any solid surface. The normal force is the reason this book remains upon the surface of the table and the reason you can sit in a chair. The magnitude of the

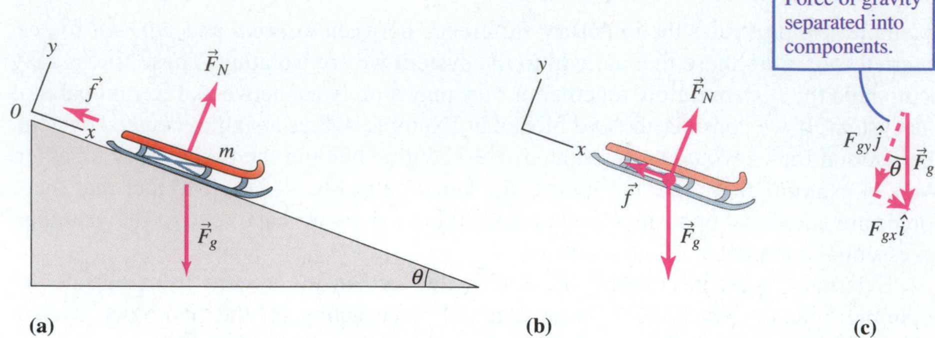

▶ **FIGURE 4–22** (a) A sled on an inclined plane, with the forces acting on it. (b) Free-body diagram for the sled. (c) The force of gravity is decomposed into components perpendicular and parallel to the plane.

normal force will always adjust itself to a situation (to a break point), and the adjustment is such as to make the normal force cancel any forces that might make the sled penetrate the hill. The normal force would be different if a child sat on the sled than for the sled alone because the force of gravity is different for the sled alone than it is for the sled with a child on it. The normal force is simple to use in practice—just as simple as gravity. In contrast to gravity, though, which is a direct and fundamental force between two massive objects (here, Earth and the sled), a normal force results from the addition of many complicated intermolecular forces within and between the materials making up the object and the surface on which it rests. Fortunately, we do not have to worry about these complications, as we are looking at a larger system that includes these forces as internal forces. We draw the normal force with its tail at the point chosen for the force of gravity.

Finally, there is a third force on the sled, included in Fig. 4–22 in the same way as the other two forces, and that is the friction force, which we label $\vec{f}$. Friction, like the normal force, is an approximation to a complicated interaction between the sled rails and the snow, and friction is also a type of contact force. Its direction is opposite to the motion at the surface, here parallel to the hill and uphill. If the sled were moving uphill, the force of friction would act downhill. Friction is always parallel to the surface and is therefore always perpendicular to any normal force at the surface.

The second law as applied to the sled, $\vec{F}_{\text{net}} = m\vec{a}$, becomes

$$\vec{F}_g + \vec{F}_N + \vec{f} = m\vec{a}. \tag{4–16}$$

A vector equation such as this one stands for three equations for the three components, so we must now break this equation into component form. To do so, we must first choose a set of axes. These axes should be placed in an inertial (nonaccelerating) frame to avoid the complications associated with noninertial frames. A convenient choice is one in which the decomposition of the force vectors is simplest. Another convenient choice is one in which the acceleration is parallel to one axis. When there are several forces and they cannot *all* point along an axis, as in this example, it is best to align the axes with as many forces as possible. A good choice for this example is shown in Fig. 4–22, where the y-axis is perpendicular to the hill and the x-axis points downhill. Then two of the three forces—$\vec{F}_N$ and $\vec{f}$—are along these axes. By including the axes in the free-body diagram (Fig. 4–22b), it becomes simpler to read off the components of the second law.

In terms of our coordinate axes, the forces $\vec{F}_N$ and $\vec{f}$ are simple:

$$\vec{F}_N = F_N\hat{j}, \qquad \vec{f} = -f\hat{i}.$$

In contrast, the force of gravity has components in both the x- and y-directions. For the breakdown of $\vec{F}_g$ into components, we have drawn a useful diagram in Fig. 4–22c. From this figure we find that

$$\vec{F}_g = (F_g \sin\theta)\hat{i} + (-F_g \cos\theta)\hat{j},$$

where F_g stands for the (positive) magnitude of the force of gravity. Because all three forces have only x- or y-components, any motion due to the forces—any acceleration, in other words—occurs entirely in the xy-plane. Any component of motion in the z-direction (perpendicular to the page) is a constant-velocity component.

Equation (4–16) can now be written as

$$(F_g \sin\theta)\hat{i} + (-F_g \cos\theta)\hat{j} - f\hat{i} + F_N\hat{j} = ma_x\hat{i} + ma_y\hat{j},$$

or in component form,

$$\text{in the } x\text{-direction: } F_g \sin\theta - f = ma_x; \tag{4–17}$$

$$\text{in the } y\text{-direction: } -F_g \cos\theta + F_N = ma_y. \tag{4–18a}$$

We can immediately simplify Eq. (4–18a) by setting $a_y = 0$ as we know that the sled does not leave the hill surface. It is this constraint that determines the magnitude of $\vec{F}_N$. With $a_y = 0$, Eq. (4–18a) becomes

$$\text{in the } y\text{-direction: } -F_g \cos\theta + F_N = 0. \tag{4–18b}$$

Equation (4–17) is the equation that describes the unknown sled acceleration a_x. It depends on the magnitude of the friction force, f. As we shall see in Chapter 5, f depends on the value of F_N, which, in turn, we can find from Eq. (4–18b).

Problem-Solving Techniques

The preparation of a free-body diagram is an important skill. We can lay out some general guidelines using Fig. 4–19 to help us identify steps:

1. Identify and isolate the object in question. Make a sketch with the object clearly labeled.
2. Identify all the forces acting on the isolated object. Draw each force on the free-body diagram as a labeled arrow; include rough approximations of the direction and magnitude of each force. These arrows should all start from a single point somewhere in the "middle" of the object—remember that until we begin to worry about the rotations of extended objects we are treating our objects as points.
3. Draw a set of coordinate axes with the origin at a fixed point of the diagram; that is, the origin should not be attached to the object itself. Choose these axes so that you can easily pick out the components of the various forces along them.
4. Remember that force equations are vector equations and they stand for a set of equations for the components. Depending on the problem, you may have to separate the forces into their components with respect to the coordinate axes you have chosen.
5. If you include an arrow that represents your guess as to the acceleration of the object, distinguish it clearly from the arrows representing forces. Remember, the acceleration is the response to a force, not a force itself. Figure 4B1–1 summarizes these steps and gives some further hints for finding the net force.

How to Draw

More on Drawing Free-Body Diagrams

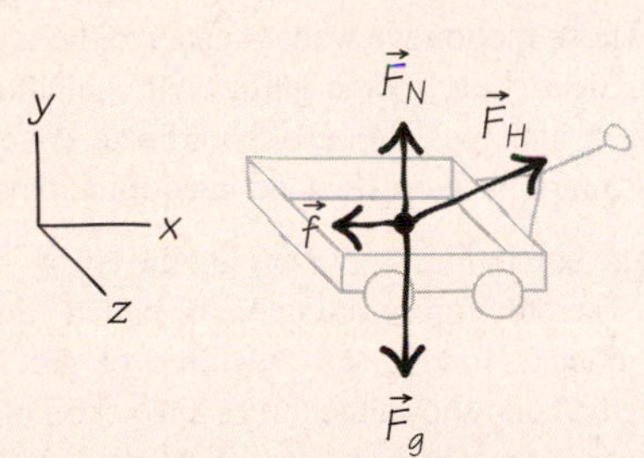

(a) Draw free-body diagram (see Fig. 4–19)

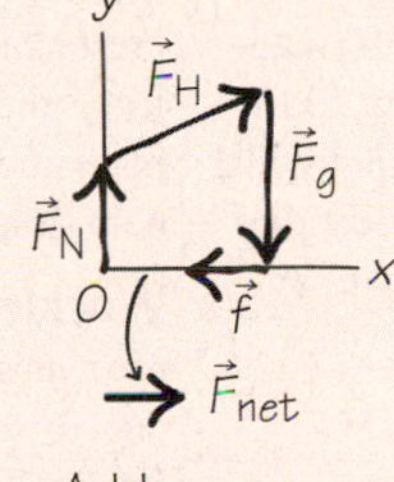

(b) Add forces, tail to head

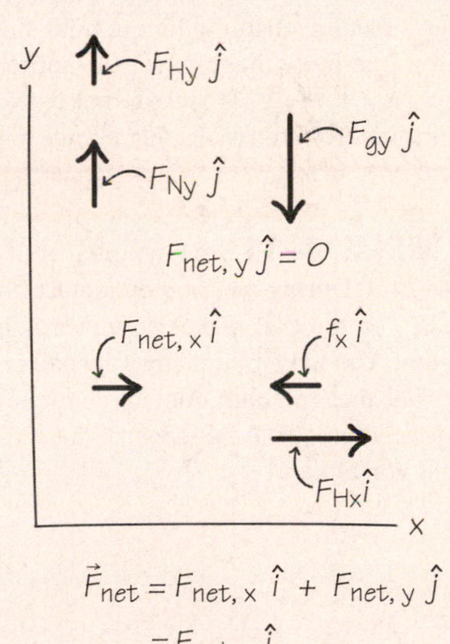

(c) Add components

▲ **FIGURE 4B1–1** More hints on drawing free-body diagrams. Parts (b) and (c) show two ways to find the net force, the first a graphical method and the second a method in which components are used.

CONCEPTUAL EXAMPLE 4–10 Two rock climbers—climber 1 and climber 2—are loosely tied together by a rope that can be assumed to be *almost* inextensible. Since the rope is loose, there is no tension. Climber 1 falls but is saved because the upper climber is at a belay point and when the rope pulls taut the fall is arrested. What are the forces that act on the climbers at various times? Which forces on the two-climber system are internal and which are external at the end of the action, when climber 1 has been saved and all is motionless?

Answer In Fig. 4–23 we sketch the situation before and after the fall. Before the fall, the downward force on each climber is the force of gravity, that is, down with magnitudes m_1g and m_2g, respectively. In each case the normal force due to the rocks acts on the hands and feet of the climbers and just cancels these. While climber 1 is falling, the only force acting on him is the force of gravity; climber 2 is unaffected by the events. Since the rope is almost inextensible, the falling climber is almost instantaneously arrested once the rope has been drawn straight. During the brief period of arrest, he experiences a large deceleration and therefore a very strong upward force. This is due to a rapid rise in the rope tension. This large tension also acts downward on climber 2. It obliges him to adjust his already secure hold in such a way that he can maintain the increased upward contact forces from the rock that must act if he is to remain stationary. Helped by the belay, we asssume that he can indeed do this, and then the forces acting on him are the downward force of gravity and downward force of the final tension of the rope, magnitude T. Because this final tension keeps climber 1 from accelerating, it is equal in magnitude to the force of gravity on climber 1, $T = m_1g$. In turn, climber 2 must be able to maintain an upward contact force from the rocks that is equal and opposite to the net downward force on him, magnitude $T + m_2g = (m_1 + m_2)g$. He does this with an appropriate "stiffness" in his hands and feet, which is where his muscles and balance come in.

The most important internal force for the two-climber system at the end of the action when the rope is taut and nothing is moving is the tension. There are also many complicated internal forces within climber 2 that are enabling him to hold on. None of these would appear in a free-body diagram for the entire system. The *net* force of gravity, $(m_1 + m_2)g$, is an external force for the system, as is the normal force from the rocks that climber 2 is holding.

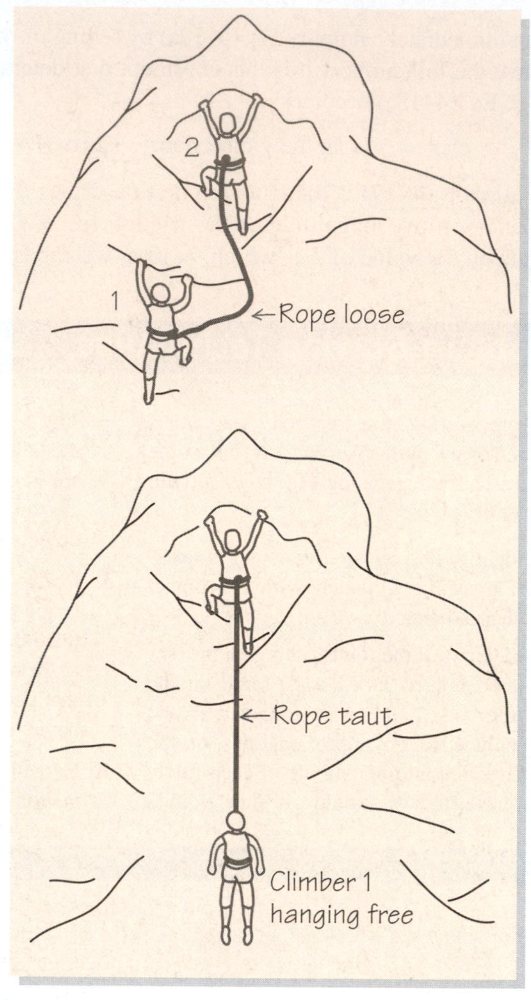

▲ **FIGURE 4–23** The falling climber is saved from falling by a second climber with the aid of the rope.

EXAMPLE 4–11 A swinging golf club strikes a golf ball (Fig. 4–24a). During a period of contact, the force of the club (magnitude F_c) on the ball makes an upward angle θ with respect to the horizontal. You may assume that the ball is no longer in contact with the tee but that the club contact continues. Draw a free-body diagram for the golf ball and specify the forces in the coordinate system that you choose.

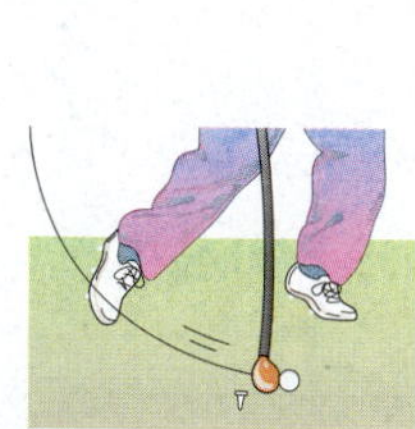

(a)

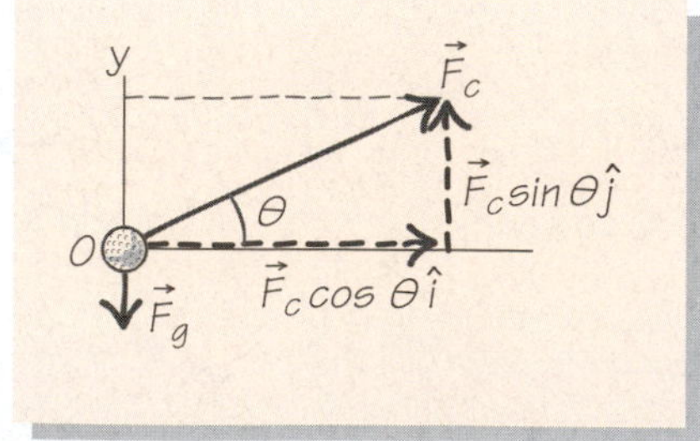

(b)

Strategy As usual, in the preparation of the free-body diagram, we must isolate the object whose motion we want to analyze (here the object is the golf ball), then identify each force acting on it, including both magnitude and direction. A judicious choice of coordinate system will help simplify the expression of Newton's second law for the object.

Working It Out The ball in Fig. 4–24a has left the tee, so there is no contact force from it. The only forces acting on the ball are the force due to the club, $\vec{F}_c$ (a contact force), and the force of gravity $\vec{F}_g$. Figure 4–24b isolates the ball and shows the forces that act on it. It also contains a coordinate system for which the forces are decomposed as

$$\vec{F}_c = (F_c \cos\theta)\hat{i} + (F_c \sin\theta)\hat{j}$$

and

$$\vec{F}_g = -mg\hat{j},$$

where m is the mass of the golf ball.

◀ **FIGURE 4–24** (a) Golf club hitting a golf ball. (b) Free-body diagram for the golf ball. The ball has already left the tee, so there is no contact force from the tee, but the ball is still in contact with the golf club at this point.

What Do You Think? In this example a net force with a horizontal component and, if $F_c \sin\theta > mg$, with an upward vertical component acts on the ball. Does this mean that during the period that the contact force acts the ball will accelerate in the horizontal direction? Furthermore, will the ball accelerate upward during the contact period?

4-6 Using Newton's Laws: Finding the Motion

We have learned how to use free-body diagrams to help us write Newton's second law for an object that is acted on by forces. This is a very important step along the way to finding the motion of that object. The object's motion is described by *solving* the equations expressing Newton's second law. It is for this reason that we call those equations the *equations of motion*. We can solve them only if we know the particular *force law*; that is, the force expressed as a function of variables such as position, velocity, time, or any other parameters of the problem. Force laws are determined in experiments that measure these forces in controlled conditions. Many of the most common forces (e.g., the force of a spring) depend only on the position of the object—and the simplest case of all occurs when the force is constant, that is, independent of the position of the object (e.g., gravity). We'll see how to solve the equations of motion for a constant force in Example 4–12. We will need to use the kinematic equations that we developed for constant acceleration in Chapters 2 and 3.

EXAMPLE 4–12 The constant net force $\vec{F}$, shown in Fig. 4–25, acts on a nugget of gold whose mass is m. (You could imagine that the nugget is in midair, on a tabletop, or whatever you like; we only ask you to assume that the vector sum of all the forces acting is constant.) What is the subsequent motion of the nugget given that its initial velocity is $\vec{v}_0$?

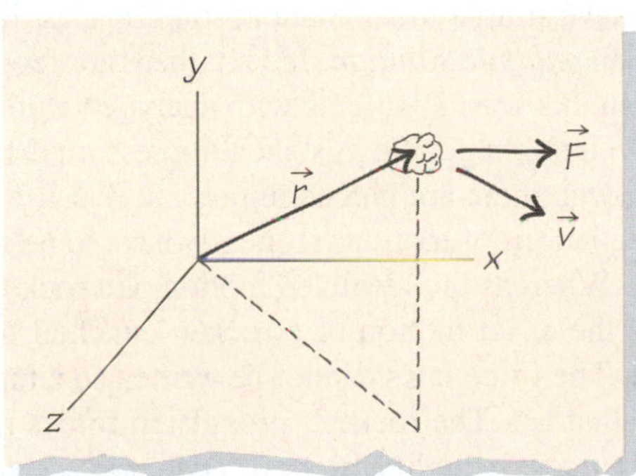

▲ **FIGURE 4–25** A gold nugget is located at a displacement $\vec{r}$ from the origin of a coordinate system and moves with velocity $\vec{v}$.

Strategy Finding the motion in a situation where Newton's second law can be applied means finding the acceleration using that law, then solving the kinematic equations of Chapters 2 and 3 that determine position and velocity given that acceleration.

Working It Out Because $\vec{F}$ is a constant vector, it points in a fixed direction. For convenience, we choose the x-axis to point in the same direction as the force, so that

$$\vec{F} = F\hat{i}. \qquad (4\text{–}19)$$

The force has no components in the y- and z-directions so the nugget of gold will not have a component of acceleration in these directions. The nugget can nevertheless have motion with constant-velocity components in these directions, and these components are the components of $\vec{v}_0$. Motion of constant velocity in the y- and z-directions means that these components of the nugget's position vector change linearly with time:

$$y(t) = y_0 + v_{0y}t, \qquad (4\text{–}20a)$$

$$z(t) = z_0 + v_{0z}t. \qquad (4\text{–}20b)$$

The quantities v_{0y} and v_{0z} are the constant components of the velocity vector in the y- and z-directions, and y_0 and z_0 are the values of the y- and z-components of the position at time $t = 0$.

For the x-direction things are different. The x-component of Newton's second law is

$$F_x = F = ma_x, \qquad (4\text{–}21)$$

or, equivalently,

$$F = m\frac{d^2x}{dt^2}. \qquad (4\text{–}22)$$

We want to find the function $x(t)$ that satisfies this equation. Equivalently, Eq. (4–21) tells us that a_x has a constant value F_x/m, and we want to find the position of the object given this constant acceleration. We have already encountered this situation in Section 2–4, and we can turn to that section for the answer. Once the *initial conditions* are stated, the answer is given by Eq. (2–21), and we restate the result in Eq. (4–23). The initial conditions are as follows: We place the location of the object at $t = 0$ to be $x = x_0$, and the x-component of the velocity at that time is $dx/dt = v_{0x}$. The x-component of the position of the nugget is then

$$x(t) = x_0 + v_{0x}t + \frac{F}{2m}t^2. \qquad (4\text{–}23)$$

This result obeys both the equation of motion [Eq. (4–22)] and the initial conditions at $t = 0$. We can verify this by direct differentiation and direct substitution.

What Do You Think? If the nugget is twice as massive, the acceleration will have half the magnitude it has here. Does this mean that the nugget goes half as far in time t?

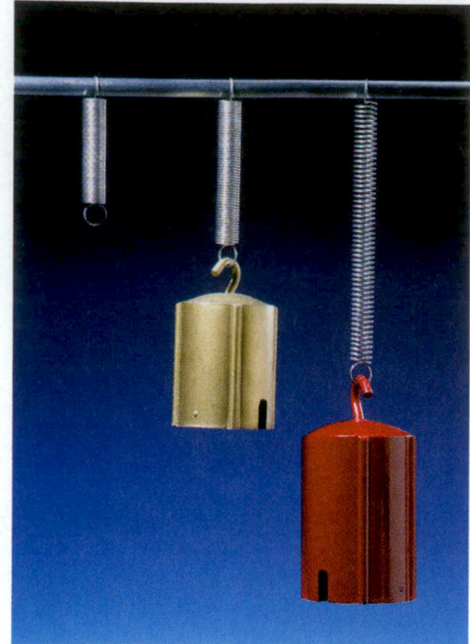

▲ **FIGURE 4–26** (a) an unstretched spring; (b) a 1 kg mass hangs from the spring stretching it out; (c) with a 2 kg mass, the spring stretches more. The amount of stretch is proportional to the weight.

The motion described in Example 4–12 is really only a review of our study of kinematics with constant acceleration from Chapters 2 and 3. What is new here is the idea that constant acceleration is associated with a constant force. We will be returning to this example in Chapter 5. It is an important example because it is applicable to the force law of gravity near Earth's surface.

When the force on an object is not constant but depends explicitly on the object's position, the solutions to the equations of motion become more complex. In fact, there are only a few cases of the force law for which we can find simple expressions for this type of motion. A spring is one example, and we shall see later that the force law for a spring that has been compressed or stretched by an amount x from a relaxed position is proportional to x, always acting in a direction opposite to the compression or stretching (i.e., if x is positive, the force is negative, and if x is negative, the force is positive). The larger the compression or stretch, the stronger the spring force (Fig. 4–26). Mathematically, the proportionality is expressed as

$$\text{spring force law: } F \propto -x.$$

We also have Newton's second law,

$$F = m\frac{d^2x}{dt^2}.$$

The equation of motion for a spring therefore has the form

$$-x \propto \frac{d^2x}{dt^2}.$$

The equation shows proportionality rather than equality because we have left off all the constants. When all the constants are properly included, we are left with a *differential equation* for x. The meaning of the equation in this case is that x must be a function of time whose second derivative is proportional to the negative of the function itself. Only certain functions will satisfy such an equation.[†]

When the equations of motion have simple mathematical solutions, as they indeed have for the spring, we say we have an *analytic solution*. In fact, there are few force and motion problems that we can solve in this way. Problems with analytic solutions populate textbooks such as this one, and you might get the mistaken impression that all problems can be solved in this way. Although there are indeed important, real-life problems that can be solved analytically, the equations of motion frequently have to be solved numerically with the aid of computers. Whereas the idealized motion of a rock thrown up in the air is simple, the problem of the exact motion of a rocket launched from Cape Canaveral is quite a different matter. The force laws cannot be written so simply in this case because of the range of forces that act. The rocket's propulsion forces need to be adjusted to compensate for varying forces, including air resistance, the acceleration of gravity, and wind forces, which all vary with altitude. Sophisticated computer calculations are needed to figure out these adjustments so that we know just how a rocket will behave as it rises into orbit.

Summary

Newton's three laws express the dynamics of motion by showing how forces acting between objects determine the subsequent motion of those objects. The first law states what happens to an object—moving or at rest—when no net force acts on it:

> **$\vec{F}_{net}$ is the vector sum of any individual forces that act on an object. When $\vec{F}_{net}$ is zero, the object moves with constant velocity.**

Forces, which are vectorial quantities, act on objects and cause them to accelerate. For a given force, this acceleration is inversely proportional to the mass m of the object in question. This is expressed in Newton's second law,

$$\vec{F}_{net} = m\vec{a}. \tag{4–6}$$

In SI, the force is measured in newtons, abbreviated N, where $1\ N \equiv 1\ kg \cdot m/s^2$.

[†] We shall see later that these solutions are sinusoidal: the object moves back and forth.

Forces act *between* objects. If objects A and B interact, that is, if there are forces acting between them, then Newton's third law states that the force on object A due to object B, $\vec{F}_{AB}$, is equal and opposite to the force on object B due to object A, $\vec{F}_{BA}$:

$$\vec{F}_{BA} = -\vec{F}_{AB}. \tag{4–15}$$

Observers in reference frames moving with respect to one another observe the motion of a given object differently. An observer who verifies that Newton's second law holds, with known or identifiable sources of forces, is said to be in an inertial frame. If a second observer moves with constant velocity relative to the first, the second observer is also in an inertial frame; if there is nonuniform relative motion, the second observer is in a noninertial frame. Observers in inertial frames agree on the forces they see acting on an object. There is no experiment they can perform to decide who is moving in an absolute sense. Observers in noninertial frames disagree on the forces that act on an object and, in effect, Newton's second law does not hold from the point of view of an accelerating observer.

If we know the nature of the forces that act on an object, then Newton's laws can help us determine the motion of the object. Conversely, the laws allow us to measure the forces acting on an object by measuring the object's motion. If we want to determine the motion of an object, we must know the forces that act on it, and knowing a force means knowing how the force due to a particular source varies with position, time, or other variables.

Newton's third law explains why forces that act within an object or within a system of objects have no effect on the motion of the object. These forces are called internal forces. Only forces external to the object or system determine an object's motion—we refer to these as the external forces.

To best use Newton's laws, we draw free-body diagrams that conceptually isolate an object or system. We include all the forces acting on it, keeping in mind that a force is a vector. Once a free-body diagram has been prepared, we choose a convenient set of axes to write the three components of the second law. These equations can be solved, either analytically or numerically, to find the object's motion.

Understanding the Concepts

1. A small but dense mass is swinging freely at the end of a light string. A very sharp knife cuts the string when the mass is at the bottom of its swing; the knife does not disturb anything else. What is the subsequent flight of the mass?
2. A baseball is hit out of the park. Sketch a trajectory and show the forces (use arrows) that act on the ball at various points along its path.
3. Someone pushes on a wall. What experiment can you propose to determine the force with which the person pushes?
4. If you were in a freely falling elevator, the contact force on you due to the floor would drop to zero. If the elevator were to accelerate rapidly upward, the contact force between you and the floor would increase. Why?
5. You are standing in an elevator that is at rest. The elevator starts moving up. In due course you reach your desired floor, and the elevator slows down and stops. Assume that you are standing on a scale during the whole trip. Describe and explain how the pointer on the scale will move from beginning to end.
6. An astronaut is working while in orbit. When the astronaut assembles a piece of equipment, will he or she notice a difference between working with components of large mass as opposed to components of small mass?
7. Two masses on a smooth flat surface are successively pushed against a compressed spring. The spring is compressed by the same amount each time. One mass is twice as heavy as the other, as measured by a vertical spring scale. True or false: The spring is released, and when the speed each mass attains is measured, they are found to be identical, just as the speeds of falling objects of different mass are the same.
8. When a satellite travels around Earth in a circular orbit, it moves at a constant speed. Does Newton's first law apply in this situation? Is the velocity constant? Is there a force present?
9. There is a well-known parlor trick in which a tablecloth is pulled sharply from beneath a dinner setting, leaving the setting in place. Why does this work?
10. Is a spaceship heading from Earth to the Moon traveling in a force-free environment? Explain. What about a spaceship traveling from Earth to Mars that is currently in a region far away from either planet?
11. An adult sits on a child's table. The table is about to break. Is it correct to say that it is the weight of the adult that is causing the table to break?
12. If you have ever ridden a bicycle with friends, you know that, even without pedaling, different cyclists descend the same hill at different speeds. Isn't this in conflict with the claim that all falling objects accelerate at the same rate under the influence of gravity?
13. From your experience with forces, which of these are contact forces: friction, the force of gravity, the normal force on an object on the floor, the force due to a magnet, the tension in a rope used in a pulley?
14. A box is placed on a table. The box's weight and the normal force on it are equal and opposite. Is this an example of Newton's third law?
15. Shortly after jumping from an airplane, a parachutist will descend with constant velocity. Why is this?
16. Using Newton's second law, devise a system (other than the compressed spring described in Section 4–2) that could be used to measure masses.
17. Determining the mass of objects in space, where the motion is that of free fall, can be difficult. How do you think it can be done?

18. Newton stated that a reference frame at rest with respect to the distant stars would be a good inertial system. Comment on this. Suggest systems that would be good inertial systems for experiments conducted (a) in a physics lab, (b) on the space shuttle, (c) on a ship at sea, and (d) on Mars.
19. A rubber ball and a golf ball have the same mass but the rubber ball has a larger radius than the golf ball. If they are identically accelerated with the same initial force, why might the golf ball go farther in the atmosphere?
20. A fellow student has drawn the free-body diagram in Fig. 4–27 for a block on a smooth ramp, with the arrows representing different forces. What, if anything, is wrong with this diagram?

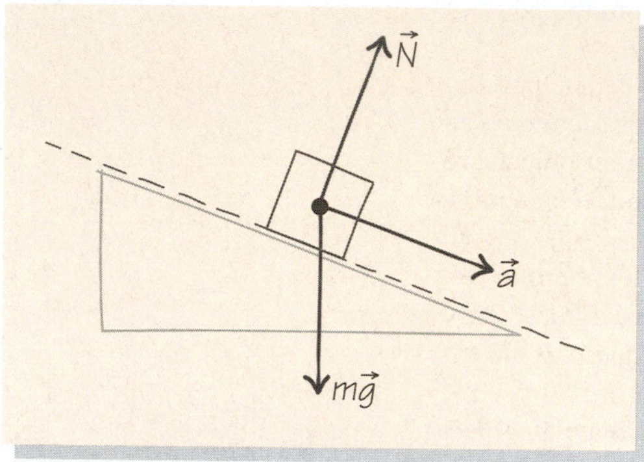

▲ **FIGURE 4–27** Question 20.

21. Describe the forces that are responsible for the acceleration of an automobile. In particular, in what way is friction between the tires and the road responsible for the acceleration? What, then, is the role of the engine?
22. If you tie a rock to one end of a piece of string and swing it in a circle, you experience a force on your hand. Why?
23. A girl stands on a merry-go-round halfway along a line from the center to the rim. She jumps straight up into the air and lands back on the merry-go-round. Analyze where she lands and how this depends on the rotation speed of the platform.
24. When a baseball hits a bat, a force must be exerted on the ball to change its direction of motion. Describe the forces between the ball and the bat. What effect does the ball have on the bat? Does the batter feel the effects of any of these forces?
25. Give three examples of noninertial forces.
26. You put your hand on a table, place a brick on top of the hand, and give the brick a smart blow with a hammer. Your experience is quite different than what it would be if the brick were not present. Why?
27. A marble is placed on the side of a bowl. It rolls down into the bowl, then continues to roll up and down the bowl's side until it finally comes to rest at the bottom. Explain this in terms of the first and second laws.
28. Describe qualitatively how you might go about constructing an accelerometer, which measures the acceleration of a ship. [*Hint*: Think about how a spring placed between your back and the back of a seat would react when an automobile in which you are sitting accelerates forward. Such devices are of great importance for "blind" navigation, which might be required for a submarine. They allow us to follow the path when we know the acceleration as a function of time and then reconstruct the velocity and the position as functions of time. (This procedure is described in Chapter 2.) It is, in fact, difficult to construct an accurate version of such a device without knowing about rotational motion.]
29. You are standing on a scale in an elevator. The elevator suddenly starts to move upward. What happens to the reading of the scale and why?
30. A diver jumps from a high platform and experiences a feeling of weightlessness. Is the force of gravity no longer acting on the diver?
31. An apple hangs from a tree, firmly attached to the branch. Does Earth exert a force on the apple even though the apple is at rest? If so, then by Newton's third law, the apple exerts a force on Earth. What keeps Earth from accelerating toward the apple in this situation?
32. A 5-kg mass is placed on a table. A professor states that the normal force on the mass due to the table is (5 kg) $(9.8\ \mathrm{m/s^2})$ and that this is a consequence of Newton's third law. Is this a sound analysis?
33. In Example 4–7, we discussed professor A, who is at rest on a train platform, and professor B, in the accelerating train, to illustrate the consequences of the noninertial nature of professor B's reference frame. How do we know that professor A's frame is inertial?
34. Consider a horse that pulls on a cart. By Newton's third law, the cart pulls on the horse with a force of equal magnitude but in the opposite direction. How can there be any motion?
35. What makes a car go forward when the engine is turned on and the transmission is engaged?
36. You are standing in a stationary bus and suddenly find yourself thrown backward. What does Newton's second law say about that?
37. A car is stationary on a flat parking lot. The force of gravity acts downward and an equal and opposite normal force acts upward. Is it correct to say that these forces are equal and opposite because of Newton's third law?
38. In a tug of war, one side is stronger than the other. Assuming that both sides exert themselves to the maximum, discuss the motion of a handkerchief tied to the rope being pulled.
39. A fellow student states that forces cause an object to move. Criticize this statement.
40. How can we be sure that an object moving with a constant velocity has no force on it? Perhaps we are observing the object from a noninertial frame of reference.
41. Baron Munchausen claimed that it is possible for a very strong man to pull himself off the ground and rise into the air by pulling on his bootstraps. Discuss this mode of liftoff in the cold light of Newton's laws.

Problems

4–1 Forces and Newton's First Law

1. (I) In applying Newton's laws, we must identify the forces acting on an object. Are there any forces acting on the following objects? If so, list them: (a) the space shuttle in Earth orbit; (b) an ice skater coasting on ice; (c) the *Voyager I* spacecraft far past the orbit of the planet Pluto.
2. (I) A skater is gliding at what is very close to constant speed on a frozen lake. (There is negligible friction between blade and ice.) A stiff but very steady wind is blowing. What are the forces acting on the skater? Do they balance?
3. (I) In a tug of war, a red ribbon tied around a point on the rope between the two teams moves with a uniform velocity of 0.1 m/s in the y-direction. One team exerts a force on the rope of 600 N in the y-direction. What force does the other team exert on the rope?
4. (I) A boat sailing in the northeasterly direction with constant speed experiences a wind force of magnitude 3×10^3 N from the

south. What is the force on the sailboat due to the resistance of the water to motion through it?

5. (I) In a classic demonstration, Otto von Guericke used 16 horses—8 on each side—to try to pull apart two hemispheres forming a sphere from which air had been evacuated. Could he as well have used only 8 horses on one side, with the other side tied to a sturdy tree?

6. (II) Three nonzero forces act on a particle at the origin of a coordinate system: $\vec{F}_1$ is in the z-direction, whereas $\vec{F}_2$ and $\vec{F}_3$ lie in the xy-plane. Can you arrange the magnitudes and directions of $\vec{F}_2$ and $\vec{F}_3$ (keeping them always in the xy-plane) so that the particle does not accelerate?

7. (II) (a) A spider is suspended from a single vertical thread; the spider has a mass $m = 30$ mg. The spider is acted upon by the force of gravity, which is directed downward and has magnitude 3.0×10^{-4} N, and by the *tension T* in the thread, a common type of contact force that always acts in the direction of the thread and is directed away from the point at which it is attached. In this case, the tension acts in the upward direction. What is the magnitude of the tension? (b) The spider is now attached to two threads of equal length that make a 120° angle with each other, as in Fig. 4–28. The spider is motionless, waiting for a victim. What is the tension in each thread? [*Hint:* The tension is a vector directed away from the attachment point. Newton's first law must be satisfied in its vector form.]

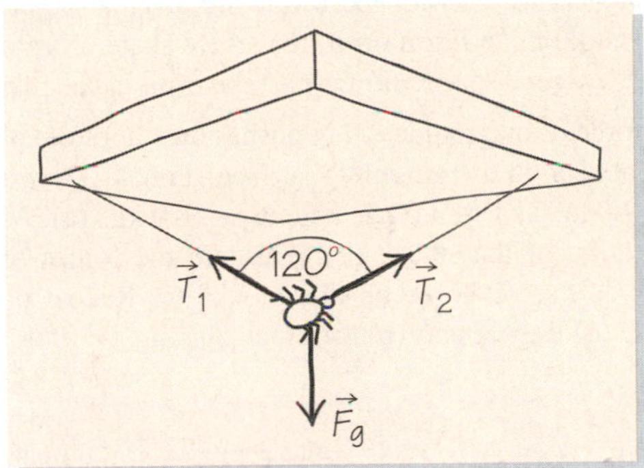

▲ **FIGURE 4–28** Problem 7.

8. (II) A very large—consider it to be infinitely large—mesh of stiff wires makes a horizontal plane and large electric charges are placed at the intersections of the wires. A mass with another charge is placed below one of the charges on the mesh. Electric charges exert forces on one another that are proportional to the strength of the charges. Each charge on the mesh repels the lone charge with a force that varies with the distance between the charges. The forces are directed along the line between the charges. Show that the net force on the lone charge is a repulsion directed straight downward.

4–2 Newton's Second Law of Motion

9. (I) A car coasts along a road with initial velocity $\vec{v}_0$. It inevitably slows down and finally comes to rest. (a) Describe why this is so. (b) An observer traveling with uniform velocity $\vec{v}_0$ starts out at rest relative to the car. What does she see with the passage of time? How does she explain what happens?

10. (I) The force of gravity on an apple (mass 0.15 kg) has a magnitude of about 1.5 N. What is the acceleration of the apple as it falls toward Earth? How large is the force of gravity on a falling SUV, mass 2500 kg, if it has the same acceleration as the apple? In each case, assume that only the force of gravity acts.

11. (I) The force of gravitation attracts two masses m_1 and m_2 to each other. If the masses are separated by a distance d, the magnitude of the force on each mass is

$$F = Gm_1m_2/d^2,$$

where G is a constant. Suppose $m_2 = 3m_1$. Make a sketch of the two masses with vectors that indicate the direction of the forces on the two bodies. Draw the lengths of the vectors to correspond to the magnitude of the two forces. Repeat the sketch but replace the force vectors with acceleration vectors.

12. (I) The forces acting on an airplane are the following:

gravity:	$F_g = 6.0 \times 10^5$ N down;
engine thrust:	$F_E = 2.0 \times 10^5$ N forward;
lift:	$F_L = 6.0 \times 10^5$ N up;
air drag:	$F_D = 1.5 \times 10^4$ N backward.

What is the net force on the airplane? In what direction, if any, is it accelerating?

13. (II) A forensic expert wants to examine the striations on a bullet fired by a gun. A bullet of mass 2.0 g is fired from the gun with muzzle speed 400 m/s into a special resistive material. The bullet is stopped in a distance of 14 cm. If we assume the negative acceleration is constant, what is the acceleration of the bullet inside the material and what force is exerted on the bullet as it accelerates?

14. (II) A spring exerts a force when it is compressed or extended. The force is proportional to the distance x by which it is compressed or stretched away from its equilibrium position. The direction of the spring's force is toward its equilibrium position. Draw diagrams with the spring compressed and extended. Choose a direction for x. Write mathematical expressions for the force in both situations.

15. (II) Five forces, all of the same magnitude F, act on an object of mass m at the origin of the coordinate system shown in Fig. 4–29. Two of the forces are aligned along the x-axis; one is oriented in the $+x$-direction and one in the $-x$-direction. Two other forces are similarly aligned with the z-axis. The fifth force points in the $+y$-direction. What are the direction and magnitude of the acceleration of the object? If we had specified the forces in a different order, would the answer have been different?

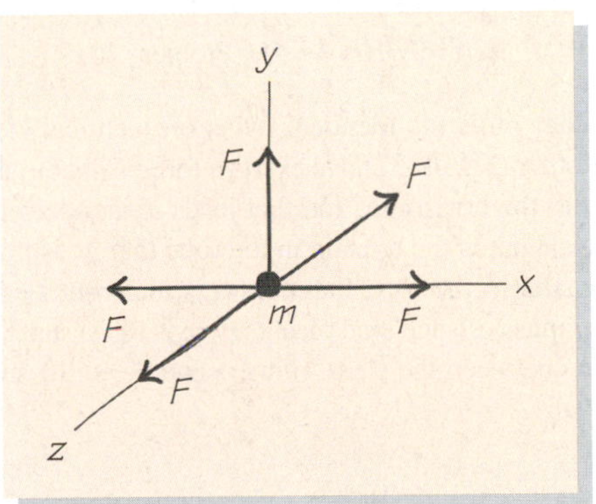

▲ **FIGURE 4–29** Problem 15.

16. (II) An object can move along a line between two points A and B separated by a distance $2L$. The object is attracted to each point by a force that is a constant c times the distance between the object and the point. Find an expression for the net force on the object. The net force tends to pull the object back to a particular point. Where is that point?

17. (II) Two forces act on an object of mass 2.5 kg: force $\vec{F}_1$ that is directed along the $+x$-direction and has magnitude 0.50 N and force $\vec{F}_2$ that points at a 45° angle in the $+y$ and $-x$ quadrant and has magnitude 2.0 N. Find the additional force, if any, such that the object will accelerate in the $+y$-direction with magnitude 1.5 m/s^2.

18. (II) A sports car of mass 720 kg accelerates from 0 to 60 mi/h in 6.7 s. What is the average force, in newtons, that the road exerts on the wheels of the car? A huge SUV of mass 2400 kg can accelerate from 0 to 60 mi/h in 9.7 s. What force does the road exert on the SUV? What would the SUV's acceleration be if it were acted on by the same external force as the sports car?

19. (II) A car of mass 1150 kg accelerates from rest to 100 km/h in 11 s. With additional streamlining, the same car undergoes acceleration to the same speed in 9.0 s. What is the difference in the force exerted by the air (the *drag force*) on the car in the two cases? For this problem, assume the drag force is constant. (This assumption is a poor one in practice.)

20. (II) A common type of contact force is that provided by a taut rope. Suppose that a taut rope is attached to an object. In what direction can the rope apply a force to the object? This force is the tension. (a) A cart being pulled horizontally by a light rope (the word "light" in this context means that you can ignore the mass of the rope in all your considerations) has a mass of 25 kg and accelerates in the horizontal direction at 2.40 m/s^2. What is the tension in the rope? If the same cart is pulled so that it accelerates at 0.65 m/s^2, what is the tension? Assume that the only force acting on the cart is the tension. (b) Suppose that the rope passes over a fixed pulley (Fig. 4–30) and the cart accelerates horizontally at 1.4 m/s^2. What is the upward force on the pulley?

▲ **FIGURE 4–30** Problem 20.

21. (II) A father pulls his identical twins on identical sleds tied one after the other (Fig. 4–31). He exerts a force $\vec{F}$ that makes an angle of 30° with the horizontal and that leads to an acceleration of the two sleds. What is the tension in the rope that he is pulling? What is the tension in the rope that connects the front sled to the rear sled? The mass of each sled plus its twin is m. Assume that there is no friction between the sleds' runners and the snow surface.

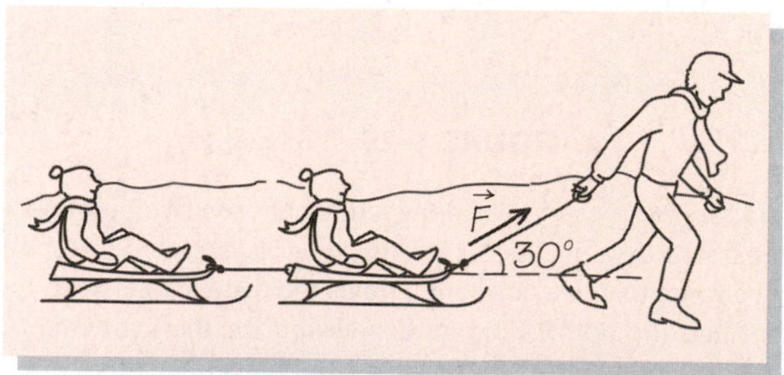

▲ **FIGURE 4–31** Problem 21.

22. (II) Electrons, mass about 10^{-30} kg, are constituents of atoms and respond to electrical forces. Such forces, of varying strengths, can be generated in the laboratory. Suppose that a constant electrical force of 5×10^{-14} N acts on an electron. What is the speed of the electron after 10^{-10} s? After 10^{-9} s?

4–3 Newton's Third Law of Motion

23. (I) A hook is screwed into a ceiling; one end of a string is tied to the hook and the other end has an 8-kg mass attached to it. Assuming that the force of gravity acting on a mass m has magnitude mg, where $g = 9.8$ m/s^2, what force does the hook exert on the string?

24. (I) A 3000-kg pickup truck pulls a 1200-kg boat on a trailer, and they are accelerating together at 1.2 m/s^2. What is the horizontal force that the truck and boat trailer exert on the road?

25. (I) A falling automobile, mass 950 kg, has an acceleration of magnitude 9.8 m/s^2 when only the force of gravity acts on it. What is the magnitude of the upward acceleration of Earth? Take the mass of Earth to be 6.0×10^{24} kg.

26. (I) The force exerted on a satellite by an astronaut is $(6.5\text{ N})\hat{i} + (3.7\text{ N})\hat{j} + (-4.7\text{ N})\hat{k}$. What is the force exerted by the satellite on the astronaut? What is the magnitude of this force?

27. (II) In Problem 1, you were asked to find the forces acting on certain objects. By Newton's third law, these original objects are the sources for forces that act on other objects. What are these other objects for each of the following original objects: (a) the space shuttle in an Earth orbit; (b) an ice skater coasting on ice; (c) the *Voyager I* spacecraft far past the orbit of the planet Pluto?

28. (II) A force of magnitude 8.0 N pushes on a horizontally stacked set of blocks on a frictionless surface (Fig. 4–32) with masses $m_1 = 2.0$ kg, $m_2 = 3.0$ kg, and $m_3 = 4.0$ kg. (a) What is the acceleration of the stack? (b) What are the forces on block 1 as well as the net force on this block? (c) Repeat part (b) for block 2. (d) Repeat part (b) for block 3.

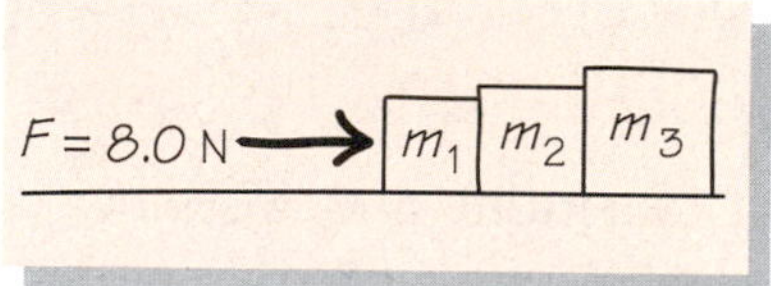

▲ **FIGURE 4–32** Problem 28.

29. (II) Two blocks connected by a spring are placed on a frictionless flat table. The blocks are pulled apart and then released. A measurement of their accelerations relative to the table shows that the acceleration of one block is twice that of the other. (a) What can you say about the masses of the blocks? (b) After the blocks compress the spring enough, they are pushed apart again. Suppose that while this is happening the spring is cut, so that the blocks are no longer connected. What can you say about the velocities of the blocks after this happens?

30. (II) Three blocks of equal mass are stacked vertically. Block 1 is on top and rests on block 2, which in turn rests on block 3, which rests on a table. There is a downward force on the table of 3 N. (a) What are the forces acting on block 3? (b) What are the forces acting on block 2? (c) What are the forces acting on block 1?

31. (II) The engine of a train pulls five cars each of mass $m = 21{,}000$ kg. During a period of acceleration, the force between the engine and the first car is 15,000 N. (a) What is the acceleration of the train? (b) Sketch the forces acting on the first car. (c) How large is the tension in the hook between the first and second cars?

32. (II) Repeat Problem 28, this time with the blocks stacked in the reverse order, that is, block 3 to the left and block 1 to the right.

33. (II) Three charges move through space with no forces acting on them except the electric forces that they exert on each other. In an appropriate coordinate system, some of the forces can be broken down as follows: The force that charge 1 exerts on charge 2 is $\vec{F}_{21} = (2\,\text{N})\hat{i} + (-3\,\text{N})\hat{j} + (1\,\text{N})\hat{k}$; the force that charge 1 exerts on charge 3 is $\vec{F}_{31} = (-3\,\text{N})\hat{i} + (2\,\text{N})\hat{j} + (-3\,\text{N})\hat{k}$. What is the total, or net, force on charge 1?

*4–4 Noninertial Frames

34. (I) An observer inside an elevator that is in free fall will see any object that was initially at rest in midair inside the elevator remain in that position. How does he explain this fact, assuming that he knows about the existence of gravity?

35. (I) Consider the situation described in Example 4–7, with $v_A = 0.0$ and $v_B = 3.0\,\text{m/s}$. The train is accelerating in the x-direction at $0.70\,\text{m/s}^2$, according to professor A. Can professor B tell that he is accelerating? Does he think that professor A is accelerating? Do our physical senses help us solve these questions? What is the acceleration of professor A according to professor B? Is this a real acceleration?

36. (II) An observer sits in a cylindrical bathtub that can rotate about a vertical axis. At first, the tub is at rest—relative to a neighboring tree for example. A little later, the tub is rotating at high speed around its vertical axis. The observer is still at rest relative to the tub, but the water level is no longer flat; it runs up on the sides of the tub. How does the observer explain this?

37. (II) A simple demonstration of the effect of a noninertial frame can be made by dangling a watch from a chain while you are taking a trip in a jet. What is the angle of the chain with respect to a vertical window edge when (a) the jet is parked at the gate, (b) the jet is accelerating with some acceleration a' along the runway just before takeoff, and (c) the jet is at cruising speed at 30,000 ft? You can try this type of experiment in an accelerating automobile.

38. (II) A mass of 2 kg lies on a horizontal table that is placed in the back of a truck (Fig. 4–33a). The mass is held in position by a string and a force of 6 N acts on it in the $-y$-direction. The truck then accelerates in the $+x$-direction. As a result, an observer in the truck sees that the string holding the mass on the table makes an angle of 20° with respect to its initial orientation, as shown in Fig. 4–33b. What is the acceleration of the truck? Describe what the observer in the truck sees and how he might interpret his observations.

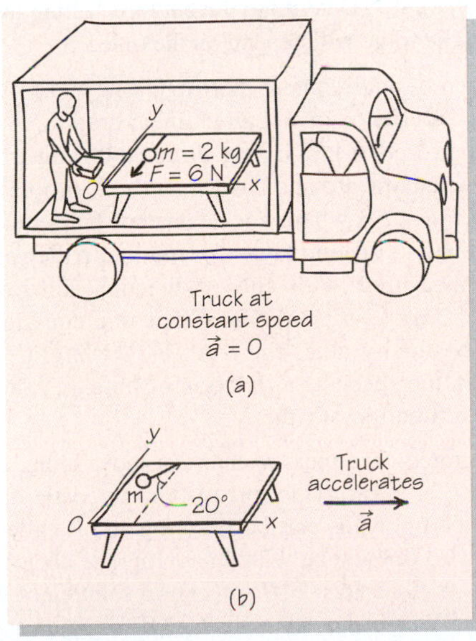

▲ FIGURE 4–33 Problem 38.

4–5 Using Newton's Laws I: Free-Body Diagrams

39. (I) A car is accelerating straight ahead on a flat road. Draw the free-body diagram for the car.

40. (I) A child slides down a water slide at an amusement park. What forces are present? Draw a free-body diagram for the child.

41. (I) Consider an astronaut sitting in a rocket just after blastoff from Cape Canaveral. Draw a free-body diagram showing the forces present on the astronaut sitting in her seat.

42. (I) Consider the arrangement shown in Fig. 4–34. The whole system is at rest. Draw all the forces, draw the free-body diagram for each mass, and give the equations of motion for each mass.

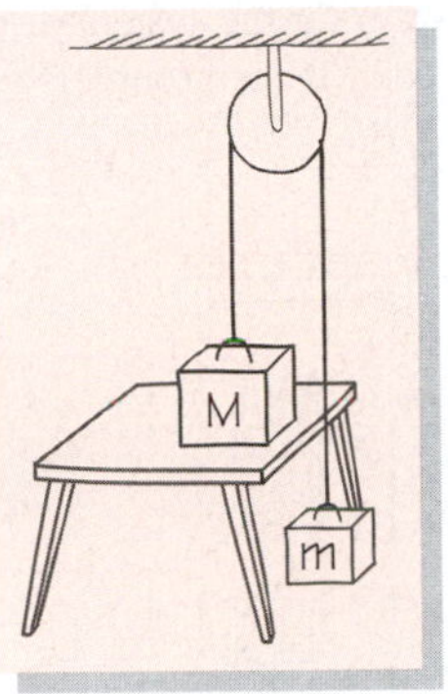

▲ FIGURE 4–34 Problem 42.

43. (I) A person throws a medicine ball into the air. Draw separate free-body diagrams for the ball and for the person (a) for the time just before the ball leaves the hands and (b) after the ball has been thrown.

44. (I) A piano sits within a large freight elevator. (a) Draw a free-body diagram for the piano alone. (b) Draw a free-body diagram for the elevator alone.

45. (II) Using a rope, a horse pulls a wagon that is initially at rest. (a) Draw a free-body diagram for the horse. Include friction, Earth's gravity, the influence of the wagon, and any other forces that may be operating. Draw the direction of the forces as accurately as possible. (b) Repeat part (a) for the wagon. (c) Repeat part (a) for Earth, including only the horse and wagon as influences. (d) Why doesn't the horse accelerate backward toward the wagon? In answering, think about the different forces acting on the horse and wagon.

46. (II) A rope spans a gap by being attached at two points on either side of the gap. The two points are at the same height, and the rope is attached in such a way that the tension in it is 50 N. (See Problem 20 for a discussion of tension.) (a) Can the rope be perfectly horizontal? You can assume here that the rope is so light that you can neglect any mass it has. (b) A mass of 2.0 kg is hung by another (light) rope tied to the gap-spanning rope at its midway point. Can the gap-spanning rope remain horizontal? (c) If your answer to part (a) is no, what is the angle the appropriate parts of the gap-spanning rope make with the horizontal?

47. (II) An engine pulls three identical railroad cars along a track. The train's acceleration is a. Draw a free-body diagram for the train as a whole, assigning masses to the cars and engine and putting in all the forces, including those between cars. Which forces are external and which are internal?

48. (II) A block of mass M sits on a rough horizontal surface. A rope inclined upward at an angle θ with the horizontal exerts a force of magnitude T on the block. The block remains stationary. Draw a free-body diagram and express all the forces on the block in terms of M, g, θ, and T.

49. (II) Consider a brick sliding down an inclined plane of 21°. Draw a free-body diagram for the brick that includes the gravity, friction, and normal forces. Choose a coordinate system and give the force components in equation form.

50. (II) Draw two free-body diagrams for a tug of war in which the strengths of the two sides are unbalanced. (The two objects in the diagram will be the two groups of contestants.) Be sure to include all forces!

51. (II) A heavy man stands on a ladder that is leaning against a rough-surfaced wall and rests on the ground (Fig. 4–35). Draw the free-body diagram for the system of man and ladder. Ignore the mass of the ladder. If the ground is very smooth, the ladder may start to slip. Why?

▲ **FIGURE 4–35** Problem 51.

52. (II) Two astronauts, Joe and Moe, are taking a space walk. They are connected to the ship only by a slack rope. They have a light rope between them and decide to have an outer space tug of war. In their space suits, Joe has a mass of 100 kg, whereas Moe has a mass of 110 kg. The loser of the tug of war is the first astronaut to cross an imaginary line midway between them. (a) Draw a free-body diagram for each astronaut. What is the constraint of Newton's third law? (b) Show that Moe wins the tug of war. (c) Miffed at his loss, Joe decides to challenge Moe to a tug of war with their feet firmly planted on Earth. Draw the free-body diagram for each astronaut. (d) Joe wins the tug of war this time. How is this possible?

53. (II) An electrically charged mass, when placed between two charged parallel plates, experiences a constant force F_e perpendicular to the plates. Suppose that such a mass moving vertically with some speed v_y enters the region between two such plates, which are oriented vertically. Draw a free-body diagram for the mass while it is between the plates. Include the force of gravity. Note that there are two possibilities for the direction of F_e; choose one or the other.

54. (II) Objects made of iron are attracted by an electromagnet when a switch is thrown activating it. Suppose that a small piece of iron is suspended by a thread in the vicinity of an electromagnet, as shown in Fig. 4–36. Draw a free-body diagram for the iron object (a) immediately after the switch is closed and the electromagnet is activated and (b) when the switch has been closed for a long time.

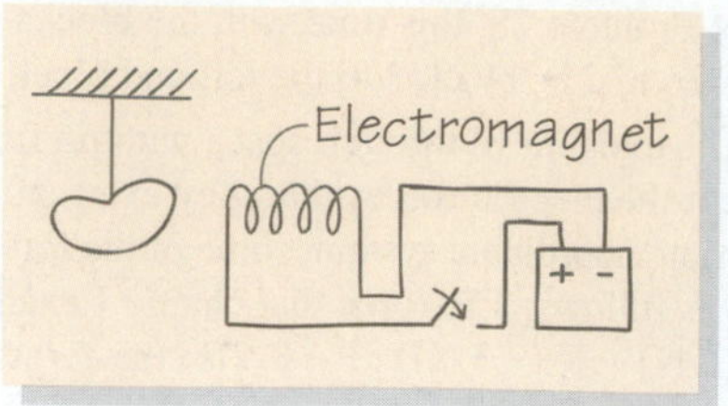

▲ **FIGURE 4–36** Problem 54.

55. (II) A person is standing in an elevator that is moving upward with a constant speed. (a) What are the forces acting on the person? (b) Suppose that the elevator is accelerating upward with an acceleration $\vec{g}$ that is equal in magnitude to the acceleration that the person would have if the only force acting on him were gravity. What are the forces acting on the person and how large are they? (c) Suppose that the cable breaks and the elevator, together with its unfortunate passenger, is falling freely. What forces are acting on the person?

4–6 Using Newton's Laws II: Finding the Motion

56. (I) An object of mass 43 g that can move only in the xy-plane is at rest at the origin of the xy-coordinate system at time $t = 0$ s. Constant forces $\vec{F}_1 = (0.071, 0, 0)$ N and $\vec{F}_2 = (0, 0.081, 0)$ N act on the object. (a) Draw the free-body diagram for the object. (b) State the initial conditions, that is, the position and velocity vectors at $t = 0$ s. (c) What are the position and velocity of the object at time $t = 1.200$ s? (d) At $t = 3.600$ s?

57. (II) An object of mass M is subject to two constant forces: $\vec{F}_1$, pointing in the x-direction, and $\vec{F}_2$, pointing in the y-direction. Show that the motion is one of constant acceleration. Find the magnitude and direction of this uniform acceleration.

58. (II) A third constant force $\vec{F}_3 = (F_{3x}, F_{3y}, 0)$ acts on the object in Problem 56 at $t = 0$ s in addition to the forces $\vec{F}_1$ and $\vec{F}_2$. (a) What must F_{3x} and F_{3y} be so that the object does not accelerate? (b) What must F_{3x} and F_{3y} be so that at $t = 5.000$ s the object is at position $\vec{r} = (1.000, 1.000, 0)$ m?

59. (III) At some time t, the displacement of an object moving in one dimension of $m = 2.0$ kg is measured to be $x = At^{3/4}$, where $A = 0.030\ \text{m/s}^{3/4}$. Determine the net force acting on this object. Note that the force will depend on the time.

60. (III) When a small sphere starting from rest falls through a liquid, it experiences a so-called linear drag force $\vec{F}_D$, in addition to the force of gravity mg directed downward. The force $\vec{F}_D$ is directed upward and has a magnitude proportional to the velocity, $F_D = bv$, where b is a constant. The equation of motion—Newton's second law—for motion in the vertical direction z (measured with the +-direction downward) reads $m\,dv/dt = mg - bv$. (a) Show that this equation is solved when the velocity takes the form $v(t) = mg(1 - e^{-bt/m})/b$. (b) Sketch the speed as a function of time. (c) Sketch the position as a function of time.

61. (III) The force on a mass m that can move along the x-axis is given by $F = -kx$. (This force is characteristic of the spring; k is a property of this particular spring and is called the spring constant.) (a) Show that, with an appropriate choice of the constant c, $x = A\cos(ct) + B\sin(ct)$is a solution of the equation of motion. What is the value of c? (b) If the object is at $x = 0$ at $t = 0$ and if its velocity at that time is v_0, what are A and B?

General Problems

62. (I) The gravitational force that the Sun exerts on Earth has a magnitude of 3.5×10^{22} N and is directed toward the Sun. What force does Earth exert on the Sun?

63. (II) A load of mass 53.2 kg is to be lifted by two ropes each going over pulleys. The ropes are separated in such a way that both ropes initially make an angle of 65° with the horizontal (Fig. 4–37). The magnitude of the forces exerted by each of the ropes is 333 N. (a) Draw a free-body diagram for the load. The only forces acting on it are the forces of the ropes, directed along the ropes, and the force of gravity, which points down and has a magnitude given by the mass of the load times the constant acceleration $g = 9.80\ \mathrm{m/s^2}$. (b) What is the net force on the load? (c) What is the initial acceleration (magnitude and direction) of the load?

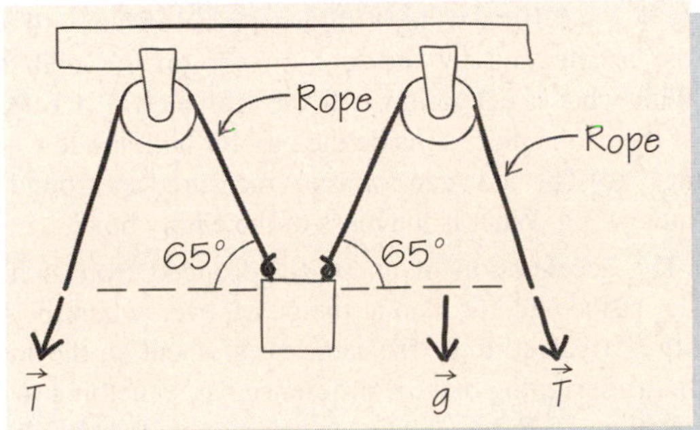

▲ **FIGURE 4–37** Problem 63.

64. (II) An object of weight $F_g = 20$ N is suspended as shown in Fig. 4–38. The rope and pulleys are very light and friction is negligible. Determine the weight of the object that maintains the equilibrium given that the angle between the rope and the horizontal is 35° on both sides of the movable pulley.

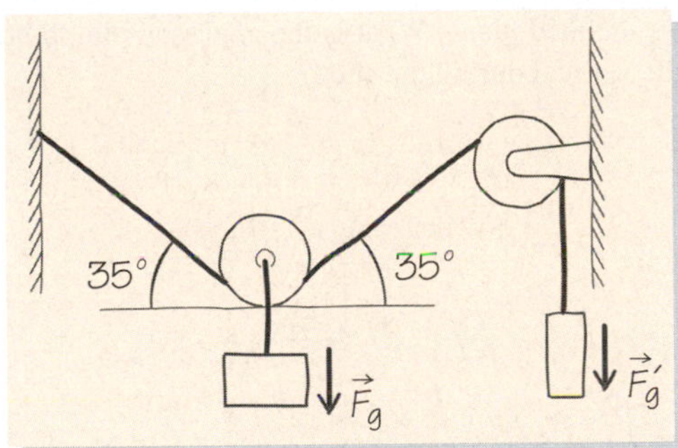

▲ **FIGURE 4–38** Problem 64.

65. (II) A window washer sits on a board supported by two ropes that go over pulleys attached to a scaffold. He holds the other ends of the ropes in his hands (Fig. 4–39). He pulls on them in such a way that he moves upward at constant speed. Draw a free-body diagram for the window washer only and give all the equations; assume that the tension in a given rope is the same everywhere in the rope, even after it wraps around a pulley. Suppose that the effect of the window washer's efforts leads to an upward acceleration of both the window washer and the board. How are the equations modified?

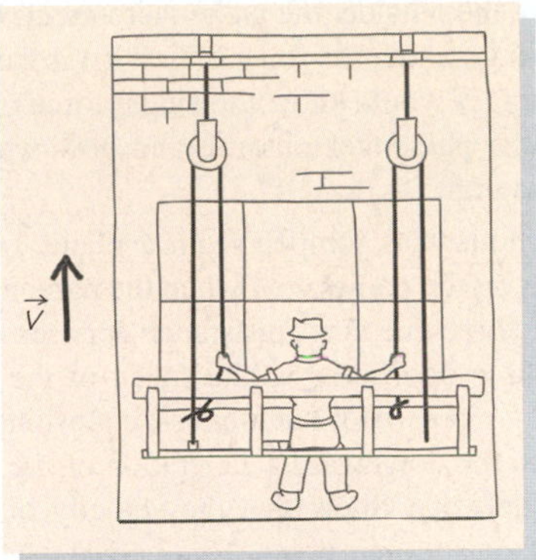

▲ **FIGURE 4–39** Problem 65.

66. (II) A dead tennis ball is dropped onto a granite slab and it bounces back to about 20 percent of the height from which it was dropped. What forces acted on the tennis ball for it to exhibit this behavior? In what direction do these forces point? Estimate their magnitude.

67. (II) A train consists of an engine and three cars tied closely together. The mass of the whole train is 1.7×10^5 kg; the engine alone has a mass of 80,000 kg and each car has a mass of 30,000 kg. Through its wheels, the engine can exert a horizontal force of 3.0×10^4 N on Earth. (a) When the engineer wishes to accelerate forward, in which direction should the engine exert its force on Earth? What is the magnitude of the force exerted by Earth on the engine then? (b) When the engineer acts to accelerate the train, what is the force on the set of three cars and what is the subsequent acceleration of these cars? (Treat the three cars as a unified whole.) (c) When the engineer acts to accelerate the train, what are all the forces on, and subsequent acceleration of, the second car of the train?

68. (II) A boat is being pulled up the middle of a canal at a steady speed by two horses, one on each side of the canal (Fig. 4–40). The ropes tying the boat to the horses each make an angle of 30° with the lengthwise direction of the canal. (a) Draw a free-body diagram for the boat. Include the friction force due to the boat sliding through the water. (b) If the force exerted on each horse by its rope is 2000 N, what is the force of friction that the boat experiences due to the water?

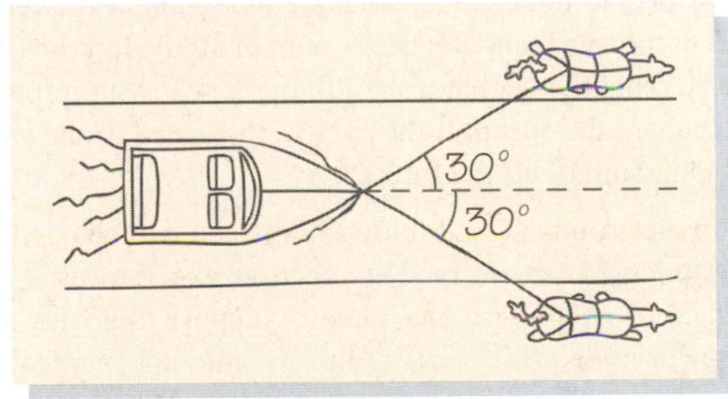

▲ **FIGURE 4–40** Problem 68.

69. (II) A passenger in a stationary elevator is acted on both by the force of gravity, which is directed downward, and by a contact force directed upward by the floor. The contact force from the floor cancels the force of gravity, and in this way there is no net force on the passenger, which explains why she measures herself to be at rest with respect to the walls. (a) What is the free-body diagram on the passenger when the elevator descends at constant velocity? (b) The elevator cable breaks. (Do not be alarmed—the emergency brake will act if you finish this problem in time.)

As seen from the outside, the passenger and elevator fall together and accelerate under the force of gravity. A free-body diagram for the passenger would show the single force of gravity acting. What does the passenger observe and how would she draw a free-body diagram for herself?

70. (II) A parachutist has jumped from a plane. (a) What are the forces acting on the parachutist when the parachute opens? Suppose that the force due to air resistance opposes the parachutist's motion and is proportional to the speed of the system; that is, $\vec{F}_{\text{air resistance}} = -A\vec{v}$. (b) What are the dimensions of the constant A? (c) Sketch the downward acceleration of the parachutist as a function of velocity. Show that the velocity of the system becomes constant at some point. This constant velocity is called the *terminal velocity*. (d) If the force of gravity on the system is 1000 N and the terminal velocity is 6.0 m/s, what is the value of the constant A?

71. (II) A parachutist experiences two forces: One is the force of gravity, which is of the form $\vec{F}_g = -mg\hat{j}$. Here, m is the mass of the parachutist and g is the acceleration due to gravity. The other force is a drag force, and, in contrast to Problem 70, it has the form $\vec{F}_d = Av^2\hat{j}$ in this problem, where the velocity of the parachutist is given by $-v\hat{j}$. (a) What are the dimensions (and units) of A? (b) Complete the equation $dv/dt = \ldots$. (c) At some point, the parachutist reaches a terminal velocity that is constant. What is it?

72. (II) An egg of mass 80 g is in a stiff box of mass 200 g surrounded on all sides by padding that keeps the egg in place. A force $\vec{F} = (1.2\text{ N})\hat{i} + (-0.08\text{ N})\hat{j}$ acts on the outside of the box, in addition to the force of gravity (which is oriented in the $-y$-direction). (a) Draw free-body diagrams for the egg–box system and for the egg. (b) Express Newton's second law as it applies to the egg. (c) What are the acceleration of the egg–box system and net force on the egg?

73. (II) A bicyclist rides a flat course at constant speed v with respect to the ground into a stiff but steady headwind of speed v_w, also with respect to the ground. (a) Draw a free-body diagram for the cyclist–cycle system. List the forces acting on this system, including the wind resistance, which pushes in his face, and the frictional forces with the ground, which propel him forward. (b) If you are observing the cyclist from a reference frame at rest with respect to the ground, what is the vector sum of all the forces? (c) If you are observing the cyclist from a frame at rest with respect to the bicycle, what are the speed of the cyclist, the speed of the wind, and the vector sum of all the forces? (d) If you are observing the cyclist from a frame at rest with respect to the air, what are the speed of the cyclist, the speed of the wind, and the vector sum of all the forces?

74. (II) An electron is placed midway between two charged parallel plates oriented vertically. The electron experiences a constant force perpendicular to the plates. Suppose that the electron, whose mass is 9.0×10^{-31} kg, is moving vertically with $v_y = 2.0 \times 10^6$ m/s and enters the region between the two plates (Fig. 4–41). This region has a vertical height of 1 cm, and while in it, the electron experiences a force of 3.0×10^{-18} N. (a) How long does the electron stay in the region? (b) After the electron has passed through the region, what is its horizontal velocity component?

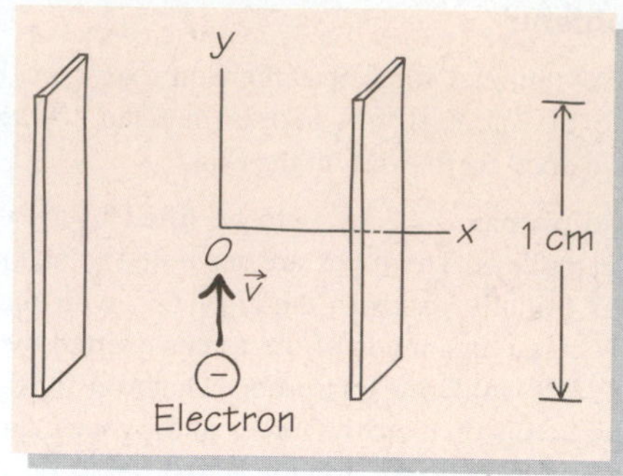

▲ **FIGURE 4–41** Problem 74.

75. (II) The air stream from a leaf blower exerts a force F_0 on the low setting and a force $2F_0$ on the high setting. When on the low setting the leaf blower accelerates a soda can at 0.20 m/s^2 and an empty box at 0.10 m/s^2. (a) With what acceleration will each object be accelerated by the leaf blower on the high setting? (b) With what acceleration will the leaf blower accelerate the box if the can is placed inside the box for both the low and high settings? (c) The soda can's mass is measured and found to have the value 4.3 g. What is the mass of the empty box?

76. (III) The acceleration of an object dropped from a height is smaller at the equator than at the poles, even when the force of gravity is assumed to be the same everywhere on the surface of Earth. (a) Assuming that the difference has something to do with the rotation of Earth, obtain an expression for the difference between the accelerations in terms of the radius of Earth and its rotation frequency. (b) Given that the radius of Earth is $R = 6.38 \times 10^6$ m and that the acceleration at the poles is $g_0 = 9.80$ m/s^2, calculate the acceleration of a falling object at the equator. (c) Calculate the acceleration of a falling object as a function of the angle of latitude in terms of g_0. (d) If experiment shows a deviation from your prediction,what effects could be responsible for that deviation?

77. (III) As shown in Fig. 4–42, a plumb bob suspended from a frame settles to a steady position as the frame slides down a frictionless inclined plane. What is the angle the plumb bob makes with the vertical during the slide?

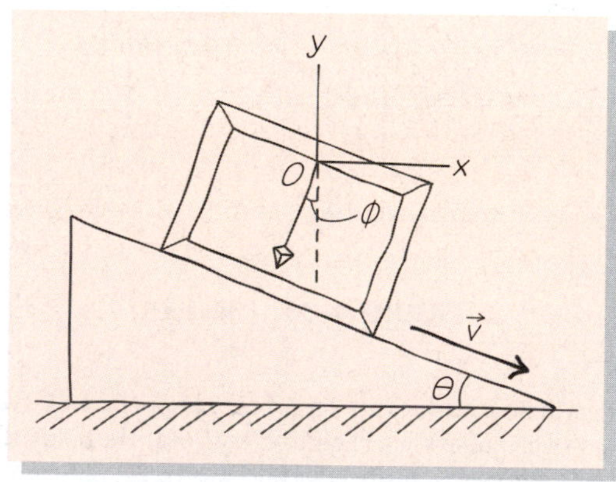

▲ **FIGURE 4–42** Problem 77.

◀ It is a contact force exerted by the starting blocks against the foot of a sprinter that accelerates him forward. An equal and opposite force is exerted on the starting block, and, because the block is attached to Earth, on Earth itself.

Applications of Newton's Laws

In this chapter we apply Newton's laws to a variety of situations in which forces act. We will consider several types of forces, including gravity, tension, normal forces, friction, and drag forces, all of which act on us and the objects around us. We shall also look at the role of forces in circular motion as well as the features of motion in rotating frames. Finally, we will look at how the forces are ultimately described in terms of more fundamental forces—in particular, ones that act at a microscopic level.

5–1 Common Forces Revisited

The forces that we encountered in Chapter 4 demand further study, as we have not yet explored all of their significant features. These forces include the force of *gravity* $\vec{F}_g$, here treated as a constant force that acts on every object, the tension $\vec{T}$ exerted by a taut rope, and the normal force $\vec{F}_N$ that keeps you from falling through your chair. We'll study friction in Section 5–2.

▲ **FIGURE 5–1** The horse follows a parabolic path as it jumps the rail.

Gravity

The parabolic form of a projectile's trajectory near Earth (Fig. 5–1) is due to the force of gravity. As we saw in Chapter 3, projectiles in the vicinity of Earth's surface travel with a constant acceleration $\vec{a}$ that points downward and has magnitude g. With Newton's second law, we can say that the force of gravity causes the projectile's constant acceleration and that the force has constant magnitude near Earth's surface and is always directed down toward the center of Earth. More precisely, the acceleration $\vec{a}$ of our projectile is given in terms of the force of gravity $\vec{F}_g$ on the projectile and the inertial mass m of the projectile as

$$\vec{a} = \frac{\vec{F}_g}{m}. \tag{5–1}$$

A very special characteristic of the force of gravity is that, at any given location, *it causes all objects to accelerate in the same way—no matter what their mass.* In other words, the right-hand side of Eq. (5–1) is independent of the mass m. The only way this can happen is for the *force itself to be proportional to the mass*, so that the factor m in the denominator cancels with another such factor in the force itself.

Let's look more carefully at the meaning of this remarkable fact. That an object has constant acceleration under the influence of gravity means only that this force takes the form

$$\vec{F}_g = m_g\vec{g}, \tag{5–2}$$

where $\vec{g}$ is a constant vector with dimensions of acceleration and points to the center of Earth.[†] The "gravitational mass" m_g is the property of the object that determines the strength of the gravitational force acting on it. However, the acceleration of *all* objects under the influence of gravity is *precisely the same*; mathematically, this statement means

$$m_g = m. \tag{5–3}$$

This relation between gravitational mass and inertial mass is remarkable because, at least until Einstein's theory of general relativity, not developed until about 1915, we had no reason to think that the force of gravity has anything special to do with the inertial mass. (See Chapter 12 for further discussion.) We recall from Chapter 4 that the inertial mass determines the *response* to a force, and tests to determine the inertial mass are possible without using gravity at all—say, by seeing how much the object accelerates when *any* known force, even one that has absolutely nothing to do with gravity, acts on it. Conversely, tests to determine the force of gravity on an object can be performed without using motion at all—say, by observing the equilibrium stretch of a spring while the object hangs from it under the influence of gravity. Nevertheless, the equality of Eq. (5–3) has been experimentally verified to a very high degree of accuracy.[‡] Thus the force of gravity has the simple form

$$\vec{F}_g = m_g\vec{g} = m\vec{g}. \tag{5–4}$$

FORCE OF GRAVITY

The acceleration $\vec{a}$ of any object at Earth's surface under the influence of gravity alone is then given by the constant vector $\vec{g}$:

$$\vec{a} = \frac{\vec{F}_g}{m} = \vec{g}. \tag{5–5}$$

As described in Chapter 3, the magnitude of $\vec{g}$ at Earth's surface is roughly

$$g = 9.80\ \text{m/s}^2. \tag{5–6}$$

Experiments show that this value varies by about 1 percent over Earth's surface, with the higher values occurring at the poles. This variation is due to irregularities in the shape and density of Earth and to Earth's rotation.

[†] The force near any astronomical object has the same form and direction, but the magnitude of g is different. We'll learn more about this in Chapter 12.

[‡] The so-called Eötvös experiments verify that the inertial and gravitational masses are the same to 1 part in 10^{12}.

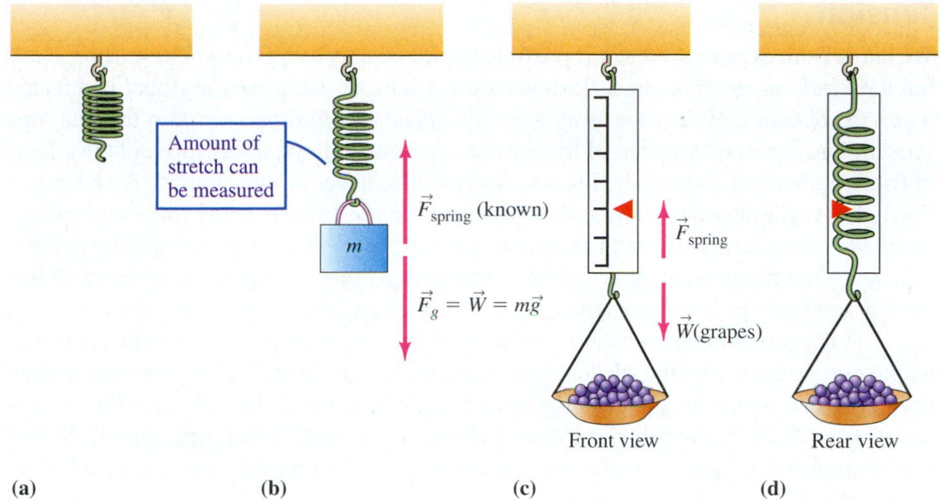

▲ **FIGURE 5–2** The weight of an object is found by attaching it to a spring, for which the amount of stretch corresponds to a given force. In (a) the spring is unstretched. (b) Calibration (the force corresponding to a given amount of stretch) is mapped out by hanging a series of known masses. (c, d) Once the calibration is done, the weight of any other object can be found.

We noted in Chapter 4 that the force of gravity on an object is commonly called the object's weight, $\vec{W}$,

$$\vec{W} = \vec{F}_g = m\vec{g}. \tag{5–7}$$

WEIGHT

We can experimentally determine the weight of an object by balancing the force of gravity against a second calibrated force such as that exerted by the stretched spring shown in its unstretched state in Fig. 5–2a and stretched, with a mass attached, in Fig. 5–2b. Once we know how much force is exerted for a given stretch, we can find the weight of any object by suspending it from the spring and observing the amount of stretch (Figs. 5–2c and d). In fact, when you step on a bathroom scale, you are indirectly measuring the force of gravity on your body in a similar way; the second force is the calibrated bathroom scale, which is in essence a compressible spring. While we might in everyday conversation sometimes use the words *weight* and *mass* interchangeably, this is not correct. Be careful not to confuse the weight of an object, measured in newtons or pounds, with its mass. We know from Chapter 4 that the mass of an object is the quantity of matter it contains, which is measured by its inertia—its resistance to any change in motion. The mass and weight of an object, though, are numerically proportional to each other through g—the weight of an object equals the object's mass times $\vec{g}$. In other words, the mass is intrinsic to the object, but its weight depends on where it is: If you went to the Moon, where a falling object falls with a different acceleration $\vec{g}_{\text{Moon}}$, a 1-kg mass would still be a 1-kg mass, but its weight would have magnitude mg_{Moon} rather than mg_{Earth}.

CONCEPTUAL EXAMPLE 5–1 Let us imagine that the inertial mass of any object is exactly 5 percent larger than its gravitational mass. How would this affect our treatment of the projectile motion?

Answer The force of gravity $\vec{F}_g = m_g\vec{g}$ can be determined by weighing the object. However the two factors cannot be determined without knowing more about gravity (this is the subject of Chapter 12). The acceleration due to the force of gravity is given by

$$\vec{a} = \frac{\vec{F}_g}{m} = \frac{m_g}{m}\vec{g} = \frac{1}{1.05}\vec{g}.$$

It is this quantity that is determined by dropping an object from a given height, and it is this acceleration that has the value 9.81 m/s^2 on Earth. Furthermore it is this quantity that enters into projectile motion. Thus nothing would be changed, *provided* that the ratio of m to m_g (here 1.05) is a universal constant, and does not change from object to object.

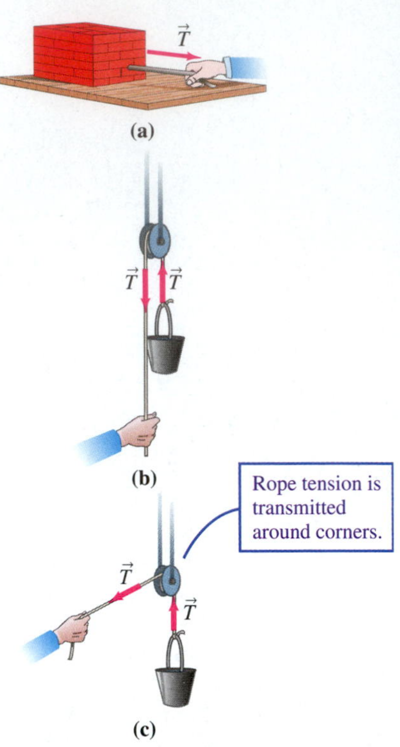

Tension

We know from experience that it is possible to pull objects using ropes. The tension $\vec{T}$ is a force exerted on an object by a flexible rope (or wire or cable or string) that is directed *along* the direction of the rope away from the object. By *flexible*, we mean that the rope goes around corners (by means of frictionless idealized pulleys, a way to avoid the effects of friction), not that it stretches. Tension always pulls, never pushes (Figs. 5–3a, b, and c). For light (negligible-mass) ropes, the *magnitude of the tension is the same everywhere along the rope*. (A rope of negligible mass—the kind of idealized rope we employ in these chapters—is a rope whose mass is small compared to other masses in the problem. When the mass of the rope is not negligible, then the tension will vary along the rope.)

The tension will adjust itself to different values according to the situation. For example, a rope from which a bucket is suspended (Figs. 5–3b and c) has less tension than does the same rope with a piano suspended from it because in the two cases the tension cancels very different weights. If we were able to look inside a taut rope, we would find that tension arises from the molecular forces that hold the rope together and give it flexibility. We do not need to know the microscopic details to be able to use tension in everyday problems.

◀ **FIGURE 5–3** (a) The tension force pulls but cannot push. (b, c) A rope of negligible mass with a given tension can maintain the magnitude of the tension even if the direction of the rope is changed. Tension can be transmitted around corners without change of magnitude because we are assuming an ideal pulley.

EXAMPLE 5–2 Consider Fig. 5–4a, which shows a fishing line hanging from a hook attached to the ceiling. This fishing line is rated as 10-lb-test line, which means that it should hold as long as the tension within it does not exceed 10 lb. A box of mass 2.0 kg is attached to the line. Find the tension in the line. Will the line hold? What will happen if a box of mass 5.0 kg is attached instead?

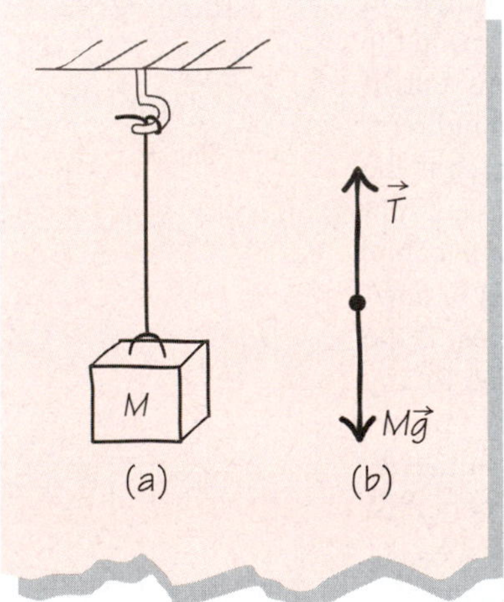

▲ **FIGURE 5–4** (a) Mass hangs from hook using 10-lb-test line. (b) Free-body diagram for a 2.0-kg package suspended from a line.

Setting It Up We are given the mass of the box, which we label as M. If we can calculate the unknown line tension's magnitude T, we can decide whether the line will break. We draw a sketch of the situation (Fig. 5–4a).

Strategy An important first step is to adapt the sketch and make it into a free-body diagram for the box (Fig. 5–4b). With the aid of this diagram we can write Newton's law for the box and use the condition that the mass is stationary—that is, that there is no acceleration, so that the net force is zero—to find T.

Working It Out The free-body diagram shows us that the tension and the force of gravity are the only forces acting on the box. Thus the net force on it is zero:

$$\vec{F}_{net} = \vec{T} - M\vec{g} = 0.$$

We solve this for T:

$$T = Mg.$$

The sign is positive because we want only the magnitude of the tension force. With $M = 2.0$ kg,

$$T = Mg = (2.0\ \text{kg})(9.8\ \text{m/s}^2) = 19.6\ \text{N}$$

$$= (19.6\ \cancel{\text{N}})\left(\frac{1\ \text{lb}}{4.45\ \cancel{\text{N}}}\right) = 4.4\ \text{lb}.$$

The line will hold. But with $M = 5.0$ kg,

$$T = (5.0\ \text{kg})(9.8\ \text{m/s}^2) = 49\ \text{N} = (49\ \cancel{\text{N}})\left(\frac{1\ \text{lb}}{4.45\ \cancel{\text{N}}}\right) = 11\ \text{lb}.$$

The line may not hold with the heavier mass suspended from it as the tension within it exceeds its breaking tension.

What Do You Think? Why do fishermen use fishing line that is rated much higher than the weight of any fish they are likely to catch? *Answers to **What Do You Think?** questions are given in the back of the book.*

How can we measure the tension within a rope at a given point? One way would be to cut the rope at the given point and insert a calibrated spring scale. The scale will stretch by an amount corresponding to the tension in the rope. This experiment could be performed at different points along the rope to show that the tension is the same throughout.

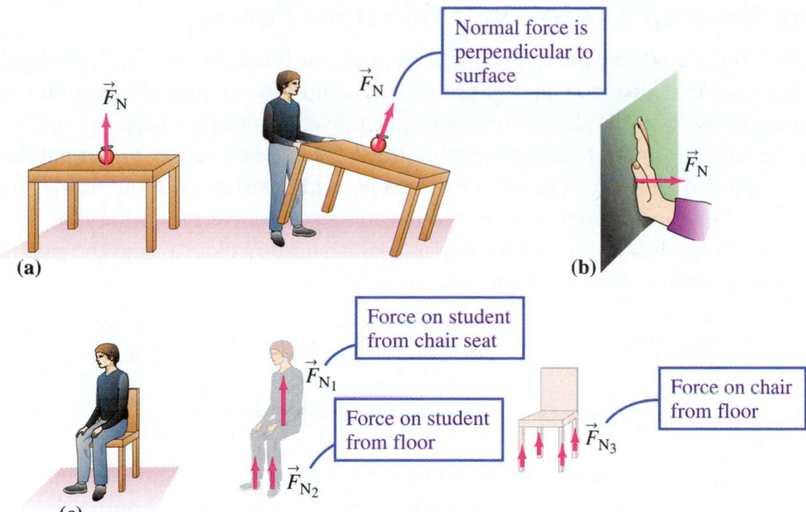

FIGURE 5–5 Some normal forces. These forces are always perpendicular to the contact surface, keeping the object on which the force acts from entering the surface.

FIGURE 5–6 (a) A scale can be used to measure the normal force on an apple. (b) We can draw the forces acting on the apple.

(a)

1) Draw gravitational force

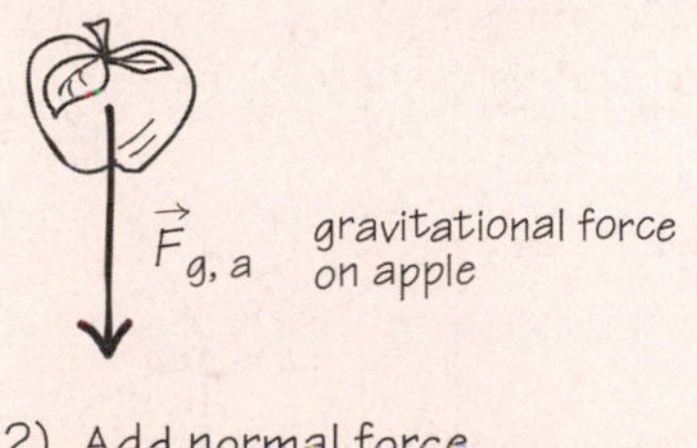

2) Add normal force

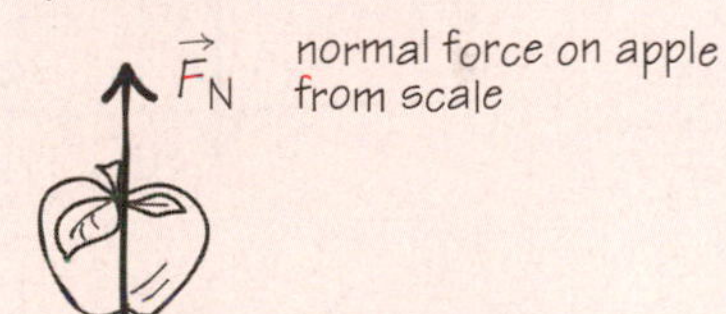

3) Draw free-body diagram

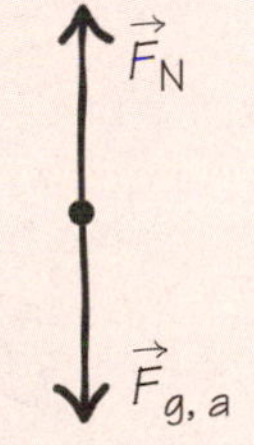

4) Apple doesn't accelerate

Therefore, $\vec{F}_{net} = 0$

$\vec{F}_{net} = \vec{F}_{g,a} + \vec{F}_N = 0$

(b)

Normal Force

Let's look at the forces acting on an apple placed on a table (Fig. 5–5). In addition to the force of gravity, a *normal force* $\vec{F}_N$ acts on the apple. In Chapter 4, we were introduced to this force and acknowledged that such a normal force must exist—without it, gravity would cause the apple to accelerate into the surface of the table. The normal force is the result of the complex interaction between the molecules of the table. We call a material that can give rise to a normal force a *solid*. The normal force on an object acts only when the object is in contact with the table. The normal forces on the objects shown in Fig. 5–5 adjust themselves to cancel the components of forces perpendicular to the surface. If the weight is too large, however, the supporting surface will collapse. The normal force pushes but never pulls. It acts perpendicular to and away from the surface at the object's point of contact. If this were not true, an apple on a table would accelerate to one side of the table or the other. In fact, whatever the tilt of the table, the apple does not move into its surface because *the normal force always acts perpendicular to and away from the surface*. Because of this property, the normal force cannot counteract any forces parallel to the surface. For example, the normal force cannot oppose the frictional force parallel to the table's surface in the case of the tilted table in Fig. 5–5b and cannot keep the apple from rolling down the table. All the normal force can do is keep the apple from penetrating the table.

We can see how the magnitude of $\vec{F}_N$ is determined in a situation by the use of Newton's first law. An apple of 0.25 kg sitting on a table must experience an upwardly directed normal force that is equal and opposite to the downward force of gravity; the magnitude of this force is $mg = (0.25\text{ kg})(9.8\text{ m/s}^2) = 2.5\text{ N}$. If a pumpkin of mass 2.5 kg sits on the table, the normal force is 25 N. As we'll see in the examples below, the normal force on the apple will differ in magnitude from the apple's weight if we tilt the table.

An ordinary bathroom scale gives us a way to read directly the magnitude of the normal force. The spring in the scale opposes the force of gravity in exactly the same way as does the normal force and therefore gives the magnitude of the normal force via a calibrated scale. In Fig. 5–6, a scale measures the normal force of 2.5 N that acts on the apple (although the force is often expressed in units other than newtons). The scale could equally well be placed on other surfaces that exert normal forces. For example, the normal force exerted on a hand by a wall can be measured directly if a scale is inserted between the wall and the hand.

There are other forces similar to the tension and the normal force. Among these we could include the support forces supplied by hooks, connection points, bearings, and so forth. The direction and magnitude of such forces are determined by the requirement that the attachment point in question does not accelerate. For example, a hook from which a 20-lb weight is suspended must supply an upward force of 20 lb if the mass is not to fall.

Applying Newton's Laws with Constant Forces

In many everyday situations, the tension and the normal force are constant. Examples involving constant tensions, constant normal forces, and gravity provide us with solvable equations of motion. We shall investigate several such examples here.

It may be helpful first to recall the motion that results when *any* constant net force $\vec{F}$ acts on an object of mass m. The object's velocity and position as a function of time are solutions to Newton's second law, $\vec{F} = m\vec{a}$, and also involve the initial conditions of the motion. In one dimension, where the acceleration, the velocity, and the position have only x-components, these solutions are

$$\text{for constant force: } x = x_0 + v_0 t + \frac{1}{2}at^2 = x_0 + v_0 t + \frac{1}{2}\frac{F}{m}t^2; \qquad (5\text{–}8)$$

$$\text{for constant force: } v = v_0 + at = v_0 + \frac{F}{m}t. \qquad (5\text{–}9)$$

The constants x_0 and v_0 refer here to the values of position and velocity at time $t = 0$ and are often given or known quantities. For a force with three constant components, we must apply the solution to each component separately. Remember that these formulas apply only when $\vec{F}$ is constant.

EXAMPLE 5–3 A coffee cup of mass 75 g is placed on a slippery (frictionless) ramp tilted at an angle 20° to the horizontal. The coffee cup starts from rest and slides down the ramp. How far down the ramp has the cup moved after 2.0 s?

Setting It Up Figure 5–7a sketches the situation. We label the given mass of the cup as m and the ramp angle as θ. We want the distance x that the cup has moved as a function of time t.

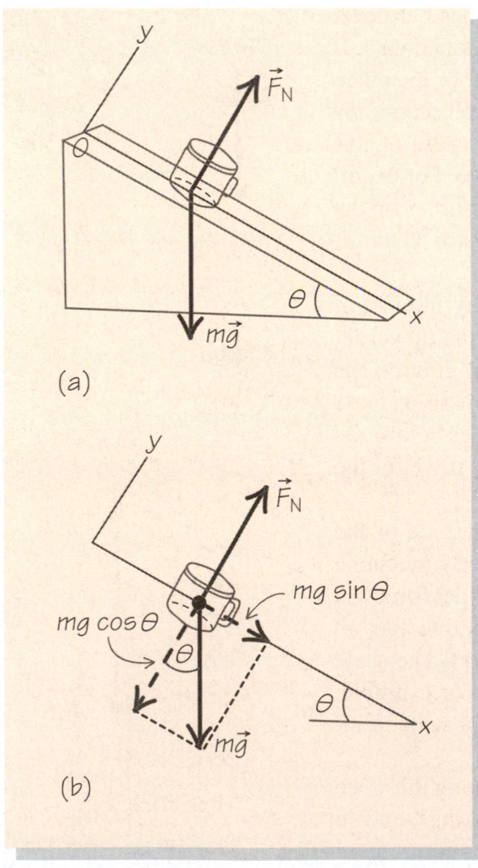

▲ **FIGURE 5–7** (a) A coffee cup on a frictionless, tilted ramp. (b) Free-body diagram for the coffee cup, including the vector decomposition of the forces.

Strategy We first prepare a free-body diagram, choose a convenient set of axes, and decompose the forces into their components along these axes (Fig. 5–7b). We then write Newton's second law, which will determine acceleration. At that point kinematic equations can be used to find the displacement. Because there is no friction, the only forces acting on the cup are gravity $\vec{F}_g$ and the normal force $\vec{F}_N$. Note that $\vec{F}_N$ is perpendicular to the surface of the inclined plane and is *not* oriented in the vertical direction.

The origin is placed at the starting point of the coffee cup. We choose the axes shown in Fig. 5–7b because we know from experience the acceleration $\vec{a}$ of the cup will be along the ramp, with no component perpendicular to the ramp. Accordingly, the x-direction points down along the ramp and the acceleration will have only an x-component a_x, $\vec{a} = a_x\hat{i}$. We must then decompose the forces acting along this set of axes and express Newton's second law. While only the component of forces along x will act to accelerate the cup, the constraint that there is *no* acceleration perpendicular to the surface may still be helpful, and so we will write Newton's law for that direction as well. (Although we may not use this in this example, writing equations for all vector components is a good habit to get into and will be necessary when friction becomes an issue.)

Finally, it is always useful to have a way to check our result. Here we can use the fact that we expect no movement if the plane is horizontal.

Working It Out With our choice of axes, the normal force has only a y-component:

$$\vec{F}_N = F_N\hat{j}.$$

The magnitude F_N is as yet unknown, but we will find it below from the requirement that the acceleration has no y-component. The force of gravity, however, has both x- and y-components (Fig. 5–7b):

$$\vec{F}_g = (mg\sin\theta)\hat{i} - (mg\cos\theta)\hat{j}.$$

Newton's second law has the vector form $\vec{F}_g + \vec{F}_N = m\vec{a}$, or

$$(mg\sin\theta)\hat{i} - (mg\cos\theta)\hat{j} + F_N\hat{j} = ma_x\hat{i}.$$

Each component of this equation is a separate equation of motion:

$$\text{for the } x\text{-component: } mg\sin\theta = ma_x; \qquad (5\text{–}10)$$

$$\text{for the } y\text{-component: } -mg\cos\theta + F_N = 0. \qquad (5\text{–}11)$$

Now we solve the equations of motion. Equation (5–10) gives

$$a_x = g \sin \theta. \qquad (5\text{–}12)$$

Equation (5–11) shows that $\vec{F}_N$ has magnitude $mg \cos \theta$.

To find how the position of the coffee cup changes with time, we use the kinematic equation for constant acceleration in one dimension: Eq. (5–8). In this equation, the cup starts at the origin, so $x_0 = 0$; moreover, $v_0 = 0$ because the cup is initially at rest. We then insert Eq. (5–12) into Eq. (5–8):

$$x = \tfrac{1}{2}a_x t^2 = \tfrac{1}{2}(g \sin \theta)t^2.$$

At $t = 2.0$ s,

$$x = (0.5)(9.8\ \text{m/s}^2)(\sin 20°)(2.0\ \text{s})^2 = 6.7\ \text{m}.$$

We find the cup has moved 6.7 m after 2.0 s. Does this seem far to you? The lack of friction is an important effect! As is typical for problems involving gravity, the answer is independent of the mass of the cup. The angle $\theta = 0°$ presents a special limit; in this case, the plane is horizontal, and there should be no acceleration whatsoever. This is ensured by the $\sin \theta$ factor in the acceleration, which is zero when $\theta = 0°$. In this limit the normal force has magnitude mg, as expected. A second limit is the case $\theta = 90°$, when $\sin \theta = 1$. Here, $a_x = g$ and $F_N = 0$, also as expected.

What Do You Think? It seems unwise to put a coffee cup on an inclined frictionless ramp. What would change in this example if we put other objects—a golf ball or a car—on the ramp?

EXAMPLE 5–4 Because of a wager, a woman wishes to lift a professional football player off his feet. The player is a large interior lineman (a tackle) with a mass of 149 kg. (He weighs 328 lb.) The woman has devised a system for the task, which is shown in Fig. 5–8a. We will assume that all pulleys, ropes, and miscellaneous gear in the apparatus have negligible mass and are frictionless. What is the magnitude of the downward force the woman must exert on the end of the rope in order to lift the lineman?

Setting It Up The given mass of the lineman is M; the mass of the woman is denoted by m. The downward force the woman exerts at the point that the lineman is lifted is just the tension T in the rope, and this is the quantity we want to find.

Strategy Even a smooth lift at constant velocity is sufficient—the lineman does not need to accelerate continuously to be lifted. This is a condition that the net force of the ropes on the lineman balances his weight. The tension in the rope, which is the force that must be supplied by the woman pulling on the rope, is the same throughout the rope.

If we examine Fig. 5–8a, we see that all four rope segments, 1 through 4, pull upward on the lineman. Figure 5–8b illustrates the external forces on the isolated system more clearly. Since it is a single massless rope passing around all the pulleys, the tension in each one is the same, namely T. The free-body diagram for the system is Fig. 5–8c. When the woman begins to pull on the rope, it acquires a tension of magnitude T, and the sum of the tensions $T_1 + T_2 + T_3 + T_4 = 4T$ increases from zero. As long as this sum is less than the weight of the lineman, Mg, he will remain on the ground. But when this sum becomes equal to Mg, there is no net force on the lineman, and any additional tension—no matter how small—will start him accelerating upward. Thus the condition to lift the lineman is

$$4T = Mg.$$

We need only plug in the numbers.

Working It Out We have

$$T = \tfrac{1}{4}Mg = \tfrac{1}{4}(149\ \text{kg})(9.80\ \text{m/s}^2) = 365\ \text{N}.$$

This is the magnitude of the smallest downward force that the woman must apply to her end of the rope. Provided that she weighs more than 365 N (82.1 lb), which corresponds to a mass $m = W/g = 37.3$ kg, she can apply this force just by hanging on the rope. (If she weighs more than 82.1 lb, then she can "partially" hang from the rope, meaning that her feet don't completely lose contact with the ground.)

The arrangement described here is called a *block and tackle* and is used, for example, to enable a single person to lift an engine from a car.

What Do You Think? Why did we make the assumption that all pulleys, ropes, and miscellaneous gear in the apparatus have negligible mass and are frictionless?

(a)

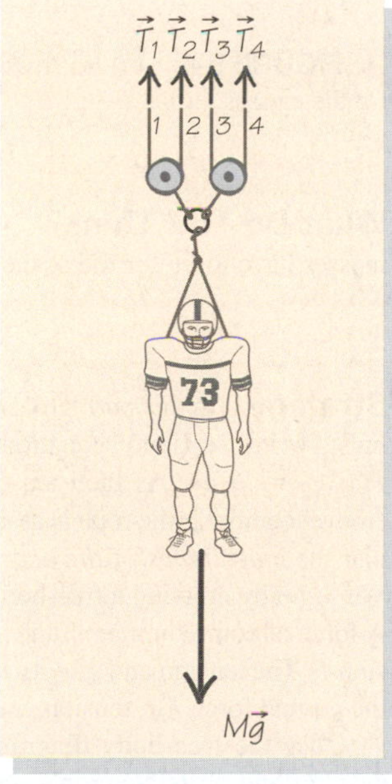

(b)

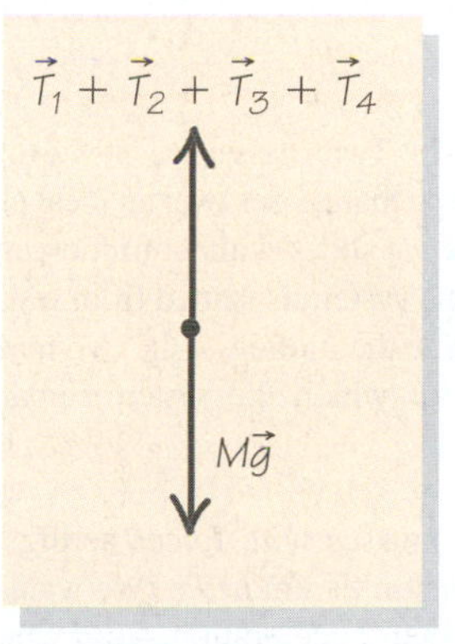

(c)

◀ **FIGURE 5–8** (a) A lineman lifted by a rope via a system of pulleys. (b) The lineman is clearly isolated, with the external forces shown. (c) Free-body diagram for the lineman.

In Examples 5–5 and 5–6 we investigate further all three forces discussed so far in this section—tension, the normal force, and gravity. These examples also introduce something new, in that two masses are involved. Each mass requires an identification of the forces acting on it, a free-body diagram, and an equation of motion.

EXAMPLE 5–5 Masses $m_1 = 1.1$ kg and $m_2 = 2.3$ kg are attached to opposite ends of a massless rope draped over a pulley (Fig. 5–9a). (This device is called an *Atwood machine.*) Mass m_2 rests on a scale that measures the normal force exerted on m_2. There is no motion. What is the reading on the scale, and what is the tension in the rope?

Setting It Up We want the rope tension, magnitude T, and the normal force, magnitude F_N, that the scale exerts on m_2.

Strategy We start with a free-body diagram for each mass (Fig. 5–9b), including a coordinate, here y, and find the equations of motion for each mass. Because there is no acceleration, these equations express the fact that net force on each mass is zero. All three forces—gravity, tension, and normal force—are vertical, so we have only one component (the y-component) to consider and we can drop the vector notation. There are two masses, hence two equations, and these should be enough to determine the two unknown forces.

Working It Out Take the upward direction as positive. The forces acting on m_1 are gravity $(-m_1g)$ and the rope tension (T), and the net force on m_1 is

$$F_{\text{net},1} = -m_1g + T = 0. \qquad (5\text{–}13)$$

Three forces act on mass m_2: $-m_2g, T$, and F_N. Notice that the same value of T acts on each mass, pulling (acting upward) in each case. The net force on m_2 is

$$F_{\text{net},2} = -m_2g + T + F_N = 0. \qquad (5\text{–}14)$$

We then can solve Eqs. (5–13) and (5–14) for the tension and the normal force (the scale reading). Solving Eq. (5–13),

$$T = m_1g = (1.1 \text{ kg})(9.8 \text{ m/s}^2) = 11 \text{ N}.$$

If we insert this value of T into Eq. (5–14), we can solve for the normal force and hence the scale reading:

$$\begin{aligned} F_N &= m_2g - T = m_2g - m_1g = (m_2 - m_1)g \\ &= (2.3 \text{ kg} - 1.1 \text{ kg})(9.8 \text{ m/s}^2) \\ &= (1.2 \text{ kg})(9.8 \text{ m/s}^2) \cong 12 \text{ N}. \end{aligned}$$

For comparison, the weight of m_2 alone is $(2.3 \text{ kg})(9.8 \text{ m/s}^2) = 23$ N.

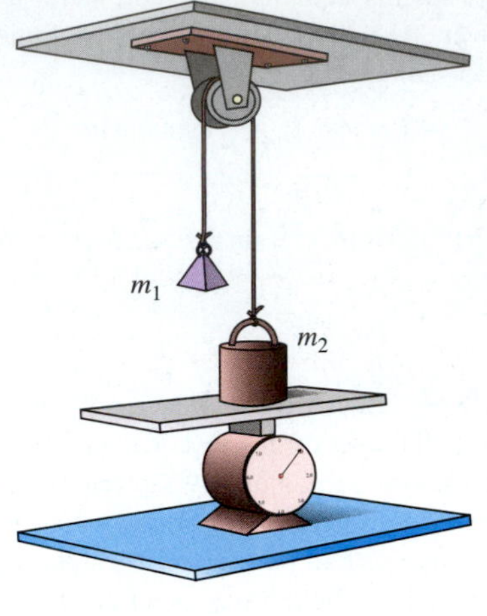

(a)

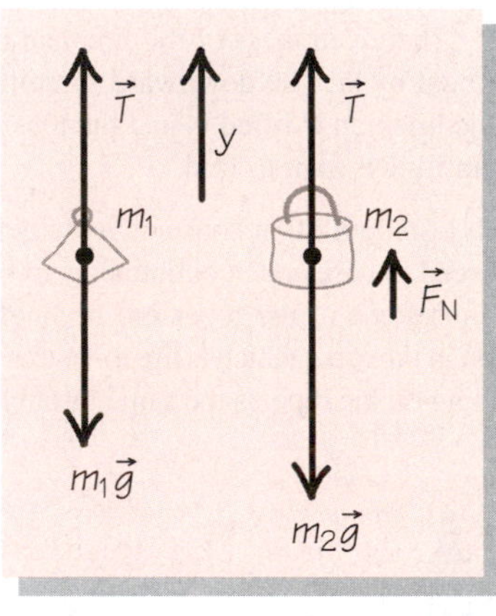

(b)

▲ **FIGURE 5–9** (a) An Atwood machine. (b) Free-body diagrams for the masses.

What Do You Think? What would the relation between the masses have to be for the scale to read zero?

EXAMPLE 5–6 Consider Fig. 5–10: Two masses $m_1 = 1.00$ kg and $m_2 = 2.00$ kg are connected by a rope that passes over an ideal pulley. Mass m_1 hangs straight down, while m_2 slides without friction on a ramp inclined at an angle θ. At $t = 0$, the system is started from rest in the position shown in Fig. 5–10a. Describe the motion of the two masses for $\theta = 25.0°$. Find the angle θ' for which the system remains motionless.

Setting It Up We are given two masses with forces acting on them. One of those forces, the rope tension, is unknown. We want to find the motion of the masses, that is, their acceleration, from which we can work out position. We also want to describe a geometry for which there is no acceleration.

Strategy In this problem, we again have two masses for which we must write separate equations of motion: $\vec{F}_{\text{net on 1}} = m_1\vec{a}_1$ and $\vec{F}_{\text{net on 2}} = m_2\vec{a}_2$. As they are connected by a rope, the two masses "move together" (the rope is assumed to be inextensible), and in particular the *magnitudes of the accelerations* are the same, $a_1 = a_2 = a$. We begin by drawing a free-body diagram for each mass (Fig. 5–10b). A force of common magnitude acts on each mass; this force is the tension $\vec{T}$. The tension and gravity $m_1\vec{g}$ act on m_1. Three forces act on m_2: the normal force $\vec{F}_N$, tension, and gravity $m_2\vec{g}$.

Once the free-body diagrams are drawn, we choose coordinate systems and write Newton's second law. Remember that the coordinate systems we choose are for our convenience and do not affect the result, and we *do not have to choose the same coordinate system*

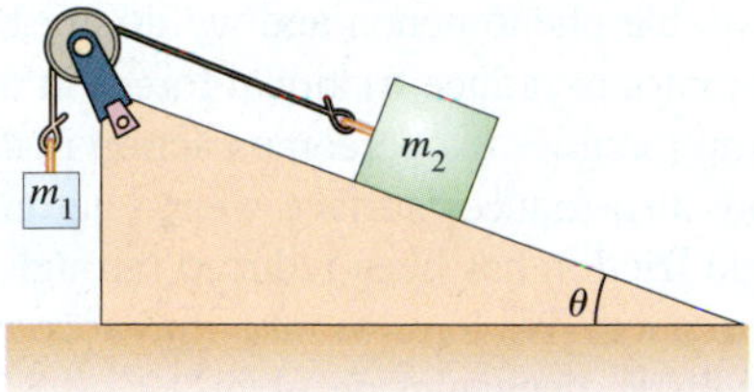

(a)

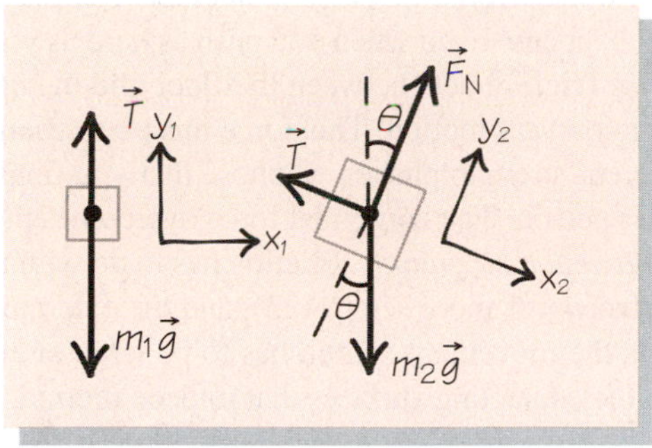

(b)

▲ **FIGURE 5–10** (a) Two masses connected by a rope via a pulley. (b) Free-body diagrams for masses m_1 and m_2.

for the two masses. In fact, choosing different systems makes this problem easier to solve (Fig. 5–10b). We will have to count equations and determine if there are enough to solve for the unknowns, in particular a.

As for finding the angle for which $a = 0$, if we had found enough equations to allow us to solve for a in the first part, we could see whether there is an angle for which $a = 0$. This is much simpler if we leave our expressions in algebraic form, a very good general rule.

Working It Out We choose the origin of each coordinate system to be at the location of the respective mass at $t = 0$. Because m_1 moves only in the vertical direction, labeled y_1, its acceleration $\vec{a}_1$ is aligned with y_1. For m_1, then, we need look only at the component of Newton's second law along y_1,

$$T - m_1 g = m_1 a_1. \qquad (5\text{–}15)$$

Although the mass m_2 has forces acting in two directions, it moves only in the direction labeled x_2 in Fig. 5–10b. Thus $\vec{a}_2$ is aligned with x_2. We must now decompose the forces acting on m_2 into components in this coordinate system. The normal force $\vec{F}_N$ is in the $+y_2$-direction and $\vec{T}$ is in the $-x_2$-direction. The third force, gravity, has two components; we can determine these components by recalling from geometry that the angle θ indicated in Fig. 5–10b is the same as the ramp angle θ in Fig. 5–10a. Then the force of gravity $m_2\vec{g}$ has x_2-component $m_2 g \sin\theta$ and y_2-component $-m_2 g\cos\theta$. Thus Newton's second law for mass m_2, $\vec{F}_{\text{net on }2} = \vec{F}_N + \vec{T} + m_2\vec{g} = m_2\vec{a}_2$, breaks down into two component equations:

$$\text{for the } x_2\text{-component: } -T + m_2 g \sin\theta = m_2 a_2; \qquad (5\text{–}16)$$

$$\text{for the } y_2\text{-component: } F_N - m_2 g\cos\theta = 0. \qquad (5\text{–}17)$$

The three equations (5–15), (5–16), and (5–17) are not enough to solve for the four unknowns T, F_N, a_1, and a_2. The needed fourth equation is the expression that states that the the two masses have accelerations of the same magnitudes,

$$a_1 = a_2 \equiv a. \qquad (5\text{–}18)$$

The single acceleration magnitude a is then substituted in Eqs. (5–15), (5–16), and (5–17), which become three equations for the three unknowns a, T, and F_N. Equation (5–17) gives the normal force,

$$F_N = m_2 g\cos\theta.$$

The tension cancels in the sum of Eqs. (5–15) and (5–16):

$$T - m_1 g - T + m_2 g\sin\theta = m_1 a + m_2 a.$$

We are left with an equation for a with solution

$$a = \frac{m_2\sin\theta - m_1}{m_1 + m_2}g. \qquad (5\text{–}19)$$

Now that we have found the accelerations we can find positions as a function of time by using the constant-acceleration equations (5–8). Each mass starts from rest at the origin of its respective coordinate system, so $v_0 = 0$ for both masses, and both y_{10} ($\equiv y_1$ at $t = 0$) and x_{20} ($\equiv x_2$ at $t = 0$) are zero. Thus Eq. (5–8) gives

$$\text{for } m_1\text{: } y_1 = \tfrac{1}{2}at^2;$$

$$\text{for } m_2\text{: } x_2 = \tfrac{1}{2}at^2.$$

The masses move only in these directions. Together with Eq. (5–19), these equations describe the motion fully.

For the case $\theta = 25.0°$, we use $\sin 25.0° = 0.423$, so

$$a = \frac{(2.00\,\cancel{\text{kg}})(0.423) - 1.00\,\cancel{\text{kg}}}{1.00\,\cancel{\text{kg}} + 2.00\,\cancel{\text{kg}}}(9.80\text{ m/s}^2) = -0.503\text{ m/s}^2.$$

Thus $y_1 = (-0.503\text{ m/s}^2)t^2/2 = x_2$. Note the minus sign in the acceleration: The sign indicates that m_1 drops and m_2 moves up the ramp. Equation (5–19) shows that the acceleration—including its sign—depends on the masses and the ramp angle.

Finally, to find the angle for which the acceleration is zero, we note that, according to Eq. (5–19), the acceleration is zero at an angle θ' for which

$$m_2\sin\theta' - m_1 = 0, \qquad (5\text{–}20)$$

or $\sin\theta' = m_1/m_2$. For this particular problem the forces will balance and acceleration will be zero for $\sin\theta' = (1.00\text{ kg})/(2.00\text{ kg})$, or $\theta' = 30.0°$.

What Do You Think? Consider the system at the angle $\theta' = 30.0°$. (a) What happens for masses $m_1 = 3.0$ kg and $m_2 = 6.0$ kg? (b) What happens for masses $m_1 = 4.0$ kg and $m_2 = 6.0$ kg? (c) What happens for masses $m_1 = 3.0$ kg and $m_2 = 5.0$ kg?

5–2 Friction

Friction is a familiar concept. It is a contact force that impedes sliding, and we experience it in all aspects of our lives. Sometimes friction is useful to us: It is friction that holds nails and screws in place (Fig. 5–11); if there were no friction between our feet and the ground, we could not walk, and if there were no friction between the wheels of a car and the road, the engine would cause the wheels to spin but there would be no forward or backward

▲ **FIGURE 5–11** Nails are held in place by the force of friction, which can be quite substantial.

motion. In other situations friction is not a desirable phenomenon and we do our best to minimize it. Even with the oil added to a car's motor to reduce frictional forces in the engine, as much as 20 percent of gasoline consumption goes to overcome friction in the engine. Lubrication reduces friction and therefore also reduces surface wear—automobile engines now tend to last longer because internal friction has been reduced through more precise manufacturing and more effective lubrication. We have to take into account the forces of friction if we are to understand any realistic mechanics problem.

Static and Kinetic Friction

Suppose you want to slide a crate full of books from one place to another. You push on it with a small horizontal force but nothing happens. Even when you push as hard as you can, the crate does not move. Why not? **Static friction** acts between the floor and the crate in the absence of motion in such a way as to *prevent* motion. This force must be variable because it balances each of your own different-strength pushes. Suppose that you finally get the crate moving with the help of another person. The combined force overcomes the static friction because *static friction has a maximum magnitude*. Friction has to do with the interlocking of microscopically rough surfaces (for more see p. 133), and the external force that overcomes the interlocking and gets the movement started has to be large enough to bend or break the tiny protuberances on the contacting surfaces that impede motion.

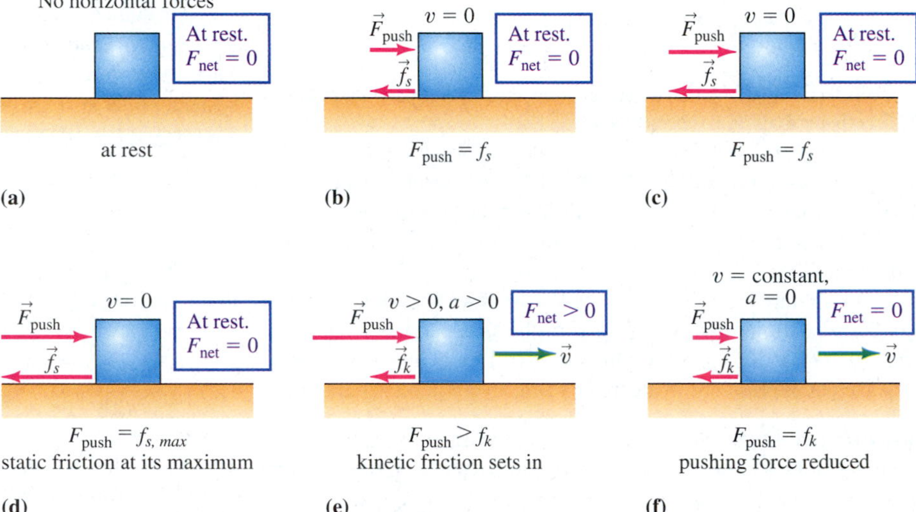

▶ **FIGURE 5–12** Sequence in which an increasing pushing force is opposed by an equal and opposite static friction force. Static friction can increase only to a certain point, after which the crate accelerates. Kinetic friction, which applies when the crate actually moves, is smaller than the maximum size of static friction, so it takes less pushing force to move the crate once its motion has begun.

Once the crate is moving, it is easier to keep it moving at a constant speed. There is still friction opposing your push, but it is now **kinetic** (or **sliding**) **friction**, that is, friction associated with motion. Experiment shows that the magnitude of kinetic friction is smaller than the maximum value of static friction. The entire sequence of getting the crate started and keeping it moving is illustrated in Fig. 5–12. Static friction acts in a direction *opposite* to the component of an applied force *along* the surface; sliding friction acts *opposite* to the direction of the velocity of a sliding object at its point or points of contact. In each case the friction force is parallel to the surface (Fig. 5–13a). Figure 5–13b illustrates the magnitude of friction when an object such as the crate starts from rest and is pushed with a steadily increasing external force. The magnitude of static friction increases as the external force increases until the "break point" arrives, at which point kinetic friction enters. This force does not vary with the external pushing force. When the applied force decreases to zero, kinetic friction continues to act until the body stops. At that point we revert to the static situation.

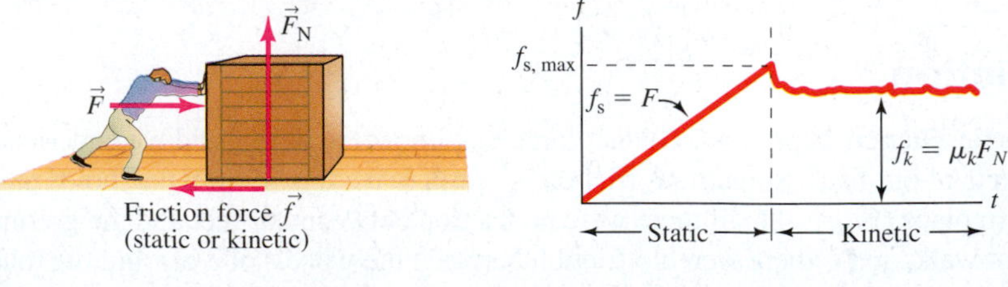

▶ **FIGURE 5–13** (a) Friction opposes the motion of the crate. (b) With a steadily increasing push, static friction will increase in magnitude until the crate starts to move, at which point kinetic friction, whose magnitude does not vary with the pushing force, takes over.

CONCEPTUAL EXAMPLE 5–7 In the discussion above we stated that a large enough force has to be applied to overcome static friction, and the applied force must be large enough to bend or break the small protuberances that keep a surface from being perfectly flat. Given the fact that measurements of the coefficient of static friction between a block of wood and a table give the same answer after many experiments, would you argue in favor of bending or breaking the surface imperfections?

Answer If static friction were due primarily to obstacles that have to be broken in order to facilitate motion, then after a few experiments all the obstacles would have been broken off, and static friction would decrease significantly. Because this does not happen, the bending of deformations must be the primary source of friction. We should add that if you actively smooth a surface by, for example, sanding it, then you are presumably doing some breaking of the protuberances on it.

Quantitative Properties of Friction

Quantitative tests on friction were made by Leonardo da Vinci some 200 years before Newton's work on dynamics. Leonardo experimented with a set of blocks of varying sizes sliding on table tops and discovered some surprising facts. He found that both static and kinetic friction are independent of the surface area of the blocks in contact with the table top. Moreover, both *static and kinetic friction are proportional to the magnitude F_N of the normal force exerted by the table top on the blocks*. The experiments that led Leonardo to his conclusions are quite simple: Take a given block and turn it so that faces of different areas are in contact with the table top. The friction force on the block is the same no matter what face is down.

The proportionality constant that relates the friction force and the normal force is the **coefficient of friction** μ. This (positive dimensionless) constant is determined experimentally. As the maximum value of static friction is generally not equal to the force of kinetic friction, we distinguish two coefficients: μ_s for static friction and μ_k for kinetic friction.

If we write the force of static friction as $\vec{f}_s$ and that of kinetic friction as $\vec{f}_k$, their magnitudes are given by

$$0 \le f_s \le \mu_s F_N; \tag{5–21}$$

STATIC FRICTION

$$f_k = \mu_k F_N. \tag{5–22}$$

KINETIC FRICTION

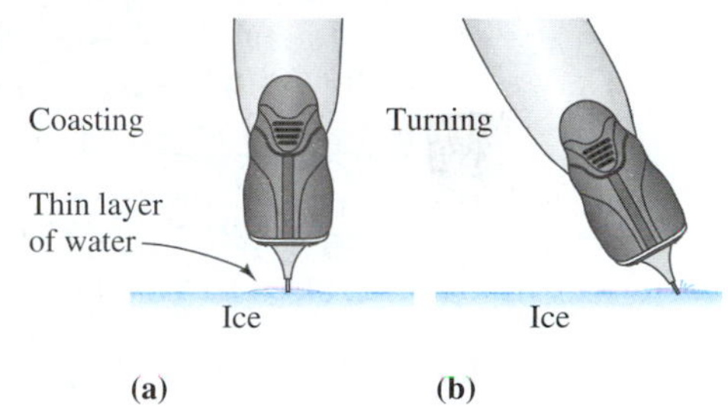

▲ **FIGURE 5–14** (a) A layer of water lubricates and decreases the action of kinetic friction for an ice skater. (b) By manipulating the blade of the skate, a skater makes static friction act in the sideways direction and can make a turn.

Equation (5–21) expresses a range because, as we have described above, static friction takes a value that depends on the external conditions. The experimental fact that the maximum value of static friction exceeds kinetic friction implies the inequality

$$\mu_s > \mu_k. \tag{5–23}$$

You experience this when you have to exert a greater force to get a crate of books moving than you have to exert to keep it moving. We also make the assumption, reasonably well satisfied by experiment, that *μ_k is independent of the relative speed of the two surfaces.*

The coefficients of friction depend on the two surfaces involved. We know from everyday experience that a basketball shoe on a basketball court involves a larger coefficient of friction than does the blade of an ice skate on a frozen lake. A lubricating material—such as sweat—between the basketball shoe and the court will drastically reduce the coefficient of friction. In ice skating, the lubricating material is a layer of liquid water between blade and ice (Fig. 5–14a); an ice skater can make static friction take over by adroit use of the blade—in Fig. 5–14b the skater is pushing off or making a turn by digging into the ice rather than gliding over it. Table 5–1 shows some typical values of coefficients of static friction. The materials listed are generally unlubricated ("dry"). Coefficients of kinetic friction can be anywhere from roughly 25 percent to 100 percent of the corresponding coefficients of static friction.

The values in Table 5–1 are meant only to be indicative—the coefficients of friction are sharply dependent on such things as the cleanliness of the surfaces, their roughness, and so forth. Two very rough objects may have a large coefficient of friction that can be reduced once the objects are smoothed. But if two objects of the same material

TABLE 5–1 • Some Coefficients of Static Friction

Materials	μ_s
Automobile brake shoes on a brake drum	1.2
Dry tire on dry asphalt	1.0
Hard steel on hard steel	0.8
Oak on oak, parallel to the grain	0.6
Book on a table	0.3
Wet tire on wet asphalt	0.2
Ice on wood	0.05
Teflon on steel	0.04

are smoothed too much and are free of dirt and oxidation as well, the coefficient of friction may rise virtually to infinity because the surfaces weld together! In effect, the molecules at the surface between the two objects can interact just as they do at the interior, and the two objects become one.

Example 5–8 illustrates one method of measuring the coefficient of static friction.

EXAMPLE 5–8 A box of mass m is set at rest on a horizontal surface; there is friction between the box and the surface. The surface is slowly raised at one side so that it becomes a ramp, making an angle θ with the horizontal. By analyzing the forces on the box, find the critical ramp angle θ_c at which the box will start to slide.

Setting It Up Figure 5–15a illustrates the situation. We include in the figure a coordinate system that simplifies the solution: The $+x$-direction is oriented down along the ramp and the y-direction is perpendicular to the ramp. The origin is placed at the initial position of the mass.

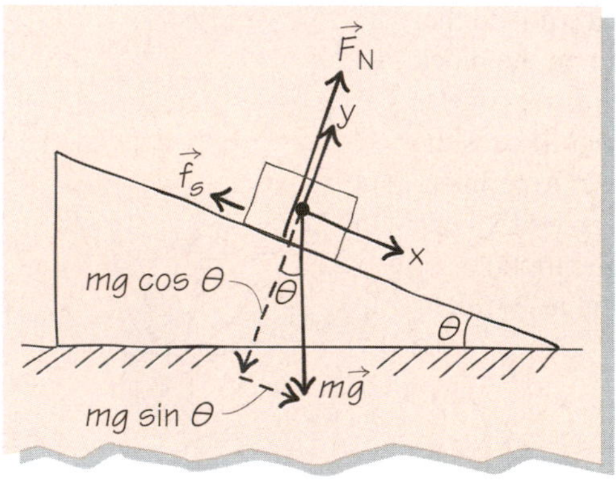

(a)

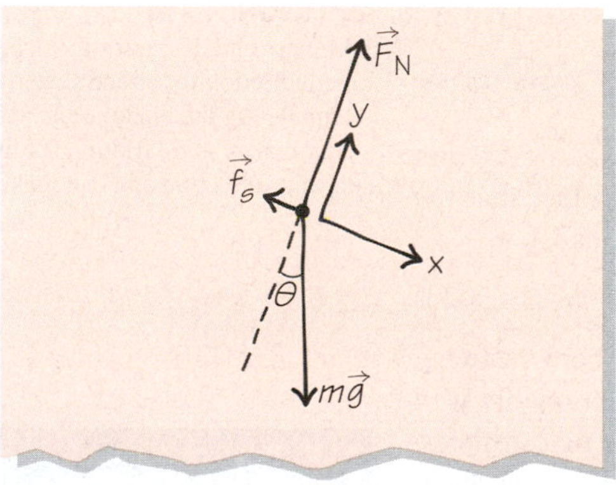

(b)

▲ **FIGURE 5–15** (a) Box of mass m experiencing a friction force on a ramp. The ramp angle θ at which the mass starts to slide yields the coefficient of static friction. (b) Free-body diagram for the box.

Strategy The free-body diagram is Fig. 5–15b. The three forces acting on the box before it starts to move are the force of gravity (vertically down), the normal force $\vec{F}_N$ (perpendicular to the surface), and the force of static friction, $\vec{f}_s$ (up the ramp). The friction force opposes the motion that would take place if there were no friction. The magnitude of the force of static friction is determined by the fact that the mass remains motionless for a sufficiently shallow ramp angle. The maximum value of static friction depends on F_N [see Eq. (5–21)], so we must also find its value. We can use the fact that there is no acceleration perpendicular to the ramp to find F_N.

As the ramp angle increases, the components of the normal and gravity forces along and perpendicular to the ramp change. As long as the box is not slipping (there is no net force—we refer to the box as being in *static equilibrium*), the friction force (along the ramp) is static friction. Once the value of f_s reaches its maximum value $\mu_s F_N$, however, the box will slip, and this determines the critical value of the ramp angle θ_c.

Working It Out We start with the box stationary, so that Newton's second law with zero acceleration reads $\vec{F}_g + \vec{F}_N + \vec{F}_s = 0$, or

$$(mg\sin\theta)\hat{i} - (mg\cos\theta)\hat{j} + F_N\hat{j} - f_s\hat{i} = 0.$$

In component form, the equations of motion are

$$\text{for the } x\text{-component: } mg\sin\theta - f_s = 0; \tag{5–24a}$$

$$\text{for the } y\text{-component: } F_N - mg\cos\theta = 0. \tag{5–24b}$$

The x-component equation relates the force of static friction to the force of gravity,

$$f_s = mg\sin\theta. \tag{5–25}$$

The y-component equation determines F_N as a function of θ:

$$F_N = mg\cos\theta. \tag{5–26}$$

As θ (and $\sin\theta$) increases, the force of friction from Eq. (5–25) that is needed to hold the box in static equilibrium also increases. Eventually, static friction reaches its maximum value, $\mu_s F_N$. Beyond that point, the box will start to slide. Setting static friction to its maximum value, $f_s = \mu_s F_N$, Eq. (5–25) determines the critical angle θ_c at which the box starts to slide. Equation (5–25) then becomes

$$\mu_s F_N = mg\sin\theta_c,$$

or, from Eq. (5–26),

$$\mu_s mg\cos\theta_c = mg\sin\theta_c.$$

Cancel the factor mg from this equation to find θ_c:

$$\frac{\sin\theta_c}{\cos\theta_c} = \tan\theta_c = \mu_s. \tag{5–27}$$

This equation tells us that if we measure the angle at which the box begins to slip, we measure μ_s, and this is in fact a useful way to determine coefficients of static friction between surfaces.

What Do You Think? Imagine that a person pushes down on the box in a direction opposite to the normal force. How does this change the frictional force? What happens to the critical angle?

Example 5–8 suggests a related experiment that will measure the coefficient of kinetic friction. Once the box begins to slip, kinetic friction acts, given by $f_k = \mu_k F_N = \mu_k mg\cos\theta$. If we use Newton's second law and apply it along the x-direction, then instead of Eq. (5–24a), we find

$$mg \sin \theta - \mu_k mg \cos \theta = ma_x. \qquad (5\text{–}28)$$

Because μ_k is smaller than μ_s, the ramp can be lowered back down, decreasing θ while the mass is still sliding, and the box will continue to slide. There is a second critical value of θ—call it θ'_c—for which the forces in Eq. (5–28) cancel and the object no longer accelerates. Instead, the object slides at constant velocity. This critical angle is given by

$$\frac{\sin \theta'_c}{\cos \theta'_c} = \tan \theta'_c = \mu_k. \qquad (5\text{–}29)$$

Thus θ'_c measures the coefficient of kinetic friction.

EXAMPLE 5–9 A professor with a light eraser (assume massless) in her hand leans against a blackboard. Her straight arm makes an angle of 60° with the horizontal, and the force $\vec{F}_{\text{prof}}$ exerted by her arm on the eraser has magnitude $F_{\text{prof}} = 50$ N. The coefficient of static friction between the eraser and the blackboard is $\mu_s = 0.15$. Does the eraser slip?

Setting It Up We sketch the situation in Fig. 5–16b, and we can use this figure as a free-body diagram for the eraser. We choose a coordinate system with x into the board and y vertically upward. We are interested in finding the force of friction between the board and the eraser. If that value exceeds the maximum value of static friction, the eraser will slip.

Strategy After drawing the free-body diagram to allow us to pick out the forces acting, we apply Newton's law of motion to the eraser under the assumption that it is not slipping. The components of that equation should allow us to solve for the magnitude of the friction force to find out if it exceeds its maximum value.

Working It Out If there were no friction (a perfectly slippery board), the eraser would slide up. Therefore, the static friction force must be down, in the $-y$-direction. As long as there's no acceleration, Newton's first law applies to the eraser, $\vec{F}_N + \vec{f}_s + \vec{F}_{\text{prof}} = 0$, and we have

$$-F_N \hat{i} - f_s \hat{j} + (F_{\text{prof}} \cos \theta)\hat{i} + (F_{\text{prof}} \sin \theta)\hat{j} = 0.$$

The two component equations are (θ is the angle with respect to x, here 60°)

$$\text{for the } x\text{-component: } -F_N + F_{\text{prof}} \cos \theta = 0;$$

$$\text{for the } y\text{-component: } -f_s + F_{\text{prof}} \sin \theta = 0.$$

The x-component equation determines F_N from the requirement that it balances the horizontal component of the professor's force $\vec{F}_{\text{prof}}$. Once we have found that $F_N = F_{\text{prof}} \cos \theta$, we can determine the *maximum* value of static friction, $f_{s,\text{max}} = \mu_s F_N = \mu_s F_{\text{prof}} \cos \theta$. When this maximum value is exceeded, the eraser begins to slip. Thus, when we substitute the maximum value of static friction into the y-component equation, we find a condition for the critical angle θ_c for which the eraser begins to slip:

$$-\mu_s F_{\text{prof}} \cos \theta_c + F_{\text{prof}} \sin \theta_c = 0;$$

$$\frac{\sin \theta_c}{\cos \theta_c} = \tan \theta_c = \mu_s.$$

Note the striking feature that the critical angle at which the eraser starts to slip is independent of the force the professor exerts! Numerical substitution yields $\tan \theta_c = 0.15$, or $\theta_c = 8.5°$. This angle is less than the 60° angle made by the arm, so the eraser indeed slips.

What Do You Think? Suppose that the eraser were not massless. Is the critical angle θ_c still independent of the professor's force?

(a)

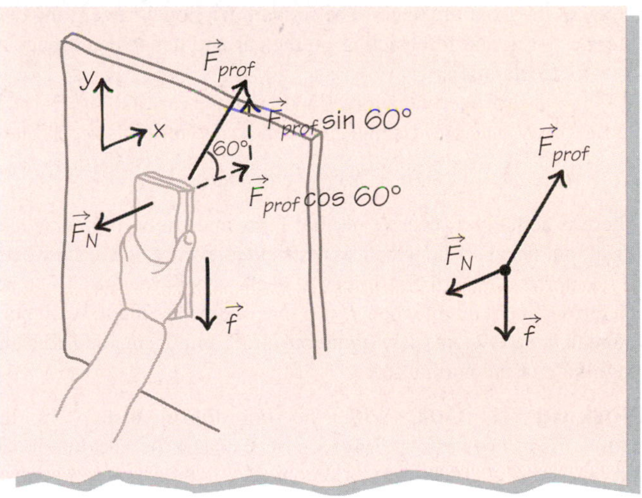

(b)

▲ **FIGURE 5–16** (a) Professor erases the blackboard. (b) Sketch of forces and free-body diagram for the eraser.

THINK ABOUT THIS . . .

FRICTION ACTS BACKWARD—HOW CAN IT ALLOW US TO WALK FORWARD?

▲ **FIGURE 5–17** A runner uses the force of static friction between shoe and ground to accelerate forward.

We have stated that friction acts against motion, but at the start of the chapter we also stated that the presence of friction allows us to walk and cars to accelerate. Friction can act to produce a positive acceleration on us. How is this possible? The answer lies in the fact that people are extended and flexible systems and for such systems static friction can act *forward on the system as a whole*, even though it will always act to *impede motion at the point of contact*. Walking or running is perhaps the most familiar example of this phenomenon (Fig. 5–17). The runner is exercising a muscle that pushes the contact foot backward, and without friction the motion of the foot at the point of contact would therefore be backward. (You can easily test this by trying to walk on a frozen puddle.) But friction opposes the backward motion at the point of contact and therefore is a force that acts in the *forward* direction on the runner's foot. With sufficient stiffness in the runner, this forward friction force acts on the entire system to move her forward. By Newton's third law, there is a corresponding backward-directed force on Earth. An automobile similarly moves forward because of static friction, as we describe more quantitatively in Example 5–10.

EXAMPLE 5–10 An automobile with four-wheel drive and a powerful engine has a mass of 1000 kg. Its weight is evenly distributed on its four wheels, whose coefficient of static friction with the dry road is 0.8. If the car starts from rest on a horizontal surface, what is the greatest forward acceleration that it can attain without spinning its wheels?

Setting It Up The situation is illustrated in Fig. 5–18a. The car mass is m and the coefficient of static friction between the wheels and the road is μ_s. The unknowns are the car's acceleration, magnitude a, and in particular maximum magnitude a_{max}.

Strategy It is static friction, magnitude f_s, that accelerates the car forward, and the automobile has its greatest forward acceleration when f_s is at its maximum. The engine of the car creates the rotational motion shown in Figure 5–18a in the wheels. If there were no friction between the wheels and the road, the wheels would simply spin. When there is no slipping, it is *static* friction that acts, and μ_s, not μ_k, enters the problem. The motion of the tires at the point of contact with the road is to the rear in the absence of friction. The force of friction opposes this rearward motion, so the direction of $\vec{F}_s$ is toward the front of the car. The forward frictional force is the *only* external horizontal force acting on the car and hence determines its forward acceleration.

Figure 5–18b shows the forces acting on the car, and Fig. 5–18c is the free-body diagram. Using it, we can write Newton's second law:

$$\vec{F}_N + m\vec{g} + \vec{f}_s = m\vec{a}.$$

(There is actually a separate normal force and friction force at each wheel, but because the weight is distributed evenly over each wheel, all the normal and friction forces at each wheel are equal.) The acceleration is maximum when f_s is a maximum. To find it, we'll need to break the Newton's law equation into components and find the magnitude of the normal force.

Working It Out With the coordinate system in the figure, $\vec{F}_N = F_N\hat{j}$, $m\vec{g} = -mg\hat{j}$, $\vec{f}_s = f_s\hat{i}$, and the acceleration is forward, so $\vec{a} = a\hat{i}$. Newton's second law is then

$$F_N\hat{j} - mg\hat{j} + f_s\hat{i} = ma\hat{i},$$

and the component equations are

$$\text{for the } x\text{-component: } f_s = ma;$$
$$\text{for the } y\text{-component: } N - mg = 0.$$

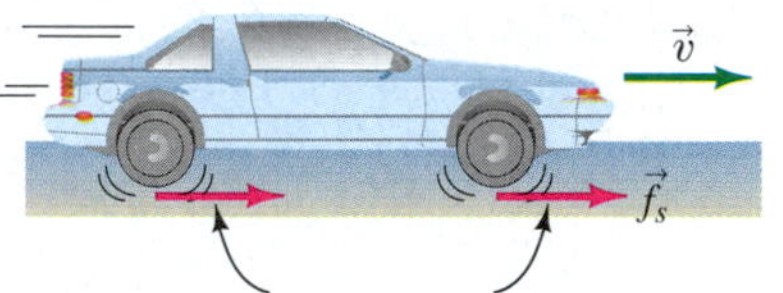

Force of static friction on rear and front tires. The same force acts on the left side tires.

(a)

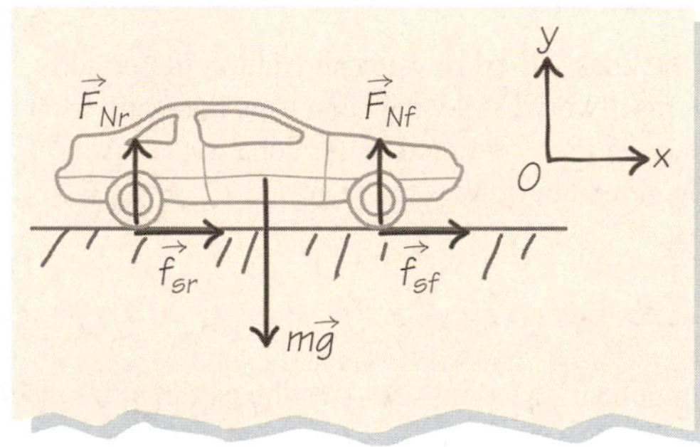

(b)

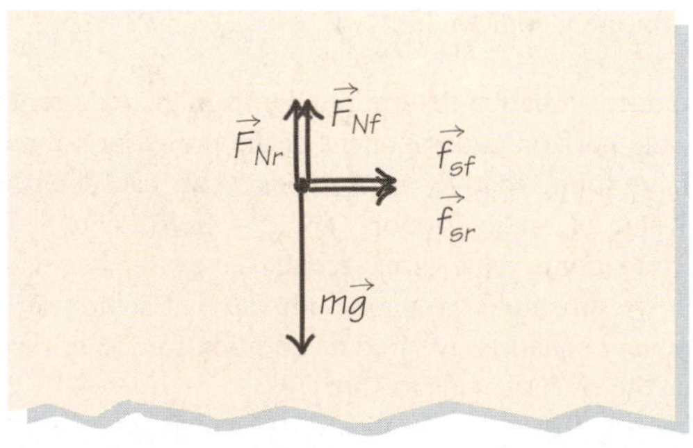

(c)

▲ **FIGURE 5–18** (a) A car accelerates forward under the influence of static friction. (b) Forces acting. (c) Free-body diagram for the car.

We find a_{max} when static friction is at its maximum value $\mu_s F_N = \mu_s mg$, where we have used $F_N = mg$ from the y-component equation. Thus

$$\mu_s mg = ma_{max}.$$

The mass of the car cancels out of this expression, leaving

$$a_{max} = \mu_s g = (0.8)(9.8\ \text{m/s}^2) \cong 8\ \text{m/s}^2.$$

This is quite a significant forward acceleration. Note that as μ_s decreases, the maximum acceleration decreases; in other words, when μ_s is zero, the automobile can only spin its wheels.

What Do You Think? Cars with front-wheel drive perform better in snow than those with rear-wheel drive. Why do you suppose this is?

THINK ABOUT THIS . . .

WHERE DOES FRICTION COME FROM?

To understand the origins of friction, let's look closely at two surfaces that rub against one another. Figure 5–19 shows a microscopic view of two such surfaces. Because of the hills and valleys present on any rough surface—and all surfaces are rough when viewed closely enough—the area on two surfaces that actually touch together is a small fraction of the area that appears to be in contact. Friction forces are due to three major effects: the interlocking of surface irregularities, the attraction between the contact points due to forces between the molecules of the two objects (the objects "adhere"), and the "plowing out" of softer materials by harder ones. The coefficient of static friction can be greater than that of kinetic friction because the materials have a longer time to "settle in" together.

This description helps us understand why the friction force is independent of the apparent surface area that is in contact while it is dependent on the normal force. The normal force is a measure of how strongly the two surfaces are pressed together; when the normal force is large, the two surfaces are pressed strongly together. The rough surfaces shown in Fig. 5–19 mesh more closely when pressed strongly together, and the *actual* surface area in contact increases. In fact, the normal force is a good measure of the actual surface contact area. Whether we place the broad side or the narrow end of a brick on a table, approximately the same surface areas are in actual contact, even if the apparent contact areas are vastly different. So the friction force comes from the interaction of the two surfaces at the atomic level and is proportional to the real area that is in contact at this microscopic level.

The study of friction, wear, and lubrication is called *tribology*, a subject of obvious importance. Despite much effort, a truly fundamental understanding of friction remains elusive. The discovery of Teflon™—a very slippery coating material that you have likely seen in frying pans—was a happy accident, not the result of a planned development program. The discoverer has stated that he was lucky he was not blown up in the process.

(a)

(b)

▲ **FIGURE 5–19** If we use both of the figures chosen: (a) The two surfaces appear smooth, (b) but a microscopic view of the contact region reveals rough surfaces.

5–3 Drag Forces

A spoon dropping through molasses, an automobile moving at highway speeds, and the space shuttle using a parachute to slow down during landing are all subject to a substantial *drag force*, which is a resistive force somewhat like friction. Drag forces act like sliding friction in that *they act in a direction opposite to that of the motion*, but they differ from sliding friction in two ways: They depend on the *speed* v of the object that is moving through the medium and there is no equivalent to a normal force to set their magnitude.

In many everyday situations, an automobile moving on a highway, for example, the drag force $\vec{F}_D$ is found by experiment to have magnitude

$$F_D = \tfrac{1}{2}\rho A C_D v^2, \qquad (5\text{–}30)$$

where ρ is the mass density (the mass per unit volume) of the medium through which the object moves, A is the maximum cross-sectional area presented by the moving object, and C_D is the *drag coefficient*. The drag coefficient is dimensionless and depends on the object's shape. A highly streamlined object might have a drag coefficient as small as 0.1, whereas a particularly awkward shape will have a drag coefficient greater than 1. The most streamlined automobiles have drag coefficients around 0.25.

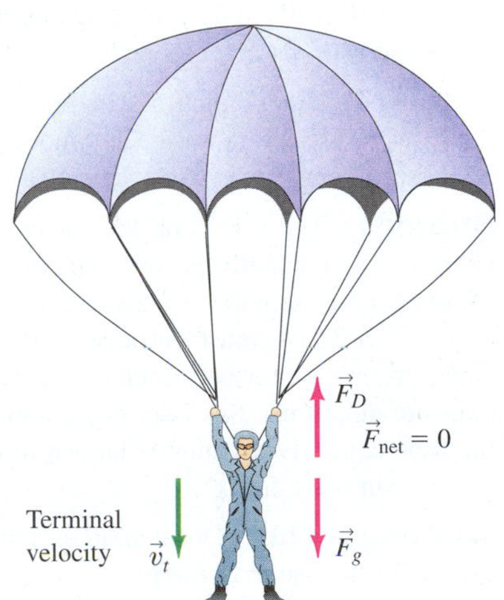

▲ **FIGURE 5–20** When the drag force and the force of gravity acting on a parachutist are equal and opposite, the parachutist has reached his terminal speed.

Terminal Speed

The fact that the drag force on an object increases with the speed has an important consequence. Consider a parachutist falling through the air (Fig. 5–20) and acted upon by both gravity and a drag force, as described by Eq. (5–30). When the parachutist first starts to fall, his speed is slow, so the drag force is small and there is an acceleration $\vec{g}$ due to gravity. As the parachutist's speed increases, so does the drag; at some point the drag will be equal and opposite to the force of gravity. Then the parachutist no longer accelerates and his speed remains constant at a maximum value. The maximum speed v_t is called the *terminal speed.* Some typical values are 5 m/s for a parachutist, 8 m/s for a falling raindrop, 40 m/s for a baseball, and 60 m/s for a skydiver (parachutist with unopened parachute).

Let's calculate the terminal velocity in the case of falling objects, where only the force of gravity and the drag force [Eq. (5–30)] act. The terminal speed is found from the condition that the forces of gravity and drag are equal and opposite: $\vec{F}_g = -\vec{F}_D$, or in terms of magnitudes

$$mg = \tfrac{1}{2}\rho A C_D v_t^2 .$$

We solve for the terminal speed:

$$v_t = \sqrt{\frac{2mg}{\rho A C_D}}. \qquad (5\text{–}31)$$

CONCEPTUAL EXAMPLE 5–11 Consider a skydiver falling through the atmosphere. How can he change his terminal speed? When several skydivers jump out of the same airplane and want to connect together, how do they all get to the same location (Fig. 5–21)?

Answer The terminal speed is attained when the magnitude of the drag force equals that of gravity. The skydiver cannot change the force of gravity on himself, but he can change the drag force and subsequent terminal speed by varying the area A and the drag coefficient C_D [see Eq. (5–31)]. Thus skydivers can make themselves compact to increase their speed through the air or spread their arms and legs to slow down. Skydivers can even change direction slightly by varying their drag characteristics, and their clothes may be loose so that they can be "deployed." For many skydivers to meet up together and form a chain or circle, some skydivers have to increase their terminal speeds and others must decrease theirs.

▲ **FIGURE 5–21** These skydivers had to change their terminal speeds and directions in order to arrive at the same location at the same time and link up.

EXAMPLE 5–12 The maximum force with which a certain automobile engine can accelerate a car is 3200 N. The density of air is about 1.2 kg/m^3, the cross-sectional area of the automobile is 3.4 m^2, and the drag coefficient is 0.50. Assuming that the drag force of Eq. (5–30) is the only force resisting the motion, what is the maximum speed of the automobile?

Setting It Up Figure 5–22a shows the forces, with Fig. 5–22b the free-body diagram for the automobile. The term v_t labels the terminal speed and that is what we wish to find.

Strategy The forces on the car are $\vec{F}_{\text{engine}}$ and $\vec{F}_{\text{drag}}$. (Friction from the road actually propels the car forward, but this occurs because of forces provided by the engine, so we will refer to this force as $\vec{F}_{\text{engine}}$.) These must be equal and opposite at terminal speed, when there is no further acceleration. Since the two forces point in opposite directions, this is in fact a condition that the force magnitudes be equal. Everything is known in the condition except for v_t, and we can solve for it.

Working It Out Our condition reads $F_{\text{engine}} = F_{\text{drag}}$, or, using Eq. (5–30) for the drag force,

$$F_{\text{engine}} = \tfrac{1}{2}\rho A C_D v_t^2 .$$

We solve this equation for the terminal speed squared:

$$v_t^2 = \frac{2F_{\text{engine}}}{\rho A C_D} = \frac{2(3200\ \text{N})}{(1.2\ \text{kg/m}^3)(3.4\ \text{m}^2)(0.50)} = 3.1 \times 10^3\ \text{m}^2/\text{s}^2.$$

The terminal speed of the automobile is then

$$v_t = 56\ \text{m/s}, \qquad \text{or } 125\ \text{mi/h}.$$

What Do You Think? In what ways would reducing the drag coefficient of automobiles be useful? Have drag coefficients for cars changed over the years?

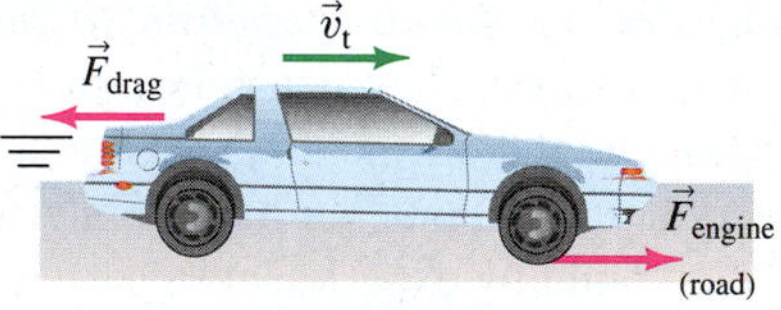

(a)

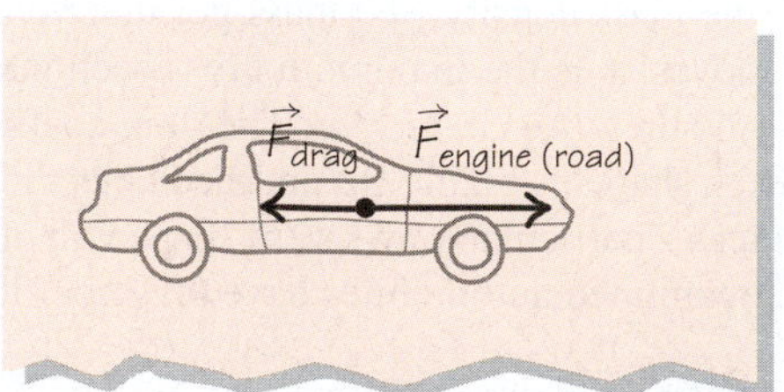

▲ **FIGURE 5–22** (a) Car moving on road. (b) Free-body diagram for an automobile that moves under the influence of the force provided by the engine and the drag of the air through which it moves.

In Example 5–12, air drag was used to determine the maximum speed of an automobile. Road friction, sometimes referred to as rolling friction, also plays an important role in opposing the motion of an automobile. Road friction is fairly constant over a large range of speeds and, as a rule of thumb, road friction has an effect equal to that of drag at about 40 mi/h. As the speed increases, the relative size of drag compared to road friction increases.

A Better Approximation to Drag Force

Precise experiments reveal that the force law for the drag force depends in a more complicated way on the medium through which an object moves and on the shape and size of the moving object than Eq. (5–30) suggests. In fact, at low speeds the drag force is proportional to v, not v^2. It is only after the speed has increased enough to produce turbulence in the medium that the v^2 term comes in. Thus, a better approximation to the drag force magnitude than Eq. (5–30) is

$$F_D = bv + cv^2. \tag{5–32}$$

The coefficients b and c contain information on the shape of the moving object as well as on the medium in which it moves. The first term always dominates for sufficiently low speeds, while the second term dominates for higher speeds.

EXAMPLE 5–13 A marble of mass 5.0 g falls into a jar of oil. The drag force on the marble is given by Eq. (5–32) with $b = 0.20$ kg/s and $c = 0.10$ kg/m. (These values are typical for real fluids such as oil and for an object the size and shape of a marble.) Find the value of the speed for which the two terms in the drag force are equal. Which of the two terms is dominant when the drag force is comparable to the force of gravity on the marble? Will the two terms ever be equal as the marble falls through the oil?

Setting It Up Figure 5–23 shows the progress of the marble as it falls through the fluid.

Strategy We need only evaluate terms in Eq. (5–32) for this situation, and this equation will also show which drag force is dominant. As before, we can recognize that the terminal speed is reached when the overall drag force matches the force of gravity. This speed may be less than or greater than the speed for which the two terms in the drag force equation are comparable. If the speed is less, then the b term always dominates in the drag; if it is greater, then the c term will dominate near the terminal speed.

Working It Out Let the value of the speed for which the two terms in the drag force are equal be v'. Then v' is determined by

$$bv' = cv'^2.$$

This equation is solved by

$$v' = \frac{b}{c} = \frac{0.20\ \text{kg/s}}{0.10\ \text{kg/m}} = 2.0\ \text{m/s}.$$

At terminal speed, the magnitudes of the force of gravity and the drag force are equal; that is, $F_D = mg$. For now, let's utilize only the first term in the drag, bv, $bv = mg$:

$$v = \frac{mg}{b} = \frac{(5.0 \times 10^{-3}\ \text{kg})(9.8\ \text{m/s}^2)}{0.20\ \text{kg/s}} = 0.25\ \text{m/s}.$$

When the values of the two terms in the drag force are equal—when $bv = cv^2$—the speed of the marble is 2.0 m/s. However, when the gravity and drag are the same size, the b term in the drag force dominates; this yields a speed of only 0.25 m/s. The marble will never reach speeds at which the c term has increased to where it is comparable to the b terms.

What Do You Think? The speeds in this example are fairly high and perhaps represent a marble falling in oil at a relatively high temperature. What happens to oil at low temperature?

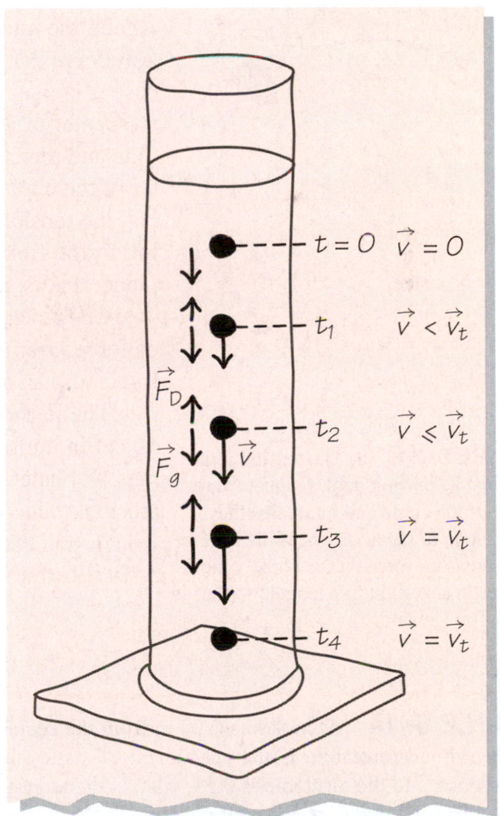

▲ **FIGURE 5–23** The marble is captured at various stages of its progress through the oil. When the marble has reached the terminal speed, the magnitudes of the drag force and gravity are equal.

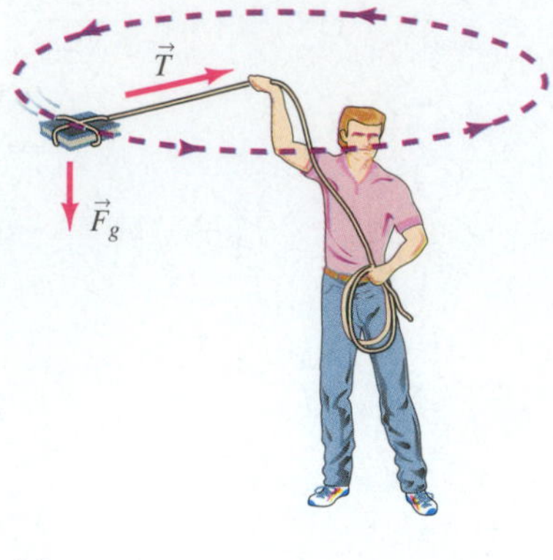

(a)

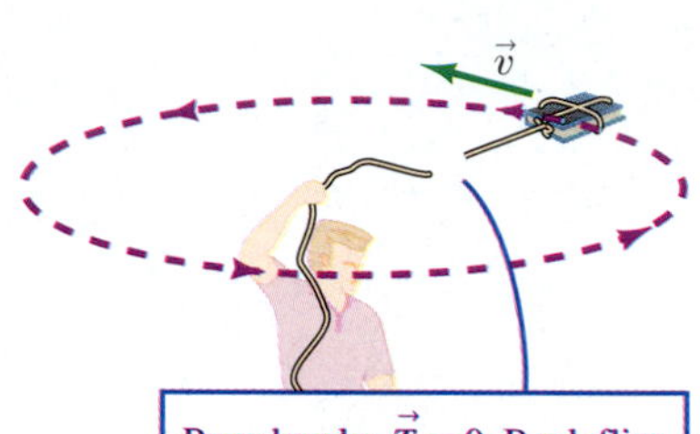

(b)

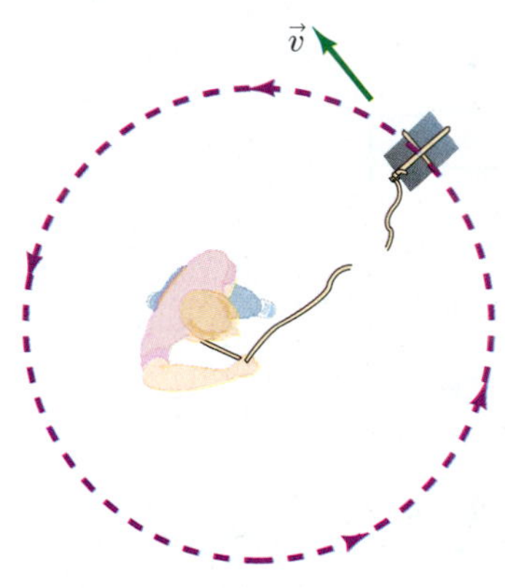

(c)

▲ **FIGURE 5–24** (a) The tension in the rope holding a whirling book points toward the center of the circle and maintains the circular motion of the book. (b) If the rope breaks, there is no force on the book, which then moves in a straight line tangent to the circle. (c) Top view of (b).

5–4 Forces and Circular Motion

If you were to tie this book to the end of a rope and swing it smoothly over your head in a nearly horizontal circle, keeping your hand as close to the center of the circle as possible (Fig. 5–24a), you would feel the rope become taut. (Because of gravity, the rope could not be perfectly horizontal, but we'll ignore gravity here.) This tautness is tension in the rope. As we discussed in Section 5–1, a rope under tension pulls on whatever it is connected to—in this case your hand at one end and the book at the other. In this situation the pull of the rope is the only horizontal force on the book, so there is a net horizontal force on the book, and the book will accelerate in the horizontal plane. This isn't surprising, though; we saw in Chapter 3 that an object in circular motion is accelerating even if it has constant speed (*uniform circular motion*). There is acceleration because the *direction* of the velocity vector is changing. If the constant speed of the uniform circular motion of the book is v and the radius of its circular motion is r, then Eq. (3–56) tells us that the acceleration vector $\vec{a}$ of the book is

$$\vec{a} = -\frac{v^2}{r}\hat{r}. \tag{5–33}$$

The unit vector $\hat{r}$ points outward from the center of the circular motion. The minus sign in Eq. (5–33) indicates that the acceleration is *toward* the center of the circle.

Newton's second law, $\vec{F} = m\vec{a}$, tells us that the force $\vec{F}$ required to keep a mass m in uniform circular motion is

$$\vec{F} = -\frac{mv^2}{r}\hat{r}. \tag{5–34}$$

NEWTON'S SECOND LAW FOR UNIFORM CIRCULAR MOTION

Again, the minus sign reminds us that the force vector points to the center of the circle. Any force that points in toward the center of a circle is called a *centripetal force*; in this case it is the rope tension that is a centripetal force. Similarly, any acceleration toward the center of a circle is a *centripetal acceleration*. Equation (5–34) tells us the magnitude and direction of the rope tension. Because the rope leads from the book to the center of the circle, the direction of the tension is indeed toward the center. If the rope were cut, the tension that provided the centripetal force would no longer be present. According to the first law, then, the book would move without acceleration in the horizontal plane; that is, in a straight line (Figs. 5–24b and c).

Centripetal forces, whether they are realized by a rope tension or any other source of force, *are not constant forces*. They are constantly changing in direction and therefore cause an acceleration that is constantly varying.

The tension of a rope is not the only force that can be centripetal and maintain an object in uniform circular motion, and we'll look at other forces in the next few examples. A planet in circular orbit about the Sun or a satellite in circular orbit about Earth are maintained in their circular motion by gravitational forces. Friction between the wheels and the road is the force that takes an automobile around a curve or, as Example 5–14 illustrates, that holds an object on a rotating turntable in its circular path.

EXAMPLE 5–14 A book sits 0.75 m from the center of a rotating playground turntable. If the coefficient of static friction between the book and the turntable is 0.55, what is the largest possible angular speed such that the book will not slide off?

Setting It Up Figure 5–25a shows the book on the turntable. We are not given the mass m of the book, so we leave it in symbolic form. (It will in fact turn out to be unimportant.)

Strategy Figure 5–25b is the free-body diagram for the book. The force maintaining the book in uniform circular motion is static friction $\vec{f}_s$ between the book and the turntable. The largest rotation frequency that is possible without the book slipping occurs when the friction force reaches its maximum value, $f_s = \mu_s F_N$.

Newton's second law, $\vec{F}_{net} = \vec{F}_N + m\vec{g} + \vec{f}_s = m\vec{a}$, must be decomposed into equations for the vertical direction and for the horizontal plane. For uniform circular motion, the horizontal motion equations are represented by a single radial component equation, Eq. (5–34). There is no acceleration component in the vertical direction; that is, the direction in which $\vec{F}_N$ and $\vec{F}_g = m\vec{g}$ act. The condition that the vertical acceleration is zero will give us F_N and hence the maximum value of friction. Once we have this value, we can use it in the equation for the horizontal component of Newton's law. We can then use this equation to determine the maximum speed, that is, the speed for which the radial acceleration just matches the maximum force of friction. Finally, we will have to translate the maximum speed to a maximum angular speed.

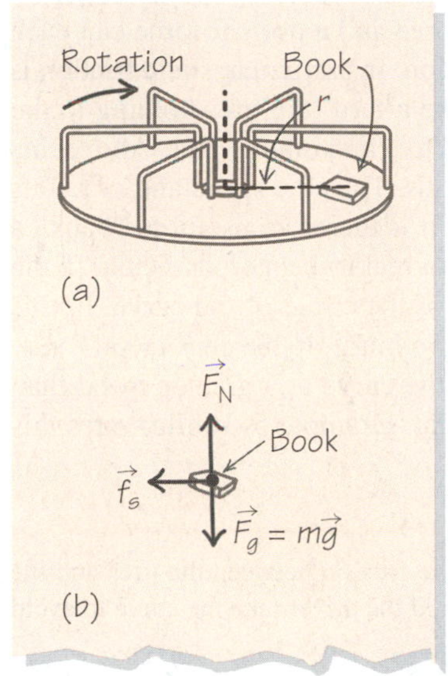

FIGURE 5–25 (a) Book rotates on playground turntable. (b) Free-body diagram for the book.

Working It Out The vertical component of Newton's second law is

$$F_N - mg = 0.$$

This gives $F_N = mg$. In turn, the maximum value of static friction is

$$f_s = \mu_s F_N = \mu_s mg.$$

The maximum radial acceleration $\vec{a}_{max}$ occurs for the maximum value of static friction. The radial component of Newton's second law, the coefficient of $\hat{r}$ in Eq. (5–34), is then

$$\mu_s mg = m\frac{v_{max}^2}{r}.$$

Note that the mass of the book cancels. We want to express this result in terms of an angular speed. From Chapter 3, we know that if the turntable is rotating with angular speed ω, then a point on the turntable that is a distance r from the center moves with speed $v = \omega r$. Thus our radial equation of motion becomes

$$\mu_s g = \frac{v_{max}^2}{r} = \frac{(\omega_{max} r)^2}{r} = \omega_{max}^2 r.$$

We solve for ω_{max}:

$$\omega_{max} = \sqrt{\frac{\mu_s g}{r}} = \sqrt{\frac{(0.55)(9.8\ \text{m}/\text{s}^2)}{0.75\ \text{m}}} = 2.7\ \text{rad/s}.$$

What Do You Think? If the angular velocity in this example exceeds the value calculated, what will happen to the book? How can you explain that based on the free-body diagram of Fig. 5–25b?

Normal forces can also act centripetally, as in Example 5–15.

EXAMPLE 5–15 A space station is an environment where the astronauts are in permanent free fall along with the station and so are in "weightless" conditions—a scale beneath their feet would register no contact force. One way to avoid weightlessness would be to give the station some rotation. Accordingly, consider a station that consists of a hollow circular tube that is rotating around its central axis, with the astronauts moving at the outermost radius (Fig. 5–26). If that radial distance is 50 m, what must be the speed v of a point on the outer wall such that a bathroom scale will read the same as it would on Earth?

Setting It Up We note that the reading of a bathroom scale on Earth is the weight mg, where m is the astronaut's mass. We are not given m.

Strategy Rather than setting up Newton's second law in what might be a rather complicated coordinate system, let's try to think through this problem to gain more insight. An astronaut of mass m who stands on a scale on the outermost wall within the space station moves in a circular path of radius R at constant speed v. He therefore accelerates toward the center of the circle (the axis) with magnitude v^2/R from Eq. (5–33). The force that causes this acceleration is the normal force $\vec{F}_N$, of magnitude mv^2/R, which is supplied by the tube's outer wall (Fig. 5–26b). The bathroom scale indicates this force, just as it would indicate the magnitude of the (upwardly directed) normal force on the astronaut if he were standing on the scale on Earth. Because the normal force on Earth balances gravity, it would have magnitude mg. Thus the scale on the space station reads as it would read on Earth provided that

$$mv^2/R = mg.$$

This can be solved for v.

(a)

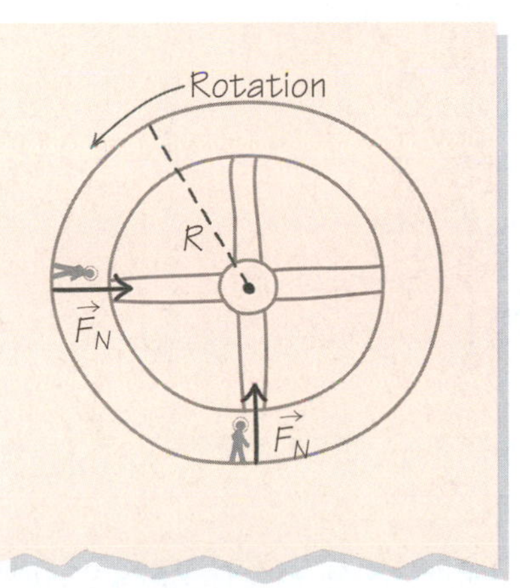

(b)

FIGURE 5–26 (a) Space station from the movie *2001: A Space Odyssey.* (b) A normal force $\vec{F}_N$ pointing to the circle's center acts on each astronaut. The presence of this force gives the astronauts the illusion of weight.

(continues on next page)

Working It Out The mass m cancels out from this equation and we can solve for the speed v:

$$v = \sqrt{gR} = \sqrt{(9.8\ \text{m/s}^2)(50\ \text{m})} = 22\ \text{m/s}.$$

This is a fairly considerable speed—about 50 mi/h. Motion such as this involves internal forces that the station must be constructed to withstand. These internal forces are proportional to v^2. The space station would have to be extremely well constructed, and hence extremely expensive, to have a large enough rotation to give the astronauts the illusion of being on Earth. Actual space stations will have to make do with much less of a normal force—if any at all. Each piece of material that they are constructed from has to be transported from Earth and so must be very light.

What Do You Think? What would happen if the astronaut tried to walk on the inside surface of the innermost wall of the rotating station? Or on the "side walls," the ones parallel to the plane of rotation?

We have now discussed how tension, a friction force and a normal force can each act as the centripetal force responsible for circular motion. In particular, static friction is the centripetal force that causes an automobile to make a turn (the tires holding to the road rather than skidding). But static friction is limited to a maximum value. Above this value sliding friction takes over and the automobile skids. Thus the speed and/or radius with which an automobile can make a turn is limited. It is therefore useful to "bank" a curve. In a banked curve, the road is tilted so that the normal force, perpendicular to the road surface, has a component pointing to the center of the circle of the curve. In this way, the normal force acts together with static friction to make up the centripetal force. The combination accelerates the automobile through the curve at a greater speed than could be achieved without skidding if the road were flat. Example 5–16 illustrates this point in the extreme case where there is no friction.

EXAMPLE 5–16 It is a dark and stormy night, and a driver advancing along an icy road must negotiate a turn (Fig. 5–27a). The road traces out an arc of a circle with radius 320 m and is banked at an angle of 5.1°. Assume that the friction between the tires and the road is zero. At what speed should the driver take the curve to avoid sliding off the road?

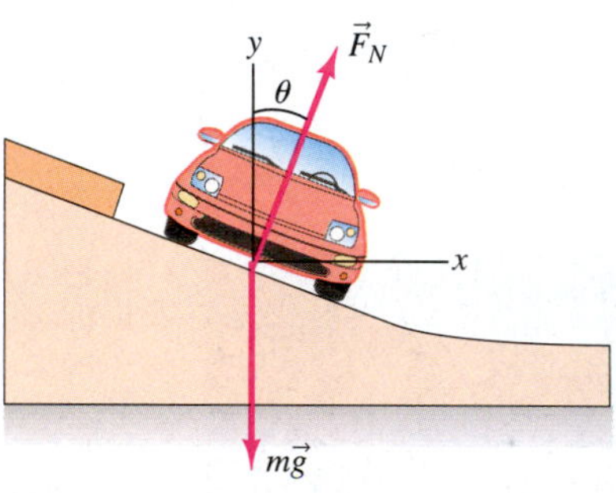

(a)

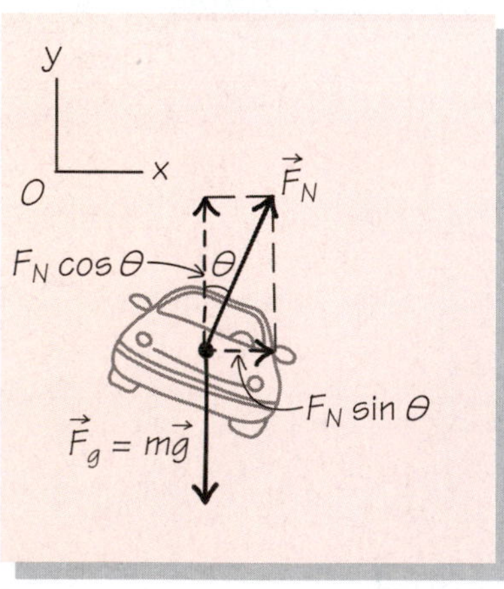

(b)

▲ **FIGURE 5–27** (a) Car making turn on icy, banked road. (b) Free-body diagram for the automobile, seen face on. The center of the circle described by the road curve is to the right.

Setting It Up The translation of the question to a quantitative setup is "What is the speed for which the car will not skid?" We denote the given radius of the circular path by r; other quantities are specified in the figure.

Strategy Figure 5–27b is a free-body diagram for the car as seen from the front together with a coordinate system. Note that here we have chosen axes that are vertical and horizontal rather than along and perpendicular to the plane. The center of the circle is to the $+x$-direction. In the absence of friction, only the normal force and gravity act on the car, and we want to write Newton's law and then break it into horizontal and vertical components. The vertical component has no acceleration, as the car stays on a horizontal path if there is no slipping. The horizontal component will have the acceleration for the circular motion of the car. These two equations should be enough to determine uniquely the speed.

Working It Out Newton's second law is $\vec{F}_{\text{net}} = \vec{F}_N + \vec{F}_g = m\vec{a}$, or in our coordinate system

$$(F_N \cos\theta)\hat{j} + (F_N \sin\theta)\hat{i} - (mg)\hat{j} = \frac{mv^2}{r}\hat{i}.$$

In component form, we have

$$\text{for the } x\text{-component: } F_N \sin\theta = mv^2/r; \tag{5–35}$$

$$\text{for the } y\text{-component: } F_N \cos\theta - mg = 0. \tag{5–36}$$

Equation (5–36) determines F_N:

$$F_N = \frac{mg}{\cos\theta}.$$

Note that, unlike other situations that we have studied, in this case the normal force is larger than the force of gravity. This result is inserted into Eq. (5–35) and solved for v:

$$\frac{mg}{\cos\theta}\sin\theta = \frac{mv^2}{r};$$
$$v = \sqrt{gr\tan\theta}.$$

With $r = 320$ m and $\theta = 5.1°$ ($\tan 5.1° = 0.089$),

$$v = \sqrt{(9.8\text{ m/s}^2)(320\text{ m})(0.089)} = 17\text{ m/s} \cong 60\text{ km/h}.$$

Only the *x*-component of F_N acts as the centripetal force to carry the car around the curve. When the automobile moves without sliding, the vertical component of F_N just balances the force of gravity, and the horizontal component provides the centripetal force, as in Eq. (5–34).

Note that this system is balanced quite delicately. For a given banking, curvature, and speed, Eq. (5–35) uniquely determines F_N. If F_N is too small—if the speed is too slow—the product $F_N\cos\theta$ cannot balance gravity, and the car slips *down* the slope and off the road. If F_N is too large—if the speed is too fast—the vertical component of the normal force will exceed mg, and the car will skid *up* the slope and off the road. Although you might have been tempted to think that this question involves the *maximum* speed with which the driver can take the curve, there is in fact only one possible speed.

What Do You Think? We treated this example under the assumption that the road was perfectly slick. Suppose instead that there was a nonzero coefficient of static friction. Without doing any calculations, how does this new element affect our arguments about how delicately balanced the system is?

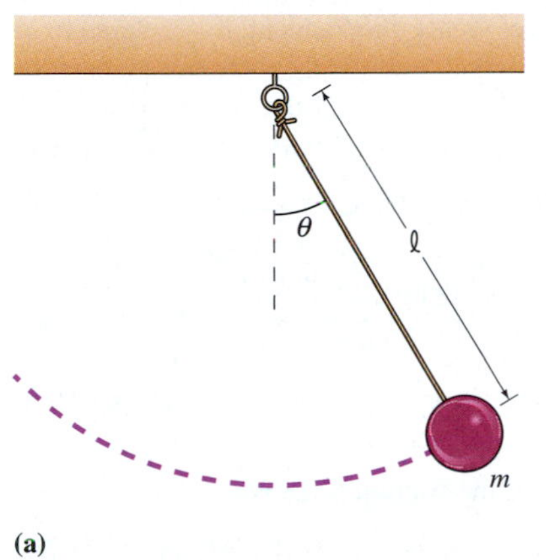

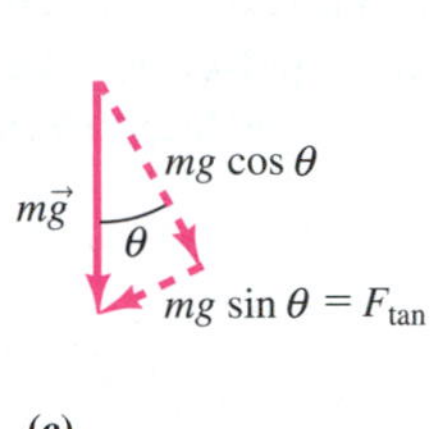

◀ **FIGURE 5–28** (a) Simple pendulum of mass m and length ℓ. The path followed by the bob traces a circular arc with the speed along the path varying with angle θ. (b) Free-body diagram for the bob of the simple pendulum. (c) Components of the force of gravity on the pendulum bob.

Circular Motion with Changing Speed

What happens in circular motion with changing angular speed? Figure 5–28a illustrates a *simple pendulum*, a point mass m (the bob) suspended from a string of negligible mass that has a length ℓ. The pendulum swings in a plane under the influence of gravity. During this swing, the mass on the end follows a circular path determined by the length of the string, but it is *not* uniform circular motion because the speed is not constant. To understand this, we draw a free-body diagram (Fig. 5–28b) for the bob, which is located at angle θ. Two forces act on it: gravity and tension. The bob accelerates according to Newton's second law, $\vec{F}_{net} = \vec{F}_g + \vec{T} = m\vec{a}$. The forces can be broken into components perpendicular to the path (radial forces) and components along the path (tangential forces). The tension is purely radial and inwardly directed (negative), whereas the force of gravity has an outward (positive) radial component $mg\cos\theta$ and a tangential component $mg\sin\theta$ (Fig. 5–28c). Thus the second law expressed in component form is

$$\text{for the radial component:}\quad -T + mg\cos\theta = -\frac{mv^2}{r};$$

$$\text{for the tangential component:}\quad mg\sin\theta = ma_{tan} = m\frac{dv}{dt}.$$

Here the tangential component of the acceleration, a_{tan}, is the rate of change of speed. *The speed of the pendulum changes because a tangential force is present.* The radial forces simply guide the mass on a circular path and *do not change its speed.* Figure 5–29 enables us to see that the pendulum changes its speed.

For *any* curving path there will generally be forces acting that have components perpendicular to the path and tangential to the path. Force components perpendicular to the path change the direction of the object's motion but not its speed. The tangential components of the net force change the magnitude of the velocity but not its direction.

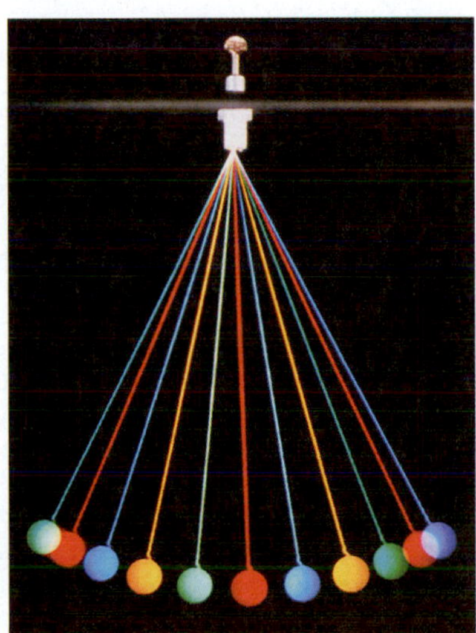

▲ **FIGURE 5–29** As the pendulum bob moves under the influence of gravity and the string tension, it speeds up to a maximum at the bottom of the swing and moves most slowly at the top. This is revealed here by an open-shutter photograph of the swing with light flashes at equal time intervals of $\frac{1}{10}$ s.

CONCEPTUAL EXAMPLE 5–17 At the beginning of this section we referred to attaching this book to a rope and swinging it in a horizontal circle. Let's consider this situation again, and again ignore the gravitational force and consider only the forces that make the book describe a circle. We know that when the book is in uniform circular motion, the force is *central*. In other words, the rope must lead back to the center of the circular motion. What forces were required to bring the book to this state?

Answer In order to start the book moving and bring its speed up to its final value, there must be a tangential force component exerted by the rope on the book through the rope tension. Thus the tension cannot be purely radial but must in addition have a tangential component. This is managed by moving the hand responsible for swinging the book away from the center of the circle—the hand itself will follow a smaller circle about the motion center. The larger the tangential component of the tension and the longer it operates, the faster the book swings. As the book swings faster, the radial tension force also increases, but the hand holding the rope must take this increasing radial component into account with a firmer hold. In reality there are forces acting such as gravity and air resistance that will slow the motion down unless a continuous effort is made to keep it up, as you would quickly realize if you were asked to keep the book swinging for a significant period of time.

EXAMPLE 5–18 A skater of mass 52 kg is swinging in a circle of radius 2.0 m at a speed 3.0 m/s. He maintains this motion by holding onto a rope attached through a frictionless bearing to a fixed pole at the center of the circle (Fig. 5–30). The tension of the rope provides the centripetal force for his circular motion, and we can regard the ice as frictionless. As he swings around the pole, however, he runs into a rough patch of ice 48 cm long, where the coefficient of friction is 0.10. (a) What is the tension T in the rope before he hits the rough patch? (b) What is his speed just after he passes the rough patch? (c) How must the tension change to keep him on the same circular path after he has passed the rough patch?

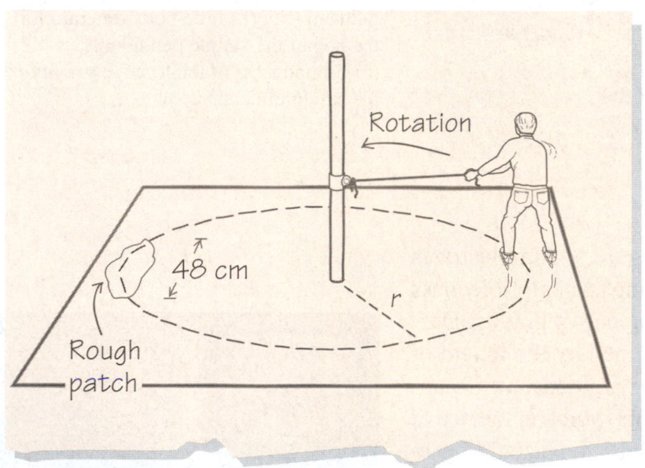

▲ **FIGURE 5–30** Ice skater rotating about pole by means of a rope.

Setting It Up The given mass, radius of motion, and speed are labeled *m, R,* and v, respectively. We label the coefficient of kinetic friction in the rough patch as μ_k. The unknown tension has magnitude T.

Strategy For part (a), we simply use the fact that tension supplies the centripetal force required for the motion. For part (b), we note that the friction of the rough patch is a constant tangential force, which slows the skater down. The patch is small enough for us to assume that the motion of the skater through the patch is a straight line. The frictional force yields the tangential acceleration (negative) through Newton's second law. We then use our kinematic formulas for one-dimensional motion to find the speed change. For part (c), we redo the exercise of part (a) but with the reduced speed.

Working It Out (a) Newton's second law for the circular motion reads

$$T = \frac{mv^2}{r} = \frac{(52\ \text{kg})(3.0\ \text{m/s})^2}{2.0\ \text{m}} = 2.3 \times 10^2\ \text{N}.$$

Compare this to the skater's weight, $mg = (52\ \text{kg})(9.8\ \text{m/s}^2) = 5.1 \times 10^2\ \text{N}$.

(b) The magnitude of the friction force is

$$f_k = \mu_k F_N = \mu_k mg = (0.10)(510\ \text{N}) = 51\ \text{N}.$$

Here f_k/m is the magnitude of the tangential acceleration, pointing oppositely to the motion (friction will slow the skater down). We then use the one-dimensional formula relating speed change to distance traveled for a constant force, Eq. (2–24):

$$v^2 = v_0^2 + 2a\,\Delta x = v_0^2 - 2\frac{f_k}{m}\Delta x$$

$$= (3.0\ \text{m/s})^2 - 2\frac{51\ \text{N}}{52\ \text{kg}}(48\ \cancel{\text{cm}})\frac{1\ \text{m}}{100\ \cancel{\text{cm}}} = 8.1\ \text{m}^2/\text{s}^2.$$

After passing the rough patch, the skater has slowed to

$$v = \sqrt{8.1\ \text{m}^2/\text{s}^2} = 2.8\ \text{m/s}.$$

(c) We simply follow the first procedure and compute the tension with the new speed:

$$T = \frac{mv^2}{r} = \frac{(52\ \text{kg})(2.8\ \text{m/s})^2}{2.0\ \text{m}} = 2.1 \times 10^2\ \text{N}.$$

This is 20 N less than the tension in the rope before the rough patch is encountered.

What Do You Think? You are a fairly good skater and decide to try the motion described in this example. You find a suitable, safe place. What should you be concerned about when performing the maneuver?

Circular Motion and Noninertial Frames

The doughnut-shaped space station in Example 5–15 illustrates how a normal force can be the centripetal force that moves an object in a circle.This system also nicely illustrates the effects of noninertial frames, first mentioned in Section 4–4. An astronaut in the space station is an observer in a noninertial frame. Seen from outside, the astronaut is in uniform circular motion and is therefore accelerating. The centripetal force responsible for this acceleration is the normal force, which can be read by a scale at the astronaut's feet. But what does the astronaut see? As he perceives it, everything around him within the station is at rest. The scale reads the usual kind of normal force, directed toward the center of the circle.

The astronaut's analysis of the situation is that he is at rest and there is no identifiable source for the force that keeps him on the scale—this force is therefore a fictitious force. It is directed outward (*centrifugal*, meaning "directed away from the center") because it balances the inward-directed normal force that the scale reads.

While the notion of a centrifugal force is a commonplace one, we should keep in mind that *it is not a force in the same sense as the forces that appear in Newton's laws*. It is the consequence of an observer being in a rotating (and hence accelerating) reference frame. We use the language of the centrifugal force to interpret the invisible hand that forces us against the door when an automobile makes a high-speed turn. Seen from outside the automobile, a passenger is accelerating centripetally. A centripetal normal force from the door of the automobile pushing on the passenger is the force responsible for moving the passenger in his curving path. The automobile passenger views things differently. The passenger perceives herself to be at rest with respect to the car, yet she feels the normal force from the door. The passenger then invents a "centrifugal force" to balance the normal force from the door.

5–5 Fundamental Forces

The forces that we have dealt with so far include the force of gravity (weight), normal forces, friction, tension, drag, and spring forces. With the exception of weight, these forces are what we can call secondary forces, the manifestation on a macroscopic level of something happening at the microscopic level. At the microscopic level, the constituents of matter interact through the fundamental forces of electricity and magnetism. The forces themselves are well understood, but there are so many atoms and molecules in any piece of matter or in any object, and they interact in such a complex way, that even with the largest computers at our disposal we would be unable to derive a spring force, say, from the electromagnetic interactions between the molecules of the spring. Thus we must be content to use the spring force—a secondary force—as a simple effective force to solve problems relevant to the everyday scale. The force of gravity is a different case. It corresponds to a more general law—the fundamental force of *universal gravitation*, first understood by Newton—but used over a limited range.

What we now refer to as the fundamental forces, the ones that underlie the secondary forces, are not necessarily the forces that were thought of as the fundamental forces 200, 100, or even 25 years ago. Our progress on the understanding of what is truly fundamental has been steady, marked especially with a series of *unifications*. Thus what were once thought to be distinct fundamental forces are now viewed as different aspects of the same fundamental force.

We now believe that there are only three fundamental forces in nature. These are the force of **universal gravitation**, the **electroweak** force, and the **strong** force (often called the nuclear force). The expression of the force of universal gravitation (see Chapter 12) was one of Newton's greatest discoveries. Albert Einstein replaced Newton's expression in his law of general relativity, making some new predictions of tiny experimental effects that are not predicted by Newtonian gravitation. Einstein also made persistent but unsuccessful efforts to unify gravitation with the other fundamental forces. The electroweak force is a relatively recent discovery, dating from a theory proposed by Sheldon L. Glashow, Abdus Salam, and Steven Weinberg in the 1960s and 1970s, that was verified experimentally in the early 1980s. This work unified two forces formerly thought to be independent and fundamental: the *weak* force, responsible mainly for some types of radioactive processes in nuclei and important in the evolution of the universe, and the forces of *electromagnetism* (Chapters 21 to 34). On the scale appropriate to the secondary forces that we deal with every day, the electromagnetism aspect of the electroweak force is dominant, and it is often convenient to refer simply to the electromagnetic force. Electromagnetism is itself the result of a nineteenth-century unification of the forces of *electricity* and *magnetism*; these two forces were previously thought to be different kinds of forces. The third fundamental force, the strong force, is responsible for holding together the nuclei of atoms. Both the strong force and the electroweak force have been the object of more recent attempts at unification, which have yet to bear fruit. Since the 1980s, there has also been a new effort to bring gravitation into the unification program. Unification continues to be a fascinating beacon that guides to a simpler description of the universe.

On the everyday scale of the secondary forces that we have discussed in this chapter, only the force of gravity is a direct aspect of a fundamental force: universal gravitation. All the other forces that we have investigated are ultimately due to the electromagnetic force, which binds atoms and molecules together into ordinary matter; for example, contact forces result from forces exerted at the atomic scale. We do not directly see the strong force on our everyday scale; it holds nuclei together so tightly that, for most practical purposes, we can think of them as indivisible lumps of matter. Only when we discuss nuclear energy or the composition of stars do strong forces come into practical play.

Within the nucleus, the typical strong force is about 100 times larger than the electromagnetic forces. The gravitational force between two protons in the nucleus is many, many orders of magnitude smaller than either the strong or the electroweak force. But as the strong force acts over such a limited range, we experience the electroweak and gravitational forces as the dominant forces at the everyday scale. The gravitational force dominates on the astronomical scale because the atoms of matter are electrically neutral, and so electromagnetic forces cancel very nearly to zero.

Summary

This chapter is devoted to describing various forces that occur in nature and to applying the problem-solving techniques developed in Chapter 4. These techniques are ones that we use throughout the book.

The force of gravity (or weight), tension, and the normal force are common forces for which it is often possible to solve the equations of motion given by Newton's second law. The force of gravity is expressed as

$$\vec{F}_g = m\vec{g}, \tag{5–4}$$

where $\vec{g}$ is a vector of constant magnitude that points down toward Earth's center.

The tension $\vec{T}$ is a variable force that is determined according to the circumstance and exerted by ropes (or wires or cables or strings). Tension always pulls on a mass in the direction of the rope; it is transmitted everywhere along the rope, taking a single constant magnitude throughout the rope when the mass of the rope is negligible.

The normal force $\vec{F}_N$ is also variable. It is directed perpendicularly to a surface and acts to cancel the effects of any other forces that might make a mass accelerate into the surface.

The friction force acts when two surfaces slide or attempt to slide across one another. Static friction $\vec{f}_s$ is variable and acts in a direction that would oppose any sliding motion that would occur if there were no friction. Its magnitude can increase up to a maximum value proportional to the magnitude of the normal force:

$$\text{static friction: } 0 \leq f_s \leq \mu_s F_N. \tag{5–21}$$

Kinetic friction $\vec{f}_k$ acts when sliding actually occurs and is also proportional to F_N:

$$\text{kinetic friction: } f_k = \mu_k F_N. \tag{5–22}$$

The constants μ_s and μ_k are the coefficients of friction, and generally

$$\mu_s > \mu_k. \tag{5–23}$$

Another type of friction, the drag force, occurs when objects move through fluids or gases. The drag most frequently encountered varies with the speed squared:

$$F_D = \tfrac{1}{2}\rho A C_D v^2. \tag{5–30}$$

Any drag force that increases with speed leads to a situation in which objects accelerating within media can be accelerated only up to a terminal speed.

Forces are responsible for accelerating objects moving in a curved path. When the motion is uniform circular motion, then Newton's second law takes the form

$$\vec{F} = -\frac{mv^2}{r}\hat{r}. \tag{5–34}$$

The vector $-\hat{r}$ is directed to the center of the circle and the force is said to be centripetal. Any of the forces we mentioned above can act as the centripetal force that keeps an object in uniform circular motion. When both the direction and the magnitude of the velocity change, then $\vec{F}$ has a component tangential to the motion. This component causes the magnitude of the velocity to change, while a force component perpendicular to the motion causes the direction of the motion to change.

All the forces of nature are ultimately described in terms of three fundamental forces: the force of universal gravitation, the electroweak force, and the strong force. The electromagnetic force, which is part of the electroweak force, is responsible for most of the secondary forces, including tension, friction, drag, normal forces, and spring forces.

Understanding the Concepts

1. A tightrope walker moves to the center of a thin wire that was initially stretched taut to a horizontal position. Why is it that the wire cannot remain horizontal?
2. We have said that it is actually the force of friction that is responsible for both the acceleration of automobiles and our ability to walk. What is the role of the engine or of the muscles in these processes?
3. An observer sees a mass hanging motionless from a vertical string. Under what circumstances is the tension in the string greater than or less than the weight of the mass?
4. What are some factors that could limit how fast a hot rod can go in a $\frac{1}{4}$-mi race?
5. How does the fact that a rope has mass complicate solving a problem about lifting a load with a pulley? What is the effect of friction in the pulley?
6. Why is it helpful for an automobile with an engine in the front to have front-wheel drive? Why is it useful to put sand in the trunk of your car in winter if your car has rear-wheel drive?
7. Why do bicyclists or motorcyclists "lean into" a curve? In explaining why, make use of the fact that we are able to balance ourselves best when we feel that the net force on us is coming from directly beneath our feet. Also think about the forces that friction must oppose.
8. How might the result of Example 5–10 vary if the weight was not evenly distributed over the wheels and/or the car was equipped with rear-wheel drive instead of four-wheel drive?
9. Suppose that a rope has tension because a mass is suspended from its end. Let's say that the rope is now cut, a spring is inserted at the cut, and we observe the stretch of the spring as a measure of the rope's tension. If the mass of the rope is negligible, does the observed tension of the rope depend on where the cut is made along the string? Does your answer change if the rope cannot be considered massless?
10. A bowl of water with floating ice cubes is placed on a scale. The ice cubes melt. Does the reading of the scale change? Why or why not?
11. What is the role of the keel, which runs along the center of the bottom of a sailboat? Some sailboats have centerboards rather than keels. These are simply large boards that can be lowered or raised in the position of a keel. What is the role of a centerboard? Why might you prefer a keel to a centerboard or vice versa?
12. Suppose the woman in Example 5–4 had just the limiting mass of 37.3 kg and the suspended lineman ate a cheeseburger, adding 500 g to his mass. Assuming the woman doesn't let go of the rope, describe qualitatively the resulting motion.
13. The block and tackle illustrated in Example 5–4 employed the same rope four times to allow a mass m_1 to lift a mass $m_2 = 4m_1$. Could you have increased the number of pulleys in that example so that the same rope is employed 100 times and a mere baby could have lifted the football lineman? What are the practical limitations in this sort of process?
14. We have referred to massless ropes and the tension in them. What physical considerations allow you to think of a rope as massless?
15. The speed of the boats (shells) used in scull racing is, to a good approximation, independent of the number of people rowing (provided that the number is larger than three or four). At first, this might appear strange: The more people there are, the larger the forward propelling force available to overcome the drag of the water. Can you explain this seeming contradiction?
16. Describe a series of experiments that would have allowed Leonardo da Vinci to decide that the force of friction depends only on the normal force on an object.
17. True or false:
    (a) Sliding friction can accelerate an object.
    (b) Sliding friction can be used to increase the speed of an object.
    (c) Static friction can accelerate an object.
    (d) Static friction can be used to increase the speed of an object.
18. At the moment a car in a loop-the-loop roller coaster is at the top of the loop (directly below the track), can there be a normal force on it? Such a force would point straight down, in the same direction as that of gravity.
19. The riders on the loop-the-loop of Question 18 experience a feeling of near weightlessness close to the top of the loop. Why? Would the coins in their pockets fall out?
20. Suppose a block of wood sits on the floor of an elevator with the elevator moving upward with a constant velocity. Will the normal force differ from what it would be when the elevator is stationary?
21. Suppose a block of wood sits on the floor of an elevator with the elevator moving upward with a positive acceleration. Will the normal force differ from what it would be when the elevator is stationary?
22. Suppose that it were possible for a ship to sail all the way around the world along a great circle. Is a centripetal force necessary to keep the ship moving in this circle? What force or forces would act centripetally?
23. Look at the pendulum photo in Fig. 5–29. Is the net force on the pendulum bob constant throughout its motion? Is there a point at which the net force equals zero? Describe the forces acting on the pendulum bob when it just stops and turns around on the right side.
24. When you sit in the passenger seat of a car that makes a tight turn to the left, you could be thrown out of the car if the door should open (and you are not belted in). Why?
25. At the beginning of Section 5–4, we spoke of swinging a book on the end of a rope in a nearly horizontal circle. Could the motion form a perfectly horizontal plane?
26. Imagine that you and your partner are on skates on a perfectly frictionless ice surface. You hold on to opposite ends of a rope and pull toward each other. What happens if you each have exactly the same mass? If you have half your partner's mass?
27. Why is it hard to run when the ground is icy?
28. You could attach the end of a stick to a mass and use the stick to pull the mass, just as with a rope. Is the associated force a tension force? If not, in what ways does it differ from tension?
29. Tarzan swings from tree to tree on a vine. At which point in his swing is the vine most likely to break and why?
30. A person on a rapidly moving Ferris wheel feels that she is about to fly off the seat when her seat reaches the top of the circular path. Why is that?

31. A hemispherical bowl is placed open end up on a table and rotates around its own vertical axis. A die is allowed to slide down from the edge into the bowl. Describe the motion of the die as seen by a hypothetical observer at the bottom of the bowl, assuming that there is no friction between the bowl and the die.

32. The person in Fig. 5–17 is moving along briskly at a constant speed. What is the net force on her? Describe the various external forces acting on her.

33. Consider Eq. (5–32) for the drag force, which has two terms. Which term, if any, dominates at low speeds and which, if any, dominates at high speeds? How would you decide the size of the domains for which one term is much larger than the other?

34. You sit on a comfy chair sinking into the seat cushion. Is a normal force acting as you do? Describe how the normal force is acting in this case.

Problems

5–1 Some Simple Constant Forces

1. (I) An 8-g bullet that travels at 500 m/s is fired into a rigidly fixed block of wood. The bullet is found 7 cm into the wood. What was the average force exerted by the wood opposing the bullet's motion? Assume that the deceleration was uniform.

2. (I) In Example 5–4, a woman lifts a football lineman (mass 149 kg) by pulling down on one end of a rope with 365 N of force. Suppose that the woman has a mass of 50 kg. If a bathroom scale were beneath her feet as she lifted the lineman, what would the scale read in pounds? Analyze the free-body diagram for the woman and remember that a scale of this type reads the upward normal force that the floor exerts on her.

3. (I) A helium balloon just manages to lift 100 kg (including the mass of the balloon, the helium it contains, and its payload) off the ground and then hovers 1 m off the ground. The upward force that maintains the balloon is *buoyancy* (we shall treat this force in Chapter 16). What is the magnitude of the buoyancy in this case?

4. (I) A woman of mass 61 kg sits in a racing car. When she depresses the accelerator, the car accelerates in a straight line to 210 km/h in 7.3 s. What are the direction and magnitude of the force she experiences? Where is the force applied?

5. (I) A hockey puck of mass 0.10 kg slides without friction on ice. In an appropriate coordinate system, its velocity $\vec{v}_1 = (1.4\ \text{m/s})\hat{i} + (3.0\ \text{m/s})\hat{j}$. A constant force $\vec{F} = (4.0\ \text{N})\hat{i}$ is then applied to the puck. After how many seconds will the puck have a speed of 6.0 m/s?

6. (II) A man of mass 80 kg is escaping a burning building using a rope that will break if the tension exceeds 600 N. (a) With what acceleration must he slide down the rope if it is not to break? (b) How far down the rope is he and what is his velocity after 5.0 s assuming he drops with the minimum acceleration of part (a)?

7. (II) A metal rod of mass 5.6 kg and length 3.5 m is suspended from the ceiling. (a) What is the tension in the rod at a distance of 2.0 m from the top? (b) What is it at a distance of 3.0 m from the top?

8. (II) A brick hangs from a string attached to the ceiling. When a horizontal force of 12 N is applied to the brick, the string makes an angle of 25° with the vertical (Fig. 5–31). What is the mass of the brick?

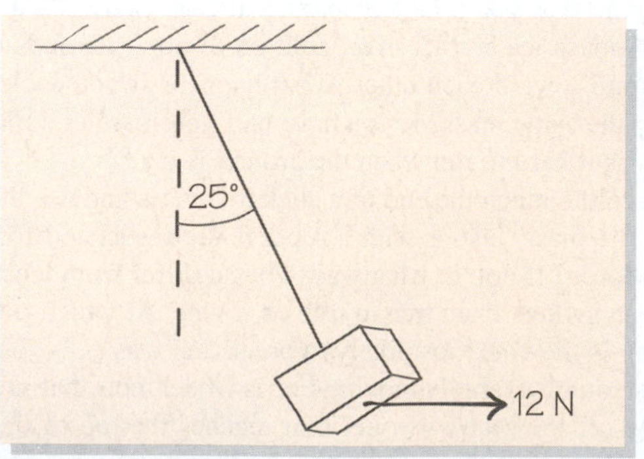

▲ **FIGURE 5–31** Problem 8.

9. (II) Figure 5–32 shows a person applying a horizontal force in trying to push a 25-kg block up a frictionless plane inclined at an angle of 15°. (a) Calculate the force needed just to keep the block in equilibrium. (b) Suppose that she applies three times that force. What will be the acceleration of the block?

▲ **FIGURE 5–32** Problem 9.

10. (II) Two blocks of mass m_1 and m_2 are placed in contact on a smooth surface with the more massive one (m_1) on the left. A force of magnitude F pointing to the right is applied to the block on the left. (a) What is the acceleration of the system? (b) What force acts on the block on the right?

11. (II) An automobile of mass 1200 kg pulls another automobile of mass 1400 kg with a tow rope. (a) In order to pull out onto a highway, the automobile must accelerate to 55 mi/h in an access lane that is only 120 m long. What must its acceleration be? (b) What is the tension in the tow rope?

12. (II) An Atwood machine consists of a massless string connecting two masses over a massless, frictionless pulley (Fig. 5–33). In this case, the masses are 1.70 and 1.65 kg. The system is released from rest with the 1.7-kg mass 2.15 m above the floor and the 1.65-kg mass on the floor. (a) What is the acceleration of the 1.7-kg mass? Of the 1.65-kg mass? (b) What is the speed of the 1.7-kg mass just before it hits the floor? (c) How long does it take the 1.7-kg mass to reach the floor?

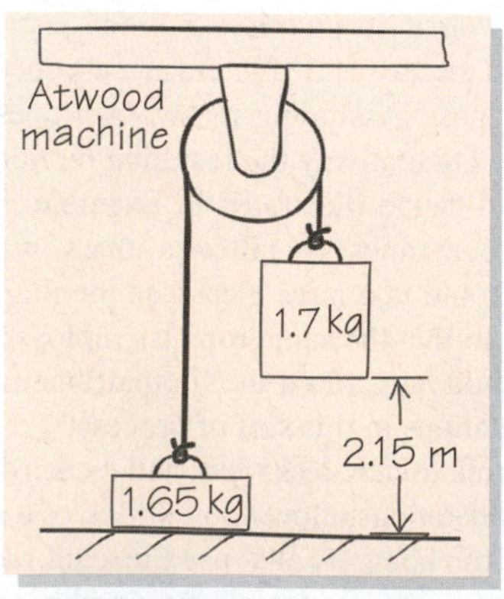

▲ **FIGURE 5–33** Problem 12.

13. (II) Two blocks of masses M and m are connected by a light rope that passes over a frictionless pulley. Mass M sits on an inclined plane with an angle of inclination of $\theta = 30°$ (Fig. 5–34). The coefficient of static friction between mass M and the inclined plane is 0.20, while $m = 3.0$ kg. Determine the largest and smallest possible values of M for which the system remains in equilibrium. Calculate the force of static friction on the block of mass M if $M = 6.0$ kg.

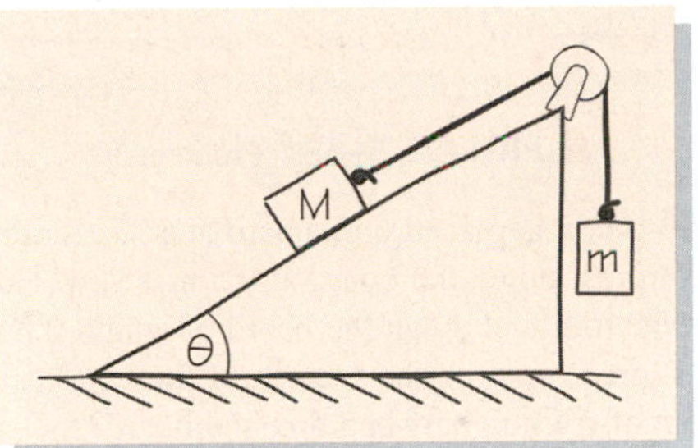

▲ **FIGURE 5–34** Problem 13.

14. (II) Consider a variation on an Atwood machine in which the masses are each on a (frictionless) incline (Fig. 5–35). The mass sliding on incline 1, m_1, is 1.50 kg, and the angle of this incline is $\theta_1 = 62°$. If the mass on the second incline, m_2, is 2.50 kg, what is the angle θ_2 so that the system does not accelerate?

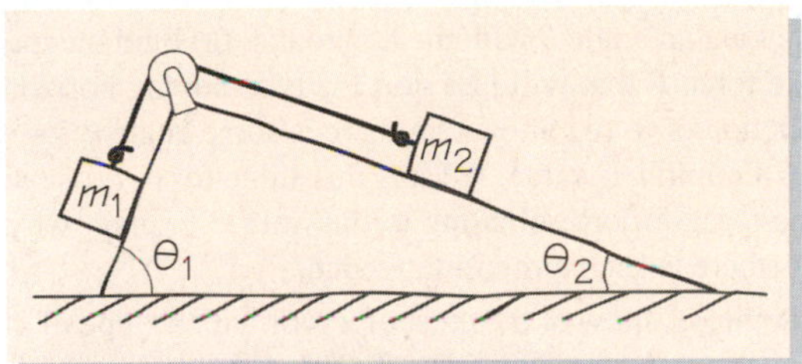

▲ **FIGURE 5–35** Problem 14.

15. (II) Suppose the woman in Example 5–4 had just the limiting mass of 37.3 kg and that the suspended lineman ate a cheeseburger, adding 500 g to his mass. Assuming the woman doesn't let go of the rope, describe quantitatively the resulting motion.

16. (II) In the text, we stated that the tension is the same everywhere in a massless rope. Here we ask you to find the tension as a function of position for a stationary hanging rope with constant mass density λ = mass/unit length. [*Hint*: Consider a particular point along the rope and find how much mass lies below that point. The tension at that point must be enough to support that mass.]

17. (III) A compound Atwood machine is constructed by replacing one of the masses of a conventional Atwood machine (see Problem 12) with the pulley of another Atwood machine (Fig. 5–36). Altogether there are three masses, two ropes, and two pulleys; the ropes and pulleys are to be considered massless and friction free. Describe the method by which the motions of the masses of this machine can be analyzed. [*Hint*: There are two independent tensions and two independent accelerations, which must be found by analyzing the equations of motion for the three masses simultaneously.] Solve these equations for arbitrary masses. What happens if $m_2 = m_3 \neq m_1$? [*Hint*: If m_1 accelerates, so does the pulley, so m_2 and m_3 share an additional acceleration.]

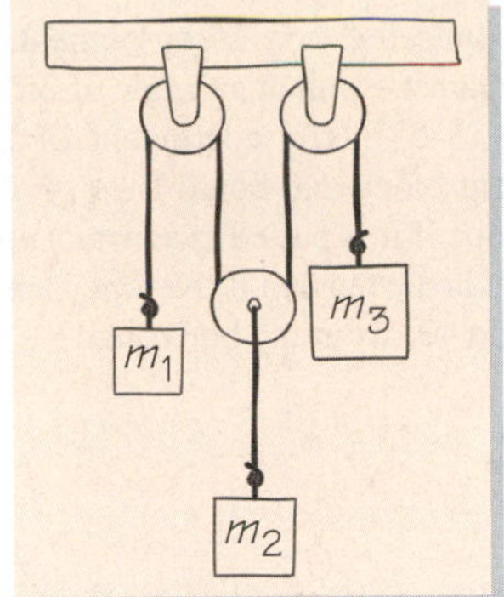

▲ **FIGURE 5–36** Problem 17.

18. (III) A double-pulley system—like that described in Problem 17—has masses $m_1 = 2.00$ kg, $m_2 = 1.20$ kg, and $m_3 = 0.800$ kg. (a) What are the accelerations of all the masses? (b) What are the tensions in all the ropes?

19. (III) Consider the three-pulley arrangement shown in Fig. 5–37. The three masses m_1, m_2, and m_3 have the values 4.00, 10.00, and 6.00 kg, respectively. All the pulleys are frictionless and the strings are massless. What are the tensions in all the strings and what are the accelerations of the masses?

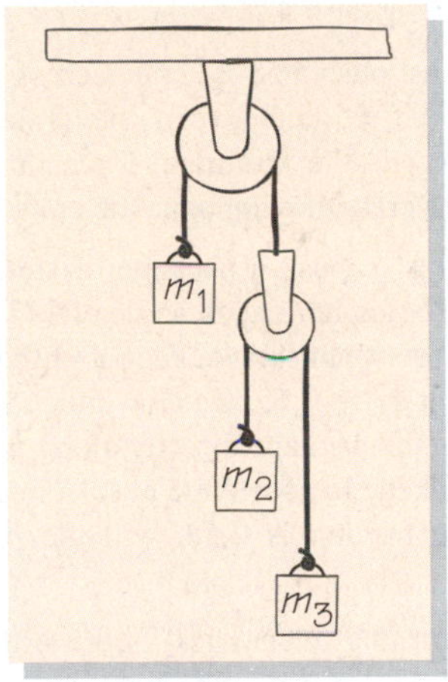

▲ **FIGURE 5–37** Problem 19.

5–2 Friction

20. (I) A boy of mass 42 kg runs into an ice-skating rink in street shoes and begins a smooth slide. He starts with a speed of 5.2 m/s and comes to a stop after sliding for 4.7 s. What is the coefficient of kinetic friction of his shoes on ice?

21. (I) A truck of mass 5000 kg accelerates on a straight road at 1.2 m/s^2. Assuming that air resistance is negligible, what is the minimum coefficient of static friction between the road and the tires? (Don't worry about differences in normal forces on each wheel and assume four-wheel drive.)

22. (I) The coefficient of static friction between a worker's shoes and the floor is 0.81, while the coefficient of static friction between the floor and a crate is 0.43. The worker, mass 80 kg, pushes the crate, which has a mass of 140 kg. What is the friction force on the worker due to friction between the shoes and the floor when the crate starts to slide?

23. (I) A car of mass 1200 kg is moving at 25 m/s. The driver suddenly sees a dog crossing the road, slams on the brakes, and manages to stop the car in 4.2 s. What is the minimum coefficient of static friction between the tires and the road? (Assume that the acceleration is constant and that there is no skidding.)

24. (I) A rope is connected to a 25-kg cement block placed on a board leaning against a wall at an angle of 25° with respect to the horizontal (Fig. 5–38). The coefficient of kinetic friction between the cement block and board is $\mu_k = 0.4$. (a) What is the tension in the rope if it is pulled at constant speed straight up the board? (b) What is the tension if the rope is pulled up at constant speed at an angle 40° from the horizontal?

▲ **FIGURE 5–38** Problem 24.

25. (I) A worker must push a 58-kg crate across a floor. The coefficient of kinetic friction between the crate and the floor is $\mu_k = 0.63$. What is the minimum force that the worker must exert to keep the crate moving once the crate starts moving?

26. (II) A man wants to push a package of shingles of total mass 15 kg up a roof being built at an angle of 27°. The coefficient of kinetic friction between the package and the roofing paper already in place is $\mu_k = 0.55$. (a) How much force does the man have to exert on the package directly along the slope of the roof to cause the package to accelerate at $0.15\ \text{m/s}^2$? (b) If the coefficient of static friction is 0.58, will the package remain on the roof?

27. (II) A person learning to snow ski will use the snowplow position; it is a rudimentary way of keeping one's skiing speed under control. Let's imagine that a beginning skier finds herself on an icy slope of 22°. Only by setting her skis in a good snowplow position, with the tips of both skis pointed inward and the inner edges dug in, is she able to keep from accelerating. Effectively, what coefficient of sliding friction is created by the snowplow?

28. (II) A crate of mass 250 kg is loaded on the back of a truck. The coefficient of static friction between the crate and the truck bed is μ_s. The truck decelerates such that it comes to a stop from a speed of 60 mi/h (26.7 m/s) in a distance of 140 m. How large must μ_s be so that the crate does not slide forward on the truck bed?

29. (II) The coefficient of static friction between a car of mass 1500 kg and an asphalt road is $\mu_s = 0.70$. (a) What is the shortest distance over which the car can accelerate from rest to a speed of 96 km/h? (b) How long will this take? (Because we neglect drag and rolling friction, the distance and time will be unusually short.)

30. (II) A block of mass $m = 2.0$ kg is placed on a horizontal surface. The coefficient of static friction between the block and the surface is $\mu_s = 0.40$. A light rope is tied to the block and thrown over a frictionless pulley (Fig. 5–39). The free end of the rope is pulled with a slowly increasing force T. At what value of T will the block start to move if the angle that the rope makes with the horizontal is 50°?

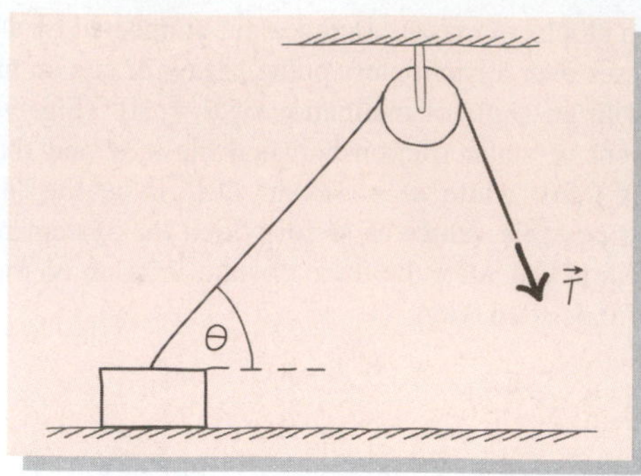

▲ **FIGURE 5–39** Problem 30.

31. (II) A 500-g box is placed on a board at a 35° incline and accelerates from rest down the board at $0.5\ \text{m/s}^2$. (a) How long does the box take to travel down the board of length 0.8 m? (b) What frictional force opposes the motion of the box? (c) What is the coefficient of friction between box and board?

32. (II) A block of mass 0.70 kg rests on top of another block of mass 0.90 kg, which rests on a frictionless surface. The coefficient of static friction between the blocks is $\mu_s = 0.45$. What is the maximum horizontal force that can be applied to the upper block so that the blocks accelerate together without the upper block sliding on the lower one? If the horizontal force is applied to the lower block instead, what is the maximum force that will give rise to the same motion?

33. (II) A 50-kg box rests on a rough horizontal surface with which the box has a large coefficient of static friction, $\mu_s = 0.75$. The box is pulled by means of a light rope with a force of magnitude F, making an angle θ with the horizontal. (a) Find the magnitude of the force F that will just start the box moving horizontally as a function of θ. (b) Show that there is some angle θ for which F takes a minimum value. What is this value for our case and what is the force F corresponding to this value? Explain why, physically, there is such a minimum value.

34. (II) A pile of snow at the crest of a roof with a slope of 40° from the horizontal starts to slide off. The distance from the crest to the edge of the roof is 8 m, and the coefficient of kinetic friction for the snow on the roof is 0.1. (a) What is the speed of the pile of snow when it reaches the edge of the roof? (b) Assuming that it is 6 m from the edge of the roof to the ground, how far out from the base of the building does the snow land?

35. (II) Consider again the professor of Example 5–9. Her mass is 55 kg and the coefficient of static friction between her and the floor is μ_1. What is the minimum value of μ_1 for which she will not slip on the floor?

36. (III) A mass m_1 rests on top of another mass, m_2, which in turn rests on a frictionless horizontal surface (Fig. 5–40). A light cord is attached to m_2, which is used to pull on it with a force F. (a) Find the acceleration of each object when the surface between the two objects is frictionless. (b) Find the acceleration of each object when the surface between the two objects is rough

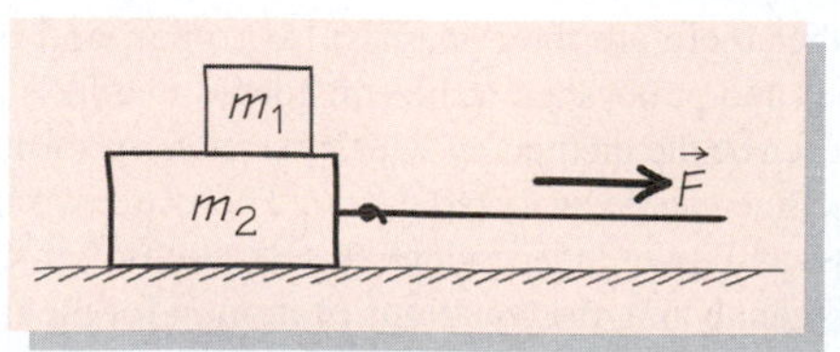

▲ **FIGURE 5–40** Problem 36.

enough to ensure that m_1 does not slide on m_2. (c) What are the magnitude and direction of the contact forces exerted by the lower object on the upper one assuming that the upper object is sliding on the lower object with a nonzero coefficient of kinetic friction μ_k? (d) Find the acceleration of each object if the surface between the two objects is such that the upper object is sliding on the lower one under the influence of kinetic friction, with a coefficient of kinetic friction μ_k.

5–3 Drag Forces

37. (I) Estimate the drag force on an automobile cruising at 65 mi/h. Assume that the drag coefficient C_D is 0.45 and that the car's cross-sectional area is 4 m^2. Take air to have a density of 1.25 kg/m^3.

38. (I) A parachute is rigged so that a parachutist of total mass 116 kg with full gear reaches the ground at a terminal speed of 4.9 m/s. Assuming that the drag force on the parachutist, moving with speed v, has a magnitude equal to kv^2, what is the value of k?

39. (I) A ball of mass 500 g is observed to reach its terminal speed of 18 m/s after being dropped from the top of a tall building. Assume that the density of air is 1.25 kg/m^3 and the drag coefficient C_D is 0.40. What is the effective cross-sectional area of the ball? You can use Eq. (5–30) here.

40. (I) A race car of mass 800 kg has a maximum acceleration from rest of 4.8 m/s^2. Assume that the car's engine is such that the force on the tires is constant and that the car's effective cross-sectional area into the air is 1.8 m^2. If the car's top speed is observed to be 90 m/s, what is its drag coefficient C_D? Take the density of air as 1.25 kg/m^3.

41. (II) A barge moving at uniform speed is pulled by two horses moving on opposite sides of a canal in which the barge floats. The ropes connecting the horses to the barge make an angle of 32° with the line of motion of the barge. If the resistance to the motion of the barge is characterized by a frictional force given by $F = -(220\ \text{N}\cdot\text{s/m})v$, where v is the speed of the barge in meters per second and the tension in each of the ropes is 74 N, what is the speed of the barge?

42. (II) A sphere of radius r_1 and mass m_1 falls through the air and is found to have a terminal speed v_1. Equation (5–30) applies. (a) What is the terminal speed v_2 for a sphere with twice the radius and the same mass density as the first sphere? (b) Generalize the result of part (a) to find the terminal speed for a sphere with a radius z times the first radius but the same mass density.

43. (II) The terminal speed of a skydiver of mass 75 kg can be controlled by the orientation of her body and can range from 40 to 60 m/s. Assume that she can change the area presented to the ground by a factor of 1.5 in going from the minimum to the maximum terminal speed and that the larger area presented to the ground slows the skydiver down. Now, how does the drag coefficient change? (Express your answer as a fraction.)

44. (II) Assume that the drag force on a parachute is given by Eq. (5–30) in the text. The effective area of the parachute is 30 m^2 and the density of air is 1.25 kg/m^3. If a 90-kg parachutist finds that his terminal speed is 6.0 m/s, what is the drag coefficient?

45. (II) A parachutist of 60-kg mass uses a parachute similar to the one used by the parachutist of the previous problem. Determine the terminal speed of this lighter person.

46. (II) A marble of mass m falls through a fluid and is subject to the drag force $\vec{F}_D = -A\vec{v}$, where $\vec{v}$ is the velocity of the marble. The marble will reach a terminal speed given by $\vec{v}_t = m\vec{g}/A$. Use dimensional analysis to estimate how long it will take to reach the terminal speed. [*Hint*: A characteristic "time" can be constructed from A, g, and m.]

5–4 Forces and Circular Motion

47. (I) An airplane of mass 2×10^4 kg executes a banked turn of radius 30 km while flying at 200 m/s. What acceleration will the passengers have as a result of the turn?

48. (I) A rock swings in a nearly horizontal circle at the end of a string whose breaking tension is 26 N. The circular path is 0.35 m in radius, and the rock's mass is 220 g. What is the maximum speed the rock can have before the string breaks?

49. (I) A man of mass 65 kg stands at the edge of a merry-go-round of radius 5.3 m. The merry-go-round turns at 6.0 rev/min. What are the magnitude and direction of the net force on the man?

50. (I) An accelerometer shows that an airplane flying at 650 km/h undergoes a vertical acceleration of 0.30 g's ($1\ g = 9.8\ m/s^2$) at a certain moment. What is the radius of curvature of the airplane's (horizontal) path at that point?

51. (II) An automobile makes a turn whose radius is 150 m (Fig. 5–41). The road is banked at an angle of 18°, and the coefficient of friction between the wheels and road is 0.3. Find the maximum and minimum speeds for the car to stay on the road without skidding up or down the banked road.

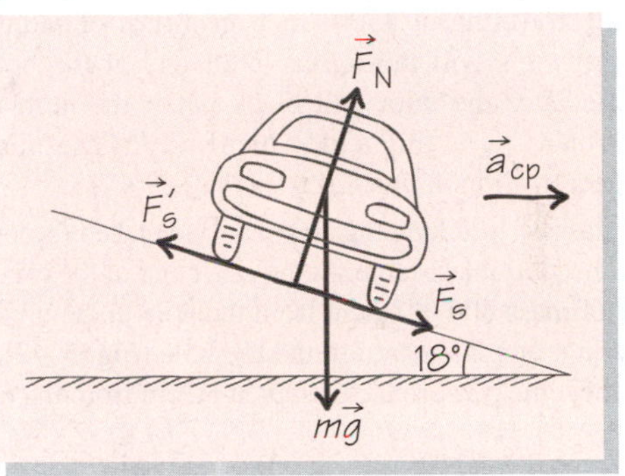

▲ **FIGURE 5–41** Problem 51.

52. (II) A merry-go-round has a circular platform that is 1 m from the central axis at its inner edge and is 5 m from the central axis at its outer edge. The ride turns at a rate of one full rotation every 10 s. A passenger holds himself to the surface with a pair of very sticky shoes and is most comfortable when he orients his body length along the line of the net force on him. Determine the angle his body makes to the vertical (a) 1 m from the axis, (b) 3 m from the axis, and (c) 5 m from the axis.

53. (II) A spring is 0.6 m long. One end is permanently attached to a pivot on a horizontal table top. The force that it exerts on a body attached to the other end is $(4.2\ \text{N/m})x$, where x is the distance that the spring is stretched beyond its normal length. Suppose a mass of 400 g is attached to the other end of the spring and the whole system is set in circular motion. (The entire motion takes place on the horizontal table top.) How far will the spring be stretched if the mass rotates with an angular velocity of 8 rad/s?

54. (II) A student carrying an accelerometer on a large roller coaster measures a value of 0.93 m/s^2 perpendicular to the track when going over the top of a circular portion of the track. At another curve the student finds the accelerometer to read 2.20 m/s^2 at the top of the curve. The student has no way of knowing how fast she is going in both situations. When she gets off the roller coaster, she estimates that the radius of curvature for the second curve is about one-half the value of the other. What was her relative speed through the second curve with respect to the first?

55. (II) The coefficient of static friction between a small stone and a horizontal turntable is measured in the following way. The stone is placed on the turntable at a distance R from the axis, and the speed of rotation is slowly increased to 33 rev/min. When the experiment is repeated for several different values of R, it is found that the stone remains on the turntable if $R < 21$ cm and that it slides off with increasing speed of rotation if $R > 21$ cm. Determine μ_s from these data.

56. (II) A Ferris wheel in an amusement park has a radius of 30 m and makes one complete turn every 75 s. Calculate the normal force that a passenger of mass 60 kg experiences through the seat of the pants (the seat bottom is parallel to the ground) when the passenger is (a) at the bottom of the path, nearest the ground, and (b) at the maximum height of the path.

57. (II) Assume that the acceleration of the Moon due to Earth's gravity is 0.0027 m/s^2. What is the velocity of the Moon with respect to Earth if the period of the Moon's motion around Earth is 28 days? Do not look up the distance between Earth and the Moon; instead, calculate it. Compare this result with the distance given in Appendix III–1.1.

58. (II) A fighter pilot makes a dive almost vertically down and pulls up while traveling at 1500 km/h in a turn of radius 1.75 km. How many g's will the fighter pilot feel at the bottom of the dive? Because the pilot will black out if the number of g's is greater than 11, is this a safe maneuver? (The number of g's is the acceleration in units of $g = 9.80$ m/s^2.)

59. (II) A mass of 1.00 kg hangs from a rope placed through a hole in a smooth, frictionless table. At the other end of the rope is attached a puck of mass 400 g, 80 cm from the hole in the table. The puck swings in a circular orbit around the hole (Fig. 5–42). With what speed must the puck rotate to keep it 80 cm from the hole?

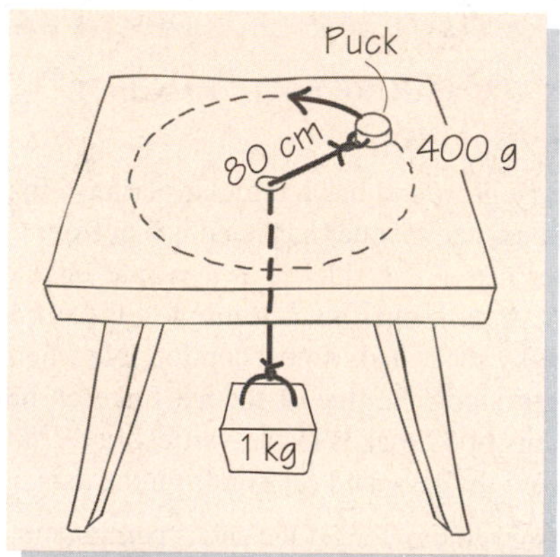

▲ **FIGURE 5–42** Problem 59.

60. (II) A mass m at the end of a string of length r moves at constant speed v in a circle on a frictionless table. (a) Find the tension in the string in terms of m, r, and v. (b) A second mass identical to the first is attached at the midpoint of the string and the two are whirled; the speed of the outer mass is again v. Draw free-body diagrams for the two masses and calculate the tensions in terms of m, r, and v.

61. (II) Consider the *conical pendulum*, a mass on the end of a massless string, with the other end of the string fixed on a ceiling. Given the proper push, this pendulum can swing in a circle at a given angle θ, maintaining the same height h throughout its swing, as shown in Fig. 5–43. (a) What is the free-body diagram for such a pendulum? (b) If the mass of the pendulum is 0.2 kg, the length of the pendulum is 0.5 m, and the angle at which it swings is $\theta = 10°$, what is the speed of the mass as it swings?

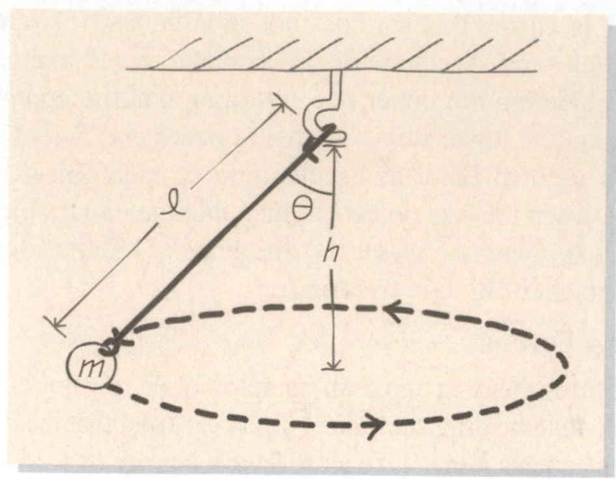

▲ **FIGURE 5–43** Problems 61 and 63.

62. (II) An object moves along a trajectory that is parallel to the unit vector $\hat{r}$ of some fixed coordinate system. Are any of the statements below true? Which ones and why or why not?
(a) Any force acting on the object is perpendicular to $\hat{r}$.
(b) If the magnitude of the velocity of the object is unchanging, the force acting on it is zero.
(c) Any force acting on the object must be aligned with $\hat{r}$.

63. (II) A heavy bob is attached to one end of a string whose other end is attached to a hook on the ceiling. The system acts as a conical pendulum (see Fig. 5–43), with the string making an angle of 30° with the vertical and the bob traveling in the horizontal plane at an angular velocity of $\pi/2$ rad/s. How long is the string?

64. (II) A small mass slides without friction in a horizontal circular path around the sides of a circular bowl. The bottom of the bowl may be described as a parabola, with the height h above the bottom varying quadratically with the distance r from the axis: $h = br^2$. The mass is observed to move in its circular path with a speed v. What is the height of the path?

65. (II) A small block slides in a horizontal circle on the inside of a conical surface, with the cone making an angle of 44° with the vertical (Fig. 5–44). Assuming that there is no friction between the block and the surface and the block slides with an angular speed of 3.8 rad/s, at what vertical height above the apex of the cone does the block slide?

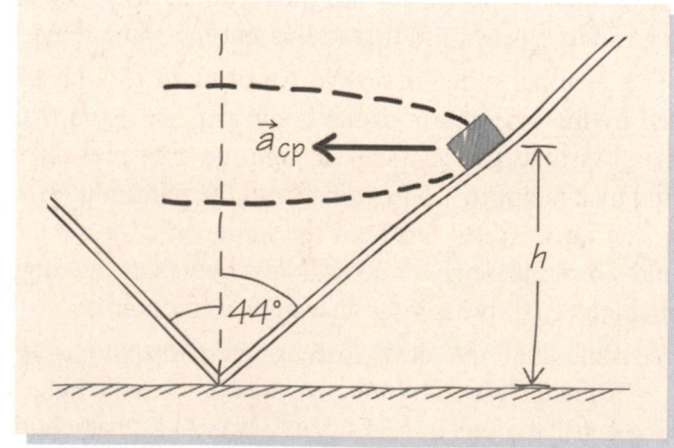

▲ **FIGURE 5–44** Problem 65.

66. (II) Two light strings 1.0 m in length are attached to a vertical support 1.0 m apart, and a mass of 5.0 kg at the end of the two strings is whirled about the vertical z-axis (Fig. 5–45, see next page). Both strings are taut, so that they and the vertical support form an equilateral triangle. The tension in the upper string is measured to be 150 N. (a) What is the tension in the lower string? (b) How much time does it take for the apparatus to make one complete circuit around the vertical support?

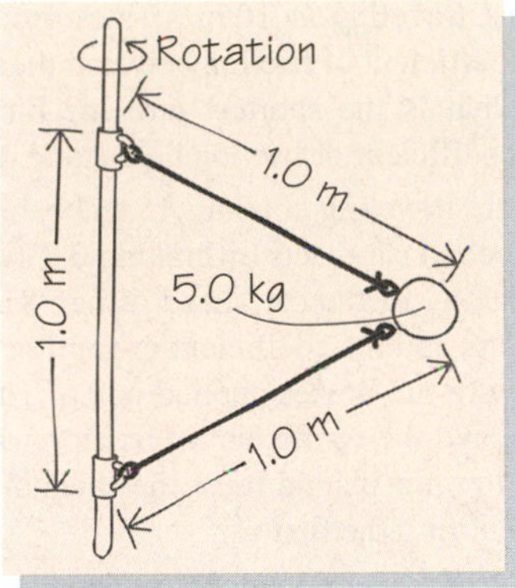

▲ **FIGURE 5–45** Problem 66.

67. (II) The road in Example 5–16 is now in better condition, and the coefficient of static friction between the tires and the road is 0.8. With what maximum constant speed can the motorist now negotiate the curve? Friction acts here both to provide a centripetal force and to help oppose the vertical forces. Is there one and only one speed with which the curve can be taken, as in Example 5–16? Why or why not?

68. (II) A train of mass 1.5×10^5 kg is traveling horizontally at 80 km/h and rounding a bend whose radius of curvature is 2 km. At the same time it is decelerating at a rate of 0.2 m/s^2. The length of the train is negligible compared with the size of the bend, and the train can be treated as a point. What net force does the track exert on the train? Give an approximate answer in which the net change in speed is small compared to the speed itself. Such questions help engineers decide how "robustly" a track must be constructed or how much to bank it.

69. (II) A small puck of mass $m = 0.1$ kg moves in a circle of radius 0.3 m on a table top; the puck is tied with a massless string to a tether at the origin. The coefficient of kinetic friction between the puck and the table top is $\mu_k = 0.25$. At $t = 0$ s, the puck has a velocity in the $+y$-direction of magnitude 8.0 m/s. (a) What is the tension in the string at $t = 0$ s? (b) What is the tension in the string at the end of one revolution?

70. (II) A pendulum hangs at rest from a hook in a ceiling of a building. The building is located at a latitude such that the radius vector from the center of Earth to the building makes an angle θ with Earth's axis of rotation. Assume that the force of gravity is the same everywhere and points directly to the center of Earth. What is the angle that the pendulum makes with the vertical because of Earth's rotation? (By definition, Earth rotates about its axis once a day.)

71. (III) A satellite of mass 3000 kg travels in a circular orbit 180 km above Earth, where the acceleration due to gravity is 5 percent smaller than on Earth's surface. Assume that, in a year, the satellite loses 5.0 km in altitude because of the drag of the extremely thin atmosphere at that altitude. What would you estimate the density of air to be at that altitude given that the effective area of the satellite is 6.0 m^2 and the drag coefficient C_D in Eq. (5–30) is 1.0?

72. (III) Consider a ball thrown outward from the center of a platform that rotates counterclockwise with uniform angular velocity ω (Fig. 5–46). An observer standing off the platform (in an inertial reference frame) will describe the ball as moving with uniform velocity in a straight line. (Ignore the effect of gravity; imagine looking down on the platform from above, so that you do not see the up-and-down motion of the ball.) The inertial observer will see that the ball reaches a horizontal distance r from the center in time $t = r/v$, where v is the speed of the ball. (a) Show that, in time t, a point at a radius r on the platform will have moved a distance $d = v\omega t^2$. (b) An observer moving with the platform will see the ball curve away to the right (as seen from the center of the platform). Show that the perceived acceleration is perpendicular to the velocity vector and that its magnitude is $2\omega v$. (c) What is the direction of the perceived acceleration of a ball thrown by a person on the platform from the rim toward the center? It is not surprising that an observer in an accelerated frame sees force-free motion that nevertheless undergoes acceleration. Under certain circumstances—for example, in the study of global wind motion on a rotating Earth—it is convenient to study motion as seen in an accelerating frame. The frame-imposed acceleration is then attributed to a purely fictitious force, which, for rotating systems like the one treated in this problem, is called the *Coriolis force*.

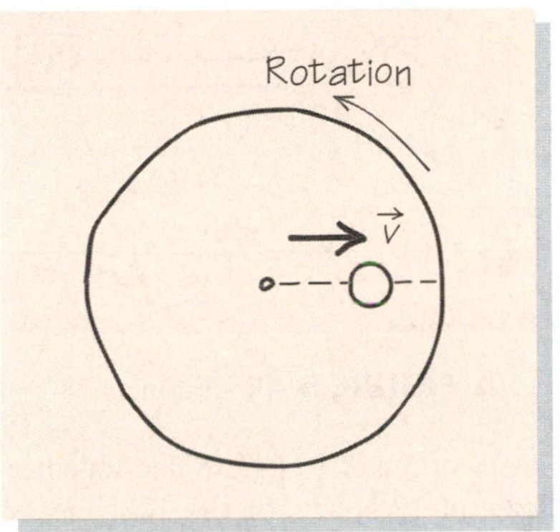

▲ **FIGURE 5–46** Problem 72.

General Problems

73. (I) A 5.0-kg sphere at the end of a 1.2-m cable swings in a horizontal circle on a frictionless surface at the rate of one revolution every 1.4 s. What is the tension in the cable?

74. (II) A small die that is placed 4 in from the center of a turntable begins to slide. The turntable is rotating at 45 rev/min. What is the coefficient of static friction between the die and the turntable?

75. (II) Three masses (from left to right: 0.3, 0.4, and 0.2 kg) are connected by light cords to make a "train" sliding on a frictionless surface. They are accelerated by a constant horizontal force $F = 1.5$ N that pulls the rightmost mass to the right. What is the tension T in the cord (a) between the 0.3- and 0.4-kg masses and (b) between the 0.4- and 0.2-kg masses?

76. (II) A string 6.95 m long is strung between two pegs (4.96 m apart) on a ceiling. A mass of 3.88 kg is attached to a point 2.96 m along the string. What are the tensions in the two segments of the string?

77. (II) Consider a system of masses connected by light ropes that pass over massless and frictionless pulleys (Fig. 5–47). (a) When

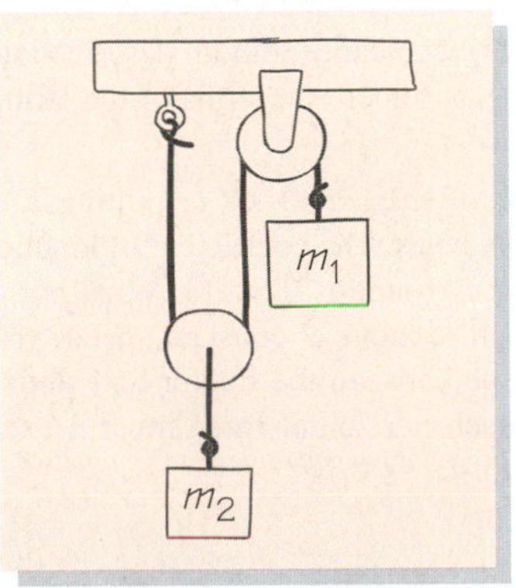

▲ **FIGURE 5–47** Problem 77.

m_1 is displaced vertically by Δx_1, what is the displacement Δx_2 of m_2? (b) For $m_1 = 1.2$ kg and $m_2 = 1.8$ kg, calculate the respective accelerations of the two masses. (c) What is the tension in the string for the masses given in part (b)?

78. (II) Masses $m_1 = 0.80$ kg and $m_2 = 1.10$ kg are connected by a taut rope. Mass m_2 is just over the edge of a ramp inclined at an angle of $\theta = 25°$, as in Fig. 5–48, and the masses have a coefficient of kinetic friction $\mu_k = 0.25$ with the surface. At $t = 0$ s, the system is given an initial speed of $v_0 = 1.2$ m/s, which starts mass m_2 down the ramp. (a) Draw the free-body diagram for each mass. (b) Solve the equations of motion to predict the motion of the system with time. Assume that the rope is long enough so that mass m_1 does not hit the pulley.

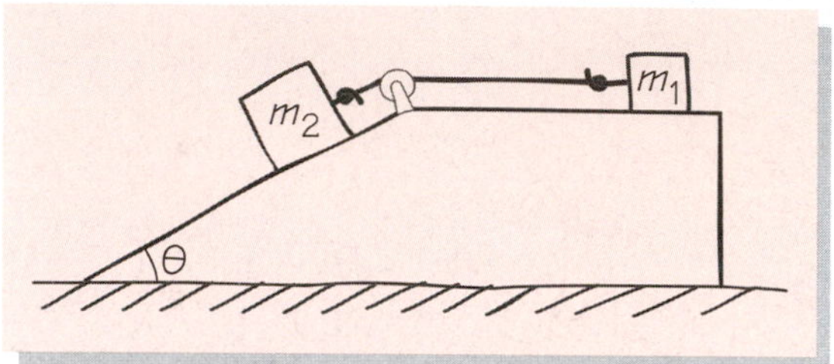

▲ **FIGURE 5–48** Problem 78.

79. (II) Two masses of 5 and 10 kg are tied together by a massless spring. A force of 12 N acts on the 10-kg mass directly away from the 5-kg mass. At a particular instant the 5-kg mass has an acceleration of 3 m/s^2 in the direction of the larger mass. What is the acceleration of the 10-kg mass?

80. (II) A light cord attached to a ball will break if the tension in the string exceeds 40 N. The ball has a mass of 150 g. A student swings the ball in a circle in the vertical plane. (a) You bet that the cord will break when the ball is at the top of the circle. Will you win the bet? (b) If the length of the cord is 1.0 m, how fast will the ball be moving when the cord breaks?

81. (II) One of the entertainments at the carnival is a rotating cylinder. The participants step in and place themselves against the interior wall. The cylinder starts to rotate more and more rapidly, and at some point the floor falls away, leaving the customers stuck like so many flies to the wall. If the cylinder were to slow down without the floor coming back up, the participants would begin to slip down. In terms of the relevant parameters, express the rotational speed ω at which this happens.

82. (II) A motorcycle moves in a horizontal circular path on the inside surface of a vertical cylinder of radius 8 m. Assuming that the coefficient of static friction between the wheels of the motorcycle and the wall is 0.9, how fast must the motorcycle move so that it stays in the horizontal path?

83. (II) Consider the conical pendulum described in Problem 61. Express the angular velocity in terms of the string angle θ and the string length ℓ.

84. (II) A tractor of mass 800 kg is pulling a sled loaded with 1450 kg of hay bales. The coefficient of kinetic friction between the sled and the ground is 0.68. (a) What horizontal force must the tractor exert to move at constant speed? (b) What is the tension in the rope between the tractor and sled? (c) If the tractor stops, how much horizontal force must it exert to get the sled moving again if $\mu_s = 0.70$?

85. (II) A bicyclist traveling at 10 m/s rides around an unbanked curve. If the coefficient of friction between the tires and the road is $\mu = 0.4$, what is the shortest turn the bicyclist can safely make? Is the coefficient of friction here static or kinetic?

86. (II) Two cars are traveling at 60 mi/h, one behind the other. The driver of the second car reacts by braking 0.8 s after she observes the sudden braking of the car ahead of her. The front car has a mass of 1200 kg, and the coefficient of friction between the tires and the road with the brakes applied is 0.8. The second car has mass 1600 kg, and the coefficient of friction with the brakes applied is 0.7. How far behind must the second car have been in order to avoid hitting the first car?

87. (II) A stunt motorcyclist rides with uniform speed on the inside rim of a vertical circular ramp of radius 8 m (Fig. 5–49). How fast must the motorcyclist travel to avoid leaving the surface at the top of the loop?

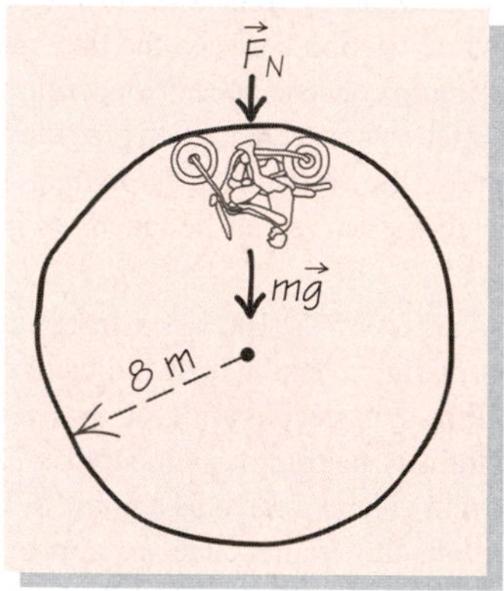

▲ **FIGURE 5–49** Problem 87.

88. (II) A pith ball is a typically small ball that is very light and can be given electrical charge—electrical charges exert forces on one another. If two such balls, labeled 1 and 2, are given charges q_1 and q_2, respectively, then the force on one due to the other is repulsive (directed along the line between them and tending to push them apart), with magnitude roughly $(9 \times 10^9) \times q_1 \times q_2 \times d^{-2}$ N, where d is their separation, measured in meters, and the unit of charge is the coulomb (C). Two tiny pith balls, mass 0.5 g, are hung from nearly the same point on the ceiling by identical strings , 60 cm long, and given identical electrical charges q. At equilibrium, the balls separate, and the strings are found to make an angle of 60° with one another. Find q.

89. (III) The new white belt of a long horizontal conveyor is moving with a constant speed $v = 3.0$ m/s. A small block of carbon is placed on the belt with zero initial velocity relative to the ground. The block will slip a bit before moving with the belt, leaving a black mark on the belt (Fig. 5–50). How long is that mark if the coefficient of kinetic friction between the carbon block and the belt is 0.20 and the coefficient of static friction is 0.30?

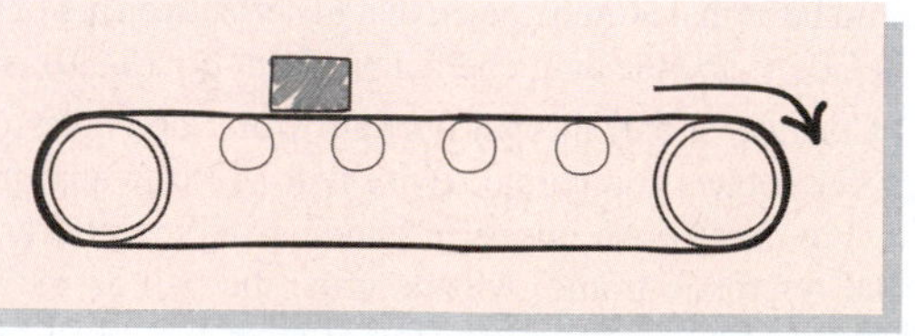

▲ **FIGURE 5–50** Problem 89.

6

As the elevator moves up and down, the force of gravity and the contact forces from the floor do work on the people inside. This work can be linked to changes in energy, and eventually to the principle of the conservation of energy.

Work and Kinetic Energy

A pitcher throws a baseball—can we predict the motion of the ball? In the previous chapters we've seen that as long as we have enough information about initial positions, velocities, and the forces that act on the ball, we can, in principle, use Newton's laws to predict everything that is predictable about the ball's motion. Nevertheless, the ball has a complex shape and spin, and we would have a hard time being certain that we had accurate information on the ball's interaction with the air. Even if we did have all the correct information, the equations would be complex and unlikely to have analytic solutions. Nowadays we can resort to numerical computation on ever more powerful computers when details of a complex motion are needed. For example, the motion of a rocket bound for Pluto is complex because it is affected by the gravitational forces due to Earth, the Sun, and other planets along the way. These forces are all calculable, but Newton's second law must be solved numerically, and such a solution has to be carried out with high precision to get a suitable trajectory.

Scientists of the eighteenth and nineteenth centuries did not have computers and instead developed powerful analytical methods to deal with certain aspects of the motion of objects. Their research led to a much deeper understanding of the dynamical properties of systems than the "black box" of computer programs can possibly provide. One of their most important discoveries was the concept of *energy*. Energy is a quantity that an object or a physical system possesses. As we shall learn in future chapters, there are many forms of energy—kinetic energy, potential energy, thermal energy, electrical energy, chemical energy, nuclear energy, and mass energy. The names of these different forms of energy were due to the circumstances in which they were studied, but the forms of energy are closely linked and the names are unimportant. When considering energy, what is important is to keep in mind the deep physical law, known as the **principle of the conservation of energy**, which states that energy cannot be created or destroyed, only transformed from one form to another. In other words, all the forms of energy are interchangeable, fundamentally one and the same. The law of the conservation of energy is one of the most fundamental of all physical laws, and we will be occupied with it in this chapter and the next.

In this chapter, we are concerned with only one form of energy: energy of motion, or more technically, *kinetic energy*. We start our study of kinetic energy through consideration of a precisely defined quantity called *work*. A force acting on an object can do work on the object when the object moves. Thus all the forces acting on a baseball moving through the air do work on it. The hand that throws the ball, gravity, and the friction, or drag, forces due to the air all do work on the baseball and in doing so cause the kinetic energy of the ball to change. The close relation between work and energy is a practical tool, useful for learning about the aspects of the baseball's motion that don't depend directly on time—for example, how fast the baseball is moving when it is at a certain location. In Chapter 7, we'll change our focus away from work and toward seeing how we can view the connection between work and energy as a conservation law† for energy. In doing so, we'll begin to expand our ideas about energy to forms other than kinetic energy.

6–1 Kinetic Energy, Work, and the Work–Energy Theorem

We begin by rethinking some simple features of motion in one dimension. Consider, for example, a bobsled with mass m being pushed by some bobsledders on flat ground (Fig. 6–1) over an icy surface. (We'll assume there is no friction between the sled and the surface.) The bobsled's motion is described by a position x and a velocity v—as the motion is strictly one dimensional, we can dispense with vector notation. When the bobsled is at rest, we assign it a value of **kinetic energy** $K = 0$. When the bobsled moves, its kinetic energy will have a nonzero value. In other words, its kinetic energy has something to do with its motion. This energy is supplied by the forces applied by the three men pushing the sled. We have already learned that if an object such as the bobsled changes its velocity from $v = 0$ to a nonzero value, it must, by definition, accelerate. For our initial analyses we'll suppose this to be a *constant acceleration* a. After moving a distance Δx, subject to this acceleration and starting from rest, the bobsled will acquire a speed given by [see, e.g., Eq. (2–24)]

▲ **FIGURE 6–1** As the bobsled is pushed, it gains kinetic energy.

$$2a\,\Delta x = v^2. \tag{6–1}$$

We also know that a net force F_{net} leads to an acceleration, and here this force is supplied by the men pushing the sled. For constant acceleration the net force will be *constant*. Thus our result may be rewritten in a different form. Multiplying by $\frac{1}{2}m$ and replacing the combination ma by F_{net} gives

$$F_{\text{net}}\,\Delta x = \tfrac{1}{2}mv^2. \tag{6–2}$$

This equation is the root of all that will follow in this chapter. On the right we have an expression that is associated with the motion of our bobsled, which we identify with the bobsled's **kinetic energy**. On the left are the quantities that produce this kinetic energy: a net force acting on the bobsled and the sled's displacement. This side of our relation is the **work** W_{net} that the net force does on the bobsled. Equation (6–2) is the starting point for all that follows in this chapter in that we will find ways to generalize work to situations where the force can vary and where the motion is more than one dimensional. Similarly, we will find ways to generalize the kinetic energy to cases where the motion is not one dimensional and the object is not treated as a point. The main point to re-

†It is important to realize from the outset that when we talk about a conservation law, we mean that there is something that stays constant in time, not the more colloquial meaning of not wasting it!

member throughout is that *when a net force acts to displace an object, it transmits (kinetic) energy to it*. The energy transmitted will be W_{net}, which is *the net work done by the net force on the accelerating object*, or briefly, the *work*. In the case of our constant force in one dimension,

$$W_{net} \equiv F_{net}\,\Delta x. \qquad (6\text{–}3)$$

CONSTANT NET FORCE IN ONE DIMENSION

The object's resulting *kinetic energy* K is given by

$$K \equiv \tfrac{1}{2}mv^2. \qquad (6\text{–}4)$$

KINETIC ENERGY

If the object is not initially at rest but has initial velocity v_0, then the results of Chapter 2 for constant acceleration show us that Eq. (6–1) is replaced by

$$v^2 - v_0^2 = 2a\,\Delta x \qquad (6\text{–}5)$$

and Eq. (6–2) is replaced by

$$W_{net} = \tfrac{1}{2}mv^2 - \tfrac{1}{2}mv_0^2 = \Delta K, \qquad (6\text{–}6)$$

WORK–ENERGY THEOREM

where for a constant force, $W_{net} = F_{net}\,\Delta x$. This equation is known as the **work–energy theorem**. It cements our interpretation that the left side of Eq. (6–2), the net work, is the kinetic energy transmitted. The work done by the net external force to accelerate the object from its initial velocity v_0 to its final velocity v equals the *change* in the object's kinetic energy from its initial value $K_0 = \frac{1}{2}mv_0^2$ to the final value $K = \frac{1}{2}mv^2$. The kinetic energy change is $\Delta K = K - K_0$. If we write Eq. (6–6) as

$$K_0 + W_{net} = K, \qquad (6\text{–}7)$$

then we can view the work–energy theorem as saying that if we start with an object with kinetic energy K_0 and do net work on it, the sum of these is the final kinetic energy. Again, the net work is the amount of additional kinetic energy that our object has been given. We will find more general versions of the work–energy theorem for motion in three dimensions and nonconstant forces.

Remember, work is something done to an object by a force as the object is displaced (Fig. 6–2); kinetic energy is something that the object possesses by virtue of having work done on it.

The *sign* of the work can be either positive or negative, and accordingly the kinetic energy can either increase or decrease. We see from the definition of work in Eq. (6–3) that the sign of W_{net} is determined by the relative sign of F_{net} and of the displacement Δx. If the net force is directed in the same direction as the displacement (they have the same sign), then the net work done by that force is positive, and it increases the kinetic energy of the object to which the force is applied. If the force acts in a direction *opposite* to the displacement (they have opposite signs), the work done is negative, and the object's kinetic energy decreases. Consider our bobsled. Suppose that the bobsledders have acted to give the sled a forward velocity and decide that they want to bring it to a halt, always on level ground. They therefore run to the front of the sled and apply a force opposite to its motion, doing work on the sled even as the sled continues to move forward (Fig. 6–3). The displacement is to the right while the pushing force is to the left. The directions of the force and the displacement are opposite, so their product (the work done) is negative, and the change in kinetic energy is negative—the sled slows down.

A cautionary note: *Work* and *energy* have technical meanings which we have been careful to spell out, but these terms also have colloquial meanings that may differ from our definitions. For example, holding a bag of groceries while waiting for a bus would ordinarily be described as work. In our definition this cannot be work because work involves displacement and the bag of groceries does not move. In a technical sense the

▲ **FIGURE 6–2** As he raises the barbell from a position near the floor to a position at his shoulder, this weightlifter does positive work. However, the weightlifter does no work in holding the barbell stationary at his shoulder.

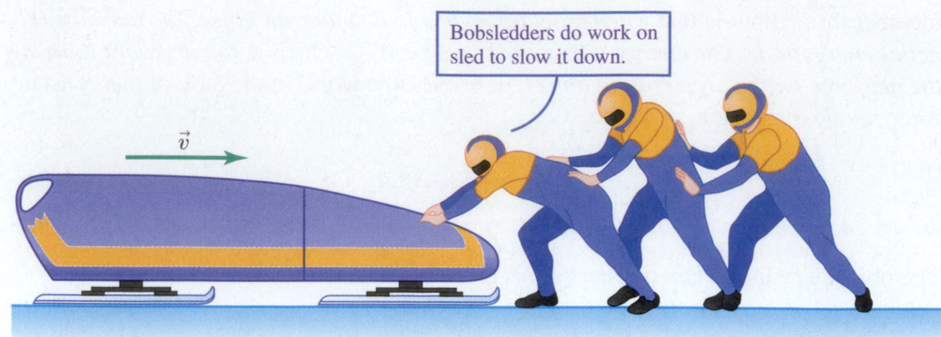

▶ **FIGURE 6–3** Bobsledders bring the sled to a stop by applying a force opposite to the displacement of the sled.

force exerted by our muscles does no work on the bag. The word *energy* also has a variety of colloquial meanings, but in physics the meaning of the word is very tightly prescribed, even if we deal with energy in many different contexts.

THINK ABOUT THIS . . .

WHAT WORK IS DONE WHEN WE LIFT AN OBJECT FROM THE FLOOR?

We remarked above that no net work is done on a bag of groceries when it is held steady—the fact that the person holding the bag may feel tired as a result is irrelevant in the definition of work, which requires a displacement. That same person may also have lifted the bag from the floor. Is there any net work involved in that case? We can break up the motion into three parts: (i) starting at floor level the bag is brought up to lifting speed in a short time (and moves a short distance in that time); (ii) the bag is raised smoothly, at a constant speed, to the desired height; and (iii) in a short period of time the bag is brought to a halt. During stage (i) the person lifting must exert an upward force exceeding the force of gravity on the bag. There is a net force, there is a displacement in the same direction, and there is positive net work done on the bag. The bag accelerates, gaining kinetic energy. During stage (ii) there is a displacement but no net force—the contact force of your hand is equal and opposite to the force of gravity—and hence no net work. The kinetic energy of the bag remains unchanged. Finally, during stage (iii) there is a net force, a (small) displacement, and net work is done. The kinetic energy changes, but the change is negative. The net work is also negative and is due primarily to gravity, which points down, even as the displacement continues upward. Over all three stages we can say that no net work is done. The positive net work done in stage (i) is canceled by the negative net work done in stage (iii): The bag started at rest and it ends at rest.

Units of Work and Energy

Work and kinetic energy must have the same dimensions. Their units correspond to these dimensions. Work is measured in $(\text{newtons})(\text{meters}) = (\text{kg}\cdot\text{m/s}^2)(\text{m}) = (\text{kg})(\text{m/s})^2$, and kinetic energy is measured in $(\text{kg})(\text{m/s})^2$; these are indeed the same. The unit of work and energy is so important that it has a special name in SI, the **joule** (J). One joule is the work done by the force of one newton pushing an object a distance of one meter along the direction of the force. Thus $1\ \text{J} = 1\ \text{N}\cdot\text{m}$.

In the cgs system, the unit of energy is the **erg**, which is the work done by a force of one dyne in moving an object a distance of one centimeter. The conversion is the following:

$$\begin{aligned} 1\ \text{erg} &= (1\ \text{dyne})(1\ \text{cm}) = (1\ \text{g}\cdot\text{cm/s}^2)(1\ \text{cm}) \\ &= (1\ \text{g}\cdot\text{cm}^2/\text{s}^2)(1\ \text{kg}/10^3\ \text{g})(1\ \text{m}^2/10^4\ \text{cm}^2) \\ &= 10^{-7}\ \text{kg}\cdot\text{m}^2/\text{s}^2 = 10^{-7}\ \text{J}. \end{aligned}$$

EXAMPLE 6–1 Consider the kinetic energies given in Table 6–1.

(a) Given that the electron mass is 0.9×10^{-30} kg, how fast is it moving in its orbit around the nucleus?

(b) If the sprinter runs 100 m in 10 s, what is her mass?

(c) Given that Earth's mass is 6×10^{24} kg, how fast is it moving around the Sun?

Strategy In each of the cases we can use Table 6–1 and the relation $K = \frac{1}{2}mv^2$ [Eq. (6–4)] and solve for the unknown.

Working It Out (a) Here the unknown is the speed v. It follows from $K = \frac{1}{2}mv^2$ and from the given values of K and m that the electron speed is

$$v = (2K/m)^{1/2} \approx 10^6 \text{ m/s}.$$

(b) In this case we know $v = 10$ m/s; hence we use $m = 2K/v^2 =$ 20 kg. This doesn't mean that we are dealing with a five-year-old prodigy sprinter! Rather, Table 6–1 only gives values to the nearest power of 10.

(c) With Earth's mass m, we find its orbital speed from $v = (2K/m)^{1/2} = 2 \times 10^4$ m/s.

TABLE 6–1 • Some Orders of Magnitude for Kinetic Energies

System	Kinetic Energy (J)
Electron in orbit around a nucleus	10^{-18}
Molecule of air at room temperature	10^{-17}
Electron in a TV tube	10^{-15}
Walking ant	10^{-8}
Falling raindrop	10^{-3}
Running human	10^3
Automobile on a highway	10^5
Cruising airplane	10^{11}
Large earthquake	10^{17}
Earth in orbital motion around the Sun	10^{33}

THINK ABOUT THIS . . .

HOW CAN WE ESTIMATE THE ENERGY OF A SNOW AVALANCHE?

Estimating a quantity always involves having some knowledge of the physical system involved. In this case, you may know that a snow avalanche typically starts in a steep mountain gulley and that it is a "shelf" of snow forming a layer over a previously frozen surface which slides down the gulley. This layer will likely have formed as the result of a snowfall on the earlier surface, so a reasonable value for its thickness is 15 cm (about 6 in), and you could guess that a plaque perhaps 50 m^2 in area breaks off and slides. The volume is thus $50 \times 50 \times 0.15 \text{ m}^3 = 375 \text{ m}^3$. You might also know that a foot of snow is equivalent to about an inch of rain, suggesting that snow is roughly 10 percent as dense as liquid water, which has a mass density of $1 \text{ g/cm}^3 = 10^3 \text{ kg/m}^3$. We therefore give our snow shelf a density of 10^2 kg/m^3, and the mass of the moving snow is the product of volume and density, around 4×10^4 kg. At this point we can use the work–energy theorem. We might take a height $h = 500$ m for our mountain valley, and we'll assume that the avalanche moves approximately vertically. The work done by gravity uses the fact that the force has magnitude $mg = (4 \times 10^4 \text{ kg})(10 \text{ m/s}^2) = 4 \times 10^5 \text{ kg}\cdot\text{m/s}^2$. We then estimate the work done by gravity as $W = F \times h = mg \times h = (4 \times 10^5)(500) \text{ J} = 2 \times 10^8$ J. This is our estimate for the kinetic energy. It is interesting to convert this estimate to a speed for the mass of snow: We have $v^2 = 2K/m = 2(mgh)/m = 2gh$. (Not surprisingly, this is the speed squared we would find from the kinematics of an object of any mass falling under constant acceleration g.) Numerically, $v^2 = 2(10)(500)(\text{m/s})^2$, or $v = 100$ m/s. This is nearly 200 mi/h, and even though it does not take into account drag due to air, it is not far off of measured values. ■

CONCEPTUAL EXAMPLE 6–2 A furniture mover pushes a sofa across a carpeted floor. The mover applies a horizontal force of magnitude F to the sofa, and he and the sofa move with uniform velocity. Discuss the notion of work and energy for this situation. What is different if the sofa is pushed with the same force on the frictionless surface of a slick floor (but the mover has nonslip soles on his shoes)?

Answer The sofa moves with uniform velocity, that is, it does not accelerate. Therefore there is no *net force* acting on it. The force applied by the mover has magnitude F, and therefore the force of kinetic friction between sofa and floor has the same magnitude and points in the opposite direction. If the net force is zero, no work is done on the sofa; equivalently, the sofa's kinetic energy is unchanging. (There is no motion in the vertical direction, and we can treat the entire problem as a one-dimensional one.) The mover does work $W = F\,\Delta x$ in displacing the sofa by Δx, but this work alone does not go into increasing the kinetic energy of the sofa; only the net work does that. Is there nevertheless a transfer of energy associated with the work done by the mover? Yes, the work done by the mover *must* involve a transfer of energy, but it does not go into the kinetic energy of an object. Instead the work is just the right amount to overcome the work done by friction in response to the sofa's motion. The work done by the mover goes into heating the sofa's feet and the floor and into abrasion. We will learn about this in more detail later.

In the case of the slippery floor (no kinetic friction), the *net* force is just the force exerted by the worker. Net work is done as the sofa moves, so the kinetic energy of the sofa changes, that is, there is an acceleration. This increased kinetic energy is given by the work–energy theorem [Eq. (6–6)].

What Do You Think? We speak of the mover's energy being used to push the sofa. What energy are we talking about? *Answers to **What Do You Think?** questions are given in the back of the book.*

EXAMPLE 6–3 In a movie stunt, a car of mass 1200 kg falls a vertical distance of 24 m starting from rest (Fig. 6–4a). What is the work done by the force of gravity on the car? Use the work–energy theorem to find the final velocity of the car just before it hits the water. (Treat the car as a pointlike object.)

Setting It Up We draw a sketch of the car falling in Fig. 6–4b. We only need the vertical y-axis, which points upward. The car, of given mass m, has a downward displacement $\Delta y = y_f - y_i = 0\text{ m} - 24\text{ m} = -24\text{ m}$ and an initial speed $v_0 = 0$. We want to find the work done by gravity on the car and the car's final speed v_f.

Strategy The force of gravity is the only force acting, so it makes the only contribution to the net work. We can calculate the work done by gravity and then use the work–energy theorem to find the final speed.

Working It Out Gravity has magnitude mg and is oriented downward, so the y-component of the net force is $F_{\text{net}} = -mg$. Therefore the net work is

$$W_{\text{net}} = F_{\text{net}}\,\Delta y = (-mg)\,\Delta y$$
$$= (1200\text{ kg})(-9.8\text{ m/s}^2)(-24\text{ m}) = 2.8 \times 10^5\text{ J}.$$

Note that this work is positive. We can now use the work–energy theorem, Eq. (6–6), to find v_f. The initial kinetic energy, K_i, is zero because the car starts from rest. Thus we have

$$\Delta K = K_f = W_{\text{net}} = mg\,\Delta y, \qquad \tfrac{1}{2}mv_f^2 = mg\,\Delta y.$$

The mass cancels and we find that

$$v_f = \sqrt{2g\,\Delta y} = \sqrt{2(-9.8\text{ m/s}^2)(-24\text{ m})}$$
$$= 22\text{ m/s}, \quad \text{about 79 km/h (nearly 50 mi/h).}$$

(We could have solved this problem in Chapter 2; we consider it here to bring in the concepts of work and kinetic energy.)

What Do You Think? If the mass of the car is doubled, is (i) the work done by gravity and (ii) the final velocity of the car (a) doubled, (b) halved, or (c) the same? (Answer this without doing any calculations.)

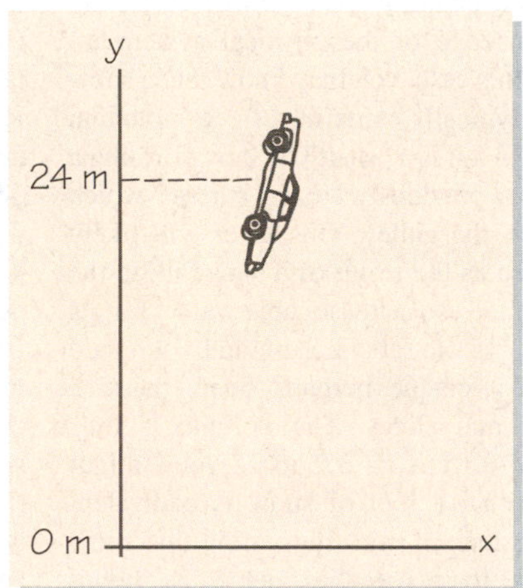

▶ **FIGURE 6–4** (a) Gravity pulls the car down. (b) The car leaves the draw bridge at $y = 24$ m.

Work Done by Individual Forces

It is possible to calculate the work done by each force acting on an object. For example, if each force is constant and all the forces act along the x-axis, then the work done on an object by a particular force $\vec{F} = F\hat{i}$ as the object moves through a displacement Δx is given by

$$W = F\,\Delta x.$$

We anticipated being able to think about the work done by individual forces in Conceptual Example 6–2 when we calculated the work done by the mover as distinct from the net work. Remember, however, that the work–energy theorem, Eq. (6–6), specifies the *net* work, that is, the work done by the *net* force. To apply it, we have to consider *all* the forces acting on an object. When a piano is slowly lowered at constant speed by a rope, the force of gravity is canceled by the rope tension and there is no acceleration. Here, the net force is zero, so the net work is also zero. This is consistent with the fact that there is no change in kinetic energy and no change in speed.

Because the net force is the sum of the individual forces acting, the net work can be decomposed into a sum of the work done by each of the individual forces. *This is a result that will hold for the most general definition of work.* It is sometimes simpler to find the work W_n done by individual forces F_n and then take the algebraic sum $\sum W_n$ to find the net work than to find the net force and calculate the work done by it. In the case of the piano lowered by a rope at constant speed, positive work is done by gravity as the piano is lowered. The tension of the rope, however, which points upward, does the same magnitude of *negative* work on the piano. The forces cancel and the net work is zero in this case.

CONCEPTUAL EXAMPLE 6–4 You are standing in a stationary elevator. Suddenly the elevator accelerates upward for some distance and then stops very suddenly. What do you expect will happen to *you*? Explain this in terms of work and energy.

Answer In a stationary elevator, I know I am subject to the force of gravity (downward), but since I am not falling, there must be an upward normal force due to the presence of the solid floor under my feet to compensate and cancel gravity. When the elevator accelerates, I accelerate with it. The forces that act on *me* are still the force of gravity and the normal force. Now, however, the normal force exceeds that of gravity, so that there is a net upward force that accelerates me. My speed increases with that of the elevator, and therefore my kinetic energy increases. This energy is supplied by the work done on me by the net force, in accordance with the work–energy theorem. When the elevator suddenly stops, I have a certain amount of kinetic energy, associated with an upward velocity. If the elevator stops quickly enough, I would leave the floor. In that case I would be subject only to the downward force of gravity. There is negative work done on me by that force as I continue to rise, and the work–energy theorem implies that my kinetic energy has to decrease as I move upward. At some point my kinetic energy goes to zero. As I fall back to the floor of the elevator, my kinetic energy increases again because the work done by the force of gravity is positive. When I hit the floor of the elevator, I experience an upward normal force larger than the force of gravity, so that the net work done is negative—the normal force points upward while my displacement as I come to rest is in the downward direction. This brings me to rest again, removing the kinetic energy that I had in coming down. (In a real situation, I would reduce the acceleration by bending my knees.) Once both the elevator and I are stationary again, the force of gravity and the normal force cancel, so that there is no further net work done on me by external forces. My kinetic energy is zero.

What Do You Think? Describe in terms of work and energy what happens to the elevator together with its passenger seen as a single system—in other words, imagine you were looking at the elevator from the outside, without being able to see the various actions within. [*Hint*: What is it that moves and stops the elevator?]

EXAMPLE 6–5 A box of books of mass 100 kg is pushed with constant speed in a straight line over a rough floor with a coefficient of kinetic friction 0.2. Find the work done by the force that pushes the box if the box is moved 3 m across the floor.

Setting It Up We draw a free-body diagram in Fig. 6–5, including an x–y coordinate system and the forces. We know the mass m of the box of books, the coefficient of kinetic friction μ_k between the box and the floor, and the distance d that the box moves along the x-axis. The box moves with constant speed. We want to find the work W done by the force F that pushes the box.

Strategy Because the box moves with a constant speed (no acceleration), there is no *net* force on the box. Therefore, we can set the net forces in both x- and y-directions equal to zero. This allows us to determine the pushing force $\vec{F}$, and we can then find the work it does on the box.

Working It Out Because the box moves with a constant velocity, the *net* horizontal force must vanish. Thus the pushing force $\vec{F}$, must be equal in magnitude but opposite in direction to the force of friction $\vec{f}$, whose magnitude is given by $f = \mu_k F_N$. The forces in the vertical direction must cancel and so $F_N = mg$, and friction has magnitude $\mu_k mg$. Hence the magnitude of $\vec{F}$ is also $\mu_k mg$. The force $\vec{F}$ is oriented along the same direction as the displacement, so that the work done by the pushing force is positive; this work is

$$W = Fd = \mu_k mgd$$
$$= (0.2)(100\ \text{kg})(9.8\ \text{m/s}^2)(3\ \text{m}) = 6 \times 10^2\ \text{J}.$$

What Do You Think? Suppose the force is unchanged but the coefficient of friction gradually increases as the box moves into a region of rougher floor. Use the work–energy theorem to explain qualitatively what happens.

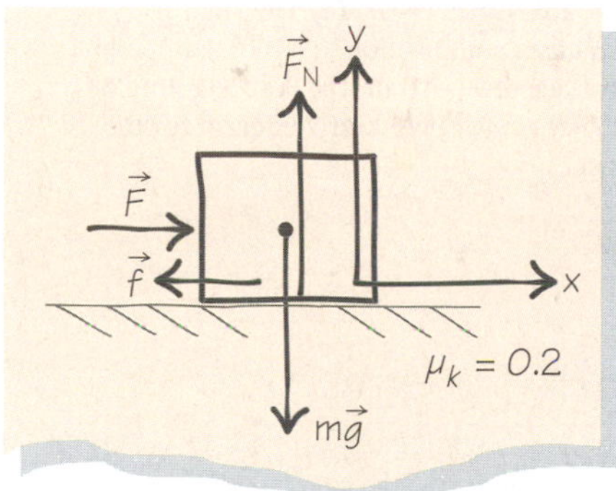

(a)

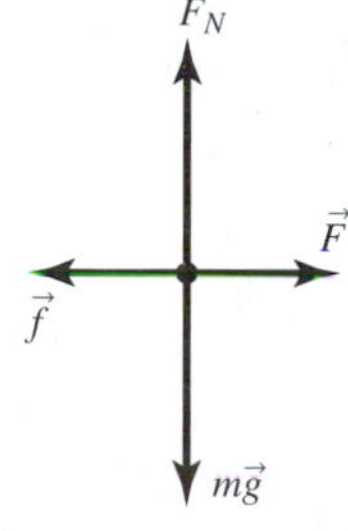

(b)

FIGURE 6–5 (a) A variety of forces act on the box of books in motion. This picture of the situation includes a coordinate system. (b) Free-body diagram for a box of books.

EXAMPLE 6–6 A crate of mass 96 kg is pushed across a horizontal floor by a force $\vec{F}$. The coefficient of kinetic friction between the crate and the floor is $\mu_k = 0.27$. The crate moves with uniform velocity. What is the magnitude of $\vec{F}$? Suppose that at some point the crate passes onto a new section of floor, where $\mu_k' = 0.085$. The pushing force on the crate is unchanged. After 1.25 m on the new section of the floor, the crate moves with a speed $v_f = 2.3$ m/s. What was the original speed v_i of the crate?

Setting It Up We illustrate the situation in Fig. 6–6a, including an x–y coordinate system. We know the crate mass m, the coefficients of kinetic friction for the two floor sections, the final speed of the crate after being pushed a distance $\Delta x = 1.25$ m in the second floor section, and the fact that the speed was constant in the first floor section. We want to find the magnitude of the pushing force and the speed of the crate in the first floor section.

Strategy We draw a free-body diagram in Fig. 6–6b and 6c and denote all the forces acting on the crate: gravity, normal force, pushing force, and friction. With a zero net force (constant speed) on the first section, we have enough information to find each of the forces there, including the pushing force. This pushing force acts in the second floor section, where we can use the work–energy theorem in a situation where the friction force has changed magnitude. This will tell us by how much the crate's kinetic energy has changed, and from that we can find the original speed.

Working It Out The free-body diagram permits us to find the pushing force, which, because the crate moves with constant speed on the first floor section, must balance the force of friction. This friction force has magnitude

$$f = \mu_k mg = (0.27)(96\text{ kg})(9.8\text{ m/s}^2) = 2.5 \times 10^2\text{ N},$$

and the pushing force $\vec{F}$ must have just this magnitude: $F = 2.5 \times 10^2$ N. We have used the cancellation of the vertical forces to find the magnitude of the normal force and hence the magnitude of the friction force.

On the new floor section, the force of friction is less, whereas the pushing force remains the same. Thus there is a (constant) net force in the direction of motion and the crate accelerates uniformly. On the new section, the force of friction has magnitude $f' = \mu_k' mg$. Therefore the net force on the crate has magnitude $F_{\text{net}} = F - f' = (\mu_k - \mu_k')mg$ and acts in the direction of motion of the crate. The net work done on the crate as it moves a distance Δx on the new section of floor is then

$$W_{\text{net}} = F_{\text{net}}\,\Delta x = (\mu_k - \mu_k')mg\,\Delta x.$$

According to the work–energy theorem, this is the increase in kinetic energy of the crate as it moves over the new section of floor:

$$W_{\text{net}} = K_f - K_i = \tfrac{1}{2}mv_f^2 - \tfrac{1}{2}mv_i^2.$$

We solve this equation for the initial speed:

$$\begin{aligned} v_i^2 &= v_f^2 - \frac{2W_{\text{net}}}{m} = v_f^2 - \frac{2(\mu_k - \mu_k')\cancel{m}g\,\Delta x}{\cancel{m}} \\ &= (2.3\text{ m/s})^2 - 2(0.27 - 0.085)(9.8\text{ m/s}^2)(1.25\text{ m}) \\ &= 0.76\text{ m}^2/\text{s}^2, \\ v_i &= 0.87\text{ m/s}. \end{aligned}$$

What Do You Think? Suppose that instead of hitting a smoother section of the floor, the coefficient of friction does not change but the floor begins to slope downward at the $x = 0$ m point. Use the work–energy theorem to explain what happens.

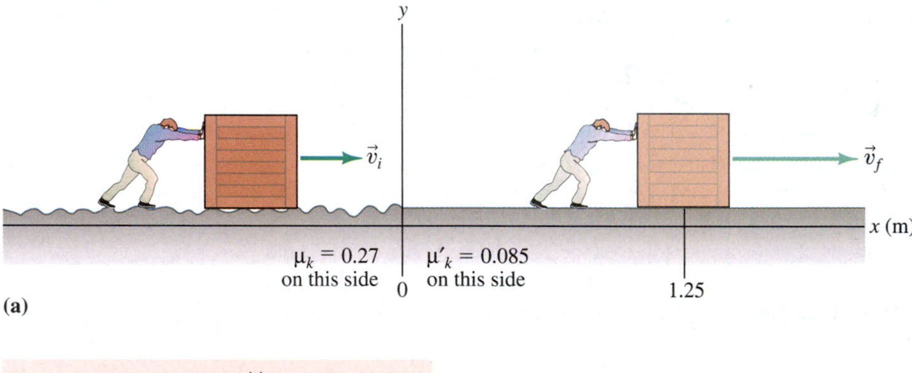

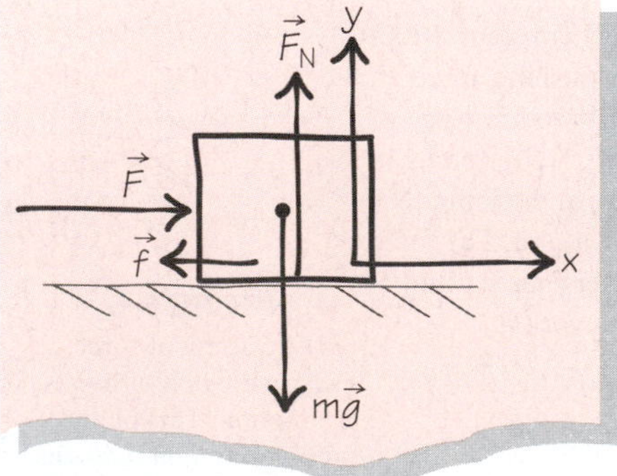

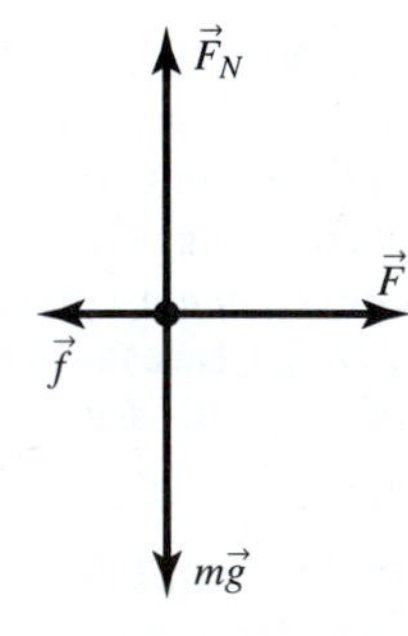

▶ **FIGURE 6–6** (a) The floor changes composition to a more slippery surface at $x = 0$ m. (b) Forces acting on the crate. (c) Free-body diagram for the crate.

THINK ABOUT THIS . . .

HOW DID THE EGYPTIANS BUILD THE PYRAMIDS?

While we don't know the fine details, we can be sure that the pyramids (Fig. 1–10), a truly immense project, were constructed with the aid of at least some of the classic "simple machines"—ramp, lever, block and tackle (see Example 5–4), screw thread, and so forth (Fig. 6–7). The expression for work contains both the displacement of the object being moved and the magnitude of the force. We can do the same amount of work by applying a larger force over a smaller distance or a smaller force over a larger distance. The simple machines are devices that do a given amount of work by employing a smaller force over a longer distance; in a sense they amplify your ability to do work by allowing you to exchange a large force applied over a short distance for a smaller force applied over a larger distance. Hoisting a several-ton block to its place in the pyramids requires an immense amount of work. It would not have been possible for the Egyptians to lift such a block vertically, but they could get it to the top of the pyramid by applying a smaller force than the weight of the block over a longer distance, thereby doing the same work on the block. This would still require many workers but is achievable by pushing the block up a sloping ramp built along the side of the pyramid, rather than over a purely vertical displacement. (We'll describe the work done as an object moves along a ramp in Section 6–2.) To help transport the stone blocks from their source, the Egyptians likely used another simple machine, the lever, in the form of a huge crowbar, to raise the blocks high enough for logs to be slid underneath. The stone blocks could then be pushed with less force than if the blocks were pushed along the ground because rolling involves less friction than sliding. The Egyptians made use of these elegant and practical force "amplifiers" thousands of years ago, and we still use them today. It is convenient to use a ramp to put a refrigerator in a truck, to use a lever in the form of a screwdriver to pry open the lid of a paint can, or to employ a block and tackle as in Example 5–4. In each case, we say we are using a "mechanical advantage," or that we are using "leverage."

▲ **FIGURE 6–7** A simple machine is a kind of force amplifier. It does the same work that a large force would do over a small distance by applying a small force over a large distance. Thus a lever lifts a large mass a height h by applying a force much less than the mass weight over a distance much greater than h. Here, we can see a lever, a ramp, a wedge, and various screw-driven devices.

6–2 Constant Forces in More Than One Dimension

We have so far discussed the concepts of kinetic energy, work, and the work–energy theorem as they apply to objects that move only in one dimension. We now want to extend these concepts to two and three dimensions. We continue to assume that the net force $\vec{F}$ is a constant, meaning that it is a vector that does not change with time and has the same magnitude and direction at every point in space. Throughout, we'll take care to emphasize that force is a vector, as are displacement, velocity, and acceleration. (For simplicity, we have dropped the subscript "net" on the net force).

The application of energy and work to two and three dimensions is a straightforward one; we know that Newton's laws include the information that motion and the effects of forces on motion are *independently applicable to the different Cartesian directions*. Let's first apply these concepts to two dimensions for simplicity. Newton's second law, $\vec{F} = m\vec{a}$, encompasses two separate equations, one for the x-direction and one for the y-direction. If we label the components of all vector quantities with the appropriate axes, then we can derive an equation like the work–energy theorem—Eq. (6–2)—for each direction:

$$F_x\,\Delta x = \tfrac{1}{2}mv_x^2 - \tfrac{1}{2}mv_{0x}^2, \tag{6–8a}$$

$$F_y\,\Delta y = \tfrac{1}{2}mv_y^2 - \tfrac{1}{2}mv_{0y}^2. \tag{6–8b}$$

Here, the displacement vector is $\Delta\vec{r} = \Delta x\,\hat{i} + \Delta y\,\hat{j}$. [Note that the quantity $\frac{1}{2}mv_x^2$ in Eq. (6–8a), say, is not the "x-component" of kinetic energy; kinetic energy involves the speed squared, and the speed involves all the components of velocity.] The velocity $\vec{v}$ and the initial velocity $\vec{v}_0$ have each been separated into their components, as has the net force $\vec{F}$.

Let's now take the sum of these two equations. The square of the *magnitude* of the velocity—speed squared (v^2)—is the sum of the x- and y-components of the velocity squared:

$$v^2 = v_x^2 + v_y^2.$$

The summed equations then form a generalization of the work–energy theorem:

$$F_x\,\Delta x + F_y\,\Delta y = \tfrac{1}{2}mv^2 - \tfrac{1}{2}mv_0^2. \tag{6–9}$$

Defining kinetic energy just as we have before—Eq. (6–5), $K = \frac{1}{2}mv^2$— the right-hand side of this equation is once again the change in kinetic energy.

The left-hand side of Eq. (6–9) generalizes the definition of the work done on an object in two dimensions. The work done easily generalizes to three dimensions:

$$\text{for constant force: } W = F_x\,\Delta x + F_y\,\Delta y + F_z\,\Delta z. \tag{6–10}$$

This definition of the work done contains the components of the two vectors $\vec{F}$ and $\Delta\vec{r}$. One takes each component of the force and multiplies it by the corresponding component of the displacement, with the resulting terms added together. This combination of two vectors occurs in situations other than this one, and because it appears so frequently, it is given a name: the **scalar product** (or **dot product**) $\vec{A}\cdot\vec{B}$ of two vectors $\vec{A}$ and $\vec{B}$. The work, then, is the scalar product of $\vec{F}$ and $\Delta\vec{r}$, namely $\vec{F}\cdot\Delta\vec{r}$. A scalar product, although it is the product of two vectors, is itself a scalar quantity. If you are not already familiar with this concept, refer to the box The Scalar Product.

From Eq. (6–10) we can now see that the work W done by a constant force $\vec{F}$ acting on an object that moves through a displacement $\Delta\vec{r}$ is

$$W = \vec{F}\cdot\Delta\vec{r} \tag{6–11}$$

Using this expression for work, the work–energy theorem takes exactly the same form as before—$W_{net} = \Delta K$. Keep in mind that Eq. (6–11) applies only as long as the net force is a constant vector. (We'll see later how to deal with a varying force.)

We can also write the work differently than Eq. (6–10). Using the general properties of the scalar product (see the box), we have

$$W = \vec{F}\cdot\Delta\vec{r} = F\,\Delta r\cos\theta, \tag{6–12}$$

where θ is the angle between the two vectors (Fig. 6–8). Equivalently, the work done is the simple product of the magnitude of force *in the direction of the displacement* and the magnitude of the displacement. Put another way, *only the component of the force along the direction of the displacement* (here $F\cos\theta$) *counts in the work.* From this fact we can conclude that a force perpendicular to the motion of an object does no work on the object. A frequently occurring example of this is the normal force $\vec{F}_N$, which is perpendicular to the surface on which an object moves; $\vec{F}_N$ does no work on that object. Another important example of a force that does no work in this way is the centripetal force responsible for uniform circular motion (see Section 6–4). Note also that the quantity $F\Delta r\cos\theta$ can be positive or negative, depending on whether the displacement is "with" or "against" the force.

Remember that, although the scalar product is formed from two vectors, it is itself a scalar. Work is always a scalar quantity.

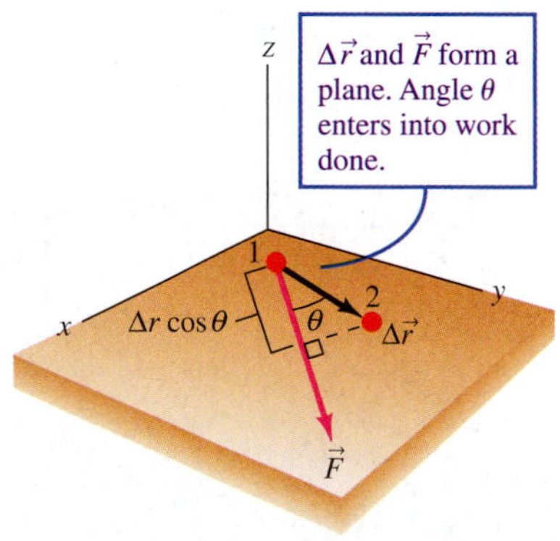

▲ **FIGURE 6–8** The two nonparallel vectors $\Delta\vec{r}$ and $\vec{F}$ always form a plane. We have labeled that plane as the xy-plane. The angle between these vectors determines the work done by the constant force on an object undergoing displacement $\Delta\vec{r}$, namely $W = F\Delta r\cos\theta$. We have indicated the component of $\Delta\vec{r}$ along the force, namely $\Delta r\cos\theta$.

Problem-Solving Techniques

The Scalar Product

In Chapter 1, we discussed the definition of vectors and the multiplication of vectors by scalars. The product $b\vec{A}$ of a scalar b and a vector $\vec{A}$ is a vector. It points in the same direction as $\vec{A}$ and has magnitude $|b|A$, where A is the magnitude of $\vec{A}$. One way to multiply two vectors—$\vec{A}$ and $\vec{B}$, for example—is the **scalar product** $\vec{A}\cdot\vec{B}$. The scalar product is a scalar quantity whose value is

$$\vec{A}\cdot\vec{B} \equiv AB\cos\theta. \qquad \text{(B1–1)}$$

Here, θ is the angle between the directions of the two vectors (Fig. 6B1–1a). The scalar product has the properties that

$$\vec{A}\cdot\vec{B} \equiv \vec{B}\cdot\vec{A} \qquad \text{(B1–2)}$$

and

$$\vec{A}\cdot(\vec{B}+\vec{C}) = (\vec{A}\cdot\vec{B}) + (\vec{A}\cdot\vec{C}). \qquad \text{(B1–3)}$$

If two vectors are perpendicular (*orthogonal*) to each other, then $\theta = 90°$ and $\cos\theta = 0$, and their scalar product is zero. If the vectors are parallel to each other, then the scalar product takes on its maximum value, that is, the product of the magnitudes of the two vectors. The scalar product of a vector with itself is the square of its magnitude, $\vec{A}\cdot\vec{A} = A^2$. The unit vectors $\hat{i}, \hat{j}$ and $\hat{k}$ along some set of orthogonal axes x, y, and z have the property that

$$\hat{i}\cdot\hat{i} = \hat{j}\cdot\hat{j} = \hat{k}\cdot\hat{k} = 1. \qquad \text{(B1–4)}$$

Because they are orthogonal to each other,

$$\hat{i}\cdot\hat{j} = \hat{j}\cdot\hat{k} = \hat{i}\cdot\hat{k} = 0. \qquad \text{(B1–5)}$$

Two vectors $\vec{A}$ and $\vec{B}$ can be decomposed into their vector components: $\vec{A} = A_x\hat{i} + A_y\hat{j} + A_z\hat{k}$ and $\vec{B} = B_x\hat{i} + B_y\hat{j} + B_z\hat{k}$. The rules in Eqs. (B1–4) and (B1–5) allow us to write the scalar product of $\vec{A}$ and $\vec{B}$ as

$$\begin{aligned}\vec{A}\cdot\vec{B} &= (A_x\hat{i} + A_y\hat{j} + A_z\hat{k}) \\ &\quad\cdot(B_x\hat{i} + B_y\hat{j} + B_z\hat{k}) \qquad \text{(B1–6)} \\ &= A_xB_x + A_yB_y + A_zB_z.\end{aligned}$$

Thus the scalar product of two vectors is the sum of the product of the components of the two vectors.

The scalar product is a scalar quantity, so it remains the same even if the axes of our coordinate system are rotated. If we consider two vectors $\vec{A}$ and $\vec{B}$, we may choose our coordinate frame in such a way that $\vec{A}$ lies along the x-axis, $\vec{A} = A\hat{i}$. The other axes can be arranged so that the vector $\vec{B}$ has only x- and y-components, $\vec{B} = B_x\hat{i} + B_y\hat{j}$. These vectors are shown in Fig. 6B1–1b, which is a view looking down on the plane formed by $\vec{A}$ and $\vec{B}$. According to Eq. (B1–6), the scalar product is then given by

$$\vec{A}\cdot\vec{B} = AB_x.$$

Thus the scalar product of two vectors may be described as the *product of the length of one vector and the projection of the other vector along the direction of the first one.* [Because $B_x = B\cos\theta$, we recover here our original definition: Eq. (B1–1).] This way of looking at things makes it clear that the orientation of coordinate axes is irrelevant to the value of the scalar product.

Scalar products are useful in many mathematical and physical manipulations. Consider, for example, the following result from analytic geometry: For a triangle whose sides have lengths a, b, c and where the angle between known sides a and b is θ,

$$c^2 = a^2 + b^2 - 2ab\cos\theta. \qquad \text{(B1–7)}$$

In vector notation, and as in Fig. 6B1–2, if two sides of a triangle are denoted by the vectors $\vec{a}$ and $\vec{b}$, then the third side is given by $\vec{c} = \vec{a} - \vec{b}$. Squaring both sides ("squaring" a vector means taking the scalar product of the vector with itself) gives

$$\vec{c}^2 = (\vec{a} - \vec{b})^2 = \vec{a}^2 + \vec{b}^2 - 2\vec{a}\cdot\vec{b}, \qquad \text{(B1–8)}$$

which implies Eq. (B1–7). In this book, in addition to all the applications associated with energy, we'll meet the scalar product again in fluid motion, electricity and magnetism, and other places.

(a)

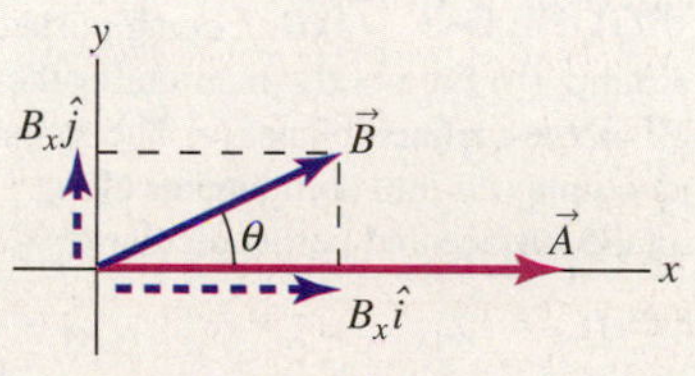

(b)

▲ **FIGURE 6B1–1** (a) Two nonparallel vectors $\vec{A}$ and $\vec{B}$ have been displaced so that their tails meet at the same point. They are oriented in space with the angle θ between them. (b) The x-axis of our coordinate system has been redefined so that $\vec{A}$ lies along the $+x$-direction. The scalar product is independent of the orientation of the axes.

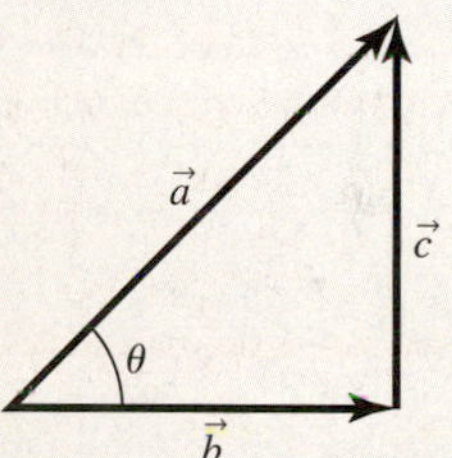

▲ **FIGURE 6B1–2** Illustrating the rule for the length of the third side of a triangle if the length of two sides and the angle between them are known.

EXAMPLE 6–7 A mover has to place a box of books of mass m in a truck, and it is too heavy to lift directly. He therefore uses a ramp that makes an angle θ with the horizontal and pushes horizontally, applying a force $\vec{F}$ to the box. Find the magnitude of $\vec{F}$ such that the box moves up the plane with acceleration $\vec{a}$. What is the work done by $\vec{F}$? (Assume that the ramp is rough, with coefficient of kinetic friction μ_k, and that the distance the box moves along the ramp is d.)

Setting it Up We illustrate the situation in Fig. 6–9a, including a suitable coordinate system.

Strategy We first draw a free-body diagram and use it to find the components of the forces along the axes. Newton's second law then allows us to find the force necessary to push the box up the ramp with acceleration $\vec{a}$. We know the displacement, so we can then find the work done by this force.

Working It Out Figure 6–9b is the free-body diagram for the box. We then separate the forces into their components in Fig. 6–9c. The component of the force along the ramp is $F \cos\theta$, so the work done by this force on the box as it moves a distance d is $W = Fd\cos\theta$.

Now we find the magnitude F. The y-components of the forces must add up to zero because there is no acceleration in the direction perpendicular to the ramp. We thus have

$$F_N - F\sin\theta - mg\cos\theta = 0.$$

From this equation, $F_N = F\sin\theta + mg\cos\theta$. There is also an acceleration of the box up the ramp, which is determined by Newton's second law applied to the x-direction:

$$\begin{aligned} ma &= F\cos\theta - mg\sin\theta - \mu_k F_N \\ &= F\cos\theta - mg\sin\theta - \mu_k(F\sin\theta + mg\cos\theta). \end{aligned}$$

This equation can be solved for F:

$$F = \frac{ma + mg\sin\theta + \mu_k mg\cos\theta}{\cos\theta - \mu_k\sin\theta}.$$

The work done by the force in accelerating the box is then

$$W = Fd\cos\theta = md\frac{a + g\sin\theta + \mu_k g\cos\theta}{1 - \mu_k\tan\theta}.$$

Notice that this is not the net work, as that involves the pushing force, friction and gravity.

What Do You Think? What happens if the ramp angle is larger? Give a qualitative answer.

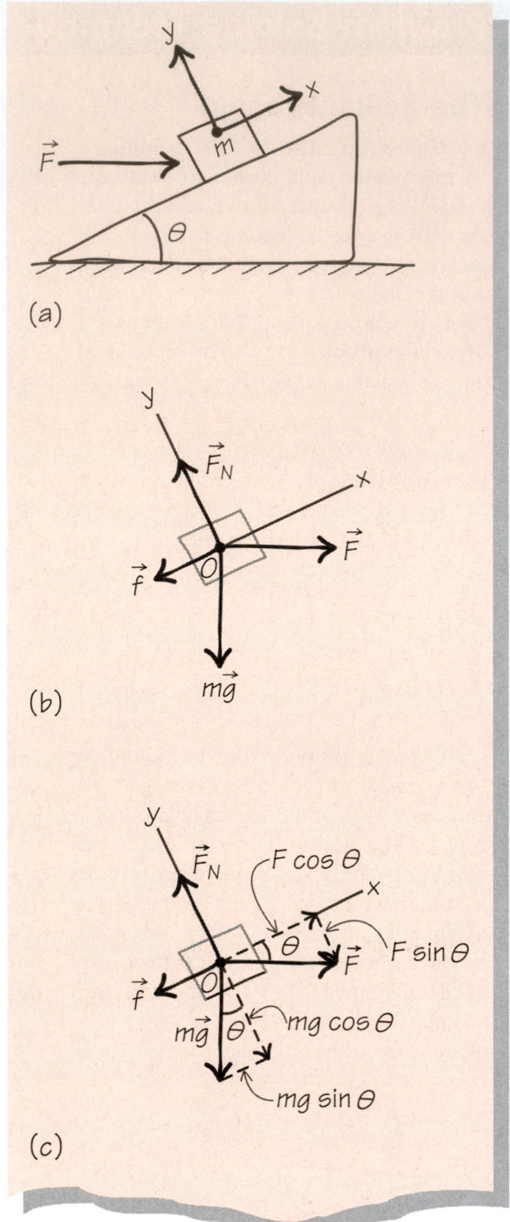

▶ **FIGURE 6–9** (a) Box being pushed up a ramp. (b) Free-body diagram for the block on the inclined plane. (c) The forces are decomposed into components along the plane surface and perpendicular to it.

6–3 Forces That Vary with Position

In our study of the work–energy theorem so far the force acting on an object has been constant. Many forces in nature and in engineering, however, such as the gravitational force or the forces exerted by springs, vary with position. It is possible to have a force whose magnitude varies from point to point, one whose direction varies from point to point, or one for which both the magnitude and direction vary from point to point. We want to generalize the definition of work to include these cases so that the work–energy theorem continues to hold.

Variable Forces in One Dimension

Let's first go back to motion in one dimension. The only change from the constant-force situation of Section 6–1 is that the force magnitude F depends on position x, so that $F = F(x)$. Figure 6–10a shows a force that varies with position and is one possible form of $F(x)$. We may adapt what we have learned so far to this situation by approximating the curve in Fig. 6–10a by a series of "steps"—their width being some interval in x—that come close to matching $F(x)$. An examination of Fig. 6–10b shows that the width of these steps is made to vary: If $F(x)$ is fairly flat in a certain region, the width of the step (its Δx) can be relatively large. If $F(x)$ changes rapidly, we must make the width of the steps in that range very small. Even if $F(x)$ in any interval differs a little from the step height across the interval, the error in treating F as a constant (e.g., taking its value in the middle of the width of the interval) is small as far as the work calculation is concerned, since if the interval Δx is small, $F\,\Delta x$ will be small.

To rephrase this more mathematically, divide the total displacement, $x_f - x_0$, into a series of small intervals across each of which, to within a small error, the force is constant. When the force is changing only slowly with position, it will remain roughly constant over a relatively large interval (see the parts of Fig. 6–10a marked "1"), whereas if the force varies considerably with a small change of position (see the part of Fig. 6–10a marked "2"), it will remain roughly constant over only a very small interval. Figure 6–10b shows how the width of the intervals would have to vary so that $F(x)$ is approximately constant over each interval. For mathematical simplicity we now take the width of the intervals to be all the same, and to ensure our assumption that the force is almost constant over this interval, we set our common interval width to the width of the smallest interval in Fig. 6–10b, as drawn in Fig. 6–10c. We call this width Δx. We have now approximated our variable force as a series of constant forces, each varying only very slightly from its value in a neighboring interval.

Let us now denote the (average) value of the force in the interval from x_0 to x_1 by F_1, the (average) value of the force in the interval from x_1 to x_2 by F_2, and so on, as in Fig. 6–11. Let us also denote the velocities at the edges of the intervals as follows: at x_0 the velocity is v_0, at x_1 it is v_1, at x_2 it is v_2, and so on. There are altogether $N = (x_f - x_0)/\Delta x$ intervals. The work–energy theorem, Eq. (6–2), applied in succession to each of these N intervals gives

$$\begin{aligned}
F_1\,\Delta x &= \tfrac{1}{2}mv_1^2 - \tfrac{1}{2}mv_0^2,\\
F_2\,\Delta x &= \tfrac{1}{2}mv_2^2 - \tfrac{1}{2}mv_1^2,\\
F_3\,\Delta x &= \tfrac{1}{2}mv_3^2 - \tfrac{1}{2}mv_2^2,\\
&\;\;\vdots\\
F_N\,\Delta x &= \tfrac{1}{2}mv_f^2 - \tfrac{1}{2}mv_{N-1}^2.
\end{aligned}$$

Here, v_N is equivalent to the final speed, so we denote it v_f. If we add all these equations, we see that all the intermediate kinetic energies cancel, leaving

$$F_1\,\Delta x + F_2\,\Delta x + F_3\,\Delta x + \cdots + F_N\,\Delta x = \sum_{i=1}^{N} F_i\,\Delta x = \tfrac{1}{2}mv_f^2 - \tfrac{1}{2}mv_0^2.$$

The right side is the change in kinetic energy between the initial point x_0 and the final point x_f. The left side of this equation is the work done in going from x_0 to x_1 plus the

(a)

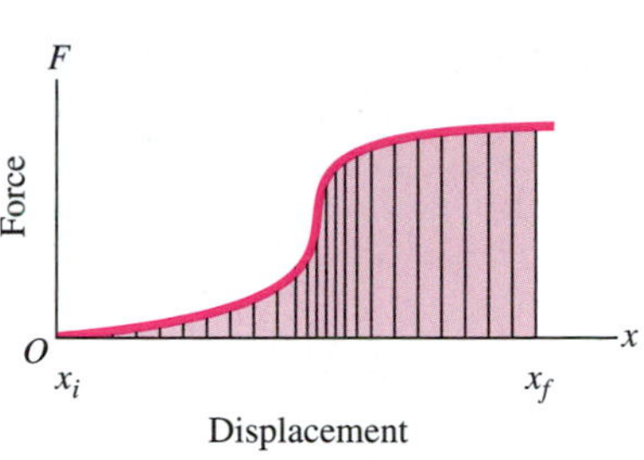

(b)

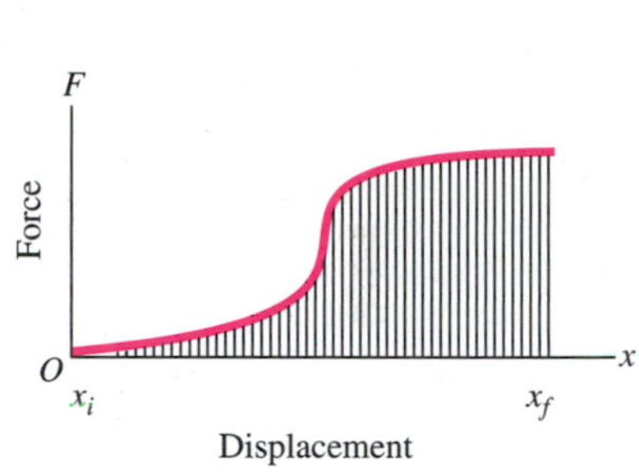

(c)

▲ **FIGURE 6–10** (a) A smoothly varying force. (b) If there is a region where the force changes more rapidly, the intervals in that region can always be made smaller so that the force can be thought of as constant in each interval. (c) Take the width of each interval to be the width of the smallest interval in (b).

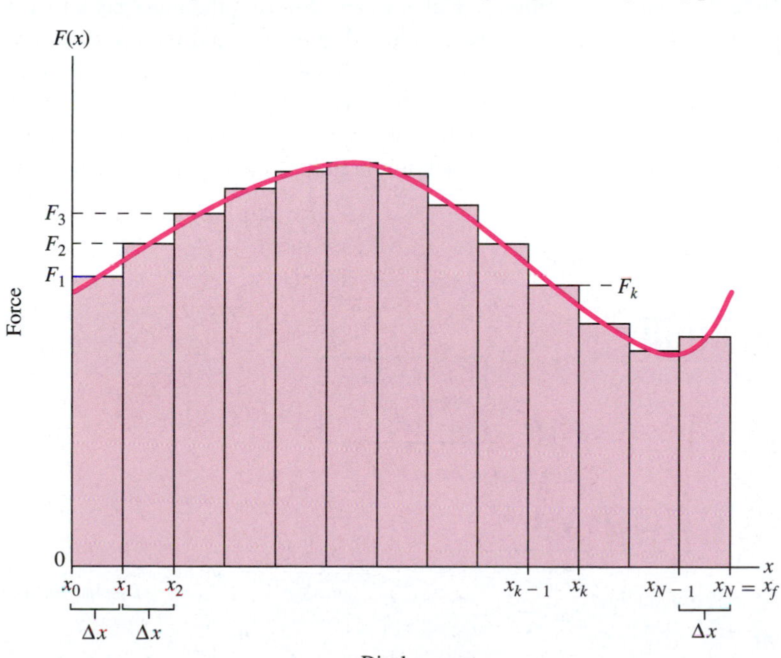

◀ **FIGURE 6–11** A variable force $F(x)$ has an approximately constant value $F(x_k) = F_j$ within the jth small interval, of width Δx, in the region between x_0 and x_N. The work done by the force on an object as the object moves between these limits is the sum of the work W_k done in each interval, for which we can use the constant-force formula for work, $W_k = F_k\,\Delta x$. This sum, W, is the integral of $F(x)$ over x, and it is equal to the area under the curve of F versus x.

work done in going from x_1 to x_2, and so on. Each of these individual terms can be interpreted as the area of the rectangle formed by the interval width and the function height in that interval. The sum of these terms is approximately the area under the curve of F versus x. As the number of terms becomes infinite—as we make the width of our intervals infinitesimally small—the approximation becomes exact. The work done over the entire distance from x_0 to x_f is the *total area under the curve of force versus position*. In other words, it is the integral of the function $F(x)$ over the interval starting at point x_0 and ending at point x_f. (For a review of integration refer to the box Integration, a Quick Review.) We may therefore write the above equation as

$$\int_{x_0}^{x_f} F(x)\,dx = K_f - K_0 = \Delta K. \tag{6–13}$$

In terms of the work–energy theorem—that is, the change in kinetic energy of a system is the net work done on the system—we can identify the left side as the net work done,

$$W = \int_{x_0}^{x_f} F(x)\,dx. \tag{6–14}$$

(Again, we have not bothered with the subscript "net.") We can also say that the work done by any one force that varies with position in one dimension has this form. In the cases that we deal with in this chapter the integral can be done explicitly, or at least numerically. In the simple case of a constant force of magnitude F_0, this form reduces to

$$W = (F_0)\int_{x_0}^{x_f} dx = (F_0)(x_f - x_0) = F_0\,\Delta x,$$

which gives us back the result of Eq. (6–3).

Work Done by a Spring

One of the most important examples of a one-dimensional variable force is the force exerted on a mass by a spring attached to it (Fig. 6–12). The force takes the form

$$F = -kx. \tag{6–15}$$

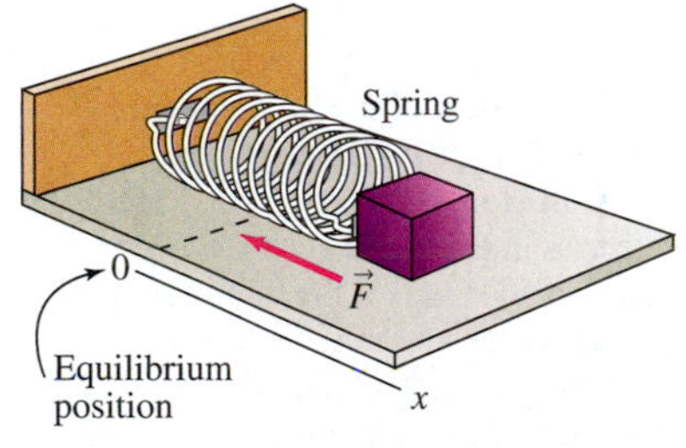

▲ **FIGURE 6–12** A spring exerts a force on an object that tends to bring the object back to the equilibrium position. Here the spring is stretched past the equilibrium position.

Here, x measures the displacement of the mass from an equilibrium position; k is a constant characteristic of the particular spring, known as the **spring constant.** This force law is known as **Hooke's law,** after its seventeenth-century discoverer, Robert Hooke. Note the sign: The spring force always acts to bring the mass back to $x = 0$. When x is positive, the mass is on the right side in Fig. 6–13a and the force points to the left, and when x is negative, the mass is on the left side of the origin and the force points to the right. The universal importance of this force is that it applies to virtually any system that has an equilibrium point in which a small movement away from that point brings in forces that tend to bring you back to the point; this is called a stable equilibrium point.

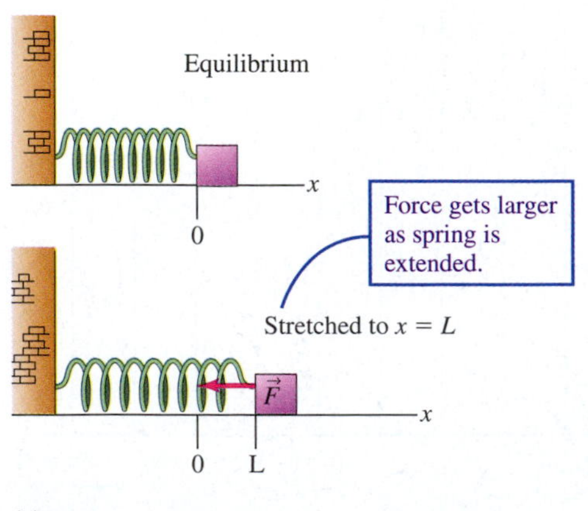

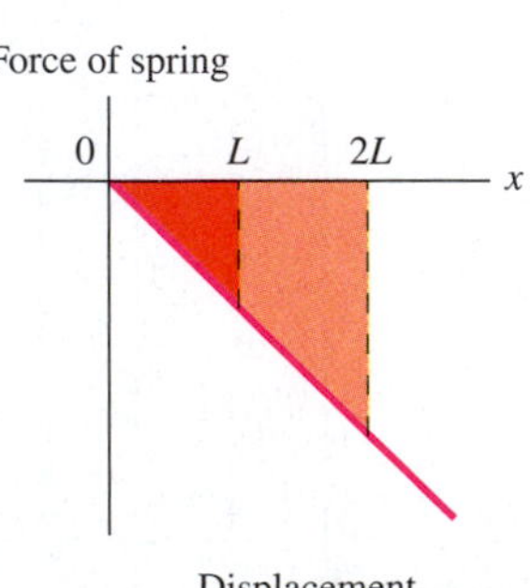

▶ **FIGURE 6–13** (a) The force exerted by a spring changes as the spring stretches (or compresses). Twice the stretch corresponds to twice the force. (b) The area under the curve of force versus x is the work the spring force performs on a mass attached to its end. The work done as the spring stretches from $x = 0$ to $x = L$ is the area shaded in red.

Examples include pendulums and the forces between the atoms that make up solids. This will be discussed in more detail in the next chapter.

Let's now calculate the work done on the mass by the spring force when the mass moves from the equilibrium position (where the spring force is zero) to the position $x = L$ (Fig. 6–13a). Note that if the mass is at rest, it cannot move away from $x = 0$ under the influence of the spring force alone; the spring force is zero there. But the mass may be acted on by other forces or may have started its motion at a different point than $x = 0$. If other forces are involved, then the work done on the mass by the spring force alone is not the *net* work. No matter; we can always calculate the work done by the spring force alone as the mass is displaced. To calculate the work done by the spring, we must integrate the (nonconstant) force over the displacement, as in Eq. (6–14):

$$W = \int_0^L (-kx)\,dx.$$

We use the general integration formula for powers from Appendix IV–8, with $p = 1$, to find that

$$W = -k\int_0^L x\,dx = -\tfrac{1}{2}kx^2\Big|_0^L = -\tfrac{1}{2}kL^2.$$

The work done by the spring on the mass varies with the square of the distance moved. What does the negative sign mean? The spring force is in the direction opposite to the displacement of the mass, so the work done by the spring is negative. This is sensible given that the force acts to bring the mass back to $x = 0$, and if the spring force were the only force acting, the mass would slow down as it goes from zero to L. Figure 6–13b illustrates how the work done is equal to the area under the curve of F versus x, here shaded in red. The area of the triangle is indeed $\frac{1}{2}kL^2$, and it is negative since F itself is negative.

Problem–Solving Techniques

Integration, a Quick Review

Here we want to remind you of some important features of the process of integration. (A more complete discussion of integration is contained in Section 2–6.) Integration of a function is the inverse of differentiation of that function. In other words, if we integrate some function $f(x)$ and then differentiate the result, we get the function $f(x)$ back again. More precisely, suppose that the functions $f(x)$ and $g(x)$ are related by

$$g(x) = \int_{x_0}^{x} f(x')\,dx'. \quad \text{(B2–1)}$$

Then

$$\frac{dg(x)}{dx} = f(x). \quad \text{(B2–2)}$$

As we saw in the subsection Variable Forces in One Dimension, we can interpret work as the area under the curve of F versus x. If the function F is negative and the displacements are positive, as in Fig. 6B2–1, the area and hence the work are negative. Or, if the displacement is negative and the force positive, the work will again be negative. In the language of integrals, this occurs because the integral changes sign when we reverse the limits on it:

$$\int_{x_f}^{x_i} f(x)\,dx = -\int_{x_i}^{x_f} f(x)\,dx. \quad \text{(B2–3)}$$

If the integral on the right-hand side is positive, then the integral on the left-hand side—which represents the work done in a displacement from x_f to x_i—is negative.

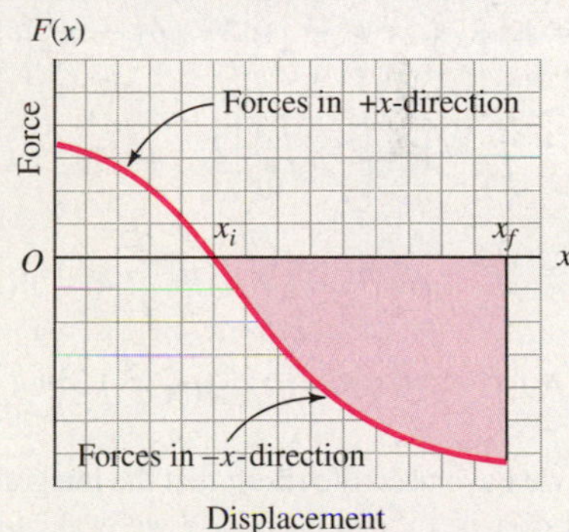

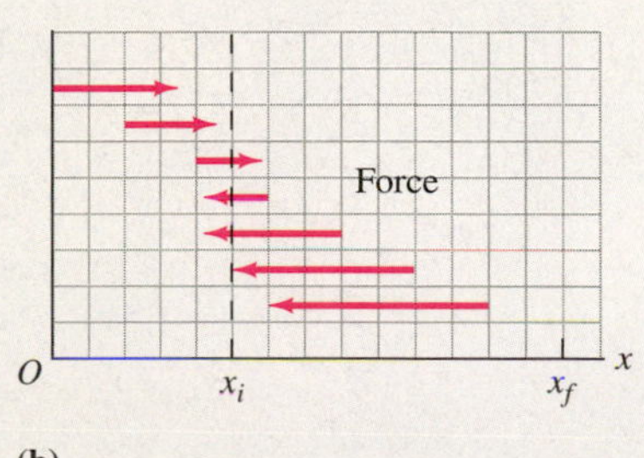

▶ **FIGURE 6B2–1** (a) In the region between x_i and x_f, the shaded area under the curve is negative, (b) corresponding to negative values for the force $F(x)$.

CONCEPTUAL EXAMPLE 6–8 What is the amount of work done by the spring force in moving from $x = 0$ to $x = L$ compared to the work done moving from $x = 0$ to $x = 2L$?

Answer We can use the graph in Fig. 6–13b for this question. The work done by the spring force as the mass moves from one value of x to another is the area under the curve of F versus x, between the two values of x. We consider the shaded triangle under the force curve (here, a straight line with slope k) from $x = 0$ to $x = L$ and a second shaded triangle under the force curve from $x = 0$ to $x = 2L$. These triangles are similar, but the second one has twice the base length and twice the height. The second triangle has four times the area of the first, so the spring force does four times as much work when the mass moves twice as far.

EXAMPLE 6–9 A worker pushes a 20-kg crate straight across a 1-m-long section of horizontal floor with a constant force of 29 N. This section of floor has the peculiarity that it becomes rougher from beginning to end, and the crate is moving at 1.2 m/s when it arrives at the start of this section. The coefficient of kinetic friction is 0.15 at the start and 0.25 at the finish, varying linearly with distance in between. What is the speed of the crate at the end of the section?

Setting It Up Figure 6–14 indicates a coordinate system in which the starting point for the section is at $x = 0$ and the end point is at $x_f = 1$ m. We let m be the given mass of the crate; v_i the given (positive) initial velocity; v_f the unknown final velocity; F, which acts along the $+x$-direction, the given pushing force; and f_k the force of kinetic friction, taking the form $f_k = -\mu_k F_N$. The minus sign in f_k shows it acts in the $-x$-direction. Finally, we know the initial and final values of μ_k, namely μ_k^i and μ_k^f, as well as the fact that μ_k varies linearly with distance over the range $x = 0$ to $x = x_f = 1$ m. This translates into a coefficient of friction that takes the algebraic form $\mu_k = \mu_k^i + (\mu_k^f - \mu_k^i)(x/x_f)$.

Strategy This type of question is tailor made for the work–energy theorem. We use it to find the final energy, hence the final speed, in terms of the calculable initial kinetic energy and the calculable net work. The net work involves the net force in the x-direction, and the net force is composed of both F and f_k. The free-body diagram in Fig. 6–6b suffices and allows us to see that $f_k = \mu_k mg$.

Working It Out Begin with the net work: $F_{net} = F + f_k = F - [\mu_k^i + (\mu_k^f - \mu_k^i)(x/x_f)]mg$. Therefore

$$W_{net} = \int_0^{x_f} F_{net}\,dx = \int_0^{x_f}\left[F - \mu_k^i mg - (\mu_k^f - \mu_k^i)\left(\frac{x}{x_f}\right)mg\right]dx$$

$$= [F - \mu_k^i mg]x_f - [(\mu_k^f - \mu_k^i)mg]\frac{x_f^2}{2x_f}$$

$$= [F - \mu_k^i mg]x_f - [(\mu_k^f - \mu_k^i)mg]\frac{x_f}{2}.$$

Here we have used the result that the integral of unity is x and the integral of x is $\frac{1}{2}x^2$; in each case we evaluate at the upper and lower limit and take the difference, and we have included the multiplicative constants as well. Numerically,

$$W_{net} = [29\text{ N} - (0.15)(20\text{ kg})(9.8\text{ m/s}^2)](1\text{ m})$$
$$- [(0.25 - 0.15)(20\text{ kg})(9.8\text{ m/s}^2)]\frac{1\text{ m}}{2} = -10\text{ J}.$$

The sign is negative because the friction force dominates. You can see that the first term in square brackets is very nearly zero. It is composed of a term involving the pushing force and a canceling term involving the initial value of kinetic friction, suggesting that the crate was moving at constant velocity over a section of floor with coefficient of kinetic friction 0.15 before it arrived at the section treated here. In any case, with negative net work, we expect the speed to decrease.

The initial kinetic energy is $K_i = \frac{1}{2}mv_i^2 = \frac{1}{2}(20\text{ kg})(1.2\text{ m/s})^2 = 14$ J. We then use the work–energy theorem in the form of Eq. (6–7),

$$K_f = W_{net} + K_i = -10\text{ J} + 14\text{ J} = 4\text{ J}.$$

This gives us a final speed according to

$v_f^2 = (2/m)K_f = (2/20\text{ kg})(4\text{ J}) = 0.4\text{ (m/s)}^2$, or $v_f = 0.6$ m/s.

What Do You Think? Does this example have anything to do with the spring force?

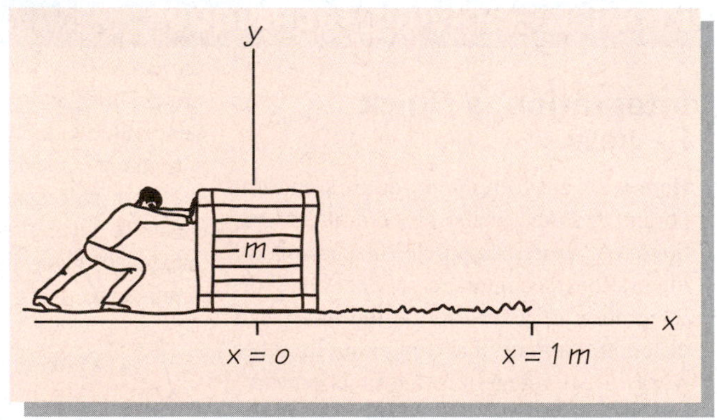

▲ **FIGURE 6–14** Coordinate system for forces acting on crate.

Forces That Vary in Both Magnitude and Direction

For many common forms of motion forces appear that vary in magnitude and/or direction: When an object moves under the influence of a force, either the force may change direction as the object moves or the object may change its direction as it moves. Think of driving your car along a steep and winding mountain road. Gravity is constant in both direction and magnitude, but you are constantly changing your direction. Thus the work done on you by gravity near one location—a scalar product of the form $m\vec{g}\cdot\Delta\vec{r}$, where $\Delta\vec{r}$ is a small local displacement at that first location—may be different in another location, in this case because the direction of the displacement changes.

Therefore we want to generalize the net work done, the quantity that appears in the work–energy theorem, to encompass the possibility either that the force varies its direc-

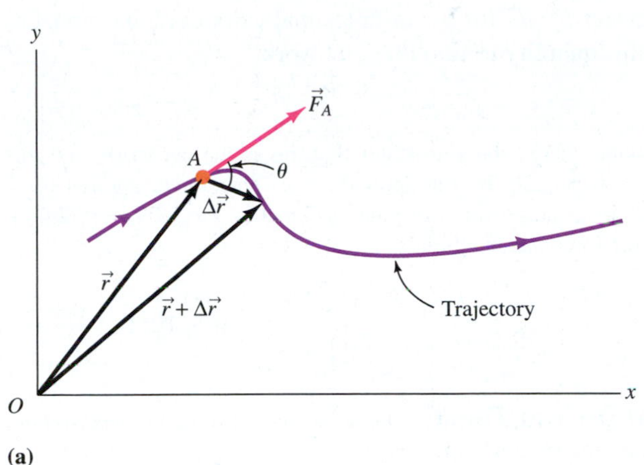

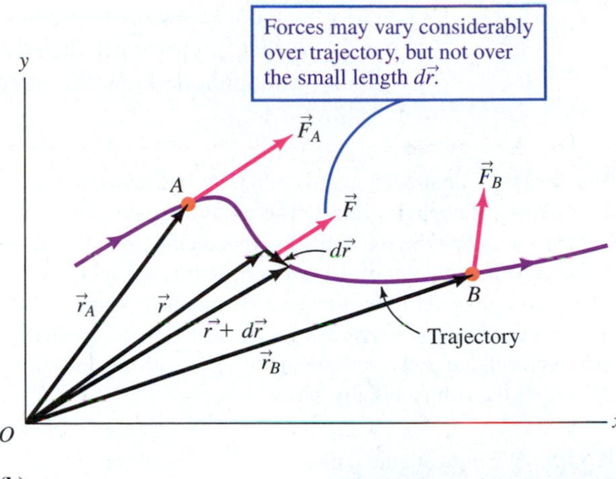

▲ **FIGURE 6–15** (a) An object moves from position $\vec{r}$ at time t to position $\vec{r} + \Delta\vec{r}$ at time $t + \Delta t$ while a force $\vec{F}_A$ acts on it. The work done on the object by the force over the displacement $\Delta\vec{r}$ is $\vec{F}_A \cdot \Delta\vec{r}$. (b) In a finite time, there is a net displacement from point A to point B.

tion from point to point or that the net displacement $\vec{r}_f - \vec{r}_i$ of the moving object results from a rather complicated *path* in space (Fig. 6–15) or both. (Here we have drawn the path in a plane, but the path could move through three-dimensional space.) The path is described with a changing position vector $\vec{r}$. At the point A, which is a distance r from the origin, the force on the object is $\vec{F}_A$. This force does not change very much as the object moves through a small displacement $\Delta\vec{r}$ from the point labeled by position vector $\vec{r}$ to an adjacent point labeled by position vector $\vec{r} + \Delta\vec{r}$ (Fig. 6–15a). As $\Delta\vec{r} \to 0$, this displacement is *tangential* to the curve at the tip of $\vec{r}$, whereas the force vector, which can be regarded as constant over a sufficiently small displacement, points in some other direction. The work done on the object in moving through the small interval is then $\vec{F}_A \cdot \Delta\vec{r}$. We find the total work done for a displacement from point A with position vector $\vec{r}_A$ to point B with position vector $\vec{r}_B$ (Fig. 6–15b) by summing the small contributions from each small interval along the path. In the limit that the small displacements go to zero, the sum takes the form of an integral:

$$W = \lim_{\Delta\vec{r}\to 0} \sum \vec{F}\cdot\Delta\vec{r} = \int_{\vec{r}_A}^{\vec{r}_B} \vec{F}\cdot d\vec{r}. \qquad (6\text{–}16)$$

This formula is the general definition of work and is consistent with all our earlier definitions. The integral that appears here is called a **line integral** because it depends not only on the beginning and ending points A and B but also, in general, on the path, or line, taken to move between these points. We shall explore the properties of this integral in Section 6–4.

With the definition of work of Eq. (6–16), *the work–energy theorem in the form given by Eq. (6–6),* $W_{net} = \Delta K$, *applies for the most general case.*

▲ **FIGURE 6–16** As long as a Ferris wheel passenger is moving in uniform circular motion, the net work done on the passenger along any segment of the arc of the circle is zero. The net force that uniformly moves the passenger in a circle is made up of a combination of gravity and normal forces. If the Ferris wheel moves with a uniform velocity, why does it need a motor?

No Work Is Done in Uniform Circular Motion

We conclude this section with an important observation: *No net work is done on a particle that undergoes uniform circular motion* (Fig. 6–16). This holds for *any* part of a circular trajectory at constant speed. Recall from Section 3–5 or 5–4 that an object undergoing uniform circular motion experiences an acceleration that is directed along the radius toward the center of the circle. Thus the force is directed in the (negative) radial direction and is always *perpendicular* to the direction of motion, which is tangential to the circular trajectory. If the infinitesimal displacement along an arc is $d\vec{s}$, as in Fig. 6–17, then the scalar product $\vec{F}\cdot d\vec{s}$ is zero because the force has no component in the direction of the displacement. Thus no work is done.

If the motion is circular with varying speed, a tangential force $\vec{F}_t$ must be present (one that is parallel to the direction of the displacement). Work is done in this case with

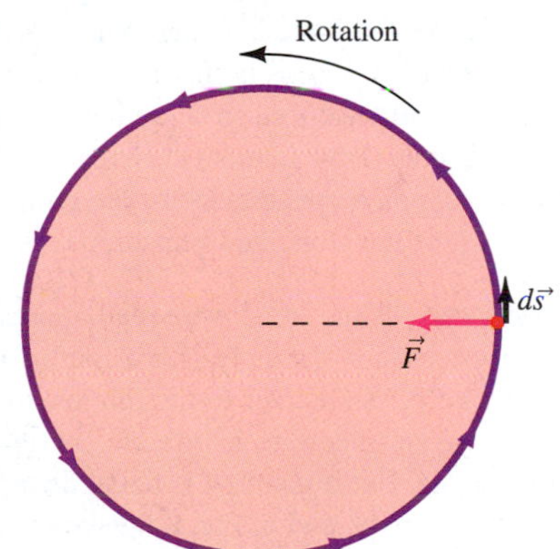

▲ **FIGURE 6–17** In the uniform circular motion of an object, the infinitesimal displacement is tangent to the circle. The force responsible for this motion is directed toward the center of the circle and is perpendicular to the displacement. The work done by this force on the object is therefore zero.

a consequent change in energy. Because the force is directed along (or opposite to) the displacement $d\vec{s}$, the scalar product $\vec{F}_t \cdot d\vec{s}$ for the infinitesimal work done has nonzero magnitude $F_t ds$. We integrate this quantity to find the total work.

EXAMPLE 6–10 A ball with a mass of 8.0 kg is attached to the end of a rod 1.5 m long of negligible mass. The rod is attached to a vertical shaft in such a way that the rod is held perpendicular to the shaft (Fig. 6–18). The attachment point of the rod and shaft is a frictionless pivot point that allows the rod to rotate about the shaft. A tangential force of constant magnitude F is applied to the ball for one-quarter turn. As a result, the ball and rod rotate about the shaft with an angular speed of 1.2 rev/s. If the ball is initially at rest, what is F?

Setting It Up We indicate the tangential force F and the velocity v in the figure. The rotation is about the z-axis, which is vertical. We know the mass m of the ball, the radius r of the circular motion of the ball (r being rod length), and the ball's angular velocity ω after starting from rest and rotating a quarter turn.

Strategy If we find the ball's final speed from the given final angular speed, we can then find the ball's final kinetic energy. Because the initial kinetic energy is zero, we know the change in kinetic energy and can use the work–energy theorem to find the corresponding net work. Given the distance over which the force acts, this allows us to find the net force itself.

Working It Out The angular speed after the force has acted is $1.2\ \text{rev/s} = (1.2\ \text{rev/s})(2\pi\ \text{rad/rev}) = 7.5\ \text{rad/s}$, so the speed of the ball at the end of the push is $v = \omega r = (7.5\ \text{s}^{-1})(1.5\ \text{m}) = 11\ \text{m/s}$. Thus

$$K_f - K_i = \tfrac{1}{2}mv_f^2 - \tfrac{1}{2}mv_i^2$$
$$= \tfrac{1}{2}(8.0\ \text{kg})(11\ \text{m/s})^2 = 4.8 \times 10^2\ \text{J}.$$

The work–energy theorem states that this is the net work, and this work is done only by the applied force. This force, applied for a quarter turn, acts over a distance $\frac{1}{4}(2\pi r) = \frac{1}{4}\pi(3.0\ \text{m}) = 2.4\ \text{m}$. Thus the force has magnitude

$$F = \frac{K_f - K_i}{\text{distance}} = \frac{4.8 \times 10^2\ \text{J}}{2.4\ \text{m}} = 2.0 \times 10^2\ \text{N}.$$

What Do You Think? Is the tangential force referred to here the only force acting on the ball?

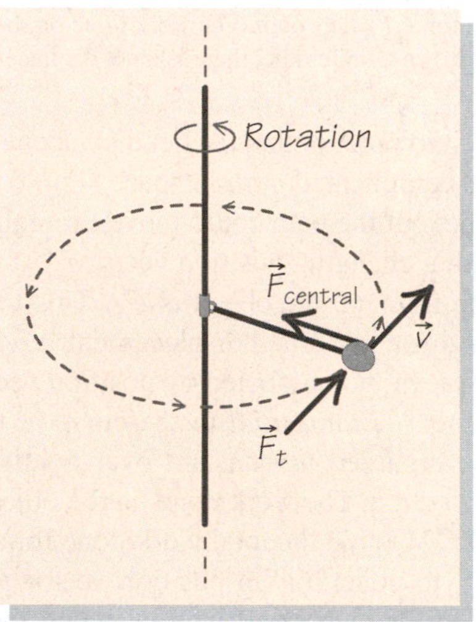

▲ **FIGURE 6–18** Forces acting on ball.

CONCEPTUAL EXAMPLE 6–11 Consider a stone twirled at the end of a length of rope. Due to the centripetal tension of the rope, the stone is undergoing nearly circular motion in a horizontal plane. The rope is slowly shortened by pulling it in, and as you may have experienced in a similar situation, the stone then moves more rapidly. How is this consistent with the fact that the force due to the rope is perpendicular to the path of the stone? [*Hint*: Sketch the path of the stone and reexamine the statement of the question.]

Answer If the stone were really undergoing circular motion with the rope tension the only (horizontal) force acting on it, this force would be perpendicular to the motion and could do no work; by the work–energy theorem, its kinetic energy and hence its speed could not change. But if the rope is being drawn in, the path of the stone is actually a spiral, and in Fig. 6–19 we see that the line from the center of rotation O to the stone is not quite perpendicular to the spiral path. Thus there is a small component of the force vector (the tension of the rope) along the in-going part of the spiral. This component of the force speeds up the stone as it moves along its path. You can see this in another way if you think of a rope shortening as it wraps around a pole. The pole must have a nonzero radius, and this means that the motion is not quite circular about the center of the pole. We'll see in Chapter 10 that we can view this as a matter of conserving angular momentum and that the conservation of angular momentum implies an increase in speed as the radius of the motion decreases.

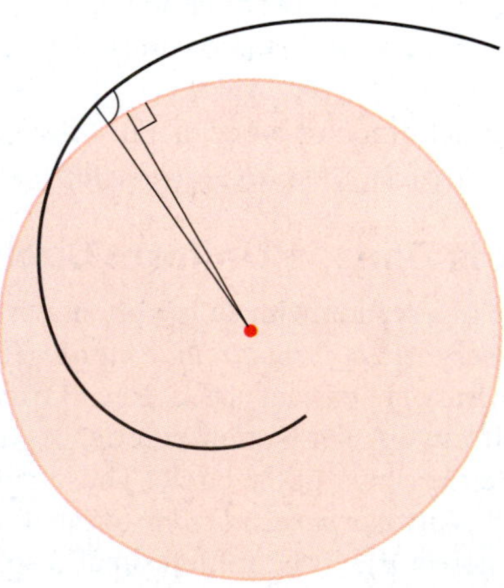

▲ **FIGURE 6–19** In spiral motion the direction from the center of the motion is not quite perpendicular to the direction of the motion, as it is in circular motion. The work done is therefore not zero.

6–4 Conservative and Nonconservative Forces

One of the most important questions that arises when an object undergoes a certain displacement concerns *path dependence*: How does the work done by a force depend on the path taken by the object during its displacement? We are not necessarily considering net force here; certainly when the work–energy theorem is applied, we must do so. But the question of path dependence is one that is best addressed in the context of individual forces. We illustrate what is involved by considering, in order, gravity, friction, and the spring force.

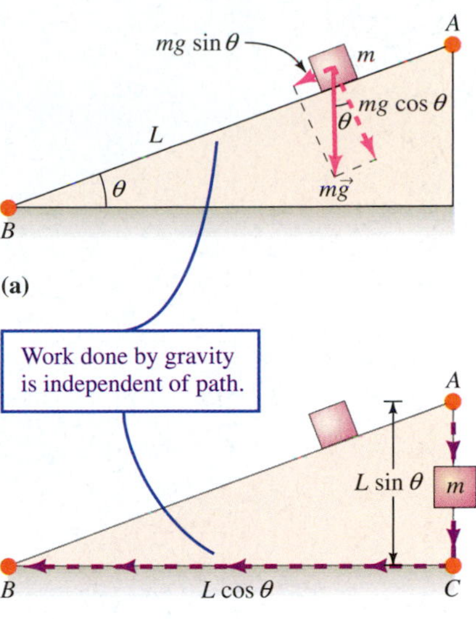

▲ **FIGURE 6–20** (a) An object of mass m moves from point A to point B on an inclined plane under the influence of gravity. Gravity does positive (or negative) work on the object as it moves down (or up) the plane. (b) The object now moves from point A to point B by a different path: a vertical motion from point A to point C followed by a horizontal movement from C to B. The work done by gravity is exactly the same as in part (a).

Gravity: Let us consider the work that gravity does when a box of books of mass m moves from the top (point A) of a ramp of length L to the bottom (point B). Figure 6–20a shows the geometry of the ramp, which makes an angle θ with the horizontal. The force of gravity acts vertically downward on the box with magnitude mg. This force can be decomposed into a normal component $mg \cos\theta$ perpendicular to the ramp and a parallel component $mg \sin\theta$ pointing along the ramp toward its bottom. When the box moves down the ramp, as in Fig. 6–20a, the motion is perpendicular to the normal component of gravity, and the work done by that component of the force is therefore zero. The work done by the force of gravity to move the object from A to B (a total length L) along the direction of motion thus involves only the component $mg \sin\theta$ of gravity *along* the ramp,

$$W = (mg \sin\theta)L. \qquad (6\text{–}17)$$

Consider now a second path from the top of the ramp to the bottom. This second displacement is achieved by moving the box vertically off the back of the ramp (from A to C in Fig. 6–20b) and then horizontally to the previous end point (from C to B in Fig. 6–20b). For the first leg, the component of the force along the motion is mg, and the distance through which the force acts is $L \sin\theta$. Thus the work done by gravity on the box over this first leg is $(mg)L \sin\theta$. The box then moves horizontally; gravity does no work on the box during this portion of the displacement because the force of gravity is perpendicular to the motion. Thus the total work done by gravity on the box for this second path is $mgL \sin\theta$, the same as that calculated in Eq. (6–17). We would find the same result for other paths: The work done by gravity depends only on the difference between the final and initial heights of the box, in this case $L \sin\theta$; it does not depend on the path by which this height difference is reached. In fact, if the height decreases by an amount h, the work done in moving through any path is mgh (see Example 6–3). In this particular case, $h = L \sin\theta$.

Friction: Let us next consider the work done by sliding friction on a coffee cup of mass m that is pushed across a rough horizontal surface. (While there are other forces besides friction acting on the cup, it is only the work done by friction that interests us here.) The friction force on the cup has magnitude $\mu_k mg$ and is directed *opposite* to the motion. Suppose that the cup in Fig. 6–21 is moved (with the aid of some external force) along the $+x$-axis from $x = 0$ to $x = L$ and then back again. For the first half of the motion, the displacement of the cup is $+L\hat{i}$, whereas the force of friction is directed to the $-x$-direction:

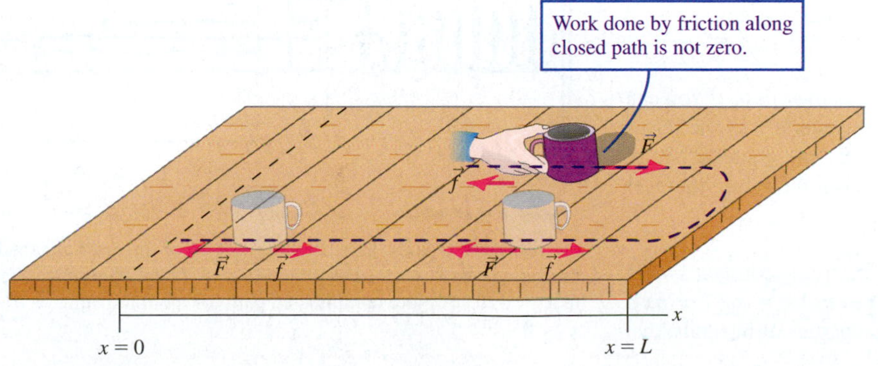

◀ **FIGURE 6–21** A cup moves along a horizontal surface from $x = 0$ to $x = L$ and back again. One of the forces acting is friction, and the work done by friction in the back-and-forth motion is not zero.

$\vec{f} = -\mu_k mg\hat{\imath}$. In the motion back to the origin, the displacement is $-L\hat{\imath}$, whereas the force of friction is directed to the $+x$-direction: $\vec{F} = +\mu_k mg\hat{\imath}$. Then the work done by friction is

$$W = W_{out} + W_{back} = (-\mu_k mg)(L) + (\mu_k mg)(-L) = -2\mu_k mgL.$$

Rather than canceling, friction does the same negative work coming back as going out, and there is work done even though the net displacement is zero. Moreover, the amount of work depends on L; that is, it depends on the path taken by the cup. The least amount of work is associated with the shortest path. In this case the starting and finishing points are the same and the shortest path is $L = 0$, that is, there is no motion at all. You can guess that if the starting and finishing points were different, the path over which friction does the least amount of work is the straight line between the points.

Spring Force: Equation (6–14), $W = \int_{x_0}^{x_f} F(x)\,dx$, is an expression for the work done that applies to any one-dimensional force that depends only on the position of the object. It is a general property of integration that one-dimensional integrals such as these depend only on the end points of the integration, not on any intermediate points. Thus the work done by such is independent of the path. This is illustrated with the spring force in the following example.

EXAMPLE 6–12 A mass m is attached to the end of a spring with spring constant k. The equilibrium position of the mass is at $x = 0$ (Fig. 6–22a). Consider two paths by which the mass can move from the point $x = L$ to the point $x = 0$: Path 1 is the direct motion (Fig. 6–22b) and path 2 is the displacement from $x = L$ to $x = a\,(a < 0)$ followed by a displacement to $x = b\,(b > L)$ and completed with a final displacement to $x = 0$ (Fig. 6–22c). Find the work done on the mass by the spring for each path.

Strategy Because we know the spring force $F = -kx$, we can find the work by using $W = \int F\,dx$ between suitable limits of integration.

Working It Out In the case of path 1, we have

$$W = \int_L^0 (-kx)\,dx = -\frac{kx^2}{2}\Big|_L^0 = -\left(0 - \frac{kL^2}{2}\right) = \frac{kL^2}{2}.$$

The work done by the spring on the mass is positive.

In the case of path 2, there are three contributions to W—corresponding to the three displacement steps described—and their sum is

$$W = \int_L^a (-kx)\,dx + \int_a^b (-kx)\,dx + \int_b^0 (-kx)\,dx$$
$$= -\frac{kx^2}{2}\Big|_L^a - \frac{kx^2}{2}\Big|_a^b - \frac{kx^2}{2}\Big|_b^0$$
$$= -\frac{ka^2}{2} + \frac{kL^2}{2} - \frac{kb^2}{2} + \frac{ka^2}{2} - \frac{k \times 0^2}{2} + \frac{kb^2}{2} = \frac{kL^2}{2}.$$

The spring does the same work over path 2 as over path 1. You can convince yourself rather quickly, given the general way the terms canceled in the work done over path 2, that the equality of the work done over the two paths would have been the same whatever the form of $F(x)$.

What Do You Think? Suppose that the spring constant k were not really constant but varied slightly with x so that $k = k_0(1 + bx)$. Give an argument, based only on general properties of integrals, that the work done by this force as the object under its influence moves from one point to another is independent of the path between the points.

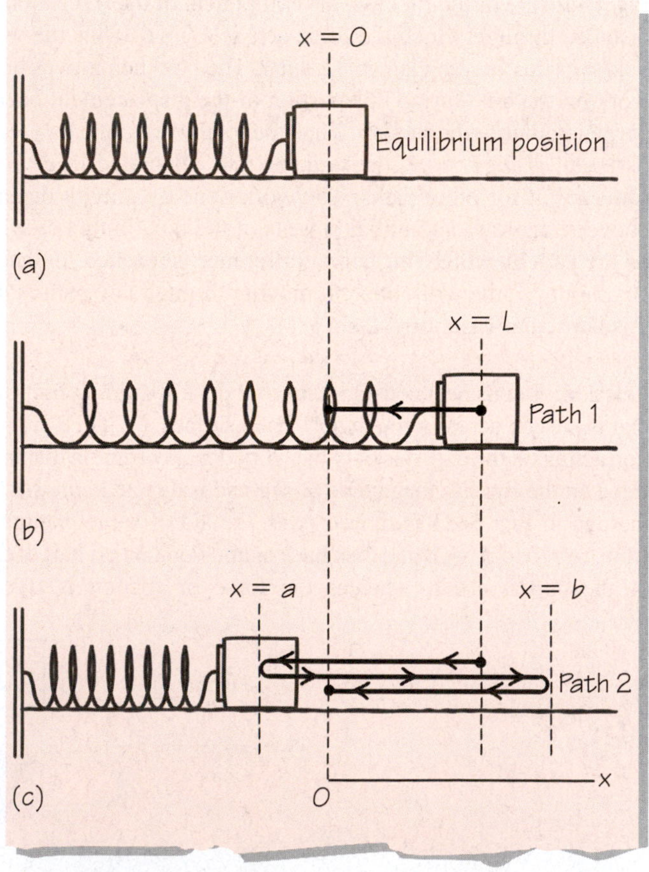

▲ **FIGURE 6–22** Equilibrium position of the mass attached to the spring. (b) The spring is pulled to $x = L$ and then moves back to $x = 0$. (c) A more complicated path for motion from $x = L$ to $x = 0$.

Conservative and Nonconservative Forces

We have seen that when an object moves from one position to another as forces act on it, the work done by gravity or the spring force is independent of the path taken, whereas the work done is dependent on the path taken if the force of friction acts on the object. We must be aware that the work, the integral of Eq. (6–16), may or may not depend on the path. We can categorize forces by whether the work they do is independent or not of the path taken. Forces for which the work is independent of the path are called **conservative forces**; those for which the work depends on the path are called **nonconservative forces**. Gravity and the spring force are conservative, whereas sliding friction is nonconservative.

The work done by a conservative force in moving an object from one position to another is independent of the path taken by the object.

We can reformulate this statement in terms of closed paths (paths that end at the same point at which they start):

The work done by a conservative force in moving an object along any closed path is zero.

Here is why these two statements are equivalent. One possible closed path is the path for which the object does not move at all, in which case W is obviously zero. But the statement that the work done is path independent means that the same amount of work is done along this path as along *any* path for which the object starts and finishes at the same point. Thus a conservative force does no work on an object moving along any closed path, and indeed this is a definition of a conservative force.

As an example, consider a bag of concrete. If we move it from the back of a truck to the ground, the force of gravity does positive work. If we then lift the bag from the ground back to the truck, gravity does negative work. The total work done by gravity in this process must be zero because the bag of concrete is back where it started—more precisely it is back at the same height at which it started.

We can now restate our conclusions about forces that act in one dimension, which were earlier stated in terms of path independence. *Any* force that can be written in the form $F = F(x)$ is a conservative force. This includes constant forces as a special case.

Conservative forces matter for two reasons. First, we shall see in Chapter 7 that we can dispense with work for these forces and instead use the notion of a *potential energy*. Second, *most of the forces we deal with are conservative*. This is true of all the fundamental forces in nature. In particular, *central forces*—those for which the force is directed along a line from a fixed center and whose magnitude depends only on the distance from the center—are conservative. The gravitational force between the Sun and the planets is an important example of a central force.

THINK ABOUT THIS . . .

FRICTION APPEARS TO BE A CONSTANT FORCE; WHY ISN'T IT CONSERVATIVE?

You might think from this discussion that kinetic (sliding) friction, which we have explicitly shown is not conservative, provides a counterexample to the idea that forces with constant magnitude are always conservative. After all, isn't friction on a horizontal surface a constant? This apparent difficulty is resolved by noting that friction is *not*, in fact, constant. Its *magnitude* is constant if the coefficient of sliding friction is constant. On a horizontal surface the magnitude of the friction force is $\mu_k F_N$, and F_N is the magnitude of gravity in simple situations. However, the *direction* of the friction force is not constant. The direction depends on the direction of motion of the object on which the force acts. When the object moves to the right, friction acts to the left; when the object moves to the left, friction acts to the right. Mathematically,

$$\vec{f} = \mu_k F_N \text{ in the direction opposite to } \vec{v},$$

which means that, strictly speaking, $\vec{f}$ depends on velocity. The work done by friction when an object moves out cannot possibly cancel the work done by friction as the object comes back along a reversed path, as was the case for the spring, because along each leg the friction does negative work. We can safely conclude that *sliding friction is nonconservative*.

The fact that we won't be able to associate a "potential energy" with nonconservative forces does not put us into any fundamental difficulty. We will simply have to deal with conservative and nonconservative forces differently. The work–energy theorem works for conservative and nonconservative forces alike, and we can always separate the work done by the two types of forces. To understand why this might be useful, consider a physical situation where there is an unknown frictional or drag force present, for example, an accelerating automobile subject to an unknown drag force from the air. We may then use the work–energy theorem to calculate the work done by this unknown force. The theorem in this case reads

$$\begin{pmatrix}\text{initial kinetic}\\ \text{energy}\end{pmatrix} + \begin{pmatrix}\text{work done by}\\ \text{known force}\end{pmatrix} + \begin{pmatrix}\text{work done by}\\ \text{unknown drag force}\end{pmatrix} = \begin{pmatrix}\text{final kinetic}\\ \text{energy}\end{pmatrix}.$$

The term *work done by unknown drag force* can be regarded as the energy lost, or **dissipated**. Rearranging, we find the work done by the unknown drag force.

$$\begin{pmatrix}\text{work done by}\\ \text{unknown drag force}\end{pmatrix} = \begin{pmatrix}\text{final kinetic}\\ \text{energy}\end{pmatrix} - \begin{pmatrix}\text{initial kinetic}\\ \text{energy}\end{pmatrix} - \begin{pmatrix}\text{work done by}\\ \text{known force}\end{pmatrix}.$$

The work done by the known force can be either positive or negative.

EXAMPLE 6–13 A mass of 50 g is placed on the end of a spring with a spring constant of 6.0 N/m. The mass is released from rest at $x = 10$ cm and moves to $x = 0$ (Fig. 6–23). Opposing the force of the spring is an unknown drag force. Find the work done on the mass by the drag force if the velocity of the mass at the equilibrium position $x = 0$ is 0.85 m/s.

Setting It Up The motion is all along one direction that we take as the x-axis. Figure 6–23 shows the mass both at the equilibrium position $x = 0$ and at the position released from rest, $x_0 = 10$ cm. We know the values of the mass m at the end of a spring and the spring constant k as well as the position of initial release from rest x_0 and the speed v_1 when the mass passes back through the equilibrium position.

Strategy We must put several elements in place to be able to use the work–energy theorem and find the work done by the drag force. The net work is the difference between the final kinetic energy K_f and the initial kinetic energy K_i; we know both K_i (which is zero because the mass starts from rest) and $K_f\left(=\frac{1}{2}mv_1^2\right)$. The net work, however, is also the algebraic sum of the work done by the spring, W_s, and the work done by the drag force, W_D. Thus, if we can find W_s, we can solve for W_D.

Working It Out The spring force $F = -kx$, so the work done by the spring on the mass as it moves from point x_0 to zero is

$$W_s = \int_{x_0}^{0} (-kx)\,dx = -\int_{0}^{x_0} (-kx)\,dx = \tfrac{1}{2}kx^2\Big|_0^{x_0} = \tfrac{1}{2}kx_0^2.$$

We now apply the work–energy theorem:

$$W_{\text{net}} = W_s + W_D = K_f - K_i,$$

or

$$\begin{aligned} W_D &= K_f - K_i - W_s = \tfrac{1}{2}mv_1^2 - 0 - \tfrac{1}{2}kx_0^2 \\ &= (0.5)(50 \times 10^{-3}\text{ kg})(0.85\text{ m/s})^2 \\ &\quad - (0.5)(6.0\text{ N/m})(0.10\text{ m})^2 = -0.012\text{ J}, \end{aligned}$$

where W_D, the work done by the unknown drag force, is negative, showing that energy is dissipated. Drag forces always dissipate energy, that is, they never add energy. This is because drag forces always act in a direction opposite to the direction of the displacement.

What Do You Think? Drag forces may arise in different ways. Give some ideas of how energy is dissipated if the mass at the end of the spring slides on a rough surface. Do the same if the spring and masses are immersed in molasses.

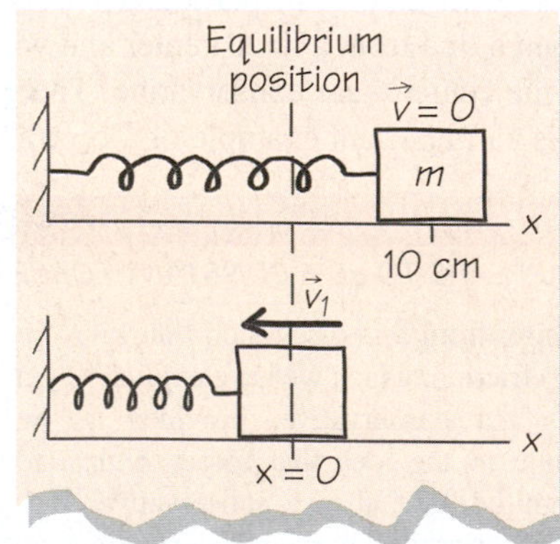

▲ **FIGURE 6–23** Coordinate system for mass on spring.

6–5 Power

Up to this point, we have said nothing about how *rapidly* work is done. When you go up a flight of steps, you are doing a certain amount of work, essentially mgh, where h is the vertical distance. But you can go up the steps very slowly or you can run up as fast as you can. In the first case you are doing the given amount of work over a long period. In the second case the rate at which you do the work is larger. **Power** P is *the rate at which work is done*:

$$P \equiv \frac{dW}{dt}. \quad (6\text{–}18)$$

DEFINITION OF POWER

It is straightforward to calculate the power for a constant force. In one dimension, we have Eq. (6–3), $W = F\,\Delta x$, where F is the constant force and Δx is the displacement from some fixed starting point. If we divide by Δt and take the limit of small Δt (so that Δx is also small), then we can recognize $\Delta x/\Delta t \rightarrow v$; on the left side we have $W/\Delta t$, that is, the work done per unit time, or power. We have shown that

$$P = Fv. \quad (6\text{–}19)$$

In two or three dimensions, the work done by a constant force is $W = \vec{F} \cdot \Delta\vec{r}$ [Eq. (6–11)]. It is easy to generalize our result for one dimension to

$$P = \vec{F} \cdot \frac{d\,\Delta\vec{r}}{dt} = \vec{F} \cdot \vec{v}. \quad (6\text{–}20)$$

Since the quantities on the right side of this expression can change with time, the power P in Eq. (6–20) is more appropriately called the **instantaneous power**.

The SI units of power are joules per second (J/s), and the unit has been given its own name, the **watt** (W for short; do not confuse this unit with the algebraic symbol W that we use for work). One watt is the power generated when a force of one newton displaces an object moving with a speed of one meter per second. Another commonly used unit of power is the **horsepower** (hp):

$$1 \text{ hp} = 550 \text{ ft. lb/s} = 746 \text{ W}.$$

Horsepower is sometimes used today as a measure of the power of automobile engines, but cars could just as well be rated in watts. You undoubtedly know that the watt is commonly used to rate the power output of lightbulbs (we study *electrical energy* in Chapter 26). A useful measure of electrical energy is based on the watt or, more conveniently, the kilowatt (kW). The **kilowatt-hour** (kWh) is the amount of work done when one kilowatt of power is generated for one hour.† Because there are 3600 s in 1 h,

$$1 \text{ kWh} = 3.6 \times 10^6 \text{ J}.$$

EXAMPLE 6–14 Early in the nineteenth century, James Watt wanted to market his newly developed steam engine to a society that until then had relied heavily on horses for mechanical work. So Watt invented a unit that made it clear how useful a steam engine could be. He conducted a demonstration in which a horse lifted water from a well over a certain period of time and called the corresponding power expended "one horsepower" (Fig. 6–24a). He could then compare his engine (favorably). Assume that water has a mass density of $1.0 \times 10^3 \text{ kg/m}^3$, that the well was 20.0 m deep, and that the horse worked for 8.0 h. How many liters of water did the horse raise from the well?

Setting It Up We sketch the situation in Fig. 6–24b. The water is pulled up vertically through the tension in the rope, with the pulley used to change the direction of the tension T supplied by the horse. We are given the mass density ρ of water, the depth Δy of the well, and the time t the horse worked. The tension in the rope, of magnitude T, is as yet unknown; it will be needed to find the work done and from that the power.

Strategy To determine the work done by the horse, we need T and the (known) distance the water traveled. Assuming the bucket of water does not accelerate, we can use the fact that there is no net force on the bucket to learn that $T = mg$. By dividing the work done by the tension (i.e., by the horse) by the time taken, we can find the power supplied by the horse, but we know this independently (it is 1 hp), so we can solve for the mass raised and hence the volume of water.

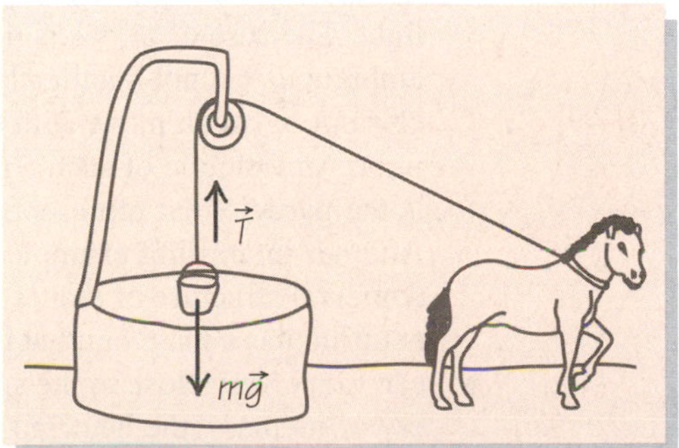

FIGURE 6–24 (a) As this nineteenth-century representation shows, James Watt's steam engine could do the same work as many horses. (b) Sketch to help in determining the power supplied by the horse.

†A typical large hydroelectric project generates a power of 10^{10} W; household energy consumption in the northern United States typically runs from 250 to 1000 kWh per month.

(continues on next page)

Working It Out The work done by the tension in lifting a mass $m = \rho V$ of water from the bottom of the well is

$$W = T\,\Delta y = mg\,\Delta y = \rho V g\,\Delta y.$$

Since this work was done over a period of time t, the power is

$$P = \frac{\text{work}}{\text{time}} = \frac{\rho V g\,\Delta y}{t}.$$

We solve for the unknown volume V:

$$V = \frac{Pt}{\rho g \Delta y}.$$

All the quantities on the right-hand side are known. The power is, by definition, 1 hp = 746 W, and

$$V = \frac{(746\ \text{W})(8.0\ \text{h} \times 3600\ \text{s/h})}{(1.0 \times 10^3\ \text{kg/m}^3)(9.8\ \text{m/s}^2)(20.0\ \text{m})} = 1.1 \times 10^2\ \text{m}^3.$$

Because there are 10^3 L in 1 m^3, the volume lifted by the horse in liters is 1.1×10^5 L.

What Do You Think? In the problem statement we used the phrase "in a certain period." So is the horsepower an instantaneous quantity?

* 6–6 Kinetic Energy at Very High Speeds

We mentioned in Chapter 5 that Newton's laws cease to be applicable in two domains. One is the domain in which speeds approach the speed of light and the other is the domain of quantum physics, which applies largely to atoms and smaller entities. The first domain is discussed in Chapter 39, whose subject is special relativity. In Chapter 39, we will show that the maximum speed attainable by any particle is the speed of light itself. We will also show that the precise expression $\frac{1}{2}mv^2$ for the kinetic energy—the Newtonian form of kinetic energy—of a particle of mass m should be replaced with a more general result,

$$K = mc^2\left(\frac{1}{\sqrt{1 - (v/c)^2}} - 1\right). \tag{6–21}$$

Here c is the speed of light, which has a value of 3×10^8 m/s. This kinetic energy appears in the same work–energy theorem that we used throughout if one also appropriately modifies how work is calculated. In this section, we explore this relativistic form for the kinetic energy.

Equation (6–21) reduces to the usual expression $K = \frac{1}{2}mv^2$ when $(v/c)^2$ is very small (see Problem 77). We call a quantity x "very small" when x can be neglected in comparison with 1, and we denote this by $x \ll 1$. For example, if $v/c = 10^{-2}$, then $(v/c)^2 = 10^{-4}$; this is indeed much less than 10^{-2} if we maintain 1 percent accuracy. Under normal conditions, v/c is much smaller than 10^{-2}. The speed v_E of Earth around the Sun is, by comparison to, say, the speed of an automobile, a large number, yet $v_E/c \cong 10^{-4}$. Molecular speeds in air are also of this order of magnitude.

For particles in cosmic-ray showers and in particle accelerators, v can be very close to c, and the factor $1/\sqrt{1 - (v/c)^2}$ can be very large. In the highest energy accelerators this factor is larger than 10^4, and the relativistic kinetic energy is very much larger than the Newtonian kinetic energy formula would indicate. We note that the formula for K becomes uncontrollably large in the limit $v/c \to 1$. It is indeed an essential tenet of the theory of relativity, which improves on Newton's laws for high-speed particles, that no information—that is, no particles—can be sent at a speed greater than the speed of light. The case of $v = c$ is delicate: With $m = 0$, the expression for K in Eq. (6–21) is ambiguous but not manifestly wrong. The theory of relativity states that massless particles not only can move with the speed of light but must *always* move with the speed of light. An example of such a particle is the elementary particle called the photon, which is the particle that represents light itself in the quantum physics description of nature. Another interesting example is the *neutrino*, which plays an important role in the fundamental structure of matter, in astrophysics, and in cosmology. Experiment shows the neutrino has a mass but that it is a very small fraction of the electron mass. This particle can move very close to the speed of light but not *at* the speed of light. Until the discovery of its mass, the neutrino was thought to move always at exactly the speed of light; once you learn that the mass is not zero, no matter how small, you know that its speed cannot ever reach the speed of light.

Summary

Work W is done by a force that acts on an object when the object moves; a moving object possesses a kinetic energy K given by

$$K = \tfrac{1}{2}mv^2. \tag{6–4}$$

Here K is a scalar quantity formed from the velocity vector, with v^2 (the speed squared) $= v_x^2 + v_y^2 + v_z^2$. The central result of this chapter is the work–energy theorem as it applies to an object under the influence of one or more forces,

$$W_{net} = \Delta K, \tag{6–6}$$

where W_{net} is the net work done by the net force on the object. Here ΔK is the *change* in the kinetic energy of the object as it moves from an initial position $\vec{r}_i$ to a final position $\vec{r}_f$ through a displacement $\Delta\vec{r} = \vec{r}_f - \vec{r}_i$, $\Delta K = K_f - K_i$. The work–energy theorem tells us that when net work—positive or negative—is done on an object, the kinetic energy increases or decreases by the amount of the net work. The net work depends on the net force acting on an object as well as its displacement and, in general, on how the displacement is made. The net work is the work done by the net force; equivalently, it is the sum of the work done by the individual forces that make up the net force. The work done by a force can be expressed in different ways according to the form the force takes. We can enumerate these forms:

For a constant net force in one dimension $\quad W \equiv F\Delta x. \quad$ (6–3)

For a constant force in three dimensions

$$W = F_x\,\Delta x + F_y\,\Delta y + F_z\,\Delta z \tag{6–10}$$

$$= \vec{F}\cdot\Delta\vec{r} \tag{6–11}$$

$$= F\,\Delta r\cos\theta, \tag{6–12}$$

where θ is the angle between the vectors $\vec{F}$ and $\Delta\vec{r}$.

For a nonconstant force in one dimension

$$W = \int_{x_i}^{x_f} F(x)\,dx. \tag{6–14}$$

For a nonconstant force in three dimensions

$$W = \int_{\vec{r}_A}^{\vec{r}_B} \vec{F}\cdot d\vec{r}. \tag{6–16}$$

In Eq. (6–16), the displacement is between points A and B. This form for work is the most general one and reduces to the other forms in the appropriate limit. It shows in particular that no work is done on an object that is in uniform circular motion. Both work and kinetic energy are measured in the SI unit of the joule (J).

In some cases, the work done when an object moves between two points depends on the path the object takes; in others, the work done is independent of the path. When the work done is path dependent, we say that the force is nonconservative; friction provides an example. When the work done is path independent, we say that the force is conservative; gravity and the spring force provide examples. Conservative forces are important because all the fundamental forces of nature are conservative.

The work–energy theorem allows us to calculate the speeds of objects when the work done by the forces is known, and it allows us to calculate the net work that must be done if a certain speed is to be achieved. This theorem is often much simpler to use for these purposes than is Newton's second law.

Power is the rate at which work is done:

$$P \equiv \frac{dW}{dt}. \tag{6–18}$$

The SI unit of power is the watt, equivalent to 1 J/s. From the definition of the instantaneous power, we find that in one dimension

$$P = Fv. \tag{6–19}$$

In two or three dimensions, we can generalize this result to

$$P = \vec{F}\cdot\vec{v}. \tag{6–20}$$

Understanding the Concepts

1. We mentioned in the introduction that the work–energy theorem might help us to analyze the motion of a baseball under the influence of both gravity and drag force from the air. How would you do so?
2. You are sitting in an automobile with a ball on your lap. The automobile stops very suddenly and the ball shoots forward out of your lap. To you, its kinetic energy has evidently changed. How would you use the work–energy theorem to analyze the situation?
3. Does it make sense to refer to a force doing negative work when an object moves under its influence? What does negative work mean? In answering, consider what happens when an object is stopped by a force.
4. It certainly seems like work to us when we hold a bag of groceries for a long period of time. Are we expending energy when we hold a bag of groceries for a long period of time? How is the answer to this question consistent with the work–energy theorem?
5. The centripetal forces that cause uniform circular motion do no work because they are perpendicular to the motion. How do such forces fit into the work–energy theorem?
6. A piano can be lifted to the third story of a building by having a crew carry it up the stairs or by using some type of pulley system. Is the same work done in both cases? Assume that friction can be neglected.
7. If the moving crew of Question 6 uses a rope and pulley, it pulls on the rope in the same direction as the force of gravity. Because the crew pulls the rope in the direction opposite to the displacement of the load, is the crew doing negative work on the load?
8. Is the force of a tennis racket acting on a tennis ball a conservative force?
9. A man pushes against the smokestack on a cruise boat. When the ship is stationary, he does no work. When the ship starts to move in the direction in which he is pushing, he appears to be doing work yet he experiences no change in the level of his exertion. Why is this? Keep in mind that the man does not fall down because the force of friction keeps his shoes from sliding backward. What is the work done on the man by the rough deck?
10. You do no net work when you walk at a constant speed. Why do you get tired?
11. A parachutist jumps from a plane and lands safely in a field. Does the net work done on the parachutist depend on the height from which he or she jumps? (The work done by gravity does depend on that height.)
12. No work is done in uniform circular motion. Suppose that you observe circular motion in which the moving object first speeds up and then slows down to its original speed. Is any net work done?
13. The work done by friction on a box that slides across a floor from one spot to another is negative. Can one conceive of a physical situation in which the work done by friction on an object is positive? [*Hint*: In an idealized (no-air-resistance) situation, the net external force on an automobile is friction between tires and road.]
14. No work is done on a bag of groceries while you are holding it stationary. Is work done on the same bag if you are holding it while you move steadily upward in an elevator? What is the difference in the two situations? Are your hands still the origin of the force that does work on the bag?
15. A car with cruise control transports you at a constant speed. The engine does work. How much of that work is done on you?
16. Discuss to what extent the following description of a conservative force is equivalent to the ones given in the text: "In the motion of objects that are subject to conservative forces, there is no energy dissipation involved."
17. One of the entertainments at a carnival is a rotating (vertical) cylinder. The participants step in and place themselves against the interior wall. The cylinder starts to rotate more and more rapidly, and at some point the floor falls away, leaving the customers stuck like flies to a wall. Is any work done on the participants? If so, what force does the work?
18. Tarzan swings from tree to tree on a jungle vine (Fig. 6–25). Is there net work done on him during the motion? If so, what forces do the work?

▲ **FIGURE 6–25** Question 18.

19. A stunt consists of one acrobat standing on the short end of a seesaw whose pivot point is not at its midpoint. A second acrobat leaps down on the long end of the board and flips the first acrobat several meters into the air. How would you determine the work done by the second acrobat in flipping the first one into the air?
20. When a dropped egg hits the ground, it abruptly loses the kinetic energy it had just before it struck the ground. Does this mean that the ground has done work on the egg? If so, what is the sign of this work?
21. Two identical twins work side by side as butchers. They use identical motions and identical hatchets. One brings a hatchet down on some very tender meat and the other on a large bone. Which one does more work per swing?
22. You tow a small child on a sled at a constant speed by pulling the sled with a rope. The rope makes an angle θ with respect to the horizontal. What forces act on the sled and which ones do work?
23. A one-dimensional force acts on an object, changing its velocity from zero to $\vec{v}$. By the work–energy theorem, the work done is $W = \frac{1}{2}mv_f^2 - \frac{1}{2}mv_i^2 = \frac{1}{2}mv^2$. An observer moving with velocity $\vec{v}$ with respect to the original system sees the initial velocity as $-\vec{v}$ and the final velocity as zero. This observer would conclude that $W = -\frac{1}{2}mv^2$. What accounts for the difference?
24. A diver plunges from a 10-m-high diving board into water. How would you determine the average force of resistance of the water that slows down and stops the diver? (Neglect the force of gravity on the diver while she is in the water—we shall see in Chapter 16 how the water's buoyancy takes care of that.)
25. A parachutist jumps off a tower. What measurements would you have to make to determine the work done by the drag force of the air during the entire fall? The drag force is a rather complicated function of the velocity of the jumper.
26. Are the following forces conservative or nonconservative? (a) Air drag on a parachute. (b) The force opposing the fall of a steel ball bearing in a beaker of water. (c) The explosive force causing a bullet to leave a rifle barrel. (d) The force of an ideal trampoline that propels you into the air.
27. How do you know that the drag forces you experience when you swim are not conservative?
28. An object can be said to have a certain kinetic energy. The work–energy theorem relates the change in kinetic energy on the object to the work that is done on it. Does this mean that the object also "has" a certain amount of work?

Problems

6–1 Kinetic Energy and Work

1. (I) An automobile of mass 10^3 kg moves at 1.0 km/h = 0.28 m/s. (a) What is its kinetic energy? (b) At what speeds must a person of mass 80 kg and a bullet of mass 10 g move to have the same kinetic energy as the automobile? (c) What would the speed of the automobile be if the kinetic energy doubled?

2. (I) A construction worker of mass 85 kg rides in an elevator up to the 15th floor, which is 42 m above the ground. The elevator travels with uniform speed. (a) What is the net work done on the worker? (b) What is the work done on the worker by the contact force of the elevator? (c) What is the work done on the worker by gravity?

3. (I) A person lifts a suitcase of mass 10 kg from the floor. Ignore the initial acceleration of the suitcase and suppose that it moves upward at a constant speed between a height $h = 0$ m and $h = 1$ m. (a) What are the forces acting on the suitcase as well as the net force? (b) What is the net work done on the suitcase? (c) Find the work done on the suitcase by the person.

4. (I) A bedroom dresser of mass 38 kg is moved from the first floor of an apartment building to the penthouse on the 44th floor, 130 m higher. (a) What is the work done on the bureau by three men in carrying it up the steps? (b) If the three men take it up on an elevator, how much work is done on the dresser by the normal force of the floor of the elevator?

5. (I) A truck carrying a 66–kg crate accelerates uniformly from rest to 63 km/h in 15 s. Calculate the work done on the crate by the truck.

6. (I) An old piano of mass 180 kg is removed from an apartment building being converted into condominiums. The previous owners found it too much trouble to remove and left it. The workmen decide the easiest thing to do is to drop it out of a double-width window to the ground 25 m below. (a) How much work do the workmen do if they just push it out the window? (b) If the men slowly lower the piano by rope, what is the work done on the piano by the rope's tension? (c) How much work does gravity do in each case?

7. (I) A man pushes a refrigerator of mass 40 kg at uniform speed for a distance of 1.5 m to the kitchen wall. The coefficient of friction between the refrigerator and the floor is $\mu_k = 0.4$. (a) How much work does the man do in moving the refrigerator? (b) What other sources of work done are there? (c) What is the net work done in this process?

8. (I) A person pulls a heavy load of mass 37 kg up the side of a building by using a frictionless pulley. The load travels up a distance of 7.5 m. Take the load to move with constant velocity and ignore any acceleration at the beginning or end of the move. (a) How much work is done on the load by gravity? (b) By the tension of the rope? (c) By the person?

9. (II) Consider the woman who lifts the huge interior lineman in Example 5–4. How much work does she do while pulling down a 2-m length of rope? What is the work done by gravity on the lineman while this is going on?

10. (II) The mass $M = 40$ kg is lifted to a height $h = 4$ m using the system of pulleys shown in Fig. 6–26. The motion is slow and the initial acceleration is negligible. (a) Find the force that must be applied at the free end of the rope. (b) Find the work done on the mass by this force. (c) Calculate the work done on the mass by gravity during the process.

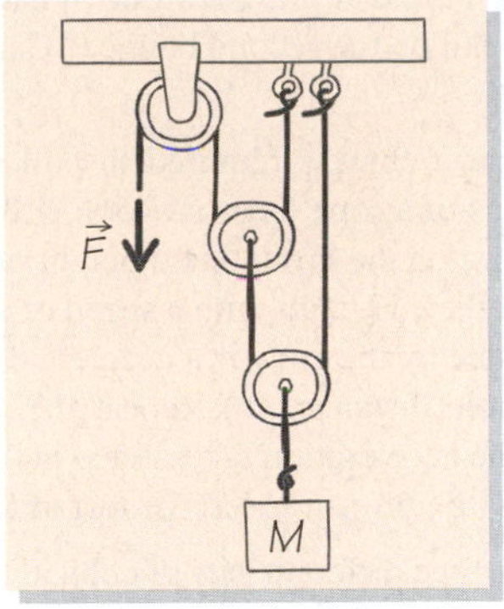

▲ **FIGURE 6–26** Problem 10.

11. (II) A baseball of mass 145 g leaves a pitcher's hand at 96.6 mi/h, but, due to air resistance, it arrives at home plate 60.0 ft away traveling at 95.3 mi/h. Assume that the magnitude of the ball's acceleration is constant and that the ball travels in a straight line (ignore gravity). How much work is done by friction during the flight of the ball?

12. (II) A ball of mass 240 g is dropped from a height of 2 m. (a) What is the work done on the ball by gravity? (b) Suppose that the ball bounces to a height of only 1.5 m. How much work is done by gravity on the ball as it moves from ground level to 1.5 m?

13. (II) A construction worker of mass 75 kg hoists a load of bricks of mass 42 kg by throwing a rope attached to the load over a pulley and letting his weight lift the load. Assuming that there is no friction, what is the work done by gravity during a 2.0-s period?

14. (II) Two masses are connected by a light string over a light, frictionless pulley, as in Fig. 6–27. The table surface is also frictionless. (a) Apply the work–energy theorem for this system to calculate the speed of the masses after the masses have moved a distance Δx starting from rest. Note that the work of the tensions drops out. (b) Use this result to obtain the acceleration of the system.

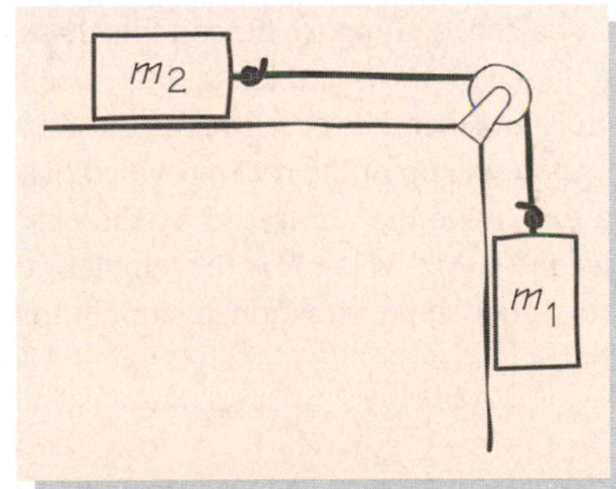

▲ **FIGURE 6–27** Problem 14.

15. (II) A waterfall of height 40 m has 200 m^3 of water falling every second. How many joules of work are done by gravity every hour? (The mass of 1 m^3 of water is 10^3 kg.)

16. (II) A ball of mass 85 g is dropped from a height of 3.00 m. It bounces back to a height of 2.75 m. Use the work–energy theorem to calculate the change in kinetic energy between the beginning of the contact with the floor and the termination of the contact, assuming that air resistance is totally negligible.

17. (II) Consider the ball of the previous problem. Assume that the energy loss on contact with the floor is proportional to the kinetic energy of the ball as it hits the floor. What will be the height reached by the ball on the second bounce? Can you generalize to the nth bounce?

18. (II) A construction crew is required to pull up a load of mass 106 kg by means of a rope thrown over a pulley. They are to lift the load from rest on the ground to a height of 4 m, and the load should arrive at the end point with a speed of 2.0 m/s. (a) Calculate the work done by the crew if it accelerates the load uniformly over the whole distance. (b) Repeat the calculation for the case in which the acceleration takes place in the first 1 m and the load is pulled with uniform speed the rest of the way.

19. (II) A child has three different sets of cubical blocks (Fig. 6–28). The first set consists of 3 blocks, each 12 cm on a side and of mass 36 g; the second set consists of 6 blocks, each 6 cm on a side and of mass 18 g; the third is a set of 12 blocks, each 3 cm on a side and of mass 9 g. For each set, what is the work the child must do to stack the blocks into a tower 36 cm high? The blocks can be treated as point objects at their *centers* in calculating the work.

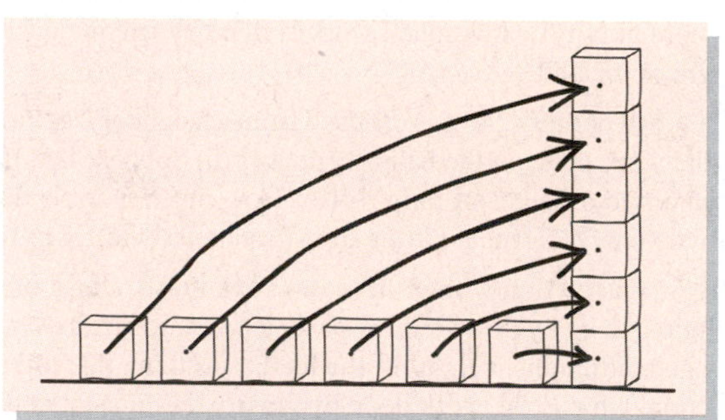

▲ **FIGURE 6–28** Problem 19.

20. (II) Archimedes is supposed to have said: "Give me a fulcrum and I will move the world." As a model (Fig. 6–29), consider a rod of length L. It rests on a sharp rock so that on the longer side the length of the rod is l_1 and on the shorter side it is l_2 (so that $L = l_1 + l_2$). A mass m_1 is placed at the end of the longer side and a mass m_2 at the end of the shorter side. The masses are such that the rod is balanced. Suppose the rod tilts through a *tiny* angle θ so that the mass m_1 is lowered just a little. Use the work–energy theorem to show that the rod will not continue to rotate as an acceleration about the tip of the rock provided that $m_1 l_1 = m_2 l_2$. [*Hint*: For a tiny angle the arc traced by the end point of a rod hinged at one end is $R\theta$, where R is the length of the rod, and the arc is, to a very good approximation, a straight line.]

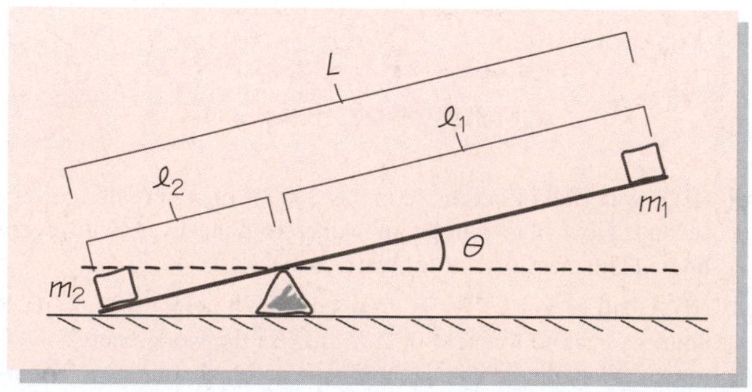

▲ **FIGURE 6–29** Problem 20.

21. (II) A mass swings on the end of a rope of length R, rising from a low point to a position $R(1 - \cos\theta)$ above that point (Fig. 6–30). Is work being done on the mass? Ignoring air resistance, what are the forces acting on the mass? Which of those forces does work on the mass, if any? Calculate the work done in any way you choose.

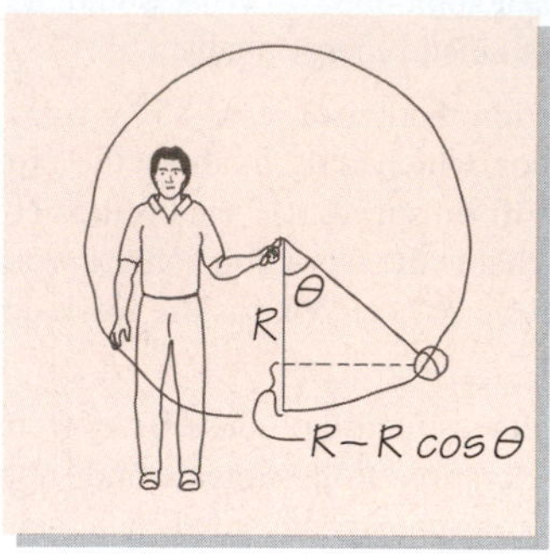

▲ **FIGURE 6–30** Problem 21.

22. (III) Find the work that must be done by a force lifting against gravity to raise a coiled rope of length L and mass M entirely off a level surface. [*Hint*: Use the method of Problem 19 and divide the rope into more and more segments.]

6–2 Constant Forces in Space

23. (I) What is the scalar product of $\vec{A} = -2\hat{i} + 3\hat{j} - 5\hat{k}$ and $\vec{B} = 5\hat{i} + \hat{j} - 2\hat{k}$?

24. (I) A force $\vec{F} = (-3.1\text{ N})\hat{i} + (2.7\text{ N})\hat{j}$ is used to displace an object of mass 17 kg by an amount $\vec{r} = (0.50\text{ m})\hat{i} + (-0.75\text{ m})\hat{j}$. What is the work done by the force on the object?

25. (I) An object of mass 0.23 kg is initially at the origin and is acted on by the sole force $\vec{F} = (0.50\text{ N})\hat{i}$. After a certain amount of time, the object is at a position $\vec{r} = (0.88\text{ m})\hat{i}$. What is the change in the object's kinetic energy?

26. (I) Show that the vector $\vec{v} = -y\hat{i} + x\hat{j}$ is always perpendicular to the vector $\vec{u} = x\hat{i} + y\hat{j}$.

27. (I) Consider two vectors, $\vec{u} = 3\hat{i} - 4\hat{j} + 7\hat{k}$ and $\vec{v} = -2\hat{i} + 3\hat{j} + z\hat{k}$. What must z be so that $\vec{u}$ and $\vec{v}$ are orthogonal?

28. (I) A person puts a suitcase of mass 11.5 kg into a van, moving the suitcase a total distance of 0.9 m: 0.6 m up and 0.3 m horizontally. How much work is done by the person?

29. (I) A man pulls a sled by a rope, moving his two daughters to the top of a 15° slope. He holds the rope parallel to the slope. If the daughters and the sled have a total mass of 43 kg and the length of the slope is 36 m, how much work does the man do on the sled, assuming that he pulls the sled with uniform velocity? Ignore all friction on the sled.

30. (I) A block of material with mass 1300 kg is used in the construction of a building. During one part of the process of setting the block in place, a complex network of cables acts on it and its motion is transformed from a horizontal motion with speed 15 cm/s to a vertical motion with speed 21 cm/s. What is the net work done on the block during this motion?

31. (II) A skier of mass 72 kg (including skis), starting from rest, slides down a slope at an angle of 18° with the horizontal. The coefficient of kinetic friction is $\mu_k = 0.12$. What is the net work done on the skier in the first 7.0 s of descent?

32. (II) Consider a vector $\vec{A}$ in the xy-plane. Its x- and y-components are A_1 and A_2, respectively. Show that any vector $\vec{B}$ in the same plane that points in a direction perpendicular to $\vec{A}$ must have components $-cA_2$ and cA_1, respectively, where the magnitude of c is the ratio of the lengths ($|c| = B/A$).

33. (II) Sketch the direction of the vector $\vec{e} = (\cos\theta)\hat{i} + (\sin\theta)\hat{j}$. Show that it has unit length and use your sketch to give an expression for the unit vectors $\vec{f}$ that are perpendicular to $\vec{e}$ and that lie in the xy-plane. How many such vectors are there?

34. (II) Consider the vector $\vec{A} = 7\hat{i} + 3\hat{j} - 6\hat{k}$. Find the most general vector in the yz-plane that is perpendicular to $\vec{A}$.

35. (II) Consider the unit vector $\hat{e} = -0.6\hat{i} + 0.8\hat{j}$. What is the magnitude of the projection of vector $\vec{A} = 3\hat{i} - 2\hat{j}$ onto the line along which $\hat{e}$ points?

36. (II) A stone is thrown from a height h_0 above a level field, leaving the hand at a 40° angle. Ignore all effects of air resistance. (a) Compute the work done by gravity as the stone follows its trajectory back to the height h_0. Recall that the motion can be divided into motion in the vertical direction and motion in the horizontal direction. (b) Show, by applying the work–energy theorem, that the speed of the stone when it reaches h_0 again is identical to the speed it had when it left the hand.

37. (II) A force $\vec{F} = (2\hat{i} - 5\hat{j})$ newtons acts on an object that moves from $\vec{r}_1 = (7\hat{i} - 8\hat{j} + 2\hat{k})$ meters to a new position $\vec{r}_2 = (5\hat{i} - 4\hat{j} + 5\hat{k})$ meters. How much work does this force do on the object?

38. (II) A 32-kg crate slides down a plane that makes an angle of 17° with the horizontal, starting from rest at the top. The speed of the crate when it reaches the bottom of the 10-m-long slide is 2.5 m/s. What is the coefficient of friction? How much work is done by the force of friction?

39. (II) A small object is forced to move within a horizontal groove aligned with the x-axis. A constant force $\vec{F} = F_x\hat{i} + F_y\hat{j}$, where each component is constant, acts on the object, making it accelerate within its groove. Find the work done on the object during the period where it moves a horizontal distance L. What is the kinetic energy at the final point assuming that the object started at rest?

6–3 Forces That Vary with Position

40. (I) A spring with spring constant $k = 12$ N/m is attached to a wall at ground level. The end of the relaxed spring is on the floor at a location that we take to be the origin. A mass of 3.0 kg is attached to the end of the spring, and the spring is stretched by 50 cm and released. How much work has the spring done on the mass by the time the mass passes through the origin?

41. (I) A small gizmo is confined to a groove that is aligned with the x-direction. A rod pulls the gizmo in the $+x$-direction. The rod is attached to an apparatus such that the pulling force is 0.3 N when the gizmo is to the left of a point in the groove we label as the origin and 0.7 N when the gizmo is to the right of the origin. What is the work done by the pulling force on the gizmo as it moves from $x = -6$ cm to $x = +7$ cm?

42. (I) A one-dimensional force on a particle is given by $F = \alpha x$, where $\alpha = -3.00$ N/m for $x < 0$ and $\alpha = +7.00$ N/m for $x > 0$ (Fig. 6–31). Calculate the work done by the force on a block when the block is moved from $x_i = -1.50$ m to $x_f = +1.50$ m.

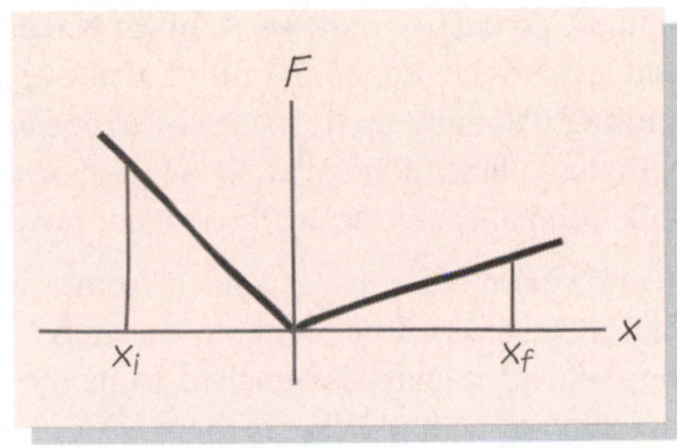

▲ **FIGURE 6–31** Problem 42.

43. (I) A one-dimensional force F depends on the position x of a particle on which it acts as $F = g_1 x - g_2 x^3$, where g_1 and g_2 are constants. What is the work done in moving the particle from the origin to $x = 2.0$ m?

44. (II) A man pushing a 50-kg crate up a slope that makes an angle of 30° with the horizontal exerts a force parallel to the plane. The coefficient of kinetic friction varies along the slope and is given by $\mu_k = \mu_1 + [(\mu_2 - \mu_1)s/L]$, where s is the distance along the slope starting at the bottom. The largest value of s is L, where $L = 10$ m is the length of the slope; $\mu_1 = 0.2$ and $\mu_2 = 0.3$. What is the work done on the crate by the force as a function of s if the force varies such that the crate is pushed at a constant speed?

45. (II) A spring gun is made by compressing a spring (assumed to be perfect) and latching it. A spring of constant $k = 60$ N/m is used and the latch is located at a distance of 7 cm from equilibrium. The pellets have mass 4 g. What is the muzzle velocity of the gun?

46. (II) An asteroid drops straight toward the Sun; the force on the asteroid due to the Sun has magnitude (constant)/r^2, where r is the distance from the center of the Sun to the asteroid, and is directed to the Sun's center. Given that the surface of the Sun is a distance R_{Sun} from its center and the asteroid starts from rest an infinite distance from the Sun, what is its kinetic energy when it reaches the Sun's surface?

47. (II) A nonstandard spring exerts a force $F = -k_1 x - k_2 x^3$ to restore itself to equilibrium, where x is the distance from equilibrium. The values of k_1 and k_2 are 5.0 N/m and 15 N/m^3, respectively. Calculate the work done to stretch the spring from 0.10 to 0.20 m.

48. (II) A rocket is scheduled to blast off from Cape Canaveral to study a neighboring solar system. Earth's gravitational force is $F = K/r^2$, where r is the distance from Earth's center and K is a negative constant. What is the minimum work the rocket engine must do so that the rocket leaves the gravitational force of Earth? Assume that the mass of the rocket does not vary in the process. (This assumption is actually very poor, but a more exact treatment must wait until Chapter 8.)

6–4 Conservative and Nonconservative Forces

49. (I) A child swings a streamer toy over her head in a nearly horizontal plane. The toy, of mass 85 g, is at the end of a massless string of length 1.5 m. She starts twirling the toy from rest while she slowly lets out the string to full length and gets the angular speed up to 2 rev/s. How much work has she done?

50. (I) A 74-g ball is tossed straight up in the air, rises to a maximum point, then falls back until it is 0.60 m below the position of the hand that tossed it up. A second ball is simply dropped from the same hand position and also lands 0.60 m below that position. What is the net work done by gravity in the two cases?

51. (I) You are in the process of moving. A 54-kg bed can be brought from ground level to the second floor (4.0 m above ground level) either by pulling it straight up by means of a rope or by dragging it up a frictionless plane inclined at 30° to the horizontal. Calculate the work done in each case by those who move the bed.

52. (I) An object of mass 600 g is suspended from a vertical spring that is attached to the ceiling. Without the mass, the spring is 25 cm long. When the mass is attached to it, the spring is extended to a length of 65 cm. What is the work done by the force of gravity during the extension of the spring?

53. (I) A constant force of 10 N pushes a particle along the x-axis. The position of the particle is represented by $x = 11\text{ m} - (2\text{ m/s})t + (0.5\text{ m/s}^2)t^2$. Find the work done by the force between $t = 0$ s and $t = 1$ s and between $t = 1$ s and $t = 2$ s. Is the force conservative?

54. (II) A force $\vec{F}$ has components $F_x = axy - by^2$, $F_y = -axy + bx^2$, where $a = 2\text{ N/m}^2$ and $b = 2\text{ N/m}^2$ (Fig. 6–32). Calculate the work done on an object of mass 4 kg if it is moved in a closed path from (x, y) values of $(0, 1)$ to $(4, 1)$, to $(4, 3)$, to $(0, 3)$, and back to $(0, 1)$ (all coordinates in meters). The path between the points is always the shortest straight one, and all the distances are given in meters.

58. (II) An object of mass m is to be moved from the top of a building of height h to a point on the ground a horizontal distance h from its original location so that the position vector may be chosen to be $h\hat{j}$ at the beginning, and $h\hat{i}$ at the end. Two possible paths are: (a) the object is lowered at constant speed by rope and, after it reaches the ground, it is moved horizontally to the final location; (b) the object is allowed to slide along a straight support that runs from the initial point to the final point. Show that the work done by the force of gravity is the same in both cases.

59. (III) Prove that a force acting in one dimension is conservative if it is a function of position only and not a function of any other information about the motion of an object under its influence. Does this include forces with constant magnitude? In view of your answer, how does the friction force manage to be nonconservative?

60. (III) A small object of mass m is moved up along a track that forms one-quarter of a circle of radius R in the vertical plane (Fig. 6–33). The object moves with a small uniform speed maintained by a tangential force (that varies with the angle) along the track. Calculate the work done by this force in moving the object through a 90° arc from the lowest point to the highest point by (a) direct use of the definition of work and (b) using the work–energy theorem.

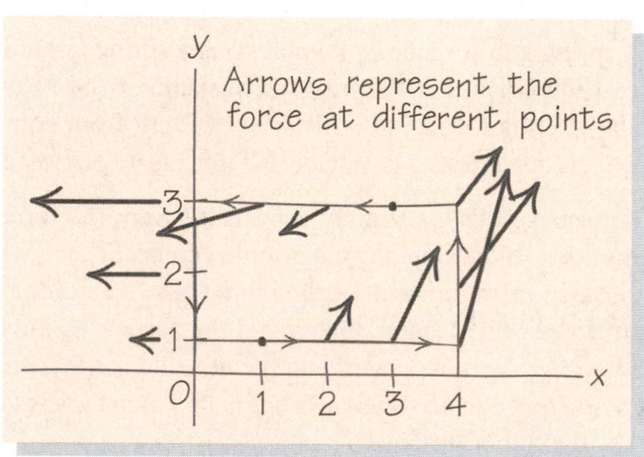

▲ **FIGURE 6–32** Problem 54.

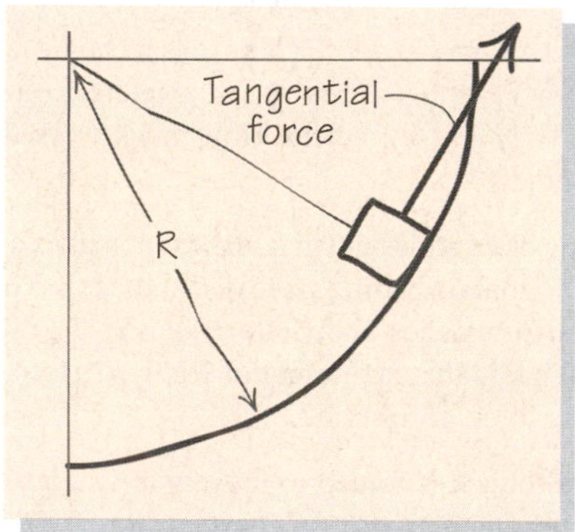

▲ **FIGURE 6–33** Problem 60.

55. (II) A force with an x-component acts on a mass with a strength that varies only with the position x of the mass on the x-axis, according to $|F(x)| = Ax^2$. The sign of the force is negative when x is positive and positive when x is negative, indicating that the force always attracts the mass toward the origin. Compute the work done by this force when $A = 1500\text{ N/m}^2$ and the mass moves along the x-axis (a) from $x = -5.0$ cm to the origin, (b) from $x = -5.0$ cm to $+5.0$ cm, (c) from $x = +5.0$ cm to $+2.0$ cm, and (d) from $x = -2.0$ cm to -5.0 cm.

56. (II) A child's playground ride consists of four seats, of mass 12 kg each, connected to a vertical axle with spokes of small mass. The seats are placed equidistant in a circle of radius 1.8 m and rotate about the vertical axle. A child of mass 21 kg sits in one of the seats, and his friend pushes the ride to accelerate him from rest to 0.6 rev/s. How much work does the friend do?

57. (II) The net force acting on a particle depends on the position of the particle on the x-axis according to the relation $F = F_0 + Cx$, where $F_0 = 5$ N and $C = -2$ N/m. The particle is initially at rest at the point $x = 0$ m when the force begins to act. (a) Calculate the work done by the force when the particle reaches x values of 1, 2, 3, and 4 m. (b) Determine any positions (other than at $x = 0$ m) where the work done is zero. (c) Is the force conservative?

61. (III) The (one-dimensional) force acting on an object of mass m is given by the expression $F(x) = C|x|$, where $|x|$ is measured in centimeters. (a) How much work is done by the force when the object is moved from $x = -4.0$ cm to $x = +4.0$ cm? Compare your result with the amount of work done if the form of the force law were $F(x) = Cx$. (b) Repeat the calculation if the object moves from $x = 0.0$ cm to $x = 8.0$ cm.

62. (III) Consider a force that acts on an object of mass m which moves in the xy-plane. The force is given by $\vec{F}(x, y) = k_1 x\hat{i} + k_2 y\hat{j}$. Calculate the work done by the object if it moves in a circle of unit radius (given by $x^2 + y^2 = 1\text{ m}^2$) starting at $x = 0$ m and $y = -1$ m and ending at a point that makes an angle of (a) 90°, (b) 180°, and (c) 360° with the original direction of the position radius vector. [*Hint*: The problem is simplified with polar coordinates r and θ, where $x = r\cos\theta$ and $y = r\sin\theta$.]

6–5 Power

63. (I) Electricity costs about \$0.08/kWh. Your monthly electric bill is \$26.00. Assuming that your only use of electricity is for light and that you keep your house lit 5 h/day, how many 100-W bulbs do you keep going?

64. (I) How much energy is used by running a fleet of one hundred 80-hp cars around the clock for one month?

65. (I) How long does it take you to climb four flights of steps? Assume that your body is 20 percent efficient and estimate the power you have to generate.

66. (I) Two engines are used to move a mass of 80 kg, starting from rest, a distance of 2.5 m in a straight line along a frictionless flat surface. Engine 1 exerts a constant force of 0.05 N, and engine 2 exerts a constant force of 0.75 N. (a) What is the work done by each engine? (b) What is the average power expended by each engine during the process?

67. (I) A test car of mass 700 kg is moving at a speed of 15 mi/h when it crashes into a wall to test its bumper. If the car comes to rest in 0.3 s, how much average power is expended in the process? (To find the average power, simply imagine Eq. (6–18) for a finite time interval: $P_{av} = \Delta W/\Delta t$, where ΔW is the work done during the time interval Δt.)

68. (I) Assume that a car of mass 1200 kg has an engine with power output of 80 hp. How long would it take to accelerate such a car to a speed of 100 km/h? (Neglect air resistance, which would make this time much larger.)

69. (II) The maximum power of a particular horse is 1 hp. With what speed can this horse pull a sled on level ground if the weight of the sled with its load is 5000 N and the coefficient of kinetic friction is $\mu_k = 0.03$? What is the maximum speed on a 5° upward incline?

70. (II) An accelerator accelerates a proton to $0.99c$, where c is the speed of light. If 6.50×10^{10} protons are accelerated every minute, how much power is expended by the accelerator, assuming 5.00 percent efficiency? [For this problem, with v of the order of c, use Eq. (6–21) for the kinetic energy. The *efficiency* is the fraction of the total power that goes into changing the kinetic energy of the protons.]

71. (II) Consider the waterfall in Problem 15. If the waterfall is used to produce electricity in a power station and the efficiency of conversion of kinetic energy of falling water to electrical energy is 60 percent, what is the power production of the station?

72. (II) An escalator moves people from one floor up to another. The height difference between the floors is 4.2 m, and the angle that the escalator makes with the horizontal is 20°. The speed of the escalator is 1.2 m/s, and it is supposed to carry a maximum of 75 passengers, with an average mass of 75 kg. How much power must be generated by the motor that runs the escalator?

73. (II) Trained athletes can exert power for their movements ranging from around 5 hp for 1 s to 0.4 hp or less for periods extending over several hours. (a) A bicyclist is limited by wind resistance, which is roughly of the form $F = Av^2$, where $A = 0.08$ kg/m. Estimate the speed a cyclist can maintain for 1 h. (b) Estimate the time it takes a weightlifter to lift 100 kg a distance of 2 m. (c) Assuming that not too much time is taken up turning the corners, estimate how fast it is possible to climb three flights of steps, a vertical distance of 12 m. You can easily try this one!

*6–6 Kinetic Energy at Very High Speeds

74. (I) A proton is accelerated from rest to a final speed of $0.85c$, where c is the speed of light. How much work is done by the accelerator on the proton given that a proton's mass is 1.7×10^{-27} kg?

75. (I) What is the kinetic energy of an electron (mass of 9.1×10^{-31} kg) moving at a speed of $0.9999c$? Of $0.9999999999c$?

76. (I) (a) How much work does it take to accelerate an electron from a speed of $0.1c$ to $0.5c$? (b) From $0.5c$ to $0.99c$? (c) From $0.99c$ to $0.999c$?

77. (II) In order to show that Eq. (6–21) reduces to the usual form for small v/c, we need an approximation for $1/\sqrt{1-x}$ for small values of x, namely, $1/\sqrt{1-x} \approx 1 + \frac{1}{2}x + \frac{3}{8}x^2$. Check this approximation by calculating both sides of the expression on your calculator. For what value of x is the approximation correct to within 10 percent? To within 1 percent?

General Problems

78. (I) A grocery store pays a monthly power bill of \$475. Electricity costs \$0.09/kWh. How many joules of energy were used in the month?

79. (I) A ball with mass 100 g is set in motion inside a bowl, with initial speed $v = 2$ m/s. The ball ultimately comes to rest due to friction. How much net work was done by the external forces acting on the ball?

80. (I) A pile driver works by lifting a large mass and dropping it to the ground. A driver used to put pilings in the ground for a tall building has a mass of 6400 kg and is raised to a height of 2.5 m in 0.50 s. (a) How much work is done by the engine each time the weight is lifted? (b) What horsepower engine must be used to run the pile driver?

81. (I) The so-called Domesday Book recorded a general census carried out in England in the year 1086. It catalogued all 6000 waterwheels in the country, each of which had a power output of roughly 2 hp. Waterwheels were the main source of nonanimal energy at the time. By contrast, the power of a Boeing 747 jet airplane at maximum thrust is approximately 1 MW, and at cruise level the power output is roughly 0.3 MW. What fraction of the total nonanimal power of eleventh-century England does a cruising 747 represent?

82. (II) A mass $M = 3$ kg moving without friction in the xy-plane starts at the point labeled by the position vector $\vec{r}_i = 0\hat{i} + 0\hat{i}$ with velocity $\vec{v}_i = (2\hat{i} + \hat{j})$ meters per second. Two forces, $\vec{F}_1 = (2\hat{i} + 7\hat{j})$ newtons and $\vec{F}_2 = (2\hat{i} - 5\hat{j})$ newtons, act on the mass as it moves in a straight line to the point labeled by the position vector $\vec{r}_f = 10\hat{i} + 5\hat{j}$ meters (Fig. 6–34). (a) How much work is done by $\vec{F}_1$ as the mass moves from $\vec{r}_i$ to $\vec{r}_f$? (b) What power is provided by $\vec{F}_1$ at the instant the mass is at $\vec{r}_i$? (c) What is the kinetic energy of the mass when it reaches $\vec{r}_f$?

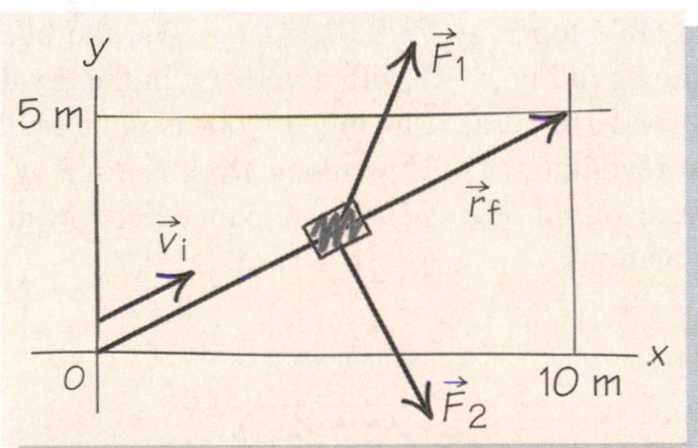

▲ **FIGURE 6–34** Problem 82.

83. (II) A rocket in outer space initially at rest is accelerated uniformly at the rate of 2.0 m/s^2 for 33 s. (Neglect gravity throughout.) (a) If the mass of the rocket, assumed here to be constant, is 1.0×10^4 kg, how much work is done by the rocket engine? (b) Suppose that the rocket decelerates at such a rate that it comes to rest in 55 s. How much work is done on the rocket during the deceleration?

84. (II) A worker pushes a box of mass $m = 25$ kg in a straight line along a rough floor. The applied force $\vec{F}$ has magnitude 85 N and acts downward at an angle $\theta = 10°$ with respect to the horizontal. The box is initially at rest at the position $x_1 = 0$ m, and it has speed $v_2 = 0.55$ m/s at position $x_2 = 3.50$ m. (a) Use the data to calculate the coefficient of friction. (b) What is the net work done? (c) How much work is done to overcome friction? (d) What is the instantaneous power generated by the worker at $x = x_2$?

85. (II) A block of mass $m = 2.6$ kg is placed on an inclined plane that makes an angle $\theta = 32°$ with the horizontal. It is given an initial speed v_0 up the ramp and slides a distance 1.3 m up the ramp before it comes to a stop. The coefficient of kinetic friction between the block and the ramp is $\mu_k = 0.25$. (a) What are the three forces acting on the block? Give the magnitude and direction of each. (b) What is the work done by each of the three forces during the motion of the block? (c) What was the initial speed v_0?

86. (II) The engine of an automobile requires 45 hp to maintain a constant speed of 80 km/h. (a) What is the resistive force against the automobile? (b) If the resistive force is proportional to the velocity, what must the engine power be to drive at a constant speed of 60 km/h? (c) At 140 km/h?

87. (II) A mass of 4.0 kg is attached to a string tied to a hook in the ceiling. The length of the string is 1.0 m, and the mass is released from rest in an initial position in which the string makes an angle of 30° with the vertical. Calculate the work done by gravity by the time the string is in a vertical position for the first time. [*Hint:* Recall that you can use any path you like to get from the initial to the final point because the work done by gravity is path independent.]

88. (II) A mass of 5 kg is accelerated by applying a force. In each case, calculate the work, in joules, that this force must do on the mass. (a) The mass is brought from rest to a speed of 0.5 m/s. (b) The mass is accelerated from 10 to 10.1 m/s. (c) The speed of the mass is increased from v to $v + 1$ meters per second. (d) If $v \gg 1$ m/s, find an approximation that simplifies the calculation in part (c).

89. (II) A puck of mass $m = 0.2$ kg moves in a circle of radius 0.8 m on a table top and is tied with a massless rope to a tether at the origin. The coefficient of kinetic friction between the puck and the table top is $\mu_k = 0.02$. At $t = 0$ s, the puck is at the point shown in Fig. 6–35 with a velocity in the $+y$-direction of magnitude 10 m/s. (a) How much work is done by the rope on the first revolution? (b) How much work is done by friction on the first revolution? (c) What is the kinetic energy at the end of one revolution?

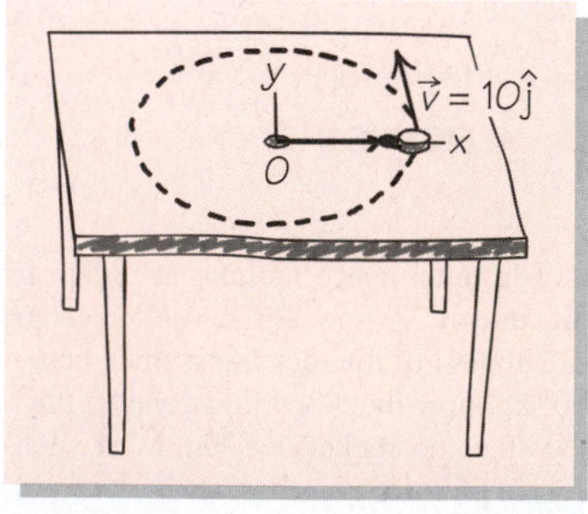

▲ **FIGURE 6–35** Problem 89.

90. (II) An automobile of mass 1100 kg is brought to a halt by applying the brakes, which lock the wheels. The coefficient of kinetic friction between the wheels and the road is 0.55. The car leaves skid marks 48 m long. (a) What is the force of friction between the car and the road? (b) What is the work done on the car by friction in bringing the car to a halt? Include the sign of the work. (c) What was the speed of the automobile when the brakes were first applied?

91. (II) A mass m is hauled from ground level up an inclined plane that makes an angle θ with the horizontal by means of a rope passing over a frictionless pulley. The mass is pulled along until it reaches a height H. The building of the Egyptian pyramids and many other ancient construction jobs used such ramps. Logs were sometimes used as rollers to reduce friction. (a) Show that if the contact between the mass and the ramp is frictionless, the work done by the tension in the rope (or, equivalently, by the person hauling the rope) is independent of the angle θ. (b) Calculate the work done by the tension in the rope as a function of the ramp angle θ if the coefficient of kinetic friction between the mass and the surface is μ_k.

92. (II) A large laser designed for nuclear fusion can produce 6×10^4 J of energy over a time period of 0.3×10^{-9} s. (a) How much power can such a laser produce during its discharge? (b) How much power is required to reenergize it over a period of 20 min?

93. (II) A pendulum of length L and mass m starts from an initial position in which it makes an angle θ_i with the vertical. Calculate the work done by the force of gravity as the mass moves from θ_i to θ_f and use your result to calculate the speed of the pendulum at the bottom of the swing. Why is it possible to ignore the tension in the string of the pendulum? [*Hint*: The work done by gravity is path independent.]

94. (II) A small dam produces electrical power. The water falls a distance of 18 m to turn a turbine. If the efficiency to produce electrical energy is only 68 percent, at what rate must water flow over the dam to produce 850 kW of electrical energy? (See Problem 70 for the meaning of efficiency.) The power system costs $3.5 million. How many years will it take for the power plant to pay for itself if electricity can be sold for $0.10/kWh? Ignore effects such as inflation and the cost of borrowing money.

95. (II) The force the Sun exerts on a planet in a circular orbit of radius x around the Sun is given by an expression of the form $F(x) = -mK/x^2$, where m is the mass of the planet and K is a constant. How much work must be done by a passing celestial body to move the planet to a radius that is 1 percent larger?

96. (II) Early twentieth-century investigations of the structure of atoms carried out by Ernest Rutherford involved the collisions of alpha particles (α particles) and gold atoms. The most interesting results concern the cases in which an α particle is scattered at a large angle from a gold nucleus. Assume that an α particle approaches a gold nucleus (at rest) head on and that the force between these two objects is repulsive, with magnitude $F = k/r^2$, where $k = 3.65 \times 10^{-26}\ \text{N}\cdot\text{m}^2$. Suppose that the α particle moves in toward the nucleus from far away under the sole influence of the repulsive force of the nucleus. What is the minimum kinetic energy the α particle must have initially so that it can get within 1.00×10^{-14} m?

97. (II) Consider the motion of an object on the xy-plane. Show that the force whose x- and y-components are respectively $F_x = (3\ \text{N/m})y$ and $F_y = 0$ is not conservative.

Chapter 7

The motion of the balls illustrates the conversion of potential energy into kinetic energy. As a ball rises, it slows, losing kinetic energy but gaining potential energy; as it drops, it speeds up, gaining kinetic energy and losing potential energy. If there is no air resistance, once it has been tossed the total energy of a ball—the sum of the kinetic and potential energies—is conserved.

Potential Energy and Conservation of Energy

In Chapter 6 we looked at the energy an object can possess by virtue of its motion—its kinetic energy. The kinetic energy changes when work is done on the object, as described by the work–energy theorem. Work can be done by both conservative and nonconservative forces, but if the work is done by *conservative forces*, the expression for work takes a form that we can relate to another kind of energy: the *potential energy*. For conservative forces, the work–energy theorem is a simple and powerful principle, the principle of the conservation of energy. We can say this another way: Consider an object in motion on a trajectory under the influence of a conservative force. At any given point the object has a total mechanical energy that consists of the sum of two terms, a potential energy and a kinetic energy. This total energy remains unchanged throughout the motion. The kinetic energy may change, the potential energy may change, but their sum remains constant. The potential energy is a function of position, and each type of conservative force is described by a different potential energy function. Since the kinetic energy depends on speed, the conservation of energy gives us a powerful and simple relation between position and speed.

The balls tossed by a juggler move under the influence of gravity, a conservative force, and clearly illustrate the principle. The initial speed with which they leave the juggler's hand sets the scene. After that, only the force of gravity acts and the conservation of energy provides us with a direct link between height and speed. As the ball rises, it loses speed and its initial kinetic energy decreases, but at the same time it gains in potential energy. At the top of the trajectory the ball has no velocity and all of its initial kinetic energy has been converted to potential energy, which is *given back* as the ball falls. As the ball falls, the potential energy decreases and the kinetic energy increases once more. The potential energy associated with gravity depends on position, in this case height.

A nonconservative force also acts on the tossed balls, air resistance. The work–energy theorem continues to hold, but the work done by the nonconservative force cannot be related to a potential energy. Does that mean that the conservation of energy is only an approximation? Superficially, that is the case. But when we look more closely, as we will in chapters to follow, we will see that it is possible to define other forms of energy associated with nonconservative forces and to extend the principle of the conservation of energy; for example, the tossed ball causes turbulence and heating of the air and ball (small, to be sure) in its passage, and friction causes surfaces to get hot, so we can often associate thermal energy with nonconservative forces. When one looks at a more fundamental level, there is an explanation of nonconservative forces that is both elegant and interesting. On a microscopic scale *there are no nonconservative forces.* When a brake pad gets hot due to the nonconservative force of friction, molecules in the material of the pad and in the surrounding air speed up, so there is microscopic energy of motion. In fact, *all* of the microscopic forces—atomic, molecular, nuclear, and so on—*are conservative.* So although on a macroscopic scale some forces are nonconservative, if we look at nonconservative forces on the microscopic scale we will see that they result from conservative forces. The principle of energy conservation, in which the sum of kinetic and potential energy is unchanged, is exact. Energy conservation is a fundamental law of physics and is also an enormously useful principle for the solution of problems.

7–1 Potential Energy and Conservative Forces

Our starting point is the work–energy theorem, which relates the work done on an object to a change in its kinetic energy, and we will concentrate on the case where the net work done comes from conservative forces. We defined conservative forces (Section 6–4) as those for which the work done in moving an object from one location to another is *independent of the path along which the object is moved.* This simple statement has enormous consequences.

The work done on a given object by a conservative force when the object moves from point A to point B depends only on those points; it does not depend on any points in between: $W = W(A, B)$. We can actually say more than this. Suppose we go from A to B, then from B to C (Fig. 7–1). In this case, the work done by our conservative force does not depend on the intermediate point B, and thus

$$W(A, C) = W(A, B) + W(B, C).$$

But this, in turn, can be true only if W has the form

$$W(P_1, P_2) = U(P_1) - U(P_2), \tag{7–1}$$

where $U(P)$ is some scalar function that depends only on position P. [You can easily check that when W has this form, then $W(A, C) = W(A, B) + W(B, C)$.] We can say that W depends on the end points alone and that it does so as a difference of two values of a function that depends only on position. If we know this function for a given conservative force, then we know how much work that force does as an object moves from one position to another. We have already worked out some cases in Chapter 6. For example, when an object moves from height y_1 to height y_2, the work done by gravity on the object is

$$W(y_1, y_2) = mg(y_1 - y_2).$$

Comparing the work done by gravity with Eq. (7–1), we find

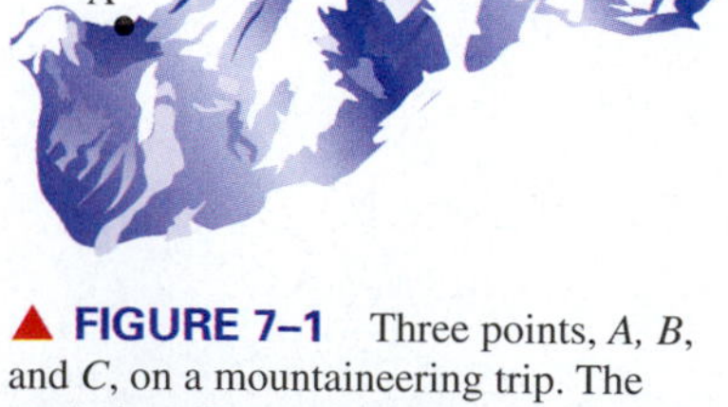

▲ **FIGURE 7–1** Three points, *A, B,* and *C*, on a mountaineering trip. The work done by gravity depends only on the end point of any segment of the trip, not on the path taken between those points.

$$U(y) = mgy + U_0, \tag{7–2}$$

POTENTIAL ENERGY OF GRAVITY

where U_0 is a constant. The presence of the constant does not affect the basic relationship shown in Eq. (7–1) because it is the same in both terms and therefore cancels in the subtraction. We call the particular function U for any given conservative force the **potential energy** for that force. Equation (7–2) gives U for gravity, and forces different from gravity (e.g., the spring force) will have other potential energy functions. Although for the moment U is just a way to express the work, we'll show below that U appears on a par with kinetic energy as a piece of a total energy.

The case of gravity is the case of a constant force. Let us now follow the idea for a general one-dimensional conservative force $F(x)$. From Chapter 6, the work done by such a force on an object that moves from point x_0 to point x is

$$W(x_0, x) = \int_{x_0}^{x} F(z)\, dz.$$

Using Eq. (7–1), we now write the work done by our force on an object moving from x_0 to x as $U(x_0) - U(x)$, or in other words

$$U(x) - U(x_0) = -W(x_0, x). \tag{7–3}$$

Note the sign: The potential energy at point x minus its value at some initial point is the *negative* of the work done by the associated force as the object moves to point x from the initial point. We can rewrite this result as

$$U(x) - U(x_0) = -\int_{x_0}^{x} F(z)\, dz. \tag{7–4}$$

This equation defines the potential energy $U(x)$ for us. Once we know the function $F(z)$, we know $U(x)$ aside from an additive number $U(x_0)$ or just U_0, independent of x. Equation (7–4) gives us U as an integral of F. We can also go backward and find F as a derivative of U. Recall that the derivative of an integral with respect to its upper limit is just the integrand evaluated at the upper limit. Applied to Eq. (7–4), we find

$$F(x) = -\frac{dU(x)}{dx}. \tag{7–5}$$

[This equation confirms that the conservative force associated with the potential energy of gravity, Eq. (7–2), is just $-mg$.] It is important to remember that $U(x_0)$ is a constant, no different from the U_0 that appeared in Eq. (7–2). The work done as an object moves between two points is a difference between U evaluated at those points, so U_0 always cancels out. We can choose U_0 to be whatever we like, and we'll see below that this is often a help in solving problems.

CONCEPTUAL EXAMPLE 7–1 An unknown conservative force does negative work on a mass. Does the potential energy of the mass increase, decrease, or remain unchanged?

Answer We identified [Eqs. (7–3) and (7–4)] the work done by our force as $U(x_0) - U(x)$, that is, as the *negative* of the change in the potential energy. That means that when positive work is done, the potential energy decreases, and when negative work is done, the potential energy increases, which gives the answer to the question. There is another way to think about this, namely that the sum of the kinetic and potential energies is conserved, or remains fixed. The conservation law states that if the kinetic energy increases, then the potential energy must decrease, and if the kinetic energy decreases, then the potential energy must increase to compensate. It is as if the potential energy were a kind of energy reservoir that you can draw on to increase the kinetic energy or refill by decreasing the kinetic energy. We already know from the work–energy theorem that when negative work is done on a mass, its kinetic energy decreases. In the language of the conservation law, this means that potential energy must increase, in accordance with what we already concluded above.

The Conservation of Energy

We are now in a position to establish that energy, consisting of potential plus kinetic energy, is conserved. We suppose that an object moves from point x_0 to point x_1 under the sole influence of a conservative force. From the work–energy theorem [Eq. (6–7)] and Eqs. (7–3) and (7–4), we have

$$K_1 - K_0 = W(x_0, x_1) = U(x_0) - U(x_1),$$

where $K_1 = \frac{1}{2}mv_1^2$ and $K_0 = \frac{1}{2}mv_0^2$. Simple rearrangement gives

$$K_1 + U(x_1) = K_0 + U(x_0). \tag{7–6}$$

This remarkable relation *expresses explicitly the principle of* ***conservation of energy***. Because our two points x_0 and x_1 can be any two points from the trajectory, Eq. (7–6) shows that $\frac{1}{2}mv^2 + U(x)$ *has the same value at every point in the trajectory of the object*. We call this quantity E, the **total mechanical energy** (or more simply, the **energy**) of the object:

$$E \equiv K + U(x) = \tfrac{1}{2}mv^2 + U(x). \tag{7–7}$$

TOTAL MECHANICAL ENERGY

We have shown that when only conservative forces act, E is *conserved* during the motion, no matter how it is transformed from potential energy to kinetic energy and back again—$\Delta E = 0$. A football, which moves to a good approximation under the sole influence of gravity, starts with a certain kinetic energy (Fig. 7–2). This kinetic energy is converted to potential energy as the ball rises to its maximum height; the potential energy is converted back to kinetic energy as the ball falls. Throughout the ball's flight, the total energy is always the same. The idea that the total energy is conserved is one of the most powerful ideas in physics as well as one of the most useful.

Applications of the Conservation of Energy

The conservation of E means that any change ΔK in the kinetic energy is compensated by an equal and opposite change ΔU in the potential energy:

$$\Delta K = -\Delta U. \tag{7–8}$$

This equation explains the origin of the term potential energy. Potential energy is energy that *has the potential* to be converted into kinetic energy (and, as we shall see later in the book, into other forms of energy).

Energy conservation is a useful tool for determining the speed of an object when only conservative forces act. If the function $U(x)$ is known and the initial conditions are given, we can determine the speed at any point x on the trajectory of an object. We simply use the fact that the total energy is constant.

We have already remarked that we can add an arbitrary constant U_0 to the potential energy $U(x)$ because this constant will cancel on both sides of Eq. (7–6). In other words, the conservation of energy involves only *changes* in the potential energy, so any

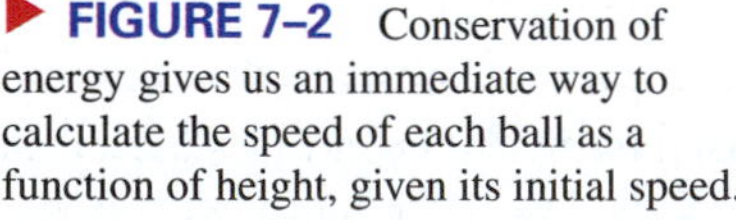

▶ **FIGURE 7–2** Conservation of energy gives us an immediate way to calculate the speed of each ball as a function of height, given its initial speed.

additive constant in the potential energy does not enter into the change. This is explicit in Eq. (7–8), which is the conservation-of-energy equation rewritten to make apparent the fact that only changes in potential energy are important, not the absolute values. The freedom to choose U_0 is equivalent to choosing a particular point where the potential is zero. To see this, consider the case of gravity and suppose that in your coordinate system you have measured height y from sea level. Then if in Eq. (7–2) you choose $U_0 = 0$, the potential energy will be $U(y) = mgy$. For this potential energy, $U(y) = 0$ at $y = 0$—we say the zero of potential energy is at $y = 0$. If instead we wanted the zero of potential energy to be at $y = 3$ m, we could choose $U_0 = -mg \times (3\text{ m})$. With this choice $U(y) = mgy - mg \times (3\text{ m})$, and $U(y) = 0$ when $y = 3$ m. The freedom to choose U_0, or equivalently to choose the point where $U = 0$, is a very useful practical tool; often the physical situation suggests a natural choice for the zero of U. This will be seen in a series of examples that follow.

CONCEPTUAL EXAMPLE 7–2 Consider the motion on a flat frictionless table of two masses connected by a spring, as in Fig. 7–3. You observe the motion at sea level, then you take the entire apparatus to the top of a tall building, $y_0 = 100$ m higher, and observe the motion again. A colleague claims that the motion must be different because you have added mgy_0 to the potential energy, so there will be less kinetic energy. Is your colleague right?

Answer The system has a potential energy that consists of two terms: One involves the interaction between the masses on the table, governed by the spring between them and associated with horizontal motion of the masses; the second is the potential energy of gravity, which we can write as Mgy, where M is the total mass of the system and y is the height above sea level of the table top in the two situations. Changing y to $y + y_0$ just corresponds to adding a constant mgy_0 to the potential energy due to gravity, which, as we have seen, changes nothing. Your colleague is wrong; it does not matter at what altitude the motion takes place. Note that this is consistent with what you would conclude from Newton's second law. Gravity is canceled by the normal force from the table top, and as long as the table top is frictionless, it will play no role whatsoever in the motion of the masses.

What Do You Think? By changing y to $y + y_0$, from what point are you measuring height? Where is the zero of the potential energy now? *Answers to **What Do You Think?** questions are given in the back of the book.*

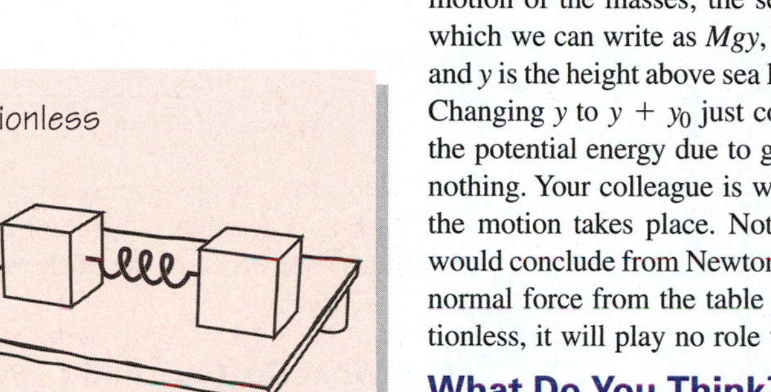

FIGURE 7–3 Two masses on frictionless table connected by a spring.

Gravity: Equation (7–2) gives the potential energy of a mass m subject to the force of gravity. When only gravity acts, the conserved total energy is

$$E = \tfrac{1}{2}mv^2 + mgy + U_0, \tag{7–9}$$

where v is the speed and y is measured upward from some origin.

Suppose we drop a ball from rest from the roof of a building of height H with zero initial velocity and ask for its speed at height y. The value of E at $y = H$, where we know $v^2 = 0$, is $E = mgH + U_0$. A convenient choice for U_0 would be $U_0 = 0$, which corresponds to a potential energy that is zero at $y = 0$ (ground level). With an initial velocity of zero, the initial value of energy is mgH whereas the value at height y is $\tfrac{1}{2}mv^2 + mgy$, where v is the speed at height y. By energy conservation,

$$mgH = \tfrac{1}{2}mv^2 + mgy.$$

We easily solve this for v^2:

$$v^2 = 2g(H - y). \tag{7–10}$$

An alternative choice for U_0 might be simpler if instead we wished to calculate how high an object will go when projected upward from the roof with an initial speed v_0. The initial value of E is $mv_0^2/2 + mgH + U_0$. Here a convenient choice might be $U_0 = -mgH$ so that the initial value of energy is $mv_0^2/2$. We have chosen the zero of potential energy at $y = H$ because then $U(y) = mgy + U_0 = mg(y - H)$, and this is zero at $y = H$. Due to energy conservation E has exactly the same value at its maximum height y_{max}, where $v = 0$, and this choice of U_0 gives

$$\frac{1}{2}mv_0^2 = \frac{1}{2}m(0)^2 + mg(y_{max} - H). \tag{7–11}$$

This is easily solved for y_{max}.

EXAMPLE 7–3 A brick is thrown straight up from 18 m above the ground by a person leaning over the edge of a roof so that the brick will reach the ground on a vertical path. Use the fact that energy is conserved to determine the speed with which the brick is thrown, given that it reaches the ground with a speed of 24 m/s.

Setting It Up We draw the sketch in Fig. 7–4. Included is a y-axis with $y = 0$ at ground level, and the height from which the ball is thrown is at $y = H$. Assuming that the clock starts when the brick is thrown, we know the height H at $t = 0$. We also know the final speed when the brick hits the ground ($y = 0$) is v_f. We want to determine the initial speed v_0.

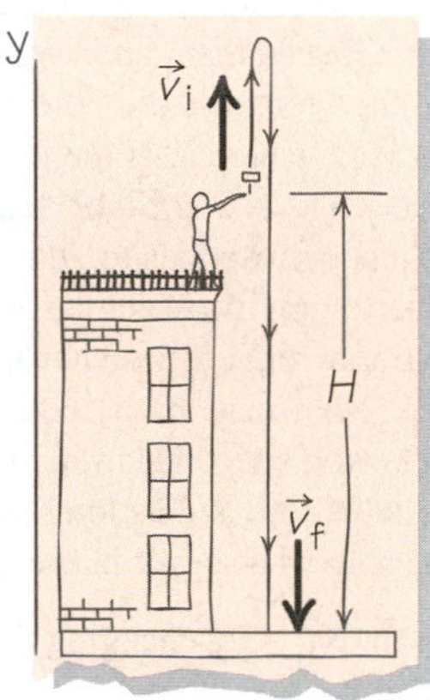

▲ **FIGURE 7–4** Ball thrown initially up falls back down to ground.

Strategy The total energy E at $t = 0$ will be the same as the energy when the brick hits the ground. Equation (7–9) provides us with an expression for the energy. We write out the energy for the two different situations and equate them—this is an equation whose only unknown is v_0, and we can solve for it.

Working It Out The initial energy is $\frac{1}{2}mv_0^2 + mgH + U_0$ and the final energy is $\frac{1}{2}mv_f^2 + U_0$. Equating these, we have

$$\frac{1}{2}mv_0^2 + mgH + U_0 = \frac{1}{2}mv_f^2 + U_0.$$

As before, U_0 cancels, and we have an expression that we can solve for the unknown v_0:

$$v_0^2 = v^2 - 2gH$$

or

$$v_0 = \pm\sqrt{(24\text{ m/s})^2 - 2(9.8\text{ m/s}^2)(18\text{ m})} \cong \pm 15\text{ m/s}.$$

Note that energy conservation determines only speed, so either sign for v_0 is possible. Other information is necessary to find the sign; in this case we know the brick was thrown upward, so the plus sign is the right one.

What Do You Think? Suppose the process were reversed and somebody at ground level were to throw the brick upward with speed 24 m/s. Can you say, without doing the calculation, what the speed of the brick is at the initial launching position?

Spring Force: An ideal spring (one with negligible mass and that exerts a force given by Hooke's law) provides another example of a conservative force for which we can calculate a potential energy. You can think of any wind-up toy, in which a turned key effectively compresses a spring, to realize that you can store energy in a stretched or compressed spring. The displacement from the equilibrium position of a mass m attached to the spring is labeled as x. The force on the mass due to the spring is a restoring force, pulling the mass back to equilibrium:

$$F(x) = -kx, \tag{7–12}$$

where k is the spring constant. The "equilibrium point" of the spring is $x = 0$, since there is no force acting on the mass when it is at that point. Equation (7–4) takes the form

$$U(x) = U(x_0) - \int_{x_0}^{x}(-kx)\,dx = U(x_0) + \frac{1}{2}kx^2 - \frac{1}{2}kx_0^2.$$

It is convenient to choose the potential energy to be zero at the equilibrium point. With this choice, we set $U(0) = 0$, and applying this to the equation above, we see that this means $U(x_0) = \frac{1}{2}kx_0^2$, leaving

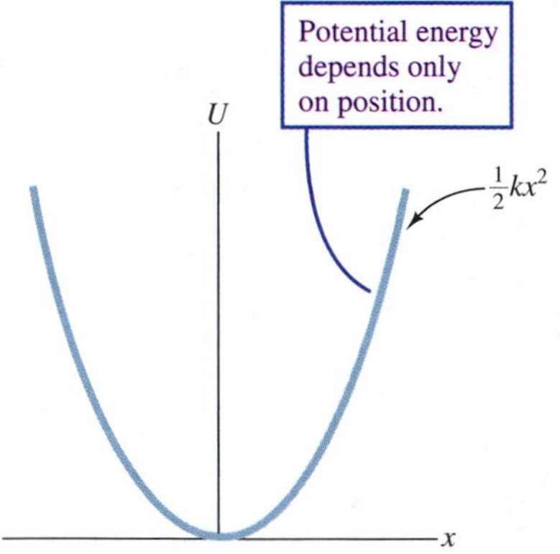

▲ **FIGURE 7–5** The potential energy function for a mass attached to a spring. The mass position $x = 0$ is the position where the spring is relaxed, and we have taken the potential energy to be zero there.

$$U(x) = \frac{1}{2}kx^2. \tag{7–13}$$

POTENTIAL ENERGY OF A SPRING

This function is illustrated in Fig. 7–5. The total conserved energy is given by the sum of the kinetic and potential energies: $E = (mv^2/2) + (kx^2/2)$.

EXAMPLE 7–4 One end of a massless spring is welded to a flat surface; the other end points upward (Fig. 7–6a). A mass of 1.0 kg is gently set down on top of the spring until the spring is compressed by 17 cm to a new equilibrium position (Fig. 7–6b). What is the spring constant?

Now, the 1.0-kg mass is removed and a 2.0-kg mass is set on top of the spring. The spring is then compressed by hand so that the end of the spring is 42 cm lower than the position of the spring with no mass on top (Fig. 7–6c). The spring is then suddenly released. What is the maximum kinetic energy of the 2.0-kg mass?

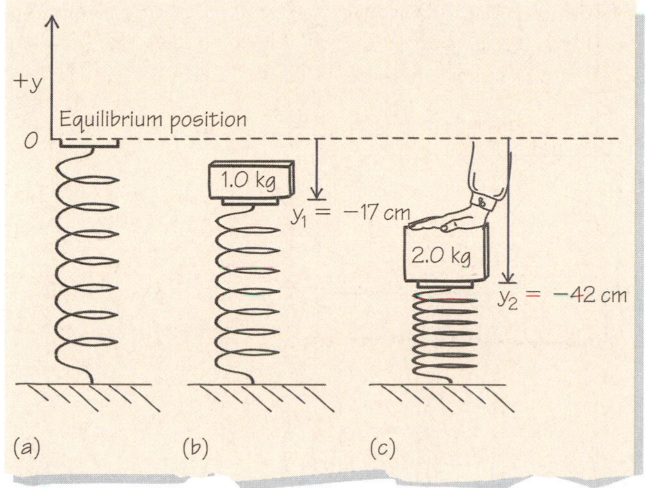

▲ **FIGURE 7–6** (a) In our coordinate system, the top of the relaxed spring is at $y = 0$. (b) When a 1.0-kg mass is placed gently on top of the spring, the new equilibrium position is $y = -17$ cm. (c) A 2.0-kg mass is attached to the end of the spring, and the mass is then pushed down to $y = -42$ cm. The mass is then released.

Setting It Up We set $y = 0$ at the equilibrium position in Fig. 7–6a, the top of the unloaded spring. The positive y-direction is up. The motion is all one dimensional, which is why we have not used vector notation. We know the spring position y_1 when we set the first mass m_1 on the spring. We want to find the spring constant k. In the second part, we start with a second mass m_2 at rest at a given height y_2, and we want to find its maximum kinetic energy in the subsequent motion.

Strategy At equilibrium for m_1, the net force is zero. With the knowledge that the only forces acting are gravity (down) and the spring force (up), setting the net force to zero is an equation that can be solved for the spring constant. For the second part, we have a known spring force, hence a known potential energy, containing terms associated with the spring and with gravity. Only these conservative forces act, so we can use the conservation of energy, which relates the values of potential and kinetic energies. The maximum kinetic energy will occur when the maximum amount of potential energy has been converted to kinetic energy, and this occurs when the overall potential energy has its minimum value. If you look at Eq. (7–5), then you see that the minimum of the potential, which occurs at $dU(x)/dx = 0$ is the place where the force $F(x) = 0$. We'll use this condition to find the minimum point.

Working It Out For the first part, we set the net force to zero. The force due to the spring acts upward, $F_{spring} = -ky_1$ (note y_1 is negative). We add the force of gravity, $-m_1g$, to obtain the net force $F_{net} = -ky_1 - m_1g$, and this must be zero. We then solve for the spring constant k,

$$k = -\frac{m_1 g}{y_1} = -\frac{(1.0\text{ kg})(9.8\text{ m/s}^2)}{-0.17\text{ m}} = 58\text{ kg/s}^2 = 58\text{ N/m}.$$

For the second question, we want the potential energies associated with each of the forces. We choose both the potential energy of gravity and the potential energy of the spring to be zero at $y = 0$. Thus

$$U_g = mgy, \qquad U_{spring} = \tfrac{1}{2}ky^2,$$

where k is the spring constant. The total energy E of the system in this situation is the sum of the kinetic energy, the spring potential energy, and the potential energy of gravity:

$$E = \tfrac{1}{2}m_2v^2 + \tfrac{1}{2}ky^2 + m_2gy.$$

We evaluate E by noting that $v = 0$ m/s at $y = y_2$. Thus

$$\begin{aligned} E &= \tfrac{1}{2}(2.0\text{ kg})(0\text{ m/s})^2 + \tfrac{1}{2}(58\text{ N/m})y_2^2 + (2.0\text{ kg})(9.8\text{ m/s}^2)y_2 \\ &= (29\text{ N/m})(-0.42\text{ m})^2 + (19.6\text{ kg}\cdot\text{m/s}^2)(-0.42\text{ m}) \\ &= -3.1\text{ J}. \end{aligned}$$

Now we look for the point y_3 where the kinetic energy is a maximum, which as described above is the point where the net force is zero, that is, where the spring force and the force of gravity cancel. We have $-ky_3 - m_2g = 0$, or

$$y_3 = -\frac{m_2 g}{k} = -\frac{(2.0\text{ kg})(9.8\text{ m/s}^2)}{58\text{ N/m}} = -0.34\text{ m}.$$

At this point, $E = K_{max} + \tfrac{1}{2}ky_3^2 + m_2gy_3$. By energy conservation, this must equal the initial value of E, -3.1 J, or

$$\begin{aligned} K_{max} &= -\tfrac{1}{2}ky_3^2 - m_2gy_3 + E = -\tfrac{1}{2}(58\text{ N/m})(-0.34\text{ m})^2 \\ &\quad - (2.0\text{ kg})(9.8\text{ m/s}^2)(-0.34\text{ m}) - 3.1\text{ J} \\ &= 0.21\text{ J}. \end{aligned}$$

Note that there is conversion not only between kinetic energy and potential energy but also between the two different kinds of potential energy. These conversions all occur within the constraint of a constant total energy.

What Do You Think? Before the mass m_1 was placed on the spring, and with our choice of the zero of potential energy, the total energy was zero. With m_1 added, the spring is compressed. Is the total energy still zero?

The Energy of Systems

The additive nature of energy makes it easy to apply it to entire systems—as long as those systems involve conservative forces. Let's look at an example involving a system composed of two masses connected by a rope.

EXAMPLE 7–5 Consider the Atwood machine introduced in Example 5–5 (Fig. 5–9). The two masses have the values $m_1 = 1.37$ kg and $m_2 = 1.51$ kg. The system is released from rest with m_2 at height $h_2 = 84$ cm from the floor. Use energy conservation to find the speed of m_2 just before it hits the floor.

Setting It Up Figures 7–7a and 7–7b show the initial and final states, respectively. We have included a (vertical) y-axis, with $y = 0$ at the floor. We have also indicated the initial positions h_1 and h_2 of the two masses.

Strategy Energy conservation relates the speed and height of the masses. Assuming no friction at the massless pulley and no air resistance, only gravity acts on this system. Gravity is a conservative force, so we can use the conservation of energy with the potential energy due to gravity. We also assume that the rope is unstretchable, so that the speeds of the two masses will be identical, one moving up as the other moves down.

Working It Out We consider the two masses tied together as a whole, choosing the potential energy of gravity to be zero at $y = 0$; that is, U takes the general form mgy. Before the system is released (Fig. 7–7a), the total energy E_{init} is

$$E_{init} = m_1gh_1 + m_2gh_2.$$

In the final state of the system, just before m_2 hits the floor (Fig. 7–7b), the system contains kinetic as well as potential energy. Each mass has the same speed, v, and therefore the final energy E_{final} is

$$E_{final} = m_1g \times (h_1 + h_2) + \tfrac{1}{2}m_1v_1^2 + m_2g \times (0) + \tfrac{1}{2}m_2v_2^2$$
$$= m_1g \times (h_1 + h_2) + \tfrac{1}{2}(m_1 + m_2)v^2.$$

By conservation of energy, $E_{init} = E_{final}$, and this gives

$$(m_2 - m_1)gh_2 = \tfrac{1}{2}(m_2 + m_1)v^2.$$

Solving for v, we find the speed of m_2 near the floor,

$$v = \sqrt{\frac{2(m_2 - m_1)gh_2}{m_2 + m_1}}$$
$$= \sqrt{\frac{2(1.51\text{ kg} - 1.37\text{ kg})(9.81\text{ m/s}^2)(0.84\text{ m})}{1.51\text{ kg} + 1.37\text{ kg}}}$$
$$= 0.90\text{ m/s}.$$

What Do You Think? Suppose that the pulley in the Atwood machine were attached to the top bar by a spring instead of being fixed to it. Would you expect the motion to be different?

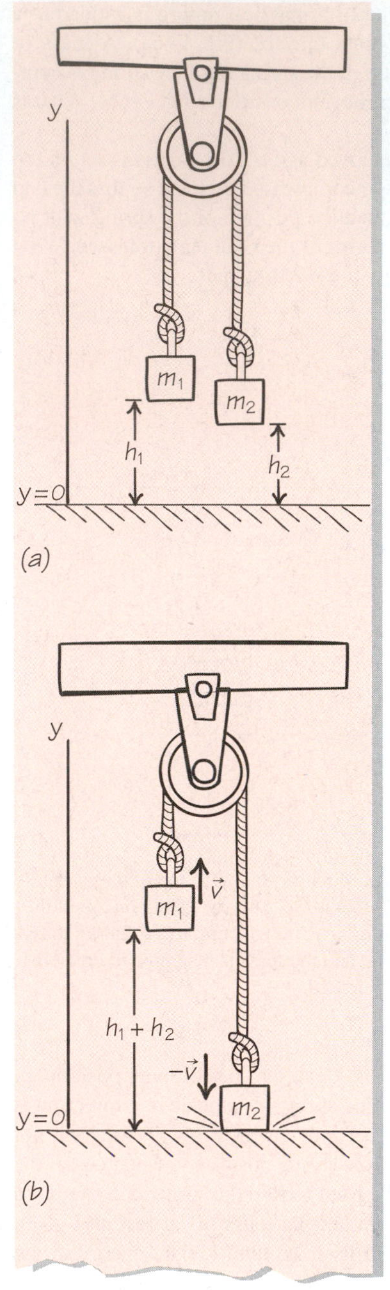

FIGURE 7–7
(a) Initial conditions for Example 7–5.
(b) The situation just before mass m_2 hits the floor.

We have just considered a few examples showing how useful energy considerations can be in analyzing a system. The extension of the energy conservation principle to include many forms of energy provides a powerful tool for the study of complicated systems. All that happens around us can to a very large extent be described in terms of energy flows; the conservation of energy allows us to understand in a very direct way the nature of energy flows, the paths by which the energy moves, and the rate at which it is transferred. Using it, we can account for energy entering or leaving a system by certain routes and at certain rates. For example, if a certain amount of electric lighting is necessary in a new building, how does the heat associated with the light bulbs affect the heating or air conditioning that must be installed? In constructing a refinery, how do the heights at which various pipes must be placed affect the pumping apparatus required? The conservation of energy is a bookkeeping tool that allows engineers to design and analyze the behavior of a system without knowing every last detail of the system's variables. The extension of the energy conservation principle to include many forms of energy is one of the central themes of this book, and we'll come back to it frequently.

7–2 Energy Conservation and Allowed Motion

Let's take a look at the energy conservation equation for the one-dimensional motion of a mass attached to an ideal spring with spring constant k (Fig. 7–8a). To do so, we will use an **energy diagram** of the system (Fig. 7–8b). To create an energy diagram, we mark

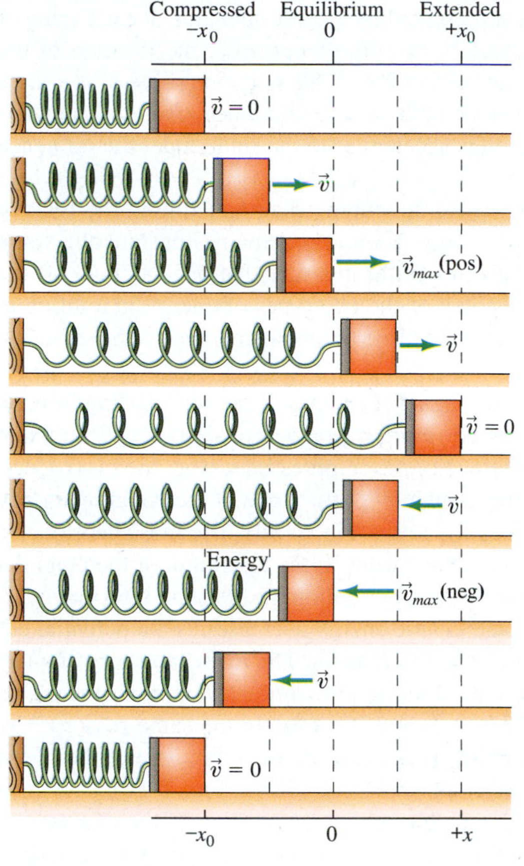

(a)

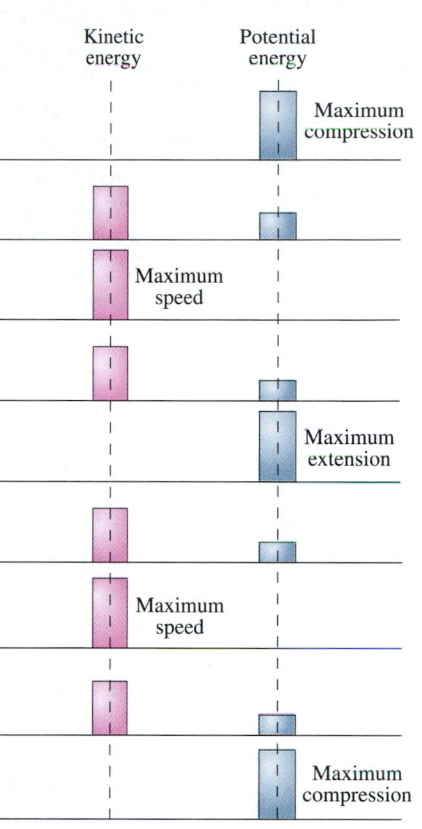

(c)

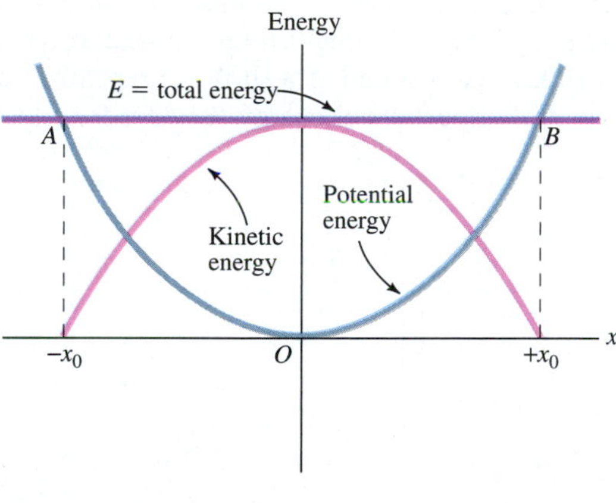

(b)

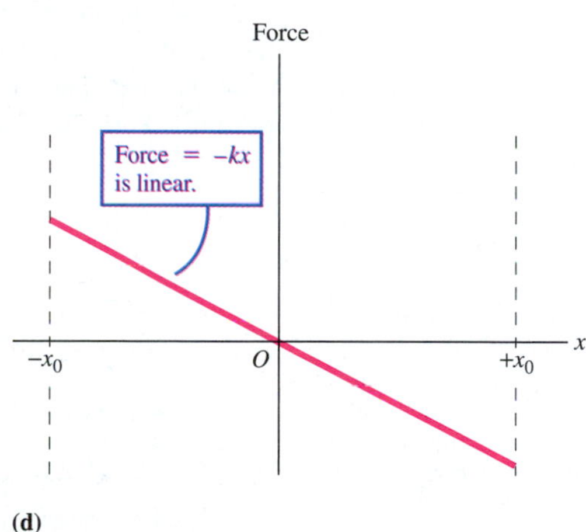

(d)

▲ **FIGURE 7–8** A mass on the end of a spring; equilibrium is $x = 0$. (a) First, the spring is compressed by an amount x_0, then released. During one period (or cycle), the mass moves through $x = 0$ to an extension $x = x_0$ and back to $x = -x_0$. (b) Energy diagram for the mass. The potential energy $U = \frac{1}{2}kx^2$ is drawn. The straight line represents the constant total energy E; the kinetic energy K is the difference between E and U. The curve of K is determined purely by the conservation of energy. Points $+x_0$ and $-x_0$ are turning points for the motion, where the kinetic energy and therefore the speed are zero. The mass cannot go outside the range $-x_0 \leq x \leq +x_0$ as the kinetic energy would be negative. (c) As the mass moves, the kinetic and potential energies associated with the mass–spring system change: One increases as the other decreases. (d) Plot of the force of a spring on a mass attached to its end.

the position of the mass (the object whose motion is of interest to us) on the horizontal axis, and then on the vertical axis we draw any of the energies: potential energy, $U = kx^2/2$; kinetic energy, $K = mv^2/2$; or total energy, $E = U + K$. Suppose that the spring is initially compressed by an amount x_0, then released (Fig. 7–8a). Initially, $v = 0$ and $x = -x_0$; at this point, then, $E = kx_0^2/2$. Because E is unchanging, this is the value of E at any value of x. We represent it in the energy diagram by a horizontal line (Fig. 7–8b). We can plot the potential energy as a function of x on the same figure, and the kinetic energy is then automatically represented by a curve such that at each value of x its value is $E - U(x)$. When the spring is initially compressed, the location of the mass is at $-x_0$. The force on the mass, plotted in Fig. 7–8d, is the negative of the slope of the potential energy, Eq. (7–5). The slope of $U(x)$ at $-x_0$ is negative, which means that the force is positive there; the mass thus accelerates to the right under the influence of the spring force.

In the subsequent motion we can, using the energy diagram, follow the relative sizes of the kinetic energy and potential energy. Their sum remains constant at a value of E, but their respective values at any given x differ in a way that one can read off the energy diagram. We can also follow this play of the two types of energy in a different way: Figure 7–8c shows the magnitude of kinetic and potential energies as the sequence of positions of Fig. 7–8a is attained.

With Fig. 7–8c in mind, let us now look again at the mass's motion, starting from its initial position on the end of the compressed spring (negative x). We already stated that under the influence of the right-directed force of the compressed spring, the mass accelerates to the right. As the mass passes the point $x = 0$, the slope of the potential energy curve changes sign, and so therefore does the force (Fig. 7–8d). The mass decelerates, and the kinetic energy, which reached its highest value at the minimum of the potential energy $(x = 0)$, falls below the horizontal line representing the constant total energy E. As we go further to the right of $x = 0$, the kinetic energy decreases as the potential energy curve rises. Finally, at $+x_0$, $U(x)$ is equal to E again. The mass cannot go any farther to the right; there, $U(x) > E$ and this would require a negative kinetic energy, an impossibility for the quantity $\frac{1}{2}mv^2$. Thus $+x_0$ and (for exactly the same reason) $-x_0$ are *turning points* for the motion. **Turning points** are those points where the speed drops to zero and the mass changes direction (points A and B in Fig. 7–8b). The mass must therefore remain between $-x_0$ and $+x_0$ throughout the motion. [If the total energy were larger, then the horizontal line of E in Fig. 7–8b would have a greater value (Fig. 7–9), with turning points corresponding to points C and D. A turning point x_0 is characterized by the fact that $E = U(x_0)$, or $K = \frac{1}{2}mv^2 = 0$.]

Let's suppose that the mass moves from $-x_0$ to $+x_0$ and then back to $-x_0$. It then restarts its motion from $-x_0$, and there is nothing to distinguish this second traversal from the first. The motion of the mass attached to the end of a spring is repetitive, or **periodic**. Periodic motion actually occurs for any potential energy function for which

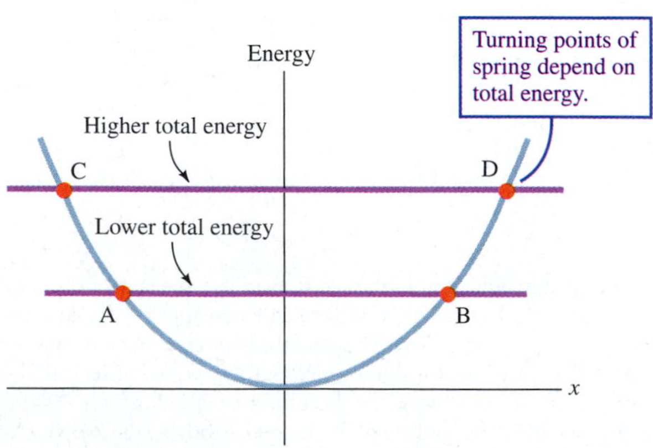

▶ **FIGURE 7–9** When the total energy E of a mass on the end of a spring increases, the turning points move to large values of position. Here the turning points A and B move out to new turning points C and D when the energy increases.

the mass is constrained to move between two boundaries. It is spring-like near *any* minimum of a potential energy function. We will consider this further in Chapter 13, and there we will see many examples.

More about Energy Diagrams

The energy diagram provides a useful visualization. Let us write $U(x) = mgh(x)$. We may think of $h(x)$ as the *height* of the mass at the point x and $U(x)$ as the *gravitational potential energy* there. The motion of a mass in that potential may be visualized as the motion of a cross-country skier (on frictionless snow) on varying terrain whose height is given by $h(x)$. The particular form taken by $h(x)$ depends on the potential in question; for example, for the spring the terrain forms a parabolic bowl with a minimum at $x = 0$ (Fig. 7–5). This form of visualization is useful because we are intuitively familiar with how a skier might move on mountains of various shapes.

Energy diagrams can give us a great deal of insight into qualitative aspects of motion. We'll start with motion in one dimension again, and instead of a skier we treat a small object initially at rest on a frictionless landscape (Fig. 7–10) with a height $h(x)$ that depends on x (a profile). If the object starts on a slope, it will slide downward, converting some of its potential energy into kinetic energy. There are places, however, at which the object starting at rest *will remain* at rest. These points, which must be those for which the force is zero, or equivalently [see Eq. (7–5)] those points for which the potential energy function is flat, are called **equilibrium** points. There are three kinds of equilibrium points (Fig. 7–10). The object starting at rest at the bottom of a valley will not move. If you give this object a tiny bit of kinetic energy, it will move up the side of the valley, but soon it will reach a turning point and slide back, subsequently oscillating about the minimum point. The equilibrium point is said to be **stable** because a displacement from the equilibrium point puts the object in a position where *restoring* forces act to bring the object back to the equilibrium point. Consider next the equilibrium point with the object starting from rest at the top of a hill in our landscape. A little kinetic energy or a starting point a little off the top puts the mass on a slope with forces acting to move the mass further downhill, away from the equilibrium point. In other words, a small departure from the equilibrium point is amplified, and we call such an equilibrium **unstable.** A third possibility is a flat region in $U(x)$, a plateau in our landscape picture. An object at rest at one point on the plateau can be displaced to any other point on the plateau, and there will still be no forces acting on it. We describe this situation as one of **neutral** equilibrium.

Careful examination of the energy diagram can yield a great deal of information about the motion of an object in an arbitrary potential $U(x)$. We illustrate the procedure of extracting this information in the following example.

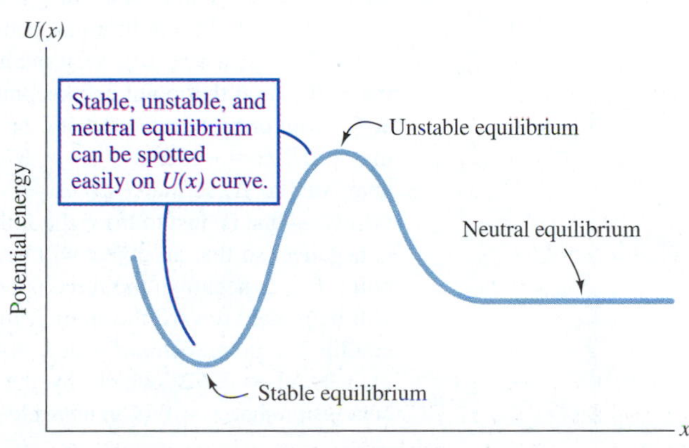

FIGURE 7–10 Energy diagram for one-dimensional motion, with the potential energy plotted as a function of x. We see three kinds of equilibrium for a mass acted on by the associated force: stable, unstable, and neutral.

EXAMPLE 7–6 Consider the potential energy given by

$$U(x) = \tfrac{1}{4}x^4 - \tfrac{1}{3}x^3 - 3x^2,$$

where the coefficients of the various powers of x have units such that U has units of joules.

(a) Sketch $U(x)$ on an energy diagram.

(b) Find the equibrium points and classify them.

(c) Find the turning points for E values of -2 and 3 J, respectively.

Setting It Up We want to find the equilibrium points corresponding to the given potential and classify them, and we also want turning points (x-values) for two values of E.

Strategy After sketching a figure that corresponds to this potential, we find the equilibrium points—defined as those at which the force vanishes—by letting $dU(x)/dx = 0$ or by seeing where the flat spots, or extrema, of $U(x)$ are found on the plot of $U(x)$ versus x. Maxima are unstable equilibrium points; minima are stable equilibrium points. The turning points are points for which the kinetic energy is zero, that is, $U = E$, and these occur at the values of x where $U = E$, which should be visible on our sketch.

Working It Out (a) Figure 7–11 is a sketch of $U(x)$ with units of x understood. (The process can be carried out with a graphing calculator or in any number of other ways.) In drawing figures such as this there are several things to keep in mind. First, if the potential is a polynomial, as here, the large x behavior, positive or negative, is dominated by the highest power of x. In this case, $U(x) \cong \frac{1}{4}x^4$ at large $|x|$, always positive. Second, the degree and form of the polynomial determines the number of zeros. Here $U(x)$ is x^2 times a second-order polynomial. That means there is a double zero at $x = 0$ (which represents an extremum at $x = 0$ but not an axis crossing—the function has no sign change just above and just below $x = 0$) and two more zeros coming from the second-order part. Nevertheless, we see from the graph that the extremum at $x = 0$ is a maximum.

(b) The equilibrium points are defined as those at which the force vanishes, or equivalently the points where $dU(x)/dx = 0$. In this case

$$\frac{dU(x)}{dx} = x^3 - x^2 - 6x = x(x + 2)(x - 3),$$

so that the equilibrium points are $x = -2$, $x = 0$, and $x = 3$. These are the extrema of the graph (the units are understood). The simplest way to classify the equilibrium points is to look at the sketch of the potential. The points $x = -2$ and $x = 3$ lie at the bottom of valleys and are therefore stable equilibrium points. The point $x = 0$ lies on top of a hill and is therefore a point of unstable equilibrium.

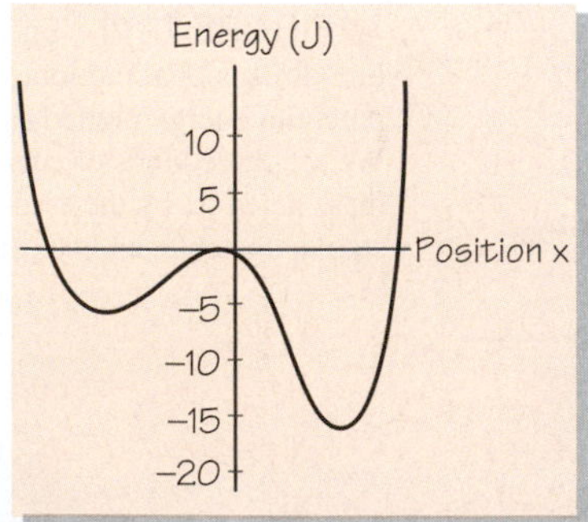

▲ **FIGURE 7–11** Plot of the potential energy function of Example 7–6.

(c) The turning points are those points at which the kinetic energy is zero, that is, $U = E$. Thus we find the turning points by solving the equations $U(x) = -2$ J and $U(x) = 3$ J, respectively, for x. Since U is a fourth-order polynomial, this can only be done numerically.

More insight is obtained by drawing horizontal lines at $E = -2$ J and $E = 3$ J on the energy diagram (Fig. 7–12). We see that for $E = -2$ J there are *four* turning points, corresponding to two different regions in which the kinetic energy can be positive. A mass moving in the left-side valley cannot go over to the right-side valley. It encounters a **potential barrier** and is trapped between the turning points in a **potential well**. The same is true of a mass that starts its motion in the right-side valley. It encounters a potential barrier on the left in the form of the potential's hill around $x = 0$. On the other hand, for $E = 3$ J, there are two turning points on the outer wings of the potential energy curve. The kinetic energy can be positive anywhere between these two turning points, and in particular, it is larger than the potential energy at the intermediate high point at $x = 0$. It can thus cross $x = 0$ with kinetic energy to spare.

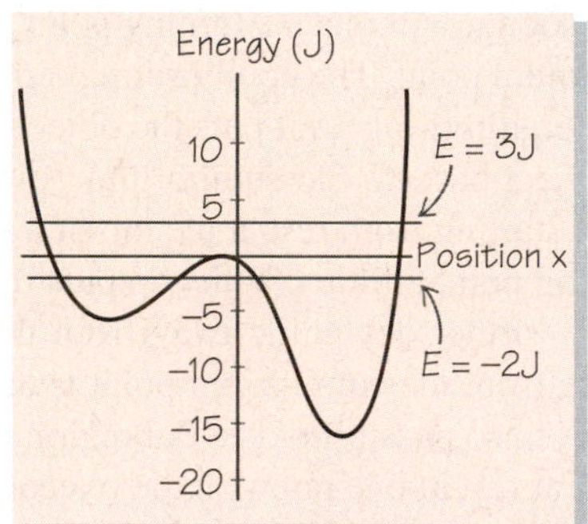

▲ **FIGURE 7–12** If the total energy is negative, an object moving under the influence of the potential will be trapped in one of two potential wells on either side of the origin. If the total energy is positive, the object is still trapped, but this time there is only a single trapping region.

Alternative Approach to part (b) Another, more mathematical way of looking at the task in part (b) is to look at the force *in the vicinity* of the equilibrium points. For example, at the point $x = -2 + \Delta$, where Δ is very small, the force is given by calculating $-dU/dx$ at that point and keeping only the term proportional to Δ (i.e., ignoring terms quadratic or higher in Δ). Here this means that $-dU/dx = -10\Delta + 7\Delta^2 - \Delta^3$ and that we ignore Δ^3 and Δ^2 compared to Δ, so that $-dU/dx \cong -10\Delta$. This tells us that for Δ positive—that is, just to the right of the equilibrium point—the force is negative, so that an object will be pulled back to the equilibrium point. For Δ negative, the force is positive, so that again the object will be pushed toward the equilibrium point. Thus we have a stable equilibrium point. Similarly, at $x = 0 + \Delta$, we get $-(\Delta)(2 + \Delta)(-3 + \Delta) \cong +6\Delta$, which, by the same arguments, leads to the conclusion that $x = 0$ is an unstable equilibrium point.

What Do You Think? Can you describe the motion of an object that has total energy $E = 0$?

THINK ABOUT THIS . . .

ARE POTENTIAL BARRIERS REALLY IMPENETRABLE?

The answer to this question would appear to be a simple yes, such barriers are indeed impenetrable. If E were less than $U(x)$, which would correspond to being "inside" the barrier, the kinetic energy would be negative, and this is impossible given the form $K = \frac{1}{2}mv^2$. It is therefore very surprising that it is a well-established fact that on a microscopic scale (when we deal with nuclei, atoms, and even large molecules) *barrier penetration*, or *tunneling*, occurs. This is a phenomenon in which a particle can appear on the other side of a potential barrier even though it does not have enough energy to do so classically. On the microscopic scale Newtonian mechanics gives an inadequate description of the dynamics of systems, and in the microscopic domain, one must use another set of physical laws, the laws that form the subject of *quantum mechanics* (Chapter 40). Quantum theory allows for the phenomenon of tunneling and also shows that it is a very improbable process on the macroscopic scale. The physics behind this phenomenon is that matter exhibits wave-like characteristics in quantum mechanics and waves behave in many ways quite differently from particles.

Tunneling through potential barriers is an important aspect of many areas of modern engineering. Of particular interest is the application to scanning tunneling microscopy. This technique, which is based on the fact that electrons can tunnel through the potential barrier that normally keeps them confined to the vicinity of nuclei, allows for a detailed study of surfaces of materials. In another very important set of applications, many modern electronic devices—transistors and other control devices—that appear in computers or other circuits depend on tunneling for their operation.

7–3 Motion in Two or Three Dimensions

As for one dimension, there exists a potential energy that is a function of position for conservative forces acting in two or three dimensions. This potential energy is again defined so that the change in its value in going from one point to another is the negative of the work done by the force between those points. (Remember, for a conservative force the work done will depend only on the starting and ending points of the motion.) We then define the potential energy as before [see Eqs. (7–3) and (7–4)]: It is the negative of the work done, or more precisely

$$U(\vec{r}) - U(\vec{r}_0) = -W(\vec{r}, \vec{r}_0) \qquad (7\text{–}14)$$

The potential energy is now a function of all three variables x, y, and z or, equivalently, of the position vector $\vec{r}$. The total energy E is given by

$$E \equiv \tfrac{1}{2}m(v_x^2 + v_y^2 + v_z^2) + U(x, y, z)$$
$$= \tfrac{1}{2}mv^2 + U(\vec{r}). \qquad (7\text{–}15)$$

It is independent of time; that is, it is a constant of the motion—in other words, as before energy is conserved. The value of the constant E is fixed by the initial value of U (at the starting point) and by the initial speed. The potential energy function $U(\vec{r})$ determines the force in a manner analogous to the one-dimensional case of Eq. (7–4), which gives $F(x) = -dU/dx$.

Rather than becoming involved in the calculus of several variables, we can gain some insight into the motion in *two dimensions* by generalizing our picture of the potential energy in one dimension as the profile of a one-dimensional landscape with hills and valleys to a picture of the potential energy in two dimensions as a surface landscape just like the landscape of Earth. Such a landscape can be represented on a topographic map. Figure 7–13 represents such a landscape, which includes peaks, ridges, lake beds, valleys, and saddles. The function $h(x, y) = U(x, y)/mg$ represents the height of the terrain above sea level. The force at the point (x, y) in a given direction is the slope of the terrain in that direction. Near the bottom of a lake, the force in all directions is one that brings a mass back to the bottom—this is what is meant by true minimum. The bottom of a lake bed is therefore a stable equilibrium point. Similarly, at the top of a peak, the terrain slopes downward in every direction, so that although an object could sit at the very top with no force on it, any small movement away from the top brings in a force that moves the object away from the top. Peaks are unstable equilibrium points. Saddles

▲ **FIGURE 7–13** The potential corresponding to a force that depends on the two space coordinates of a plane exhibits many of the features of Earth's surface and is equivalent to a topographic map of Earth's surface.

(mountain passes) present a new situation, as they do not occur in one-dimensional potentials. Saddles occur where the slope is downward on the two sides of a pass and upward along the ridge of the pass. In general, then, the motion can be more complicated than in one-dimensional motion. In one-dimensional motion an object, starting somewhere on a slope with an initial positive kinetic energy, will move up or down, and if the initial kinetic energy is large enough, it will pass over a potential barrier (the total energy is above the top of the barrier). In two dimensions, an object on a slope with some initial kinetic energy can move in a variety of directions, some of which will take the object across ridges and some of which won't. In particular, if the initial velocity of the object is *along a contour line*, that is, a line of constant elevation in our landscape, the potential energy of the object will remain constant. We call a line of constant values of potential energy an **equipotential**. An object will move along an equipotential with constant kinetic energy (constant *speed*).

We can draw an important conclusion from this discussion. We have a unique landscape described by the scalar function $U(x, y)$, and by using our slope argument, we can find the components of the vector force in different directions. For conservative forces these comes from a single function $U(x, y)$ and therefore are related. We can describe motion more economically with a potential energy than with three components of a force.

Potential Energy for Projectile Motion

In Section 7–1, we derived and applied the expression $U(y) = mgy + U_0$ [Eq. (7–11)] for pure vertical motion under the influence of gravity. We can now show that this expression is valid also for arbitrary projectile motion under the influence of gravity.

The force of gravity is conservative in that the work does not depend on the path taken—even when that path contains horizontal components, as we saw when we studied the motion of a mass on a ramp in Section 6–4. Therefore gravity is conservative for motion in more than one dimension. To find the potential energy when an object can move in more than one dimension, we must evaluate the work done when gravity acts. But the work, and hence the potential energy change, depends only on the change in height of the object and not on the object's horizontal motion. For example, $W = mgL \sin\theta$ [Eq. (6–22)] is the work done by gravity in moving a mass a total length L along a ramp that makes an angle θ with the horizontal. But the quantity $L \sin\theta = h$ is the height through which the mass drops (see Fig. 6–20). The work done is thus mgh, which is just the expression that holds when we consider vertical motion only. We conclude that *the potential energy function $U(y) = mgy + U_0$ applies to general projectile motion under the influence of gravity.*

EXAMPLE 7–7 A golfer hits a bad shot and the ball leaves the tee with an initial speed of 28 m/s at an angle of 84° with respect to the horizontal. A bee is cruising innocently at a height of 37 m when it has the bad luck to meet the golf ball. What is the speed of the ball when it hits the bee? Ignore all effects of air resistance.

Setting It Up We sketch the situation in Fig. 7–14 using an xy-coordinate system with the origin at the golf tee. The height where the ball and bee meet is $y = h$. We know v_i, the initial speed of the ball, and the initial elevation angle. We want to find the speed v_f of the golf ball when it hits the bee at the known altitude $y = h$.

Strategy Conservation of energy tells us that the initial energy of the struck ball has to equal the energy when the collision occurs, and this will relate speed and height. We let the potential energy be zero at ground-level height ($h = 0$ m), so $U(h) = mgh$.

Working It Out The golf ball has an initial energy $E_i = K_i + U(0) = \frac{1}{2}mv_i^2 + mg \times 0 = mv_i^2/2$, where m is the mass of the ball. The energy of the ball when it meets the bee at height h is $E_f = K_f + U(h) = \frac{1}{2}mv_f^2 + mgh$. Conservation of energy then reads

$$E_i = \frac{1}{2}mv_i^2 = E_f = \frac{1}{2}mv_f^2 + mgh.$$

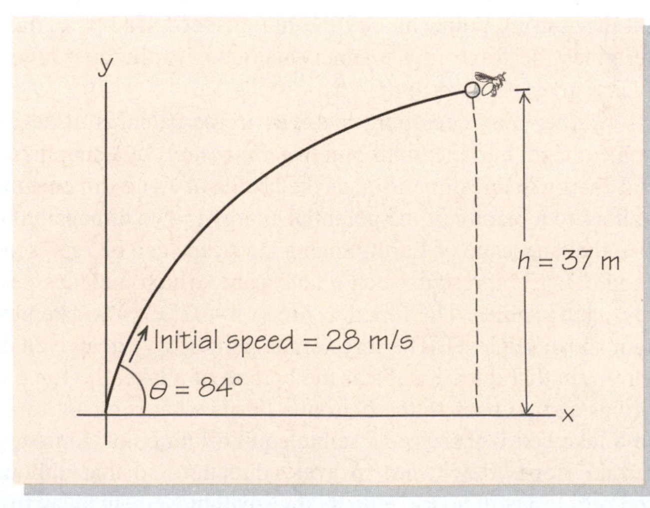

▲ **FIGURE 7–14** Golf ball hits bee at height of 37 m.

Cancel the factor m and solve for v_f:

$$v_f^2 = v_i^2 - 2gh$$
$$= (28\ \text{m/s})^2 - 2(9.8\ \text{m/s}^2)(37\ \text{m})$$
$$= 59\ \text{m}^2/\text{s}^2,$$

and thus

$$v_f = 7.7\ \text{m/s}.$$

The golf ball is fairly close to the top of its trajectory and has slowed down considerably. Note that we do not need to use the detailed information about the initial angle of the ball or calculate the angle at which the ball is moving when it hits the bee. That is part of the utility of working with the conservation of energy.

What Do You Think? Can we decide with this approach whether the golf ball hit the bee while the ball was still rising or when it had begun to fall back to the ground?

Central Forces

There is a most interesting case that is important in nature: a force with a magnitude that depends only on the radial distance from a fixed point and has a direction aligned along the corresponding radius vector. For example, the force of gravitational attraction on Earth due to the Sun depends on the distance between Earth and the Sun and is aligned along the line between the two objects. The fundamental law of the attraction between particles with electrical charge—the force that allows atoms to exist—is also of this nature. Such forces are called **central forces** and *all central forces are conservative*. Such a force takes the form

$$\vec{F}(\vec{r}) = F(r)\left(\frac{\vec{r}}{r}\right) = F(r)\hat{r}, \tag{7–16}$$

where $\hat{r}$ is a unit vector pointing away from the origin. This force is reminiscent of a one-dimensional force in that it depends on only one space coordinate. The potential energy function of a central force depends only on the radial distance r; the expression for the force in terms of a potential energy function $U(r)$ is

$$\vec{F} = -\frac{dU(r)}{dr}\hat{r}. \tag{7–17}$$

Example 7–8 illustrates how we can apply Eq. (7–17).

EXAMPLE 7–8 The potential energy describing the gravitational interaction (the *gravitational potential energy*) between two point masses such as Earth and the Moon, with masses m_1 and m_2, respectively, is given by the expression

$$U(r) = -\frac{Gm_1m_2}{r},$$

where G is a constant and r is the distance between the masses. Calculate the force experienced by the Moon due to the presence of Earth. Is the force attractive or repulsive?

Setting It Up We define the unit vector $\hat{r}$ to be in the direction from Earth (taken to be located at the origin) to the Moon (Fig. 7–15).

Strategy This is a straightforward application of Eq. (7–17), which gives the force in terms of the potential.

Working It Out From Eq. (7–17) and the fact that $(d/dr)(1/r) = -1/r^2$, we find

$$\vec{F} = -\frac{Gm_1m_2}{r^2}\hat{r}.$$

The force is proportional to $1/r^2$ (an inverse-square form), and it is directed opposite to the position vector $\vec{r}$ from Earth to the Moon. It therefore acts to pull the Moon toward Earth and is an *attractive* force. This force represents the centripetal force that allows the Moon to orbit Earth in a nearly circular orbit (see Chapter 12 for a more complete treatment of the subject).

What Do You Think? What choice has been made for the location(s) where the potential energy vanishes? Why is this a reasonable choice?

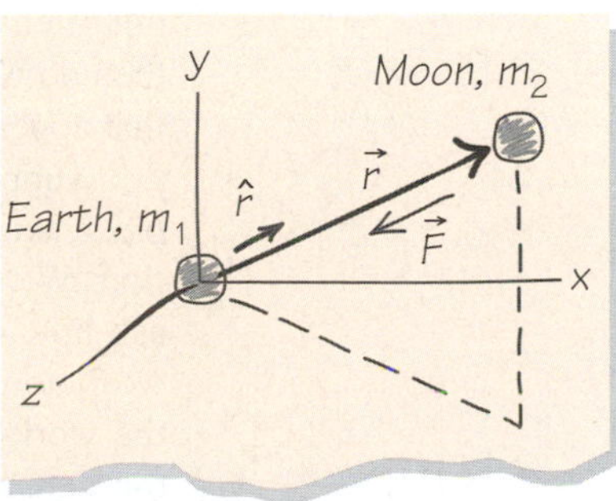

▲ **FIGURE 7–15** The vector $\vec{r}$ pointing radially outward from the origin (the location of mass m_1) is the position vector of mass m_2. The force vector on m_2 acts inward radially.

7–4 Is Energy Conservation a General Principle?

At first glance, it may appear that energy conservation is not all that useful, as many physical systems involve nonconservative forces that dissipate energy such as friction or drag. In the presence of friction, mechanical energy, the sum of the potential and kinetic energies, is not conserved. But as we discussed briefly at the end of Section 7–1, the energy that is lost because of the presence of friction can be identified through experiment as **thermal energy**, that is, the type of energy that raises the temperature of two objects when they are rubbed together. There are still other forms of energy that we will identify in later chapters, such as the energy in sound, in light, in various electrical systems, and in the interaction between molecules. When thermal and indeed all the forms of energy are measured and included as still other components of the total energy of a large system, *the total energy is then conserved*. However, there is a catch. Whereas potential energy can be converted to kinetic energy and vice versa, thermal energy is not a form of potential energy because it is not completely retrievable—it cannot be fully converted to kinetic energy of a mass, for example. A system that contains thermal energy differs from a system for which kinetic and potential energies can be freely converted back and forth. The ways in which mechanical and thermal energy can be converted back and forth are important, and we shall study this in detail in our discussion of thermal phenomena (Chapters 17 to 20).

To illustrate that some forms of energy are not completely retrievable, consider two situations. In one we drop a rubber ball on a stone floor, and in the other we drop a glass. In each situation energy is conserved, but mechanical energy is conserved only in the first case. In the absence of air resistance, the ball will hit the ground and return to the hand—mechanical energy is conserved. What happens to the mechanical energy of the glass? The glass at first merely acquires kinetic energy through loss of potential energy, but when it hits the floor, it shatters. Shards fly all over the place, sliding to a stop under the influence of friction, and your eardrums vibrate because they hear the crash. The glass could bounce up into your hand only if all the initial potential energy it contained could be retrieved and transformed back into kinetic energy. What would it take to retrieve that energy? You would have to run the movie of the crash backward, and you can imagine that getting the energy back into each shard at the right time and the right place so that the glass is reconstructed is effectively impossible. The thermal energy given to the stone floor when the shards slid to a stop would have to be extracted. Even the energy that left as sound would have to be recovered and put back into the glass!

Nevertheless, when we look closely enough, *all the fundamental forces in nature are conservative*, and conservation of energy always holds. ***The conservation of energy is one of the most fundamental principles of physics***. It has received experimental support from all fields of physics, and we shall encounter energy conservation in a variety of applications, including thermodynamics, electromagnetism, relativity, and quantum physics.

Energy Conservation and Nonconservative Forces

Although it is important to realize that all fundamental forces are conservative, we do not often operate at the level of the fundamental forces in our macroscopic world. How then do we deal with the nonfundamental and nonconservative forces, such as friction, that appear in everyday situations?

Suppose both conservative and nonconservative forces act on an object. For example, a skier descending a mountain moves under the influence of both gravity and friction. We can first construct the potential energy as if there were no friction and then ask how much energy is lost due to friction; the loss of energy is attributable to any work done by friction. We can do this because we can divide the work as it appears in the work–energy theorem into a part associated with a potential energy (the work of the conservative forces) and another part that is the work done by nonconservative forces. Thus, if we write W_{nc} as the work done by the net nonconservative force and U as the potential energy of the net conservative force, the work–energy theorem $W_{net} = W_{nc} + W_{conservative} = \Delta K$ becomes

$$W_{nc} = \Delta(K + U) = \Delta E. \tag{7–18}$$

Here, we have continued to think of E as the total mechanical energy $K + U$, but E is no longer conserved.

For our skier, if friction is ignored, we have $E = K + mgy =$ constant, where $K = mv^2/2$ is the kinetic energy at a height y above some zero level. When friction is introduced, the initial and final energies E_i and E_f, respectively, differ, with $E_f < E_i$. The deficit is due to the work done by friction. We have

$$W_{\text{friction}} = E_f - E_i = (K_f + mgy_f) - (K_i + mgy_i). \qquad (7\text{–}19)$$

The work done by friction is negative, so that the final energy is less than the initial energy.

EXAMPLE 7–9 A ball of mass 10 kg is attached to a 5.0-m-long wire that swings freely from a support. We have a *simple pendulum* (Fig 7–16a). The ball is pulled aside so that the wire makes an angle of 31° to the vertical and it is released from rest. After 10 swings, the maximum angle that the ball reaches is 25° from the vertical. What is the work done by air resistance and any other nonconservative forces acting on the ball during these 10 swings?

Setting It Up The illustration (Fig. 7–16a) shows the starting point for the motion; Fig. 7–16c includes labeling for the angle θ of the pendulum from the vertical and the vertical distance y from the minimum position of the pendulum's swing. We know the ball's mass m, the pendulum length L, and the maximum angle θ_i at time $t = 0$. After 10 swings, the new maximum angle θ_f is given. We want the work done by nonconservative forces during the 10 swings.

Strategy We compute directly the change in the total mechanical energy E (kinetic energy plus potential energy) associated with the conservative forces, in this case gravity, that was lost over the 10 swings. We are given information at moments where E is entirely in the form of potential energy, so that the energy change can be expressed entirely in terms of the difference of two potential energies. This difference must equal the work done by nonconservative forces, in this case air resistance on the ball.

Working It Out The forces acting on the ball are specified in a free-body diagram (Fig. 7–16b). This diagram includes the forces of gravity, the tension of the wire, and the air resistance. Tension acts in a direction perpendicular to the motion, so it does no work and plays no role in any energy considerations. Gravity is conservative and there is a potential energy associated with it. We will first express conservation of energy as if air resistance were not present. If we take the zero of potential energy at the minimum point of the swing and measure the height y from that level, then $U(y) = mgy$. Figure 7–16c shows that

$$y = L(1 - \cos\theta),$$

where L is the wire's length. There is no kinetic energy at the top of the swing, so the total energy is

$$E = U(y_{\max}) = mgy_{\max} = mgL(1 - \cos\theta_{\max}).$$

This expression will apply to the initial and final situations with the maximum angle given by θ_i and θ_f, respectively. The *loss* of energy over the 10 swings is then

$$\begin{aligned} E_i - E_f &= mgL(1 - \cos 31°) - mgL(1 - \cos 25°) \\ &= mgL(\cos 25° - \cos 31°) \\ &= (10\text{ kg})(9.8\text{ m/s}^2)(5.0\text{ m})(\cos 25° - \cos 31°) \\ &= 24\text{ J}. \end{aligned}$$

This energy is lost by virtue of the nonconservative forces; equivalently, it must match the work done by these forces, here air resistance.

What Do You Think? You are told that during the time that the pendulum keeps on swinging it acts as a clock. But a good clock does not slow down. How is it possible that the pendulum described here could make a good clock?

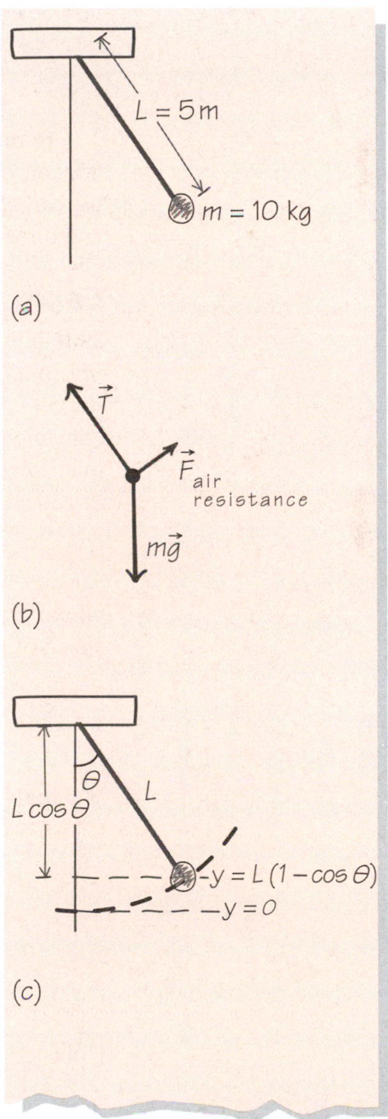

▲ **FIGURE 7–16** (a) Simple pendulum. (b) Force diagram for the mass. (c) Geometry for the determination of the potential energy.

THINK ABOUT THIS . . .

HOW DOES OUR SOCIETY PRODUCE AND USE ENERGY?

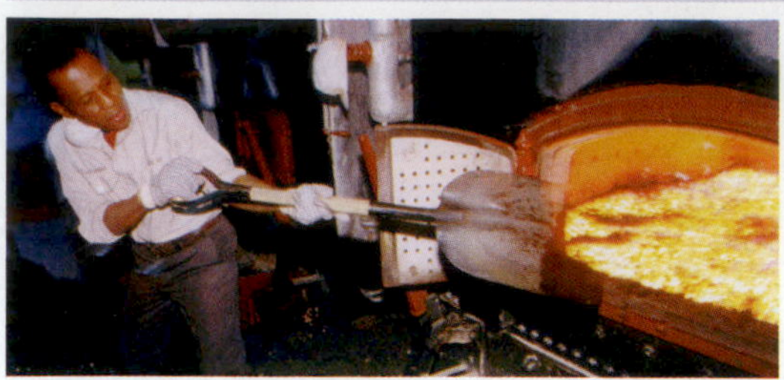

▲ **FIGURE 7–17** When fuel is burned, chemical energy is converted into thermal energy as well as electromagnetic energy—light.

There is a great deal of justifiable concern in the world about producing energy, using energy, and wasting energy. When we talk about energy production, we have in mind electric power generators of various kinds, including coal- or oil-burning plants, nuclear reactors, hydroelectric plants, solar energy plants, and windmills. In each case we are not producing energy but *converting* it. Coal or oil contain chemical energy, a form of potential energy associated with the chemical bonds that tie atoms together into the molecules of the fuel (Fig. 7–17). The burning, which is a chemical reaction, breaks some of these bonds and this releases energy, generally in the form of thermal energy. Some of that energy may be used to convert water to steam, and the pressure of the steam may be used to turn turbines that can generate electrical energy. In nuclear reactors the potential energy is contained in the nuclei that undergo fission—sometimes referred to as *nuclear energy*—which generates rapidly moving particles whose kinetic energy is again available for further conversion. In hydroelectric plants, water in a reservoir is allowed to flow downhill, converting gravitational potential energy to kinetic energy of the water, some of which can turn the wheels of turbines. Similarly, windmills convert the kinetic energy of moving air.

The electrical energy "produced" is not used up when, for example, it lights buildings. But part of the energy is wasted in the sense that even if we put effort into it, the thermal energy given up by a light bulb could only partially be captured and reconverted into energy that could be reused. Whatever use we make of it, energy is always conserved. What we can do is try to ensure that the energy is used as efficiently as possible and the processes we employ have as little deleterious effect on the environment as possible. ■

In this discussion and example, we used the energy loss to find the work done by nonconservative forces. But, of course, we could use this approach in a different way. If we can calculate the work done by friction—we can do this if we know its magnitude and direction and the distance over which it acts—then we can *predict* the energy loss.

Energy Conservation as One of the Underpinnings of Physics

Within this chapter we have seen that energy conservation can be a powerful shortcut to information that can be obtained only laboriously from the equations of motion. It is important to understand that the statement that E in Eq. (7–15) is a constant is entirely equivalent to Newton's laws. It represents a more fundamental formulation of dynamics than Newton's second law, even though the two are equivalent. This is not so easy to see in general, but is quite transparent for motion in one dimension. We write

$$E = \tfrac{1}{2}mv^2 + U(x), \tag{7–20}$$

where E is a constant and both x and v are functions of time; they are the position and speed of the object in motion. Because energy is conserved, we have

$$\frac{dE}{dt} = 0. \tag{7–21}$$

We apply this to Eq. (7–20) using the chain rule when needed:

$$\frac{dE}{dt} = \frac{1}{2}m\frac{dv^2}{dt} + \frac{dU(x)}{dt} = \frac{1}{2}m\frac{dv^2}{dv}\frac{dv}{dt} + \frac{dU(x)}{dx}\frac{dx}{dt}$$
$$= \frac{1}{2}m(2v)\frac{dv}{dt} + \frac{dU(x)}{dx}v = v\left[m\frac{dv}{dt} + \frac{dU(x)}{dx}\right] = 0.$$

Unless $v = 0$ at all times, the contents of the square brackets must be zero; that is,

$$m\frac{dv}{dt} = -\frac{dU(x)}{dx} = F(x). \tag{7–22}$$

This is just our original formulation of Newton's second law.

Physicists and mathematicians studied Newton's equations almost from the time of their discovery. But it is interesting to note that the principle of energy conservation—formulated by Julius Robert Mayer (1842), Hermann von Helmholtz (1847), and James Prescott Joule (1850)—grew instead out of experimental research by engineers on the properties of heat and the conversion of mechanical energy into thermal energy.

Summary

Conservative forces can be specified in terms of a single scalar function called the potential energy. In one-dimensional motion, the potential energy $U(x)$ depends on one coordinate only and is defined by

$$U(x) - U(x_0) = -\int_{x_0}^{x} F(x')\,dx', \tag{7–4}$$

where $U(x_0)$ $(= U_0)$ is an arbitrary constant and x_0 is an arbitrary point. In other words, the potential energy $U(x) - U(x_0)$ is the negative of the work done by the force as the object on which it acts moves from x_0 to x. The force is given in terms of U by

$$F(x) = -\frac{dU(x)}{dx}. \tag{7–5}$$

Two important examples of potential energy are associated with gravity and with the spring force. For gravity,

$$U(y) = mgy + U_0, \tag{7–2}$$

where y is the height above ground level. For the spring,

$$U(x) = \tfrac{1}{2}kx^2, \tag{7–13}$$

where x is the distance from the equilibrium point of the spring and the zero of the potential is chosen at the equilibrium point $x = 0$.

For central forces (forces aligned in the radial direction), the potential energy depends only on r, and

$$\vec{F} = -\frac{dU(r)}{dr}\hat{r}. \tag{7–17}$$

For conservative systems, the total mechanical energy—the sum of the kinetic and potential energies—does not change during the motion; thus, the total energy is a constant whose value may be determined at any time during the motion:

$$E = \tfrac{1}{2}mv^2 + U(\vec{r}). \tag{7–15}$$

The fact that E is constant means that any change in the kinetic energy is compensated by an equal but opposite change in the potential energy during the motion. *The total energy is conserved.* The utility of this conservation law is enormous: We can determine the speed when the position of an object is known. Energy diagrams provide us with a systematic way to approach problems that involve energy conservation. With them, we can understand useful concepts such as turning points, potential energy barriers, and stable and unstable equilibria.

Nonconservative (dissipative) forces such as friction or air resistance appear in the macroscopic world. They spoil energy conservation, but since the work–energy theorem always holds, one can use it to find the energy dissipated through friction or drag forces, even if one cannot calculate this energy loss directly. Dissipative forces do not exist on the microscopic level; they are effective forces that come from the action of the microscopic laws. Thus at a deeper level the conservation of energy always holds. Energy is neither created nor destroyed; it can only change from one form to another.

Understanding the Concepts

1. Are drag forces such as air resistance conservative?
2. Discuss why it is possible for the total mechanical energy to be negative, even if the kinetic energy cannot be negative.
3. The potential energy of gravity can be given the form mgh, where h is the height above sea level. That means that if one goes into a deep mine whose bottom lies below sea level, the potential energy becomes negative. What is the meaning of this negative potential energy?
4. If we add a constant term to the potential energy of an object, why doesn't this change the object's motion?

5. Explain why a rubber ball seems never to bounce back to its original height when dropped from rest. What happens to the energy? Explain why the ball can bounce back to a much greater height when it is thrown down.
6. The force $F = -kx$ due to a one-dimensional spring is conservative. Is it still conservative if k is allowed to vary with time? In thinking about this question, it will be helpful to use some of the definitions of conservative forces given in Chapter 6.
7. Is energy conserved when you ingest sugar and then go out and exercise? In what ways, and in which of the various processes for converting food into muscular activity, might the energy conservation principle come into play?
8. For which of the following forces can one write a potential energy function? In each case, justify your answer. (a) The force exerted by an ideal spring. (b) The drag force exerted by the air on an object falling through it (Chapter 5). (c) The normal force exerted by an object sitting on a table top. (d) A force that is the same everywhere in space.
9. Does a real spring exert a truly conservative force? What experiment might you perform to check the answer to this question?
10. Suppose you are given the components of a two-dimensional force in the x- and y-directions as functions of x and y. Can you think of a way of figuring out whether the force is conservative?
11. How, if at all, does experiment demonstrate that the gravitational force is conservative?
12. How would you figure out the energy lost to friction for a skier going down a mountain?
13. A golf ball is dropped with zero initial velocity from a height of 20 m. It bounces on a concrete pad and rises to a maximum height of 19 m. Are any of the forces that the ball experiences during the motion nonconservative?
14. Consider motion with a potential energy of the shape given in Fig. 7–5. Let $U(x)$ describe the height of a smooth slide above the ground. In the absence of friction the motion is easily predicted to repeat with a uniform repetition time. Discuss the motion, starting from the right or the left side, when a small amount of friction is present.
15. When an object slides down a frictionless, curved slope there are normal forces as well as gravitational forces present. How can we assert, without knowing the shape of the slope, that we are dealing with a conservative system of forces?
16. When a force acts in two dimensions, are there two separate laws for the conservation of energy that correspond to motion in each direction?
17. The potential energy is zero at a given point. Is the force necessarily zero or nonzero at that point?
18. An object moves on a rough inclined plane. Can you still use the concept of energy conservation to relate the height and the speed? How?
19. A ball tethered to a vertical rod is struck by a bat every time it passes by a certain point. Every time it goes around the rod, it goes faster, so that on moving back to the starting point it has a different kinetic energy. Why are the forces involved not conservative?
20. The motion of the pendulum of a grandfather clock is slowed by air resistance and bearing friction, yet such clocks can run without stopping for years. What types of mechanisms are typically used to supply the necessary energy?
21. For which, if any, of the following systems does it make sense to say that the energy of the system is conserved? (a) A community of animals living at the bottom of the ocean near a source of hot water and minerals. (b) Earth as a whole. (c) A sealed ecosystem under a huge glass dome, such as has been built in Arizona. (d) A perfectly sealed cylinder of hot gas that can drive a piston.
22. You are watching fireworks. You see a glowing point streak into the air. This is followed by a loud noise and, about a second later, by the appearance of many scattered dots of light that sink in graceful curves before disappearing. Account for all of the energy supplied and spent at each stage of the spectacle.
23. A steel marble is bounced off the top step of a staircase. It rises to 1 m above the step, then falls onto the second step, bounces off, falls onto the third step, and so on, for 39 steps. Assuming that there is no air resistance or other source of energy loss, how high will the marble rise after the last bounce? (Do not do 39 calculations to answer this question!)

Problems

7–1 Energy Conservation

1. (I) An unknown constant force F pushes a 10-kg body from rest on the ground vertically upward. At a height of 2.0 m, the velocity of the object is $\vec{v} = (2.4\ \text{m/s})\hat{j}$. (a) Find the change in the potential energy associated with gravity. (b) What is the net work done, and what is the work done by the unknown force?
2. (I) Consider the expression $U = mgy$ for an object of mass m under the influence of gravity, where y is measured from the ground up. Express U in terms of z, where z is measured in a downward direction from a rooftop 30 m above the ground.
3. (I) A baseball pitcher throws a ball at 95 mi/h off a roof that is 80 m above the ground. How high will the ball be when it is traveling at 120 mi/h?
4. (I) A rock is thrown straight down into a deserted quarry from the edge, which is 45 m above the bottom. The rock has a speed of 42 m/s when it reaches the bottom. What was the rock's initial speed? Ignore air resistance.
5. (I) The spring of a toy gun launches rubber-tipped projectiles with a spring constant of 5 N/m. The spring is compressed by 7 cm with the projectile in place. How much kinetic energy is imparted to the projectile?
6. (II) The potential energy of an archery bow is measured to be $U(x) = bx^2 + cx^3$, where x is the distance the bow string is pulled back from its equilibrium position. When an archer pulls the string a distance x, what force does the archer exert on the string and what force does the string exert on the archer?
7. (II) A package of mass 5.0 kg is subject to a constant force of 8.0 N pointing in the $+x$-direction. (a) Calculate the potential energy of the package as a function of its position x, defining it such that $U(x)$ at $x = 0.0$ m is zero. (b) Assuming that the package has a velocity of 2.0 m/s at $x = -1.0$ m, calculate the total mechanical energy of the package. (c) What is the speed of the package at $x = 3.0$ m?
8. (II) A spring has a spring constant of 16.5 N/m and obeys Hooke's law. How far must the spring be pulled back if its potential energy is to be 3.77 J? What is the mass of a ball at the end of the spring if the maximum speed of the ball is observed to be 1.71 m/s when the spring is released?

9. (II) The energy of a harmonic oscillator (a mass moving on the end of a spring) is given by $E = \frac{1}{2}mv^2 + \frac{1}{2}kx^2$. Plot contours of constant E on a graph in which x is measured along one axis and v is measured along the perpendicular axis. Choose the parameters $E = 16.0$ J, $m = 2.0$ kg, and $k = 8.0\ \text{J/m}^2$. Such a plot is called a *phase plot*; the motion of a system is restricted to the curve corresponding to the energy E.

10. (II) Sketch the potential energy for a 5-kg mass that can move between the ground and a height of 10 m. (a) Assume that zero potential energy is at the ground, (b) at a height of 10 m, and (c) at 4 m.

11. (II) A block of mass 0.528 kg slides with uniform velocity of 3.85 m/s on a horizontal frictionless surface. At some point, it strikes a horizontal spring in equilibrium. If the spring constant is $k = 26.7$ N/m, by how much will the spring be compressed by the time the block comes to rest? What is the amount of compression if the surface is rough under the spring, with coefficient of kinetic friction $\mu_k = 0.411$?

12. (II) An archery bow acts much like a spring displaced from equilibrium when the bow is drawn. Suppose that an archer displaces the string from equilibrium by 47 cm and exerts a force of 65 N. (a) What is the "spring constant"? (b) What is the speed of an arrow of mass 40 g that leaves the bow as the string reaches the equilibrium position? [*Hint*: Use the expression for energy for a mass moving under the influence of a spring.]

13. (II) A cannonball of mass 15 kg is dropped from rest from a height of 6.0 m. It falls onto a large vertically oriented spring that is compressed from its relaxed position when the cannonball lands on it. The spring has spring constant $k = 10^4$ N/m. What is the maximum compression of the spring? How much would the spring be compressed if a man of mass 60 kg jumped onto it from a height of 1.5 m?

14. (II) A mass $m = 0.70$ kg slides along the x-axis of a horizontal frictionless surface with speed $v_x = 2.2$ m/s. It runs into a relaxed spring oriented along the x-axis. This spring had previously been observed to stretch by 3.8 cm when it was oriented vertically with the mass suspended from it. (a) What is the maximum compression of the spring when the mass runs into it? (b) The mass rebounds as a result of having compressed the spring. What is its velocity when it leaves contact with the spring?

15. (II) A spring with spring constant $k = 200$ N/m is used as a launcher for a small block whose mass is 10 g. The block is placed against the compressed spring in a horizontal arrangement on a smooth horizontal surface. The spring, with the block, is compressed 5 cm and then released. (a) Find the speed of the block just as it leaves the spring. (b) The block encounters a rough surface as it leaves the spring. How much work does friction do in bringing the block to an eventual stop? (c) The block slides a distance of 3.5 m before stopping. What is the coefficient of kinetic friction between the block and surface?

16. (II) A person of mass 70 kg jumping on flat ground can raise his center of mass by 1.0 m, whereas the same person jumping on a trampoline, which acts as a type of spring, can raise his center of mass by 2.5 m. What is the potential energy contained in the trampoline at the bottom of the person's 2.5-m jump? Do you need to make an assumption about the mass of the trampoline's elastic surface?

17. (II) Over a large enough height difference, the gravitational acceleration g is not constant but decreases with altitude. If we could approximate this change by $g \to g_0 - g'y$, where $g' \ll g_0$, where y is the height above sea level, what would be the potential energy of a mass under the influence of gravity? Choose the zero of the potential energy where $y = 0$.

7–2 Energy Conservation and Allowed Motion

18. (II) For a conservative one-dimensional force, show that the sign of the slope of the potential energy function at a position x determines the direction, positive or negative, in which the force acts.

19. (II) When two atoms on a line are far apart, there is no force between them. As they start to move closer, there is an attraction between them, which, at very close distances, turns into a strongly repulsive force. Sketch the potential energy as a function of the distance between the atoms.

20. (II) Figure 7–18 shows the force $F(x)$ that acts on a particle moving along the x-axis. (a) Plot the potential energy of the particle as a function of x. (b) The particle starts its motion at $x = -0.5$ m, with zero initial velocity. How far to the right will the particle travel?

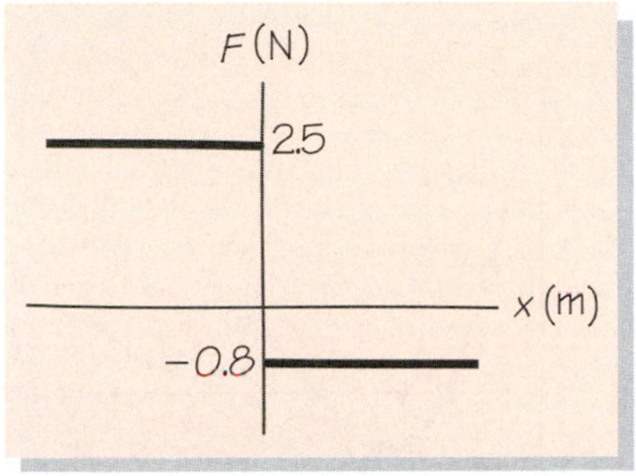

▲ **FIGURE 7–18** Problem 20.

21. (II) Consider the force described in Problem 20. Is there a stable equilibrium point associated with this force? If so, find the turning points that correspond to a given initial total energy.

22. (II) The potential energy for a mass $m = 1.0$ kg moving in one dimension is given by $U(x) = (2.0\ \text{J})\sin \pi x$. The mass starts at $x = 0$ with an initial velocity $v = +0.71$ m/s. (The plus sign means the motion is in the positive x-direction.) Describe the subsequent motion of the mass. Suppose the initial velocity were +3.0 m/s. What would the subsequent motion look like?

23. (II) Consider a force $\vec{F}(x)$ acting along the x-axis that is opposite to the spring force (Hooke's law). In other words, the force has the single vector component $F(x) = +kx$. A mass m is under the influence of this force and of no other forces. Suppose that the mass is placed at rest just to the right of the origin. (a) Which way will the mass move, if at all? (b) Find the speed of the mass as a function of its distance from the origin. (c) Repeat the problem but assume that the mass had been placed just to the left of the origin.

24. (II) Consider the energy diagram in Fig. 7–19. (a) What are the limits of motion for energies E_1 and E_2? Redraw the figure and label it as necessary. (b) Describe the circumstances under which

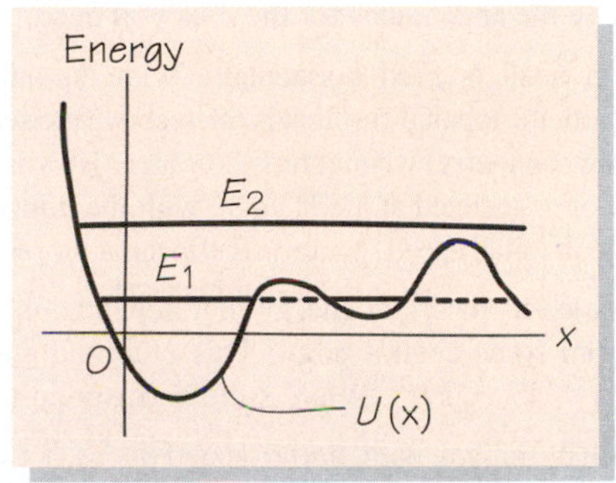

▲ **FIGURE 7–19** Problem 24.

the particle is always at rest. (c) Find the energies and positions for which motion within turning points is possible. (d) Find the equilibrium positions on your drawing. Are they stable or unstable?

25. (II) Draw a one-dimensional potential energy $U(x)$ diagram with the following characteristics: (a) The particle can never reach negative x. (b) There are three regions in x where the particle can move within turning points. (c) The particle can never reach infinity. (d) The particle has unstable equilibrium positions at 1 and 2 nm.

26. (II) Consider the potential energy $U(x)$ shown in Fig. 7–20. (a) What is the sign of the force at positions 1 through 6? (b) Which positions have the most positive, most negative, and zero force? (c) Find the equilibrium positions and indicate whether they are stable or unstable.

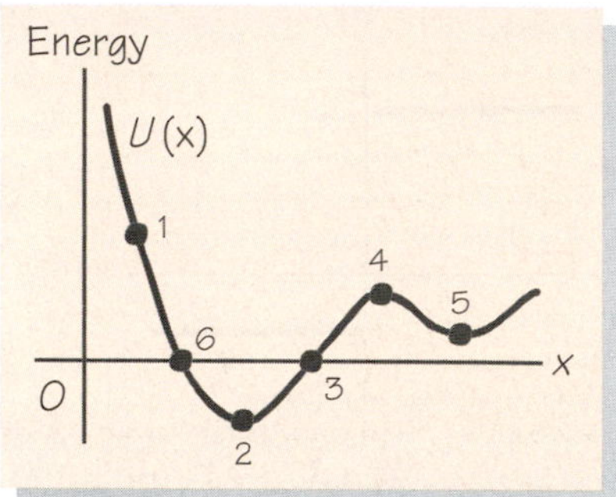

▲ **FIGURE 7–20** Problem 26.

27. (II) The potential energy of an object constrained to move in the x-direction is given by $U(x) = \alpha x^4 + \beta x^2$, where $\alpha = 26\ \mathrm{J/m^4}$ and $\beta = -3.0\ \mathrm{J/m^2}$. Find the equilibrium points and state whether they are stable or unstable.

28. (II) A particle is moving in a potential well described by the potential energy

$$U(x) = -\frac{1.5\ \mathrm{J\cdot m^2}}{1.2\ \mathrm{m^2} + (x + 0.80\ \mathrm{m})^2} - \frac{1.5\ \mathrm{J\cdot m^2}}{1.2\ \mathrm{m^2} + (x - 0.80\ \mathrm{m})^2},$$

where x is measured in meters. (a) Sketch the shape of the potential energy for $-4\ \mathrm{m} < x < 4\ \mathrm{m}$. (b) The speed of a particle of mass 0.50 kg at $x = -2$ m is 3.2 m/s. Can the particle reach $x = 1.5$ m?

7–3 Motion in Two or Three Dimensions

29. (I) A projectile fired from a gun leaves the barrel at a speed of 500 m/s. The gun is placed 180 m above a level plain. Use energy conservation to calculate the speed of the projectile when it is 16 m above the plain. Neglect all drag effects.

30. (I) What is the gravitational potential energy of the Earth–Moon system? See the appendices for the data you need.

31. (I) A cannonball is fired horizontally with an initial speed of 125 m/s from the top of a cliff that is 68 m above the sea. What is the speed of the cannonball when it hits the water? How is that changed if the cannon is inclined at a 32° angle with the horizontal without any change in initial speed? Ignore all effects of air resistance.

32. (I) A particle has potential energy that depends only on the distance r from some central point. This potential energy has the form $U(r) = U_0 - k/r^2$. What is the corresponding force law?

33. (II) A ball is thrown with initial speed v_0 in a trajectory that makes an initial angle θ with the ground. Air resistance is small. (a) By using the principle of conservation of energy, show that the speed of the ball when it reaches the height h above the ground is $v_h = \sqrt{v_0^2 - 2gh}$, independent of the angle of the throw. (b) Use the result of part (a) to find the initial speed v_0 that is required if a vertically thrown ball is to reach height H just before turning back. (c) Use the results of part (a) to find the initial speed v_0 required so that a ball thrown at 45° reaches a maximum height H.

34. (II) An object of mass m is subject to two forces: One force acts only in the x-direction and is due to a spring of spring constant k; the other force acts only in the y-direction and is due to a spring with the same spring constant k. (a) What is the potential energy? (b) Use your result to show that the net force is proportional to the distance from the equilibrium point and is directed towards it. (c) Using your result from part (b), show that one possible motion is uniform circular motion in the xy-plane. [*Hint*: Make a simple choice for equilibrium point.]

35. (II) A ball of mass m is thrown with a speed v at an angle θ from the horizontal off the top of a tall building at height h. (a) Show that, in general, it is not possible to tell whether the angle θ is above or below the horizontal when a ball hits the ground just by measurement of the speed of the ball when it hits the ground. (b) By knowing the horizontal distance that the ball travels, show that the ambiguity can be resolved.

36. (II) A skier slides down a hill starting with zero velocity at a height of 43 m above the bottom of the hill. The shape of the terrain is shown in Fig. 7–21. What is the velocity of the skier on top of the second, intermediate hill, whose height is 37 m? What is the skier's velocity at the bottom of the hill? Neglect all frictional effects. Is this neglect reasonable?

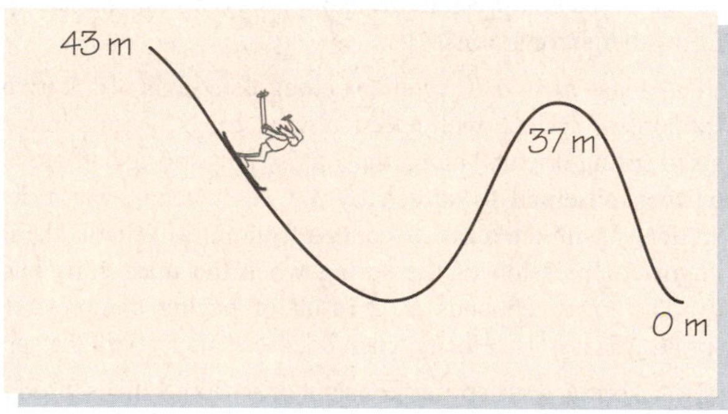

▲ **FIGURE 7–21** Problem 36.

37. (II) Suppose that the skier of Problem 36 reaches the bottom of the hill at a speed of 23 m/s. Assuming that the skier, including equipment, has a mass of 75 kg, how much work is done by the resistive forces of friction and drag?

38. (II) A ski jumper starts from rest and follows, with its several ups and downs, the rather bumpy ski jump shown in Fig. 7–22. The

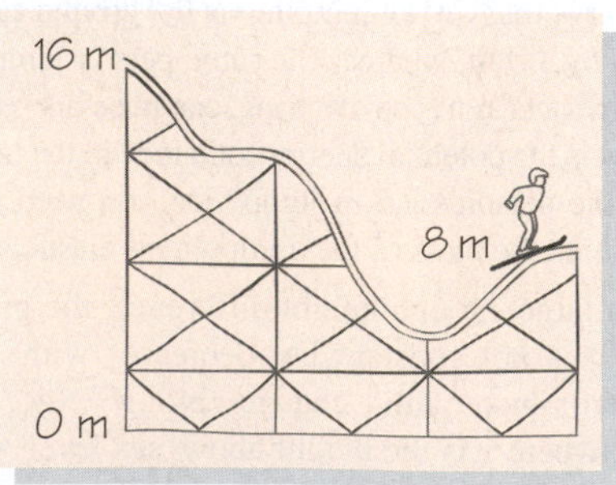

▲ **FIGURE 7–22** Problem 38.

jump track starts at a height $h = 16$ m above the eventual landing point, and the jumper leaves the track while he moves horizontally at a height above the landing point that is exactly one-half h. Assuming that the effects of air resistance are negligible (in real ski jumping they are not), what is the horizontal distance of the edge of the track from the landing point?

39. (II) A particle of mass $m = 30$ g slides inside a bowl whose cross section has circular arcs at each side and a flat horizontal central portion between points a and b of length 20 cm (Fig. 7–23). The curved sides of the bowl are frictionless, and for the flat bottom the coefficient of kinetic friction $\mu_k = 0.21$. The particle is released from rest at the rim, which is 10 cm above the flat part of the bowl. (a) What is the speed of the particle at a? (b) What is the speed of the particle at b? (c) Where does the particle finally come to rest?

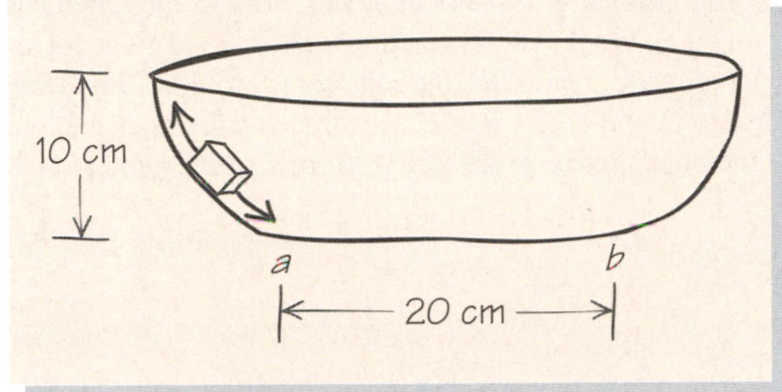

▲ **FIGURE 7–23** Problem 39.

40. (II) A toy car of mass M slides down a frictionless track that makes a circular loop of radius R at the bottom (Fig. 7–24). Suppose that the car starts from rest at a height H, with $H > 2R$. (a) What is the car's speed at the bottom of the circle? (b) At the top of the circle? (c) What is the force exerted by the track at the top of the circle? (d) What is the minimum value of H such that the car goes around the loop without falling off under the force of gravity?

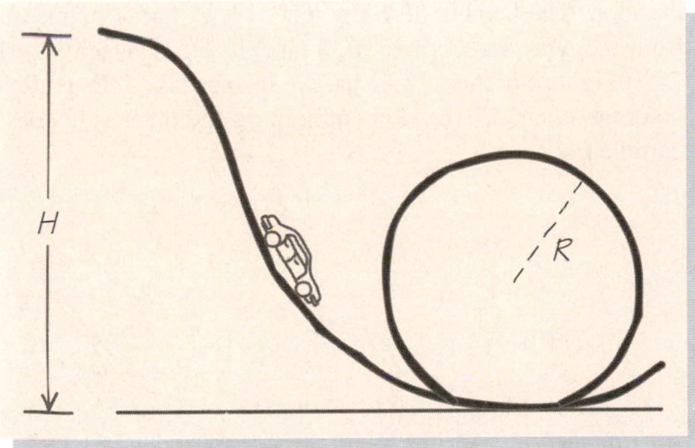

▲ **FIGURE 7–24** Problem 40.

41. (II) A block is constrained to move without friction along the x-axis. The block is attached to a spring of equilibrium length L. The other end of the spring is fixed at a point $x = 0$, $y = h$ (Fig. 7–25), where $h < L$. (a) What is the potential energy of the system? (b) What is the net force acting on the block? Sketch both $U(x)$ and the force.

42. (II) The potential energy of two atoms separated by a distance r may be written in the form $U(r) = U_0 \times [(r_0/r)^{12} - 2(r_0/r)^6]$ (Fig. 7–26). Find the separation r at which there is no force between the atoms. What is the magnitude of the potential energy there?

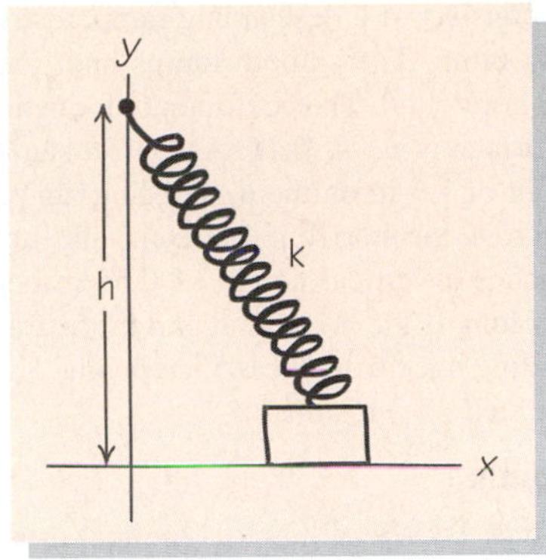

▲ **FIGURE 7–25** Problem 41.

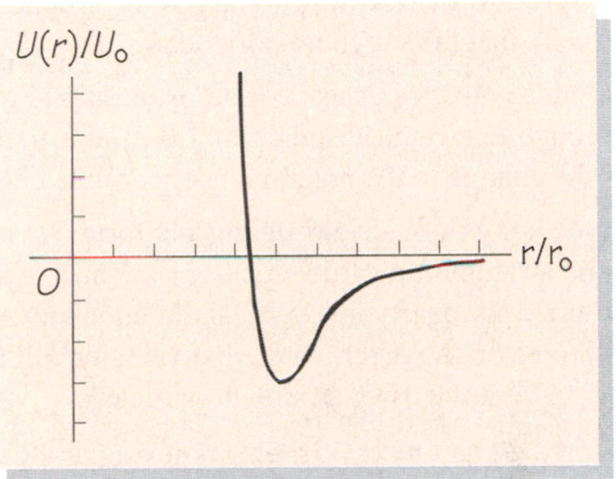

▲ **FIGURE 7–26** Problem 42.

43. (II) The potential energy of a satellite of mass m moving in a circular orbit of radius r about a planet of mass M is given by $U(r) = -GMm/r$, where G is a universal constant. (a) Calculate the force on the satellite. (b) From the equation $\vec{F} = m\vec{a}$, calculate the kinetic energy for a particle in this orbit. (c) What is the total energy?

44. (II) The potential energy of a particle moving in the xy-plane is given by $U(x, y) = a_1x^2 + a_2xy + a_3y^2$, where $a_1 = 3\ \text{J/m}^2$, $a_2 = -14\ \text{J/m}^2$, and $a_3 = 1.5\ \text{J/m}^2$. Calculate the force vector.

45. (III) The ski jumper of Problem 38 has survived the jump and prepares for another jump on a track of a different design: The track takes the jumper through a vertical distance H from its start to its takeoff point, but the jumper can choose the angle at which he leaves the takeoff point. The lip at the edge can be varied so that the jumper leaves the jump at any angle—from taking off horizontally to leaving vertically upward. Show that the angle θ with the horizontal that leads to the maximum-distance jump onto a horizontal plane a distance D below the lip is given by $\sin^2\theta = 1/[2 + (D/H)]$. Ignore all friction and drag effects. (A real landing area is, in fact, sloped downward to allow the jumper to make a smooth landing.)

7–4 Is Energy Conservation a General Principle?

46. (I) A 10-g Ping-Pong ball is dropped with zero initial velocity from a height of 1.0 m, and it bounces back to a height of 0.90 m. What is the work done by the nonconservative forces in this process?

47. (I) A parachutist jumps off a training tower that is 85 m high. She starts at rest and reaches the ground with a vertical speed of 5.0 m/s. How much work was done by the drag forces acting on her given that her mass is 75 kg?

48. (II) A track consists of a descending ramp, a straight track, and an ascending ramp. The smooth ramps both make an angle of 20° with the horizontal. The coefficient of kinetic friction on the horizontal surface is $\mu_k = 0.18$. An object starts from rest at a vertical height of 1.3 m on the descending ramp. It slides down the ramp, across the horizontal stretch, and up the ascending ramp. It reaches a vertical height of 0.55 m before coming to rest. (a) How long is the horizontal part of the track? (b) The object starts sliding back from the 0.55 m height. How far along the horizontal stretch does it slide?

General Problems

49. (I) Assume that 1 kWh of electric energy costs 12 cents. Estimate the cost of lighting a three-room apartment per day.

50. (I) A rock falls off the edge of a cliff moving initially in the horizontal direction with speed 2 m/s. The cliff is 20 m high. How does the speed of the rock, as it reaches the ground, differ from that of a rock that falls with no horizontal velocity?

51. (I) A furniture mover pushes a crate of mass 60 kg up a rough slope through a vertical distance of 1.0 m at a uniform speed. What is the change in the potential energy of the crate?

52. (I) A diver jumps off a rigid diving platform 5.0 m above the water with an initial upward velocity of 2.2 m/s. Assuming that his takeoff is very nearly in a vertical direction and that there are no drag forces on the diver, with what velocity will the diver hit the water? (Treat the diver as a point particle.)

53. (I) The force on an object of mass m moving along the x-axis is given by $F(x) = -ax + bx^2$, where $a = 3$ N/m and $b = 0.2$ N/m^2. (a) Calculate the potential energy function $U(x)$, letting $U(x) = 3$ J at $x = 0$ cm. (b) Sketch $U(x)$ as a function of x from $x = 0$ m to $x = 4.0$ m in steps of 0.5 m.

54. (II) A massless spring hangs vertically in equilibrium with no mass at its end. When a 2.0-kg mass is connected to the bottom, the new equilibrium position is 5.0 cm lower. The mass is then pulled down and released. It is observed that the speed of the mass is 2.1 m/s when the mass passes the original equilibrium position (before the mass was attached). How far down were the mass and spring pulled together when released?

55. (II) Are the following forces conservative or nonconservative? (a) The force $\vec{F}(x) = (ax + bx^3 + cx^4)\hat{i}$. (b) The force $\vec{F}(x, y) = Ax^2\hat{i} + Bxy\hat{j}$.

56. (II) A ball of mass 50 g falls on a hard surface from a height of 1.0 m. After each bounce it reaches 95 percent of the height from which it falls. How much energy is dissipated at the first bounce, at the second bounce, and at the third bounce? Can you generalize to the nth bounce?

57. (II) A mass moving in the xy-plane is subject to a force in the x- and y-directions whose components are given by $F_x = 2Ax^2y$ and $F_y = Axy^2$. Calculate the work done in going in a counterclockwise direction around a square that has corners at $(x, y) = (0,0), (1,0), (1,1), (0,1)$. Can you write a potential energy function that gives rise to this force?

58. (II) An anharmonic spring exerts a force, as a function of displacement from equilibrium, given by $F = -(0.9\text{ N/m})x - (1.4\text{ N/m}^3)x^3$ (Fig. 7–27). This force acts on an object of mass 1.25 kg and displaces it from $x = 0$ m to $x = 3$ m. Consider the object being displaced along two different paths: (a) directly from $x = 0$ m to $x = 3$ m and (b) from $x = 0$ m to $x = -2$ m, then to $x = 7$ m, and back to $x = 3$ m. Show that the work done by the force is the same for both paths.

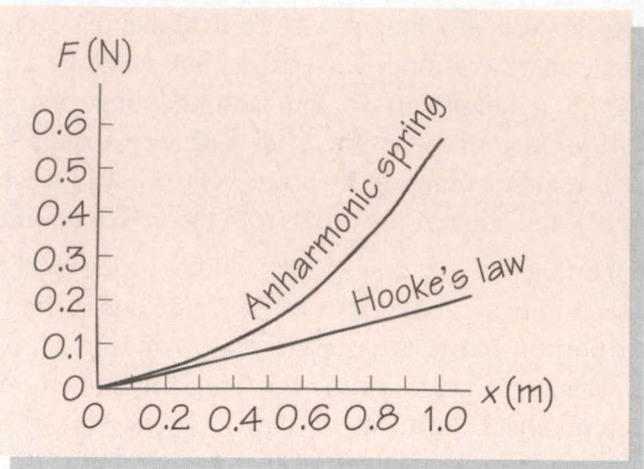

▲ **FIGURE 7–27** Problem 58.

59. (II) A conservative force does 2 J of work in moving a particle from point A to point C via path ABC (Fig. 7–28). The force does -1 J of work to move the particle from D to F, 3 J for E to B, 1 J for E to F, and 1 J for B to C. How much work does the force do as the object moves from C to A, from A to E, from D to C?

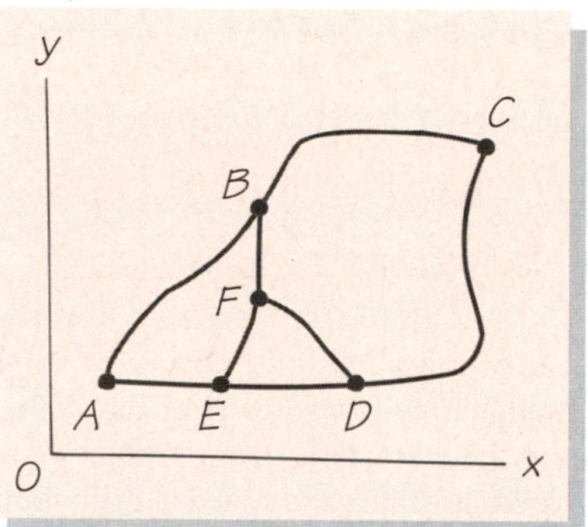

▲ **FIGURE 7–28** Problem 59.

60. (II) A particle of mass 50 g leaves, from rest, point a on a loop-the-loop. The heights of the points a, b, c, and d as measured from the table level are 10, 0, 8, and 12 cm, respectively (Fig. 7–29). Ignore friction. (a) What are the speeds of the particle at points b, c, and d? (b) How high up on the other side does the particle rise?

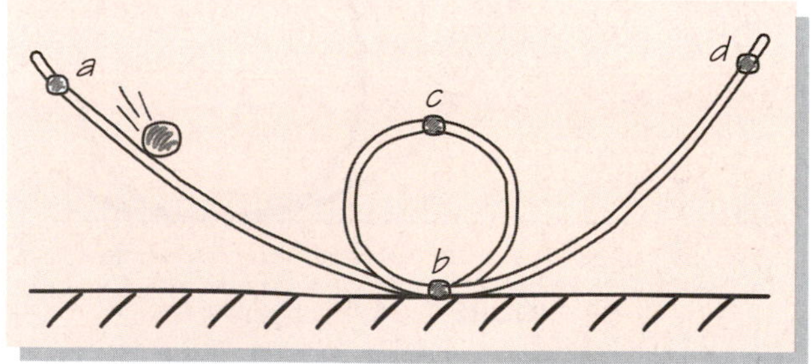

▲ **FIGURE 7–29** Problem 60.

61. (II) A skier skis from rest from a vertical height $h_1 = 18$ m over two successively lower hills of vertical heights $h_2 = 15$ m and $h_3 = 7$ m (Fig. 7–30, see next page). The summit of the third hill fits a circle of radius h_3 centered at height 0 m. Friction with the snow and air resistance are negligible. (a) Find her speeds at x_1, x_2, and x_3. (b) Does the skier leave the surface at x_3? If not, what should h_1 be so that she just leaves the surface at x_3?

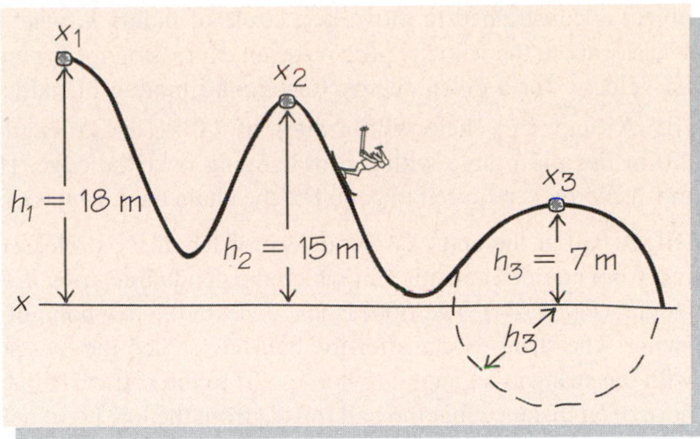

▲ **FIGURE 7–30** Problem 61.

62. (II) A pile driver works by lifting a large mass and dropping it to the ground. The mass is 1300 kg, and it is raised to a height of 6 m above the pile for each stroke. The pile driver encounters resistance of a constant force of 2.5×10^6 N on each stroke. Use a combination of energy conservation and the work–energy theorem to determine how many strokes it takes to drive a pile 5 m into the ground. For which forces involved is it possible to use energy conservation, and for which forces is this not possible?

63. (II) The mass of a simple pendulum of length $L = 1$ m is released with the string originally in a horizontal position. (a) Calculate the speed of the mass at its lowest position. (b) What is the speed when the string makes an angle of 45° with the vertical? (c) Determine the tension in the string in both positions if the mass is 0.2 kg.

64. (II) A simple pendulum of length $L = 1.0$ m and mass 0.20 kg is released from the horizontal position. When the mass is at its lowest point, the string hits a nail a distance h above the mass, so that the mass loops around the nail (Fig. 7–31). How large can h be so that the string of the pendulum remains taut even when the mass loops to a point right above the nail? [*Hint*: The string is taut as long as there is tension in the string.]

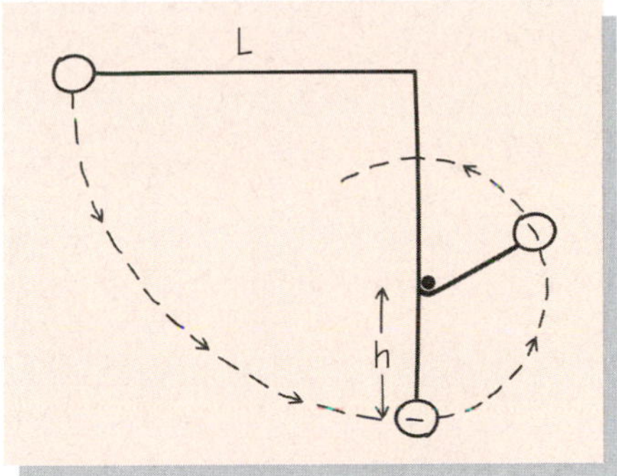

▲ **FIGURE 7–31** Problem 64.

65. (II) An electron is attracted to a proton (the latter is much heavier than the former) with a central force whose magnitude is given by Coulomb's law, $F = C/r^2$, where $C = 2.3 \times 10^{-28}$ kg·m^3/s^2. (a) Is this a conservative force? (b) Write an expression for the potential energy associated with this system. (c) An electron is very far away from a proton and starts from rest. It falls straight toward the proton under the influence of the force. What is the speed of the electron when it is 1.2×10^{-12} m from the proton?

66. (II) A batter hits a baseball and the baseball leaves the bat making a 46° angle with the ground. Air resistance has negligible effect on the trajectory of the baseball, which travels a total horizontal distance of 130 m. (a) What is the speed of the baseball just after it leaves the bat? (b) Use the conservation of the total energy of the baseball to calculate the maximum height to which the baseball rises. (c) What is the speed of the baseball when it has first risen to half its maximum height? (d) When it falls back to half its maximum height?

67. (II) A 3.0-kg block is held against a spring with spring constant $k = 25$ N/cm, compressing the spring 3 cm from its relaxed position. When the block is released, the spring expands and pushes the block upward along a rough surface inclined at a 20° angle (Fig. 7–32). The coefficient of kinetic friction between the block and the surface is $\mu_k = 0.1$. What is the work done on the block (a) by the spring as it extends from its compressed position to its equilibrium position? (b) By friction while the block moves 3 cm as in part (a)? (c) By gravity during the same motion? (d) What is the speed of the block when the spring reaches its equilibrium position? (e) If the block is not attached to the spring, how far up the incline will it slide before it comes to rest? (f) Suppose that the block is attached to the spring so that the spring is extended when the block slides past the equilibrium point. By how much will the spring be extended before the block comes to rest?

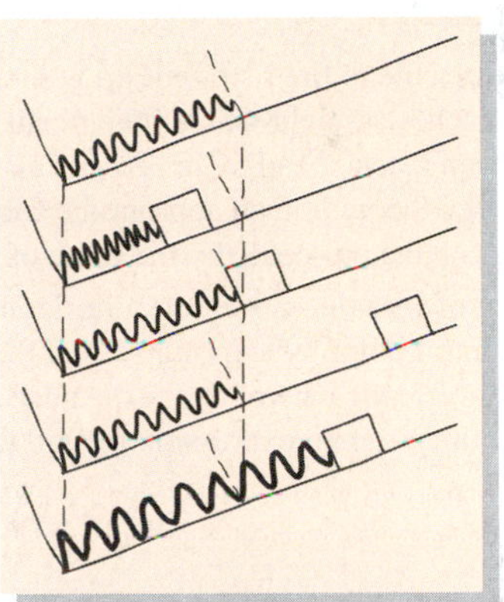

▲ **FIGURE 7–32** Problem 67.

68. (II) A block of wood of mass 1.8 kg is placed on a horizontal table and attached by a massless rope of length 31 cm to a vertical axis that passes through the table. The rope is initially swung around the axis, causing the wood to have a tangential velocity of 3.5 m/s. After one revolution the wood is observed to have a speed of only 2.1 m/s. (a) How much work has friction done on the block of wood during the first revolution? (b) What is the coefficient of kinetic friction between the block of wood and the table? (c) What is the potential energy of the block of wood at the beginning, after one revolution, and when the block of wood comes to rest? (d) How many revolutions does the block of wood make before stopping?

69. (II) Two blocks of mass $m_1 = 5.0$ kg and $m_2 = 2.0$ kg are supported by the system of light frictionless pulleys and massless strings shown in Fig. 7–33 (see next page). Mass m_1 is at rest at a height $h = 0.8$ m above the ground when the system is released. Use the conservation of energy to determine the speed of m_1 when it hits the ground.

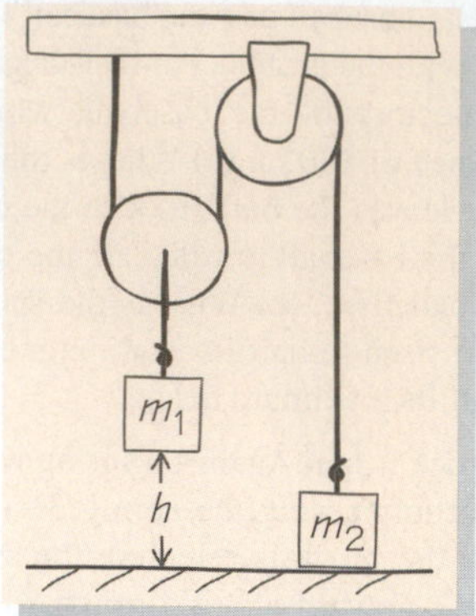

▲ **FIGURE 7–33** Problem 69.

70. (II) A representation of the nuclear force between two nucleons (neutron or proton) is given by the *Yukawa potential energy function* $U(r) = -Ae^{-kr}/r$, where $1/k$ has the approximate value of 10^{-15} m and A is a constant. (a) Plot $U(r)$ versus r in steps of 0.2×10^{-15} m up to 2.4×10^{-15} m. Plot U in units of $A \times 10^{15}$. (b) At what distance is the potential energy a minimum? (c) Determine the force $F(r)$. (d) Determine the force at $r = 0.1 \times 10^{-15}$ m and 10×10^{-15} m.

71. (II) Given the relationship between the force and the potential energy, show that Newton's third law is satisfied if the potential energy has the form $U(x_1 - x_2)$ for two particles located at x_1 and x_2, respectively.

72. (II) A 5 kg projectile is fired straight up with an initial speed of 30 m/s and reaches a height that is 90 percent of the height that would have been reached had there been no air resistance. Apply the work–energy theorem in an appropriate form in order to find the work done on the projectile by the force of the air resistance.

73. (II) An object of mass m moves on a horizontal table. It is attached to a central point by a spring so that the radial force acting on the mass is $-kr$, where r is the distance of the mass to the central point. (a) Write an expression for the total energy. If the object is constrained to move in a circle of radius R, what can you say about the velocity? (c) Write an expression for the angular velocity for a given energy for circular motion of radius R.

74. (III) A (uniform) chain with a mass of 7.0 kg and a length of 2.0 m lies on a table with 0.6 m hanging over the edge. How much energy is required to get all of the chain back on the table?

75. (III) A ball at the end of a pendulum of length L is released at rest from an initial position in which the pendulum string is horizontal (Fig. 7–34). The floor is just beneath the low point of the swing. The string is cut after the ball has passed the low point with the string at an angle $\alpha = \theta - 90°$ to the vertical. Find the horizontal distance that the ball travels from the low point before it bounces on the floor.

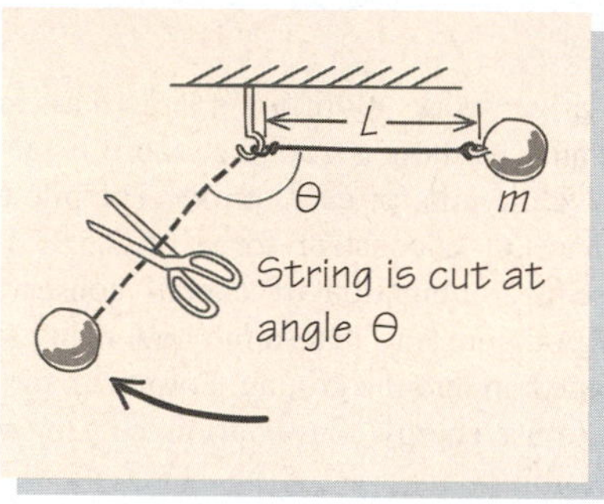

▲ **FIGURE 7–34** Problem 75.

76. (III) The shape of the surface of water in a bucket is determined by the condition that the potential energy per unit mass, as determined in the frame in which the water is at rest, is constant at all places on the surface. Assume that the bucket is set into rotation at an angular speed ω (introduced in Chapter 3) about its vertical central axis and that the water rotates with the bucket. Write a formula that gives the shape of the water surface. [*Hint*: Note that the normal force of the liquid must account for the centripetal acceleration.]

8

The two soccer (football) players collide inelastically with each other, but collide almost elastically with the ball. We will study collisions in this chapter and learn that linear momentum is always conserved in such collisions.

Linear Momentum, Collisions, and the Center of Mass

In the world around us collisions are commonplace—think of raindrops colliding with the ground or ocean waves with the shore, the collision of a golf club with the ball, or the gentle collision of a mother taking a baby in her arms. Behind these everyday scenes are the incessant motion of air molecules and their collisions. Even the distant interactions between stars in the galaxy can be thought of as collisions—slow ones, for the most part. Behind the complexity of these collisions we can find a great simplification, the conservation of a new quantity called the momentum. The momentum is simply formed as the product of the mass and velocity of a particle, and its importance is associated with the fact that Newton's laws are very simply restated in terms of it. In particular, Newton's third law then shows us that momentum is conserved within an isolated system of particles—a system free from net external forces, such as when two hockey pucks sliding on ice collide—no matter how complex the internal interactions of the particles making up the system. In other words, the total momentum of such a system is constant. The conservation of momentum is enormously useful for understanding the behavior of colliding objects, and there are good practical reasons for wanting to understand collisions—much of our information about the world on the atomic scale and below and about the structure of materials comes from observing collisions in one form or another.

As we study the momentum of a system, we will learn that there is a particular point of the system—the *center of mass*—which moves in an especially simple way. For an isolated system, the center of mass moves without acceleration. When external forces act on the system, the center of mass accelerates according to Newton's second law just as a point object does.

8–1 Momentum and Its Conservation

Newton's second law, $\vec{F} = m\vec{a}$, describes how forces change the motion of objects. In previous chapters this law has been expressed in terms of the mass and the acceleration of an object. Another form of the second law is applicable even if the mass changes, as for an airplane when it consumes fuel. This more general form of the second law is

$$\vec{F}_{\text{net}} = \frac{d(m\vec{v})}{dt}. \tag{8–1}$$

The combination mass times velocity, $m\vec{v}$, is called the **linear momentum**, or just **momentum**, of an object. We denote this quantity by $\vec{p}$:

$$\vec{p} \equiv m\vec{v}. \tag{8–2}$$

LINEAR MOMENTUM

The momentum of an object is a vector whose direction is that of the velocity. Its dimensions are those of a mass times a velocity, namely, $[MLT^{-1}]$; in SI, the units of momentum are kilogram-meters per second.

In terms of momentum, the second law [Eq. (8–1)] has the general form

$$\vec{F}_{\text{net}} = \frac{d\vec{p}}{dt}. \tag{8–3}$$

NEWTON'S SECOND LAW

The kinetic energy of an object can also be expressed in terms of the momentum:

$$K = \frac{1}{2}mv^2 = \frac{p^2}{2m}, \tag{8–4}$$

where p is the magnitude of the momentum. From Eq. (8–3) it can be seen that when a large net force acts on an object, the object's momentum will change rapidly and a small net force will result in a slow momentum change.

CONCEPTUAL EXAMPLE 8–1 The same net force acts on a table tennis ball and a bowling ball. Compare the rates at which their momenta change.

Answer This is a bit like the trick question "Which weighs more, a pound of feathers or a pound of nails?" We did not ask for the rate at which the velocity changes—the acceleration—which will be significantly less for the bowling ball than for the table tennis ball. We asked for the rate of change of momentum, and that is precisely given by the force acting [Eq. (8–3)]. The rate of momentum change is the same for each ball since the net force acting on each is the same.

Conservation of Momentum

When objects exert a force on one another, we say they interact. Let's consider the interaction between objects 1 and 2 in both parts of Fig. 8–1. The two objects may be in contact, as in a collision of two billiard balls (Fig. 8–1a), or they may exert a force on each other at a distance, as in the gravitational attraction between Earth and the Moon, or they may be connected by a spring (Fig. 8–1b). Let $\vec{F}_{12}$ denote the force exerted on object 1 by object 2 and let $\vec{F}_{21}$ denote the force exerted on object 2 by object 1 (the first subscript always labels the object that is acted *upon*). Then, Newton's third law states that

$$\vec{F}_{12} = -\vec{F}_{21}. \quad (8\text{–}5)$$

When it is expressed in terms of momentum, Newton's second law [Eq. (8–3)] tells us that the rate of change of each object's momentum is the force acting on it:

$$\frac{d\vec{p}_1}{dt} = \vec{F}_{12}, \quad (8\text{–}6)$$

$$\frac{d\vec{p}_2}{dt} = \vec{F}_{21} = -\vec{F}_{12}. \quad (8\text{–}7)$$

Addition of these two equations leads to

$$\frac{d\vec{p}_1}{dt} + \frac{d\vec{p}_2}{dt} = \vec{F}_{12} - \vec{F}_{12} = 0;$$

$$\frac{d(\vec{p}_1 + \vec{p}_2)}{dt} = 0. \quad (8\text{–}8)$$

As a consequence,

for zero net external force: $\vec{p}_1 + \vec{p}_2 = \text{a constant}.$ (8–9)

CONSERVATION OF MOMENTUM

We will see below that this result is not confined to two objects. In words, Newton's third law implies that

the sum of the momenta of an isolated system of objects is a constant, no matter what forces act between the objects making up the system.

This is called the **principle of conservation of momentum**. Like the principle of conservation of energy, the conservation of momentum is important both as a general principle and as a powerful tool for solving problems.

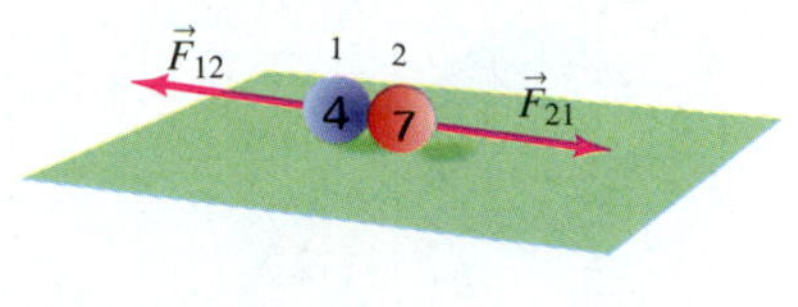

(a)

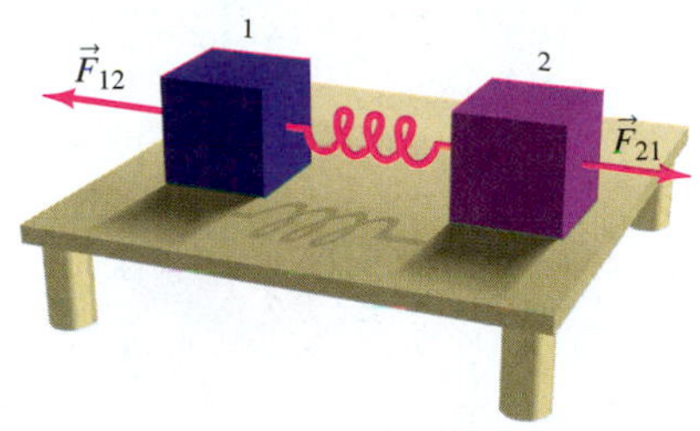

(b)

▲ **FIGURE 8–1** (a) Two billiard balls at the moment of collision and (b) two masses connected by a spring. In each case, the objects exert forces on one another. Here $\vec{F}_{12}$ is the force exerted on object 1 by object 2, and $\vec{F}_{21}$ is the force exerted on object 2 by object 1. In these collisions, the forces are contact forces although, in general, physical contact is not necessary for two objects to exert forces on one another.

EXAMPLE 8–2 A cue ball moves with a velocity of 1.20 m/s in the $+y$-direction on a billiard table and strikes an equally massive ball initially at rest (Fig. 8–2a). The cue ball is deflected so that its velocity has a component of 0.80 m/s in the $+y$-direction and a component of 0.56 m/s in the $+x$-direction (Fig. 8–2b). What is the velocity of the struck ball immediately after the collision?

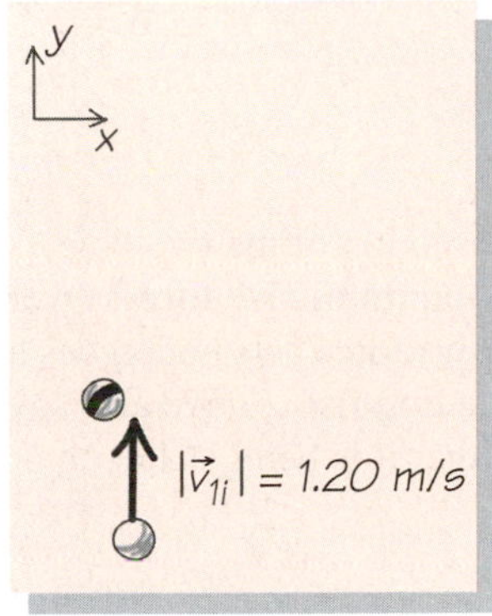

Before

(a)

y
x
$\vec{v}_{1f}$
$\vec{v}_{2f}$
0.80 m/s
+ 0.56 m/s

After

(b)

▲ **FIGURE 8–2** Two colliding billiard balls. (a) Before the collision. (b) After the collision.

Setting It Up The figures label the initial and final velocities of the cue ball as $\vec{v}_{1i}$ and $\vec{v}_{1f}$, respectively, and the final velocity of the struck ball as $\vec{v}_{2f}$. The given velocities are $\vec{v}_{1i} = (1.20\ \text{m/s})\hat{j}$ and $\vec{v}_{1f} = (0.56\ \text{m/s})\hat{i} + (0.80\ \text{m/s})\hat{j}$.

Strategy We can find the initial and final momenta of the sum of the two balls and use the conservation of momentum, which will relate the velocities in question. The only unknown is $\vec{v}_{2f}$.

Working It Out If m is the mass of each ball, the initial momentum is $\vec{p}_i = m\vec{v}_{1i}$ because the struck ball has an initial velocity of zero. Then the conservation of momentum reads

$$m\vec{v}_{1i} = m\vec{v}_{1f} + m\vec{v}_{2f}.$$

This simplifies to $\vec{v}_{1i} = \vec{v}_{1f} + \vec{v}_{2f}$ or $\vec{v}_{2f} = \vec{v}_{1i} - \vec{v}_{1f}$. Numerically,

$$\vec{v}_{2f} = \vec{v}_{1i} - \vec{v}_{1f} = (1.20\ \text{m/s})\hat{j} - [(0.56\ \text{m/s})\hat{i} + (0.80\ \text{m/s})\hat{j}]$$
$$= (-0.56\ \text{m/s})\hat{i} + (0.40\ \text{m/s})\hat{j}.$$

This corresponds to a final speed of $v_{2f} = \sqrt{(v_{2f,x}^2 + v_{2f,y}^2)} = 0.69\ \text{m/s}$.

What Do You Think? In this example we specified some information about the final state (after the collision) as well as all the information about the initial state. If no information had been supplied about the final state, could the statement that kinetic energy is conserved, if it were true here, have allowed a full solution? *Answers to What Do You Think? questions are given in the back of the book.*

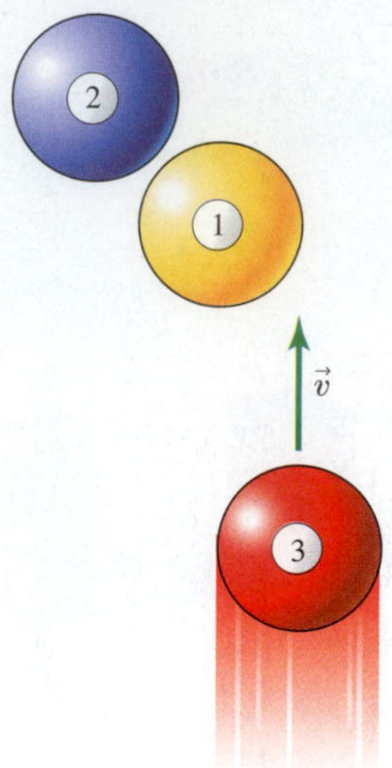

▲ **FIGURE 8–3** System of three interacting objects. Any time one ball touches another, there is a force between those two balls.

Conservation of Momentum for a System of Many Objects: The conservation of momentum is not confined to a system of two interacting objects. Suppose that there are three objects in a system on which no external forces act. [For example, you could imagine three billiard balls on a table, the 1 ball and the 2 ball quite close to one another and the 3 ball coming in to interact with both of them (Fig. 8–3).] Then the total force $\vec{F}_1$ on object 1 is given by the sum of the forces on object 1 due to objects 2 and 3:

$$\vec{F}_1 = \vec{F}_{12} + \vec{F}_{13}. \tag{8–10a}$$

Similarly, the total force $\vec{F}_2$ on object 2 is given by the sum of the forces on object 2 due to objects 1 and 3; so

$$\vec{F}_2 = \vec{F}_{21} + \vec{F}_{23}, \tag{8–10b}$$

and similarly,

$$\vec{F}_3 = \vec{F}_{31} + \vec{F}_{32}. \tag{8–10c}$$

(Note that in Fig. 8–3 the 3 ball may not ever directly hit the 2 ball, so that in this case $\vec{F}_{23}$ and $\vec{F}_{32}$ would be zero.) Adding these three equations and using $\vec{F}_{12} = -\vec{F}_{21}$, $\vec{F}_{13} = -\vec{F}_{31}$, and $\vec{F}_{23} = -\vec{F}_{32}$ yield

$$\begin{aligned}\frac{d\vec{p}_1}{dt} + \frac{d\vec{p}_2}{dt} + \frac{d\vec{p}_3}{dt} &= \vec{F}_1 + \vec{F}_2 + \vec{F}_3 \\ &= \vec{F}_{12} + \vec{F}_{13} + \vec{F}_{21} + \vec{F}_{23} + \vec{F}_{31} + \vec{F}_{32} = 0.\end{aligned} \tag{8–11}$$

Consequently, the sum of the momenta of the three objects is constant throughout the motion:

$$\vec{P} = \vec{p}_1 + \vec{p}_2 + \vec{p}_3 = \text{a constant}. \tag{8–12}$$

We can easily extend this demonstration to N interacting objects and prove that the sum of the objects' momenta is constant throughout the motion.

8–2 Collisions and Impulse

What happens when objects *collide*? The word "collision" evokes the image of an action with a short, sharp contact, such as the collision between two billiard balls or two automobiles. We can more formally think of a collision between objects as an interaction between them—a set of forces—that is limited in time. (Of course, the word "limited" allows us a lot of leeway.) We will want to think of the colliding objects as otherwise isolated in order to be able to apply the conservation of momentum to the situation. Before we do so, it will pay us to study the idea of briefly acting forces, the kind that occur in collisions, in more detail.

Impulsive Forces

We'll suppose that during a collision the force that alters the motion of the two objects is active for only a short time Δt. We refer to such a force as an **impulsive force**. Over the time Δt the momentum of the object on which the impulsive force acts undergoes a momentum change $\Delta\vec{p}$. We'll call this momentum change the **impulse** $\vec{J}$. (We also say that an object receives or gives an impulse according to whether it is being acted on or is the source of the impulsive force, respectively.) We write

$$\vec{J} \equiv \Delta\vec{p} = \vec{p}_f - \vec{p}_i. \tag{8–13}$$

A little calculus allows us to express the impulse for an impulsive force in terms of that force. The change in momentum is the integral of the rate of change of momentum between the initial and final times that the force acts, and from Newton's second law, the rate of change of momentum is the (impulsive) force $\vec{F}$ that acts on the object:

$$\vec{J} = \int_{t_i}^{t_f} \left(\frac{d\vec{p}}{dt}\right) dt = \int_{t_i}^{t_f} \vec{F}\, dt. \qquad (8\text{–}14)$$

IMPULSE AND FORCE

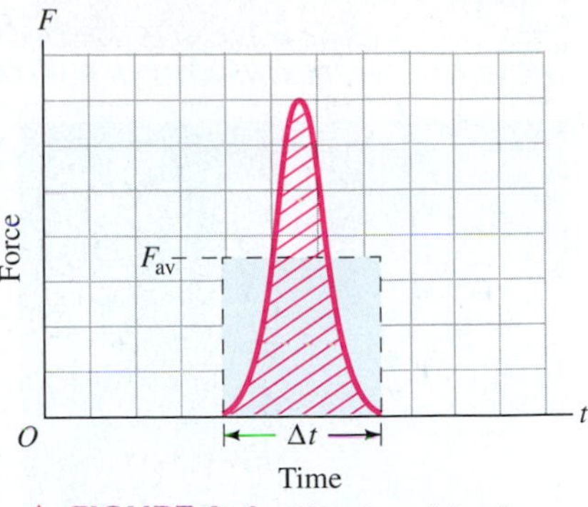

▲ **FIGURE 8–4** If an impulsive force acts on an object, its momentum will be changed. That force acts over a time period Δt, varying with time as it acts. The same change in momentum is produced by another force that is constant over Δt and takes the actual force's average value over that time period.

We have used the fact that the force is zero outside the time interval $\Delta t = t_f - t_i$.

Because the integral of force over time is the area under a curve of force versus time, we can also represent the impulse as the product of the time interval Δt and a quantity that we may call the *average force* $\vec{F}_{av}$ (Fig. 8–4). The average force refers to the average value of the force over the time interval Δt. In this case, the impulse or change in momentum can be written in the form

$$\Delta\vec{p} = \vec{J} = \int_{t_i}^{t_f} \vec{F}\, dt = \vec{F}_{av}\, \Delta t. \qquad (8\text{–}15)$$

Remember that this equation is a *vector* equation. In Example 8–3, however, the vector aspect does not play a crucial role.

EXAMPLE 8–3 The magnitude of the average force exerted by a bat on a baseball during the time of contact (a period of 2.00×10^{-3} s) is 6660 N. The mass of the baseball is 0.145 kg and its speed is 33.5 m/s just before the bat collides with it. What is the velocity of the ball when it leaves the bat? Assume that the ball leaves the bat along the same line of direction from which it is pitched.

Setting It Up Figure 8–5 illustrates this problem. The x-axis is horizontal and to the right. We know the average force magnitude F_{av} and the time Δt over which it is applied. We are also given the ball's mass m and its initial speed v_0. The motion is one dimensional, in a direction that we label as the x-axis, with the original direction of the ball—the direction of the pitch—in the $+x$-direction. We want to find the final velocity of the ball.

Strategy Because we know both F_{av} and Δt, we can find the ball's impulse—the change in its momentum. Then we can use the value of the impulse to find the ball's final momentum knowing its initial momentum. We can then use its final momentum to find its final speed.

Working It Out The momentum change $\vec{p}_f - \vec{p}_i$ of the ball is given by Eq. (8–15), with $\vec{p}_i = mv_0\hat{i}$. The direction of the force (and hence of the impulse) is in the $-x$-direction, so from Eq. (8–15),

$$\vec{p}_f = \vec{p}_i + \vec{F}_{av}\,\Delta t = (mv_0 - F_{av}\,\Delta t)\,\hat{i}.$$

The final velocity is the final momentum divided by m:

$$\vec{v}_f \equiv v_f\hat{i} = \frac{\vec{p}_f}{m} = \frac{mv_0 - F_{av}\,\Delta t}{m}\,\hat{i} = \left(v_0 - \frac{F_{av}\,\Delta t}{m}\right)\hat{i}.$$

Thus the final velocity is oriented along the x-axis, with x-component

$$v_f = +33.5\text{ m/s} - \frac{(6660\text{ N})(2.00 \times 10^{-3}\text{ s})}{0.145\text{ kg}} = -58.4\text{ m/s}.$$

The minus sign indicates that the ball moves in the negative direction, back toward the pitcher.

What Do You Think? Which sort of racket will allow you to return a tennis ball with more velocity: an ordinary strung tennis racket or a solid wooden paddle of the same shape and weight?

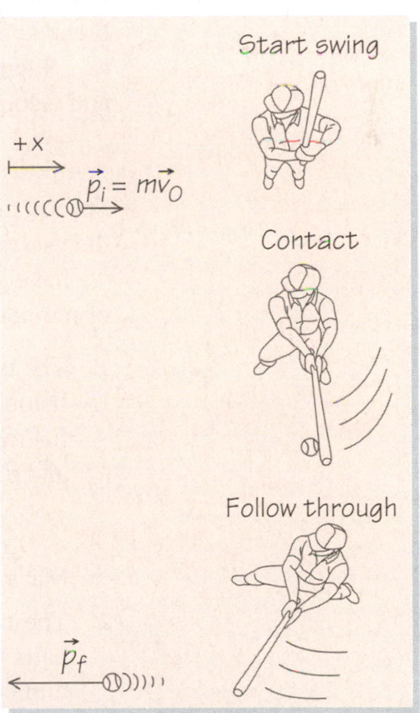

▲ **FIGURE 8–5** The direction of the impulse imparted to a ball hit by a bat is to the left, in the $-x$-direction.

THINK ABOUT THIS . . .

HOW ARE THE FOUNDATIONS FOR TALL BUILDINGS MADE?

▲ **FIGURE 8–6** The pile driver is one of the devices that enables us to build large structures.

One way of constructing foundations (Fig. 8–6) is to drive vertical "piles" into the soil down to bedrock and then anchor the building to them. These piles are often made of steel. A pile driver is used to force them into the ground, and momentum and impulse are key to the operation of this device. In Fig. 8–6, there is a weight sitting within a cylinder that encloses the top of the pile. The weight is lifted by exploding fuel and then drops back onto the pile. During the fall, the weight gains momentum. When it reaches the top of the pile, it is brought to rest in an exceedingly short time (determined by the compressibility of the weight and the pile, and the distance the pile moves). Thus the impulse to the dropped cylinder is large, and by Newton's third law there is an equal and opposite impulse to the pile. In turn, the pile delivers an impulse to the rock and soil beneath it, breaking the rock and soil and allowing the pile to move down until friction and normal forces bring it back to a stop. The action is repeated, driving the pile further down with each repetition. ■

Classification of Collisions

Attempts to understand collisions were carried out by Galileo and his contemporaries. The description of collisions in one dimension were formulated by John Wallis, Christopher Wren (best remembered today as an architect), and Christian Huygens in 1668, and the principle of the conservation of momentum plays a central role in understanding their results.

We can recap our previous discussion on the collision of two objects as follows: The two objects move freely before the collision—no net forces act on either of them—and each has its own constant momentum. During the brief interaction, their individual momenta change because each object experiences an impulsive force due to the other object. After the collision, the two objects are again free but have momenta that differ from those they had before the collision. However, the impulses of the two objects are equal and opposite because the forces each exerts on the other are equal and opposite, so that the change in the momentum of one object is equal and opposite to the change in momentum of the other. In other words, *the sum of their momenta is unchanged*. This feature, the conservation of total momentum of the isolated system, provides a governing constraint. It will hold even if the objects stick together or, at the other extreme, break apart into many pieces.

Suppose that initially object 1 has mass m_1 and velocity $\vec{v}_1$ and object 2 has mass m_2 and velocity $\vec{v}_2$. The total initial momentum is given by

$$\vec{p}_{\text{init}} = m_1\vec{v}_1 + m_2\vec{v}_2. \tag{8–16}$$

Several distinct and interesting possibilities for what the final state can look like present themselves. Figure 8–7 shows these cases and we enumerate them below. In the figure, we have *drawn* the collision in one dimension, although we'll express the conservation of momentum in more general form.

1. The two masses hit each other and stick together, coalescing into one, as in the collision of two blobs of putty or a comet colliding with a planet. Figures 8–7a and b illustrate the before and after for this case. If the mass of the single final object is M and its velocity is $\vec{v}$, momentum conservation reads

$$M\vec{v} = m_1\vec{v}_1 + m_2\vec{v}_2. \tag{8–17}$$

   Mass conservation gives us the additional information[†] that $M = m_1 + m_2$.

2. The two masses can remain distinct and unchanged, as in the collision of billiard balls (Fig. 8–7c). We label the final velocities $\vec{v}_1'$ and $\vec{v}_2'$, respectively, and momentum conservation takes the form

$$m_1\vec{v}_1' + m_2\vec{v}_2' = m_1\vec{v}_1 + m_2\vec{v}_2. \tag{8–18}$$

[†] Because special relativity plays an important role in nuclear or subnuclear collisions, the conservation of mass per se is only an approximation, and sometimes a very bad one, in those cases.

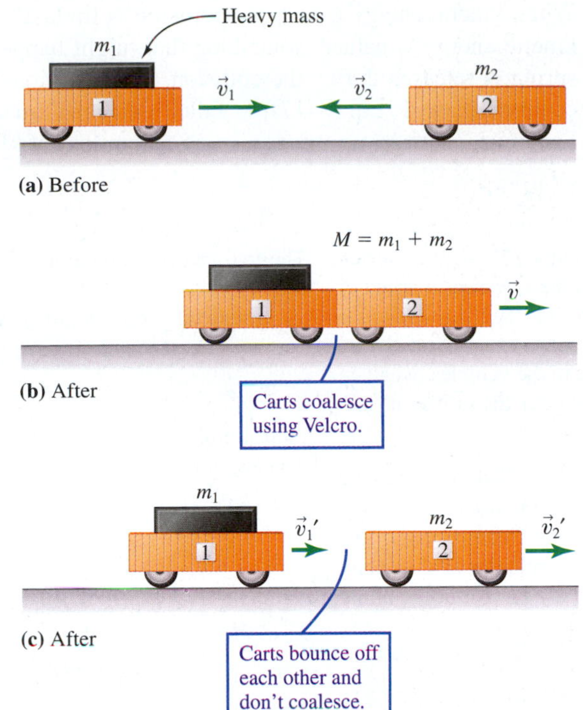

FIGURE 8–7 A collision between two objects, as shown before the collision in part (a), can have several outcomes. The two objects shatter into several pieces (not shown). (b) The two masses combine as the objects coalesce into one object. (c) Two objects leave the collision, each with the mass of the original objects respectively, drawn for the case of object 1 continuing to move in the original direction.

3. Mass can be transferred from one object to the other such that after the collision one would have two objects with masses m_3 and m_4. As an example, a carbon atom can collide with a molecule of carbon dioxide to make two carbon monoxide molecules. We label the final velocities of masses m_3 and m_4 as $\vec{v}_3$ and $\vec{v}_4$, so that momentum conservation reads

$$m_3\vec{v}_3 + m_4\vec{v}_4 = m_1\vec{v}_1 + m_2\vec{v}_2. \qquad (8\text{–}19)$$

In this case mass conservation would add the information that $m_3 + m_4 = m_1 + m_2$.

4. One or both of the objects can shatter into several pieces (a more complex possibility that we won't deal with in detail).

The vector equations above are indispensable tools for understanding collisions. However, even given all the information about the initial state, they are generally not enough to determine everything about the final state. Only in the case of coalescence [Eq. (8–17)] are the three vector equations sufficient to determine the three components of the single final velocity. In all the other cases, more information is required to understand the details of the motion, and this is usually information about the energy of the objects.

Energy Considerations in Collisions

The degree to which the kinetic energy is conserved in a collision provides us with another piece of information that we can use to help us understand the process. Although for the reasons described in Chapter 7 the *total* energy is always conserved in a collision, some of the energy may be *dissipated* in ways that make it lost to us. For example, when friction is present, some of the energy goes into heating. To take another example, in a collision between two cars, some energy goes into the crumpling of metal, and as a result there is less energy available for motion—the kinetic energy after the collision. If the kinetic energy decrease is known (as in Example 8–4 below), we can use this information to help us calculate the details of the motion after a collision.

The degree to which kinetic energy is conserved provides another way to classify collisions. Kinetic energy is conserved if all the initial kinetic energy of the two colliding objects goes into the kinetic energy of the objects present after the collision, and we call the collision **elastic**. (We are assuming the objects retain their identities as well, as in case 2 above, although that case also includes the possibility that energy is lost.)

When kinetic energy is lost, the collision is **inelastic**—we would also use this term if kinetic energy is gained, something that might happen if the equivalent of an internal spring is released during the collision. The situation in which two objects collide and coalesce [case 1, Eq. (8–17)] is called **perfectly inelastic** because, as we shall see in Section 8–6, it corresponds to the maximum loss of kinetic energy.

EXAMPLE 8–4 A 14,000-kg truck and a 2000-kg car have a head-on collision. Despite attempts to stop, the truck has a speed of 6.6 m/s in the $+x$-direction when they collide and the car has a speed of 8.8 m/s in the $-x$-direction. If 10 percent of the initial total kinetic energy is dissipated through damage to the vehicles, what are the final velocities of the truck and the car after the collision? Assume that all motion takes place in one dimension.

Setting It Up Figure 8–8 shows the collision, with the $+x$-direction to the right. The given masses of the truck and car are M and m, respectively. The known initial velocity component of the truck is V_i and that of the car is v_i. We want to find the final velocity components of the truck V_f and the car v_f if 10 percent of the initial kinetic energy is lost.

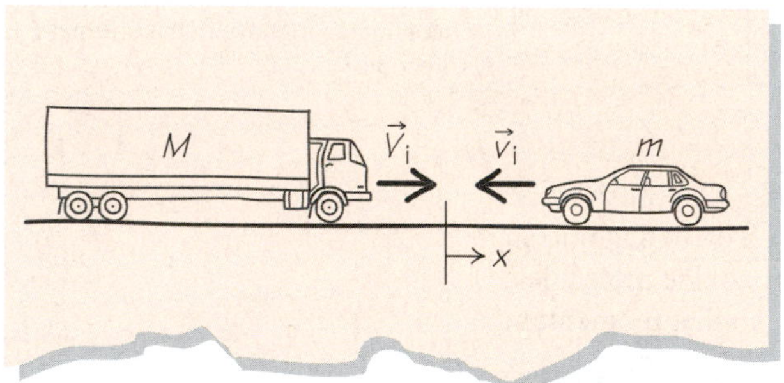

▲ FIGURE 8–8 Head-on collision between large truck and small car.

Strategy We can use the conservation of momentum, which, because the motion is one dimensional, is a single equation for the x-component of velocity. The energy information we have is that the final mechanical energy (all kinetic) is 90 percent of the initial energy. Both the initial momentum and initial energy are easily calculated from the given information. Thus our two conditions should be enough to solve for the two unknowns.

Working It Out Momentum conservation reads

$$MV_f + mv_f = MV_i + mv_i \equiv p_i.$$

The initial total energy is the sum of the kinetic energies of the two vehicles, $K_i = (MV_i^2/2) + (mv_i^2/2)$. The sum of the final kinetic energies is 90 percent of this quantity, so

$$K_f = (MV_f^2/2) + (mv_f^2/2)$$
$$= 0.9[(MV_i^2/2) + (mv_i^2/2)] \equiv 0.9K_i.$$

The two centered equations above are two algebraic equations for the two unknowns V_f and v_f.

Momentum conservation directly yields

$$V_f = \frac{p_i - mv_f}{M}.$$

Substituting this value into the energy relation, we have a quadratic equation for v_f:

$$\frac{1}{2}M\left(\frac{p_i - mv_f}{M}\right)^2 + \frac{1}{2}mv_f^2 - 0.9K_i = 0;$$

$$\frac{1}{2}M\frac{p_i^2}{M^2} - \frac{1}{2}M\left(\frac{2p_imv_f}{M^2}\right) + \frac{1}{2}M\frac{m^2v_f^2}{M^2} + \frac{1}{2}mv_f^2 - 0.9K_i = 0;$$

$$(m^2 + Mm)v_f^2 - 2mv_fp_i + p_i^2 - 2(0.9)MK_i = 0.$$

This equation has solutions

$$v_f = \frac{2mp_i \pm \sqrt{(2mp_i)^2 - 4(m^2 + Mm)[p_i^2 - 2(0.9)MK_i]}}{2(m^2 + Mm)}.$$

When numbers are inserted for the two possible solutions represented by the $\pm$ sign, the solution with a minus sign gives a negative velocity for the car and a positive velocity for the truck. This means that the car continues its motion to the left, going "through" the truck, while the truck similarly goes "through" the car. Since this is not possible, the correct solution has the plus sign. Inserting numbers, we find

$$v_f = 17 \text{ m/s} \quad \text{and} \quad V_f = 2.9 \text{ m/s}.$$

The truck continues in the $+x$-direction with a speed less than its initial speed; the car has completely reversed its direction and is moving even faster than its initial speed. The car has "bounced" from the much more massive truck, much like a tennis ball against a tennis racket. Analysis of this type is used by crash-scene investigators. Adding information such as stopping distance under friction, they can learn, for example, the speeds and directions of the vehicles just before the crash.

What Do You Think? Assuming that they are protected from injury directly associated with the automobile collapsing on them, which would be better for the occupants of the car, an elastic or an inelastic collision?

8–3 Perfectly Inelastic Collisions; Explosions

Perfectly Inelastic Collisions

There is a range of possible collisions from elastic to inelastic to perfectly inelastic, according to how much kinetic energy is lost. The simplest is the case of *perfectly inelastic* collisions in one dimension, which are those in which the objects coalesce as a result of the collision. For example, an asteroid hitting Earth would be a perfectly inelastic collision. Momentum conservation for these collisions are described by Eq. (8–17).

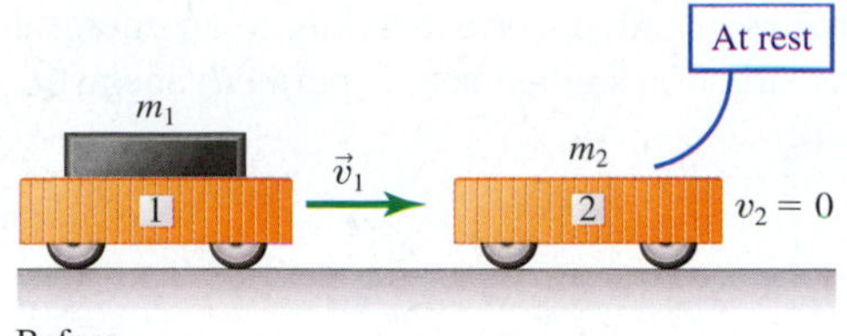

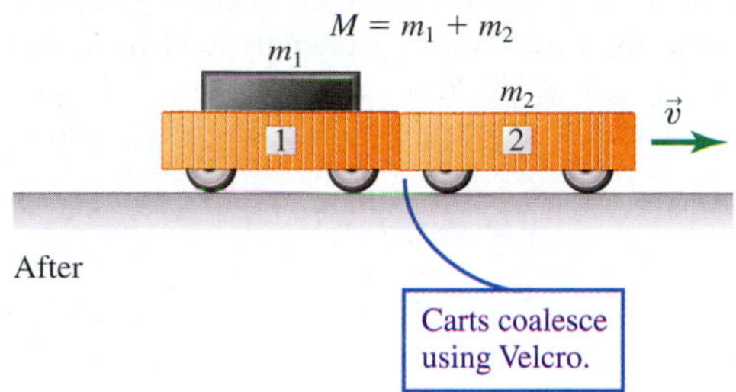

FIGURE 8–9 A cart with mass m_1 and velocity $\vec{v}_1$ collides with a cart of mass m_2 at rest. The two carts stick together and move on together with velocity $\vec{v}$.

Mass conservation implies that $M = m_1 + m_2$. We can divide Eq. (8–17) by this factor. The velocity of the coalesced object then becomes (in one-dimensional motion)

$$v = \frac{m_1 v_1 + m_2 v_2}{M}. \tag{8–20}$$

Let's analyze this result for some special cases.

If one of the objects (m_2) is at rest $(v_2 = 0)$ and the other (m_1) runs into it (Fig. 8–9), then

$$v = \frac{m_1}{M} v_1. \tag{8–21}$$

If $m_1 \gg m_2$, the "composite" object will move with a velocity nearly equal to that of the initially moving object; a car colliding with a bug does not slow down very much. In contrast, when $m_1 \ll m_2$, as when a stationary athlete catches a ball, we get the opposite effect; that is, the athlete will recoil with only a low velocity, just the fraction $m_1/(m_1 + m_2) \cong m_1/m_2$ of the velocity of the ball.

Next, consider the case of a head-on collision in which the two objects have equal and opposite velocities $(v_2 = -v_1)$. In this case, Eq. (8–20) becomes

$$v = \frac{m_1 - m_2}{m_1 + m_2} v_1. \tag{8–22}$$

In the special case that $m_1 = m_2$, the two objects have equal and opposite *momenta* because

$$m_1 v_1 + m_2 v_2 = m_1 v_1 + m_1 v_2 = m_1(v_1 + v_2) = 0. \tag{8–23}$$

In that case, the final momentum must be zero and thus $v = 0$, as Eq. (8–22) verifies. The objects collide and come to rest.

Energy Loss in Perfectly Inelastic Collisions: Let's find the change in energy for the collision described above. Before the collision, the total energy E_i is the sum of the kinetic energies of the two objects. The final energy E_f is the kinetic energy of the composite object of mass $M = m_1 + m_2$. The change in energy $\Delta E = E_f - E_i$ is, using Eq. (8–20),

$$\begin{aligned}
\Delta E &= \frac{1}{2}Mv^2 - \left(\frac{1}{2}m_1 v_1^2 + \frac{1}{2}m_2 v_2^2\right) \\
&= \frac{1}{2}\frac{M(m_1 v_1 + m_2 v_2)^2}{M^2} - \left(\frac{1}{2}m_1 v_1^2 + \frac{1}{2}m_2 v_2^2\right) \\
&= \frac{1}{2}\frac{m_1^2 v_1^2 + 2m_1 m_2 v_1 v_2 + m_2^2 v_2^2 - M(m_1 v_1^2 + m_2 v_2^2)}{M} \\
&= \frac{1}{2}\frac{m_1 m_2(-v_1^2 - v_2^2 + 2v_1 v_2)}{M} = -\frac{1}{2}\frac{m_1 m_2}{M}(v_1 - v_2)^2.
\end{aligned} \tag{8–24}$$

The right-hand side of Eq. (8–24) is always negative, corresponding to an energy loss, or to an inelastic collision. As to why the situation studied here is *perfectly* inelastic, see the Think About This box in Section 8–6.

EXAMPLE 8–5 A dog who jumps into the interior of a stationary ice boat is moving at $v_1 = 32$ km/h when he enters the boat (Fig. 8–10), and his landing on the boat can be regarded as a collision. The dog's mass is 14 kg and that of the boat plus boater is 160 kg. You can assume all the motion is horizontal. (a) Assuming that the ice surface is frictionless, what is the velocity of the boat after the collision? (b) What is the ratio of the energy loss to the initial energy?

▲ **FIGURE 8–10** A dog jumping into the boat changes the boat's momentum.

Setting It Up We specify that the motion is in the x-direction by adding an axis to Fig. 8–10. We know the dog's mass m_1 and that of the boat and person, m_2. We also know the dog's velocity before the collision, or equivalently its x-component v_1, as well as the boat's ($v_2 = 0$). We want the velocity x-component v of the boat after the collision and the ratio of the energy loss ΔE to the initial kinetic energy K_i.

Strategy This is a perfectly inelastic collision, so conservation of momentum applied to the collision that occurs when the dog enters the boat will determine the boat's velocity after the collision. With all speeds known, we are left with a direct calculation of the kinetic energy before and after the collision to find the needed energy ratio.

Working It Out (a) The initial momentum has only the x-component,

$$p_i = m_1 v_1.$$

The final momentum is given by $p_f = Mv$, where $M = m_1 + m_2$. Equating p_f and p_i, we find $Mv = m_1 v_1$, or

$$v = \frac{m_1 v_1}{M} = \frac{(14\text{ kg})(32\text{ km/h})}{174\text{ kg}} = 2.6\text{ km/h} = 0.72\text{ m/s}.$$

(b) The initial energy is the kinetic energy of the dog, $K_i = \frac{1}{2}m_1 v_1^2$. The final energy is again all in the form of kinetic energy:

$$K_f = \frac{1}{2}Mv^2 = \frac{1}{2}M\left(\frac{m_1 v_1}{M}\right)^2 = \frac{1}{2}\left(\frac{m_1}{M}\right)m_1 v_1^2 = \frac{m_1}{M}K_i.$$

Thus the energy loss is given by

$$\Delta E = K_i - K_f = K_i - \frac{m_1}{M}K_i = K_i\left(1 - \frac{m_1}{M}\right),$$

and the ratio of the energy loss to the initial energy is

$$\frac{\Delta E}{K_i} = 1 - \frac{m_1}{M} = \frac{M - m_1}{M} = \frac{m_2}{M},$$

a number less than 1. The energy has decreased. Numerically,

$$\frac{\Delta E}{K_i} = \frac{160\text{ kg}}{174\text{ kg}} = 0.92.$$

What Do You Think? Kinetic energy is lost. Which of the following is correct? (a) It went into gravitational energy. (b) It went into the dog and boater "giving and stretching" as the dog lands. (c) It went into melting the ice.

EXAMPLE 8–6 A 10-g bullet is fired in the $+x$-direction into a stationary block of wood that has a mass of 5.0 kg. The speed of the bullet before entry into the wood block is 500 m/s. What is the speed of the block just after the bullet has become embedded? What distance will the block slide on a surface with a coefficient of friction equal to 0.30?

Setting It Up We have drawn the situation in Fig. 8–11. Positive x is to the right. We label the known mass of the bullet as m_1, its known initial velocity as $\vec{v}_1 = v_1\hat{i}$, and the known total mass of block and bullet as M. We want to find the distance d the block will slide on a surface for which there is a known coefficient of kinetic (sliding) friction μ_k.

Strategy We can use conservation of momentum to find the velocity of the block and bullet immediately after the collision, then Newton's second law to find the friction-induced negative acceleration of the block. Knowing the acceleration and the initial condition, kinematic relations will give us the distance d the block moves.

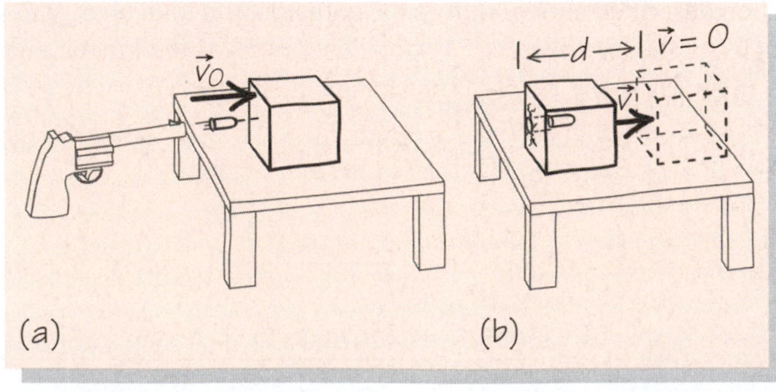

▲ **FIGURE 8–11** A bullet fired into a stationary block of wood moves the block.

Working It Out The x-component of the initial momentum is $p_i = m_1 v_1$, while the momentum immediately after the collision is Mv. Momentum conservation gives

$$m_1 v_1 = Mv;$$

$$v = \frac{m_1}{M} v_1 = \frac{10\ \text{g}}{5010\ \text{g}}(500\ \text{m/s}) = 1.0\ \text{m/s}.$$

We now turn to the problem of finding the acceleration given the value v of the initial speed of the bullet–block composite. The normal force N on the block from the table has magnitude Mg, so the force of friction between block and table is $-\mu_k N = -\mu_k Mg$. (The minus sign indicates that friction points to the left, along the $-x$-direction.) The friction force has constant magnitude and leads to a constant acceleration a of the block, according to Newton's second law, namely $F_{net} = Ma$ reads $-\mu_k Mg = Ma$, or

$$a = -\mu_k g.$$

The negative sign means the block slows down, traveling a distance d before it stops (Fig. 8–11b). Because the acceleration is uniform, we can use the relation $v_f^2 - v_i^2 = 2ad$ [from Eq. (2–24)]. With $v_f = 0$ and the initial speed $v_i = v$ above, we have

$$d = -\frac{v_i^2}{2a} = \frac{1}{2}\frac{v^2}{\mu_k g} = \frac{1}{2}\frac{(1.0\ \text{m/s})^2}{(0.30)(9.8\ \text{m/s}^2)} = 0.17\ \text{m}.$$

What Do You Think? In what way could we have used the fact that the mass of the bullet is much less than the mass of the block?

Explosions

Imagine that we were to film a perfectly inelastic collision in a frame of reference in which the total momentum is zero. In this reference system, the two objects approach each other and merge, leaving a composite object at rest. If we ran the film in reverse, it would look like a film of an explosion. The "initial" object of mass $M = m_1 + m_2$, at rest, breaks up into two objects, m_1 and m_2, with their total momentum equal to zero,

$$m_1 \vec{v}_1 + m_2 \vec{v}_2 = 0. \tag{8–25}$$

Energy conservation tells us that an explosion is possible if there is an initial potential energy U within the "unexploded" system that can be converted into kinetic energy. For the case we are referring to here, this could be as simple as a compressed spring between two masses that is then released. We'll have

$$U = \tfrac{1}{2} m_1 v_1^2 + \tfrac{1}{2} m_2 v_2^2. \tag{8–26}$$

An explosion can involve many more than two objects in the final state. In the explosion of, say, dynamite, the potential energy is stored in its molecules, in the form we call chemical energy. To take another example, Fig. 8–12 shows the remnants of a stellar explosion far from Earth. In these more complicated cases, we will always have the overriding simplicity that the initial momentum of the system before it explodes is the same as the sum of the momentum of all the fragments after the explosion. Let's next take a look at an explosion that occurs when an unstable atomic nucleus disintegrates—a nuclear decay.

▲ **FIGURE 8–12** A small portion of the Cygnus Loop supernova blast wave passes through clouds of interstellar gas. The collision heats and compresses the gas, which causes the glow. Such images taken by the Hubble Space Telescope reveals the structure of the interstellar medium.

EXAMPLE 8–7 One type of polonium nucleus (symbol ^{210}Po), with mass 3.49×10^{-25} kg, can decay into an α particle (actually a helium nucleus), mass 6.64×10^{-27} kg, and a certain type of lead nucleus (symbol ^{206}Pb), mass 3.42×10^{-25} kg:

$$^{210}\text{Po} \rightarrow \alpha + {}^{206}\text{Pb}.$$

In this process, the final decay products have a kinetic energy of 8.65×10^{-13} J if the polonium nucleus decays at rest, the situation we consider here. What are the speeds of the α particle and the lead nucleus?

Setting It Up We show the decay in Fig. 8–13. With a decay from rest into two bodies, the momenta of the two decay products must go off back to back because that is the only way they can add to zero. We call this direction the x-axis. We label the known masses of the alpha particle, the polonium nucleus, and the lead nucleus as M_α, M_{Po}, and M_{Pb}, respectively, and we denote the known kinetic energy in the final state as Q. We want to find the speeds v_α and v_{Pb} of the alpha particle and lead nucleus, respectively.

Strategy Conservation of (one-dimensional) momentum is a single equation involving both speeds, as is the known final energy value. Therefore we have two equations for the two unknowns.

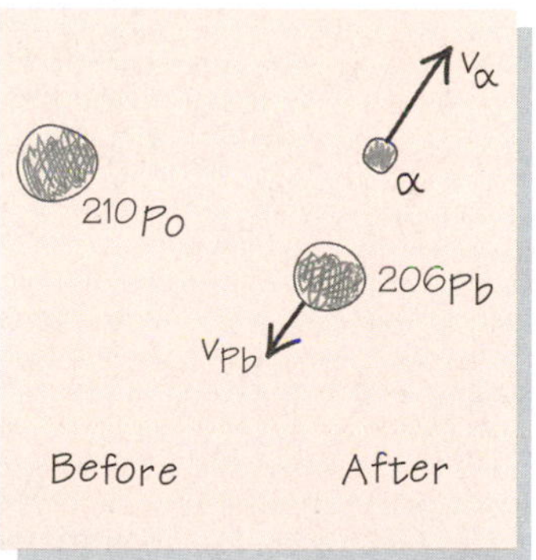

▲ **FIGURE 8–13** Decay of a nucleus. Before: Original nucleus. After: Two outgoing fragments that result from the decay (explosion).

Working It Out Conservation of momentum and the expression for the kinetic energy read, respectively,

$$M_\alpha v_\alpha = M_{Pb} v_{Pb};$$
$$Q = \tfrac{1}{2} M_\alpha v_\alpha^2 + \tfrac{1}{2} M_{Pb} v_{Pb}^2.$$

These two equations can be solved for the two variables v_α and v_{Pb}.

(continues on next page)

We find that the speeds are

$$v_\alpha = \sqrt{\frac{2Q}{M_\alpha(1 + M_\alpha/M_{Pb})}},$$

$$v_{Pb} = \sqrt{\frac{2Q}{M_{Pb}(1 + M_{Pb}/M_\alpha)}}.$$

Substitution of the known numerical values gives $v_\alpha = 1.60 \times 10^7$ m/s and $v_{Pb} = 3.10 \times 10^5$ m/s. The speed of the α is about 5 percent of the speed of light, and this is where special relativity begins to play a role—we have ignored that here of course.

This decay is a form of radioactivity; the fragments could conceivably be dangerous. A relevant element is the energy carried by the two fragments, and you should be able to calculate the kinetic energy of each one.

What Do You Think? An object at rest explodes into two fragments of unequal mass. Which of the following is true? (a) The lighter fragment could have the same speed as the heavier fragment but most often moves more quickly. (b) The lighter fragment always moves off more quickly. (c) The lighter fragment can move off more slowly, depending on the details of the explosion.

THINK ABOUT THIS . . .

HOW DOES A JET ENGINE WORK?

▲ **FIGURE 8–14** A jet engine propels an airplane through the operation of the conservation of momentum.

Conservation of momentum is one way to understand the operation of a jet engine (Fig. 8–14). Air is brought into the front of the engine by intake fans. The air, containing oxygen molecules (O_2), is then mixed with fuel. Among chemical reactions that occur during the ensuing combustion, there is the production of two water molecules (H_2O) for each O_2. This doubles the volume of that part of the oxygen from the air that combines with the hydrogen. Other combustion reactions leave the number of molecules and hence the volume of the air unchanged, and some part of the air, principally the nitrogen, undergoes no chemical reaction. Because of the water-producing reaction, the net effect is that a bigger volume of gas must leave the engine than entered it. To be able to keep up the continuous action of the engine, the outgoing gas must therefore leave with a velocity greater than the incoming velocity; the combustion provides the necessary kinetic energy to enable this to happen. Furthermore, the mass of the departing gas is larger, since the mass of the fuel used in combustion has been added to it. Thus the outgoing gas has substantially larger momentum in the backward direction than the incoming gas. To conserve the momentum of the entire system, the airplane gains momentum in the forward direction. The force that makes the escaping gas accelerate out the back has its third law partner in a forward force on the airplane. ■

8–4 Elastic Two-Body Collisions in One Dimension

Let's continue to work in one dimension. As usual, the word "velocity" will mean the velocity *component* in the direction of motion; this can be positive or negative. In an *elastic collision*, there is no mass transfer from one object to another. Further, *all the kinetic energy in the initial state goes into kinetic energy in the final state.* If the final velocities of objects 1 and 2 are denoted by v_1' and v_2', then, in addition to the momentum conservation equation for one dimension,

$$m_1v_1 + m_2v_2 = m_1v_1' + m_2v_2', \tag{8–27}$$

we have the energy conservation equation

$$\tfrac{1}{2}m_1v_1^2 + \tfrac{1}{2}m_2v_2^2 = \tfrac{1}{2}m_1v_1'^2 + \tfrac{1}{2}m_2v_2'^2. \tag{8–28}$$

With this information we can find the final velocities of the colliding objects if their initial velocities are known. We rewrite the momentum conservation equation (8–27) as

$$m_1(v_1 - v_1') = -m_2(v_2 - v_2'). \tag{8–29}$$

We use the fact that $v_1^2 - v_1'^2 = (v_1 - v_1')(v_1 + v_1')$ and $v_2^2 - v_2'^2 = (v_2 - v_2')(v_2 + v_2')$ to rewrite the energy conservation equation (8–28) in the form

$$\tfrac{1}{2}m_1(v_1 - v_1')(v_1 + v_1') = -\tfrac{1}{2}m_2(v_2 - v_2')(v_2 + v_2'). \tag{8–30}$$

Dividing both sides of Eq. (8–30) by the two sides of Eq. (8–29) leads to the equation

$$v_1 + v_1' = v_2 + v_2'. \tag{8–31}$$

If we use the letter u to denote the *relative velocity* of the two colliding objects, then

$$u_i = v_1 - v_2 \quad \text{and} \quad u_f = v_1' - v_2'.$$

Using these quantities, Eq. (8–31) can be written in the form

$$u_i = -u_f. \tag{8–32}$$

Equation (8–32) states that *when the collision is elastic, the relative velocity of the colliding objects changes sign but does not change magnitude.* A simple way to remember this result is that the relative velocity behaves like the velocity of a perfectly elastic rubber ball hitting a brick wall.

We may solve Eq. (8–31) for one of the unknown variables, v_2', for example,

$$v_2' = v_1 - v_2 + v_1',$$

and substitute this value into the momentum conservation equation (8–27). We then have

$$m_1 v_1 + m_2 v_2 = m_1 v_1' + m_2(v_1 - v_2 + v_1'),$$

which may be rewritten in the form

$$(m_1 + m_2)v_1' = (m_1 - m_2)v_1 + 2m_2 v_2;$$

$$v_1' = \frac{m_1 - m_2}{m_1 + m_2}v_1 + \frac{2m_2}{m_1 + m_2}v_2. \tag{8–33}$$

A similar calculation leads to the formula

$$v_2' = \frac{2m_1}{m_1 + m_2}v_1 + \frac{m_2 - m_1}{m_1 + m_2}v_2. \tag{8–34}$$

These equations are complicated and it is useful to consider two special cases that simplify them.

1. *Object 2 is initially at rest.* We set $v_2 = 0$, so Eqs. (8–33) and (8–34) become

$$v_1' = \frac{m_1 - m_2}{m_1 + m_2}v_1 \tag{8–35a}$$

and

$$v_2' = \frac{2m_1}{m_1 + m_2}v_1. \tag{8–35b}$$

Let's consider the following situations (in all of which object 2 is initially at rest):

a. The objects have equal masses (Fig. 8–15a). In this case, $v_1' = 0$ and $v_2' = v_1$. The two objects in effect change roles: The moving object comes to rest and the object that was initially at rest moves with the initial velocity of the first object. This effect can be seen vividly in hard billiard shots along a line.

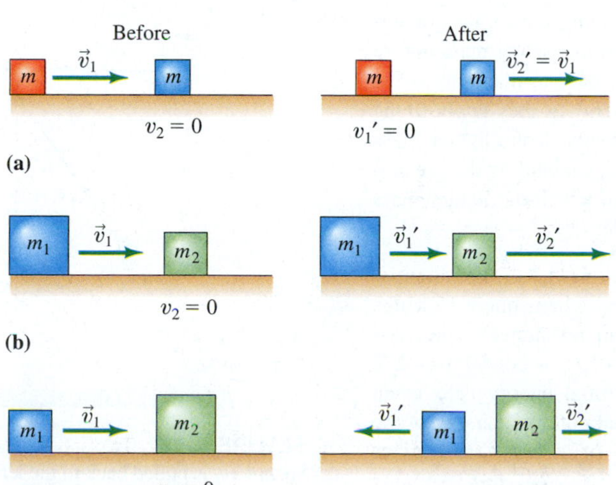

FIGURE 8–15 Two objects collide elastically in one dimension, with the second object initially at rest. (a) If the two masses are equal, the objects simply exchange velocities. (b) If $m_1 \gg m_2$, both objects move off to the right. (c) If $m_2 \gg m_1$, mass m_1 reverses its direction and m_2 moves slowly off to the right.

b. Mass $m_1 \gg$ mass m_2 (Fig. 8–15b). In this case, Eqs. (8–35) yield $v_1' \cong v_1$ and $v_2' \cong 2v_1$. The velocity of the moving object decreases a little, while the object that was at rest picks up almost twice the velocity of the incoming object.

c. Mass $m_2 \gg$ mass m_1 (Fig. 8–15c). In this case, Eqs. (8–35) yield $v_1' \cong -v_1$ and $v_2' \cong (2m_1/m_2)v_1$. The moving object very nearly reverses its velocity, while the object initially at rest recoils with a very small velocity. In the limit that m_2 approaches infinity, the recoil velocity can be neglected and the final velocity of the first object is equal and opposite to its incident velocity. This is just what happens when a tennis ball is bounced off a wall.

2. *The initial total momentum is zero.* The two objects approach each other with velocities such that the initial total momentum is zero, $m_1v_1 + m_2v_2 = 0$. Thus

$$v_2 = -\frac{m_1}{m_2}v_1. \tag{8–36}$$

When this value is substituted into Eq. (8–33), we find

$$v_1' = \frac{m_1 - m_2}{m_1 + m_2}v_1 + \left(\frac{2m_2}{m_1 + m_2}\right)\left(-\frac{m_1}{m_2}\right)v_1 = \left(\frac{m_1 - m_2 - 2m_1}{m_1 + m_2}\right)v_1 = -v_1. \tag{8–37}$$

The initial total momentum was zero and so, by momentum conservation, the final total momentum $(m_1v_1' + m_2v_2')$ is also zero and

$$v_2' = -\frac{m_1}{m_2}v_1' = \frac{m_1}{m_2}v_1 = -v_2. \tag{8–38}$$

Therefore, in the case where the total momentum is zero, the velocities of the objects are unchanged in magnitude but they change sign. In effect, under these circumstances, each of the objects acts as if it hit an infinitely massive brick wall.

EXAMPLE 8–8 Two spheres with masses of 1.0 and 1.5 kg hang at rest at the ends of strings that are both 1.5 m long. These two strings are attached to the same point on the ceiling. The lighter sphere is pulled aside so that its string makes an angle $\theta_i = 60°$ with the vertical. The lighter sphere is then released and the two spheres collide elastically. When they rebound, what is the largest angle with respect to the vertical that the string holding the lighter sphere makes?

Setting It Up We specify in Fig. 8–16a an initial angle θ_i and in Fig. 8–16b a final angle θ_f. We know the values of the light mass m_1 and the heavy mass m_2. The string length L and the initial angle θ_i are also known. We want the rebound angle, θ_f, of the lighter mass after the collision.

Strategy Conservation of energy, including the initial gravitational potential energy, gives us the speed of the lighter mass before the collision. The collision is elastic, so we have available to us the conservation both of momentum and of energy as they apply to the collision. These will give us the recoil velocity of the lighter mass and we can then apply conservation of energy, including the gravitational potential energy, to find the height to which the lighter mass rises. Geometry then gives us the final angle.

Working It Out We begin by calculating the initial potential energy, which is converted to kinetic energy when sphere 1 swings down to the minimum point (neglecting air resistance). The mass m_1 is raised a distance $L(1 - \cos\theta_i) = L(1 - \cos 60°) = L/2$ above the minimum point. With the potential energy zero when the spheres are hanging vertically, the initial potential energy of the system is $U_i = m_1gL/2$. Conservation of energy then gives $\frac{1}{2}m_1v_1^2 = \frac{1}{2}m_1gL$, or

$$v_1 = \sqrt{gL}.$$

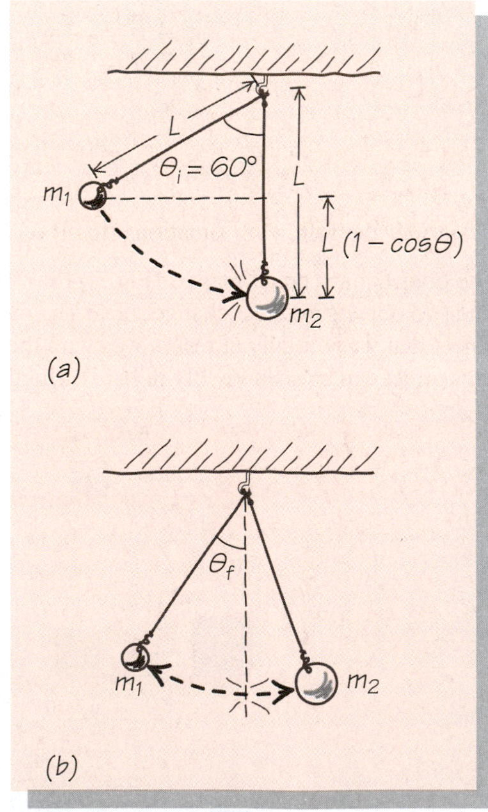

▲ **FIGURE 8–16** Two spheres hang from strings of equal length. (a) Sphere m_1 is pulled back to the left at an angle θ_i and released. It collides with sphere m_2, which is at rest. (b) The balls recoil, with sphere m_1 reaching a maximum height characterized by the angle θ_f.

We now treat the collision, which takes place at the bottom of the swing, where $\vec{v}_1$ is horizontal, $\vec{v}_1 = v_1\hat{i}$. (In fact, the collision itself is a one-dimensional one, with all motion aligned with the x-axis, so we'll drop the explicit vector notation and write all our equations for the collision in terms of the x-components of the vectorial quantities momentum and velocity.) The initial momentum is

$$p_i = m_1 v_1 + m_2 v_2 = m_1 v_1.$$

To find the velocities v_1' and v_2', we use conservation of momentum, which states $p_f = p_i$, or

$$m_1 v_1' + m_2 v_2' = m_1 v_1. \qquad (8\text{–}39)$$

The collision is elastic, so we can also use Eq. (8–32), which is a consequence of energy conservation and which states that the initial relative velocity and the final relative velocity are equal in magnitude but opposite in sign. With

$$v_{\text{rel, initial}} = v_1 - v_2 = v_1,$$

we have a final relative velocity of

$$v_{\text{rel, final}} = v_1' - v_2' = -v_{\text{rel, initial}} = -v_1. \qquad (8\text{–}40)$$

Equations (8–39) and (8–40) are two simultaneous equations that can be solved for v_1' and v_2'. We are interested only in the final velocity of the lighter sphere, and the solution for this quantity is

$$v_1' = \frac{m_1 - m_2}{m_1 + m_2}\sqrt{gL}.$$

With m_2 larger than m_1, this quantity is negative, indicating that the lighter sphere recoils back to the left.

We next find the height to which m_1 recoils. The kinetic energy right after the collision, $m_1 v_1'^2/2$, is converted into gravitational potential energy $U = m_1 gh$ as the sphere rises by h. At the top of the recoil motion, all the kinetic energy is converted to potential energy. As Fig. 8–16b shows, the height risen is $L(1 - \cos\theta_f)$, so we have

$$m_1 gL(1 - \cos\theta_f) = \tfrac{1}{2}m_1 v_1'^2.$$

Numerically,

$$1 - \cos\theta_f = \frac{v_1'^2}{2gL} = \frac{(m_1 - m_2)^2}{(m_1 + m_2)^2}gL\frac{1}{2gL} = \frac{(m_1 - m_2)^2}{2(m_1 + m_2)^2}$$
$$= \frac{(1.0\text{ kg} - 1.5\text{ kg})^2}{2(1.0\text{ kg} + 1.5\text{ kg})^2} = 0.020,$$

or $\theta_f = 11°$.

What Do You Think? Describe the same process in the case that the two masses are equal.

8–5 Elastic Collisions in Two and Three Dimensions

In one dimension, the possible motion of colliding objects is limited. When collisions are no longer restricted to lie along a line, as when billiard balls collide on a billiard table, the vector nature of the mathematical equations becomes important. We work here with collisions in which the identities of the two objects are preserved and in which kinetic energy is conserved.

The law of conservation of momentum [Eq. (8–9)] for the collision of two objects of masses m_1 and m_2, with initial velocities $\vec{v}_1$ and $\vec{v}_2$ and final velocities $\vec{v}_1'$ and $\vec{v}_2'$, reads

$$m_1\vec{v}_1 + m_2\vec{v}_2 = m_1\vec{v}_1' + m_2\vec{v}_2'. \qquad (8\text{–}41)$$

The energy conservation law is

$$\tfrac{1}{2}m_1 v_1^2 + \tfrac{1}{2}m_2 v_2^2 = \tfrac{1}{2}m_1 v_1'^2 + \tfrac{1}{2}m_2 v_2'^2. \qquad (8\text{–}42)$$

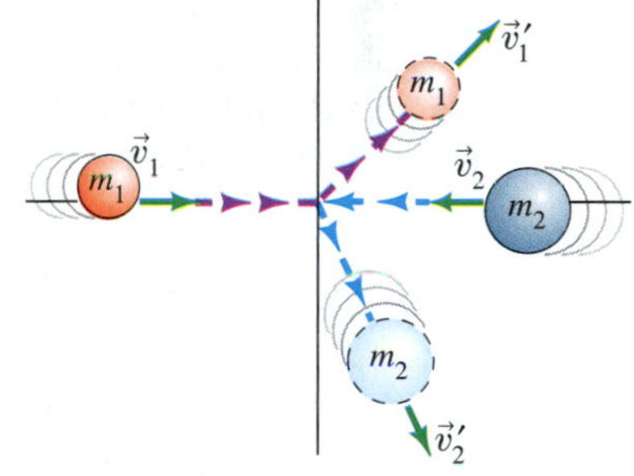

▲ **FIGURE 8–17** An object of mass m_1 and velocity $\vec{v}_1$ collides with another object of mass m_2 and velocity $\vec{v}_2$ in two dimensions.

Let's first consider collisions in two dimensions, where everything happens in a plane, such as the xy-plane shown in Fig. 8–17. The collision of billiard balls on a billiard table represents this case, but with equal masses. Given information about the initial motions, we want to find the magnitudes of the final velocities and the angles they make with the x-axis. Equivalently, we want the x- and y-components of the final velocities of objects 1 and 2. There are four unknowns but only three equations. [Equation (8–41) is a vector equation and actually comprises two equations: one for the x-components and one for the y-components.] Therefore, for a given set of initial velocities, *there is no unique solution for the final velocities* and the final objects can move in a variety of directions with a variety of speeds. Nevertheless, the three equations do impose substantial constraints.

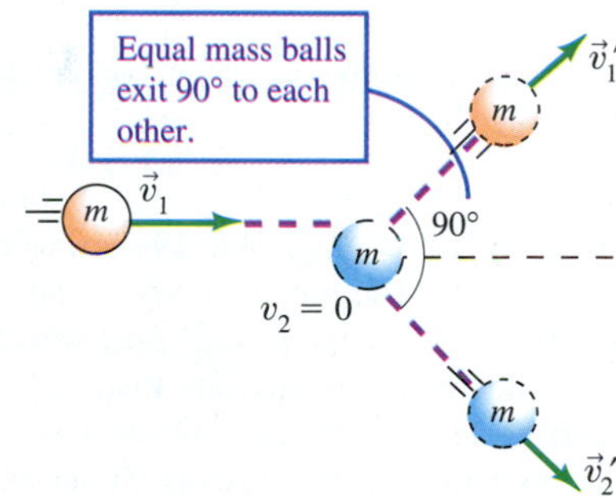

▲ **FIGURE 8–18** An object of mass m_1 and velocity $\vec{v}_1$ collides elastically with an object of the same mass at rest. The angle between the final velocities is 90°.

An interesting example of these constraints occurs when the two masses have identical values m and one of the objects is initially at rest. This is very literally the billiards case, in which the projectile is the cue ball and the target is initially stationary. Figure 8–18 shows the geometry—we have chosen object 2 to be initially at rest, $v_2 = 0$. After canceling a common factor of m, we take the square of Eq. (8–41) (the square of a vector equation $\vec{A} = \vec{B}$ implies $\vec{A}\cdot\vec{A} = \vec{B}\cdot\vec{B}$). With $v_2 = 0$, we find

$$v_1^2 = v_1'^2 + 2\vec{v}_1'\cdot\vec{v}_2' + v_2'^2. \qquad (8\text{–}43)$$

We can also cancel a common factor of $\frac{1}{2}m$ from Eq. (8–42), leaving, with $v_2 = 0$,

$$v_1^2 = v_1'^2 + v_2'^2. \tag{8–44}$$

Comparing Eqs. (8–43) and (8–44), we conclude that $\vec{v}_1' \cdot \vec{v}_2' = 0$. In other words, the final velocity vectors $\vec{v}_1'$ and $\vec{v}_2'$ are perpendicular to one another. The fact that the angle between the velocities of the outgoing cue ball and the recoiling target ball is a right angle is a fact well known to billiards players (Fig. 8–18). (If spins come in, then this is no longer necessarily true.) It is also worthwhile noting what Eq. (8–44) tells us: The final velocities form the sides of a right triangle whose hypotenuse has magnitude $|v_1|$.

CONCEPTUAL EXAMPLE 8–9 There are two near extremes that describe billiard shots. In the first, the cue ball strikes the target ball nearly head on and the target ball moves rapidly in nearly the original direction of the cue ball. In the second, the cue ball barely grazes the target ball and the cue ball continues with nearly the same speed in its original direction. Qualitatively, what happens to the cue ball in the first case and what happens to the target ball in the second case?

Answer Carried to the limit, these are one-dimensional collisions. In the limit of the first collision, the cue ball stops dead, and in the limit of the second collision the cue ball just misses the target ball, which therefore does not move. The answer to our question must be close to these cases. The billiard balls have equal masses, and the target ball is originally at rest, so that we also have "right angle rules," in which the final velocities obey a right triangle rule [Eq. (8–44)] and the billiard ball motions after the collision make a right angle. Thus in the first case the cue ball moves off slowly at nearly a right angle to the original direction of motion. In the second case the target ball moves slowly off to the side at nearly a right angle to the original direction of motion.

EXAMPLE 8–10 Two billiard balls of equal mass m approach each other along the x-axis; one is moving to the right with a speed of $v_1 = 10\text{ m/s}$ and the other is moving to the left with a speed of $v_2 = 5\text{ m/s}$. After the collision, which is elastic, one of the balls moves in the direction of the y-axis (Fig. 8–19). What are the velocities $\vec{v}_1'$ and $\vec{v}_2'$ of the balls after the collision?

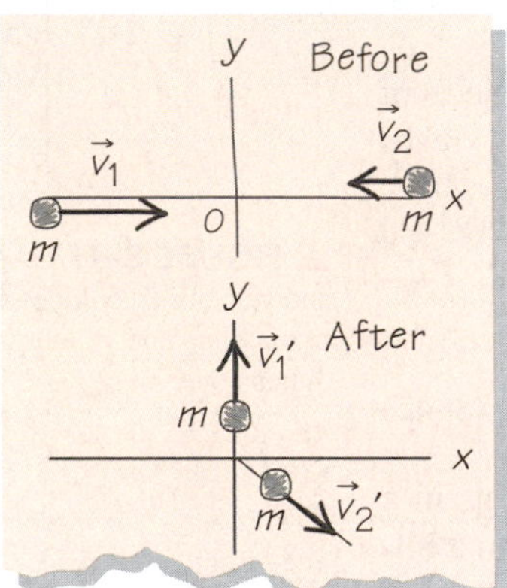

▲ **FIGURE 8–19** Two billiard balls approach, collide, and change directions.

Setting It Up Figure 8–19 shows the situation together with the x- and y-axes. The initial speeds $\vec{v}_1$ and $\vec{v}_2$ are specified in the problem, as is the direction of the final velocity $\vec{v}_1' = v_1'\hat{j}$ of one of the balls—this ball has only one final velocity component. We do *not* know the value of the mass m, which will in fact cancel. We want the final velocities of the two balls, $\vec{v}_1'$ and $\vec{v}_2'$.

Strategy If we use the momentum conservation equation and divide by the common factor of m, we deal with a conservation of velocities. Because the collision is elastic, we can also use the conservation of kinetic energy, from which a common factor of m also cancels. The unknown mass m will cancel from the problem. We will want to count equations to make sure we have enough information to find the unknowns, which are the three final velocity components, one for ball 1 and two for ball 2. Assuming there is enough information, we can algebraically solve for the velocity components.

Working It Out With the cancellation of the mass factors, the momentum and energy conservation equations read

$$\vec{v}_1 + \vec{v}_2 = \vec{v}_1' + \vec{v}_2', \qquad v_1^2 + v_2^2 = v_1'^2 + v_2'^2.$$

The balls move in the xy-plane, so the first equation stands for two component equations and the second equation provides a third. There are three unknowns, the single component of $\vec{v}_1'$ and the two components of $\vec{v}_2'$. The problem is thus solvable. Writing $\vec{v}_2' = v_{2x}'\hat{i} + v_{2y}'\hat{j}$, momentum conservation reads

$$(10\text{ m/s})\hat{i} + (-5\text{ m/s})\hat{i} = v_1'\hat{j} + v_{2x}'\hat{i} + v_{2y}'\hat{j}.$$

We now separately equate the coefficients of $\hat{i}$ and of $\hat{j}$; that is, we use the fact that momentum conservation holds in the x-direction and in the y-direction separately. We find

$$v_{2x}' = (10\text{ m/s}) + (-5\text{ m/s}) = 5\text{ m/s} \quad \text{and} \quad v_{2y}' = -v_1'.$$

The energy conservation equation reads similarly

$$(10\text{ m/s})^2 + (-5\text{ m/s})^2 = v_1'^2 + (v_{2x}'^2 + v_{2y}'^2),$$
$$100\text{ m}^2/\text{s}^2 + 25\text{ m}^2/\text{s}^2 = v_1'^2 + 25\text{ m}^2/\text{s}^2 + v_1'^2.$$

Thus we have $100\text{ m}^2/\text{s}^2 = 2v_1'^2$, or $v_1' = \sqrt{50}\text{ m/s}$. We also have $v_{2y}' = -\sqrt{50}\text{ m/s}$.

Although the collisions of real billiard balls are quite elastic, this is not the case for many other real collisions. A high-speed photograph of a baseball meeting a baseball bat (Fig. 8–20) shows that the ball undergoes significant deformation, which, even though the deformation is not permanent, usually means that there is inelasticity, that is, some energy is lost in the collision.

Problem-Solving Techniques

Collision problems involve an interaction between objects that occurs in a limited space (and a limited time). At some early time, the objects do not exert any forces on one another; at some later time, they again exert no forces on one another. Typically, we need to find some parameters of the final (or initial) motion—the final velocity of one of the objects, for example. The following steps can be useful:

1. We must identify the relevant interaction between the objects involved in a collision. We must therefore be able to assume that the objects are isolated or that during the collision external forces are small compared to the impulsive forces.
2. We must identify the objects involved before the collision and those involved after the collision. The objects that result from a collision may not be the same as the objects before the collision, for example, explosions or collisions in which the colliding objects coalesce.
3. We must identify the quantities that are known and the quantities that are to be found. In particular, a count of the number of known and unknown quantities, including the number of vector components, is helpful.
4. Remember that the conservation of momentum, which is a vector relation, is always applicable if there is no external force on the colliding system. If the motion is in one, two, or three dimensions, the conservation of momentum gives one, two, or three relations among the momenta (or velocities), respectively.
5. If the collision is known to be elastic, the conservation of energy provides an additional equation that involves the speeds of the objects. If the collision is not elastic, then an equation for the energy is available only if information about the energy loss is available.
6. The number of equations that include the unknowns must match the number of unknowns if the problem is to be solved.

We have thus far concentrated on two-dimensional elastic collisions. What about elastic collisions in three dimensions? If you observe the collision from a frame of reference known as the center-of-mass frame of the two colliding objects (we'll discuss the meaning and use of this inertial frame in Section 8–6), then the incoming objects come in along a line. They would leave along another line in that frame, and since two intersecting lines determine a plane, the collision is two dimensional in this frame. The results established for elastic collisions in two dimensions are thus directly applicable to three dimensions.

▲ **FIGURE 8–20** The softball undergoes a significant deformation during the brief time period that an impulsive force due to the bat acts on it.

8–6 Center of Mass

In a system of many parts, including the two-piece systems we have dealt with to this point in this chapter, there is a particular point with special properties, called the *center of mass*. As we shall see, the importance of this point is that it moves under Newton's second law as if the total mass of the system were concentrated there. In this section we'll learn how to calculate the position of the center of mass and to understand its properties.

In Section 8–1 we saw that in the absence of external forces the sum of the momenta of the N particles that make up a system (an extended object) is unchanging. (There we actually worked this through for $N = 3$, but the procedure we used is easily generalized.) In other words, if $\vec{P}$ is the total momentum of the system, then in the absence of external forces

$$\vec{P} = m_1\vec{v}_1 + m_2\vec{v}_2 + \cdots + m_N\vec{v}_N = \text{a constant.} \tag{8–45}$$

Here we have labeled the particles, or pieces, that make up the system with subscripts. The momentum of the system is a constant vector—one that does not change in magnitude or direction as a function of time.

We can use this result to locate the center of mass of the extended object. If we also assume that the particle masses do not change as a function of time, Eq. (8–45) can be written as

$$\frac{d}{dt}(m_1\vec{r}_1 + m_2\vec{r}_2 + \cdots + m_N\vec{r}_N) = \text{a constant.} \tag{8–46}$$

We divide the quantity in parentheses by the total mass, $M = m_1 + m_2 + \cdots + m_N$, to obtain the position vector $\vec{R}$ of the center of mass:

$$\vec{R} \equiv \frac{m_1\vec{r}_1 + m_2\vec{r}_2 + \cdots + m_N\vec{r}_N}{M}. \tag{8–47}$$

CENTER OF MASS

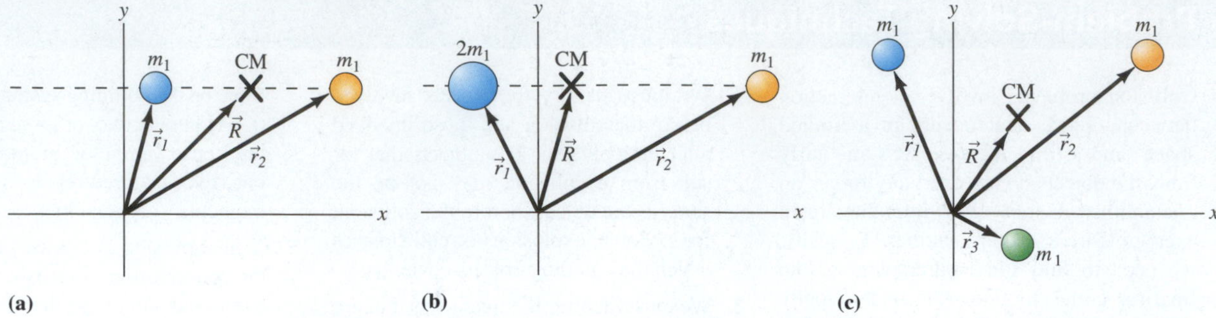

▲ **FIGURE 8–21** Three examples of the center of mass (CM). (a) Both balls have equal masses. (b) One ball has twice the mass of the other. (c) Three balls of equal mass.

From its definition, the center-of-mass vector $\vec{R}$ has Cartesian components X, Y, and Z given by

$$X = \frac{m_1x_1 + m_2x_2 + \cdots + m_Nx_N}{M}, \tag{8–48a}$$

$$Y = \frac{m_1y_1 + m_2y_2 + \cdots + m_Ny_N}{M}, \tag{8–48b}$$

$$Z = \frac{m_1z_1 + m_2z_2 + \cdots + m_Nz_N}{M}. \tag{8–48c}$$

Figure 8–21 illustrates some examples of the center of mass for two and three pieces. The following two examples illustrate the calculation of the position of the center of mass. There are some simple ideas that we can extract from the figure and the examples. In particular, we mention here that if the system is made from two pieces, the center of mass will lie along the line between them and a more massive piece will tend to "pull" the center of mass toward it.

EXAMPLE 8–11 Two pointlike masses are placed on a massless rod that is 1.5 m long. The masses are placed as follows: 1.6 kg at the left end and 1.8 kg 1.2 m from the left end. (a) What is the location of the center of mass? (b) By moving the 1.8-kg mass, can you arrange to have the center of mass in the middle of the rod?

Setting It Up Figure 8–22 illustrates this one-dimensional problem. We align the rod with length L along the x-axis, and we place its left end at the origin. The 1.6-kg mass m_1 is placed at $x_1 = 0$ m and the 1.8-kg mass m_2 is placed at $x = 1.2$ m.

Strategy Both parts of this example ask us to calculate in a straightforward way the location of the center of mass in a one-dimensional system. Equation (8–48a) provides the necessary formulation.

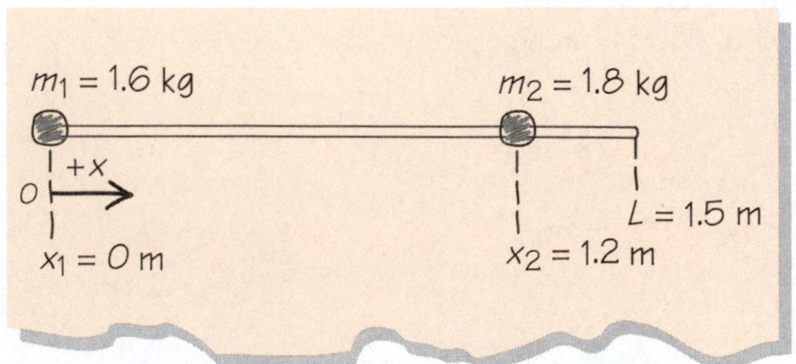

▲ **FIGURE 8–22** A rod with two pointlike masses is aligned along the x-axis, with its left end at the origin.

Working It Out (a) The location of the center of mass is given by Eq. (8–48a):

$$X = \frac{m_1x_1 + m_2x_2}{m_1 + m_2} = \frac{(1.6\text{ kg})(0\text{ m}) + (1.8\text{ kg})(1.2\text{ m})}{1.6\text{ kg} + 1.8\text{ kg}}$$
$$= 0.64\text{ m}.$$

(b) The midpoint of the rod is $x_{\text{mid}} = 0.75$ m. Let's place m_2 at a new position x_2 and ask if there is a solution for x_2 along the rod ($x_2 < L$) such that $X = x_{\text{mid}}$. We have

$$X = x_{\text{mid}} = \frac{m_1x_1 + m_2x_2}{m_1 + m_2} = \frac{m_1(0\text{ m}) + (m_2x_2)}{M} = \frac{m_2x_2}{M}.$$

We can solve this equation for x_2:

$$x_2 = \frac{Mx_{\text{mid}}}{m_2} = \frac{(1.6\text{ kg} + 1.8\text{ kg})(0.75\text{ m})}{1.8\text{ kg}} = 1.4\text{ m}.$$

This value is indeed less than L, so if we place the 1.8-kg mass at this point, the center of mass is at the midpoint of the rod.

What Do You Think? The center of mass of the Earth–Moon system is (a) on a line that connects their centers, relatively close to the Moon or even within it; (b) on a line that connects their centers, relatively close to Earth or even within it; (c) along the line that runs through their centers, near Earth's center, on the side away from the Moon; or (d) impossible to tell because each of the two bodies is not a point.

EXAMPLE 8–12 Add a third mass of 2.3 kg to the masses of Example 8–11 at the point shown in Fig. 8–23. Find the center of mass.

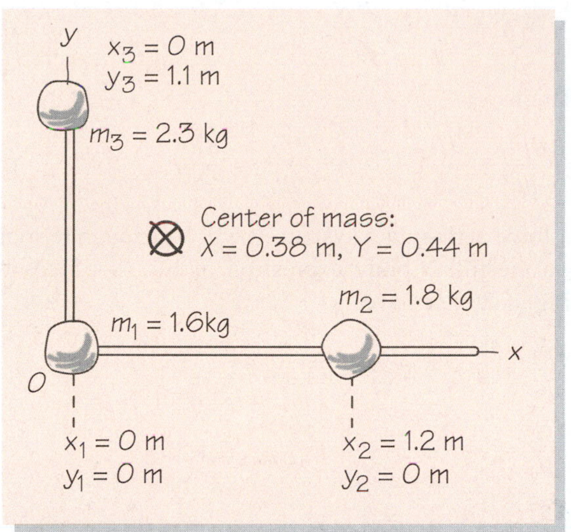

▲ **FIGURE 8–23** The location of the center of mass is marked with a circled X.

Setting It Up With the first two masses aligned with the x-axis, we add a y-axis to the previous sketch (See Fig. 8–23) and place the new mass m_3 in it at the coordinates $(x_3, y_3) = (0, 1.1\text{ m})$.

Strategy Again we have a simple calculation of the center of mass, except that the problem has become a two-dimensional one. Equations (8–48) give both X and Y.

Working It Out With $M = 1.6\text{ kg} + 1.8\text{ kg} + 2.3\text{ kg} = 5.7\text{ kg}$, we have, from Eqs. (8–48),

$$X = \frac{m_1 x_1 + m_2 x_2 + m_3 x_3}{M}$$

$$= \frac{(1.6\text{ kg})(0\text{ m}) + (1.8\text{ kg})(1.2\text{ m}) + (2.3\text{ kg})(0\text{ m})}{5.7\text{ kg}}$$

$$= 0.38\text{ m},$$

$$Y = \frac{m_1 y_1 + m_2 y_2 + m_3 y_3}{M}$$

$$= \frac{(1.6\text{ kg})(0\text{ m}) + (1.8\text{ kg})(0\text{ m}) + (2.3\text{ kg})(1.1\text{ m})}{5.7\text{ kg}}$$

$$= 0.44\text{ m}.$$

This point is indicated in Fig. 8–23.

What Do You Think? Could you have alternatively first found the center of mass of the pair 1 and 2, then combined that with object 3 to find the center of mass of the three-body system?

Center of Mass Motion in the Absence of External Forces

The velocity $\vec{V}$ of the center of mass is the time derivative of its position $\vec{R}$ [Eq. (8–47)]:

$$\vec{V} \equiv \frac{d\vec{R}}{dt} = \frac{m_1\vec{v}_1 + m_2\vec{v}_2 + \cdots + m_N\vec{v}_N}{M}. \tag{8–49}$$

The numerator is the sum of the momenta of the individual pieces, so we may write this result as

$$\vec{V} = \vec{P}/M. \tag{8–50}$$

This is of course the same relation between momentum, velocity, and mass obeyed by a point object.

We have already seen that the total momentum of an object is constant in the absence of external forces. Equation (8–50) shows that *in the absence of external forces the center of mass moves with constant velocity*. No matter how complicated the motion of the constituent objects, the motion of its center of mass is constant-velocity motion, corresponding to the motion of a point mass in the absence of net force on it (Fig. 8–24).

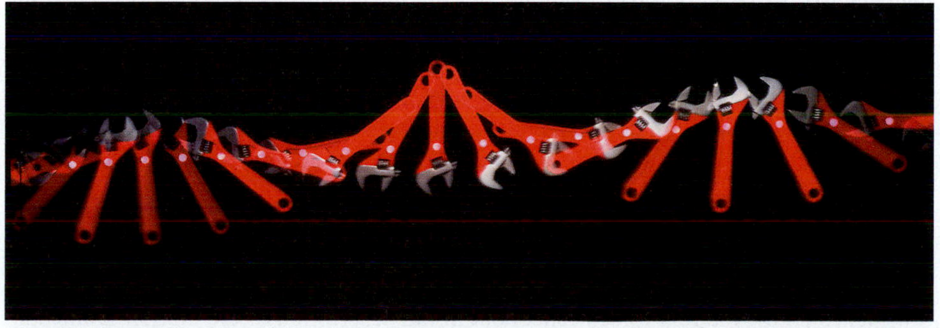

▲ **FIGURE 8–24** Note that the center of mass of the complex object (wrench) moves in a straight line (denoted by the white dot) with constant velocity if the net force on the object is zero.

Center of Mass Motion in the Presence of External Forces

What happens when there are external forces present in addition to interparticle forces, the ones that hold a extended object together, rigidly or otherwise? We begin by writing the acceleration of the center of mass. By taking a derivative of the velocity [Eq. (8–50)], we find the acceleration $\vec{A}$ of the center of mass:

$$\vec{A} \equiv \frac{d\vec{V}}{dt} = \frac{d^2\vec{R}}{dt^2}. \tag{8–51}$$

Multiplying by M gives the relation

$$M\frac{d^2\vec{R}}{dt^2} = \frac{d\vec{p}_1}{dt} + \frac{d\vec{p}_2}{dt} + \cdots + \frac{d\vec{p}_N}{dt} = \frac{d\vec{P}}{dt}. \tag{8–52}$$

Let's now consider an external force acting on a system. It will be enough to look at a three-body example rather than the full N-body expression of Eq. (8–52). With the three-object system shown in Fig. 8–25, we have

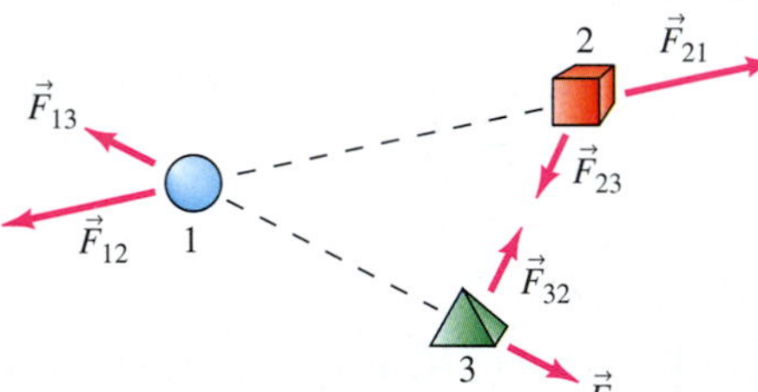

▲ **FIGURE 8–25** The internal forces in a three-object system. These forces are labeled by both the object that causes them and the object on which they act. According to Newton's third law, these forces, whether they are attractive or repulsive, are equal and opposite in pairs. The external forces are not shown.

$$\frac{d\vec{p}_1}{dt} = \vec{F}_{12} + \vec{F}_{13} + \vec{F}_{1,\text{ext}},$$

$$\frac{d\vec{p}_2}{dt} = \vec{F}_{21} + \vec{F}_{23} + \vec{F}_{2,\text{ext}},$$

$$\frac{d\vec{p}_3}{dt} = \vec{F}_{31} + \vec{F}_{32} + \vec{F}_{3,\text{ext}}.$$

Here $\vec{F}_{i,\text{ext}}$ is the external force on object i and the $\vec{F}_{ij}$ are the internal forces (see Section 8–1). If we add these equations, using Newton's third law—according to which $\vec{F}_{12} + \vec{F}_{21} = 0$, $\vec{F}_{13} + \vec{F}_{31} = 0$, and $\vec{F}_{23} + \vec{F}_{32} = 0$—we find that all the internal forces cancel out,

$$\frac{d\vec{p}_1}{dt} + \frac{d\vec{p}_2}{dt} + \frac{d\vec{p}_3}{dt} = \frac{d\vec{P}}{dt} = \vec{F}_{1,\text{ext}} + \vec{F}_{2,\text{ext}} + \vec{F}_{3,\text{ext}} = \vec{F}_{\text{tot,ext}}. \tag{8–53}$$

Equation (8–52) and the extension of Eq. (8–53) to N objects give the general result

$$\vec{F}_{\text{tot,ext}} = \frac{d\vec{P}}{dt} = M\frac{d^2\vec{R}}{dt^2}. \tag{8–54}$$

For any system with total mass M in which there are both internal and external forces, we have shown that *the center of mass moves like a single point mass of mass M subject to the total external force on the system.* In other words, the system obeys Newton's second law

$$\vec{F}_{\text{tot,ext}} = M\vec{A}, \tag{8–55}$$

EQUATION FOR LINEAR MOTION OF THE CENTER OF MASS

where $\vec{A}$ is the acceleration of the center of mass. The internal forces play no role in the the center-of-mass motion. For example, despite the complicated motion of a twisting, somersaulting diver, the diver's center of mass moves with the same, simple parabolic motion as that of a rock thrown from the diving board with the same initial velocity as the diver's initial center-of-mass velocity.

A bowling ball rolling down an inclined plane consists of some 10^{27} molecules, all interacting through electric forces. Despite the immense complexity of the internal interactions, it is nevertheless possible to treat a bowling ball as a point mass (or as a simple rigid object, as we shall see in Chapters 9 and 10) and to study its motion without being aware of the underlying atomic structure of the ball or the laws that govern the motion of atoms. In deriving Eq. (8–54), we have seen why this is true: Newton's third law makes all the internal forces cancel out in the description of the motion of the center of mass. Thus, when we say we treat the bowling ball as a single object, we mean we describe the motion of the center of mass. This simplification of the overall motion of

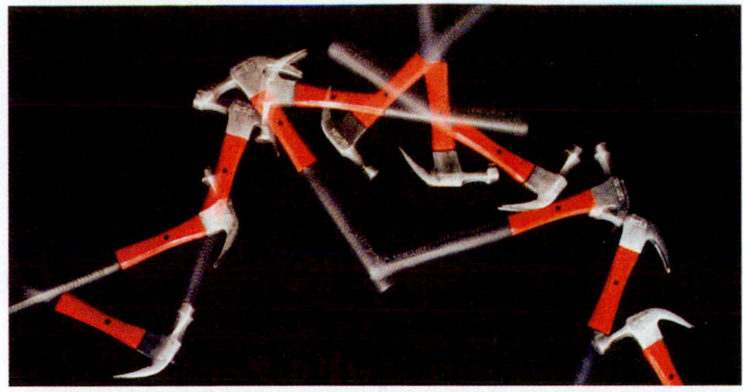

FIGURE 8–26 The center of mass of the hammer (denoted by the small black dot) moves as a point mass under the influence of an external force. In this case, the external force is gravity so the center of mass has a parabolic trajectory.

an object to the motion of its center of mass is a crucial feature of our ability to describe the world around us. In fact, the idealized "point particle" motion that we discussed in the early chapters of this book can now be seen as an exact description of the center-of-mass motion of real objects. And, as we shall see in Chapters 9 and 10, this way of looking at things will allow us to break down the motion of an extended system into motion of the center of mass and, in a highly independent way, motion of the pieces of the system about the center of mass.

Figure 8–26 shows that the motion of a system's center of mass is simple even though there is a rather complex motion of parts of the system. In the next subsection we describe an important practical property of the center of mass: Assuming that our system has no net external force on it, a coordinate system with its origin at the center of mass is an inertial frame within which it is especially simple to treat collisions (interactions) of the constituents of the system.

Conservation of Momentum in Different Inertial Frames

The center of mass of a system with no external forces acting on it moves with constant total momentum, a result that generalizes Newton's first law to systems. As we saw in Chapter 4, Newton's first law is closely associated with the equivalence of inertial frames. Our generalization of the first law is accordingly a generalization of the idea of inertial frames.

We can start with a frame of reference whose origin is placed at the center of mass and moves with it. In this frame the center of mass is at rest—this is called the **center-of-mass frame**—and the total momentum is zero [Eq. (8–50)]. We can then go to another inertial frame by adding any constant velocity to the entire system (i.e., to each part of the system). In this way one can move back and forth between these inertial frames, observing the behavior of the system in the different frames. We can use this fact to great advantage in problem solving for collisions because calculations involving colliding objects may be simpler in one inertial frame than in another. In particular, the center-of-mass frame presents simplifications.

An example best illustrates how to use this technique.

EXAMPLE 8–13 Two objects of equal mass collide in what we call the laboratory frame. In this frame, one of them is moving with momentum $\vec{p} = p\hat{i}$ and the other is at rest. Assuming that you have the same two distinct objects after the collision, describe this collision in the center-of-mass frame and use your result to find the general form of the final momenta in the laboratory frame.

Setting It Up We sketch what is described in Fig. 8–27a, including xy-coordinates. We are asked here first to describe the two-body-to-two-body collision in the center-of-mass frame, then to take this general description and transform it to the laboratory frame.

Strategy We have objects moving in one inertial frame and want to see how they look in another. As we described above, moving between such frames involves a shift by a constant velocity; in other words, we can add a constant overall velocity $\vec{v}$ to every individual particle in the problem. Although in this problem all the masses are the same, that will not always be true. In general, if the particles have initial momenta $\vec{p}_1 = m_1\vec{v}_1$ and $p_2 = m_2\vec{v}_2$, respectively, then, as viewed from the new frame, their momenta will be $m_1(\vec{v}_1 + \vec{v})$ and $m_2(\vec{v}_2 + \vec{v})$, respectively. The center-of-mass (or, as we see below, the center-of-momentum) frame is obtained if we can find $\vec{v}$ such that the sum of the new momenta is zero. This condition reads $m_1(\vec{v}_1 + \vec{v}) + m_2(\vec{v}_2 + \vec{v}) = 0$, and it is satisfied with the choice

$$\vec{v} = -\frac{m_1\vec{v}_1 + m_2\vec{v}_2}{m_1 + m_2}.$$

(continues on next page)

We can see that this is indeed the center-of-mass frame by comparison with Eq. (8–49), which shows that we are viewing the collision from the point of view of an observer moving along with the center of mass of the colliding objects. But for our purposes the most important feature is that this is the frame in which the particles come in along a straight line with equal and opposite momenta, and must leave with equal and opposite momenta along another straight line. This fact tells us as much as we can know about the final state. Particle 1 will have a final center-of-mass frame momentum $m_1\vec{v}'_{c2}$ and the momentum of particle 2 will be $m_2\vec{v}'_{c2} = -m_1\vec{v}'_{c1}$ in the center-of-mass frame.

Finally we take these most general center-of-mass frame final momenta back to the laboratory frame. Thus the momentum of particle 1 in the laboratory frame will be $m_1(\vec{v}'_{c1} - \vec{v})$, and so forth.

Let us now return to the special conditions of this problem, that $m_1 = m_2 = m$ and that particle 2 is initially at rest. If we call $\vec{p}$ the initial laboratory momentum of particle 1, then by substitution into the equations above we see immediately that the center-of-mass momentum of particle 1 is $\vec{p}/2$ and that of particle 2 is $-\vec{p}/2$. For the equal-mass case (and *only* for this case), we can go from laboratory frame to center-of-mass frame by adding to the individual momenta a constant momentum, $-\vec{p}/2$, and we go back from the center-of-mass frame to the laboratory frame by adding to the individual momenta a constant momentum, $+\vec{p}/2$.

Working It Out We suppose we have made the transformation described above to the center-of-mass frame. As above the final momenta in this frame have the form $\vec{p}'$ and $-\vec{p}'$ (Fig. 8–27c). Without information about the conservation of kinetic energy, we cannot say much more about the magnitude p' of these momenta, and any angle is possible. The only requirement is that in the center-of-mass frame the final objects go off back to back.

To go back to the laboratory frame, we add the momentum $+\vec{p}/2$ to every part of the system. Thus in the laboratory frame the final momenta must be of the form $\vec{p}' + \vec{p}/2$ and $\vec{p}' - \vec{p}/2$, respectively (Fig. 8–27d).

What Do You Think? We stated that any angle is possible for the final objects in the center-of-mass frame. Is that true also for the laboratory frame? If not, what determines the restriction?

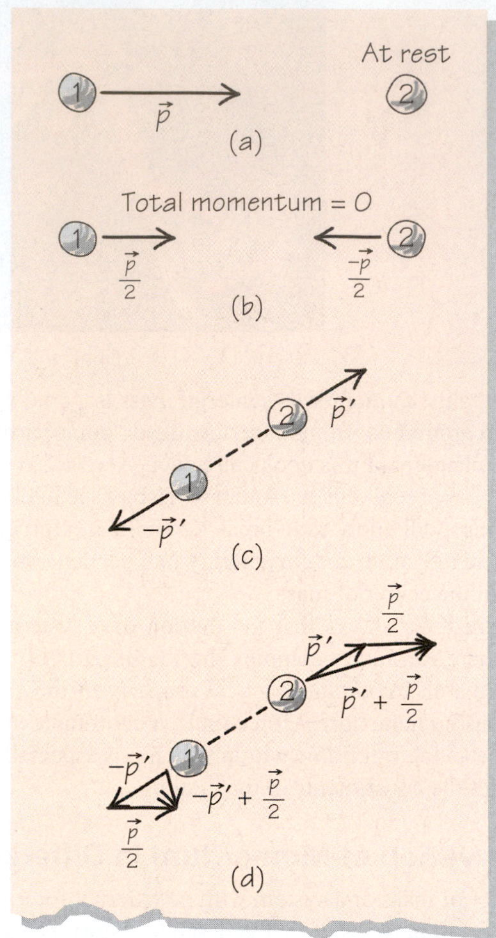

▲ **FIGURE 8–27** (a) View before collision from the frame in which ball 2 is at rest. (b) View before collision from the frame in which the total momentum is zero. (c) View after collision from the frame in which the total momentum is zero. (d) View after collision from the frame in which ball 2 was originally at rest.

THINK ABOUT THIS . . .

WHY ARE COALESCING COLLISIONS TERMED "PERFECTLY INELASTIC"?

Now that we have learned about the center-of-mass frame, we can understand why the term perfectly inelastic is appropriate in the case that two colliding objects "stick" and form a single final object. In the center-of-mass frame, this object has zero total momentum; hence it is at rest. The collision is correctly termed *perfectly inelastic* because in the center-of-mass frame there is *no* final kinetic energy. *All the energy of motion* in the initial state of the system goes into the "sticking together" of the objects. If the final state consists of more than one object, however slowly separating, there is some kinetic energy in the center-of-mass frame and the inelasticity will be less than "perfect."

Center of Mass of a Continuous Mass Distribution

We know that if we look closely enough matter is composed of discrete masses in the form of atoms. But to a detector that is not sufficiently acute to see atoms, such as our eyes, a solid object looks and behaves like a continuum of matter; that is, a *continuous mass distribution*. We may define the center of mass of a continuous distribution of

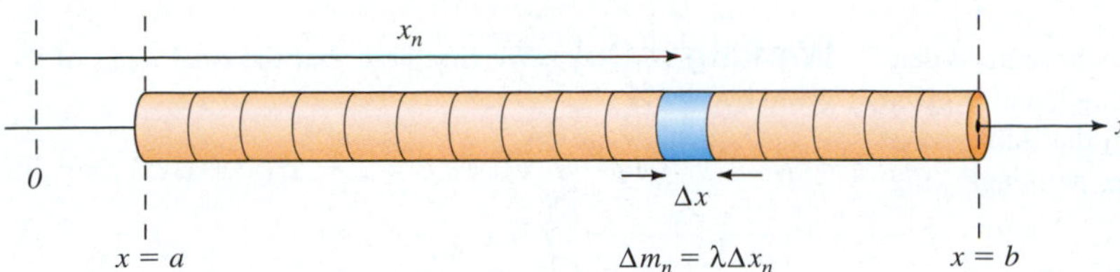

▲ **FIGURE 8–28** A continuous mass is divided into N tiny mass segments of equal length Δx (where N is large). This technique allows us to find the center of mass. The nth segment, x_n, has mass Δm_n.

mass by a simple generalization of the center-of-mass vector equation, Eq. (8–48). Consider a thin rod of mass M lying along the x-axis (Fig. 8–28). We may take one end to lie at $x = a$ and the other at $x = b$. Now let's divide the interval between a and b into a large number N of tiny segments. The segments are of equal length, Δx, but they are not necessarily of equal mass, because the rod need not have a uniform mass distribution. The position of the nth segment is labeled x_n, and we call the mass of this tiny segment Δm_n. We may write this mass in terms of the local mass density:

$$\Delta m_n = \left(\frac{\Delta m_n}{\Delta x}\right)\Delta x. \tag{8–56}$$

The masses of all the segments add up to the total mass of the rod, so that we have

$$\Delta m_1 + \Delta m_2 + \cdots + \Delta m_n = \left(\frac{\Delta m_1}{\Delta x} + \frac{\Delta m_2}{\Delta x} + \cdots + \frac{\Delta m_n}{\Delta x}\right)\Delta x = M. \tag{8–57}$$

If we were to use infinitesimal intervals, this equation would read

$$M = \int_a^b \left(\frac{dm}{dx}\right) dx. \tag{8–58}$$

According to Eq. (8–48a), we have

$$\begin{aligned} X &= \frac{\Delta m_1 x_1 + \Delta m_2 x_2 + \cdots + \Delta m_n x_n}{M} \\ &= \frac{1}{M}\left[\left(\frac{\Delta m_1}{\Delta x}\right)x_1 + \left(\frac{\Delta m_2}{\Delta x}\right)x_2 + \cdots + \left(\frac{\Delta m_n}{\Delta x}\right)x_n\right]\Delta x. \end{aligned}$$

As $N \to \infty$, this equation becomes an integral:

$$X = \frac{1}{M}\int_a^b \left(\frac{dm}{dx}\right)x\, dx.$$

It is usual to write the mass density dm/dx of a one-dimensional object as λ. If λ is a constant, it can come out of the integral; if it is not a constant, $\lambda = \lambda(x)$, then it must remain under the integral. (You could easily imagine that this happens because the thickness of our "one dimensional stick" varies along its length.) In this notation, the position of the center of mass is

$$X = \frac{1}{M}\int_a^b x\lambda(x)\, dx. \tag{8–59}$$

Equation (8–59) allows us to calculate the center of mass of a one-dimensional object, as in Example 8–14.

EXAMPLE 8–14 Consider a rod of length L whose mass density (mass per unit length dm/dx) varies along its length and is given by $\lambda = C(1 + ax^2)$, where x is the distance from the light end of the rod and C is a constant with dimensions of mass per length (Fig. 8–29). Calculate the center of mass of the rod.

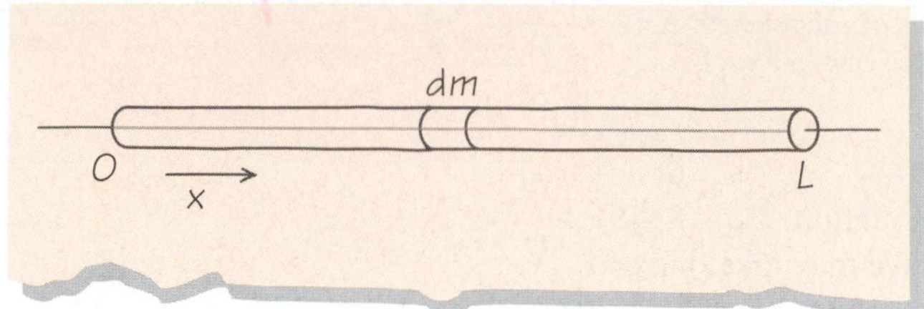

▲ **FIGURE 8–29** A rod has varying density along its length.

Setting It Up In Fig. 8–29 we align the rod with the x-axis as shown.

Strategy This is a straightforward application of the center-of-mass formula for an object of varying density. Equation (8–59) provides us with the tool to find the center-of-mass position. To use it, we must also calculate the total mass M of the rod.

Working It Out We first note that the total mass of the rod is given by

$$M = \int_0^L \lambda(x)\,dx = \int_0^L C(1 + ax^2)\,dx$$
$$= C\left(x + \frac{ax^3}{3}\right)\Bigg|_0^L = C\left(L + \frac{aL^3}{3}\right).$$

Now, from Eq. (8–59),

$$X = \frac{1}{M}\int_0^L xC(1 + ax^2)\,dx = \frac{1}{M}C\left(\frac{x^2}{2} + \frac{ax^4}{4}\right)\Bigg|_0^L$$
$$= \frac{1}{M}C\left(\frac{L^2}{2} + \frac{aL^4}{4}\right) = \frac{C[(L^2/2) + (aL^4/4)]}{C[L + (aL^3/3)]}$$
$$= \left(\frac{L}{2}\right)\frac{1 + (aL^2/2)}{1 + (aL^2/3)}.$$

If $a = 0$, then the rod is uniform, and the center of mass is at $L/2$, as expected. In the case of nonuniformity, the center of mass is closer to the more massive end. Note that the parameter C of the mass density cancels.

What Do You Think? In all the examples we have thus far looked at, including this one, the center of mass lies within the object. Can you think of an object for which the center of mass lies outside the object?

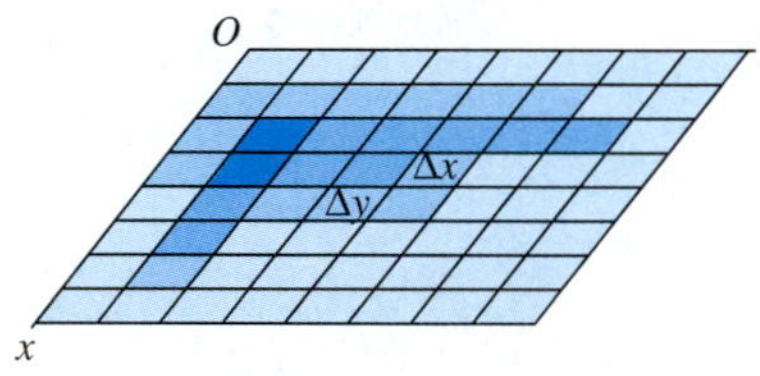

▲ **FIGURE 8–30** We have placed a two-dimensional object in the xy-plane and divided it up into squares, each of area $\Delta x\,\Delta y$. The mass of a square at a given point is the two-dimensional density σ times the area. The shading of the squares in this figure represents the value of σ, with dark blue representing high density and light blue representing low density; we see the density varying from point to point.

Continuous Objects in Two and Three Dimensions: Suppose that instead of a one-dimensional object we deal with a two-dimensional one such as a sheet of metal with an area A. We divide the object into tiny elements of area $\Delta A = \Delta x\,\Delta y$ and mass Δm (Fig. 8–30). Suppose that the two-dimensional mass density, or mass per unit area, for the element at the point (x, y) is $\sigma(x, y)$. This quantity is analogous to the one-dimensional mass density λ; σ has dimensions of mass per unit area. If we take the area element small enough so that σ is a constant within that area, then the mass of that area element is

$$\Delta m = \sigma\,\Delta x\,\Delta y.$$

Remember, the value of σ can vary from place to place, as the varying colors within Fig. 8–30 indicates. Thus σ depends on x and y. We can find the total mass by summing the masses over all the elements, and, in the calculus limit, this summation takes the form†

$$M = \lim\sum\Delta m = \lim\sum\sigma\,\Delta x\,\Delta y = \int_{\text{Surface}}\sigma(x, y)\,dx\,dy. \qquad (8\text{–}60)$$

Similarly, we find the coordinates of the center of mass by summing over the contributions of the elements:

$$MX = \lim\sum x\,\Delta m = \int_{\text{Surface}} x\sigma(x, y)\,dx\,dy, \qquad (8\text{–}61\text{a})$$

$$MY = \lim\sum y\,\Delta m = \int_{\text{Surface}} y\sigma(x, y)\,dx\,dy. \qquad (8\text{–}61\text{b})$$

For a three-dimensional object, we must know its three-dimensional mass density $\rho(x, y, z)$, measured in mass per unit volume. The dependence on x, y, z indicates that the mass density can vary from point to point. The total mass is

$$M = \int_{\text{volume}}\rho(x, y, z)\,dx\,dy\,dz, \qquad (8\text{–}62)$$

†The subscript "Surface" on the single integral sign denotes a two-dimensional integration over the area A; a subscript "volume" similarly indicates an integration over the full three-dimensional space.

while the center of mass is at the position

$$M\vec{R} = \int_{\text{volume}} \vec{r}\rho(x, y, z)\, dx\, dy\, dz. \qquad (8\text{–}63)$$

While the integrals above look difficult, they are normally done as a sequence of single integrations and therefore present nothing fundamentally new. The following example illustrates this.

CONCEPTUAL EXAMPLE 8–15 A square sheet of metal of area 1 m^2 sits as shown (Fig. 8–31a). Its mass density (mass per unit area) is constant. Find the location of the center of mass.

Answer There is a two-dimensional integral to be performed here. We approach it by taking the square and breaking it up into a series of vertical strips (Fig. 8–31b). We find the center of mass of each strip, then add the effect of each strip as if it were entirely concentrated at its own center of mass (see the subsection below for more on this point). Each strip is like the rod of Example 8–14, but with a constant (linear) mass density, so the center of mass of each strip is at its center, that is, at $y = \frac{1}{2}$ m (Fig. 8–31c). (We could have calculated this as a single integration.) When we look at this arrangement, the center of mass of each vertical strip forms a horizontal strip of uniform linear mass density, and the center of mass of that strip (again, by a single integration) is at its center, that is, at $x = \frac{1}{2}$ m. The center of mass is therefore at $(x, y) = \left(\frac{1}{2}\text{ m}, \frac{1}{2}\text{ m}\right)$. We have broken our double integration down into a series of two single integrations.

What Do You Think? Could you have found this answer even more quickly?

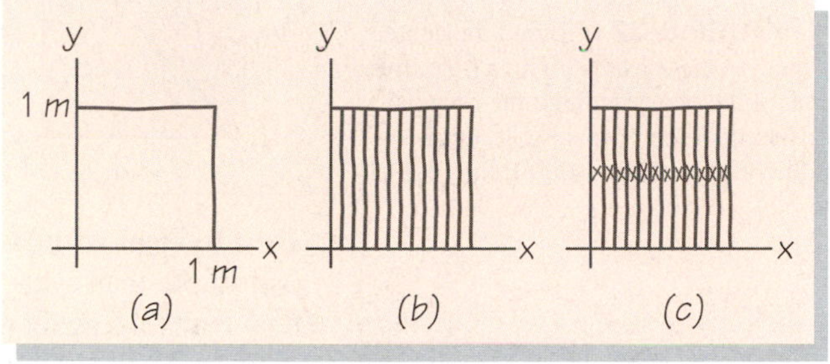

▲ **FIGURE 8–31** To find the center of mass of a sheet of metal, as in (a), we can divide the metal sheets into vertical strips, as in (b), find the center of mass of each strip (a one-dimensional problem), then replace each strip by all of its mass located at the strip's center of mass, and treat this as a horizontal one-dimensional line of masses, as in (c).

A Few Tips for Finding the Center of Mass

We have described the calculation of the center-of-mass of systems composed of discrete point masses and of continuous systems. In either case there are a few simple ideas to keep in mind that will help you to do the calculations—we have already used these in our calculations above, although we did not single them out.

Symmetry Helps: The center of mass of two equal masses lies midway on the line between them, the center of mass of a square flat uniform plate lies at the center of the square, and the center of mass of a sphere lies at its center if the distribution of its mass is spherically symmetric (no direction from the center different from any other direction).

How Subsystems Enter: You may have a system that you can think of as composed of subsystems, themselves distributions of mass. The center of mass of the whole system can be found by first finding the center of mass of the subsystems and then combining these into a final center of mass as if each subsystem were a particle (a point mass) with the total subsystem mass located at the subsystem's center of mass. (We used this in Conceptual Example 8–15 when we treated each strip as if it were a particle located at its own center of mass.) This result follows because of the simple linear nature of the calculation of the center-of-mass position and can be easily understood with a three-particle example in one dimension: Three particles with masses m_1, m_2, and m_3 are placed along the x-axis at positions x_1, x_2, and x_3, respectively. The center-of-mass position is

$$X = \frac{m_1 x_1 + m_2 x_3 + m_3 x_3}{m_1 + m_2 + m_3}.$$

Now we consider the "two-body" system consisting of particle 1 alone and particles 2 and 3 taken as a subsystem located at X_{23}, the center of mass of the 2–3 subsystem. We have $X_{23} = (m_2 x_2 + m_3 x_3)/(m_2 + m_3)$, and the expression of the full center of mass in this view is

$$X = \frac{m_1 x_1 + (m_2 + m_3) X_{23}}{m_1 + (m_2 + m_3)}.$$

It is no trouble whatever to verify that our two expressions are identical, and this result easily generalizes.

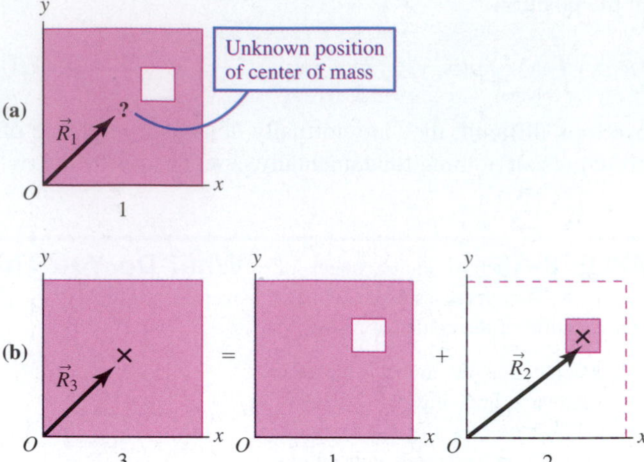

▶ **FIGURE 8–32** To find the center of mass of an object (1) with a hole in it, think of the object without the hole (3) as consisting of the original object plus another object that would fill the hole (2).

How to Deal with Holes: Sometimes it is easier to think of an object in terms of what isn't there than what is. We can use the paragraph above, How Subsystems Enter, to deal with this; again we'll proceed by example. Consider the flat plate with a hole cut in it, as in Fig. 8–32a. We'll label this system 1. We can think of this distribution entering as one piece of a full plate, without the hole, labeled as 3 in Fig. 8–32b; we find the center of mass of plate 3 by combining plate 1 with the material from the hole, labeled as 2. We know the center of mass of plate 3 and plate 2—by symmetry these points are at their centers and are marked $\vec{R}_2$ and $\vec{R}_3$ in Fig. 8–32b. We have

$$\vec{R}_3 = \frac{m_1\vec{R}_1 + m_2\vec{R}_2}{m_1 + m_2},$$

and we can solve this to find the unknown center of mass of the system with the hole,

$$\vec{R}_1 = \frac{(m_1 + m_2)\vec{R}_3 - m_2\vec{R}_2}{m_1}.$$

This technique is quite useful.

▲ **FIGURE 8–33** An astronaut can get a small push back toward the space shuttle by throwing a hammer in the opposite direction.

*8–7 Rocket Motion

A rocket is a system that undergoes a kind of continuous explosion (see Section 8–3). To understand rocket motion, imagine an astronaut who finds himself or herself at rest with respect to the vehicle but separated from it. No amount of bodily contortions will change the fact that the astronaut's center of mass remains at rest in the absence of external forces. However, the astronaut could, for example, throw a hammer in a direction away from the space vehicle. Then he or she will move toward the space vehicle and safety (Fig. 8–33). Momentum conservation lies behind this motion. The total momentum of the astronaut–hammer system is initially zero, and this remains true after the hammer is thrown. A rocket moves through the same principle. The ejected rocket fuel acts as a series of thrown hammers. The presence or absence of an atmosphere has nothing to do with the basic principle of operation.

A rocket is propelled by the ejection of hot gases from burning chemical fuel. By the "rocket" we mean the forward section containing the payload and any remaining unburned fuel. Two important parameters describe the rocket (Fig. 8–34). First there is the

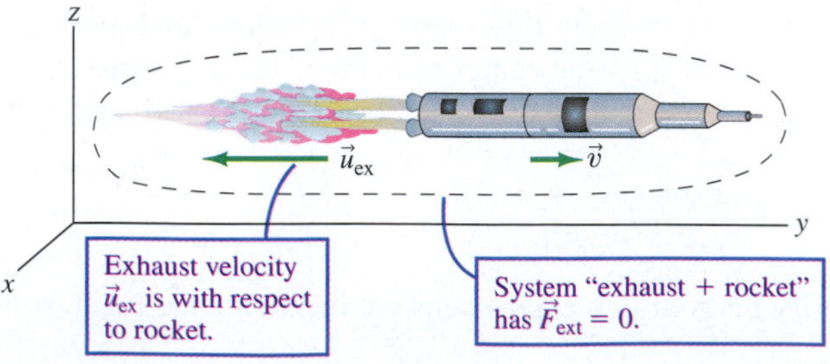

▶ **FIGURE 8–34** A rocket expels gases that have a speed u_{ex} with respect to the rocket. If we know the rate at which the gases are expelled, conservation of momentum allows us to find the velocity $\vec{v}$ of the rocket.

exhaust speed of the hot gases, u_{ex}, which by definition is positive. Keep in mind that the exhaust speed involves a burning reaction within the rocket and that it has a given value *with respect to the rocket*. Second, there is the rate of change of mass dm/dt of the rocket, that is, the rate at which the rocket expels the exhaust gases. Don't confuse these parameters; you could have two rockets that each burn fuel by the same reaction and for which u_{ex} is identical, but one could burn it at a very different rate than the other and therefore expel exhaust at a very different rate. In our discussion below, we'll suppose that all the motion is in one direction. If this direction is along a particular axis, any "velocity" is the component of the velocity along that axis. That component could be positive or negative.

Suppose now that the initial mass of the rocket with all its fuel is m_0; at some later time t, its velocity with respect to some ground-based (inertial) observer is v while its mass has been reduced to m. In the next small time interval Δt, the observer sees that a small bit of exhaust gas, mass Δm, has been ejected backward relative to the rocket; the observer sees this bit of gas moving with velocity $v - u_{ex}$, while he sees the rocket itself now traveling at a velocity $v + \Delta v$ and with a decreased mass $m - \Delta m$.

The conservation of momentum requires that the momentum change of the rocket be matched by the momentum carried away by the gas. In the absence of external forces, the total momentum at time t must equal the total momentum at time $t + \Delta t$. We have $P(t) = mv$, while $P(t + \Delta t) = (\Delta m)(v - u_{ex}) + (m - \Delta m)(v + \Delta v)$. The last two terms are the momentum of the exhaust gas and the rocket, respectively. Thus $P(t) = P(t + \Delta t)$ reads

$$\begin{aligned} mv &= (\Delta m)(v - u_{ex}) + (m - \Delta m)(v + \Delta v) \\ &= (\Delta m)v - (\Delta m)u_{ex} + mv + m\,\Delta v - (\Delta m)v - (\Delta m)(\Delta v), \end{aligned}$$

or, neglecting the doubly small term $(\Delta m)(\Delta v)$,

$$m\,\Delta v - (\Delta m)u_{ex} = 0. \qquad (8\text{–}64)$$

Divide Eq. (8–64) by the time interval Δt and take the limit $\Delta t \to 0$. Rearranging, we find $u_{ex}(dm/dt) = m(dv/dt)$. Now we are looking for an equation for the rocket. But the quantity dm/dt refers to the (positive) rate at which the ejected fuel mass increases. We can get an equation for the rocket itself by noting that the rate at which the ejected fuel mass changes is the negative of the rate at which the rocket mass changes. Thus if we change the sign of the dm/dt term we can interpret dm/dt as the rate of mass change of the rocket itself:

$$-u_{ex}\frac{dm}{dt} = m\frac{dv}{dt}. \qquad (8\text{–}65)$$

This equation looks just like Newton's second law $F = ma$. The left side is the force term, the **thrust** of the rocket, a positive quantity. It contains both of our rocket parameters.

Equation (8–65) may be written in a form that leads directly to a solution for the speed of the rocket as a function of its remaining mass by dividing both sides by mu_{ex}:

$$-\frac{1}{m}\frac{dm}{dt} = \frac{1}{u_{ex}}\frac{dv}{dt}. \qquad (8\text{–}66)$$

We now use

$$\frac{d(\ln m)}{dt} = \left(\frac{1}{m}\right)\left(\frac{dm}{dt}\right). \qquad (8\text{–}67)$$

[This formula follows from the calculus results in Appendix IV–7. We use $d(\ln x)/dx = 1/x$. We must replace x by m and use the chain rule of calculus.] Equations (8–66) and (8–67) imply

$$\frac{d}{dt}\left(\ln m + \frac{v}{u_{ex}}\right) = 0.$$

The quantity in parentheses must then be a constant,

$$\ln m = -\frac{v}{u_{ex}} + \text{a constant.} \tag{8–68}$$

Equation (8–68) describes the rocket's speed as a function of its mass. Assuming the rocket starts from rest at $t = 0$, we have at that time $m = m_0$ and $v = 0$. That means that the constant in Eq. (8–68) is given by $\ln m_0$. Equation (8–68) is then rearranged to give

$$v = u_{ex}[\ln m_0 - \ln m] = u_{ex} \ln\left(\frac{m_0}{m}\right), \tag{8–69}$$

ROCKET SPEED

where in the last step we have used the fact that $(\ln a) - (\ln b) = \ln(a/b)$. This is a fundamental equation in rocket propulsion.

What does Eq. (8–69) tell us? First of all, we note that the speed v is proportional to u_{ex}. This dependence is to be expected because u_{ex} is the only quantity with the dimensions of speed that appears in the problem. Next, we observe that the coefficient of u_{ex} must be a dimensionless quantity, and therefore it must be a function of m/m_0, which is the fraction of the rocket mass left over after time t. Because the logarithm is a very slowly varying function of its argument, the most productive way to get a sizable speed is by making u_{ex} large. Chemical rockets can generate gas exhaust speeds up to a maximum of about 4000 m/s, and thus the typical speeds of rockets are also in the range of 10^3 m/s. Much higher exhaust speeds can be achieved with nuclear reactions, but rockets that use nuclear energy for propulsion in this way present enormous environmental problems. A different way to increase the speed of the rocket would be to carry more fuel, that is, increase m_0/m. But large values of m_0/m help only moderately; for example, $\ln 10 = 2.3$, so changing the mass ratio by a factor of 10 improves the speed by a factor of 2.3.

Rocket Motion in the Presence of Gravity

The previous discussion involves a rocket free of all external forces. When gravity acts on the rocket, Eq. (8–69) is modified. In Problem 92 we outline the steps that lead to the conclusion that the velocity (assumed one-dimensional as before, here vertical) includes a new term, $-gt$, in addition to the term proportional to $u_{ex} \ln(m_0/m)$ due to the fuel exhaust. (The new term corresponds to free fall under the influence of gravity, as we already know from Chapter 2. Thus Eq. (8–69) is modified to

$$v = u_{ex} \ln(m_0/m) - gt. \tag{8–70}$$

EXAMPLE 8–16 A Saturn V rocket (the vehicle that sent humans to the Moon) of mass 2.5×10^6 kg takes off from Earth in a vertical direction. It burns fuel at a uniform rate of 1.6×10^4 kg/s for a duration of 2 min. If the exhaust speed of the gas is given by 3.0 km/s, what is the speed of the rocket immediately after the combustion ceases?

Setting It Up We know the rocket's initial mass m_0 as well as the rate dm/dt at which it loses mass through the burn; we also know the burn time t and the exhaust speed u_{ex}.

Strategy This is a direct application of the equation that describes rocket speed, Eq. (8–70). All the parameters of the right side of that equation are known, except that we need to calculate the final mass, a calculation that involves a known starting mass and a given rate of mass loss for a given time.

Working It Out The mass m in Eq. (8–70) is the final mass of the rocket after the gas has been burned,

$$m = m_0 - \frac{dm}{dt}t = 2.5 \times 10^6\ \text{kg} - (1.6 \times 10^4\ \text{kg/s})(120\ \text{s})$$
$$= 0.58 \times 10^6\ \text{kg}.$$

We also have

$$\ln\left(\frac{m_0}{m}\right) = \ln\left(\frac{2.5 \times 10^6\ \text{kg}}{0.58 \times 10^6\ \text{kg}}\right) = \ln\left(\frac{2.5}{0.58}\right) = 1.5.$$

Thus, from Eq. (8–70),

$$v_f = (3.0 \times 10^3\ \text{m/s})(1.5) - (9.8\ \text{m/s}^2)(120\ \text{s})$$
$$= 3.2 \times 10^3\ \text{m/s}.$$

Our calculations are correct only near the surface of Earth. Earth's gravitational force on the rocket decreases as the rocket moves away from the surface, so in effect g decreases as the rocket moves away.

What Do You Think? If you were interested in sending a rocket such as this far far away, would it help to launch it from the Moon rather than from Earth?

*8–8 Momentum Transfer at High Energies

High-energy physics is the study of the properties of the elementary particles that make up matter at the most fundamental level. These properties are studied with the aid of high-energy accelerators. There, projectile particles such as protons or electrons are accelerated to high speeds and allowed to collide with target particles of the same type. The results of such collisions are then analyzed by means of sophisticated detectors in which the energies and momenta of the particles that emerge after the collisions are precisely measured. The patterns revealed in these measurements are used to probe the underlying structure of the colliding particles.

THINK ABOUT THIS . . .

HOW DO WE KNOW THAT ATOMS HAVE A NUCLEAR STRUCTURE?

The fact that the atom has a massive central structure was discovered in experiments carried out by Ernest Rutherford in the first decade of the twentieth century. These experiments involved the scattering of projectiles known as α particles (later these were discovered to be themselves the nuclei of helium atoms) from atoms. Imagine a BB gun fired at two targets, one of cotton candy and the other of cotton candy but with a marble implanted at its center. A BB shot (the projectile) will hardly be deflected by the first target. Most of the time, the same will be true for the second target. Occasionally, however, the projectile will penetrate the second target close to its center and, upon colliding with the marble, will be deflected through a large angle. An analysis of the collisions in the two cases would allow us to deduce that the second target had a hard, compact center (the marble). The pattern of many projectile deflections would even allow us to say whether the marble is spherical or cubical. Rutherford's experiments were similar, with the nucleus playing the role of the marble in the cotton candy. His projectile was occasionally deflected through large angles, and analysis of the pattern revealed a dense small center, the atomic nucleus. ■

There is a benefit in increasing the speed—or momentum—of the projectile in scattering experiments, a benefit that the rules of classical physics could not have predicted. For collisions of atomic or subatomic particles, the rules of *quantum mechanics* apply. This is a field of study developed in the first part of the twentieth century, with some surprising consequences. One consequence of quantum mechanics is that high momenta (or high energy) are *required* to study the regions close to the center of a target. More precisely, the **Heisenberg uncertainty relation**, discovered by physicist Werner Heisenberg, describes a relation between the magnitude of the momentum change, $\Delta\vec{p}$, of a target particle in a collision and the size of the region around the center of the target that can be studied. Let us denote by r_{coll} the smallest radius of a spherical region around the center of a target that can be studied in a collision. Then the uncertainty relation is

$$|\Delta\vec{p}|r_{\text{coll}} \cong \hbar, \tag{8–71}$$

where $\hbar \cong 10^{-34}\ \text{J}\cdot\text{s}$ is Planck's constant divided by 2π. According to Eq. (8–71), the larger the value of $\Delta\vec{p}$, the smaller r_{coll} can be.

Experiments with elementary particles—like the BB gun fired at cotton candy—have been carried out and have revealed much about the structure of these particles. We have already mentioned the structure of the atom itself in the Think About This box. More recently, experiments involving protons colliding at momenta that could probe down to distances of 10^{-16} m revealed that the proton, which is itself a component of the nucleus, contains "pellets" known as *quarks*. When a next generation of accelerators is built, we will discern even more detail. Judging by the enormous difference between the physics of atoms and the physics of nuclei, we can expect some surprises as we probe more and more deeply.

Summary

The momentum (or linear momentum) of an object of mass m moving with velocity $\vec{v}$ is defined by

$$\vec{p} \equiv m\vec{v}. \tag{8–2}$$

The rate of change of momentum with time is given by Newton's second law:

$$\vec{F} = \frac{d\vec{p}}{dt}, \tag{8–3}$$

where $\vec{F}$ is the net force acting on the object. The momentum of a collection of objects with momenta $\vec{p}_i$ (where $i = 1, 2, \ldots, N$) is their vector sum, and the net external force on the system determines the rate of change of the total momentum. In particular, if the only forces present are the internal forces exerted by the objects on each other, then the total momentum does not change with time; that is, momentum is conserved. In the case of two objects, this relation is

$$\text{for zero net external force: } \vec{p}_1 + \vec{p}_2 = \text{a constant.} \tag{8–9}$$

One consequence of Eq. (8–3) is that the impulse $\vec{J}$, the change in momentum of an object influenced by a force that acts over a limited time (an impulsive force), is an integral of the force over time:

$$\Delta\vec{p} = \vec{J} = \int \vec{F}\,dt = \vec{F}_{\text{av}}\,\Delta t. \tag{8–15}$$

The impulse does not involve the detailed time dependence of the force.

Collisions involve the (generally brief) interactions between pieces of an isolated system. Since the system is isolated, the total momentum is conserved, and the vector equations that express this conservation law is one set of constraints on the kinematics of the particles involved in the collision. We can also distinguish collisions by whether energy is or is not dissipated in the process. Accordingly, collisions may be elastic, inelastic, or perfectly inelastic. In elastic collisions—the collisions of billiard balls is a close approximation—the sum of the kinetic energies is conserved. In inelastic collisions the sum of kinetic energies in the initial state, before the collision, differs from the sum of kinetic energies in the final state. The change is generally a decrease due to the presence of dissipative forces such as friction, or by deformations, but an increase in kinetic energy is also possible, as in explosions. The maximum amount of kinetic energy is lost in perfectly inelastic collisions, where the objects that collide merge into a single object.

The center of mass of a collection of objects with masses m_i and position vectors $\vec{r}_i$ $(i = 1, 2, \ldots, N)$ is the mean position of a system's mass, which is defined by

$$\vec{R} \equiv \frac{m_1\vec{r}_1 + m_2\vec{r}_2 + \cdots + m_N\vec{r}_N}{M}. \tag{8–47}$$

In the absence of external forces, the center of mass moves with uniform velocity $\vec{V}$, given by

$$\vec{V} \equiv \frac{d\vec{R}}{dt} = \frac{m_1\vec{v}_1 + m_2\vec{v}_2 + \cdots + m_N\vec{v}_N}{M} = \frac{\vec{P}}{M}. \tag{8–49, 8–50}$$

The total mass of the system is $M = m_1 + m_2 + \cdots + m_N$. The center of mass for a continuous object may be calculated in terms of the mass density. For a one-dimensional object with mass density $\lambda(x)$, it is given by

$$X = \frac{1}{M}\int_a^b x\lambda(x)\,dx, \tag{8–59}$$

where the mass is related to $\lambda \equiv dm/dx$ by

$$M = \int_a^b \lambda(x)\,dx. \tag{8–58}$$

For a two-dimensional object with mass density $\sigma(x, y)$, the corresponding expressions are

$$X = \frac{1}{M}\int_{\text{Surface}} x\sigma(x, y)\,dx\,dy \tag{8–61a}$$

and

$$Y = \frac{1}{M}\int_{\text{Surface}} y\sigma(x, y)\,dx\,dy, \tag{8–61b}$$

where

$$M = \int_{\text{Surface}} \sigma(x, y)\, dx\, dy. \tag{8–60}$$

Similar expressions hold for a three-dimensional object with mass density $\rho(x, y, z)$.

When a net external force $\vec{F}_{\text{net}}$ acts on a system, the center of mass continues to move in a simple way, as described by Newton's second law: The position of the center of mass $\vec{R}$ obeys the equation

$$\vec{F}_{\text{tot, ext}} = \frac{d\vec{P}}{dt} = M\frac{d^2\vec{R}}{dt^2} = M\vec{A}. \tag{8–54, 8–55}$$

Here $\vec{A}$ is the acceleration of the center of mass.

The concept of momentum conservation is useful for the description of rocket motion. For a rocket that is not influenced by external forces, whose mass is initially m_0 and whose mass at time t is m, the speed is

$$v = u_{\text{ex}} \ln\left(\frac{m_0}{m}\right). \tag{8–69}$$

Here, u_{ex} is the exhaust speed of the gas relative to the rocket.

Understanding the Concepts

1. A comet enters the solar system, is deflected, and then leaves the solar system in a direction different from the direction by which it entered. Is momentum conserved in this situation?
2. A common toy consists of a series of five balls that touch and form a line; each ball is suspended as a pendulum. When ball 1 is pulled away and then released, it strikes the line and ball 5 rises in a motion like the reverse of ball 1. How can you explain this motion?
3. Two pucks collide on a table top and eventually come to rest. Does this mean that their center of mass comes to rest? If so, does this violate the principle that the center of mass of an isolated system moves with constant velocity?
4. Describe a physical object for which the center of mass is not actually inside the object.
5. A diver leaps off the diving board, performing a difficult series of maneuvers. Can he cause his center of mass to perform a midair loop?
6. You have two masses, labeled 1 and 2, moving in space, and their momenta are equal and opposite, $\vec{p}_1 + \vec{p}_2 = 0$. Pick out the statements that are necessarily true.
   (a) The two masses are going to collide.
   (b) The masses could be moving in two planes that are parallel to one another.
   (c) The masses are moving parallel to one another but not necessarily along the same line.
   (d) The masses are moving along the same line.
7. The center of mass of a championship-level high jumper passes below the bar even though the jumper passes above the bar. How is this possible?
8. A particle collides with another particle at rest. If there are two particles observed to come out of the collision, do the momenta of the two final state particles and the initial incoming particle have to lie in a plane? Why or why not? What if the final system consists of three particles?
9. A piece of machinery is modeled as a thin circle of iron, radius R, lying in the xy-plane and centered at the origin, with a second point mass on the x-axis at the point $x = R$. Is the center of mass within the circle? How would you find the position of the center of mass of this system with the least calculation?
10. A vase falls to the floor and shatters. Is momentum conserved in the collision? What objects need to be taken into account in describing the conservation of momentum?
11. As a tennis racket hits a tennis ball, the racket continues to move forward. Is this consistent with the conservation of momentum?
12. If a tennis player wants more power, should he or she choose a racket with more or less tension on the strings? If the player wants more control? [*Hint*: Think about how long the ball is on the strings in each case.]
13. The mass on the end of a pendulum swings in part due to the effects of gravity on the mass. Does momentum have to be conserved in the interaction of this mass with Earth? If so, does this mean that Earth moves back and forth along with the mass?
14. A very small child and a large adult want to use a see-saw, a board with a pivot at its midpoint. By comparison to the position of the child, should the adult sit (a) closer to the midpoint, (b) farther from the midpoint, or (c) at the same distance from the midpoint? Justify your answer.
15. We have said that an impulsive force acts for only a short time. What decides whether a time is "short" or not?
16. Cricket players catch balls hit as hard as baseballs but they do not use padded mitts. How do they avoid injuring their hands?
17. An amusing experiment involves the use of a golf ball (or any other small, light ball) and a large ball. If you keep the large ball stationary on the ground and drop the small ball onto it, the small ball will bounce back to almost the height from which you dropped it. If, however, you take both balls and, holding the smaller ball just above the larger one, drop them together, you will find that the small ball might hit the ceiling with some force! Try this and explain what is going on.
18. If you have the misfortune to be in an automobile collision, you are better off in a more massive car (all other things being equal). Why?

19. In films with naval battles, especially involving pirate ships, the guns that fire always roll back. Why is that?

20. If you have the misfortune to be in an automobile collision, you are better off in a car that tends to crumple on impact rather than a car that holds together stiffly. Why?

21. A closed railroad car is at rest on a flat stretch of track. A cannon located inside the car at the front end points to the rear. What is the motion of the railroad car when the cannon is fired and the shell is absorbed by the rear wall?

22. A large, closed crate contains many pigeons that sit on the floor of the crate. A sudden noise makes them all fly up and hit the top of the crate at the same time. Will they be able to lift the crate off the ground?

23. Drag forces act when a parachute opens. The parachute and its load will slow down to a terminal velocity. Why is the drag force velocity dependent? Think of the parachute as colliding with a lot of tiny air molecules.

24. Suppose two identical billiard balls moving with equal and opposite velocities along a line have a head-on collision. How will this look to an observer moving with a uniform velocity equal to that of one of the billiard balls?

25. A billiard ball strikes the cushion of a billiard table at an angle of 45°. Assuming that there is no energy loss (and no effects due to the spin of the ball), what can you say about the angle at which the billiard ball bounces off? What would happen if there were some energy loss?

Problems

8–1 Momentum and Its Conservation

1. (I) Calculate the magnitudes of the momenta of (a) a 40-g arrow traveling at a speed of 110 km/h; (b) a 145-g baseball traveling at a speed of 35 m/s; (c) a 72-kg sprinter running at 22 mi/h; (d) a 95-kg tackler running 100 m in 12.5 s.

2. (I) A 1.65-kg mass falls vertically downward from a roof 27.5 m high. What is the momentum of the object after 1.6 s given that the initial velocity is zero?

3. (I) Calculate the magnitudes of the momenta of (a) a man of mass 70 kg, running 6 m/s; (b) a freight car of 100,000 kg, moving 60 m/s; (c) a car of 1100 kg, moving 25 mi/h; (d) a proton moving at 2×10^5 m/s; (e) a feather of 10 g in an airless container that has fallen 10 cm due to gravity.

4. (II) An object of mass m is constrained to move in a circle of radius R by a central force F. (a) What is the magnitude of the momentum of the object? (b) Suppose that the force has magnitude $F = Kv$, where K is a constant and v is the speed of the object. Calculate R in terms of K and the momentum of the object. (This kind of force acts on a charged object in a uniform magnetic field.)

5. (II) A 7-kg rifle is used to fire a 10-g bullet that travels with a speed of 700 m/s. (a) What is the speed of recoil of the rifle? (b) How much energy does it transmit to the shoulder of the person using the rifle as it stops?

6. (II) Two objects, of masses m and M, respectively, move in circular orbits that have the same center in such a way that they are always at opposite sides of a diameter (Fig. 8–35). If the force that gives rise to this motion is a force of attraction between the two objects acting along a line joining them, using momentum conservation and Newton's second law, (a) show that they move with the same angular speed (see Chapter 3) and (b) calculate the ratio of the radii of the two circular orbits.

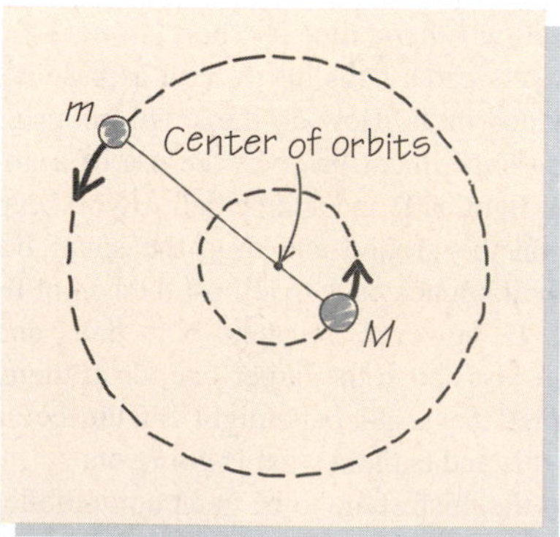

▲ **FIGURE 8–35** Problem 6.

7. (II) Two objects are moving in the xy-plane. The first, with mass 2.4 kg, has a velocity $\vec{v}_1 = (-2.0\text{ m/s})\hat{i} + (-3.5\text{ m/s})\hat{j}$; the second object, with mass 1.6 kg, has velocity $\vec{v}_2 = (1.8\text{ m/s})\hat{i} + (-1.5\text{ m/s})\hat{j}$. (a) What is the total momentum of the system? (b) If the system observed at a later time shows that the 2.4-kg object has $\vec{v}_1' = (2.5\text{ m/s})\hat{i}$, what is the velocity of the 1.6-kg object? (c) Consider again the initial situation. Now suppose that there has been a mass transfer so that the first object now has a mass of 2.1 kg. The total mass is conserved. What is $\vec{v}_1'$ if the velocity of the second object is $(-2.5\text{ m/s})\hat{j} + (1.3\text{ m/s})\hat{k}$? (d) Calculate the sum of the kinetic energies in the initial configuration and in the configurations of parts (b) and (c). Compare and discuss them.

8. (III) (a) A billiard ball collides head on with a second billiard ball that is touching a third billiard ball placed directly behind it. Describe the subsequent motion of all three billiard balls. Assume that all collisions are elastic. (b) A moving billiard ball hits the two balls at rest precisely between the two touching balls at rest. The velocity of the moving ball is perpendicular to the line between the centers of the two balls at rest. Again, all collisions are elastic. Describe the subsequent motion now.

8–2 Collisions and Impulse

9. (I) A baseball of mass 0.15 kg moving horizontally with a momentum of 4.0 kg · m/s is struck head on by a baseball bat with an impulse of 10 N · s. What is the speed of the baseball after it is struck?

10. (I) If a 32-g arrow moving at 160 km/h penetrates a block of wood suspended by a rope, what impulse is delivered to the block? What velocity will the wood block acquire if its mass is 3.5 kg? Ignore the mass of the arrow compared to that of the wood block.

11. (I) A 145-g baseball traveling at a speed of 36 m/s hits a bat and moves back along its incoming trajectory with a speed of 45 m/s. What is the impulse delivered to the ball by the bat? If the duration of the bat–ball collision is 7.0×10^{-4} s, what is the average force exerted by the bat on the ball during this period?

12. (I) A 440-g ball is dropped from a height of 3.2 m onto a hard floor and bounces back to exactly that height. (a) What is the impulse received by the ball? (b) What is the average force on the ball during the 0.008 s during which, an independent measurement shows, it was in contact with the floor?

13. (I) A fire rescue unit uses a tightly woven net to catch an 80-kg person who jumps out of a burning building from a height of 11 m. What is the impulse transmitted to the net? If the net sinks 70 cm as it slows down the jumper, what is the average force exerted on the jumper by the net?

14. (II) (a) Estimate the possible impulse received by a nail when it is hit by a hammer. (b) Estimate the possible impulse received by a football when it is kicked in a kickoff. The ball has mass of about 250 g and travels a horizontal distance of some 50 m. Note the difference between parts (a) and (b); in the latter, you have additional information on the momentum the ball receives.

15. (II) A high jumper of mass 55 kg clears 6 ft 7 in. What is the impulse transmitted to the jumper by the ground? (It has been found that the center of mass of the high jumper remains *below* the bar. In this problem, assume that the center of mass starts at 3 ft 10 in and reaches the height 6 ft 7 in.)

16. (II) A communications satellite of mass 850 kg can be ejected from the cargo bay of the space shuttle by means of springs. A particular satellite is ejected at 0.45 m/s. (a) What impulse does the spring provide? (b) If the spring operates over a time period of $\Delta t = 0.85$ s, what average force does the spring provide?

17. (II) A ball of mass 260 g is dropped from a height of 2.0 m. It hits the ground and rebounds to a height of 1.4 m. Assuming that the ball is in contact with the ground for 0.004 s, what is the average force exerted on the ball during the contact?

18. (II) A golf ball of mass 0.05 kg placed on a tee is struck by a golf club. The speed of the golf ball as it leaves the tee is 100 m/s. (a) What is the impulse? (b) If the time of contact between the club and the ball is 0.02 s, what is the average force? (c) If the force decreases to zero linearly with time during the 0.02 s, what is the value of the force at the beginning of the contact?

19. (II) A catcher catches a 45-m/s pitch and his glove recoils 0.25 m. The mass of the ball is 0.14 kg. Assume that the deceleration of the ball during the catching time, Δt, is constant and equal to some a_{av}. (a) Find the magnitude of the average force F_{av} that the catcher exerts. (b) How much work does he do? (c) Find Δt.

20. (II) Superman rushes to save Lois Lane, who has fallen (has been pushed?) from a window 65 m above a crowded street. Superman swoops down in the nick of time, arriving when Lois is 1.0 m above the street and stopping her just at ground level. Lois has a mass of 52.5 kg. Ignore air resistance throughout. (a) What is the impulse that Lois receives as Superman catches her? (b) If the force that Superman supplies in stopping Lois is constant, how long does it take for Lois to come to rest? (c) What is the average force Superman applies to Lois? Compare this to the force of gravity on her. Draw your own conclusions about these last-minute rescues.

21. (II) A ball of mass 150 g is dropped from a height of 60 cm onto the first step at the top of a staircase. Each step is 20 cm above the next. The ball bounces perfectly elastically, but it has a small horizontal velocity, so it hits the second step on the next bounce, and then the third, and so on. Assume that the size of each step is such that the ball always bounces onto the next step down. What is the impulse transmitted to the nth step?

22. (II) A golfer playing on a level course hits a ball that leaves the tee at a 30° upward angle and hits the ground 175 m away. What was the impulse given to the ball by the swinging club? If you need the information, you can estimate the mass of the golf ball.

23. (II) Two apparently identical balls of the same mass 120 g are dropped from rest 1.5 m above a floor. One of them is a "superball," and when this ball hits the floor, it bounces up such that its speed immediately after leaving the floor is equal to its speed just before it hits. The other ball hits the floor with a thud and just stops there. What is the impulse received by the floor in each case?

24. (II) A brother and sister live near a lake. In the middle of winter, when the lake is solidly and smoothly frozen, they play catch by sliding a rock back and forth between them across the ice. Ignore all friction. The mass of the two children is 40 kg each, and the rock has a mass of 2.0 kg. Each time the rock is given a push by the child who has just caught it, it leaves the hand with a speed of 1.0 m/s (relative to the hand). (a) The children are initially at rest, and the sister first slides the rock toward her brother. What is the initial motion of the sister after she releases the rock? (b) The brother receives the rock. What is his initial velocity after he has received it? (c) The brother then slides the rock to his sister. What is his velocity now? (d) The sister catches the rock. What is her velocity now? (e) Will either child ever have a speed such that the rock will never reach the other player? In other words, can the game go on forever?

8–3 Perfectly Inelastic Collisions; Explosions

25. (I) A 250-g cart moves on an air track (a one-dimensional system in which motion is friction free) at 1.2 m/s. It collides with and sticks to another cart of mass 500 g, which was moving in the opposite direction at 0.80 m/s before the collision. What is the velocity of the composite cart after the collision?

26. (I) An object of mass m_0 with speed v_0 hits another object of mass m at rest. The two masses stick together in the process. Find the fractional change in kinetic energy.

27. (I) A snowball of mass 400 g is thrown with a speed of 10 m/s and hits the loosely placed head of a snowman. If the snowball sticks to the head, which is of mass 5 kg, with what initial speed will the now enlarged head recoil?

28. (I) A fireworks rocket is shot straight up in the air. Its "payload," mass 3.80 kg, explodes just as it reaches its peak height and separates into two fragments before eventually exploding again into spiraling colors. The first fragment, of mass 1.1 kg, heads straight down with an initial speed of 15 m/s. What is the velocity of the second fragment immediately after the explosion?

29. (I) Two objects each move with speed v in opposite directions along a line. They meet and have a perfectly inelastic collision. After the collision, the composite object moves along the same line with a speed of $v/2$. What is the ratio of the masses of the two objects?

30. (I) Two automobiles have a perfectly inelastic collision. The automobiles, which are identical models except for the color and which contain only drivers of identical masses, meet at an intersection. Each was moving with speed 12 mph, one coming from the south, the other from the east. What is the velocity of the final composite object?

31. (II) An asteroid of diameter 1.5 km and density of 7.1×10^3 kg/m^3 traveling at 25 km/s hits the Moon in a completely inelastic collision. Estimate the maximum fractional change in velocity of the Moon in its orbit as a result of this impact.

32. (II) An ^{241}Am nucleus at rest emits an alpha particle (a ^{4}He nucleus). The energy released during the process is 6 MeV $= 9.6 \times 10^{-13}$ J. (a) What are the speeds of the alpha particle and the remaining ^{237}Np nucleus? (b) What is the kinetic energy of the ^{237}Np nucleus? (The masses of the nuclei may be taken to be $241m_0$, $4.0m_0$, and $237m_0$, respectively, where $m_0 = 1.66 \times 10^{-27}$ kg.)

33. (II) In a *ballistic pendulum*, a bullet of mass m and speed v embeds in a block of mass M suspended by a string (Fig. 8–36). The block and bullet, of total mass $m + M$, then move as a pendulum and the maximum height h that it reaches can easily be measured. This apparatus can be used to measure the speed of the bullet. (a) What fraction of the bullet's kinetic energy is lost in the collision? (b) Give a formula for measuring v in terms of m, M, g, and h.

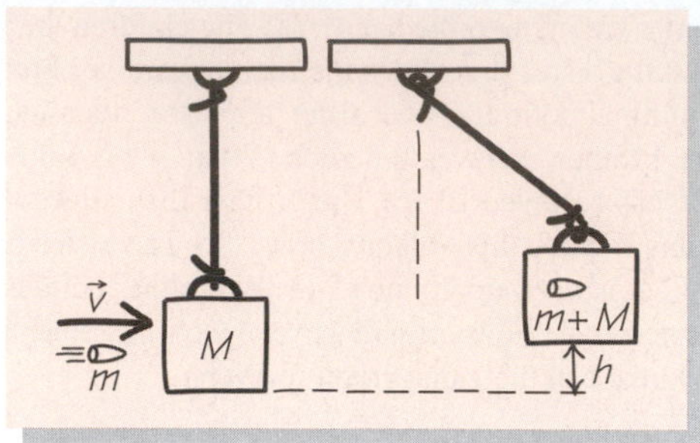

▲ **FIGURE 8–36** Problem 33.

34. (II) A bullet, mass 18 g, strikes a ballistic pendulum (see Problem 33) whose target block has a mass of 1.8 kg. The block is observed to rise to a height of 4.5 cm. What was the bullet's speed?

35. (III) Two persons, one of mass m_1 and the other of mass m_2, are sitting initially at rest on a bobsled of mass M on a frozen lake. Answer the following questions for a frame fixed to the lake. (a) What is the final velocity of the bobsled if the person of mass m_1 jumps off the rear of the bobsled with horizontal velocity v followed a few seconds later by the second person, who also jumps off in the same way with the same horizontal velocity with respect to the sled? (b) What is the velocity of the bobsled if the person of mass m_2 jumps off first? (c) What is the velocity if both jump off at the same time?

8–4 Elastic and Inelastic Collisions

36. (I) An object with velocity $(1.4\ \text{m/s})\hat{i}$ and mass 0.30 kg collides with an object whose velocity is $(-2.5\ \text{m/s})\hat{i}$ and whose mass is 0.15 kg. The motion takes place in one dimension. (a) What are the final velocities of the objects if the collision is elastic? (b) What is the total initial kinetic energy in the collision?

37. (I) On an air track, a 0.4-kg mass m_1 moves at 3.0 m/s in the positive direction. It approaches a stationary mass m_2 of 0.8 kg. They collide and, after the collision, the velocity of mass m_2 is 1.6 m/s in the positive direction. (a) What is the velocity of mass m_1 after the collision? (b) Is this collision elastic or inelastic? If the latter, what percentage of the maximum possible kinetic energy loss occurs?

38. (I) A bullet of mass 70 g is moving horizontally at a speed of 450 m/s when it strikes a 2.6-kg block at rest. (a) The block is made of wood; the bullet penetrates and stops. What is the initial speed of the bullet–block combination? (b) The block is made of a very hard steel, as is the bullet, and the collision is perfectly elastic. What is the resulting velocity of the block? [*Hint*: Use the approximation that the mass of the bullet is much less than the mass of the block.]

39. (II) A block of mass 126 g is moving along the $+x$-axis with a speed of 0.875 m/s. Just ahead of it is a 9.66-kg mass moving in the same direction with the same speed. At some point, the large mass hits a wall and bounces off the wall perfectly elastically (Fig. 8–37). What is the return speed of the small mass after its elastic collision with the large mass?

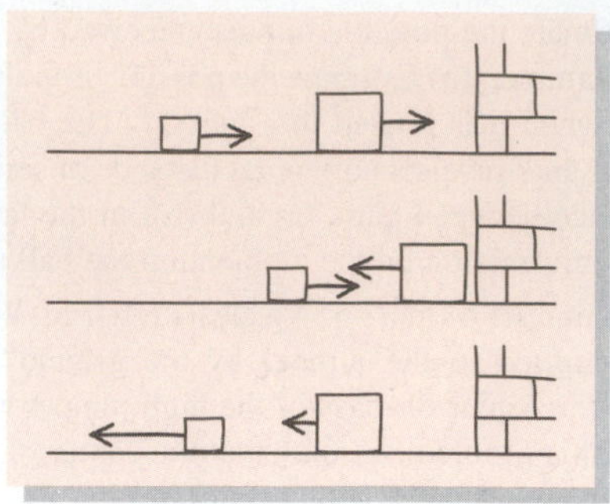

▲ **FIGURE 8–37** Problem 39.

40. (II) A machine gun in automatic mode fires 20-g bullets with $v_{\text{bullet}} = 300$ m/s at 60 bullets/s. (a) If the bullets enter a thick wooden wall, what is the average force exerted against the wall? (b) If the bullets hit a steel wall and rebound elastically, what is the average force on the wall?

41. (II) Two spheres of masses 400 and 600 g, respectively, are suspended from the ceiling by massless strings 1.00 m long. The lighter sphere is pulled aside through an angle of 80° and let go. It swings and collides elastically with the second sphere at the bottom of the swing. How high will the lighter of the spheres swing (in terms of an angle)?

42. (II) A pendulum of mass $m = 0.37$ kg and length $l = 0.95$ m is released from a horizontal position. At the bottom of its swing, it collides elastically with a mass $M = 0.56$ kg. (a) What is the velocity of mass M right after the collision? (b) To what height above its low point does mass m rebound?

43. (II) A superball has collisions that are nearly perfectly elastic. A superball of mass M is dropped from rest from a height h (where $h \gg$ the size of the superball) together with a smaller marble of mass m; the marble is initially just a little above the top of the superball and remains right over it throughout the fall. The superball hits the floor first and immediately rebounds elastically, colliding with the marble. (a) What is the speed of the superball and the marble just before the superball hits the floor? (b) Just after the superball rebounds from the floor but before it hits the marble? (c) What is the velocity of the marble after the superball hits it in the head-on collision? (d) How high does the marble go after its collision with the superball, assuming the marble has stayed in line with the superball so that all the motion is vertical? (e) What is the answer to part (d) in the limit $M \gg m$?

44. (II) Consider a perfectly elastic collision in one dimension between a ball of mass M and another of mass m, where $M \gg m$. The light ball is initially at rest, and the heavy ball has a given initial velocity v_i and final velocity v_f. What is the fractional velocity change $(v_i - v_f)/v_i$ of the large mass? Use $M \gg m$ to find an approximate expression.

8–5 Elastic Collisions in Space

45. (I) A ball of mass 1.2 kg moves along the positive x-axis with a speed of 2.4 m/s. Another ball (mass 0.80 kg) moves along the negative x-axis with a speed of 3.6 m/s. After colliding with each other, the lighter ball moves at a speed of 1.8 m/s along a line that makes an angle of 60° with the positive x-axis. What are the speed and direction of the heavier ball? Is the collision elastic?

46. (I) Two objects of equal mass approach each other with equal but opposite velocities along the x-axis. After a collision, one particle has velocity $v_1\hat{i} + v_2\hat{j}$. What is the velocity of the other particle?

47. (I) A billiard ball with velocity $\vec{v} = (2.50\text{ m/s})\hat{i}$ strikes a stationary billiard ball of the same mass. After the collision, the first billiard ball has velocity $\vec{v}_1 = (0.50\text{ m/s})\hat{i} + (-1.00\text{ m/s})\hat{j}$. What is the velocity of the second ball? Is the collision elastic?

48. (I) Suppose there are two billiard balls of equal mass; one is at rest and one is moving with a speed of 3.5 m/s. They collide elastically. After the collision, one of the two balls is measured to be moving with speed 2.3 m/s. What is the speed of the other ball?

49. (I) A billiard ball moving at 3.0 m/s collides with another billiard ball at rest. The balls move off at right angles to one another. If the first ball continues with a speed of 1.5 m/s, what is the speed of the ball that was initially at rest?

50. (II) Two objects with masses 2.0 and 3.0 kg move toward each other, both with speeds $v_0 = 5.0$ m/s. They collide head on and stick together. (a) Calculate their final velocity. (b) Calculate the amount of kinetic energy lost during the process. (c) Suppose the two masses approach each other at 90° before the collision (e.g., along the x- and y-axes). What will be the kinetic energy loss in this case?

51. (II) In a target shooting game, wooden blocks are thrown into the air and shot in flight. A block of 0.80 kg has a speed of 10 m/s at the top of its trajectory when it is hit by a bullet from below at an angle of 60° from the horizontal (Fig. 8–38). The mass of the bullet is 5.0 g and its speed is 550 m/s when it hits the block. The bullet is embedded in the block. What is the velocity of the block immediately after impact?

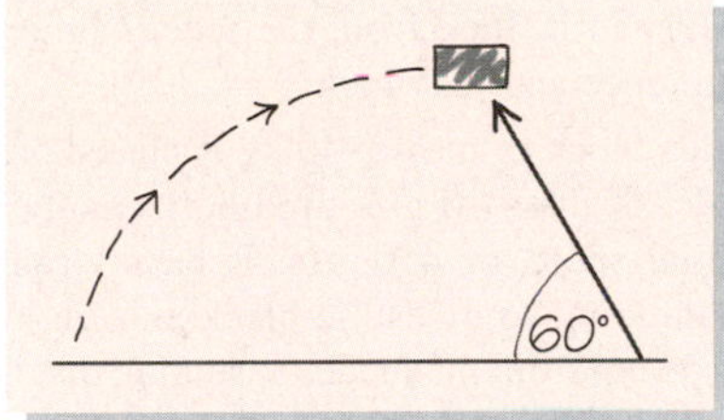

▲ **FIGURE 8–38** Problem 51.

52. (II) An air puck with mass 0.15 kg and velocity $(-1.7\text{ m/s})\hat{i} - (2.0\text{ m/s})\hat{j}$ on a frictionless table collides with a second air puck of mass 0.22 kg and velocity $(3.6\text{ m/s})\hat{i}$ (Fig. 8–39). As a result of the collision, the first air puck comes to rest. What is the kinetic energy of the second air puck after the collision?

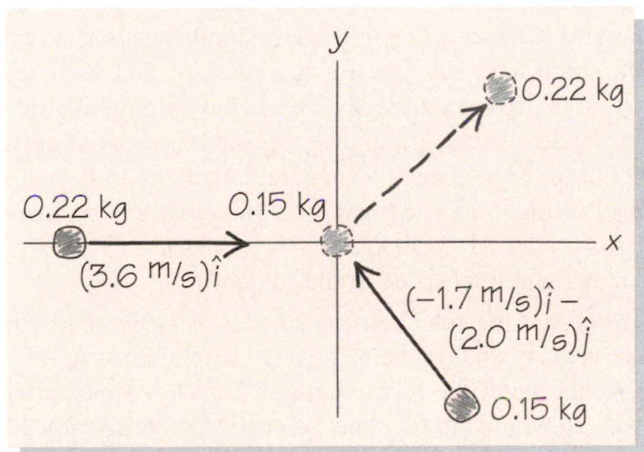

▲ **FIGURE 8–39** Problem 52.

53. (II) In a court hearing, a police expert reconstructs an accident in the following way (Fig. 8–40): A sports car of mass $m_1 = 1000$ kg collided with a parked pickup truck of mass $m_2 = 1500$ kg. From the skid marks, it is estimated that the speed of the pickup immediately after the collision was 21.6 m/s at an angle of 33.7° with the direction of the road. The sports car left a mark at an angle of 60° with the road but did not stop, leaving the scene badly damaged. Can the expert determine from these data if the sports car was speeding and its driver fully responsible?

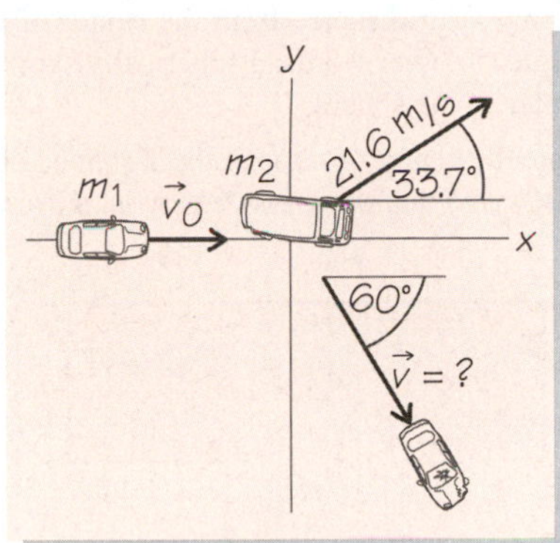

▲ **FIGURE 8–40** Problem 53.

54. (III) A proton of mass m and kinetic energy K scatters elastically from an alpha particle at rest (with mass four times that of the proton). The proton is deflected at 30° from its original direction. (a) At what angle does the alpha particle recoil? (b) Draw a momentum diagram of the collision. (c) What are the final energies of the two particles?

8–6 Center of Mass

55. (I) A massless, rigid bar that is 2.24 m long connects two tiny spheres (of negligible radii compared to the length of the bar) of masses 895 and 478 g, respectively. Locate the system's center of mass.

56. (I) Point masses of 0.15, 0.40, and 0.25 kg are located in the xy-plane at the origin, at $\vec{r}_2 = (35\text{ cm})\hat{j}$, and at $\vec{r}_3 = (15\text{ cm})\hat{i} + (58\text{ cm})\hat{j}$, respectively. Where is their center of mass?

57. (I) Redo Example 8–12 by placing the origin at the location of mass m_3, which was added to the original system.

58. (I) What is the velocity of the center of mass of the system in Problem 8, both before and after the collision (both parts), if the speed of the moving billiard ball is v?

59. (I) Two children of masses 25 and 30 kg, respectively, stand 2.0 m apart on skates on a smooth ice rink. The lighter of the children holds a 3.0-kg ball and throws it to the heavier child. After the throw the lighter child recoils at 2.0 m/s. With what speed will the center of mass of the two children and the ball move?

60. (II) A uniform iron bar that is 0.75 m long with mass 7.5 kg is placed along the y-axis. From the midpoint of that bar a 0.25-m-long bar of negligible mass extends perpendicular to it in the x-direction; a 1.5-kg point mass is placed at its end. Where is the center of mass of this system? [*Hint*: First calculate the center of mass of the iron bar, $\vec{R}_1$, then treat that bar as a point mass of mass 7.5 kg at $\vec{R}_1$.]

61. (II) A mallet forms a symmetric T-shape. The top of the T is a uniform iron block of mass 4.0 kg. The wooden handle is uniform, 1.2 m long, and has a mass of 1.8 kg. Where is the mallet's center of mass? (See the hint in Problem 60.)

62. (II) Where is the center of mass of a uniform, L-shaped iron rod of sides 1.1 and 0.25 m, respectively? (Assume that the rod is so narrow that the dimensions of the outer bend are the same as those of the inner bend of the L.)

63. (II) By using arguments based on symmetry, show that the center of mass of a uniform, thin, spherical shell made of any material of uniform density is at the center of the shell. Using this result, show that the center of mass of any sphere whose density varies arbitrarily only with distance from the center of the sphere but not with angle (in other words, a spherically symmetric density) is at the center of the sphere.

64. (II) Calculate the center of mass of the T-shaped object shown in Fig. 8–41. Assume that the mass density is uniform.

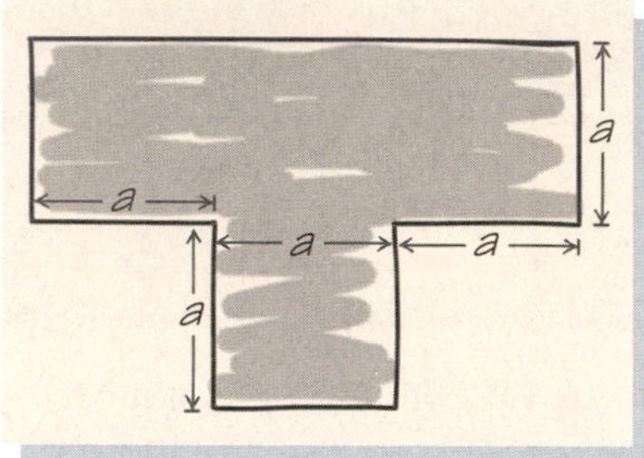

▲ **FIGURE 8–41** Problem 64.

65. (II) Find the center of mass of a flat, semicircular object of radius R. Assume that the density is uniform.

66. (II) A cylinder 30 cm in diameter and 1.00 m long has a hole drilled in it. The hole is 10 cm in diameter, 30 cm long, and it is symmetrically aligned along the axis of the cylinder. Where is the center of mass of the drilled cylinder?

67. (II) A set of four point masses are arrayed on the xy-plane. A mass of 1 kg is placed at the origin, two similar masses are at the points $(x, y) = (1\text{ m}, 0)$ and $(0, 1\text{ m})$, respectively, and a fourth mass of 2 kg is at the point $(1\text{ m}, 1\text{ m})$. Where is the center of mass? Suppose that the fourth mass were 1 kg rather than 2 kg. Find, without a detailed calculation, the location of the center of mass in this case.

68. (III) Show that the center of mass of a planar object can be found by dividing the object into a convenient number of sections of known mass M_i, finding the center of mass of each section, and then treating each section as a collection of point masses M_i located at the center of mass of that section.

69. (III) Calculate the center of mass of a stick extending from $x = 0$ to $x = L$; assume that the density of the mass distribution is given by $\lambda(x) = K/(x^2 + a^2)$, where K is a constant with dimensions $[ML]$.

70. (III) Find the center of mass of a flat triangular object of uniform density with sides of lengths a, b, and c.

8–7 Rocket Motion

71. (I) The thrust of a rocket is 10×10^6 N. If the speed of its engine's exhaust gas is 3.0 km/s, how fast is the rocket's mass changing?

72. (I) An orbiter must fire its thruster rockets to increase its speed. If the mass of the orbiter is about 20,000 kg and the exhaust gas velocity is 2600 m/s, how much mass must be discarded to increase the speed from 5800 to 5900 m/s?

73. (I) After ejecting a communication satellite, the space shuttle must make a correction to account for the change in momentum. One of the thrusters with $u_{ex} = 10^3$ m/s is used to increase the orbital velocity by 10 m/s. What percent of the mass of the space shuttle must be discarded?

74. (I) A rocket that uses a fuel with an exhaust speed of 2700 m/s is shot straight up and reaches a final speed $v = 3400$ m/s when the fuel is exhausted, which happens 120 s after the launch. What fraction of the initial mass of the rocket with all its fuel is the mass of the portion of the rocket that remains? Do not ignore gravity, but assume the force of gravity to be the same throughout the burn period.

75. (II) The *burnout velocity* is the final velocity of a rocket when all the fuel is burned away. If m_0 is the total initial mass of the rocket and m_{fuel} is the mass of the fuel, show that the burnout velocity is given by

$$v = u_{ex} \ln\left(\frac{m_0}{m_0 - m_{fuel}}\right) - gt,$$

where g is assumed to be constant. If the ratio m_{fuel}/m_0 is 70 percent, $t = 90$ s, and u_{ex} is 2800 m/s, what is the burnout velocity?

76. (II) The rocket considered in Problem 75 carries human passengers. Consequently, its initial upward acceleration may not exceed $2.5g$. What is the largest initial burnout rate (the mass expelled over unit time) that satisfies this condition? (The mass of the rocket is 60,000 kg.)

77. (II) A wooden block of mass 0.40 kg is placed on a fence 1.0 m high. A bullet of mass 10 g is fired horizontally into the block. If the speed of the bullet is 450 m/s, how far will the block, with the bullet in it, land from the fence? Ignore any friction between the fence and the block.

78. (II) A wooden block of mass 0.40 kg is placed on a fence 1.0 m high. A bullet of mass 10 g is fired horizontally into the block with an initial speed of 450 m/s. It passes right through the block, and in doing so gives the block enough kick so that the block lands 55 cm from the fence. Assuming that the land behind the fence is completely flat, how far from the fence will the bullet land? Ignore any friction between the fence and the block and any air drag on the block or bullet.

79. (II) To test certain aspects of rocket propulsion, a student stands initially motionless on frictionless ice. He has 100 watermelon seeds that he can spit horizontally in the $-x$-direction, thereby propelling himself in the $+x$-direction. The speed he can give one seed with respect to himself is u, but if he tries to spit two seeds simultaneously, he can give them only speed $u/2$. More generally, if he tries to spit n seeds simultaneously, he can give them speed u/n. The student has mass M and each seed has mass m. (a) To maximize his speed after spitting all 100 seeds, is the student better off spitting out all 100 seeds at once, spitting out the seeds one at a time, or something in between? Assume that the ratio m/M and even $100\,m/M$ can be taken to be small. (b) Let $M = 50$ kg, $m = 1$ g, and $u = 3$ m/s. What is the maximum final speed of the student?

80. (III) Because the final velocity of rockets is limited by the exhaust velocity and by the mass ratio of total mass to the rest of the rocket minus the fuel, engineers have developed multistage rockets in which the fuel tank of each stage is discarded before the next stage is ignited. Consider a two-stage rocket for which m_0 is the total mass of the rocket at liftoff, m_1 is the mass of the fuel in the first stage, m_1' is the mass of the first stage that is discarded, and m_2 is the mass of the second-stage fuel. Find an equation that gives the final velocity of the rocket; assume that gravity is constant.

General Problems

81. (I) A 48-kg figure skater (skater 1) moving at 6 m/s is picked up by her 82-kg partner (skater 2), who is skating in the same direction at 9 m/s. They then coast off together. Ignore all friction. (a) What is the velocity of their center of mass before he picks her up? (b) After he picks her up? (c) What were the velocities of skaters 1 and 2 in the center-of-mass frame before he picked her up? (d) During a brief period of 0.05 s, they crash into the barrier and collapse together. What average force is exerted on the barrier?

82. (II) A railroad car of mass 16 Mg ($1\ \text{Mg} = 10^3\ \text{kg}$) is released from rest in a railway switchyard and rolls to the bottom of a slope 1.8 m below its original height. At the low point, it collides with and sticks to another car of 8 Mg. The two cars roll off together up another slope to a position a height h above the low point, where they come to a stop. Ignoring the effects of friction, find h.

83. (II) A cannon of mass 800 kg shoots a 5-kg projectile at an angle of 35° from the horizontal. The projectile has an initial speed of 800 m/s. The cannon is on wheels. (a) Draw a diagram showing the momentum vectors. (b) With what velocity does the cannon recoil along the horizontal ground? (c) What happens to the recoiling cannon's component of momentum perpendicular to the ground?

84. (II) A small bag of sand is suspended by a thread of length $L = 1.2$ m. A bullet of mass $m = 8.0$ g moving with speed $v_0 = 600$ m/s in the horizontal direction hits the bag and leaves with a speed $v = 250$ m/s (Fig. 8–42). The bag swings to a maximum angle of $\theta = 40°$. What is the mass of the sandbag?

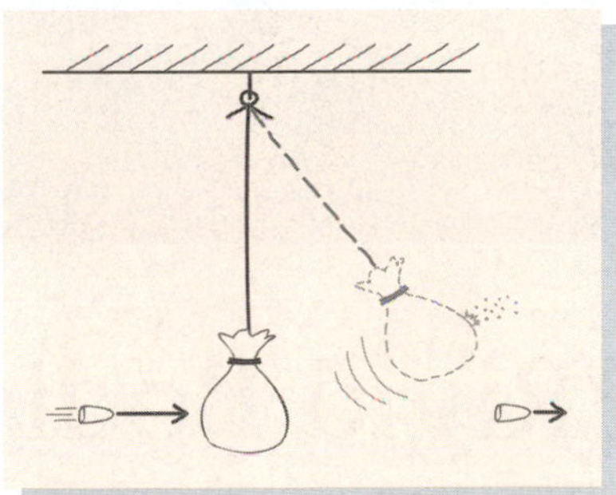

▲ **FIGURE 8–42** Problem 84.

85. (II) Two masses slide without friction down the sides of a hemispherical bowl. They each start with zero velocity at the lip, which is a height h above the bottom. They move in opposite directions and collide elastically at the bottom of the bowl. Suppose that the masses are m and M, respectively, with $m < M$. (a) By how much will the lighter mass overshoot the lip after the collision? (b) How high will the masses move if they coalesce on contact?

86. (II) A boxcar of length 9 m and height 3.0 m is at rest on frictionless rails. Inside the boxcar, whose empty mass is 4200 kg, a tank containing 1800 kg of water is located at one end. The tank is 1.5 m long and 2.0 m high. At some point, the walls of the tank start to leak and the water fills the floor of the boxcar uniformly. Assuming that all the water stays in the boxcar, describe how the breakage of the tank manifests itself to an outside observer.

87. (II) A steer of mass 500 kg runs at 30 kph past a cowboy of mass 75 kg, who jumps on in an attempt to bring down the steer. Assuming that the cowboy is not dragging his feet, what is the speed of the steer after the cowboy jumps on?

88. (II) In the Olympic Games of the year 2020, a new sport is introduced—cyclo-shooting. A bicyclist moving in the $+x$-direction fires a gun in the y-direction in an attempt to hit a target lined up with the position $x = 0$ m. The mass of the cyclist plus the bicycle is 65 kg, and the cyclist is originally moving with speed 35 km/h. The gun shoots a projectile, with mass 15 g, at a speed of 1600 km/h. The target is a distance 30 m from the bicycle track. (a) At what x-value should the cyclist fire in order to hit the target? Ignore air resistance. (b) Immediately after she fires, the cyclist moves in a new direction due to the recoil. What is the new value of her velocity, both magnitude and direction?

89. (II) A sphere of styrofoam has radius R. A cavity of radius $R/2$ centered a distance $R/2$ directly above the center of the sphere is hollowed out and filled with a solid material of density five times the density of styrofoam (Fig. 8–43). Where is the center of mass of the new sphere?

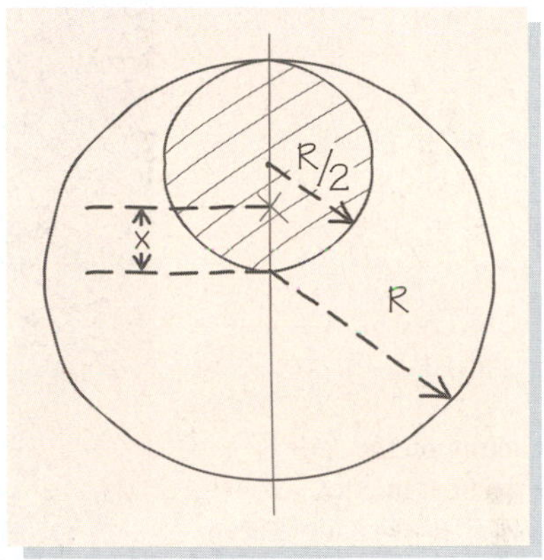

▲ **FIGURE 8–43** Problem 89.

90. (III) Two identical spherical billiard balls each with a radius of 3 cm move toward each other with velocities $(1.5\ \text{m/s})\hat{i}$ and $(-1.1\ \text{m/s})\hat{i}$. The center of one of them moves along the $+x$-axis ($y = 0$ m), and the center of the other one moves in the $-x$-direction, 1.0 cm below the x-axis ($y = -1.0$ cm). Assuming that the collision is perfectly elastic, what are the final velocities of the two billiard balls? Pretend that the objects slide without friction. Note that the impulse is directed along the line that joins the two centers at the moment of collision.

91. (III) Two students sit on opposite ends of a 4-m-long sled that is initially at rest on frictionless ice. Each student has a mass of 65 kg; the sled's mass is 30 kg. The student at one end slides a 3-kg object on the sled across to the other at a uniform speed of 6 m/s relative to the sled (the object moves friction free on the sled). (a) What is the sled's speed relative to the ice before the second student catches the object? (b) After the second student catches the object? (c) Over what distance does the sled move while the object slides across? (d) Over what distance does the center of mass move while the object slides across?

92. (III) When a rocket moves vertically in the presence of gravity, the motion is different than when gravity is not present. If you go back to the derivation of Eq. (8–65), you will see that the starting point of that derivation, the condition that $P(t + \Delta t) - P(t) = 0$, where P is the total momentum of the rocket plus its exhaust gases, would be replaced by $P(t + \Delta t) - P(t) = -F_g\,\Delta t$. Here F_g is the magnitude of the force of gravity and the minus sign in front of F_g signifies that the "up" direction is positive in our one-dimensional derivation. Follow the steps that led to Eq. (8–65) to find the equation of motion when gravity is included [the analog of Eq. (8–65)] and show that the solution of the equation of motion is Eq. (8–70).

Chapter

9

▶ The wind acting on the sails is trying to rotate the boat in one direction; the weight of the sailors acts to rotate it in the opposite direction. The description and dynamics of rotations are an important element in understanding the behavior of extended systems such as this sailboat.

Rotations of Rigid Bodies

To this point our discussion of dynamics has treated objects from crates to planets as if they were all pointlike (particles). Our study of an extended object, or of a system of particles, has reached the point where we can show that under the influence of a net external force the center of mass moves as if the entire system were concentrated at that single point. But for an extended object that rotates there is more to its motion than just the motion of its center of mass. Consider the wheels of a skateboard, a tossed blackboard eraser, the flow of water down the drain of a bathtub, or the motion of Earth. All these systems rotate, or spin, about an axis, as do many other objects from galaxies to molecules, and we must be able to analyze this sort of motion if we are to understand the dynamics of our world.

In this chapter, we study the rotational motion of **rigid bodies**, which are objects of fixed form that do not distort or deform as they move. Blackboard erasers, basketballs, Earth, and compact disks are, at least to a good approximation, all rigid bodies, and all can have a rotational aspect to their motion. (The water flowing down a drain is not a rigid object, but a rotation is an important part of its description—the motion of these more complicated nonrigid systems will be explored more fully in Chapter 10.) Here we start with a simple description (kinematics) of the rotational motion of rigid objects and then continue with the causes of their rotations (dynamics). Even though rotations appear to encompass many new concepts, these motions are in fact described by the application of Newton's laws to collections of particles. No new physical laws are involved.

9–1 Simple Rotations of a Rigid Body

Motions of a Rigid Body

Suppose we have a rigid body that has an axis within it about which the body can rotate and that is permanently fixed in space. What kind of motions are possible? The characteristic feature of a rigid body is that any point P of the object remains at a fixed distance from any other point and hence at a fixed distance from the axis. To see the effect of this fact, consider a clock hand with an axis passing through the end of the hand at the clock center and perpendicular to the face. If this axis is fixed, the possible motion corresponds to each point of the hand moving in circles about the fixed axis (Fig. 9–1). The particular constant distance of our point is the radial distance R to the center. In simple language, the clock hand rotates about this axis.

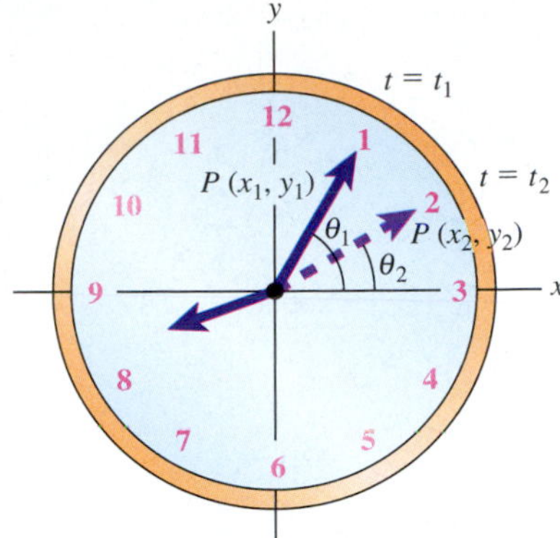

▲ **FIGURE 9–1** A flat object in the xy-plane rotates about a fixed axis perpendicular to the plane. Here, the object is the hand of a clock, which rotates through the angle $\theta_2 - \theta_1$. The fixed point is at the center of the clock face.

In Fig. 9–1, we have established a set of fixed axes with their origin at the center. At some time t_1, point P has coordinates (x_1, y_1); at a later time, point P has moved to where its coordinates are now (x_2, y_2). *Both* x and y change with time, and this is complicated. Things are simpler if we work with polar coordinates. We can describe point P on the hand with an angle θ measured from the x-axis and a radial distance R from the origin. In this description, *the only variable that changes with time is the angle* θ. Our point P has undergone an *angular displacement*.

Perhaps you are wondering if the fact that the clock hand moves in a plane has a special role. Consider then a more complicated rigid object, with the fixed axis again passing through it, as in Fig. 9–2. Any point in the object—here, a pot on a potter's wheel—that lies along the axis is fixed in space. A point P that is off the axis will always be at a fixed perpendicular distance from the axis. The motion of the pot consists of rotations about the axis. And as the discussion of the clock hand shows, to describe rotations about a fixed axis, we need only specify a single angle. Other examples of this limited motion are provided by Earth's rotation about its axis or by a turning merry-go-round.

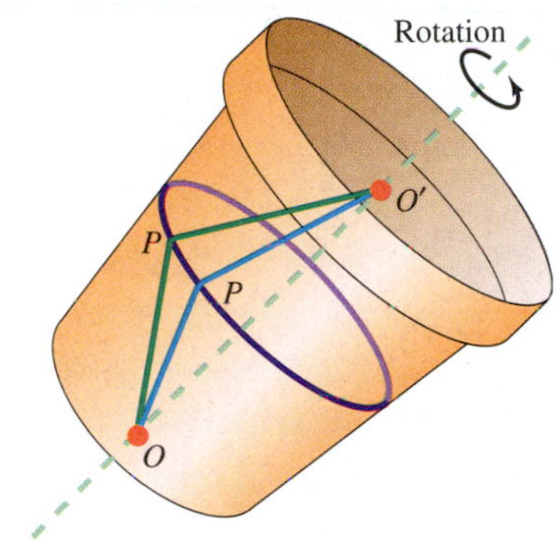

▲ **FIGURE 9–2** A three-dimensional object—a clay pot on a potter's wheel—rotates about the axis defined by the line between the two fixed points O and O'. The point P remains the same distance from these points or from any point on the axis as the pot rotates about the axis.

To repeat, what distinguishes a rigid body is that when it moves through a certain angle in the rotational motion about an axis, all points in the rigid body turn about the axis by exactly the same angle. Even if the motion is not uniform, it is still just circular motion. We need only one variable—the angle of circular rotation θ.

Angular Velocity and Angular Acceleration

To describe the motion mathematically, let's orient the z-axis along the rotation axis (Fig. 9–3a). Consider the position vector $\vec{r}$ from the origin to point P. Take the projection of $\vec{r}$ on the xy-plane to go to point P' and let $\theta(t)$ be the angle this projection makes with the x-axis at some time t (Fig. 9–3b). This angle, which in Section 3–5 was denoted ϕ, is measured in *radians* (2π rad $= 360°$). Figure 9–3b also shows that at a later time $t + \Delta t$ the projection point P' has moved from angle $\theta(t)$ to angle $\theta(t + \Delta t)$.

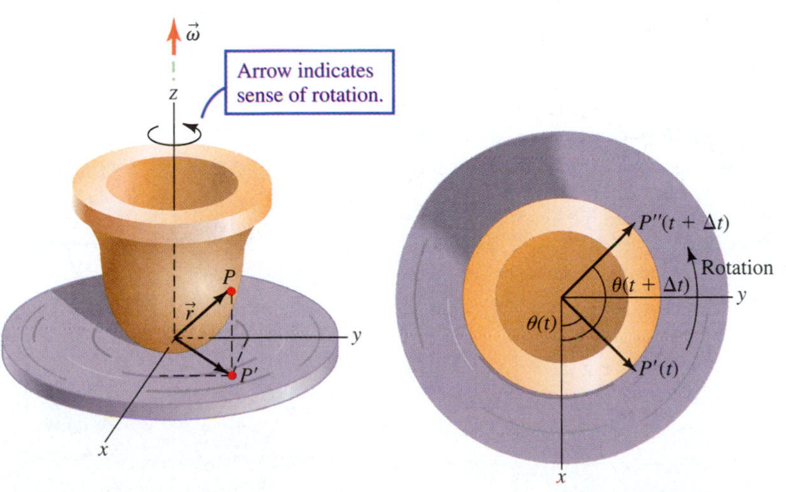

◀ **FIGURE 9–3** (a) A pot on a potter's wheel rotates about the z-axis. (b) The pot as viewed looking down the z-axis. The angle of rotation θ, which varies with time, is defined to be the angular displacement from the x-axis of some point—any point—in the pot. It is positive when the pot rotates counterclockwise.

The average angular velocity, ω_{av}, over a finite time interval Δt is obtained from the change in the angle over that time interval:

$$\omega_{av} = \frac{\theta(t + \Delta t) - \theta(t)}{\Delta t}. \tag{9–1}$$

Every point in our rigid body moves with the same average angular velocity. It will equally be true that every point in the rigid body will move with the same *instantaneous angular velocity* (or just the *angular velocity*) $\omega(t)$, a quantity that we define by taking the limit $\Delta t \to 0$ of Eq. (9–1) [see also Eq. (3–46), where we discuss the magnitude, or angular speed]:

$$\vec{\omega} = \frac{d\theta}{dt}\hat{\omega}. \tag{9–2}$$

This expression is a vectorial one, and we did not use the term "velocity" casually: The angular velocity is a vector. Its direction, which is contained in the unit vector denoted $\hat{\omega}$ in Eq. (9–2), is defined conventionally but unambiguously as follows. We start by saying that it is aligned with the axis of the rotation. This leaves an ambiguity; if the axis of the rotation is vertical, is the angular velocity up or down? We break this ambiguity by the following rule: If *the fingers of your right hand curl along with the rotation [that is, in the direction of the angle change that appears in the numerator of Eq.* (9–2)], *your thumb will give the direction of the angular velocity*. This is a so-called *right-hand rule* (Fig. 9–4). As Fig. 9–4 shows, the movement of the rotating object is in the xy-plane with a counterclockwise motion as seen from above. With the right-hand rule, the extended thumb of the right hand points in the $+z$-direction (along the axis of rotation), and this is the direction of the angular velocity $\vec{\omega}$.

If we look back at Fig. 9–1, we see that for a rotation by $\Delta\theta$, the arc distance s traced out by a point a distance r from the axis of rotation is given by $s = r\,\Delta\theta$; similarly the linear speed of that point will be $v = r\omega$.

The units that describe angular velocity are *radians per second* (rad/s). Because a radian (the ratio of an arc length to the radius of the arc) is a dimensionless measure, the dimensions of ω are $[T^{-1}]$.

Suppose that the angular velocity is a *constant;* that is, $\omega(t) = \omega_0$ in some fixed direction that defines an axis. Then the angle θ that describes rotation around the axis changes linearly with time:

$$\theta = \theta_0 + \omega_0 t, \tag{9–3}$$

where θ_0 is the value in radians of the angle θ at time $t = 0$. A full rotation is made over a time T, which is the *period* of the motion. The angle θ goes from θ_0 to $\theta_0 + 2\pi$ in time T, so $\omega_0 T = 2\pi$, or

$$T = \frac{2\pi}{\omega_0}. \tag{9–4}$$

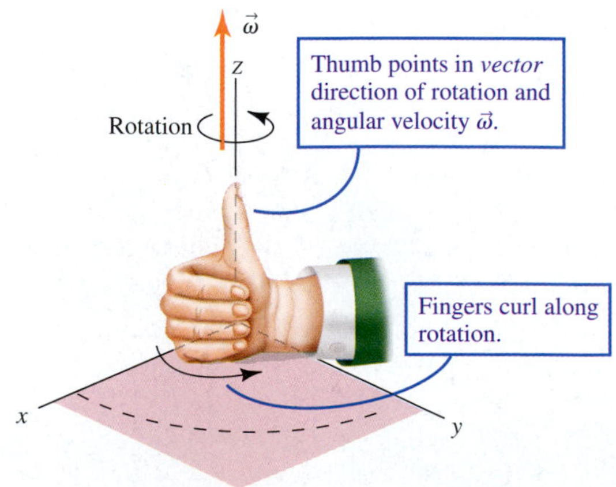

FIGURE 9–4 Right-hand rule. The thumb of the right hand points along the axis of rotation (the z-axis) when the fingers curl in the direction of rotation (from the x-axis to the y-axis). This will also specify the direction of the angular velocity vector.

The *frequency* f measures the number of times per second that point P returns to its original position in uniform rotation:

$$f = \frac{1}{T} = \frac{\omega_0}{2\pi}. \tag{9–5}$$

FREQUENCY AND PERIOD

For nonuniform motion, we also speak of an instantaneous frequency, $f(t) = \omega(t)/2\pi$. For more on these relations you might want to refer back to Section 3–5.

Angular Acceleration

If the angular velocity is not constant, we can define an average angular acceleration:

$$\vec{\alpha}_{\text{av}} = \frac{\vec{\omega}(t + \Delta t) - \vec{\omega}(t)}{\Delta t}. \tag{9–6}$$

The instantaneous angular acceleration, or more simply the **angular acceleration**, is defined as the instantaneous rate of change of the angular velocity:

$$\vec{\alpha} = \frac{d\vec{\omega}}{dt} = \frac{d^2\theta}{dt^2}\hat{\omega}, \tag{9–7}$$

The last step applies when the angular velocity changes in magnitude but not direction. Note that this equation has the same structure as the equation for linear motion,

$$\vec{a} = \frac{d\vec{v}}{dt} = \frac{d^2\vec{r}}{dt^2}.$$

The direction of the angular acceleration vector is specified by the rate of change of the angular velocity vector, exactly as the linear acceleration vector is the rate of change of the linear velocity vector. As for the angular velocity, the angular acceleration is the same for all the points in the rigid body. The angular acceleration has units of radians per second squared, and its dimension is $[T^{-2}]$, again because radians are dimensionless.

Constant Angular Acceleration: Provided that the angular acceleration and the angular velocity are parallel (this is the case for most situations we'll deal with here), a constant angular acceleration means that the angular velocity has a constant time derivative and an unchanging direction. (If we understand that the equations below should contain a unit vector in the relevant direction, we can now dispense with the over-arrow notation. All the vector information will be contained in the *sign* of the respective quantities.) With constant angular acceleration the angular velocity changes linearly with time; that is,

$$\omega = \omega_0 + \alpha t, \tag{9–8}$$

where ω_0 is the angular velocity at $t = 0$. The angle is then given by

$$\theta(t) = \theta_0 + \omega_0 t + \tfrac{1}{2}\alpha t^2. \tag{9–9}$$

This solution corresponds to the familiar linear motion equation $x(t) = x_0 + v_0 t + \frac{1}{2}at^2$, valid for constant linear acceleration a. Just as the position of a car moving on a line and undergoing constant linear acceleration changes quadratically with time, so the angle of the wheels of that car change quadratically with time if the wheel undergoes constant angular acceleration.

We may eliminate the time from Eqs. (9–8) and (9–9) to find that

$$\omega^2 = \omega_0^2 + 2\alpha(\theta - \theta_0), \tag{9–10}$$

which is the analog of the linear motion expression $v^2 = v_0^2 + 2a(x - x_0)$.

EXAMPLE 9–1 A pulley is rotating at a frequency of 32 rev/min. A motor speeds up the wheel so that 30 s later the frequency of the motion is 82 rev/min. (a) What is the average angular acceleration in radians per second squared during that period? (b) How far will a point 0.30 m from the center of the pulley have traveled during the acceleration period assuming that the acceleration is constant?

Setting It Up The rotating pulley is shown in Fig. 9–5, with the direction of rotation shown. We label the given initial and final rotational frequencies as f_i and f_f and the time interval Δt. We want the initial and final angular speeds ω_i and ω_f, the average angular acceleration α_{av}, and the distance D traveled by a point a given radial distance r from the center of the pulley under the assumption that the angular acceleration is constant.

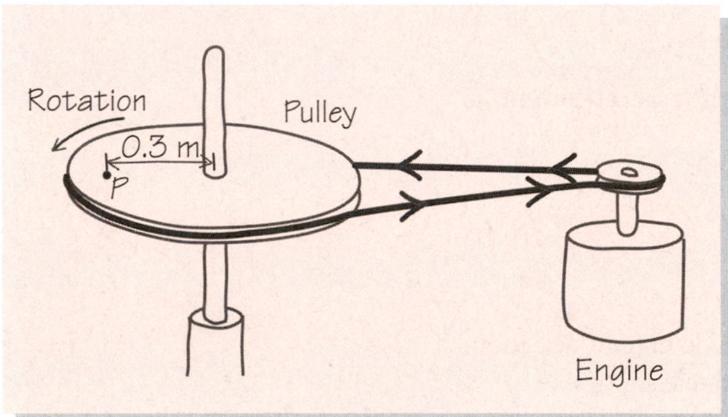

▲ **FIGURE 9–5**

Strategy We first will need to change the frequencies to angular speeds. This is accomplished by recalling that each revolution consists of an angular displacement of 2π radians. Thus we multiply frequency by 2π to find angular speed. We can then use this in the definition of average angular acceleration [Eq. (9–6)] to find the average angular acceleration. As for the distance traveled by a point, we can find the angular displacement under the assumption of constant angular acceleration using Eq. (9–9), then use the relation between angular change, radius, and arc length to find the distance traveled.

Working It Out (a) A rotation frequency of $f_i = 32$ rev/min corresponds to an angular speed of $\omega_i = (32\ \text{rev/min})(2\pi\ \text{rad/rev}) = 64\pi\ \text{rad/min}$, or $64\pi\ \text{rad}/60\ \text{s} = 3.4\ \text{rad/s}$. Similarly, after 30 s the angular velocity ω_f is $(82\ \text{rev/min})(2\pi\ \text{rad/rev})(1\ \text{min}/60\ \text{s}) = 8.6\ \text{rad/s}$. Thus the average angular acceleration is

$$\alpha_{av} = \frac{\omega_f - \omega_i}{\Delta t} = \frac{(8.6 - 3.4)\ \text{rad/s}}{30\ \text{s}} = 0.17\ \text{rad/s}^2.$$

(b) We calculate the angular displacement $\Delta\theta$ during the period by using Eq. (9–9), which applies when there is uniform acceleration. In our case $\omega_i = 3.4$ rad/s and $\alpha_{av} = 0.17\ \text{rad/s}^2$, so Eq. (9–9) gives

$$\begin{aligned}\Delta\theta = \theta - \theta_i &= \omega_i t + \tfrac{1}{2}\alpha_{av}t^2 \\ &= (3.4\ \text{rad/s})(30\ \text{s}) + \tfrac{1}{2}(0.17\ \text{rad/s}^2)(30\ \text{s})^2 \\ &= 180\ \text{rad}.\end{aligned}$$

The distance traveled is therefore

$$D = r\,\Delta\theta = (0.30\ \text{m})(180\ \text{rad}) = 54\ \text{m}.$$

The unit "radian" is deleted from the final value because radians are dimensionless.

What Do You Think? How does the answer to part (a) depend on the details of the speed-up? *Answers to* ***What Do You Think?*** *questions are given in the back of the book.*

Acceleration of a Point in a Rotating Rigid Body

Consider the potter's wheel of Fig. 9–6, and, more specifically, a point within it located at a distance r from the axis of rotation. The point moves in a circle; its velocity has a magnitude of $r\omega$ and a direction that is tangential to the circle of radius r. We learned in Section 3–5 that a point moving in a circle of radius r with speed v undergoes an acceleration in the radial direction toward the center of the circle—we refer to the point's radial acceleration. The magnitude of that acceleration is v^2/r. In terms of the angular speed $\omega = v/r$, this acceleration has magnitude

$$a_{radial} = \frac{v^2}{r} = \omega^2 r. \tag{9–11}$$

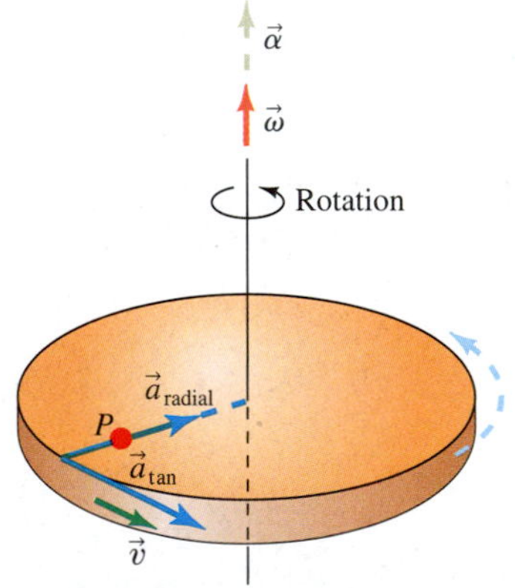

▲ **FIGURE 9–6** A rotating potter's wheel undergoes an angular acceleration. Any point P must have an acceleration with both a centripetal component (this component keeps the point on its circular path) and a tangential component.

If the potter's wheel had a changing angular speed, or in other words an angular acceleration α, then our given point would also have a linear acceleration in the tangential direction, a tangential acceleration, of magnitude

$$a_{tan} = \alpha r. \tag{9–12}$$

Figure 9–6 illustrates these acceleration components for a point in a rotating potter's wheel. If a point in a rigid body accelerates, there must be forces involved. Both the radial acceleration and any tangential acceleration occur at least in part through the interatomic forces that fix the relative positions of the atoms of the rigid body. We'll discuss this in more detail later.

CONCEPTUAL EXAMPLE 9–2 A lump of clay rotates on a potter's wheel as the potter speeds it up. Consider two points, P_1 close to the rotation axis and P_2 far from the rotation axis. Which of these points has the larger radial acceleration a_{radial} and which the larger tangential acceleration a_{tan}?

Answer As we have tried to emphasize, the angular variables are common for all points in a rigid object; all points move through the same angle in a given time, all points move with the same angular velocity, and all points have the same angular acceleration. Therefore, to make a comparison of the accelerations of different points within a rotating object, it is best to write their accelerations as much as possible in terms of angular speed and acceleration. For a_{radial}, Eq. (9–11) provides us with the necessary expression: The angular speed is the same for the two points, so $a_{radial} \propto r$. For the tangential acceleration, we can use Eq. (9–12): The angular acceleration is the same for the two points, so $a_{tan} \propto r$. Point P_2, which sits at a larger radial distance, has both the larger radial and tangential acceleration components.

What Do You Think? Which of the two points has the larger angular acceleration?

9–2 Rotational Kinetic Energy

Except for points on the rotation axis, every point of a rotating object such as that shown in Fig. 9–7a moves, so the object has **rotational kinetic energy** K. To analyze it, we take the hamburger package in Fig. 9–7a, which rotates with angular velocity ω about an axis that we label as the z-axis. In Fig. 9–7b we divide the package into elements labeled by the indices i, with each element having mass Δm_i. The total kinetic energy of the object is given by

$$K = \sum_i K_i = \frac{1}{2}\sum_i \Delta m_i v_i^2.$$

Because each mass element is rotating about the z-axis with angular velocity ω, the speed of an element is given by $v_i = R_i\omega$. The length R_i is the *perpendicular distance* between the mass element and the rotation axis, as shown in Fig. 9–7b. Substitution of this value of v_i into the expression for the kinetic energy gives

$$K = \frac{1}{2}\Big(\sum_i \Delta m_i R_i^2\Big)\omega^2. \qquad (9\text{–}13)$$

The factor ω has come out of the sum because it is the same for all the elements of the object. The kinetic energy of our rotating object is thus proportional to ω^2. We may write

$$K = \tfrac{1}{2}I\omega^2, \qquad (9\text{–}14)$$

KINETIC ENERGY OF ROTATION

where

$$I \equiv \sum_i \Delta m_i R_i^2 \qquad (9\text{–}15)$$

ROTATIONAL INERTIA

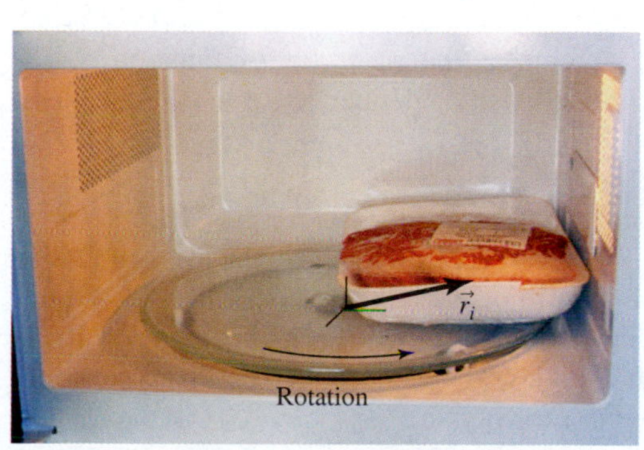

(a)

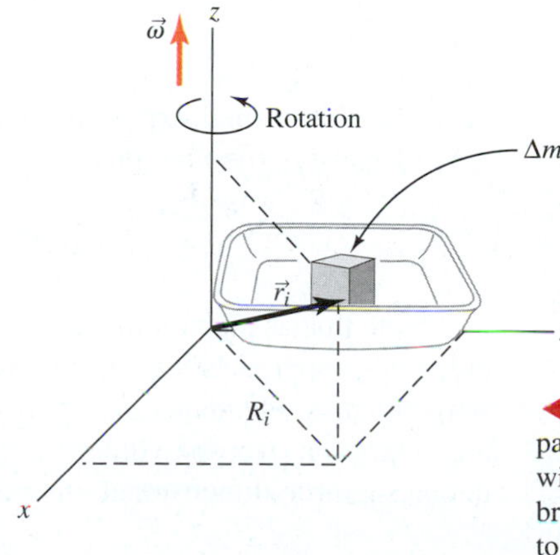

(b)

FIGURE 9–7 (a) The hamburger package has kinetic energy associated with its rotation on the turntable. (b) We break the package into elements in order to analyze the kinetic energy in terms of rotational variables.

The quantity I is a property of the rigid object called the **rotational inertia** or the **moment of inertia**. It has dimensions of $[ML^2]$ and is measured in units of kilogram-meters squared in SI. From its definition, we see that I sums the contributions of each element of the object; for a single element or single particle, then,

$$I = mR^2, \tag{9–16}$$

where R is the perpendicular distance from the single element to the axis of rotation. According to Eq. (9–15), the rotational inertia can be defined with respect to any axis we choose (we refer to the reference axis), not just the axis of rotation. Moreover, *I depends on the axis from which R_i is measured.*

Rotational inertia plays the same part in rotational motion as mass does in translational (linear) motion. It differs from the mass in that it depends on the choice of a particular axis. The kinetic energy of a particle in translational motion is one-half its mass times the velocity squared; the kinetic energy of an object rotating about a fixed axis is one-half its rotational inertia times its angular velocity squared; that is, $K_{\text{translation}} = mv^2/2$ and $K_{\text{rotation}} = I\omega^2/2$. Just as the mass measures the resistance of an object to changes in velocity, we shall see that the rotational inertia measures the resistance of an object to changes in its angular velocity.

THINK ABOUT THIS . . .

CAN YOU RUN A CAR USING ROTATIONAL KINETIC ENERGY?

▲ **FIGURE 9–8** This 14-foot flywheel was used in a steam pumping engine to pump water in the Southend Waterworks Company in Great Britain. It was installed in 1931, but is no longer operational.

A *flywheel* is a rotating disk with a large rotational inertia and the ability to reach a large angular velocity; a flywheel thereby stores energy, and various means can be used to recover this energy. Flywheels (see Fig. 9–8) have a long history; for instance, from the start of the industrial revolution flywheels were used to accumulate energy from turning waterwheels, then that energy was distributed through systems of belts to machine tools. In their early configuration, San Francisco's cable cars used three flywheels 15 ft in diameter with a mass of 25 tons to keep the cables moving steadily as cable cars gripped and ungripped the cables. By the 1950s Zurich was running a bus system in which energy to run the buses came from flywheels mounted within them—the buses' flywheels were about $1\frac{1}{2}$ m in diameter, had a mass of 1500 kg, and turned at 3000 rpm. In the Zurich buses, the flywheel powers an electric generator and the electricity runs electric motors used to drive the wheels. Periodically the flywheel is sped up by connection to the local electrical grid at the stops. In a search for nonpolluting automobiles, the possibility that a rapidly spinning flywheel in an automobile could power that car is a very real one. Of course, the electricity that would have to be used to bring the flywheel up to speed would have to be generated somewhere, but the pollution of that plant could at least be localized. ■

We have defined the rotational inertia by treating an object as composed of discrete elements. Most real objects are more conveniently thought of as continuous, and a treatment of this type requires us to take a limit in which the elements have infinitesimal volume and infinitesimal mass. In this case, the sum becomes an integral over the volume of the object. We can write this continuum expression by labeling the mass element by the differential notation dm, and thus

$$I = \int R^2\, dm. \tag{9–17}$$

To make this formal expression useful, we employ the *mass density* function ρ. The mass density $\rho(\vec{r})$ at the location $\vec{r}$ of the ith element is defined as

$$\rho(\vec{r}) \equiv \lim_{\Delta V_i \to 0} \frac{\Delta m_i}{\Delta V_i} = \frac{dm}{dV}.$$

Note that as the volume ΔV_i of the ith element approaches zero, the mass of that element also approaches zero, leaving a finite ratio ρ. Having defined the density, we can write the mass element as $dm = \rho\, dV$. The total mass is found from the mass density by an integration over the volume V (the single integral sign with the subscript "Volume" denotes a three-dimensional integration over the volume):

$$M = \sum_i \Delta m_i \to \int dm = \int_{\text{Volume}} \rho\, dV. \tag{9–18}$$

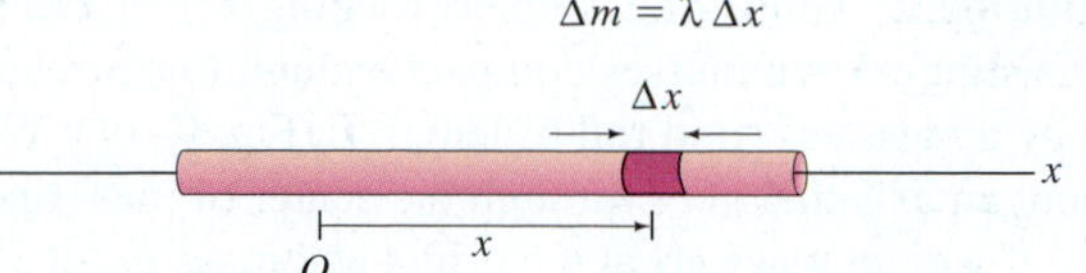

◀ **FIGURE 9–9** A one-dimensional object is aligned with the x-axis. A segment of width Δx is a displacement x from the origin.

In this language the rotational inertia of a continuous object is

$$I = \int R^2\,dm = \int_{\text{Volume}} \rho R^2\,dV. \tag{9–19}$$

The integrals referenced here may look difficult because they refer to such a general situation. But in many particular cases of interest in this book—for example, the solid cylinder below—we can use symmetry to reduce the integrals to simple one-dimensional integrals. At worst, the integrations can be formulated as a series of simple one-dimensional integrals (see Example 9–7).

An equation like Eq. (9–19) applies to one-dimensional (linear) objects or two-dimensional (flat) objects. In these cases, we replace the mass per unit volume ρ by a mass per unit length $\lambda \equiv dm/dx$ or a mass per unit area $\sigma \equiv dm/dA$, respectively. Thus

$$\text{for a one-dimensional object: } dm = \lambda\,dx;$$
$$\text{for a two-dimensional object: } dm = \sigma\,dA;$$
$$\text{for a three-dimensional object: } dm = \rho\,dV.$$

The densities λ, σ, and ρ may vary from point to point of the object.

To evaluate the rotational inertia about an axis perpendicular to a one-dimensional object, we can align our object with the x-axis, and

$$I = \int \lambda x^2\,dx. \tag{9–20}$$

The distance x is measured from the reference axis, which passes through the origin, as shown in Fig. 9–9, and the integral is carried out over the length of the object. When the object is two dimensional, we have

$$I = \int_{\text{Surface}} \sigma r^2\,dA. \tag{9–21}$$

We have taken the reference axis perpendicular to the object and integrated over its entire area (Fig. 9–10). The quantity r^2 is the square of the distance of the integration element to the axis about which I is measured.

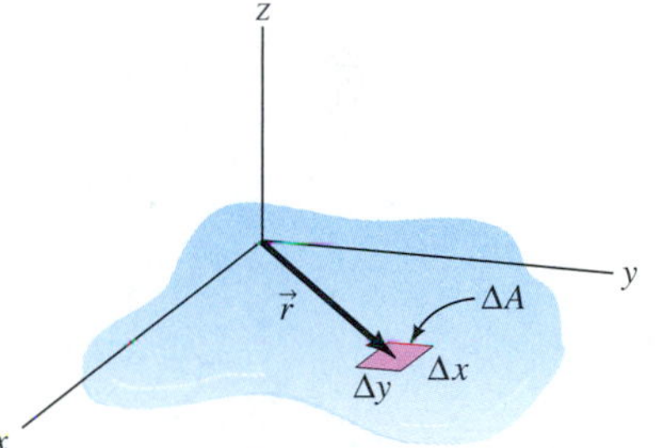

▲ **FIGURE 9–10** A two-dimensional object is aligned in the xy-plane. A segment of area ΔA is a position $\vec{r}$ from the origin.

9–3 Evaluation of Rotational Inertia

A description of the rotational motion of an object requires us to evaluate its rotational inertia. For continuous objects, this evaluation involves integrations. We will see how this works in a series of examples. Table 9–1 contains a summary of the results as well as rotational inertias of other objects for which the calculations are not carried out in this section.

TABLE 9–1 • Rotational Inertias of Simple Solids

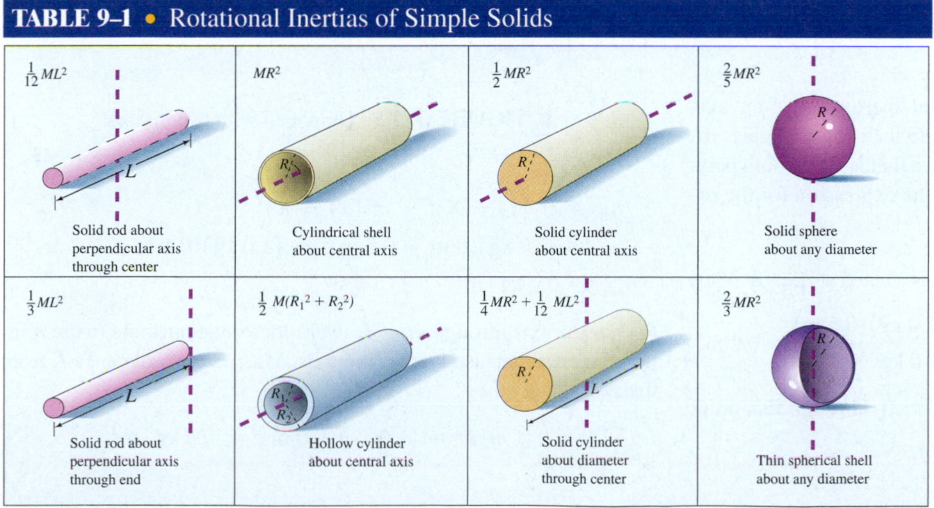

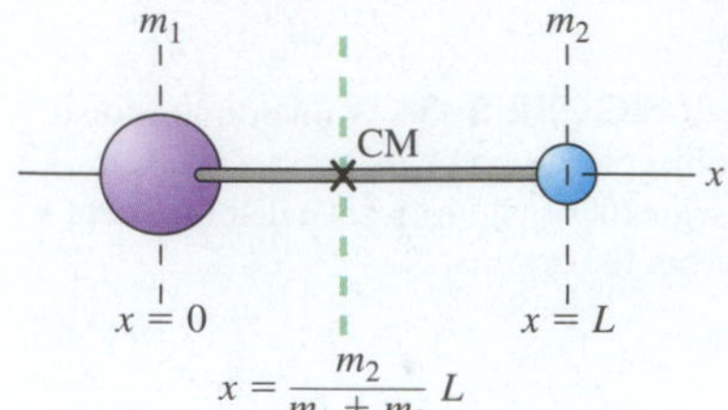

▲ **FIGURE 9–11** A dumbbell consists of two point masses, m_1 and m_2, connected by a rigid massless rod of length L. The object is aligned along the x-axis, with m_1 at the origin. The axis of rotation is perpendicular to the rod, passing through the center of mass.

Dumbbell: Almost the simplest rotating object that we can contemplate is a dumbbell consisting of two masses compact enough to be treated as points, m_1 and m_2, connected by a massless rigid rod of length L (Fig. 9–11). We'll find here the rotational inertia about an axis that goes through the center of mass and is perpendicular to the rod.

If we put mass m_1 at the origin and mass m_2 at $x = L$, Eq. (8–48a) tells us that the location X of the center of mass is

$$X = \frac{m_1x_1 + m_2x_2}{m_1 + m_2} = \frac{(m_1)(0) + (m_2)(L)}{m_1 + m_2} = \frac{m_2L}{M},$$

where M is the total mass. Our reference axis is therefore perpendicular to the rod and passes through the point $x = m_2L/M$. As measured from the center of mass, the coordinates of m_1 and m_2 are $-m_2L/M$ and $L - (m_2L/M) = m_1L/M$. As in Fig. 9–11, the rotational inertia about the axis passing through the center of mass and perpendicular to the axis of the dumbbell is therefore

$$I = m_1\left(-\frac{m_2L}{M}\right)^2 + m_2\left(\frac{m_1L}{M}\right)^2 = \frac{(m_1m_2^2 + m_2m_1^2)L^2}{M^2}$$

$$= \frac{m_1m_2L^2(m_2 + m_1)}{(m_1 + m_2)^2} = \frac{m_1m_2}{m_1 + m_2}L^2. \qquad (9\text{–}22)$$

If the masses are equal, then $m_1m_2/(m_1 + m_2) = m/2$; if $m_2 \gg m_1$, then it is a good approximation to write $m_1m_2/(m_1 + m_2) = m_1$, the mass of the light point mass. These results are easy to understand physically. In the case of equal masses m, the center of mass is the midpoint of the rod, a distance $L/2$ from each mass, and the rotational inertia about this axis is then $I = 2m(L/2)^2 = mL^2/2$. In the case of one very light mass, $m_1 \ll m_2$, to a good approximation the center of mass lies very near to the mass m_2, and m_2 no longer contributes to the rotational inertia about this point. The lighter mass m_1 is a distance L from the axis, and $I = m_1L^2$.

Note finally that if the reference axis is the line connecting the two point masses, then each mass lies directly on the axis. Their respective distances to the axis is zero, and the rotational inertia about this axis is zero. This axis also passes through the center of mass!

EXAMPLE 9–3 A dumbbell consists of point masses 2.0 and 1.0 kg attached by a rigid massless rod of length 0.6 m. Calculate the rotational inertia of the dumbbell (a) about an axis perpendicular to the rod of the dumbbell and going through the center of mass and (b) about an axis perpendicular to the rod of the dumbbell and going through the 2.0-kg mass.

Setting It Up Figure 9–12 serves as a suitable sketch. The masses are treated as points. The rod lies along the x-axis, and the rotation axes referred to in the question are also drawn. With the coordinate system of the sketch, the larger mass m_1 is at the origin, $x_1 = 0$ m, and the smaller mass m_2 is at $x_2 = L$, L being the known length of the rod.

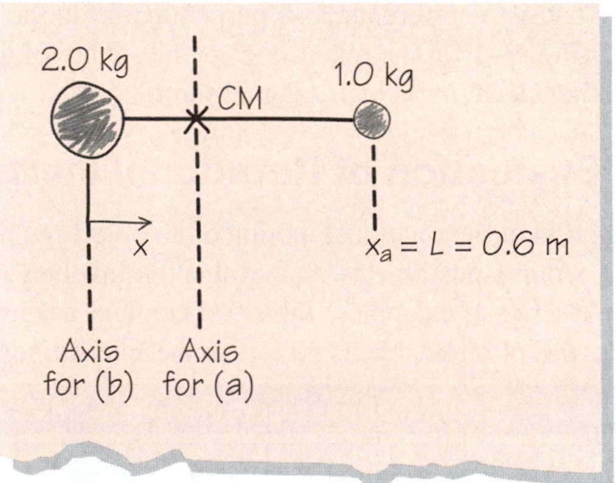

▲ **FIGURE 9–12** Dumbbell with two masses.

Strategy If we are to find a rotational inertia about an axis through the center of mass, we must first find the center-of-mass location; we can do this with Eq. (8–48a). Once that is done, both parts of the problem are simple substitution into the expression for the rotational inertia, Eq. (9–16).

Working It Out (a) The center of mass is located at [Eq. (8–48a)]

$$X = \frac{m_1x_1 + m_2x_2}{m_1 + m_2} = \frac{(2.0\text{ kg})(0\text{ m}) + (1.0\text{ kg})(0.6\text{ m})}{2.0\text{ kg} + 1.0\text{ kg}} = 0.2\text{ m}.$$

Now we find the rotational inertia about the axis through this point perpendicular to the dumbbell rod:

$$I_{cm} = m_1(x_1 - X)^2 + m_2(x_2 - X)^2$$
$$= (2.0\text{ kg})(0\text{ m} - 0.2\text{ m})^2 + (1.0\text{ kg})(0.6\text{ m} - 0.2\text{ m})^2$$
$$= 0.2\text{ kg}\cdot\text{m}^2.$$

(b) For the axis through m_1, m_1 itself does not contribute to the rotational inertia because it is on the axis. Mass m_2 is a distance L from the axis. Thus

$$I = m_2L^2 = (1.0\text{ kg})(0.6\text{ m})^2 = 0.4\text{ kg}\cdot\text{m}^2.$$

CONCEPTUAL EXAMPLE 9–4 In the discussion of the dumbbell to this point, we assumed that the masses were point masses. Would the calculation of the position of the center of mass or of the rotational inertia change if the masses were not point masses but were instead uniform spheres of some radius?

Answer The mass distribution changes when the masses are no longer pointlike. This has no effect on the calculation of the center of mass. The center of mass is linear in the positions of the elements involved, and as we established in Chapter 8, we could first find the center of mass of each sphere, which would be at the center of the sphere, then use those as a two-pointlike-mass input to the overall center of mass. Thus the result for the center of mass position is unchanged. The rotational inertia would change, however, because that quantity involves distances squared. This is most easily understood with a simplified example: Place an object of mass M at the right end of a massless rod of length L and find the rotational inertia about an axis perpendicular to the rod and through the rod's left end. If the object is pointlike, $I = ML^2$. But if instead the object were composed of a pointlike mass $M/2$ a distance $L - \Delta L$ from the left end and a pointlike mass $M/2$ a distance $L + \Delta L$ from the left end, with $\Delta L \ll L$, we would find

$$I = (M/2)(L - \Delta L)^2 + (M/2)(L + \Delta L)^2 = M[L^2 + (\Delta L)^2].$$

This differs from the single-pointlike-mass case. The fact that it differs by an amount that is second order in the "size" of the object tells us that it is a good approximation to treat a real dumbbell as a dumbbell with pointlike masses at the end.

Thin Cylinder: The thin-walled cylinder (a cylindrical shell) shown in Fig. 9–13 has radius R, height h, and total mass M distributed uniformly around the ring. Let's find the rotational inertia about the central axis of the cylinder. The word "thin" means that we can consider all of the mass to be the same distance R from the axis. Thus, the rotational inertia is simply

$$I = MR^2. \tag{9–23}$$

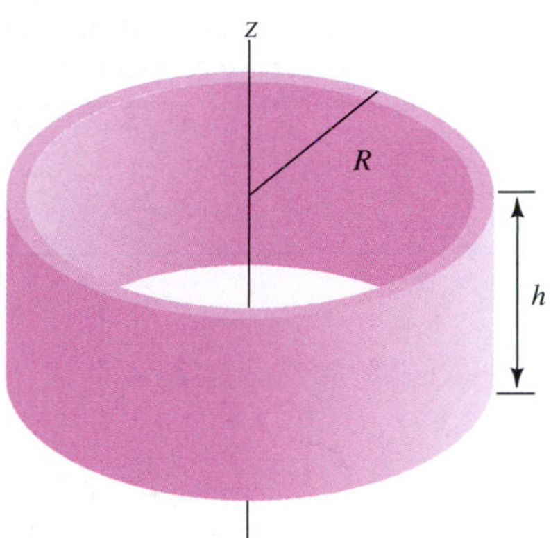

▲ **FIGURE 9–13** A thin cylinder of mass M has height h and radius R. We want to find its rotational inertia about the z-axis, which is the symmetry axis of the cylinder.

Solid Cylinder: We can very often simplify the calculation of the rotational inertia of an object by breaking the object up into appropriate pieces. The rotational inertia of a uniform solid cylinder about its central axis can be calculated by breaking the solid cylinder into a series of concentric thin cylinders and using our result for the thin cylinder. We start with the mass density. The solid cylinder has exterior radius R, height h, and mass M, so its mass density ρ is

$$\rho = \frac{M}{V} = \frac{M}{\pi R^2 h}. \tag{9–24}$$

Now we divide the solid cylinder into thin cylindrical shells. The jth shell is a distance r_j from the axis and has (small) thickness Δr and mass Δm_j (Fig. 9–14). The cylindrical shell labeled j in Fig. 9–14 contributes an amount $\Delta m_j r_j^2$ to the rotational inertia I of the solid cylinder. We sum these contributions to find I:

$$I = \sum_j (\Delta m_j) r_j^2. \tag{9–25}$$

Just as we remarked in the discussion leading to Eq. (9–19), we can carry out this sum once we find the mass of each shell. The mass of a given volume is ρ times that volume; for the thin shell labeled j, the volume is approximately the shell thickness times the shell circumference times the height, $\Delta V_j = (\Delta r)(2\pi r_j)h$. Thus $\Delta m_j = \rho(\Delta r)(2\pi r_j)h$, and

$$I = \sum_j (\Delta m_j) r_j^2 = \sum_j \rho(\Delta r)(2\pi r_j) h r_j^2 = 2\pi h\rho \sum_j (\Delta r) r_j^3.$$

In the limit that the cylindrical shells are very thin, the sum, which includes shells from $r = 0$ to $r = R$, becomes an integral:

$$I = 2\pi h\rho \int_0^R r^3\, dr = 2\pi h\rho \frac{R^4}{4}.$$

When we substitute Eq. (9–24) for ρ, we find

$$I = 2\pi h \frac{M}{\pi R^2 h} \frac{R^4}{4} = \frac{1}{2} MR^2.$$

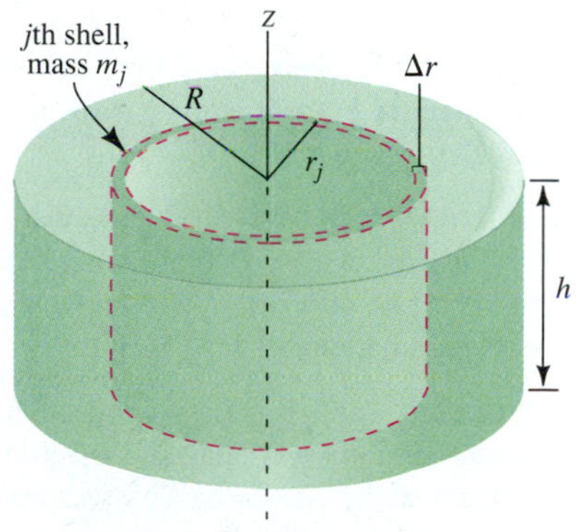

▲ **FIGURE 9–14** To find the rotational inertia of a thick cylinder of mass M, height h, and radius R about its symmetry axis, we break it into a series of thin concentric cylinders. We know the rotational inertia of the thin cylinders from Eq. (9–23) and can sum these inertias to find the rotational inertia of the solid cylinder.

Note that we have reduced the volume integration (a three-dimensional integration) to a one-dimensional one. How did we do this? All the mass at a given distance—meaning at all angles—contributed equally. In other words, symmetry has allowed us to perform the angle integration "automatically." In addition, there is no dependence on height, so the integration over height was also done automatically. This sort of simplification is very often possible when the object involved has some symmetry.

Note that the value of I is less than that of a thin cylinder of the same radius and mass. The weighted sum for the solid cylinder includes mass pieces that give a smaller contribution because they are closer to the axis. Note too that the result is independent of the height and applies for a thin disk as well as a tall cylinder.

CONCEPTUAL EXAMPLE 9–5 Will the rotational inertia of a thick cylinder about its central axis be (a) less than, (b) greater than, or (c) the same as the rotational inertia of a thin cylinder of the same outer radius, length, and total mass?

Answer (a) There is relatively more of the mass of the thin cylinder at the outer radius, and this "large radius" material is weighted more heavily in the calculation of the rotational inertia. Indeed, we can use a dimensional analysis to write the rotational inertia about their axes of a set of cylinders of the same outer radius, height, and mass in the form $I = CMR^2$. Here C is a dimensionless parameter that runs from 1 (all mass concentrated at outer radius, as in the thin cylinder) to 0 (all mass concentrated along the central axis). By the reasoning here, we can conclude that

$$C_{\text{thin cylinder}} > C_{\text{thick cylinder}} > C_{\text{solid cylinder}}.$$

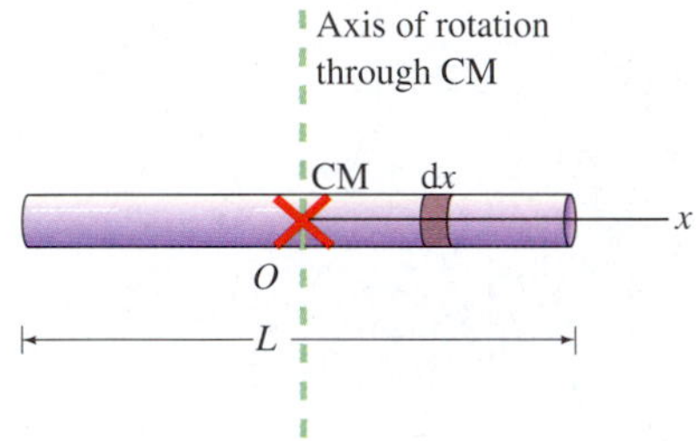

▲ **FIGURE 9–15** Calculating the rotational inertia about an axis perpendicular to the rod passing through the center of mass. This is at the midpoint if the rod's density is constant.

Rod: Next we consider the rotational inertia of a uniform rod of length L about an axis that is perpendicular to it and passes through its center of mass (Fig. 9–15). As this is a thin rod, it can be approximated as a one-dimensional mass distribution, and Eq. (9–20) can be used. The center of mass is the midpoint of the rod, and we will place the origin of our coordinate system at this point. The mass density is the total mass divided by the rod's length, $\lambda = M/L$. Consider the mass dm contained in a slice of the rod of length dx at a distance x from the axis. We have $dm = \lambda\, dx$, so the rotational inertia about the axis is given by

$$I = \int_{-L/2}^{L/2} x^2 \lambda\, dx = \lambda \int_{-L/2}^{L/2} x^2\, dx = \lambda \frac{x^3}{3}\bigg|_{-L/2}^{L/2} = \frac{ML^2}{12}.$$

If λ is not a constant but is symmetric about the midpoint, so that the midpoint is still the center of mass, then λ is replaced by $\lambda(x)$, and the rotational inertia about the center of mass is the integral

$$I = \int_{-L/2}^{L/2} x^2 \lambda(x)\, dx.$$

EXAMPLE 9–6 Find the rotational inertia of a uniform thin rod of length L about an axis perpendicular to the rod and passing through one end of it.

Setting It Up We sketch the situation in Fig. 9–16. The x-axis is along the rod with $x = 0$ at the end that is the site of the rotation axis. The rod, of length L and mass M, has a constant mass density (mass per unit length) $\lambda = M/L$.

Strategy This is a direct application of the expression for the rotational inertia of a continuous object in one dimension, Eq. (9–21). The fact that the rotation axis passes through one end of the rod makes it useful for the origin of our x-axis to pass through that end as well.

Working It Out With the rod running along the x-axis from $x = 0$ to $x = L$,

$$I = \int_0^L x^2 \lambda\, dx = \lambda \int_0^L x^2\, dx = \frac{\lambda}{3} x^3 \bigg|_0^L = \frac{\lambda L^3}{3} = \frac{1}{3} ML^2. \quad (9\text{–}26)$$

The mass density is a constant, which explains why it came out of the integration.

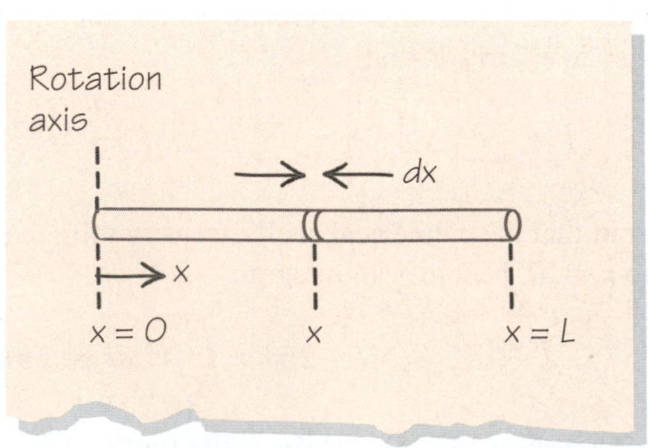

▲ **FIGURE 9–16** The axis of rotation passes through the end at $x = 0$ and is perpendicular to the rod.

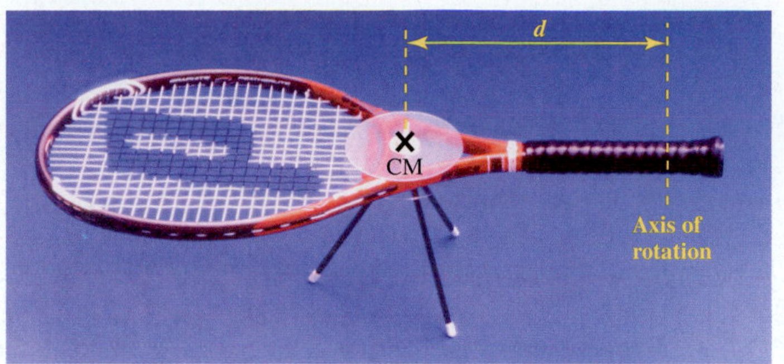

◀ **FIGURE 9–17** To find the rotational inertia of any three-dimensional object about a given axis of rotation, first find the rotational inertia about the parallel axis through the center of mass. The parallel-axis theorem gives a general relation between these two rotational inertias. Here, the object in question is a tennis racket.

Parallel-Axis Theorem

We have worked out the rotational inertia for several simple systems and for several specific axes. Yet how useful are such exercises, as after all there is a different answer for every axis? There is, however, a natural set of axes for every object, and these are axes that go through the center of mass of the system. It may be sufficient to calculate the rotational inertia of objects about these axes because the rotational inertia about a set of other axes can be obtained from the **parallel-axis theorem**. This theorem, whose proof we outline below, states that there is a simple relation between the rotational inertia of an object about an axis through the object's center of mass and the rotational inertia about any other parallel axis (Fig. 9–17). More precisely, the parallel-axis theorem states that

$$I_{pa} = I_{cm} + Md^2, \tag{9–27}$$

PARALLEL-AXIS THEOREM

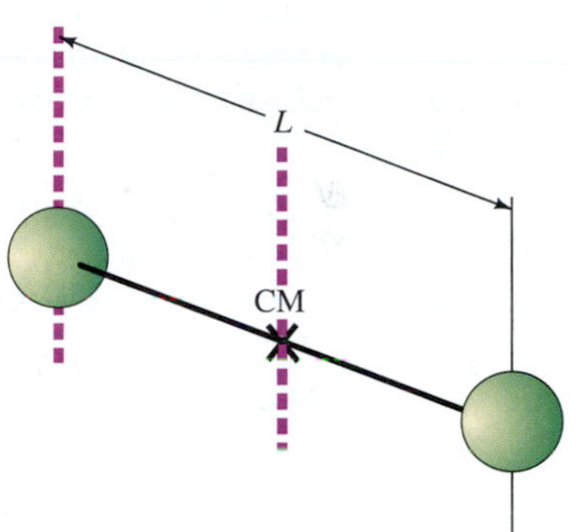

▲ **FIGURE 9–18** Two parallel axes, one through the center of mass and the other through one end, can be drawn through this dumbbell of equal masses. By using the rotational inertia for the center of mass axis and then the parallel-axis theorem, we can easily find the rotational inertia about the axis through one end.

where I_{cm} and I_{pa} are the rotational inertias between the axis through the center of mass and the second axis, respectively, M is the total mass of the object, and d is the perpendicular distance between the axis through the center of mass and the parallel axis.

As a simple example, let's take a dumbbell with equal masses m and consider two parallel axes each perpendicular to the connecting rod: One passes through the center of mass and one through one of the masses (Fig. 9–18). For the axis through the center of mass, we have $I_{cm} = m(L/2)^2 + m(L/2)^2 = mL^2/2$, while direct calculation for the axis through the end gives $I_{end} = mL^2$. What does the parallel-axis theorem give for I_{end}? The distance d in Eq. (9–27) is $L/2$, while $M = 2m$, so the parallel-axis theorem gives $I_{end} = I_{cm} + M(L/2)^2 = mL^2/2 + (2m)(L/2)^2 = mL^2$, the same result given by the direct calculation.

The theorem shows that the rotational inertia of any object is smallest if the axis of rotation goes through the object's center of mass ($d = 0$). This means that the kinetic energy of rotational motion of an object for fixed angular velocity ω is smallest if the rotation is about an axis containing the center of mass.

You can get an idea of how a proof of the parallel-axis theorem works by looking at a two-dimensional object (Fig. 9–19). The reference axes in question will be perpendicular to the page. We have set the origin of the coordinate system at the position O of the center of mass, and the second axis passes through the point P; we have chosen the x-axis along the line OP. A particular element is picked out with mass dm; the rotational inertia about the axis through P is then $I_P = \int r'^2\,dm$. But $r'^2 = d^2 + r^2 - 2r'd\cos\theta$, so that

$$I_P = \int (r^2 + d^2 - 2rd\cos\theta)\,dm = \int r^2\,dm + d^2\int dm - 2d\int r\cos\theta\,dm.$$

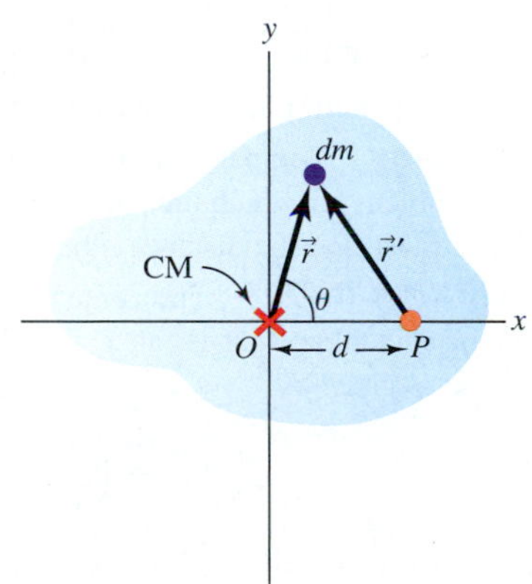

▲ **FIGURE 9–19** For establishing the parallel-axis theorem, we set two axes perpendicular to the page, one through the center of mass at O and the other through the point P.

The first term on the right is just the rotational inertia about the axis through the center of mass, I_{cm}. The second term is Md^2. The factor $r\cos\theta$ in the integral of the third term is x, and the integral $\int x\,dm$ is by definition the x-component of the center of mass. But the center of mass is at the origin, so this term is zero. We are left with Eq. (9–27), the parallel-axis theorem.

EXAMPLE 9–7 Consider a uniform, thin, rectangular sheet of metal of mass M with width a and length b. Calculate the rotational inertia about an axis that is perpendicular to the sheet and passes through one of the corners.

Setting It Up Choose the corner through which the axis passes as the origin and line up the width along the x-axis and the length along the y-axis, as shown in Fig. 9–20.

Strategy This problem is another direct application of the calculation of a rotational inertia for a continuous object. The object is two dimensional, so Eq. (9–21) is relevant. That equation is expressed in terms of the two-dimensional mass density (mass per unit area) σ, which in this case is uniform.

Working It Out To apply Eq. (9–21), we note that $0 \le x \le a$ and $0 \le y \le b$, that r^2 in the integral is $x^2 + y^2$, and that the two-dimensional uniform mass density σ is

$$\sigma = \frac{M}{\text{area}} = \frac{M}{ab}.$$

Also, the area element dA in Eq. (9–21) can be written as $dx\,dy$—we have broken the area down into little rectangles each of area width times height. To sum over the whole area (integrate), we take consecutive steps in which we first fix x and sum the rectangle heights dy over all y-values and then sum the rectangle widths dx over all x-values. Thus

$$I_{\text{corner}} = \int_0^a dx \int_0^b dy\, \sigma(x^2 + y^2)$$

$$= \frac{M}{ab}\int_0^a dx \int_0^b dy(x^2 + y^2)$$

$$= \frac{M}{ab}\left(\int_0^a x^2\,dx \int_0^b dy + \int_0^a dx \int_0^b y^2\,dy\right)$$

$$= \frac{M}{ab}\left[\left(\frac{a^3}{3}\right)b + \left(\frac{b^3}{3}\right)a\right]$$

$$= \frac{1}{3}M(a^2 + b^2).$$

Note that this is a case where we did not directly reduce a multidimensional integral to a one-dimensional integral, as we did, for example, in the calculation of the rotational inertia of the solid cylinder. However, note that the integration procedure above is in fact a sequence of one-dimensional integrals.

What Do You Think? Does it matter if the sheet is thick rather than thin?

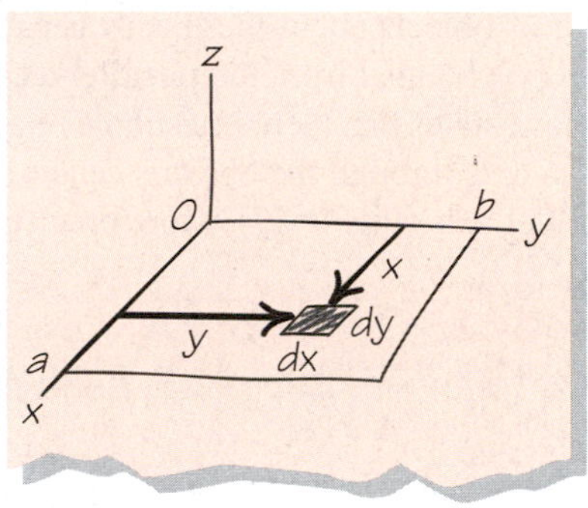

▲ **FIGURE 9–20** We calculate the rotational inertia about the z-axis, which passes through one corner.

EXAMPLE 9–8 Use the result of Example 9–7 to calculate the rotational inertia of the same sheet of metal about an axis that is perpendicular to the plane of the sheet and passes through the sheet's center of mass.

Setting It Up Figure 9–21 sketches the situation. We include the same coordinate system as in Fig. 9–20.

Strategy Comparison of Figs. 9–21 and 9–20 suggests that this is a problem for which the parallel-axis theorem is applicable. To use it, we must find the distance d between the two axes.

Working It Out The center of mass is located at the midpoint of the rectangle, that is, at $x = a/2$, $y = b/2$. The distance to the corner is given by

$$d^2 = \left(\frac{a}{2}\right)^2 + \left(\frac{b}{2}\right)^2.$$

Thus, by the parallel-axis theorem, $I_{\text{corner}} = I_{\text{cm}} + Md^2$:

$$I_{\text{cm}} = I_{\text{corner}} - Md^2 = \frac{1}{3}M(a^2 + b^2) - M\left[\left(\frac{a}{2}\right)^2 + \left(\frac{b}{2}\right)^2\right]$$

$$= M(a^2 + b^2)\left(\frac{1}{3} - \frac{1}{4}\right) = \frac{1}{12}M(a^2 + b^2).$$

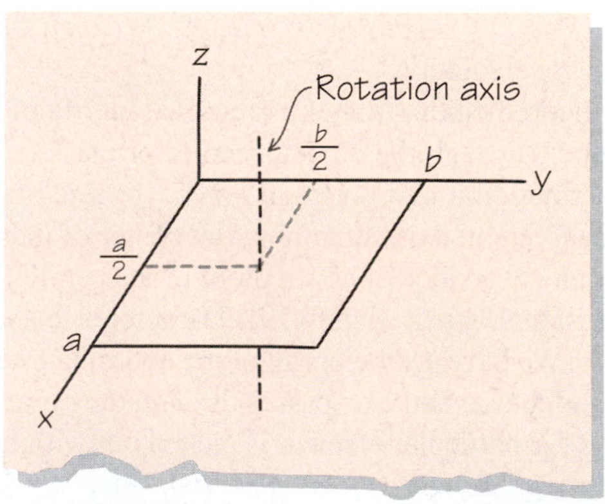

▲ **FIGURE 9–21** As for Example 9–7, except that we calculate the rotational inertia about an axis passing through the center of the plate.

What Do You Think? Why is the answer to this example much smaller than that of the previous example?

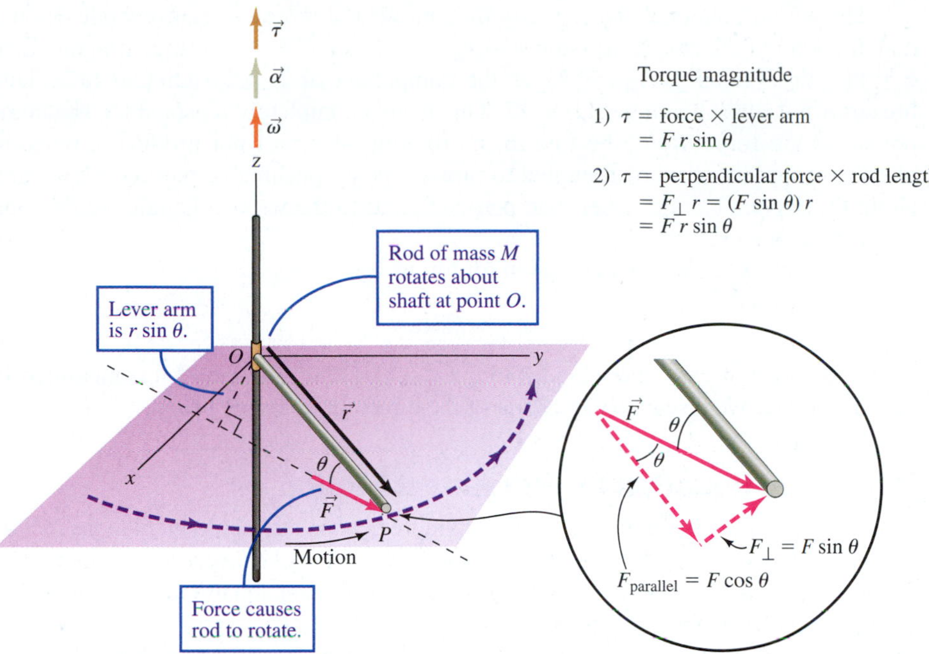

▲ **FIGURE 9–22** A rod is attached to a shaft and can rotate in the plane perpendicular to the shaft. The shaft acts as a rotation axis. A force lying in the plane in which the rod can move is applied to the rod, causing a rotation. The effectiveness of the force in causing a rotation depends on the strength of the force, on where along the rod it is applied, and on its direction, specified by the angle θ. The lever arm is $r\sin\theta$.

9–4 Torque

We have described rotations; now, let's turn to the question of what causes them. The analogies we have established between rotational motion and the motion of point particles (we refer to this as *linear motion*) will be useful here. Newton's second law describes the dynamics of linear motion by the equation $\vec{F} = m\vec{a}$. We'll see here that for rotational motion there is a dynamical equation for angular acceleration with a quantity analogous to force called **torque**, symbol $\vec{\tau}$. Just as a force provides a push or a pull, torque provides a twist. Thus it takes a torque to accelerate the rotation of a wheel or to turn a screw.

It is useful to keep a definite example in mind as we go through our discussion, the rotation of a rod attached at a point O to a shaft that can rotate (Fig. 9–22). The rod, which has length r and mass M, makes a right angle with the shaft. The motion is entirely in the plane perpendicular to the shaft. Suppose now that a force $\vec{F}$ acts in that plane on the end of the rod, at point P in Fig. 9–22. This force causes the system to undergo rotational motion, and in what follows we describe the effectiveness of the force in bringing about changes in rotational motion by making a suitable definition of the torque.

How effective a force is in setting an extended system (such as that of Fig. 9–22) into rotational motion about a point O depends on two features. First, a larger force will be more effective in exerting torque than would an otherwise similar force of smaller magnitude, so we expect the torque to be proportional to the force. Second, a force is most effective at producing a torque if it is applied at a point P in Fig. 9–22 that is *far* from the rotation axis (i.e., far from point O in Fig. 9–22) and is at *right angles* to the line between the rotation axis and the point of application of the force (line OP in Fig. 9–22). In other words, the torque τ is also reasonably proportional to $r_\perp \equiv r\sin\theta$ in Fig. 9–22, where $\vec{r}$ is the vector from point O to point P and θ is the angle between $\vec{r}$ and $\vec{F}$. The factor $r\sin\theta$ is known as the **lever arm** or **moment arm**. You can see from Fig. 9–22 that the lever arm is found by extending the line along which the force acts and measuring the closest distance of this line to the rotation axis. Plumbers who use long wrenches to free stubborn pipes are quite familiar with the effectiveness of a large lever arm (Fig. 9–23). Pulling the proportionality to both F and $r_\perp$ together, we would write $\tau = r_\perp F$.

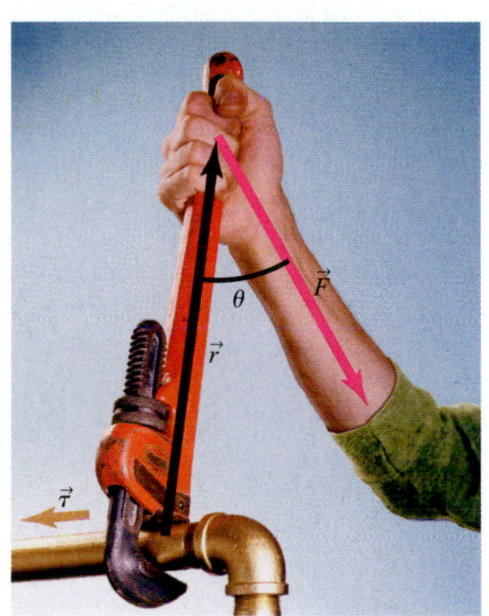

▲ **FIGURE 9–23** A plumber taking advantage of a long lever arm. A long wrench is more effective than a short one for freeing a stubborn pipe joint.

There is still another way to verify that our idea of torque is a correct one. Rather than thinking of the trial expression $\tau = r_\perp F = (r \sin\theta)F$, we arrange it in the form $\tau = r(F \sin\theta) = rF_\perp$, where $F_\perp$ is the component of $\vec{F}$ perpendicular to $\vec{r}$. This breakdown is also shown in Fig. 9–22. This is also completely reasonable: The component of the force *along* the line to the rotation axis will not produce a twist. A plumber's apprentice who attempted to turn a pipe by pushing or pulling the wrench along the wrench handle rather than perpendicular to the wrench handle would soon be out of a job!

To summarize, our trial expression for torque is

$$\tau = rF \sin\theta. \tag{9–28}$$

The dimensions of this expression for torque are $[ML^2T^{-2}]$, and in SI it is measured in newton-meters. We discuss the question of the direction of torque below.

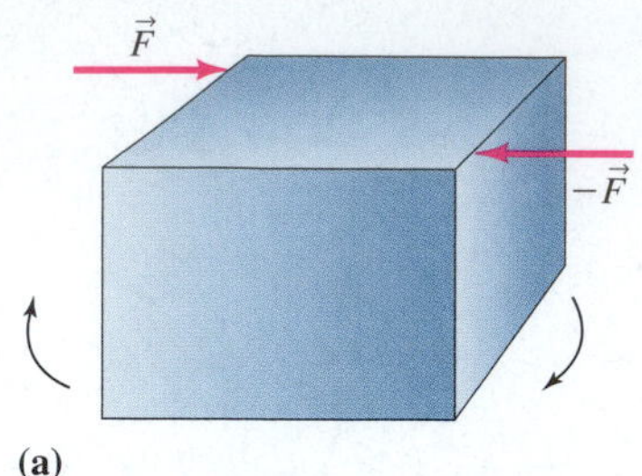

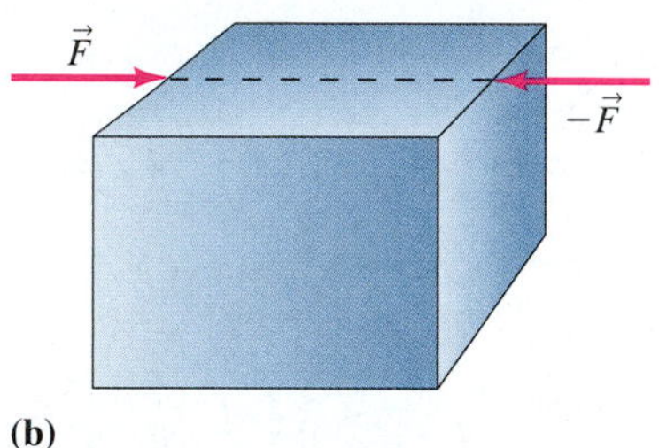

▲ **FIGURE 9–24** (a) A pair of equal and opposite forces that do not act along the same line exert a torque on an object. (b) There is no torque only if the same pair of forces act along the same line.

Dynamical Equation for Rotational Motion

To this point, our expression for torque is no more than an attempt to include the factors we know intuitively are effective at producing a twist. To make this concept a more useful one, we want to connect it *quantitatively* to a dynamical equation for rotations. We do this by applying Newton's second law.

Both internal and external forces can act on a given element of an object. In linear motion, internal forces do not cause any acceleration because they come in pairs that by Newton's third law cancel. In the same way, these pairs of canceling internal forces cannot cause a twist because they are aligned with each other, so we shall consider only external forces. And as Figs. 9–24a and b show, a pair of equal and opposite external forces can produce a twist if they are not aligned.

Suppose, then, that a net external force $\vec{F}_i$ acts on an element of a rod labeled by i (Fig. 9–25); this is similar to the situation described in Fig. 9–22. We suppose that the force is perpendicular to the rod; this is good enough because any component along the rod has zero lever arm and is ineffective at producing a twist. The direction of the force is therefore purely *tangential.* The ith element is a distance r_i from the rod's attachment point to the fixed shaft and has mass Δm_i. The torque on element i about the origin therefore has magnitude $\tau_i = r_i F_i$. At the same time, we know from the second law that F_i produces an instantaneous tangential acceleration a_i of element i whose magnitude is $a_i = F_i/\Delta m_i$. Because the acceleration a_i is tangential, we can use Eq. (9–12), $a_i = \alpha r_i$. In other words, because the force is perpendicular to the rod, the instantaneous acceleration of the mass element is associated with an *angular* acceleration of the mass element. Because of internal forces within the (rigid) rod, the entire rod undergoes the same angular acceleration, so there is no need for a subscript i on the quantity α. Combining, we have Newton's second law:

$$F_i = \Delta m_i \, r_i \alpha;$$

hence, the torque on the element i has magnitude

$$\tau_i = r_i F_i = \Delta m_i \, r_i^2 \alpha.$$

The *net* torque τ comes from summing the torques on each element and therefore has magnitude

$$\tau = \sum_i \Delta m_i \, r_i^2 \alpha = \Big(\sum_i \Delta m_i \, r_i^2\Big)\alpha.$$

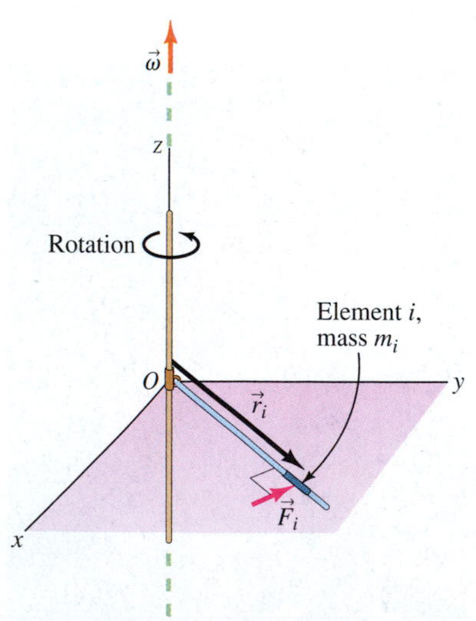

▲ **FIGURE 9–25** A force of magnitude F_i acts on an element i of a rod attached to a rotating shaft. The force is perpendicular to the rod.

The quantity in parentheses is the rotational inertia I about the shaft [Eq. (9–15)]. We have found the dynamical relation we were looking for:

$$\tau = I\alpha. \tag{9–29}$$

EQUATION FOR ROTATIONAL MOTION

When there is an angular acceleration about an axis perpendicular to the plane of rotation, it is dynamically determined by the torque about the same point according to Eq. (9–29). This equation is analogous to Newton's second law, which determines the linear acceleration of an object in terms of the object's inertial mass and the force acting on the object. The major difference is that all the quantities in Eq. (9–29)—torque, angular acceleration, and rotational inertia—refer to a given axis.

It is worth emphasizing that we have not introduced any new laws of physics here. All the laws of rotational motion follow from Newton's original laws.

Newton's second law is a vector relation and we have already stated that the angular acceleration is a vector. Where is the vector aspect of the torque? Figure 9–26 shows that we can use a right-hand rule to define correctly the direction of the torque: If the fingers of the right hand are aligned along the line perpendicular to the rotation axis and extending to the point where the force is applied (in other words, along the vector $\vec{r}$) and then curled in the direction of the force, the thumb points in the direction of the torque vector $\vec{\tau}$. The procedure is illustrated in Fig. 9–27, and we can see that the torque is always in the same direction as the angular acceleration. The vector equation of motion takes the form

$$\vec{\tau}_{\text{net}} = I\vec{\alpha}, \tag{9–30}$$

where, as usual, the word "net" means the vector sum of all the individual torques acting.

In this chapter, the vector nature of torques will not play a very important role, because we are restricting ourselves to rotations about a single axis. It is important, however, to become used to thinking of the dynamics of rotational motion in terms of vectors. In the more general treatment given to rotations in Chapter 10, the vector nature of torque is critical.

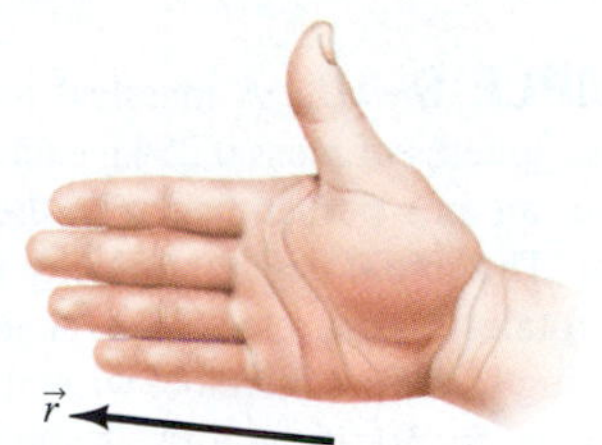

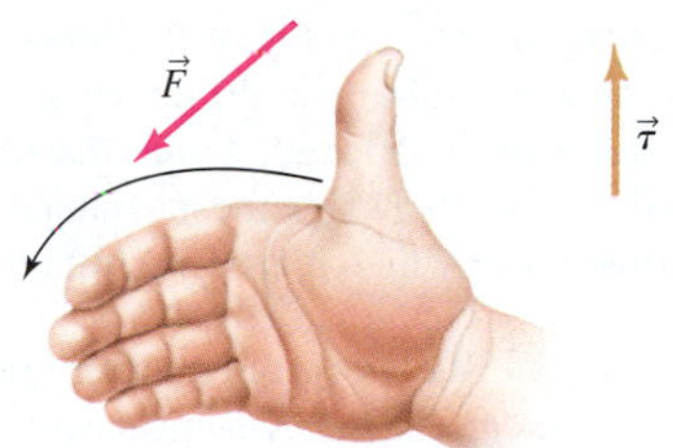

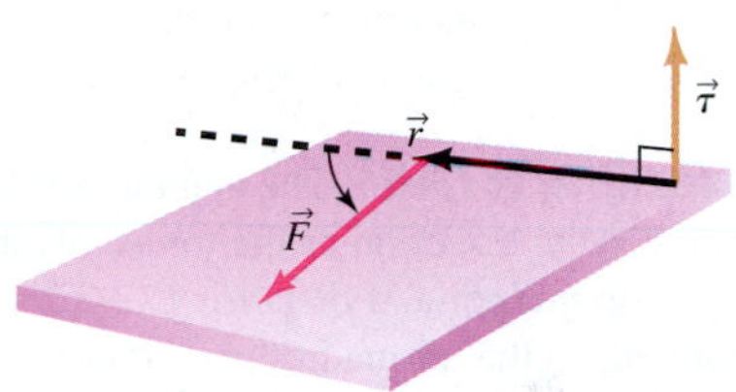

▲ **FIGURE 9–26** The direction of the torque can be determined by a right-hand rule applied to the vectors $\vec{r}$ and $\vec{F}$. The fingers of the right hand are aligned along $\vec{r}$ and then curled toward $\vec{F}$, with the angle between $\vec{r}$ and $\vec{F}$ always taken to be less than 180°. The torque is then along the thumb.

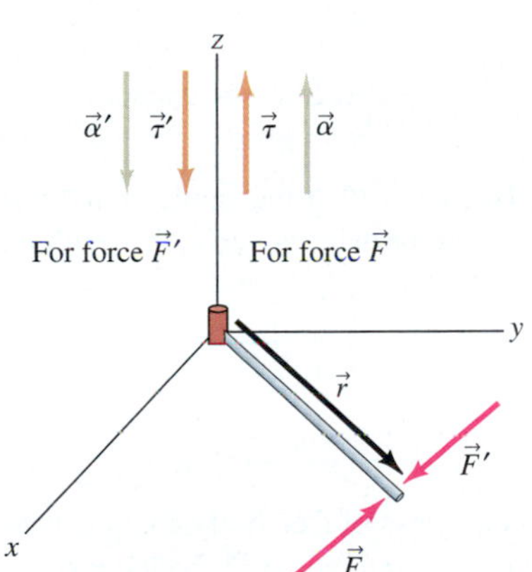

◀ **FIGURE 9–27** The direction of the angular acceleration of the rod, $\vec{\alpha}$, and of the torque on the rod, $\vec{\tau}$, depends on the direction of the force. When the force is in one direction $(\vec{F})$, the torque $\vec{\tau}$ and angular acceleration point along $+z$; when the force is in the other direction $(\vec{F}')$, the torque $\vec{\tau}'$ and angular acceleration point along $-z$.

CONCEPTUAL EXAMPLE 9–9 Look at Fig. 9–28. What is the direction of the torque on the large ship about its center of mass? Supposing that the pushing force applied by the tugboat is at its maximum, is the magnitude of this torque also a maximum?

Answer The torque will make the large ship turn in a clockwise fashion as viewed from above. Using a right-hand rule for the sense of this rotation, the right thumb will point down to Earth's center, and this is the direction of the torque. The force is applied at the very end of the ship, and this is as far from the ship's center of mass as one can get. In addition, the force is apparently applied at right angles to the line from the center of mass. Thus the torque has maximum magnitude.

▲ **FIGURE 9–28** Tugboat rotating ship.

EXAMPLE 9–10 A massless rod of length 0.83 m connects two small spheres of mass 0.25 kg each. The rod is constrained to rotate about an axis perpendicular to the rod and passing through its midpoint. The initial angular velocity has magnitude 2.1 rad/s, and the rod rotates counterclockwise. A tangential force of magnitude 9.6 N acting in the counterclockwise directon is applied to one of the spheres (Fig. 9–29). If the force is applied for 2.0 s, what is the final angular speed of the rod?

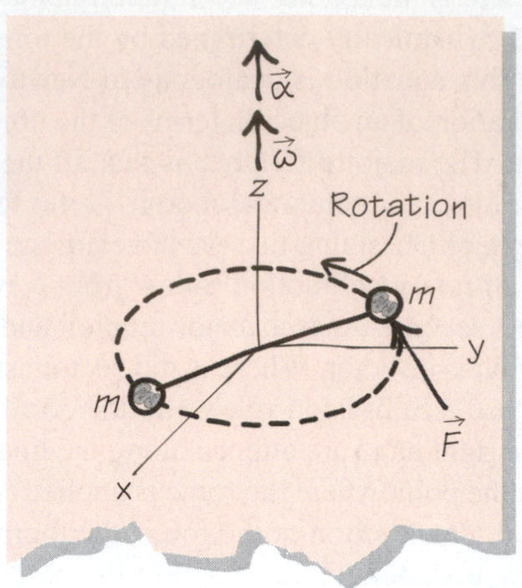

▲ **FIGURE 9–29** Force produces torque which causes angular acceleration.

Setting It Up As Fig. 9–29 shows, we place the z-axis along the rotation axis, with $+z$ in the direction of the initial angular velocity. We know the rod length L, the masses m, and the initial angular speed ω_0. We also know the magnitude F of the tangential force and the time t over which it acts. We want the final angular speed ω.

Strategy We are given enough information to find the torque. With it, we can use the dynamical equation relating torque and angular acceleration, and because the torque is constant, so is the angular acceleration—we can solve for its value from the dynamical equation. The kinematics of constant angular acceleration will then lead us to a final angular velocity linearly changing with time.

Working It Out The angular acceleration about the axis points in the same direction as the torque about that axis. The rotation axis is at the midpoint, a distance $L/2$ from the point where the force is applied, so the torque has magnitude

$$\tau = (L/2)F.$$

Equation (9–29) then gives an angular acceleration in terms of the torque, constant in magnitude for a constant-magnitude torque, $\alpha = \tau/I$. We have just found τ, but we also need the rotational inertia of the two-sphere system about the axis,

$$I = m\left(\frac{L}{2}\right)^2 + m\left(\frac{L}{2}\right)^2 = \frac{mL^2}{2}.$$

Thus we have

$$\alpha = \frac{\tau}{I} = \frac{LF/2}{mL^2/2} = \frac{F}{mL}.$$

With a constant angular acceleration, the angular speed has a linear dependence on the time,

$$\omega = \omega_0 + \alpha t = \omega_0 + \frac{F}{mL}t.$$

At time $t = 2.0$ s, this gives

$$\omega = 2.1 \text{ rad/s} + \frac{9.6 \text{ N}}{(0.25 \text{ kg})(0.83 \text{ m})}(2.0 \text{ s}) = 95 \text{ rad/s}.$$

What Do You Think? True or false: If the applied force had half the magnitude but were applied for twice the time, the resulting final angular speed would be the same.

Free-Body Diagrams Revisited

The discussion of torque suggests that to extend the usefulness of free body-diagrams to situations with rotations of extended objects we need to take into account the location at which forces act. A description of appropriately extended free-body diagrams and advice on how to approach problems that involve rigid-body motion are given in the Problem-Solving Techniques box.

Problem-Solving Techniques

1. Identify and isolate the rigid body on which external forces act.
2. Identify the external forces that act on the object as well as where they act.
3. Prepare an extended free-body diagram: All forces should be included, as well as *information on where they act* on the rigid body. Where the force acts is of paramount importance in understanding torques. Note that gravity acts as if it were applied to the center of mass of the object [see Eq. (9–36) and the following text].
4. Identify a single convenient axis, which will normally be the axis about which the rigid body rotates.
5. Find the torques about this axis that result from the acting forces. The direction of positive torque should be clearly understood using the right-hand rule as necessary.
6. Express the dynamical equations of motion that correspond to the *net* torque. This step generally requires knowledge (or calculation) of the rotational inertia about the axis of rotation. The solutions of the dynamical equations describe the rotational motion of the rigid body.
7. The net force governs the linear motion of the center of mass.

Let us look at an example. A bucket of water of mass m is connected to a rope of negligible mass (Fig. 9–30a). The rope is wrapped around a pulley of mass M, radius R, and rotational inertia I about its axis. The pulley is free to rotate without friction about its axis, which is horizontal and fixed to the wall. The bucket is released from rest and drops, making the pulley turn. We want to use extended force diagrams to find the equations that would allow you to find the angular velocity of the pulley after the bucket has fallen for a time t.

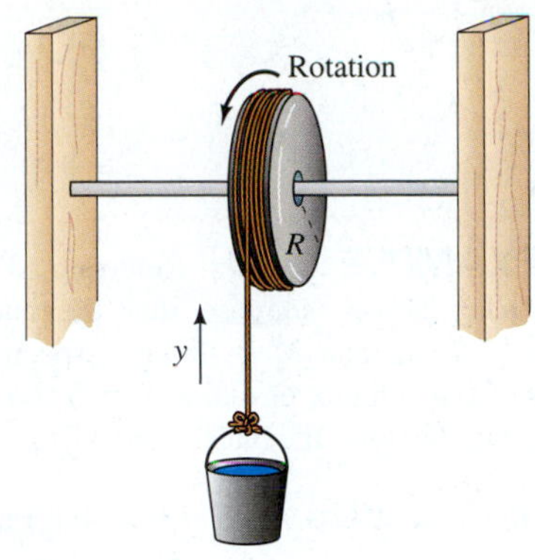

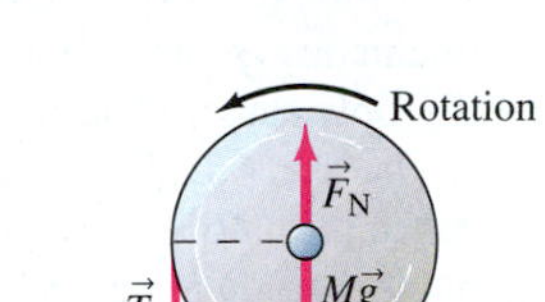

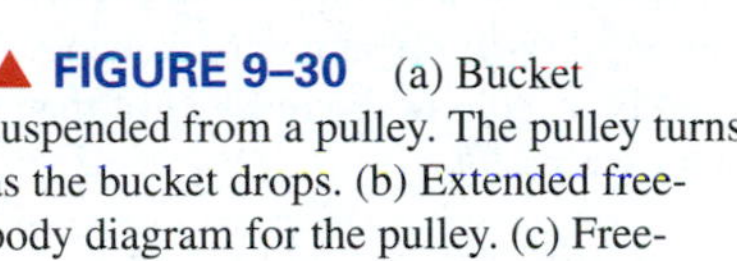

▲ **FIGURE 9–30** (a) Bucket suspended from a pulley. The pulley turns as the bucket drops. (b) Extended free-body diagram for the pulley. (c) Free-body diagram for the bucket.

From Fig. 9–30a we see that there are two objects to consider, the bucket and the pulley. We will therefore identify the forces on and draw extended free-body diagrams for both of these objects. Figure 9–30a also reminds us that there is a connection, namely an unstretchable rope, between the pulley and the bucket, which results in the angular acceleration of the pulley being kinematically connected to the linear acceleration of the bucket.

We next identify all the forces on *each* object and draw corresponding extended force diagrams. Once that is done, we can express Newton's second law and any torque equations. Finally, we will use the fact that the rotational motion of the pulley is connected to the linear motion of the bucket because of the unstretchable rope connecting them. Once we have written all the corresponding expressions, we can count equations and unknowns to ensure that we can carry through the solution.

The forces acting on the pulley are a reaction force that keeps the pulley in place, F_N, at its axis, the force of gravity, magnitude Mg, and the tension of the rope, magnitude T. These appear in the extended free-body diagram for the pulley, Fig. 9–30b. The forces on the bucket are the rope tension and the force of gravity on the bucket, magnitude mg, and the extended free-body diagram for the bucket is Fig. 9–30c, although the "extended" free diagram for the bucket is really just an ordinary free-body diagram.

With the free-body diagrams drawn, we can write the dynamical equations. All the forces are vertical, so there is a single (y-component) Newton's law equation for each object—this axis is identified in Fig. 9–30a. There is a torque equation only for the pulley, and if we calculate angular quantities about the pulley axis, we can see that only the rope tension produces a torque. This is also only a single equation as the angular velocity in Fig. 9–30b will point along the axis of rotation—in our drawing the right-hand rule shows that it will be out of the page. Explicitly we recognize that the center of mass of the pulley does not accelerate, as it is fixed to the wall, so the net force acting on it must be zero:

$$\text{second law for pulley: } F_N - T - Mg = 0. \tag{9–31}$$

The torque acting on the pulley has magnitude TR; hence

$$TR = I\alpha, \tag{9–32}$$

with α the pulley's angular acceleration magnitude and I the given rotational inertia of the pulley about its axis. As the torque is directed out of the page, α will also be directed out of the page, corresponding to an increasing angular speed. Finally, for the bucket, there is an (as-yet-unknown) acceleration a, which we assume is downward by writing the acceleration term of the second law as $-ma$:

$$\text{second law for bucket: } T - mg = -ma. \tag{9–33}$$

(The minus sign on the right-hand side will take care of itself; if the equations were to give us a value for a that is negative, that would mean that the actual acceleration is upward.)

We now have three dynamical equations but four unknowns: F_N, T, a, and α. We have not yet brought in a last element, a *kinematic* relation between a and α. This relation comes from the fact that if the bucket moves at a speed v, then any point on the rope moves at the same speed, and thus the outer rim of the flywheel moves at that speed. This means that the pulley rotates about its axis at an angular speed ω such that

$$v = \omega R.$$

If we take a time derivative of this equation, we find

$$a = \alpha R. \tag{9–34}$$

Equations (9–31) to (9–34) are enough information to solve for the unknown a, which will be a constant. We can then use the fact that the speed v increases linearly with time, and once we have v, we have $\omega = v/R$.

EXAMPLE 9–11 Referring to the discussion of the falling bucket above, suppose that the bucket and pulley masses are $m = 12$ kg and $M = 88$ kg, respectively, and that the pulley is a uniform cylinder of radius $R = 0.50$ m. Find the angular velocity of the pulley after the bucket has fallen for 5 s.

Setting It Up See Fig. 9–30 for a suitable diagram.

Strategy This is a matter of solving Eqs. (9–31) to (9–34) for one of the quantities (e.g., the angular acceleration of the pulley), then using kinematics to obtain the angular speed as a function of time. (Actually only three of the equations are needed.) We'll also need to find the rotational inertia of the pulley, contained in Eq. (9–32). As for virtually all problems, we first solve algebraically, plugging in numbers only at the end.

Working It Out First the rotational inertia of the pulley: The cylinder is a uniform solid, so $I = \frac{1}{2}MR^2$ (see Table 9–1). Next we solve our equations for the desired unknown, here the angular acceleration α, a process that can be done in a number of ways. The simplest is to solve Eq. (9–33) for T, $T = m(g - a)$. Now we substitute $a = \alpha R$ in this result and insert it into the torque equation (9–32):

$$TR = m(g - \alpha R)R = I\alpha.$$

This equation can be immediately solved for α, namely,

$$\alpha = \frac{mgR}{mR^2 + I} = \frac{mgR}{mR^2 + (M/2)R^2} = \frac{mg}{[m + (M/2)]R}$$

This angular acceleration is constant, implying that the angular velocity changes linearly with time:

$$\omega = \omega_0 + \alpha t = \omega_0 + \frac{mgt}{[m + (M/2)]R}.$$

By the initial conditions of the problem, $\omega_0 = 0$ rad/s, and after 5 s,

$$\omega = \frac{(12\text{ kg})(9.8\text{ m/s}^2)(5\text{ s})}{[(12\text{ kg}) + (88\text{ kg}/2)](0.50\text{ m})} = 21\text{ rad/s}.$$

What Do You Think? The fact that both the angular velocity vector of the pulley and its angular acceleration vector are directed out of the page means that the angular velocity is (a) constant in time, (b) changing quadratically with time, (c) changing in magnitude while unchanging in direction, or (d) none of the above.

CONCEPTUAL EXAMPLE 9–12 Consider once again the arrangement of the falling bucket that turns the pulley shown in Fig. 9-30. Will the angular frequency increase more quickly or less quickly if more of the mass M of the pulley is concentrated near the outer radius R?

Answer If more of the mass M of the pulley is concentrated near the outer radius R, the rotational inertia of the pulley about its axis will be larger. The increased rotational inertia for torque is like increased mass for force: In each case the motion in response is more "sluggish." For a given tension in the rope the angular acceleration will be smaller. We can therefore conclude that the angular speed will increase more slowly if the mass of the pulley is concentrated at the edge. In equations, the rotational inertia of the pulley will take the general form CMR^2, where $C = 0$ corresponds to all the mass concentrated at the pulley axis, $C = \frac{1}{2}$ is a uniform solid cylinder, and $C = 1$ corresponds to all the mass concentrated at the outer radius R. You can trace through the algebra of the previous example and see that ω increases more slowly for $C = 1$.

Gravity and Extended Objects

Gravity acts on every point of an object at once. Here, we argue that the torque on an extended object due to gravity can be computed as if the entire mass of the object were concentrated at its center of mass. It is enough to work with an easily generalized example, a rod of length L and mass M fixed to a pivot at its upper end (Fig. 9–31a). We divide the rod into a series of pieces labeled by the subscript i. The pieces, which have

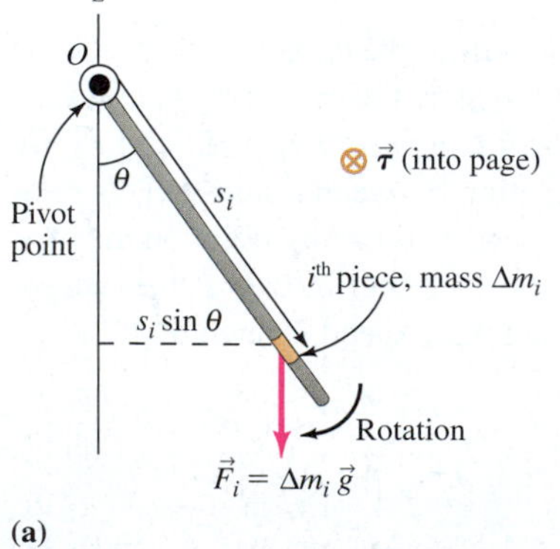

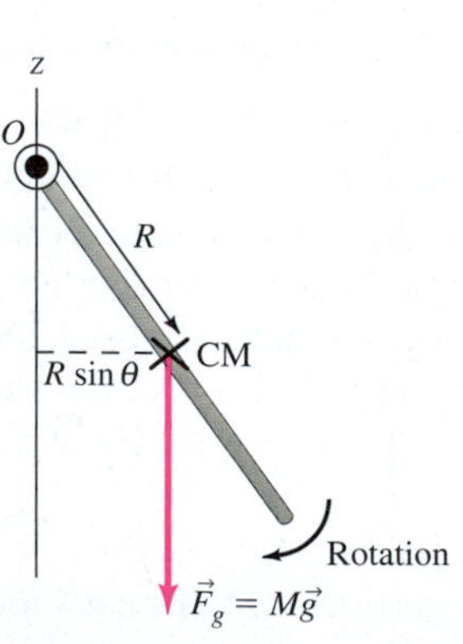

▶ **FIGURE 9–31** The force of gravity rotates a rod about the point O. (a) The differential element of force $\vec{F}_i$ acts on a differential mass Δm_i located a distance s_i down the rod. The circled cross indicates that the torque τ is directed into the plane of the page. (b) The net effect of the sum of the torques acting on each piece is a net torque that is the same as would be present if the entire mass of the rod were concentrated at the rod's center of mass.

mass Δm_i, are small enough so that we can consider each of them to be a given distance s_i down the rod. We will compute the net torque due to gravity about the pivot point assuming the rod makes an angle θ to the vertical. The net torque τ is the vector sum of the torques on each piece, τ_i. All the torques are directed into the page, so this sum is the simple algebraic sum of the magnitudes τ_i. In turn, the moment arm for the force of gravity on the piece labeled by i is $s_i \sin\theta$, so the torque τ_i is

$$\tau_i = (s_i \sin\theta)\,\Delta m_i\, g.$$

The net torque then has magnitude

$$\tau = \sum_i \tau_i = \sum_i (s_i \sin\theta)(\Delta m_i\, g) = \left[\sum_i (s_i)(\Delta m_i)\right](\sin\theta)g. \qquad (9\text{–}35)$$

By definition (see Chapter 8), the quantity in brackets is MR, where R is the distance of the center of mass from the reference point. Thus the net torque due to gravity is (Fig. 9–31b)

$$\tau = R(Mg)\sin\theta. \qquad (9\text{–}36)$$

Compare this result with Eq. (9–28), which is our general form for the torque. *The torque due to gravity on an extended object of total mass M may be represented by the torque due to gravity acting on a particle of mass M located at the object's center of mass.* This important and very useful result is a general one that holds for any object, not just a rod, as you could prove by breaking up an arbitrary object into pieces as we did for the rod.

We could add as a corollary that gravity cannot exert torque on an object about the object's center of mass. This can be helpful in problem solving.

The same reasoning can be applied to energy considerations. *When an extended object of mass M moves in any way, we can find the change in its gravitational potential energy by finding the change in height, Δh, of the center of mass. The change in gravitational potential energy is then $Mg(\Delta h)$.*

9–5 Angular Momentum and Its Conservation

We have uncovered a series of analogues between rotation and linear motion—the analogous descriptive variables appear in analogous positions in expressions for energy and the dynamical equation for rotational motion is analogous to Newton's second law. The analogue between linear and rotational motion can be of further use to us. In our original formulation of Newton's second law, the quantity $m\, d\vec{v}/dt$ appears. In Chapter 8 we found it useful to define the linear momentum $\vec{p} \equiv m\vec{v}$ and to formulate a more general form of Newton's second law:

$$\vec{F} = \frac{d\vec{p}}{dt}.$$

Similarly, it is useful to define the **angular momentum** $\vec{L}$ of a symmetrical object that rotates about its symmetry axis (or an axis parallel to the symmetry axis) with angular velocity $\vec{\omega}$ by

$$\vec{L} \equiv I\vec{\omega}. \qquad (9\text{–}37)$$

ANGULAR MOMENTUM

Here, I is the rotational inertia of the object with respect to the rotation axis. We see here that angular momentum is a vector aligned with $\vec{\omega}$. For uniform rotational motion about an axis, the angular momentum does not change in either magnitude or direction.

To gain some idea of the possible range of angular momenta in nature, you might try calculating the angular momentum of Earth's motion about its axis of rotation given that its mass is 6×10^{24} kg and its radius is 6.4×10^{6} m. With a uniform mass density,

$I = 2/5MR^2 = 10^{38}\ \text{kg} \cdot \text{m}^2$. The value of ω comes from the known period T, 1 day, namely $\omega = 2\pi/T$. After conversion of T to seconds, you find $\omega = 7.3 \times 10^{-5}$ rad/s. Combining, the magnitude of Earth's angular momentum is $L = I\omega = 7 \times 10^{33}\ \text{kg} \cdot \text{m}^2/\text{s}$. (In fact, the actual value of the angular momentum is about 20 percent less than this, and this is because the distribution of mass is not uniform but is more concentrated at the center, reducing I from the value we found by 20 percent.) At another extreme there is a smallest nonzero value of angular momentum that occurs in nature, about $10^{-34}\ \text{kg} \cdot \text{m}^2/\text{s}$. This is a significant fact associated with quantum mechanics that we shall discuss in the next chapter and elsewhere.

In terms of angular momentum, the dynamical equation for rotational motion takes the more general form

$$\vec{\tau} = \frac{d\vec{L}}{dt}. \tag{9–38}$$

EQUATION FOR ROTATIONAL MOTION

Because angular momentum is so fundamental to rotational motion, it is useful to write the energy in terms of it. We have for a simple rotation

$$K = \frac{1}{2}I\omega^2 = \frac{1}{2}\frac{(I\omega)^2}{I} = \frac{L^2}{2I}. \tag{9–39}$$

Angular momentum shares with momentum (linear momentum is simply momentum) the important property that it is independent of time for a system that is "left alone," that is, a system on which there is no (net) torque due to external forces. (Note that the external torque may be zero even when the external force is not zero, depending on where the external force is applied and on its direction. Conversely, it is possible to have a net torque even though the net force is zero.) When the net torque on an object is zero, its angular momentum is independent of time—the angular momentum is *conserved.* For rigid bodies, the rotational inertia is constant, and so the conservation of angular momentum tells us that the angular velocity is constant in time. The conservation of angular momentum in more complicated systems is a powerful tool. The following example shows how collisions can be analyzed with the help of this conservation law. In Chapter 10 we'll show in much more detail how we can use the conservation of angular momentum.

EXAMPLE 9–13 An engineer works with a turntable that accepts a stack of disks. The engineer starts with a model in which a freely rotating turntable, friction free, of rotational inertia I_1 about the axis passing through its center and perpendicular to the center is rotating with an angular speed ω_i when a disk of rotational inertia I_2 about the same axis is dropped from rest onto the turntable, with its center on the turntable's center. Friction between disk and turntable brings them to a common final angular speed ω_f. What is ω_f?

Strategy We recognize that this is a colliding system upon which no external torques act so that angular momentum is conserved. The initial angular momentum is associated with the turntable—the disk initially has none. As for the final angular momentum, we recognize that when the turntable and the disk rotate together about a common axis, their net rotational inertia is the sum of their individual rotational inertias about the same axis, $I_1 + I_2$. We can then set the initial angular momentum to the final angular momentum and solve for ω_f.

Working It Out The initial angular momentum is $I_1\omega_i$ and the final angular momentum is $(I_1 + I_2)\omega_f$. We set them equal and easily solve for ω_f:

$$\omega_f = \frac{I_1\omega_i}{I_1 + I_2}.$$

Note that this is less than ω_i: The disk speeds up but the turntable slows down.

What Do You Think? Is energy conserved in the process described here?

Parallels between Rotational and Linear Motion

Throughout this chapter, we have emphasized the similarities and the differences between rotational motion and linear motion. The major differences are that the vectors that describe the parameters of rotational motion are all measured with respect to an axis, and a right-hand rule applies to specify their direction. The similarities, or parallels, between these two types of motion are many. In fact, in Chapter 10 we'll accumulate a few more than we already have, and there we'll summarize the analogues in the form of a table. Although the description of rotational motion may appear to be complicated, keep in mind that no fundamentally new physical laws are involved. Everything we have discussed to this point is derived from application of Newton's laws to extended systems of point masses.

THINK ABOUT THIS . . .

CAN ANGULAR MOMENTUM BE TRANSFERRED IN COLLISIONS?

In Chapter 8 we described how linear momentum is transferred in collisions. Angular momentum can similarly be transferred. Billiard players certainly know how to put spin on a billiard ball when the cue stick collides with it. If you have an air table available, you can easily see a demonstration of the effect with pucks that have "sticky" sides. Another good example is in a context completely familiar to you. In the normal operation of an automobile, angular momentum contained in the turning engine is transferred to the wheels when a clutch, or its equivalent in an automatic transmission (Fig. 9–32), is engaged—the engagement is a kind of soft inelastic collision. (Indeed, the very word "transmission" describes the process of transfer of the engine's behavior to the wheels.) The transmission shafts and the rear wheels, which start at rest before the clutch begins its engagement, have rotational inertia, and as the clutch is engaged, some of the engine's angular momentum is transferred to the shafts and wheels. The angular momentum of the engine is accordingly reduced. The total energy of the engine/transmission-wheel system is also reduced. This is analogous to the completely inelastic collision of a moving object with a stationary one, in which momentum is conserved but kinetic energy is reduced.

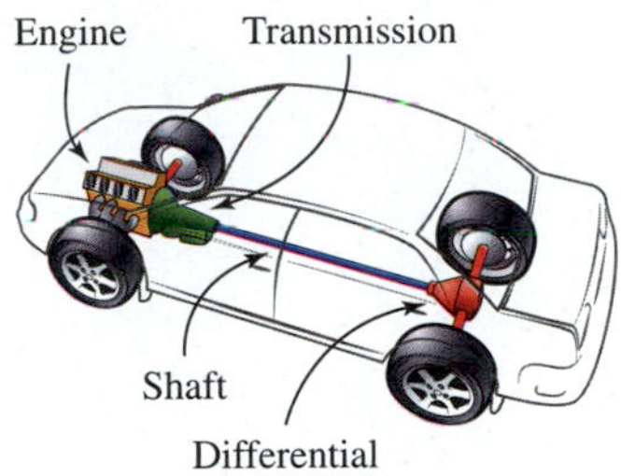

▲ **FIGURE 9–32** An automobile transmission is a device to transmit the rotational motion of the engine to a rotational motion of the wheels.

9–6 Rolling

Kinematics of Rolling

Our discussion so far has dealt with the kinematics of pure rotation. *Rolling* is a type of motion that is a combination of rotational and linear motion. A rolling wheel undergoes pure rotation about its axis, but to an observer on the ground the axis is moving (Fig. 9–33); as a consequence, the observer sees a point on the wheel's rim undergo motion that is a kind of connected linear and rotational motion. Our first task is to describe rolling mathematically and to discover in just what way the connection between the linear and rotational motion is realized.

Consider a wheel of radius R that rolls without slipping or skidding in a straight line on a horizontal surface (Fig. 9–34a). The center of the wheel moves at uniform speed v. In rolling without slipping, the connection between the speed v and the magnitude of the angular velocity, ω, of the wheel's rotation is

$$v = R\omega. \quad (9\text{–}40)$$

ROLLING MOTION CONNECTION

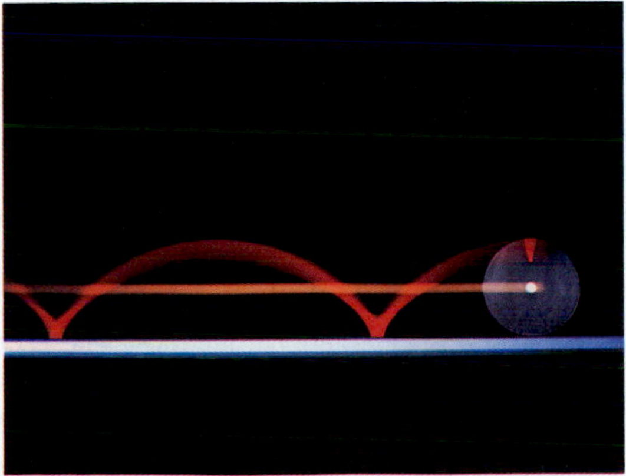

◀ **FIGURE 9–33** Rolling cylinder. A small light at the center and one at the edge of the cylinder show that the center of mass moves linearly, whereas a point along the edge has rotational motion and traces out a *cycloid*.

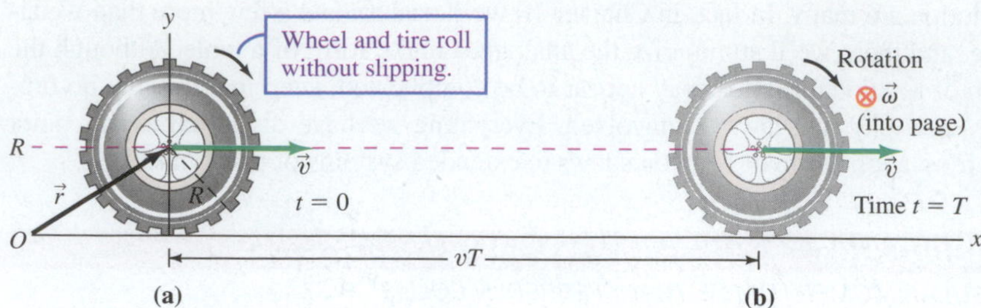

▶ **FIGURE 9–34** (a) A wheel of radius R rolls without slipping in the x-direction. Its plane lies in the xy-plane and its center is at the position $\vec{r}$. (b) If the center of the wheel is at $x = 0$ at $t = 0$ and moves at speed v along the $+x$-direction, then the angular speed about the center has magnitude $\omega = v/r$.

A simple way to see this result is to imagine a chalk mark made on the rim of the wheel that rolls on a surface. The wheel makes chalk marks on its path as it rolls, and the distance between chalk marks made in one revolution is $2\pi R$ (the wheel's circumference). If the time taken to move this distance is T, then the center of mass will have traveled a distance vT (Fig. 9–34b). The distances are equal, so $vT = 2\pi R$, or $v = 2\pi R/T$. But $2\pi/T$ is the angular speed ω, and we end up with Eq. (9–40). You can easily see that if there is slipping then Eq. (9–40) no longer holds, by thinking of the extreme in which there is no friction whatever between the wheel and the surface: The wheel will never "grip" and there will be no rotation at all.

CONCEPTUAL EXAMPLE 9–14 The motion of a rolling wheel may be viewed as a rotation about the contact point with the road (Fig. 9–35) in that the point of contact of the wheel is *instantaneously* at rest. If the linear speed of the wheel hub is v (the speed of the hub is the speed of the vehicle), find the angular speed of the wheel about the contact point.

Answer The angular velocity about that point is determined by the fact that the wheel hub, at a distance R from the point of contact, has speed v (Fig. 9–35). Thus the instantaneous angular speed of the hub about the contact point is $\omega = v/R$. The wheel is rigid, so *every* point on the wheel has the same angular speed about the contact point.

What Do You Think? Could you use your answer to find the linear speed of the top of the wheel?

(a)

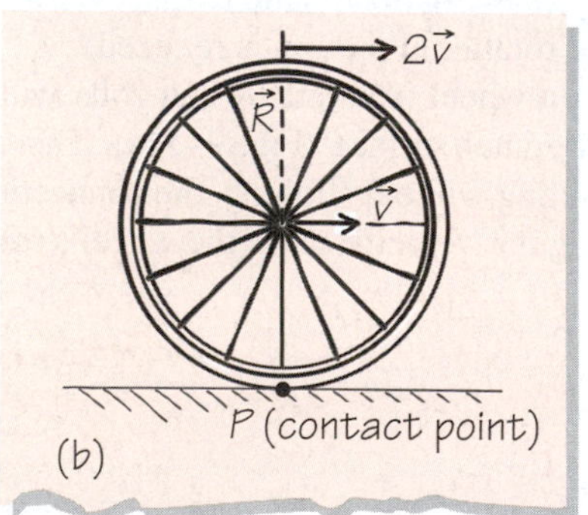

▶ **FIGURE 9–35** We work in the frame of reference in which the point of contact between the rolling bicycle wheel and the road is at rest. In this frame, the center of the wheel moves with velocity $\vec{v}$ and the top of the wheel with velocity $2\vec{v}$. Both points rotate about the contact point.

Energy in Rolling

With respect to the point of contact, the motion of the wheel is a *pure* rotation about the point of contact, and the total kinetic energy is

$$K = \tfrac{1}{2} I_{\text{contact}} \omega^2, \tag{9–41}$$

where I_{contact} is the rotational inertia of the wheel about an axis through the point of contact. By the parallel-axis theorem, the rotational inertia I_{contact} may be written in terms of the rotational inertia about the central axis through the center of mass:

$$I_{\text{contact}} = I_{\text{cm}} + MR^2, \tag{9–42}$$

where M is the mass of the rolling object. Thus the total kinetic energy of the rolling object is

$$K = \frac{1}{2}(I_{cm} + MR^2)\omega^2 = \frac{1}{2}I_{cm}\omega^2 + \frac{1}{2}Mv^2, \quad (9\text{–}43)$$

where we have used $v = R\omega$. Equation (9–43) expresses an important result: *The kinetic energy of an object that rolls without slipping is the sum of the kinetic energy of rotation about its center of mass ($I_{cm}\omega^2/2$) and the kinetic energy of the linear motion of the object as if all the mass were at the center of mass.*

EXAMPLE 9–15 Calculate the total kinetic energy of a uniform solid ball of mass 2 kg and radius 10 cm that rolls without slipping on a flat surface at a speed of 0.8 m/s.

Setting It Up We have a solid (and constant-density) sphere of known mass M and radius R; it rolls with a uniform speed v (its center of mass moves with that speed). We want the total kinetic energy, that is, the linear plus rotational kinetic energy.

Strategy The kinetic energy for a rolling object is given by Eq. (9–43). First we find—it will suffice to look in Table 9–1—the rotational inertia of a uniform solid sphere about an axis through its center, then use the fact that $v = \omega R$ for a rolling object. The problem can then be solved by inserting the values into Eq. (9–43).

Working It Out From Table 9–1, the rotational inertia of a solid ball about an axis through its center is $\frac{2}{5}MR^2$. From Eq. (9–43), the total kinetic energy is then

$$K = \frac{1}{2}\left(\frac{2}{5}MR^2\right)\omega^2 + \frac{1}{2}Mv^2 = \frac{1}{5}Mv^2 + \frac{1}{2}Mv^2 = \frac{7}{10}Mv^2$$
$$= 0.7(2\text{ kg})(0.8\text{ m/s})^2 = 0.9\text{ J}.$$

(In the first step we have replaced ω by v/R.) The kinetic energy associated with the rotation is 40 percent of the energy associated with the linear motion.

What Do You Think? Would you expect the percentage of energy in rotation to be larger for larger I assuming the mass and radius of the rolling object stays the same?

Dynamics of Rolling

Consider an object of mass M and radius R that can roll without slipping down a plane that makes an angle θ with the horizontal. The object is symmetric about the central axis but is otherwise unrestricted; for example, it could be a solid cylinder, a hollow cylinder, or a sphere. In Fig. 9–36 we choose the shape to be a solid cylinder, but our analysis does not depend on this choice. Let us use the dynamical equations of motion to find the angular velocity of the object after its center of mass has traveled a distance ℓ down the incline, starting from rest.

The forces are included in Fig. 9–36, which thereby serves as a free-body diagram. The force of gravity, $M\vec{g}$, may be separated into a component normal to the plane, $Mg\cos\theta$, and a component parallel to the plane, $Mg\sin\theta$. Both components act on the center of mass of the rolling object, as we learned in Section 9–4. There is also the contact force $\vec{F}_N$, normal to the plane and acting at the point of contact, which just cancels the component of gravity normal to the plane. Finally, there is the force of friction, $\vec{f}$, which acts at the point of contact and points back up along the plane. It is *static* friction that acts here; there is no relative motion of the rolling object and the ramp at the point of contact. The existence of the static friction is crucial for there to be rolling. Without it the surface would not "grip," and, round or not, the object would slide rather than roll. However, remember that up to some maximum the magnitude of static friction takes a value determined by the dynamics.

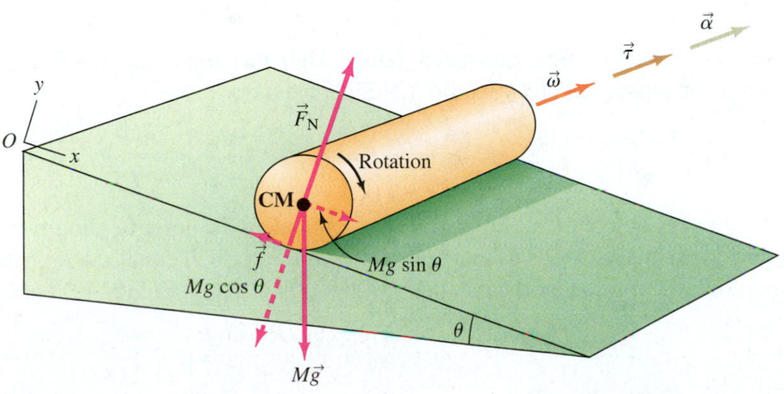

FIGURE 9–36 A cylinder of radius R rolls down an inclined plane. Static friction causes the cylinder to roll without slipping. The rotational motion can be analyzed about any axis, but a convenient one passes through the geometrical center of the cylinder and another is formed by the line of contact between the cylinder and the plane. The forces acting are friction $\vec{f}$, the normal force $\vec{F}_N$, and gravity $M\vec{g}$. We have decomposed the force of gravity into components along and perpendicular to the inclined plane.

To be able to apply the equations of motion to the rolling object, we want to find the net torque acting on the object. We will use the axis through the center of the object as a reference axis, as shown in Fig. 9–36. Friction is the *only* force of all the forces acting on the object that exerts a torque about the center of mass. Both the normal force and gravity have no lever arm through this point, gravity because it effectively acts at that point and the normal force because if you extend the line along which it acts (perpendicular to the surface) that line passes through the center of mass.

The net force along the plane determines the linear acceleration, magnitude a in this direction. According to Newton's second law,

$$Ma = Mg \sin\theta - f. \tag{9–44}$$

Because it is static friction that is involved, its magnitude is not uniquely determined until we include the dynamical equation of rotation. The torque due to friction has magnitude

$$\tau = fR. \tag{9–45}$$

By the right-hand rule, the torque's direction is along the axis of rotation $\vec{\omega}$ in Fig. 9–36. The dynamical equation that determines the rate of change of angular momentum about the axis of the cylinder comes from Eq. (9–29), with the torque given by Eq. (9–45):

$$I\alpha = \tau = fR. \tag{9–46}$$

This equation implies that the friction force is given by $f = I\alpha/R$. Substituting into Newton's second law, Eq. (9–44), we find

$$Ma = Mg \sin\theta - \frac{I\alpha}{R}. \tag{9–47}$$

Finally, we use Eq. (9–40), $v = R\omega$, to supply a relation between a and α. The time derivative of that equation immediately gives us $a = R\alpha$. Substitution into Eq. (9–47) then gives us a single equation for the angular acceleration:

$$M\alpha R = Mg \sin\theta - \frac{I\alpha}{R},$$

with the algebraic solution

$$\alpha = \frac{MgR \sin\theta}{MR^2 + I}. \tag{9–48}$$

Only geometric factors appear on the right, so the angular acceleration is a constant. With a constant angular acceleration, the linear dependence of angular velocity on time is

$$\omega = \omega_0 + \frac{MgR \sin\theta}{MR^2 + I}t. \tag{9–49}$$

Here, the angular velocity at $t = 0$ is ω_0. If the object starts from rest, then $\omega_0 = 0$. Similarly, the linear acceleration is constant, implying a center-of-mass velocity that increases linearly with time according to

$$v = \omega R = v_0 + \frac{MgR^2 \sin\theta}{MR^2 + I}t. \tag{9–50}$$

We can also determine the speed gained after the object has rolled a total distance ℓ down the plane. According to Eq. (2–25d), we have

$$v^2 = v_0^2 + 2a\ell = v_0^2 + 2\frac{MgR^2 \sin\theta}{MR^2 + I}\ell. \tag{9–51}$$

For an object of radius R and mass M that is symmetric about a central axis, dimensional analysis tells us that the rotational inertia will equal CMR^2, where C is a pure number determined by the geometry of the object. Thus Eq. (9–49) becomes

$$\omega = \omega_0 + \frac{(g/R) \sin\theta}{1 + C}t,$$

or equivalently

$$v = v_0 + \frac{g \sin \theta}{1 + C} t. \tag{9–52}$$

Similarly,

$$v^2 = v_0^2 + 2a\ell = v_0^2 + 2\frac{g \sin \theta}{1 + C}\ell. \tag{9–53}$$

It is instructive to compare this last result with the speed gained by a point object that moves down the same ramp without friction: $v^2 = v_0^2 + 2g\ell \sin \theta$. The rolling object moves less rapidly because *some of the potential energy goes into rotational motion.* If there is some slipping, then this same reasoning about energy tells us that we end up in an intermediate situation, with speed greater than in the case of no slipping but less than in the case of pure sliding (no friction at all). Note also the striking result that *the speed of the center of mass is independent of both M and R*. We are familiar with the speed independent of mass—that is the typical behavior under gravity. The lack of dependence on R, which you may find surprising, is something you easily test.

Let's look at a few special cases. The smallest possible value of C is zero, corresponding to an object with all its mass concentrated at its central axis; this is just the case of a sliding object since there is no rotational energy in this case. The case $C = \frac{1}{2}$ is a solid cylinder (see Table 9–1). The case $C = 1$ corresponds to a thin hollow cylinder with all its mass concentrated at the outside. Still larger values of C are possible if we consider an object like a yo-yo, where the rolling takes place on the inner cylinder. Thus the factor $1 + C$ increases from a minimum value of 1, and the rate at which ω, or v, increases with time for symmetric rolling objects is always less than that of a sliding object.

EXAMPLE 9–16 A ramp of length 1.5 m is set at an angle of 5.0° to the horizontal. Two objects are initially at rest at the top of the ramp and simultaneously start rolling without slipping. Object 1 is a solid cylinder of mass 0.65 kg and radius 4.7 cm. Object 2 is a hollow, thin-walled tube of the same radius and of mass 0.13 kg. How much time does it take for each object to arrive at the bottom of the ramp?

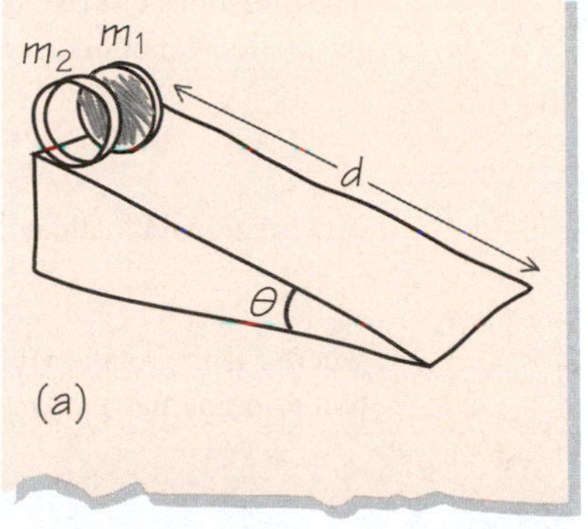

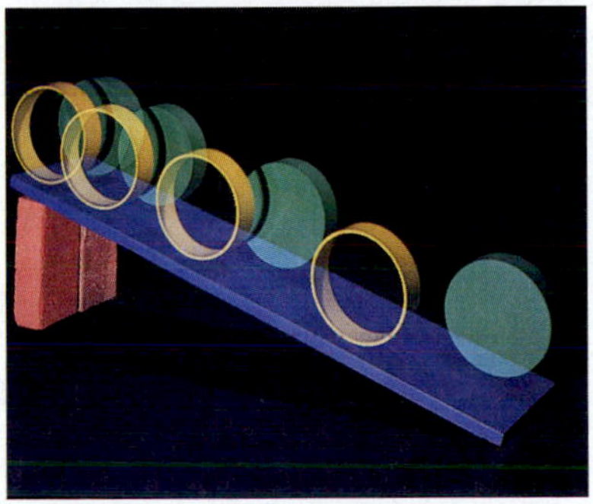

(b)

◀ **FIGURE 9–37** The rate at which a rolling object accelerates depends on its rotational inertia. In a race between a solid cylinder and a hollow cylinder, both starting from rest and rolling down an inclined plane, the solid cylinder comes in first.

Setting It Up We sketch the ramp and the rolling objects in Fig. 9–37a. The coordinate system is positioned at the top of the ramp with the x-axis down the ramp. The ramp length is d and its angle is θ. The solid roller has mass M_1 and the hollow one mass M_2; both have the same radius R. All five of these variables are known.

Strategy Because there is no slipping, we can use the direct connection between linear motion and angular motion of the rolling objects. We can also use directly the results of the discussion above, but with different values of I for the two different objects. We therefore know the values of the linear acceleration in each case. This acceleration is constant, so kinematics can give us the travel time.

Working It Out The time required to travel a given straight-line distance d under constant acceleration a when the initial velocity is zero is expressed by the constant-acceleration equation $d = at^2/2$, or $t = \sqrt{2d/a}$. Next we need to find the value of the acceleration. Equation (9–48) with $a = \alpha R$ gives

$$a = \frac{MgR^2 \sin \theta}{MR^2 + I},$$

where M is the object's mass. For the case of the solid cylinder, $I = MR^2/2$. The center of the solid cylinder then has linear acceleration

$$a_1 = \frac{g \sin \theta}{1 + \frac{1}{2}} = \frac{2}{3} g \sin \theta.$$

The hollow cylinder has a rotational inertia about its axis of $I = MR^2$; thus, its linear acceleration has magnitude

$$a_2 = \frac{g \sin \theta}{1 + 1} = \frac{1}{2} g \sin \theta.$$

(continues on next page)

In each case, the linear acceleration is independent of the mass and the radius, so this information is not needed. The only relevant difference between the solid cylinder and the hollow tube is that they have different rotational inertias because their masses are distributed differently. Numerically,

$$a_1 = \tfrac{2}{3} g \sin\theta = (0.67)(9.8\ \text{m/s}^2)(\sin 5.0°) = 0.57\ \text{m/s}^2;$$
$$a_2 = \tfrac{1}{2} g \sin\theta = (0.5)(9.8\ \text{m/s}^2)(\sin 5.0°) = 0.43\ \text{m/s}^2.$$

The descent time of the solid and hollow cylinders respectively are

$$t_1 = \sqrt{\frac{2d}{a_1}} = \sqrt{\frac{2(1.5\ \text{m})}{0.57\ \text{m/s}^2}} = 2.3\ \text{s}$$

and

$$t_2 = \sqrt{\frac{2(1.5\ \text{m})}{0.43\ \text{m/s}^2}} = 2.6\ \text{s}.$$

It is easy to try this experiment and confirm that a solid cylinder reaches the bottom of a ramp faster than a hollow one, as Fig. 9–37b shows.

What Do You Think? How would a thick cylinder of the same radius as the thin cylinder do if it became the third entry in this race? What if the thick cylinder has radius $2R$?

Summary

In this chapter we have explored the fact that a rigid body can undergo rotations as well as linear motion. Rotational motion can be described using quantities that are analogous to the quantities used to describe linear motion. The linear motion of an object can be described by specifying the motion of the center of mass. A rotation about a fixed axis can be described with a single angle θ. The angular velocity $\vec{\omega}$ describes the rate of change of this angle, while the angular acceleration $\vec{\alpha}$ describes the rate of change of the angular velocity:

$$\vec{\omega} = \frac{d\theta}{dt}\hat{\omega}, \qquad (9\text{–}2)$$

$$\vec{\alpha} = \frac{d\vec{\omega}}{dt}. \qquad (9\text{–}7)$$

The directions of these vectors are determined by a right-hand rule. The period T and frequency f of rotational motion are related by

$$f = \frac{1}{T} = \frac{\omega}{2\pi}. \qquad (9\text{–}5)$$

The rotational kinetic energy of a rigid body rotating about an axis is

$$K = \tfrac{1}{2} I\omega^2, \qquad (9\text{–}14)$$

where I is the rotational inertia of the rotating object with respect to that axis. For a discrete system of point masses Δm_i that is a perpendicular distance R_i from an axis,

$$I \equiv \sum_i \Delta m_i R_i^2. \qquad (9\text{–}15)$$

In particular, for a single mass m that is a perpendicular distance from the axis of rotation, the rotational inertia is

$$I = mR^2. \qquad (9\text{–}16)$$

The definition in Eq. (9–15) can be extended to continuous objects: for an object with mass density ρ,

$$I = \int_{\text{Volume}} \rho R^2\, dV, \qquad (9\text{–}19)$$

where R^2 is the perpendicular distance of an internal point of the object to the axis.

Rotational inertia plays a role in rotational motion analogous to the role played by mass in linear motion. A useful tool for its evaluation is the parallel-axis theorem. This theorem states that the rotational inertia of an object about a given axis is

$$I = I_{\text{cm}} + Md^2. \qquad (9\text{–}27)$$

In this equation, I_{cm} is the rotational inertia about an axis that goes through the center of mass and is parallel to the given axis, M is the mass of the object, and d is the perpendicular distance between the axes.

Torque τ is the cause of changes in rotational motion and is analogous to force, the cause of changes in linear motion. For a rigid body rotating about a fixed axis with rotational inertia I about that axis, the equation of motion is analogous to Newton's second law:

$$\vec{\tau}_{\text{net}} = I\vec{\alpha}, \qquad (9\text{–}30)$$

where $\vec{\alpha}$ is the angular acceleration. With constant torque, the angular velocity grows linearly with time.

The net torque is expressed in terms of the net force by the expression

$$\tau = rF \sin \theta, \tag{9–28}$$

where r is the distance from the point of application of the force to the axis of rotation and θ is the angle between the net force and r. The combination $r \sin \theta$ is the lever arm, or perpendicular distance from the axis to the line along which the force acts—the line of action. The direction of the torque is specified by a right-hand rule. The torque due to gravity acts as if it were applied to the center of mass.

The technique of free-body diagrams for solving linear motion can be extended to rotational motion by specifying the point of application of the force.

The equation of motion takes an alternative form

$$\vec{\tau} = \frac{d\vec{L}}{dt}, \tag{9–38}$$

where $\vec{L}$ is the angular momentum—a quantity analogous to the linear momentum in linear motion. For rotations of a symmetric, rigid body about a symmetry axis, the angular momentum is

$$\vec{L} = I\vec{\omega}. \tag{9–37}$$

When the net torque is zero, the angular momentum is constant; thus, the angular momentum is conserved for isolated systems.

When an object rolls, linear motion is combined with rotational motion. Objects with radius R that roll without slipping have an angular velocity about their axes, ω, that is related to the speed v of the center of mass of the object:

$$v = R\omega. \tag{9–40}$$

The kinetic energy of a rolling object is the sum of its rotational kinetic energy about its axis of rotation, $I_{cm}\omega^2/2$, and the kinetic energy of its linear motion:

$$K = \tfrac{1}{2}(I_{cm} + Mr^2)\omega^2 = \tfrac{1}{2}I_{cm}\omega^2 + \tfrac{1}{2}Mv^2. \tag{9–43}$$

Understanding the Concepts

1. A record turntable rotates in the clockwise sense when seen from above. In what direction is the angular velocity of the turntable? If you had used a left-hand rule rather than a right-hand rule, which direction would you have chosen for the direction of the angular velocity vector? Is the choice between a right-hand or a left-hand rule purely conventional?
2. In the discussion of rolling without slipping, we saw that the kinetic energy of the rolling object is greater than the kinetic energy associated with the linear motion alone. Are there cases of rolling, with or without slipping, for which this is not true?
3. In a juggling act a juggler spins a plate rapidly in its plane and then supports it on a stick, walking around with it, and hardly paying attention to the exact angle that the stick makes with the vertical. How does the trick work?
4. Suppose that you have a set of spherical objects with the same total mass but different radial distributions of mass. Which objects have the larger rotational inertias about an axis through the center: the spheres with more mass at the center or the spheres with more mass toward the outer surface?
5. A paddle wheel that propels a Mississippi River boat dips down into the water. Discuss the direction of motion of the portion of the paddle wheel under the water according to an observer on the shore and then according to an observer on the boat.
6. If you were to hold an arm out stiffly at a right angle while grasping a rock, the effect of the force of gravity on the rock would translate to a torque on you about, say, your center of mass. Does this mean you would rotate?
7. A ball starts from rest down the inside of a parabolic bowl and rolls without slipping. At the bottom, the surface is a frictionless surface. The ball then moves up the other side of the bowl to a certain height. Is this height higher than, lower than, or the same as the height from which the ball started?
8. Solid and hollow cylinders both roll from rest down an inclined plane. Explain the difference in speeds at the bottom.
9. You have two cylindrical cans, one filled with congealed wax and the other with water. The cans are identical, and the total masses of the two cans with their contents are the same. If you roll the two down an inclined plane, which will reach the end first?
10. Devise a method to determine the rotational inertia of a sphere whose density can vary with the distance from its center.
11. Two skaters holding on to opposite ends of a rope circle a common point between them. In order to double their angular velocity about that point, should they lengthen or shorten the distance between them? By how much?
12. You have a flat outline of the continental United States cut out in $\frac{1}{4}$-in plywood. Devise a method to find the center of mass of this outline.
13. When canoeing in rapidly flowing water, it is possible for the opposite ends of the canoe to become pressed against rocks such that the canoe is aligned across the flow of the stream. Why is this bad?
14. When a landing airplane first touches the runway, a puff of smoke comes from the wheels; the smoke stops after a moment and the airplane can roll to a halt. Describe these events in terms of the relevant forces acting.

15. What limits the amount of energy a flywheel could store in its rotational motion?
16. A solid and a hollow cylinder are rolled from rest down an inclined plane. At the bottom, the center of the solid cylinder is moving more rapidly, even if the two cylinders have the same radius and mass. Is this a violation of the conservation of energy?
17. Why did the wheels used to steer large sailing ships have large radii?
18. The accuracy of guns improved greatly when their barrels were rifled, making the bullets emerge with a rapid rotation about their axes. Can you explain why this might improve accuracy?
19. When you switch to a longer wrench, you have not suddenly become stronger. So why is it easier to loosen a pipe with a long wrench than with a short wrench?
20. We stated that the rotational inertia of a dumbbell about an axis that runs through the rigid rod connecting the two masses is zero. What approximations (or idealizations), if any, are made in coming to this conclusion?
21. Attach one end of a given spring to a fixed pivot point on a frictionless table. Attach a mass to the other end and start the mass in circular motion about the pivot point (Fig. 9–38). What happens when the revolution time is decreased by some external means and why?
22. When a wheel of radius R rolls without slipping, the relationship between the speed of the center of the wheel, v, and the angular speed, ω, is $v = R\omega$. How would this relationship change if there were some slipping?
23. A driver presses on the accelerator, increasing the angular speed of the driven wheels by a factor of 2. Assuming that there is no slipping, by what factor does the linear speed of the automobile increase?

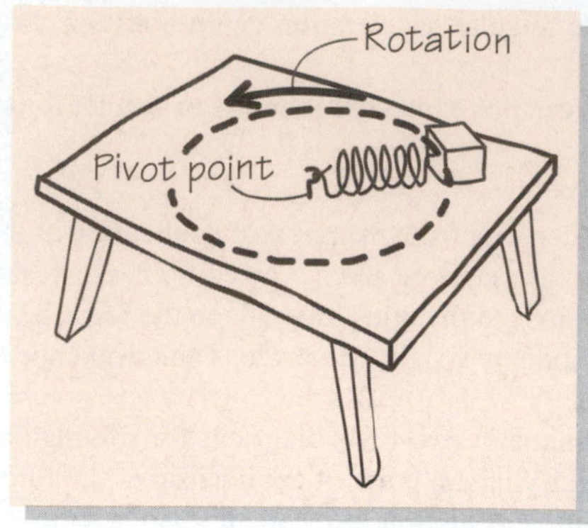

▲ **FIGURE 9–38** Question 21.

24. Suppose that you swing a rock at the end of a string in a circle in a vertical plane. What happens to the rock if the string suddenly breaks? Does it suddenly lose its angular momentum?
25. You have two wheels with rotational inertias in the ratio 2 : 1. Initially, the wheels are at rest. If the same torque is applied to both of them for the same length of time, what will be the ratio of their angular momenta at the end of that time interval?
26. You have two wheels with rotational inertias in the ratio 2 : 1. Initially, the wheels are at rest. If the same torque is applied to both of them for the same length of time, what will be the ratio of their rotational kinetic energies at the end of that time interval?
27. A wheel rotates about a central axis. When the angular speed increases beyond a certain value, the wheel breaks. Why?

Problems

9–1 Simple Rotations of a Rigid Body

1. (I) When a phonograph turntable is switched off, it comes to rest from its original $33\frac{1}{3}$ rev/min in 5.15 s. What is the magnitude of the angular deceleration assuming that it is constant while the turntable comes to rest?
2. (I) How many revolutions does the turntable of Problem 1 make between the moment it is switched off and the moment it stops?
3. (I) A skater does a pirouette at the rate of 2.0 rev/s and then stops within $\frac{3}{4}$ rev. Assume that the angular deceleration is constant and calculate its magnitude.
4. (I) A turntable slows down from 15 rev/s to rest in 45 s. What is the average angular acceleration of the turntable?
5. (I) What is the angular velocity of Earth's rotation about its axis? What is the angular velocity of Earth in its orbital motion around the Sun?
6. (I) A carousel has a 7-m radius and requires 8 s for a single revolution at full speed. A carousel pig sits at a distance of 3 m from the axis and a carousel horse sits at a distance of 6 m. (a) What is the period T for a single revolution of the pig? (b) Of the horse? (c) What is the angular frequency of the motion of the pig? (d) Of the horse? (e) What is the velocity of the pig? (f) Of the horse? (g) What is the centripetal acceleration of the pig in its motion around the axis? (h) Of the horse?
7. (II) A centrifuge whose maximum rotation rate is 10,000 rev/min can be brought to rest in 4.00 s. (a) What is the average angular acceleration of the centrifuge? (b) What is the distance that a point on the rim travels during the deceleration time assuming that the radius of the centrifuge is 8 cm and that the acceleration is uniform?
8. (II) A CD turntable is rotating at 313 rev/min in a clockwise direction, viewed from above. (a) What is its angular velocity, both direction and magnitude, in radians per second? (b) The turntable is switched off and comes to rest in 0.55 s. What is the average value of the angular acceleration, direction, and magnitude during that period?
9. (II) A carousel initially at rest has an angular acceleration of 0.4 rad/s^2 and accelerates for 5 s. It then rotates at a constant angular velocity for 30 s before slowing down at the same rate with which it accelerated. (a) What is the average acceleration during the first 20 s? (b) How many total revolutions does it make? (c) How far does a child sitting on a horse 3 m from the center travel?
10. (II) A thread is wrapped around a cylindrical spool of radius 1.5 cm whose central axis is fixed on a support (Fig. 9–39). The

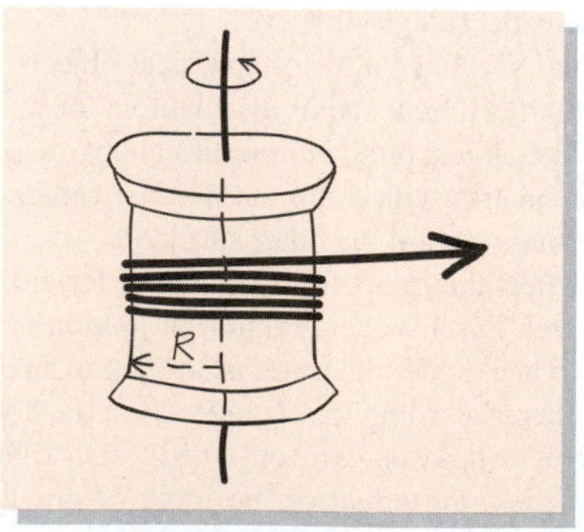

▲ **FIGURE 9–39** Problem 10.

thread is pulled off at a constant rate, causing the spool to spin at a constant rate; it takes 2 s to pull off 3 m of thread. What is the angular velocity of the spool while the thread is pulled off?

11. (II) A more careful measurement of the unwrapping of the thread in Problem 10 shows that the spool accelerates from rest at a steady rate in the 2 s it takes to pull off 3 m of thread. (a) Give a formula for the position of the hand that pulls the thread as a function of time. What is the value of the constant (linear) acceleration? (b) What is the value of the angular acceleration of the spool? (c) Give a formula for the magnitude of the angular velocity as a function of time.

12. (II) A vacuum pump is connected to its electric motor by a belt drive (Fig. 9–40). The motor rotates at a rate of 1200 rev/min and the diameter of the motor shaft is 1.5 cm. How large should the pulley be if it is designed for a speed of 33 rev/s?

▲ FIGURE 9–40 Problem 12.

13. (II) The path of the tip of a needle on a phonograph record may be described by the formula $r = r_0 - (\theta\rho/2\pi)$, where r_0 is the outer starting radius, θ is the angle (in radians) that a fixed radial line in the record makes with the needle arm, and ρ is the spacing between grooves. Assuming that the record turns at a rate of $33\frac{1}{3}$ rev/min and that the radial distance traveled by the needle is 9 cm in 20 min, what is ρ?

14. (II) The angular acceleration of a wheel starting from rest has magnitude $C_1t + C_2t^3$, where $C_1 = 48\ \text{rad/s}^3$ and $C_2 = -9.5\ \text{rad/s}^5$. (a) What is its angular velocity at 3.0 s? (b) How many revolutions has the wheel made after 2.0 s? (c) When will the wheel be at rest again?

15. (II) By marking a point on its edge, the angle θ that measures the rotation of a turntable is found to increase with time as t^3 during a certain time interval. How did the angular acceleration change with time, if at all, during that interval?

16. (II) The angular velocity of a wheel that can rotate in the xy-plane around a fixed axle oriented along the z-axis varies with time as $\vec{\omega} = (w_1t - w_2t^2)\hat{k}$, where w_1 and w_2 are constants. Assuming that the wheel starts at $t = 0$ with a mark on its edge sitting on the positive x-axis, where is the mark the next time the wheel comes to rest?

17. (III) A rigid solid undergoes rotational motion about an axis. Its angular velocity has magnitude $\omega = \alpha t$, where α is constant. As the angular speed ω increases with time, the period T decreases. (a) Show that the rate of change of the period is described by the equation $dT/dt = -2\pi/\alpha t^2$. (The period is infinite at $t = 0$ because the rotation has not yet started, and a measurement of the time for 1 rev at that rate would be infinite!) (b) Show that the change in the period between $t = t_1$ and $t = t_2$ is given by

$$T(t_2) - T(t_1) = \frac{2\pi}{\alpha}\left(\frac{1}{t_2} - \frac{1}{t_1}\right).$$

9–2 Rotational Kinetic Energy

18. (I) A metal ball of mass 350 g at the end of a 1.75-m long wire rotates with an angular speed of 85 rev/min ($85 \times 2\pi$ rad/min). What is the rotational kinetic energy of the ball?

19. (I) Two identical balls are spinning on a flat surface. Ball 1 has three times the angular speed of ball 2. What is the ratio of their kinetic energies?

20. (I) Measurements of the amount of energy used by an electric motor to speed up a wheel from rest show that to bring the wheel from rest to an angular speed of 3.7 rad/s the motor expends 7600 J. In a second use of the motor and wheel, the motor expends 9200 J to bring the wheel up from rest to an unmeasured angular speed. The motor and wheel have practically negligible amounts of friction or other type of damping. What is the unmeasured angular speed?

21. (II) A string is wrapped around a cylindrical spool of radius 1 cm. The axis of the spool is fixed. A length of string of 0.8 m is pulled off in 1.5 s at a constant tension of 20 N. What is the rotational inertia of the spool?

22. (II) What is the rotational kinetic energy of a dumbbell consisting of two equal (compact) masses of 1.5 kg each connected by a massless rod of length 0.62 m when the dumbbell rotates about an axis through the center of and perpendicular to the rod at 36 rev/min? What is the rotational kinetic energy if the dumbbell rotates with the same angular velocity about a parallel axis through one of the masses?

23. (II) A ball of mass 0.75 kg is attached by a 1.5-m-long rope to the top of a rod. The ball swings in a circle at the rate of 25 rad/s with the rope making an angle of 30° with the vertical. What is the rotational kinetic energy of the ball? What is it when the angle is 60°?

9–3 Evaluation of Rotational Inertia

24. (I) A pipe made of aluminum with a density of $2.7\ \text{g/cm}^3$ is a right cylinder 16 cm long whose outer diameter is 5.0 cm and whose inner diameter is 4.0 cm. What is the rotational inertia about the central axis of the pipe? Note that the rotational inertia of the thick cylinder can be expressed as the rotational inertia for a solid cylinder of radius R_2 minus the rotational inertia of the solid cylinder of radius R_1.

25. (I) What is the rotational inertia of a uniform 4.0-kg iron rod 0.25 m long about (a) an axis through its center point and perpendicular to the rod and (b) an axis through an end point and perpendicular to the rod?

26. (II) A dumbbell consists of two point masses 0.10 and 0.15 kg connected by a (nearly massless) rod of length 35 cm. Find the rotational inertia of this object about an axis that is parallel to the rod and a distance 10 cm away from it.

27. (II) A neutron star has a constant density of $6 \times 10^{17}\ \text{kg/m}^3$ and a mass five times that of our Sun. Compare its rotational inertia with that of Earth (assume constant density). In both cases the reference axis is an axis through the center of the sphere; Table 9–1 gives the rotational inertia for such an axis.

28. (II) Mass m_1 sits at the point $(x, y, z) = (0\ \text{m}, 0\ \text{m}, 0\ \text{m})$, and mass m_2 at point $(0\ \text{m}, 1\ \text{m}, 0\ \text{m})$. (a) Where is the center of mass? (b) What is the rotational inertia about an axis through the center of mass and parallel to the z-axis? (c) Parallel to the y-axis? (d) Parallel to the x-axis? (e) What is the rotational inertia about an axis through the origin and along the z-axis? (f) Verify the parallel-axis theorem for this system using the results of parts (b) and (e).

29. (II) Find the rotational inertia about the symmetry axis of a thick cylinder of mass M. Take the inner radius to be R_1 and outer radius to be R_2.

30. (II) Calculate the rotational inertia of a section of a right circular cylinder of radius R that subtends an angle θ_0 at the origin (Fig. 9–41) when the reference axis is at the origin and perpendicular to the section.

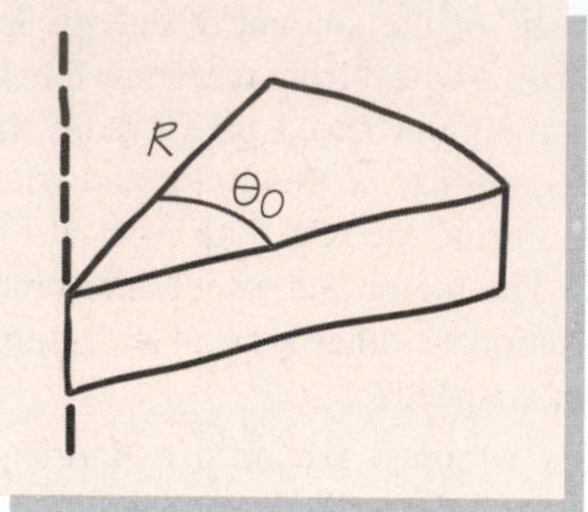

▲ **FIGURE 9–41** Problem 30.

31. (II) Calculate the rotational inertia about the central axis of the solid cone of mass M illustrated in Fig. 9–42, of opening half-angle α and height H.

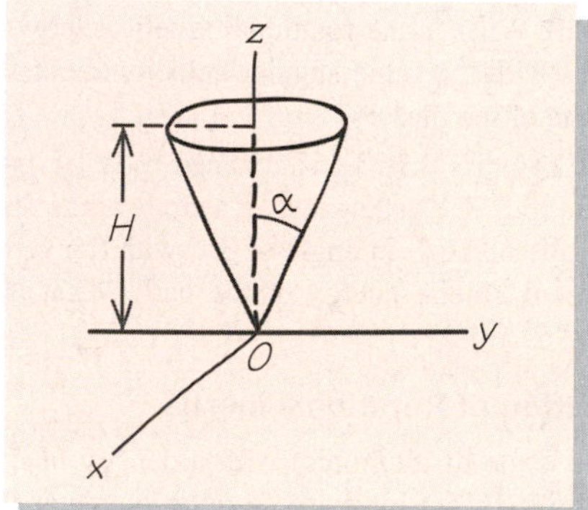

▲ **FIGURE 9–42** Problem 31.

32. (II) A thin stick of length $L = 1.6$ m is denser at one end than at the other: Its mass density is $\lambda = (0.40\ \text{kg/m}) - (0.070\ \text{kg/m}^2)x$, where x measures the distance from the heavier end of the stick. The stick rotates with period $T = 1.1$ s about an axis perpendicular to the stick through the heavy end. Determine the rotational kinetic energy of the stick.

33. (II) Calculate the rotational inertia of a sphere of radius R and mass M about an axis through the center of the sphere; assume that the density is not uniform but is given by ρ_1 for $0 \le r \le R_1$ and by ρ_2 for $R_1 \le r \le R$ (Fig. 9–43).

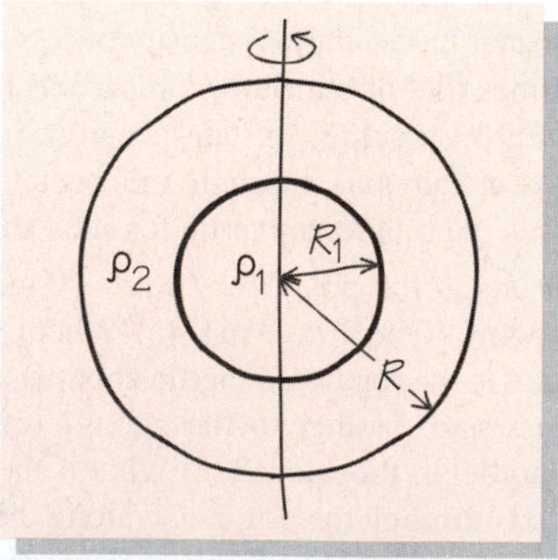

▲ **FIGURE 9–43** Problem 33.

34. (III) Use Eq. (9–19) to show that the rotational inertia about an axis through the center of a uniform, constant-density sphere of mass M and radius R is $\frac{2}{5}MR^2$.

35. (III) In Problem 89 of Chapter 8, we described a styrofoam sphere of radius R. A cavity of radius $R/2$ centered a distance $R/2$ directly above the center of the sphere was hollowed out and filled with a solid material of density five times the density of styrofoam. In that problem, the location of the center of mass of the composite sphere was determined. What is the rotational inertia of the composite sphere about a horizontal axis through the center of mass? Express your result in terms of the total mass M of the composite sphere and its radius R. (*Hint*: You may view the mass as consisting of a large sphere of radius R and density ρ and a small sphere, off-center, of radius $R/2$ and density 4ρ. Use the parallel-axis theorem.)

36. (III) A square of mass m is made of thin wire with sides of length a. Calculate the rotational inertia about the axes shown in Fig. 9–44.

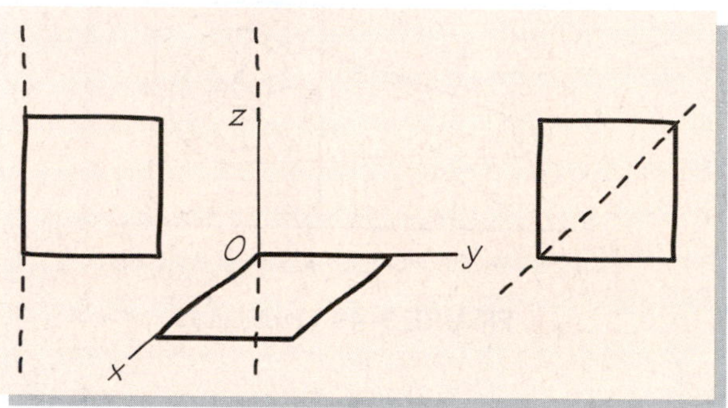

▲ **FIGURE 9–44** Problem 36.

9–4 Torque

37. (I) An instruction manual calls for a bolt to be tightened to 20 ft · lb. All you have is a 10-in wrench. How much force do you need to apply to the end of the wrench to tighten the bolt as required?

38. (I) A plumber of mass 74 kg just loosens a rusted-in bolt with the help of a 45-cm-long wrench. He places the wrench in a horizontal position and applies torque by hanging from the end of the wrench. What is the torque applied?

39. (I) The assistant of the plumber in Problem 38 meets a similar situation but aligns the wrench at an angle of 50° from the vertical (Fig. 9–45). What must his mass be in order for him to use the same technique to loosen the same bolt?

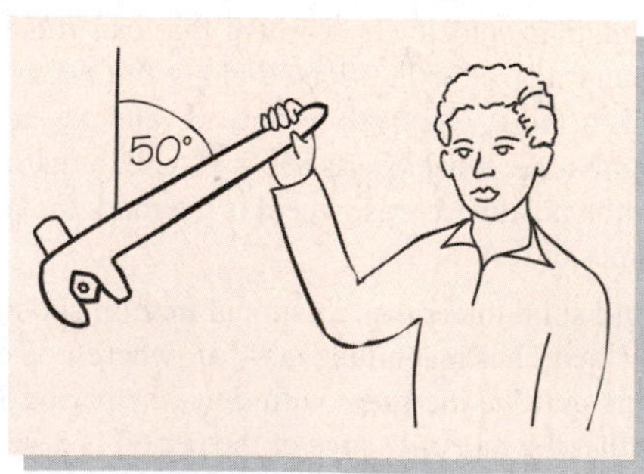

▲ **FIGURE 9–45** Problem 39.

40. (I) A flywheel of rotational inertia $I = 53\ \text{kg} \cdot \text{m}^2$ rotates with angular speed 4.0 rad/s. A tangential force of 6.5 N is applied at a distance of 0.36 m from the center in such a way that the angular speed decreases. How long will it take for the wheel to stop?

41. (II) A uniform rod of length L lies along the x-axis. A force F_{1y} is applied to one end of it and a force $-F_{2y}$ is applied to the other end of it. How large is the torque on the rod about its center of mass?

42. (II) A uniform rod 1.1 m long with mass 0.7 kg is pivoted at one end, as shown in Fig. 9–46, and released from a horizontal position. Find the torque about the pivot exerted by the force of gravity as a function of the angle that the rod makes with the horizontal direction.

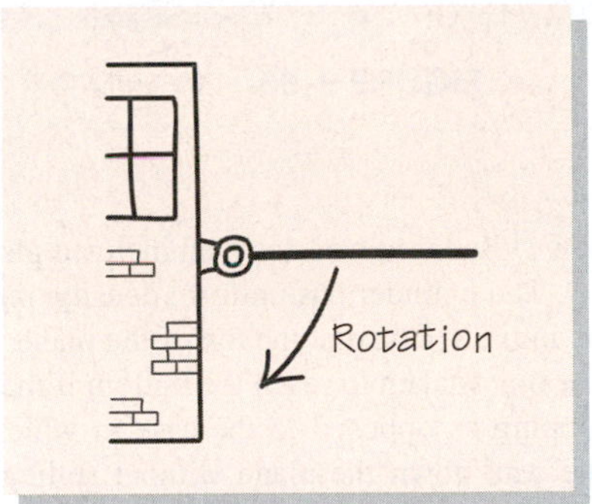

▲ **FIGURE 9–46** Problem 42.

43. (II) A seesaw pivots as shown in Fig. 9–47. (a) What is the net torque about the pivot point? (b) Give an example for which the application of three different forces and their points of application will balance the seesaw. Two of the forces must point down and the other one up.

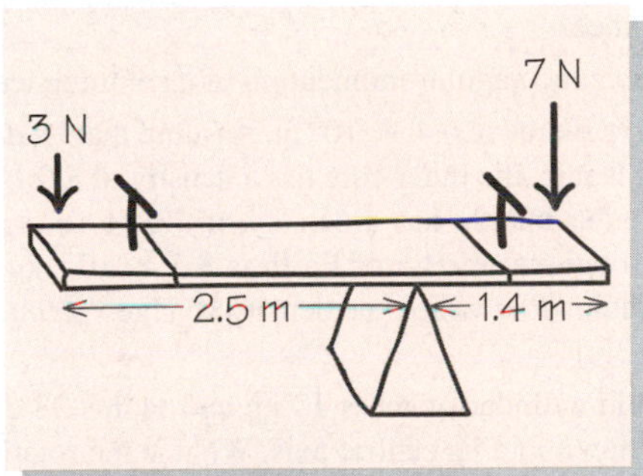

▲ **FIGURE 9–47** Problem 43.

44. (II) A two-dimensional object placed in the xy-plane has several forces acting on it. Find the torques about points A and B in Fig. 9–48.

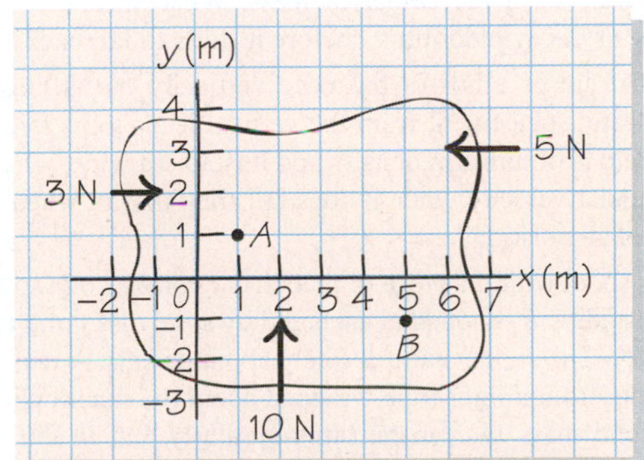

▲ **FIGURE 9–48** Problem 44.

45. (II) An aluminum casting is finished with a band sander. The sandpaper is stretched by two cylindrical rollers, one of them driven by a motor (Fig. 9–49). The rollers have diameter of 16 cm. How large is the torque that has to be applied to the driven cylinder if the coefficient of kinetic friction between aluminum and sandpaper is 1.2 and the force applied to push the casting against the sandpaper is 4 N?

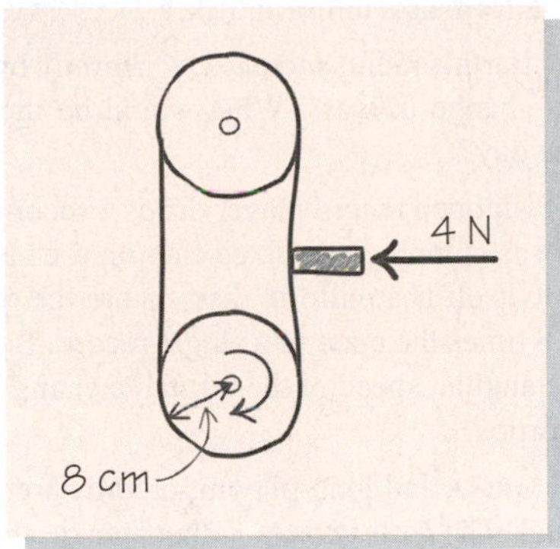

▲ **FIGURE 9–49** Problem 45.

46. (II) A massless rope is wrapped around a hollow cylinder of radius 12 cm whose central axis is fixed in a horizontal position. A mass of 4.0 kg hangs from the rope and, starting from rest, moves 180 cm in 2.0 s. What is the mass of the cylinder?

47. (II) A wheel of radius 24.6 cm whose axis is fixed starts from rest and reaches an angular velocity of 4.15 rad/s in 2.68 s due to a force of 13.4 N acting tangentially on the rim. (a) What is the rotational inertia of the wheel? (b) What is the change in the angular momentum during the 2.68 s? (c) How many revolutions does the wheel make? (d) How much rotational kinetic energy does the wheel have after 2.68 s?

48. (III) A motorcycle has a mass of 500 kg, the wheels have a diameter of 60 cm, and the centers of the wheels are separated by 1.5 m. Assuming that the weight is distributed uniformly over the wheels, that the wheels roll without sliding, and that the coefficient of static friction between the wheels and the road is $\mu = 0.5$, calculate the torque about the center of the front wheel exerted by the forces between the road and the wheels when there is a maximum braking.

9–5 Angular Momentum and Its Conservation

49. (I) A student sits on a piano stool and holds the axle of a bicycle wheel that rotates with angular velocity of magnitude 4π rad/s, pointing upward. The wheel's axis of rotation goes through the axis of the stool, which is at rest. The rotational inertia of the wheel about its axis is 1.2 kg·m^2, and the rotational inertia of the student and stool about the stool's axis is 8 kg·m^2. The student suddenly flips the shaft of the wheel so that its angular velocity points down. How fast and in what direction will the student and stool rotate? Ignore friction.

50. (I) Two identical tops spin with angular velocities 35π rad/s up and 25π rad/s down, respectively, about vertical axes on a table. The tops bump into one another and separate. After the collision, one of the tops has an angular velocity of 30π rad/s in its original direction. What is the angular velocity of the other top?

51. (I) A spherically symmetric celestial object rotates at 3.24592 rev/s. Through some mechanism that does not involve an application of external torque, a change of total mass, or a change of shape, its radius decreases rapidly and uniformly. As a consequence, the rate of revolution changes to 3.24608 rev/s. What is the fractional change in the radius?

52. (II) A child of mass 25 kg stands at the edge of a rotating platform of mass 150 kg and radius 4.0 m. The platform with the child on it rotates with an angular speed of 6.2 rad/s. The child jumps off in a radial direction. (a) What happens to the angular speed of the platform? (b) What happens to the platform if, a little later, the child, starting at rest, jumps back onto the platform? (Treat the platform as a uniform disk.)

53. (II) Suppose Earth's radius increased *uniformly* by 0.001 percent without any change in mass. What would be the change in the length of the day?

54. (II) An old-fashioned record player drops a second record onto a first one that is spinning at a fixed rate on a coasting turntable. The turntable itself is a uniform disk, as are the records, and the turntable is 8 times the mass of a single record. By what percentage does the angular speed of the turntable change when the second record drops?

55. (II) Compact disks and long-playing records are made from the same material. The former have a diameter of about 12 cm; the latter, about 32 cm. When in use, records spin at $33\frac{1}{3}$ rev/min and compact disks spin at, say, 400 rev/min. What is the ratio of the angular momentum of a compact disk in use to that of a record? Assume that a compact disk has half the thickness of a record.

9–6 Rolling

56. (I) Bicycle racers sometimes use solid wheels in order to cut down the drag force between the air and the spokes of an ordinary wheel. This can be an important effect because drag forces rise quite rapidly as the speed of an object through the air increases. If the radius of a wheel is 35 cm and the speed of the bicycle relative to the ground is 22 mi/h (9.8 m/s), what is the speed relative to the ground of the end of the spoke closest to the rim for (a) a spoke leading to the contact point with the ground, that is, a spoke pointing vertically down? (b) A spoke pointing vertically up? (c) A spoke that is horizontal and points forward? (d) A spoke that is horizontal and points backward?

57. (II) A cylindrical shell starting from rest rolls down an inclined plane that makes an angle of 20° with the horizontal. How far will the shell travel in 4 s? How far would a solid cylinder travel in the same time?

58. (II) A homogeneous cylinder of mass 1.20 kg and diameter 25 cm rolls down an inclined plane that makes an angle of 25° with the horizontal. What is the speed of the axis of the cylinder by the time the cylinder has rolled 1.00 m down the plane?

59. (II) The homogeneous cylinder of the preceding problem rolls down an inclined plane that makes an angle of 12.5° with the horizontal. How much further does it move down the ramp before it reaches the same speed that it did in rolling 1.00 m in the preceding problem?

60. (II) The following objects all roll without slipping and have uniform density, mass M, and radius R and the speed of the center of mass in each case is v. Find the ratio of the rotational kinetic energy to the total kinetic energy for (a) a solid cylinder, (b) a hollow cylinder, and (c) a solid sphere.

61. (II) Figure 9–50 shows a disk attached to an axle placed on an incline made of two parallel bars. The radius of the disk is 0.12 cm and its mass is 0.8 kg; the radius of the axle is 0.020 m and its mass is 0.10 kg, not including the part inside the disk. Calculate the acceleration of the system if the incline makes an angle of 5° with the horizontal and the axle rolls without slipping.

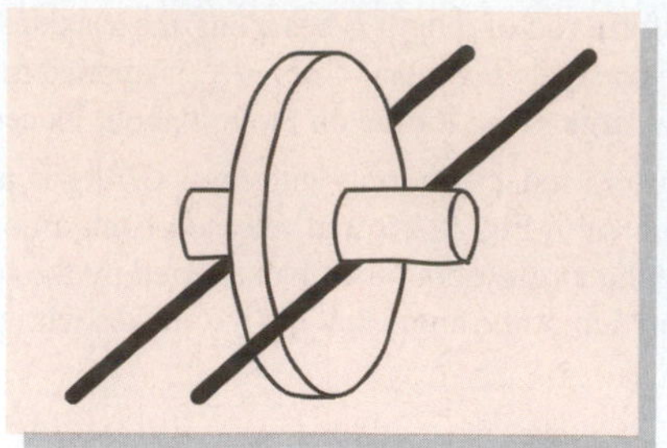

▲ **FIGURE 9–50** Problem 61.

62. (II) A hollow cylinder moves down an inclined plane of length ℓ and angle θ. The cylinder has uniform density, mass M, and radius R. It is initially at rest at the top of the plane. Calculate and compare the times taken to reach the bottom if the cylinder rolls without slipping as opposed to the case in which the cylinder slips all the way down the plane without rolling. In the latter case, the coefficient of friction must be zero.

General Problems

63. (I) A basketball player shoots a desperate last shot, spinning the ball at an angular speed of 15 rad/s to give it "action." The ball is shot with an initial velocity of 4.1 m/s at an elevation angle of 45° and leaves the player's hands 1.7 m off the floor. Unfortunately, the shot misses the backboard, the rim, and the net. How many revolutions has the ball made when it hits the floor? Ignore air resistance.

64. (I) *Estimate* the angular momentum of a spinning ice skater.

65. (II) Earth's radius is 6.4×10^6 m. Assume that its density is not uniform; that is, the inner core has a density of $8.0 \times 10^3\ \text{kg/m}^3$ and the outer mantle has a density of $3.0 \times 10^3\ \text{kg/m}^3$. Given that the rotational inertia of Earth is $8.3 \times 10^{37}\ \text{kg}\cdot\text{m}^2$, calculate the radius R at which the density changes. [*Hint*: Go back to Problem 33.]

66. (II) A solid cylinder of mass 17 kg and radius 33 cm rotates at 300 rev/min about its central axis. What is the rotational kinetic energy of this motion? Suppose that a 0.8-kg mass is attached at one point on the rim of the cylinder. If the additional mass can be treated as a point mass and the rotational speed is unchanged, what is the percentage change in the rotational kinetic energy?

67. (II) A solid uniform cylinder of mass M and radius R is projected up an incline of angle θ. It rolls without slipping from an initial speed v_0 of the center of mass. What distance s does the center of the cylinder travel before it starts to fall back?

68. (II) The lid of a box is balanced vertically on its hinges. The slightest displacement from the vertical leads to its falling shut. If the lid is of uniform density and it is 30 cm wide, what will be its angular velocity when it does fall shut, that is, when it is in a horizontal position?

69. (II) A spool within a piece of machinery consists of a solid cylinder of radius R_1 capped at each end by solid disks of a larger radius R_2. The spool can turn freely about its axis. A mass m_1 can be hung from a light rope wrapped about the inner cylinder and a second mass m_2 can be hung similarly, but in the opposite sense, from one of the disks (Fig. 9–51, see next page). What is the value of m_2 such that the spool will not turn? How would your answer change if the central cylinder were hollow, with internal struts for support?

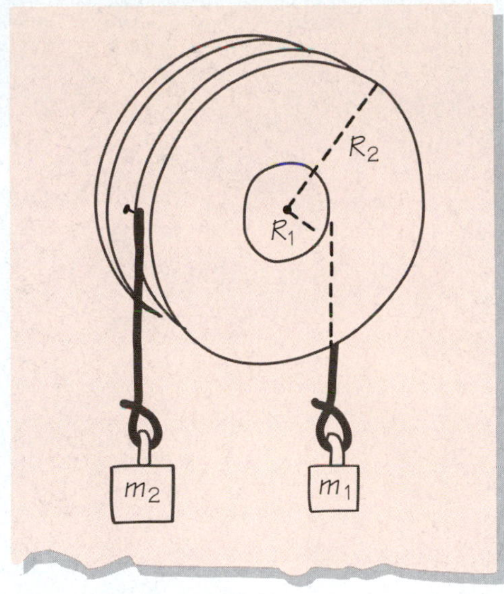

▲ **FIGURE 9–51** Problem 69.

70. (II) When a bicycle rider accelerates, he must accelerate his own and his bicycle's linear motion as well as the angular motion of the wheels. Suppose that the cyclist has mass 55 kg; the bicycle (not counting wheels), 8.0 kg; and both wheels together, 1.8 kg. Assume that the wheels, each of radius 30 cm, have all their mass concentrated in the (thin) rim. (a) At 25 km/h, what fraction of the kinetic energy of the rider plus his bicycle is in linear motion and what fraction is in rotational motion? (b) Suppose that the cyclist loses 3.0 kg on a diet. What percentage of the original force is required to accelerate the system uniformly from 0 to 25 km/h in 10 s? (c) Suppose that instead of going on a diet the cyclist replaces his wheels with ones of total mass 1.2 kg. Now what is the percentage of the original force required to accelerate the system uniformly from 0 to 25 km/h in 10 s?

71. (II) A hollow cylinder of radius 15 cm and mass 3.0 kg rolls without slipping at a constant speed of 1.6 m/s. (a) What is its angular momentum about its symmetry axis? (b) What is its rotational kinetic energy? (c) What is its total kinetic energy?

72. (II) A pulsar (the remnant of a star after a supernova explosion) has a mass of 3.7×10^{30} kg and a radius of 17 km. It rotates with a period of 0.15 s. (a) Assuming a spherical shape and constant density, what is the kinetic rotational energy of the pulsar? (b) If the period changes by 1 part in 10^8 in 1 yr, what is the rate of energy loss of the pulsar?

73. (II) A yo-yo has mass M and external radius R. The central stem has negligible mass and a radius r. The string is pulled horizontally on the lower side with a constant force F, while the yo-yo rests on a rough horizontal surface (Fig. 9–52). What is the maximum value of F for which the yo-yo will roll without slipping assuming that the coefficient of static friction between the yo-yo and the surface is μ?

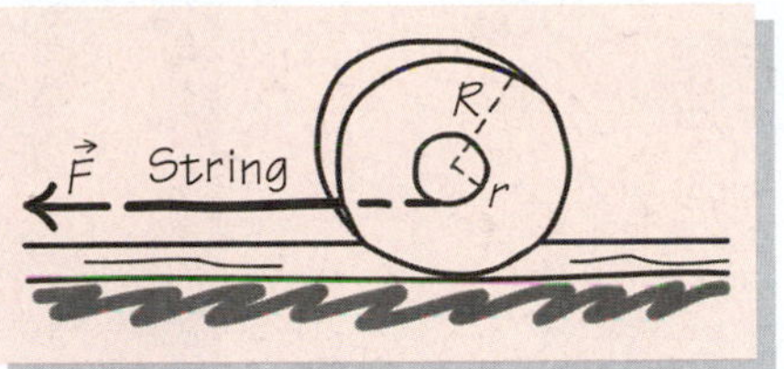

▲ **FIGURE 9–52** Problem 73.

74. (II) A cylinder of known mass M, radius R, and rotational inertia I is placed on an inclined plane with angle θ (Fig. 9–53). A string is wound around the cylinder and pulled up with a tension T parallel to the inclined plane. The coefficient of static friction is large enough to prevent slipping. (a) Find the tension T_0 needed to keep the cylinder in equilibrium. (b) Find the acceleration of the cylinder if the tension is known and is different from T_0.

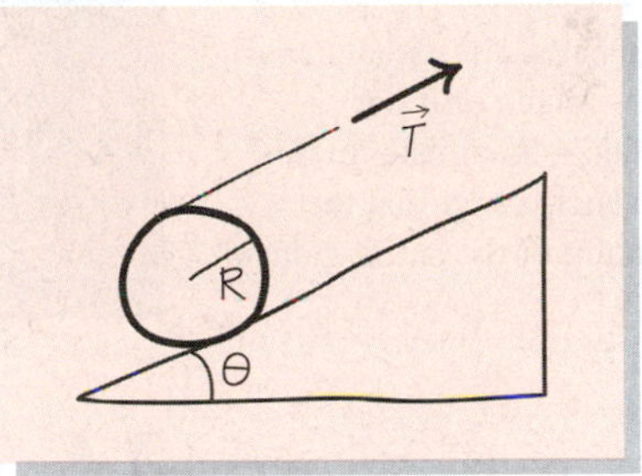

▲ **FIGURE 9–53** Problem 74.

75. (II) Suppose that the yo-yo described in Problem 73 is released from rest while the upper end of the cord is held steady. What is the tension in the cord during the yo-yo's downward motion?

76. (II) A thin rod of mass M and length ℓ is lying on a frictionless table. It is given an impulse of 4.5 N · s by a force of magnitude F on one end at an angle of 35° to the rod. How far will the center of mass of the rod travel when the rod has completed 12 rev?

Chapter 10

The complicated and violent swirl of stars and gas—much of it not glowing or hidden—that make up this galaxy move in such a way that the angular momentum of the entire galaxy is conserved.

More on Angular Momentum and Torque

While the rigid body rotations we studied in Chapter 9 are certainly observable in the world, many other real physical systems, such as the collections of stars that make up galaxies, a whirlpool, or a figure skater executing a complex maneuver, are not rigid, and their motions are equally interesting. Just as the concept of linear momentum and its conservation provides us with important simplifications for systems in linear motion, angular momentum and its conservation are extremely useful concepts for systems of particles that do not form rigid bodies. In this chapter we deal in more detail with the vectorial nature of rotational motion, angular momentum, and torque. This study will allow us to extend and expand the correspondence between the dynamics of linear motion and rotational motion.

10–1 Generalization of Angular Momentum

In Section 9–5 we attributed angular momentum to rigid systems in pure rotation. Here we shall see that we can define this quantity for any system, and that, as for the rigid object, angular momentum is conserved if the system is isolated. Our technique for doing this is to define angular momentum for the simplest possible system—the point particle moving freely in space—and then to extend the definition, essentially by superposition, to a larger system composed of pointlike constituents. As in Chapter 9, the angular momentum of even a point mass moving in space must be defined with respect to a fixed origin.

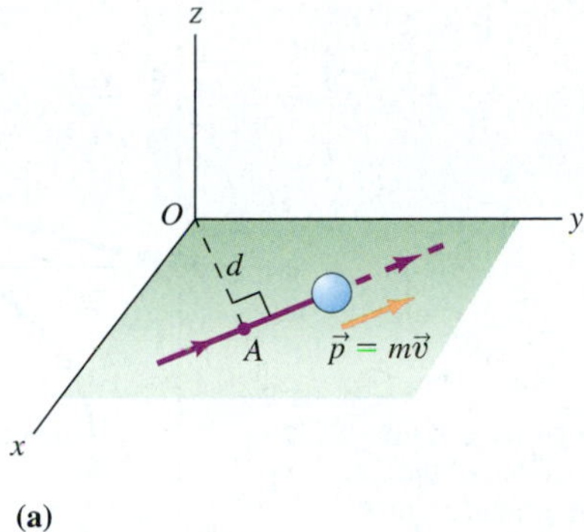

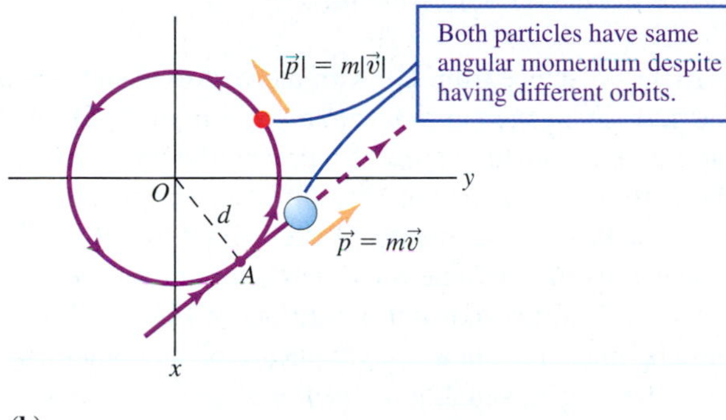

FIGURE 10–1 (a) A point mass m moves freely in a plane with velocity $\vec{v}$. (b) To find its angular momentum with respect to a point O we match its motion at point A, when it is closest to O, to a particle of the same mass moving in a circle about O.

What may be surprising is the idea that a free particle—one with no forces on it—has something to do with rotations. Such an object moves along a straight line with constant linear momentum (or just momentum) $\vec{p}$; as for example a hockey puck sliding along the ice. (Strictly speaking, the puck has no *net* force on it, but because, as we shall see, it also has no net torque, it behaves as the truly free object.) Consider our free particle, together with a point that we label as the origin O. As long as this point is not on the line formed by the trajectory, it and the trajectory form a plane, which we label the xy-plane (Fig. 10–1a). Our aim here is to define the particle's angular momentum with respect to the point. We will start by insisting that the angular momentum $\vec{L}$ that we assign to the particle is constant, because there are no forces, and thus no torques, acting on the particle. As the particle moves, its point of closest approach to O is a distance d from the origin. At this point, which we label A on Fig. 10–1, the line of motion is perpendicular to the line from the origin; further, the particle's velocity at point A is *instantaneously* identical to that of a second particle of the same mass moving in a circular path in the xy-plane about point O (Fig. 10–1b). We already know how to find the angular momentum of the second particle: Its angular momentum is in the $+z$-direction and has magnitude

$$L = I\omega = (md^2)(v/d) = mvd = pd.$$

Now, since the motion of the free particle is indistinguishable at the point of closest approach to the motion of the second particle moving in a circular path, we assign the free particle the instantaneous value of the angular momentum of the second particle. And since the angular momentum of the free particle does not change, this is the angular momentum that we assign to the freely moving particle for all times:

$$\vec{L} = pd\,\hat{k}. \tag{10–1}$$

As Fig. 10–2 illustrates, the freely moving particle may at some time be at point P, a distance r from the origin. Here, the momentum makes an angle θ with respect to the line from O. Geometry shows that $d = r\sin\theta$ so that

$$\vec{L} = pd\,\hat{k} = (pr\sin\theta)\hat{k}. \tag{10–2}$$

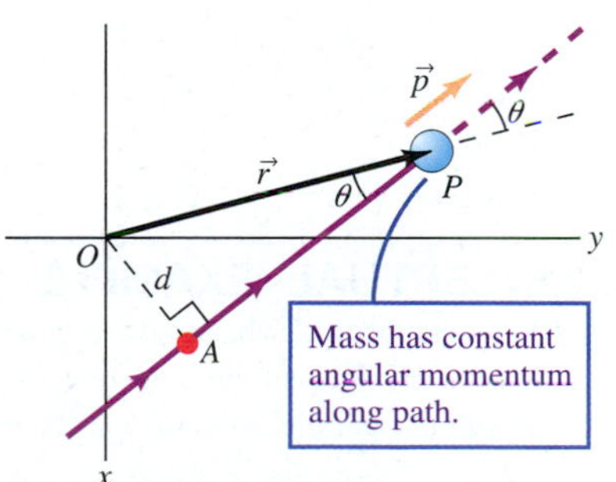

FIGURE 10–2 At a later time mass m has moved to point P. Its angular momentum, however, is the same as it was in Fig. 10–1.

Equation (10–2) *is the basis for a more general definition of the angular momentum of a point particle* with respect to a point O. If our particle has linear momentum of magnitude p and the vector from O to our particle's location has magnitude r (i.e., r is the

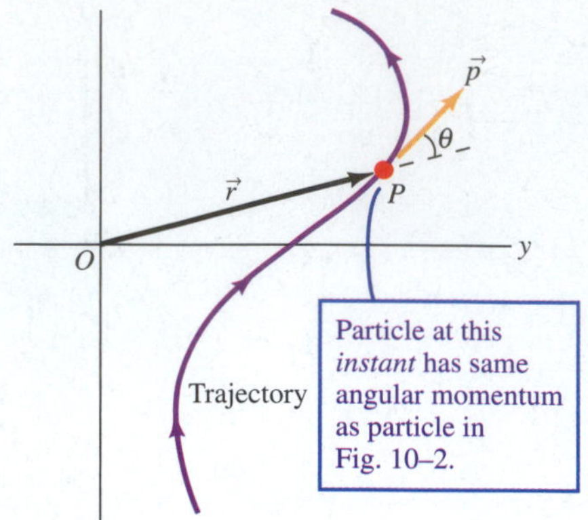

FIGURE 10–3 A particle of the same mass and velocity as that of the free particle in Fig. 10–2 at the same point P. Its angular momentum at that point is the same as that of the freely moving particle.

particle's distance from O), with the vectors $\vec{r}$ and $\vec{p}$ making an angle θ, then we assign the particle an angular momentum as in Eq. (10–2). (Note that we now determine angular momentum with respect to a point rather than an axis.) This is our definition whether the particle is free or not. The difference between a free particle and one on which a torque acts is that for the free particle, Eq. (10–2) is the angular momentum for *all times*; whereas for the particle on which a torque acts, Eq. (10–2) is only the *instantaneous* value of $\vec{L}$, at the moment picked out in Fig. 10–2. It does not matter if the particle is moving in a circle, in a straight line, or on any other trajectory (Fig. 10–3).

The torque, suitably defined, will determine how the angular momentum changes. Generally, we'll want to define torque with respect to the same point O that serves for the definition of angular momentum. Note that a force may act, changing the linear momentum, but if that force exerts no torque on the particle about point O, Eq. (10–2) remains the constant value of $\vec{L}$.

It is worthwhile to emphasize what our definition entails in terms of vectors, and to understand that a right-hand rule is involved in the determination of the direction of angular momentum. We have a vector $\vec{r}$ from the origin (the reference point) to our particle, which has momentum $\vec{p}$. The direction of the angular momentum is perpendicular to both $\vec{r}$ and $\vec{p}$ (two vectors always form a plane, and you can always find the direction perpendicular to the plane), pointing along the direction of the thumb of the right hand if the fingers curl from $\vec{r}$ to $\vec{p}$. You should check that this is indeed how Eq. (10–2) works. Below we'll find a more convenient way to say this.

We arrived at our result by thinking of a freely moving particle as momentarily equivalent to an object in circular motion. Therefore it is not surprising that it is possible—although we'll forgo the exercise—to start from Eq. (10–2), apply it to the point-like components that make up a rigid object, all rotating about a fixed axis with angular speed ω, and recover the results of Chapter 9, namely

$$\vec{L} = I\vec{\omega}. \tag{10–3}$$

Here I is the rotational inertia about the rotation axis, and both $\vec{L}$ and $\vec{\omega}$ are aligned with the rotation axis according to the right-hand rule. In other words, our new definition of the angular momentum about a point—Eq. (10–2)—reduces to the earlier form, Eq. (9–37), for the angular momentum about an axis in the case of a rotating rigid body.

CONCEPTUAL EXAMPLE 10–1 A powerful rocket is fired straight up. What is its angular momentum with respect to Earth's center? Answer the same question about a little ball that a baby tosses initially parallel to Earth's surface. Ignore Earth's rotation.

Answer To find the angular momentum of an object with respect to some reference point, we can extrapolate the line of its momentum and find the closest approach of that line to the reference point. That is the moment arm d for the angular momentum. In this case, "straight up" means a line that, if extrapolated back toward Earth's center, passes directly through the center. Thus the moment arm is zero, and so is the angular momentum of the rocket with respect to Earth's center. For the case of the ball, which we might suppose has mass m and an initial speed v, the motion is perpendicular to the line from Earth's center, a distance R_E from the center, where Earth's radius is R_E. The moment arm is R_E, and the angular momentum of the ball with respect to Earth's center is mvR_E.

EXAMPLE 10–2 In an engineering design, a light but stiff rod of length R is attached at an angle θ to a shaft along the z-axis; it is used to rotate a mass M about the shaft (Fig. 10–4a). The mass moves with speed v. Describe the angular momentum of the mass with respect to the attachment point of the rod.

Setting It Up The origin of the coordinate system, placed at the point of attachment of the rod to the shaft, is the reference point for the angular momentum. In Fig. 10–4b we have drawn a side view with the velocity of the mass into the page; the page is in the yz-plane. This second drawing will be useful for understanding the geometry of the situation.

Strategy In identifying the vectors $\vec{R}$ from the origin O to the mass M and $\vec{p}$—the momentum vector, which points along the direction of the velocity—we can work out both the magnitude and the direction of the angular momentum. $\vec{L}$ must be perpendicular to each, as determined by a right-hand rule, and the magnitude L is the product of the magnitudes of $\vec{R}$ and $\vec{p}$ times the sine of the angle between them, as in Eq. (10–2).

Working It Out From our discussion of the direction of the angular momentum, perpendicular to both $\vec{R}$ and $\vec{p}$, we can draw the angular momentum vector (Fig. 10–4b). Note in particular that $\vec{L}$ makes the angle θ with the horizontal. $\vec{R}$ and $\vec{p}$ are perpendicular to each other, hence the angular momentum has magnitude $L = Rp = RMv$. In addition, Fig. 10–4b makes it evident that $\vec{L}$ has both z-component and a radial component toward the shaft, given by

$$L_z = L\sin\theta = L\frac{d}{R} = (RMv)\frac{d}{R} = dMv;$$

$$L_{\text{radial}} = -L\cos\theta = -L\frac{\sqrt{R^2 - d^2}}{R} = -\sqrt{R^2 - d^2}Mv.$$

The minus sign in L_{radial} indicates it points to the rotation axis.

We can anticipate from our discussion in Chapter 9 that because the angular momentum is not constant (the radial component rotates so as to point to the shaft), there must be a torque about point O. This classic engineering problem tells us that the attachment to the shaft must be appropriately constructed to supply the torque. We'll describe such a torque in Conceptual Example 10–4 in Section 10–2.

What Do You Think? How would the result change if the attachment point were in the plane formed by the rotating mass? *Answers to* ***What Do You Think?*** *questions are given in the back of the book.*

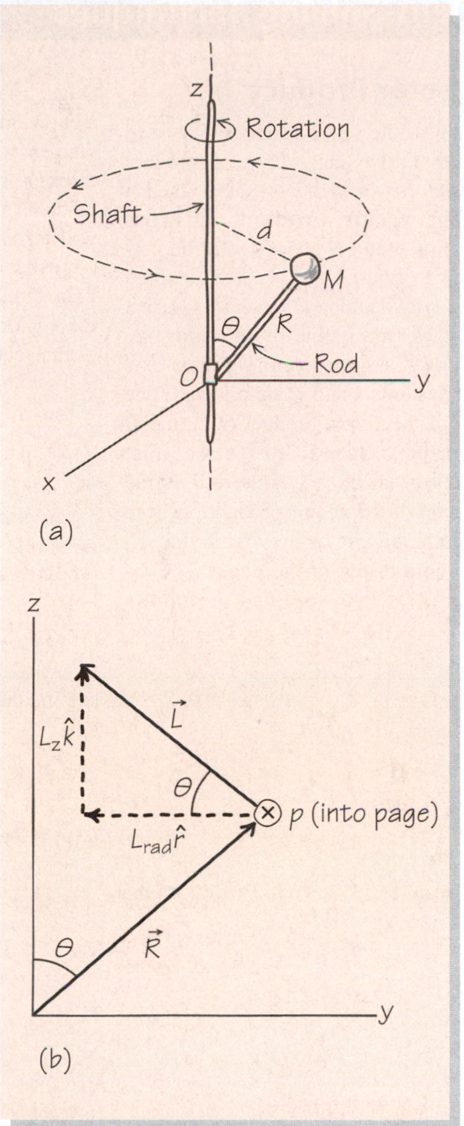

▲ **FIGURE 10–4** (a) A rod attached at the origin with a mass at its end rotates about the z-axis. In (b) we draw a side view of (a).

Angular Momentum as a Vector Product

Vector products ("The Vector Product", p. 284) are useful for many of the quantities and relations involved in rotation. One of the most important examples is the angular momentum itself about some origin O of a point mass with momentum $\vec{p}$, which is given by

$$\vec{L} \equiv \vec{r} \times \vec{p}, \tag{10–4}$$

ANGULAR MOMENTUM OF A POINT MASS

where $\vec{r}$ is the position vector of the point mass with respect to O (Fig. 10–5). We can reduce this result to that previously given in Eq. (10–2) and described there. The magnitude of $\vec{L}$ is $rp\sin\theta$, where θ is the angle between $\vec{r}$ and $\vec{p}$, and $r\sin\theta$ is the perpendicular distance d of Eq. (10–2). The vector product described in the box has the property that the angular momentum vector $\vec{L}$ defined in Eq. (10–4) is perpendicular to *both* $\vec{r}$ and $\vec{p}$, with the direction given by the right-hand rule (see Figs. 10–6b, c). Note that $\vec{L}$ is zero when $\vec{r}$ and $\vec{p}$ are parallel; that is, when the straight-line extension of the vector $\vec{p}$ passes through the reference point O. Example 10–3 illustrates the use of Eq. (10–4).

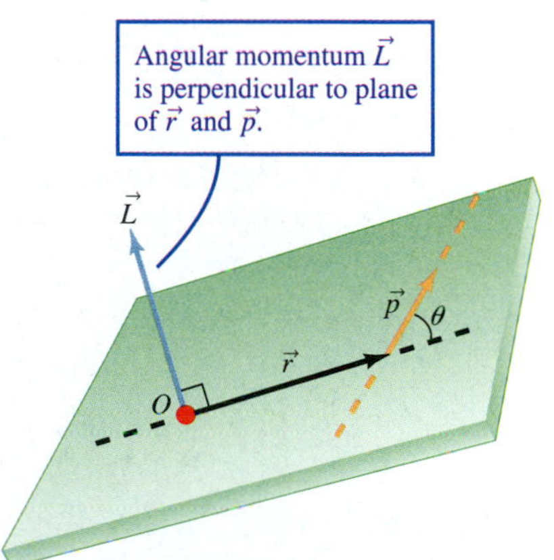

▲ **FIGURE 10–5** Vector relation for the angular momentum $\vec{L}$ of a point particle about a point O.

Problem-Solving Techniques

The Vector Product

In addition to the scalar product discussed in Chapter 6, we can define a product of two vectors—$\vec{A}$ and $\vec{B}$—that is itself a vector. The **vector product** (or **cross product**) of $\vec{A}$ and $\vec{B}$ is defined to be perpendicular to both $\vec{A}$ and $\vec{B}$; it is denoted $\vec{A} \times \vec{B}$. Any two nonparallel vectors form a plane that we can define by choosing the x-axis along $\vec{A}$ and then defining the y-axis so that $\vec{B}$ has only x and y components. The direction of the vector product of $\vec{A}$ and $\vec{B}$ is then perpendicular to this plane (Fig. 10–6a). A right-hand rule is used to define the direction of the vector product because there is an ambiguity as to whether the direction is into or out of the plane. The direction of $\vec{A} \times \vec{B}$ is specified as follows (Figs. 10–6b, c): Point the fingers of your right hand in the direction of $\vec{A}$ and curl toward $\vec{B}$ according to the rule that the angle through which the fingers are curled to reach $\vec{B}$ must always be less than 180° (it will be hard to get the fingers to follow the opposite of this rule!). The direction of your thumb then indicates the direction of the vector product $\vec{A} \times \vec{B}$. The magnitude of the vector product is given by

$$|\vec{A} \times \vec{B}| = AB \sin\theta, \qquad \text{(B1–1)}$$

where θ is the angle (less than 180°, so that the sine is positive) between $\vec{A}$ and $\vec{B}$.

An immediate consequence of the right-hand rule is that

$$\vec{A} \times \vec{B} = -\vec{B} \times \vec{A}. \qquad \text{(B1–2)}$$

Thus the vector product is not commutative. It also follows from Eq. (B1–1) that *the vector product of two vectors that are parallel (or antiparallel) to each other is zero:* When $\theta = 0°$ or $180°$, $\sin\theta = 0$. The vector product attains its maximum magnitude when $\vec{A}$ and $\vec{B}$ are perpendicular to each other because then $\sin\theta = 1$. Contrast this to the scalar product, which is zero when the two vectors are perpendicular and which has a maximum magnitude when they are parallel.

The unit vectors $\hat{i}$, $\hat{j}$, and $\hat{k}$ (Fig. 10–6d) along the x-, y-, and z-axes, respectively, obey the relations

$$\hat{i} \times \hat{j} = -\hat{j} \times \hat{i} = \hat{k}, \qquad \text{(B1–3a)}$$

$$\hat{j} \times \hat{k} = -\hat{k} \times \hat{j} = \hat{i}, \qquad \text{(B1–3b)}$$

$$\hat{k} \times \hat{i} = -\hat{i} \times \hat{k} = \hat{j}. \qquad \text{(B1–3c)}$$

Also $\hat{i} \times \hat{i} = \hat{j} \times \hat{j} = \hat{k} \times \hat{k} = 0$. We can expand the vectors $\vec{A}$ and $\vec{B}$ into their components and find

$$\begin{aligned}\vec{A} \times \vec{B} &= (A_x\hat{i} + A_y\hat{j} + A_z\hat{k}) \times (B_x\hat{i} + B_y\hat{j} + B_z\hat{k}) \\ &= A_xB_y(\hat{i} \times \hat{j}) + A_xB_z(\hat{i} \times \hat{k}) + A_yB_x(\hat{j} \times \hat{i}) + A_yB_z(\hat{j} \times \hat{k}) \\ &\quad + A_zB_x(\hat{k} \times \hat{i}) + A_zB_y(\hat{k} \times \hat{j}).\end{aligned}$$

With the results of Eqs. (B1–3), we find that

$$\vec{A} \times \vec{B} = (A_yB_z - A_zB_y)\hat{i} + (A_zB_x - A_xB_z)\hat{j} + (A_xB_y - A_yB_x)\hat{k}. \qquad \text{(B1–4)}$$

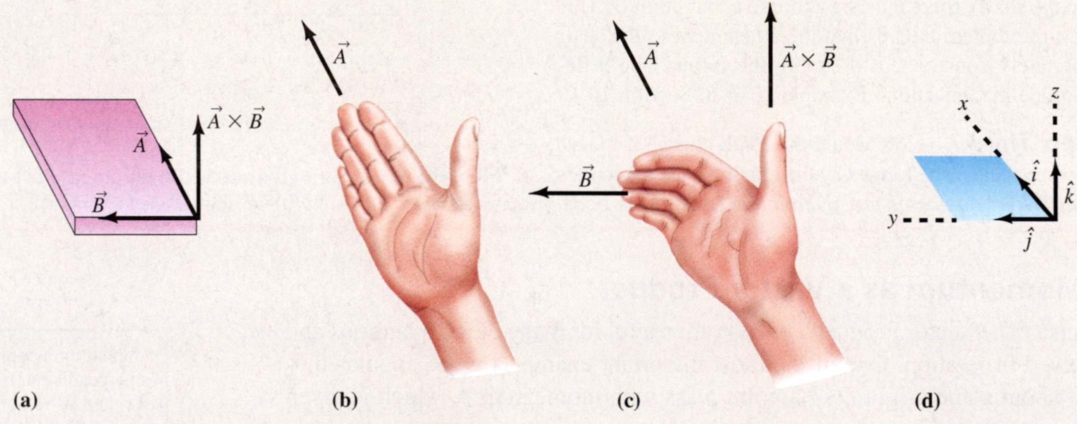

▲ **FIGURE 10–6** (a) Two vectors $\vec{A}$ and $\vec{B}$ and their vector product $\vec{A} \times \vec{B}$. (b), (c) A right-hand rule specifies the direction of the vector product. (d) The process is broken down to vector products between Cartesian unit vectors.

EXAMPLE 10–3 A tether ball is tied by a rope of length R to a central pole. As the ball whirls around the pole in a (nearly) horizontal plane, the rope gradually winds around (or unwinds from) the pole, shortening (or lengthening) the amount of rope between the ball and the pole (Fig. 10–7). The motion is spiral rather than circular: At the time t, when the ball is at a distance nearly equal to R and rotating around the pole with angular speed nearly equal to ω, the distance is more precisely $r = R - A\omega t$, where A is a constant with dimensions of length and $A\omega t$ is small compared to R. Find the ball's angular momentum with respect to the pole at time t.

Setting It Up The plane of the rope is the xy-plane, and z is along the pole. We denote by θ the angle made by the rope as it goes around the pole (Fig. 10–7). We know the ball's motion in detail and are asked to compute its angular momentum.

Strategy Because the tether ball does not follow a circular path, we must use the more general form $\vec{L} = \vec{r} \times \vec{p}$ for the angular momentum. We first write out the x- and y-coordinates of the ball as a function of time and use these to calculate the velocities and thus the momentum. Finally, we can use the calculation of the linear momentum vector to find the angular momentum.

Working It Out We have

$$x = r\cos\theta = (R - A\omega t)\cos(\omega t);$$
$$y = r\sin\theta = (R - A\omega t)\sin(\omega t).$$

The momentum components of the ball are therefore

$$\begin{aligned} p_x &= Mv_x = M\frac{dx}{dt} \\ &= M[(R - A\omega t)(-\omega\sin(\omega t)) + (-A\omega)\cos(\omega t)]; \\ p_y &= Mv_y = M\frac{dy}{dt} \\ &= M[(R - A\omega t)(\omega\cos(\omega t)) + (-A\omega)\sin(\omega t)]. \end{aligned}$$

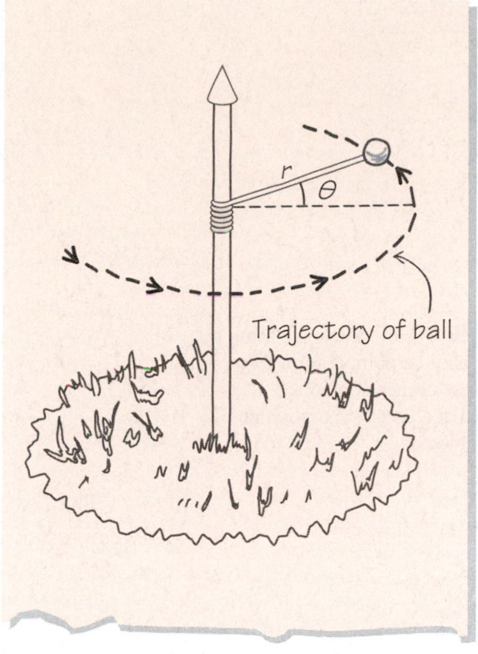

▲ **FIGURE 10–7** As the rope winds around the pole, it shortens.

We can now find $\vec{L}$ using Eq. (10–4). From Eq. (B1–4) in "The Vector Product" box on page 284, we have

$$\begin{aligned} \vec{L} &= (xp_y - yp_x)\hat{k} \\ &= M\begin{Bmatrix} [(R - A\omega t)\cos(\omega t)][(R - A\omega t)(\omega\cos(\omega t)) + (-A\omega)\sin(\omega t)] \\ -[(R - A\omega t)\sin(\omega t)][(R - A\omega t)(-\omega\sin(\omega t)) + (-A\omega)\cos(\omega t)] \end{Bmatrix}\hat{k} \\ &= M(R - A\omega t)^2\,\omega[\cos^2(\omega t) + \sin^2(\omega t)]\hat{k} \\ &= M(R - A\omega t)^2\,\omega\hat{k}. \end{aligned}$$

$\vec{L}$ always lies in the z-direction because the motion is, by assumption, always in the xy-plane. But the magnitude of the angular momentum changes with time—albeit slowly ($R \gg A\omega t$). As a check, note that if $A = 0$ (pole diameter zero), the angular momentum has the familiar magnitude corresponding to uniform circular motion of a point mass, namely $MR^2\omega = MvR$.

What Do You Think? The rope holding the ball suddenly breaks cleanly. By how much is the angular momentum of the ball changed from the moment just before the rope breaks to the moment just after?

10–2 Generalization of Torque

Torque, introduced in Chapter 9, has a magnitude that is given by a lever arm times the magnitude of the applied force, and a direction specified by a right-hand rule. The lever arm is the distance to the reference axis times the sine of the angle between the force and the line to the axis. This has an immediate expression in terms of a vector product: The torque $\vec{\tau}$ with respect to any reference point O due to a force $\vec{F}$ applied at some point P of an object is

$$\vec{\tau} = \vec{r} \times \vec{F}, \tag{10–5}$$

THE TORQUE

where $\vec{r}$ is the position vector from O to the point of application (P) of the force (Fig. 10–8). The object in question can be an extended system or it can even be a point mass located at P. The reference point O about which the torque is defined can be located inside or outside the object. Aligning the fingers of the right hand along $\vec{r}$ and curling them toward $\vec{F}$, the torque is in the direction of the thumb. In Fig. 10–8 this is upward.

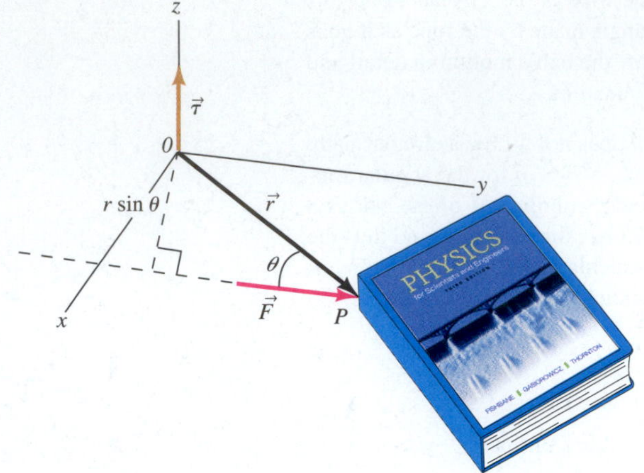

▶ **FIGURE 10–8** Vector relation for the torque about a point O for a force applied to the corner of a book, i.e., applied at a specific point on an extended object.

The test as to whether this is correct is to see whether the dynamical relation between torque and angular momentum, Eq. (9–39), continues to be satisfied:

$$\vec{\tau} = \frac{d\vec{L}}{dt}. \tag{10–6}$$

DYNAMICAL EQUATION FOR ROTATION

To check that Eq. (10–6), which is the rotational analog of Newton's second law, is indeed satisfied, we take Eq. (10–4), which is our generalized expression for the angular momentum of a point mass, and compute its rate of change. The rate of change of angular momentum can be found by using the chain rule of calculus (Appendix IV–7):

$$\frac{d\vec{L}}{dt} = \frac{d(\vec{r} \times \vec{p})}{dt} = \left(\frac{d\vec{r}}{dt} \times \vec{p}\right) + \left(\vec{r} \times \frac{d\vec{p}}{dt}\right). \tag{10–7}$$

The first part of Eq. (10–7) drops out because it is of the form $\vec{v} \times (m\vec{v})$, and the vector product of two parallel vectors is zero. For the second part of Eq. (10–7) we use Newton's second law in the form $d\vec{p}/dt = \vec{F}$ and obtain the equation

$$\frac{d\vec{L}}{dt} = \vec{r} \times \vec{F} = \vec{\tau},$$

which is just the desired Eq. (10–6).

We argued in Section 10–1 that it was possible to show that our expression for the angular momentum of a point object, $\vec{L} = \vec{r} \times \vec{p}$, reduces to the form $I\vec{\omega}$ for symmetric rigid bodies. Similarly, $\vec{\tau} = d\vec{L}/dt$ reduces to the form $I\vec{\alpha}$.

We can also use Eq. (B1-4) to express torque in terms of the components of $\vec{r}$ and $\vec{F}$. To do this, write $\vec{r} = x\hat{i} + y\hat{j} + z\hat{k}$. Then Eq. (B1–4) immediately leads to

$$\vec{\tau} = \vec{r} \times \vec{F} = (yF_z - zF_y)\hat{i} + (zF_x - xF_z)\hat{j} + (xF_y - yF_x)\hat{k}. \tag{10–8}$$

CONCEPTUAL EXAMPLE 10–4 In Example 10–2 we studied a mass on the end of a light stiff rod attached at an angle to a vertical axis and rotating about it. The angular momentum calculated with respect to the attachment point O of the rod had a constant vertical component but a changing radial component (Fig. 10–9a). To change the radial component of angular momentum there must be a torque about point O. Where does that torque come from, what is the direction of the torque, and how does all this affect the engineering design of the attachment of the rod to the shaft?

Answer We start with an extended free-body diagram (Fig. 10–9b) and find the net force on the mass. Only gravity and a contact force $\vec{N}$ from the rod act on the mass; the vertical component of the contact force merely cancels gravity, allowing the mass to stay in a horizontal plane. The net force on the mass is therefore the horizontal component of the contact force, which is radial and must point inward to make the mass move in its circular path; we'll label it as $\vec{N}_{\text{radial}}$. As we see below, this force will supply the needed torque.

We now verify that there is a torque due to this net force about point O. In Fig. 10–9c, which picks out the radial component of the contact force, the cross product $\vec{r} \times \vec{N}_{\text{radial}}$ would be out of the page. This follows from the fact that both $\vec{r}$ and $\vec{N}_{\text{radial}}$ are in the page plane and the cross product is perpendicular to the plane made by the two of them, together with use of the right-hand rule. You can verify by

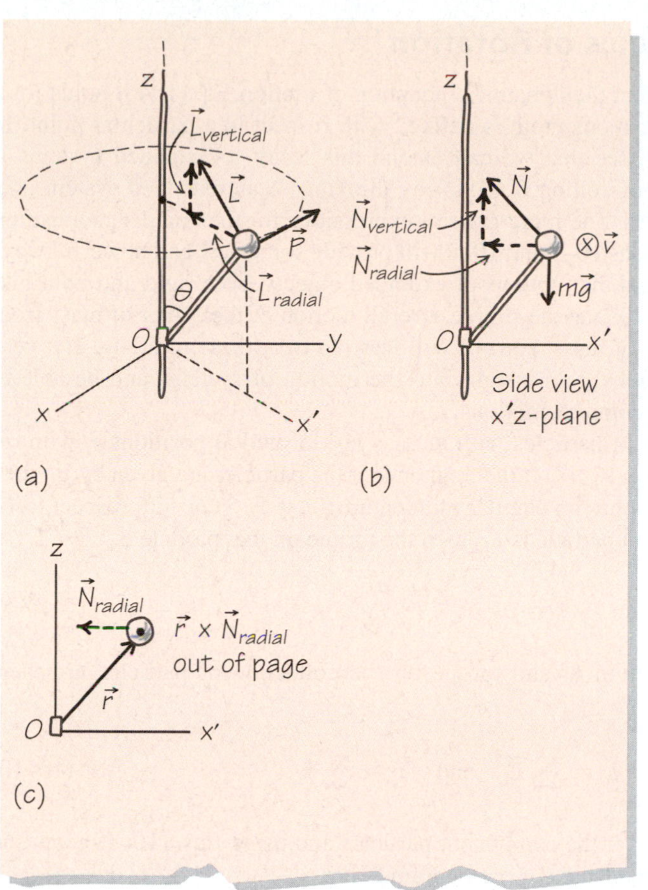

repeating the exercise of Example 10–2 that this is indeed the direction of the change in the radial component of mass's angular momentum on this part of the trajectory.

How does this affect the engineering design for the attachment point? The point is fixed. But the normal force from the rod acting on the mass means that the mass exerts a normal force on the rod, and this in turn means that the shaft must supply a horizontal (and vertical) force on the rod to cancel properly the force on the rod due to the mass. In particular, the vertical force from the shaft must be fixed, while the horizontal force from the shaft will have to change with time, and the attachment must be constructed strongly enough to accommodate this.

What Do You Think? How does your answer about torque change if the attachment point is in the plane of the motion ($\theta = 90°$)—that is, the rod is perpendicular to the shaft?

◀ **FIGURE 10–9** In part (a) we are reminded of the results of Example 10–2. (b) A free-body diagram for the mass. The direction of the contact force from the rod is only approximate here, but it must have an upward vertical component to cancel gravity and it must have an inward component to enable the mass to move in a circle. (c) The torque supplied by the radial component of the normal force.

EXAMPLE 10–5 A stone of mass $m = 1$ kg is dropped from an outstretched arm of length $\ell = 0.8$ m (Fig. 10–10). Find the net torque on the stone, taking the reference point O to be the person's shoulder, at the moment the stone has dropped exactly $d = 1$ m. Ignore air resistance.

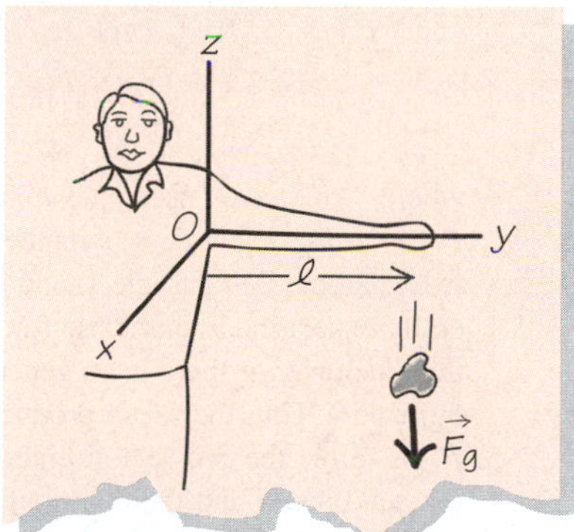

▲ **FIGURE 10–10** The origin of the coordinate system is placed at the shoulder. $\vec{F}_g$ is the force of gravity on the falling stone.

Setting It Up The coordinate system chosen (Fig. 10–10) sets the rock's motion parallel to the z-axis. The origin of this coordinate system is point O, so we want the torque on the stone about the origin.

Strategy The net torque is determined by the net force, and in this case the only force acting is gravity, $\vec{F} = -mg\,\hat{k}$. Given this information, we can find the net torque by the "lever arm" method described in Section 9–4 or by the equivalent, but more formal, vector product expression of Eq. (10–5). We'll do both.

Working It Out 1 We extend the line of the force in Fig. 10–10 and observe that the distance of closest approach to O is the arm length ℓ; this is the lever arm. The magnitude of the torque is the product of lever arm and force magnitude, i.e., ℓmg, with direction given by the right-hand rule to be in the $-x$-direction, that is, into the page. *The answer is independent of the distance the rock has fallen.* Numerically, the magnitude is $\tau = (0.8\text{ m})(1\text{ kg})(9.8\text{ m/s}^2) = 8\text{ N}\cdot\text{m}$.

Working It Out 2 In the more formal approach, we use the definition of the vector product in terms of components. We have, from Eqs. (10–5) and (B1–3b),

$$\vec{\tau} = \vec{r} \times \vec{F} = \ell\hat{j} \times (-mg\,\hat{k}) = -\ell mg(\hat{j} \times \hat{k}) = -\ell mg\;\hat{i},$$

the same result as in the first method.

What Do You Think? The torque is constant in time and independent of the distance the stone has fallen. This is because (a) there is no time dependence in the force; (b) the moment arm is independent of time; (c) the moment arm is independent of the distance fallen; (d) all of the above.

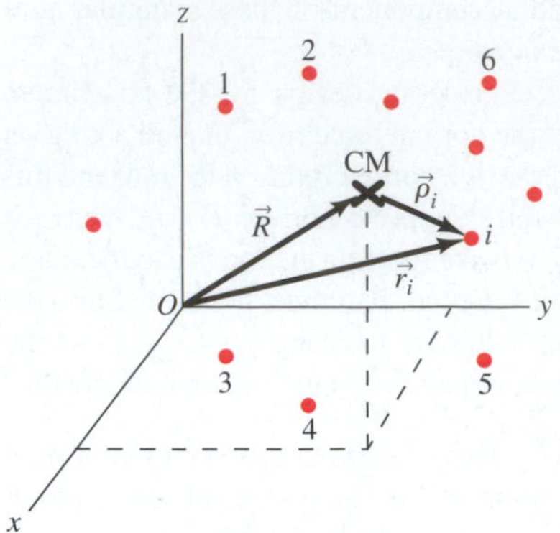

▲ **FIGURE 10–11** A system of many particles labeled by the index i. The position of particle i with respect to an origin O is $\vec{r}_i$, and the displacement of particle i from the center of mass is $\vec{\rho}_i$.

10–3 The Dynamics of Rotation

In Section 10–2 we saw that the dynamical equation of motion Eq. (10–6) holds for a point mass whose angular momentum is defined with respect to a particular point. In this section we shall first see that we can extend this result to extended systems—whirlpools, spinning divers, rolling objects—by thinking of an extended system as a collection of point particles. The more general expressions for the angular momentum and torque set out in Sections 10–1 and 10–2 will provide some new and powerful ways to think about rotations and the motions of extended objects, both rigid and nonrigid. We will see how we can separate the simple, overall motion of the center of mass from rotations about the center of mass, and we will learn to treat these rotations in a very general way. We will thereby gain insight into the motion of systems and be able to solve problems involving complex motions.

Consider a collection of particles with masses m_i, located at positions $\vec{r}_i$ with respect to a point O, as in Fig. 10–11. If the momenta of the particles are given by $\vec{p}_i$, then the particle with the label i has an angular momentum $\vec{L}_i = \vec{r}_i \times \vec{p}_i$ with respect to O. If the force acting on the ith particle is $\vec{F}_i$, then the torque on that particle is

$$\vec{\tau}_i = \vec{r}_i \times \vec{F}_i = \frac{d\vec{L}_i}{dt}. \tag{10–9}$$

The total angular momentum of and torque on the collection of particles are then, respectively,

$$\vec{L} = \sum_i \vec{L}_i \quad \text{and} \quad \vec{\tau} = \sum_i \vec{\tau}_i. \tag{10–10}$$

By summing Eq. (10–9) over the component particles and using Eqs. (10–10), we find the rate of change of the total angular momentum to be

$$\vec{\tau} = \frac{d\vec{L}}{dt}. \tag{10–11}$$

This is the same as Eq. (10–6) and shows that we can expand the rotational version of Newton's second law to any extended system.

THINK ABOUT THIS . . .

WHAT IS THE ROLE OF INTERNAL FORCES IN THE TORQUE ON A SYSTEM?

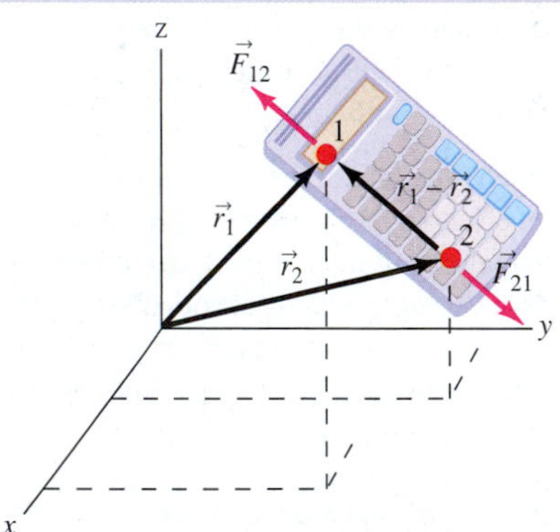

▲ **FIGURE 10–12** Two mass points, labeled 1 and 2, within an extended object, and the internal forces they exert on one another.

In Chapter 8 we were able to establish that internal forces play no role in how the linear momentum of an extended system changes under the influence of forces. Similarly, only external torques—torques associated with external forces—determine how a system changes its angular momentum. It will be enough to look at just two particles (or two small mass regions), with labels 1 and 2; our result can then be generalized to include more pieces of the larger systems. For the torque on particles 1 and 2, respectively, we have

$$\vec{\tau}_1 = \vec{r}_1 \times (\vec{F}_1^{\text{ext}} + \vec{F}_{12})$$

and

$$\vec{\tau}_2 = \vec{r}_2 \times (\vec{F}_2^{\text{ext}} + \vec{F}_{21}),$$

where $\vec{F}_{ij}$ is the (internal) force exerted on particle i due to particle j (Fig. 10–12) and $\vec{F}_i^{\text{ext}}$ is the external force on particle i.

The total torque is the sum of the two terms above. The internal forces contribute to the total torque through the terms

$$(\vec{r}_1 \times \vec{F}_{12}) + (\vec{r}_2 \times \vec{F}_{21}) = (\vec{r}_1 \times \vec{F}_{12}) - (\vec{r}_2 \times \vec{F}_{12}) = (\vec{r}_1 - \vec{r}_2) \times \vec{F}_{12},$$

where we have used Newton's third law, $\vec{F}_{12} = -\vec{F}_{21}$. Now $\vec{r}_1 - \vec{r}_2$ is the vector leading from particle 2 to particle 1, and we would generally expect the interparticle forces to lie along this line (even if there is no general rule, we require this). Thus the vector product $(\vec{r}_1 - \vec{r}_2) \times \vec{F}_{12}$ is zero—the two vectors entering are parallel to each other—and the contribution to the total torque from the internal forces is zero. In other words, *only the external forces appear in the total torque* $\vec{\tau}$ that is to be used in Eq. (10–11). Just as for linear motion, we can find the rotational motion of extended bodies without knowing the details of how their individual parts interact with each other. ■

The Role of the Reference Point

Throughout we have emphasized that rotational quantities—angular momentum, torque, rotational inertias—are defined with respect to a reference point. While some reference points—for example, the center of mass itself—are more "natural" choices than others, there is no physical principle that demands one choice over another. To put this another way, physical motion does not depend in any way on the choice of this point. What is true is that rotational problems can often be solved more easily by the choice of a good reference point; the right choice may, for example, simplify the geometry of a situation. Here we find some results that allow us to vary the reference point for rotational quantities. After we present the results, we illustrate their usefulness with an example.

Let us first recall the fact [Eq. (9–37)] that for a rigid body with rotational inertia I about an axis through the center of mass, the dynamical equation for rotation takes the form

$$\vec{\tau} = \frac{d\vec{L}}{dt} = I\frac{d\vec{\omega}}{dt} = I\vec{\alpha}. \tag{10–12}$$

The quantities in this equation all are assigned relative to the specified axis, or as we would now say, with respect to the location of the center of mass itself.

We now extend the "center of mass" results with a series of three relations that allow us instead to consider rotations about a second point A. These results apply to nonrigid as well as rigid systems. Let $\vec{R}$ be a vector from point A to the center of mass. Then, as we prove in a separate optional section, the following three results hold:

1. The angular momentum about A, which we call $\vec{L}_A$, is the sum of two terms: (a) the angular momentum $\vec{R} \times \vec{P}$ about A of a point mass that is carrying the whole mass (M) of the object, as though it were located at the center of mass and moving with the velocity $\vec{V}$ of the center of mass; and (b) the angular momentum $\vec{L}_{\text{cm}}$ of the object about the center of mass:

$$\vec{L}_A = (\vec{R} \times \vec{P}) + \vec{L}_{\text{cm}}. \tag{10–13}$$

Here, $\vec{P} = M\vec{V}$.

2. The total torque about A is the sum of two terms: (a) the torque about A due to the total external force applied to the center of mass; and (b) the torque about the center of mass:

$$\vec{\tau}_A = (\vec{R} \times \vec{F}_{\text{tot}}) + \vec{\tau}_{\text{cm}}. \tag{10–14}$$

3. The rate of change in each term of the total angular momentum in Eq. (10–13) is equal to the corresponding term of the total torque; that is,

$$\vec{R} \times \vec{F}_{\text{tot}} = \frac{d}{dt}(\vec{R} \times \vec{P}) \quad \text{and} \quad \vec{\tau}_{\text{cm}} = \frac{d}{dt}\vec{L}_{\text{cm}}. \tag{10–15}$$

*How to Get Equations (10–13) Through (10–15) for Angular Momentum and Torque

We break our extended object into a set of discrete pieces labeled with the subscript i. The center of mass position $\vec{R}$ with respect to an origin O was defined in Eq. (8–47) by

$$\vec{R} = \frac{1}{M}\sum_{i=1}^{N} m_i\vec{r}_i,$$

where $M = \sum_i^N m_i$ is the total mass, N is the number of pieces, and $\vec{r}_i$ is the position of the ith piece. We introduce the vector $\vec{\rho}_i$ (Fig. 10–11), which is the position vector of particle i as measured from the center of mass:

$$\vec{r}_i = \vec{R} + \vec{\rho}_i.$$

The angular momentum with respect to point O may now be written in the form

$$\vec{L} = \sum_i \vec{r}_i \times \vec{p}_i = \sum_i (\vec{R} + \vec{\rho}_i) \times \vec{p}_i = (\vec{R} \times \sum_i \vec{p}_i) + (\sum_i \vec{\rho}_i \times \vec{p}_i)$$

$$= (\vec{R} \times \vec{P}) + (\sum_i \vec{\rho}_i \times \vec{p}_i),$$

where $\vec{P} = \sum \vec{p}_i$ is the total momentum of the system. The term $\sum \vec{\rho}_i \times \vec{p}_i$ is the angular momentum $\vec{L}_{cm}$ about the center of mass. We have thus proved Eq. (10–13).

There is a similar decomposition of the total torque:

$$\vec{\tau} = \sum_i \vec{r}_i \times \vec{F}_i = \sum_i (\vec{R} + \vec{\rho}_i) \times \vec{F}_i = (\vec{R} \times \sum_i \vec{F}_i) + (\sum_i \vec{\rho}_i \times \vec{F}_i)$$

$$= (\vec{R} \times \vec{F}_{tot}) + (\sum_i \vec{\rho}_i \times \vec{F}_i).$$

The second term in the final equality is the torque about the center of mass, and so we have demonstrated Eq. (10–14).

To derive the third result, we take the rate of change of the $\vec{R} \times \vec{P}$ term in Eq. (10–13). We find that it is the same as the first term of Eq. (10–14):

$$\frac{d(\vec{R} \times \vec{P})}{dt} = \left(\frac{d\vec{R}}{dt} \times \vec{P}\right) + \left(\vec{R} \times \frac{d\vec{P}}{dt}\right) = \left(\frac{1}{M}\vec{P} \times \vec{P}\right) + (\vec{R} \times \vec{F}_{tot})$$

$$= \vec{R} \times \vec{F}_{tot}.$$

This in turn implies that the rate of change of the total angular momentum about the center of mass is equal to the torque about the center of mass due to all the forces:

$$\frac{d\vec{L}_{cm}}{dt} = \vec{\tau}_{cm}.$$

Why It Is Useful to Be Able to Choose the Reference Point

The results above are of practical importance because they provide alternative ways to approach problems involving both linear and rotational motion. We can choose convenient points of reference for torque, angular momentum, and rotational inertia. This is illustrated with a cylinder rolling down an incline, a problem we first examined in Chapter 9 (Fig. 10–13a). There we used a point on the symmetry axis of the cylinder as a reference. Let's now choose our reference point to be point O, the point of contact between the cylinder and the plane. Why choose this point? The answer is that it is easiest to compute net torque about a point through which the largest number of individual forces act. As the extended free-body diagram (Fig. 10–13b) shows, such forces have a zero lever arm for this point and so they do not contribute to the torque about the point. In this case, only gravity produces a torque about the contact point; moreover, only the component of gravity that is parallel to the plane—of magnitude $mg \sin \theta$—contributes.

EXAMPLE 10–6 A cylinder, mass M and radius R_0, whose mass density is symmetric about its axis, rolls down a ramp making an angle θ to the horizontal. (You can think for definiteness of any wheel; this is certainly a problem of practical interest.) Assuming its angular speed ω about its axis at $t = 0$ is ω_0, find the angular speed at a later time t.

Setting It Up Figure 10–13 illustrates the problem. While a rotational inertia for the cylinder will enter the problem, we don't have the precise mass distribution, so it will have to remain as an unknown quantity.

Strategy In keeping with the express aim of this example, we compute all angular quantities with respect to the instantaneous

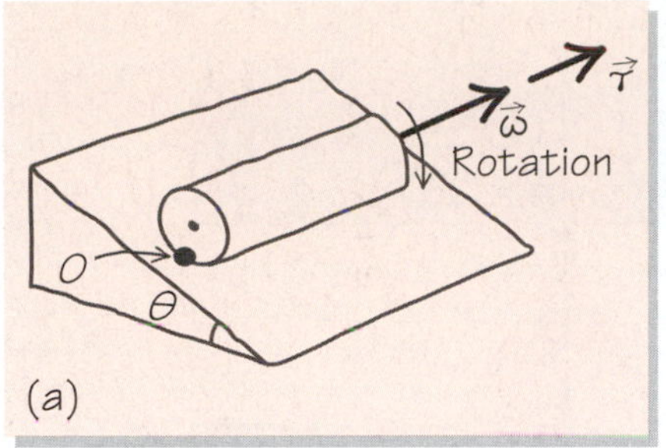

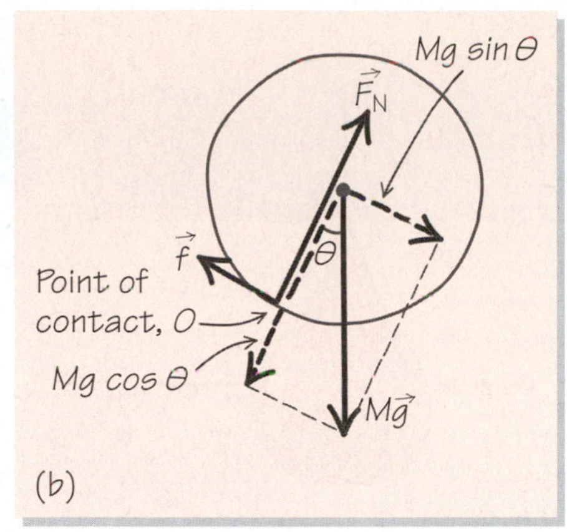

FIGURE 10–13 (a) A cylinder rolls without slipping down an inclined plane. (b) Force diagram for the rolling cylinder.

contact point O between the cylinder and the ramp (Fig. 10–13b). As described prior to the example, with this reference point only gravity produces a torque, and this is easily calculable. The angular momentum in this situation is a little more complex, and we turn to the technique described above and expressed by Eq. (10–13). This expression utilizes the momentum $\vec{P}$ of the center of mass, as well as the angular momentum about the symmetry axis of the cylinder, which has magnitude $L = I_{cm}\omega$. Finally, once we know the torque and hence the rate of change of the two terms in the angular momentum about the axis A, we will have the angular acceleration and can use kinematic relations to find angular speed as a function of time.

Working It Out Using the right-hand rule, we see that the torque due to gravity about point O points into the page and has magnitude

$$\tau = (Mg\sin\theta)R. \tag{10–16}$$

Equation (10–13) gives us the angular momentum about O. This expression contains two terms. First, there is the angular momentum of the total mass as though that mass were placed at the object's center of mass, $\vec{R}\times\vec{P} = \vec{R}\times(M\vec{V})$. $\vec{R}$ and $\vec{V}$ are perpendicular to each other, and the magnitude of $\vec{R}$ is R_0, so this term has magnitude

$$MVR_0 = M(\omega R_0)R_0 = MR_0^2\omega. \tag{10–17}$$

Second, there is the angular momentum about the symmetry axis passing through the center of mass, magnitude $L = I_{cm}\omega$. This is in the same direction as the $\vec{R}\times\vec{P}$ term above, so the magnitude of the total angular momentum about the point of contact is the sum of Eq. (10–17) and $I_{cm}\omega$:

$$L = (MR_0^2 + I_{cm})\omega. \tag{10–18}$$

Equations (10–16) and (10–18) are the torque and angular momentum, respectively, about the same reference point. The dynamical equation $dL/dt = \tau$ is therefore applicable:

$$(MR_0^2 + I_{cm})\frac{d\omega}{dt} = MgR_0\sin\theta;$$

$$\alpha = \frac{d\omega}{dt} = \frac{MgR_0\sin\theta}{MR_0^2 + I_{cm}}.$$

This equation states that the magnitude of angular acceleration α is constant, so the angular speed increases linearly with time:

$$\omega(t) = \omega_0 + \alpha t = \omega_0 + \frac{MgR_0\sin\theta}{MR_0^2 + I_{cm}}t. \tag{10–19}$$

This is the desired answer.

Comparison with the Chapter 9 treatment of this problem is useful. For a symmetric rolling object, $I_{cm} = CMR_0^2$, where C is a numerical constant (Section 9–3). Then Eq. (10–19) tells us that the linear speed $v = \omega R_0$ of the cylinder center is

$$v(t) = v_0 + \frac{g\sin\theta}{1 + C}t. \tag{10–20}$$

This result is the same one found in Eq. (9–52). In Chapter 9 we used the center of mass as a reference point, while here we use the contact point of the rolling cylinder and the plane. Compare this example with the discussion of Eq. 9–52, and consider which, if any, is the preferable approach. Having a choice of reference point is always valuable.

Angular Impulse

The angular impulse for the motion of extended systems is useful in the same way as the linear impulse for the motion of point masses. The linear impulse, $\Delta\vec{p} \equiv \vec{J} = \vec{F}\,\Delta t$, describes the change in momentum when a force $\vec{F}$ acts for a short duration Δt. Similarly, the dynamical equation for rotational motion tells us how the change in the angular momentum of a system, $\Delta\vec{L} \equiv \vec{J}_\tau$ (the **angular impulse**), is described by a torque $\vec{\tau}$ acting for a duration Δt:

$$\Delta\vec{L} \equiv \vec{J}_\tau = \vec{\tau}\,\Delta t. \tag{10–21}$$

Here, both the torque and angular momentum are measured with respect to the same reference point.

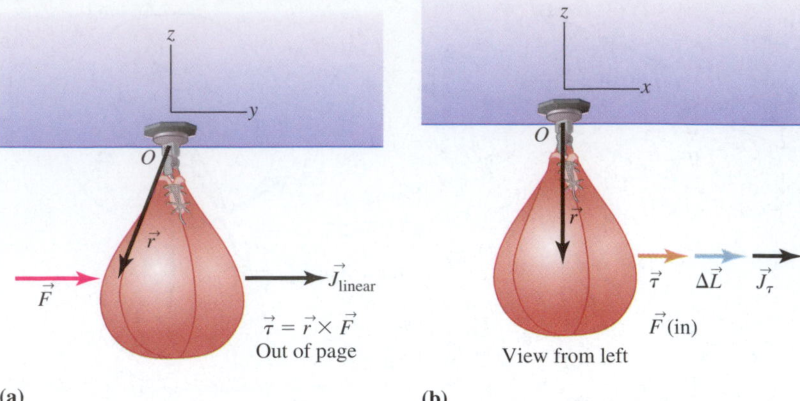

▶ **FIGURE 10–14** The angular impulse delivered to a punching bag. In parts (a) and (b) we draw two views of the situation.

We can illustrate how forces lead to angular impulses with the punching bag of Fig. 10–14. Let the reference point for torque and angular momentum be O. The force, $\vec{F}$, is applied briefly at $\vec{r}$. The torque is then $\vec{\tau} = \vec{r} \times \vec{F}$; multiplying by Δt, we see that the angular impulse is given by

$$\vec{J}_{\tau} = \vec{\tau}\,\Delta t = \vec{r} \times (\vec{F}\,\Delta t) = \vec{r} \times \vec{J}. \tag{10–22}$$

EXAMPLE 10–7 A law-court expert is called upon for a case in which a stationary automobile is struck from the side. To model this situation, he treats the automobile as a uniform rod of mass M and length ℓ at rest on a frictionless surface. An impulse of magnitude $F\,\Delta t$ is applied at right angles at a distance $\ell/6$ from one end of the rod (Fig. 10–15). Describe the subsequent motion of the rod.

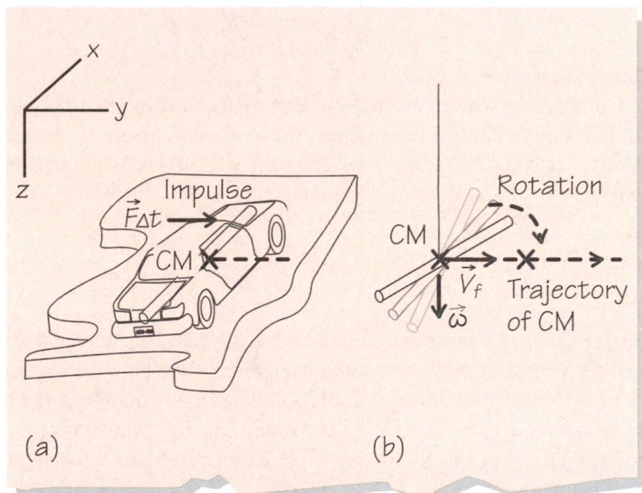

▲ **FIGURE 10–15** The automobile is modeled by a uniform rod. (a) Before: An impulse is given to the stationary rod a distance $\ell/6$ from the top. (b) After: The rod moves with a combination of a linear center-of-mass motion and a rotation about the center of mass.

Setting It Up In Fig. 10–15 the rotation axis goes through the center of mass. Let us label the rod's center of mass momentum and velocity as $\vec{P}$ and $\vec{V}$, respectively, and the angular momentum and angular speed about its center of mass as $\vec{L}$ and $\vec{\omega}$, respectively. We want to find these quantities after the collision (subscript f).

Strategy The motion may be decomposed into motion of the center of mass and rotation about the center of mass. The initial momentum of the center of mass is zero, and therefore its final momentum is given by the linear impulse. We can translate information of the momentum change to details of the speed change, through the mass of the automobile. The angular impulse is equal to the change in angular momentum in time Δt. The rotational inertia of the rod about the rotation axis will then allow us to relate the change in angular momentum to the change in angular speed.

Working It Out With $P_i = 0$,

$$\Delta\vec{P} = \vec{P}_f - \vec{P}_i = \vec{P}_f = \vec{F}\,\Delta t,$$

and the final center of mass speed is

$$\vec{V}_f = \frac{\vec{P}_f}{M} = \frac{\vec{F}\,\Delta t}{M}.$$

Similarly, the angular impulse is equal to the change in angular momentum in time Δt. The initial angular momentum is zero, and the distance from the point of application of the force to the center of mass is $(\ell/2)-(\ell/6)$. Thus

$$\Delta\vec{L} = \vec{L}_f - \vec{L}_i = I\vec{\omega}_f = (F\,\Delta t)\left(\frac{\ell}{2} - \frac{\ell}{6}\right)\hat{k} = \frac{(F\,\Delta t)\ell}{3}\hat{k},$$

pointing down into the street. $I = M\ell^2/12$ (see Table 9–1), so the angular speed can be found:

$$\frac{M\ell^2\omega_f}{12} = \frac{(F\,\Delta t)\ell}{3} \quad \text{or} \quad \omega_f = \frac{4}{M\ell}(F\,\Delta t) = \frac{4}{\ell}V_f$$

The final linear and angular motions are shown in Fig. 10–15.

What Do You Think? Suppose the car was struck at its center rather than off center. Would the center of mass move off with (a) more (b) less or (c) the same velocity as in the example?

The fact that angular impulse, torque, and angular momentum are vectors has some consequences that are not very intuitive, as we can see in the following conceptual example. Keep in mind as you look at this example that the laws that govern rotational motion follow directly from the laws for linear motion.

CONCEPTUAL EXAMPLE 10–8 A bicycle wheel is rotating about an extended axle (Fig. 10–16), with its initial angular momentum $\vec{L}_i$ aligned horizontally in the xy-plane, and with one end of the axle attached to a pivot about which it can move freely. The other end of the axle is struck by a downward hammer blow, receiving an impulse $\vec{J} = -J\hat{k}$ (Fig. 10–16a). What is the immediate response of the wheel to the impulse, assuming that gravity can be neglected in the short run?

Answer The change in the wheel's angular momentum is the angular impulse. We can find this quantity using the fixed end (pivot point) of the axle as a reference point. If the vector from the fixed end to the point where the impulse is applied is $\vec{\ell}$, then the angular impulse is

$$\Delta\vec{L} = \vec{J}_\tau = \vec{\ell} \times \vec{J}.$$

By a right-hand rule, the direction of this vector lies in the xy-plane as shown (Fig. 10–16b). Even though the blow comes from above, *the free end of the wheel tends to rotate back toward the y-axis!* The wheel moves in a direction perpendicular to the direction of the blow. The result we have found here is a surprising consequence of the vector nature of angular momentum and torque. This counterintuitive motion is echoed by a similar phenomenon in the motion of rotating objects (gyroscopes) under the effect of gravity (Section 10–8).

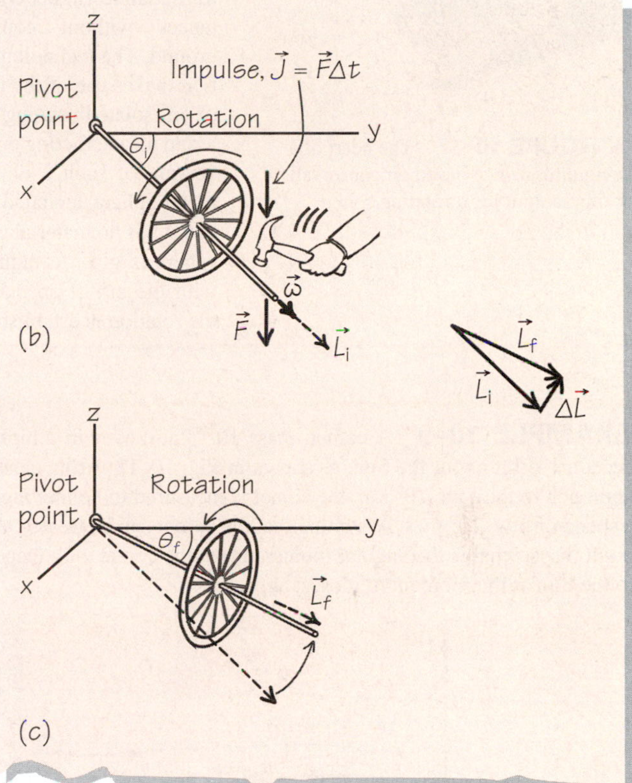

▶ **FIGURE 10–16** (a) Rotating wheel precesses about string. (b) A rapidly rotating bicycle wheel is spinning with its angular velocity, $\vec{\omega}$, and its initial angular momentum, $\vec{L}_i$, aligned as shown. One end of the axis is fixed on a pivot point, and the other end is given an impulse $\vec{J} = -J\hat{k}$. (c) When a corresponding angular momentum change, $\Delta\vec{L}$, is added to the initial angular momentum, the result is that the free end of the wheel tends to rotate toward the y-axis.

10–4 Conservation of Angular Momentum

The general dynamical equation for rotational motion is $\vec{\tau}_{\text{net}} = d\vec{L}/dt$, where the torque always refers to *external* forces. If there is no net external torque on a system—rigid or otherwise—the angular momentum of the system is *conserved*,

$$\frac{d\vec{L}}{dt} = 0, \qquad (10\text{–}23)$$

meaning that the angular momentum is constant during the motion of the system. This fact has a variety of powerful consequences that we shall explore in this section.

Angular Momentum and Central Forces

Central forces—forces directed along a line from a given source—provide examples of forces with no torque about a special point. In this case, the point is the source of the force itself. For example, the Sun is responsible for a central force on the planets, and this gravitational force is directed toward the Sun itself. This force exerts no torque on the planets about the Sun, and the angular momentum of the planets as they orbit the Sun is therefore constant. Because both gravitation and electrostatic forces (such as the force holding electrons to nuclei in atoms) are central, central forces are of particular importance in the physical world. Example 10–9 illustrates how to find angular momentum for an object under the influence of a central force.

THINK ABOUT THIS . . .

HOW DOES A SUBMARINE NAVIGATE 'SILENTLY'?

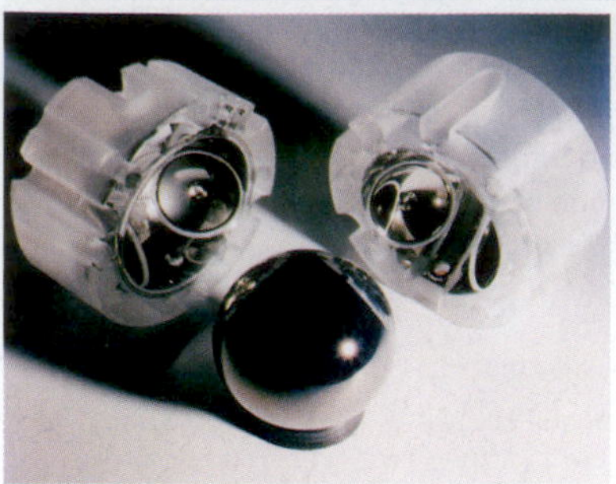

▲ **FIGURE 10–17** The heart of a navigating device based on conservation of momentum for a rotating sphere.

Nuclear submarines can navigate with great accuracy without communication with anyone or anything above the surface, an obvious necessity given what submarines do. Similarly, even before the advent of the Global Positioning System or other satellite-based communication, transatlantic flights could fly across great distances without communication with the ground. The technology used for silent navigation makes use of the fact that the rotation axis of an isolated rotating object (one with no external torques acting on it) will not change. One version of such a device consists of a small metal sphere levitated by a magnetic field so that it has no material contact (Fig. 10–17). The sphere is given a high initial angular velocity with the aid of special magnetic fields. Once the rotation is established, the sphere's rotation axis remains aligned in a constant spatial direction. As the sphere does not touch anything, there are no mechanical bearings, and so there are no vestigial torques that can cause the sphere's axis to rotate. The sphere is engraved with black and white marks that can be read with the aid of photocells. When the submarine changes direction, the position of these marks relative to a casing attached to the submarine will change. The photocells and an associated computer keeping track of the marks will observe the sphere's rotation axis move relative to the submarine and can measure the exact degree of the motion. The measurement is a way to keep track of any changes in the submarine's direction of motion throughout the voyage. Such devices can provide positioning within meters over great distances. ■

EXAMPLE 10–9 A comet, mass 10^{15} kg, moves in a highly eccentric orbit about the Sun, as shown in Fig. 10–18. At its closest approach to the Sun, 10^6 km, the comet is measured to be moving at a speed of 6×10^6 m/s. Assuming that the comet can be treated as a point mass, what is the angular momentum of the comet with respect to the Sun at the moment of closest approach?

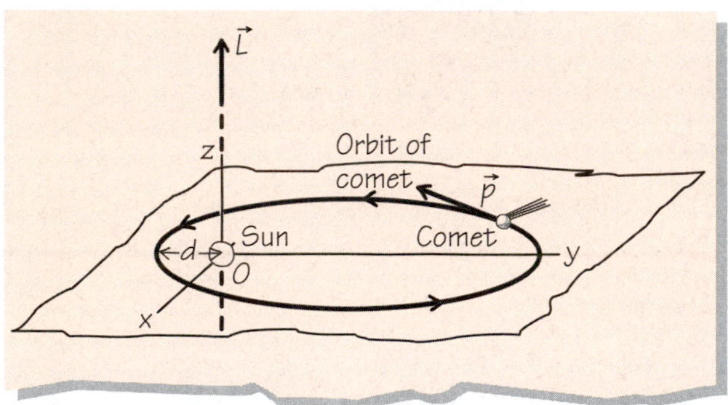

▲ **FIGURE 10–18** The direction of the angular momentum is found from a right-hand rule to be up out of the plane of the orbit, in the $+z$-direction.

Setting It Up We know the comet's mass M, the distance d of its closest approach to the Sun, and its speed v at that point.

Strategy The angular momentum is given by the linear momentum $p = Mv$ times the distance of closest approach, i.e., magnitude $L = Mvd$, so the problem can be solved through simple substitution. We can also use a right-hand rule to find the direction of the angular momentum.

Working It Out The angular momentum has magnitude

$$L = Mvd = (10^{15}\text{ kg})(6 \times 10^6\text{ m/s})(10^6\text{ km})[(10^3\text{ m})/(1\text{ km})]$$
$$= 6 \times 10^{30}\text{ kg}\cdot\text{m}^2/\text{s}.$$

The direction of the angular momentum is given by a right-hand rule. In Fig. 10–18, this direction is up out of the plane of the orbit, in what is labeled the $+z$-direction. (Below we'll see why angular momentum is conserved for central forces, so that the angular momentum at the moment the comet is closest to the Sun is its angular momentum throughout its orbital motion.)

What Do You Think? Suppose the comet of the example has an orbit that is much less eccentric (i.e., more nearly circular) with speed at closest approach of 12×10^6 m/s and a distance of closest approach 0.5×10^6 m. Does the comet have (a) more angular momentum, (b) the same angular momentum, or (c) less angular momentum than in the example?

The torque on a particle is zero if the force and the displacement vector $\vec{r}$ from the point of reference (the origin) to the point of application of the force are *parallel* (or antiparallel), because then the vector product between $\vec{r}$ and $\vec{F}$ is zero. This applies for central forces, so *for central forces the angular momentum about the origin is conserved*, as in Example 10–9.

Consider an object moving under the influence of a central force, and take the origin of the coordinate system to be at the source of this force. For example, let us think about the comet of Example 10–9 moving under the influence of the Sun. Let the initial position and velocity of the comet be $\vec{r}_0$ and $\vec{v}_0$, respectively. The initial direction of the angular momentum is given by $\vec{r}_0 \times \vec{v}_0$, that is, perpendicular to the plane formed by $\vec{r}_0$ and $\vec{v}_0$. Because the angular momentum is constant and its direction does not change, *the comet's motion is always confined to the initial plane formed by* $\vec{r}_0$ and $\vec{v}_0$. To obtain the magnitude of the angular momentum, consider the path shown in Fig. 10–19a.

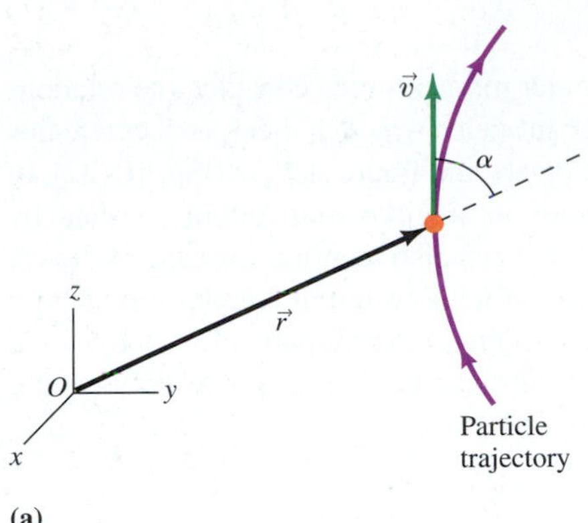

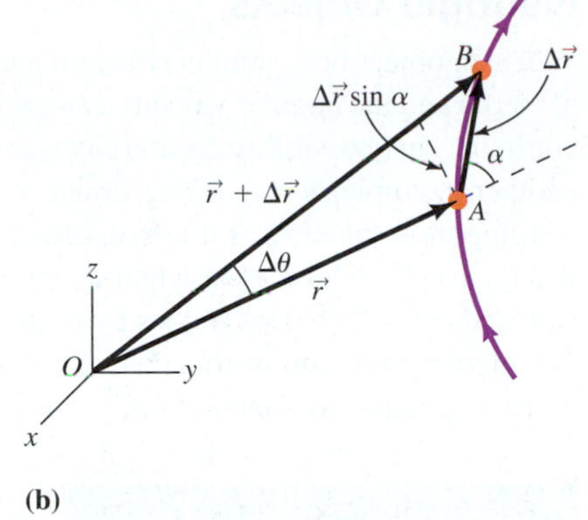

FIGURE 10–19 The trajectory of a particle with constant angular momentum. (a) The velocity at any given time is tangent to the trajectory and makes an angle α with the position vector. (b) In a small time interval, the particle moves from point A to point B and is displaced by $\Delta\vec{r}$. The angle $\Delta\theta$ is approximately $\Delta\theta = (|\Delta\vec{r}|\sin\alpha)/r$.

At any given moment t, the comet is at position $\vec{r}$ and is moving with a velocity $\vec{v}$ tangent to the trajectory. The magnitude of the angular momentum is given by

$$L = |\vec{r} \times \vec{p}| = |\vec{r} \times m\vec{v}| = rmv \sin\alpha, \tag{10–24}$$

where α is the the angle between $\vec{r}$ and $\vec{v}$. If the object's position changes from $\vec{r}$ to $\vec{r} + \Delta\vec{r}$ in a small time interval Δt as in Fig. 10–19b, and if we replace $\vec{v}$ by $\Delta\vec{r}/\Delta t$, we find

$$L = mr\frac{|\Delta\vec{r}|\sin\alpha}{\Delta t} = mr\frac{r\,\Delta\theta}{\Delta t}. \tag{10–25}$$

We have used the geometric relation $|\Delta\vec{r}|\sin\alpha = r\,\Delta\theta$. In the limit $\Delta t \to 0$, Eq. (10–25) becomes

$$L = mr^2\frac{d\theta}{dt} = mr^2\omega, \tag{10–26}$$

where ω is the instantaneous angular velocity about the origin. Thus $r^2\omega$ is constant throughout the motion when, as here, the angular momentum is constant. For uniform circular motion both r and ω are constant, and there is nothing new in Eq. (10–26)—this is simply the circular motion described in Section 3–5. If r is not constant, then Eq. (10–26) can be interpreted as follows: Consider the time interval Δt and the geometry of Fig. 10–19b. For infinitesimal displacements Δr, the area of the triangle OAB is given by

$$\Delta A = \frac{1}{2}(r\,\Delta\theta)r = \frac{1}{2}r^2\,\Delta\theta, \tag{10–27}$$

from which it follows that $dA = \frac{1}{2}r^2\frac{d\theta}{dt}dt = \frac{1}{2}r^2\omega dt$. In other words,

$$\frac{dA}{dt} = \frac{1}{2}r^2\omega = \frac{L_0}{2m}, \tag{10–28}$$

where we have used Eq. (10–26) and L_0 is the constant value of the angular momentum. Whatever the trajectory of a particle moving under the influence of a central force (which depends on the detailed form of the central force), it follows from angular momentum conservation that *the rate at which the radius vector sweeps out an area is constant*. In the context of the gravitational force (Chapter 12), this is known as **Kepler's second law**. With that force a comet moves in an elliptical orbit with the Sun located at one of the focal points; this orbit sweeps out equal areas in equal lengths of time (Fig. 10–20). The comet must move rapidly when it is near the Sun compared with its speed at large distances from the Sun.

Another example of a particle that moves with no torque on it is a freely moving particle such as a puck sliding on ice. The reference point is any point whatever, and this motion too satisfies the condition of Eq. (10–28), i.e., its position vector, as measured from any point, sweeps out each area at a uniform rate. This is demonstrated in Fig. 10–21.

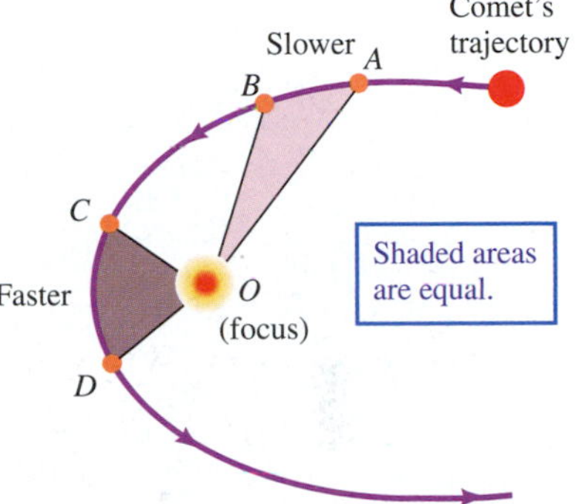

FIGURE 10–20 If a particle moves with constant angular momentum, then its trajectory sweeps out equal areas in equal times. If the motion of a comet from A to B takes time Δt and the motion from C to D takes the same amount of time, then the particle must move more quickly in traveling from C to D.

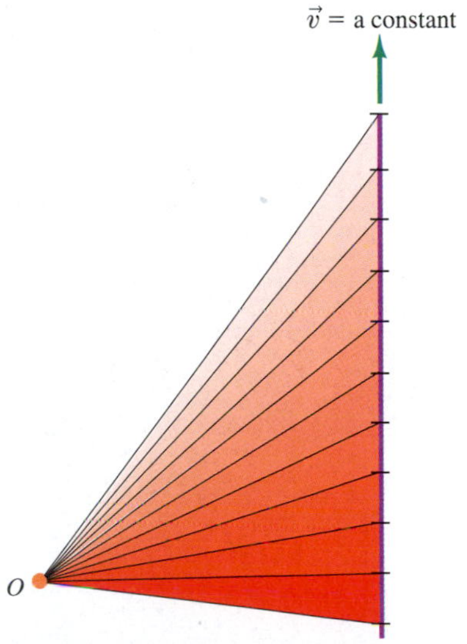

FIGURE 10–21 The triangles that a uniformly moving particle sweeps out with respect to any point O have equal areas.

Nonrigid Objects

When an object or system is not rigid and its angular momentum is constant, the rotational inertia and the angular velocity can each change in such a way that their product remains constant. This possibility commonly occurs with divers and figure skaters (Fig. 10–22). In another example, water going down a drain holds its angular momentum constant by swirling more quickly as it approaches the drain. We can also mention the case of the cat in Fig. 10–22, which even when dropped back down with no initial angular momentum can land on its feet. This it does by rotating and extending different parts of its body while the angular momentum remains zero (as it must—the cat has no means to do anything about its angular momentum once it has been dropped).

(a) (b)

▲ **FIGURE 10–22** (a) A cat can maneuver in midair while conserving angular momentum. (b) Although a figure skater spins more slowly when her arms are extended than when her arms are close to her body, her angular momentum remains constant.

EXAMPLE 10–10 In a common classroom experiment, a student sits on a spinning stool with weights in each hand. An idealized version of this experiment is as follows: a solid cylinder of diameter 0.5 m and mass 50 kg is oriented vertically and spins freely about its axis with a period of 3 s. Two massless rods are attached horizontally to the cylinder, with their ends 1 m from the surface, and there is a mass 2 kg at the end of each rod (Fig. 10–23a). The rods are drawn into the cylinder by an internal mechanism until the 2-kg masses are at the surface of the cylinder (Fig. 10–23b). What are the initial and final angular velocities?

Setting It Up We know the diameter d and mass M of the spinning cylinder, as well as the length D of each of the attached rods; the masses m at the end of each cylinder are also given, as is the initial period of rotation T. We want to find the initial and final angular speeds, ω_0 and ω_f, respectively.

Strategy The angular momentum $L = I\omega$ about the symmetry axis of the cylinder is constant because there are no external torques. (The force pulling in the "arms" is *internal* and *radial*, each property by itself sufficient to ensure that there is no torque about the central axis.) We can calculate the initial and final rotational inertia I_0 and I_f by simple geometry, and we can find ω_0 from its kinematic connection with period. The condition that L is conserved then determines ω_f. Physically, we can anticipate as a check that as the masses are pulled in, I decreases; therefore, ω must increase.

Working It Out The conservation of angular momentum reads

$$I_0\omega_0 = I_f\omega_f,$$

or

$$\omega_f = \omega_0\frac{I_0}{I_f}.$$

I_0 is the rotational inertia of a solid cylinder ($MR^2/2$ from Table 9–1) of mass $M = 50$ kg and radius $R = d/2 = 0.25$ m, plus the rotational inertia of the two masses of 2 kg a distance $D + R = 1.25$ m from the axis:

$$I_0 = \tfrac{1}{2}MR^2 + 2m(D + R)^2.$$

I_f is found the same way, except the two masses are a distance R from the axis:

$$I_f = \tfrac{1}{2}MR^2 + 2mR^2 = [(M/2) + 2m]R^2.$$

The initial angular velocity is found from the period, $\omega_0 = 2\pi/T = 2\pi/(3\text{ s}) \cong 2$ rad/s. Thus

$$\begin{aligned}\omega_f &= \omega_0\frac{(MR^2/2) + 2m(D + R)^2}{[(M/2) + 2m]R^2}\\ &= (2\text{ rad/s})\frac{(0.5)(50\text{ kg})(0.25\text{ m})^2 + 2(2\text{ kg})(1.25\text{ m})^2}{(0.5)(50\text{ kg})(0.25\text{ m})^2 + 2(2\text{ kg})(0.25\text{ m})^2}\\ &= 9\text{ rad/s}.\end{aligned}$$

The angular velocity has increased by a factor of 4.5. The movement of the end weights into the axis is remarkably effective at decreasing the rotational inertia! You can perform this demonstration in class (Figs. 10–23c and 10–23d).

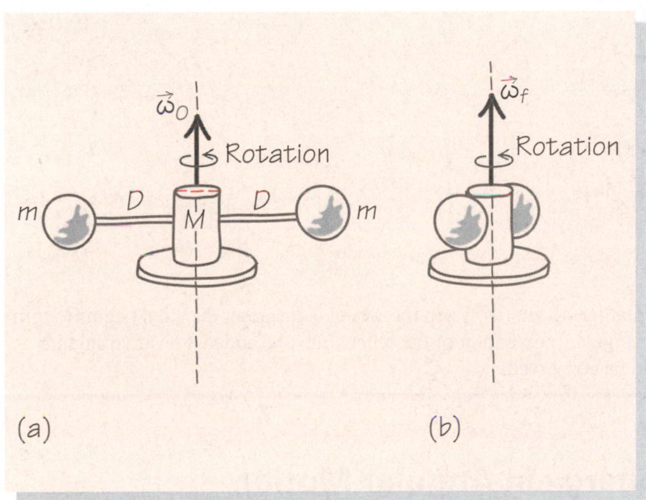

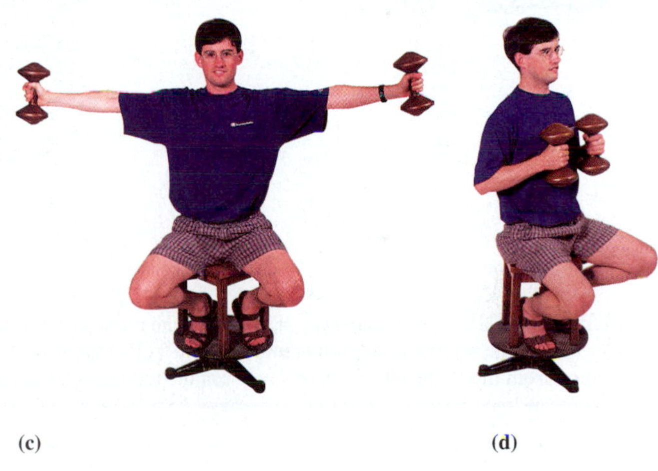

▲ **FIGURE 10–23** (a) An object rotates with masses at the end of a rod. (b) When the masses are brought in closer to the rotation axis, the conservation of angular momentum requires the rotation to speed up. (c), (d) A demonstration of this.

An Astrophysical Example: A supernova occurs when a massive star stops burning because it has used up its fuel and undergoes a collapse due to gravitational forces, gaining enough energy in the process to blow off most of its material in a stupendous explosion (Fig. 10–24). What is left behind is a *neutron star*: an extremely dense sphere of matter with a mass of several solar masses and a diameter of perhaps 10 km (compared with 10^6 km for the original star). If the original star is rotating, as most stars do, at some rate on the order of several days, and if the outer layers are blown off nearly radially, then the neutron star will have a good part of the original angular momentum, and because it has so much mass packed into a tiny sphere, its angular speed will be very large. Many neutron stars so produced are known to have periods on the order of milliseconds.

▲ **FIGURE 10–24** At the center of the outwardly moving remnants of the Crab Nebula supernova, a neutron star spins rapidly.

CONCEPTUAL EXAMPLE 10–11 Suppose that a spinning bicycle wheel with its angular velocity pointing up (the wheel spins counterclockwise as seen from above) is handed to a student sitting on a stool that can freely rotate about the vertical axis (Fig. 10–25a) but is otherwise attached to the ground. The student turns the wheel upside down, so that its angular velocity points downward. Which of the following is true? (a) The student will be forced to the right and off the stool by an angular impulse; (b) the stool (with the student) will also pick up a downward angular velocity; (c) the stool (with the student) will pick up an upward angular velocity; (d) the wheel is isolated from the student and nothing happens to the student.

Answer The answer is (c). An upward angular velocity corresponds to an upwardly directed angular momentum. Because the system of student, stool, and wheel is isolated as far as vertical rotations are concerned, the z-component of angular momentum is conserved. Before the wheel is turned, the total angular momentum is oriented up. After the wheel is turned, the wheel has an angular momentum oriented down, so the student and stool must have enough angular momentum oriented up to keep the total angular momentum unchanged. That means the student-stool spins counterclockwise—the original direction of rotation of the wheel (Fig. 10–25b). This new spin direction is generated to conserve the total angular momentum (Fig. 10–25c).

If you trace the motion through, you can see that there was a moment when the wheel was oriented with its angular momentum in the horizontal direction. At that moment the only way the entire system could have no net angular momentum in the horizontal plane is for the student and stool to rotate over! The attachment point of the stool will in fact exert a compensating torque to keep things upright—in other words, the system is not truly isolated from its surroundings for this situation.

What Do You Think? Suppose the student now flips the wheel back to its starting orientation. What will the result be?

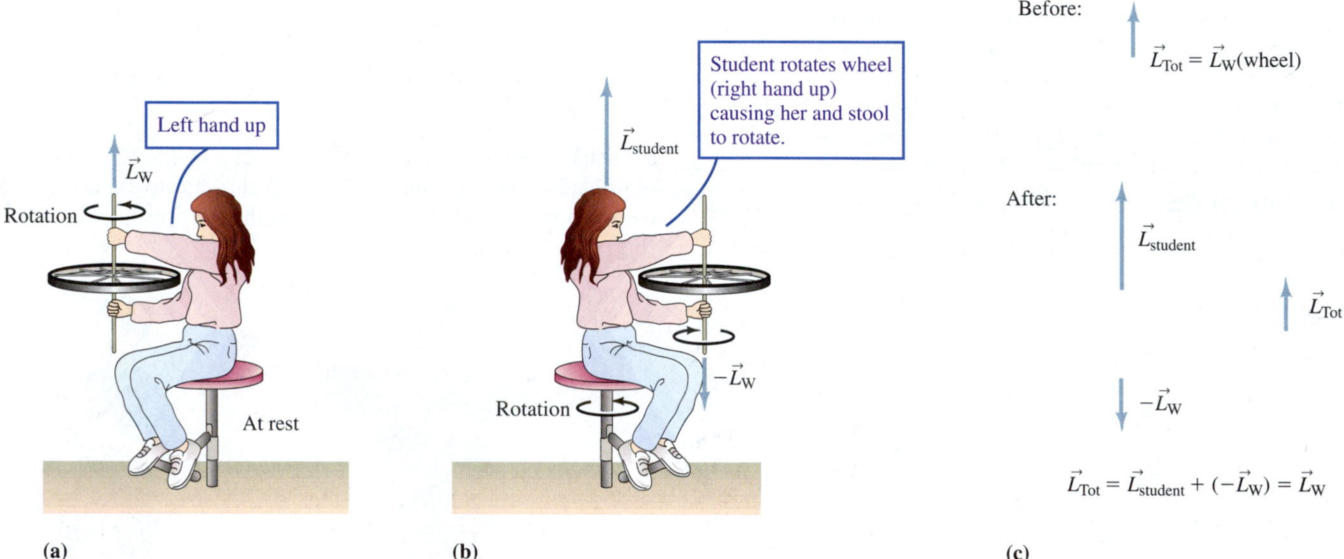

▲ **FIGURE 10–25** (a) A student on a stool that can rotate holds a freely rotating wheel. (b) When the wheel is inverted, the stool begins to turn in order to hold the total angular momentum constant. (c) To the now negative angular momentum of the wheel must be added a positive angular momentum from the student. Only in this way can the total angular momentum be conserved.

10–5 Work and Energy in Angular Motion

Rotating systems of particles, even those that are nonrigid, have energy, and work has to be done on them to change their energy. For example, the motor must do work to start an airplane's propeller, and gravity does work on a ball that rolls down a hill. Both of these examples are symmetric, rigid objects, the subject of Chapter 9. There we saw that if the object has rotational inertia I and angular speed ω about the axis of rotation, its energy is

$$K = \frac{1}{2}I\omega^2. \qquad (10\text{–}29)$$

The angular velocity is a vector, which can be expressed as

$$\omega = \omega_x \hat{i} + \omega_y \hat{j} + \omega_z \hat{k} \qquad (10\text{–}30)$$

and

$$\omega^2 = \vec{\omega} \cdot \vec{\omega} = \omega_x^2 + \omega_y^2 + \omega_z^2. \qquad (10\text{–}31)$$

If a torque is present, the rotational kinetic energy of the body will change because the angular speed changes. We calculate the rate of change of energy, that is, the *instantaneous power* that must be supplied by whatever changes the energy,

$$\frac{dK}{dt} = \frac{1}{2}I\frac{d}{dt}\omega^2 = I\frac{d\vec{\omega}}{dt}\cdot\vec{\omega} = I\vec{\alpha}\cdot\vec{\omega} = \vec{\tau}\cdot\vec{\omega}, \quad (10\text{–}32)$$

where we have used Eq. (10–12), $I\vec{\alpha} = \vec{\tau}$. Equation (10–32) is the analog of the linear motion equation for power that we saw in Chapter 6: $\frac{dK}{dt} = \vec{F}\cdot\vec{v}$.

The Work–Energy Theorem for Rotations

It is fairly straightforward to derive a work–energy theorem for rotational motion *about a fixed axis*. To define the work, we first follow the analog with linear motion once more. Just as the infinitesimal work in linear motion is defined as $dW = \vec{F}\cdot d\vec{x}$, the infinitesimal work done in rotating a rigid body through an infinitesimal angle $d\theta$ about the axis is defined to be

$$dW \equiv \vec{\tau}\cdot d\vec{\theta}. \quad (10\text{–}33)$$

INFINITESIMAL WORK DONE BY TORQUE

The infinitesimal angle is a vector whose direction is defined by a right-hand rule: If the fingers curl in the direction of the infinitesimal angle, the thumb direction gives the vector direction. When the torque points along the axis of rotation, this reduces to the form

$$dW = \tau\, d\theta. \quad (10\text{–}34)$$

Exactly as in Chapter 6, the test as to whether this is a reasonable definition for the work is that it leads to an appropriate work–energy theorem. To verify this, we use $\tau = I\, d\omega/dt$ and $d\theta = \omega\, dt$, and hence

$$W = \int_{\theta_0}^{\theta} \tau\, d\theta = \int_0^t I\frac{d\omega}{dt}\omega dt = \int_0^t I\frac{1}{2}\frac{d\omega^2}{dt}dt$$

$$= \frac{1}{2}I\int_{\omega_0^2}^{\omega 2} d\omega^2 = \frac{1}{2}I(\omega^2 - \omega_0^2) = K - K_0. \quad (10\text{–}35)$$

The work done is indeed the change in kinetic energy. We have found the usual work–energy theorem for rotational motion about a fixed axis when the torque is directed along the axis.

The Energy of an Extended Object in Motion

We next consider the energy of an extended object that may or may not be rigid. First, we break up the object into a number of individual pieces with masses m_i located at $\vec{r}_i$ relative to some origin. We introduce the center of mass position $\vec{R}$ and the coordinates $\vec{\rho}_i$ relative to the center of mass according to $\vec{r}_i = \vec{\rho}_i + \vec{R}$, as in Fig. 10–11. When we take a derivative with respect to time, we have

$$\vec{v}_i = \frac{d\vec{r}_i}{dt} = \frac{d\vec{R}}{dt} + \frac{d\vec{\rho}_i}{dt} = \vec{V} + \vec{u}_i, \quad (10\text{–}36)$$

where $\vec{u}_i \equiv d\vec{\rho}_i/dt$ are the velocities of the masses m_i with respect to the center of mass and $\vec{V}$ is the velocity of the center of mass. We can then show that the kinetic energy of the system takes the form

$$K = \sum_i \frac{1}{2}m_i\vec{v}_i^2 = \frac{1}{2}MV^2 + \frac{1}{2}\sum_i m_i\vec{u}_i^2. \quad (10\text{–}37)$$

(We do not prove this result here.) The total kinetic energy splits into two parts: the kinetic energy of the total mass of the object moving with the velocity of the center of mass, and the kinetic energy of the motion relative to the center of mass. By now this should be a familiar type of split.

The kinetic energy of Eq. (10–37) takes on a familiar form for a *rigid* body. In that case, the vectors $\vec{\rho}_i$ have a fixed magnitude, and only the angular variable changes. For rotations with angular speed ω about an axis through the center of mass, the speed of any point that is a radial distance r from the axis is given by $v = \omega r$. Thus $u_i = \omega \rho_i$, and $u_i^2 = \omega^2 \rho_i^2$. We can substitute this expression into Eq. (10–37) to find that

$$K = \frac{1}{2}MV^2 + \frac{1}{2}I_{cm}\omega^2. \tag{10–38}$$

Equation (10–38) shows that *the total kinetic energy of a rigid body consists of the kinetic energy of the total mass moving with the velocity of the center of mass together with the rotational kinetic energy of the object rotating about an axis passing through the center of mass*. We had already derived this result [see Eq. (9–43)] for the special case of rolling objects. Now we see that it is more general. This result is useful in calculations involving a combination of rotational and linear motion, as Example 10–12 shows.

EXAMPLE 10–12 A spool of thread of total mass M, rotational inertia I about its axis, and radius R falls and unwinds under the force of gravity. By using energy considerations, find the speed of the spool's center of mass after it has unwound a length h of thread.

Setting It Up Figure 10–26 is a sketch of the situation.

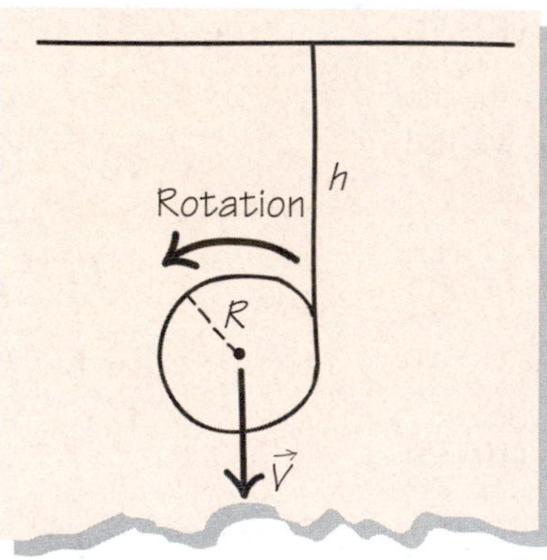

▲ **FIGURE 10–26** A spool of thread unwinds as it falls.

Strategy This example asks explicitly for the use of the conservation of energy. The falling object is subject to a conservative force, gravity, so that it has both a potential energy and a kinetic energy, the sum of which is conserved. As the spool falls it loses potential energy and gains kinetic energy. This situation is different from that of a falling object in that the kinetic energy consists of both rotational and linear terms. Because the thread unwinds, the motion is effectively that of a rolling object, and the center of mass speed v and the angular speed ω of the spool about the symmetry axis are related by $v = R\omega$.

Working It Out We begin with the total kinetic energy:

$$K = \frac{1}{2}Mv^2 + \frac{1}{2}I_{cm}\omega^2.$$

With the constraint $v = R\omega$,

$$K = \frac{1}{2}MR^2\omega^2 + \frac{1}{2}I_{cm}\omega^2 = \frac{1}{2}(MR^2 + I_{cm})\omega^2.$$

For the potential energy of gravity, we set the zero of the potential energy at the initial height of the spool. Thus if the spool falls a distance h, the potential energy is $-Mgh$ and the spool will have gained a kinetic energy Mgh. With the spool starting from rest, the kinetic energy after the spool has fallen a distance h is

$$K = Mgh.$$

Thus

$$\tfrac{1}{2}(MR^2 + I_{cm})\omega^2 = Mgh.$$

We can solve this expression for ω and hence for v:

$$v = R\omega = R\sqrt{\frac{2Mgh}{MR^2 + I_{cm}}}.$$

Alternative Approach A simple application of the parallel-axis theorem shows that the quantity in parentheses in the expression for the kinetic energy is the rotational inertia about the point at which the thread separates from the spool. So we can find the kinetic energy by treating the motion as pure rotation about that point.

What Do You Think? Imagine two spools with the same mass and radius but different mass distributions: Spool A is solid, while spool B is hollow. Which of these two spools would have a larger speed after dropping as in the example a distance h?

10–6 Collecting Parallels Between Rotational and Linear Motion

Through both Chapters 9 and 10 we have emphasized two things: First, there are thus far no new laws of physics to describe rotational motion. Second, there are a series of analogies between the expressions for linear and rotational motion. Of course, these two things are not separate. The analogies appear because aggregate systems are made up of individual pieces that separately obey the laws of linear motion discovered by Newton. At this point it is useful to gather all the analogies into Table 10–1. Study of this table is well worthwhile.

TABLE 10–1 • Analogies Between Linear and Rotational Motion

Linear Motion	Rotational Motion
Infinitesimal linear displacement: $d\vec{r}$	Infinitesimal angular displacement: $d\vec{\theta}$
Velocity: $\vec{v} = \dfrac{d\vec{r}}{dt}$	Angular velocity: $\vec{\omega} = \dfrac{d\theta}{dt}\hat{\omega}$
Acceleration: $\vec{a} = \dfrac{d\vec{v}}{dt}$	Angular acceleration: $\vec{\alpha} = \dfrac{d\vec{\omega}}{dt}$
Momentum: $\vec{p} = m\vec{v}$	Angular momentum: $\vec{L} = I\vec{\omega} = \vec{r} \times \vec{p}$
Force: $\vec{F} = \dfrac{d\vec{p}}{dt}$	Torque: $\vec{\tau} = \dfrac{d\vec{L}}{dt} = \vec{r} \times \vec{F}$
Impulse: $\Delta\vec{p} = \vec{F}\,\Delta t$	Angular impulse: $\Delta\vec{L} = \vec{\tau}\,\Delta t$
Kinetic energy: $\frac{1}{2}mv^2$	Kinetic energy: $\frac{1}{2}I\omega^2$
Work: $\int \vec{F}\cdot d\vec{r}$	Work: $\int \vec{\tau}\cdot d\vec{\theta}$
Power: $\vec{v}\cdot\vec{F}$	Power: $\vec{\omega}\cdot\vec{\tau}$

*10–7 Quantization of Angular Momentum

In systems with atomic dimensions or smaller, the effects of quantum physics become important. One quantum effect is that *angular momentum can have only certain discrete values*—we say that it is *quantized*. Consider an electron that is orbiting the nucleus of an atom, under the influence of the central force between the electron and the nucleus. This force is of the same form as the gravitational force, and if the atom were a classical system (i.e., described with the Newtonian physics we have studied to this point) the electron would behave like a planet orbiting the Sun—in particular, it would have an angular momentum $\vec{L}$ that points along an axis perpendicular to the plane in which it orbits. In Newtonian physics, angular momentum can have any magnitude. Niels Bohr proposed in 1913 that this is in fact incorrect and that the angular momentum can have only a component in the direction perpendicular to the plane of motion (here the z-direction) given by

$$L_z = n\hbar. \tag{10–39}$$

In this quantum mechanical expression, n has *only* the integer values $0, \pm1, \pm2, \ldots,$ and $\hbar$ is Planck's constant, h, divided by 2π, with value $\hbar \simeq 10^{-34}\ \mathrm{J\cdot s}$.

An angular momentum of $10^{-34}\ \mathrm{J\cdot s}$ is not one that is within our common experience; a "typical" macroscopic angular momentum is closer to $1\ \mathrm{J\cdot s}$ than $10^{-34}\ \mathrm{J\cdot s}$. Perhaps you can get an idea of just how small this value of angular momentum is by imagining a uniformly rotating solid wheel of radius 1 cm and mass 20 g—a rotational inertia of $\frac{1}{2}MR^2 = 10^{-6}\ \mathrm{kg\cdot m^2}$. An angular momentum of magnitude $L = 10^{-34}\ \mathrm{J\cdot s}$ would mean an angular speed of $\omega = L/I = 10^{-28}\ \mathrm{s^{-1}}$, which would correspond to a full revolution of the wheel in a time $T = 2\pi/\omega \cong 2 \times 10^{22}$ yr, or about a factor of 10^{12} larger than the lifetime of the universe! To take a second example, and one that is relevant to the real world, imagine a mass m in circular motion of radius of 10^{-10} m about a center and moving with a speed of 1% of the speed of light ($v \cong 3 \times 10^6\ \mathrm{m/s}$). If the angular momentum of the mass about the center had magnitude $L = 10^{-34}\ \mathrm{J\cdot s}$, then the mass would be $m = L/(vR) \cong 3 \times 10^{-31}$ kg. This is nearly the electron mass, and the example corresponds to the atom! In the atomic and molecular worlds, angular momenta of order $\hbar$ really do appear, and Bohr's quantization rule has a very noticeable effect.

We have enough evidence to believe that angular momenta is really quantized, and Eq. (10–39) must apply to macroscopic as well as microscopic systems. So why don't we notice that angular momentum is restricted to certain values in everyday systems? The answer is that the incredibly small step between allowed values is too small to detect in such systems. The typical macroscopic value for angular momentum of 1 J · s is 10^{34} units of $\hbar$. The quantization is not observable in our typical macroscopic system, because its observation would require a determination of the radius or the angular velocity to an accuracy of one part in 10^{34}, an accuracy far beyond the capacities of our most refined instruments. In molecular or atomic systems, however, angular momenta take on values of one or several units of $\hbar$. The difference between, say, 5 and 6 units of $\hbar$ represents 10 to 20 percent of the whole. The fact that the angular momentum is quantized—it has only the discrete values dictated by quantum mechanics—is easily detectable in experiments that measure the energies of the molecule, such as those described next.

Quantization of Energy

In Eq. (9–43) we showed that the kinetic energy of a system with an angular momentum could be expressed in terms of the angular momentum. As we shall see in our more detailed work later, this can be extended to include the total energy. But if the angular momentum is quantized, then we are forced to conclude that *the allowed values of energy of a system such as an electron orbiting an atomic nucleus are quantized as well.*

When the system has one of its possible values of energy, corresponding to different values of n, it is said to be in an **allowed state**. The possible values of energy can then be labeled by the particular value of n—as E_n—and we then refer to a particular **energy level**. These energy levels are detectable through a central fact of our physical world: Atoms emit electromagnetic radiation (light) when an electron makes a transition from one energy level to another one of lower energy. The energy of the radiation is just the difference between the final and initial energy levels—energy is conserved in the transition process. But if the final and initial levels are quantized, then so is the energy of the radiation. We add one more ingredient to this recipe: The energy of electromagnetic radiation is associated with the frequency of that radiation, and our eyes perceive different frequencies as different colors.

(a)

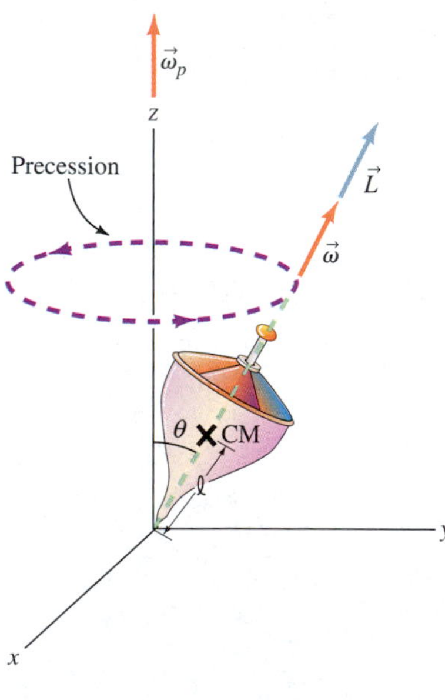

(b)

▲ **FIGURE 10–27** (a) Simple precession. (b) The precession of a top is governed by Eq. (10–41), where ℓ is the distance between the pivot point and the top's center of mass.

Putting all this together, we conclude that as a result of the quantization of angular momentum, light is emitted by atoms in only certain, characteristic frequencies. The frequency of any such radiation is easily measurable by a variety of techniques. It was in fact the observation that atoms and molecules emit light with discrete frequencies that provided one of the keys to the development of quantum mechanics.

The consequences of the discreteness of atomic (and molecular) energies are momentous. For example, there is a minimum energy—an *energy gap*—required to move an atomic system such as a hydrogen atom up the "quantum ladder" from a lowest energy level to higher energy levels. It is the relatively large size of the energy gap that explains the stability of atoms. It is difficult to excite a hydrogen atom. The fact that even large molecular systems have energy gaps is crucial to the stability of biological systems. If there were no energy gaps, even the slightest perturbation would change molecular systems in important ways.

*10–8 Precession

If you have ever played with a spinning top, you will recall that its motion is quite complex. The top spins about its symmetry axis, but if the axis is not oriented to start in the vertical direction, the top of the symmetry axis sweeps out a circle about a vertical through the point where the tip touches the ground, a phenomenon called **precession** (Fig. 10–27). Still another type of motion is a complex up-and-down bobbing motion of the symmetry axis while it rotates about the vertical, a movement called *nutation* (Fig. 10–28). The movement depends on how the motion starts and also on the mass distribution of the top; some tops will actually flip over. While the general motion of tops and their close cousins—gyroscopes—is beyond the scope of this book, we can show you how precession arises. It is helpful to precede this discussion with an example that reminds us of how a torque on an already spinning object can have some rather unintuitive consequences.

FIGURE 10–28 The interplay of gravity, angular momentum, and torque lead to perturbations on the precession of a top, resulting in the up-and-down bobbing motion called nutation.

EXAMPLE 10–13 In making a science-fiction movie about an asteroid that collides with Earth, a studio constructs a model Earth attached firmly to a support fixed at the South Pole and spinning rapidly about the (vertical) south-to-north axis with angular speed ω pointing out of the North Pole. A model asteroid approaches horizontally and strikes a grazing blow with the model Earth just at the North Pole. Describe the new motion of the model Earth.

Setting It Up We show the collision in Fig. 10–29a, with a coordinate system in which the north direction is to $+z$ and the asteroid approaches from the $+y$-direction. The model Earth spins with initial angular velocity $\vec{\omega}$ and angular momentum $\vec{L}_i$. We want to describe the change in both of these quantities as a result of the collision.

Strategy Because the model Earth is fixed at its pivot at the South Pole, we calculate the torque on the model Earth about this point. The collision delivers a linear impulse $\vec{J}$ to the model Earth. We can then use Eq. (10–22) to find the angular impulse $\vec{J}_\tau$ about the support point. This angular impulse describes the change in angular momentum, $\Delta\vec{L}$, of model Earth. The new angular momentum will be the old angular momentum plus this change.

Working It Out Given the direction of the asteroid's motion, the linear impulse delivered to model Earth is $\vec{J} = -J\hat{j}$. Then from Eq. (10–22) the angular impulse about the support point is $\vec{J}_\tau = \vec{r} \times \vec{J}$, where $\vec{r}$ is the vector from the fixed point to the point where the impulse is delivered. Because $\vec{r}$ is in the $+z$-direction, the vector product $\vec{r} \times \vec{J}$ is in the $+x$-direction (Fig. 10–29b). But this describes the change in angular momentum, $\Delta\vec{L}$, produced by the angular impulse. The initial angular momentum $\vec{L}_i$ is in the $+z$-direction, and when $\Delta\vec{L} = (+\Delta L)\hat{i}$ is added to this, the new angular momentum $\vec{L}_f$ corresponds to Earth having tilted toward the $+x$-axis (Fig. 10–29b). Note that this is *not* the direction from which the model asteroid comes.

What Do You Think? Will the magnitude of the model Earth's angular momentum change?

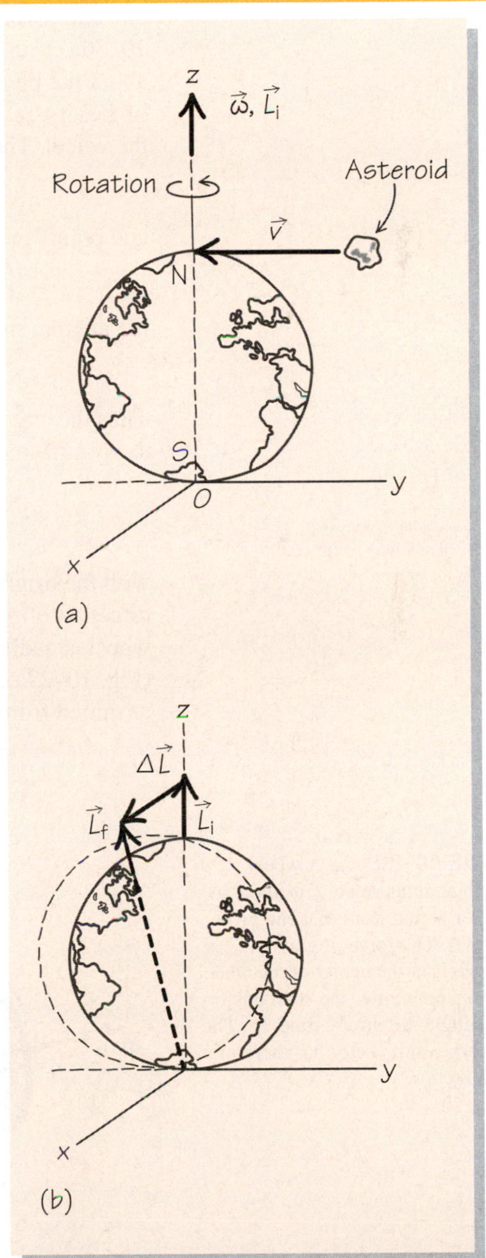

FIGURE 10–29 (a) An asteriod is imagined to hit Earth near the North Pole. (b) The collision of the model asteroid along the $-y$-direction causes the spinning model Earth to tilt toward the $+x$-direction.

Torque on a Spinning Top

In Example 10–13, there is a torque, and hence a change in angular momentum, which is perpendicular to an existing angular momentum. As a result the existing angular momentum changes its direction. This is reminiscent of something that we have seen in linear motion, where there are two rather different ways that linear momentum can change in response to forces. If the net force is parallel to the linear momentum, the momentum will change in magnitude but not direction, and this means a change in speed. But another possibility, which is realized when the acting net force is perpendicular to the momentum, is that the momentum can change *direction* rather than magnitude, as in circular motion. The same is true for rotational motion. If the applied torque is parallel to the angular momentum, then we have changes in the magnitude of the angular momentum, and this represents the majority of the cases we have studied. But if the torque is perpendicular to the existing angular momentum, then we can make that angular momentum change direction without changing its magnitude. With the net torque supplied by gravity, this is what makes a top precess.

To see how this works, consider a top (here in the form of a wheel) of rotational inertia I rotating with angular velocity ω about a horizontal shaft of length ℓ, which has one end pivoted at point A and the other end free. The system is shown at $t = 0$ in Fig. 10–30a. If the sense of rotation of the wheel is as shown, then the angular momentum about the pivot initially has magnitude $L = I\omega$ and points in the x-direction. The force of gravity, magnitude Mg and pointing in the $-z$-direction, acts on the center of mass of the wheel. The torque about point A due to this force has magnitude

$$\tau = Mg\ell \tag{10–40}$$

and points in the $+y$-direction by the right-hand rule. As

$$\Delta\vec{L} = \vec{\tau}\,\Delta t,$$

the angular momentum after a short time interval Δt is given by

$$\vec{L} = I\omega\hat{i} + Mg\ell(\Delta t)\hat{j}. \tag{10–41}$$

Thus the angular momentum vector has rotated slightly in a counterclockwise direction, as seen from above. As Fig. 10–30b shows, the new angular momentum makes an angle

$$\Delta\theta = \frac{Mg\ell\Delta t}{I\omega} \tag{10–42}$$

with the original direction. This is the angle of the shaft, and so we have shown that the direction of the shaft changes with time in a steady way. When the shaft of the spinning wheel steadily rotates with the pivot point fixed, we say that the shaft *precesses* (Fig. 10–27a). The angular velocity of the precession of the angular velocity vector determined from Eq. (10–42) is

$$\omega_p = \frac{\Delta\theta}{\Delta t} = \frac{Mg\ell}{I\omega} \tag{10–43}$$

▶ FIGURE 10–30 (a) A wheel rotates with angular velocity $\vec{\omega}$ about a massless horizontal shaft that can pivot about point A. The force of gravity, $\vec{F}_g = m\vec{g}$, acts at the center of the wheel. (b) As seen from above, the shaft has rotated an angle $\Delta\theta$ after a time Δt. The angular momentum vector has therefore changed direction.

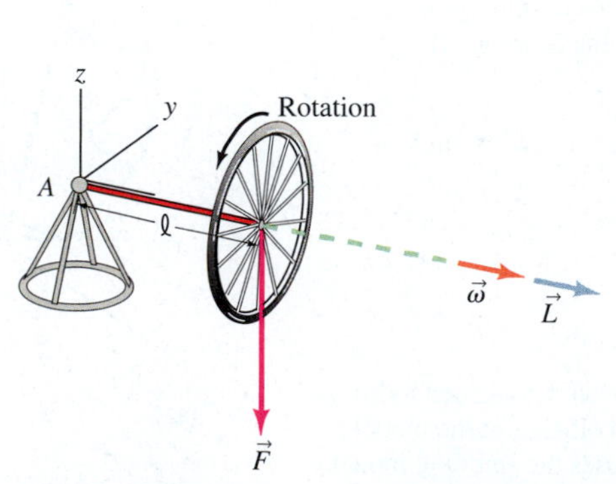

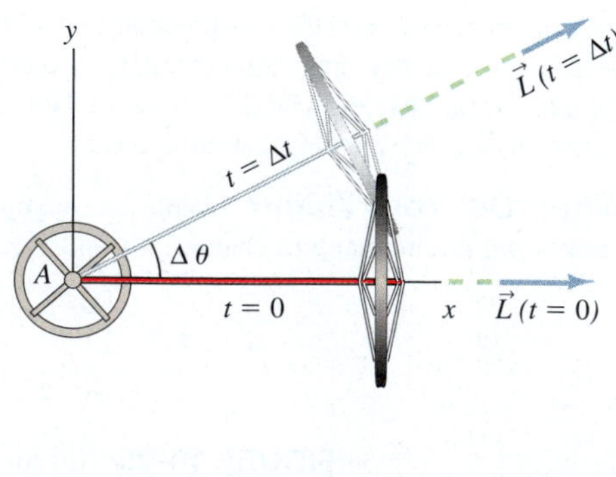

and is known as the *angular frequency of precession*. The precession frequency is, in fact, a vector that points in the $+z$-direction due to the right-hand rule. The precession frequency turns out to be independent of the angle the initial value of $\vec{L}$ makes with the vertical (see Problem 45). This formula applies for any top—not just a rotating wheel—if ℓ is the distance of the top's center of mass from the support point.

Our description of the precession of the wheel's shaft is only approximately correct. A more elaborate treatment involving energy considerations would show how to explain the up-and-down bobbing motion of the shaft called nutation (Fig. 10–28). The effect is small if the angular velocity of the wheel is large or, more precisely, if $\omega_p \ll \omega$. Indeed, the nutation becomes more important when the top slows down, and the wobble exaggerates as the top finishes its motion. Correctly applied, the laws of rotational motion described in this chapter are sufficient to account for all these features.

Summary

A particle moving with momentum $\vec{p}$ has an angular momentum about a point given by

$$\vec{L} = \vec{r} \times \vec{p}, \tag{10–4}$$

where $\vec{r}$ is the position vector from the point to the particle. This result involves the vector product of the vectors $\vec{r}$ and $\vec{p}$. If θ is the angle between these two vectors, the vector product has magnitude $rp \sin \theta$, and its direction—perpendicular to both $\vec{r}$ and $\vec{p}$—is determined by a right-hand rule. The angular momentum of a system of particles, rigid or otherwise, about some point is a sum over terms like Eq. (10–4).

Extended systems obey a dynamical equation that follows from Newton's second law. It states that the rate of change of angular momentum about some reference point O is given by

$$\vec{\tau} = \frac{d\vec{L}}{dt}, \tag{10–11}$$

where $\vec{\tau}$ is the net torque on the system. This net torque is a sum of terms of the type

$$\vec{\tau} = \vec{r} \times \vec{F}, \tag{10–5}$$

where $\vec{r}$ is the vector from the reference point to the point where one of the pieces, $\vec{F}$, of the net force is applied.

There is a simple way to relate the motion of a system whose rotational aspects are described about the center of mass O to a description about some other point A.

(1) The angular momentum consists of two parts. The angular momentum about A is the sum of the angular momentum about A of a point mass carrying the whole mass of the object, as if it were located at the center of mass and moved with the velocity of the center of mass, plus the angular momentum of the object about the center of mass,

$$\vec{L} = (\vec{R} \times \vec{P}) + \vec{L}_{\text{cm}}. \tag{10–13}$$

(2) Similarly, the total torque about A is the sum of the torque about A due to the total external force applied to the center of mass and the torque about the center of mass,

$$\vec{\tau}_A = (\vec{R} \times \vec{F}_{\text{tot}}) + \vec{\tau}_{\text{cm}}, \tag{10–14}$$

and (3) the rate of change of each term of the total angular momentum equation is equal to the corresponding term of the total torque.

In the absence of net torque, the angular momentum of a system is constant throughout the motion. This fact is an important tool for the study of the motion of nonrigid extended systems. The torque will naturally be zero in the absence of external forces, but it also is zero when only central forces are present. The conservation of angular momentum as applied to the motion of a single particle moving under the influence of a central force implies in particular that the areas swept out by position vectors from the source of the force to the moving object in equal times are equal.

The kinetic energy of rotational motion is given by $K = I\omega^2/2$, and the change in K equals the work done—just as in linear motion. For rotational motion in which the torque lies along the angular velocity, this relation (the work–energy theorem) reads

$$W = \int_{\theta_0}^{\theta} \tau d\theta = K - K_0, \tag{10–35}$$

where K_0 is the initial rotational energy at the angle θ_0. In many problems both rotational and linear motion are present. The total energy of a rigid body may be written as the sum of the linear kinetic energy, calculated as if all the mass of the object were concentrated at the center of mass, and the rotational energy of the object about the center of mass:

$$K = \frac{1}{2}MV^2 + \frac{1}{2}I_{cm}\omega^2. \tag{10–38}$$

Quantum physics tells us that angular momentum is quantized in the form

$$L_z = n\hbar = n\frac{h}{2\pi}, \tag{10–39}$$

where h is Planck's constant and $n = 0, \pm1, \pm2, \ldots$. The consequences of this quantization are important at the atomic scale and include the result that not all energy values are permitted for atoms and molecules.

Many subtle effects are associated with the effect of torque on angular momentum, and we illustrated one of them: the precession of angular momentum for a spinning top. If the top is spinning while inclined from the vertical, then gravity produces a torque on the top that causes its axis to precess about the vertical with a precession frequency

$$\omega_p = \frac{Mg\ell}{I\omega}. \tag{10–43}$$

Understanding the Concepts

1. Angular momentum is sometimes given in units of joule-seconds $(\mathrm{J \cdot s})$. Is this a correct SI unit for angular momentum?
2. A comet is heading at high speed straight into the center of the Sun. Why is the angular momentum of the comet with respect to the Sun zero?
3. Why is a cyclist more stable on a rapidly moving bicycle than on one that is almost stationary?
4. When a quarterback throws a football, he tries to put quite a bit of spin on it. Since some of the energy put into the ball then goes into rotational motion, this means that there is less energy for translational motion. Why is it nevertheless done?
5. In taking a fast corner on a bicycle, it is safer if you crouch as low as possible. Why is that?
6. Films from Skylab show the astronauts reorienting themselves by spinning their arms as they float weightlessly. Is this consistent with the conservation of angular momentum?
7. Relief from the heat is often provided by large fans that are enclosed in rectangular structures. If you try to move one of these structures while the fan is on, you will encounter some difficulties. Explain what these might be.
8. A long, flexible, heavy bar can be very useful to a tightrope walker. Why?
9. Why is it easier on your back for you to lift heavy objects by bending your knees, keeping your back straight, and straightening your knees, rather than by picking up the object as you bend over from your waist with your legs straight?
10. An astronaut floating in a space station holds the axle of a rotating wheel. When the astronaut is vertical and the axle points away from her, she suddenly rotates the axle to point up. What happens to the astronaut?
11. A diver prepares for a complex set of midair maneuvers. In springing off an elastic diving board, the diver wishes to acquire angular momentum relative to her center of mass. What should she strive for in the takeoff?
12. In Example 10–3, we studied a tether ball and found that as the rope tying it to the pole winds, the angular momentum decreases. Where does the corresponding torque come from?
13. When a cue hits a billiard ball off center, the ball will move as well as spin. What is the sense of rotation, and how does this follow from considerations of the angular impulse?
14. It is possible to tell whether an egg is hard-boiled without cracking it open by setting it in rotation on a table. Discuss what you would find, and why. Compare the behavior of the hard-boiled egg to that of a raw one.
15. If you were to tie a rock to a rope and swing it in a horizontal circle above your head, you could rather easily do so without spinning around yourself. Is this a violation of the conservation of angular momentum?
16. The center of mass of an object accelerates as the result of an impulse (a brief force). If the object is extended, the impulse may also be an angular impulse. Is it possible to have an angular impulse without a linear impulse?
17. A comet falls straight toward the Sun, under the influence of gravitation, a central force, having started at rest with respect to the Sun. What is the angular momentum of the comet with respect to an axis through the Sun?
18. A diver executes a series of midair maneuvers. To do so, is it necessary for the diver to give herself some angular momentum about her center of mass?
19. We saw that the torque–angular momentum relation is a consequence of Newton's laws. Is it also true in Newtonian mechanics that the conservation of angular momentum follows from Newton's laws?
20. You are given a stool of known rotational inertia that can rotate with minimal friction. You are also given a stopwatch, a very light meter stick, and two known masses that can slide along the meter stick. How would you use this apparatus to measure the rotational inertia about the rotational axis of a person sitting on the stool?
21. The propeller of a single-engine airplane rotates clockwise, as seen from the cockpit. The plane makes a slow turn to the right. What else happens?
22. If only one set of handbrakes on your bicycle works when you descend a steep downhill slope, which set would you prefer?

23. In a conceptual example, we discussed a student, initially at rest, on a stool free to rotate about a vertical axis. The student holds a spinning wheel with angular momentum oriented up. What would have happened if instead of reversing the wheel completely, the student gave the wheel a turn such that its spin axis became horizontal?

24. A woman stands on the edge of a freely rotating platform. She walks toward the center along a radius. Will the speed of rotation of the platform change? If so, in what way, and what is the source of the torque?

25. A horizontal platform is rotating at a certain speed. A boy jumps onto the platform from an overhanging tree branch. He lands with both feet straddling the center and remains standing. Will the platform speed up, slow down, or neither?

26. To prevent rolling, a boat can be stabilized by attaching a large flywheel to the sides (Fig. 10–31). The attachment point is above the waterline. Both ω and the rotational inertia, I, about the axis are large. (a) A wave hits the boat on the side. The wave would tend to rock the boat or rotate it about its longitudinal axis without the flywheel. With the flywheel installed, how does the boat react to this wave? (b) A wave comes straight at the bow (the front). It would lift the bow of the boat without the flywheel. With the flywheel, what happens?

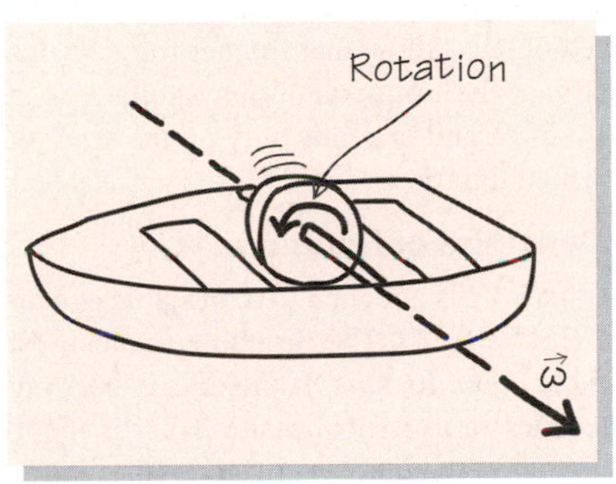

▲ **FIGURE 10–31** Question 26.

27. In an industrial machine a cylinder is spinning without friction about its (fixed) axis, angular speed ω_i (Fig. 10–32). A small movement of its axis is allowed so that it comes into contact with a second identical cylinder, also free to spin about its fixed axis but initially at rest. As a result of friction between the surfaces, both end up spinning with equal and opposite angular velocity, magnitude ω_f. The net final angular momentum about, say, the axis of the first cylinder is therefore zero. What happened to the angular momentum?

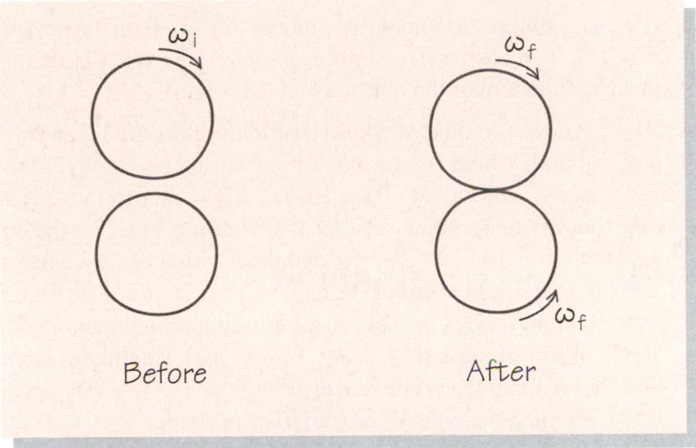

▲ **FIGURE 10–32** Question 27.

28. When jugglers perform juggling acts, the props they use are frequently put into a spinning motion when they are tossed into the air. Why is this a good idea?

29. A baseball is thrown horizontally, hits a rough floor, and bounces. The ball is spinning in the direction of its motion (like a rolling bicycle wheel). Which of the following is true? When the ball leaves the floor, it does so at (a) the same angle at which it struck the floor; (b) a larger angle with respect to the floor; (c) a smaller angle with respect to the floor.

30. Think about the demonstration in which someone spinning on a stool pulls in his or her arms and speeds up because of angular momentum conservation. In the process, does the energy of rotation decrease, remain constant, or increase?

Problems

10–1 Generalization of Angular Momentum

1. (I) An airplane of mass 2000 kg located 100 km north of New York City is flying 200 km/h in an easterly direction. (a) What is its angular momentum with respect to New York City? (b) What if it is flying in a northeasterly direction?

2. (I) What is the angular momentum about the origin of a particle of mass 270 g at position $\vec{r} = (0.1\hat{i} - 0.5\hat{j} + 0.2\hat{k})$ m, moving with a velocity of $\vec{v} = (12\hat{i} - 7\hat{j} - 3\hat{k})$ m/s?

3. (I) You are standing on the corner of Main Street and Elm, watching the cars on Main pass at a steady 10 m/s. You stand 5 m from the line of traffic. You watch a red convertible, whose mass is about 1000 kg, from the moment it is one block away, a distance of 200 m. Treat the car as pointlike, and assume that Main Street is straight. (a) What are the car's angular momentum, magnitude, and direction, with respect to you when it is one block away? (b) What is its angular momentum with respect to you when it passes your position on the corner?

4. (I) Assuming that each object is pointlike, how large is the angular momentum (a) of Earth about the Sun; (b) of the Moon about Earth? (c) Compare these results to the angular momentum of Earth (no longer pointlike) about its own axis; assume constant density.

5. (I) A bicycle travels east. The mass of the wheel, 1.8 kg, is uniformly distributed along the rim, with the mass of the hub and spokes negligible. If the radius of the wheel is 18 cm and the wheel rotates at the rate of 4.2 rev/s, what is the direction and the magnitude of the angular momentum of the wheel about its axis? (Neglect the width of the rim.)

6. (II) Calculate the angular momentum about the origin of a particle of mass m moving along the trajectory $y = ax + b$ with uniform speed v.

7. (II) A rock of mass 60 g is thrown with initial horizontal speed $v_x = 25$ m/s off a building from a height of 30 m. Calculate the angular momentum of the rock about the line along the edge of the roof as a function of time.

8. (II) A unicycle has a wheel of mass 1.5 kg, rotational inertia $0.28\ \text{kg}\cdot\text{m}^2$ about the axle, and radius 0.38 m. What is its angular momentum with respect to a point on the road if the wheel rolls without slipping with an angular velocity of 2.5 rad/s?

9. (II) The position vector of an object of mass m subject to two constant forces that act at right angles is given by $\vec{r} = \left(\frac{1}{2}at^2\right)\hat{i} + (vt)\hat{j} + \left(\frac{1}{2}bt^2 - wt\right)\hat{k}$. Calculate the angular momentum of this object about the origin.

10. (II) An object of mass m moves in a path given by $\vec{r} = (x_0 + \rho\cos[\omega t])\hat{i} + (y_0 + \rho\sin[\omega t])\hat{j}$. What is the angular momentum of the object about the origin?

11. (II) Consider two objects whose position vectors are given by $\vec{r}_1$ and $\vec{r}_2$ and whose momenta are given by $\vec{p}_1 = m_1\vec{v}_1$ and $\vec{p}_2 = m_2\vec{v}_2$, respectively (Fig. 10–33). Show that in the special case that the center of mass of the two bodies is at rest at the origin (that is, $\vec{P} = \vec{p}_1 + \vec{p}_2 = 0$, and the position of the center of mass is $\vec{R} = 0$), the sum of the angular momenta of the two objects about the center of mass equals the angular momentum of a single object of mass $\mu = m_1m_2/(m_1 + m_2)$, rotating in circular motion about the origin at a distance $r = r_2 - r_1$. The quantity μ is called the *reduced mass*. [*Hint*: Introduce $\vec{r} \equiv \vec{r}_2 - \vec{r}_1$, and express $\vec{p}_1$ in terms of $d\vec{r}/dt$.]

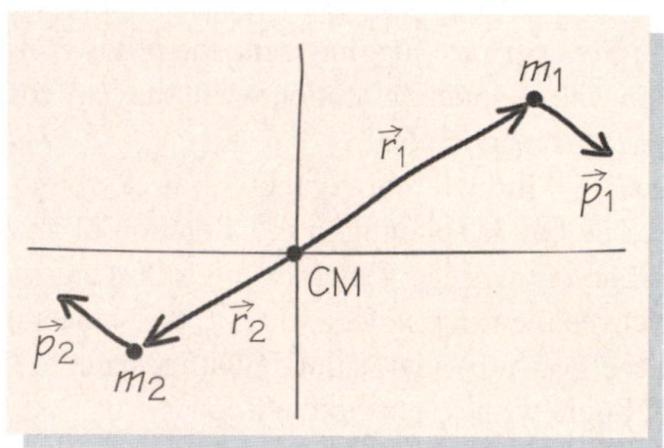

▲ **FIGURE 10–33** Problem 11.

12. (II) Earth is not a point object but is a sphere with a rotational inertia of $9.8 \times 10^{37}\ \text{kg}\cdot\text{m}^2$ about its axis. Assume that Earth's axis of rotation is parallel to the axis of the orbital motion of Earth around the Sun and that the Sun is so massive that it can be considered to be fixed. Calculate (a) the rotational inertia of Earth about the axis of its orbital motion around the Sun; (b) the total angular momentum of Earth about that same axis. (c) Calculate the fractional difference between your result for part (b) and the angular momentum you would find for part (b) if Earth were pointlike; explain why the difference is small.

13. (II) A square, 20 cm on the side, is made of very light sticks. Four identical masses of $m = 0.1$ kg form the corners of the square. The square rotates with an angular velocity of 8 rad/s about an axis perpendicular to its plane through the center of the square. (a) Calculate the rotational inertia of the system about the rotation axis and use it to find the angular momentum about this axis. (b) Use the general definition of angular momentum to calculate the angular momentum of each mass with respect to the center of the square, and add these up. Compare the results of (a) and (b).

14. (II) Consider the square studied in Problem 13. Calculate the angular momentum of each particle about a point on the axis of rotation 14 cm below the plane of the square. Compare the total angular momentum calculated in this way with the results of Problem 13.

15. (II) Three identical masses m are attached to the corners of an equilateral triangle of sides d. Calculate the angular momentum (a) if the triangle rotates at an angular velocity ω about the center of mass around an axis perpendicular to the plane of the triangle; (b) if the triangle rotates with angular velocity ω about one of its sides; (c) if the triangle rotates about an axis going through one of its vertices to the midpoint of the opposite side.

10–2 Generalization of Torque

16. (I) A construction worker of mass 72 kg stands at the end of a 3.4-m-long (massless) horizontal mast attached to a building. What is the magnitude of the torque exerted on the hinge that holds the mast fixed in position?

17. (I) What is the torque about the origin on a particle positioned at $\vec{r} = (3\hat{i} - \hat{j} - 5\hat{k})$ m, exerted by a force of $\vec{F} = (2\hat{i} + 4\hat{j} + 3\hat{k})$ N?

18. (I) A flagpole 2.2 m long is attached to a building. It makes an angle of 20° with the horizontal. A mass of 18 kg is suspended from the end. What is the torque acting on the point of attachment to the building due to the suspended mass?

19. (I) A 1.0-m-long massless stick lies on a table. A force of 200 N is applied to one end of the stick for 0.1 s at an angle of 45° to the stick, and in a direction pointing away from the center of the stick. What is the torque about the other end of the stick during the brief period that the force is acting?

20. (II) Consider two forces $\vec{F}$ and $-\vec{F}$ that act at different points on an extended object. Show that the net force of this combination is zero and that the torque about any point P is independent of the location of P, and depends only on the separation of the two points at which the forces act.

10–3 The Dynamics of Rotation

21. (I) A point mass M is attached to a turntable at a distance R from the center (Fig. 10–34). The turntable rotates with constant angular speed ω about its axis. If the axis is horizontal so that the turntable rotates in a vertical plane, what is the torque that the force of gravity on the mass exerts about the axis as a function of time? Assume that the mass is at the topmost position at $t = 0$.

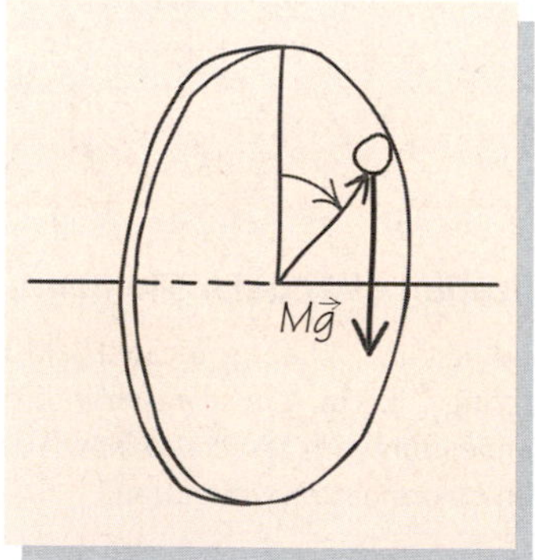

▲ **FIGURE 10–34** Problem 21.

22. (II) The position of a ball of mass m thrown from a building is given by $\vec{r} = (v\cos\theta)t\hat{i} + [(v\sin\theta)t - gt^2/2]\hat{j}$, measured from the point from which the ball was thrown. What is the torque about the origin that the force of gravity exerts on the ball?

23. (II) A ball of mass m slides at speed v on a frictionless horizontal surface and bounces elastically from a wall. The initial path of the ball makes an angle θ with the wall (Fig. 10–35, see next page). Find the initial and final angular momenta of the ball about the point A. What causes the change in angular momentum?

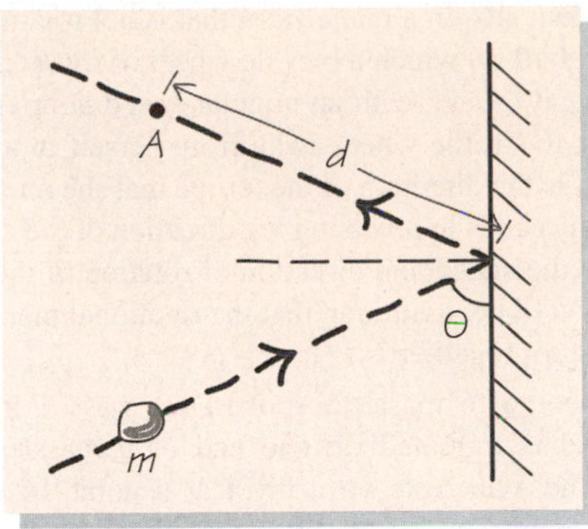

▲ **FIGURE 10–35** Problem 23.

24. (II) A pulley system is used to lift a heavy mass. How much force must be applied to lift the object in Fig. 10–36 at a steady speed? Neglect friction at the axle.

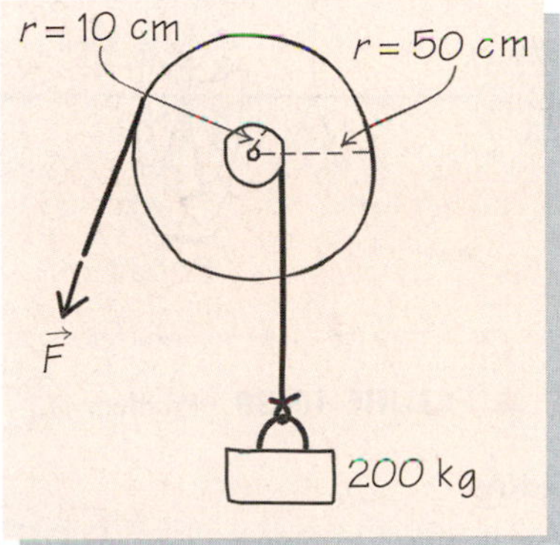

▲ **FIGURE 10–36** Problem 24.

25. (II) What is the vector product of $\vec{A} = 2\hat{\imath} - 4\hat{\jmath} + 5\hat{k}$ and $\vec{B} = \hat{\imath} + 3\hat{\jmath} - 2\hat{k}$?

26. (II) Show that the magnitude of the vector product of two vectors is the area of the parallelogram for which the two vectors form adjacent sides.

10–4 Conservation of Angular Momentum

27. (I) A playground merry-go-round of diameter 3.4 m and rotational inertia 120 kg · m^2 is pushed with no one on it by three children to an angular speed of 2.5 rad/s. Two of the children, of mass 25 kg each, jump on the edge of the merry-go-round, coming radially in. What is the new angular speed?

28. (I) A uniform disk rotating without friction about its (vertical) central axis, with total mass 38 kg and radius 1.7 m, acts as a turntable. Its angular speed is $\omega = 0.075$ rad/s. A person of mass 71 kg jumps straight down onto the rotating turntable. The person lands 0.9 m from the axis. What is the new angular speed of the turntable?

29. (I) A firetruck, mass 6000 kg, passes a parked car on a straight street at $t = 0$ with speed 15 m/s. A physics graduate student finds that, 10 s later, the angular momentum of the firetruck with respect to the parked car is twice its value at $t = 0$. What is the speed of the truck at $t = 10$ s?

30. (I) A skater twirls at 0.7 rev/s with her arms extended and holds a 3-kg mass in each hand; each mass is 0.8 m from the axis of rotation. She pulls the masses in along the radial direction until they are 0.4 m from the axis of rotation. Assuming that the rotational inertia of the arms is negligible and that the rotational inertia of the skater without the masses is 2.3 kg · m^2, what is the speed of rotation after the masses have been pulled in?

31. (II) A bug of mass $m = 2.0$ g walks around a horizontal turntable, which may be viewed as a uniform cylinder of mass $M = 0.24$ kg. If both the turntable and the bug are initially at rest, how much does the turntable rotate relative to the ground while the bug makes one full circle relative to the turntable?

32. (II) A small mass of 17 g slides down a frictionless slope starting from rest at 1.1 m above the ground level. When the slope reaches the bottom, it levels off, and the mass strikes the bottom of a vertical uniform bar of mass 0.2 kg and length 20 cm, pivoted at its midpoint, and sticks to it (Fig. 10–37). With what angular speed will the bar start its rotation?

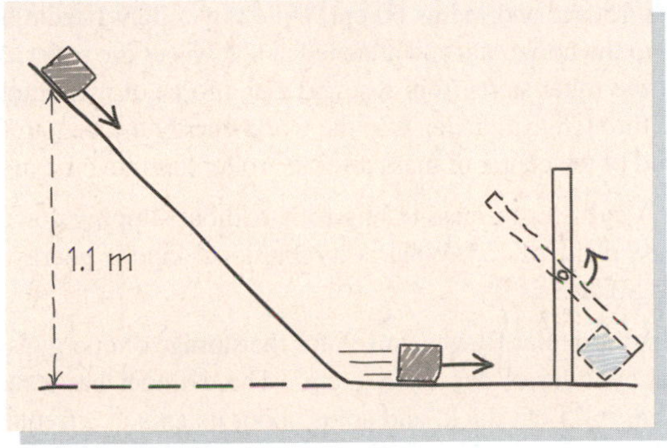

▲ **FIGURE 10–37** Problem 32.

33. (II) A gob of clay, mass 100 g, falls from rest a distance 75 cm before striking and sticking to the edge of a wheel free to rotate about a horizontal axis through its center (Fig. 10–38). The wheel can be approximated as a solid disk of mass 10 kg and radius 50 cm. What is the angular speed of the wheel with the gob of clay attached?

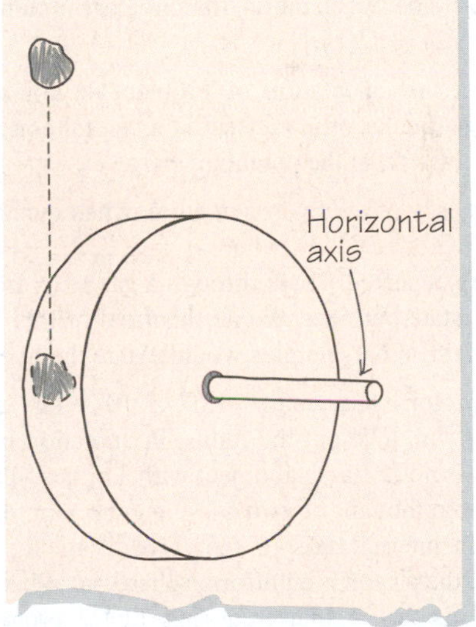

▲ **FIGURE 10–38** Problem 33.

10–5 Work and Energy in Angular Motion

34. (I) A flywheel has radius 1.2 m and mass 680 kg, almost all of which is concentrated on the rim. It is spinning at $\omega = 4.5$ rad/s about its axis when a torque is applied along the axis, producing an angular acceleration of 0.3 rad/s^2. How much time does it take for ω to increase to 6 rad/s, and how much work is done by the torque during this time?

35. (I) An airplane engine develops 240 hp while turning the propeller at 3200 rev/min. What is the torque exerted on the propeller axis by the engine?

36. (II) Repeat the problem of the cylinder rolling down the inclined plane treated in Section 9–6, but this time use energy techniques to learn the speed of the cylinder as a function of distance traveled.

37. (II) The work done to slow down a spool used in a manufacturing process during a certain period is 1200 J. The spool is fixed on an axis and has rotational inertia of 0.033 kg $\cdot$ m^2 about that axis. The spool is rotating at an angular speed of 490 rad/s before the specified period. What is the spool's angular speed at the end of the period?

38. (II) A lawn roller consists of a cylinder—essentially solid—of mass 150 kg and radius 60 cm, pulled by a light handle oriented 30° to the horizontal and attached at the axis of the roller. Assume that the roller starts from rest and that a force of magnitude 55 N acts through the handle. Use the work–energy theorem to find the speed of the center of mass after the roller has moved 2 m.

39. (II) A cylinder of mass 0.2 kg rolls without slipping down an inclined plane of 15°. What is its rotational kinetic energy after it rolls 80 cm?

40. (II) A particular flywheel used for the storage of energy is a solid steel cylinder of density 8 g/cm^3. The cylinder has a radius of 1.2 m, is 45 cm thick, and spins about its axis at a frequency of 260 rev/min. (a) What is the rotational inertia of the flywheel? (b) What is the work the flywheel can do in being brought to a halt if there is no energy loss to frictional forces?

*10–6 Quantization of Angular Momentum

41. (II) Consider an object of mass m that moves in a circular orbit caused by a central force given by $F = -kr$. Suppose that the Bohr quantization condition is applied to this motion. What are the allowed quantized radii, velocities, and kinetic energy values? [*Hint*: The acceleration for circular motion is v^2/r, and $E = (mv^2/2) + U(r)$.]

42. (II) Repeat the calculation of Problem 41 for an object that moves in a circular orbit caused by a central constant attractive force, obtained from the potential energy $U(r) = Cr$.

43. (II) The energy of the hydrogen atom, when quantized, is given by $E_n = -(13.6\text{ eV})/n^2$, where $n = 1, 2, 3\ldots$ Particles of energy 2.0 eV repeatedly pass through a gas of hydrogen atoms in the $n = 1$ state but never excite them out of this lowest-energy state. Explain. What energies would excite the hydrogen atoms?

44. (II) A proton has mass 1.67×10^{-27} kg and "radius" 1.3×10^{-15} m. (We put the radius in quotation marks because the proton is not a classical object with a radius, like a baseball.) It also can be thought of as having angular momentum with respect to an internal axis of 0.5×10^{-34} kg $\cdot$ m^2/s. Take as a model that the proton is a uniform sphere. (a) What, according to the model, is the angular frequency of the proton's rotational motion? (b) What is the speed with which the outermost portion of the proton, on its equator, moves? Compare this to the speed of light, 3×10^8 m/s. (c) What is the energy associated with the rotational motion of the proton?

*10–7 Precession

45. (II) By repeating the derivation of the precession frequency of a top, but with the top making an angle ϕ with respect to the vertical, show that the precession frequency is *independent* of the angle ϕ.

46. (II) A student sits on a piano stool that is not rotating. She holds a vertical shaft on which a bicycle wheel of rotational inertia I is mounted and rotates with an angular speed ω oriented upward. She wants to tilt the wheel away from herself in a radial direction. What is the direction of the torque that she must exert? Suppose she succeeds in reversing the direction of the shaft by 180°. What will the speed and direction of rotation of the student and the piano stool be, assuming that the rotational inertia of the student and stool together is I^*?

47. (II) A wheel with massless spokes has mass 1 kg and radius 10 cm and is mounted on one end of a massless axle (Fig. 10–39). The axle rests on a pivot at a point 16 cm from the mounting point and 10 cm from the wheel. At the other end a mass of 0.8 kg is attached. The wheel spins at an angular frequency of 10 rad/s. What is the rate of precession?

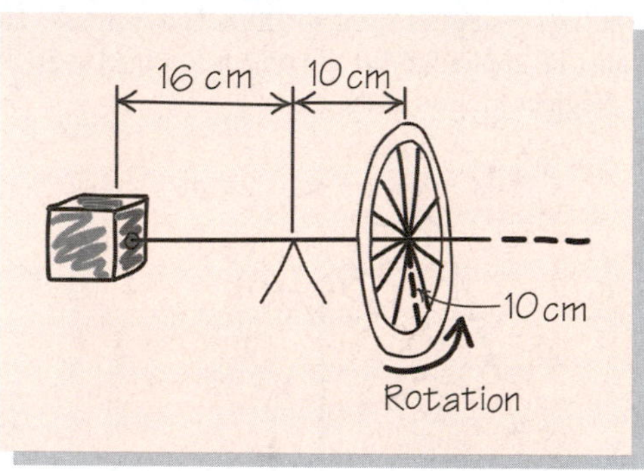

▲ **FIGURE 10–39** Problem 47.

General Problems

48. (I) A 110-g ball is thrown out of a second-story window 6.0 m above the ground. The initial velocity of the ball is horizontal, with magnitude 4.5 m/s. What is the angular momentum of the ball, as a function of time, about the point on the ground directly below the window?

49. (II) A solid cylinder of mass 0.85 kg and radius 4.2 cm initially at rest rolls down a plane inclined at 28° with the horizontal and 1.5 m long. Use energy conservation to calculate the angular velocity of the cylinder at the bottom of the ramp; assume that all the kinetic energy of the cylinder is in rolling motion (that is, there is no sliding).

50. (II) A constant-density cylinder of mass 0.5 kg and radius 4 cm can rotate freely about an axis through its center. It has thread wound around an attached axle of radius 0.5 cm that also runs through its center (Fig. 10–40). The thread is attached to a mass of 1 kg, which slides down an inclined plane with an acceleration of 0.1 m/s^2. What is the coefficient of kinetic friction between the block and plane?

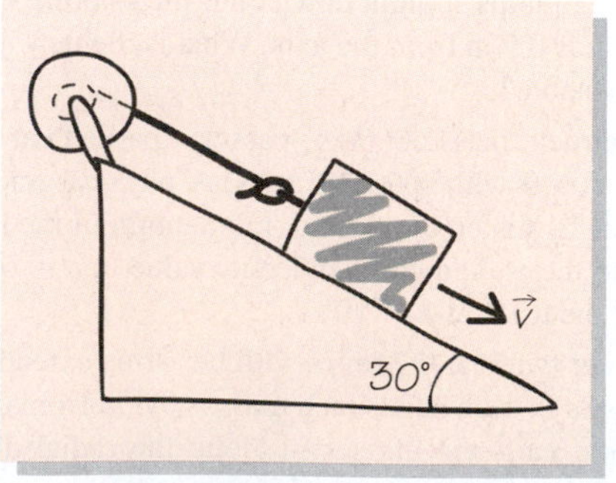

▲ **FIGURE 10–40** Problem 50.

51. (II) A child of mass 32 kg stands at the center of a platform of radius 2 m and rotational inertia 450 kg · m^2. The circular platform rotates about a frictionless shaft with angular speed of 0.8 rad/s. The child walks in a radial direction until he reaches the rim. What will the angular velocity of the platform be when that happens? What is the change in energy of the platform plus child? Identify the source of the work responsible for the change in rotational kinetic energy.

52. (II) A wooden board 4 m long and of mass 20 kg lies on a frictionless surface. A mass of 30 kg slides along the frictionless surface, strikes the board at right angles near one of its ends, and stays attached to it (as in a perfectly inelastic collision). Describe the subsequent motion of the board and mass together.

53. (II) A door hangs on frictionless hinges. A ball of mass 35 g moving with a velocity of 45 m/s strikes the door at the edge opposite to the hinged side and bounces back with a velocity of 35 m/s. Assuming that the door has a mass of 3.0 kg and is 85 cm wide, what will be the angular velocity acquired by the door?

54. (II) A bullet of mass 15.0 g and velocity 350 m/s passes through a wheel at rest (Fig. 10–41). The wheel is a solid disk of mass 3.0 kg and radius 18 cm. The bullet passes through the wheel at a perpendicular distance of 14 cm from the center, and the bullet's final velocity is 270 m/s. What are the wheel's angular velocity, angular momentum, and kinetic energy? Is energy of motion conserved?

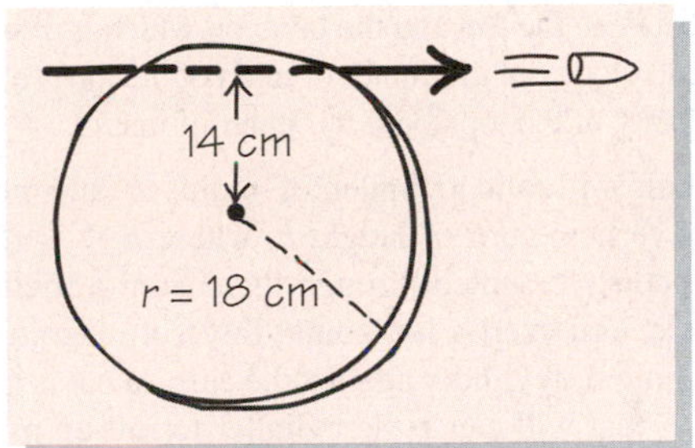

▲ **FIGURE 10–41** Problem 54.

55. (II) Figure 10–42 shows the Atwood's machine treated in the problems of Chapter 5, with a rope of negligible mass 2 m long. Earlier the pulley was treated as massless. Now suppose that the pulley can be approximated by a solid disk of radius 0.1 m and mass 2 kg. The system is released from rest with the 4 kg mass 1.5 m from the floor and the 1 kg mass on the floor. (a) What is the speed of either block just before the 4 kg mass hits the floor? (b) How long does it take the 4 kg mass to reach the floor?

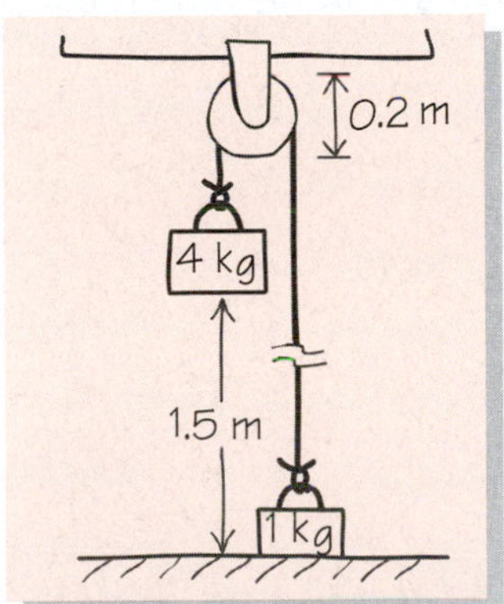

▲ **FIGURE 10–42** Problem 55.

56. (II) Two weights of mass 1.2 kg and 0.85 kg, respectively, are connected by a massless string that passes over a pulley. The pulley is a hollow cylinder of radius 18 cm and mass 0.45 kg. What are (a) the acceleration of the system and (b) the time that it takes for the larger mass to descend a distance of 1.6 m if the weights start from rest?

57. (II) A ball of mass M, radius R, and uniform density falls from rest from the top of a hemispherical bowl (Fig. 10–43). The left side of the bowl is frictionless, but the right side has a large coefficient of friction with the ball, and for all practical purposes the ball immediately rolls without slipping. How far up the right side of the bowl does the ball reach? Explain your answer in light of our assumptions about the very short distance in which sliding changes to rolling.

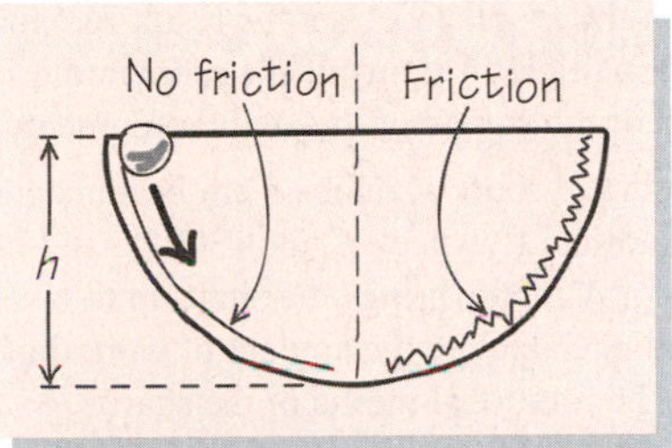

▲ **FIGURE 10–43** Problem 57.

58. (II) A solid door of mass $M = 15$ kg and width $\ell = 120$ cm is hit at a right angle by a mud ball of mass $m = 0.3$ kg, which, as Fig. 10–44 shows, hits the door at the edge with speed $v = 12$ m/s and sticks. (a) What is the rotational inertia of the door about the hinges? (b) What is the angular velocity of the door after having been struck? (c) What fraction of the initial energy does the moving door–mud ball system retain?

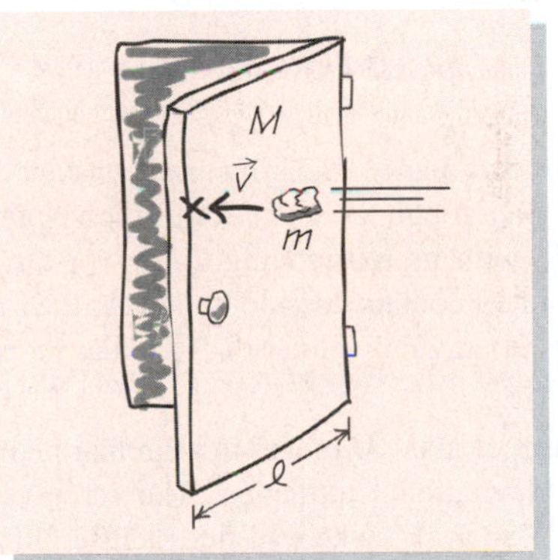

▲ **FIGURE 10–44** Problem 58.

59. (II) A 75-kg bank robber is escaping on a 450-kg motorcycle, runs out of gas, and thereafter coasts (friction free) at 80 km/h. As he passes under an overpass, a 75-kg policeman drops vertically onto the back of the seat and hangs on. Ignore the impulse due to the sudden change of the force of friction on the road on the translational and rotational motion. (a) Find the final velocity of the motorcycle. (b) What fraction of the initial kinetic energy (motorcycle plus robber) is lost? Ignore the rotational energy in the wheels. (c) Redo part (b) but include the effect of the rotational energy of the wheels, which each have a rotational inertia about their axes of 3 kg · m^2 and a radius $r = 0.5$ m.

60. (II) Electric power is used to speed up a centrifuge whose rotational inertia is 1.2 kg · m²; 1.6 kW of power were used to make the centrifuge accelerate at a steady rate from rest to 17,000 rev/min. If the electricity use was 100 percent efficient, how much time is required to speed up the centrifuge?

61. (II) A point mass $m = 0.2$ kg is attached to a string, which passes through a hole in a table and rotates in a circle of radius $r = 0.8$ m with an angular velocity of 40 rad/s. What mass M must be attached to the end of the string under the table to maintain this motion? Suppose that mass M is slowly increased by an amount that makes it descend a distance 0.1 m. What is the amount of the increase of M? What will the new angular velocity of the point mass be? [*Hint*: Use angular momentum conservation.]

62. (II) Show that $(\vec{r} \times \vec{p}) \cdot (\vec{r} \times \vec{p}) = r^2p^2 - (\vec{r} \cdot \vec{p})^2$. [*Hint*: It is convenient, without any loss of generality, to assume that both $\vec{r}$ and $\vec{p}$ lie in the xy-plane.] Use this result to express the kinetic energy of a particle in terms of the momentum in the radial direction and of the square of the angular momentum.

63. (II) A cylindrical shaft of radius 5 cm is connected by a band to a solid cylindrical flywheel of mass 300 kg and of radius 0.35 m (Fig. 10–45). A motor brings the shaft up to a rotational rate of 1400 rev/min. Calculate the amount of work done by the motor, neglecting the rotational inertia of the shaft.

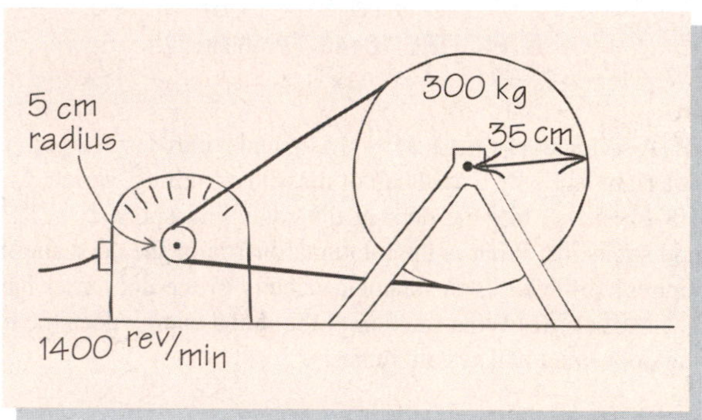

▲ **FIGURE 10–45** Problem 63.

64. (II) A thin rod of mass M, length ℓ, and constant density is standing on end on a rough table that forms the xy-plane. The rod begins to fall, with its top moving in the $+x$-direction, but as it falls, its point of contact does not move. As the rod hits the table, what are its (a) angular velocity, (b) angular momentum, and (c) kinetic energy?

65. (II) An object of mass M moves in a circular planar orbit about a center of gravitational attraction. The force of attraction has magnitude $F = K/r^2$, where r is the radius of the circle, and it is directed toward the center. Calculate (a) the velocity, (b) the radius, (c) the period, T, and (d) the acceleration of the object, all in terms of the angular momentum L, M, and K.

66. (II) An object of mass m moves in a plane with its path described by the radius vector $\vec{r} = \hat{i}A \cos \omega_1 t + \hat{j}B \sin \omega_2 t$. Calculate the angular momentum about the origin. In what direction will the angular momentum point? Under what circumstances will the angular momentum be constant?

67. (II) A hurricane is a vast swirl of Earth's atmosphere. Using your knowledge of the size of such storms, the depth of the atmosphere, the speed of the winds, the density of air, and so forth, estimate the kinetic energy contained as well as the angular momentum. Compare your estimate of the angular momentum with that of Earth itself (see Section 9–5).

68. (III) A putty ball of mass $m = M/5$ is thrown with velocity $\vec{v} = v\hat{i}$ and hits the top of the thin rod in Problem 64 as the rod stands vertically. If the putty ball makes a completely inelastic collision, and if again the point of contact between the rod and the table does not move, what are the angular velocity, angular momentum, and kinetic energy of the system as it hits the table?

69. (III) A particular top can be approximated as a solid cylinder of mass 100 g and radius 2 cm. A string of negligible mass and length 1 m is wound around the top, which is started by pulling horizontally on the string with a constant force of magnitude 0.6 N. The top starts from rest at point O, and the string is pulled off. Ignore all friction between the top and the table on which it moves. (a) What is the final velocity of the center of mass of the top? (b) the final angular velocity of the top about its center of mass?

70. (III) A uniform solid cylinder of radius R and mass M rests against a vertical curb of height h, where $h < R$ (Fig. 10–46). The cylinder is mounted through its axis on a frictionless horizontal axle. You exert a horizontal force of magnitude F on the axle, pushing the cylinder against the curb. What is the minimum value of F that will cause the cylinder to roll up over the curb?

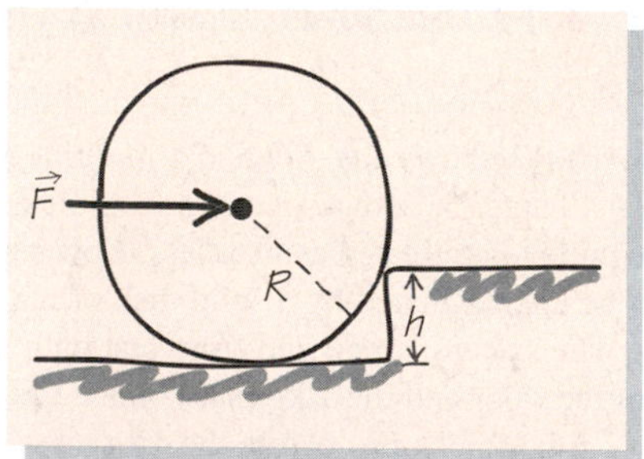

▲ **FIGURE 10–46** Problem 70.

◀ The tightrope artist Jay Cochrane, seen here walking towards the tower of Casino Niagara with the American Falls in the background, uses the large bar to control the position of his center of mass and minimize any bad effects of the torque acting on the system formed by him and the bar. Cochrane completed the 70 m walk more than 40 stories high without a net in less than ten minutes.

Statics

In the previous two chapters, we have seen how extended objects or systems can have a rich array of motions, both linear and rotational. In this chapter, we are going to look at rigid objects—or at least systems that are approximately rigid—that are *motionless*. This is the area of physics known as **statics**. Although an object may be motionless, this does not mean that there are no forces or torques acting upon it. In fact it is likely that the object is motionless *because* forces and torques act upon it. Consider the tightrope walker in the opening photograph and the rock in Fig. 11–1. What forces and torques do you think are acting upon them? Statics has its most notable use in the field of structural engineering, in the design of bridges, buildings, and other structures. Statics also plays an important part in the analysis of the role of muscle, tendon, and bone in living systems. We can also use this opportunity to begin our study of the properties of solids, as rigid objects are formed from them. What we shall see is that solids themselves are not truly rigid, and the knowledge of how solids deform in response to forces acting on them is one step in understanding in what ways extended systems are truly rigid. In fact, well-engineered structures are never truly rigid, and part of the genius of good design lies in the degree to which one can avoid the overbuilding that true rigidity would require.

▲ **FIGURE 11–1** Among the forces and torques acting on this rock are those due to contact forces and to gravity. This static marvel has remained balanced for a very long time, suggesting that the equilibrium is stable.

11–1 Static Conditions for Rigid Bodies

As long as a building does not move, we can say that the net force and the net torque on it is zero. But a building is not an indivisible thing. Its components exert torques and forces on one another, and even one end of a heavy beam can be said to exert forces on

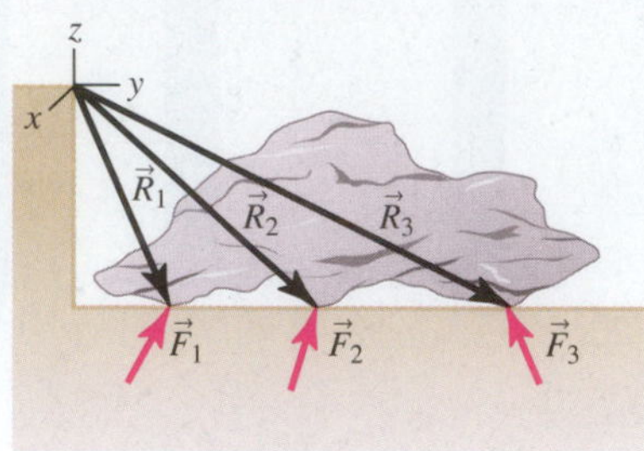

(a)

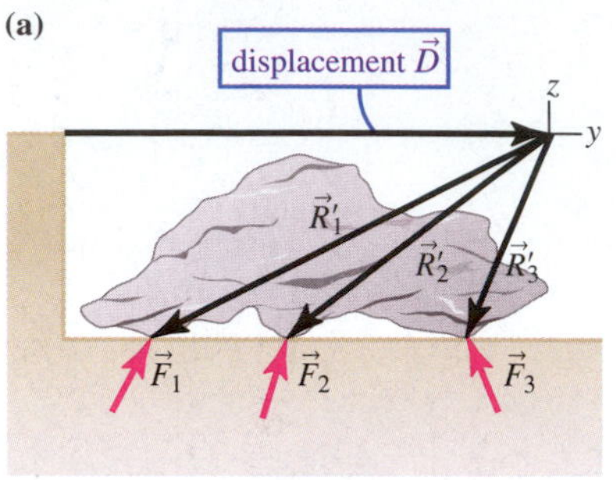

(b)

▲ **FIGURE 11–2** (a) A series of forces act at several places on this object, each at a different position measured with respect to an origin. To keep the figure clear, we have left off gravity. (b) The same forces act but the origin that measures their positions of action has shifted.

a neighboring part of the beam. The complete analysis of a structure, with all its internal forces, can be quite difficult. The problem becomes much simpler if we can think of the building as *rigid*. Under this approximation, it is only the external forces that determine whether the structure will stand or tip. This may be an approximation, but it is often a reasonable one in practice—in other words, many of the structures that we are interested in come close to being rigid.

We know that external forces acting on a rigid object have two effects: First, no matter where they are actually applied on the object, their vectorial sum produces a linear acceleration of the center of mass. Second, depending on where they are applied, they may produce torques that act to rotate the entire object. In the previous two chapters, we were interested in the resulting motion. In this chapter, we will look at the conditions for *no center-of-mass acceleration and no angular acceleration about any point*; although it is irrelevant from the point of view of the dynamical equations, we will also normally have *no center-of-mass velocity and no angular velocity about any point*. The object in question is then said to be in equilibrium. This is a term that we first met in Chapter 7, where we looked at the question of stable, unstable, and neutral equilibrium in terms of the potential energy. In statics, we normally are in situations where we want to avoid an unstable equilibrium, and certainly an engineer who is designing a bridge will carry through an analysis which will reveal the nature of the equilibrium conditions and ensure that the structure is in a stable equilibrium. In this chapter we won't look systematically into this question, although we will have occasion to remark on it from time to time.

Suppose that we have a set of external forces $\vec{F}_i$, each member of the set labeled by the index i, that act at various points described by the position vectors $\vec{R}_i$ on a rigid body of mass M (Fig 11–2a). Then the full dynamical equations for the object are

$$\vec{F}_{\text{net}} = \sum_i \vec{F}_i = M\vec{A} \tag{11–1}$$

and

$$\vec{\tau}_{\text{net}} = \sum_i (\vec{R}_i \times \vec{F}_i) = \frac{d\vec{L}}{dt} = \vec{\alpha}. \tag{11–2}$$

The points at which the forces act have position vectors $\vec{R}_i$ measured with respect to any convenient origin. Here, $\vec{A}$ is the acceleration of the center of mass due to the net force $\vec{F}_{\text{net}}$; $\vec{L}$ is the angular momentum of the object whose change is due to the net torque, $\vec{\tau}_{\text{net}}$, about the point chosen as the origin. We can recall, apropos of the torque equation, that a torque has a magnitude given by the product of the force and the *lever arm*, also known as the moment arm. Recall from Section 9–4 that to find the lever arm about a certain point O, extend the line of the force, and the closest, or perpendicular, distance between the extension line and point O is the lever arm. The direction of the torque is given by a right-hand rule. These review remarks are summarized in Figs. 9–22 and 9–26.

We can now write down the conditions for statics. When the linear acceleration $\vec{A} = 0$ in Eq. (11–1) the net force is zero:

$$\sum_i \vec{F}_i = 0. \tag{11–3}$$

FORCE EQUILIBRIUM CONDITION

and when the angular acceleration $\vec{\alpha} = 0$ in Eq. (11–2), the net torque is zero:

$$\sum_i (\vec{R}_i \times \vec{F}_i) = 0. \tag{11–4}$$

TORQUE EQUILIBRIUM CONDITION

The key to statics then is that the vector sums of forces and torques acting on an object are zero. We shall sometimes refer to these two equations as the equilibrium conditions.

CONCEPTUAL EXAMPLE 11–1 Rock climbing is a demanding activity, and it is important for a climber to be able to rest from time to time. Consider the climber in Fig. 11–3 resting in a "chimney." What are the forces that keep the climber in equilibrium?

Answer Gravity is certainly one force acting. The other forces acting on the climber at rest are either contact forces that act perpendicular to the rock surfaces or friction that acts along those surfaces. As the chimney walls are vertical, the normal forces will not oppose gravity. However, that does not mean they are not present and that they are not important to this problem. In particular, the maximum magnitude of static friction is proportional to the normal force, and friction can act vertically in this case and in a direction that cancels gravity. By pushing with his feet and back into the walls, the normal forces at the walls are increased, and that has the positive effect of increasing the friction forces on the climber. These forces will cancel gravity and allow the climber to be in equilibrium. The technique of pushing the walls within cracks such as this is instinctive and effective.

▲ **FIGURE 11–3** A rock climber is resting in a chimney. Note that the forces acting on him are in equilibrium and allow him to be at rest.

The Condition of No Torque Is Independent of the Choice of Reference Point

When we say that there is no angular acceleration [Eq. (11–4)], to which reference point are we referring? We can show that if an object has no angular acceleration about any one point, then, as long as the object is not in linear acceleration, it will not have an angular acceleration about any other point. Thus Eq. (11–4) applies to *any* choice of origin when Eq. (11–3) holds.

In order to prove this important result, we reconsider the system of Fig. 11–2a but with an origin displaced by $\vec{D}$ from the old origin (Fig. 11–2b). We'll find the net torque about this new origin due to the forces, labeled by an index i, shown in the figure; we'll suppose these are the only forces acting. We also assume that the vector sum of these forces is zero, so that there is no linear acceleration. If the force labeled i is applied at point $\vec{R}_i$ with respect to the original origin, then it is applied at the point $\vec{R}_i'$ in the new system:

$$\vec{R}_i' = \vec{R}_i - \vec{D}.$$

The condition for no rotational acceleration is

$$0 = \sum_i (\vec{R}_i \times \vec{F}_i) = \sum_i [(\vec{R}_i' + \vec{D}) \times \vec{F}_i] = \sum_i (\vec{R}_i' \times \vec{F}_i) + \sum_i (\vec{D} \times \vec{F}_i)$$
$$= \sum_i (\vec{R}_i' \times \vec{F}_i) + (\vec{D} \times \sum_i \vec{F}_i).$$

But there is no net force—the sum over the forces in the second term of the right-hand side is zero. Thus

$$\sum_i (\vec{R}_i \times \vec{F}_i) = \sum_i (\vec{R}_i' \times \vec{F}_i) = 0,$$

and this equation shows that if there is no torque about one origin, then there is no torque about any other origin.

Because a static object has no net torque about any point, we can place the point about which we calculate torques wherever the calculation will be easiest. A good choice leads to considerable simplification, as the calculation of the torques about some points may be trivial—the torque about the point where a force is applied, or anywhere along the line of that force, is zero because the lever arm is zero for that force.

CONCEPTUAL EXAMPLE 11–2 Consider a hard ball being pushed against a curb by a horizontal force $\vec{F}$ that is applied at precisely curb height (Fig. 11–4a). The ball itself has a radius equal to the curb height. Will this force cause the ball to climb over the curve?

Answer A force can make the ball climb the curb if it causes the ball to rotate about the point P, and this can happen only if there is a net torque about P. In the present case, the contact force of the curb point, $\vec{F}_C$, and the external force, $\vec{F}$, are both directed so that extensions of them pass through point P. These forces therefore produce no torque about P, as the extended free-body diagram in Fig. 11–4b shows. Thus the external force cannot make the ball climb the curb, and the ball is static.

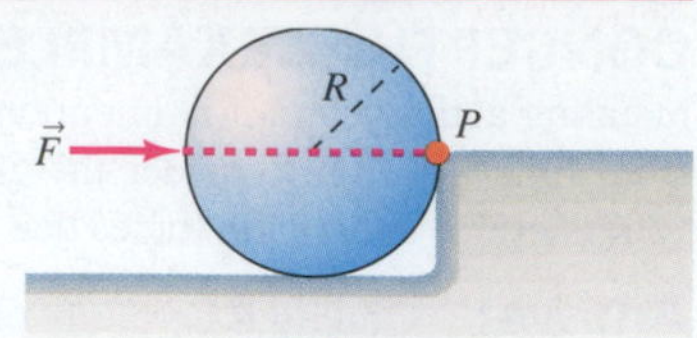

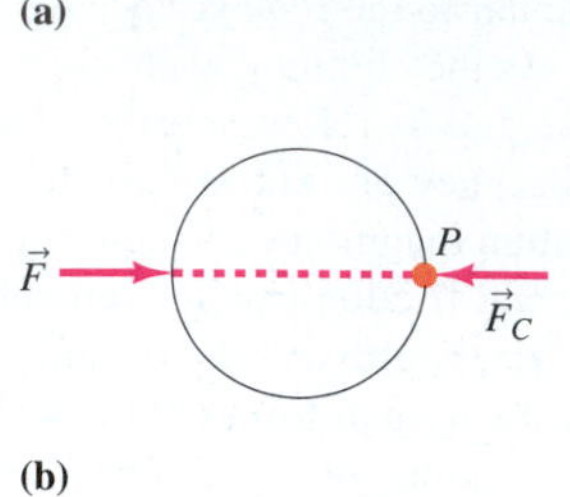

▶ **FIGURE 11–4** (a) The ball has a radius equal to the height of the curb, and a horizontal force $\vec{F}$ is applied at exactly curb height. (b) Extended force diagram (we have ignored the vertical forces). The forces shown can have no torque about point P.

11–2 Gravity and Rigid Bodies

Gravity acts in all the static situations we consider in this chapter, and it is an important element in the analysis of large structures, so it will pay us to study it more carefully. The mass of a large structure is distributed, and gravity therefore acts all across the structure. In Section 9–4 we showed that the result of an analysis of this situation is quite simple: *Gravity acts as though it were applied to a concentrated (point) mass equal to the total mass M at the center of mass of the extended object.* Thus, gravity produces a torque as though the gravitational force acted on the center of mass, which is sometimes called the *center of gravity* in this context.

EXAMPLE 11–3 Consider a rectangular book with a uniform mass density and a length L. The book is lying on a table with one side parallel to the table edge, hanging off that edge by an amount ℓ (Fig. 11–5a). How large can ℓ be before the book rotates off the edge and falls?

Setting It Up Figure 11–5b is a free-body diagram for the problem and includes a coordinate system with its origin at the table edge.

Strategy In the extended free-body diagram of Fig. 11–5b, gravity, $m\vec{g}$, acts on the book's center of mass, which is at a point midway along its length. If the book falls off the table, it will do so by rotating about the table edge. The force equilibrium condition, Eq. (11–3), can be satisfied through the cancellation of gravity by a normal force, $\vec{F}_N = -m\vec{g}$. For the torque equilibrium condition, Eq. (11–4), we want to calculate the torque about an axis passing through the origin along the edge of the table. The book will not fall off the table if we can maintain the torque equilibrium condition. Note that just at the point of rotating off the table, the normal force acts *at* the table edge (Fig. 11–5c).

Working It Out As long as the center of mass of the book lies above the table surface, we can satisfy the torque equilibrium condition, Eq. (11–4), if $\vec{F}_N$ acts at the same point as the force of gravity, as shown in Fig. 11–5b. If, however, the book's center of mass is beyond the edge of the table, then the second equilibrium condition can no longer be satisfied. The normal force will be acting at the edge of the table; at this point, the lever arm for the torque of the normal force is zero. The force of gravity will have a torque for rotations about the edge. Thus the equilibrium condition can no longer be satisfied, and the book falls (Fig. 11–5c) by rotating off the edge of the table. We conclude that the equilibrium conditions *can* be satisfied if the center of mass of the book lies above the table surface, and the largest possible value of ℓ is $L/2$.

What Do You Think? If you observe a book that is resting more than halfway over the table, what can you conclude? *Answers to **What Do You Think?** questions are given in the back of the book.*

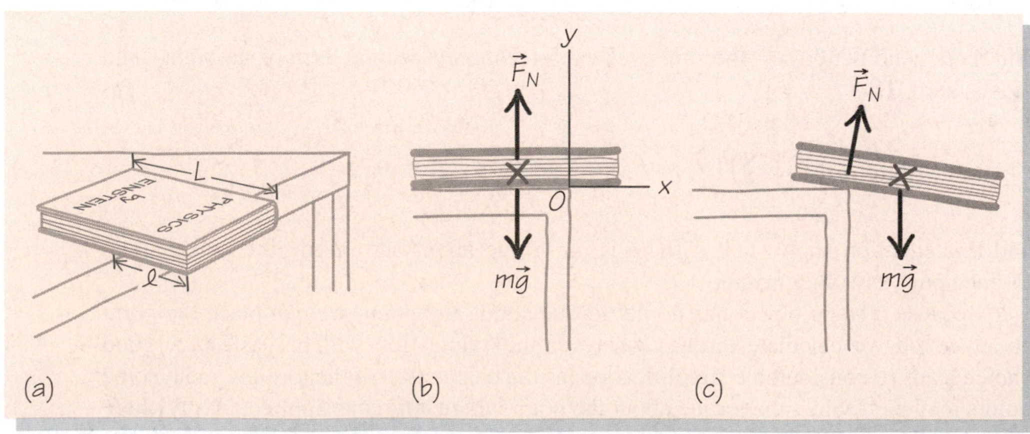

▶ **FIGURE 11–5** (a) A book rests on a table top with a length ℓ hanging off. (b) The center of mass of the book is over the table surface. (c) The center of mass of the book is beyond the table edge.

THINK ABOUT THIS . . .

IS CEMENT AN INDISPENSABLE ELEMENT OF CONSTRUCTION?

An interesting generalization of Example 11–3 involves the static conditions for a stack of several books of mass m on a table, shown in Fig. 11–6. If you stack enough books, it is possible to make a static pile with the topmost book lying as far from the table as you like (see Problem 13). Even with a stack of only four books, the topmost book can be made to lie completely off the table edge. While it is static, the extreme case is in unstable equilibrium, in the sense that a fly walking too far to the edge could cause the pile to fall over. These results can be applied to other constructions; for example, if building blocks of stone are used judiciously it is possible to build stable structures without the need for cement (usually used in the form of mortar between bricks and stones). A crucial ingredient for such structures is the keystone (Fig. 11–7), which, in effect, allows the joining of the topmost edges of two stacks of books. With the keystone, neither pile need be stable by itself; with it the structure is truly stable—the equilibrium is a stable one—in the sense that the structure will continue to stand even when the external forces change a little, as in an earthquake. Mortar often represents the weak point of a structure, and many ancient building structures have lasted precisely because good use was made of the possibility of stable equilibrium without mortar. Among these are numerous Egyptian, Greek, and Roman monuments.

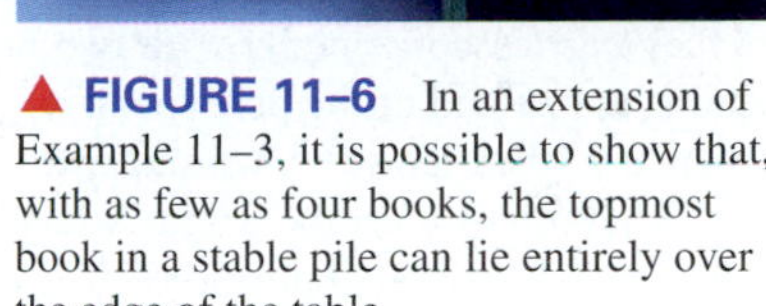

▲ **FIGURE 11–6** In an extension of Example 11–3, it is possible to show that, with as few as four books, the topmost book in a stable pile can lie entirely over the edge of the table.

◀ **FIGURE 11–7** The arch is a stable architectural element that has been used since ancient times, here at the Roman ruins in Djemila, Algeria.

CONCEPTUAL EXAMPLE 11–4 You have a cutout map of the United States made of thick cardboard. Use the fact that gravity effectively acts on the center of mass of the map to devise an experimental method to determine its center of mass. [*Hint*: First show that because gravity effectively acts on the center of mass, a suspended object will be in stable equilibrium when the center of mass lies directly below the suspension point. Then use this fact to devise your method of finding the center of mass.]

Answer We know that a simple pendulum has its lowest potential energy when the pendulum bob lies directly below the support. This is therefore a stable equilibrium point. An analysis in terms of torque about the point of suspension shows that only gravity exerts a torque, and that torque has the effect of rotating the bob back to the equilibrium point. Knowing that gravity acts on a suspended object as if all the mass were concentrated at the center of mass, the same analysis shows that only gravity exerts a torque about the suspension point and that the suspended object has a stable equilibrium with its center of mass directly below the suspension point.

We can now find the center of mass of our map with the aid of a plumb bob. The map is suspended twice by hanging it from two holes punched in it in different locations (Fig. 11–8). A plumb bob is dropped from each suspension point and a vertical line is drawn on the object (Figs. 11–8a, b). The center of mass lies at the intersection of the two lines. Hanging the object from a third point, as in Fig. 11–8c, is a good check on the technique.

What Do You Think? You have a little floor stand with a vertical shaft coming to a point, and you want to locate the center of mass of the map by a different method, this time with the map in a horizontal orientation. How would you proceed?

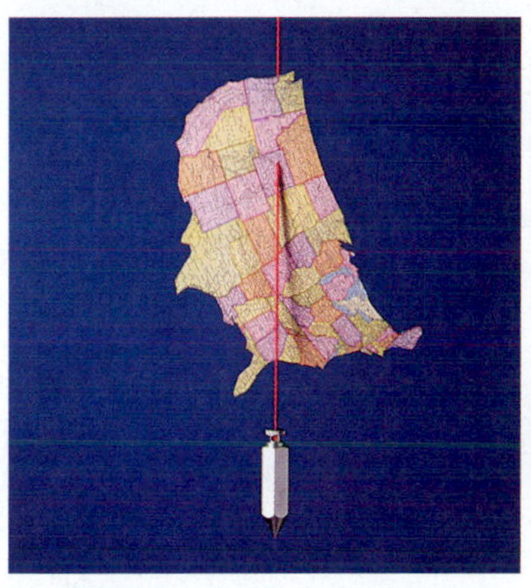

(a)

(b)

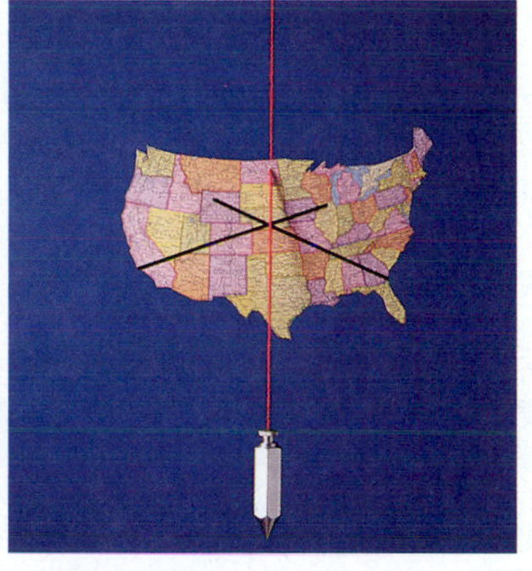

(c)

▲ **FIGURE 11–8** Here's a technique to find the center of mass of this cutout map of the United States. (a) Suspend the map from any point and drop a plumb line from that point. The center of mass lies somewhere along the line. (b) The same is true for a second plumb line dropped from a second point, and the intersection of the two lines is the location of the center of mass. (c) The plumb line dropped from a third point does indeed pass through the center of mass.

11–3 Applications of Statics

How many beams of a given size are required to support a roof of a given weight? Can the beams be reduced in size and still safely support the roof? Where should the beams be located? The equations for statics are often used to determine the forces ("loads") on such components, which in turn are used to find the location and size of components needed for construction on various scales, or to determine the limits of safe construction. In Examples 11–5 and 11–6, we look at considerations like these for some everyday situations.

EXAMPLE 11–5 In order to handle a hot pizza of mass 1.0 kg, a cook slides a light (i.e., of negligible mass) but stiff spatula of length 2.00 m from the right hand to the center of the pizza, which he then carries horizontally with two hands (Fig. 11–9a). His right hand is at one end of the pole (point A); his left hand is at point B, a distance 0.50 m farther down the pole; and the pizza is at the other end. What forces must the worker's hands exert on the pole so that it maintains its horizontal position?

Setting It Up The origin of our coordinate system is placed at point A (at the right hand). The figure labels the distance ℓ between the hands, the total length L of the pole, and the pizza mass m, as well as the unknown forces due to the hands.

Strategy The external forces acting on the pole are gravity and contact forces from each hand, and our strategy will be to use the static conditions to find the contact forces. In the free-body diagram for our extended object (Fig. 11–9b), we assumed that the hands will exert upward forces. But, as is usual with free-body diagrams, it is not necessary that the forces drawn correspond precisely in direction and magnitude to the forces to be determined—these values are determined through the static conditions for these forces. Having identified the forces and drawn the free-body diagram, our goal is simply to solve the static condition equations for those forces. Point A is a simple point about which to evaluate the torque because one of the forces runs through it.

Working It Out A net force on the pole of zero, Eq. (11–3), reduces to one equation for the vertical components of the force. A net torque on the pole of zero, Eq. (11–4), is also one equation because the torques about point A are all into or out of the page. With two equations we can find the two unknown forces exerted by the hands. Dropping the vector indications, our static conditions are

$$\text{Force: } F_A + F_B - mg = 0$$
$$\text{Torque: } \ell F_B - Lmg = 0.$$

Note that F_B and mg must exert torques in opposite directions.

We solve these equations for the unknown forces:

$$F_B = mg\left(\frac{L}{\ell}\right);$$

$$F_A = mg - mg\frac{L}{\ell} = mg\left(\frac{\ell - L}{\ell}\right).$$

Because $\ell < L$, F_A is negative: Hand A must in fact exert a downward force. In effect, point B, which lies between the two ends of the pole, acts as a pivot point about which the pole may rotate. The force at A must point in the *same* direction as the force of gravity on the mass to allow the torque to be zero. Note the coefficients in the forces that must be exerted by the hands. For example, F_B is much larger than mg if L is much larger than ℓ. It is difficult to hold a mass at the end of a long horizontal pole!

Inserting numbers, we find

$$F_A = (1.0\,\text{kg})(9.8\,\text{m/s}^2)\frac{0.50\,\text{m} - 2.00\,\text{m}}{0.50\,\text{m}} = -29\,\text{N};$$

$$F_B = (1.0\,\text{kg})(9.8\,\text{m/s}^2)\frac{2.00\,\text{m}}{0.50\,\text{m}} = 39\,\text{N}.$$

What Do You Think? Assume that the distance ℓ is fixed. If the pole were shorter, would the forces exerted by the pizza cook have to be (a) smaller, (b) larger, or (c) unchanged?

(a)

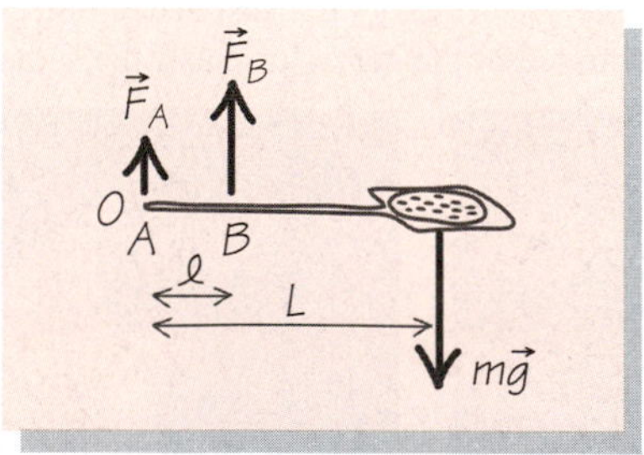

(b)

FIGURE 11–9 (a) A cook holds a pizza with a 2-m long spatula. (b) Extended force diagram. The object on which the forces act is the massless rod with the pizza at the end. The problem solution shows that the actual direction of $\vec{F}_A$ points down.

Problem-Solving Techniques

Suggestions for Statics Problems

1. After clear identification of the object in question, prepare an extended free-body diagram for it. This will include all the forces and where they act. Follow the first three steps described in the Problem-Solving Techniques box in Chapter 9.
2. Choose an origin and axes. In particular, the torques due to the forces may be simpler to compute about some origins than about others; an origin through which more than one force passes is good. Further, the more forces are aligned with coordinate axes, the easier the analysis.
3. Be clear about the direction you have chosen for positive torque.
4. Write down the conditions for statics that follow from steps 1 through 3—Eqs. (11–3) and (11–4).

Problem-Solving Techniques *(continued)*

5. Count the resulting equations as well as the unknown forces to be determined to make sure that the number of equations corresponds to the number of unknowns and the problem does indeed have a unique solution.
6. In solving the equations for statics, carry through an algebraic rather than a numerical solution as far as possible. In this way, you can make checks in various limits or for special cases and see whether the results are reasonable.

EXAMPLE 11–6 A crane whose cabin and engine are effectively fixed to Earth is used to lift a 5300-kg mass (Fig. 11–10a). The 10.0-m-long arm of the crane is supported at its base, at point B, by a strong but friction-free pivot, and at its top, at point A, by a supporting cable. The arm and the supporting cable make angles of 45° and 32°, respectively, with the horizontal. The mass is lifted by a line from a point on the arm 0.52 m from point A. Ignoring the mass of the arm (reasonable compared to the 5-ton mass being lifted) and assuming the suspended mass is not accelerating, compute the tension in the supporting cable.

Setting It Up We place the origin of the coordinate system origin at the bottom of the crane arm, point B (Fig. 11–10b). (In the discussion of strategy we'll see why this is a good choice.) We know the mass m that is to be lifted, the arm length L, the length ℓ from the end of the arm to where the cable lifting the heavy mass is attached, the angle θ_T between the horizontal and the supporting cable, and θ_{arm} that the crane arm makes with the horizontal. These quantities are all labeled in Fig. 11–10b. We want the tension magnitude, T, in the supporting cable.

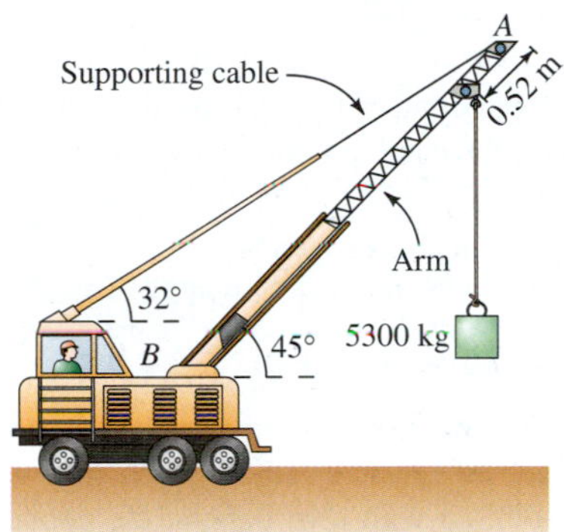

(a)

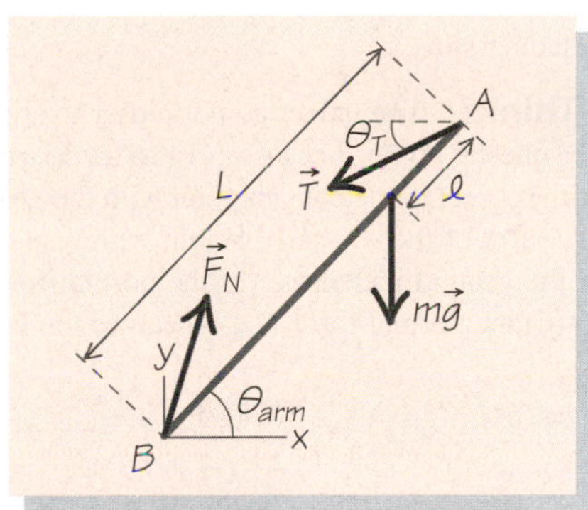

(b)

▲ **FIGURE 11–10** (a) Example 11–6. (b) Extended force diagram. The object on which the forces act is the arm of the crane. The relevant angle between the gravitational force and the position vector from point B along the crane's arm is given by $\theta_{\text{arm}} + 90°$. The relevant angle between the tension and the position vector from point B along the crane's arm is given by $-\theta_{\text{arm}} + 180° + \theta_T$. The sines of these angles enter into the expression for the torque about point B due to the forces of gravity and tension, respectively.

Strategy Figure 11–10b is an extended free-body diagram for the arm, which is the system in equilibrium. The forces acting on the arm are (1) the tension, $\vec{T}$, acting at A along the cable; (2) the weight of the suspended mass; and (3) an unknown contact force, $\vec{F}_N$, at the bottom pivot point. The contact force points in a direction to be determined, although it lies within the plane of the page. We must choose a reference point for the torque equilibrium equation. If this point is at the application point of any of the forces, then there will be no torque due to that force. Since the contact force has unknown direction, it will be easiest to choose the reference point to be point B, where the crane arm is attached (Fig. 11–10b). We set the origin there. We'll also want to count the equilibrium equations, Eqs. (11–3) and (11–4), to ensure that there are enough to allow us to solve for the magnitude of $\vec{T}$.

Working It Out From the free-body diagram (Fig 11–10b), we see that the force equation has two vector components:

$$x\text{-component: } F_{Nx} - T\cos\theta_T = 0, \tag{11–5}$$

$$y\text{-component: } F_{Ny} - T\sin\theta_T - mg = 0. \tag{11–6}$$

The forces all produce torques perpendicular to the page, so the torque equation has a single vector component perpendicular to the page:

$$-(L-\ell)mg\sin(\theta_{\text{arm}} + 90°) + LT\sin(-\theta_{\text{arm}} + 180° + \theta_T) = 0. \tag{11–7}$$

There are three equations: two for the x- and y-components of the net force and one for the torque. There are three unknowns: the two components of $\vec{F}_N$ and the tension magnitude T. The number of equations match the number of unknowns, and the problem has a solution.

Without $\vec{F}_N$ in the torque equation, Eq. (11–7) alone is sufficient to solve for T:

$$\begin{aligned} T &= \left(\frac{L-\ell}{L}\right)\frac{\sin(\theta_{\text{arm}} + 90°)}{\sin(-\theta_{\text{arm}} + 180° + \theta_T)}mg \\ &= \left(\frac{10.0\text{ m} - 0.52\text{ m}}{10.0\text{ m}}\right)\frac{\sin(45° + 90°)}{\sin(-45° + 180° + 32°)}(5300\text{ kg})(9.8\text{ m/s}^2) \\ &= 1.5\times10^5\text{ N} \end{aligned}$$

A useful check on this result is that if $\ell = L$, the mass is hung directly from point B and, from physical considerations, we expect no tension to be required; this is indeed the case. It is also true that if $\theta_T = \theta_{\text{arm}}$—so that the cable is also attached at the pivot point at B—then there is no way that T can provide a torque of the opposite sign from the torque coming from the weight, and the arm cannot be stabilized. Indeed, we cannot solve for T when the angles are equal.

What Do You Think? We refer to a crane effectively fixed to Earth at the start of this example. What might happen if it is not fixed to Earth?

An analysis like that of Example 11–6 could be useful if you were to design a crane and had to choose the size of the supporting cable. Notice that for this type of crane the tension in the cable must be much greater than the weight itself. For safety it would also be necessary to solve for the contact force to ensure that the pivot itself is sufficiently strong.

As an exercise, choose some other point—say, the other end of the arm—to be the origin and verify that the answer is independent of the choice of origin (see Problem 24).

EXAMPLE 11–7 A ladder of length 3.0 m is leaning against a wall at an angle of 58° as in Fig. 11–11a. Its eight rungs are spaced 0.33 m apart. The ladder's mass is insignificant compared to the 85-kg mass of a window washer who is climbing the ladder. The coefficient of static friction between the rubber feet of the ladder and the floor is 0.51, while you can ignore friction between the top of the ladder and the wall. Is the ladder safe from slipping if the window washer climbs to the seventh rung?

Setting It Up The extended free-body diagram, Fig. 11–11b, sets a coordinate system with origin at the base of the ladder. We know the ladder length L, the angle θ the ladder makes with the ground, the coefficient of static friction μ_s between the ladder and the floor, the mass m of the window washer, and the spacing d between rungs. Some of these are labeled in Fig. 11–11b. We want to know if static friction will be large enough to maintain static equilibrium when the mass m is centered at a given spot.

Strategy The extended object on which the forces act is the ladder along with the washer. We start by drawing the extended free-body diagram. The forces acting on this system are the normal forces $\vec{F}_{floor}$ and $\vec{F}_{wall}$, the friction force $\vec{f}$ between floor and ladder, and, given that the ladder is much lighter than the window washer, gravity acting on the washer. A crucial piece of information in solving this problem is that the static friction force takes on its *maximum value* just before slipping. Finally, to apply the torque equation, we need a reference point, and the chosen origin at the ladder base is a useful one because two forces (friction and the normal force of the floor) act here, and neither will exert torque about that point. (The answer will not in the end depend on the choice of reference point for torque, but we want to make the calculation as simple as possible.) The forces act in the plane of the page, and the torque is perpendicular to the page, so we'll have three (equilibrium) equations for the two normal forces and friction. We compare the calculated force of static friction with its maximum value,

$$f_{max} = \mu_s F_{floor}. \tag{11–8}$$

If the calculated magnitude of friction exceeds this maximum value, then the ladder will slip.

Working It Out Placing the washer at the nth rung of the ladder (from the bottom), and noting that friction acts to the left, the statics equilibrium condition for the forces are, in the x- and y-directions, respectively,

$$F_{wall} - f = 0, \tag{11–9}$$

$$F_{floor} - mg = 0. \tag{11–10}$$

For the torque, Fig. 11–11b shows that the lever arm for the force of gravity about the origin is $nd \cos\theta$, whereas the lever arm for the normal force of the wall is $L \sin\theta$. These tend to rotate the system in opposite directions. The torque equilibrium condition is thus

$$mgnd \cos\theta - F_{wall} L \sin\theta = 0. \tag{11–11}$$

Equations (11–9) and (11–11) can be solved for the contact force at the wall and the friction, and we find for friction

$$f = \frac{mgnd \cot\theta}{L}. \tag{11–12}$$

Equation (11–10) gives the contact force at the floor directly. The condition that the friction force at the floor not exceed the maximum value of static friction is

$$\frac{mgnd \cot\theta}{L} \le \mu_s F_{floor} = \mu_s mg,$$

$$nd \le \mu_s L \tan\theta.$$

The higher the washer climbs, the closer the ladder comes to slipping. The right-hand side of the inequality is $\mu_s L \tan\theta = (0.51)(3.0\text{ m})(\tan 58°) = 2.4\text{ m}$, so with $d = 0.33\text{ m}$, the condition can be written as

$$n \le (2.4\text{ m})/(0.33\text{ m}) = 7.3.$$

At the 8th rung the ladder slips.

What Do You Think? The ladder comes closer to slipping as the window washer climbs higher because (a) the lever arm of the washer's weight with respect to the contact point with the ground decreases; (b) the lever arm of the washer's weight with respect to the contact point with the ground increases; (c) the normal force F_{floor} becomes too large; (d) the normal force F_{wall} becomes too large.

(a)

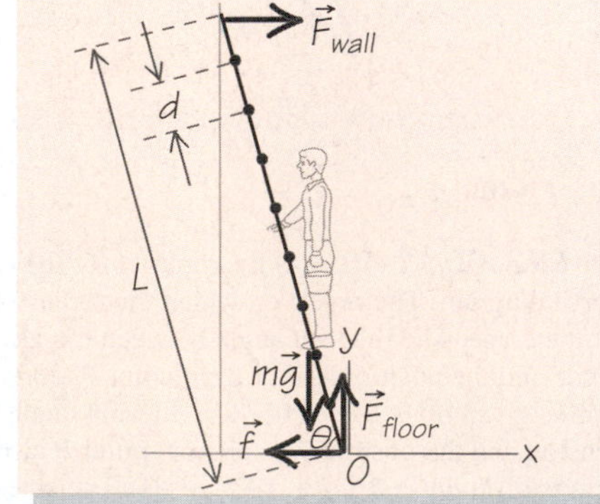

(b)

▶ **FIGURE 11–11** (a) The window washer is standing vertically, such that his center of mass is over the rung on which he stands. (b) Extended force diagram for the ladder. The lever arm for the force of gravity about the origin is $nd \cos\theta$, and the lever arm for the normal force of the wall is $L \sin\theta$.

EXAMPLE 11–8 The biceps muscle is responsible for bending your arm. It acts through a kind of lever system (Fig. 11–12a), with an upward force where the biceps attaches to the bone and a downward force at the attachment point of the triceps. The elbow–hand distance a, and x, the distance from the biceps attachment point to the elbow, are important parameters. If a book of mass M is held in your hand with your forearm horizontal and your upper arm vertical, what upward force does the biceps have to exert on the forearm bones (radius and ulna) to remain in equilibrium? Ignore the mass of the forearm bones, and assume the hand-forearm forms a single system.

Setting It Up The extended force diagram of Fig. 11–12b, discussed further below, includes specification of the parameters a and x. We label the magnitude of the desired biceps force F_B.

Strategy The three vertical forces on the forearm are the force $\vec{F}_H$ from the upper arm acting at the elbow, the force $\vec{F}_B$ of the biceps on the forearm, and the weight of the mass, each included in the extended free-body diagram in Fig. 11–12b. For the torque, we choose the reference point to be the elbow, although the hand would do just as well. For this situation there is one condition for equilibrium from the force equilibrium equation (all forces are oriented vertically), and one from the torque equilibrium equation. These two conditions should be enough to solve for the two unknown force magnitudes, including F_B.

Working It Out We have for the torque equilibrium condition

$$0 = (Mg)a - F_B x.$$

We see that this condition alone determines F_B, and we have no need to write the force equilibrium condition. (That is because the force F_H exerts no torque when the reference point is the elbow.) We immediately solve:

$$F_B = Mg\frac{a}{x}.$$

The force exerted by the contracting muscle must be a large factor greater than the mass's weight. Some typical values for an arm are $a = 30\text{ cm}$ and $x = 4\text{ cm}$, for which the factor is $(30\text{ cm})/(4\text{ cm}) = 8$.

What Do You Think? Consider two people of the same mass, height, and muscle strength. One of them can do pull-ups much more easily than the other. What is a possible explanation?

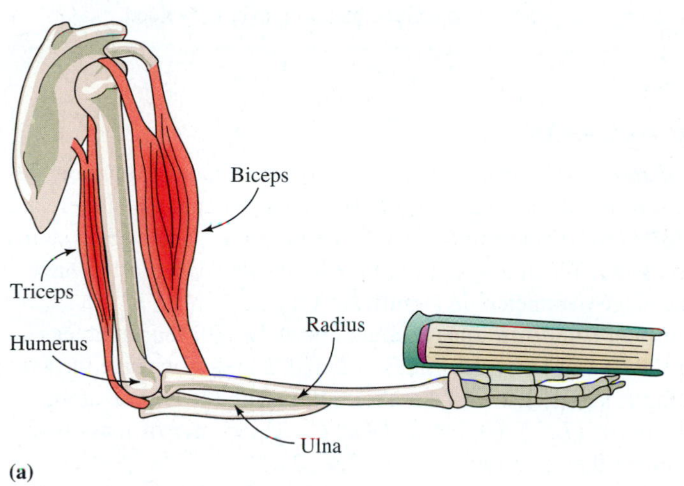

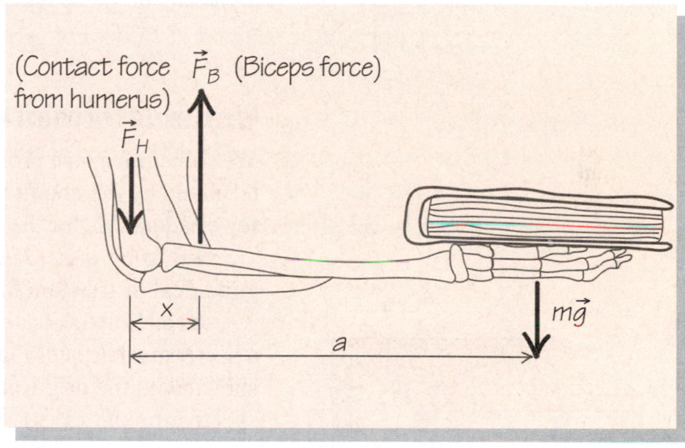

▲ **FIGURE 11–12** (a) The bicep muscle bends the arm at the elbow. (b) Extended force diagram for forearm/book.

Statics is an important element in understanding the mechanical aspects of living things such as ourselves, and the structure treated in Example 11–8 is representative of the study of this sort of question. Living animals are remarkably adaptable, many creatures being capable of a wide variety of activities. When we apply equilibrium conditions to flexible objects such as animals, we approximate them as a collection of rigid parts (bones) that have variable orientations (Fig. 11–13).

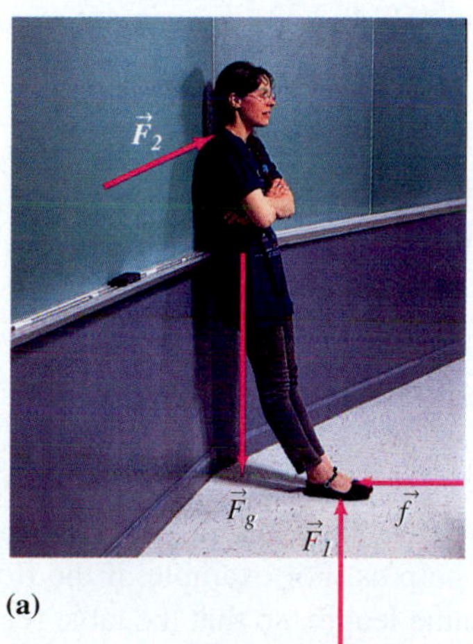

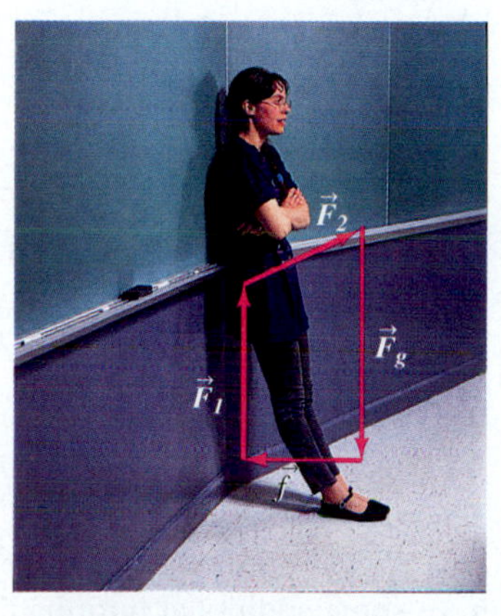

◀ **FIGURE 11–13** Biological systems can change the relative orientations of their different parts to allow them to be stable. (a) At least four different external forces act on the professor. (b) These forces add to a net force of zero.

THINK ABOUT THIS . . .

WHY DON'T CONSTRUCTION CRANES TIP OVER?

▲ **FIGURE 11–14** Construction cranes are a familiar feature of the urban landscape.

A common crane used at construction sites (Fig. 11–14) consists of a flimsy-looking tower constructed from three or four planes arranged so that the tower has a triangular or square cross section. Each plane consists of a lattice of triangular forms (this is called a geodetic construction). At the top of this tower is a horizontal boom that can lift the large masses involved in the construction. The tower can rotate and the boom can move the mass radially inward or outward, all in order to position the object at the correct location; the object is finally set down. (While these towers are tall, they do not have to be excessively so; they are gradually moved up as a skyscraper under construction gets taller.) The strength of such a tower is insufficient to withstand the torque on it due to a large mass at the end of the boom, so there is a balance weight at the far side of the crane that moves in an opposite direction from the lifting equipment and the object being moved. This exerts a torque on the tower equal and opposite to the torque from the weight of the large mass. The motion of the counterweight is controlled by sensors at the top of the tower, which ensure that the boom produces no net torque at the top of the tower, and this in turn ensures that the tower does not tip over.

Another technique is applied to the portable cranes that are carried by trucks. These cranes have sensors that measure the downward force of the crane and truck body on each wheel. If this force reaches a lower limit, indicating that a wheel is lifting off the ground, movement of the crane boom is halted, with operator controls being overridden. Thus the danger that the crane might tip over is avoided.

Underdetermined Systems

We sometimes face problems in static systems where the forces cannot be uniquely determined by the conditions of force and torque equilibrium. This occurs when there are fewer equations for the forces to be determined than there are forces, and such systems are said to be *underdetermined*. When a system is underdetermined, we often have a good deal of freedom to change parameters in significant ways.

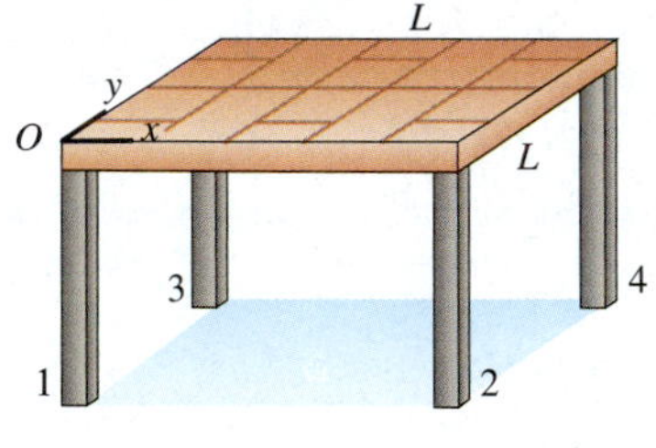

(a)

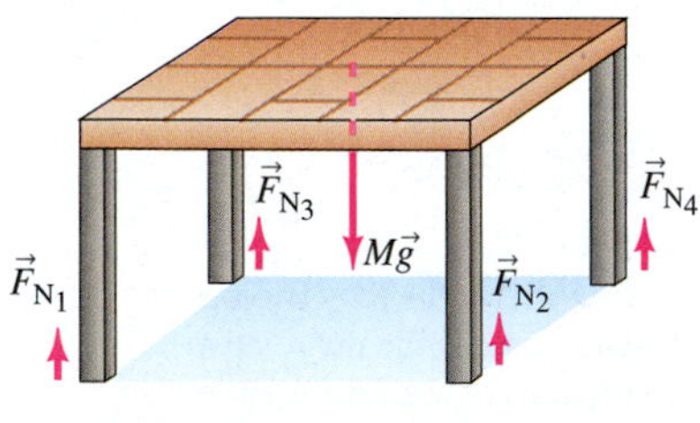

(b)

▲ **FIGURE 11–15** (a) A uniform four-legged table. (b) Extended force diagram for the table.

As an illustration, consider a uniform square table of mass M with four light legs on a horizontal surface (Fig. 11–15a). The top has sides of length L and, if we take the axes shown with the origin at leg 1, legs 1, 2, 3, and 4 make contact with the ground at the respective points $(x, y) = (0, 0)$, $(L, 0)$, $(0, L)$, and (L, L). The center of mass of the table is at its geometric center, the point $(x, y) = (L/2, L/2)$.

The forces acting on the table are the four contact forces $\vec{F}_{N1}, \ldots \vec{F}_{N4}$ acting upward on the legs, and the force of gravity, $M\vec{g}$, acting on the center of mass. Figure 11–15b is the free-body diagram. The forces all act in the vertical, or z-, direction. We calculate the torque about the origin. The torque vectors due to the forces are perpendicular to these forces and therefore lie in the xy-plane. Our static conditions thus consist of one force equation (for the z-component) and two torque equations (for the x- and y-components). These equations are, respectively,

$$F_{N_1} + F_{N_2} + F_{N_3} + F_{N_4} - Mg = 0,$$

$$LF_{N_3} - \frac{L}{\sqrt{2}}Mg \sin 45^\circ + \sqrt{2}LF_{N_4} \sin 45^\circ = 0,$$

$$-LF_{N_2} + \frac{L}{\sqrt{2}}Mg \sin 45^\circ - \sqrt{2}LF_{N_4} \sin 45^\circ = 0. \qquad (11\text{–}13)$$

With $\sin 45^\circ = 1/\sqrt{2}$, the last two (torque) equations simplify to

$$F_{N_3} - \frac{1}{2}Mg + F_{N_4} = 0, \qquad (11\text{–}14)$$

$$F_{N_2} - \frac{1}{2}Mg + F_{N_4} = 0. \qquad (11\text{–}15)$$

We have only three equations to determine the four unknown forces $\vec{F}_{N_i}$. We could have some additional information to help us. For example, if the floor were rigorously flat and if every leg had exactly the same length, so that the table really was symmetric, then we could say that the contact forces must be the same. But lacking this information, we cannot solve uniquely for the individual contact forces. The flip side of the coin is that there is a good deal of freedom in our ability to change parameters and still have a stable table. One leg—any leg—could be a millimeter shorter than the others and the table would still be stable against falling over even if it wobbled at that point. Indeed, we could even remove any one leg completely and still have a table in equilibrium. Then Eqs. (11–13) through (11–15) give three equations for the three remaining unknown contact forces. But a four-legged table is in stable equilibrium—it remains in equilibrium if there are small changes in parameters like the length of the legs—while a three-legged table is in unstable equilibrium for some motions, meaning that a small change could make a big difference in the situation. If a teacup were placed just off the center of the table toward the corner with the missing leg, the table would tip over in that direction.

In engineering language, we might say that an underdetermined system is a "robust" system. But we could also say that it is an overbuilt system in that it could be made lighter and simpler in various ways. Which is more desirable is a matter of budget and tolerance for instability. You certainly wouldn't put a square table with three legs in your dining room!

11–4 Solids and How They Respond to Forces

To this point we have assumed that all the objects we have dealt with are rigid. Of course that is only an approximation, and not just because there is some give between different elements that make up a structure. Even the solid materials that make up our "rigid" objects deform in various ways when a force is applied. While solids may appear at first glance to resist any kind of deformation, close examination reveals that they exhibit responses to external influences. When compressed, their volumes change slightly; when put under tension or heated, they stretch.

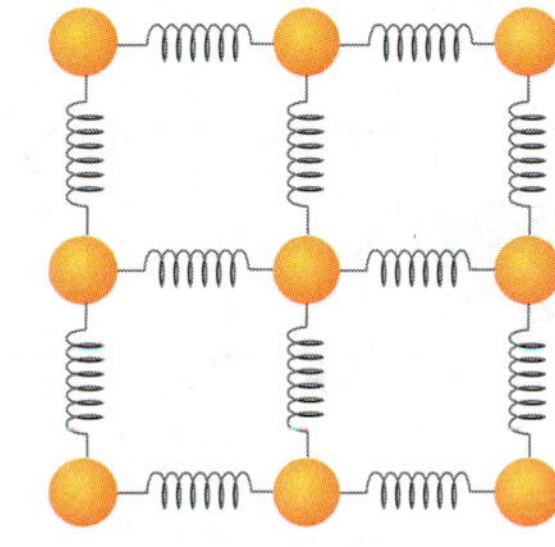

▲ **FIGURE 11–16** Schematic diagram of atoms connected by bonds in a lattice structure.

The explanation for this behavior lies in the underlying atomic structure of the solid, an extremely important ingredient in our understanding of the physical world. Solids are aggregates of atoms or molecules for which the interatomic forces lead to an organized, three-dimensional grouping of atoms, called a *lattice structure*. Figure 11–16, a schematic two-dimensional representation of such a structure, depicts the interatomic forces as springs. Because the lattice structure is one of stable equilibrium, the use of spring forces is at least approximately correct. A **crystal** is an object whose atoms have been arranged in a lattice structure. Different crystals are determined by the nature of the interatomic forces between atoms or molecules and by the shapes and orientations of these constituents, that is, how they fit together; atoms can be arranged to form crystal structures in many different ways. Figure 11–17 shows the three-dimensional structure of one particularly simple type of crystal structure, the *simple cubic lattice*.

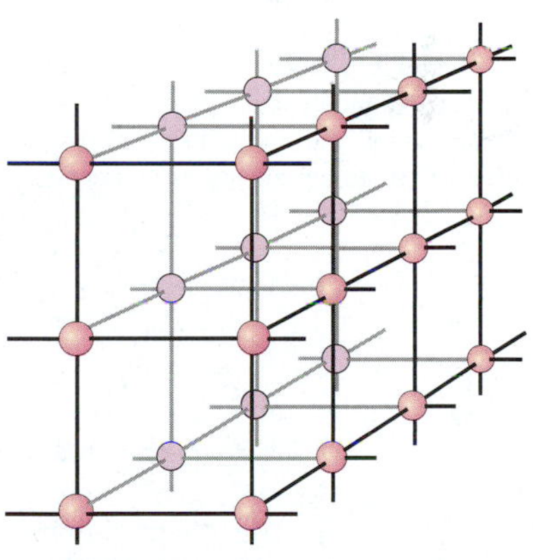

▲ **FIGURE 11–17** Sketch of a simple cubic lattice.

Even one kind of simple atom can form different crystals—both graphite and diamonds, two solids with very different properties, are formed of carbon atoms alone. If geometrically complicated molecules rather than simple atoms make up a crystal, or if several different kinds of atoms are present, the number of possible crystal structures and, perhaps more important, how the crystal structure influences the bulk appearance and behavior, increases rapidly. The way different atoms assemble into crystals depends also on the conditions under which they are formed. The formation processes for diamond and graphite are quite different. In a certain sense, the formation of a crystal by atoms is like a three-dimensional jigsaw puzzle with identical pieces, *which has more than one solution*!

As important as crystalline structure is to the macroscopic behavior of a solid, so are its lattice imperfections, called **defects** or dislocations. Defects are closely associated with the way a crystal grows. Figure 11–18 shows a simple two-dimensional crystal that is easy to grow and observe: a raft of bubbles that float on the surface of a liquid. Figure 11–18a is a perfect crystalline form, whereas a defect has been introduced in Fig. 11–18b. Can you recognize it? A more general type of defect is associated with grains and grain boundaries. A **grain** is a region in a solid where the crystal structure is perfect. Chance plays a role in the organization of atoms when crystal growth starts, and as an object cools and changes over from a liquid to a solid, a lattice may start to form in one region with one

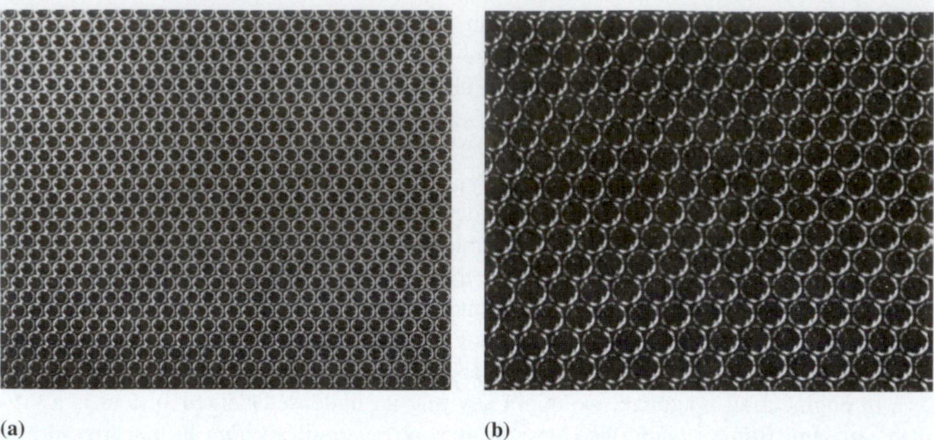

▶ **FIGURE 11–18** (a) A perfect crystalline raft of bubbles. Bubbles will arrange themselves in crystalline patterns (in this case, two-dimensional), just like the atoms in real solids. (b) A defect in a raft of bubbles.

orientation and in another in another region. Once a lattice starts growing, it serves as a template for the lining up of other atoms, but if there are two growth centers with different orientations, these will not combine to a single crystal. Rather, there will be two grains. The boundaries where these grains meet involve special kinds of defects. Figure 11–19a shows such a boundary. A real solid is made up of many grains (Fig. 11–19b). Another class of defects occurs when single lattice sites are empty or are occupied by an impurity (an atom chemically different from those that make up the bulk of the crystal), or when impurities occupy spaces between lattice sites. Defects are related to a crystal's color, luminescence, transport properties, and mechanical properties. The study of the nature of defects is a major field of research and has taught us much about the behavior of solids.

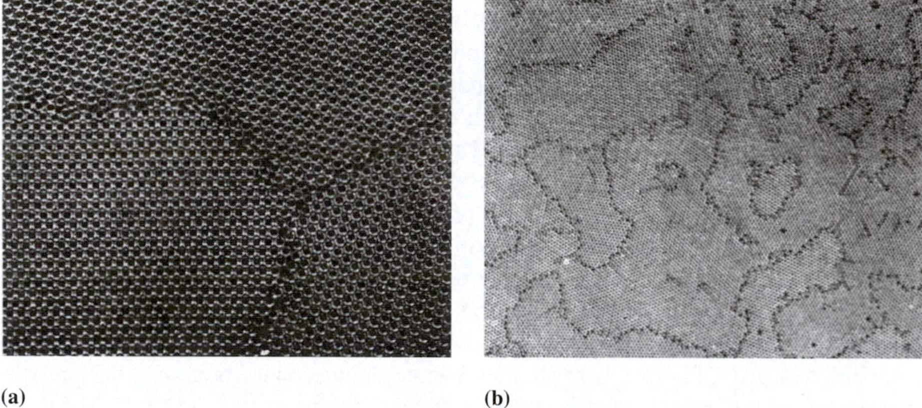

▶ **FIGURE 11–19** (a) A close look at the grain boundaries in a crystal, as illustrated by a raft of bubbles. (b) An overview of numerous grains in a crystal.

Our simple picture of lattice structure illustrates why a solid is not truly rigid. Figure 11–20 suggests how a cubic structure might react to compression (Fig. 11–20a), to stretching (Fig. 11–20b), and to a *shear* (Fig. 11–20c). Just how hard it is to

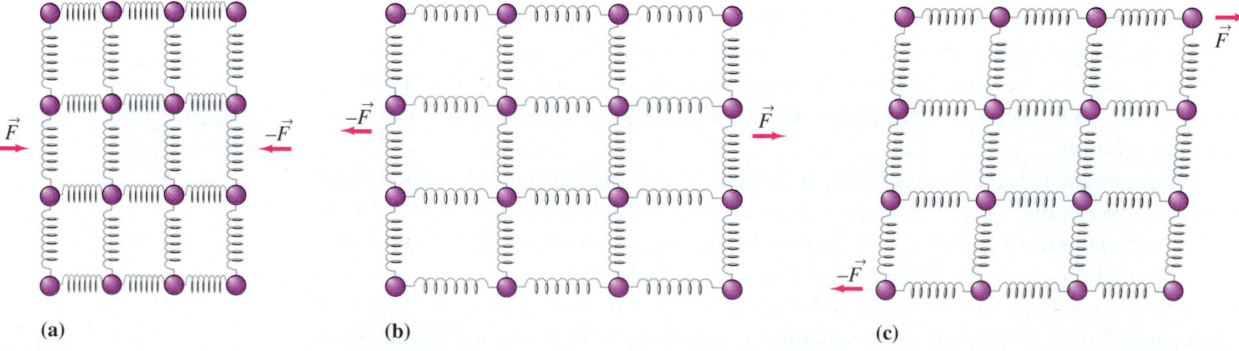

▲ **FIGURE 11–20** Schematic diagram of lattice distortions: (a) compressed in the horizontal direction; (b) stretched in the horizontal direction; (c) subjected to shear forces, which act in opposite directions at the top and bottom.

compress, stretch, or shear a solid depends on its interatomic forces, sometimes referred to in shorthand as the interatomic *bonds*. In the final analysis, it is these forces, along with the grain and defect structure, that differentiate steel from a pie crust. (For more on this, see the "Think About This" feature on the strength of solids on p. 327.)

In principle, all properties of solids can be described by describing the atomic structure of solids, but in practice it is simpler to describe the properties of solids using observed macroscopic behavior. This is the point of view we take in the remainder of this section, where we develop empirical relations about the deformations of solids. Remember that even if these rules are sometimes called "laws," they are more properly rules of thumb, valid only under restricted conditions.

Stresses and Strains

Stress is a measure of the force that acts on a solid object, whereas **strain** is a measure of deformation, the object's response to that stress. Our aim here is to connect these quantities. To do so, we shall assume that the applied force is small in some sense, which implies that the response is correspondingly small. (Another type of response, more properly described as catastrophic, is associated with large forces. When a heavy hammer strikes a piece of ice, the ice shatters. Or a girder may buckle if it is loaded too much. This aspect the behavior of materials is as important as the small response to small forces, and we'll discuss it later.)

We define the stress on a solid object, S, as the external force per unit area that is exerted on it:

$$S \equiv \frac{F}{A}, \qquad (11\text{–}16)$$

STRESS DEFINED

where F is the magnitude of the force that acts perpendicularly to an area A of the object. (Actually we are oversimplifying here. We really should include the possibility that the force is not perpendicular to the surface in question—see our discussion of shear below. However, what we have done is adequate for the kind of discussion carried out in this chapter.) As an example of this, we might have a force that acts on a solid rod along its length and tends to stretch it (Fig. 11–21)—this particular force is a *tension*. When the rod is subjected to a force along its length and tends to shorten it, you have a compression. These are the same sorts of forces but with opposite sign, and the usual convention is that tension corresponds to a positive stress, and compression to a negative stress. Stresses of this type are called *longitudinal stresses*. You will quite commonly have this sort of situation to deal with in the beams that make up a structure. Stress has the same dimensions as pressure and has SI units of newtons per square meter. As this unit is rather small for ordinary engineering applications, a more practical unit for stress is the meganewton per square meter (MN/m^2).

The *response* to the stress associated with either a tension or a compression is called the *compressional strain, e*. It is defined as *the fractional change in the length of the solid object along the direction of the compressional force*:

$$e \equiv \frac{\Delta L}{L}. \qquad (11\text{–}17)$$

COMPRESSIONAL STRAIN DEFINED

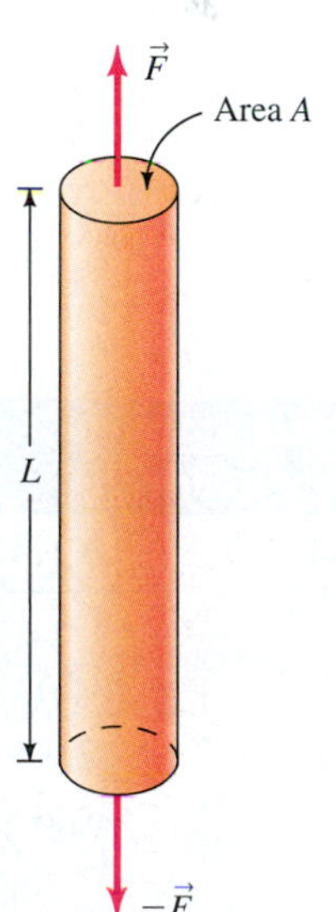

▲ **FIGURE 11–21** Equal but opposite forces applied evenly to opposite ends of a piece of material cause the material to stretch.

Note that e is dimensionless. It is negative if the stress is due to compression, and positive if it is due to tension.

The result of a tension on the rod in Fig. 11–21 is more than just an increase in length; when it stretches, the bar also *shrinks* in its lateral dimensions—the width h and depth w (or the radius if the rod has a circular cross section). If the force is a compression, the bar will instead bulge in the lateral directions. The fractional shrinking or bulging is proportional to the compressional strain by an amount characteristic of the material, and it is described by the positive constant called *Poisson's ratio*, σ, after Siméon Poisson, who worked in the first part of the 19th century:

$$\frac{\Delta h}{h} = \frac{\Delta w}{w} = -\sigma e. \tag{11–18}$$

A typical value of σ for solid material is 0.3, and this constant must in fact always be less than 0.5. You can show (see Problem 60) that if it were to exceed 0.5, the volume of the solid would increase when a uniform pressure acts *in* on it, and this would be quite at odds with our experience.

If an external force F of constant magnitude is applied everywhere perpendicular to the entire surface of a solid object, as for example for rocks below Earth's surface, the volume V of the object changes. The volume decreases if the force is inward and increases if the force is outward. The *volume stress, p*, is defined as stress was defined in Eq. (11–16): $p \equiv F/A$, where A is now the *total* surface area of the solid. For solids the volume stress can equally well refer to an outward force on the body, as when an object is hollow and contains a gas of high pressure. The accompanying response is a *volume strain*, e_V, defined by the fractional change in volume due to a volume stress: $e_V \equiv \Delta V/V$. If Poisson's ratio were small, the volume strain would be just a special case of the compressional strain; a bar that stretches while its cross-sectional area A remains constant has a volume $V = AL$, and when such a bar is stressed longitudinally, the change in volume is $\Delta V = A\,\Delta L$. When these quantities are substituted in the definition for e_V, we obtain Eq. (11–17) for e.

There is a third kind of stress and its corresponding strain that we want to mention: shear stress and shear strain. Suppose that a uniform force acts *along* a face of a solid object, as in Fig. 11–22a. A tangential force like this is called a *shear force*. If the bottom surface of the solid is fixed (by some kind of contact force—glue, say), the result is a distortion of the object (Fig. 11–22b). The quantity F/A is the *shear stress* on this solid, and the *shear strain* is the fractional amount by which the upper surface moves, $\Delta L/L$. The volume of an object is essentially unchanged under shear stress, in contrast to what happens under a longitudinal stress.

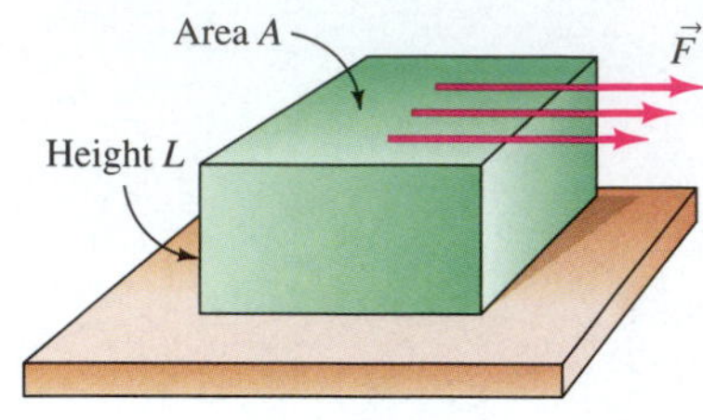

(a)

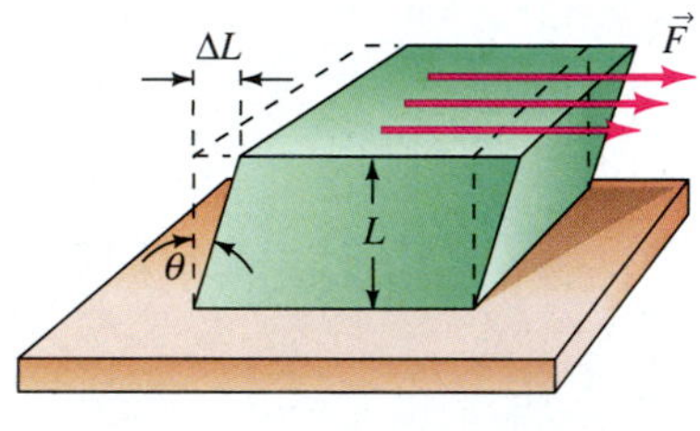

(b)

▲ **FIGURE 11–22** (a) An object fixed along its bottom surface has a force applied. (b) Shear stress from uniform force distorts the object.

The Relation Between Stress and Strain

Now we turn to the question of "laws" that relate a stress and its corresponding strain. As early as the seventeenth century, Robert Hooke realized that to a good approximation stress and compressional strain are proportional to each other. A clear understanding of this proportionality was first provided by Thomas Young in 1800, who wrote down the relation

$$\frac{S}{e} = Y. \tag{11–19}$$

LINEAR RELATION BETWEEN STRESS AND STRAIN

This equation applies to the relation between (small) longitudinal stresses (compression or tension) and the corresponding strain. The parameter Y is called **Young's modulus**, or the *elastic modulus*. Its value, which is a measure of the "stiffness" of a material, varies with the material. Y is analogous to the *spring constant, k*, in the spring force $F = -kx$. In fact, Young's modulus is a constant only for *small* strains because the atomic bonds behave just like simple springs, and we know that if a spring is stretched too much it will no longer exhibit a force that varies linearly with the amount of stretch. We all know about materials for which the linear relation between stress and strain fails: bread dough or licorice, materials that certainly don't show a decreased strain when the stress is decreased, i.e., they don't go back to their original shapes when the force "lets up." Because e is dimensionless, Young's modulus has the same dimensions as S and therefore has units of force per area—meganewtons per square meter, for instance. Table 11–1 lists values of Young's modulus for various solid materials. For small stresses and strains, the value of Young's modulus is generally independent of whether the material is under tension (positive strain) or under compression (negative strain). However, for some materials this is not a very good rule, as we shall discuss in the "Think About This" feature below.

TABLE 11–1 • Young's Moduli for Various Solids

Material	Y (MN/m²)
Rubber	7
Wood	14,000
Concrete	17,000–30,000
Bone	9,000–21,000
Glass	70,000
Aluminum	73,000
Steel	210,000
Diamond	1,200,000

EXAMPLE 11–9 A vertical steel rod of length 2.000 m and diameter 2.0 cm is fixed at the top and has a 9500-kg mass hanging from its lower end. Given that Young's modulus for this particular type of steel is 250,000 MN/m^2, calculate the elongation of the rod. What is the strain?

Setting It Up We have sketched the situation in Fig. 11–23.

Strategy The vertical force due to the hanging mass acts across the cross section of the rod; the force is a tension, so we can find the stress using Eq. (11–16). Given the stress, Eqs. (11–17) and (11–19) express the amount of stretch ΔL of the rod of length L and Young's modulus Y in terms of the stress.

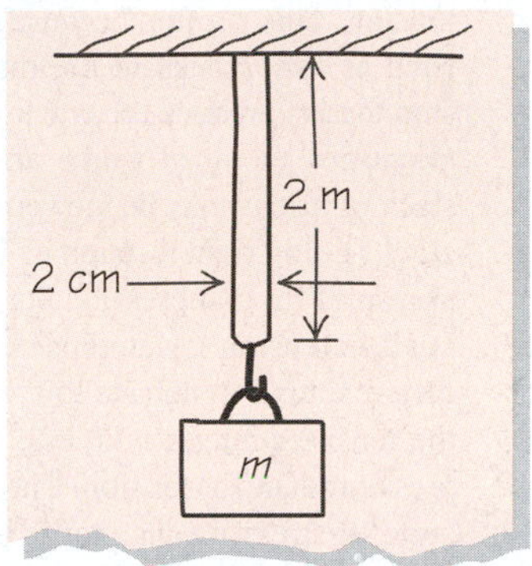

▲ **FIGURE 11–23** A large mass causes a slight elongation to a steel rod supporting its weight.

Working It Out The magnitude of the tension on the bar is mg, so that according to Eq. (11–16) the stress is $S = mg/A$, where A is the cross-sectional area of the rod. We then use Eqs. (11–17) and (11–19) to find

$$\Delta L = eL = \frac{S}{Y}L = \frac{mg}{AY}L.$$

With $A = \pi r^2$, where $r = \frac{1}{2}(2.0\text{ cm}) = 1.0\text{ cm}$ is the radius of the cross section,

$$\Delta L = \frac{mg}{\pi r^2 Y}L = \frac{(9.5 \times 10^3\text{ kg})(9.80\text{ m/s}^2)}{3.14(1.0 \times 10^{-2}\text{ m})^2(2.5 \times 10^{11}\text{ N/m}^2)}(2.000\text{ m})$$
$$= 2.4 \times 10^{-3}\text{ m}.$$

The 2.4-mm elongation gives a strain of

$$\frac{\Delta L}{L} = \frac{2.4 \times 10^{-3}\text{ m}}{2.000\text{ m}} = 1.2 \times 10^{-3}.$$

What Do You Think? We assumed the cross section was constant over the stretch. Was that a good approximation in this case?

The relation between volume stress and volume strain takes exactly the same form as the relation between ordinary stress and strain, for small strains. The proportionality constant B = (volume stress)/(volume strain), which is analogous to Young's modulus, is called the **bulk modulus**. We have

$$B \equiv -\frac{F/A}{\Delta V/V} = -\frac{p}{e_V}. \tag{11–20}$$

The minus sign means that an *inward* pressure on a solid, which is positive, implies a *decrease* in volume, $\Delta V < 0$. There is a relation between Young's modulus and the bulk modulus that must involve Poisson's ratio. Only two of these three constants are independent of one another (see Problem 59), and since many solid materials have nearly the same Poisson ratio, the bulk modulus of different materials tends to track the Young's moduli, listed in Table 11–1. Finally, there is a *shear modulus* G, which is given by the ratio of shear stress to shear strain. Liquids have a shear modulus of zero, meaning that they have an arbitrarily large shear strain for even small shear stresses. Liquids do not resist shear at all.

THINK ABOUT THIS . . .

WHAT MAKES A SOLID STRONG?

An important property of solids is their strength or, more specifically, their *tensile strength*, which is also known as *fracture stress*. This quantity is the stress required to break a material into two pieces—by pulling it along an axis, for example. The tensile strength of different materials varies a great deal, and this is a criterion in the selection of materials for a variety of tasks, such as in the manufacture of household goods, or in civil engineering. Steel may break when the tension is 3000 MN/m^2, whereas a piece of concrete may be pulled apart by a stress of only 5 to 10 MN/m^2. The tensile strengths for a variety of materials are listed in Table 11–2. The values of tensile strength are a factor of 100 to 1000 times smaller than the values of Y (Table 11–1) for the same materials. This huge difference reflects the importance of even tiny cracks and imperfections in the lattice structure of a solid.

(continues on next page)

THINK ABOUT THIS . . . (cont.)

TABLE 11–2 • Tensile Strengths for Various Solids

Material	Tensile Strength (MN/m^2)
Steel piano wire	3000
Steel	400–1500
Cast iron	70–250
Aluminum (pure)	70
Aluminum alloys	140–550
Copper	140
Titanium alloys	700–1400
Spruce, along grain	100
Spruce, across grain	3
Glass	700–170
Brick	5
Cotton	350
Spider silk	240
Human tendon	100
Rope	80

The tensile strength is a kind of *critical stress*. We may define a corresponding *critical strain* as $(\Delta L/L)_c$, such that

$$\text{tensile strength} \equiv Y\left(\frac{\Delta L}{L}\right)_c.$$

TENSILE STRENGTH DEFINED

The factor of 100 to 1000 difference between Young's modulus and the tensile strength for a given solid means that a fractional extension of a rod by 1 percent, or even by 0.1 percent, will break the rod. This is because the pervasive existence of cracks weakens the material. When a stretched rod breaks, the ultimate reason is that the interatomic forces between atoms are overcome. Without cracks, the stress on a section of rod between two layers of atoms is carried by many bonds, but when cracks are present, the stress is concentrated on fewer bonds. The force F in the expression F/A for the stress is the same, but the real area A over which the force acts is very much smaller than the apparent area. This means that the stress experienced by the material is much larger than the stress obtained by using the apparent area. For extremely thin—micrometer-size—fibers, which are nearly perfect crystals, measurements give a tensile strength of around $0.2Y$. This value is what we would expect upon evaluating $(\Delta L/L)_c$ from our understanding of interatomic forces. The discrepancy between the values of the tensile strength for bulk materials and tiny perfect crystals illustrates the importance of imperfections in the structure of solids.

In some materials the failure under compression is not the same as the failure due to tension. This is often because of imperfections, such as tiny cracks in the material. Compression tends to reduce the cracks, whereas tension magnifies them. (To take an extreme case, a stack of bricks may be viewed as one brick with cracks going right through it. The stack behaves as a solid for compression but has no resistance at all to tension.) Materials such as cast iron or concrete tend to behave this way, and for them the tensile strength is in effect larger for compression than for tension. The response of concrete to stress means that it is relatively weak under tension, and this is remedied by the addition of rebar (iron rods that form reinforcing bars) to make *reinforced concrete*. The rods have a high tensile strength under tension but can catastophically buckle under compression. Reinforced concrete works because its two components have compensating properties against failure and is therefore strong under both tension and compression.

Summary

An extended object is static when the net external force and the net external torque about any origin are both zero:

$$\sum_i \vec{F}_i = 0 \tag{11–3}$$

and

$$\sum_i (\vec{R}_i \times \vec{F}_i) = 0. \tag{11–4}$$

In order to apply these conditions, it is necessary to know both what the external forces acting on an object are and *where* they act. Gravity acts as though it were applied to the center of mass of a rigid body. We can say that the weight of an object acts on its center of mass.

The static conditions are most often used to learn under what conditions a structure can be held static or to learn what forces various components of a structure must endure. Some structures are underdetermined, which means that it is not possible to determine uniquely the forces acting through the consideration of the static conditions alone.

Solids consist of atoms arranged in regular crystalline structures. They are held in this structure by the interatomic forces, which can be considered springlike. As a consequence, solid bodies may stretch or be compressed when subjected to external forces, and are not truly rigid. For longitudinal (pulling or pushing) forces, the force per unit area, or stress, S, is related to the fractional change in length, or strain, e, by Y, Young's modulus:

$$\frac{S}{e} = Y. \tag{11–19}$$

A similar relation applies to volume stress, p, and volume strain, e_V, which are related by the bulk modulus, B:

$$B = -\frac{p}{e_V}, \tag{11–20}$$

and finally there is still another relation for shear stress and strain. Beyond the small deformation that occurs for small stresses, solids break and buckle. The real tensile strength of solids, which is a measure of the amount of strain at which they fracture, is smaller than one would expect with a description based purely on interatomic forces and a perfect crystal structure. That is because of the many imperfections that occur in the crystalline structure of solids. The nonrigidity of solids is evidently reflected in the fact that real structures are never perfectly rigid.

Understanding the Concepts

1. It is a common (and quite useful) piece of advice for those who are learning mountaineering techniques to "stand away" from the mountain on steep slopes rather than to follow one's instincts and "hug" the mountainside. Explain why this is so, using your knowledge of torques and forces on a climber standing on the steep slope of a mountain.
2. A baby pulls straight down with all his might on the flush handles of the closed drawer of a bureau (Fig. 11–24). Can he cause the bureau to tip over?

▲ **FIGURE 11–24** Question 2.

3. Why does a rope from which an object is suspended line up with the vertical?
4. Is it possible for an object not to be in equilibrium even though the net force on the object is zero? If you answered yes, give at least one example.
5. You are sitting quietly in a porch rocker that is suspended by chains from the ceiling. Are you in a stable or unstable equilibrium?
6. Consider Example 11–7. Does it help (in the sense of allowing the washer safely to climb higher) if the ladder is massive? You may want to think about this by imagining the ladder is *very* massive.
7. Bridges and buildings are not really rigid. Does this mean that nothing we have said in this chapter is relevant to them? How can an object be "approximately" rigid?
8. Look at the rock structure in Fig. 11–1. If your task were to topple the rock with a single minimum force, where would you apply it and why?
9. A massive bar is maintained horizontal by three point supports placed at three specified positions along the length of the bar. True or false: You can determine the contact forces at each of the three supports.
10. Is the height of the four-legged table discussed in Section 11–3 irrelevant to the calculation of the equilibrium conditions for the table?
11. A motorcycle with its weight equally distributed over the wheels is resting on level ground with its front wheel bumped up against a curb, perpendicular to the curb. The motorcycle is driven by its rear wheel. Describe the equations that determine the motion of the motorcycle. Can the cycle climb the curb?
12. A pendulum suspended from the roof of an accelerating rail car makes a nonzero angle with the vertical. The pendulum is not swinging. Is this a case of stable equilibrium?
13. For some objects (U-shaped objects, for example), the center of mass is outside the object itself. For such objects, can we still think of gravity as though it acts on the center of mass?
14. The doughnut-shaped space station in Stanley Kubrick's film *2001: A Space Odyssey* rotates about its axis of symmetry (see Fig. 5–27a). Is the rotating station in stable equilibrium, unstable equilibrium, or neither?
15. Suppose you firmly hold the shoelaces of your shoe, one at an angle of 30° with the horizontal, the other at an angle of 60° with the horizontal. You then push straight down with your foot. One of the shoelaces breaks. Which one will it be?
16. Does the method of finding the center of mass of a flat object, as discussed in this chapter, work even if the density of the object is not constant?
17. You want to hold a beam with its end against a frictionless wall. Is it possible to do so by running a rope from the far end of the beam to any attachment point on the wall?
18. When a cart has to be pushed uphill, it is better to exert one's force by pushing on the top of the wheel rather than at the axle of the wheel. Why is that?
19. Can a ladder placed against a rough vertical wall remain standing when the floor is so smooth that there is no friction between it and the feet of the ladder?
20. Identical twins are placed at opposite ends of a seesaw pivoted about its midpoint. No forces other than those due to the twins, the pivot, and gravity act on the seesaw. What determines the inclination to the horizontal made by the seesaw when it is balanced? Is the equilibrium stable, unstable, or neutral?
21. Suppose that the seesaw of Question 20, still pivoted about its midpoint, has a sharp downward bend at the midpoint. If there is an equilibrium, will it be stable, unstable, or neutral?
22. Why is cement unsuitable as a construction material for a boiler?
23. In Chapter 5 of the Book of Exodus, the Israelites in captivity complain to Pharaoh that they are being asked to make bricks without straw. Why was it a good idea to put straw in the clay that was allowed to dry in the hot sun?
24. Why does the amount of stretch of a bar on which a tension acts depend on the overall length of the bar? [*Hint*: Think of breaking the bar into two pieces of equal length and of how much each piece would stretch under the same force.]
25. If it is possible to make diamonds from graphite by using high pressure, how would it be possible to make graphite from diamonds?
26. On a macroscopic level, solids are distinguishable from liquids by their resistance to shear. Is this distinction likely to be a sharp one?
27. What kind of crystal structure might have little resistance to shear forces in some directions and much resistance in other directions? Would such a crystal have direction-dependent resistance to stretching?

Problems

11–2 Gravity and Rigid Bodies

1. (I) A uniform board of mass 80 kg and length 3.6 m is placed on top of a pivot 1.2 m from one end. What mass must be put at that end to allow the board to balance?

2. (I) A 20-kg board 2.5 m long is supported in a horizontal position at the two ends. A 70-kg worker stands 1.2 m from one end. What forces are exerted by the board on the two support points?

3. (I) Two workmen each carry one end of a 2.2-m-long ladder of mass 24 kg. The ladder is tapered so that its center of mass is 0.9 m from the wider end. What are the forces exerted by the ladder on the two workmen?

4. (I) A rail of length 3.0 m and mass 8.0 kg runs horizontally between two scales; a bowling ball of mass 5.5 kg is allowed to roll at a steady speed of 0.15 m/s from the left scale to the right scale (Fig. 11–25). During the time the ball is moving, how do the readings on the two scales change?

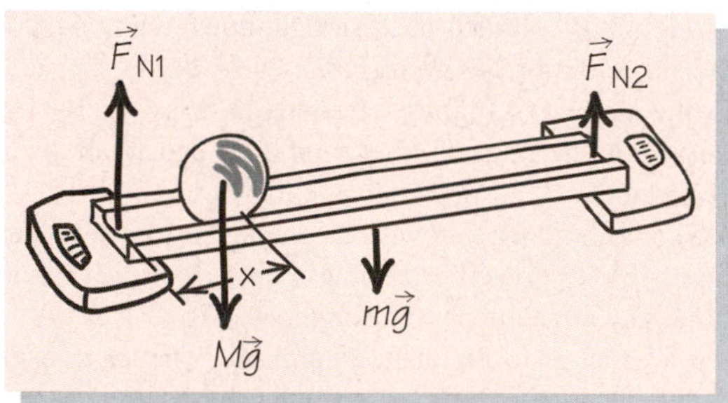

▲ **FIGURE 11–25** Problem 4.

5. (I) A projectile has broken into two parts. At time t_0, the two parts are located at the following points: m_1 at $(x, y, z) = (3, 0, 0)$ and m_2 at $(0, 0, 3)$, where all distances are measured in meters. The mass m_1 is twice the mass m_2. What is the location at time t_0 of the point that follows a parabolic trajectory (assuming that there is such a point)?

6. (II) Two people of unequal strength must carry a uniform beam of length L while holding it horizontal. The weaker of the two holds the beam at one end. (a) How far from the other end must the stronger person hold the beam in order to support three-quarters of the weight? (b) Is there a way in which the stronger person can carry the beam at one end and still support more than half the weight of the beam?

7. (II) A playground seesaw is balanced at its midpoint. Two children, weighing 25 kg and 40 kg, respectively, want to balance on the seesaw. If the children are separated by a distance of 2.8 m, how far from the pivot point will the lighter child sit?

8. (II) Consider a seesaw whose total mass is 8 kg and total length is 3.50 m. Suppose the seesaw is placed off center on the pivot point so that the pivot point is 24 cm from the center of the seesaw (Fig. 11–26). How far from the center will the children of Problem 7 have to sit if the lighter child sits on the longer part of the seesaw? (Their separation is still 2.80 m.)

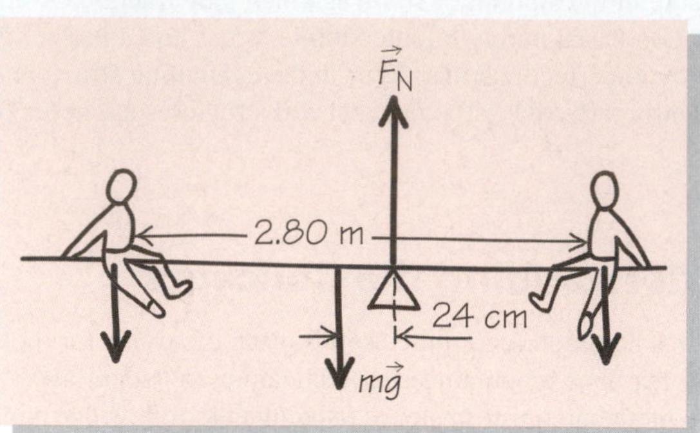

▲ **FIGURE 11–26** Problem 8.

9. (II) In order to hang a load of mass $M_1 = 30$ kg from the horizontal, flat roof of a building, a plank of length $L = 2.4$ m is placed on the roof (Fig. 11–27). One end is held in place with a chunk of concrete of mass $M_2 = 15$ kg, and the other supports the load M_1 with a light rope. How far can the end of the plank reach without tipping over? Neglect the mass of the plank.

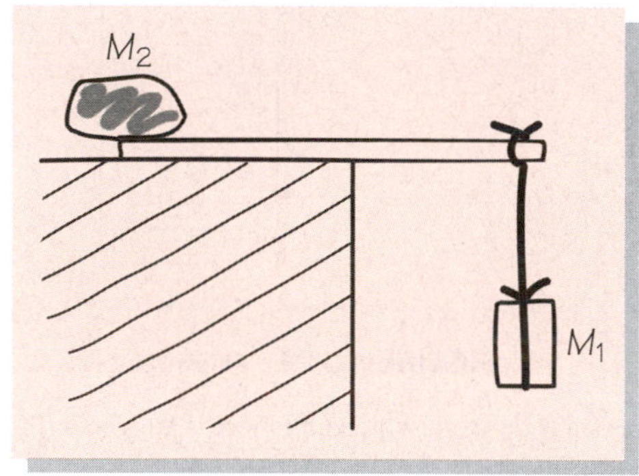

▲ **FIGURE 11–27** Problem 9.

10. (II) A uniform beam of mass 80 kg and length 3.0 m rests on two pivots, one at the left edge and one 2.4 m from the left edge. How far to the right of the right pivot can a mass of 150 kg be placed without the beam tipping?

11. (II) A rectangular piece of plywood (60 cm × 120 cm) lies in the horizontal xy-plane. The surface mass density of the plywood is 3 kg/m^2. Calculate the torque about one of the corners due to the force of gravity.

12. (II) A uniform book of mass 1.0 kg is placed such that 60% of the book is hanging over the edge of a table, with a paperweight of mass m is centered on top of the 40% of the book that is on the table. What is the minimum mass of the paperweight such that the book doesn't fall off the table?

13. (II) Two books are stacked at the edge of the table, with their lengths perpendicular to the table edge. If the width of each book is L, how far out from the table edge can the top book's extreme edge be placed without the books falling down? How does this generalize to three books?

11–3 Applications of Statics

14. (I) A door 90 cm by 195 cm of mass 14 kg hangs on two hinges: One is attached to the bottom of one side of the door, and the other to the top of the same side. What are the horizontal forces exerted on the door by the hinges?

15. (I) A football player is at the top of a pushup. The angle that the (rigid) torso makes with the floor is 25°. His arms are perpendicular to his torso and his center of mass is located at a point 3/8 of the distance from the shoulders to the feet. Assuming that the mass of his head can be neglected, what is the force, in terms of the player's weight, along his arms?

16. (I) A student wants to place a flower pot on a board that juts out from a window so it will get more sunlight (Fig. 11–28). The flower pot has a mass of 3.5 kg and needs to be 50 cm from the windowsill. The student can only place a nail into the sill 4 cm from the edge. Neglect the mass of the board and find out how much force the nail must exert to hold the board in place. What is the nature of the force "exerted" by the nail?

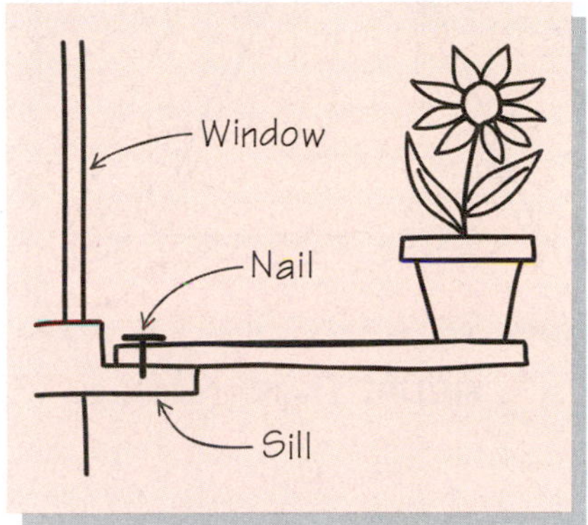

▲ **FIGURE 11–28** Problem 16.

17. (I) A uniform rod, mass 12 kg and length 1.5 m, rests on two points, one at its left end and one at the center point. What are the contact forces on the rod at these points? Comment on the stability of the situation.

18. (II) An 8.5-m extension ladder of mass 26 kg is propped up against a wall, touching at a point 8.0 m above the level ground. A man of mass 75 kg climbs 7 m up the ladder to repair a window. The ladder rests against a frictionless wall, but the ground has friction. Determine all the forces on the ladder.

19. (II) Consider a ladder of mass 10 kg and length 4 m, leaning against a vertical wall at an angle of 30° with the vertical. The coefficient of friction between the ladder and the floor is $\mu_s = 0.40$, and there is no friction between the wall and the ladder. A man of mass 80 kg climbs up the ladder. (a) How high can he climb before the ladder begins to slip? (b) Work out your calculation by taking the torques about the three points *A*, *B*, and *O* in Fig. 11–29, and show that the resulting equations are independent of the choice of point.

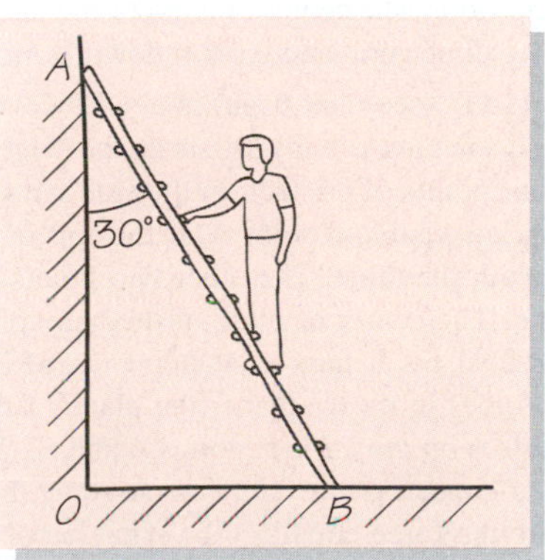

▲ **FIGURE 11–29** Problem 19.

20. (II) Consider the ladder in Example 11–7 (see Fig. 11–11a). The 3.0-m-long ladder is placed against a frictionless wall, making the same 58° angle to a different horizontal surface; this time, much to his dismay, the same 85-kg window washer finds himself starting to slip when he steps to the second rung. What is the coefficient of static friction between the ladder and floor?

21. (II) Using a uniform strut, a rigid brace hinged at the floor, a person holds a 30-kg engine in equilibrium while it is being repaired. The strut has a mass of 12.5 kg. A smooth rope passes over a pulley at the end of the strut (Fig. 11–30). (a) What is the force exerted on the rope by the person? (Specify the direction of this force by calculating the angle θ that the rope makes with the horizontal.) (b)What forces are exerted by the strut?

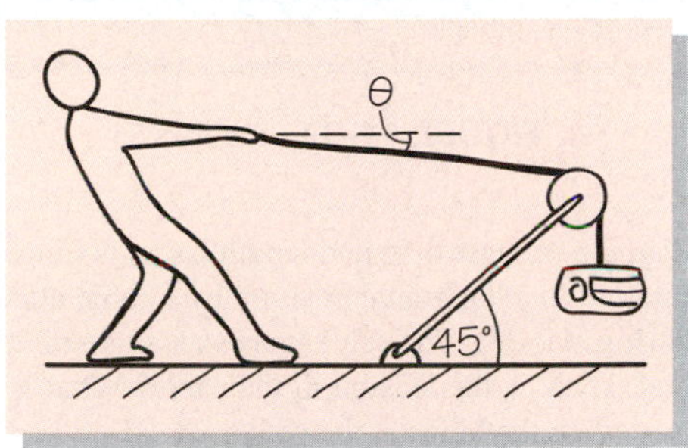

▲ **FIGURE 11–30** Problem 21.

22. (II) A piece of plywood leans against a wooden wall with which it has a coefficient of static friction of $\mu_s = 0.28$. (a) If the coefficient of static friction between the board and the floor is 0.35, what is the minimum angle that the board can make with the floor yet still not slip? (b) What happens if the coefficient of friction between the board and the floor is zero?

23. (II) A lawn mower of mass *m* is at rest on a rough slope, coefficient of static friction μ_s between wheels and ground, as shown in Fig. 11–31. Find the largest slope angle θ for which the mower will not (a) slip down the slope and (b) tip over. You should be able to show that what determines whether the mower tips or slides is the same for any angle.

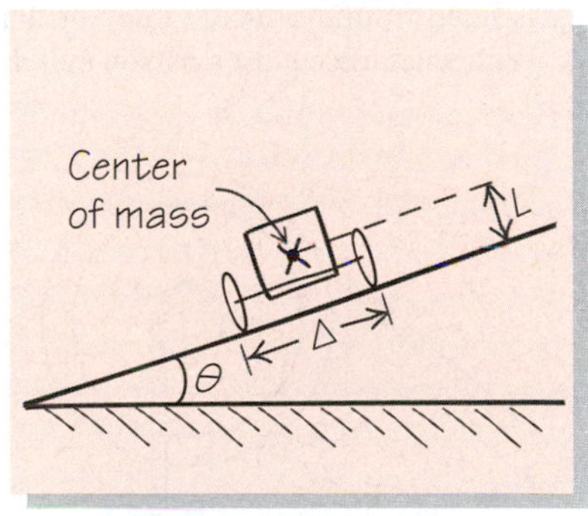

▲ **FIGURE 11–31** Problem 23.

24. (II) Re-solve the problem posed in Example 11–6, this time using as a reference point for the torque equation the high end of the crane arm. You should find the same result as in the example solution.

25. (II) A stepladder consisting of two ladders of mass M and length L is held together by a crossbar attached to the midpoints of the two ladders (Fig. 11–32). What force is exerted on the crossbar by each ladder if the length of the crossbar is $L/2$? Assume that any friction between the ladders and the floor is negligible.

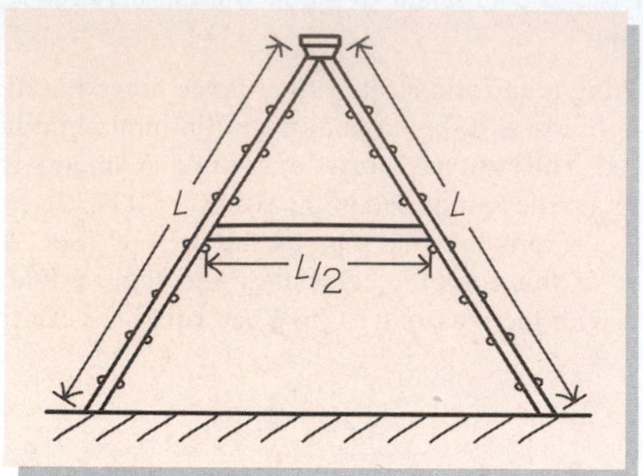

▲ **FIGURE 11–32** Problem 25.

26. (II) A flagpole of mass 6 kg and length 2.4 m is hinged at a wall and supported in a horizontal position by a cable attached to the free end (Fig. 11–33). The cable makes an angle of 25° with the horizontal. What is the tension in the cable? What is the vector force exerted on the hinge at the wall?

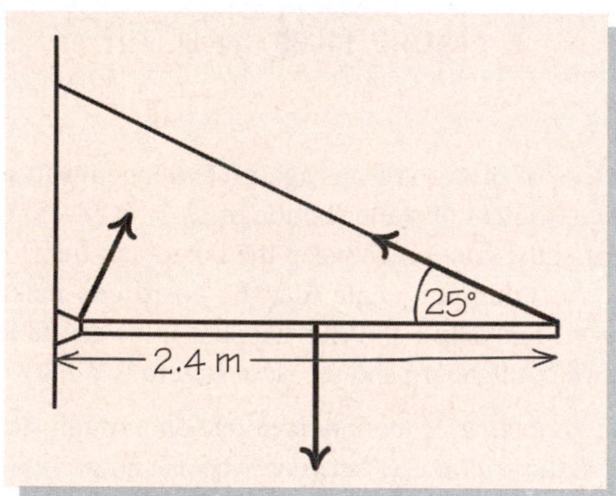

▲ **FIGURE 11–33** Problem 26.

27. (II) Two pulleys are mounted on the same axis. A rope is attached to the large pulley of diameter 20 cm. A car engine of mass 300 kg is hung from the small pulley of diameter 8.0 cm (Fig. 11–34). With what force must a person pull the rope to hold up the engine?

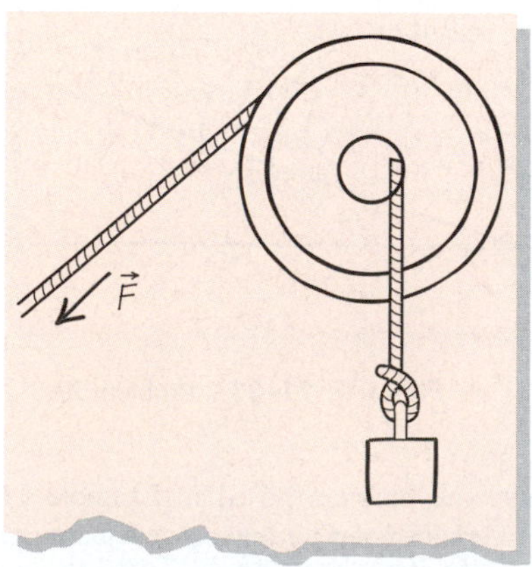

▲ **FIGURE 11–34** Problem 27.

28. (II) A frictionless pivot joins two uniform boards of the same mass m and length L, which are then placed to form a symmetrical "tent" on a rough horizontal ground surface (Fig. 11–35). The coefficient of static friction between the board ends and the ground is μ. What is the largest angle θ that each board makes with the vertical to the ground such that the arrangement will not slip?

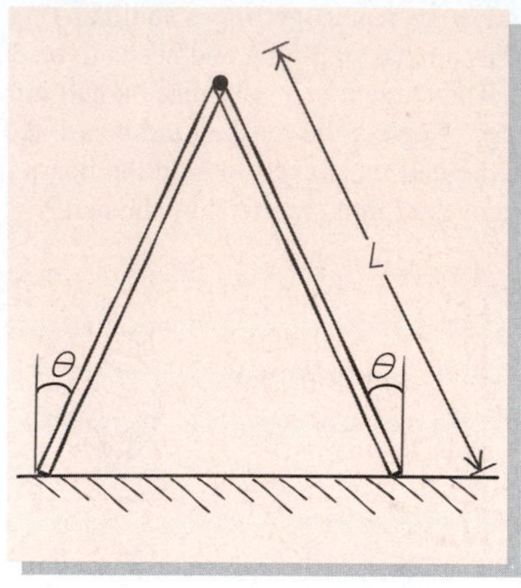

▲ **FIGURE 11–35** Problem 28.

29. (II) A desk of height 0.82 m, length 1.54 m, and mass 43 kg is pushed across a horizontal floor at a steady speed with a horizontal force $\vec{F}$ applied at the top (Fig. 11–36). The coefficient of kinetic friction between the legs and the floor is $\mu_k = 0.45$. What is the friction force at each leg, and what is $\vec{F}$? Assume that the two right legs each support the same forces, as do the left legs.

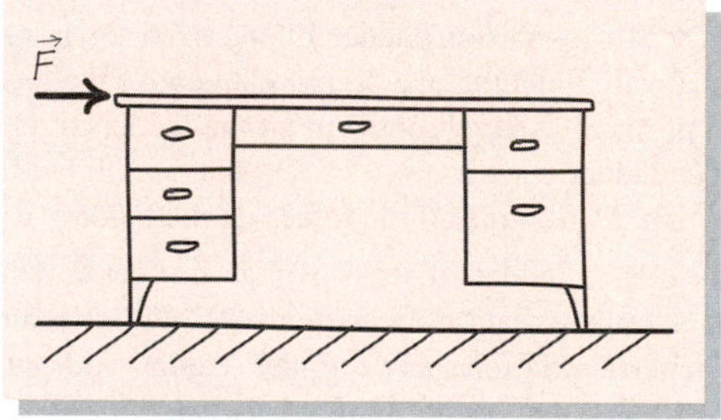

▲ **FIGURE 11–36** Problem 29.

30. (II) Repeat Problem 29, but this time assume that the desk is being pushed down a slope of 3.5°. The force applied is parallel to the sloping floor. The center of mass of the desk is in the middle of its long dimension and 0.38 m down from its top surface.

31. (II) Figure 11–37 (see next page) is a side view of a seat used for babies. It enables a baby to sit at the edge of a table by means of four points of contact; in the side view shown, two of these points are visible—point A at the top of the table, and point B beneath the table. The other two points are aligned directly behind. The center of mass of the baby plus the seat can be approximated by a mass m at point C. (a) Taking into account only forces in the xy-plane (the plane of the figure), calculate the forces *on the table* at points A and B. (Remember that there are four contact points, not two. Assume that the symmetric legs share the force equally.) (b) What happens for $\ell_2 \to 0$? for $\ell_1 \to 0$? (c) Work out numerical values for the forces for $m = 10$ kg, $\ell_1 = 20$ cm, and $\ell_2 = 30$ cm.

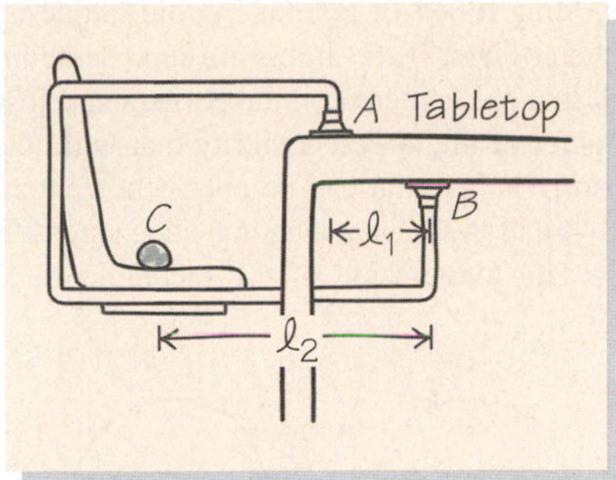

▲ **FIGURE 11–37** Problem 31.

32. (II) Consider the baby seat of Problem 31. The table has a total mass $M = 18$ kg, all concentrated in the (uniform) top, and the center of the table is 70 cm from the vertical extension of point P. For the geometry of the baby seat and the tabletop shown in Fig. 11–38, what mass m of the baby plus the seat will cause the table to tip over by a rotation at point P? Can the possible movements of a baby seriously destabilize the situation?

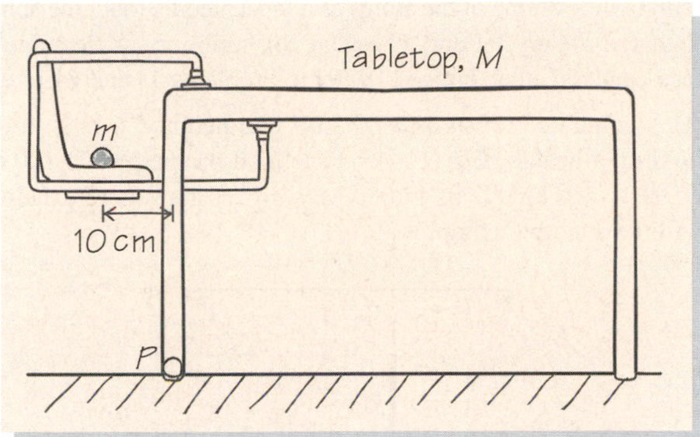

▲ **FIGURE 11–38** Problem 32.

33. (II) A trap door 1.8 m square with mass 20 kg is hinged at one edge and is attached to a rope at the opposite edge (Fig. 11–39). The trap door makes an angle of 55° with the horizontal, and the rope is perpendicular to the door. (a) What is the tension in the rope? (b) What is the force vector acting on the door at the hinge?

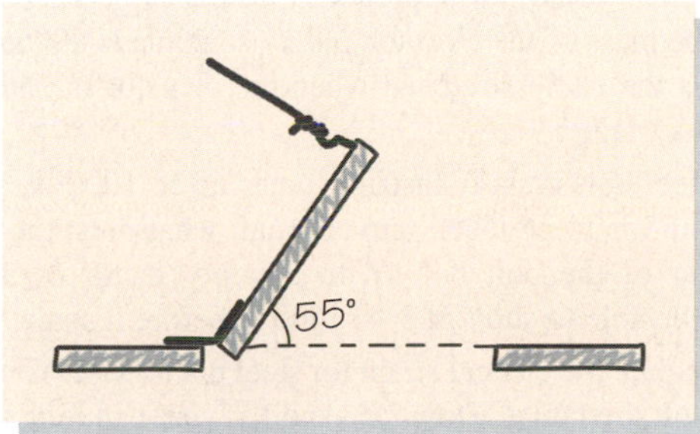

▲ **FIGURE 11–39** Problem 33.

34. (II) A beam of mass 12 kg is hinged at the top, and its bottom end is pulled to the side by a horizontal rope. The beam makes an angle of 15° with the vertical. What is the tension in the rope, and what is the force exerted on the beam at the suspension point?

35. (II) A variant of the crane in Example 11–6 is shown in Fig. 11–40. The pivot at point B is frictionless, and the beam, of length 3.00 m, has mass 100 kg. The rope makes an angle of 30° with the horizontal and can withstand a tension of 10,000 N before breaking. If this arrangement is used to lift masses from the point shown on the beam, what is the maximum mass that can be so raised?

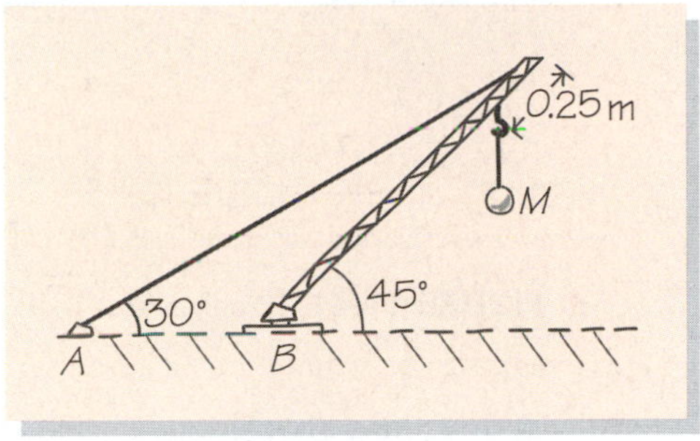

▲ **FIGURE 11–40** Problem 35.

36. (II) A uniform door is attached to the wall by two hinges. Draw a force diagram for the door and describe the various forces that can act. Describe the conditions for static equilibrium, and find the forces that are solvable. Is this an underdetermined system? If so, which forces are underdetermined?

37. (II) Consider the four-legged table discussed in Section 11–4, with leg 1 slightly shorter than the others. Suppose that a tea cup were placed on the diagonal between legs 1 and 4, at location

$$(x, y) = \left(\frac{L}{2} + x, \frac{L}{2} + x\right).$$

(a) Find a satisfactory solution to the magnitude of the contact forces on the legs if $x > 0$. Neglect the masses of the legs. (b) Repeat part (a) for $x < 0$.

38. (II) A solid block of mass M, height h, and base width w stands on a rough floor. By applying a large enough force F perpendicular to one side of the block, it is possible to topple it: the block will rotate on one edge so that the edge remains in contact with the floor without sliding. Through what critical angle must the box be rotated before it falls over? What is the work that must be done to tip it over?

39. (II) One way to keep a hollow cylinder of length L, mass M, and radius R from rolling down a rough (coefficient of static friction μ) inclined plane of angle θ is to exert a torque on it about its center that would tend to rotate the cylinder in the direction opposite from its rotation sense as it rolls. How large should this torque be in terms of the parameters given here?

40. (II) A large, spherical satellite is held in place in the bay of a space station with six ropes: four are attached and equally spaced around the equator and the other two are attached at the poles. (a) What equations describe the situation in which the satellite is held motionless with respect to the station? (b) Is this an underdetermined system? (c) Is the system underdetermined when all the ropes are under identical tension?

41. (III) A seaside tower is supported by a cable (Fig. 11–41, see next page). A horizontal wind that increases with height often blows from the left, exerting a horizontal force to the right on the tower. The horizontal force on a unit length of the tower increases with the height h according to force/unit length $= \alpha h$, where h is the height from ground level, and, if h is measured in meters, $\alpha = 50\ \text{N/m}^2$. The total height of the tower is 20 m. What is the tension in the cable?

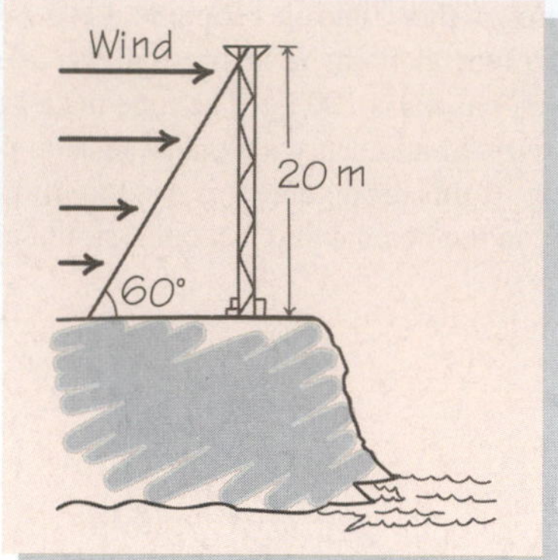

▲ **FIGURE 11–41** Problem 41.

42. (III) The mythical Greek king Sisyphus, pushing a large round rock up a mountain, wishes to take a rest and supports the rock with a horizontal rope attached to the top of the rock (Fig. 11–42). The coefficient of static friction between the rock and the slope is $\mu_s = 0.6$. (a) What is the largest value of θ for which this method of support is possible? (b) If the mass of the rock is 1088 kg, what tension must the rope be able to support on this maximum slope? (c) How does the tension vary as a function of θ for angles less than the maximum angle? [*Hint*: It is easiest to take moments about the point of contact.]

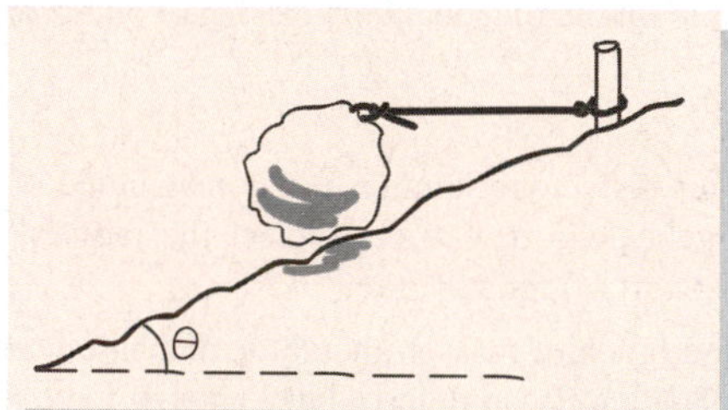

▲ **FIGURE 11–42** Problem 42.

11–4 Solids and How They Respond to Forces

43. (I) The crystal structure of copper is face-centered cubic; i.e., the structure is built of cubes of edge length a, with copper atoms at the corners as well as the face centers of the cube (Fig. 11–43). The lattice length $a = 0.361$ nm. (a) Calculate the diameter of a copper atom. (Assume that atoms are described by spheres that are centered on the corners and centers, and that they just touch). (b) Show that the structure can also be described as a stacking of triangular layers of atoms, perpendicular to the maximum diagonal of the cube.

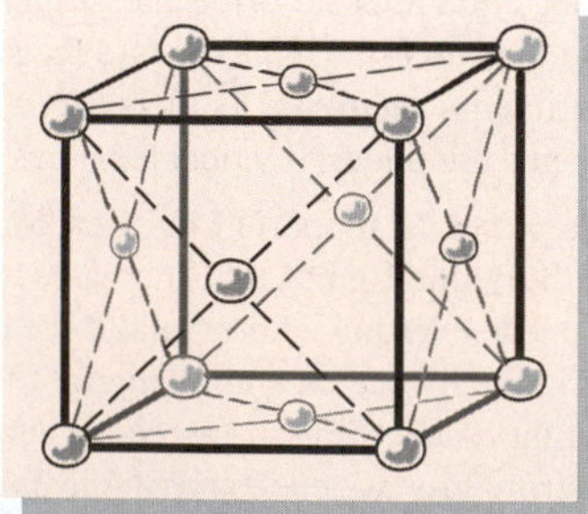

▲ **FIGURE 11–43** Problem 43.

44. (I) The building block of a simple cubic lattice is a cube with atoms at the corners. If the atoms are represented by spheres of diameter d that are centered on the corners and just touch, what is the diameter of the largest impurity that will not displace any of the existing atoms that can be put at the center of the cube? [Figure 11–44 shows what a diagonal slice across the cube looks like. *Hint*: The longer side of the rectangle is $\sqrt{2}$ times the shorter side.]

▲ **FIGURE 11–44** Problem 44.

45. (II) The volume filling of a crystal structure can be defined as the ratio of the volume of the atoms as represented by touching spheres to that of the crystal unit. Calculate this ratio for simple cubic and face-centered cubic lattices. [Refer to Problems 43 and 44.]

46. (I) A concrete pier of area $0.85\ \text{m}^2$ and height 3.6 m is built to hold up a bridge (Fig. 11–45). The load on the pier is 100 tons $(0.91 \times 10^5\ \text{kg})$. $Y = 17{,}000\ \text{MN/m}^2$ for the concrete. How far will the concrete compress?

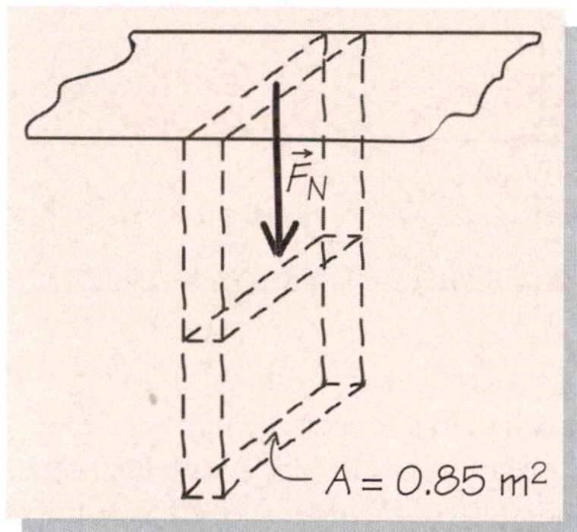

▲ **FIGURE 11–45** Problem 46.

47. (I) A mine elevator is supported by a steel cable 2.0 cm in diameter. The mass of the elevator and its contents is 800 kg. By how much is the cable stretched when the elevator is 250 m below Earth's surface?

48. (I) A steel bolt must withstand forces up to 1000 N. If a safety factor of 3 is to be taken into account, what must the minimum diameter of the bolt be? (With a safety factor of 3, the bolt should be able to support 3×1000 N before it snaps.)

49. (I) Calculate the critical strain for steel piano wire. You will need to consult the tables. How far would a wire 1 m long stretch before breaking?

50. (I) Assume that the piano wire of Problem 49 is 0.85 mm in diameter. Calculate the weight it could hold before fracturing.

51. (II) A crate of mass 30 kg slides across the ground. The coefficient of kinetic friction between the crate and the ground is 0.3. The physical contact area between the crate and the ground (which is less than the area of the whole crate) is $0.35\ \text{m}^2$. Calculate the shear stress on the crate.

52. (II) A steel wire of diameter 5.0 mm and length 2.0 m stretches 0.30 mm when a load of 60.0 kg is hung from it. What is its Young's modulus? How much mass can the wire hold before it may fracture? Use the tensile strength of steel from the tables.

53. (II) The 3-km-long cables on a large suspension bridge are stretched from their equilibrium length by 3 m. Estimate the change in the equilibrium separation between any two adjacent atoms along the cable.

54. (II) A steel bar of length 3.5 m is placed in a structure where it is subject to extreme stress under tension. The area of the bar is 25 cm^2, and it has room to stretch by 3 mm. It cannot stretch any more than this without butting up against a much stronger part of the structure. Does the bar break before it stretches the 3 mm? Use the Young's modulus and tensile strength of the tables.

55. (II) A uniform beam of length 2.0 m and mass 10 kg is freely pivoted at one end about a point fixed to the wall (Fig. 11–46). It is held in a horizontal position by a steel cable of diameter 2.0 mm. The cable makes an angle of 30° with the horizontal. A load of 30 kg is suspended from the end of the beam. What will be the angle of the beam relative to the horizontal? [*Hint*: The distortion is small, so you may calculate the tension in the cable without taking into consideration the slight change in angle.]

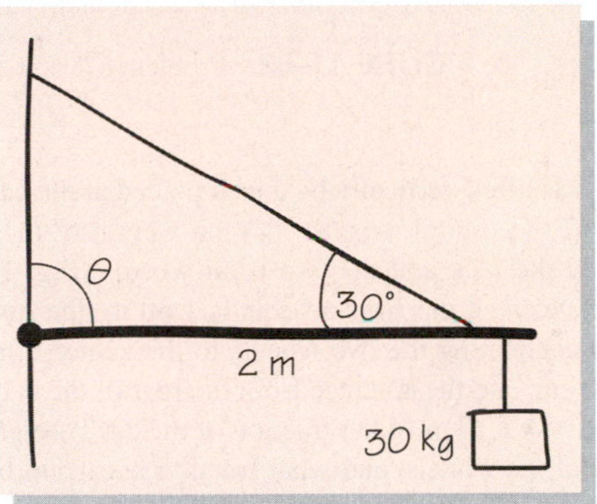

▲ **FIGURE 11–46** Problem 55.

56. (II) A 0.5-m-long piece of metal is compressed longitudinally by 0.1 mm. Its Poisson's ratio is 0.32. Calculate its volume strain.

57. (II) Poisson's ratio for steel is 0.3. What is the new diameter of the steel rod used in Example 11–9 after the rod has been stretched?

58. (III) According to our discussion of Poisson's ratio, a bar of length L and cross-sectional area A under compression at its ends expands at the sides. Find the amount of pressure that must be exerted on the sides to keep them fixed when there is a compression F on the two ends. Find the total amount of compression in the longitudinal direction under these conditions, and show that a bar with constrained sides is stiffer under compression than the same bar with unconstrained sides. (Keep the terms up to order σ^2 only.)

59. (III) Let the strain on a rod along the direction of applied stress be e_1. Because the stretching produces a slight reduction in the diameter of the rod, we may speak of an induced transverse strain, given by $e_{tr} = -\sigma e_1$, where σ is Poisson's ratio (Fig. 11–47). Recall that Y is defined in terms of longitudinal stress and strain. A cube of the rod's material immersed in a liquid, so that the pressure on all sides is p, suffers a volume change given by $\Delta V/V = 3p(1 - 2\sigma)Y$. (a) Derive this result. (b) Show that when Poisson's ratio is not zero, the relation between Young's modulus and the bulk modulus under uniform pressure is given by

$$B = \frac{1}{3}\frac{Y}{1 - 2\sigma}.$$

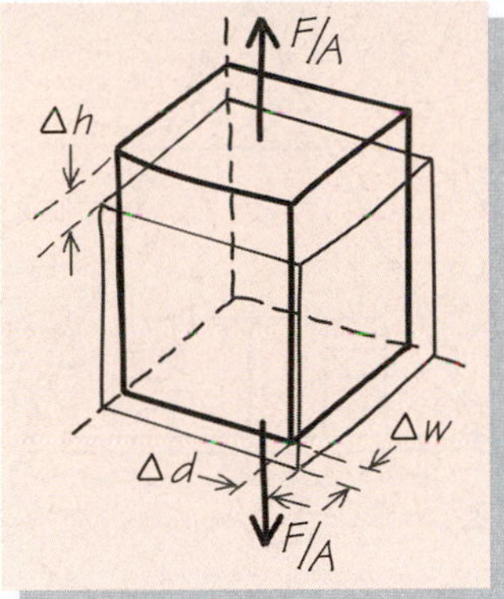

▲ **FIGURE 11–47** Problem 59.

60. (II) By using the results of Problem 59, show that Poisson's ratio, σ, must be less than 0.5 because, if it exceeds 0.5, the total volume of the solid would increase when a uniform pressure is applied to it.

61. (III) *Compressibility* is defined by $-\Delta V/pV$, the reciprocal of the bulk modulus. Using the equation for $\Delta V/V$ in Problem 59, calculate the compressibility of silver, given that $\sigma = 0.38$ and $Y = 7.9 \times 10^4\ \text{MN/m}^2$.

62. (III) Show by analogy with a simple spring, for which the force is $F = -kx$, that a volumetric potential energy associated with strain—that is, potential energy per unit volume of a wire or rod of length L and cross-sectional area A—is given by $u = \frac{1}{2}$ (stress)(strain). [*Hint*: Recall the work–potential energy connection.]

General Problems

63. (II) A 30-cm-wide shelf is supported at the wall by a cable at each end of the shelf placed at 45° to the wall (Fig. 11–48). A 20-kg sack of potatoes is placed and centered on the shelf. What is the tension of the cable if we ignore the mass of the shelf?

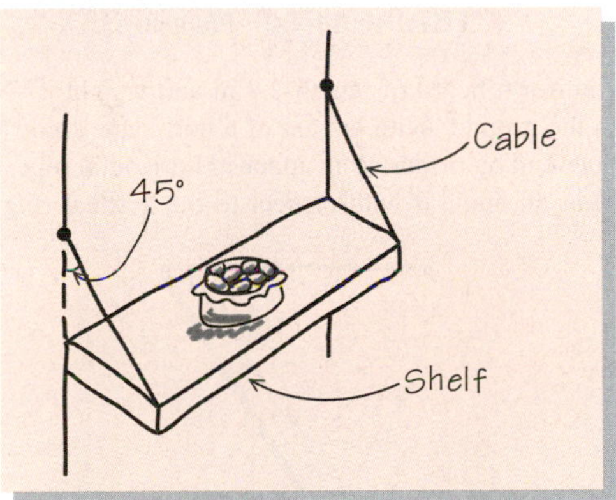

▲ **FIGURE 11–48** Problem 63.

64. (II) The cross section of an A-frame house is shown in Fig. 11–49 (see next page). The total height of the apex is 5.0 m, and the 1.5-m-long crossbeam is two-thirds of the way up the roof line. The crossbeam must support two roof beams, each of mass

3000 kg. (a) Does the crossbeam push the roof beams out or does it pull them in? (b) What is the force exerted by the roof beams on the crossbeam?

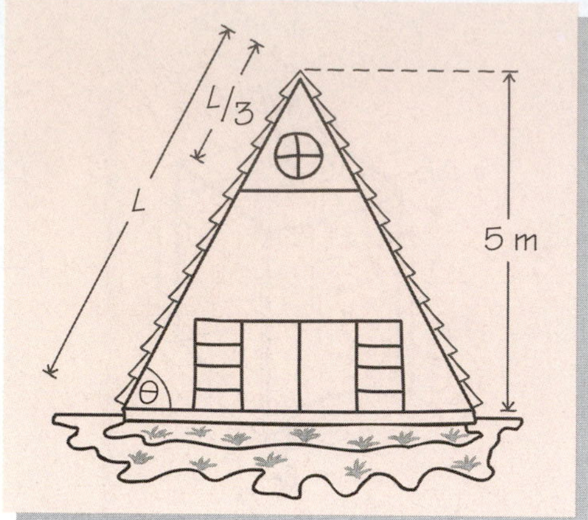

▲ **FIGURE 11–49** Problem 64.

65. (II) A uniform beam of length L and mass M is freely pivoted at one end about an attachment point in a wall. The other end is supported by a horizontal cable also attached to the wall, so that the beam makes an angle θ_0 with the horizontal (Fig. 11–50). (a) What is the tension in the cable? (b) The cable snaps. What is the angular acceleration of the beam about its pivot point immediately afterward? (c) What is the angular velocity of the beam as it falls through the horizontal position?

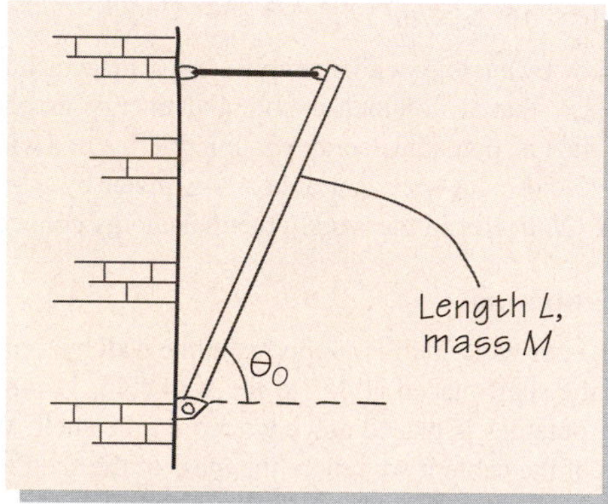

▲ **FIGURE 11–50** Problem 65.

66. (II) A uniform board of length 2.4 m and weight 47 N has one end on the ground. With the aid of a horizontal force applied at the upper end by means of an attached horizontal rope, the board is held at an angle θ with respect to the vertical (Fig. 11–51). The coefficient of static friction between the end of the board and the ground is $\mu_s = 0.32$. What is the range of angles the board can make with the vertical and still be in static equilibrium? How does the tension in the rope vary with the angle within the angle's possible range?

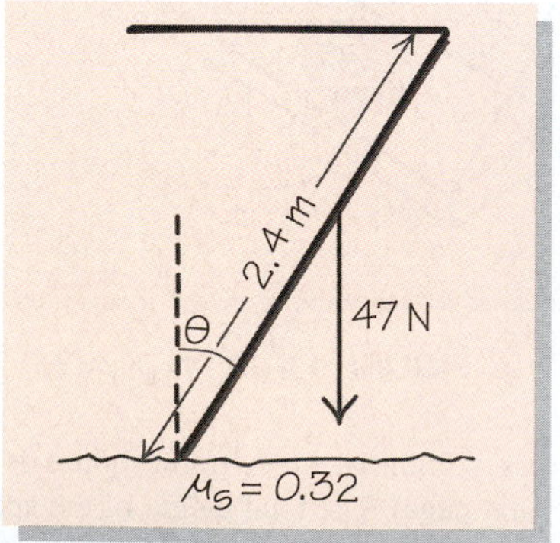

▲ **FIGURE 11–51** Problem 66.

67. (II) A sign is to be constructed from a piece of plywood in the shape of a 30° right triangle of mass 15 kg. It is to be attached to a wall, as shown in Fig. 11–52. The lower attachment point is a frictionless pivot, and the upper point is a rope that can be reeled in or out to make the bottom of the sign and the rope itself horizontal. What is the tension in the rope?

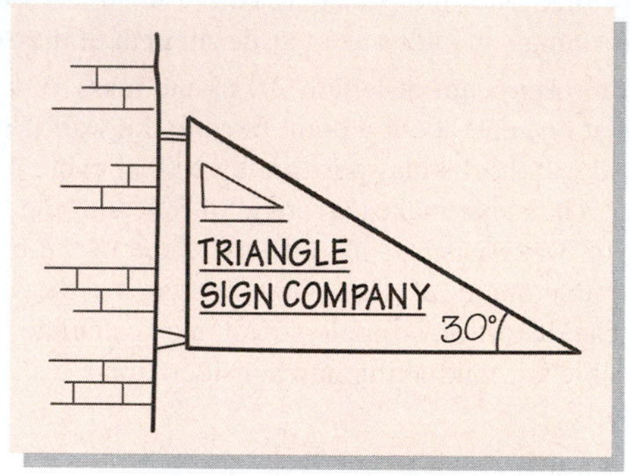

▲ **FIGURE 11–52** Problem 67.

68. (II) A car is lifted vertically by a jack placed at the car's rear end 40 cm off the central axis, so that the weight of the car is supported by the jack and the two front wheels (Fig. 11–53). The distance between the front wheels is 1.60 m, the distance from the axis connecting the two wheels to the center of mass of the car is 80 cm, and the distance from the rear of the car to the center of mass is 2.10 m. What fraction of the car's weight is carried by each of the wheels, and what fraction is carried by the jack? (Note that the weight on the wheels will *not* be symmetrically distributed.)

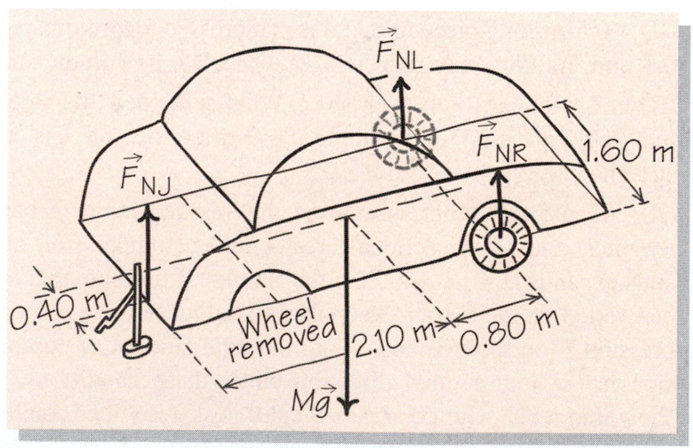

▲ **FIGURE 11–53** Problem 68.

69. (II) A roller of radius 30 cm and mass 80 kg is pulled by a force F applied horizontally to the axle. How large must F be in order for the roller to climb a step 15 cm high?

70. (II) Consider the roller of Problem 69. Assume that the force F can be applied in any direction. What direction minimizes the magnitude of F necessary to make the roller climb the step?

71. (II) A chest of drawers 58 cm wide, 1.6 m long, and 1 m high is pulled by a horizontal rope attached at a height of 58 cm to the midpoint of the long side. The force is such that the chest moves with uniform velocity. (a) Express the force on the legs on the side with the rope, and the force on the legs on the opposite side, in terms of the coefficient of kinetic friction μ_k. (b) For what value of μ_k will the chest topple over?

72. (II) A wooden box of uniform density and dimensions $h = 1.4$ m, base 0.3 m $\times$ 0.3 m, and mass 50 kg stands on a rough surface with coefficient of static friction $\mu = 0.70$. When a large enough force is applied perpendicularly to one side, the box will either slide or topple over, depending on the height above the ground that the force is applied to. What is the highest point at which you can apply the force to slide the box rather than toppling it?

73. (II) A spherical nut is placed between the handles of a nutcracker (Fig. 11–54). As the handles are closed, the nut slides away from the hinge until the angle α decreases to the point at which $\tan(\alpha/2) = \mu$, where μ is the coefficient of friction between the nut and the inner surface of the nutcracker. Show that no matter what the force on the handles is, the nut will not slide any farther.

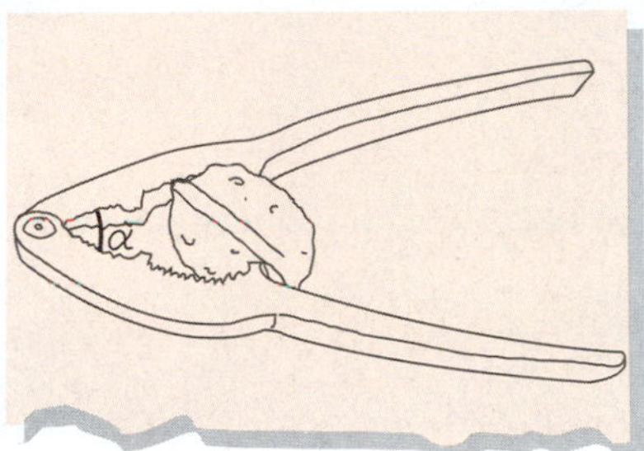

▲ **FIGURE 11–54** Problem 73.

74. (II) A box 0.6 m long and 1.5 m tall is placed on a flatbed truck. The truck accelerates at a rate of 1.2 m/s^2. Will the box topple over?

75. (II) A cylinder of mass M and radius R rests on an inclined plane (Fig. 11–55). It is held in place by a horizontal string that is attached to the edge of the cylinder. If the angle that the plane makes with the horizontal is θ, and the coefficient of static friction between the cylinder and the plane is μ_s, what is the smallest value of μ_s that will maintain this position as an equilibrium position?

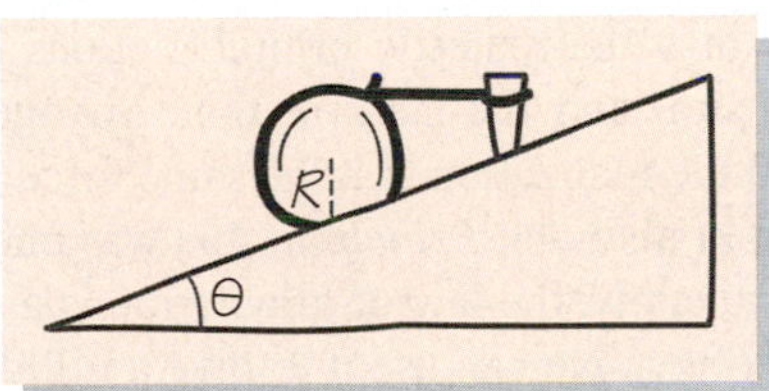

▲ **FIGURE 11–55** Problem 75.

76. (II) A centrifugal governor consists of light rods pivoted at points A, B and C, as shown in Fig. 11–56. They are loaded with masses M, m, and m respectively, and the whole apparatus rotates about the vertical axis. Depending on the angular velocity, the mass M slides up or down, and its position can be used to control the flow of steam in a steam engine. Derive a relationship between the angular velocity and the angle the arms make with the vertical.

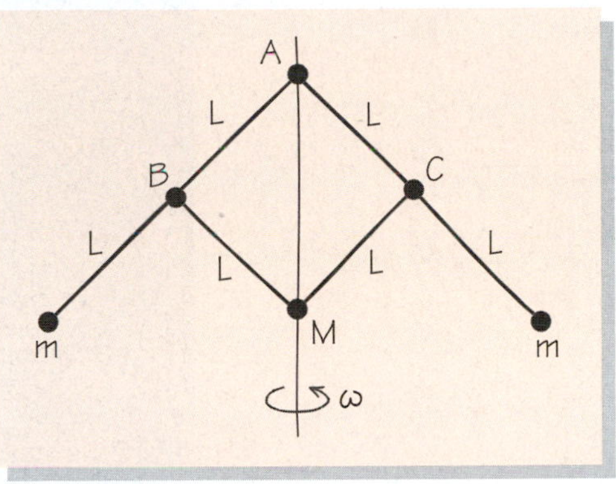

▲ **FIGURE 11–56** Problem 76.

77. (II) A rubber tether ball is attached by a string to the wall as shown in Fig. 11–57, with the string tangential to the ball. The mass of the ball is 150 g, the diameter is 12 cm, and the string is 90 cm long. The string is attached at a position 11 cm from the wall. What is the smallest possible value of the coefficient of static friction between the ball and the wall such that the ball will not fall?

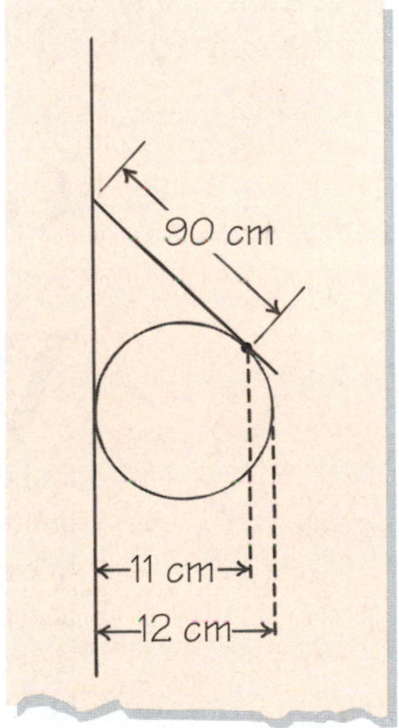

▲ **FIGURE 11–57** Problem 77.

78. (III) Three steel ball bearings of mass m and radius 1.33 cm are placed in a closed vertical tube of diameter 4.00 cm that is closed on the bottom. Find the forces on the middle ball bearing.

79. (III) Two steel ball bearings of mass m and radius 2 cm are placed in a vertical tube of diameter 6 cm that is closed on the bottom. Find all the forces on the two balls.

Chapter 12

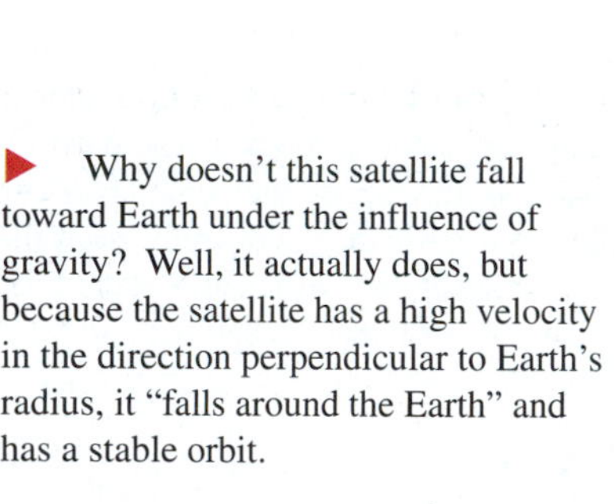

▶ Why doesn't this satellite fall toward Earth under the influence of gravity? Well, it actually does, but because the satellite has a high velocity in the direction perpendicular to Earth's radius, it "falls around the Earth" and has a stable orbit.

Gravitation

While the forces that we see and experience every day are ultimately derived from the more basic forces that govern the interactions of matter, in most cases the basic force itself is well hidden. Examples include friction, tension, and contact forces, which are all ultimately explained by the forces that hold the components of matter together—the electromagnetic force—although it would be very difficult to explain all the characteristics of these "effective" forces from first principles. There is, however, one important exception, and that is gravity. Earth's gravity, as we experience it every day, is a simple manifestation of the force of universal gravitation; as far as we know today, gravitation is one of the truly fundamental forces. Gravitation governs not only the motion of a falling apple but also the majestic orbital motions of the Moon around Earth, the planets around the Sun, and the stars in their voyage around the galaxy. In 1687 Isaac Newton published the realization that the same force governs both the falling apple and the orbiting Moon in his book *Principia*; this was one of the great intellectual leaps in human history. Fortunately, the law of gravitation is a force law for which the equations of motion can be solved; we can use it to give a full description of the motion of astronomical bodies. Gravitation is conservative, and we can use potential energy to help analyze many situations involving the gravitational force.

From the start, the study of the motion of celestial objects was motivated by philosophical and esthetic considerations ("the harmony of the spheres"), but there are also good, practical reasons to want to understand gravity and gravitation. The measurement and understanding of local variations of the strength of gravity has taught us a lot about the Earth's interior structure and is an important prospecting tool for oil and minerals.

Satellites, whose orbits are determined by the gravitational force, are an indispensable part of the world's communications system, and we rely on the satellites of the Global Positioning System (GPS) for navigation. Indeed, the GPS relies for its functioning not just on Newtonian gravitation but on Einstein's general theory of relativity, a subject that had once been thought to have no practical consequences. Finally, much of our exploration of the solar system relies on understanding complicated orbital motions. From the earliest times, people have looked into the sky and considered the Sun, Moon, and stars. This interest in observing our celestial surroundings has led to a deep understanding of gravitation and its consequences.

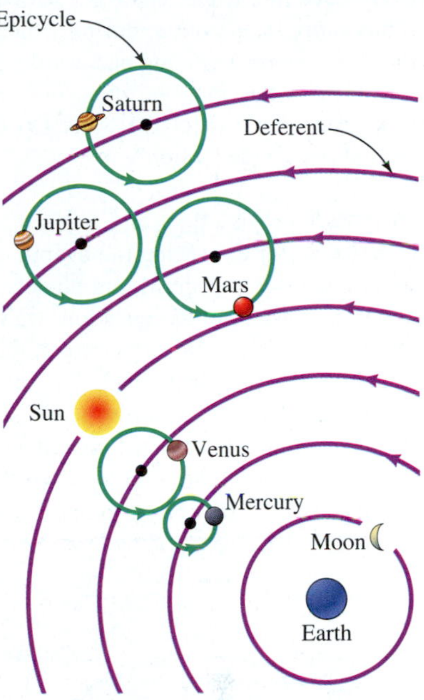

▲ **FIGURE 12–1** Planetary motion based on the Ptolemaic system, which was the generally accepted framework for celestial phenomena from the second century to the 1600s. According to Ptolemy, Earth is the center of the world system. Planets move in small circular paths called epicycles, and the centers of the epicycles move around Earth on large circles called deferents.

12–1 Early Observations of Planetary Motion

The earliest astronomical observations of the night sky led ancient peoples to divide the points of light they saw into two classes: the so-called fixed stars, which move each day in nearly perfect circles around Earth, and the planets (in Greek, the "wanderers"), which move in what appear superficially to be complicated, erratic patterns in the night sky. The first interpretations of these observations placed Earth at the center of the universe, in a **geocentric** frame. Earth was pictured as being surrounded by a rotating spherical shell in which the stars were fixed. The more complicated motion of the planets, Sun, and Moon were explained by placing the planets upon moving, transparent shells within the outer, fixed-star shell. These shells had to move in complicated ways if the observations were to be consistent with observation. In the second century A.D., Claudius Ptolemy made the most detailed formulation of these notions, and to explain the various paths taken by planets, he had to depart from the simple spherical shell description and construct paths of planetary motion that form circles that themselves lie on circles around Earth (Fig. 12–1). His theory was based on a culturally imposed belief that circular motion is in some sense "perfect" and that superposed circular motions embodied that perfection. As his *epicycles* (the circles on circles used to describe the movements of the planets) gave a reasonable description of the motion of the planets, the idea proved serviceable for over 1400 years.

The Copernican Picture

In 1543 Nicolaus Copernicus introduced a revolutionary and controversial view of the motion of the planets in a **heliocentric** frame, with the Sun rather than Earth at the center of the solar system. Copernicus continued to insist on describing all motions with circles, and because the true motions of the planets about the Sun are not circles, epicycles were still needed in the Copernican description to accommodate the observations of planetary motion (Fig. 12–2). The epicycles meant that the Copernican description was about as unwieldly as the Ptolemaic description, so in the light of continued philosophical objections, there was no compelling reason to accept the Copernican view; the Copernican hypothesis was not accepted for almost a century. During this transitional period the construction of more refined instruments (protractor-like instruments called quadrants, as there were still no telescopes) allowed Tycho Brahe to improve the knowledge of planetary **orbits**, or paths, to an accuracy of less than half a minute of arc by the end of the sixteenth century.

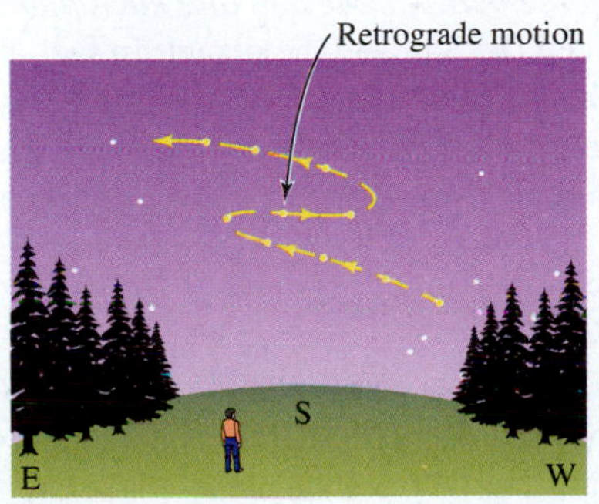

◀ **FIGURE 12–2** The Sun and Moon move west to east relative to fixed stars from night to night, but some planets do a retrograde "loop" from east to west every once in a while. The yellow dashed line traces a planetary position relative to the stars over a period of several months. The simplest Ptolemaic universe does not easily explain the retrograde motion of Mars shown here; "epicycles" are required (Fig. 12–1). It is simpler to explain this motion with the Copernican description.

CONCEPTUAL EXAMPLE 12–1 As Earth rotates, Mars moves across the sky from east to west, as do the Sun, the Moon, the other planets, and the stars. Most of the time Mars moves across the sky a little more rapidly than the background stars, but every once in a while, and for a period of a few months, Mars exhibits *retrograde motion*, during which it moves more slowly than the stars, in effect moving backward with respect to them. How can you explain this motion, given our understanding that Earth and Mars each orbit about the Sun? Do the Sun and the Moon ever exhibit retrograde motion?

Answer The effect of Earth's daily rotation about its axis is responsible for our seeing the overall east-to-west motion, and this motion applies to Sun, Moon, planets, and stars alike. In thinking about the answer to the question, we'll subtract out this aspect of the motion. Earth also orbits about the Sun, and there is an additional moving platform effect. The stars are so distant that they remain fixed. The planets move with respect to the stars because they themselves are orbiting the Sun and we are close enough to them to clearly observe that motion. Most of the time we see them moving "forward" in their orbits against the backdrop of the fixed stars, meaning each day they will be a little farther along their paths than they were the previous day. But if Earth in its orbit overtakes them, as happens with Mars, then they will appear to move "backward" against the stars. For a useful analogy, think about yourself in a car on a highway. The distant landscape hardly seems to move, while you see other cars on the highway move against the landscape because of the relative motion between you and them. Cars moving in your direction move forward against the landscape. The analogy to retrograde motion of a planet occurs when you pass another car. From your point of view the car you overtake will move backward with respect to the landscape. A similar thing happens with Mars seen from Earth. Earth has a solar orbit inside that of Mars, and Earth moves faster in its orbit than does Mars. As we on Earth overtake Mars on an inside track, we see it move backward against the fixed background. The Moon and Sun have no such motion with respect to us, because Earth orbits the Sun, and the Moon orbits Earth. While this retrograde motion is simple in the Sun-centered picture, it is explicable, though in a more complex way, in the Ptolemaic system with epicycles. Indeed, a glance at Fig. 12–2 reveals possible retrograde motion for Mars, and the occurence of retrograde motion is not enough to kill the Ptolemaic picture.

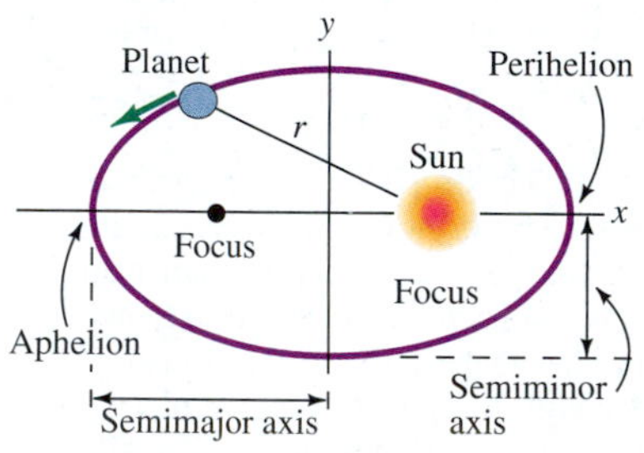

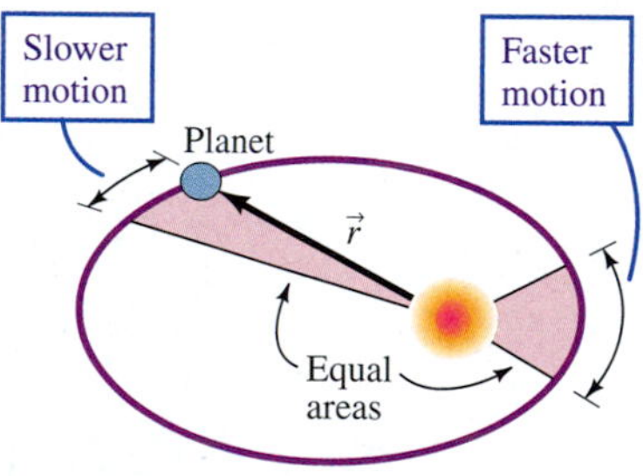

▲ **FIGURE 12–3** (a) Kepler's first law specifies that the planets move in elliptical orbits with the Sun at one focus. The perihelion is the distance of closest approach and the aphelion is the farthest distance between a planet and the Sun. (b) Kepler's second law specifies that the radius vector from the Sun to a planet sweeps out equal areas in equal times. The two shaded areas shown are equal if the same time is taken to sweep out each of them. This means that the planet moves more quickly over a part of the orbit closer to the Sun than over a more distant part.

Kepler's Laws

After Brahe's death in 1601, his assistant Johannes Kepler inherited the data that Brahe had accumulated. Kepler spent some 20 years analyzing these data, looking for mathematical regularities. In a crucial breakthrough he concluded that the idea of circular orbits should be discarded and replaced with elliptical orbits. **Kepler's laws** summarize Kepler's most important conclusions:

1. Planets move in planar elliptical paths with the Sun at one focus of the ellipse (Fig. 12–3a).
2. During equal time intervals the radius vector from the Sun to a planet sweeps out equal areas (Fig. 12–3b).
3. If T is the time that it takes for a planet to make one full revolution around the Sun, and if R is half the major axis of the ellipse (R reduces to the radius of the planet's orbit if that orbit is circular), then

$$\frac{T^2}{R^3} = C, \tag{12–1}$$

where C is a constant *whose value is the same for all planets*.

Kepler's laws were so simple that once they were known it was no longer possible for scientists to cling to pre-Copernican ideas.

We found the origin of Kepler's second law in Chapter 10: Eq. (10–28). The law follows from the conservation of angular momentum. We saw in Chapter 10 that the angular momentum of a body under the influence of a central force is conserved. Here angular momentum conservation, and hence the "equal area in equal times" rule, is a consequence of the fact that *the gravitational force between the Sun and each planet is central*; that is, the force acts along the line between the Sun and the planet. In fact, Kepler's second law can be taken as evidence that the gravitational force is central. Conservation of angular momentum also implies that the trajectories of planets must lie in a plane—the plane that is perpendicular to the direction of the fixed angular momentum vector.

12–2 Newton's Inverse-Square Law

Newton recognized the importance of Kepler's laws. He realized that since the planets do not move in straight lines, they must be subject to a net force. He concluded from Kepler's second law that the net force on a planet must point from the planet to the Sun—it must be

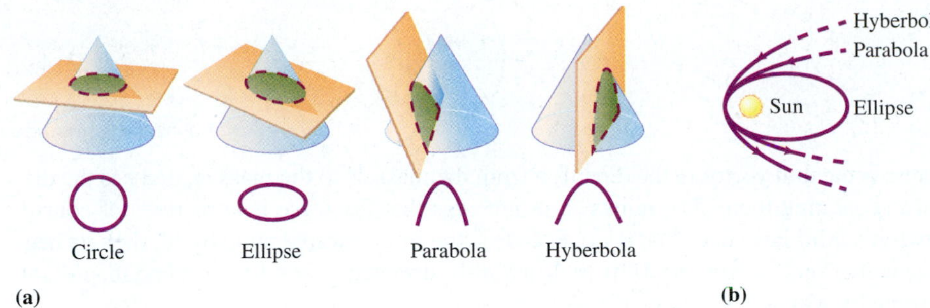

FIGURE 12–4 (a) Slicing a cone with a plane produces conic sections. The conic sections are the familiar geometric curves shown here. (b) The orbits of astronomical objects such as planets can follow all the curves drawn in (a).

central. He also found that the elliptical paths described by Kepler can be explained by an *inverse-square force law*: If the source of the force is at the origin O and the mass on which the force acts is at position $\vec{r}$, then the force has magnitude k/r^2, where k is a constant.

Newton showed a more general result as well: The possible paths of a mass under the influence of a force of this form had to be **conic sections**. Conic sections are the shapes obtained if a cone is sliced by a plane. As can be seen in Fig. 12–4a, if a cone is sliced by a plane parallel to the cone's base, the conic section is a **circle**. If the plane is tilted a little, the conic section is an **ellipse**. If the slice is cut parallel to the slope of the cone, the conic section turns out to be a **parabola**. If the slice is steeper than the slope of the cone, the conic section is a **hyperbola**. The planets from which Kepler discovered his first law move in closed elliptical orbits, but there are known astronomical bodies that move along parabolic and hyperbolic trajectories as well (Fig. 12–4b).

By postulating a central inverse-square force law, Newton was able to explain not only Kepler's first two laws but Kepler's third law as well. Here we'll see how Kepler's third law follows for a circular orbit of radius R. From above, the force has magnitude k/R^2, and we know from Chapter 3 that the acceleration for an object moving with speed v in a circle of radius R is v^2/R. (We are assuming the Sun is fixed in space, an assumption justified as we'll see later by its large mass.) We can now apply Newton's law of motion $F = ma$ for the case of a planet of mass m, so that F is replaced by k/R^2 and ma is replaced by mv^2/R:

$$\frac{k}{R^2} = \frac{mv^2}{R}. \tag{12–2}$$

In terms of the *period*, T, of the planet's orbital rotation, we have $v = \text{circumference}/T$, so that $T = 2\pi R/v$. Squaring, we have

$$T^2 = \frac{4\pi^2 R^2}{v^2}.$$

If we now substitute $v^2 = kR/mR^2 = k/mR$ from Eq. (12–2), we find

$$T^2 = \frac{4\pi^2 R^2}{k/mR} = \left(\frac{4\pi^2 m}{k}\right)R^3. \tag{12–3}$$

This is just the statement of Kepler's third law as it applies to a circle, with the additional condition that the constant $C = (4\pi^2 m/k)$ in Eq. (12–1) should be the same for all planets, which it will be if k is proportional to the mass of the planet.

The Law of Universal Gravitation

The constant k just described involves the force between two objects; by Newton's third law, there is a symmetry between these objects as they appear in the force law. If k is proportional to one of the masses, it must also be proportional to the other one. Then k must have the form GmM, where m is the mass of one object, M is the mass of the other, and G is a proportionality constant known as the **gravitational constant**. In 1686 Newton put all this together into what is now known as **Newton's law of universal gravitation**: The force of gravitation acting on a point mass of mass m due to another point mass of mass M is an attractive force with an inverse-square form,

$$\vec{F} = -\left(\frac{GmM}{r^2}\right)\hat{r}. \tag{12–4}$$

LAW OF UNIVERSAL GRAVITATION

Here $\hat{r}$ is the unit vector in the direction from the mass M to the mass m, and r is the distance separating them. The minus sign indicates that the force is attractive. Of course, Newton's third law states that if Eq. (12–4) is the force exerted on m by M, then its negative is the force exerted on M by m. Mass M is attracted to m with the same magnitude of force by which m is attracted to M.

We can now return briefly to Kepler's third law, specializing to circular orbits. According to Newton, the constant C in Kepler's third law is given by

$$C = \frac{4\pi^2 m}{k} = \frac{4\pi^2 m}{GmM} = \frac{4\pi^2}{GM}. \tag{12–5}$$

Kepler's third law, Eq. (12–3), thus takes the form

$$T^2 = \frac{4\pi^2 R^3}{GM}. \tag{12–6}$$

This is a remarkable equation. It provides a quantitative realization of Kepler's third law†, and it shows that its analog with a different value of M applies equally for the motion of satellites and moon around a planet as well. In fact, once we know G, we can use it to obtain the mass of a planet using the period of one of its moons and the moon's distance from the center of the planet.

The Gravitational Constant: The constant G characterizes the strength of the gravitational force and must be determined from experimental data. Its dimensions are $[L^3M^{-1}T^{-2}]$, and its units in SI are $\text{N}\cdot\text{m}^2/\text{kg}^2$. The constant G in Eq. (12–4) cannot be measured independently unless the masses of the two objects involved are known; for this reason, G cannot be measured from astronomical objects such as Earth, the Moon, or the Sun. The value of G was first determined by Henry Cavendish in 1798 in an experiment whose basic idea is shown in Fig. 12–5. Two masses m at the end of a rod of

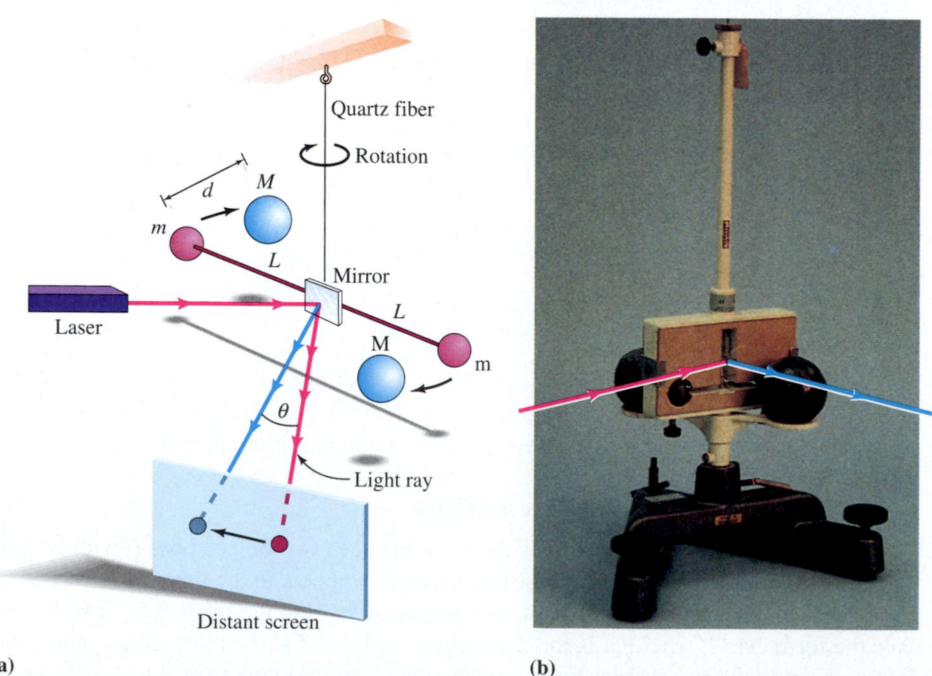

▶ **FIGURE 12–5** (a) In experiments modeled after those of Cavendish, the attraction of the smaller masses to the larger ones twists the fiber from which the rod is suspended. A light shining on a mirror on the rod indicates the small rotation by reflection on a distant screen. Switching the positions of the large masses reverses the rotation. (b) The apparatus used in a Cavendish-type experiment.

† Although we have expressed it for a circular orbit, Newton's mathematical techniques were powerful enough to show it holds for elliptical orbits with R being the semimajor axis of the ellipse. Most of the planetary orbits are close enough to circular for us to approximate them that way.

negligible mass and length $2L$ are suspended by a wire or a quartz fiber. Large masses M are placed near the masses m, with their centers of mass separated by a distance d. The magnitude of the torque about the point of suspension due to the force of gravitation on the masses is counterbalanced by a second torque due to the resistance of the fiber to twisting and proportional to the angle of deflection θ. When this resistance is known, the torque due to the gravitational force, and hence G, can be determined. The most recent measurements lead to the value

$$G = 6.673 \times 10^{-11}\ \mathrm{N\cdot m^2/kg^2}, \qquad (12\text{–}7)$$

with an uncertainty of about 0.06 percent. The magnitude of the gravitational force between two 10-kg masses separated by 10 cm is only about 7×10^{-7} N. We only sense the force of gravity when large masses such as the Earth are involved. The Cavendish experiment is difficult because the gravitational force is weak, making G one of the least accurately measured of all the fundamental physical constants.

EXAMPLE 12–2 Calculate the mass of the Sun, assuming that Earth's orbit around the Sun is circular, with radius $r = 1.5 \times 10^8$ km.

Strategy Kepler's third law in its extended form, Eq. (12–6), relates the given distance of Earth from the Sun, the period of Earth's orbit, and the mass of the Sun. We know for ourselves that the period of the orbit is one year. We can thus directly solve Eq. (12–6) for M_S.

Working It Out Equation (12–6) gives

$$M_S = \frac{4\pi^2 r^3}{GT^2}.$$

Given the value of G, the radius of the orbit, r, and the period, T, which is 365 days $= (365\ \mathrm{d})(24\ \mathrm{h/d})(60\ \mathrm{min/h})(60\ \mathrm{s/min}) = 3.15 \times 10^7$ s, we find that

$$M_S = \frac{4\pi^2(1.5 \times 10^{11}\ \mathrm{m})^3}{(6.67 \times 10^{-11}\ \mathrm{N\cdot m^2/kg^2})(3.15 \times 10^7\ \mathrm{s})^2}$$
$$= 2.0 \times 10^{30}\ \mathrm{kg}.$$

This is a factor of 3×10^5 larger than the mass of Earth.

What Do You Think? Mars is farther from the Sun than is Earth. Assuming that its orbit is also circular, is Mars's year (a) longer than Earth's, (b) shorter than Earth's, or (c) could it be either, depending on additional information? *Answers to **What Do You Think?** questions are given in the back of the book.*

EXAMPLE 12–3 Calculate how high above Earth's equator a satellite must be to stay above the same point at all times.

Setting It Up Visualization of this problem is particularly helpful. Figure 12–6 shows a satellite in circular orbit above the equator with Earth *turning beneath it* in its daily rotation. With this figure we can understand immediately that a satellite will be stationary above a particular point if its period is the same as Earth's—24 hours.

Strategy We'll start with the assumption, implicit in the problem statement, that the orbit is circular, with radius r to be determined. As described above, we want a period of 24 hours. Given the period, we can find the distance r using the explicit form of Kepler's third law given in Eq. (12–6).

Working It Out We solve Eq. (12–6) for the radius r as a function of period,

$$r^3 = \frac{GM_{\mathrm{Earth}}T^2}{4\pi^2}.$$

Numerically the orbital radius is

$$r = \left[\frac{(6.67 \times 10^{-11}\ \mathrm{m^3/kg\cdot s^2})(5.976 \times 10^{24}\ \mathrm{kg})(24.0 \times 3600\ \mathrm{s})^2}{4\pi^2}\right]^{1/3}$$
$$= 4.22 \times 10^7\ \mathrm{m} = 26{,}200\ \text{miles}.$$

This is 22,300 miles above the surface of Earth, or seven times Earth's radius.

What Do You Think? Is it possible to place a satellite that remains permanently above Washington, D.C.?

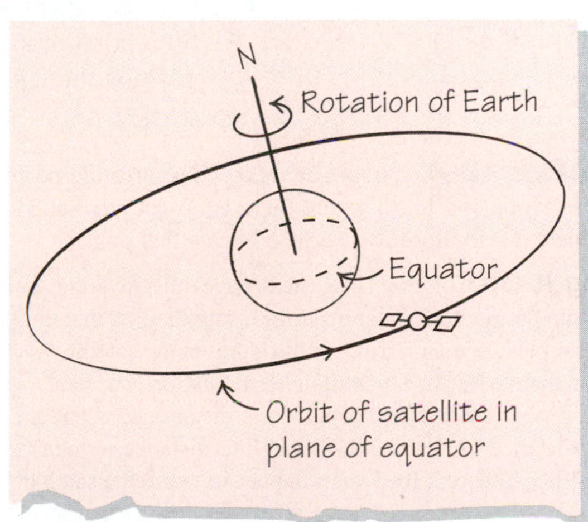

▲ **FIGURE 12–6** A sketch of a satellite in an orbit whose plane is the plane of the equator. Earth rotates as the satellite orbits.

The Potential Energy Associated with Newton's Gravitational Force

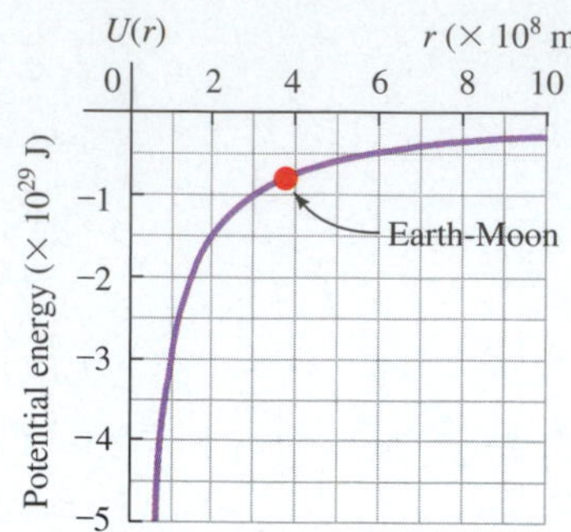

▲ **FIGURE 12–7** The potential energy of the gravitational force is negative and has a $1/r$ dependence. We define it to be zero at $r = \infty$.

The gravitational force is central and depends only on the distance of the influenced object from the force center. It is therefore *conservative* and can be derived from a potential energy function (Section 7–1). We know the value of having such a function—the conservation of energy principle is very useful. We can show here that the potential energy of a system of two point masses interacting with each other through the gravitational force, Eq. (12–4), is

$$U(r) = -\frac{GmM}{r}. \qquad (12\text{–}8)$$

GRAVITATIONAL POTENTIAL ENERGY

A graph of this function is shown in Fig. 12–7, where we have used numbers appropriate to the Earth/Moon system, with the point corresponding to the actual lunar orbital radius also shown. As always, we are free to choose the point where the potential energy is zero, and we have chosen the potential energy to be zero at infinity, the point where the force itself is zero.

To derive Eq. (12–8), we recall the definition of potential energy from Chapter 7, Eq. (7–4):

$$U(r) - U(\infty) = -\int_{\infty}^{r} \vec{F}(\vec{r}\,')\cdot d\vec{r}\,',$$

where we have chosen the point x_0 in Eq. (7–4) at ∞. The force points from the location of mass m to the origin (the location of M), and as any integration path gives the same result (the force is conservative), we can choose the path to go directly along a radial direction, so that $\hat{r}\,'\cdot d\vec{r}\,' = dr'$. We thus obtain

$$U(r) - U(\infty) = -\int_{\infty}^{r} \frac{-GmM}{r'^2}\,dr' = -\frac{GmM}{r'}\bigg|_{\infty}^{r} = -\frac{GmM}{r}.$$

By choosing $U(\infty) = 0$, Eq. (12–8) follows.

The Superposition Principle

Newton's gravitational force, like all other forces, is vectorial and obeys the **principle of superposition**: When several objects of varying masses exert gravitational forces on a point mass m, the total force is the (vectorial) sum of the individual forces. This is equivalent to the fact that potential energies are *additive*: If we have two masses M_1 and M_2, then the potential energy U of m in the presence of the two masses is the *sum* of the potential energies U_1 and U_2. If the source of the gravitational force or potential energy for a point mass is an extended continuous object, we can apply the principle of superposition through integration.

EXAMPLE 12–4 A satellite is to be sent to the position between the Moon and Earth where there is no net gravitational force on an object due to those two bodies. Locate that point.

Setting It Up Figure 12–8 (next page) displays the position of the satellite between Moon and Earth. Using the fact that the Moon is much less massive than Earth, we have drawn the satellite as closer to the Moon than to Earth; a quantitative calculation is needed to confirm this. We denote the known Earth–Moon distance as d and the Earth–satellite distance as x; the Moon–satellite distance is then $(d - x)$. We want the value of x for which the net force on the satellite is zero.

Strategy The forces on the satellite due to the Moon and Earth are respectively $\vec{F}_M$ (to the right) and $\vec{F}_E$ (to the left). Each of these can be expressed as functions of the unknown x. We set the magnitudes of the two forces equal and solve for x.

Working It Out The equality of the force magnitudes reads

$$F_M = F_E$$
$$\frac{GmM_E}{x^2} = \frac{GmM_M}{(d-x)^2}.$$

The satellite mass m cancels, and, on rearranging, we have

$$(d-x)^2 = x^2\alpha,$$

where the ratio $\alpha = M_M/M_E = 0.0123$ (see Appendix III–1.1). This quadratic equation for x has the following solution with x between 0 and d:

$$x_0 = \frac{1-\sqrt{\alpha}}{1-\alpha}d = \frac{1-\sqrt{0.0123}}{1-0.0123}d = 0.900\,d.$$

Zero net force occurs nine-tenths of the way to the Moon.

What Do You Think? How would you approach the problem of finding out whether the zero point determined in this example is a stable point of equilibrium? (A stable point is one for which the forces are such that a little movement away from the point would push you back to that point.)

▲ **FIGURE 12–8** The satellite is placed at a position on the line between Earth and the Moon and is affected by gravitational forces from both bodies.

THINK ABOUT THIS . . .

HOW DO WE KNOW EARTH'S MASS?

The force of local gravity is just Newton's law of universal gravitation at a (nearly) fixed distance $r = R_E$. (You may be wondering how this can be, since Earth is not a point object to us. In fact, as we will discuss in detail in Section 12–4, the gravitational force due to a spherically symmetric object—Earth is a good approximation of one—acts as if the mass of the object were concentrated at its center.) Given this fact we can determine Earth's mass, M_E, from the size of the acceleration due to gravity, g, and Earth's radius, R_E. The force exerted on an object of mass m on Earth's surface by Earth's gravitational attraction is given by

$$F = \frac{GmM_E}{R_E^2} = mg.$$

We thus obtain

$$M_E = \frac{gR_E^2}{G} = \frac{(9.80\ \text{m/s}^2)(6.37 \times 10^6\ \text{m})^2}{6.673 \times 10^{-11}\ \text{N}\cdot\text{m}^2/\text{kg}^2} = 5.96 \times 10^{24}\ \text{kg},$$

where we have used the measured values of g, R_E, and G. This mass determination shows that Earth has an average density of 5.5 g/cm^3. ■

12–3 Planets and Satellites

The gravitational force [Eq. (12–4)] determines the trajectories (or orbits) followed by astronomical bodies and other properties of their motion. The orbit of an object subject to the gravitational force due to a single source can be derived analytically from Newton's second law. This is certainly not true for *every* force law! We have already remarked that Newton showed that these orbits were conic sections.

It was in thinking about orbits that Newton was able to cement his revolutionary idea that an apple and the Moon obey the same force law. What is surprising about Newton's idea is that the motion of an apple and the Moon look, superficially, so different. But Newton was able to reconcile this apparent difference with the simple thought experiment shown in Fig. 12–9, taken directly from Newton's work. If you shoot a cannonball horizontally from a mountaintop, it will fall to the ground on a parabolic path, the motion we would expect from an apple. But if you increase the initial speed of the cannonball, it will go farther, eventually going far enough that the surface of the spherical Earth "falls away" beneath it. At a sufficiently high speed, the path of the falling cannonball will never catch up with Earth's surface falling away, and the motion will be a circular orbit. The Moon in its orbit, Newton reasoned, is "falling" the same way as the cannonball. This is precisely how the satellite can continue to orbit the Earth in the chapter opener photo.

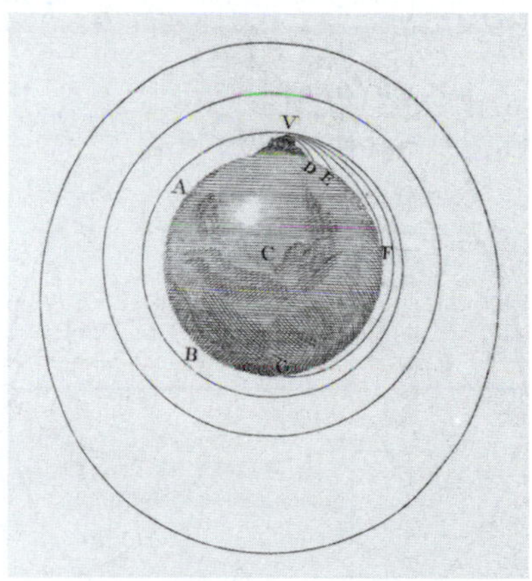

▲ **FIGURE 12–9** Newton's cannon shoots a projectile horizontally from a mountaintop. If the projectile has just the right speed, it will attain a circular orbit at the mountaintop's height. This figure is from Newton's work *A Treatise of the System of the World*, published posthumously.

In the remainder of this section, we'll describe various aspects of orbital motion.

Escape Speed

It is simplest to approach the orbit of a single object—we refer to such objects as satellites—from the point of view of energy. The energy is the sum of the kinetic and potential energies and is thus given by

$$E = K + U = \frac{1}{2}mv^2 - \frac{GmM}{r}. \tag{12–9}$$

The **escape speed**, v_{esc}, of a projectile launched from Earth's surface is the minimum speed with which the projectile must leave the surface in order to leave the vicinity of Earth forever—that is, travel an infinite distance from Earth. In other words, the escape speed is the speed at $r = R_E$ that gives zero speed at infinite r. (If the initial speed at R_E is greater than the escape speed, then the projectile moves away with nonzero speed even when it is arbitrarily far.) An object given the escape speed will have zero kinetic energy at $r = \infty$, and the potential energy is defined to be zero at that point, so the total energy will be $E = 0$ at infinity. Because the total energy is conserved, we can set $E = 0$ everywhere; then the condition that $E = 0$ at the surface is the condition that determines v_{esc}. Let the mass of the projectile be m and Earth's mass be M_E. At Earth's surface, $r = R_E$, and

$$E = \frac{1}{2}mv_{esc}^2 - \frac{GM_Em}{R_E} = 0.$$

If we solve this equation for v_{esc}, the factor m cancels, and we find that

$$v_{esc} = \sqrt{\frac{2GM_E}{R_E}}. \tag{12–10}$$

With the known values of G, M_E, and R_E, the escape speed from Earth's surface is $v_{esc} = 1.12 \times 10^4 \text{ m/s} = 11.2 \text{ km/s}$.

CONCEPTUAL EXAMPLE 12–5 The escape speed is independent of the direction in which the object leaves Earth's surface. Why is this?

Answer The simple response to this question is that Eq. (12–10) contains only radial quantities, not angular ones. The potential energy associated with a conservative central force such as gravity depends *only* on radial distance from the center. The kinetic energy depends only on speed. Thus although the object may leave the surface in different directions, its speed depends on its initial speed and its radial distance. Physically we can understand this result as follows: When one leaves the surface, the angular aspect appears important—certainly you can tell if your takeoff is vertical or at a 20° angle to the vertical or at any other angle. But when one is very far away, that aspect becomes irrelevant. You look back and, whatever angle you took off at, the distant Earth looks like a point object from which you have gone out in a radial direction.

What Do You Think? Did it require the *Apollo* astronauts a higher escape speed to leave the Earth originally or to leave the Moon on the return trip?

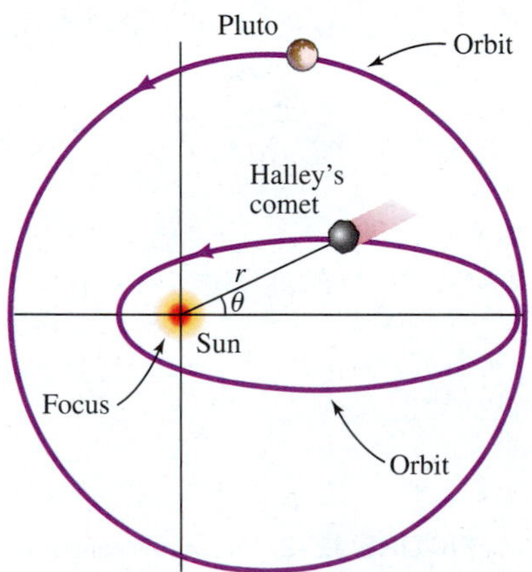

▲ **FIGURE 12–10** For negative total energies, the value of r is limited and the orbit is elliptical (or circular). In this figure we see two elliptical orbits, one not very elongated (that of Pluto) and one quite elongated (that of Halley's comet).

Types of Orbits

We saw earlier that the orbits, or trajectories, of objects (satellites) under the influence of the gravitational force can take one of four forms: circular, elliptic, hyperbolic, or parabolic (Fig. 12–4). A related way of classifying orbits is through the sign of the total energy. Consider Eq. (12–9) once more:

$$E = K + U = \frac{1}{2}mv^2 - \frac{GmM}{r}.$$

Why are the form of the orbit and the sign of the energy closely related? If the total energy E is positive, then the object's kinetic energy is greater than its potential energy at any radius r, and as r becomes infinite, $|v| > 0$. This corresponds to an orbit that never closes, and it is, in fact, a hyperbolic trajectory. When the energy of the object is exactly zero, we have the special case of a parabolic orbit, also an open trajectory. The object starts at infinity with a speed of zero, swings past the source of the gravitational force (for example, the Sun) in a parabolic orbit, and slows down as it moves away from the source, ending up with zero velocity when r becomes infinitely large. This is exactly how we determined the escape speed. For negative energies the potential energy is always greater than the kinetic energy of the object. Then r cannot become too large, because if it did, then v^2 would have to be negative, which is impossible. In that case the orbit is limited and is elliptical (Fig. 12–10) or circular. As applied to the Sun, the planets all have slightly elongated elliptical orbits; Halley's comet (Fig. 12–11), like many other comets, follows a very elongated elliptical orbit. The comet's potential energy is greatest at the farthest point from the Sun, and its kinetic energy least at this point. Conversely, at the nearest point to the Sun the comet will have its greatest velocity and kinetic energy and its least (most negative) potential energy. Circular orbits are the simplest of all, as Example 12-6 illustrates. Only the direction of the motion changes while the potential energy and kinetic energy remain the same throughout. In summary, when $E < 0$, the satellite has a bound orbit, and when $E \geq 0$, the object follows an unbound orbit.

▲ **FIGURE 12–11** Halley's comet, shown here in a recent passage near Earth, has an elongated elliptical orbit with a period of 76 yr.

EXAMPLE 12–6 Imagine that a small moon of Jupiter, of mass m, moves in a circular orbit of radius r under the influence of the gravitational force due to Jupiter, mass $M \gg m$ (Fig. 12–12). Calculate the total energy of the moon as a function of r.

Strategy Equation (12–9) gives the total energy, but it is a function of both v and r. To eliminate the speed, we recall that the acceleration (which is centripetal) has magnitude v^2/r for a circular orbit and that the force (also centripetal) has magnitude GmM/r^2. We can then find v as a function of r by equating the force on m with m multiplied by the acceleration. Once we know the speed as a function of radius, we can calculate the energy as a function of radius.

Working It Out The second law for the moon is

$$\frac{GmM}{r^2} = \frac{mv^2}{r}.$$

(All directions are centripetal.) We solve for the speed squared,

$$v^2 = \frac{GM}{r}.$$

We can then use this in the expression for the energy, Eq. (12–9),

$$E = \frac{1}{2}mv^2 - \frac{GmM}{r} = \frac{1}{2}m\frac{GM}{r} - \frac{GmM}{r} = -\frac{1}{2}\frac{GmM}{r}.$$

The total energy of the orbiting moon is just one half the potential energy for a circular orbit. It is negative, as is appropriate for a closed orbit. Nothing we did depends on the fact that this is a Jovian moon, and it is generally true that for a circular orbit the total energy is one half the potential energy.

What Do You Think? Which parameter best characterizes the difference between bound and unbound orbits? (a) speed, (b) kinetic energy, (c) potential energy, (d) total energy.

▲ **FIGURE 12–12** Here Europa is a distance r from Jupiter's center. The innermost moon, Io, can be seen by its bright, brown-yellow surface against Jupiter's background. This photo was taken by Voyager 1 in 1979 as it passed Jupiter on its way out of our solar system.

EXAMPLE 12–7 A satellite moves in a circular orbit around Earth, taking 90.0 min to complete 1 revolution. We are given the following information: Earth's radius is $R_E = 6.37 \times 10^6$ m; the distance from the Moon to Earth is 3.84×10^8 m; the Moon's orbit is circular; the period of the Moon's rotation about Earth is 27.32 d; and Earth's gravitational force acts as if all of Earth's mass were concentrated at its center. With this information, calculate the height of the satellite above Earth.

Setting It Up We label the known periods of Moon and satellite as T_M and T_S, respectively. The Moon's orbit radius is R_{ME}. All these are given in the problem statement. Let the satellite orbit radius be R_S; it determines the height of the satellite above Earth through $h = R_S - R_E$.

Strategy Once again, Kepler's third law comes into play, but this time applied to Earth as the force center. Both the Moon and a satellite in circular orbit around Earth behave like planets in circular orbit around the Sun. Here the simpler form of Kepler's third law is useful [Eq. (12-1)], that the cube of the radius of the orbit is proportional to the square of the orbital period. In other words, the ratio $(\text{radius})^3/(\text{period})^2$ is the same for both the Moon and the satellite.

Working It Out The ratio equality reads

$$\frac{R_S^3}{T_S^2} = \frac{R_{ME}^3}{T_M^2}.$$

This equation can be solved for R_S,

$$R_S = R_{ME}\left(\frac{T_S}{T_M}\right)^{2/3}.$$

The period of the satellite must be in the same units as that of the Moon. We thus have $T_S = 90.0 \text{ min} = 1.5 \text{ h} = (1.5 \text{ h})/(1 \text{ d}/24 \text{ h}) = 0.0625$ d. This gives us the numerical result

$$R_S = (3.84 \times 10^8 \text{ m})\left(\frac{0.0625}{27.32}\right)^{2/3} = 6.67 \times 10^6 \text{ m}.$$

The height is then

$$h = R_S - R_E = (6.67 - 6.37) \times 10^6 \text{ m}$$
$$= 3 \times 10^5 \text{ m} = 300 \text{ km}.$$

Many satellites use this orbit, which is one that lies just above Earth's atmosphere. Small air drag makes the orbit long-lasting, and it takes less energy to put a satellite in a lower than a higher orbit.

What Do You Think? When the radius increases, so does the period. Does the speed also increase?

Properties of Noncircular Orbits: The hyperbolic, parabolic, and elliptical orbits corresponding to different signs of total energy are closely related; as stated in Section 12–2, they are all *conic sections* (see Fig. 12–4). Let's focus our attention on the solar system, although we should keep in mind that our results will be applicable to, say, satellites orbiting Earth as well. The Sun, which is very heavy in comparison to the planets, plays a special role in the orbits. For closed orbits in particular, the Sun sits at one *focus* of the ellipse, a fact noted by Kepler in his first law and illustrated in Fig. 12–3a. The *semimajor axis* in Fig. 12–3a is denoted as a, and the *semiminor axis* as b. The equation describing an ellipse is

$$\frac{x^2}{a^2} + \frac{y^2}{b^2} = 1. \tag{12–11}$$

The point where the planet makes its closest approach to the Sun $(r = r_{min})$ is called the *perihelion*; the point where the farthest distance between a planet and the Sun is attained $(r = r_{max})$ is the *aphelion*. (For Earth satellites the corresponding words are perigee and apogee.) From Fig. 12–3a, we have

$$\text{semimajor axis} = a = \tfrac{1}{2}(r_{min} + r_{max}). \tag{12–12}$$

The *eccentricity* e of the orbit is a dimensionless measure of the elongation of the orbit and is proportional to the difference between r_{max} and r_{min}:

$$e = \frac{r_{max} - r_{min}}{2a}. \tag{12–13}$$

Note that when $r_{max} = r_{min}$, the orbit reduces to a circular one of radius a and the eccentricity is zero. The most extreme possible orbit corresponds to $r_{min} = 0$, when $r_{max} = 2a$, and

$$e_{extreme} = \frac{2a}{2a} = 1.$$

The eccentricity therefore varies between zero and one. The role of these quantities in the shape of the possible elliptical orbits is detailed in Fig. 12–13, which also includes the possible non-elliptical orbits.

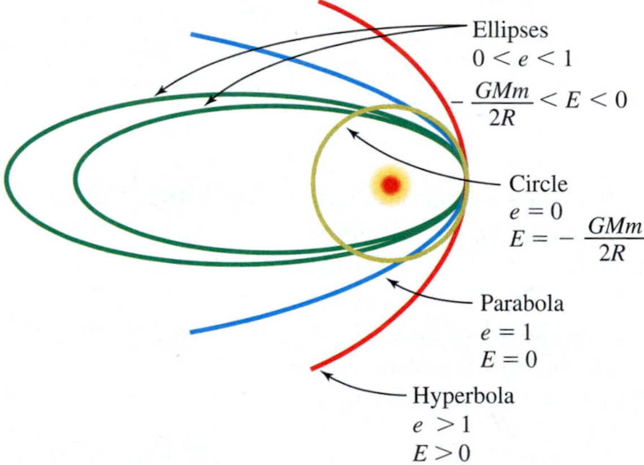

FIGURE 12–13 The set of possible orbits for objects whose distance of closest approach to the Sun are the same. Two ellipses are included, and the more elongated of these has the larger value of eccentricity.

EXAMPLE 12–8 Consider the motion of a comet in an elliptical orbit around a star. The eccentricity of the orbit is given by 0.80, and the distance between the perihelion and the aphelion is 1.0×10^{10} km. (a) Find the distances of nearest and farthest approaches of the comet. (b) If the speed of the comet is 81 km/s at perihelion, what is its speed at aphelion?

Setting It Up We show the motion of the comet about the star in Fig. 12–14 (next page). We are given the eccentricity, which we label by e, as well as the sum of the aphelion and perihelion distances, which we label by $2a$. For part (a) we are asked to find the distances r_{max} and r_{min} labeled in the figure.

Strategy Part (a) is simply a matter of solving equations that relate the unknowns, r_{max} and r_{min}, to the knowns, $2a$ and e. Equations (12–12) and (12–13) apply; we solve them for r_{max} and r_{min} in terms of a and e. For part (b) we note that at both perihelion and aphelion the orbit is at right angles to the vector $\vec{r}$ from the star to the comet. The angular momentum is therefore just the product of r with mv, where v is the speed. Because the angular momentum is conserved in motion under the influence of gravitation, we must have

$$mv_{perihelion}r_{min} = mv_{aphelion}r_{max}. \tag{12–14}$$

Since after part (a) everything in this equation except $v_{aphelion}$ is known, we can solve for it.

Working It Out (a) If we add Eq. (12–12) to a times Eq. (12–13), r_{min} cancels, and

$$r_{max} = a + ae = a(1 + e). \tag{12–15}$$

In turn,

$$r_{min} = 2\left(a - \frac{r_{max}}{2}\right) = 2\left[a - \frac{a(1 + e)}{2}\right] = a(1 - e). \tag{12–16}$$

Numerically, with $e = 0.80$ and $2a = 1.0 \times 10^{10}$ km,

$$r_{max} = \tfrac{1}{2}(1.0 \times 10^{10}\ \text{km})(1.80) = 9.0 \times 10^{9}\ \text{km};$$
$$r_{min} = \tfrac{1}{2}(1.0 \times 10^{10}\ \text{km})(0.20) = 1.0 \times 10^{9}\ \text{km}.$$

(b) We solve Eq. (12–14) to find the speed at aphelion:

$$v_{aphelion} = v_{perihelion}\frac{r_{min}}{r_{max}} = (81\ \text{km/s})\left(\frac{1.0 \times 10^{9}\ \text{km}}{9.0 \times 10^{9}\ \text{km}}\right)$$
$$= 9.0\ \text{km/s}.$$

We used the conservation of angular momentum to relate the speed at perihelion to the speed at aphelion in this example. The use of energy conservation would also allow us to find the star's mass (see Problem 38).

What Do You Think? A comet comes in toward the Sun, comes very close to the Sun's surface, and moves out again. Is this enough information to deduce that the comet is in an elliptical orbit?

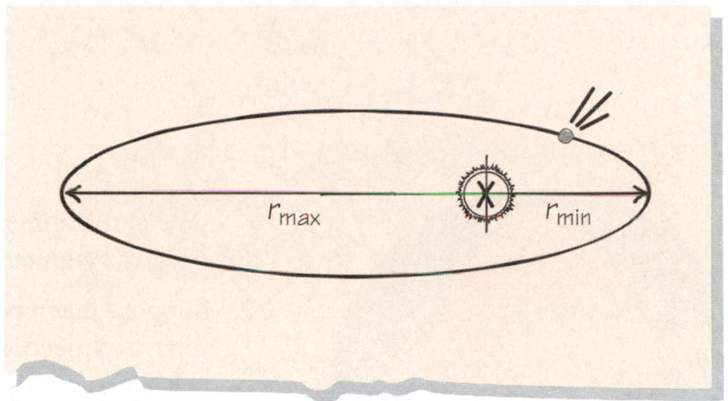

▲ **FIGURE 12–14** A comet's orbit, with the minimum and maximum distances from the star (perihelion and aphelion, respectively).

12–4 Gravitation and Extended Objects

We have viewed the planets and the Sun on the scale of the entire solar system and have treated them as point masses. But they are not really pointlike. This problem is especially evident if we think about how the gravitational force from Earth acts on an apple. To an apple, Earth most certainly is an extended object, yet we treated the motion of the apple as if all the mass of Earth were concentrated at Earth's center. We can study how extended objects behave by using the principle of superposition (Sec. 12–3). Generally in this section we consider spherical objects, the (approximate) form of many large astronomical objects.

The Gravitational Force Due to a Spherically Symmetric Object

We claimed in Section 12–2 that when an extended object is spherically symmetric, the force it exerts on a point mass outside the extended object is the same as the force that would be exerted if the entire mass of the extended object were concentrated at its center. It is this assertion that allowed us to relate the acceleration due to gravity at Earth's surface, g, to Earth's mass and radius, and thereby weigh Earth itself (see the feature Think About This ... How do we know Earth's mass?). In fact, for any spherically symmetric object[†] of mass M and radius R, the acceleration due to gravity at its surface is

$$g = \frac{GM}{R^2}. \tag{12–17}$$

Thus we would expect the acceleration of gravity to be different at the surfaces of the different planetary objects, such as the Moon.

Many mass distributions are spherically symmetric, or nearly so. For such systems the mass density depends only on the distance from the center of the distribution. The mass density at the center could vary with radial distance, but it must not depend on angle. Earth conforms well to these requirements (Fig. 12–15). Earth's core is denser than the outer layers—it most likely consists primarily of iron. (The small deviations from spherical symmetry come from the fact that, because of its rotation, Earth bulges a little in the equatorial region; there is an additional distortion that makes the planet look a little pear-shaped. In addition, there are local regions of slightly greater or lesser density, due for example to the presence of ore bodies.)

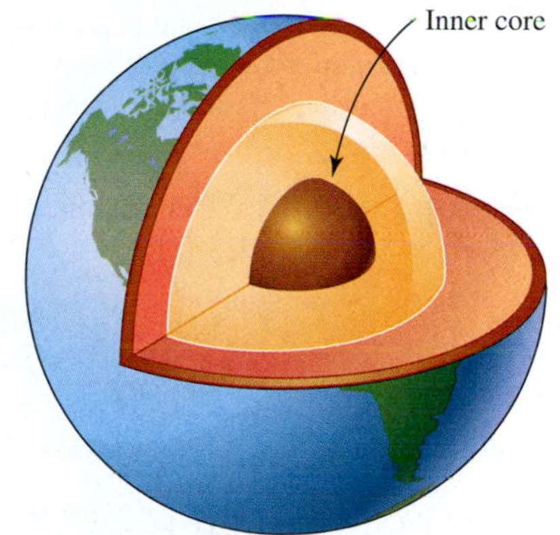

▲ **FIGURE 12–15** A view of Earth's internal structure. The inner core is substantially denser than the rest.

[†]This means its density varies only with distance from the center.

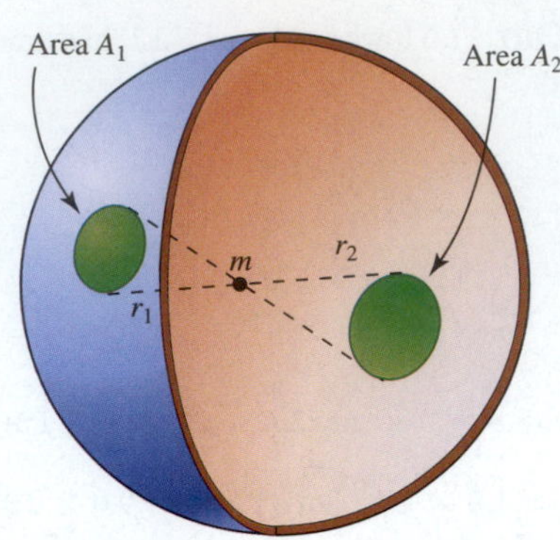

▶ **FIGURE 12–16** The gravitational force on an object of mass m inside a spherical shell. Regions from opposite sides of the shell exert equal and opposite forces on the object, and the result is that there is no net force on the mass m.

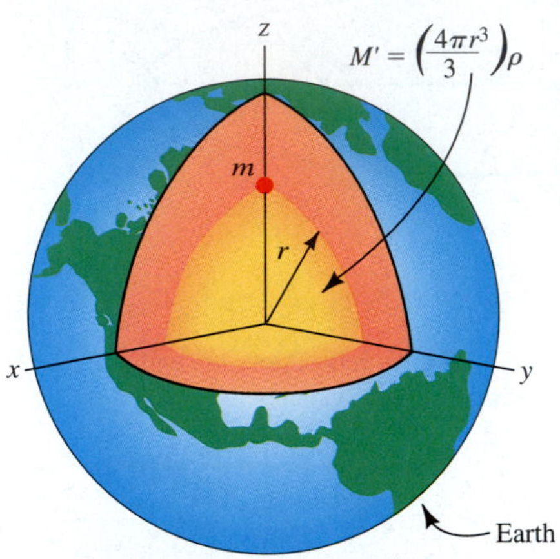

▲ **FIGURE 12–17** The gravitational force on an object with mass m located inside Earth depends on the total amount of mass M' inside the sphere whose density is ρ and whose radius r is the distance of mass m from the center.

We can summarize the principal conclusions about the gravitational force due to spherically symmetric systems as follows.

1. Suppose that a point mass is *outside* a spherically symmetric object. The gravitational force experienced by the point mass is identical to the force that would arise if the whole mass of the spherical object were concentrated at its center. In other words, the force exerted on the point mass is the same as the force that would be exerted on it if the uniform spherical object (total mass M) were a point mass M located at the center of the sphere.
2. Suppose that a point mass is somewhere *inside* a thin spherical shell of constant density (Fig. 12–16). Then there is *no* gravitational force on the mass. This conclusion holds for a point mass inside an arbitrarily thick shell as well, as long as the mass density of the shell depends *only* on the distance from the geometric center of the shell. We'll describe the reasoning behind this surprising result later.

These two important facts show that the gravitational force on a point mass m within Earth at a distance r from the center, for example, would be due to a mass M' concentrated at Earth's center; where M' is the mass contained within Earth only up to the radius r (Fig. 12–17).

EXAMPLE 12–9 Suppose that a tunnel is drilled through our planet along a diameter. Assume that Earth's mass density is uniform and is given by ρ. Describe the force on a point mass m dropped into the hole as a function of the distance of the mass from the center.

Setting It Up As in Fig. 12–18a, the tunnel passes through Earth's center. In addition to the quantities that are directly given, we can assume we know either Earth's mass M_E or its radius R_E (given one, we can find the other using ρ).

Strategy The gravitational force on the point mass m is due only to the mass $M(r)$ of the material contained within a radius r, where r is the distance from the point mass m to Earth's center. The force is attractive, toward the center. We must calculate $M(r)$ in order to find the force, and that mass is the density ρ times the volume of the sphere of radius r.

Working It Out The magnitude of the force at a radius $r \le R_E$ is

$$F(r) = -\frac{GmM(r)}{r^2}.$$

Since the volume of a sphere of radius r is $(4/3)\pi r^3$, we have

$$M(r) = \left(\frac{4\pi r^3}{3}\right)\rho.$$

Thus, as is sketched in Fig. 12–18b,

$$F(r) = -\frac{Gm}{r^2}\left(\frac{4\pi\rho r^3}{3}\right) = -\left(\frac{4\pi Gm\rho}{3}\right)r \quad \text{for} \quad r \le R_E.$$

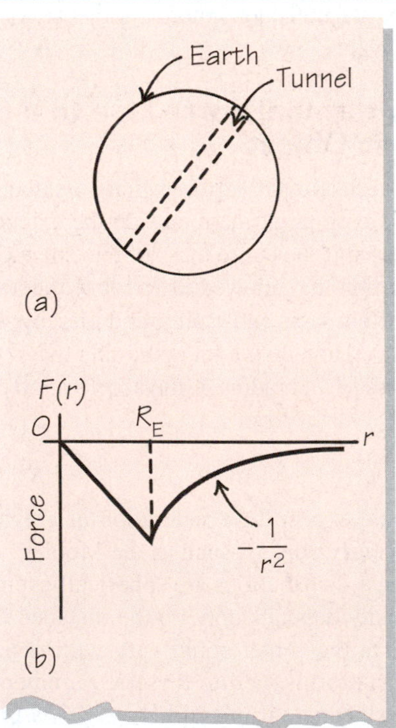

▲ **FIGURE 12–18** (a) A tunnel passes through Earth's center. (b) A graph of the variation of the force with r shows that the force of gravity increases linearly from Earth's center out to the surface (assuming constant density), and dies away with the inverse square of distance from Earth's center at distances from the center greater than Earth's radius.

As we shall describe in Chapter 13, the fact that $F \propto r$ means that inside Earth the point mass acts as if it were moving under the influence of a spring, or as a pendulum. This motion is oscillatory, and the point mass moves from one end of the tunnel to the other and back.

What Do You Think? Suppose Earth's density were not uniform but rather more mass was concentrated at the center, which is in fact the case. The force on the object as it moved through the tunnel would (a) be unchanged; (b) vary linearly with r but with a different coefficient; (c) fall off more rapidly than r as $r \to 0$; (d) fall off less rapidly than r as $r \to 0$, and perhaps even grow.

The two summarizing conclusions about the force of gravity for spherically symmetric objects will not be derived mathematically until Chapter 23, where we give a simple but indirect proof. Direct proof would follow from integrating the contributions of the force from different pieces of our extended object, but it is an unnecessary diversion at this point. However, we can gain a physical picture of why a spherically symmetric shell that surrounds a point mass exerts no force on the point mass.[†]

Figure 12–16 shows the point mass placed off center within a thin spherical shell. Consider a double cone making an opening angle θ at the point mass and, in particular, the gravitational effects of the two circular sections that the cone cuts on the shell. The point mass is a distance r_1 from each point in the area on the left, A_1, and a different distance r_2 from each point in the area on the right, A_2. If the density of the shell is ρ and its thickness is τ, then the force F_1 that attracts the point mass to the left-hand area is

$$F_1 = \frac{Gm}{r_1^2}\rho\tau A_1,$$

and the force F_2 that attracts the point mass to the right-hand side is

$$F_2 = \frac{Gm}{r_2^2}\rho\tau A_2.$$

Now we note that areas A_1 and A_2 are proportional to r_1^2 and r_2^2, respectively, so that

$$\frac{A_1}{r_1^2} = \frac{A_2}{r_2^2}.$$

This means that the two forces are independent of the distance of the point mass to the circular sections and the forces cancel because they pull in opposite directions. In effect, as the point mass moves closer to one side, the inverse-distance-squared factor in the force increases, but the amount of mass seen decreases by the same factor; thus, the pull of every sector of the shell is the same and cancels.

This curious effect is very special to the inverse-square law—a law that also holds for electric charges. Indeed, it was Benjamin Franklin who noticed that there is no force on an electric charge surrounded by a shell of the opposite charge, and it was on the basis of this observation that Joseph Priestley in the eighteenth century first suggested that electric forces obey the inverse-square law. We will consider electrical charge further in Chapter 21.

Dark Matter

Astronomers have inferred the presence of what is known as *dark matter* by using the properties we described above. Dark matter gets its name from the fact that it emits no radiation and hence cannot be observed with telescopes. The discovery of dark matter has had important implications for our understanding of the overall structure of the universe. To understand how dark matter is revealed, consider a mass m outside of a spherically symmetric mass distribution of total mass M and moving in circular motion at a distance r from the center of the distribution. For that motion, the acceleration is centripetal, with magnitude v^2/r, and because of our result that the net gravitational force from the distribution is the same as if it were all concentrated at the center, Newton's second law reads

$$\frac{mv^2}{r} = \frac{GmM}{r^2}$$

[†]Strictly speaking, our arguments hold only for small values of θ. But generalization is possible.

or

$$v \propto \frac{1}{\sqrt{r}}.$$

This result is now tested for clouds of visible ionized hydrogen that circulate around galaxies, outside the visible ("shining") distribution of matter—glowing stars and visible dust—in those galaxies. If the only matter making up the galaxies were the visible matter, we would expect the measured speed of the hydrogen clouds to drop off with distance from the galaxy as $1/r^{1/2}$. But instead the speeds of the hydrogen clouds orbiting about the galaxy remain constant with distance from the center of the galaxy. This only makes sense if there is matter that goes beyond the visible boundaries of the galaxy and which influences the orbit of the hydrogen clouds. More precisely, suppose that the *total* (not just the visible) galactic mass enclosed in a sphere of radius r is $M(r)$, and assume that the mass is symmetrically distributed. Then the orbital motion of a cloud at a distance r from the center of the sphere is governed by that mass,

$$\frac{mv^2}{r} = \frac{GmM(r)}{r^2},$$

so that

$$v \propto \sqrt{\frac{M(r)}{r}}.$$

If the speed v is a constant for large r, as the observation of the hydrogen clouds indicates, we can conclude that $M(r)$, grows linearly with r. But any material at the distance of the glowing hydrogen clouds themselves, which are outside the radius of the visible galaxy, is not luminous—it does not shine. Although the details are somewhat different, studies of clusters of galaxies also lead to the conclusion that dark matter must be present to explain the motions of galaxies within a cluster. Altogether the various pieces of orbital evidence tell us that the mass of shining matter is roughly a factor of 10 smaller than the mass of dark matter. What exactly is this dark matter? No one knows at this point, and this question holds center stage in much current research.

How *g* Varies with Altitude

Equation (12–17), together with the result that a symmetric, spherical Earth behaves like a centered point mass, implies that g varies with altitude. Suppose that we measure altitude, h, from sea level, and that Earth's radius at sea level is R_E. From Eq. (12–17) we find

$$g(h) = \frac{GM}{(R_E + h)^2} = \frac{GM}{R_E^2[1 + h/R_E]^2}. \qquad (12\text{–}18)$$

Now, if the ratio $h/R_E \ll 1$ (and even for the top of Mount Everest, $h/R_E \cong 1.5 \times 10^{-3}$), then we can use the approximation $1/(1 + x)^2 \cong 1 - 2x$ (for $x \ll 1$; see Appendix IV–10), so $g(h) \cong (GM/R_E^2)[1 - 2h/R_E)]$, or

$$\frac{g(h)}{g(0)} \cong \frac{(GM/R_E^2)[1 - (2h/R_E)]}{(GM/R_E^2)} = 1 - \frac{2h}{R_E}. \qquad (12\text{–}19)$$

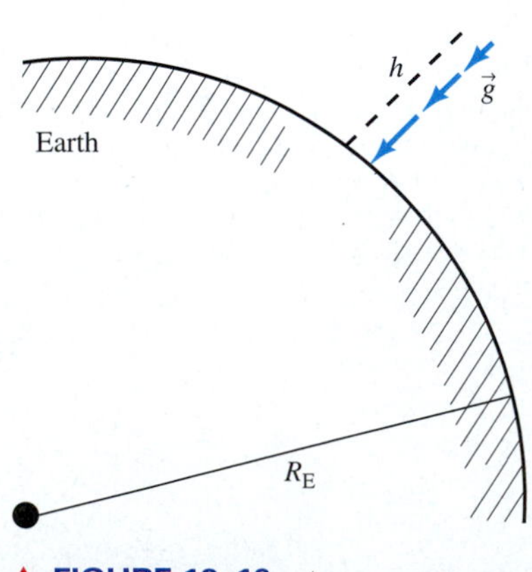

▲ **FIGURE 12–19** As we move away from the surface, the value of g decreases.

To get a feeling for this effect, we can see from Eq. (12–19) that, at the top of the Sears Tower in Chicago, with $h = 443$ m, g is 99.99 percent of its value at sea level; even at the top of Mount Everest, with $h = 8848$ m, g is 99.74 percent of its value at sea level (Fig. 12–19).

Other effects modify Eq. (12–17). Earth is not a perfectly uniform sphere. Not only is it not spherical but it contains lumps of higher or lower density that affect g. Measurements of g are routinely made to many significant figures, more than enough to reveal all the effects we have discussed as well as one associated with Earth's rotation (Conceptual Example 12–10).

CONCEPTUAL EXAMPLE 12–10 Earth's rotation has an effect on g. Which of the following statements are true, if any? (a) The effect is largest at the equator, where g increases over its zero-rotation value; (b) The effect is largest at the poles, where g decreases; (c) The effect is largest at the equator, where g decreases; (d) The effect is largest at the poles, where g increases.

Answer (c) The effect is due to the fact that a rotating frame is an accelerating frame, with fictitious forces present (see Chapter 5). In other words, there is a "centrifugal force" that makes it appear to the observer in the rotating frame that he or she is being pushed outward. If the frame rotates with angular velocity ω, then the acceleration that produces the effect is proportional to $\omega^2 r$. Thus the effect is absent at the poles, where the r factor is 0 and is largest at the equator, for which $r = R_E$. Since the "force" is outward, it tends to cancel the acceleration of gravity—g is decreased. Quantitatively, it is not hard to calculate, using the $\omega^2 r$ value, that at the equator g is 99.57 percent of what it would be if Earth did not rotate.

(a)

(b)

FIGURE 12–20 Alma beach in the Fundy National Park, New Brunswick, Canada, at (a) high and (b) low tide. The Bay of Fundy has some of the largest tidal differences in the world.

Tidal Forces

Why do the oceans on Earth experience two tides each day? To answer this, we need to consider how gravity acts on an extended object. If the object it acts on is large, the strength of the gravitational force will be different on different parts of the object, and when different parts of the object experience different forces, they will accelerate differently. This fact can cause the object to distort in shape or even fall apart if there are no compensating internal forces that hold the object together. We say that such distortions are due to **tidal forces**, a term we apply to the *difference* of gravitational forces across an extended object. Newton, who first correctly developed the idea of tidal forces, showed that just as the Moon attracts the water nearest it more strongly than Earth as a whole, making a bulge toward the Moon, so the Moon also attracts Earth as a whole more strongly than the water on the *far* side of Earth, leaving behind a second bulge on the far side. The result is two high tides per day (Fig. 12–20). We can make this argument a little more quantitative using the same reasoning that gave us Eq. (12–19).

While we are going to apply tidal forces to understand ocean tides on Earth, there are other astronomical situations where tidal forces play an important role. A revealing if fanciful example occurs if we apply Eq. (12–18), the variation of g with height, to an object of height $h = 2$ m moving on a grazing orbit under the influence of a *neutron star*, an object of very high surface gravity. We take realistic numbers for the star, $M = 2 \times 10^{30}$ kg (one solar mass) and $R = 10$ km $= 10^4$ m. Then the difference between g at the top of the object farthest from the star and g at the bottom, 2 m nearer to the star, is

$$\frac{GM}{R^2[1 + (h/R)]^2} - \frac{GM}{R^2} \cong -\frac{GMh}{R^3} \cong 27 \times 10^7\ \mathrm{m/s^2} = 2.7 \times 10^7\ g_{\mathrm{Earth}}.$$

No object that we know of could remain intact under this sort of differential acceleration. The tidal forces are just too strong.

Let us turn now to ocean tides on Earth. These are correlated, though imperfectly, with the position of the Moon, and we'll concentrate on the Moon's role. (The Sun is important too, although we'll ignore it at first.) We begin by finding the tidal forces on Earth due to gravitational attraction to the Moon (Fig. 12–21). Let the distance from the Moon, mass M_M, to Earth's center O be D and Earth's radius be denoted by R_E. Then the acceleration experienced by a piece of matter at Earth's center is in the direction of the Moon, with magnitude

$$a_O = \frac{GM_M}{D^2}.$$

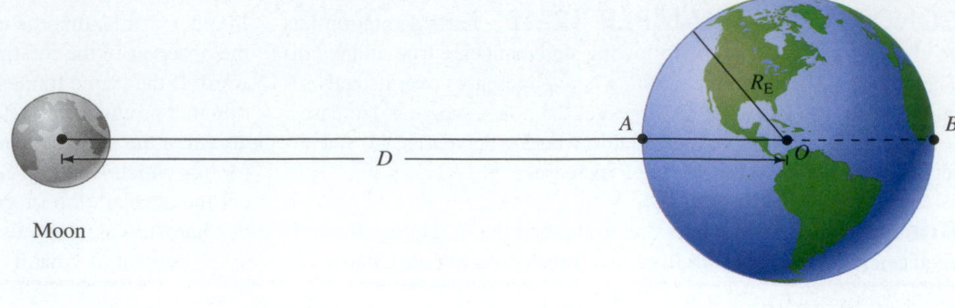

▶ **FIGURE 12–21** A sketch helpful in calculating tidal forces on Earth due to the Moon.

The acceleration experienced by a mass at point A on Earth's surface (Fig. 12–21) is in the same direction but has magnitude

$$a_A = \frac{GM_M}{(D - R_E)^2} = \frac{GM_M}{D^2\left(1 - \dfrac{R_E}{D}\right)^2}.$$

We can now use the fact that $R_E \ll D$, and approximate this just as we did for Eq. (12–19):

$$a_A = \frac{GM_M}{D^2}\frac{1}{\left(1 - \dfrac{R_E}{D}\right)^2} \cong \frac{GM_M}{D^2}\left(1 + \frac{2R}{D}\right) = \frac{GM_M}{D^2} + 2\frac{GM_M R_E}{D^3}.$$

The first term on the right is just a_O, and the second term shows that a_A is a little larger in magnitude than the acceleration of a point at Earth's center—the Moon is a little closer and has a little larger effect. We can interpret the second term as the acceleration *relative to Earth's center* of the piece of matter closest to the Moon; it is in a direction toward the Moon. In other words, the piece of matter at A experiences a differentially larger force that tends to move it toward the Moon, away from Earth's center. Similarly, the acceleration experienced by a mass bit at point B has magnitude

$$a_B = \frac{GM_M}{(D + R_E)^2} \cong \frac{GM_M}{D^2} - 2\frac{GM_M R_E}{D^3}.$$

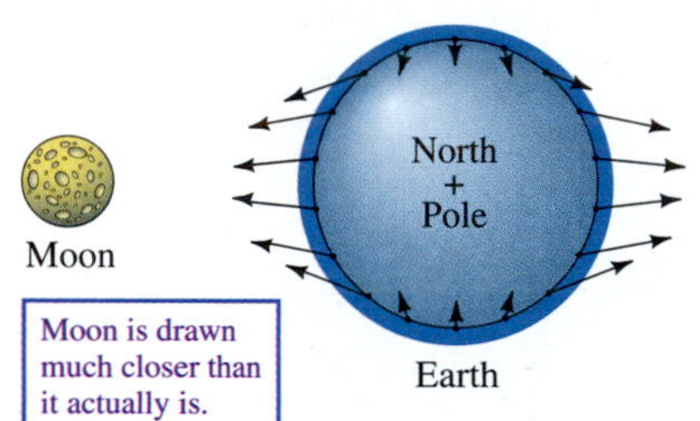

(a)

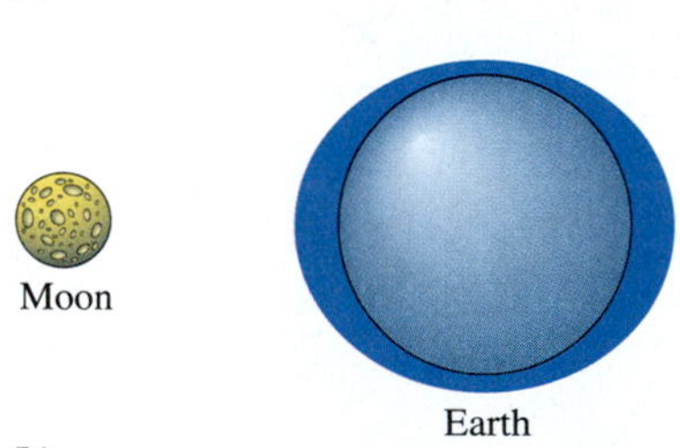

(b)

▲ **FIGURE 12–22** (a) The arrows indicate how the forces from the Moon at Earth's surface differ from the average value of the force on Earth. The arrows along the plane perpendicular to the line connecting Earth and the Moon should be much smaller than are drawn here; the tidal forces there are about $1/60$ of the tidal forces along the line from Earth to the Moon. (b) The resulting water distribution has two bulges.

A mass bit at point B, the point farthest from the Moon, will have a smaller acceleration than that of Earth toward the moon, that is, it will accelerate *away* from Earth's center and from the Moon with an acceleration relative to Earth's center of magnitude $2(GMR/D^3)$. This acceleration is interpreted by someone measuring things with respect to Earth's center as due to a force tending to move the mass bit away from the Moon. Of course, in applying this to oceans on Earth, we should make the same calculation for every spot on Earth. We could make a map of the differential forces—the tidal forces—and we would find that tidal forces occur everywhere on Earth's surface (Fig. 12–22a), and they all tend to reinforce a piling up of water at points A and B (Fig. 12–22b).

What is the numerical size of the corrections? The accelerations due to the tidal forces are of order $2R/D$ of the Earth's acceleration due to the Moon, and this is approximately a factor $1/30$. The Sun also contributes to tidal forces. The Sun is about 2.7×10^7 times more massive than the Moon, but its distance from Earth is also a factor 3.9×10^2 larger. Since the correction to the acceleration is proportional to M/D^3, the tidal effects due to the Sun are $(2.7 \times 10^7)/(3.9 \times 10^2)^3 \cong 0.45$ times smaller than those due to the Moon.[†] This is still significant, and it shows up most clearly when

[†] The actual gravitational force from the Sun is almost 200 times that from the Moon—it is the tidal forces due to the Sun that are less than those due to the Moon.

the Sun, Moon, and Earth are aligned, when the tides are about 50% higher than average (spring tides), and when the line between Earth and the Sun is perpendicular to the line between Earth and the Moon, when the tides are about 50% lower than average (neap tides).

If water could flow arbitrarily easily, the high tides would always be directly under the Moon. But water is viscous, and a kind of friction acts within it to oppose its flow. For this reason the high tides lag the Moon's position. The friction has the additional effect of slowing down the rate of Earth's rotation. Since angular momentum is conserved (there are no external torques), the Moon must accordingly move away from Earth. This increase in the Moon's orbital distance from Earth is about 0.5 m per year. Lunar ranging—the measurement of the distance between the Moon and Earth by means of reflection of laser light from reflectors placed on the Moon combined with extremely high precision time measurements—confirms detailed calculations of the effect.

Ultimately Earth will slow its rotation until one side always faces the Moon, much like the Moon presents only one of its faces to Earth. At that point there will be no more tides and no further slowing. One can show that this will occur when the Moon has receded to a distance 1.44 times its present distance to Earth, a very long time at a rate of 0.5 m per year. One could then use Kepler's third law to estimate the length of the lunar month. Because Earth will always be presenting the same face to the Moon, this will also be the length of the day!

THINK ABOUT THIS . . .

WHAT DETERMINES HOW BIG MOUNTAINS CAN GET?

Mount Everest and the other very tall mountains on Earth are about 10 km high. Is this an accident? Why shouldn't the tallest mountains be 100 km, or 1 km high? In an entertaining demonstration of how to pull in knowledge from many areas of physics, the late Victor Weisskopf, who was a very talented teacher, argued that 10 km is no accident. Although we can't explain every step of the argument at this point, it is nonetheless well worth thinking about. Tectonic forces—the movements of great crustal plates at Earth's surface—cause uplift of parts of the crust. Think then about a block of matter of height h. If this block gets too high, then its mass will be so large that it will cause plastic deformation of the base: The base will melt and run out, and the block will settle back down. In effect, the melting of a layer of thickness d is equivalent to moving a layer of thickness d from the top of the mountain to the bottom, and we know the energy involved is proportional to gh. In order to convert this to a maximum height, we must realize that if this number is big enough, it will cause the rock to melt. We have to know how much energy is involved in raising the rock temperature to the melting point, about 2000°C, then melting it, and this can be obtained from a knowledge of its atomic structure or looked up in tables. When these numbers are brought in, the maximum height h on Earth is 14 km, a number very close to the measured values.

If Mars is made of the same basic material as on Earth, then this argument allows us to estimate the maximum height of Mars mountains: $g_{\text{Mars}}h_{\text{max Mars}}$ must be the same as $g_{\text{Earth}}h_{\text{max Earth}}$, and we need only find g_{Mars}. The result of this exercise (see Problem 56) is a maximum height about 30 km, and this also corresponds very well with the observed highest mountain, about 25 km.

12–5 A Closer Look at Gravitation

When we have applied the law of gravitation, it has always been in the context of the attraction exerted on one object by another with a much larger mass object. We have considered planets or comets moving around the Sun, tennis balls falling on Earth's surface, and so on. We assumed that the massive object was at rest and that the orbits being studied (for example, the ellipses) were those of the light object. We know that just as Earth exerts a force on the Moon, the Moon exerts a force on Earth. The Moon moves in an orbit

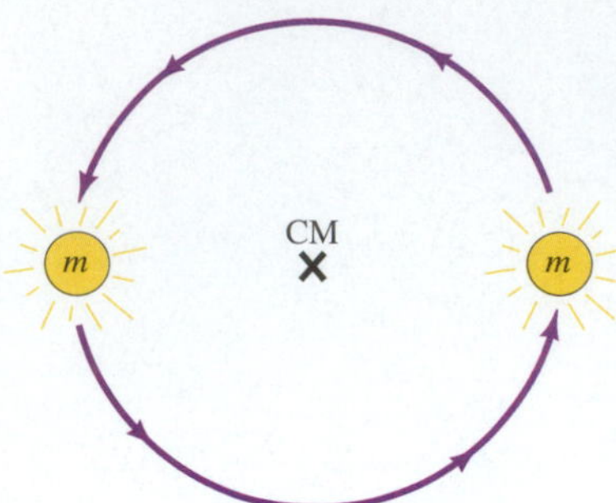

▲ **FIGURE 12–23** In a double star system with stars of equal mass, the center of mass is midway between the stars. Each star orbits around this point.

around Earth, but does Earth move around the Moon? The answer to this question is yes. When two objects move under the influence of the gravitational forces they mutually exert on each other, they move about their common center of mass. This is a property we first observed in the context of momentum. If we regard the two objects as an isolated system, then there is no external force on that system, and its center of mass remains unaccelerated. When one of the two objects is much heavier than the other, the center of mass is closer to the heavier object and may even be within that object. In the Earth–Sun system, the Sun's motion is a very small orbit about a point very close to the center of (and well within) the Sun. Cases in which the center of mass is well separated from the two objects in question occur in double star systems with stars of comparable masses. These stars each move in an orbit around a point between them (Fig. 12–23).

Effects of Other Objects

The superposition principle tells us that the orbit of any planet is affected not only by the Sun but also by the presence of all other planets—although to a much smaller extent. The largest effects are due to the most massive planets, Jupiter and Saturn. We can write the potential energy of a planet of mass m in the form

$$U = -Gm\left(\frac{M_S}{r_S} + \frac{M_1}{r_1} + \frac{M_2}{r_2} + \cdots\right), \tag{12–20}$$

▲ **FIGURE 12–24** An orbit that does not close on itself. The orbit is said to precess.

where the terms in parentheses represent the contribution of the Sun and those of the other planets. Thus the net force on a planet is no longer a pure $1/r^2$ force directed exactly at the Sun; there are small corrections to it. As a consequence, the orbits are no longer exact ellipses that close on themselves, but instead, the orbits *precess* (Fig. 12–24), which means that the perihelion gradually moves, and the orbit never actually repeats. Almost all the major figures in nineteenth-century mathematics worked on the problem of computing orbits subject to perturbations from additional masses; in fact, accurate orbits were computed even before the advent of large-scale computing machines. Two interesting historical events are worth mentioning in this connection. First, the calculation of the orbit of Uranus, with the inclusion of all the perturbations, did not fit the observed orbit. In 1845 both John Adams (an undergraduate at Cambridge University) and Urbain Le Verrier in France calculated the potential effects of a hypothetical new planet and published their results. Adams's work was ignored, whereas Le Verrier was more successful in mounting a search for the new planet, which culminated in the discovery of the planet Neptune in 1846. Second, the calculation of the precession of the perihelion of the planet Mercury was also carried out to great accuracy by Le Verrier. The result of the comparison of observation and theory left a discrepancy in the rate of precession of the perihelion that amounted to only 43″ of arc per century (out of an observed total of some 5600″ of arc per century). Both theory and experiment were so good that there was no doubt of the existence of this discrepancy, but explanation of the discrepancy had to await Einstein's theory of gravitation.

Equality of Inertial and Gravitational Masses

The parameter m in $\vec{F} = m\vec{a}$ describes a property of an object that is properly called the *inertial mass* [see Eq. (5–1)]. It is a constant that characterizes the object, and it appears as a coefficient of the acceleration in response to *any* force. The parameter m that appears in the expression for the gravitational force that is exerted on that object is the *gravitational mass*, and there is no a priori reason why the inertial and gravitational masses should be equal.

The equality of inertial and gravitational masses to one part in 10^{11} has been demonstrated by Robert Dicke and Vladimir Borisovitch Braginsky. Newton had already measured the equality of these quantities to an accuracy of one part in 10^3, and Loránd von Eötvös carried out measurements in the period from 1890 to 1922 to an accuracy of one part in 10^9. The equality of inertial and gravitational masses made a great impression on Einstein and led toward his formulation of the equivalence principle, a cornerstone of the general theory of relativity (see Section 12–6).

Is Newton's Law of Gravitation Right over All Distances?

We know to a good degree of accuracy that interactions between celestial objects obey an inverse-square law, but these phenomena occur on scales ranging from hundreds of kilometers to much greater distances. The Cavendish experiment and its modern versions verify the inverse-square law at distances ranging from less than centimeters to meters. But there remain large gaps in the range of distances for which the law has been well tested, and these are all the targets of current experiment. Such experiments are often motivated by theoretical thinking, and there are some current ideas that suggest the law may fail in some as-yet untested distance range. Even without the theoretical motivation, Newton's predictions are subject to the same experimental verification as are those of any other scientist, and the passage of 300 years does not mean that we should stop asking questions about the validity of scientific ideas.

*12–6 Einstein's Theory of Gravitation

As accurate as it is for astronomical scales, Newton's theory of gravitation was superseded in 1915 by a still more accurate description: Albert Einstein's theory of gravitation. The Einstein theory, also known as the **general theory of relativity**, reduces to Newton's theory for objects that move with a speed $v \ll c$, where c is the speed of light, and for gravitational potential energies small compared to mc^2. These conditions are satisfied except in extreme circumstances, such as in regions very close to masses on the order of stellar masses and larger, depending on the size of the star, or for measurements at a precision only possible in the 20th century, so Newton's theory has always been adequate in virtually all its applications.

Einstein's theory arose out of his attempts in 1915 to combine Newton's theory with the special theory of relativity (see Chapter 39). While working on this problem, Einstein had what he described as "the happiest moment of my life" when he realized the importance of the fact that a freely falling person does not feel (and has no way to measure) his or her own weight. This idea was generalized in 1907 to form the **equivalence principle**. According to this principle, no experiment can distinguish between the following two situations: (1) a physical system at rest that is subject to a uniform gravitational force; and (2) a physical system that is uniformly accelerating in the absence of gravity. A simple example can help illustrate what is implied. Suppose an observer stands in an elevator and experiences a force on his feet. He can interpret this effect as being due to an upward acceleration of the elevator. Because the observer, of mass M, experiences an acceleration, a, he must be subject to an upward force of magnitude $F = Ma$. In fact, this force is the contact force that the elevator floor exerts on his feet. A second interpretation is that the elevator is at rest, but there is a uniform gravitational force acting to pull him downward. He again feels the upward normal force, F_N, of the floor and, because the elevator and thus the observer are at rest, that force must just cancel the gravitational force. Thus $F_N = Mg$, where g is the acceleration due to gravity. Notice that we have used the same mass M in both descriptions. Strictly speaking, the mass M in $F = Ma$ is the *inertial* mass, whereas it is the *gravitational* mass in the force-identity equation $F_N = Mg$. The gravitational and inertial masses were observed by Newton to be identical, but he was unable to draw deep conclusions from this observation. Einstein's equivalence principle states that if g has the same numerical value as a, there is no way for the observer to distinguish the two cases; thus, the inertial mass and the gravitational mass are *required* to be equal.

Predictions of the Equivalence Principle

Light Falls under the Influence of Gravity: Light falls just like matter does. To see this, let's consider the elevator again. A beam of light shines across the elevator in a horizontal direction from one side just as the elevator accelerates upward (Fig. 12–25). Because it takes light some time to travel across the elevator, it will hit the opposite wall closer to the floor than a horizontal line parallel to the floor would indicate. More important, for constant acceleration, the amount by which the elevator moves upward is proportional to the square of the time. That means that a *series* of measurements made by an occupant of

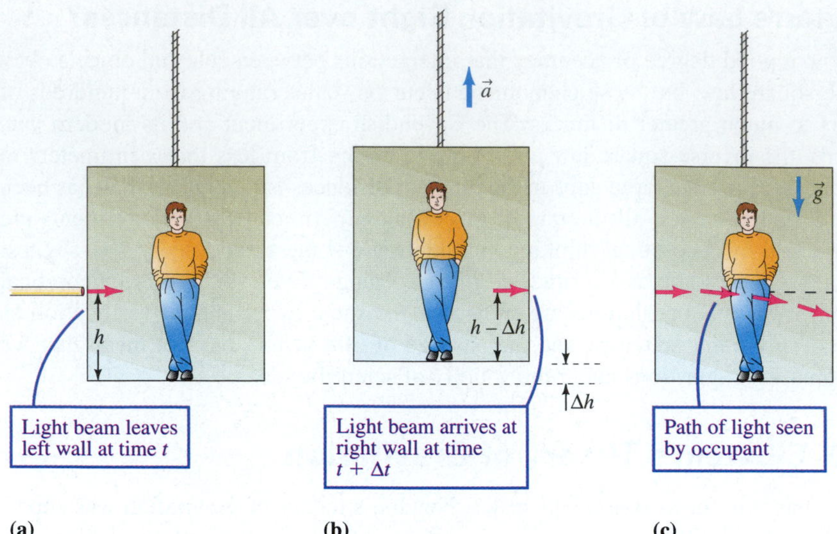

▶ **FIGURE 12–25** (a) A beam of light shines across an elevator at a time t when the elevator starts to rise. (b) At time $t + \Delta t$ the light hits the opposite side of the elevator at a point lower than it would have hit had the elevator not been moving. (c) The observer in the elevator cannot know whether the elevator is undergoing an upward acceleration $\vec{a}$ or whether a local gravitational acceleration of the same magnitude is present. We conclude from thought experiments like this that light must be bent when gravitational forces are present.

the elevator that located the light beam relative to the floor would show that the light beam follows a parabolic path, just as a falling object would under the effect of gravity. If we are to interpret this observation from the point of view that the elevator and its contents feel a gravitational force, then the observed deflection of the light must be due to the gravitational force. Light, in this sense, does not behave any differently than matter, except that it moves faster. The first observation of the deflection of light due to gravitation was made during a solar eclipse in 1919. At such a time, pairs of stars whose light passes very close to the Sun become visible. The angular spread between the stars when their light comes around the two sides of the Sun can be compared with their observed angular spread when they are seen away from the Sun (Fig. 12–26). The confirmation of the predicted effect brought Einstein's theory to the public's attention.

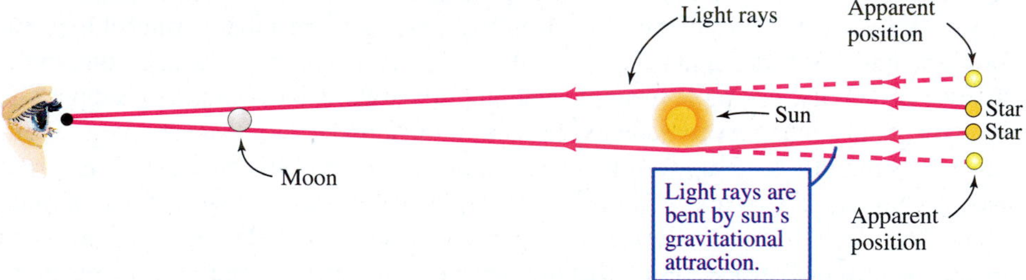

▶ **FIGURE 12–26** During a solar eclipse, it is possible to observe the light of two stars on either side of the Sun. Because the Sun's gravity bends the light, the stars appear to be farther apart than they actually are.

Gravitational Lenses: Astronomers today routinely observe what is known as *gravitational lensing*. In an example, consider the observation of two identical images of *quasars* (extraordinarily bright sources of light) very close to one another as seen from Earth (Fig. 12–27). Because there is good evidence that quasars are billions of light-years away, any two quasars are literally quite disconnected from one another; thus, it is highly improbable that any two could be as nearly identical as the members of the observed pair appear to be. The correct explanation is that there is only one quasar, but its light passes close to an extremely massive object, which bends light going

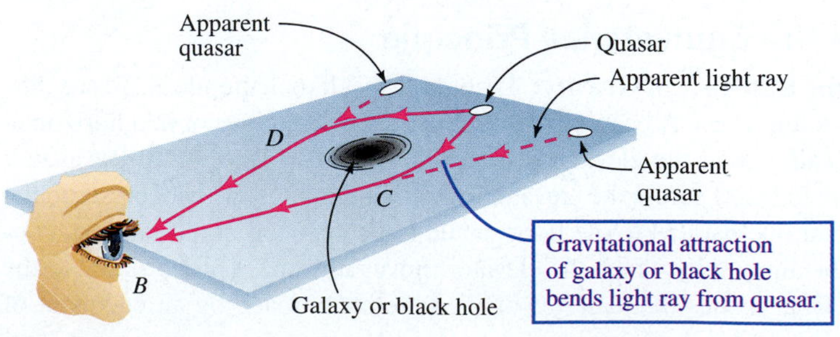

▶ **FIGURE 12–27** A double image results when light from a single quasar is bent as it passes on either side of an object with strong gravity, such as a galaxy or a large black hole. The effect is greatly exaggerated here.

around it just as a lens would; this gives rise to a double image; the effect is said to be due to a *gravitational lens*. An even more spectacular demonstration of this effect is seen in a ring image of the source (Fig. 12–28). (Can you see how the ring is produced?) To produce such a large effect, the light must be bent by a galaxy whose mass equals that of about 300 billion suns. Observation of lensing is so routine that it is now used as a tool to learn about the otherwise invisible intermediate objects that cause the effect.

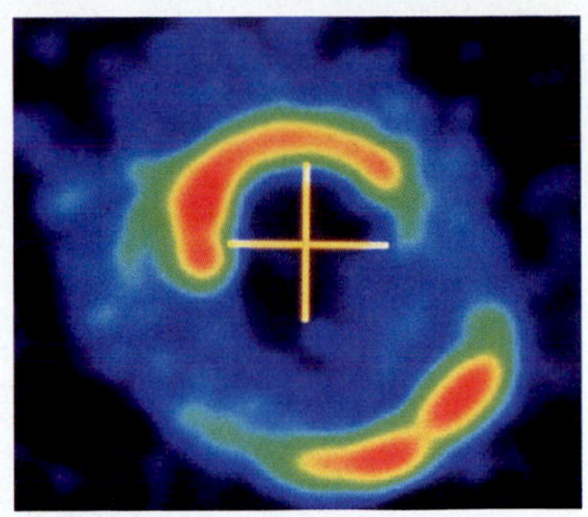

▲ **FIGURE 12–28** This photograph, taken by the Hubble Space Telescope, of a massive, compact galactic cluster illustrates gravitational lensing. The lensing effects are evident in the arclike pattern. They are caused when light from an object far beyond the cluster passes near the cluster, which magnifies, distorts, and brightens that light on its way to our eyes. The image tells us a great deal both about the matter far beyond the cluster and about the cluster itself.

Black Holes: The fact that light falls when it passes near masses implies that it is possible to imagine a large enough mass, localized in a small enough region, for which the speed of light is smaller than the escape speed from the surface of the mass. Such a mass would not be directly visible, because light or matter could not escape from it. It would manifest itself only through the gravitational force it exerts. Such a mass forms a *black hole*. Astonomers have accumulated impressive evidence for both small and large black holes. The "small" ones have masses of one or several solar masses. For example, there are pairs of stars only one of which can be seen directly. The second star is invisible, but its presence and properties can be deduced from the motion of the visible star and other features of the system. Some such pairs emit X-rays copiously, and the characteristics of the X-rays indicate that the invisible second star is in fact a black hole and the X-rays come from matter falling into it. The strong gravitational force necessary to produce such dramatic effects can come only from a black hole. Even more spectacularly, many galaxies, including our own, contain at their centers black holes with a mass of millions of Suns.

Precession of Planetary Orbits: The equivalence principle forms the foundation of the general theory of relativity, which is the full theory of gravitation. The mathematical application of the theory leads to subtle corrections to Newton's gravitational force law. The corrections to the Newtonian form predict, among other things, that the perihelion of Mercury, even in the absence of other planets, should precess by 43″ of arc per century, an amount that is in agreement with observation when the effects of other planets on the motion of Mercury are taken into account. Other predictions of the general theory of relativity are being confirmed by recent experiments.†

Summary

Astronomical observations led Kepler to three laws for planetary motion:

1. Planets move in planar elliptical paths with the Sun at one focus of the ellipse.
2. During equal time intervals, the radius vector from the Sun to a planet sweeps out equal areas.
3. If T is the time that it takes for a planet to make one full revolution around the Sun, and if R is half the major axis of the ellipse (R reduces to the radius of the orbit of the planet if that orbit is circular), then

$$\frac{T^2}{R^3} = C, \tag{12–1}$$

   where C is a constant whose value is the same for all planets.

Newton showed that these laws are a consequence of a law of universal gravitation, which states that any two point masses m and M, separated by a distance r, will attract each other with a force that is along the radius vector connecting the masses. The force on mass m due to mass M has the inverse-square form

$$\vec{F} = -\left(\frac{GmM}{r^2}\right)\hat{r}, \tag{12–4}$$

where $\hat{r}$ is the unit vector pointing from mass M to mass m. The constant G has been measured, and its value is $G = 6.673 \times 10^{-11}\ \text{N}\cdot\text{m}^2/\text{kg}^2$. The gravitational force is conservative and may be derived from a potential energy:

$$U(r) = -\frac{GmM}{r}. \tag{12–8}$$

†There is a wealth of literature on this fascinating subject, easily uncovered with a search on the Internet.

Masses that exert gravitational forces are not always pointlike. The most important case of an extended object is a spherically symmetric mass distribution, such as Earth or the Sun. In this case the gravitational force is the same as if all the mass of the extended object were concentrated at the center of the spherical distribution. Another consequence of the formula given in Eq. (12–8) is that the gravitational force exerted on an object anywhere inside an arbitrarily thick spherical shell, with a mass distribution that depends only on the distance from the center, is zero.

The gravitational force between two objects acts on both objects; for example, not only does Earth revolve around the Sun but the Sun also revolves around Earth. The tides are explained when this is taken into account in the Earth–Moon system. More precisely, when a gravitational force acts between two objects, both objects revolve around the center of mass of the two-object system.

The Newtonian theory of gravity is a limiting case of a more accurate and fundamental theory of gravity: Einstein's general theory of relativity. That theory is based on the equivalence principle, which states that no experiment can distinguish between a uniform acceleration and the effects of a uniform gravitational force. One consequence is the equality between the inertial mass (m_i in $F = m_i a$) and the gravitational mass (the masses that appear in the law of universal gravitation). Another consequence is the fact that light falls just like ordinary matter when it is subject to gravity.

Understanding the Concepts

1. What are some forces, other than those mentioned in this chapter, that are not fundamental? From what fundamental forces are they derived?
2. Any projectile fired with enough initial speed will eventually escape Earth, regardless of the direction in which it is fired. How do you reconcile this statement with the fact that the height above Earth to which a cannonball will rise depends on the angle at which it is fired?
3. To a good approximation, the Sun and the Moon both move with respect to Earth in the plane of Earth's equator. Does this mean there would be little or no tide at the North Pole if there were surface water at the North Pole?
4. The European satellite launch area is in French Guiana, because less energy is required to launch rockets into orbit from there than from a point in Europe. Explain why this is so.
5. How can Earth's rotational motion be used to minimize the fuel needed to boost a satellite into a given orbit around the planet? How can Earth's orbital motion be used to minimize the fuel needed to boost a satellite into a given orbit around the Sun?
6. The space shuttle orbits at an altitude of about 400 km. At that distance the acceleration of gravity due to Earth is very close to its surface value. Why then do the astronauts float around inside the space shuttle?
7. The satellites of Jupiter follow Kepler's third law: The square of their periods, divided by the radius of their orbits cubed, is a constant. Is this the same constant as for the planets moving around the Sun?
8. In our discussion of the tidal forces near a neutron star [Section 12–4], we had our unfortunate subject passing a neutron star in a close orbit. Why did we do this rather than having him or her simply stand on the surface? [*Hint*: What would the value of g itself be on the surface?]
9. Describe the path of a celestial object whose angular momentum with respect to the Sun is zero.
10. What gravitational force would a 1-kg piece of tungsten feel at the center of Earth?
11. If the gravitational force were a central force proportional to $1/r^3$ rather than $1/r^2$, the planetary orbits would no longer be closed (unless they are circular). Would the planets still sweep out equal areas in equal times?
12. We often hear that Earth satellites burn up when they leave their orbit and return to Earth. Why don't satellites burn up as they go up into orbit?
13. The same side of the Moon always faces Earth. What does this tell us about the rotational motion of the Moon? As we saw in our discussion of tides, there is a reason for this.
14. The acceleration of Earth due to the Moon is about 175 times smaller than that due to the Sun. How can it be that Earth's acceleration due to the Sun is so much larger than its acceleration due to the Moon yet the tidal forces due to the Moon are larger than those due to the Sun?
15. When astronauts float in the bay of their spacecraft, they are in outer space, orbiting the Earth. Is gravity acting on them?
16. Scientists and engineers have proposed that a base on the Moon would be useful for interplanetary launches. This would be advantageous compared to a launch from Earth's surface or from an orbit close to Earth because (a) there is no air on the Moon; (b) it would cost less net energy; (c) one wouldn't need to take as much fuel to the Moon's orbit distance in one launch.
17. Very careful measurements of the orbits of satellites can help teach us about Earth's internal structure. How can we use such measurements to study regions with a mass density that is higher or lower than that of Earth's average mass density?
18. If Earth were a perfect sphere, would you weigh more or less at the equator than at the poles?
19. Earth is not a perfect sphere. What types of observations might we use to learn this fact?
20. When a satellite is in circular orbit around Earth, there is a direct relationship between its angular momentum and the radius of the orbit. Suppose that such a satellite collides elastically with a meteor that was heading directly toward the center of Earth. Because the impulse is in a radial direction, the angular impulse is zero. What happens? Draw some diagrams, using the conservation laws that you know. [*Hint*: If you get stuck, review Kepler's first law.]
21. Consider the apparatus in Fig. 12–29 (see next page). Assume that the friction is normal sliding friction and that the rope and pulley are ideal. The motion can depend on (a) M_1 and M_2 separately; (b) the ratio M_1/M_2; (c) neither; the motion is independent of either mass; (d) the product $M_1 \times M_2$.

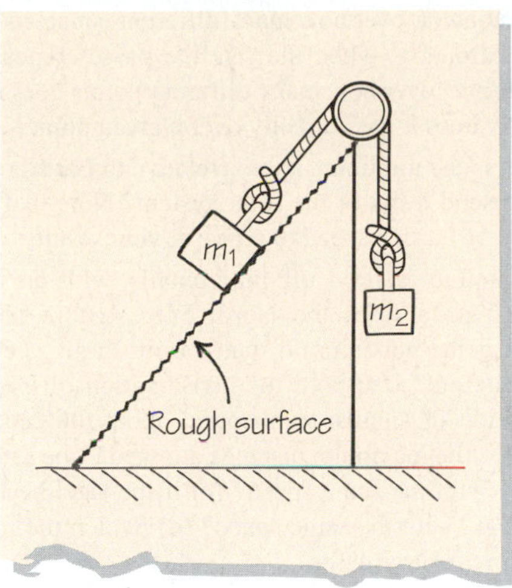

▲ **FIGURE 12–29** Question 21.

22. In the text we spoke of friction between Earth and its shell of water as slowing down the rotation frequency of Earth. Would this still occur if the Moon were not present?
23. The "day" on Jupiter is 9h 50 min at the equator, and 9h 55 min at its poles. What does this tell you about the rigidity of the surface of Jupiter?
24. Is it possible for friction from the tides to slow down Earth's rotation and still conserve Earth's angular momentum (with its oceans)? The origin of the tides involves the gravitational force due to the Moon on Earth's oceans, so Earth and its oceans do not form an isolated system.
25. A satellite is in circular orbit around Earth. How much work is done on the satellite by the gravitational force of Earth during one orbit?
26. Assume that Earth is perfectly spherical and that its density depends only on the distance from its center. A large asteroid makes a close pass to Earth. Could such an asteroid (which is nevertheless small compared to Earth itself) change the rate of Earth's rotation without actually colliding?
27. You are in a spaceship very far away (say, 100 Earth radii) from Earth. Could you move to a region where the gravitational force due to Earth is less by ejecting some material from the spaceship in the proper direction?
28. In Section 12–3 we described Newton's thought experiment demonstrating how a cannonball shot horizontally off a mountaintop could attain a circular orbit if its initial speed increases to some large-enough value. Describe the motion of the cannonball if its initial speed exceeds this value.
29. Edgar Rice Burroughs's character Tarzan discovers that Earth is hollow. Tarzan finds an entry and discovers a whole new civilization, with modernistic buildings, people walking around or driving modernistic vehicles, and so on, in Earth's interior. What is wrong with this picture?
30. Spring tides (especially high tides) occur both when the Sun and Moon are aligned on the same side of Earth and when they are aligned on opposite sides of Earth. How do you explain these facts?

Problems

12–1 Early Observations of Planetary Motion

1. (I) Use the data listed in the appendices to calculate C in Eq. (12–1).
2. (I) In planetary tables we find that Jupiter's satellites Io and Europa each follow nearly circular orbits: Io's orbit has a mean radius of 422,000 km and Europa's orbit has a mean radius of 671,400 km. The period of Io is 152,854 s. What is Europa's period?
3. (II) Consider an object of mass m, moving in a circular orbit, subject to a central attractive force whose magnitude is given by $F(r) = h/r^3$. (a) What are the dimensions of h? (b) Show that the angular momentum for the motion is uniquely determined by h and m. (c) What is the resulting relation between period and radius analogous to Kepler's third law for this force?
4. (III) Angular momentum is conserved for a radial, or central, force. Show that the orbits due to a radial force lie in a plane.

12–2 Newton's Inverse-Square Law

5. (I) A man of mass 95 kg is dancing with his wife, who has a mass of 68 kg. Assume that each person's mass is concentrated at their respective centers of mass, which are separated by 48 cm. (a) What is the gravitational attraction between them? (b) Which person has the greater gravitational attraction toward the other?
6. (I) Calculate the gravitational attraction between a proton and an electron in a hydrogen atom if the radius of the atom is 0.6×10^{-10} m. The masses can be found in Appendix II.
7. (I) A Cavendish experiment involves the force between two spheres of 1 kg each whose centers are separated by 40 cm. Using the known value of G, find the gravitational force between these spheres. Compare this force to the weight of a fly.
8. (I) What is the acceleration due to gravity on the surface of (a) the Moon ($R = 1.74 \times 10^3$ km, $m = 7.35 \times 10^{22}$ kg); (b) Mars ($R = 3.40 \times 10^3$ km, $m = 6.42 \times 10^{23}$ kg); (c) Jupiter ($R = 7.14 \times 10^4$ km, $m = 1.90 \times 10^{27}$ kg); (d) the Sun ($R = 6.96 \times 10^5$ km, $m = 1.99 \times 10^{30}$ kg)?
9. (I) A satellite orbits Earth in 90 minutes. What is the radius of its motion around the center of Earth?
10. (I) What is the period of a satellite circling the Moon at a height of 90 km above the Moon's surface?
11. (I) What is the period of a satellite circling Earth at a height of 300 km above Earth's surface?
12. (I) What is the surface gravity (the value of g) on a spherical asteroid of diameter 30 km and density 5400 kg/m^3?
13. (I) Two identical satellites move in circular orbits around Earth. One has twice the kinetic energy of the other. The radius of the faster one's orbit is three Earth radii. What is the radius of the slower one's orbit?
14. (II) A weight lifter can lift 138 kg on Earth. What mass could the same weight lifter lift on (a) the Moon, (b) the Sun (use the data in Problem 8)?
15. (II) The height achieved in a jump is determined by the initial vertical velocity that the jumper is able to achieve. Assuming that this is a fixed number, how high can an athlete jump on Mars if she can clear 1.85 m on Earth?
16. (II) There is a point on the line joining two astronomical bodies where there is no gravitational force on a rocket. Find this point for (a) the Earth–Sun system, (b) a binary system of stars, one of which is one solar mass and the other two solar masses.

12–3 Planets and Satellites

17. (I) Pluto has the most eccentric orbit of all the planets, with $e = 0.25$. Its semimajor axis is 39.5 AU. What is $(r_{max} - r_{min})/(r_{max} + r_{min})$ for Pluto's orbit?

18. (I) The semimajor axis of Earth's orbit is 149.6×10^6 km, while its eccentricity is 0.017. What is the maximum distance between Earth and the Sun as Earth traces out its orbit?

19. (I) Determine the escape speed of an object from the Sun's surface.

20. (I) What are the escape speeds on the surface of (a) the Moon, (b) Mars, and (c) Jupiter? (Use the data in Problem 8.)

21. (I) Consider the asteroid of Problem 12. What is the escape speed from the surface of that satellite?

22. (I) The radius of a neutron star is 750 times smaller than Earth's radius, and its mass is 1.8×10^5 times larger than Earth's mass. What is the escape velocity from the surface of a neutron star? (Ignore the fact that, at high speeds, one should not really use $mv^2/2$ for the kinetic energy.)

23. (II) If the asteroid of Problem 12 rotates with an angular speed ω about an axis, material on the equator will have a tendency to be thrown off. How slowly must it rotate so that material that is not attached permanently just barely stays on the surface?

24. (II) A rocket is sent vertically upward from Earth's surface with an initial speed of 6.8 km/s. How far above Earth's surface will it go before falling back? Ignore atmospheric friction.

25. (II) Astronomers discover a meteorite at a distance of 80,000 km from the center of Earth. The meteorite is moving directly toward Earth with a velocity of 2000 m/s. What will be the velocity of the meteorite when it hits Earth's surface? Ignore all drag effects.

26. (II) Determine the minimum energy needed to allow an unmanned rocket of mass 3800 kg to leave the Moon's surface and arrive at a point very far away. Ignore the effect of Earth.

27. (II) Calculate the distance from Earth's center to a satellite in circular orbit with a period that is (a) one-third the Moon's period; (b) three times the Moon's period; (c) one-thousandth of the Moon's period.

28. (II) Using the information in Example 12–3, determine a satellite's time of revolution about Earth if it is in a circular orbit 370 km above Earth. How would your answer change if the mass of the satellite were to double?

29. (II) Determine the velocity with respect to Earth of a satellite of mass 500 kg in a circular orbit 200 km above Earth's surface. What is its kinetic energy? What is its angular momentum?

30. (II) A geosynchronous communications satellite orbits Earth, always positioned above the same point on the equator. (a) What is the period and angular velocity of the satellite? (b) What is the radius of the orbit? (c) Show that Kepler's third law applies to the orbits of the satellite and the Moon.

31. (II) The mass of Mars is 6.42×10^{23} kg, and its radius is 3393 km. What is the period of a satellite in a circular orbit 95 km above the surface of Mars?

32. (II) The Moon goes around Earth once in 27.3 d. What is the distance between the Moon and Earth?

33. (II) Use the mass of the Sun, given in Problem 8, to estimate the distance from Earth to the Sun. Assume that the orbit is circular.

34. (II) Consider a satellite on a circular polar orbit, one whose plane is the same as a plane made by a great circle passing through the North and South Pole. What is the altitude of the orbit with a 24-hour period? When the satellite passes repeatedly over the equator, over how many different equatorial points does it pass? Similarly, when the satellite passes repeatedly over a given latitude, over how many different points does it pass? How frequently does it pass directly over a given point?

35. (II) What is the minimum speed (relative to Earth) required for a rocket to send it out of the solar system? Note that you need to make use of Earth's speed to arrive at your result.

36. (II) A satellite is fired off horizontally with an initial speed $v_0 = 10.5$ km/s from the North Pole. Ignore air resistance. (a) What is the maximum distance from Earth's center attained by the satellite? Use both the conservation of energy and the conservation of angular momentum about the center of Earth. (b) What is the maximum distance attained if the satellite is fired vertically with the same speed? (c) if the satellite is fired at an angle of 41° with the same speed? (d) Sketch the motion of the satellite in the three cases.

37. (II) A small package is fired off Earth's surface with a speed v at a 45° angle. It reaches a maximum height h above the surface at $h = 6370$ km, a value equal to Earth's radius itself (Fig. 12–30). What is its speed when it reaches this height? Ignore any effects that might come from Earth's rotation.

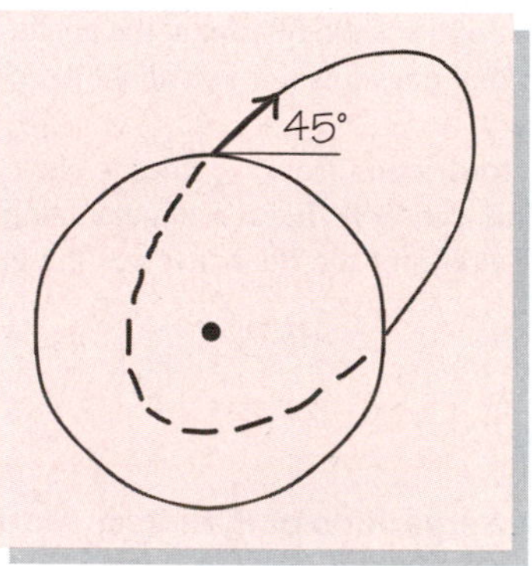

▲ **FIGURE 12–30** Problem 37.

38. (II) In Example 12–8 we used the conservation of angular momentum to find the speed of a comet at perihelion and aphelion around a star. Use the fact that the energy at these two points must be equal to find the mass of the star.

39. (II) An object of mass 3×10^{15} kg approaches the solar system (Fig. 12–31). When it is very far away—where the gravitational potential energy can be neglected in comparison with its kinetic energy—the object moves with a velocity of 12 km/s in a straight line. By straight-line extrapolation, the closest this line would come to the Sun is 3×10^8 km. The point of the object's nearest approach to the Sun is characterized by the fact that the radius vector from the object to the Sun is perpendicular to

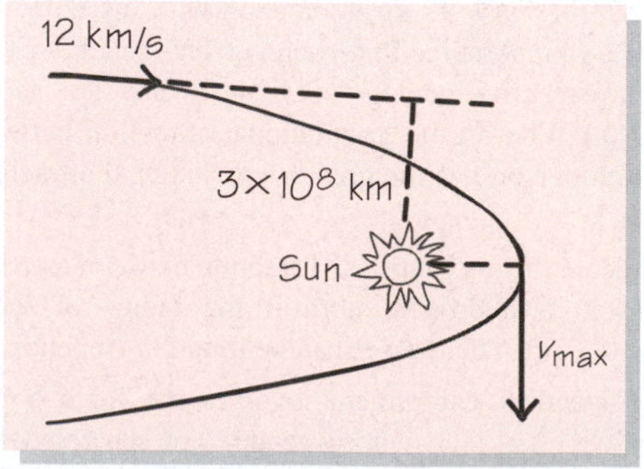

▲ **FIGURE 12–31** Problem 39.

the tangent to the path at that point. (a) Sketch the orbit of the object. (b) Use conservation of energy and of angular momentum to calculate the velocity of the object at the point of nearest approach. (c) Calculate the distance of nearest approach.

40. (II) A satellite is in a circular orbit of radius R_1 around Earth. Small rockets aboard the satellite change its direction so that it has an elliptical orbit. The change causes the satellite to lose half its orbital angular momentum, but the total energy remains constant. In terms of R_1, what are the perigee and apogee distances of the new orbit with respect to the center of Earth?

41. (II) A satellite of mass 300 kg is in circular orbit 2000 km above Earth's surface $(M_E = 6 \times 10^{24}\text{ kg}, r_E = 6370\text{ km})$. (a) What is the orbital speed of the satellite? (b) What is the angular momentum of the satellite? (c) A rocket engine is fired, reducing the speed of the satellite to half its initial value, but leaving the direction of motion unchanged. What is the new angular momentum? (d) Does the satellite crash as a result of the maneuvers in part (c)? Explain your answer.

42. (III) The distance of closest approach of Halley's comet to the Sun is 8.9×10^{10} m. Its period is 76 yr. What is the nature of its orbit? Calculate the following: (a) semimajor axes; (b) eccentricity; (c) aphelion distance (farthest distance from Sun).

43. (III) A satellite of mass 2000 kg is in circular orbit about Earth at a distance of 300 km above the surface. (a) What is the speed of the satellite in its orbit? (b) What is the angular momentum of the satellite about the center of Earth? (c) What is the total energy of the satellite in its orbit? (d) Controllers back on Earth wish to move the satellite to a new orbit 500 km above the surface. They propose to do this by briefly firing a rocket engine on the satellite for several seconds in the direction of the center of Earth; that is, the force on the satellite is directly away from Earth's center. What is the torque on the satellite about the center of its orbit? (e) Can the new orbit be circular?

12–4 Gravitation and Extended Objects

44. (I) What is the approximate difference between the value of g at sea level and the value on top of a 14,000-ft-high mountain? Assume that Earth has a constant density, and Earth's radius is the radius at sea level.

45. (I) The right-hand-side of Eq. (12–19) is an approximation. The exact form, given in Eq. (12–19), is $g(h)/g(0) = R_E^2/(R_E + h)^2$. Verify the accuracy of the approximation by calculating the ratio $g(h)/g(0)$ for $h = 10{,}000$ m (the altitude of a cruising passenger jet) according to the approximate and the exact forms.

46. (I) How much does the acceleration of gravity due to Earth decrease from sea level to (a) the 37,000-foot elevation flown by a jet airplane, and (b) the position of zero net gravitational attraction on the line between Earth and the Moon (see Example 12–4)?

47. (II) An object of mass m falls freely toward a large sphere of radius R and mass M from a great distance, starting from rest. The object has no angular momentum with respect to the large sphere. It arrives at the large sphere at a spot where there is a small hole and passes within. The large sphere turns out to be hollow. How long does it take for the object to make the trip from one side of the large sphere to the other?

48. (II) A deep hole in Earth reaches a depth of one half of Earth's radius (Fig. 12–32). How much work is done when a 1-kg mass is slowly lifted from the bottom of the hole to Earth's surface?

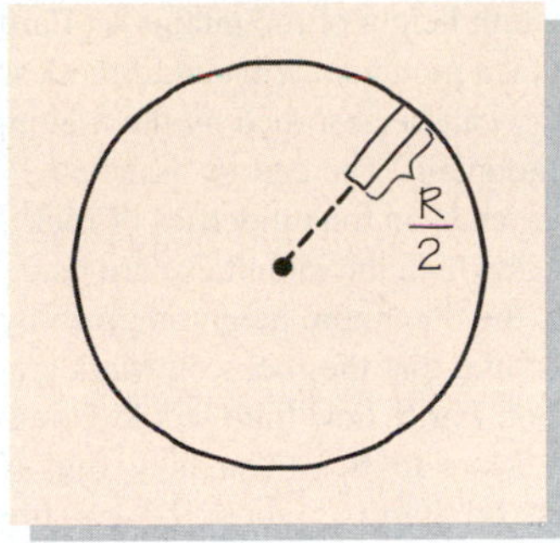

▲ **FIGURE 12–32** Problem 48.

49. (II) Jupiter is about 5.2 times as far from the Sun as is Earth, and its diameter is 11.2 times that of Earth. From what you know about the Sun's effect on tides on Earth, what can you say about tides on Jupiter due to the Sun?

50. (II) How big is the acceleration of Earth due to the Moon compared to that due to the Sun? Use the astronomical tables and estimate the average acceleration of Earth due to Jupiter, compared to that due to the Moon.

51. (II) What is the speed of the mass in the tunnel through Earth in Example 12–9 as it passes through the center of Earth?

52. (III) Rather than a tunnel through Earth's center, as in Example 12–9, consider a tunnel drilled along a chord of Earth, meaning that it passes a perpendicular distance d away from the center of Earth (Fig. 12–33). Find the potential energy of a mass placed in such a tunnel as a function of (a) its distance r from the center of Earth and (b) its distance x from the midpoint of the tunnel.

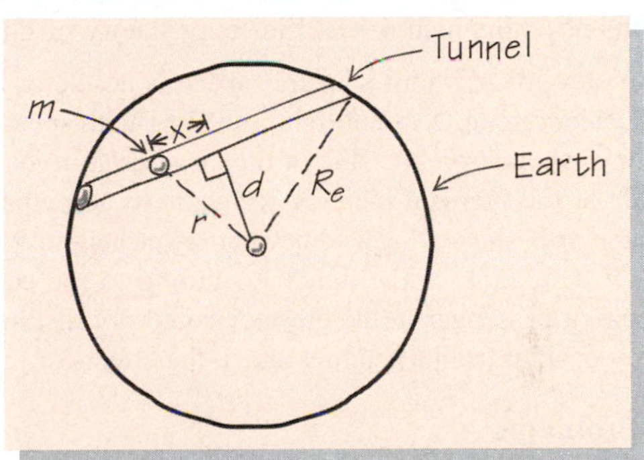

▲ **FIGURE 12–33** Problem 52.

12–5 A Closer Look at Gravitation

53. (I) The mean Earth–Sun separation is 1.50×10^8 km, Earth's radius is 6.37×10^3 km, the mass of the Sun is 1.99×10^{30} kg, and Earth's mass is 5.98×10^{24} kg. Given these data, locate the center of mass of the Earth–Sun system.

54. (I) At a certain moment, the Sun, Jupiter, and Saturn are lined up, with Jupiter between the Sun and Saturn. Suppose the orbits of Jupiter and Saturn were circular, with mean distances from the Sun of 7.78×10^8 km and 1.42×10^9 km, respectively. The mass of Saturn is 0.029 percent of that of the Sun. What is the ratio of the gravitational force on Jupiter due to Saturn to the gravitational force on Jupiter due to the Sun, at the specified moment?

55. (II) Some delicate gravitational measurements must be accurate to one part in 10^{11}. Suppose a 20-ton truck drives by the laboratory at a distance of 20 m. By what angle will this deflect a pendulum from the vertical (defined as the equilibrium position in the absence of the truck). Should one worry about trucks near the laboratory?

56. (II) The maximum height of mountains on Earth can be estimated by noting that a mountain will sink if the resultant decrease in potential energy can be absorbed by the melting of the rock underneath the mountain. The energy associated with melting is a number that depends on the properties of rock. The result is that for any planet in which the mountains are made of the same material as Earth, the maximum height is given by $gh = C$, a constant. (a) Assuming that the rocks on Mars are more or less the same as those on Earth, how high would you expect the highest mountains on Mars to be? (You may find it interesting that Mount Everest on Earth is about 10 km high, while Mount Olympus on Mars is about 25 km high.) (b) The highest mountains on the Moon, where g is about $1/6$ of g_{Earth}, are about 2 km in height. What does this tell you about the Moon?

57. (II) A binary star system consists of two stars, each of mass M, orbiting around their common center of mass with radii R from the center of mass. Determine the period of revolution.

*12–6 Einstein's Theory of Gravitation

58. (I) Fighter pilots are able to withstand accelerations up to $7g$ for a short period. A jet dives toward Earth and then pulls up in a parabolic orbit. Draw a force diagram showing the various forces on the pilot at the bottom of the orbit. If the dive was at night, could the pilot tell the difference between an increased value of gravitational force and the effect of the contact forces on him at the bottom of the dive?

59. (II) A very sharply defined laser beam, directed horizontally, enters a hotel room at height h. At what height does the light beam hit the opposite wall, which is 8 m from the first wall? Compare the difference in heights to the size of an atom. Is this a feasible experiment with which to test Einstein's theory of gravitation?

60. (II) An elevator of width w in free space is accelerated upward with acceleration g. A ray of light, traveling with speed c, enters through a pinhole on one side of the elevator, at right angles to the side at the moment the elevator starts to accelerate. It will strike the opposite wall at a somewhat lower height. What is the angle of deflection of the light? According to the equivalence principle, a passenger in the elevator could not distinguish this bending of light from a bending due to the effects of gravity.

General Problems

61. (I) A spaceship of the future is cylindrical in shape, with a radius of 60 m. In order to simulate terrestrial gravity on the inside surface of the cylinder, the spaceship is made to rotate about its axis. What is the angular velocity of the spaceship about its axis?

62. (I) (a) What is the acceleration g due to gravity on a planet with the same density as Earth but with 1.6 times the radius? (b) The orbital period and radius of Jupiter's moon Ganymede are 7.16 d and 660,000 mi, respectively. What is the period of the moon Io, whose orbital radius is 262,000 mi? (c) Planets A and B are both in circular orbits around a star. Planet A has two-thirds the orbital speed of planet B. What is the radius of A in terms of the radius of B?

63. (II) The Little Prince (a character in a book by Antoine de Saint-Exupery) lives on the spherically symmetric asteroid B-612 (Fig. 12–34). The density of asteroids, including B-612, is $5.2 \times 10^3\ \text{kg/m}^3$. Assume that the asteroid does not rotate. The Little Prince noticed that he felt lighter whenever he walked quickly around his asteroid. In fact, he found that he became weightless and started to orbit the asteroid like a satellite whenever he speeded up to 2 m/s. (a) Estimate the radius of the asteroid from these data. (b) What is the escape speed for the asteroid? (c) Suppose that B-612 does rotate about an axis such that the length of the day there is 12 h. Can the Little Prince take advantage of this rotation when he wants to orbit his asteroid?

▲ **FIGURE 12–34** Problem 63.

64. (II) A neutron star has a mass of 5.4×10^{30} kg and a radius of 12 km. (a) Calculate the acceleration due to gravity at the surface of the neutron star. (b) What is the difference between the gravitational forces acting on the top and the bottom of a tiny dumbbell held vertically on the surface (that is, with one end on the surface of the neutron star and the other 1 mm above the surface)? The dumbbell consists of two 1 g point masses connected by a massless connector of length 1 mm.

65. (II) Suppose that, instead of a $1/r^2$ dependence, an attractive central force varied with distance as $1/r^n$. (a) Would such forces support a circular orbit? (b) Find the resulting relation between period and radius analogous to Kepler's third law for this force.

66. (II) Consider a cluster of galaxies that fills a sphere of radius R and average mass density ρ. (There are so many galaxies that you can assume uniform density.) There is a galaxy of mass M at the edge of this sphere (Fig. 12–35). (a) Write an expression for the energy of the galaxy. (b) In the big-bang model of the origin of the universe, the velocity of the galaxy is directed radially outward from the center of the sphere; the galaxy's speed is $v = HR$, where $H = (15\ \text{km/s})/(10^6\ \text{ly})$ is the Hubble parameter. For what critical density ρ_c of the large cluster will the galaxy be able to escape to infinity with a final velocity of zero?

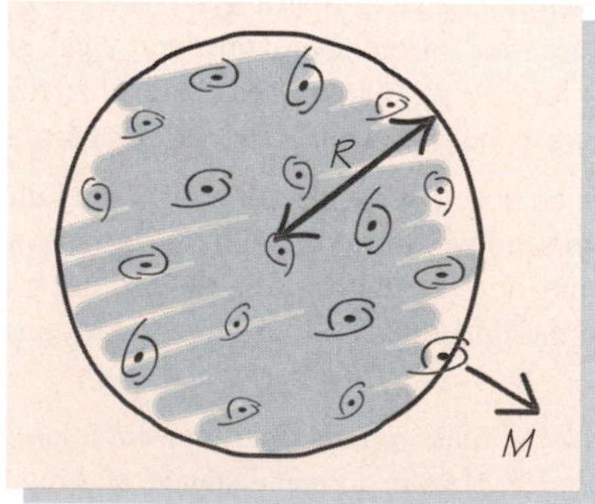

▲ **FIGURE 12–35** Problem 66.

67. (II) A satellite in low circular Earth orbit is subject to a very small constant friction force, f, due to the thin atmosphere. As it spirals in, it slowly decreases its radius. Find the decrease in radius per revolution under the assumption that the orbit is approximately circular with radius r. Find the changes in potential energy, total energy, and kinetic energy per orbit. [*Hint*: If you find the kinetic energy increasing, you are on the right track!]

68. (II) The electric force between an electron and a (much more massive) proton is attractive and of magnitude e^2/r^2, where $e^2 = 2.3 \times 10^{-28}\ \text{N}\cdot\text{m}^2$. The electron circles the proton in a circular orbit of total energy $E = -1.6 \times 10^{18}$ J. (We have taken zero potential energy to be at infinite separation.) (a) What is the radius of the orbit? (b) What is the period of the orbit? (c) If another electron were in another circular orbit around the proton, with an orbit radius three times as large as the first, what would its total energy and period be?

69. (II) An astronaut of mass 115 kg (including equipment) finds himself drifting away from his orbiting space ship at 0.05 m/s. He throws a 3-kg wrench in the direction of his drift and comes to rest relative to the ship 1 m from its surface. The ship is a sphere of radius 12 m and mass 10^5 kg. (a) At what speed does he throw the wrench? (b) How many hours must he wait for the gravitational attraction of the ship to pull him to its surface, assuming that the force of gravity is approximately constant in the region of interest?

70. (II) Suppose that an object of mass m is placed at the point at which the gravitational attraction of the Moon is just canceled by that of Earth; further, suppose that the object is displaced by a small distance x along a line perpendicular to the line connecting the centers of Earth and the Moon (Fig. 12–36). What are the magnitude and direction of the net force on the object as a function of x? Calculate your answer by using the approximation $(r^2 + x^2)^n = r^{2n}[1 + (nx^2/r^2) + \cdots]$, valid for $x^2/r^2 \ll 1$.

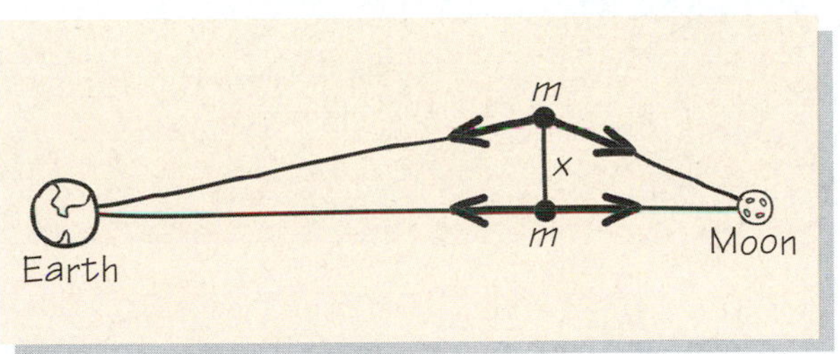

▲ **FIGURE 12–36** Problem 70.

71. Jupiter's moon Io has a radius of 1815 km, a mass 4.7×10^{-5} that of Jupiter, and an orbital radius about Jupiter that is 5.95 times Jupiter's own radius (Fig. 12–37). Given that Jupiter's mass is 318 times that of Earth and that its radius is 26 times larger than Earth's, calculate the acceleration due to gravity on Io at (a) the point nearest Jupiter and (b) the point farthest from Jupiter.

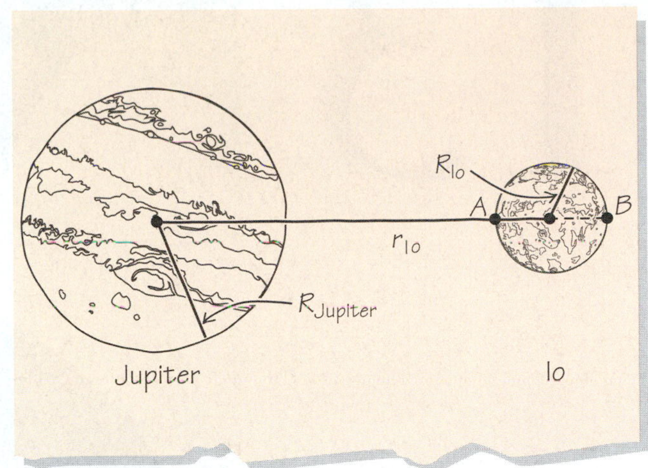

▲ **FIGURE 12–37** Problem 71.

72. (III) Astrologers claim that a person's life is influenced by the position of the planets at the moment of that person's birth. To check whether this influence could be due to gravity, compare the following two quantities: the change in the gravitational force on a baby in a hospital due to the change in the position of Jupiter from one day to the next, and the change in the gravitational force due to the presence or absence of a 4-ton truck parked near the hospital at a distance of 75 m. Jupiter has a mass of 1.90×10^{27} kg; its mean distance from the Sun is 0.78×10^9 km, and its period is 11.9 yr. Assume a circular orbit for Jupiter, and a circular orbit of radius 1.5×10^8 km for Earth. Choose the region of closest approach of the two planets for convenience.

73. (III) Three stars, each the mass of the Sun, form an equilateral triangle. Each moves in a circular orbit about the center of mass of the system because of the gravitational force exerted by the other two stars. (a) Is such an arrangement possible? (b) If so, what is the period of the motion, assuming that the side of the triangle is an Earth–Sun distance? (c) Is the system stable?

74. (III) When the first nuclear weapons were detonated, concern was expressed in some quarters that a huge nuclear chain reaction would be set up, blowing Earth to pieces. Show that the energy that would be required to disassemble Earth completely into pieces totally separate from each other is $\frac{3}{5}GM_E^2/R_E$. [*Hint*: Imagine that layers of Earth are peeled off one by one, like layers of an onion.]

Chapter 13

A backyard swing provides an example of oscillatory motion. Such motion occurs everywhere in the physical world, from vibrations in molecules to oscillations in the shape of the Sun.

Oscillatory Motion

Rhythmic motion—also known as periodic motion—is a common occurrence in the physical world. The very concept of time arises from the observation that certain motions, such as the human heartbeat and the cycling of the seasons, repeat themselves in a reliable and regular way. An important class of periodic motions involves what are known as restoring forces, forces that act to bring an object back to an equilibrium point. As we have already seen in Chapter 7, such restoring forces have potential energy functions with minima at the equilibrium point. Objects in this sort of motion oscillate, and oscillatory motion is the central subject of this chapter. The most basic type of oscillatory motion is omnipresent in nature: **simple harmonic motion**. This motion occurs when the strength of the restoring force is directly proportional to the object's displacement from the equilibrium point. Everyday examples are the motion of a mass on the end of a spring and the motion of a pendulum. The position of an object in simple harmonic motion varies with time as a sine or a cosine. While the spring force is an example that we will use repeatedly, simple harmonic motion is of universal importance because virtually any small oscillatory motion about a stable equilibrium point is simple harmonic motion.

We'll also see the effects of dissipative forces in this chapter—which not surprisingly cause the motion to progressively die out—and the effects of an oscillatory driving force. The presence of the driving force illustrates the remarkable feature known as resonance, in which the motion can become catastrophically large if the frequency of the driving force is just right.

13–1 The Kinematics of Simple Harmonic Motion

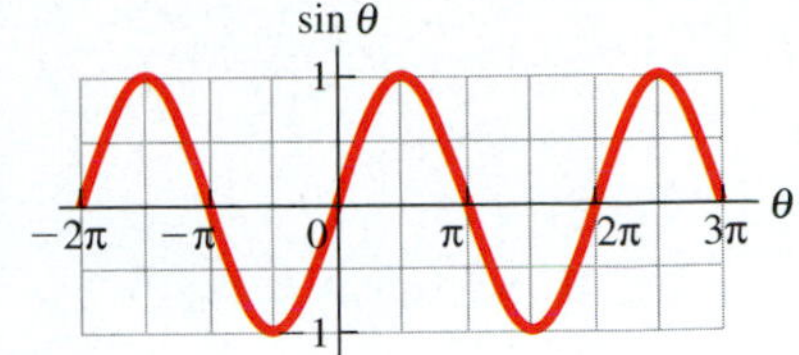

(a)

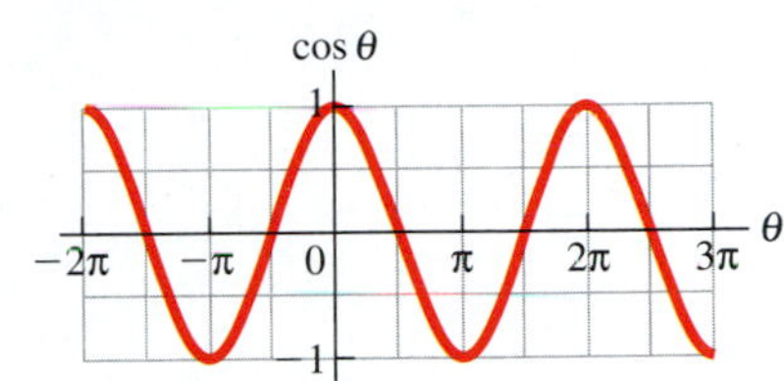

(b)

▲ **FIGURE 13–1** Plots of (a) sin θ and (b) cos θ, both as a function of θ.

Simple harmonic motion, which describes the small repeating motion followed by a mass on the end of a spring or a pendulum, is a simple form of **oscillatory** motion. The word "harmonic," signifying agreement and accord, reveals that humankind have always seen beauty in this motion. In the back-and-forth of simple harmonic motion, the position $x(t)$ of an object is of the form $\sin(\omega t)$ or $\cos(\omega t)$, where the coefficient ω is the **angular frequency**. Both sines and cosines repeat themselves periodically as time t passes. The trigonometric functions are functions of a dimensionless argument, an angle measured in radians (or, sometimes, degrees). Thus the coefficient of the time must have the dimensions $[T^{-1}]$. We'll see later that the angular frequency ω is a fundamental property of the motion, determined by the inertia of the moving objects and the restoring force acting on them.

How do we figure out whether the motion of a mass on the end of a spring is described by a sine or by a cosine? Let's look at a graph of $\sin\theta$ versus θ next to a graph of $\cos\theta$ versus θ (Fig. 13–1). Both functions repeat every time the angle θ changes by 2π rad. When $\theta = 0$, the sine function is zero, whereas the cosine function is $+1$, but this is only a matter of placing the axis. Indeed, the functions are *identical* if the origin of the θ axis is shifted. We can specify such a shift of θ by an angle we call the **phase**, δ. By what angle δ would θ have to be shifted so that the $\sin\theta$ curve in Fig. 13–1a is coincident with the $\cos\theta$ curve of Fig. 13-1b? If δ is chosen properly, the function $\sin(\omega t + \delta)$ can represent $\sin(\omega t)$, $\cos(\omega t)$, or anything in between. The phase simply makes explicit the "starting" point for harmonic motion. Both sine and cosine have the same shape, but displaced, and the phase sets the amount of displacement.

Another quantity that characterizes oscillatory motion is how far the moving object gets from the equilibrium position before it turns around. In the case of a point mass in simple harmonic motion in, say, the x-direction, the motion is symmetric from one side to the other, and the maximum distance of displacement to the right of the equilibrium point equals the maximum distance of displacement to the left. We call this distance the **amplitude**, A. It is by definition positive. The sine function is dimensionless and varies between -1 and $+1$. But $x(t)$ has dimensions of length. To express $x(t)$, we therefore have to multiply the harmonic sine (or cosine) function by a constant with dimensions of length, and this constant is the amplitude A described above. The resulting expression for the position of an object in simple harmonic motion is

$$x(t) = A\sin(\omega t + \delta), \tag{13–1a}$$

SIMPLE HARMONIC MOTION

and we can immediately confirm that A describes the magnitude of the maximum excursion away from the point of zero displacement (Fig. 13–2). An alternative form of this expression turns out to be very useful. We can use the basic trigonometry rule $\sin(x + y) = \sin x \cos y + \cos x \sin y$ to rewrite $\sin(\omega t + \delta)$ as

$$x(t) = [A\cos\delta]\sin(\omega t) + [A\sin\delta]\cos(\omega t).$$

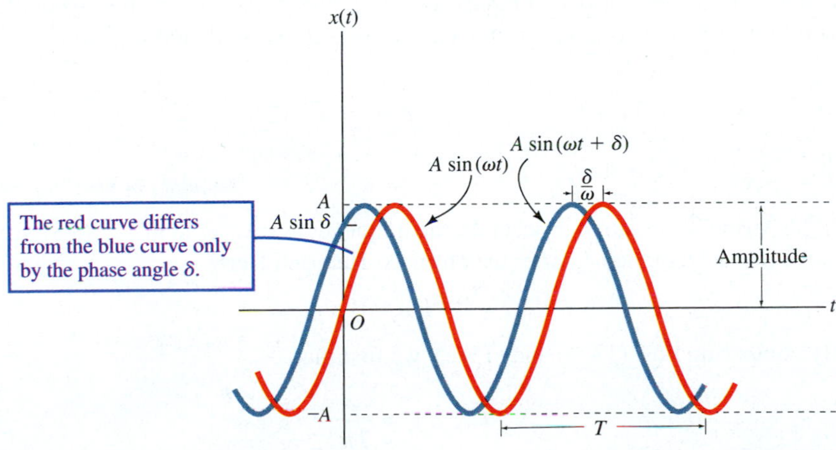

◀ **FIGURE 13–2** In simple harmonic motion the phase, δ, corresponds to a sliding of the curve of displacement versus time to earlier or later times. The amplitude and period of the motion are also shown.

The original quantities A and δ are two constants that characterize the motion, and we can think of the two quantities in square brackets as combinations of these two constants that equally well characterize the motion. In other words, simple harmonic motion can alternatively be expressed as

$$x(t) = a_1 \sin(\omega t) + a_2 \cos(\omega t). \tag{13–1b}$$

Comparing the intermediate step with Eq. (13–1a), we find relations that can be used to connect the constants a_1 and a_2 to the constants A and δ,

$$a_1 = A \cos \delta, \; a_2 = A \sin \delta.$$

Inverting, we can express A and δ in terms of a_1 and a_2:

$$A^2 = a_1^2 + a_2^2, \; \tan \delta = a_2/a_1.$$

Which of these two forms, Eq. (13–1a or b), is more convenient depends on the circumstances, and we'll sometimes use one and sometimes the other.

Properties of Simple Harmonic Motion

Three independent parameters appear in simple harmonic motion and describe the motion: the amplitude A, the phase δ, and the angular frequency ω. The amplitude and the phase are determined by specifying the position $x(t)$ at $t = 0$ and the maximum magnitude of $x(t)$. It follows in this case from Eq. (13–1a) that $x(0) = A \sin \delta$, while $|x_{\max}| = A$. These two equations give A and δ in terms of $x(0)$ and $|x_{\max}|$. Or one may know the position $x(t)$ and velocity $v(t) = dx/dt$ at an initial time $t = 0$. In this case we say that A and δ are determined by the **initial conditions** for the motion. It follows from Eq. (13–1a) that $x(0) = A \sin \delta$. We can also use Eq. (13–1a) to find the velocity $v(t) = dx/dt = A\omega \cos(\omega t + \delta)$ [see Eq. (13–7)], so that $v(0) = A\omega \cos \delta$. The two expressions $x(0)$ and $v(0)$ are enough to specify both amplitude and phase provided that ω is known. A similar analysis can be done for Eq. (13–1b), in which the two constants a_1 and a_2 are determined by initial conditions. The fact that ω needs to be known here suggests that we should turn to that constant next.

The angular frequency ω is a measure of the repetition time for the motion, i.e., the time for one full cycle of the motion. We call this time the **period** T. The sine function repeats itself either when the angle increases by 2π rad (see Fig. 13–1) or, because δ is a constant, when ωt increases by 2π. Thus the period satisfies $\omega T = 2\pi$. We can solve for the period:

$$T = \frac{2\pi}{\omega}. \tag{13–2}$$

PERIOD OF SIMPLE HARMONIC MOTION

Thus the value of the angular frequency ω determines the period. In Chapter 3, where we described uniform circular motion, we defined the **frequency**, f, as the number of full oscillations per unit time, or equivalently the inverse of the period. A period of 5 s means a frequency of one complete repeat of the motion every five seconds, while a period of 0.5 s means a repeat frequency of two per second, and so forth:

$$f = \frac{1}{T}. \tag{13–3}$$

FREQUENCY OF SIMPLE HARMONIC MOTION

If the period is measured in seconds, the frequency is measured in s^{-1}. In SI, the unit s^{-1} is the **hertz** (Hz), named after the physicist Heinrich Hertz:

$$1 \text{ Hz} = 1 \text{ s}^{-1}. \tag{13–4}$$

By comparing Eqs. (13–2) and (13–3), we find that

$$f = \frac{\omega}{2\pi}. \tag{13–5}$$

CONCEPTUAL EXAMPLE 13–1 Your classmate states that if the acceleration of a mass acted on by a spring is proportional to the displacement from the equilibrium point of the mass, then the farther the mass gets from the equilibrium, the larger the acceleration, and the mass will soon be accelerating so much that it will be in the next county in a few minutes. Is he right? How would you correct him?

Answer It is indeed true that the acceleration is proportional to the displacement, but as a look at Eq. (13–8) verifies, *there is a crucial minus sign* in the relation. This sign keeps the motion within bounds. If the displacement is to the right of the equilibrium point, the acceleration is to the left, tending to send the mass back to the left; if the displacement is to the left, the acceleration is to the right, tending to send the mass back to the right. Without the minus sign, your classmate is correct. In that case, the position is an exponential function of time rather than oscillatory.

We will see in Section 13–2 that the angular frequency ω can be identified with the angular speed, a quantity we have already defined and used in Sections 3–5 and 9–1 in connection with circular motion. Inversion of Eq. (13–2) or (13–5) gives

$$\omega = \frac{2\pi}{T} = 2\pi f. \tag{13–6}$$

When the position is specified as a function of time, the velocity and the acceleration are determined by taking successive derivatives. As a consequence of Eq. (13–1a), we have (see Appendix IV–7)

$$v(t) = \frac{dx}{dt} = \frac{d}{dt}[A\sin(\omega t + \delta)] = \omega A\cos(\omega t + \delta). \tag{13–7}$$

One further derivative gives the acceleration as a function of time:

$$a(t) = \frac{dv}{dt} = -\omega^2 A\sin(\omega t + \delta) = -\omega^2 x(t). \tag{13–8}$$

The acceleration is proportional to the displacement. Since we will argue that virtually all stable equilibrium situations, from the back and forth of a rocking chair to the oscillation of a spider on his web in the breeze, are associated with simple harmonic motion; thus, the proportionality of the acceleration and the displacement is a universal property of motion near equilibrium.

Relations Among Position, Velocity, and Acceleration in Simple Harmonic Motion

In Fig. 13–3 we plot the position, velocity, and acceleration of an object in simple harmonic motion over two full periods, starting with $x(t) = A\sin(\omega t)$. (For convenience, the phase has been taken to be zero. The relations discussed here are not affected by the phase.)

The photo in Fig. 13–4 represents the up-and-down motion of a ball on a spring, presented so you can follow the ball's vertical position as a function of time. This motion matches the motion described in Fig. 13–3. In Fig. 13–3a the object is at the origin at $t = 0$. As we see in Fig. 13–3b, the velocity at $t = 0$ is maximum in magnitude and is positive, while Fig. 13–3c shows that at this time the acceleration is zero, so that the velocity is not changing. After one-quarter of the period ($\omega t = \pi/2$), the object has moved to the right-hand extreme of its motion and is ready to turn around. The velocity is zero at this turnaround point, but the acceleration has actually reached a maximum in magnitude and is negative, indicating that the velocity will be turning to the left and will become

◀ **FIGURE 13–4** A photograph of the simple harmonic motion of the mass on the end of a spring.

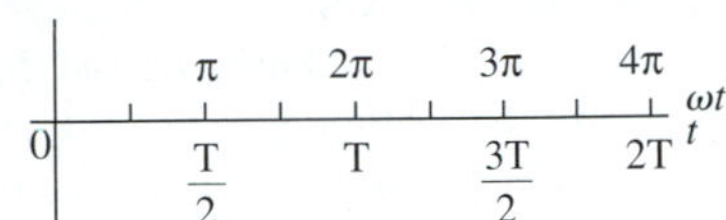

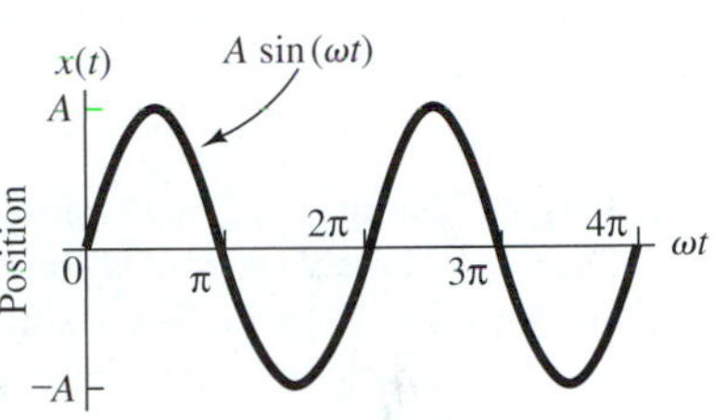

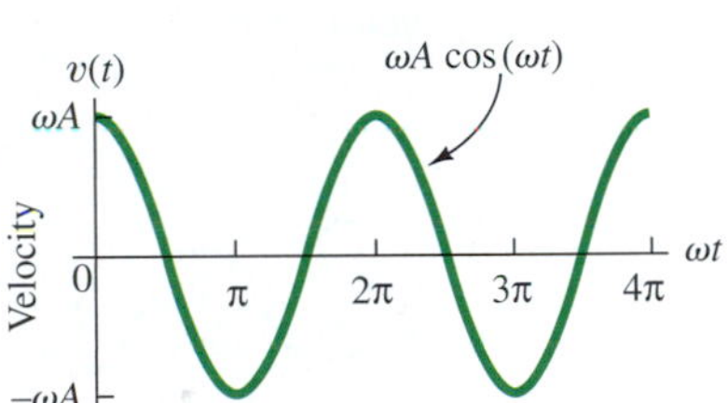

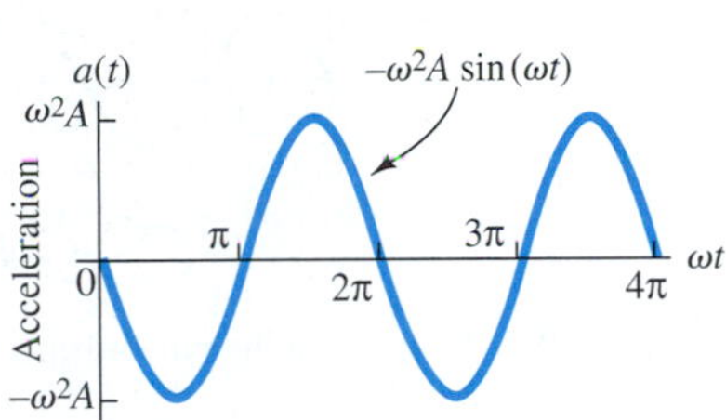

▲ **FIGURE 13–3** Starting with (a) the graph of position $x(t) = A\sin(\omega t)$, a single derivative gives (b) the velocity $v(t) = \omega A\cos(\omega t)$. One further time derivative gives (c) the acceleration $a(t) = -\omega^2 A\sin(\omega t)$. We have in each case plotted two cycles, or periods, for zero phase; the curves repeat after every period. Note also, as the upper scale indicates, the correspondence between the number of periods and ωt as a multiple of 2π.

negative. (Think of a ball thrown in the air; at the maximum height—the turnaround point—the velocity is zero even if the acceleration is nonzero and directed toward Earth.) After one-half the period ($\omega t = \pi$), the object once again passes through the origin, this time moving to the left. The acceleration is again zero. The three-quarter mark ($\omega t = 3\pi/2$) is at another turnaround, characterized by a maximum negative value of x—the object is at its left-hand extreme—and zero velocity. The acceleration is maximum and positive, meaning that the velocity is becoming positive, and the object will subsequently move back to the right. Finally, after one full period ($\omega t = 2\pi$), the object has come back to its starting point, moving to the right through the origin with its largest positive velocity and zero acceleration. The situation at $t = 2\pi/\omega$ is identical to what it was at $t = 0$.

EXAMPLE 13–2 A cork floating on a pond moves in simple harmonic motion, bobbing up and down over a range of 4 cm. The period of the motion is $T = 1.0$ s, and a clock is started at $t = 0$ s when the cork is at its minimum height. What are the height and velocity of the cork at $t = 10.5$ s?

Setting It Up We draw a graph of the motion in Fig. 13–5, which is along a z-axis whose origin is the midpoint of the motion. The maximum value of z is $z_{max} = 2$ cm, and the minimum value is $z_{min} = -2$ cm, which is the location at $t = 0$.

Strategy We must find an expression for position and velocity as a function of time given the information in the problem, and then evaluate these at $t = 10.5$ s. For position, we'll use the general form of Eq. (13–1b), which requires two constants and knowledge of ω. We are given T, and that will determine ω directly. With Eq. (13–1b) we can find the velocity by taking the derivative of the position. To evaluate constants of our expressions, we can use the facts that at $t = 0$, $z = z_{min}$ and $v = 0$.

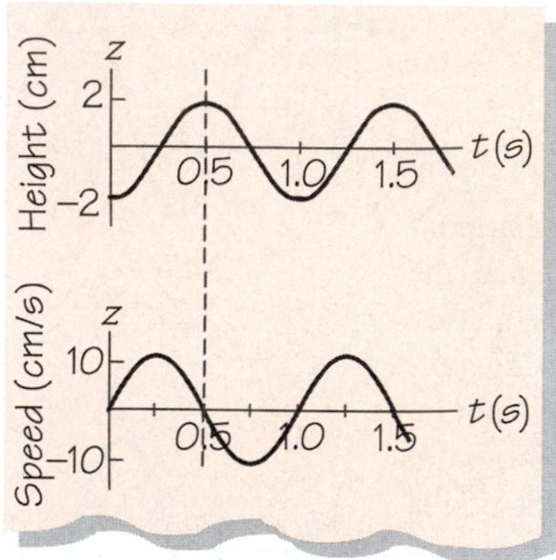

▲ **FIGURE 13–5** The height and speed of a bobbing cork in a pond.

Working It Out We know the period, T, and from Eq. (13–6), $\omega = 2\pi/T$. The motion (position) takes the general form

$$z(t) = a_1 \sin(\omega t) + a_2 \cos(\omega t).$$

With a single derivative, we also get the velocity:

$$v(t) = a_1\omega \cos(\omega t) - a_2\omega \sin(\omega t).$$

To find the constants a_1 and a_2, we use the initial conditions. As stated above, these read, $z = z_{min}$ and $v = 0$ at $t = 0$. The second equation above is simple to apply: v can only be 0 at $t = 0$ if the constant $a_1 = 0$. Applying this, the condition that $z = z_{min}$ at $t = 0$ then gives immediately $a_2 = z_{min}$. Finally, $\omega = 2\pi/T = 2\pi/(1\text{ s}) = 2\pi$ rad/s. In summary,

$$z(t) = a_2 \cos(\omega t)$$

and

$$v(t) = -a_2\omega \sin(\omega t)$$

$$\text{with } a_2 = -2\text{ cm and } \omega = 2\pi\text{ rad/s}.$$

The second part of Fig. 13–5 shows the velocity of the cork.

It is straightforward to plug $t = 10.5$ s into these expressions. We can also employ some simple reasoning to make a shortcut to the numerical answer. Since both z and v repeat themselves every period, the values of z and v at 10.5 s are the same as at 0.5 s (0.5 period). Moreover, after half a period, the cork moves from the bottom of the motion to the top, i.e., z will move from z_{min} to z_{max} and the velocity will once again be zero as the motion of the cork turns around. Thus

$$\text{for } t = 10.5\text{ s},\ z = z_{max} = +2\text{ cm} \quad \text{and} \quad v = 0\text{ m/s}.$$

What Do You Think? For what value(s) of z does the acceleration of the cork have maximum magnitude? For what value(s) of z does the acceleration have minimum magnitude? ***Answers to What Do You Think? questions are given in the back of the book.***

▲ **FIGURE 13–6** The relation between uniform circular motion and simple harmonic motion is evident in the piston-linkage connection on the train wheel and the resulting motion.

13–2 A Connection to Circular Motion

In Chapter 3 we discussed another kind of periodic motion: uniform circular motion. The photograph in Fig. 13–6 of the wheels and driving piston of a steam engine suggests that circular motion has a simple connection to harmonic motion, and we next demonstrate this connection in more detail. Figure 13–7 shows uniform circular motion for a point moving in the xy-plane a constant distance R from the origin. The motion is described by an angle θ, measured from the x-axis, that varies linearly with time:

$$\theta = \omega t + \delta. \tag{13–9}$$

The phase, δ, is just the value of θ at time $t = 0$.

If we were to look at a side view of the uniform circular motion of a pin stuck on a rotating turntable, we would see the pin oscillate in simple harmonic motion. Figure 13–7

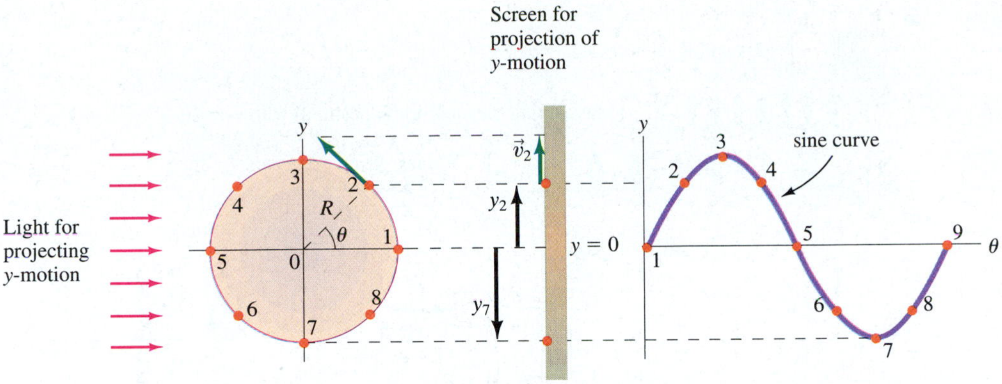

◀ **FIGURE 13–7** Uniform circular motion in the xy-plane, and its projection onto the y-axis. The projection represents simple harmonic motion, easily visible in a plot of y versus $\theta = \omega t + \delta$, as on the right.

indicates the *projection* of the circular motion on the y-axis, but you could easily project onto both x and y. Simple trigonometry gives us these projections:

$$x = R\cos\theta = R\cos(\omega t + \delta); \tag{13–10}$$

$$y = R\sin\theta = R\sin(\omega t + \delta). \tag{13–11}$$

Thus uniform circular motion corresponds to simple harmonic motion in both the x- and y-directions. A cosine rather than sine appears in x, but as we discussed above, this is just the standard form with a different phase. We can use the trigonometric identity $\sin[\theta + (\pi/2)] = \sin\theta\cos(\pi/2) + \cos\theta\sin(\pi/2) = \cos\theta$ to replace the cosine in Eq. (13–10) with a sine function, and we thereby obtain

$$x = R\sin\left(\omega t + \delta + \frac{\pi}{2}\right). \tag{13–12}$$

Both the x- and y-motions are now in the standard form of Eq. (13–1a). The two motions have a phase that differs by exactly $\pi/2$ (90°), and the sign of this phase difference specifies the direction—clockwise or counterclockwise—of the corresponding uniform circular motion (see Problem 22).

13–3 Springs and Simple Harmonic Motion

Having described simple harmonic motion—the kinematics—and armed with our knowledge of Newton's second law, we now can turn to the cause of the motion. Springs give rise to simple harmonic motion. Let's restrict ourselves to one-dimensional motion and dispense with vector notation. The spring force on a mass displaced by x from the equilibrium position of the spring is a restoring force linearly dependent on x, the form known as *Hooke's law*:

$$F = -kx. \tag{13–13}$$

This form is valid provided the spring is not overly stretched or compressed, in which case it loses its "springiness" and distorts—this is why we have spoken about "small" motions about the equilibrium point. Here k is the *spring constant*. It is the minus sign in Eq. (13–13) that indicates that the force is a *restoring force*. A displacement in the $+x$-direction gives rise to a force that acts in the $-x$-direction and vice versa. Figure 13–8a shows a series of possible starting points for the motion. Let us choose the third one, where the mass is released at $t = 0$ from an extended position. The resulting motion is shown in Figure 13–8b over a complete period of the motion. Newton's second law provides us with the connection between the force and the acceleration; namely, $F = -kx = ma$. Thus the acceleration of a mass on the end of a spring is proportional to its displacement, with a minus sign:

$$a = -\frac{k}{m}x. \tag{13–14}$$

An acceleration proportional to the position, with a minus sign, is just the kinematic characteristic that we found in Section 13–1 for simple harmonic motion. Comparison of

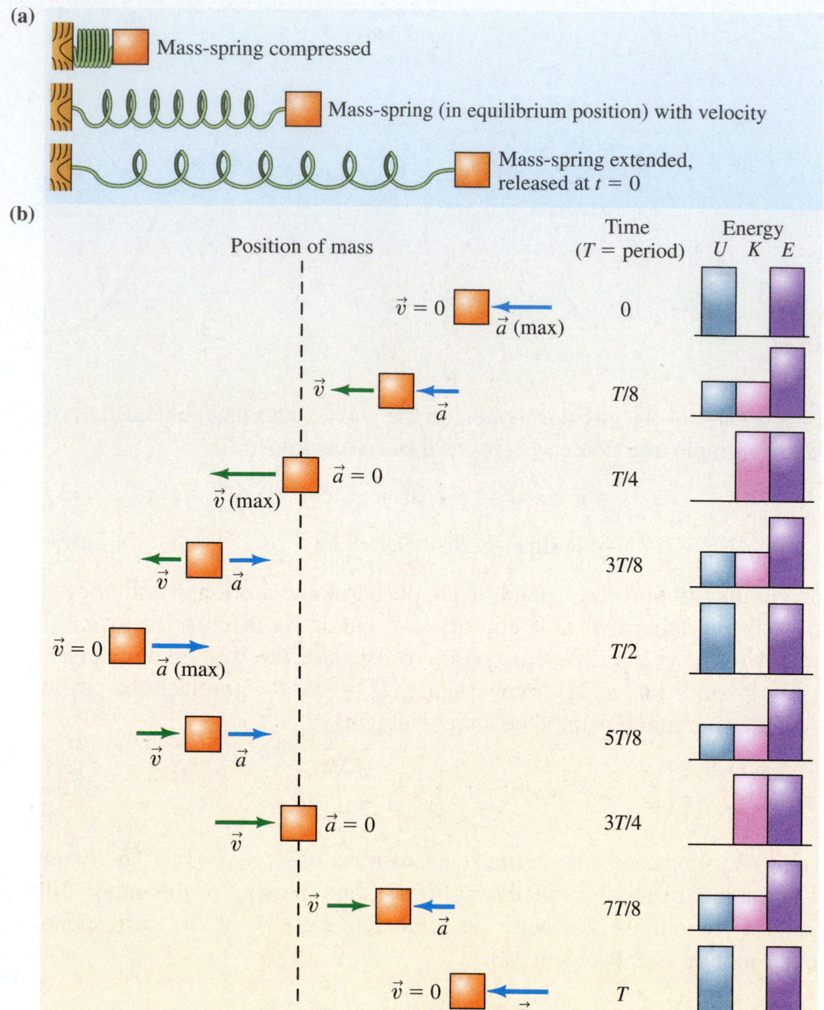

FIGURE 13–8 (a) Some possible starting points for the motion of a mass on the end of a spring. (b) The simple harmonic motion of the mass when it is released from the stretched position. The speed is lowest (and the acceleration is highest) when the displacement from equilibrium is a maximum, and the speed is highest (and the acceleration is lowest) when the displacement is a minimum. We can also see the play between kinetic and potential energy; one is large where the other is small.

Eqs. (13–8) and (13–14) yields the important result that the angular frequency is determined by the mass and the spring constant:

$$\omega^2 = \frac{k}{m}; \tag{13–15}$$

$$\omega = \sqrt{\frac{k}{m}}. \tag{13–16}$$

ANGULAR FREQUENCY FOR MASS ON A SPRING

In turn, Eqs. (13–2) and (13–3) give the period and the frequency of the oscillations:

$$T = 2\pi\sqrt{\frac{m}{k}} \quad \text{and} \quad f = \frac{1}{2\pi}\sqrt{\frac{k}{m}}. \tag{13–17}$$

Remarkably, *the period of the motion is independent of the amplitude*. The same is then true for the frequency.

The spring is the prototype of dynamical systems moving back and forth about a stable equilibrium—virtually all such systems exhibit simple harmonic motion. All these systems reduce to a mass on the end of a spring, in that the *form* of the force is the same as that of the spring, a restoring force linear in some variable.

CONCEPTUAL EXAMPLE 13–3 The spring constant k of a mass-spring system is doubled. By what factor does m have to change so that (a) the acceleration at $x = 0$ is unchanged; (b) the acceleration at $x = A$ [A is the original amplitude] is unchanged; (c) the velocity at $x = A$ is unchanged; (d) the period of the motion is unchanged?

Answer (a) The acceleration is proportional to x; at $x = 0$, the acceleration remains zero, regardless of the values of k or m.

(b) From Eq. (13–14), the original acceleration at $x = A$ is $a = -(k/m)A$. If k is doubled, doubling m will leave a unchanged.

(c) The velocity at the extremes of the motion—i.e., $x = A$—is zero, and this is independent of the values of k or m.

(d) The period is inversely proportional to the angular frequency, which is in turn a function of k/m. So double m to leave the period unchanged.

EXAMPLE 13–4 A mass $m = 0.50$ kg moves along the x-direction under the influence of a spring with spring constant $k = 2.0$ N/m. The origin of the x-axis is at the equilibrium point of the mass. At $t = 0$ s, the mass is at the origin and moving with a speed of 0.50 m/s in the $+x$-direction. (a) At what time t_1 does the mass first arrive at its maximum extension? (b) What is this maximum extension?

Setting It Up We note specifically that we are given initial conditions, in this case the position and velocity at $t = 0$ s.

Strategy We can give a description of the position at all times, then substitute specific times. The motion as a function of time is given by either of the two Eqs. (13–1)—we'll use Eq. (13–1a) here. With k and m known, we can find the angular frequency, ω. And the initial conditions will be sufficient to find the two remaining parameters A and δ in Eq. (13–1a). In part (a) we want the time to go from the origin to the maximum extension, and this is just a quarter period—for that we need only ω. For part (b) the parameter A *is* the maximum extension.

Working It Out From Eq. (13–16), the angular frequency, ω, is

$$\omega = \sqrt{\frac{k}{m}} = \sqrt{\frac{2.0\ \text{N/m}}{0.50\ \text{kg}}} = \sqrt{4.0\ \text{s}^{-2}} = 2.0\ \text{rad/s}.$$

(a) The time to go from the equilibrium position to the maximum extension is $T/4$:

$$t_1 = \frac{1}{4}T = \frac{1}{4}\frac{2\pi}{\omega} = \frac{1}{4}\frac{2\pi\ \text{rad}}{2.0\ \text{rad/s}} = \frac{1}{4}3.1\ \text{s} = 0.78\ \text{s}.$$

(b) We use the information about x and v at $t = 0$ s to find the amplitude. Writing $x(t) = A\sin(\omega t + \delta)$, we have $x(t = 0) = A\sin\delta = 0$. This implies that $\delta = 0$. We use this, in turn, for the value of v at $t = 0$, $v(t = 0) = A\omega\cos(0) = A\omega$. (The argument of the cosine is zero because both t and δ are zero.) Thus

$$A = \frac{v(t = 0)}{\omega} = \frac{0.50\ \text{m/s}}{2.0\ \text{rad/s}} = 0.25\ \text{m},$$

which is the maximum excursion of the mass from the origin.

What Do You Think? If the speed at $t = 0$ were doubled, then the time to reach the maximum extension would be (a) doubled (b) the same (c) halved.

Additional Constant Forces

Suppose we start with a spring force and we add a constant force to it that acts along the same line. How different is the motion of an object under the influence of both these forces from the motion with the spring force alone? The answer is, remarkably little. The only thing that changes is the equilibrium point. As we have seen, the original (one dimensional) spring force always takes the form $F_{\text{spring}} = -k(x - x_0)$ (the sign takes into account the vector nature in one dimension), here aligned with the x-axis. The quantity $x - x_0$ is the displacement of the mass from its equilibrium point at $x = x_0$. The period of this spring, or indeed any spring, is independent of the equilibrium point. Now imagine adding (also acting along the x-axis) a constant force F_c. We can *always* write F_c in the form

$$F_c = kx_1,$$

where k is the same spring constant as for the original spring and $x_1 \equiv F_c/k$. That means the net force takes the form

$$F_{\text{net}} = F_{\text{spring}} + F_c = -k(x - x_0) + kx_1 = -k(x - [x_0 + x_1]).$$

This is again a spring force, with the same spring constant as the original spring force. Thus the motion will have the same frequency, but a shifted equilibrium point, $x_0 + x_1$ instead of x_0.

This behavior is exhibited by a mass hanging vertically from a spring. The supplementary constant force is that of gravity. The frequency of the simple harmonic motion will be the same whether the spring is hanging vertically or not. For the hanging case, and assuming the spring itself is much less massive than the mass attached to its end, the equilibrium position will be lowered by an amount Δy proportional to the additional weight of the mass, as in Fig. 13–9. More precisely, we have

$$mg = k\,\Delta y, \quad \text{or} \quad \Delta y = mg/k.$$

The harmonic motion is measured from the new equilibrium position.

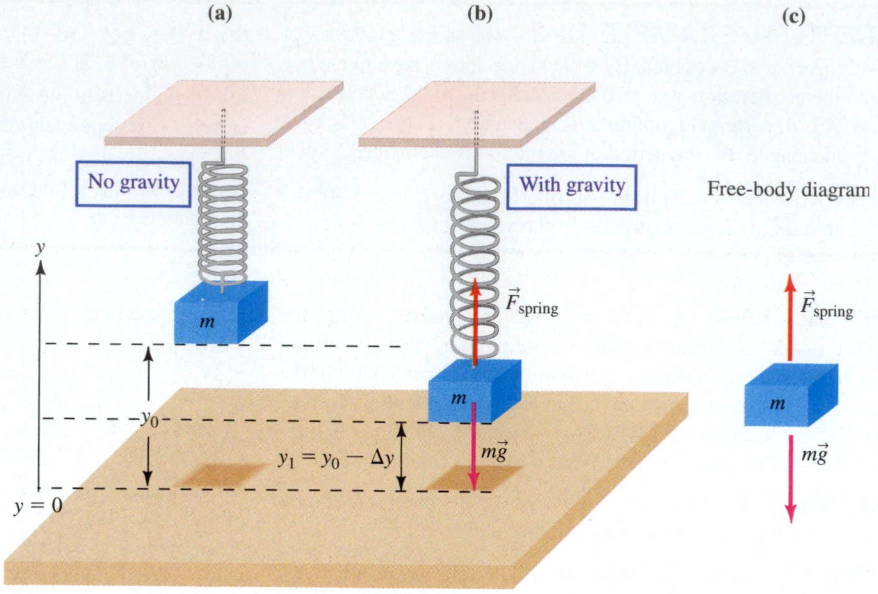

(a) (b) (c)

No gravity
With gravity
Free-body diagram

FIGURE 13–9 A mass on the end of a spring is suspended vertically. (a) If its equilibrium length would place it at height y_0 in the absence of gravity, then (b) it will be stretched an additional amount, Δy, to a new equilibrium position, y_1, under the influence of gravity. (c) Free-body diagram for the mass.

13–4 Energy and Simple Harmonic Motion

We examined energy considerations for the spring force in Chapter 7, where we found that the work done by a spring force in moving a mass from one position to another is independent of the path taken by the mass. That means that the spring force is conservative and has a potential energy function $U(x)$ associated with it. The total energy E (the sum of kinetic energy, K, and potential energy) is *conserved* throughout any motion.

In Section 7–1 we computed the potential energy $U(x)$ of an object attached to a spring and found

$$U(x) = \frac{1}{2}kx^2. \qquad (13\text{–}18)$$

POTENTIAL ENERGY FOR MASS ON A SPRING

In Eq. (13–18) zero potential energy has been chosen at the equilibrium position of the spring, $x = 0$. The kinetic energy is simply

$$K = \frac{1}{2}mv^2. \qquad (13\text{–}19)$$

Because both x and v are known for simple harmonic motion from Eqs. (13–1) and (13–7), the variation in time of U and K can be plotted. If we write the argument $\omega t + \delta$ as θ, we have

$$U = \frac{1}{2}kA^2 \sin^2 \theta, \qquad (13\text{–}20)$$

and using $\omega^2 = k/m$ [Eq. (13–15)],

$$K = \frac{1}{2}mA^2 \omega^2 \cos^2 \theta = \frac{1}{2}kA^2 \cos^2 \theta. \qquad (13\text{–}21)$$

Figure 13–10 is a plot of the potential and kinetic energy functions as θ varies between 0 and 2π, which corresponds to a complete cycle. Both $\sin^2 \theta$ and $\cos^2 \theta$ vary between 0 and 1; when $\sin^2 \theta$ is a minimum, $\cos^2 \theta$ is a maximum and vice versa. Thus U and K each vary between 0 and $kA^2/2$. Suppose that an object attached to a spring starts at the origin and moves to the right, motion you can follow on the graphs of Fig. 13–10. At the origin the potential energy is zero and K is a maximum. As the mass moves to the right, it slows until it has reached its turnaround point at one-quarter cycle, where the velocity and hence K are zero. Because x is at its maximum here, U is also a maximum. The mass now moves to the left, gaining speed until the speed is a maximum as it passes

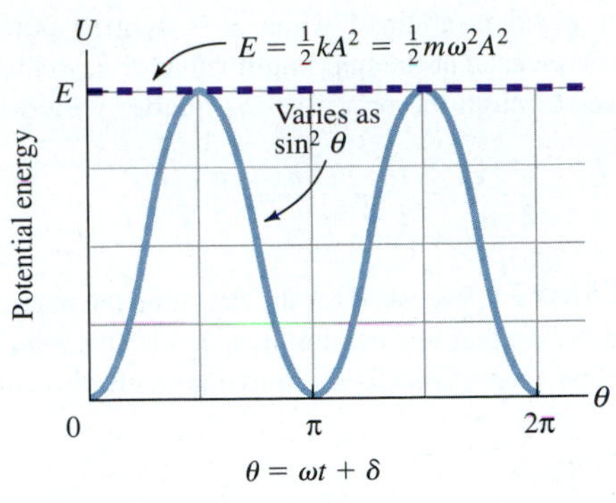

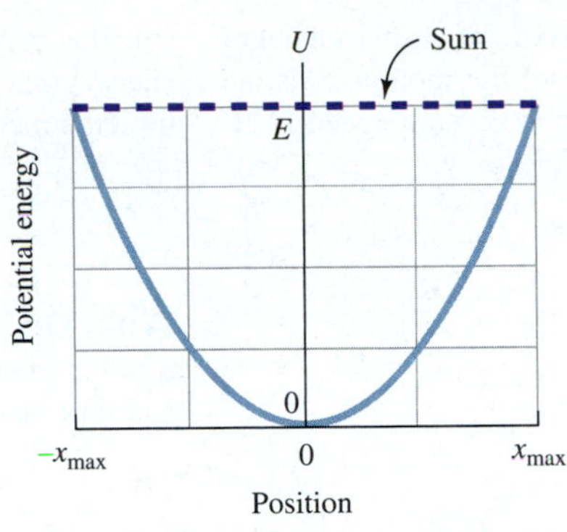

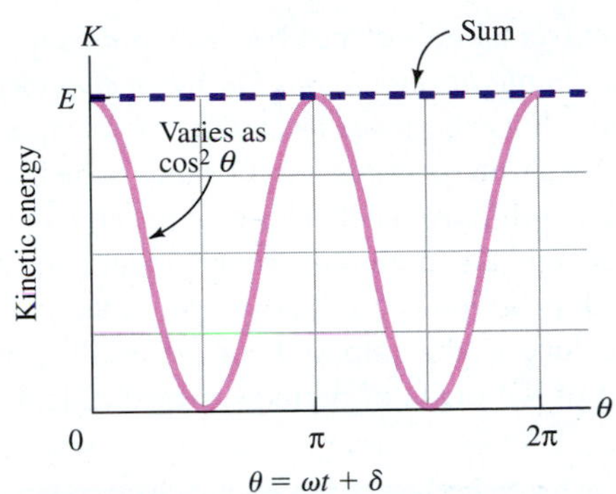

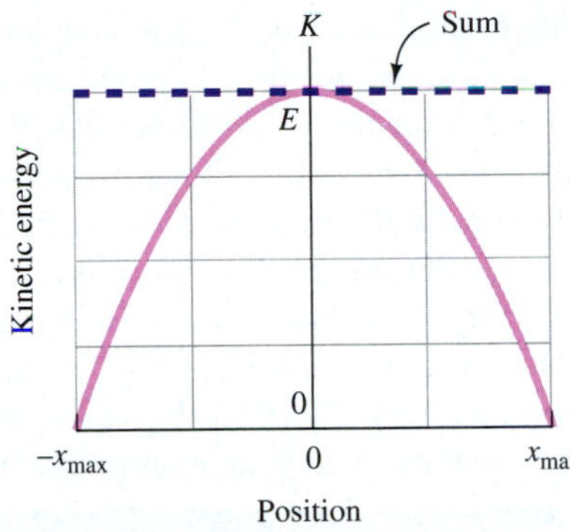

FIGURE 13–10 The potential energy and the kinetic energy of a mass in simple harmonic motion plotted over one cycle (a) as a function of θ, with the origin at the equilibrium point, and (b) as a function of displacement x. When one is a maximum, the other is a minimum, and their sum, the total energy, is conserved.

through the origin once more. Here, after one-half cycle, K is a maximum and U is a minimum. Finally, at the left-hand turnaround point, K is a minimum and U is a maximum. *The energy flows back and forth between U and K.*

The Total Energy

The total energy, $E = U + K$, must be constant. We have (again, $\theta = \omega t + \delta$)

$$E = \frac{1}{2}kA^2 \sin^2\theta + \frac{1}{2}kA^2\cos^2\theta$$

$$= \frac{1}{2}kA^2[\sin^2\theta + \cos^2\theta]. \quad (13\text{–}22)$$

Because the sum of $\sin^2\theta$ and $\cos^2\theta$ is unity for any θ, E is indeed constant in time:

$$E = \frac{1}{2}kA^2. \quad (13\text{–}23)$$

TOTAL ENERGY OF MASS ON A SPRING

The dependence of energy on the square of the amplitude is typical of simple harmonic motion.

EXAMPLE 13–5 A mass m attached to a spring of spring constant k is stretched a length X from its equilibrium position and released with no initial motion. (a) What is the maximum speed attained by the mass in the subsequent motion? (b) At what time is this speed first attained?

Strategy For part (a) the conservation of energy is a useful tool. Initially all the energy is potential, and the maximum speed occurs later, when all the potential energy is converted to kinetic energy. Once we know the maximum kinetic energy, we also know the maximum speed. For part (b) we are asked about time, and we need more information than energy alone can supply. However, we can use our knowledge that in spring motion the potential energy is zero when the mass passes through the origin, and that time is one-quarter period later than the time it is at a maximum extension, which in this case is the starting point of the motion.

Working It Out (a) Just before the mass is released from rest at a position $x = X$, all of its energy is potential energy; that is, the total energy is

$$E = \frac{1}{2}kX^2.$$

(continues on next page)

This agrees with Eq. (13–23) because the maximum displacement of the motion is, by definition, the amplitude of the motion. E is the value of the energy at all times. When the maximum speed is attained, all the energy is in the form of kinetic energy:

$$\frac{1}{2}mv_{max}^2 = E = \frac{1}{2}kX^2.$$

We solve for v_{max}:

$$v_{max} = \sqrt{\frac{k}{m}}X = \omega X.$$

(b) The maximum speed is attained when $x = 0$ (zero potential energy). The mass is released at the maximum value of x, so the first time the mass passes through the origin is one-quarter period later:

$$t = \frac{T}{4} = \frac{1}{4}\frac{2\pi}{\omega} = \frac{1}{4}2\pi\sqrt{\frac{m}{k}} = \frac{\pi}{2}\sqrt{\frac{m}{k}}.$$

What Do You Think? We asked for the first time the maximum speed is attained, implying that this maximum speed is attained a second time at least. How many times is the maximum speed attained?

It's Not Just About Springs

The motion described in this chapter is of universal importance because *almost all systems that are in stable equilibrium exhibit simple harmonic motion when they depart slightly from their equilibrium position.* Everyday experience bears this out in a qualitative way. For example, a marble nudged a little from its stable equilibrium at the bottom of a bowl rolls back and forth, a child's swing will move back and forth through the stable equilibrium position when it is disturbed, and an automobile rocks up and down on its worn shock absorbers. It is obvious that this motion is oscillatory, and as we'll argue below, it is also simple harmonic motion as long as the amplitude of the oscillations is small enough. Table 13–1 gives a sampling of the range of periods of mechanical systems that move in simple harmonic motion.

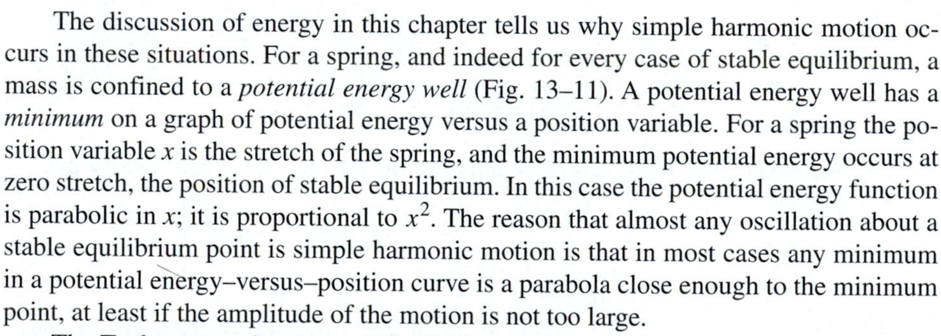

TABLE 13–1 • Periods of Mechanical Systems in Simple Harmonic Motion

Mechanical System	Period (s)
Sloshing of water in a tidal basin or large lake	10^2 to 10^4
Large structures (bridges, buildings)	>1
Strings or air columns of musical instruments	5×10^{-2} to 10^{-4}
Piezoelectric crystals, ultrasound generators	10^{-5} to 5×10^{-1}
Vibrations in molecules	10^{-14}

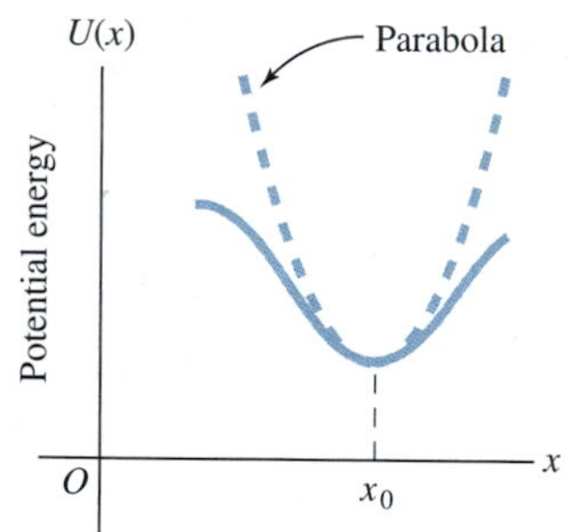

▲ **FIGURE 13–11** A potential well, in which potential energy has a minimum at $x = x_0$. This point is a point of stable equilibrium. The dashed line is a parabola that matches the minimum of the well.

The discussion of energy in this chapter tells us why simple harmonic motion occurs in these situations. For a spring, and indeed for every case of stable equilibrium, a mass is confined to a *potential energy well* (Fig. 13–11). A potential energy well has a *minimum* on a graph of potential energy versus a position variable. For a spring the position variable x is the stretch of the spring, and the minimum potential energy occurs at zero stretch, the position of stable equilibrium. In this case the potential energy function is parabolic in x; it is proportional to x^2. The reason that almost any oscillation about a stable equilibrium point is simple harmonic motion is that in most cases any minimum in a potential energy–versus–position curve is a parabola close enough to the minimum point, at least if the amplitude of the motion is not too large.

The Taylor expansion (Appendix IV–8) is a very general mathematical result that allows us to see why the minimum of a potential energy well forms a parabola and explains why simple harmonic motion is universal near equilibrium. Suppose we apply the Taylor expansion to a potential energy function near a minimum. Let's label the position of the minimum as the origin, $x = 0$. Then the Taylor expansion says that

$$U(x) = \{U(0)\} + x\{U'(x)|_{x=0}\} + (x^2/2)\{U''(x)|_{x=0}\} + \cdots$$

where we have labeled differentiation with respect to x with a prime. The quantities in curly brackets in this expression are constants, and the variable x no longer appears in them. The constant $U(0)$ plays no physical role, and as we know, we can always replace it by 0. (This is implicit in the expression $U = \frac{1}{2}kx^2$ that applies for the spring itself.) The first derivative of U at $x = 0$ is zero because that is a minimum point. Thus, if we keep the first nonzero term in the Taylor expansion—and this is a good approximation if x remains small, so our result refers to small oscillations—we find

$$U(x) \cong (x^2/2)U''(0). \qquad (13\text{–}24)$$

[Here we have used the notation $\{U''(x)|_{x=0}\} = U''(0)$]. This is indeed in the form of a spring force, with $U''(0)$ playing the role of the spring constant. Thus the force takes the general form near the equilibrium point:

$$F(x) = -\frac{dU}{dx} \cong -U''(0)x. \tag{13–25}$$

The force is proportional to the displacement and in a direction opposite to the displacement. It is the familiar linear restoring force of the spring.

We can conclude that almost all stable equilibrium behavior is simple harmonic motion close to the equilibrium point. (The "almost" is present as it is conceivable that a force might have a potential for which the term $U''(0)$ is zero. This requires, however, a restoring force of a very special form, and for these cases you would have to go to the term of order (x^3) in U to find the leading term. Figure 13–11 shows how a minimum on a potential energy curve can be approximated by a parabola, the dashed curve in Fig. 13–11.

EXAMPLE 13–6 A mass m on a frictionless table is attached to two pegs by springs with spring constants k_1 and k_2, respectively. The mass can move along the straight line between the pegs. The separation between the pegs has been arranged so that each spring is in its relaxed position, neither stretched or compressed, when the mass is placed at an equilibrium position. What is the motion of the mass when it is displaced from this position? In particular, assuming the motion is periodic, what is the period?

Setting It Up In Fig. 13–12 we show in part (a) the mass at equilibrium, at the point $x = 0$ where there is no force on the mass from either spring; in part (b) the mass is displaced to position $x \neq 0$ as indicated. We measure x positive to the right. The physical situation implies a point of equilibrium at $x = 0$. When the mass moves away from $x = 0$, the forces tend to send it back to that point, so it is stable in this position. We want to show that when the mass is displaced from $x = 0$, the net force is a linear restoring force, and then find the period of the harmonic motion.

Strategy The motion is one-dimensional, along the line between the pegs. We find the net force acting on the mass, which is a force composed of the forces from the two springs. From the general discussion of stable equilibrium, we expect that the net force will be proportional to the displacement x, and the coefficient will give us the *net, or effective, spring constant*. From this we can deduce the period of the motion. To calculate the net force, we simply add the two forces, taking into account their signs.

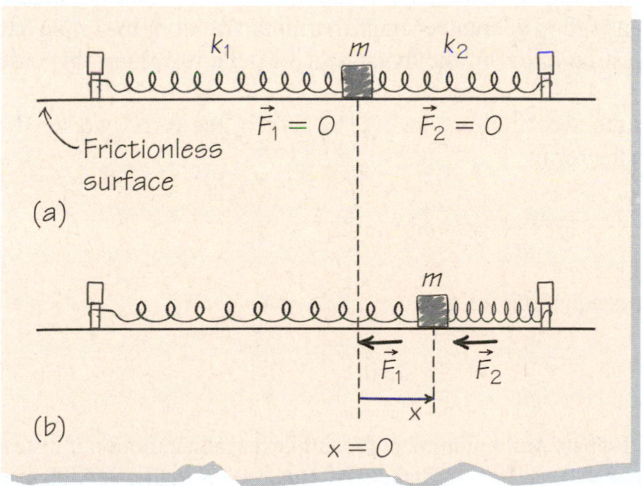

▲ **FIGURE 13–12** In (a) the mass is at its equilibrium position (no net force acts on it). In (b) the mass is no longer in the equilibrium position, and it feels a force from both springs.

Working It Out We let positive values of force be to the right, which takes care of the vector aspect of this problem. From Fig. 13–11 we see that for the displacement shown, the force from the left-hand spring is $F_1 = -k_1 x$, while the force from the right-hand spring is similarly $F_2 = -k_2 x$. [You can check that the signs are correct: With x positive (to the right), the left-hand spring is stretched and its force is to the left, while the right-hand spring is compressed and its force is also to the left.] Adding, the net force on the mass is

$$F_{net} = F_1 + F_2 = -(k_1 + k_2)x.$$

Thus the two springs together act as a single spring with effective spring constant

$$k_{eff} = k_1 + k_2.$$

The motion is simple harmonic motion, with period

$$T = 2\pi\sqrt{\frac{k_{eff}}{m}} = 2\pi\sqrt{\frac{k_1 + k_2}{m}}.$$

Alternative Strategy A different strategy utilizes the potential energy in the two springs. For the displacement of the figure, the potential energy in springs 1 and 2 are

$$U_1 = \frac{1}{2}k_1 x^2 \quad \text{and} \quad U_2 = \frac{1}{2}k_2 x^2,$$

respectively—we have chosen the zero of potential energy at $x = 0$ for each spring. The total potential energy is then

$$U = U_1 + U_2 = \frac{1}{2}(k_1 + k_2)x^2.$$

This is a harmonic oscillator potential energy for a spring with spring constant $k_1 + k_2$, so we get a period corresponding to spring constant $(k_1 + k_2)$, the same result we obtained using forces.

What Do You Think? Suppose that the initial separation between the pegs were larger than that of the problem. Would there still be harmonic motion for movement on the line between the pegs?

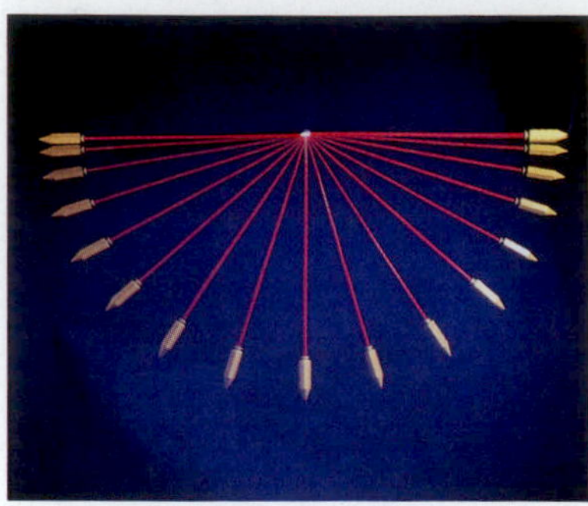

▲ **FIGURE 13–13** The simple pendulum, illuminated by a strobe light at equal time intervals. Half of a complete cycle is imaged. The pendulum bob moves faster near the bottom of its swing and more slowly near the ends. Here the motion is periodic but not harmonic.

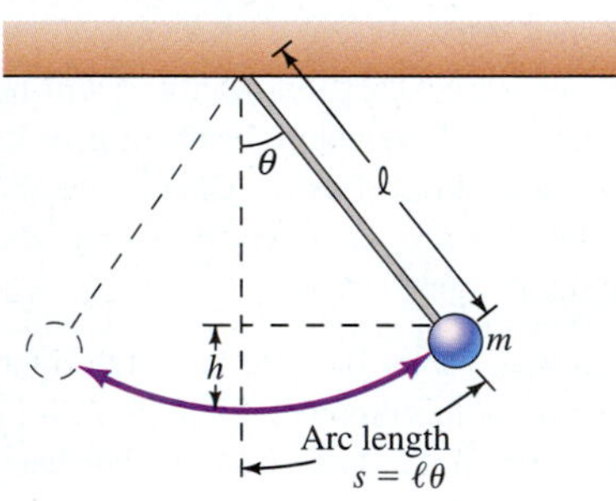

(a)

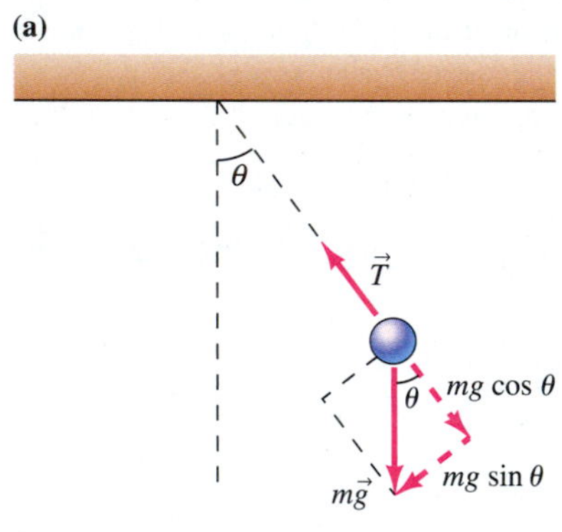

(b)

▲ **FIGURE 13–14** (a) Geometry of the simple pendulum. (b) Force diagram for the simple pendulum. The tension, $\vec{T}$, is along the radial direction, but $m\vec{g}$ has components along both the radial direction and tangent to the arc of the circle traced out by the motion of the mass.

13–5 The Simple Pendulum

A particularly important example of simple harmonic motion is the ordinary pendulum, most commonly observed in some clocks. For centuries, such clocks were the most accurate way to measure time, and the pendulum clock, now more a curiosity or a work of art than a scientific instrument, was crucial in setting standards for time. When we idealize the form of the pendulum to a point mass suspended from a massless string of length ℓ, as in Fig. 13–13, we have the **simple pendulum**. The mass moves along the arc of a circle traced out by the end of the taut string (Fig. 13–14a).

Let's first look at how Newton's second law applies. Suppose that the string makes an angle θ with the vertical. The force diagram for the mass, Fig. 13–14b, includes the force of gravity, $m\vec{g}$, and the tension of the string, $\vec{T}$. The tension is perpendicular to the path of the motion. Its only role in the motion is to constrain that motion to lie along the arc of a circle of radius ℓ. The position s of the mass along the arc of the circle is given by

$$s = \ell\theta, \tag{13–26}$$

where s is measured from $\theta = 0$. The angle θ varies with time, and we wish to determine just how it varies.

To obtain the velocity along the arc of the circle, we differentiate s with respect to time. Because ℓ is a constant, we find

$$v = \frac{ds}{dt} = \ell\frac{d\theta}{dt}. \tag{13–27}$$

The tangential acceleration, which is the component of the total acceleration along the arc of the circle, is in turn associated with changes in the magnitude of this velocity:

$$a = \frac{dv}{dt} = \ell\frac{d^2\theta}{dt^2}. \tag{13–28}$$

This component of the total acceleration (which should *not* be confused with the centripetal component of the acceleration) is due to the tangential force component

$$F_t = -mg\sin\theta. \tag{13–29}$$

The sign of this force component is important. It is negative when θ is itself positive—when the mass is on the right side of the vertical—and in the positive direction when the mass is on the left side of the vertical. This means that *the force of gravity always acts to bring the mass back to the vertical.* This is enough to ensure that the motion will be oscillatory but not enough to guarantee simple harmonic motion: In simple harmonic motion the force must be *linear* in the dynamical variable itself—in this case, the angle.

Using Eq. (13–28) for the acceleration and Eq. (13–29) for the force, we see that Newton's second law takes the form

$$m\ell\frac{d^2\theta}{dt^2} = -mg\sin\theta.$$

Canceling the mass from this equation, we get

$$\ell\frac{d^2\theta}{dt^2} = -g\sin\theta. \tag{13–30}$$

Equation (13–30) would satisfy our requirement for simple harmonic motion if instead of $\sin\theta$, the angle θ itself appeared on the right-hand side. But $\sin\theta$ is in fact very close in value to θ when θ is small. To see this, refer to Fig. 13–15a for $\sin\theta$ and θ, which gives

$$\sin\theta = \frac{x}{\ell} \quad \text{and} \quad \theta = \frac{s}{\ell}, \tag{13–31}$$

where s is the arc length corresponding to angle θ. We can see from Fig. 13–15b that x comes closer and closer to the value of s as θ becomes smaller. Thus, for small θ, $\sin\theta \cong \theta$. This can also be seen from the Taylor expansion for the sine function:

$$\sin\theta = \theta - \frac{\theta^3}{3!} + \frac{\theta^5}{5!} - \cdots. \quad (13\text{–}32)$$

When θ is small, the terms of order θ^3, θ^5, and so forth can be ignored, justifying our approximation. For example, when $\theta = 0.2$ rad (about 11°), the difference between $\sin\theta$ and θ is about 1 percent, but when $\theta = 0.1$ rad (about 6°), the difference is only about 0.1 percent.

The small-θ approximation demonstrates that a pendulum will have true simple harmonic motion only for small excursions. With the small-excursion approximation $\sin\theta \cong \theta$ taken as an equality, Eq. (13–30) becomes for small θ

$$\ell\frac{d^2\theta}{dt^2} = -g\theta. \quad (13\text{–}33)$$

As long as the approximation is good, this equation is *precisely* the equation for simple harmonic motion, as we can see if we compare it to Newton's second law for the spring, $md^2x/dt^2 = -kx$. We need change only some variable names and constants to go from the equation of simple harmonic motion for a mass-spring system to that for a pendulum. Using horizontal arrows to indicate name substitutions,

$$x \rightarrow \theta, \quad k \rightarrow g, \quad \text{and} \quad m \rightarrow \ell \quad (13\text{–}34)$$

summarizes the changes needed. The solution to the motion of the simple pendulum for small angles is then taken directly from the solution for the motion of the spring:

$$\theta = \theta_0 \sin(\omega t + \delta) \quad \text{with} \quad \omega = \sqrt{\frac{g}{\ell}}. \quad (13\text{–}35)$$

The quantities θ_0 and δ are determined from initial conditions. θ_0 is the amplitude of the angular motion, the maximum angle attained. Again, it is the dynamics of the motion that determines the angular frequency ω—you cannot adjust it by a change of initial conditions. The period and frequency of the pendulum's motion come from Eqs. (13–2) and (13–3):

$$T = 2\pi\sqrt{\frac{\ell}{g}} \quad \text{and} \quad f = \frac{1}{2\pi}\sqrt{\frac{g}{\ell}}. \quad (13\text{–}36)$$

Just as for a mass on an ideal spring, *the period of the small-amplitude pendulum is independent of the amplitude.*

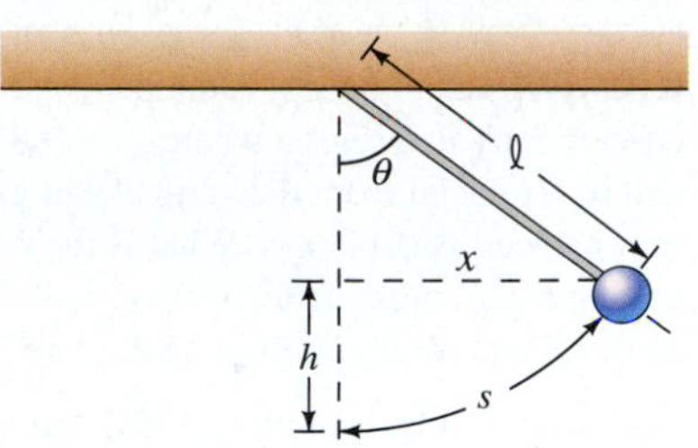

(a)

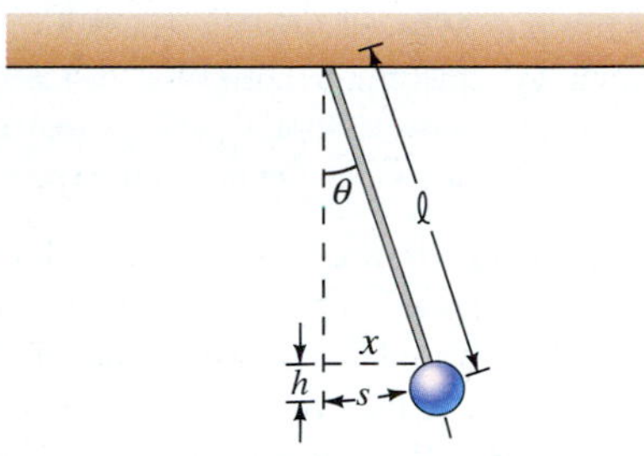

(b)

▲ **FIGURE 13–15** Comparing the horizontal displacement x of the mass of a simple pendulum with the arc length s traced out by the motion. These two lengths differ by a larger percentage (a) when θ is large than (b) when θ is small.

THINK ABOUT THIS . . .

WHY WERE PENDULUM CLOCKS SO USEFUL?

Why did the pendulum play such a central role in the development of time standards? The meaning of standards is that they are easily reproducible—two persons can have two different clocks and still agree on what constitutes a given time period. The pendulum clock fills this bill because its period is independent of amplitude as long as we stick to small amplitudes, so that the first term of Eq. (13–32) dominates, and $\sin\theta \cong \theta$. The fact that $\sin\theta$ is a series in odd powers rather than all powers of θ means that the first correction to $\sin\theta \cong \theta$ is two powers of θ down rather than just one, and $\sin\theta \cong \theta$ is therefore an especially good approximation. The result of all this is that one can construct pendulum clocks that keep the same time without worrying that their amplitudes have to match. One only needs to make sure that the length of different pendula are the same, and this can easily be arranged by screw mechanisms below the mass to allow small adjustments to the length. Once two pendulum clocks have pendula of the same length, one does not have to worry about whether they run at the same amplitude. Similarly, a given clock will keep consistent time even if perturbations cause its amplitude to change (although the driving mechanism is meant to hold the amplitude constant). The time is *not* consistent, however, if there are effects, in particular changing temperatures, that change the length of the pendulum, and indeed clockmakers struggled to construct pendulums whose length would not change under varying conditions.

EXAMPLE 13–7 A simple pendulum 2.00 m long is suspended vertically in a region where $g = 9.81\ \mathrm{m/s^2}$. The point mass at the end is displaced from the vertical and given a small push, so its maximum speed is 0.11 m/s. What is the maximum horizontal displacement of the mass from the vertical line it makes when at rest? Assume that all the motion takes place at small angles.

Setting It Up Figure 13–15b can help us with the geometry of this situation. In the figure we have labeled the pendulum length ℓ, the angle θ of the swing, and the horizontal displacement x of the mass, with $x = 0$ at the bottom of the swing. We can also use it to solve x in terms of θ. We have drawn a sequence of the motion in Fig. 13–16. We want the maximum horizontal distance $x = x_{max}$ reached by the swinging mass, assuming always that the swing covers only small angles. We are given the maximum speed v_{max}.

Strategy θ varies harmonically, with a known angular frequency ω. By calculating v, we find a formula in which the amplitude appears. We have information on the maximum velocity v_{max}, which occurs at the bottom of the swing ($\theta = 0$), and this will allow us to find the amplitude of the swing. Knowing the amplitude, we can find the maximum value of θ and, using geometry, the maximum value of x.

Working It Out We know $\theta = \theta_0 \sin(\omega t + \delta)$, where ω is the angular frequency. The arc length s traced by the mass at the end of the string is given by $s = \ell\theta$, where ℓ is the string's length. Thus s also varies harmonically: $s = A \sin(\omega t + \delta)$, where $A = \ell\theta_0$. The velocity is the rate of change of s (understood to be in the tangential direction), so we write $v = ds/dt$. Thus we have

$$v = \frac{d}{dt}[A \sin(\omega t + \delta)] = A\frac{d}{dt}[\sin(\omega t + \delta)]$$
$$= A\omega \cos(\omega t + \delta).$$

From this expression, we see that v varies harmonically, with amplitude $A\omega$. The maximum magnitude of v, v_{max}, is thus $A\omega = 0.11\ \mathrm{m/s}$. We can find ω from Eq. (13–35):

$$\omega = \sqrt{\frac{g}{\ell}} = \sqrt{\frac{9.81\ \mathrm{m/s^2}}{2.00\ \mathrm{m}}} = 2.21\ \mathrm{rad/s}.$$

Thus from $v_{max} = A\omega$,

$$A = \frac{v_{max}}{\omega} = \frac{0.11\ \mathrm{m/s}}{2.21\ \mathrm{rad/s}} = 0.050\ \mathrm{m}.$$

As we can see from Fig. (13–15b), this is to a good approximation also the horizontal displacement of the mass as long as the angle is small. Our result for maximum displacement, 5.00 cm, is indeed small compared to the length, so our small-angle approximations are good.

What Do You Think? If the mass had been pushed a little harder, the maximum angle would have been larger, and at some point you could no longer trust the small-angle approximation. How would you go about deciding whether the small-angle approximations still hold at the larger angle?

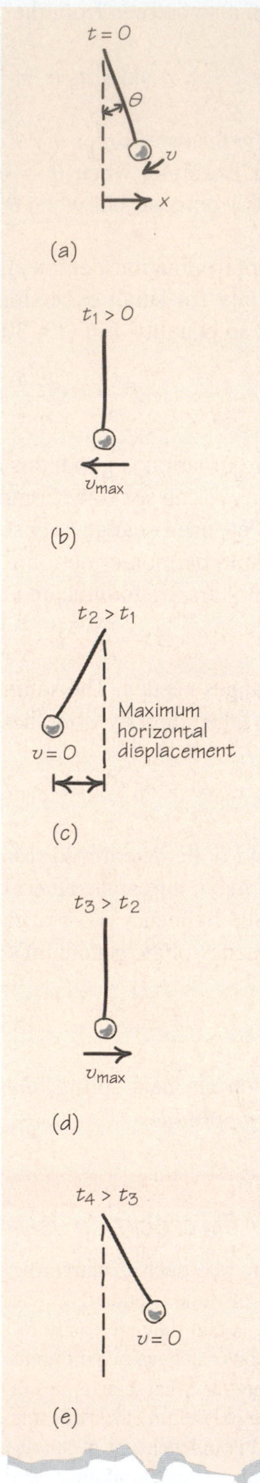

▲ **FIGURE 13–16** Sequence of motion for a simple pendulum.

The Energy of the Simple Pendulum

Let's now consider the simple pendulum from the point of view of energy. The kinetic energy, K, which is a function of θ, is found by expressing the speed as a function of θ. We can use Eq. (13–27):

$$K(\theta) = \frac{1}{2}mv^2 = \frac{1}{2}m\ell^2\left(\frac{d\theta}{dt}\right)^2. \tag{13–37}$$

The only force with a component along the motion of the mass is gravity. This force is conservative, and to find the associated potential energy function, U, we can express the height h gained by the mass in terms of θ. From Fig. 13–15a, we see that $h = \ell - \ell \cos\theta$, so that

$$U = mgh = mg\ell(1 - \cos\theta). \tag{13–38}$$

Here we have taken zero potential energy to be at the bottom of the swing, at $\theta = 0$. In a small-angle approximation, $\cos\theta \cong 1 - (\theta^2/2)$, and

$$U(\theta) \cong mg\ell\left(1 - 1 + \frac{1}{2}\theta^2\right) = \frac{1}{2}mg\ell\theta^2. \tag{13–39}$$

Comparing Eqs. (13–37) and (13–39) for the kinetic and potential energies with their counterparts for the spring, Eqs. (13–19) and (13–18), we see that they have the same form, with just the name changes of Eq. (13–34).

CONCEPTUAL EXAMPLE 13–8 Two teams of students are each given a stopwatch and a simple pendulum whose length, around 1 m, has been precisely measured. They are asked to measure the acceleration of gravity, g. One team pulls their pendulum back a small amount and measures the time for 1 oscillation. The second team releases the pendulum from the same point as the first and lets it swing for a minute, carefully counting the number of oscillations as the amplitude drops to a low value. Which team will come up with the more accurate value of g?

Answer We can assume that the initial displacement in each case is small enough so that the period is correctly given by Eq. (13–36). In that case the precision of each team is limited by the ability of the member with the stopwatch to start and stop the watch at the beginning or end of a full swing. Reflex times for stopwatches are on the order of a tenth of a second, and we'll take this as a measure of the uncertainty Δt in the measurement of a time period. The percentage error made by the first team in its measurement of the period T is $\Delta t/T$, and we can estimate from experience [or from Eq. (13–36)] that a pendulum with length of around 1 m has a period of around 2 s. Thus the percentage error in the time measurement is roughly 5%. That translates into a 10% error in the measurement of g, as g depends on T^2 (see Section 1–3). If the second team keeps track of the number of oscillations, which is just an integer number N and not likely subject to an error, they will have a fractional error in their time measurement of $\Delta t/(NT)$. Over a minute, say, the pendulum will make 30 swings, so the error is $(5\%)/30$, or only about 0.2%. The error in g is diminished by the same factor of 1/30. It is often more accurate to make a cumulative measurement.

13–6 More About Pendulums

The Physical Pendulum

A pendulum is not limited to a massless string with a pointlike mass at the end of it. When a suspended, swinging object has some other form, we call it a *physical pendulum.* Any object can be suspended from any point on it to act as a physical pendulum. For example, when you are walking, the leg that is not in contact with the ground swings and acts as a pendulum. Such examples illustrate the point that oscillatory motion is a general characteristic of motion about a stable equilibrium. We can approach such problems by recasting them as a spring problem, through analysis of either the potential energy or the dynamical equation, and then use the results we have already found for the spring.

▲ **FIGURE 13–17** A swinging sign is an example of a physical pendulum. The center of mass of the oscillating sign executes simple harmonic motion about the stable equilibrium position, $\theta = 0$.

From previous chapters, we have built up the necessary tools to handle the physical pendulum: We study the torque on it, τ, and the corresponding angular acceleration. Figure 13–17 illustrates a physical pendulum allowed to pivot through some horizontal axis, called y, which defines the vertical plane of the swinging. We take the y-axis to be into the page and through the oscillation point. We need to know the rotational inertia, I, about the pivot axis, the total mass M of the object, and the distance r from the center of mass to the pivot axis. The stable equilibrium point for this object is $\theta = 0$, when the center of mass hangs directly below the axis. When $\theta \neq 0$, only the force of gravity, which acts on the center of mass, exerts a (restoring) torque. From Eq. (9–29) the equation of motion governing the behavior of the object is

$$\tau = I\alpha, \tag{13–40}$$

where α is the angular acceleration about the pivot point. The magnitude of α is $d^2\theta/dt^2$.

The torque is given by $\vec{\tau} = \vec{r} \times \vec{F}$, where $\vec{r}$ is the vector from the pivot to the center of mass and $\vec{F}$ is the force of gravity acting on the center of mass. According to the right-hand rule, this torque is along the $+y$-axis (when θ is positive, as shown); the angular acceleration $\vec{\alpha}$ must therefore also be along the $+y$-axis, which corresponds in Fig. 13–17 to an angular acceleration that brings the object back to equilibrium. (Both $\vec{\tau}$ and $\vec{\alpha}$ would change signs if the object were drawn with θ on the other side of the vertical axis.) The magnitude of the torque is

$$\tau = rF \sin\theta, = rMg \sin\theta, \tag{13–41}$$

so the equation of motion, Eq. (13–41), becomes

$$Mgr \sin\theta = -I\frac{d^2\theta}{dt^2}. \tag{13–42}$$

Equation (13–42) has exactly the same form as the equation governing the motion of the simple pendulum, which is Eq. (13–30). If the motion is restricted to small angles, then $\sin\theta$ can be replaced with θ itself and, as for the simple pendulum, simple harmonic motion follows: $\theta = \theta_0 \sin(\omega t + \delta)$. To find ω, it is necessary only to replace ℓ and g for the simple pendulum with I and Mgr, respectively, for the physical pendulum. Thus, from Eqs. (13–35) and (13–36),

$$\omega = \sqrt{\frac{Mgr}{I}}, \tag{13–43a}$$

$$T = 2\pi\sqrt{\frac{I}{Mgr}}. \tag{13–43b}$$

Because I is always M multiplied by some length squared, M will cancel from Eqs. (13–43). *The period of a physical pendulum is independent of its total mass—only how the mass is distributed matters.* We can verify that the period of the physical pendulum as expressed in Eq. (13–43b) reduces to the simple pendulum when the swinging object is a point mass m on a massless string of length ℓ. In that limit, $I = m\ell^2$, $r = \ell$, $M = m$, and T reduces to the appropriate value,

$$T = 2\pi\sqrt{\frac{m\ell^2}{mg\ell}} = 2\pi\sqrt{\frac{\ell}{g}}.$$

EXAMPLE 13–9 A thin, uniform rod of mass M and length L swings from its end as a physical pendulum. What is the period of the oscillatory motion for small angles? Find the length ℓ of the simple pendulum that has the same period as the swinging rod.

Setting It Up Figure 13–18 illustrates the situation.

Strategy Equation (13–43b) gives the period of a physical pendulum swinging in small angles and thus is directly applicable here. Its use requires us to know the rotational inertia of a rod about an axis through its end. This quantity was calculated in Eq. (9–26) and is listed in Table 9–1. For the second part, we compare the period for our pendulum with that of a simple pendulum of length ℓ and solve for ℓ.

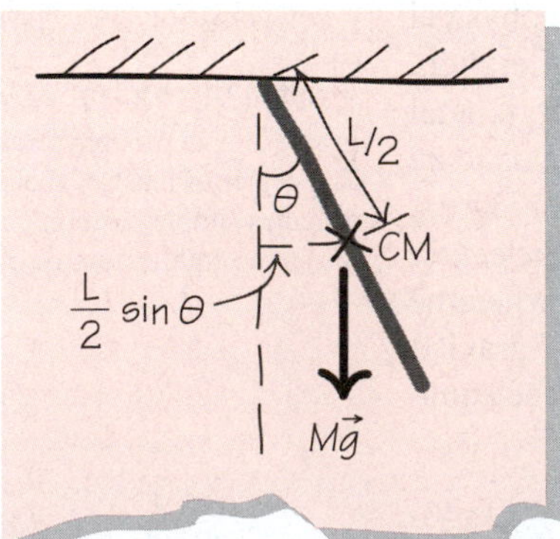

▲ **FIGURE 13–18** A thin, uniform rod swings as a physical pendulum.

Working It Out From Chapter 9 the rotational inertia of our rod is

$$I = \frac{1}{3}ML^2.$$

As we can see in Fig. 13–18, the distance of the center of mass from the end is $r = L/2$, and Eq. (13–43b) then gives the period of the motion as

$$T_{\text{rod pendulum}} = 2\pi\sqrt{\frac{I}{Mgr}} = 2\pi\sqrt{\frac{ML^2/3}{MgL/2}} = 2\pi\sqrt{\frac{2}{3}\left(\frac{L}{g}\right)}.$$

The length ℓ of a simple pendulum of the same period is determined by

$$T_{\text{simple pendulum}} = 2\pi\sqrt{\frac{\ell}{g}} = 2\pi\sqrt{\frac{2}{3}\left(\frac{L}{g}\right)};$$

$$\ell = \frac{2}{3}L.$$

What Do You Think? If a toy monkey is attached to the bottom of the rod, does the period (a) increase; (b) decrease; (c) remain unchanged; or (d) change in a way that cannot be determined from the information given?

13–7 Damped Harmonic Motion

Almost all physical systems, including masses, springs, and other oscillating systems, are affected by friction or drag (resistive) forces. These forces tend to remove energy from a moving system and thereby slow it down, or damp its motion. The universality and importance of these forces is evident in the world around us: Without "pumping," a child on a swing slows down to a stop over a few minutes, and a marble in a bowl will not oscillate indefinitely. When energy is lost in this way, the motion is said to be *damped*. From Eq. (13–23) we can see that a decrease in energy implies a decrease in amplitude. If a clock pendulum is to maintain a given amplitude, energy must be supplied from, say, a wound spring or hung weights.

What are the *quantitative* effects on a mass-spring of these nonconservative, or dissipative, forces? We can answer in the case of a drag force $\vec{F}_d$ that is proportional to velocity (Fig. 13–19):

$$\vec{F}_d = -b\vec{v} = -b\frac{d\vec{x}}{dt}, \tag{13–44}$$

where b is the *damping coefficient* (or *damping parameter*). The minus sign indicates that this force is always opposite to the direction of motion. For this case the equations of motion are solvable in analytic form, and we can use these solutions as a guide to the behavior of other damped systems. For a one-dimensional system, the equation of motion is

$$-kx - b\frac{dx}{dt} = m\frac{d^2x}{dt^2}. \tag{13–45}$$

By finding the function $x(t)$ that satisfies this equation, we will have found the position of a mass on a spring with damping. To solve differential equations such as Eq. (13–45), we attempt trial solutions that are educated guesses and see if they work. As a guess to the solution of Eq. (13–45), we keep a sinusoidal component but also incorporate a decreasing term due to damping. Our trial solution is

$$x = Ae^{-\alpha t}\sin(\omega' t + \delta), \tag{13–46}$$

where A, δ, and α are constants to be determined. We use a frequency ω' rather than ω_0, the frequency in the absence of drag, because we want to allow for the possibility that the frequency is changed by the damping. (In this context and in the context of the next section, ω_0 is called the **natural frequency**. As we have emphasized throughout, any

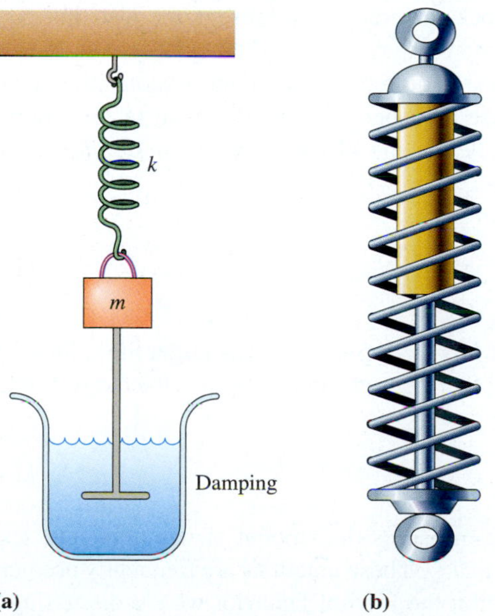

FIGURE 13–19 (a) A damped oscillator. (b) An automobile shock absorber is a damped harmonic oscillator.

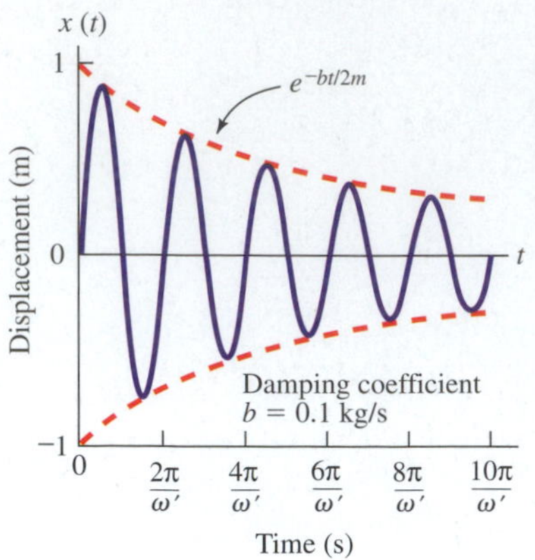

(a)

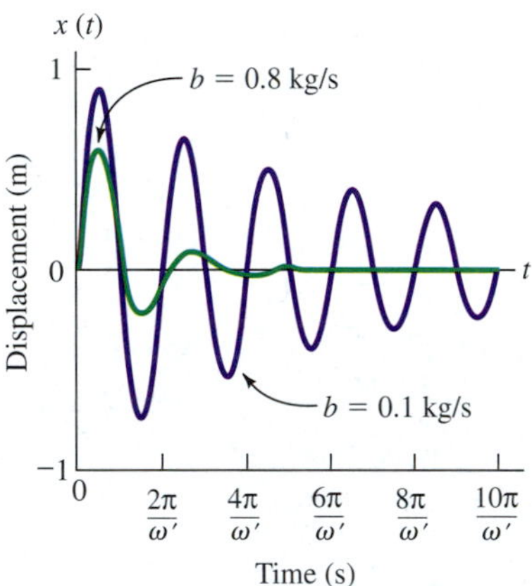

(b)

▲ **FIGURE 13–20** Damping of simple harmonic motion by a drag force $\vec{F} = -b\vec{v}$. Plotted is the function $x(t) = Ae^{-bt/2m}\sin(\omega' t)$ versus t, with $\omega' = \sqrt{k/m - b^2/4m^2}$. The values of the parameters are those chosen in the text.

physical system will have a natural frequency for motion near a stable equilibrium point.) We check whether the trial solution is a good one by seeing if the equation of motion, in this case Eq. (13–45), is satisfied when the trial solution is plugged into it. This procedure is left to Problem 77. The result is that the trial solution is satisfactory provided that

$$\alpha = \frac{b}{2m} \tag{13–47}$$

and that

$$\omega' = \sqrt{\frac{k}{m} - \frac{b^2}{4m^2}} = \sqrt{\omega_0^2 - \frac{b^2}{4m^2}}. \tag{13–48}$$

Thus *the damping factor α and the modified angular frequency ω' are determined by the equations of motion.* (A and δ remain undetermined.) It is easy to check that Eq. (13–46) reduces to standard simple harmonic motion when the damping coefficient $b = 0$.

If we plot our solution [Eq. (13–46)], we see that the falling exponential function forms a kind of *envelope*, marked in Fig. 13–20a by dashed lines. The damping thereby modulates what would otherwise be simple harmonic motion, such that the amplitude of the motion decreases as time goes on. The argument of the exponential is directly proportional to b; that is, to the size of the drag force. Figure 13–20b shows the motion described by Eq. (13–46) for two values of b, one larger than the other. Here we have taken $k = 1$ N/m, $m = 1$ kg, and $A = 1$ m, and in Fig. 13–20b the two values of b are $b = 0.1$ kg/s, and $b = 0.8$ kg/s. In each case the decrease of the amplitude with time is clearly visible, but the motion for $b = 0.1$ kg/s is weakly damped compared to that for $b = 0.8$ kg/s. For both these systems the oscillations continue to some extent and the motion is said to be **underdamped.** The difference between ω' and the angular frequency $\omega_0 = \sqrt{k/m}$ of the spring without drag is harder to see.

As b increases, the angular frequency ω' decreases (and hence the period of the motion increases) until, from Eq. (13–48), $\omega' = 0$ when $b^2 = 4mk$. We refer to this value of b as the *critical* value b_c,

$$b_c = \sqrt{4mk} = \sqrt{4m^2\omega_0^2}. \tag{13–49}$$

When $b = b_c$ the system is said to be **critically damped**; there are no oscillations at all, as is the case when $b > b_c$ and the system is **overdamped**.

Damping is quite frequently introduced to systems that would otherwise oscillate in undesirable ways. Automobile shock absorbers are present for just this purpose (Fig. 13–19).

The quantity $b/2m$, which appears multiplying t in the exponential factor in Eq. (13–46), has dimensions of inverse time because the argument of an exponential must be dimensionless. We therefore define the **lifetime** (or the *mean life*), τ, of the damped oscillator by

$$\tau \equiv \frac{m}{b}. \tag{13–50}$$

In terms of τ, the exponential envelope has the form $e^{-t/2\tau}$. The larger the value of τ, the slower the exponential falloff. Still another nomenclature employs the dimensionless ***Q* factor**, defined by

$$Q \equiv \omega_0 \tau. \tag{13–51}$$

Because Q is proportional to τ, it too measures the amount of damping. The less the damping, the larger are the values of τ and Q. These notations are frequently used in connections with electric circuits that exhibit the kind of behavior we are discussing here.

EXAMPLE 13–10 A gong struck with a hammer can be modeled as a damped harmonic system, with its frequency of oscillation measured by the musical tone emitted and the "loudness" of its sound measured by the amplitude squared (see Chapter 14). A certain gong is struck, and after 9.0 s the loudness has dropped to 0.85 times the original loudness. How much more time will have elapsed before the loudness is 0.25 times the original loudness?

Setting It Up We are given information on the rate of falloff of the loudness, but not a direct value for the lifetime τ characteristic of the system. If, however, we can find τ, then we would know all we need about the loudness at any time t.

Strategy In the damped systems we are studying, the amplitude decrease is governed by the exponential factor $e^{-t/2\tau}$, hence the loudness varies with time as $e^{-t/\tau}$. The information on the loudness at 9.0 s will allow us to find τ; once we have that, the loudness as a function of all time is determined.

Working It Out Denoting the loudness as $L(t)$, we have $L(t) = L(0)\,e^{-t/\tau}$. At 9.0 s, $L = L(0) \times (0.85)$, and therefore

$$e^{-(9.0\text{ s})/\tau} = 0.85.$$

We invert to find τ. The inverse of the exponential is the natural logarithm, ln (see Appendix IV–6). If we take the natural log of both sides of this equation, we have

$$\ln\left[e^{-(9.0\text{ s})/\tau}\right] = -\frac{9.0\text{ s}}{\tau} = \ln(0.85) = -0.16$$

and hence $\tau = (-9.0\text{ s})/(-0.16) = 55\text{ s}$.

Knowing τ, we want to find the time for which $L(t)/L(0) = e^{-t/\tau} = 0.25$. We insert $\tau = 55\text{ s}$ and take the natural log of both sides:

$$\ln\left[e^{-t/(55\text{ s})}\right] = -\frac{t}{55\text{ s}} = \ln(0.25) = -1.4,$$

or $t = (1.4) \times (55\text{ s}) = 76\text{ s}$.

What Do You Think? Can we tell anything about the period of oscillation from the data given?

13–8 Driven Harmonic Motion

Systems such as a mass on the end of a spring may be subject to external forces. We saw the effect of a constant force (gravity) on an oscillator in Section 13–5. Another common situation involves a repeating external force (we refer to a driving force), as occurs, for example, in any pendulum clock or in a child being pushed on a swing. The simplest example of such a force is a *harmonic driving force*, in which the driving force varies sinusoidally with time (Fig. 13–21); we can treat this case in a straightforward manner. The resulting motion of the mass is called **driven harmonic motion**. (Damped harmonic motion is undriven.)

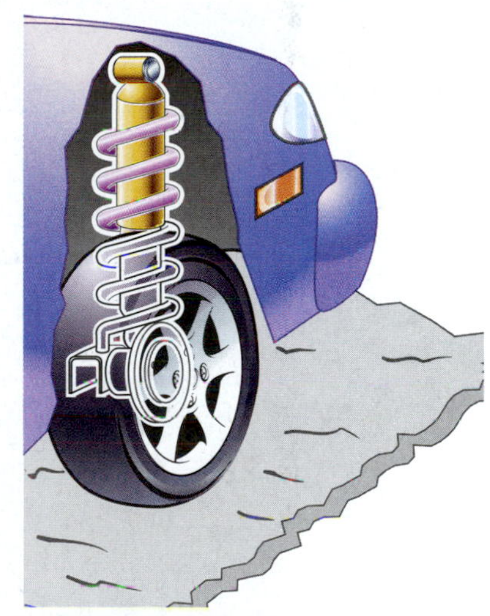

▲ **FIGURE 13–21** A car passing over a rough road is an example of driven, damped harmonic motion. The washboard road provides the force, the shock absorber provides the damping, and the spring provides the harmonic motion.

The most important consequences of driven harmonic motion involve **resonances**. In resonance phenomena, the amplitude of simple harmonic motion grows enormously when the frequency of the driving force matches the natural frequency of the oscillating system. You may have read about the possibility of a bridge being driven to large oscillations and collapsing if soldiers march over it in step with the right rhythm. Less spectacular examples are familiar: the trampoline jumper who, by timing her jumps, can make them much more effective; the swinging child who gets a regular push or who "pumps" in a regular fashion; or the coffee that sloshes out over the edge of the cup when you walk at just the wrong pace. Resonance phenomena are even more important in microscopic situations, as in the action of a microwave oven, where microwave radiation drives the electrons of water molecules with a natural frequency of the molecular system. Resonances also occur in acoustical phenomena, and many musical instruments are driven in harmonic resonance in order to produce their notes. Finally, electric circuits often behave like driven harmonic systems even though they are not mechanical.

It is easy to demonstrate the resonance phenomenon for yourself on a swing or with the simple aid of a loose rubber band tied to a mass such as a kitchen utensil. By extending the suspended band and releasing it, you can get a good idea of the natural frequency of the system and observe the amplitude steadily reduced by the damping. If you then tie the band to one hand, you can drive the system by moving your hand up and down in an approximation of harmonic motion—first with a frequency less than, then greater than, and finally equal to the natural frequency. At resonance, when the hand frequency equals the natural frequency of the system, the oscillation amplitude is dramatically large.

Equations of Motion for Driven Harmonic Motion

To see how resonance arises, consider a mass subject to a spring force, a drag force proportional to the speed, and an external harmonic driving force F whose time variation is determined by the angular frequency ω, where $F(t) = F_0 \sin(\omega t)$. F_0 is a type of amplitude for the driving force. It is useful to recall the natural frequency of the unencumbered mass on the spring, $\omega_0 = \sqrt{k/m}$. All motion is in the x-direction, so the x-component of Newton's second law, $\vec{F} = m\vec{a}$, is

$$-kx - b\frac{dx}{dt} + F_0 \sin(\omega t) = m\frac{d^2x}{dt^2}. \qquad (13–52)$$

Equation (13–52) appears to be more complicated than the equation of motion with a drag force alone, but if we consider the equation further, we can see that simplification is possible. By definition, the driving force has gone on and will go on forever—we mean, of course, for a long time—and we would expect that eventually the mass would have to move as the driving force dictates—that *the motion of the mass is simple harmonic motion with the frequency of the driving force.* Any transient effects due to friction will have long since died out exponentially. Imagine, for example, a spring with a mass on its end suspended from a harmonically moving hand. In time the mass will move with the frequency of your hand motion, even if the motion of the mass and your hand are not in phase. Therefore we expect that after long times the solution to the equation of motion is

$$x = A \sin(\omega t + \delta). \qquad (13–53)$$

This solution can be verified by substitution into Eq. (13–52), and the amplitude A and phase δ are *determined* in this substitution. This is a complicated exercise in algebra, which we forgo. In particular, the physically interesting amplitude of oscillation is

$$A = \frac{F_0}{\sqrt{m^2(\omega^2 - \omega_0^2)^2 + b^2\omega^2}}. \qquad (13–54)$$

AMPLITUDE OF HARMONICALLY DRIVEN SYSTEM

This amplitude indeed displays the remarkable property of *resonance*: *It is peaked when the driving frequency*, ω, *nears the natural frequency*, ω_0. In fact, if there were no damping ($b = 0$), the amplitude would become infinite when $\omega = \omega_0$. This is not a realistic physical situation, because it corresponds to the spring being stretched to infinite length. A real spring will snap rather than accept an infinite stretch; in other words, some form of damping will ultimately occur. But it does illustrate that, at resonance, the response of a harmonic system to a driving force can be catastrophically large.

THINK ABOUT THIS . . .

HOW CAN WE REDUCE THE WIND-DRIVEN SWAY OF SKYSCRAPERS?

The engineering of modern skyscrapers uses driven harmonic motion to *reduce* a natural harmonic movement. This is a kind of "anti-resonance" phenomenon. Think about the child on a swing getting a regular push. You know that not only should the driving force have the same frequency as the natural frequency but that it should be applied with a phase that gives a kick when the mass is ready to take the kick as a positive help to its motion. This means that the phase of the driving force should match that of the motion. But if we want to inhibit the motion of an object, we can instead give a regular kick when the kick tends to hinder the natural motion. For example, if you push the child backward just when he or she is coming through at the bottom of the swing, the motion will be inhibited. An application of this idea occurs in skyscrapers, which without a "tuned mass damper" can sway alarmingly in the wind. The damper is a large mass placed within the building that can be moved back and forth with a phase that tends to damp the movement of the building. The first such damper in the United States was placed within the 280-m-tall Citicorp building in New York. A 400-metric ton mass of concrete is slid back and forth within the building, ensuring a relative stability. Today many tall buildings use dynamical damping of this type.

Properties of Resonance

If we plot amplitude as a function of ω for several values of b (Fig. 13–22), we can clearly see the effect of resonance near the natural frequency ω_0. We choose $F_0 = 0.01$ N, $\omega_0 = 1$ rad/s, and $m = 1$ kg. The damping coefficient b is given the values 0.01, 0.05,

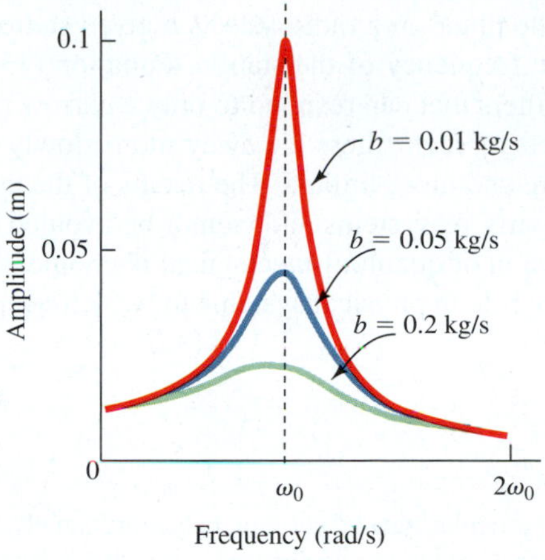

FIGURE 13–22 The amplitude of a harmonic system under the influence of a harmonic driving force with frequency ω. The natural frequency of the harmonic system is $\omega_0 = 1$ rad/s. There is a drag force with damping coefficient b. Resonance occurs at the peak (near $\omega = \omega_0$) and is strongest when b is smallest.

and 0.2 kg/s. From the plot, we can see that the position of the peak amplitude shifts slightly from $\omega_0 = 1$ rad/s when b increases from a small value. From Eq. (13–54) the peak amplitude generally occurs at

$$\omega_{max}^2 = \omega_0^2 - \frac{1}{2}\left(\frac{b^2}{m^2}\right). \tag{13–55}$$

That the peak is less sharp for larger b is sometimes expressed as the fact that the *width* of the peak is a measure of b. More precisely, it is possible to compute the *total width at half-maximum* $\Delta\omega$ of the peak, which we find by evaluating the amplitude where it is one half its peak value and measuring the spread of frequencies to which this corresponds, $\Delta\omega$ (Fig. 13–23). For small b, this width can be shown to be

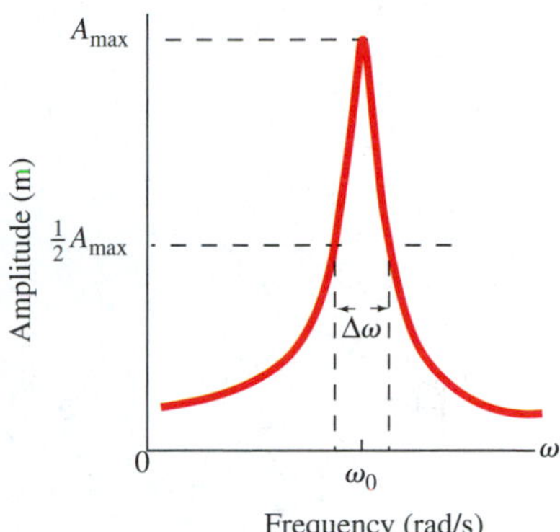

FIGURE 13–23 The width at half-maximum of a resonant peak near ω_0 is $\Delta\omega$, and is a measure of the damping coefficient of the system.

$$\Delta\omega \cong \frac{2b}{m}. \tag{13–56}$$

Thus the sharpness $\Delta\omega$ of the resonance peak is a direct measure of the damping coefficient, b, divided by m: The smaller the damping, the sharper the resonance peak.

*Resonance and Uncertainty

We have seen that the amplitude for a driven, damped harmonic oscillator has a resonance peak. We have also seen that the undriven damped harmonic oscillator has a characteristic decay time. These phenomena are closely related, and this relation has important consequences for our ability to construct systems with resonances, such as radio tuners or filters for screening electronic noise.

Equation (13–46) can be used to find the rate at which the energy is dissipated in an undriven damped oscillator. The energy is proportional to the amplitude *squared* and therefore falls as $e^{-t/\tau}$, where $\tau = m/b$ is the lifetime of the motion. The width in frequency of the driven harmonic oscillator is given in Eq. (13–56), and we see that it is inversely proportional to τ. From Eqs. (13–50) and (13–56) we have

$$\tau\,\Delta\omega \text{ is on the order of 1.} \tag{13–57}$$

This equation is known as the **uncertainty principle**: It expresses the impossibility of measurements that are arbitrarily precise in both time and frequency. Strictly speaking, we have derived it only for a special damping force. But it actually represents a very general property. It states that if the damping time of a harmonic oscillator is large, then the resonance width is small, and vice versa. The weaker the damping of a harmonic oscillator, the more sharply it responds to, or selects, a harmonic driving force of the right frequency. What is the significance of this result? One of the important uses of the resonance phenomenon, both in mechanical systems and in electric circuits, is that it allows us to select, or filter, certain frequencies by letting these frequencies act as a driving force on our selector. For example, the tuner on a radio selects a given station by letting in only frequencies near the central frequency of the station. Equation (13–57) sets strong limits on our ability to design filters that can respond to only a narrow range of driving frequencies. The transient effects of such filters die away more slowly with time as the frequencies they select are more and more limited. The results of the uncertainty principle reflect a fundamental property of systems and cannot be avoided with clever design. We'll see later in our discussion of quantum-mechanical phenomena that the uncertainty principle plays an important role in physical systems for which quantum mechanics is important.

Summary

Simple harmonic motion is characterized by a particular type of periodic behavior; namely, sinusoidal time dependence of the position of a moving object. In one dimension this dependence can be written as

$$x(t) = A\sin(\omega t + \delta). \tag{13–1a}$$

The amplitude, A, measures the maximum displacement from equilibrium. The phase, δ, is the angle by which the motion is shifted from $x = 0$ at $t = 0$; the angular frequency, ω, measures how rapidly the motion repeats. Alternatively, the motion can be written as

$$x(t) = a_1 \sin(\omega t) + a_2 \cos(\omega t). \tag{13–1b}$$

The period, T, and frequency, f, of the simple harmonic motion are related to ω by

$$T = \frac{2\pi}{\omega} \tag{13–2}$$

and

$$f = \frac{1}{T}. \tag{13–3}$$

In simple harmonic motion the velocity and the acceleration are also sinusoidal. In particular, the acceleration is proportional to the displacement but opposite in sign:

$$a = -\omega^2 x. \tag{13–8}$$

Uniform circular motion is closely related to simple harmonic motion. It is just the result of simultaneous simple harmonic motion of equal amplitudes and frequencies in the x- and y-directions, 90° out of phase.

Because acceleration is proportional to displacement in simple harmonic motion, the force that leads to this motion is a restoring force proportional to the displacement. The prototype of this is the force exerted on a mass by a spring, Hooke's law:

$$F = -kx, \tag{13–13}$$

where k is the spring constant and x is the displacement of the mass from the equilibrium position of the spring. The resulting angular frequency of the motion of a mass m subject to this force is

$$\omega = \sqrt{\frac{k}{m}}. \tag{13–16}$$

The period of the motion is independent of the amplitude.

The potential energy of a mass subject to a spring force is

$$U(x) = \frac{1}{2}kx^2. \tag{13–18}$$

It oscillates with time as $\sin^2(\omega t + \delta)$, whereas the kinetic energy oscillates with time as $\cos^2(\omega t + \delta)$, and the total energy, E, which is their sum, remains constant with time:

$$E = \frac{1}{2}kA^2. \tag{13–23}$$

The simple pendulum, consisting of a mass on the end of a light string of length ℓ, also undergoes simple harmonic motion when it is not allowed to swing too far. In this case, it is the angle with respect to the vertical that varies sinusoidally, and the period of the motion is independent of the mass as well as the amplitude:

$$T = 2\pi\sqrt{\frac{\ell}{g}}. \quad (13\text{–}36)$$

Almost all small departures from stable equilibrium situations exhibit simple harmonic motion. Two examples are a bead rolling back and forth across the bottom of a bowl and a physical pendulum. The period of a physical pendulum depends on the geometric factor contained in its rotational inertia about the axis through which it swings, but does not depend on the object's mass.

The presence of additional velocity-dependent drag, or resistive forces, causes the amplitude of a particle that moves under the influence of springlike forces to decrease, or damp. When a system that by itself would move in simple harmonic motion is driven by a force with sinusoidal time dependence, the system moves with the frequency of the driving force. The amplitude of the resulting motion of the system shows resonant behavior when the frequency of the driving force equals the natural frequency of the system. The width of the resonance peak is inversely related to the exponential rate of falloff of the undriven system due to damping—a result known as the uncertainty principle.

Understanding the Concepts

1. Not all periodic motion is harmonic. Sketch the pattern of motion of a ball that falls on a flat surface and bounces back, in the idealized case of a perfectly elastic bounce.
2. Describe an experiment to determine the spring constant of a spring.
3. If the amplitude of simple harmonic motion is doubled, what happens to the maximum kinetic energy?
4. Will a pendulum clock lose time or gain time when it is taken from sea level to the top of a mountain? Ignore all damping effects.
5. In the previous question we suggested that all damping effects be ignored. Suppose that you could not ignore damping effects due to the air, which becomes less dense as you gain in altitude. Why might you find it difficult to predict whether a pendulum clock gains or loses time when it is taken from sea level to the top of a mountain?
6. In Section 13–3 we showed that when a constant force is added to a spring force and the two forces are aligned in the same direction, the motion remains periodic and with the same period; only the equilibrium point shifts. Is the same true when the constant force is aligned with a different direction than that of the spring force?
7. A damped harmonic oscillator driven by a harmonic external force maintains a steady oscillatory motion. Is energy lost to friction in the motion? If so, what keeps the oscillator moving?
8. The length of a simple pendulum is doubled, and the mass at the end is halved. What happens to the period?
9. Grandfather clocks with long pendulums have a screw device at the end to adjust the period and keep the clock in time. Sometimes it is necessary to make a winter-summer adjustment to account for changes in pendulum length associated with temperature changes. If the clock is running slow, how should the length of the pendulum be changed?
10. What is the phenomenon that allows you to increase the amplitude of your motion when you swing on a swing?
11. Suppose you stand on a swing instead of sitting on it. Will your frequency of oscillation increase or decrease?
12. It can be shown that if the frequency, the energy, or the amplitude of a harmonic oscillator changes very slowly then, although the energy E is no longer constant, the ratio E/ω, where ω is the angular frequency, does not change. Suppose you have a pendulum whose maximum angle of deflection is θ_0. If the bob at the end of the pendulum consists of fine sand in a spherical container, and the sand leaks very slowly out the bottom, what will happen to the motion?
13. Make use of the assertion in Question 12 to discuss what happens to the period and amplitude of a pendulum in which the length is very slowly decreased by a factor of 1.5.
14. We say the period of a physical pendulum doesn't depend on the pendulum's mass. Does it depend on the distribution of the mass? In what way?
15. Discuss if and why each of the following is an example of simple harmonic motion, damped harmonic motion, and/or driven harmonic motion: (a) leaves blowing on a tree limb; (b) children seesawing; (c) a child swinging on a playground swing; (d) a car bouncing up and down after hitting a large pothole in the road; and (e) water sloshing back and forth in a tub.
16. Consider the pendulum of Example 13–9. Suppose you could adjust the place along the rod from which the rod could be hung, and that you hung it from a spot very near to the center of mass of the rod than in the example. Would the period be shorter or longer?
17. A mass is suspended from two ropes of equal length, attached to different points on a ceiling. Would small oscillations about the point of stable equilibrium represent simple harmonic motion?
18. A simple pendulum acts as a *conical pendulum* when the mass moves in a horizontal circle (Fig. 13–24). What force keeps it moving in a circle? Does the total energy consist of potential and kinetic energies that each vary in this case?

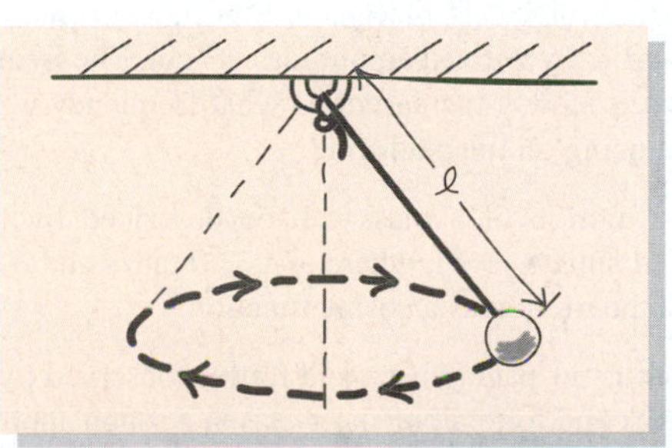

▲ **FIGURE 13–24** Question 18.

19. Consider a pendulum with *large*-angle oscillations. Will the motion still be periodic? Will the period be longer or shorter than for a small-angle oscillation?
20. Suppose that there were a harmonic driving force on a mass attached to a horizontal spring and that very little damping occurs. What could supply the energy that would allow the amplitude to become enormously large near resonance?
21. What are you doing when you adjust your bounce up and down on a diving board to get a big boost for your dive?
22. Suppose you studied a pendulum in a place where a steady horizontal wind blew in the direction parallel to the swing. How would the period of the pendulum differ from its value in a protected place?
23. A large sled is sliding down a snowy hill. Resting on the sled are a mass and a spring; one end of the spring is attached firmly to the sled, and the other end is attached to the mass. There is little friction between the mass and the sled, and the mass is set into oscillation. Will the sled accelerate smoothly down the hill?

Problems

13–1 The Kinematics of Simple Harmonic Motion

1. (I) What phase, δ, is necessary if $\sin(\theta + \delta) = \cos(\theta)$? This can be shown graphically, with Fig. 13–1, or by using a trigonometric identity for the sine of the sum of two angles.

2. (I) A simple harmonic motion along the x-direction has the following properties: The maximum amplitude $= 0.04$ m, the time between the maximum and minimum x-values $= 0.30$ s, and $x = 0.01$ m for $t = 0.10$ s. Find the period, the angular frequency, and the general equation of motion.

3. (I) A spring has the speed $v = 0.4\sin(\omega t + \pi)$ m/s, where $\omega = 2.00$ rad/s. Plot x, v, and a as functions of time for three periods of motion.

4. (I) The angular frequency of a mass on the end of a spring in simple harmonic motion is 3.827 rad/s. What is the period of the motion?

5. (I) The amplitude of the motion of a mass attached to a spring is $A = 2.84$ m, while the maximum speed of the mass is 4.36 m/s. What is the period of the motion?

6. (I) The maximum speed of a mass attached to a spring is $v_{max} = 0.371$ m/s, while the maximum acceleration is 1.05 m/s^2. What is the maximum displacement of the mass?

7. (I) What is the position of the mass in Problem 6 at time $t = 1$ s, given that the mass is precisely at the origin ($x = 0$ m) at $t = 0$ s?

8. (II) The expression for the position of a simple harmonic oscillator is given by $x(t) = B\cos(\omega t) + C\sin(\omega t)$. (a) Show that this can be written in the form given in Eq. (13–1a), and express the constants B and C in terms of the constants A and δ. (b) Express B and C in terms of the position x_0 and the speed v_0 at time $t = t_0$. (c) What is the maximum speed in terms of B and C?

9. (II) A harmonic oscillator operates at a frequency of 813.52 Hz. What is the amplitude for which the maximum acceleration is 183.25 m/s^2?

10. (II) A small object is placed on a horizontal platform that vibrates vertically with an amplitude of 3 cm. The frequency of the vibration is slowly increased. At what frequency will the object start bouncing on the platform?

11. (II) The motion of a mass can be described by the function $x(t) = A\sin(\omega t + \delta)$, where $\omega = 2.0$ rad/s and $\delta = 0.40$ rad. Express the motion as a cosine function.

12. (II) A professor pacing back and forth is observed by the class to move back and forth along the x-axis in a rough approximation to simple harmonic motion. Relative to the center of the classroom, the motion is between the two extremes $+3$ m and -3 m. The professor was at $x = -0.3$ m at $t = 0$ s, and the motion is observed to repeat itself six times in 90 s. What are the amplitude, phase, period, frequency, and angular frequency of the motion?

13. (II) A spring has spring constant 0.50 N/m and a 0.20 kg mass on its end, which has a maximum speed of 2.0 m/s. (a) What are the angular frequency and period of the system? (b) What is the amplitude of the motion?

14. (II) A sailing ship rolls sideways in simple harmonic motion, with the period given by $T = 5.0$ s (Fig. 13–25). The tip of a 25-m mast travels a maximum of 2.5 m from the vertical position. What is the speed of the tip of the mast at the instant it is in a vertical position?

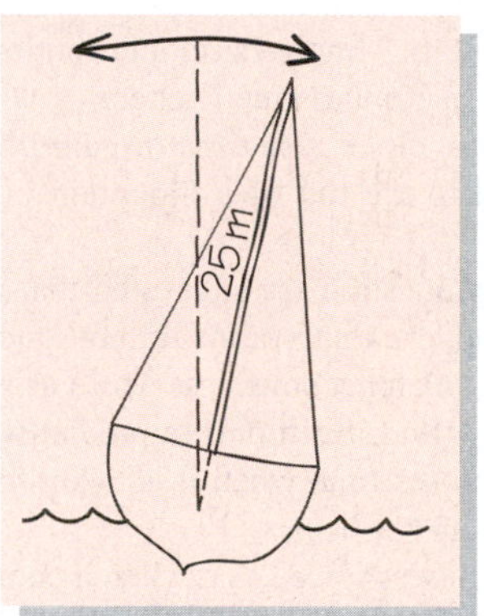

▲ **FIGURE 13–25** Problem 14.

15. (II) A particle undergoing simple harmonic motion travels a total distance of 6.98 cm during one cycle of 1.71 s. (a) What is the average speed of the particle? (b) What are its maximum speed and acceleration?

16. (II) A mass $m = 0.35$ kg is attached to an ideal spring of spring constant $k = 12$ N/m. All motion takes place in the (horizontal) x-direction, and the equilibrium position of the mass is defined to be $x = 0$ m. The mass is then displaced to $x = 0.040$ m and released from rest. (a) Write an expression for the function $x(t)$ that describes the subsequent motion. (b) At what position x does the maximum positive acceleration of the mass occur? (c) What is the magnitude of the maximum acceleration? (d) What is the maximum speed of the mass?

17. (II) An object oscillates with an angular frequency of 3.0 rad/s. Its initial displacement from equilibrium is $+3.0$ cm, and the initial velocity is 5.0 cm/s in the direction of the equilibrium point. (a) Find the displacement as a function of time. (b) How soon will the displacement be $+3.0$ cm again? (c) At what times will the object move with a speed of 5.0 cm/s? (d) Sketch x and v as functions of t.

13–2 A Connection to Circular Motion

18. (I) Find the phase angle δ for uniform circular motion when $x = -R$ and $y = 0$ at $t = 0$.

19. (I) A particle moves in the xy-plane so that its x and y coordinates are described by Eq. (13–10) and Eq. (13–11); namely,

$$x = R\cos(\omega t + \delta), \quad y = R\sin(\omega t + \delta).$$

Show that the distance of the particle from the origin is a constant, and find that constant.

20. (I) When a certain uniform circular motion in the xy-plane is projected onto the x- and y-axes, this projection gives $x(t) = R\sin[\omega t + \delta - (\pi/2)]$ and $y(t) = R\sin(\omega t + \delta)$. (a) Show that the uniform circular motion is clockwise. (b) What would happen if instead of an extra phase $-\pi/2$ in $x(t)$, there was an extra phase $+\pi/2$ in $y(t)$?

21. (I) Write down a formula that describes the motion of Earth around the Sun as seen by a distant observer in the plane of Earth's orbit. Assume a circular orbit.

22. (II) A small object is placed at the outer edge of a turntable of diameter 8 in. The turntable rotates clockwise at 45 rev/min. (a) What is the projection of the object's motion on the x-axis, assuming that at $t = 0$ the projection of the motion places the object at $x = 0$? (The point $x = 0$ corresponds to the center of the record.) (b) Give the amplitude, angular frequency, and largest speed of the projection of the motion. (c) What is the acceleration of the projection of the motion on the x-axis? (The answer to this last part is another way to see why the acceleration in uniform circular motion is centripetal.)

23. (II) Show that if the motion along the x- and y-axes of a mass moving in the xy-plane is $x(t) = R_1\cos(\omega t + \delta)$ and $y(t) = R_2\sin(\omega t + \delta)$, then this motion traces out an ellipse of axis lengths R_1 and R_2. (An ellipse is the curve described by $[x^2/a^2] + [y^2/b^2] = 1$, where a and b are the axis lengths.)

24. (II) An object on a frictionless table is attached to a peg by a spring. How should you start it off, if you want it to move in a circle? What should you do to make it move in a straight line? (Pretend that the object can go right through the peg anchoring one end of the spring.)

13–3 Springs and Simple Harmonic Motion

25. (I) A mass is attached to a vertical spring. When it is pulled down 6.0 cm and released, it starts upward with an acceleration of 40 cm/s^2. What is the period of the motion? With what speed does it pass the equilibrium point?

26. (I) A 4.0 kg ball is suspended motionless under the influence of gravity from a spring with a force constant of $k = 300$ N/m. (a) How much is the spring stretched? (b) The same spring is sent to the Moon and suspended in the same way. By how much is the spring stretched in this case?

27. (I) A student has a spring with $k = 200$ N/m and wants to build a horizontal mass-spring system with period 1.0 s. What mass should the student use at the end of the spring?

28. (I) The spring on a scale is compressed by 2.45 cm when a 60-kg man stands on it. What is the spring constant of the spring?

29. (I) The period of a 45-g mass attached to the end of a spring is 3.1 s. Find the spring constant of the spring.

30. (I) A mass moves along the x-axis under the influence of a spring whose equilibrium position is at the origin. The mass moves between limits of -3.5 cm and $+3.5$ cm, and the period of its motion is 1.3 s. (a) What is the period if the mass is doubled? Halved? (b) What is the period if the amplitude is increased to 4.0 cm? Decreased to 1.0 cm?

31. (II) The plate at the base of a floor sander moves back and forth in simple harmonic motion at a frequency of 20 oscillations/s, and the amplitude of the motion is 0.80 cm. If the mass of the oscillating plate is 1200 g, what is the maximum value of the driving force?

32. (II) A spring is placed in a vertical position by suspending it from a hook at its top. A similar hook on the bottom of the spring is measured to be 45 cm above a tabletop. A mass of 180 g, of negligible size, is then suspended from the bottom hook, which is now found to be 20 cm above the tabletop. The mass is then pulled down a distance of 15 cm and released (Fig. 13–26). Find (a) the spring constant, k; (b) the angular frequency, ω; (c) the position of the bottom hook after 5 s.

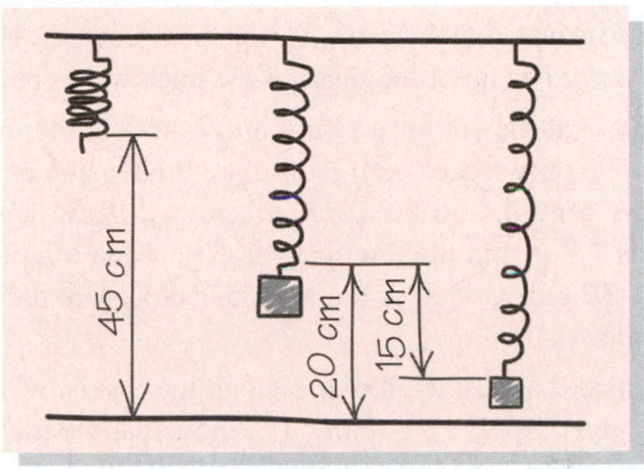

▲ **FIGURE 13–26** Problem 32.

33. (II) A spring is suspended from the ceiling. A mass of 25 g is attached to it, and the spring stretches by 12 cm. Ignore damping. What is the period of the oscillation of a 75-g mass attached to the spring?

34. (II) Two identical springs of spring constant k are attached end to end to make one longer spring. Show that this new spring has spring constant $k/2$. The springs are said to be attached *in series*. The case of n springs attached in series gives a spring n times as long, with a spring constant $k_n = k/n$.

35. (II) A small object of mass $m = 0.060$ kg is held in place by two springs (Fig. 13–27). The one acting on the left has spring constant $k_1 = 100$ N/m; the one acting on the right has spring constant $k_2 = 200$ N/m. The object is moved away from its equilibrium position by 1.0 cm to the right and released at time $t = 0$. What is the displacement of the object as a function of time?

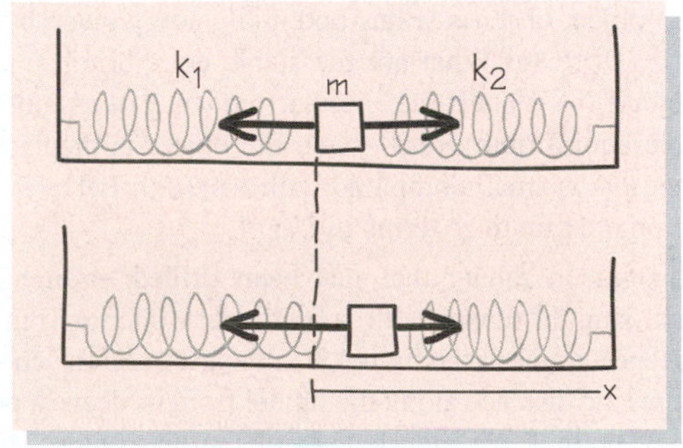

▲ **FIGURE 13–27** Problem 35.

13–4 Energy and Simple Harmonic Motion

36. (I) What is the energy of a 1.2-kg mass that moves with amplitude 14 cm on a flat, frictionless table and is attached to a spring of spring constant 375 N/m?

37. (I) A trampoline acts like a spring of spring constant 1200 N/m. A mass is placed on the trampoline. The trampoline surface is depressed by a total of 30 cm. What is the period of oscillation of the system if this mass is pushed beyond the equilibrium displacement?

38. (I) A mass of 1.2 kg, attached to a spring, is in simple harmonic motion along the x-axis, and its period is $T = 2.5$ s. If the total energy of the spring and mass is 2.7 J, what is the amplitude of the oscillation?

39. (I) A spring with a 1 kg fish at its end and with spring constant $k = 2$ N/m is compressed 3 cm from equilibrium and then released. Use the conservation of energy to find the maximum speed of the fish.

40. (I) A puck of mass 350 g moves horizontally with speed $v = 0.88$ m/s on a frictionless surface toward the end of a relaxed spring for which $k = 140$ N/m, and compresses the spring. By how much is the spring compressed? How would your answer change if the mass of the puck were doubled?

41. (I) A mass at the end of a string moves in a circle in a vertical plane. The only forces acting on it are the tension of the string, which is central, and the force of gravity. If the length of the string is 1.2 m and the angular velocity at the top of the circle is $\omega = 2.2$ rad/s, what is the angular velocity at the bottom of the circle?

42. (II) Consider a mass m, moving along the x-axis, with potential energy given by $U(x) = \frac{1}{2}m\omega^2x^2$. Show that the motion of this mass is simple harmonic motion with angular frequency ω by using $dE/dt = 0$, where E is the (constant) total energy of the object.

43. (II) A point mass m on a turntable that rotates with angular frequency ω is located a distance d from the center of the turntable. Show that the energy of the point mass is the sum of the energies of the harmonic motions in the x- and y-directions.

44. (II) A mass m of 0.50 kg is attached to the end of a spring and released from rest at $t = 0$ s from an extended position $x_{max} = 13$ cm. All other forces acting on the spring cancel, so that the motion of the mass is due entirely to the effect of the spring. After its release, the mass's speed drops back to zero for the first time after 0.45 s. Find the maximum speed of the mass.

45. (II) A mass m attached to the end of a spring is released from rest at $t = 0$ s from an extended position x_{max}. The mass $m = 0.2$ kg, and $k = 1$ N/m. After 0.5 s, the speed of the mass is measured to be 1.5 m/s. Calculate x_{max}, the maximum speed of the motion, and the total energy.

46. (II) An object of mass m has potential energy given by $U(x) = U_0(x^2 - a^2)^2$. (a) What are the stable equilibrium positions of the object? (b) What is the angular frequency of its motion about a stable equilibrium point, when the object is displaced by an amount that is small compared with a? [*Hint*: Let $x = x_{eq} + z$, and keep only up to z^2 terms in $U(z)$.]

47. (II) Imagine a tunnel that has been drilled through Earth: a smooth, straight tunnel with a frictionless interior (Fig. 13–28). The deepest point of the tunnel is at depth d, and the coordinate x measures the distance along the tunnel from its deepest point to an arbitrary point P a distance ℓ from Earth's center. The known parameters are the mass and radius of Earth, M_E and R_E, respectively, as well as G. Earth is assumed to have uniform density. (a) What is the total mass of that portion of Earth that lies within the distance ℓ in terms of ℓ, M_E, R_E, and G? (b) What is the gravitational force, in direction and magnitude, acting on a ball of mass m at point P? (c) What is the *total* force, in direction and magnitude, acting on the ball *as a function of* x and of the constants of the problem? Why is there no net force acting perpendicular to the tunnel? (d) What is the period of the motion if the ball is released at rest at an entrance to the tunnel? Ignore air resistance. (e) What is the period of a satellite in circular orbit around Earth at a radius equal to Earth's radius?

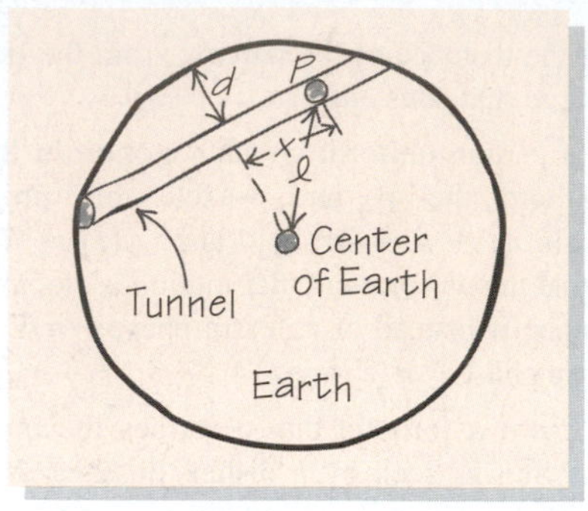

▲ **FIGURE 13–28** Problem 47.

48. (III) A bead of mass $m = 40$ g slides without friction at the bottom of a bowl whose bottom traces out the arc of a circle of radius $R = 15$ cm. If the bead moves in a plane that passes through the bottom of the bowl, its position can be described by an angle θ measuring its angular displacement from the bottom. (a) Express the potential energy as a function of θ for very small angles θ. [*Hint*: The potential energy is a linear function of height. Reexpress the energy in terms of θ, and use an approximation suitable for small θ.]

At $t = 0$ s the mass is released from rest from a very small angle $\theta_0 = 0.1$ rad. Parts (b)–(e) refer to this initial condition. (b) Give an expression $\theta(t)$ describing the subsequent motion. In particular, what is the frequency of small oscillations at the bottom of the bowl? (c) What is the total energy of the system? (d) the velocity of the bead at $t = 0.1$ s? (e) the acceleration of the bead at $t = 0.2$ s?

49. (III) A small metal block of mass m is placed on a smooth horizontal table and constrained to move along a frictionless, rectilinear groove. The block is attached to one end of a spring (of spring constant k) whose other end is fastened to a pin P (Fig. 13–29). Let length ℓ be the equilibrium length of the spring and the perpendicular distance to the groove. The spring is now pulled a distance x_0 from the equilibrium position and released. Show that, if the displacement along the groove $x \ll \ell$, the restoring force on the block is proportional to x^3, so the motion is not simple harmonic (although it will still be periodic). [*Hint*: For $x \ll \ell$, $\sqrt{x^2 + \ell^2} \simeq \ell + (x^2/2\ell)$.]

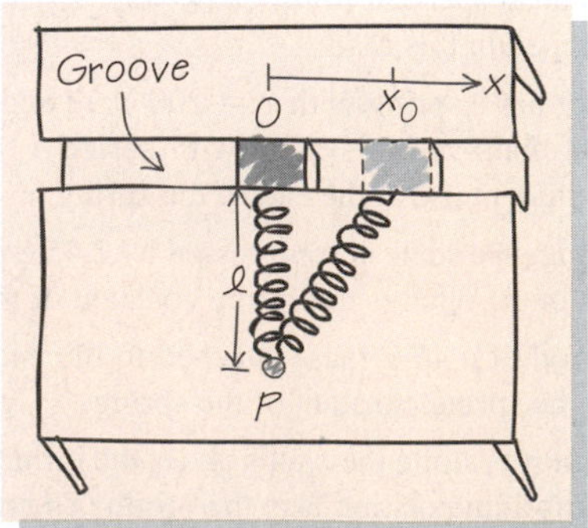

▲ **FIGURE 13–29** Problem 49.

13–5 The Simple Pendulum

50. (I) A simple pendulum 1.20 m long is suspended in a location where g is 9.82 m/s^2. What is the period of the pendulum?

51. (I) A simple pendulum has a frequency of 0.342 Hz. The length of the pendulum string is 2.12 m. What is the local value of g?

52. (I) A thin wire 3.88 m long is attached to the ceiling of a lecture room. A lead ball is attached at the bottom. The wire is displaced by an angle of 0.055 rad and released. Express the angular displacement of the lead ball as a function of time, given that $g = 9.80$ m/s^2.

53. (II) You need to measure the height of a room. You have a watch but no meter stick. You also have your human brain, not to mention an opposable thumb. A long pendulum with a point mass at the end extends from the ceiling to the floor and has a period $T = 3.0000$ s. (a) What is the height of the room? (b) You take the same pendulum to the top floor of a skyscraper and measure a period $T = 3.0002$ s. What is the height of the skyscraper?

54. (II) The difference in temperature between summer and winter causes the length of the pendulum in a clock to change by one part in 30,000. What time-difference error will this make in one week?

55. (II) A small lead ball of mass 2 kg is suspended at the end of a light string 1 m in length. A small peg, 0.5 m below the suspension point, catches the string in its swing (Fig. 13–30). The ball is set swinging through small angles. (a) What is the period of this pendulum? (b) The ball is started swinging on the side that does not catch the peg, at an initial height 0.05 m above the low point. How high does it rise on the side where the peg restricts the pendulum length to 0.5 m?

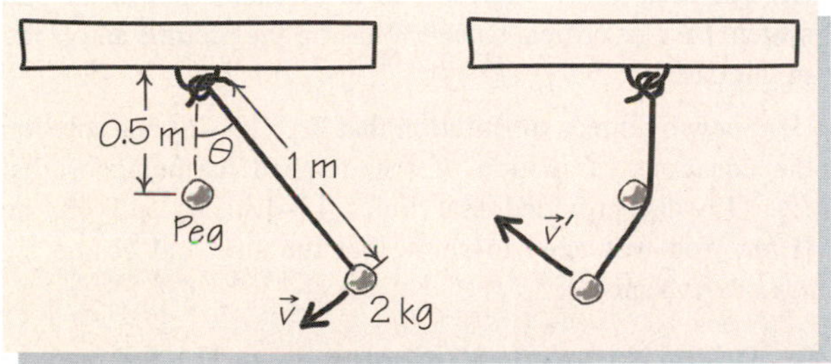

▲ **FIGURE 13–30** Problem 55.

56. (II) A mass m moves as a simple pendulum, the massless string holding it making an angle θ with the vertical $\theta(t) = \theta_0 \sin(\omega t)$. It receives a tangential momentum impulse of magnitude J, to the left ($-\theta$-direction) when the pendulum is at the angle $\theta = \theta_0$ (i.e., at $t = T/4$). Find the new amplitude and phase of the motion.

57. (II) A simple pendulum of length L with a bob of mass m is undergoing small oscillations of maximum angle α. Express the total energy of the pendulum in terms of these parameters. Suppose the maximum angle were not small. Find in that case a corresponding expression for the total energy.

13–6 More About Pendulums

58. (I) A small door to allow a dog to pass in and out of the house has a mass of 0.45 kg and is 28 cm wide and 40 cm tall. The door is hinged along the top of the 28-cm width. What is the frequency of oscillation of the door?

59. (I) A student wants to build a pendulum out of a circle of plywood as shown in Fig. 13–31. The circle has a radius of $R = 10$ cm and the plywood has a mass of 200 g. What is the period of the motion?

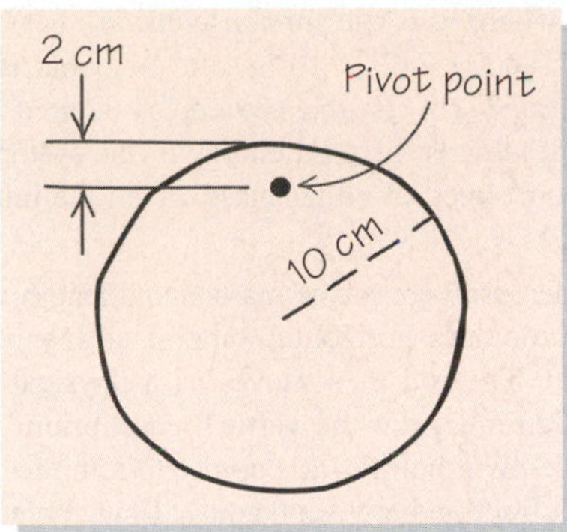

▲ **FIGURE 13–31** Problem 59.

60. (II) A uniformly dense book of dimensions 6 cm × 15 cm × 21 cm has a mass of 1.8 kg. You pick it up close to one corner with a pair of calipers and let it swing. What is the angular frequency? You can do this experiment with your own textbook.

61. (II) A thin, uniform rod of mass M and length L oscillates with small amplitude in a vertical plane about a pin that pierces the rod a distance y from the upper end. Calculate the period of the motion as a function of y. [*Hint*: You need to use the parallel-axis theorem to calculate the rotational inertia to be used in Eq. (13–43b).]

62. (II) A straight wire 80 cm long is bent in the middle into an L-shape and balanced with the two ends down on a knife edge. With what frequency will it oscillate about its equilibrium position?

63. (II) A thin uniform disk of mass M and radius R hangs from a nail driven straight into the disk at a distance ℓ from the center. (a) What is the rotational inertia of the disk about the nail? (b) What is the equation of motion for small oscillations of this pendulum about the point where the nail is driven in? [*Hint*: Use Newton's law for the torque about the point in question.] (c) What is the period, T, of oscillations about the point of suspension? (d) What is T in the limit where ℓ goes to zero?

64. (II) A uniform circular disk of radius R and mass M is attached at its center to one end of a massless, rigid rod of length L. The other end of the rod is attached to the ceiling and pivots freely about that point. The system thus makes a physical pendulum. What is the period of small oscillations of the pendulum?

65. (II) A *torsion pendulum* consists of a dumbbell suspended from its center by a wire that resists being twisted (Fig. 13–32). The dumbbell has length ℓ and has masses m on each end. It remains in the horizontal plane as it twists back and forth about the equilibrium position, $\theta = 0$ rad. For small oscillations about this equilibrium position, the torque exerted by the wire has the form

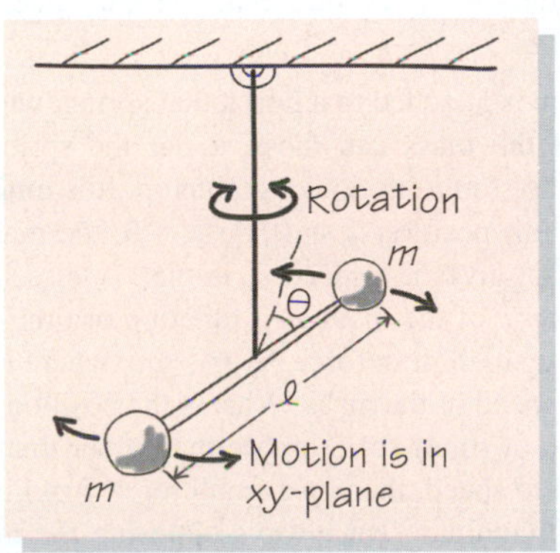

▲ **FIGURE 13–32** Problem 65.

$\tau = -\gamma\theta$, where γ is the *torsion constant*. (a) For $m = 80$ g, $\ell = 30$ cm, and $\gamma = 2 \times 10^5\ \text{g}\cdot\text{cm}^2/\text{s}^2$, what is the period of the oscillation? (b) If the system is started from rest at $\theta = 0.1$ rad, what is the total energy of the system? (c) What is the maximum speed of either mass, given the initial conditions of part (b)?

66. (II) A slender, uniform rod of mass m and length ℓ can be pivoted on a frictionless horizontal support at any point along the rod's length. The rod then moves as a physical pendulum for small oscillations about the vertical equilibrium position. Suppose that the pivot point is at a point $z\ell$ from the end of the rod, where z is a fraction between 0 and 1. Using an equation for the angular acceleration of the rod as a function of the angle θ, which measures the departure from the vertical, find the period of small-angle oscillations. Check the answer by specifying $z = 0.5$, where the period should become large.

13–7 Damped Harmonic Motion

67. (I) Consider the gong struck in Example 13–10. What is the loudness of the sound (as a fraction of the original loudness) after 2 min? How much time does it take for the sound to have a loudness 1/10,000 of the original loudness? Your ear is a very good detector of loudness, and you would have no trouble hearing the gong for this entire time.

68. (I) A mass on a spring with a natural angular frequency $\omega_0 = 3.6$ rad/s is placed in an environment in which there is a damping force proportional to the speed of the mass. If the amplitude is reduced to 0.35 times its initial value in 12.9 s, what is the angular frequency of the damped motion?

69. (I) The damping coefficient of a damped harmonic oscillator can be adjusted. Two measurements are made: First, when the damping coefficient is zero, the angular frequency of motion is 3880 rad/s. Second, a static measurement shows that the effective spring constant of the system is 184 N/m. To what value should the damping coefficient be set in order to have critical damping?

70. (II) Suppose we have a mass m attached to a spring of mass constant k. The mass slides on a rough surface with coefficient of sliding friction μ. The mass starts from rest with the spring extended in the positive direction to a distance A. Describe the subsequent motion. How large must μ be so that we get an "overdamped" situation? You can assume that static friction plays no role here.

71. (II) Using your knowledge of the approximate mass of an automobile as well as the frequency with which you feel it oscillate when you go over a bump, estimate the value of the effective spring constant of the suspension. Judging how long it takes for the oscillations to die out, estimate the value of the damping constant b.

72. (II) A mass m is attached to a horizontal spring, with spring constant k, and the mass can move under the spring's influence along the x-axis on a frictionless tabletop. It is initially at rest at the equilibrium position $x = 0$. At $t = 0$, the mass is struck a brief blow that gives it a speed v_0 in the $-x$-direction. (a) What is the position x of the mass as a function of time? (b) Suppose that there is a small drag force $f_D = -bv$, where b is a constant and v is the speed of the mass. What is the position x of the mass as a function of time? (c) Suppose that rather than a drag force proportional to speed, there is a small force f of kinetic friction, constant in magnitude but always opposing the motion, acting on the object. How, qualitatively, does the motion differ from that described in part (b)?

73. (II) A harmonic oscillator with natural period $T = 2.0$ s is placed in an environment where its motion is damped, with a damping force proportional to its speed. The amplitude of the oscillation drops to 70 percent of its original value in 5.0 s. What is the period of the oscillator in the new environment?

74. (II) Consider a damped harmonic oscillator. The damping, proportional to the speed, is sufficiently weak so that it is a good approximation to view the amplitude as constant over the duration of a cycle. What is the energy of the oscillator at time t if its original energy at time $t = 0$ is E_0? [*Hint*: For one cycle, the usual energy formula involving the square of the amplitude can be used.]

75. (II) A spring with $k = 12.0$ N/m and an attached bob oscillates in a viscous medium (Fig. 13–33). A given maximum, of +6.0 cm from the equilibrium position, is observed at $t = 1.5$ s, and the next maximum, of +5.6 cm, occurs at $t = 2.5$ s. What will the position of the bob be at 3.0 s and at 4.8 s? What was its position at $t = 0$ s?

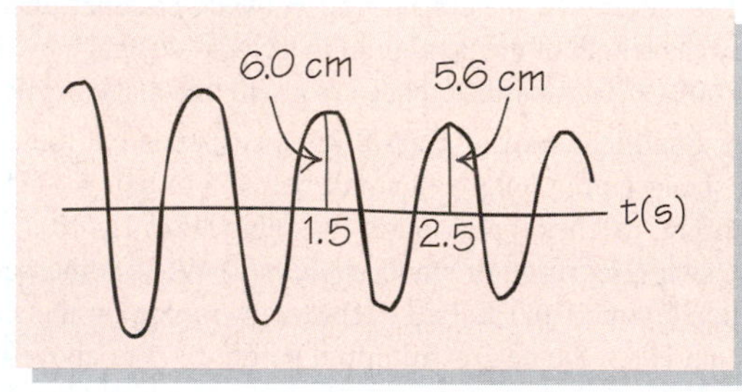

▲ **FIGURE 13–33** Problem 75.

76. (II) For what mass m of the bob in Problem 75 will the spring system have its critical value? What are the lifetime and Q factor in this case?

77. (II) Show by direct substitution that Eq. (13–46) is a solution to the equations of motion of the damped harmonic oscillator, Eq. (13–45), provided that Eqs. (13–47) and (13–48) hold. [*Hint*: You will need to show that the sine and cosine terms *separately* vanish.]

13–8 Driven Harmonic Motion

78. (I) A long, flexible, and very light strip of metal, such as a saw blade, is clamped to the edge of a table, extending horizontally over the edge like a miniature diving board. When a gob of putty is placed on the end, the strip sags and comes to a new equilibrium position with its tip a distance 0.6 cm below its original position. If the end of the strip is then lightly tapped, its tip will oscillate in simple harmonic motion. With what frequency should the tip be tapped to make the strip oscillate with maximum amplitude?

79. (I) A mass of 0.5 kg is suspended from a spring, which stretches by 8 cm. The support from which the spring is suspended is set into sinusoidal motion. At what frequency would you expect resonant behavior?

80. (II) Show that resonance occurs for driven harmonic motion when $\omega = \sqrt{\omega_0^2 - (b^2/2m^2)}$. The resonant frequency occurs when the amplitude has a maximum as a function of frequency.

81. (II) A particular spring has a spring constant of 86 N/m and a mass of 0.548 kg at its end. When the spring is driven in a viscous medium, the resonant motion occurs at an angular frequency of 12.2 rad/s. What is (a) the damping parameter due to the viscous medium? (b) The lifetime of the system? (c) The sharpness of the resonance peak?

82. (II) Consider the driven, damped harmonic oscillator discussed in Section 13–8. (a) Calculate the power dissipated by the damping force. (b) Calculate the average power loss, using the fact that the average of $\sin^2(\omega t + \delta)$ over a cycle is one-half. [*Hint*: Recall that $P = Fv$.]

General Problems

83. (II) When you bounce up and down on a diving board of negligible mass, you find yourself and the board bouncing with the maximum amplitude when your bounce frequency is once every 1.2 s. How much will the board deflect vertically when you stand on the end without bouncing? Assume that the damping that must be present is small.

84. (II) Consider simple harmonic motion of a mass on the end of a spring, $x = (0.35\text{ m})\sin(\omega t + \delta)$. At $t = 0$ s the position is -0.080 m, and the velocity is 2.1 m/s in the $-x$-direction. The total energy of the motion is 6 J. What is the value of the (a) phase, δ; (b) frequency, f; (c) acceleration at $t = 0$ s; (d) spring constant, k; (e) mass, m?

85. (II) A mass m on the end of a spring oscillates with angular frequency ω. The mass is removed, the spring is cut in two, and the mass is reattached. What is the new angular frequency?

86. (II) When a 100-g mass is placed gently onto a foam-rubber mattress, the mass sinks 4.0 cm into it. Suppose a 200-g mass is dropped from a height of 30.0 cm onto the same slab of foam rubber. How far will this mass sink into the mattress?

87. (II) A pendulum whose period is exactly 1 s is taken to the Moon, whose mass is 7.35×10^{22} kg and whose radius is 1738 km. What is the period there?

88. (II) A spring of equilibrium length 30 cm has one end anchored, and a mass is attached to the other end. The mass is set in uniform circular motion in a plane. The angular frequency of the rotational motion is two-thirds the natural angular frequency of the spring, ω_0. (a) By how much is the spring extended by the motion? (b) Derive the general result for $\omega = \alpha\omega_0$.

89. (II) In the spectacular sport of bungee jumping, a light elastic cord (a bungee cord) is tied tightly around the ankles of someone who jumps from a bridge of height H (the other end of the cord is attached to the bridge). The length of the cord is calculated so that the jumper, of mass m, will not quite reach the surface of the water below the bridge before he or she springs back up. Suppose that the cord behaves like a spring of spring constant $10\,mg/H$, where g is the acceleration due to gravity. (a) How long must the cord be so that a jumper just touches the water before being pulled back up? Neglect the height of the jumper and any effects due to friction. (b) Friction damps the up-and-down motion of the jumper that results after the initial jump. How far above the water would the jumper be when the oscillations have ceased? Express your answer as a fraction of H.

90. (II) A block of mass $m = 0.80$ kg is dropped onto a horizontal platform supported by a spring with spring constant $k = 3.5$ N/m. The mass of the platform and spring are negligible compared with m. The speed of the block when it hits the platform is 1.7 m/s. (a) Find the maximum compression of the spring. (b) How long does it take to reach the lowest point after the block hits the platform?

91. (II) The following systems, illustrated in Fig. 13–34, exhibit simple harmonic motion. What is the period of each motion? (a) A toy tightrope walker, with geometry as shown, whose body has mass much less than that of the barbell weights, each of mass 50 g. The toy sways from side to side in harmonic motion. (b) A mass m attached to two parallel springs, each of spring constant k. Ignore gravity. (c) A spring with spring constant k and a mass m attached to each end. (d) A mass m hanging vertically from a spring, of spring constant k, under the influence of gravity.

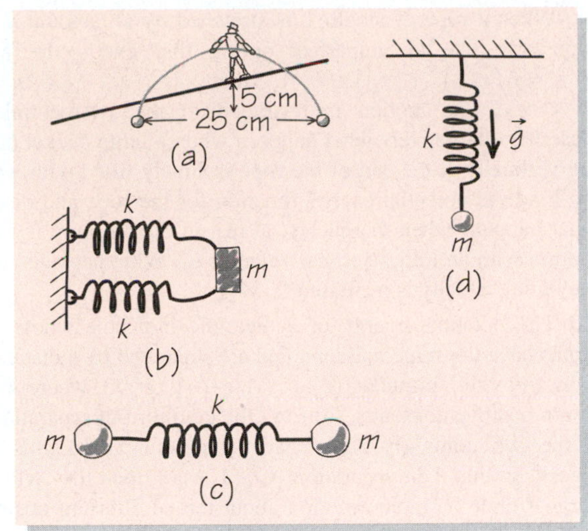

▲ **FIGURE 13–34** Problem 91.

92. (II) A simple pendulum of mass $m = 0.28$ kg and length $L = 0.65$ m is attached to a cart of mass $M = 1.0$ kg (Fig. 13–35). The mass of the pendulum support is negligible. The cart can roll freely on a horizontal surface. At $t = 0$, the pendulum bob is released from rest when the string makes an angle of 10° with the vertical. Assume that the resulting motion of the cart relative to the ground is simple harmonic motion. Determine the amplitude of the motion of the cart. Why would you expect the cart to move in simple harmonic motion?

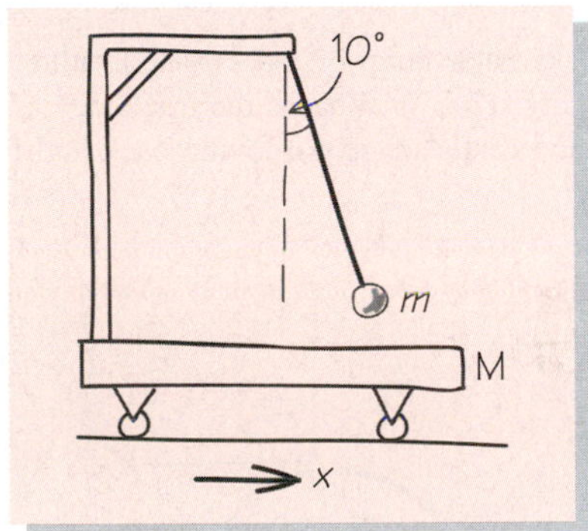

▲ **FIGURE 13–35** Problem 92.

93. (II) A bar 5.0 m long with a mass of 12 kg has a sharp bend (totaling 14°) at its midpoint (Fig. 13–36). The bar rests on a pivot placed under the bend. Twins of mass 32 kg each are seated at opposite ends of the bar. What is the period of small oscillations of this modified seesaw?

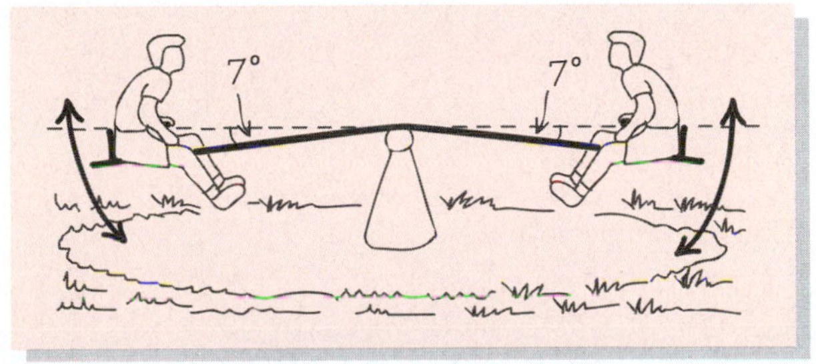

▲ **FIGURE 13–36** Problem 93.

94. (II) When a rope of length L is stretched by an amount x, the rope acts like an imperfect spring that exerts the force $F = -f_1(x/L) - f_2([x/L]^2)$, where $f_1 > 0$, $f_2 > 0$, and $f_1 \gg f_2$. The motion is restricted to the x-direction. (a) Describe qualitatively what happens when a large bucket hanging vertically at the end of the rope is slowly filled with water. (b) Find the potential energy function for the rope and discuss what happens when the bucket, at rest in equilibrium, is set in motion with an initial vertical velocity v_0. What happens when the initial velocity is increased?

95. (II) The potential energy of a diatomic molecule whose two atoms have the same mass, m, and are separated by a distance r is given by the formula $U(r) = (A/r^2) - (e^2/r)$, where A and e^2 are positive constants. (a) Find the equilibrium separation r_0 of the two atoms. (b) Show that if the atoms are slightly displaced, so that their separation is $r_0(1 + x)$, then they will undergo simple harmonic motion about the equilibrium position. Calculate the angular frequency of the harmonic motion. Use

$$(1 + x)^{-2} = 1 - 2x + 3x^2 - \ldots \quad \text{and}$$
$$(1 + x)^{-1} = 1 - x + x^2 - \ldots \quad \text{for} \quad x \ll 1.$$

(Remember the reduced mass.)

96. (II) A small mass m is placed midway on the line between two large masses M. The large masses are separated by a distance L. (a) What gravitational force does the small mass experience? (Assume that the system is isolated from other masses.) (b) The small mass is displaced a distance y in a direction perpendicular to the line between the masses. What is the direction of the net force on that mass? (c) Evaluate the magnitude of the net force for $y \ll L$. Give an expression that incorporates the inequality. (d) What is the motion of m when it is displaced from the midway point?

97. (II) A thin, circular hoop of mass m and radius R hangs on a small peg (Fig. 13–37). What is the frequency of small oscillations with the center of mass moving back and forth along the x-direction?

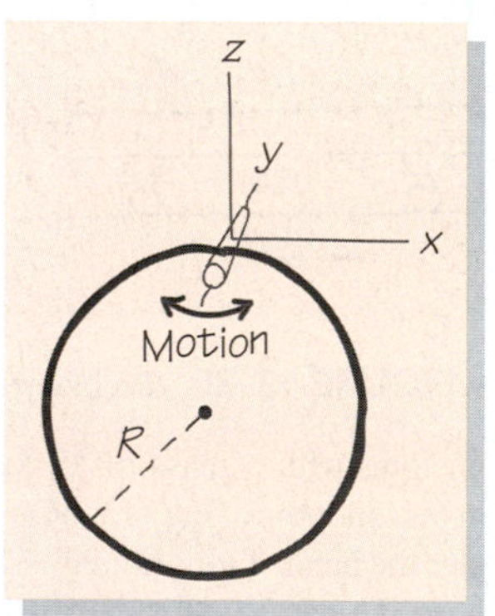

▲ **FIGURE 13–37** Problem 97.

98. (II) A machine gun fires bullets in a direction perpendicular to the plane of a target 180 m away from the gun. The bullets leave the muzzle at a speed of 370 m/s. The machine gun is set up on a platform that is vibrating with a frequency of 4.0 Hz and an amplitude of 8.50 cm in a horizontal direction parallel to the plane of the target. Describe the distribution of bullet impact points on the target. The bullets are fired with a frequency much greater than 4 Hz.

99. (III) A solid, uniform cylinder of mass m and radius R is fitted with a frictionless axle along the cylinder's long axis. A horizontal spring (spring constant k) is attached to this axle (Fig. 13–38). Under the influence of the spring, the cylinder rolls back and forth without slipping on a horizontal surface. What is the frequency of this motion?

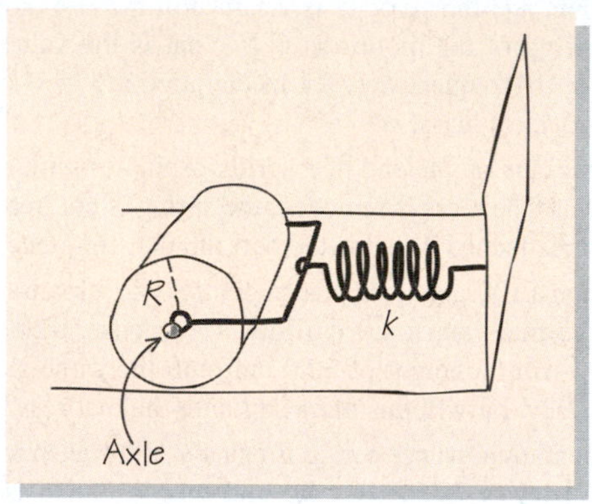

▲ **FIGURE 13–38** Problem 99.

The Canadian goose is generating a series of waves that spread in circles from where the goose sits. These waves represent a kind of organized disturbance on the water surface that are induced by a primary disturbance at the center of the spreading circles. It is only the disturbance that spreads; there is no flow of water away from the goose.

Waves

Waves are familiar to all of us. We can see ocean waves by means of light waves. Music may originate as waves on piano strings, be transported as waves of electrical current in a stereo system, form waves on the cone of a loudspeaker, and reach our ears as sound waves in air, which in turn form waves within the mechanical structure of our ears and are carried as waves of electrical impulses to our brains. Waves may appear as traveling waves, which move in some direction, as ocean surf moves toward a beach. They may appear as standing waves, like the vibrations of a guitar string. They may appear in one-dimensional systems, such as a Slinky spring; two-dimensional surfaces, such as the surface of a pond; or three-dimensional media, such as the air around us.

Consider a struck bell. To a very good approximation it is a rigid body, yet its ringing shows that it is vibrating just as a guitar string does. Even what look like the hardest of objects distort, if only by small amounts, in response to internal and external forces—indeed, a bell made of the hardest steel may ring the best. The resulting waves are generated within continuous media. In Chapters 14 and 15 we study waves on a guitar string, on the sea, or in the air, all of which involve continuous media. Eventually we shall encounter (in Chapter 34) electromagnetic waves such as light waves, in which no medium at all is required!

14–1 Types of Waves

In Chapter 13 we emphasized that when a point mass is displaced from its stable equilibrium position, the displacement gives rise to a restoring force, which leads to oscillatory motion. Moreover, the motion is generally harmonic as long as the displacement is small. The same physical principle applies to extended systems, but this time disturbances can be transmitted from one part of the system to another through internal restoring forces. A simple example of this is a system composed of a series of identical pendulums hanging in line side by side and connected to their neighbors on each side by identical springs (Fig. 14–1a). If the pendulums are at rest, we have a stable equilibrium situation. Let us now displace one of the pendulums from its vertical position and let it swing in a direction perpendicular to the line of pendulums (Fig. 14–1b). What we observe is that the stretched spring tends to displace the neighboring pendulums from their equilibrium position, and they too start to swing, with their motion lagging the originating pendulum. These pendulums in turn affect their neighbors—both the originating

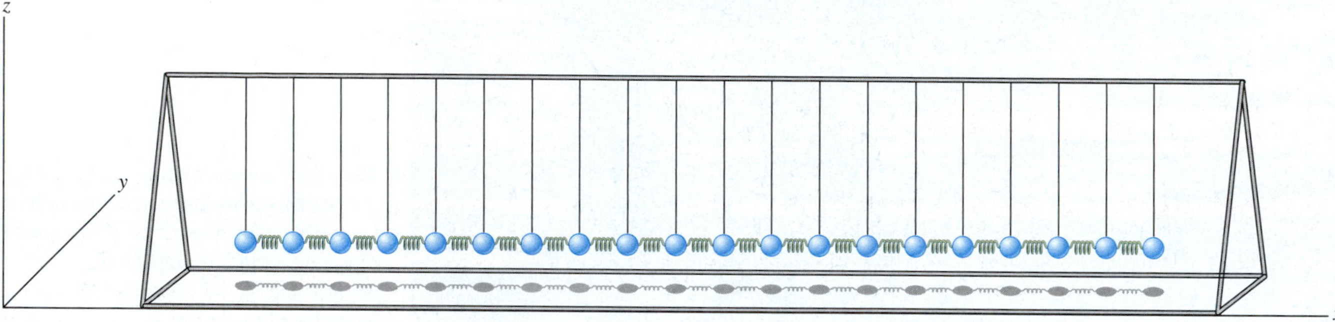

(a) Pendulum at rest. Each bob mass is connected with its nearest neighbor by a spring.

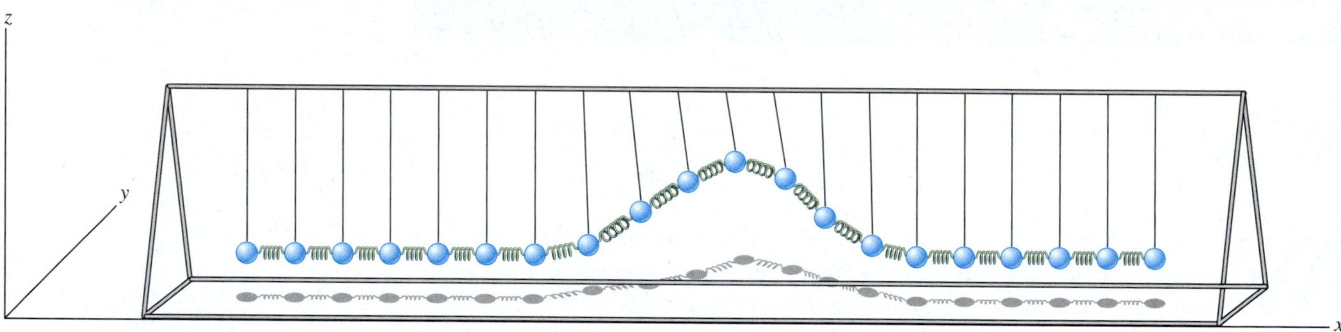

(b) A central bob mass is pulled in the *y* direction, pulling a group of neighboring masses along.

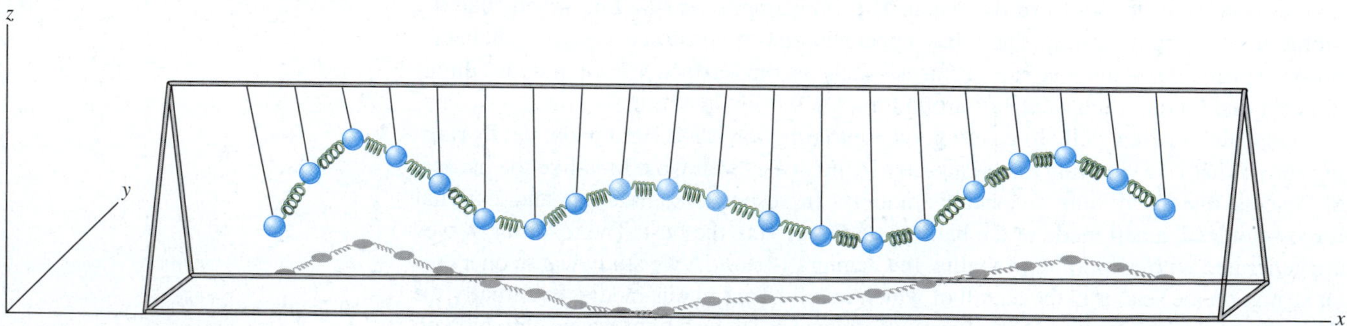

(c) As the central mass swings back the neighboring masses induce pulses of motion going in both directions away from the initiating disturbance.

▲ **FIGURE 14–1** Each pendulum in this line of pendulums swings in a plane parallel to the *yz*-plane. The line itself is along the *x*-axis. The pendulums are coupled by springs, and an initial swing of any given pendulum will be transmitted from one pendulum to the next, propagating a disturbance along the line.

pendulum and their other neighbors. What one eventually observes is a wave of displacements that travels from the originating pendulum in both directions.

This little toy models motions that can occur along a taut rope. Imagine that you are holding one end of the rope. If you rapidly displace your end of the rope to the side a little, the displacement affects the adjacent part of the rope, which in turn disturbs and displaces its neighboring part of the rope. Each section of the rope that is displaced acts on the adjacent sections to pull them in the direction of its own displacement, and by Newton's third law the adjacent sections act back on the displaced section to pull it back into alignment with the rest of the rope—the forces on the displaced section are *restoring forces*. The result is that the sideways displacement effectively travels down the rope, section by section, and gives rise to a traveling **transverse wave**.

How, beyond the generalities we have discussed so far, is the series of pendulums connected by springs a model for the rope? Below we'll show quantitatively how tension within the rope acts like the springs connecting the pendulums to provide the necessary restoring forces. At a deeper level, springlike atomic forces between neighboring atoms or molecules in the rope provide restoring forces analogous to those of the springs between the pendulums.

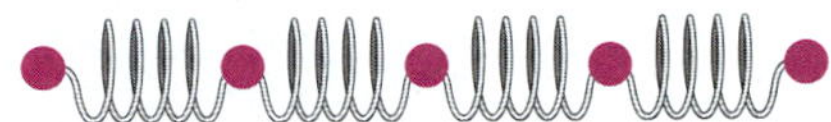

▲ **FIGURE 14–2** The line of masses connected by springs can propagate a longitudinal disturbance. A displacement of the first mass along the x-direction, which is the direction of the line, will propagate from mass to mass through the springs.

The motions that generate a moving disturbance can also occur in a *longitudinal* direction (along the line of the system). A series of point masses connected by springs and lying in a line (Fig. 14–2) presents a simple illustration of how such a **longitudinal wave** is transmitted. Again one needs to look no further than Newton's laws to understand the wave. A displacement of one of the masses to the right causes the spring on the left to stretch and the one on the right to compress. These act to displace the neighboring masses, and again a *traveling wave* of compressions and extensions (rarefactions) results (Fig. 14–3). This system is a model for longitudinal waves in general, including sound waves in air and longitudinal waves that travel through solids (also referred to as sound).

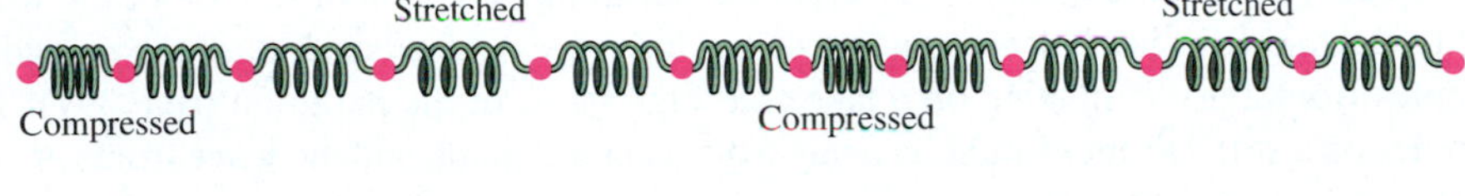

(a)

Motion

Longitudinal waves

(b)

◀ **FIGURE 14–3** (a) The point masses of the system illustrated in Fig. 14–2 are displaced longitudinally (along the equilibrium line). Some springs are compressed, some are stretched. (b) Longitudinal waves on a Slinky spring.

Keep in mind another feature of the propagating disturbances that we have described here. In transverse waves, like those that can be generated on a rope, all the actual motion is to the side (transverse). In longitudinal waves, like sound, no material is displaced a distance farther than the back-and-forth displacements of the system elements. Yet the disturbance, or wave, travels along the medium for what may be very long distances indeed. The sound wave that an earthquake generates within the Earth may involve actual atomic motions that are limited to motions on the order of several atomic spacings, yet the disturbances that the sound represents may propagate all the way to Earth's opposite side. And it is not just a "disturbance" that is traveling; we'll see below that a traveling wave carries energy and momentum.

Standing waves represent another manifestation of propagating disturbances, but in an apparently very different form from traveling waves. A harp string under tension is fastened at both ends and can be set into transverse motion with a pluck. An organized, or regular, pattern of displacements from the equilibrium position of the string results, and the pattern is *periodic*. Just as for the simple harmonic motion we have already enountered, the motions of a standing wave are not just periodic but are *harmonic* when the displacements are not too large (Figure 14–4). The vibrating harp string is an example of a transverse standing wave, but longitudinal standing waves, such as sound in a closed organ pipe, are also possible. Standing waves have a distinct property—*only certain oscillation frequencies are allowed*. This is why a plucked harp string or a given organ pipe has a certain pitch.

▲ **FIGURE 14–4** The vibrating strings of a harp form an example of transverse standing waves, which can be established on strings fixed at both ends.

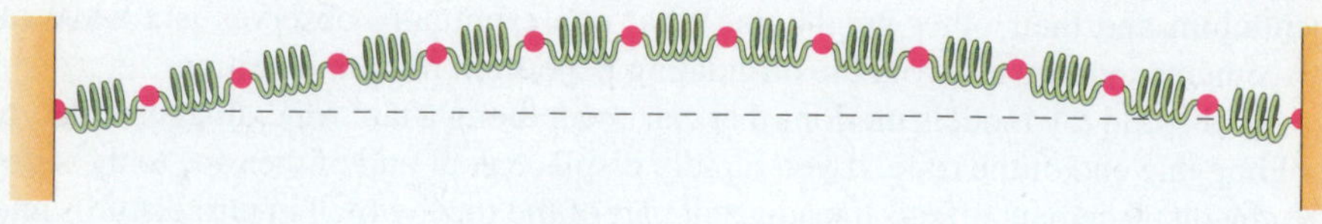

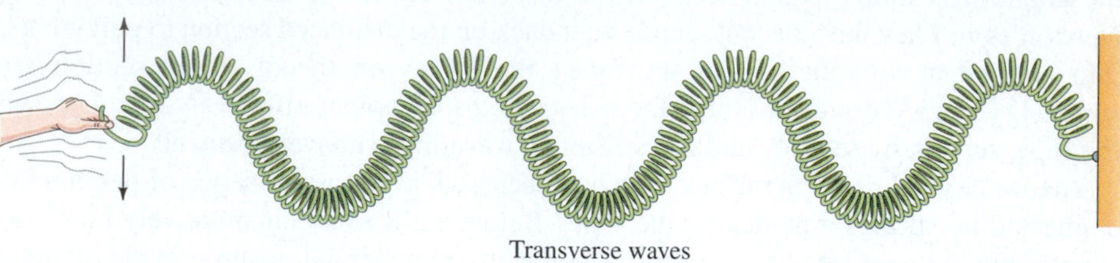

▶ **FIGURE 14–5** (a) The point masses of the system illustrated in Fig. 14–2 are displaced transverse to the equilibrium line (dashed). The stretching of a spring acts to bring the system back to the equilibrium line. (b) Transverse waves on a Slinky spring.

So far we have been discussing waves in a one-dimensional system, such as the displacement of a spring (Fig. 14–5) from its straight-line equilibrium shape, or the compression and rarefaction in the closeness of the spirals in a horizontally laid out Slinky (Fig. 14–3), and it is such waves that we will be mostly dealing with in this chapter and in Chapter 15. However, two- and three-dimensional systems also support waves. For example, standing waves can be generated on a drumhead—a two dimensional surface—in the same way that they can be generated on a string. These standing waves are transverse, consisting of an organized harmonic motion transverse to the drumhead (Fig. 14–6). Traveling waves in two-dimensional systems form **wave fronts**, which are lines that are formed by a series of propagating crests (or any other identifiable part of the wave)—we'll see later that these fronts are perpendicular to the direction of motion of the wave. This is a familiar phenomenon that we see in the waves at the beach or in the spreading ripples from a goose floating on a lake (see Fig. 14–7, or the photo on page 397). The ripples on a still lake move radially away from a central point, and the wave fronts are expanding circles. Just as for one-dimensional traveling waves in ropes or wires, water waves travel, but the actual movement of the water is entirely local (limited), as is easily seen by putting a cork in a pond of water (Fig. 14–8). (This figure illustrates that the local water motion, which is what the cork traces out for us, is a combination of transverse and longitudinal.) For a three-dimensional system such as air or Earth's interior, the wave fronts formed by traveling waves move out as spheres from a central point.

We begin with a description of the dynamical equation that leads to waves. Newton's second law relates the acceleration of a small portion of the medium to the forces that arise from its displacement from equilibrium. The propagation of a disturbance through the system is governed by the dynamical equation, which in this context

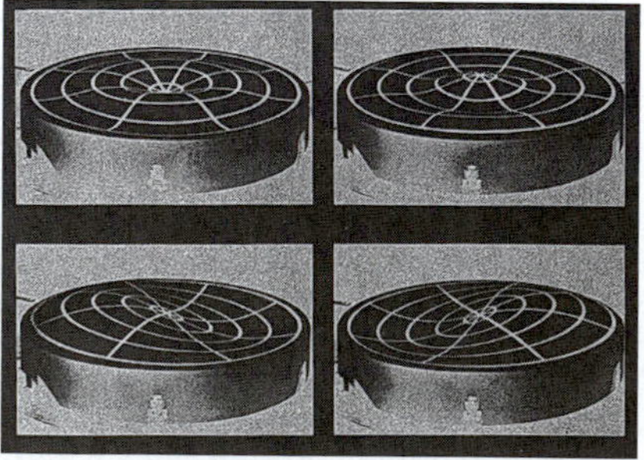

▲ **FIGURE 14–6** Standing waves on a drumhead. The *nodes*, points where the drumhead is not in motion, form lines rather than individual points as they did for the string. The lines shown are not necessarily nodes. Can you determine the nodes for each pair of vibrations?

▲ **FIGURE 14–7** The spreading waves resulting from a disturbance of a water surface form circles centered on the disturbance.

is called a **wave equation**. We will see that the name is in fact justified: The solutions to the dynamical equations properly describe waves. The wave equation in its various forms can be used to describe the mechanical waves of this chapter and the next; it will reappear when we discuss electromagnetic waves in Chapter 34. In particular, the wave equation will allow us to understand how to find the speed with which waves travel and the discrete vibration frequencies of standing waves.

FIGURE 14–8 A cork floats on the ocean surface, where the water is not too shallow, as a wave passes. The X marks the crest of the passing wave, the dashed circle marks the path followed by the cork, and the arrow marks the direction of motion of the cork at the particular point in the wave's passage.

14–2 The Wave Equation

Let us apply Newton's second law to a string under tension, neglecting the forces due to gravity in this analysis. Figure 14–9 shows a rope displaced transversely from its equilibrium position along the horizontal x-axis. We write the sideways displacement as the function $z(x, t)$. It changes with time, and it changes from place to place along the x-axis. It is important to note here that there are two space coordinates involved—z, the transverse displacement, and x, the distance along the rope—and that to understand wave motion, you need to keep them distinct in your thinking. Because z varies along the rope, we cannot treat the entire rope, but only a tiny segment of it. We denote the length of the segment by Δl. We assume that the rope is uniform with a mass per unit length μ. Thus the mass of the element is $\mu\,\Delta l$. Since the transverse position of the rope segment is described by z, its (transverse) acceleration is

$$a = \frac{d^2 z(x, t)}{dt^2}. \tag{14–1}$$

x labels the point on the rope where the transverse displacement occurs, and in this expression we differentiate with respect to t while holding x fixed. The conventional way of denoting this is to replace d^2/dt^2 by the mathematically correct $\partial^2/\partial t^2$. The *partial derivative* sign ∂ indicates only that the function whose derivative is taken depends on more than one independent variable, and that for the partial derivative with respect to, say, t, all the remaining variables (here x) are regarded as constants.

Let us now consider the net force $\vec{F}$ acting on the element of the rope. We see in Fig. 14–9 that all the forces that act on our element come from the tension of the rope. The force that acts on the left side of the element at x has two components. If the angle that the rope makes with the horizontal direction at x is θ_1, then the horizontal component of this force is $-T\cos\theta_1$ and the vertical component of the force is $-T\sin\theta_1$, where T is the magnitude of the tension. The minus signs are present because we define the positive directions as that of the positive x-axis, and the vertical pointing upward in our figure. The force that acts on the right side of the rope, at $x + \Delta x$, has a horizontal component that acts to the right, and therefore is $+T\cos\theta_2$, where θ_2 is the angle the

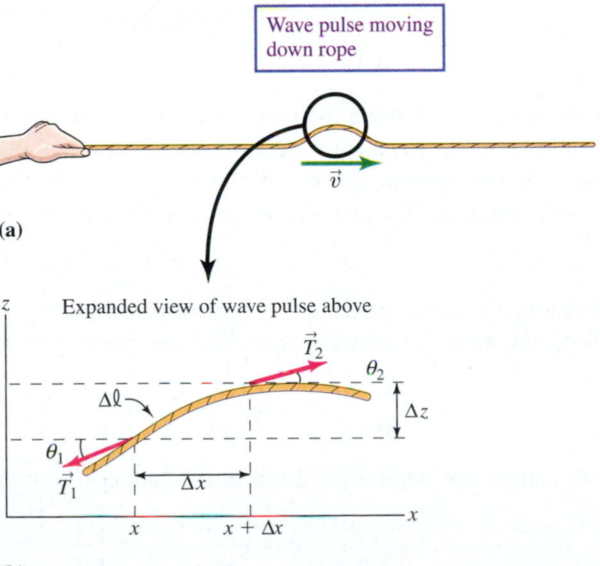

FIGURE 14–9 A string has been transversely displaced. We have isolated a segment of length $\Delta\ell$, as well as the tension forces acting on it. Due to the curvature, the direction through which the tension acts is different on each side. The magnitudes of $\vec{T}(x)$ and $\vec{T}(x + \Delta x)$ are the same.

rope makes at $x + \Delta x$. The vertical component acts in the upward direction and it is $+T \sin \theta_2$. Before we write down the net force on the element, it is helpful to relate the angle θ to the displacement at point x and to make some approximations. The first observation is that if we make the rope segment small enough, the angles θ_1 and θ_2 won't differ by much, and unless we are dealing directly with their difference, we can just call them θ. Next we note that $\tan \theta$ is the slope of the piece of rope at x.

$$\tan \theta = \frac{dz(x,t)}{dx}. \tag{14–2}$$

As in Eq. (14–1), the derivative is with respect to just one variable, in this case x (t is fixed here), so that we should really replace dz/dx by $\partial z/\partial x$. Next we note that the displacement is *small* and the angle θ itself is small (the rope is *nearly* flat), and we may replace $\tan \theta$ by $\sin \theta$ and $\cos \theta$ by unity. With these approximations in hand, let's look at the net force. The horizontal component of the net force is

$$F_x = T \cos \theta_2 - T \cos \theta_1 = T(\cos \theta_2 - \cos \theta_1) = T[\cos \theta(x + \Delta x) - \cos \theta(x)].$$

We have here recognized that θ_2 is θ at $x + \Delta x$ while θ_1 is θ at x. If we divide the right hand side by Δx and multiply it by the same quantity, and use the fact that

$$\frac{d \cos \theta}{dx} = \frac{\cos \theta(x + \Delta x) - \cos \theta(x)}{\Delta x},$$

we see that

$$F_x = T\,\Delta x \frac{d \cos \theta}{dx} = T\,\Delta x\left(-\sin \theta \frac{d\theta}{dx}\right). \tag{14–3}$$

This is *doubly* small, since θ (hence $\sin \theta$) as well as its derivative are small. The horizontal component of the net force is thus negligible compared to the vertical component of the force (see below). This confirms our assertion that the motion is really up and down, and we can drop this term. Using the exact same techniques, the vertical component of the net force is upward,

$$F_{\text{vert}} = T \sin \theta_2 - T \sin \theta_1 = T[\sin \theta(x + \Delta x) - \sin \theta(x)] = T\,\Delta x \frac{d \sin \theta}{dx}. \tag{14–4}$$

Using Eq. (14–2) and the small-θ approximation, we have

$$\sin \theta \cong \tan \theta = \frac{\partial z(x,t)}{\partial x}. \tag{14–5}$$

Thus

$$F_{\text{vert}} = T\,\Delta x \frac{d}{dx}\frac{\partial z(x,t)}{\partial x} = T\,\Delta x \frac{\partial^2 z(x,t)}{\partial x^2}. \tag{14–6}$$

(We have again written the *partial* derivative: We are differentiating with respect to x, while t remains fixed.) We are now close to our final result. We write Newton's second law for the vertical motion and use Eqs. (14–6) for the force, Eq. (14–1) for the acceleration, and the fact that the mass of the piece of rope is

$$\mu\,\Delta l \cong \mu\,\Delta x,$$

because for small angles $\Delta l = \Delta x/\cos \theta \cong \Delta x$. Both the force and the mass of the segment are proportional to Δx, which we cancel, leaving

$$T\frac{\partial^2 z(x,t)}{\partial x^2} = \mu \frac{\partial^2 z(x,t)}{\partial t^2}, \quad \text{or} \quad \frac{\partial^2 z(x,t)}{\partial x^2} = \frac{\mu}{T}\frac{\partial^2 z(x,t)}{\partial t^2}. \tag{14–7}$$

We can go one step further. Look at the dimensions of the ratio μ/T:

$$\left[\frac{\mu}{\text{Tension}}\right] = \frac{[M/L]}{[ML/T^2]} = \left[\frac{T^2}{L^2}\right] = \frac{1}{[\text{velocity}^2]}. \tag{14–8}$$

This dimensional argument suggests that the combination μ/T will play a role as a speed, and accordingly we define

$$v \equiv \sqrt{\frac{T}{\mu}}. \tag{14–9}$$

Equation (14–7) now takes its final form, the **wave equation**.

$$\frac{\partial^2 z(x,t)}{\partial x^2} = \frac{1}{v^2}\frac{\partial^2 z(x,t)}{\partial t^2} \tag{14–10}$$

THE WAVE EQUATION

This equation may be viewed as a generic equation for the motion of all small displacements away from the equilibrium position in a one-dimensional continuous medium. That includes not only traveling waves but standing waves as well. It, or something very much like it, governs the wavelike behavior of many different systems that we described above.

What is the meaning of the parameter v that appears in the wave equation? We shall see in Section 14-4 that it is *the speed with which disturbances, or waves, propagate along the rope*. Here we will just note that this quantity depends on *internal* properties of the medium, in this case the tension and mass density of a rope.

14–3 Periodic Waves

If the hand in Fig. 14–9 shakes the rope just once, then the resulting disturbance will propagate down the rope. If instead the hand regularly—periodically—repeats the motion, then a continuous, periodic wave will be generated (Fig. 14–10). Such periodic disturbances are the ones that will interest us the most in this chapter. To establish that these periodic disturbances are indeed what occur, we will want to show that they satisfy the wave equation. In other words, we want to investigate periodic solutions of the wave equation—that is, solutions that repeat every time interval τ, the *period* of the motion†. Mathematically this is expressed by the condition that

$$z(x,t) = z(x,t+\tau). \tag{14–11}$$

If you were to look at this wave through a narrow vertical slit located at some point x, you would see the rope rising and falling with a definite period. Figure 14–10 reveals a displacement along the rope with regularly spaced peaks and troughs, with consecutive peaks separated by a distance which we label λ and which we refer to as the wavelength. This observation is expressed mathematically as

$$z(x,t) = z(x+\lambda,t). \tag{14–12}$$

Our waves are periodic in both time and space.

We know a very simple set of periodic functions the trigonometic functions, and we have seen how they enter into motion in Chapter 13. In that chapter we saw that a function periodic in time could have, for example, the form $\cos \omega t$. The quantity ω, which has dimensions of inverse time (so that the argument of the cosine is dimensionless), is

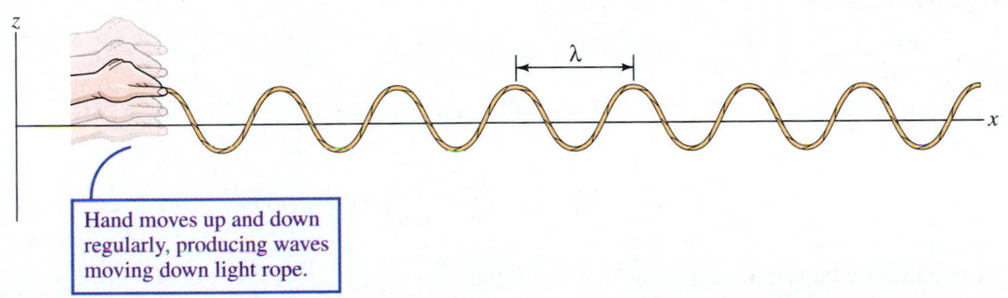

FIGURE 14–10 A periodic wave on a rope has a disturbance that repeats at regular intervals. The distance over which the pattern repeats is the wavelength λ.

†We use τ rather than T, as in Chapter 13, to avoid confusion with the tension T, which plays a role in many of our illustrations of wave motion.

the angular frequency familiar from Chapter 13. The **period** τ comes from the requirement that $\cos \omega t = \cos \omega(t + \tau) = \cos(\omega t + \omega\tau)$, or

$$\omega\tau = 2\pi. \tag{14–13}$$

The period is thus given in terms of ω by

$$\tau = 2\pi/\omega. \tag{14–14}$$

Note that in terms of the **frequency** f, which is the inverse of the period ($f = 1/\tau$), we have

$$\omega = 2\pi f. \tag{14–15}$$

We had already argued above how you can interpret the period or frequency for the traveling wave on a rope described above: The period is the time it takes for a complete back-and-forth motion at any given point on the rope, and the frequency measures how often this motion repeats per unit time.

All the remarks above hold true for $\sin \omega t$ as well, and therefore for any linear combination of $\sin \omega t$ and $\cos \omega t$.

Functions that repeat in space can be treated in exactly the same way. Consider, for example, $\cos kx$. The quantity k, which has dimensions of inverse length (so that the argument of the cosine is dimensionless), is analogous to ω. We call it the **wave number**. Our function is of course repetitive: $\cos kx = \cos k(x + \lambda) = \cos(kx + k\lambda)$ provided that

$$k\lambda = 2\pi. \tag{14–16}$$

This establishes that the repetition length λ, which we call the **wavelength**, is expressed in terms of k by

$$\lambda = 2\pi/k. \tag{14–17}$$

Now we combine the periodicity in time and in space to find a wave that is periodic in both. Take, for example,

$$z(x, t) = A \cos kx \cos \omega t. \tag{14–18}$$

In this form, or in other combinations of trigonometric functions of the arguments kx and ωt, the value of $z(x, t)$ runs between $+A$ and $-A$. The quantity A is therefore called the **amplitude**. The most general such combination is

$$z(x, t) = A \cos kx \cos \omega t + B \cos kx \sin \omega t + C \sin kx \cos \omega t + D \sin kx \sin \omega t. \tag{14–19}$$

This is a sum of different waves of the type shown in Eq. (14–18) and sketched in Fig. 14–10. Each one is characterized by a different amplitude A, B, C, or D. (There is no simple way of characterizing an overall amplitude.) It is not hard to check that Eq. (14–19) will satisfy the wave equation, *although only for special values of* k *and* ω. If, for example, we take just the last term and substitute it into the wave equation in the form

$$\frac{\partial^2 z}{\partial x^2} - \frac{1}{v^2}\frac{\partial^2 z}{\partial t^2} = 0,$$

we find

$$\begin{aligned}\frac{\partial^2(\sin kx \sin \omega t)}{\partial x^2} - \frac{1}{v^2}\frac{\partial^2(\sin kx \sin \omega t)}{\partial t^2} \\ &= -k^2 \sin kx \sin \omega t - \left[\left(-\frac{\omega^2}{v^2}\right)\sin kx \sin \omega t\right] \\ &= \left(-k^2 + \frac{\omega^2}{v^2}\right)\sin kx \sin \omega t = 0.\end{aligned}$$

This can only be satisfied if k and ω are related:

$$k^2 = \frac{\omega^2}{v^2}. \tag{14–20}$$

Using Eqs. (14–15) and (14–16), we see that this is equivalent to

$$\lambda f = v. \tag{14–21}$$

The same conclusion is reached with each of the terms in the more general solution form, Eq. (14–19). Thus Eq. (14–21) is a general relation between the wavelength, the frequency, and v, which as we shall see in Section 14–4 is *the speed with which waves travel*.

EXAMPLE 14–1 A disturbance along a long string aligned initially along the x-axis has a wave form that along a small part of it takes the form $z(x, t) = a(x - vt)^3$, where $z(x, t)$ is the transverse displacement of the string from the x-axis. Here a is a small constant. (Figure 14–11 illustrates what is meant here: The solid part of the curve is the one we are concentrating on, and that part is cubic in the variable $x - vt$.) (a) Does this part of the disturbance satisfy the wave equation? (b) If so, describe the position of the spot on the string whose displacement is zero.

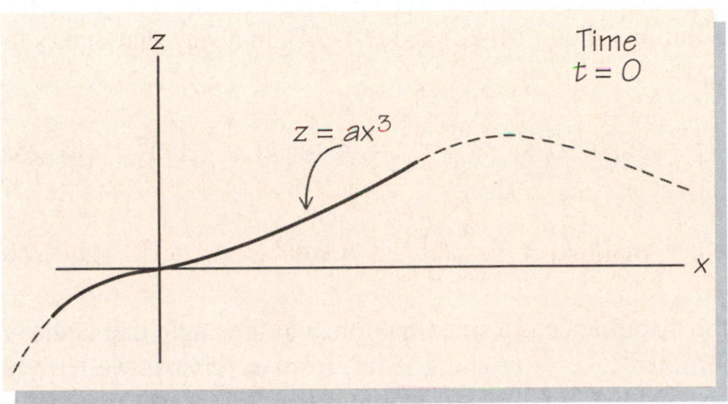

▲ **FIGURE 14–11** A disturbance which over a limited range has the form $z(x, t) = a(x - vt)^3$ plotted vs x at $t = 0$.

Setting It Up We regard Fig. 14–11 as a snapshot at $t = 0$.

Strategy For part (a) we need only take derivatives and check that Eq. (14–10) is satisfied. For (b) we look for the position x as a function of t for which $z = 0$; this requires setting $z = 0$ and solving the resulting expression to find x as a function of t.

Working It Out (a) We have

$$\frac{\partial^2}{\partial x^2}[a(x - vt)^3] = \frac{\partial}{\partial x}[3a(x - vt)^2] = 6a(x - vt)$$

$$\frac{\partial^2}{\partial t^2}[a(x - vt)^3] = \frac{\partial}{\partial t}[3a(-v)(x - vt)^2] = 6a(-v)^2(x - vt).$$

Taking the first term minus $(1/v^2) \times$ the second term, we find zero, so the wave equation is satisfied.

(b) The transverse displacement is zero at one spot only, when $x - vt = 0$. This point is at $x = 0$ when $t = 0$, and *it moves at a speed* v, $x = vt$. This verifies our claim that the wave equation allows solutions in which the disturbance propagates with speed v, as we will see in more detail in later sections.

What Do You Think? Does this displacement function really satisfy all the conditions that we insisted on to establish the wave equation? *Answers to* ***What Do You Think?*** *questions are given in the back of the book.*

14–4 Traveling Waves

Transverse Waves

We have seen that the wave equation has solutions that are periodic in both space and time for a transverse displacement on a rope—by transverse we mean that the disturbance is perpendicular to the rope's alignment. If we call the rope's alignment direction the x-axis and the direction perpendicular to the rope the z-axis, the fully periodic solution $z(x, t)$ is described by combinations of the products of trigonometic functions in x and in t, as in Eq. (14–19). At this point we want to show that some solutions of this type represent trigonometric shapes (sines or cosines) that move along the rope with a fixed speed, and we refer to these waves as **traveling waves**.

We first consider a solution formed by two terms from Eq. (14–19), with equal and opposite coefficients:

$$z(x, t) = A \sin kx \cos \omega t - A \cos kx \sin \omega t.$$

If we use the trigonometric identity $\sin(a - b) = \sin a \cos b - \cos a \sin b$, we see that the above solution has the form

$$z(x, t) = A \sin(kx - \omega t). \tag{14–22}$$

We can now show that this represents a traveling wave. Consider first what is happening at time $t = 0$. We have $z(x, 0) = A \sin kx$, which is drawn in Fig. 14–12a. This function has a *node* (the technical name for a zero of the function) at $x = 0$. Where is this node a short time Δt later? The node is located for any time at the zero of the argument in the solution in Eq. (14–22). So if our node is displaced by Δx, we can find that value by requiring that

$$k\,\Delta x - \omega\,\Delta t = 0.$$

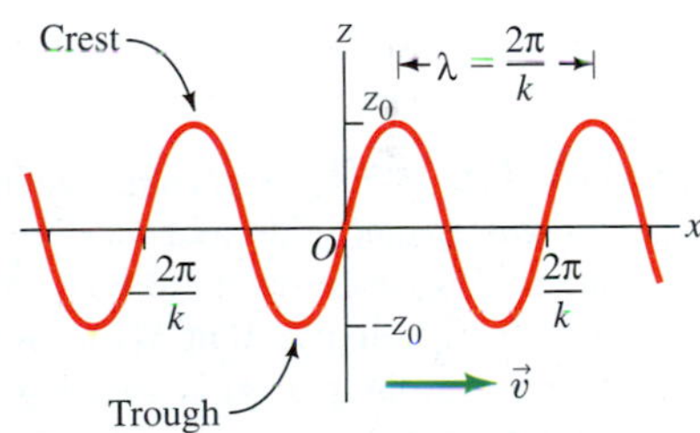

(a) $t = 0$

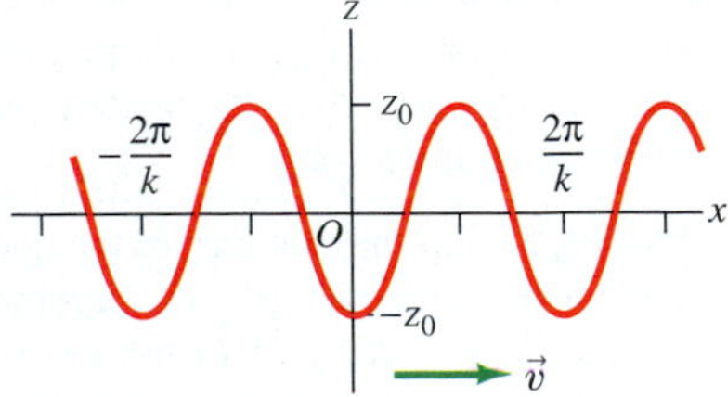

(b) $t = \Delta t$

▲ **FIGURE 14–12** (a) The harmonic traveling wave of Eq. (14–22) is plotted at the fixed time $t = 0$. (b) As time passes, this pattern moves to the right.

This gives for the new location $\Delta x = (\omega/k)\,\Delta t$, a little to the right. This same reasoning works for *any* point on the sine-curve, hence the entire distortion of the rope looks like the drawing of Fig. 14–12b, which is just the $t = 0$ curve (Fig. 14–12a) displaced an amount Δx to the right. Equivalently, the speed with which the node (or any other point) has moved is

$$\frac{\Delta x}{\Delta t} = \frac{\omega}{k} = \frac{2\pi f}{2\pi/\lambda} = \lambda f = v. \qquad (14\text{–}23)$$

Thus the constant v that appears in the wave equation is the *speed* of the traveling wave, just as stated at the end of Section 14–2. This wave travels in the positive x direction, since x increases as t increases.

We can easily construct waves that travel in the negative x direction. A solution made up of a different combination of terms in Eq. (14–19) will do the trick, namely

$$A\cos kx \sin\omega t + A\sin kx\cos\omega t = A\sin(kx + \omega t). \qquad (14\text{–}24)$$

(We have again used a trigonometric identity.) Here the argument of the sine function will remain the same as t increases only if Δx *decreases*. The wave moves to the left.

We can write our two solutions, Eqs. (14–22) and (14–24), in a way that makes the role of the velocity evident:

$$A\sin(kx - \omega t) = A\sin\left[k\left(x - \frac{\omega}{k}t\right)\right] = A\sin[k(x - vt)] \qquad (14\text{–}25a)$$

$$A\sin(kx + \omega t) = A\sin\left[k\left(x + \frac{\omega}{k}t\right)\right] = A\sin[k(x + vt)]. \qquad (14\text{–}25b)$$

In each case, the shape of the disturbance is a sine function, and the angle that is the argument of the sine is proportional to $x - vt$ and $x + vt$, respectively. As we have already described, these correspond to right- and left-moving harmonic waves of speed v. We can also see by following the argument above that the particular shape of the wave (its **waveform**), here a sine function, is completely independent of the discussion that shows that the waveform moves to the right or the left. The advancing of the shape, which is what characterizes traveling waves, depends only on the fact that the argument of the waveform is $x - vt$ or $x + vt$. We'll confirm this in more detail at the end of this section. Finally, we emphasize again that the value of v is determined by internal properties of the wave medium—here the tension and mass density.

EXAMPLE 14–2 A person standing at the narrow entrance to a harbor sees regular sinusoidal water waves moving into the harbor. He counts 42 wave crests in 1.0 min, and he estimates the distance between the crests to be 3.0 m. Write an expression for the height of the wave where the person is standing. What are the wavelength, wave number, frequency, angular frequency, and speed of these waves?

Setting It Up Figure 14–13 labels an appropriate axis as well as the wavelength λ. Note that the maximum height (amplitude) of the waves is not given, so we simply treat that as an unknown parameter h_0, where h_0 labels the maximum wave height above a steady background water level.

Strategy For the first part of the question we can use any of the trigonometric expressions with argument $kx - \omega t$ (right-moving waves, as in the sketch). Since we are asked only for the wave height where the person is standing, we substitute $x = 0$. We can find the wave number k directly in terms of the wavelength λ. The frequency involves counting waves per second, which is given information, and the speed is always the product of the frequency and the wavelength.

Working It Out Let $h(x, t)$ be the height of the waves above the average water level. Then using, say, Eq. (14–22), $h(x, t) = h_0\sin(kx - \omega t)$. This function at the observer's position is

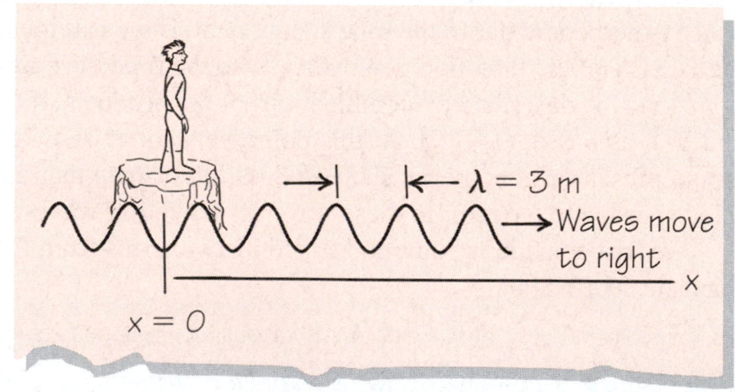

▲ **FIGURE 14–13**

$$h(x, t) = -h_0\sin(\omega t).$$

The wavelength is the distance between crests, $\lambda = 3.0$ m. The wave number, k, is given in terms of λ by Eq. (14–16):

$$k = \frac{2\pi}{\lambda} = \frac{2\pi}{3.0}\,\text{m}^{-1} = 2.1\ \text{m}^{-1}.$$

The frequency of the waves is $f = 42$ per min, or

$$f = (42\ \text{min}^{-1})\frac{1\ \text{min}}{60\ \text{s}} = 0.70\ \text{s}^{-1}.$$

The angular frequency, ω, and the frequency, f, are related by Eq. (14–15):

$$\omega = 2\pi f = (2\pi\text{rad})(0.70\ \text{s}^{-1}) = 4.4\ \text{rad/s}.$$

Finally, the speed v is given by Eq. (14–21),

$$v = \lambda f = (3.0\ \text{m})(0.70\ \text{s}^{-1}) = 2.1\ \text{m/s}.$$

What Do You Think? Suppose the observer sees waves in the same system but with a wavelength of 2.0 m. Can he state with confidence whether the frequency of the waves will be less or greater than that of the 3.0-m waves she has observed?

CONCEPTUAL EXAMPLE 14–3 We have several wires available with different mass densities. Tests indicate that for the particular construction of these wires the maximum tension that the wires can support before breaking increases as the square of the linear mass density. Which wire can produce a traveling wave of the greatest speed?

Answer We already know that the speed of traveling waves on the wire is given by Eq. (14–9), $v = (T/\mu)^{1/2}$. Thus for a given wire of linear mass density μ, you get the greatest speed when the tension is highest. We therefore set the tension in each wire at its largest value, namely $T = \kappa\mu^2$, where κ is a constant with the dimensions of $(\text{N m}^2/\text{kg}^2)$—if the tension were higher, the wire would break. The maximum wave speed for each wire is then

$$v = (T/\mu)^{1/2} = (\kappa\mu^2/\mu)^{1/2} = (\kappa\mu)^{1/2}.$$

We conclude that the greatest wave speed can be achieved with the wire of largest linear mass density.

EXAMPLE 14–4 A very long wire is held under tension by suspending weights from an end that passes over a pulley. The speed of transverse traveling waves on the wire is 51 m/s. The wire is replaced by another with three times the mass density, and twice as much weight is suspended from the end of the new wire. What is the speed of the traveling waves now?

Setting It Up With Fig. 14–14 it is clear that the tension T in the wire equals the suspended weight mg. We are given neither the initial tension nor mass density, but since the question only involves the velocities, we only need the ratio of these quantities. We are given the initial speed v_1 and want the final wave speed v_2.

Strategy We can write down the expression for v_1 in terms of T and μ, Eq. (14–9). The expression for v_2 involves the same expression with T multiplied by 2 and μ multiplied by 3.

Working It Out Equation (14–9) gives $v_1 = (T_1/\mu_1)^{1/2}$. The second speed v_2 is

$$v_2 = \sqrt{\frac{2T_1}{3\mu_1}} = \sqrt{\frac{2}{3}}\sqrt{\frac{T_1}{\mu_1}} = \sqrt{\frac{2}{3}}v_1.$$

Numerically, $v_2 = (0.67)^{1/2}\ (51\ \text{m/s}) = 42\ \text{m/s}$. Increasing the tension tends to increase the speed, whereas increasing the mass density tends to decrease the speed. In this case, the mass density is increased by a larger factor than is the tension, so the speed decreases.

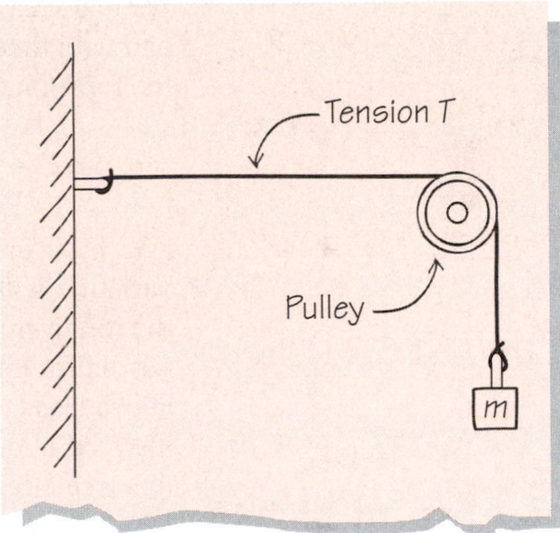

▲ **FIGURE 14–14** A wire is held under tension by the gravitational force mg.

What Do You Think? How would you adjust the tension and/or mass density so that the wave speed drops to zero? What is the physics behind your answer?

Longitudinal Waves

A Slinky spring provides us with a familiar example of a longitudinal traveling wave. Figure 14–3 showed regions of rarefaction and compression along the coil that move along the system. The actual motion within the spring is along the same line as the spring itself, which is also the line along which the wave propagates. A traveling wave like this is generated by a hand moving longitudinally back and forth at the end of the spring. Another example of a longitudinal wave is a wave of compression and rarefaction traveling in a metal rod—we'll refer to such waves as **sound** waves. We can treat this case quantitatively using the stress and strain properties of the rod. (You may want to review Section 11–4.) We first derive a wave equation for this kind of wave in the same way as we did for transverse waves on a string. We'll find an analogous equation, with v determined this time by a mass factor analogous to the mass density of the string and a force factor analogous to the tension in the string.

Area A

x

$\vec{F}_{\text{stress}}$

A force called a *stress* is applied to section of rod.

x $x + \Delta x$

(a)

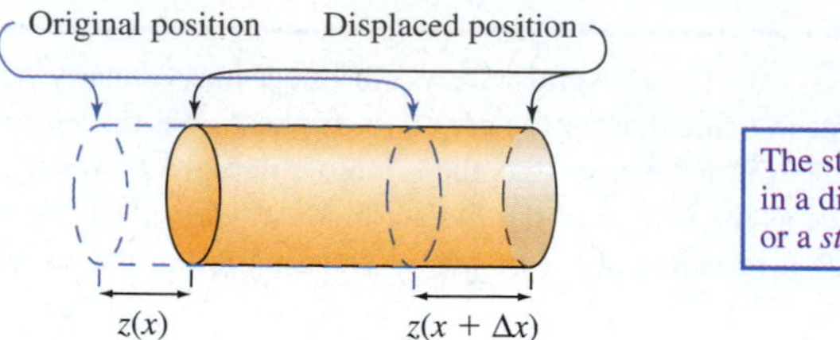

(b)

FIGURE 14–15 The (a) stresses on and (b) resulting elongation of a section of a rod lying between the positions x and $x + \Delta x$.

Start with a rod lying along the x-axis of cross-sectional area A and concentrate on a segment of the rod between x and $x + \Delta x$ (Fig. 14–15a). Just as the net force on a string segment originates with tension from one side and from the other, the net stress S (force per unit area) on our segment comes from the right and the left. It will vary with position and with time. Since generally the force on the right is $F(x + \Delta x) = AS(x + \Delta x)$ and the force on the left is $F(x) = AS(x)$, the net force on the segment is

$$\Delta F = F(x + \Delta x) - F(x) = A[S(x + \Delta x) - S(x)] = A\left[\frac{\partial S(x)}{\partial x}\right]\Delta x.$$

(We have immediately gone to a partial derivative notation, anticipating independent variation with time.) Let's denote the longitudinal displacement of the material within the rod from its equilibrium position by $z(x, t)$. (Note that we use the same symbol z as we did for transverse waves to represent the actual displacement of material.) The force above leads to an acceleration and hence displacement of the material. The mass Δm of material in our segment is the mass density ρ multiplied by the volume, $\Delta m = \rho A\,\Delta x$; the acceleration of the material is given by the second time derivative of the displacement $z(x, t)$, $a = \partial^2 z(x, t)/\partial t^2$. Newton's second law, $(\Delta m)a = \Delta F$, reads

$$\rho A\,\Delta x \frac{\partial^2 z}{\partial t^2} = A\left(\frac{\partial S}{\partial x}\right)\Delta x$$

or

$$\rho \frac{\partial^2 z}{\partial t^2} = \frac{\partial S}{\partial x}. \tag{14–26}$$

We now relate the stress S (a force factor) to the strain e (a fractional displacement of the material). What is the strain here? The net change in the length Δx of our little segment of the rod is the difference between the displacements measurable at the two ends of the section, $z(x + \Delta x) - z(x)$ (Fig. 14–15b). Thus the strain is given by the ratio of the difference of displacements to the original length, Δx:

$$e = \frac{z(x + \Delta x, t) - z(x, t)}{\Delta x} = \frac{\partial z}{\partial x}.$$

Equation (11–19) states that $S = eY$, where Y is Young's modulus, a characteristic constant of the material:

$$S = Y\frac{\partial z}{\partial x}.$$

When this result is substituted into Eq. (14–26), we find that

$$\rho\left(\frac{\partial^2 z}{\partial t^2}\right) = Y\left(\frac{\partial^2 z}{\partial x^2}\right). \tag{14–27}$$

This result has the form of the wave equation, Eq. (14–10), and the speed can be identified as

$$v = \sqrt{\frac{Y}{\rho}}. \tag{14–28}$$

If you compare this to the speed for transverse waves on a string, Eq. (14–9), you will see that the form, as well as the kind of factors that appear, are entirely analogous.

We can get some feeling for the speed of sound waves in solids by calculating v for a steel rod. For steel, $\rho = 7.9 \times 10^3\ \text{kg/m}^3$, while Young's modulus is $Y = 200{,}000\ \text{MN/m}^2$. Therefore

$$v = \sqrt{\frac{Y}{\rho}} = \sqrt{\frac{2.0 \times 10^{11}\ \text{N/m}^2}{7.9 \times 10^3\ \text{kg/m}^3}} = \sqrt{0.25 \times 10^8\ \text{m}^2/\text{s}^2} = 5.0 \times 10^3\ \text{m/s}.$$

THINK ABOUT THIS . . .

WHY ARE LONGITUDINAL WAVES IN SOLIDS CALLED SOUND?

We think of sound in terms of our hearing, and this type of sound does involve waves, propagating in air from the source of the disturbance to our ears. The method by which they propagate is not identical to the way "sound" propagates in a solid. But the dynamics of a sound wave in air has certain similarities to the longitudinal waves in solids described above, and that is why we use the same name for both. Both are longitudinal waves of compression and rarefaction. In the case of solids, the dynamics that keeps the disturbance moving has to do with the atoms of the solid, densely packed, interacting through what for all practical purposes are little springs—actually electrical forces at the atomic level. In air the disturbance is propagated through collisions of fairly sparse moving molecules (see Chapter 19). The number we found above for the numerical value of the speed of sound in steel is, typically for sound in dense media, much larger than the speed of sound in air. It is in excellent agreement with experimental data. We can get an idea of the variety of values of sound waves from Table 14–1, which gives the speed of longitudinal waves—sound—in various media.

TABLE 14–1 • The Speed of Sound Waves in Various Media†

Medium	Speed of Sound (m/s)
Hydrogen	1284
Air	330
Liquid mercury (20°C)	1450
Methyl alcohol	1189
Water	1402
Polyethylene	920
Lead	1210
Silver	2700
Aluminum	5000
Beryllium	12,870

† *Longitudinal waves. Temperature = 0°C, pressure = 1 atm unless otherwise stated.*

How the Wave Speed Depends on the Medium

Once we have established that waves traveling through a certain medium obey a wave equation, we can try to determine the wave speed, which, as we've said before, depends on properties of the medium. For the string $[v = (T/\mu)^{1/2}]$, the tension T acts as a restoring force that tends to bring the string back to its equilibrium position. The mass density μ describes the reaction of the string to this restoring force. For sound waves in solids $[v = (Y/\rho)^{1/2}]$, the Young's modulus is the "spring constant" of a stress, and the mass density ρ describes how "nimbly" the solid responds to this force. Generally, this suggests that the speed of mechanical waves is a function of an internal force factor (an elastic restoring force) divided by a mass factor characteristic of the system.

A dimensional analysis helps us to extend this analysis. The tension is a force and has dimensions $[MLT^{-2}]$. The mass density (mass per unit length) has dimensions $[ML^{-1}]$. Therefore

$$\left[\frac{\text{tension}}{\text{mass density}}\right] = \left[\frac{MLT^{-2}}{ML^{-1}}\right] = [L^2T^{-2}].$$

Thus if v is a function of the ratio of the restoring force factor to the mass factor, *the particular function must be a square root*:

$$\text{wave speed} = \sqrt{\frac{\text{restoring force factor}}{\text{mass factor}}}. \tag{14–29}$$

WAVE SPEED

If we restrict ourselves to bulk (macroscopic) properties of the string, Eq. (14–9) is the *only* dimensionally correct relation for wave speed. Indeed, the *only* bulk properties of the string that we know are the tension, the total mass, the length, and the thickness; the first three of these are figured into the wave speed. (For the true mass density, or mass per unit *volume*, the thickness of the string would also enter.) As we discussed in Chapter 1, dimensional analysis does not permit a calculation of constant factors such as 2 or π, but it will give the system's dependence on the bulk properties, and it will give a correct estimate of orders of magnitude.

The Speed of Sound in Air

We can use Eq. (14–29) to estimate the speed of sound waves in a gas (such as air). We must find a restoring force factor and a mass factor for the gas. We would not expect the total amount of gas to enter, because the speed is not dependent on the total amount of gas. That leaves us with only one choice for the mass factor, namely the mass density ρ_0. The restoring force factor is the pressure p_0 within the gas. We will look at this quantity in detail in Chapter 16, but here we can describe pressure as follows. Within the gas, molecules are constantly in motion. If a tiny membrane is placed into the gas, it is bombarded on each side by many molecules. When the molecules collide elastically with the membrane, they transfer momentum to it. A transfer of momentum is equivalent to a force, so we can characterize the gas by a force per unit area, the **pressure**, that acts on any surface in contact with the gas. Pressure has dimensions of force per area, or $[MLT^{-2}][L^{-2}] = [ML^{-1}T^{-2}]$, whereas density has dimensions of mass per volume, or $[ML^{-3}]$. The ratio p_0/ρ_0 has dimensions $[L^2T^{-2}]$, a speed squared. We thus estimate for the speed of sound in a gas

$$v_{\text{sound}} \cong \sqrt{\frac{p_0}{\rho_0}}. \tag{14–30a}$$

This result was first found by Newton and is appropriately called Newton's formula.

A more detailed calculation yields a better value for the speed of sound,

$$v_{\text{sound}} = \sqrt{\frac{\gamma p_0}{\rho_0}}. \tag{14–30b}$$

Here γ is a constant with the approximate value 1.4 for air—we'll see in Chapters 17–20 that γ has something to do with the thermal properties of the gas. (Thermal properties enter because local compression and rarefaction of a gas result in local heating and cooling.) The numerical correction to our estimate due to the factor γ is of the order of 20%. For normal atmospheric conditions at sea level, $p_0 \cong 1.0 \times 10^5\ \text{N/m}^2$, $\rho_0 \cong 1.3\ \text{kg/m}^3$, and

$$v_{\text{sound}} \cong 330\ \text{m/s} = 740\ \text{mi/h}, \tag{14–31}$$

or about 1000 ft/s.

Shear Waves in Solids

Just as small longitudinal displacements within a solid lead to longitudinal waves, small transverse displacements within a solid lead to transverse waves. This is because solids resist shear forces, and the wave speed of transverse waves—also called *shear waves*—is described by using the shear modulus, G, of Section 11–4. (This correctly implies that bulk gases or fluids do not support transverse waves because they do not resist shear forces—there is no restoring force for shear displacements.) Application of Eq. (14–29) leads to

$$v_{\text{shear}} = \sqrt{\frac{G}{\rho}}. \tag{14–32}$$

In most cases, the shear modulus is lower than Young's modulus, and the speed of shear waves is less than that of longitudinal waves.

THINK ABOUT THIS . . .

HOW DO WE KNOW WHAT'S INSIDE EARTH?

Both sound and shear waves can travel through Earth. When they are initiated by earthquakes, we refer to them as *seismic waves*. This phenomenon is the basis for an important set of tools used in geological exploration and research. The varying arrival times of the seismic waves from an earthquake at a seismograph—a device whose vibrations detect seismic waves—gives information about the location of the earthquake. More important, the study of these waves has allowed us to build up a picture of the inside of the Earth. One of the ingredients in the analysis is the fact that at a boundary between two media, waves are reflected or transmitted, a subject we'll be discussing in Chapter 15. In the case of Earth, these boundaries are places where the composition or state of the internal material changes. Seismic waves from an earthquake can follow many paths, reflecting from various internal layers and boundaries between different materials (Fig. 14–16), and information on Earth's internal structure comes from the arrival time and strength of waves generated by distant earthquakes. We have learned that Earth has an overall structure that can be divided into four main parts: crust, mantle, outer core, and inner core. We know for another example that Earth's outer core is liquid because shear waves do not pass through it. The search for petroleum also relies heavily on evidence from wave propagation within Earth. In this case the waves are generated by explosions judiciously set off in the vicinity of the area to be explored. This tool can reveal in detail the structure of the interior deep below the surface.

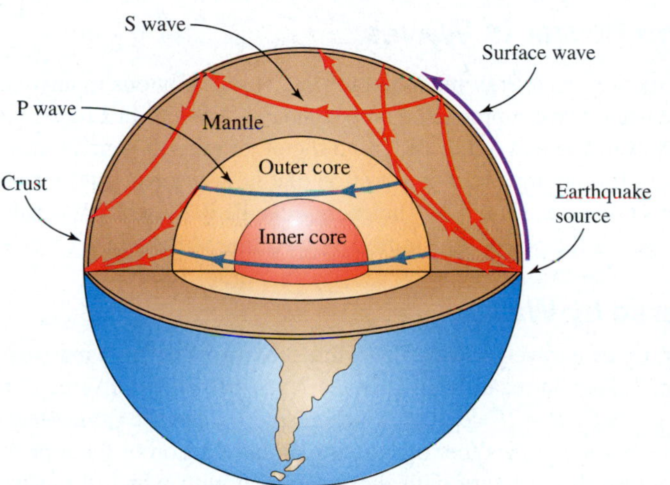

◀ **FIGURE 14–16** Several types of waves are possible in solids. Seismic waves, produced in events such as earthquakes, include all types. The P waves marked here stand for primary, geologists' notation for the sound waves, while the S waves stand for secondary, geologists' notation for shear waves. These waves travel at different speeds; moreover, as we have remarked, fluids do not support shear waves. Hence P and S waves convey different information.

The Propagation of Arbitrary Waveforms

We have thus far concentrated on traveling *harmonic* waves, of generic form

$$z(x, t) = A\sin(kx - \omega t) = A\sin k\left(x - \frac{\omega}{k}t\right) = A\sin k(x - vt).$$

We will now show that a waveform of *any shape* can be propagated provided it has argument $x - vt$:

$$z(x, t) = f(x - vt). \tag{14–33}$$

To prove this, we need to show that this function is a solution of the wave equation, which is a simple mathematical exercise (see Problem 15). Here $f(x - vt)$ is any function we like; it describes the shape of whatever waveform is propagating.

We'll illustrate the procedure with a simple waveform, one that we call a *pulse*, as in Fig. 14–17. (This is a term normally reserved for a nonrepetitive waveform of limited

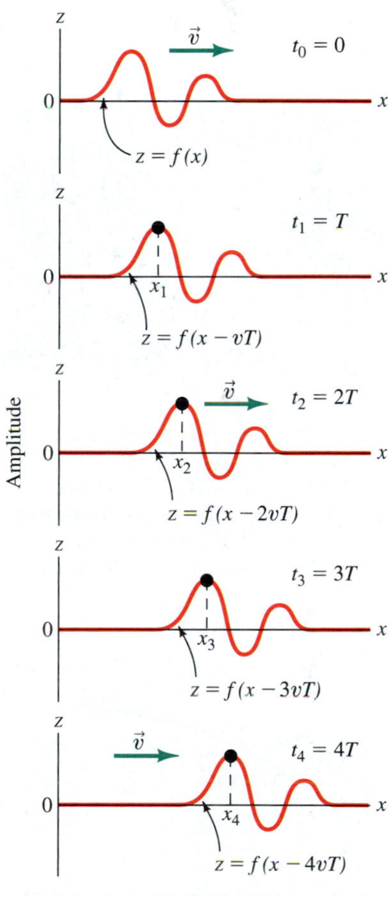

▲ **FIGURE 14–17** The curve that is formed by the pulse represents some function f of the variable $x - vt$. As time goes on (here $t_2 > t_1$), the curve moves to the right with speed v.

duration.) First we single out a particular point on the function, such as a peak, and mark this point with a black dot. At time t_1 this point is at an x-value of x_1. If t_1 increases to t_2, the argument of f will have the same value—the point will still be at the black dot—only if x increases to compensate for the decreased value of the term $-vt$. In order for the point to stay at the same argument, we must have

$$x_2 - vt_2 = x_1 - vt_1,$$

which can be solved for x_2:

$$x_2 = vt_2 + x_1 - vt_1 = x_1 + v(t_2 - t_1). \tag{14–34}$$

Eq. (14–34) describes a point moving to the right with speed v. Since the argument applies for any point of the pulse, all points of the pulse move in the same way—the waveform $f(x - vt)$ moves as a whole. Similarly, $f(x + vt)$ represents a waveform moving to the left. In summary, $f(x - vt)$—or $f(x + vt)$—*represents a waveform that moves to the right—or left—with speed* v.

CONCEPTUAL EXAMPLE 14–5 Suppose you laid out a very long rope with a small bulge in it (as in Fig. 14–17), poured water on it, and then let it freeze. You have a frozen waveform, which you then arrange to be carried at a uniform velocity past an observer. Mathematically this is identical to a traveling pulse. But is it really a traveling wave? Is a "wave" of cheering, popular at football games, a traveling wave?

Answer No, to both questions. In the first case, matter is moving in the direction of propagation. With a real wave, a bit of colored string attached to some part of the rope would stay at some level, then rise and fall, but always stay at the same lateral position relative to the observer (e.g. right in front of him). With the fake frozen wave, the string would be moving along with the fixed velocity v. We can add that in the fake wave, there are no forces acting to accelerate portions of the rope, only static forces that hold it rigid. As to the second question, the stadium "wave" does not have the defect of the moving frozen rope: Matter does not travel along the stadium wave. Nevertheless, it is not like the waves treated here in the sense that no power is carried by the wave. No forces, other than moral persuasion, act between neighboring elements.

▲ **FIGURE 14–18** Waves carry energy, a fact well known to anyone who has spent time at the seashore, or, especially, who has tried to surf them.

14–5 Energy and Power in Waves

In waves, masses are in motion, so energy must be present. This is obvious to anyone who has stood within the water waves approaching a beach (Fig. 14–18). In a traveling wave, energy is delivered from one end of the system to the other even though no mass is transferred in the motion. It is an important observation that in waves *energy is transmitted without the transport of mass*. Moreover, the energy can be delivered at a steady rate, and this leads us to look at the power transmitted by traveling waves.

The Power Delivered by Waves

Consider right-moving transverse traveling waves on a uniform rope that has a mass per unit length μ and tension T. As you recall from Chapter 7, the power P delivered at a point x along the rope is given by $P = \vec{F}\cdot\vec{v}_{\text{transverse}}$, where $\vec{v}_{\text{transverse}}$ is the (transverse) velocity of the rope at that point and $\vec{F}$ is the force acting on the portion of the rope to the right of x. (Don't confuse this velocity with the velocity $\vec{v}$ with which the wave propagates!) Only the transverse component of the force enters into the scalar product. To find the transverse component of the force, refer to Fig. 14–19, which shows that the transverse component of force at x acts downward (*negative* sign) with magnitude

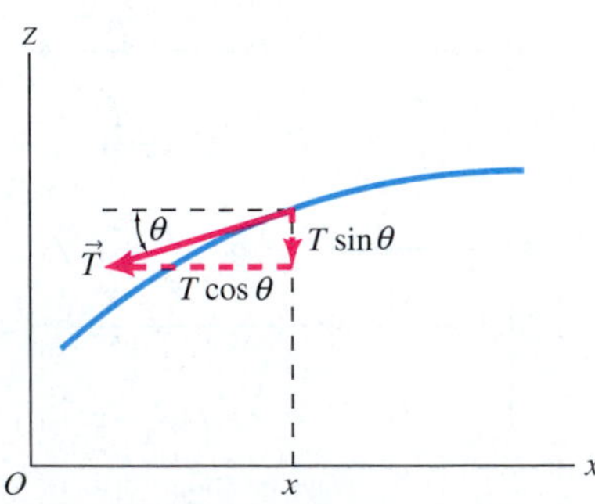

▲ **FIGURE 14–19** A segment of a distorted string is subject to a tension force from either side whose direction depends on the distortion. We are interested in this case in the vertical component.

$$T\sin\theta \cong T\tan\theta = T\frac{\partial z}{\partial x}.$$

We have used the small distortion approximation here—the angle θ is small—and we have remembered to use the partial-derivative notation again, as a reminder that z in fact depends on two variables, x and t. The transverse velocity of the rope at x is given by

$$v_{\text{transverse}} = \frac{\partial z}{\partial t},$$

where the sign of the time derivative determines whether the sign of the velocity is positive (moving up) or negative (moving down). Thus the power delivered to the rope at point x is

$$P = \vec{F}\cdot\vec{v}_{\text{transverse}} = -T\frac{\partial z}{\partial x}\frac{\partial z}{\partial t}. \tag{14–35}$$

It is simple to calculate the derivatives involved here for a harmonic traveling wave, $z = A\sin(kx - \omega t)$. We find

$$P = Tk\omega A^2\cos^2(kx - \omega t) = \mu v\omega^2 A^2\cos^2(kx - \omega t). \qquad (14\text{–}36)$$

To find the last form, we have used Eq. (14–9), which gives $T = \mu v^2 = \mu(\omega/k)v$, where v is the wave speed.

The power in Eq. (14–36) oscillates with time, and it is useful to think instead about the time average of the power delivered. We denote this average with angular brackets. To find its value, we need the average value taken by the cosine squared over a long time period, namely 1/2. Thus

$$\langle P\rangle = \tfrac{1}{2}\mu\omega^2 A^2 v. \qquad (14\text{–}37)$$

Energy Transport

There is no mechanism for energy loss in a rope such as we have described. (Of course, if we have to take into account the fact that the rope vibrates within the air and that there is air resistance, there is energy loss. We ignore this possibility among others.) Thus the power supplied by whatever generates the waves in the first place is transmitted from one piece of the rope to another. This power is the rate P of energy transport through the rope. We can treat this issue by thinking of an energy density u—an energy per unit length—in the vibrating rope. This energy density must also take the form of a traveling wave, which travels down the rope. How is the energy density u related to the power P? We note that the quantity $v\,\Delta t$ is the length of wave that moves past a given point in a brief time interval Δt. All the energy in that length, namely the energy density times the length, or $uv\,\Delta t$, will be delivered; the rate of energy delivery P is this quantity divided by Δt. In other words, $uv = P$, or

$$u(x, t) = P/v = \mu\omega^2 A^2\cos^2(kx - \omega t). \qquad (14\text{–}38)$$

We have shown that the power delivered and the energy density on the rope are each quantities that travel with the medium's wave speed. These quantities are always positive. Moreover, *they are proportional to the squares of both the amplitude and the frequency of the wave*. Although we have derived our results for traveling waves on a rope, the quadratic dependence of the rate of energy transport, of power, or of energy density on amplitude and frequency is a general feature of *all* waves.

EXAMPLE 14–6 A long rope of mass density 150 g/m is tied at one end. A person holds the other end with a horizontal force of 95 N so that the equilibrium position of the rope is horizontal. The person then shakes the held end of the rope up and down such that sinusoidal traveling waves are generated with a frequency of $1.0\ \text{s}^{-1}$ and an amplitude of 5.0 cm. What average power must be delivered by the hand that is shaking the rope?

Setting It Up Figure 14–20 shows the wave produced when the rope is shaken up and down. We are given all the parameters that enter into the calculation of average power, namely the tension $T = 95$ N, the mass density $\mu = 0.15$ kg/m, the frequency $f = 1.0$ Hz, and the amplitude $A = 0.050$ m.

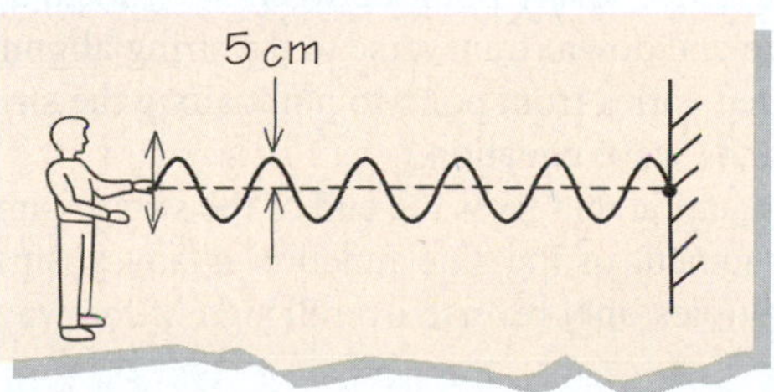

▲ **FIGURE 14–20** A person shaking a rope produces traveling waves.

Strategy The average power delivered to the rope is the average power carried by the rope in its wave motion. Thus Eq. (14–37) can be used to calculate the power. In this expression only the angular frequency and the wave speed are not given directly. The frequency gives the angular frequency, while we find the wave speed from the given tension and the given mass density using Eq. (14–9), and only the numerical details need to be worked out.

Working It Out To apply Eq. (14–37), we need $\omega = 2\pi f = 6.3$ rad/s, and $v = (T/\mu)^{1/2} = [(95\ \text{N})/(0.15\ \text{kg/m})]^{1/2} = 25$ m/s. Then Eq. (14–37) gives

$$P = \tfrac{1}{2}(0.15\ \text{kg/m})(6.3\ \text{s}^{-1})^2(0.050\ \text{m})^2(25\ \text{m/s}) = 0.19\ \text{W}.$$

This is a tiny fraction of 1 hp, easily delivered by the hand holding the rope.

What Do You Think? In stating this problem, we made the unspoken assumption that gravity does not act on the rope. Since the rope has mass, it is subject to gravity. Would you expect inclusion of the force of gravity to lead to a significant adjustment to the solution?

THINK ABOUT THIS . . .

WHY DOES A DISTANT SHOUT SOUND FAINTER THAN A NEARBY ONE?

Our results for the energy density, together with the conservation of energy, determine how the amplitude (wave height) of a periodic wave originating at a point, in this case the shouter's mouth, varies with distance from that point. For a one-dimensional system, the amplitude and the average energy density remain unchanged as the wave propagates—you might say that the energy has nowhere else to go. For waves in two and three dimensions the energy spreads out as it moves away from the source. We can visualize this if we think of the two-dimensional case of a toe periodically dipped into a pond. The moving toe is the source of the energy contained in the ripples that spread outward in circles from the toe. If a given ripple has radius r, the energy in the ripple is spread evenly over the circle, which has length $2\pi r$. Thus the total energy density falls off as $1/r$, while the amplitude falls off as $r^{-1/2}$ because the energy density is proportional to the square of the wave's amplitude. Similarly, a shout in open air is the source of waves that spread outward from the source as a sphere. Sound energy is spread over the surface of area $4\pi r^2$ of a sphere of radius r. Thus the energy density at a point decreases as $1/r^2$, and the amplitude of the sound waves from the sound decreases as its square root—that is, as $1/r$.

14–6 Standing Waves

When you pluck a guitar string, it is the string as a whole that vibrates, with the vibration largest in the middle of the string and smallest—indeed, zero—at the stopping fret and the bridge. As we remarked in Section 14–1, this doesn't seem much like the traveling waves we have been discussing. In fact, as we shall now see, it is a very closely related phenomenon.

Let us return to the very general harmonic solution to the wave equation, Eq. (14–19), which we rewrite here for convenience,

$$z(x, t) = A \cos kx \cos \omega t + B \cos kx \sin \omega t + C \sin kx \cos \omega t + D \sin kx \sin \omega t. \tag{14–19}$$

Can we find solutions that describe the vibrating guitar string by choosing the parameters A, B, C, and D correctly? If the string runs from $x = 0$ to $x = L$, such solutions would give the string's ends fixed at $x = 0$ and $x = L$, that is, $z(0, t) = z(L, t) = 0$ *for all time*. The requirement that $z(0, t) = 0$ means that $A \cos \omega t + B \sin \omega t = 0$ for all times, and this is only possible if $A = B = 0$, and leaves us with

$$z(x, t) = (C \cos \omega t + D \sin \omega t) \sin kx.$$

The term $C \cos \omega t + D \sin \omega t$ can always be written in the form $z_0 \cos(\omega t + \phi)$. To do this, simply write, as is always possible, $C = z_0 \cos \phi$ and $D = -z_0 \sin \phi$. Then

$$C \cos \omega t + D \sin \omega t = z_0 \cos \phi \cos \omega t - z_0 \sin \phi \sin \omega t = z_0 \cos(\omega t + \phi).$$

Our wave now has the form

$$z(x, t) = z_0 \cos(\omega t + \phi) \sin kx. \tag{14–39}$$

We must still enforce the requirement that the end of the string at $x = L$ does not vibrate, i.e. $z(L, t) = 0$ for all time, or by comparison with Eq. (14–39),

$$\sin kL = 0. \tag{14–40}$$

This is a condition on the k-values that we'll investigate immediately below. We call the form Eq. (14–39) a **standing wave**, since at any point between $x = 0$ and $x = L$, the guitar string moves harmonically up and down (transverse to the string alignment) with time, with an amplitude $z_0 \sin kx$ that varies from point to point along the string. Nothing "propagates" in this solution to the wave equation.

What does Eq. (14–40)—the equation that fixes the end of the string—imply? This equation can be satisfied if the argument of the sine function is any positive integer multiple of π (negative integer multiples only reverse overall sign of the wave), or if k takes on the discrete values

$$k_n = \frac{n\pi}{L} \quad \text{for } n = 1, 2, 3, \ldots. \tag{14–41}$$

(The case $n = 0$ gives just a flat or undistorted string—we say the string is *unexcited.*) This is a condition on the wavelength of the wave, since $k = 2\pi/\lambda$, and Eq. (14–41) then implies that the allowed wavelengths also carry an index n:

$$\lambda_n = \frac{2\pi}{k_n} = \frac{2}{n}L \quad \text{for } n = 1, 2, 3, \ldots. \tag{14–42}$$

When the string is set into motion with values of k that correspond to a certain n-value, we say the string is *excited* into the wave labeled by n. Equation (14–42) can be rewritten as $L = \frac{1}{2}n\lambda_n$, providing a convenient way to recall the possible values of the wavelength. For a standing wave *the total length of a string (or any one-dimensional medium with fixed ends that supports waves) can be divided into an integral number of half-wavelengths*. The allowed waves are called **modes**—possible vibration patterns—of the standing waves, and they are labeled by the value of n. $n = 1$ labels the *fundamental mode* (see Fig. 14–21), also called the *first harmonic*. $n = 2, 3, \ldots$ are called the second, third, ... harmonics. Note that for odd values of n, the number of wavelengths that can fit on the string is a whole number plus one half, while for even values of n, a whole number of wavelengths fits into the string. For $n > 1$ there are *nodes*, locations along the x-axis where the string rests motionless with no transverse displacement. Although these nodes are not caused by a physical clamping of the string as at the end points, they otherwise behave as if the string were fixed at these points. Not counting the end points, there are $n - 1$ of these nodes—for example, one node in the $n = 2$ case of Fig. 14–21. We can also see the presence of *antinodes*, locations where the displacement is a maximum. How many nodes and antinodes are there in the standing wave with $n = 4$ in Fig. 14–21?

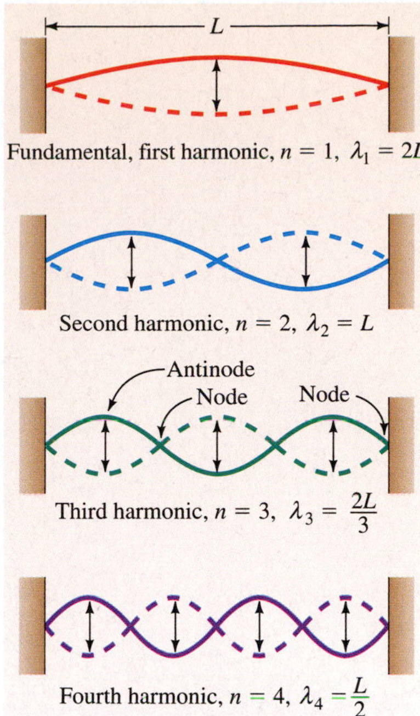

▲ **FIGURE 14–21** Some modes of transverse standing waves on a string with fixed ends. Higher modes—those characterized by larger values of n—have more nodes and antinodes. The arrows indicate the extremes of the motion as the string moves back and forth.

CONCEPTUAL EXAMPLE 14–7 Consider two strings, both fixed at each end, one twice as long as the other. Do they have any allowed wavelengths for standing waves in common?

Answer If the length of the shorter string is L, then for the shorter string the possible wavelengths are $\lambda_1 = 2L/n_1$, and for the longer string, length $2L$, we have $\lambda_2 = 4L/n_2$. The question then becomes: For what values of n_1 and n_2 can we have $\lambda_1 = \lambda_2$? The quickest way to see this is to eliminate L by writing a ratio

$$\frac{\lambda_2}{\lambda_1} = \frac{4L}{n_2} \times \frac{n_1}{2L} = \frac{2n_1}{n_2}.$$

This ratio must equal 1. We thus see that for every *even* value of n_2 there is an n_1, half its value, that leads to the equality of the wavelengths. The n values are integers, so this is the only solution.

EXAMPLE 14–8 Standing waves can be excited on a guitar string of length 1.00 m whose ends are fixed at $x = 0$ and $x = L$. Find the two smallest values of the wave number and the two longest wavelengths allowed for standing waves. Sketch the space dependence of these waves.

Strategy The possible wave numbers are given by Eq. (14–41). The allowed values of k grow with n, so the smallest wave numbers will correspond to $n = 1$ and $n = 2$, and we will use these values in our calculation. The wavelength is inversely proportional to the wave number, so the two smallest wave numbers are also the two longest wavelengths.

Working It Out The smallest values of k and the largest values of λ are the two cases $n = 1$ and $n = 2$ of Eqs. (14–41) and (14–42), respectively. These are

$$k_1 = \pi/L = 3.14\ \text{m}^{-1} \quad \text{and} \quad k_2 = 2\pi/L = 6.28\ \text{m}^{-1}$$

and

$$\lambda_1 = 2L = 2.00\ \text{m} \quad \text{and} \quad \lambda_2 = 2L/2 = 1.00\ \text{m}.$$

The space dependence of the waves is given by the function $\sin(k_n x)$ times an arbitrary amplitude z_0. These are sketched in Fig. 14–22.

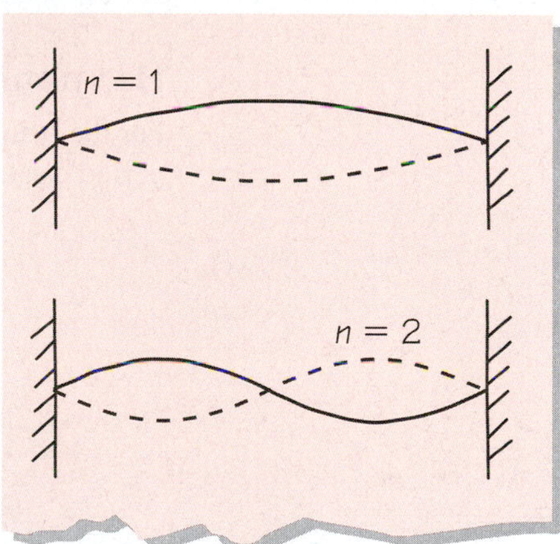

▲ **FIGURE 14–22** Standing waves on a guitar string.

What Do You Think? Suppose the guitar string is additionally clamped at $x = 20$ cm, for example by a finger against the fret. How do you think the results will change to the left of the clamping point? To the right of the clamping point? Will there be certain wavelengths such that the pattern looks like a single string?

The Frequencies of Standing Waves

The waves we have described above cannot have arbitrary wavelengths, and we'll now see that they cannot have arbitrary frequencies. The relation between frequency and wavelength follows from the wave equation, Eq. (14–7). If we substitute Eq. (14–39) into the the wave equation rewritten in the form $\frac{\partial^2 z(x,t)}{\partial x^2} - \frac{1}{v^2}\frac{\partial^2 z(x,t)}{\partial t^2} = 0$, we find for the left side

$$\frac{\partial^2}{\partial x^2}\sin kx\cos(\omega t+\phi) - \frac{\mu}{T}\frac{\partial^2}{\partial t^2}\sin kx\cos(\omega t+\phi)$$

$$= -k^2\sin kx\cos(\omega t+\phi) - \frac{\mu}{T}[-\omega^2\sin kx\cos(\omega t+\phi)]$$

$$= \left(-k^2 + \frac{\mu\omega^2}{T}\right)\sin kx\cos(\omega t+\phi).$$

This can equal 0 (the wave equation is satisfied) only if $k^2 = \mu\omega^2/T$ or, in terms of the wavelength and the wave speed $v = (T/\mu)^{1/2}$ (which here has no apparent connection to an actual speed),

$$\omega = \frac{2\pi}{\lambda}\sqrt{\frac{T}{\mu}} = \frac{2\pi v}{\lambda}. \tag{14–43a}$$

Equivalently this gives a condition on the period $\tau = 2\pi/\omega$ of the oscillations,

$$\tau = \lambda\sqrt{\frac{\mu}{T}} = \frac{\lambda}{v}. \tag{14–43b}$$

What do Eqs. (14–43a, b) tell us about the motion of the string?

1. For a particular string, the standing waves with smaller wavelengths also have smaller oscillation periods.
2. For a given wavelength and string length, a less dense string vibrates more rapidly than a denser one.
3. For a given wavelength and string length, a string under more tension vibrates more rapidly than a string under less tension.

Harmonics

For the string with fixed ends, $\lambda_n = (2/n)L$, and the frequency, angular frequency, and period can have the discrete values

$$f_n = \frac{n}{2L}\sqrt{\frac{T}{\mu}}, \tag{14–44a}$$

$$\omega_n = \frac{n\pi}{L}\sqrt{\frac{T}{\mu}}, \tag{14–44b}$$

$$\tau_n = \frac{2L}{n}\sqrt{\frac{\mu}{T}}. \tag{14–44c}$$

The frequency f_1, the lowest frequency, is known as the **fundamental frequency**, and $\lambda_1(= 2L)$ is the **fundamental wavelength**. The frequency $f_n = nf_1$ is called the ***n*th harmonic** of the string, and the full set is called the harmonic series; Fig. 14–21 depicts some harmonics. (We will return to these topics in the next section.)

EXAMPLE 14–9 The time taken to complete 100 full vibrations (100 periods) in the fundamental mode of a piano wire is 0.5 s. The wire length is 2 m, and the total mass of the wire is 25 g. What is the tension on the wire? What value should the tension have in order to halve this time?

Setting It Up We have the length L and total mass M of the wire as well as the period τ_1 of the fundamental mode.

Strategy The period, which in this case is 1/100th of 0.5 s, of any mode is directly related to the tension and mass density [Eq. (14–44c)]. Although it is not given directly, the mass density is $\mu = M/L$.

Working It Out Equation (14–44c) gives

$$\tau_1 = 2L\sqrt{\frac{\mu}{T}}.$$

This may be solved for the tension, T:

$$T = \frac{4\mu L^2}{\tau_1^2} = \frac{4L^2}{\tau_1^2}\frac{M}{L} = \frac{4ML}{\tau_1^2}.$$

With $L = 2.0$ m, $\tau_1 = (0.50\text{ s})/100 = 5.0 \times 10^{-3}$ s, and $M = 0.025$ kg

$$T = \frac{4(0.025\text{ kg})(2.0\text{ m})}{2.5 \times 10^{-5}\text{ s}^2} = 8.0 \times 10^3\text{ N},$$

equivalent to the average weight of ten men.

Since the period is proportional to $T^{-1/2}$, we can halve the period if the tension is increased by a factor of 4, to 3.2×10^4 N.

What Do You Think? Suppose the frequency of a particular piano note is to be shifted upward by 1%. Will the piano tuner in Fig. 14–23 (a) increase the tension by 1%? (b) decrease it by 1%? (c) decrease it by 2% or (d) increase it by 2%? Do all the strings in a piano differ only in the tensions?

▲ **FIGURE 14–23** This piano tuner adjusts the natural frequencies of the strings on this piano by adjusting their tensions.

THINK ABOUT THIS . . .

WHAT DO STANDING WAVES LOOK LIKE IN TWO AND THREE DIMENSIONS?

Traveling waves in two and three dimensions, such as ocean waves or sound waves in air, all behave according to the appropriate wave equation. Just as for waves on a string, if the medium supporting them has boundary conditions that restrict the displacement, standing waves will occur. Drumheads are an example of standing waves on a two-dimensional surface. The boundary conditions for a drumhead require that the vibrations vanish on the circular rim of the drum. Figure 14–6 shows a few of the modes possible on a circular drumhead. The modes on an elliptical drumhead differ from those on a circular drumhead, as the boundary conditions are different. Your microwave oven, which sets up standing waves within its interior volume, provides an example of three-dimensional standing waves. The boundary conditions determining the allowed values of k and the dynamics determining the allowed values of ω are more complicated when the system is a surface or a volume rather than a one-dimensional system like the string, but the common feature of all these systems is that the allowed wavelengths and frequencies are discrete and can be labeled by integers.

Energy and Standing Waves

There is no energy flow in a standing wave such as the vibrations of a guitar string, as there is no continuing source of energy input into the string. Nevertheless, there are internal forces leading to (changing) transverse velocity, and the motion contains energy; the motion simply does not translate into an energy flow. To understand why, we look to the formula for power. As we saw in Eq. (14–35), the power is given by $P = -T(\partial z/\partial x)(\partial z/\partial t)$. For a traveling wave of the form $z(x, t) = A\sin(kx - \omega t)$, we find $P = A^2Tk\omega\cos^2(kx - \omega t)$. This is always positive and is itself a traveling wave. But for a standing wave of the form $z(x, t) = A\sin kx\cos\omega t$, we get

$P = A^2Tk\omega[\cos kx \sin kx] \times [\cos \omega t \sin \omega t] = (ATk\omega/4) \sin 2kx \sin 2\omega t$. This is another standing wave. While it does not vanish, it oscillates in time, and since the time average of $\sin 2\omega t$ vanishes, the average power in the standing wave is zero.

14–7 More About Sound

In Section 14–4 we described sound waves in a metal bar and, more briefly, in air. Sound waves in air are generated by the mechanical motion of some object in contact with the air—the vibration of vocal cords, an oscillating violin string, or the collapsing hot-air column formed by lightning. A traveling sound wave transports energy through the air, which can then be picked up by another mechanical system such as an eardrum. In this section we look more closely at these waves.

The Nature of Sound Waves

Let's look at sound in air on a microscopic level. The molecules in air are widely spaced so the intermolecular forces that act as springs in solids or liquids play a very limited role here. Except for brief, random collisions with each other, air molecules move freely. How, then, can a wave propagate? Consider a pipe oriented along the x-axis as in Fig. 14–24. At one end of the pipe is a speaker. When the speaker diaphragm moves to the right, the air molecules within a thin slice of the pipe to the right of the diaphragm are subject to a force that acts along the $+x$-direction. These air molecules acquire a net momentum in the $+x$-direction in addition to their random motion. Some of the molecules in the original slice will then move to an adjacent region, increasing the number of molecules there and creating a region of *compression*, where the molecular density is higher. When the diaphragm moves back to the left, fewer molecules remain in this region, which corresponds to a region of *rarefaction* where the molecular density is lower. We thus have adjacent regions of compression and rarefaction, as in Fig. 14–24. In the region of compression, there are more molecules and therefore more molecular collisions. The air molecules that bounce out of the region of more collisions will fill in the rarefied region. The molecules to which the net momentum had been transferred will move in the $+x$-direction, so the region of compression moves to the right in the form of a pulse. If the diaphragm undergoes periodic motion, regions of compression and rarefaction are propagated in the $+x$-direction as a periodic wave in air. Figure 14–24 shows the molecular pattern of a sound wave in air. The displacement of the molecules is aligned with the propagation direction of the wave, so sound waves in air are *longitudinal* waves.

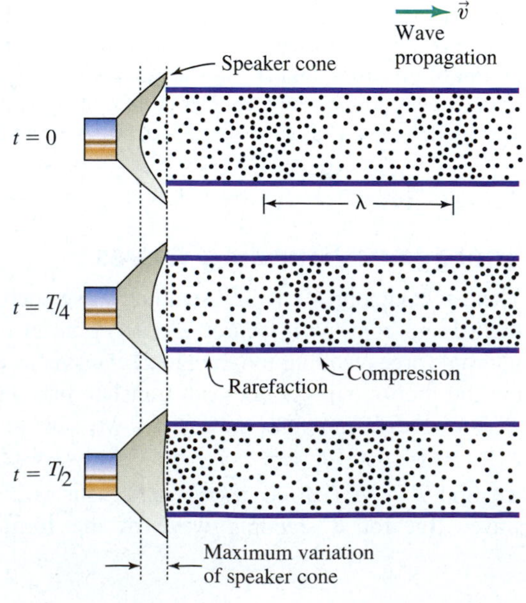

FIGURE 14–24 When a piston—here a speaker diaphragm—at one end of an air-filled pipe moves along the $+x$-direction, there is a force on the air molecules to the right of the piston, initially supplied by collisions of molecules with the moving piston, and subsequently by collisions between molecules. If the piston moves back and forth regularly, the result is alternate regions of higher and lower density propagating along the pipe, which are longitudinal waves.

There is no net movement of molecules in the direction of the wave. It is rather the alteration of the density from the average (or equilibrium) density that propagates. If ρ is the density of air (or any other gas), then sound waves may be described in terms of $\Delta\rho = \rho - \rho_0$, the deviation from the equilibrium density ρ_0. A harmonic sound wave traveling in the $+x$-direction would be described by

$$\Delta\rho = A_\rho \sin(kx - \omega t), \tag{14–45}$$

where A_ρ is the maximum deviation of the density from ρ_0. Generally, $A_\rho \ll \rho_0$, so we may speak of a nearly constant mass density, even in the presence of sound waves. The speed of the sound waves obeys the relation $v = \omega/k$, and it is given by Eq. (14–30b), $v = \sqrt{\gamma p_0/\rho_0}$, where ρ_0 is the equilibrium density of the gas, p_0 is the pressure, and γ is a numerical constant equal to 1.4 in air.

Standing Sound Waves

Standing sound waves can be generated if the necessary boundary conditions exist. For example, a pipe that is much narrower than the wavelength of a wave will quite easily generate standing waves that we can view as one dimensional. If the pipe is closed at one end, then there can be no displacement of gas molecules across that end. The closed end of the pipe is then like the fixed end of a string. If both ends are closed, we have the equivalent of a string fixed at both ends. As Fig. 14–25 indicates in a schematic way, the wavelength for these standing waves takes on only the discrete values allowed by Eqs. (14–41) and (14–42). The frequencies are determined in terms of the wavelengths by Eq. (14–21),

$$v_{\text{sound}} = \lambda f. \tag{14–46}$$

As for a vibrating string, the discrete integer n labels the modes [Eqs. (14–44)]. When $n = 1$, we have the fundamental frequency (discussed in Section 14–6); when $n > 1$, we speak of the nth harmonic.

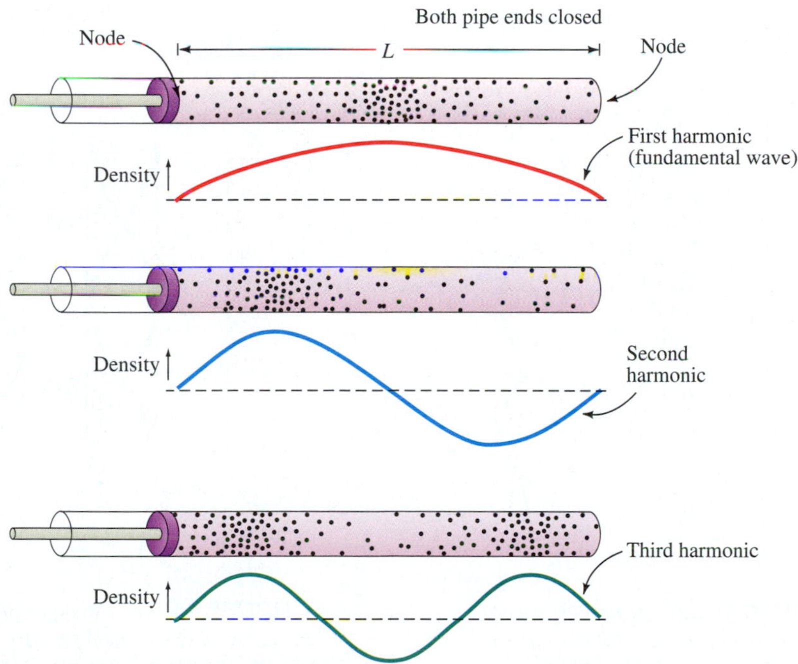

▲ **FIGURE 14–25** There is a node of molecular displacement at each end of a closed pipe. Thus the longest possible wavelength in this situation is the one that has one half of the standing sound wave within the pipe. The longest wavelength represents the fundamental frequency, corresponding to the first harmonic. The next two harmonics are also shown here. The curves represent the amplitudes of the density wave which, you should keep in mind, is a wave in the departure of the density from its equilibrium value described by the horizontal dashed line.

EXAMPLE 14–10 A pipe closed at both ends has length $L = 4.0$ m. Find the fundamental frequency for standing sound waves, as well as the frequencies of the next two modes.

Setting It Up Figure 14–25 shows the fundamental and the next two harmonics.

Strategy As Eq. (14–42) states, the fundamental wavelength, λ_1, for a system with two closed ends is twice the length of the system: $\lambda_1 = 2L$. Higher modes have shorter wavelengths, enumerated in this case by $n = 2, 3$: $\lambda_2 = \lambda_1/2$, $\lambda_3 = \lambda_1/3$. The relation between frequency and wavelength is given by Eq. (14–46). Thus the frequencies in question are determined by

$$f_n = \frac{v_{\text{sound}}}{\lambda_n} = \frac{v_{\text{sound}} n}{2L}.$$

Working It Out Taking v_{sound} from Eq. (14–30b), we find

$$f_1 = \frac{330 \text{ m/s}}{2(4.0 \text{ m})} = 41 \text{ Hz}, \; f_2 = 2f_1 = 83 \text{ Hz}, \; f_3 = 3f_1 = 124 \text{ Hz}.$$

Figure 14–25 shows how the standing wave of fundamental frequency fits within the pipe—it has a node at each end, but none in between.

What Do You Think? The standing waves in this example run from end to end of the pipe. Do we have to assume that the pipe is infinitely narrow to justify this? If not, are there standing waves that go across the pipe?

Other boundary conditions are possible for standing sound waves. Consider the organ pipe in Fig. 14–26. When one end of a pipe is open, the pressure at that end takes on the value of the pressure of the exterior. There is no variation of the pressure there, which corresponds to a pressure node. This can be shown to mean that the displacement has an amplitude maximum (or *antinode*) rather than a node at the open end. Figure 14–27 shows a similar pipe, but this time open at both ends. When both ends are open, there is an antinode at each end (Fig. 14–27), in which the first few harmonics are drawn.

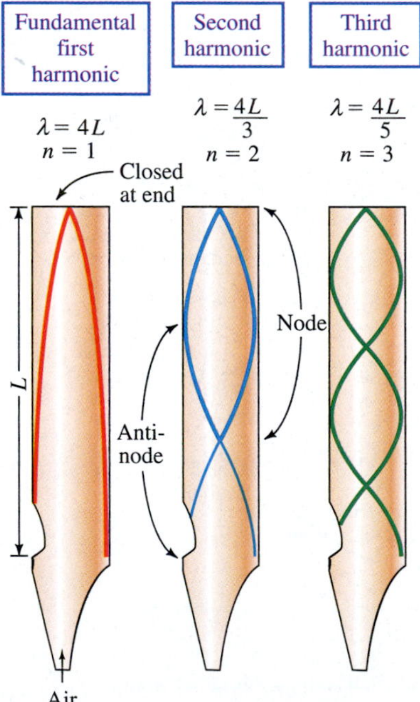

▲ **FIGURE 14–26** A standing sound wave in an organ pipe open at one end has a maximum at the open end. In this situation, one-quarter of the standing wave with the longest possible wavelength—the fundamental—fits within the pipe. The next harmonic is also shown. The curves represent the amplitudes of the density wave.

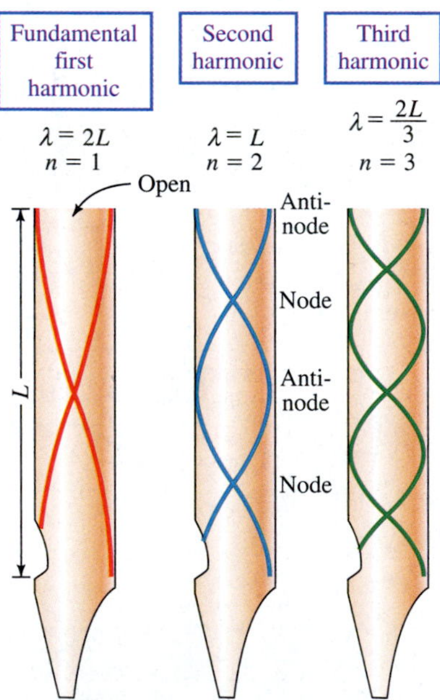

▲ **FIGURE 14–27** An organ pipe open at each end has standing waves with antinodes at each end, hence a different harmonic series than for the organ pipe closed at one end.

EXAMPLE 14–11 An organ pipe with the dimensions of the pipe in Example 14–10 is closed at one end and open at the other. Find the fundamental frequency and next harmonic.

Setting It Up We are asked to find frequencies just as in the previous example. What is different about this situation? Here the boundary conditions have been changed, and we must take this into account.

Strategy We must have a node at one end and an antinode at the other. The longest possible standing wave that fits this description is easily drawn (Fig. 14–26). The figure shows that the wavelength is four times the length of the pipe, $\lambda_1 = 4L$. This is double the wavelength of the fundamental mode when both ends are closed. Equivalently, the corresponding frequency is half the fundamental frequency for both ends closed.

Working It Out The fundamental frequency is

$$f_1 = \frac{v_{\text{sound}}}{\lambda_1} = \frac{v_{\text{sound}}}{4L} = \frac{330\ \text{m/s}}{4(4.0\ \text{m})} = 21\ \text{Hz}.$$

The next harmonic is not the harmonic with a frequency twice the fundamental frequency; such a wave would have a node at the open end. Rather, it is the harmonic with three times the fundamental frequency, also illustrated in Fig. 14–26. For this wave, three-quarters of one wavelength fits into the pipe. Mathematically, notice that at L the magnitude of the solution $\sin kx$ must have its maximum value, so that $\sin kx = \pm 1$. Thus $kx = \pi/2,\ 3\pi/2,\ 5\pi/2 \ldots = (2n + 1)\pi/2$, with $n = 0, 1, 2, \ldots$

What Do You Think? If you take an empty bottle and blow across the open end, you get a deep sound. Why? Suppose you were to blow across a cardboard tube with both ends open. What would you hear?

The Excitation of Standing Sound Waves: We can excite standing waves in an organ pipe if we vibrate the air within at the standing wave frequency. One way to do this is to direct externally generated sound waves of that frequency to the pipe. This process must also go in the other direction—the standing waves within the pipe must generate traveling sound waves of the same frequency if the members of an audience are to perceive the standing waves as sound. The couplings between these different types of waves normally involve rather complicated mechanisms. The many different ways that standing waves in musical instruments are excited (bow, single and double reed, mouthpiece, and so forth) indicates the rich variety of physical phenomena involved.

In each case, the excitation of waves involves a transfer of energy. This transfer of energy, from the energy of an external source—such as the air forced across a clarinet mouthpiece—to the energy of the standing wave within the tube of a clarinet, takes place in ways closely related to the *resonance* phenomenon discussed in Chapter 13. If a tuning fork is placed near the end of an organ pipe and if the frequency of the tuning fork is the same as the fundamental frequency of the pipe, strong standing waves will rapidly be established within the pipe. Resonance of this type is responsible for the fact that when a trumpet plays a note whose frequency matches the natural frequency of a delicate goblet, standing waves can be induced in the goblet strong enough to make it break (Fig. 14–28).

▲ **FIGURE 14–28** The glass breaks because its natural frequencies are resonant with a frequency (or with several) of the sound wave, which allows energy to be transferred easily from the sound to the glass.

Hearing Sounds

The combination of ear and brain acts as a very sensitive detector of sound with many interesting properties. The most basic quantities it detects are pitch and loudness. *Pitch* is a measure of frequency, and the human ear can hear frequencies ranging from about 20 Hz to 20,000 Hz. *Loudness* is a measure of the power carried in a wave.

Because sound waves in air form two-dimensional fronts (Fig. 14–29), it is appropriate to define the average power per unit area rather than the total power in the wave.

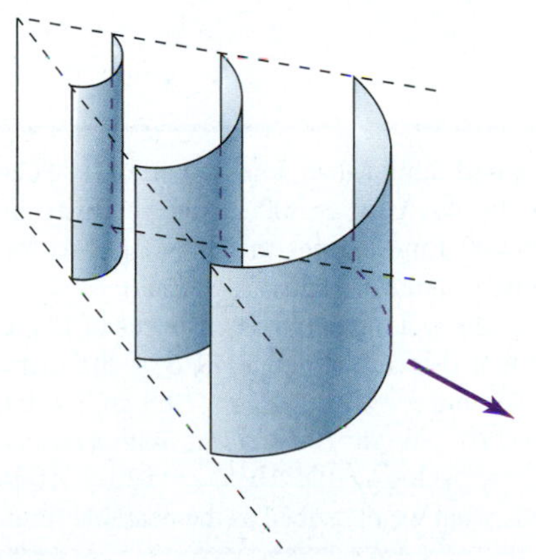

◀ **FIGURE 14–29** Waves generated in a three-dimensional medium form two-dimensional fronts. A front is defined as traced out by, say, maxima of a harmonic wave. In one-dimensional waves, the front would be a point moving along the medium and in two-dimensional waves the front is a line.

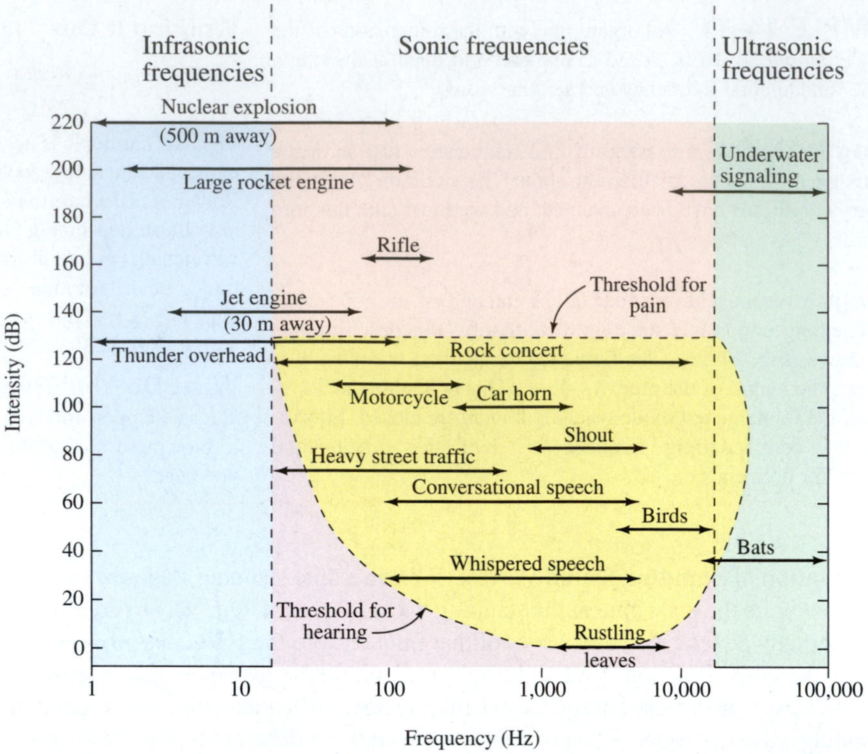

FIGURE 14–30 Typical noise levels at the source. We have here plotted intensity, measured in decibels, as a function of frequency. The frequency axis is divided into three parts based on the central, audible, portion of the spectrum—each is plotted on a logarithmic scale.

This average power per unit area is called the **intensity**, I. The intensity measures energy per unit time per unit area and has SI units of watts per square meter $(\mathrm{W/m^2})$. Intensity, like power, is proportional to the square of both the amplitude and the frequency. Intensity itself is not a particularly convenient measure physiologically; a measure better suited to the human ear/brain combination is the *decibel scale*, in which the measure of intensity is a dimensionless quantity, β, with units of decibels (dB). It is defined by

$$\beta = 10 \log_{10}\left(\frac{I}{I_0}\right), \tag{14–47}$$

where

$$I_0 = 10^{-12}\ \mathrm{W/m^2}.$$

A logarithmic scale such as this easily covers a large range of intensities (Fig. 14–30). The smallest detectable intensity is I_0 itself, for which $\beta = 0$, and the largest intensity still perceived as sound is about $I = 1\ \mathrm{W/m^2}$, or $\beta = 120$ dB. Higher intensities are not perceived as sound, but rather as something painful. The range of intensities detectable as sound by the ear–brain combination is thus an impressively large factor of 10^{12}, as Figure 14–30 shows. Curiously enough, the range of sound intensities detectable by the human ear is about the same as the range of light intensities detectable by the human eye.

CONCEPTUAL EXAMPLE 14–12 A football fan sits in the upper deck of a large football stadium and has a portable sound device to cheer for his team. The device emits its sound equally in all directions, and it has the property that the sound is loud enough that someone sitting 300 m away on the other side of the stadium can hear it above the crowd noise of 80 dB. Why do you not want to be sitting 1 m in front of the fan?

Answer The intensity I is proportional to $1/D^2$ where D is the distance from the device (see "Think About This. . . . Why does a distant shout sound fainter than a nearby one?" in Section 14–5), so that whatever the intensity at 300 m, the intensity at 1 m will be $(300)^2$ i.e. 90,000 times larger than the intensity of the ambient noise in a football stadium.

We can work this out quantitatively in terms of β. At 300 m the intensity is such that β is larger than 80. If $\beta = 80$, then $I/I_0 = 10^{12} I = 10^{\beta/10} = 10^8$, and $I = 10^{-4}\ \mathrm{W/m^2}$. At 1 m, the intensity is then $I = (300)^2 \times (10^{-4}\ \mathrm{W/m^2}) = 9\ \mathrm{W/m^2}$, with a corresponding decibel level of $\beta = 10 \log_{10}(9 \times 10^{12}) = 120 + 10 \log_{10}(9)$. This number exceeds what we described as the bearable limit of sound.

14–8 The Doppler Effect

Traveling waves move with a finite velocity. It is therefore possible for an observer who measures the waves (the receiver) to move relative to them or for the generator of traveling waves (the source) to move relative to the waves it produces. The movement of the source or of the receiver affects the measured frequencies. An extreme case illustrates the effect: A receiver moves along with the wave crests and measures a wavelength, just as if he or she were stationary with respect to the medium, by extending a yardstick from one crest to another. But this receiver measures no frequency at all, for the frequency measures the rate at which crests pass the observer.

It is not hard to see the result of motion of receiver or source in the case of sound waves. The perception of a sudden change in the pitch of an automobile horn (a source) as a vehicle passes is a common experience. In 1842 Christian Doppler gave an explanation for this phenomenon, as well as for a corresponding phenomenon for light. The **Doppler effect** is the shift in frequency and wavelength of waves that results from a source moving with respect to the medium, receiver moving with respect to the medium, or even a moving medium. We'll treat the various possibilities in order.

Case A: A Moving Source: Consider a point source of traveling waves, such as a siren wailing at some frequency f_0 (period $\tau_0 = 1/f_0$). These waves spread away from the source at a wave speed v that is characteristic of the medium alone and is unaffected by the source's motion. If the source is stationary with respect to the medium, these waves are symmetric on all sides of the source, and the wavelength is $\lambda = v/f_0$ [from Eq. (14–46)]. But if the source moves with velocity $\vec{v}_s$ with respect to the medium, then the wave fronts will be squeezed together in the direction of $\vec{v}_s$ and spread apart in the opposite direction (Figs. 14–31a, b). The larger circles in Fig. 14–31a and b correspond to emission at some earlier time, when the source was farther back in the direction $-\vec{v}_s$. Thus the wavelengths are shorter along the direction $+\vec{v}_s$ and longer in the opposite direction.

Consider, for simplicity, the source to be moving directly toward one observer and directly away from another observer (Fig. 14–31c). For the case that the source moves toward the observer, during one period τ_0 it moves a distance $v_s\tau_0 = v_s/f_0$. The wavelength is then decreased by this amount:

$$\lambda' = \lambda - \frac{v_s}{f_0} = \frac{v - v_s}{f_0}. \tag{14–48}$$

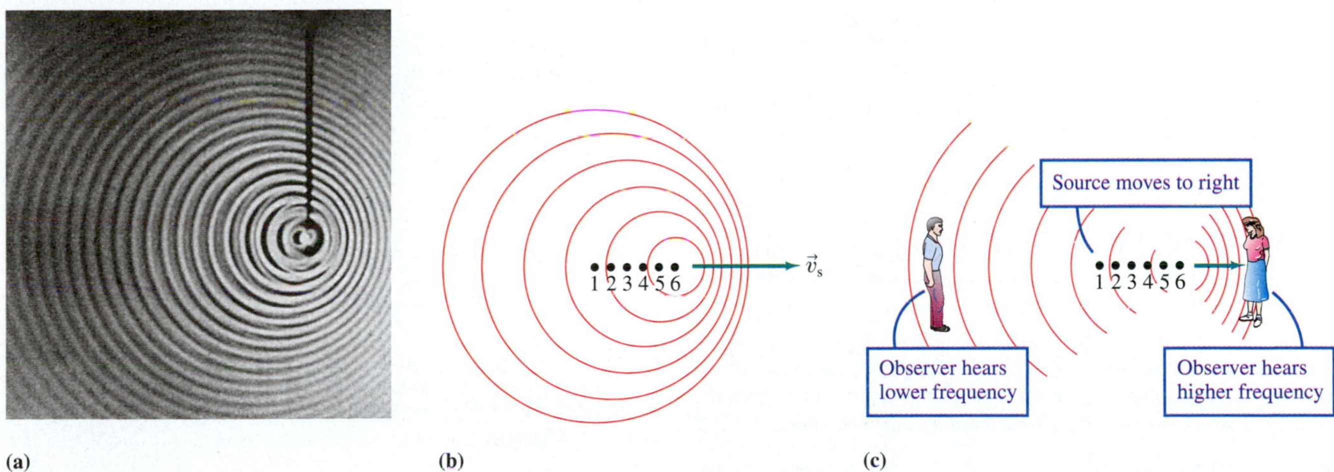

▲ **FIGURE 14–31** (a) The Doppler effect is seen in the waves of a ripple tank. Here the "dipper," the device that generates the circular waves by rhythmically touching the surface, is moving slowly to the right, so that successive centers of the spreading ripples are farther to the right. (b) The dots along the direction of the velocity $\vec{v}_s$ of the source represent the source position when the wave centered on them was emitted. Dot 1 corresponds to the outermost wave, dot 2 to the second from the outermost wave, and so forth. (c) Two observers at rest on either side of the moving source detect the waves, and observe different frequencies.

Because the wave speed remains unaffected throughout, the frequency with which the observer receives the waves becomes

$$f' = \frac{v}{\lambda'} = f_0\left(\frac{v}{v - v_s}\right) = \frac{f_0}{1 - (v_s/v)}. \tag{14–49}$$

When the source moves toward the observer, the wavelength *decreases* and the frequency *increases*—the pitch will be higher in the case of sound. For the case that the source moves away from the observer, we simply reverse the sign of v_s:

$$\lambda' = \lambda + \frac{v_s}{f_0} = \frac{v + v_s}{f_0}; \tag{14–50}$$

$$f' = \frac{f_0}{1 + (v_s/v)}. \tag{14–51}$$

When the source moves away from the observer, the wavelength *increases* and the frequency *decreases*—for sound, the pitch drops. A manifestation of this phenomenon is familiar to anyone who has heard a siren drop to a lower pitch as the vehicle with the siren approaches, then passes.

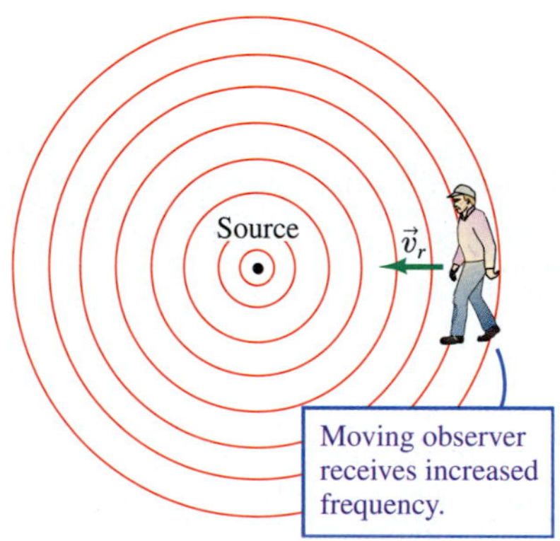

▲ **FIGURE 14–32** A receiver (an observer) moving with velocity $\vec{v}_r$ receives waves from a stationary source. In the situation shown, the observer hears an increased frequency.

Case B: A Moving Observer: Figure 14–32 depicts an observer moving with velocity $\vec{v}_r$ with respect to a wave-carrying medium, within which a stationary source is exciting waves with frequency f_0. We have illustrated the case of an observer moving directly *toward* the source. Such an observer will measure exactly the same wavelength λ as if he or she were at rest, but the observer sees the wave crests pass by more quickly. The observer measures a *modified wave speed*

$$v' = v + v_r. \tag{14–52}$$

The wavelength and wave speed provide the modified wave frequency that the moving observer measures:

$$f' = \frac{v'}{\lambda} = \frac{v + v_r}{\lambda} = f_0 + \frac{v_r}{\lambda} = f_0\left(1 + \frac{v_r}{v}\right). \tag{14–53}$$

When an observer moves *toward* the source of traveling waves, the wavelength is unchanged, the wave speed *increases*, and the frequency *increases*. We reverse the sign of v_r and repeat the derivation to learn that when an observer moves *away from* a wave source, the wavelength is again unchanged and the wave speed and the frequency both *decrease*:

$$v' = v - v_r; \tag{14–54}$$

$$f' = \frac{v - v_r}{\lambda} = f_0 - \frac{v_r}{\lambda} = f_0\left(1 - \frac{v_r}{v}\right). \tag{14–55}$$

EXAMPLE 14–13 While moving along a straight road, you first approach, and then pass, a stationary car just at the side of the road. The car's horn is being continuously sounded. Having perfect pitch, you recognize that the frequency you hear approaching the car is 853 Hz, while you hear a note of 741 Hz as you recede from the car. Use the value 330 m/s for the speed of sound in air to find the frequency the occupant of the car hears and your own speed.

Setting It Up We call your speed v_r, f_0 the frequency of the horn at rest, f_+ the frequency as you approach the horn, and f_- the frequency as you recede from it.

Strategy The numbers can be found with a proper application of the Doppler effect. As you move toward the car, f_+ is increased over f_0 according to Eq. (14–53), $f_+ = f_0\left(1 + \frac{v_r}{v}\right)$, where v is the speed of sound in air. Similarly, as you move away, you hear a decreased frequency $f_- = f_0\left(1 - \frac{v_r}{v}\right)$. These are two equations for the two unknowns f_0 and v_r.

Working It Out We solve the two equations by taking their sum and difference:

$$f_+ + f_- = 2f_0$$
$$f_+ - f_- = 2f_0(v_r/v) = (f_+ + f_-)(v_r/v).$$

The first equation immediately gives us f_0 as the average of the two frequencies,

$$f_0 = \tfrac{1}{2}(f_+ + f_-) = \tfrac{1}{2}(853\text{ Hz} + 741\text{ Hz}) = 797\text{ Hz}.$$

This is the frequency the car's occupant hears. The second equation gives us v_r:

$$v_r = v\frac{f_+ - f_-}{f_+ + f_-} = (330 \text{ m/s})\frac{853 \text{ Hz} - 741 \text{ Hz}}{853 \text{ Hz} + 741 \text{ Hz}} = 23.2 \text{ m/s}.$$

What Do You Think? Suppose you are driving a car toward the base of a massive cliff, and you honk your horn, which has a frequency f_0. A few seconds later you hear an echo—the sound reflected by the cliff. Do you expect to hear a higher or a lower frequency? How would you calculate it, knowing your speed? You may assume that a sound wave does not change its frequency when it is reflected.

Case C: A Moving Source and a Moving Observer: If all motion is in the same line, which we label the x-axis, then we can simply combine the results of cases A and B. Any velocity, including that of the wave itself, is positive if it is to the right and negative if it is to the left. With this reminder about the sign of the velocity, we drop the absolute-value signs. The effect of the moving source is to change the wavelength but not the wave speed, and the effect of the moving observer is to change the wave speed but not the wavelength. This can be expressed as

$$\lambda' = \frac{v - v_s}{f_0};$$
$$v' = v - v_r. \tag{14–56}$$

The modified frequency f' is then

$$f' = \frac{v'}{\lambda'} = \left(\frac{v - v_r}{v - v_s}\right)f_0. \tag{14–57}$$

Source and observer approach one another when v_s and v have the same sign and v_r has the opposite sign. In this case, f' is *increased* over f_0. When the source and observer are moving away from each other, v_r and v have the same sign and v_s has the opposite sign. The perceived frequency is *decreased*. Equations (14–56) and (14–57) include cases A and B as special cases. Note that Eq. (14–57) is *not* symmetric between the source and the observer. If we know the relative speed, we can tell what is moving—the source or the observer—by measuring the at-rest and in-motion frequencies.

When the speeds of the observer and the source are *small* compared to v, as is often the case when the Doppler effect applies to sound waves, then Eq. (14–57) is *approximately*

$$f' \cong \left(1 + \frac{v_s - v_r}{v}\right)f_0. \tag{14–58}$$

In this approximation, the Doppler shift (the change in frequency) depends only on the relative velocity between source and observer, and it is no longer possible to say who is moving—the source or the observer.

Case D: A Moving Medium: We can consider a fourth possibility: that the medium is itself in motion with respect to some fixed reference frame. Waves generated in a stream provide an example. The source is fixed with respect to the bank, and the stream is moving with speed v_m (Fig. 14–33). Let's take a look at how an observer on the bank downstream

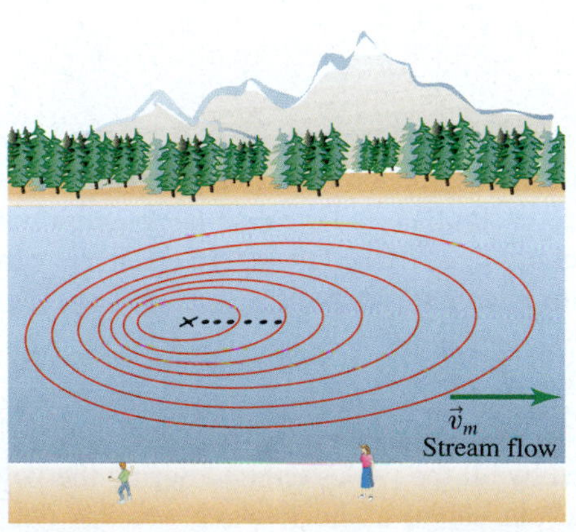

FIGURE 14–33 Traveling waves are generated at a fixed point, here marked with a cross, in a moving medium. The spreading waves move downstream; we can describe their positions by thinking of their centers as having moved downstream.

from the source sees the waves. There are two effects: First, the observer sees the crests moving by at an increased speed $v' = v + v_m$. Second, the wavelength is also increased because a downstream crest moves an additional distance before a second crest is emitted. From Fig. 14–33, it is easy to deduce that the wavelength increases from λ to $\lambda' = \lambda[1 + (v_m/v)]$. The frequency measured by the observer is $f' = v'/\lambda' = (v + v_m)/\lambda[1 + (v_m/v)] = (v/\lambda)[(v + v_m)/(v + v_m)] = v/\lambda = f$.

Similarly, an observer on the bank upstream from the source will see both the wave speed and the wavelength decreased, and in the same ratio, $[1 + (v_m/v)]$, so the frequency $f' = v'/\lambda'$ is unchanged. In other words, because both the speed and the wavelength are changed in the same way by the movement of the medium, *the frequency is unchanged by the movement of the medium*. The fact that the frequency is unchanged is the reason why it is possible to give outdoor concerts even under windy conditions. The intensity of the sound may be decreased if the wind is blowing in the wrong direction, but at least the pitch remains true.

Uses of the Doppler Shift

You are well aware of the use of the Doppler shift in police work—many speeders are caught this way. A radar beam (electromagnetic wave) is bounced from a moving car. The reflected wave is coming from a moving source, and the device's receiver translates the shifted frequency to the car's speed. Doppler radar can also be used in meteorological studies, where frequency shifts in radar waves reflected from moving objects such as raindrops convey a great deal of information about the dynamics of storms. Probably the most important use of the Doppler shift is in astronomy, where the measurement of the light emitted by distant stars exhibits a Doppler shift according to the motion of the star. The detection of the shift is greatly facilitated by the fact that certain characteristic discrete frequencies are emitted in starlight. By observing that the speed and direction with which stars and galaxies move has a correlation with their distance, Edwin Hubble was able to show that the universe is expanding.

*14–9 Shock Waves

When a source of traveling waves moves with a speed exceeding the wave speed, we encounter a new situation in which *shock waves* can form. This can be seen from Eqs. (14–49) or (14–57), which show that when $v_s = v$, the observed frequency becomes infinite. When $v_s = v$, the waves emitted directly in front of the source pile up on top of one another, so the wavelength is zero. Thus the time separation between the waves is zero, which corresponds to an infinite frequency. When v_s exceeds v, the source outruns the motion of the waves themselves; thus, instead of the squeezed-together wave fronts that occur for $v_s < v$ (see Fig. 14–31), the wave fronts are actually left behind.

The geometry of this situation can be analyzed as follows (Fig. 14–34). Suppose that the wave source is at point x_0 at $t = 0$ and emits a crest. In a time t_2, the source has traveled to x_2, a distance $v_s t_2$ from x_0. In this same time, the crest has moved out to a circular wave front a distance $v t_2$ from x_0. The perpendicular to the line drawn from x_2 tangent to this circle makes an angle θ_0 with respect to the line perpendicular to the motion of the source, and

$$\cos\theta_0 = \frac{v t_2}{v_s t_2} = \frac{v}{v_s}. \tag{14–59}$$

All the wave fronts emitted as the source moves lie along this tangent, which results in a piling up of wave fronts that reinforce each other and create a large-amplitude wave front. For instance, consider the crest emitted at t_1 when the source is at x_1, a distance $v_s t_1$ from x_0. The geometry of the figure gives

$$\cos\theta_1 = \frac{v(t_2 - t_1)}{v_s(t_2 - t_1)} = \frac{v}{v_s}.$$

Thus $\theta_1 = \theta_0$ and the tangent is formed by all the waves emitted during the voyage of the source. This line—a linear wave front called a **shock wave**—moves out with a

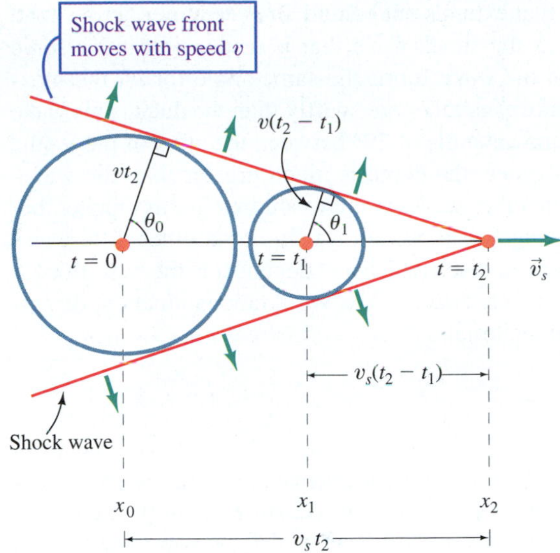

◀ **FIGURE 14–34** When the wave speed v is less than the source speed v_s, a linear wave front forms along the red lines. This front is called a shock wave. The source rhythmically emits waves that form outgoing circles in two dimensions or spheres in three dimensions; their crests are marked in blue in the figure. At $t = 0$ the source is at the center of the larger circle, at a later time t_1 at the center of the smaller circle, and at a still later time t_2 about to emit another wave crest. Successive circles form the continuous line of the shock wave.

speed v, maintaining the angle given by Eq. (14–59) (Fig. 14–35). When the wave is propagating in a three-dimensional medium, such as air, a cone is formed instead. We should stress that *there is only one such wave front*. A cone whose apex is at any location on the trajectory of the source represents just an earlier or later location of the cone that appears in Fig. 14–34. The wave front will pass an observer located on the ground just once.

When $v < v_s$, Eq. (14–59) determines without ambiguity the angle of the shock wave. The ratio v_s/v is called the *Mach number*; the larger the Mach number, the closer θ_0 comes to 90°—that is, the more acute is the angle of the line or cone. The sonic boom that occurs whenever an airplane moves faster than the speed of sound is a manifestation of the arrival of a shock wave of sound at an observer's ear. A second type of shock wave is associated with the movement of an electrically charged particle through some medium such as a gas or Plexiglas. The speed of light in this medium can be less than the speed of light in a vacuum, and a charged particle, whose speed cannot exceed that of light in a vacuum, can nevertheless move through the medium faster than the speed of light in that medium. Light waves then propagate along a shock-wave front whose angle is given by Eq. (14–59). These waves are called *Cherenkov radiation*, after Pavel Cherenkov (Fig. 14–36). When the speed of light in a given medium is known, measurement of the angle of the shock wave gives the speed of the charged particle. This is a useful measurement tool in experiments on particle physics.

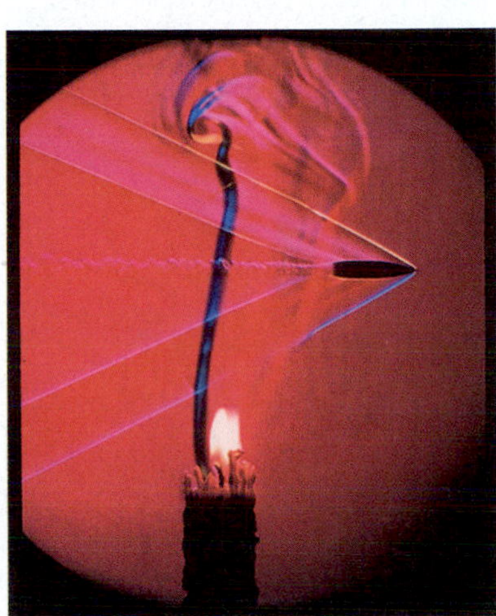

▲ **FIGURE 14–35** A bullet passing through the hot gases above a candle flame produces a shock wave because its speed is faster than the speed of sound in air. The ratio of these speeds determines the angle of the shock wave.

◀ **FIGURE 14–36** The blue glow is Cherenkov radiation. It is produced when a charged particle moves through a medium at a speed greater than the speed of light in the medium. In this case the medium is the water of a nuclear reactor and the particles are the product of the radioactivity of the reactor elements.

▲ **FIGURE 14–37** The wakes of these motorboats on the Connecticut River are a consequence of the fact that the speed of a water wave depends on its wavelength.

The *wake*, or track (Fig. 14–37) that extends out behind an ocean liner, a row boat, or a paddling duck is not the same as the shock wave that is responsible for a sonic boom. Upon observation, we see that the wake forms the same angle for *all* these examples, even though the ocean liner moves much more swiftly than the duck. This wake forms a V with the boat at the point and an angle of 39° between the arms of the V (the linear wave fronts). A wave in deep water (the depth is much greater than the wavelength) behaves differently from all the other waves we have studied in this chapter, because *its speed depends on its wavelength.* The wake of a ship, and other differences between that wake and the shock waves discussed here, depend on the variation of speed with wavelength, a phenomenon called *dispersion.* We'll talk again about dispersion when we deal with light waves in materials.

Summary

Extended bodies will often respond elastically to external and internal forces, and as a result they can support the organized motions we call waves. Waves can be classified as traveling waves, in which limited internal motion propagates progressively with some definite speed and in some definite direction through the medium, or as standing waves, in which the motion of the entire medium has the same sinusoidal time dependence. Both traveling and standing waves are further classified into either longitudinal waves or transverse waves. In transverse waves the motion within the medium is perpendicular to the direction in which the waves are organized. In longitudinal waves the motion is parallel to the wave direction.

The dynamics that allow waves to propagate are described by the wave equation, which for the particular case of transverse waves on a string of mass density μ under tension T takes the form

$$\frac{\partial^2 z(x,t)}{\partial x^2} = \frac{\mu}{T}\frac{\partial^2 z(x,t)}{\partial t^2}. \qquad (14\text{–}7)$$

Here the string is aligned with the x-axis, and z describes the transverse displacement of a point on the string. More generally, wave motion can be described by the wave equation, generically of the form

$$\frac{\partial^2 z(x,t)}{\partial x^2} = \frac{1}{v^2}\frac{\partial^2 z(x,t)}{\partial t^2}. \qquad (14\text{–}10)$$

The quantity v is the speed with which disturbances propagate.

For *harmonic* traveling waves, the motion of any point in the medium is described by

$$z(x,t) = A\sin(kx - \omega t). \qquad (14\text{–}22)$$

The parameters of this wave are the amplitude A, the angular frequency ω, and the wave number k. The wavelength, λ, the length over which a wave repeats itself, is related to the wave number by Eq. (14–16), $\lambda = 2\pi/k$. The parameters of frequency f and period τ are also used to describe wave motion. These parameters are all related by $\lambda f = \lambda/\tau = \omega/k$, from Eq. (14–21). The speed of the traveling wave is

$$v = \lambda f = \frac{\lambda}{\tau} = \frac{\omega}{k}.$$

This speed, here described in terms of its kinematic relation to the parameters of $z(x,t)$, is fixed by the wave equation. For example, comparison of Eqs. (14–7) and (14–10) shows that for a string of mass density μ under tension T, the speed is

$$v = \sqrt{\frac{T}{\mu}}. \qquad (14\text{–}9)$$

Waves carry energy, and traveling waves carry energy and momentum along the direction of their motion, even if the medium has no net motion in this direction. The power delivered by the wave is proportional to the square of the amplitude A and of the angular frequency ω. Its average value is

$$\langle P\rangle = \tfrac{1}{2}\mu\omega^2A^2v. \qquad (14\text{–}37)$$

Traveling sound waves in solids, liquids, or gases are longitudinal waves. The speed of sound in air is given by

$$v_{\text{sound}} = \sqrt{\frac{\gamma p_0}{\rho_0}}, \tag{14–30b}$$

where γ is a constant close to 1, and p_0 and ρ_0 are the pressure and the density of the medium, respectively. For normal atmospheric conditions at sea level, the speed of sound in air is roughly 330 m/s.

For standing waves, v is once again fixed by dynamical variables such as tension and mass density. The possible values of k and λ are determined by the boundary conditions of the medium and are generally discrete. The frequency of such waves is then determined from $f\lambda = v$, so that the angular frequency is

$$\omega = 2\pi f = 2\pi \frac{v}{\lambda}.$$

We see that the wave speed v affects the properties of standing waves even though they are not propagating.

The wave speed for mechanical waves takes the general form

$$v = \sqrt{\frac{\text{restoring force factor}}{\text{mass factor}}}. \tag{14–29}$$

The Doppler effect describes what happens to the frequency and wavelength of traveling waves emitted by a source moving with velocity $\vec{v}_s$ and detected by a receiver moving with velocity $\vec{v}_r$. If v is the speed of a wave within the medium, then the frequency perceived by the receiver is

$$f' = \left(\frac{v - v_r}{v - v_s}\right) f_0. \tag{14–57}$$

Here v_r and v_s are positive (negative) when the corresponding motion is along (against) the motion of the wave.

Understanding the Concepts

1. Give some examples of transverse waves and longitudinal waves and discuss the nature of the mechanical motion in each wave.
2. Meteorologists use something called "Doppler radar" to study the motion of clouds, tornadoes, and so on. What do you think is going on?
3. When a gong is sounded, it produces sound waves in air. If the same gong were struck in a vacuum, would it continue to vibrate for less time, for the same amount of time, or for more time than it does when struck in the same way in air? Explain your reasoning.
4. In old cowboy movies, you have probably seen someone place his ear on railroad rails to tell if a train is coming. Why is this done, and how does it work?
5. You have two ropes—one of them half the length and twice the mass density of the other—both under the same tension. The ropes are firmly attached to walls at each end. Compared to those for the shorter rope, the allowed wavelengths for the fundamental and first harmonic modes ($n = 1$ and 2, respectively) for the longer rope are (a) both doubled; (b) $n = 1$ doubled, $n = 2$ the same; (c) the same.
6. Would you expect the speed of sound (a) in hydrogen and (b) in argon to be greater than or less than the speed of sound in oxygen if all gases are under the same pressure?
7. When a lecturer in a classroom demonstration takes a lungful of helium and speaks, his or her voice is comically higher in pitch. Why is that?
8. There is a biological reason why a church does not need an organ pipe longer than 4.0 m. What is that reason? There are much longer pipes. What is their effect?
9. Captain Kirk is somewhere between galaxies. When a gong sounds in a neighboring spaceship, Kirk reacts to the sound. What is wrong with this scenario?
10. Suppose that a string supports a standing wave with a node at a particular location. How would things change if the point of the node were clamped physically?
11. In the expression for $h(x, t)$ in Example 14–2, the wave height at the observer's position varies sinusoidally, with a minimum (a trough) at $t = 0$. How would you write a sinusoidal function that would allow you to set the wave height at an arbitrary value (between $+h_0$ and $-h_0$) at $t = 0$?
12. What prevents sound waves from being transverse waves in a dilute gas such as air? In solids, transverse "sound" waves are generally possible. What can we learn about the nature of the intermolecular forces in such a solid?
13. Are the nodes actually motionless when a longitudinal standing wave is established on a Slinky?
14. Suppose the supports that hold the two ends of a piano wire are just simply pulled apart, so that the length of the wire increases by a small amount. How will this affect the fundamental period?
15. The speed of deep-water waves of wavelength λ is given by the formula $v = \sqrt{g\lambda/2\pi}$. When there is a storm somewhere on the ocean, it manifests itself by large breakers (breaking waves) at the shore. Initially you observe that the breakers arrive every 20 s at the shore. Somewhat later you observe that they are coming every 10 s. How do you interpret this?

16. A child concocts a telephone with two cans that are open at one end and connected by a string attached through a small hole to the closed ends. When the string is made taut by pulling on the cans at opposite ends, two children talking into the open ends can carry out a soft conversation over rather large distances. How does this happen?
17. The two supports that hold the end of a piano wire are pulled apart. If you wanted the fundamental frequency (pitch) of the wire to remain unchanged, what should you do about the tension, if anything?
18. A common toy consists of a series of five balls that touch and form a line; each ball is suspended as a pendulum. When ball 1 is pulled away and then released, it strikes the line and ball 5 rises in a motion like the reverse of ball 1. If there is a small gap between the balls at equilibrium, then a series of collisions occurs in motion, each with an exchange of velocities, until the last ball swings up. The phenomenon does not change even if the balls touch when at rest. If we describe the phenomenon as a collision of the first ball with an object four times its mass (this is how we describe the four touching balls), we do not get the observed motion of the balls. Why is the last description inadequate?
19. Suppose a rope on which standing waves are generated has its already substantial tension increased by a factor of 4. Do you expect that the velocity will be *exactly* doubled? If not, in which direction will it depart from doubling?

Problems

14–4 Traveling Waves

1. (I) A traveling wave is described by a displacement function $y = A\sin(ax - bt)$, where $a = 0.30\ \text{m}^{-1}$ and $b = 0.020\ \text{s}^{-1}$. Plot the displacement versus t at both $x = 1.0$ m and $x = 12$ m.
2. (I) A wave is described by $y = A\sin(kx + \omega t)$, where $k = 3.0\ \text{m}^{-1}$, $\omega = 2.0\ \text{s}^{-1}$, and $A = 0.40$ m. Determine (a) the wavelength, (b) the frequency, (c) the maximum amplitude, (d) the period, and (e) the wave number.
3. (I) Consider a long length of the wire used in Example 14–4. The wire is taut under a tension of 55 N. What is the mass density of the wire?
4. (I) The French train known as the TGV travels at 300 km/hr. What is the frequency with which the cars of this train pass a given point at that speed, given that the cars are 30 m in length?
5. (I) A harmonic wave with wavelength 0.27 m moves with speed 13 m/s. Find the wave number k and the frequency f of this wave.
6. (I) A rope of density μ_1 is attached to another rope, of density μ_2 (Fig. 14–38). A traveling wave of speed v_1 is sent down the first rope. What will the speed in the second rope be?

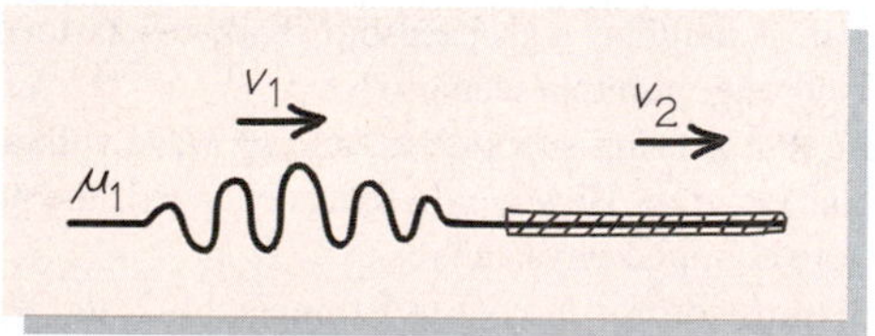

▲ **FIGURE 14–38** Problem 6.

7. (I) The values of Young's modulus, Y, for aluminum and copper are $Y_{Al} = 7.0 \times 10^{10}\ \text{N/m}^2$ and $Y_{Cu} = 11 \times 10^{10}\ \text{N/m}^2$. Their mass densities are $2.70 \times 10^3\ \text{kg/m}^3$ and $8.96 \times 10^3\ \text{kg/m}^3$, respectively. Calculate the speed of elastic wave propagation in these two media.
8. (I) A mountain climber of mass 68 kg hangs from a nylon rope 28 m below the top of a cliff. The total mass of the rope is 1.3 kg; assume that this mass is small enough so that the variation in equilibrium tension along the rope can be ignored. How long does it take a wave pulse to travel up the rope from the man to the top of the cliff?
9. (I) The speed of sound in a material of density $2.7 \times 10^3\ \text{kg/m}^3$ is measured to be 5.1×10^3 m/s. What is Young's modulus for that material?
10. (I) The shear modulus of copper is $G = 4.2 \times 10^{10}\ \text{N/m}^2$. Given that the density of copper is $8.93 \times 10^3\ \text{kg/m}^3$, what is the speed of a shear wave in copper?
11. (I) The density of pure gold is $19.3\ \text{g/cm}^3$, and the speed of sound in a small sample is measured to be 2.04×10^5 cm/s. What is Young's modulus for gold in dynes/cm^2?
12. (I) Given that the ratio of densities for steel and lead is 7.86 to 11.3, and that the ratio of Y values is 200,000 MN/m^2 to 16,000 MN/m^2, what is the ratio of the speeds of sound in these two materials?
13. (II) Use Eq. (14–22) for the traveling wave $z(x, t)$ and the wave equation to show that $Tk^2 = \mu\omega^2$.
14. (II) A perfect triangular pulse travels along a string with a velocity of 4 m/s. The shape of the pulse at $t = 0$ is shown in Fig. 14–39. Plot the displacement and velocity of the segment at $x = 3$ m as a function of time. Is such a pulse physically possible?

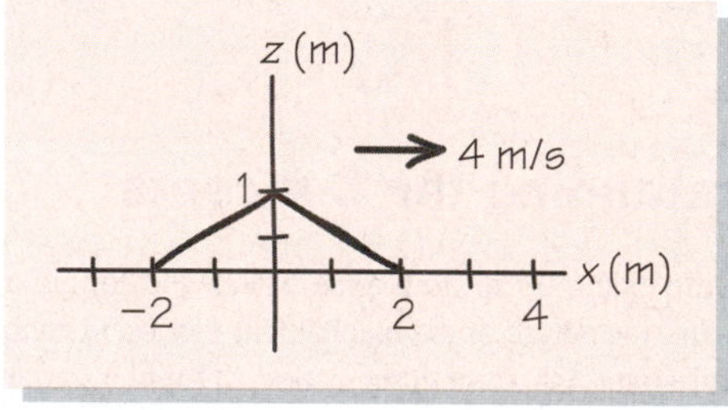

▲ **FIGURE 14–39** Problems 14, 16.

15. (II) Show by substitution that the general form of a traveling wave, $f(x - vt)$, satisfies the wave equation. [*Hint*: Let $s = x - vt$ and use the chain rule for differentiation.]
16. (II) Between $x = 0$ cm and $x = 10$ cm, a very long wire (so long that the boundary effects do not matter) is distorted from its straight-line equilibrium form into a triangular shape like that of Fig. 14–39, with the peak 4.0 cm above the original position. The tension and density of the wire are such that waves travel along the wire at a speed of 400 m/s. (a) What is the shape of the wire at the initial time $t = 0$? (b) What is the shape of the wire at an arbitrary time t, assuming that the wave travels in the $-x$-direction?
17. (II) A scuba diver has to send a message to save an oil platform in the Gulf of Mexico. He realizes that sound travels faster in steel than in air or water. With only one second left for the message to be received on the platform, he finds a steel mooring cable that is 5.0 km long, which runs directly to the platform. Will the message reach the platform in time? (The density of steel is $7.9 \times 10^3\ \text{kg/m}^3$ and $Y = 210{,}000\ \text{MN/m}^2$ for steel.)
18. (II) Young's modulus for Earth can be approximated by $5.5 \times 10^5\ \text{MN/m}^2$, and the corresponding shear modulus by $2.0 \times 10^5\ \text{MN/m}^2$. A first signal of an earthquake arrives at a

detection station at $t = 0$ min, and a second signal arrives 2.0 min later. Assume that the successive signals are those of longitudinal waves and shear waves, respectively, that follow the same straight-line path through the planet. Assume further that the average density of Earth is 5.5 g/cm^3. How far from the station did the earthquake take place?

19. (II) Earthquake waves within Earth are of two types: S (transverse) waves, which can be supported within Earth's mantle but not in the liquid part of the core, and P (longitudinal) waves, which propagate in both the mantle and the core (see Fig. 14–16). The speeds of S waves and P waves in the mantle are approximately 4.0 km/s and 8.0 km/s, respectively. The speeds of these waves can be calculated from the Young's and shear moduli. Take the density of Earth's mantle to be 5.0 g/cm^3. Calculate the Young's and shear moduli. If the difference between the time of arrival of the two waves is 27 min, how far away was the earthquake? Ignore Earth's thin crust, and assume that the waves travel in a straight line.

20. (II) A steel rod of length 0.80 m is rigidly supported at one end. Calculate the frequencies of the lowest three longitudinal modes, and the lowest three transverse (shear) modes, given that $Y = 2.1 \times 10^{11}$ N/m^2 and $G = 8.1 \times 10^{10}$ N/m^2.

21. (II) Consider a closed loop chain of mass density μ kg/m, which is spinning at a high angular velocity ω till it moves like a hoop. Find the tension in the chain in terms of μ, ω, and the radius R.

22. (II) Assume that experiment has shown that the velocity of waves in shallow water is independent of the wavelength of these waves. Use dimensional analysis to show that the speed of these waves is proportional to the square root of the depth D. Express the speed in terms of the relevant constants. [*Hint*: The relevant force here is that due to gravity.]

23. (II) Assume that the expression for the velocity of shallow-water waves in terms of g and depth D that you derived in Problem 22 is exact. A speed boat that travels close to the speed of such waves tends to build up a large wave just at its bow and thus will have trouble going any faster. At what speed in kilometers per hour will this happen in a harbor in which the water is 5.0 m deep?

24. (II) The considerations of Problem 22 apply to all water waves for which the wavelength is much larger than the depth. For tidal waves (or *tsunami:* waves caused by underwater earthquakes), the wavelengths are enormous, many hundreds of kilometers. What is the speed of a tidal wave in water that is 5.0 km deep?

25. (III) When n springs—each of spring constant k and equilibrium length ℓ—are attached in a line, the resulting system is a spring with spring constant $k_n = k/n$ and length $L = n\ell$. If there is a mass m at the connection point between each spring (the springs themselves are assumed to be massless), then the resulting system can support longitudinal, or compressional, waves (see Fig. 14–3b). By imagining an arbitrarily long series, you can see that the speed of these waves is independent of the total length of the spring, and hence of n. Use the available constants k, ℓ, and m to estimate the speed of longitudinal waves. This system is an excellent model for compressional waves on a real spring. [*Hint*: Dimensional analysis can be useful here.]

26. (III) A set of tiny beads is connected in a line by springs, as in Problem 25. The equilibrium length of the springs is 3.0 cm. The beads have mass 5.0 g each, and the spring constant of the springs is 20 N/m. Longitudinal waves can be generated on the string of beads by moving the end of the string along the direction of the string. Estimate the speed of such longitudinal pulses. [*Hint*: Dimensional analysis can be useful here.]

14–5 Energy and Power in Waves

27. (I) A wave of wavelength 0.70 m is sent moving down a rope of density 12 g/cm, under a tension of 800 N. The amplitude of the wave is 5.0 cm. What is the average power transported by the rope? Suppose that the wavelength is doubled; what is the average power transported by the rope now?

28. (I) Calculate the kinetic energy and potential energy contained in one wavelength of a harmonic traveling wave on a string whose mass density is μ and whose tension is T.

29. (I) The power delivered per unit area by a sound wave is given by an expression similar to that of Eq. (14–36). In this case, μ is replaced by the density, ρ, of the gas in which the sound propagates: $P/\text{area} = \text{intensity} = \frac{1}{2}\rho\omega^2 A^2 v$, where A is the amplitude of the displacement of air molecules from their average positions. Given an 8600 Hz sound wave with an intensity of 5.0×10^{-7} W/m^2, what is A? The density of air is 1.30 kg/m^3, and the speed of sound in air is 330 m/s.

30. (II) A sound wave has an intensity of 1.7×10^{-6} W/m^2. Given that the speed of sound in air is 330 m/s, what is the energy density of the sound wave?

31. (II) Obtain an expression analogous to Eq. (14–36) for the rate of energy transport in a left-moving wave. In particular, show that the power remains positive.

32. (II) Harmonic water waves of a given amplitude are generated on a water surface by rhythmic dipping of a pointer at one spot. The waves thus generated spread in circular wave fronts. If R is the distance from the generation point, show that conservation of energy implies that the amplitude A must vary with R, and that this variation is of the form $A \propto 1/\sqrt{R}$.

33. (III) We derived the power or rate of energy transport for a sinusoidal wave [Eq. (14–36)]. Obtain an expression for the power delivered by an arbitrary right-moving wave of the form $f(x - vt)$.

14–6 Standing Waves

34. (I) The third harmonic of a certain piano wire ($\mu = 0.010$ kg/m, $T = 500$ N) has a frequency of 600 Hz. What is the wavelength of the fundamental?

35. (I) A guitar string has end points 54 cm apart. What are the three largest wavelengths of standing waves this string can support? Sketch the standing waves for these wavelengths.

36. (I) A string that is fastened at the two end points, $x = \pm 60$ cm, oscillates in its fundamental mode (no nodes except at the walls). The period of the oscillation is 1 s, the displacement at $x = 0$ cm at $t = 0.0$ s is 2.0 cm, and at $t = 0.1$ s, 3.0 cm. Express $z(x, t)$ for the displacement.

37. (I) A standing wave is formed on a string for which three and one half wavelengths fit into the total length of the string, which is 2.7 m. The wave has a period of 0.10 s. If the string has a mass density of 220 g/m, what is the tension in the string?

38. (I) The tension on a wire is 2840 N. The wire is 2.6 m long and the fundamental frequency is 310 Hz. What is the mass of the wire?

39. (I) Two wires of the same length and under the same tension have fundamental frequencies that are in the ratio 3 : 1. What is the ratio of their masses?

40. (II) A string with fixed ends is in a standing wave that vibrates in five segments (Fig. 14–40, see next page). The frequency of this mode is 120 Hz. (a) Find the fundamental frequency of the string. (b) The tension in the string is reduced by a factor of 9. What is the new fundamental frequency?

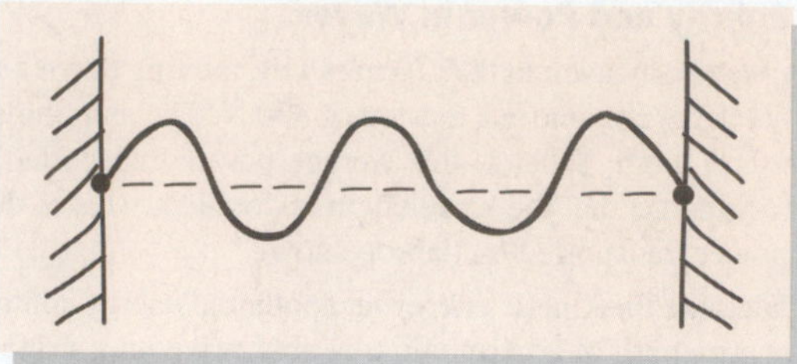

▲ **FIGURE 14–40** Problem 40.

41. (II) A steel string under tension vibrates in its lowest mode at a frequency of 1200 Hz. A string of the same material and length but three times the thickness vibrates in its lowest mode at a frequency of 400 Hz. What is the ratio of the tensions of the two strings?

42. (II) Figure 14–41 shows a pulse at $t = 0$ traveling to the right with a speed of 0.50 m/s. (a) Sketch, properly scaled, the shape of the rope at $t = 4$ s. (b) Plot the displacement of the point at $x = 5.0$ m as a function of time.

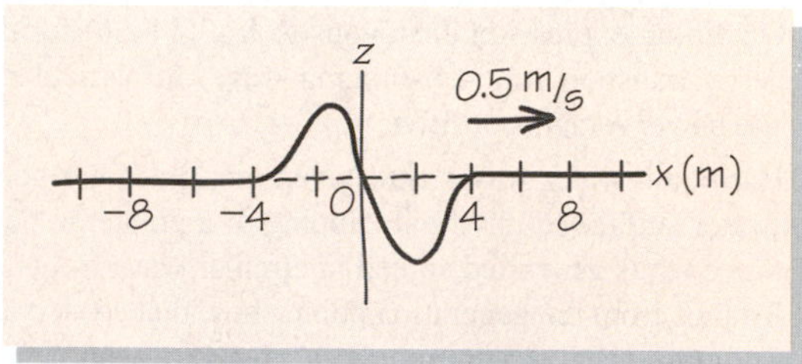

▲ **FIGURE 14–41** Problem 42.

43. (II) A rope has a mass density of 40 g/cm and is under tension of 200 N. The rope is fixed at one end and is connected to a rod at the other end that sets up standing waves by moving slightly. The rope is 1.0 m long. Determine the three lowest possible frequencies and the corresponding angular frequencies and periods.

44. (II) A string with length L and linear density μ is under tension T. One end is fixed; the other end remains horizontal while it moves periodically up and down. Standing waves with an antinode (a point of maximum displacement) can be induced in this way, as in Fig. 14–42. Find the possible wavelengths and frequencies.

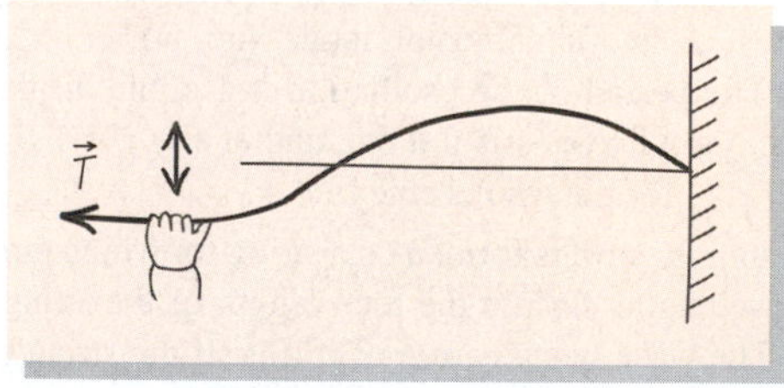

▲ **FIGURE 14–42** Problem 44.

45. (II) The mass density of copper is 8.92×10^3 kg/m^3. What is the tension on a copper wire 1.00 mm in diameter and 60.0 cm long, fastened at each end, when the frequency of vibration in the $n = 3$ mode is 870 Hz?

14–7 More About Sound

46. (I) Sound waves in air travel at 330 m/s, whereas light travels at 3.0×10^8 m/s. You can determine how far away a lightning strike is by counting the number of seconds between the time you see the lightning flash and the time you hear the thunderclap. How far are you from the lightning bolt for each 1 s difference between the flash and the thunderclap?

47. (I) Using the information in Table 14–1, compute the densities of air and hydrogen.

48. (I) A telephone receiver is at your left ear, with the sound of the conversation at a level of 58 dB. At the same time, your little sister's screams, at a level of 93 dB, enter your right ear. What is the ratio of the sound intensities that enter your two ears?

49. (I) A series of noises have intensities 8.3×10^{-4} W/m^2, 3.5×10^{-6} W/m^2, and 7.2×10^{-9} W/m^2, respectively. What is the sound level of each in decibels?

50. (I) A sound suddenly appears to the ear to be twice as intense as usual. How has the amplitude of the pressure wave (the magnitude of the maximum pressure in the wave) changed?

51. (II) A sound source emits sound with equal intensity in all directions. What is the power output of the source if the intensity of the sound measured at 5.0 m from the source is 83 dB?

52. (II) Two sounds have levels of 35 dB and 53 dB, respectively. What is the ratio of the power of the two noises? What is the ratio of the amplitudes of the pressure waves they make?

53. (II) In an ideal fluid, the speed of sound is given by $v = \sqrt{B/\rho}$, where B is the *bulk modulus*, a measure of how easy it is to compress the fluid. Sound waves in a fluid are only longitudinal because an ideal fluid cannot bounce back from the application of transverse forces, sometimes called shearing forces. Using the information given in Table 14–1, calculate the bulk modulus of water.

54. A source of sound waves is operating in midair, far from any reflecting surfaces. If, 1.0 m from the source, the sound intensity level is 120 dB, what is the intensity 1000 m from the source? What is the total power output of the source?

55. (II) Warm air has a higher density than cold air, and the density at Earth's surface can be taken to be inversely proportional to the temperature, measured on the Kelvin scale K (see Chapter 17). In this scale water freezes at close to 273K. Given that the speed of sound at 273K is 330 m/s, what is its speed at room temperature, roughly 295K?

14–8 The Doppler Effect

56. (I) A musician with perfect pitch—the capacity to identify pitch by ear—has run a red light and is being pursued by a police car whose siren emits at the single frequency $f = 400$ Hz. The musician's car is moving east at 60 mi/h (27 m/s), and the police car follows at 80 mi/h. The speed of sound is 330 m/s, and the air is still. What does the musician say the frequency of the siren is?

57. (I) A horn emits sound at a frequency of 160 Hz. What frequency is heard by an observer moving away from the source at 26.0 m/s?

58. (I) A child is blowing a whistle while riding her bicycle toward you on the sidewalk. You have set up an accurate audio receiver that measures the frequency of the whistle sound to be 793 Hz. You estimate the bicycle to be moving at 7 mph (3.1 m/s). What would you measure the frequency of the whistle to be at rest?

59. (I) A whistle emits sound with a frequency of 333 Hz. The whistle is mounted on a train that travels at 140 km/h in the $-x$-direction toward an observer at rest. (a) What frequency is heard by the observer? (b) Suppose that the whistle is at rest and the observer moves in the $+x$-direction at 140 km/h. What frequency is heard by the observer?

60. (II) A truck is traveling at 60 mi/h (27 m/s) down the interstate highway where you are changing a flat tire. As the truck

approaches, the truck driver sounds an air horn, which has a frequency of 185 Hz. (a) What frequency do you hear? (b) Suppose that you were driving your car at 50 mi/h while the truck driver blows his air horn after he passes you. What frequency would you hear?

61. (II) The frequency of light emitted by a star moving away from us (the observer) with speed v will be *red-shifted* (its frequency will be lowered) by an amount properly described by the treatment in Section 14–7, provided $v/c \ll 1$, where c is the speed of light ($c = 3.0 \times 10^8$ m/s). How fast is a galaxy receding from us if the observed frequency is 0.95 times the natural frequency of the light?

62. (II) You are moving southward on a divided highway and, heading northward on the other side, a fire engine passes with its (one-note) siren screaming. You are close enough to each other as you pass to assume that you and the fire engine are moving on the same line. Your speedometer reads 45 mi/h, and being a trained musician, you recognize that the frequency you hear before you pass the fire engine is 910 Hz and 680 Hz after you pass. Use the value 330 m/s for the speed of sound in air to find the frequency emitted by the siren and the speed of the fire engine.

63. (II) An ambulance siren is the source of sound emitted with a frequency of 1600 Hz. The ambulance is driving toward the base of a large cliff at 95 km/h (Fig. 14–43). What is the frequency of the reflected sound wave heard at the ambulance?

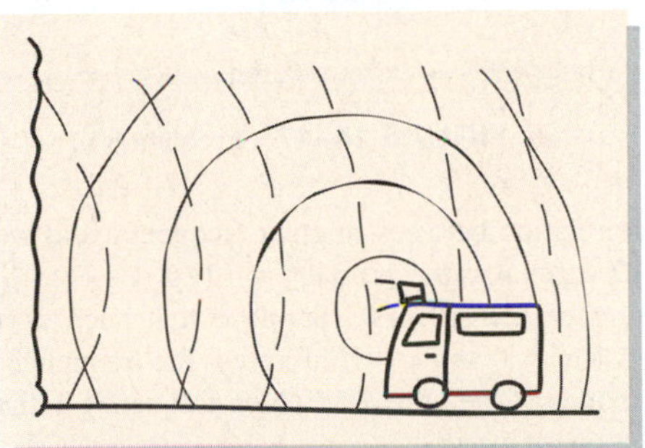

▲ **FIGURE 14–43** Problem 63.

64. (III) Show that for $v \ll c$, where c is the speed of a wave in a certain medium, the frequency shift for an observer at rest and a source moving away at speed v is the same as that for a source at rest and the observer moving at speed v in the opposite direction.

65. (III) A tuning fork with frequency 440 Hz is dropped from a tower of height 100 m by an enterprising student. She can hear the fork sounding as it falls. Given that the speed of sound in air is 330 m/s and that $g = 10$ m/s^2, what frequency does the student hear as a function of time? Do not forget that it takes time for sound to travel back to the student.

*14–9 Shock Waves

66. (I) An airplane moving at three times the speed of sound (Mach 3) flies horizontally and at a very low altitude. It emits sound continuously with a frequency that the pilot hears as 300 Hz. (a) What angle does the plane's shock wave cone make with the vertical? (b) What is the frequency of the sound heard by a stationary observer after the plane has passed? (Assume that the plane is moving directly away from the observer.)

67. (I) The angle made by the shock wave associated with the sonic boom of a low-flying airplane is $\theta_0 = 70°$. What is the speed of the airplane?

General Problems

68. (I) Two children stretch a rope between two trees. In spite of their efforts, the tension turns out to be only 20 N. The mass density of the rope is 48 g/cm. If they strike the rope hard with a stick, how fast does the resulting wave travel to the other tree?

69. (I) The human voice system, which includes the vocal cords, the mouth, and the lips, is a device that can change its shape and size and so is capable of emitting sound with a good deal of complication. There is nevertheless a fundamental frequency. Assume the vocal cavity is approximately 14 cm long and determine the fundamental frequency of the human voice. Assume a model where the vocal tract is open at both ends (larynx and mouth) and use 350 m/s as an approximation to the speed of sound at the human body temperature.

70. (II) A string is stretched between a pulley and a wave generator consisting of a plate vibrating up and down with small amplitude and frequency 120 Hz (Fig. 14–44). A standing wave with two intermediate nodes is generated when the string has a mass of 215 g attached to it. How large a load is needed to generate standing waves with one and with four nodes?

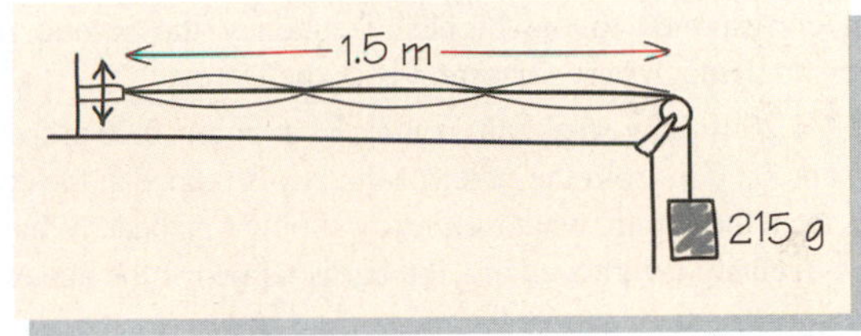

▲ **FIGURE 14–44** Problem 70.

71. (II) A string is fastened to a wall at $x = 0$ cm and is free to move on another wall at $x = 60$ cm, with the constraint that it is always perpendicular to the wall there (Fig. 14–45). What are the permissible values of the wavelength?

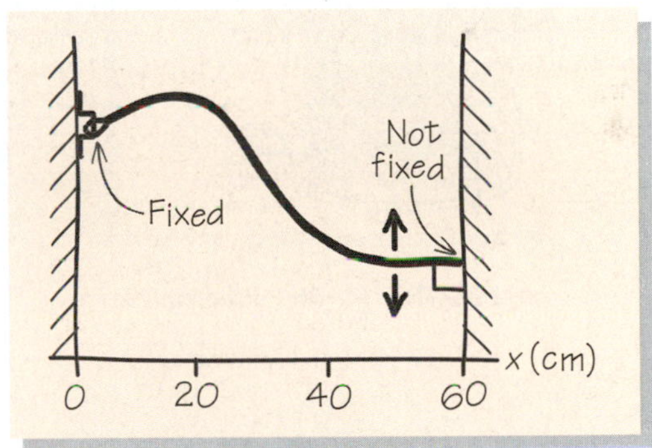

▲ **FIGURE 14–45** Problem 71.

72. (II) Transverse traveling waves are generated on a uniform wire of constant tension and a given mass. By what factor will the required power be increased or decreased if (a) the given mass of wire is doubled in length and the angular frequency remains constant? (b) the amplitude is doubled and the angular frequency is halved? (c) both the wavelength and the amplitude are doubled? (d) the given mass of wire is doubled in length and the wavelength is halved?

73. (II) The sound intensity at a rock concert reaches the pain threshold of 120 dB. Normal conversation has an intensity of 55 dB. What is the ratio of the power delivered to the ear at a rock concert compared to that delivered by a typical conversation?

74. (II) The left-hand end of a long, taut string is moved harmonically up and down with amplitude 0.14 m. This motion produces a wave of wavelength $\lambda = 2.5$ m that travels in the $+x$-direction at speed $v = 245$ m/s. (a) What is the frequency, f, of the traveling wave? (b) What is the maximum transverse velocity of a point along the string? (c) What is the maximum transverse acceleration of a point along the string?

75. (II) The density of neon is approximately 0.90 kg/m^3 at standard atmospheric pressure and temperature. The value of γ for neon is 5/3. A loudspeaker emits a note with a frequency of 400 Hz in air. What will the frequency and wavelength of that note be if the loudspeaker is put into a chamber filled with neon?

76. (II) A wire is attached to two walls 1.0 m apart. The wire oscillates in such a way that a point on the wire 25 cm from one end is held fixed at its equilibrium position. What are the possible wavelengths for the first three allowed modes?

77. (II) A tuning fork with frequency $f = 512$ Hz is held by someone who swings it vigorously in a circle in the horizontal plane. The radius of the circle is 1.0 m, and the frequency of revolution is 3.0 rev/s. (a) What are the maximum and minimum frequencies that a second person would hear? (b) Which part of the rotation corresponds to the highest frequency the second person hears, and which part corresponds to the lowest?

78. (II) A 1.20-m-long steel bar is clamped at points 0.20 m from one end and 0.60 m from the other. If the bar is tapped at the end farther from the clamp, what frequency should one hear? What if the 0.60-m clamp is removed and the bar is tapped at the free end?

79. (II) A tuning fork with a frequency of 440 Hz is held just above the top of a uniform tube containing water (Fig. 14–46). The tube can excite the column of air above the water, whose level can be changed by a spigot at the bottom of the tube. As the water is drained out, the sound intensity of the fork is enhanced when the air column has a length of 0.60 m and again when the air column has a length of 1.0 m. Use these data to find the speed of sound in air.

▲ **FIGURE 14–46** Problem 79.

80. (III) A square sheet of metal with sides of length L is clamped on all four sides. It is convenient to put these sides at $x = 0$, $x = L$, and $y = 0$, $y = L$ (Fig. 14–47). The vertical displacement at any point on the sheet is given by $z(x, y, t)$. and it satisfies the wave equation

$$\frac{\partial^2 z}{\partial x^2} + \frac{\partial^2 z}{\partial y^2} = \frac{1}{v^2}\frac{\partial^2 z}{\partial t^2}.$$

Show that a standing wave of the form $z(x, y, t) = A \sin(kx) \sin(qy) \cos(\omega t)$ satisfies the wave equation, and find the values of k and q that satisfy the boundary conditions and the constraints of the wave equation.

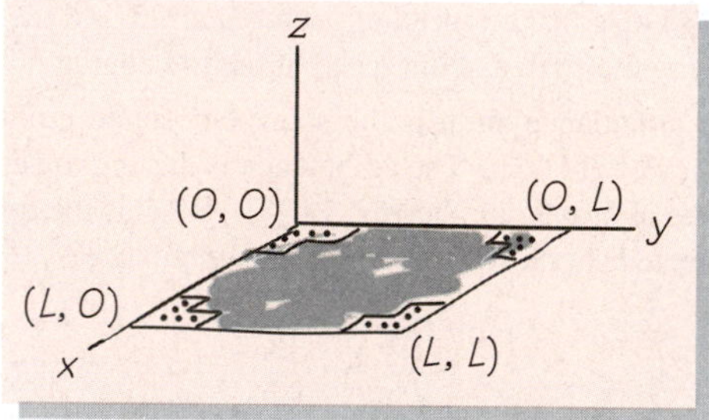

▲ **FIGURE 14–47** Problem 80.

81. (III) The relation between angular frequency and wave number for a stiff wire takes the form $\omega^2 = (T/\mu)k^2 + \alpha k^4$, where α is a measure of the stiffness. Suppose that such a stiff wire is clamped at $x = 0$ and $x = L$. Express the frequency for the two lowest modes of the wire. For which frequency is the stiffness a more important effect?

Chapter

◀ The water waves behind the ship display an important characteristic: They add or subtract algebraically, creating interference patterns.

Superposition and Interference of Waves

The harmonic waves discussed in Chapter 14 are idealizations. Waves in nature or in technology are rarely as simple as harmonic waves. These include water waves, sound in air, waves in material, waves within or on musical instruments, the waves generated by a doctor's ultrasound machine, or the vast gamut of electromagnetic waves that include light. The surface of the ocean is a chaotic welter of waves moving in various directions. A guitar string is in fact set into a very complicated motion when it is plucked. A single shake of the end of a rope sends a single pulse rather than a sine function traveling down the rope.

Nevertheless, we have the tools to understand these more complex waves in terms of simple harmonic waves: Harmonic waves can be combined, or superposed, to produce the full variety of waves observed in nature. Superposed waves, which sometimes reinforce and sometimes (partially or totally) cancel each other, are said to interfere. This interference may produce regular patterns of reinforcement and cancellation; such patterns are seen most easily in light waves as the colors in oil spots or in the pattern of ocean waves as they enter a harbor.

Our goal in this chapter is to understand the different ways in which waves can combine. We will see how the standing waves we encountered in Chapter 14 are formed by the superposition of traveling waves, how the pulsing phenomenon known as beats arises, how there can be well-organized interference patterns in space, and how waves of any shape can be formed by the superposition of harmonic waves.

15–1 The Superposition Principle

The waves that appear in nature are complex. Nevertheless, our work on harmonic waves is relevant because most elastic media that undergo small displacements in response to forces exhibit **linearity**. Linearity means that if two waves can travel through a medium, then a third wave whose form is the algebraic sum of these waves is possible. Thus if a stretched rope supports two different waves in the form $z_1(x, t)$ and $z_2(x, t)$, then it can also support the wave that is described by $z(x, t) = z_1(x, t) + z_2(x, t)$. In other words, *the sum of two solutions of the wave equation for the system is also a solution*. When linearity applies, the sum of any number of waves can add to form another wave, and a wave multiplied by any constant is also an allowed wave. Of course, when we say the "sum" or "multiple" of waves, we are talking about the displacement involved, the quantity we have often called z that describes the wave. This could be a transverse displacement on a rope, a longitudinal displacement in a solid, or the pressure change or density change in a gas. This fact is known as the **superposition principle**.

Let us see how a sum of two solutions of the wave equation can also be a solution to that equation. We saw in Chapter 14 that for transverse waves on a string the wave equation reads [Eq. (14–10)]

$$\frac{\partial^2 z}{\partial x^2} = \frac{1}{v^2}\frac{\partial^2 z}{\partial t^2}, \tag{15–1}$$

where x is the distance along the string and $z(x, t)$ is the transverse displacement of the string at point x and time t. Equation (15–1) is *linear* because the function z appears linearly; that is, as z rather than as z^2 or as z to some other power. If $z_1(x, t)$ and $z_2(x, t)$ are two different solutions satisfying some linear equation such as the wave equation, then the sum $z_1(x, t) + z_2(x, t)$ also satisfies that equation. Superposition works only for waves of small amplitude because when a medium is stretched or distorted too far, the restoring force is no longer purely springlike, and the dynamical equation becomes nonlinear.

Interference

The superposition principle allows us to add sinusoidal waves of different amplitudes, wavelengths, and frequencies, each of them moving in different directions. These sums are *algebraic* sums, complete with the possibilities of algebraic cancellations (Fig. 15–1). This cancellation is known as **interference**, and the result of the interference is a rich palette of patterns called **interference phenomena**.

A simple and striking example of interference is the case of two transverse harmonic waves that each travel to the right on a rope. Let the waves have the same amplitude and be 180° (π rad) out of phase. The two waves have the respective forms

$$z_1(x, t) = z_0 \sin(kx - \omega t) \tag{15–2}$$

and

$$z_2(x, t) = z_0 \sin(kx - \omega t + \pi) = -z_0 \sin(kx - \omega t). \tag{15–3}$$

The superposition of these two transverse waves adds to zero! This is an extreme case of **destructive interference**, which is interference that occurs at any point where two waves have the opposite sign. Although interference may be *fully* destructive (waves adding to zero at a given point), keep in mind that less than total cancellation is possible. **Constructive interference** occurs when the waves add in such a way that they reinforce. If, for example, our two waves z_1 and z_2 had exactly the same phase, we would find a summed solution with the same x- and t-dependence of either wave but with amplitude $2z_0$. Here too, less than maximal levels of constructive interference are possible.

Coherence

What conditions are necessary for waves from two different sources (for example, the sound waves coming from two loudspeakers) to exhibit an interference pattern? If the two sources put out a series of waves with different frequencies and with phases that change randomly, then there is no *consistent* interference pattern. Any pattern would change so

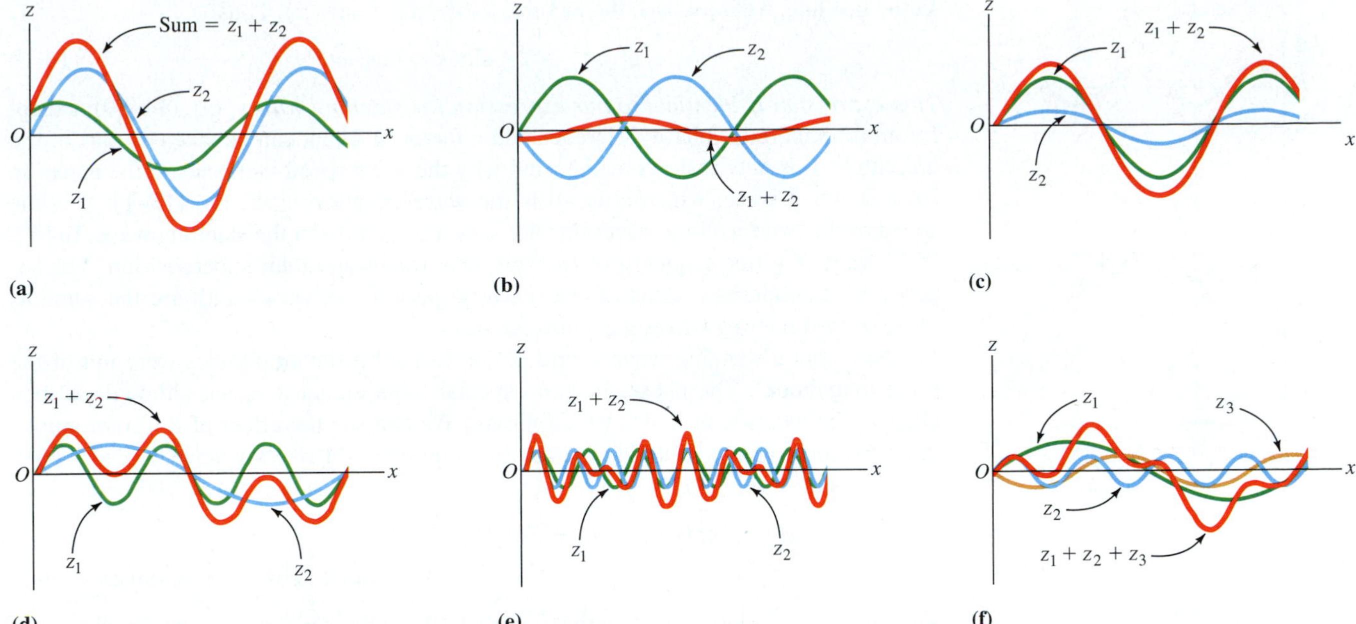

▲ **FIGURE 15–1** The superposition of (a) two waves of the same frequency, with a slight phase difference between them; (b) two waves of the same amplitude and frequency, with a phase difference of nearly 180° between them; (c) two waves of the same frequency and phase but different amplitudes; (d) two waves of the same amplitude but frequencies that differ by a factor of 3; (e) two waves of the same amplitude but slightly different frequencies, resulting in beats; (f) three waves of different amplitudes and frequencies.

rapidly that we could not recognize it. *An interference pattern occurs only when the waves have a definite, stable relation between their frequencies and phases.* The waves are then said to be **coherent**. When, in contrast, we have waves without a definite, stable relation between their frequencies and phases, the waves are **incoherent**. Sources such as piano strings or lightbulbs or the human voice that are not point sources are generally incoherent with other sources such as other piano strings or other voices. If the form of vibration of the piano string is complex, as it generally is, even waves coming from different parts of the string will not generally be coherent with each other. You won't observe interference by listening to a rock band! This is discussed further at the end of Section 15–4.

15–2 Standing Waves Through Interference

Standing waves can be viewed as the result of the superposition of traveling waves moving in opposite directions. Imagine a very long string with two waves that move in opposite directions but are otherwise identical, each a solution of the wave equation:

$$\text{for the left-moving wave: } z_l(x, t) = A\sin(kx + \omega t); \tag{15–4a}$$

$$\text{for the right-moving wave: } z_r(x, t) = A\sin(kx - \omega t). \tag{15–4b}$$

We have taken the wavelengths, frequencies, phases, and amplitudes to be equal. (You may want to review the meaning and relations between these quantities at this point; see Section 14–3.) According to the principle of superposition, the resultant displacement is

$$z_1(x, t) = z_r(x, t) + z_l(x, t) = A[\sin(kx - \omega t) + \sin(kx + \omega t)]. \tag{15–5}$$

If we use the trigonometric identity 5.3 of Appendix IV–4, with $a = kx - \omega t$ and $b = kx + \omega t$, then we obtain

$$\sin(kx - \omega t) + \sin(kx + \omega t)$$
$$= 2\sin\left[\frac{(kx - \omega t) + (kx + \omega t)}{2}\right]\cos\left[\frac{(kx - \omega t) - (kx + \omega t)}{2}\right]$$
$$= 2\sin(kx)\cos(\omega t).$$

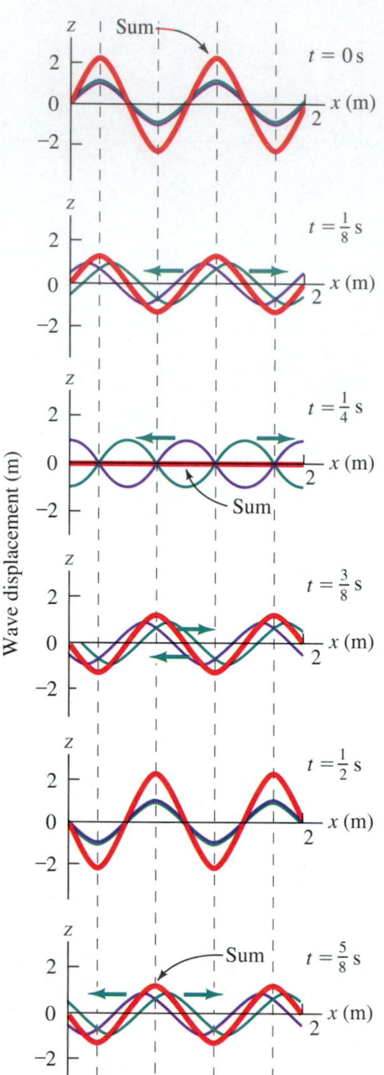

In the last line, we have used the fact that $\cos(-a) = \cos(a)$. Thus

$$z(x, t) = 2A \sin(kx) \cos(\omega t). \tag{15–6}$$

This expression is identical to our expression for standing waves, Eq. (14–39), except for an (unimportant) phase difference and a factor of 2 that can be taken as part of the amplitude. It is now easier to understand why the wave speed v appears in the equation for a standing wave, which after all is the wave equation itself, Eq. (15–1): It is the speed of the two traveling waves that are superposed to form the standing wave. In Fig. 15–2 we plot a time sequence of two traveling waves and their superposition. The superposition produces a standing wave whose period and wavelength are the same as those of the traveling waves that compose it.

Note that a standing wave would not be formed if the amplitudes were not of the same magnitude. (The phases have no special requirement to meet, although as they change, the location of nodes would move.) We can see the effect of different amplitudes by, for example, doubling the amplitude in the right-moving wave [Eq. (15–4b)]. Then the sum would give

$$A[\sin(kx - \omega t) + \sin(kx - \omega t) + \sin(kx + \omega t)]$$
$$= 2A \sin kx \cos \omega t + A \sin(kx - \omega t).$$

The last term is a leftover wave that travels to the right. This means that creating pure standing waves is not a trivial matter. Of course, standing waves are created on the strings of musical instruments all the time—this is because, as we'll see below, in the reflection of waves from a place where the string is fixed, such as a fret, the conditions for a standing wave are automatically satisfied.

◀ **FIGURE 15–2** Superposition of two traveling waves produces a standing wave, plotted in red. Shown are the waves $\sin(kx - \omega t)$, moving rightward, and $\sin(kx + \omega t)$, moving leftward. The amplitude of each wave is 1 m, the wavelength is 1 m, and the period is 1 s. Thus, $k = 2\pi \text{ m}^{-1}$ and $\omega = 2\pi$ rad/s. Arrows indicate the direction of motion of the two traveling waves. Time progresses in $\frac{1}{8}$-s intervals starting at $t = 0$ s and ending at $t = \frac{5}{8}$ s. Note that the peaks of the standing wave remain at exactly the same x-values.

EXAMPLE 15–1 Two traveling waves move in opposite directions in a one-dimensional medium and form a standing wave. The angular frequency of that standing wave is 2.9 rad/s, and the velocity of each of the traveling waves is 1.8 m/s. For comparison purposes you also have a string attached tightly to two walls whose separation can be adjusted. In particular, they are adjusted so that the largest wavelength of the standing waves (lowest mode) that can be supported on your string is the same as the wavelength of the superposition of the two traveling waves. What is that separation?

Setting It Up The medium with the traveling waves is not specified, but we can imagine it too is a string. This system and the comparison string are both shown in Fig. 15–3. We are given the angular frequency ω as well as the speed v of the traveling waves that form the standing wave in the medium. The wavelength λ, which is labeled in the figure, is unknown; we can find the wall separation L in terms of it.

Strategy We first must find the wavelength of the standing wave set up within the medium, given that we know the angular frequency and the wave speed. For this we can use the relation $v = f\lambda = (\omega/2\pi)\lambda$. We then want to adjust the separation L of the walls so that the lowest, or fundamental, mode (longest wavelength) of the string connected to the walls has wavelength λ. For that we need only recall that the longest wavelength on a string connected at its two ends is twice the string length, Eq. (14–42) with $n = 1$. Thus we set $L = \lambda/2$.

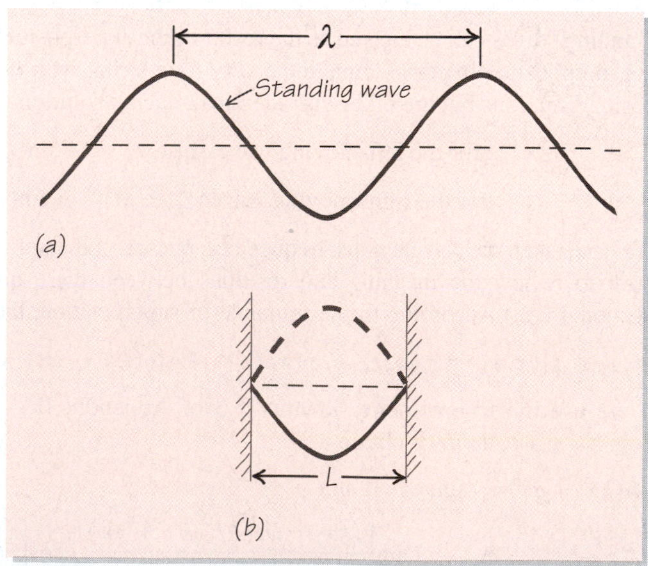

▲ **FIGURE 15–3** Part (a) shows the standing wave on the one-dimensional system, while (b) shows the string attached to walls.

Working It Out We find

$$\lambda = 2\pi v/\omega = (2\pi)(1.8\ \text{m/s})/(2.9\ \text{rad/s}) = 3.9\ \text{m}.$$

In turn,

$$L = \lambda/2 = 2.0\ \text{m}.$$

What Do You Think? Is the frequency of the wave on the string necessarily that of the standing wave in the medium? *Answers to **What Do You Think?** questions are given in the back of the book.*

Interference of Incident and Reflected Waves

In order to understand the production of standing waves, it is helpful to consider an *incident*, or incoming, harmonic traveling wave that encounters a wall. To carry this wave, we can stretch a long rope aligned with the x-axis attached to a wall at the point $x = 0$ and extending to the left. Since the string is tied to the wall at that point, the displacement $z(x, t)$ is always zero at $x = 0$. A harmonic incident wave is initiated somewhere on the far left and arrives at $x = 0$. When the wave arrives at the wall it is *reflected*. The energy initially carried to the right bounces back and travels back down the string to the left as a reflected wave. The reflected wave has the same frequency f as the incoming wave, as you can see by thinking about the reflection process: A peak of the incident wave will produce a peak of the reflected wave, and it will take the same period for the reflected wave to produce its next peak as it did for the next peak of the incident wave to arrive at the wall. Also, the rigid wall absorbs no energy, so the reflected wave must have the same amplitude as the incident wave. Finally, the reflected wave moves with the same speed as the original incident wave because the wave speed v is a property of the string. These are just the conditions that give us Eq. (15–5), and they lead to Eq. (15–6): The reflected wave interferes with the rightward-moving incident wave to produce a standing wave with the same frequency and wavelength. This is why the idea that a standing wave on a string is a superposition of traveling waves is more than merely a mathematical construct.

One feature of the reflection that creates a standing wave deserves special notice. The standing wave formed on the string must have a node at the fixed end $x = 0$. This means that the incident wave and its reflection must *cancel* at $x = 0$ (and hence form a node at that point). This requires that *the sinusoidal wave that reflects from a fixed end has a displacement with a sign opposite to that of the incoming wave and is therefore inverted.* Whether reflected waves are inverted or not is an interesting subject and one that we'll take up again in Section 15–5.

15–3 Beats

Beats occur when two waves of nearly the same frequency are superposed; a long-wavelength (low-frequency) interference pattern is produced. Musicians use beats to tune their instruments, and the phenomenon is of scientific importance because it provides a sensitive technique for measuring frequency (or period) differences. A simple visual analogue to beats can be constructed with the help of two combs with slightly different tooth spacings; the slightly different spacings are analogous to slightly different periods. In Fig. 15–4b, two combs are superposed. Where the teeth lie directly on

(a)

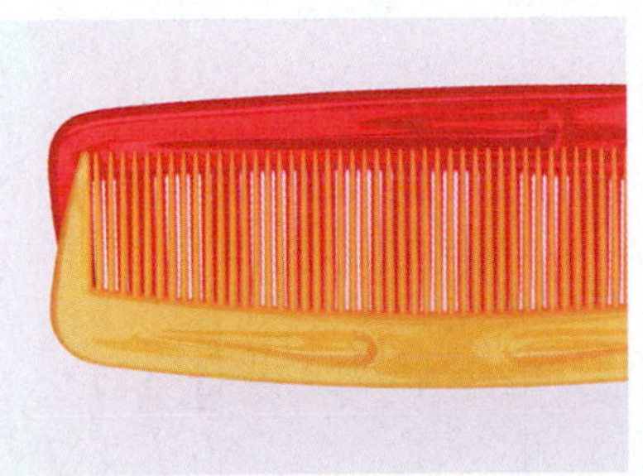

(b)

FIGURE 15–4 An analogy indicating beats. (a) Two combs have slightly different spacings. (b) When superposed, the combs produce an interference pattern known as a moiré pattern.

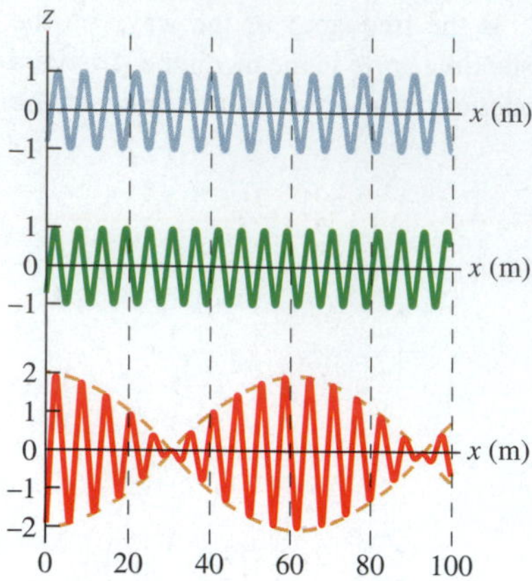

▲ **FIGURE 15–5** Two traveling waves of identical amplitude, $z_0 = 1$ m, and nearly the same frequency and wavelength, at $t = 1$ s. The blue traveling wave has $\omega = 1$ rad/s and $k_1 = 1\ \text{m}^{-1}$. The green traveling wave has $\omega_2 = 0.9$ rad/s and $k_1 = 1.1\ \text{m}^{-1}$. Beats are visible in their superposition (the red wave), formed by the envelope shown as a dashed line. The vertical lines trace the regions of constructive and destructive interference.

top of one another, the spaces are visible. Some distance away, the teeth will lie out of phase and no spaces are visible, farther still and the spaces will once again be visible. The superposed combs exhibit a pattern that repeats over a much greater distance than the separation of the teeth. The large spacing of the pattern observed when the two combs are superposed is equivalent to the large period of the interference pattern observed when waves with two slightly different periods are superposed.

Consider two traveling waves of slightly different frequencies, as in the blue and green waves in Fig. 15–5. The two waves have equal amplitudes of 1 m. The first wave has $\omega_1 = 1.00$ rad/s and $k_1 = 1.00\ \text{m}^{-1}$, and the second has $\omega_2 = 0.90$ rad/s and $k_2 = (1/0.90)\ \text{m}^{-1} = 1.11\ \text{m}^{-1}$. To the eye, the two waves are barely distinguishable from each other. Yet their superposition has a striking visual signature: *A very different wave—one with a long wavelength—is formed* (the red wave in Fig. 15–5 and the close-up view of Fig. 15–6).

Here is how this pattern is formed: Both waves start out together at $x = 0$ m, in phase, and hence interfere constructively. The amplitude of the superposed wave is therefore large at this point. Because the wavelengths of the two waves differ slightly, they begin to develop a phase difference as we move along x, and at some point along x (at about 30 m), they are almost perfectly out of phase. Whereas they started out with maximum on maximum or minimum on minimum, they now are such that the maximum of one is at the same point as the minimum of the other. At this point they interfere destructively, and the amplitude of the superposed wave is small. The cycle repeats itself as the waves slowly come together to interfere constructively again, at about 63 m, as for the combs of Fig. 15–4.

Calculation of the Beat Frequency

To analyze this situation more quantitatively, start with the two waves

$$z_1(x,t) = z_0 \sin(k_1 x - \omega_1 t) \quad \text{and} \quad z_2(x,t) = z_0 \sin(k_2 x - \omega_2 t). \qquad (15\text{–}7)$$

The angular frequencies and wave numbers differ a little. A convenient way to account for this fact is to deal with the sum and the difference of the frequencies. The frequency difference will be a small quantity. If $\omega_1 > \omega_2$, and $k_1 > k_2$, we define

$$\delta\omega \equiv (\omega_1 - \omega_2) \quad \text{and} \quad \Omega \equiv \frac{1}{2}(\omega_1 + \omega_2), \qquad (15\text{–}8)$$

and similarly,

$$\delta k \equiv (k_1 - k_2) \quad \text{and} \quad K \equiv \frac{1}{2}(k_1 + k_2). \qquad (15\text{–}9)$$

It is easy to show that the ratios $\delta\omega/\delta k$ and Ω/K are given by the wave speed $v = \omega_1/k_1 = \omega_2/k_2$ in the medium. Both $\delta\omega$ and δk are positive and small.

We can now reexpress the original frequencies and wave numbers in terms of their sums and differences:

$$\omega_1 = \Omega + \frac{1}{2}\delta\omega \quad \text{and} \quad \omega_2 = \Omega - \frac{1}{2}\delta\omega, \qquad (15\text{–}10)$$

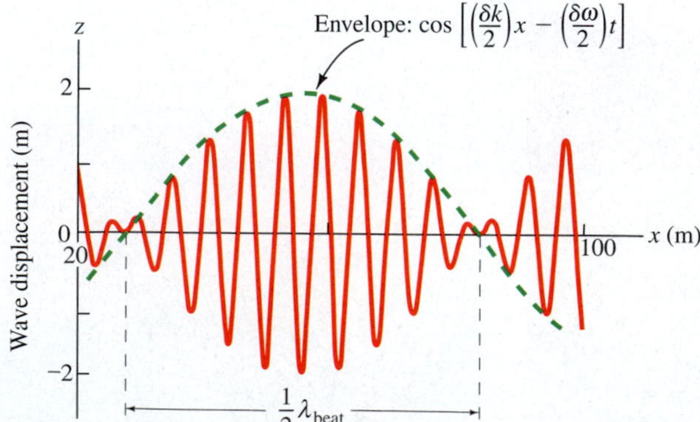

▶ **FIGURE 15–6** An enlarged look at the superposition of the red wave of Fig. 15–5. The dashed envelope is a traveling wave of wavelength $\lambda_{\text{beat}} = 126$ m.

$$k_1 = K + \frac{1}{2}\delta k \quad \text{and} \quad k_2 = K - \frac{1}{2}\delta k. \qquad (15\text{–}11)$$

When Eqs. (15–10) and (15–11) are substituted into the waves of Eq. (15–7), we find that

$$\begin{aligned} z_1(x,t) &= z_0 \sin\left[\left(K + \frac{\delta k}{2}\right)x - \left(\Omega + \frac{\delta\omega}{2}\right)t\right] \\ &= z_0 \sin\left\{(Kx - \Omega t) + \frac{1}{2}[(\delta k)x - (\delta\omega)t]\right\}, \end{aligned} \qquad (15\text{–}12)$$

$$\begin{aligned} z_2(x,t) &= z_0 \sin\left[\left(K - \frac{\delta k}{2}\right)x - \left(\Omega - \frac{\delta\omega}{2}\right)t\right] \\ &= z_0 \sin\left\{(Kx - \Omega t) - \frac{1}{2}[(\delta k)x - (\delta\omega)t]\right\}. \end{aligned} \qquad (15\text{–}13)$$

To sum these two terms and find the resultant superposition, we expand the trigonometric functions with identity 5.3 of Appendix IV–4, and we use the fact that $\cos(-a) = \cos(a)$. The result is

$$z(x,t) = z_1(x,t) + z_2(x,t) = 2z_0 \sin(Kx - \Omega t)\cos\left[\left(\frac{\delta k}{2}\right)x - \left(\frac{\delta\omega}{2}\right)t\right]. \qquad (15\text{–}14)$$

Equation (15–14) states that the sum of the original two waves can be reformulated as a *product* of two traveling waves. The wavelength and frequency of the first of these traveling waves, which is the sine term in Eq. (15–14), are given by the *average* of the corresponding quantities for the original two waves. The envelope formed by the second traveling wave is the dashed curve in Fig. 15–5. This envelope is a wave traveling with the same speed as the sine wave but with a much longer wavelength. The envelope wave is called a *beat* of the original two waves. The coefficient of x in the cosine factor in Eq. (15–14), $\delta k/2$, gives the wave number of the beats, whereas the coefficient of the time, $\delta\omega/2$, gives the angular frequency. In turn, we can use Eqs. (15–8) and (15–9) to find the wavelength and frequency of the beats:

$$\frac{1}{\lambda_{\text{beat}}} = \frac{k_{\text{beat}}}{2\pi} = \frac{k_1 - k_2}{4\pi} = \frac{1}{2}\left(\frac{1}{\lambda_1} - \frac{1}{\lambda_2}\right); \qquad (15\text{–}15)$$

$$f_{\text{beat}} = \omega_{\text{beat}}/2\pi = (\omega_1 - \omega_2)/4\pi = \tfrac{1}{2}(f_1 - f_2). \qquad (15\text{–}16)$$

BEAT WAVELENGTH AND FREQUENCY

The beat has a wave number and frequency that are one half the *differences* of the original wave numbers and frequencies. Because this difference is small by assumption, the wave number of the envelope is small and the wavelength is *large*. Similarly, the frequency of the beat is *small*. In deriving this result, we tacitly assumed $f_1 > f_2$, so that f_{beat} was positive. Note, however, that the beat factor in Eq. (15–14) is a cosine factor, and the cosine is unchanged if its argument changes sign. This means that if $f_2 > f_1$, f_{beat} would be given by $(f_2 - f_1)/2$ rather than $(f_1 - f_2)/2$: The beat frequency is always positive.

Figure 15–6 is an enlarged version of the superposed waves plotted in Fig. 15–5. Recall that the original two waves have $k_1 = 1.0\ \text{m}^{-1}$ and $k_2 = 1.1\ \text{m}^{-1}$. From Eq. (15–15), the beat wavelength is then

$$\lambda_{\text{beat}} = \frac{4\pi}{|(k_1 - k_2)|} = \frac{4\pi}{|(1.00\ \text{m}^{-1} - 1.10\ \text{m}^{-1})|} = 126\ \text{m}.$$

Only a little over half a wavelength is plotted in Fig. 15–6, and we can indeed see that $\lambda_{\text{beat}}/2 \cong 63$ m. The frequency of the beat is, in this case,

$$f_{\text{beat}} = (\omega_1 - \omega_2)/4\pi = (1.00\ \text{rad/s} - 0.90\ \text{rad/s})/4\pi = 0.0080\ \text{Hz}.$$

Beats in Sound: When we hear beats of sound, the combination of ear and brain has a definite physiological response. In addition to the note of average frequency, we hear a second note with the pitch of the beat frequency if the beat frequency exceeds 25 Hz or so. If the beat frequency is less than this, however, we hear regular pulses in amplitude. Because the ear interprets loudness as proportional to the wave intensity, which is in turn proportional to the amplitude squared, the ear hears a pulse every time the beat passes through either a maximum or minimum. The result of this is that *the frequency of the perceived pulse is twice the beat frequency.*

CONCEPTUAL EXAMPLE 15–2 Two tuning forks labeled as identical actually differ in frequency by approximately 0.5%. By determining the beat frequency, can you determine precisely the difference in frequencies? Can you determine which tuning fork has the higher frequency? Suppose you had some electrical tape. Knowing that winding the tape around the upper end of the prongs of a tuning fork reduces the frequency, could you then determine which tuning fork has the higher frequency?

Answer Since the beat frequency is one half of the frequency difference, an accurate measurement of the beat frequency determines the frequency difference accurately. However, the beat frequency only determines the absolute value of the frequency difference, so that this measurement alone cannot determine which tuning fork has the higher frequency.

The tape can be used to determine which tuning fork has the higher frequency. Suppose the two tuning forks are distinguished by color, say blue and green. Suppose we put the electrical tape on the blue one. If the blue one has the lower frequency, then the tape lowers it further, and the beat frequency increases. If the blue one has the higher frequency, then the tape reduces the difference and that will be noticed in a measurement of the beat frequency.

What Do You Think? Why would putting electrical tape on the end of the tuning fork prongs reduce the frequency?

EXAMPLE 15–3 Two sinusoidal waves traveling along a string with equal amplitudes and with periods of 19 and 20 seconds, respectively, are observed to have a maximum displacement at a particular point and time (say $x = 0$ and $t = 0$). After how many seconds will the resultant displacement be zero at $x = 0$? When will this occur for the second time?

Setting It Up We sketch the two waves as a function of time at the fixed point $x = 0$ (Fig. 15–7). They coincide at $t = 0$, where they are each a maximum. The sketch extends to beyond $t = 5$ s, the time when the wave with 20-s period crosses the axis, while the wave with 19-s period will have crossed it at $t = 4.75$ s.

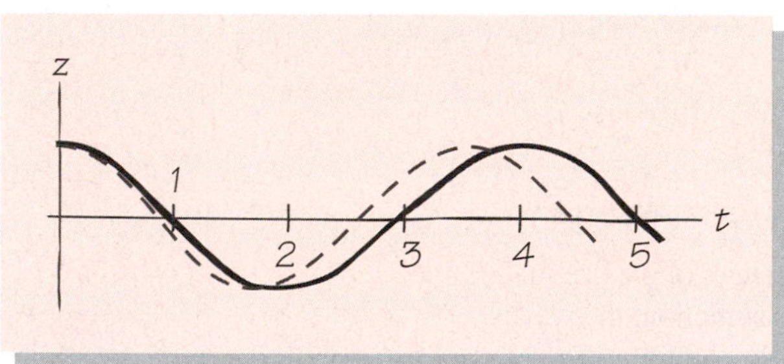

▲ **FIGURE 15–7** Two traveling waves have the same amplitude, but slightly different frequencies.

Strategy We deal at a given space point ($x = 0$) with the superposition of two traveling waves with slightly different periods and equal amplitudes. The first step is to write expressions for the two waves at $x = 0$, taking into account the fact that they are both at their maximum at $t = 0$. We then add the two waves and use trigonometric identities to express the sum as a product of single waves, one involving a beat frequency and one involving an "average" frequency, just as in the text above. Because the oscillation in the "average" frequency is faster, we expect that it is the average frequency term that will vanish first. At that point it is a matter of numerically identifying the time when the harmonic function involving the average frequency is zero. We then need to convert periods to angular frequencies ω and then work out the details.

Working It Out A harmonic wave at a particular point is a function only of t. If it takes its maximum value at $t = 0$, it must be proportional to $\cos \omega t$. We have two waves, and since their amplitudes are equal, we describe them by $A \cos \omega_1 t$ and $A \cos \omega_2 t$, respectively.

Now we work out the sum, breaking out sums and differences of the frequencies:

$$A(\cos \omega_1 t + \cos \omega_2 t) = A\left[\cos\left(\frac{\omega_1 + \omega_2}{2}t + \frac{\omega_1 - \omega_2}{2}t\right) + \cos\left(\frac{\omega_1 + \omega_2}{2}t - \frac{\omega_1 - \omega_2}{2}t\right)\right]$$
$$= 2A\cos\left(\frac{\omega_1 + \omega_2}{2}t\right)\cos\left(\frac{\omega_1 - \omega_2}{2}t\right).$$

The first cosine term contains the average frequency, the second the beat. Our expression will be zero when either $\frac{\omega_1 + \omega_2}{2}t = \frac{\pi}{2}$ or $\frac{\omega_1 - \omega_2}{2}t = \frac{\pi}{2}$, whichever happens earlier. These correspond respectively to $t_{1,\text{av}} = \frac{\pi}{\omega_1 + \omega_2}$ or $t_{1,\text{beat}} = \frac{\pi}{\omega_1 - \omega_2}$. As must be the case, $t_{1,\text{av}}$ is smaller. The sum of the two waves will again be zero at *either* $t_{1,\text{beat}}$ or at the second zero of the average frequency factor $t_{2,\text{av}}$, whichever occurs first. We have $\frac{\omega_1 + \omega_2}{2}t_{2,\text{av}} = \frac{3\pi}{2}$ or $t_{2,\text{av}} = \frac{3\pi}{\omega_1 + \omega_2}$. We will find which occurs first numerically, but since the frequencies of the two waves are almost the same, we might expect the beat period to be very large, and $t_{2,\text{av}}$ to be first.

We start with the relation $\omega = 2\pi f = 2\pi/T$ to calculate $\omega_1 = 2\pi/(19\text{ s}) = 0.33\text{ rad/s}$ and $\omega_2 = 2\pi/(20\text{ s}) = 0.31\text{ rad/s}$. We substitute these into our expressions for $t_{1,\text{av}}$, $t_{2,\text{av}}$, and $t_{1,\text{beat}}$. We find that these are, respectively, 4.9 s, 15 s, and 1.6×10^2, and the beat period is much larger than the times associated with the "average" wave.

What Do You Think? Would these results be changed if at $x = 0$ each wave had their maxima at $t = 5$ s rather than 0 s? Would they change if each wave had a $t = 0$ maximum at a different space point than $x = 0$?

15–4 Spatial Interference Phenomena

▲ **FIGURE 15–8** Two sources that produce waves in a ripple tank form an interference pattern, visible as regions of constructive and destructive interference radiating out from the sources.

Sound waves, light waves, and water waves all exhibit *interference patterns* in space.† *Any* waves that propagate in two or three dimensions can show spatial interference phenomena provided that the interfering waves are coherent. For example, if a laser beam is separated into two beams by passing it through two holes (to ensure two laser light sources of the same wave characteristics), it is possible to produce a regular series of intense spots where the two beams interfere constructively. A shallow tank of water—what is often called a ripple tank—is another useful tool for demonstrating interference patterns, in this case of water waves (Fig. 15–8). Spatial interference phenomena are behind many precision distance measurements, as we'll see when we study the interference of light waves (Chapters 37 and 38).

To see how spatial interference phenomena arise, imagine a single source of harmonic traveling waves with wavelength λ and frequency f, such as a harmonically driven loudspeaker, and two pipes that carry the sound waves away from the loudspeaker (Fig. 15–9). The pipes fix the path that the waves can follow, and the two paths do not necessarily have the same length even if they end up at the same place. At the source, i.e., the beginning of the respective paths, the waves are in phase. For example, if there is a wave crest at the start of path 1, then the same crest is at the start of path 2. These two crests each travel down their paths at the same speed. If the paths are the same length, $L_1 = L_2$, then they will both arrive at the observation point P at the same time. They will still be in phase at P, and they will interfere constructively. Under these circumstances, the adjacent troughs will arrive together, as will the next crests, and so forth. The interference will always be constructive.

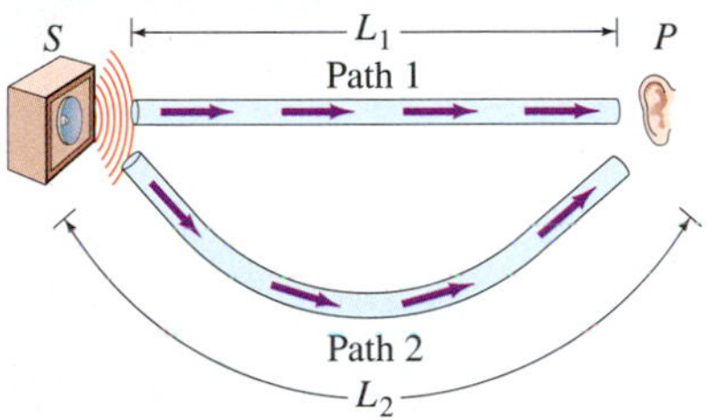

▲ **FIGURE 15–9** Unless they propagate on a string or on other one-dimensional media, waves from a source S may follow different paths to get to a given observation point P. Here two pipes of different lengths determine two different paths between S and P.

If the path lengths are different, however, then these same crests will not arrive at the same time. Whether the interference is constructive or destructive depends on the *difference* in path length, $\Delta L = L_2 - L_1$. Suppose path 2 is longer, so that ΔL is positive. If ΔL is one full wavelength, $\Delta L = \lambda$, then the second crest will arrive at P along path 1 just when the first crest arrives at P along path 2. The two waves will be in phase and the interference will be constructive. If $\Delta L = \lambda/2$, then when the crest arrives at P from path 2, it meets a trough from path 1. In this case the interference is destructive, and no disturbance is recorded at P. The situation can be summed up as follows:

$$\text{for constructive interference: } \Delta L = n\lambda,\ n = 0, \pm 1, \pm 2, \ldots \qquad (15\text{–}17)$$

$$\text{for destructive interference: } \Delta L = \left(n + \tfrac{1}{2}\right)\lambda,\ n = 0, \pm 1, \pm 2, \ldots. \qquad (15\text{–}18)$$

CONDITIONS FOR INTERFERENCE

The effect just described also occurs with two sources at different distances from an observation point. Imagine two separate sources emitting harmonic waves of the same wavelength in a given medium. We could for example make the same rhythmic disturbance at two different points on a water surface. At each point, a series of spreading ripples is created that interfere with one another. In Fig. 15–10, we have two separated sources, S_1 and S_2, each producing waves with the same wavelength λ and frequency f. We assume that these waves are *in phase*; that is, a crest always emerges from S_1 at the same moment that a crest emerges from S_2. These waves are coherent. An observation point P is located a distance L_i from source S_i; the path difference is $\Delta L = L_2 - L_1$. The conditions for constructive interference are simply that the crests, or the troughs, of the waves from the two sources arrive at P together. For this two-source situation, the same conditions respectively apply for constructive and destructive interference as before; namely, Eqs. (15–17) and (15–18). Since different places in space are different distances from the sources, there will be constructive interference at some points in space and destructive interference at others.

†Not to be confused with beats, in which we are primarily interested in time dependence at a particular point.

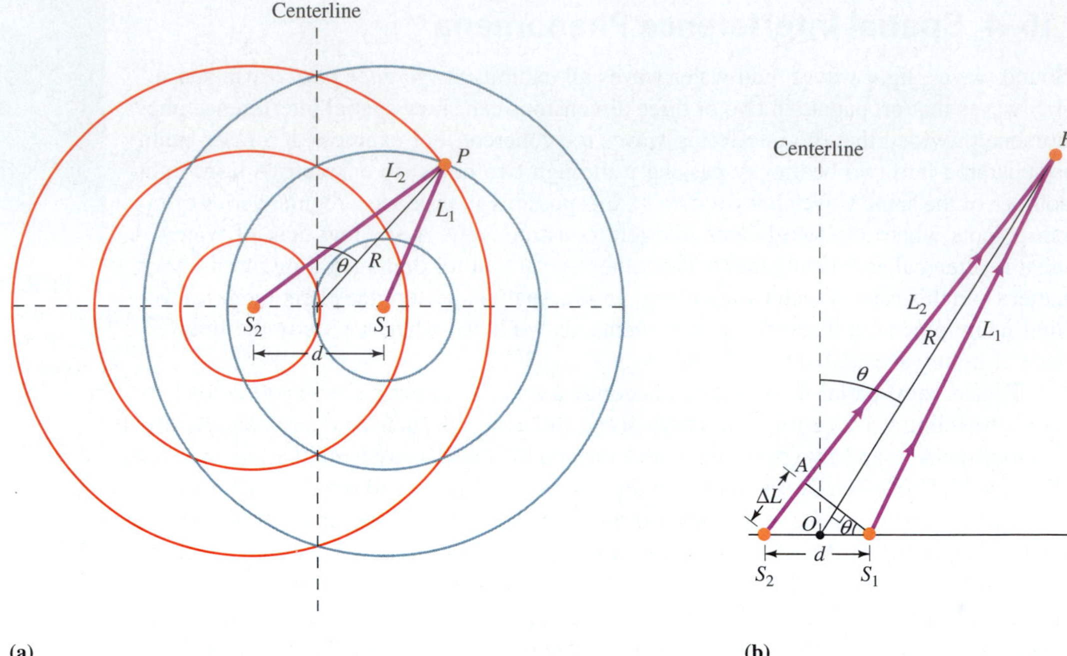

▲ **FIGURE 15–10** (a) The water waves generated by two coherent sources in phase spread away in circles from these sources. Whether the interference is constructive or destructive at observation point P depends on the difference in path lengths. The circles are (arbitrarily) drawn on the wave crests. (b) The geometry necessary to determine the path difference $\Delta L = L_2 - L_1$ and hence for the pattern of interference maxima and minima. The observation point P is far away compared to the separation of the two sources, d.

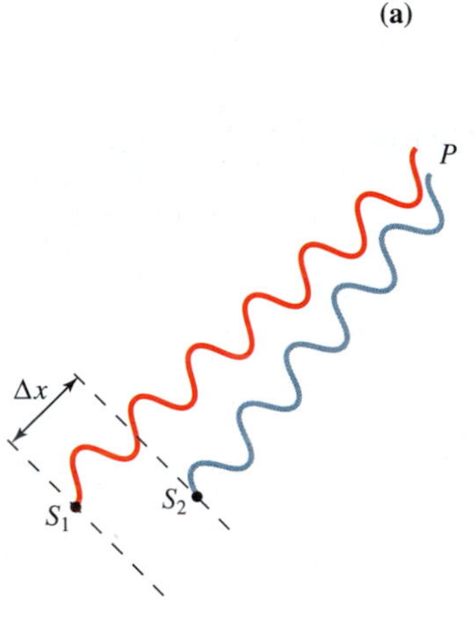

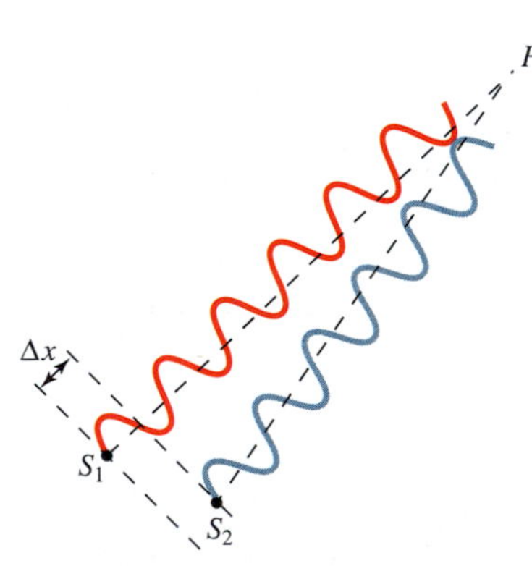

▲ **FIGURE 15–11** The conditions for maxima and minima in the interference of coherent waves of the same wavelength emitted by two sources. (a) Maxima correspond to the difference in path length equal to an integer multiple of λ, while (b) minima occur when the path difference is an integral multiple of λ plus $\lambda/2$.

Locations of Maxima and Minima

The observation point P can be anywhere that the waves propagate. Some locations will have constructive interference, some destructive, and some points will have an intermediate level of interference, depending only on the distances from the sources. The locations where the interference is constructive are said to exhibit interference *maxima*, and the locations where the interference is destructive exhibit interference *minima*. Finding the spatial pattern of the maxima and minima is an exercise in geometry—of the type sketched out in Fig. 15–11.

We can get a sense of patterns of this type with a visual analogue to wave interference. Figure 15–12a shows a set of finely spaced gray concentric circles around a point representing a wave source. Think of the gray circles as simplified crests and the clear areas between the circles as simplified troughs. When two such wave sources are present, as they are in Fig. 15–12b (too closely spaced to see easily in that figure), we can think of constructive interference as an overlap of the gray areas (crests) with themselves, giving very dark areas, or of the clear areas (troughs) with themselves, leaving a clear area; we can think of destructive interference as a uniform gray where the gray areas and the clear areas overlap. In fact, bands of maxima and minima are clearly visible in the figure, even if the centers of the waves are not.

To find the patterns of maxima or of minima defined by Eqs. (15–17) and (15–18), the necessary geometric information is the wavelength, λ, and the separation of the two sources, d. The geometry is shown in Fig. 15–10. We have put an origin, O, midway between the sources and measured the distance R from point P to the origin, as well as the angle θ between the center line and the line from the origin to P. The center line is the line perpendicular to and bisecting the line of the sources. Note that at the angle $\theta = 0$, where $L_1 = L_2$, we have $\Delta L = 0$; *there is a maximum everywhere along the center line*. The regions of the other maxima and minima are simple to see when $R \gg d$; that is, when P is very far from the pair of sources. This case is illustrated in Fig. 15–10b, which shows, to a good approximation, that all three lines—the line from S_1 to P (S_1P), the line from the origin to P, and the line from S_2

(a)

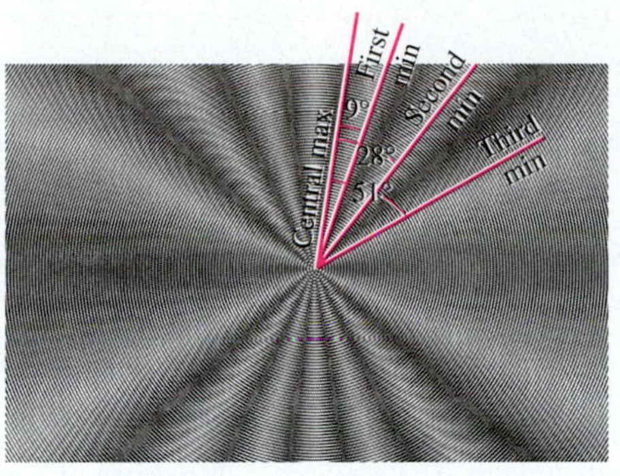

(b)

FIGURE 15–12 (a) Concentric circles representing waves propagating out from a source. (b) Overlay of the circles that spread from two adjacent sources. Regions of constructive and destructive interference, marked as max (maxima) and min (minima), are clearly visible. We have measured $\lambda = 5$ mm and $d = 16$ mm on the original of this reproduction, so the ratio $\lambda/d = 0.32$. The first minimum should occur at $\sin\theta_1 = \lambda/2d = 0.16$, or $\theta_1 = 9°$, and this is the value of the angle found in the figure.

to P (S_2P)—make the same angle θ with the center line. Then ΔL can be measured by drawing a line from S_1 perpendicular to S_2P. This line meets S_2P at point A in the figure. S_2S_1 is perpendicular to the center line, and AS_1 is perpendicular to S_2P, so the angle formed by lines AS_1 and S_2S_1 is θ. Because $AP = S_1P$, we have $\Delta L = L_2 - L_1 = S_2A$, and $\sin\theta = \Delta L/d$, or

$$\Delta L = d\sin\theta. \qquad (15\text{–}19)$$

As a result of this approximation, for $R \gg d$ we can restate Eqs. (15–17) and (15–18) for the maxima and minima in terms of the angle θ by using Eq. (15–19):

$$\text{for maxima: } \sin\theta = n\frac{\lambda}{d},\ n = 0, \pm1, \pm2, \ldots; \qquad (15\text{–}20)$$

$$\text{for minima: } \sin\theta = \left(n + \frac{1}{2}\right)\frac{\lambda}{d},\ n = 0, \pm1, \pm2, \ldots. \qquad (15\text{–}21)$$

These equations determine a set of angles for which there are interference maxima and minima. Positive θ values are to the right of the center line and negative values are to the left. The array of angles is symmetric about the center line, so we can concentrate on one side. On the right side $n = 0$ is a maximum at $\theta = 0$, which is along the center line itself. As we increase θ, the first minimum occurs at $\sin\theta = \lambda/2d$. The second maximum is at $\sin\theta = \lambda/d$, the second minimum at $3\lambda/2d$, and so forth. In Fig. 15–12b these angles are marked on the pattern made by overlaying two sets of concentric circles, one set slightly displaced from the other. In the situation shown here $\lambda = 5$ mm and $d = 16$ mm. The first three minima should then be at $\sin\theta_1 = 0.16$, $\sin\theta_2 = 0.47$, and $\sin\theta_3 = 0.78$, or $\theta_1 = 9°$, $\theta_2 = 28°$, and $\theta_3 = 51°$, in agreement with Fig. 15–12b.

Any unchanging phase difference between two harmonic waves—coherent waves—will lead to an interference pattern. Above, the two sources have the same frequency and no phase difference between them. But suppose we change things a little. Source S_1 could consist of a sharp object dipping into water with some fixed frequency, and S_2 consists of a similar object dipping into the water with the same frequency but a time interval τ later. This changes the relative phase of the two waves (by $\omega\tau$), but the relative phase is still fixed, so that we still have coherence. Let us now change source S_2 as follows. Its frequency is still the same as that of S_1, but the sharp object dips into the water in different positions around what was previously the fixed position of S_2. This means that for each wave, the interval ΔL is slightly different. If ΔL were much smaller than the wavelengths involved, then there would still be an interference pattern, but a blurred one. If ΔL is of the order of the wavelength or more, then the interference pattern is gone: There are still points of total constructive or destructive interference, but these vary from wave to wave, and no steady pattern of the type seen in Fig. 15–8 will appear. This is an example of incoherence—no interference pattern appears when incoherent waves superpose.

Example 15–4 describes the interference pattern that occurs when two waves of the same wavelength have phases differing by π.

EXAMPLE 15–4 Two loudspeakers, S_1 and S_2, are separated by 0.50 m. These speakers form sound waves of the same amplitude at a frequency of 4400 Hz. The amplifier emits the two waves 180° out of phase. A set of chairs is arranged in a semicircle 30.0 m from the midpoint of the two speakers. Find the amplitude of the wave at the chair on the perpendicular bisector (center line) of the line between the speakers. At what distance to the right of this central chair is there a first maximum in sound intensity?

Setting It Up The situation is sketched in Figure 15–13. We have marked on the figure the known speaker separation d and the radius R of the semicircle of chairs surrounding them. The speakers emit with known frequency f. The unknown distance D where a maximum occurs is also marked.

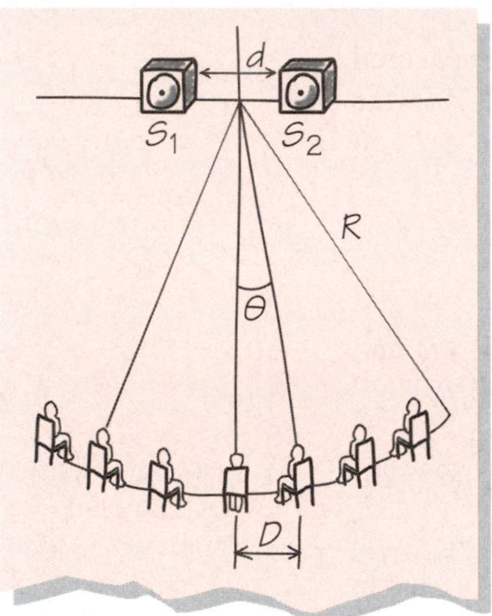

▲ **FIGURE 15–13** The geometry is the same as that illustrated in Fig. 15–10b.

Strategy The first step is a geometrical one, the calculation of the difference in the lengths of the paths from the two speakers to a point on the semicircle. This can be done exactly, but in the approximation that $(d/R)^2 \ll 1$ we can treat the chairs not far from the center line *as if* they were on a straight line, perpendicular to the center line, and we can approximate the sine of the small angle θ by θ alone. We then need to translate that difference in path length into a statement about destructive/constructive interference. Because the two speakers emit the waves 180° out of phase, the rules in Eqs. (15–20) and (15–21) are interchanged: *Minima become maxima* and vice versa. Thus, along the center line, for which the path difference ΔL is zero, the interference is now destructive. We want the smallest angle θ for which ΔL implies constructive interference, and that translates geometrically into D.

Working It Out With the phase change taken into account, the rules are

$$\text{for minima:}\quad \sin\theta = n\frac{\lambda}{d},\ n = 0, \pm1, \pm2, \ldots;$$

$$\text{for maxima:}\quad \sin\theta = \left(n + \tfrac{1}{2}\right)\frac{\lambda}{d},\ n = 0, \pm1, \pm2, \ldots.$$

The first maximum ($n = 0$) is at an angle given by

$$\sin\theta = \frac{\lambda}{2d}.$$

If the semicircle of chairs is a distance R from the speakers, then $\theta = D/R$. For small θ, $\sin\theta \cong \theta$, and the value of D for which the first maximum occurs is

$$\sin\theta = D/R = \lambda/2d, \quad\text{or}\quad D = (\lambda/2d)R = (v/2fd)R.$$

We have used the relation $\lambda f = v$, where v is the wave speed. The speed of sound is roughly 330 m/s, so

$$D = \frac{(330\ \text{m/s})}{2(4400\ \text{Hz})(0.50\ \text{m})}(30.0\ \text{m}) = 2.3\ \text{m}.$$

We see that $D \ll R$, so our small-angle approximation is justified.

What Do You Think? Suppose the listener in the chair at which the first maximum is heard rolls radially away from the midpoint of the speakers at a rate of 3 m/s. What properties of the sound will have changed when he reaches a distance of 60 m?

CONCEPTUAL EXAMPLE 15–5 The spatial interference pattern of maxima and minima that we have studied forms lines corresponding to particular angles from a center line. These angles depend on the wavelength λ of the interfering waves and the separation d of the sources. When these angles are small, we say the pattern is *narrow*; when the angles are large, we say the pattern is *broad*. If you had a particular pattern and you wanted to make it broader, and you could control both λ and d, what would you do?

Answer The angular distance between adjacent angles is given by $\sin\theta_{n+1} - \sin\theta_n = (\lambda/d)$. Thus increasing λ and/or decreasing the source separation d make this distance increase. We say that the pattern *spreads* (or is broadened). The pattern *shrinks*, or *narrows*, as λ decreases and/or d increases. For example, as the separation of our two loudspeakers decreases, the pattern of interference maxima and minima spreads in angle.

What Do You Think? By how much can the pattern spread?

15–5 Pulses

When an incident ocean wave meets a wall, the wave rebounds and forms a reflected wave that moves back toward the sea. If the wall is low, then part of the incident wave can pass over the top of it and continue as a transmitted wave. This demonstrates that waves can both reflect from, and be transmitted across, boundaries. A simple lecture demonstration of reflection is obtained by attaching a long rope to a wall and holding the other end fairly taut flipping the wrist. The result is a bump traveling to the wall and

THINK ABOUT THIS . . .

HOW DO NOISE-CANCELING HEADPHONES WORK?

The noise-canceling headphones worn by workers in helicopters or in other very noisy environments (Fig. 15–14) make use of the interference phenomena. They contain both speakers and a microphone located close to the entrance of the ears. Sound received at the microphone is amplified and supplied to the speaker with a phase difference, such that the sound from the speaker cancels the original signal as completely as possible. Maximum cancellation is set to occur within the ear, nominally at the surface of the eardrum. Communication from the radio or intercom of the aircraft is not subject to cancellation and is therefore audible to the crew members.

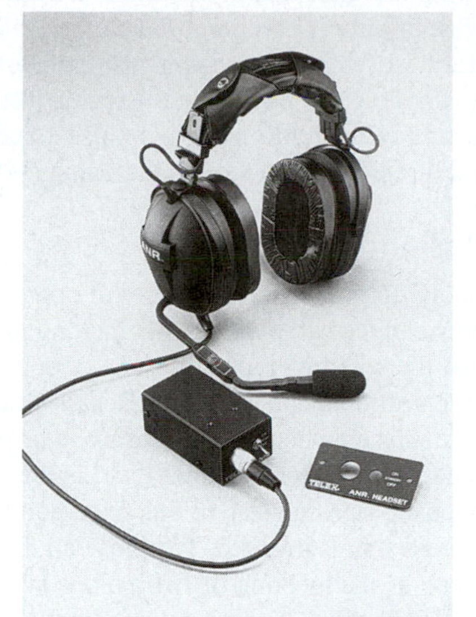

▶ **FIGURE 15–14** An electronic noise cancellation headset such as this one can reduce noise in the low-frequency range by some 10–15 dB over conventional headsets.

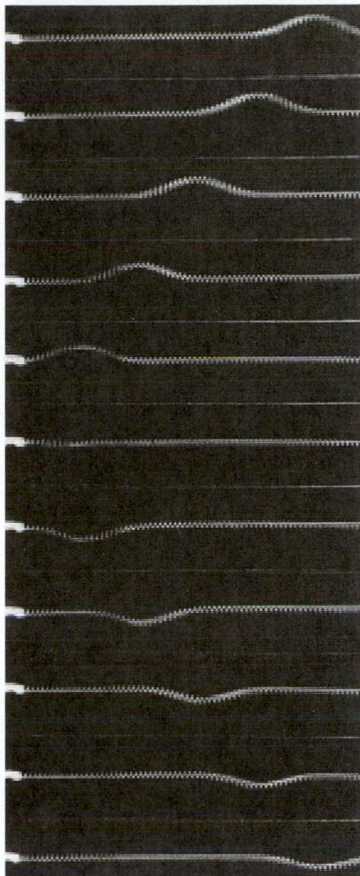

▲ **FIGURE 15–15** A pulse is generated on a light spring on the right side and moves to the left until it encounters a wall, where the spring is attached. The displacement at this point is zero, and the pulse is then reflected from the wall towards the right with an inverted pulse as explained below.

then traveling back to the other end (Fig. 15–15). If the held rope is connected to another rope of different mass density, with the second rope connected to the wall, then both the reflection and transmission are visible at the two-rope boundary.

The bump generated in the ropes is a **pulse**, and the discussion of reflection and transmission is somewhat easier to visualize in terms of pulses. What are pulses? Pure harmonic waves (cosines or sines), or any truly periodic waves, are idealizations in that they are infinitely long. Pulses are disturbances that have a beginning and an end, whether short or not. For us to be able to deal with pulses we need only be sure that they obey the wave equation. This is not much of a limitation, basically only requiring for us that the amplitudes involved are not too great. Aside from that, we saw in Chapter 14 that *any* function of the form $F(x \pm vt)$ satisfies the wave equation. The wave equation is *linear*, i.e., the sum of two solutions is also a solution. So the sum of two pulses is another acceptable pulse.

We are interested in pulses because they are the way waves occur in the physical world. Communication of any kind—the passage of information—is by means of pulses. The sound of a drum, a flash of light, and a ripple caused by a pebble dropped into a pond are all pulses. In Section 15–6 we will see that pulses are formed from the superposition of harmonic waves, but in this section we will use the superposition of pulses to provide us with insight into wave interference.

Collisions Between Pulses

Consider a pulse moving to the right along a string (Fig. 15–16). In the figure the velocity of the transverse motion is represented by vertical arrows. There is upward transverse motion along the leading edge of the pulse. The transverse motion vanishes at the peak, and there is downward transverse motion along the trailing edge until the string again comes to rest after the pulse has passed.

Consider next the superposition of two pulses of identical shape on the string: Pulse 1 is positive and moves to the right; pulse 2 is negative and moves to the left. Figure 15–17 illustrates a sequence in which the pulses approach each other, superpose, then continue on. At the moment when their respective peaks would be at exactly the same position, $t = 0$ s, the superposition of the two pulses gives a flat string. This situation is in fact no more peculiar than that of a ball that has been thrown vertically and is momentarily at rest at the peak of its flight. Even if the ball is momentarily at rest, its acceleration remains constant. Similarly, even though the string is flat momentarily,

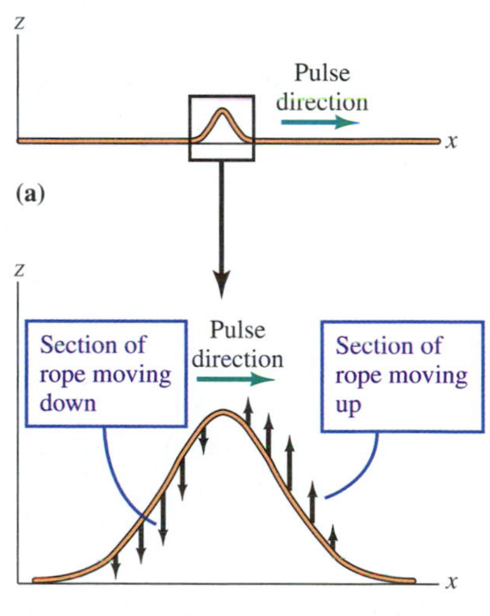

▲ **FIGURE 15–16** A pulse moving to the right (horizontal arrow) on a string. The vertical arrows indicate the velocities of elements of the string.

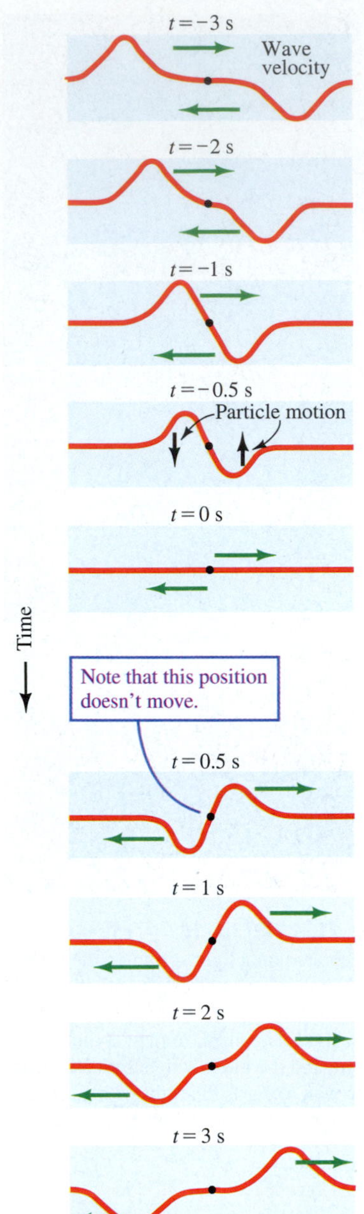

▲ **FIGURE 15–17** The superposition of two pulses, one moving to the right and one to the left. The center point between the two waves, at the dot, remains motionless throughout.

it may still be moving. Not only do the displacements of the pulses superpose but so do their time derivatives, which measure their transverse velocities. The velocities of the string are indicated in sequence in Fig. 15–17. When the peaks superpose, the leading edge of pulse 1 is superposed on the trailing edge of pulse 2. Both edges are moving upward, so there is a region just right of the center of the superposed peaks moving upward rapidly. Similarly, the region just left of the center of the superposed peaks, consisting of the trailing edge of pulse 1 and the leading edge of pulse 2, is moving downward rapidly. Even though the string is flat for a brief moment, it is not at rest!

Reflection

In our discussion of two pulses of opposite sign that meet and interfere, the central point between the two pulses—marked on Fig. 15–17 by a dot—never moves at all. It is clear that the colliding pulses must be of opposite sign and equal magnitude in order to leave a central point undisplaced. Recall now that there is also a single point that remains undisplaced when an incident wave or pulse reflects from the fixed end of a string (or rope). The stationary point there is the point fixed to the wall. Because only pulses of opposite sign can lead to this fixed point, we then argue that the pulse that reflects from a fixed wall is inverted. To phrase all this in another way, we think of the reflection of a pulse as the meeting of two pulses, with one pulse coming from the other side of the wall. The string and the inverted pulse or wave on the other side of the wall are purely imaginary, just an aid to picturing the form of the reflected wave. Figure 15–18 shows a sequence of a pulse reflecting from the fixed end of a string. The pulse moving away from the boundary is the *reflected* pulse.

Suppose that instead of being fixed at the wall, the end of a rope under tension is *free to move in the vertical direction*. This could be arranged by fixing the rope to a support that allows it to slide freely, as in Fig. 15–19a. The analogous sequence of reflection is shown in Fig. 15–19b. As before, the reflected pulse matches the incident pulse

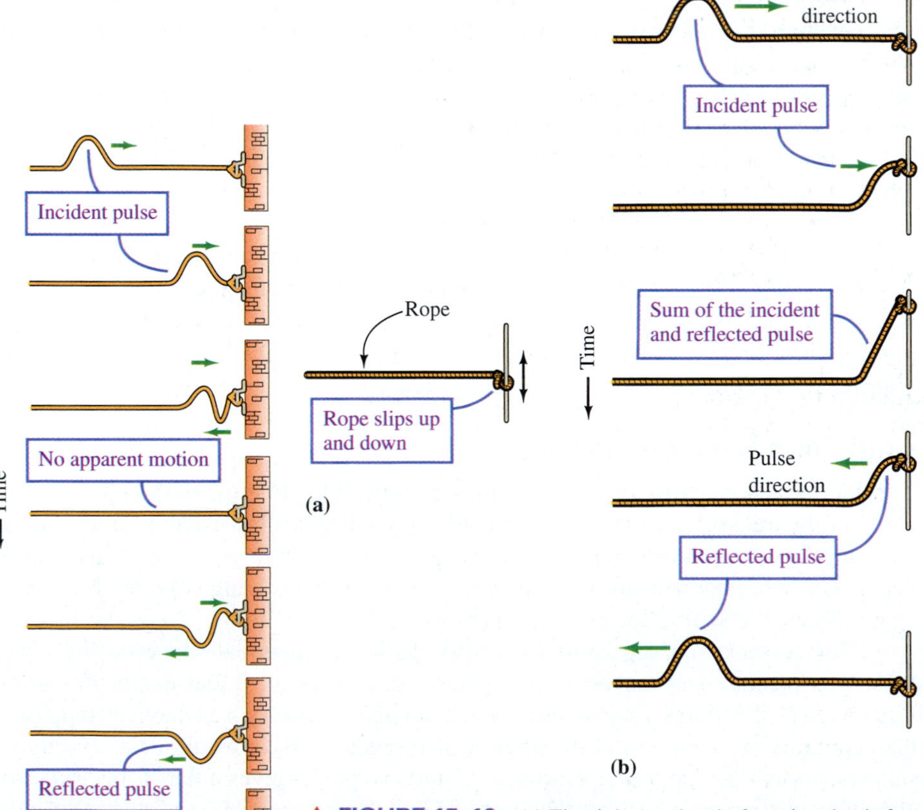

▶ **FIGURE 15–18** An incident pulse reflecting from the fixed end of a string is inverted. If the pulse is symmetric, then at some point the string is flat, because the reflected pulse cancels the incoming pulse by destructive interference.

▲ **FIGURE 15–19** (a) The right-hand end of a string that is free to slide vertically. (b) The reflection of an incident pulse is upright when the end of the string is free.

in shape. This time, however, the reflected pulse must be right-side up rather than inverted because the end of the string has been thrown upward by the incident pulse. In this case, we can also visualize an (imaginary) traveling wave on the far side of the slide support moving to the left. We see from the sketch that when the two pulses meet and combine at the slide support, the pulse height at the support is twice as high as the original amplitude. (What has happened here is that the free end gets a vertical impulse when the pulse arrives. Because there is no mass to be accelerated farther down the string—there is no string to the right of the spot—the end can move an additional amount.) Just as the fixed end of the string was an argument that the reflected wave was inverted, the large size of the displacement at the end of the string in this case is an argument that when the end of the string is not fixed, the reflected wave is right-side up. This is easily confirmed by experiment.

The conclusions we have drawn about whether reflected pulses are inverted or not can also be used to establish the form of reflected harmonic waves, as we see in the following example.

EXAMPLE 15–6 A harmonic sound wave travels in the $+x$-direction. We can describe this wave by using the quantity z that measures the longitudinal displacement of a layer of air from its equilibrium position. z will be a function of the longitudinal position x and will take the form $z_i = A_i \cos(kx - \omega t)$. (We have managed to set the clock so that there is no additional phase here.) The wave encounters a brick wall at $x = 0$, and because no air can cross the brick wall, the boundary condition $z = 0$ at $x = 0$ holds. What is the form of the reflected wave, and what is the form of the superposition of the incident and reflected waves?

Strategy First of all, we need to construct a generic form of the reflected wave. This will have a form similar to the incident wave, with the change from $(kx - \omega t)$ to $(kx + \omega t)$, with a different amplitude A_r and with an unknown phase. We next impose the condition that $z = 0$ at $x = 0$. The expectation is that this condition, which must hold *at all times*, will fix the reflected amplitude and phase. Once we have established the form of the reflected wave, we can study how it adds to the incident wave.

Working It Out The reflected wave has form $z_r = A_r \cos(kx + \omega + \phi_r)$. The boundary condition stipulates that the superposition of the two waves (the net wave) must be zero at $x = 0$: $z_i + z_r = 0$ at $x = 0$ or, for all t,

$$A_i \cos(k \times 0 - \omega t) + A_r \cos(k \times 0 + \omega t + \phi_r)$$
$$= A_i \cos(-\omega t) + A_r \cos(\omega t + \phi_r) = 0.$$

We can expand the second cosine by using the identities in Appendix IV–4—here, $\cos(\omega t + \phi) = \cos \omega t \cos \phi - \sin \omega t \sin \phi$. Then we separate the coefficients of $\sin \omega t$ and $\cos \omega t$ that result:

$$(A_i + A_r \cos \phi_r) \cos \omega t + (-A_r \sin \phi_r) \sin \omega t = 0.$$

If this equation is to hold for all t, the coefficients of both $\sin \omega t$ and $\cos \omega t$ must be zero. For example, if the expression is to be zero at $t = 0$, where $\sin \omega t$ is zero automatically, the coefficient of $\cos \omega t$ must be zero; if the expression is to be zero at $t = \pi/2\omega$, where $\cos \omega t$ is zero automatically, the coefficient of $\sin \omega t$ must be zero. Thus

$$A_i = -A_r \cos \phi_r \quad \text{and} \quad 0 = A_r \sin \phi_r.$$

The second equation has as its most interesting solution that $\phi_r = 0$. Then the first equation will give $A_r = -A_i$.

Pulling things together, the reflected wave takes the form

$$-A_i \cos(kx + \omega t).$$

This is like the incident wave, except propagating to the left and inverted. The sum of the incident and reflected wave is therefore

$$z_{total} = z_i + z_r = A_i[\cos(kx - \omega t) - \cos(kx + \omega t)]$$
$$= 2A_i \sin kx \sin \omega t.$$

The important results to note are that (a) $z = 0$ at $x = 0$, as required, and (b) the conditions that determine the reflected amplitude and phase *do not depend on the value of* ω.

What Do You Think? How is this wave different, if at all, from the standing waves we saw in Chapter 14?

Transmission

Reflections of waves and pulses are familiar to us through sound echoes and the bouncing of water waves from boundaries. Less familiar, but equally important, is the passage of waves and pulses through boundaries that separate different media. This is the process of *transmission*. The transmission of waves occurs when sound waves penetrate from one medium to another—for example, from water to air, and when light waves pass between air and glass (Chapter 35).

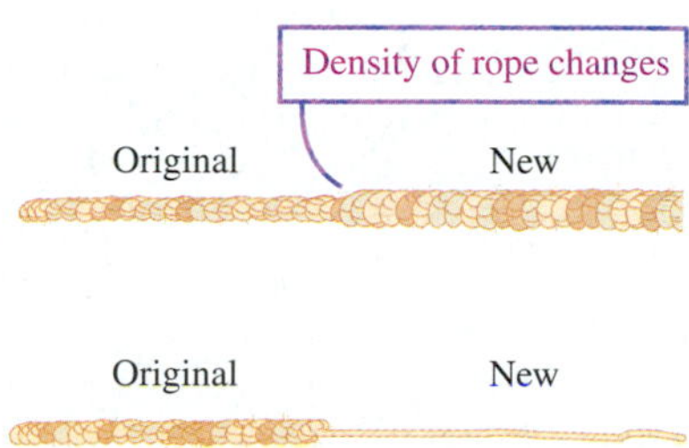

▲ **FIGURE 15–20** A string is attached to a heavier string (top) and to a lighter string (bottom).

In this chapter we will discuss transmission in terms of harmonic waves or pulses moving along strings. Up to this point, we have supposed that one end of a string was either tightly fixed or completely free. Imagine instead that our string is attached to a new string that is heavier or lighter than the original one (Fig. 15–20). If the new string is sufficiently heavy, then it acts like the wall; if the new string is sufficiently light, then

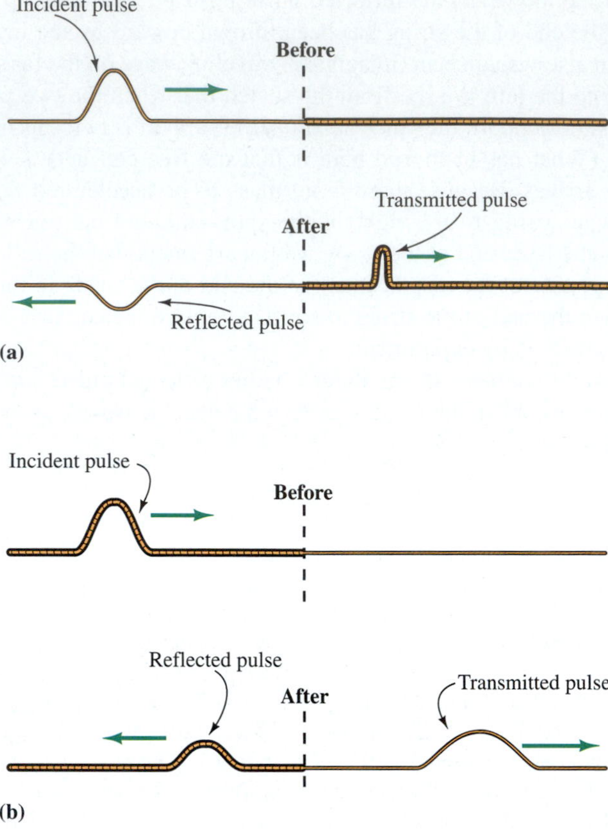

▲ **FIGURE 15–21** (a) A pulse is incident on a heavy string from a lighter one. The reflected pulse is reduced in amplitude and inverted. The transmitted pulse has an amplitude less than that of the incident pulse and moves more slowly than either the incident or reflected pulse. (b) A pulse is incident on a light string from a heavier one. The reflected pulse is reduced in amplitude and right-side up. The transmitted pulse has an amplitude less than that of the incident pulse and moves more rapidly than either the incident or reflected pulses.

it serves as a device that allows the end of the original string to move freely. The situations in Fig. 15–21 are between these extremes, and there we observe both a wave reflected back along the original string and a transmitted wave carried by the new string. We can most easily understand this by considering a pulse traveling along the first string toward the boundary.

The Amplitudes of Reflected and Transmitted Pulses: Figure 15–21a shows what happens when a pulse is incident on a heavier string and Fig. 15–21b shows what happens when a pulse is incident on a lighter string. Consider first the *reflected* pulse. In view of our discussion of the string with fixed or free ends, it is not surprising that when the new string is heavier than the original string, the reflected pulse is inverted, whereas when the new string is lighter, the reflected pulse is right-side up. The amplitude of the reflected pulse decreases to zero in the limit where the new string has a mass density equivalent to that of the original string because, in that case, there is no boundary. This behavior is both smooth and reasonable. What can we say about the transmitted pulse other than that it has the same sign as that of the incident pulse? To say more, we use energy conservation.

Consequences of the Conservation of Energy: If we have an incident pulse that is partly reflected and partly transmitted, then the sum of the energies in these two pulses equals the energy of the incident pulse. If there is no transmitted pulse, as is the case when the new string is either infinitely heavy or infinitely light, then the energy in the reflected pulse matches the energy in the incoming pulse. The transmitted and reflected waves must share the energy of the incident wave and each therefore has an amplitude reduced from that of the incident wave. We'll illustrate the quantitative consequences in Example 15–7.

The Change in Velocity with a Change in Medium: Consider again two different strings attached to each other. The tension throughout the two strings is uniform, as there is no horizontal acceleration. The speed, however, is not uniform. With $v = \sqrt{T/\mu}$, where μ is the mass density of the string, we see that when the new string is heavier (per unit length), the velocity of a harmonic wave is smaller; when it is lighter, the velocity is larger.

We'll state without proof that if the wave speed in the medium containing the incident wave is greater than the wave speed in the medium containing the transmitted wave, then the reflected wave is inverted but is otherwise similar in shape to the incident wave. The transmitted wave is always right-side up, but its shape will be stretched if the wave speed is higher in the second medium and compressed if the wave speed is lower in the second medium (Fig. 15–21). This information is further summarized in Table 15–1.

TABLE 15–1 • Properties of Reflected Waves and Transmitted Waves†

A Wave in Medium 1 Is Incident on a Boundary That Separates Medium 1 from Medium 2	Reflected Wave	Transmitted Wave
Medium 1 is less dense than medium 2 (wave speed in 1 is higher than in 2)	Inverted	Right-side up, Compressed
Medium 1 is denser than medium 2 (wave speed in 1 is lower than in 2)	Right-side up	Right-side up, Stretched

† *See Fig. 15–21.*

EXAMPLE 15–7 A string with mass density μ_1 is attached at $x = 0$ to a second string, with mass density μ_2. The strings are under a common tension T. An incident harmonic traveling wave of amplitude A_i arrives on the first string. A reflected harmonic wave of amplitude A_r results on the first string, and a transmitted harmonic wave of amplitude A_t results on the second string (Fig. 15–22). The strings behave elastically so there is no mechanism for energy loss. Use the fact that energy is conserved to relate the three amplitudes.

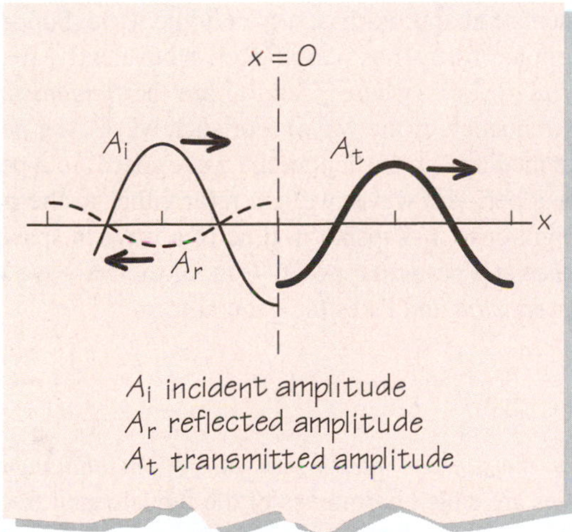

▲ **FIGURE 15–22**

Setting It Up Figure 15–22 labels the amplitudes of the incident, reflected, and transmitted wave. The reflected wave is drawn as having changed sign, which is appropriate if $\mu_2 > \mu_1$, but nothing in what follows depends on that.

Strategy We can solve this using energy conservation alone. However, we can equally well deal with the energy carried per unit time by the three waves—that is, the power—and as we know how to write expressions for the average power in a traveling wave, Eq. (14–37), this is an effective way to proceed. We will therefore write down expressions for the incident, reflected, and transmitted power. Generically the average power is given by $P = \mu\omega^2 A^2 v/2$, where μ is the mass density, ω the angular frequency, A the amplitude, and v the velocity of the harmonic wave. The statement of energy conservation is equivalent to the statement that the incident power is equal to the sum of the reflected and the transmitted power. We also need to use the condition that the frequencies are the same for all three waves—they are the same because they simply represent a counting of the rate at which crests reach and leave the boundary. (If the frequency changed, wave crests would be lost or would pile up somewhere.) When we apply the energy conservation condition, we find a relation for the amplitudes.

Working It Out The statement that the average incident power is equal to the sum of the average reflected and transmitted power is

$$\frac{\mu_1\omega_i^2 A_i^2 v_1}{2} = \frac{\mu_1\omega_r^2 A_r^2 v_1}{2} + \frac{\mu_2\omega_t^2 A_t^2 v_2}{2}.$$

Here we have noted that the speed of both the incident and reflected waves is v_1, whereas the speed of the transmitted wave is v_2. In particular, $v_1 = \sqrt{T/\mu_1}$ and $v_2 = \sqrt{T/\mu_2}$. The (identical) frequencies and the factors of $\sqrt{T}$ and of $\frac{1}{2}$ cancel, and we are left with the relation between the three amplitudes:

$$\sqrt{\mu_1}A_i^2 = \sqrt{\mu_1}A_r^2 + \sqrt{\mu_2}A_t^2.$$

Note that when there is no transmitted amplitude—either because there is a wall at $x = 0$ or because the end of the first string is free—then the relation is $A_i^2 = A_r^2$. Because this relation depends on the *squares* of the amplitudes, it tells us nothing about whether the reflected wave is right-side up (positive A_r) or inverted (negative A_r).

What Do You Think? What other conditions do you think would be available for the determination of the amplitudes?

THINK ABOUT THIS . . .

HOW ARE REFLECTED AND TRANSMITTED WAVES OF USE TO US?

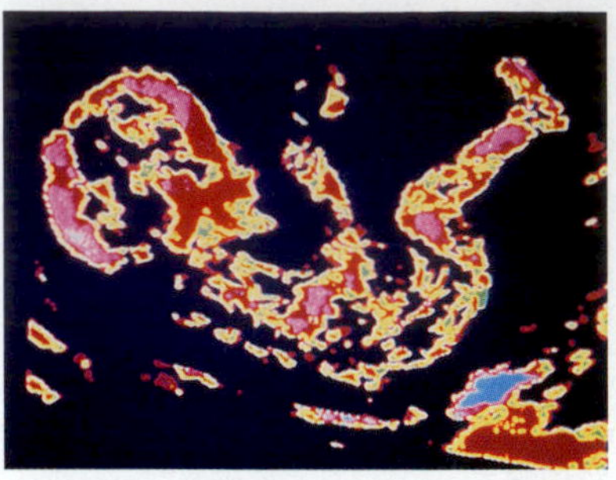

▲ **FIGURE 15–23** The colors in this ultrasound image of a human fetus are the result of computer processing of the waves that reflect back to the source/receiver.

Our eyes primarily detect light reflected from the objects around us, and it is with the information in these reflected light waves that we interpret the world around us. The light arrives at Earth's surface having traveled through space from the Sun and having been transmitted into the atmosphere. Many technologies use the phenomena of reflection and transmission. For example, geophysicists use sound waves to explore Earth's interior (see the Chapter 14 "Think About This... How do we know what's inside Earth?"). We can emphasize here the important role of reflection and transmission across rock boundaries. Doctors use ultrasound (sound of very high frequencies) to make interior images of the human body (Fig. 15–23), and the same techniques are used for investigating imperfections in materials. Radar and sonar devices emit electromagnetic waves or sound waves, then detect the arrival of their reflections, allowing us to learn about the reflecting object. Finally, Doppler shifts of reflected waves tell us about the speed of the reflecting object.

*15–6 Fourier Decomposition of Waves

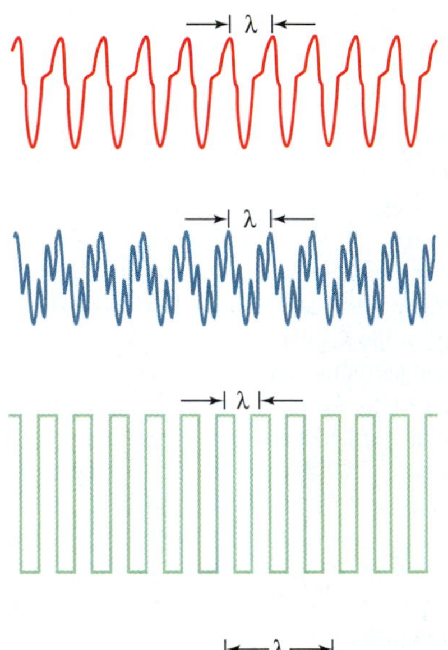

▲ **FIGURE 15–24** Some nonharmonic periodic waves.

We have studied two types of waves: single pulses, which are not periodic phenomena, and harmonic (or sinusoidal) waves, which are. In another important category we might put periodic waves that are not harmonic (Fig. 15–24). The superposition principle, together with a remarkable mathematical result called *Fourier's theorem*, shows that the waves in these three categories are very closely related. The substance of the theorem is (1) *any periodic (not necessarily sinusoidal) wave can be approximated by a superposition of purely harmonic (or sinusoidal) waves with different frequencies in the form of a sum*, and
(2) *any pulse can be approximated by a superposition of purely harmonic waves in the form of an integral.*

Fourier's theorem does more than just state a possibility; it also lays out the rules for finding what the harmonic components are that make up a wave and the relative weights of these components—what we call the *decomposition* of the wave. Even better, it shows how only a few harmonic components can approximate any periodic wave. Suppose that the wave we wish to fit with harmonic forms has period T or, equivalently, frequency $f = 1/T$, and angular frequency $\omega = 2\pi f$ (where f and ω are the *fundamental* frequency and *fundamental* angular frequency of the wave). For such waves, the period in time is related by properties of the medium (in particular, the wave speed) to a period in space—the wavelength. Thus, for a periodic wave, we can refer either to the periodic time dependence at a particular point or to a snapshot in time of a wave in space. Let's look at the periodic time dependence at a particular point, $f(t)$, of the wave we want to fit. The fit is known as a *Fourier expansion* and takes the general form

$$f(t) = \sum_{n=1}^{\infty} A_n \sin(n\omega t + \phi_n). \tag{15–22}$$

The only frequencies that enter into the sum are integer multiples of the fundamental frequency, $f_n = nf$. These frequencies are called harmonics of the fundamental frequency, with second harmonic referring to $n = 2$, third harmonic referring to $n = 3$, and so on.

If, instead of concentrating on the time dependence, we take a snapshot of a periodic wave $g(x)$ in space, then that wave, however complex in form, has a fundamental wavelength λ or a fundamental wave number $k = 2\pi/\lambda$. The Fourier expansion of our wave takes the form

$$g(x) = \sum_{n=1}^{\infty} A'_n \sin(nkx + \theta_n). \tag{15–23}$$

Here, the higher harmonics have wave numbers that are integer multiples of the fundamental wave number. The Fourier expansion tells exactly how to calculate the constants in Eqs. (15–22) and (15–23) with a set of rules. While we won't detail the expansion rules here, we do emphasize that these rules are not at all ambiguous.

If the infinite sum in Eq. (15–22) or (15–23) is truncated (cut off), the resulting finite sum forms an approximation to the original waveform. Such approximations are quite useful. We may want to learn about the important frequencies in a particular waveform so that we can suppress or enhance certain modes. For example, in electronic sound equipment, we may want to boost or suppress certain components of a note without changing its pitch, thus changing the quality of the sound.

Fourier Decomposition of a Triangular Wave

To demonstrate Fourier decomposition, let's try to fit the triangular wave illustrated in Fig. 15–25 with harmonic functions. (Such waves might be used in electronics applications where they are generated precisely by the superposition of harmonic waves.) The triangular wave has a period T and an angular frequency $\omega = 2\pi/T$. The rules for the constants that appear in the expansion require the phases ϕ_n to equal $\pi/2$. This means that only cosines appear in the expansion. This is reasonable because the triangular waveform is symmetric about $t = 0$, and the cosine also has this symmetry. The cosine is thus a "natural" expansion function for the triangular wave. Cosines that reach negative maxima at $t = \pm T/2$ would make it easy to build the sharp negative trough of the triangular wave at these points. This feature is shared by $\cos(\omega t)$, $\cos(3\omega t)$, $\cos(5\omega t)$, and so forth, but *not* by $\cos(2\omega t)$, $\cos(4\omega t)$, and so on. The calculation of A_n bears this out: $A_2, A_4, \ldots$ all turn out to be zero. When the expansion rules are applied, the exact Fourier expansion of our wave is

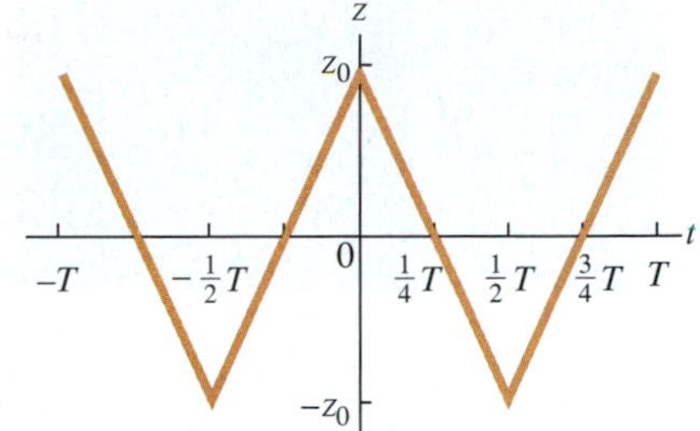

▲ **FIGURE 15–25** A triangular wave of period T. The time dependence of the transverse displacement z of a string is plotted at some fixed point in space.

$$z(t) = \frac{8}{\pi^2} z_0 \sum_{n=\text{odd}} \frac{1}{n^2} \cos(n\omega t)$$
$$= 0.81 z_0 \left[\cos(\omega t) + \frac{1}{9}\cos(3\omega t) + \frac{1}{25}\cos(5\omega t) + \cdots\right]. \qquad (15\text{–}24)$$

Figure 15–26 illustrates this Fourier expansion. Figure 15–26a shows the individual terms $z_1 = 0.81 z_0 \cos(\omega t)$ and $z_2 = (0.81/9) z_0 \cos(3\omega t)$. Figure 15–26b shows how first one term and then two terms of Eq. (15–24) match the triangular wave.

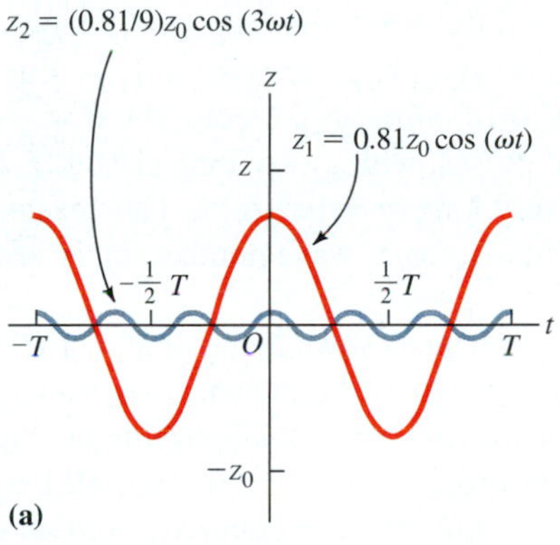

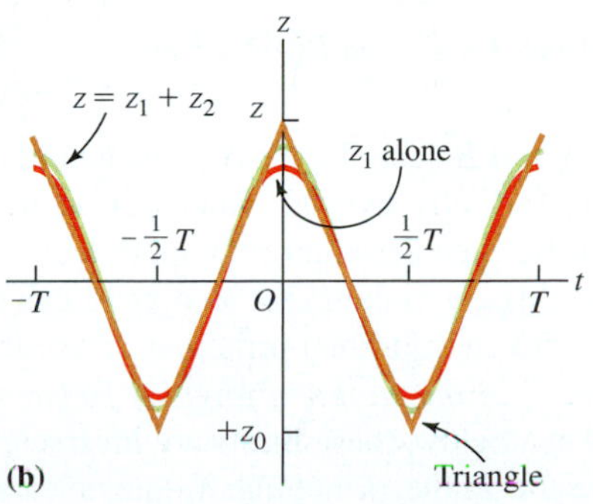

◀ **FIGURE 15–26** Fourier analysis of the triangular waveform of Fig. 15–25. (a) The first two contributing harmonics, $z_1 = 0.81\, z_0 \cos(\omega t)$ and $z_1 = (0.81/9)\, z_0 \cos(3\omega t)$. (b) The triangular waveform, the expression z_1 alone, and an approximation containing the first two terms of Eq. (15–24). The approximation with z_1 alone is already a good one, and the two-term expression is an even better one.

The $\cos(\omega t)$ term alone does reasonably well in approximating the triangular wave. By the time the second term is included, the approximation is at worst good to 10%.

The Fourier theorem is of great importance in engineering and physics because of its relevance to the solution of many different problems. In addition to the applications in vibrations and wave motion that are simply sketched out here, heat-flow problems and the solutions of a great class of problems in electricity and magnetism across the entire gamut of electrical engineering require the Fourier expansion.

Fourier Analysis and the Ear

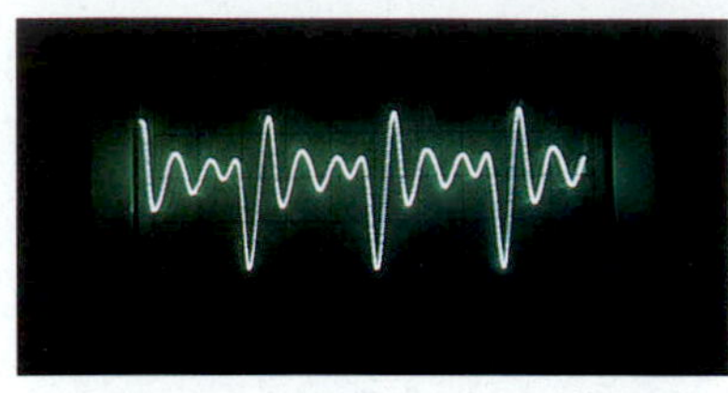

(a) Trumpet

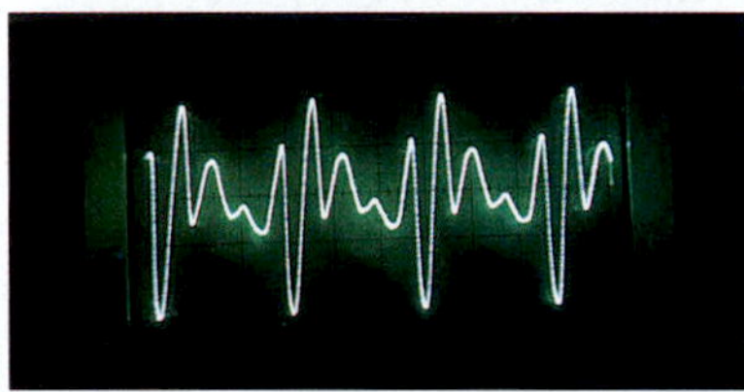

(b) Synthesized trumpet

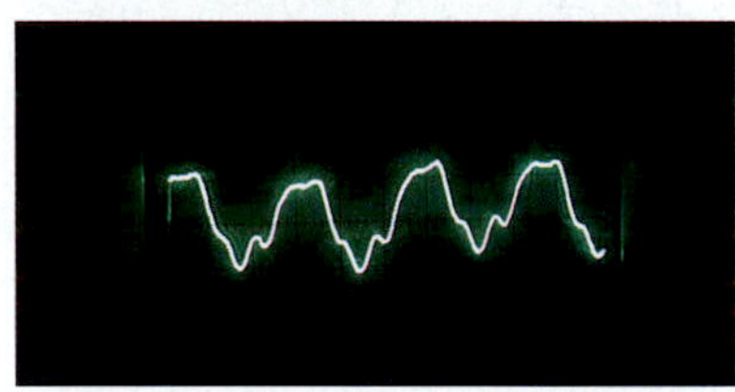

(c) Guitar

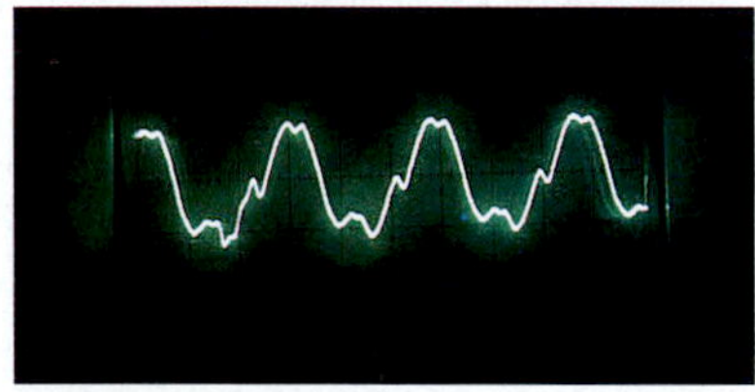

(d) Synthesized guitar

▲ **FIGURE 15–27** The characteristic sounds of different instruments are associated with frequencies above that of the sound of the "note" being played. These higher frequencies are known as overtones, and their presence makes the waveform of the sound different than the purely harmonic sine or cosine form. A synthesizer electronically generates the characteristic pattern of overtones to trick the ear. Shown in Figs. (a) to (d) is the note of F, played, by a trumpet, a synthesized trumpet, a guitar, and a synthesized guitar, respectively.

Our ears work very well as Fourier analysis devices. When the note A is produced by an electronic "tuning fork," a pure harmonic wave of period $T = 1/(440\text{ Hz}) = 2.27 \times 10^{-3}$ s emerges. The same note played on a piano, a violin, a trombone, an oboe, or a saxophone contains a harmonic wave of the same fundamental frequency. But each instrument emits its own complicated waveform (Fig. 15–27). In the language of Fourier analysis, this means that a note played by any of those instruments contains vast numbers of differing harmonic components. The ear is capable of picking out the fundamental frequency; we can all agree when the same note is played by a piano and an electric guitar. It is the higher components by which the ear differentiates between different instruments or between different voices. Even different violins have very different higher harmonic contents, a very subtle but important matter in the construction of violins. Although the physical differences between Stradivarius violins and others of lesser value are so slight that attempts to duplicate "Strads" fail, the sounds of Strads are generally more appealing than those of ordinary violins. A Fourier analysis shows that better violins have a far greater content of higher harmonics. Our ears can even distinguish between individual violins made by the same master!

*15–7 Pulses and the Uncertainty Principle

When we treat pulses with Fourier's theorem, we discover a fundamental relation between the width of the pulse and the frequencies of the periodic waves of which it is composed. This relation is known as the **uncertainty principle**.

The Fourier Decomposition of Pulses

In Section 15–6 we used Fourier's theorem to talk about general periodic waves. At first sight, pulses would seem to have little to do with such periodic waves. But the last drawing in Fig. 15–24 suggests a way to think of pulses as periodic. Here we see a periodic wave that is a series of short, repeated pulses, with a very long period, T, between the pulses or, equivalently, separated by a very long wavelength, λ. This means that the fundamental angular frequency, ω (or the fundamental wave number, k), is very small, because $\omega = 2\pi/T$ (or $k = 2\pi/\lambda$).

A true pulse in time is like the periodic sequence shown in the limit that $T \to \infty$, or $\omega \to 0$. Similarly, a true pulse in space is like the periodic sequence shown in the limit that $\lambda \to \infty$, or $k \to 0$. In the Fourier expansion, the sum is taken over integer multiples of angular frequencies (wave numbers). Therefore, in the $\omega \to 0$ limit, *all* frequencies enter into the sum and *the sum in the Fourier expansion is converted into an integral.* More concretely, the statement of the theorem is

$$P(t) = \int_0^\infty d\omega A(\omega)\cos\omega t + \int_0^\infty d\omega B(\omega)\sin\omega t$$

for any pulse function $P(t)$. The functions A and B describe the relative amounts of different frequencies. While the integral runs formally over all values of ω, it is really $A(\omega)$ and $B(\omega)$ that determine the range of frequencies entering the integral.

Just what can we say about this range? Although the exact answer to this question depends on the shape of the pulse, we can state an important principle. *If a pulse is limited in time (or space), then it is also limited in the range of angular frequencies (or wave numbers) that enter into it.* (As we'll see below, these limits are inversely related; the narrower the range in time, the larger the range in angular frequencies and vice versa.) In other words, the angular frequencies that contribute are in a narrow range, and we can estimate that range by the following argument. The different frequencies that contribute to a pulse interfere constructively in the region where the pulse is strong and interfere destructively in the region where the pulse drops to zero. Suppose that a pulse lasts a time Δt; we say it has a **width** in time of Δt. Suppose also that the contributing frequencies go from a minimum angular frequency ω_{min} to a maximum ω_{max}, with $\Delta\omega = \omega_{max} - \omega_{min}$. We want these waves to be out of phase at the beginning of the pulse, to be in phase at the center of the pulse, and then to be out of phase once more at the end of the pulse. If in time span Δt the number of periods of the waves of frequency ω_{max} differs by just one from the number of periods of the waves of frequency ω_{min}, then waves that are out of phase at the beginning of the pulse will be in phase at the center and out of phase at the end. Now the number of periods of a wave of angular frequency ω contained in a time span Δt is

$$\frac{\Delta t}{T} = \Delta t \frac{\omega}{2\pi}.$$

Our condition is then $\frac{\Delta t}{2\pi}(\omega_{max} - \omega_{min}) \cong 1$, or $(\Delta t)(\Delta\omega) \cong 2\pi$. In fact, this approximate equality holds for some pulses, and a more detailed mathematical study shows that for others it is an inequality that is usually stated as

$$(\Delta t)(\Delta\omega) \geq 2\pi. \tag{15–25}$$

This equation is a **reciprocal relation**; if Δt is large, $\Delta\omega$ is small, and vice versa. A very sharp pulse in time contains a very broad range of harmonic components. Similarly, a pulse that is very broad in time will have only a limited range of angular frequencies.

A more commonly used reciprocal relation follows if we use the frequency $f = \omega/2\pi$ in place of the angular frequency, ω:

$$(\Delta t)(\Delta f) \geq 1. \tag{15–26}$$

RECIPROCAL RELATION FOR TIME AND FREQUENCY

It is also worthwhile to see how the reciprocal relation works when we study a pulse in space rather than time. Here, the breakdown of the pulse is in terms of a continuous range of wave numbers (or wavelengths) rather than angular frequencies. Suppose that the pulse has width Δx in space. Then the same method of estimating the range of wavelengths, $\Delta\lambda$ of the contributing harmonic waves, can be used. The condition would then be

$$\Delta x\left(\frac{1}{\lambda_{min}} - \frac{1}{\lambda_{max}}\right) \cong 1$$

or, in terms of the wave number, $k = 2\pi/\lambda$, $(\Delta x)(\Delta k) \cong 2\pi$. Again, the approximate equality is actually an inequality,

$$(\Delta x)(\Delta k) \geq 2\pi. \tag{15–27}$$

RECIPROCAL RELATION FOR SPACE AND WAVE NUMBER

Here, $\Delta k = k_{max} - k_{min}$ is the range of wave numbers that must enter into the construction of a pulse whose width in space is Δx. Again, the smaller Δx is (i.e., the narrower the pulse), the larger the range of wave numbers.

Uncertainty Relations

In the context of the microscopic world of atoms, the reciprocal relations we have discussed are called **uncertainty relations**; the fact that such relations come into play is the *uncertainty principle*. In atomic physics quantum mechanics is important, and the key idea in quantum mechanics is that all matter has wavelike characteristics. Although the words "uncertainty principle" are often used only in connection with quantum mechanics, the principle really applies to any wavelike phenomenon. The trade-off in the context of quantum mechanics involves the position and speed of microscopic particles. If you measure the position of a particle at any given time with high precision, then you can know the speed—or the momentum—only with poor precision, and vice versa. The momentum is closely related to a wave number in quantum mechanics, which we'll discuss further in Chapter 40.

THINK ABOUT THIS . . .

DOES THE RECIPROCAL RELATION HAVE PRACTICAL CONSEQUENCES?

There is a very real place where the reciprocal relation is relevant: communication by signal construction. Pure harmonic waves cannot convey information, because they have neither a beginning nor an end. Signals and information transfer are necessarily in the form of pulses. For example, all movement of information within and between computers is done with sequences of pulses in binary form. Rapid communication of information requires that the pulses carrying the information be very narrow. If not and, in particular, if pulses overlap to make wider pulses, information is lost in the confusion between the pulses. According to the reciprocal relation, this means that a very broad spectrum of frequencies must go into the construction of these narrow pulses: The narrower the pulse, the broader the range of frequencies must be. But this costs money—in a very literal sense. The energy density and power in harmonic waves of frequency ω are proportional to ω^2, as discussed in Section 14–5. The reciprocal relation forces a trade-off: A sharper pulse, which is more effective in transferring information, is more expensive.

Summary

The algebraic sum, or superposition, of two solutions of the wave equation is also a solution. The collection of physical phenomena that follow from this fact fall under the general heading of wave interference. When two waves that are superposed have opposite algebraic signs at a particular point or a particular moment of time, they interfere destructively. When they have the same sign, they interfere constructively.

Standing waves can be considered to be the result of the superposition of traveling waves moving in opposite directions. Standing waves on a string can be decomposed into the superposition of traveling waves with their own reflections from the ends.

When two waves with nearly equal frequencies, f_1 and f_2, are superposed, the result is that the original wave is modulated by a wave of much smaller frequency and larger wavelength called a beat. The beat frequency is

$$f_{\text{beat}} = \tfrac{1}{2}(f_1 - f_2). \tag{15–16}$$

Two waves of the same wavelength are coherent if there is a constant phase difference between them. If the distances that the waves travel differ by ΔL, an interference pattern that depends only on ΔL and on the phase difference is traced out. If the phase difference at the source is zero, the interference pattern is

$$\text{for constructive interference: } \Delta L = n\lambda,\ n = 0, \pm 1, \pm 2, \ldots; \tag{15–17}$$

$$\text{for destructive interference: } \Delta L = \left(n + \tfrac{1}{2}\right)\lambda,\ n = 0, \pm 1, \pm 2, \ldots. \tag{15–18}$$

Suppose that two coherent wave sources of wavelength λ and zero phase difference are separated by a distance d that is much smaller than the distance to the observation point. Then there is an interference pattern that is a function of the angle θ made with the perpendicular bisector of the line separating the wave sources:

$$\text{for maxima: } \sin\theta = n\frac{\lambda}{d},\ n = 0, \pm 1, \pm 2, \ldots; \tag{15–20}$$

$$\text{for minima: } \sin\theta = \left(n + \frac{1}{2}\right)\frac{\lambda}{d},\ n = 0, \pm 1, \pm 2, \ldots. \tag{15–21}$$

The pattern described by these equations broadens (the angle θ corresponding to a particular minimum or maximum increases) as the wavelengths involved increase and/or the separation between the sources decreases. This is a very general feature of interference patterns.

Waves reflect from boundaries and are transmitted past boundaries that separate one wave-carrying medium from another. A reflected wave is either inverted or right-side up, compared to the original wave, according to whether the medium from which the wave reflects is denser or less dense than the original medium.

Fourier's theorem shows that any wave, both periodic and nonperiodic (pulses), can be approximated as closely as we like by a superposition of purely harmonic waves with different frequencies. This result allows for the construction of arbitrary waveforms from harmonic waves and for the analysis of physical systems in terms of harmonic waves.

A pulse can also be treated by Fourier analysis. If a pulse has a time span of Δt, then the range of frequencies, Δf, that comprise that pulse is limited:

$$(\Delta t)(\Delta f) \geq 1. \tag{15–26}$$

There is a similar relation between the spatial extent of the pulse, Δx, and the range in wave numbers, Δk, that enter into the pulse's decomposition:

$$(\Delta x)(\Delta k) \geq 2\pi. \tag{15–27}$$

These relations hold whenever wave phenomena are present and explain why sharp information-carrying signals are expensive to generate.

Understanding the Concepts

1. A standing wave on a string can be constructed from two traveling waves that move in opposite directions. Can a traveling wave be constructed from the sum of two standing waves?
2. When transverse positive and negative pulses that have the same symmetric shape and size but travel in opposite directions meet, is it necessary that there be a moment when the string or wire on which they move is flat? If so, how do the pulses "know" to continue moving on the string?
3. Two people on each end of a long rope send off a wave pulse. If both wave pulses are on the same side of the rope, describe what happens when the pulses meet. What about when the pulses are on opposite sides? What happens when one pulse is oriented at a 90° angle to the second pulse?
4. If a higher harmonic than the fundamental one is excited on the string attached between the two walls, will it be the frequency or the velocity that changes?
5. Listening to the beat frequency between an unknown and a standard (known) tuning fork tells you only about the difference between the two frequencies. The unknown frequency could be larger or smaller than the known frequency. With a piece of chewing gum at your disposal, can you devise a method to determine the unknown frequency?
6. A wave with truly sharp edges, such as the triangular wave discussed in Section 15–7, can never exist on a string in nature. What might limit how sharp the changes can be?
7. Consider a sinusoidal traveling wave that moves down a sequence of successively lighter strings tied together end to end. The wave starts at the end containing the heaviest string. Qualitatively describe what happens to the frequency and wavelength of the wave by the time it reaches the lightest end of the system.
8. We saw what interference pattern two speakers close to one another make near the line perpendicular to and bisecting the line that connects the speakers. Will there be an interference pattern even far from this center line, where the small angle approximation that we used to derive the pattern is not valid?
9. A pebble dropped into still water produces a pulse. The speed of water waves of different wavelengths differ from one another. Why does a dropped pebble produce a series of spreading concentric waves rather than a single wave front?
10. How do you reconcile the principle of conservation of energy with the observation that, in one region, the wave medium is not in motion when waves interfere destructively?
11. In Section 15–1 we referred to two waves that move to the right and cancel; they give no wave at all. Yet each wave alone has an energy density. What happens to the energy? [*Hint*: What are the implications of the superposition principle for the energy density?]
12. Musicians often use beats to tune to some standard, such as an oboe. Can a trumpet player tell from the beat frequency whether one particular note is sharp (at a higher frequency) or flat (at a lower frequency) compared to the same note played on the oboe?

Problems

15–1 The Superposition Principle

1. (I) Show that the superposition of the two traveling waves $\sin(kx - \omega t)$ and $\cos(kx - \omega t)$ can be written in the form $A\sin(kx - \omega t + \phi)$, and find A and ϕ.

2. (I) Two traveling waves have the forms $z_1 = A\sin(kx - \omega t)$ and $z_2 = A\sin(kx - \omega t + \pi/2)$, respectively. Sketch the superposition of these two waves.

3. (I) Work out the result of the superposition of the two waves of Problem 2 in the form $z = z_1 + z_2 = A\sin(kx - \omega t + \delta)$.

4. (II) Consider the superposition of two harmonic waves, $\psi(x, t) = A\sin(kx - \omega t) + 3A\cos(kx - \omega t + \theta)$. (a) For what values of θ will this superposition result in the maximum constructive interference? (b) the maximum destructive interference? (c) If this superposition is written in the form $B\sin(kx - \omega t + \alpha)$, express B and α in terms of A and θ.

5. (II) Two sinusoidal waves with the same frequency travel down a long rope (Fig. 15–28, see next page). The waves have amplitudes 3 cm and 2 cm, respectively, but the second wave trails the first wave by a phase of $\pi/2$. Determine the amplitude of the resultant motion.

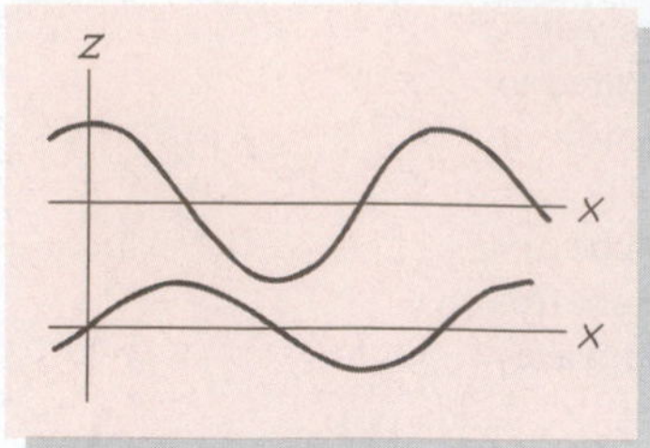

▲ **FIGURE 15–28** Problem 5.

6. (II) Four sinusoidal waves have the same frequency, but the first two waves have amplitudes twice that of the other two waves. The phases of the four waves are 0, $\pi/2$, π, and $3\pi/2$, respectively. Plot the resulting superposed wave, and describe its motion.

7. (II) Two sine waves, $A\sin(\omega t + kx)$ and $A\sin(\omega t + kx + \delta)$, combine to form a sine wave with the same amplitude A. For what values of δ of the second wave is this possible?

15–2 Standing Waves Through Interference

8. (I) Two waves are described by $\psi_1 = (2.5\text{ cm})\sin(kx - \omega t)$ and $\psi_2 = (2.5\text{ cm})\sin(kx + \omega t)$, where $k = 3\pi/2\text{ cm}^{-1}$ and $\omega = 2\pi\text{ s}^{-1}$. (a) Show that the addition of the two waves produces a standing wave. (b) Determine the amplitude and frequency of the resulting standing wave.

9. (I) Consider traveling waves moving in opposite directions: one has the form $z_r(x,t) = z_0\sin(kx - \omega t)$ and the other has the form $z_l(x,t) = z_0\cos(kx + \omega t)$. Do these waves add to a standing wave? If so, what is the expression for the standing wave, and what are the locations of the nodes? Note that $\cos\theta$ can be written as $\sin[\theta + (\pi/2)]$.

10. (II) Show by a specific construction that two standing waves can add up to give a traveling wave. What standing wave would you have to add to $A(\cos kx)(\cos\omega t)$ to get a wave traveling to the left?

11. (II) Two traveling waves that move in opposite directions along the x-axis interfere to produce a standing wave of the form $6\sin kx\cos\omega t$, where $k = 3\pi\text{ m}^{-1}$ and $\omega = 22\pi\text{ s}^{-1}$. (a) What are the frequencies and speeds of the traveling waves? (b) What are the forms of the traveling waves?

12. (II) One harmonic of an organ pipe has a single node at 3/4 of the length of the pipe and an antinode at the open end (see Fig. 14–26). Write an expression to describe this mode as a standing wave. Find two traveling waves that produce the same result. The length of the pipe is 0.80 m and the speed of sound is 330 m/s.

13. (II) Show that the sum of a wave that travels to the right and a wave that travels to the left with the same amplitude, wave number, and frequency is a standing wave, independent of phase. [*Hint*: Consider $\psi(x,t) = A\sin(kx - \omega t) + A\sin(kx + \omega t + \theta)$; define $u \equiv kx + \theta/2$ and $v \equiv \omega t + \theta/2$, and use the appropriate identities.]

14. (II) The most general forms of two traveling waves that move in opposite directions along the x-axis with the same amplitude, wavelength, and frequency are $z_r(x,t) = z_0\sin(kx - \omega t + \phi_1)$ and $z_l(x,t) = z_0\sin(kx + \omega t + \phi_2)$. What is the form of the standing wave that results from their superposition?

15. (II) A harmonic wave traveling to the right along the x-axis has its maximum amplitude at the time $t = 0$ s at the point $x = 0$ m. Superimposed on this is a wave traveling to the left with equal amplitude, wavelength, and frequency, such that the net standing wave has nodes at $x = 3$ m and at $x = 6$ m. What is the largest possible value of the wavelength?

15–3 Beats

16. (I) You have a tuning fork of unknown frequency and a tuning fork of frequency 552.0 Hz. Combined, the two produce a pulse frequency ($f_{\text{pulse}} = 2f_{\text{beat}}$) of 1.8 Hz. What are the possible values of the frequency emitted by the first tuning fork?

17. (I) You have a tuning fork of frequency 512 Hz and look for another tuning fork of smaller frequency so that there is a perceived beat frequency of 6 Hz between them. What frequency should the other tuning fork have?

18. (I) Two violinists play their (identical) A-strings, nominally 440 Hz. A beat is heard at a pulsing of 2 Hz. What is the magnitude of the fractional difference in wave speed, $(v_1 - v_2)/v_1$, for the two strings? [*Hint*: The length L of a violin string is the same from violin to violin.]

19. (I) Two strings of the same material on a pair of guitars are in tune with a frequency of 414 Hz. The string of one of them is tightened slightly, and you hear a beat frequency of 8 Hz. What is the frequency of the tightened string?

20. (I) Two speakers are wired so that they emit sound coherently. Because of a manufacturing error, the same electrical impulse that leads one to emit sound of wavelength 27.10 cm leads the other one to emit sound of wavelength 26.71 cm. What is the pulse frequency ($f_{\text{pulse}} = 2f_{\text{beat}}$)?

21. (II) Two identical piano wires have a frequency of 512 Hz when under the same tension. As a consequence of slippage, the tension of one of the wires changes slightly, so that its frequency decreases slightly. If 4 beats/s are heard when the wires both vibrate, what is the fractional change in the tension of the lower-frequency string?

15–4 Spatial Interference Phenomena

22. (I) Two speakers that emit waves with the same wavelength λ are placed at $x = L$ and $x = -L$, respectively. Assuming that the speakers emit waves with equal amplitudes A and that the phase of the wave on emission is zero at $t = 0$ in both cases, what is the form of the traveling waves that travel toward a microphone located between them at $x = 0$?

23. (I) Find the waveform picked up by the microphone in Problem 22; that is, the superposition of the two waves arriving at the microphone.

24. (I) Two speakers separated by 80 cm are fed with the same pure sinusoidal wave of 750 Hz. A microphone is placed 4.0 m away along the center line between the speakers (Fig. 15–29). It picks up an intensity maximum. It is then moved slowly out to the side. (a) How far out to the side, in meters, does the microphone have to be moved to pick up a first minimum? (b) How much farther does the microphone now have to be moved to pick up a second maximum? (c) How do the answers to parts (a) and (b) change if the speakers are now separated by just 10 cm? The approximation $\sin\theta \cong \theta$ is not valid here, although the angles of the lines from the speakers to the microphone are nearly equal.

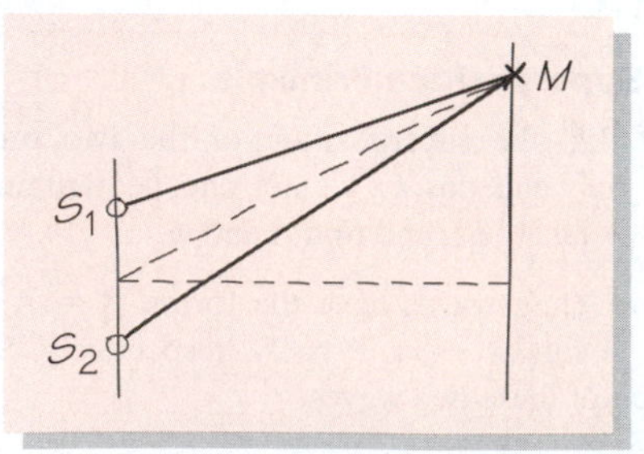

▲ **FIGURE 15–29** Problem 24.

25. (I) Two coherent sources are 3.0 cm apart and make harmonic ripples of the same frequency (see Fig. 15–8). Consider the ripples along a straight line L_1 parallel to the line L_2 that connects the sources and 40.0 cm away from line L_2. If the distance between the central maximum at L_1 and the next maximum on L_1 is 8.0 cm, what is the wavelength of the ripples?

26. (I) Sound of frequency 600 Hz is emitted in phase from two speakers 1.4 m apart (Fig. 15–30). A microphone can move along a line parallel to the line that joins the two speakers and is 10.0 m from the base of the speakers; it measures sound intensity, which can then be plotted versus distance. The interference produced is constructive in the center. Where is the first minimum in the sound intensity curve along the track of the microphone?

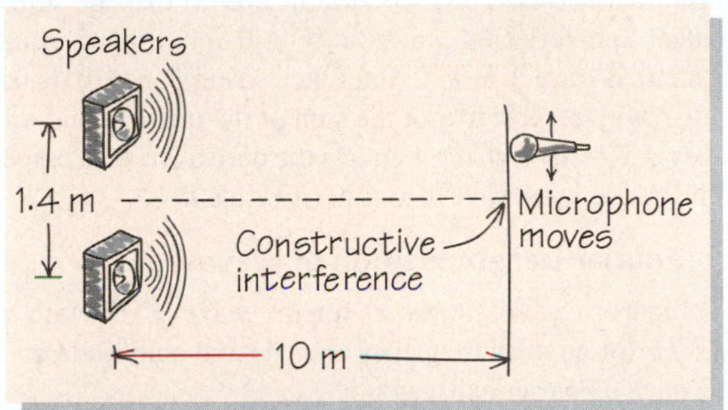

▲ **FIGURE 15–30** Problem 26.

27. (II) Two coherent wave sources 5 cm apart make ripples in a water tank (see Fig. 15–8). The wavelength of the waves is 3 cm. (a) Determine the angle between the perpendicular bisector of the line segment connecting the two wave sources and the direction of the first noncentral maximum. (b) What is the direction of the first maximum if the two sources operate with opposite phases? (c) How will the answers to (a) and (b) change if the wavelength is changed to 6 cm?

28. (II) Harmonic light waves are emitted by a source equidistant between two narrow vertical slits 0.18 mm apart in a screen (Fig. 15–31). The light waves detected at a screen parallel to the screen containing the slits but 35 cm away have the usual maximum at the point opposite the midpoint of the two slits, and the first minimum occurs at a point 0.6 mm from the maximum. What is the wavelength of the light?

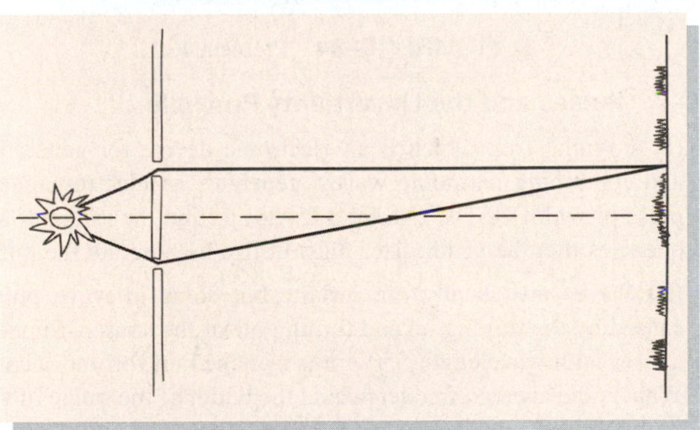

▲ **FIGURE 15–31** Problem 28.

29. (II) Coherent waves emerge from two slits separated by a distance $2a$. A screen is placed at a distance L away from the slits, and the first interference minimum is found at a distance d from the center point. Use the small-angle approximation to show that the relationship between the wavelength (λ) and the other lengths is given by $\lambda = 4ad/L$.

30. (II) Two tuning forks vibrating in phase with frequency 440 Hz are placed 85 cm apart. A detector is moved in a circle of radius 540 cm around the center of the tuning forks (Fig. 15–32). Plot the approximate variation of the square of the net amplitude as a function of angle. Neglect the small variation in the magnitude of the amplitude that results from the sources not being at the same distance from the detector. Take the velocity of sound to be 330 m/s.

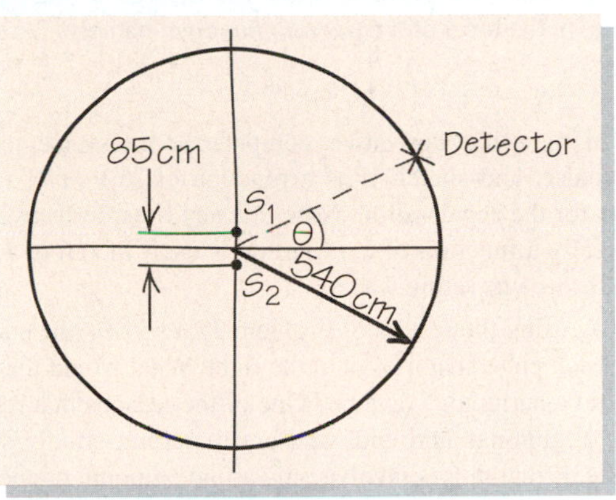

▲ **FIGURE 15–32** Problem 30.

31. (II) An oscillator, located at $(x, y) = (-2\text{ cm}, 0\text{ cm})$, that dips in and out of water in a tank produces a water wave of the form $\psi_1 = (3\text{ cm})\sin(\pi r_1 - \pi v t)$, where r_1 is the distance from the oscillator; a second oscillator, located at $(x, y) = (+2\text{ cm}, 0\text{ cm})$, produces a water wave of the form $\psi_2 = (3\text{ cm})\sin(\pi r_2 - \pi v t)$, where r_2 is the distance from the second oscillator. (a) Where on the line between the sources (the x-axis) are the interference maxima located? (b) What is the form of the wave along the line $x = 0$ cm? (Take all times in seconds.)

32. (II) Two tuning forks that vibrate coherently (in phase) at 512 Hz are placed 0.35 m apart on a horizontal surface. A short circular wall of radius 8 m, centered on the midpoint between the sources, is placed around them. Describe the interference pattern of maxima and minima at the wall.

33. (II) Suppose that two in-phase sources of surface waves in water generate water waves that spread out in circles with equal amplitudes. (a) Show that the intensity of the disturbance or the energy generated in the water is, at a maximum, four times the intensity for each wave taken separately. (b) Similarly, show that there is no intensity at a minimum. (c) Show that the intensity at any given point, if there were no wave interference, is twice the intensity from each wave. (d) Do the different answers to parts (a), (b), and (c) mean that the conservation of energy principle must be abandoned for waves? Explain your answer.

15–5 Pulses

34. (I) A student pulls a rope taut and shakes it once, sending a wave pulse down the rope. The pulse is described by a *Gaussian function*,

$$z(x, t) = z_0 \exp\left[-\frac{(x - (1\text{ m/s})t)^2}{(0.1\text{ m})^2}\right].$$

Where is the center of the pulse at time t?

35. (I) At $t = 0$, a sawtoothed pulse has the shape $z = 0$ for $x < -a$ and for $x > 0$; $z = k(x + a)$ for $-a < x < 0$. Sketch this pulse. Suppose that the pulse moves at speed v in the $+x$-direction. What is the algebraic expression for the moving pulse?

36. (II) Consider a train of pulses of the type described in Problem 35. Here, at $t = 0$, the displacement is $z = 0$ for $x < -a + nb$ and for $x > nb$; $z = k(x + a - nb)$ for $-a + nb < x < nb$; n is any integer—positive, negative, or zero—and $b > a$. (a) Sketch this wave train of pulses. (b) Suppose that the entire train of pulses moves at speed v in the $+x$-direction. What is the algebraic expression for the train? (c) What are the period and wavelength of this nonharmonic wave?

37. (II) Consider a transverse wave that moves to the right on a string, in the form of a *Gaussian function*, namely

$$z(x, t) = z_0 e^{[-(x-vt)^2/\alpha^2]}.$$

By taking a time derivative, compute the transverse velocity of this pulse, and sketch it as a function of $(x - vt)/\alpha$. Do the same for the acceleration. Note that any time derivative is automatically a function of $x - vt$ and is itself therefore a quantity that forms a traveling wave.

38. (II) By using the results of Problem 37, *estimate* the energy in a Gaussian pulse that moves to the right. What would the total energy in a harmonic wave be? One of the reasons that real waves have beginnings and ends and are therefore, strictly speaking, pulses, is that pulses involve only a finite amount of energy.

39. (II) Two transverse pulses travel along a string at speed v in opposite directions. The pulse that moves to the right approximates a square with sides of length ℓ projecting above the equilibrium level of the string. The pulse that moves to the left approximates a square with sides of length $\ell/2$ projecting below the equilibrium level. Assume that the superposition principle holds, and describe the approximate shape of the string as the pulses collide.

40. (I) A pulse shaped like an upright two-dimensional pyramid, or an isosceles triangle, travels to the left along a string under tension (Fig. 15–33). The left-hand end of the string is free to move up and down. Sketch a sequence of drawings that show how the pulse reflects from the free end.

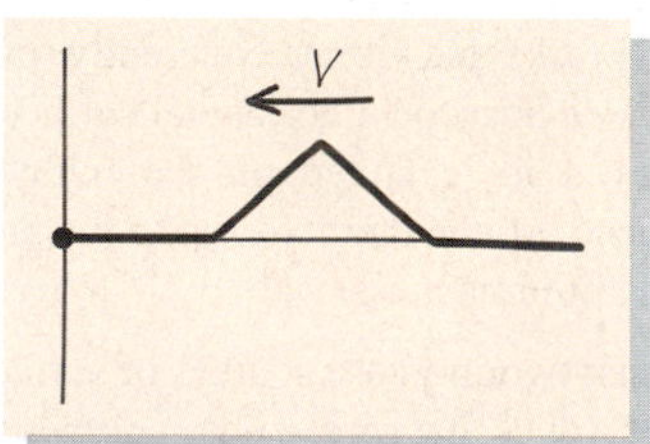

▲ **FIGURE 15–33** Problem 40.

41. (I) A string with linear mass density μ_1 is attached to another string, with linear mass density μ_2. A wave with amplitude A that travels along the μ_1 string will at the attachment point give rise to a reflected wave with amplitude B and a transmitted wave with amplitude C, where

$$B = \frac{\sqrt{\mu_2} - \sqrt{\mu_1}}{\sqrt{\mu_2} + \sqrt{\mu_1}} A, \; C = \frac{2\sqrt{\mu_1}}{\sqrt{\mu_2} + \sqrt{\mu_1}} A.$$

For $\mu_1 = 5.0 \times 10^{-2}$ kg/m and $\mu_2 = 3.8 \times 10^{-2}$ kg/m, what are B and C in terms of the amplitude of the incoming wave?

42. (I) Use the formulas in Problem 41 to show that the average power of the reflected wave plus the average power of the transmitted wave add to the average power of the incident wave.

43. (I) A string with mass density μ_1 is attached to a string with mass density μ_2. The amplitude of a reflected pulse is 85 percent of the amplitude of the incident pulse and is inverted. What is the ratio of μ_1 to μ_2? (See Problem 41.)

44. (II) When a symmetric pulse is incident on the fixed end of a rope, the reflected and incident pulse interfere, and there is a moment when the rope is completely flat. Is this a violation of the conservation of energy? (The energy of waves is discussed in Chapter 14.) Where is the energy at the moment the rope is flat?

45. (III) A string with mass density μ_1 is attached at the point $x = 0$ to a string with mass density μ_2. A tension T is maintained in both strings. A transverse wave of the form $A\cos(kx - \omega t)$, moving to the right, is generated far to the left in the first string. Use the fact that the string is unbroken at $x = 0$ and that no sharp kinks can develop there to obtain the forms of the reflected and transmitted waves. [*Hint*: The stated boundary conditions are sufficient to determine all the parameters of the reflected and transmitted waves. These conditions are (a) that the sum of the incident and reflected waves at $x = 0$ on string 1 equals the transmitted wave at $x = 0$ (unbroken string), and (b) that the derivative with respect to x of the sum of the incident and reflected waves at $x = 0$ on string 1 equals the derivative with respect to x of the transmitted wave at $x = 0$ (no kinks).]

*15–6 Fourier Decomposition of Waves

46. (III) Figure 15–34 shows a square wave $f(t)$ with period $T = 2\pi$ (or angular frequency $\omega = 1$) and amplitude $= 1$ unit. The wave's Fourier expansion is

$$f(t) = \frac{4}{\pi}\left[\sin(\omega t) + \frac{1}{3}\sin(3\omega t) + \frac{1}{5}\sin(5\omega t) + \cdots\right].$$

Plot an approximation to the square wave that corresponds to the first, the first two, and the first three terms in this series. (It helps in this to have access to computer software or to a calculator that can plot graphs.)

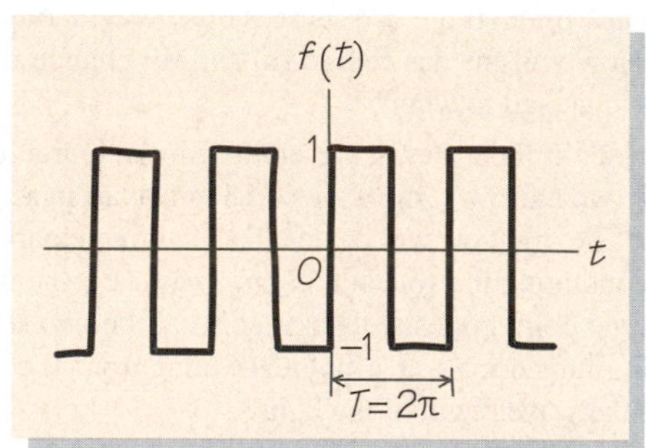

▲ **FIGURE 15–34** Problem 46.

*15–7 Pulses and the Uncertainty Principle

47. (I) A synthesizer, which is an electronic device for generating and combining harmonic waves, generates an electromagnetic pulse of width 10^{-6} s. Use Eq. (15–26) to find the range of frequencies that the synthesizer must utilize to generate the pulse.

48. (II) Waves are usually not infinite but come in wave pulses caused by the turning on and turning off of the source. Suppose a laser with wavelength 550 nm is switched on for one second. What is the average frequency and the width of the range of frequencies in this pulse?

General Problems

49. (I) Two harmonic waves are given by $y_1 = A\cos(kx - \omega t)$ and $y_2 = A\sin[kx - \omega t + (\pi/3)]$, where $k = 5\pi\ \text{m}^{-1}$, $\omega = 800\pi\ \text{s}^{-1}$, and $A = 4.0$ cm. (a) What is the frequency of each wave? What is the wavelength of each wave? What is the speed of each wave? (b) What is the amplitude of the wave that is the superposition, $y_1 + y_2$, of these two waves?

50. (I) Two rectangular pulses 4 cm long and of amplitude 3 cm but opposite sign are traveling down a string from opposite directions. At $t = 0$ s, the pulses approach each other with their centers 16 cm apart. The wave speed on the string is 1 cm/s. (a) Describe and sketch the shape of the string at $t = 0, 2, 5, 8, 10$, and 15 s. (b) What has happened at $t = 8$ s? What is the energy at this time?

51. (I) Two violinists play their (identical) A-strings, nominally 440 Hz. A beat is heard at a pulsing of 2 Hz. What is the fractional difference in tensions $(T_1 - T_2)/T_1$ of the two strings? [*Hint*: The length of a violin string is the same from violin to violin; when a violinist tunes his or her instrument, the string tension is changed.]

52. (II) The police are busy on the day that you hear two police cars sounding their one-note sirens as you wait at a light (Fig. 15–35). The police cars are moving at 67.1 mi/h (30.0 m/s) and their drivers hear their own sirens emitting at a frequency of 650 Hz. One police car passes you while the other is still approaching. Do you hear a beat? If so what is it?

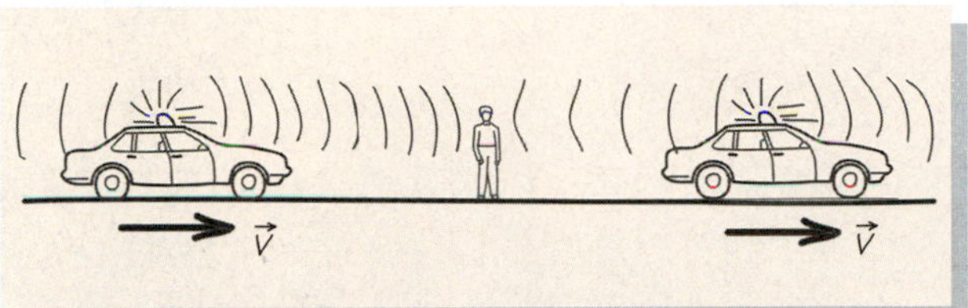

▲ **FIGURE 15–35** Problem 52.

53. (II) Consider two sources of sound that both emit waves with a frequency of 1800 Hz. These sources are placed on a turntable of radius 50 cm that rotates with an angular speed of 120 rev/min. Calculate the pulse frequency ($f_{\text{pulse}} = 2f_{\text{beat}}$) that results from the opposite Doppler shifts of the two sources, heard at a large distance from the turntable as a function of time. Ignore the fact that the sources are not quite equidistant from the detector of the sound. The speed of sound in air is 330 m/s.

54. (II) A guitar string has a fundamental frequency of 240 Hz. Another apparently identical guitar string is stretched beside the first one. (a) During successive stretching, each time tightening the second guitar string, the beat frequencies are 10, 5, and 1 Hz, consecutively. During the tightening, the beat frequency never passes through zero. What are the fundamental frequencies of the second guitar string for each case? (b) By what percentage should the tension be increased for the final adjustment to make the guitar strings have identical frequencies?

55. (II) A loudspeaker (source) that emits sound waves uniformly in all directions is 10 m from a microphone (receiver). Both the source and the receiver are 20 m from a smooth, rigid wall that reflects the sound waves from the source (Fig. 15–36). As the frequency of the waves from the source is smoothly increased from zero, there is a first interference maximum detected by the receiver and subsequently a series of further maxima. List the frequencies of the three maxima that occur above 80 Hz. Assume that sound reflects from the walls as would a ball in an elastic collision.

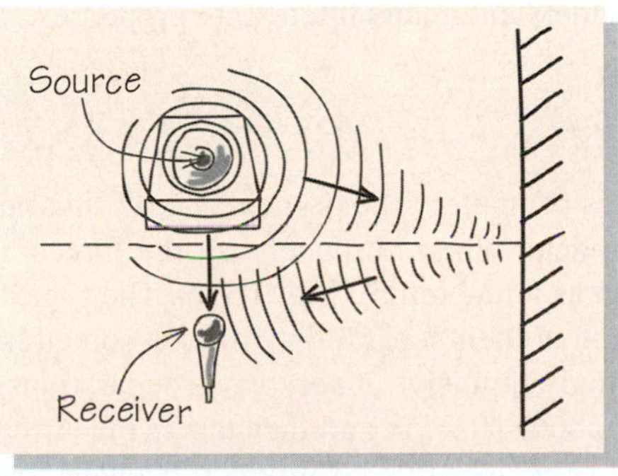

▲ **FIGURE 15–36** Problem 55.

56. (II) Consider two wave sources on a water surface. They generate identical waves of $\lambda = 1$ cm. What possible values of the separation of the sources, d, can you take to ensure that there is no disturbance to the side of the sources (that is, on the extension of the line joining the sources)? When this happens, is there a disturbance on that same line but *between* the sources? If so, describe it; if not, explain why not. Neglect the change in amplitude as a function of distance from the source; that is, treat this as a one-dimensional problem.

57. (II) The speed of the propagation of light in different media depends on the properties of the media. The formula is given by $v = c/n$, where $c = 3.0 \times 10^8$ m/s and n is the *index of refraction* ($n > 1$ for all media). Consider light of frequency 0.50×10^{15} Hz entering from empty space, with $n = 1.00$, into glass, with $n = 1.52$. What is the wavelength of the light in empty space? What is the wavelength of the light in glass?

58. (II) Waves are generated in phase in a water tank. The waves move along the x- and y-axes, respectively, with equal amplitude, so that the total wave is given by $\psi(x, y, t) = \sin(kx - \omega t) + \sin(ky - \omega t)$. Show that the pattern of interference maxima is given by the line $x - y = 2n\pi/k$ (where $n = 0, \pm 1, \pm 2 \ldots$).

59. (II) Which of the following equations are linear? Explain your reasoning:

(a) $\dfrac{\partial^2 \psi}{\partial t^2} = v^2 \dfrac{\partial^2 \psi}{\partial x^2}$

(b) $\dfrac{\partial \psi}{\partial t} = -K \dfrac{\partial^2 \psi}{\partial x^2}$

(c) $\dfrac{\partial^2 \psi}{\partial t^2} = a\psi^2 + b\dfrac{\partial \psi}{\partial x}$

(d) $\dfrac{\partial \psi}{\partial t} = 2\psi \dfrac{\partial \psi}{\partial x} + 3\psi$

(e) $\dfrac{\partial^2 \psi}{\partial t^2} = -k \ln \psi$

60. (II) Two sources emit sound coherently and in phase at a frequency of 1250 Hz with the same intensity. They are 55.0 cm apart on a horizontal surface. A circular wall of radius 220 cm (centered on the midpoint between the sources) is placed around them. Describe systematically the interference pattern of maxima and minima at the wall.

61. (III) Two waves of the form $z_1(x, t) = z_0 \sin(k_1 x - w_1 t)$ and $z_2(x, t) = z_0 \sin(k_2 x - w_2 t)$, respectively, propagate at speed v in a nondispersive medium. The frequency difference $\Delta f = f_1 - f_2$ is small. What function describes the envelope of the resulting superposition of the two waves? Express this envelope as a wave, and give its frequency and wavelength.

62. (III) Sinusoidal waves are generated on an infinitely long rope. One wave, $y_1(x, t)$, moves to the left and has amplitude y_0, wave number k_1, and angular frequency ω_1; the other, $y_2(x, t)$, moves to the right with amplitude y_0, wave number k_2, and angular frequency ω_2. Each wave has the same phase. (a) Express $y_i(x, t)$ for each of the waves. (b) Assuming that the rope is a nondispersive medium, what is ω_2 in terms of k_1, k_2, and ω_1? In parts (c) and (d), assume that $k_2 = k_1 + \delta k$ and $\omega_2 = \omega_1 + \delta\omega$. Take the speed of traveling waves on the rope to be v. (c) What is $\delta\omega$ in terms of δk and v? (d) Superpose these waves, assuming that δk and $\delta\omega$ are small. Use the notation $k_{\text{av}} = (k_1 + k_2)/2$ and $\omega_{\text{av}} = (\omega_1 + \omega_2)/2$. Show that the result of the superposition is two waves, one moving at the low speed $v_l = v\,\delta k/2k_{\text{av}}$ and wavelength $2\pi/k_{\text{av}}$, and one moving at the high speed $v_r = 2vk_{\text{av}}/\delta k$ and the long wavelength $4\pi/\delta k$.

Chapter 16

The remote controlled submarine robot, *Jason Junior*, is peeking through a window of the *Titanic* in 1986 during the investigation of the sunken ship. *Jason Junior* was attached to the three-man submersible *Alvin* (which was just outside the *Titanic* at a depth of 4000 m) by a 200 foot cable and eventually entered *Titanic* and explored extensively.

Properties of Fluids

Waves on the sea, smoke rings, and a whirlpool of water flowing down a drain are demonstrations of the wide variety of phenomena visible in fluid flow. Although fluids can support waves, as we have seen, a little observation suggests that the response of fluids to forces goes well beyond wave motion. The physical laws involved are those we have already discussed, but applied in a new, more complex setting. Our job here will be to develop descriptive and dynamical tools for handling both static fluids and fluids in motion. Conservation laws for momentum, energy, and mass will be very helpful in this task. With these tools and a series of approximations that turn out to be surprisingly useful, we can arrive at a description of the behavior of fluids that allows us to understand many interesting properties.

16–1 States of Matter

Let us start with a review of the states of matter. The particular form that matter takes, whether solid, gaseous, or liquid, is determined ultimately by the forces between its molecules, as well as conditions such as temperature (Fig. 16–1). The typical force between a pair of molecules as a function of their separation will look something like Fig. 16–2. Generally these forces are strongly repulsive at very small separations. The force becomes attractive at a separation of about 10^{-10} m and then tapers off virtually to zero at a distance of a few times 10^{-10} m. How does this information allow us to understand the three states of matter?

◀ **FIGURE 16–1** The three states of matter, which for water are ice, liquid water, and water vapor, are very different on the macroscopic scale. These differences are reflections of differences on the microscopic scale. In solids, the molecules are arranged in ordered arrays called crystals. In liquids, the spacing of the molecules is not very different than in solids, but the order of the lattice is lost. The molecules are close enough together in each case that intermolecular forces are important. In gases the molecules are so widely spaced that their interactions come into play only very briefly in collisions.

Gases: If molecules are, on average, sufficiently far apart, then the forces between them are small. In effect, their only interaction occurs when the molecules occasionally collide with each other. The molecules behave as a **gas**. We will study gases further in Chapters 17 through 20 and only note here that gases are easily compressed and that the shape of a body of gas is easily deformed.

Solids: If molecules are sufficiently close together that the forces due to neighboring molecules are significant, matter behaves in a way that depends on the detailed properties of the intermolecular forces. In some cases the forces act to organize the molecules into a **lattice**, which is a regularly repeating three-dimensional structure in which the molecules or atoms can move only by small displacements (Fig. 16–3). This is a typical **solid**: a substance whose structure resists forces that would deform its shape. We described the lattice structure as well as some properties of solids in Chapter 11.

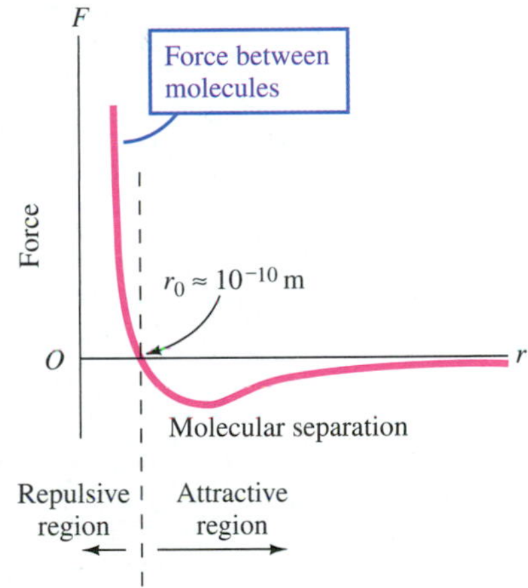

▲ **FIGURE 16–2** Qualitative behavior of the force between a pair of molecules as a function of their separation, r. Near $r = r_0$, the point at which the force is zero, the force is springlike (restoring). For $r < r_0$, the force is strongly repulsive; for $r > r_0$, it is attractive.

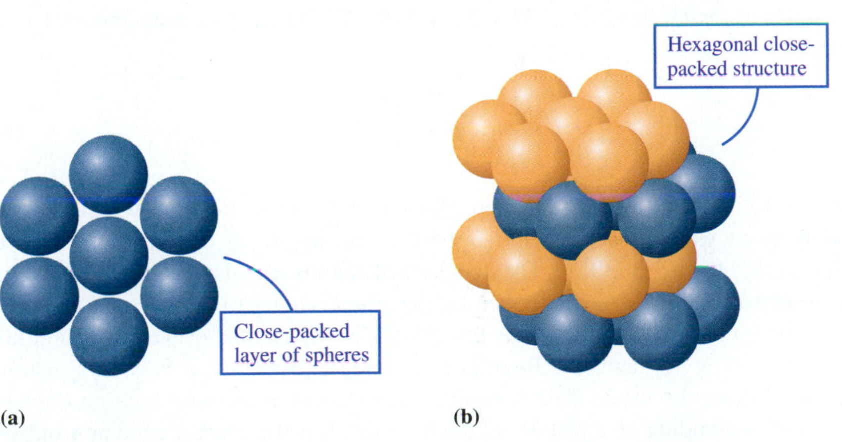

▲ **FIGURE 16–3** A lattice structure characteristic of solids. Here we show a hexagonal close-packed structure in which (a) one layer is fit together and (b) the layer above is nested into the depressions in the first layer. The forces between the molecules are springlike for small displacements from the equilibrium positions.

Liquids: In liquids the molecules are closely spaced and the intermolecular forces between the molecules at their typical spacing are strong, but for a variety of reasons they do not lead to an organized lattice structure. For such aggregates of molecules, there is no resistance to deformation. Unlike the molecules of gases, the molecules of a liquid are close enough to each other so that the repulsive intermolecular forces are responsible for an aggregate that resists compression. As a result, liquids have a definite volume but an easily changeable shape. A major difference between gases and liquids is in their compressibility.

The distinction among the three forms of matter is not absolute. If the forces are large enough, even rock will flow, and gases also share many properties with liquids. A collection of molecules can form a solid, a liquid, or a gas, depending on external circumstances. Still other states are possible in the right circumstances. For example, in one type of *liquid crystal* rod-shaped molecules line up in such a way that seen from end on they are ordered like a crystalline solid, whereas seen from the side there is no such ordering. This is a state of matter that in some ways looks like a solid and in others looks like a liquid. We say that matter appears in different **phases** or states; the particular phase depends on external conditions. Water is solid at low temperatures, but at atmospheric pressure it can become a liquid at 0° C and a gas at 100° C. At very high temperatures, such as those found within the Sun, the molecules themselves may dissociate into negatively charged electrons and positively charged ions. That kind of matter has very different properties because electrical forces between the constituents play an important role. Matter that is electrically dissociated into positive ions and negatively charged electrons is called a *plasma*. This state of matter is one that we don't directly deal with very often, and we won't discuss it further.

We shall define a **fluid** to be a substance that is not resistant to the action of shear forces (see Section 11–4). Essentially a fluid undergoes deformation without elastic forces to bring it back to its original shape. This description involves how a material responds to external forces and applies to both liquids and gases. Therefore, except for phenomena in which compressibility is important, our description of the behavior of fluids *applies to gases as well as to liquids.*

16–2 Density and Pressure

We start our study of fluids with **hydrostatics**—the study of fluids at rest. Density and pressure are the two important variables for such fluids, and in this section we'll look at them in some detail.

Density

A quantity of matter has a mass M and a volume V. The average **mass density** ρ of the matter is

$$\rho = \frac{M}{V}. \tag{16–1}$$

Table 16–1 gives densities of some typical substances. Densities vary with temperature; the density of water at sea level is $0.9998 \times 10^3\ \text{kg/m}^3$ at 0°C, but it decreases to $0.9584 \times 10^3\ \text{kg/m}^3$ at 100°C. The densities of liquids can also change under a change of the pressure on them ("compression" if the effect is to try to decrease the volume), although the effect is often small. The fractional density change of water in a cylinder is only about 1.5×10^{-8} when a force per unit area of $3000\ \text{N/m}^2$ acts on a piston that acts to compress the water. This is equivalent to the force per unit area produced by a 75-kg person standing on a piston of area $0.25\ \text{m}^2$. It is therefore a good approximation to treat water as **incompressible**. Finally, we can define the **specific gravity** of a substance as the ratio of the mass density of that substance to the mass density of water; thus, for example, the specific gravity of glycerin is 1.26.

TABLE 16–1 • Mass Densities

Material	Phase	Density (kg/m^3)	Material	Phase	Density (kg/m^3)
Hydrogen (0°C, 1 atm)	Gas	0.090	Liquid mercury	Liquid	13.6×10^3
Air at sea level	Gas	1.29	Sun's core	Plasma	1.6×10^5
Chlorine (0°C, 1 atm)	Gas	3.21	Cork	Solid	0.25×10^3
Ethyl alcohol	Liquid	0.79×10^3	Applewood	Solid	0.745×10^3
A good olive oil	Liquid	0.92×10^3	Ice	Solid	0.917×10^3
Water	Liquid	1.00×10^3	Quartz	Solid	2.65×10^3
Glycerin	Liquid	1.26×10^3	Steel alloys	Solid	7.6×10^3–8.9×10^3
Earth's core	Liquid	9.5×10^3	Iridium	Solid	22.4×10^3

EXAMPLE 16–1 A neutron star is a star that is much smaller than our Sun and has the density of an atomic nucleus. A typical neutron star has a radius of 10 km, but a mass of 2×10^{30} kg (the mass of the Sun). How much would a 1 cm^3 volume of such a star weigh under the influence of Earth's gravity?

Strategy We use the given radius R to find the volume V, which together with the given mass M allows us to find the density. For the weight of 1-cm^3 of a neutron star on Earth's surface, we use the density we have calculated to find the mass of this volume and hence the weight.

Working It Out The density is

$$\rho = \frac{M}{V} = \frac{M}{4\pi R^3/3} = \frac{2 \times 10^{30}\ \text{kg}}{4\pi(10^4\ \text{m})^3/3} = 0.5 \times 10^{18}\ \text{kg/m}^3$$
$$= 0.5 \times 10^{12}\ \text{kg/cm}^3.$$

Thus a volume $V_{cube} = 1\ \text{cm}^3\ (10^{-6}\ \text{m}^3)$ of the material of a neutron star has a mass $m = \rho V_{cube} = 0.5 \times 10^{12}$ kg, and it weighs $mg = (0.5 \times 10^{12}\ \text{kg})g = (0.5 \times 10^{12}\ \text{kg})(9.8\ \text{m/s}^2) = 0.5 \times 10^{13}$ N. For comparison, 1 cm^3 of water on Earth's surface has a mass of 1 g and weighs 0.01 N. A neutron star is incredibly dense!

What Do You Think? If the radius of a solid sphere doubles, and its mass also doubles, then is the density (a) less than (b) unchanged from (c) greater than the original density? *Answers to What Do You Think? questions are given in the back of the book.*

Pressure

Pressure is a measure of the force acting on the surface of an object within a fluid, even a fluid at rest. Swimmers know from experience that liquids do exert such forces. If you dive into deep water, your eardrum may hurt or even rupture due to the pressure of the water on it. Why then doesn't a sheet of glass placed under water shatter? It is because there is as much pressure on one side of it as on the other, and indeed it is a characteristic of the pressure in a static fluid that it acts upon every surface, perpendicular to the surface.

As we see more precisely below, the pressure is defined as the force per unit area. We can in principle determine the force exerted by a fluid by using the spring-balance system shown in Fig. 16–4. The plate on the spring balance is in contact with the fluid, while the spring attached to the other side of the plate is inside an enclosure where there is a vacuum, so that there is no force on the plate from the inside due to any pressure. Only the spring force resists the pressure from the fluid, and the force from the fluid is therefore measured by observing the compression of the spring. Our device would allow us to define the average pressure p on any surface of area A as the *force per unit area* acting perpendicular to that surface:

$$p \equiv \frac{F}{A}. \tag{16–2}$$

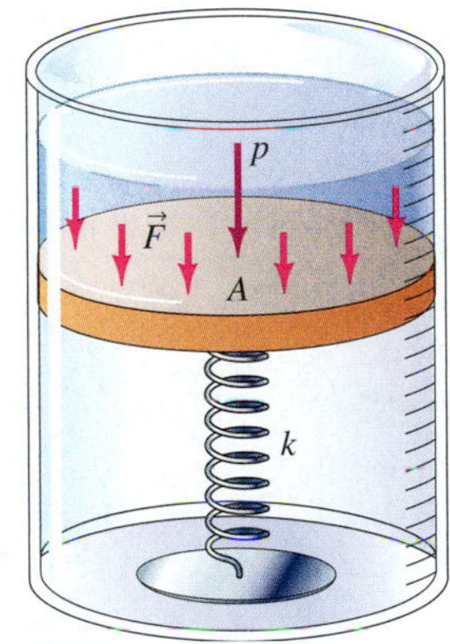

▲ **FIGURE 16–4** A device that would allow the measurement of pressure in a liquid. The fit between the piston and cylinder walls allows the piston to move but is tight enough so that a near vacuum can be maintained on the side with the spring. The compression of the spring provides a measurement of the force, from which the pressure is determined.

The device described above is not a very realistic one. The **manometer** shown in Fig. 16–5 is a realistic way to do the measurement. As we shall see later in this chapter, the difference in the mercury levels in the two sides of the U-shape is sensitive to the pressure in the fluid compared to the pressure in the atmosphere. With such an instrument one could see how pressure varies with position, and since it does vary, a more precise local definition of the pressure is in order. We consider a small area ΔA located at point

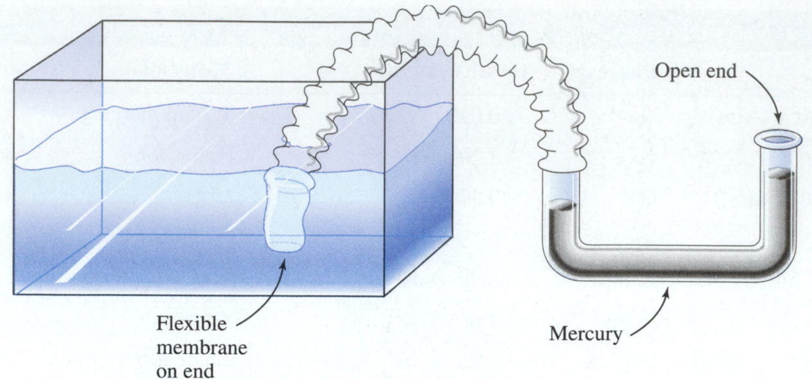

▶ **FIGURE 16–5** A more realistic device to measure pressure. The end of the tube is covered by a flexible membrane that is pushed inward if the pressure in the fluid increases. This pushes the column of mercury higher on the right side of the U-shaped tube than on the left side.

$\vec{r}$, and take the limit $\Delta A \to 0$. If we denote the force perpendicular to this area as ΔF, we find

$$p = \lim_{\Delta A \to 0} \frac{\Delta F}{\Delta A} = \frac{dF}{dA}. \tag{16–3}$$

PRESSURE

This is the pressure at point $\vec{r}$. Pressure is measured in SI units as *pascals* (Pa), named after Blaise Pascal, who studied pressure in the mid-seventeenth century:

$$1 \text{ Pa} = 1 \text{ N/m}^2. \tag{16–4}$$

Other common units for pressure are pounds per square inch (lb/in^2): $1 \text{ lb/in}^2 = 6.9 \times 10^3$ Pa; the atmosphere (atm), which is the atmospheric pressure at sea level: $1 \text{ atm} = 1.01 \times 10^5 \text{ Pa} = 14.7 \text{ lb/in}^2$; the bar: $1 \text{ bar} = 10^5$ Pa; and the torr: $1 \text{ torr} = 133.32$ Pa.

If you took your manometer and moved the membrane about, you would discover two important facts about pressure in static fluids: First, pressure varies only with depth, increasing as one goes deeper. As we shall explain in the next section, this is a consequence of the presence of gravity. Second, you would learn that *when a fluid is at rest, the pressure at any given point must be the same in all directions.* The microscopic origin of pressure is the bombardment of the surface on which the pressure acts by the individual molecules of the substance (Chapter 19). The pressure is the same in all directions because there is no preferred direction for the random motion of molecules. For fluids at rest, the pressure is then simply a positive scalar quantity. Note also that the pressure at a static interface between two fluids, say a layer of oil on water, is the same on both sides of the interface; if it were not, there would be a net force at the interface and a resulting motion of the interface.

16–3 Pressure in a Fluid at Rest

The gravitational force causes the pressure of fluids at rest to be larger at ocean depths than at Earth's surface. In order to establish exactly how pressure varies with depth for an incompressible fluid, consider a liquid of fixed density ρ at rest in a container. Imagine isolating a very thin cylinder of that liquid (Fig. 16–6). (The top, base, and walls of the cylinder are purely imaginary; their sole purpose is to help us visualize the forces acting on an element of the fluid.) The cylinder's top, of area A, is horizontal and lies a distance y below the surface of the liquid. The bottom of the cylinder is at a depth $y + \Delta y$. Note that we measure the depth y downward from the surface, so it increases as we go down. We denote the pressure at the top of our cylinder by p and the pressure at the bottom by $p + \Delta p$. We want to find an expression for Δp.

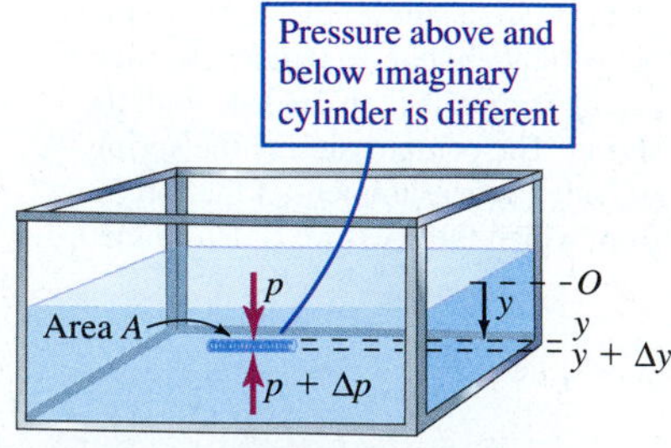

▲ **FIGURE 16–6** An imaginary thin cylinder of fluid is isolated to show the forces that act to keep it in equilibrium.

The downward force on the cylinder, which is applied at the top surface by pressure from the surrounding fluid, is pA; the upward force, which is applied at the bottom surface by the pressure there, is $(p + \Delta p)A$. Additionally, there is a downward force on the cylinder due to gravity; namely, the weight of the liquid inside the cylinder. That force has a magnitude of $(\Delta m)g$, and the mass of the cylinder Δm is the density times the volume of the cylinder. Thus, the downward force of gravity is $\rho(A\Delta y)g$. All together, we find

$$F_{\text{up}} = (p + \Delta p)A,$$
$$F_{\text{down}} = pA + (\Delta m)g = pA + \rho(A\Delta y)g.$$

Because the cylinder does not accelerate, the net force must cancel:

$$F_{\text{net}} = F_{\text{up}} - F_{\text{down}} = (p + \Delta p)A - pA - \rho(A\Delta y)g = 0.$$

We cancel the pA terms, divide by A, and rearrange terms to find that

$$\frac{\Delta p}{\Delta y} = \rho g.$$

In the limit $\Delta y \to 0$, this result becomes

$$\frac{dp}{dy} = \rho g. \qquad (16\text{–}5)$$

According to Eq. (16–5), the pressure change is positive for positive Δy. Pressure therefore increases with increasing depth (Fig. 16–7). In physical terms this phenomenon is a measure of the increased amount, and hence increased weight, of the liquid above any given point as the depth of the point increases. There is a greater weight of water above a diver at an ocean depth of 50 m than above a diver at a depth of 10 m.

Equation (16–5) looks just like the formula for the constant acceleration g of an object acted on by the constant force of gravity, $dv/dt = g$. When we studied projectile motion in Chapters 2 and 3, we saw that this equation has a solution $v = v_0 + gt$. If the fluid is incompressible (as water is, but gases are generally not), then the density does not depend on the depth y, and we can directly compare to write the solution to Eq. (16–5), which gives the pressure as a function of depth:

$$p = p_0 + \rho g y. \qquad (16\text{–}6)$$

Alternatively, we can integrate Eq. (16–5) to find this result. The constant p_0 is the value of the pressure at the surface, where $y = 0$; for a body of fluid such as the ocean, this is just atmospheric pressure. The pressure increase with depth y is equal to the weight per unit area of the overlying fluid. The total mass of the fluid above the cylinder is $\rho V = \rho A y$, so its weight is $\rho A y g$, and thus the weight per unit area is $\rho y g$.

The variation of atmospheric pressure with depth (from the top of the atmosphere) is not at all like that of water, Eq. (16–6). The difference is that gases are compressible, and this changes the form of the equation obeyed by the pressure, and with it its variation with depth. The atmospheric pressure will be treated in Chapter 19.

▲ **FIGURE 16–7** Divers experience the increased pressure of increased water depth. To equalize the pressures of the surrounding water and the sinus cavities, a diver blows air into his or her sinuses.

CONCEPTUAL EXAMPLE 16–2 Look at Fig. 16–8: the level of the liquid in all the connected vessels is the same. Why?

Answer The derivation that led us to Eq. (16–6) is completely independent of the horizontal location of the imaginary cylinder in the water tank of Fig. 16–6. Equation (16–6) itself shows that the pressure in a fluid depends *only* on the depth—not the horizontal position. In our series of vessels the pressures at the fluid surfaces are all equal to atmospheric pressure (the surface is in contact with the atmosphere and is therefore at the same pressure). According to Eq. (16–6), a given pressure corresponds to a given height, and so the level of the top of the fluid will be the same for each vessel.

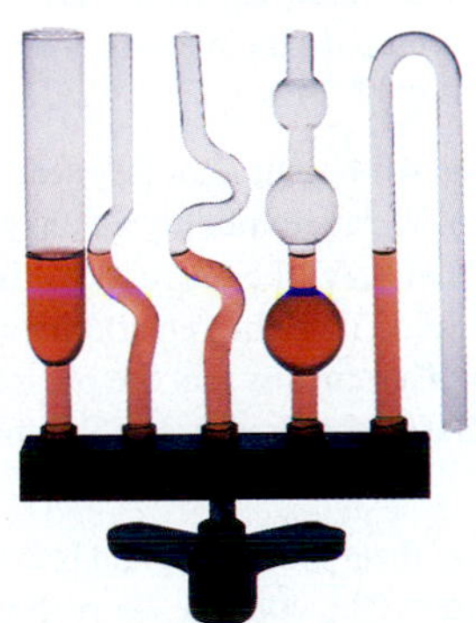

◀ **FIGURE 16–8** The surface pressure is equal in each individual vessel.

We can see from Eq. (16–6) that if p_0 is changed as a result of some external effect, p is changed by exactly the same amount. In other words, *a change in pressure applied to any point in a fluid at rest results in the same pressure change at every point in the fluid.* This statement is commonly known as **Pascal's principle**.

EXAMPLE 16–3 Many automobile repair shops have a hydraulic jack (Fig. 16–9), which is a device used to lift cars. This device has two connecting vertical cylindrical columns filled with fluid, a wider one that sits beneath the car to be lifted and a narrower one that is pushed down on by an operator. In this problem the wider column has a radius of 20.0 cm and lifts a car of mass 1000 kg. The narrower column has a radius of 2.5 cm, and the operator pushes down on a piston in it by applying air pressure from a pump. (a) What pump pressure is needed to lift the car? (b) If the car is lifted up a distance of 2.00 m, what downward distance must the piston in the narrower cylinder move?

▲ **FIGURE 16–9** A schematic view of the hydraulic lift used in garages.

Setting It Up Figure 16–9 labels two of the variables with which we are going to deal, both given, the areas A_1 and A_2 of the wider and narrower columns, respectively. We denote as d_1 the given vertical distance that the car, mass m, moves. We are going to find the pump pressure p that will allow the car to be raised and the distance d_2 that the piston in the narrow cylinder must move.

Strategy (a) We can lift the car with an upward force of magnitude mg; this has to be supplied by the fluid pressure p, and since this pressure acts over an area A_1, the upward force is pA_1. That will determine what p has to be. (This pressure is in addition to the pressure that is already present and allows the car to remain stationary.) By Pascal's principle this additional pressure p can be applied at the narrower cylinder—at the pump—and it will be transmitted throughout the fluid. (b) There are various ways to approach this part. In one method we note that the car has acquired a potential energy mgd_1 in being raised. According to the work–energy theorem, this energy must come from the work done at the narrower tube. If the force acting at the narrower tube is $F_2 = pA_2$, the work done is d_2F_2. We match this to the energy gain of the car to give us an equation for d_2. In another method, we use the fact that the fluid is incompressible, so the same volume of fluid d_1A_1 that has been added to the wide cylinder must come from the smaller cylinder. Since the volume in the smaller cylinder is d_2A_2, equating d_2A_2 to d_1A_1 gives us an equation for d_2. We'll use the second method, which is very simple.

Working It Out We have $pA_1 = mg$, or $p = mg/A_1$. In terms of the given radius $R_1 = 0.20$ m of the wider tube,

$$p = \frac{mg}{\pi R_1^2} = \frac{(1000\ \text{kg})(9.8\ \text{m/s}^2)}{\pi(0.20\ \text{m})^2} = 7.8 \times 10^4\ \text{Pa}.$$

This is the necessary pump pressure, a number 80% of the ambient atmospheric pressure. (The ambient pressure alone cannot lift the car as it acts downward on the topmost surfaces of the car and upward on the bottom surfaces, and so forth, and the net force from the ambient pressure is zero. The pump pressure here is really an additional pressure, and it acts in one particular direction because of the orientation of the plate that supports the car.)

(b) We set $d_2A_2 = d_1A_1$ and solve for d_2:

$$d_2 = \frac{A_1}{A_2}d_1 = \frac{\pi R_1^2}{\pi R_2^2}d_1 = \frac{(20.0\ \text{cm})^2}{(2.5\ \text{cm})^2}(2.0\ \text{m})$$
$$= 8^2 \times (2.0\ \text{m}) = 1.3 \times 10^2\ \text{m}.$$

In the narrower tube there will be some sort of ratcheting mechanism which allows the vertical motion to be carried out in many strokes. Many air pumps have such mechanisms.

What Do You Think? Why do bicycle and car tires have such thin valves?

EXAMPLE 16–4 A column of mercury ($\rho_{\text{Hg}} = 13.6 \times 10^3$ kg/m^3) in a vertical tube that is open at the bottom rests in an open pool of mercury. The column is sealed off at the top after all the air has been evacuated from the tube, creating a vacuum above the mercury. What is the height of the column of mercury?

Setting It Up In Fig. 16–10 we see the height H of the mercury column that is asked for in this problem. We are explicitly given only the density of mercury, and more information will be required to find the height of the mercury column. The clue is in the remark that the pool of mercury is open at the bottom. That means that the pressure of the atmosphere acts on it, and so that is the pressure in the mercury at the level of the pool.

Strategy The fact that the tube above the mercury column is evacuated means that there is no pressure above the column. (In reality, a little mercury will have evaporated, but the pressure that it exerts is negligible.) However, the surface of the open pool, and therefore in the interior of the column at the same height, is that of the atmosphere—call this pressure p_0. If the mercury column has a cross section A, then there is an upward force at the bottom, p_0A, that counteracts the weight of the mercury in the column, mg. We can express the mass m of mercury in the column as the density of mercury times the column volume V, and $V = HA$. In this way, H is determined.

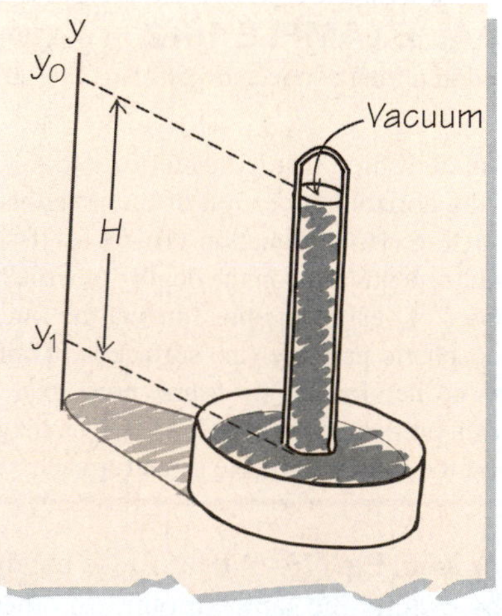

▲ **FIGURE 16–10** A column of mercury balanced by atmospheric pressure.

Working It Out The expression for the balancing of the forces is $p_0 A = mg$. But the volume of the column of mercury is $V = HA$, hence the mass is $m = \rho_{Hg} V = \rho_{Hg} HA$, and we have

$$p_0 A = \rho_{Hg} HAg.$$

The area cancels from both sides. We solve the remaining equation for H:

$$H = \frac{p_0}{\rho_{Hg} g}.$$

Numerically,

$$H = \frac{1.01 \times 10^5 \text{ Pa}}{(13.6 \times 10^3 \text{ kg/m}^3)(9.8 \text{ m/s}^2)} = 0.76 \text{ m}.$$

This example describes the working principle of a **barometer**, a device used to measure atmospheric pressure. One atmosphere of pressure will support a column of mercury that is 760 mm high.

What Do You Think? What would the height of the mercury column be if the vacuum at the top of the tube were instead a fraction of 1 atm?

EXAMPLE 16–5 A U-shaped tube of uniform cross section is open at both ends. First, amyl bromide is poured into the tube. (Amyl bromide is a fluid that does not mix with water.) Then water is carefully added to one side of the U, so there is water at the top of one column of fluid and amyl bromide is at the top of the other. The two columns of fluid meet at a point that we label $h = 0$, as in Fig. 16–11. What is the ratio of the heights of the fluids above $h = 0$ in each column? The density of water is $\rho_W = 1.00 \times 10^3 \text{ kg/m}^3$ and that of amyl bromide is $\rho_{ab} = 1.26 \times 10^3 \text{ kg/m}^3$.

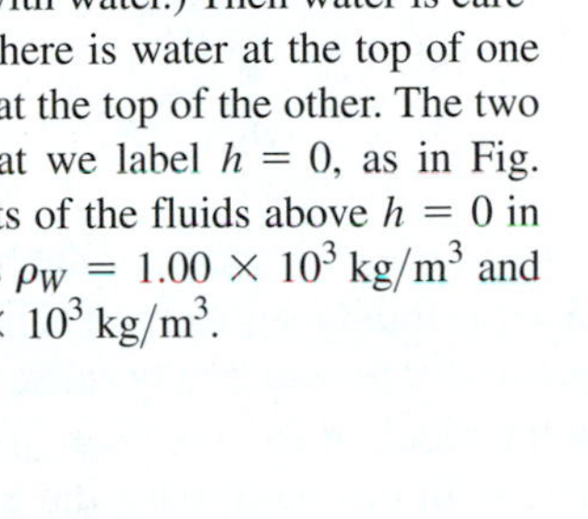

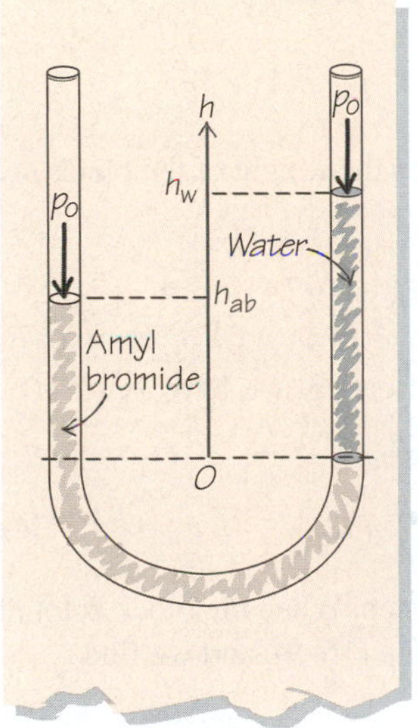

▲ **FIGURE 16–11**

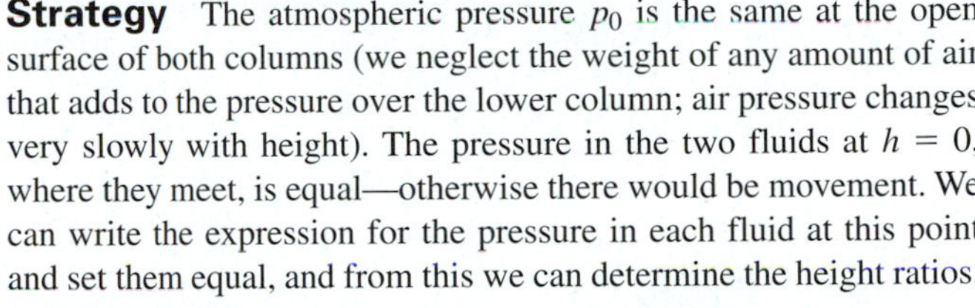

Strategy The atmospheric pressure p_0 is the same at the open surface of both columns (we neglect the weight of any amount of air that adds to the pressure over the lower column; air pressure changes very slowly with height). The pressure in the two fluids at $h = 0$, where they meet, is equal—otherwise there would be movement. We can write the expression for the pressure in each fluid at this point and set them equal, and from this we can determine the height ratios.

Working It Out We measure heights from the point where the fluids meet and denote h_{ab} as the height of the amyl bromide column and h_w as the height of the water column. The pressures p_{ab} and p_w at $h = 0$ in the respective columns is the air pressure at the top plus the pressure due to the weight of the fluid above,

$$\text{for the amyl bromide, } p_{ab} = p_0 + \rho_{ab} h_{ab} g;$$

$$\text{for the water, } p_w = p_0 + \rho_w h_w g.$$

Setting these pressures equal to one another, we have

$$\rho_{ab} h_{ab} g = \rho_W h_W g.$$

We rearrange this equation to find

$$\frac{h_{ab}}{h_W} = \frac{\rho_W}{\rho_{ab}} = \frac{1.00 \times 10^3 \text{ kg/m}^3}{1.26 \times 10^3 \text{ kg/m}^3} = 0.794.$$

The water column is higher than the amyl bromide column. This makes sense physically because amyl bromide is denser than water; the shorter column of amyl bromide weighs the same as the taller column of water. Note that this result is independent of the amount of either fluid.

What Do You Think? What would happen if we used heptyl ether, another fluid that does not mix with water and has a mass density of $0.81 \times 10^3 \text{ kg/m}^3$?

16–4 Buoyancy and Archimedes' Principle

A cork floats in water while a brick sinks. Why? Experience tells us that this has something to do with the density of the material (and not the mass; a small brick sinks just like a large one). Let's consider the situation in Fig. 16–12a, a block of material of volume $V = Ah$ and density ρ (the mass is thus $m = \rho V$) partly immersed in water, which has a density of ρ_W. What forces do you think are acting on the block in this equilibrium position? In particular, focus on the lower surface of the partly submerged block, as in Figure 16–12b. The upward force is due to the pressure of the liquid, as given by Eq. (16–6): $p = p_0 + \rho_W g y$, where y is the vertical submerged distance of the block. The upward force on the block is therefore

$$F_{up} = pA = p_0 A + \rho_W g y A.$$

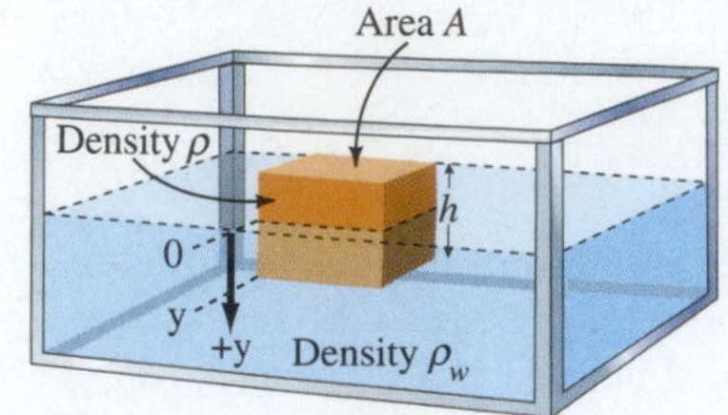

(a)

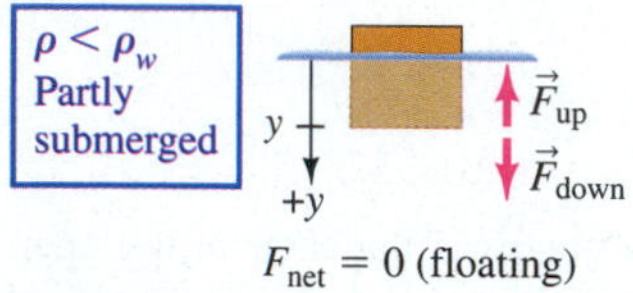

(b)

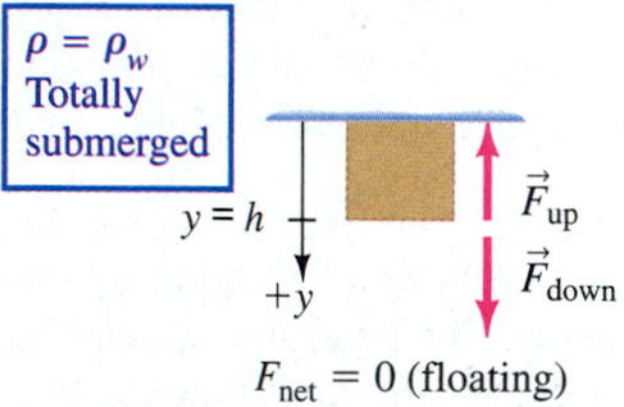

(c)

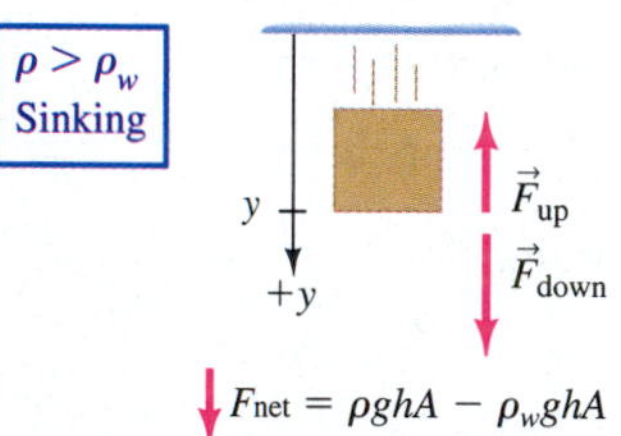

(d)

▲ **FIGURE 16–12** (a) A block of material of density ρ is partly immersed in water, density ρ_w. The volume of the block is $V = Ah$, and the volume of water displaced by the block is $V_d = Ay$. (b) Force diagram of the forces acting on the lower surface of the block, which is partly immersed, as in part (a). The net force is $F_{net} = 0$, and the block floats. (c) When $\rho = \rho_w$, the block is totally submerged, and the block floats because $F_{net} = 0$. (d) When $\rho > \rho_w$, the block sinks because $F_{net} > 0$.

The downward force has two components: that due to the atmospheric pressure p_0 above the block, and the weight of the block. The first downward component is $F_{atm} = p_0 A$, and the second is the force of gravity on the block, $F_g = mg = (\rho Ah)g$. Thus the total downward force is

$$F_{down} = p_0 A + \rho ghA.$$

As the block is in equilibrium—floating but not moving—the net force will be zero. Therefore,

$$F_{net} = 0 = F_{down} - F_{up} = p_0 A + \rho ghA - p_0 A - \rho_W gyA = \rho ghA - \rho_W gyA. \tag{16–7}$$

Dividing by gA, we obtain

$$\text{for a floating object: } \rho h = \rho_W y;$$

$$\frac{\rho}{\rho_W} = \frac{y}{h}. \tag{16–8}$$

If $\rho < \rho_W$, then $y/h < 1$ and only a fraction of the block is submerged. In the limiting case that $\rho = \rho_W$, the block is totally submerged and $y = h$ (Fig. 16–12c). In this case, the block would float just under the surface because the upward and downward forces cancel. If $\rho > \rho_W$, then the block must sink (Fig. 16–12d) because the two forces can no longer cancel completely. In this case, once the block is fully submerged $(y \geq h)$, the net (downward) force is

$$F_{net} = \rho ghA - \rho_W ghA > 0. \tag{16–9}$$

We can interpret the difference between the weight of the block ρghA and F_{net} as an *upward* **buoyant force**:

$$F_{buoy} = F_g - F_{net}. \tag{16–10}$$

When the block is partly submerged and floating, we have $F_{net} = 0$, and Eq. (16–10) gives

$$\text{for partial submergence: } F_{buoy} = \rho ghA - 0 = \rho ghA = \rho_W gyA,$$

where we have used Eq. (16–8) in the last step. When the block is totally submerged (at any depth where $y \geq h$), F_{net} is given by Eq. (16–9), and we find

$$\text{for total submergence: } F_{buoy} = \rho ghA - (\rho ghA - \rho_W ghA) = \rho_W ghA.$$

If we think of $V_{submerged}$ as the volume of the object beneath the water ($V_{submerged} = yA$ or hA, according to whether the object is partly or totally submerged), then we can combine our results into the single expression (Fig. 16–13)

$$F_{buoy} = \rho_W g V_{submerged}. \tag{16–11}$$

THE BUOYANT FORCE

The buoyant force opposes the downwardly directed force of gravity, ρghA, on the block. For a floating body, the buoyant force and the force of gravity cancel exactly. When $\rho > \rho_W$, the buoyant force opposes the force of gravity even though it does not cancel the force of gravity completely; a sinking object descends with less acceleration than it would under the influence of gravity alone. When a metal sphere is immersed in

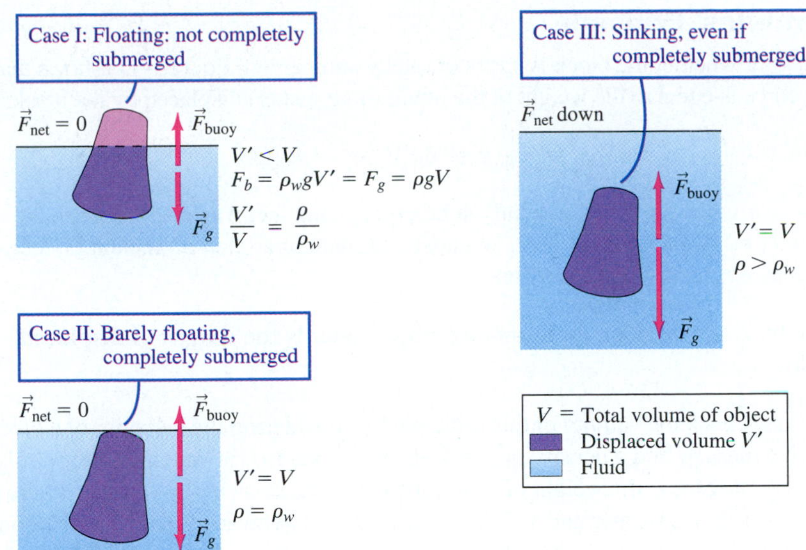

▲ **FIGURE 16–13** By comparing the buoyant force F_{buoy} to the force of gravity F_g, we determine whether a body sinks, just floats, or pops up.

a bucket of water and suspended from a spring balance, the balance will indicate a lower weight than it would for the same sphere in air. The difference in these weights is the buoyant force.

CONCEPTUAL EXAMPLE 16–6 Think about the hot-air balloon in Fig. 16–14a and the steel ball floating in mercury in Fig. 16–14b. Explain both phenomena.

Answer Above we saw that if a material is less dense than the medium in which it is is submerged, the (upward) buoyant force on it exceeds the force of gravity acting downward. The object will float or, if completely submerged, it will rise. This describes the hot-air balloon: Hot air is less dense than cooler air, and the balloon is completely submerged in the cooler air, so the net force on the balloon is upward. More precisely, we can use Eq. (16–7) with $h = y$, ρ_w replaced by the density of cool air and ρ equal to the density of hot air (very nearly the density of the entire balloon). The net upward force is thus $F_{net} = Vg(\rho_{cool} - \rho_{hot})$.

The steel ball is in a situation like that of Fig. 16–13a, floating on the surface of a medium denser than it is. Perhaps that seems odd because we think of steel as being quite dense. That is true, but mercury is a liquid of extraordinarily high density. The steel ball floats easily because its density, 7.8 g/cm^3, is less than that of mercury, 13.6 g/cm^3.

(a)

(b)

◀ **FIGURE 16–14** (a) A hot air ballon is floating in air. (b) The heavy steel ball floats in mercury.

Archimedes' Principle

Regardless of whether the block is partly or totally submerged, Eq. (16–11) shows that the buoyant force is equal to the weight of the liquid (here, water) displaced by the block:

$$W_{\text{liq}} = m_{\text{liq}}g = \rho_{\text{liq}}V_{\text{liq}}g = \rho_{\text{liq}}yAg,$$

where $y = h$ when the block is totally submerged. This fact was discovered in the third century B.C. by Archimedes, a great scientist and mathematician of ancient Greece, and is known as **Archimedes' principle**:

> **The buoyant force on an immersed object equals the weight of displaced fluid.**

We may check the validity of this principle by considering the behavior of a block of iron, with a mass M and a density $\rho_{\text{Fe}} = 7.86 \times 10^3\ \text{kg/m}^3$, in water. The upwardly directed buoyant force is the weight of water displaced, $F_{\text{buoy}} = mg = \rho_W Vg$, where m is the mass of displaced water and V is its volume. The downward force of gravity on the iron block is $F_g = Mg = \rho_{\text{Fe}}Vg$. Because $\rho_{\text{Fe}} > \rho_W$, the net force is downward and the block sinks. The net downward force—effectively the weight of iron in water—is

$$F_{\text{net}} = F_g - F_{\text{buoy}} = \rho_{\text{Fe}}Vg - \rho_W Vg = \left(1 - \frac{\rho_W}{\rho_{\text{Fe}}}\right)\rho_{\text{Fe}}Vg$$
$$= \left(1 - \frac{\rho_W}{\rho_{\text{Fe}}}\right)F_g. \qquad (16\text{–}12)$$

We can check the magnitude of the net force, and verify Archimedes' principle, by weighing an object in water (Fig. 16–15). Archimedes is said to have discovered this technique when he was asked to ascertain that a crown was made entirely of gold without marring the crown; he did this by weighing the crown in air and in water, thereby measuring its density [Eq. (16–12)]. It is said that he found the density of the crown to be less than that of gold, so that the gold in the crown must have been diluted with another metal.

The crown's owner would have not been very happy if Archimedes had been required to melt the crown down and recast it as a rectangular block to make his test! Although our calculations for buoyancy and Archimedes' principle were carried out for a uniform rectangular block, the results do not depend on the shape or the unifor-

(a)

(b)

(c)

▲ **FIGURE 16–15** By Archimedes' principle, the magnitude of the buoyant force on the sphere, F_{buoy}, equals the weight of the water the sphere displaces. The scale measures the difference of the magnitudes of the force of gravity on the sphere and the buoyant force on it. (a) No water is displaced; the scale measures the weight of the sphere. (b) Half the volume of the sphere is displaced. (c) The entire volume of the sphere is displaced. The buoyant force in this case is about 10 N.

mity. An object of any shape can be thought of as being composed of small rectangular blocks. Similarly, an object of varying density can be treated as a collection of many elements, each of constant density. The end result is that all that matters is the weight of the liquid that is displaced by the object.

EXAMPLE 16–7 A metal sphere weighs 29.4 N in air and 18.5 N in water. What is its average density?

Setting It Up In contrast to the situation of Fig. 16–14, the sphere sinks. When its weight is measured while it is beneath the surface of the water, there is a nonzero reading (Fig. 16–15c). We label the weight of the metal sphere in air $F_g = mg$ and in water W_w.

Strategy Equation (16–12) applies directly; it gives the weight of the sphere in water. Since W_w is proportional to F_g, it is simplest to eliminate the weight in air by taking the ratio W_w to mg. The resulting relation can be solved for the unknown average density ρ_{sphere} of the sphere.

Working It Out We have

$$\frac{W_w}{F_g} = \frac{\left(1 - \dfrac{\rho_w}{\rho_{\text{sphere}}}\right)mg}{mg} = \left(1 - \frac{\rho_w}{\rho_{\text{sphere}}}\right).$$

We solve for ρ_{sphere}:

$$\rho_{\text{sphere}} = \rho_w \frac{1}{1 - \dfrac{W_w}{mg}} = \rho_w \frac{1}{1 - \dfrac{18.5}{29.4}}$$

$$= \rho_w \times 2.70 = 2.70\ \text{g/cm}^3.$$

At this point you could also solve for the unknown mass of the sphere: Knowing the sphere density, the expression for W_w would give us the mass.

What Do You Think? If the metal sphere had been weighed in salt water rather than fresh water and given the same weight reading, would the density of the metal sphere have been (a) greater, (b) smaller, (c) the same, (d) can't tell. (Salt water is denser than fresh water.)

EXAMPLE 16–8 A rectangular tub made of a thin shell of cement has length $\ell = 1$ m, width $w = 80$ cm, depth $d = 60$ cm, and mass $M = 200$ kg. The empty tub floats in a lake. How many people of mass $m = 80$ kg each can stand in the tub before it sinks?

Setting It Up The tub is shown in Fig. 16–16. The depth of the bottom of the tub is y, and we denote the unknown number of passengers by N.

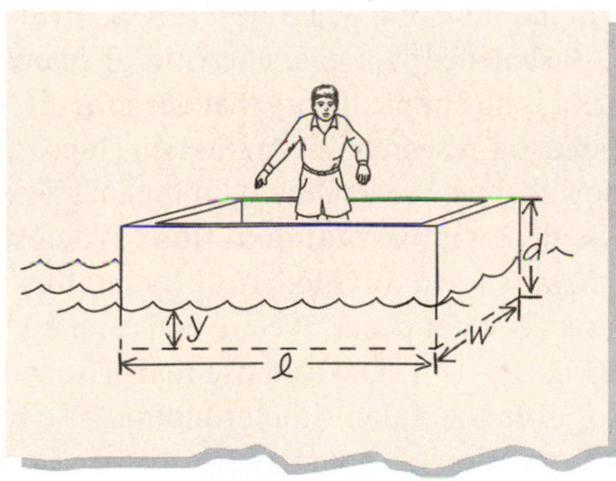

▲ **FIGURE 16–16** A tub sinks to a water depth y. When $y > d$, the depth of the tub, the tub sinks.

Strategy With the tub submerged to a distance y, we can find the volume $\ell w y$ of displaced water and then the weight of the displaced water. This will allow us to find the buoyant force. The buoyant force will cancel the weight of the tub with its passengers up to the point that the immersed depth is equal to the tub's height, $y = d$. This is the point at which the boat is barely floating. Beyond this the tub will fill with water, increasing its average density, and it will sink to the bottom. So we set the buoyant force at $y = d$ to the total weight, and this relation will determine that total weight and in turn the maximum number of passengers N_{max}.

Working It Out With the tub submerged a vertical distance y, the mass of the displaced water is $\rho(\ell w y)$, where ρ is the water density, $1\ \text{g/cm}^3 = 10^3\ \text{kg/m}^3$. Then the weight of the displaced water is the mass times g, and this is the buoyant force,

$$F_{\text{buoy}} = \rho(\ell w y)g.$$

This force cancels the total weight of the tub and N passengers, $(M + Nm)g$. The equality of the buoyant force and the total weight,

$$\rho(\ell w y)g = (M + Nm)g,$$

is what determines y. We see from this relation that y increases as N does (the tub moves lower in the water), until the critical value $y = d$ is reached, at which point any additional mass sinks the tub. We thus set $y = d$ to find N_{max}. With $\rho(\ell w d)g = (M + N_{\text{max}}m)g$, we solve for N_{max}:

$$N_{\text{max}} = \frac{\rho \ell w d - M}{m}.$$

Numerically,

$$N_{\text{max}} = \frac{(10^3\ \text{kg/m}^3)(1\ \text{m})(0.80\ \text{m})(0.60\ \text{m}) - 200\ \text{kg}}{80\ \text{kg}} = 3.5.$$

With three people the tub does not sink; with four people she goes down. Cargo ships were built out of concrete during both World Wars because of the high demand for steel. The older ships also served a useful purpose when they were deliberately sunk during the Normandy invasion of World War II to serve as a breakwater that would improve conditions for landing ships.

What Do You Think? If this tub had been weighed in water, what would the scale read?

16–5 Fluids in Motion

The marvelous fluid flow patterns observed in nature go far beyond the particle and rigid body motions we have studied to this point. **Hydrodynamics** describes fluids in motion, including the flow of streams, the eddies of oars in a pond, mudslides, smoke rings, ocean waves and tides, arterial blood flow, Jupiter's (and Earth's) atmosphere, the rotational flow of galaxies, and nuclear fission. Numerical calculations performed by supercomputers are necessary to solve the most complex types of behavior, but a remarkable variety of phenomena can be described with a very simplified set of equations. This simplified set involves a number of assumptions that we describe here.

We assume, as we have already discussed, that the fluid under consideration is incompressible. This does not mean that we exclude phenomena that involve gases, such as airflow; but it does mean that we will discuss these phenomena only under circumstances in which compressibility does not play a role.

We also assume that we are dealing with a fluid at constant temperature. (We will learn more about temperature and heat flow in Chapters 17 through 20.) Phenomena such as the **convection** of fluids—in which a liquid in the bottom of a vessel is heated, rises, cools, and falls in a circulating pattern—will not be considered.

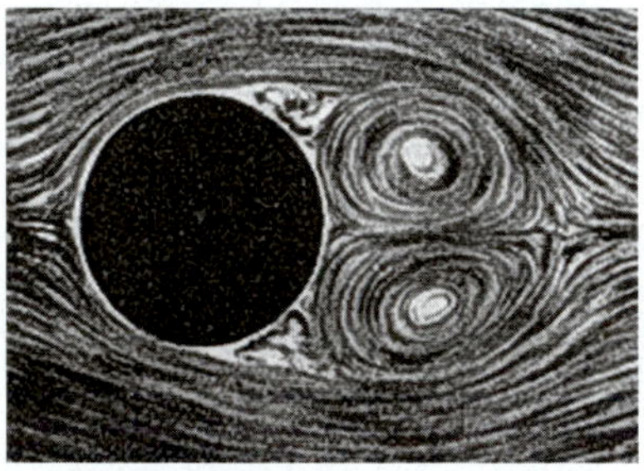

We assume **steady flow**, in which the velocity $\vec{v}$ and the pressure p of a fluid at a given point do not depend on time. The velocity and pressure may vary from point to point, but we assume that any changes are smooth ones.

▲ **FIGURE 16–17** Streamlines for a cylinder pulled steadily through a fluid, showing circulation in the fluid in the form of the two vortices on the right.

The description of how velocity can change from point to point deserves a more detailed discussion and leads us to some further simplifying assumptions. Consider a fluid in motion. We could represent the direction of the velocity at some point by an arrow at that point, with the length of the arrow proportional to the magnitude of the velocity. For steady (time-independent) flow, there is a *local velocity* $\vec{v}(\vec{r})$. If we start at some point, then a tiny element of the fluid is at that point, and it will move in the direction of the velocity vector. An instant later the element is at an adjacent point, which has a slightly different velocity vector. This vector dictates the element's direction and speed of motion in the next instant and determines the next step of the motion. In this way, the motion of a tiny element of fluid can be traced. (That we can follow such an element is an idealization that approximates reality if the element is small enough.) The line of motion traced out by our element is called a **streamline**. Flow described by streamlines is called **laminar flow**; in laminar flow each element of fluid travels along a smooth, well-defined path. Under certain circumstances fluid flow ceases to be laminar and becomes turbulent. **Turbulent flow**, as in the flow of a white-water river, involves rapidly fluctuating velocities and pressures. Turbulence is a characteristic of many flow patterns that we observe in nature, but there is no simple theory that describes it. We'll look further at turbulence, an exciting area of research in engineering and physics, in Section 16–9, but we'll assume for now that we are dealing with laminar flow.

We will also assume here that we deal with **irrotational flow**. When streamlines close in on themselves, such as those shown in Fig. 16–17, the flow is called **rotational**, and in irrotational flow this does not occur. We are consequently unable to discuss here vortices, or whirlpools (Fig. 16–18), such as those that arise in water running out of a bathtub, smoke rings, or even the much simpler motion of coffee stirred by a spoon in a cup. A mathematical expression of the statement that the flow is irrotational is that the **circulation**, defined as the integral of the velocity along any *closed* path in the fluid, is zero.

▲ **FIGURE 16–18** The water moves faster near the center of the vortex than it does far from the center. The water is in rotational flow.

To simplify matters further, we make one more assumption: We neglect the analogue of friction in the fluid. That there is such an analogue is obvious from simple observations. The flow of honey is different than the flow of water. The friction that enters here is an internal friction which is called **viscosity**. In neglecting viscosity, we are neglecting internal energy losses, and the conservation of energy can then be used profitably.

To summarize, in the next two sections we make the following assumptions about fluid flow:

1. The fluid is incompressible.
2. The temperature does not vary.

3. The flow is steady, so that the velocity and pressure do not depend on time.
4. The flow is laminar rather than turbulent.
5. The flow is irrotational, so there is no circulation.
6. There is no viscosity in the fluid.

In Section 16–9 we will look at some of the important differences between this idealized flow and the flow in real fluids. In particular, we'll come back to the subjects of viscosity and turbulence.

16–6 The Equation of Continuity

For a fluid in motion, we can focus on a group of adjacent streamlines, which we say form a *bundle*, and follow them (Fig. 16–19). (We must work with a bundle because a single streamline is just a one-dimensional line—it contains no mass, hence no energy, no momentum, and so forth.) By their definition, streamlines cannot cross. If they did, the crossing point would be a point where the fluid has two different velocities simultaneously, and that is impossible. And because streamlines never cross, the bundle, which forms a thin tube of nonuniform cross section known as a **streamtube**, has the property that no fluid passes from the inside to the outside of it. We can then choose a streamtube thin enough so that the velocity across a section drawn at right angles to the tube is the very nearly the same across the whole cross section. The volume of fluid that crosses the section of area A_1 (Fig. 16–19) in a given time t is the volume of a cylinder whose base is A_1 and whose length is $\ell_1 = v_1 t$, where v_1 is the speed across that section; the volume of fluid is thus $v_1 t A_1$. If the density of the fluid at this section is ρ_1, then the mass of fluid crossing the section in the given time t is

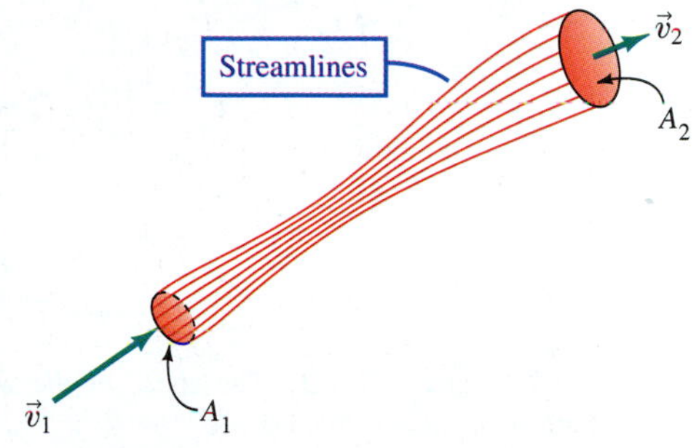

▲ **FIGURE 16–19** Flow contained in a bundle of streamlines (streamtube). The velocity perpendicular to the surface of area A_1 is $\vec{v}_1$, and that perpendicular to the surface of area A_2 is $\vec{v}_2$.

$$m_1 = \rho_1 v_1 t A_1. \tag{16–13}$$

Similarly, the mass of fluid crossing a section of area A_2 at right angles to the bundle of streamlines with speed v_2 and density ρ_2 in the same time t is

$$m_2 = \rho_2 v_2 t A_2.$$

But because fluid neither enters nor leaves the tube by the side walls, the *conservation of mass* implies that $m_1 = m_2$, or, if we cancel the time t,

$$\rho_1 v_1 A_1 = \rho_2 v_2 A_2. \tag{16–14}$$

EQUATION OF CONTINUITY

This equation, the **equation of continuity**, is true even for fluids whose densities can vary with position. Note that $m/t = \rho v A$ is the mass flow per unit time.

When we restrict ourselves to incompressible flow, we have $\rho_1 = \rho_2$, and Eq. (16–14) simplifies to the very useful equation

$$v_1 A_1 = v_2 A_2. \tag{16–15}$$

CONSERVATION OF RATE OF FLOW, OR FLUX

Since $\rho v A$ is the mass flow per unit time, for constant density the combination vA is the *volume flow per unit time*—usually we call this just the rate of fluid flow. Equation (16–15) is a conservation law for the rate of flow. It states that the product of speed and area of any bundle of streamlines is the same anywhere along the bundle. The speed of fluid flow thus varies inversely with the streamtube cross section A. The rate of flow—the product (velocity perpendicular to an area) × (area)—is also known as the **flux**, Φ,

$$\Phi \equiv vA. \tag{16–16}$$

Equation (16–15) describes the **conservation of flux**. Perhaps you have observed the conservation of flux while entering or leaving a building on a windy day: The wind speed is greatest in the smallest doorway.

EXAMPLE 16–9 Water flows at a rate of 0.25 L/s through a garden hose of inside diameter 2.0 cm. What is the speed of the water in the hose? The nozzle of the hose is a circular opening of diameter 1.0 cm. What is the speed of the water when it emerges?

Setting It Up Figure 16–20 indicates the area A_{hose} of the hose and the area A_{noz} of the nozzle. We denote the known flow rate as Φ. We want the water speeds v_{hose} in the hose and v_{noz} in the nozzle—the latter is also the speed just outside the nozzle.

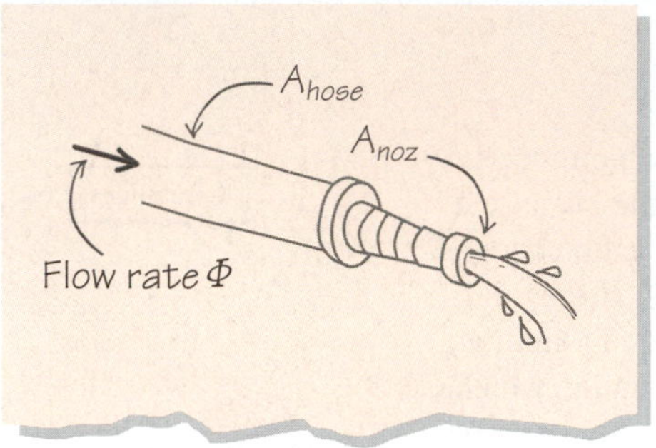

▲ **FIGURE 16–20** The speed of the water must be larger in the narrow opening of the nozzle.

Strategy The given flux Φ is the product vA. The fact that flux is conserved allows us to say that product vA is the same at the nozzle as in the hose. We can then solve for v for either the hose or nozzle in terms of Φ and the appropriate area.

Working It Out We have immediately $v_{\text{hose}} = \Phi/A_{\text{hose}}$ and $v_{\text{noz}} = \Phi/A_{\text{noz}}$, where Φ is the same in the two sections. We can note that since the nozzle is half the diameter of the hose, the area of the nozzle is one quarter the area of the hose, so that the speed in the nozzle is four times higher than in the hose. This will save a little numerical work. The quantitative results are

$$v_{\text{noz}} = \frac{(0.25\ \text{L/s}) \times (10^3\ \text{cm}^3/\text{L})}{\pi(0.50\ \text{cm})^2} = 3.2 \times 10^2\ \text{cm/s} = 3.2\ \text{m/s},$$

and $v_{\text{hose}} = \frac{1}{4}v_{\text{noz}} = 0.87\ \text{m/s}$.

What Do You Think? If the nozzle were of a type that adjusts the diameter of the hose when twisted, should the nozzle diameter be reduced or increased to increase the distance reached by the water stream?

16–7 Bernoulli's Equation

In principle, the motion of a fluid can be understood by dividing the fluid into small elements and applying Newton's second law to each of them. Alternatively, providing that only conservative forces are present, we can use the conservation of energy. If there is no viscosity in a fluid, and hence no energy dissipation, and if the fluid is incompressible, it is straightforward to formulate the conservation of energy for fluids. This is a simple and important tool for the study of fluid motion. In the application here, we will use the conservation of energy in the form of the work–energy theorem and will end with an equation that relates speed, pressure, and height for an ideal fluid.

We again consider a streamtube—a bundle of streamlines of fluid—this time under the influence of gravity. Concentrate on the fluid forming a plug between two points along the tube: point 1, which has height h_1 and cross-sectional area A_1 and point 2, which is at height h_2 and has area A_2 (Fig. 16–21). The velocity perpendicular to the cross sections will have magnitudes v_1 and v_2, respectively, and we take our tube to be thin enough so that the velocity does not vary across the slices. The fluid to the left of the plug exerts pressure p_1 on the plug to the right, while the fluid to the right of the plug exerts pressure p_2 on the plug to the left. There will be, say, flow to the right if the force to the right, p_1A_1, is larger than the force to the left, p_2A_2.

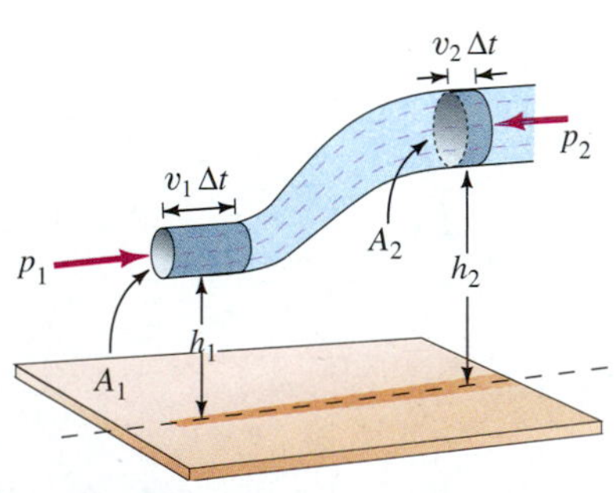

▲ **FIGURE 16–21** A streamtube of fluid at different heights. In time Δt, the fluid moves to the right under the influence of a difference in pressures between p_1 and p_2. In effect, the shaded slice in region 1 is transferred to the shaded slice in region 2.

Let's examine what happens during a time interval dt. The plug of fluid that starts out in the region between A_1 and A_2 at $t = 0$ will advance—in Fig. 16–21 we have supposed the movement is to the right. The distance by which it advances at point 1 is $v_1\,dt$, and the distance by which it advances at point 2 is $v_2\,dt$. We calculate the net work done in advancing the fluid in the tube by looking at the forces acting at either end. The work done by the force of the fluid to the left of point 1 is the force there times the displacement. The force is p_1A_1, so this work is

$$W_1 = (p_1A_1)(v_1dt). \tag{16–17}$$

The work done by the force of the fluid to the right of point 2 can be calculated similarly, but there is a minus sign because the force (and thus the pressure) acts to the left while the displacement is to the right, so that $\vec{F}\cdot\vec{s} = -Fs$. Thus

$$W_2 = -(p_2A_2)(v_2\,dt).$$

The net work done on the streamtube plug between regions 1 and 2 is then

$$W_{\text{net}} = W_1 + W_2 = (p_1 - p_2)vA\,dt, \tag{16–18}$$

where in the last line we have made use of Eq. (16–15), which expresses the constancy of the flux vA at any point along the tube: $vA = v_1A_1 = v_2A_2$.

We have now determined the amount of work done in moving the fluid in the plug during the time dt. To apply the work–energy relation, we next find the change in energy. If we examine Fig. 16–21, we can see that in the time interval dt, the effect of the motion is to transfer a small slice of fluid from region 1 to region 2. This transfer has two effects on the energy: The potential energy changes as the slice is lifted from h_1 to h_2, and the kinetic energy changes as the speed changes from v_1 to v_2. The change in potential energy is

$$\Delta U = mg\Delta h = (\rho vA\,dt)g(h_2 - h_1). \tag{16–19}$$

Again we have used the constancy of vA. The change in kinetic energy is

$$\Delta K = \frac{1}{2}(\rho vA\,dt)(v_2^2 - v_1^2). \tag{16–20}$$

Because $W_{\text{net}} = \Delta K + \Delta U$, we find, from Eqs. (16–18) through (16–20),

$$(p_1 - p_2)vA\,dt = \frac{1}{2}(\rho vA\,dt)(v_2^2 - v_1^2) + (\rho vA\,dt)g(h_2 - h_1).$$

After the factor $vA\,dt$ is divided out, we are left with

$$p_1 - p_2 = \frac{1}{2}\rho(v_2^2 - v_1^2) + \rho g(h_2 - h_1).$$

This equation can be rearranged into a conservation law, with one side dependent on the quantities defined in region 1 and the other side dependent on the quantities defined in region 2:

$$p_1 + \tfrac{1}{2}\rho v_1^2 + \rho gh_1 = p_2 + \tfrac{1}{2}\rho v_2^2 + \rho gh_2. \tag{16–21}$$

In other words,

$$\text{along a streamline: } p + \tfrac{1}{2}\rho v^2 + \rho gh = \text{a constant.} \tag{16–22}$$

BERNOULLI'S EQUATION

Our derivation deals with the motion of one particular streamline. But it can be shown in general that *for irrotational flow, the constant in Eq. (16–22) is the same for all streamlines* within a fluid. Equation (16–22) is **Bernoulli's equation**, derived in 1738 by Daniel Bernoulli, one of a remarkable family of scientists and mathematicians. In Section 16–8 we discuss applications of this formula.

Problem-Solving Techniques

Bernoulli's equation expresses conservation of energy for a fluid; as such, its use is similar to the use of all conservation laws. Keep in mind the following points:

1. When we say the right-hand side of Eq. (16–22) is a constant, we mean that it takes on the *same* value for all points within the fluid. If we know the value of $p + \frac{1}{2}\rho v^2 + \rho gh$ at one point, then that is the value it takes at all points.
2. In its typical application, we can find the value of the constant in Bernoulli's equation because we know the value of all the variables at some point. Then we can use the value of the constant to find the unknown value of a variable at another point.
3. When one of the three terms in the expression $p + \frac{1}{2}\rho v^2 + \rho gh$ is the same at all points, then we can ignore it and concentrate on using Bernoulli's equation to learn about the other variables. For example, in Section 16–8 we first apply Bernoulli's equation to the case of a fluid at rest. In that case, v is the same everywhere ($v = 0$), and we can use Bernoulli's equation as a conservation law for the quantity $p + \rho gh$.

16–8 Applications of Bernoulli's Equation

Bernoulli's equation is useful for describing a variety of phenomena. First, we observe that if a fluid is at rest, so that $v = 0$ in Eq. (16–22), then

$$p + \rho g h = \text{a constant.} \tag{16–23}$$

This is just the hydrostatic relation between pressure and height that we obtained in Eq. (16–6).

The Bernoulli Effect

For flow at constant height, the conservation of energy [Eq. (16–22)] implies that

$$p + \tfrac{1}{2}\rho v^2 = \text{a constant.} \tag{16–24}$$

Thus the pressure must be lower in a region in which a fluid is moving faster. You can observe this phenomenon—known as the **Bernoulli effect**—in a simple experiment. If you blow air across the top of a single sheet of paper (Fig. 16–22), the pressure above the paper is reduced. However, the pressure is unchanged below the paper; thus, the paper rises. The Bernoulli effect also enters into the operation of a chimney. When the wind blows over the top of a chimney, the pressure is decreased at the top by the Bernoulli effect, and the air in the tube rises. This is an important element in a chimney's "draw."

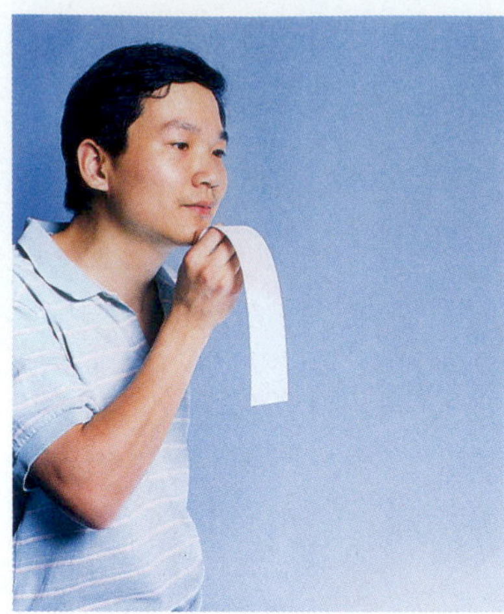

(a)

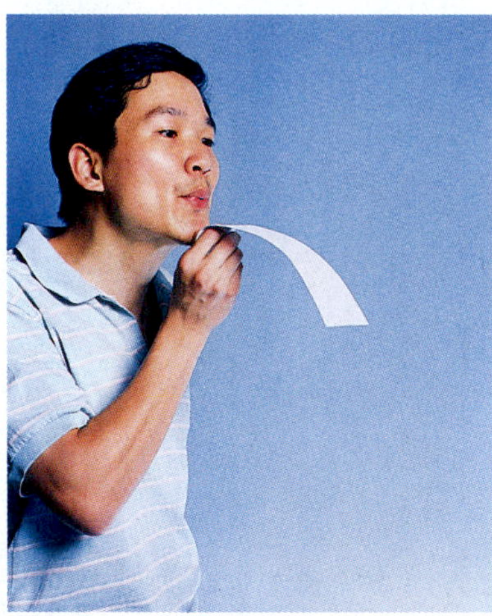

(b)

▲ **FIGURE 16–22** By blowing over the top of the paper, the pressure above the paper is reduced according to Bernoulli's principle. Below the paper the pressure is unchanged, and there is a net force that makes the paper rise.

THINK ABOUT THIS . . .

HOW IS FLUID SPEED IN A CLOSED PIPE MEASURED?

How does an engineer running a refinery know the speed of the fluids that pass through the pipes? Equation (16–24) is applied in the *Venturi flowmeter*, a device designed for just this purpose (Fig. 16–23a). Figure 16–23b is a simplified version of the flowmeter. The large pipe, with cross-sectional area A_1, is constricted smoothly to a smaller area A_2. A U-shaped tube containing a different fluid connects the two segments of pipe. The speeds in the larger and smaller sections of the pipe are v_1 and v_2, respectively; similarly, the pressures in the two sections are p_1 and p_2, respectively. Then Eq. (16–24) tells us that

$$p_1 + \frac{1}{2}\rho v_1^2 = p_2 + \frac{1}{2}\rho v_2^2;$$

or

$$p_1 - p_2 = \frac{1}{2}\rho(v_2^2 - v_1^2).$$

The continuity equation, Eq. (16–15), tells us that $v_2 = v_1 A_1/A_2$, so

$$p_1 - p_2 = \frac{1}{2}\rho v_1^2\left[\left(\frac{A_1}{A_2}\right)^2 - 1\right]. \tag{16–25}$$

The difference in pressures is determined from the difference in heights, h, of the two columns of fluid in the U-shaped tube (see Section 16–3). Equation (16–25) is then used to find speed v_1 in terms of this pressure difference.

(a) (b)

▲ **FIGURE 16–23** (a) The Venturi flowmeter. (b) The speed v_1 of a fluid in a pipe is measured by the difference in heights, h, of column 1 and column 2 of a different fluid. The number of measuring tubes that tap the pipe through which the fluid flows can be varied according to the measurement desired.

EXAMPLE 16–10 A giraffe needs a strong heart because of its long neck. Suppose that the difference in height between the aortic valve (the place where arterial blood comes out of the heart) and the head of a giraffe is 2.50 m, and that the artery leading from near the aortic valve to the head has constant cross section all the way to the head. What is the minimum pressure at the aortic valve? Blood is an incompressible fluid with density $\rho = 1.0\ \text{g/cm}^3$.

Setting It Up See Fig. 16–24 for a sketch. We will use Bernoulli's equation with constant speed in order to relate pressure at the heart to pressure at the head and the height change, and for that we will need to know the minimum value of pressure at the giraffe's head. For the survival of the giraffe this value must be more than $p = 0$ and we would have to ask a veterinarian what the true minimum at the head is. For our purposes we'll just assume that $p = 0$ at the head is the smallest physical value allowed.

Strategy Because blood is incompressible and the arterial area is constant, the equation of continuity (flux conservation) implies that the speed of the blood will be the same at the head as at the heart. Bernoulli's equation in the form of Eq. (16–23) (the form appropriate to constant speed) is a relation between the heart height and pressure and the head height and pressure. Only the heart pressure is unknown in this expression, and we can solve for it.

Working It Out Let's use subscripts 1 and 2 for the heart and head, respectively. Then, with $p_2 = 0$, Eq. (16–23) reads $p_1 + \rho g h_1 = \rho g h_2$, which can be solved for p_1:

$$p_1 = \rho g(h_2 - h_1).$$

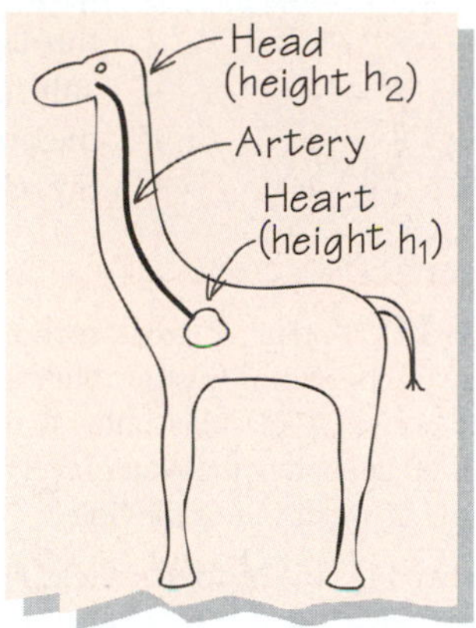

▲ **FIGURE 16–24** A giraffe's long neck requires a strong heart for arterial blood flow.

Numerically, $p_1 = (1.0\ \text{g/cm}^3)(10^2\ \text{cm/m})^3(10^{-3}\ \text{kg/g})(9.80\ \text{m/s}^2)(2.50\ \text{m}) = 2.5 \times 10^4\ \text{Pa}$. Various factors, in particular the loss of energy due to friction between the blood and the arterial walls and the desire to have a larger value than 0 for the head pressure, make this a low estimate. Comparing this to the *peak* output pressure of the human heart, 1.6×10^4 Pa, we can see that the giraffe's long neck requires a strong heart.

What Do You Think? What would be the effect on the giraffe if the artery diameter narrowed down as it approached the brain?

Fluid Motion with Constant Speed

In the Bernoulli effect, height is constant but the speed of the flowing fluid is not. But we can easily imagine reversing this and confronting constant speed but changing height. To fix the speed, we need only confine our moving fluid to a pipe of unchanging cross section. If the fluid is incompressible, then the equation of continuity, Eq. (16–15), shows that the speed cannot change. However, the pipe could perfectly well move from one height to another, and in this situation the pressure will change. For this case Bernoulli's equation is $p + \rho g h =$ a constant, identical to the static case expressed by Eq. (16–23) (see Example 16-10).

Flow from a Tank

Another interesting phenomenon of practical importance in fluid flow is the speed of flow of a liquid from an opening in a tank (Fig. 16–25a). If we make a small hole in the side of an open tank of water at a depth h below the surface, we can apply Bernoulli's equation, Eq. (16–22), to the fluid at different locations in the tank. We denote the height above the hole by the variable y, so that the height at the water surface is $y = h$ and at the opening $y = 0$. Equation (16–22) states that the value of $p + \frac{1}{2}\rho v^2 + \rho g y$ is the same at point 1 (on the upper surface, where the pressure $p = p_0$) and point 2 (just outside the hole, where $p = p_0$ because the sides of the stream of fluid flowing from the tank are in equilibrium with the atmosphere). The speed v_2 at point 2 is to be determined. We equate the values of $p + \frac{1}{2}\rho v^2 + \rho g y$ at the two points and obtain

$$p_0 + \tfrac{1}{2}\rho v_1^2 + \rho g h = p_0 + \tfrac{1}{2}\rho v_2^2 + \rho g \times 0.$$

We can neglect the small velocity v_1 of the liquid at the top of the tank. We solve for the speed at point 2 (the hole) to find that

$$v_2 = \sqrt{2gh}, \tag{16–26}$$

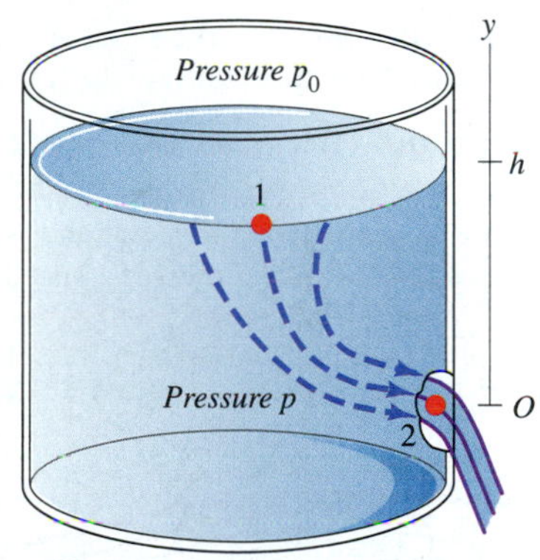

(a)

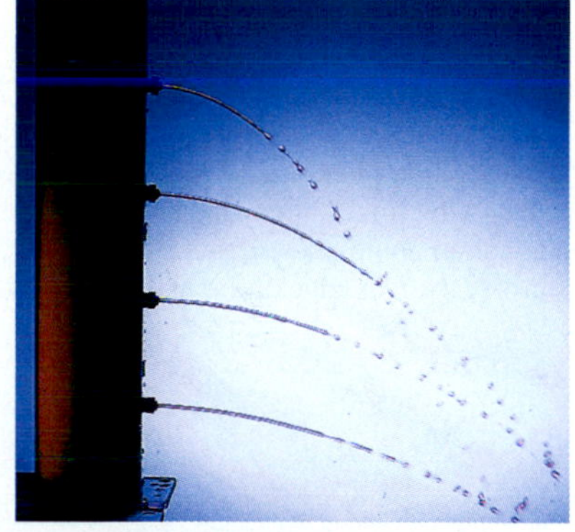

(b)

▶ **FIGURE 16–25** The escape of fluid through a hole in a tank. (a) The curved lines are streamlines for real fluid flow. (b) The pressure at the bottom of the tank is greater than the pressure at the top, so a stream of water leaving the tank has greater speed if it leaves from a hole that is lower on the tank.

as is consistent with Fig. 16–25b. This is just the speed of a single particle falling through a height h under the influence of gravity, as in Eq. (2–27). We find the same result for the falling liquid as for a falling object because they both are consequences of the conservation of energy. To find the rate of fluid flow from the hole,† we use the expression for the flux, Eq. (16–15), namely $\Phi = vA$.

EXAMPLE 16–11 A tank of cross-sectional area $0.07\ m^2$ is filled with water. A tightly fitting piston, with a total mass 10 kg, rests on top of the water. A circular hole of diameter 1.5 cm is opened at a depth of 60 cm below the water level of the tank. What is the initial rate of flow of water out of the hole?

Setting It Up See Fig. 16–26 for a sketch. The sketch labels the tank area A, the hole area a, the piston mass m, and the hole depth h; all three quantities are given in the question. We'll also need to assume that the atmospheric pressure p_0 does not change from the top of the tank to the place where the hole is cut.

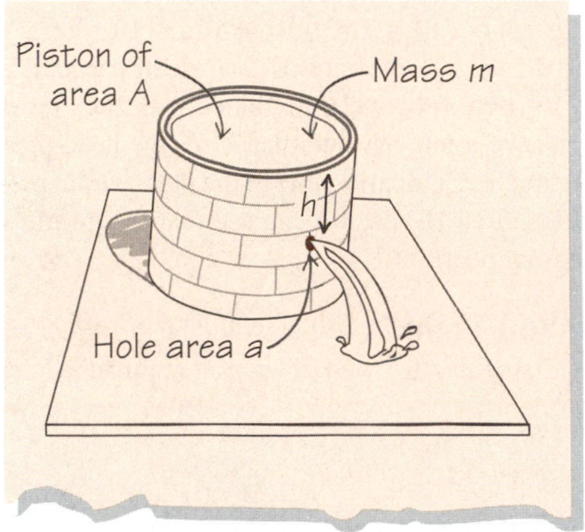

▲ **FIGURE 16–26** The speed of the water flowing out of the small hole depends on the height h of the water above it.

Strategy This is a typical situation where Bernoulli's equation, which relates pressure, speed, and height, is useful: We know some of these quantities at one point and want to learn about them at another point. Here, the pressure, speed, and height are known at the top, as are the pressure and height at the hole. We can then use Bernoulli's equation to learn about the speed at the hole. Then given the water speed v, we can find the flow rate va.

Working It Out At surface level, the total pressure $P = p_0 + p_1$, where p_0 is the atmospheric pressure and p_1 is the pressure due to the piston resting on the water, $p_1 = mg/A$. The water at the surface, which we set at height h, is essentially stationary. At the hole, which here is at height 0, the pressure just outside the hole is p_0, as in the discussion above. An application of Bernoulli's equation at the top of the water and at the opening leads to the equality

$$p_0 + p_1 + \rho g h = p_0 + \tfrac{1}{2}\rho v^2.$$

We cancel p_0 from both sides of the equation and solve for v^2:

$$v^2 = \frac{2(p_1 + \rho g h)}{\rho} = \frac{2\left(\dfrac{mg}{A} + \rho g h\right)}{\rho} = 2g\left(\frac{m}{\rho A} + h\right).$$

In turn, the flow rate is $\Phi = va = v\pi(d/2)^2$, where d is the hole diameter. If $\rho = 10^3\ kg/m^3$, we find

$$v^2 = 2(9.8\ m/s^2)\left[\frac{10\ kg}{(10^3\ kg/m^3)(0.07\ m^2)} + 0.6\ m\right] = 15\ m^2/s^2,$$

or $v \cong 4\ m/s$, and $\Phi = (4\ m/s)\pi(7.5 \times 10^{-3}\ m)^2 = 7 \times 10^{-4}\ m^3/s = 0.7\ L/s$.

What Do You Think? Let's assume that we want the water coming out of the hole to land as far away as possible. How would you determine where to put the hole? Discuss, qualitatively, how far the stream would land according to whether the hole is near the top or bottom of the tank.

*Lift

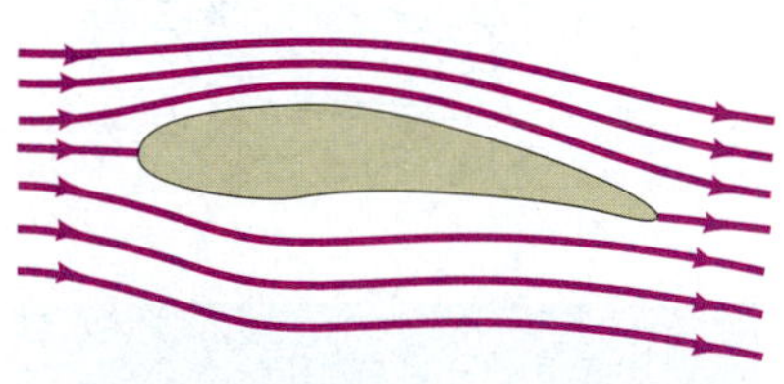

▲ **FIGURE 16–27** Schematic version of streamlines over an airplane wing section. There is a streamline that ends on the front of the wing at the stagnation point and a streamline that starts at the rear edge of the wing.

The upward force that keeps a heavy airplane in the sky is an example of **lift**. The Bernoulli effect—the fact that when the speed of a fluid increases, pressure in the fluid decreases—helps us to understand the origin of lift. However, there are some subtleties, and circulation, viscosity, and turbulence all play an important role in lift. Thus it is necessary to relax some of the simplifying assumptions listed at the end of Section 16–5 to provide a correct description of this phenomenon. Consider the cross section of a wing shown in Fig. 16–27. We visualize the situation from the reference frame of the wing, so that we think of the air moving toward the wing, which is at rest, rather than the wing moving through stationary air. (This is, in fact, just what happens in a wind tunnel in which problems such as this one are studied.) Air approaches the wing with some speed v along a set of streamlines. Some of the streamlines go below the wing, some go above. The wing is designed so that the air speed above the wing is greater than that below the wing, which can happen only if there is circulation in the clockwise direction‡. Circulation occurs only around the wing.

†A more careful treatment replaces the rate of flow with the expression kvA, where the coefficient k is less than 1. Its correct value for a simple small hole is 0.62. In the examples and problems, we use $k = 1$ for simplicity.

‡No air actually moves forward under the wing; rather, the speed beneath the wing is less than the speed above the wing.

The flow elsewhere is irrotational, which means that the constant appearing in the Bernoulli equation, Eq. (16–22), is the same for all the streamlines that flow near the wing, below or above. From Eq. (16–24) we see that the pressure on top of the wing, p_t, must therefore be less than the pressure below, p_b, and there is a net upward force, or lift, on the wing. The net upward pressure, $p_{net} = p_b - p_t$, is determined from Eq. (16–24):

$$p_b + \tfrac{1}{2}\rho v_b^2 = p_t + \tfrac{1}{2}\rho v_t^2,$$

where here ρ is the air density. We thus find that

$$p_{net} = p_b - p_t = \frac{1}{2}\rho(v_t^2 - v_b^2) = \rho\left(\frac{v_t + v_b}{2}\right)(v_t - v_b) \cong \rho v(v_t - v_b), \quad (16\text{–}27)$$

where v is the average of the top and bottom speeds. Because the difference between the top speed and the bottom speed is small, both are nearly equal to the average speed, which we call the *airstream speed* v.

The expression in Equation (16–27) can be used to calculate the lift. If the wings have a width w and a tip-to-tip span S, then their total area is wS (we are ignoring the fuselage), and the lift is given by

$$L = wSp_{net} = wS(p_b - p_t) \cong wS\rho v(v_t - v_b).$$

Experiments show that for speeds much slower than the speed of sound, $v_t - v_b$ is proportional to the airstream speed v, with a proportionality constant K (in general, $K \ll 1$), which depends on the wing design and the wing angle relative to the airplane direction. Thus we can finally write the lift as

$$L \cong wS\rho Kv^2. \quad (16\text{–}28)$$

A typical value of the constant K is 0.12, so for a small plane with a total wing area of 30 m^2 and an air speed of 200 km/h = 56 m/s and with air density of about 1.3 kg/m^3, the lift is given by

$$L = (0.12)(30\ m^2)(1.3\ kg/m^3)(56\ m/s)^2 = 1.6 \times 10^4\ N.$$

Thus the mass that can be lifted is $(1.6 \times 10^4\ N)/(9.8\ m/s^2) \cong 1600$ kg. The fact that air is compressible plays little role if the relative speed of the wing and the air is much smaller than the speed of sound in air, $v_s \cong 330$ m/s. (The design and operation of the wings of supersonic aircraft presents a different set of problems.)

To give the flow around the wing a nonzero circulation, it is necessary to have some turbulence. The turbulence is induced by the viscosity of the air and further enhanced by the placement of small protuberances (ailerons) on the wings.

THINK ABOUT THIS . . .

WHY DO RACING CARS HAVE WINGS?

In contrast to airplanes, racing cars are designed so that the air beneath the car moves faster than the air above it, and a downward pressure is exerted on the car. This downward pressure is produced or augmented by both front and rear wings, sometimes called *spoilers*. These act like upside-down airplane wings, producing a reverse lift that pushes the car vertically down onto the road. The increased contact force with the road is very useful to a racing car, because static friction between the wheels and the track is proportional to the contact force. Increased friction allows more torque to be supplied to the wheels without them spinning, and in this way larger accelerations and speeds are possible. Moreover, the increased friction allows better steering control.

*16–9 Real Fluids

Our discussion has mostly been limited to idealized kinds of flow of idealized fluids. Some of our assumptions apply only approximately, at best, to real fluids. Nonviscous flow, for example, has been called the flow of "dry water." Here we give some qualitative notions of how real fluids behave under ordinary circumstances.

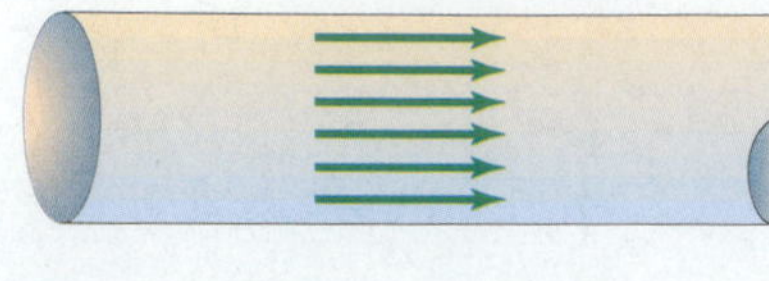

(a)

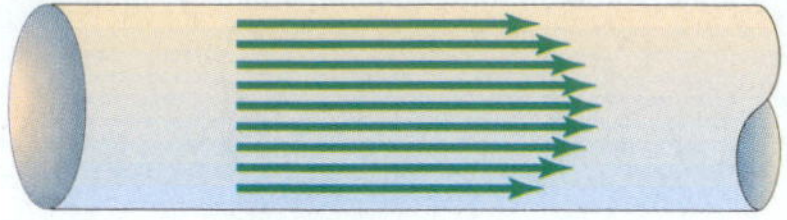

(b)

▲ **FIGURE 16–28** Velocity profiles for laminar flow in a pipe: (a) idealized fluid flow that does not take into account viscosity; (b) realistic fluid flow (Poiseuille flow).

Viscosity

Figure 16–28a is a picture of the velocity profile of pipe flow for fluids idealized as we have described. For real fluids the liquid near the surface of the pipe is at rest, and the velocity profile is actually parabolic (Figure 16–28b). Such flow is referred to as **Poiseuille flow**, after Jean Poiseuille, who studied blood circulation. The reason for this effect is the internal friction of the fluid, or *viscosity*. Viscosity comes from intermolecular collisions. The molecules of a fluid have a random motion, to be described in Chapter 19. If two adjacent layers of fluid flow with slightly different speeds, the random sidewise intrusion of some slower molecules into the faster stream will tend to slow down the faster stream, whereas intrusion of faster molecules into the slower stream will tend to speed up the slower stream.

Viscous flow can be analyzed by taking two glass plates with a film of fluid between them and sliding one plate over the other. The easier it is to slide the plates, the less viscous is the fluid between them. To quantify this effect, take the smaller plate to have an area A; it is separated from the other plate by a distance y (Fig. 16–29). If the upper plate is to move with a uniform velocity $\vec{v}$, a force must be applied—just as if a brick is made to slide along a rough surface at constant speed. This force, F, is proportional to A and inversely proportional to y. In addition, the force increases linearly with speed (like the damping force on a mass at the end of a spring or like air drag). A formula that reflects these factors will have a proportionality constant η that will differ in different materials,

$$F = \frac{\eta v A}{y}. \qquad (16\text{–}29)$$

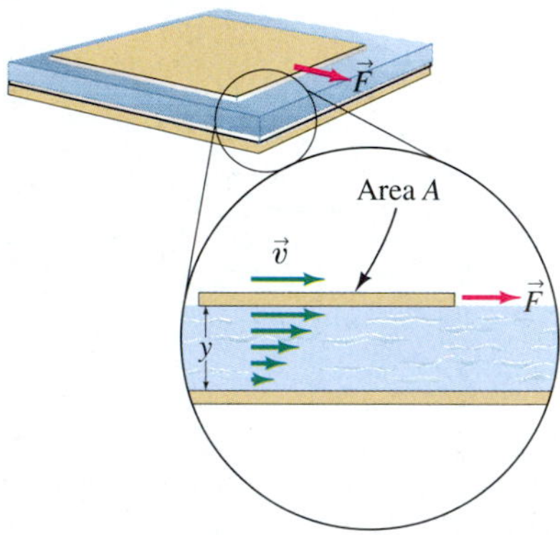

▲ **FIGURE 16–29** Two glass plates have a film of fluid between them; one slides against the other. The top plate moves with speed v. Fluid near the top plate moves with the same speed as that plate, while fluid near the bottom plate is at rest.

The coefficient η is the **coefficient of viscosity**. It is measured in newton seconds per square meter or equivalently kilograms per meter second. Typical values are $10^{-3}\ \text{kg}/(\text{m}\cdot\text{s})$ for water and $830 \times 10^{-3}\ \text{kg}/(\text{m}\cdot\text{s})$ for glycerin, both at 20°C. We mention the temperature because, as anyone who has poured syrup over pancakes knows, viscosity is strongly temperature-dependent.

Turbulence

A fluid flow marked by complex and chaotic movement is referred to as turbulent. At the onset of turbulence, the whole notion of neatly separated streamlines ceases to make sense (Fig. 16–30). A dimensionless parameter known as the **Reynolds number**, *Re*,

▶ **FIGURE 16–30** Turbulence, an important feature of fluid flow, is seen in the steam and smoke from the eruption of Mount Augustine in Alaska.

determines when turbulent flow occurs, and this parameter contains the fluid speed. It is defined as

$$Re = vL\left(\frac{\rho}{\eta}\right), \tag{16–30}$$

where ρ is the density of the fluid, v is its speed, L is a length associated with the flow, such as the pipe's diameter for flow in a pipe, and η is the viscosity coefficent. When the Reynolds number increases past about 2000 to 3000, the flow becomes turbulent. For water flowing through a pipe 1 cm in diameter, $Re \cong (10^4\ \text{s/m})v$ so the flow becomes turbulent when v reaches a mere 0.3 m/s. Turbulence is very common, but fortunately a small amount of turbulence does not change the applicability of Bernoulli's equation for short distances of flow (just as a small amount of friction does not disturb energy conservation over short times), so Eq. (16–22), although not exact, is still a very good and useful approximation.

THINK ABOUT THIS . . .

OF WHAT USE ARE THE DIMPLES ON A GOLF BALL?

When an object such as a golf ball moves through the air, there is drag on it. We can identify two different sources of drag. There is drag associated with turbulence, in which energy is lost in churning the air. There is also a type of drag known as pressure drag associated with laminar (smooth) flow, in which to satisfy continuity equations the air flow leaves a region *behind* the moving object where there is a lower pressure than the pressure before it. Pressure drag can be more important than turbulent drag at high speeds. The dimples on a ball increase the turbulence around the ball, and while this means more turbulent drag, it has a second important effect: it decreases the size of the region with reduced pressure behind the ball and thereby decreases the pressure drag. Experiment shows that this is a winning strategy: a dimpled ball will travel farther. Moreover, the increased turbulence due to the dimples can also increase a lift associated with the spin of the golf ball. The dimples may look cute, but they also have a solid positive effect on the flight of the ball.

Summary

Liquids are distinguished from gases in that the molecules that form liquids are closer together and impose strong forces on each other. Liquids are distinguished from solids in that the molecules in liquids do not form a rigid lattice. Both liquids and gases deform in response to external forces and therefore flow as fluids. The density of a material of mass M and volume V is unchanging in incompressible fluids:

$$\rho = \frac{M}{V}. \tag{16–1}$$

The fluids we study in this chapter are for the most part incompressible.

Pressure, measured in SI in pascals, expresses the way that fluids transmit forces. The pressure in a fluid is the (pushing) force per unit area on any area A:

$$p \equiv \frac{F}{A}; \tag{16–2}$$

and is perpendicular to A. When a fluid is at rest, the pressure at any given point is the same in all directions. For an incompressible fluid in Earth's gravity, the pressure increases with the depth y according to

$$p = p_0 + \rho g y, \tag{16–6}$$

where p_0 is the pressure at $y = 0$. A change in pressure applied to a fluid at rest is transmitted without change to every part of the fluid, a statement known as Pascal's principle. This principle is the basis of hydraulic lifts and brakes.

For an object immersed in a fluid, there is an upward buoyant force equal to the weight of the displaced fluid,

$$F_{\text{buoy}} = \rho g V_{\text{submerged}}. \tag{16–11}$$

This is Archimedes' principle.

Nonviscous (friction-free) fluids in smooth (nonturbulent) motion follow paths traced out by streamlines. The conservation of mass for a fluid moving in this way is stated as the continuity equation,

$$\rho_1 v_1 A_1 = \rho_2 v_2 A_2. \tag{16–14}$$

Here ρ and v are the density and speed at some point in a tube of streamlines; A is the corresponding cross section of the tube. If the fluid is incompressible, the (constant) density cancels, and the resulting expression is the conservation of flow rate, or flux:

$$v_1 A_1 = v_2 A_2. \tag{16–15}$$

The conservation of energy for a moving fluid under the influence of gravity is Bernoulli's equation:

$$\text{along a streamline: } p + \tfrac{1}{2}\rho v^2 + \rho g h = \text{a constant}. \tag{16–22}$$

When the fluid is at rest, this reduces to Eq. (16–6); when the fluid flow is horizontal, this reduces to

$$p + \tfrac{1}{2}\rho v^2 = \text{a constant}. \tag{16–24}$$

This equation is the basis of the Venturi flowmeter, which measures the speed of fluids in pipes, as well as the basis for understanding the flow of fluids from holes and the lift on wings.

It is possible to go quite far in understanding fluids with a number of simplifications. The relaxation of the simplifying assumptions brings in rotational flow, viscosity, and turbulence. The behavior of real fluids remains a field of active study.

Understanding the Concepts

1. How does a hot-air balloon get off the ground?
2. We weigh a brick twice with a bathroom scale. The first time, the scale and the brick are inside a tub of water, and the second time, they are outside the tub. Water can enter the scale. Does the scale indicate a difference in weight in the two cases?
3. One accurate method of measuring the fat content of your body is to measure your weight twice: once while you are immersed in a tank of water, and once when you are out of the water. Explain how such a method might work.
4. The most exciting white water for canoeing often occurs when the river narrows. Why is this?
5. If an empty balloon is weighed on a scale, a certain value is obtained. If a child now blows up the balloon and waits for several minutes so that the temperature of the air in the balloon is the same as the room temperature, what value will the scale give for the balloon's weight?
6. Suppose that a certain fluid has no viscosity. Will the acceleration of objects falling under gravity within this fluid be independent of the mass of the objects?
7. You have two objects in a bathtub—one floating high up, the other just barely floating. Does the buoyant force cancel gravity in each of these cases?
8. Water-storage tanks for communities normally are placed high off the ground rather than underground. This is obviously a safety hazard, so why is it done?
9. Mercury is poisonous. Why, then, is it typically used in barometers to measure atmospheric pressure?
10. The "antilift" effect for a racing car, the opposite of the lift effect on airplanes, is said to be so strong that a car could race on an upside-down track. Is this a plausible statement? Make some rough estimates of the surface area of a racing car, its mass, and its velocity, and assume that the difference between the air speeds under the car and over the car is about 20 m/s.
11. In Example 16–10 we suggested that the minimum pressure at the giraffe's head must be $p = 0$. But why couldn't the minimum value of pressure be negative? Comment on the signs of the three terms in Bernoulli's equation, Eq. 16–21.
12. One way to get the water out of a stopped-up sink is to use a long piece of rubber tubing. You put one end in the water and suck on the other end until you have filled the tubing. You then close off your end, bring that end to a point below the bottom of the sink, and reopen that end (Fig. 16–31). The water will flow out through the tubing until the sink is empty. You have made use of the *siphon mechanism* (see Problem 73). Explain how it works.

▲ **FIGURE 16–31** Questions 12, 13, and Problem 73.

13. The siphon shown in Fig. 16–31 and described in Question 12 is limited in the height h_0 that it can overcome. What determines this limit?
14. Suppose mercury were less dense than it is. Would the height of a barometer column be (a) taller or (b) shorter?
15. Do sailboats make use of any of the aspects of Bernoulli's laws, such as those we described for airplanes? In particular, do sails need to be curved in order to work?
16. A common classroom demonstration employs a table tennis ball in a funnel. An air hose is connected to the tube of the funnel and air is blown through. If the funnel is held upside down, the ball does not fall out. Why?

17. Which weighs more, a ton of Styrofoam or a ton of lead? Which of the two has the larger volume? How might you estimate the density of each?
18. People find it very easy to float in the Dead Sea. Why?
19. The point of a cone-shaped bottle (pointed end up) is removed to provide an opening. The pressure of the liquid at the very bottom of the bottle is the sum of the atmospheric pressure and the weight of the liquid above. If we consider a part of the bottom that is not directly below the opening, we might argue that there is less pressure because, above that point, there is less liquid and no atmosphere. This argument violates Pascal's principle. What has the argument left out?
20. In Example 16–11 we described water flowing from a hole cut in a tank. What happens to the flow with the passage of time? In particular, is the rate of flow constant? Is the distance from the tank to the landing point of the stream constant?
21. Suppose that you put a mixture of oil and water into a centrifuge that rotates at high speeds (the oil is less dense than the water and does not mix with it). The two liquids separate. Why, and which component will be farthest from the center of the centrifuge?
22. An old proposal for transcontinental travel along a fixed latitude is the following: Take a balloon high into the stratosphere. Wait until Earth has rotated the desired distance beneath you, then descend. Comment on the feasibility of this mode of transport.
23. A snorkel is a breathing tube meant to be used while you swim under water; it runs from your mouth to the water surface. Why is the length of a snorkel limited?
24. A pressure of 8 kPa across the eardrum is enough to cause serious pain, and 17 kPa will rupture the eardrum. What fraction of an atmosphere is this? Given your answer, why is it that merely stepping outside is not dangerous?
25. In the course of a lazy summer morning, ice cubes floating in a pitcher of water melt. What happens to the water level in the pitcher? (Ignore evaporation.)
26. Suppose that you put a kilogram weight and a jar half-filled with water on a kitchen scale. The scale reads 2.5 kg. Now you place the kilogram weight *inside* the jar, and no water spills out. You might argue that because the kilogram weight weighs less in water than in air, the scale would read less than 2.5 kg, but the reading remains 2.5 kg. What is wrong with your argument?

Problems

16–2 Density and Pressure

1. (I) Jupiter has a radius $R = 7.14 \times 10^4$ km, and the acceleration due to gravity at the surface is $g_J = 22.9\ \mathrm{m/s^2}$. Use these data to calculate Jupiter's average density.
2. (I) The numbers given in Problem 1 have the following values for some of the other planets: Venus: $R = 6.05 \times 10^6$ m, $g = 8.86\ \mathrm{m/s^2}$; Mars: $R = 3.39 \times 10^6$ m, $g = 3.73\ \mathrm{m/s^2}$; Uranus: $R = 2.54 \times 10^7$ m, $g = 9.12\ \mathrm{m/s^2}$; Neptune: $R = 2.48 \times 10^7$ m, $g = 12.0\ \mathrm{m/s^2}$. Calculate the densities of these bodies.
3. (I) The density of a nucleus is about $2 \times 10^{17}\ \mathrm{kg/m^3}$. The amount of water in a large lake is $10^{13}\ \mathrm{m^3}$. If this amount of water were compressed to nuclear density, how many liters of water would there be?
4. (I) A platinum sphere has a diameter of 0.96 cm. What is the diameter of an aluminum sphere of the same mass, given that the densities of platinum and aluminum are $21.4 \times 10^3\ \mathrm{kg/m^3}$ and $2.70 \times 10^3\ \mathrm{kg/m^3}$, respectively?
5. (I) Atoms of an imaginary two-dimensional solid are arranged in a square lattice. The distance of neighboring atoms is a. Assume that the solid is acted upon by a shear force and deformed by a *small* angle f without any change in the nearest neighbor distance (Fig. 16–32). Calculate the change in the distance between atoms situated at opposite corners of a square.

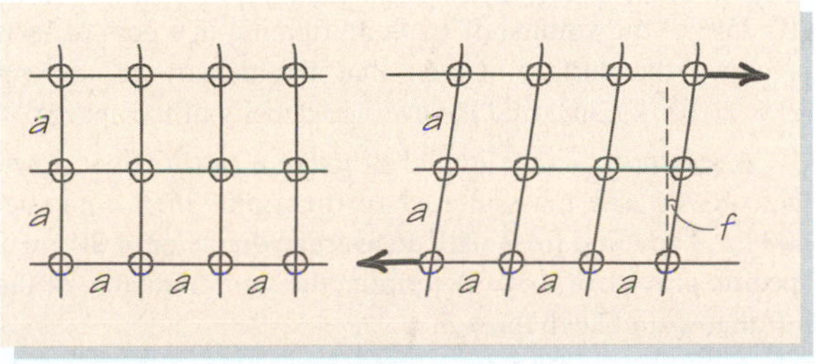

▲ **FIGURE 16–32** Problem 5.

6. (I) The density of mercury is 13.6 times that of water. Compare the height of a column of water to that of a column of mercury, assuming that the pressure exerted by the weights of the respective liquids at the bottom of each column are the same.
7. (I) A circus clown stands on a pair of stilts that each have a square cross section of 4.0 cm per side. If the mass of the clown plus the stilts is 68 kg, what pressure is exerted on the floor?
8. (II) A wedge of mass 15 N and opening angle $\theta = 55°$ is floating on the surface of water in the symmetrical position shown in Fig. 16–33. Calculate the force of the water that acts on each of the two surfaces.

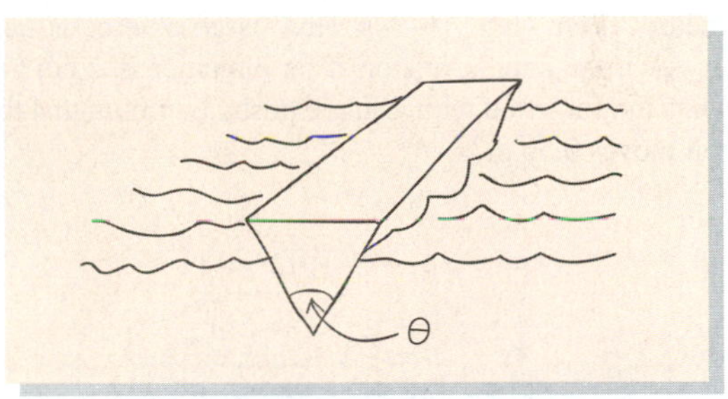

▲ **FIGURE 16–33** Problem 8.

9. (II) A hollow stainless steel sphere of radius 20 cm is evacuated so that there is a vacuum inside. (a) What is the sum of the magnitudes of the forces that act to compress the sphere? (b) There is a circular hole of diameter 4 cm on the side of the sphere for access to the inside. Calculate the force needed to pull a flat plate off the hole when the sphere is evacuated. Do you think that you could remove such a plate by pulling on it?
10. (II) During a hurricane, the atmospheric pressure changes dramatically. Explain why it is recommended that house windows be kept slightly open during a hurricane. What is the net force on a wall that is 300 $\mathrm{ft^2}$ in area when the pressure on one side is 14.7 $\mathrm{lb/in^2}$ and the pressure on the other is 14.0 $\mathrm{lb/in^2}$?

11. (II) A hollow metal pyramid with a square base that is 15 cm on each side has a mass of 1.8 kg. What is the pressure exerted by the pyramid on the table on which it stands? Suppose that the temperature in the room rises so that the metal expands. Will the pressure increase or decrease as a result of the expansion?

16–3 Pressure in a Fluid at Rest

12. (I) Using the densities in Table 16–1, what is the height of a fluid column in a barometer that has water as its fluid, in an air pressure of 1 atm? One that has alcohol? (Neglect the vapor pressure of the fluid above the column.)
13. (I) What is the pressure at depths of 1 m, 10 m, 100 m, and 10 km under the surface of the ocean? Take $\rho = 1.03 \times 10^3\ \mathrm{kg/m^3}$ for the density of sea water and $p_0 = 1.01 \times 10^5$ Pa for the atmospheric pressure at the ocean surface. (Ignore the fact that, to this accuracy, density varies with depth.)
14. (I) Recreational scuba divers rarely go deeper than 20 m. What is the water pressure at this depth? By what factor does your result differ from sea-level water pressure?
15. (I) What is the pressure 1 cm below the surface of a column of mercury exposed to air? 3 cm below the surface?
16. (I) An oceanic research vehicle operates at a depth of 1.20 km under water. Assuming that the pressure inside the vehicle is 1.0 atm, what is the force on a small window of dimensions 10 cm × 10 cm?
17. (II) Assume that a spaceship lands on Venus with the cabin pressure maintained at 1 atm. What would be the height of a column of mercury in a barometer in the spaceship on Venus? Repeat for a spaceship landing on Neptune. (Use the data given in Problem 2.)
18. (II) Tests in pressure test chambers show that mental processes are affected when the gauge pressure exceeds $35\ \mathrm{lb/in^2}$. *Gauge pressure* is the pressure in excess of atmospheric pressure. This is a well-known effect that limits the depth at which scuba divers should breathe pure air. In sea water, whose density is $1.03\ \mathrm{g/cm^3}$, to what depth should a diver be limited?
19. (II) A hydraulic jack is used to lift a 1200-kg car on a piston of diameter 30 cm (Fig. 16–34). How large a force is needed to push down the smaller piston if its diameter is 2 cm? By how much is the car lifted with a single push, assuming that the small piston moves 0.50 m?

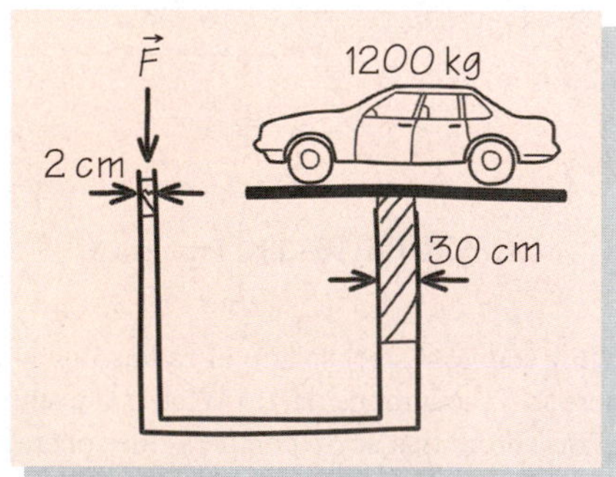

▲ **FIGURE 16–34** Problem 19.

20. (III) The vertical wall of a small dam is 32 m long and it holds back a body of water that is 5 m deep (Fig. 16–35). Calculate the force acting on the wall due to the pressure of the water. Determine the effective point of action of this force. [*Hint*: Recall the method used to calculate the center of mass of a nonuniform object in Chapter 8.]

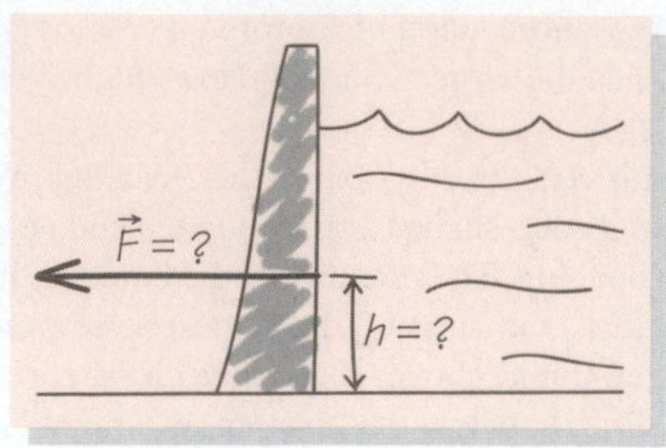

▲ **FIGURE 16–35** Problem 20.

16–4 Buoyancy and Archimedes' Principle

21. (I) A bathysphere (a spherical, watertight research facility) 2.6 m in diameter has a mass of 9400 kg. It is released from a submarine at a depth of 20 m below the surface. Will the bathysphere sink or rise?
22. (I) Use the densities in Table 16–1 to show that ice floats in water. What fraction of the volume of a floating ice cube will be above the surface?
23. (I) A cubical box, contents unknown, floats in water; 16 percent of the volume of the box is above water. What is the average density of the box and its contents?
24. (I) Over 10,000 icebergs are calved from Greenland each year. These icebergs are a serious danger to Atlantic shipping because most of the iceberg remains under water, invisible. If 10% of the iceberg sticks out of the water, what is an iceberg's density?
25. (II) A supertanker can carry $2.2 \times 10^5\ \mathrm{m^3}$ of oil with a density of $0.88\ \mathrm{g/cm^3}$. When fully loaded its mass is 2.3×10^8 kg, and the dimensions of its hull are approximately 400 m long, 60 m wide, and 38 m high. Given that the density of seawater is $1.03\ \mathrm{g/cm^3}$, how deeply is the hull submerged in the water?
26. (II) A hemispherically shaped bowl is floating on the surface of water; the mass of the bowl is 0.6 kg (Fig. 16–36). Water is poured into the bowl; when $4.3 \times 10^3\ \mathrm{cm^3}$ of water has been added, the bowl just sinks. What is the outer radius of the bowl?

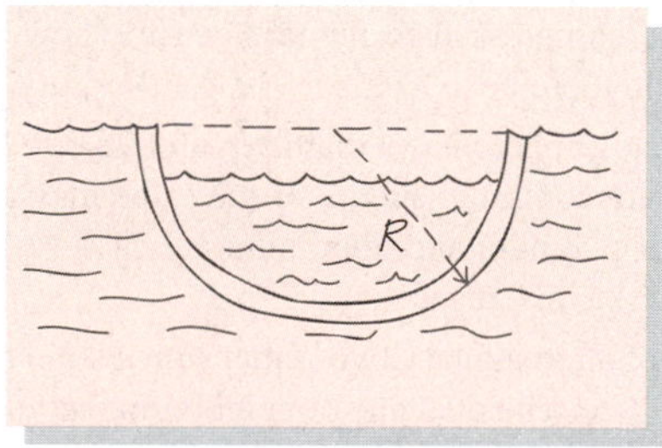

▲ **FIGURE 16–36** Problem 26.

27. (II) Find the volume of the smallest block of ice which when floating on salt water will carry a person of mass 80 kg. Assume that the density of salt water is $1.03 \times 10^3\ \mathrm{kg/m^3}$. You will have to look up the density of ice.
28. (II) 75% of the volume of a sphere floating in a pool of mercury is above the surface. Given that the density of mercury is $13.6 \times 10^3\ \mathrm{kg/m^3}$, find the average density of the sphere.
29. (II) A group of Scouts intend to make a raft to float down the Ohio River. The mass of four of them plus their equipment is 400 kg. They find trees with an average diameter of 20 cm and a specific gravity of 0.80. Determine the minimum area of the log raft that would keep them dry.
30. (II) A temporary bridge is supported on floating empty drums that are 1.0 m in diameter and 1.8 m long. The design criterion is

that the drums may only be submerged to 2/3 of their diameter under a maximum load of 5000-kg trucks, 7.5 m long, that follow each other bumper to bumper in a single line. How far apart should the drums be placed? Neglect the weight of the bridge and of the drums.

31. (II) A valuable preserved biological specimen is weighed by suspending it from a spring scale. It weighs 0.45 N when it is suspended in air and 0.081 N when it is suspended in a bottle of alcohol. What is its density?

32. (II) Consider a spherical balloon filled with helium, with a density of 0.18 kg/m^3. The density of air is 1.3 kg/m^3. What must the radius of the balloon be in order to lift a load of 130 kg (including the mass of the balloon)?

33. (II) Olive oil floats on water. Take ρ_1 to be the density of the oil and ρ_2 to be the density of the water. Consider an oil–water interface across which a bouillon cube of density ρ_3 floats; use the geometry shown in Fig. 16–37. We know that for the oil to float on the water, ρ_1 must be less than ρ_2. What is the condition on ρ_3 so that the cube floats at the oil–water interface? How much of the cube will be in the water?

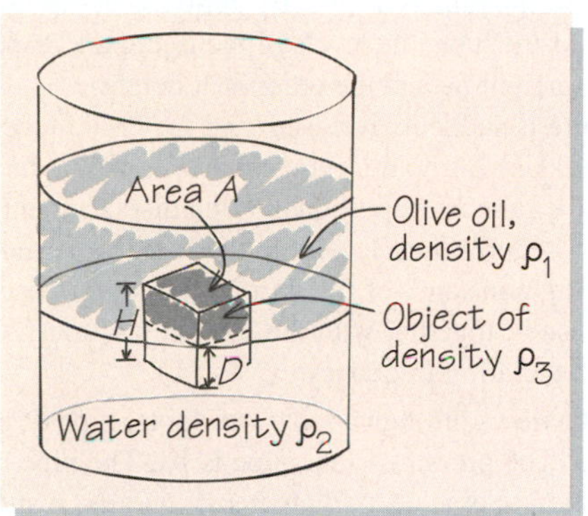

▲ **FIGURE 16–37** Problem 33.

34. (II) A beaker of mass 350 g contains 2.8 kg of water and rests on a bathroom scale. A 5.2-kg block of aluminum, of specific gravity 2.7, is suspended from a spring scale and completely submerged in the water. Find the readings on both scales.

35. (II) Balloons of mass 3.5 g can be filled with helium to make spheres of radius 21 cm. How many of these balloons must a child of mass 32 kg hold in order to float off Earth's surface? Assume that helium has a density of 0.18 kg/m^3.

36. (II) An air-filled lead balloon of radius $R = 0.1$ m is totally submerged in a tank of water (Fig. 16–38). What is the thickness t of the lead skin of the balloon if the balloon neither rises nor sinks? The density of lead is 11.3×10^3 kg/m^3.

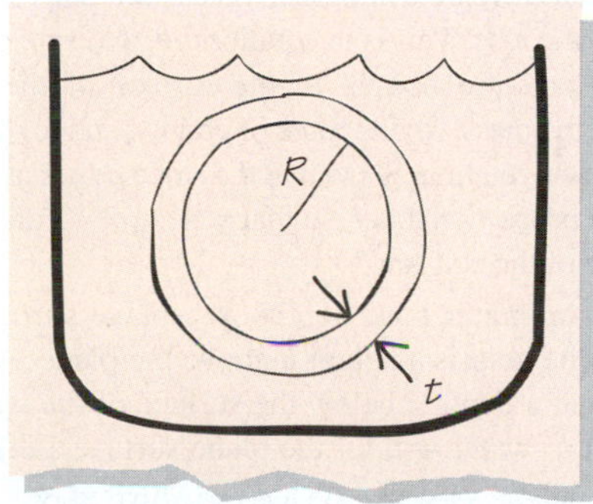

▲ **FIGURE 16–38** Problem 36.

37. (II) (In this problem we ask you to draw the figure that is described here.) When a body floats it is because the downward force of gravity just cancels the upward force of buoyancy. This alone is not enough; one also wants to know if the floating is stable, that is, whether the floating object might tip over. In thinking about whether, say, a tilting boat will right itself, it is important to note that gravity acts as a single downward force passing through the center of mass of the body (denote this point by M) while the buoyancy acts as a single upward force *passing through the center of mass of the displaced fluid* (a point denoted by B). In equilibrium B and M lie on the vertical, with B usually below M. To determine stability, we look at the body tilted. Denote the new center of mass of the displaced fluid by C. Draw a vertical line up through C. If this line intersects MB below M then the torques due to the forces of gravity and the force of buoyancy act in the same direction, and the body will topple. (a) Draw a diagram to show that a wooden log will not float vertically in water. (b) Draw a diagram to show why having several people stand up in a canoe is likely to tip it over.

38. (III) A sphere of radius R of material with an average density of 0.82 g/cm^3 is immersed in a body of water. What is the height of that portion of the sphere that projects above the surface?

16–6 The Equation of Continuity

39. (I) Water flows at a speed of 1.3 m/s through a hose of diameter 1.5 cm and emerges from a nozzle of radius 0.50 cm. With what speed does the water emerge? How much water emerges from the nozzle per second?

40. (I) Water flows down a channel 1.0 m deep and 0.5 m wide at a rate of 2 metric tonnes/s. At some point the channel widens to 0.8 m. How fast does the water flow in the wider channel?

41. (I) A steady stream of automobiles moves along the eastbound side of a highway in a region where there is neither exit nor entrance. Where two eastbound lanes are available, the automobiles are spaced so that there are 66 cars per mile in each lane, and the average speed is 45 mi/h. In a region to the west, where three eastbound lanes are available, the automobiles are spaced to give 35 cars per mile in each lane. What is the average speed in the region where three eastbound lanes are available?

42. (II) A lawn sprinkler is connected to a garden hose of inside diameter 2.5 cm that has a water-flow rate of 0.5 L/s. The sprinkler has four nozzle openings, each with a diameter of 7 mm. (a) Calculate the speed of the water that emerges from the nozzles. (b) How could the nozzles be arranged to cause the sprinkler to rotate? What principle could you use to calculate the torque on the sprinkler? What are the limitations, if any, on the speed of rotation of the sprinkler?

43. (II) Students would like to spray water 18 m across their neighbor's yard onto a patio. They have a hose of inside diameter 1.5 cm that can spray a distance of only 1.5 m. What size nozzle do they need?

44. (II) When water drains out of a hole in the bottom of a tank, the stream radius contracts as the speed of the water in the stream increases. Assume that the stream starts with zero water velocity at the top of the tank, at a height H above the hole, and ignore the various instabilities that will eventually cause the stream to break up. Find the distance below the hole at which the radius of the stream is 80 percent of the hole radius.

16–8 Applications of Bernoulli's Equation

45. (I) A bilge pump is used to pump water out of a leaking ship (Fig. 16–39). The pump's hose has a diameter of 3.0 cm, and the pump moves water through the hose up and out of a porthole 5.0 m above the waterline at a speed of 4.0 m/s. Calculate the power of the pump.

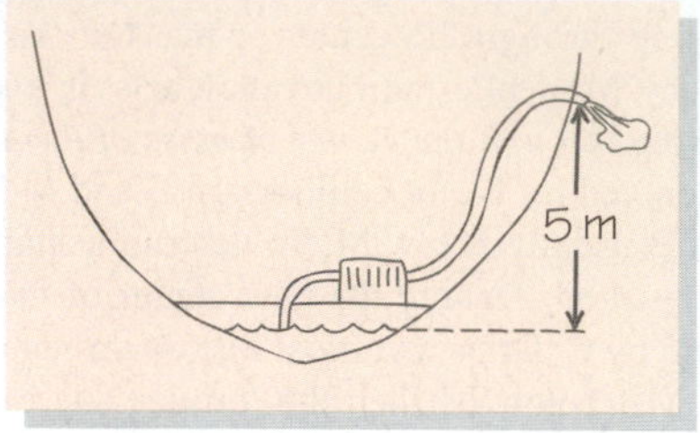

▲ **FIGURE 16–39** Problem 45.

46. (I) What pressure must a pump generate to get a jet of water to leave a nozzle at a speed of 8 m/s? Can this jet reach a height of 8 m above the nozzle? If so, what angle should the nozzle make with the horizontal so that the jet reaches 8 m at its maximum height?

47. (I) Wind gusts at 60 mi/h past a ship's cabin in which the air is at rest and the pressure is 1 atm. What are the pressure outside the cabin and the net pressure on the walls past which the wind blows?

48. (I) A stream of water sprays like a fountain vertically from a small hole in a pipe, reaching a height of 2.4 m. Assuming that the water in the pipe is static, what is the gauge water pressure in the pipe? (See Problem 8.)

49. (I) A folded piece of notebook paper can exhibit lift in the air flow from a fan of a window air conditioner. A plastic straw is taped perpendicular to the plane of the paper through the paper's middle, and a thread runs through the straw allowing the paper to move up and down the thread. The folded paper has an area of 300 cm^2. The air speed is about 4 m/s and the proportionality constant K in the approximate lift formula $KA\rho v^2$ is taken to be $K = 0.14$. Here, A is the surface area of the wings, ρ is the air density, and v is the air speed. How much lift is possible? Given the mass of the paper and straw to be 8 g, will the paper rise?

50. (II) A fountain shoots a vertical jet of water at a rate of 0.33 L/s to a height of 1.6 m. (a) What is the initial speed of the jet, and what is the radius of the hole out of which the jet passes? (b) What pressure must the pump of the fountain supply? (Assume that it sits just below the emerging jet.) (c) At height 1.2 m, what is the speed of the jet, and what is the radius of the column of water? Ignore effects of turbulence; in particular assume that the jet does not break up.

51. (II) Heavy rains have flooded your favorite professor's basement to a depth of 15 cm. This basement is 7.5 m × 12 m in area. You have a hose of diameter 1.2 cm and can run the hose from the water in the basement to a level 3 m below the level of the basement floor, into a low portion of the yard. When the hose is filled with water, it acts as a siphon (see Question 9) and empties the basement. (a) What is the speed of the stream of water leaving the hose? (b) Approximately how long does it take to empty the basement?

52. (II) A liquid of density 1.05×10^3 kg/m^3 flows in a horizontal pipe. The cross-sectional area in one part of the pipe is 42 cm^2. When the liquid enters another part of the pipe with cross-sectional area 56 cm^2, the pressure as measured by a gauge is 0.4×10^4 Pa higher than it was in the first part. Calculate the velocities of the liquid in the two parts of the pipe.

53. (II) A large horizontal pipe of diameter 10 cm contains water flowing with a speed of 0.3 m/s. The pipe branches into four pipelets, each of diameter 2 cm. The four pipelets each run horizontally after an initial height change of 3.5 m (Fig. 16–40). (a) What are the speeds of the fluid in the horizontal portion of each pipelet? (b) If the pressure in the large pipe is $p = 2.5$ atm, what is the pressure in the pipelets?

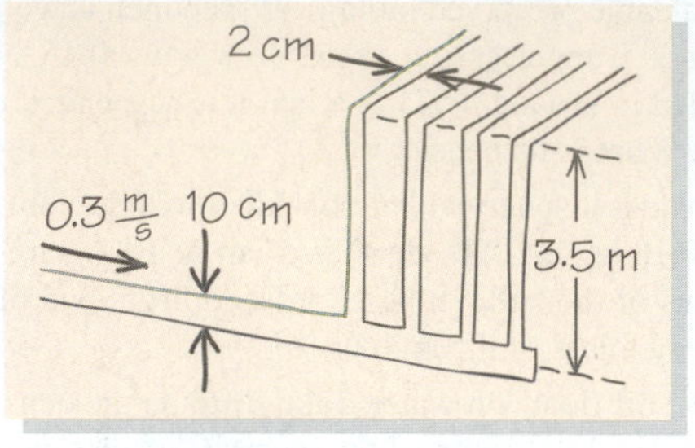

▲ **FIGURE 16–40** Problem 53.

54. (II) A uniform glass tube with internal cross-sectional area A is formed into a U-shape and partly filled with a volume V of an incompressible fluid of density ρ. The two ends of the tube are open to the atmosphere. At equilibrium, the level in each vertical segment of the tube is $h = 0$. By applying additional pressure to one end of the tube, the levels of the liquid are displaced, so one side is at height h_0 and the other is at height $-h_0$. The additional pressure is suddenly released, and the liquid level starts to oscillate. (a) Use Bernoulli's equation to show that the force on the fluid in the tube is proportional to the displacement of the fluids from their equilibrium levels, so the force is harmonic. (b) Find the "spring constant" of the harmonic motion described in part (a) and use it, together with the mass in motion (ρV), to calculate the oscillation frequency.

55. (II) Water flows through a large horizontal pipe of diameter D at speed v_0. The pressure in the pipe is p_0. The pipe branches into two horizontal pipelets of diameter d_1 and d_2, respectively, both at the same height as the large pipe. Write the equations that would allow you to find the pressures p_1 and p_2 and the speeds v_1 and v_2 in the two pipes. Show that one of these four quantities must be known to be able to solve for the others.

56. (II) A vertical cylinder of height 1.0 m and radius 10 cm is filled with water and placed at the edge of a table. A 1 cm diameter hole is drilled in the side wall near the base. Water starts flowing out, forming a stream that passes over the edge of the table. Neglecting any turbulence, calculate the velocity of the water as it leaves the cylinder. What will be the diameter of the stream of water at a point 1.0 m below the opening?

57. (III) The surface of a bucket of water centered on a rotating turntable will form a parabola with its minimum on the rotation axis. Prove that this is the case by the consideration of a small mass element at the surface, located at the point (y, r), where y is the height from the eventual minimum on the axis and r is the radial distance from the axis. This is in equilibrium, and subject to (a) the atmospheric pressure normal to the surface of the liquid at the location of the mass, (b) the force of gravity, and (c) the forces associated with the rotation. Show that the slope dy/dr at the location of the mass is proportional to r, so that $y = y_0 + (\text{constant})r^2$. What is the value of the constant?

58. (III) An open water tank stands on a plane surface. The water surface in the tank is a height h above the plane. A small hole is opened up at a depth y below the surface of the water. (a) Show that the jet of water will hit the plane surface a distance D from the tank, where $D = \sqrt{4y(h - y)}$. (b) Show that the hole should be placed at a depth $y = h/2$ for the jet to cover a maximum horizontal distance.

*16–9 Real Fluids

59. (II) The retarding force on a sphere of radius R that moves with speed v through a fluid of viscosity η is proportional to η and depends on R and v, in the form $R^a v^b$. Use dimensional analysis to determine the powers a and b.

60. (III) The result of the solution of Problem 59, the retarding force on a sphere of radius R that moves with speed v through a fluid of viscosity η, is $F = 6\pi\eta R v$, known as *Stokes's law.* Such a sphere, with mass density ρ_m, falls through glycerin, whose density is ρ. The forces on the sphere are the retarding Stokes's law force, the buoyant force, and the force of gravity. (a) Derive a formula for the value of the velocity v_t for which the net force is zero. (b) How does v_t depend on the radius of the sphere? (c) If the viscosity of glycerin is $8.3\ \mathrm{kg/(m\cdot s)}$ and its density is $\rho = 1.26 \times 10^3\ \mathrm{kg/m^3}$, what is the terminal velocity v_t of a sphere of iron, which has $\rho = 8.5 \times 10^3\ \mathrm{kg/m^3}$ and a diameter of 5 cm? Of an iron sphere with a radius of 1 mm?

General Problems

61. (II) A rectangular container is divided into two parts by a movable vertical partition. This partition is sealed so that no water can seep from one side to the other. Initially, the wall is held in the middle of the 30-cm length of the container. One side is filled with 2 kg of water to a depth of 10 cm; the other side is attached to a spring in equilibrium (unstretched, uncompressed). The spring constant is 180 N/m. The partition is now released and the spring compresses. What is the position of the partition at maximum compression? Assume that the process takes place without turbulence so that no energy is dissipated.

62. (II) A hollow sphere of radius 0.08 m and total mass 0.35 kg is placed in water. One very small lead pellet of mass 0.08 kg is attached to the inside surface of the sphere. When the sphere is floating in equilibrium, the lead pellet is oriented on the bottom. If the sphere is rotated slightly, so that the pellet is moved off to the side, and then released, the sphere will rock, and the pellet will move like a pendulum. Calculate the period of small oscillations for this motion. Assume that the center of the sphere does not move.

63. (II) A depth of about 100 ft (≈ 30 m) is as deep as a scuba diver should go without special precautions because of an effect that causes disorientation. What is the water pressure at a depth of 30 m? Does a scuba diver take in the same mass of air in each breath at 30 m that he or she would at a depth of 5 m? Explain.

64. (II) A cylindrical bucket of fluid of density ρ sits at the center of a turntable that rotates with some fixed angular velocity, ω. Show that the pressure variation in the fluid at a horizontal distance r from the axis is given by $p = p_{\mathrm{axis}} + \frac{1}{2}\rho\omega^2 r^2$. [*Hint*: If you place yourself within the rotating fluid and at rest relative to it, the problem is one of hydrostatics. The only force in the horizontal direction is a fictitious force away from the axis of the rotation, due to the fact that your frame of reference is not inertial. This force is the centrifugal force.]

65. (II) Late in the seventeenth century, King Louis XIV of France had a series of fountains constructed at Marly, some 20 km from his chateau at Versailles. These fountains were fed from the Seine River, whose altitude is some 150 m below that of Marly, through a remarkable series of pumps. The tallest of these fountains was the Grand Jet, which is calculated to have risen a height of 37 m from its base. (a) At what speed does this jet leave the orifice at the fountain's base? (b) The rate of flow in the Grand Jet is measured to be $0.051\ \mathrm{m^3/s}$. Compare this to an estimate of the rate of flow from a garden hose. (c) Given the rate of flow in part (b), calculate the area of the orifice from which the Grand Jet comes. (d) What is the pressure of the water just behind the orifice from which the Grand Jet comes? Beyond the crude calculations outlined here, the height to which a jet of water will rise can be affected by the addition of a short length of conical pipe above the level of the orifice at the base.

66. (II) A cylindrical tank is filled with water to a height h_1 above its base. Unfortunately, someone has cut a hole in the bottom through which fluid escapes in a vertical stream into your room. (a) What is the speed of the escaping stream at the hole? (b) You insert a flexible hose into the hole. If the other end of the tube is held at the same height as the base of the tank and is directed horizontally, what is the speed of the water as it leaves the hose? (c) You hold the tube, whose end is still at the height of the base of the tank, so that the stream is directed straight up. How high does the escaping stream shoot?

67. (II) An inventor proposes to you the following perpetual motion machine (a machine that can do work "forever" without your having to add energy to it) (Fig. 16–41). A cylindrical tower of height h filled with water rests on the ground. Blocks of wood are inserted at the bottom of the tower. They float to the top, where they are removed. The potential energy that they acquired in floating to the top can be converted into work. They can, for example, be dropped down for re-use, with the kinetic energy they acquire in falling used to turn the wheels of a turbine. When they reach the bottom, the heavier blocks are again inserted, and the process is repeated. How can you prove to the mad inventor that at best this machine can do no net work? [*Hint*: Calculate the work required to insert a block, which requires in effect that a mass of water whose volume is that of the block be raised to the top. Compare this to the energy the block acquires in being accelerated to the top under the influences of gravity and the buoyant force. Neglect any drag caused by the water.]

Do not buy this machine! Perpetual motion machines are impossible to construct (see Chapters 18 and 20). A close analysis of the details of such machines always reveals this fact.

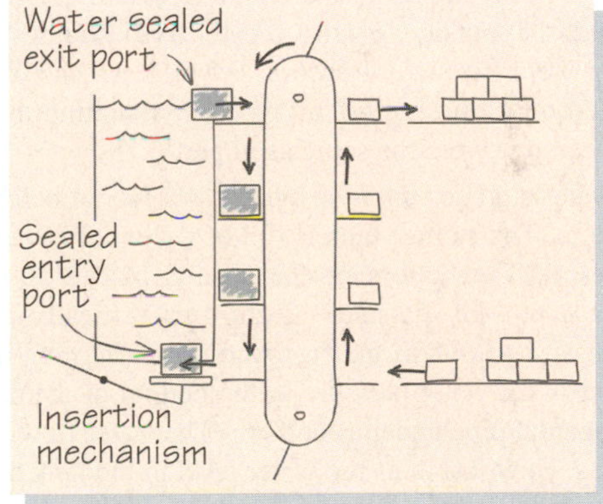

▲ **FIGURE 16–41** Problem 67.

68. (II) Consider the machine discussed in Problem 67. Why does the machine still fail if you replace the entry port by an exchange chamber (Fig. 16–42, see next page) with two sliding doors, so that the insertion of the blocks does not require that work be done?

69. (II) A cubical block of wood 0.20 m on each side floats in water, with 8.0 cm of the block above the surface of the water (Fig. 16–43). (a) What is the density of the wood? (b) Suppose that the block of wood is pushed down below the equilibrium level by a distance Δ. What is the magnitude of the restoring force that pushes the wood back toward the equilibrium level? (c) In general, if dissipation of energy is ignored, the block will overshoot the equilibrium level and rise to a height Δ above that level, and is then subjected to a downward force. How large is that force?

▲ **FIGURE 16–42** Problem 68.

(d) Recall from Chapter 13 that a force that is proportional to the displacement Δ from equilibrium (of the form $F = -k\Delta$) and acts on an object of mass m leads to oscillations with the frequency $f = (k/4\pi^2 m)^{1/2}$. Calculate that frequency for the block that bobs in the water.

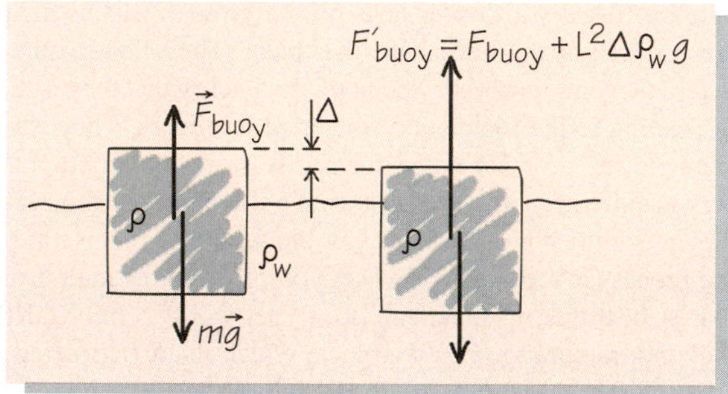

▲ **FIGURE 16–43** Problem 69.

70. (II) The ticket-window opening of a movie theater, where money goes into the cashier's hands and tickets emerge, is a rectangle 12 cm × 15 cm. A breeze of speed 3 m/s comes out of this opening. Sometimes, as a result, paper money goes flying. (a) What is the difference in air pressure between the inside and the outside of the theater? (b) Suppose that the opening were the size of a door, 0.9 m × 2.1 m. What would the speed of the breeze be in this case, assuming the air pressure difference were the same as in part (a)?

71. (II) A water-storage tank is an upright cylinder of height 8 m and radius 6 m. The storage tank is full of water but is vented to the atmosphere. The bottom of the tank is placed 30 m off the ground. A pipe of diameter 12 cm runs vertically down from the tank and goes 1 m underground before turning horizontal. (a) What is the water pressure in the bottom of the tank? (b) in the horizontal pipe underground? (c) The water flow in the pipe is 80 L/s. How fast does the water level in the tank drop? (d) A saboteur drills a hole of diameter 6 mm near the bottom of the tank. How fast does the water shoot out? (e) What volume flow of water is lost out the hole?

72. (II) A *Pitot tube* is a device for measuring flow velocities (Fig. 16–44). It is as useful for measuring the flow velocity of blood in an artery as for measuring the air speed of a jet airplane. The fluid passes the opening at point B, creating a pressure p_B inside. The fluid enters the opening of the tube and comes to a halt, creating a pressure p_A. Apply Bernoulli's equation as needed to show that the flow velocity is given by $v = \sqrt{2(p_A - p_B)/\rho} = \sqrt{2gh\rho_\ell/\rho}$, where ρ is the density of the fluid being measured, ρ_ℓ is the density of the fluid shown in the device, and h is the difference in height of the liquids.

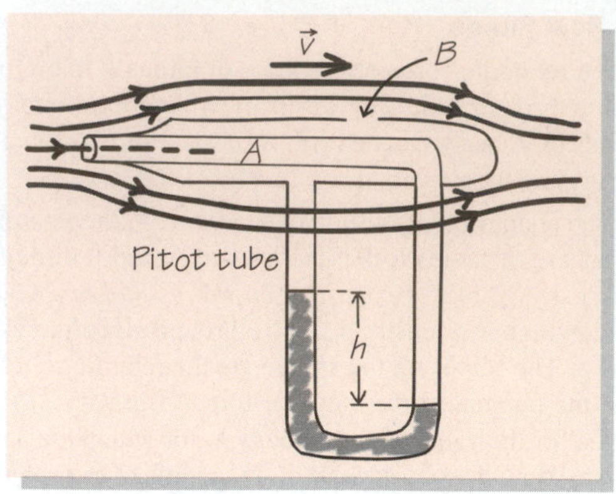

▲ **FIGURE 16–44** Problem 72.

73. (II) With a siphon, it is possible to transfer fluids from one container to another. Figure 16–31 shows a siphon that must be filled with fluid to start the transfer. If the speed with which the fluid leaves the higher container is zero, and if the fluid has a fixed density ρ, find an expression for the speed of the fluid as it enters the lower container, in terms of the heights h_0 and h_1.

74. (III) A tank of surface area A is filled with water to a height x above the position of a hole of surface area σ, and the water flows out. Assume throughout that the fluid speed is zero at the top surface. (a) How much water flows out of the hole in the first time interval Δt after the hole is opened? (b) How large is the drop in the water level, Δx, in that time interval? (c) Use the results of parts (a) and (b) in the limit of small Δt to express a relation of the form $dx/dt = f(x)$. This is a differential equation for the height as a function of time. Take care to get the right signs, noting that x decreases with time, so that Δx is a negative quantity. (d) Show that this equation is solved by the formula

$$x(t) = \left[\sqrt{x_0} - \frac{1}{2}\left(\frac{\sigma}{A}\right)t\sqrt{2g}\right]^2.$$

(e) How long will it take for the water to drop to the level of the hole?

75. (III) The ice-cream cone shown in Fig. 16–45 is filled with melted ice cream of density 1.2 g/cm^3. The cone has a diameter of 6 cm at the larger end and is 10 cm long. Find the pressure at the bottom of the cone. If a small hole of diameter 1 mm is opened at the bottom, the ice cream starts to run out. Ignoring the viscosity of the melted ice cream, find the amount of time it takes the ice cream to run out. Assume that the fluid speed is zero at the top.

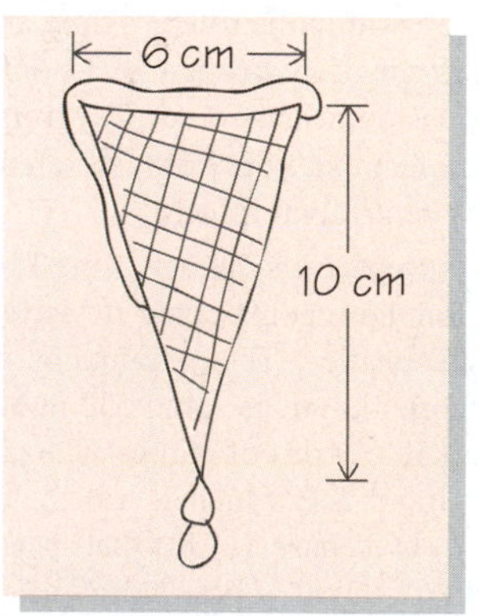

▲ **FIGURE 16–45** Problem 75.

◀ That the properties of materials depend on temperature is no surprise. Whether we are referring to the steel being worked in this photograph or to the properties of the air over the polar ice cap, we would fail to pick out essential features if we did not recognize the role of temperature.

Temperature and Ideal Gases

Many properties of the physical world depend on temperature. For example, the compound water behaves quite differently at temperatures below its freezing point than it does at temperatures above its boiling point. Temperature has something to do with energy, as we can see from the fact that hot steam can lift a piston, and the hotter the steam, the better it is able to perform the work. How much energy does hot steam contain, and how does that energy get into the steam? How can we quantify "hotness" or "coldness?" To answer these questions, we will consider both the macroscopic properties and microscopic behavior of gases, liquids, and solids, focusing primarily on gases. These properties lead us to a good definition of temperature. We will see how temperature relates to energy, how that energy is like the mechanical energy we have earlier studied, and how efficiently that energy can be used.

The historical development of thermal physics was based on the properties of bulk matter—such as density, pressure, volume, and temperature—before atomic theory was well established. The study of thermal phenomena received its strongest impetus from engineers who studied the steam engines that powered the Industrial Revolution of the 19th century. Their treatment of thermal phenomena on the macroscopic scale is known as **thermodynamics**. Nineteenth-century physicists also learned how these macroscopic phenomena could be understood in terms of the behavior of the atomic constituents of matter. This work, known as **statistical mechanics**, brings in an entirely new idea—that it is the statistical behavior of these constituents that is reflected in thermal behavior in the macroscopic world. All these topics will be discussed in Chapters 17 through 20 under the general heading of thermal physics. Thermal phenomena are pervasive in the physical world, and our understanding of them has had a profound effect on the way we live.

17–1 Temperature and Thermal Equilibrium

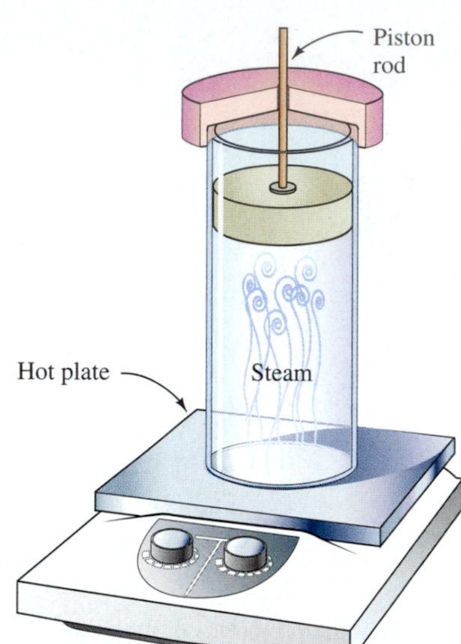

▲ **FIGURE 17–1** A bottle of hot steam with a piston at one end provides us with a system that can do work by virtue of the thermal properties of the steam. In particular, the hotter the steam, the farther it can move the piston.

We call a physical system a **thermal system** whenever we are interested in its temperature-dependent properties. By virtue of the fact that it is hot, steam can lift a piston and do work on its surroundings, so that a piston and a cylinder full of hot steam (Fig. 17–1) form a good example of a system that is worth keeping in mind as we proceed, even if we will be more interested in the work that thermal systems do in later chapters. We all have an intuitive feeling for temperature—for example, we can usually tell when one system is hotter than another or, in other words, that its temperature is higher. Our aim here is to develop a more precise notion of temperature.

Central to the idea of temperature is the notion of thermal equilibrium. Suppose that we pour cold water into a bucket of hot water. Experience shows that after a short time the water becomes lukewarm throughout. Until that point, we cannot say that the water has a particular temperature. When the temperature and temperature-dependent quantities are no longer changing, we can say that **thermal equilibrium** has been reached (Fig. 17–2). At this point the temperature becomes a variable that characterizes the whole bucket of water; that is, the whole thermal system. Similarly, the gas in a container eventually comes to thermal equilibrium if its container is in thermal equilibrium with its surroundings, and two containers of gas at different temperatures will come to thermal equilibrium if they are allowed to mix. One mechanism for this process is collisions between the constituent molecules of the gas, but as we shall see there are other mechanisms and mixing is not necessary. We shall see in some detail that the temperature is a measure of how fast the molecules of the gas are moving, and in collisions between fast-moving and slow-moving molecules, the fast-moving ones will generally slow down and the slow-moving ones speed up, thus bringing the ensemble to a common temperature.

▲ **FIGURE 17–2** The vendor's stand is not in thermal equilibrium with its environment. The steam is evidence of an interaction with the surroundings that would eventually cool the stand down to where it would be serving frozen hot dogs.

If two thermal systems are in thermal equilibrium with one another, then *they have the same temperature*. We can add that if two systems are each in thermal equilibrium with a third, then they are in thermal equilibrium with one another. The observation that if system A is in thermal equilibrium with system B, and if system B is in thermal equilibrium with system C, then A is in thermal equilibrium with C is sometimes called the **zeroth law of thermodynamics.**

If we put an ice cube into a bathtub full of hot water, the temperature of the tubful of water changes very little; if we add a teaspoon of hot water to a swimming pool, the temperature change is virtually undetectable. These observations allow us to define a **thermal reservoir** (or **heat bath**) as a thermal system so large that it maintains a constant temperature when it interacts with other thermal systems. A thermal reservoir is of importance because it allows us to bring a given (smaller) system to a predetermined temperature, the temperature of the reservoir, simply by bringing the system into **thermal contact** with the reservoir.

Thermal contact is established between thermal systems in three important ways. We can illustrate them with two metal plates in an enclosure, one red hot and the other cold. Thermal contact by *conduction* occurs when the two plates actually touch. "Something" passes between them, which with foreknowledge we can describe as the agitation of the molecules of the hot plate passing through collisions to the agitation of the molecules of the cold plate. When you sit on a cold metal bleacher seat at a football game, there is thermal contact by conduction between you and the seat. Thermal contact by *convection* occurs when there is a fluid (a liquid or a gas) between the plates. The fluid acts as a third thermal system whose temperature increases in the region near the hot plate through its own thermal contact with the plate. The fluid then circulates in currents to the colder plate, increasing its temperature. The reason for this circulation has to do with thermal properties of the conducting fluid; for example, as we saw in Chapter 16, hot gas is less dense than cold gas, so that it tends to rise. If you live in a house with radiators, then you are warmed by the air that starts near the radiator and then circulates in the room. Similarly, if your house has hot air heat, the fact that hot air rises explains why heating ducts are placed on or near the floor. Finally, in thermal contact by *radiation*, a hot plate emits electromagnetic waves—the "radiation" in question—while a cold plate absorbs them. (We'll say a good deal more about electromagnetic radiation in the context of thermal physics in Section 17–5.) This form of thermal contact operates even with the plates separated by some distance and even across a vacuum. You experience this form of thermal contact if you tan on the beach or if you stand in the vicinity of a fire.

In contrast to the descriptions above, if two fluids at different temperatures are separated by a double-walled, silvered glass flask with a vacuum between the two glass surfaces, the fluids will take a very long time to reach the same temperature. The vacuum prevents thermal contact by conduction or convection, and the silvered surfaces of the glass walls reflect radiation back to the heat source and prevent thermal contact by radiation. The liquids are **thermally isolated** from one another. Don't confuse thermal contact with mechanical contact. If the pistons in an automobile engine were not in mechanical contact with the drive train, the car could not go very far, but the manufacturer goes to some length to keep the hot cylinders and their contents out of close thermal contact with most of the remainder of the vehicle.

THINK ABOUT THIS . . .

IS IT EASY TO MAKE A POT OF TEA IN A SPACE STATION?

That depends on the method you use. Aside from the fact that you would have to make sure the pot was closed to keep the water from drifting out, the technique you use for boiling water would have to change from what you do at home. On Earth the heating of a pot of water proceeds as follows: The water at the bottom becomes heated first, by conduction, and then a process of convection begins. The liquid moves in a circular pattern. The hot water rises, because on heating it expands, and since the density of the hot water is lower, there is a buoyant force. The hot water then pushes the cold water from above it to the bottom and the process continues until the water is uniformly heated. On a space station, the fact that the density of hot water is less than that of cold water makes no difference, because without an "up" or "down" as on Earth, there is no buoyancy. So some external stirring mechanism is needed to help you boil the water. Or you could wait for the entire pot to heat by conduction, a slower process. We might guess that microwaving a container of water would be a good way to proceed, since with a good microwave oven the heating of the water occurs uniformly throughout the volume.

Thermometers

A **thermometer** is a device that measures the temperature of a thermal system quantitatively, generally by coming to thermal equilibrium with that system. A thermometer must satisfy several criteria. It should have a minimal effect on the system being measured. (A small room thermometer brought inside from the cold will have this property.) The thermometer must make good thermal contact with the system being measured. Like a meter stick for measuring length, the thermometer should have a reproducible and easily read scale.

There are several properties of matter that are suitable for thermometry. Ordinary fever thermometers rely on the fact that fluids, such as mercury, expand and contract when their temperatures change. We can therefore make a thermometer by marking numbers on a column of mercury that measure the mercury's volume. Other properties, such as the pressures and volumes of gases, the volumes of liquids or solids, reflective properties of liquid crystals, electrical and magnetic properties of solids, and the color of light emitted by hot objects, all vary with temperature and can be used as thermometers (Fig. 17–3). Not all of the properties of matter that depend on temperature, however, are suitable for making thermometers: The brittleness of many materials depends on temperature, but we would not make a thermometer that works by seeing how easy it is to smash a material with a hammer! Table 17–1 lists some useful thermometers.

(a)

(b)

FIGURE 17–3 (a) This thermometer contains a (red) fluid that expands at higher temperatures and contracts at lower temperatures. (b) This digital thermometer uses the temperature dependence of electronic properties of materials. It is calibrated in the Celsius scale, in which the temperature of ice melting in pure water is 0 degrees.

TABLE 17–1 • Some Thermometers

Thermometer	Physical Property Measured	Comments
Ideal gas	Pressure and volume of dilute gas	See Section 17–2
Mercury bulb	Expansion or contraction of fluid	Good where fluid does not change phase
Bimetallic strip	Difference in expansion of two metals	
Resistance	Electrical resistance	See Chapter 26
Thermocouple	Electrical voltage across different metals	Most widely used thermometer in industry
Paramagnetic	Magnetic properties of matter	Useful at ultracold temperatures; see Section 31–5
Optical pyrometer	Color of emitted light	Useful at high temperatures; see Section 17–5

CONCEPTUAL EXAMPLE 17–1 A friend proposes that you check the temperature of boiling water that you have prepared in order to make some hot tea by dumping 10 large ice cubes of a calibrated size in it and seeing how long it takes for the ice cubes to melt. What is wrong with this thermometer?

Answer This is indeed a thermometer, and assuming that the initial temperature of the ice is known, a measurement of the melting time would indeed give you a way to measure the initial temperature of the water. However, this thermometer would change the temperature of the very system it is supposed to measure. By the time you complete your measurement, your water would be too cold to make your tea!

17–2 Ideal Gases and Absolute Temperature

Consider a gas in a container of fixed volume with a gauge that can read the pressure of the gas (Fig. 17–4). If we dip the container into thermal reservoirs at various temperatures, we find that as the gas becomes hotter, its pressure increases. The pressure correspondingly decreases when the gas becomes colder.

We can alternatively arrange to allow the volume to vary while keeping the pressure fixed, as in Fig. 17–1, with masses of various sizes sitting on the top of the piston. We find that the hotter gas has a larger volume for a fixed pressure, while the colder gas has a smaller volume for that same pressure. In fact, if we take all this into account, we find the following relation between the pressure p, the volume V, and the amount of gas n:

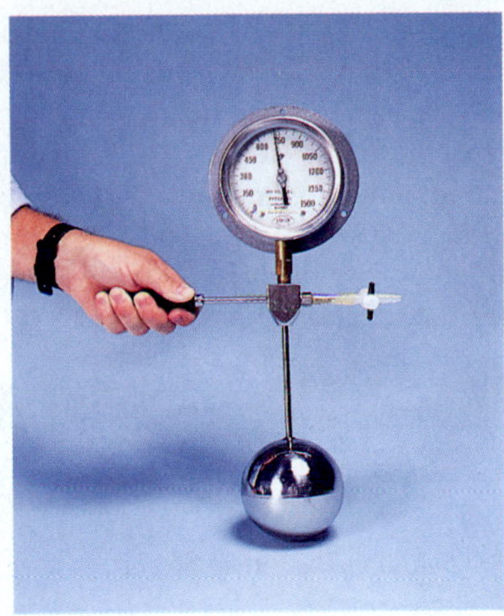

▲ **FIGURE 17–4** A container of gas of fixed volume with means to measure the pressure of the gas. This system forms a constant volume thermometer whose pressure is a measure of the temperature of the system with which the gas is in thermal contact.

The product of pressure and volume divided by the amount of gas in the bottle increases or decreases according to whether the thermal reservoir in contact with the bottle is hotter or colder, respectively. We accordingly define temperature T as proportional to the product of pressure and volume, divided by the number of moles, that is, $T \propto pV/n$.

The *constant volume gas thermometer* that we are describing here comes extremely close to being a universal thermometer because experiment shows that the proportionality constant is the same for all gases as long as their *densities are sufficiently low.* Dilute (or low-density) gases, which we call **ideal gases**, will enable us to give an unambiguous definition of temperature, a definition that can be employed by anyone who follows the simple set of instructions that we'll give below. In other words, we will have a thermometer that meets the requirements we have laid out. Just how dilute the gas needs to be is simply a question of the accuracy demanded of the thermometer. The lower the density of the gas, the more 'ideal' the behavior of the gas. Gas thermometers accurate enough for laboratory use may contain gas hundreds of times less dense than air at sea level.

Figure 17–5 plots the pressure of a fixed amount of a dilute gas at fixed volume against our temperature scale. The dashed part of the curve, near where T and p drop to zero, is an extrapolation and is drawn as a straight line.

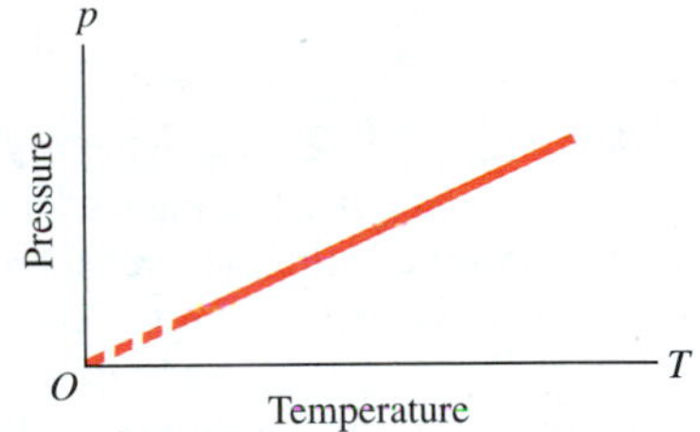

▲ **FIGURE 17–5** As p varies in an ideal gas of constant volume, T is defined by drawing a straight line.

We have not yet set our temperature scale. To do so, work with a constant-volume ideal gas thermometer and define the temperature $T = 0$ as the point where the pressure, p, would become *zero* (if that were possible), so

$$\text{for constant volume: } p = (\text{a constant}) \times T. \tag{17–1}$$

The point $T = 0$ is sometimes called **absolute zero**. Although experiments have come quite close to absolute zero, this temperature cannot in fact be reached and it corresponds to an extrapolation. We need to specify the constant in this relation, and we do so by defining T at some other point. This second point, called a *defined point*, was chosen in 1968 by international agreement to be the **triple point** of water. The triple point of water is the state at which water vapor, liquid water, and frozen water coexist. This can only happen at a single pressure p_{tp} (= 4.58 mm of mercury) and temperature; that temperature is defined as 273.16K. The letter K stands for kelvin, and the temperature scale so defined is the **Kelvin scale**, after Sir William Thomson, Lord Kelvin, who did important work on thermal phenomena and on electricity and magnetism during the mid- to late nineteenth century.) To summarize, temperature in the Kelvin scale, T, is defined by a constant-volume ideal-gas thermometer as

$$T = \left(\frac{p}{p_{tp}}\right) 273.16\text{K}, \tag{17–2}$$

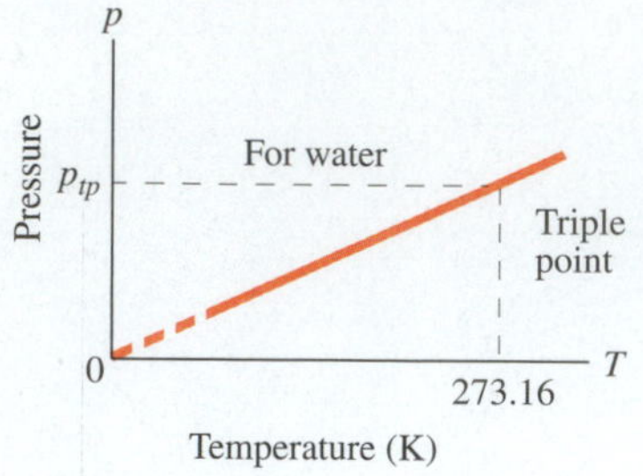

▲ **FIGURE 17–6** The Kelvin temperature scale is defined by a constant-volume ideal-gas thermometer. The two fixed points are $T = 0\text{K}$, located where p extrapolates to zero, and $T = 273.16\text{K}$, located at the pressure of the triple point of water, where $p = p_{tp}$.

where p is the pressure of the ideal gas in the thermometer. Figure 17–6 illustrates Eq. (17–2). The assignment of 273.16K as the triple point makes the Kelvin scale compatible with the Celsius scale, to be described below.

Table 17–2 gives us a feeling for the Kelvin scale by listing some temperatures that occur in nature. We call your attention to the numbers that refer to the origin of our universe in the so-called Big Bang, which occurred some 13 billion years ago. No one was around at that time to take notes, but remarkably these values can be estimated with some confidence from laboratory data.

TABLE 17–2 • Temperatures Occurring in Nature

Physical System	Temperature (K)
Lowest temperature reached in the lab	$\cong 10^{-9}$
Temperature of background radiation that fills the universe	3
Liquid helium at 1 atm	4.2
Coldest recorded outdoor temperature on Earth	185
Average surface temperature on Mars	218
Freezing point of water	273.15
Typical sea-level temperature on Earth	300
Body temperature	310
Hottest recorded outdoor temperature	331
Boiling point of water	373
Melting point of gold	1335
Surface of the Sun	6000
Interior of the Sun	10^7
Helium fusion reaction	10^8
Interior of hottest stars	10^9
Universe at 3 min after the Big Bang	5×10^8
Universe at 1 s after the Big Bang	10^{10}
Universe at 10^{-6} s after the Big Bang	10^{14}
Universe at 10^{-12} s after the Big Bang	10^{16}

EXAMPLE 17–2 Estimate the pressure in a hypothetical closed bottle of air that, starting with sea-level temperature and pressure, has been heated to the temperature of the surface of the Sun. Express your answer in atmospheres.

Setting It Up "Typical" sea-level temperature and pressure are 300K and 1 atm, respectively. As Table 17–2 shows, the temperature at the Sun's surface is 6000K.

Strategy We assume that the gas in the bottle is ideal over the temperature range involved. According to Eq. (17–1), the ratio of the two temperatures in question—300K and 6000K—is the same as the ratio of the two pressures.

Working It Out If we refer to the temperature and pressure on Earth and the Sun with the subscripts E and S, respectively, we have

$$\frac{p_E}{T_E} = \frac{p_S}{T_S};$$

$$p_S = p_E\left(\frac{T_S}{T_E}\right) = 1\ \text{atm}\left(\frac{6000\text{K}}{300\text{K}}\right) = 20\ \text{atm}.$$

Ideal-gas thermometers are not very practical at high temperatures, as it is not easy to construct a suitable container for temperatures much above 2000K, even at atmospheric pressure.

What Do You Think? According to Fig. 17–5, there also seems to be a difficulty in producing ideal-gas thermometers at low temperatures and pressures. Why is this so? *Answers to* ***What Do You Think?*** *questions are given in the back of the book.*

Other Temperature Scales

By changing both the slope and the intercept in Fig. 17–6, we can construct other temperature scales. Equivalently, if two points on the linear curve are assigned to special temperatures, a new scale emerges. The temperature in the **Celsius scale**—once known as the centigrade scale—is denoted by t_C and measured in units of °C. It assigns the value $t_C = 0°C$ to the freezing point of water at 1 atm of pressure (the *ice point*) and $t_C = 100°C$ to boiling water at 1 atm. From Fig. 17–7, we see that

$$t_C = T - 273.15, \tag{17–3}$$

where T is the temperature in the Kelvin scale.† Note that the slopes of the Kelvin and the Celsius scales are exactly the same.

Other scales differ from the Kelvin scale in slope as well as in zero level. The **Fahrenheit scale**, in which the temperature t_F is measured in °F, is an example. Different scales are useful for different purposes, and the Fahrenheit scale was invented with humans in mind. The ice point of water is 32°F, the temperature of the human body is about 100°F, and the steam point of water is 212°F. The relation between the Fahrenheit scale and the Kelvin scale is

$$t_F = \frac{9}{5}T - 459.67, \tag{17–4}$$

with T again measured in kelvins. We can easily find the relation between any two scales (Figs. 17–7 and 17–8).

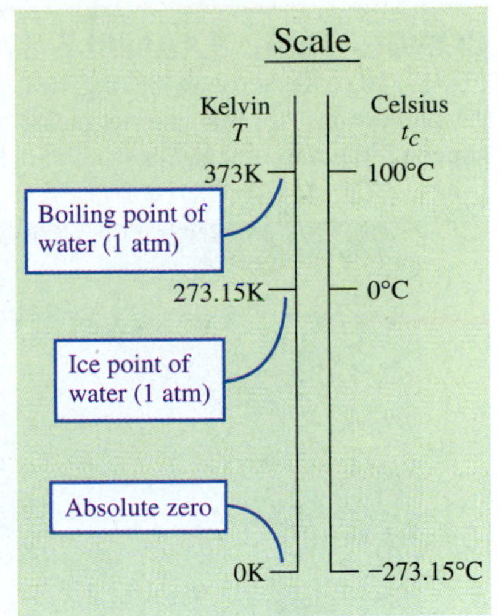

▲ **FIGURE 17–7** Comparison of the Kelvin and Celsius temperature scales.

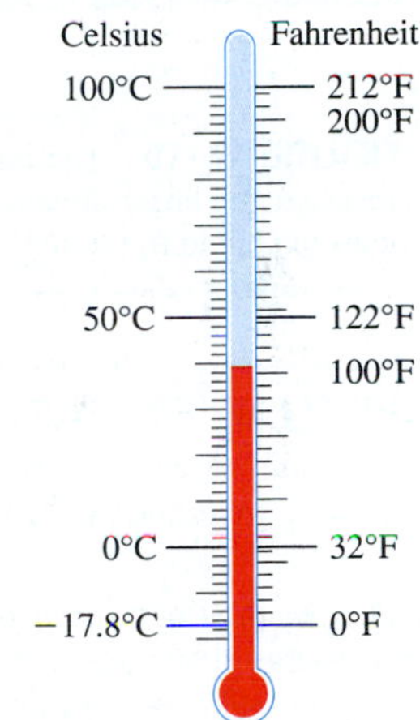

▲ **FIGURE 17–8** Comparison of the Celsius and Fahrenheit temperature scales.

17–3 Thermal Expansion

Observation shows that a solid expands when its temperature increases. A simple demonstration of this fact is shown in Fig. 17–9. Empirically, this expansion is described by the **coefficient of thermal expansion**, α, defined by

$$\alpha \equiv \frac{1}{L}\frac{dL}{dT}. \tag{17–5}$$

The quantity $\alpha\Delta T$ is the fractional change $\Delta L/L$ in the length L of a solid due to a change ΔT in its temperature. In principle, the coefficient α is a function of temperature T, but tends to vary quite slowly with T. The analogous **coefficient of volume expansion**, β, is simply related to α. It gives the fractional change in the volume V due to a temperature change ΔT: $\beta\Delta T = \Delta V/V$, or

$$\beta \equiv \frac{1}{V}\frac{dV}{dT}. \tag{17–6}$$

We can show that $\beta = 3\alpha$ by considering a cube of volume $V = L^3$. We have

$$\beta = \frac{1}{V}\frac{dV}{dT} = \frac{1}{L^3}\left(3L^2\frac{dL}{dT}\right) = \frac{3}{L}\frac{dL}{dT} = 3\alpha. \tag{17–7}$$

This result applies as long as the material is *isotropic*; that is, its properties are the same in any direction within the material. Unless a uniform solid is a single crystal, with consequent preferred directions, we would normally expect it to be isotropic. Thermal expansion also occurs in liquids, which are more correctly described as isotropic than are solids. The volume expansion coefficient β is usually given when liquids are involved. Both β and α are measured in units of K^{-1}. Table 17–3 lists some thermal expansion coefficients.

▲ **FIGURE 17–9** The expansion joint in this bridge allows the material of the bridge to undergo expansion in hot weather and contraction in cold weather.

TABLE 17–3 • Coefficients of Thermal and Volume Expansion (at 20°C)

Material	$\alpha(K^{-1})$	Material	$\beta(K^{-1})$
Aluminum	23.0×10^{-6}	Ethyl alcohol	11.2×10^{-4}
Copper	16.7×10^{-6}	Gasoline	9.5×10^{-4}
Iron	19×10^{-6}	Mercury	1.82×10^{-4}
Pyrex glass	3.2×10^{-6}	Water	2.07×10^{-4}
Fused quartz	0.42×10^{-6}		
Steel	10.5×10^{-6}		

†The difference of 0.01 between 273.16 in Eq. (17–2) and 273.15 here is the difference between the temperature of water at its triple point and water at its ice point.

CONCEPTUAL EXAMPLE 17–3 The unheated ball in Fig. 17–10 passes through the ring; heated it will not, because its linear dimensions—in this case its radius—have increased. What will happen if the ring is heated instead of the ball?

Answer In thermal expansion, *all* the linear dimensions increase, including the dimensions of the holes. For the ring, both inner and outer radii are linear dimensions. The holes—increase in size just as if they were made of the material in question, and the ball will go through.

(a)

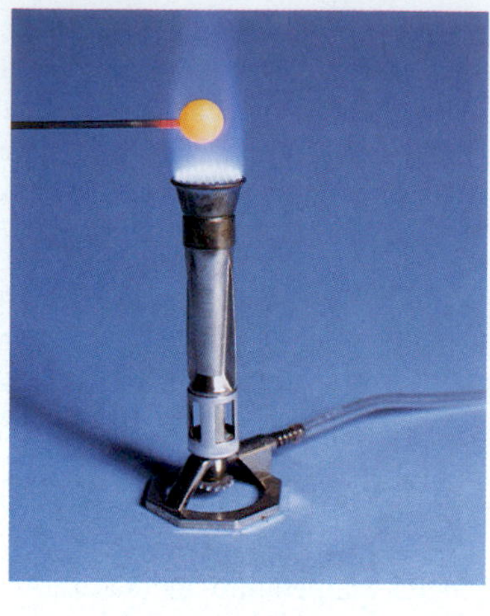

(b)

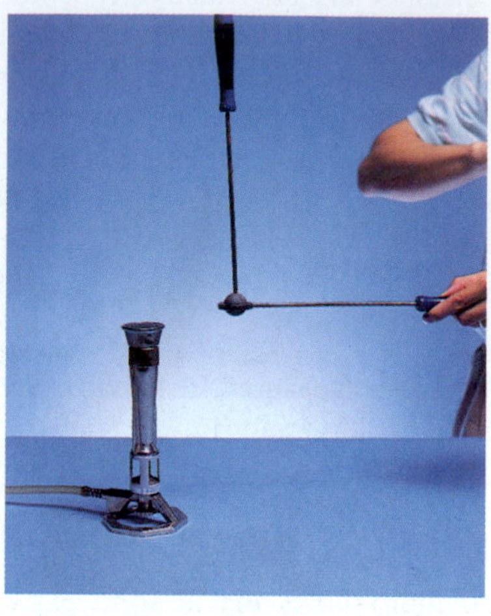

(c)

▲ **FIGURE 17–10** Thermal expansion: The linear dimensions of a material expand proportionally with the temperature increase. In this instance, the linear dimension of the metal ball is its radius. (a) It just fits through a ring at room temperature, (b) is heated, and (c) becomes too big to fit through the ring.

CONCEPTUAL EXAMPLE 17–4 We learn by experience that we can loosen a metal lid that is stuck on a glass jar by pouring hot water over the lid. Why does this work?

Answer As the lid's temperature rises, it expands. Movement occurs where the glass and metal are stuck and the lid releases (Fig. 17–11). In fact, you could dip the entire system (lid and glass container) in hot water, and the different expansions of the metal and the glass will lead to the same result. Note that the same difference in the thermal expansion is used in making a good seal: lids are placed on jars when the contents are hot.

▶ **FIGURE 17–11** By pouring hot water over this lid, you can take advantage of thermal expansion to free it from the jar.

EXAMPLE 17–5 A steel bridge is 600 m long. How much allowance must be made for linear expansion between the temperatures of +40°C and −40°C?

Setting It Up We know the values of L, ΔT, and the coefficient of expansion α for steel; we are interested in the length change ΔL.

Strategy This is a straight application of Eq. (17–5), which gives ΔL in terms of L and ΔT.

Working It Out We take α from Table 17–3: $10.5 \times 10^{-6}\,\mathrm{K}^{-1}$ for steel (α varies little with temperature in the range +40°C to −40°C). Eq. (17–5) rearranged in the form $\Delta L = \alpha L \Delta T$ then gives

$$\Delta L = (10.5 \times 10^{-6}\,\mathrm{K}^{-1})(600\,\mathrm{m})(80\mathrm{K}) = 5 \times 10^{-1}\,\mathrm{m} = 50\,\mathrm{cm}.$$

Expansion joints that allow for this much thermal expansion are distributed at regular intervals along bridges (Fig. 17–9). If there are N joints, then each joint need accommodate a length for $\Delta L/N$.

What Do You Think? Would it be more pleasant to drive over the steel bridge in this example at +40°C or −40°C?

EXAMPLE 17–6 You want to fill an aluminum container with gasoline in the dead of winter, at a temperature of $-20°C$. The filled container is to be stored into the middle of summer, when the temperature may reach $T = +40°C$. How much unfilled space must you leave in the container to allow for expansion when summer comes around?

Setting It Up We know from Table 17–3 that $\beta_{gas} = 9.5 \times 10^{-4}\,K^{-1}$ is the coefficient of volume expansion for gasoline. The coefficient of volume expansion for aluminum can be found from α using Eq. (17–7); from the table $\beta_{alu} = 3\alpha_{alu} = 6.9 \times 10^{-5}\,K^{-1}$. We want to find the difference in the fractional volume changes of the gasoline and the container.

Strategy Both the gasoline and the aluminum of the container will expand when T increases. We calculate and compare the fraction volume change $\Delta V/V = \beta \Delta T$ for both. With $\beta_{gas} > \beta_{alu}$ the gasoline expands more than the container, and the difference is the fraction that would be lost if the container were full; it is thus the fraction that should be left unfilled. If β_{gas} had turned out to be smaller than β_{alu}, the fractional volume increase of the gasoline would be smaller than that of the container, in which case there would be no need for expansion space.

Working It Out The temperature change ΔT is positive, $\Delta T = +60°C = +60K$. Then for the gasoline

$$\Delta V/V = \beta_{gas}\Delta T = (9.5 \times 10^{-4}\,K^{-1})(60K) = 0.057.$$

For the container the volume change is

$$\Delta V/V = \beta_{alu}\Delta T = (6.9 \times 10^{-5}\,K^{-1})(60K) = 0.004.$$

The container volume change is negligible compared to the nearly 6% volume change of the gasoline. You should fill the container only to the 94% level in winter if you want to avoid a dangerous situation.

What Do You Think? Can you safely fill the container to the brim in summer?

17–4 The Equation of State of Gases

A thermal system such as a gas in a bottle is described by its pressure p, temperature T, volume V, and the amount (or mass) of the gas. We have already discussed the first three variables. A useful unit for the amount of material in a thermal system is the **mole** (abbreviated mol), which represents either an amount of gas with a certain mass (according to the chemical species of the material) or a count of the number of molecules. (You should be familiar with this unit from your chemistry studies.) One mole of a given gas is defined to be the amount of gas with a mass in grams equal to the *atomic* (or *molecular*) *weight* of the gas. For example, 1 mol of helium gas has a mass of 4 g; the atomic weight of helium is 4 units of atomic mass. One mole of gas *always* contains *Avogadro's number* of molecules—$N_A = 6.022 \times 10^{23}$ molecules. The total number of molecules in a container, N, can be written in terms of the number of moles, n, as

$$N = nN_A. \tag{17–8}$$

Variables such as p, T, V, and n describe thermal systems, and we refer to them as **thermodynamic variables**.

CONCEPTUAL EXAMPLE 17–7 In an experiment, a tiny known volume V of oil is placed on calm water and subsequently spreads over an area A, forming a film of oil one molecule thick (a monolayer). The oil has a known density ρ and molecular weight M. How can you use this information to (a) estimate the size of a molecule of the oil, and (b) estimate Avogadro's number N_A. What additional assumptions are necessary for your estimates? (This experiment was apparently first performed by Benjamin Franklin.)

Answer (a) If we make the additional assumption that the oil molecules in the monolayer are as close as they are in the original drop, then we estimate the thickness t of the monolayer as follows: The original drop forms a "pancake" of volume At equal to the original volume V. Thus $t = V/A$. If we now assume further that the molecules themselves are compact—that we can, for example, treat them as little cubes of size l—then we can identify $l = t = V/A$. (b) For an estimate of N_A, we note that the number N of our molecules is given by $N = V/l^3$. But we have another way to get N: If $m = \rho V$ is the (known) mass of the drop, then $N : N_A = m : M$. (Remember, N_A is the number of molecules in a mass M of the material.) In other words, $N_A = N(M/m)$.

Ideal gases—think of dilute gases here—obey *Boyle's law*, discovered in 1662 by Robert Boyle. Boyle's law states that for a dilute gas held at constant temperature, the product of pressure p and volume V divided by the number of moles n is a constant:

$$\text{for constant } T\text{: } \frac{pV}{n} = \text{a constant}. \tag{17–9}$$

BOYLE'S LAW

We previously defined temperature as a linear function of pressure if volume is constant. We can combine this result [Eq. (17–1)] with Boyle's law into the **ideal gas law**:

$$pV = nRT, \qquad (17\text{–}10)$$

IDEAL GAS LAW

where R is a proportionality constant. The ideal gas law is an example of an **equation of state**, a relation between the thermodynamic variables of a thermal system. Such a relation holds for any thermodynamic system. The ideal gas law is the equation of state for an ideal gas. Measurements of ideal gases give

$$R \cong 8.314\frac{\mathrm{N \cdot m}}{\mathrm{mol \cdot K}} = 8.314\ \mathrm{J/mol \cdot K}. \qquad (17\text{–}11)$$

R is known as the **universal gas constant**. If we count the number of molecules N of a gas rather than the number of moles n, we can use Eq. (17–8) in order to rewrite Eq. (17–10) as

$$pV = NkT, \qquad (17\text{–}12)$$

IDEAL GAS LAW

where the constant k, **Boltzmann's constant**, is given by

$$k = \frac{R}{N_A} \cong 1.381 \times 10^{-23}\ \mathrm{J/K}. \qquad (17\text{–}13)$$

Boltzmann's constant, named for Ludwig Boltzmann, one of the important figures in nineteenth-century thermal physics, provides the bridge between temperature and energy. To see how, note that the quantities pV/n and pV/N, from Eqs. (17–10) and (17–12) respectively, have the dimensions of [force/length^2][length^3] = [force][length], which is the same as work or, equivalently, energy. *The dimensions of RT and kT are those of energy.* This is already reflected in the units of R and k [Eqs. (17–11) and (17–13)]. The significance of this observation is that temperature, as defined by the ideal gas law, must be closely related to the energy of a thermal system. This *thermal* (or *internal*) energy is the subject of Chapter 18.

CONCEPTUAL EXAMPLE 17–8 Before you drive through West Texas on Interstate 10 toward El Paso in the summer, you are advised not to pressurize your automobile tires to the maximum allowed pressure. Why?

Answer First, you should know that the air temperature is likely to be high and that people drive very fast on the wide-open roads of Texas. The tires will get quite hot. However, tires are constructed with steel belts so that they do not expand in volume past a certain amount. If the volume stays constant, and the temperature increases considerably, then the pressure in the tires will increase substantially, and they may become unsafe against a blowout. This is good advice for driving anywhere in summer.

EXAMPLE 17–9 Calculate the volume occupied by 1.0 mol of an ideal gas at 20°C and at atmospheric pressure. Use this volume to calculate the mass density of air, whose molecular weight M is 29 g/mol.

Strategy The ideal gas law gives us the volume for n moles of gas, $V = nRT/p$. We then use the M of air ($= 29$ g/mol) to find the density of air, $\rho = nM/V$. Here we'll want to set $n = 1.0$.

Working It Out The temperature T is 20°C = (20 + 273)K = 293K, and T must be in K for the ideal gas law. The volume is

$$V = \frac{nRT}{p} = \frac{(1.0\ \mathrm{mol})(8.3\ \mathrm{J/mol \cdot K})(293\mathrm{K})}{1.0 \times 10^5\ \mathrm{Pa}} = 2.4 \times 10^{-2}\ \mathrm{m}^3.$$

As for the mass density of air,

$$\rho = \frac{29\ \mathrm{g}}{2.4 \times 10^{-2}\ \mathrm{m}^3} = \frac{(29\ \mathrm{g})(10^{-3}\ \mathrm{kg/g})}{2.4 \times 10^{-2}\ \mathrm{m}^3} = 1.2\ \mathrm{kg/m}^3.$$

What Do You Think? Where does the molecular weight 29 g/mol used in this example come from?

Changing the Thermodynamic Variables of a Gas

Thermal systems change as the thermodynamic variables change. We say that a *thermal transformation* or, alternatively, a *thermal process*, occurs. For example, when the cylinder of gas in Section 17–2 was put in contact with a hotter thermal reservoir, the gas' temperature changed; so does the gas' volume in the set-up in Fig. 17-1. In this transformation the temperature, the volume, and the pressure all changed subject to the constraints of the equation of state. There are many different ways to change a thermal system. For example, in an **isothermal** transformation the temperature remains fixed while the pressure and volume both change. In an **isobaric** transformation the pressure is held fixed while both temperature and volume change. Such transformations are most easily shown on diagrams in which one variable is plotted against another. For example, in a p–V diagram, pressure is plotted against volume. When such a diagram shows various *phases* of a material—whether it is solid, liquid, or gas—it is referred to as a *phase diagram.* Figure 17–12 is a p–T phase diagram for water. We'll be looking at phase diagrams throughout, concentrating on phase changes—transitions across phase boundaries—in Chapter 18. Careful: the line that separates phases on a phase diagram is not a transformation!

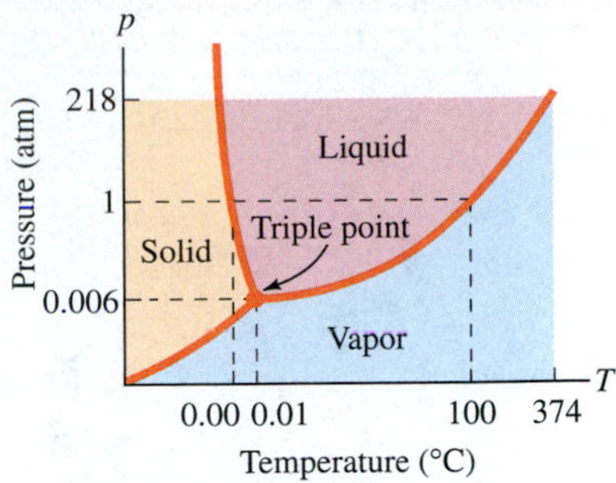

▲ **FIGURE 17–12** Phase diagrams are useful for the different phases of a material, such as those for water, shown here: liquid, solid (ice), and vapor (steam). The intersection of these phases is the triple point of water.

CONCEPTUAL EXAMPLE 17–10 Referring to Fig. 17–12, what changes can be made to water to allow you to go from solid ice to liquid water without any change in temperature? How might this be useful for understanding ice skating?

Answer Consider the almost vertical line near 0.00°C on the phase diagram in Figure 17–12. In the region of 1 atm, increasing the pressure—moving up along the constant temperature line, say at 0°C—takes you from the solid phase (ice) to the liquid one. It is the fact that the slope of the line describing the solid-liquid phase separation is negative on the p–T diagram that allows this to happen. As an ice skater moves over the ice, the weight of his or her body increases the pressure under the blade on the ice. If the temperature is not too low, this increased pressure will be enough to melt the ice, forming a film of water between the ice skate and ice, in effect lubricating the motion of the blade over the ice. A narrow blade makes the pressure higher and the effect more pronounced. A glance at the diagram shows that this effect is more difficult to realize at lower temperature.

EXAMPLE 17–11 Exactly 1 mol of an ideal gas is taken through the sequence of changes shown on the p–V diagram in Fig. 17–13. The change $A \to B$ is an isothermal transformation. In $B \to C$ the transformation is isobaric, that is, the gas is compressed in volume in such a way that the pressure remains fixed. (a) If $p_A = 5.00$ atm and $V_A = 8.00$ L, what is the temperature T_A? (b) If $V_B = 40.0$ L, what is p_B? (c) If $T_C = T_A/8$, what is V_C?

Setting It Up Figure 17–13 provides a very useful picture of what is happening. Diagrams such as this one are very important in problems involving thermal transformations.

Strategy We can apply the equation of state for the ideal gas, $pV = nRT$, with $n = 1$ mol. This gives us T_A immediately.

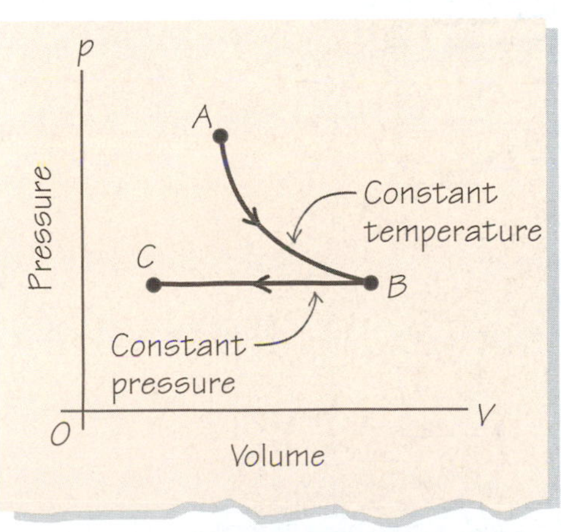

▲ **FIGURE 17–13**

Then, since $T_B = T_A$, we can use the ideal gas law to find p_B. Finally, since $p_C = p_B$ and T_C is given, we can again use the ideal gas law to find V_C.

Working It Out (a) We know p_A and V_A; the temperature T_A is then given by

$$T_A = \frac{p_A V_A}{nR} = \frac{(5.00\ \text{atm})(8.00\ \text{L})}{(1\ \text{mol})(8.314\ \text{N}\cdot\text{m/mol}\cdot\text{K})}.$$

We make some unit conversions in order to cancel units:

$$T_A = \frac{(5.00\ \text{atm})(1.01 \times 10^5\ \text{Pa/atm})(8.00\ \text{L})(10^{-3}\ \text{m}^3/\text{L})}{(1\ \text{mol})(8.314\ \text{N}\cdot\text{m/mol}\cdot\text{K})} = 486\text{K}.$$

(b) $T_B = T_A$ from the conditions stated in the problem. We know V_B and T_B, and hence we can compute p_B:

$$p_B = \frac{nRT_B}{V_B} = \frac{nRT_A}{V_B} = \left(\frac{nR}{V_B}\right)\left(\frac{p_A V_A}{nR}\right) = p_A\left(\frac{V_A}{V_B}\right)$$

$$= (5.00\ \text{atm})\left(\frac{8.00\ \text{L}}{40.0\ \text{L}}\right) = 1.00\ \text{atm}.$$

(c) Since we know $p_C = p_B$ and we know T_C, we can calculate V_C. It is given by

$$V_C = \frac{nRT_C}{p_C} = \frac{nR}{p_C}\left(\frac{T_A}{8}\right) = \frac{nR}{8p_C}\left(\frac{p_A V_A}{nR}\right) = \frac{p_A V_A}{8p_C} = \frac{p_A V_A}{8p_B}$$

$$= \frac{(5.00\ \text{atm})(8.00\ \text{L})}{8(1.00\ \text{atm})} = 5.00\ \text{L}.$$

What Do You Think? Suppose you had followed a straight line from A to C on the diagram of Fig. 17–13. What would have been constant: *p, V, or T*?

THINK ABOUT THIS . . .

HOW DO WE GET THE FIZZ INTO DISPENSED DRINKS?

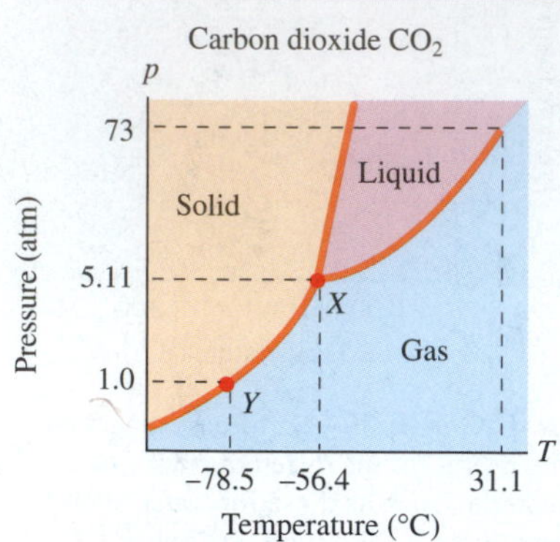

◀ **FIGURE 17–14** Phase diagram for carbon dioxide. At a pressure of 1 atm, only the solid and gas phases exist.

A fizzy drink from a carbonated beverage dispenser is made by mixing a "syrup" of sugar and flavorings with water and carbon dioxide as it is dispensed. It is the carbon dioxide that gives the drink its fizz. But how is the carbon dioxide handled? At a pressure of 1 atm, carbon dioxide exists only in the solid phase (at low temperatures) and in the gas phase (at normal temperatures). This is visible in the p–T phase diagram for carbon dioxide (Fig. 17–14). Both forms are inconvenient to use; the gas is bulky, and the solid is difficult to incorporate in mixing systems. Carbon dioxide is therefore transported and delivered at pressures greater than 40 atmospheres, pressures at which the room-temperature phase is liquid. This liquid can then be conveniently stored, transported efficiently through pipes, controlled by automatic valves, and so forth. When it is needed for incorporation into soft drinks near room temperature, it is released to one atmosphere of pressure, at which point it changes to the gas phase, becoming the bubbles in your drink.

How Close Do Real Gases Come to Being Ideal?

When is a real gas "sufficiently dilute" to obey the ideal gas equation of state? Of course, this is a question without a definite answer, because there is no sharp line dividing ideal gases from nonideal ones, just a matter of tolerance to approximation. But we can approach this question by asking how well Boyle's law is satisfied as we vary the density of a gas. Consider a series of identical closed tubes. A pressure gauge can measure the pressure of any gas in the tubes. Put different amounts (a different number of moles, n) of the same gas in each tube and immerse them in the same thermal reservoir. If Boyle's law is satisfied exactly, the (measured) ratio pV/n will be the same in each tube, independent of the pressure in the tube. One can also imagine a set of thermal reservoirs with different temperatures. In that case the more stringent test for ideal gas behavior is that the ratio pV/nT is the same in each tube.

Experiment shows that for monatomic gases,† such as helium, the ratio pV/nT is independent of the pressure over a wider range of pressures and therefore fits the ideal gas limit better than do polyatomic gases such as O_2, or carbon dioxide, or ammonia. However, *all* gases give precisely the same value of pV/n when this quantity is extrapolated to *zero density* (Fig. 17–15). All gases are indeed ideal in this limit. A more appropriate definition of temperature than that presented in Eq. (17–2) is obtained by *extrapolating the pressure ratio* in Eq. (17–2) to zero density ρ:

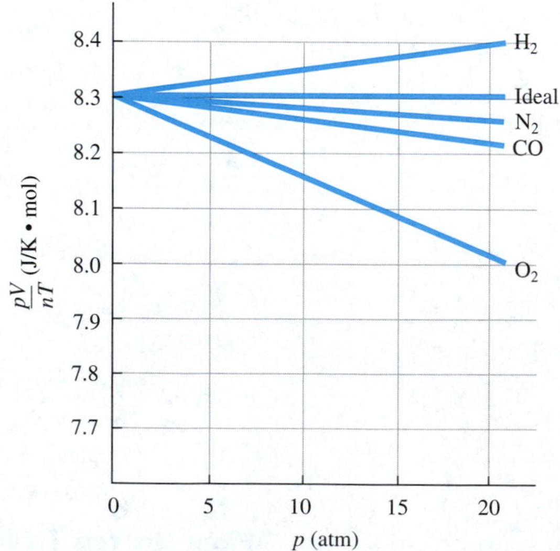

▶ **FIGURE 17–15** As the pressure (and hence the density) of any gas drops to zero, the ratio pV/nT goes to the same constant, which is the universal gas constant. This behavior characterizes the ideal gas.

†The molecule of a monatomic gas consists of one atom; the molecule of a polyatomic gas consists of several.

$$T = (273.16\text{K}) \lim_{\rho \to 0} \frac{p}{p_{tp}}. \tag{17–14}$$

KELVIN TEMPERATURE SCALE DEFINED

Equation (17–14) provides us with the desired universal definition of temperature because it holds in the domain where *every* gas is ideal.

The van der Waals Equation of State

In 1873 Johannes D. van der Waals proposed an equation of state that describes the properties of many real gases better than the ideal gas law. His formula, known as the *van der Waals equation*, takes the form

$$\left[p + a\left(\frac{n}{V}\right)^2\right]\left(\frac{V}{n} - b\right) = RT. \tag{17–15}$$

The constants a and b are both positive and specific to a given gas. The van der Waals equation includes a correction for intermolecular forces through the constant a and a correction for finite molecular size through the constant b. In contrast, the ideal gas law can be derived at the molecular level through assumptions that the gas molecules are noninteracting point particles. (See Chapter 19 for more on this.) We can verify that the van der Waals equation reduces, as it must, to the ideal gas law when the gas is very dilute: In the limit as $n \to 0$, the term $a(n/V)^2$ is small compared to p, and the term V/n is large compared to b. Thus, in this limit, Eq. (17–15) reduces to $pV/n = RT$.

The van der Waals equation of state is a considerable empirical improvement over the ideal gas law for polyatomic gases that are about as dilute as the air we breathe. It describes all gases reasonably well, even when a gas is nearly a liquid; that is, at very low temperatures and at densities approaching those of real liquids.

*17–5 Blackbody Radiation

Most of us have enjoyed an evening before a roaring blaze that slowly dies down during the course of the evening. As the fire dies, its color changes from brilliant white to bright yellow, to orange, and finally to a dull red. The change is more marked if you look into a cavelike hole within the embers, a spot where the temperature is steadier because it is protected from the vagaries of drafts. This phenomenon is the basis of a thermometer called the *optical pyrometer* (Table 17–1). This thermometer works because the color or, rather, the range of colors emitted by a hot object is characteristic of the temperature of that object. This thermometer works best for relatively hot objects with temperatures in the range of several hundred kelvins.

The light emitted by hot objects is **electromagnetic radiation**. Radiation exists in the form of waves, and all frequencies are possible in principle. One small range of frequencies corresponds to visible light, and different frequencies within this range are perceived as different colors, but even if we cannot literally see the radiation at other frequencies, we often use it (radio, infrared, microwave, infrared, X-rays, etc). All matter, at any temperature, absorbs and emits electromagnetic radiation, and across the full range of frequencies. The basic physical emission mechanism is that atoms go into *excited states* when the temperature of the matter they comprise is raised. They then radiate energy when they return to their normal states. For example, the atoms heated by an electric current in a lightbulb filament radiate visible light, from red to orange to yellow to green to blue to violet, as the frequency of the radiation increases from 0.5×10^{15} Hz to approximately 1.0×10^{15} Hz. The filament also radiates a great deal of energy in the form of longer wavelength radiation, and that is why the bulb is hot as well as brilliant.

In some ways, radiation in an enclosure acts like a dilute gas in a bottle; in particular, the radiation has a temperature just as a gas does. However, there is an important difference: Radiation achieves thermal equilibrium through the fact that it is absorbed and emitted by the atoms of the walls of the enclosure rather than through collisions among its components (as do the molecules that make up a dilute gas). The radiation in the enclosure at thermal equilibrium at a given temperature is a mix of *different*

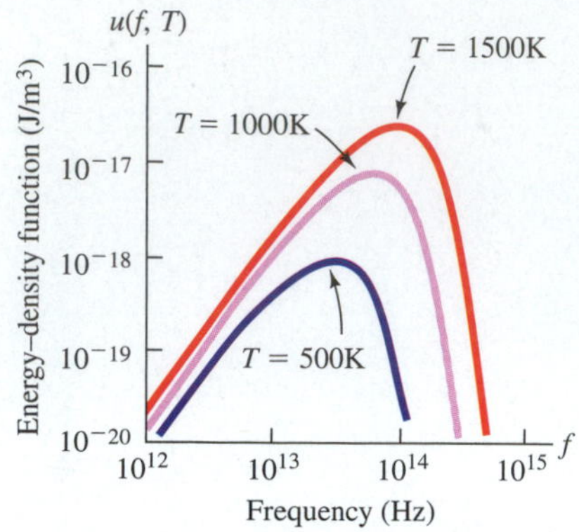

▲ **FIGURE 17–16** The energy density of electromagnetic radiation in thermal equilibrium at the indicated temperatures. The maximum energy density occurs for increasing frequency as the temperature increases. The maximum energy density increases very rapidly with T for a given f; note that this is a log–log plot.

frequencies, and what is important here is that in the mix the proportion of high frequencies to low frequencies rises with temperature. The average frequency of the radiation associated with a white-hot object is greater than the average frequency associated with the object when it is glowing red. We can give a function that describes the radiation at *different* frequencies associated with a given temperature. Such a function is denoted by $u(f, T)$, where *$u(f, T)\, df$ is the radiation energy per unit volume, or energy density, with a frequency between f and $f + df$.* We say that the radiation is at a given temperature when it is in thermal equilibrium, meaning that the radiation interacts with an object or system at that temperature over a long period and that the object and the associated radiation is isolated from its surroundings. Radiation of this type is known as **blackbody radiation**, and the object that emits and absorbs such radiation is known as a blackbody. Figure 17–16 shows this function for different values of temperature. *The measurement of $u(f, T)$ provides a measurement of T.*

We sometimes refer to this radiation as being "emitted" when we look into the enclosure; that is, we can see the radiation that escapes from the "bottle" of radiation. This picture most exactly describes your looking deeply into a furnace through a small window or a campfire through a little opening in the pile of wood. If the window is large, then so much radiation escapes to the outside that the system may cool down, which is why we refer to a small window. But equivalently, if we have a way to maintain a high temperature, then we just say that the blackbody radiation is emitted from an object at that temperature. An example is provided by the filament of a lightbulb, which would quickly cool if we didn't maintain an electric current through it that holds it at a white-hot temperature.

The form of the energy–density function $u(f, T)$ was a subject of intense study in the latter part of the 19th century. The work of Gustav Robert Kirchhoff, Wilhelm Wien, Lord Rayleigh, and James Jeans set the stage for the breakthrough of Max Planck, who determined in 1900 that

$$u(f, T) = \frac{8\pi h}{c^3}\frac{f^3}{e^{hf/kT} - 1}. \qquad (17\text{–}16)$$

This formula, known as the **Planck formula**, perfectly fits the data on the energy density of radiation for a wide range of frequencies and temperatures. In addition to Boltzmann's constant, k [introduced in Eq. (17–13)], and the Kelvin temperature, T, the formula contains the speed of light, $c = 3 \times 10^8\ \text{m/s}$, which is the speed of electromagnetic waves of *every* frequency. A new fundamental constant also appears in this formula. The new constant h, called *Planck's constant*, was found by fitting the formula to the observed energy density, and its value is $h = 6.625 \times 10^{-34}\ \text{J}\cdot\text{s}$. This is an awfully small number, and many of the effects associated with it are visible only on an atomic scale, but its presence in the formula for blackbody radiation shows that it can perfectly well have a macroscopic effect. Note that the quantity hf has dimensions of energy as does the quantity kT. (Indeed, these quantities must have the same dimensions because their ratio is the argument of the exponential function.)

The discovery of the Planck formula was the first step on the path to the basic theory of matter and radiation, called *quantum theory*. This theory is based on a set of ideas that revolutionized the physical sciences in the twentieth century. (The other great revolution in 20th-century physics—Einstein's theory of relativity—also has its origin in the investigation of electromagnetic radiation.)

An interesting consequence of the Planck formula is that the total power radiated through a small hole in an enclosure containing radiation in thermal equilibrium at temperature T is proportional to the fourth power of the temperature. The **Stefan–Boltzmann formula**, which was discovered experimentally by Josef Stefan in 1879 and inferred from thermodynamics by Boltzmann in 1884 before the discovery of the Planck formula, states that the radiated power (energy per unit time) per unit area emitted through the hole is given by $P(T)/A$, usually written as

$$E(T) = \sigma T^4, \qquad (17\text{–}17)$$

where $\sigma = 5.67 \times 10^{-8}\ \text{W}/(\text{m}^2\cdot\text{K}^4)$. The T dependence in Eq. (17–17) can be obtained by integrating Eq. (17–16) over all frequencies f (see Problem 76).

EXAMPLE 17–12 A surface that is "white hot" emits about 10 times more power than a "red hot" surface of the same area. What does this tell us, quantitatively, about the relative temperatures?

Strategy Power and temperature are related through the Stefan–Boltzmann formula, Eq. (17–17). By setting the ratio of the powers to 10, we have an equation for the ratio of the temperatures.

Working It Out From Eq. (17–17),

$$\frac{P(\text{white hot})}{P(\text{red hot})} = 10 = \frac{\sigma T_{\text{white}}^4}{\sigma T_{\text{red}}^4} = \left(\frac{T_{\text{white}}}{T_{\text{red}}}\right)^4$$

This implies that the temperature of the "white hot" surface exceeds that of the "red hot" surface by a factor of

$$\frac{T_{\text{white}}}{T_{\text{red}}} = 10^{1/4} = 1.78.$$

Let's examine the Planck formula for the energy–density function, Eq. (17–16). If the frequencies are small, $hf \ll kT$, then the exponential function in $u(f, T)$ has a small argument, and we can use the expansion

$$\text{for small } x\text{: } e^x \cong 1 + x + \cdots$$

In this case, the denominator of $u(f, T)$ is

$$e^{hf/kT} - 1 \cong 1 + \frac{hf}{kT} - 1 = \frac{hf}{kT},$$

and the function $u(f, T)$ takes the approximate form

$$\text{for } hf \ll kT\text{: } u(f, T) \cong \frac{8\pi h}{c^3} f^3 \frac{kT}{hf} = \frac{8\pi f^2}{c^3} kT. \tag{17–18}$$

This expression is independent of Planck's constant and is thus independent of quantum physics. It was, in fact, first found by Lord Rayleigh in 1900 by using a derivation based on classical physics. The energy density in Eq. (17–18) rises with increasing frequency as f^2 and linearly with temperature.

If, however, the frequencies are large, $hf \gg kT$, then the exponential argument in $u(f, T)$ is large and is the exponential function itself. The denominator then becomes

$$e^{hf/kT} - 1 \cong e^{hf/kT}.$$

(Here we have used the fact that for $hf/kT \gg 1$, $\exp(hf/kT) \gg 1$.) Then the energy density is approximately

$$\text{for } hf \gg kT\text{: } u(f, T) \cong \frac{8\pi h}{c^3} f^3 e^{-hf/kT}. \tag{17–19}$$

Because the exponential function falls with f much faster than the polynomial factor f^3 grows, $u(f, T)$ decreases as f grows in this range of f. Note that h plays a crucial role in this part of the energy–density function, indicating that quantum physics is important.

As we have seen, the energy–density function increases with f for $hf \ll kT$ and decreases with f for $hf \gg kT$. Such a function has an intermediate maximum, which we can estimate as occurring at about $hf \cong kT$. The underlying physics of the high-frequency side of the curve of the energy–density function in Fig. 17–16 is quantum physics because it involves Planck's constant, but the low-frequency side can be explained by classical physics. The maximum moves to higher values of f as T increases. We can therefore say that the importance of quantum physics decreases as the temperature grows, and this is generally true.

EXAMPLE 17–13 The surface of the Sun is observed to emit light most strongly at a frequency of 2×10^{14} Hz. Estimate the temperature of the Sun's surface (Fig. 17–17).

Setting It Up We can safely assume the radiation is blackbody; thus we want to find the blackbody temperature that gives a maximum in the energy–density function $u(f_0, T)$ at a given frequency f_0.

Strategy As we argued above, the frequency f_0 for which u is a maximum is given by $hf_0/kT \cong 1$. We can solve this for T. (*Alternatively*, we could go beyond this estimate and calculate the maximum of the energy–density function, Eq. (17–16), by taking

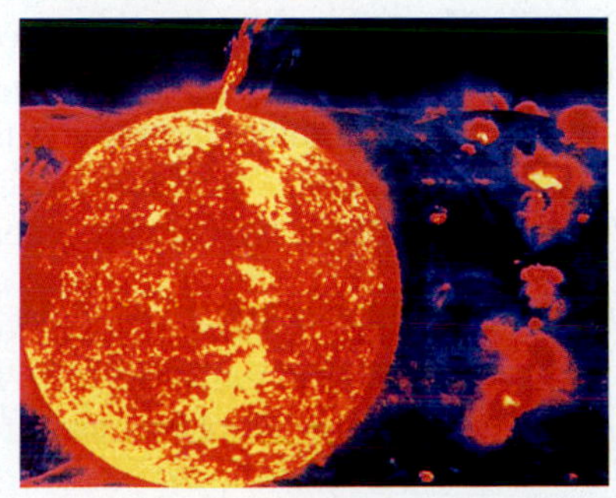

◀ **FIGURE 17–17** The temperature of the Sun's surface is not constant but varies somewhat in active areas, such as regions with Sun flares.

(continues on next page)

the derivative of u with respect to f and setting the derivative to zero. The simpler estimate will be enough here.)

Working It Out We have

$$T \cong \frac{hf_0}{k} = \frac{(6.63 \times 10^{-34}\,\text{J}\cdot\text{s})(2 \times 10^{14}\,\text{Hz})}{1.38 \times 10^{-23}\,\text{J/K}} = 10^4\text{K}.$$

The more precise alternative solution described above gives a solar surface temperature of 6000K (see Problem 80).

What Do You Think? The interior of the Sun is at a much higher temperature than the surface. What happens to the radiation, characteristic of the higher internal temperatures, that must be present in the interior?

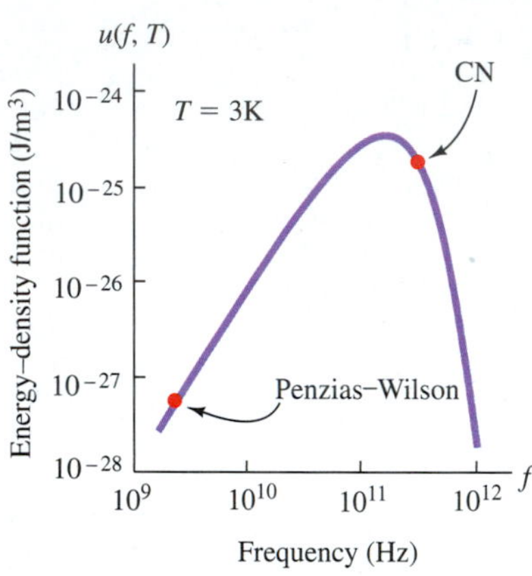

▲ **FIGURE 17–18** The energy density of electromagnetic radiation in thermal equilibrium at $T = 3\text{K}$. The Penzias–Wilson frequency measurement and the frequency associated with interstellar cyanogen, labeled CN, are marked by red dots.

The Discovery and Measurement of the Background Radiation of the Universe

In 1964 Arno Penzias and Robert Wilson used a large radio antenna to study the radio waves emitted within our galaxy. They found a puzzling anomaly, which they at first thought might be a problem with their antenna: A background electromagnetic radiation that comes from no particular source is present. Their antenna was sensitive only to the particular frequency $f = 4.08 \times 10^9$ Hz, but later measurements over other frequencies showed that this background radiation was consistent with the Planck formula.

Penzias and Wilson were not aware of work by Ralph Alpher, George Gamow, and Robert Herman in the late 1940s, or of work by James Peebles that was contemporary with their own, in which a cosmological, or large-scale, model of the universe predicted the existence of this radiation. As described by this model, the early universe expanded from an initial catastrophic event—the Big Bang. As the universe expanded, it cooled, maintaining thermal equilibrium between electromagnetic radiation and matter up to a point, some 100,000 years after the Big Bang, when the temperature was about 3000K. At that point, the model shows that electromagnetic radiation and matter decoupled; they were no longer in thermal equilibrium. However, and quite remarkably, as the universe continued to expand, the Planck formula still described the radiation, but with a temperature that dropped as the universe expanded. *A measurement of the present-day temperature of the radiation provides us with a measurement of the rate of expansion of the universe; hence, it provides us with a measurement of the time that has elapsed since the radiation was no longer in thermal equilibrium with matter.* Presently the temperature of the radiation, as characterized by the Planck formula, is about 3K, and this radiation is the oldest directly observable relic of the explosion in which the universe is formed.

A measurement of a single frequency is not a measurement of an entire spectrum! Penzias and Wilson's frequency measurement occurs at one value of frequency on the classical, or low-frequency, side of the Planck formula, as Fig. 17–18, the energy–density function for $T = 3\text{K}$, indicates. Since 1964 many other Earth-based measurements have been made by radio antennas sensitive to a great variety of frequencies. Because Earth's atmosphere absorbs radiation of high frequency, these measurements are all made on the classical side of the Planck formula. Fortunately, one other technique can make a measurement on the quantum, or high-frequency, side of the peak of energy density, and measure the amount of light emitted by interstellar cyanogen, the CN molecule. This light is emitted with a strength that can be explained only if the cyanogen is in the presence of background radiation of a certain temperature. The frequency measurement of CN, first made in 1941, is marked on Fig. 17–18. Today we have a series of satellite measurements of the background radiation. Measurements can be made with extraordinary precision across a broad range of the spectrum, including across the maximum of the energy–density curve. We show the results in Fig. 17–19; a temperature of about 2.7K fits the curve perfectly. (The actual temperature can be determined to six significant figures!) Still other measurements reveal that the temperatures in different directions differ ever so slightly from one another. Some regions are somewhat hotter, while others are cooler. These temperature inhomogeneities may well be relics of the instabilities that led to the formation of the great clusters of galaxies present today. Indeed, the precise nature of the inhomogeneities provides further evidence for the Big Bang as well as other cosmological (universe-wide) features. The Big Bang model of the universe, which predicts the presence of this background blackbody radiation, has a remarkable amount of experimental support.

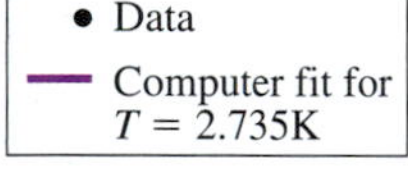

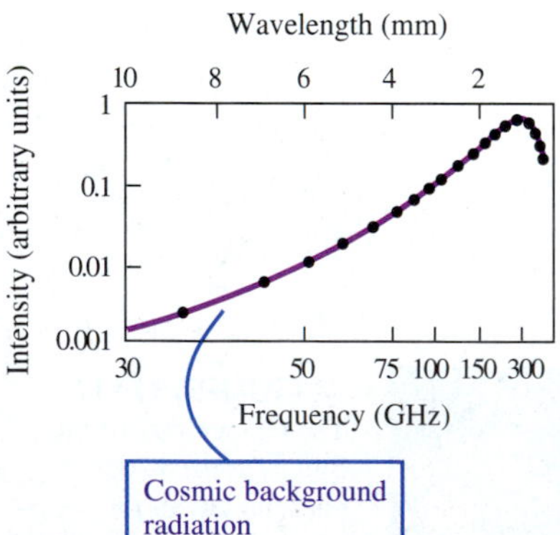

▲ **FIGURE 17–19** Satellite-based measurements of the cosmic background radiation show remarkable agreement with the Planck formula. The data shown for $T = 2.735\text{K}$ fit the smooth blackbody curve very well indeed.

Summary

The temperature of a system in thermal equilibrium is defined on the Kelvin scale by the behavior of the pressure, p, of a dilute gas in a constant-volume gas thermometer:

$$T = (273.16\text{K}) \lim_{p\to 0} \frac{p}{p_{tp}}, \quad (17\text{–}14)$$

where p_{tp} is the pressure of the gas in the thermometer at the triple point of water (where the gaseous, liquid, and solid phases coexist); thus 273.16K is chosen to be the temperature at the triple point. The other fixed point of the Kelvin scale is built into the thermometer: $T = 0\text{K}$, absolute zero, is the point where the pressure extrapolates to zero. Two other temperature scales in common use are the Celsius scale and the Fahrenheit scale. The Celsius temperature is given in terms of the Kelvin temperature by

$$t_C = T - 273.15. \quad (17\text{–}3)$$

Any sufficiently dilute gas behaves as an ideal gas, meaning that its thermodynamic variables are related by the ideal gas law, an equation of state:

$$pV = nRT = NkT, \quad (17\text{–}10,\ 17\text{–}12)$$

where the universal gas constant is

$$R \cong 8.314\ \text{J/mol}\cdot\text{K} \quad (17\text{–}11)$$

and Boltzmann's constant is

$$k = \frac{R}{N_A} \cong 1.381 \times 10^{-23}\ \text{J/K}, \quad (17\text{–}13)$$

with N_A Avogadro's number. n and N measure the quantity of gas, in moles or in numbers of molecules, respectively. The quantities kT and RT have dimensions of energy.

Real gases are ideal only in the limit that they are very dilute. The small departures from ideal behavior are accurately reflected for many real gases by the empirical van der Waals equation of state,

$$\left[p + a\left(\frac{n}{V}\right)^2\right]\left(\frac{V}{n} - b\right) = RT. \quad (17\text{–}15)$$

Materials expand when heated, and they are characterized by a coefficient of thermal expansion, α, defined by

$$\alpha \equiv \frac{1}{L}\frac{dL}{dT}. \quad (17\text{–}5)$$

Electromagnetic radiation displays the characteristics of a thermal system. The total power emitted per unit area by radiation in equilibrium with another thermal system at temperature T, which is known as blackbody radiation, is given by the Stefan–Boltzmann formula,

$$P(T)/A = E(T) = \sigma T^4, \quad (17\text{–}17)$$

where $\sigma = 5.67 \times 10^{-8}\ \text{W/(m}^2\cdot\text{K}^4)$. Blackbody radiation contains a range of frequencies f that have an energy per unit volume (or energy density) per unit frequency, given by

$$u(f, T) = \frac{8\pi h}{c^3}\frac{f^3}{e^{hf/KT} - 1}, \quad (17\text{–}16)$$

a result known as the Planck formula. The constant h is Planck's constant.

Electromagnetic radiation with this frequency range is observed to occur throughout the universe at a characteristic temperature of 3K. This is strong evidence for the Big Bang theory of the origin of the universe.

Understanding the Concepts

1. Estimate to within an order of magnitude the number of molecules in a breath of air.

2. Why is mercury used in most thermometers rather than a more common liquid such as water?

3. If a scuba diver takes the same volume of air for each breath and breathes at a constant rate, why is the tank of air used up much faster at greater water depths?

4. A student proposes to make a steel sphere in which helium gas can be inserted at various pressures. The student will insert the sphere into various temperature baths (boiling water, ice water, liquid nitrogen, dry ice-methanol) and measure the pressure at each temperature. Then he or she will change the amount of helium in the sphere and do it again. If the student plots pressure vs. temperature for these separate amounts of helium, where, if anywhere, should they intersect?

5. How can a triple point of water exist? If ice, water, and vapor all existed together, would the ice cool the vapor, making it water? Or would the vapor and water melt the ice?
6. Why is it advisable to measure the pressure in automobile tires when the car has not been driven for some time?
7. Devise an experiment to prove the experimental fact that hot objects radiate more energy per unit area than cold objects do.
8. It is well known that chameleons can change their skin color. In this chapter we spoke about the different colors associated with light emitted from objects of different temperature. Is this how a chameleon changes its color—by changing its temperature?
9. Why is it important to check the pressure in automobile tires before the winter season?
10. All gases obey the ideal gas law when they are made sufficiently dilute. Is this true when the temperature is lowered sufficiently?
11. In recent years an attempt has been made to convert Americans to the use of the metric system, so that, for example, two sets of tools would not be necessary for a mechanic who works on both American and European cars. Would it be of equal practical importance to convert Americans from the use of Fahrenheit to Celsius temperature scales?
12. When we observe a fire, we are observing radiation that has escaped the region of the hot fire. Is this a mechanism for cooling the fire? In what way could a fire be enclosed so that it remains hot?
13. In the experiment shown in Fig. 17–10, it often happens that the ball gets stuck in the ring while performing the experiment. Why do you think this happens? Could you dislodge the ball by heating up either the ball or the ring?
14. Suppose that we have no idea that gases consist of molecules. Could we still use the low-pressure limit of a gas to define temperature?
15. Substances in thermal contact reach thermal equilibrium after some time. Can you think of any experimental test that will indicate just when that time will have arrived?
16. Water expands when it freezes. You also have heard that 90% of icebergs are hidden beneath the surface. Why is the latter statement true? Can you estimate by how much water expands on freezing?
17. If the molecules of gases move around a lot, why do we never see a container of gas shake?
18. The pressure of a gas produces an outward force on the walls of a container. Why is there no force at an angle to the walls?
19. The Chamber of Commerce of a town plagued by temperature extremes devises a way to avoid bad publicity: It defines a new temperature scale, the G scale, such that (a) −40°F is defined to be +20°G; (b) +120°F is defined to be −40°G; and (c) the scale is linear. Is there anything wrong with this scale?
20. If a doughnut-shaped piece of a solid that expands under heat is heated, does the hole expand or shrink?
21. Water is one of the relatively rare substances that becomes less dense rather than more dense when it freezes. Describe what might happen to the ocean if ice were denser than liquid water.

Problems

17–1 Temperature and Thermal Equilibrium

1. (II) Which of these pairs of systems are in thermal equilibrium with one another? (a) A roast reaching the rare stage and the oven in which it sits. (b) The point of a meat thermometer in the roast of part (a) and the roast. (c) A sunbather getting a tan and the air around him. (d) A bather in a hot tub and the water around her. (e) An ice cube in a glass of water at 0°C. (f) Molten iron ore in a blast furnace and the walls of the furnace.

2. (II) You must measure the different temperatures of a series of bottles that contain 1 L of an unknown fluid. Which one of the following techniques would make a suitable thermometer, and why or why not? (a) The time it takes 1L of ice cubes dumped into the fluid containers to melt. (b) The melting time for ice cubes sealed in full 1-L plastic bags that are dumped into the bottles. (c) The melting time for ice sealed in full 1-mL plastic bags that are dumped into the bottles. (d) All of the above techniques, but the fluids in the bottles are gently stirred during the measurement.

3. (II) Thermodynamic variables are either extensive or intensive. A variable is *extensive* if, when two identical thermodynamic systems are combined into one, the variable of the combined system is double its original value in each system; otherwise, it is *intensive*. Which of the following variables are extensive, and which are intensive: (a) volume, (b) temperature, (c) pressure, (d) number of moles?

4. (II) A physicist needs to measure room temperature. Not having anything better at hand, she decides to make use of the thermal expansion of water to construct a thermometer (Fig. 17–20). She calibrates her apparatus to a dependable thermometer at 14°C and 38°C, and divides the range between these temperatures into 24 equal intervals. What will be the true temperature when her equipment reads 28°C? She knows that the volume of water can be described in this temperature range by the formula

$$V = V_0 \times (1 - 2.525 \times 10^{-4} + 4.98 \times 10^{-6}T + 4.94 \times 10^{-6}T^2),$$

where T is the temperature in °C.

▲ **FIGURE 17–20** Problem 4.

17–2 Ideal Gases and Absolute Temperature

5. (I) At what temperature values are the following scales the same: (a) the Fahrenheit and the Celsius, (b) the Celsius and the Kelvin, (c) the Fahrenheit and the Kelvin?

6. (I) The title of Ray Bradbury's science-fiction novel *Fahrenheit 451* refers to the temperature at which paper ignites. What is that temperature on the Celsius scale?

7. (I) An oral fever thermometer ranges from 35.5°C to 42.5°C. What is the corresponding range on the Fahrenheit scale?

8. (I) An ideal gas in a container is heated from 17°C to 120°C. By what factor does the pressure increase?

9. (I) A very dilute gas confined to a closed 1-L container is put on the burner of a stove and the temperature is raised from 271K to 349K. By how much does the pressure, which is originally at 1Pa, change?

10. (I) If an ideal gas thermometer could be constructed to measure the temperature of a hot star (see Table 17–2), by what factor would the pressure in the thermometer differ from p_{tp}?

11. (II) (a) Determine the temperature of the surface of Venus (730K) in the Celsius and the Fahrenheit scales. (b) Determine the temperature of the boiling point of liquid nitrogen (77K) in the Celsius and the Fahrenheit scales. (c) A comfortable room temperature is 75°F. Determine this temperature in the Kelvin and Celsius scales. (d) Heat pumps become less efficient than furnaces for heating homes when the temperature drops below 36°F. New laws may require that all temperatures be posted in the Celsius and the Kelvin scales as well as the Fahrenheit scale. What is this temperature in these other systems?

12. (II) In 1701 Isaac Newton proposed a linear temperature scale in which ice water has a temperature of 0°, whereas the human body in good health is assigned a temperature of 12°. In this scale, what is the temperature of someone with hypothermia (chilling)? What is the temperature of boiling water?

13. (II) A temperature scale—the Reaumur scale—developed in the early eighteenth century sets the ice point at 0°R and the boiling point of water at 80°R. Give the conversion formula from the Reaumur scale to the Kelvin scale and to the Fahrenheit scale.

14. (II) For the new temperature scale described in Question 19, what are the general conversion rules from °G to °F and vice versa? What is the conversion rule from °C to °G?

15. (II) The Rankine scale for temperature, named for William Rankine, is used in engineering applications. It has units of °R and is defined by $t_R = 9T/5$, where T is the temperature in the Kelvin scale. Find the value of t_R for the temperatures 100°C, 4.2K, 6000K, 32°F, −30°F, and −25°C.

16. (II) A bottle that has a freely sliding vertical piston and contains an ideal gas is used as a thermometer for water: The bottle is submerged in a water bath (whose temperature is to be measured) until thermal equilibrium is reached (Fig. 17–21). When the bottle is in contact with ice water, the volume inside the bottle is 0.35 L. The bottle is removed from the ice water and put in a second bath. This time, the enclosed volume is 0.38 L. What is the temperature of the second bath, in °C?

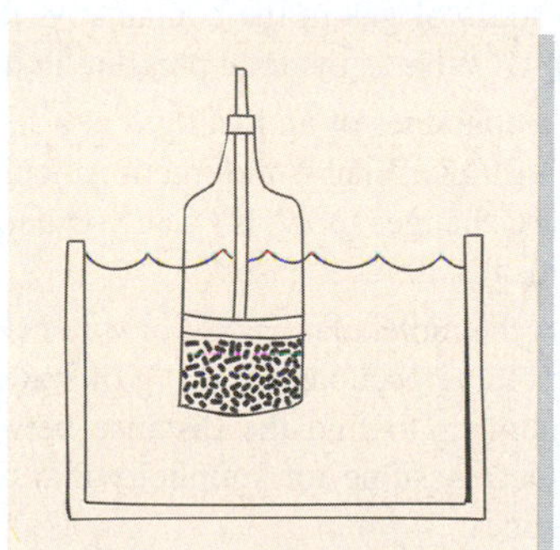

▲ **FIGURE 17–21** Problem 16.

17. (II) One Sunday morning a family takes an automobile trip to Grandma's. At the start of the trip, the temperature is 288K (15°C), and the gauge pressure in the tires is 32 lb/in^2 (psi). (The gauge pressure is the excess over 14.5 psi, the exterior air pressure.) After an hour's ride over an interstate highway, the gauge pressure in the tires is 38 psi. What is the temperature of the air in the tires, assuming that air behaves as an ideal gas? Neglect any changes in volume of the tires.

17–3 Thermal Expansion

18. (I) By how much will a steel rod of length 1.5 m and diameter 1.75 cm elongate if it is heated by 100°C?

19. (I) How much farther than a steel rod will a 1 m long aluminum rod expand when each undergoes a temperature increase of 40°C?

20. (I) The coefficient of thermal expansion for Invar is $0.70 \times 10^{-6}\,\text{K}^{-1}$. (Invar is a steel alloy.) Given that an Invar pendulum clock keeps perfect time at a room temperature of 20°C, how much time will the clock gain or lose per day when it is in a room at 30°C? (For the time-keeping properties of a pendulum, see Chapter 13.)

21. (I) The area of the circle formed by a circular loop of metal is found to increase by 1.6% when the temperature changes by 10°C. What is the coefficient of thermal expansion of the metal?

22. (II) When the temperature of mercury changes by ΔT, the fractional volume change, $\Delta V/V$, is given by $\Delta V/V = \beta \Delta T$, where $\beta = 1.8 \times 10^{-4}\,\text{K}^{-1}$ is the coefficient of thermal expansion. Consider a thermometer containing a bulb attached to a thin, cylindrical capillary tube. If the bulb and capillary tube contain 0.2 cm^3 of mercury, how large must the diameter of the capillary tube be so that a 1K change in temperature corresponds to a 2 mm change in the mercury level in the capillary tube? (Neglect the change in volume of the glass bulb when the temperature is raised by 1K.)

23. (II) A copper bowl of volume 1500 cm^3 is filled to the brim with water. Assume that the coefficient of volume expansion is $5.10 \times 10^{-5}\,\text{K}^{-1}$ for copper and $2.07 \times 10^{-4}\,\text{K}^{-1}$ for water. How much water will spill out when the temperature of the system is raised from 20°C to 50°C?

24. (II) Show that the area A of a plate whose coefficient of thermal expansion is α undergoes a fractional change $dA/A \cong 2\alpha\, dT$ when the temperature changes by a small amount dT.

25. (II) Look at the experiment done in Fig. 17–10. (a) How hot do we have to heat the brass ball of 2.028 cm diameter at room temperature to increase the diameter by 0.004 cm, so it will not pass through the ring? (b) What is the coefficient of volume expansion β for brass? (c) By what fraction does the volume expand for this temperature change?

26. (II) A metal bar is 1.75 m long with a coefficient of thermal expansion of $1.34 \times 10^{-5}\,\text{K}^{-1}$. It is rigidly held between two fixed beams. When the temperature rises, the metal bar takes on the shape of the arc of a circle (Fig. 17–22). What is the radius of curvature of the circle when the temperature rises by 40°C? [*Hint*: Use the small-angle approximation $\sin\theta \cong \theta - \theta^3/6$.]

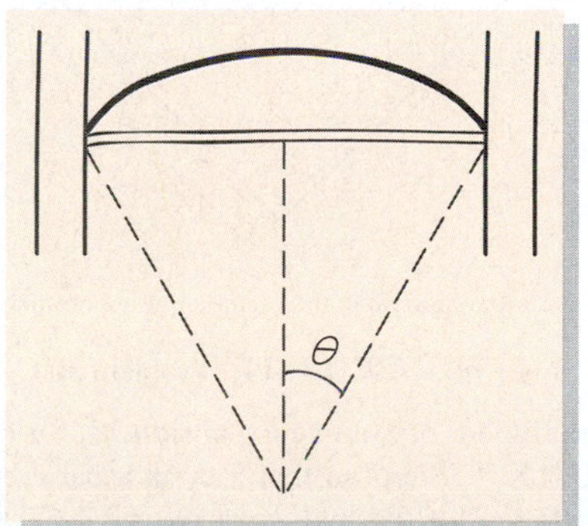

▲ **FIGURE 17–22** Problem 26.

27. (I) In the design of a mercury thermometer, it is desirable for each °C change to correspond to a 1 cm expansion of the metal. What is the cross section of the mercury column if the mercury at the lowest reading on the thermometer is contained in a bulb whose volume is 0.7 cm^3? Ignore any possible expansion of the glass tube.

28. (II) In space, where heat transfer by air is missing, large temperature differences can develop. A space telescope (Fig. 17–23) consists of a main mirror and an instrument unit, held together by titanium rods. The optical axis of the mirror (perpendicular to the mirror, through its center) hits the middle of the receiving window of the instrument unit. Suppose that the temperature of two neighboring rods increases by 20°C due to direct sunlight, with the temperature of the other two remaining unchanged. How far will the optical axis of the mirror move on the instrument unit, given that the rods are 5 m long and that $\alpha = 8.5 \times 10^{-6}\ K^{-1}$?

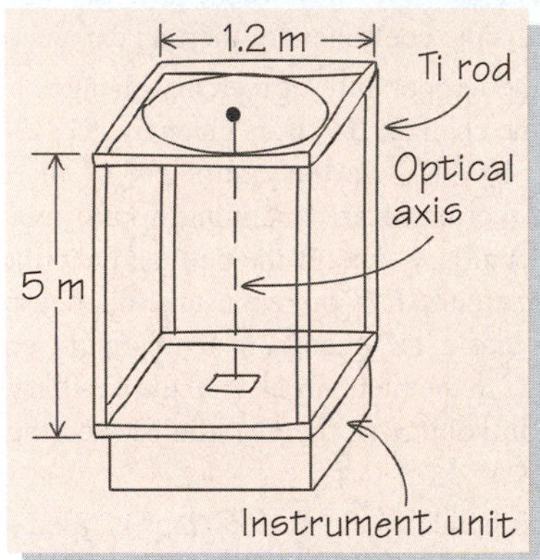

▲ **FIGURE 17–23** Problem 28.

29. (II) As we know from Chapter 13, the time expressed by a grandfather clock depends on the length of the pendulum. (a) Which is the best material for the pendulum of a clock to be used outdoors: aluminum, copper, or steel? (See Table 17–3.) (b) If the temperature of the clock ranges from -20°C to $+30$°C over the course of a year, and if the clock ticks with frequency $f = 1.00000$ Hz at 5°C, what is the yearly frequency range for each of the three materials?

30. (III) A strip of brass 1.5 mm thick and 12 cm long is glued back-to-back to a strip of Invar with the same dimensions. When the temperature is increased by ΔT, the composite strip tends to curl into an arc of a circle (Fig. 17–24). Find the radius of curvature of this circle for $\Delta T = 8$°C, given that the coefficient of thermal expansion is $19 \times 10^{-6}\ K^{-1}$ for brass and $0.7 \times 10^{-6}\ K^{-1}$ for Invar.

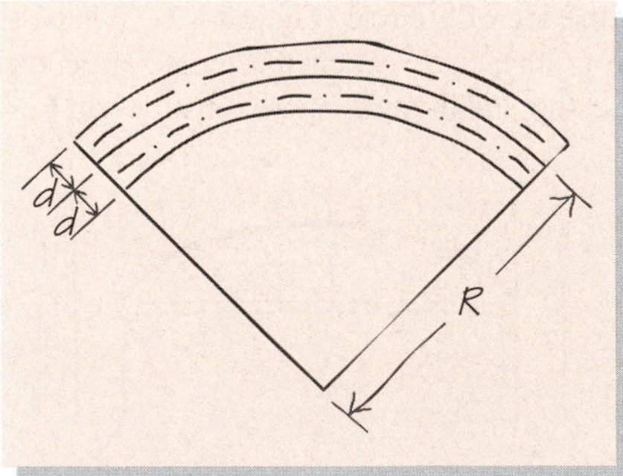

▲ **FIGURE 17–24** Problem 30.

31. (II) The coefficient of volume expansion, β, for ethyl alcohol is $1.12 \times 10^{-3}\ K^{-1}$. Suppose that a cylindrical copper container, with a coefficient of thermal expansion of $1.67 \times 10^{-5}\ K^{-1}$, is filled to the very brim with ethyl alcohol at 5°C. What percentage of the alcohol will spill if the temperature is raised to 25°C?

17–4 The Equation of State of Gases

32. (I) A container holds gas at a pressure of 1.0 atm and a temperature of 300K. Half the gas leaks out while the temperature is raised to 340K. What is the pressure in the container?

33. (I) A dilute gas in a container fitted with a piston occupies a volume of 2500 cm^3. If the pressure is increased by 50% and the temperature in kelvin is decreased by 15%, what volume will the gas occupy?

34. (I) A 1000 L container is filled with an ideal gas at 1.2 atm and 310K. How many moles are in the container? How many gas molecules?

35. (I) What is the volume occupied by three moles of an ideal gas at 30K and 10^{-7} atm in the upper atmosphere?

36. (I) What is the mass density of steam at a temperature of 100°C and a pressure of 1 atm, given that the molecular weight (the number of grams of 1 mol of the material) of water is 18 g/mol?

37. (I) What is the mass density of air (average molecular weight 29 g/mol), at a pressure of 0.31 atm and a temperature of -43°C? (These conditions correspond roughly to those at the top of Mount Everest.)

38. (I) A helium balloon is filled with 150,000 cm^3 of helium gas at *STP* (standard temperature and pressure: 0°C and 1 atm). How many moles and molecules of helium (molecular weight 4.0 g/mol) are contained in the balloon?

39. (I) A container of ideal gas whose volume can be changed by the movement of a plunger is placed in a large bath that maintains the temperature of the gas at a fixed value of 295K. The plunger is pulled out so that the volume is increased by a factor of 1.78. By what factor is the pressure changed in the vessel, if at all?

40. (I) The atomic weight A of an elemental substance is closely related to the mass of the atoms of that substance. Atoms consist of electrons, of negligible mass, and a total of A protons and neutrons, which each have nearly the same mass M. The total mass of an atom is therefore approximately AM, and N_A atoms have a total mass AMN_A. By the definition of Avogadro's number, N_A, this mass is A g. Given the value of N_A, what is M in grams?

41. (I) A container at 0°C holds a mixture of gases: 80% nitrogen (N_2 molecular weight 28 g/mol) and 20% oxygen (O_2 molecular weight 32 g/mol) by weight. Assuming that for a dilute gas each constituent behaves as if it alone occupied the volume, what is the partial pressure due to each constituent of the mixture, given that the total mass of gas in the container is 8.0 g and that the volume is 4.0 L? What is the total pressure in the container?

42. (I) Consider a container of an ideal gas at a fixed pressure. The temperature is 290K. What is the fractional change in volume if the temperature changes to 296K? Can you derive a general formula that holds for all ideal gases?

43. (II) Given that the molecular weight of water (H_2O) is 18 g/mol and that the volume occupied by 1.0 g of water is $10^{-6}\ m^3$, use Avogadro's number to find the distance between neighboring water molecules. Assume for simplicity that the molecules are stacked like cubes.

44. (II) The lowest pressure achievable with a rotary vane pump is on the order of $10^{-2}\ N/m^2$. What is the number density of molecules of the gas in a chamber that holds gas at this pressure and at a temperature of 20°C?

45. (II) The pressure of an ideal gas in a closed container is 0.60 atm at 35°C. The number of molecules is 5.0×10^{22}. (a) What are the pressure in pascals and the temperature in kelvins? (b) What is the volume of the container? (c) If the container is heated to 120°C, what is the pressure in atmospheres?

46. (II) Freon has van der Waals values of $a = 1.08\ \text{J}\cdot\text{m}^3/\text{mol}$ and $b = 10^{-4}\ \text{m}^3/\text{mol}$. (a) Calculate the pressure inside a 0.40 L container of 1 mol of freon at 15°C. (b) What is the percentage difference between this pressure and the pressure you would have found had Freon been ideal? (c) What would your answers be if there were 100 mol inside the container?

47. (II) The Dieterici equation of state is

$$p\left(\frac{V}{N} - b\right) = kT\exp\left(-\frac{aN}{kTV}\right).$$

(a) Find the equation of state to which this reduces if $a = b = 0$. (b) By expanding the exponential function, find the equation of state to first order in a. (c) Sketch the isothermal curve on a p–V diagram, together with the corresponding curve for the ideal gas. In making your sketch, you can assume a and b are small.

48. (II) The molecular weight of H_2 is 2.0 g/mol. Consider H_2 gas in a container at STP (see Problem 38). (a) What volume is occupied by 30 g? (b) What mass and volume are occupied by 0.50 mol and 10 kmol?

49. (II) Use the ideal gas law to calculate the volume occupied by 1 mol of ideal gas at 1 atm pressure and 0°C. Given that the average molecular weight of air is 28.9 g/mol, calculate the mass density of air, in kg/m^3, at the above conditions.

50. (II) Use the result of Problem 49 and Avogadro's number for the number of molecules in 1 mol, 6.02×10^{23}, to estimate the number of molecules in your lecture hall.

51. (II) What is the mass density of helium gas (atomic weight 4.00 g/mol) at 1 atm pressure and a temperature of 8K?

52. (II) A child takes 20 puffs of air to blow up a spherical balloon. The diameter of the balloon is 23 cm. The temperature is 23°C, and the exterior atmospheric pressure is 1 atm. (a) How many molecules are in each of the child's puffs? (b) If the surrounding air and the balloon are heated to 28°C, what size will the balloon be?

53. (II) In an experiment, a vacuum of 10^{-10} atm is achieved in a bottle. If the bottle is at room temperature (30°C), what is the number of molecules in the bottle per cubic centimeter?

54. (II) At altitudes above 14,000 ft, pilots must breathe air with enriched oxygen because the density of the atmosphere is decreased. The pressure of air is only about 0.26 atm at a typical height for flying of 32,000 ft. If the airplane is not pressurized, what must the fraction of oxygen in the air be in order for a pilot to breathe the same amount of oxygen as at sea level?

55. (II) An ideal gas is contained in a tank at 120 atm of pressure and 263K. (a) If half the mass of gas is drawn off and the temperature then rises by 50K, what is the new pressure? (b) Suppose instead that the temperature first rises by 50K and then half the mass of the gas is drawn off. What is the new pressure?

56. (II) For safety reasons, compressed gases for laboratory use are often stored outdoors and transferred to where they are needed via a system of tubes and valves (Fig. 17–25). Suppose that a 150 L tank contains hydrogen at 25 atm at −5°C at the beginning of an experiment. If 190 L are used at 25°C and 1 atm pressure, what is the pressure of the remaining hydrogen? The temperature outside does not change during the measurement.

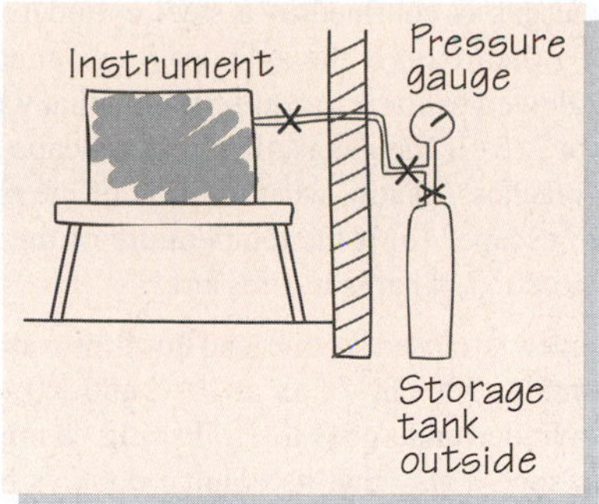

▲ **FIGURE 17–25** Problem 56.

57. (II) The cylinder of a manual bicycle tire pump is 35 cm long; its inside diameter is 2.5 cm. When the piston is pulled up, the valve to the tire is closed and air at atmospheric pressure enters the cylinder through the gap between the piston and the wall of the cylinder, which acts like another valve (Fig. 17–26). When the piston is pushed down, the air in the cylinder is compressed. The valve to the tire opens when the pressure in the tire is reached, and air is transferred to the tire during the rest of the stroke. The last 5% of the volume of air remains in the cylinder and the hose. Suppose that you are pumping up a tire that already contains air at 2.5 atm gauge pressure. How much air is transferred to the tire at this pressure? What is the volume of this air at atmospheric pressure? Sketch a p–V diagram of the process. Assume that the temperature is constant.

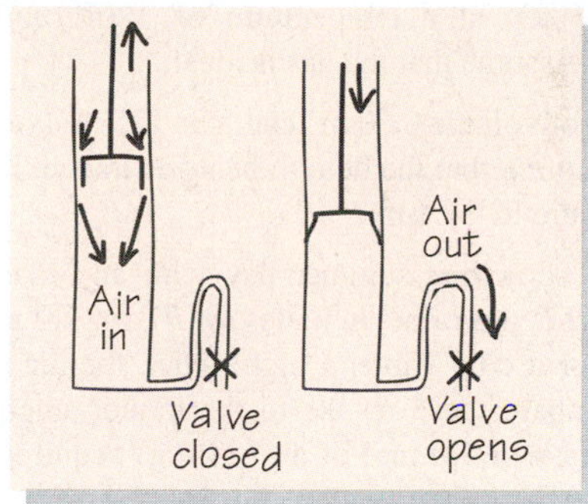

▲ **FIGURE 17–26** Problem 57.

58. (II) Figure 17–27 represents the variation of the volume and pressure of gas contained in a cylinder with a movable piston. The curved sections of the curve are isotherms. Calculate the temperature, pressure, and volume at the points 1, 2, 3, and 4 given that $p_1 = 1.00$ atm, $V_1 = 1.00$ L, $T_1 = 0.0$°C, $V_2 = 3.5V_1$, and $p_4 = 3p_1$.

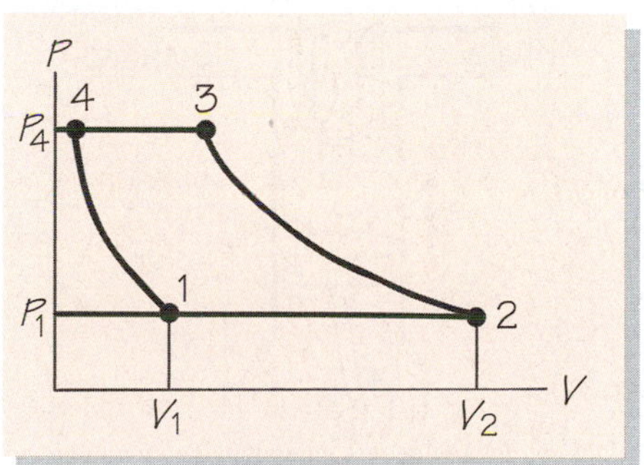

▲ **FIGURE 17–27** Problem 58.

59. (II) The specific volume of a gas is its volume per unit mass; that is, the inverse of the density. Compute the specific volume and density of oxygen gas, O_2, at $T = 15$°C and 1.0 atm of pressure, assuming that oxygen gas is ideal.

60. (II) An ideal gas is confined in a steel cylinder at 25°C and a pressure of 7.0 atm. (a) If the cylinder is surrounded by boiling water and allowed to come to equilibrium, what will the pressure of the gas be? (b) If the gas is allowed to escape until the pressure again reaches 7.0 atm, what fraction of the original gas, by weight, will escape? (c) If the temperature of the remaining gas now drops to 20°C, what is its pressure?

61. (II) A cylinder is closed at one end by a movable piston. The cylinder contains 300 cm^3 of air at 20°C and 1.0×10^5 Pa pressure. The cylinder undergoes the following changes: (a) the gas is heated in such a way that its volume doubles but its pressure remains constant; (b) the volume is kept constant and the temperature is changed until the pressure increases by 30 percent; (c) the gas cools, and the piston position is adjusted to keep the pressure constant until the initial volume is reached. Calculate the volume, pressure, and temperature after each step.

62. (II) The law of Joseph Gay-Lussac and Jacques Charles (1802) states that any ideal gas undergoes the same fractional increase in the product of pressure, p, and volume, V, when the gas is brought from one temperature to another. What is the fractional increase in pV for an ideal gas brought from the temperature of ice melting at 1 atm of pressure to the temperature of water boiling at the same pressure?

63. (II) Calculate the volume of 24 g of the gaseous form of ethyl ether, $C_4H_{10}O$, at a temperature of 120K and pressure of 0.080 atm. Assume that the gas is ideal.

64. (II) A tank of volume 0.80 m^3 contains 125 mol of helium gas at 25°C. Assuming that the helium behaves like an ideal gas, what is the pressure in the tank?

65. (II) It is a scorching summer day with an air temperature of 308K. What is the value, in joules, of RT for 1.0 mol of air? Repeat this for a cold winter's night when the air temperature is 258K. Compare these results to the kinetic energy that a ball with the mass of 1.0 mol of air (28.9 g) acquires when it falls 1.0 m under the influence of gravity.

66. (II) At the beginning of a compression stroke, the cylinder in a combustion engine contains 1.0 L of air at atmospheric pressure and a temperature of 20°C (Fig. 17–28). At the end of the stroke, the air has been compressed to a volume of 60 cm^3, and the total pressure is 35 atm. What is the temperature of the air when it is so compressed?

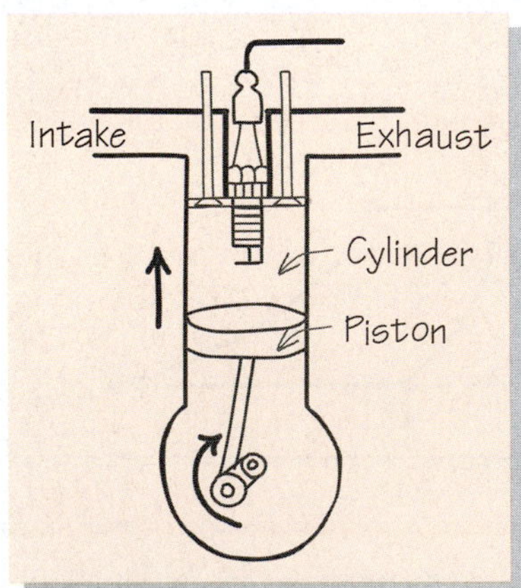

▲ **FIGURE 17–28** Problem 66.

67. (II) The steam point of water is the point where water vapor (steam) and boiling water coexist at a pressure of 1.00 atm. When a constant-volume gas thermometer calibrated to read a temperature of 273.16K at the triple point of water is used to measure the temperature of the steam point, the result depends slightly on the pressure of the gas in the thermometer because the gas is not precisely ideal (Fig. 17–29). In particular, if oxygen is used in the thermometer, the temperature of the steam point is determined to be 373.35K when the gas pressure is 0.40 atm, and 373.25K when the pressure is reduced to 0.20 atm. What is the temperature of the steam point?

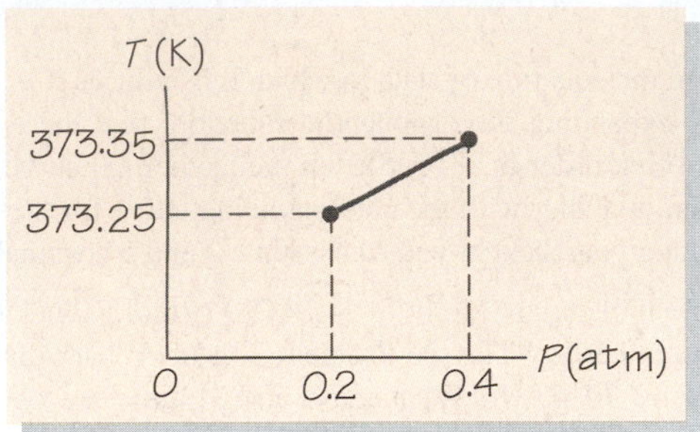

▲ **FIGURE 17–29** Problem 67.

68. (II) Carbon dioxide obeys the van der Waals equation of state with $a = 5.96 \times 10^6$ atm·cm^6/mol^2 and $b = 98.6$ cm^3/mol. We know that 0.30 moles of CO_2 occupies 2.5×10^3 cm^3 at 1.7 atm of pressure. (a) Find the temperature of the gas. (b) Redo your calculation, but this time assume that CO_2 is an ideal gas. What is the percentage difference between the two answers?

69. (II) There is a certain temperature T_B (the *Boyle temperature*) for which a van der Waals gas behaves as if it were ideal. (a) Show that an expression for this temperature is $T_B = (a/bR)[1 - (bn/V)]$. (b) The constants a and b of the van der Waals equation of state are $a = 0.140$ m^6·Pa/mol^2 and $b = 4.00 \times 10^{-5}$ m^3/mol for argon gas. What is the value of the Boyle temperature of argon gas when the gas is so dilute that the term bn/V in the expression for the Boyle temperature can be ignored?

70. (III) An isotherm of 1 mol of a gas that obeys the van der Waals equation of state and is plotted on a p–V curve falls monotonically for large T. As T is decreased and new isotherms are plotted, the curve begins to show a kink, and at a temperature known as the *critical temperature*, T_c, there is a flat spot in the curve, where the slope is zero (Fig. 17–30). At this point, $p = p_c$ and $V = V_c$. Find T_c, p_c, and V_c in terms of the constants a and b.

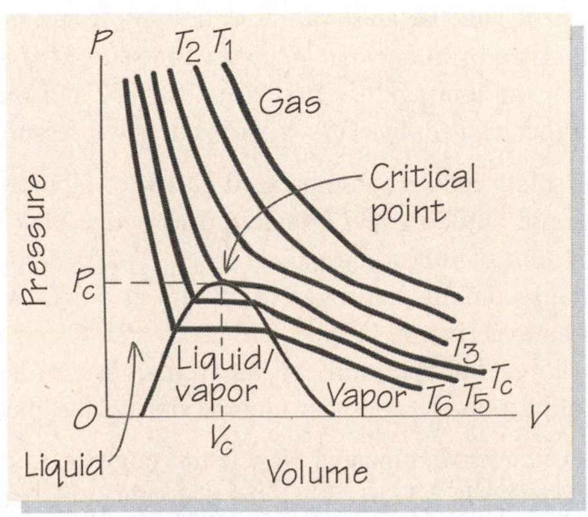

▲ **FIGURE 17–30** Problem 70.

*17–5 Blackbody Radiation

71. (I) According to the Stefan–Boltzmann formula, the energy of radiation emitted by a blackbody per unit area per unit time is proportional to T^4, where T is the temperature in the Kelvin scale. Compare the energy emitted by a tungsten filament at 3200°C and the same filament at room temperature (25°C).

72. (I) The intensity of radiation from a small source decreases with the square of the distance from the source. Consider the energy of radiation that reaches your face after being emitted by solid aluminum at room temperature (25°C). At what distance will this energy of radiation be the same as the energy emitted by the same amount of molten aluminum (800°C) that reaches your face when you are 20 m from the aluminum? (Your answer will surprise you. Can you explain why? A look at Fig. 17–16 should help.)

73. (I) The dominant frequency of radiation emitted by an object is related to its temperature by $hf \cong kT$. Find the dominant radiation frequency emitted by (a) an object in interstellar space at 3K; (b) a body of water at 280K; (c) an electric stove heating unit (800K); (d) melting tantalum (3000K).

74. (I) The human eye is most sensitive to yellow light with a wavelength of approximately 550 nm. What is the temperature of an incandescent bulb filament that radiates most of its energy as a blackbody at this wavelength?

75. (II) The temperature of your skin is about 35°C. Calculate the wavelength at which the Planck radiation curve has a maximum for this temperature, and therefore the wavelength at which your body radiates the most energy. Use the more acccurate estimate indicated in Example 17–13.

76. (II) Show that the integral over all frequencies of the Planck formula given by

$$U(T) = \int_0^\infty \left(\frac{8\pi hf^3}{c^3}\right)\frac{1}{e^{hf/kt} - 1}df$$

gives a result that is of the form (a constant) $\times\, T^4$. [*Hint*: Change variables from f to hf/kT.] The energy emitted per unit area per unit time, $P(T)$, is proportional to $U(T)$, and thus $P(T)$ is also proportional to T^4, as in the Stefan–Boltzmann formula, Eq. (17–17).

77. (II) The surface of the Sun is at a temperature of 6000K. At what rate is energy radiated from the whole surface of the Sun, given that the radius of the Sun is $R = 6.95 \times 10^8$ m?

78. (II) Assume that the radiation emitted from the Sun moves radially outward from the Sun and that no radiation is absorbed between the Sun and Earth. How much energy in the form of radiation will fall per second on an area of 1 m^2 on Earth, if that area is perpendicular to the straight-line path of the radiation? The distance from the Sun to Earth is 1.5×10^{11} m.

79. (II) The average surface temperature of Earth is 290K. How much energy per second is radiated by Earth's surface, assuming that Earth simulates a blackbody. Compare this result with the amount of radiative energy that reaches Earth from the Sun. Use the result of Problem 78.

80. (III) The energy density per unit frequency in the frequency range from f to $f + df$ in blackbody radiation is $u(f, T)\, df$, where $u(f, T)$ is given by Eq. (17–16). An alternative way to express the blackbody radiation is to give $u'(\lambda, T)\, d\lambda$, the energy density per unit wavelength in the wavelength range from λ to $d\lambda$. (a) Use the fundamental wave relation $c = \lambda f$, where c is the speed of light, to show that $u(f, T)\, df = u[(c/\lambda), T]c\, d\lambda/\lambda^2$, so $u' = u[(c\lambda), T]c/\lambda^2$. (b) Assuming that the temperature is fixed, use the results of part (a) to find an equation for the λ_{max} for which $u'(\lambda, T)$ has a maximum. (c) The result of part (b) is a transcendental equation. Solve it for $\lambda_{max}T$. For what range of temperatures does λ_{max} fall within the visible spectrum, $\lambda \cong 450$ nm to 650 nm?

81. (III) Find the constant coefficient of T^4 in the Stefan–Boltzmann formula, Eq. (17–17), given that the relation between $P(T)$ and the total energy density $U(T)$ calculated in Problem 76 is $P(T) = cU(T)/4$, where c is the speed of light. [*Hint*: An appropriate entry in a table of integrals is

$$\int_0^\infty [x^3/(e^x - 1)]\, dx = \pi^4/15.]$$

General Problems

82. (II) Some astronomers estimate that toward the center of the galaxy, the average interstellar gas is molecular hydrogen, at a density of roughly 1 molecule per cubic centimeter and a temperature of between 10 and 20K. Estimate the pressure of this gas. Compare this with atmospheric pressure (1 atm = 760 torr) and very good laboratory vacuums (10^{-10} torr) that are achieved only with some difficulty.

83. (II) Use Archimedes' principle (Section 16–4) to estimate how much helium is needed for a balloon to lift a payload of 230 kg (including the mass of the balloon) in air at 1.0 atm pressure and 10°C. You will need the molecular weight of helium (4.00 g/mol) and that of air (29 g/mol).

84. (II) A constant-volume gas thermometer is placed in contact with ice water, and the pressure in the thermometer is observed to be 0.76×10^5 N/m^2. If the same thermometer is then placed in a pot of boiling water, the pressure rises to a new value of 0.90×10^5 N/m^2. The thermometer is calibrated in °C. What temperature will this thermometer read if it has a pressure reading of 0.84×10^5 N/m^2?

85. (II) Suppose some measurable property Z of a material you have is sensitive to temperature and that you wanted to make a thermometer with a Celsius scale based on it. You therefore write $T_C = a + bZ$. How do you establish the values of a and b? Would your scale necessarily coincide with the Celsius scale that you would get from an ideal gas?

86. (II) The human eye can detect colors over the electromagnetic wavelength range of about 300 to 700 nm. (a) Use the results of Section 17–5 to calculate what temperatures correspond to the maximum energy densities ($hf \cong kT$) for this range of frequencies? (b) The answer to Problem 80 is $\lambda_{max}T = 2.9 \times 10^{-3}$ m·K. Calculate the temperatures using this relation. (c) What is the percentage difference, if any, for the ends of the range in the two calculations?

87. (II) A cylindrical vessel with a tight but movable piston is placed in a vertical position so that the piston, whose area is 70 cm^2, is subject to atmospheric pressure. When the gas in the vessel is heated from a temperature of 20°C to 80°C, a 0.5 kg mass must be placed on top of the piston to hold the piston at the position it occupied at the lower temperature (Fig. 17–31, see next page). What is the volume of 0.2 mol of the gas?

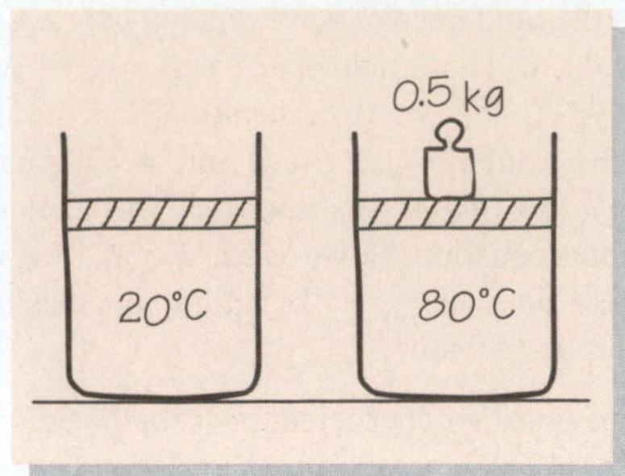

▲ **FIGURE 17–31** Problem 87.

88. (II) An air bubble of volume 8 cm^3 rises from the bottom of a lake 15 m deep. The temperature at the bottom is 8°C. The bubble rises to the surface, which is at 20°C and at atmospheric pressure. Assuming that the bubble is in thermal equilibrium with the surroundings at all times, calculate the volume of the bubble at the surface.

89. (II) In Chapter 14 we stated that the speed of sound in air is given by $v_{sound} = \sqrt{\gamma p/\rho_0}$, where the constant $\gamma \cong 1.4$ for air near STP (see Problem 38), p is the pressure, and ρ_0 is the density of the air. Reexpress the speed of sound in terms of the temperature. What is the difference in the speed of sound at the coldest and hottest times of day in a desert?

90. (II) A bottle containing air is closed with a watertight yet smoothly moving piston. The bottle with its air has a total mass of 0.30 kg. At the surface of a body of water whose temperature is a uniform 285K throughout, the volume of air contained in the bottle is 1.5 L (Fig. 17–32). Recall that the pressure of water increases with depth below the surface, D, as $p = p_0 + \rho g D$, where p_0 is the surface pressure and $\rho = 1.0$ kg/L. The bottle is submerged. (a) What is the volume of the air in the bottle as a function of depth? (b) Calculate the buoyant force on the bottle as a function of depth. (c) At what depth do the buoyant force and the force of gravity cancel? (d) Is the depth at which the force on the bottle is zero, calculated in part (c), a stable or an unstable equilibrium?

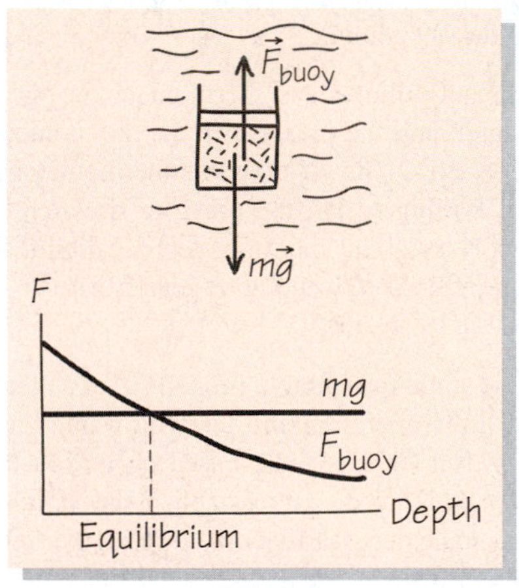

▲ **FIGURE 17–32** Problem 90.

91. (II) Suppose that, instead of the Kelvin scale, an alternative temperature scale—call it the Kelvin′ scale, with the units K′—were defined in which the zero level remains unaffected but the temperature at the triple point of water is taken to be 500K′. Calculate the value of Boltzmann's constant in units of J/K′.

92. (II) Consider the radiation emitted from a furnace through a hole of area 1 cm^2. Assume that all the radiation is absorbed by a beaker that contains 42 g of water. Given that it takes 4.2 J of energy to raise the temperature of 1 g of water by 1°C, how long will it take for the water to be heated by 8°C if the furnace temperature is 1600K? Assume that there are no energy losses due to the beaker.

93. (III) The van der Waals equation of state, Eq. (17–15), can be put into a form that is an expansion in powers of the density. Dividing by RT and then by $[(V/n) - b]$ in Eq. (17–15), we can write

$$p + \frac{\alpha(n/V)^2}{RT} = \left(\frac{V}{n} - b\right)^{-1} = \frac{n}{V}\left(1 - b\frac{n}{V}\right)^{-1}.$$

(a) Show that, for $bn/V < 1$, this equation has the approximate form

$$\frac{p}{RT} = \frac{n}{V} + b\left(\frac{n}{V}\right)^2 - \frac{\alpha}{RT}\left(\frac{n}{V}\right)^2.$$

(b) The result in part (a) is of the general form

$$\frac{p}{RT} = \frac{n}{V} + B_2(T)\left(\frac{n}{V}\right)^2,$$

where, in this case, the function $B_2(T)$ is $B_2(T) = b - (a/RT)$. This is an example of a *virial expansion* for the equation of state, and the function $B_2(T)$ is known as a *virial coefficient.* Show that $B_2(T)$ starts out negative as T increases from some small value, passes through zero, and then becomes positive. The value of T at which $B_2(T)$ is zero is the Boyle temperature, where a real gas satisfying the van der Waals equation behaves as though it were ideal (see Problem 69).

Part of a large plant near Geyserville, in northern California, that uses thermal energy from within Earth to do work and generate electric power. While such a plant does not require a separate fuel source that must be replenished, any electric energy that it has produced has depleted at least that much thermal energy from the underground thermal source.

Heat Flow and the First Law of Thermodynamics

Steam contains thermal energy, and as a result steam can push a piston in a cylinder or turn a turbine and can thereby do work. The observation that we can transfer thermal energy to a mechanical system and do work is important. In this chapter we will see how thermal energy is related to other forms of energy with which we are familiar and how it can be used to do mechanical work. We will see that as the temperature of a thermal system increases, so does the amount of thermal energy in the system.

Changes in the energy of thermal systems (thermal energy) and its relation to changes in the energy of the systems' surroundings is summarized in two important laws of physics—the first and second laws of thermodynamics—which we examine here and in Chapter 20, respectively. The first law of thermodynamics describes thermal energy as just another form of energy—like kinetic energy or potential energy. The principle of the conservation of energy that we developed in Chapter 7 is now expanded to include the thermal energy of a system. This energy can be accounted for in mechanical form: It is the kinetic and potential energy of the system's microscopic atomic constituents (Chapter 19). In other words, the first law of thermodynamics is nothing more than the conservation of energy principle applied to thermal systems. The second law of thermodynamics states limitations on how much of a system's thermal energy can be converted to other forms of energy.

It is difficult to overestimate the importance of the conservation of energy as a scientific principle. The discovery that thermal energy is just another form of energy, and the consequent expansion of the principle of conservation of energy, constitutes one of the truly important scientific advances of the 19th century.

18–1 Changes in Thermal Systems

As we saw in Chapter 17, thermal systems in equilibrium, such as a container of hot steam, are described by only a few thermodynamic variables. For a gas these variables are the temperature T, the pressure p, the volume V, and the number of moles n (or the number of molecules, N). Thermodynamic variables describe the *state* of the thermal system, and an equation of state relates these variables. For example, the equation of state for an ideal gas [Eqs. (17–7) and (17–9)] is $pV = nRT = NkT$, where R is the universal gas constant and k is Boltzmann's constant. But other equations of state are equally possible and can all be handled the same way. If we assume that n is fixed, the remaining three variables are connected by an equation of state, and we can very usefully describe the state as a point on a two-dimensional plot, such as a p–V diagram, a p–T diagram, or a V–T diagram. Changes, or transformations, in the state of a thermal system can be represented as curves on such diagrams—at least if thermal equilibrium is established at every stage of the transformation. We have already described and used such curves in Section 17–4 (see Example 17–11).

To maintain thermal equilibrium at each step of a transformation, changes must be made slowly. Just how slowly depends on the internal dynamics that lead to equilibrium. For example, it is not difficult to compress a gas in such a way that it maintains thermal equilibrium at each stage. To see why, consider a piston compressing a gas. When molecules bounce from the inward-moving piston, they are speeded up (thereby gaining energy) and distribute their additional energy to other molecules through collisions. The time required for the effect of the push to be evenly spread through the gas is characterized by the size of the container divided by the *speed of sound* in the gas. That is because the speed of sound is a measure of the spread of an organized disturbance through the gas through successive collisions. The speed of sound is large compared to typical mechanical motion, such as the movement of the piston; equivalently, the intermolecular collision time is short compared to the time for mechanical motion.

Reversible and Irreversible Processes

A change in which thermal equilibrium is maintained throughout, such as the sufficiently slow compression of a piston in an insulated cylinder, or the cooling of a bottle of soda in a refrigerator, is called a **reversible process**. Such changes can always be reversed by changing the external conditions in such a way that the thermodynamic variables return to their original values following the same path, only reversed, on the graph of thermal variables. As an example, the volume of a cylinder of gas held at a fixed temperature is increased from V_1 to V_2 in a controlled way when a piston slowly moves out of the cylinder with an external force balancing the pressure, as in Fig. 18–1a. If the piston is slowly pushed back in, as in Fig. 18–1b, the external conditions are reversed, and the thermodynamic variables of the gas recover their initial values. These changes are reversible, and Fig. 18–1c shows the transformation as a curve on a p–V diagram. Because the thermal system remains in equilibrium throughout a reversible process, reversible processes always connect equilibrium states.

Conversely, a gas expanding to fill a vacuum through an open stopcock (Fig. 18–2a) provides an example of an **irreversible process**, a process in which the thermal system's changes cannot be retraced. The expansion from V_1 to V_2 occurs rapidly and without control. Nothing you do with the stopcock will cause the released gas to reverse itself and reconcentrate within the original volume. We can mark the original and final points on a p–V diagram (Fig. 18–2b), but we cannot mark the points in between, because the system is not in thermodynamic equilibrium during the expansion. The particular irreversible process described here, in which the gas expands freely, is known as **free expansion**. Other examples of irreversible thermal transformations include the melting of ice cubes in lemonade and the heating of a brake shoe as it rubs against the brake drum. Unless we state otherwise, in this chapter *we consider only reversible changes of state.*

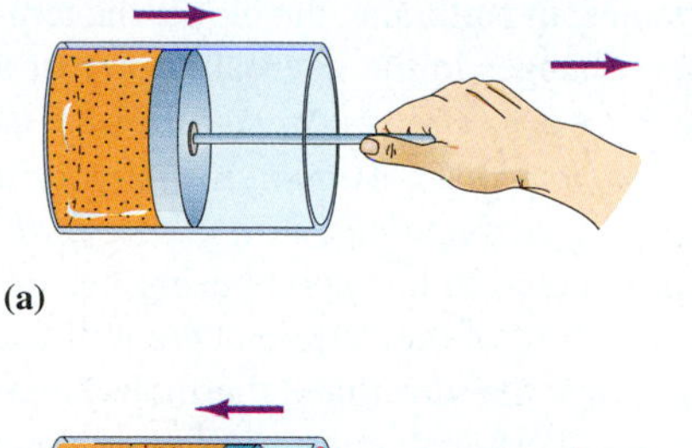

(a)

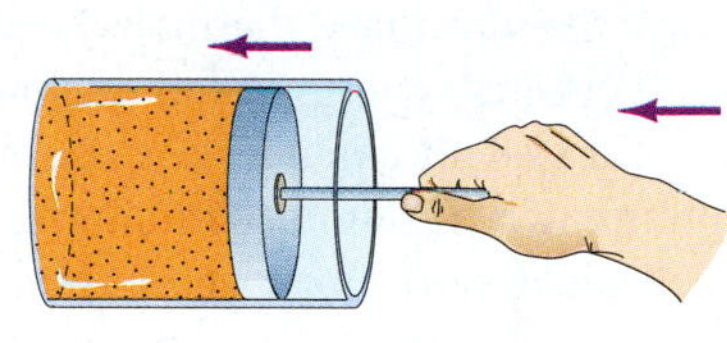

(b)

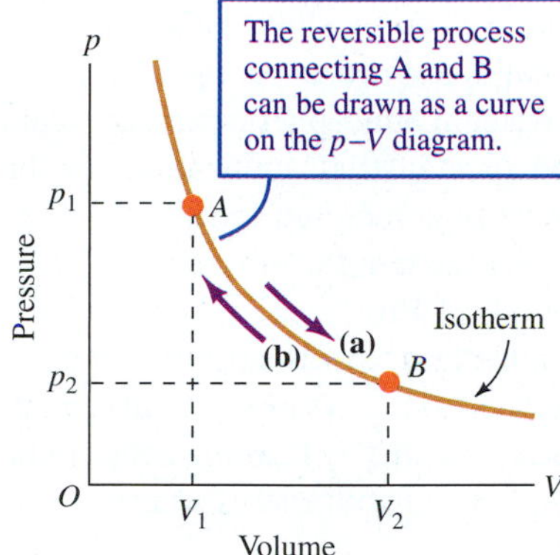

(c)

◀ FIGURE 18–1 The volume of a gas in a cylinder is reversibly changed when a piston is (a) slowly pulled or (b) slowly pushed. The temperature is held fixed during the transformations. (c) Constant-temperature reversible transformation of a gas between V_1 and V_2 on a p–V diagram. The curve is an *isotherm.*

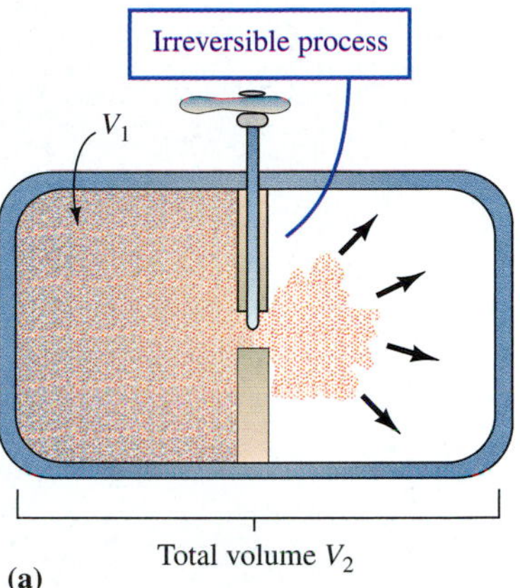

(a)

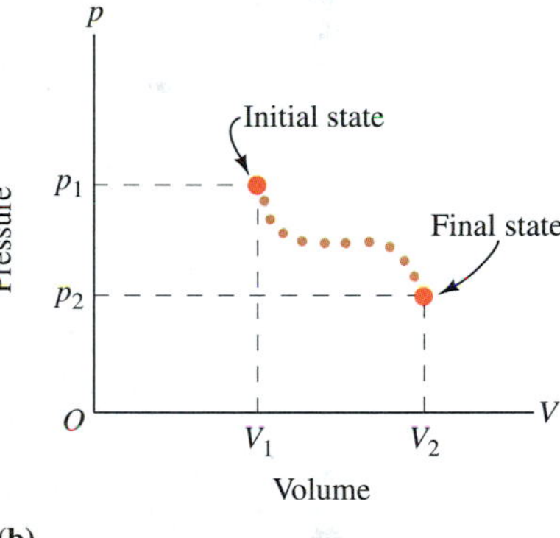

(b)

▶ FIGURE 18–2 (a) A gas initially confined to the left-hand side of a container expands to fill the container's total volume when a stopcock is opened. (b) Irreversible expansion of a gas from V_1 to V_2. The path is not well defined, and the dots serve only to indicate that there has been some irreversible change.

CONCEPTUAL EXAMPLE 18–1

A gas in contact with a thermal reservoir at temperature T is expanded reversibly from V_1 to V_2. The trace of points made by this change on a p–V diagram is an isotherm. Describe the isothermal curve if the gas is ideal.

Solution Because we have a gas in contact with a thermal reservoir, the temperature T must remain constant. Therefore, the ideal gas law, $pV = nRT$, reveals that the product of p times V must be a constant, because nRT is constant. We can see that as the volume increases from V_1 to V_2, the pressure must decrease from p_1 to p_2 in order to keep the product pV constant as shown in Figure 18–1c. The curve that corresponds to pV = a constant is a hyperbola. Figure 18–3 shows a family of hyperbolas, each curve corresponding to a different constant temperature.

What Do You Think? What would such a curve look like on a p–T diagram? *Answers to **What Do You Think?** questions are given in the back of the book.*

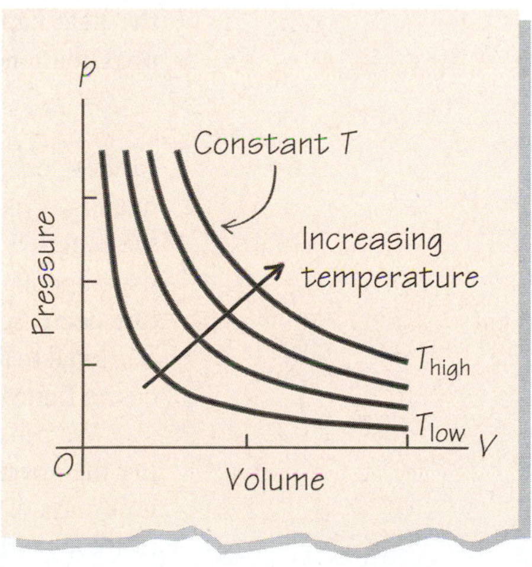

▲ FIGURE 18–3 An isotherm for an ideal gas is a hyperbola, $p \propto TV^{-1}$. The isotherms corresponding to different temperatures are indicated.

18–2 Heat Flow

We suggested in Section 17–3 that the temperature of a system is closely related to its **internal energy**, or **thermal energy**. We will learn here that the internal energy of a thermal system is a function of the thermodynamic variables; in particular, the higher the temperature of a system, the higher the internal energy. Changes in the internal energy of a thermal system can manifest themselves in only a few ways: *The system can transfer its internal (thermal) energy by changing the temperature (or phase) of another system, or it can use its internal energy to do mechanical work on its surroundings, or it can do both.*

When the temperature of a thermal system in contact with a neighboring system changes, we say that there has been a **heat flow** into or out of the system.† We will use the term *heat flow* rather than *heat* in order to discourage the idea that a thermal system has a certain amount of "heat." Heat flow is more like work than energy: Work is done *on* or *by* a mechanical system, and we do not speak of a system containing a certain amount of work. It would be equally misleading to think of a system as containing a certain amount of "heat." Heat flow into, and/or mechanical work done on, a system may each contribute to raising the temperature of a thermal system. We employ the symbol Q for heat flow or, often, ΔQ.

As far as units, we start with the historical unit of heat flow, the calorie (cal). (We are going to have to learn some further physics before we can sensibly relate this to SI units with which we are familiar.) Provisionally, 1 cal is defined to be the amount of heat flow required to raise the temperature of 1 g of water at atmospheric pressure from 14.5°C to 15.5°C. Another unit for heat flow used in industrial applications is the *British thermal unit* (Btu), which is defined as the heat flow required to raise the temperature of a 1-lb weight of water by 1°F, averaged over the temperature range 32°F to 212°F. The British thermal unit is related to the calorie by 1 Btu = 252.02 cal.

A major aim of this chapter is to establish the fact that heat flow is a transfer of energy and that thermal systems have an internal energy related to the system's thermal properties. In the remainder of this section we will set the stage for this by learning about some properties of heat flow that are independent of its identification with energy transfer.

Heat Capacity

When there is a heat flow to or from a thermal system, the temperature of the system changes. The connection between heat flow and temperature change is the **heat capacity**, C. A *small* heat flow, dQ, *into* a thermal system will change the temperature by a small amount dT, with a proportionality constant that is the heat capacity C:

$$dQ = C\,dT. \tag{18–1a}$$

This can also be written for a *finite* heat flow ΔQ and a *finite* temperature change ΔT if the heat capacity does not depend on the temperature—a situation that is a good approximation for many physical situations:

$$\Delta Q = C\Delta T. \tag{18–1b}$$

In these equations the heat flow, whether large or small, can be positive or negative. The quantity C is always positive; thus, the sign of the temperature change indicates the sign of the heat flow. Equations (18–1a, b) state that there is a positive heat flow into a system when its temperature increases and there is a negative heat flow into the system when its temperature decreases. The value of the heat capacity depends on the particular material and is proportional to the amount (mass) of material. It may also depend on the temperature, in which case a finite heat flow is related to a finite temperature change through an integral.

We can eliminate the dependence of heat capacity on the amount of material by defining the **specific heat**, c, as the heat capacity of 1 g of the material—the heat capacity per unit mass of the material—and the **molar heat capacity**, c', as the heat capacity of 1 mol of the material. If the mass in grams of a thermal system consisting of n moles is m, then

$$C = mc = nc', \tag{18–2}$$

†We will see later that we can also have a heat flow without a temperature change when the system does work or undergoes a phase change.

and if A is the molecular weight (the mass of 1 mol, in grams), then

$$c' = Ac. \tag{18–3}$$

The units of heat capacity are cal/K; those of specific heat are cal/g · K; and of molar specific heat, cal/mol · K.

CONCEPTUAL EXAMPLE 18–2 You have a substance with a large specific heat—water is an example. For a given positive heat flow out of a given amount of the substance, the temperature will (a) rise a lot; (b) rise a little; (c) fall slightly; (d) fall a lot.

Answer If we remove energy (heat flow) from a substance with a large specific heat capacity, the temperature of the substance will fall only slightly. The substance has a "large capacity" for holding its temperature. (The word "slightly" should take into account the amount of the substance.) Water has a relatively large specific heat, which is why large bodies of water have strong moderating effects on the climate in their vicinity.

Path Dependence of Heat Flow

The heat capacity is defined in Eq. (18–1) in terms of a temperature change. But the conditions for a temperature change are not unique—for example, the volume or the pressure, or neither, can be held constant during the temperature change. The heat capacity generally depends on how the temperature change is made and on the values of the thermodynamic variables p, V, and T. This dependence is indicated by writing $C = C(p, T)$ or $C = C(V, T)$. We cannot use Eqs. (18–1a, b) to find a net heat flow unless we know more about how either pressure or volume changes as temperature changes. We say that the heat flow *depends on the path* along which temperature changes.

Let us examine two cases of special interest: In the first case the volume is kept constant, and in the second the pressure is held constant. For these cases, we label the heat capacity with the subscript V or p, and Eq. (18–1a) becomes

$$\text{for constant volume: } dQ = C_V\,dT; \tag{18–4}$$

$$\text{for constant pressure: } dQ = C_p\,dT. \tag{18–5}$$

Generally, C_V differs from C_p, as we will see later when we find C_V and C_p for an ideal gas. If these quantities are independent of temperature, then the generalizations of the finite change Eq. (18–1b) applies. Indeed, the heat capacities do turn out to be independent of temperature for monatomic ideal gases, which describe at least roughly all gases that are not too hot, and that will simplify things. Note that the subscript must be used on c and c' as well. (The volume change with temperature for liquids is insignificant compared with that of gases; for solids, pressure is not a useful variable. Therefore, the subscript is not used for these cases.)

One can show (to see why, see Section 18–7) that the heat capacity for constant pressure is larger than the heat capacity for constant volume:

$$C_p \geq C_V. \tag{18–6}$$

This inequality holds even for nonideal gases.

EXAMPLE 18–3 An ideal gas undergoes transformations that take it from point A to point B on a p–V diagram by the two different paths, α and β, shown in Fig. 18–4. Points A and B have the same temperature, $T = T_0$ (they lie on an isotherm). Assuming that C_V and C_p are constants, what is the heat flow into the gas (a) for path α (through point 1 at temperature T_1) and (b) for path β (through point 2 at temperature T_2)?

Strategy Figure 18–4 labels the various thermodynamic variables. We can use Eqs. (18–4) and (18–5) to find the heat flow along the various legs of the transformation. But because both C_V and C_p are constants, we can use these equations in "Δ" form, that is, we can use them for finite heat flows and finite temperature changes.

(continues on next page)

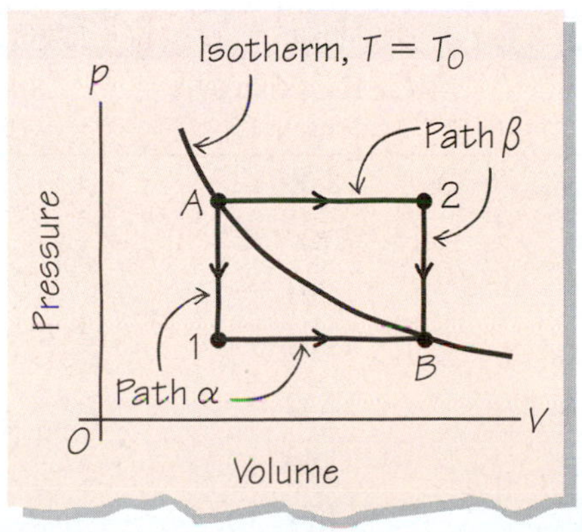

▲ **FIGURE 18–4** An ideal gas undergoes a transformation between points A and B by two different paths.

Working It Out (a) Along path α, there are two contributions: from point A to point 1 (constant volume), and from point 1 to point B (constant pressure). Thus

$$\Delta Q_\alpha = C_V\,\Delta T_{A\to 1} + C_p\,\Delta T_{1\to B}$$
$$= C_V(T_1 - T_0) + C_p(T_0 - T_1) = (T_0 - T_1)(C_p - C_V).$$

(b) For path β, there are also two contributions: from point A to point 2 (constant pressure), and from point 2 to point B (constant volume).

$$\Delta Q_\beta = C_p\,\Delta T_{A\to 2} + C_V\,\Delta T_{2\to B}$$
$$= C_p(T_2 - T_0) + C_V(T_0 - T_2) = (T_2 - T_0)(C_p - C_V).$$

The factor $(T_0 - T_1)$ is not equal to the factor $(T_2 - T_0)$. Thus the two answers are different, and the heat flow depends on the path that is followed.

What Do You Think? There is no temperature difference between points A and B. Why isn't the heat flow zero?

Calorimetry

The measurement of heat capacities is called **calorimetry**. Calorimetry, or more properly *adiabatic calorimetry*, is based on the fact that *when a system is mechanically and thermally isolated, heat flow is conserved within the system.* (A system is said to be *mechanically isolated* if it cannot do mechanical work on its surroundings.) A heat flow into one part of such a system must be matched with a corresponding heat flow out of another part. In such circumstances, heat flow is present only as long as there is a temperature difference between parts of the system; it ceases when equilibrium is established throughout the system.

Our definition of the calorie shows that the specific heat for water at 15°C and at 1 atm of pressure is

$$c_{H_2O} = 1\ \text{cal}/(\text{g}\cdot{}^\circ\text{C}).$$

The measurement of the specific heats of other substances is based on that of water, as in Example 18–4.

EXAMPLE 18–4 A block of iron (mass 100 g) is heated in an oven to 500K and then plunged into a closed, thermally insulated container of 500 g of water at 292K. The block and the water come to an equilibrium temperature of 297K. What is the specific heat of iron, c_{Fe}? Assume that the specific heats of water and iron do not vary significantly over the temperature ranges in question.

Strategy There is a positive heat flow ΔQ_{water} into the water—it is positive because the iron is initially hotter than the water and the water temperature increases. Similarly the iron cools, so the heat flow into it, ΔQ_{Fe}, is negative. The entire system of iron and water together is isolated, so that

$$\Delta Q_{system} = 0 = \Delta Q_{water} + \Delta Q_{Fe}.$$

We next relate the heat flows to the specific heats. There are two things to keep in mind: First, the heat capacities C are independent of temperature here, so we can use Eq. (18–1b) for finite temperature shifts. These temperature changes are all specified in the problem statement. Second, we are interested in the specific heats c, so we can use the relation $C = mc$, where m is the amount of material, also specified. These two facts allow us to write $\Delta Q = mc\Delta T$ for both water and iron in the equation above, and that is enough to solve for c_{Fe}.

Working It Out We have

$$\begin{aligned}0 &= \Delta Q_{water} + \Delta Q_{Fe}\\ &= m_{water}c_{water}(\Delta T)_{water} + m_{Fe}c_{Fe}(\Delta T)_{Fe}\\ &= (500\ \text{g})(1\ \text{cal/g}\cdot\text{K})(297\text{K} - 292\text{K}) + (100\ \text{g})c_{Fe}(297\text{K} - 500\text{K})\\ &= 2500\ \text{cal} + (-2.03\times 10^4\ \text{g}\cdot\text{K})c_{Fe}.\end{aligned}$$

We solve for c_{Fe} to find that

$$c_{Fe} = \frac{2500\ \text{cal}}{2.03\times 10^4\ \text{g}\cdot\text{K}} = 0.12\ \text{cal/g}\cdot\text{K}.$$

What Do You Think? What if we had used an aluminum block instead of iron in this experiment? Would the final system temperature be (a) higher, (b) lower, (c) no change? (Look at Table 18–1 for the specific heat of aluminum.)

TABLE 18–1 • Molar Heat Capacities and Specific Heats of Some Materials at Constant Volume†

Material	Molar Heat Capacity (cal/mol·K)	Specific Heat (cal/g·K)	Material	Molar Heat Capacity (cal/mol·K)	Specific Heat (cal/g·K)
Helium gas	3.00	0.75	Silicon	4.78	0.17
Argon gas	30.00	0.75	Aluminum	5.82	0.22
N_2 gas	4.94	0.176	Copper	5.85	0.092
O_2 gas	4.97	0.155	Iron	6.00	0.107
SO_2 gas	7.52	0.118	Lead	6.32	0.031
Diamond	1.46	0.12	Water	18.0	1.00
Beryllium	3.85	0.43	Mercury	6.7	0.033

† *Gases at standard temperature and pressure.*

Example 18–4 shows how the specific heat of iron (or of many materials) can be measured. Calorimetry is used most often in the opposite sense: Given the specific heats, temperature changes are calculated in situations where there is a heat flow from one system to another. Table 18–1 gives the molar heat capacities and specific heats *at constant volume* for a variety of materials at room temperature. In this table, the gases are at standard pressure. In Chapter 19 we will see that the values in Table 18–1 can be explained in terms of the microscopic constituents and structure of the material.

THINK ABOUT THIS . . .

WHAT ARE THE ORIGINS OF THE TERMS CALORIE, CALORIMETRY, CALORIC?

The prefix "calor" is derived from the Latin word *calor*, meaning heat in its colloquial sense. The French scientist Lavoisier coined the term "caloric" in his caloric theory of 1787; this theory proposed that a massless substance (a fluid) flows from one material to another when an object is heated. The fluid was called the *caloric*. The unit of heat flow was later called the *calorie*. Caloric was thought never to be created or destroyed, and all substances had caloric that could be absorbed or released. The term calorimetry refers to the science of measurements associated with heat flow; the term is still used today even though we know the idea of the caloric is wrong. A calorimeter is a device used to measure heat flow.

Count Rumford (Benjamin Thompson) played an important role in discrediting the caloric theory. While he was supervising the boring of cannon barrels in Bavaria in 1798, Rumford observed and quantified the continual heating of the metal (Fig. 18–5). He reasoned that if the heating were due to the transfer of a fluid—as proposed by the caloric theory—then the fluid eventually would be used up and no more heating should occur. He found that heat was caused by the process of friction and that there was a continual heat flow as long as the boring continued. He even used the hot chips of metal that were bored out to boil water!

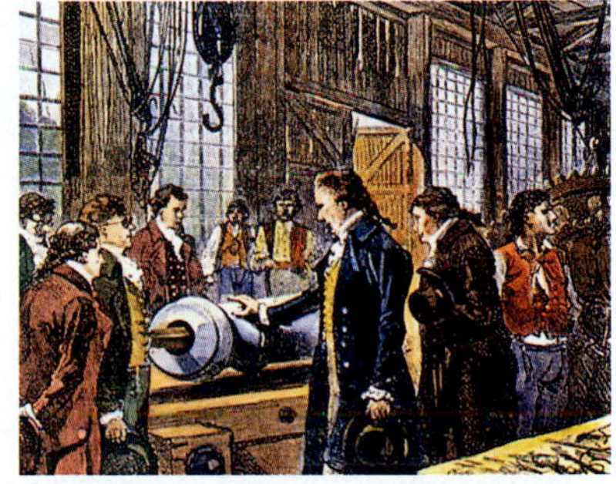

▲ **FIGURE 18–5** Count Rumford observed that the temperature of cannon barrels rose steadily as they were being hollowed out. He concluded that the caloric theory could not be correct.

Phase Changes and Heat Flow

When there is a **change of phase** in a system, there is no temperature change even though there is a heat flow. For example, a solid state may change to a liquid state by melting, or a gaseous state may change to a liquid state by condensation (Fig. 18–6). It takes a certain heat flow to convert 1 g of ice at its melting temperature of 0°C to liquid water at 0°C (Fig. 18–7). The heat flow necessary to bring about the transformation is called the **latent heat of fusion**, L_f. Measurements show that $L_f = 79.6$ cal/g for ice.

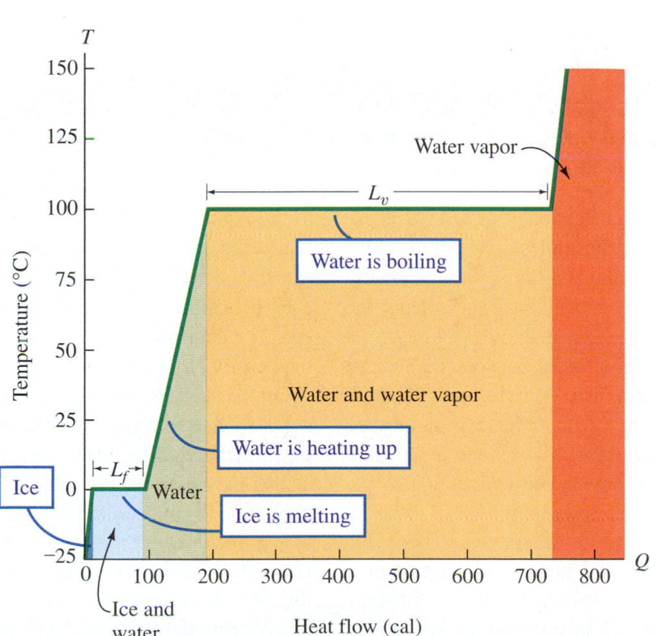

▲ **FIGURE 18–7** Phase changes for 1 g of water. Note that the temperature remains constant as heat flow is added during melting and vaporization.

(a)

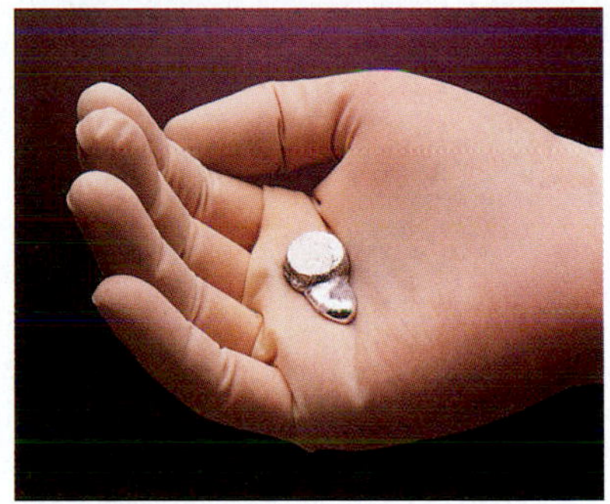

(b)

▲ **FIGURE 18–6** (a) Mist forms when water droplets condense from water vapor. (c) Gallium melts at the temperature of the human body. When it does so, it absorbs heat flow from its surrounding environment.

During the melting process, there is no change of temperature and, to a good approximation, there may be little change of pressure or volume. The heat flow goes into changing the molecular arrangement of the substance (see Fig. 16–1). Similarly it takes a certain heat flow to convert 1 g of water at its vaporization temperature, 100°C (at 1 atm pressure), to 1 g of steam. The heat flow necessary for this is called the **latent heat of vaporization**, L_v, whose value for water is 540 cal/g. Most of this heat flow goes into overcoming the intermolecular forces that hold the molecules close together and make a liquid much denser than a gas. Some latent heats are given in Table 18–2.

TABLE 18–2 • Some Latent Heats

Material	Melting Temperature (K)	Latent Heat of Fusion (kJ/mol)
H_2	14	0.12
O_2	54	0.44
Water (H_2O)	273	6.0
Ethyl alcohol (CH_3CH_2OH)	159	4.8
Mercury	234	2.3
Lead	600	4.8
Copper	1357	13

Material	Boiling Temperature† (K)	Latent Heat of Vaporization (kJ/mol)
H_2	20	0.92
O_2	90	6.8
Water (H_2O)	373	41
Ethyl alcohol (CH_3CH_2OH)	351	39
Mercury	630	59
Lead	2023	178
Copper	2839	300

† *Boiling takes place at atmospheric pressure.*

THINK ABOUT THIS . . .

WHY ARE LATENT HEATS OF VAPORIZATION SIGNIFICANTLY LARGER THAN LATENT HEATS OF FUSION?

As we described in Section 16–1, the arrangement of atoms and molecules of a substance differs between the solid and liquid phase. Broadly speaking, solids are crystalline arrangements; the atoms or molecules that make up solids form a lattice structure. The forces between the nearest neighbors are springlike. With the input of a little energy, which we can anticipate is associated with a heat flow, the atoms vibrate about their equilibrium positions on the lattice, and as the energy input increases, the amplitude of these oscillations increases. At larger separations the forces become weaker, and the melting point is reached when the oscillations become of large enough amplitude to take the atoms beyond the point where the forces are springlike. At this point, which is where the substance has gone to the liquid phase, the atoms are no longer tied to a particular neighbor and can move in random directions. In the liquid phase there are still, however, bonds between neighboring atoms, and the density of liquids remains essentially the same as the density of the corresponding solid. So in this simplified picture the energy needed to melt a solid is determined only by the requirement that the *directional order* be destroyed—that the molecules leave their neat lattice arrangement. In vaporization the molecules form a gas, and (as long as we are not near the critical point) the distance between them increases enormously. To achieve this, it is necessary to *break the bonds* between neighboring atoms or molecules, and this takes much more energy than a rearrangement without actual separation. ■

CONCEPTUAL EXAMPLE 18–5 One of the games that children used to play was to take a toy tin soldier, put it in a spoon over a flame till the soldier melted, and then drop it into a pan of water. The shape of the solid mass was then used to cast shadows or to "read," like tea leaves. The temperature of the water naturally rose. Would it have risen more, less, or the same if the tin soldier had been heated to the point of almost melting and then dropped in the water?

Answer The just-melted tin and the not-quite-melted tin are at very nearly the same temperature. Nevertheless, they have quite different effects on the water. The water temperature rises because there is heat flow to it from the hot tin. But when a liquid solidifies (freezes), as does the melted tin, there is heat flow in the form of a latent heat of fusion. The not-quite-melted soldier does not have this additional thermal energy. There is thus more heat flow in the case of the melted tin soldier than in the case of the solid tin soldier, and the temperature of the water will rise higher when the melted soldier is dropped in rather than the not-quite-melted one.

EXAMPLE 18–6 One hundred grams of ice at 0°C is dropped into 200 g of liquid water at 49°C. The system is thermally isolated (see Chapter 17). After a period of time, the ice is entirely melted, leaving 300 g of water at 6°C. Assume that the specific heat of liquid water has the constant value $c = 1\ \text{cal/g}\cdot{}^\circ\text{C}$. Calculate the latent heat of fusion of water.

Setting It Up As in Fig. 18–8, we label the mass of ice as m_1, its initial temperature as T_i, the initial mass of liquid water as m_2, and its initial temperature as T_i'. The final temperature of the system is T_f. All these quantities are given.

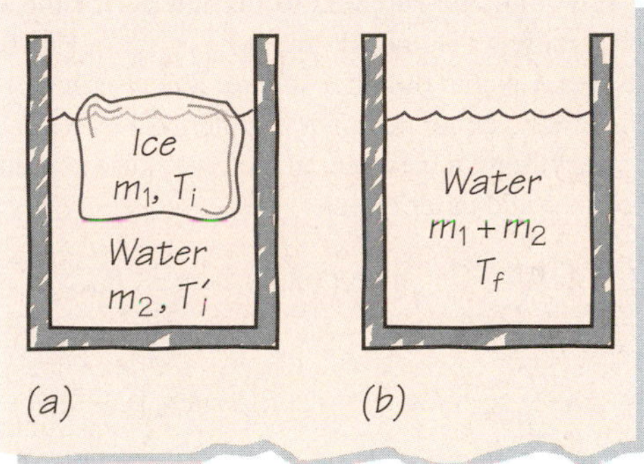

▲ **FIGURE 18–8** Ice dropped into water changes the temperature of the system.

Strategy With the system thermally isolated, the temperature cannot change as a result of thermal contact with the surroundings. Any heat flow must occur within the system between the water and the melting ice, and the net heat flow to the outside must be zero. We can summarize this condition as

$$\Delta Q_1 + \Delta Q_2 + \Delta Q_3 = 0,$$

where ΔQ_1 is the heat flow needed to melt the ice, ΔQ_2 is the heat flow necessary to raise the temperature of an amount m_1 of liquid water (from the melted ice) to the T_f, and ΔQ_3 is the heat flow necessary to cool an amount m_2 of liquid water (the initially present liquid) to T_f. The specific heat of water is assumed to be a constant, so we can use the finite heat-flow version of Eq. (18–1). By definition the latent heat of fusion, L_f, is the heat flow necessary to melt 1 g, so if m_1 is measured in grams, we have

$$\Delta Q_1 = m_1 L_f,\ \Delta Q_2 = c m_1 (T_f - T_i),\ \Delta Q_3 = c m_2 (T_f - T_i').$$

The only unknown is L_f, so to solve for it, we use the condition that $\Delta Q = 0$ for the system.

Working It Out Let's calculate ΔQ_2 and ΔQ_3:

$$\Delta Q_2 = (1\ \text{cal/g}\cdot{}^\circ\text{C})(100\ \text{g})(6^\circ\text{C} - 0^\circ\text{C}) = 600\ \text{cal}$$
$$\Delta Q_3 = (1\ \text{cal/g}\cdot{}^\circ\text{C})(200\ \text{g})(6^\circ\text{C} - 49^\circ\text{C}) = -8600\ \text{cal}.$$

Now we have

$$m_1 L_f + \Delta Q_2 + \Delta Q_3 = 0$$

or

$$L_f = \frac{-\Delta Q_2 - \Delta Q_3}{m_1} = \frac{-600\ \text{cal} + 8600\ \text{cal}}{100\ \text{g}} = 80\ \text{cal/g}.$$

You can find $\Delta Q_1 = L_f \times (100\ \text{g}) = 8.00 \cdot 10^3$ cal. This is much larger than the value associated with raising the temperature of the melted ice; we have $L_f/c = 80$. Latent heats are typically quite significant.

What Do You Think? Describe what would be different if the latent heat of fusion of water were (a) much larger, (b) much smaller than it is. Think about lakes in Minnesota or about iced drinks for your answer.

18–3 Heat Flow in Materials

When one end of a solid beam of length L is maintained at a temperature T_1 and the other end is kept at a lower temperature T_2, there will be a steady heat flow from the higher-temperature end to the lower-temperature end. Experiment shows that the rate of heat flow is proportional to the cross section A of the beam and to the temperature difference $\Delta T = (T_1 - T_2)$ between the two ends; it is inversely proportional to the length of the beam L. This rate is thus described by an equation of the type

$$\frac{\Delta Q}{\Delta t} \propto A \frac{\Delta T}{L}. \qquad (18\text{–}7)$$

Heat flow is only linearly dependent on the temperature difference ΔT if ΔT is small. This poses no difficulty if we think about the rate of heat flow for a thin slice of length

TABLE 18–3 • Thermal Conductivities for Various Materials

Material	κ (kW/m·K)
Aluminum	0.21
Copper	0.39
Lead	0.035
Silver	0.42
Water (at 0°C)	5.65×10^{-4}
Water (at 20°C)	5.99×10^{-4}
Air	2.6×10^{-5}
Asbestos	0.80×10^{-4}
Brick	6.3×10^{-4}
Glass	10.5×10^{-4}

Δx rather than for the entire beam at once. In that case ΔT is small and the linear relation between ΔT and the rate of heat flow is valid. In this "thin-slice limit" we replace L by Δx and rewrite Eq. (18–7) in derivative form as

$$\frac{dQ}{dt} = -\kappa A \frac{dT}{dx}. \tag{18–8}$$

The coefficient κ is the **thermal conductivity**, a constant that varies from material to material and may have a temperature dependence. The SI units of κ are W/(m·K), or J/(m·s·°C). (Note that a difference of 1°C is identical to a difference of 1K. The appearance of watts in the units of κ anticipates that heat flow is an energy transfer.) We present some typical values of thermal conductivity in Table 18–3. When κ is large, the material is said to be a good thermal conductor, and when κ is small, the material is a poor thermal conductor. The minus sign in Eq. (18–8) indicates that if the *temperature gradient* dT/dx is negative (i.e., the temperature is lower for larger values of x), there is positive heat flow toward larger x. In other words, heat flow goes from a hot region to a cold region. The differential equation Eq. (18–8) applies to liquids and gases as well as to solids, and Table 18–3 lists values of κ for materials of all three states.

EXAMPLE 18–7 An igloo, a hemispherical enclosure built of ice ($\kappa = 1.67$ J/m·s·°C), has an inner radius of 2.50 m. The thickness of the ice is 0.30 m. At what rate must thermal energy be generated to maintain the air inside the igloo at 5°C when the outside temperature is −40°C? Ignore all thermal energy losses through the ground or by air currents.

Setting It Up Given the configuration and dimensions of an igloo, as in Fig. 18–9, we want the rate of heat flow out of it for a given inside-outside temperature difference. This much heat flow has to be supplied from a heater to keep the inside temperature steady.

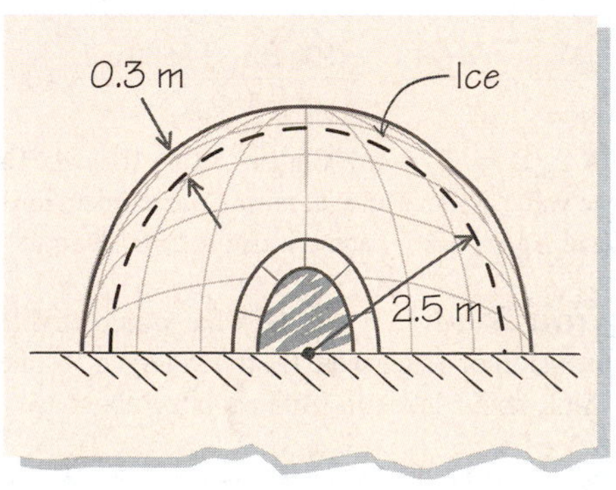

▲ **FIGURE 18–9**

Strategy The right-hand side of Eq. (18–8) tells us the rate at which heat flow must be supplied to the interior of the igloo. The temperature gradient is given by $(T_{in} - T_{out})/(\text{ice thickness})$. The surface area A is that of half a sphere. Because the ice thickness is small compared to either the inner or outer radius, we could use either radius or anything in between; in fact, we'll use a radius halfway between the inner and outer radius.

Working It Out The appropriate radius is $R_{med} = 2.50\text{ m} + 0.15\text{ m} = 2.65\text{ m}$. Then

$$\frac{dQ}{dt} = -\kappa A \frac{dT}{dx} = -\kappa A \frac{\Delta T}{\Delta x} = -\kappa(2\pi R_{med}^2)\frac{T_{in} - T_{out}}{\text{ice thickness}}$$

$$= -2\pi(1.67\text{ J/m·s·°C})(2.65\text{ m})^2 \frac{(5°\text{C}) - (-40°\text{C})}{0.30\text{ m}}$$

$$= -1.1 \times 10^4\text{ J/s} = -11\text{ kW}.$$

This igloo requires an 11 kW heater just to keep things at a chilly 5°C. Note the use of units involving joules rather than calories. For comparison, a room-size kerosene heater puts out some 5 kW.

What Do You Think? Purely from a thermal point of view, why is it best to make igloos hemispherical?

Thermal Resistance

The **thermal resistance**, or **R value**, is used as an alternative to the thermal conductivity rating; it is inversely proportional to thermal conductivity. For a slab of a given material of thickness L, R is defined by

$$R \equiv \frac{L}{\kappa}. \tag{18–9}$$

Whereas κ depends only on the type of material, *R depends also on the thickness of the material*. Therefore, R characterizes the effectiveness of a thermal barrier of given thickness. A large R value corresponds to a small thermal conductivity and/or a large thickness; a piece of material with such characteristics is a good *thermal insulator*, whereas a piece of material with a low R value is a poor insulator. The R value has SI units of $m^2 \cdot K/W$. When it is used in connection with insulation in construction in America, however, it is often quoted in units of $ft^2 \cdot h \cdot °F/Btu$, where $1\ ft^2 \cdot h \cdot °F/Btu = 0.18\ m^2 \cdot K/W$. The R value of a 6-in piece of fiberglass insulation is $22\ ft^2 \cdot h \cdot °F/Btu$, compared to $0.32\ ft^2 \cdot h \cdot °F/Btu$ for a 3/8-in piece of plasterboard.

In terms of R, the relation between thermal energy flow and temperature difference from Eqs. (18–8) and (18–9) is

$$\frac{\Delta Q}{\Delta t} = \frac{1}{R} A\, \Delta T, \tag{18–10}$$

where A is the area across which thermal energy is transported. Because R refers to a given material of a given thickness, it is interesting to study the effective R values of different pieces of material combined in different ways—after all, that is how a house is constructed.

Thermal Resistance in Series: Let's first consider combining two conducting materials of the same area A *in series*, as shown in Fig. 18–10. Suppose that the left-hand piece has an R value of R_1 and the right-hand piece R_2 and that the temperature varies from T_c (cold) on the left to T_h (hot) on the right. We want to find the effective R value of the combined piece, R_{eff}. When thermal energy flows at a steady rate through the two materials, as shown in Fig. 18–10, then energy conservation implies that the rate of transport of thermal energy across one piece must equal the rate of transport across the second piece, and hence also the rate across the combined piece. Application of this idea (see Problem 25) implies that

$$\text{for two solids in series: } R_{eff} = R_1 + R_2. \tag{18–11}$$

This result generalizes for many slabs of material that are combined in series.

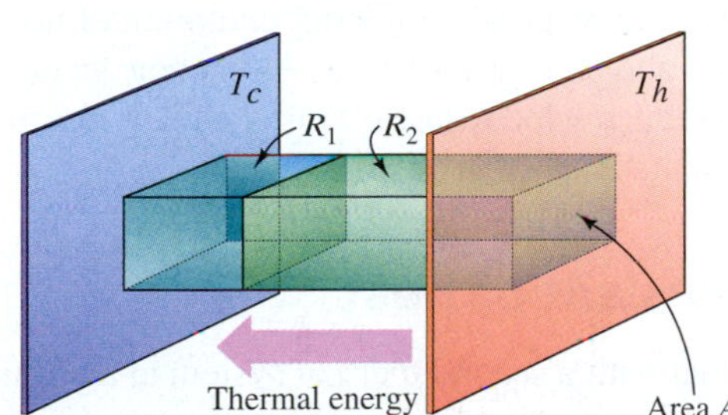

FIGURE 18–10 Thermal resistance, or R values, for two materials placed "in series"—one behind the other. Thermal energy passes through the materials as shown.

Thermal Resistance in Parallel: Now suppose that instead of being placed in series, two slabs are placed *in parallel*, as shown in Fig. 18–11. In this case the temperature difference on the right-hand and left-hand sides of both slabs is the same, and the total rate of transport of thermal energy is the sum of the transport rates through the two slabs. In Problem 27, this fact gives the result

$$\text{for two solids in parallel: } \frac{1}{R_{eff}} = \frac{1}{A_1 + A_2}\left(\frac{A_1}{R_1} + \frac{A_2}{R_2}\right). \tag{18–12}$$

This result generalizes for more than two slabs placed in parallel.

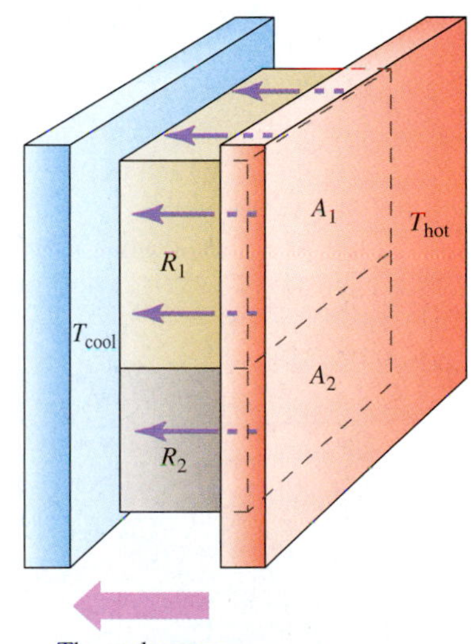

FIGURE 18–11 Thermal resistance, or R values, for two materials placed "in parallel." Thermal energy passes through the materials as shown.

EXAMPLE 18–8 A hole in a wall has been temporarily patched on the inside with a 1 sq-ft piece of plywood for which $R = 1\ ft^2 \cdot h \cdot °F/Btu$. The rest of the wall is insulated with fiberglass insulation for which $R = 22\ ft^2 \cdot h \cdot °F/Btu$. What is the R value of the repaired wall, whose total area is $100\ ft^2$?

Setting It Up The sketch of this situation, Fig. 18–12, shows that we have two materials placed in parallel. We know A_1 and R_1, the area and R value of the plywood, and A_2 and R_2, the area and R value of the fiberglass, respectively. We want R_{eff}, the R-value of the wall formed as in the figure.

(continues on next page)

Strategy This is an application of the formula for materials in parallel; Eq. (18–12) applies. We require only a numerical evaluation.

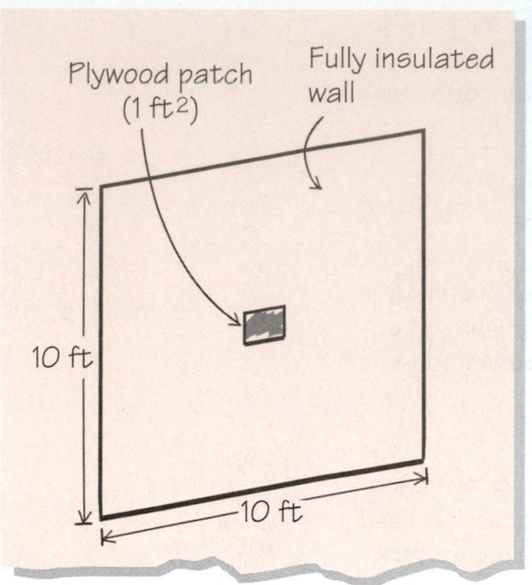

▲ **FIGURE 18–12** Hole in wall is patched with plywood.

Working It Out We have

$$\frac{1}{R_{\text{eff}}} = \frac{1}{A_1 + A_2}\left(\frac{A_1}{R_1} + \frac{A_2}{R_2}\right)$$

$$= \frac{1}{100\ \text{ft}^2}\left(\frac{1\ \text{ft}^2}{1\ \text{ft}^2 \cdot \text{h} \cdot {}^\circ\text{F/Btu}} + \frac{100\ \text{ft}^2 - 1\ \text{ft}^2}{22\ \text{ft}^2 \cdot \text{h} \cdot {}^\circ\text{F/Btu}}\right)$$

$$= \frac{1}{18\ \text{ft}^2 \cdot \text{h} \cdot {}^\circ\text{F/Btu}}.$$

Thus

$$R_{\text{eff}} = 18\ \text{ft}^2 \cdot \text{h} \cdot {}^\circ\text{F/Btu}.$$

Note that because the R values already take the thickness of the materials into account, the plywood is not necessarily the same thickness as the fiberglass insulation. In fact, plywood is only a fraction of an inch thick, whereas the fiberglass is 6 in thick. The rest of the thickness may be taken up by air, but because this air will be continuous with the outside environment, we have ignored its insulating properties.

What Do You Think? Having 1% of the area in plywood changed the R value by almost 20%. If instead of 1% the plywood area were 10%, the R value would have changed by about a factor of 3. Why such a big effect?

In Example 18–8 a hole in a wall that is patched with little insulation and takes up only 1% of the area of the wall reduces the R value by some 18%. This effect is typical of thermal insulators combined in parallel. In particular, if one of the elements of a set of insulators combined in parallel has a very small R value, Eq. (18–12) shows that the effective R value is very small. Small holes can be very costly sources of thermal energy loss. Indeed, the best window has a thermal resistance of about one-tenth of normal wall and ceiling insulation. Thus once wall and ceiling energy losses become small, the only way to conserve energy further is to reduce window area. This is already practiced in many commercial buildings in cold climates.

18–4 The Mechanical Equivalent of Heat

Placing a thermal system in thermal contact with a second thermal system at a different temperature is one way to produce a temperature change in the first system, via heat flow in this case. For example, we can heat water by placing it in a pot on a hot stove. In this case a heat flow is responsible for the temperature change. In the first half of the nineteenth century the major discovery was made that the temperature of a thermal system can also be raised by performing mechanical work on that system. For example, when an object is rubbed so that work is done on it by friction, its temperature is raised, just as in Count Rumford's late 18th century observation (Fig. 18–5). Such observations and the realization of their significance took place through a slow buildup of data rather than through a single breakthrough experiment. The history is thus rather tangled, with many participants. Rumford argued that Lavoisier's caloric theory could not be correct. Rumford's result was not widely accepted at the time, but in the mid-19th century the careful experiments of James Joule established with precision that a given amount of mechanical work done on a thermal system produces a temperature rise that corresponds to a given heat flow, as defined by Eq. (18–1).

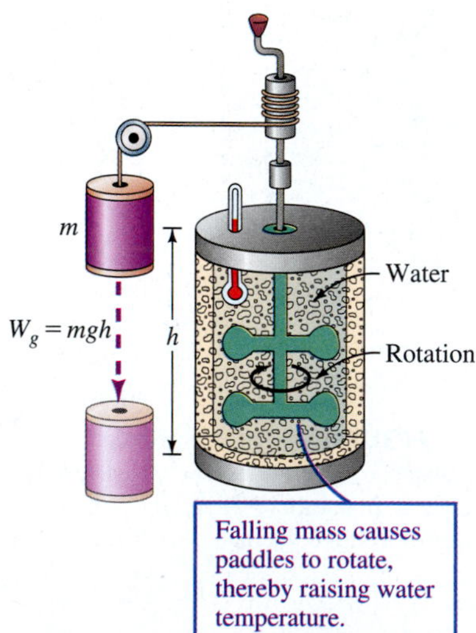

▲ **FIGURE 18–13** Schematic diagram of the apparatus used in Joule's classic experiments. Churning paddles raise the temperature of a thermally isolated container of water.

Joule's Experiments

In an idealized version of Joule's experiments, the temperature of a thermally isolated quantity of water is measurably raised by a set of paddles that churn the water (Fig. 18–13). We know that a drag force is present to resist the motion of the turning paddles, and work must be performed to keep them rotating. The amount of work done can be measured precisely by attaching the paddles to a falling mass that keeps the paddles moving, and then measuring the distance the mass falls.

When the paddles have done a certain amount of work on the water, the temperature of the system (water plus paddles) will have risen by a definite amount. And once the temperature has been raised, *there is no way to tell* whether the temperature increase occurred because work has been done on the system or because of heat flow into it. A series of experiments of this type establishes that there is a specific temperature rise for a specific amount of work; therefore, there is a precise equivalent between work done on a system, W, and the heat flow into it, ΔQ. Because *no* experiment can be performed that can tell whether a given temperature change was produced by mechanical work or by heat flow, we can assert that *heat flow, like work, is an energy transfer*. This result is known as the **mechanical equivalent of heat**. It means that the unit of heat flow, the calorie, must be convertible to the unit of energy—that is, there is a direct relation between the SI unit of work, the joule, and the unit of heat flow, the calorie, and Joule's experiment allows one to find the numerical relation:

$$1 \text{ cal} = 4.185 \text{ J}. \qquad (18\text{–}13)$$

THE MECHANICAL EQUIVALENT OF HEAT

This relation may seem to be a simple unit conversion; however, it is far more. Until Joule's experiment, the calorie was unrelated to any other units familiar to us. Once his work was completed, one could say that the calorie has the dimensions of energy and that the SI unit of heat flow is the joule. Equation (18–13) thus states the crucial result of a basic experiment. It should not be confused with, say, the statement that 1 Btu is 252.00 cal, which is merely the relation between different units.

Today the *numerical definition* of the calorie is taken to be Eq. (18–13); it is no longer based on a measurement of the temperature change of water, as it was in Section 18–2. Such a definition is possible only because of the physical equivalence between heat flow and work. The calorie continues to be a widely used unit.

CONCEPTUAL EXAMPLE 18–9 Two *very* thin-walled (but unbreakable and undentable!) containers are dropped from a tall building. One of them holds water, the other an equal mass of mercury. Which of the liquids inside will be warmer after the container hits the ground?

Answer When the containers land, *kinetic energy* is converted into thermal energy in each case. The mass of each container is the same, so each has the same kinetic energy just before landing. But since the specific heat of water is higher than that of mercury (by a factor of 30; see Table 18–1), the water will undergo a smaller temperature change as a result. Note that the temperature does not rise gradually as the container falls; the conversion to thermal energy takes place rapidly when the container hits the ground and the fluid is strongly shaken.

What Do You Think? Would the heating be the same if the containers were flung from the height of about 1 m, with a velocity comparable to the one they reach in falling from the building?

EXAMPLE 18–10 The paddles of the apparatus shown in Fig. 18–13 are driven by a mass m of 0.50 kg falling at a constant speed. The mass of the thermally isolated water in the apparatus is 250 g. How much is the temperature of the water raised if the driving mass falls a distance $h = 2.0$ m? Ignore the mass of the paddles.

Strategy According to the work–energy theorem, the work W done on the water is the change in energy of the falling mass m. Since the mass falls at constant speed, that change is the change in the potential energy of m, namely mgh. But this work is equivalent to a heat flow, $\Delta Q = W$, and if we note that ΔQ is related to the temperature change ΔT by the specific heat (which we assume constant) according to Eq. (18–1), we can solve for ΔT.

Working It Out The specific heat of water has the constant value $c_{H_2O} = 1.00 \text{ cal/g} \cdot {}^\circ\text{C}$, so Eq. (18–1) reads $m_{H_2O} c_{H_2O} \Delta T = mgh$, or

$$\Delta T = \frac{mgh}{m_{H_2O} c_{H_2O}} = \frac{(0.50 \text{ kg})(9.8 \text{ m/s}^2)(2.0 \text{ m})}{(250 \text{ g})(1.00 \text{ cal/g} \cdot {}^\circ\text{C})}(4.185 \text{ cal/J}) = 0.16^\circ\text{C}.$$

What Do You Think? It looks as though you could reverse the entire process: You turn the crank in Fig. 18–13 in reverse and thereby raise the weight. Will the thermal process also reverse, leaving you with cooler water?

18–5 Work Done by Thermal Systems

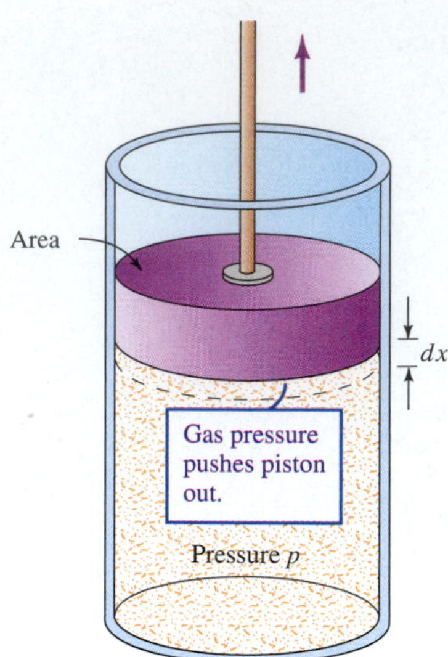

▲ **FIGURE 18–14** The volume of a cylinder increases when a piston of area A moves out a distance dx under the influence of the pressure, p, of the gas contained in the cylinder.

In the last section, we saw that work can be done on a thermal system. By a simple sign reversal, we can think of this as work done by the thermal system on its surroundings—negative work, in the case of the churning paddles of Joule's experiments. Thermal systems, particularly gases, which can expand greatly, can also do *positive* work on their surroundings.

Consider a piston of area A pushed out by the pressure p of the gas contained in a cylinder (Fig. 18–14). By the definition of pressure, the magnitude of the force F on the piston is pA. Moreover, if the piston moves out, this force acts in the direction of the displacement, which we take to have infinitesimal magnitude dx. The infinitesimal work done by the gas in moving the piston is

$$dW = \vec{F} \cdot d\vec{x} = pA\,dx. \tag{18–14}$$

Because $A\,dx = dV$ (the infinitesimal change in the volume of the gas), this expression can be rewritten as

$$dW = p\,dV. \tag{18–15}$$

Equation (18–14) is specific to the linear motion of a piston with the geometry of Fig. 18–14, but Eq. (18–15) is a more general form that holds when a gas changes its volume by dV in pushing against its surroundings in any direction. Note the sign carefully: dW, the work the gas does *on* its surroundings, will be positive or negative according to whether dV is positive or negative.

As for the heat flow, we must specify the way in which the finite volume change is made in the integration of Eq. (18–15). Thus, *the net work done by the gas depends on the path in a p–V, p–T, or V–T diagram*. The integral expressing the work done in going from some state 1 to another state 2 is

$$W = \int_{V_1}^{V_2} p\,dV. \tag{18–16}$$

This integral is defined only when the path $1 \rightarrow 2$ is specified, meaning that how p is changing as we follow the path has to be specified.

If the path of the transformation $1 \rightarrow 2$ is drawn on a p–V diagram, as in Fig. 18–15a, then the work done by the gas has a simple interpretation: *It is the area under the curve of p versus V*, which is shaded in the figure. The sign of the work done by the gas is determined by whether V_2 is larger than V_1, as in Fig. 18–15a, or smaller than V_1, as in Fig. 18–15b. In Fig. 18–15b, the net work done *by* the gas is negative; in other words, there is positive work done *on* the gas to compress it. All this follows the normal rules of calculus, where $\int_{x_1}^{x_2} F(x)\,dx$ is the area under a curve of $F(x)$ versus x.

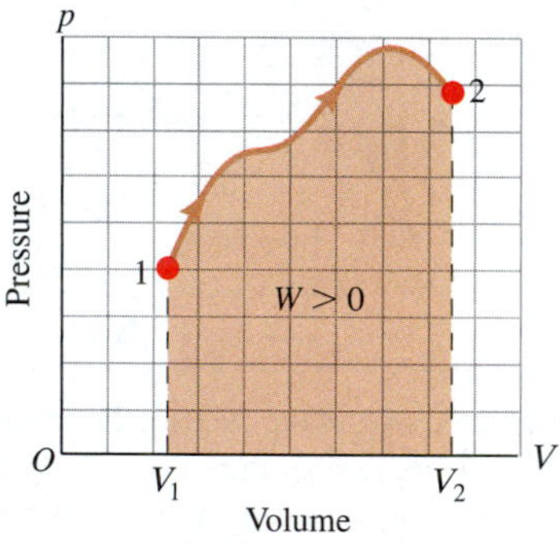

(a)

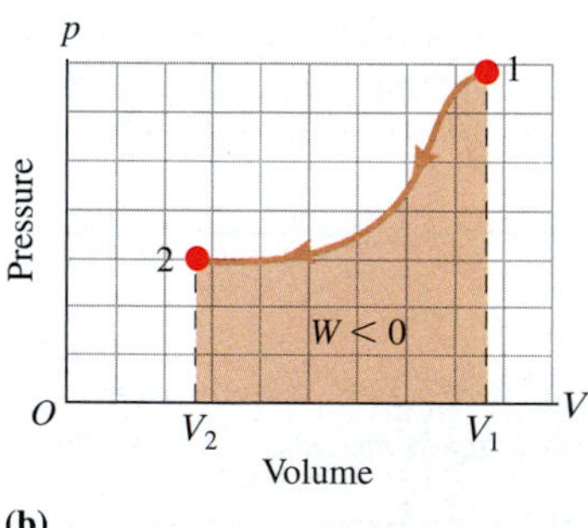

(b)

▲ **FIGURE 18–15** The work done by a gas on its surroundings (like that of the piston–cylinder system of Fig. 18–14) in moving reversibly from point 1 to point 2 is the area under the curve of p versus V. (a) If the volume V_2 is greater than V_1 (the gas expands), the work the gas does on its surroundings is positive. (b) If the volume V_1 is greater than V_2, the gas is compressed, and the work it does on its surroundings is negative.

If we think of the work done by a thermal system as the area under a curve of p versus V, the fact that the work done depends on the path of the change becomes evident. The path describes the curve between two states, 1 and 2, and the area under different curves may certainly differ.

Cyclic Transformations

The operation of the pistons in an automobile engine is repetitive.† This is an example of a **cyclic transformation**: a change in which a thermal system follows a path that takes it back to its original state. The importance of such processes rests on the fact that all engines work in cycles. The net work done by the thermal system in a cyclic transformation can be calculated by starting anywhere along the path shown on a p–V diagram (Fig. 18–16a). (Note that we are assuming that all the changes treated here are reversible ones; the very fact that we are tracing the path indicates that.) Let's divide the cycle of Fig. 18–16a into two parts, as in Fig. 18–16b and c. The first part is the change $B \rightarrow A$,

†If you are not familiar with the details of such an engine, the only important aspect here is the repetitive nature of an engine's action. But you can also peek ahead to Chapter 20 if you are curious.

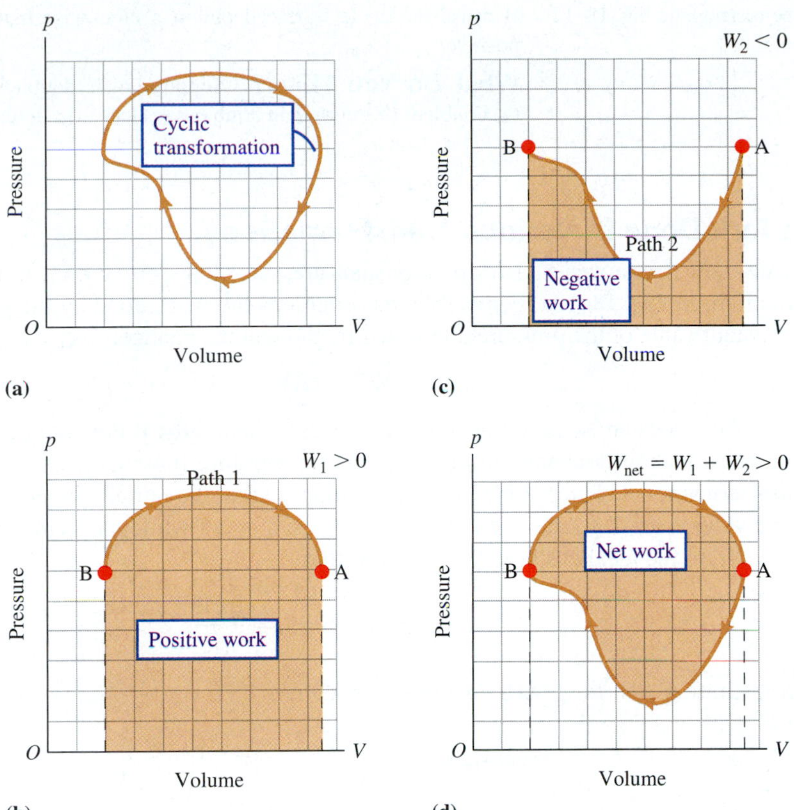

◀ **FIGURE 18–16** In a closed cycle, the work done by a gas on its surroundings is the area enclosed by the curve of p versus V. To show this, the full cycle is broken into two paths.

along path 1 (Fig. 18–16b), and the second part is $A \to B$, along path 2 (Fig. 18–16c). The net work W_{cycle} adds algebraically (Fig. 18–16d) and is thus given by

$$W_{\text{cycle}} = \underset{\text{path 1}}{W_{B \to A}} + \underset{\text{path 2}}{W_{A \to B}}.$$

As we saw in Chapter 6 during our discussion of work, $\int_{x_1}^{x_2} f(x)\,dx = -\int_{x_2}^{x_1} f(x)\,dx$, when the two integrals are taken over the same path. Thus the integral over path 2, which goes from larger to smaller values of x, is the negative of the area under that path. The net work is then the area under path 1 minus the area under path 2:

$$W_{\text{cycle}} = \text{area enclosed by cyclic path.} \tag{18–17}$$

Because the cycle is traced in a clockwise direction, the work done by the gas along path 1 is larger than the work done on the gas along path 2. The net work done on the surroundings is positive; if this cycle were traced in a counterclockwise direction, the net work done would be negative.

EXAMPLE 18–11 A thermal system is taken around the cycle shown in Fig. 18–17. Along the paths $A \to B$ and $C \to D$, the volume is held fixed at V_1 and V_2, respectively. From $B \to C$ and from $D \to A$, the pressure is held fixed at p_2 and p_1, respectively. What is the work done by the thermal system in tracing out one cycle?

Strategy We can solve the total work around the cycle by calculating the work over each step using Eq. (18–16) and summing the work done in each step, or we can do what is suggested in Eq. (18–17) and evaluate the area enclosed by the p–V diagram. We shall use the second method, which is simpler because it summarizes the steps of the first method. Nevertheless, it is useful to make one comment on the first method. Two steps of the cycle, the ones from $A \to B$ and the ones from $C \to D$, are constant volume segments. For these legs the differential element $dV = 0$, and the integral is zero, as there is no area beneath a vertical line on the p–V diagram.

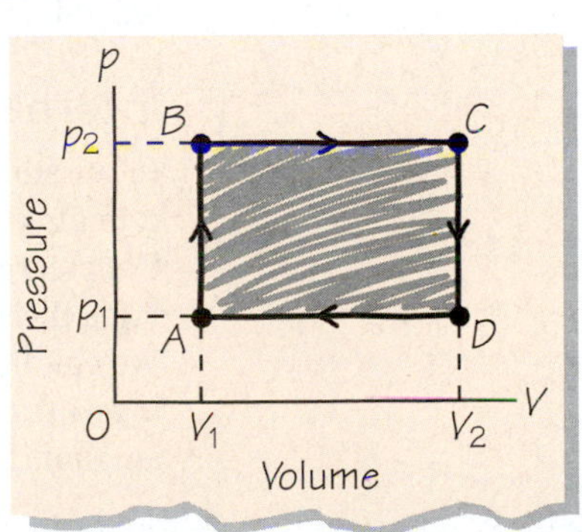

▲ **FIGURE 18–17** A thermal system undergoes a cyclic transformation.

(continues on next page)

Working It Out The area enclosed by the rectangle in Fig. 18–17 is the width $(V_2 - V_1)$ times the height $(p_2 - p_1)$:

$$W_{\text{cycle}} = (p_2 - p_1)(V_2 - V_1).$$

Because the cycle is traced out in a clockwise manner, W_{cycle} is positive.

What Do You Think? Suppose p_2 is doubled. The work is (a) doubled; (b) more than doubled; (c) less than doubled.

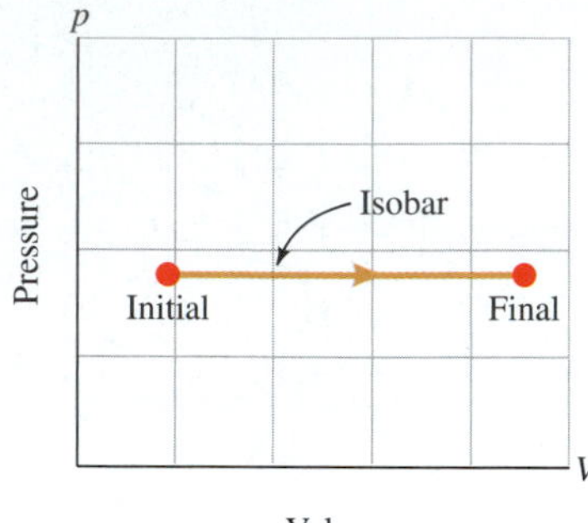

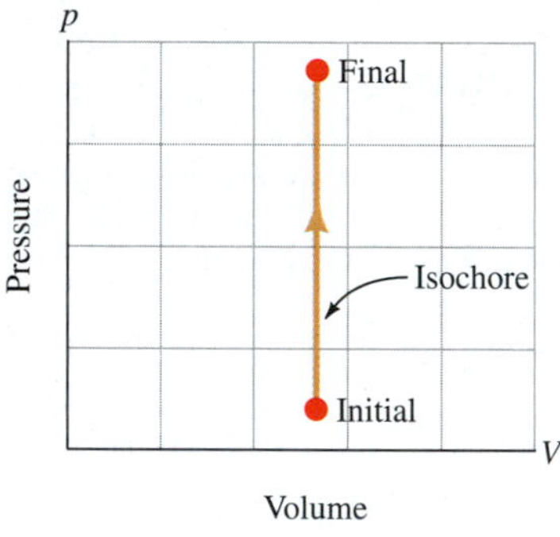

▲ **FIGURE 18–18** (a) An isobar, which represents a constant-pressure transformation. (b) An isochore, which represents a constant-volume transformation.

Work Done in Various Transformations

Constant Pressure: An isobaric (constant-pressure) transformation is drawn on a p–V diagram in Fig. 18–18a. For such transformations the work done by the gas is just the constant value of the pressure multiplied by the volume change,

$$W = p\Delta V. \tag{18–18}$$

How does an isobaric transformation occur? Generally, if the volume of a gas is to increase without a change in pressure, the temperature must increase. Let's calculate this temperature change for an ideal gas. Suppose that the initial temperature, pressure, and volume are T, p, and V and that V is increased by dV while p is held constant. It follows from the ideal gas law, $T = pV/nR$, that

$$\text{for constant } p\text{: } dT = \frac{p\,dV}{nR}. \tag{18–19}$$

Substituting T/V for p/nR, Eq. (18–19) becomes

$$\text{for constant } p\text{: } dT = \frac{T}{V}dV \quad \text{or} \quad \frac{dT}{T} = \frac{dV}{V}. \tag{18–20}$$

Conversely, if the heat flow is such that the fractional increase in temperature is equal to the fractional increase in volume, as in Eq. (18–20), then for an ideal gas the pressure must be constant, and the heat flow is $C_p\,dT$.

Constant Volume: When the volume is kept fixed, the curve of the transformation is said to be an **isochore** (Fig. 18–18b). Thermal systems do no work—positive or negative—in an isochoric transformation because there is zero area under the curve on a p–V diagram. The pressure and temperature both change, however, so such a transformation must be accompanied by a heat flow, just as for the isobaric and the isothermal transformations. In this case the heat flow is $dQ = C_V\,dT$.

Adiabatic: Finally, there is a particular curve that will be followed if the gas does work while it is thermally isolated from its environment. Imagine a cylinder of gas enclosed in a thermal insulator, such as Styrofoam, with only the mechanical movements of a piston connecting it to its surroundings. Because of the thermal isolation, there is no heat flow dQ into or out of the system. When $dQ = 0$, the temperature can change only if work is done on the system. Reversible transformations of a thermal system in which there is no heat flow to the system are called **adiabatic** transformations. In Section 18–8 we will explore the precise curve followed by adiabatic transformations for ideal gases.

Internal Energy

In an adiabatic transformation a thermal system has no thermal contact with its surroundings but does work on them. The work–energy theorem, as formulated in Chapters 6 and 7, can be applied to the system, which we can accordingly view as having a thermal (or internal) energy that can change if work is done on or by the system. We call this thermal energy† U. In particular, when a system does work on its surroundings in an adiabatic transformation, *it changes its internal (thermal) energy* by the amount of work it does:

$$\text{for an infinitesimal adiabatic transformation: } dU = -dW; \tag{18–21}$$

$$\text{for a finite adiabatic transformation: } U_B - U_A = -W_{A\rightarrow B}. \tag{18–22}$$

†Don't confuse U with a potential energy!

In other words, any changes in a thermal system's internal energy are matched by mechanical work done if a transformation is adiabatic. If we want to study how the internal energy of a thermal system changes when we go from one thermodynamic state to another, we should look at adiabatic transformations because then we can avoid the complications of heat flow. *The adiabatic process allows us to learn just how the internal energy depends on the thermodynamic variables.*

The thermal (internal) energy is a function of p and V, p and T, or T and V. The thermal energy is associated with the energy of the molecules of a thermal system; it is simply a function of the thermodynamic variables of the system, not of how those variables might have changed. Changes in internal energy, like changes in the potential energy of a conservative mechanical system, are *not* path dependent. This contrasts sharply with work or heat flow, which *are* path dependent.

In Eq. (18–21) both dU and dW are written as differentials. The symbol dU has its usual meaning as a differential, in that a finite change in internal energy is

$$\int_A^B dU = U_B - U_A, \tag{18–23}$$

which is independent of the path between A and B. There is no equivalent expression for work, because the integral of dW *does* depend on the path taken. In a strict mathematical sense, dW is not a true differential, but there is no need for a more complicated notation, provided we keep in mind that the integral of dW is path dependent.

The minus signs on the right-hand sides of Eqs. (18–21) and (18–22) occur because W always denotes the work done *by* the system *on* its surroundings. When the system does positive work in an adiabatic transformation, its internal energy U decreases; when the work done is negative, the internal energy increases.

Let's now reconsider the Joule experiments that determined the mechanical equivalent of heat (Section 18–4). The water whose temperature is raised by the churning paddles is thermally isolated. The transformation is irreversible; it is not possible to "unchurn" the water and use it to raise the fallen weight. Nevertheless, the (positive) net work done by the paddles does indeed determine the (positive) change in the internal energy of the water. *How* a change is made is irrelevant for the internal energy, because the internal energy is purely a function of state. Thus Eq. (18–22) holds for both reversible and irreversible transformations of thermally isolated systems.

18–6 The First Law of Thermodynamics

The **first law of thermodynamics** is a statement of the conservation of energy for thermal systems. It is essential in understanding any thermal system—particularly those in which fuel is burned and work is done, such as engines or power plants. We have seen that heat flow and work are equivalent ways to change a system's internal, or thermal, energy. The first law simply recognizes this by stating that the change in the internal energy of a thermal system in going from one state to another is the negative of the work that the system does on its surroundings, plus the heat flow into the system, during that transformation. In other words, if we label our initial and final states by A and B, respectively,

$$\Delta U = U_B - U_A = -W_{A\to B} + Q_{A\to B}. \tag{18–24}$$

FIRST LAW OF THERMODYNAMICS

This equation is the primary result of this chapter. The minus sign in front of W indicates that W is the work done *by* the thermal system, and the plus sign in front of Q indicates that Q is the heat flow *into* the system. In Fig. 18–19 we give another view of the meaning of these terms.

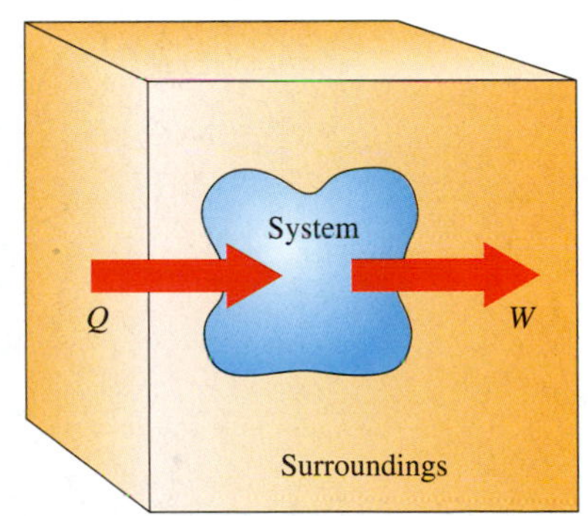

▲ **FIGURE 18–19** In the first law of thermodynamics, W represents the work done by the system, while Q represents the heat flow into the system. Either term can be negative or positive, depending on the transformations involved.

We should mention that there is no universal agreement as to the sign convention for work W. Engineers and physicists tend to use the convention stated here; engineers and physicists are primarily interested in what the system does to its surroundings—for example, how much work an engine is capable of doing. But chemists prefer to say that W is positive when it is done *on* the system, a convention that makes the sign in Eq. (18–24) + instead of −. That is because chemists are more interested in the thermal system itself than its effects on its surroundings.

Because U is a function of state only, ΔU is the same whether the transformation $A \rightarrow B$ is reversible or irreversible. Similarly, the work done by the system can be measured by observing its effect on its surroundings—for example, how high the system lifts a mass—independently of whether the transformation $A \rightarrow B$ is reversible or irreversible. The heat flow $Q_{A \rightarrow B}$ can be determined from the first law of thermodynamics if the transformation is irreversible or determined independently by making a calorimetric measurement.

The first law of thermodynamics is often written as

$$\Delta U = -\Delta W + \Delta Q \tag{18–25a}$$

or in a differential form as

$$dU = -dW + dQ. \tag{18–25b}$$

While both dW and dQ depend on the path, dU is independent of path. Remarkably, the path dependence of the two separate terms cancels in the difference $-dW + dQ$.

The First Law of Thermodynamics in Closed Cycles

Engines operate and perform work in closed cycles. Thus it is of practical importance to be able to apply the first law of thermodynamics to a closed cycle. There is no change in internal energy over a cycle, because the system comes back to its original state. The work done by the thermal system over the cycle must then match the heat flow into the system, by the first law:

$$Q_{\text{cycle}} = W_{\text{cycle}}. \tag{18–26}$$

Here Q_{cycle} and W_{cycle} are the net heat flow into, and net work done, by the system over the entire cycle.

CONCEPTUAL EXAMPLE 18–12 Suppose that a thermal system follows the cycle $A \rightarrow B \rightarrow C \rightarrow D \rightarrow A$, as drawn in Fig. 18–17 and treated in Example 18–11. What is the total heat flow into the system after one cycle is completed?

Answer To repeat the argument above, the internal energy change over one cycle is zero because the initial point is the same as the final point. The fact that the system passes through a complicated succession of intermediate states is irrelevant. Then we apply Eq. (18–26), $Q_{\text{cycle}} = W_{\text{cycle}}$, together with the results of Example 18–11, where the work done for this cycle is calculated:

$$Q_{\text{cycle}} = (p_2 - p_1)(V_2 - V_1).$$

What Do You Think? Suppose that instead of the rectangular path of Fig. 18–17, an arbitrary closed curve were followed. What can you say about the sign of the heat flow into the system?

How the Energy Changes in Constant-Volume and Constant-Pressure Transformations

No work is done in reversible changes under constant volume ($dW = p dV$, and $dV = 0$), so Eq. (18–25) becomes

$$\text{for constant volume: } dU = dQ_V = C_V\, dT. \tag{18–27}$$

We have emphasized the fact that the volume is constant by the subscript on dQ. No subscript is necessary for dU, because the change in U is independent of the path taken. Similarly, we can express the first law of thermodynamics for reversible changes under constant pressure as

$$\text{for constant pressure: } dU = -dW_p + dQ_p = -p dV + C_p\, dT. \tag{18–28}$$

Equations (18–27) and (18–28) can both be integrated over paths with constant V and constant p, respectively, to find the net internal energy change under a finite volume or pressure change. Keep in mind that in such an integration the heat capacities may depend on temperature, so they cannot be removed from under the integral sign.

18–7 Internal Energy of Ideal Gases

Joule performed an experiment in 1845 to study the properties of U for ideal gases. He demonstrated that *for ideal gases, the internal energy is a function of temperature alone*: $U = U(T)$. The equation of state adds to this by saying that U is equivalently a function only of the product pV.

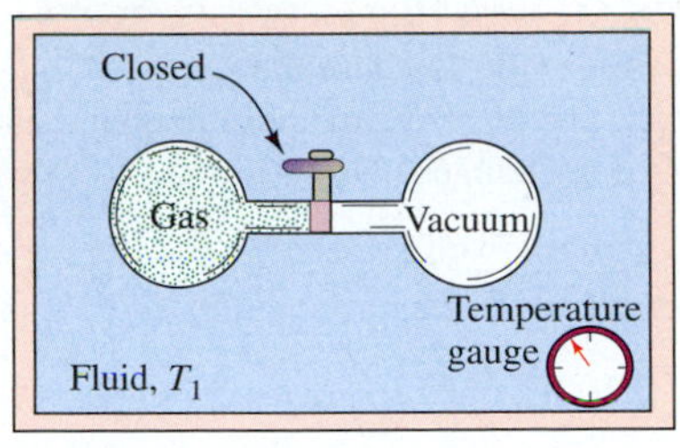

(a)

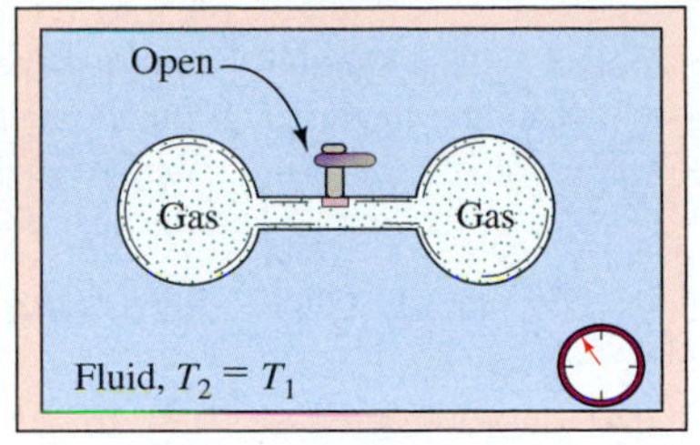

(b)

FIGURE 18–20 Schematic diagram of Joule's experiment. When Joule opened the stopcock to allow free expansion, he observed no temperature change for a dilute (ideal) gas.

Joule's experiment is as follows: Fill one part of a two-part container with a dilute sample of gas (if it is sufficiently dilute, we can be sure it is ideal). The other part of the container is evacuated, and the two parts are connected by a tube with a stopcock (Fig. 18–20a). The container as a whole is sealed and does no work on the outside world. The container is submerged in a thermally isolated container of water or some other fluid whose temperature can be monitored. This fluid, initially at temperature T_1, acts as a calorimeter: If its temperature changes because of some change in the state of the gas in the container, then there has been a heat flow into or out of the gas.

The stopcock is now opened, and the gas undergoes free expansion into the previously empty part of the container. The crucial feature of free expansion is that the gas does no work, because it does not cause mechanical movement in its surroundings. The new temperature of the fluid, T_2, is then measured. When the gas is sufficiently dilute, the experiment shows that $T_2 = T_1$ (Fig. 18–20b). *The temperature of an ideal gas undergoing free expansion remains constant.* This result means that there has been no heat flow into the gas. There is no work done and no heat flow in free expansion. Thus, according to the first law of thermodynamics, the internal energy, U, remains unchanged. But because the volume of the gas does change, U must be independent of volume. Therefore, for an ideal gas, U is a function only of temperature:

$$U_{\text{ideal}} = U(T). \tag{18–29}$$

ENERGY OF AN IDEAL GAS

This result follows directly from experiment, but it can also be derived from the second law of thermodynamics, which we will explore in Chapter 20.

For real gases—which are not perfectly ideal—there is a deviation from the result of Eq. (18–29), and temperature changes during free expansion. In other words, U is a function of density as well as temperature. We typically ignore this effect.

Let's now couple the fact that the internal energy of an ideal gas depends only on the temperature with Eqs. (18–27) and (18–28). Because U depends *only* on T, the integral of Eq. (18–27) gives

$$U(T_B) - U(T_A) = \int_{T_A}^{T_B} C_V \, dT. \tag{18–30}$$

This equation must hold for *all* values of the volume because the left-hand side of the equation is independent of volume. Hence Joule's experiment shows that C_V must itself be independent of volume. Equation (18–30) expresses only a *change* in energy, so it is equivalent to the form

$$U(T) = \int_0^T C_V \, dT + \text{a constant}. \tag{18–31}$$

The constant is arbitrary because only *changes* in U matter, not the actual values of $U(T)$ and $U(0)$. We therefore take the constant to be zero. Finally, we can use the experimental fact that over a wide temperature range, C_V is independent of temperature. We can remove C_V from behind the integral sign and find

$$\text{for } C_V \text{ independent of temperature: } U(T) = C_V T. \tag{18–32}$$

The internal energy of an ideal gas is simple indeed, as long as C_V is constant with temperature.

EXAMPLE 18–13 An ideal gas in contact with a thermal reservoir at temperature $T = T_0$ does work by expanding against a piston. That work is 10 J. How much heat flow goes into the gas during its expansion?

Setting It Up The path, which for an ideal gas making an isothermal transformation is a hyperbola, is already drawn in Fig. 18–1c.

Strategy The first law of thermodynamics gives us the heat flow, Q, in terms of the work W done by the system and the change in internal energy, ΔU: $Q = \Delta U + W$. Because the gas is ideal, U is a function only of temperature, and therefore in an isothermal transformation such as this one, U is unchanged. This allows us to solve for the heat flow Q.

Working It Out With $\Delta U = 0$, we have

$$Q = W = 10\text{ J}.$$

What Do You Think? Where does the heat flow come from?

A Relation Between C_p and C_V for Ideal Gases

We can express C_p in terms of C_V. Equations (18–27) and (18–28) (respectively $dU = C_V\,dT$ and $dU = -p\,dV + C_p\,dT$) are two differential expressions for the energy change, dU. Because Eq. (18–28) specifies constant pressure, the term $p\,dV$ that occurs on the right-hand side can also be written for an ideal gas as $p\,dV = nR\,dT$. Then both Eqs. (18–27) and (18–28) involve only terms proportional to dT on the right-hand side. Because the energy shift dU is independent of whether the infinitesimal transformation is made at constant V or at constant p, the two expressions for dU must be equal:

$$C_V\,dT = -nR\,dT + C_p\,dT.$$

We cancel the common factor dT from this equation to find that

$$C_p = C_V + nR, \qquad (18\text{–}33)$$

RELATION BETWEEN C_p AND C_V

which is a result that applies to ideal gases. As C_V is a function of T only, C_p is also; and to the extent that C_V is in fact constant with T, so is C_p. Equation (18–33) is supported by experiment for ideal gases and is a good approximation for real gases. To go further than this relation—actually to compute quantities such as C_V—it is necessary to consider the microscopic dynamics of gases. This calculation will be made in Chapter 19.

18–8 More Applications for Ideal Gases

In order to understand the behavior of engines or, more generally, of gaseous thermal systems, we want to express the changes that occur in terms of the thermodynamic variables. In principle, this is possible as long as the transformations are reversible. Better, if we can approximate real gases by ideal gases, then all the transformations we have discussed can be expressed mathematically, as can the corresponding work done, heat flow, energy changes, and related quantities. Some of these mathematical expressions have already been given. For example, the shape of the p–V curve that represents an isothermal transformation at $T = T_0$ is a hyperbola, $p = nRT_0/V$. The relation $C_p = C_V + nR$ [Eq. (18–33)] is a consequence of the fact that the gas is ideal.

Of the four reversible transformations we discussed earlier—constant-volume (isochoric), constant-pressure (isobaric), constant-temperature (isothermal), and adiabatic transformations—the first two are simple to treat. Let's now look at the other two in more detail.

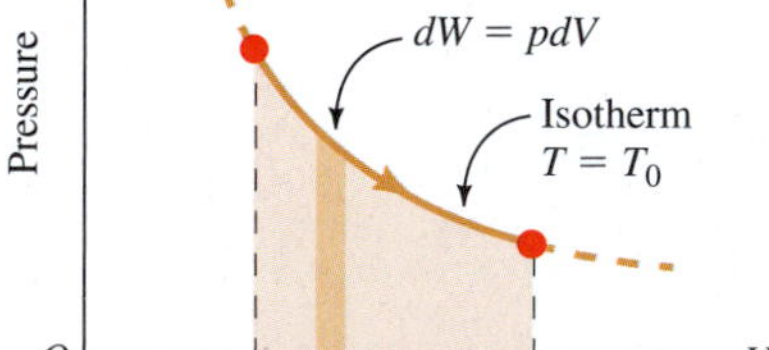

▲ **FIGURE 18–21** An ideal gas expands from V_1 to V_2 while its temperature is held fixed, and in doing so does work on its surroundings.

Isothermal Transformations of an Ideal Gas

When the temperature is constant, the work done by an ideal gas is simple to compute. We see from Fig. 18–21 that

$$W = \int_{V_1}^{V_2} p\,dV = \int_{V_1}^{V_2} \frac{nRT_0}{V}\,dV = nRT_0 \int_{V_1}^{V_2} \frac{dV}{V}. \qquad (18\text{–}34)$$

The integral of $1/x$ is the natural logarithm, $\ln x$ (see Appendix IV-6). Because $\ln x_1 - \ln x_2 = \ln(x_1/x_2)$, we have, for the work done by an ideal gas in an isothermal transformation:

$$W = nRT_0 \ln\left(\frac{V_2}{V_1}\right). \tag{18–35}$$

As expected, W is positive when the gas expands $(V_2 > V_1)$ and negative when the gas contracts $(V_2 < V_1)$.

EXAMPLE 18–14 Suppose that 0.12 mol of an ideal gas, in contact with a thermal reservoir that holds the temperature fixed at $T_0 = 9.8°\text{C}$, has an initial volume of 1.3 L and does 14 J of work. What is the final volume, and what is the final pressure?

Strategy We can use Eq. (18–35) to find the final volume V_f because every other quantity in the equation is known. (We'll need to convert T_0 to the Kelvin scale.) Then we can use the ideal gas law, $pV = nRT_0$, to find the final pressure of the gas.

Working It Out To start, we have $T_0 = 9.8 + 273.15 \cong 283\text{K}$. Then from Eq. (18–35)

$$\ln\left(\frac{V_f}{V_i}\right) = \frac{W}{nRT_0} = \frac{14\ \text{J}}{(0.12\ \text{mol})(8.314\ \text{J/mol}\cdot\text{K})(283\text{K})} = 0.050,$$

or

$$\frac{V_f}{V_i} = \exp(0.050) = 1.05;$$

$$V_f = 1.05\,V_i = 1.05 \times (1.3\ \text{L}) = 1.4\ \text{L}.$$

The final pressure is given by the equation of state:

$$p_f = \frac{nRT_0}{V_f} = \frac{(0.12\ \text{mol})(8.314\ \text{J/mol}\cdot\text{K})(283\text{K})}{(1.4\ \text{L})(10^{-3}\ \text{m}^3/\text{L})}$$

$$= 2.0 \times 10^5\ \text{Pa} = 2.0\ \text{atm}.$$

The pressure has decreased from an initial value of $p_i = p_f(V_f/V_i) = 2.2$ atm.

What Do You Think? Did (positive) thermal energy come out of the reservoir, or did it go into the reservoir, or was there no thermal energy flow at all?

The internal energy of an ideal gas is unchanged in an isothermal transformation. Thus by the first law of thermodynamics, Eq. (18–35) also expresses the heat flow into the gas during an isothermal transformation.

Adiabatic Transformations of an Ideal Gas

In an adiabatic expansion, the first law of thermodynamics becomes $dU = -dW$. Because dW is positive, dU is negative. U is proportional to T, so the temperature will fall in an adiabatic expansion and rise in an adiabatic compression. When a gas expands in a thermally isolated container, it cools; when it is compressed, it heats up (as you can feel when you compress a bicycle pump).

The curves that describe this transformation are derived in the next subsection. They are given by

$$pV^\gamma = p_0 V_0^\gamma = \text{a constant}, \tag{18–36}$$

where (p_0, V_0) are the initial values of pressure and volume, respectively, of the ideal gas that undergoes an adiabatic transformation to new values (p, V). The constant γ is given by

$$\gamma \equiv \frac{C_p}{C_V} = \frac{C_V + nR}{C_V} = \frac{c'_V + R}{c'_V}, \tag{18–37}$$

where c' is the molar heat capacity, from Eq. (18–2). We have used Eq. (18–33), which relates C_p to C_V. Because $C_p > C_V$, γ is greater than unity (a result that is true even for nonideal gases). In terms of V and T, the transformation follows the curves

$$TV^{\gamma-1} = \text{a constant}. \tag{18–38}$$

Compare Eq. (18–36) with the curve that describes an isothermal transformation, $pV = $ a constant, or $p \propto V^{-1}$. Figure 18–22 contains an isotherm and an adiabatic curve for a monatomic ideal gas, for which $\gamma = 1.67$. The isothermal curve is a hyperbola, $p \propto V^{-1}$. The adiabatic curve $p \propto V^{-\gamma}$ falls more steeply than the hyperbola as V increases, because $\gamma > 1$.

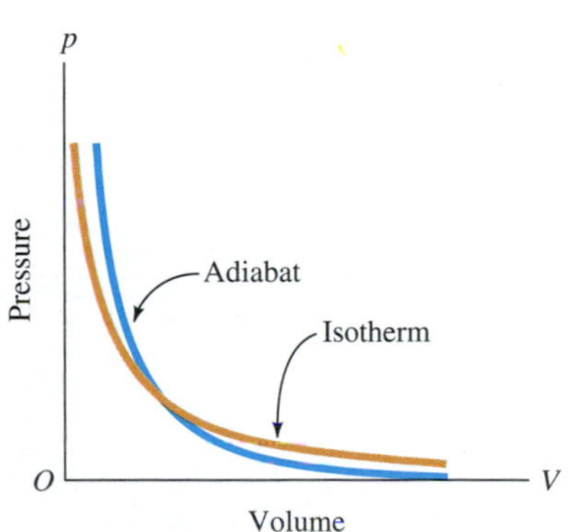

▲ **FIGURE 18–22** For ideal gases, adiabatic curves $(p \propto V^{-\gamma})$, called adiabats, are steeper than isotherms $(p \propto V^{-1})$ on a p–V diagram.

*How to Derive the Adiabatic Transformation of an Ideal Gas on a *p*–*V* Diagram

To derive the curves that are described by Eq. (18–36), we start with the first law of thermodynamics in its differential form [Eq. (18–25)], with the heat flow, dQ, set to zero: $dU = -dW$. The infinitesimal work done is still given by $dW = p\,dV$. We also know that dU is proportional to the temperature change, dT. In particular, dU can be calculated by differentiating Eq. (18–31) with respect to the temperature, T. Because C_V is constant for an ideal gas, we have $dU/dT = C_V$, or

$$dU = C_V\,dT.$$

Thus the adiabatic transformation satisfies

$$C_V\,dT = -p\,dV. \tag{18–39}$$

All three thermodynamic variables appear here, but because of the ideal gas law, $nRT = pV$, only two are independent. To find the curve of p versus V, we eliminate T in favor of p and V. The ideal gas law implies that

$$nR\,dT = p\,dV + V\,dp.$$

(See Appendix IV–7.) Equation (18–39) can thus be rewritten as

$$C_V\frac{p\,dV + V\,dp}{nR} = -p\,dV.$$

Solving for $V\,dp$, we find

$$V\,dp = -\frac{C_V + nR}{C_V}p\,dV. \tag{18–40}$$

The numerator on the right-hand side of Eq. (18–40) is the heat capacity at constant pressure, C_p, as Eq. (18–33) shows. Thus the right-hand side may be written as $-(C_p/C_V)p\,dV = -\gamma p\,dV$, with γ defined in Eq. (18–37).

With the definition of γ, we rewrite Eq. (18–40) in the form

$$\frac{dp}{p} = -\gamma\frac{dV}{V}.$$

Each side of this equation may be integrated separately, with the lower limits on the integrals determined by the initial state (p_0, V_0, T_0). Thus

$$\int_{p_0}^{p}\frac{dp}{p} = -\gamma\int_{V_0}^{V}\frac{dV}{V}.$$

Each integral gives a logarithm, and

$$\ln\left(\frac{p}{p_0}\right) = -\gamma\ln\left(\frac{V}{V_0}\right).$$

Because $A\ln x = \ln(x^A)$, it follows that

$$\ln\left(\frac{p}{p_0}\right) + \ln\left[\left(\frac{V}{V_0}\right)^{\gamma}\right] = 0.$$

We use $\ln x_1 + \ln x_2 = \ln(x_1x_2)$ to find

$$\ln\left[\left(\frac{p}{p_0}\right)\left(\frac{V}{V_0}\right)^{\gamma}\right] = 0.$$

We now take the exponential of both sides of the equation to find

$$pV^{\gamma} = p_0V_0^{\gamma} = \text{a constant}.$$

This is Eq. (18–36). By using the ideal gas law, we can show that the corresponding equation in terms of T and V is Eq. (18–38).

Variation of Atmospheric Temperature with Height

If you have taken a long-distance commercial flight, you know that the air high up is quite a bit colder than the air at ground level. Adiabatic transformations explain the variation of atmospheric temperature with height rather well. Air is transported from high altitude to low altitude and vice versa by large atmospheric wind currents. Air is a sufficiently good thermal insulator, so the transport of air between different altitudes is, to a good approximation, adiabatic. When air rises, the pressure decreases, so that the air undergoes adiabatic expansion; its temperature therefore falls. When air is carried to lower altitudes, it undergoes adiabatic compression, and its temperature rises.

If we assume that air is an ideal gas, we can quantify these concepts. Suppose that pressure and temperature are related by an adiabatic transformation. The adiabatic curve for p versus T is derived from, for example, Eq. (18–36) with $V \propto T/p$. For the p-versus-T adiabatic curve, this gives the equation $p^{1-\gamma}T^{\gamma}$ = a constant. If we raise both sides to the power $1/\gamma$ and solve for T, we find that

$$T = (\text{a constant})p^{(\gamma-1)/\gamma}. \tag{18–41}$$

If we know how pressure varies with height, then Eq. (18–41) can tell us how temperature varies with height. But how does pressure vary with height? In Chapter 16 we found that pressure decreases linearly with increasing height for an *incompressible* fluid, such as water. Air, which is assumed to obey the ideal gas law in this case, is certainly not incompressible, but we can use the methods of Section 16–3 to find an equation for the way pressure varies with height. There we considered an imaginary horizontal slab cut out of the fluid, with its base at height h, where the pressure is p. The top of the slab, where the pressure is $p + dp$, is at height $h + dh$. Note that we expect dp to be negative because pressure should decrease with height. According to Eq. (16–5),

$$\frac{dp}{dh} = -\rho g,$$

where ρ is the density of the fluid. From this equation, we see that as h decreases, p increases due to the increasing weight of the overlying fluid (in this case, air). For water, ρ is constant. For air, ρ is not constant but is given by

$$\rho = \frac{m}{V} = \frac{nM}{V} = M\left(\frac{p}{RT}\right),$$

where M is the mass of 1 mol of air (its molecular weight). The equation for pressure as a function of height becomes

$$\frac{dp}{dh} = -\left(\frac{Mg}{RT}\right)p. \tag{18–42}$$

Rather than solving this equation directly (see Problem 61), we can convert it into an equation for temperature as a function of height (see Problem 77). This is done by using the adiabatic curve, Eq. (18–41). Temperature is thus found to change linearly with height as

$$T = T_0 - \left(\frac{mgh}{R}\right)\left(\frac{\gamma - 1}{\gamma}\right). \tag{18–43}$$

Because $\gamma > 1$, temperature does indeed drop with increasing height—as anyone who has been to the mountains can appreciate.

How does Eq. (18–43) compare with observation? The molecular weight of air is approximately 0.029 kg/mol, $\gamma_{\text{air}} \cong 1.4$, and $g \cong 9.8\ \text{m/s}^2$. Then if we take the derivative of T [Eq. (18–43)] with respect to h, we find that

$$\frac{dT}{dh} = -\frac{(0.029\ \text{kg/mol})(9.8\ \text{m/s}^2)}{8.3\ \text{J/°C}\cdot\text{mol}}\left(\frac{1.4 - 1}{1.4}\right) = 0.010\text{°C/m}.$$

We have estimated that temperature drops about 10°C/km. The experimental result is approximately 6.5°C/km. The main sources of error in our calculation are the neglect of the effects of water condensation and, closer to the surface, the effects of local topography, such as mountain ranges. In the vicinity of flat deserts and high in the atmosphere, the value calculated is a fairly good approximation.

Summary

Reversible changes of state are changes that are made sufficiently slowly that the temperature, T, the pressure, p, and the volume, V—the thermodynamic variables that describe a thermal system such as a gas—are well defined at all times. The three variables are related by the equation of state of the gas. Reversible transformations, in contrast with irreversible transformations, can be traced out on a plot of any one of the three variables versus either of the other two. In this chapter we dealt with isothermal, isobaric, isochoric, and adiabatic changes.

When the temperature of a system changes, then the internal energy of the system will have changed. This may happen because the system has done work on its surroundings—for example, it pushed a piston and raised a weight—and/or because there has been a heat flow into the system. Infinitesimal heat flow into a system, dQ, is related to the infinitesimal temperature change of the system, dT, by the heat capacity, C, which depends not only on the system's composition but on how the temperature change is made:

$$\text{for constant volume: } dQ = C_V\, dT; \tag{18–4}$$

$$\text{for constant pressure: } dQ = C_p\, dT. \tag{18–5}$$

It is generally true that

$$C_p \geq C_V. \tag{18–6}$$

Because dQ depends on how a transformation is made, the net heat flow in a finite transformation is path dependent. Heat flow is measured in calories. Calorimetry, the measurement of heat capacities, is based on the fact that heat flow is conserved in a mechanically and thermally isolated system.

The specific heat of a material, c, is the heat capacity of 1 g of the material; its molar heat capacity, c', is the heat capacity of 1 mol of the material.

Solids conduct thermal energy in a characteristic way. The rate of transport of thermal energy depends on the temperature drop per unit length and the area of the material through which the thermal energy flows. The thermal conductivity, κ, is defined by

$$\frac{dQ}{dt} = -\kappa A \frac{dT}{dx}, \tag{18–8}$$

where the left-hand side represents the flow of thermal energy per unit time, A is the area, and dT/dx is the temperature drop per unit length. The reciprocal of κ is related to the thermal resistance, or R value, of the insulating material, with

$$R \equiv \frac{L}{\kappa}. \tag{18–9}$$

where L is the thickness of the material. The R values of two insulating slabs of the same area are additive when the slabs are connected in series, whereas the quantities area/R are additive when the slabs are connected in parallel.

Heat flow, like work, is an energy transfer. Any transformation of a thermal system caused by a heat flow can be made equally well by doing work on the system. This means that heat has an exact mechanical equivalent and that the calorie can be expressed as an energy unit:

$$1 \text{ cal} = 4.185 \text{ J}. \tag{18–13}$$

A gas undergoing an infinitesimal transformation does an infinitesimal amount of work on its surroundings, given by

$$dW = p\,dV. \tag{18–15}$$

As for heat flow, the net work done in a finite transformation is path dependent. If the path is adiabatic $(dQ = 0)$, the work done by the system is the negative of the internal energy change,

$$\text{for an infinitesimal adiabatic transformation: } dU = -dW; \tag{18–21}$$

$$\text{for a finite adiabatic transformation: } U_B - U_A = -W_{A\to B}. \tag{18–22}$$

An adiabatic transformation is a reversible transformation in which there is no heat flow between a system and its surroundings.

The internal energy (thermal energy), U, is a function of state. Any changes U undergoes in thermal transformations are path independent. The first law of thermodynamics specifies that the change in the internal energy of a thermal system in going from state A to state B is the negative of the work it does on its surroundings, plus the heat flow into the system:

$$\Delta U = U_B - U_A = -W_{A \to B} + Q_{A \to B}. \qquad (18\text{–}24)$$

The work in this equation, W, is the work done *by* the thermal system, and Q is the heat flow *into* the system. The differential form of the first law of thermodynamics is written as

$$dU = -dW + dQ. \qquad (18\text{–}25)$$

The internal energy of ideal gases is a function of temperature alone:

$$U_{\text{ideal}} = U(T). \qquad (18\text{–}29)$$

To the extent that C_V is independent of temperature, the dependence is linear,

$$\text{for } C_V \text{ independent of temperature: } U(T) = C_V T. \qquad (18\text{–}32)$$

It is also true that for an ideal gas

$$C_p = C_V + nR. \qquad (18\text{–}33)$$

The curve of an adiabatic transformation can be calculated for an ideal gas by using the ideal gas law. It is given on a p–V diagram by

$$pV^{\gamma} = \text{a constant}, \qquad (18\text{–}36)$$

where

$$\gamma \equiv \frac{C_p}{C_V} = \frac{C_V + nR}{C_V} = \frac{c'_V + R}{c'_V}. \qquad (18\text{–}37)$$

This equation can be used to predict the variation of temperature with height in the atmosphere.

Understanding the Concepts

1. When a bicycle tire is pumped up, the end of the pump near the valve feels hotter. What is happening?
2. A gas expands adiabatically, doing 500 J of work. By how much does the internal energy of the gas change?
3. We stated that the system of Example 18–6 was thermally isolated. Shouldn't we have added that it is mechanically isolated as well?
4. If a bucket of water is carried uphill, raising the water's potential energy, will its temperature change?
5. Which of the following are reversible processes: (a) the slow inflation of a balloon with a bicycle pump; (b) the heating up of a drill bit used to bore a hole in a log; (c) the slow stretching of a wire by an external force, carried out at a constant temperature?
6. Thermopane windows are windows made of two glass panes with air between them. Explain why they are better thermal insulators than a simple window pane of twice the thickness.
7. You want to make a reversible transformation between 2 points A and B on a p–V diagram. Assume that $V_B > V_A$. On what path is the least amount of work done by the gas?
8. To survive a cold night in the woods, you are advised to make a thick bed of dry leaves to sleep on and a thick cover of dry leaves to sleep under. Why does this work?
9. Why are cross-country skiers told to wear several layers of clothing rather than a single, thick coat?
10. Why is free expansion not an example of an adiabatic process?
11. Would it make sense for you to shake a container of soup in order to heat it? Would the fact that the soup is in an insulating container change your answer?
12. Should the cooling fluid for an engine have a large or a small heat capacity?
13. What use can you think of for fluids with very large specific heats? With very small specific heats?
14. You glue together two strips of metals that have different coefficients of thermal expansion, back-to-back. What will happen to such a composite strip when it is heated or cooled? Can you propose some technological use for such strips?
15. When a mass m_1 of water at temperature T_1 is mixed with a mass m_2 of water at temperature T_2, with $T_1 > T_2$, the temperature change of the hotter mass, ΔT_1, is related to the temperature change of the colder mass, ΔT_2, by $\Delta T_1/\Delta T_2 = -m_2/m_1$. In an experiment in which water and mercury are mixed, it was found that 100 g of mercury acts like 3.3 g of water. Explain both the formula and the observation about mercury.
16. A well-insulated container is divided into two separate compartments by an adiabatic wall. One compartment contains gas at a pressure p_1 and temperature T_1; the other contains gas at a pressure p_2 and temperature T_2. What happens when the partition is removed?
17. Why is the latent heat of vaporization for water so much larger than the latent heat of fusion?

18. When a freeze threatens, how can spraying fruit trees with water protect the fruit from frost damage?
19. A cyclic transformation shows up as a closed curve in a p–V plot. Suppose a cyclic transformation differs from the first one only in that the value of V is double at every step. How are the values of the net work related for the two transformations?
20. Why does perspiring during exercise help you to cool off?
21. For an ideal gas, the heat capacities are independent of temperature, and it is easy to relate temperature change to heat flow. Would it be impossible to relate temperature change to heat flow if the heat capacities varied with temperature?
22. Can all thermodynamic transformations be plotted on a graph of one thermodynamic variable versus another?
23. A change in a thermal system is reversible if (a) there is no friction and (b) the change is carried out slowly. Explain these constraints.
24. In the evening, after a hot summer day, why is it much cooler in the suburbs than downtown?
25. In Example 18–10 you have an apparatus that is capable of measuring ΔT to 0.02°C. If you want a better than 20% accuracy measurement, you need at least a 0.10°C temperature increase. Would you be better off with mercury instead of water in the experiment? (See Table 18–1.)
26. The center of a continent has more temperature extremes than do coastal areas at the same latitude. What effect contributes significantly to this?
27. People who drive cars in cold climates observe that the car's engine sometimes overheats if they keep the engine idling. This does not happen while they are driving, even though the engine puts out much more power then. Why?

Problems

18–1 Changes in Thermal Systems

1. (II) Are the following thermal transformations reversible or irreversible? Show the changes on a diagram when one is requested. (a) The sudden release of a gas from one portion of a container, of volume V_1, into a second portion, previously under vacuum. The total final volume, the volume of the whole container, is V_2 (p–V diagram). (b) As in part (a), but rather than a sudden release, the release is made a little at a time until there is no further flow of gas (p–V diagram). (c) The compression of a gas in thermal contact with a heat reservoir by loading weights, a little at a time, onto a piston connected to the gas (V–T diagram). (d) The slow heating of a gas by the release, a little at a time, of a warmer gas into the original sample of gas (p–T diagram). (e) The slow heating of a gas by placing it in thermal contact with a series of ever hotter thermal reservoirs (p–T diagram). (f) The slow rusting of a nail. (g) The release of electrical energy as lightning. (h) The release of chemical energy in the operation of a rechargeable battery to lift a weight.

18–2 Heat Flow

2. (I) (a) Using Table 18–1, calculate the heat capacity of a 5.7-kg aluminum rod. (b) Assuming that heat capacity is constant with temperature, how much heat flow is required to raise the temperature of the rod from 14°C to 58°C?
3. (I) A water heater heats 0.8 L of water/min from 20°C to 50°C. If the efficiency of the heater is 80%, how much power is consumed by the heater?
4. (I) The latent heat of fusion of oxygen (O_2) is 0.44 kJ/mol. Given that the atomic weight of molecular oxygen is 32 g/mol, find the latent heat of fusion in units of (a) kJ/kg and (b) cal/g.
5. (I) Consider an ideal gas initially in a state (p, V) and finally in a state $(p, 2V)$. If the transformation occurs at a constant pressure, what is the heat flow into the gas in terms of p and V?
6. (II) A 16 kg lead brick is at room temperature, 25°C. The melting point of lead is 327°C. Using the specific heat for lead, given in Table 18–1, calculate the heat flow to the brick needed to bring the brick to the melting temperature.
7. (II) A 300 cm^3 glass is filled with 100 g of ice at 0°C and 200 g of water at 25°C. (a) Characterize the content of the glass after equilibrium has been reached. Neglect heat transfer to and from the environment. (b) Repeat your calculations for 50 g of ice and 250 g of water.
8. (II) In Example 18–4 an experiment in calorimetry was described in which the assumption is made that the heat capacity of water and iron are constant over the temperature range of the experiment. Describe a series of experiments that would establish whether or not these heat capacities are indeed constant.
9. (II) The quantity 0.60 kg of thermally isolated water at an initial temperature of 291K serves as a calorimeter. Into it can be placed 1.20 kg of metal that has been heated to 373K. Find the specific heats and molar heat capacities of the metals if the equilibrium temperature is (a) 316K; the metal is aluminum. (b) 305K; the metal is iron. (c) 297K; the metal is lead.
10. (II) A calorimeter consists of 400 g of water at 25°C. A 500 g piece of copper at 100°C is thrown into the water, and the equilibrium temperature is found to be 28.3°C. What are the specific heat of copper in cal/g·K and the molar heat capacity, given that 1 mol of copper has a mass of 63.5 g?
11. (II) The quantity 200 g of thermally isolated water at an initial temperature of 293K serves as a calorimeter. Into it are placed, *successively*, 200 g of iron, then 200 g of silver, each of which has been heated to 400K. Using the known specific heats of iron and silver, find the final equilibrium temperature. (Use $c_{Ag} = 0.0557$ cal/g·K and $c_{Fe} = 0.112$ cal/g·K.)
12. (II) An axe head consisting of 1.8 kg of iron is left outdoors one cold winter's night and is brought indoors when the outside temperature is a brisk 240K. The room into which it is brought is initially at a nice, comfortable 293K and 1.0 atm of pressure. The volume of the room, which is well insulated, is 38 m^3. Assuming that the axe head comes to thermal equilibrium with the air in the room, by how much is the temperature of the room lowered? (Ignore the thermal interaction with furniture, walls, and so forth. Use $c_{air} = 0.172$ cal/g·K, $c_{Fe} = 0.112$ cal/g·K, and 28.8 g/mol for the molecular weight of air.)
13. (II) A thermally isolated system consists of 1 mol of helium gas at 100K and 2 mol of a solid at 200K; the gas and the solid are separated by an insulating wall of negligible mass. If the wall is removed, what is the equilibrium temperature of the system? Assume that at the given temperatures, the heat capacity of the solid is given by $3R/2$ per mole.

18–3 Heat Flow in Materials

14. (I) A single window pane of thickness 6.0 mm has an area of 1.0 m^2. The inside and outside temperatures are 73°F and 13°F,

respectively. Calculate the rate of thermal energy loss through the window. (See Table 18–3.)

15. (I) Compare the rate of thermal energy transfer across a piece of asbestos that is 1.0 m^2 in area and 1.0 cm thick with that of a piece of glass of the same dimensions for a temperature difference of 60°C.

16. (I) The glass from a basement window is replaced by a sheet of asbestos that is 0.80 cm thick. The window is 60 cm wide and 35 cm high, and there is no air leakage around the edges. Given that the heat loss through the surface is 0.006 J/s, what is the temperature difference between the outside and the inside of the basement?

17. (I) A brick wall 5.0 m by 5.0 m in size and 0.10 m thick separates the inside of a house from the outside, where the temperature is 285K (12°C). A heater supplies energy inside at a rate of 0.35 kW. What is the inside temperature? Ignore all surfaces except the brick wall in question.

18. (I) Your refrigerator can be thought of as a box with six sides of total area 2.5 m^2. The effective R value of the walls is 1.5 $m^2 \cdot K/W$. The temperature inside is 5°C, while the temperature outside is 25°C. Calculate the rate of heat loss.

19. (I) A house of dimensions 36 ft × 21 ft × 9 ft is built of bricks that are 9.0 in thick. The inside and outside temperatures are 76°F and 25°F, respectively. Calculate the rate of thermal energy loss through the brick wall. (See Table 18–3.)

20. (II) The R value of brick is about 1.1 $ft^2 \cdot h \cdot °F/Btu$ across the brick's width. A dog who expends 57 W of energy lives in a brick doghouse of total effective area 1.6 m^2. How much temperature variation between the inside and outside of the doghouse can the dog withstand if only its body heat warms the house? Assume that the door locks out drafts and that the heat conduction through the door is identical to that through the walls.

21. (II) Pine wood 1 in thick has an R value of 1.3, compared with 3.0 for fiberglass insulation of the same thickness. (The units of R are $ft^2 \cdot h \cdot °F/Btu$.) Calculate the rate of thermal energy loss of a 6-in thickness of these same materials across a 4.0 ft × 8.0 ft section of wall for a temperature difference across the wall of 35°F. If 10 percent of the wall section is made of wood studs and the space between studs is filled with fiberglass, calculate the effective R value of the wall section (Fig. 18–23).

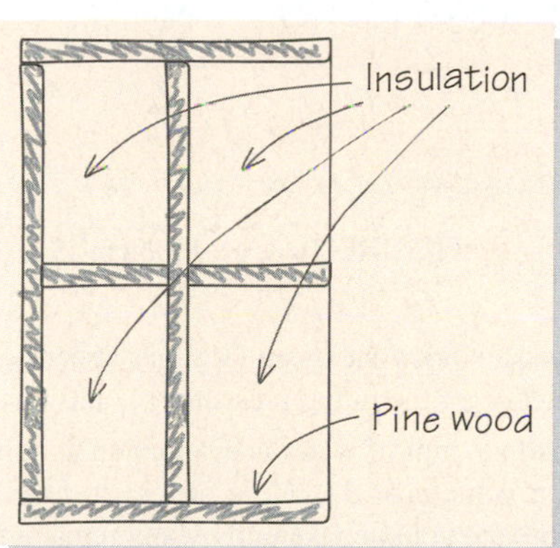

▲ **FIGURE 18–23** Problem 21.

22. (II) Consider a wall of thickness L and area A. Part of the area, A_1, consists of a material with thermal conductivity κ_1 and the remaining area, $A - A_1$, consists of a material with thermal conductivity κ_2 (Fig. 18–24). Calculate how much thermal energy flows through each part of the wall, and calculate an effective thermal conductivity for the wall as a whole.

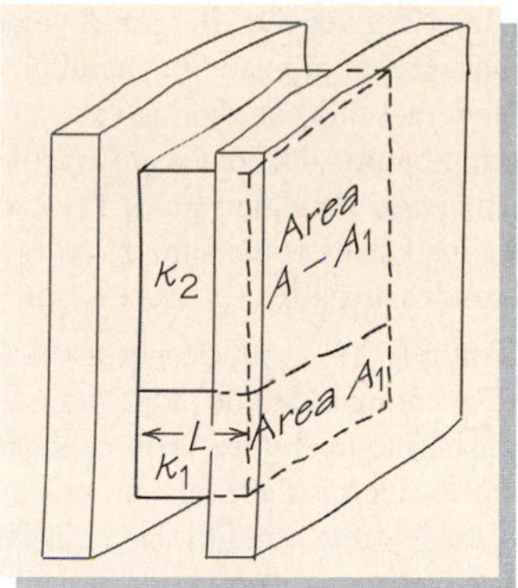

▲ **FIGURE 18–24** Problem 22.

23. (II) A wall of area A consists of two slabs: one of thickness L_1 and thermal conductivity κ_1, the other of thickness L_2 and thermal conductivity κ_2, as shown in Fig. 18–25. The temperatures are T_1 and T_2, respectively, on the left-facing side and the right-facing side of the wall. (a) Calculate the effective thermal conductivity of the wall in terms of the parameters given. (b) Calculate the temperature at the interface of the two slabs.

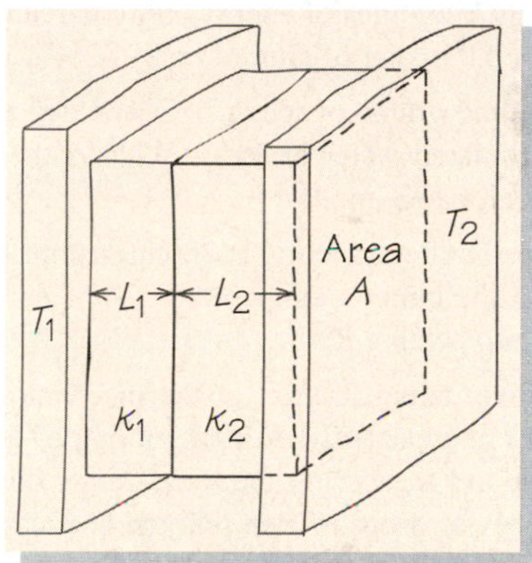

▲ **FIGURE 18–25** Problem 23.

24. (II) Two pieces of material are arranged in series as in Fig. 18–10. If the temperature to the left is T_1 and the temperature to the right is T_2, find the temperature between the pieces, T_{int}, as a function of T_1, T_2, and the R values of the two pieces.

25. (II) Equation (18–11) states that when two slabs of conducting material are combined in series, as pictured in Fig. 18–10, their R values are additive. Show that this is so by using the principle that the rate of flow of thermal energy is the same for each slab and for the two slabs combined:

$$\frac{dQ}{dt} = \frac{1}{R_1}A(T_1 - T_{int}) = \frac{1}{R_2}A(T_{int} - T_2)$$
$$= \frac{1}{R_{eff}}A(T_1 - T_2).$$

26. (II) A trapper lives in a cave. At the cave's entrance, the trapper has constructed an exterior wall from lumber 6.0 cm thick. The wood has a thermal conductivity, κ, of 0.35 $W/m \cdot K$. Assume that all thermal energy loss occurs through the wall, which forms a rectangle of dimensions 2.5 m × 2.0 m. This wall, together with a stove, is capable of maintaining the cave's interior at a temperature of 8°C when the outside temperature is −25°C. What is the power output of the stove?

27. (II) Equation (18–12) gives the effective R value for two slabs of conducting material combined in parallel, as pictured in Fig. 18–11. Show that this equation is valid by using the principle that the temperature difference on opposite sides of both slabs is the same, and that the rate of flow of thermal energy across the combined slabs is the sum of the rate of flow of thermal energy across each slab.

28. (III) Repeat Example 18–7 but do not assume that the igloo's thickness is small compared to its radii. You may assume that the surfaces of constant temperature form concentric hemispheres. [*Hint*: Break up the hemisphere of ice into thin hemispherical shells and find the R value of all these hemispheres combined in series. Note that the area of the hemispherical shells is not constant, so Eq. (18–8) must be modified appropriately.]

29. (III) A long, cylindrical hot-water pipe of radius R_1 is at temperature T_1. It is wrapped with insulation of thickness α and of thermal conductivity κ; the outer surface of the insulation is at temperature T_2. Derive an expression for the rate of heat loss per unit length of insulated pipe. [*Hint*: The temperature within the insulation varies with radial position r. Break the insulation into thin cylindrical shells and find the areas of these shells as a function of their radii.]

18–4 The Mechanical Equivalent of Heat

30. (I) How many kilojoules of energy are required to raise the temperature of a 3.0 kg bar of aluminum by 32°C?

31. (I) One hundred grams of ice at 0°C is added to 500 grams of thermally insulated water at 24°C. What is the temperature of the water after the ice melts?

32. (I) A 50 g lead bullet traveling at 200 m/s stops in a target. How much would the bullet's temperature rise if 80% of the energy were to go into heating it?

33. (I) A primitive demonstration of the mechanical equivalent of heat consists of repeatedly dropping a bag of lead shot from a given height and measuring the temperature change in the bag after a number of drops. Given that the heat capacity of lead is 128 J/kg · K, how many times would the bag have to be dropped from a height of 2.0 m to change its temperature by 5°C? Neglect all heat flows from the lead.

34. (II) Suppose that a kilogram of water drops, starting from rest, from a height of 13 m. When the water hits the ground suppose that half of the kinetic energy it attains in falling is converted into thermal energy within the water. By how much will the temperature of the water rise? Neglect the effect of air resistance.

35. (II) Julius Robert Mayer, who first enunciated the principle of the conservation of energy, estimated the mechanical equivalent of heat by considering a quantity of air confined in a cylinder by a movable piston. If the volume is fixed, 0.172 cal must be supplied to raise the temperature of 1 g of air by 1.0°C. If the piston is not fixed but exerts a constant pressure of 1 atm ($= 1.01 \times 10^5\ \text{N/m}^2$), 0.241 cal are needed to raise the temperature of 1 g of air by 1°C. If the volume change experienced by 1 g of air when the temperature changes by 1°C at constant pressure is $2.83 \times 10^{-6}\ \text{m}^3/\text{g} \cdot \text{K}$, what is the relation between the calorie and the joule?

36. (II) The amount of solar energy delivered to a horizontal surface in Washington, D.C., averaged over a full year, is 160 W/m^2. Assuming a 10% efficiency for absorption and conversion to usable heating, how many liters of water can be heated from 0°C to 20°C in a 24 h period by a 3 m^2 solar panel?

37. (II) Angel Falls, in Venezuela, has the highest free waterfall in the world, at 2,648 ft. The water at the top of the falls is at a temperature of 10°C. (a) What is the temperature of the water in the pool at the base of the falls? As a first approximation, ignore the effects of air resistance, so that there is no terminal velocity. Also, ignore the effects of evaporation. [*Hint*: The kinetic energy that the water gains in falling, thereby losing potential energy, goes into turbulent motion at the base of the falls.]The exact terminal velocity is a complicated function of the precise configuration of the waterfall, such as whether it forms a tight tube or whether it is a broad, dispersed spray. (b) Repeat part (a) for Niagara Falls, which is approximately 50 m high. (c) Return to Angel Falls. Discuss the effect of air resistance and the fact that there is a terminal velocity. Would you expect the water to be heated as much as in part (a)? (d) It takes a positive heat flow into a liquid, the latent heat of vaporization, to cause it to evaporate. What is the sign of the temperature change in the waterfall, when you take into account the effect of evaporation of the falling droplets?

38. (II) Work is done on a thermally isolated container of 1.5 kg of water under 1 atm of pressure, initially at a temperature of 295K. Assume that the specific heat of water does not vary with temperature, and that churning paddles do work on the water at a rate of 40 W. How long must the paddles continue to turn to bring the water to its boiling point, 373K?

18–5 Work Done by Thermal Systems

39. (I) The volume of a gas in a cylinder slowly increases by a factor of 3 while its pressure decreases from 9×10^5 Pa to 1×10^5 Pa, as shown in Fig. 18–26. How much work is done on the piston by the gas during the expansion?

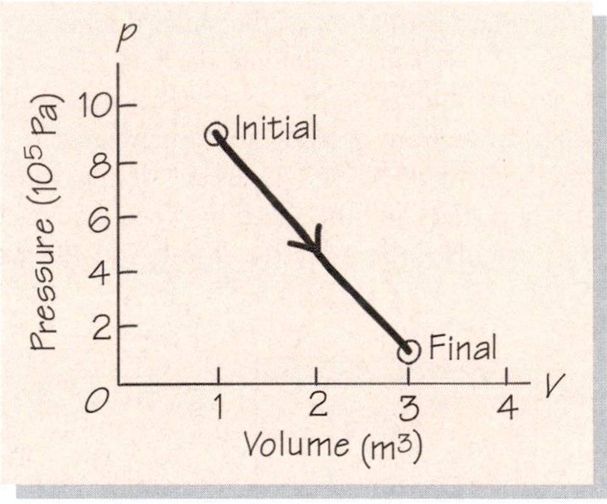

▲ **FIGURE 18–26** Problem 39.

40. (I) Express the work done by an ideal gas at constant temperature T_0 as a function of the initial pressure, p_i, and final pressure, p_f.

41. (I) If a small amount of water is evaporated at atmospheric pressure until a volume of 3.0 m^3 is occupied, how much work is done? Ignore the volume taken up by any remaining liquid water.

42. (I) Gas in a cylinder is allowed to expand from a volume of $0.85 \times 10^{-3}\ \text{m}^3$ to a volume of $0.95 \times 10^{-3}\ \text{m}^3$, while the pressure changes linearly with the volume from an initial pressure of 4.8 atm to a final pressure of 4.2 atm. What is the work done by the gas?

43. (II) Half a mole of an ideal gas is compressed isothermally at 293K so that the pressure is increased from 1.0 atm to 3.6 atm. What are the initial and final volumes of the gas, and how much work is done by the gas during the compression?

44. (II) Two liters of nitrogen gas at an initial temperature of 320K expand at a constant pressure of 5 atm against a piston, thereby doing work to raise a total mass of 100 kg. The gas is allowed to expand to a volume of 5.0 L. (a) How high can the mass be lifted? (b) What is the final temperature of the gas?

45. (II) A gas is compressed from V_1 to V_2 by reversibly following the path shown in Fig. 18–27, $p = p_0 - \beta V$, where β is a constant. (a) What is the dimension of β? (b) Determine the work done by the gas during the transformation by integrating the infinitesimal work along the path, $dW = p\,dV$. (c) Check your answer by evaluating the area under the p–V curve in the figure.

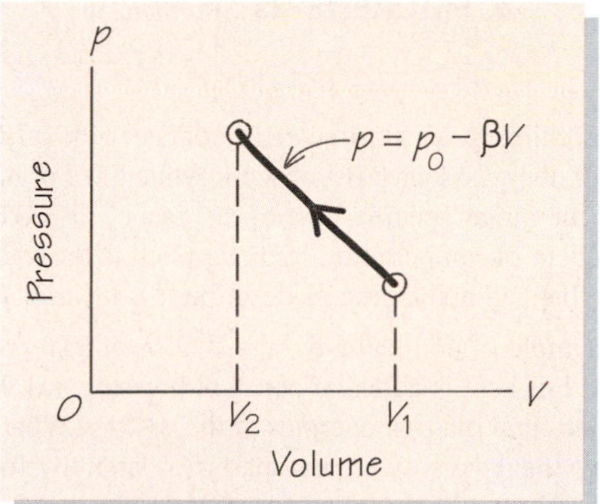

▲ **FIGURE 18–27** Problem 45.

46. (II) Figure 18–28 shows a reversible path for a gas being taken from state A to state B. Both states have the same pressure. (a) What is the work done in taking this path? (b) Draw a second reversible path that connects the same points by which no work is done. If the gas is ideal, to what minimum temperature is the gas taken on this path? (Real gases cannot be taken to this temperature.) (c) Draw a third path in which the work done is greater than the work done in following the original path.

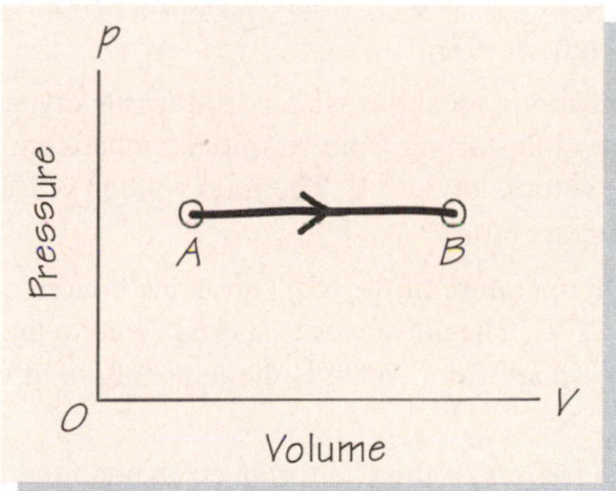

▲ **FIGURE 18–28** Problem 46.

47. (II) Two moles of an ideal gas are carried around the thermodynamic cycle shown in Fig. 18–29. The cycle consists of (1) an isothermal expansion $A \rightarrow B$ at a temperature of 700K, with the pressure at A given by $p_A = 8.0$ atm; (2) an isobaric compression $B \rightarrow C$ at 3.0 atm; and (3) an isochoric pressure increase $C \rightarrow A$. What work is done by the gas per cycle?

48. (III) An external force compresses 0.10 mol of an ideal gas in thermal isolation. The curve of this transformation on a p–V diagram is $pV^{1.4}$ = a constant. The gas initially has a volume $V_i = 1.6$ L and a temperature of 275K. When the compression is

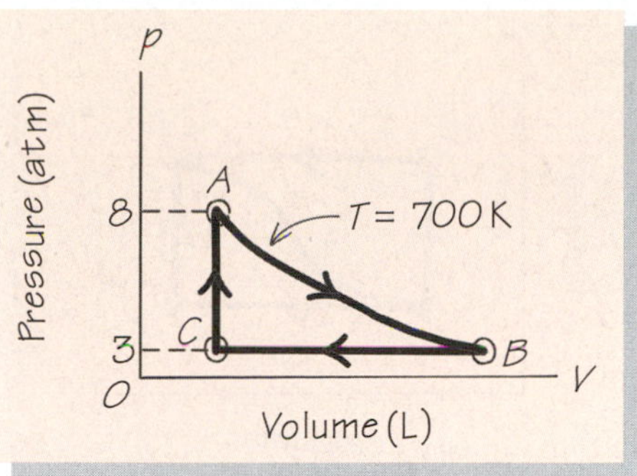

▲ **FIGURE 18–29** Problem 47.

finished, the temperature has increased to 380K (Fig. 18–30). (a) What are the final volume and final pressure? (b) How much work was done on the gas to compress it?

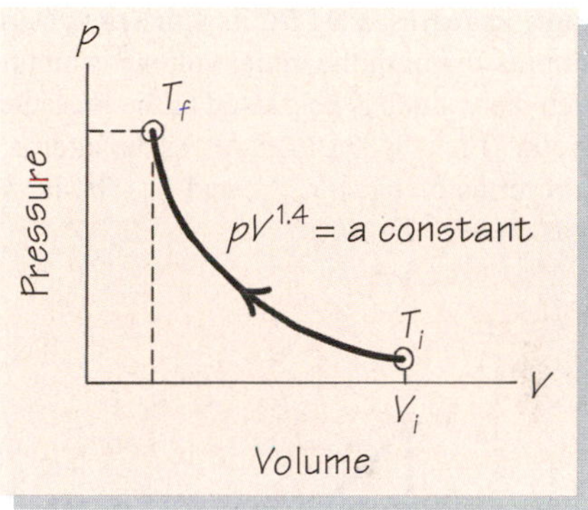

▲ **FIGURE 18–30** Problem 48.

18–6 The First Law of Thermodynamics

49. (I) A certain engine follows a closed thermodynamic cycle in which the heat flow into the system is measured to be 633 cal. Assuming that all transformations of the cycle can be approximated as reversible ones, what is the net work done, in joules, by the engine in one cycle?

50. (I) An ideal gas undergoes a reversible transformation from an initial state ($p_i = 1.0$ atm, $V_i = 350\ \text{cm}^3$) to a final state ($p_f = 2.0$ atm, $V_f = 175\ \text{cm}^3$). The transformation is carried out in two steps: The gas is compressed at constant pressure, and when the final volume is achieved, the pressure is increased while the volume is held fixed. Use the first law of thermodynamics to calculate the heat flow into the gas during this transformation. Assume (as is shown in Section 18–7) that, for an ideal gas, the internal energy U is a function of T only.

51. (II) Consider the p–V diagram shown in Fig. 18–31. If the system is taken from point A to point B via path $A \rightarrow C \rightarrow B$, 40,000 cal of heat flows into the system, and the system does 20,000 cal of work. (a) How much heat flows into the system along path $A \rightarrow D \rightarrow B$ if the work done is 7000 cal? (b) When the system returns from point B to point A along the curved path, the work done is 15,000 cal. Does the system absorb or liberate heat, and how much?

52. (II) The latent heat of vaporization for water at 1 atm and 100°C is 540 cal/g. How much of that heat is converted into the mechanical work needed to change the volume of 1 g of water into 1 g of steam? [*Hint*: Calculate the volume of 1 g of steam.]

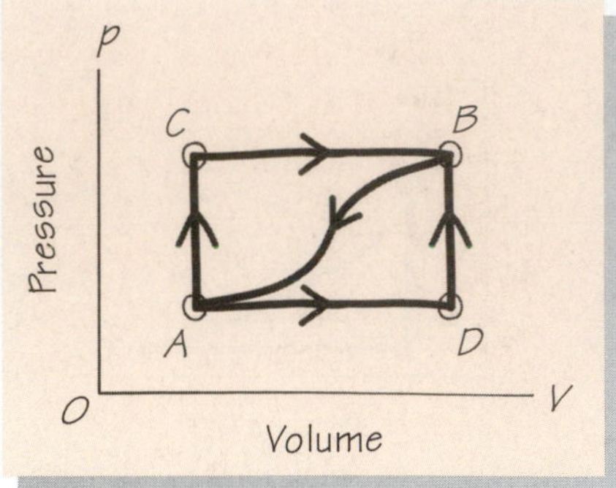

▲ **FIGURE 18–31** Problem 51.

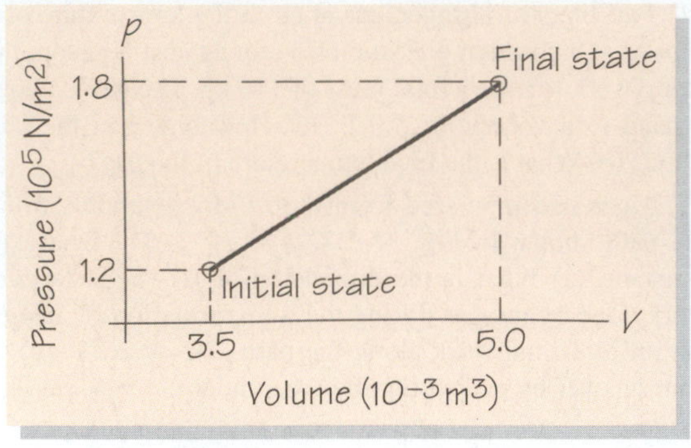

▲ **FIGURE 18–33** Problem 55.

18–7 Internal Energy of Ideal Gases

53. (I) A thermal system consisting of 1 mol of an ideal gas forms a cycle that contains a leg AB in which pressure decreases linearly as the volume increases, a leg BC in which the gas is compressed at constant pressure until the initial volume is attained, and a leg CA in which the volume is held fixed as the pressure increases to its initial value (Fig. 18–32). (a) Express the internal energy at A, B, and C in terms of p_A, p_B, V_A, and V_B. (b) How much work does the gas do over the cycle?

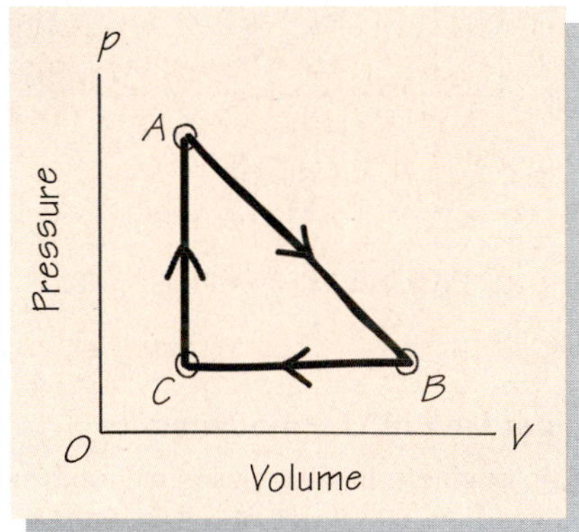

▲ **FIGURE 18–32** Problem 53.

54. (II) An ideal monatomic gas, with $c'_V = 3R/2$, undergoes a change of state. On a plot of pressure versus volume, it follows straight lines from an initial point (p_0, V_0) to $(p_0, 2.5V_0)$ to $(1.5p_0, 2.5V_0)$. (a) Draw the p–V plot. (b) By how much does the internal energy change? (c) How much work is done by the gas on its surroundings? (d) How much heat is added to the gas?

55. (II) Five grams of N_2 gas (molecular weight 28 g/mol) are allowed to expand from an initial state $(p_i, V_i) = (1.2 \times 10^5\ \text{N/m}^2, 3.5 \times 10^{-3}\ \text{m}^3)$ to a final state $(p_f, V_f) = (1.8 \times 10^5\ \text{N/m}^2, 5.0 \times 10^{-3}\ \text{m}^3)$ along a path that forms a straight line on a p–V plot (Fig. 18–33). (a) What are the initial and final temperatures? (b) What is the change in the internal energy of the gas? (c) What is the work done by the gas? (d) What is the heat flow into or out of the gas during the process?

56. (II) One mole of an ideal gas in contact with a thermal reservoir at 350K is under 10 atm of pressure. The pressure is slowly increased to 25 atm. (a) What are the initial and final volumes of the gas? (b) How much work is done *on* the gas? (c) What is the change in internal energy of the gas? (d) What is the heat flow into the gas?

57. (II) According to a calorimeter in contact with 0.20 mol of an ideal gas, the gas does 8.0 J of work while 5.0 J of heat is added to it. If the molar specific heat of the gas, c'_V, is 20.8 J/mol·K, independent of temperature, by how much is the temperature of the gas changed in the process described? Is it warmed or cooled?

58. (II) One mole of an ideal gas, $\gamma = 5/3$, is at STP (0°C, 1 atm), and 350 J of heat is added at constant pressure. (a) What are the initial and final internal energies of the gas? (b) What is the work done by the gas? (c) Repeat parts (a) and (b) for the same amount of heat added at constant volume.

18–8 More Applications for Ideal Gases

59. (I) How much work does it take to compress 30 mol of an ideal gas at a fixed temperature of 15°C to half its volume?

60. (I) One mole of ideal gas undergoes an adiabatic transformation in which its temperature changes by 25K. How much work is required to bring about this transformation, given that the constant-volume molar heat capacity of the gas is $c'_V = 30.00$ cal/mol·K?

61. (I) Equation (18–42) is an equation for pressure as a function of height for the atmosphere, which is treated as an ideal gas. Show that, if T is independent of height, this equation has the solution $p(h) = p_0 e^{-Mg(h-h_0)/RT}$, where p_0 is the pressure at some reference height $h = h_0$.

62. (I) A monatomic ideal gas with $\gamma = 1.67$ undergoes an adiabatic compression starting from an initial temperature of 248K and an initial volume of 3.01 L. The final volume is 2.22 L. What is the final temperature?

63. (II) The temperature of the (dry) air at the bottom of a mountain valley is 26°C. The air moves as a brisk wind to the ridge 600 m above the valley floor. What is the temperature of the air on the ridge?

64. (II) The wildfires of the California coast are often fed by desert winds, which heat up to about 100°F in the desert at an altitude of 500 m, then rush down to sea level. What is the temperature of the air in these winds at sea level?

65. (II) Consider two thermal states of an ideal gas, A and B, together with the possible reversible paths that connect them. Is there an adiabatic path that connects these states when (a) $V_A = V_B$? (b) $p_A = p_B$? (c) $T_A = T_B$?

66. (II) An ideal gas is contained in a cylinder with a movable piston of cross sectional area 300 cm^2. The external pressure is held constant at 1.01×10^5 Pa. An external force lifts the piston from a height of 30 cm to a height of 60 cm. What is the work done by the applied force on the gas if the process takes place isothermally?

67. (II) An ideal gas expands adiabatically from an initial state p_0, V_0 to a final state p_f, V_f. Such an expansion is characterized by the p–V curve pV^γ = constant, where $\gamma > 1$ is a constant. Show that the work done by the gas is $W = (\gamma - 1)^{-1}(p_0V_0 - p_fV_f)$.

68. (II) A diatomic ideal gas such as air, for which $\gamma = 1.4$, expands adiabatically to 40 times its original volume. (a) By what factor does the temperature change? (b) By what factor does the pressure change?

69. (II) One liter of gas characterized by $\gamma = 1.4$ is allowed to expand adiabatically to twice its volume. How does this pressure change compare with the pressure change that would occur for an isothermal expansion?

70. (II) How much work is done by 1.00 L of helium when it expands adiabatically from sea-level pressure (1.00 atm) to the point where the pressure is 0.10 atm? Helium is monatomic, with $c'_V = 3R/2$.

71. (II) A particular adiabatic transformation of an ideal gas occurs as the pressure increases by a factor of 5 while the volume decreases by a factor of 3. What is the ratio of C_V to C_p? How much work is done on the gas during the compression, in terms of the initial pressure and volume? By how much does the internal energy of the gas change?

72. (II) Use the result of Problem 61 to find the pressure at the height of Mt. Everest, which is about 10 km. Find also the temperature at this height. How much volume does 1.0 mol of air occupy? Calculate the density of air at this altitude and compare it with the density of air at sea level. The molecular weight of air is 29.2 g/mol.

73. (II) A gasoline engine of volume 25 cm^3 intakes air at 25°C and 1.0 atm. The engine compresses the air adiabatically to 30% of the original volume. Find the final pressure and temperature and how much work is performed on the gas. You can assume that the air is ideal.

74. (II) The compression ratio (the ratio of the maximum volume to the minimum volume) of the cylinder of a diesel engine is 15:1. The working gas is air (for which $\gamma = 1.4$), which enters the cylinder at room temperature, approximately 300K. The compression is so rapid that there is no heat flow through the cylinder walls, and the compression can be said to be adiabatic. To what temperature is the air heated?

75. (II) In Problem 17–57 the operation of a bicycle pump was studied on the assumption that the air in the pump is in thermal equilibrium with the environment. Repeat the analysis for the case when the pump is pushed down so quickly that no heat exchange occurs and the process is adiabatic. Assume that the temperature of the environment is 22°C. (The real process actually lies between the very slow isothermal and the very rapid adiabatic case).

76. (III) Consider the *Carnot cycle* shown in Fig. 18–34. One mole of an ideal gas with volume V_A at temperature T_1 undergoes an isothermal expansion to a volume V_B. This is followed by an adiabatic expansion to a volume V_C and then an isothermal compression at temperature T_2 to volume V_D; the cycle is closed by an adiabatic compression ending with volume V_A at temperature T_1. You will need to know the form of the transformation curve followed on each leg, and the heat flow to the gas for each leg. For the adiabatic legs, you know the heat flows, but for the isothermal legs, you may want to find the work done directly and use your knowledge of the energy change of an ideal gas when the temperature is constant. You also need to know the starting and ending points of the curves. What is the net heat, expressed in terms of the given variables, supplied to the gas during the cycle?

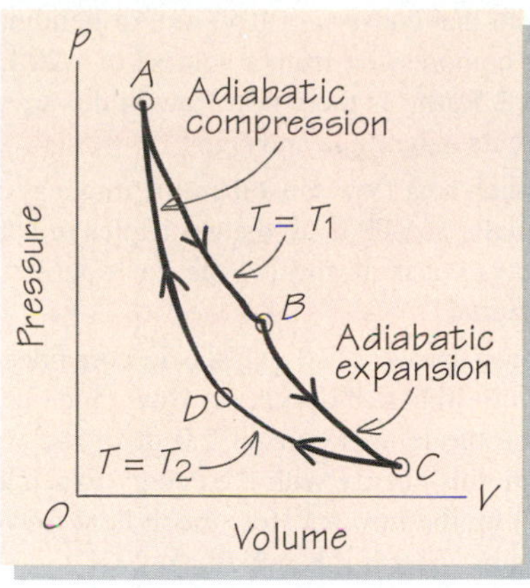

▲ **FIGURE 18–34** Problem 76.

77. (III) In Eq. (18–43) we state that by following an adiabatic curve, the temperature of Earth's atmosphere drops linearly with height h from the surface, and give the coefficient of h. Here this result is derived. The starting points are Eq. (18–41), which expresses the relation between T and p in an adiabatic transformation of an ideal gas, as well as Eq. (18–42), which expresses the pressure of the atmosphere as a function of height. (a) Take the derivative with respect to height of Eq. (18–41), using the chain rule of differentiation (see Appendix IV–7) for the right-hand side, to show that

$$\frac{dT}{dh} = \text{a constant}\left(\frac{\gamma - 1}{\gamma}\right)p^{[(\gamma-1)/\gamma]-1}\frac{dp}{dh}.$$

(b) Show that this result can be rewritten as

$$\frac{dT}{dh} = \left(\frac{\gamma - 1}{\gamma}\right)Tp^{-1}\frac{dp}{dh}.$$

(c) Substitute Eq. (18–42) for dp/dh into the result derived in part (b) to find that

$$\frac{dT}{dh} = -\left(\frac{Mg}{R}\right)\left(\frac{\gamma - 1}{\gamma}\right).$$

(d) Show that Eq. (18–43) is the solution of the differential equation derived in part (c).

78. (III) An experiment shows that the amount of work done is 88.9 J when a certain ideal gas is compressed adiabatically from an initial pressure of 1.00 atm and an initial volume of 1.00 L to a final volume of 0.50 L. In a second experiment, the same amount of gas is again compressed from 1.00 L to 0.50 L, but this time it is in thermal contact with a thermal reservoir at $T = 20$°C, and 189 J of work is required for the compression. (In the second experiment, the initial pressure is not necessarily 1 atm.) (a) How many moles of gas are there? (b) What is γ for the gas? [*Hint*: Calculate the work done in the adiabatic compression for several values of γ between 1.2 and 2.] (c) What is C_V for the gas?

General Problems

79. (I) Five liters of oxygen at one-half atmospheric pressure and 273K is heated at constant pressure until its volume has tripled. It is then compressed isothermally back to 5 L. (a) Draw the process on a p–V diagram. (b) What is the final pressure of the gas? (c) How much work is done by the gas during the entire process?

80. (II) A system that consists of 0.40 mol of helium gas undergoes an isobaric compression from a volume of 1.20 L to 0.70 L at a pressure of 2.5 atm. Is there heat flow in this transformation? If so, what are its magnitude and sign?

81. (II) How much heat flow must there be into a system of 10 L of oxygen initially at STP if (a) the gas triples in volume while the pressure stays constant and (b) the gas doubles in pressure at constant volume?

82. (II) A camper carries a 5.0 gal plastic container of water up a 40-m-high hill to her campsite. (a) How much heat flow does it take to raise the temperature of 2.0 qt of the water by 65°C? (b) Compare this energy with the energy expended to carry the entire 5 gal up the hill. (c) How much heat flow will it take to boil the 2 qt of water if it is initially at 23°C?

83. (II) Consider the adiabatic and isothermal processes that each pass through the point (p_0, V_0) on a p–V diagram (Fig. 18–35). Which of the processes has the greater slope at this point? Compute the ratio of the slopes at this point.

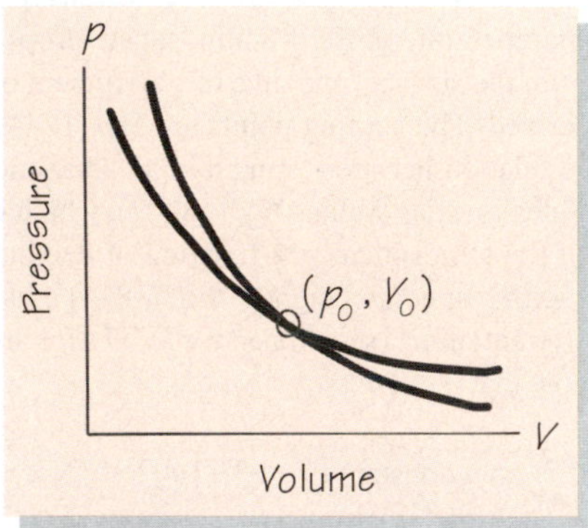

▲ **FIGURE 18–35** Problem 83.

84. (II) A 1-mol sample of an ideal gas for which $c'_V = 3R/2$ follows the closed thermodynamic cycle shown in Fig. 18–36. There are three legs in the cycle: an isothermal expansion, $A \rightarrow B$; an isobaric compression, $B \rightarrow C$; and a constant-volume increase in pressure, $C \rightarrow A$. The temperature during the isothermal leg is $T = 340\text{K}$, $p_A = 1.8$ atm, and $p_B = p_C = 1.2$ atm. What are (a) V_A? (b) V_B? (c) The work done by the system during leg $A \rightarrow B$? (d) Q for leg $A \rightarrow B$? (e) T_C? (f) the net work done by the system over the complete cycle?

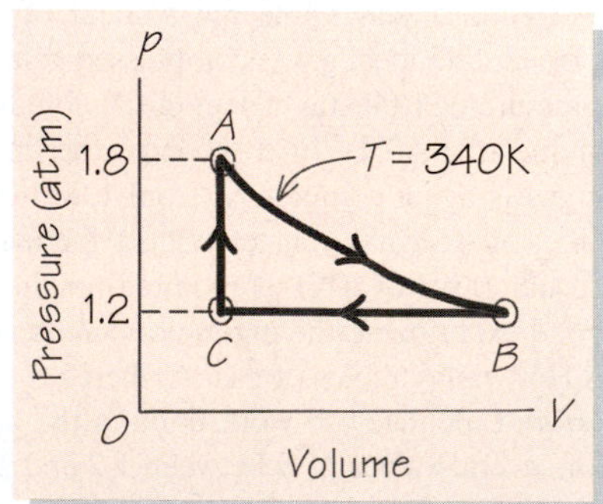

▲ **FIGURE 18–36** Problem 84.

85. (II) The energy in the form of solar radiation that falls on Earth, after reflection is taken into account, is 956 J/m^2 s. The effective area that faces the Sun is that of a disk whose radius is the same as that of Earth. In equilibrium Earth radiates at a rate of σT^4 where σ is given in Section 17–5. T is the temperature of Earth. Under equilibrium conditions, what is T?

86. (II) Consider a van der Waals gas (Section 17–4). Find the work done by such a gas in an isothermal expansion from V_0 to V_f. Is this more or less than the work done by an ideal gas in an isothermal transformation from V_0 to V_f at the same temperature? (The signs of the constants a and b for the van der Waals gas are discussed in Section 17–4.)

87. (II) A 100-g piece of copper is heated from 0°C to 100°C at atmospheric pressure. What is the change in its internal energy?

88. (II) A pan contains 300 g of water at 23°C. (a) What heat flow does it require to raise the temperature of the water to 100°C and vaporize it at 1 atm? (b) How much work is done by the water as it changes to the gaseous phase? (c) What is the total change in internal energy of the water?

89. (II) A nuclear power plant requires 1000 MW (megawatts) of cooling. If environmental concerns limit the temperature rise of the water to 12°C, how much water flow is required to cool the plant?

90. (II) Heat flow is transferred to 0.30 mol of an ideal gas at a temperature of 500K and pressure of 2.5 atm, resulting in an isothermal expansion of the volume by a factor of 5. How many calories are involved in the heat flow?

91. (II) One mole of N_2 gas initially at a temperature of 300K and pressure of 1 atm is compressed isothermally until its volume is reduced by a factor of 10. (a) What is the final pressure of the gas? (b) What is the work done by the gas? (c) How much heat flow is transferred?

92. (II) An electric immersion heater is placed in a coffee cup to heat the water to boiling. The cup initially contains 0.18 L of water at 72°F. The heater is rated for 400 W. Neglect heat loss from the water and the cup. How long does it take to heat the water up to boiling? How long after this does it take to boil the water away completely?

93. (II) The internal energy of a van der Waals gas (Section 17–4) is

$$U = \frac{3}{2}nRT + \frac{an^2}{V} + \text{constant}.$$

(a) Find C_p as a function of V and T. (b) Find C_V as a function of V and T. (c) What is the value of the parameter $\gamma \equiv C_p/C_V$ for this gas? Check that your result reduces properly in the low-density limit.

94. (II) A thermally isolated container contains 1.0 L of water at 20°C. A 100-g block of aluminum at 80°C is placed into the container. (a) What is the final temperature of the water? Next 500 g of ice at −20°C is placed in the container. (b) What is the final temperature of the water, iron, and ice (if any)? $c_{\text{ice}} = 0.467\ \text{cal/g} \cdot \text{K}$.

95. (II) A glass contains 150 cm^3 of water at 70°F. Four ice cubes of 25 g each of temperature 10°F are dropped into the water. Neglecting heat-flow loss, what is the final temperature of the water? The specific heat of ice in this temperature range is $2.04\ \text{J/g} \cdot \text{K}$.

96. (II) The temperature of espresso coffee (mostly water) can be increased by blowing 100°C steam into it. How much steam (in grams) is needed to heat up a 20-cm^3 cup of espresso from 50°C to 80°C? What is the volume of this quantity of steam, assuming that the steam is an ideal gas?

97. (II) One mole of an ideal gas is carried around the thermodynamic cycle shown in Fig. 18–37. The cycle consists of an isothermal expansion at a temperature of 400°C with an initial pressure of 5 atm, leading to a doubling of the initial volume. This is followed by a pressure drop of a factor of 2 at constant volume, and then by an isobaric compression until the initial volume is restored. The cycle is completed by a constant-volume pressure increase to the initial value of 5 atm. Calculate the values of p, T, and V not given, and the work done by the gas during one cycle.

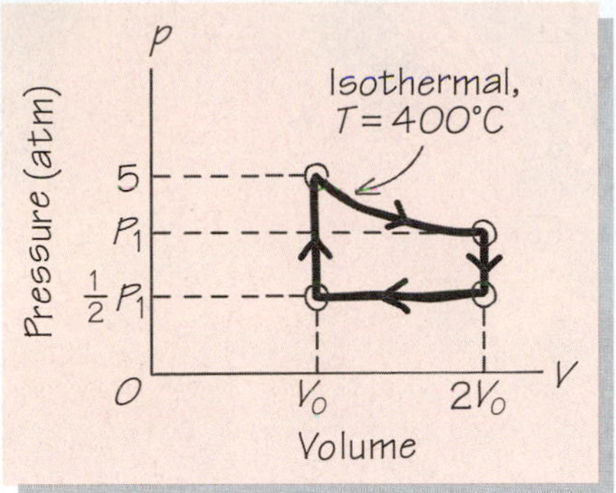

▲ **FIGURE 18–37** Problem 97.

98. (II) In experiments carried out between 1759 and 1762, Joseph Black started with a glass cup of mass 32 g, containing 467 g of water at 88°C. Black took a piece of ice of mass 404 g at 0°C and put it in the cup. The final equilibrium temperature was measured to be 12°C. How many calories are needed to melt 1 g of ice; that is, convert it to water at 0°C? That quantity is the latent heat of fusion. What would the equilibrium temperature be if no energy were required to convert ice to water at 0°C? Ignore the heat capacity of the glass.

99. (II) An ideal gas for which $c'_V = 5R/2$ is carried around a cycle $a \rightarrow b \rightarrow c \rightarrow a$. There are 2.4 mol of the gas in the cycle. The expansion $a \rightarrow b$ is a straight line on a p–V diagram with $T_b = T_a = 540\text{K}$, $b \rightarrow c$ is a constant-pressure segment with $p = 1.8$ atm, and $c \rightarrow a$ is a constant-volume segment at 15 L. (a) What is p_a? (b) What is V_b? (c) What is the work done in segment $a \rightarrow b$? (d) What is T_c? (e) What is the change in internal energy in segment $c \rightarrow a$? (f) What is the net work done during the entire cycle?

100. (II) The trapper of Problem 18–26 tires of looking at the blank entrance wall and decides to cut a window in it. The window will have a sheet of mica 1.0 mm thick with a thermal conductivity of 0.85 W/m · K. How large a window can he construct if he keeps the same stove and plans to keep the temperature inside the cave at 5°C?

101. (II) Cold air blows through a crack in the wall of the previous problem, so the trapper covers the crack with a piece of wood 5 mm thick. The thermal conductivity of the wood is 25×10^{-2} W/m · K. If the area of the crack is 6 cm^2 and the temperature difference between outside and inside is 30°C, what is the heat loss through the covered crack?

102. (III) A cyclic process with 0.40 mol of ideal gas is represented by a circle on the p–V diagram (if the appropriate scale is chosen) (Fig. 18–38). Calculate (a) the amount of heat transferred to the gas between the minimum and maximum volumes, as given in the figure; and (b) the largest and smallest internal energies during the cycle, given that $c'_V = 5R/2$.

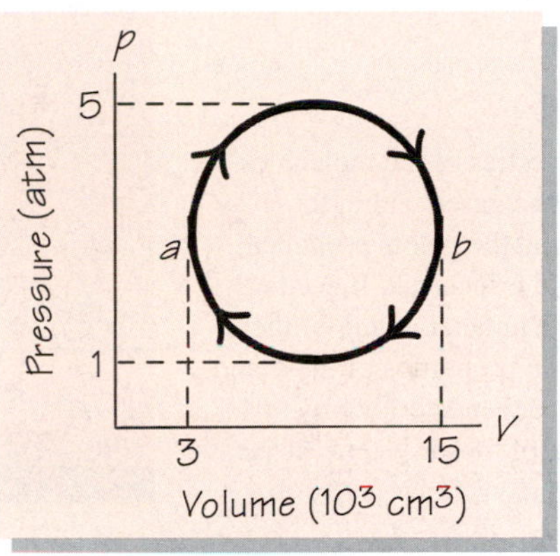

▲ **FIGURE 18–38** Problem 102.

Chapter

19

▶ The properties of the molecules that make up the gases within these lights determine the colors produced when electrical effects act. Indeed, a more complete understanding of the thermodynamic properties of these and other systems depends critically on an understanding of these systems at the molecular level.

The Molecular Basis of Thermal Physics

If we could watch a film of the molecular motion of a dilute gas, we would see the molecules rushing around at different velocities, spaced widely compared to their sizes, and bouncing elastically off each other and off the walls of the container in a gigantic three-dimensional game of billiards. Huge numbers of molecules make up even the smallest measurable quantity of a gas. The behavior of such a vast number of molecules can be expressed only in terms of averages. The **kinetic theory** of ideal gases, which is a statistical treatment of the large ensemble of molecules that make up a gas, is the focus of this chapter. We will use kinetic theory to understand the behavior of ideal and nonideal gases. Along the way, we will learn about the velocities of molecules and their statistical distributions, and about how such quantities as energy and momentum can move through a gas—transport phenomena. The kinetic theory of gases is remarkable because it explains so much of the thermodynamics we have already studied with the simplest of assumptions.

Statistical techniques for the study of large collections of particles have been the subject of intensive work in physics, chemistry, astronomy, and other fields of physical science. Early work was done by James Clerk Maxwell in the 1850s and by Ludwig Boltzmann and Josiah Willard Gibbs in the latter half of the 19th century. Research in this area falls under the general heading of **statistical physics**, a vast field that also includes many aspects of the physics of liquids and solids. Statistical techniques allow us to explain a great many of the physical properties observed in bulk matter. Statistical physics has been used to explain new and often unexpected phenomena: phase changes, superconductivity, complex crystalline structure, and chaotic behavior, to name just a few.

19–1 A Microscopic View of Gases

A volume of air the size of a birthday balloon contains some 10^{23} molecules. This is an easy number to calculate, given that the air around you is at least approximated by an ideal gas, so that the number of molecules of a volume V of a gas at standard temperature and pressure is $N = pV/kT$, where k is Boltzmann's constant. Equivalently, the number of moles is pV/RT, where R is the universal gas constant. As we know from Chapter 17, the ratio R/k gives us Avogadro's number N_A, some 6.02×10^{23} molecules, the number of molecules in 1 mol of any gas. The properties of individual molecules throw additional light on this number. Atoms and molecules consist of electrons bound to nuclei by electrical forces. Nuclei consist of protons and neutrons. These have almost equal masses, about 1.67×10^{-27} kg $= 1.67 \times 10^{-24}$ g. The atomic weight A of a type of molecule is the total number of protons and neutrons, so that the mass of a single molecule, in grams, is $A \times (1.67 \times 10^{-24}\ \text{g})$. Now one mole of a substance of atomic weight A has, by definition, a mass of A grams, so to find the number of molecules N_A in one mole, we divide the mass A g of a mole by the mass of one molecule: $N_A = (A\ \text{g})/(A \times 1.67 \times 10^{-24}\ \text{g}) = 6.02 \times 10^{23}$ molecules.

Avogadro's number is incomprehensibly large. But just how crowded are the molecules? It follows from the ideal gas law that, at standard temperature and pressure, the volume of 1 mol of gas is 22.4 L $= 22.4 \times 10^{-3}\ \text{m}^3$. This means that, on average, each molecule occupies a volume of $(22.4 \times 10^{-3}\ \text{m}^3)/(6 \times 10^{23}) \cong 4 \times 10^{-26}\ \text{m}^3$. Using a radius r of 10^{-10} m for a typical molecule, we calculate the molecular volume to be $(4/3)\pi r^3 \cong 4 \times 10^{-30}\ \text{m}^3$. A molecule therefore takes up only $(4 \times 10^{-30}\ \text{m}^3)/(4 \times 10^{-26}\ \text{m}^3) = 10^{-4}$ of the volume available to it! The cube root of this number, namely about $1/20$, gives the ratio of the typical molecular radius to the mean spacing between molecules. If we were to scale molecules to the size of, say, sheep, with a radius of 0.5 m, the molecules would be spaced some 10 m, or 30 ft, apart.

As a gas becomes more dilute, molecular collisions occur less frequently. A dilute (or ideal) gas is a collection of independent molecules that move about with differing velocities and relatively rare collisions. However rare, collisions play an important role. They are the means by which gases come to thermal equilibrium. When a hot gas (consisting of rapidly moving molecules) is mixed with a cold gas (composed of more slowly moving molecules), elastic collisions between fast molecules and slow molecules slow down the fast molecules and speed up the slow ones (see Chapter 8). The entire system comes to equilibrium at an intermediate temperature.

Our treatment of gases will be based on a statistical picture, in which averages of quantities such as speed appear rather than their values for individual molecules. Why do averages tell us anything at all about a gas? After all, if we toss 10 coins ("a set of tosses") many times, the average number of heads in each set of tosses is 5, but we have very little confidence that we will actually get heads 5 times in just 10 tosses. (Try it!) It is only because the number of molecules we deal with is so large that we can be confident that average quantities really do represent the behavior of gases. Suppose that the number of coin tosses (N) is very large. The average number of heads is $0.5N$. But more important, in any given trial (a set of N tosses), the probability that we will get heads more than the average number of times depends on N. For example, the probability that we will get heads more than 51% of the time is about 38 in 100 for $N = 100$. This probability drops to about 2 in 100 for $N = 10^4$, and is on the order of 1 in 10^{88} for a million tosses! It is generally true that for systems with a large number of independent components (for example, the number of tosses in a trial), large deviations from the average value of a variable are unlikely. The molecules in a gas can be viewed as moving independently, and when a gas that contains 10^{24} or so molecules in thermal equilibrium is sampled, the chances that the measured value of any randomly varying quantity will differ from its average value are tiny indeed.

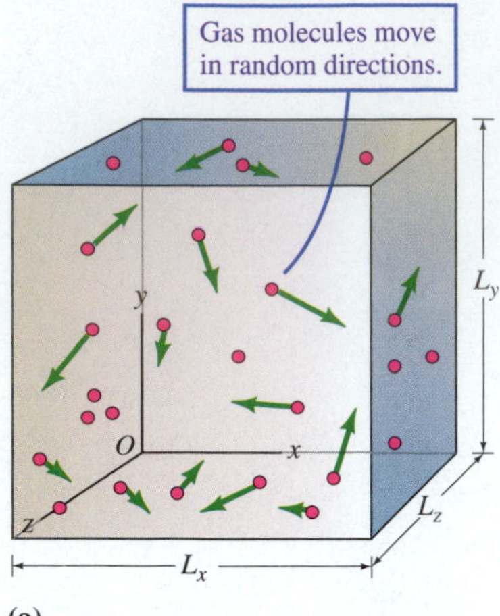

(a)

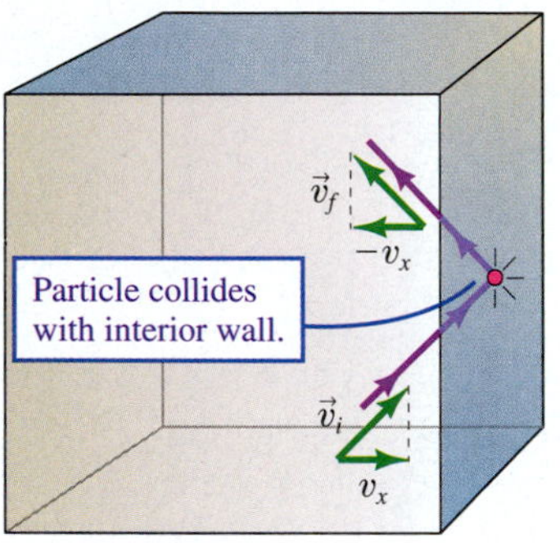

(b)

▲ **FIGURE 19–1** (a) The dimensions and orientation of a box that contains a gas. (b) A typical molecule, whose x-component of velocity is v_x, has an elastic collision with the right-hand wall. The x-component of the velocity changes its sign in the collision, but the y- and z-components are unchanged.

19–2 Pressure and Molecular Motion

Consider a container of dilute gas that consists of N independently moving molecules each of mass m. In kinetic theory these molecules move with a variety of speeds in a variety of directions, and pressure is due to molecular collisions with the walls of the container. Each time a molecule hits and rebounds from the wall, there is a momentum transfer to the wall, and since a rate of momentum transfer is a force, there is accordingly a force on the wall. To calculate the pressure we need to calculate the *average* rate of momentum transfer.

Some Average Values in a Gas

Suppose that the dilute gas in the container above is in thermal equilibrium at temperature T. The container has volume V and is oriented in a Cartesian coordinate system, as shown in Fig. 19–1a. The molecules in the box are moving in random directions and *the average velocity of the molecules is zero* because as many move in one direction as move in any other. In other words, the average x-, y-, or z-components of the velocity are zero. We will use angular brackets, $\langle\ \rangle$, to indicate the average value, so

$$\langle v_x \rangle = \langle v_y \rangle = \langle v_z \rangle \quad \text{so that} \quad \langle \vec{v} \rangle = 0. \tag{19–1}$$

While the average velocity may be zero, the same is not true for the average *speed*. The speed is by definition positive, and its average value is likewise positive. We will concentrate on the average value of the velocity squared $\langle v^2 \rangle$, or more precisely on $\sqrt{\langle v^2 \rangle}$, a quantity we call the **root-mean-square (rms) speed**. Note that $\langle v^2 \rangle$ is *not* the same as the square of the average value of the velocity: $\langle v^2 \rangle \neq \langle \vec{v} \rangle^2$; in fact, the latter quantity is zero because $\langle \vec{v} \rangle$ is zero.

We can relate $\langle v^2 \rangle$ to the internal energy of a dilute gas by kinetic theory. Forces between molecules are important only when the molecules are close together, and this happens relatively rarely in dilute gases. Thus the internal energy, U, consists mainly of the kinetic energies of the molecules. In other words, U is given by

$$U = N\langle K \rangle = N\left(\frac{1}{2}m\langle v^2 \rangle\right), \tag{19–2}$$

where $\langle K \rangle$ is the average kinetic energy per molecule and N is the total number of molecules. This equation provides us with a link to the temperature of the gas, because, as we learned in Chapter 18, the internal energy of an ideal gas is a function of temperature only.

The average of a sum of terms is the sum of the average of those terms, so

$$\langle v^2 \rangle = \langle v_x^2 + v_y^2 + v_z^2 \rangle = \langle v_x^2 \rangle + \langle v_y^2 \rangle + \langle v_z^2 \rangle. \tag{19–3}$$

The gas as a whole is not moving, so there must be an equivalence among the three directions. The average value of the x-component of the velocity squared must be the same as that of the y-component and that of the z-component. Thus

$$\langle v_x^2 \rangle = \langle v_y^2 \rangle = \langle v_z^2 \rangle. \tag{19–4}$$

Using this relation in Eqs. (19–3) and (19–2), it is possible to relate the average of the components of the velocity squared to the thermal energy of the gas:

$$\langle v^2 \rangle = 3\langle v_x^2 \rangle;$$

$$\langle v_x^2 \rangle = \frac{1}{3}\langle v^2 \rangle = \frac{2}{3}\frac{U}{mN}. \tag{19–5}$$

Next we will relate $\langle v_x^2 \rangle$ to the pressure of the gas.

The Origin of Pressure

Pressure arises from the multiple collisions the molecules of a gas have with the walls that contain the gas. For a quantitative calculation, we compute the momentum transfer to a wall due to a single collision, and then find the number of molecules that strike the wall per unit time. We use this to find the average momentum transfer to the wall per unit time, which is the force on the wall. Pressure is then the force per unit area. Figure 19–1b shows a molecule colliding elastically with the right-hand wall of the box of Fig. 19–1a. The wall is oriented parallel to the yz-plane, so only the x-component of the velocity changes. If the velocity before the collision is

$$\vec{v}_i = v_x\hat{i} + v_y\hat{j} + v_z\hat{k},$$

then the velocity after the collision is

$$\vec{v}_f = -v_x\hat{i} + v_y\hat{j} + v_z\hat{k}.$$

The momentum change of the molecule, $\Delta\vec{P}_{\text{mol}}$, is

$$\Delta\vec{P}_{\text{mol}} = m\vec{v}_f - m\vec{v}_i = -2mv_x\hat{i};$$

the momentum transfer to the wall, $\Delta\vec{P}$, is the negative of this:

$$\Delta\vec{P} = 2mv_x\hat{i}. \quad (19\text{–}6)$$

This sign is reasonable: The wall is pushed to the right by the collision.

We have found the momentum transfer from a single molecule. As we continue, we think of this molecule as an "average" molecule; it has the average x-component of velocity, which we continue to write as v_x for now. The next question is: How many collisions with the wall occur per unit time? Consider Fig. 19–2a: The number of molecules with an x-component of velocity of magnitude v_x that strike an area A in a time interval dt is the number of molecules contained in an imaginary cylindrical volume whose base, which is against the wall, is the area A and whose length is $v_x\,dt$. Molecules that travel toward this area but are farther away than the cylinder height will not reach the wall in the infinitesimal time interval dt. Note also that by making the time interval short enough, we can safely assume that the molecules do not collide with each other before they reach the wall. If the number of molecules per unit volume, the number density, is N/V, then the total number of molecules in our cylindrical volume is $(N/V)(v_x\,dt)(A)$. Thus the number of collisions with the wall in time dt is

$$\#\text{ of collisions} = N_{\text{coll}} = \frac{1}{2}\frac{N}{V}(v_x\,dt)A. \quad (19\text{–}7)$$

The factor $1/2$ is present because only half of the molecules are moving to the right; the other half are moving to the left. Only the ones that move to the right contribute to the pressure in this case. The number of collisions must be multiplied by the individual collision momentum transfer ΔP_x, the x-component of Eq. (19–6), to find the total momentum dP_x transferred in the time interval dt:

$$dP_x = (2mv_x)\left(\frac{N}{2V}\right)(v_x\,dt)A = mv_x^2\frac{N}{V}A\,dt.$$

In turn, the total momentum transfer per unit time—the force exerted on area A—is

$$\frac{dP_x}{dt} = F_x = mv_x^2\frac{N}{V}A. \quad (19\text{–}8)$$

The pressure on the wall, p, is the force per unit area:

$$p = \frac{F_x}{A} = mv_x^2\frac{N}{V}. \quad (19\text{–}9)$$

Finally, recall that we have employed the average x-component of velocity squared for each molecule. We make this explicit by employing the notation $\langle v_x^2\rangle$ rather than v_x^2. Then using Eq. (19–5), we find

$$p = m\langle v_x^2\rangle\frac{N}{V} = m\left(\frac{2}{3}\frac{U}{mN}\right)\frac{N}{V} = \frac{2}{3}\frac{U}{V}; \quad (19\text{–}10)$$

$$pV = \frac{2}{3}U. \quad (19\text{–}11)$$

PRESSURE IN TERMS OF ENERGY

This derivation is an important one. We have used the microscopic properties of a gas to find a relation between macroscopic thermodynamic variables.

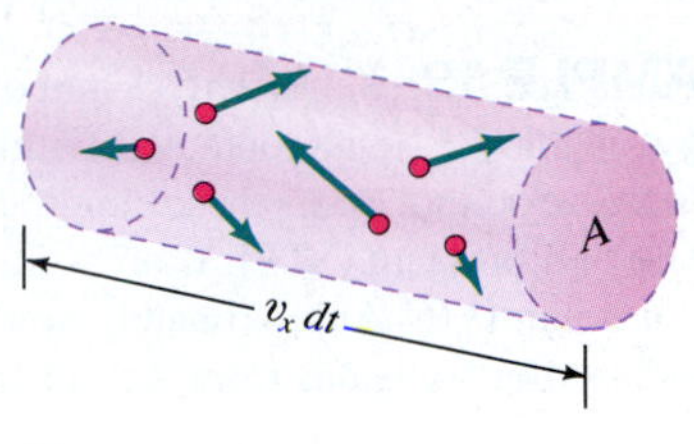

(a)

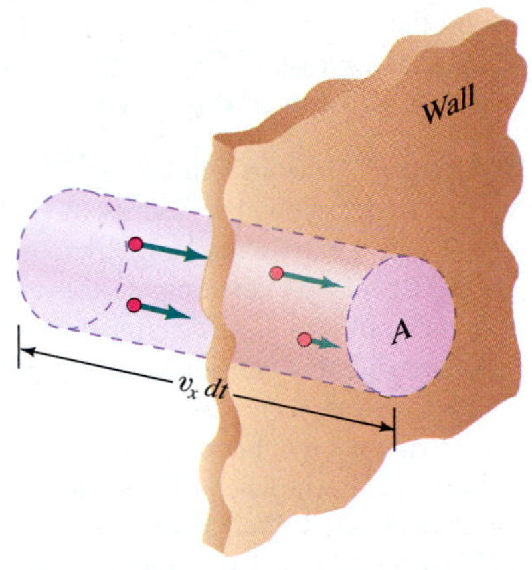

(b)

▲ **FIGURE 19–2** (a) All the molecules that will collide with an area A of a wall in a time interval dt are contained in a cylinder of area A and length $v_x\,dt$. (b) Only the molecules in the cylinder whose x-component of the velocity is directed towards the wall will contribute to the pressure. The figure only shows the x-component of the velocity.

EXAMPLE 19–1 Exactly 1 mol of helium gas has an internal energy of 3600 J. It is contained within a cube of sides 0.50 m. (a) Compare helium under these conditions with air at standard temperature and pressure (STP). If air is ideal at STP, can the same be said for helium? (b) Approximately how many times per second do the walls suffer collisions from the molecules of the helium gas?

Setting It Up For part (a) we want a "comparison"; we find the physically measurable quantities for air at STP and the same quantities for the helium. Specification of STP means specification of temperature and pressure. For part (b) we must pass from macroscopic properties to the microscopic ones.

Strategy We start with air at STP (0°C; 1 atm). Given p and T, we can find the number density N/V from the ideal gas relation $pV = NkT$. Next we turn to helium. Given the internal energy, we can use Eq. (19–11) to compute the pressure; given the volume and the number of moles, we can find the number density. If this is smaller than the number density for air at STP, the helium will be an ideal gas and we can then compute the temperature using the ideal gas law. To find the number of collisions, we know from Eq. (19–7) that in time Δt the number of collisions with one wall of area A is $(N/2V)(v_x\,\Delta t)A$. We can find an appropriate value for v_x from the rms speed, which is calculable from the total energy. We must take into account the fact that there are six sides to the box.

Working It Out (a) We start with the number density of air. Rearranging the ideal gas law,

$$\text{air: } \frac{N}{V} = \frac{p}{kT} = \frac{1.01 \times 10^5\ \text{Pa}}{(1.38 \times 10^{-23}\ \text{J/K})(273\text{K})} = 2.68 \times 10^{25}\ \text{m}^{-3}.$$

We can find the number density of helium from the fact that there are 6.02×10^{23} molecules (1 mol) in the given volume of $(0.5\ \text{m})^3$:

$$\text{helium: } \frac{N}{V} = \frac{6.02 \times 10^{23}}{(0.50\ \text{m})^3} = 4.8 \times 10^{24}\ \text{m}^{-3}.$$

This is about one sixth that of air at STP, so we can treat the helium as ideal.

We use Eq. (19–11) for the helium pressure: $p = \frac{2}{3}\frac{U}{V} = \frac{2}{3}\frac{3600\ \text{J}}{(0.50\ \text{m})^3} = 1.9 \times 10^4$ Pa, which is about 0.2 atm. The temperature can be computed from the ideal gas law:

$$T = \frac{pV}{nR} = \frac{(1.9 \times 10^4\ \text{N/m}^2)(0.50\ \text{m})^3}{(1\ \text{mol})(8.31\ \text{J/mol})} = 2.9 \times 10^2\text{K}.$$

This is in the range of room temperature. Our container of helium has a temperature close to that of the air at STP, is about six times less dense, and has a pressure one fifth that of air.

(b) The collision rate (# of collisions per second) with a wall of area A perpendicular to the x-axis is $N_{\text{coll}} = (N/2V)v_x A$; for v_x we take $\langle v_x^2\rangle^{1/2}$, and we can use (symmetry) $\langle v_x^2\rangle = \langle v^2\rangle/3$. Finally, we can find $\langle v^2\rangle$ from $U = N\left(\frac{1}{2}m\langle v^2\rangle\right)$. Pulling this all together, we have

$$\frac{N_{\text{coll}}}{\Delta t} = 6\frac{N}{2V}A\sqrt{\langle v^2\rangle/3} = 6\frac{N}{2V}A\sqrt{\frac{2U}{3mN}}.$$

The factor 6 takes into account the number of walls. The quantity mN, the mass of one molecule times the number of molecules, is the total mass of a gas, M. There is 1 mol of helium, so the mass is $4.0\ \text{g} = 4.0 \times 10^{-3}$ kg. Thus

$$\frac{N_{\text{coll}}}{\Delta t} = \frac{6(6.02 \times 10^{23})}{2(0.50\ \text{m})^3}(0.50\ \text{m})^2\sqrt{\frac{2(3600\ \text{J})}{3(4.0 \times 10^{-3}\ \text{kg})}}$$

$$= 2.8 \times 10^{27}\ \text{collisions/s}.$$

What Do You Think? If the internal energy is doubled, how will the wall collision rate change? *Answers to* ***What Do You Think?*** *questions are given in the back of the book.*

19–3 The Meaning of Temperature

The microscopic picture of a gas provides us with a simple and enlightening interpretation of temperature. The fact that the energy is the sum of the kinetic energy of all the molecules, $U = \frac{1}{2}Nm\langle v^2\rangle$ [Eq. (19–2)], on the one hand, and that the energy is related to the pressure, $U = (3/2)pV$ [Eq. (19–11)], on the other, provides a link to temperature. From the ideal gas law $pV = nRT = NkT$, where here n is the number of moles of the gas, R is the universal gas constant, and k is Boltzmann's constant. Thus

$$U = \frac{3}{2}nRT = \frac{3}{2}NkT. \qquad (19\text{–}12)$$

Equations (19–2) and (19–12) together give a microscopic interpretation of temperature:

$$kT = \frac{2}{3}\frac{U}{N} = \frac{2}{3}\langle K\rangle. \qquad (19\text{–}13)$$

TEMPERATURE AS ENERGY

The temperature of an ideal gas is a measure of the average kinetic energy of the constituents. The number of molecules has canceled from this expression, so T is (correctly) independent of the amount of gas.

Figure 19–3 gives another view of the close ties between energy and molecular motion, showing at the microscopic level how the change in energy of a system is associated with the work done by the system.

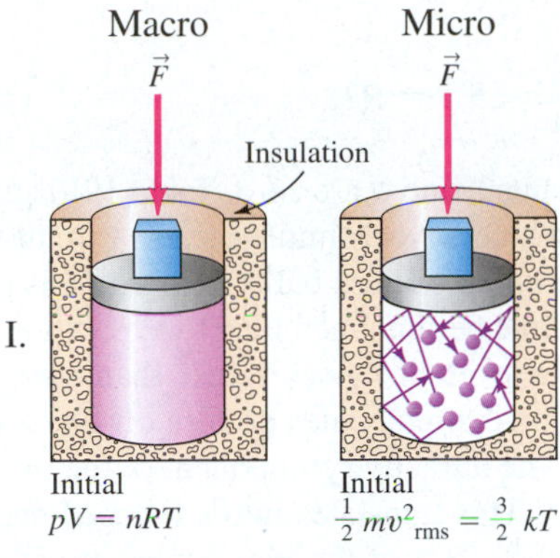

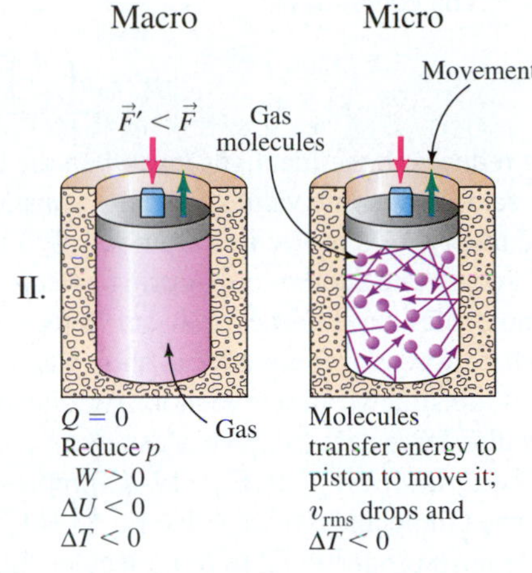

◀ **FIGURE 19–3** Macroscopic and microscopic views of the work done by an insulated gas. Molecules colliding with the recoiling piston lose energy. Thus the temperature drops.

CONCEPTUAL EXAMPLE 19–2 You have two identical bottles, one containing neon gas and the other containing the same number of atoms of argon, each in thermal equilibrium with the same thermal reservoir. Both gases can be regarded as ideal. Atoms of argon have twice the mass of atoms of neon. How does the average kinetic energy of the neon atoms differ from that of the argon atoms? How do the measurable thermodynamic variables T, p, and V compare?

Answer Both bottles of gas are at the same temperature (this is what equilibrium with the same thermal reservoir means). We can conclude from this that since the kinetic energy per atom is precisely a measure of temperature [Eq. (19–13)], it is the same for the two gases. The volumes of the two gases are the same, since the bottles containing them are identical. It follows from the ideal gas law, rewritten as $p = NkT/V$, that with N, T, and V the same for the two gases, the pressure will be the same.

EXAMPLE 19–3 1.00 mol of helium gas is contained in a large volume at 31.5K. What is the internal energy of the gas? What is the average value of the x-component of molecular velocity squared, $\langle v_x^2 \rangle$?

Strategy Because the gas is in a large volume, it is dilute and behaves as an ideal gas. We may therefore use Eq. (19–12) in the form $U = (3/2)nRT$ to find the internal energy. Every quantity on the right-hand side of the equation is known. As for $\langle v_x^2 \rangle$, that is dependent on the temperature but independent of the number of molecules: We can use $m\langle v^2 \rangle/2 = 3kT/2$, then $\langle v_x^2 \rangle = \langle v^2 \rangle/3$. Finally, the mass m of a helium atom is obtained by noting that 1 mole $= N_A$ atoms have a mass of 4.00 g, that is, $m = (4.00 \times 10^{-3}\ \text{kg})/N_A$.

Working It Out For the internal energy we have

$$U = (3/2)nRT = (3/2)(8.31\ \text{J/K})(31.5\text{K}) = 393\ \text{J}.$$

For the square of the x-component of the velocity we have

$$\langle v_x^2 \rangle = (1/3)(3kT/m) = (kT)[N_A/(4.00 \times 10^3\ \text{kg})].$$

We can calculate this or alternatively recognize that $N_A k = R = 8.31\ \text{J/mol}\cdot\text{K}$ to get

$$\begin{aligned}\langle v_x^2 \rangle &= RT/(4.00 \times 10^{-3}\ \text{kg}) \\ &= (8.31\ \text{J/mol}\cdot\text{K})(31.5\text{K})/(4.00 \times 10^{-3}\ \text{kg}) \\ &= 6.5 \times 10^4\ \text{m}^2/\text{s}^2.\end{aligned}$$

The square root of this quantity is about 250 m/s. Even at temperatures as low as 31.5K, the helium atoms are moving quite rapidly.

What Do You Think? A few argon atoms are added to your gas. After a little while, will they, on average, move (a) faster than, (b) slower than, or (c) at the same speed as the helium atoms?

THINK ABOUT THIS . . .

WHAT DOES ABSOLUTE ZERO MEAN AT THE MOLECULAR LEVEL?

Absolute zero, the zero of the Kelvin temperature scale, is the point at which the pressure drops to zero. In the microscopic view of an ideal gas, the temperature is zero when the average kinetic energy of the ideal gas is zero. Pressure vanishes because the molecules no longer move around and bounce against the walls.

Interpretation of the van der Waals Gas

Knowing how the microscopic properties of an ideal gas relate to its macroscopic behavior enables us to understand the origin of the van der Waals equation of state, which we encountered in Section 17–4, Eq. (17–15). This equation describes gases away from the dilute gas limit more precisely than does the ideal gas equation of state. As you may recall, the van der Waals equation takes into account the fact that the gas molecules are not point particles and also provides a correction for intermolecular forces—neither of these effects are included in the ideal gas law. Here we can see in more detail how these corrections are made.

The equation is

$$\left[p + a\left(\frac{n}{V}\right)^2\right]\left(\frac{V}{n} - b\right) = RT. \qquad (17\text{–}15)$$

It reduces to the ideal gas law when the constants a and b are zero. Table 19–1 gives us a sample of some values for these constants, which have a simple meaning in the kinetic theory. Gas molecules are not truly like pointlike billiard balls. Billiard balls exert a force on one another only during a very brief time period when they touch, whereas real molecules have forces that act between them over distances larger than their radii. These forces are repulsive at short range, so that the molecules bounce off one another, but are slightly attractive at longer distances. The attractive component of the intermolecular force acts to make a gas more compact. This translates into a reduced pressure: The term $a(n/V)^2$ in Eq. (17–15) represents the effect of the long-range attraction between molecules. If we solve Eq. (17–15) for p, we see it is reduced by $a(n/V)^2$. As for the term proportional to b, molecules, like billiard balls, take up some space. This term is present because of the strong repulsion between molecules at a characteristic radius. By appearing as a term subtracted from V, the constant b measures the volume unavailable for the motion of molecules because it is space already occupied by other molecules—it is the volume taken up by one mole of molecules. Example 19–4 illustrates how we can use our interpretation of b to estimate the size of molecules.

TABLE 19–1 • Some a and b Coefficients that Appear in the van der Waals Equation of State

	a (Pa · m^3)	b (m^3/mol)
Helium	3.46×10^{-3}	23.71×10^{-6}
Neon	2.12×10^{-2}	17.10×10^{-6}
Hydrogen	2.45×10^{-2}	26.61×10^{-6}
CO_2	3.96×10^{-1}	42.69×10^{-6}
Water vapor	5.47×10^{-1}	30.52×10^{-6}

EXAMPLE 19–4 Measurements show that nitrogen gas obeys the van der Waals equation of state with the constant $b = 3.94 \times 10^{-5}$ m^3/mol. What is the size of a nitrogen molecule?

Strategy We argued above that the parameter b represents the volume occupied by one mole of molecules. We can therefore find the volume of a single molecule by dividing b by Avogadro's number N_A. Setting the result equal to $4\pi R^3/3$ allows us to calculate the radius R of the molecule. (The problem statement gives no indication of the shape of the molecule, so we will assume it is spherical.)

Working It Out The volume of a single molecule is

$$V = \frac{b}{N_A} = \frac{3.94 \times 10^{-5}\ \text{m}^3/\text{mol}}{6.02 \times 10^{23}\ \text{molecules/mol}}$$
$$= 0.654 \times 10^{-28}\ \text{m}^3/\text{molecule}.$$

We now set this equal to $4\pi R^3/3$, and solve for R:

$$R = \left(\frac{3V}{4\pi}\right)^{1/3} = \left(\frac{3 \times 65.4 \times 10^{-30}\ \text{m}^3}{4 \times 3.14}\right)^{1/3} = 2.50 \times 10^{-10}\ \text{m}.$$

This radius is in fair agreement with more direct atomic measurements.

What Do You Think? Table 19–1 shows that all of the b values are within a factor of two of each other. What does this tell you about molecules? Why are the a values all positive?

19–4 Probability Distributions

Just as every student in a class does not receive the average grade on a given test, the molecules in a gas do not each have the average velocity or the average position. Instead, each molecule has its own velocity and position. Two functions describe the probability that a molecule has a particular velocity or a particular position. The **velocity distribution function** describes how many molecules have one velocity, how many have another, and so forth. The **position distribution function** similarly describes how the molecules are distributed in space. We need to know something about these distributions to further develop our understanding of the thermal properties of gases.

We all have a certain familiarity with the notions of probability and averages. Perhaps you have seen films or even played games of chance such as roulette. In this section, we will sharpen these notions so that we can use them in the analysis of gases and other thermal systems.

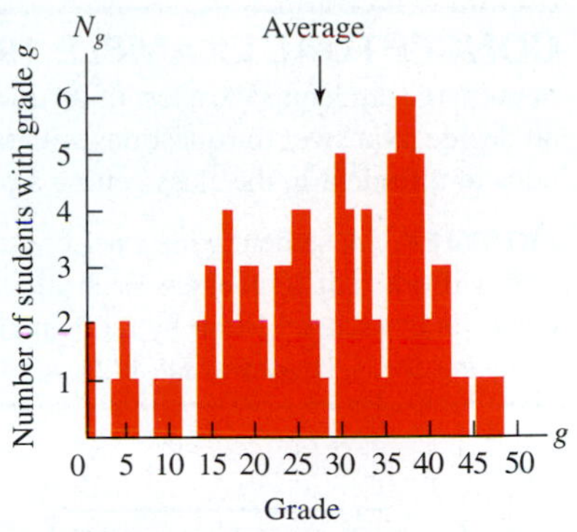

▲ **FIGURE 19–4** Bar graph of quiz grades in a class.

A Case Study: The Distribution of a Set of Test Grades

Consider a quiz taken by a class of 100 students. The grades can run from 0 to a maximum score of 50. The number of students who receive each grade is shown in Fig. 19–4, where the number of students with grade g, N_g, is plotted as a function of the grade g. The total number of students here is $N = 100$, and if we add up all the N_g values, we get all of the students taking the quiz:

$$\sum_g N_g = N. \tag{19–14}$$

Given such a distribution of grades, we can find the average grade. We take the number of people with a given grade, multiply it by the grade, add these products for each grade, and divide by the total number of students. We denote the average grade by $\langle g \rangle$. According to the above technique, we find

$$\langle g \rangle = \frac{1}{N}\left(\sum_g gN_g\right). \tag{19–15}$$

In the case illustrated, $\langle g \rangle$ is 28. We could also ask for the average of the *square* of the grade $\langle g^2 \rangle$ given by

$$\langle g^2 \rangle = \frac{1}{N}\left(\sum_g g^2N_g\right). \tag{19–16}$$

In the case illustrated, $\langle g^2 \rangle = 908.6 = (30.1)^2$. Note that $\langle g^2 \rangle \neq \langle g \rangle^2$, analogous to the remark we made about molecular velocities in Section 19–2. The quantity $\langle g^2 \rangle$ contains different information than does $\langle g \rangle$.

What is the probability that any one student, chosen at random, has a grade of 38 points? This probability is the number of students with a grade of 38 divided by the total number of students—in this case, 6/100. More generally, the probability of finding a student who received a grade g is

$$P_g = \frac{1}{N}N_g; \tag{19–17}$$

we refer to P_g as the **probability distribution** for the grades. It can run from a minimum of 0 to a maximum of 1, with 1 being the proverbial "sure thing." As a consequence of Eq. (19–14), the probability distribution is such that

$$\sum_g P_g = 1. \tag{19–18}$$

That the sum of probabilities is 1 is known as a **normalization condition**. It expresses the fact that it is a "sure thing" that *something* will happen, in this case that the student will get some grade. As we see from Eqs. (19–15) and (19–16), averages are obtained from weighted sums over P_g:

$$\langle g \rangle = \sum_g gP_g, \tag{19–19}$$

$$\langle g^2 \rangle = \sum_g g^2P_g. \tag{19–20}$$

We can ask one more type of question: If we pick one student at random, what are the odds that his or her grade lies between 15 and 19? The probability of finding that the student has a grade between 15 and 19 is the number of students with grades in that range divided by the total number of students. We write this as follows:

$$P(15 \le g \le 19) = \frac{1}{N}\sum_{g=15}^{19} N_g = \sum_{g=15}^{19} P_g.$$

For our example, this is equal to $13/100 = 0.13$. We will see that this type of question is especially relevant to quantities that vary continuously.

CONCEPTUAL EXAMPLE 19–5 A group of 100 totally unprepared students take a test of 20 true-or-false questions, and they all decide to answer the questions with random guesses. What are the odds of a student in the class getting a perfect score?

Answer A student with a perfect score has chosen the right answer with each of 20 guesses. Each guess is independent of a previous one (this is what we mean by random guesses), and thus the odds of choosing the right answer is $(1/2) \times (1/2) \times (1/2) \times \ldots$ with 20 factors. We know that $2^{10} = 1024$, so that this number is $(1/1024)^2$, or approximately 10^{-6}. One in a million students would have a perfect score; equivalently, the odds that any one student has a perfect score are one in a million. To find out the probability that there is one perfect score in the class, we remark that if a perfect score were to appear, it could be student 1, or student 2, or any of the students. As you likely know, in "or" situations, probabilities add, so the odds that there will be a perfect score at all is $100 \times 10^{-6} = 10^{-4}$.

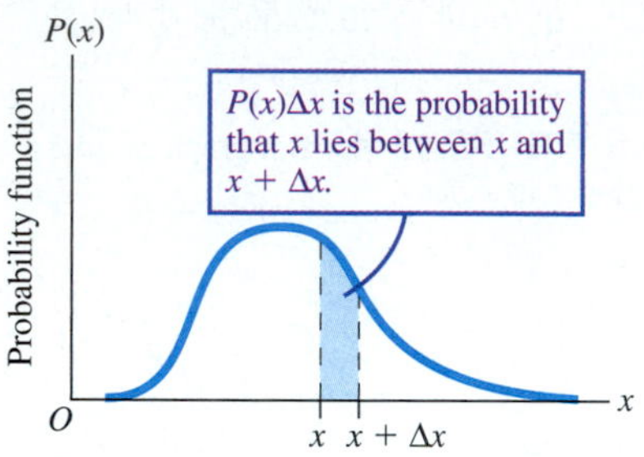

▲ FIGURE 19–5 The probability $P(x)\,\Delta x$ associated with a continuous distribution is the area beneath the curve between x and $x + \Delta x$.

Continuous Distributions

Some distributions do not involve discrete quantities, such as the grades on an exam, but instead involve continuous quantities, such as the heights of students. Much of the analysis of this case follows the discussion above of the discrete case, with one exception: We must recognize that the probability that any student will have a *particular* height x—say, $x = 1.73000055\ldots$ m, with an infinite number of digits—is zero. Instead, we must measure the probability $P(x)\,\Delta x$ that the outcome lies within some small interval Δx around the value x (Fig. 19–5). In our example, this could be the probability that a student has a height within an interval $\Delta x = 0.01$ m around $x = 1.73$ m. The normalization condition now expresses the fact that all students must have some height, so that the sum of all probabilities adds to 1:

$$\Sigma P(x)\,\Delta x = 1. \tag{19–21}$$

A sum appears here because—even if x takes on a continuous range of values—the intervals Δx have divided the region into a *finite* number of segments.

The concept we have described can be refined by making the interval smaller and smaller, thereby finding the probability that an outcome near a particular x-value occurs with more and more accuracy. Calculus notation is appropriate in this limit. The quantity Δx is written as dx, and the sum becomes an integral whose upper and lower limits are the upper and lower values of the variable in question, x_2 and x_1:

$$\int_{x_1}^{x_2} P(x)\,dx = 1. \tag{19–22}$$

This is the normalization condition in the case of continuous variables.

The averages of functions of x are found from formulas similar to those of the discrete case. For example,

$$\langle x^n \rangle = \int_{x_1}^{x_2} x^n P(x)\,dx. \tag{19–23}$$

EXAMPLE 19–6 Suppose that all heights between 1.5 m and 1.8 m are equally likely among a group of students and that no students are outside this height range. Find the probability distribution and use it to calculate the average height.

Strategy Since sizes form a continuum, we must take the distribution to be continuous. The first step is to determine $P(x)$. where we use x to denote the height, within the distribution. The data tells us that $P(x) = 0$ for $x < 1.5$ m and for $x > 1.8$ m. In between, the probability of finding any given height is denoted by the constant P. Its value will be found using the normalization condition Eq. (19–22). The calculation of $\langle x \rangle$, the average height, is then a simple matter of working out an integral like that of Eq. (19–23).

Working It Out The normalization condition reads

$$1 = \int P(x)\,dx = P\int_{1.5\text{ m}}^{1.8\text{ m}} dx$$
$$= P \times (1.8\text{ m} - 1.5\text{ m}) = (0.3\text{ m})P.$$

We can solve this for P, $P = 3.33\text{ m}^{-1}$. The average height is then

$$\langle x \rangle = \int xP(x)\,dx = P\int_{1.5\text{ m}}^{1.8\text{ m}} x\,dx = P\left[\frac{x^2}{2}\right]_{1.5\text{ m}}^{1.8\text{ m}}$$
$$= (3.33\text{ m}^{-1})(0.5)((1.8\text{ m})^2 - (1.5\text{ m})^2) = 1.65\text{ m}.$$

As all heights are equally likely, this is a uniform distribution, so it is not surprising that the average lies right in the middle of the interval between 1.5 m and 1.8 m.

What Do You Think? Suppose the probability distribution $P(x)$ were still 0 for $x < 1.5$ m or > 1.8 m, but otherwise symmetrically U shaped, with the bottom (the least likely height) at 1.65 m. What would you expect the average to be in this case?

19–5 The Velocity Distribution of Gases

In this section, we extend our ideas about probability distributions to molecules in a gas. The velocity distribution function $F(\vec{v})$ is a probability distribution for the velocities of the gas molecules. Velocities form a continuum, so we use a probability function appropriate for continuous variables. The velocity distribution function for ideal gases was first described by James Clerk Maxwell in 1859, a time when the molecular model of matter was by no means universally accepted. As was true of the height distribution of Section 19–4, the probability of finding any particular velocity $\vec{v}$ is zero. Therefore we start with a number distribution $N(\vec{v})$ such that

$$N(\vec{v})\,d^3v = \text{number of gas molecules with a velocity between } \vec{v} \textit{ and } \vec{v} + d\vec{v}.$$

What this means is that we describe the probability that a velocity will be somewhere in the neighborhood of $\vec{v}$. This "neighborhood" is a box of volume

$$d^3v = dv_x\,dv_y\,dv_z, \tag{19–24}$$

centered about the tip of the vector $\vec{v}$ (Fig. 19–6). The total number of molecules is N, so that

$$\int N(\vec{v})\,d^3v = N. \tag{19–25}$$

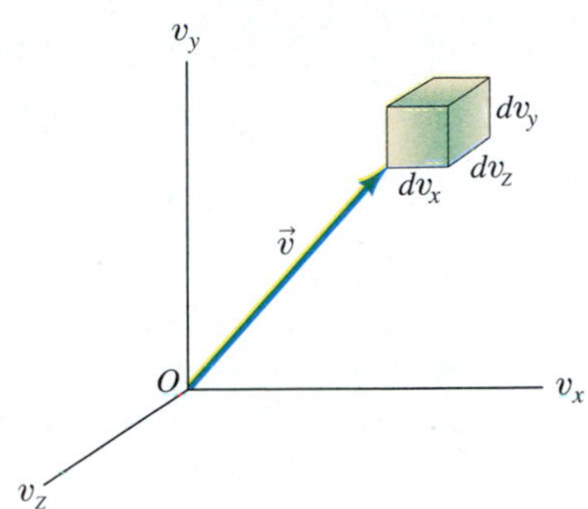

▲ **FIGURE 19–6** A molecule has a certain probability of having a velocity $\vec{v}$ within the "volume" $d\vec{v} = d^3v = dv_x\,dv_y\,dv_z$. Here we show the small box d^3v within which the velocity vector is located.

The distribution $N(\vec{v})$ leads directly to a *probability* distribution $F(\vec{v})$:

$$F(\vec{v}) = \frac{1}{N}N(\vec{v}). \tag{19–26}$$

The meaning of this function is that

$$F(\vec{v})\,d^3v = \text{probability that a gas molecule velocity is between } \vec{v} \text{ and } \vec{v} + d\vec{v}.$$

From $F(\vec{v})$ we can calculate the averages for continuously varying quantities. For example, the average of the velocity squared, which is relevant to the internal energy and the temperature, is

$$\langle v^2 \rangle = \int v^2 F(\vec{v})\,d^3v. \tag{19–27}$$

To determine the probability distribution $F(\vec{v})$, we must make some *physical* assumptions. The model of molecules darting about and occasionally colliding with other molecules suggests that a given molecule may have any velocity. This leads to a fundamental physical postulate: *Any way in which a gas's total energy and total momentum (which is zero) can be shared among the molecules is equally likely.* Once this postulate is accepted, we are in the same position as someone asking for the probability of getting any particular number between 100 and 600 when throwing 100 dice. What counts is the *number of ways* of getting that number. There is only one way of getting 100 or 600 (throwing 100 ones, or 100 sixes, respectively), but there are many ways of getting 400; thus the probability of getting 400 is much greater than the probability of getting 100 or 600. When this analysis is done for molecules whose individual kinetic energy is $\frac{1}{2}mv^2$ and whose average kinetic energy is $\frac{1}{2}m\langle v^2\rangle = 3/2\,kT$, the distribution function is found to be

$$F(\vec{v}) = \left(\frac{m}{2\pi kT}\right)^{3/2} e^{-mv^2/2kT}. \tag{19–28}$$

IDEAL GAS VELOCITY DISTRIBUTION

The prefactor in this equation ensures that this velocity distribution function is properly normalized, $\int F(\vec{v})\,d^3v = 1$, meaning that the sum over all probabilities is 1. $F(\vec{v})$ is a function only of the magnitude of the velocity (that is, the speed), not the direction. This expresses the physically reasonable fact that there is no preferred direction in a container of gas in which the gas as a whole is not moving—that is, all directions are equally likely. It is possible to make a direct experimental verification of Eq. (19–28) by cutting a tiny hole in a container of gas and measuring the different speeds of the molecules that leak out.

The Average of the Velocity Squared

Here we look at an important concrete example of how F is used: Given F, we calculate $\langle v^2 \rangle$. We have

$$\langle v^2 \rangle = \int v^2 F(\vec{v})\, d^3v = \left(\frac{m}{2\pi kT}\right)^{3/2} \int v^2 e^{-mv^2/2kT} d^3v. \qquad (19\text{–}29)$$

This integral appears to be a very tough one. It is multidimensional, and moreover it has some parameters in it that at first glance look hard to deal with. But we can extract the essential from this integral with very little work by using a technique known as **scaling**. This very useful technique allows us to extract the dependence of the integral on physical parameters of the problem. The idea is to change the integration variable to a dimensionless one, removing the dimensional dependence to a factor. In particular when there is an exponential function (or any transcendental function), the argument of the function is dimensionless. Accordingly, we change the integration variable in Eq. (19–29) from $\vec{v}$ to the dimensionless vector $\vec{u}$ by the definition

$$\vec{v} \equiv \vec{u}\sqrt{\frac{2kT}{m}}. \qquad (19\text{–}30)$$

This definition means that each component of the vector $\vec{v}$ is transformed; Eq. (19–30) stands for three equations. With this change the argument of the exponential function is simply $-u^2$, and u itself is dimensionless. This can be seen in another way: The integration variable v is not dimensionless. Both kT and mv^2 have the dimensions of energy, so v^2 has the dimensions of kT/m and v those of $(kT/m)^{1/2}$. By comparing this fact with Eq. (19–30), we see right away that u is indeed dimensionless.

In terms of our new variable, the volume element d^3v is

$$d^3v = \left(\frac{2kT}{m}\right)^{3/2} d^3u.$$

The factor $\sqrt{2kT/m}$ is cubed because d^3v stands for $dv_x\, dv_y\, dv_z$, and there is a factor for each variable. We also replace the factor v^2 in the integration of Eq. (19–29) with $u^2(2kT/m)$. Pulling everything together, we have

$$\langle v^2 \rangle = \left(\frac{m}{2\pi kT}\right)^{3/2} \left(\frac{2kT}{m}\right)^{5/2} \int u^2 e^{-u^2} d^3u = \frac{2kT}{m}\frac{1}{\pi\sqrt{\pi}} \int u^2 e^{-u^2} d^3u. \qquad (19\text{–}31)$$

The remaining integral on the right contains no quantities with dimension; it is a pure number. But even without knowing this number (which is indeed a hard integral), Eq. (19–31) shows that $\langle v^2 \rangle$ is a dimensionless constant times $2kT/m$. We anticipated that in our discussion of Eqs. (19–11) and (19–13) which combined with the ideal gas law led to

$$\langle v^2 \rangle = \frac{2U}{Nm} = \frac{3kT}{m}. \qquad (19\text{–}32)$$

Thus Eq. (19–31) is in accord with Eq. (19–32)—at least in terms of dimensions. To finish the evaluation, we note with the help of a table of integrals that:

$$\int u^2 e^{-u^2} d^3u = \frac{3}{2}\pi\sqrt{\pi}.$$

When this is inserted into the right side of Eq. (19–31), we find that the result of Eq. (19–32) is confirmed.

With Eq. (19–32) in hand, we can get a feel for molecular speeds from the rms speed v_{rms}:

$$v_{\text{rms}} = \sqrt{\langle v^2 \rangle} = \sqrt{\frac{3kT}{m}}. \qquad (19\text{–}33)$$

For helium gas, which has an atomic mass $m = 6.7 \times 10^{-27}$ kg, the rms speed is 556 m/s, 1362 m/s, and 2486 m/s for $T = 50$K, $T = 300$K, and $T = 1000$K, respectively. Contrast this with argon gas, which has a molecular mass 10 times greater than helium. The corresponding rms speeds are $\sqrt{10}$ times smaller: 176 m/s, 431 m/s, and 786 m/s.

One can also use the distribution F [Eq. (19–28)] to calculate the mean *speed* of the molecules. The expression will be similar to the rms speed, since dimensional analysis shows that it has to be of the form $[\text{energy/mass}]^{1/2}$, which must be $(\text{constant})(kT/m)^{1/2}$. The fact that the speed is inversely proportional to the square root of the mass implies that lighter molecules (faster ones) escape in greater quantity through a tiny hole than heavier ones (slower ones). This is the practical basis for the separation of uranium isotopes.† The uranium isotope ^{238}U is more common, but the relatively rare isotope ^{235}U is more useful for nuclear reactors such as those used for generating electric power—it is more fissionable, and therefore more useful for the creation of chain reactions. These isotopes cannot be separated by chemical means because they have the same chemical properties. However, the fact that ^{238}U is a bit heavier than ^{235}U means that when they are mixed in a gas—uranium hexafluoride, UF_6, is used—the ^{238}U component moves, on average, at a lower speed. The UF_6 gas is allowed to escape through many tiny holes from one container at a given temperature into another container that is initially empty. The percentage of ^{235}U in the gas that has escaped is slightly higher than that in the original container. Thousands of repetitions of this process of isotope separation lead to significantly increased percentages of ^{235}U. This method was used during World War II in the Manhattan Project.

THINK ABOUT THIS . . .

WHY IS THERE SO LITTLE HYDROGEN IN EARTH'S ATMOSPHERE?

The most common element in the universe is hydrogen, yet there are virtually no H_2 molecules in Earth's atmosphere. It would seem reasonable that the material out of which the planets condensed contained a lot of hydrogen. Jupiter, for example, consists largely of hydrogen, as does the Sun. The explanation lies in the escape speed, 11.2 km/s for objects on Earth (see Section 12–3). This number can be compared to the rms speed of molecules in a gas at temperature T, $v_{rms} = \sqrt{3kT/m}$, where m is the molecular mass. With a T value of 290K, we get for air ($m = 29.2 \times 1.67 \times 10^{-27}$ kg, an average coming mainly from oxygen and nitrogen molecules) a value of $v_{rms} = 0.5$ km/s. For H_2 m is a factor of 14.6 smaller, so that $v_{rms} = 1.9$ km/s. Both numbers are small compared with 11.2 km/s, but since the number of molecules with some speed v is proportional to $e^{-v^2/2kT} = e^{-3/2(v/v_{rms})^2}$, there are many more hydrogen molecules at speeds >11.2 km/s than there are molecules of the other components of air moving with this speed. Thus the upper layers of the atmosphere will lose their faster (H_2) molecules much more rapidly than they lose their slower (O_2 or N_2) air molecules. It is a slow process, but ultimately all the hydrogen molecules get "filtered out." ■

19–6 The Maxwell–Boltzmann Distribution

An ideal gas may consist of molecules that have just one atom (monatomic) such as helium (He) or argon (Ar), or it may consist of molecules with two (diatomic) or more atoms, such as nitrogen (N_2), oxygen (O_2), or water (H_2O) (Fig. 19–7). Is there a difference between these types of ideal gases? The answer came in the last part of the 19th century from the work of Ludwig Boltzmann.

The velocity distribution of a monatomic ideal gas [Eq. (19–28)] has the general form

$$F(\vec{v}) = \frac{1}{Z}e^{-mv^2/2kT} = \frac{1}{Z}e^{-K/kT}, \tag{19–34}$$

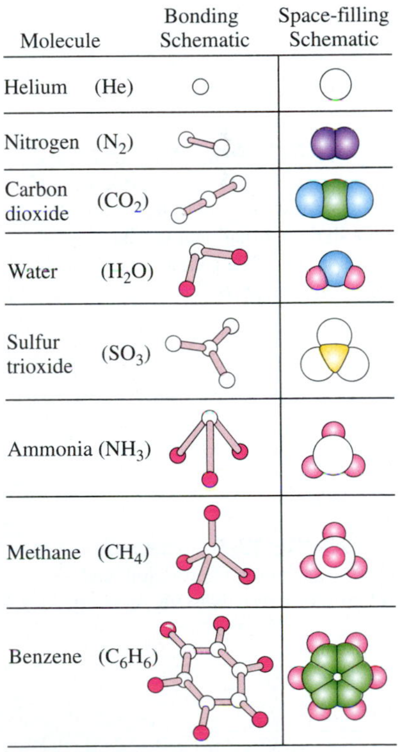

Molecule		Bonding Schematic	Space-filling Schematic
Helium	(He)		
Nitrogen	(N_2)		
Carbon dioxide	(CO_2)		
Water	(H_2O)		
Sulfur trioxide	(SO_3)		
Ammonia	(NH_3)		
Methane	(CH_4)		
Benzene	(C_6H_6)		

▲ **FIGURE 19–7** Molecules come in different shapes, as shown here in two schematic representations.

†Elements are characterized by their chemical properties, which depend on the number of electrons they carry, and hence on the number of (positively charged) protons in their nuclei. *Isotopes* of an element differ in the number of (electrically neutral) neutrons in the nucleus of the element, and thus differ in atomic weight. For uranium, the atomic weights 235 and 238 refer to the total number of protons plus neutrons.

where K is the kinetic energy ($K = mv^2/2$); the factor $1/Z$ takes into account the normalization. Boltzmann generalized the velocity distribution to a distribution that describes the probability that any one molecule has a given energy E. He discovered that in the original velocity distribution, Eq. (19–34), *the total energy of the molecule, E, should replace the kinetic energy of a point mass, K*:

$$F = \frac{1}{Z}e^{-E/kT}. \qquad (19\text{–}35)$$

MAXWELL–BOLTZMANN DISTRIBUTION

This result was obtained under the assumption that *all* the variables describing molecular motion, such as velocity and angular velocity, have equal probability of taking on *any* value, subject to the constraint that energy must be conserved, very much in line with the assumptions that led to Eq. (19–28). This distribution function is known as the **Maxwell–Boltzmann distribution**. Equation (19–35) is one of the most frequently used formulas in the physics of the behavior of matter. This distribution function depends on all the dynamical variables that enter into the energy of a molecule. It is only for free, pointlike molecules that the energy depends on speed alone. The energy of a rotating molecule with rotational inertia also depends on the angular velocities of rotation about the various axes. Later in this section we will see how this affects the specific heat of gases.

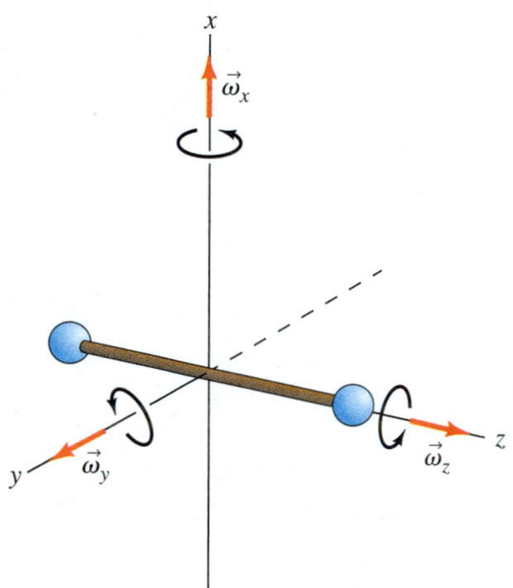

▲ **FIGURE 19–8** A dumbbell-like molecule can rotate about three axes. If the masses at the end of the rod are pointlike, the rotational inertia about the z-axis is zero. The rotational inertia about the x- and y-axes is I.

The Energy Distribution for Diatomic Molecules

To see the effect of molecular structure on the thermal properties of a gas, let's consider a gas composed of diatomic molecules. To describe a symmetric diatomic molecule, such as O_2, we use a classical model in which the two atoms act as though they are connected by a rigid rod. The rotational inertia has the same value, I, about each of the two axes that pass through the center of the rod and are perpendicular to it (Fig. 19–8). (We can assume that the rotational inertia about the molecular axis is so small that we can ignore it.) The energy of the molecule is a sum of the overall translational kinetic energy from linear motion, or translation (Fig. 19–9a), and of the rotational motion of the atoms about the center of mass (Fig. 19–9b; see Chapters 9 and 10). The rotational energy is

$$E_{\text{rot}} = \frac{1}{2}I\omega_x^2 + \frac{1}{2}I\omega_y^2, \qquad (19\text{–}36)$$

where ω_x and ω_y are the components of the angular velocity along the x- and y-axes, respectively (Fig. 19–8). The ω_z term is missing from Eq. (19–36) because the rotational inertia about the z-axis is zero. The probability distribution F now refers to the probability that the velocity of a molecule lies in a region d^3v about the velocity vector $\vec{v}$ *and* that the molecule's angular velocity lies in a region $d^2\omega = d\omega_x\, d\omega_y$ about the angular velocity vector ω whose components are ω_x and ω_y. The total energy of a molecule is now $E = K + E_{\text{rot}}$, and, from Eq. (19–35), the Maxwell–Boltzmann distribution is

$$F = \frac{1}{Z}e^{-(mv^2/2kT)-(I\omega_x^2/2kT)-(I\omega_y^2/2kT)}. \qquad (19\text{–}37)$$

Since not only the molecular velocity but also the rotation speed can vary, the average energy $\langle E\rangle$ takes the form

$$\langle E\rangle = \tfrac{1}{2}m\langle v_x^2\rangle + \tfrac{1}{2}m\langle v_y^2\rangle + \tfrac{1}{2}m\langle v_z^2\rangle + \tfrac{1}{2}I\langle\omega_x^2\rangle + \tfrac{1}{2}I\langle\omega_y^2\rangle. \qquad (19\text{–}38)$$

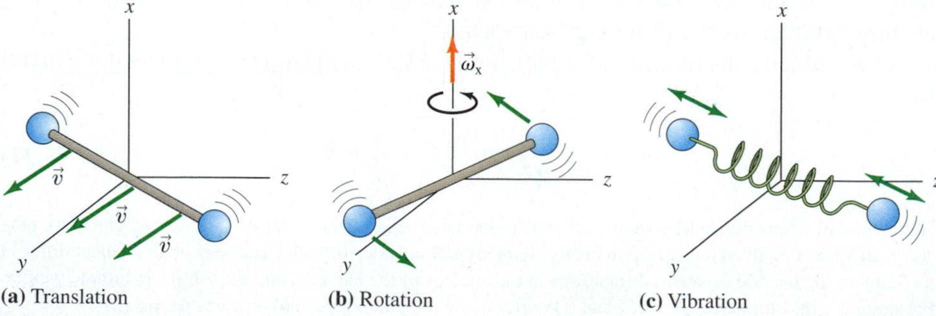

▶ **FIGURE 19–9** Schematic diagram of (a) translation, (b) rotation, and (c) vibration of a diatomic molecule.

The average is found by taking the integral of $E \times F$, and that integration is over d^3v, $d\omega_x$, and $d\omega_y$. The integrals are all of the form that lead to Eq. (19–31), and evaluation of each of the five terms in Eq. (19–38) is precisely $kT/2$. Thus for our rigid-rod diatomic molecule,

$$\langle E \rangle = \frac{5}{2}kT. \tag{19–39}$$

The total internal energy of the gas is then $U = N\langle E \rangle$, where N is the number of molecules. This result should be compared to the monatomic case, in which there are only the three terms of the kinetic energy of linear motion and $\langle E \rangle = \frac{3}{2}kT$ [Eq. (19–12)].

There is yet another aspect to diatomic molecules that we have not yet mentioned. If the diatomic molecules are not rigidly attached to each other, the bond between them acts like a spring with spring constant k (Fig. 19–9c). There are then two additional vibrational contributions to the energy, of the form $E_{\text{vib}} = (mV^2/2) + (kx^2/2)$. Here, V is the speed of the atoms of the molecule relative to the molecule's center of mass and x is the displacement of the vibrating atoms from an equilibrium position. Each of these terms also contributes $kT/2$ to the average energy and, for these molecules, $\langle E \rangle = \frac{7}{2}kT$.

We mention here in passing that the different average energies for molecules of different structure will be reflected in different values for the heat capacity. Thus these effects are directly translated to the macroscopic scale. We'll explore this in more detail below.

Equipartition

The discussion above shows that the total average energy per molecule depends on how many independent motions a molecule can have. If a molecule acts like a point mass, all it can do is move in the x-, y-, and z-directions. In this case, the energy has the three terms, proportional to v_x^2, v_y^2, and v_z^2, corresponding to linear motion in the x-, y-, and z-directions, respectively. If, in addition, the molecule is diatomic with a rotational inertia about axes x and y, we have two new terms in the energy—this time proportional to ω_x^2 and ω_y^2—making five terms in all. If vibration is possible, then there are two more terms, this time proportional to the squares of the relative speed of the atomic constituents and to their separation; thus there are in total seven such terms. Every term in the energy expression that is quadratic in an independent dynamical variable designates a **degree of freedom**. Generally, *the contribution of each degree of freedom to the average energy of a molecule is $kT/2$.* This result is called the **equipartition theorem**. If s is the number of degrees of freedom, then

$$\langle E \rangle = \frac{s}{2}kT. \tag{19–40}$$

EQUIPARTITION

Remember, $\langle E \rangle$ may contain contributions associated with translation, rotation, and vibration.

CONCEPTUAL EXAMPLE 19–7 Suppose you have a gas of rigid diatomic molecules constrained to move in just two spatial dimensions, for example between two plates spaced a molecular diameter apart. What would you expect the value of $\langle E \rangle$ to be?

Answer We may visualize this by thinking of the molecules as a collection of matches on a table forming the xy-plane, as in Fig. 19–10. The velocity of the center of mass can have only x- and y-components, and the integration that gives the average value of energy is over only two rather than the usual three translational degrees of freedom. Rotations can only occur when the nuclei rotate about their centers of mass in the xy-plane, that is, there are rotations only about a z-axis. This means there is only one rotational degree of freedom. The total number of degrees of freedom is three, just as for a monatomic gas. Therefore the average energy $\langle E \rangle$ is $3 \times kT/2$, the same result as for a three-dimensional monatomic gas.

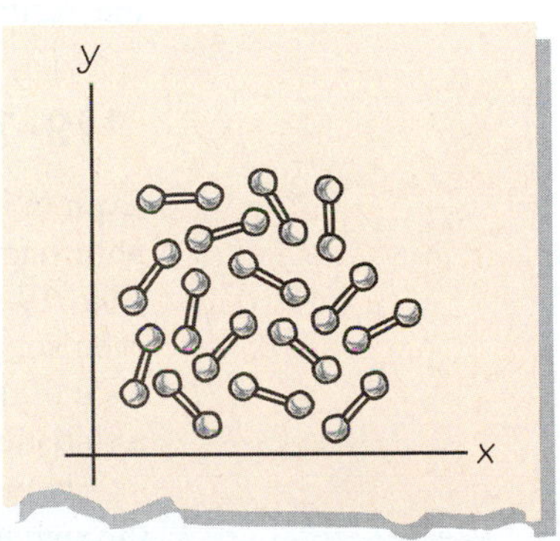

▲ **FIGURE 19–10** Diatomic molecules move in xy-plane.

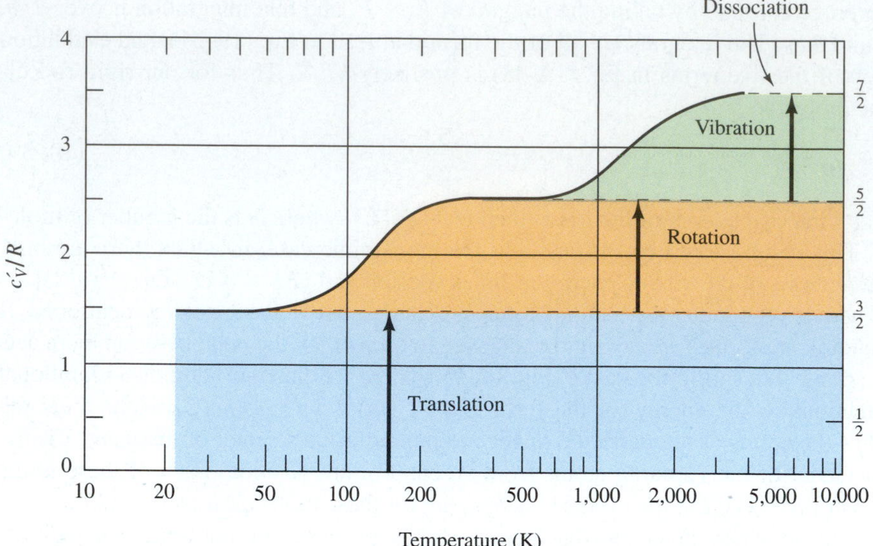

▶ **FIGURE 19–11** The constant-volume molar heat capacity divided by the gas constant, c'_V/R, for H_2 as a function of temperature.

A dilute gas—even one whose molecules have a complicated structure—continues to obey the ideal gas law. The internal energy is

$$U = N\langle E\rangle = \frac{s}{2}NkT. \tag{19–41}$$

The internal energy continues to be linearly dependent on temperature, as for any ideal gas. The term by which the temperature is multiplied in Eq. (19–41) is the constant-volume heat capacity of the gas, C_V. The heat capacity therefore depends on the number of degrees of freedom of the molecules through the relation

$$C_V = \frac{s}{2}nR. \tag{19–42}$$

Molecular hydrogen (H_2), for example, should have a *molar heat capacity*, c' or heat capacity per mole ($n = 1$), of $(7/2)R$ at constant volume: $(3/2)R$ is from the overall motion of the center of mass, $(2/2)R$ is from rotations about two axes, and $(2/2)R$ is from the kinetic and potential energies of the springlike bond that connects the two hydrogen atoms. How does this compare with experimental results? Figure 19–11 plots c'_V/R where c'_V is the constant-volume molar heat capacity, as a function of temperature. At lower temperatures, hydrogen gas behaves like a monatomic system, with a molar heat capacity of $(3/2)R$, which explains why we can treat H_2 at room temperature as a monatomic gas. The full factor of $(7/2)R$ comes in only at high temperatures. The steplike nature of the curve suggests that, somehow, the effects enter one at a time. The reason for this cannot be found in the realm of classical physics, but instead requires quantum mechanics. Quantum physics sets certain minimum temperatures for the excitation of the different degrees of freedom.

*19–7 Collisions and Transport Phenomena

Even in a dilute ideal gas, molecules follow a tortuous path in their container, colliding with one another and the walls, changing in direction and speed with each collision (Fig. 19–12). At STP an air molecule undergoes billions of collisions per second, but even so, the collisions are brief, and most of the time the molecule is free of the influence of other molecules. The *average* distance a molecule travels between collisions is a statistical quantity that can be calculated just as the average speed can be calculated.

It is by means of these collisions that molecules can carry physical properties through a gas. If there is local heating, that extra thermal energy can be carried to other parts of the gas through successive collisions. If some molecules with a particular odor or color are introduced in one location, they will spread through the gas through colli-

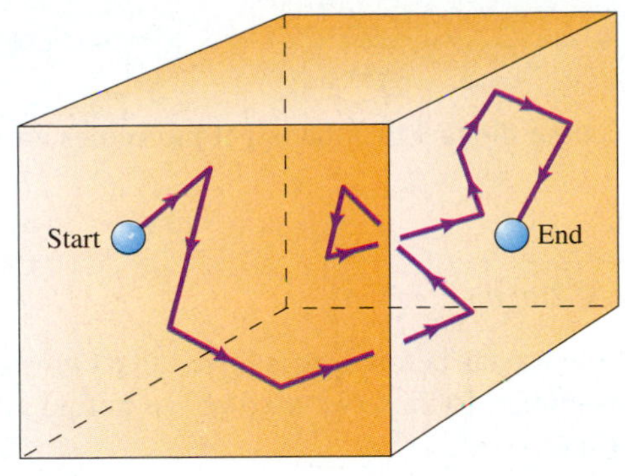

◀ **FIGURE 19–12** A molecule follows a tortuous zigzag path because of its multiple collisions with other molecules.

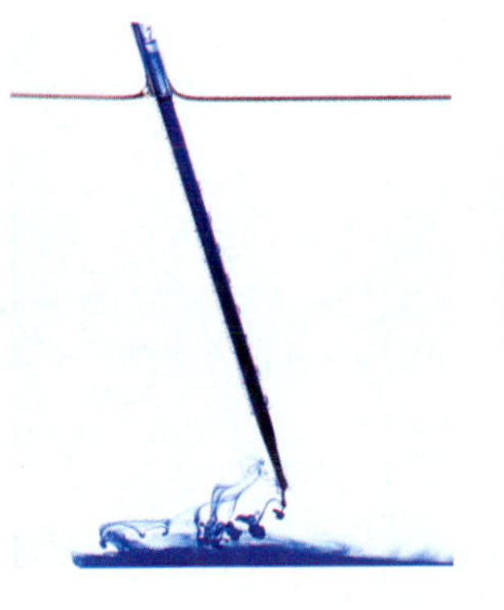
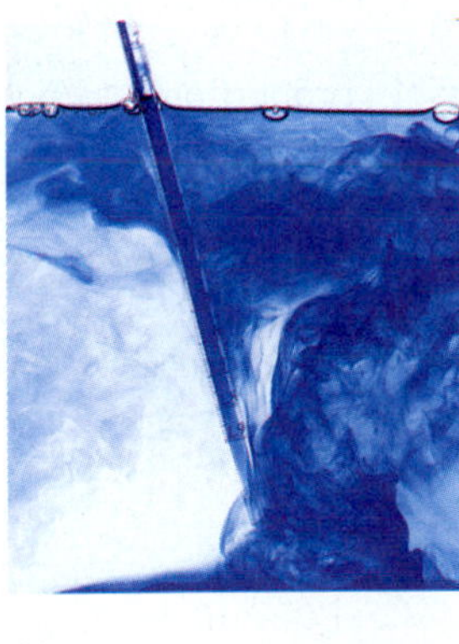
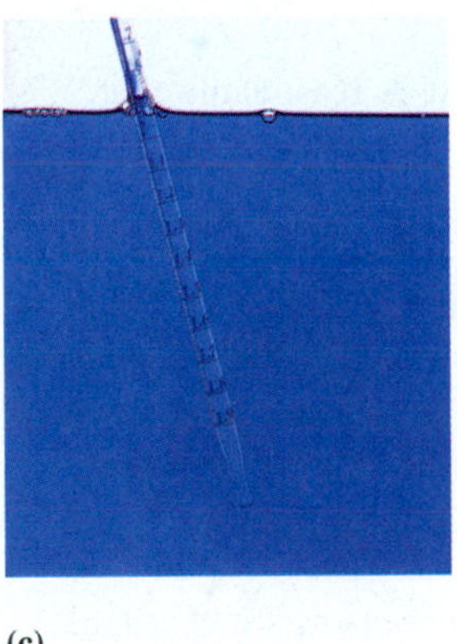
(a) (b) (c)

◀ **FIGURE 19–13** (a) Dye is placed with the help of a pipette at the bottom of a flask of water. (b) A combination of convection and diffusion has moved the dye into other regions of the flask. (c) Dye has diffused throughout the flask of water to make a uniform distribution. While stirring can speed the process, diffusion will make this happen if given enough time.

sions, even without wind currents (convection). The movement of such properties is called **transport** (of thermal energy, odor, and so forth), and, more particularly, the movement of molecules by random collisions is called **diffusion**. Figure 19–13 illustrates transport in a liquid. Ordinary thermodynamics has nothing to say about transport phenomena, but kinetic theory can explain them.

Collisions and Molecular Movement in a Gas

For a given gas density, the distance that a molecule travels before it has a collision depends on the size of the molecules. If molecules were infinitely small, they would never encounter each other; if they were very large, they would always be colliding. Molecules of diameter D collide when the path of the center of one molecule lies within an area πD^2 presented by the second molecule (Fig. 19–14). This area is the **collision cross section**, σ.

$$\sigma = \pi D^2. \tag{19–43}$$

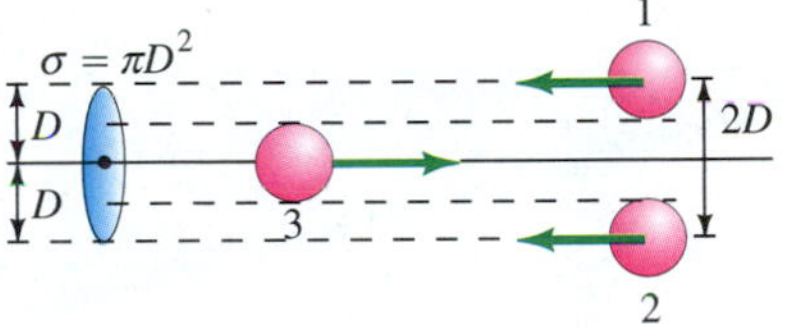

▲ **FIGURE 19–14** If the projected centers of two molecules of diameter D that move together are less than or equal to a distance D apart, the molecules will collide. An area $\sigma = \pi D^2$ of one molecule, shown in blue, is available for collisions with another molecule.

Even if the molecules interact differently than billiard balls, there is some effective distance D that characterizes the collision, and the collision cross section still has the form of Eq. (19–43).

Consider now a molecule that moves with speed v and sweeps out an area σ. If no other molecules are present, then in time t the molecule travels a distance $d = vt$, sweeping out a volume $V = \sigma d = \sigma vt$. However, our molecule is not alone: Within the volume V, there are $N = nV$ target molecules, where n is the number density of molecules (not the number of moles). The traveling molecule therefore suffers N collisions. Even though the path is bent with each collision, the volume V remains unchanged if the speed is not changed by the effect of multiple collisions, and, on average, it is not. Because the number of collisions in time t is N, there is on average a collision every t/N seconds. This is the **mean collision time**, τ. Since we are averaging over many molecules, we replace v with some average value. It is reasonable to replace it with v_{rms} so that

$$\tau = \frac{t}{N} = \frac{t}{nV} = \frac{t}{n\sigma v_{\text{rms}}t} = \frac{1}{n\sigma v_{\text{rms}}}. \tag{19–44}$$

A more precise calculation takes into account the fact that the target molecules are also moving, and Eq. (19–44) is modified to

$$\tau = \frac{1}{\sqrt{2}n\sigma v_{\text{rms}}}. \tag{19–45}$$

The average distance a given molecule travels before it is involved in a collision is the **mean free path**, λ. Using arguments similar to those we used above to find τ, λ is the rms speed times the mean collision time:

$$\lambda = \tau v_{\text{rms}} = \frac{1}{\sqrt{2}n\sigma}. \tag{19–46}$$

It is reasonable that λ is inversely proportional to both the density and the collision cross section.

EXAMPLE 19–8 What is the mean free path in air at sea level when the temperature is 300K? By how much does the mean free path change when the temperature drops to 275K? Take $r \cong 10^{-10}$ m as a typical molecular radius, and treat the atmosphere as ideal.

Strategy To find the mean free path we need to assemble all the quantities that enter into the expression for it, Eq. (19–46). The collision cross section σ can be found with the given molecular radius using Eq. (19–43). The number density is obtained from the ideal gas law written in the form $pV = NkT$, from which we get the number of molecules per unit volume as $n = N/V = p/kT$. All that remains is to put in the numbers. For the second part of the problem, we notice that since $n \propto 1/T$, the mean free path $\lambda \propto T$. We can then get the mean free path at 275K by taking ratios.

Working It Out We write $\sigma = \pi(2r)^2$ and $n = p/kT$. We have $r = 10^{-10}$ m, $p = 1.0$ atm $= 1.0 \times 10^5$ Pa and $T = 300$K, so that

$$\lambda = \frac{1}{\sqrt{2}n\sigma} = \frac{1}{\sqrt{2}}\frac{kT}{p\pi(2r)^2}$$
$$= \frac{1}{\sqrt{2}}\frac{(1.38 \times 10^{-23}\text{ J/K})(300\text{K})}{(1.0 \times 10^5\text{ N/m}^2)\pi(2 \times 10^{-10}\text{ m})^2} = 2.3 \times 10^{-7}\text{ m}.$$

For the second part of the problem (the mean free path at 275K), the linear dependence tells us that it is 90% of that at 300K:

$$\frac{\lambda(275\text{K})}{\lambda(300\text{K})} = \frac{275}{300} = 0.9.$$

What Do You Think? If you slowly increase the pressure by slowly moving a piston into a cylinder containing the gas, will the mean free path of the molecules increase or decrease?

The Random Walk and Diffusion

Molecules move through a gas by diffusion. How far does a molecule move, *on average*, from its initial position in a given amount of time? This problem is similar to a classic problem in mathematics called the **random walk**, known historically as the drunkard's walk. In this problem, a drunkard starts at a lamppost and takes steps that are equal in length but random in direction (Fig. 19–15). Compare Fig. 19–15 to Fig. 19–12, a typical molecular path in a gas. If the length of the drunkard's step is replaced by the mean free path, and if the time between steps is replaced by the mean collision time, then—aside from the fact that the molecule moves in three dimensions rather than two—the problems are very similar.

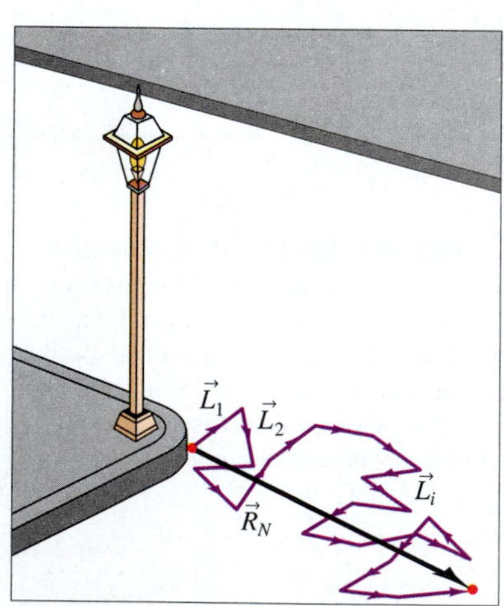

▲ **FIGURE 19–15** The random walk (a drunkard's walk around a lamppost).

We can find the average displacement of a molecule after N steps as follows. Let the successive displacements of the molecules be $\vec{L}_1, \vec{L}_2, \ldots \vec{L}_N$. These have random directions, but their magnitudes are all the same value, L. After N steps the net displacement of a molecule is $\vec{R}_N = \vec{L}_1 + \vec{L}_2 + \cdots + \vec{L}_N$. Squaring this quantity, we find that

$$R_N^2 = L_1^2 + L_2^2 + \cdots + L_N^2 + 2\vec{L}_1 \cdot \vec{L}_2 + 2\vec{L}_1 \cdot \vec{L}_3 + \cdots + 2\vec{L}_{N-1} \cdot \vec{L}_N.$$

We want the average value of this quantity. The directions of the vectors $\vec{L}_i$ are random, so all the dot products, such as $\vec{L}_1 \cdot \vec{L}_2$, have an average over time of zero, whereas the terms L_i^2 all equal L^2. We are left with

$$\langle R_N^2 \rangle = NL^2. \tag{19–47}$$

To connect this result to the properties of the gas, we replace L^2 by the mean free path squared. The time interval between steps is the mean collision time, τ, so after N steps, a time $t = N\tau$ has elapsed. N can thus be replaced by t/τ. Therefore, Eq. (19–47) states that after time t, a molecule will have moved on average a distance squared given by

$$\langle r^2 \rangle = \frac{t}{\tau}\lambda^2. \qquad (19\text{–}48)$$

It is typical of random-walk problems that the displacement *squared* is linear in time, or equivalently that the displacement is proportional to the square root of the time.

This result (or the slightly modified result of a more detailed calculation) as it applies to molecules can be verified experimentally. Molecules that have been marked in one way or another (for example, by using some radioactive molecules) can be traced as they move. Even better, the actual motion of this random walk is observable when particles larger than molecules, such as smoke particles, are added to a gas. This motion is called **Brownian motion**. The observation of Brownian motion and its explanation in kinetic theory was a "smoking gun" that established once and for all the atomic picture of atoms.

THINK ABOUT THIS . . .

HOW DOES THE MOLECULAR PICTURE EXPLAIN THERMAL ENERGY CONDUCTION IN MATERIALS?

In Section 17–1 we discussed the various methods of thermal contact, which we subsequently learned involves the flow of thermal energy. We have learned here that thermal energy is the energy of motion of the constituents of matter and so we can now look at thermal conductivity in materials from a molecular point of view. It is useful to think about gases and solids separately.

Gases

In gases, thermal energy movement by conduction is a transport phenomenon of the type discussed in this section. Molecules that move more rapidly because they are in a region of higher temperature collide with molecules in a neighboring region, giving the adjacent molecules more kinetic energy and consequently more thermal energy. This model accounts satisfactorily for such quantities as the thermal conductivity κ (see Chapter 18).

Table 18–3 shows that air, like all gases, has a small thermal conductivity. So why is a house built of double walls with air between them poorly insulated? The reason is that convection, which involves large-scale movement ("wind currents"), is the main form of heat transport in fluids. Thermal energy is transported efficiently by convection. Fiberglass insulation is placed in walls so that air is trapped, unable to flow, or convect, easily.

Solids

It is a fact that metals conduct thermal energy better than nonmetals do. This is because in some sense metals contain a gas of electrons, and the thermal conduction mechanism available to gases is also available to metals. Within the atoms of metals, the outermost electrons that orbit the nucleus are attached very loosely (they are *weakly bound*) to the nucleus, and it takes very little energy to make them move almost freely through the material. The practically free electrons of metals are efficient in transferring energy from one end of a piece of metal to the other. The same electrons also transfer electric charge (Chapter 26), and thermal and electrical conductivity in metals are closely related. (The analogy between metals and gases only goes so far, and there is no analog to convection in metals.)

Nonmetals also conduct thermal energy, but less well than metals do. The outermost electrons of the atoms of nonmetals are attached tightly to their nucleus or to other atoms, and are therefore not readily available to conduct thermal energy. Both metals and nonmetals can move energy in the form of waves (Chapters 14 and 15), but purely elastic waves traverse a material without depositing energy along the way. Thus thermal energy cannot be distributed through material this way. What is required to do so is a set of impurities or defects in the medium, because when an elastic wave meets such obstacles it can indeed transfer some of its energy to them. ■

Summary

Molecules are sparse in gases under normal conditions. The kinetic theory of ideal gases is a statistical description in which molecules bounce randomly and relatively infrequently off each other and off the walls of their container. This theory accurately predicts many thermodynamic properties.

The internal energy, U, is given in this model by the number of molecules, N, times the average energy of a molecule:

$$U = N\langle K\rangle = N\left(\frac{1}{2}m\langle v^2\rangle\right), \tag{19–2}$$

where $\langle v^2\rangle$ is the average of the velocity squared. The pressure of a gas arises from multiple elastic collisions between the walls and the molecules. Pressure is related to internal energy by

$$pV = \frac{2}{3}U, \tag{19–11}$$

where V is the volume of gas. In the ideal gas law, pV is proportional to temperature, so temperature can be interpreted as a measure of the internal energy, or of the average energy of molecules:

$$kT = \frac{2}{3}\frac{U}{N} = \frac{2}{3}\langle K\rangle. \tag{19–13}$$

Probability distributions describe the probability that various outcomes will be represented in a large statistical sample. For gases, $F(\vec{v})\,d^3v$ describes the probability that a given molecule will have a velocity $\vec{v}$ in the range from $\vec{v}$ to $\vec{v} + d\vec{v}$. Averages are computed from the probability distribution by integration. For example, the average of v^2 is

$$\langle v^2\rangle = \int v^2 F(\vec{v})\,d^3v. \tag{19–27}$$

The probability distribution $F(\vec{v})$ for the velocity of the molecules of a gas is

$$F(\vec{v}) = \left(\frac{m}{2\pi kT}\right)^{3/2} e^{-mv^2/2kT}. \tag{19–28}$$

Boltzmann generalized the velocity distribution to the cases in which molecules have internal energy as well as a center-of-mass kinetic energy, or in which the molecules are subject to external forces. The Maxwell–Boltzmann distribution is given by

$$F = \frac{1}{Z}e^{-E/kT}, \tag{19–35}$$

where E is the total energy of a molecule. Every variable that appears quadratically in the expression for the energy of a single molecule—such variables are called degrees of freedom—contributes $\frac{1}{2}kT$ to the average energy. If the number of degrees of freedom is labeled s, then the average energy of a molecule is

$$\langle E\rangle = \frac{s}{2}kT. \tag{19–40}$$

This result is the equipartition theorem. The internal energy is then $U = N\langle E\rangle$; in addition to the energy itself, the heat capacity, which is related to how fast U changes with temperature, is also determined, completing the link to thermodynamics.

Transport phenomena describe the motion of molecules, momentum, energy, and so forth through a gas as a result of multiple molecular collisions. Molecules have a collision cross section, σ, which describes the area available for collisions with another molecule. If the number of molecules per unit volume is n, then the mean time τ between collisions, called the mean collision time, is

$$\tau = \frac{1}{\sqrt{2}n\sigma v_{\mathrm{rms}}}. \tag{19–45}$$

The average distance a given molecule travels before it suffers a collision is the mean free path, λ:

$$\lambda = \tau v_{\mathrm{rms}} = \frac{1}{\sqrt{2}n\sigma}. \tag{19–46}$$

As a result of multiple collisions with other molecules, an individual molecule within a gas will undergo a random walk. In a time t, the molecule moves—diffuses—a distance that is proportional to the square root of t.

Understanding the Concepts

1. Consider an ideal gas in an insulated cylinder outfitted with a piston. The piston is moved slowly, so that the volume of the gas is decreased. What happens to the temperature of the gas? How do you explain your answer in terms of the motion of molecules?
2. On average, which will travel faster: the oxygen molecules or the nitrogen molecules in your room?
3. In Example 19–6 the calculated quantity P is 3.3 m^{-1}. Is this a probability? If so, how can it be greater than one?
4. In our discussion of pressure, we used the fact that the collisions of molecules with the walls are elastic. If there are diatomic molecules, they could be set into rotation by the collision, and thus their translational kinetic energy after the collision could be smaller than before the collision. How can these statements be reconciled? What is the effect on the pressure?
5. We made the connection between temperature and the average kinetic energy of a molecule without taking into account the large number of intermolecular collisions estimated in Section 19–1. Why do you think this approximation holds?
6. Why would more massive molecules in a gas at a given temperature have smaller average speeds?
7. In an extremely dilute gas, the rate of interatomic collisions is very small. As the gas becomes less dense, the rate of interatomic collisions becomes much less than the collision rate with the container's walls. Can we still make a connection between temperature and the average kinetic energy of a molecule? What are the possible problems in making the connection in this limit?
8. Very few hydrogen molecules are present in Earth's atmosphere today. Is this fact in conflict with the possibility of their abundance a long time ago? In explaining your answer, ignore the possibility that hydrogen is removed by chemical reactions.
9. Can we assign a temperature to a single molecule? Explain.
10. Why, from the point of view of kinetic theory, does the air near a hot stove become heated?
11. In Section 19–1 we stated that the typical distance a molecule travels before it suffers a collision in 1 mol of a gas at standard temperature and pressure is roughly 3×10^{-7} m. This will be much less than the size of a container of gas. How is it that the approximation that each molecule in the cylinder collides freely with the wall, without worrying about intermediate collisions between molecules, leads to a satisfactory result for the pressure, Eq. (19–11)?
12. Suppose you place a thin chip of paint of dimensions 1 cm by 1 cm in a gas. Do the collisions knock it to one side? Which side?
13. At standard temperature and pressure, a molecule of one mole of an ideal gas has a mean free path of about 3×10^{-7} m, some 100 times greater than the average intermolecular spacing. How is it possible for the mean free path to be greater than the average spacing?
14. You have a container of a dilute gas at a certain temperature and inject a small quantity of a different type of molecule into that container. What processes determine how long it takes the new molecules to come to thermal equilibrium with the original molecules?
15. If you were to consider helium gas and air at the same temperature, which gas would have a higher root-mean-square *momentum*?
16. Gravity keeps the atmosphere of Earth close to the surface. Why, then, is there no atmosphere on the Moon, which also exerts a gravitational force?
17. In Newtonian mechanics, energy is added to a system when work is done on it. In the language of classical mechanics, how is energy added to a gas in a container when a piston is pushed in and the container's volume is changed?
18. We have a container of helium gas and another of nitrogen gas. Both have molecules with the same average speed. What is the ratio of their temperatures?
19. Given our explanation of the origin of pressure, why is there no net force on a pane of glass placed in a container of gas, even though there may be a net force on a wall of the container?
20. Will the air just above the surface of hot water in a bowl be at the same temperature as the water? If so, what molecular mechanism is responsible for the equality of the two temperatures?
21. Why will a cup of hot water cool off much more rapidly when air is blown across its surface than when the air above the cup is stationary?
22. The smoke from the burning end of a cigarette rises; the flame of a candle points upward. Why?
23. In our discussion of the distribution of heights we claimed that in a group of persons all heights are equally likely. How would you check this premise? Can you know this with certainty, given that the number of people in the world is finite?
24. When a candle burns, carbon combines with the oxygen in air to form carbon dioxide, a compound used in fire extinguishers to put out fires. Why does the candle not snuff itself out? Could you burn a candle in a satellite in orbit?
25. Does temperature depend on the rotational kinetic energy of molecules?

Problems

19–1 A Microscopic View of Gases

1. (I) We can define a volume ratio R_V of a gas to be the ratio of the volume taken up by the molecules to the total volume of the container; similarly, we can define a spacing ratio R_L to be the ratio of the linear size of a molecule to the mean spacing between molecules. Compute R_V and R_L for a gas with molecules of diameter 2×10^{-10} m. Assume that the gas obeys the ideal gas law for the following values of temperature and pressure: (a) $T = 300\text{K}$, $p = 1.0$ atm; (b) $T = 5\text{K}$, $p = 1.0$ atm; (c) $T = 300\text{K}$, $p = 10^{-8}$ atm; (d) $T = 5\text{K}$, $p = 10^{-8}$ atm.
2. (I) For an ideal gas at 0°C and 1.0 atm, the ratio of the typical molecular radius (10^{-10} m) to the intermolecular spacing is about 1/20. What is this ratio in liquid water (H_2O, molecular weight 18 g/mol)?
3. (II) Assume that the carbon and hydrogen atoms that make up oil behave like spheres of diameter 10^{-8} cm, and that the fundamental molecular component of oil, CH_2, would take up a corresponding area of 10^{-15} cm^2 on a water surface. How many such components are there in 0.1 L of oil if the density of oil is 90 percent of the density of water? What would the area of a one-molecule-thick layer of this amount of oil on water be? (This type of experiment, said to have been performed by Benjamin Franklin, among others, is sometimes used to compute the size of molecules.)

19–2 Pressure and Molecular Motion

4. (I) A rubber ball, confined to move in the x-direction, bounces elastically back and forth between two walls. The ball, which has a mass of 85 g, moves at a speed of 11 m/s, and the spacing of the walls is such that the ball strikes the right-hand wall 20 times in a ten-second period. What is the average force on the right-hand wall due to the ball during that ten-second span?

5. (I) Grains of sand of mass 3×10^{-3} g each, fall from a height of 0.8 m on a sticky surface at a rate of 50 grains per second per cm^2. What pressure does this shower of sand exert on the surface, assuming that air resistance can be neglected?

6. (II) Use dimensional analysis to estimate the number of air molecules that strike the 12-in $\times$ 12-in screen of a television set during a 1-h program.

7. (II) One mol of a monatomic ideal gas is placed in a chamber under 5 atm pressure. The volume of the chamber is 5000 cm^3. (a) What is the internal energy of the gas? (b) What is the temperature of the gas? (c) Assuming that the mass of a molecule of the gas is 3.36×10^{-26} kg, what is the value of $\langle v^2 \rangle$ of a gas molecule? (d) What is the root-mean-square (rms) velocity?

8. (II) A cubic box of volume 0.080 m^3 contains helium gas at 0.85 atm pressure and 22°C. (a) How many total collisions do the molecules make with the walls per second? (b) What is the total internal energy of the gas? (c) If the temperature is doubled, what happens to the rate of collisions?

9. (II) Consider a gas constrained to move in a plane of area A rather than a volume in three dimensions. The pressure in that case is a force/unit length exerted at the edge. What is the relation between pressure and internal energy in this case?

10. (II) Consider a spherical satellite of mass M and radius R, moving with speed v through a dilute gas, the upper atmosphere. We shall take v to be much smaller than v_{rms} so that we can ignore the distribution in velocities of the molecules of the gas. Assuming that the gas molecules are stationary, and that the satellite is slowed down by collisions with the gas molecules, obtain an expression for the change of velocity of the satellite as a function of time in terms of m, the molecular mass, n, the number density of molecules in m^{-3}, and the characteristics of the satellite. [*Hint*: The area presented by the satellite to the stationary molecules is just the same as if the satellite were a disc of radius R, oriented perpendicular to the motion.]

11. (II) The apparatus shown in Fig. 19–16 is designed to demonstrate aspects of the kinetic theory of gases. It consists of a transparent cubical box, 20 cm on each side, containing 100 steel balls of diameter 5 mm. The density of steel is 7.8 g/cm^3. The bottom of the box vibrates so that the steel balls bounce around. The top of the box consists of a movable piston of mass 1 kg. What is the rms speed of the steel balls if the top of the box is in dynamic equilibrium with the "gas" of steel balls? Ignore gravity in your treatment of the motion of the steel balls. Is this approximation justified?

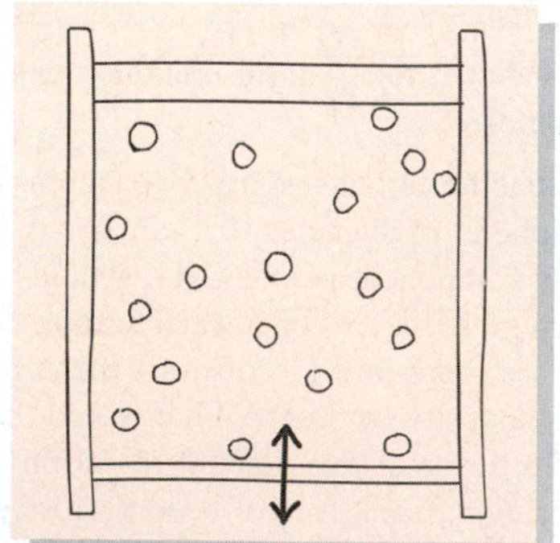

▲ **FIGURE 19–16** Problem 11.

12. (II) When particles of mass 1.3×10^{-17} kg are suspended in a liquid at room temperature, they are observed to have an rms speed of 0.030 m/s. Use this to determine Avogadro's number. (You will need to look up the gas constant R.)

13. (II) The rms speed of galaxies in a large part of the visible universe is roughly 100 km/s. The number density of these galaxies is $3 \times 10^{-20}/\text{ly}^3$, and the average mass of a galaxy is 3×10^{41} kg. What is the pressure of a gas of such galaxies?

14. (II) Molecules are confined to move in a plane, for example, by putting them between two glass plates separated by a distance that is small compared to the mean free path. What is the relation between the pressure and the internal energy? What is the relation between internal energy and temperature for an ideal gas?

19–3 The Meaning of Temperature

15. (I) Consider the mole of helium gas with an internal energy of 3600 J that we discussed in Example 19–1. (a) What is the temperature of the gas? (b) If 1/1000 of this amount of gas had the same internal energy as the total gas, what would its temperature be? (c) What would the internal energy of 1/1000 of the original sample of gas be, at its original temperature?

16. (I) A gas is heated from 120K to 470K. What is the change in the rms speed?

17. (I) What is the average kinetic energy per molecule of a gas at room temperature, 293K? How fast would a baseball of mass 0.10 kg be moving to have this same kinetic energy?

18. (I) An ideal gas is contained in a vessel of volume 3.0×10^{-4} m^3 that is under a pressure of 1.2 atm. What is the internal energy of the gas?

19. (I) The temperature of hydrogen in a dilute plasma—a state of matter in which atoms are broken apart into electrons and nuclei—inside a nuclear fusion reactor needs to be on the order of 20 million K in order to initiate nuclear fusion. What is the rms speed of the hydrogen nuclei in the plasma?

20. (II) A mixture of nitrogen gas and sulfur dioxide gas is in equilibrium at 27°C. What is the rms speed of the sulfur dioxide molecules (SO_2, molecular weight 64 g/mol), and that of the nitrogen molecules (N_2, molecular weight 28 g/mol)?

21. (II) Calculate the rms speed at 300K of the principal components of air: O_2 molecules (molecular weight 32 g/mol), N_2 (molecular weight 28 g/mol), CO_2 (molecular weight 44 g/mol) and H_2 (molecular weight 2.0 g/mol).

22. (II) The rms speed of 1 mol of argon atoms (atomic weight 40 g/mol) in a box is 680 m/s. (a) What is the temperature inside the box? (b) What is the internal energy? (c) If the box has a volume of 10^{-3} m^3, what is the pressure? Treat the gas as ideal.

23. (II) Estimate v_{rms} for hydrogen atoms (a) on the surface of the Sun, where the temperature is 6000K; (b) on the surface of the Moon, where the temperature at one point is 150K.

24. (II) A proton of kinetic energy 3×10^7 eV comes to a stop because of molecular collisions in a thin tube that contains 1.0×10^{-3} mol of oxygen at STP, as in Fig. 19–17. By how much is the temperature in the tube increased at equilibrium?

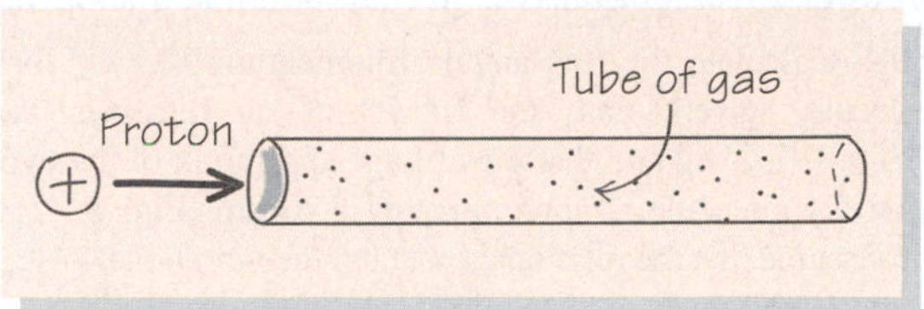

▲ **FIGURE 19–17** Problem 24.

25. (II) A "gas" of water droplets 10^{-6} m in diameter is in equilibrium with air molecules at 300K. Given that the density of water is $1.0 \times 10^3\ \text{kg/m}^3$, what is the value of v_{rms} for the droplets?

26. (II) A relativistic gas is one in which a significant fraction of the constituents have speeds that are some finite fraction of the speed of light, $c \cong 3 \times 10^8$ m/s. Suppose that the rms speed of the constituents of a gas of atomic hydrogen is just 1% of c. What is the temperature? The energy that binds the electron and the proton together into a hydrogen atom is 2.18×10^{-18} J. Would this gas break down into a gas of electrons and protons? In the interior of stars, where temperatures comfortably exceed the temperature you have calculated, atoms cannot exist; they are broken into their components.

27. (II) If their kinetic energy greatly exceeds the average gravitational potential energy, galaxies in the sky may be viewed as an ideal gas. Given that the rms speed of galaxies is 100 km/s and the average mass is 3×10^{41} kg, what is the temperature of a gas of galaxies?

19–4 Probability Distributions

28. (I) A lottery that costs \$1 to play has a jackpot of \$10,000,000. The state that runs the lottery is obligated to tell the players that the probability of winning is one in 50,000,000. Is it a good idea from a financial point of view to play this lottery? Answer this question by calculating the ratio of your outlay to your revenue under the assumption that you play the game an unlimited number of times.

29. (I) Nine cars are measured at a given spot on a highway to have speeds v of 52.3, 54.5, 57.0, 57.2, 57.9, 63.6, 63.6, 68.1, and 82.2 mi/h. Calculate (a) the average speed, and (b) the rms speed.

30. (I) Take a deck of cards. Draw cards from that deck, one by one, without putting any cards back in the deck. What is the probability that the first card is a spade? What is the probability that the second one is also a spade? What is the probability that the third one is also a spade? (You may extend this procedure to calculate the probability that when you draw thirteen cards, they are all spades, by multiplying all the thirteen probabilities together).

31. (I) Four dice are tossed. What are the probabilities of getting (a) four sixes; (b) three sixes and a five?

32. (II) A cage of 50 Ping-Pong balls has air blowing through it to keep the Ping-Pong balls moving. Five Ping-Pong balls have the number 0 painted on them, five have the number 1, and so forth, finishing with five Ping-Pong balls painted with the number 9. The apparatus is used to choose the numbers for a lottery, and the balls are not returned to the cage after they are chosen. (a) The first drawing is for three balls. Your number is 186. Before any balls are drawn, what is the probability that you will win? (b) The first ball drawn is 1. What are your chances now? (c) The second ball drawn is 8. Now what is the probability that you will win?

33. (II) Consider the distribution of grades g discussed in Section 19–4. (a) Plot the distribution of the numbers D for the students, where D is defined as the deviation of the grade from the average. (D can be positive or negative.) (b) Calculate the average value of D^2. (c) Show that $\langle D \rangle = 0$ and $\langle D^2 \rangle = \langle g^2 \rangle - \langle g \rangle^2$.

34. (II) Find an expression for the probability of getting 13 spades in a bridge hand. (In bridge, the entire deck of 52 cards is dealt to four players.)

35. (II) In a true-or-false test, a correct answer is awarded +1 point and an incorrect answer is awarded −1 point. In a test involving 87 students, the following grade distribution is found (Fig. 19–18):

−50 to −30:	3 students;
−30 to −10:	18 students;
−10 to +10:	29 students;
+10 to +30:	22 students;
+30 to +50:	15 students.

(a) Calculate the average grade. (b) Calculate $\langle D^2 \rangle$, where D is defined in Problem 33. (c) Suppose that the distribution was such that every student has a grade between +10 and +30. With this information, what would the average be?

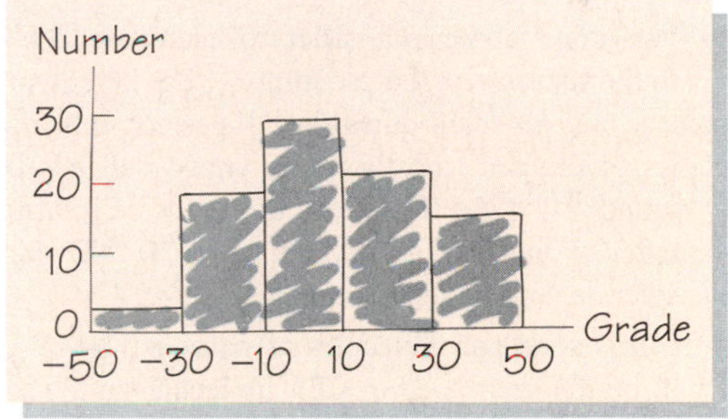

▲ **FIGURE 19–18** Problem 35.

36. (II) Suppose that the distribution of grades (the number of test takers, N, with a grade x) in a national test is approximated by a continuous curve of the form (Fig. 19–19)

$$N(x) = 25x \quad \text{for } 0 < x < 60$$
$$= 6000 - 75x \quad \text{for } 60 < x < 80$$
$$= 0 \quad \text{for } 80 < x < 100.$$

(a) What is the total number of students who took the test? (b) What is the average grade?

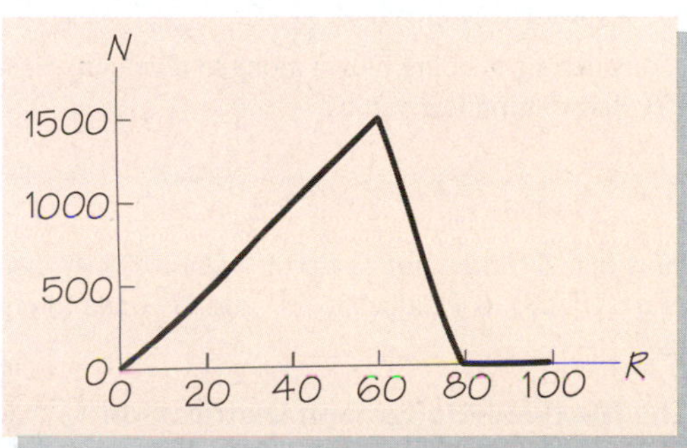

▲ **FIGURE 19–19** Problem 36.

37. (III) A way to find the value of π (*Buffon's method*) is the following: Drop a needle of length L repeatedly onto a grid of parallel lines separated by a distance L (Fig. 19–20). Show that the probability of the needle falling on a line is $2/\pi$. [*Hint*: Take the lines as parallel to the x-axis. Calculate the average value of the component of the needle's length along the y-axis.]

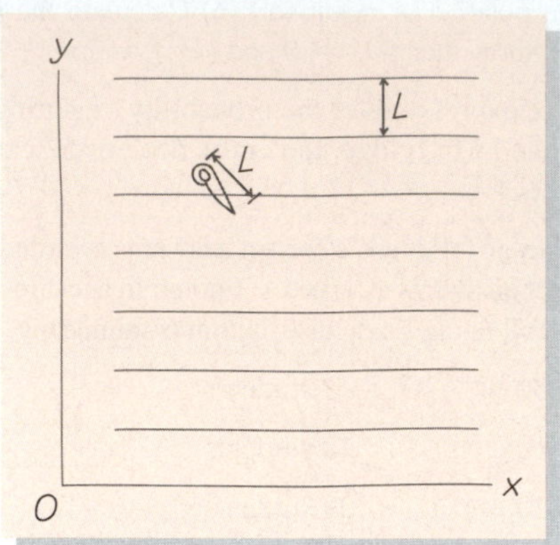

▲ **FIGURE 19–20** Problem 37.

19–5 The Velocity Distribution of Gases

38. (I) What is the rms speed of a helium atom in the vapor above liquid helium at 20K? (The system is enclosed in a perfectly insulating bottle.)

39. (I) Oxygen atoms come in a variety of isotopes: ^{16}O, ^{17}O, and ^{18}O (with the superscript representing A). The latter two isotopes are present only in small quantities. A gas consists of diatomic molecules of oxygen (mostly $^{16}O_2$, with small admixtures of ^{16}O–^{17}O and ^{16}O–^{18}O). Calculate the ratios of the rms speeds for the latter two to that of the dominant (^{16}O–^{16}O) form. Does your answer depend on temperature?

40. (I) If the rms speed of molecules of gaseous H_2O is 200 m/s, what will be the rms speed of CO_2 molecules at the same temperature? Assume that both of these are an ideal gas.

41. (II) Calculate $\langle v \rangle$, the average speed for an ideal gas. In doing this calculation with the help of the velocity distribution $F(\vec{v})$, use the fact that for a spherically symmetric integrand, you may replace d^3v by $4\pi v^2\, dv$. You will need the integral

$$\int_0^\infty z^3 e^{-z^2}\, dz = \tfrac{1}{2}.$$

[*Hint*: This problem illustrates the use of dimensional analysis as an intermediate step.]

42. (II) Use the result of Problem 41 to find the numerical value of the average speed for an ideal gas of CO_2 at 550K, and compare it with v_{rms} at that temperature.

43. (III) Consider a gas being blown along at a velocity $\vec{u}$, so that its velocity distribution is given by

$$F(\vec{v}) = \frac{1}{Z} e^{-m(\vec{v}-\vec{u})^2/2kT}.$$

(a) Show that Z is the same as for the usual velocity distribution, as in Eq. (19–28). (b) Calculate $\langle \vec{v} \rangle$ and $\langle v^2 \rangle$ and interpret your results.

19–6 The Maxwell–Boltzmann Distribution

44. (I) The value of c_p' for carbon monoxide (CO) is 29.2 J/mol · K. What value do you expect for c_V' for the diatomic CO molecule?

45. (I) One mole of hydrogen gas is determined to have an internal energy of 1.87×10^4 J at 900K. How many degrees of freedom are there for the hydrogen molecule? Which degrees of freedom are likely to be available?

46. (I) A gas of N_2 molecules is in equilibrium at a temperature of 450K. Calculate the rms angular momentum $I\sqrt{\omega^2}$ for an N_2 molecule, given that the mass of a nitrogen atom is 2.33×10^{-26} kg and that the average separation of the two atoms in an N_2 molecule is approximately 0.70 nm (Fig. 19–21). (You may assume that the rotational—but not the vibrational—degrees of freedom participate in the equipartition theorem.)

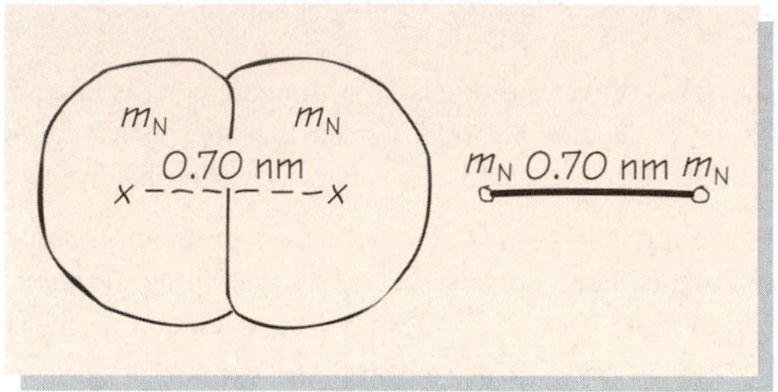

▲ **FIGURE 19–21** Problem 46.

47. (II) The following data can be found in a table of thermodynamic properties of gases:

$$\text{Argon (Ar): } c_p = 0.214 \text{ cal/g}\cdot\text{K};\ c_V = 0.074 \text{ cal/g}\cdot\text{K}$$
$$\text{Oxygen (O}_2\text{): } c_p = 0.219 \text{ cal/g}\cdot\text{K};\ c_V = 0.157 \text{ cal/g}\cdot\text{K}$$
$$\text{Water (H}_2\text{O): } c_p = 0.445 \text{ cal/g}\cdot\text{K};\ c_V = 0.335 \text{ cal/g}\cdot\text{K}$$
$$\text{Carbon dioxide (CO}_2\text{): } c_p = 0.201 \text{ cal/g}\cdot\text{K};$$
$$c_V = 0.156 \text{ cal/g}\cdot\text{K}$$

What can you infer about the shape of the molecules from these data?

48. (II) Leaving out any contributions due to internal motion, what is the average molecular energy for a dilute gas of methane (see Fig. 19–7) at temperature T according to the Maxwell–Boltzmann distribution?

49. (II) A container has 1.3 mol of argon gas and 2.0 mol of nitrogen gas at room temperature. (a) What is the total internal energy of the gas? (b) What is the constant volume specific heat of the mixture?

50. (II) A gas of diatomic molecules is in equilibrium at a temperature of 350K. Assuming that the rotational inertia of the molecule [as used in Eq. (19–36)] is 2.4×10^{-40} kg · m^2, what is the rms angular speed of the molecules? What is the mean angular momentum of the molecules?

51. (III) Write down an integral for the probability that a molecule is found to have a velocity larger than 90 percent of v_{rms}. Show that the probability is independent of the temperature. Show that this is not the case if we ask for a velocity larger than some fixed number; for example, the escape velocity from a planet.

52. (III) A crystal can be thought of as a collection of N atoms arranged in a lattice, with all the atoms connected by springs (Fig. 19–22 see next page). Thus we can think of each atom as a mass at the end of three springs, aligned respectively along the x-, y-, and z-directions. Using this model of the lattice, count the number of degrees of freedom for each atom, then use the equipartition theorem to compute the molar heat capacity of the crystal at temperature T.

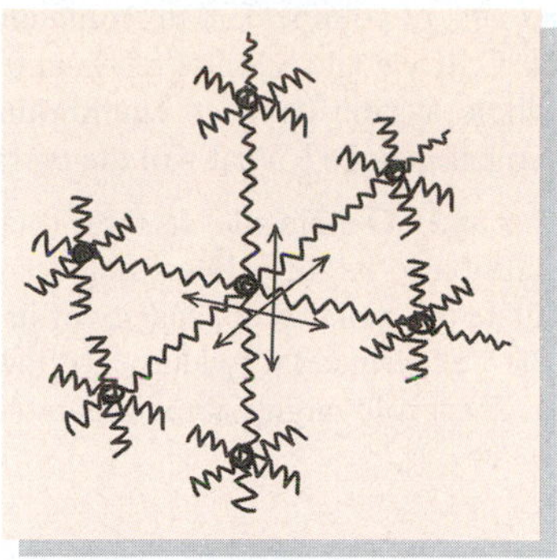

▲ **FIGURE 19–22** Problem 52.

*19–7 Collisions and Transport Phenomena

53. (I) For a collision between two molecules of radii R_1 and R_2, respectively, show that the collision cross section is $\sigma = \pi(R_1 + R_2)^2$ (Fig. 19–23).

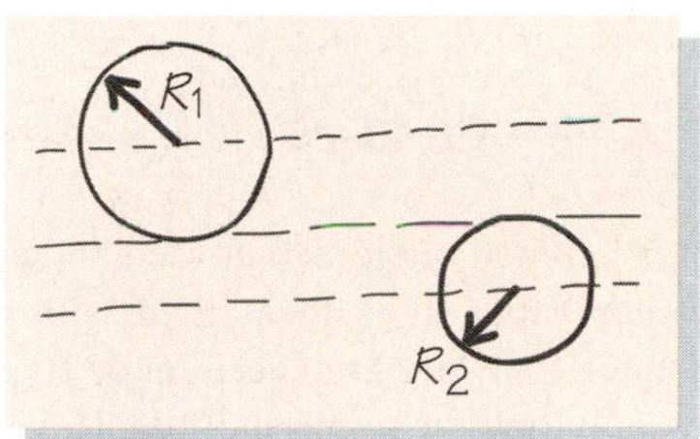

▲ **FIGURE 19–23** Problem 53.

54. (I) Compute the pressure of a container of air molecules of diameter 2.5×10^{-10} m if the temperature is 273K and the mean free path is (a) 1 mm, (b) 1 μm. The gas should be treated as ideal.

55. (II) One can treat the flow of gas through a vacuum hose as if the gas were a fluid if the diameter of the hose is much larger than the mean free path. If, however, the mean free path is much larger than the diameter of the hose, the motion of the individual free molecules has to be considered. Consider a hose of diameter 1 cm. At what pressure is the mean free path of air at 300K shorter than 0.01 cm? Longer than 100 cm? Take the diameter of an air molecule to be 3×10^{-10} m.

56. (II) Using a reasonable value for the collision cross section or, more simply, a model of a gas in which the molecules are like tiny billiard balls, make an estimate of the density at which the mean free path becomes equal to the mean spacing between molecules.

57. (II) Ten tennis balls are rolling around randomly on a tennis court (Fig. 19–24). Estimate the mean free path for collisions.

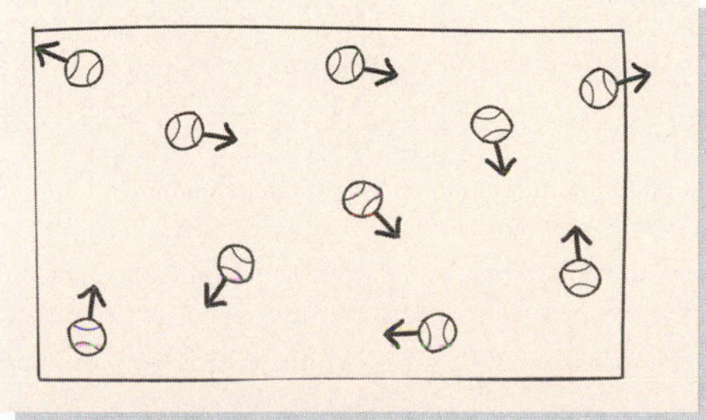

▲ **FIGURE 19–24** Problem 57.

58. (II) A student walks across a field that is 60 m $\times$ 40 m. Estimate the student's mean free path if there are (a) 10, (b) 100, (c) 300 other students on the field.

59. (II) A cubic box of sides 10 cm, contains helium atoms (diameter $\cong 1.0 \times 10^{-10}$ m) at a pressure of 3×10^4 Pa. If the mean free path in the box is 4×10^{-4} m, how many helium atoms are in the box? How many are there if the mean free path is 40 cm? What will the pressure be if the temperature is held fixed?

60. (III) According to one theory, planets are formed as a consequence of near collisions between stars. The number density of stars in our part of the galaxy is 10^{-51} m^{-3}. The velocity of the Sun relative to these stars can be taken to be 25 km/s, and the size of the stars is roughly 10^9 m. If we assume that a passage of two stars at a distance of 100 radii is close enough to set off the tidal forces that give rise to planet formation, what is the probability that the Sun would have acquired a planetary system in 10 billion yr?

General Problems

61. (I) Argon and methane (CH_4) gas are mixed together at 400K. What is the ratio of the rms speeds of the argon and methane atoms? What happens to this ratio if the temperature is halved?

62. (II) The density as well as the temperature of material in intergalactic space is rather uncertain. Suppose that the density and temperature were one hydrogen atom per cubic meter and 3K, respectively. (a) What is the pressure? (b) What is the mean free path if the diameter of a hydrogen atom is 10^{-10} m?

63. (II) The acceleration due to gravity on the surface of Mercury is 3.5 m/s^2, and Mercury's radius is 2.4×10^6 m. (a) What is the escape speed of a particle on Mercury? (b) Suppose that the atmosphere of Mercury were pure H_2 gas. What would the temperature be so that, if the molecules had the velocity distribution given by Eq. (19–28) for that temperature, the rms speed of the H_2 molecules matched the escape speed? Qualitatively, what is the effect on the temperature of the remaining gas? (c) If the temperature were less than the result of part (b), would there be a similar effect, and why or why not? (d) Suppose that Mercury's atmosphere has two or more molecular components. What happens to the composition of the atmosphere over time as a result of the effects discussed here?

64. (II) Some molecules (magnetic dipoles) carry a (vector) property $\vec{\mu}$ called the magnetic dipole moment. When such a dipole is placed in a magnetic field $\vec{B}$ it has an energy $-\vec{\mu} \cdot \vec{B}$. When a collection of N molecules with this property are placed in a box within which there is a magnetic field (meaning that at every point in the box the magnetic field has the magnitude and direction of the given $\vec{B}$), they each have an energy $-\vec{\mu} \cdot \vec{B}$. If the collection of molecules are at temperature T, what is the most likely orientation of $\vec{\mu}$? What is $P(30°)/P(60°)$, where $P(\theta)$ is the probability of finding one of them making an angle θ with the given magnetic field?

65. (II) In the quantum mechanical description of nature, electromagnetic radiation is carried in units called photons. These come with different frequencies. The equilibrium distribution of electromagnetic radiation in a cavity can be thought of as a collection of photons of different frequencies, appearing and disappearing as they are emitted or absorbed by the walls. For a given cavity volume and temperature, the probability of finding n photons of a given frequency f is $P_f(n) = x^n(1 - x)$, where $0 < x < 1$. (a) Is $P_f(n)$ normalized correctly? (b) What is $\langle n \rangle$, the average number of photons of frequency f? (c) What is the fractional spread $\dfrac{\sqrt{\langle n^2 - \langle n \rangle^2 \rangle}}{\langle n \rangle}$? Does this ever go to zero?

66. (II) The surface of the Sun contains both hydrogen and helium atoms at 6000K. (a) What is the average kinetic energy of each type of atom? (b) What is the rms speed of each? (c) Which is most likely to escape from the Sun?

67. (II) The regions in which the beams of accelerated particles circulate in accelerators have high vacuums (10^{-9} torr) to allow the particles to circulate freely. (a) What are the mean free paths of air molecules inside the accelerator at room temperature? (b) By how much does the mean free path decrease if the pressure inside increases to 10^{-6} torr?

68. (II) Suppose that molecules were confined to move in a plane. Calculate the velocity distribution $F(\vec{v})$, assuming that it is of the general form $(1/Z)e^{-mv^2/2kT}$. In other words, calculate Z so that $\int_{\text{surface}} F(v)\, d^2v = 1$.

69. (II) Brownian motion is the random walk followed by individual molecules as well as by larger objects such as smoke particles. The Brownian motion of smoke particles in a room can be observed with the aid of a microscope. If the mass of the smoke particles is typically 3×10^{-15} kg, what is the rms speed of the particles at room temperature?

70. (II) Peter and Paul are standing in opposite corners of a large room, 14 m apart. The air in the room is still. Peter holds a small vial full of H_2S while Paul holds a vial containing Cl_2. They open their vials at the same time (Fig. 19–25). Paul smells H_2S after 3.5 min. When will Peter smell the Cl_2?

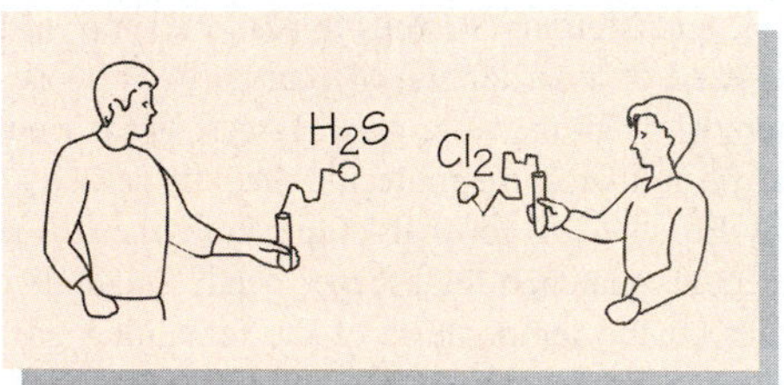

▲ **FIGURE 19–25** Problem 70.

71. (II) Calculate an expression for the pressure of a gas of massless particles, using the following facts: Massless particles always move with the speed of light, so that $v = c$; the energy of a massless particle, E, is related to its momentum magnitude P by $E = Pc$. Use this expression to show that for a gas of massless particles, $pV = U/3$. This result is applicable to the blackbody radiation discussed in Chapter 17.

72. (II) The relation $pV = 2U/3$ is correct only for a monatomic gas. Show that the more general relation is $pV = (\gamma - 1)U$, where $\gamma = c'_p/c'_V$.

73. (II) Consider a tank of volume 0.30 m^3 containing 2.5 mol of helium gas at 20°C. If we add 1.0 mol of O_2 at 0°C to the helium and let the whole system come to equilibrium, what will the equilibrium temperature be? What will the pressure be?

74. (II) Assume that an H_2O molecule has three rotational degrees of freedom in addition to its three linear degrees of freedom (Fig. 19–26). Calculate the molar heat capacity of its vapor. Compare this with the experimental value at low pressures of 0.48 cal/g · K. What role might vibrational motion play?

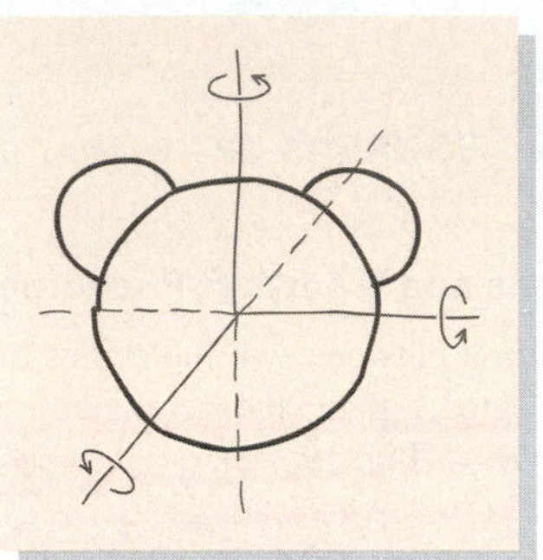

▲ **FIGURE 19–26** Problem 74.

75. (II) A vessel is pumped out to a high vacuum and then heated to get rid of water vapor. It is then filled with dry helium and pumped out again to a very high vacuum of 10^{-11} torr, so that it now contains only helium atoms at 300K. Helium atoms have a diameter of about 10^{-10} m. (a) What is the density of helium atoms? (b) the mean collision time? (c) the mean free path?

76. (II) The *law of partial pressures* states that the total pressure of a mixture of gases is the sum of the pressures that each gas would exert if it were present alone in the same amount as in the mixture. Use the kinetic theory to demonstrate this fact. Could this law be true if the gases were not ideal?

77. (III) The escape speed of a particle from Earth is 11 km/s. Assuming that the oxygen molecules at Earth's surface are at a temperature of 300K, how would you use the velocity distribution to calculate the probability that a particular molecule at the surface would escape Earth's gravitational attraction if the molecule were to suffer no further molecular collisions? Express this probability in the form of an integral, and manipulate the integral into a form that involves the ratio $v_{\text{rms}}/v_{\text{escape}}$, but do not attempt to perform the integration. [*Hint*: This problem is analogous to finding the probability that a given student in a class has a grade of 35 or more on a test.]

Chapter 20

The engines that power racing cars such as these are highly sophisticated, a world away from the first steam engines of the industrial revolution. Yet both types of engines are subject to fundamental limitations on their possible efficiency.

The Second Law of Thermodynamics

The first law of thermodynamics does not specify the direction of thermal-energy transfer. Nevertheless, we know from experience that the spontaneous transfer of thermal energy always proceeds from a hotter system to a cooler one. For example, if you put ice cubes in hot water, the ice cubes melt and the water cools off—nothing would be more surprising than to see the water becoming hotter and the ice cubes colder. This directionality in nature is enshrined in the second law of thermodynamics. The directionality of the spontaneous transfer of thermal energy can be understood from a molecular point of view by using the notions of probability developed in Chapter 19. These notions will allow us to develop the idea of order in a thermal system and to see that a spontaneous process proceeds from a more ordered to a less ordered state.

The second law has an important practical consequence: An engine, which is a device that takes thermal energy and uses it to do work, has a limit to how much of the thermal energy can be used in this way. (This limit is one that goes beyond the limits imposed by the first law of thermodynamics, which states that energy is conserved.) In this chapter we'll explore the meaning of the limit of how much thermal energy can be used to do work. What does this limit have to do with the direction of spontaneous processes? The answer to this question is provided by the concept of entropy, a new state variable for thermal systems, which quantifies the degree of order in thermal systems in thermodynamic terms.

20–1 Beyond Energy Conservation

The first law of thermodynamics extends the conservation of energy to thermal systems. We saw in Chapter 18 that we can construct engines, which are cyclic systems that do work on their surroundings—they may lift weights, or make vehicles go forward—by using thermal energy from thermal reservoirs. (These reservoirs may quite literally be heat baths, or they come in the form of hot gases that have been brought to high temperatures by burning fuel.) However, the first law cannot explain certain facts about thermal systems. For example, it is a fact that it is impossible to devise an engine that extracts all the thermal energy from a thermal reservoir and turns that energy into work. If we had such an engine, we could use it to power New York City—run its subway, light all its buildings, and provide heat in winter—while doing nothing more than cooling the Atlantic Ocean a little. Yet the existence of such an engine is completely consistent with the first law of thermodynamics.

There is a second set of facts about thermal systems that looks superficially different and that is also not explained by the first law. An example is the fact that when ice cubes are put into hot tea, the ice cubes melt and the tea cools. The first law of thermodynamics requires only that the heat flow into the ice cubes match the heat flow out of the hot tea, and heat flow could proceed in either direction. Nevertheless, lukewarm tea never *spontaneously* turns into hot tea with a couple of ice cubes floating in it. Another example is the observation that gas released into an evacuated chamber through a small opening spontaneously spreads throughout the available space—free expansion (Sections 18–1 and 18–7). We never see the gas spontaneously concentrating in a corner of the chamber, leaving most of the chamber evacuated.

These experimental phenomena can be summarized as follows:

1. *Not all the thermal energy in a thermal system is available to do work.*
2. *Thermal systems spontaneously change only in certain ways, and in particular, spontaneous heat flow always goes from a body at higher temperature to a body at lower temperature.*

These facts are expressions of the **second law of thermodynamics**. We will look in more detail at the second law and how it applies to point 1 in the next section, and how it applies to point 2 in Section 20–5. We can nevertheless make a few pertinent remarks that lean on the microscopic picture of thermal behavior, in which molecules undergo random motion. It is this chaos that lies behind the second law. The only way to get all the thermal energy out of a gas in a cylinder would be to get all the molecules moving in the same direction such that they can collide with and lift a weight together. But this is a very unlikely occurrence in a thermal system made of so many constituents. Similarly, the directionality of spontaneous processes has to do with probability—as we shall see, the directionality is determined by what configuration of molecules is the most probable one. Keep this very rough set of ideas in the back of your mind as you go through this chapter.

Engines and Refrigerators

In Sections 18–5 and 18–6 we described engines in terms of a set of thermal transformations that form a cycle. We emphasized that even though over one cycle the thermal system ends up where it starts, net work is done on the surroundings. To do this work, the engine has to get energy from someplace, and this comes in the form of thermal energy extracted from the equivalent of thermal reservoirs. An engine is thus a device that cyclically transforms thermal energy into mechanical energy. For example, in an automobile engine the thermal energy of burning fuel is converted into the energy of the automobile's motion. So is an automobile engine really like the abstract-sounding set of transformations we described in Chapter 18? The answer is yes. You can trace a set of transformations that describe the operation of an automobile engine on a p–V diagram—the engine does consist of cylinders in which gases expand against pistons and so forth. There is one difference between a real automobile engine and the systems

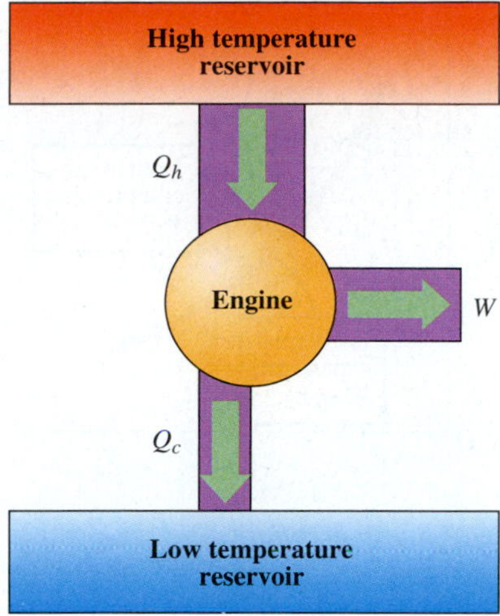

▲ **FIGURE 20–1** Abstract representation of an engine, which takes energy—here represented by a system at high temperature—to perform work. We shall learn in this chapter that it is not possible to convert thermal energy entirely into work, so there is some thermal energy left over, which is "dumped" at cooler temperatures.

we described: The automobile engine, even though it is well lubricated, has friction in it, and there is some irreversibility in its operation—some energy is lost this way. The engines we'll deal with here, at least to start, are idealized in that they are described by a set of reversible transformations, but below we will certainly address the issue of friction.

There is one feature of the cycles that we dealt with in Chapter 18 that we did not emphasize very much: There are parts of the cycle where there is positive heat flow *into* the system, and other parts of the cycle where there is positive heat flow *out of* the system. In every engine the conversion of thermal energy into mechanical energy is accompanied by the emission of exhaust gases, which carry off some thermal energy. This means that only part of the thermal energy of the burning fuel is converted into mechanical energy (Fig. 20–1).

We can restate the points above by recognizing that a useful engine exhibits two key features:

1. *An engine must work in cycles if it is to be useful.* We can understand this best by considering why a *noncyclic* operation is less useful. Consider an ideal gas confined to a cylinder with a movable piston that undergoes isothermal expansion. It is held at constant temperature through contact with a thermal reservoir; for example, it could literally be sitting in a large container of hot water. If there is a mass on top of the upward-oriented piston, the gas does work as it expands. To satisfy the first law, there must be a heat flow from the reservoir to the gas, and this means that thermal energy from the reservoir is converted into mechanical energy. But is this system an engine? It is not, because the process is not cyclic. The volume of the cylinder cannot increase indefinitely, because the pressure, which is inversely proportional to the volume, will decrease until at some point the expanding gas will no longer be able to push the piston. You will have an ineffectual container of gas that takes up a lot of room! To have a functioning engine, *the gas must be recompressed and the piston restored to its original position, ready to do work again.*
2. *A cyclic engine must include more than one thermal reservoir.* To understand this, consider the cylinder and piston again. If the piston were returned to its original position while the cylinder remained in thermal contact with the original thermal reservoir, then all the work that the gas did while it expanded would be needed to recompress the gas. In this scenario the original expansion is reversed, and nothing has been gained. In Chapter 18 we learned that the area enclosed by a closed path on a p–V diagram is the work done during the cycle. In this case the area enclosed is

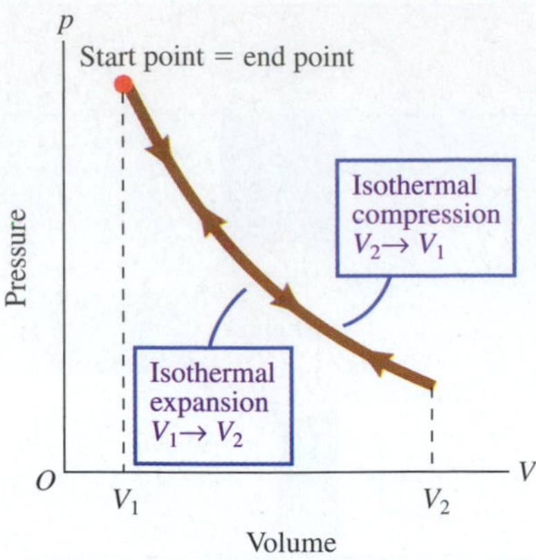

▲ **FIGURE 20–2** A cycle in which a gas expands isothermally and is then compressed isothermally at the same temperature. The work done is zero.

zero, so the net work done by the system is also zero (Fig. 20–2). For finite work to be done over a cycle, thermal contact with the original thermal reservoir must be broken, and *temperatures other than the temperature of the original reservoir must come into play.* Here we would want a second large container of water, this time with cold water, so that the gas is compressed at a temperature that is lower than the temperature at which it expanded. If we do this, the pressure will be lower, and less work will be needed for the compression than was produced in the expansion. You will have done net work, and you will be back where you started, ready to repeat the operation. Note that in the process thermal energy is taken from the hot reservoir and dumped to the colder reservoir. We'll come back to this very important theme later in the chapter.

A young engineer, Sadi Carnot, gave an early but perfectly correct statement of the second law of thermodynamics in 1824. Working during the Industrial Revolution, he was interested in the very practical problem of improving the efficiency of steam engines. He found that there are limits to the possible efficiency of any engine. (It is interesting that his discovery of the second law was made even before the first law was understood!) Accordingly, we now look at engine efficiency.

Efficiency: Engines have an **efficiency**, which is a measure of what fraction of the heat flow from the hotter thermal reservoir (such as a burning fuel) is converted into work (Fig. 20–1). If the work done in one complete cycle is W, then we define the efficiency, η, as the ratio of the work done to the total positive heat flow supplied by the burning fuel, Q_h:

$$\eta \equiv \frac{W}{Q_h}. \tag{20–1}$$

DEFINITION OF ENGINE EFFICIENCY

According to the *first* law of thermodynamics, this efficiency can run from 0 to 1. When the efficiency is 1, or 100%, all the thermal energy taken from the thermal reservoir is converted to mechanical work; this is allowed by the conservation of energy (the first law). We will see, however, that an efficiency of 100% is impossible according to the second law. Figure 20–3 illustrates how the efficiency is calculated. This is an important quantity to be able to calculate, and we'll illustrate it by example in several places below.

Finally let us turn to refrigerators, which at first sight might seem to have nothing to do with engines. In fact, they are very closely related: A refrigerator is an engine run in reverse. Since the path on a p–V diagram of an idealized engine is reversible, we can indeed run it in reverse. Work will be done on the system by the surroundings, and thermal energy taken from a colder reservoir and released to a hotter reservoir. The motor in a refrigerator is even called a compressor; it compresses the gas of the idealized version

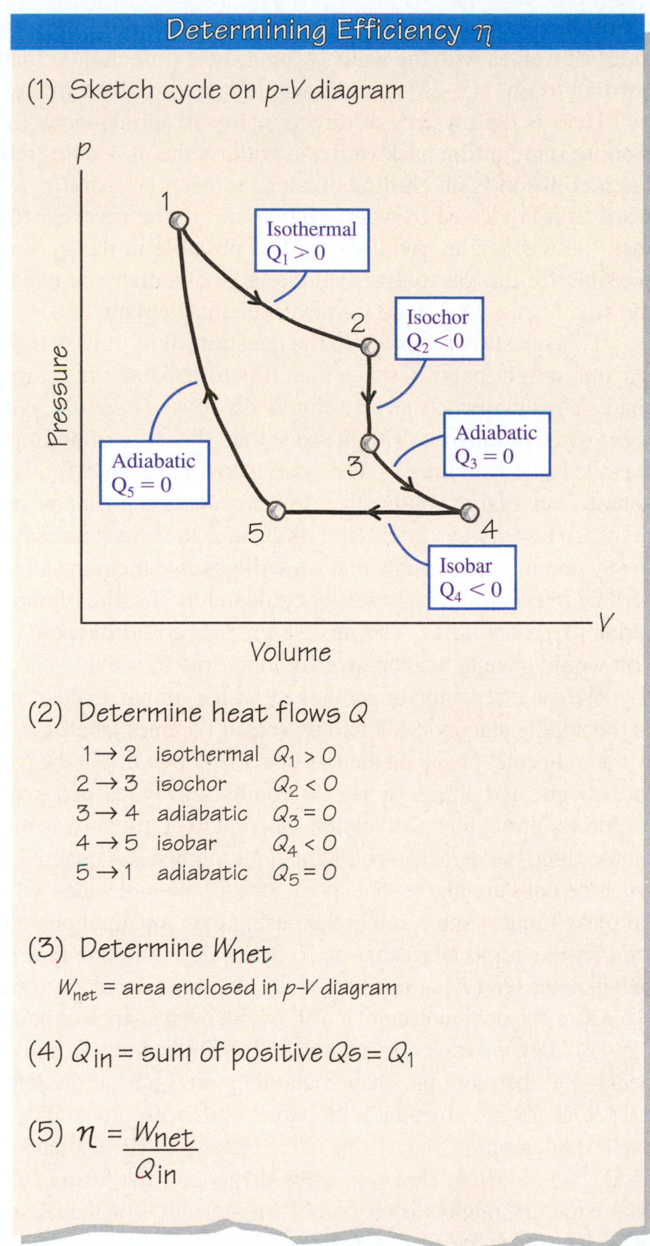

FIGURE 20–3 An illustration of the technique for calculating efficiency.

of the refrigerator and thereby does work on the system. The colder reservoir is the icebox—by removing thermal energy from it, you make it still colder or maintain its temperature when your little brother stands in front of the open door. Where is the hotter reservoir to which thermal energy is added? It is in the coils at the back of the refrigerator, and it ends up warming the air in the room in which the refrigerator sits.

Spontaneous Processes

The directionality of spontaneous processes is illustrated by the two examples we have mentioned—hot tea with ice cubes, and free expansion. Let's concentrate on free expansion, which we can discuss directly in terms of molecules. When a valve is opened and gas escapes from a small tank into a large, empty chamber, the gas spreads and fills the chamber until the density and temperature are uniform throughout that chamber. Now imagine a film of this process made on a scale in which individual molecules can be seen. The molecules move rapidly and collide frequently in the small tank. When the valve is opened, molecules that reach the opening escape. They enter a vast space, empty of molecules, and the first batch of molecules encounter nothing in their way. Their first collisions will be with the walls of the chamber. As more and more molecules escape into the large chamber, they will begin to collide with other molecules as well as with the walls. After a short time their velocities will be distributed according to Eq. (19–28), and the whole system will be in thermal equilibrium.

Here is the mystery of directionality in spontaneous free expansion. Suppose we were to run the film backward and witness the molecules returning to the smaller tank. Each collision is an elastic collision, so each collision is as convincing viewed backward as it is viewed forward; each action in the reversed film is perfectly possible. In fact, the whole film, run backward, is possible in theory, and so in theory it should be possible for the gas to evacuate the large chamber spontaneously and concentrate into the small tank. So why do we never see this happen?

This question is similar to the question, Why it is so much harder to parallel park a car into a tight parking space than it is to leave the parking space and get back on the road? The answer to that question is obvious: *There are many fewer ways of getting it right* when pulling into a tight space than there are of pulling out into a wide boulevard. It is the same with the gas. There are many more ways for the molecules of the gas to be spread over a large volume than to concentrate themselves into a small one. Saying that there are fewer ways for a set of molecules to be in state B than in state A is another way to say that it is improbable that we will observe them in state B. In this way, we have described irreversible processes, or equivalently the direction of spontaneous processes, in terms of probabilities. The reverse process could happen, but it is so improbable that you would have to wait an awfully long time to see it.

We can discuss this quantitatively with a simple model. Consider a box with a partition in the middle that divides it into two equal volumes labeled L (left) and R (right). If initially the molecules are all on the left side, then upon lifting the partition, molecules can move to the right, and after a short time equilibrium is reached, with the molecules distributed uniformly throughout the volume. There will be roughly as many molecules on the left as on the right, but there can be fluctuations that we can quantify. We could ask, for example, what the odds are that at some point 80% of the molecules will be on one side or the other. To make things simple, rather than using the typical number of order of $N_A = 6 \times 10^{23}$, we consider a gas consisting of 10 molecules. Any one molecule has an equal chance of being on the left (L) or on the right (R)—that is, the chance of being in, say, L is $1/2$. Consider now the next molecule: it also has an even chance of being on the left. We now make a *physical assumption*: The molecules are not tied together in any way—they are independent—and therefore the chance that they are both on the left is $(1/2) \times (1/2)$. In fact LL,LR,RL,RR are all equally probable, each with probability $(1/2)^2$. With 10 molecules each configuration, consisting of a total of 10 L's and R's, has a probability of $(1/2)^{10} = 1/1024$. However, every division of the form $(10{,}0)$, $(9{,}1)$, $(8{,}2)$, $(7{,}3)\ldots$, where the first number is the number of molecules on the left and the second is the number of molecules on the right, is not equally likely. That is because the number of ways of getting them is different. Let us work out the number of ways of getting the configurations $(10{,}0)$, $(9{,}1)$, $(8{,}2)$, and so on. The number of ways in which we can have all 10 molecules on the left and none on the right is just one. It consists of the assignment $LLL\ldots L$ (10 terms). If we ask for the number of ways of having 9 on the left and 1 on the right, we can pick any one of the 10 molecules to go to the right, and therefore there are 10 *different* ways of having the division $(9{,}1)$. For the division of $(8{,}2)$ we have 10 ways of picking one molecule to go on the right, and then, since there are only 9 left, 9 ways of picking the second one. That would give us 10×9 ways. However, with this way of figuring things, we have overcounted. We have counted as different possibilities two different orders for choosing the two molecules: M_1 first and M_2 second, and the other way around. This is overcounting because we are interested only in the fact that there are 2 molecules on the right, and not in the order of their arrival. Thus we must divide by 2 for the 2 orders, and the number of ways of getting $(8{,}2)$ is not (10×9) but rather $(10 \times 9)/2 = 45$. For the order $(7{,}3)$, we can pick out the three molecules in sequence: there are 10 ways of picking the first, 9 ways of picking the second, and 8 ways of picking the third. Here again we have overcounting. If we label the molecules with 1,2,3, then they are counted as arriving in all the possible 6 sequences: (123), (132), (231), (213), (312), (321). We must therefore divide by 6, and we get $(10 \times 9 \times 8)/6 = 120$ for the $(7{,}3)$ division. It is not very hard to see what the whole picture yields [Fig. 20–4]:

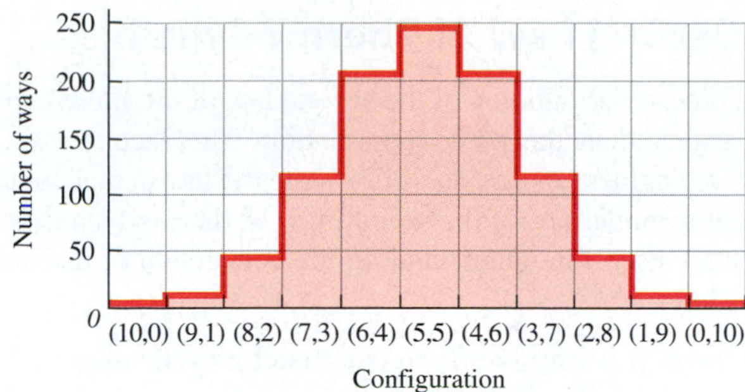

▲ **FIGURE 20–4** The number of ways, and hence the relative probability, that in a collection of 10 molecules the configuration in a box is (# on left, # on right). For example, there are 120 ways to get 7 on the left and 3 on the right. We are assuming that the "choice" of whether a molecule goes left (or right) is independent of the other molecules.

Division	# of ways	Probability P
(10,0) or (0,10)	1	1/1024
(9,1) or (1,9)	10	10/1024
(8,2) or (2,8)	$(10 \times 9)/(2 \times 1) = 45$	45/1024
(7,3) or (3,7)	$(10 \times 9 \times 8)/(3 \times 2 \times 1) = 120$	120/1024
(6,4) or (4,6)	$(10 \times 9 \times 8 \times 7)/(4 \times 3 \times 2 \times 1) = 210$	210/1024
(5,5)	$(10 \times 9 \times 8 \times 7 \times 6)/(5 \times 4 \times 3 \times 2 \times 1) = 252$	252/1024

The probability results above show that with even as few as 10 molecules, the odds of equal division with a fluctuation of 1 [i.e. $(5, 5) + (6, 4) + (4, 6)$] compared with the outer extreme with a fluctuation of 1 [i.e. $(10, 0) + (9, 1) + (0, 10) + (1, 9)$] are in the ratio of $(252 + 210 + 210)$ to $(1 + 10 + 10 + 1) = 672/22 = 30.1$ to 1. There are many more ways of finding nearly equal numbers on both sides than there are ways of finding almost all the molecules on one side or the other. When we get to numbers as large as 10^{24} molecules, this effect is much bigger. In fact, the ratio of the odds becomes more like 10^{12} than 30.1. This means that once equilibrium is established, there is effectively no going back.

Our simple model suggests that the directionality of spontaneous processes is determined by what takes you to the state with the highest probability. There may be subsequent fluctuations around the most probable state, but they are relatively small when the numbers of molecules is high. The irreversibility of a spontaneous process such as free expansion is not guaranteed but it is an almost sure bet.

EXAMPLE 20–1 Three dice are thrown. Assuming honest dice, what is the probability of getting a total of 3? What is the probability of getting a total of 10?

Strategy When it lands, each die gives 6 possible outcomes, and each outcome is equally likely. Thus the probability of any one given outcome on one of the dice is 1/6. Since the three dice are totally independent, any single outcome has probability $(1/6)^3 = 1/216$. (Equivalently, there are 216 different outcomes for the set of three dice.) The difference between the three dice giving a total of three and a total of 10 lies in the number of ways of getting these totals with three dice. The simplest way of getting at these numbers is straight enumeration.

Working It Out There is only one way of getting a total of 3, and that is by the dice reading (111). This means that the probability of getting a total of 3 is 1/216.

Let's now enumerate ways of getting a total of 10: We list the triplets that can add up to 10 as follows: (136), (145), (226), (235), (244), (334). For an entry in this list in which all three digits differ, there are six different ways to get them. For example, if square brackets mean that we are keeping track of the order, we can get (136) as [136], [163], [316], [361], [613], or [631]. When two of the integers are the same, then there are only three ways of getting them—for example, we can get (226), by the throws [226], [262], or [622]. Thus in the list of the number of ways to get 10 we have $6 + 6 + 3 + 6 + 3 + 3 = 27$ different throws. Thus the probability of getting 10 is 27/216, 27 times more probable than the probability of a 3.

What Do You Think? If you deal with the triplet (136), do you really have to list all six possibilities to convince yourself that there are six? *Answers to **What Do You Think?** questions are given in the back of the book.*

20–2 The Second Law of Thermodynamics

Here we'll give a precise formulation of the second law of thermodynamics, using some of the facts we gathered in the preceding section. The second law was discovered through the study of engines, and we shall first state the law in that context. There are in fact two equivalent formulations of the second law of thermodynamics associated with engines, and both are useful in understanding the conversion of thermal energy to mechanical energy. They are:

> **The Kelvin form. It is impossible to construct a cyclic engine that converts thermal energy from a body into an equivalent amount of mechanical work without a further change in its surroundings.**

For example, there is no way to extract thermal energy from the ocean and use that energy to run an electrical generator without further effect on the surroundings of the system, such as a heating of the atmosphere. In the example discussed in Section 20–1 (a gas expanding against a piston without recompression), thermal energy from a thermal reservoir is indeed converted into work, but the engine is not cyclic. In an engine that takes thermal energy from a thermal reservoir cyclically and converts that energy to work, the additional effect is typically a heat flow from the engine into its lower-temperature surroundings, as in point 2 (page 574) of Section 20–1. This brings us to the second formulation of the second law of thermodynamics:

> **The Clausius form.**[†] **It is impossible to construct a cyclic engine whose only effect is to transfer thermal energy from a colder body to a hotter body.**

In other words, there is no cyclic engine that can start with a pot of cold water, make it still colder by taking thermal energy from it, and use that energy to warm up a second pot of water *with no further effect*. In this case, the further effect is associated with the fact that the apparatus *must* take some mechanical energy from its surroundings (such as an electric motor could provide).

We will use mainly the Kelvin form, which was first stated by Lord Kelvin around 1850. Detailed analysis, however, shows that the two formulations of the second law of thermodynamics are entirely equivalent.

THINK ABOUT THIS . . .

HOW DO PERPETUAL MOTION MACHINES WORK?

The search for a *perpetual motion machine* has stimulated thought for centuries. The simplest version of such a machine is an engine that does more work than the energy that goes into it. (Analysis of a seemingly attractive example of such a machine can be found in Problem 16–67.) The discovery of the first law of thermodynamics, and the underlying law of conservation of energy, explained why no such machine (a "perpetual motion machine of the first kind") ever worked. Such knowledge has not stopped "inventors" from trying to violate this law, though the U.S. Patent Office no longer even considers granting patents on such contraptions. Inventors also attempt to construct machines that do not violate energy conservation but have the property that they take energy from a reservoir at a lower temperature and transfer it to one at a higher temperature, without requiring further energy input. Such a machine could take in thermal energy from the ocean and, for free, use it to heat a city on the shore. This is a very attractive idea. Unfortunately, it violates the second law of thermodynamics. It is this law that forbids the construction of this type of machine, also known as a "perpetual motion machine of the second kind."

We can construct wind-power generators that extract energy from the wind and convert it to useful power. The second law is followed in this case because not *all* of the energy in the air flux that hits a windmill can be converted to work in the form of the turning blades. In other words, a windmill does not have a 100% efficiency. Effectively the wind that flows in acts as a hot reservoir, and the outgoing air, which has been stirred up by its passage past the blades, acts as a cold reservoir. The role of such reservoirs in the operation of the second law is described in Section 20–3.

[†] Rudolf Clausius is best known for introducing the concept of entropy.

20–3 The Carnot Cycle

We argued earlier that thermal reservoirs at two (or more) different temperatures are necessary for the operation of a reversible thermodynamic cycle that performs work; that is, a reversible engine. A minimal version of such an engine—one that requires exactly two such reservoirs—at temperatures T_h and T_c, respectively, was invented by Sadi Carnot in 1824. The reversible cycle followed by this engine—known as a **Carnot cycle**—is of great importance in thermodynamics because it is the most efficient engine one can construct that operates between any two temperature extremes. While the Carnot cycle is not used for real engines in common use, some real engines at least approximately match it. By learning the efficiency of the Carnot cycle, we learn the best we can hope to accomplish in engine efficiency.

The Carnot cycle, demonstrated in Fig. 20–5 with a gas-filled cylinder and a piston, consists of four reversible steps. Figure 20–5a is a p–V diagram of these four steps, and Fig. 20–5b depicts the cycle schematically. Step I (AB in Fig. 20–5a) is an isothermal expansion: The expansion takes place with gas in the cylinder in thermal contact with a thermal reservoir—say, a kitchen hot plate—at temperature T_h. In step II (BC) the cylinder of gas has been insulated from thermal contact with its surroundings so that there is no heat flow ($\Delta Q = 0$; the transformation is adiabatic). The volume slowly expands further, while the pressure as well as the temperature of the gas decrease. The temperature of the gas is monitored throughout this adiabatic expansion. When the temperature has dropped to T_c (the temperature of a second, colder, thermal reservoir—for example, a bathtub full of cold water), it is placed in thermal contact with this second reservoir (step III). During this stage (CD), an isothermal compression of the gas at T_c

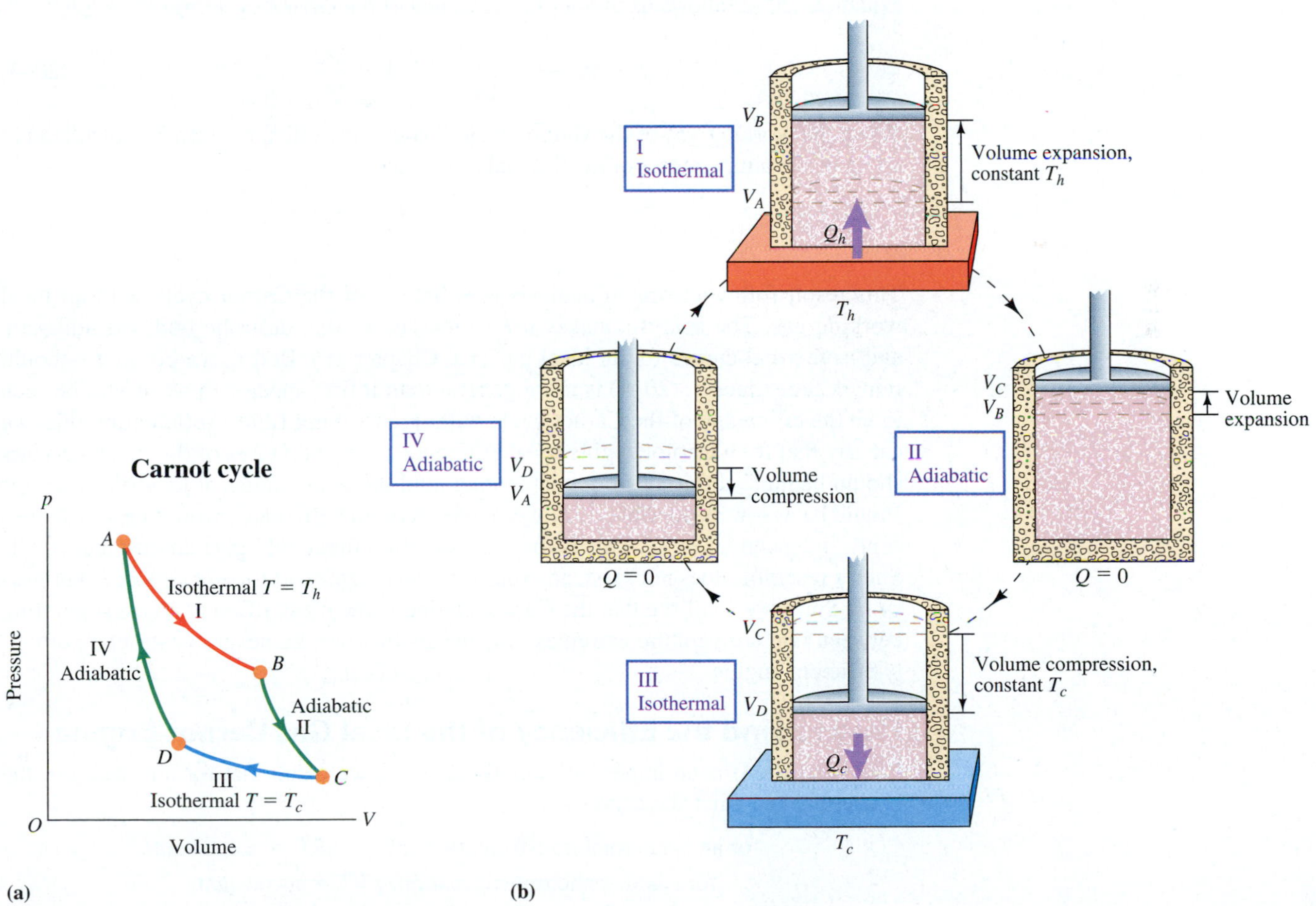

▲ **FIGURE 20–5** (a) The Carnot cycle, which consists of two adiabatic and two isothermal transformations. (b) How the cycle might be realized.

takes place. The final pressure of step III is determined such that when the gas is removed from contact with the bathtub and again placed in thermal isolation, the adiabatic compression of step IV (DA) increases the temperature and pressure of the gas until it returns to the starting point, A.

In Fig. 20–5a the isothermal legs are drawn as hyperbolas, and the adiabatic legs follow the curve $p = \text{(a constant) } V^{-\gamma}$ for $\gamma > 1$—as is appropriate for an ideal gas (see Chapter 18). However, the working fluid does not have to be an ideal gas.† *Any* gas or other compressible fluid is suitable, as isothermal or adiabatic changes are possible for any substance. The crucial features of the Carnot cycle are independent of whether the working fluid is an ideal gas or not.

The work W done *by* the gas in tracing out a complete Carnot cycle is the area enclosed by the curve in the p–V diagram. The cycle shown is clockwise, so W is positive. There is also heat flow associated with the Carnot cycle, although not during steps II and IV, which are by definition adiabatic. During step I, there is a positive heat flow Q_h from the hotter thermal reservoir into the thermal system. (For an ideal gas, $\Delta U = 0$ during an isothermal process, and then $Q_h = W_{A\to B}$, the positive work done by the gas in that expansion.) Similarly, there is a negative heat flow $-Q_c$ from the colder thermal reservoir to the system during step III. In other words, there is a positive heat flow Q_c from the thermal system to the colder reservoir. Thus the Carnot engine absorbs a heat flow Q_h during step I, gives up a heat flow Q_c during step III, and does a total amount of mechanical work W. (This is shown schematically in Fig. 20–1, without making precise the steps.) The first law of thermodynamics can be applied to the entire cycle. Because the engine returns to its original state, the internal energy is unchanged, a statement we could make even if the working fluid is not an ideal gas. Then, according to the first law,

$$0 = W - Q_h + Q_c. \tag{20–2}$$

Equation (20–2) allows us to find the efficiency of the Carnot cycle, $\eta_C = W/Q_h$:

$$\eta_C = \frac{Q_h - Q_c}{Q_h} = 1 - \frac{Q_c}{Q_h}. \tag{20–3}$$

When the working gas of the Carnot cycle is ideal, the efficiency can be calculated in terms of the temperatures of the thermal reservoirs:

$$\eta_C = 1 - \frac{T_c}{T_h}. \tag{20–4}$$

This result follows from an analysis (see below) of the Carnot cycle with an ideal working gas. The analysis makes use of the curves that describe both the adiabatic and isothermal curves for an ideal gas (see Chapter 18). Before we go on we should remark here that Eq. (20–4) is more general than it first appears to be: It will be seen to be the efficiency of the Carnot cycle with *any* working fluid. Anticipating this, we can say that for a Carnot engine to be efficient, the temperatures of the two reservoirs should be quite different from one another, and the temperature of the cold reservoir should be as low as possible. If T_c were to be zero, the efficiency would be 1 (100 percent), independent of T_h. This is the only situation that could give an efficiency of 1. Such a reservoir does not exist, and thus *a perfectly efficient Carnot engine is not possible*. Since we shall see that the Carnot engine is the most efficient engine operating between two temperature extremes, this means that we can never construct a perfectly efficient engine.

How to Find the Efficiency of the Ideal Gas Carnot Engine

Equation (20–4) is an important one. To derive it, we recall that for an ideal gas, the isothermal and adiabatic curves are

$$\text{for an isothermal transformation: } pV = nRT = \text{a constant},$$

$$\text{for an adiabatic transformation: } pV^{\gamma} = \text{a constant},$$

where the quantity $\gamma = C_p/C_V$ is greater than 1.

†The "working fluid" is the material that expands and contracts, allowing the engine to do work.

Refer now to Fig. 20–5a. For an ideal gas the thermal energy is a function of temperature only, so the thermal energy, U, is constant during the isothermal steps (I and III). For these steps, $\Delta U = 0$, and the first law of thermodynamics, $\Delta U = Q - W$, implies that

$$\text{for step I: } Q_h = W_{A\to B},$$

and

$$\text{for step III: } Q_c = -W_{C\to D}.$$

There is a minus sign in the second equation because we have defined Q_c as the heat flow *from* the thermal system to the cold reservoir, whereas in the first law of thermodynamics the heat flow is always *to* the thermal system.

The work done by the gas during the isothermal steps is

$$W_{A\to B} = \int_{V_A}^{V_B} p\,dV = \int_{V_A}^{V_B} \frac{nRT_h}{V}\,dV = nRT_h \ln\left(\frac{V_B}{V_A}\right);$$

$$W_{C\to D} = \int_{V_C}^{V_D} p\,dV = \int_{V_C}^{V_D} \frac{nRT_c}{V}\,dV = nRT_c \ln\left(\frac{V_D}{V_C}\right).$$

Using these results, we find the efficiency of the cycle:

$$\eta_C = 1 - \frac{Q_c}{Q_h} = 1 - \frac{-W_{C\to D}}{W_{A\to B}} = 1 + \frac{nRT_c \ln(V_D/V_C)}{nRT_h \ln(V_B/V_A)}$$

$$= 1 - \frac{T_c \ln(V_C/V_D)}{T_h \ln(V_B/V_A)}. \tag{20–5}$$

We have not yet used any information about the shape of the adiabatic curves in this expression. This information will relate the various volumes. Points B and C lie on the same adiabatic curve, so $p_B V_B^\gamma = p_C V_C^\gamma$. Similarly, for points A and D, $p_A V_A^\gamma = p_D V_D^\gamma$. The ratio of these two equations gives

$$\frac{p_B V_B^\gamma}{p_A V_A^\gamma} = \frac{p_C V_C^\gamma}{p_D V_D^\gamma}. \tag{20–6}$$

Next, we substitute for the ratio of the pressures on each side of this equation. Because A and B lie on an isotherm, $\dfrac{p_B}{p_A} = \dfrac{nRT_h/V_B}{nRT_h/V_A} = \dfrac{V_A}{V_B}$, and similarly, because C and D also lie on an isotherm, $\dfrac{p_C}{p_D} = \dfrac{nRT_c/V_C}{nRT_c/V_D} = \dfrac{V_D}{V_C}$. We insert these ratios into Eq. (20–6) to find that

$$\frac{V_B^{\gamma-1}}{V_A^{\gamma-1}} = \frac{V_C^{\gamma-1}}{V_D^{\gamma-1}}.$$

The same power, $\gamma - 1$, appears on both sides of this equation, so the arguments must be equal:

$$\frac{V_B}{V_A} = \frac{V_C}{V_D}.$$

This result considerably simplifies the expression for efficiency. The arguments of the logarithms in Eq. (20–5) are the same, so the logarithms cancel and we finish with Eq. (20–4).

EXAMPLE 20–2 In a Carnot engine with an ideal gas, calorimetry reveals that the heat flow Q_h from the hot thermal reservoir in one cycle equals 38 J, and the heat flow Q_c to the cold thermal reservoir is 28 J. The temperature of the cold reservoir is 290K. (a) What is the mechanical work done during one cycle? (b) What is the efficiency of the cycle? (c) What is the temperature of the hot reservoir?

Strategy For part (a) we find the mechanical work done during one cycle by finding the area enclosed by the cycle on a p–V diagram. This evaluation can in principle be done by integration, but only with great difficulty because all the parameters of the cycle are not yet known. A much simpler way to compute the work done is to use the first law of thermodynamics, which, for a closed cycle, states that the work done plus the heat flow out of a thermal system equals the heat flow into the system: $W + Q_c = Q_h$, from Eq. (20–2), so that $W = Q_h - Q_c$.

For part (b) we can use the expression for the efficiency in terms of the heat flows, given in Eq. (20–3). Once we have the efficiency, we can solve part (c): We express the efficiency in terms of the temperatures and solve this for the temperature of the hot reservoir, $T_h = \frac{T_c}{1 - \eta}$. Everything on the right hand side of this is known.

Working It Out (a) We have

$$W = Q_h - Q_c = 38\text{ J} - 28\text{ J} = 10\text{ J}.$$

(b) Equation (20–3) expresses the efficiency directly in terms of the given heat flows:

$$\eta = 1 - \frac{Q_c}{Q_h} = 1 - \frac{28\text{ J}}{38\text{ J}} = 0.26.$$

(Remember, this the *most* efficient engine you can have that operates between these two temperature extremes.)

(c) We have

$$T_h = \frac{T_c}{1 - \eta} = \frac{290\text{K}}{1 - 0.26} = 3.9 \times 10^2\text{ K}.$$

What Do You Think? What single parameter change will increase the efficiency of an ideal Carnot engine? If the efficiency cannot change, how could you make sure to get more work done by the gas in the engine?

CONCEPTUAL EXAMPLE 20–3 Describe the operation of a Carnot cycle run in reverse.

Answer Running a Carnot engine—or any reversible engine—in reverse means that the direction of the transformation in each step is reversed, so that the path runs counterclockwise on the p–V diagram. For example, step I in Figure 20–5a becomes an isothermal compression rather than an isothermal expansion. The work done during one cycle is then negative. Positive work must be supplied to the cycle from the exterior to make it run—while heat flow Q_c is taken from the cold reservoir and heat flow Q_h goes to the hot reservoir. We'll see in the subsection on heat pumps and refrigerators below how an engine run in reverse is useful to us.

The Importance of the Carnot Engine

The second law of thermodynamics implies two important results; the second of these explains why any discussion of engine efficiency must start with discussion of the Carnot engine.

1. *All Carnot cycles that operate between the same two temperatures have the same efficiency.* In particular, the efficiency of a Carnot cycle *does not depend on the use of an ideal gas*. The expression for the efficiency of a Carnot engine with an ideal gas, Eq. (20–4), is the expression for a Carnot engine with any working fluid.
2. *The Carnot engine is the most efficient engine possible that operates between any two given temperatures.* (We will demonstrate this in Section 20–4.) This fact is the reason why the Carnot cycle is so important. In a sense it provides us with a target for the best possible engine.

We can demonstrate the first result with a standard method: we show that we would violate the second law of thermodynamics if it were not true. Suppose that we have two Carnot engines, A and B, that operate between the same two temperatures. Further, suppose that somehow the efficiency $\eta_A = W_A/Q_{hA}$ of cycle A were 75 percent, whereas the efficiency $\eta_B = W_B/Q_{hB}$ of cycle B were 74 percent. Let's adjust the amount of working gas in the cycles so that $Q_{hA} = Q_{hB}$, say, 100 J. This does not change the efficiencies. The work done by the two engines, which is the product of efficiency and heat flows, would then be different; namely, $W_A = (75\%)(100\text{ J}) = 75\text{ J}$, whereas $W_B = (74\%)(100\text{ J}) = 74\text{ J}$. Cycle A does more work than cycle B by an amount $\Delta W = W_A - W_B = 1\text{ J}$. We can now connect these two cycles in a way that violates the second law. Like any Carnot engine, engine B is reversible. We connect it to engine A as in Fig. 20–6 and run it in reverse. (In Conceptual Example 20–3 we described what is meant by running an engine in reverse.)

We can divide the 75 J of work done by engine A in Fig. 20–6 into two parts: We use 74 J to run engine B and 1 J is left over. An equivalent amount of heat flow goes into the hot reservoir as is taken out, so this reservoir plays no role. Thus the net effect of this combination engine is to employ only one reservoir to extract 1 J of work with each

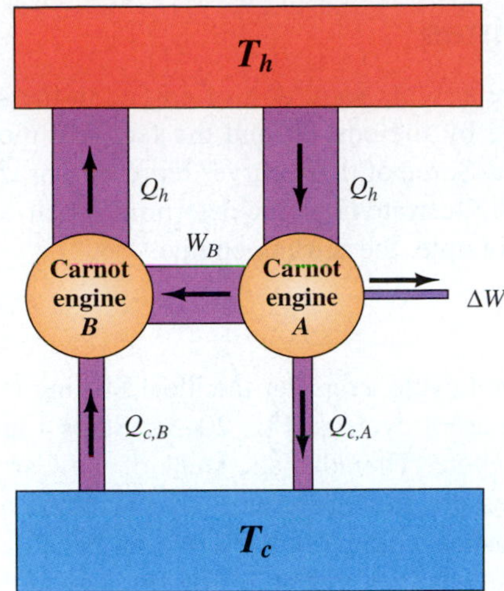

FIGURE 20–6 Carnot engines A and B are connected together, with B run in reverse. If engine A is more efficient than engine B, the net effect of the combination is to circumvent the Kelvin form of the second law of thermodynamics.

cycle. But such an arrangement violates the Kelvin form of the second law of thermodynamics. This result is quite general, and it is a remarkable one because Carnot engines can be constructed with many different working fluids.

We have shown that the efficiency of *any* Carnot cycle that operates between temperatures T_c and T_h is given by Eq. (20–4), whether or not the gas is ideal:

$$\eta_C = 1 - \frac{T_c}{T_h}. \tag{20–7}$$

EFFICIENCY OF ANY CARNOT CYCLE

We have found a universal expression for the efficiency of a Carnot cycle in terms of the temperatures of the thermal reservoirs. Note that the temperature scale here is the ideal-gas temperature scale—the Kelvin scale.

What happens if instead of the Carnot engine, with its idealized reversible cycle, we have an engine with some element of irreversibility? For example, there might be friction between the piston and the cylinder walls. We could reasonably expect that such an engine is less efficient than the Carnot engine that operates between the same two temperatures. We can, in fact, prove that this must be the case (see Problem 19).

CONCEPTUAL EXAMPLE 20–4 Consider a Carnot cycle with an ideal gas. Instead of sketching the steps in the cycle on a p–V diagram, sketch the steps on a U–V diagram, where U is the internal energy.

Answer For an ideal gas we have the relation $U(T) = C_V T$, with C_V independent of temperature, so that, aside from a scale change, plotting the steps on a U–V diagram is equivalent to plotting the steps on a T–V diagram (Fig. 20–7). The first step AB is a horizontal line at T_h extending from V_A to V_B. Similarly, step III is a horizontal line at T_c extending from V_C to V_D. Step II is an adiabatic expansion. This is described by the curve pV^γ = constant. We are able to express p in terms of T and V since we are dealing with an ideal gas. Assuming one mol, we have $p = RT/V$. Thus the adiabatic curve is represented by $TV^{\gamma-1}$ = a constant. In fact, the constant is given by $T_h V_B^{\gamma-1}$. This allows us to determine V_C when the temperature is T_c. Step IV is the same kind of curve. Here $TV^{\gamma-1} = T_h V_A^{\gamma-1}$. This allows us to determine V_D at the lower temperature.

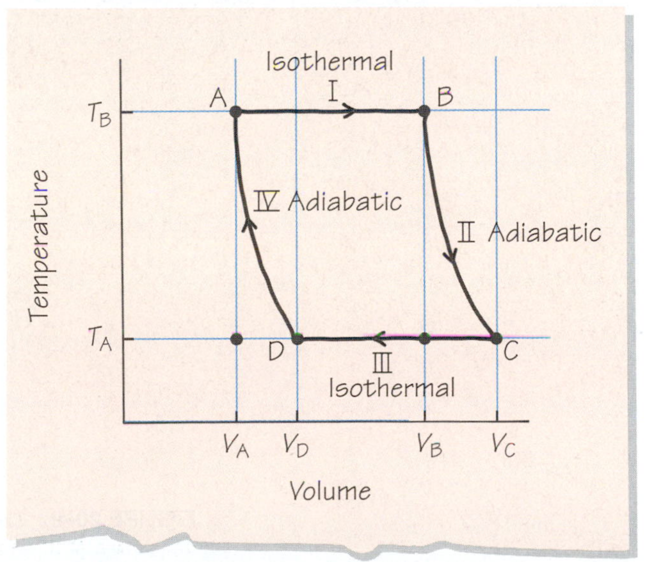

FIGURE 20–7 The Carnot cycle plotted on a T–V diagram.

20–4 Other Types of Engines

Any closed curve in the p–V diagram represents some sort of engine in the sense that the enclosed area represents work done by the engine, and the fact that the curve is closed implies that it can work in a cycle. Some of these curves represent, or at least approximate, real, functioning engines. To illustrate how one determines their efficiency, we consider a particularly interesting example, the Stirling engine.

The Stirling Engine

As an example of a reversible non-Carnot cycle, consider the ideal Stirling cycle (Fig. 20–8), and compare this cycle to the Carnot cycle in Fig. 20–5a. Steps I and III are isothermal steps, just as in the Carnot cycle. The adiabatic steps of the Carnot cycle, however, are replaced by changes in pressure and temperature that maintain a constant volume. No work is done during the constant-volume steps. However, because the temperature changes, the internal energy also changes so that, by the first law of thermodynamics, there is a heat flow to or from the working gas of the engine during the constant-volume steps. This heat flow occurs at decreasing temperatures during step II and at increasing temperatures during step IV. It is as though a series of ever colder thermal reservoirs enter into step II and ever hotter thermal reservoirs enter into step IV. Increasingly, hotter fuel must be used throughout step IV. Under these conditions, we note that the efficiency of an engine is really a measure of the work done for the total fuel burned; thus the denominator of the efficiency should be the entire positive heat flow to the thermal system, which we write as Q_{pos}.

In the subsection immediately below we work out the efficiency of the ideal Stirling engine, η_S, when the working fluid is 1 mol of an ideal gas. We find that

$$\eta_S = \frac{W}{Q_{\text{pos}}} = \frac{R(T_h - T_c)\ln(V_2/V_1)}{RT_h \ln(V_2/V_1) + C_V(T_h - T_c)}. \tag{20–8}$$

This result would be the corresponding Carnot efficiency if it were not for the term proportional to C_V in the denominator (due to step IV). Because this extra term is positive, the denominator of η_S is greater than the denominator of η_C, and thus *the efficiency of the Stirling cycle is less than that of the Carnot cycle*.

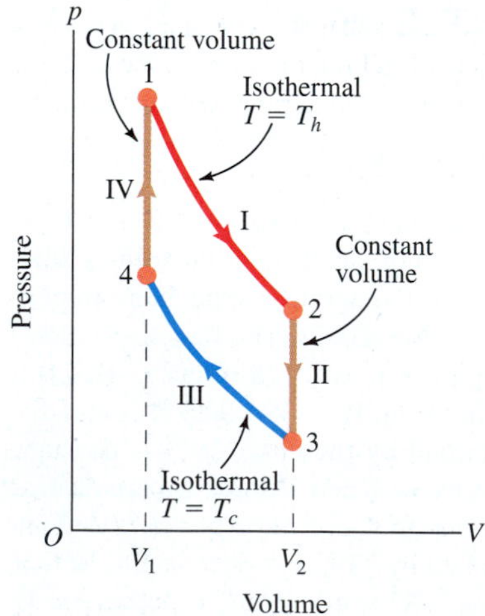

▲ **FIGURE 20–8** The ideal Stirling cycle. There is a positive heat flow to the engine in steps I and IV, and a heat flow from the engine in steps II and III. This cycle was described by Robert Stirling almost 20 years before Carnot discussed his own ideas. The engine Stirling patented in 1816 may be useful in outer space and under other conditions and has become the subject of active engineering work.

How to Get the Efficiency of the Stirling Engine

We assume that the working gas in a Stirling engine is ideal. We need the total positive heat flow $Q_{pos} = Q_I + Q_{IV}$ as well as W, the work done during one cycle (Fig. 20–8). During step I, the temperature is constant and $\Delta U = 0$. Thus, by the first law of thermodynamics, the heat flow to the engine is the work done during the transformation:

$$Q_I = W_I = \int_{V_1}^{V_2} p dV = RT_h \int_{V_1}^{V_2} \frac{dV}{V} = RT_h \ln\left(\frac{V_2}{V_1}\right).$$

According to the first law, the heat flow to the engine during step IV is the change in internal energy $U(T_h) - U(T_c)$ because no work is done by the engine during this portion of the cycle. Thus

$$Q_{IV} = U(T_h) - U(T_c) = C_V \times (T_h - T_c).$$

Combining these two equations, we see that the total heat flow to the engine is

$$Q_{pos} = Q_I + Q_{IV} = RT_h \ln\left(\frac{V_2}{V_1}\right) + C_V \times (T_h - T_c).$$

The total work done in the cycle is the area enclosed by the cycle on the p–V diagram:

$$W = W_I + W_{III} = RT_h \ln\left(\frac{V_2}{V_1}\right) + RT_c \ln\left(\frac{V_1}{V_2}\right) = RT_h \ln\left(\frac{V_2}{V_1}\right) - RT_c \ln\left(\frac{V_2}{V_1}\right)$$
$$= R(T_h - T_c) \ln\left(\frac{V_2}{V_1}\right).$$

The efficiency of this cycle, W/Q_{pos}, is then given by Eq. (20–8).

Other Engines and Their Efficiencies

In addition to the Stirling cycle there are other reversible cycles that represent idealized (friction ignored) versions of real engines. A short list would include the Diesel cycle (Fig. 20–9a), which as you know is used in many different settings; the Otto cycle (Fig. 20–9b), which is very close to an automobile engine; and the Brayton cycle (Fig. 20–9c). Even ignoring friction, these cycles are inherently less efficient than the Carnot cycle for given temperature extremes. With friction, which is always present in real engines, the efficiency is reduced still further. A more detailed treatment of these cycles and their efficiency can be found within the end-of-chapter problems.

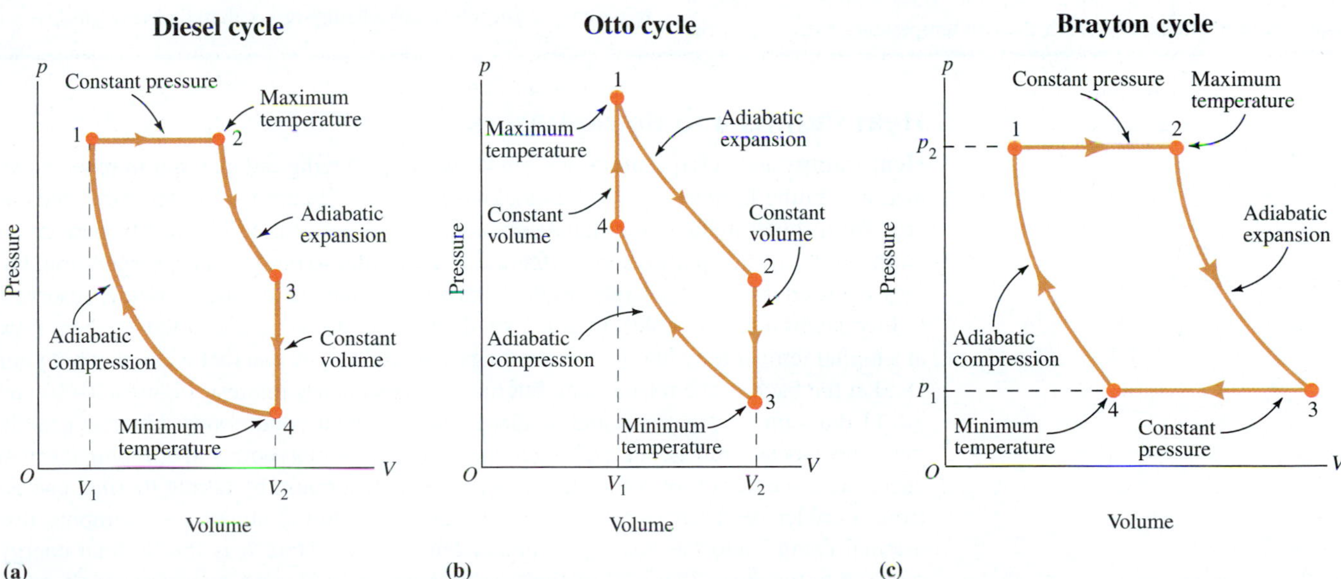

▲ **FIGURE 20–9** (a) The Diesel cycle, for the engine patented by Rudolf Diesel in 1892. (b) The Otto cycle, for the engine designed by Nikolaus Otto in 1876. (c) The Brayton cycle, named for George B. Brayton.

There is a way to decompose an arbitrary reversible cycle into a series of Carnot cycles operating at intermediate temperatures that allows one to prove that any non-Carnot cycle (even if it is reversible) that operates between two temperature extremes T_h and T_c is less efficient than the Carnot cycle that operates between those same temperatures. While we won't work this out in detail, we point out that a Carnot cycle is more efficient if the two temperatures are farther apart. An arbitrary reversible cycle is approximated by Carnot cycles operating at intermediate temperatures and therefore involves a decrease in efficiency.

THINK ABOUT THIS . . .

WHY IS IT SO HARD TO MAKE AUTOMOBILE ENGINES THAT DON'T POLLUTE?

From the end of the 1960s the U.S. government has been requesting that the efficiency of internal combustion engines be improved and air pollution reduced. Automobile engines of that period used an excess of fuel and the residue of unburnt hydrocarbons was a significant pollution source. The pollution problem was alleviated by adding air to the products of combustion as they left the engine's cylinder, thereby burning them. But, how is engine efficiency improved? The easiest way to accomplish this is to increase the maximum temperature at which the fuel is burned. But a higher temperature leads to a secondary reaction: the oxidation of the nitrogen in the added air, and the resulting oxides of nitrogen add very significantly to "acid rain" problems. Thus the government-mandated increase in efficiency resulted in a new pollution problem. The solution to this inherent conflict of efficiency and pollution is a complex one, involving recycled gases and catalysts, and it is not yet completely resolved. This area presents problems whose solution involves physics, chemistry and engineering. In the meantime, although we do have more efficient engines than in recent decades, some of us choose to drive oversized vehicles, and we are back in an era of vehicles that consume excessive amounts of fuel. ■

CONCEPTUAL EXAMPLE 20–5 The efficiency of a certain engine is half that of a Carnot engine operating between the same temperature extremes. Does that mean that there is an element of irreversibility in the working of the engine?

Answer An engine with an efficiency smaller than that of the Carnot cycle can differ from the Carnot cycle in two ways. First, the path traced out in a p–V diagram may not have the form of the Carnot cycle (two isothermal transformations and two adiabatic transformations), even though the path is reversible. These engines involve heat flows at temperatures between the two temperature extremes of the Carnot cycle and are in principle always less efficient than the Carnot cycle. This makes sense if we consider the fact that Carnot cycles themselves are more efficient when the two temperatures between which they work are larger. When one brings in intermediate temperatures, as for these "other" engines, the efficiency drops. Second, real engines generally involve some friction (that is, some irreversibility), and as you might expect this leads to less efficiency. But without knowing more details about the engine we cannot say which of these effects is the reason for the loss of efficiency, so we cannot definitely say whether or not there is irreversibility.

Heat Pumps and Refrigerators

Heat pumps and **refrigerators** are simply work-producing engines run in reverse. An engine running in reverse—counterclockwise on a p–V diagram—transfers thermal energy from colder to hotter thermal reservoirs. For this to occur, work must be performed *on* the system. Heat pumps and refrigerators are similar to one another in operation, but they are used differently. A refrigerator removes thermal energy from a colder reservoir (which keeps a freezer cold). In doing so, the refrigerator is responsible for a heat flow at a higher temperature into its surroundings, something you can feel when you put your hand at the back of the refrigerator, but that is of secondary interest. Figures 20–10 and 20–11 diagram a refrigerator and its close cousin, the air conditioner, which is used in the same way as a refrigerator, taking thermal energy from a room and dumping thermal energy to the exterior. In contrast, a heat pump heats a house by taking thermal energy from a colder reservoir (for example, cool ground below a house) and dumping this thermal energy into the house at a higher temperature. Here it is the thermal energy released to the house that is of interest. In all these cases, it is necessary to supply additional energy through the performance of work on the thermal system.

We can analyze heat pumps or refrigerators by the same method that we used to analyze heat engines. Because of the different ways these devices are used, their

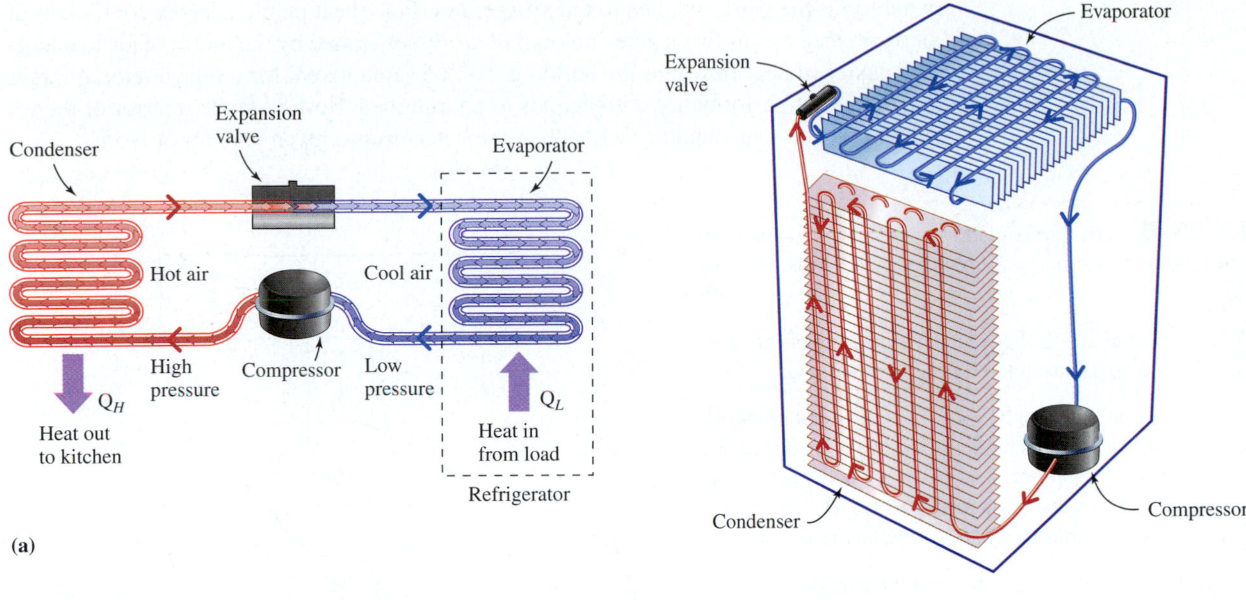

▲ **FIGURE 20–10** (a) Diagram of how a refrigerator operates. A compressor, which is typically operated by electrical energy, does work on the working fluid (freon or an environmentally acceptable substitute) by compressing it, thereby raising its temperature. That thermal energy is released by thermal contact to the exterior of the refrigerator at the coils, and the working fluid is allowed to expand (at the expansion valve) to reduce its temperature. Again thermal contact with the interior of the refrigerator at the evaporator then cools the interior of the refrigerator, and the cycle repeats. (b) The condenser coils are normally on the back of the refrigerator and the compressor coils inside where we want it to be cold.

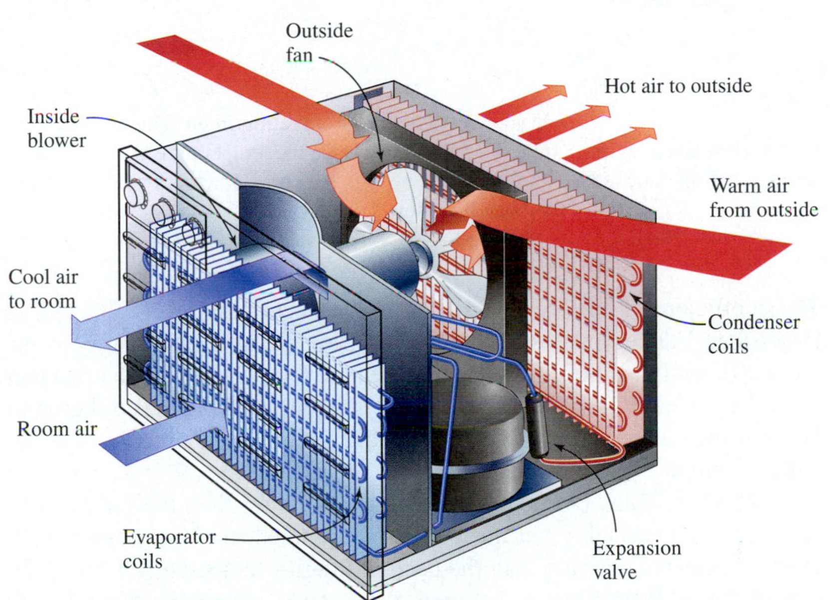

◀ **FIGURE 20–11** Diagram of a room air conditioner, whose operation follows the steps described for the refrigerator.

efficiencies are defined differently. For a heat pump, where the total positive heat flow rejected from the pump into a building, Q_{rej}, is of interest, we use the *coefficient of performance COP*, which is defined as

$$COP \equiv \frac{Q_{\text{rej}}}{W}, \tag{20–9}$$

where W is the total work that must be supplied in one cycle. For a refrigerator, where we are interested in the total positive heat flow absorbed from the interior of the refrigerator, Q_{abs}, the coefficient of performance K_{ref} is defined by

$$K_{\text{ref}} \equiv \frac{Q_{\text{abs}}}{W}, \tag{20–10}$$

where W is the work supplied to the refrigerator. For a heat pump, a larger coefficient of performance means that a given amount of work performed by the motor leads to a larger transfer of heat flow into the building. By the same token, for a refrigerator, a larger coefficient of performance corresponds to a larger heat flow from the interior of the refrigerator to its surroundings when the motor performs a given amount of work.

EXAMPLE 20–6 The Carnot cycle shown in Fig. 20–5a is run in reverse and used as a heat pump. Show that the COP is given by $T_h/(T_h - T_c)$.

Setting It Up Figure 20–12 is a schematic drawing of the operation, with W, Q_c, and Q_h all taken to be positive.

Strategy All formulas given previously for the operation of the Carnot cycle can be used, except that the signs must be changed because the cycle runs in reverse. A figure such as Fig. 20–12 helps us to figure out these signs. It is also necessary to interpret the total heat flow rejected, Q_{rej}, as the heat flow Q_h transferred to the thermal reservoir at T_h.

Working It Out By the first law of thermodynamics,

$$W = Q_h - Q_c,$$

so COP is given by

$$COP = \frac{Q_h}{Q_h - Q_c} = \frac{1}{1 - (Q_c/Q_h)},$$

according to Eq. (20–9). The ratio Q_c/Q_h is, in fact, the ratio of temperatures for any Carnot cycle—see Eq. (20–4) and the accompanying discussion. It is not important that the signs of the heat flows are reversed because we use only the ratio of these quantities. With $Q_c/Q_h = T_c/T_h$, we find

$$COP = \frac{1}{1 - (T_c/T_h)} = \frac{T_h}{T_h - T_c}. \quad (20\text{–}11)$$

What Do You Think? The coefficient of performance, as defined in Eq. (20–10), for a Carnot cycle run as a refrigerator (see Problem 26) is given by

$$K_{ref} = \frac{T_c}{T_h - T_c}. \quad (20\text{–}12)$$

If your refrigerator is operating in an igloo, where the air temperature is only a little above the temperature you want inside the refrigerator, do you need to (a) perform a lot of work, or (b) little work for a given heat flow?

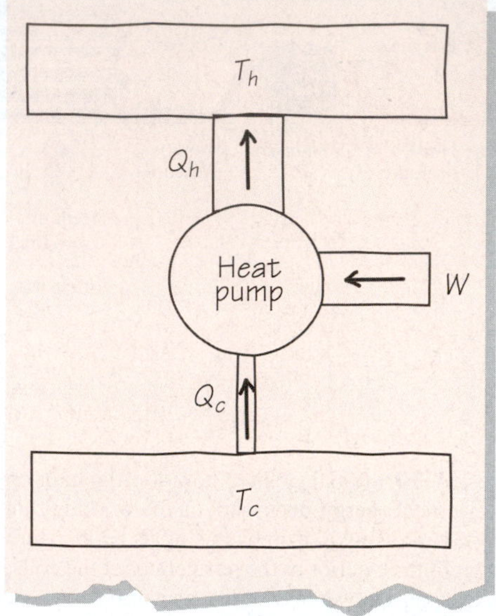

▲ **FIGURE 20–12** The Carnot cycle operated as a heat pump.

The coefficient of performance of the Carnot heat pump, Eq. (20–11), is a number larger than 1 and it increases, or improves (you are using the energy to run the pump more efficiently), if the difference between the temperatures of the two thermal reservoirs is reduced: It is easier to transfer thermal energy from the cold ground to a warm house if the temperature difference is small. A heat pump used for heating a house in winter might operate between the temperatures $T_c = -5°\text{C} = 268\text{K}$ and $T_h = 45°\text{C} = 318\text{K}$. (Although 45°C is much hotter than the interior of a house, if the house is reasonably but not perfectly insulated, the temperature of the exhaust must be somewhat higher than the desired interior temperature.) The COP of an ideal Carnot cycle that operates between these temperatures is then $T_h/(T_h - T_c) = (318\text{K})/(50\text{K}) = 6.36$. In contrast, a commercial heat pump that operates between the temperatures 45°C and −5°C has a COP that is only about half this value.

20–5 Entropy and the Second Law

In Section 20–1 we argued that the direction of spontaneous processes as well as the ability to extract thermal energy and use it to do work had something to do with the random nature of thermal energy and with the relative probability of certain states. Here we will find a way to measure the degree of disorder: the entropy. We will in fact start with a thermodynamic approach and use the Carnot cycle to identify entropy as a new thermodynamic variable. This variable is like internal energy in that it is a function of state. On a macroscopic level, we will see that the entropy is a measure of the energy unavailable to do useful work. By understanding entropy on a microscopic level, we will make the connection back to disorder and the role of probability.

Entropy as a Thermodynamic Variable

In a Carnot cycle, the efficiency, defined by $1 - (Q_c/Q_h)$, is given by Eq. (20–7): $\eta_C = 1 - (T_c/T_h)$. Thus Q_c/Q_h must equal T_c/T_h [see Eqs. (20–3) and (20–4)]. This relation can be expressed as

$$\frac{-Q_c}{T_c} + \frac{Q_h}{T_h} = 0. \tag{20–13}$$

Recall that Q_c is the heat flow from the Carnot engine to the cold thermal reservoir. The expression $-Q_c$ in the first term of Eq. (20–13), minus sign and all, is then the heat flow to the engine at the cold temperature, just as Q_h is the heat flow *to* the engine at the high temperature. The implication of all of this is that for the entire Carnot cycle, made of a series of steps on the p–V diagram,

$$\sum_i \frac{Q_i}{T_i} = 0, \tag{20–14}$$

where the index i labels the steps and where we always regard the heat flow as the heat flow *to* the engine. We will now show that this is a more general property of thermal systems going through reversible transformations, not just a particular property of the Carnot cycle. Our demonstration lacks in generality only in that it assumes that we are operating with an ideal gas. Consider the integral $\int_A^B \frac{dQ}{T}$ between any two points A and B on a p–V diagram. Then, working with one mole of gas,

$$\int_A^B \frac{dQ}{T} = \int_A^B \frac{dU + pdV}{T} = \int_A^B \frac{C_V\,dT}{T} + \int_A^B R\frac{dV}{V}$$
$$= (C_V \ln T_B + R \ln V_B) - (C_V \ln T_A + R \ln V_A). \tag{20–15}$$

In the first step we have used the first law of thermodynamics; to evaluate the temperature integral we have used the fact that C_V is independent of T. Equation (20–15) implies that the value of the integral depends only on the initial and final values of the variables, and not on the path along which the integral is evaluated. This property of the integral in fact holds even if the working fluid is not an ideal gas. This implies that there is a state function S which, like the internal energy U, depends only on the state of the gas, and not on how it reached that state. This state function is called the **entropy** of the system. It is precisely defined by

$$S(B) - S(A) = \int_A^B \frac{dQ}{T}. \tag{20–16}$$

This immediately leads to the generalization of (20–15),

$$\oint \frac{dQ}{T} = 0, \tag{20–17}$$

CLAUSIUS' THEOREM

for a reversible cycle. The circle placed on the integral is a reminder that the integral is taken over a cycle, i.e. a closed path. This result is known as **Clausius' theorem.** A mathematical way of expressing our result is to say that dS, defined by

$$dS = \frac{dQ}{T}, \tag{20–18}$$

is a perfect differential for any reversible transformation: its integral between two points is the difference of a function of each point, so that the integral over a closed path trip, for which the initial and final points are identical, vanishes. This statement also applies to dU but not to dQ.

EXAMPLE 20–7 Calculate the entropy change of 1.0 mol of an ideal gas that undergoes an isothermal transformation from an initial state of pressure 1.50 atm and volume 500 cm^3 to a final state of pressure 0.90 atm. (Being able to find the entropy change in ideal gases will help us later to understand the connection between entropy and chaos on the microscopic scale.)

Setting It Up Figure 20–13 includes the isothermal curve for this transformation and labels the initial and final points i and f as well.

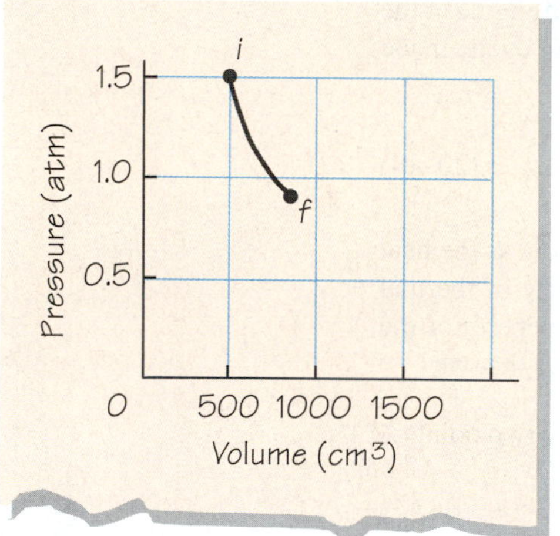

▲ **FIGURE 20–13** An isothermal change of state.

Strategy The specified path is one of constant temperature, so the integral of dQ/T along the specified path is $(1/T)$ times the integral of dQ along the path. By the first law we can write $dQ = dW + dU$, and since for an ideal gas thermal energy is a function of temperature only (that is, $dU = 0$ along this path), we have $dQ = dW = pdV$. For an ideal gas $pV = nRT$ so that $p = nRT/V$ and therefore $dS = \frac{1}{T}dQ = \frac{1}{T}\left(\frac{nRT}{V}dV\right) = nR\frac{dV}{V}$. This can be integrated between the limits. We are lacking the final volume V_f, but since the temperature is fixed, we can use $p_iV_i = p_fV_f$ to replace V_f with known quantities.

Working It Out The change in the entropy ΔS is

$$\Delta S = nR\int_{V_i}^{V_f}\frac{dV}{V} = nR\ln\frac{V_f}{V_i}.$$

Using $V_f/V_i = p_i/p_f$ this reads

$$\Delta S = nR\ln\frac{p_i}{p_f} = (1.0\ \text{mol})(8.3\ \text{J/mol}\cdot\text{K})\ln\frac{1.5\ \text{atm}}{0.90\ \text{atm}} = 4.2\ \text{J/K}.$$

Note that we did not need the value of the initial volume of the gas to get the answer.

What Do You Think? What is the entropy change if the transformation from the initial state to the final state is accomplished in two steps: a first step at constant volume as the pressure drops from 1.50 atm to 0.90 atm, followed by an isobaric transformation with the volume increasing from 500 cm^3 to the final volume of the example?

How Entropy Changes for Irreversible or Spontaneous Processes

Irreversible processes (processes with friction, for example) lead to increased inefficiency in any otherwise reversible engine. We'll now see what friction or other irreversibilities imply for heat flows to such a cycle. We start by looking again at efficiency, which for an arbitrary cycle is

$$\eta = \frac{W}{Q_{\text{abs}}} = 1 - \frac{Q_{\text{rej}}}{Q_{\text{abs}}},$$

where Q_{rej} is the total heat flow rejected by the cycle and Q_{abs} is the total heat flow absorbed. We have used the relation $W = Q_{\text{abs}} - Q_{\text{rej}}$ from the first law of thermodynamics. Each of these heat flows is positive.

The efficiency is reduced if $Q_{\text{rej}}/Q_{\text{abs}}$ gets bigger, and this can happen in two ways. First, reduce the denominator Q_{abs}, the heat flow *to* the system. Second, increase the numerator Q_{rej}, the heat flow *out* of the system. But an algebraic increase of Q_{rej} is the same as an algebraic decrease of $-Q_{\text{rej}}$, and $-Q_{\text{rej}}$ can be interpreted as heat flow *to* the system. Both ways of reducing the efficiency boil down to reducing the heat flow *to* the system. In other words, anything that causes the *algebraic sum* of the heat flows into the system to decrease will make the engine less efficient. But this is exactly the effect of irreversibilities that occur anywhere along the cycle. For example, in a real automobile engine, there is heat flow to the exhaust, but there is additional heat flow out (less heat flow in) due to the heating of the pistons and other parts of the engine. Broadly speaking, all irreversibilities cause the algebraic heat flow to the system to be reduced. More heat flow out is equivalent to less heat flow in. That means less energy to do work, and hence less efficiency.

What is the effect of a reduction in the heat flow to the system on Clausius' theorem, Eq. (20–17)? In that integral, dQ represents the differential flow to the system, so the integral will be be decreased. We can say that *for an irreversible cycle* (one with any element of irreversibility),

$$\oint \frac{dQ}{T} < 0. \tag{20–19}$$

This result is known as **Clausius' inequality**.

Equations (20–17) and (20–19) can be combined into

$$\oint \frac{dQ}{T} \leq 0. \tag{20–20}$$

CLAUSIUS' THEOREM AND INEQUALITY

The equality, Clausius' theorem [Eq. (20–17)], applies for a reversible path; the inequality applies when the transformations include any irreversibility.

Now we can use Equation (20–20) to learn about entropy changes in irreversible transformations. Consider the cycle shown in Fig. 20–14, in which leg $A \rightarrow B$ is irreversible. From Eq. (20–20),

$$0 > \oint \frac{dQ}{T} = \int_{\substack{A \\ \text{irrev}}}^{B} \frac{dQ}{T} + \int_{\substack{B \\ \text{rev}}}^{A} \frac{dQ}{T}.$$

The second term on the right-hand side of this equation is given by $S(A)–S(B)$ because it is taken over a reversible path. When we move this quantity to the left-hand side, we find that

$$S(B) - S(A) > \int_{\substack{A \\ \text{irrev}}}^{B} \frac{dQ}{T}. \tag{20–21}$$

Thus *the difference in entropy between two points is greater than the integral of* dQ/T *over an irreversible change.*

How do we compute the increase in entropy when there is an irreversible transformation from thermodynamic state A to thermodynamic state B? We have seen that entropy is a function of state, so we can compute the change $S(B)–S(A)$ by any means we like; that is, *we can use a reversible path to find the change in entropy as long as that path connects the two given states.* We will illustrate this procedure in Section 20–6.

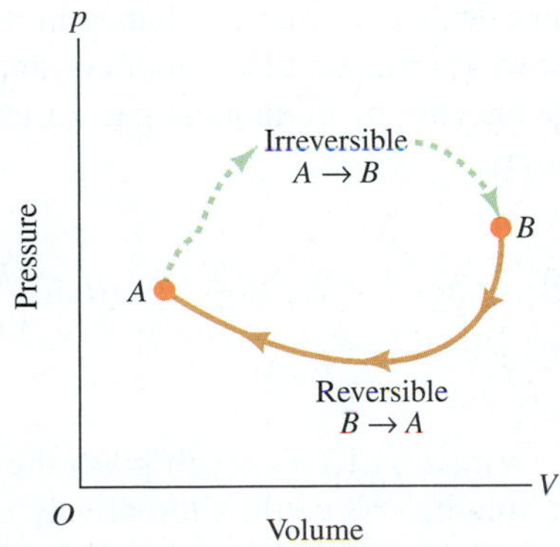

▲ **FIGURE 20–14** A cycle that contains an irreversible path from A to B and a reversible path from B back to A.

The Entropy of an Isolated System Never Decreases

In any reversible process in which there is no heat flow—an adiabatic process—the change in entropy will be zero. If there is no heat flow, but there are irreversible processes, then the entropy increases. That is because with no heat flow, the right-hand side of Eq. (20–21) is zero, and the equation reads $\Delta S > 0$. As an example, consider the free expansion of a gas from a small container into a bigger one, the entire system enclosed in Styrofoam. There is no heat flow into or out of the system, but the entropy increases. More generally a *spontaneous process in an isolated system increases the system's entropy*. Because the universe—our entire surroundings—is in contact with no other system, we say that irreversible processes increase the entropy of the universe. A consequence of this property of entropy is that the state of maximum entropy is a stable one. Once one has arrived at such a state, no further spontaneous changes can occur. (Or, more precisely, no more macroscopic changes occur—there will always be very small fluctuations about the equilibrium state, ones that are invisible on the macroscopic scale.) In other words, for an isolated thermal system *the state of maximum entropy is the state of stable equilibrium*. This is still another way to state the second law of thermodynamics. Below we are going to connect the entropy with the degree of disorder, with the direction of spontaneous processes determined by an increase in disorder. Thus what we have concluded here is that for an isolated thermal system the stable equilibrium state is the state of maximum disorder.

Sometimes you will hear people say that the existence of life is in contradiction with the second law, because living systems take thermal energy from the Sun and turn it into the order of a beautiful flower or a thinking human race that can organize parts of the entire planet. This reasoning is incorrect; a living being is not an isolated system. It is the whole system of Sun plus living creatures on Earth that is better approximated as isolated, and the entropy of this system is indeed increasing.

CONCEPTUAL EXAMPLE 20–8 Suppose you had never heard of engines, and only knew the second law of thermodynamics in the form $\Delta S \geq 0$. Is this enough to show that it is impossible to construct a device that transfers thermal energy from the ocean and converts it into an equivalent amount of work (which could be used, for example, to run the New York subways) without any other effect?

Answer The system in this case is the ocean together with wherever the thermal energy taken from it is used or dispersed. Let's denote the thermal energy taken from the ocean by Q. This implies that the reservoir, the ocean, absorbs thermal energy $-Q$. The entropy change of the ocean is therefore $\Delta S = -Q/T$. This is negative and if it were the only entropy change would violate the second law in the form $\Delta S \geq 0$. There must be entropy change somewhere else in the system, and hence there must be a heat flow somewhere else—the entire heat flow Q cannot be converted to work. The impossibility of taking heat flow and turning it into an equivalent amount of work is the Kelvin form of the second law (Section 20–2).

20–6 Entropy and Ideal Gases

When a gas is ideal, we can calculate its entropy change in reversible transformations. We demonstrated this when we used a combination of constant-volume and isothermal transformations to calculate the entropy of an ideal gas, resulting in Eq. (20–15). That result generalized to n moles is

$$S(B) - S(A) = C_V \ln\frac{T_B}{T_A} + nR\ln\frac{V_B}{V_A}. \tag{20–22}$$

ENTROPY OF AN IDEAL GAS

[We have used $\ln a - \ln b = \ln(a/b)$.] This result gives the entropy change in terms of the temperature and the volume. We could alternatively express it in terms of the pressure and volume by using the equation of state, which allows us to write $T_B/T_A = (p_BV_B)/(p_AV_A)$. If we additionally use $\ln(p_BV_B/p_AV_A) = \ln(p_B/p_A) + \ln(V_B/V_A)$, we find

$$S(B) - S(A) = C_V \ln\frac{p_B}{p_A} + (nR + C_V)\ln\frac{V_B}{V_A} = C_V \ln\frac{p_B}{p_A} + C_p \ln\frac{V_B}{V_A}, \qquad (20\text{–}23)$$

where we recall that $C_p = C_V + nR$ [Eq. (18–33)].

We used reversible paths to calculate the change in entropy. But because entropy is a state function, the same result obtains for *any* transformation, reversible or irreversible between the same states. In particular, we may use Eq. (20–22) to calculate the change of entropy in a free expansion of a gas at a fixed temperature, an irreversible process. Equations (20–22) and (20–23) can be used for any process whose end points can be put on any graph, be it a p–V plot, a T–V plot, or a p–T plot.

CONCEPTUAL EXAMPLE 20–9 (a) Plot the steps of a Carnot cycle on an S–T plot. (b) What is the significance of the area enclosed by the closed path thus formed in the S–T plane?

Answer (a) We shall follow the steps ABCD as in Fig. 20–5a. The S–T graph is shown as Fig. 20–15. The first step, AB, is an isothermal transformation. With our choice of vertical T axis, this is a horizontal line at $T = T_h$. The step BC is adiabatic. This means that there is no heat flow into the system and therefore the entropy does not change, and so the step is represented by a vertical line corresponding to a constant value of S—call it S_B. The next step is again isothermal so that the transformation CD is a horizontal line in the direction of decreasing S at the horizontal level of T_c. The last step, DA, is again *isentropic* (no change in entropy), so that it is represented by a vertical line at S_A.

(b) Although it is easy to find the area directly—it is just $(T_h - T_c) \times (S_B - S_A)$—the interpretation follows from an integral representation. The area under the upper horizontal line is equal to $\int_A^B T\,dS = \int_A^B T\frac{dQ}{T} = \int_A^B dQ$. The integral is taken along the T_h isotherm, and the result is the total heat flow from the higher temperature reservoir, that is Q_h. From this we must subtract the same integral along the T_c isotherm, and by the same token, the result is Q_c. The total area is therefore $Q_h - Q_c = W$, the work done. This is the same result that one finds by calculating the enclosed area in the p–V plane.

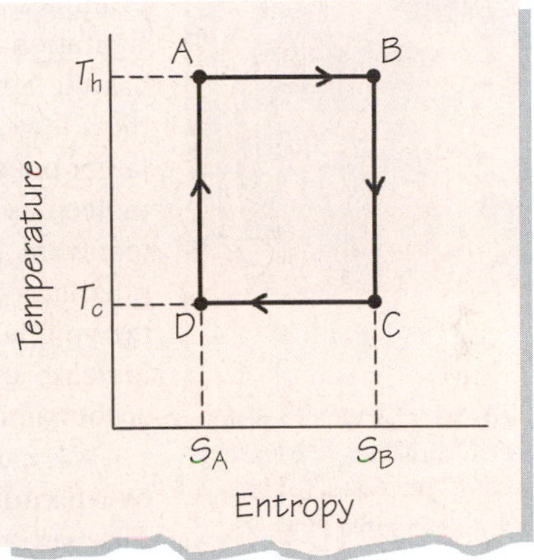

▲ **FIGURE 20–15** The Carnot cycle on an S–T plot. The area of the cycle is W, the work done.

EXAMPLE 20–10 You have 3.0 mol of an ideal gas in thermal contact with a thermal reservoir at $T = 300\text{K}$. The gas undergoes free expansion (no work is done) from 12 L to 24 L. What is the entropy change of the gas and of the universe? Repeat the exercise for an expansion from 12 L to 120 L.

Strategy The entropy change of the gas, being a state function, can be computed when the end states are known, as they are here, even if the transformation is irreversible. In this case, we can use Eq. (20–22) with fixed T. Because $\ln(1) = 0$, only the ratio of volumes is involved. What about the entropy change of the universe? To determine the entropy change in the rest of the surroundings, we must determine the heat flow to the thermal reservoir. The gas does no work in its expansion, so the heat flow out is equal to the change in internal energy. But this change is zero because the gas is ideal and because its internal energy depends only on the (unchanging) temperature. There is thus no heat flow to the thermal reservoir, and its entropy change, $\Delta Q/T$, is zero. The entropy change of the universe is therefore the entropy change of the gas. The fact that it is positive indicates that the transformation involved—in this case, free expansion—is irreversible.

Working It Out From Eq. (20–22) with $T_B = T_A$, we find $\Delta S = nR\ln(V_B/V_A)$, so that numerically $\Delta S = (3.0\text{ mol})(8.3\text{ J/mol}\cdot\text{K})\ln[(24\text{ L})/(12\text{ L})] = (25\text{ J/K})\ln 2 = 17\text{ J/K}$. The entropy change of the gas, which is also the entropy change of the universe, is positive. (The numerical result would have been the same had the question dealt with an isothermal leg of a Carnot cycle. In that case there would be a compensating $-\Delta Q/T$ contribution of entropy to the reservoir, so that there would be no net change to the entropy of the entire system of gas plus reservoir, as would be appropriate to a reversible transformation in a completely isolated system.)

When the expansion is from 12 L to 120 L, the entropy change is as just described, except that ln(2) is replaced by ln(10), and $\Delta S = 25\ln(10)\text{ J/K} = 58\text{ J/K}$. The natural logarithm changes fairly slowly with its argument, so, compared to the ratio of volume changes in the two cases, the difference in entropy changes is not very great.

What Do You Think? Suppose you had two containers, each of volume V. One contains one mole of CO_2 and the other contains one mole of N_2O. Suppose you now connect the two containers by a tube, open the stopcock on the tube, and let the gases mix. Do you think the entropy will increase or stay the same?

(a)

(b)

▲ **FIGURE 20–16** (a) When a plate breaks, it is more disordered—the entropy of the "plate system" increases. (b) In urban renewal, the demolition of an abandoned building raises its entropy very quickly!

20–7 The Meaning of Entropy

The entropy of a system and all its surroundings—the universe—increases when an irreversible process takes place. In effect, increased entropy corresponds to a lost heat flow, and hence a lost ability to do work. The loss of heat flow represents energy that can never be turned into work because, by the nature of irreversible processes, the process cannot be reversed to recapture this heat flow without making other changes. Of course, the energy is not really lost; energy is always conserved. But it has been turned into the random motion of the microscopic constituents of the thermal system and cannot be sufficiently well "organized" to allow all of it to go into, say, lifting a weight. By organization we mean order; if all the molecules were moving in precisely the same direction, for example, they would be well ordered. They might have the same total energy as a disorganized gas, but *all* of their energy could be used to lift a weight. Any degree of irreversibility reduces our ability to do work because it increases the disorder. Let's consider this idea more closely.

We began our discussion of spontaneous processes in Section 20–1 with two examples. In one, a gas expands into an empty space, its molecules move from a configuration that has a low probability (all the molecules are in one corner) to one that has the largest possible probability (the molecules are spread uniformly throughout the allowed volume). Initially, all the molecules are confined in a small box inside a larger one. This initial state is more *ordered* than the final state because the individual molecules are more localized. A loss of order is associated with a change of the system from a state of low probability to one of higher probability. Equation (20–22) applied at a fixed temperature illustrates how entropy increases for an ideal gas when the volume increases. From this perspective entropy is a measure of disorder, and the increase of the entropy in an irreversible process is the increase in disorder that accompanies such processes (Fig. 20–16).

We can illustrate this idea quite concretely by mixing two gases initially separated by a partition (Fig. 20–17). Each gas starts out with the same temperature and pressure. The partition is opened, and the gases in the two parts mix. Before the partition is opened, the degree of order is greater: All the molecules of each gas are on their respective sides. After the partition is opened, the disorder is greater: Any one molecule is somewhere in the greater volume, rather than being on one side or the other. It is more probable that each gas will spread separately throughout the greater volume than that it will stay on its respective side. Thus, we would expect the entropy to increase when the partition is opened, which is in line with the idea that the mixing of the two sides is an irreversible process. This increased entropy is the **entropy of mixing**, illustrated quantitatively in the example below.

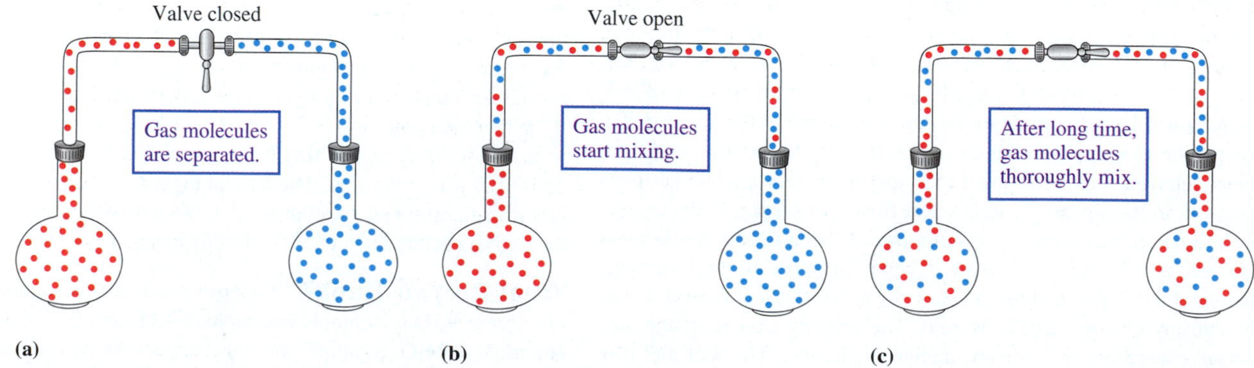

▲ **FIGURE 20–17** (a) Two gases are initially unmixed. (b) A valve is opened, and each gas diffuses into the other's flask. (c) The final state is more disordered than the initial state. The system's entropy has increased.

EXAMPLE 20–11 A quantity of 0.20 mol of argon gas contained in a volume of 5.0 L mixes with 0.50 mol of neon gas contained in a volume of 12.5 L, making a total volume of 17.5 L. Both gases are at the same temperature and can be regarded as ideal. Each volume is thermally isolated. What is the change in entropy?

Setting It Up We draw the situation in Fig. 20–18, showing the situation before and after mixing.

Strategy The gases start at the same temperature, and they will be at that same temperature after expansion. Only their volumes will have changed. Each will then have an entropy change given by Eq. (20–22) with the temperature term absent, $\Delta S = nR\ln(V_f/V_i)$. This applies for the two gases separately, and to find the total entropy change we add the results.

Working It Out We have

$$\Delta S_{\mathrm{Ar}} = (0.20\ \mathrm{mol})(8.3\ \mathrm{J/mol\cdot K})\ln[(17.5\ \mathrm{L})/(5.0\ \mathrm{L})]$$
$$= (1.7\ \mathrm{J/K})\ln 3.5 = 2.1\ \mathrm{J/K}$$

and

$$\Delta S_{\mathrm{Ne}} = (0.50\ \mathrm{mol})(8.3\ \mathrm{J/mol\cdot K})\ln[(17.5\ \mathrm{L})/(12.5\ \mathrm{L})]$$
$$= (4.2\ \mathrm{J/K})\ln 1.4 = 1.4\ \mathrm{J/K}$$

The entropy of mixing is the sum of the increases of entropy of the two gases, so that

$$\Delta S = \Delta S_{\mathrm{Ar}} + \Delta S_{\mathrm{Ne}} = 3.5\ \mathrm{J/K}.$$

What Do You Think? Is it possible to arrange things so that the two gases, originally in the separate containers described above, spontaneously pass into a common container whose volume is 3.0 L?

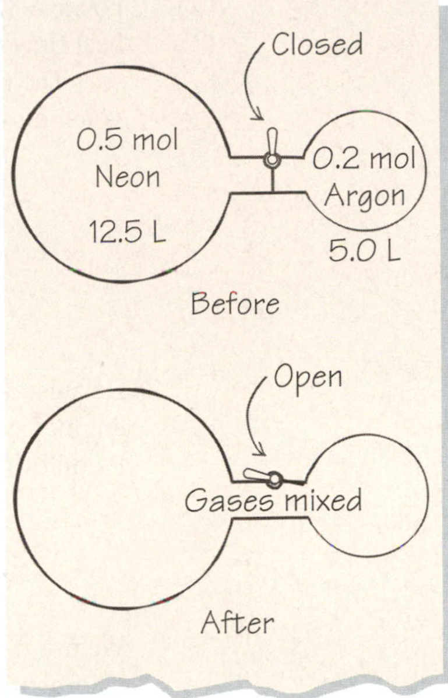

▲ **FIGURE 20–18** The neon and argon gases mix when the valve is opened. What happens to the entropy?

The second example of a spontaneous process described in Section 20–1 was the fact that ice cubes melt when they are placed in hot tea. We also interpret this case, and more generally the fact that systems with different temperatures tend to the same intermediate temperature when they are placed in thermal contact in terms of an increase in disorder. In this case, the more ordered system is the cooler one. On average, the velocity vectors of the molecules of steam at a lower temperature are confined to a smaller range than are those of steam at a higher temperature. The heated system is more disordered because, on average, the velocity vectors spread over a greater range. Equation (20–22), this time applied for changing temperature but constant volume, describes how entropy increases for an ideal gas when temperature changes. More advanced treatments of statistical physics quantify the idea that entropy is a measure of disorder. We have already described how the thermodynamic approach connects entropy change to the second law. Here we claim that entropy change is associated with an increase in disorder. We draw the conclusion that the second law follows from the interpretation of entropy as disorder. In fact, a more rigorous treatment than we can perform here confirms this: *All the consequences of the second law of thermodynamics follow from the treatment of entropy as a measure of disorder.*

Engines and Entropy

The principle that entropy never decreases is a way of formulating the second law of thermodynamics that emphasizes the directionality of thermal processes. The application of this formulation to engines is less obvious because the entropy in a reversible cycle does not change. Here (see also Conceptual Example 20–8) we show how the entropy formulation of the second law applies to engines and leads to our earlier formulations of the second law.

Consider a thermal reservoir at temperature T_h. Can we construct an engine that, in a single cycle, extracts thermal energy through a (positive) heat flow Q_h from that reservoir to lift a weight and does nothing else? Let's suppose that we could do so. There is no change in the entropy of the weight that has been lifted and there is no change in the engine after a full cycle. Thus, in the isolated system that consists of the weight, the engine, and the reservoir, the only entropy change is that of the reservoir, as given by $\Delta S = -Q_h/T_h$. This entropy change is negative and therefore violates the entropy formulation of the second law. We deduce that in order to satisfy the entropy formulation of the second law, there must be a second thermal reservoir (at temperature T_c) such that the net entropy change is positive. If this second reservoir is to have a positive entropy change, then there must be a heat flow Q_c into it. This gives rise to an entropy change $\Delta S' = Q_c/T_c$.

The isolated system, which now consists of the weight, the engine, and the two reservoirs, must satisfy the entropy formulation of the second law, namely $\Delta S + \Delta S' \geq 0$, or

$$-\frac{Q_h}{T_h} + \frac{Q_c}{T_c} \geq 0;$$

$$\frac{Q_c}{Q_h} \geq \frac{T_c}{T_h}.$$

Because the work done to lift the weight is $W = Q_h - Q_c$ and the efficiency of the engine is defined as $\eta \equiv W/Q_h = (Q_h - Q_c)/Q_h = 1 - (Q_c/Q_h)$, the entropy formulation of the second law leads to

$$\eta \leq 1 - \frac{T_c}{T_h}.$$

We are now on familiar ground. The equality applies for the reversible Carnot cycle, where there is no heat flow except at the two temperature extremes. In this case, ΔS_{tot} is zero, which requires that $(Q_c/Q_h) = (T_c/T_h)$, and the efficiency is $\eta = 1 - (T_c/T_h)$. For cycles with irreversibility, or for cycles such as the Stirling cycle in which there is heat flow at intermediate temperatures, the inequality sign holds. Thus, the entropy formulation of the second law implies that the Carnot engine is the most efficient type of engine.

We have now connected the two apparently dissimilar aspects of the second law of thermodynamics: the impossibility of constructing a perfectly efficient engine and the inevitable direction of spontaneous processes.

Order, Entropy, Probability, and the Arrow of Time

The title of this subsection carries all the catchwords that are associated with the second law of thermodynamics. Entropy as a state function was introduced through our study of engines, and the quantitative definition as given in Eq. (20–16) allowed us to learn how it changed in irreversible processes. There are other irreversible processes that do not involve thermodynamics. An example is a deck of cards arranged in order [4 aces, 4 kings, 4 queens, . . . 4 threes and 4 twos]. We would call this an arrangement that has a great deal of *order* in it. Shuffle the deck. The order will be lost. Shuffle it over and over again, and different arrangements will appear, but order, as defined by the initial arrangement, will not reappear. We may say that the process of shuffling a deck leads to an *irreversible* loss of order. Can we apply the idea of entropy to such a process? One would expect that it should be possible, because the shuffling of the deck of cards bears some resemblance to what happens to molecules initially confined to a small box. When the lid of the small box is opened and the molecules are allowed to escape into a larger box that contains the small one, the system becomes much more disorderly. One would have a harder time finding a particular molecule in the larger box than in the confines of a smaller one. Furthermore, the process is irreversible: It is extremely improbable that all the molecules would go back into the small box. Common sense tells us that if the volume of the small box is 1/10 of the large one, then roughly 10% of the molecules might be found in the small one, but never 100% of them. What do the cards and the molecules have in common? In this context it is the notion of probability. We already saw in Chapter 19 that the kinetic theory of gases relies on probabilities through distributions which allow us to calculate the temperature as the average kinetic energy of molecules and relate pressure to the average momentum transfer to the walls of a container.

In fact, Boltzmann discovered a deep relationship between entropy and probability. We can give a hint of how this relationship works out by considering the very simple model of the small open box inside a large box, with molecules in the large box. If the small box has volume V_A and the large box containing it has volume V_B, the chance of a particular molecule being in the small box is V_A/V_B. A second molecule, independent of the first one, will have the same chance V_A/V_B of being found in the small box. This means that the chance of both of them being in the small box is $(V_A/V_B)^2$. With three molecules, the chance of all three being in the small box is $(V_A/V_B)^3$ and so on. Thus with n moles of gas, nN_A molecules, the probability of all of them being in the small box is $(V_A/V_B)^{nN_A}$. We are using the fact that probabilities for independent events multiply. But as we saw in Example 20–11, entropies add. How can we bring these facts together? If we have the relation $P_{12} = P_1 \times P_2$, then we can make this into an additive relation by taking logarithms of both sides: $\ln(P_{12}) = \ln(P_1) + \ln(P_2)$.

Apply this to the full set of molecules. The natural logarithm of $P = (V_A/V_B)^{nN_A}$ yields

$$\ln(P) = nN_A \ln(V_A/V_B).$$

Let us compare this with the change in entropy [Eq. (20–22)] in letting an ideal gas escape spontaneously (that is, without a change in temperature) from a volume V_A into a larger volume V_B. We found

$$S(B) - S(A) = nR \ln(V_B/V_A).$$

Taking into account that $R = kN_A$, where k is Boltzmann's constant and comparing the entropy calculation with the logarithm of the probability, we have the relation

$$\Delta S = k \ln(1/P) = -k \ln P. \tag{20–24}$$

This relation between entropy and probability is in fact quite general, and it allows us to extend the notion of entropy to all processes in which irreversible changes occur.

The *arrow of time* is the name given to the universal appearance of directionality in processes in the world. This directionality is associated with the higher probability of a disordered state relative to an ordered state. Figure 20–16 dramatizes this directionality. A few boxes of TNT and an abandoned building represent a much more ordered state than the one that appears after the explosion. As we mentioned when we first discussed spontaneous processes, the indications of an arrow of time all around us is not in contradiction with the reversibility of the fundamental laws of nature at a microscopic level.

The elementary forces between the microscopic constituents of matter are, with one very tiny exception discussed in Chapter 45, indifferent to the direction of time, or *invariant under time reversal*. In other words, aside from the exception, this invariance applies to all the fundamental forces in nature, including the electromagnetic forces that govern the collisions of gas molecules. Nevertheless, we have seen that there is a preferred direction of time—an **arrow of time**—in which dropped eggs break rather than reassemble and leap up. Spontaneous processes, in which entropy increases, proceed in one direction and not the other.

We have learned that the reason is statistical. It is not against any physical laws for a broken egg to reassemble itself. It is just so unlikely that it would never happen in many lifetimes of the universe. The overwhelming majority of collisions between gas molecules lead to more disordered situations. The arrow of time expresses not a definite preferred direction of time in a fundamental sense but an overwhelming likelihood that out of a random sequence of microscopic interactions, less ordered situations will result. Irreversibility means only that the spontaneous reversal of the "irreversible" process is so unlikely as not to be worth considering.

Summary

The second law of thermodynamics limits the efficiency of cyclic engines and governs the direction of spontaneous processes in thermal systems. Cyclic engines do work in repeated cycles by extracting a heat flow from hotter external systems, converting part of that heat flow into work, and rejecting a heat flow to cooler thermal systems. The efficiency of such an engine is defined by

$$\eta \equiv \frac{W}{Q_h}, \tag{20–1}$$

where W is the work done in one complete cycle and Q_h is the total positive heat flow the cycle extracts from its sources of fuel.

Two equivalent ways to state the second law of thermodynamics are the Kelvin form and the Clausius form. Each of these emphasizes the role of engines and thermodynamic processes. The Kelvin form states that it is impossible to convert thermal energy from a body at a given temperature into an equivalent amount of mechanical work with no other effect; the Clausius form states that it is impossible to transfer thermal energy from a colder body to a hotter body with no other effect.

In the Carnot cycle, which involves four reversible steps that occur at just two temperatures, a working fluid undergoes alternate adiabatic and isothermal transformations. It is the most efficient cycle possible that operates between two temperatures T_h and T_c. No matter what the nature of the working fluid, the efficiency of the Carnot engine is

$$\text{for any Carnot cycle: } \eta_C = 1 - \frac{T_c}{T_h}. \tag{20–7}$$

The Carnot cycle that operates between two temperatures is more efficient than any other cycle that operates between the same temperature extremes. Any element of irreversibility also decreases the efficiency.

When the reversible cycles of work-producing engines are run in reverse, they act as refrigerators or heat pumps. Work is done on both of these machines while they extract a heat flow from a cold reservoir and release a heat flow to a hot reservoir.

The thermodynamic variable, entropy, S, measures the amount of disorder in a thermal system. Like thermal energy, it is a function of state, and only *changes* in entropy have physical significance. Entropy changes are path independent. For infinitesimal reversible transformations,

$$dS = \frac{dQ}{T}. \tag{20–18}$$

For an ideal gas, entropy can be expressed as a function of T and V as

$$S(B) - S(A) = C_V \ln\frac{T_B}{T_A} + nR \ln\frac{V_B}{V_A}. \tag{20–22}$$

Note that entropy decreases as temperature decreases.

The path independence of entropy changes is equivalent to the fact that, for a reversible cycle,

$$\oint \frac{dQ}{T} = 0, \tag{20–17}$$

a result known as Clausius' theorem. More generally, we include the possibility of an element of irreversibility by modifying Clausius' theorem to Clausius' inequality,

$$\oint \frac{dQ}{T} \leq 0. \tag{20–20}$$

The $<$ sign holds when there is irreversibility. If the change $A \rightarrow B$ is irreversible, the following relation holds:

$$S(B) - S(A) > \int_{\substack{A\\ \text{irrev}}}^{B} \frac{dQ}{T}. \tag{20–21}$$

In particular, for transformations of thermally isolated systems, we find that $\Delta S \geq 0$.

A spontaneous process in an isolated system increases the system's entropy, and the state of maximum entropy is the state of stable equilibrium. The increase of entropy expresses the inevitable increase in the disorder of thermally isolated systems.

Understanding the Concepts

1. If the ocean is viewed as a reservoir at one temperature and the atmosphere as a reservoir at another temperature, when can thermal energy flow spontaneously from the ocean to the air?
2. When the brakes are applied, a car decelerates and then stops. A film of the process run in reverse obviously looks fake. Does the process in reverse violate the second law of thermodynamics? If so, in what way?
3. The historian Herman Daly has called the second law of thermodynamics "the law of random, ravage, and rust." Is he right?
4. Why is it not possible for an engine—even a Carnot engine—to have an efficiency of one?
5. In Russia the heat exhaust from nuclear power plants is sometimes used to heat local cities. Does this increase the efficiency of the "engine"? If so, how?
6. Suppose that a gas obeys the van der Waals equation of state. What is the efficiency of a Carnot engine that uses this gas as a working fluid in terms of the temperatures of the hot and cold reservoirs?
7. In what climates are heat pumps most likely to be used?
8. A professor spends a couple of hours straightening up his office—putting books away, filing papers, throwing old exam papers away, and so on. He has created a less probable configuration out of a more probable configuration. Has he violated the second law of thermodynamics?
9. You have a small bottle initially filled with a mysterious gas X in a tightly sealed air-filled room. Gas X is not a component of air. Both gases are initially at 1 atm of pressure. The bottle turns out to be permeable to gas X, but perfectly nonpermeable to any gas in air. What will happen?
10. Imagine a microscopic device at the boundary between two containers of a gas that could sense the arrival of a particularly fast molecule heading from container A toward container B and could open a tiny door to let that molecule through. The same device, which is known generically as a *Maxwell demon*, could sense the arrival of a particularly slow molecule heading from container B toward container A and could let the molecule pass through. How would such a device violate the second law of thermodynamics? Is such a device feasible?†
11. Give examples of things you have observed recently that exemplify the second law of thermodynamics. Discuss the entropy change associated with them.
12. Would heat pumps that use an underground water reservoir be more efficient than those that use the outside air as a thermal reservoir if the pumps are employed to (a) heat in the winter? (b) cool in the summer?
13. Are electric resistive heaters 100% efficient?
14. List and discuss some processes not already mentioned that will not violate the first law of thermodynamics but will violate the second law.
15. When gasoline is burned in the chamber of a piston, chemical energy is turned into mechanical energy that is used to propel an automobile. How is this process consistent with the second law of thermodynamics?
16. If you were to run a film of an egg that drops to the floor, you would have little trouble determining if the film were running backward or not. Yet if the film were in extreme closeup, so that you were observing the interaction of molecules in the dropped egg, you would be hard pressed to say if the film were running forward or backward. How are these two statements consistent?
17. Is it possible to cool a house by leaving a refrigerator door open? What would be the net effect if you were to leave the door open?
18. Turbocharged engines take hot exhaust gas and send it back to the engine to heat up the intake air passing into the engine. Is this a satisfactory way to increase the efficiency of an engine?
19. Is the entropy of a gas an extensive quantity (such as volume) or an intensive quantity (such as pressure or temperature)? (See Problem 17–3.)
20. Biological systems use energy (ultimately from the Sun) in organizing highly ordered organisms, such as humans. Does this mean that biological systems violate the second law of thermodynamics?

Problems

20–1 Beyond Energy Conservation

1. (I) An engine produces 3.8×10^5 J of mechanical energy accompanied by a total heat flow of 7.6×10^5 J into the engine. What is its efficiency?
2. (I) Suppose that you toss a pair of dice. Make a list of all the possible outcomes, and calculate the probability of getting totals of 2, 3, 4, . . . 12, respectively.
3. (II) Six tokens, numbered one to six, are put into a hat. The tokens with even numbers are white; the ones with odd numbers are black. Suppose you draw four tokens at random, without putting them back into the hat. (a) List all the possible outcomes, without regard to order (for example, 1234 and 2134 do not count as separate outcomes). (b) What is the probability of drawing the tokens with 1, 2, 3, 4 on them? (c) What is the probability of drawing two white and two black tokens?
4. (II) N molecules of a gas are placed in the left half of a chamber and kept from entering the right half by a membrane. The membrane is then removed and the molecules distribute themselves uniformly throughout the whole volume (Fig. 20–19). What is the probability that, at some time during their subsequent random motion, they will all collect in the left half of the

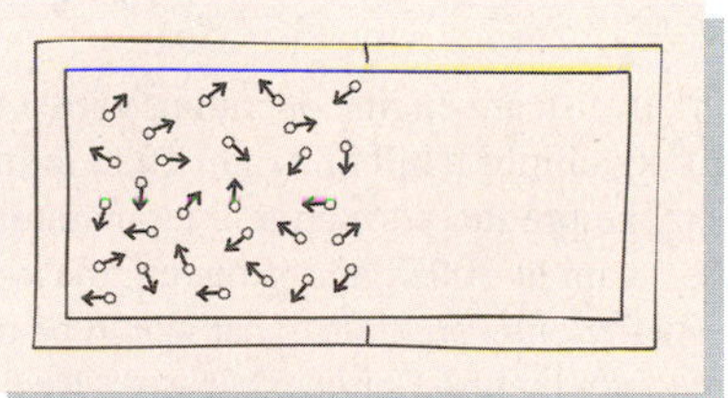

▲ **FIGURE 20–19** Problem 4.

†For an interesting discussion of Maxwell demons, see Charles H. Bennett, "Demons, Engines, and the Second Law," Scientific American, Nov. 1987, pp. 108–116.

chamber, for the cases $N = 2, 100, 10^6$, and 10^{23}? [*Hint*: This is equivalent to the probability of getting nothing but heads in 10^{23} coin tosses. To get an idea of this number in scientific notation, use the fact that $\frac{1}{2} \cong 10^{-0.3}$.]

5. (II) It is difficult to carry out experiments with just a few molecules. Here is an analogue to the problem of 10 molecules and on whether they are on the left or the right side of a container. Assume that you have 10 fair coins. Each coin gets tossed to produce a series of 10 heads or tails. Compare the probability that you end up with 10 heads (a maximally ordered array) with the probability of getting five heads and five tails in any order.

6. (II) You have a small 1-L bottle initially filled with a mysterious gas X in a tightly sealed air-filled room whose dimensions are 3 m × 3 m × 3 m. Gas X is not a component of air. Both gases are initially at 1 atm of pressure. The bottle turns out to be permeable to gas X but perfectly opaque to any gas in air. What will be the equilibrium pressure in the bottle, if any?

20–3 The Carnot Cycle

7. (I) What is the maximum efficiency of a steam engine whose boiler (the hottest-temperature reservoir in contact with the engine) is at 160°C and whose condenser (the coldest-temperature reservoir in contact with the engine) is at 35°C?

8. (I) A heat engine absorbs heat from a thermal reservoir at 550K, but the engine's maximum efficiency is only 38%. What must the temperature of the reservoir be to raise the efficiency to 41%?

9. (I) A heat engine receives heat from a thermal reservoir at 750K and transfers it to a thermal reservoir at 340 K. The efficiency of the engine is only 70% of the maximum possible. (a) What is the engine's actual efficiency? (b) How much work does the engine produce for each calorie of heat input?

10. (I) Temperature differences in the ocean have been proposed as a possible energy source in the tropics. Surface water at 29°C could act as a hot reservoir, and deep water at 3.0°C could serve as a cold reservoir. Ammonia gas could be a working fluid in a heat engine that runs between the two thermal reservoirs. What is the maximum efficiency of such an engine?

11. (II) A Carnot engine absorbs thermal energy from a reservoir at 450°C. Its efficiency is 0.55, and the work delivered by the engine is 5.0 kWh. (a) What is the temperature of the cold reservoir? (b) How much thermal energy flows into the system, and how much is rejected?

12. (II) What is the maximum efficiency of a heat engine that operates between 430°C and 125°C? If this engine generates 9800 J of mechanical energy, how many calories does it absorb from the hot reservoir, and how many calories does it transfer into the cold reservoir?

13. (II) A locomotive steam engine operating with a total train mass of 9.9×10^6 kg climbs a hill of height 1000 m in 10 minutes. If the steam engine produces steam at a temperature of 390K and exhausts the steam at 300K, and if the engine were at the theoretically best efficiency possible, what would be the rate, in mass per unit time, at which steam is exhausted? Real steam engines are much less efficient than the theoretical maximum.

14. (II) An athlete metabolizes energy at a rate such that his power output is 0.35 horsepower. (a) Assuming that the efficiency of his "engine" is 25%, how much thermal energy is dissipated? (b) If the dissipation is primarily through the evaporation of sweat, how much sweat is evaporated per minute?

15. (II) An inventor claims to have built an engine that takes in 3.0×10^8 J of thermal energy at 450K, rejects 1.4×10^8 J of thermal energy at 250K, and delivers 1.0×10^8 J of work in 1 h of cyclic operation. Is there anything wrong with this claim?

16. (II) A power plant generates 800 MW of electric power. At what rate does the plant generate waste heat if its efficiency is 28%? Assuming that the plant operates between 480°C and 80°C, what is the maximum efficiency possible?

17. (II) An ideal diatomic gas is used in a Carnot engine with thermal reservoirs at 77 K and 300 K. What is the ratio of the maximum volume to the minimum volume if the pressure drops by a factor of 2 during the isothermal expansion? [*Hint*: For a diatomic gas, $\gamma = 1.4$.]

18. (II) Prove that two reversible adiabatic paths cannot intersect. Do this by assuming that they do, constructing a cycle that consists of the two paths and another reversible transformation, and showing that such a cycle would violate the second law of thermodynamics (Fig. 20–20). (This result is simply shown for an ideal gas. Your proof should apply to a nonideal gas as well.)

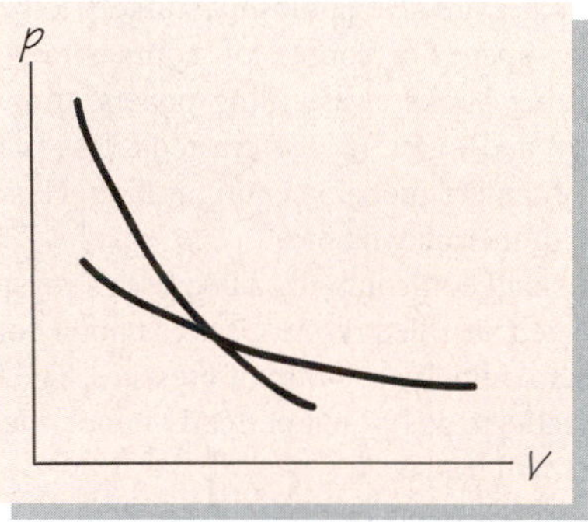

▲ **FIGURE 20–20** Problem 18.

19. (III) Any element of irreversibility makes a Carnot engine and, in fact, any engine, less efficient. Let engine A be a pseudo-Carnot engine, meaning that it has some irreversibility somewhere along the cycle; let engine B be an ideal Carnot engine—the two arranged in size so that they exhaust the same heat flow Q_c. If engine A is more efficient than engine B, then by combining the two engines such that A runs normally and B runs backward, as in Figure 20–6, show that you arrive at a contradiction of the Kelvin form of the second law of thermodynamics. Now that you have shown that engine A must be less efficient than engine B, why can the preceding argument not be reversed, running the Carnot cycle forward and the pseudo-Carnot cycle backward to get a contradiction of the second law?

20. (III) Show that if it is possible to build an engine that violates the Kelvin form of the second law of thermodynamics, it is also possible to build an engine that violates the Clausius form of the second law. [*Hint*: Consider a Carnot engine driven by the Kelvin-violating engine.]

20–4 Other Types of Engines

21. (I) An air conditioner is rated at 8000 Btu/h. Assuming maximum possible efficiency, an exhaust temperature of 38°C, and an interior temperature of 22°C, what is the electrical power consumption? (The work done on the air conditioner is supplied by electricity.)

22. (I) What is the coefficient of performance for an ideal Carnot heat pump operated between outside air at 40°C and a deep water well at 8°C? Would the coefficient of performance be

larger or smaller if your heat pump followed any other type of cycle?

23. (I) Two refrigerators can each be approximated as Carnot cycles run in reverse. Each refrigerator works with the same cold temperature reservoir $T_c = 273$K. Refrigerator 1 evacuates to a hot reservoir at temperature $T_h = 293$K, while refrigerator 2 evacuates to a hot reservoir at a temperature of $T_h = 303$K. Compare the coefficients of performance of these refrigerators. (Real refrigerators will have coefficients of performance substantially less than these.)

24. (I) A heat pump acts like a refrigerator that cools the outside of a house and gives rise to a heat flow to the inside of the house. If the outside temperature is +3°C and the inside temperature is 21°C, what is the maximum amount of British thermal units a pump can deliver per kilowatt-hour of electric energy supplied? (The work done on the pump has electric energy as its source.)

25. (I) Consider the diesel cycle in Fig. 20–9a. In which portions of the cycle is there heat flow into or out of the engine? What is the work done in the transformation $2 \rightarrow 3$ in terms of T_2 and T_3?

26. (II) An "ideal refrigerator," a Carnot engine that operates in reverse, has a compressor as its engine. It does work W to extract heat Q_c from a cold reservoir at temperature T_c and dumps heat Q_h into a hot reservoir at T_h. Use the known efficiency of a Carnot engine that operates between T_h and T_c and the conservation of energy for one cycle to show that the coefficient of performance $K_{ref} \equiv Q_c/W$ for the compressor is given by $K = T_c/(T_h - T_c)$. Note that K_{ref} is large when the cold and hot reservoirs have nearly the same temperature.

27. (II) A refrigerator with coefficient of performance $K_{ref} = 5.1$ (see Problem 26) gives rise to a heat flow out of the cooling compartment at a rate of 400 cal/min. What is the required power of the motor that operates this refrigerator?

28. (II) To compensate for a given energy loss when a refrigerator is opened too often, the refrigerator's compressor must do work. Take $T_c = 2.0$°C and $T_h = 29$°C, and suppose that the refrigerator's coefficient of performance is 35% of its ideal value. What is the rate at which the compressor does work; that is, what is the power consumption for a given rate of heat loss dQ/dt?

29. (II) The specifications of a freezer claim that it can remove 80 cal/s heat flow from the compartment at −20°C and release 94 cal/s into the room at 25°C while using 60 W of electrical energy to drive the compressor. Can you trust this statement?

30. (II) In order to make an ice cube from water, a refrigerator must extract 1400 cal of heat at the temperature of the freezer, −12°C. The room temperature is 30°C. What minimum work must an ideal refrigerator do?

31. (II) You can buy a 5-lb bag of ice for about $2. Estimate the fraction of this price that pays for the electric energy needed to run the ice-making machine. Assume that the inside of the machine is at −10°C, that the outside temperature is 30°C, that the initial temperature of the water is 20°C, and that the coefficient of performance is 30% of the ideal. Electrical energy costs $0.10/kWh.

32. (II) The temperature inside a Carnot refrigerator placed in a room at 68°F is 38°F. The heat flow from the refrigerator into the room is 0.95×10^8 J/h. How many watts are needed to operate the refrigerator?

33. (II) A heat engine operates on the cycle shown in Fig. 20–21, with heat absorbed or ejected on each leg as indicated. Its operating fluid is an ideal gas with heat capacity at constant volume C_V. (a) What are temperatures T_3 and T_4 at points 3 and 4, respectively, in terms of the indicated pressures and volumes?

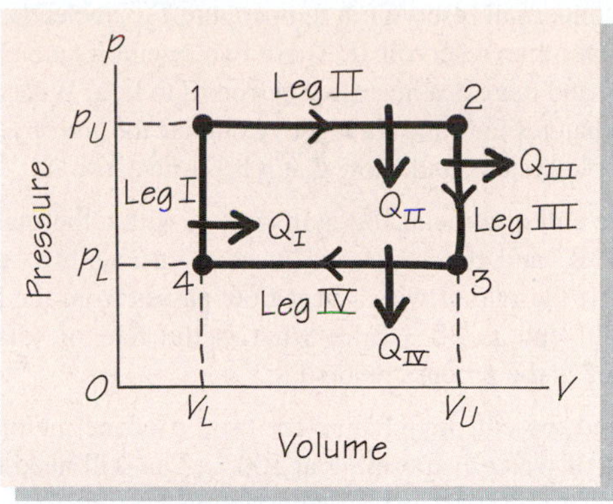

▲ **FIGURE 20–21** Problem 33.

(b) What is the internal energy change along leg IV, ΔU_{IV}? Include the sign. (c) What is Q_{IV} in terms of ΔU_{IV} and the indicated pressures and volumes? (d) What is the efficiency of the engine in terms of the indicated pressures and volumes and C_V?

34. (II) Consider the ideal Stirling cycle shown in Fig. 20–8, working between a maximum temperature T_h and a minimum temperature T_c, and a minimum volume V_1 and a maximum volume V_2. Suppose that the working gas of the cycle is 0.1 mol of an ideal gas with $c'_V = 5R/2$. (a) What are the heat flows to the cycle during each leg? Be sure to give the sign. For which legs is the heat flow positive? (b) What work is done by the cycle during each leg? Again, be sure to include the sign. (c) If, in the definition of the efficiency of this cycle, $\eta = W/Q_{pos}$, where Q_{pos} is the total positive heat flow to the engine, what is the efficiency of the cycle when $T_h = 700$K and $T_c = 400$K? $V_1 = 0.5$ L and $V_2 = 1.5$ L. Compare this efficiency to the efficiency of a Carnot cycle that operates between the same temperature extremes.

35. (II) The Otto cycle, represented in Figure 20–9b, runs between minimum and maximum volumes V_1 and V_2 and minimum and maximum temperatures T_1 and T_2, respectively. (a) Show that the efficiency is given by $\eta = 1 - (T_2 - T_3)/(T_1 - T_4)$. (b) Show that if the working fluid of the cycle is an ideal gas, the efficiency of this cycle can alternatively be written as $\eta = 1 - (V_1/V_2)^{\gamma-1}$, where $\gamma = C_p/C_V$. The efficiency of this cycle is thus independent of the temperatures between which it operates, depending instead only on γ and on geometry. The ratio V_2/V_1 is the compression ratio. A typical compression ratio is 8, and $\gamma = 1.4$, which gives a predicted efficiency of 56%.

36. (III) The Brayton cycle is represented by Fig. 20–9c. Show that when its working fluid is an ideal gas, the cycle's efficiency is given by the same formula as that for the Otto cycle in the previous problem. In this case, V_1 and V_2 are not the extreme volumes of the engine but are the volumes at the start and finish of the adiabatic expansion step, respectively.

37. (III) Find the efficiency of the diesel cycle illustrated in Fig. 20–9a in terms of $\gamma = C_p/C_V$ and the temperatures at points 1, 2, 3, and 4. Assume that it contains 1 mol of an ideal gas. (Rudolf Diesel tried to produce an engine based on the Carnot cycle. He was not able to produce such an engine, but he came close: A diesel engine nearly matches a Carnot engine in efficiency. Diesel's engine is one of the most efficient in use today.)

20–5 Entropy and the Second Law

38. (I) Consider a large, isolated thermal system in two parts. One part is a thermal reservoir at temperature T_1, labeled I, and the other part is a thermal reservoir at temperature T_2, labeled II. Reservoir I is colder than reservoir II. These two reservoirs are briefly connected, and there is a heat flow Q from II to I. (a) What are the entropy changes for I and for II? (b) Compute the entropy change for the whole system, and show that it is positive.

39. (I) One end of a metal rod is in contact with a thermal reservoir at 1273K, and the other end is in contact with a reservoir at 293K. If the rate at which heat flow passes from the hot end to the cold end is 30 J/min, what is the rate of total entropy change? Is the process reversible?

40. (II) Find the entropy change for (a) 1 g of ice melting at 0°C; (b) 1 g of water evaporating at 100°C. You will need to look up the heats of melting and of vaporization.

41. (II) One kilogram of iron at 80°C is dropped into 0.5 L of water at 20°C. Given that the specific heat of water is 1 cal/g·K and that of iron is 0.107 cal/g·K, calculate the final equilibrium temperature of the system and the increase in entropy.

42. (II) One hundred and twenty grams of water at 100°C is changed reversibly into steam at 100°C. (a) Describe such a process. (b) What is the change in entropy of the water and of the universe? (c) How would part (b) change if the process were irreversible?

43. (II) A boater dumps 8.7 kg of ice from his cooler into the lake at the end of a fishing trip. The ice is at −3°C, and the lake is at 18.7°C. What is the change in entropy of the new lake–ice cube system after the ice cubes come to equilibrium with the lake?

44. (II) Calculate the entropy change of the universe when a 1-kg block of copper at a temperature of 200°C is put in a lake at a temperature of 4°C. The total heat capacity of this block of copper is 375 J/K.

45. (II) Calculate the change in entropy of the universe if 0.3 kg of water at 70°C is mixed with 0.2 kg of water at 15°C in a thermally insulated container. [*Hint*: Even though this process is irreversible, the change in entropy can be calculated by devising a reversible process that leads from the given initial state to the expected final state.]

46. (III) Thermodynamic tables often give the specific heat of materials in the form of a list of coefficients *a, b, c, d* in the expression $c_p = a + bT + cT^2 + d/T$. How can you use these data to find the entropy change of the material as it undergoes a transformation from T_i to T_f?

20–6 Entropy and Ideal Gases

47. (I) Five moles of an ideal gas is in thermal isolation and undergoes free expansion from 35 L to 100 L. What is the entropy change of the gas? of the universe?

48. (I) The heat capacity at constant volume of a sample of gas is determined to be 22 J/K. If this amount of gas is heated in a constant-volume container from 273K to 840K, what is the change in entropy?

49. (I) One mole of an ideal gas expands at constant pressure from an initial volume of 250 cm^3 to a final volume of 650 cm^3. What is the change in entropy, assuming that the molar specific heat is $c'_V = 3R/2$?

50. (I) One mole of ideal gas with molar specific heat $c'_V = 3R/2$ is heated from 273K to 350K at constant volume. What is the change in entropy?

51. (I) An ideal gas undergoes an isothermal compression. Does the entropy increase?

52. (II) Ten moles of a monatomic gas held in a tank at 17 atm escape from the tank isothermally at 27°C. What is the change in the entropy of the universe?

53. (II) One mole of a monatomic ideal gas at an initial pressure of 30 atm and a temperature of 600K is allowed to undergo a rapid (adiabatic) free expansion from a small vessel into a vessel of 50 times greater volume. Find the change in temperature and the increase in entropy.

54. (II) One mole of a monatomic ideal gas is carried in two steps from (p_1, V_1) to $(p_1, 5V_1)$ to $(0.3p_1, 5V_1)$ by reversible thermodynamic processes that are straight lines on a *p–V* diagram (Fig. 20–22). (a) What is the overall change in the gas's internal energy? (b) What is the overall work done by the gas? (c) How much net heat flow is added overall? (d) What is the net change in the entropy?

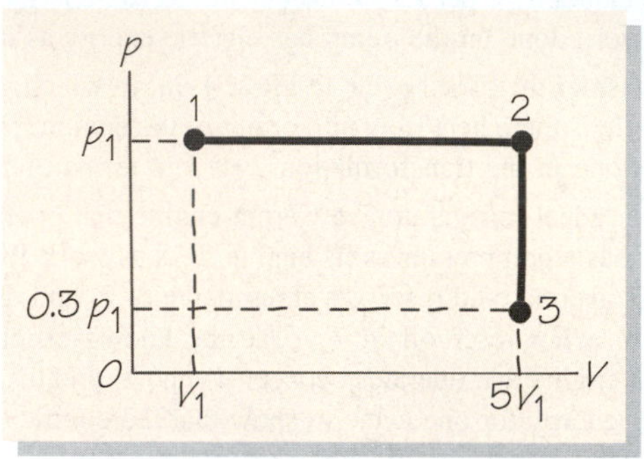

▲ **FIGURE 20–22** Problem 54.

55. (II) A gas obeys the well-known equation of state $pV =$ (a constant) T. The gas expands, doubling in volume. (a) Plot p versus V when the expansion is isobaric and when it is isothermal. (b) What work is done by the gas on its surroundings for both cases above? (c) What is the entropy change of the gas for both cases above?

56. (II) Show that in terms of temperature and pressure, the entropy of an ideal gas with constant heat capacities is $S(T, p) = C_p \ln(T/T_0) - nR \ln(p/p_0) + S(T_0, p_0)$.

57. (III) Dilute helium gas is taken through the following reversible cycle on a *p–V* diagram: *AB* is an isobaric compression; *BC* is an adiabatic expansion; *CD* is an isothermal expansion; *DA* is a constant-volume pressure increase (Fig. 20–23). Express the heat flow, work done on the system, change in internal energy, change in temperature, and change in entropy for each leg of the cycle in terms of the quantities T_A, T_B, T_C, and p_A. Pay attention to the signs.

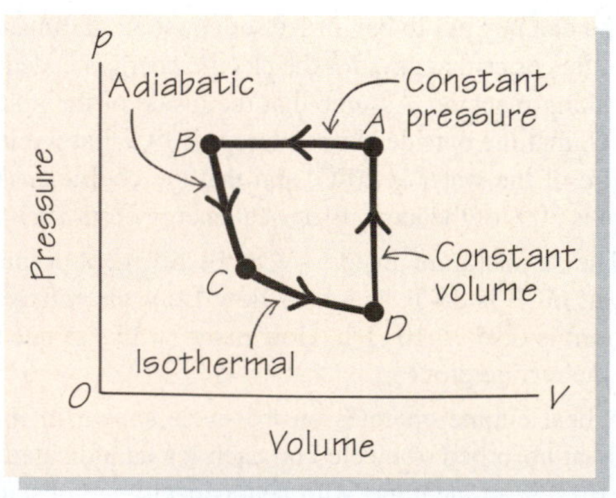

▲ **FIGURE 20–23** Problem 57.

20–7 The Meaning of Entropy

58. (I) A total of 12 molecules are in a large box, which we mentally divide into two halves, a left half and a right half. Any one molecule is equally likely to be in either of the two halves. What is the probability that all 12 molecules will be found in the left half? How does your answer change if there are 120 molecules?

59. (II) An ideal gas in a cylinder with a movable piston is placed in thermal contact with a thermal reservoir, and the gas is slowly compressed. How does the entropy of the gas in the cylinder change? How does the entropy of the reservoir change? Is the answer to these questions in conflict with the second law of thermodynamics? Explain your answer.

60. (II) A box of volume 3.0 L contains 1.8 L of He in one side and 1.2 L of O_2 in the other, separated by a partition. The temperature is 290K, and the pressure in both sides of the box is 1.2 atm. The partition is removed. (a) Describe the new equilibrium configuration. (b) What is the change in entropy?

61. (II) A thermally isolated container of volume $3V_0$ is divided by a partition. In one portion, of volume V_0, there are n mol of a monatomic ideal gas at temperature T_0. The other portion is empty. The partition is then broken, and the gas expands to fill the entire container (Fig. 20–24). (a) What is the final temperature of the gas? (b) The change in entropy of the gas? (c) The change in entropy of the universe? (d) Suppose that the gas is originally confined to volume V_0 by means of a piston instead of a partition. The piston is then slowly withdrawn until a final volume of $3V_0$ is obtained. What is the change in entropy of the gas?

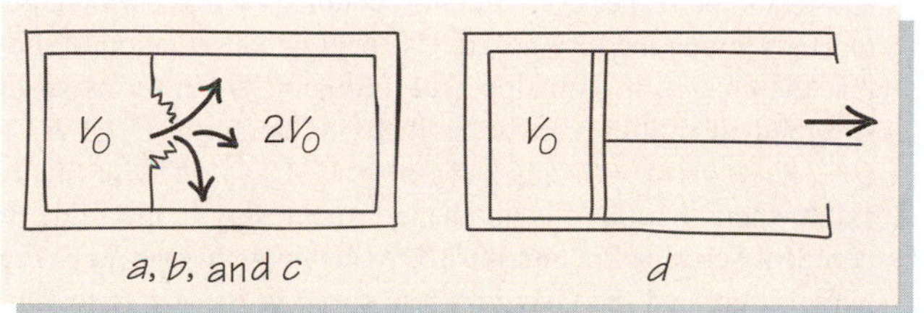

▲ **FIGURE 20–24** Problem 61.

General Problems

62. (II) A nuclear power plant requires 1450 MW of cooling. If environmental concerns limit the temperature rise of the cooling water to 9°C, how much water flow is required to cool the heat generated by the plant? If the water is at an initial temperature of 287K, what is the entropy change per kilogram of water?

63. (II) Fifty grams of oxygen gas at 320K do 80 J of work while 40 cal of heat flow is absorbed by the gas. (a) What is the change in internal energy? (b) In the temperature of the gas? (c) In entropy, assuming that the change is isobaric?

64. (II) Calculate the entropy change when 1 kg of water is cooled from 100°C to 0°C. Calculate the entropy change when the water turns to ice at 0°C (Fig. 20–25). The latent heat of fusion is 80 cal/g.

65. (II) Sketch on an S–T diagram the following thermodynamic reversible transformations: (a) An adiabatic transformation at constant volume, involving a pressure change. (b) A rough picture of the Stirling cycle.

66. (II) You have a box of volume V divided by an imaginary partition (molecules pass right through it) into two parts, volumes V_1 and V_2. You start out with N_1 molecules of gas g_1 and N_2 molecules of gas g_2, all spread evenly through the box. What is the probability that you find all the g_1 molecules in V_1 and all the g_2 molecules in V_2? Work out the numbers for $V_1 = V/3$ and $V_2 = 2V/3$ while $N_1 = N_2 = 100$. Repeat, but for $N_1 = N_2 = 1000$.

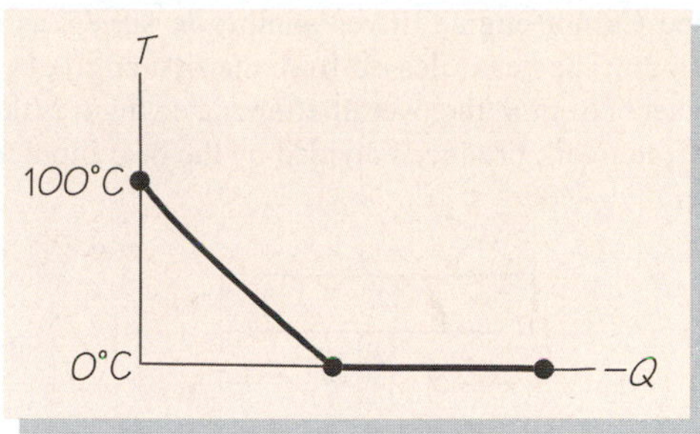

▲ **FIGURE 20–25** Problem 64.

67. (II) Two moles of helium at STP are compressed in an isothermal process to a pressure of 2 atm. (a) What is the final volume? (b) the final temperature? (c) the change in entropy?

68. (II) A Carnot engine for which the working fluid is 1 mol of an ideal gas with $c_V' = (5/2)R$ operates between two thermal reservoirs of temperatures $T_h = 550\text{K}$ and $T_c = 350\text{K}$. When the gas is in contact with the hot reservoir, its minimum volume is $V_1 = 0.70$ L, and when it is in contact with the cold reservoir, its maximum volume is $V_3 = 8.0$ L. (a) What is the efficiency of this engine? (b) For one cycle, what is the net energy change of the gas? (c) The net entropy change of the gas? (d) The net change in the entropy of the universe? (e) What is V_2, the maximum volume when the gas is in contact with the hot reservoir?

69. (II) Three moles of oxygen at STP undergoes an adiabatic compression to a final pressure of 3.0 atm. (a) What is the final volume? (b) The final temperature? (c) The change in entropy?

70. (II) One liter of nitrogen at STP is adiabatically compressed to a volume of 0.50 L when a membrane breaks and the nitrogen gas undergoes free expansion back to 1.0 L. (a) What is the final pressure? (b) The final temperature? (c) The change in entropy?

71. (II) A heat engine operates between the temperatures of 400K and 300K, doing 100 J of work. This work is used to run a refrigerator between the same temperatures. (a) Calculate the heat transfer between each reservoir and the engine or the refrigerator, assuming that both of them have ideal efficiency/coefficient of performance. (b) Repeat the calculation if the heat engine's efficiency is only 80% of the ideal. (c) Repeat the calculation for the case of an ideal heat engine but for a refrigerator whose coefficient of performance is 80% of the ideal.

72. (II) An inventor proposes that you invest in a newly developed engine. It purports to absorb 140,000 Btu at a temperature of 700°C, and reject 60,000 Btu at a temperature of 400°C. Should you invest?

73. (II) For one mole of a van der Waals gas, the energy change dE for a given temperature change dT and volume change dV takes the form

$$dE = C_V\,dT + \frac{a}{V^2}dV,$$

where in this problem we'll assume that C_V has no temperature dependence. Your gas undergoes free expansion from V_i, T_i to V_f, T_f. Does the temperature (a) increase, (b) decrease, or (c) remain unchanged? [*Hint*: In free expansion the energy is unchanged. This energy has the form $E(T, V) = C_V \int_{T_0}^{T} dT + \int_{V_0}^{V} \frac{a}{V^2}\,dV$.]

74. (II) One Carnot engine drives another in series, as shown in Fig. 20–26. The heat released from the first engine is absorbed by the second. Find the overall efficiency, which is determined by the total work produced divided by the heat input to the first engine.

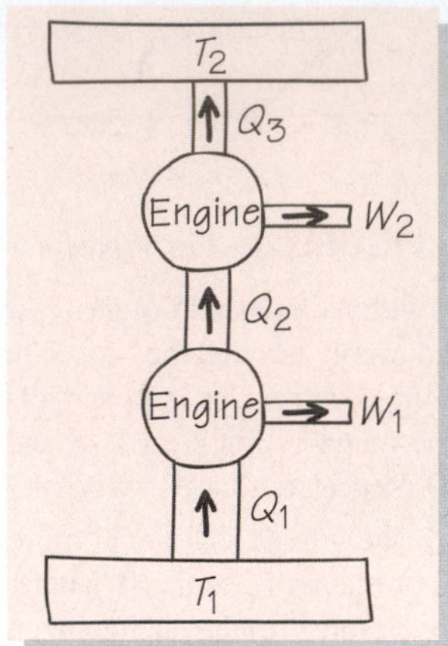

▲ **FIGURE 20–26** Problem 74.

75. (II) A coal-fired, 300-MW generating plant operates between 750K and 400K, with an efficiency 60% of the maximum efficiency possible. At what rate is waste heat flow produced? Suppose that water is used to carry off the waste heat flow, and it does so by being heated by 12°C. How much water must flow through the machinery per second?

76. (II) Consider the reversible thermodynamic cycle of the following form (Fig. 20–27): (i) One mole of a monatomic ideal gas at temperature T_a and pressure p_a is allowed to expand isobarically until the volume has increased by a factor r. (ii) The gas is then cooled so that its absolute temperature is reduced by a factor of 2, with the cooling performed at constant volume. (iii) The gas is then compressed isobarically to its original volume. (iv) Finally, the gas is heated to its starting temperature, T_a, at constant volume. (a) What is the mechanical work done by the engine? (b) The heat transfer in the four legs of the cycle? (c) The efficiency of the cycle?

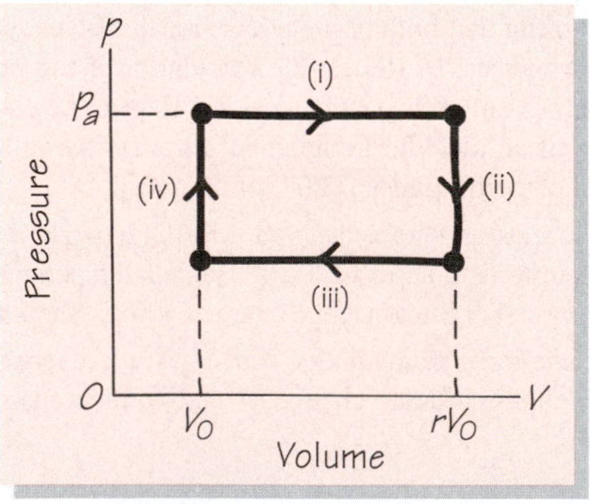

▲ **FIGURE 20–27** Problem 76.

77. (II) A heat engine operates in a reversible cycle with the following steps (Fig. 20–28): (i) An ideal gas characterized by $\gamma = 1.4$ expands adiabatically, so that it cools from an initial temperature of 330°C to 30°C; (ii) it is compressed isothermally until it reaches its initial volume; and (iii) it is then heated at constant volume until it reaches its initial temperature. Calculate the efficiency of the cycle and compare it with the Carnot efficiency that corresponds to these temperatures.

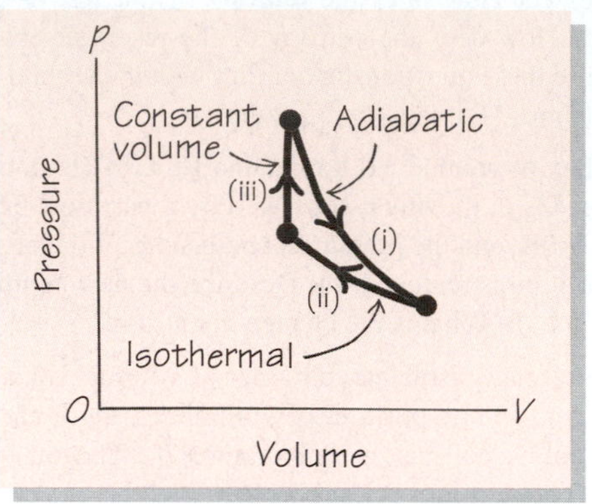

▲ **FIGURE 20–28** Problem 77.

78. (II) Suppose that there is steady flow of a mass M of gas through a turbine. Gas enters the turbine at an elevation z_1, pressure p_1, internal energy U_1, and speed v_1, and leaves at an elevation z_2, pressure p_2, internal energy U_1, and speed v_2. Thermal energy Q is supplied to the turbine. Work W is done by the turbine, and V_1 and V_2 are the respective volumes occupied by mass M on entering and leaving the turbine. (a) Use energy conservation to show that the change in enthalpy (see Problem 79) in terms of the specified quantities can be written in the form $H_2 - H_1 = Q - W + \frac{1}{2}mv_1^2 + Mgz_1 - \left(\frac{1}{2}mv_2^2 + Mgz_2\right)$. (b) Use this result to show that there is no change in enthalpy in the Joule experiment described in Section 18–7 (in which there is no change in elevation, and the change in the speed of the gas is so small that $v_2^2 - v_1^2$ can be neglected).

79. (II) A useful quantity called enthalpy, H, is defined by $H = U + pV$. Calculate the enthalpy of an ideal gas as a function of temperature.

80. (III) Three reversible paths, legs i, ii, and iii, connect points (p_2, V_2) and (p_1, V_1) in Fig. 20–29. (a) Calculate $\int_1^2 dW$ and $\int_1^2 dQ$ for the three paths. (b) Calculate $\int_1^2 dQ/T$ for the three paths.

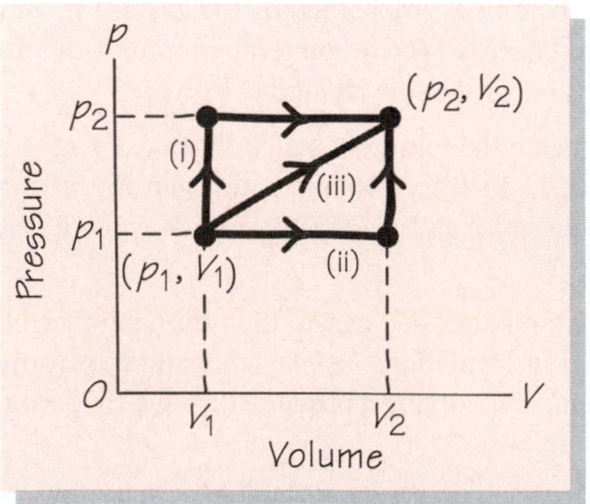

▲ **FIGURE 20–29** Problem 80.

81. (III) The Brayton cycle is used to extract work from a high-temperature, gas-cooled nuclear reactor with helium (a monatomic ideal gas) as the working fluid. The helium enters a compressor, and its pressure is raised adiabatically. It is then heated at constant pressure to a high temperature in the reactor core. The energy is supplied by the fission of uranium or plutonium. The helium then enters a turbine, in which it is allowed to expand adiabatically. It then passes through a precooler, in which heat is rejected at constant pressure until it reaches the initial conditions. Suppose that helium enters the compressor at 30°C and a pressure of 20 atm. It is compressed to 50 atm and then heated to 1200°C in the reactor core (Fig. 20–30). (a) What are the temperatures at the end of compression and expansion? (b) What are the heat flows per kilogram of helium in each cycle? (c) How much net work is done per kilogram of helium in each cycle?

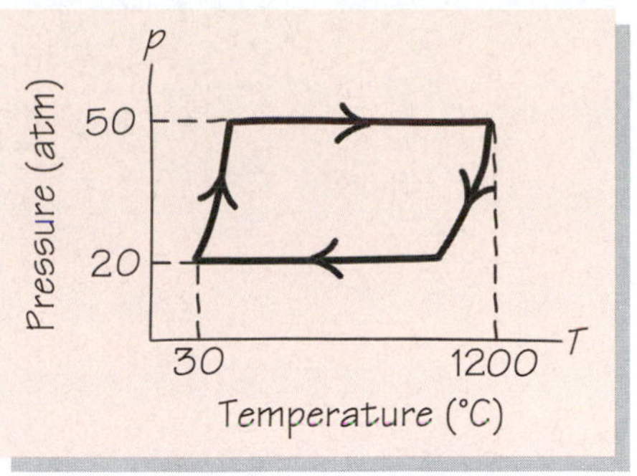

▲ **FIGURE 20–30** Problem 81.

Chapter

◀ Investigations of the nature of lightning were important in the understanding of electrical phenomena. Here, Benjamin Franklin is represented performing a kite experiment, whose description he published between 1751 and 1753. The painting is inaccurate in that Franklin flew the kite before lightning struck, he did not stand on the open, and a key was attached to the string, which was then charged by the clouds. In fact, whether Franklin actually performed the experiment is not known.

Electric Charge

Electricity, together with magnetism, governs virtually all that we see of the physical world. Electromagnetic forces control the structure of atoms and all materials. Light and other electromagnetic waves are pervasive. The understanding of these forces is one of the great success stories of science. In this chapter, we introduce electric charge, a property of atomic constituents, and we discuss the fundamental law of the interaction of two charges at rest, Coulomb's law. This force law is as fundamental as the universal law of gravitation. The interaction between charges has the same space dependence as gravitation, but the force described by Coulomb's law can be either attractive or repulsive.

21–1 Charge—a Property of Matter

Over the last century, we have become more and more dependent on electricity in our everyday lives, and most people are aware that electric charges exist. The experimental evidence for electric charges and the understanding of charge developed over a long time period. We begin with a very brief sketch of this development.

A Brief History of the Study of Electricity and Magnetism

The word *electricity* has its roots in the Greek word for "amber" (*electrum*), and the first written mention of the curious effects of rubbed amber dates from the fifth century B.C. It was not until the 1700s that the critical discovery that electric forces can be either repulsive or attractive was made. The idea developed that a quantity (which we now call electric charge) is associated with electric forces. Among the many important names associated with these discoveries are Stephen Gray, Charles Dufay, and Benjamin Franklin.

Benjamin Franklin is best known for his exploitation of the existing idea that electrical phenomena were associated with a kind of fluid contained in matter. He surmised that repulsion and attraction were the result of an excess or deficiency of the fluid. Although we now know that view is not quite right, implicit in Franklin's model is what we can recognize as the phenomenon of the conservation of charge. For example, if the "fluid" were to flow out of an object, it would leave behind a deficiency. Franklin introduced the terms "positive" and "negative" for the two types of charge and also set the standard sign convention in which the electron, the actual particle that moves in conductors, has negative charge. Franklin was known for his spectacular (and dangerous) experiments with lightning, which he recognized as an electrical effect. Franklin and his friend Joseph Priestley, as well as Henry Cavendish, are linked with the discovery that *the fundamental force between electric charges is proportional to the inverse square of the distance between them.* This law was confirmed more directly by John Robison and then by Charles Coulomb in the mid- and late eighteenth centuries, respectively. This inverse-square law is now known as **Coulomb's law**.

The nature of magnetism and its relation to electricity became clearer starting around 1820, primarily through the work of Hans Christian Oersted, André-Marie Ampère, and Michael Faraday. James Clerk Maxwell completed the unification of electricity and magnetism in the 1860s: Electricity and magnetism were aspects of a single fundamental set of phenomena, electromagnetism. This subject will occupy us for many chapters to come.

The Significance of Electric Forces

In this chapter we introduce a second basic force of nature. To the law of universal gravitation we add knowledge of the electrical interaction, as represented by the Coulomb force. Gravitational forces have an inverse-square dependence on the distance between interacting pointlike objects, and as we shall see, the same is true for the electric force. Both forces are proportional to the product of a characteristic attribute of the two objects—mass for the gravitational force and electric charge for the Coulomb force.

On the cosmic scale, gravity looms large. It is the force that keeps Earth rotating around the Sun, and the Moon rotating around Earth. The reasons why gravitation dominates electric forces on the astronomical scale are twofold. First, astronomical bodies have a great deal of mass. Second, astronomical bodies are almost exactly charge neutral, so the electric forces between them are relatively small. On anything less than an astronomical scale, however, the electric forces are normally much larger than the gravitational ones; apart from the direct effects of Earth's gravity, our everyday experience depends far more on the electric force than on the gravitational one.

As we shall see through study of the hydrogen atom, the electric force dominates the gravitational force on a microscopic scale. Even though a full explanation requires the inclusion of magnetic forces and quantum physics, we can now state that the electric force is responsible for

1. electrons binding to a positive nucleus, forming a stable atom;
2. atoms binding together into molecules;
3. atoms or molecules binding together into liquids and solids;
4. all chemical reactions; and
5. all biological processes.

The electric force is also responsible for such nonfundamental forces as friction and other contact forces. Electric energy fuels our homes, starts our cars, and runs our factories. We can say that electrical forces are dominant in the behavior of matter as we know it.

Matter and Electric Charge

In most of our discussions to this point, we have characterized bulk matter—and the atoms that make up matter—by a single attribute: mass. When we probe the structure of atoms more deeply, we find that atoms are made up of electrons and nuclei, which are characterized by an additional attribute, **electric charge** (usually labeled q). Electric charges exert (electric) forces on one another that are proportional to the product of their charges, just as masses exert gravitational forces on one another that are proportional to the product of their masses. Charges come in two types, termed positive and negative, and have the fundamental property that opposite charges attract and like charges repel. Electrons and nuclei carry opposite charges, and as we will describe in more detail later, it is the attractive electrical forces between them that hold the atom together. The set of phenomena associated with the forces between stationary charges form the subject of **electrostatics**.

A correct description of atoms requires quantum mechanics. But we can give a qualitative classical picture—the remarks about charge in this picture carry over to a quantum-mechanical picture. An atom's electrons (labeled by e) each carry the same unit of negative charge, $q_{\text{electron}} = -e$. (Note that the same symbol e stands for the electron and a charge quantity.) The electrons are bound in shell-like regions around the much heavier nucleus, which consists of electrically neutral neutrons (labeled n) and positively charged protons (labeled p). The proton charge is equal in magnitude but opposite in sign to that of the electron, $q_{\text{proton}} = +e$. When the number of protons in an atom equals the number of electrons, the atom is **electrically neutral**—it has no net charge.

Chemical elements differ in the number of protons in the nuclei of their atoms. With no ionization, the number of protons is the same as the number of electrons surrounding the nucleus. Chemical properties are closely linked to the behavior of the atom's electrons. The electrons that are, on average, closest to the nucleus are more strongly bound to the nucleus and are more difficult to dislodge from the atom and less available to interact with other atoms. The outermost electrons—farther away from the nucleus—are attracted less strongly and are more easily dislodged. The ease with which the outermost electrons can be removed from the atom, as well as how strongly an atom attracts other electrons, greatly influences the properties of the elements and how they interact chemically with one another. An atom that has lost one or more electrons and is thus left with a positive charge is called a *positive ion*, and a *negative ion* is an atom that has gained extra electrons. Bulk matter is typically electrically neutral or very nearly so. Even if there are chemical phenomena that produce positive and negative ions within the matter, these will appear in equal numbers to produce a neutral bulk.

Although bulk matter is overall neutral, it will have electrical properties that are associated with the outer electrons of its atoms. If the outer electrons of atoms in bulk matter are *weakly bound* to their respective nuclei, they behave as though they are almost free, and can move through the material almost unimpeded. Such materials—metals for the most part—are said to be good **conductors** of electricity; the term refers to how easily charge can move through the material. A certain class of materials—**superconductors**—have electrons that, in effect, move with *no* inhibition when the material is made sufficiently cold. The electrons of most nonmetallic solids do not travel easily; such solids, including rubber, glass, and plastics, are **insulators**. Silicon, germanium, and a large number of synthetic combinations of materials are substances that we can make into insulators or conductors by controlling either their electrical properties or their temperature. Such substances are called **semiconductors**, and they play an important role in electronic devices—transistors and so forth.

We can transfer charges back and forth between different materials by rubbing them together, allowing electrons to move from one material to the other. For example, when we rub a Teflon rod with fur, electrons are transferred from the fur to the Teflon rod (see Fig. 21–1a, b). The fur then has an excess positive charge: It has lost some electrons. The Teflon rod to which the electrons have been carried will have an excess of electrons and is now negatively charged. When charge is carried from one object to another in this way, the objects are said to be **charged by friction**. Both objects will be charged. We can similarly transfer electric charge by rubbing a glass rod on silk. The glass acquires a positive charge, because it transfers electrons to the silk, which in turn acquires an equal but opposite (negative) charge (Fig. 21–1c).

(a)

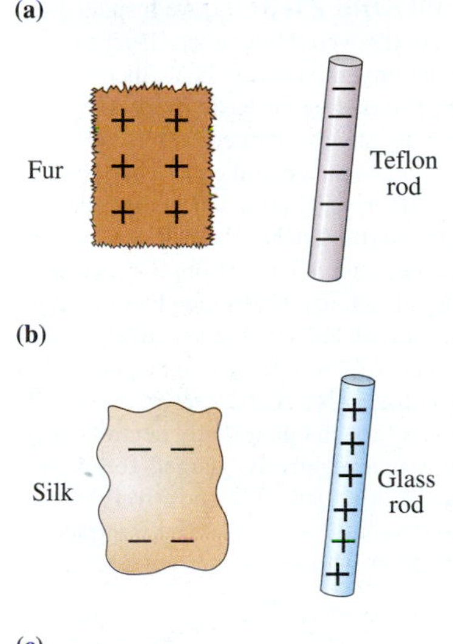

▲ **FIGURE 21–1** When Teflon is rubbed against fur (a, b) and glass is rubbed against silk (c), electric charge is transferred.

THINK ABOUT THIS. . .

IF WE ACCUMULATE A CHARGE, WHERE DOES IT EVENTUALLY GO?

When we walk across a carpet on a dry winter day, we may slowly accumulate a charge through the rubbing of our shoes on the carpet fibers. We specify that the air must be dry because if it is not, then this charge is easily transferred to water molecules in the air and carried off. And your shoes should be good insulators, because otherwise the charge will travel back to the carpet. Without these possibilities, you may later touch a radiator or another person. Your accumulated charge is then shared with the second object, sometimes in sudden and shocking fashion. (The passage of the charge may even be visible as a spark, an interesting subject on its own that we'll discuss in Chapter 25.) But just what is the object with which you are sharing your charge if you touch the radiator? The answer is that the metal pipes of the radiator may form a very long path, ultimately connecting you to Earth as a whole. If the carpet itself touches Earth through a sequence of conductors, you got your charge from Earth, and you are simply returning that charge to Earth. Even if the carpet is not well connected with Earth, by sharing your accumulated charge with Earth as a whole, your portion of it will be very small—Earth is large! By touching an object—here, it is you—to Earth, we say that the object is **grounded**. When a negatively charged object is grounded, electrons flow from the object to the ground, leaving the object (very nearly) neutral. If, instead, an object has an excess positive charge, then electrons flow from the ground and neutralize the object. The electric shock associated with a carpet is small, but there is potential for more damaging electric shocks around the home from the electricity supply. We therefore tend to ground any objects that might become highly charged, so that any excess charge will travel into Earth and not to you (Fig. 21–2). ■

▲ **FIGURE 21–2** The large copper rod that is being pounded into the ground will serve as the electrical ground for the household electrical service box.

Evidence That Charges Are of Two Types

A small mass will react to forces more visibly than a large mass, so we use small (initially neutral) masses to study the effects of forces between charges. We can do this by using small balls made of cork and coated with a conducting paint that allows charge to move around easily on the surface. Such a ball is hung by a thin insulating thread (Fig. 21–3a). If we touch a negatively charged Teflon rod to a cork ball, the ball is immediately repelled by the rod (Fig. 21–3b). If we touch the negatively charged Teflon rod to *two* suspended (neutral) cork balls, the balls will then strongly repel each other (Fig. 21–3c). Similar behavior occurs between two cork balls that have been touched by a positively charged glass rod, as they will both have acquired a positive charge. However, if we touch the Teflon rod to one cork ball and the glass rod to one cork ball, the oppositely charged balls attract each other (Fig. 21–3d).

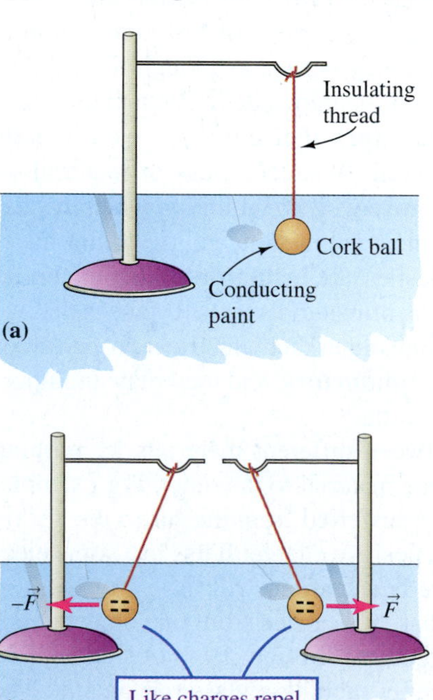

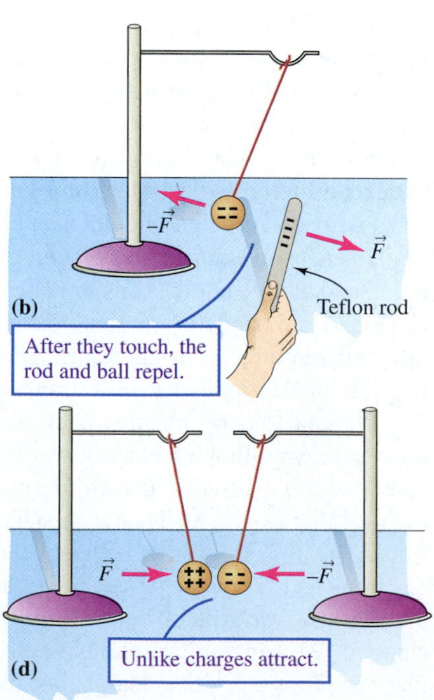

▶ **FIGURE 21–3** (a) An insulated cork ball covered with a thin layer of conducting paint can indicate the presence of small electric charges. (b) A negatively charged Teflon rod approaches the neutral coated cork ball, which is initially attracted to the rod. After the rod touches the ball, the ball becomes charged and strongly repels the charged rod. (c) If we touch two initially neutral cork balls with a negatively charged Teflon rod, the two balls repel each other: Like charges repel. (d) If we touch one initially neutral cork ball with a negatively charged Teflon rod and a similar ball with a positively charged glass rod, the two balls attract each other: Unlike charges attract.

We conclude from these experiments that the electric charges on the Teflon and glass rods are different, and that

Like charges repel, and unlike charges attract.

For example, when the Teflon rod touches the cork ball, some of the rod's negative charge is transferred to the ball. Now both the ball and the rod have a negative charge. The ball, which has been charged by conduction, immediately jumps away from the rod. Our other observations are similarly explained by the rule that like charges repel and unlike charges attract.

Another, more subtle, effect is also present. Before the negatively charged Teflon rod actually touches the neutral cork ball, the ball is *attracted* to the rod, not repelled by it. How can we explain this initial attraction? Because we have coated the cork ball with conducting paint, there are mobile electrons on the surface of the ball. When the negatively charged Teflon rod comes near, the mobile electrons are repelled and move to the far side of the cork ball (Fig. 21–4). That leaves an equal amount of excess positive charge on the area of the ball near the rod. If we additionally infer that the electric force is stronger when the charges are closer, those positive charges are attracted to the rod more strongly than the negative charges on the other side of the ball are repelled. In other words, when the positive charges on the cork ball are closer than the ball's negative charges to the Teflon rod, the *net* force is attractive. We call this phenomenon, in which charges within an object are redistributed due to the presence of external charge, **charge polarization**. Figure 21–5 illustrates an 18th-century demonstration of charge polarization. The fact that electrical forces between interacting charges weaken with distance—the inference that we had to make to explain the initial attraction of the ball and rod—is of great importance, and we shall return to it.

▲ **FIGURE 21–4** The neutral cork ball is initially attracted to the charged Teflon rod because some electrons on the ball move to the far side due to the repulsive force from the rod. The positive charges on the ball are on average closer to the rod, so the attractive force on them due to the rod is greater than the repulsive force on the redistributed electrons.

◀ **FIGURE 21–5** An eighteenth-century experiment on static electricity by Stephen Gray. The boy, suspended in air, carries a net charge (positive, let's say). As a result, charge polarization is induced in electrically neutral bits of paper near him. Negative charges on each paper bit tend to move toward the boy, leaving positive charges on the part of the paper bit farthest from the boy. The result is a net attractive force between the boy and each bit of paper.

Charge by Induction

An experiment closely related to the charge polarization experiments explains how initially neutral conductors can obtain a *charge by induction*, or an *induced charge*. Consider two neutral metal spheres, each standing on an insulated post and in side-by-side contact, so that they form a single conductor (Fig. 21–6a). If we bring a negatively charged Teflon rod close to one sphere, mobile electrons in the spheres move to the opposite side of the far sphere, leaving opposite charges on the two spheres (Fig. 21–6b). The spheres have a total charge of zero, but one is positive and the other negative. While the Teflon rod is still near, we separate the two spheres, leaving them oppositely charged (Fig. 21–6c). If we now remove the Teflon rod, the charges induced by the rod will remain on the two metal spheres (Fig. 21–6d). The spheres have been **charged by induction**. These charges can be transferred to two coated cork balls, by bringing a ball in contact with each sphere. The cork balls attract, demonstrating that the charges are opposite in sign. Note that only conductors can be charged by induction.

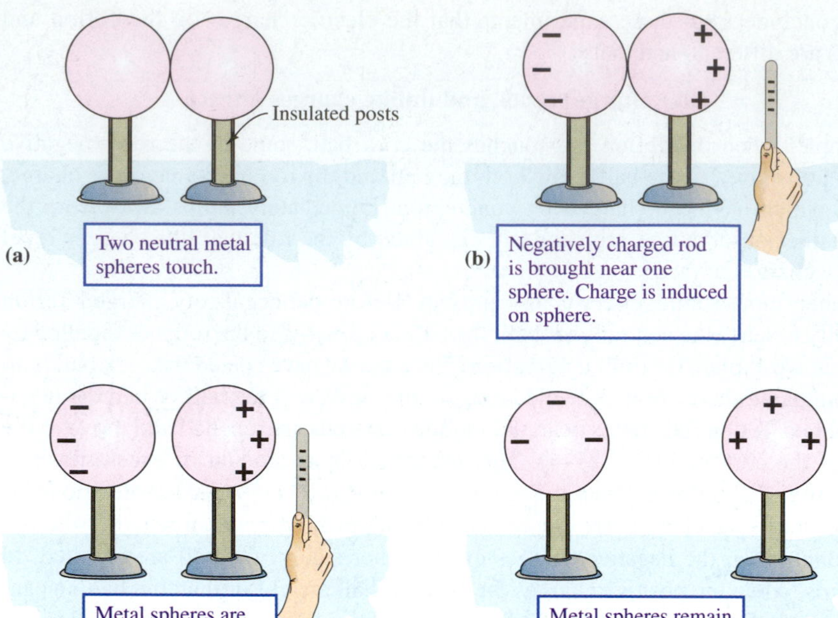

FIGURE 21–6 (a) Two neutral metal spheres on insulated posts touch. (b) A negatively charged Teflon rod polarizes the metal spheres. (c) If the metal spheres are separated while the Teflon rod is nearby, the spheres are charged oppositely. (d) When the Teflon rod is removed, the two metal spheres are still charged oppositely. Note that the total charge of the two spheres remains zero throughout.

Units of Charge

While the amount of charge on an electron is fixed, the numerical value of this amount of charge depends on how the scale for charge is defined. The SI unit of charge is called the **coulomb** (C). We can define the value of the coulomb by, say, specifying the magnitude of the force between two objects, each carrying 1 C of charge, and separated by a distance of 1 m. We'll assume that a procedure of this type has been carried out [see below Eq. (21–7)].

The magnitude of the charge on the electron—the smallest charge that can be isolated in nature—has been measured to high precision. An approximation sufficient for our purposes is

$$e \cong 1.60 \times 10^{-19}\ \text{C}. \tag{21–1}$$

The mass and charge of the neutron, proton, and electron are given in Table 21–1 and also in the tables behind the front cover.

TABLE 21–1 • Mass and Charge of Atomic Constituents

	Mass (kg)	Charge (C)
Neutron, n	1.675×10^{-27}	0
Proton, p	1.673×10^{-27}	1.602×10^{-19}
Electron, e^-	9.11×10^{-31}	-1.602×10^{-19}

EXAMPLE 21–1

A glass rod rubbed with silk has a charge of $+110\ \text{nC}\ (110 \times 10^{-9}\ \text{C})$. By how many electrons is the rod deficient?

Setting It Up We have made an unspoken (and reasonable) assumption that the rod has started off electrically neutral. Thus the rod acquires a positive charge through either an excess of positive charge or a deficiency in negative charge. In thinking in terms of a deficiency, we are saying that the positive charge has come from transferring electrons from the rod.

Strategy Because we know that each electron has a charge of magnitude e, we can find the number of transferred electrons by dividing the remaining charge by e.

Working It Out We have

$$\text{transferred electrons} = \frac{\text{net charge}}{\text{charge magnitude of each electron}}$$
$$= \frac{110 \times 10^{-9}\ \text{C}}{1.6 \times 10^{-19}\ \text{C/electron}}$$
$$= 6.9 \times 10^{11}\ \text{electrons}.$$

What Do You Think? Could we detect the number of missing electrons by the change in mass of the charged rod? You can find the mass of the electron in Table 21–1. *Answers to **What Do You Think?** questions are given in the back of the book.*

EXAMPLE 21–2 The largest American Eagle gold coin has a mass of 28.4 g. The atomic number of gold—the number of protons in the nucleus of an atom of gold—is 79, and thus the number of electrons in a neutral gold atom is also 79. The atomic mass of gold is 197, which means that 1 mol of gold has a mass $m_{\text{Au}} = 197$ g. How many electrons are contained in one pure gold coin? What is the total negative charge contained in the coin?

Setting It Up The only piece of information not specifically contained here is the fact that a mole of any element contains Avogadro's number, $N_A = 6.02 \times 10^{23}$, of atoms.

Strategy Avogadro's number N_A is the number of atoms in 1 mol of gold; the number of atoms in a given mass m of the gold is the ratio $(m)/(m_{\text{Au}})$ times N_A. We multiply this number by 79 electrons/gold atom to find the total number of electrons in the coin. Finally, we multiply the resulting number by the charge of one electron to find the total negative charge.

Working It Out The number of gold atoms in a mass of $m = 28.4$ g is

$$\frac{m}{m_{\text{Au}}} N_A = \frac{(28.4\text{ g})}{(197\text{ g/mol})}(6.02 \times 10^{23}\text{ atoms/mol}) = 8.68 \times 10^{22}\text{ atoms}.$$

The total number of electrons is then

$$\text{number of electrons} = (79\text{ electrons/atom})(8.68 \times 10^{22}\text{ atoms}) = 6.86 \times 10^{24}\text{ electrons}.$$

The total charge of these electrons is

$$\text{total electron charge} = (\text{number of electrons})(\text{charge per electron}) = (6.85 \times 10^{24}\text{ electrons})(-1.60 \times 10^{-19}\text{ C/atom}) = -1.1 \times 10^{6}\text{ C}.$$

What Do You Think? (a) If every electron were removed from the coin, what would be the charge left behind? (b) If one electron were removed from each gold atom, what would be the charge of the coin?

The Electroscope

The *electroscope* is a device used to detect excess free charge (Fig. 21–7a). There are two ways to use the electroscope. First, when charge is directly transferred to the electroscope by touching the metal ball at the top of the electroscope with a charged rod, the gold leaf inside separates from the vertical metal stem. This follows because the charge that is transferred to the electroscope is distributed throughout it, including to the leaf and stem; these each carry a charge of the same sign and hence repel one another, as in Fig. 21–7b. The leaf moves away from the stem until the vertical component of the electrostatic repulsion is balanced by the force of gravity on the leaf. Addition of more charge moves the leaf still more, and the angle made by the leaf is a measure of the amount of charge involved.

We also find that if we bring a charged rod near the metal ball at the top of the uncharged electroscope, the gold leaf still separates. What is happening? The overall charge on the electroscope is zero, and since the rod never actually touches the electroscope, the overall charge remains at zero. However, when the charged rod (positively charged, say) is brought close to the metal ball at the top of the electroscope, a negative charge is induced on the ball, as electrons are attracted by the positive charge on the rod. The positive ions collect on the leaf and stem, leaving them both positively charged. Again the leaf and stem repel (Fig. 21–7c). (An example of how this approach can be made more quantitative is given in Problem 10.)

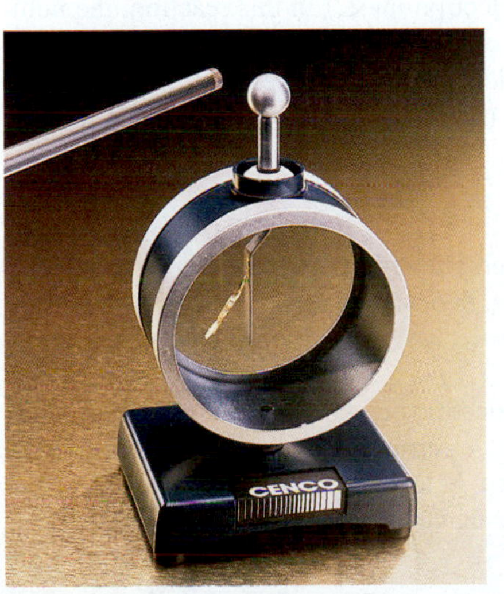

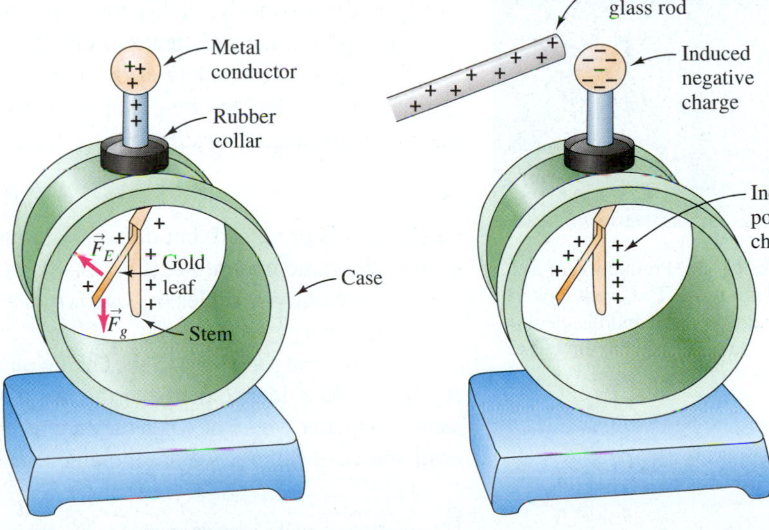

FIGURE 21–7 (a) An *electroscope*, a device that detects the presence of charge. (b) When free charge is added to the metal conductor, the gold leaf and the vertical metal stem repel, causing the gold leaf to move away from the stem. (c) When a charged object is brought close to the metal ball at the top, a charge is induced on the ball, leaving a charge of the opposite sign on the stem and leaf. They again repel.

THINK ABOUT THIS. . .

HOW DO SMOKE DETECTORS WORK?

The smoke detectors common in many homes depend on the separation of charges. They contain a small radioactive source centered within a cylinder that is open at one end. The source steadily emits alpha (α) particles (nuclear constituents) with enough energy to knock electrons from air molecules with which the α particles collide. The outer cylinder contains a negative charge, and the positive ions produced when the air molecules are broken up are attracted to it. The rate at which the ions arrive at the outer cylinder is measured electronically. When there is a house fire, large organic molecules such as those contained in bacon fat or pinewood resin enter the cylinder. These large molecules are an easy target for the α particles, and as electrons are more loosely bound in these molecules, they are also more likely to be removed from the molecule in the collision. The increased number of ions arriving at the cylinder sets off the alarm. The only disadvantage to this system is that any large molecule in the air—paint solvents or ordinary cooking smoke, for example—will set off the alarm. Many detectors have on/off switches for such cases.

21–2 Charge Is Conserved and Quantized

The simple experiments described in Section 21–1 strongly suggest that *charge is conserved.* Further experiments show that the **conservation of charge** is a fundamental physical law: *Net* charge is the same before and after any interaction. Moreover, charge conservation is *local*. That means that if we have a big box, the charge is conserved not only in the box as a whole but in any subvolume, down to as small as we can measure with precision. If there is less charge in a subvolume than before, then that amount of charge will have crossed the boundary on its way out.

Evidence of Charge Conservation

The reactions of subatomic particles such as nuclei or their constituents allows us to test charge conservation at a fundamental level. Let's look at some of these reactions. One of the reactions between atomic nuclei that takes place in a nuclear reactor is[†]

$$n + {}^{235}_{92}\mathrm{U} \rightarrow {}^{143}_{56}\mathrm{Ba} + {}^{90}_{36}\mathrm{Kr} + 3n + \text{energy}.$$

Here, the total number of protons (92) is the same on both "sides" of the reaction.

Even when the number of electrons or protons changes during a reaction, the total charge remains unchanged. Thus, another reaction that can take place in a nucleus is *electron capture*,

$$e^- + p \rightarrow n + \nu,$$

where ν stands for a neutral particle called the *neutrino*. (The neutrino, unlike the neutron, has a mass much smaller than even the electron mass.) In this reaction, the numbers of both protons and electrons change, but charge is still conserved.

Other particles, called *positrons*, are identical to electrons except for the *sign* of the charge, and are denoted by e^+. In the reaction (γ is the symbol for the photon, a package of electromagnetic radiation)

$$\gamma + p \rightarrow p + e^+ + e^-,$$

▲ **FIGURE 21–8** Production of an electron–positron pair. The event took place in a magnetic field, and the electron and positron spiral in opposite directions in this field as they lose energy.

an electron is produced, but then only in partnership with a positron, whose charge has exactly the same magnitude (Fig. 21–8). In fact, in observed reactions involving the so-called elementary particles, *no one has ever seen a single case of net charge appearing or disappearing*.

Is it possible for a little of the charge on an electron or a proton to wear off, like paint? Again, all the evidence points to the fact that the electron and the proton charges are the same, no matter where or when they are measured. In looking at quasars (distant and powerful sources of light), we are looking at matter that existed billions of years ago (it has

[†]The superscript on the element symbol is the atomic mass, the sum of the numbers of protons and neutrons in one atom; the subscript is the number of protons.

taken that long for the light to reach Earth). Observations of the color of the light that quasars emit suggest that, to a very high accuracy, the properties of their atoms are identical to the properties of atoms here on Earth. This implies that the charge of electrons and protons are not only identical but have remained constant over billions of years.

CONCEPTUAL EXAMPLE 21–3 A new theory makes the unusual proposition that the equality of the magnitude of the electron and proton charge was not always true, so that perhaps a billion years ago they differed by one part in a billion. What sort of arguments could you advance to check whether this is true or not?

Answer We have already mentioned a couple of effects that a change in magnitude of both the electron and proton charges would make: differences in the color (spectrum) of the light emitted by atoms, for example. Such effects follow when, even though changed, the electron and proton each carry charge of *identical* magnitude. Here we are asked what would happen if the magnitudes were not the same, and that has even more dramatic consequences. If the relative numbers of electrons and protons were unchanged, it would mean that the entire universe would have a net charge. Indeed, each atom would also have a net charge, all of the same sign, and unless the difference in charge were extremely tiny, the repulsive force between atoms would overwhelm the attractive gravitational forces that allow matter to clump into galaxies and stars. Bulk matter would not form in the way that we know it, at least not on the time scale of the universe.

Charge Quantization

We have already indicated that charges appear to be organized in discrete amounts. The magnitude of this minimum amount of charge is that of one electron. Greater charges are always multiples of these values. The facts that, within experimental accuracy, charge occurs in integral multiples of the electron charge, known as **charge quantization**, and that charges are never observed with values smaller than the electron charge were first established in 1909 through the pioneering experiments of Robert Millikan.

In summary, we can say that

Charge is conserved absolutely

and that

Free charge is quantized in positive or negative integral multiples of *e*.

THINK ABOUT THIS. . .

DO QUARKS CARRY CHARGE THAT IS LESS THAN THE ELECTRON CHARGE?

In 1964, Murray Gell-Mann and George Zweig proposed that protons and neutrons are composed of even more fundamental particles, called **quarks**, whose charges are either $2e/3$ or $-e/3$. Thus the proton, say, contains three quarks whose total charge adds up to $+e$: two charge $(+2e/3)$ quarks and one charge $(-e/3)$ quark. There is strong experimental evidence that quarks really do make up particles such as protons, but strangely enough, and for reasons that are only partially understood, quarks cannot be isolated—they cannot be removed and separated from a proton in the way that a hydrogen atom can be separated into an electron and a proton. Despite many searches, quarks or any other freely moving object carrying fractional electron charges have never been observed. Most physicists now believe that only combinations of quarks possessing a net charge that is an integer multiple of e can ever be isolated and independently observed. We refer to any charge that can be isolated as **free charge**. ■

21–3 Coulomb's Law

Encouraged by Benjamin Franklin, Joseph Priestley concluded in the mid-eighteenth century from Franklin's and his own experiments that the electric force between two charged objects varies as the inverse square of the distance between the objects. Priestley made this deduction after he observed that there is no charge on the inside surface of a closed or nearly closed metal vessel—all the charge is on the outside surface—and that the force on a charged object placed inside such a vessel is zero. This is like the phenomenon we discussed in Chapter 12: There is no gravitational force on an object inside a uniform spherical shell of matter. As we argued in Chapter 12, this result is a direct consequence of the $1/r^2$ nature of the force law. By analogy with gravitation, Priestley argued that the electric force responsible for his observations must have a $1/r^2$ dependence.

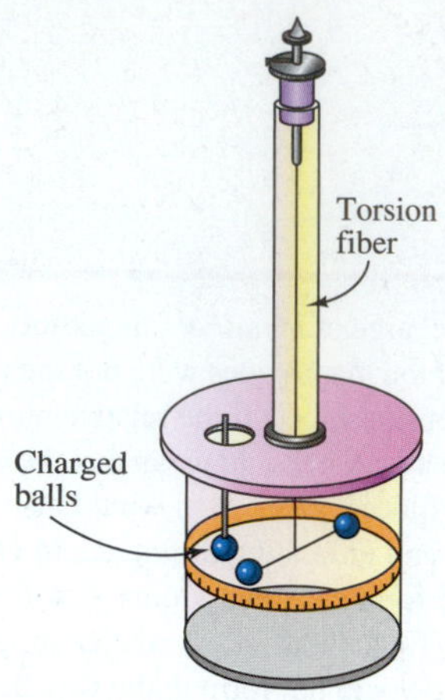

▲ **FIGURE 21–9** Coulomb's torsion balance, used to verify the inverse-square form of the force between electric charges.

In 1785, Charles Coulomb directly determined the force law for electrostatics. He performed the relevant experiments with a torsion balance similar to the one Henry Cavendish would use in 1798 to measure the gravitational constant, G (Fig. 21–9). The role played by massive balls in the Cavendish experiment (see p. 342) is here played by charged ones. Coulomb showed that the electrostatic force is central—directed on the line between the charges—and varies as

$$F \propto \frac{1}{r^2}, \tag{21–2}$$

where r is the distance between the centers of the charge sources. By changing the charge on the balls, Coulomb inferred that the force is proportional to the product of the charges q_1 and q_2 on the balls:

$$F \propto q_1 q_2. \tag{21–3}$$

To demonstrate the results of Eq. (21–3), we can ground one cork ball, neutralizing it, and charge another identical ball, giving it net (unknown) charge q. After we touch the two balls together, they each have a charge of $q/2$. Then we measure the force between these two balls. Next, we ground one ball again to neutralize it, and touch the balls together once more. Thus, each has a charge of $q/4$, and we measure the force between them to have decreased by a factor of 4 for the same amount of separation. This set of results is consistent with Eq. (21–3): In the first case, $F \propto (q/2)(q/2) = q^2/4$; in the second, $F \propto (q/4)(q/4) = q^2/16$.

Combining Eqs. (21–2) and (21–3) gives us a first view of Coulomb's law, the electrostatic force law. The magnitude of the force is

$$F = \frac{k|q_1 q_2|}{r^2}, \tag{21–4}$$

where k is a proportionality constant. The force is attractive when the charges have opposite sign and repulsive when they have the same sign. Moreover, the force obeys superposition: The force on a given charge from a collection of charges is the sum of the forces due to each charge in the collection.

The constant k plays the same role that the constant G plays in Newton's law of universal gravitation. The magnitude of k depends on the units used for charge—here we use a system of units that is consistent with the SI. It is then possible to define the coulomb by assigning a value to k:

$$k = \frac{1}{4\pi\varepsilon_0}, \tag{21–5}$$

where ε_0 is known as the *permittivity of free space*. (We shall see later that the value of ε_0 follows directly from the defined value of the speed of light, so in this sense ε_0 is itself defined.) To four significant figures, the permittivity is

$$\varepsilon_0 \cong 8.854 \times 10^{-12}\ \mathrm{C^2/N \cdot m^2}. \tag{21–6}$$

The value of k (to four significant figures) follows from Eqs. (21–5) and (21–6):

$$k = 8.988 \times 10^9\ \mathrm{N \cdot m^2/C^2}. \tag{21–7}$$

Now that we have assigned a value to k, we can tentatively define the coulomb. From Eqs. (21–4) and (21–7), we say that

When the force between two identical charges separated by 1 m is equal to the numerical value of k in newtons (8.988×10^9 N), these charges are each 1 C.

The definitive definition of the coulomb will come after we have discussed the forces between electric currents in Chapter 29.

CONCEPTUAL EXAMPLE 21–4 An electrically neutral object can be divided into a piece with charge $+q$ and another piece with charge $-q$. If these pieces are widely separated, they attract one another. Is there any way to break a neutral object into two pieces that repel each other?

Answer If an electrically neutral object is divided into two charged pieces, charge conservation demands that the two pieces are oppositely charged with charges of equal magnitude. If these two pieces are separated by distances large compared to their size, then they will look like point charges to one another and so attract. However, things can be different when the pieces are separated by relatively short distances, because then by distributing the charges properly within the pieces, we can use the distance dependence of Coulomb's law to produce a net repulsion. For example, imagine that we have one large and one small piece—the small piece with charge of $+q$ and the large piece with overall charge $-q$. However, we have distributed the charge in the large piece as in Fig. 21–10. According to Coulomb's law, the effect on a given charge of a nearby charge is much bigger than the effect of distant charge. The effect of having the positive charge concentrated at the side of the large piece that is closest to the positively charged small piece means that the overall force between the pieces is repulsive—the two pieces repel at short distances, even though at large distances they attract. While this scenario might be difficult actually to prepare, it reminds us that the distance dependence of the force between charges, with its potentially very large $1/r^2$ factor, can have important effects.

What Do You Think? Consider the larger piece with its charge arranged as in Fig. 21–10. If charge were free to flow within this piece, is this the way the charge would arrange itself?

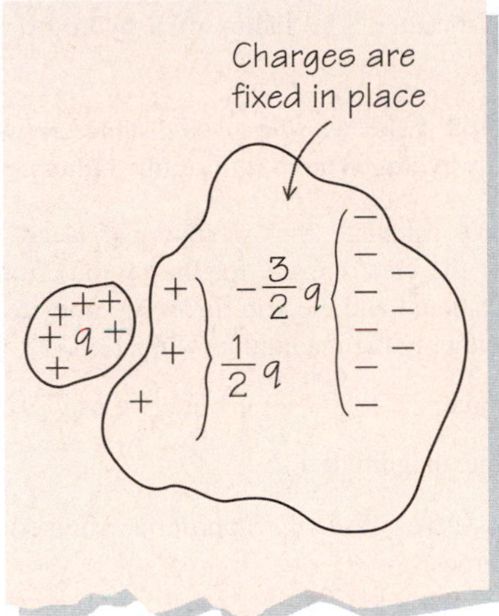

▲ **FIGURE 21–10** Two oppositely charged objects can repel at short distances if it is possible to rearrange the charge within one of them. Here the net charge $-q$ in the right-hand object has been shifted so that there is a positive piece very close to the positively charged object on the left. The magnitude of Coulomb forces falls off rapidly with distance, so that this arrangement could produce repulsion over short distances.

The electric force between point charges, the **Coulomb force**, has a direction and is described by a vector. We write Coulomb's law as

$$\vec{F}_{12} = \frac{1}{4\pi\varepsilon_0}\left(\frac{q_1 q_2}{r_{12}^2}\right)\hat{r}_{12}, \qquad (21\text{–}8)$$

COULOMB'S LAW

where $\vec{F}_{12}$ is the force exerted on point charge q_1 due to point charge q_2 when they are separated by a distance r_{12}. Newton's third law tells us that the force exerted on point charge q_2 due to point charge q_1 is then $\vec{F}_{21} = -\vec{F}_{12}$. The unit vector $\hat{r}_{12}$ is directed from q_2 to q_1 along the line between the two charges (Fig. 21–11). Note that if q_1 and q_2 have opposite signs, Eq. (21–8) indicates that the force is attractive, along $-\hat{r}_{12}$. But rather than remembering the subscripts on $\vec{F}$ and the unit vector $\hat{r}$, it is easier to remember that like charges repel and unlike charges attract.

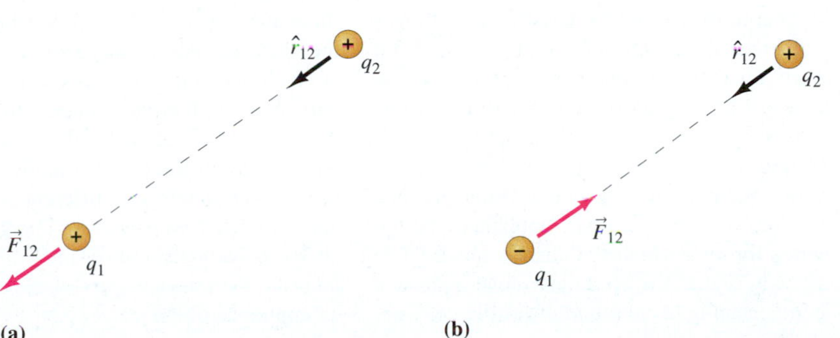

◀ **FIGURE 21–11** $\vec{F}_{12}$ is the force on q_1 due to q_2. The force is in the direction (a) $\hat{r}_{12}$ for like charges, and (b) $-\hat{r}_{12}$ for opposite charges.

EXAMPLE 21–5 Compare the electric force and the gravitational force between the single proton and single electron in a hydrogen atom. Assume a purely classical model of the hydrogen atom, in which the electron moves in a circular orbit around the proton, which is at the atom's center. The radius of a hydrogen atom is about 5×10^{-11} m.

Setting It Up Label as r the given distance between the proton and electron in a hydrogen atom (the atomic radius).

Strategy We calculate the gravitational force, obtaining the masses (m_e for the electron, m_p for the proton) from Table 21–1. Both the gravitational and electric forces are attractive in this case, and we calculate only the magnitudes. Equation (12–4) gives us the gravitational force, $F_g = \dfrac{Gm_em_p}{r^2}$, while we use Eq. (21–8) to find the electric force, magnitude F_E.

Working It Out With the appropriate numerical values for masses and charges,

$$F_g = \frac{(6.67 \times 10^{-11}\ \text{N}\cdot\text{m}^2/\text{kg}^2)(9.11 \times 10^{-31}\ \text{kg})(1.67 \times 10^{-27}\ \text{kg})}{(5 \times 10^{-11}\ \text{m})^2}$$
$$= 4 \times 10^{-47}\ \text{N}.$$

$$F_E = \frac{(9 \times 10^9\ \text{N}\cdot\text{m}^2/\text{C}^2)(1.6 \times 10^{-19}\ \text{C})^2}{(5 \times 10^{-11}\ \text{m})^2} = 9 \times 10^{-8}\ \text{N}.$$

$$\frac{F_E}{F_g} = \frac{9 \times 10^{-8}\ \text{N}}{4 \times 10^{-47}\ \text{N}} \cong 2 \times 10^{39}.$$

This calculation of the ratio could more easily have been carried out directly because it is independent of r — the common factor $1/r^2$ would cancel in the ratio. We have shown that on the atomic scale, the electric force is much greater than the gravitational force and that we are justified in ignoring gravitation at the atomic level.

What Do You Think? If the force due to gravitation is so much smaller than that of the electric force, why do we even notice it?

EXAMPLE 21–6 Two small cork balls are both charged to 40 nC and placed 4.0 cm apart. What is the magnitude of the electric force between them? Each cork ball has a mass of 0.46 g. Compare the magnitude of the electric force between them to the weight of one of the balls. What will happen if the balls are arranged vertically within a tube, with one ball placed 4 cm above the other?

Setting It Up Denote each charge by Q, each mass by m, and the separation of the balls by d. Figure 21–12 shows the cork balls (in the arrangement appropriate for the last question).

Strategy We use Coulomb's law, Eq. (21–8), to find the electric force between the balls, magnitude F_E, and then use $W = mg$ to find the weight.

Working It Out The electric force has magnitude

$$F_E = \frac{kQ^2}{d^2} = \frac{(9 \times 10^9\ \text{N}\cdot\text{m}^2/\text{C}^2)(40 \times 10^{-9}\ \text{C})^2}{(4.0 \times 10^{-2}\ \text{m})^2}$$
$$= 9.0 \times 10^{-3}\ \text{N}.$$

The weight of each cork ball is

$$W = mg = (0.46 \times 10^{-3}\ \text{kg})(9.8\ \text{m/s}^2) = 4.5 \times 10^{-3}\ \text{N}.$$

We conclude that if the balls start in a vertical orientation as in Fig. 21–12, separated by 4 cm, the (repulsive) electric force would lift the upper cork ball. The electric force would balance the gravitational force only if the balls started out a little farther apart—you could easily calculate that distance.

What Do You Think? What would have happened if the experiment with the vertical placing were carried out as above, but each ball had twice the mass?

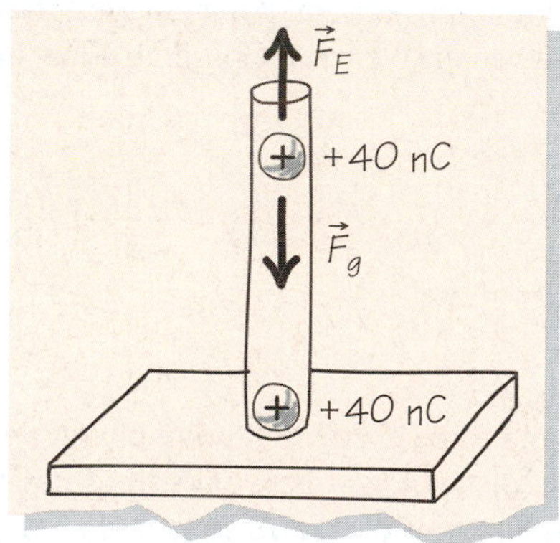

▲ **FIGURE 21–12** A small object can be suspended in space when equal but opposite gravitational and electric forces act on it.

THINK ABOUT THIS. . .

ARE THERE COULOMB FORCES BETWEEN HOME APPLIANCES?

A charge of 1 C is huge: Two such charges one meter apart exert a force of 9×10^9 N on each other. (By comparison, the force on you due to Earth's gravity is on the order of just 500 N.) In a typical household appliance, 1 C or more may move through the appliance every second, so why doesn't this produce enormous forces between appliances or the wires through which the charge moves? The answer is that the electrons move against a compensating background of positive charge, the stationary ions that are left behind when the electrons move away from their "parent" atoms in a conductor. Wires as well as the elements of appliances through which the electrons move are actually electrically neutral. Any Coulomb force is negligible. As we will see in later chapters, a different set of effects are associated with moving charges, even in an electrically neutral system, and these effects are the relevant ones for the operation of an appliance or device. ■

21–4 Forces Involving Multiple Charges

What happens if multiple charges are present? Experiment shows that the **principle of superposition** applies: The force on any one charge due to a collection of other charges is the vector sum of the forces due to each individual charge. In this respect, the Coulomb force is again like the gravitational force. The superposition principle allows us to find the force due to a set of charges on another charge or, for that matter, on another set of charges.

As an example of how superposition is applied, consider four charges, numbered 1, 2, 3, 4 (Fig. 21–13). The total force on, say, charge q_2 is the *vector sum* of the forces due to the other individual charges, q_1, q_3, and q_4:

$$\vec{F}_{2,\,\text{total}} = \vec{F}_{21} + \vec{F}_{23} + \vec{F}_{24}. \tag{21–9}$$

If there are N charges—$q_1, q_2, \ldots, q_N$—all acting on a charge q, the total force $\vec{F}$ on charge q is the vector sum of the individual forces $\vec{F}_i$ on charge q due to charge q_i:

$$\vec{F} = \sum_{i=1}^{N} \vec{F}_i = \frac{q}{4\pi\varepsilon_0}\sum_{i=1}^{N}\frac{q_i}{r_i^2}\hat{r}_i \tag{21–10}$$

The vector $\hat{r}_i$ is the unit vector from charge q_i to charge q. We have moved the common factor $q/4\pi\varepsilon_0$ out of the sum.

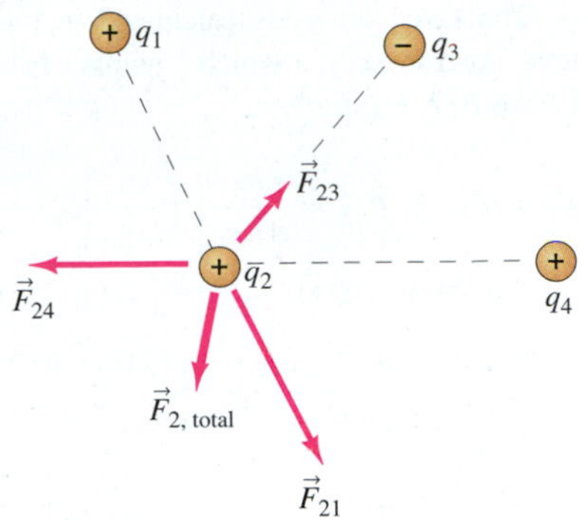

▲ **FIGURE 21–13** The superposition principle applies for multiple charges. The total force on charge q_2 is the vector sum of the individual forces on q_2 due to charges q_1, q_3, and q_4.

Problem-Solving Techniques

It is helpful to keep the following techniques in mind when looking at problems involving electric forces on a given charge in the presence of several other fixed charges or continuous distributions of charges:

1. Draw a clear diagram of the situation. Be sure to distinguish between the fixed external charges and the charges on which the forces must be found. The diagram should contain coordinate axes for reference.
2. Do not forget that the electric force that acts on a charge is a vector quantity; when many charges are present, the net force is a vector sum. In calculations, it is usually simplest to use unit vectors in a Cartesian coordinate system.
3. Search for symmetries in the distribution of charges that give rise to the electric force. When symmetries are present, the net force along certain directions will be zero. For example, if a point charge is midway between two identical charges, we know without performing any calculations that the net force on it will be zero.

EXAMPLE 21–7 Consider three point charges $q_1 = q_2 = 2.0$ nC and $q_3 = -3.0$ nC placed at the vertices of the triangle shown in Fig. 21–14. Find the net forces on q_1 and q_3, assuming that only Coulomb forces act.

Strategy The force on q_1, say, is due to the presence of charges q_2 and q_3. We find the vector forces on q_1 due to each of the charges q_2 and q_3 separately, then add them vectorially to find the net force on q_1. Coulomb's law depends on distance, and we will take these distances from Fig. 21–14. We can then do a similar calculation for the force on q_3.

Working It Out The force on q_1 is

$$\vec{F}_1 = \vec{F}_{12} + \vec{F}_{13} = \frac{q_1}{4\pi\varepsilon_0}\left[\left(\frac{q_2}{r_{12}^2}\right)\hat{r}_{12} + \left(\frac{q_3}{r_{13}^2}\right)\hat{r}_{13}\right].$$

From Fig. 21–14, we can deduce that $\hat{r}_{12} = -\hat{i}$ and $\hat{r}_{13} = -\hat{j}$. Thus

$$\vec{F}_1 = (9.0 \times 10^9\ \text{N}\cdot\text{m}^2/\text{C}^2)(2.0 \times 10^{-9}\ \text{C}) \times \left[\frac{(2.0 \times 10^{-9}\ \text{C})}{(2.0\ \text{m})^2}(-\hat{i}) + \frac{(-3.0 \times 10^{-9}\ \text{C})}{(2.0\ \text{m})^2}(-\hat{j})\right]$$
$$= (-9.0 \times 10^{-9}\ \text{N})\hat{i} + (14 \times 10^{-9}\ \text{N})\hat{j}.$$

The direction of force $\vec{F}_1$ is shown in Fig. 21–14.

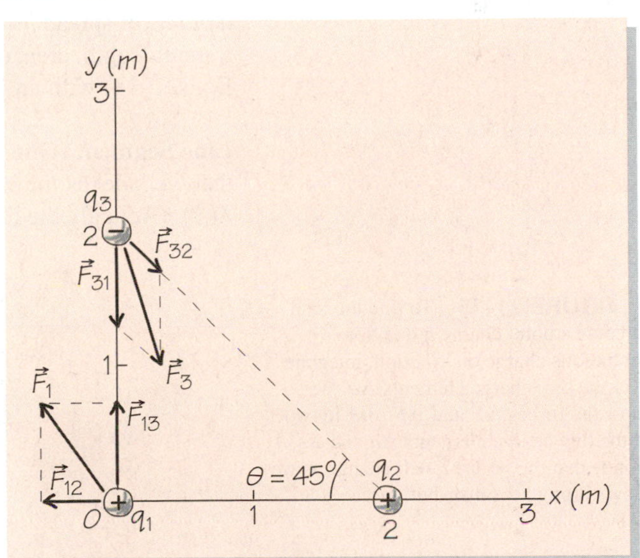

▲ **FIGURE 21–14** The positions of three point charges are indicated. Charges q_1 and q_2 are positive, while q_3 is negative. Forces $\vec{F}_{12}$ and $\vec{F}_{13}$ on charge q_1, and their resultant, $\vec{F}_1$, as well as forces $\vec{F}_{31}$ and $\vec{F}_{32}$ on charge q_3, and their resultant, $\vec{F}_3$, are drawn.

(continues on next page)

The force on q_3 is calculated in much the same way, with the unit vector $\hat{r}_{32}$, which points from q_2 to q_3, given by $(-\cos\theta)\hat{i} + (\sin\theta)\hat{j}$:

$$\vec{F}_3 = \vec{F}_{31} + \vec{F}_{32} = \frac{q_3}{4\pi\varepsilon_0}\left[\left(\frac{q_1}{r_{31}^2}\right)\hat{r}_{31} + \left(\frac{q_2}{r_{32}^2}\right)\hat{r}_{32}\right]$$

$$= (9.0 \times 10^9\,\mathrm{N\cdot m^2/C^2})(-3.0 \times 10^{-9}\,\mathrm{C})$$

$$\times\left[\frac{(2.0 \times 10^{-9}\,\mathrm{C})}{(2.0\,\mathrm{m})^2}\hat{j} + \frac{(2.0 \times 10^{-9}\,\mathrm{C})}{(2.0\,\mathrm{m})^2 + (2.0\,\mathrm{m})^2}((-\cos\theta)\hat{i} + (\sin\theta)\hat{j})\right].$$

The angle θ is 45°, or $\pi/4$ rad, so $\vec{F}_3$ becomes

$$\vec{F}_3 = (-14 \times 10^{-9}\,\mathrm{N})\hat{j} + (4.8 \times 10^{-9}\,\mathrm{N})\hat{i} - (4.8 \times 10^{-9}\,\mathrm{N})\hat{j}$$
$$= (4.8 \times 10^{-9}\,\mathrm{N})\hat{i} - (19 \times 10^{-9}\,\mathrm{N})\hat{j}.$$

What Do You Think? If q_1 and q_2 are fixed at the given locations, can we place q_3 at some point such that the net force on it is zero? If the point exists, is it a stable equilibrium point?

Continuous Distributions of Charges

The fact that charge is quantized will have no physical consequence when we deal with charges that are much larger than e. Such charges are composed of large numbers of electrons or protons. We can normally treat a large collection of point charges as a *continuous distribution* of charge. This is entirely analogous to thinking of a large object as a continuous distribution of mass, even though we know it is made of individual atoms. The techniques for treating continuous charge distributions will be very similar to the techniques we developed for treating continuous mass distributions.

We consider first the interaction of a point charge q with a large continuous charge distribution (Fig. 21–15). The force on q due to the tiny element of volume shown, which contains charge Δq and is a distance r from q, is

$$\Delta\vec{F} = \frac{q}{4\pi\varepsilon_0}\frac{\Delta q}{r^2}\hat{r}.$$

In turn, the net force on q is the sum over the forces due to the elements Δq:

$$\vec{F} = \sum\Delta\vec{F} = \sum\frac{q}{4\pi\varepsilon_0}\frac{\Delta q}{r^2}\hat{r} = \frac{q}{4\pi\varepsilon_0}\sum\frac{\Delta q}{r^2}\hat{r}. \quad (21\text{–}11)$$

We have to keep in mind that this is a vector sum.

At this point, we have not been very specific about how actually to do the sum. Just as for masses we found a more concrete way to proceed by working with a mass density, here it is useful to use the idea of a *charge density*. We separate the cases according to whether the continuous distribution of charge is distributed *along a line*, spread *over a plane*, or spread *throughout a volume*. In each case, we replace the finite charge Δq in a small length, area, or volume with the infinitesimal charge dq, and replace the sum in Eq. (21–11) with an integral. Let us look at each of these distributions in more detail:

Line Segment (One Dimension): If the charge distribution is distributed along a line that we label as the x-axis, we denote the *linear charge density* (charge/unit length) by $\lambda(x)$. (Actually the line of charge does not have to be a straight line; x really serves as a

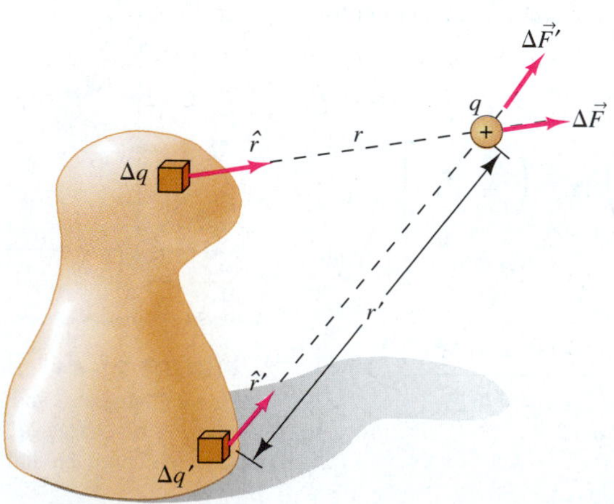

FIGURE 21–15 To find the total force on a point charge q due to a continuous charge distribution, integrate over the tiny charge elements Δq. We show the forces $\Delta\vec{F}$ and $\Delta\vec{F}'$ due to two of the tiny charge elements Δq and $\Delta q'$. Notice that the vector $\hat{r}$ will change as we move through the distribution.

way to label where you are along the distribution here.) The charge on an infinitesimal length dx of the line is

$$\text{for a charged line: } dq = \lambda(x)\,dx. \tag{21–12}$$

Note that λ can be a function of x; that is, the charge density can vary along the line. The force can be on the point charge q is then (Fig. 21–16)

$$\vec{F} = \frac{q}{4\pi\varepsilon_0}\int \hat{r}'\frac{\lambda(x)\,dx}{r'^2}. \tag{21–13}$$

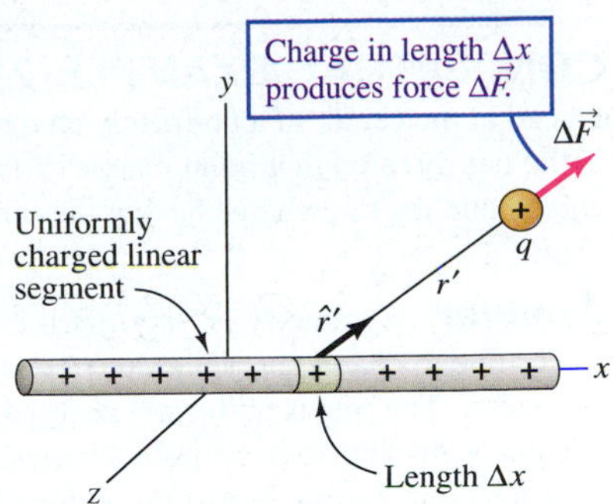

▲ **FIGURE 21–16** A one-dimensional charge distribution and the infinitesimal force on a point charge due to an infinitesimal piece of it.

Notice the meaning of the integration: We move along the line of charge, and each point along the line is a different distance r' and at a different direction $\hat{r}'$ from the charge q. Later we look at some examples to see how this works in practice. The line segment need not be straight. We must follow it wherever it goes in the integration of Eq. (21–13)—we will in the more general case refer to the integration as a line integral, as we did in our treatment of work (Chapter 6).

Surface (Two Dimensions): Here the charge is distributed across a surface. We denote the *surface charge density* (charge/unit area) by σ. The density σ could be a constant (uniform charge density), or it could vary from point to point on the surface. The charge on an infinitesimal area dS of the surface is (Fig. 21–17)

$$\text{for a charged surface: } dq = \sigma\,dS. \tag{21–14}$$

The force on the point charge q is (Fig. 21–17)

$$\vec{F} = \frac{q}{4\pi\varepsilon_0}\int_{\text{surface}} \hat{r}'\frac{\sigma\,dS}{r'^2}. \tag{21–15}$$

Here, we are integrating over all the elements of the surface, as indicated by the subscript of the integral sign. In practice, such integrals can be done by working out one-dimensional integrals.

Volume (Three Dimensions): When the charge is distributed through a volume, we write the *volume charge density* of the distribution as $\rho(\vec{r}')$, which means that the infinitesimal charge dq contained in the infinitesimal volume dV is

$$dq = \rho\,dV. \tag{21–16}$$

In terms of the charge density of the continuous charge distribution, the *net* force due to the volume element shown in Fig. 21–15 is

$$\vec{F} = \frac{q}{4\pi\varepsilon_0}\int_{\text{volume}} \hat{r}'\frac{\rho\,dV}{r'^2}. \tag{21–17}$$

The integration is over the entire volume of the charge distribution, and that is why we have used the subscript. Again, such integrals often involve simpler one-dimensional integrals in practice.

In each of these cases, the argument of the charge distribution ρ is the vector displacement $\vec{r}'$, as it is the vector displacement from an element of the charge distribution to the point charge on which the force acts that is important. We may, however, have a *uniform* charge distribution, in which charge is distributed evenly throughout a region. In that case, the linear charge density λ is the total charge on the line divided by the length of the line, the surface charge density σ is the total charge on the surface divided by the area of the surface, and the volume charge density ρ is the total charge in the three-dimensional region divided by the volume of the region. All three quantities are constants that can be removed from the integral for the net force. Keep in mind that a uniform charge distribution is not possible with a conductor, within or on which charges are free to move.

The integrals that express the force may be simple to perform, particularly if there is some symmetry in the distribution. Without the symmetry, it may be hard to find an analytical answer for the integral, but numerical integration using a computer is always possible.

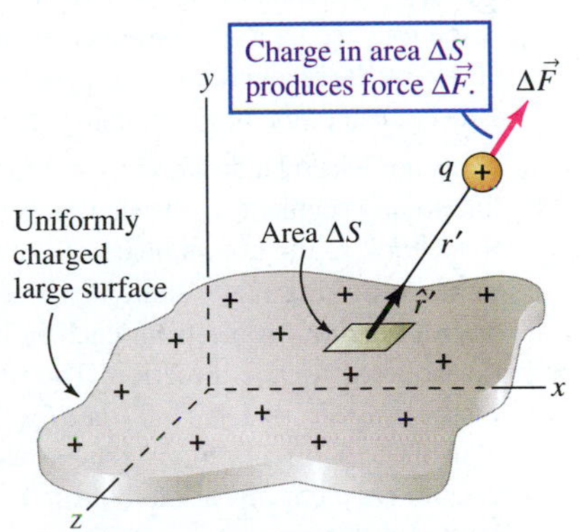

▲ **FIGURE 21–17** A two-dimensional charge distribution and the infinitesimal force on a point charge due to an infinitesimal piece of it.

CONCEPTUAL EXAMPLE 21–8 A point charge q is placed at the center of a uniformly charged ring (Fig. 21–18a). What is the net force on that point charge? Analyze this for both the point charge and the ring charge having the same sign and having different signs.

Answer Symmetry is very often a useful tool in dealing with forces due to charge distributions. In this case there is a good deal of symmetry. The ring is uniformly charged, and every point on the ring is equally distant from the point charge. There will always be equal and opposing forces due to the charged ring on either side of the charge q, as in Fig. 21–18b where we have identified areas of the ring that exert equal and opposing forces on point charge q. It does not matter whether q is positive or negative, because it will be repelled equally or attracted equally in all directions in the plane of the ring. The net force is zero, and no complex mathematical calculation is needed.

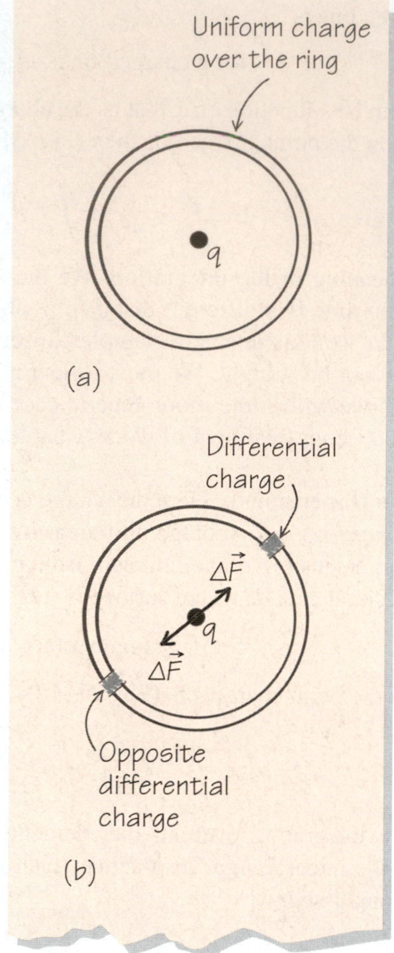

FIGURE 21–18 (a) The ring is uniformly charged. (b) Sections on opposite sides of the ring exert forces on the charge at the center that cancel.

EXAMPLE 21–9 Find the force on a point charge q_1 located on the axis of a uniformly charged ring of total charge Q. The radius of the ring is R, and q_1 is located a distance L from the center of the ring.

Setting It Up The geometry of the ring and point charge is shown in Fig. 21–19a, with appropriate labeling.

Strategy We want to find the force on q_1 due to a differential charge dq on the ring and then add the effects of all the parts of the ring. This is simplified by the recognition of the symmetry of this situation. Consider a small segment of the ring containing charge dq (Fig. 21–19a). *All* such segments are located a distance $r' = \sqrt{L^2 + R^2}$ from charge q_1, and the line to *any* segment on the ring makes the angle θ with the x-axis.

Next, look at the components of the force on q_1. Because every segment of the ring is the same distance r' from q_1, the *magnitude* of the infinitesimal force from each infinitesimal slice is the same. This is not true for the direction. The force from segment dq at the top of the ring ($z = 0, y = R$) is $d\vec{F}_{dq}$, and this force has components in the $+x$-direction and the $-y$-direction (Fig. 21–19b). The force from segment dq' at the bottom of the ring ($z = 0, y = -R$) is $d\vec{F}_{dq'}$, and this force has components in the $+x$-direction and the $+y$-direction. If the magnitude dq equals the magnitude dq', the y-components of the force will cancel each other while the x-components of the force will add. The y-components are the components perpendicular to the axis of the ring. This cancellation will hold for every perpendicular component of the force because we can always consider the charge elements in pairs. Thus we need compute only the component F_x by adding the (identical) infinitesimal components dF_x from each little element.

Working It Out The x-component from the element shown in Fig. 21–19a is

$$dF_x = \frac{q_1}{4\pi\varepsilon_0}\frac{dq}{r'^2}\cos\theta = \frac{q_1}{4\pi\varepsilon_0}\frac{\cos\theta}{(R^2 + L^2)}\,dq.$$

The net force has only an x-component and is the sum over the infinitesimal x-components:

$$F_x = \int dF_x = \int \frac{q_1}{4\pi\varepsilon_0}\frac{\cos\theta}{(R^2 + L^2)}\,dq.$$

At this point we see the symmetry come into play again: the coefficient of dq in the expression for dF_x is the same for every element and can therefore be placed outside the integral sign. Thus

$$F_x = \frac{q_1}{4\pi\varepsilon_0}\frac{\cos\theta}{(R^2 + L^2)}\int dq = \frac{q_1 Q}{4\pi\varepsilon_0}\frac{\cos\theta}{(R^2 + L^2)}.$$

We have used the fact that $\int dq = Q$, the entire charge. Finally, from trigonometry we find

$$\cos\theta = \frac{L}{\sqrt{R^2 + L^2}},$$

so

$$F_x = \frac{q_1 Q}{4\pi\varepsilon_0}\frac{L}{(R^2 + L^2)^{3/2}}. \qquad (21\text{–}18)$$

A check is always desirable, and we can immediately find one: When the point charge q_1 is very far from the ring, the ring should appear as a distant point of total charge Q, and the force should take on the Coulomb form $q_1 Q/(4\pi\varepsilon_0 L^2)$; this is indeed the limit of Eq. (21–18) when $L \gg R$.

What Do You Think? What is the force on q_1 when $L = 0$? (This is a second check on the result.)

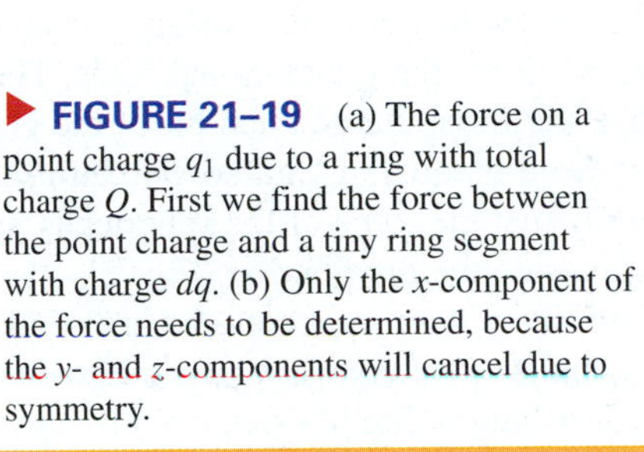

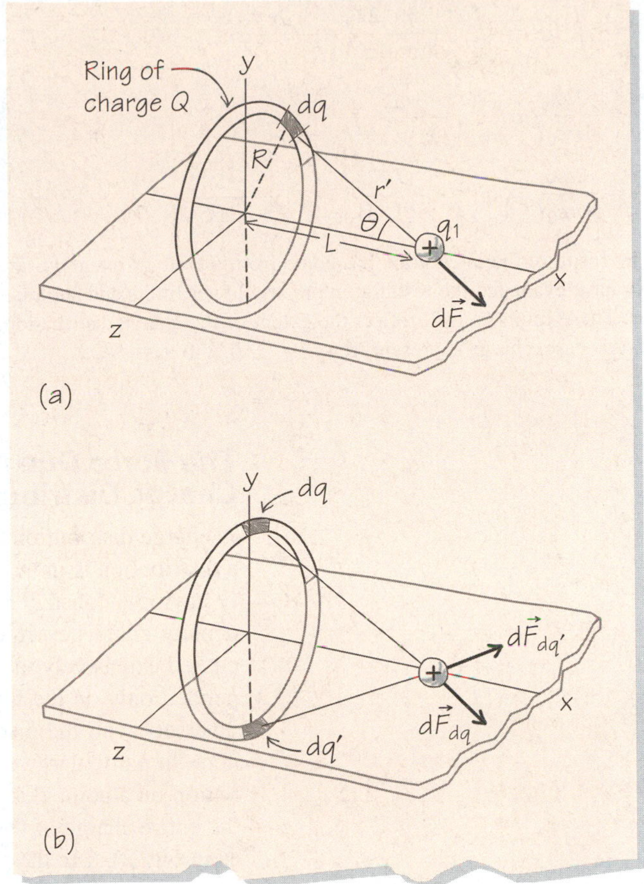

▶ **FIGURE 21–19** (a) The force on a point charge q_1 due to a ring with total charge Q. First we find the force between the point charge and a tiny ring segment with charge dq. (b) Only the x-component of the force needs to be determined, because the y- and z-components will cancel due to symmetry.

EXAMPLE 21–10 A straight rod of length L is aligned along the x-axis, with the ends at $x = \pm L/2$. The total charge on the rod is zero but the charge density is not; it is given by $\lambda(x) = 2\lambda_0 x/L$ (positive to the right of the origin, negative to the left). Find the force on a charge q located at a point $x = R$ on the x-axis, to the right of the right-hand end of the rod.

Setting It Up The geometry of the rod and of the charge on it is shown in Fig. 21–20a. The point charge for which we want to find the acting force is also indicated, at $x = R > L/2$.

Strategy We start by dividing the rod into elements that can be treated as points as far as figuring the force they exert on charge q. Thus we consider a thin slice of the rod with charge dQ located at point x, with thickness dx; this slice will have charge $dQ = \lambda(x)\,dx$. This is drawn in Fig. 21–20b. We find the force between each element of this type and q and then add these forces in the form of an integral. Although the symmetry in this case is not very marked, the integral will turn out to be an elementary one.

Working It Out With $dQ = \lambda\,dx = \dfrac{2\lambda_0}{L}x\,dx$, the infinitesimal force exerted by this charge on charge q is given by

$$d\vec{F} = \frac{q}{4\pi\varepsilon_0}\frac{2\lambda_0}{L}x\,dx\frac{1}{(R - x)^2}\hat{\imath}.$$

This force is aligned with the x-direction, as the unit vector indicates. The summation over the force due to the different slices to give the total force will also be aligned this way. This summation takes the form

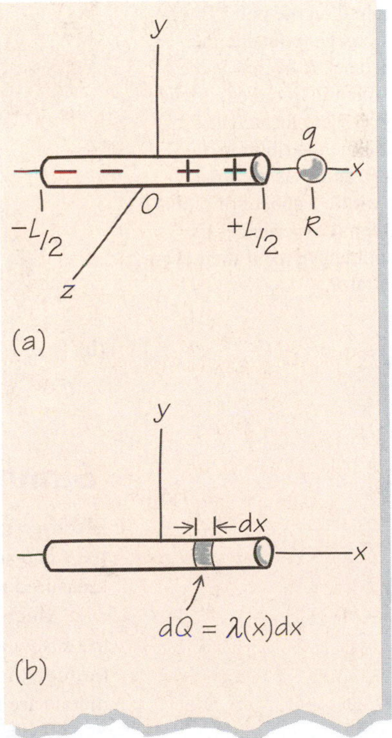

▲ **FIGURE 21–20** (a) A nonuniform charge density. (b) Isolating a section of the charge on the rod.

(continues on next page)

$$\vec{F} = \int d\vec{F} = \int_{-L/2}^{L/2} \frac{q}{4\pi\varepsilon_0} \frac{2\lambda_0}{L} \frac{x\,dx}{(R-x)^2}\hat{i}$$

$$= \frac{2q\lambda_0}{4\pi\varepsilon_0 L}\hat{i}\int_{-L/2}^{L/2} \frac{x\,dx}{(R-x)^2}$$

$$= \frac{2q\lambda_0}{4\pi\varepsilon_0 L}\hat{i}\left\{\ln\left[\frac{R-(L/2)}{R+(L/2)}\right] + R\left[\frac{1}{R-(L/2)} - \frac{1}{R+(L/2)}\right]\right\}.$$

The result of the integration has come from a table of integrals, and we have evaluated the result at upper and lower limits and subtracted. These limits, $\pm L/2$, reflect the extent of the charge distribution. In this case, charge is present from $x = -L/2$ to $x = L/2$.

If we drew a numerical plot of the factor in curly brackets in the expression for the force, we would see that this force is positive—if the charge q is positive it acts to push it to the right. This is sensible because the right side of the rod is positively charged, and the point charge is closer to the right side of the rod. The space variation of the coulomb force makes the repulsion from the closer part of the rod more important than the attraction from the farther part of the rod.

What Do You Think? Assuming the charge q is positive, what is the force on it when it is placed very close to the right end of the rod?

The Force Due to a Spherically Symmetric Charge Distribution

A charge distribution that is *spherically symmetric* is often quite easy to analyze. Such a distribution is in the form of a sphere centered at, say, point P, with the charge density having a constant value at a given (radial) distance from P. (Notice that the charge density could nevertheless vary with the radial distance from P.) This case was discussed extensively in Chapter 12 for the gravitational force. The results for gravity depended only on the fact that the force due to each bit of the charge distribution varies inversely with the distance from it squared, and so we can use the Chapter 12 results here. In particular, we can say that the force of the spherically symmetric charge distribution on a point charge q outside the distribution (Fig. 21–21a) is the same as though the entire charge of the distribution were concentrated at P (Fig. 21–21b). Moreover, if as in Fig. 21–21c the point charge q is inside any part of the distribution, then the force on q due to the part of the distribution that lies outside q is zero (Fig. 21–21d).

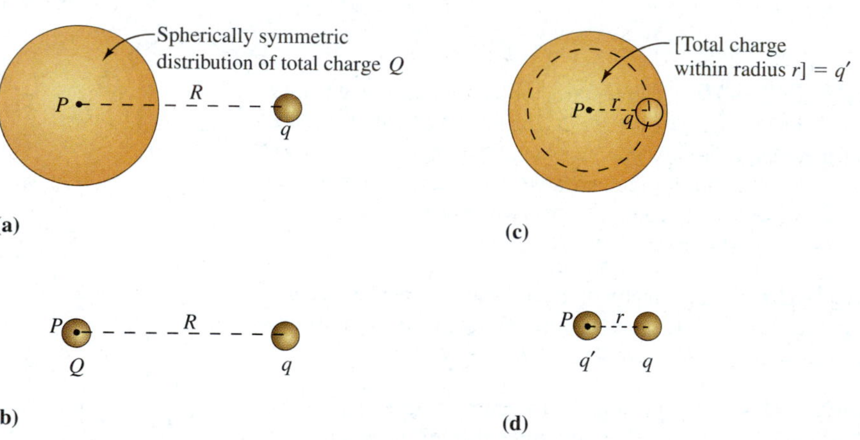

FIGURE 21–21 (a) A spherically symmetric charge distribution of total charge Q is centered on the point P. The force on a point charge q outside the distribution a distance R from P is the same as (b) the force it would experience if a point charge Q were located at P. (c) If q lies inside the distribution a distance r from P, and q' is the total charge that lies within a sphere of radius r centered on P, then it experiences the same force it would have (d) if there were a point charge q' at P.

Summary

Electric charge occurs in two forms, which we label as positive and negative charge. Charges of the same sign repel each other, and charges of unlike sign attract each other. In SI units, charge is measured in coulombs.

Much of the behavior of materials under the influence of electric forces is characterized by the ease with which electrons are dislodged from their constituent atoms and molecules and move through the material. Metals are normally good conductors of electric charge, whereas most nonmetals are not and are called insulators.

The basic electric charge is that of the electron. The electron has a charge of $-e$, and the proton has a charge of $+e$, with $e = 1.602 \times 10^{-19}$ C. Electric charge in matter is quantized in multiples of e. Charge is conserved in all interactions, meaning that the net charge before an interaction is the same as the net charge after the interaction.

The electric (Coulomb) force exerted by the point charge q_2 on the point charge q_1, when these are separated by a distance r_{12} is given by Coulomb's law:

$$\vec{F}_{12} = \frac{1}{4\pi\varepsilon_0}\left(\frac{q_1 q_2}{r_{12}^2}\right)\hat{r}_{12}, \tag{21–8}$$

where the factor $1/(4\pi\varepsilon_0)$ sets the units of charge.

The principle of superposition applies when multiple charges are present. The Coulomb forces on a point charge q due to all other charges add together vectorially. For continuous charge distributions, this addition takes the form of an integration, and the force of such a distribution on q depends on the charge distribution. For charges distributed on a line, over a surface, or through a volume, the force on the point charge q due to the distribution is, respectively,

$$\vec{F} = \frac{q}{4\pi\varepsilon_0}\int \hat{r}'\frac{\lambda(x)\,dx}{r'^2}, \tag{21–13}$$

$$\vec{F} = \frac{q}{4\pi\varepsilon_0}\int_{\text{surface}} \hat{r}'\frac{\sigma\,dS}{r'^2}, \tag{21–15}$$

and

$$\vec{F} = \frac{q}{4\pi\varepsilon_0}\int_{\text{volume}} \hat{r}'\frac{\rho\,dV}{r'^2}. \tag{21–17}$$

Here λ, σ, and ρ are the one-, two-, and three-dimensional charge densities, respectively.

On all but the astronomical scale, electric forces tend to be much stronger than gravitational forces. The electric force is responsible for making atoms, molecules, solids, and liquids stable, and all chemical reactions and biological processes are a result of electrical interactions.

Understanding the Concepts

1. Two identical positive charges are placed on a table and fixed there. Find all the places on the table where the net force on a test charge due to the two charges is zero.
2. Particles of opposite charge attract with an inverse-square law. Are there analogues of Kepler's laws for a system composed of such a pair, and what are they?
3. A balloon rubbed on a sweater and placed on a wall will often stay on the wall for a while. Explain how this happens.
4. When you walk across a carpet, you often pick up enough electric charge to cause a spark when you touch a doorknob. In climates that are dry in winter, this phenomenon is much more common in the winter than in the summer. Why?
5. Two metallic spheres on insulating stands are placed on an air-track. The mass of one sphere is five times larger than the other, and the charges are both positive in the ratio 3:1. The two objects are held at rest and then let go. What determines how far the two objects each move in a short time interval? How would you find the ratio of the distances that they travel in that interval?
6. By using the apparatus discussed in Section 21–1, how could you determine what charge you accumulate by walking across a wool rug?
7. Atoms consist of negatively charged electrons bound to the positively charged nucleus by the Coulomb force. The electrons are rearranged when two different chemicals are brought together. Would you expect the electrons that are closer to the nucleus or the ones that are farther from the nucleus to be more involved in chemical reactions?
8. Neutrons and protons are believed to be made of two types of charged particles called quarks, having charge $-1/3\,e$ and $2/3\,e$, as mentioned in Section 21–2. List the possible combinations of only three quarks that make up neutrons and protons.
9. When we unpack boxes, we often find that the "peanuts" used for cushioning stick to our hands, and it is difficult to shake them off. Why?
10. Some materials lose electrons easily by rubbing, so why are many of the objects around us not charged at all times?
11. *Earnshaw's theorem* states that a point charge cannot be in stable equilibrium while purely electrostatic forces act on the point charge. Consider a ring that is uniformly positively charged, with a positive charge at the center. It appears that the center charge suffers an identical repulsive force from every direction. How can the theorem be true?
12. How is the existence of a battery, which sends negative charges out of one of its contact points, consistent with the conservation of charge?
13. You have a cork ball with a charge of -4.8×10^{-19} C and three uncharged cork balls. Can you devise a method of touching cork balls together in sequence that will give a charge of -0.8×10^{-19} C to one of the balls?
14. You are given objects with two different charges. Can you determine whether the charges on these objects attract or repel? Can you determine with no further information whether the charges on the objects are positive or negative?
15. We spoke of generating a spark on a winter's day when we touch a conducting line to Earth and become grounded. Automobile tires are such good insulators that a car body is not connected to Earth by a conductor. How do you explain the spark that occurs when you touch a car door after you have rubbed the car upholstery?
16. Suppose that the electric charge of a fundamental particle such as an electron depends on the speed v of the particle, so that $e = e_0[1 + (\kappa v^2/c^2)]$, where e_0 is the particle's "rest charge," c is the speed of light, and κ is some tiny number. Discuss ways in which you might measure κ. Is there any experimental reason why κ must be small, if not zero?
17. Does the modification of the electric charge proposed in Question 16 necessarily violate the principle that it should not be possible to detect the absolute velocity of an object by means of any experiment?

18. The color of the light emitted by quasars is evidence that the charge on electrons has not changed over billions of years. Is saying that the charge on electrons and protons is unchanged equivalent to the statement that charge is conserved?

19. Suppose that electrons had charge $-e$ and protons had charge $+e(1 + \delta)$, with δ very small. Would there necessarily be an additional repulsive $1/r^2$ force between the Moon and Earth, for example, that could overpower the gravitational attraction between these bodies?

20. Consider a uniform, spherical positive charge distribution. A negative charge is placed at the center. Discuss the net force on that point charge. Discuss what happens to the point charge if it is placed a bit off center.

21. Describe what happens to $\vec{F}_1$ and $\vec{F}_3$ of Example 21–7 if the charge q_2 is doubled.

22. How would we know if at Alpha Centauri, the nearest star system nearest to us, the electric force had a $1/r^3$ dependence rather than the $1/r^2$ dependence on Earth?

Problems

21–1 Charge—A Property of Matter

1. (I) A cork ball is charged to $+1$ nC. How many fewer electrons than protons does the ball have?

2. (I) A uranium atom has undergone a violent collision that has stripped off 21 of its electrons. What is the charge of the resulting atom? If a uranium nucleus contains 92 protons and 146 neutrons, what is the charge of the nucleus?

3. (I) What is the total charge of all the electrons in 1 g of CO_2?

4. (I) Three identical metallic spheres are connected by wires, and a charge Q is placed on one of them. The wires are then removed. One of the spheres is then connected by wire to the ground. That wire is then removed. This particular sphere is then connected by a wire to one of the other spheres. What is the charge on each of the spheres when the process is completed?

5. (I) If you could remove 1 electron in 10^{13} from the gold coin of Example 21–2, how many would you be removing per atom, on the average?

6. (I) How many protons are contained in the gold coin of Example 21–2?

7. (II) A cork ball that is covered with conducting paint and charged to -4×10^{-10} C is touched by an identical but uncharged cork ball; the balls then separate. This second cork ball is then touched by a third uncharged cork ball, and they separate. What is the charge of each ball at the end, and how many excess electrons does each ball have?

8. (II) A cork ball covered with conducting paint is charged to -1.04×10^{-13} C. You have three similar but uncharged cork balls. Describe a method by which to produce a cork ball with a charge of -0.13×10^{-13} C. Do you need all three extra balls? Explain.

9. (II) An aluminum ball of mass 0.1 g is given a negative charge of 1 μC. What is the fractional increase in the number of electrons the ball contains?

10. (II) Two cork balls of mass 0.2 g hang from the same support point by massless insulating threads of length 20 cm (Fig. 21–22). A total positive charge of 3.0×10^{-8} C is added to the system. Half this charge is taken up by each ball, and the balls spread apart to a new equilibrium position. (a) Draw a free-body diagram for each cork ball. (b) What is the tension in the threads before the charge is added, and what is it after? (c) What is the value of angle θ in the figure? This device is a type of electroscope, or *electrometer*, a meter that measures electric charge. Angle θ measures the amount of charge on the balls if we can be sure that the charge is divided between them equally. This constraint is circumvented when the electrometer is made of a single strip of conducting material draped at its midpoint over a hook; the charge is then distributed over the strip equally, and half the strip repels the other half.

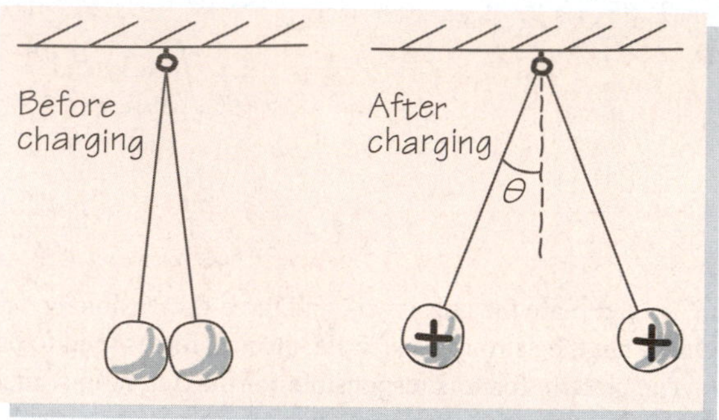

▲ **FIGURE 21–22** Problem 10.

11. (II) Silicon is the most abundant material on Earth's surface. (a) Assume that Earth is made of silicon (28 g/mol), and calculate the total number of negative charges contained within Earth. (b) When we neutralize a cork ball that has a charge of 1 μC by grounding it to Earth, what fractional change are we making in the total negative charge contained within Earth?

21–2 Charge Is Conserved and Quantized

12. (I) One possible result of the high-energy collision of two protons is the reaction $p + p \rightarrow X + p$. What is the electric charge of particle X?

13. (I) *Antiparticles* have the same mass as their counterpart particles but have an opposite charge. For example, the antiparticle of an electron, e^-, is the positron, e^+. Most antiparticles are denoted by a bar over the particle, so $\bar{p}$ is the antiparticle of the proton, and it has a charge of $-e$. Which of the following reactions satisfy the conservation of charge:
(a) $p + \bar{p} \rightarrow e^+ + e^- + e^+ + e^- + 2n$;
(b) $e^+ + e^- \rightarrow 2p + n + 2\gamma$,
(c) $e^+ + e^- \rightarrow e^+ + e^- + p + \bar{p} + 2\gamma$;
(d) $n + p \rightarrow e^- + p + \bar{p}$

14. (I) How much charge is contained in 6.5×10^{-4} g of electrons?

15. (II) The electric charge of an object is independent of the object's motion. Suppose that this were not true, but that the charge of a particle such as an electron or a proton that moves at speed v has the form $e = e_0[1 + (v^2/c^2)]$, where e_0 is the particle's charge when at rest and $c \cong 3 \times 10^8$ m/s is the speed of light. What would the net charge on a hydrogen atom be, assuming that the atom consists of a proton at rest and an electron orbiting the proton at average speed $v \cong (1/137)c$?

21–3 Coulomb's Law

16. (I) How far apart must two protons be for the Coulomb force on each other to be the same as the weight of one proton on Earth's surface?

17. (I) A proton is believed to consist of two "up" quarks of charge $+2/3\ e$ and one "down" quark of charge $-1/3\ e$. Assume that all three quarks are equidistant from each other at the distance of 1.5×10^{-15} m. What are the electrostatic forces between each pair of the three quarks?

18. (I) Two small balls, each of mass 16 g, are each charged with +8.5 nC. What distance apart must they be if the force on one of them has the same magnitude as the weight of that ball?

19. (I) Two identically charged sodium ions separated by 4.5×10^{-9} m have a force between them of 1.1×10^{-11} N. What is the charge of each ion, and how many electron charges does this represent?

20. (I) Two small cork balls have the same charge. When their centers are placed 2 cm apart, the force between them is observed to be 0.18 N. What is the cork balls' charge? Why do we have to assume that the size of the cork balls is small compared to 2 cm?

21. (I) Two tiny cork balls, both of mass 0.10 g, each have just one electron charge, $q = -1.6 \times 10^{-19}$ C. They are separated by 15 cm, which is much greater than their sizes. What is the ratio of the magnitudes of the Coulomb force between them to the gravitational force they exert on each other? Why is this result so different from that of Example 21–5?

22. (II) The experiment of Cavendish to determine the gravitational constant (see Chapter 12) relies on the measurement of a force of about 7×10^{-7} N between two masses separated by a distance of 0.1 m. One possible source of error is a small electric charge on the balls. Assuming the charges are equal, what is the magnitude of the largest allowed charge, if the force is to be measured to at least a 0.05 percent accuracy?

23. (II) Suppose that we were to measure a charge in some new unit, which we will call the esu, so defined that Coulomb's law reads, in magnitude, $F = q_1q_2/r^2$, and so that $F = 1$ dyne (10^{-5} N) when $q_1 = q_2 = 1$ esu and $r = 1$ cm. (a) How many esu are there in 1 C? (b) What is the charge of the electron in esu? (The esu is an actual unit, the *electrostatic unit*.)

24. (II) An electron and a proton attract each other with a $1/r^2$ electric force, just like the gravitational force. Suppose that an electron moves in a circular orbit about a proton. (a) If the period of the circular motion is 24 h, what is the radius of the orbit? (b) If the period is 4×10^{-16} s, as it is in a hydrogen atom, what is the radius of the orbit?

25. (II) A charge q is split into two parts, $q = q_1 + q_2$. In order to maximize the repulsive Coulomb force between q_1 and q_2, what fraction of the original charge q should q_1 and q_2 have?

26. (II) An alpha particle (a helium nucleus, composed of 2 protons and 2 neutrons) is directed onto a particular tungsten nucleus (^{184}W, with 74 protons and 110 neutrons). The alpha particle stops and turns around at a distance of 6.0×10^{-12} m from the tungsten nucleus (Fig. 21–23). Ignore the effects of electrons, and treat the alpha particle and tungsten nucleus as pointlike. What is the Coulomb force on the alpha particle at its closest approach to the nucleus?

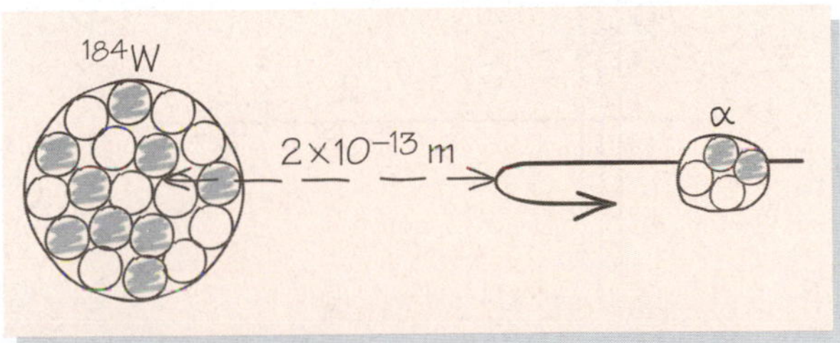

▲ **FIGURE 21–23** Problem 26.

27. (II) An electron orbits in uniform circular motion about a much heavier—and therefore nearly stationary—proton at a distance of 3×10^{-10} m. (a) What are the magnitude and direction of the Coulomb force exerted on the electron by the proton? (b) What is the speed of the electron in its circular orbit? (c) What is the frequency of the circular orbit? (d) Calculate the spring constant of a spring with an electron mass at its end and the frequency of part (c).

28. (II) Two pointlike objects are placed 8.75 cm apart and are given equal charge. The first object, of mass 31.3 g, has an initial acceleration of 1.93 m/s^2 toward the second object. (a) What is the mass of the second object if its initial acceleration toward the first is 5.36 m/s^2? (b) What is the charge of each object?

29. (II) Two cork balls, each of mass 0.20 g, are hung by insulating threads 20.0 cm long from a common point. The cork balls are given an equal charge by a Teflon rod. The balls repel and deflect as shown in Fig. 21–24. What charge q was given to each cork ball? Assume uniform charge.

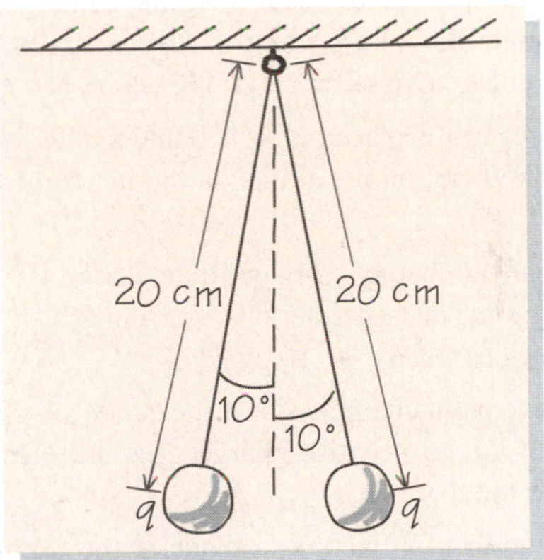

▲ **FIGURE 21–24** Problem 29.

30. (II) Astronomical data tell us that Earth's radius is 6.4×10^6 m, that its mass is 5.98×10^{24} kg, that the Moon's mass is 7.36×10^{22} kg, and that the mean Earth–Moon separation is 3.8×10^8 m. Suppose that, instead of being electrically neutral, as we believe, Earth and the Moon each have an excess positive charge of 8.5×10^{15} C. (a) What is the magnitude of the electrical repulsion between Earth and the Moon? (b) What is the ratio of this repulsive force to the attractive gravitational force? (c) If the charge on Earth were distributed uniformly throughout its volume, what would the excess charge density be, in coulombs per cubic meter (C/m^3)? (d) Assume that the excess positive charge is due to excess protons, which have an electric charge of 1.6×10^{-19} C. Calculate the density of protons, in units of protons per cubic meter, that corresponds to the conditions in part (c). (e) Earth's mean density is 5.52×10^3 kg/m^3, and a proton has mass 1.67×10^{-27} kg. Protons account for about half of Earth's mass. Compute the density of all protons in Earth, and compare this to your answer in part (d).

31. (II) Three unknown charges q_1, q_2, and q_3 exert forces on each other. When q_1 and q_2 are 15.0 cm apart (q_3 is absent), they attract each other with a force of 1.4×10^{-2} N. When q_2 and q_3 are 20.0 cm apart (q_1 is absent), they attract with a force of 3.8×10^{-2} N. When q_1 and q_3 are 10.0 cm apart (q_2 is absent), they repel each other with a force of 5.2×10^{-2} N. Find the magnitude and sign of each charge.

32. (II) An electron has a mass of 0.9×10^{-30} kg and a charge of -1.6×10^{-19} C. Earth's mass is 6×10^{24} kg, and its radius is 6.4×10^{6} m. Suppose that Earth has a net negative charge Q at its center. (a) How large would Q have to be for the charge repulsion on an electron to cancel the gravitational attraction at Earth's surface? (b) Suppose that this net charge is due to a discrepancy between the positive proton charge and the negative electron charge. Assume that half of Earth's mass is due to protons, each of which has a mass of 1.6×10^{-27} kg (the rest is neutrons, assumed to be neutral; electrons do not contribute much to the mass). What is the size of the charge discrepancy, compared to the electron charge?

33. (III) Use the similarity between Coulomb's law and the law of universal gravitation to calculate the distance of closest approach between a point charge of $+10^{-6}$ C, which starts at infinity with kinetic energy of 1 J, and a fixed-point charge of $+10^{-4}$ C. Assume that the moving-point charge is aimed straight at the fixed-point charge. [*Hint*: The similarity to gravity consists of using the notions of potential energy and energy conservation.]

21–4 Forces Involving Multiple Charges

34. (I) A charge $+4q$ is placed at $x = 0$, and a charge $+7q$ is placed at $x = 5$ units. Is there a point on the x-axis at which the net force on a charge Q is zero, and if so, where is it?

35. (I) A charge $-5q$ is placed at $x = 0$ and a charge $+3q$ is placed at $x = 10$. Where, on the x-axis, is the net force on a charge Q zero?

36. (I) Six identical charges of magnitude 3.6×10^{-8} C are placed on a straight line at 3-cm intervals, starting at $x = 0$. What is the force on the charge at $x = 15$ cm?

37. (I) A positive point charge q sits at the center of a circle of radius R on which a total negative charge Q is uniformly distributed. What is the net force on q?

38. (II) A charge of q is fixed on a plane at the origin $(0, 0)$ of an xy-coordinate system, and a charge $3q$ is fixed at $(3\text{ cm}, -3\text{ cm})$. Where must a charge of $-2q$ be placed at rest for it to be in equilibrium (that is, so that it remains at rest)? Is the equilibrium stable?

39. (II) What is the total force on each of the three quarks in Problem 17 due to the other two quarks?

40. (II) Three negative charges of magnitude 0.6 μC are placed at the corners of an equilateral triangle of sides 18 cm. What is the net force on a charge of 1.5 μC placed at the midpoint of one of the sides?

41. (II) Four positive charges $+q$ sit in a plane at the corners of a square whose sides have length d, as in Fig. 21–25. A negative charge, $-q$, is placed in the middle of the square. (a) What is the net force on the negative charge? (b) Is the equilibrium point at the center a stable equilibrium for motion of the negative charge in the plane of the square? (c) for motion of the negative charge perpendicular to the plane of the square?

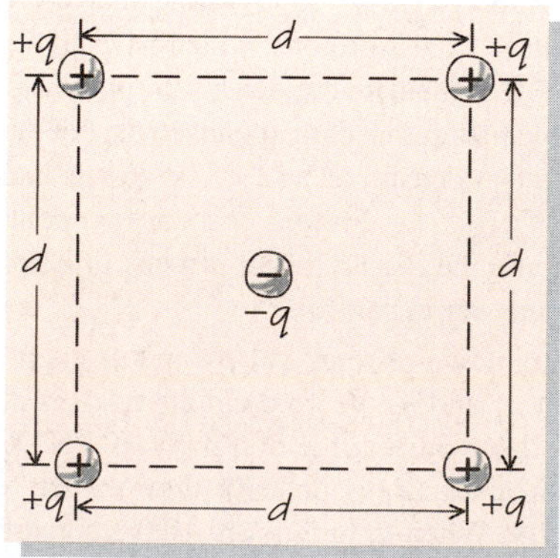

▲ **FIGURE 21–25** Problem 41.

42. (II) Calculate the force between two identical dipoles consisting of dumbbells with equal and opposite charges q and $-q$ at the end of a rigid rod of length $2d$. The dipoles are parallel, as shown in Fig. 21–26, and they are a distance x apart. Derive a first-order approximation for $d \ll x$. [*Hint*: Use $(1 + y)^n \cong 1 + ny + \cdots$ for $y \ll 1$.]

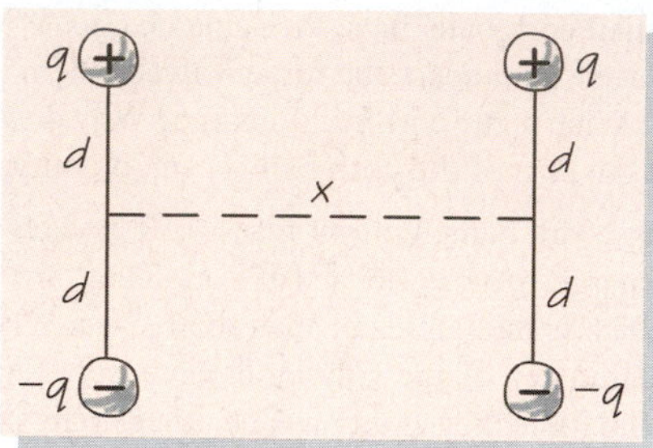

▲ **FIGURE 21–26** Problem 42.

43. (II) Charges q, $2q$, $-4q$, and $-2q$ (q is positive) occupy the four corners of a square of sides $2L$, centered at the origin of a coordinate system (Fig. 21–27). (a) What is the net force on charge q due to the other charges? (b) What is the force on a new charge Q placed at the origin?

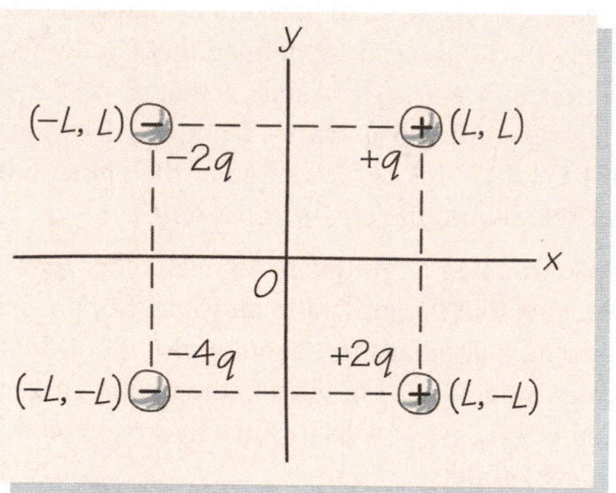

▲ **FIGURE 21–27** Problem 43.

44. (II) A charge Q is distributed uniformly along a rod of length $2L$, extending from $y = -L$ to $y = L$ (Fig. 21–28). A charge q is placed on the x-axis at $x = D$. (a) In what direction is the force on q, given that Q and q have the same sign? (b) What is the

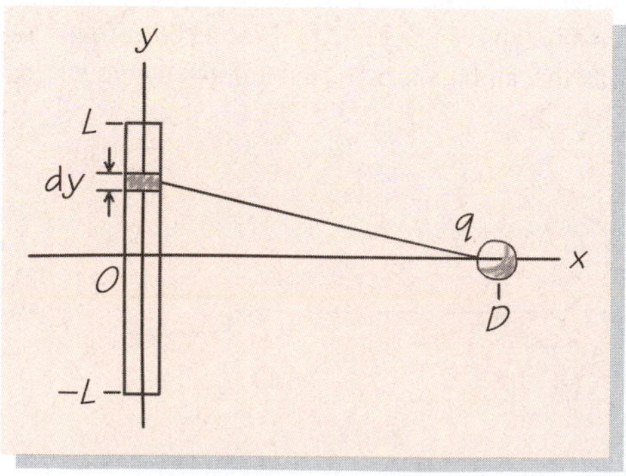

▲ **FIGURE 21–28** Problem 44.

charge on a segment of the rod of infinitesimal length dy? (c) What is the force vector on charge q due to the small segment dy? (d) Express an integral that describes the total force in the x-direction. (e) Compute the integral in order to find the total force in the x-direction.

45. (II) A charge is spread uniformly along the y-axis, stretching infinitely far in both the positive and negative directions. The charge density (charge per unit length) on the y-axis is λ. Find the force on a point charge q placed on the x-axis at $x = x_0$.

46. (II) A charge is spread uniformly along the y-axis from $y = 0$ to $y = +\infty$. The charge density on the y-axis is λ (Fig. 21–29). Find the force on a point charge q placed on the x-axis at $x = x_0$.

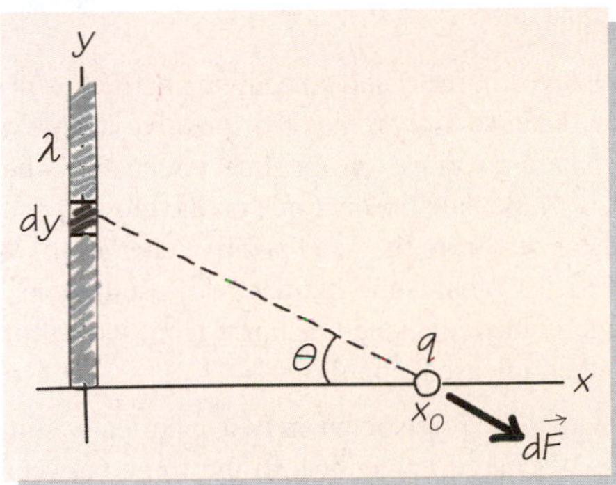

▲ **FIGURE 21–29** Problem 46.

47. (II) A long, thin rod of length L that contains a uniform distribution of charge Q points away from a point charge q. The nearest part of the rod is a distance d from the point charge. What is the electric force exerted on the charge q by the rod?

48. (II) Two uniformly charged rings of radii 25 cm and 40 cm respectively are placed parallel to each other, with a common axis. Each carries a charge of 2.2×10^{-4} C. and their centers are 100 cm apart. Where, along the common axis, should a charge q be placed so that the net force on it is zero?

49. (II) In the problem above, but now with the two rings having the same radius, the charge q is constrained to move only along the common axis. Is the position of equilibrium one of stable or unstable equilibrium? Does this depend on the sign of the charge q? [*Hint*: You may need the mathematical result that for $d \ll D$, $(D^2 + ad + bd^2)^n = D^{2n} + nadD^{2n-2}$ + terms that can be neglected.]

50. (II) A charge Q is distributed uniformly over a thin ring of radius R. The ring is oriented in the xy-plane, with its center at the origin. Find the force on a charge q located at the origin, and discuss the stability of its motion in the xy-plane. How does this compare with the case of a point charge placed at the center of a sphere whose surface is uniformly charged?

51. (II) Use the results of Example 21–9 to calculate the force on a positive point charge of magnitude 0.65 μC located 5 cm above the center of a uniformly charged solid plate of radius 8 cm that carries a total positive charge of 1.6 μC. [*Hint*: Break the disk into concentric rings, use the results of Example 21–9 for each ring, and sum over the forces due to the rings.]

52. (II) Calculate the force exerted on a charge q by an infinite plane sheet with surface charge density (charge per unit area) σ. [*Hint*: Break up the plane into concentric rings centered below the charge, use the results of Example 21–9 for the force from each ring, then sum over the forces due to the rings.]

53. (II) Two rigid plates of equal size, made of different plastics, are rubbed against each other. This results in equal and opposite charges on the two plates. How large are these charges if it takes 0.1 N to separate the two plates? The area of each plate is 0.05 m^2, and the charge distribution may be assumed to be uniform. [*Hint*: Use the result of Problem 52.]

54. (II) Consider an infinite vertical sheet that carries a charge density of $+1.2 \times 10^{-6}$ C/m^2. A cork ball of mass 8 g is suspended by a string 50 cm long at a distance of 55 cm from the charged sheet. What is the string orientation (a) if a charge $q = 0.8 \times 10^{-8}$ C is placed on the cork ball? (b) if instead a charge $q = -3 \times 10^{-8}$ C is placed on the ball?

55. (II) A total charge of 0.75 μC is distributed uniformly over a thin, semicircular wire of radius 5.0 cm. What is the force on a charge of 0.30 μC located at the center of the circle?

56. (II) A succession of $n + 1$ alternating positive and negative charges q are located along the x-axis at the points $x = 0$, $x = d$, $x = 2d, \ldots, x = nd$. An isolated charge Q is placed as shown in Fig. 21–30 at the point $x = D$ a very long distance away from the origin ($D \gg nd$). (a) Write a general expression for the electric force on charge Q. (b) Approximate your result, using the condition $D \gg nd$. Keep only leading and next-to-leading terms. [*Hint*: Use $(1 + x)^{-2} \cong 1 - 2x$ for $x \ll 1$.]

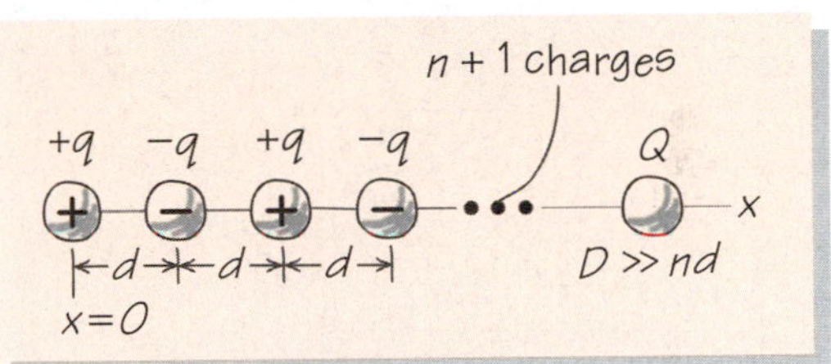

▲ **FIGURE 21–30** Problem 56.

57. (II) Charges $+q$, $-q$, $+q$, and $-q$ are placed along the x-axis, at positions $x = 0$, $x = 1$ cm, $x = 2$ cm, and $x = 3$ cm, respectively. What is the force on a charge $Q = +3q$ placed at the point $(x, y) = (1.5 \text{ cm}, y_0)$, where y_0 is a variable?

58. (III) Consider a charge of $e = 1.6 \times 10^{-19}$ C distributed uniformly over a sphere of radius $R = 0.5 \times 10^{-10}$ m. Place a point particle of charge $-e$ (-1.6×10^{-19} C) at the center of that sphere. Suppose that charge is displaced by a distance r (with $r < R$). Use the information given in Fig. 21–21 to show that the point charge will oscillate about the center of the sphere. Write down an expression for the frequency of oscillation in terms of R, the mass of the point charge m and the charge e. [*Hint*: If the acceleration for harmonic motion is given by $ma = -kr$, then the angular frequency of oscillation is $\omega = \sqrt{k/m}$.]

59. (III) What is the force per unit area between two infinite, uniformly charged plates with a surface charge density of $+10^{-5}$ C/m^2 and -10^{-5} C/m^2, respectively, when the distance between the plates is 10 cm? What if the distance between the plates is doubled? [*Hint*: You may use the result of Problem 52.]

General Problems

60. (II) A cone of height h whose radius at the open end is R carries a total charge Q. Assuming that the charge is uniformly distributed over the surface, what is the charge density in C/m^2?

61. (II) Earth has a net charge of about 6×10^5 C. Assume the charge is evenly spread on the surface of the Earth. A cork ball of

10-g mass is hanging from a thin thread so the cork ball is 10 cm from the Earth's surface. What charge would the cork ball have to have to just barely rise up due to the electric force repulsion from the Earth?

62. (II) How much charge $+Q$ should be distributed uniformly over a square, horizontal plate of dimensions 60 cm $\times$ 60 cm if a ball of mass 1.5 g and charge 0.8 μC is to remain suspended 1 mm over the surface of the plate? Take gravity into account in this problem. How would your answer change if the ball were to be suspended 2 mm over the plate? *Qualitatively*, how would your answer change if the ball were to be suspended 1 m over the plate?

63. (II) A single charge $q_1 = +2 \times 10^{-8}$ C is fixed at the base of a plane that makes an angle θ with the horizontal direction. A small ball of mass $m = 0.5$ g and charge $+2 \times 10^{-8}$ C is placed in a smooth, frictionless groove in the plane that extends directly to the fixed charge (Fig. 21–31). It is allowed to move up and down until it finds a stable position $\ell = 8$ cm from the fixed charge. What is θ?

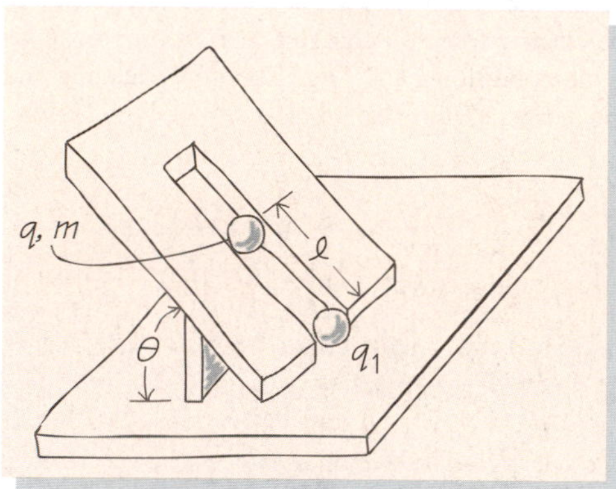

▲ **FIGURE 21–31** Problem 63.

64. (II) The nucleus of an iron atom contains 26 protons within a sphere of radius 4×10^{-15} m. What is the Coulomb force between two protons at opposite sides of this nucleus? The answer to this problem illustrates that the force that holds the nucleus together against the Coulomb repulsion of its constituents must be strong indeed.

65. (II) An electron moves in a circular planetary orbit around a proton. (a) If the centripetal force is the attractive Coulomb force, what is the speed of the electron in terms of the charge e and the radius of the circular orbit? (b) What is the angular momentum, L, of the electron in the orbit? (c) Express the speed in terms of e and L. (d) Express the radius of the orbit in terms of e and L. (e) Express in terms of e and L the time it takes for the electron to go around the circle once. (f) Evaluate all these quantities, given that $L = 1.05 \times 10^{-34}$ kg $\cdot$ m^2/s. This corresponds to a simplified version of the hydrogen atom.

66. (II) Suppose that the proton charge were slightly larger than the electron charge, so that $q_{\text{proton}} = (1 + \delta)e$ and $q_{\text{electron}} = -e$, where $0 < \delta \ll 1$. (a) Given that there are approximately 1.25×10^{57} protons (and electrons) in the Sun, and approximately 1.15×10^{44} protons and electrons in Earth, what is the upper limit on δ set by the fact that the resultant Earth–Sun electric repulsion cannot be large enough to cancel the attraction due to gravity? The mass of the Sun is approximately 2×10^{30} kg, that of Earth is approximately 6×10^{24} kg, and $G = 6.7 \times 10^{-11}$ N $\cdot$ m^2/kg^2. Assume that the number of protons is equal to the number of electrons in the Sun as well as on Earth.

67. (II) Two fixed positive charges q are separated by a length ℓ. A third positive charge q of mass m is constrained to run on a line between the two fixed charges. (a) When the third charge is placed a distance x from the left-hand fixed charge, what is the net force on the third charge? Where is this force zero? In other words, where is the equilibrium point? (b) What is the net force as a function of the displacement of the third charge from the equilibrium point of part (a)? (c) For *small* values of the displacement from the equilibrium point, the third charge behaves as if a spring were acting on it. What is the value of the oscillation frequency?

68. (II) A positive charge q and a negative charge $-\alpha q$ ($\alpha > 1$) are fixed at a distance ℓ apart. Another positive charge q of mass m is constrained to move on the line connecting the two fixed charges. (a) Calculate the net force on the moving charge when it is at a distance x from the fixed positive charge. (b) Where is the force zero? (c) What is the frequency of oscillations if the moving charge is moved a small distance from its equilibrium position and then released, and if $\alpha = 40$?

69. (II) Show that the force between two spherically symmetric distributions of charge is identical to the force between two point charges that are located at the geometric center of each distribution and have the same total charge. [*Hint*: Use the fact that the force on a point charge due to distribution 1 is the same as if distribution 1 were concentrated at its center; then use similar reasoning for distribution 2, and then use Newton's third law.]

70. (III) Two rods, each of length $2L$, are placed parallel to one another a distance R apart. Each carries a total charge Q, distributed uniformly over the length of the rod. Write down an integral for the magnitude of the force between the rods, but do not evaluate it. Without working out any integrals, can you determine the force between the rods for $R \gg L$?

71. (III) Consider an infinite number of identical point charges q located at equally spaced points on the x-axis at the locations $x = na$ (n takes on integer values that range from $-\infty$ to $+\infty$) (Fig. 21–32). (a) Write an expression for the force on a charge Q, located at $x = 0$ and $y = R$, due to all the point charges q, and show the direction of the net force. (b) Take the limit of your result when the intercharge spacing $a \to 0$ and the charge $q \to 0$ such that $q/a = \lambda$ (a fixed charge density). Show that your expression can be written as an integral, and use dimensional analysis to determine the R-dependence of the force on charge Q.

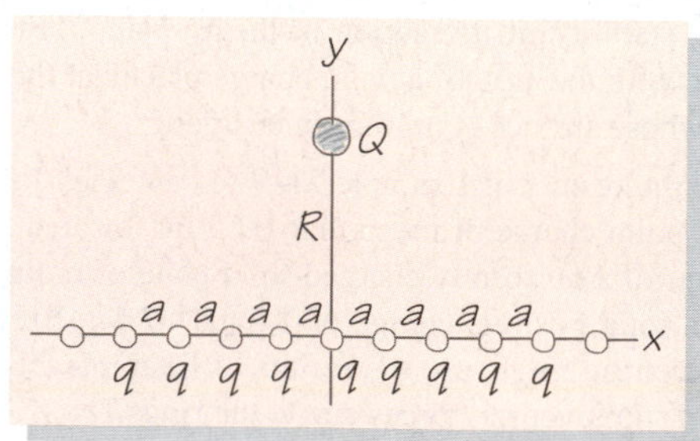

▲ **FIGURE 21–32** Problem 71.

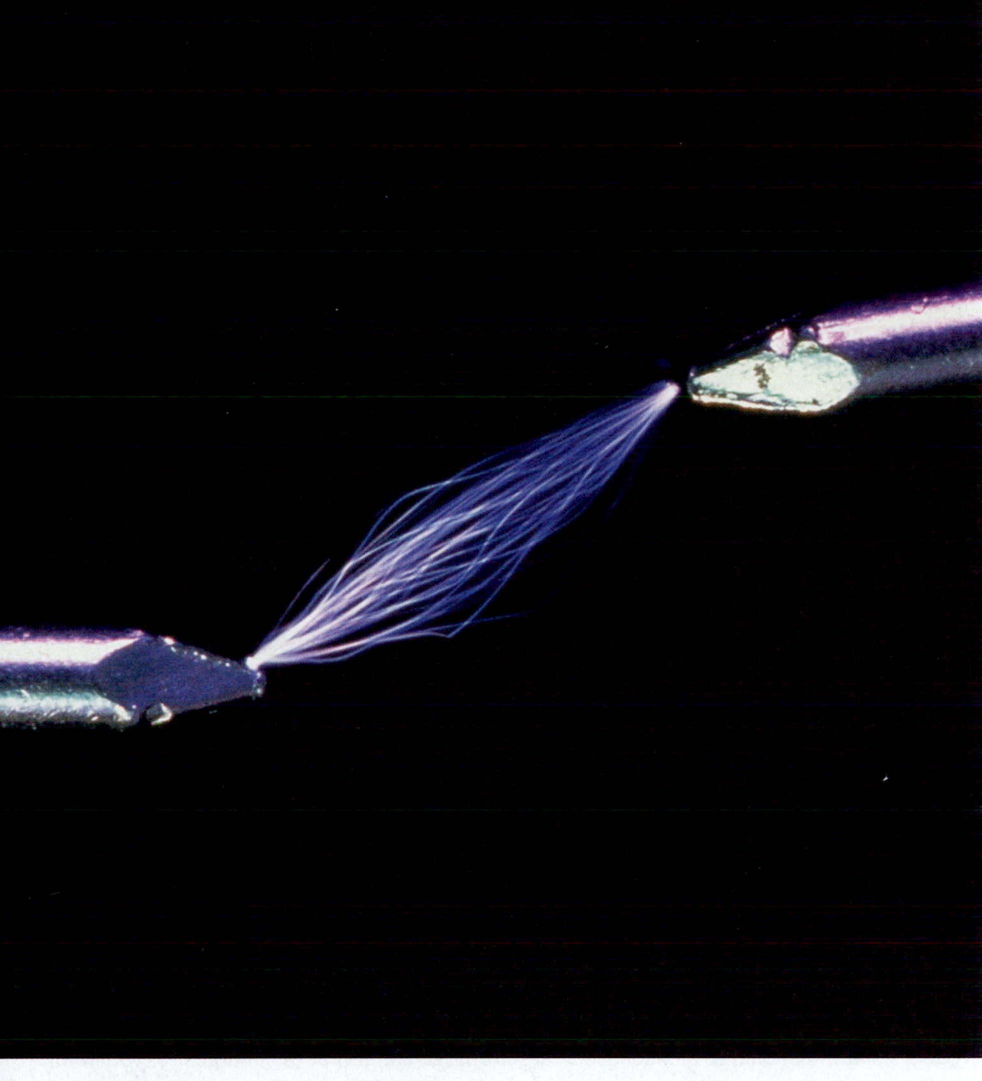

Chapter 22

An electric field forms between two charged objects, here two nails. The field is particularly strong in regions where the charged objects have sharp points, strong enough to allow charge to jump between the two objects. The charge ionizes atoms in the air, forming a series of sparks.

Electric Field

Charges exert forces on one another over large distances and across empty space. This idea of "action at a distance" poses both conceptual and technical difficulties. Action at a distance suggests that the object responsible for the force on a second object somehow reaches out, measures the distance to the second object, and then acts. Michael Faraday set out a better way to look at action at a distance: The first object influences the surrounding space by setting up a *field* around itself that is present whether there is a second object or not. When the second object is located at a given point, the field at this point acts on that object. This important idea can be developed quantitatively and, like any really good idea, leads to further ideas that go far beyond the original concept in utility and insight. In this chapter, we introduce and develop the concept of an electric field produced by static charges; here and in future chapters we will learn how it can be useful both as a practical tool and as a basis for a deeper understanding of electromagnetism.

22–1 Electric Field

It is useful to think of a distribution of charges as giving rise to an **electric field**, which acts on any other charge placed in that field. We can detect the electric field at any particular point by placing a small positive **test charge** q_0 at that location and seeing if it experiences a force. A test charge is only a probe: It does not *produce* the electric field that we are trying to measure; the field is due to other charges. The electric field $\vec{E}$ can

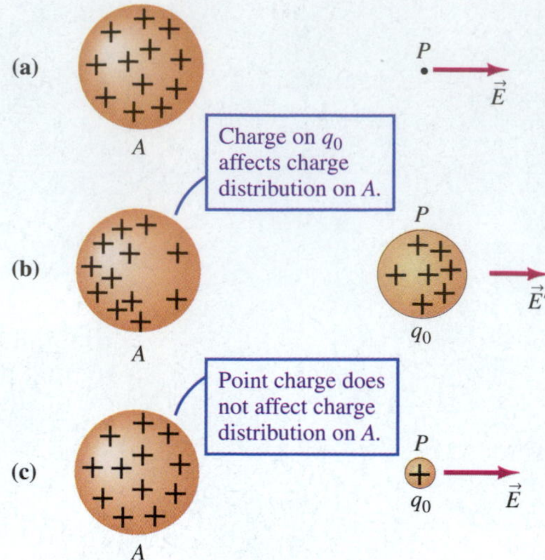

▲ **FIGURE 22–1** (a) An electric field exists at a point P due to charges on the sphere A. (b) A test charge q_0 that is too large will cause a redistribution of the charges on sphere A. A different electric field, $\vec{E}'$, is produced at P by sphere A, because the charges on A have been redistributed. (c) If we make the test charge q_0 small enough, it will hardly affect the charges on sphere A. The electric field produced by A at point P is now the same as in part (a). In each case the electric field is due to the charge on sphere A.

be defined by measuring the magnitude and direction of the electric force $\vec{F}$ acting on the test charge. The definition of the field is

$$\vec{E} \equiv \frac{\vec{F}}{q_0}. \tag{22–1}$$

We use a *small* test charge q_0 because a larger charge can affect the charges responsible for the electric field, perhaps making them move (Fig. 22–1b). This would affect the field itself. Thus, we more properly define the electric field by (Fig. 22–1c).

$$\vec{E} \equiv \lim_{q_0 \to 0} \frac{\vec{F}}{q_0} \tag{22–2}$$

DEFINITION OF ELECTRIC FIELD

Both the electric force and the electric field are vectors. We know a vector such as the electric field completely when we know both its magnitude and its direction *at every point in space*: $\vec{E} = \vec{E}(\vec{r})$.

From the definition in Eq. (22–2), the SI units of electric field are newtons per coulomb (N/C). Table 22–1 gives the magnitudes of the electric fields in various physical situations.

TABLE 22–1 • Values of Electric Fields (N/C)	
Interplanetary space	10^{-3}–10^{-2}
Atmosphere at Earth's surface in clear weather	100–200
Value sufficient to cause electrical breakdown in dry air	3×10^6
Just outside large sphere of a Van de Graaff accelerator	10^6
In the Fermilab particle accelerator	1.2×10^7
In atoms within the radius of an electron orbit	10^9
In the electromagnetic radiation of the most intense laser	10^{12}
Outside a uranium nucleus, at a distance from the center of twice the nucleus's radius	5×10^{20}

The Electric Field of a Point Charge

The simplest example of an electric field is the field associated with a point charge, q_1. Consider two point charges, q_1 and q_0, located a distance r apart. The Coulomb force on q_0 due to q_1 is given in Eq. (21–8),

$$\text{for a point charge: } \vec{F}_{01} = \frac{q_1 q_0}{4\pi\varepsilon_0 r^2}\hat{r}_{01}. \tag{22–3}$$

Here $\hat{r}_{01}$ is the unit vector pointing from q_1 to q_0. If q_0 is small, we use it as a test charge, and we can then use Eqs. (22–1) and (22–3) to find the electric field due to q_1:

$$\text{for a point charge: } \vec{E}_1 = \frac{\vec{F}_{01}}{q_0} = \frac{q_1}{4\pi\varepsilon_0 r^2}\hat{r}_{01}. \tag{22–4}$$

The value of the test charge has canceled, so the limiting process in Eq. (22–2) introduces no complications. Equation (22–4) specifies that $\vec{E}_1$ is in the same direction as $\vec{F}_{01}$. Figure 22–2 shows the direction of $\vec{E}_1$ as determined by moving our test charge to various points a distance r away from q_1. This field is radial (Fig. 22–2a). At this point we drop the subscripts on $\hat{r}_{01}$ and q_1, and we use the radial unit vector $\hat{r}$ (measured from the point charge q) to specify completely the electric field $\vec{E}_{pt}$ due to q:

$$\vec{E}_{pt} = \frac{q}{4\pi\varepsilon_0 r^2}\hat{r}. \tag{22–5}$$

FIELD OF A POINT CHARGE

The electric field points *away* from a positive charge, as shown in Fig. 22–2b. When the charge is negative, the electric field has the same magnitude but is opposite in direction. The electric field due to a negative charge points *toward* that charge, as in Fig. 22–2c.

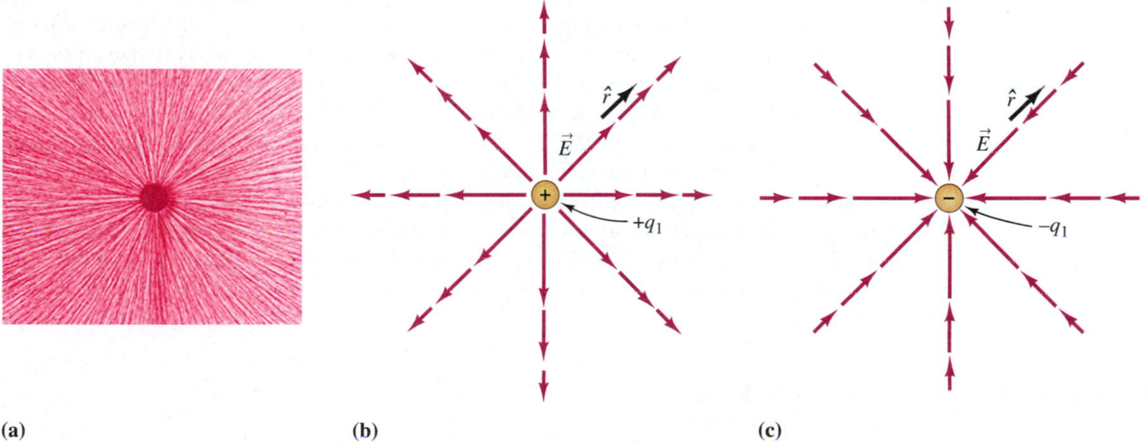

▲ **FIGURE 22–2** (a) Threads floating in oil become aligned with the electric field of this point charge. (b) The direction of the electric field $\vec{E}$ due to a point charge q_1 is radial. The charge is positive, and the field points away from it. (c) When the charge is negative the field points toward it.

CONCEPTUAL EXAMPLE 22–1 As you approach the charge in Fig. 22–2c, the magnitude of the electric field is large, and the arrows representing the field will become larger and larger, eventually overlapping each other and even passing through the (negative) charge. Does this pose a difficulty?

Answer This is a difficulty only for the artist. Remember that the arrows are just a visual representation, an aid in understanding. The tail of the arrow is placed on the position where you want to characterize the electric field. The magnitude of the electric field goes as $1/r^2$, so it can become quite large close to the point charge. But the length of the arrow has nothing to do with the scale of the drawing of charge positions.

The Usefulness of the Field Concept

Once we know the electric field $\vec{E}_{pt}$ produced by a point charge q, we can find the force on any point charge q' placed in that field by using Eq. (22–1); that is,

$$\vec{F} = q'\vec{E}_{pt}. \tag{22–6}$$

More important, *any* distribution of charges—not simply a point charge—produces an electric field throughout space. We use the subscript "ext" (for external) on the field to emphasize that this field is present *independent* of the charge q' on which the force acts, and once we know $\vec{E}_{ext}$, *the force on any point charge q' in the field* is the generalization of Eq. (22–6):

$$\vec{F} = q'\vec{E}_{ext}. \tag{22–7}$$

FORCE ON A POINT CHARGE IN AN EXTERNAL FIELD

The field concept has practical uses. We can calculate (or measure with a test charge) the field from a charge distribution once and for all, and then use Eq. (22–7) to find the effect of that distribution on any other charges within the field—we'll see numerous examples of this.

THINK ABOUT THIS. . .

HOW DOES THE FIELD CONCEPT HELP?

Why don't we deal with forces between charges instead of fields? In the chapter introduction, we mentioned the role the field plays in resolving the conceptual difficulties of action at a distance. We shall see in Chapter 24 that the field carries energy. To preserve the important idea of energy conservation, the field is a *necessary* concept. But the real power of the field concept appears when the field arises from *accelerating charges*. Even if these charges are limited to a small region (for example, within the arms of an antenna), the electromagnetic fields they produce spread through all of space at the speed of light. The supernova known as 1987A took place approximately 163,000 years ago; electric fields caused by the violent motion of many charges within and around the exploding star reached Earth on February 23, 1987. These traveling fields caused electrons in terrestrial antennas to move; this was the signal that supernova 1987A had occurred. This description of the process is easy to grasp; more, it is really the only practical way that we have to describe the process.

The notion of a field is useful in many areas of physical science. A mass distribution gives rise to a gravitational field analogous to the electric field of a charge distribution. We employ a velocity field in hydrodynamics; this field describes the velocity $\vec{v}$ at all points where fluid flow occurs, such as in the pipes of a city water system. In thermal physics, a temperature field describes the temperature at all points in a room. In this case, there is no directionality to the field; temperature forms what is called a scalar field instead of a vector field.

EXAMPLE 22–2 Find the electric field due to a point charge $+1.4\ \mu C$ at a distance of 0.10 m from the charge. What is the force on a second charge $-1.2\ \mu C$ that is placed 0.10 m from the first charge?

Setting It Up Figure 22–3a shows the electric field at point P due to a charge q. In Fig. 22–3b, we place a second charge q' at point P.

Strategy Equation (22–5) allows us to find the electric field $\vec{E}_{pt}$ at point P due to the point charge q. We then use Eq. (22–6) to find the force on the second point charge q' placed in this electric field.

Working It Out From Eq. (22–5):

$$\vec{E} = \frac{q}{4\pi\varepsilon_0 r^2}\hat{r} = \frac{(9.0 \times 10^9\ \mathrm{N\cdot m^2/C^2})(1.4 \times 10^{-6}\ \mathrm{C})}{(0.10\ \mathrm{m})^2}\hat{r}$$
$$= (1.3 \times 10^6\ \mathrm{N/C})\hat{r}.$$

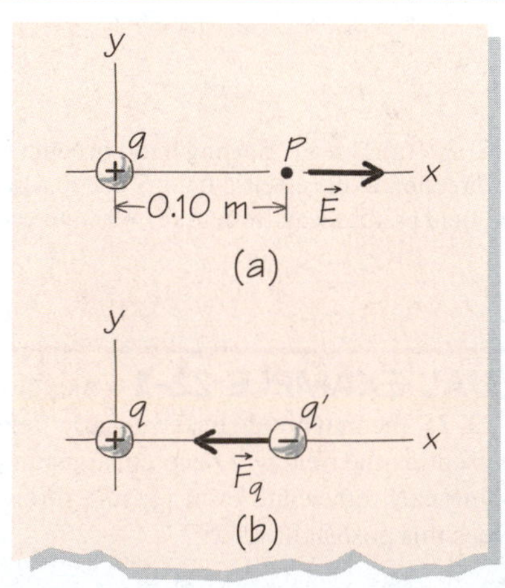

▲ **FIGURE 22–3**

The field is directed radially outward from the position of q (Fig. 22–3b). If q had been negative, the field would point radially inward rather than radially outward.

For the force, Eq. (22–6) gives

$$F = |q'|E = (1.2 \times 10^{-6}\ \mathrm{C})(1.3 \times 10^{6}\ \mathrm{N/C}) = 1.6\ \mathrm{N};$$
$$\vec{F} = (-1.6\ \mathrm{N})\hat{i}.$$

Note that the direction of the force is in the opposite direction of the electric field at point P, that is, it is along $-\hat{i}$. The opposite charges attract (Fig. 22–3b). We could of course use Coulomb's law for the force between the two charges instead, Eq. (21–8).

What Do You Think? What would be the change in the electric field and force if q' were a positive charge instead of a negative one? *Answers to **What Do You Think?** questions are given in the back of the book.*

Superposition

In Section 21–4 we described how electric forces involving multiple charges superpose. The consequence of this is that if more than one point charge is responsible for producing a net electric field, that field is determined by the principle of superposition. The superposition principle states that the net electric force on an object is the vector sum of the forces due to individual point charges. Therefore *the net electric field is the vector sum of the fields of individual charges present*. The net force exerted on our test charge q_0 due to all the other charges in the region is

$$\vec{F}_{\text{net}} = \vec{F}_{01} + \vec{F}_{02} + \vec{F}_{03} + \cdots = \sum_i \vec{F}_{0i}. \qquad (22\text{–}8)$$

Thus

$$\vec{E}_{\text{net}} = \frac{\vec{F}_{\text{net}}}{q_0} = \frac{\vec{F}_{01}}{q_0} + \frac{\vec{F}_{02}}{q_0} + \frac{\vec{F}_{03}}{q_0} + \cdots \qquad (22\text{–}9)$$

$$= \vec{E}_1 + \vec{E}_2 + \vec{E}_3 + \cdots = \sum_i \vec{E}_i. \qquad (22\text{–}10)$$

In Eq. (22–10), $\vec{E}_2$, for example, is the electric field due solely to the charge q_2 at the point in space where we have placed q_0. Using Eq. (22–5),

$$\vec{E}_{\text{net}} = \frac{1}{4\pi\varepsilon_0}\sum_i \frac{q_i}{r_i^2}\hat{r}_i. \qquad (22\text{–}11)$$

FIELD OF A GROUP OF POINT CHARGES

In this equation, r_i is the distance from the ith charge q_i to the point at which the field is evaluated, and the unit vector $\hat{r}_i$ is directed from the position of the ith charge to the point where the field is evaluated.

EXAMPLE 22–3 Consider three charges placed on a line: $q_1 = +2\ \mu\mathrm{C}$ at $x_1 = -2$ cm, $q_2 = +3\ \mu\mathrm{C}$ at $x_2 = +4$ cm, and $q_3 = -2\ \mu\mathrm{C}$ at $x_3 = +10$ cm. Find the electric field at point A, the origin of the coordinate system.

Setting It Up We sketch the configuration of charges in Fig. 22–4.

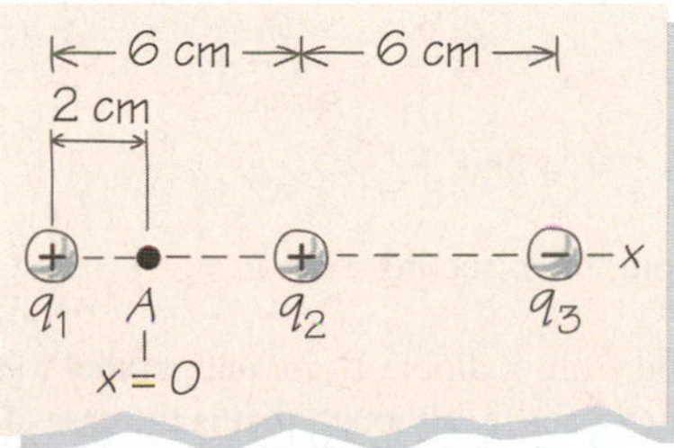

▲ **FIGURE 22–4** The electric field at point A is due to three charges. The distance x is measured from point A.

Strategy We solve this by a straightforward application of Eq. (22–11), the superposition of electric fields from point charges. Although it is generally important to remember that the required sum is vectorial, all the positions lie along a straight line in this case. The electric field at point A is

$$\vec{E}_A = \vec{E}_1 + \vec{E}_2 + \vec{E}_3,$$

where $\vec{E}_j$ (j = 1, 2, or 3) is the field at point A due to charge q_j.

Working It Out Application of Eq. (22–11) gives

$$\vec{E}_A = \frac{1}{4\pi\varepsilon_0}\left[\frac{q_1}{x_1^2}(+\hat{i}) + \frac{q_2}{x_2^2}(-\hat{i}) + \frac{q_3}{x_3^2}(-\hat{i})\right]. \qquad (22\text{–}12)$$

We must pay careful attention to signs. The unit vectors $\pm\hat{i}$ in parentheses indicate the direction of the unit vector $\hat{r}_j$, from the position of charge q_j to point A, but we must also include the signs of the individual charges. For example, the direction of $\vec{E}_3$ is $+\hat{i}$ because the negative sign of charge q_3 multiplied by $(-\hat{i})$ gives a direction $(+\hat{i})$. Putting the numbers into Eq. (22–12) gives

(continues on next page)

$$\vec{E}_A = (9 \times 10^9\ \text{N}\cdot\text{m}^2/\text{C}^2)\left[\frac{(2 \times 10^{-6}\ \text{C})}{(0.02\ \text{m})^2}\hat{i} + \frac{(3 \times 10^{-6}\ \text{C})}{(0.04\ \text{m})^2}(-\hat{i}) + \frac{(-2 \times 10^{-6}\ \text{C})}{(0.10\ \text{m})^2}(-\hat{i})\right]$$
$$= (3 \times 10^7\ \text{N/C})\hat{i}.$$

The net electric field at point A is in the $+x$-direction, or toward the right.

What Do You Think? If point A were halfway between q_2 and q_3, in what direction would the electric field point?

Electric Dipoles and Their Electric Fields

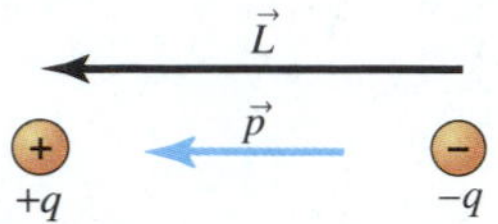

▲ **FIGURE 22–5** An electric dipole consists of equal but opposite charges separated by a distance L. The electric dipole moment $\vec{p}$ is directed from the negative charge to the positive charge.

The field of any single charge decreases as $1/r^2$. What happens if we have two charges, q_1 and q_2, spaced by a fixed distance L? Generally the field due to these charges will behave as $(q_1 + q_2)/r^2$ for $r \gg L$. Suppose now that the charges were equal and opposite. If these two charges were to sit precisely on top of one another, the two $1/r^2$ contributions to the field would exactly cancel to a zero electric field. But with the two charges separated by a finite distance, the field is not exactly zero, yet there is a partial cancellation and the field must fall off faster than $1/r^2$ for $r \gg L$. This arrangement—two charges, $+q$ and $-q$, of equal magnitude but opposite sign, that are separated by a fixed distance L—is an **electric dipole** (Fig. 22–5). We shall see that the field of the electric dipole decreases as $1/r^3$. This field depends only on the product qL, which is called the **electric dipole moment** of the neutral pair $(+q, -q)$; it is denoted by the letter p. We make $p = qL$ a vector through the displacement vector $\vec{L}$ directed from $-q$ to $+q$ (Fig. 22–5). The electric dipole moment $\vec{p}$ is then defined as

$$\vec{p} \equiv q\vec{L}. \tag{22–13}$$

ELECTRIC DIPOLE MOMENT DEFINED

The vector $\vec{p}$ points from the negative charge to the positive charge.

EXAMPLE 22–4 Find the electric field of the electric dipole shown in Fig. 22–5 at a point P that lies along the perpendicular axis that bisects the line between the two charges and is at a large distance r $(r \gg L)$ from each charge.

Setting It Up The sketch in Fig. 22–6 shows the point P, with xy-coordinates $(0, y)$.

Strategy We again use the superposition principle for point charges to find the net electric field $\vec{E} = \vec{E}_1 + \vec{E}_2$ at P, where the field $\vec{E}_1$ is due to the charge $+q$ and the field $\vec{E}_2$ is due to the charge $-q$. Since P is equidistant from both charges, the magnitudes of the two fields are the same, but $\vec{E}_1$ points away from $+q$, whereas $\vec{E}_2$ points toward $-q$. As usual, adding vectors means adding the components.

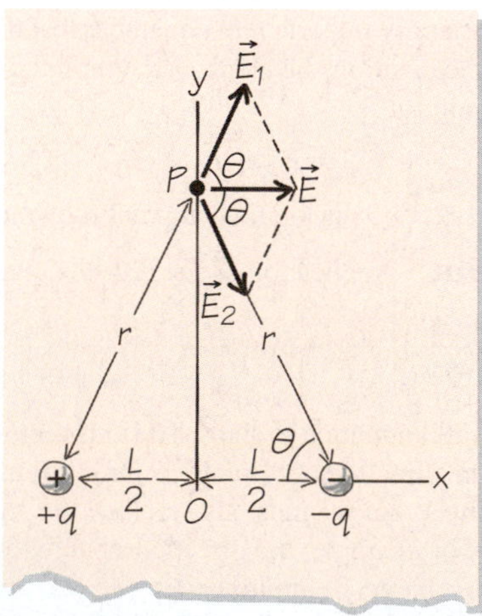

▲ **FIGURE 22–6** The net field at point P is parallel to the direction from $+q$ to $-q$.

Working It Out The y-components of $\vec{E}_1$ and $\vec{E}_2$ exactly cancel each other, and we are left with a net x-component toward the right that is twice the x-component of the field due to either charge:

$$\vec{E} = E_x\hat{i} = (E_{1x} + E_{2x})\hat{i} = 2E_{1x}\hat{i},$$

where

$$E_{1x} = \frac{q}{4\pi\varepsilon_0 r^2}\cos\theta.$$

From Fig. 22–6, we see that $\cos\theta$ is given by $\cos\theta = \dfrac{L/2}{r} = \dfrac{L}{2r}$.

Thus the total electric field of the dipole along the perpendicular bisector is

$$\vec{E} = \left(\frac{2q}{4\pi\varepsilon_0 r^2}\right)\left(\frac{L}{2r}\right)\hat{i} = \frac{qL}{4\pi\varepsilon_0 r^3}\hat{i}. \tag{22–14}$$

We made no approximations here, and Eq. (22–14) is correct along the perpendicular bisector even when the distance from the charge pair is not large. The electric field decreases with r as $1/r^3$. This is the partial cancellation we spoke of earlier. Using Eq. (22–13) for the electric dipole moment, we can write Eq. (22–14) as

$$\vec{E} = -\frac{\vec{p}}{4\pi\varepsilon_0 r^3}. \tag{22–15}$$

If $r \gg L$, then $r \cong y$ and

$$\text{along the bisecting axis: } \vec{E} \cong -\frac{\vec{p}}{4\pi\varepsilon_0 y^3}. \tag{22–16}$$

The electric field from a dipole is *not* antiparallel to the dipole moment everywhere in space, although that is the case along the bisecting axis [Eqs. (22–15) and (22–16)].

What Do You Think? Is there any place in an electric dipole field that the value depends on only q or L and not on the product?

THINK ABOUT THIS . . .

WHEN DO ELECTRIC DIPOLES APPEAR IN NATURE?

Electric dipoles can be divided into two types. External fields (due to external charge distributions) frequently induce charge separations in electrically neutral molecules and materials, leading to an excess of positive (or negative) charge on one side (and the other) and hence to an **induced electric dipole moment** (Fig. 22–7). There are also examples in nature of charge configurations with **permanent electric dipole moments** (dipole moments that are not induced by external fields). Many molecules—water is an excellent example (Fig. 22–8)—have a structure with electrons distributed preferentially in certain regions, and so have permanent electric dipole moments. In cases such as common salt (NaCl) and hydrochloric acid (HCl), electrons cluster preferentially around one atom, giving that atom a negative charge. The other atom is left with a positive charge, so such molecules have a permanent electric dipole moment. The effects of electric dipole fields have great importance at the molecular level, where they can influence, for example, boiling and melting points. At this scale permanent dipole moments will typically dominate any induced dipole moments. For example, a water molecule has a permanent dipole moment $p \cong 6 \times 10^{-30}\,\mathrm{C \cdot m}$, whereas a hydrogen atom in the rather strong field $E = 3 \times 10^{6}\,\mathrm{N/C}$ acquires an induced dipole moment of $p \cong 3 \times 10^{-34}\,\mathrm{C \cdot m}$. ■

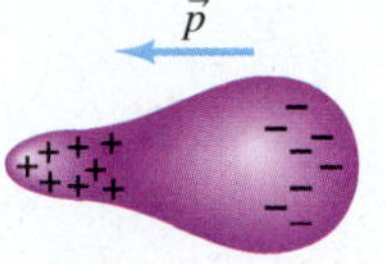

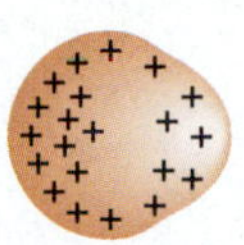

▲ **FIGURE 22–7** A nearby charge can induce a polarized charge, and hence an electric dipole, on a neutral object.

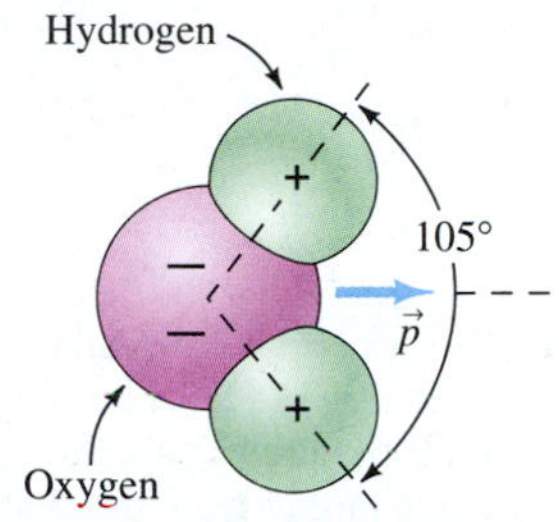

▲ **FIGURE 22–8** The water molecule, H_2O, is a permanent electric dipole. The electrons of both hydrogen atoms are shared with the oxygen atom, creating a strong electric bond that holds the molecule together (this is known as covalent bonding).

22–2 Electric Field Lines

We have already seen that we can map out the electric field by moving a test charge around in space. Unfortunately, the field is awkward to use in a visual sense. It is not easy to draw a vector at each point of a region of space that represents the magnitude and direction of the electric field at that point. The electric field can be more clearly visualized in terms of **electric field lines**. Their use was introduced by Michael Faraday around the middle of the nineteenth century, even before the concept of the electric field itself was clearly understood. Faraday used the phrase "lines of force." Electric field lines are continuous lines in space determined by the electric field, and in principle they fully describe the electric field. Keep in mind as we proceed that although the electric field itself has physical meaning, electric field lines are simply an aid to picturing the electric field and how a charge would react when placed in that field.

Electric field lines are determined according to two simple rules:

1. Electric field lines are drawn so that the tangent to the field line at each point specifies the direction of the electric field $\vec{E}$ at that point. This rule relates the *direction* of the electric field lines to the direction of the electric field. Note that we do not expect the electric field to change abruptly across any region of space that does not contain charge, so that the electric field lines in such regions are very nearly parallel to one another.
2. The *density* in space of electric field lines around a particular point is proportional to the strength of the electric field at that point (Fig. 22–9). What this means is that we take a small area oriented perpendicularly to the (nearly parallel) field lines and count the number of electric field lines that cross this small area; the line density is this number divided by the area. Thus the density is the number of lines *per unit area*. We'll explain below how we "count" the number of field lines.

Properties of Electric Field Lines

Let's draw the electric field lines of a positive point charge q. We know that the electric field points radially away from a positive charge at every location in space, so the lines are radial, pointing outward from the charge (Fig. 22–9). The field has the same magnitude at each point on a sphere centered on the charge, so the field lines are distributed uniformly about the charge. Finally, the magnitude of the electric field decreases with the distance r from the charge as $1/r^2$, so the density of lines falls off as $1/r^2$. And this is exactly how the density changes in Fig. 22–9 as long as we insist that electric field lines start (or stop, in the case of a negative charge) on the charge itself.

To establish this last point, suppose we draw N lines originating on our charge q (Fig. 22–9), and no lines appear (or disappear) as we move out. These lines radiate outward from the charge, and the number of lines that cut a sphere of radius R centered on the

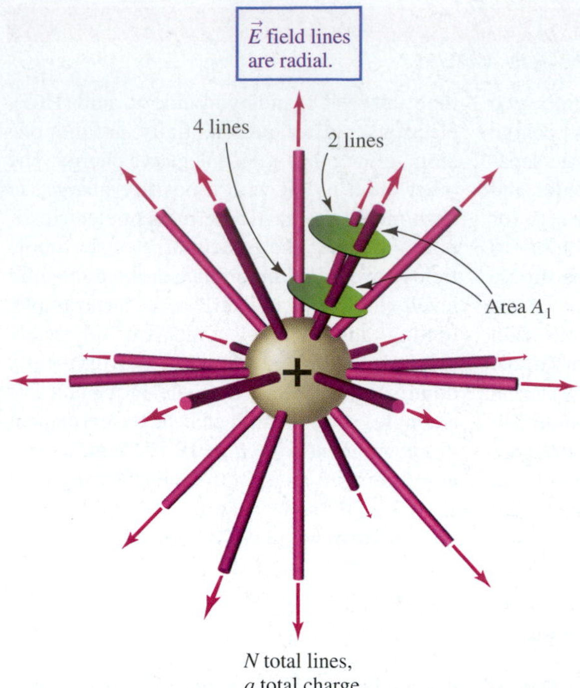

▶ **FIGURE 22–9** Representation of the radial electric field lines from a point charge. Fewer field lines pass through the same-size area farther from the charge. Note that field lines extend out to infinity.

charge is this same number N. Since the lines are evenly distributed over the sphere, the density of the lines is $N/(\text{sphere area}) = N/(4\pi R^2)$. Thus the density of the lines is proportional to $1/R^2$, and we have obtained the correct field strength dependence on R.

We can now relate the number of lines coming out of our charge to the magnitude of the charge: we take N to be any number we like! But once we have done so, the number of lines that leaves (or arrives at) any other charge is determined. For example, the number of field lines leaving a positive charge $q/2$ is $N/2$ and, more generally, the number of field lines leaving a positive charge q_i is $N_i = (q_i/q)N$. This will ensure that point 2 is satisfied: The field line density will be *proportional to* the strength of the electric field.

From the rules that determine field lines and the discussion above we can make a useful list of properties of field lines:

Property One: *Lines can start or terminate only on charges, never in empty space.* This property guarantees the connection between the strength of the field and the density of the electric field lines. We showed this above only for a single point charge, but with the help of the superposition principle, we can show that this is true in general.

Property Two: The electric field lines of a point charge go off to infinity, and by superposition this will be true for any *localized* collection of charges with a net charge—by "localized" we mean that the distribution is confined to a finite space. At distances that are large compared with the size of the charge distribution, the net charge of the distribution appears to be localized at a point, and electric field lines will be distributed evenly over a distant sphere centered on the distribution. If the net charge of a localized distribution is zero, then the field lines will curve back, not reaching to infinitely large distances.

Property Three: Electric field lines originate on, and run outward from, positive charges. They run toward, and terminate on, negative charges. This reflects the fact that electric charges are the sources of electric fields, which point away from positive charges and toward negative charges.

Property Four: *No two field lines ever cross, even when multiple charges are present.* They cannot cross because the electric field has a definite magnitude and direction at any point in space. If two or more electric field lines were to cross at some point, then the direction of the electric field at that point would be ambiguous.

Drawing Electric Field Lines

The properties above are a helpful guide for drawing the field lines associated with a given set of charges or a continuous charge distribution. Symmetry is another tool—and often a powerful one—for this process. A point charge looks the same when viewed from any direction. It has spherical symmetry, and the field lines follow the only direction that respects this symmetry—namely, they are radial. Similarly, if we are dealing with a long line of charge, there is a symmetry around the line, and the field lines must project radially outward from the line of charge, perpendicular to a cylinder that surrounds the line.

A Warning: We often draw field lines on a flat page, and since field lines are in three-dimensional space, the flat page presents certain limits. For example, you might draw Fig. 22–10 as the representation of the field lines for an isolated charge. *If you use such a drawing for determining the field strength, it should be done with care.* Figure 22–10 shows a circle of radius r centered on a positive charge. You cannot simply count the field lines that cross a particular circumference of this circle to find the field strength, as that would be equivalent to dividing by a length, not an area. If you did divide the (fixed) number of lines by the circumference $2\pi r$, you would find a field strength that falls as $1/r$, not the correct $1/r^2$. Nevertheless, planar drawings of electric field lines like Fig. 22–10 remain useful for visualizing the field and its effect on other charges.

▲ **FIGURE 22–10** Electric field lines due to a point charge $+q$. Note the number of field lines that cross the circle (sphere) at radius r.

Some Examples

The easiest way to demonstrate the usefulness of electric field lines is to look at a couple of examples. Figures 22–11a and 22–11b show the electric field lines on a plane that passes through two positive charges of equal magnitude. The field lines all extend to infinity. The field lines that approach each other between the two positive charges appear to repel because two field lines cannot cross. If we were to place a positive test charge q' in Fig. 22–11b, the field lines at the charge's location would immediately show us the direction of the force on the charge (and likewise the acceleration).

Figures 22–12a and 22–12b depict the field lines of an electric dipole. The charges have equal magnitude, $\pm q$, so an equal number of field lines are attached to them, and every field line that originates on $+q$ terminates at $-q$. Near each charge the field lines are purely radial, but every line must eventually deviate from the radial direction in order to loop back to the other charge. Notice that the field lines in Fig. 22–12b are consistent with the field $\vec{E}$ determined in Example 22–4 (compare Figs. 22–12b and 22–6).

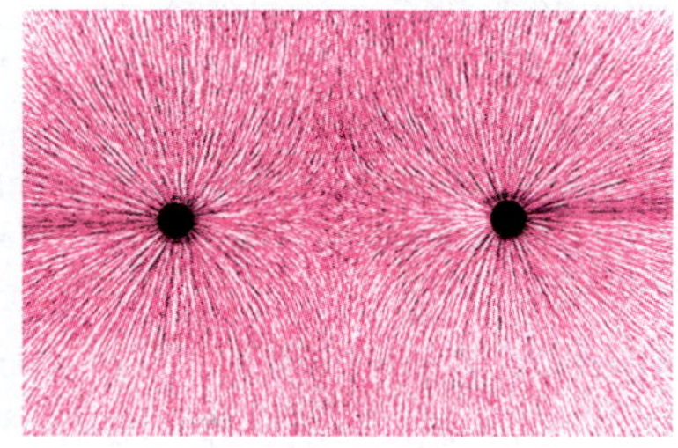

(a)

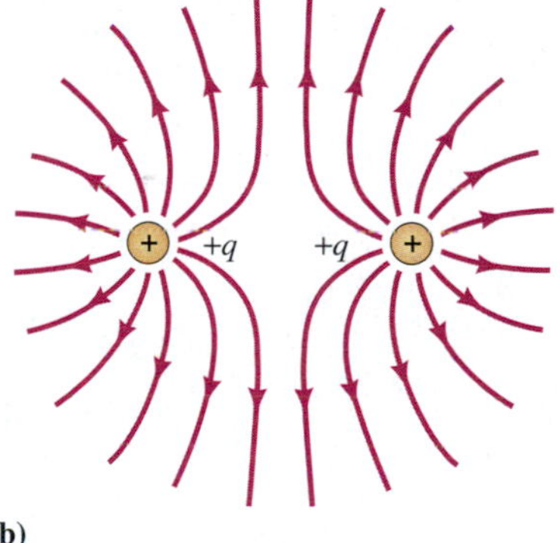

(b)

▲ **FIGURE 22–11** (a) The electric field lines due to two point charges $+q$, shown by threads in oil. (b) Schematic diagram of the field lines, which go off to infinity and appear to repel each other.

(a)

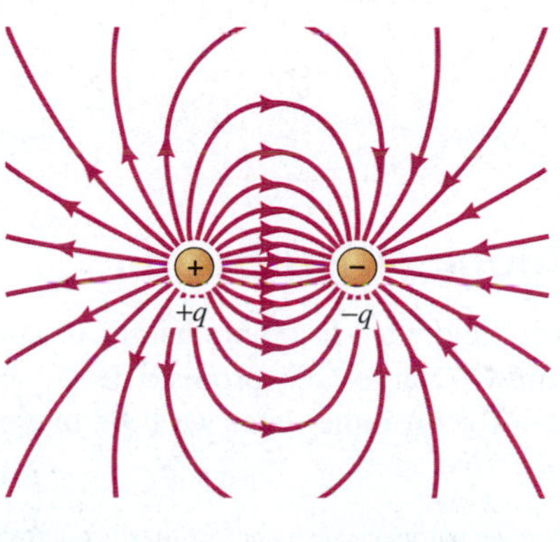

(b)

◀ **FIGURE 22–12** (a) The electric field lines due to point charges $+q$ and $-q$, a dipole, as indicated by threads in oil. (b) Schematic diagram of the field lines, all of which begin on $+q$ and end on $-q$; those that appear to head toward infinity do continue far from the charge, but eventually loop around and end on $-q$.

CONCEPTUAL EXAMPLE 22–5 The electric field lines of an isolated point charge go to infinity. How many field lines go to infinity for the electric dipole?

Answer Remember *Property Two:* Electric field lines always start on positive charges and terminate on negative charges. For an isolated point charge (say positive), the electric field lines go to infinity because there is no negative charge for them to end on. An electric dipole has one positive charge of a given magnitude and one negative charge of the same magnitude. Every line that starts on the positive charge will end on the negative charge, and so there will be none "left over" to go to infinity. We cannot see all the field lines doing this in Fig. 22–12, but if we made the figure large enough, we would see them all curve around and finally terminate on the negative charge!

EXAMPLE 22–6 Draw the electric field lines for a system of two charges, $+2q$ and $-q$, separated by a fixed distance.

Setting It Up We sketch a 2-dimensional representation of the field due to the two charges $+2q$ and $-q$ in Fig. 22–13a. Twice as many field lines leave the charge $+2q$ as end at the charge $-q$.

Strategy Arbitrarily close to each charge the field lines will be radial and uniformly spread over the area surrounding the charge. With twice as many lines coming from $+2q$ as go into charge $-q$, we can take half of the lines from the charge $+2q$ and connect them to the lines going into charge $-q$. The remaining lines go off to infinity.

Working It Out We choose, arbitrarily, to show 24 lines coming from $+2q$ so that 12 lines will go into $-q$ and 12 will go off to infinity. The final sketches are presented in Figs. 22–13b and 22–13c.

What Do You Think? What happens to the remaining 12 lines that emerge from $+2q$ that don't terminate on $-q$?

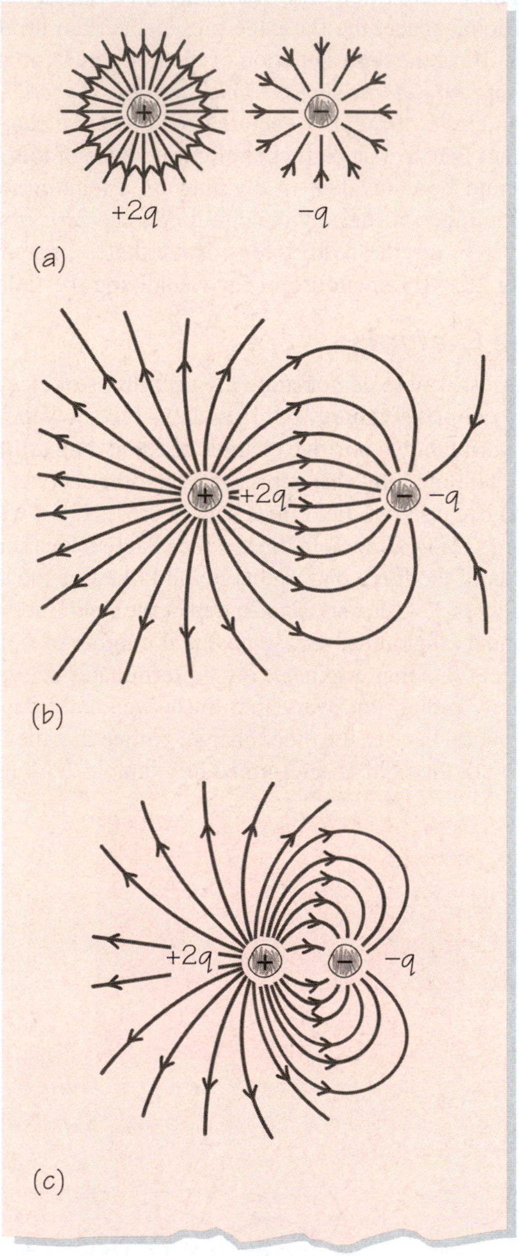

▶ **FIGURE 22–13** (a) The electric field lines close to the $+2q$ and $-q$ point charges are those of a point charge. (b) Half the electric field lines that emerge from $+2q$ end up on $-q$. (c) Far from the point charges, the electric field lines are those of a point charge $+q$. In this view you can begin to see the lines spread out to form a radial distribution.

22–3 The Field of a Continuous Distribution

We have thus far concentrated on electric fields due to point charges or collections of point charges. But *continuous* distributions of charge[†] also produce fields, and such distributions are very important in practice—for example, later we'll be interested in the

[†]Only point charges exist in nature; however, when we put many such charges together across a region, we will have a distribution that looks continuous to all but the sharpest measuring instrument.

field produced by charged capacitors. We will consider charges that are distributed *uniformly* throughout a region in space, whether a line, a surface, or a volume. We will also emphasize distributions where there is symmetry. For charge distributions that are not uniform or symmetrical, the problem of determining the resulting electric field can be more complex.

In considering forces on point charges due to charge distributions in Chapter 21, we have already set up a general framework for calculating electric fields due to line, surface, and volume distributions. Consider the calculation of the electric field at point P due to the charge distribution shown in Fig. 22–14. We divide the charge distribution into tiny elements, each of charge Δq. We first find the electric field $\Delta\vec{E}$ at point P that is due to a tiny charge element Δq, whose distance from P is r:

$$\Delta\vec{E} = \frac{\Delta q}{4\pi\varepsilon_0 r^2}\hat{r}. \tag{22–17}$$

Here, $\hat{r}$ is the unit vector pointing away from the charge element. Superposition applies, and the total electric field at P is found by summing the infinitesimal fields $\Delta\vec{E}$:

$$\vec{E} = \sum \Delta\vec{E}. \tag{22–18}$$

In the language of calculus, Eq. (22–17) becomes

$$d\vec{E} = \frac{dq}{4\pi\varepsilon_0 r^2}\hat{r}, \tag{22–19}$$

whereas the summation that gives the total electric field becomes an integral over the entire charge distribution:

$$\vec{E} = \sum_{\lim \Delta q \to 0} \Delta\vec{E} = \int d\vec{E} = \frac{1}{4\pi\varepsilon_0}\int \hat{r}\frac{dq}{r^2}. \tag{22–20}$$

The formal expression of Eq. (22–20) is identical to the one we found for the force on a point charge q due to a charge distribution, Eq. (21–11), divided by q. This is exactly how we define the electric field.

We can follow the development in Chapter 21 further to more specifically find the electric field due to a charge distribution along a line in terms of the linear charge density λ, the electric field due to a charge spread over a surface in terms of the surface charge density σ, and the electric field due to a charge distribution over a volume in terms of a volume charge density ρ (Fig. 22–15). In each case, the electric field is simply the expression for the force on the charge q divided by q; that is, Eqs. (21–13), (21–15), and (21–17) divided by q, respectively.

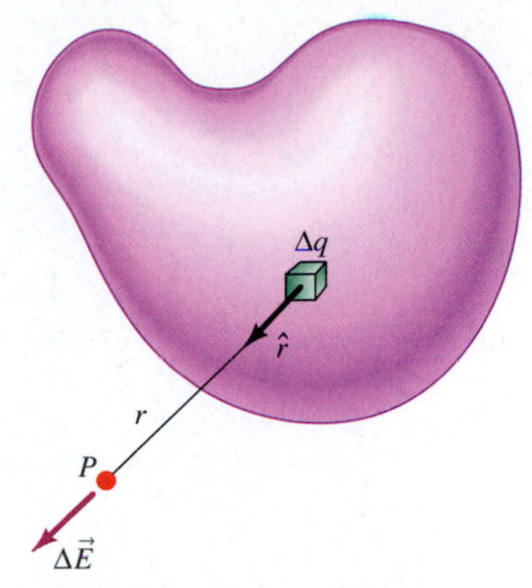

▲ **FIGURE 22–14** To find the electric field due to a continuous charge distribution, add all the electric fields $\Delta\vec{E}$ due to the charge elements Δq.

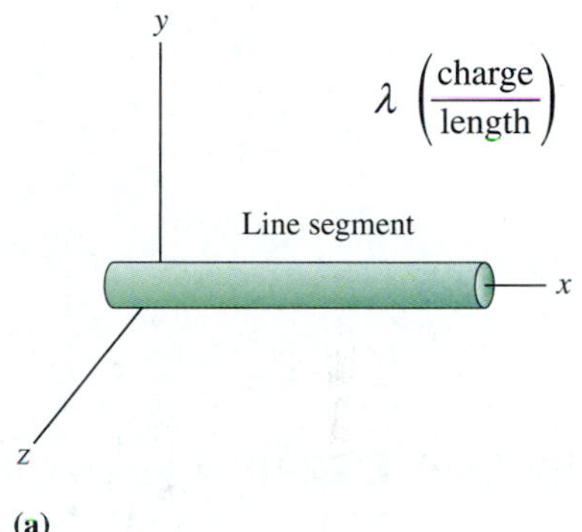

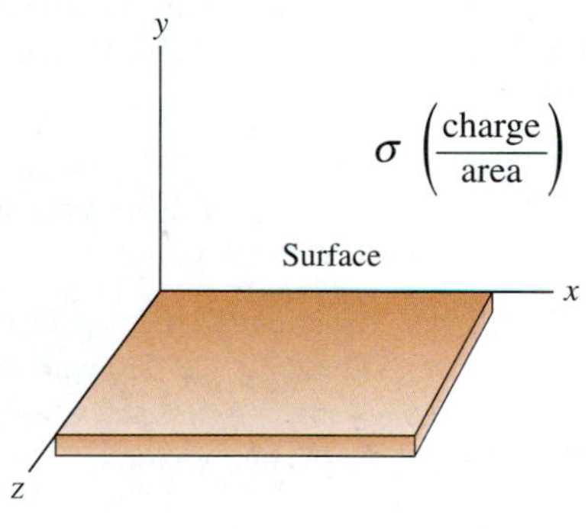

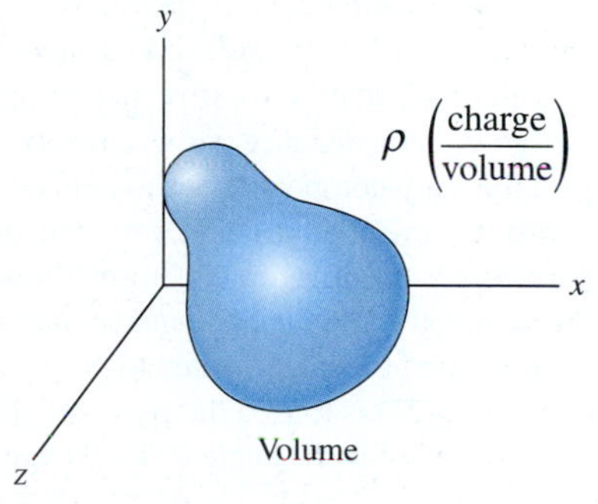

▲ **FIGURE 22–15** (a) One-dimensional, (b) two-dimensional, and (c) three-dimensional charge distribution. The charge density is labeled λ, σ, and ρ, respectively.

Constant Charge Densities

An important simplification for calculating the electric field due to a charge distribution occurs when the charge distribution, whether one-, two-, or three-dimensional, is *uniform*. This means that the charge densities are constants that can be removed from beneath the integrals expressing the electric field. Moreover, we can express the charge density in terms of the total charge Q contained in the distribution and the size of the distribution. In particular, if charge Q is distributed uniformly along a line segment of length L on a line, the linear charge density λ is

$$\lambda \equiv \frac{Q}{L}. \tag{22–21}$$

Dividing Eq. (21–13) by the test charge q, we have

$$\vec{E} = \frac{\lambda}{4\pi\varepsilon_0}\int \hat{r}\frac{dx}{r^2}. \tag{22–22}$$

Here, r is the distance to the point P at which the field is to be calculated, while $\hat{r}$ is a unit vector from the segment within the sum to point P. The integral is over a *one*-dimensional distribution.

For a charge Q distributed uniformly on a surface of area A, the surface charge density σ is

$$\sigma \equiv \frac{Q}{A}. \tag{22–23}$$

Then

$$\vec{E} = \frac{\sigma}{4\pi\varepsilon_0}\int_{\text{surface}} \hat{r}\frac{dS}{r^2}. \tag{22–24}$$

The integral in this case is over a *surface*.

Finally, if charge Q is distributed uniformly throughout a volume V, the volume charge density ρ is

$$\rho \equiv \frac{Q}{V} \tag{22–25}$$

and the electric field is

$$\vec{E} = \frac{\rho}{4\pi\varepsilon_0}\int_{\text{volume}} \hat{r}\frac{dV}{r^2}. \tag{22–26}$$

This time the integral is over a *volume*.

The three integrals in Eqs (22–22), (22–24), and (22–26) are formal, and to understand how they work in practice, it is best to look at some examples.

EXAMPLE 22–7 A straight insulating rod of length $2L$ carries a uniform linear charge density λ. Determine the electric field at point P, a distance R from the rod along the perpendicular bisector. First find the field in the limit that the rod is much longer than R $(L \gg R)$. Then find it for a distance very far from the rod $(R \gg L)$.

Setting It Up Figure 22–16a illustrates the situation. We include the origin of a coordinate system at the midpoint of the rod, which is aligned with the y-axis.

Strategy Equation (22–22) applies because the charge distribution is linear. We use Eq. (22–22) with dx replaced by dy (we are integrating along the rod) and with $\hat{r} = (\cos\theta)\hat{\imath} + (\sin\theta)\hat{\jmath}$ (Fig. 22–16a) for the unit vector. Equation (22–22) becomes

$$\vec{E} = \frac{\lambda}{4\pi\varepsilon_0}\int_{-L}^{L}[(\cos\theta)\hat{\imath} + (\sin\theta)\hat{\jmath}]\frac{dy}{r^2}. \tag{22–27}$$

Before attempting the integration, we want to recognize any symmetry. The charge $dq = \lambda\, dy$ at a distance y *below* the x-axis gives rise to a field $d\vec{E}$ that is a mirror image of the field $d\vec{E}$ due to another charge dq' at a distance y *above* the axis (Fig. 22–16b). Thus we expect the net y-component of the field on the perpendicular bisector to vanish by symmetry. Here, we shall demonstrate this formally by performing the integration; normally, we would take advantage of the symmetry to reduce the mathematical calculation.

It is often true that the key to performing integrations such as that of Eq. (22–27) is to find the right variables. In this case, the simplest variable to use is the angle θ. To change from dy to $d\theta$, we recognize

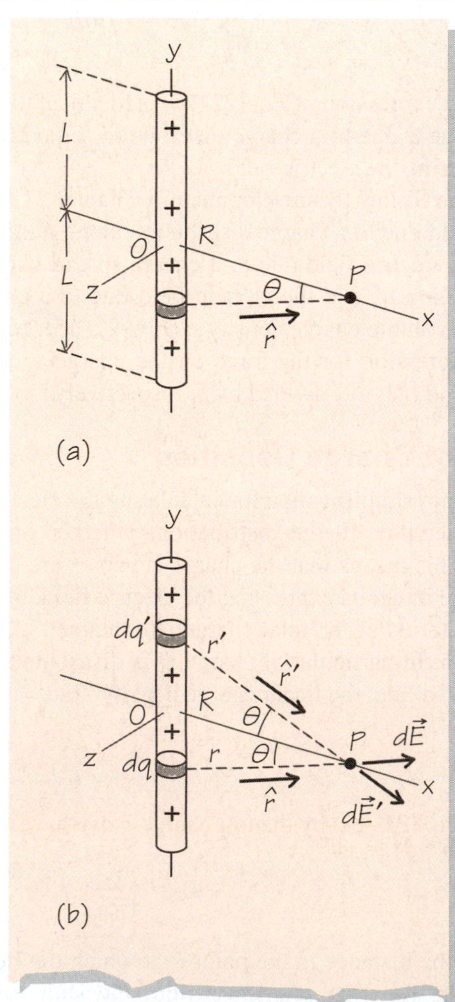

FIGURE 22–16 (a) Geometry of the situation described. (b) The y-components of electric field due to matching pieces above and below the origin cancel, whereas the x-components add.

that y, r, and $\hat{r}$ depend on θ. We must find the dependence of both y and r on θ. We have

$$\tan\theta = \frac{y}{R} \qquad (22\text{–}28)$$

and

$$\cos\theta = \frac{R}{r}. \qquad (22\text{–}29)$$

From Eq. (22–28), we obtain

$$dy = R d(\tan\theta) = R\sec^2\theta\, d\theta = \frac{R}{\cos^2\theta}\, d\theta.$$

With Eq. (22–29), the combination dy/r^2 that appears in Eq. (22–27) is

$$\frac{dy}{r^2} = \frac{1}{r^2}\frac{R}{\cos^2\theta}\, d\theta = \frac{1}{r^2}\frac{R}{(R/r)^2}\, d\theta = \frac{1}{R}\, d\theta.$$

Now we turn to the integral itself. We use the algebra above; the factor $1/R$ is a constant and comes out of the integral, leaving

$$\vec{E} = \frac{\lambda}{4\pi\varepsilon_0 R}\int_{-\theta_0}^{\theta_0}[(\cos\theta)\hat{i} + (\sin\theta)\hat{j}]\, d\theta.$$

The limits $-\theta_0$ and θ_0 are the maximum values of θ, corresponding to the two ends of the line of charge.

Working It Out In this case our strategy has helped us a great deal, because the integral over $d\theta$ is simple. Let's consider each of the two terms in turn. When we integrate the second term $(\sin\theta)$, we obtain a value proportional to $\cos\theta_0 - \cos(-\theta_0) = 0$. Therefore, the coefficient of $\hat{j}$, which is the y-component of the field, is zero, as we argued it must be by symmetry.

The coefficient of $\hat{i}$ is the x-component,

$$E_x = \frac{\lambda}{4\pi\varepsilon_0 R}\int_{-\theta_0}^{\theta_0}\cos\theta\, d\theta = \frac{\lambda}{4\pi\varepsilon_0 R}\sin\theta\Big|_{-\theta_0}^{\theta_0} = \frac{\lambda}{2\pi\varepsilon_0 R}\sin\theta_0. \qquad (22\text{–}30)$$

We can use $\sin\theta_0 = L/\sqrt{L^2 + R^2}$ if desired.

We next take the requested limits. For a rod with length $L \gg R$, $\sin\theta_0 \cong 1$, and the component E_x given by Eq. (22–30) becomes in this limit

$$\text{for } L \gg R: \quad E_x = \frac{\lambda}{2\pi\varepsilon_0 R}. \qquad (22\text{–}31)$$

Equation (22–31) gives the electric field for an almost infinitely long rod (or for a point very close to a finite rod). The direction of the field is perpendicular to the rod.

For the case of $R \gg L$, $\sin\theta_0 \cong L/R$, and in this limit Eq. (22–30) becomes

$$\text{for } R \gg L: \quad E_x = \frac{\lambda L}{2\pi\varepsilon_0 R^2} = \frac{Q}{4\pi\varepsilon_0 R^2}, \qquad (22\text{–}32)$$

where $Q = 2\lambda L$ is the total charge on the rod. In this case $(R \gg L)$, we have obtained the point-charge result, because a rod of finite length looks like a point when it is viewed from large distances.

What Do You Think? The rod is a collection of point charges, and the field from a point charge falls off as $1/R^2$. So how can the field of the rod fall off as $1/R$, as in Eq. (22–31)?

EXAMPLE 22–8 Find the electric field at a distance L from an infinite plane sheet with a uniform surface charge density σ. (This situation relates to capacitors, which are important elements in electric circuits.)

Setting It Up We sketch the plane in Fig. 22–17a, where we place the sheet in the xz-plane. The point P where we want to find the field is on the y-axis.

Strategy Again, the main challenge is to find the simplest way to perform the integral of Eq. (22–24). For this we are going to use a technique that is useful in a variety of contexts. To integrate the effect of the entire plane, we break it up into pieces for each of which the field is easy to calculate. Here we break up the plane into a series of thin concentric circular regions centered around a point below P (Fig. 22–17a). These circles have a width Δr that is so small that the area of the circular regions is $2\pi r\,\Delta r$. To see why this is a good choice, take a look at Fig. 22–17b, which shows the contributions to the field at P from two segments on opposite sides of one of these concentric circles. You can see that their contributions to the field in the xz-plane *cancel* and that their contributions to the y-component of the field *add*. This is true all the way around the circle, so that we can first say that the field at P will have only a y-component. Moreover, the contribution dE_y to the y-component of the field of each of these segments is the same. Call the charge on a little segment of the ring dq. Then from Fig. 22–17b we have $dE_y = \dfrac{dq}{4\pi\varepsilon_0}\dfrac{1}{d^2}\cos\theta$. Every factor in this expression is the same all the way around the ring, so that when we sum the contributions around the ring, the net field y-component ΔE_y from the ring is

$$\Delta E_y = \frac{\Delta Q}{4\pi\varepsilon_0 d^2}\cos\theta = \frac{\sigma(2\pi r\,\Delta r)}{4\pi\varepsilon_0 d^2}\cos\theta.$$

Here, ΔQ is the total charge on the ring, which in the second term is expressed as the surface-charge density times the area of the ring. Finally, we note that $\cos\theta = L/d$, so that

$$\Delta E_y = \frac{\sigma(2\pi r\,\Delta r)}{4\pi\varepsilon_0 d^2}\frac{L}{d} = \frac{\sigma L}{2\varepsilon_0}\frac{r\,\Delta r}{(r^2 + L^2)^{3/2}}.$$

(We have used the fact that $d = \sqrt{r^2 + L^2}$.) In thinking about the direction of $\Delta\vec{E}$, symmetry has been an important guide. We have now found the field from a single ring. To find the net field, we must now sum over the rings, and this is a straightforward integration.

Working It Out The net field is the sum over the fields from the set of rings of all different radii r from 0 to ∞; these rings cover the entire plane. This sum is a single integral over the radii r. We have

$$E_y = \sum \Delta E_y = \sum \frac{\sigma L}{2\varepsilon_0}\frac{r\,\Delta r}{(r^2 + L^2)^{3/2}} = \frac{\sigma L}{2\varepsilon_0}\int_0^\infty \frac{r\, dr}{(r^2 + L^2)^{3/2}}.$$

A table of integrals gives us

$$\int_0^\infty \frac{r\, dr}{(r^2 + L^2)^{3/2}} = \frac{-1}{(r^2 + L^2)^{1/2}}\Bigg|_{r=0}^{r=\infty} = \frac{1}{L}.$$

(continues on next page)

Thus the field due to the entire plane has magnitude $\sigma/2\varepsilon_0$ and is oriented in the y-direction:

$$\vec{E} = \frac{\sigma}{2\varepsilon_0}\hat{j}, \qquad (22\text{–}33)$$

FIELD OF AN INFINITE UNIFORMLY CHARGED PLANE

where more generally the direction is perpendicular to the plane. The electric field is *constant* in both magnitude and direction: The field $\vec{E}$ does not even depend on how far the point is from the plane (Fig. 22–17c).

In reality, we cannot have planes of infinite extent. The result above holds for finite planes when the point where the field is specified is at a distance from the finite plane that is much less than the distance to the edge of the plane.

What Do You Think? The field does not depend on how far the point is from a plane, because (a) all charged planes have the same field; (b) the infinite, uniformly charged plane looks the same from any distance; (c) this is only an approximation, and very close to the plane the field is actually larger. Which of the above statements is true?

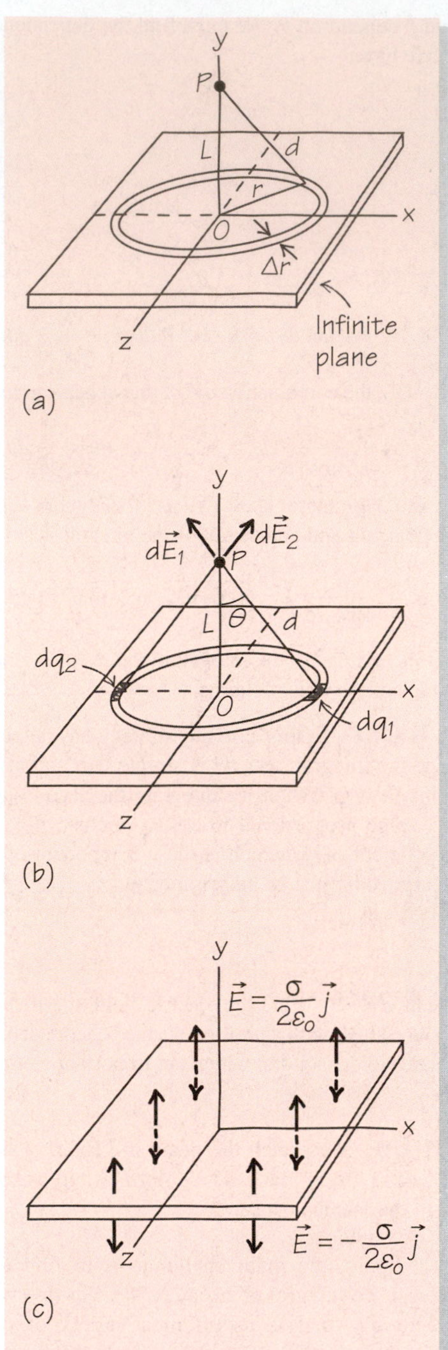

FIGURE 22–17 (a) To find the electric field at a point P a distance L above an infinitely charged plane, break up the plane into concentric circles about the axis passing through P. (b) The field due to one such circle has no horizontal component at P because the horizontal contributions from opposite points on the circle cancel; moreover, the vertical component is easy to find because every point of the circle is equidistant from P. (c) The net field after summation from all circles is uniform and perpendicular to the plane.

The preceding instance illustrates how a formal two-dimensional integration can be reduced to a single integration. There are often several ways to approach this type of problem. For instance, we could have divided our plane into narrow straight strips. The field due to a strip was found in Example 22–7, and we could have used that result. The effect of the entire plane is then found by summing over all the strips that make up the plane. Once again, the final integral would have been a single integral, with the integration variable being the differing distances of the strips from point P.

CONCEPTUAL EXAMPLE 22–9 Suppose that a very large sheet with uniform positive charge density is placed on an insulating stand parallel to Earth's (horizontal) surface. What will a charged cork ball placed a small distance above the plate do?

Answer We are told the sheet is very large and that the cork is only a small distance from the plate, so that to the cork the plate appears to be infinite and we can use the result of Example 22–6. We have a constant electric field, magnitude $E = \sigma/2\varepsilon_0$, that points upward where the cork is. Accordingly, a vertically oriented force of magnitude qE will act on the cork, where q is the charge on the cork. If q is positive, the cork will be repelled from the sheet, and if q is negative, the cork will be attracted to the sheet. For q negative, the cork will be accelerated toward the sheet. However, if q is positive, which way the cork moves depends on whether the electrical force is larger than or smaller than its weight.

EXAMPLE 22–10 Table 22–1 indicates there is an electric field near Earth's surface of about 150 N/C that points vertically down. Assume this field is constant around Earth and that it is due to charge evenly spread on Earth's surface. What is the total charge on the Earth?

Setting It Up We draw a diagram in Fig. 22–18.

Strategy Because the electric field points toward the Earth's center, the responsible charge is negative. To find the value of the charge, we can treat Earth's surface as a plane (see Question 24) and use Eq. (22–33) for the electric field. Given the field's magnitude, we can find the charge density σ, and because we know Earth's surface area, we can then find the total charge Q.

Working It Out From Eq. (22–33),

$$\sigma = 2\varepsilon_0 E = 2(8.85 \times 10^{-12}\,\mathrm{C^2/N\cdot m^2})(150\,\mathrm{N/C}) = 2.7 \times 10^{-9}\,\mathrm{C/m^2}.$$

The total charge Q is found by multiplying this charge density by Earth's surface area, $4\pi R^2$:

$$Q = \sigma 4\pi R^2 = (2.7 \times 10^{-9}\,\mathrm{C/m^2})(4\pi)(6.37 \times 10^6\,\mathrm{m})^2 = 1.4 \times 10^6\,\mathrm{C}.$$

While this may seem like a lot of charge, don't forget that it is spread out over Earth's entire surface. Washington, D.C., has an area of 68 square miles ($1.8 \times 10^8\,\mathrm{m^2}$), which would mean that it contains only a half coulomb.

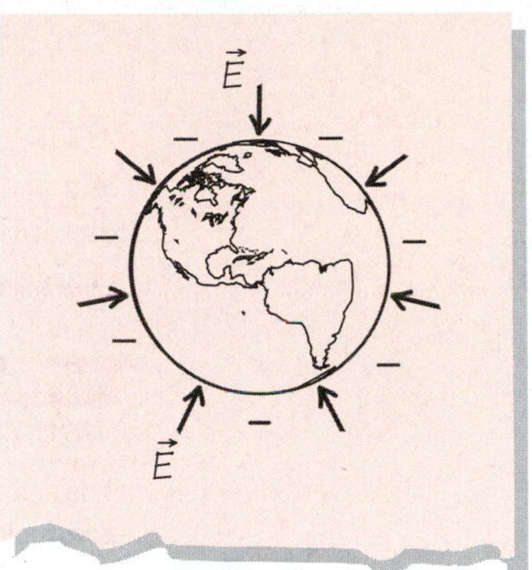

▲ **FIGURE 22–18**

What Do You Think? Suppose the Moon had a similar total charge on its surface. As a result, (a) the Moon and Earth would fly apart; (b) the Moon's surface charge density would be less than that of Earth's; (c) there would be very little effect on the Earth–Moon system; (d) the electric field at the Moon's surface would be the same as that of Earth's. Which of these statements are true?

The Electric Field Between Two Uniformly Charged Planes with Opposite Charge

Example 22–8 shows that the electric field for a positively charged plane of uniform surface charge density σ is uniform and perpendicularly directed away from the plane (Fig. 22–19a). If the plane were negatively charged, the field would be similar but would be directed *toward* the plane (Fig. 22–19b). What happens if we place the two planes, oppositely charged but with the same magnitude of charge density σ, parallel to each other? (This forms a capacitor.) As shown in Fig. 22–19c, the fields outside the

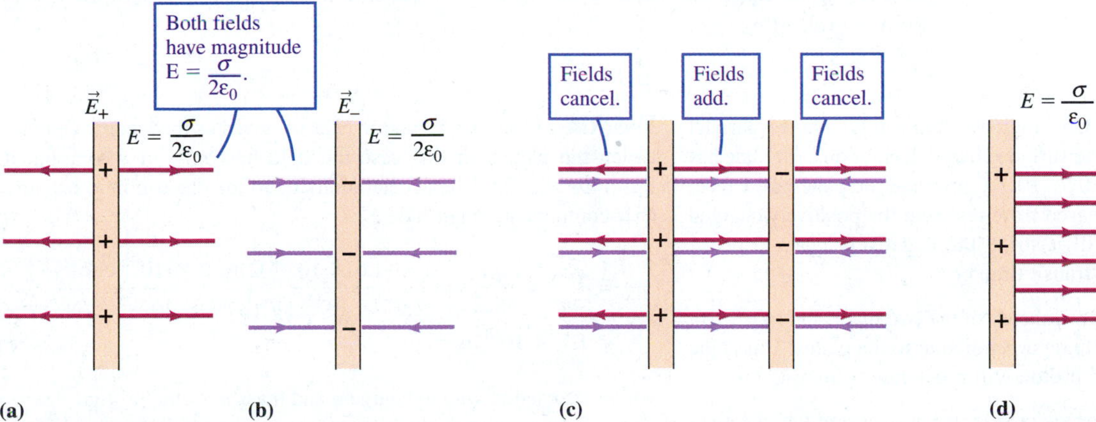

▲ **FIGURE 22–19** (a) The electric field due to a positively charged plane is directed away from the plane; (b) that due to a negatively charged plane is directed into the plane. (c) With two parallel planes carrying equal but opposite charge, the electric field cancels to zero outside the planes but is additive inside. (d) The field inside is σ/ε_0 and is directed from the positive plane to the negative plane.

parallel planes will exactly cancel each other, but the fields between the planes are additive. The resulting field is shown in Fig. 22–19d. For two parallel, oppositely charged planes, the electric field is zero everywhere except between the planes, where the field has magnitude

$$E = \frac{\sigma}{\varepsilon_0} \tag{22–34}$$

ELECTRIC FIELD BETWEEN PARALLEL PLANES OF OPPOSITE UNIFORM CHARGE DENSITY

and is directed from the positively to the negatively charged plane. (Remember that the direction of the electric field is always the direction of the force on our positive test charge q_0.)

THINK ABOUT THIS. . .

HOW DO WE KNOW THAT CHARGES COME IN MINIMUM SIZE UNITS?

The existence of a minimum charge is an experimental result. The determination of this fact and of the value of the minimum charge was carried out by Robert Millikan at the beginning of the twentieth century. The *electron* had been discovered in 1897 by the British physicist J. J. Thomson—the method of discovery will be explained in a later chapter—but all that was known about the electron at that time was the ratio q/m, the charge-to-mass ratio. Millikan's experiment of 1910, the result of several years of preparation, involves spraying of very small oil drops (size 0.1–1.0 μm) into a cavity. These oil drops fall with a very small constant velocity under the influence of gravity, the buoyancy of air, and the air's viscosity. The drops were allowed to fall through a hole into a region between two parallel horizontal plates, oppositely charged, so that there was a known, constant electric field pointing downward. In formation some of the drops acquired small electric charges, and any drops that were negatively charged were subject to a force qE that would arrest their fall, or lead to a constant upward velocity. By studying a large number of such drops, Millikan determined that the charges on the drops came in multiples of a charge $|e| = 1.6 \times 10^{-19}$ C. This smallest charge is the charge of the electron. See Problem 66 for a quantitative treatment of Millikan's experiments.

22–4 Motion of a Charge in a Field

We have been concerned thus far with finding the electric field of a given collection of charges. Let's turn now to the force that charged particles will experience in an external electric field. Newton's second law becomes

$$\vec{F} = q\vec{E}_{\text{ext}} = m\vec{a}, \tag{22–35}$$

where a particle of mass m and charge q has an acceleration $\vec{a}$ due to a given external electric field $\vec{E}_{\text{ext}}$. We then solve Newton's second law as usual. Example 22–11 demonstrates this.

EXAMPLE 22–11 Consider two oppositely charged parallel plates. The magnitude of the surface charge density on each plate has a constant value of $\sigma = 1.0 \times 10^{-6}\ \text{C/m}^2$, and the plates are 1.0 cm apart. (a) If a proton is released from rest near the positively charged plate, with what speed will it strike the negatively charged plate? (b) What will the proton's transit time be?

Setting It Up We show a sketch of the two plates in Fig. 22–20, with the x-axis to the right and perpendicular to the plates. This is the direction toward which the proton will accelerate from rest.

Strategy Given the surface charge density, we can find the electric field and hence the force on the proton; the force will be constant. This then becomes a problem in one-dimensional constant acceleration kinematics in which we determine the proton's speed and transit time.

Working It Out Equation (22–34) gives us the electric field between the plates. In this case the field has only an x-component. From Eq. (22–35), the acceleration due to the electric field has only an x-component, magnitude a_x:

$$a_x = \frac{qE_x}{m} = \frac{q\sigma}{m\varepsilon_0} = \frac{(1.6 \times 10^{-19}\ \text{C})(1.0 \times 10^{-6}\ \text{C/m}^2)}{(1.67 \times 10^{-27}\ \text{kg})(8.85 \times 10^{-12}\ \text{C}^2/\text{N}\cdot\text{m}^2)} = 1.1 \times 10^{13}\ \text{m/s}^2, \tag{22–36}$$

where we use the known charge q and mass m of the proton.

(a) Now we turn to the kinematics, starting with the speed. From Section 2–5 we have $v^2 - v_0^2 = 2a_x x$. With the initial speed v_0 zero,

$$v^2 = 2a_x x. \tag{22–37}$$

Equation (22–36) gives us a_x, while the distance traveled between the plates is $x = 1.0$ cm. Thus

$$v^2 = 2(1.1 \times 10^{13}\ \text{m/s}^2)(1.0 \times 10^{-2}\ \text{m}) = 2.2 \times 10^{11}\ \text{m}^2/\text{s}^2;$$
$$v = 4.7 \times 10^5\ \text{m/s}.$$

(b) The transit time is the final speed divided by the acceleration:

$$t = \frac{v}{a_x} = \frac{4.7 \times 10^5\ \text{m/s}}{1.1 \times 10^{13}\ \text{m/s}^2} = 4.3 \times 10^{-8}\ \text{s}.$$

The plates accelerate protons and thus represent a charged-particle accelerator.

What Do You Think? What would happen to an electron released from the same position as the proton?

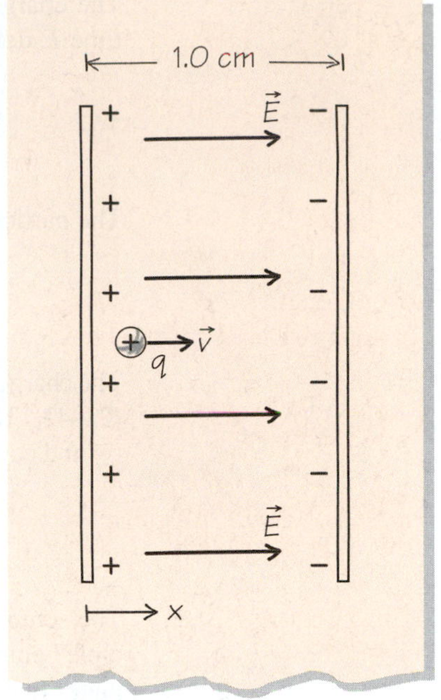

FIGURE 22–20 A charge $+q$ moving between parallel plates.

Deflection of Moving Charged Particles

Let's consider what happens when we inject a negatively charged particle (an electron, for example) into a region of uniform $\vec{E}$ between two plates (Fig. 22–21; we ignore edge effects). The particle has initial velocity $\vec{v}_0$ perpendicular to $\vec{E}$, and since it carries negative charge, it will be deflected upward in Fig. 22–21. From Eq. (22–35) the acceleration vector is

$$\vec{a} = a_x\hat{i} + a_y\hat{j} = \frac{qE}{m}\hat{j}. \tag{22–38}$$

Note that the x-component of the acceleration is zero. Because the initial velocity is only in the x-direction ($\vec{v}_0 = v_0\hat{i}$), the velocity vector becomes

$$\vec{v} = v_x\hat{i} + v_y\hat{j} = v_0\hat{i} + \left(\frac{qE}{m}t\right)\hat{j}. \tag{22–39}$$

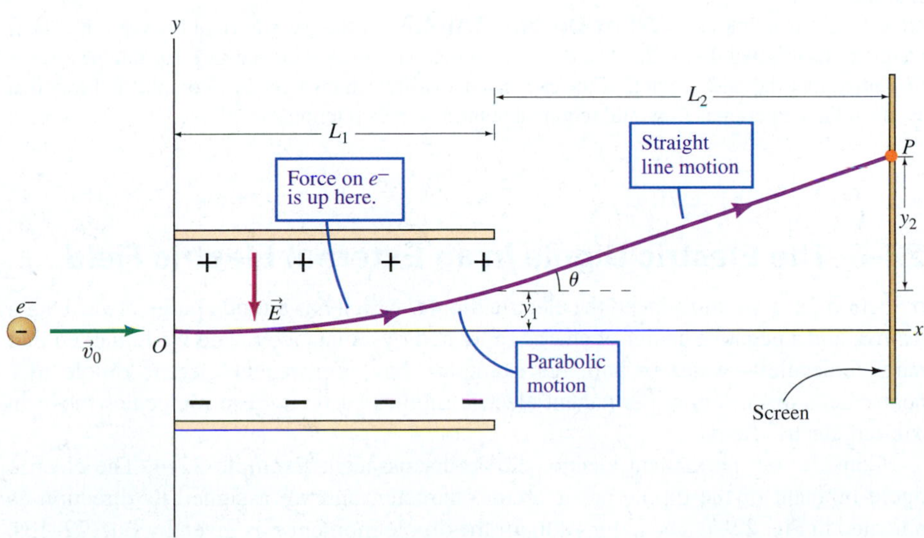

FIGURE 22–21 An electron passing between plates that deflect vertically.

The charged particle travels a horizontal length L_1 between the charged plates in the time T, determined by

$$x = v_0 T = L_1; \tag{22–40}$$

$$T = L_1/v_0. \tag{22–41}$$

The particle's deflection in the y-direction is then

$$y_1 = \frac{1}{2}a_y t^2 = \frac{1}{2}\frac{qE}{m}T^2 = \frac{1}{2}\frac{qE}{m}\frac{L_1^2}{v_0^2}. \tag{22–42}$$

The charged particle emerges from the plates at a position (x, y) given by Eqs. (22–40) and (22–42). The charged particle is then free from the influence of any force (ignoring gravity) and continues past the plates in a straight line at an angle θ from its initial direction:

$$\tan\theta = \frac{v_y}{v_x} = \frac{(qE/m)(L_1/v_0)}{v_0} = \frac{qEL_1}{mv_0^2}. \tag{22–43}$$

The setup described here allows control of a beam of charged particles and is the principle behind the picture tubes of televisions. In a television, there is also a second set of plates that controls deflection in the third direction, and a beam of electrons can then be directed anywhere on the screen.

EXAMPLE 22–12 An electron moving horizontally at a speed $v_0 = 3 \times 10^6$ m/s enters the region between two horizontally oriented plates of length $L_1 = 3$ cm. A fluorescent screen is located $L_2 = 12$ cm past these plates. Find the electron's total vertical deflection on the screen from its initial direction if the electric field between the plates points downward with a magnitude of $E = 10^3$ N/C.

Setting It Up Figure 22–21 again illustrates the situation, with the x-axis along the initial direction of the electron and the y-direction perpendicular. The electron will be deflected along the y-direction.

Strategy A constant electric field will lead to a constant acceleration, and we must apply the kinematic equations that describe constant acceleration. These equations are worked out above. Let us label as y_1 the vertical deflection of the electron between the plates, and as y_2 the additional vertical deflection of the electron after it has passed between the plates, where it continues in a straight path until it strikes the screen. We use Eq. (22–42) to find the deflection y_1 of the electron while it is between the plates. After the electron leaves the region between the plates, it travels at an angle to its original direction, given by Eq. (22–43) as $\tan\theta = qEL_1/mv_0^2$, and the additional deflection will be $y_2 = L_2 \tan\theta$. Finally, the total deflection is $y = y_1 + y_2$.

Working It Out We have from Eq. (22–42) $y_1 = \frac{1}{2}\frac{qE}{m}\frac{L_1^2}{v_0^2}$, and from Eq. (22–43) $y_2 = L_2 \tan\theta = \frac{qEL_1L_2}{mv_0^2}$. The net deflection is then

$$y = y_1 + y_2 = \frac{1}{2}\frac{qEL_1^2}{mv_0^2} + \frac{qEL_1L_2}{mv_0^2} = \frac{qEL_1}{mv_0^2}\left(\frac{1}{2}L_1 + L_2\right).$$

Numerical evaluation with a minus sign for E ($\vec{E}$ points downward) gives

$$\begin{aligned} y &= \frac{(-1.6 \times 10^{-19}\,\text{C})(-10^3\,\text{N/C})(3 \times 10^{-2}\,\text{m})}{(9.11 \times 10^{-31}\,\text{kg})(3 \times 10^6\,\text{m/s})^2} \\ &\quad \times \left[\frac{1}{2}(3 \times 10^{-2}\,\text{m}) + (12 \times 10^{-2}\,\text{m})\right] \\ &= 8 \times 10^{-2}\,\text{m}. \end{aligned}$$

Of this 8 cm, the deflection $y_1 = 1$ cm, and the deflection $y_2 = 7$ cm.

What Do You Think? If the electric field between the plates doubles, the vertical deflection (a) is unchanged; (b) doubles; (c) is half of the calculated amount; (d) changes in a complicated way that would require a numerical reevaluation.

22–5 The Electric Dipole in an External Electric Field

In Section 22–1 we introduced the electric dipole, which has a total charge of zero but a positive and a negative center of charge separated by a distance L, and we remarked that many molecules—water or salt, for example—have permanent electric dipole moments. Let's discuss how permanent electric dipoles such as these molecules move in external electric fields.

Consider the permanent electric dipole discussed in Example 22–4. The electric dipole moment of the dipole has a vector character, and we assigned its direction as indicated in Fig. 22–7. The expression for the dipole moment $\vec{p}$ is given by Eq. (22–13).

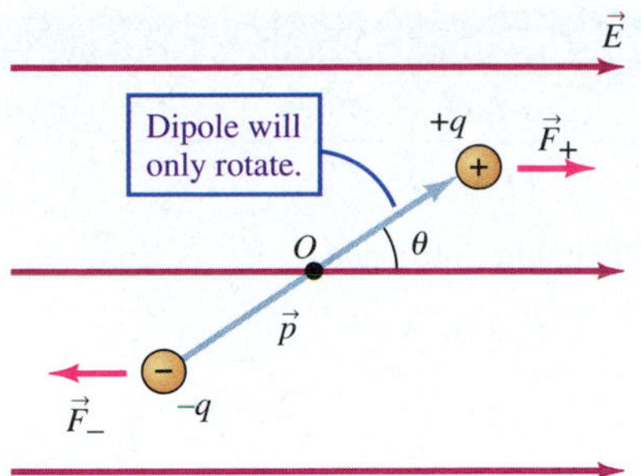

FIGURE 22–22 A dipole placed in a uniform external electric field experiences no net force but will experience a torque.

If we place the electric dipole in a uniform external field (Fig. 22–22), then the forces on $+q$ and $-q$ are, respectively,

$$\vec{F}_+ = q\vec{E},$$
$$\vec{F}_- = -q\vec{E} = -\vec{F}_+.$$

We notice that the two forces are equal and opposite and therefore cancel. *There is no net force on the dipole.*

There is, however, a torque that tends to rotate the dipole. To calculate the torque and the corresponding rotation, we must choose a reference point, and it is convenient to choose this point to be the midpoint of the dipole, which is located at point O in Fig. 22–22. The actual motion will be independent of the choice of reference point. The torque, τ, about a point due to a force that acts on another point that is a displacement $\vec{r}$ away is given by Eq. (10–6):

$$\vec{\tau} = \vec{r} \times \vec{F}, \qquad (22\text{–}44)$$

where $\vec{r}$ is measured from point O. The resulting torque from the force on each charge is then clockwise, with magnitudes

$$\tau_+ = \left(\frac{L}{2}\right)qE\sin\theta, \qquad \tau_- = \left(\frac{L}{2}\right)qE\sin\theta,$$

where the subscripts + and − refer to the charges. Because both τ_+ and τ_- are clockwise rotations, the total torque is also clockwise, with magnitude

$$\tau = \tau_+ + \tau_- = qLE\sin\theta. \qquad (22\text{–}45)$$

We can represent this expression for the torque on a dipole as the vector product of $\vec{p}$ and $\vec{E}$:

$$\vec{\tau} = \vec{p} \times \vec{E}, \qquad (22\text{–}46)$$

TORQUE ON A DIPOLE DUE TO A FIELD

which gives both the torque's magnitude $(pE\sin\theta)$ and direction (into the page in Fig. 22–22).

The maximum torque $(\tau = pE)$ occurs when $\vec{p}$ and $\vec{E}$ are perpendicular $(\theta = \pi/2)$. The torque is zero when $\vec{p}$ and $\vec{E}$ are parallel $(\theta = 0)$ or antiparallel $(\theta = \pi)$. The torque tends to rotate the electric dipole until $\vec{p}$ is parallel to $\vec{E}$. The position $\theta = 0$ corresponds to a stable equilibrium, but the position $\theta = \pi$ is one of unstable equilibrium because a small deviation will cause the dipole to rotate toward $\theta = 0$.

Suppose the dipole is initially aligned along some angle $\theta \neq 0$ with respect to an electric field. What will the motion look like? This question is most easily answered when we look at the potential energy U of the dipole in the field (see the paragraph below). There we shall see that *for small angles,* U is a constant + a term proportional to θ^2. This corresponds to the behavior of an ordinary pendulum, or a spring, and the motion is oscillatory about $\theta = 0$. As the dipole rotates toward $\theta = 0$, it gains kinetic energy and passes through $\theta = 0$ to the other side. The torque, however, then becomes counterclockwise, and the dipole slows down, stops, returns to $\theta = 0$, and passes through it again to the original side. Table 22–2 illustrates a time sequence for a rotating dipole in an electric field. Without a mechanism to dissipate the dipole's energy, the oscillations will continue.

TABLE 22–2 • An Electric Dipole Rotating in a Uniform Electric Field

Electric Field	Torque, τ	Angular Velocity, ω
Initial condition	Maximum, into page	Zero
	Decreasing, into page	Increasing, into page
	Zero	Maximum, into page
	Changed direction, out of page, increasing	Decreasing, into page
	Maximum, out of page	Zero
	Decreasing, out of page	Changed direction, increasing, out of page

The Energy of a Dipole in an External Electric Field

When a dipole rotates under the influence of an external electric field, the field does work on it. From this work we can find the potential energy of the dipole in the field. To find the work done by the field as the dipole rotates from an initial angle θ_0 to a final angle θ, we use the form $W = \int_{\theta_0}^{\theta} \vec{\tau} \cdot d\vec{\theta}$. We refer to Fig. 22–22 to see that for a counterclockwise rotation, the infinitesimal angle change is out of the page, while the torque $\vec{\tau} = \vec{p} \times \vec{E}$ is into the page. Thus the scalar product $\vec{\tau} \cdot d\vec{\theta} = -\tau \, d\theta = -(pE \sin\theta)\, d\theta$, and the work done by the field is

$$W = \int_{\theta_0}^{\theta} (-pE \sin\theta)\, d\theta = -pE(\cos\theta_0 - \cos\theta). \qquad (22\text{–}47)$$

Now, as we know from Chapter 7, the potential energy $U(\theta)$ at point θ minus its value $U(\theta_0) = U_0$ at some initial point θ_0 is the *negative* of the work done by the associated force as the object moves from θ_0 to θ. Thus we have

$$U - U_0 = pE(\cos\theta_0 - \cos\theta). \qquad (22\text{–}48)$$

We are free to choose the constant U_0, and we choose it such that $U_0 = 0$ at $\theta_0 = \pi/2$. Thus the potential energy at angle θ is given by

$$U = -pE\cos\theta. \qquad (22\text{–}49)$$

Notice that Eq. (22–49) is consistent with our choice for the zero of the potential energy, because $U_0 = -pE\cos\theta_0$, which is zero when $\theta_0 = \pi/2$.

Equation (22–49) can be written more compactly by using the scalar dot product of $\vec{p}$ and $\vec{E}$:

$$U = -\vec{p} \cdot \vec{E}. \qquad (22\text{–}50)$$

POTENTIAL ENERGY OF A DIPOLE IN A FIELD

Above, we discussed the stability of the equilibrium of the dipole in an external field. We can see directly from Eq. (22–50) that the orientation in which $\vec{p}$ is aligned with $\vec{E}$ is a point of stable equilibrium because U has a minimum there. In contrast, U has a maximum when $\vec{p}$ is antiparallel to $\vec{E}$, and therefore this is a point of unstable equilibrium.

CONCEPTUAL EXAMPLE 22–13 What is the motion, if any, of an electric dipole placed in a *nonuniform* external electric field?

Answer A possible setup of this situation is sketched in Fig. 22–23. We have used a cork ball that has a dipole moment induced by the field itself. If the field is nonuniform, the forces on the two charges in the dipole will generally differ in magnitude. Then, in addition to a torque, there will be a net force on the dipole. The resulting motion would be a combination of linear acceleration and rotation. The details of the motion depend on the particular electric field. While we have used the tip of a charged rod in this figure, we could equally well have placed a dipole in the region of a point charge, as long as the point charge field lines "spread" significantly over the size of the dipole.

The effect of a nonuniform electric field on an induced dipole explains the attraction of a neutral cork ball coated with conducting paint to a Teflon rod rubbed against fur (Fig. 22–23). The charged Teflon rod induces an electric dipole on the cork ball, which is in the nonuniform electric field of the rod. Another example of the action of a nonuniform electric field on an induced dipole is the attraction between small bits of paper and a comb that has just been charged by passing it through hair. The bits of paper have induced dipole moments and are attracted to the comb in its nonuniform field.

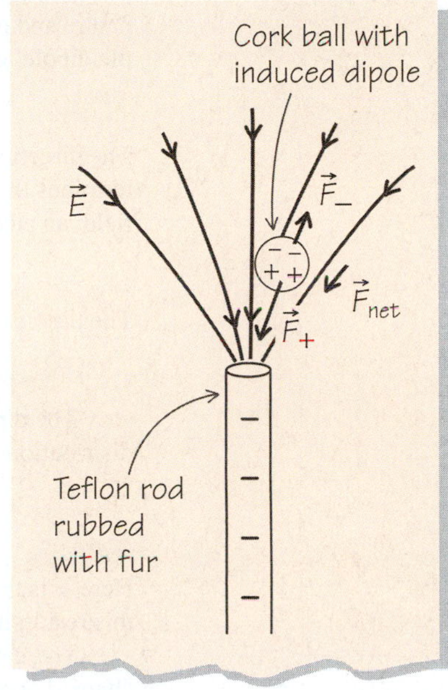

▲ **FIGURE 22–23** A dipole placed in a nonuniform external electric field can experience a net force. In this case the external electric field induces a dipole in the cork ball, which then experiences a force due to the electric field. This effect can be understood by using Coulomb's law.

Summary

Charge distributions set up electric fields in the space around them. The electric field vectors can be mapped out by moving a small, positive test charge q_0 around in this field. The field $\vec{E}$ is defined as the force $\vec{F}$ on this test charge, divided by q_0:

$$\vec{E} \equiv \lim_{q_0 \to 0} \frac{\vec{F}}{q_0} \qquad (22\text{–}2)$$

The electric field has units of N/C (or V/m). The force on a point charge q' in a given external electric field is

$$\vec{F} = q'\vec{E}_{ext}. \qquad (22\text{–}7)$$

Electric field lines aid in the visualization of the direction and magnitude of the electric field produced by various charge configurations. We can summarize the rules for drawing them and their properties by saying that

- Electric field lines are drawn so that the tangent to the field line specifies the direction of the electric field itself.
- The density in field lines per unit area is proportional to the magnitude of the electric field.
- Electric field lines start and end on electric charges, never in empty space. In particular, they start on positive charges and terminate on negative charges.
- Electric field lines never cross.

From Coulomb's law, the electric field due to a point charge q is

$$\vec{E} = \frac{q}{4\pi\varepsilon_0 r^2}\hat{r}. \tag{22–5}$$

Electric fields obey the principle of superposition:

$$\vec{E}_{\text{net}} = \vec{E}_1 + \vec{E}_2 + \vec{E}_3 + \cdots = \sum_i \vec{E}_i, \tag{22–10}$$

where $\vec{E}_i$ labels the field of the components (normally point charges) that make up a charge distribution.

In its simplest form, an electric dipole consists of a positive charge q separated by a distance L from a negative charge $-q$. Such a configuration, or any configuration that is electrically neutral but has an imbalance of positive and negative charge from one side to another, occurs often in nature and produces an electric field. This field decreases with distance r as $1/r^3$ and, for the simple dipole, is proportional to the magnitude of the electric dipole moment $\vec{p}$, given by

$$\vec{p} = q\vec{L}. \tag{22–13}$$

The direction of $\vec{L}$ (and of $\vec{p}$) is from the negative charge to the positive charge. This direction determines the angular dependence of the electric dipole field. In addition to producing an electric field, an electric dipole experiences a torque in a uniform external electric field:

$$\vec{\tau} = \vec{p} \times \vec{E}. \tag{22–46}$$

The dipole in the external field has a potential energy of

$$U = -\vec{p}\cdot\vec{E}. \tag{22–50}$$

The dipole is a configuration of two point charges, but we also deal with continuous charge distributions. The electric field due to a continuous charge distribution is

$$\vec{E} = \frac{1}{4\pi\varepsilon_0}\int \hat{r}\frac{dq}{r^2}. \tag{22–20}$$

Here, r is the distance of a charge element dq from the point where the field is measured. To use this result, it is necessary to know how dq varies throughout space.

The electric field due to an infinitely long wire is radial and perpendicular to the wire. A charged plane, infinite in area, with a charge per unit area of σ has an electric field that is uniform and directed perpendicular to the plane, with

$$E = \frac{\sigma}{2\varepsilon_0}. \tag{22–33}$$

In later chapters we'll have ample opportunity to apply the electric field concept.

Understanding the Concepts

1. In older movies, you may see gasoline trucks dragging metal chains along the road. Why?
2. Why can't two electric field lines cross?
3. We have introduced the concept of an electric field. Why might it be useful to introduce an analogous gravitational field? In what ways would this field be like an electric field, and in what ways would it be different?
4. An inflated rubber balloon is charged by rubbing it with fur. Explain what happens when that balloon is placed against (a) a metal wall; (b) an insulating wall.
5. Two charges of opposite polarities are close together. Five times as many field lines leave one charge as end up on the other charge. What is the ratio of charges?
6. Electric field lines originate from positive charges and terminate at negative charges, as exemplified by the field lines due to a dipole. Does this statement contradict the depiction of field lines due to a single positive point charge?
7. If in Example 22–3 there were a point (or points) on the line connecting the three charges at which the electric field is zero, where, qualitatively, would it (or they) be?
8. A pair of equal and opposite charges forms a dipole, and the electric field of a dipole is not zero. But if we were to look at a dipole from very far away, the two charges would appear to be on top of one another and to cancel; that is, we would see no charge, and hence we should see no field. How do you reconcile these statements?
9. In Example 22–10, we asked how many electrons the charge on Earth represents. How do we know that the charge responsible for Earth's field is a negative one?
10. Can the electric dipole induced on a spherical conducting ball cause the ball to rotate? How about the electric dipole induced on a long rod?
11. After you comb your hair, the comb can often attract small pieces of paper. The act of combing may induce an electric charge on the comb, but the combing does not itself affect the paper. What accounts for the attraction?
12. Is it possible for electric field lines to go all the way to infinity?
13. Explain how the water molecule, H_2O, acts as an electric dipole (see Fig. 22–8), given that there are *two* spatial regions (around the H atoms) with negative charge.

14. We called the arrangement in Example 22–11 an accelerator; how could we get the accelerated particle out in order to use it?
15. Explain why the electric-field-line technique would not be useful for a point charge if Coulomb's experiments had shown the electric force to decrease as $1/r$ or as $1/r^{2+\delta}$.
16. The internal motion of a liquid can be described by a velocity field, which is the velocity vector of the element of fluid at a given point. In what ways is this field like an electric field, and in what ways is it different?
17. Can you invent an arrangement of charges whose electric field would be radially directed into a point in some region of empty space? The correct answer has implications for the stability of charges placed in static electric fields.
18. Suppose that a small electric dipole ($+q$ and $-q$) is placed somewhere on a line that is perpendicular to, and bisects, a second (fixed) dipole ($+Q$ and $-Q$), as seen in Fig. 22–24. If the small dipole is free to pivot about its center, what will it do?

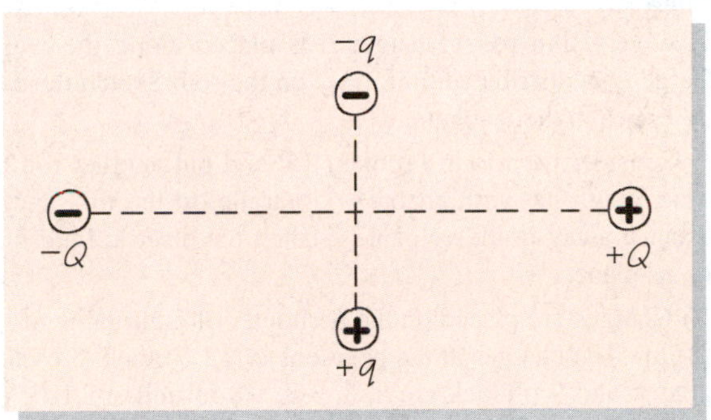

▲ **FIGURE 22–24** Question 18.

19. We have charges $+q$ at $x = \pm a$ and charge $-2q$ at $x = 0$. Is it useful to think about this configuration as two electric dipoles placed back-to-back? If so, the dipole moments for the two dipoles are opposite; is the distant field on the y-axis therefore exactly zero?
20. Consider a large number of identical dipoles centered in the xy-plane and pointing in the z-direction, distributed with uniform density. What is the electric field in the limit that the dipoles form a continuous distribution?
21. A large, flat, positively charged plate (of uniform charge density) is placed on the ground. A positively charged pellet, starting from rest, is released from above the plate. Ignore all air resistance. Qualitatively describe the motion of the pellet, according to the height from which it is dropped.
22. Suppose that a positively charged pellet is dropped from above onto the north pole of a large, positively charged sphere (of uniform charge density). Disregarding air resistance, and any small instabilities that would make the pellet move away from the vertical, describe the motion of the pellet.
23. Suppose in Example 22–3 q_1 were at $x_1 = -1$ mm rather than -1 cm. What useful approximation could you make to find the field at A?
24. In Example 22–10, we found Earth's charge by treating its surface as a plane rather than thinking of it as a sphere. Why is this a very good approximation?

Problems

22–1 Electric Field

1. (I) A 5-μC charge is located at $(x, y) = (5\text{ cm}, 0\text{ cm})$. Determine the electric field at (2 cm, 4 cm).
2. (I) Calculate the electric field at the origin due to the following distribution of charges: $+q$ at $(x, y) = (a, a)$, $+q$ at $(-a, a)$, $-q$ at $(-a, -a)$, and $-q$ at $(a, -a)$ (Fig. 22–25).

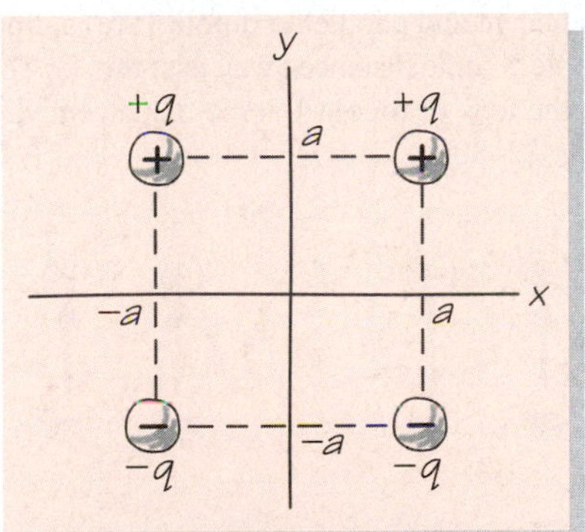

▲ **FIGURE 22–25** Problem 2.

3. (I) Find the electric field due to the nucleus of an atom of gold ($Z = 79$) at a point 1 nm from the nucleus. Treat the nucleus as a point. What is the force on an electron at that point?
4. (II) Charges of $+3\ \mu$C, $-5\ \mu$C, $+7\ \mu$C, and $-2\ \mu$C are located at the four corners of a square 4 cm on each side (Fig. 22–26). Calculate the electric field at the center of the square. [*Hint*: You may want to use a coordinate system with axes along the diagonals.]

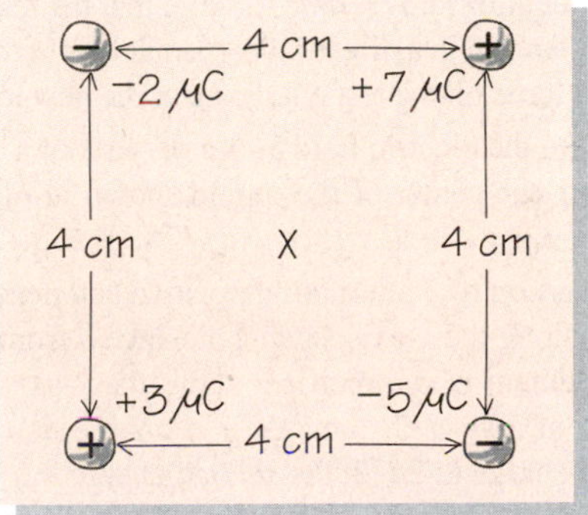

▲ **FIGURE 22–26** Problem 4.

5. (II) Five charges are located at five of the corners of a regular hexagon with sides of 10 cm, as shown on Fig. 22–27 (see next page). Find the electric field at the sixth corner of the hexagon.
6. (II) Charges $q_1 = 1.5\ \mu$C and $q_2 = -3.5\ \mu$C are a distance 22 cm apart. (a) Calculate the electric field of q_1 at the position of q_2; (b) the force acting on q_2, and (c) the total electric field at the midpoint between q_1 and q_2.
7. (II) A charge $-q$ is located at $y = -\ell/2$, and a second charge $+q$ is located at $y = +\ell/2$ (Fig. 22–28 see next page). (a) What is the electric field at the origin? (b) If the charge at $-\ell/2$ were instead $+q$, what would the electric field be at the origin? (c) For part (b), what would the electric field be in the entire xz-plane specified by $y = 0$?

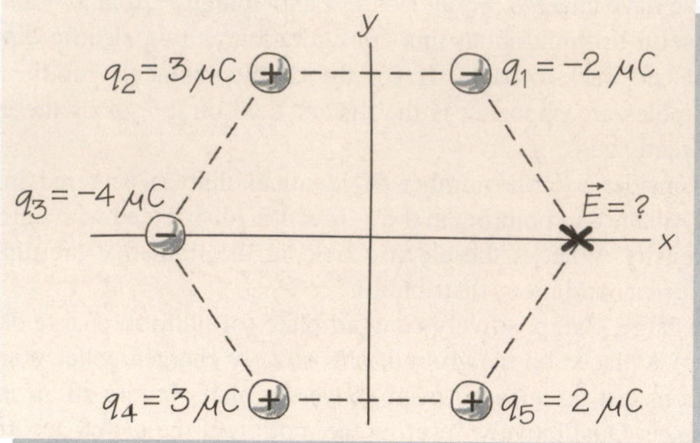

▲ **FIGURE 22–27** Problem 5.

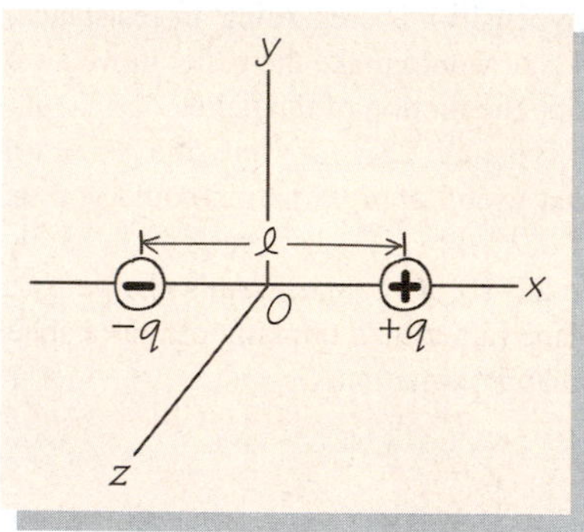

▲ **FIGURE 22–28** Problem 7.

8. (II) Identical positive charges Q are placed at $x = a$ and $x = -a$, respectively. (a) What is the electric field at $x = 0$? (b) Suppose that a positive test charge q_0 is placed at $x = 0$. Will it be in stable or unstable equilibrium? [*Hint*: Assume that the test charge is displaced a distance δ in a direction perpendicular to the x-axis. What will the net force on the test charge be at the new location?]

9. (II) Calculate the electric field *along the axis* of a dipole at a distance r from the center of the dipole shown in Fig. 22–5. Work out the field for $r \gg L$.

10. (II) A succession of n alternating positive and negative charges q are placed along the x-axis, each a distance d from its neighbors. The arrangement is symmetrical about the y-axis, with the first $+q$ charge at $x = d/2$, the first $-q$ charge at $x = -d/2$, the second $-q$ charge at $3d/2$, the second charge $+q$ at $-3d/2$, and so forth (Fig. 22–29). What is the field at a distant point $y = Y$ (where $Y \gg nd$) on the y-axis?

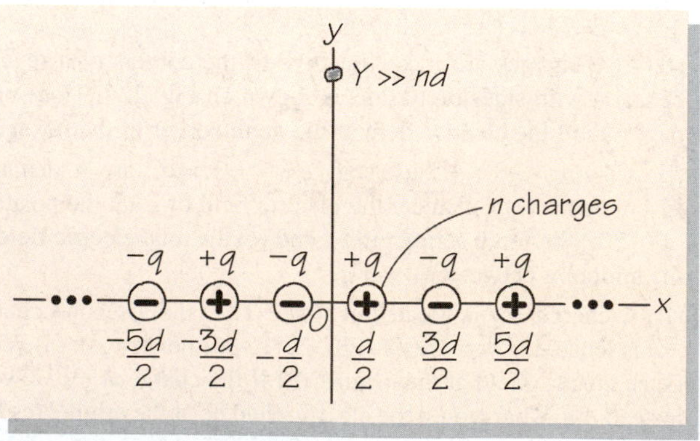

▲ **FIGURE 22–29** Problem 10.

11. (III) Suppose that the positive test charge in Problem 8 is constrained to move along the x-axis only. Will $x = 0$ be a stable equilibrium position? If it is, then the test charge should oscillate about $x = 0$ for small enough displacements. If that were the case, what would the frequency of oscillation be for a test charge of mass m? [*Hint*: Assume that the charge is displaced to a point $x = \delta$, where $\delta \ll a$, and calculate the magnitude and the direction of the electric field there. Use the approximation $1/(a + \delta)^2 = (1/a^2) - (2\delta/a^3) + \cdots$, valid for $\delta \ll a$.]

22–2 Electric Field Lines

12. (I) Draw the electric field lines due to charges of $+0.6\,\mu$C and $+1.8\,\mu$C located 15 cm apart.

13. (I) A pair of parallel plates have equal and opposite uniform charge distributions. The field is represented by parallel lines drawn with a density N per m^2. The charge density on the plates is tripled. How should the density of electric field lines be changed?

14. (I) A very thin rod of length L is placed along the x-axis. A charge Q is distributed uniformly on the rod. Sketch the electric field lines in the xy-plane.

15. (I) Consider the rod in Problem 14, and put another rod of the same length but with charge $-Q$ parallel to the first and some distance away in the xy-plane. Sketch the electric field lines in the xy-plane.

16. (II) Charges are placed on the perimeter of a turntable of radius 0.15 m. The charges at the positions of 12 o'clock, 3 o'clock, 6 o'clock and 9 o'clock are 8, 8, −8, −8 in units of 10^{-5} C, respectively. Draw the field lines for this arrangement. What is the electric field at the center of the circle?

17. (II) Six equal charges of alternating opposite sign (+, −, +, −, +, −) are aligned along the x-axis, each separated from its neighbor by a distance d. Draw the configurations of field lines.

18. (II) Consider charges q placed along the x-axis at $x = na$, with $n = 0, \pm 1, \pm 2, \pm 3, \ldots$. Sketch the electric field lines.

19. (II) The field lines due to an electric dipole $\vec{p}$ (dipole 1) are shown in Fig. 22–12b; by definition, the direction of $\vec{p}$ points from $-q$ to $+q$. Sketch the field lines for the combination of this dipole and (a) a dipole $-\vec{p}$ adjacent and parallel to dipole 1; (b) a dipole $\vec{p}$ adjacent and parallel to dipole 1; (c) a dipole $-\vec{p}$ on the axis of dipole 1 some distance away past the $+q$ charge; (d) a dipole $\vec{p}$ on the axis of dipole 1 some distance away past the $+q$ charge (Fig. 22–30).

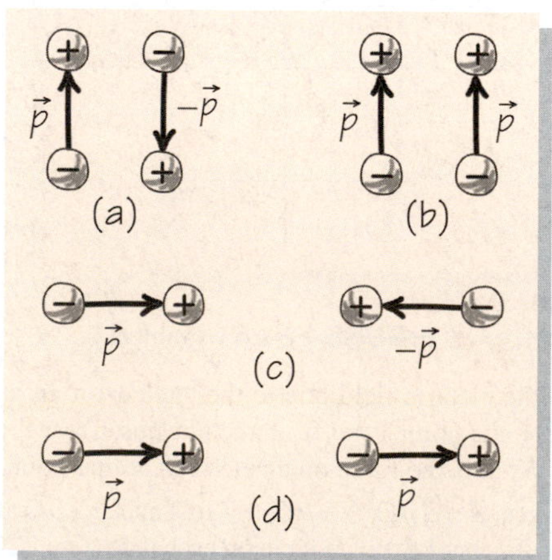

▲ **FIGURE 22–30** Problem 19.

20. (II) Charges q, q, and $-q$ form an equilateral triangle whose sides each have length 12 cm, with the negative charge at the top. Sketch the electric field lines. What is the magnitude of the electric field 0.080 cm from and directly above $-q$; what is it at a distance of 35 m from and directly above $-q$?

22–3 The Field of a Continuous Distribution

21. (I) Calculate the electric field due to an infinitely long, thin, uniformly charged rod with a charge density of 0.3 μC/m at a distance 20 cm from the rod. Assume that the rod is aligned with the x-axis.

22. (I) A thin, uniformly charged rod with a total charge of 6 μC and length 25 cm is placed along the z-axis, centered at the origin. Find the electric field at $(x, y, z) = (6 \text{ cm}, 0 \text{ cm}, 0 \text{ cm})$ and $(0 \text{ cm}, 6 \text{ cm}, 0 \text{ cm})$.

23. (I) Sketch the electric field lines between a point charge Q and a uniformly charged, flat square of area L^2 and total charge $-Q$. The point charge is located a distance L above the center of the plane.

24. (I) Consider two infinite plane sheets of insulator with uniform surface charge densities σ_1 and σ_2, respectively. The two sheets are parallel to each other and a distance L apart. What is the force on a point charge Q placed midway between the sheets?

25. (II) A negative charge is distributed uniformly on a long cylindrical shell. Sketch the field lines both inside and outside the shell. Do not include the ends of the cylinder.

26. (II) Consider positive charges distributed uniformly with a charge density λ on a circle of radius R. (a) Use symmetry arguments to deduce the direction of the electric field at a point in the plane of the circle but outside the circle. (b) What is the magnitude of the electric field at a distance L along the axis of the circle for $L \gg R$?

27. (II) A rod with a uniform negative charge is bent into a semicircle. Make a rough sketch of the electric field lines in the plane of the rod.

28. (II) Two infinite plates with a uniform charge density of 1.2 $\mu\text{C/m}^2$ are placed along the yz-plane with one plate passing through $x = 2$ cm and the other through $x = -2$ cm. Determine the electric field at $(x, y, z) =$ (a) (0 cm, 0 cm, 0 cm); (b) (8 cm, 0 cm, 0 cm); (c) (8 cm, 1 cm, 2 cm).

29. (II) Two large, flat, vertically oriented plates are parallel to each other, a distance d apart. Both have the same uniform positive charge density σ. What is the electric field in the space around and between them?

30. (II) The axis of a hollow tube of radius R and length L is aligned with the x-axis; the tube's left-hand edge is at $x = 0$, as shown in Fig. 22–31. It carries a total charge q distributed uniformly along its surface. By integrating the result for a field due to a hoop of charge along the axis of the hoop (see Example 22–8), find the electric field along the x-axis due to the tube as a function of x.

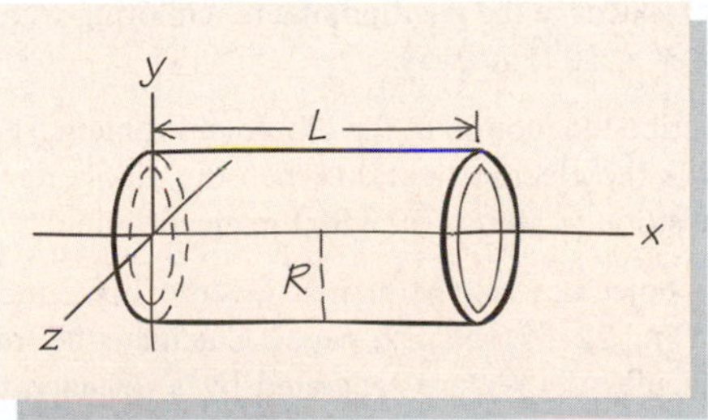

▲ **FIGURE 22–31** Problem 30.

31. (II) A thin, circular disk of radius R is oriented in the xy-plane with its center at the origin. A charge Q on the disk is distributed uniformly over the surface. (a) Find the electric field due to the disk at the point $z = z_0$ along the z-axis. (b) Find the field in the limit $z_0 \to \infty$. (c) Find the field in the limit that $R \to \infty$. Are the limits of parts (b) and (c) the same?

32. (II) Consider a thin, uniformly charged rod 18 cm long that is bent into a semicircle. The total charge on the rod is 0.36 μC. What are the magnitude and direction of the electric field at the center of the semicircle?

33. (II) A rod 30 cm long is charged uniformly with a charge density of 15 μC/m. A charge of 3 μC is placed 30 cm from the midpoint of the rod along a line perpendicular to the rod. Calculate the electric field at a point halfway between the point charge and the center of the rod.

34. (II) Determine the electric field from the large plane in Example 22–8 by assuming the plane is made of a series of charged rods, using the results of Example 22–7.

35. (II) Consider a spherical shell of radius R uniformly charged with a total negative charge $-Q$. Starting at the surface of the shell going outward, there is a uniform distribution of positive charge in space such that the electric field at the radius $(R + h)$ vanishes, where $R \gg h$. What is the positive charge density? [*Hint*: With $R \gg h$ you can assume that $(R + r)^n \cong R^n + nrR^{n-1}$ for $0 \leq r \leq h$.]

36. (III) A total charge Q is distributed uniformly over a rod of length L. The rod is aligned on the x-axis, with one end at the origin and the other at the point $x = L$ (Fig. 22–32). Calculate the electric field at a point $(0, D)$, and compare this result with the field at the point $(L/2, D)$.

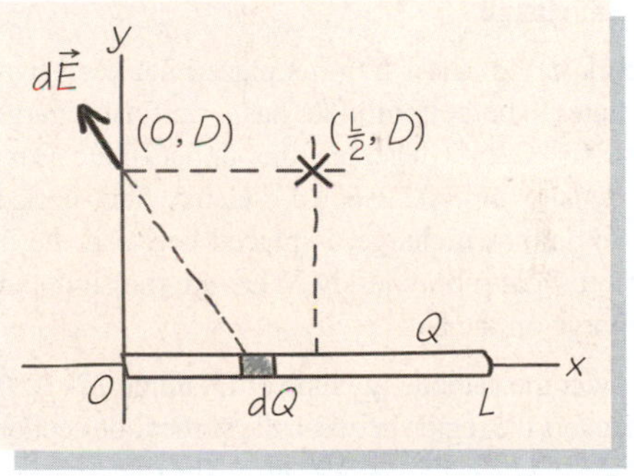

▲ **FIGURE 22–32** Problem 36.

37. (III) Consider a point at a height z_0 directly above the midpoint of a square with sides of length $2L$. The (nonconducting) square carries a uniform charge density σ. (a) Use the method of Example 22–8 to write an integral for the electric field at z_0. (b) How does the integral simplify in the limit $L \to \infty$? (c) $z_0 \to 0$?

22–4 Motion of a Charge in a Field

38. (I) An infinite plate carries a uniform charge density $\sigma = 2.17 \times 10^{-6}\ \text{C/m}^2$. A pellet of mass 0.555 g is placed at rest 0.175 m from the plate. The pellet carries a negative charge $q = -1.08 \times 10^{-6}$ C. What is its speed when it reaches the plate? Ignore all forces except the electrostatic attraction.

39. (I) A small object, mass 120 mg, is observed to undergo an acceleration of magnitude 4.6 m/s^2 when it is placed in a constant electric field of magnitude 850 N/C. What is the charge on the object?

40. (I) A sheet of uniform charge density 6.1×10^{-9} C/m^2 is placed at $z = 0$ in the xy-plane. An electron with zero initial velocity is placed at $z = 0.45$ m. What will be its velocity after 17.5 ns?

41. (I) A large, flat plate with unknown, uniform charge density σ is placed on a horizontal tabletop. A cork ball of mass 0.83 g, carrying a charge 8.5×10^{-7} C, is placed at rest above the plate and remains at rest. What is σ?

42. (I) An alpha particle approaches a gold atom head on, stops, and turns around at a distance of 10^{-11} m from the nucleus. What is the electric field due to the gold nucleus at this point? Ignore the effects of the gold atom's orbiting electrons. What is the acceleration of the alpha particle when it is stopped? An alpha particle is a helium nucleus, composed of two protons and two neutrons.

43. (II) Consider an infinite wire with uniform charge density λ along the z-axis. A negatively charged particle moves in a circle in the xy-plane centered on the wire. Calculate the particle's speed, and show that the speed is independent of the radius of the circle. Ignore all forces except those due to the wire.

44. (II) A negative charge $-q$ is restricted to move in a plane in which there is a continuous line of positive charge and a charge density λ. The negative charge, of mass m, can pass the line of positive charge freely. What is the equation of motion for the negative charge?

45. (II) A positive charge q can travel in a circular orbit about a negatively charged line with uniform charge density λ. Show that the period of the orbit is proportional to the radius of the orbit. Compare this to the dependence of the period of a circular orbit on the radius of the orbit for a point charge that interacts with another point charge.

46. (II) A cork ball of mass 5.6 g is placed between two large horizontal plates. The bottom plate has a uniform charge density of $+1.6 \times 10^{-6}$ C/m^2, whereas the upper plate has a uniform charge density of -0.22×10^{-6} C/m^2. The cork ball, which carries an unknown charge, is placed between the plates and is observed to float motionlessly. What are the sign and magnitude of the charge on the ball?

47. (II) Consider the cathode-ray tube of Example 22–12. This time an electron enters the region between the vertical-deflection plates with a total speed of $v_0 = 5.0 \times 10^6$ m/s. The direction is such that the velocity has a vertical component $v_{0y} = +2.0 \times 10^5$ m/s. Find the total vertical deflection of the electron when it reaches the screen.

48. (II) A beam of electrons is accelerated by passing it through a region between two large charged parallel plates (Fig. 22–33). Calculate the charge density on the plates if the electrons accelerate from 1.4×10^6 m/s to 3.0×10^7 m/s between the plates.

49. (II) A cork ball of mass 5 g, carrying a charge of $-2\ \mu$C, is suspended from a string 1 m long above a horizontal, uniformly charged plate of charge density 1 μC/m^2. The ball is displaced from the vertical by a small angle and allowed to swing. Show that the ball moves in simple harmonic motion, and calculate the angular frequency of that motion.

50. (III) A proton moves at speed $v = 5 \times 10^5$ m/s in the $+x$-direction and enters a certain region. An electric field in the region also is oriented in the $+x$-direction. The field's strength drops linearly with x: At the beginning of the region, $x = 0$ m, the field strength is 500 N/C; at $x = 3$ m, the field strength is zero. How much time does it take for the proton to traverse this region? [*Hint*: The equation of motion will be more familiar in terms of the variable $x' = x - 3$.]

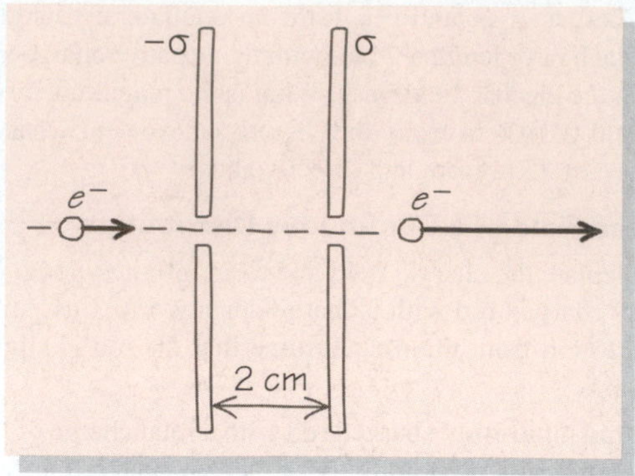

▲ **FIGURE 22–33** Problem 48.

22–5 The Electric Dipole in an External Electric Field

51. (I) An electric dipole consists of two opposite charges of magnitude 2 μC placed 10 cm apart (Fig. 22–34). The dipole is placed in a uniform electric field of 10 N/C along the x-axis, with the direction of $\vec{p}$ at an angle of $+45°$ from the x-axis in the xy-plane. Determine the torque on the dipole.

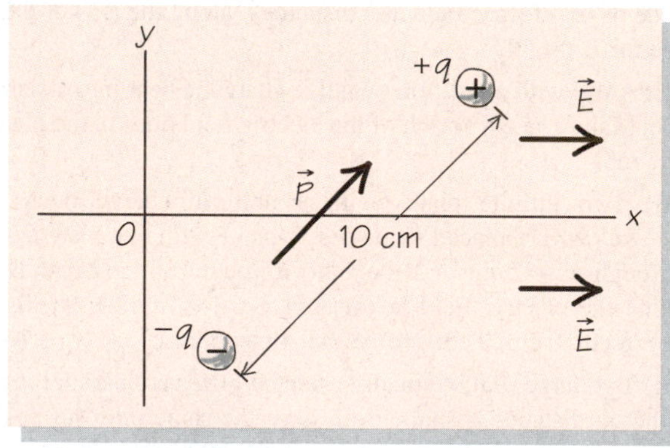

▲ **FIGURE 22–34** Problem 51.

52. (I) The magnitude of the two opposite charges that form an electric dipole is increased by a factor of 5 while the separation between the charges is tripled. What is the change in magnitude of the torque on the dipole in a uniform electric field?

53. (I) A water molecule has a permanent electric dipole moment of magnitude 6×10^{-30} C·m. Estimate the size of the electric field it produces at the position of a neighboring water molecule, which is 3×10^{-9} m away.

54. (II) Describe the motion of the dipole in Problem 51. How much work does the electric field do when the dipole moves from its initial position to alignment with the electric field?

55. (II) Two molecules with permanent electric dipole moments $\vec{p}$ are aligned (Fig. 22–35, see next page). Calculate the force between the molecules if they are separated by a distance that is large compared with the dimension of the dipoles. [*Hint*: The relation $(1 + x)^{-2} \cong 1 - 2x + 3x^2 - \cdots$ for small x is useful.]

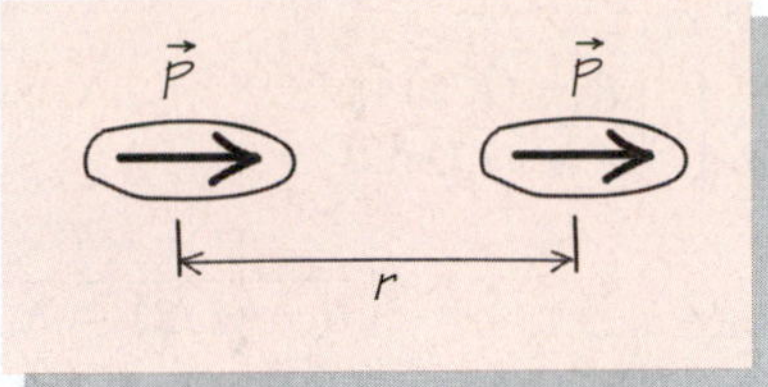

▲ **FIGURE 22–35** Problem 55.

56. (II) A molecule of lithium fluoride (LiF) has a permanent dipole moment. The molecule is placed in a uniform electric field of strength 10^4 N/C, and the difference between the maximum and minimum potential energies of the molecule in this field is 4.4×10^{-25} J. What is the electric dipole moment of the LiF molecule?

General Problems

57. (II) A point charge $-q$ is fixed at the center of a hollow spherical conductor of charge $+q$. Draw the electric field lines both inside and outside the sphere.

58. (II) A point charge $+q$ is fixed at the center of a hollow spherical conductor also of charge $+q$. Draw the electric field lines both inside and outside the sphere.

59. (II) Draw the electric field lines for a point charge $+q$ near an infinitely long, positively charged wire.

60. (II) A cork ball of radius 1.2 cm and a charge of +3.5 nC is covered with conducting paint. What is the electric field strength just outside the surface? A nickel nucleus, with a radius of 5×10^{-15} m, has a positive charge of $28e$. What is the electric field strength just outside the surface of the nucleus?

61. (II) Two infinitely long, uniformly charged rods, with charge densities of λ and $-\lambda$, respectively, are lined up parallel to each other and separated by a distance R. What are the magnitude and direction of the electric field due to the two rods at points that lie (a) on a line joining the two rods, and (b) along a perpendicular bisector of that line? Draw a figure to show the configuration, and use symmetry.

62. (II) What is the force per unit length that one of the two rods in Problem 61 exerts on the other?

63. (II) Two uniformly charged infinite plates with charge densities $-5\ \mu\text{C/m}^2$ and $3\ \mu\text{C/m}^2$ are placed at right angles, the first one along the xz-plane, the second along the yz-plane. A test particle of mass 1 g and charge 1×10^{-7} C is placed a distance of 1 m from both planes; that is, its initial position is $(x, y, z) =$ (1 m, 1 m, 0 m). What is the location of the test particle after a short time t (before it hits a plate)?

64. (II) Two infinite lines of charge density $5\ \mu\text{C/m}$ are parallel to the z-axis. One line passes through $(x, y) =$ (0 cm, 1 cm); the other, through $(x, y) =$ (0 cm, −1 cm) (Fig. 22–36). Find (a) the electric field at the origin; (b) the force on a 0.5-μC charge at the origin; (c) the force on a 6-μC charge located at $(x, y, z) =$ (4 cm, −3 cm, 0 cm).

65. (II) A proton with kinetic energy of 2×10^6 eV is fired perpendicular to the face of a large metal plate that has a uniform surface charge density of $\sigma = 8.0 \times 10^{-6}\ \text{C/m}^2$. (a) Calculate the magnitude and direction of the force on the proton. (b) How much work must the electric field do on the proton to bring it to rest? (c) From what distance should the proton be fired so that it stops right at the surface of the plate?

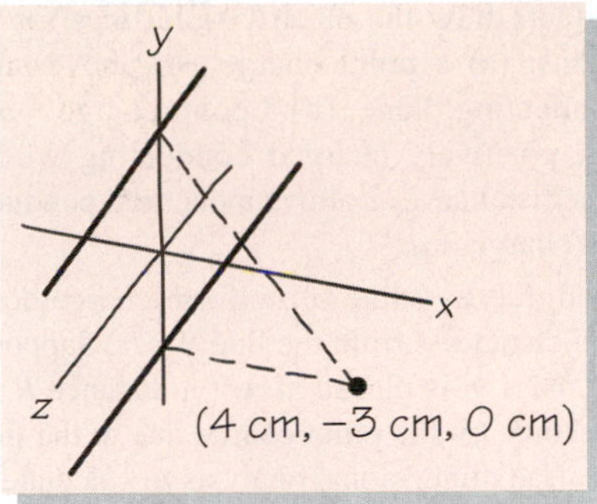

▲ **FIGURE 22–36** Problem 64.

66. (II) The electric charge with the smallest magnitude that can be isolated is the charge on the electron or the proton. In 1909, Robert A. Millikan developed a classic method to measure this charge, known as the *oil drop experiment*. Millikan was able to place charges on tiny droplets of oil, which would fall at a given terminal velocity under the influence of gravity and air drag. By placing these droplets between parallel, horizontal charged plates, as in Fig. 22–37, the electric field between the plates produces a force on the charged droplet that is directed upward and can partly cancel the gravitational force. If the mass and size of the droplet are known, then, by seeing how fast droplets fall with and without the electric field, the charge can be measured.

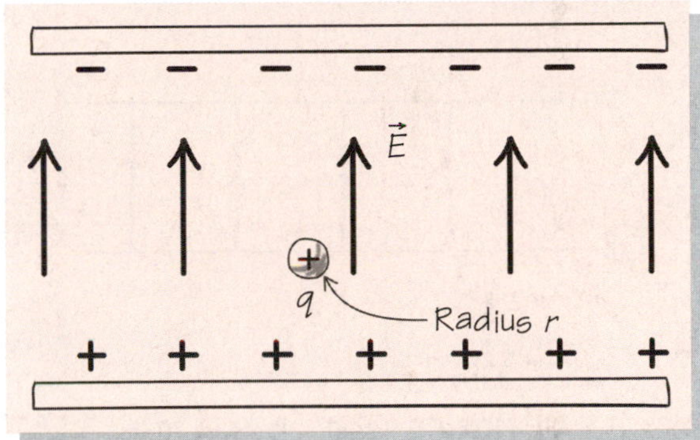

▲ **FIGURE 22–37** Problem 66.

The drag force on a droplet of radius r that falls at a steady speed v through air is also directed upward and is given by *Stokes's law*, $F_{\text{drag}} = 6\pi\eta r v$, where η is the viscosity of air. (a) Show from Newton's second law that the terminal velocity v_0 of the *uncharged* drop is $v_0 = (2/9)r^2\rho g/\eta$, where ρ is the density of the oil and g is the acceleration due to gravity. (b) Suppose that the charge on the drop, q, is positive and that the field is directed vertically upward, as in the figure, so that the electric force points up. Show by using Newton's second law that the charge is given by

$$q = \frac{18\pi(v_0 - v_1)}{E}\sqrt{\frac{v_0\eta^3}{2\rho g}},$$

where v_1 is the terminal velocity when the electric field E is imposed. (c) Take the minimum charge as 1.6×10^{-19} C, the oil's density as $0.85\ \text{g/cm}^3$, and the radius of the droplet as 2.0×10^{-4} cm. The droplet has the minimum charge. Find the value of E that will hold the droplet stationary between the plates.

67. (II) We will learn in Chapter 23 that the electric field near a conductor *must be perpendicular to the conducting surface.*

Using this fact, draw the electric field lines for the following configurations: (a) a point charge $+q$ above an infinite, uncharged conducting plane; (b) a point charge $-q$ near an infinitely long, positively charged conducting wire; (c) a point charge $+q$ a distance $L/2$ above a charged conducting plane of area L^2 and charge $+q$.

68. (II) The field due to a line of uniform charge density λ varies with a radial distance r from the line as $1/r$. Suppose that a point charge q of mass m is placed at rest a distance R from the line, and that the force on the point charge due to the field of the line is attractive. Use dimensional analysis to calculate how the time it will take for the charge to drop to the charged line depends on λ, q, m, R, and ε_0.

69. (II) Consider two thin insulating rods of equal length (0.20 m) placed parallel to each other, and 0.18 m apart. Each of the rods has a charge 2.2×10^{-4} C and -2.2×10^{-4} C placed at opposite ends, with the positive charges on the two rods nearest to each other. What is the force that one of the rods experiences due to the presence of the other rod? What torque about an axis through its midpoint is experienced by one of the rods due to the presence of the other?

70. (II) An electron of speed 3×10^6 m/s enters a region of constant electric field at an angle of 40° as shown in Fig. 22–38. How far away from where the electron enters will it strike the bottom plate?

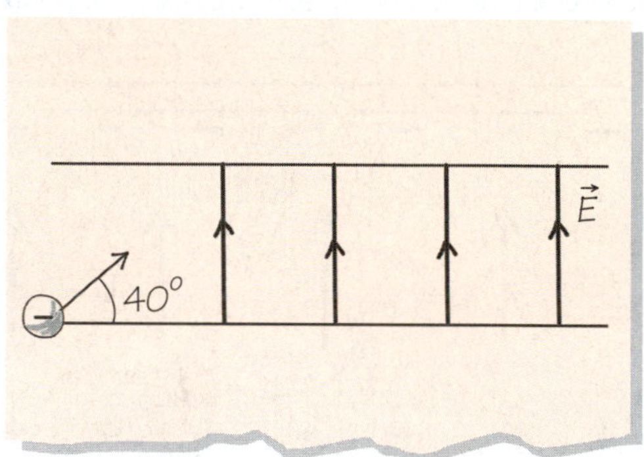

▲ **FIGURE 22–38** Problem 70.

71. (III) Consider the straight, nonuniformly charged rod of length L aligned along the x-axis, with the ends at $x = \pm L/2$, in Example 21–11. We showed there that the force on a charge q located at a point $x = R$ on the x-axis, to the right of the right-hand end of the rod, is

$$\vec{F} = \frac{q\lambda_0}{2\pi\varepsilon_0 L}\left\{\ln\left[\frac{R-(L/2)}{R+(L/2)}\right] + R\left[\frac{1}{R-(L/2)} - \frac{1}{R+(L/2)}\right]\right\}\hat{\imath}.$$

Show that for $R \gg L$, the force reduces to that of a dipole acting on q, $\vec{F} \cong (q\lambda_0 L^2/12\pi\varepsilon_0 R^3)\hat{\imath}$. What is the dipole moment? [*Hint*: Use the approximate forms $(1-x)^{-1} \cong 1 + x + x^2 + \cdots$ and $\ln(1+x) \cong x - (x^2/2) + (x^3/3) - \cdots$, both appropriate for $x \ll 1$.]

72. (III) The field of an electric dipole decreases as $1/r^3$ when the distance of a given point to the dipole, r, is much larger than the separation between the charges. The only way to arrange two charges with a total charge of zero is to form a dipole. There are, however, many ways to arrange four charges with a total charge of zero in a compact pattern. An arrangement with an electric field that behaves at great distances as $1/r^4$ is an *electric quadrupole*. (a) For four charges aligned with alternating sign (such as $(+ - - +)$ so that the combination acts like dipoles of opposite orientation) along an axis, show that the field on the axis perpendicular to the line of charges decreases as $1/r^4$, where r is much larger than any separation distance within the quadrupole. [*Hint*: Use the approximation $(r^2 + \delta^2)^{-3/2} \cong \frac{1}{r^3} - \frac{3\delta^2}{2r^5} + \cdots$ (good when $\delta \ll r$) for each of the four charges; do not forget the sign of the charge in determining the field.] (b) Sketch the field for the arrangement shown in Fig. 22–39, using the field-line technique.

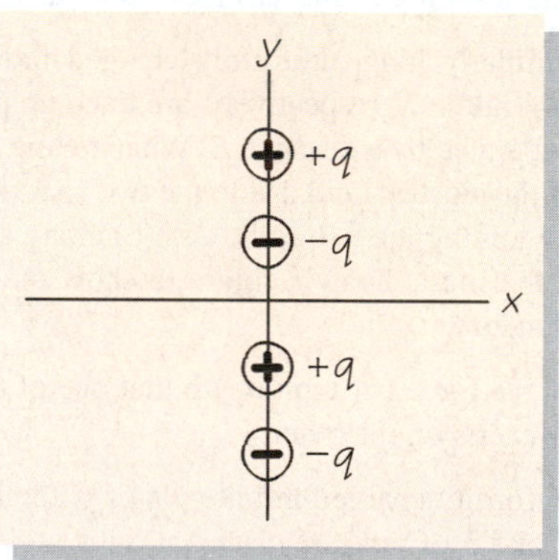

▲ **FIGURE 22–39** Problem 72.

Sparks from a high-voltage electrostatic generator at the Boston Museum of Science do not harm the operator who is sitting inside a grounded Faraday cage.

Gauss' Law

The concept of the electric field allows us to study the interaction between two charges as a two-step process: one involving the interaction of a charge with an electric field, the other involving the determination of the electric field due to electric charges. We saw in Chapter 22 that this division simplifies the problem of calculating interactions between charges. Gauss' law, the subject of this chapter and one of the fundamental laws of electromagnetism[†], gives us a new way to think about the electric field due to charge distributions. This law contains all the information in Coulomb's law and will allow us to go further. When there is symmetry in the charge distribution, Gauss' law is a powerful tool for the direct evaluation of the electric field. Gauss' law also gives us valuable insight into the behavior of conductors.

23–1 What Does Gauss' Law Do?

For a simple insight into Gauss' law, consider the electric field lines associated with a charge Q. We suppose that the charge gives rise to N electric field lines. (The choice is arbitrary, although once it is made we must use it as a normalizing factor—for example, a charge $2Q$ will give rise to $2N$ lines.) We can think of the field lines as a collection of N rigid wires that emerge radially from a point—the location of the charge—and extend to infinity. One way of finding the value of Q would be to imagine that we could somehow make a sphere of tissue appear at any fixed radius around the charge and count the holes

[†] Karl Friedrich Gauss, a great mathematician of the nineteenth century, worked on celestial mechanics, electromagnetism, optics, and the theory of errors.

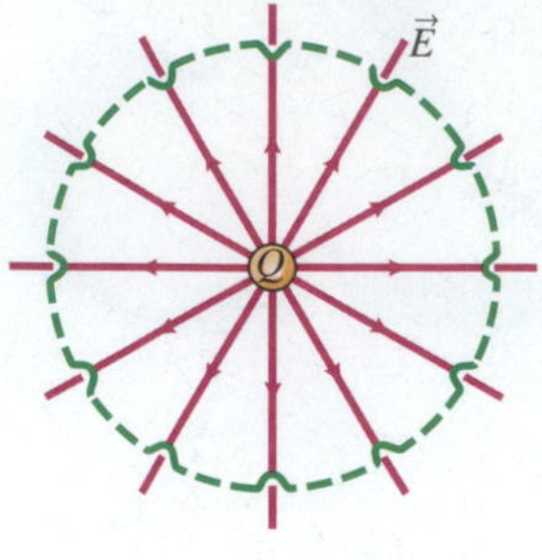

(a)

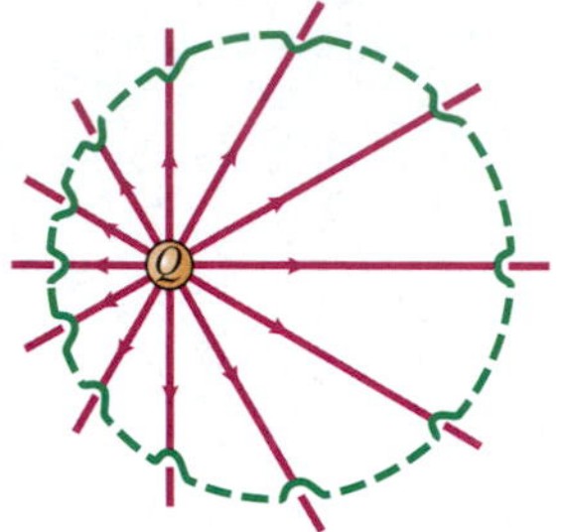

(b)

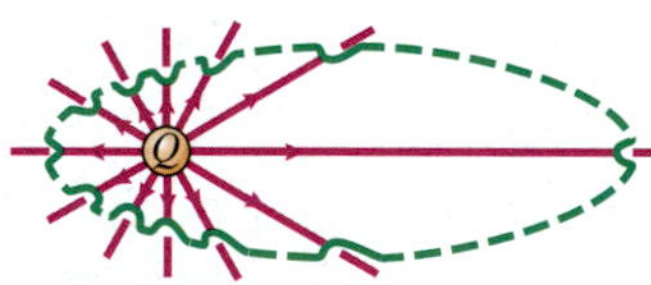

(c)

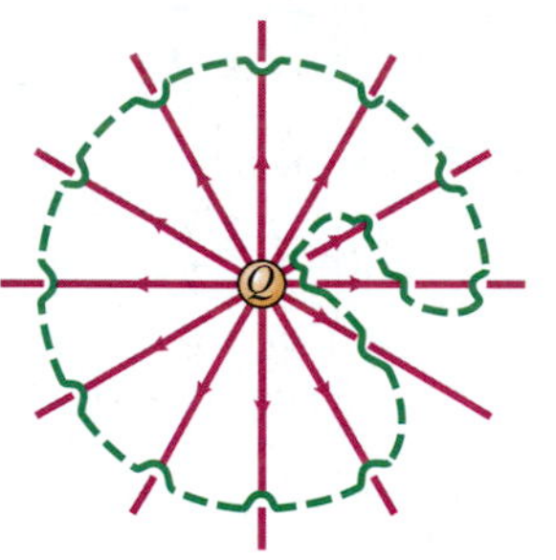

(d)

▲ **FIGURE 23–1** (a) Twelve wires representing field lines pass through this two-dimensional representation of a three-dimensional tissue forming a sphere with a charge Q at its center. (b) If the sphere is displaced the number of wires (field lines) passing through it remains 12, at least as long as the charge is somewhere within. (c) The tissue does not have to form a sphere; it can be another shape that still encloses the charge, and the count of wires that pass through it is the same. (d) If the surface enclosing the charge has folds, the counting looks as though it fails. But if one counts the number of times a line enters the surface from the outside, and assigns this number a minus sign, the algebraic total of the number of wires through the closed surface surrounding the charge is the same whatever the shape, including the case illustrated here, in which the surface has a fold.

that the wires make. There will be N holes in the tissue. Figure 23–1a represents this situation in two dimensions, not three, with the holes on the two-dimensional circle. (You should, however, be thinking about this in three dimensions.) For purposes of illustration, we have also chosen to draw 12 field lines in this figure, i.e. we have supposed that $N = 12$, but this has no fundamental significance. Suppose now that the charge were not at the center of the sphere of tissue, but somewhere else within it (Fig. 23–1b). The number of holes would still be the same, even if they are clustered more on the part of the surface that is closer to the charge. If the tissue still formed a closed surface but no longer made a sphere (Fig. 23–1c), the number of holes would again be the same. Now imagine a fold in the closed surface of tissue, as in Fig. 23–1d. This time there would be more holes in the surface, $12 + 4$ in Fig. 23–1d. But two of the four extra holes are caused by a wire going *into* the surface, and if we keep track of holes made by wires going *into* the surface and assign a negative count to such holes then in Fig. 23–1d, the number of holes would be $12 - 2 + 2 = 12$, the same as in all the other cases. The distinction between going out of the closed surface and into the surface will be an important one. To summarize, what we have seen is that we can determine the charge Q by counting the net number of "holes" made by the "field lines" through a closed surface of any shape that surrounds the charge, including a minus sign for field lines going into the surface.

Using the fact that we count lines going into the surface from the outside with a minus sign, then we see that if the charge inside our closed surface were a negative one, $-Q$ instead of $+Q$, then the electric field lines will go toward the negative charge, and the number of holes in our tissue would be -12, the sign of the number of holes correctly accounting for the charge's sign. Without a distinction about whether the field lines were ingoing or outgoing, we cannot tell the sign of Q.

The situation we have described is the essence of Gauss' law: We count "holes" created by "wires"—by this we mean the passage of field lines through an imaginary closed surface that surrounds any charge or set of charges, with a rule that field lines going out of the surface make positive holes, and field lines going into the surface make negative holes. The net number of field lines that pass through the closed surface is a measure of the charge enclosed. The charge Q does not have to be a point charge. It can be an extended charge, the only proviso being that the imaginary surface must enclose all of the charge. This is really all there is to Gauss' law, which as we'll see in the next section expresses a relation between the charge contained within a closed surface and an appropriate generalization to the number of "holes" made in the surface by the electric field.

Drawing field lines and literally counting the passage of the lines through a surface is not a practical possibility, and we now need to provide a mathematical expression for this process. We will do this by replacing the idea of holes made by rigid wires with a calculation of what is known as the *flux* of electric field, or, more simply, electric flux.

Electric Flux

We'll stay for a moment with the idea of electric field lines as represented by rigid wires, but this time we'll consider a sheet of tissue that is not closed. Rather than the field of a point charge, consider the electric field due to an infinite plane of uniformly distributed charge. As we showed in Example 22–8, the field lines are uniformly distributed (let's say with N lines per unit area of the plane) and are perpendicular to the infinite sheet of charge (aligned with the field *direction* $\hat{E}$, where the hat indicates that this is a unit vector that specifies the direction, but not the size, of the electric field). Suppose we now somehow place a flat piece of tissue, area A, above the plane and parallel to it. The wires are perpendicular to the piece of tissue, and we'll take the charge density on the plane such that there will be NA holes in the tissue (Fig. 23–2a). Now, if instead of the tissue aligning parallel to the charge plane, we could make it appear in a tilted position (Fig. 23–2b), there would be fewer holes—think of fewer field line crossings—for the simple reason that the *perpendicular* area presented to the wires by the square of tissue is smaller. We can in fact give a mathematical description to the idea that our surface makes an angle to the sheet of charge by defining the *normal* to the surface. More precisely, the normal vector $\hat{n}$ will be a unit vector *perpendicular* to the surface. In Fig. 23–2b we draw it in the direction away from the sheet of charge. Suppose now $\hat{n}$ and $\vec{E}$ make an angle θ (Fig. 23–2b). Then you can see from the figure that the area presented to the wires coming from the surface is the original area A multiplied by $\cos\theta$. A compact way of stating this is that the number of holes will be given by $NA\cos\theta = NA(\hat{n}\cdot\hat{E})$. When the tissue is aligned parallel to the plane, then $\hat{n}$ and $\vec{E}$ are parallel, and the expression for the number of holes—again, the number of field lines that cross the surface—reduces correctly to NA.

At this point, we can say that the number of field lines crossing the surface is proportional to the surface area, the orientation of the surface, and the field strength. The orientation factor is in the scalar product $\hat{n}\cdot\hat{E}$. The field strength factor is in the number of lines N, which you will recall is an arbitrary normalization of the field strength. We can take care of the factor N and remove any arbitrariness by replacing N with the magnitude of the field itself, E. Thus we replace the factor $(\hat{n}\cdot\hat{E})N$ with $\hat{n}\cdot\vec{E}$. Finally, the area A remains. But we have chosen in this discussion a large flat area—the flatness makes the normal the same. For a more general surface, there will be various curvatures, and the normal will vary from point to point. Thus, rather than A, we work with an area dA that is sufficiently tiny so that we can treat it as flat and there is no ambiguity in the value of $\vec{E}$ over it.

We have now found our equivalent to the number of holes through a piece of tissue of area dA: It is the **electric flux** $d\Phi$ through the area dA,

$$d\Phi \equiv (\hat{n}\cdot\vec{E})\,dA.$$

We have written this expression in differential form as a reminder that the area to which it refers is small; the field $\vec{E}$ is the field at the location of the surface element. Now just as the total number of holes in our tissue is the sum of the number of holes through the entire surface, we will want to define the electric flux Φ_S across a surface S as the sum over the fluxes through the elements that make it up:

$$\Phi_S = \int_S (\hat{n}\cdot\vec{E})\,dA. \tag{23–1}$$

Finally, we noted above that Gauss' law will concern *closed* surfaces. We indicate that the surface being summed over is closed with a circle on the integral sign,

$$\Phi = \oint_S \hat{n}\cdot\vec{E}\,dA. \tag{23–2}$$

There is another item to deal with: For an open surface, it is not clear from which side of the surface the normal $\hat{n}$ points. In Fig. 23–2b we chose to draw it in one of two possible directions, but this choice was in no way compelled by anything in the situation; for an open surface, the direction of $\hat{n}$ is inherently ambiguous. For a closed surface, we can decide this in a completely unambiguous way: The direction of $\hat{n}$ is defined as pointing toward the *outside* of the closed surface. With this choice, the flux through a closed surface containing a positive charge is positive because $\vec{E}$ points outward as $\hat{n}$ does, and $\hat{n}\cdot\vec{E}$ is positive. For a negative charge, $\vec{E}$ points inward and $\hat{n}\cdot\vec{E}$ is negative (Fig. 23–3).

We simplify the notation by combining the magnitude of the infinitesimal area dA with the associated normal direction, writing $\hat{n}\,dA = d\vec{A}$. Thus

$$\Phi = \oint_S \vec{E}\cdot d\vec{A}. \tag{23–3}$$

This, then, is the mathematical generalization of counting the number of field lines crossing a closed surface.

We have now established that when electric flux through a closed surface is not zero, there is a net charge within the surface. The surface, which can be picked at will just as long as it encloses the charge, is an imaginary construct generally known as a **Gaussian surface**. We are going to use such surfaces to help us calculate values for the electric field, and we shall see that surfaces that have a special symmetry in relation to the charge distribution are especially useful. For example, for spherically symmetric charge distributions, a concentric spherical Gaussian surface is particularly useful, and for charges uniformly distributed along a straight line, a Gaussian surface in the shape of a cylinder, with the line charge forming the axis, is useful.

We note in passing that the use of the term *electric flux* comes from an analogy with the flow of water through a surface, a concept we met in Chapter 16, and an analogy between electric flux and the flux of water can be made.

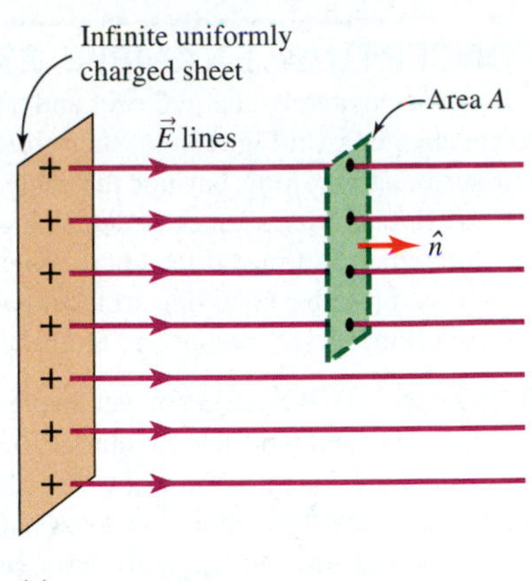

▲ **FIGURE 23–2** A plane of area A is placed in a uniform electric field. In (a) the plane is aligned perpendicular to the field. (b) The perpendicular to the plane makes an angle θ with respect to the field.

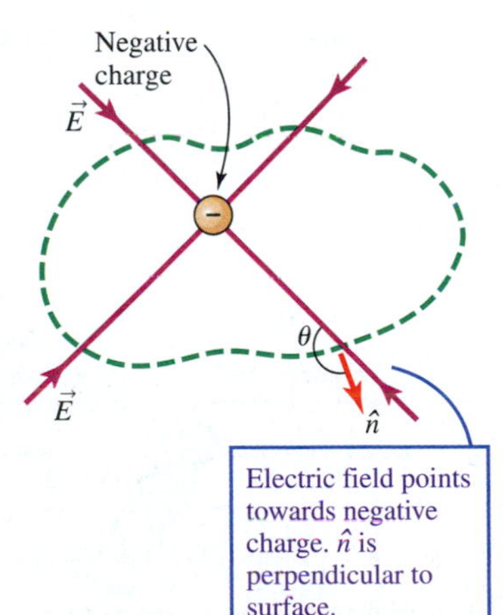

▲ **FIGURE 23–3** If the charge forming the field is negative, the angle between the perpendicular to the surface and the field is larger than 90°.

CONCEPTUAL EXAMPLE 23–1 Consider the combination of a positively charged ring and a plate with an equal and opposite charge, as in Fig. 23–4a, then think about a Gaussian surface that surrounds the ring but not the plate (Fig. 23–4b). Suppose the net flux through that surface is determined to be Φ_0. (a) What is the flux through the surface if the charge on the ring is tripled? (b) What is the flux through a Gaussian surface that encloses both the ring and the plate (Fig. 23–4c) before and after the tripling of the ring charge?

Answer (a) We have not yet made an explicit connection between the flux and the enclosed charge, but we do know that the number of field lines is *proportional* to the charge that gives rise to them and that the flux through a closed surface is also proportional to the number of field lines emerging from (or entering) the enclosed charge. From this we see that the tripling of the charge on the ring will result in the tripling of the flux through a Gaussian surface enclosing the ring only: If the initial flux due to the ring is Φ_0, the final flux is $3\Phi_0$.

(b) Initially the net charge of ring and plate is zero, so that none of the electric field lines go off to infinity. The total charge enclosed by the Gaussian surface is zero. Although field lines may cross a Gaussian surface on their way out, they return, and by our rule of counting "in" lines as yielding negative flux, the total flux is zero. When the ring charge is tripled without a change in the plate charge, the total flux through a Gaussian surface may be calculated as 3Φ due to the ring and $-\Phi$ due to the plate, so that the flux goes from an initial value of 0 to a final value of $2\Phi_0$.

(a)

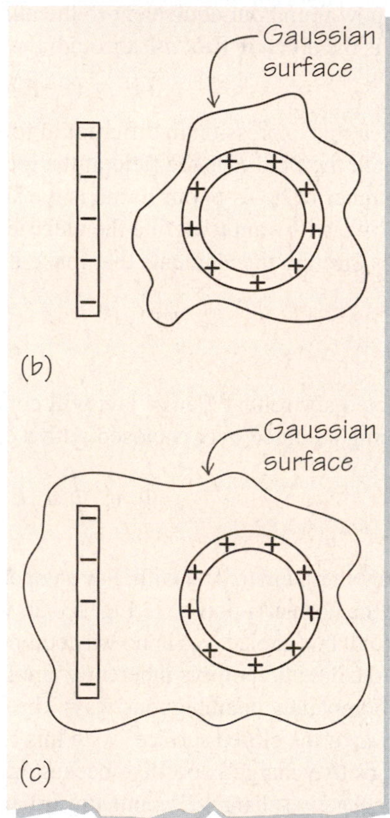

▶ **FIGURE 23–4** (a) The electric field lines due to a charged conducting cylinder placed close to an oppositely charged conducting plate, as shown by threads in a shallow dish of oil. (b) A Gaussian surface surrounds the cylinder. (c) A Gaussian surface surrounds both the cylinder and the plate.

Let us look at a calculation of electric flux through an *open* surface, where there is an ambiguity: Whereas for a closed surface there is a clear distinction between a normal to the surface $\hat{n}$ pointing out and pointing in, for an open surface this is subject to definition. The choice, once made, must be consistently preserved throughout the calculation. To take an example, we can look again at the large uniformly charged plane, with a flat square surface of area A oriented as in Fig. 23–2b. With the unit vector oriented as in that figure and a positive charge on the plane, the factor $\vec{E}\cdot\hat{n}$ in the flux is $+E\cos\theta$. This is constant over the surface, so that the flux is just $(E\cos\theta)A$. On the other hand, if we had initially chosen the normal in the opposite direction, as in Fig. 23–5, then the sign of the flux would be reversed. For an open surface such as this planar one, there is no way to decide what the "right" direction for the normal is. For a closed surface, the use of the word "outside" is unambiguous, as is the choice of direction for the normal.

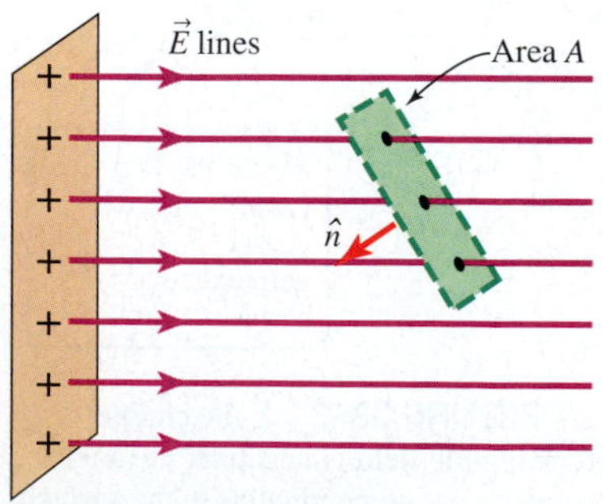

◀ **FIGURE 23–5** The perpendicular to an open Gaussian surface has an ambiguous orientation. Here we have placed it in the opposite direction to that chosen for Fig. 23–2b. This means the flux has an ambiguous sign for an open surface. This is not the case for a closed surface.

CONCEPTUAL EXAMPLE 23–2 You have a plane sheet of charge, infinite in extent, with a uniform positive charge density. You want to consider the flux through a flat, open surface. Are there circumstances in which the flux through the surface is zero?

Answer Yes. Aside from the possibility that the charge on the plane is zero, so that $E = 0$, the surface could be aligned so that the electric field is parallel to the surface. In other words, $\hat{n}$ is perpendicular to the direction of E, and therefore $\hat{n} \cdot \vec{E} = 0$.

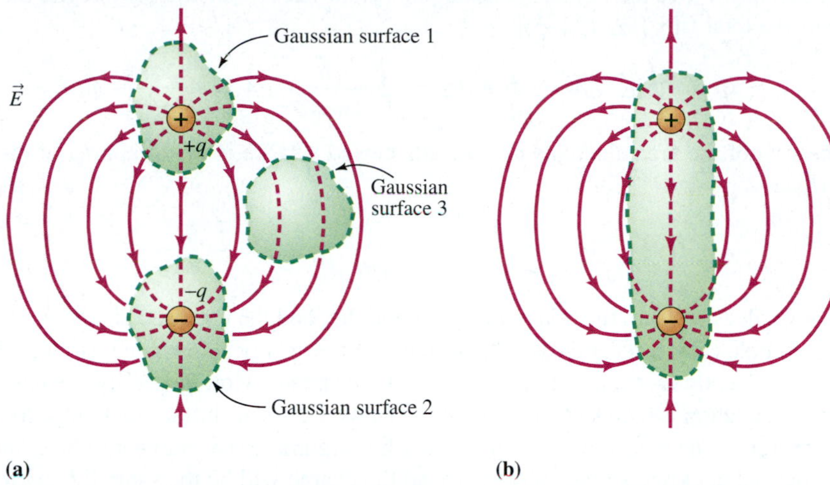

◀ **FIGURE 23–6** (a) Three Gaussian surfaces (shown dashed to remind you that they are imaginary) in the electric field of a dipole. For surface 1, which surrounds the $+q$ charge, the electric flux is positive; for surface 2, which surrounds the $-q$ charge, the flux is negative; and for surface 3, which surrounds no charge, the flux is zero. (b) A Gaussian surface surrounding both charges. The flux through this surface is proportional to the net charge and is therefore zero.

23–2 Gauss' Law

Gauss' law describes the relation between a charge and the electric flux through a closed surface—a Gaussian surface—that surrounds that charge. To start, we can most easily show that the flux through a Gaussian surface is zero if no *net* charge is enclosed by the surface by using a field-line argument. No matter how many field lines we employ, it will be true that if the enclosed charge adds to zero, then as many field lines go back into the surface as leave it. This follows by virtue of the fact that outgoing lines contribute a positive sign and ingoing lines a negative sign. For example, Fig. 23–6 shows the electric field due to a dipole as described in Chapter 22. Imagine a series of Gaussian surfaces of any convenient shape placed wherever we choose. For example (Fig. 23–6a), if we place an imaginary Gaussian surface (surface 1) around charge $+Q$ of a dipole, all the electric field lines exit the Gaussian surface, and the total electric flux is positive. If we place a second Gaussian surface (surface 2) around charge $-Q$, all the electric field lines enter the Gaussian surface, and the electric flux is negative. Any Gaussian surface, such as surface 3, that surrounds *neither* charge has no net electric flux through it because the same number of electric field lines enter and exit such a surface. If the Gaussian surface surrounds *both* charges (Fig. 23–6b), then the number of field lines that enter and exit the surface is again equal, and the total flux is zero. We conclude that

The electric flux through a closed surface that encloses no net charge is zero.

To go further, we must find the relation between the electric flux Φ and the enclosed charge. We can do so by thinking about a point charge, then using superposition. Figure 23–7 shows an (imaginary) Gaussian sphere of radius R centered on a point charge q. The centered sphere is chosen because the electric field has constant magnitude at a fixed distance from a charge, and it will be easy to find the flux through the sphere. We use Eq. (23–3) to find the electric flux that passes through the Gaussian surface. The electric field due to a point charge q was found to be [Eq. (22–5)]

$$\vec{E} = \left(\frac{q}{4\pi\varepsilon_0 r^2}\right)\hat{r}.$$

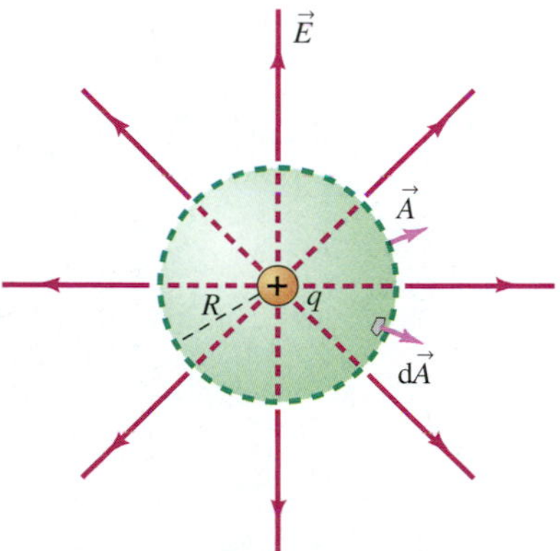

▲ **FIGURE 23–7** A simple choice for a Gaussian surface for a point charge q is a sphere of radius R.

The electric field points in the radial direction and is directed outward for positive q. Because the direction of the infinitesimal area $\hat{n} \cdot dA = d\vec{A}$ for a small area on the sphere also points outward in the radial direction, $\vec{E} \cdot d\vec{A} = E\,dA$. Because the electric field has the constant value $q/(4\pi\varepsilon_0 R^2)$ everywhere on our sphere, the infinitesimal electric flux through the infinitesimal area dA is

$$d\Phi = E\,dA = \frac{q}{4\pi\varepsilon_0 R^2}\,dA.$$

We can now move the field E (constant on the surface A) outside the integral that represents the total flux [Eq. (23–3)]:

$$\Phi = \oint \vec{E}\cdot d\vec{A} = \oint E\,dA = \oint \frac{q}{4\pi\varepsilon_0 R^2}\,dA = \frac{q}{4\pi\varepsilon_0 R^2}\oint dA.$$

The sum of the area elements dA over the closed surface is the total area of the closed surface, $A = 4\pi R^2$. Thus,

$$\Phi = \frac{q}{4\pi\varepsilon_0 R^2}A = \frac{q}{4\pi\varepsilon_0 R^2}4\pi R^2 = \frac{q}{\varepsilon_0}. \tag{23–4}$$

This result establishes the relation between the flux and the enclosed charge for a particularly simple case: a point charge at the center of the spherical Gaussian surface. The relation is evidently independent of the radius of the sphere. More generally, we have argued that this relation is independent of the shape of the Gaussian surface and of the location of the charge inside it; this follows from a field-line argument. The number of field lines passing through a closed surface placed around the charge will be the same for any form and any location as long as it encloses the charge. Then it must also be true that the electric flux through any Gaussian surface around the charge is the same for a surface of any form and any location as long as it encloses the charge, and *that value is given by Eq. (23–4).*

Using superposition, we can easily generalize Eq. (23–4) to the case of multiple point charges and then continuous charge distributions. Start with an assembly of point charges q_i that add up to a net charge Q. From the superposition principle, we know that the total electric field $\vec{E}$ is the sum of the fields $\vec{E}_i$, due to point charges q_i. The total flux Φ through a Gaussian surface S enclosing a net charge is the sum of the fluxes Φ_i due to the charges q_i:

$$\Phi = \sum_i \Phi_i = \oint_S \sum_i \vec{E}_i\cdot d\vec{A} = \frac{1}{\varepsilon_0}\sum_i q_i = \frac{Q}{\varepsilon_0}.$$

It is clear from the way we derived this result that it also holds for a continuous distribution of charge—the only difference is that the sum over point charges is replaced with an integral, but that integral remains the net charge Q.

The general statement of **Gauss' law** is therefore

$$\oint_S \vec{E}\cdot d\vec{A} = \frac{Q}{\varepsilon_0}. \tag{23–5}$$

GAUSS' LAW

The closed surface is *any* Gaussian surface that surrounds the *net* charge Q. The case in which the net charge is zero is included here—either because no charge whatsoever is enclosed by S or because there is an equal amount of positive and negative charge.

Coulomb's Law and Gauss' Law

In Chapter 22 we used Coulomb's law to determine the electric field of a point charge, and this in turn has led us to Gauss' law. Actually, the statement of Gauss' law is more general, and we may reverse the procedure to show that Gauss' law implies Coulomb's law. To do so, we center a Gaussian sphere on a point charge q (Fig. 23–7). The electric field $\vec{E}$ of the charge is assumed to be unknown. Gauss' law tells us only that the electric flux integrated over the surface of the sphere is q/ε_0. This is insufficient to determine the field, because the flux through any tiny surface element of the sphere depends on the value of the field in that region. We can, however, use a symmetry argument. There is no preferred direction for the field of a point charge. The only configuration of field around a charge that does not favor some particular direction is a radial field. The surface element $d\vec{A}$ of a Gaussian sphere is also radial. Let's assume that $\vec{E}$ is parallel to $d\vec{A}$ at all locations (the other option is antiparallel). It then follows that

$$\vec{E}\cdot d\vec{A} = E\,dA.$$

Moreover, symmetry—that is, the absence of a preferred direction—also implies that $\vec{E}$ will have the same magnitude E everywhere on the centered sphere. We can then remove E from the integral that expresses the total flux through the sphere:

$$\oint \vec{E}\cdot d\vec{A} = \oint E\,dA = E\oint dA = EA = E(4\pi r^2) = \frac{q}{\varepsilon_0},$$

where r is the radius of the Gaussian sphere. The last term in this equality is just Gauss' law. The equation can be solved for the magnitude of the electric field:

$$E = \frac{q}{4\pi\varepsilon_0 r^2}.$$

This result is consistent with Eq. (22–5). Because E is positive, we correctly chose the direction of $\vec{E}$ to be radially outward for a positive charge. The symmetry of the situation tells us only that the electric field must be radial: either outward or inward. Gauss' law determines the orientation of $\vec{E}$ to be radially outward. Coulomb's law follows directly from our result for the electric field: We put another charge, q', in the electric field and use the relation $\vec{F} = q'\vec{E}$.

Gauss' law does not require us to use any particular surface. This has practical importance, because the flux through one surface may be much easier to calculate than the flux through another.

CONCEPTUAL EXAMPLE 23–3 Find the electric flux through the Gaussian surfaces in Fig. 23–8: (a) a cube of sides L that surrounds the point charge q; (b) a sphere of radius R that surrounds the charge q, which is off center; (c) a sphere of radius b that surrounds the charges $-2q$ and $+q$.

Answer We are asking for quantitative answers, yet this is in fact a conceptual example; this illustrates the power of Gauss' law. Although the expression for the flux looks as if it involves a formidable integral, Gauss' law tells us that it is just equal to the net charge inside the surface, divided by ε_0. We can therefore immediately answer the questions.

(a) We use Gauss' law here, rather than doing a direct integration of the electric field over the cube. The total electric flux is simply q/ε_0.

(b) It does not matter that the Gaussian sphere is off center. The total electric flux is still q/ε_0.

(c) The total net charge Q enclosed by the Gaussian surface is $-2q + q = -q$, and the total electric flux through the Gaussian surface is $-q/\varepsilon_0$. We do not need to concern ourselves with the positions of the two charges within the cube.

What Do You Think? Suppose the cube in Fig. 23–8a contained three more point charges of magnitude $2q$, $-7q$ and $4q$. What would the flux through the sides of the cube be? ***Answers to What Do You Think? questions are given in the back of the book.***

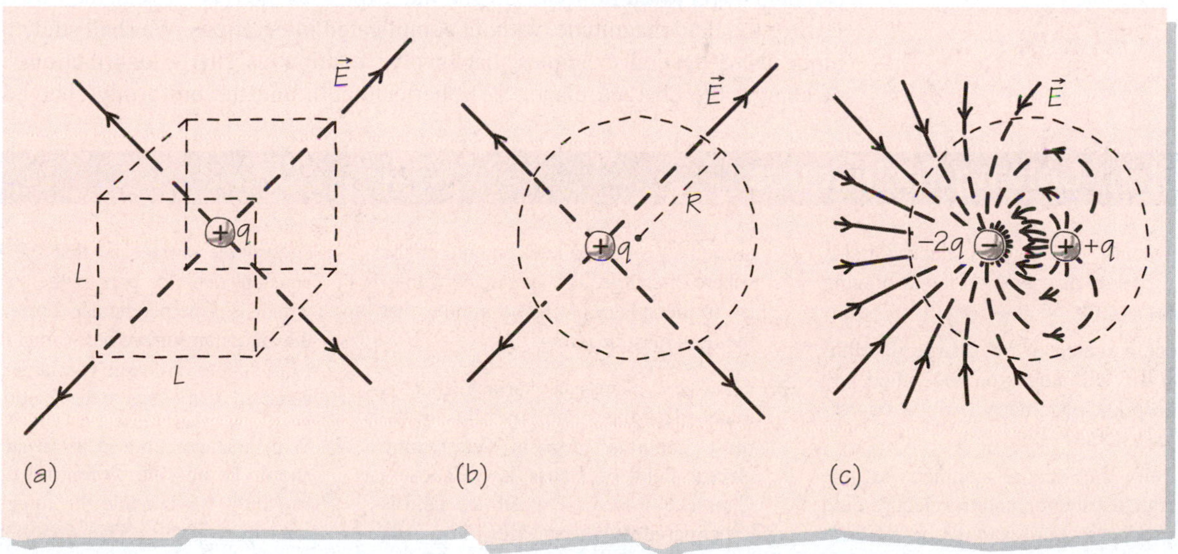

▲ **FIGURE 23–8**

EXAMPLE 23–4 Consider a point charge $q = 1$ mC placed at a corner of a cube of sides 10 cm. Determine the electric flux through each face of the cube.

Setting It Up The situation is sketched in Fig. 23–9a.

Strategy We spoke about the utility of symmetry in solving problems with the help of Gauss' law. Here we'll use the symmetry of the situation, which involves the sides joining at the corner at which the charge resides. You can see from Fig. 23–9a that for these sides, $\vec{E} \cdot \hat{n} = 0$, since the normal is perpendicular to the surfaces while the electric field goes off in a spherically symmetric pattern and lies in the sides. In other words, the electric field that originates at the charge is tangential to the surface of these three sides. This means there is no flux through these sides. The electric flux through each of the remaining three faces of the cube must be equal by symmetry. We'll refer to these sides with the label F.

To find the flux through each of the sides F, we can use a technique that puts the single charge in the middle of a larger cube. It takes seven other similarly placed cubes to surround the point charge q completely (Fig. 23–9b). The charge is at the dead center of the new, larger cube, so the flux through each of the six sides of the large cube will now have an electric flux of one sixth of the total flux. The large sides of the cube consist of four smaller squares, one of which is in fact one of the sides F, so given that the total structure is completely symmetric, the flux through a side F is one fourth of the flux through the large side.

Working It Out The total flux is q/ε_0, so that the flux through each of the sides of the large cube is $q/6\varepsilon_0$, and one quarter of that, $q/24\varepsilon_0$, goes through each of the far sides of the small cube. Numerical evaluation gives

$$\Phi_F = \frac{q}{24\varepsilon_0} = \frac{1 \times 10^{-3}\,\text{C}}{24(8.85 \times 10^{-12}\,\text{C}^2/\text{N}\cdot\text{m}^2)} = 5 \times 10^6\,\text{N}\cdot\text{m}^2/\text{C}.$$

What Do You Think? Could you use this technique to determine the flux through each of the sides if a charge q were placed at each corner of the original cube?

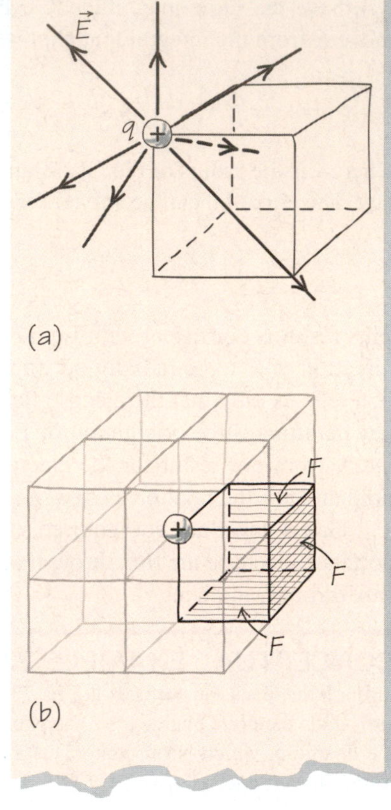

▶ **FIGURE 23–9** (a) A charge q is placed at the corner of a cube. (b) By surrounding the charge with a series of cubes such that the charge is at the center of a larger cube, we have created an arrangement sufficiently symmetric to be able to solve for desired flux values.

23–3 Using Gauss' Law to Determine Electric Fields

Gauss' law is a fundamental law in its own right. It is also a powerful tool for the determination of electric fields in situations where there is a high degree of symmetry. With enough symmetry the electric field will be constant over a simple surface, and this can then be removed from the integral that expresses the flux. We can then solve Gauss' law for the field magnitude, without complicated integrations. We shall study this technique using several examples that involve continuous charge distributions: the line of charge, the charged plane, the spherical shell, and the uniformly charged sphere.

Problem-Solving Techniques

To use Gauss' law to find electric fields given a charge distribution, the following steps are helpful:

1. Make a sketch of the charge distribution—it will help you recognize any appropriate symmetry. Add a coordinate system.
2. Identify any spatial symmetry of the charge distribution and the electric field it produces. For example, a uniform sphere of charge has spherical symmetry because it looks the same from all around any other sphere centered on the sphere of charge. The spherical symmetry of the sphere of charge implies that the field must be radial.
3. Choose a Gaussian surface that is matched to the symmetry. This is the most important step in determining electric fields by Gauss' law. Choose a Gaussian surface for which the field is either parallel to the surface ($d\Phi_E = 0$) or perpendicular to the surface ($d\Phi_E = E\,dA$) at various locations; further, choose the surface so that the field is constant over the part of the surface to which it is perpendicular. For example, the Gaussian surface best suited to a uniform sphere of charge is a larger sphere centered on the charge distribution.
4. With surfaces chosen as in step 3, it should be possible to remove the electric field from inside the integral that expresses the flux. Then Gauss' law becomes an algebraic expression for the magnitude of the field.

These cases were discussed in Chapter 22 and (for gravitation) in Chapter 12. We shall see that Gauss' law determines the fields briefly and simply. In Section 23–4 the real power of solution by Gauss' law is revealed when we discuss conductors and their electric fields, thereby learning about fields in situations that are entirely new.

In Examples 23–5 through 23–8, we use these techniques together with Gauss' law, Eq. (23–5), to determine the field. The method of solution that will be used in all of these examples involves adjusting the Gaussian surface to the charge distribution.

EXAMPLE 23–5 *Line of charge.* Determine the electric field due to an infinitely long, straight charged rod with positive, constant charge density λ.

Setting It Up Figure 23–10a illustrates the situation. We have oriented the rod along the z-axis.

Strategy To find the appropriate Gaussian surface, we want to see what symmetry tells us about the direction and magnitude of the electric field lines. These lines must leave the positively charged rod and, to be symmetric, the electric field lines must extend away from the rod radially in the xy-plane (Fig. 23–10b). The electric field lines cannot have a component along the z-direction (the direction along which the rod lies), because the symmetry prevents a choice between the $+z$- or $-z$-direction. Moreover, again by symmetry, the magnitude of the field must be the same on every point of a circle centered on the rod. Thus the field magnitude can depend only on the radial distance from the rod. The Gaussian surface that takes advantage of the symmetry is a closed cylinder whose axis coincides with the line charge. We choose it to be of radius r and height h (Fig. 23–10c). We apply Gauss' law for this surface and use the symmetry to extract the value of the field. Note that for application of Gauss' law we'll need to know that the total charge inside the cylinder is $q = \lambda h$, where λ is the charge per unit length.

Working It Out The flux through the cylinder is

$$\Phi = \int_{\text{top surface}} \vec{E} \cdot d\vec{A} + \int_{\text{bottom surface}} \vec{E} \cdot d\vec{A} + \int_{\text{side surface}} \vec{E} \cdot d\vec{A}.$$

The flux through the top and bottom surfaces is zero. That is because the electric field is parallel to these surfaces, so the surface element $d\vec{A}$ is perpendicular to the field. As for the cylindrical side surface, the electric field is perpendicular to that surface, so that $\vec{E} \cdot d\vec{A} = E\,dA$. This expression must be integrated over the curved surface of the cylinder to find the flux through the side of the cylinder. But we have chosen the cylindrical surface so that the electric field has constant magnitude over the surface. The field magnitude can therefore be removed from the integral. Thus

$$\Phi = \int_{\text{side surface}} \vec{E} \cdot d\vec{A} = E \int_{\text{side surface}} dA = E(2\pi rh).$$

Here we use the fact that the surface area of the curved part of the cylinder is the circumference $2\pi r$ multiplied by the height.

The enclosed charge is λh, so Gauss' law reads

$$E(2\pi rh) = \frac{\lambda h}{\varepsilon_0},$$

or

$$E = \frac{\lambda}{2\pi\varepsilon_0 r}. \tag{23–6}$$

The arbitrary height h has canceled. In SI units, the charge density is in coulombs per meter. We can thus check that $\varepsilon_0 E$ has the units of coulombs per square meter. Compare the ease with which we obtained Eq. (23–6) with the direct integration technique [Eq. (22–31)].

What Do You Think? What can you say about the external field due to an infinitely long cylinder whose surface is uniformly charged?

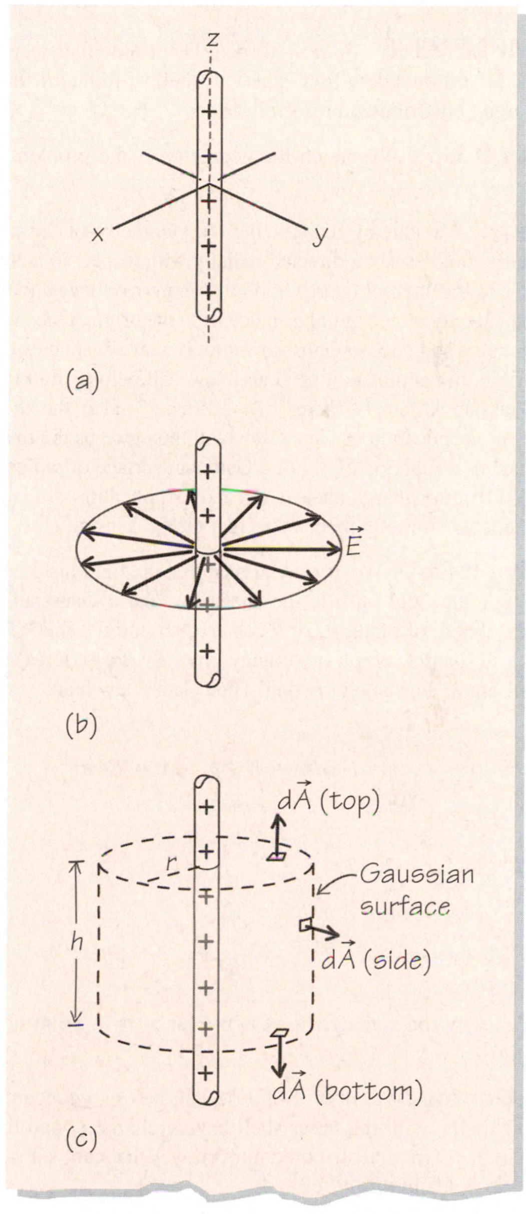

▲ **FIGURE 23–10** (a) A line of charge is oriented along the z-axis. (b) By symmetry, the direction of the electric field $\vec{E}$ is radial in the xy-plane. (c) The best Gaussian surface to use to determine the electric field of a line charge is a cylinder. The directions of the areas $d\vec{A}$ for the various surfaces of the cylinder are shown.

THINK ABOUT THIS. . .

COULD WE USE GAUSS' LAW TO FIND THE FIELD OF A FINITE LINE OF CHARGE?

Gauss' law holds for *any* distribution of charge, but for a line of charge of finite length the symmetry that allows us to determine the direction of $\vec{E}$ and remove it from the flux integration is not present. If the ends of the line are in view, they provide a guide to tell us where we are along the wire—for example, we can look to see that we are close to one end or the other. The symmetry along the wire is lost. This loss of symmetry has two consequences: First, the electric field will have a component along the wire; and second, the magnitude of the field will vary *along* the line. In fact, from far enough away, a finite charged line is indistinguishable from a point charge, and the electric field will point in a direction normal to a Gaussian sphere centered on the line charge whose radius is much larger than the line length. Gauss' law is always true but not always useful. ■

EXAMPLE 23–6 *Spherical shell.* Determine the electric field both inside and outside a thin spherical shell of radius R that has a total charge Q distributed uniformly on it.

Setting It Up We sketch the geometry of the problem in Fig. 23–11a.

Strategy We start by recognizing the symmetry of the problem: Any electric field must be directed radially with respect to an origin at the center of the shell of charge and must moreover have a magnitude that depends only on the radial distance from the origin. Once we know this symmetry and choose Gaussian surfaces that are spheres centered at the origin, the application of Gauss' law will help us find the field. For a Gaussian surface of radius $r > R$ (Fig. 23–11a), the charge enclosed is Q. Application of Gauss' law will then give us the magnitude of the field as a function of r. For a Gaussian surface of radius $r < R$ (Fig. 23–11b), the charge enclosed is zero. Application of Gauss' law with spherical symmetry then shows that the field inside must be zero.

Working It Out We already argued that the field inside the shell is zero. For the field outside the shell, we take a Gaussian surface forming a sphere of radius $r > R$ centered around the shell. Then the argument in the discussion of strategy gives $\vec{E} \cdot d\vec{A} = E\,dA$, because $\vec{E}$ and $d\vec{A}$ are in the same direction. Then Gauss' law reads

$$\frac{Q}{\varepsilon_0} = \oint_{\text{surface}} \vec{E} \cdot d\vec{A} = E \oint_{\text{surface}} dA = E4\pi r^2. \qquad (23\text{–}7)$$

It follows that for $r \geq R$,

$$E = \frac{Q}{4\pi\varepsilon_0 r^2}. \qquad (23\text{–}8)$$

This is exactly the same field as is produced by a point charge at the origin.

What Do You Think? Suppose you had *two* concentric thin spherical shells, with the inner shell having charge Q and the outer shell charge $-Q$, both distributed uniformly. What can you say about the electric fields in this case?

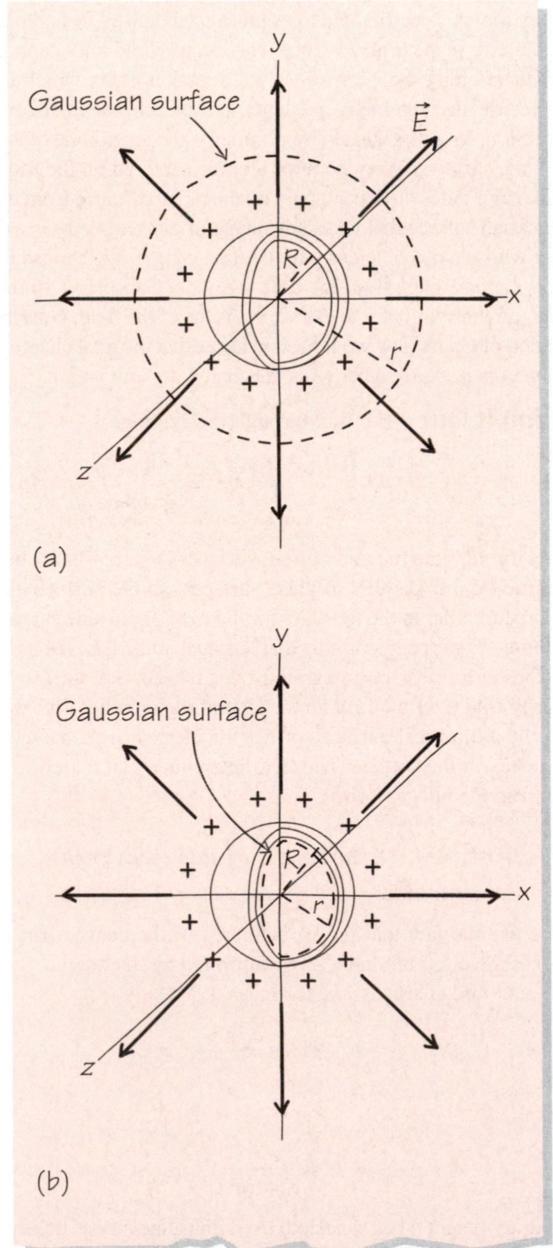

▶ **FIGURE 23–11** (a) The best Gaussian surface to determine the electric field outside a uniformly charged spherical shell is a sphere, because the symmetry is spherical. (b) The best Gaussian surface to determine the electric field inside a uniformly charged spherical shell is a sphere inside the shell.

EXAMPLE 23–7 *Solid sphere.* Find the electric field outside and inside a solid, nonconducting sphere of radius R that contains a total charge Q uniformly distributed throughout its volume.

Setting It Up The charge distribution is sketched in Fig. 23–12a.

Strategy We again make use of the symmetry of the problem and take for our Gaussian surface a concentric sphere of radius r. By symmetry, the electric field is radial and uniform over the surface of the sphere, so that we can again use Gauss' law,

$$\frac{Q_{\text{enclosed}}}{\varepsilon_0} = \oint_{\text{sphere}} \vec{E}\cdot d\vec{A} = E \oint_{\text{sphere}} dA = E4\pi r^2.$$

Here, E is the field at a distance r from the center. The quantity Q_{enclosed} requires a little thought. In working it out, we note that if our Gaussian surface is outside the sphere (Fig. 23–12a), with $r > R$, the enclosed charge is the total charge. But if our Gaussian surface is inside the sphere (Fig. 23–12b), with $r < R$, we must *calculate* the charge included within the sphere of radius r, and this is done by using the fact that the charge density is given.

Working It Out We have for this case the general result

$$E = \frac{Q_{\text{enclosed}}}{4\pi\varepsilon_0 r^2}.$$

For the field *outside* the solid sphere, the charge enclosed by a Gaussian sphere at $r > R$ is just Q and, just as for the field for a point charge or for the region outside a spherical shell,

$$E = \frac{Q}{4\pi\varepsilon_0 r^2}. \quad (23\text{–}9)$$

Inside the solid sphere, however, the situation is different (Fig. 23–12b). The enclosed charge can be calculated in terms of the *uniform* charge density, which we denote by ρ:

$$Q_{\text{enclosed}} = (\text{volume of sphere}) \times \rho = \frac{4\pi r^3}{3}\rho.$$

The density is determined by the condition that the charge Q is uniformly distributed throughout a sphere of radius R. This means that

$$\rho = \frac{Q}{(\text{volume})} = \frac{Q}{(4\pi/3)R^3}.$$

Putting all this together, we get within the sphere

$$E = \frac{Q_{\text{enclosed}}}{4\pi\varepsilon_0 r^2} = \frac{1}{4\pi\varepsilon_0 r^2} \times \frac{4\pi r^3}{3}\rho = \frac{1}{4\pi\varepsilon_0 r^2} \times \frac{4\pi r^3}{3} \times \frac{Q}{(4\pi/3)R^3}$$

$$= \frac{Q}{4\pi\varepsilon_0}\frac{r}{R^3}. \quad (23\text{–}10)$$

The electric field due to a solid sphere has the radial dependence displayed in Fig. 23–12c. As symmetry demands, the field is zero at the center of the sphere. The field increases linearly with r up to the radius of the sphere and then decreases inversely as the square of r. The fields in Eqs. (23–9) and (23–10) match at the point $r = R$, as the field is continuous.

What Do You Think? Suppose a point negative charge $-q$ is inserted at rest inside the uniformly charged sphere, a distance r from the center. How do you expect it to move?

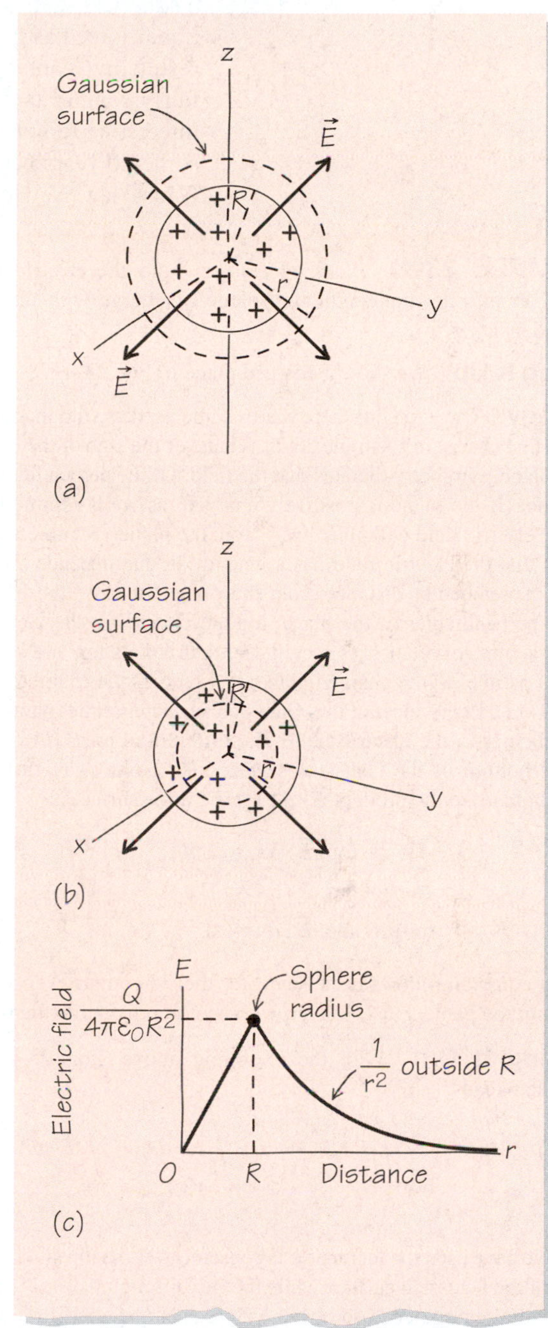

▲ **FIGURE 23–12** (a) The best Gaussian surface to determine the electric field outside a uniformly charged, nonconducting sphere is a concentric spherical surface. The symmetry is spherical. (b) The best Gaussian surface to determine the electric field inside a uniformly charged, nonconducting sphere is a Gaussian sphere inside the solid sphere. Only the charge inside the Gaussian sphere contributes to the electric field at r. (c) The electric field due to a uniformly charged, nonconducting sphere as a function of the distance from the center of the sphere.

From the two previous examples, we can draw the general conclusion that for a spherically symmetric charge distribution, the field at a radius r outside the charge distribution is that of a point charge at the center whose magnitude is the total charge within the sphere of radius r. We have seen that this is easy to prove by using Gauss' law. It holds not only for thin shells and solid spheres but indeed for *any* distribution of charge whose charge density varies only with the radius and is therefore spherically symmetric.

We noted in Chapter 12 that these same results hold for the force of gravity due to a spherical shell of matter. The mathematical problem is identical because the gravitational force has the same inverse-square form as the Coulomb force. We gave only the results in Chapter 12—without derivation—because the direct integration technique is fairly complicated. The Gauss' law derivation provided here is a very simple one. It is interesting to note that Newton delayed the publication of his theory of gravitation by some 20 years because of his lack of a simple proof of these results. If he had known Gauss' law, Newton would have saved a lot of time!

EXAMPLE 23–8 *Plane of charge*. Find the electric field outside an infinite, nonconducting plane of charge with uniform charge density σ.

Setting It Up We show a charged plane in Fig. 23–13.

Strategy The geometry here shares some aspects with that of the infinite line charge in Example 23–5. Whatever the sign of the charge on the plane, symmetry dictates that the field will be perpendicular to the plane. (If the plane is positively charged, as we'll assume here, then the electric field will point away from the plane.) Symmetry also dictates that the electric field has a magnitude that depends at most on the perpendicular distance from the plane. Because the electric field is perpendicular to the plane, a good choice for the Gaussian surface is any solid (such as a cylinder) that has its top and bottom (area A) parallel to the charged plane, that pierces the charged plane (Fig. 23–13). Every facet of this Gaussian surface is either parallel or perpendicular to the electric field. The differential areas $d\vec{A}$ for the top and bottom of the Gaussian surface also point away from the charged plane, so the product $\vec{E}\cdot d\vec{A}$ for the three surfaces is

$$\text{for the top: } \vec{E}\cdot d\vec{A} = E\,dA;$$
$$\text{for the bottom: } \vec{E}\cdot d\vec{A} = E\,dA;$$
$$\text{for the side: } \vec{E}\cdot d\vec{A} = 0.$$

The last equation follows because $d\vec{A}$ for the side points everywhere parallel to the plane, but $\vec{E}$ is everywhere perpendicular to the plane.

Working It Out With the reasoning above, Eq. (23–5) for Gauss' law reads

$$\frac{Q}{\varepsilon_0} = \oint \vec{E}\cdot d\vec{A} = \int_{\text{top surface}} \vec{E}\cdot d\vec{A} + \int_{\text{bottom surface}} \vec{E}\cdot d\vec{A} + \int_{\text{side}} \vec{E}\cdot d\vec{A}$$
$$= 2EA,$$

where we have used the fact that E is constant over the top and bottom area A of the Gaussian surface while the flux through the side is zero.

The total charge enclosed by the Gaussian surface is the charge on the plane within the surface. Because the charge density is σ and the area enclosed is A, we must have $Q = \sigma A$. The previous equation becomes

$$\frac{Q}{\varepsilon_0} = \frac{\sigma A}{\varepsilon_0} = 2EA,$$

from which we get the result

$$E = \frac{\sigma}{2\varepsilon_0}. \qquad (23\text{–}11)$$

Equation (23–11) is the same result that we found with much more difficulty by direct integration in Chapter 22 [Eq. (22–33)]. Note that E is independent of the distance from the plane.

What Do You Think? Does your result tell you anything about the field outside a *finite* uniformly charged plane?

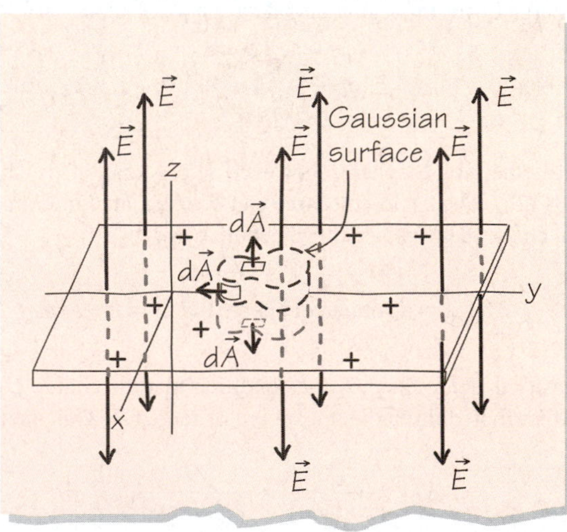

▲ **FIGURE 23–13** A convenient Gaussian surface for a uniformly charged infinite plane can be any shape whose sides are perpendicular to the plane and whose top and bottom are parallel to the plane.

23–4 Conductors and Electric Fields

A good conductor, such as silver, copper, or aluminum, has a large number of "free" electrons, which can move within the (electrically neutral) material. Any electric field that may appear inside the metal due to the presence of an external electric field will cause these electrons to move. In less than a microsecond, they rearrange themselves into a configuration that cancels the electric field inside the material. If any field whatsoever remained inside the material, it would cause the electrons of the conductor to move until they reached equilibrium. *Conductors in electrostatic equilibrium have no internal static electric field.*

The mechanism for the field cancellation within conductors (think of metals) is illustrated in Fig. 23–14. A metal is placed in a spatially constant and static external field that points to the right (Fig. 23–14a). Some electrons in the metal move to the left side of the conductor, which leaves a deficiency of electrons on the right side of the conductor. The arrangement of excess electrons on the left and a deficiency of electrons on the right forms a new, internal electric field that points to the left. This internal field will precisely cancel the external field, with the result that there is no net field within the conductor (Fig. 23–14b).

The fact that there are no static electric fields within conductors has implications for the behavior of conductors when charges are put on or near them, or when they are placed in external electric fields, and this behavior can be determined using Gauss' law. Let's consider what happens when an excess charge is added to a conductor. We might guess that because the components of this excess charge repel each other, and because they can move freely within the conductor, they move as far apart as possible. Figure 23–15 shows such a conductor as well as a Gaussian surface just inside the metal surface. If we apply Gauss' law to this surface, we find that because there is no field, there is no flux, and hence there is no net charge inside the metal. Where is the excess charge? *In electrostatic equilibrium, all excess charge is on the outside surface of a conductor.*

We can establish that the remark above is true even if our conductor has one or more cavities within it. Imagine such a cavity, filled with a nonconducting medium such as air or even a vacuum (Fig. 23–16a). Suppose that there is no excess charge within the cavity. Could charge accumulate on the surface of the cavity—an *interior* surface of the conductor? A Gaussian surface surrounding the cavity, but drawn within the conductor, has no electric flux through it because there is no static field within any conductor. Thus there is no net charge within that Gaussian surface. We have thereby shown that there can be no net charge on the interior surface of the conductor. *Any excess charge placed on a conductor, even if the conductor contains nonconducting cavities, moves to the outside surface of the conductor*, provided there is no charge within the nonconducting cavities.

We must modify our reasoning when there is charge within nonconducting cavities in the conductor. Suppose that such a cavity contains a charge $+Q$ (Fig. 23–16b). Again, draw a Gaussian surface within the metal to surround the cavity. Because there is no field inside the metal, the net charge enclosed must be zero. In this case, a charge of $-Q$ will be induced on the *inner* surface of the metal, that is, on the cavity surface. This induced negative charge keeps the electric field zero *inside the conductor*.

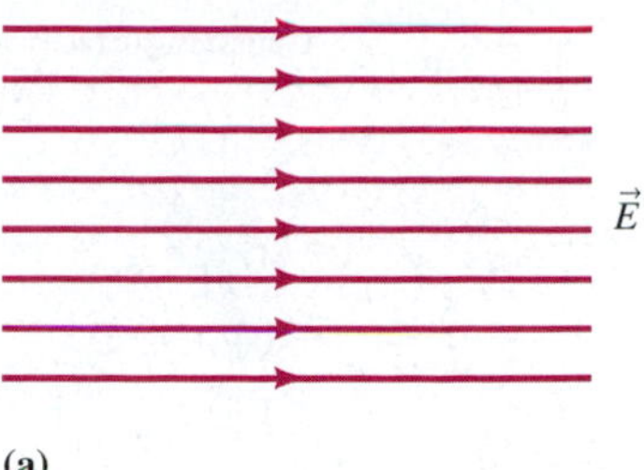

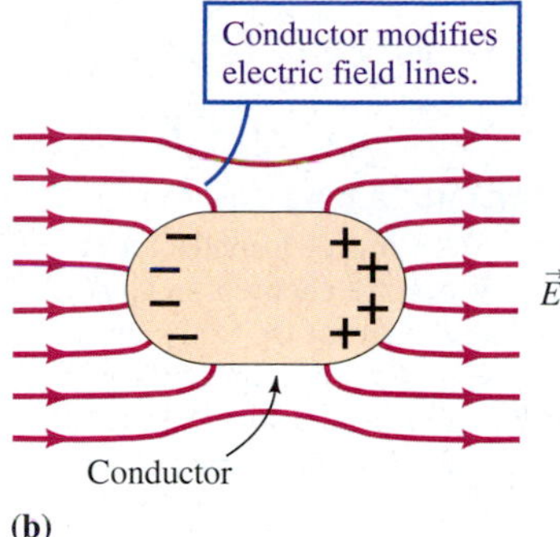

▲ **FIGURE 23–14** An uncharged conductor in an external electric field. (a) The electric field before a conductor is introduced. (b) Charges are induced on the surface of the conductor such that the electric field inside the conductor is zero. The induced charges modify the field outside the conductor, so the field no longer has its original form.

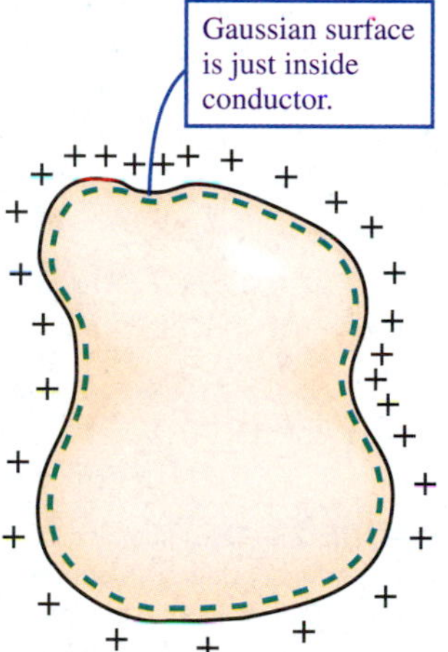

▲ **FIGURE 23–15** To find where excess charge placed on a conductor of arbitrary size and shape goes, choose a Gaussian surface just inside the surface. There is no field within the conductor, hence no flux through our surface, so all the excess charge is on the outside surface of the conductor.

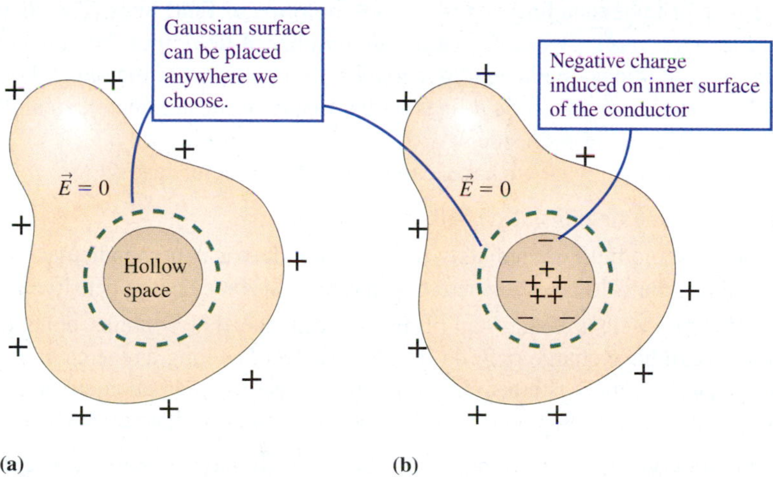

▲ **FIGURE 23–16** (a) A nonconducting cavity inside a conductor; the cavity contains no charge. Any net charge on the conductor must be on the outer surface of the conductor. (b) If we place a charge inside the hollow space, an induced charge will appear on the inside surface of the conductor, such that the electric field within the conducting material is zero. A Gaussian surface drawn just outside the cavity illustrates these results.

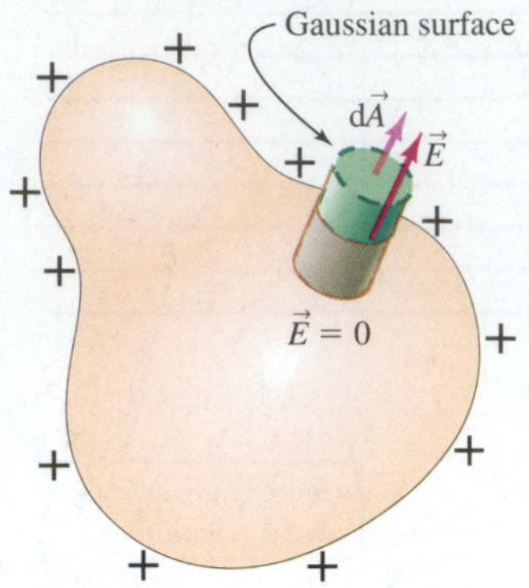

▲ **FIGURE 23–17** To find the electric field outside a conductor of arbitrary size, we choose a small right circular cylinder for the Gaussian surface. The only part of the cylinder through which there is a nonzero electric flux is the outside end of the cylinder.

Electrostatic Fields Near Conductors

We can draw two important conclusions about electrostatic fields around metals or other conductors from this discussion. First, *the electric field immediately outside a conductor must be perpendicular to the conductor's surface.* If there were a parallel component, then there would be a force on charges resting on the surface; the charges would react and move along the surface and we would not have the static situation we assumed. The charges would readjust themselves until there was no parallel component. Second, by using Gauss' law, we can find the value of this perpendicular electric field at a point near the surface in terms of the charge density at that point on the surface. Consider the conductor shown in Fig. 23–17, with a tiny Gaussian surface whose side is perpendicular and whose top is parallel to the conductor's surface. We take the Gaussian surface small enough so that the surface charge density σ is constant within it, even though σ may vary over the conductor. We refer to σ only at the point where the Gaussian surface is erected. The electric field is zero inside the metal surface and outside it is parallel to the side of the Gaussian surface. Thus, the only contribution to the flux comes from the top of the Gaussian surface. If the Gaussian surface is small enough, $\vec{E}$, which is perpendicular to the top surface, can be regarded as constant over it, and

$$\frac{Q}{\varepsilon_0} = \oint \vec{E} \cdot d\vec{A} = EA,$$

where A is the area of the top of the Gaussian surface. The total charge Q enclosed by the Gaussian surface is σA, so that in the previous equation we set $Q = \sigma A$. We see that the area cancels and we get, *just outside the surface,*

$$\vec{E} = \frac{\sigma}{\varepsilon_0}. \tag{23–12}$$

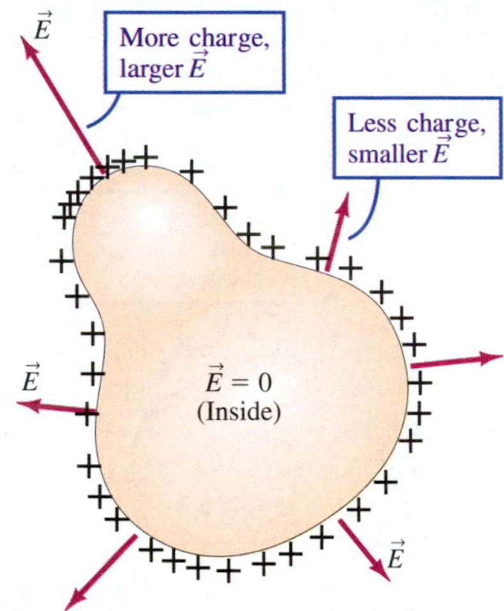

▲ **FIGURE 23–18** The electric field in and around a conductor in equilibrium. The electric field inside the conductor is zero, and just outside the conducting surface it must be perpendicular to the surface. The magnitude of the electric field varies according to the surface charge density σ, which may not be constant everywhere on the surface.

From the way we found it, we see that this result holds only near the conductor's surface. The magnitude of the field will vary around the surface of the conductor as σ does, and the field direction will always be perpendicular to the conductor near its surface (Fig. 23–18). Whether this result is useful or not depends on whether we know the charge density. We can check our result by considering a conductor that is a sphere of radius R and total charge Q. In this case, symmetry demands that the charge is spread evenly over the surface, and

$$\sigma = \frac{Q}{\text{area}} = \frac{Q}{4\pi R^2}.$$

For the field just outside the sphere, Eq. (23–12) would then give $E = Q/4\pi\varepsilon_0 R^2$, which agrees with our earlier result, Eq. (23–9).

The field just outside a conductor $(E = \sigma/\varepsilon_0)$ is twice as large as the field of a nonconducting charged plane with the same charge density $(E = \sigma/2\varepsilon_0)$, Eq. (23–11). A simple way of understanding this is to think in terms of field lines. The charge on a surface of area dA, $\sigma\, dA$, gives rise to a certain number of field lines. For a nonconducting plane, the field lines divide equally between the two sides of the plane. For a thick conducting plane, there are no field lines on the inside of the conductor, so all the field lines must emerge on the open side.

We can summarize what we have learned about conductors as follows:

1. The electrostatic field inside a conductor is zero.
2. The electrostatic field immediately outside a conductor is perpendicular to the surface and has the value σ/ε_0, where σ is the local surface charge density.
3. A conductor in electrostatic equilibrium—even one that contains nonconducting cavities—can have charge only on its outer surface, as long as the cavities contain no net charge. If there is a net charge within the cavity, then an equal and opposite charge will be distributed on the surface of the conductor that surrounds the cavity.

We can add one more important result. Suppose we have a charge-free cavity in a metal. We know that there is no field within the metal and, moreover, no *net* charge on the inner surface of the metal surrounding the cavity. But even for nonsymmetric situations, it can be shown that as long as there is no charge within the cavity, *the electric field is zero everywhere within the cavity.*

We'll discuss fields around conductors further in Chapter 24.

23–5 Are Gauss' and Coulomb's Laws Correct?

It is one of the characteristics of science to be eternally skeptical of yesterday's experiment. It is not so much that yesterday's experiment is wrong, although that certainly can happen; rather, a more accurate experiment can be done with more modern apparatus.

The equivalence of Gauss' and Coulomb's law makes it clear that testing one tests the other. Gauss' law is one of the cornerstones of our understanding of electricity and magnetism, and we must therefore ask just how well it is known and how it can be tested as precisely as possible. Many such tests rely on measurements of the $1/r^2$ behavior of Coulomb's law—you will recall that the equivalence of the two laws rests on this precise dependence. In this way, the errors implicit in measurements of Coulomb's law set limits on our knowledge of Gauss' law; these limits have been continually improved up to the present time. There are also more direct ways to look at Gauss' law, and we'll study a particularly sensitive technique for testing Gauss' law in this section.

Testing Gauss' Law with a Null Experiment

In 1773 Henry Cavendish, who you will recall from Chapter 12 measured the gravitational constant, made an early test of what later became known as Gauss' law. He placed one conducting sphere inside another and connected the two by a wire. After placing a charge on the apparatus, he disconnected the wire and looked for any charge that remained on the inner sphere. To the accuracy of his experiment, he found none. Cavendish's experiment now goes under the general name of the *Faraday "ice-pail" experiment*, after Michael Faraday, who presumably literally used an ice pail (as a conducting container) in a version of the experiment that he performed. This experiment is the basis for many of the modern high-precision tests of Gauss' law and—because Gauss' law is equivalent to Coulomb's law—of Coulomb's law.

For a simple version of the Faraday ice-pail experiment we require an electroscope (the free-charge detector introduced in Chapter 21). We also need a hollow metal container with a hole in the top, as in Fig. 23–19a, and a small metal ball on the end of an insulated rod that can be used to introduce charge to the inside of the container. The electroscope is attached to the outside of the container and thus indicates whether there is charge on the outside.

Next a positive charge, $+Q$, is placed on the small metal ball, and the ball is inserted through the small hole into the hollow container without touching it (Fig. 23–19b). Gauss' law states that there is no net charge *inside* the nearly closed metal container; therefore, a charge of $-Q$ is induced on the inside surface of the container. (The hole can be made smaller and smaller until its presence does not matter.) Because the metal container is neutral, a charge of $+Q$ must then be induced on its outside surface, and the electroscope indicates this charge. If the ball is moved around, there is no change whatsoever in the electroscope, consistent with Gauss' law—it makes no difference where in the cavity the charge is. The metal ball is subsequently touched to the interior of the hollow container (Fig. 23–19c). If Gauss' law is correct, the charge on the ball neutralizes the $-Q$ charge induced on the inside surface, leaving the $+Q$ charge on the outside surface. The electroscope indicates this result by not changing at all. When the metal ball is removed from the container, the container's outer surface remains charged (Fig. 23–19d). By touching the metal ball to another electroscope, we can verify that it carries no charge.

The description of this experiment shows why it is potentially so precise: If Gauss' law is correct, there is *no change* in the position of the gold leaf when the inner surface is touched. Equivalently, Cavendish's experiment tests for the *absence* of charge on the inner of two spheres. Experiments such as Coulomb's, the ones that try to measure departures from a $1/r^2$ fall-off of the force between two charges, look for small changes in comparison with larger effects. Such experiments are inherently less precise than experiments such as Cavendish's, which look for small changes in comparison with *no* effect. Experiments that test for small change versus no change are called **null experiments**. It is far easier to make a precise test of Gauss' law than of Coulomb's law because a null experiment can be done.

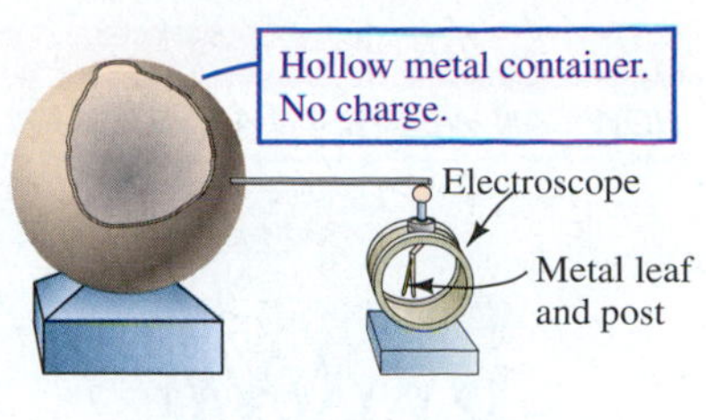

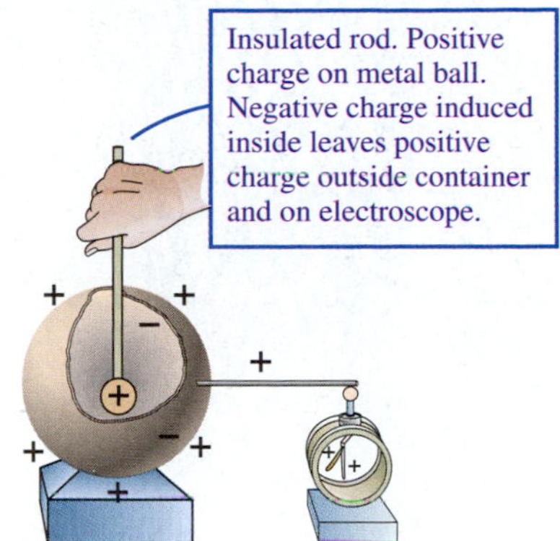

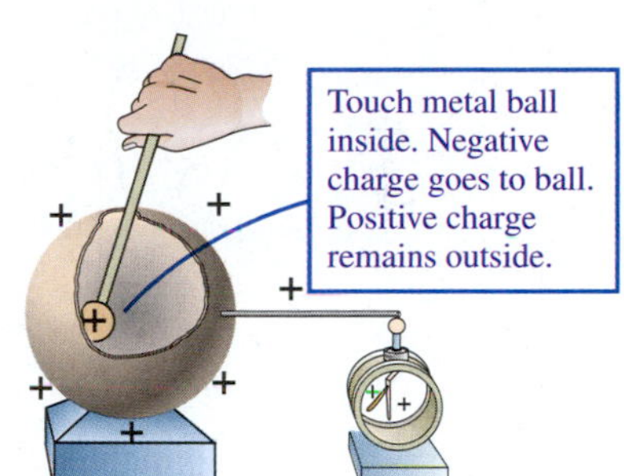

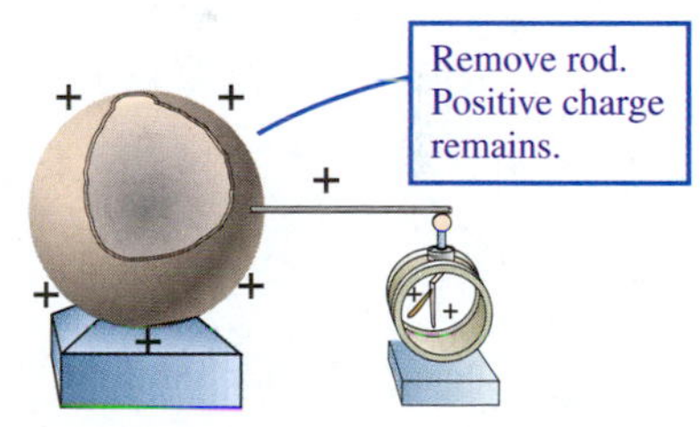

▲ **FIGURE 23–19** An electroscope is attached to the outside surface of a hollow conducting sphere to show the presence of charge. (a) No charge is present, and the gold leaf hangs down. (b) A charged metal ball on the end of an insulated rod is placed inside the sphere, and charge is induced. (c) If the metal ball touches the inside surface of the hollow conductor, all the charge passes to the outside surface. The electroscope's gold leaf indicates no change in the charge on the outside of the hollow conductor. (d) When the insulated metal ball is removed, the charge remains on the outside of the hollow conducting sphere, with no charge remaining on the metal ball.

THINK ABOUT THIS. . .

HOW CAN WE SHIELD A REGION FROM ALL ELECTRIC FIELDS?

Shielded regions or rooms in laboratories are often necessary so that electronic measurements are unaffected by outside electrical interference (Fig. 23–20). Such shielded rooms, known as *Faraday cages* and formed by enclosing a region with copper screens or sheets, rely on the fact that there are no electric fields within charge-free cavities. The enclosure is simply a cavity within a metal—in this case the conducting material is the copper that forms the screens. As long as there is no net charge inside of the enclosure, there is no electric field within it due to any external effects. If there were a net charge inside, charge would be induced on the inside of the copper screens, forcing the electric field in the copper to be zero, and there would be an electric field inside the enclosure.

Faraday cages occur beyond the laboratory. The interior of your car is a safe place in the event of nearby lightning, as any charge on the vehicle will go to the metal outer surface, but for the same reason, your car radio does not work as well when the car is located within the "cage" formed by a metal bridge.

◀ **FIGURE 23–20** By Gauss' law, there is no static electric field in an empty cavity in a metal. To the extent that this radio receives signals with only relatively slowly varying electric fields, it will not work very well within its cage. (*Copyright Jim Krider/Arizona State University.*)

*Coulomb's Law Holds over Small and Large Distances

Table 23–1 gives a summary of the accuracy to which Coulomb's law is known through experiments of the ice-pail type. It is characterized by expressing a *deviation* from Coulomb's law in the form

$$F \propto \frac{1}{r^{2\pm\delta}}.$$

When $\delta = 0$, the inverse-square law is exact; the smaller the limit on δ, the closer the law is known to be an inverse-square law. Table 23–1 expresses what is known experimentally about the possible value of δ. The limits on δ that are placed by the most recent experiments are astonishing.

TABLE 23–1 • Experimental Measurements of Deviation from an Inverse-Square Force Law[†]: Force $\propto 1/r^{2\pm\delta}$

Investigators	Date	Maximum δ
Robison	1769	0.06
Cavendish	1773	0.02
Coulomb	1785	0.10
Maxwell	1873	5×10^{-5}
Plimpton and Lawton	1936	2×10^{-9}
Williams, Faller, and Hill	1971	3×10^{-16}

[†]For more information on this subject, see A. S. Goldhaber and M. M. Nieto, "The Mass of the Photon," *Scientific American*, p. 86, May 1976.

This is not the end of the story, however. First, the experiments that we have listed in Table 23–1 test the laws only over a distance of about 1 m. Yet the laws of electrodynamics are supposed to hold in atomic systems and over galactic distances. Second, other evidence about the framework of the laws of physics suggests strongly that *a deviation from Coulomb's law of the form* $1/r^{2+\delta}$ *is not possible*. Instead, a way to characterize a deviation from Coulomb's law is with the *approximate* form

$$F \propto \frac{e^{-\mu r}}{r^2},$$

where e is the exponential constant 2.78... and μ is a constant. If Coulomb's law is correct, the parameter $\mu = 0$. The exponential function decreases with r over a distance that depends on μ. The larger μ is, the faster the exponential decreases, and the larger the violation of Coulomb's law. Any violation is, we now know, more properly expressed by limits on μ. We can determine limits on μ, and hence tests of the accuracy of Coulomb's law, from the previously reported experiments. The experiment of Williams, Faller, and Hill, for example, implies that μ is smaller than $6 \times 10^{-8}\ \text{m}^{-1}$. These limits can be extended by observing the space dependence of Earth's magnetic field and also of Jupiter's magnetic field, as measured by the spacecraft *Pioneer 10*. Although we have not yet studied magnetism, we can say that the limits on μ found thereby are indeed those associated with Gauss' law. In addition to being direct, the planetary measurements give values of μ that are smaller by an order of magnitude or more than those given by the laboratory experiments; they have the further advantage of testing Gauss' law out to large distances.

Finally, how well do we know Gauss' law at short distances? The colors of light given off by excited hydrogen atoms are very sensitive indicators of the Coulomb force at distances on the atomic scale, about 10^{-10} m. The accuracy with which Gauss' (and therefore Coulomb's) law is known is comparable to the accuracy of the experiments of Plimpton and Lawton (see Table 23–1); that is, to about one part in 1 billion. Even down to nuclear distances—about 10^{-15} m—experiments indicate consistency with the basic theory that leads to Coulomb's law.

Summary

The electric flux due to the electric field $\vec{E}$ that intersects a surface S is

$$\Phi_S = \int_S (\hat{n} \cdot \vec{E})\, dA \qquad (23\text{–}1)$$

where we must find the normal to the surface at each point to carry out the integration. Gauss' law relates the electric flux through a Gaussian surface—an imaginary *closed* surface—to the total charge enclosed by the surface, Q:

$$\oint_S \vec{E} \cdot d\vec{A} = \frac{Q}{\varepsilon_0}. \qquad (23\text{–}5)$$

Gauss' law is equivalent to Coulomb's law for static situations; it is indeed one of the fundamental equations of electromagnetism.

Gauss' law is a powerful tool for determining electric fields due to charge distributions with a high degree of symmetry. It can be used to derive in simple fashion the electric fields due to a straight-line charge or due to a conducting plane. For a general spherically symmetric charge distribution centered at the origin of a coordinate system, Gauss' law gives a simple derivation of the field at a distance r from the origin. If q is the total charge contained within a Gaussian sphere of radius r, then the electric field at r is the same as that of the field of a point charge q at the origin, $E = q/(4\pi\varepsilon_0 r^2)$.

Using Gauss' laws, we learned that conductors have the following properties:

1. The electrostatic field inside a conductor is zero.
2. The electrostatic field immediately outside a conductor is perpendicular to the surface and has the value σ/ε_0, where σ is the local surface charge density (which is not necessarily constant).
3. If there are no nonconducting cavities containing charge, a conductor in electrostatic equilibrium can have charge only on its outer surface.

Gauss' law (and its equivalent, Coulomb's law) has been subjected to many experimental tests since the mid-eighteenth century. The inverse-square law dependence on distance has been verified to a precision that ranges from one part in 10^9 to one part in 10^{16} over distances between 10^{-10} m and 10^9 m.

Understanding the Concepts

1. A temperature field is defined when the temperature of every point of a region of space is specified. Is it possible to compute a flux associated with this field?
2. In the text we refer to the Faraday ice-pail experiment and discuss one version of it in detail. The discussion concerns a sphere with a hole cut in it, and we speak of the inside and outside of this open sphere (Fig. 23–19). Yet an open sphere does not have a clear inside and outside because, unlike a closed, hollow sphere, it can be deformed continuously to a plane. Why is it possible to talk of the inside and outside of an open sphere, and why does the open sphere behave like a closed, hollow sphere (a cavity) in the experiment?
3. Consider a surface enclosing no electric flux. Discuss whether the electric field is zero everywhere on the surface.
4. Use Gauss' law to show that electric field lines must be continuous and must originate from and end on charges.
5. Describe the way in which Gauss' law would fail if the field of a point charge were to decrease as $1/r$ rather than as $1/r^2$.
6. Suppose the charge in Example 23–4 were located in the middle of the face of a cube. What symmetric arrangement would you set up, if any, to calculate the flux through the remaining sides of the cube?
7. If a large, thin, flat plate is positively charged, the field extends in both directions from the plate and has a magnitude of $\sigma/2\varepsilon_0$. If a second plate of equal but opposite charge is placed parallel to the first plate, the field around the first plate extends only toward the second plate and has a magnitude of σ/ε_0, where σ is exactly the same as before. How do you reconcile this second case with Gauss' law?
8. Consider an electric field $\vec{E}$ that is zero at every point on a closed surface S. Does this mean that there are no charges within this surface? Give an example for which there are charges inside a surface while $\vec{E} = 0$ on the surface.
9. What is the force on a charge Q that is just inside a shell of uniformly distributed charge?
10. Analyze Gauss' law as it applies to the flow of fluids. How would you formulate it? Suppose that in a certain region there are no sources of fluid, only sinks. What would you learn from Gauss' law? How does the possibility of evaporation of a fluid affect Gauss' law? What would be its counterpart in the electric Gauss' law?
11. You are sitting in your spaceship, inside an interior cubicle made entirely of aluminum. Your arch enemy shoots a trillion coulombs of charge at you with his coulomb gun. Do you survive?
12. What would Gauss' law look like for the gravitational field, which is defined by force/unit mass of a test body?
13. Charge is distributed uniformly on a wire forming a circle that is surrounded by a torus (doughnut) for which the wire serves as an axis. Does the symmetry allow us to say anything about the electric field due to the charge on the circular wire?
14. A positive point charge and a negative point charge of equal magnitude are fixed on the surface of a conductor of arbitrary shape. What, if anything, can be said about the resulting electric field lines?
15. A region in space has a uniform electric field. What can we say about whether or not any charges are inside the region?
16. You have a charge at the origin of some coordinate system, and a hemispherical surface whose center of curvature is located at the origin. The charge is tripled. How does the flux through the hemispherical surface change?
17. To derive the electric field of an infinitely long line of charge, we used a Gaussian surface in the form of a right cylinder centered on the line. Why does the use of such a surface not allow us to find the field of a line of charge of *finite* length?
18. What would be the flux if both the charge and the length of the sides were doubled in Example 23–4?
19. Suppose that the electric field in some region is known to have only x- and y-components and that the components depend only on x and y, not on z. What can you deduce about the charge distribution that gives rise to this field?
20. You have a probe that measures the electric field at any point in space. For a region in which you know independently that the charge density is constant, how can you use the probe to measure that charge density?
21. Would the result of Example 23–6, which finds the field of a thin spherical shell of uniformly distributed charge, be the same if the distribution of charge weren't spherically symmetric?

Problems

23–1 What Does Gauss' Law Do?

1. (I) An infinitely large, nonconducting, thin plate carries a uniform charge density σ. (a) What is the electric flux through a circle of radius R placed parallel to the plane? (b) What is the flux through that circle if the plane of the circle is tilted at a 30° angle with respect to its original orientation (Fig. 23–21)?

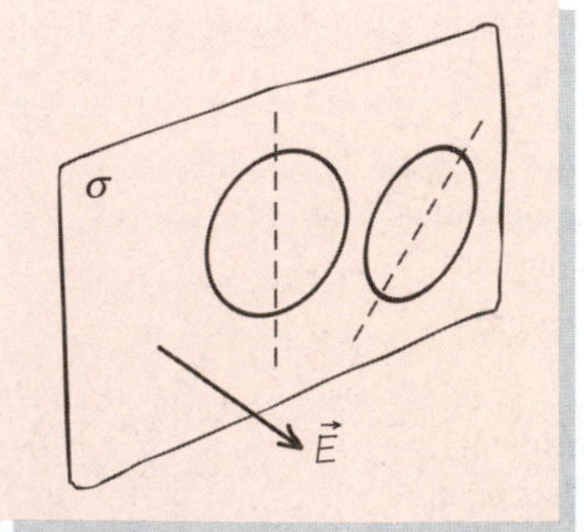

▲ **FIGURE 23–21** Problem 1.

2. (I) A region of space contains a constant electric field of magnitude 1325 N/C. A wire frame forming a square 0.27 m on a side is placed in the region, oriented so that the perpendicular to the plane of the square makes an angle of 48° with the field. What is the magnitude of the electric flux through the frame?
3. (I) The electric field due to an infinitely long, straight line of charge with uniform charge density λ points straight away from the line and has magnitude $E = \lambda/(2\pi\varepsilon_0 r)$, where r is the distance from the wire. Calculate the flux of this electric field through a right cylinder of height h and radius R, concentric with the charged line. Repeat the calculation for a cylinder of radius $2R$.
4. (I) The electric field in a certain region of space points in the z-direction and has magnitude $E = 5xz$, where x and z are measured from some origin. Calculate the flux of that field through a square perpendicular to the z-axis; the corners of the square are at $(x, y, z) = (-1, -1, 1)$, $(-1, 2, 1)$, $(2, 2, 1)$, and $(2, -1, 1)$. (All fields are measured in N/C, all distances in m.)

5. (I) An electric field has the components $E_x = 5x$, $E_y = -3y$, and $E_z = 4z$. Calculate the electric flux through the sides of a unit cube, whose corners are $(x, y, z) = (0, 0, 0)$, $(1, 0, 0)$, $(1, 1, 0)$, $(0, 1, 0)$, $(0, 0, 1)$, $(1, 0, 1)$, $(1, 1, 1)$, and $(0, 1, 1)$. (All fields are measured in N/C, all distances in m.)

6. (I) An electric field of 150 N/C points in the x-direction. A wire loop that is 4 cm^2 in area is placed so that its plane is perpendicular to the x-axis. (a) What is the electric flux through the loop? (b) If the loop is rotated about the y-axis so that the normal to the loop makes an angle of 25° with the x-axis, what is the flux through the loop now? (c) How does the flux change if the angle is increased to 335°?

7. (II) An electric field that is constant in direction is perpendicular to the plane of a circle of radius R. This electric field has a magnitude of $E_0(1 - r/R)$ at a distance r from the center of the circle. Calculate the electric flux through the plane of the circle.

8. (II) By direct calculation (that is, without using Gauss' law), find the flux of a constant electric field $\vec{E}$ through a hemispherical surface of radius R whose circular base is perpendicular to the direction of the field. Your result should be the same as the flux through the top surface of a cylinder whose circular base, of radius R, is oriented perpendicular to the field direction (Fig. 23–22). [*Hint*: The area of an infinitesimal strip at a latitude θ and a thickness $R\,d\theta$ is $2\pi R^2 \sin\theta\,d\theta$; θ varies from 0 at the North Pole to $\pi/2$ at the equator.]

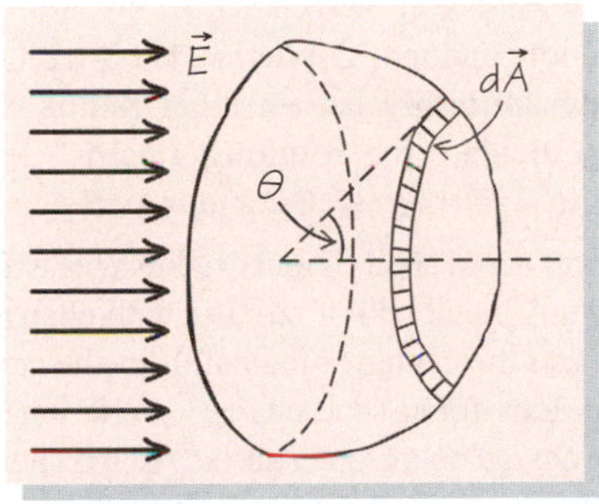

▲ **FIGURE 23–22** Problem 8.

9. (II) A point charge q is placed in the middle of a cylindrical surface of radius R, height $2h$. Find the electric flux through the surface by direct integration. [*Hint*: Use the angle θ in Fig. 23–23 as your variable of integration. Only the upper half of the cylinder is shown in the figure.]

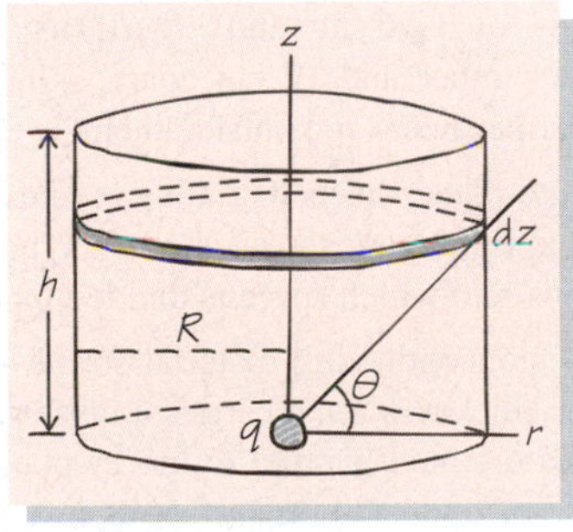

▲ **FIGURE 23–23** Problem 9.

10. (II) A charge q is placed just above the center of a horizontal circle of radius r, and a hemisphere of this radius is erected about the charge. Compute the electric flux through the closed surface that consists of the hemisphere and the planar circle (Fig. 23–24). Do not use Gauss' law.

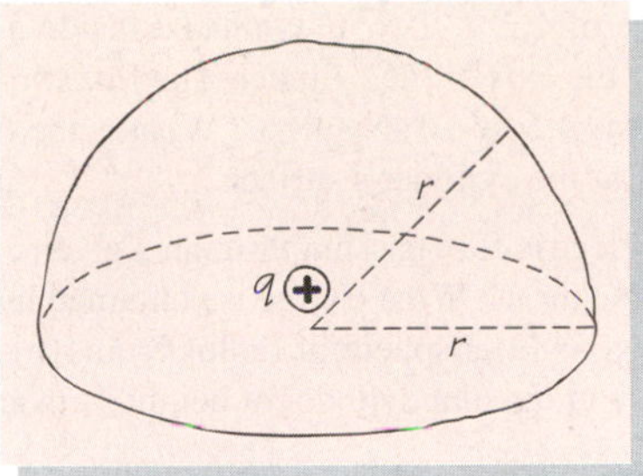

▲ **FIGURE 23–24** Problem 10.

11. (III) Consider an infinitesimal parallelepiped located at the point (x, y, z) with sides dx, dy, and dz along the x-, y-, and z-axes (Fig. 23–25). Show that the electric flux of the electric field given by $\vec{E} = E_x\hat{i} + E_y\hat{j} + E_z\hat{k}$ through the surface that bounds this volume is given by

$$\Phi = \left(\frac{\partial E_x}{\partial x} + \frac{\partial E_y}{\partial y} + \frac{\partial E_z}{\partial z}\right) dx\,dy\,dz.$$

The quantity in parentheses (the coefficient of $dx\,dy\,dz$) is called the *divergence* of the vector field $\vec{E}$.

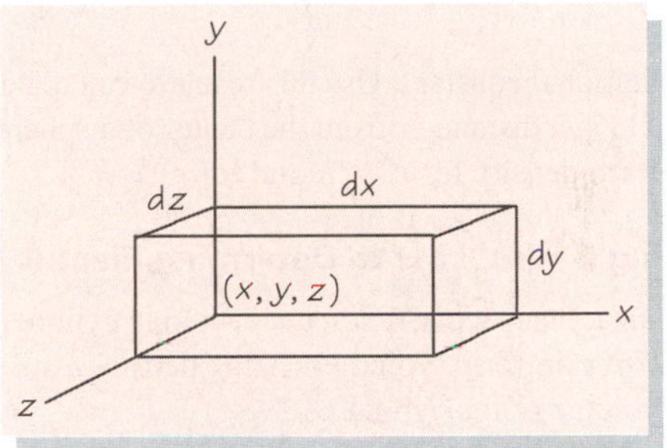

▲ **FIGURE 23–25** Problem 11.

23–2 Gauss' Law

12. (I) The flux through a closed surface surrounding a single charge is $-5.7 \times 10^{-5}\ \text{N}\cdot\text{m}^2/\text{C}$. What is the value of the charge?

13. (I) A charge of 10^{-3} C is distributed uniformly on the surface of a sphere of radius 1 cm. Calculate the total electric flux through a concentric sphere (a) just within the charged surface, and (b) just outside the charged surface.

14. (I) A 120-nC point charge is placed just inside the center of one face of an imaginary Gaussian cube. What is the flux that passes through all six faces of the cube?

15. (I) Consider the charge distribution shown in Fig. 23–26, where $q = 1\ \mu\text{C}$. Draw a spherical Gaussian surface centered on the origin with a radius of 5 cm. (a) What is the electric flux through the Gaussian sphere? (b) Do any of the electric field lines from the three charges pierce the Gaussian surface?

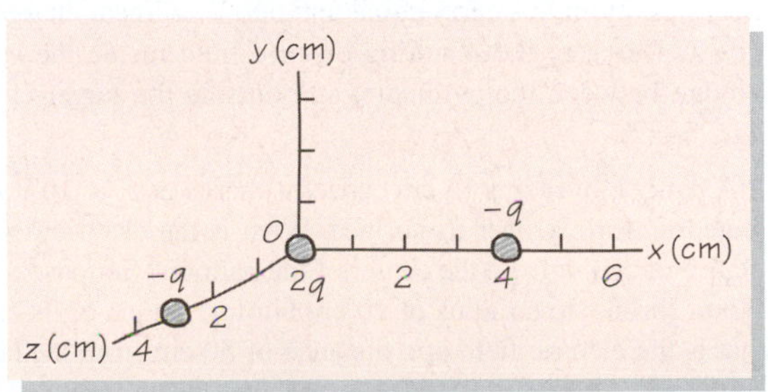

▲ **FIGURE 23–26** Problem 15.

16. (I) A charge of 1.2×10^{-4} C is placed inside a cylinder at the midpoint of the axis of the cylinder. The flux through one end of the cylinder is $4.5 \times 10^6\ \text{N}\cdot\text{m}^2/\text{C}$. What is the flux through the curved part of the cylinder's surface?

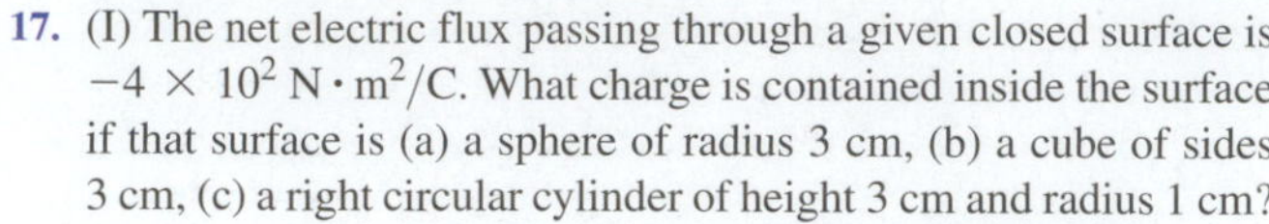

17. (I) The net electric flux passing through a given closed surface is $-4 \times 10^2\ \text{N}\cdot\text{m}^2/\text{C}$. What charge is contained inside the surface if that surface is (a) a sphere of radius 3 cm, (b) a cube of sides 3 cm, (c) a right circular cylinder of height 3 cm and radius 1 cm?

18. (II) A 420-μC charge is placed at the center of a cube of sides 8 cm. Determine the electric flux through each of the sides.

19. (II) A given region has an electric field that is a sum of two contributions: a field due to a charge $q = 5 \times 10^{-8}$ C at the origin, plus a uniform field of strength $E_0 = 3000$ N/C in the $-x$-direction. Calculate the flux through each side of a cube with sides of length 20 cm that are parallel to the x-, y-, and z-directions; the cube is centered at the origin.

20. (II) The *gravitational field* $\vec{g}$ due to a point mass M may be obtained by analogy with the electric field by writing an expression for the gravitational force on a test mass, and dividing by the magnitude of the test mass, m. Show that Gauss' law for the gravitational field reads $\Phi = \oint_S \vec{g}\cdot d\vec{A} = -4\pi GM$, where G is the gravitational constant. Use this result to calculate the gravitational field at a distance r from the center of a sphere of radius R and uniform density for $r > R$ and for $r < R$.

23–3 Using Gauss' Law to Determine Electric Fields

21. (I) Calculate the electric field outside a long cylinder of finite radius R with a uniform (volume) charge density ρ spread throughout the volume of the cylinder.

22. (I) Use Gauss' law to show that the electric field outside a large, thin, nonconducting plate with uniform charge density σ is given by $E = \sigma/2\varepsilon_0$.

23. (I) Charge is distributed on a long, straight rod with uniform density $\lambda = 6.5 \times 10^{-8}$ C/m. Compare the magnitude of the field 1 cm from the rod to the field 1 cm from a point charge $q = 6.5 \times 10^{-8}$ C.

24. (II) An infinitely long cylinder of radius R carries a uniform (volume) charge density ρ. Calculate the field everywhere inside the cylinder.

25. (II) On a clear day in Nebraska, the electric field just above the ground is 110 N/C and points toward the ground. Our planet Earth is a reasonable conductor and contains no electric field. How much net charge is contained on the surface of a 60-acre corn field (1 acre $\cong$ 4000 m^2)?

26. (II) Two long, thin cylindrical shells of radii r_1 and r_2, respectively, are oriented coaxially (one cylinder is centered inside the other). The cylinders carry equal and opposite linear charge densities λ. Describe the resulting electric field inside the smaller cylinder, between the cylinders, and outside the larger cylinder (Fig. 23–27).

27. (II) A balloon of radius 15 cm carries a charge of 5×10^{-7} C distributed uniformly over its surface. What is the electric field at a distance of 50 cm from the center of the balloon? Suppose that the balloon shrinks to a radius of 10 cm but loses none of its charge. What is the electric field at a distance of 50 cm from the center?

28. (II) A thin, cylindrical copper shell of diameter 6.0 cm has a thin metal wire of diameter 0.10 mm along its axis. The wire and the shell carry equal and opposite charges of 8.5×10^{-9} C/cm, distributed uniformly. Calculate the electric field in the region between the wire and the cylinder, and the magnitude of the electric field at the surface of the wire and at the inner surface of the cylinder.

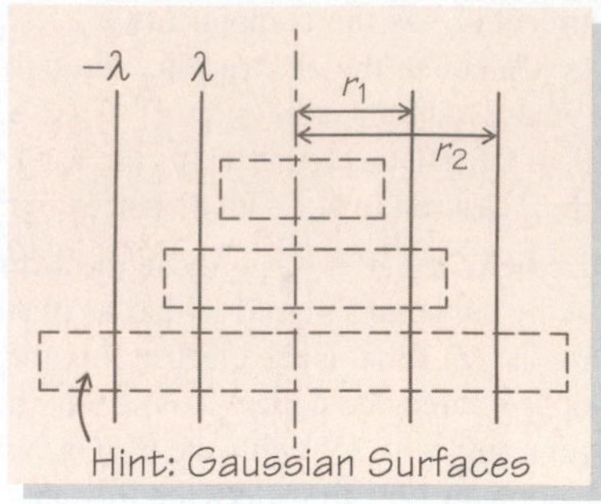

▲ **FIGURE 23–27** Problem 26.

29. (II) A long, cylindrical shell of inner radius r_1 and outer radius r_2 carries a uniform volume charge density ρ. Find the electric field due to this distribution of charge everywhere in space.

30. (II) A Teflon rod of radius 4.0 mm and height 7.0 cm is being charged uniformly over its cylindrical surface. How much charge can the rod hold before the surrounding air breaks down electrically, which happens when the electric field in air is 2.0×10^6 N/C? Ignore the likelihood of breakdown at the sharp edges.

31. (II) A thick, nonconducting spherical shell with a total charge of Q distributed uniformly has an inner radius R_1 and an outer radius R_2. Calculate the resulting electric field in the three regions $r < R_1$, $R_1 < r < R_2$, and $r > R_2$.

32. (II) A spherical metal shell of inner radius R is isolated and carries no net charge. A metal ball of radius r with charge q is suspended inside it, so that the center of the ball is at the center of the shell. Write expressions for (a) the charge on the inner surface of the shell; (b) the charge on the outer surface of the shell, (c) the electric field at a distance d from the center, where $r < d < R$.

33. (II) Consider two infinite parallel charged plates with surface charge densities of σ_1 and σ_2 respectively. (a) What is the electric field in the three regions on the far sides of the plates and between the plates? (b) What are the fields if an infinite uncharged metallic plate is inserted between and parallel to the two charged plates? Is it enough to just use Gauss' law? What else is needed?

34. (II) Two infinite-plane nonconducting, thin sheets with uniform surface charges of 3 μC/m^2 and -1 μC/m^2, respectively, are parallel to each other and 12 cm apart. What are the electric fields between the sheets and outside them?

35. (II) Two infinite-plane sheets that are just like those of Problem 34 are placed at right angles to each other. What are the fields in the four regions into which space is divided by the planes?

36. (II) A slab of nonconducting material forms an infinite plane. The slab has a thickness t and carries a uniform positive charge density ρ. It is oriented parallel to the xy-plane, with its upper surface at $z = t/2$ and its lower surface at $z = -t/2$. Use Gauss' law to find the electric field both above and below the surface, as well as at an arbitrary value of z in the interior of the slab.

37. (II) Consider a solid sphere of radius 3 cm that carries a negative charge of 2 μC distributed uniformly. The sphere is placed concentrically in a spherical shell of radius 8 cm that has a positive charge of 5 μC distributed uniformly over it. Calculate the electric field as a function of radius r for $0 < r < 15$ cm.

38. (III) Charge is distributed throughout a sphere with the charge density given by $\rho = \rho_0$ for $r < a$, $\rho = \rho_0(r - R)/(a - R)$ for $a < r < R$, and $\rho = 0$ for $R < r$ (Fig. 23–28). Calculate the flux through the spherical surfaces at $r = a$, $r = R$, and $r = 10R$, and calculate the corresponding electric fields at these radii.

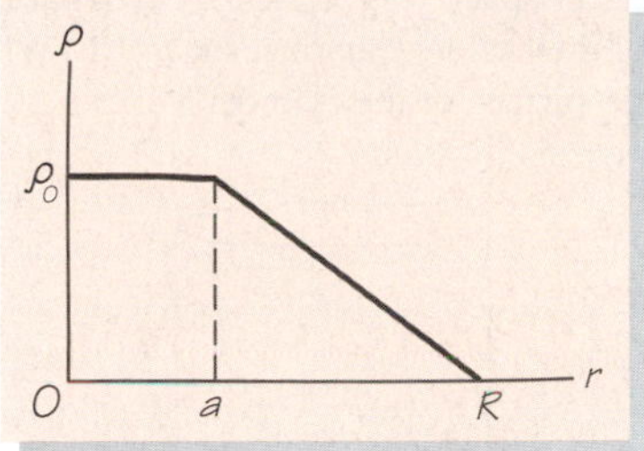

▲ **FIGURE 23–28** Problem 38.

39. (III) Consider the charge distribution given in Problem 38. Plot the charge density, the flux through a concentric shell of radius r, and the electric field as a function of r. Use $R = 3a$.

23–4 Conductors and Electric Fields

40. (I) Two large, thin, metallic plates are placed parallel to each other, separated by 11 cm. The top plate carries a uniform charge density of 6.5 $\mu C/m^2$, while the bottom plate carries a uniform charge density of $-4.8\ \mu C/m^2$. What is the electric field halfway between the plates?

41. (I) Two concentric metallic shells—conductors—have radii of R and $2R$, respectively. A charge q is placed on the inner shell, and a charge $-2q$ is placed on the outer shell. What are the electric fields in all of space due to the two shells?

42. (I) Two oppositely charged, parallel metal plates give rise to a field of 3×10^6 N/C between them. The plates are square and have dimensions 0.1 m × 0.1 m. How much charge must there be on each plate? Assume that the charge distribution and electric field are uniform, as if the plates were infinite in size. This will be a good approximation if the distance between the plates is much smaller than 0.1 m.

43. (I) Charge is placed on a large spherical surface. What is the maximum surface charge density that avoids electrical breakdown in air ($E_{max} = 3 \times 10^6$ N/C)?

44. (I) A metal sphere of radius 15 cm is concentrically surrounded by a thin spherical metal shell whose inner radius is 25 cm. The electric flux through a concentric spherical Gaussian surface at a radius of 40 cm is $1.6 \times 10^7\ N \cdot m^2/C$, and that through a concentric spherical Gaussian surface at a radius of 18 cm is $0.80 \times 10^7\ N \cdot m^2/C$. What is the ratio of the charges on the inner and outer spheres?

45. (I) What is the ratio of the charge densities on the inner and outer spheres in Problem 44?

46. (I) A solid copper cube is placed in a constant electric field that points in the $+x$-direction. The faces of the cube are parallel to the xy-, yz-, and xz-planes, and one corner is at the origin. Draw the field lines as they would be observed looking down on the cube toward the xy-plane. Show at least two electric field lines starting or stopping on each of the four sides of the cube perpendicular to the xy-plane.

47. (II) A metal sphere of radius a is surrounded by a metal shell of inner radius b and outer radius R. The flux through a spherical Gaussian surface located between a and b is Q/ε_0, and the flux through a spherical Gaussian surface just outside radius R is $2Q/\varepsilon_0$ (Fig. 23–29). What are the total charges on the inner sphere and on the shell? Where are the charges located, and what are the charge densities?

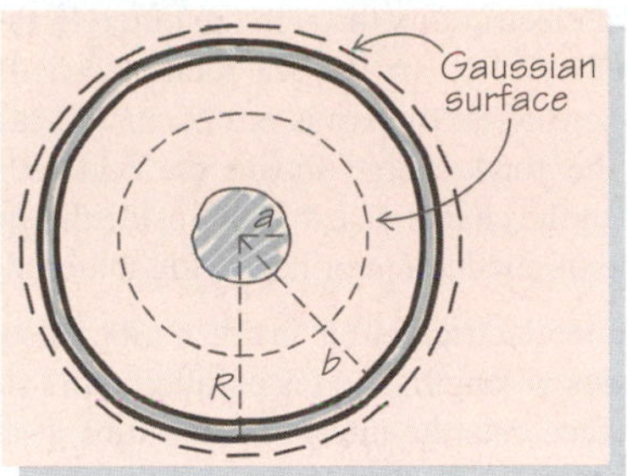

▲ **FIGURE 23–29** Problem 47.

48. (II) The electric field near Earth's surface on a given day is 100 N/C, pointing radially inward. If this were true everywhere on the Earth's surface, what would the sign and magnitude of the total charge on Earth be? If Earth is treated as a conductor, where is the charge located? What is the charge density?

49. (II) A point charge q is placed a distance $L/2$ over the center of a conducting square plate of area L^2. (a) Draw the electric field lines on both sides of the plate, which has charge $-q$. (b) Repeat part (a) for a charge on the plate of $2q$.

50. (II) The center of a solid conducting sphere of radius 18 cm and charge 380 μC is placed 15 cm above and away from the center of a flat, horizontal conducting square plate of area 1000 cm^2 and charge 15 μC. Draw the electric field lines.

General Problems

51. (II) Consider a cube of sides a located at the origin (Fig. 23–30). Suppose that an electric field is present and given by $\vec{E} = bx^2\hat{i}$, where b is a constant. Calculate the flux through each side of the cube, and use this to find the charge within the cube.

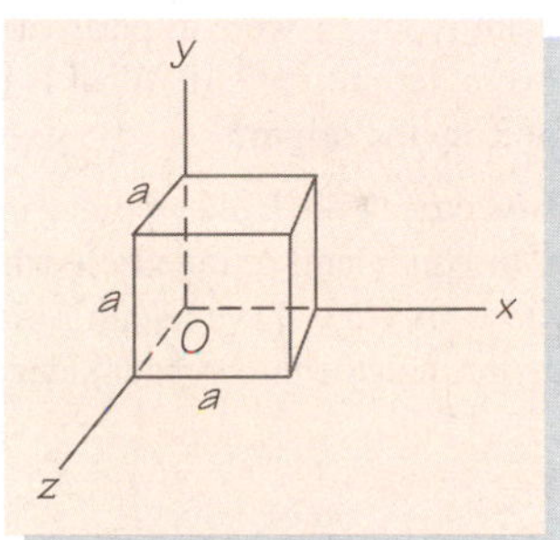

▲ **FIGURE 23–30** Problem 51.

52. (II) Consider a solid sphere of radius R with a charge Q distributed uniformly. Suppose that a point charge q of mass m, with a sign opposite that of Q, is free to move within the solid sphere. Charge q is placed at rest on the surface of the solid sphere and released. Describe the subsequent motion. In particular, what is the period of the motion, and what is the total energy of the point charge? [*Hint*: Recall the properties of the motion for which the force varies linearly with the distance from a fixed point and is a restoring force.]

53. (II) Consider a point charge Q at the center of a Gaussian sphere of radius R. The sphere has a cap "sliced off," with the area of the cap forming a fraction of 0.067 of the total area of the sphere. (This corresponds to making the slice at a latitude of 60°, measured from the equatorial plane). The cap is replaced by a flat surface. (a) What is the total flux through the flat surface? (b) What is the magnitude of the electric field at the boundary of the flat surface?

54. (II) The total electric flux through the outer surface of a uniform solid sphere is $17.1\ \mathrm{N\cdot m^2/C}$. The total electric flux through the top hemisphere of the sphere is $8.9\ \mathrm{N\cdot m^2/C}$. (a) What can you say about the total charge inside the sphere? (b) What can you say about the charge distribution inside the sphere? (c) Is the sphere made of conducting or insulating material?

55. (II) A constant electric field is inside a tube of square cross section with sides of length L and is parallel to the sides of the tube. A plane surface cuts the interior of the tube at an angle θ (Fig. 23–31). Show by explicit calculation that the flux through this surface is independent of the angle θ. How would you show this without explicitly calculating the flux through the surface?

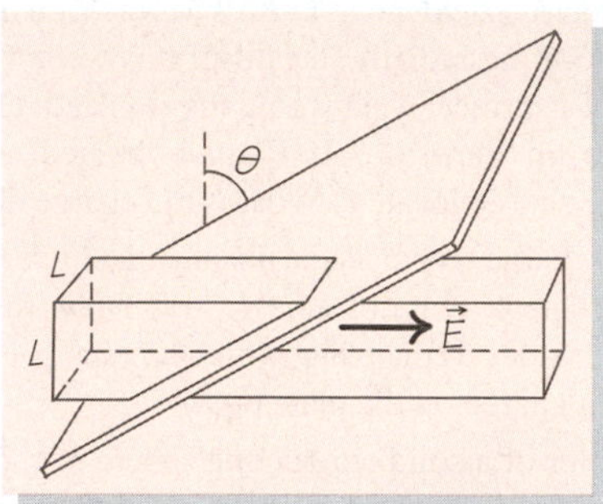

▲ **FIGURE 23–31** Problem 55.

56. (II) A conducting sphere of radius 25 cm is centered at the origin of a coordinate system, as is a surrounding conducting shell of radius 75 cm. The inner sphere has a charge density of $16\ \mu\mathrm{C/m^2}$ over its surface, and the outer sphere has a uniform charge density half that large. (a) Find the electric field at a distance 50 cm from the origin; (b) at a distance 70 cm from the origin. (c) How would your answers to parts (a) and (b) change if the outer shell were not present? (d) What is the electric field at a distance 1.5 m from the origin?

57. (II) A constant electric field E that points in the $+z$-direction passes through an equilateral tetrahedron whose base is in the xy-plane and whose six edges have length L (Fig. 23–32). Calculate the total flux through the three upper sides of the tetrahedron.

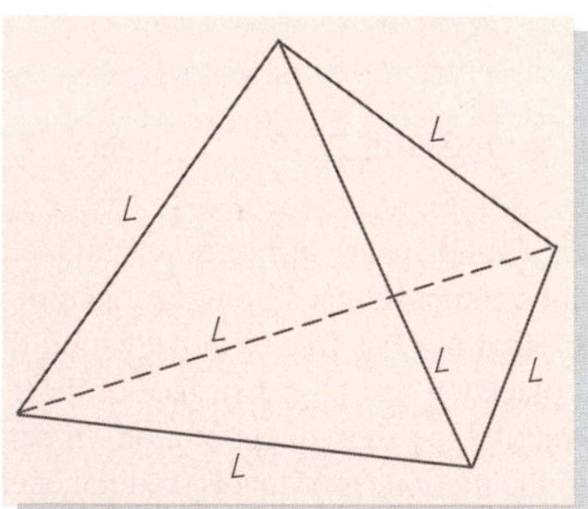

▲ **FIGURE 23–32** Problem 57.

58. (II) How should the charge density of a sphere of radius R vary with the distance from the center of the sphere to give a radial field of constant magnitude within the sphere? What happens at the origin, and why?

59. (II) A certain experiment requires an electric field that points symmetrically away from an axis and has a constant magnitude. Describe the charge distribution capable of creating such a field.

60. (II) A right solid conducting cylinder has a charge of -0.55 mC. Inside the cylinder a $+0.12$-mC charge rests at the center of a hollow spherical space (Fig. 23–33). (a) What is the charge on the surface of the hollow spherical space? (b) What is the charge on the outside surface of the cylinder?

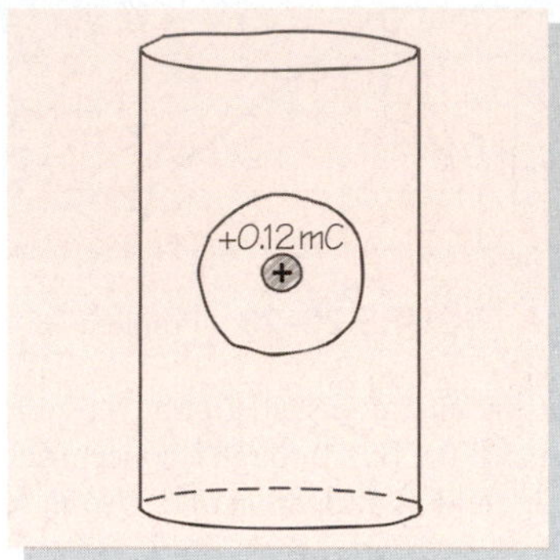

▲ **FIGURE 23–33** Problem 60.

61. (II) A conductor has a surface oriented in the yz-plane that marks the boundary of a region in which there is an electric field oriented in the $+x$-direction. The strength of this field increases linearly as x increases from $x = 0$ m to $x = 0.5$ m. At the beginning of the region, at $x = 0$ m, the field strength is 0; at $x = 0.5$ m, the field strength has increased to 3000 N/C. Describe the distribution in the x-direction of the charge that produces this field.

62. (III) A nonconducting sphere of radius R is charged uniformly with charge density ρ. Use Gauss' law to show that the electric field inside the sphere at a point P whose displacement vector from the sphere's center is $\vec{r}$ is given by $\vec{E} = (\rho/3\varepsilon_0)\vec{r}$. A small sphere centered at the point whose displacement from the origin is $\vec{a}$ is cut out of the sphere (Fig. 23–34). Use the superposition principle to calculate the electric field inside the cavity. [*Hint*: The cavity can be created by inserting in the original sphere a sphere of opposite charge density, $-\rho$, and radius b, centered at $\vec{a}$.]

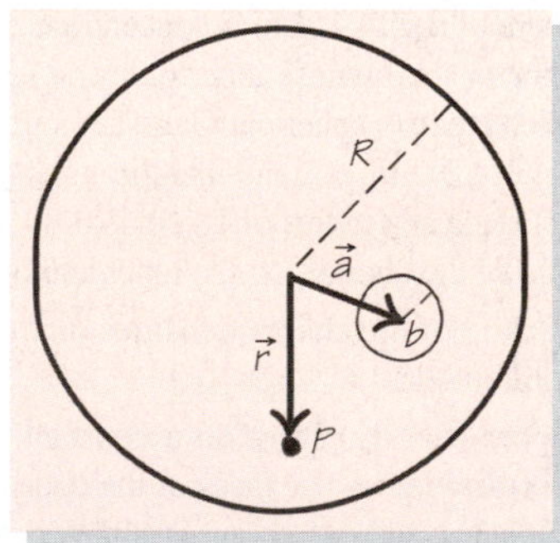

▲ **FIGURE 23–34** Problem 62.

63. (III) Use Gauss' law to show that a test charge in the electric field due to any given static charge distribution cannot be in stable equilibrium. [*Hint*: At an equilibrium point, the net electric field must be zero. What must the fields in the vicinity of that point be so that the equilibrium is stable?]

24 Chapter

The discharge of lightning bolts provides an impressive demonstration that there is energy in electric fields. Lightning forms when a significant electric potential difference exists between Earth and the clouds or between different clouds.

Electric Potential

We know that there is a potential energy associated with conservative forces, that this potential energy is a term in the total energy of systems, and that the conservation of energy is a powerful tool for solving problems. The electric, or Coulomb, force is a conservative force. Thus we expect a collection of charges to have a potential energy, the subject of this chapter. Many of the results we develop here are similar to those for gravitation (Chapter 12) because the gravitational force and Coulomb's law have the same form.

Electric force concerns the interaction of a charge distribution and a second charge and is the product of the second charge and the electric field due to the distribution. Similarly, the electrical potential energy is the potential energy of the system made up of the charge distribution interacting with a second charge. And just as it is useful to remove the second charge from the electric force and deal with the electric field due to the distribution, it is useful to remove the second charge from the electrical potential energy and deal with the *electric potential*, or just potential, a property of the charge distribution alone.

There is one more important point to keep in mind. One of the advantages of using the potential energy rather than the force is that the potential energy is a scalar quantity, while force is a vector. If we know the potential energy, we may find the force by appropriate derivatives. Similarly, the potential is a scalar quantity, easier to handle in many respects than the electric field, which is a vector quantity, and if we know the electric potential of a charge distribution, we can derive the electric field due to that distribution from it.

24–1 Electric Potential Energy

The concept of a potential energy is extremely useful. For example, in thinking about gravity we learned that a mass m at a height h (much less than Earth's radius) above Earth's surface has a potential energy that can be written as $U(h) = mgh$. This helps us to determine the object's speed at any other height if we know its speed at one height. *Any* conservative force has a potential energy associated with it. This potential energy is a function of position, and it can be converted to kinetic energy in accordance with the conservation of mechanical energy. The total energy is $E = K + U$, where K is the kinetic energy. Conservation of energy means that the change in E is zero, so $\Delta E = 0 = \Delta K + \Delta U$, or $\Delta K = -\Delta U$. Any change in U will be matched by an equal but opposite change in K.

Let us now look at Coulomb's law. The electric force on charge q_0 due to charge q, separated by a distance r, is

$$\vec{F} = \frac{qq_0}{4\pi\varepsilon_0}\frac{1}{r^2}\hat{r}, \tag{24–1}$$

where $\hat{r}$ is the unit vector that points radially outward from the position of q to the position of q_0. Recall that the gravitational force between a mass m_0 and a mass m separated by a distance r is

$$\vec{F} = -Gmm_0\frac{1}{r^2}\hat{r}. \tag{24–2}$$

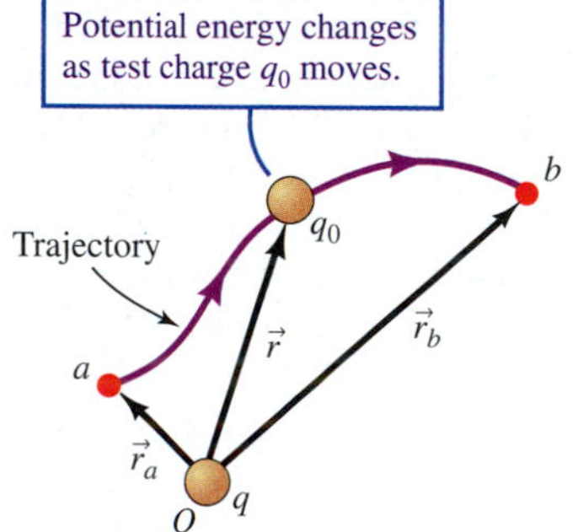

▲ **FIGURE 24–1** When a test charge q_0 moves from point a to point b in the presence of a charge q that is fixed in place, the potential energy of the system changes.

This is just the same form as Coulomb's law, although the gravitational force is always attractive, whereas the electric force is attractive or repulsive according to whether qq_0 is negative or positive. Each force is conservative, so a potential energy is associated with each force. This potential energy must of course take the same form for both cases.

Recall that only *changes* in potential energy have meaning. From Chapter 7, we know that we can express the change in potential energy of our system as the charge q_0 (or, in the case of gravitation, the mass m_0) moves from an initial point a at position $\vec{r}_a$ to a final point b at position $\vec{r}_b$ through a displacement (Fig. 24–1) as

$$\Delta U = U_f - U_i = U(\vec{r}_b) - U(\vec{r}_a) = -\int_{\vec{r}_a}^{\vec{r}_b} \vec{F}\cdot d\vec{s}. \tag{24–3}$$

The integral in this expression is a line integral whose value is *independent of the path of integration* between points a and b—that is precisely why Eq. (24–3) involves only a difference in the value of the potential energy at the two end points.

Let's now evaluate the change in electric potential energy for the point charge q at the origin and the point charge q_0 when q_0 moves from point a to point b. We start with the simplest situation (Fig. 24–2a), in which point a is on the same radius as point b.

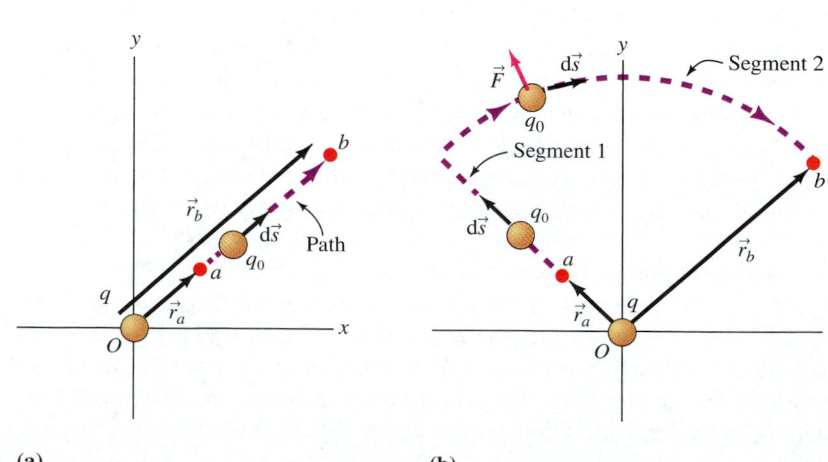

▶ **FIGURE 24–2** The change in potential energy of the system of two charges q and q_0 when the charge q_0 moves from point a to point b, in terms of a path-independent line integral. (a) Charge q_0 moves along a radius. (b) The two points are not along the same radius. The path is here taken to run radially outward to the radius of point b, and then follow the circumference at that radius.

The path from a to b is along the dashed line shown in Fig. 24–2a, and as the Coulomb force points outward along this radial direction, we have for our path

$$\vec{F} \cdot d\vec{s} = F\, dr.$$

Then, from Eq. (24–3), the potential energy change when charge q_0 moves from a to b is

$$\Delta U = -\int_{r_a}^{r_b} F\, dr = -\int_{r_a}^{r_b} \frac{qq_0}{4\pi\varepsilon_0 r^2}\, dr$$

$$= -\frac{qq_0}{4\pi\varepsilon_0}\int_{r_a}^{r_b} \frac{dr}{r^2} = -\frac{qq_0}{4\pi\varepsilon_0}\left(\frac{-1}{r}\right)\Bigg|_{r_a}^{r_b} = \frac{qq_0}{4\pi\varepsilon_0}\left(\frac{1}{r_b} - \frac{1}{r_a}\right). \qquad (24\text{–}4)$$

What if charge q_0 moves between two points that do not lie on the same radius, as in Fig. 24–2b? In this case, we follow the dashed path shown. [Remember, the result of the integration in Eq. (24–3) is path independent.] For segment 1, which runs outward radially from a to a distance r_b from the origin, the result is identical to Eq. (24–4). For segment 2, which follows a circumference at a distance r_b from the origin, the integral is zero because the force is perpendicular to the path segment $d\vec{s}$ everywhere. The result for the change in potential energy is still given by Eq. (24–4).

Let's look more closely at what Eq. (24–4) tells us. Suppose first that the charges move closer together ($r_a > r_b$). If the charges repel (qq_0 is positive), the change in potential energy is positive. This is like moving a mass *up* a mountain. If the charges attract (qq_0 is negative), the system loses potential energy when the charges move closer together. This is like moving a mass *down* the mountain. As with any potential energy, electric potential energy can be converted into kinetic energy. If there are no additional forces acting, then like-sign charges slow down—or lose kinetic energy—when they move closer together. Similarly, charges of opposite sign speed up—or gain kinetic energy—when they move closer together. We draw a similar set of conclusions when the charges move farther apart ($r_a < r_b$). Charges that repel lose electric potential energy and, if there are no other forces, gain kinetic energy. Opposite charges (which attract) gain electric potential energy when they move farther apart and lose kinetic energy in the absence of other forces.

Equation (24–3) shows that the change in electric potential energy is given by the difference of two functions, $U(r_b)$ and $U(r_a)$. We can therefore choose the zero of the potential energy function to be at whatever value of r we like. It is convenient and natural to choose zero potential energy to be at infinity. We can do this if we let $r_a \to \infty$ and let r_b take on a general value r in Eq. (24–4):

$$\Delta U = U(r) - U(r_a)\Bigg|_{r_a\to\infty} = \frac{qq_0}{4\pi\varepsilon_0}\frac{1}{r}.$$

We then say that the potential energy of a charge q_0 a distance r from charge q is the difference in potential energy between that point and infinity. When we reverse the roles of q and q_0, the potential energy of q at a distance r from q_0 is again $qq_0/4\pi\varepsilon_0 r$. We can then say that the **electric potential energy** $U(r)$ for a system of two point charges q and q_0 separated by a distance r is

$$U(r) = \frac{qq_0}{4\pi\varepsilon_0}\frac{1}{r}. \qquad (24\text{–}5)$$

It is indeed true that $U(r) = 0$ in the limit $r \to \infty$. Thus the system has no potential energy when the two charges are infinitely far apart. Note that the potential energy of the two charges depends *only* on the distance r between them and on the magnitudes and signs of the charges. Equation (24–5) has the same form as Eq. (12–9), calculated in Chapter 12 for the gravitational potential energy.

CONCEPTUAL EXAMPLE 24–1 How much work is done by the electrical force when a point charge is brought from infinity to rest at a distance r from a fixed charge of the opposite sign? What is the meaning of the sign of your result?

Answer The work done is given by $W = \int_{\infty}^{r} \vec{F} \cdot d\vec{s}$, where $\vec{F}$ is the electrical force between the charges. The quantity is the negative of the corresponding change of electric potential energy, as in Eq. (24–3), namely $-[U(r) - U(\infty)]$. Since the zero of the potential energy is at infinity, this is just $-U(r)$, with U given by Eq. (24–5). For charges of opposite sign, $-U(r)$, and therefore the work done, is positive. This is a sensible result, because when the electrical force is the only force acting the work done by it is the change in kinetic energy, and that change would indeed be positive for charges of opposite sign, which attract one another.

EXAMPLE 24–2 One measure of the strength of the nuclear forces that hold the constituents of the nucleus together is provided by the comparative strength of the repulsive Coulomb force between the (positively charged) protons in the nucleus. Find the electrostatic potential energy between two protons that are separated by the average separation between the nuclear constituents, roughly 2×10^{-15} m.

Strategy This is a straightforward application of Eq. (24–5), an expression for the electrostatic potential energy. This expression assumes that the potential energy is zero at infinity.

Working It Out The proton charge is equal and opposite to that of the electron, $+e = 1.6 \times 10^{-19}$ C. Then

$$U = \frac{(e)^2}{4\pi\varepsilon_0 r} = \frac{(9 \times 10^9\,\mathrm{N \cdot m^2/C^2})(1.6 \times 10^{-19}\,\mathrm{C})^2}{(2 \times 10^{-15}\,\mathrm{m})} = 10^{-13}\,\mathrm{J}.$$

This is a typical energy value on the nuclear scale and is about 10^5 times larger than the energy of a proton and electron that make up a hydrogen atom (see Example 24–4). This energy can be converted into kinetic energy in nuclear reactions (see Chapter 44).

24–2 Electric Potential

A point charge q is the source of an electric field $\vec{E}$ that exists in the surrounding space. The electric field affects any charge q_0 introduced into that space through a force $\vec{F}$ on q_0 given by $\vec{F} = q_0\vec{E}$. We saw in Section 24–1 that the introduction of a charge q_0 a distance r from q gives rise to the potential energy $U(r)$ of Eq. (24–5). If we write $U(r) = q_0 V(r)$, we can make a statement analogous to the statement about the electric field: The charge q is the source of an *electric potential* (or just *potential*) $V(r)$, within which any charge q_0 a distance r from q will have potential energy $U(r) = q_0 V(r)$. Strictly speaking q_0 should be a small test charge, so that its presence does not disturb charge q or any other charge distribution that gives rise to the electric potential. This description tells us how to define the **electric potential** due to a charge distribution:

$$V(\vec{r}) \equiv \lim_{q_0 \to 0} \frac{U(\vec{r})}{q_0}, \tag{24–6}$$

where $U(\vec{r})$ is the potential energy of the test charge q_0 in the presence of the charge distribution. The potential $V(\vec{r})$ is independent of the test charge q_0, just as the electric field, defined by $\vec{E} = \vec{F}/q_0$, is independent of the test charge.

The Electric Potential of a Point Charge

Let's calculate the electric potential of the simplest possible system: one point charge. Consider a test charge q_0 separated by a distance r from a single point charge q. As Eq. (24–5) shows, the potential energy of the system is $U(r) = q_0 q/4\pi\varepsilon_0 r$, and hence $U/q_0 = q/4\pi\varepsilon_0 r$. We have found *the electric potential of a point charge q at a distance r from the charge*:

$$V(r) = \frac{q}{4\pi\varepsilon_0 r}. \tag{24–7}$$

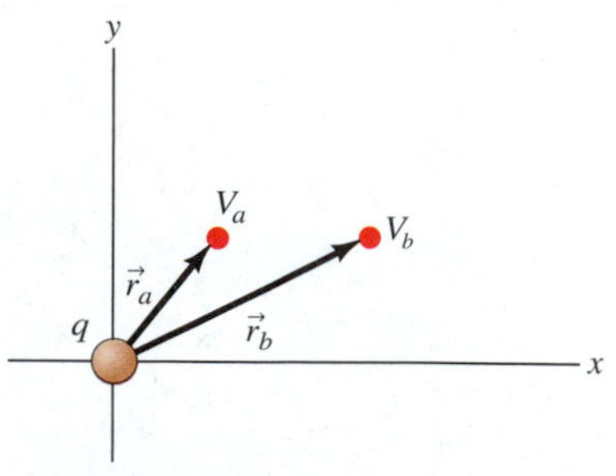

▲ **FIGURE 24–3** The potential at two different locations in space. Only the potential *difference* has physical meaning.

In Eq. (24–7), we have assumed that zero potential energy is at infinity and, as a consequence, we have taken the electric potential due to a charge q to be zero at infinity. To emphasize this point, we might say that Eq. (24–7) is the potential of a single charge *with respect to infinity.*

As for potential energy, the only physically relevant feature of the potential is how it differs between two points. The *electric potential difference* due to the charge q between the points a and b at locations $\vec{r}_a$ and $\vec{r}_b$ is given by (Fig. 24–3):

$$\Delta V = V_b - V_a = \frac{U_b - U_a}{q_0} = \frac{q}{4\pi\varepsilon_0}\left(\frac{1}{r_b} - \frac{1}{r_a}\right). \tag{24–8}$$

Here, we have abbreviated V as a function of r_a, or $V(r_a)$, as V_a, and so forth.

We can obtain another formulation of the electric potential difference by using Eqs. (24–3) and (24–8) and substituting $\vec{F} = q_0\vec{E}$:

$$\Delta V = \frac{U_b - U_a}{q_0} = -\int_{r_a}^{r_b} \vec{E}\cdot d\vec{s}. \qquad (24\text{–}9)$$

Here, the electric potential difference is expressed as a path-independent integral over an electric field. The electric field in Eq. (24–9) is not necessarily the electric field of a point charge. Equation (24–3) is the potential energy change when a test charge q_0 moves from point a to point b in the field of *any* charge distribution. Thus Eq. (24–9) is a general expression for the electric potential difference between two points. Any charge distribution produces an electric field, and an electric potential is associated with any charge distribution.

Recall (Section 7–1) that the change in the potential energy of a system is equal to the negative of the work done by the system in moving an object from point a to point b. Equivalently, $U_b - U_a$ is the work done by an external agent to move the object. By the parallel relation between force and field we can then interpret Eq. (24–9) to mean that

> **The electrical potential difference $V_b - V_a$ is the work per unit charge that must be done by an external agent to move a test charge from point a to point b without changing its kinetic energy.**

If there is no external agent, then a change in potential, which corresponds to a change in potential energy of the test charge, must be accompanied by a corresponding change in the kinetic energy of the test charge.

With knowledge of the electric potential $V(\vec{r})$ due to a charge distribution and the magnitude of a test charge q_0 we immediately have the potential energy $U(\vec{r})$ of the system composed of the distribution and the test charge q_0 placed at the position $\vec{r}$:

$$U(\vec{r}) = q_0V(\vec{r}). \qquad (24\text{–}10)$$

In the absence of other forces, this equation tells us that a positive test charge q_0 in the presence of an electric potential will move toward lower values of the potential because the potential energy decreases in that way. The charge speeds up as it moves to lower potentials.

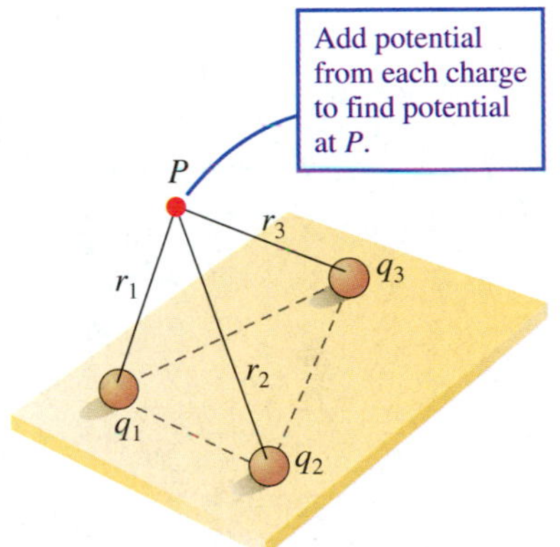

▲ **FIGURE 24–4** The superposition principle determines the potential at point P due to multiple charges. We simply add the potential due to each of the charges.

The Electric Potential of Charge Distributions

The electric field obeys the superposition principle. Therefore the electric potential of a system of charges can also be determined from the superposition principle. The superposition principle states that the electric field of a collection of charges is the sum of the electric fields of each charge. Thus *the electric potential at a point P due to n point charges* $q_1, q_2, \ldots, q_n$ (Fig. 24–4 shows three charges) at distances $r_1, r_2, \ldots, r_n$ from point P is

$$V_P = \frac{q_1}{4\pi\varepsilon_0 r_1} + \frac{q_2}{4\pi\varepsilon_0 r_2} + \cdots + \frac{q_n}{4\pi\varepsilon_0 r_n} = \frac{1}{4\pi\varepsilon_0}\sum_{i=1}^{n}\frac{q_i}{r_i}, \qquad (24\text{–}11)$$

where r_i is the distance from point charge q_i to point P. The electric potential due to a collection of charges is the *scalar* sum of the potentials due to single charges. This scalar sum is much easier to perform than the vector sum that expresses the electric field due to a collection of point charges, illustrating in another way the usefulness of the concept of potential.

The calculation of the electric potential due to a continuous charge distribution is also straightforward. We first find the electric potential dV at a point P due to a small charge dq that is part of an arbitrary charge distribution (Fig. 24–5). Because electric potential is a scalar quantity, the addition of all the tiny potentials dV is given by scalar integration. Thus *the potential due to a continuous charge distribution* takes the symbolic form

$$V = \int dV = \frac{1}{4\pi\varepsilon_0}\int\frac{dq}{r}. \qquad (24\text{–}12)$$

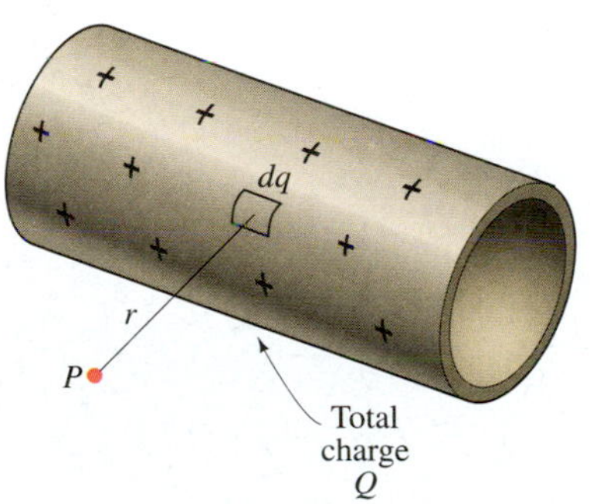

▲ **FIGURE 24–5** To find the potential at point P due to a continuous charge distribution, here a cylinder, integrate over the contribution from the differential charges dq as if each dq were a point charge.

The integration must be done over the entire charge distribution. In Section 24–5 we discuss techniques for the calculation of V in specific situations.

Units of Electric Potential

The dimension of electric potential is energy per charge; thus the SI unit is joules per coulomb (J/C). Because electric potential is used frequently, it has a separate name in the SI: the **volt**(V):

$$1\ \mathrm{V} \equiv 1\ \mathrm{J/C}. \tag{24–13}$$

(Don't confuse the roman letter V that symbolizes the unit with the italic V that is the usual symbol for the function electric potential.) It is named after Alessandro Volta, who did research at the beginning of the nineteenth century on the nature of electric energy. Note that electric potential has the dimensions of electric field times length, so the dimensions of electric field must be the dimensions of potential divided by length (V/m):

$$1\ \mathrm{N/C} = 1\ \mathrm{V/m}. \tag{24–14}$$

The Potential Energy of a System of Charges

Equation (24–10) gives the potential energy $U(r) = q_0V(r)$ of a test charge q_0 placed in the electric potential of a charge distribution. If the charge distribution is a collection of charges, then the electric potential V_P is given by Eq. (24–11), and the potential energy of the test charge is $U(r) = q_0V_P$. But it would be incorrect to call this the potential energy of the entire system of charges $q_0, q_1, q_2, \ldots, q_n$, because the product q_0V_P just represents the work that needs to be done to bring charge q_0 in from infinity. It does *not* take into account the work that must be done to bring the charges $q_1, q_2, \ldots, q_n$ in from infinity. To calculate the potential energy of a collection of three charges, for example, we assemble them one by one. To bring the first charge, q_1, in to the point P_1 requires no work by the external agent if the kinetic energy of the charge is unchanged. To bring the second charge, q_2, in from infinity to the point P_2 does require work because of the potential due to q_1. For our two charges, the work the external agent must do to bring q_2 in from infinity—the potential energy—is given by

$$U_{12} = q_2V_1 = \frac{q_1q_2}{4\pi\varepsilon_0 r_{12}}, \tag{24–15}$$

where r_{12} is the distance between charges q_1 and q_2.

What happens if we bring a third charge, q_3, in from infinity? We must calculate the additional work done by an external force to bring q_3 in. This work is given by the product of q_3 and electric potentials V_1 and V_2 due to q_1 and q_2 in place. Thus, the additional contribution to the potential energy of the system is

$$U_{13} + U_{23} = \frac{q_1q_3}{4\pi\varepsilon_0 r_{13}} + \frac{q_2q_3}{4\pi_0 r_{23}}, \tag{24–16}$$

where r_{13} and r_{23} are the distances between q_3 and q_1, q_3 and q_2, respectively. The total potential energy U of the system is the sum of U_{12}, U_{13}, and U_{23}:

$$U = \frac{1}{4\pi\varepsilon_0}\left(\frac{q_1q_2}{r_{12}} + \frac{q_1q_3}{r_{13}} + \frac{q_2q_3}{r_{23}}\right). \tag{24–17}$$

This can be generalized to any number of charges, and the resulting formula for the *electric potential energy of the system* is a simple generalization of Eq. (24–17):

$$U = \frac{1}{4\pi\varepsilon_0}\sum_{i<j}\frac{q_iq_j}{r_{ij}}, \tag{24–18}$$

where r_{ij} is the distance between the locations of the charges q_i and q_j. The sum over i and j includes all charge pairs in the system, and the inequality $i < j$ avoids the counting of pairs more than once. We can eliminate that restriction by writing the equivalent expression

$$U = \frac{1}{2}\sum_{\substack{i,j\\i\neq j}} \frac{q_i q_j}{4\pi\varepsilon_0 r_{ij}}.$$

Now the sum is unrestricted, except that we omit the case $i = j$, which is not in the original sum, Eq. (24–18). Thus we can rewrite Eq. (24–18) as

$$\begin{aligned} U &= \frac{1}{2}q_1 \sum \frac{q_j}{4\pi\varepsilon_0 r_{1j}} + \frac{1}{2}q_2 \sum \frac{q_j}{4\pi\varepsilon_0 r_{2j}} + \frac{1}{2}q_3 \sum \frac{q_j}{4\pi\varepsilon_0 r_{3j}} + \cdots \\ &= \frac{1}{2}q_1V_1 + \frac{1}{2}q_2V_2 + \frac{1}{2}q_3V_3 + \cdots, \end{aligned} \tag{24–19}$$

where V_1 is the electric potential due to all the other charges at the location of charge q_1, and so on. It should be stressed that the potential energy of q_1 in a given potential V_1 is still q_1V_1; this means that q_1V_1 can be converted into the kinetic energy of the particle that carries charge q_1. This potential energy must be distinguished from the potential energy of the *entire* charge configuration, Eq. (24–18) or (24–19). The potential energy of the entire charge configuration is the energy that would be made available if *all* the charges that appear in the problem were to move to infinity.

In Examples 24–3 and 24–4, we illustrate calculation techniques for the electric potential energy and the electric potential when two or more point charges are involved.

EXAMPLE 24–3 In an experiment to investigate the effects of electricity, Benjamin Franklin could well have placed two point charges, $q_1 = 2.0\ \mu\text{C}$ and $q_2 = -4.0\ \mu\text{C}$, at some distance apart (points P_1 and P_2, respectively) (Fig. 24–6). (a) Find the electric potential at points a and b due to these two point charges. (b) Find the potential difference between points b and a. (c) How much energy would Franklin have had to supply to bring a third charge, of magnitude 3.0 μC, in from infinity to point b?

Setting It Up The only quantity that requires new labeling here is the third charge [part (c)], which we shall call q_3.

Strategy For part (a) we can use Eq. (24–11) to determine the electric potential. We'll require various distances; for example, to find the potential at point a, we'll need the distance from point a to points P_1 and P_2. These distances are a matter of geometry. Part (b) requires taking the difference of the potentials we found in part (a). Finally for part (c), we note that the work that must be done to bring q_3 in from infinity to point b is equal to the change in the potential energy of the system. But we know from part (a) the electric potential of the original system of two charges at point b, so we use Eq. (24–10), $U_b = q_0V_b$, to find the potential energy of the new charge at point b.

Working It Out (a) The distance from point a to point P_1 is $r_{1a} = 2$ m, and the distance from point a to point P_2 is $r_{2a} = \sqrt{(2.0\text{ m})^2 + (3.0\text{ m})^2} = 3.6$ m (note that we work with two significant figures here). The electric potential V_a at point a is then

$$\begin{aligned} V_a &= \frac{1}{4\pi\varepsilon_0}\left(\frac{q_1}{r_{1a}} + \frac{q_2}{r_{2a}}\right) \\ &= (9.0\times10^9\ \text{N}\cdot\text{m}^2/\text{C}^2)\left(\frac{2.0\times10^{-6}\ \text{C}}{2.0\ \text{m}} + \frac{-4.0\times10^{-6}\ \text{C}}{3.6\ \text{m}}\right) \\ &= 1.0\ \text{kV}. \end{aligned}$$

The SI units of potential are volts. A check is always useful—in this case, the unit combination is $\text{N}\cdot\text{m/C} = \text{J/C} = \text{V}$.

For the electric potential at point b, the distance from charge q_1 to b is $r_{1b} = 2.0$ m; similarly, $r_{2b} = 1.0$ m. Therefore, the potential V_b is

$$\begin{aligned} V_b &= \frac{1}{4\pi\varepsilon_0}\left(\frac{q_1}{r_{1b}} + \frac{q_2}{r_{2b}}\right) \\ &= (9.0\times10^9\ \text{N}\cdot\text{m}^2/\text{C}^2)\left(\frac{2.0\times10^{-6}\ \text{C}}{2.0\ \text{m}} + \frac{-4.0\times10^{-6}\ \text{C}}{1.0\ \text{m}}\right) \\ &= -27\ \text{kV}. \end{aligned}$$

(b) The potential difference $V_b - V_a = -27\ \text{kV} - 1\ \text{kV} = -28\ \text{kV}$. Thus the electric potential is higher at point a than at point b.

(c) We have

$$\begin{aligned} U_b = q_3V_b &= (3.0\ \mu\text{C})(-28\ \text{kV}) \\ &= (3.0\times10^{-6}\ \text{C})(-28\times10^3\ \text{V}) = -8.4\times10^{-2}\ \text{J}. \end{aligned}$$

The work done is in the SI units of joules.

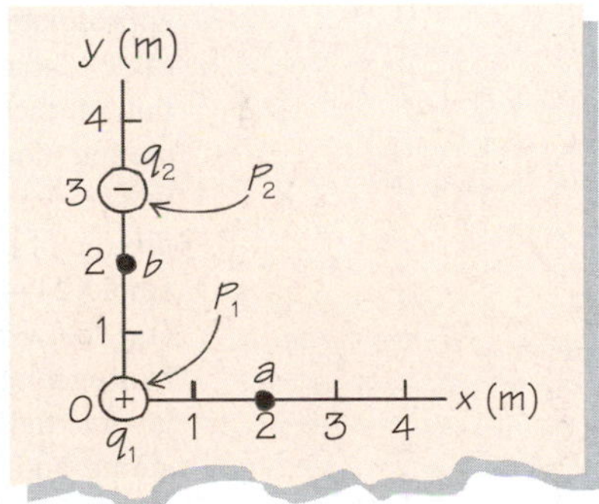

▲ **FIGURE 24–6**

What Do You Think? True or false: The sign of our result for the work that must be done in part (c) is wrong. ***Answers to What Do You Think?** questions are given in the back of the book.*

EXAMPLE 24–4 The classical model of a hydrogen atom in its normal, unexcited configuration has an electron that revolves around a proton at a distance of 5.3×10^{-11} m. What is the electric potential due to the proton at the position of the electron? Determine the electrostatic potential energy between the two particles. This energy is relevant to understanding the chemical activity of atoms.

Setting It Up We sketch this situation in Fig. 24–7.

Strategy The electric potential V_P due to the proton can be found by using Eq. (24–7), and we can subsequently find the electrostatic potential energy by using Eq. (24–15), $U = (-e)V_P$, where $-e$ is the electron charge.

Working It Out We have

$$V_P = \frac{+e}{4\pi\varepsilon_0 r} = \frac{(9.0 \times 10^9\ \mathrm{N \cdot m^2/C^2})(1.6 \times 10^{-19}\ \mathrm{C})}{5.3 \times 10^{-11}\ \mathrm{m}} = 27\ \mathrm{V}.$$

In turn,

$$U = (-e)V_P = (-1.6 \times 10^{-19}\ \mathrm{C})(27\ \mathrm{V}) = -4.3 \times 10^{-18}\ \mathrm{J}. \quad (24\text{–}20)$$

What Do You Think? What does the sign of U tell you about whether or not you must supply positive energy to separate the electron from the proton?

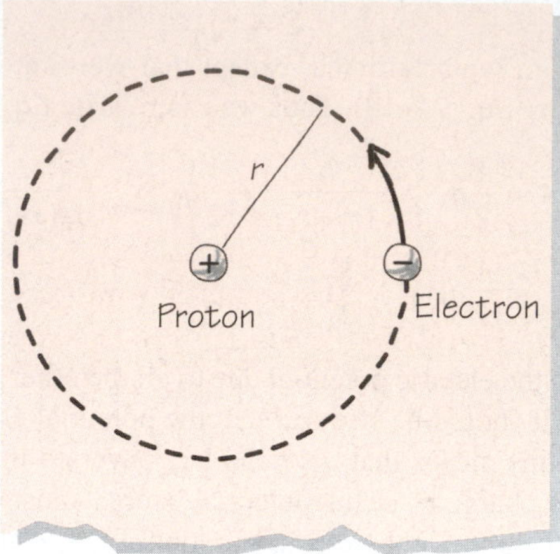

▲ **FIGURE 24–7** A simplistic representation of an electron that orbits a proton in the hydrogen atom.

The Electron-Volt

We have seen that the energy of a system is determined by multiplying charge by electrostatic potential. Because the charge on an electron is such a frequently used quantity, a useful unit of energy is that of the charge magnitude of an electron (or proton) times 1 V. We call this unit of energy an **electron-volt** (eV). An electron-volt is simply the energy an electron gains when it is accelerated though a potential difference of one volt. The electron-volt is not an SI unit. The relation between the electron-volt and the SI unit joule is

$$1\ \mathrm{eV} = (1.6 \times 10^{-19}\ \mathrm{C})(1\ \mathrm{V}) = 1.6 \times 10^{-19}\ \mathrm{J}.$$

The electron-volt is especially valuable for calculations in atomic, nuclear, and particle physics.

THINK ABOUT THIS. . .

IN WHAT REALM OF PHYSICS IS THE ELECTRON-VOLT A NATURAL UNIT?

As the name indicates, this is a unit of energy appropriate to an object with a charge comparable to that of an electron, moving in a potential of the magnitude of volts. It is in the atomic domain that we deal with particles whose charges are those of a single electron or proton, of the order of 10^{-19} C. We saw in Example 24–4 that the electrostatic potential energy between the proton and electron of the hydrogen atom is −27 eV. Chemical reactions involve the rearrangement of electrons among atoms, and in accordance with the result of Example 24–4, involve energies of the order of fractions of electron-Volts. This can translate into temperatures that are involved in chemical reactions, because energy can be expressed that way via the relation energy $= k_B T$, where k_B is Boltzmann's constant. As an example, consider the temperature at which paper burns, 451°F. This corresponds to 506K, and therefore to an energy of $E = (1.38 \times 10^{-23}\ \mathrm{J/K})(506\mathrm{K})/(1.6 \times 10^{-19}\ \mathrm{J/eV}) = 4.4 \times 10^{-2}\ \mathrm{eV}$. This number is much smaller than the ionization energy of hydrogen—the energy that must be supplied to remove the electron entirely—because rearrangements of electrons cost much less energy than stripping them off entirely.

In nuclear reactions the forces are much stronger than electrical ones, and distances are much shorter than typical atomic dimensions, and the natural unit is a million electron-volts, the MeV. For example, a typical "ionization energy" for nuclei—the energy required to remove a constituent particle from a nucleus—is 8 MeV, while rearrangement energies are in the hundreds of kilo-electron volts (keV). ■

EXAMPLE 24–5 Calculate the electric potential due to an electric dipole whose dipole moment has magnitude p at an arbitrary point Q. Work in the limit in which the distance of point Q from the dipole is much larger than the dipole length.

Setting It Up We draw the dipole in Fig. 24–8, which also contains some necessary geometrical and label information. In particular, the charges are $\pm q$, and they are separated by a distance $\ell = p/q$. The distances to the point P from the charges $\pm q$ are r and $r + \Delta r$, respectively.

Strategy A dipole consists of two pointlike charges, so Eq. (24–11)— superposition—determines the potential, which will be zero at infinity. We can see from the figure that point Q is specified in part by the angle between $\vec{p}$ and the line between the $-q$ charge and Q, and as the figure shows, Δr can be expressed in terms of that angle and the separation between the charges in the limit in which $r \gg \Delta r$, a limit that is equivalent to the limit that the distance to the dipole is much greater than the linear size of the dipole.

Working It Out Equation (24–11) gives

$$V = \frac{q}{4\pi\varepsilon_0 r} + \frac{-q}{4\pi\varepsilon_0(r + \Delta r)} = \frac{q}{4\pi\varepsilon_0}\left(\frac{1}{r} - \frac{1}{r + \Delta r}\right)$$

$$= \frac{q}{4\pi\varepsilon_0}\frac{(r + \Delta r) - r}{r(r + \Delta r)} = \frac{q}{4\pi\varepsilon_0}\frac{\Delta r}{r(r + \Delta r)}. \qquad (24\text{–}21)$$

From the figure we see that in the large r limit the dotted line is normal to the longer line, so that

$$\Delta r = \ell \cos\theta = \frac{p\cos\theta}{q}. \qquad (24\text{–}22)$$

When this result is substituted into Eq. (24–21), we find

$$V = \frac{p\cos\theta}{4\pi\varepsilon_0}\left[\frac{1}{r(r + \ell\cos\theta)}\right] = \frac{pq\cos\theta}{4\pi\varepsilon_0}\left[\frac{1}{r(qr + p\cos\theta)}\right], \qquad (24\text{–}23)$$

where in the last step we have multiplied and divided by q.

It is a mathematical exercise to calculate the electric dipole potential of Eq. (24–23) in the large r limit. While Eq. (24–23) is rather complicated, it takes a simple approximate form far from the dipole, when $r \gg \ell$. The easiest place to make an approximation is in Eq. (24–21), where we can use the fact that $\Delta r \ll r$. The numerator has one power of Δr, and if we want a calculation that is correct to first order in the small quantity Δr, we can drop the Δr in the denominator. From that point on we make our substitutions and end up with

$$\text{for } r \gg \ell\text{: } V = \frac{p\cos\theta}{4\pi\varepsilon_0 r^2}. \qquad (24\text{–}24)$$

[Another way of getting the same result is to note that the condition $r \gg \ell$ is equivalent to $qr \gg q\ell = p$, so that we can drop the second term in the denominator of Eq. (24–23).] Because of the large r approximation, we can now measure θ from anywhere between the two charges of the dipole. Note that the potential of the dipole for distant points decreases as $1/r^2$, as compared to the $1/r$ dependence for a point charge, just as the electric field of the dipole falls off as $1/r^3$ compared to the $1/r^2$ dependence of the point charge.

This example is of some importance because, as we mentioned in Section 22–1, the dipole charge distribution occurs repeatedly in nature.

What Do You Think? Our picture of the hydrogen atom in Example 24–4 consists of a charge $+e$ and $-e$ a certain distance apart. Does this neutral atom give rise to a dipole field, and if not, why not?

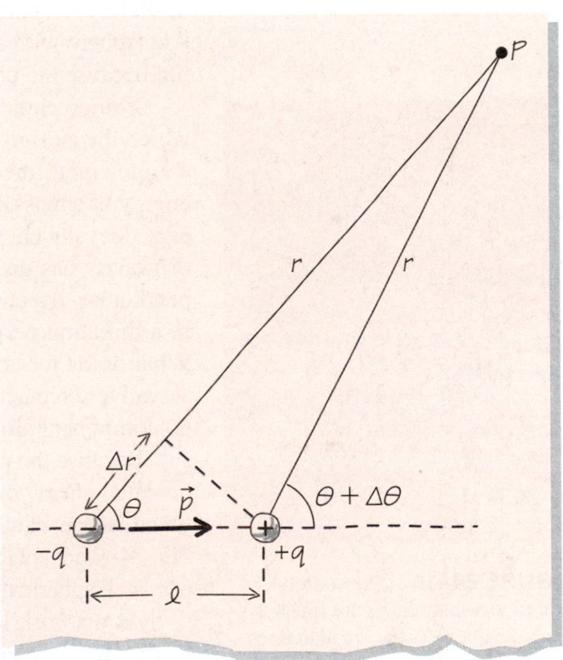

▲ **FIGURE 24–8** Using geometry to find the potential at a point P for an electric dipole. The dipole moment $p = q\ell$.

24–3 Equipotentials

Regions for which the electric potential of a charge distribution has constant values are called **equipotentials**. They are particularly interesting and worth investigating—we shall see, for example, that the surfaces of conductors form equipotentials. Suppose that a system of charges produces a certain potential. The positions in space that have the same electric potential form surfaces in three dimensions and lines in two dimensions. We say that the places where the potential has a constant value form **equipotential surfaces** in three dimensions or **equipotential lines** in two dimensions. As an example, consider the equipotential surfaces formed by a point charge. The electric potential is proportional to $1/r$ and has a constant value at any fixed radial distance from the charge.

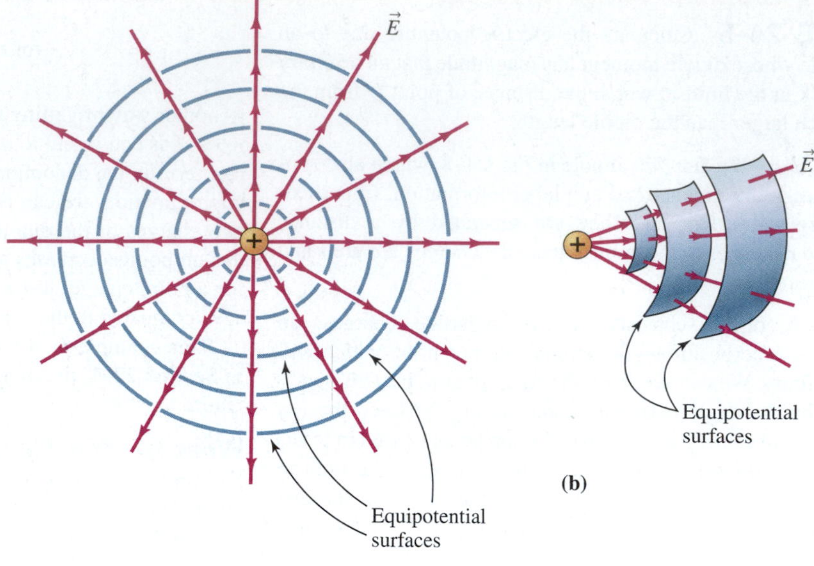

▶ **FIGURE 24–9** (a) The equipotential surfaces for a point charge in a two-dimensional representation. (b) The full three-dimensional representation; the equipotential surfaces are spheres centered on the charge.

Therefore, a sphere centered on the charge forms an equipotential surface (Fig. 24–9). Any sphere with a different radius centered on the charge forms a different equipotential because the potential varies only with the radius of the sphere.

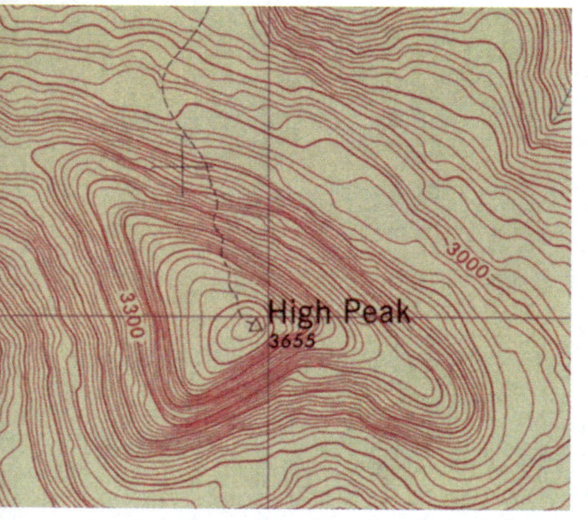

▲ **FIGURE 24–10** The contour lines on topographic maps are lines of constant elevation. These are also lines of constant gravitational potential energy. The force of gravity has no component *along* contour lines, only perpendicular to them.

Equipotentials are analogous to contour lines on a topographic map—lines for which the elevation from sea level is constant (Fig. 24–10). The lines on such maps are equipotential lines for the potential energy of local gravity. The gravitational potential energy of a mass depends only on the mass's elevation, so the gravitational potential energy does not change when a mass moves along a contour line. Consequently, the force of gravity has no component along contour lines. Gravity only has a component perpendicular to a contour line; a ball that starts on a particular contour line will accelerate in a direction perpendicular to the line, or what we would call straight down the hill. What holds for contour lines holds for any equipotential surface or line, and any conservative force acts in a direction perpendicular to the equipotential because it can have no component along the equipotential.

Because the potential has exactly the same value along an equipotential, so does the potential energy of a test charge. No work is done when the test charge moves at constant speed on an equipotential surface or line. The equipotentials for the point charge in Fig. 24–9 are spheres centered on the charge, and a test charge can move freely about any one such spherical surface without work being done by the electric field.

As no work is done by the electric force when a test charge moves on an equipotential, we can understand why the electric field does not have a component along an equipotential surface. If it did, then that component of the electric field would do work to move a charge on the equipotential surface, which is not possible. Thus

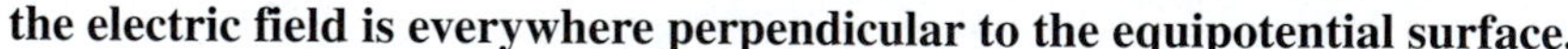

the electric field is everywhere perpendicular to the equipotential surface.

Furthermore, because all the charge on a conductor in equilibrium resides on the surface, a potential difference between two points on the surface would be quickly equalized by a flow of free charge, so

the surface of a conductor is an equipotential.

The same reasoning shows that in static equilibrium the entire conductor will be at that same electric potential—there is no field within the conductor, so it takes no work to move a charge through it.

Electric Field Lines from Equipotentials and Vice Versa

The fact that the electric field and the equipotentials are everywhere perpendicular to each other is helpful in finding equipotential surfaces if the field is known, and in finding the electric fields if the equipotentials are known. We illustrate this process for some charge

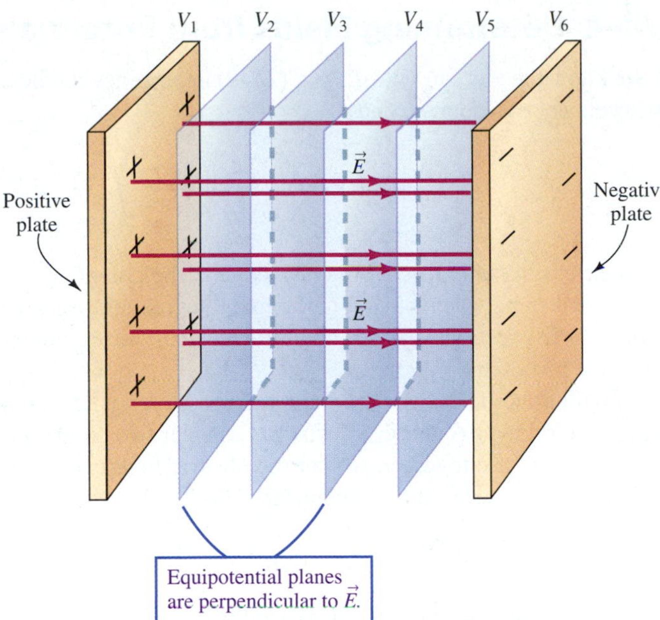

FIGURE 24–11 The electric field lines (burgundy) and the equipotentials (blue) for two oppositely charged parallel plates.

configurations for which the fields are known. We have already seen how the equipotential surfaces of a point charge are perpendicular to the (radial) electric field. For another example, this time starting from known field lines, consider two oppositely charged plates, for which the field lines are as in Fig. 24–11. The equipotential surfaces are perpendicular to these lines, and hence form planes parallel to the charged plates. If the charged plates are conductors, then they must also be equipotential surfaces. Thus we have a series of n equipotential planes $V_1, V_2, V_3, \ldots, V_n$ between and including the two charged plates.

CONCEPTUAL EXAMPLE 24–6 Sketch the equipotentials for an electric dipole given the field lines shown in Fig. 24–12a.

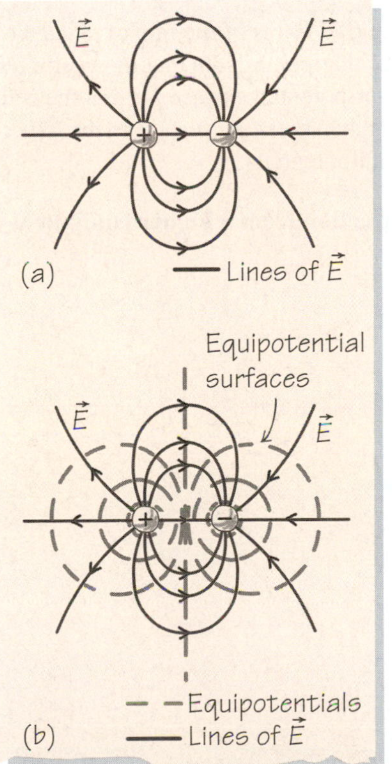

FIGURE 24–12 (a) The electric field lines for an electric dipole. (b) The equipotentials are drawn as dashed lines.

Answer Starting from Fig. 22–12a, we draw our equipotential surfaces everywhere perpendicular to the electric field lines, arriving at the dashed lines shown in Fig. 24–12b. These lines represent the locations in the plane of the page where the equipotential surfaces cut the page. You can visualize them in three dimensions as a series of nested but not concentric closed surfaces—each is a sort of flattened sphere—that encircles each charge. By sketching in this way we can visualize what the equipotential surfaces are like, even without using the algebraic form of the electric dipole potential derived in Example 24–5. A computer-generated drawing of the dipole potential is shown in Fig. 24–13.

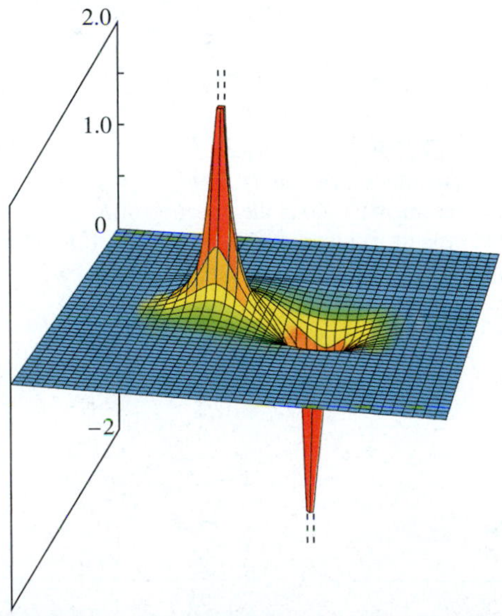

FIGURE 24–13 The dipole potential. Height represents the potential here.

24–4 Determining Fields from Potentials

If we know the electric field $\vec{E}$, Eq. (24–9) determines the potential difference $V_b - V_a$ between any two points a and b:

$$V_b - V_a = \int_{r_a}^{r_b} dV = -\int_{r_a}^{r_b} \vec{E}\cdot d\vec{s}.$$

Because electrostatic forces are conservative, the potential difference is independent of the path taken between a and b in the line integral, and we can choose a path for convenience. In this section we'll see how we can invert this relation and find the electric field given the potential.

Finding the field given the potential is analogous to finding the force between objects if their potential energy is known. Consider two equipotentials, labeled by b and a, that are very close together, so that the electric field can be considered constant in the region between them. In that region Eq. (24–9) takes the form

$$V_b - V_a = dV = -\vec{E}\cdot d\vec{s}, \tag{24–25}$$

where $d\vec{s}$ is an infinitesimal displacement vector pointing from equipotential a to equipotential b. (Actually, this relationship is true for any $d\vec{s}$ connecting the equipotentials, even a large displacement, as long as the electric field can be treated as constant along the displacement.)

It is simplest to take our infinitesimal path as in Fig. 24–14, with $d\vec{s}$ perpendicular to the two equipotential surfaces. As the electric field also points in that direction, Eq. (24–25) reads

$$dV = -E\,ds.$$

Equivalently,

$$E = -\frac{dV}{ds}. \tag{24–26}$$

This equation gives the magnitude of the electric field in terms of the rate of change of V in a direction perpendicular to the equipotential at that point. Note that the vector $d\vec{s}$ pointed from the lower-potential surface to the higher-potential surface, so that the sign in Eq. (24–26) shows that the field points from the higher-potential surface to the lower-potential surface. In summary, for closely spaced equipotentials,

> **the electric field points along the shortest direction from a higher equipotential to a lower one.**

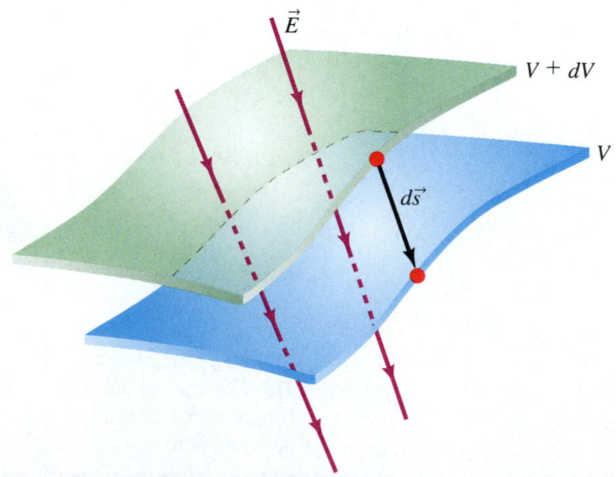

▶ **FIGURE 24–14** Two equipotentials differ by dV. The displacement $d\vec{s}$ along the direction of $\vec{E}$ between the equipotentials is perpendicular to them.

How the Potential Determines the Field in Cartesian Coordinates

For a different point of view, suppose that an arbitrary displacement vector $d\vec{s}$ is decomposed into Cartesian coordinates:

$$d\vec{s} = dx\,\hat{i} + dy\,\hat{j} + dz\,\hat{k}.$$

Here $\hat{i}$, $\hat{j}$, and $\hat{k}$ are the unit vectors in the x-, y- and z-directions, respectively. Then the scalar product in Eq. (24–25) takes the form

$$dV = -\vec{E}\cdot d\vec{s} = -E_x\,dx - E_y\,dy - E_z\,dz, \tag{24–27}$$

where we have separated the field $\vec{E}$ into its Cartesian components. In general, the potential depends on all three space coordinates, $V = V(x, y, z)$. The change in V in going from an initial position $\vec{r} = x\hat{i} + y\hat{j} + z\hat{k}$ to a new position $\vec{r} + d\vec{s} = (x + dx)\hat{i} + (y + dy)\hat{j} + (z + dz)\hat{k}$ is

$$dV = \frac{\partial V}{\partial x}dx + \frac{\partial V}{\partial y}dy + \frac{\partial V}{\partial z}dz. \tag{24–28}$$

Note the use of the partial derivatives here; this is necessary because V depends on all three Cartesian coordinates. Recall that partial derivatives are simple to use: The partial derivative with respect to x means that y and z are held fixed while the ordinary derivative with respect to x is taken. To illustrate, if $V = xz^2$, then $\partial V/\partial x = z^2$, $\partial V/\partial y = 0$, and $\partial V/\partial z = 2zx$. We can equate the coefficients of dx, dy, and dz in Eqs. (24–27) and (24–28):

$$E_x = -\frac{\partial V}{\partial x}, \quad E_y = -\frac{\partial V}{\partial y}, \quad E_z = -\frac{\partial V}{\partial z}.$$

Equivalently, *the electric field vector is given in terms of derivatives of the electric potential* by

$$\vec{E} = -\hat{i}\frac{\partial V}{\partial x} - \hat{j}\frac{\partial V}{\partial y} - \hat{k}\frac{\partial V}{\partial z}. \tag{24–29}$$

THE FIELD IN TERMS OF THE POTENTIAL

Equation (24–29) gives the Cartesian components of the electric field in terms of the potential. We have found a way to express a particular vector, the electric field, in terms of the derivatives of a scalar, the electric potential. It can be simpler to calculate the electric field of a charge distribution by first finding the potential and then differentiating, rather than doing a vector integration to get the field.

EXAMPLE 24–7 Use the electric potential of a point charge q to find its electric field.

Setting It Up Specifying the electric field means finding both its magnitude and its direction.

Strategy From the discussion above, we know that the direction of the electric field is perpendicular to the equipotential surfaces, pointing from a region of higher to lower potential. Here the potential is a function only of the radial distance from the charge, $V = q/4\pi\varepsilon_0 r$. The equipotential surfaces are therefore spheres at a constant distance from the charge. This means that the electric field lines point in the radial direction, and they point from higher potential (smaller r) to lower potential (larger r), that is, outward. We can find the magnitude of the field by using Eq. (24–26).

Working It Out We know already that the direction is the *outward radial direction*, and the magnitude is

$$E = -\frac{dV}{dr} = -\frac{q}{4\pi\varepsilon_0}\frac{d}{dr}\left(\frac{1}{r}\right) = \frac{q}{4\pi\varepsilon_0 r^2}.$$

Although we already know this answer from previous discussion, the technique is useful in contexts where we do not already know the answer!

24–5 The Potentials of Charge Distributions

We rarely deal with the electric field and potential of a single point charge. More often, we have collections of charges spread over regions of space, as when charges spread over the surface of a metal, or when the field of a complicated ionic molecule determines its chemical or biological behavior. We must therefore be able to find the potentials of continuous charge distributions. These charge distributions may not be simple ones, and we must develop strategies for calculating the corresponding electric potentials. In this section, we first summarize the underlying techniques, then illustrate them with a series of examples.

The qualitative shapes of equipotential surfaces due to a charge distribution are most easily found by graphical techniques. For quantitative calculations, we have learned two different ways to determine the electric potential of a charge distribution:

1. If the electric field is known, then Eq. (24–9) can be used to determine the potential:

$$\Delta V = V_b - V_a = -\int_{r_a}^{r_b} \vec{E} \cdot d\vec{s}.$$

2. If the electric field is not known, we may calculate the potential directly by using one of various forms:

for one point charge, Eq. (24–7): $$V = \frac{q}{4\pi\varepsilon_0 r};$$

for many point charges, Eq. (24–11): $$V = \frac{1}{4\pi\varepsilon_0}\sum_i \frac{q_i}{r_i};$$

for a continuous charge distribution, Eq. (24–12): $$V = \frac{1}{4\pi\varepsilon_0}\int \frac{dq}{r}.$$

In a direct calculation of electric potential, we must decide the location of zero potential. In fact, the convention that zero potential is at infinity is already implicit in Eqs. (24–7), (24–11), and (24–12), and is almost always the most convenient choice for a charge distribution that does not extend all the way to infinity. If a potential *difference* is calculated directly, no decision need be made about the zero level.

Examples

Let's first look at the relation between electric field and potential for two parallel conducting plates (a *parallel-plate capacitor*, a circuit device to be revisited in Chapter 25), each brought to different potentials (Fig. 24–15a). We suppose that the plates are close enough together or large enough that we can ignore the distortions of the field near the edges, as shown in Fig. 24–15b. In Fig. 24–15a, the left-hand plate is at a lower potential than the right-hand plate. The electric field between parallel plates is known to be constant, and it runs from regions of higher potential to lower potential—from right to left in this case. In Eq. (24–9), we take the path to be a straight line from the left to the right plate, so that $\vec{E}$ is antiparallel to $d\vec{s}$. We will let the direction from left to right define the x-axis, with the left plate at $x = 0$. Then the potential difference between the plates is given in terms of the field E between the plates and the separation ℓ of the plates by

$$\Delta V = V_{\text{right}} - V_{\text{left}} = -\int_{\text{left}}^{\text{right}} \vec{E} \cdot d\vec{s} = +E\int_0^{\ell} dx = E\ell,$$

and the electric field between parallel plates has magnitude

$$E = \frac{\Delta V}{\ell}. \tag{24–30}$$

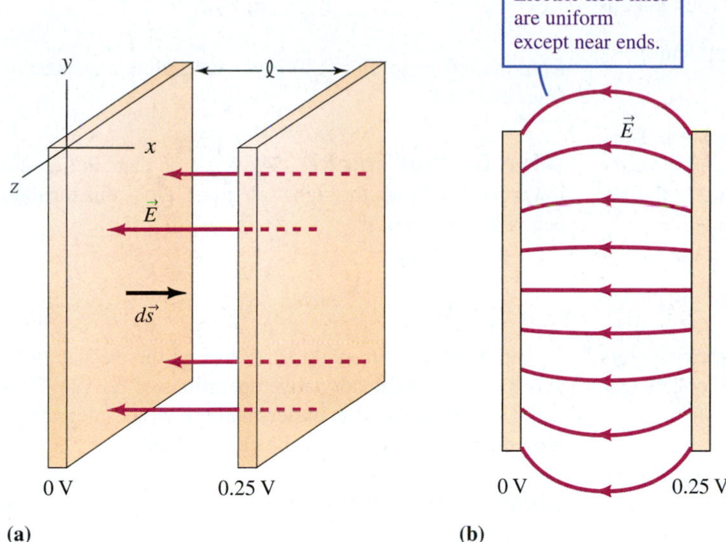

▲ **FIGURE 24–15** (a) Two parallel plates, viewed from the side, with a potential difference between them. A differential displacement $d\vec{s}$ is indicated. (b) The fields do show distortions at the edges for a real (not infinitely large) pair of plates.

Equation (24–30) and the preceding discussion yield an important practical result:

> **The magnitude of the constant electric field between parallel conducting plates is the potential difference between the plates divided by the distance between the plates. It points from the higher to the lower potential plate.**

This result gives the electric field between two parallel-plane conducting plates whose potential difference is ΔV. We can also use it to find the equipotential surfaces associated with *any* constant field. These surfaces are planes perpendicular to the field, and the potential difference for a plane a distance ℓ from a reference equipotential changes *linearly* with ℓ, $\Delta V = E\ell$. Note the sign: The potential decreases along the direction in which the electric field points.

EXAMPLE 24–8 Two parallel metal plates carrying equal and opposite charges have area $A = 225\ \text{cm}^2$ and are separated by 0.50 cm. There is a potential difference of 0.25 V between them. Find the numerical value of the electric field. What is the charge density and total charge on each plate? Draw the equipotential surfaces at 0.10 V and 0.20 V.

Setting It Up Figure 24–15a labels axes and places the plates within them. We call ΔV the given potential difference between the plates and ℓ the known separation.

Strategy Equation (24–30) directly applies here and allows us to calculate the magnitude of the electric field. The field points from the higher to the lower potential plate. Given the electric field, we can calculate the charge density, since $E = \sigma/\varepsilon_0$, and given the area, we can calculate the total charge on each plate.

Working It Out

$$E = \frac{\Delta V}{\ell} = \frac{0.25\ \text{V}}{0.0050\ \text{m}} = 50\ \text{V/m},$$

and it points from right to left in Fig. 24–16. Since the electric field between the plates is σ/ε_0,

$$\sigma = E\varepsilon_0 = (50\ \text{V/m})(8.85 \times 10^{-12}\ \text{C}^2/\text{N}\cdot\text{m}^2)$$
$$= 4.4 \times 10^{-10}\ \text{C/m}^2.$$

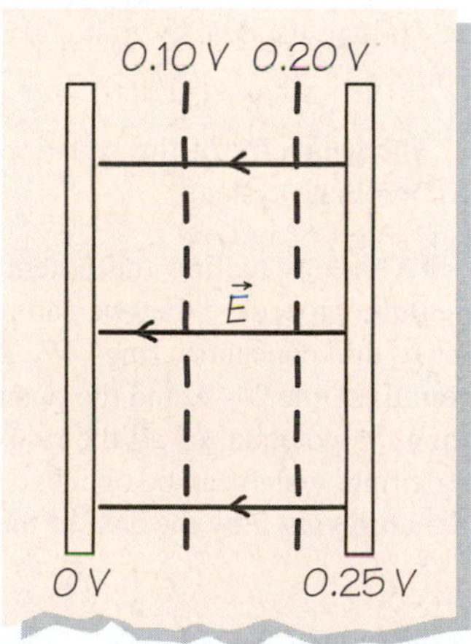

▲ **FIGURE 24–16** The electric field for parallel plates points from right to left when the potential of the left-hand plate is lower than that of the right-hand plate.

(continues on next page)

The answer is in coulombs per square meter—SI units. Given the area of the plates, the magnitude of the total charge on each plate is

$$Q = \sigma A = (4.4 \times 10^{-10}\ \text{C/m}^2)(225\ \text{cm}^2)(10^{-4}\ \text{m}^2/\text{cm}^2)$$
$$= 1.0 \times 10^{-11}\ \text{C}.$$

The electric field is constant between the parallel plates. Thus, from Eq. (24–30), we find that at a distance d from the left plate the potential differs by an amount $\Delta V = Ed$ from its value at the left plate, $d = 0$. The equipotential surface for 0.10 V is then

$$d = \frac{\Delta V}{E} = \frac{0.10\ \text{V}}{50\ \text{V/m}} = 0.20\ \text{cm}$$

from the left plate. For 0.20 V, we determine a distance of 0.40 cm (Fig. 24–15b).

What Do You Think? The left-side plate in Fig. 24–15 is at a lower potential than the right-side plate. Does this mean that it has a negative charge on it?

EXAMPLE 24–9 *The Charged Ring.* Find the electric potential due to a uniformly charged ring of radius R and total charge Q at a point P on the axis of the ring.

Setting It Up The charge distribution is shown in Fig. 24–17, together with point P and a distance labelled r.

Strategy Symmetry plays an important role in the solution, with each point on the ring the same distance $r = \sqrt{R^2 + x^2}$ from the point P. The potential due to an infinitesimal section of the ring that carries charge dq is given by $dV = \dfrac{dq}{4\pi\varepsilon_0}\dfrac{1}{\sqrt{R^2 + x^2}}$. An integration of this yields the answer—the integration is simple because of the fact that each point on the ring is the same distance from P.

Working It Out The integral expressing the potential is

$$V = \int dV = \int \frac{dq}{4\pi\varepsilon_0}\frac{1}{\sqrt{R^2 + x^2}} = \frac{1}{4\pi\varepsilon_0}\frac{1}{\sqrt{R^2 + x^2}}\int dq$$
$$= \frac{Q}{4\pi\varepsilon_0}\frac{1}{\sqrt{R^2 + x^2}}. \qquad (24\text{–}31)$$

In the second to last step, we have removed the constant distance r from the integral, leaving an integral over dq, which gives the total charge Q.

We could use our result for V to find the electric field along the axis by applying the derivative operations of Eq. (24–29) (see Problem 43). This method is easier than the direct integration technique presented in Chapter 22 for the electric field.

What Do You Think? What is the potential at the center of the ring? Interpret your answer.

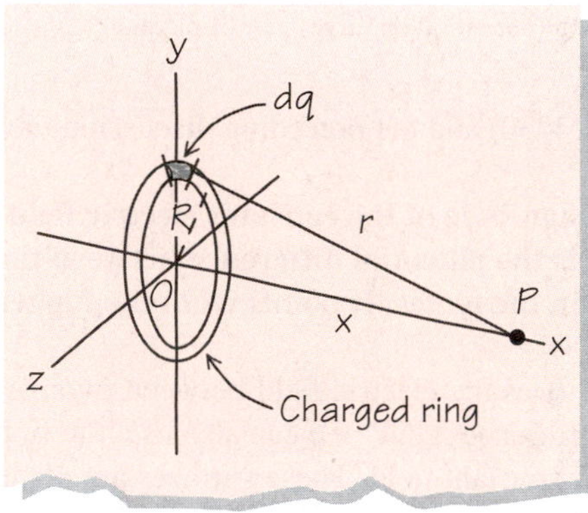

▲ **FIGURE 24–17** Geometry to find the potential at a point P on the axis of a charged ring of radius R by using a differential charge dq.

EXAMPLE 24–10 *The Charged Disk.* Find the electric field due to a thin, flat, uniformly charged disk of radius R and total charge Q at a point P along its axis, by first calculating the electric potential at this point.

Setting It Up The geometry of the situation is shown in Fig. 24–18, including a coordinate system.

Strategy The first task is to find the potential. Figure 24–18 contains in a nutshell the procedure that we can use: We divide the disk up into a series of thin concentric rings. We know the potential for a single ring from Example 24–9, and the potential of the disk is obtained by summing the potential of all the rings. We first require the charge on a single ring, which can be obtained by calculating the charge density and multiplying it by the area of the strip.

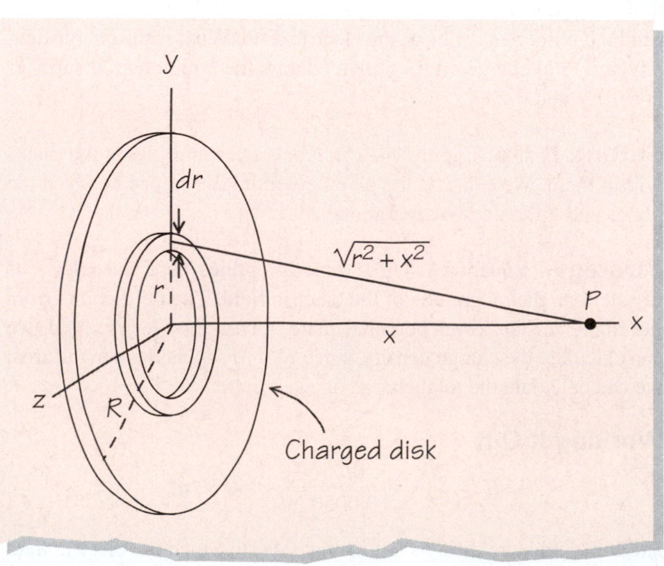

▶ **FIGURE 24–18** Geometry to find the potential at a point P on the axis of a charged disk of radius R. The potential due to the ring of radius r and width dr is first found and then integrated.

For the second part of the problem we can find the field given the potential by taking the appropriate derivative of the potential. Symmetry is helpful here: You can see that along the axis the electric field can only have an x-component. The reason is that a component in a direction perpendicular to the x-axis coming from any point on any of the concentric rings will have an equal and opposite component coming from the point on the ring at the opposite side of that ring.

Working It Out The charge density is $\sigma = Q/\pi R^2$. The area of a ring with inner and outer radius r and $r + dr$ respectively is $dA = \pi(r + dr)^2 - \pi r^2 \cong 2\pi r\, dr$ for infinitesimal dr and the charge on the ring is then $\sigma(2\pi r\, dr)$. From Example 24–9, the potential due to the ring on the axis is

$$dV = \frac{\sigma(2\pi r\, dr)}{4\pi\varepsilon_0}\frac{1}{\sqrt{r^2 + x^2}}.$$

The potential for the whole disk is obtained by summing the potentials from all the rings, i.e. integrating this result from $r = 0$ to $r = R$:

$$\begin{aligned} V &= \int dV = \int_0^R \frac{\sigma}{4\pi\varepsilon_0}\frac{1}{\sqrt{r^2 + x^2}}(2\pi r\, dr) \\ &= \frac{\sigma}{2\varepsilon_0}\int_0^R \frac{r\, dr}{\sqrt{r^2 + x^2}} \\ &= \frac{\sigma}{2\varepsilon_0}\sqrt{r^2 + x^2}\bigg|_0^R = \frac{\sigma}{2\varepsilon_0}\left(\sqrt{R^2 + x^2} - x\right) \\ &= \frac{Q}{2\pi\varepsilon_0 R^2}\left[\sqrt{R^2 + x^2} - x\right]. \end{aligned} \qquad (24\text{–}32)$$

In turn, the only component of the electric field is

$$E_x = -\frac{\partial V}{\partial x} = -\frac{Q}{2\pi\varepsilon_0 R^2}\left(\frac{x}{\sqrt{R^2 + x^2}} - 1\right).$$

What Do You Think? The result for the x-component of the field for $x \gg R$ is (a) $\sigma x/(2\varepsilon_0)$; (b) $\sigma/(2\varepsilon_0 x)$; (c) $Q/(4\pi\varepsilon_0 x)$; (d) $(Q/4\pi\varepsilon_0 x^2)$?

EXAMPLE 24–11 *The Charged Line.* Find the electric potential as a function of the radial distance R from an infinite charged line of uniform charge density λ.

Strategy We found the field for this charge configuration earlier [Eq. (22–32)], and we can here use Eq. (24–9) to find the potential from the electric field. Since the electric field has only a radial component, we integrate along a radial direction in Eq. (24–9), that is, $d\vec{s} = d\vec{r}$ in Fig. 24–19a.

Working It Out Equation (24–9) becomes

$$\Delta V = -\int E_r\, dr = -\frac{\lambda}{2\pi\varepsilon_0}\int\frac{dr}{r}.$$

The potential difference depends on the end points of the integration. Let zero potential be at $r = a$, so that

$$\Delta V = V_R - V_a \equiv V = -\frac{\lambda}{2\pi\varepsilon_0}\int_a^R\frac{dr}{r} = -\frac{\lambda}{2\pi\varepsilon_0}\ln r\bigg|_a^R,$$

or

$$V = -\frac{\lambda}{2\pi\varepsilon_0}\ln\frac{R}{a}. \qquad (24\text{–}33)$$

Note that it is not possible to set zero potential at infinity in this case because the logarithm is infinite at $a = \infty$. Physically, this is because the line itself reaches to infinity. We graph the potential of Eq. (24–33) in Fig. 24–19b, assuming that the charge of the line is positive.

What Do You Think? It is not possible to do this calculation directly by putting the charge along the z-axis, then calculating the potential by integrating the potential $\lambda\, dz/\left(4\pi\varepsilon_0\sqrt{R^2 + z^2}\right)$ due to a charge $\lambda\, dz$ a distance $\sqrt{R^2 + z^2}$ away from the point P. Why?

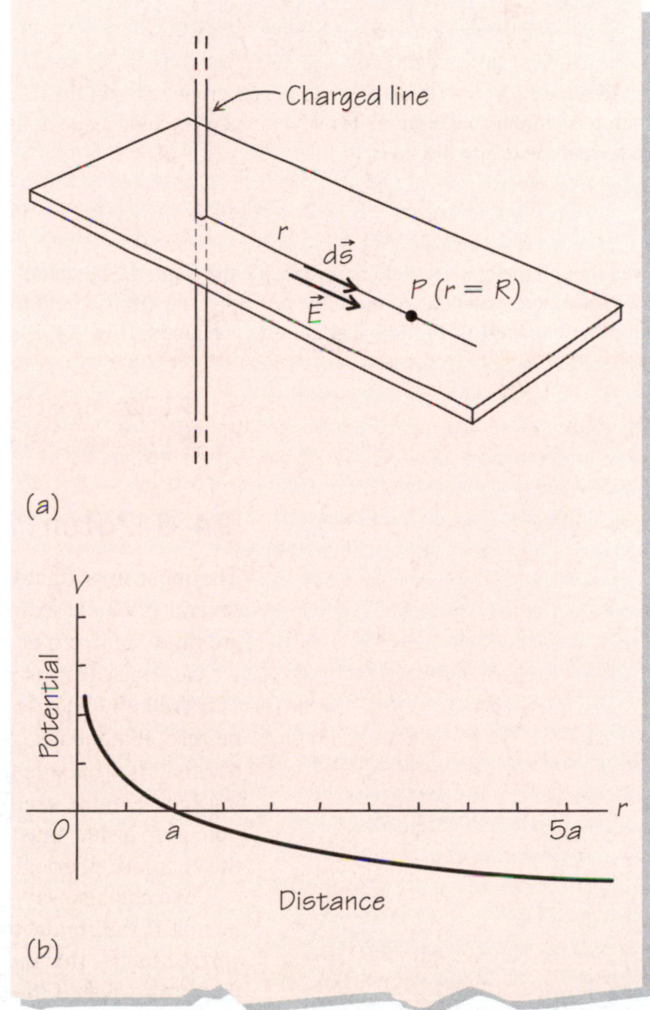

▶ **FIGURE 24–19** (a) An infinite charged line has a radial electric field. To find the potential at the point P, we consider a displacement $d\vec{s}$ in the direction of the electric field $\vec{E}$ and use the known expression for the electric field. (b) The resulting potential, defined to be zero at $r = a$, goes to positive infinity at $r = 0$ and continues to negative infinity for large r.

EXAMPLE 24–12 *The Charged Spherical Shell.* Find the potential for a uniformly charged spherical shell of total charge Q and radius R at positions both inside and outside the shell. Set zero potential at infinity.

Strategy As in the previous example, we can best work backwards from the electric field to the potential. We already know the electric field of the spherical shell from Example 23–4 and can use Eq. (24–9) to determine the electric potential, or more precisely a potential difference. But as zero potential is set at infinity, the potential $V(r)$ is the potential difference $\Delta V = V(r) - V(\infty)$. From Example 23–6, we have

$$\text{outside a spherical shell, } r > R: \qquad E = \frac{Q}{4\pi\varepsilon_0 r^2};$$

$$\text{inside a spherical shell, } r < R: \qquad E = 0.$$

We must integrate these to find the potential.

Working It Out The electric field is purely radial, so by choosing a radial path from ∞ to a radial position r, we have in Eq. (24–9) $\vec{E}\cdot d\vec{s} = E\,dr$. Then for a point *outside* the spherical shell,

$$V(r) = V(r) - V(\infty) = -\int_{\infty}^{r} E\,dr = -\frac{Q}{4\pi\varepsilon_0}\int_{\infty}^{r}\frac{dr}{r^2}$$

$$= \frac{Q}{4\pi\varepsilon_0}\left(\frac{1}{r} - \frac{1}{\infty}\right) = \frac{Q}{4\pi\varepsilon_0 r} \qquad (24\text{–}34)$$

If r is *inside* the shell, we have

$$V(r) = -\int_{\infty}^{R} E_{\text{outside}}\,dr - \int_{R}^{r} E_{\text{inside}}\,dr.$$

Because $E_{\text{inside}} = 0$, the second integral drops out, and the integration is similar to the previous one with zero potential again at infinity, namely inside the shell,

$$V(r) = \frac{Q}{4\pi\varepsilon_0 R} = \text{a constant.} \qquad (24\text{–}35)$$

Even though the electric field is zero inside the shell, the potential is not (when zero potential is at infinity). We plot the electric field for the spherical shell in Fig. 24–20a and the potential in Fig. 24–20b.

What Do You Think? How is the fact that the field is zero inside the shell consistent with a potential that is not zero?

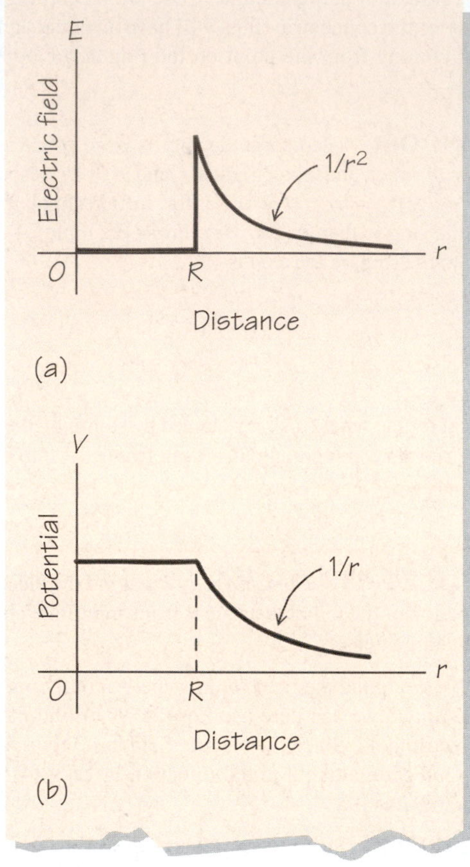

▲ **FIGURE 24–20** (a) The electric field and (b) electric potential for a spherical shell of radius R. Even though the electric field is zero inside the shell, the potential has a constant value equal to that on the shell's surface.

We have now calculated the electric field and potential for several different distributions of charge. We summarize these results in Table 24–1.

24–6 Potentials and Fields Near Conductors

The most important cases of continuous charge distributions are those on metals; for several practical examples you can glance ahead to Section 24–7. These distributions are rarely uniform because charges are free to move on and within metals. Despite this, we can learn a surprising amount about the electric potentials near metals.

We know that, in electrostatics, the electric field inside a conducting material must be zero, that the net charge on a conductor lies on the outside surface, and that the electric field just outside that surface must be normal to the surface. Moreover, the conducting surface must itself be an equipotential and, as there is a zero electric field inside, the potential inside must have the same value as it has at the surface. Example 24–12, the charged spherical shell, provides an illustration of these features.

We can also say something about properties of the electric potential outside a conductor. If the conductor is charged, the fact that the electric field is perpendicular to the surface means that the equipotentials near the surface will be parallel to the surface, and this is true even if the conductor is uncharged. Consider, for example, the constant electric field shown in Fig. 24–21a due to two parallel plates (not shown). If we place an uncharged conductor of arbitrary size in this electric field, the field around the conductor

TABLE 24–1 • Electric Fields and Potentials for Various Charge Configurations

Charge Configuration	Magnitude of Electric Field	Electric Potential	Location of Zero Potential
Point charge	$\dfrac{q}{4\pi\varepsilon_0 r^2}$	$\dfrac{q}{4\pi\varepsilon_0 r}$	∞
Infinite line of uniform charge density λ	$\dfrac{\lambda}{2\pi\varepsilon_0 r}$	$-\dfrac{\lambda}{2\pi\varepsilon_0}\ln\dfrac{r}{a}$	$r = a$
Parallel, oppositely charged plates of uniform charge density σ, separation d	$\dfrac{\sigma}{\varepsilon_0}$	$\Delta V = -Ed = -\dfrac{\sigma d}{\varepsilon_0}$	anywhere
Charged disk of radius R, along axis at distance x	$\dfrac{Q}{2\pi\varepsilon_0 R^2}\left(\dfrac{\sqrt{R^2+x^2}-x}{\sqrt{R^2+x^2}}\right)$	$\dfrac{Q}{2\pi\varepsilon_0 R^2}\left(\sqrt{R^2+x^2}-x\right)$	∞
Charged spherical shell of radius R	$r \geq R$: $\dfrac{Q}{4\pi\varepsilon_0 r^2}$	$r \geq R$: $\dfrac{Q}{4\pi\varepsilon_0 r}$	∞
	$r < R$: 0	$r \leq R$: $\dfrac{Q}{4\pi\varepsilon_0 R}$	∞
Electric dipole	Along bisecting axis only, far away: $\dfrac{p}{4\pi\varepsilon_0 r^3}$	Everywhere, far away: $\dfrac{p\cos\theta}{4\pi\varepsilon_0 r^2}$	∞
Charged ring of radius R, along axis	$\dfrac{Qx}{4\pi\varepsilon_0(R^2+x^2)^{3/2}}$	$\dfrac{Q}{4\pi\varepsilon_0\sqrt{R^2+x^2}}$	∞
Uniformly charged nonconducting solid sphere of radius R	$r \geq R$: $\dfrac{Q}{4\pi\varepsilon_0 r^2}$	$r \geq R$: $\dfrac{Q}{4\pi\varepsilon_0 r}$	∞
	$r < R$: $\dfrac{Qr}{4\pi\varepsilon_0 R^3}$	$r < R$: $\dfrac{Q}{8\pi\varepsilon_0}\left(3 - \dfrac{r^2}{R^2}\right)$	∞

will be greatly modified (Fig. 24–21b). Charge will be induced on the outside of the conductor in equilibrium, thereby forcing the electric field to be normal to the conducting surface and, again, *the equipotential surfaces near a conductor of arbitrary shape are parallel to the conductor's surface.*

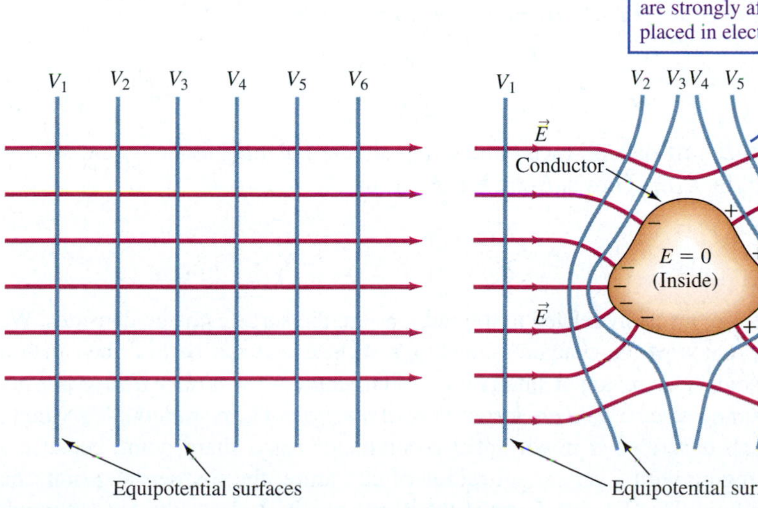

FIGURE 24–21 (a) A uniform electric field before an uncharged conductor is placed in the field. (b) Afterward, the electric field is changed dramatically, with no electric field inside the conductor. Induced charges, which make the electric field inside the conductor zero, appear on the outside surface of the conductor. These charges affect the electric field outside the conductor.

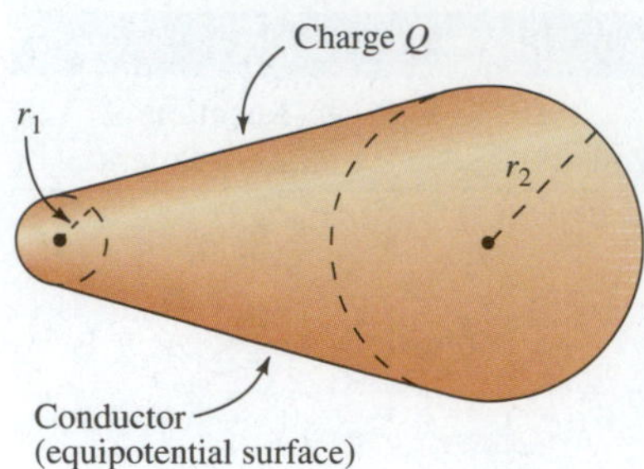

▲ **FIGURE 24–22** A conductor of irregular shape can be modeled by spheres of radii r_1 and r_2 at its ends.

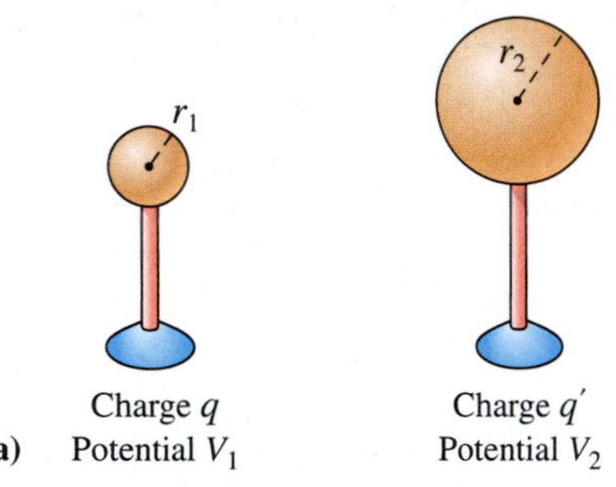

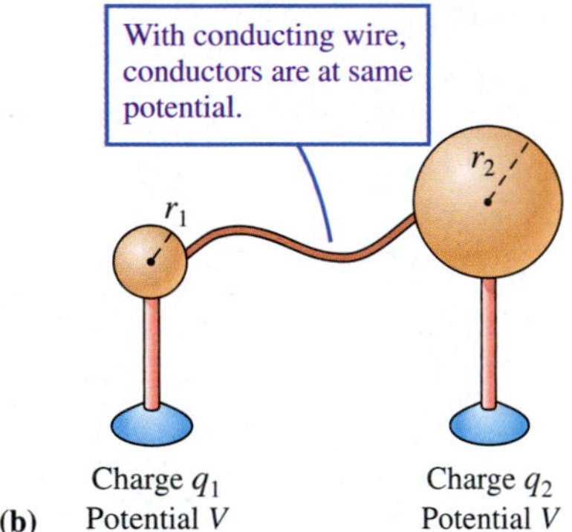

▲ **FIGURE 24–23** (a) Two conductors initially are at different potentials. (b) If the conductors are connected by a wire, charge must flow to make the potential equal everywhere throughout the connected conductors.

There is a last important result to consider: Charge density on a conductor will vary if the surface has an irregular shape. In the next subsection we shall see how the concept of potential allows us to say more about the charge density and hence the fields near irregularly shaped charged conductors.

The Role of Sharp Points on Conducting Surfaces

Many real conductors have irregular shapes and the shape of a conductor affects the electric field in its vicinity. Consider the irregular conductor shown in Fig. 24–22. To simplify its analysis, we can think of the two ends shown as approximated by sections of spheres with radii r_1 and r_2, respectively. The left region is more sharply curved than the right region, so $r_1 < r_2$. We model this conductor with a two-step process; for the first step, we consider the field and potential for the two spherical conductors shown in Fig. 24–23a. The sizes of these spheres match the two ends of our irregular conductor (Fig. 24–22). The charges q and q' are placed on the two spheres, and the electric potentials at the spheres are, respectively,

$$V_1 = \frac{q}{4\pi\varepsilon_0 r_1} \quad \text{and} \quad V_2 = \frac{q'}{4\pi\varepsilon_0 r_2}.$$

If we now connect the two spheres by a long conducting wire (Fig. 24–23b), charge will flow between the two spheres, and the entire system will come to the same potential. This potential is

$$V = \frac{q_1}{4\pi\varepsilon_0 r_1} = \frac{q_2}{4\pi\varepsilon_0 r_2},$$

where q_1 and q_2 are the equilibrium charges on the two spheres. Since the entire connected object is at a single potential, these charges and radii must be related by

$$\frac{q_1}{r_1} = \frac{q_2}{r_2}.$$

In this discussion we did not use the fact that the spheres are connected by a thin wire, only that they are connected; this is the information that ensures that the entire system is at a single potential. Therefore this calculation will apply to our irregular conductor (Fig. 24–22). In particular, the connected spheres form a model for the relative size of the electric fields at the two ends of the conductor.

Let us now relate what we have done to the surface charge density σ. For a sphere of radius r this quantity is the charge q on the sphere divided by the surface area: $\sigma = q/(\pi r^2)$, or $q = \sigma\pi r^2$. For our two connected spheres, therefore, the equation $q_1/r_1 = q_2/r_2$ is

$$\frac{\sigma_1 \pi r_1^2}{r_1} = \frac{\sigma_2 \pi r_2^2}{r_2};$$

$$\sigma_1 r_1 = \sigma_2 r_2. \tag{24–36}$$

The electric field E_i just outside each conducting sphere has magnitude σ_i/ε_0, so that we can replace σ_i by $\varepsilon_0 E_i$. Thus $\varepsilon_0 E_1 r_1 = \varepsilon_0 E_2 r_2$, or

$$\frac{E_1}{E_2} = \frac{r_2}{r_1}. \tag{24–37}$$

The electric fields are inversely related to the radii, as are the surface charge densities. We can say that at a region of the charged conductor's surface that can be inscribed with a sphere of a small radius—a region of small radius of curvature—*the surface charge density and the corresponding electric field are larger than at a region of large radius of curvature.*

The effect just described is important if a conductor has a sharp point, because a sharp point is a region with a very small radius of curvature; the sharper the point, the more intense the field (Fig. 24–24). Even if the conductor is at a low electric potential, there will be intense fields near sharp points.

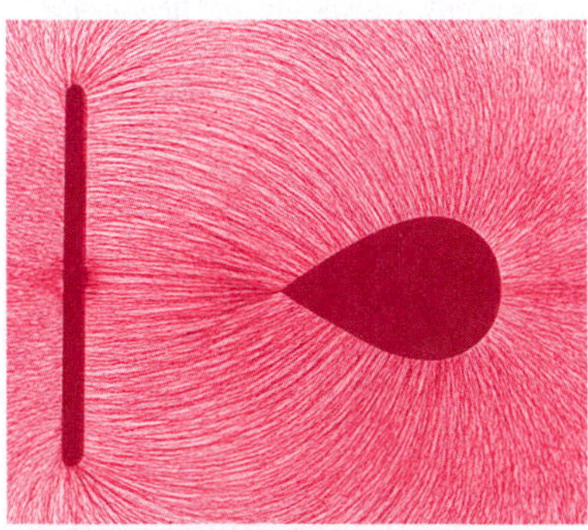

▲ **FIGURE 24–24** The electric field near small radii of curvature can be quite large, as seen for the point of this charged object.

THINK ABOUT THIS. . .

HOW DO LIGHTNING RODS WORK?

Observations show that there is an electric potential over much of the land and seas of Earth that increases with height at a rate of about 100 V/m; that is, there is an electric field that points downward and has a nearly constant strength of about 100 V/m. (To be more exact, the magnitude of the field decreases slowly till it is almost zero at about 50–100 km above the surface of Earth.) Depending on weather conditions, the electric field can be very much larger—a lightning strike is preceded by the presence of large fields. Air is a fairly good insulator; what, then, conducts electricity in air? Molecules of nitrogen and oxygen are occasionally *ionized*—electrons are removed from them—by collisions with incoming *cosmic rays*, energetic particles that come from beyond the solar system. The resulting ratio of ions (positive and negative) to molecules that are not ionized is about 1 in 10^{16}, and it is these charged particles that conduct electricity in the atmosphere.

We know that there are large electric fields near the pointed parts of a charged conductor, and ions of the opposite charge are attracted to this part of the conductor while ions of the same charge are repelled. The positive and negative ions may be strongly accelerated, and ionize other neutral molecules in their path. In this way the conductivity of the air near the pointed conductor increases measurably, and a flow of charge acts to cancel the field. On a small scale this gives rise to sparks (see Fig. 24–25); a more massive charge transfer produces a *corona discharge*. Sailors long ago saw the glow of corona discharge at the pointed tops of their masts and spars and dubbed the phenomenon *St. Elmo's fire* (see Fig. 24–26). More generally we speak of *dielectric breakdown*.

In electric storms clouds acquire a large charge by mechanisms that are not entirely understood,† and this gives rise to high fields extending over large regions. An ionization cascade can develop from small seeds of ionized material in the atmosphere and result in massive discharges from clouds to Earth or from cloud to cloud. *Lightning* (see the chapter-opening photo) is an example of such a large-scale discharge. When lightning strikes, the passage of electric charge heats the air along its path. This causes a rapid expansion of the air, and surrounding air rushes back into the partial vacuum, creating the sound that we term *thunder*.

If there is a pointed, grounded conductor (lightning rod) beneath the highly charged cloud, then the strong field at the point acts to lead the charged particles to the rod. More important than providing a convenient path for a lightning stroke, the steady discharge from the rod will serve to decrease the potential difference between the rod and the clouds and discourage the formation of a catastrophic discharge in the vicinity.

†A detailed discussion of electricity in the atmosphere may be found in Chapter 9, Vol. II of *Lectures on Physics*, by R. P. Feynman.

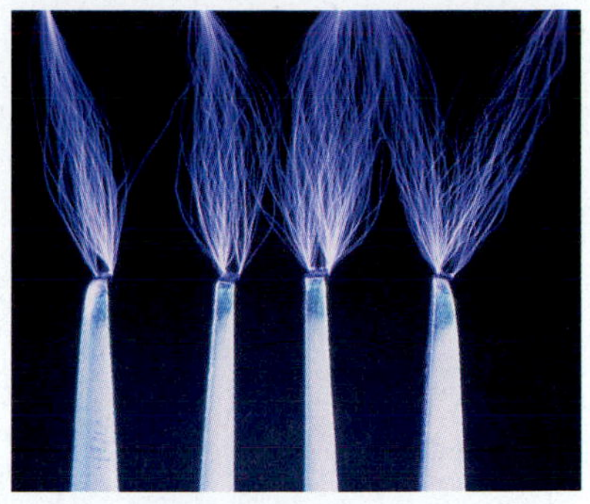

▲ **FIGURE 24–25** The electric field at the tips of these fork tines is large enough to cause a discharge from them into the air in the form of many small sparks.

▲ **FIGURE 24–26** St. Elmo's fire at the end of the masts.

24–7 Electric Potentials in Technology

Electrostatics represents only a small portion of our study of electromagnetism, but it has important applications. We briefly mention a few of them here to indicate how the understanding of basic principles can be put to good practical use.

The Van de Graaff Accelerator

If we place a charge anywhere in a conductor, the charge will move to the outside surface, and the field inside the conductor will be zero. Robert Van de Graaff took advantage of this concept in 1931 to build an *accelerator*: an apparatus that produces highly energetic charged particles. Such particles are useful for microscopic probes of matter and as cancer treatments. Van de Graaff used a device similar in concept to the apparatus shown schematically in Fig. 24–27. An insulated belt (or chain) continuously brings charge to the inside of a hollow conductor, which then moves to the outside surface of the conductor. The electric potential on the spherical conducting surface increases as charge flows to its surface ($V = q/4\pi\varepsilon_0 R$). An *ion source* produces charged atoms whose sign is such as to be repelled from the region of high potential and thus accelerated. Such devices are called **Van de Graaff accelerators** or **Van de Graaff generators** (Fig. 24–27), and the beams they or other accelerators produce play an important role in modern technology—for example, such beams are used to make microcircuits.

▲ **FIGURE 24–27** (a) Schematic diagram of a simple Van de Graaff accelerator. Charge is sprayed on the rotating belt at the bottom and taken off at the top. The charge goes to the outside surface of the conductor, and the potential continues to build up to high values. The symbol in the bottom right indicates that the base of the accelerator has been grounded. (b) The children touching this Van de Graaff generator are brought to a high electric potential. The individual hairs behave like the leaves of an electroscope.

The Field-Ion Microscope

The phenomenon that large electric fields occur at sharp points on a conductor is carried to its extreme in **field-ion microscopy**. The high electric fields involved in this technique allow us to produce images of individual atoms in the crystalline structure of the sharp point, or tip, of a metal. A fine tip of the crystalline material is prepared, commonly by dipping a mechanically formed tip in a substance that dissolves atoms off the end of the tip. These tips can be as small as 200 nm across, depending on the particular metal and the crystal being prepared. On the 200-nm scale, such a tip looks smooth, but it is still very rough at the atomic level. The tip is then introduced into a vacuum, and a large positive potential of several kilovolts is applied (Fig. 24–28). The end of the tip is thereby smoothed off even further as protruding groups of atoms of

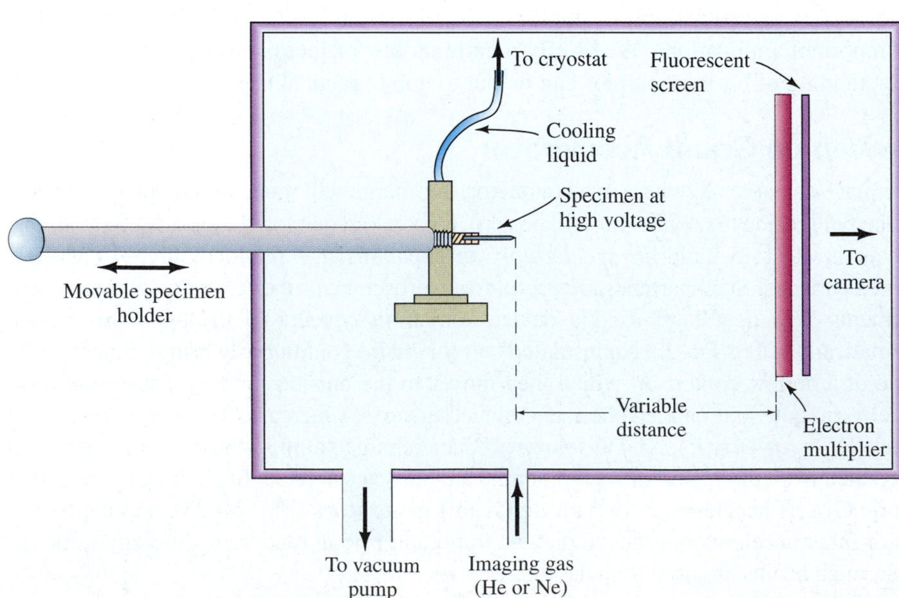

▶ **FIGURE 24–28** Schematic diagram of a field-ion microscope. The cryostat maintains a steady, low temperature in the chamber.

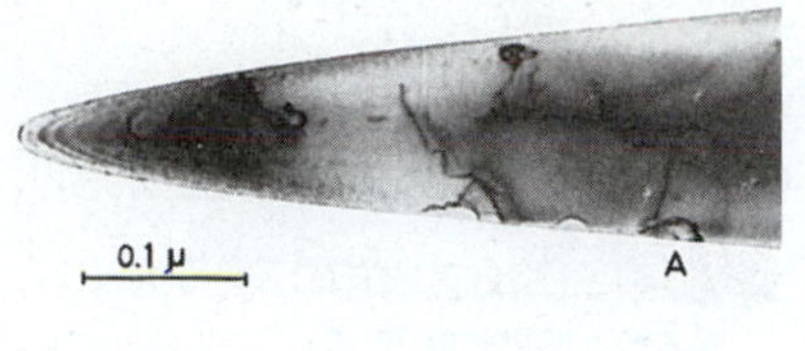

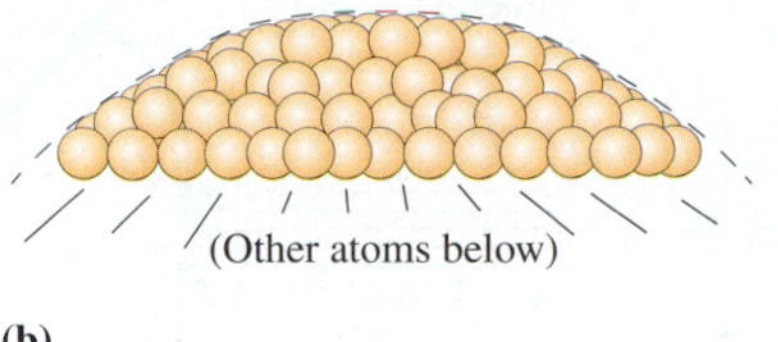

FIGURE 24–29 (a) Magnified iron tip. (b) The spheres represent individual atoms in a field-ion tip.

the metal itself are driven off in the form of positive ions by the large fields. This smoothing process leaves a tip like those in Figs. 24–29a and 24–29b. Oranges stacked in layers into a semipyramidal shape provide a familiar example of the possible structure of such a tip.

In the next stage, a dilute gas such as helium or neon—called the *imaging* gas—is introduced into the chamber that contains the tip. The positive potential on the tip is increased, again to several kilovolts, until the gas atoms just begin to be ionized. This happens only where the field is largest: right above individual atoms of the tip, where the field is strong enough to ionize the gas. The gas ions are then driven away from the tip, following the electric field lines out from the tip atoms to a grounded screen, where the impinging gas ions leave a visible trace. The image formed corresponds to the position of the individual atoms of the tip, which thus become visible in a picture such as Fig. 24–30. Field-ion microscopy is useful for observing crystal structures and the effects of impurities and defects in crystals. Even some noncrystalline materials can be investigated this way. Learning about the structure of materials is certainly one of the important steps in our ability to use them in applications.

FIGURE 24–30 This field-ion micrograph of a platinum crystal allows the atomic structure of the crystal to be determined from the many overlapping geometrical patterns. The magnification is 200,000.

Xerography

Photocopying machines take advantage of electrostatics in several steps of **xerography**, or photoreproduction. The process is illustrated in Figure 24–31. It begins with a positively charged plate coated with photoconducting material (a photoconductor is a good conductor in light but not in the dark, because light releases conducting electrons within it) such as selenium (Fig. 24–31a). Light reflected from the original to be copied passes through a lens onto the charged plate, where the dark areas remain charged, but charges flow to the plate underneath at the areas that receive light (Fig. 24–31b). The resulting image of the dark areas is represented by the remaining charges. Negatively charged toner (a black powder) is added to the positively charged plate, leaving the original dark areas with black toner on the plate (Fig. 24–31c). In the next step, paper that has also been positively charged is placed over the plate and attracts the black, negatively charged toner (Fig. 24–31d). Heat is used to fuse the toner (and thus the image) to the paper.

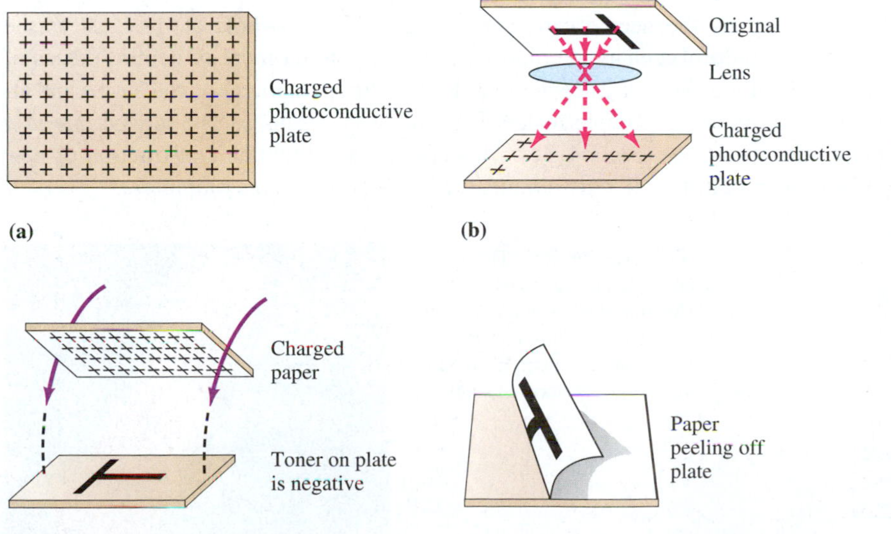

FIGURE 24–31 Schematic diagram of xerography. (a) A positively charged photoconductive plate. (b) Light from the white areas on the original neutralizes the positive charges on the plate. (c) The negatively charged toner is attracted to the positive charge. (d) The positively charged paper picks up the toner. Heat seals the toner on the paper, (e) which is then peeled off the plate.

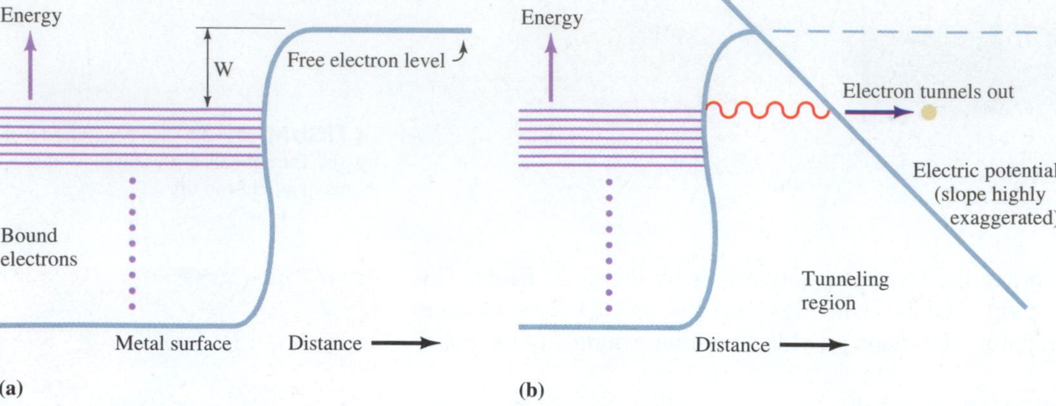

▲ **FIGURE 24–32** (a) Electrons in a metal take on a range of energies up to a highest level. This level is still negative compared to the energy of a free electron. It takes an energy *W*—typically 4 to 5 eV—to sufficiently raise the electron energy to allow the electron to leave the metal. The potential energy curve is drawn in blue. (b) When an external potential is applied to the metal, that potential plus the existing potential yields an effective potential barrier. Electrons can tunnel through this barrier.

Electric Potentials and Quantum Engineering

Electrons are trapped within a metal by electrostatic attractions to their parent ions. These forces can be represented in classical physics by a potential barrier that the electrons cannot cross. Figure 24–32a is an energy diagram of the potential energy of an electron as well as the electron's (constant) total energy. Classically, the electron cannot enter the region where its total energy is less than its potential energy.

A positively charged object brought near the metal surface pulls on the electrons. The positively charged object has an electric potential with respect to the metal, and an electron has a potential energy due to the external object (Fig. 24–32b). When the potential energy due to the external object is added to the original potential energy that holds the electron within the metal, the barrier is, in effect, reduced. Even if the external potential is too weak to lower the maximum potential energy below the electron's total energy and allow the electrons to escape classically, penetration through a barrier (*barrier tunneling*) is possible in quantum physics (Chapter 40). The tunneling is enhanced when the barrier is lowered. By tunneling, the electrons can emerge on the other side of the barrier and escape the electrostatic attraction of the parent ion, though many of them are absorbed on the way. This effect is utilized in the *scanning tunneling microscope* (Fig. 24–33). A weak positive potential is placed on an ultrafine tungsten needle. The needle scans the surface of a sample and provides the necessary potential to help electrons escape the sample by making it easier for them to penetrate the barrier. These electrons are attracted to the needle and form a flow of charge (*electric current*) through it, whose magnitude depends on the distance between the needle and the surface. This effect is used in two ways.

1. A feedback mechanism that constantly repositions the needle can be set up so that the current is constant. The distance between the needle tip and the sample's surface is therefore constant. The repositioning can be measured, and the topography of the surface is thereby mapped (Fig. 24–34).
2. The potential on the needle can exert a slight pull on whole atoms. Just as the nonuniform field of a charged comb induces a dipole moment of neutral pieces of paper and attracts them, the needle tip induces a dipole moment on the atoms and attracts them out of the sample material. In this way, atoms can be moved *one at a time* to new positions (Fig. 24–35). This effect may allow the construction of new molecules and ultrasmall logic circuits (switching circuits in computers).

▼ **FIGURE 24–33** Schematic diagram of a scanning tunneling microscope. The fine-tipped needle in the scanning head comes to within 1 nm of the sample; this distance is the tunnel gap. The tunneling current across this gap holds the gap distance constant as the tip scans the sample surface and thereby provides a map of that surface. The base voltage leads to the tunneling current, which can be used to form the driving voltage. The driving voltage moves the tip by means of piezoelectricity, a phenomenon in which a voltage applied to crystals distorts them in a predictable way.

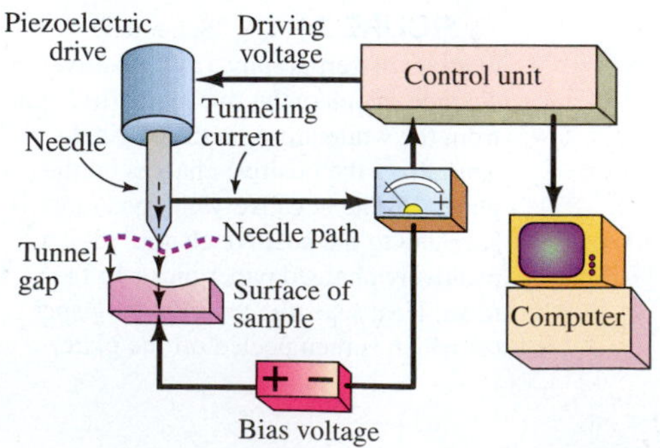

▶ **FIGURE 24–34** This is a scanning tunneling microscope (STM) image of a chain of cesium atoms (reddish yellow) on a gallium arsenide surface (blue). Each bump in this image represents a single atom.

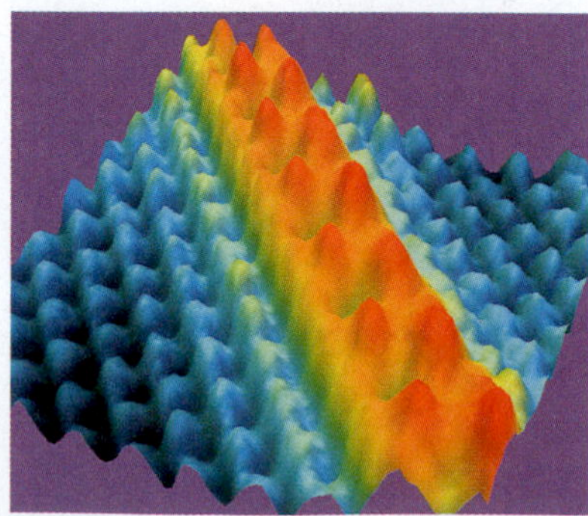

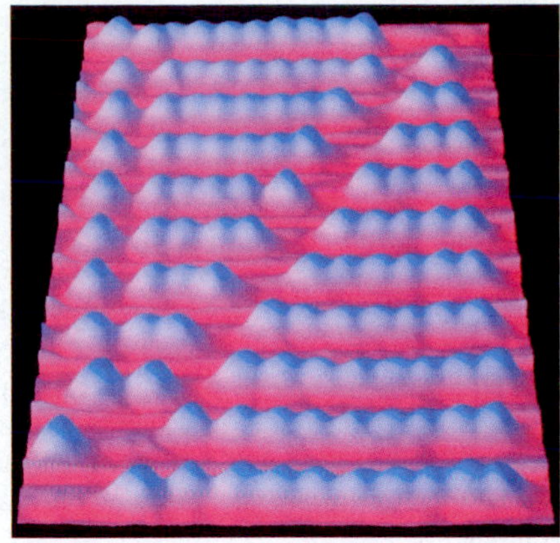

FIGURE 24–35 Individual atoms, whose size is less than a nanometer, have been moved one at a time to form the pattern shown. The vertical scale of the figure is enhanced by a factor of ten for a more dramatic effect.

The combination of electrostatics and quantum mechanics (through the use of the wavelike behavior of electrons) is an important tool in what is aptly called *quantum engineering*.

Summary

The Coulomb force is conservative, so a potential energy—electric potential energy—is associated with it. If a test charge q_0 moves from point a to point b in the presence of a point charge q at the origin, the change in potential energy is given by

$$\Delta U = \frac{qq_0}{4\pi\varepsilon_0}\left(\frac{1}{r_b} - \frac{1}{r_a}\right). \tag{24–4}$$

The electric potential difference due to any charge distribution between points a and b is defined as the change in potential energy divided by the magnitude of a test charge q_0:

$$\Delta V = \frac{U_b - U_a}{q_0} = -\int_{r_a}^{r_b} \vec{E}\cdot d\vec{s}. \tag{24–9}$$

Here $\vec{E}$ is the electric field due to the charge distribution. The integral in Eq. (24–9) is independent of the path between the end points. The potential difference $V_b - V_a$ is the work done per unit charge by an external agent in moving a test charge from point a to point b with no change in kinetic energy. The potential is independent of the test charge.

The electric potential can be determined by the following methods, in addition to graphical methods:

1. If the electric field is known, then Eq. (24–9) may be used.
2. If the electric field is not known, it is generally easier to calculate the potential directly by using one of these forms:

for one point charge:
$$V(r) = \frac{q}{4\pi\varepsilon_0 r}; \tag{24–7}$$

for many point charges:
$$V_P = \frac{1}{4\pi\varepsilon_0}\sum_{i=1}^{n}\frac{q_i}{r_i}; \tag{24–11}$$

for a continuous charge distribution:
$$V = \frac{1}{4\pi\varepsilon_0}\int\frac{dq}{r}. \tag{24–12}$$

In each of these cases, zero potential is chosen to be at infinity.

The SI unit of electric potential is the volt (V); 1 V = 1 J/C. A useful unit of energy for atomic and subatomic systems is the electron-volt (eV); 1 eV = 1.6×10^{-19} J.

If the potential is known, then the electric field can be determined in terms of derivatives of the potential:

$$\vec{E} = -\hat{i}\frac{\partial V}{\partial x} - \hat{j}\frac{\partial V}{\partial y} - \hat{k}\frac{\partial V}{\partial z}. \tag{24–29}$$

The electric field between two parallel plates is constant and is given by the potential difference divided by the distance between the plates:

$$E = \frac{\Delta V}{\ell}. \tag{24–30}$$

The electric field and potential for several charge configurations are given in Table 24–1.

Equipotential surfaces are surfaces at a fixed potential. The electric field is perpendicular to equipotentials. The surfaces of conductors form equipotentials, and the potentials inside conductors in equilibrium are everywhere the same as the potential on the surface. Electric fields just outside conductors are inversely proportional to the radius of curvature, so there are high electric fields near sharp points on conductors even if the conductors are at low potentials.

Understanding the Concepts

1. How many joules are in 1 V · C?
2. An infinite plane is uniformly charged with positive charge of density σ. How would you use a known negative test charge to measure σ?
3. How would you create an electric field inside the hollow space of a spherical metal shell that is constant in magnitude and direction in a small region of the interior space?
4. In good weather, the electric field in the lower atmosphere is approximately 100 V/m, pointing downward. What happens when a 3-m metal rod is planted in the ground?
5. When an electric field moves a charge by doing work on it, what is the source of the energy to do the work? Where did this energy come from originally?
6. In describing the potential difference as the work per unit charge to move a test charge, we added the phrase "without changing its kinetic energy" (see the boldface statement on p. 687). Why is this important?
7. For a dipole $\theta = 0$ is a point on the axis, while $\theta = \pi/2$ is a point on the perpendicular bisector of the line between the two charges. Without looking at the equations, and given that one is at the same distance from the dipole center, at which angle is the field larger, and why?
8. Are the children in Fig. 24–27 grounded?
9. Will a conductor always be an equipotential? If not, under what circumstances will that occur?
10. Using Eq. (24–29), explain why changing the location of zero potential does not affect the value of the electric field.
11. A small Van de Graaff generator can be used as a lecture demonstration device. If a person touches the dome, his or her hair stands up (see Fig. 24–27b). Explain why. Why should the person stand on an insulated mat during this demonstration?
12. Earth is typically defined to be at zero potential with respect to infinity. Does this mean that Earth can have no net charge? If Earth does have a net charge, can it still be at zero potential?
13. If we know the electric potential at a certain point, do we also know the electric field? What can we know about the electric field if we know the electric potential at two points arbitrarily close to one another?
14. Examine Fig. 24–24. In what regions do the field lines seem to be denser? In what regions do you expect dense lines, and why?
15. Is the electric potential energy of a system of point charges independent of the order in which the system is assembled?
16. Why are there so many curved surfaces on the Van de Graaff generator?
17. How do we really know that electric forces are conservative?
18. In the potential associated with a point charge, we chose zero potential to be infinitely far from the charge. What would change in our predictions about electric charges if we had chosen the potential to be zero at $r = 10^{-10}$ m from the charge?
19. If we start with point charges, for each of which zero potential is at infinity, is it possible for a superposition of charges to have zero potential other than at infinity?
20. The potential of a configuration of point charges is zero at certain points. Does this mean that the force on a test charge is zero at these points?
21. Is it possible to arrange charges so that the potential is zero over a small but finite region?
22. Consider Fig. 24–10. What is the easiest way off High Peak? Think in particular about whether or not your path is perpendicular or not to the contour lines, and how the spacing between the contours affect your choice. Describe the relevance of your reasoning to a possible set of equipotentials for a two-dimensional distribution of charge.

Problems

24–1 Electric Potential Energy

1. (I) Two protons are separated by a nuclear diameter of 5×10^{-15} m. What is their mutual electrostatic energy?
2. (I) What is the electrical potential energy between the nucleus of a uranium atom (92 protons) and a single electron located 3×10^{-12} m from the nucleus?
3. (I) A charge of 7.0×10^{-7} C is fixed at the origin of a coordinate system. A charge of 3.0×10^{-6} C is placed on a raisin of mass 0.30 g. The raisin is then brought from far away to a point 20 cm from the origin. What is the electric potential energy of the system?
4. (I) Suppose that the raisin of Problem 3 is released from rest from its position 20 cm from the origin. If no other forces act on the raisin, where will it move? What will its final kinetic energy be?
5. (I) A 3-μC charge is brought in from infinity and fixed at the origin of a coordinate system. (a) How much work is done? (b) A second charge, of 5 μC, is brought in from infinity and placed 10 cm away from the first charge. How much work does the electric field of the first charge do when the second charge is brought in? (c) How much work does the external agent do to bring the second charge in if that charge moves with unchanging kinetic energy?
6. (I) Charges $q_1 = 2.0 \times 10^{-5}$ C and $q_2 = -9.0 \times 10^{-5}$ C are placed at rest 0.50 mm apart. How much work must be done by an outside agent to move these charges slowly and steadily until they are 0.05 mm apart?
7. (II) A positive charge of magnitude 3.0×10^{-6} C is placed 5.0 cm above the origin of a coordinate system, and a negative charge of the same magnitude is placed 5.0 cm below the origin, both on the z-axis. What is the potential energy of a positive charge of magnitude 0.20×10^{-6} C placed at the position $(x, y, z) = (30\text{ cm}, 0\text{ cm}, 50\text{ cm})$? at (30 cm, 0 cm, 0 cm)?
8. (II) Repeat the calculation of Problem 7 for the case that (a) both charges on the z-axis are positive and the third charge is negative; (b) the signs and magnitudes of all charges are the same.
9. (II) A charge of 1.5 μC is placed at the point $x = 12$, $y = 25$, $z = 0$ (all distances given in centimeters). Calculate the work done in bringing a charge of -3 μC from $x = 12$, $y = 60$, $z = 50$ to the point $x = 12$, $y = 50$, $z = -25$, assuming that the charge is moved at a steady speed.
10. (II) Use potential energy arguments to show that charges of the same sign cannot form a system with a closed circular orbit.

24–2 Electric Potential

11. (I) Two equal charges of -4 mC are placed along the y-axis at -3 mm and 4 mm, respectively. Where is the electric potential zero?

12. (I) Two charges are placed along the x-axis; 3 μC at 14 cm, and -4 μC at 15 cm. Find those points along the x-axis where the potential is zero.

13. (I) The electrostatic potential of a single unknown charge 2.5 mm from that charge is 0.12 V. (The value of the potential at infinity is zero.) What is the value of the charge?

14. (I) A proton moves from point A to point B under the sole influence of an electric field, losing speed as it does so from $v_A = 8 \times 10^5$ m/s to $v_B = 5 \times 10^4$ m/s. What is the potential difference between the two points?

15. (I) An external force steadily moves a point charge of $+3.0 \times 10^{-7}$ C from a negatively charged to a positively charged plate. The plates are large and parallel, and the negatively charged plate is at a potential of $+3.0$ kV, whereas the positively charged plate is at a potential of $+17$ kV. How much work does the external force do?

16. (I) Consider two very long coaxial cylinders that carry opposite charges. The interior cylinder, negatively charged, is at a potential of $+16$ kV, whereas the exterior cylinder, positively charged, is at a potential of $+27$ kV. An external force steadily moves a point charge of -3×10^{-8} C from the negatively to the positively charged cylinder (Fig. 24–36). How much work does the external force do?

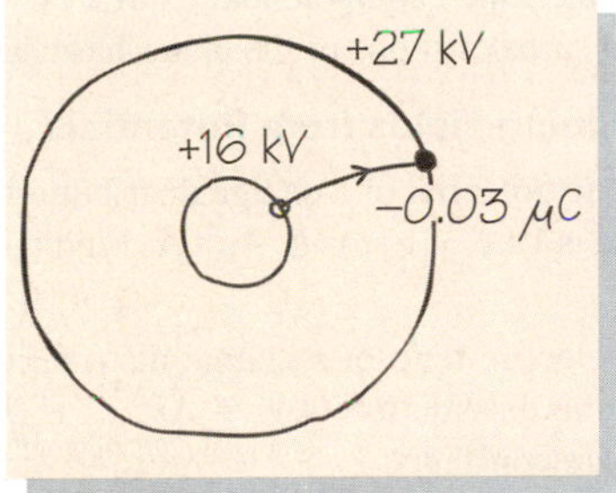

▲ **FIGURE 24–36** Problem 16.

17. (II) Three charges are at rest on the z-axis, $q_1 = 2$ mC at $z = 0$ m, $q_2 = 0.5$ mC at $z = 1$ m, and $q_3 = -1.5$ mC at $z = -0.5$ m. What is the potential energy of this system?

18. (II) Charges $+q$, $-q$, $+q$, and $-q$ are placed on successive corners of a square in the xy-plane. Plot all the locations in the xy-plane where the potential is zero.

19. (II) Consider two charges of $24 \times 10^{-2} \mu$C and $-10 \times 10^{-2} \mu$C, respectively, at opposite ends of a diameter of a circle of radius 25 cm (Fig. 24–37). (a) What is the potential on a point of the circle that is 30 cm from the positive charge? (b) How much work is required to bring a charge of -0.2 μC from infinity to that point on the circle?

20. (II) The origin of a coordinate system is at the intersection point of the perpendicular bisectors of the sides of an equilateral triangle of sides 3 cm. Calculate the potential at the origin due to three identical charges of 0.5 μC placed at the corners of the triangle.

21. (II) Consider a square of sides 14 cm. Charges are placed on the corners of the square as follows: 2 μC at (0 cm, 0 cm); -3 μC at (0 cm, 14 cm); 5 μC at (14 cm, 14 cm); $+3$ μC at (14 cm, 0 cm). What is the potential at the point (30 cm, 30 cm)?

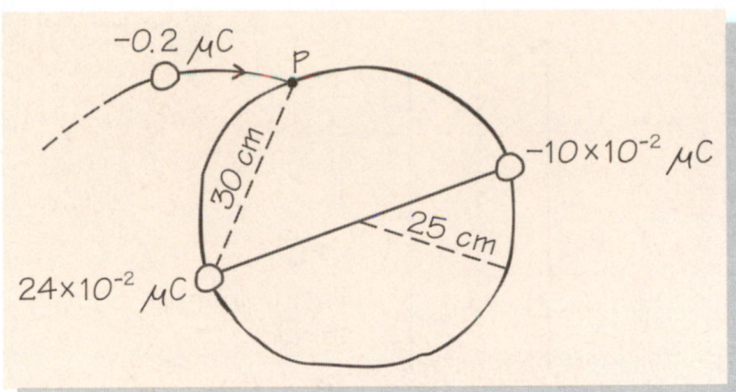

▲ **FIGURE 24–37** Problem 19.

22. (II) A 5-μC charge is fixed at $(x, y) = (15\text{ mm}, 20\text{ mm})$, a -3-μC charge is fixed at (15 mm, 30 mm), and a -2-μC charge is fixed at (25 mm, 20 mm) (Fig. 24–38). What is the potential energy of the system? Does the order in which the charges are brought in from infinity matter?

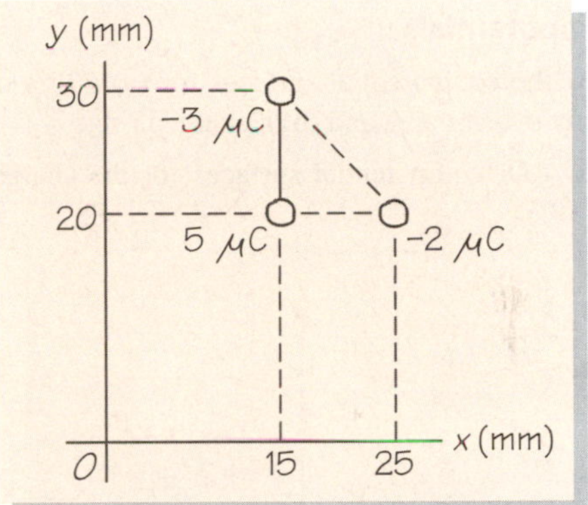

▲ **FIGURE 24–38** Problems 22 and 23.

23. (II) Calculate the electric potential at the origin due to the three charges considered in Problem 22.

24. (II) Two parallel conducting plates are brought to a potential difference of 600 V, and a small pellet of mass 2 mg carrying a charge of 3×10^{-7} C accelerates from rest at one plate. With what speed will it reach the other plate?

25. (II) Charges of $+12$ μC and -20 μC are placed along the y-axis at positions $y = +5.0$ cm and $y = -9.0$ cm respectively. (a) What is the potential at $y = 0$ and $x = 12.0$ cm? (b) What is the potential at $y = 0$, $x = 0$? (c) What is the electric field at that point?

26. (II) Two positive charges of magnitude 2.5 μC and 7.5 μC are placed a distance of 0.80 m apart. What is the electric potential along a line joining them? Where along that line will the electric field vanish? If a test charge is placed at that point, will it be in stable or unstable equilibrium as far as its motion along the line is concerned? Does your answer depend on the sign of the test charge?

27. (II) Figure 24–39 shows the cross section of a very large insulating slab that is uniformly charged to a charge density of 10^{-5} C/m^3. The thickness of the slab is 2 cm. (a) Determine the electric field of the charge on the slab at points A, B, and C. (b) Calculate the potential at points B and C, assuming that it is zero at A. (c) Plot the electric field and the potential as a function of distance from the center of the slab.

28. (II) A charge Q is distributed uniformly over the surface of a spherical shell of radius R. How much work is required to move these charges to a shell with half the radius? The charges are again distributed uniformly.

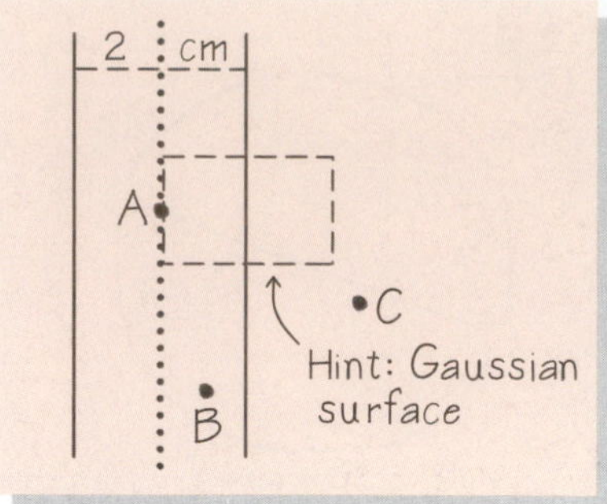

▲ **FIGURE 24–39** Problem 27.

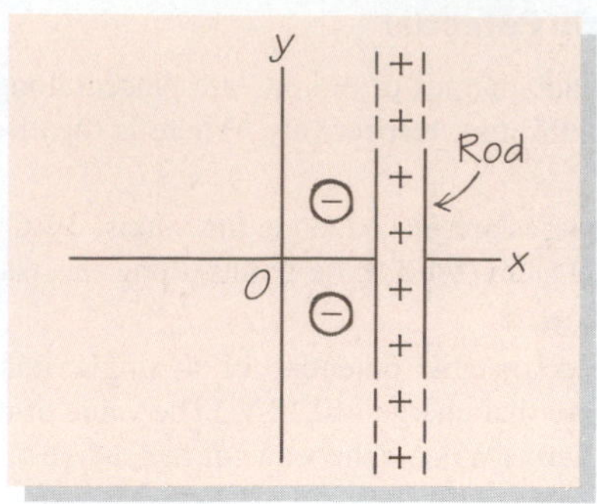

▲ **FIGURE 24–42** Problem 33.

29. (III) Calculate the potential inside and outside a sphere of radius R and charge Q, in which the charge is distributed uniformly throughout. [*Hint*: The additive constant for the potential inside the charged sphere must be chosen so that the two potentials, inside and outside, agree at $r = R$.]

24–3 Equipotentials

30. (I) Draw the equipotential surfaces for (a) a thin disk charged uniformly over its area and (b) a charged ring.

31. (I) Draw four equipotential surfaces for the charges shown in Fig. 24–40.

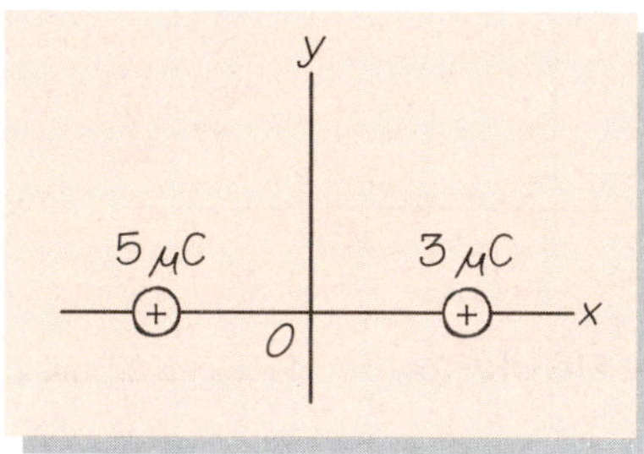

▲ **FIGURE 24–40** Problem 31.

32. (I) Sketch the equipotential surfaces for the charges shown in Fig. 24–41. Assume that the rod is an insulator.

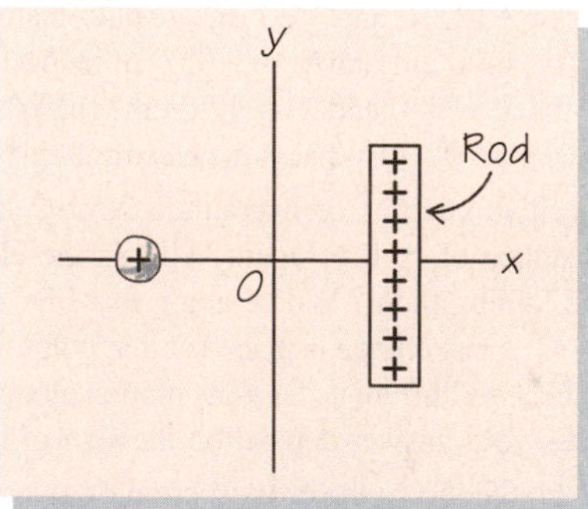

▲ **FIGURE 24–41** Problem 32.

33. (I) Sketch the electric fields and the equipotentials for the charge distribution shown in Fig. 24–42. Assume that the rod (of infinite length) is an insulator.

34. (I) Three metallic spheres are placed on the corners of an equilateral triangle. The radii of the spheres are $1/4$ of the side of the triangle. All the spheres are at the same potential V_0. Sketch the equipotentials for this system.

35. (II) A uniformly charged metal rod is placed parallel to an infinite, uncharged metal plate. Sketch the equipotentials in a plane perpendicular to the plate and to the rod, and in a plane perpendicular to the plate but parallel to the rod.

36. (II) Sketch the equipotentials in the xy-plane due to an infinite number of identical point charges q that lie on a line and are separated by a distance a, so that the coordinates of the point charges are $x_n = na$ and $y_n = 0$, where $n = 0, \pm 1, \pm 2, \pm 3, \ldots$.

37. (II) Two charges of equal magnitude but opposite sign are separated by a distance L. Sketch the equipotentials. What equipotential surfaces will have a potential of zero when the separate potentials for the two charges are chosen to be zero at infinity?

38. (II) Two infinite plates, each charged uniformly with charge density σ, are placed at right angles to each other and are almost touching. What are the equipotential surfaces? What are the equipotential surfaces if one of the plates has charge density $-\sigma$?

24–4 Determining Fields from Potentials

39. (I) The electric potential of a charge distribution within some region of space is $V(x, y, z) = Q/4\pi\varepsilon_0 x$. Find the electric field in this region.

40. (I) Find the electric field of a charge distribution if the electric potential of the distribution is $V = Ax^2y^2 + Byz^2 + C$, where A, B, and C are constants.

41. (I) In a certain region of space, the electric potential due to a charge distribution varies only with x, changing according to $V = a_0 + a_1 x$ where $a_0 = 12.7\text{ V}$, $a_1 = -6.68\text{ V/m}$, and x is in meters. Find the electric field, magnitude, and direction in this region.

42. (II) Starting from the solution in Example 24–9 of the potential due to a uniformly charged ring, use the derivative operations in Eq. (24–29) to find the electric field along the axis of the ring.

43. (II) Find the electric field far away along the bisecting axis of an electric dipole from the potential given in Eq. (24–24).

44. (II) Consider charge distributed in an infinitely long cylinder of radius R whose axis forms the z-axis. The charge distribution depends only on the distance r from the z-axis. The potential is given for $r < R$ by $V(r) = (Q/2\pi\varepsilon_0)[A(r/R) + B(r/R)^2 + C]$, where A, B, and C are constants. What is the electric field within the rod? What is the value of C if the potential is defined to be zero on the cylinder's surface?

45. (II) The potential $V(r)$ of a spherically symmetric charge distribution is given by $V(r) = (Q/4\pi\varepsilon_0 R)[-2 + 3(r/R)^2]$ for $r < R$ and $V(r) = Q/4\pi\varepsilon_0 r$ for $r > R$ (Fig. 24–43, see next page). Calculate the electric field.

46. (II) Use the results of Problem 45 and Gauss' law applied to Gaussian surfaces at various radii to calculate the charge distribution that gives rise to the potential given in that problem.

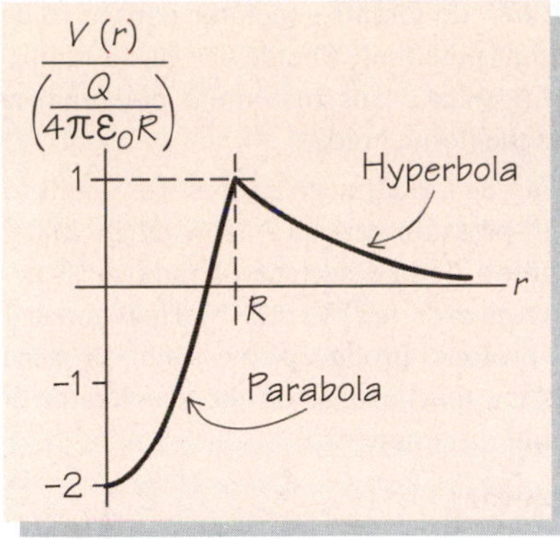

▲ **FIGURE 24–43** Problem 45.

47. (III) The potential in the xy-plane due to a certain charge distribution is given by

$$V(x, y) = \frac{Q}{4\pi\varepsilon_0 L} \times \left[\arctan\left(\frac{y}{x - a_0}\right) - 2\arctan\left(\frac{y}{x}\right) + \arctan\left(\frac{y}{x + a_0}\right)\right],$$

where L and a_0 are constant lengths. Show that the electric field at distances $x \gg a_0$, $y \gg a_0$ is proportional to a_0^2, and find its dependence on x and y. Express your answer in terms of r, the distance to the origin, and θ, the angle that the line from the origin to the point (x, y) makes with the x-axis.

24–5 The Potentials of Charge Distributions

48. (I) Two large, metal, parallel plates have a potential difference of 200 V, and the electric field between them has magnitude 7×10^3 V/m. What is the separation distance between the plates?

49. (I) The voltage along the axis of a uniformly charged ring of radius 10 cm is 5 V at a point 15 cm from the center of the ring. How much charge is on the ring?

50. (I) In fair weather, there is a constant electric field near Earth's surface whose magnitude is roughly 100 V/m, directed downward. (a) Find the potential associated with this field. (b) What is the most convenient point to choose for zero potential? (c) How does the potential energy of a test charge near Earth compare in form with the potential energy of gravity? (d) How much negative charge would have to be placed on a person of mass 50 kg to have the electric force balance the force of gravity?

51. (II) Find the potential as a function of the perpendicular distance R from an infinite line of uniform charge density by using Gauss' law and Eq. (24–9).

52. (II) Charges are distributed with uniform charge density λ along a semicircle of radius R, centered at the origin of a coordinate system. What is the potential at the origin?

53. (II) A rod that is 20 cm long is given a uniformly distributed charge of 2 μC (Fig. 24–44). Calculate the potential at a point P, which is a distance of 10 cm from the end of the rod, assuming that $V = 0$ at infinity.

54. (III) Find an expression for the electric potential at all points due to a rod of length L and uniform charge density λ, using Eq. (24–12). The rod is oriented along the z-axis, with its center at the origin (Fig. 24–45). Show that at distances much greater than L from the rod, the potential reduces to that of a point charge $Q = \lambda L$ at the origin.

55. (III) A charge $3q_0$ is placed on the x-axis at the point $x = x_0$ (where x_0 is positive), and a second charge, $-q_0$, is placed on the x-axis at the point $x = -x_0/2$. (a) What is the potential on the x-axis of this distribution of charges? Assume that zero potential for a point charge is at infinity. (b) Show that your result for part (a) can be approximated for large x by a term proportional to $1/x$, plus a term proportional to $1/x^2$, plus higher powers of $1/x$. (c) Show that the expansion of part (b) is that of a point charge at the origin, plus an electric dipole oriented along the x-axis and centered at the origin, plus other terms. Find the strength of the point charge as well as the dipole moment of the electric dipole. (d) How large must x be so that the approximation of a point charge plus a dipole comes within 1 percent of the exact answer? [*Hint*: Use the approximation $(1 + z)^k = 1 + kz + \frac{1}{2}k(k - 1)z^2 + \cdots$ good for $z \le 1$.]

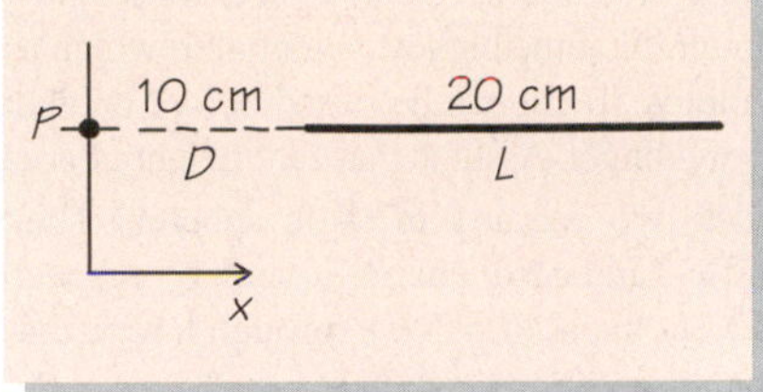

▲ **FIGURE 24–44** Problem 53.

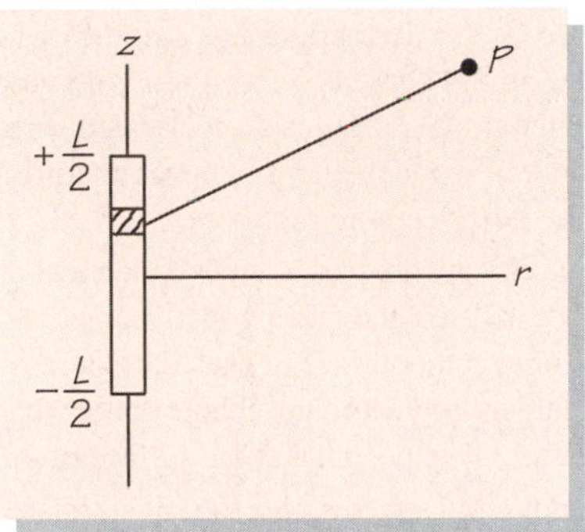

▲ **FIGURE 24–45** Problem 54.

24–6 Potentials and Fields Near Conductors

56. (I) A thin disk of radius 2.8 cm carries a total charge of 6.0×10^{-8} C spread evenly over its surface. What is the minimum work required to bring a charge $q = 3.2 \times 10^{-7}$ C at rest from infinity to a distance of 8.8 cm from the disk along its axis?

57. (I) A thin ring of radius 24 cm carries a uniformly distributed charge of 3.5×10^{-7} C. A negative charge $q = -8.5 \times 10^{-8}$ C is placed on the axis of the ring 28 cm from the plane of the ring (Fig. 24–46). How much work must an external agent do to move the charge slowly and steadily to a distance 85 cm away, also on the axis?

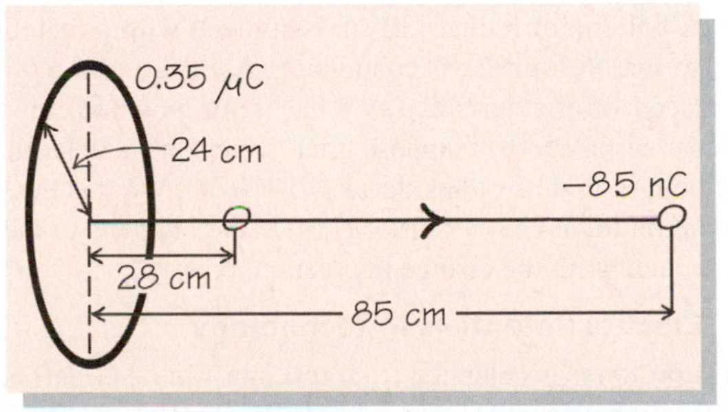

▲ **FIGURE 24–46** Problem 57.

58. (I) An electric field of 2.8×10^6 V/m is sufficiently large to cause sparking in air. Find the highest potential to which a spherical conductor of radius 30 cm can be raised before breakdown occurs in the air surrounding it. Assume that zero potential is taken at infinity.

59. (I) Consider two charged metallic spheres. The spheres have radii r_1 and r_2, and carry charge q_1 and q_2, respectively. What is the amount of charge that flows through a wire that is brought in and connected to the two spheres?

60. (II) A spark plug has a gap of 2 mm. What must be the potential difference across the gap in order for it to ionize a bridge across the gap? Assume air fills the gap.

61. (II) A Van de Graaff accelerator has a dome of radius 0.61 m. If the potential on the dome is 5.5 million volts, how much charge has accumulated? If a proton is accelerated through this potential to ground, how much energy does it acquire? Calculate how fast it will be going (ignore relativity).

62. (II) The same charges are placed on two identical drops of mercury. The drops are isolated and take perfectly spherical shapes, and the electric potential at the surface of each drop is 70 V. The drops coalesce into a larger drop with a net charge double that of either smaller charge. What is the potential at the surface of this larger charge?

63. (II) Two conducting spheres of different sizes are connected by a thin conducting wire. The radius of the larger sphere is three times that of the smaller sphere. If a total charge Q is placed on this apparatus, what fraction of Q sits on each sphere?

64. (II) Concentric metal shells, perfect conductors, have radii R and $1.5R$, respectively. A charge q is placed on the inner shell, and a charge $-3q$ is placed on the outer shell. (a) What are the electric fields in all space due to the two shells? (b) What is the potential difference between the two shells? (c) If a thin, perfectly conducting wire now joins the two shells, how does the charge redistribute itself?

65. (II) A cloud is made up of raindrops, each carrying a positive charge of 16×10^{-19} C. There are 1.2×10^{10} raindrops per cubic meter. When the electric field at the surface of the cloud builds up to 3.2×10^6 V/m there will be electrical breakdown and lightning will be seen. What will be the radius of the cloud, assuming it is spherical, when this occurs?

66. (II) Two metallic spheres of radii 0.05 m and 0.08 m respectively are placed far away from each other. They are then connected by a thin wire and a charge of 40 μC is placed on the system. Assuming that we can neglect the charge on the wire, how will the charge be distributed between the two spheres? What is the potential of the system?

67. (II) Two spherical conductors of radii 20 mm and 100 mm are connected by a thin wire and carry charges q_1 and q_2, respectively. If the wire is cut and the centers of the spheres are 250 mm apart, there is a repulsive force of 3.5 N between them. Use this information to calculate (a) q_1 and q_2 and (b) the electric fields at the surfaces of the conductors when they are connected by the wire.

68. (II) A balloon of radius 430 cm is sprayed with a metallic coating so that the surface is conducting. A charge of 1.5×10^{-5} C is placed on the surface. (a) What is the potential on the balloon's surface? (b) Suppose that some air is let out of the balloon, so that its radius shrinks to 310 cm. What is the new potential on the balloon's surface? (c) What happens to the energy associated with the change in potential energy?

24–7 Electric Potentials in Technology

69. (I) A proton is accelerated from rest in a Van de Graaff accelerator through a potential of 5.5×10^6 V. (a) What energy does the proton have—in electron-volts and joules? (b) What is the proton's final speed?

70. (I) A small Van de Graaff generator is used to demonstrate the effects of high potential. The device has a radius of 41 cm and stands in air. What is its maximum potential, and how much charge does the dome hold?

71. (II) Early Van de Graaff accelerators were built to operate in air without high-pressure gases. (a) How much voltage could an accelerator with a domed surface of radius 1.3 m produce if air breaks down at 3×10^6 V/m? (b) How much kinetic energy could the protons produced by such an accelerator have? (c) What is the total charge on the accelerator dome when the maximum field is attained?

General Problems

72. (I) We have a high-voltage power supply capable of producing 15,000 V, and we want to ionize the air molecules between parallel plates. What plate separation will give us electrical breakdown?

73. (I) The potential at a point x due to a thin, flat, uniformly charged disk of radius R and total charge Q at a point on the disk's axis a distance x away from the disk is given in Eq. (24–32). Consider two such identical disks, one in the yz-plane at $x = -a$, and another in the yz-plane at $x = +a$, both centered on the x-axis (Fig. 24–47). What is the potential at an arbitrary point on the x-axis between the two disks?

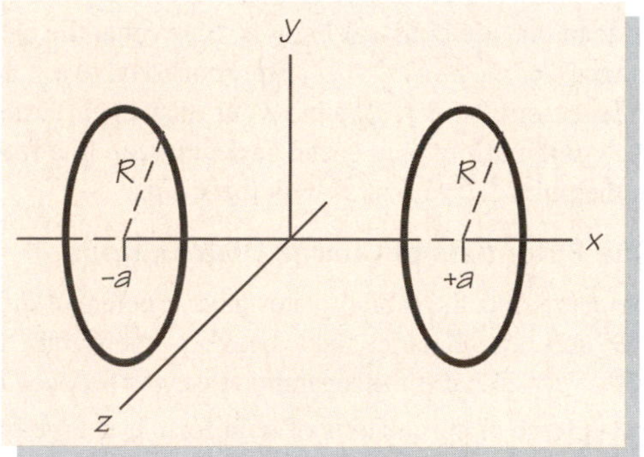

▲ **FIGURE 24–47** Problem 73.

74. (I) A metallic ring of radius R, carrying charge Q, has an associated electric potential at a distance x from the center of the ring along its axis given in Table 24–1. What is the potential at some point x due to two rings, one carrying charge Q and the other $-Q$, both in the yz-plane, centered on the x-axis, one located at $x = a$, the other at $x = -a$?

75. (II) What is the electric field at a point x on the axis due to the two charged rings described in the previous problem for $x \gg a$? [*Hint*: Use the approximation $(x^2 + 2ax + b^2)^{-1/2} \cong x^{-1}(1 - a/x + (3a^2 - b^2)/2x^2)$.]

76. (II) Write an expression for the total energy of two point charges—one positive and of magnitude Q, fixed at the origin; the other negative and of magnitude q and mass m, located at a point a distance r from the origin. Suppose the charge q, instead of being stationary, moves in a circular orbit of radius r around the charge Q. Assuming that the Coulomb attraction is responsible for the centripetal acceleration, calculate the energy. Why is the angular velocity of this motion constant?

77. (II) A nonconducting sphere of radius R carries a charge $+Q$ distributed uniformly throughout its volume. What is the potential energy of a point charge $-q$ a distance r $(r < R)$ from the center of the sphere? Show that if there is a hole drilled through the sphere so that the point charge can move through it, then the point charge oscillates as though it were attached to a spring. Find the effective spring constant.

78. (II) Three electrons are located along the x-axis at positions $-6\ \mu\text{m}$, $0\ \mu\text{m}$, and $6\ \mu\text{m}$, respectively. How much energy was required to move each of the electrons in turn from infinity? Does the order in which they were moved matter?

79. (II) A salt crystal consists of an array of positive Na and negative Cl ions, both carrying an elementary charge of magnitude e. Assume that a small "seed" crystal consists of four ions, forming a square of side 0.25 nm (Fig. 24–48). Find the electric force acting on one of the sodium atoms due to the other atoms of the seed and the work needed to remove this ion from the seed. Give your result in electron volts.

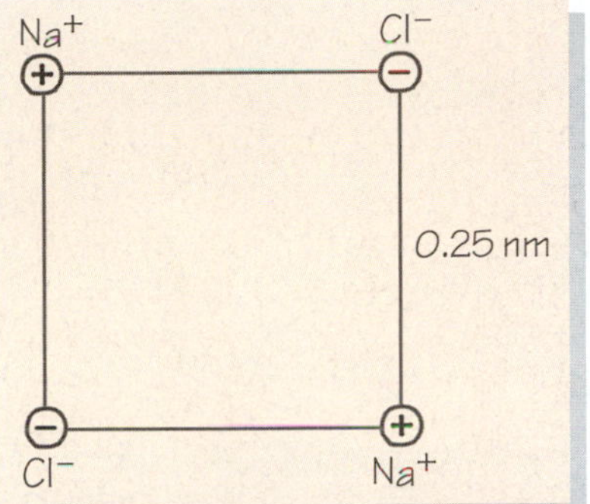

▲ **FIGURE 24–48** Problem 79.

80. (II) Calculate the potential at the point $P(x, y)$ due to the dipole in Fig. 24–49, which consists of a charge $+q$ placed at $(0, a)$ and a charge $-q$ placed at $(0, -a)$, and use this potential to calculate the electric field at point P.

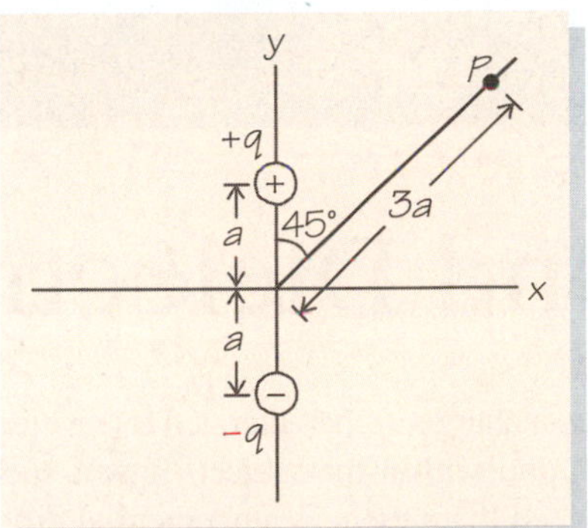

▲ **FIGURE 24–49** Problem 80.

81. (II) Find the electric field along the axis of a uniformly charged ring of radius R and total charge Q by taking the appropriate derivatives of the potential found in Example 24–9. Set up the problem by using the direct integration techniques presented in Chapter 23. Compare the difficulties of the two ways of calculating the electric field.

82. (II) A positron (charge $+e$ and same mass as that of the electron) approaches a proton (charge $+e$) head-on. As a result of the repulsion, the positron turns around a distance $r_0 = 6.5 \times 10^{-10}$ m from the proton. What is the kinetic energy of the positron when it is very far from the proton? You may assume that the proton motion can be neglected.

83. (II) An electron is moving in the field of a helium nucleus (atomic number $Z = 2$). What is the change in the electron's potential energy when it moves from a circular orbit of radius 3×10^{-10} m to one of radius 2×10^{-10} m? What is the change in kinetic energy? in the total energy of the electron? Energy conservation is not violated in this process because it can be carried away by radiation that is emitted during the change of orbits.

84. (II) Four charges $+Q, -Q, +Q$ and $-Q$ are placed at a corners of a square of side L. What is the electric potential magnitude at a point P which lies on one of the diagonals of the square at a distance R from the center of the square, with $R \gg L$. [*Hint*: Use the binomial expansion $(R^2 \pm x^2)^n = R^{2n} \pm nR^{2n-2}x^2 + \dfrac{n(n-1)}{2}R^{2n-4}x^4 \pm \cdots$, good for $x < R$.] The field's dependence on R is characteristic of an *electric quadrupole*.

85. (II) Two concentric metal shells of radii r_1 and r_2 respectively carry charge q_1 and q_2 respectively. Assuming $r_1 < r_2$, what is the potential in the range $0 \le r \le \infty$?

86. (II) An electric dipole fixed in space consists of a charge $+q$ at the point $x = -0.2$ m and a charge $-q$ at the point $x = +0.2$ m, where $q = 5\ \mu\text{C}$. A test charge $q_0 = 3\ \mu\text{C}$ is steadily moved from the point $x = +0.6$ m to the point $x = -0.4$ m by following a semicircular path of radius 0.5 m that takes the test charge through the y-axis (Fig. 24–50). How much work is required to move the test charge?

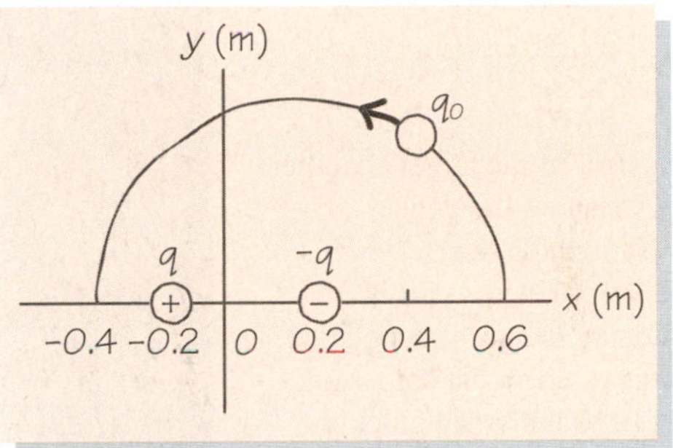

▲ **FIGURE 24–50** Problem 86.

87. (II) Two identical cork balls of charge $2.0\ \mu\text{C}$ are suspended from the same point by thin threads 0.80 m long. (a) Calculate the mass of the cork balls if the threads each make a 30° angle with the vertical. (b) Calculate the potential energy of the system of two balls due to the presence of charges and to the presence of gravity as a function of the angle θ the threads make with the vertical. Choose zero gravitational potential energy to correspond to $\theta = 0$.

88. (II) A large, square plane with sides of length L, parallel to the yz-plane and located at x_1, has charge density σ_1. A similar plane, located at x_2, has charge density σ_2. How much work must be done to bring the second plane to within a distance a of the first one? Neglect end effects; that is, calculate the fields as though the planes were infinite.

89. (III) An infinitely long cylinder of radius R is filled with uniform charge density ρ. Calculate the potential inside and outside the cylinder.

90. (III) The inner radius of a spherical dielectric shell is 16 cm, and the outer radius is 45 cm. The shell carries a charge of 5.0×10^{-6} C, distributed uniformly. Sketch the shape of the potential for all values of r, the distance from the center of the shell, and evaluate it at the center and at the inner and outer radii.

91. (III) A solid sphere of radius R has uniform charge density ρ. Calculate the total potential energy by calculating the energy required to bring a spherical shell of thickness dr and charge density ρ from infinity to a distance r from the sphere's center in the potential due to a uniformly charged sphere of radius r.

Chapter 25

▶ The interior of the target chamber of the NOVA laser at Lawrence Livermore National Laboratory. This project aims to produce controlled nuclear fusion by means of depositing large amounts of energy at the target; the energy can be delivered quickly because it is stored in large capacitor banks.

Capacitors and Dielectrics

Any conducting object that carries a charge is characterized by an electric potential that is constant everywhere on and within that object. If two such conductors have a potential difference between them then, as any potential difference is able to accelerate charges, the system effectively stores energy. A capacitor is a device that can maintain a potential difference, storing energy by storing charge. The relation between the amount of charge a capacitor stores and the potential difference it maintains depends on the geometry of the capacitor. The storage of charge is also affected by the presence of insulating (nonconducting) material—dielectric material. In this chapter, we study the role capacitors play in electrical circuits, where they are indispensable elements, how they can be used directly as energy storage devices, as well as how geometry and dielectric materials affect the properties of capacitors. We also extend our fundamental knowledge about the behavior of matter by thinking about the microscopic structure of dielectrics and how that microscopic behavior shows itself at the everyday level.

25–1 Capacitance

A pair of conductors, whether separated by empty space or by an insulating material, forms the simplest type of **capacitor**. Capacitors store separated equal and opposite charges. In their most common and useful form, capacitors are made from two conductors with charge $+Q$ on one conductor and charge $-Q$ on the other. In Chapter 24 we saw that the potential difference[†] between the conductors is linearly dependent on this

[†] We will also use the terms "potential," "voltage drop," or "voltage" rather than "potential difference."

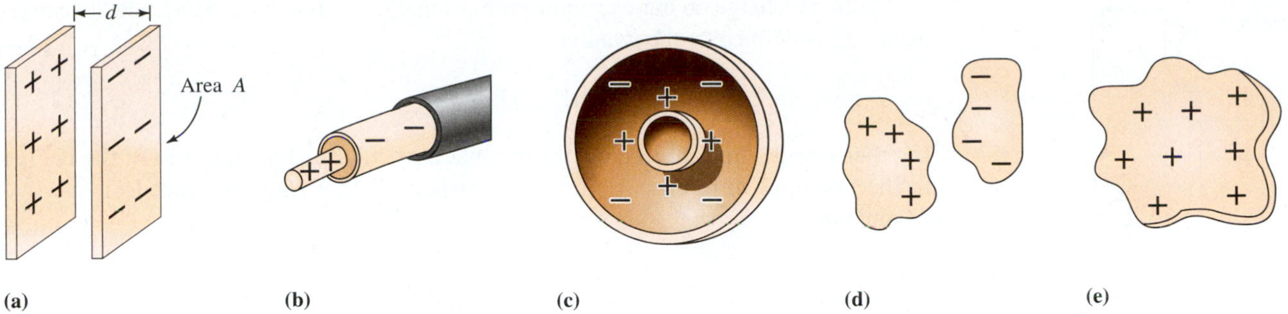

▲ **FIGURE 25–1** Various kinds of capacitors: (a) parallel-plate; (b) coaxial cable; (c) spherical (two hollow conducting spheres); (d) conductors of arbitrary size and shape; (e) isolated conductor infinitely far from second conductor.

charge; that is, $V \propto Q$. Thus, if we double the charge, we double the potential difference between the two conductors, so that the ratio of Q to V between two conductors is constant. We call this constant ratio Q/V the **capacitance** C. It depends on the shape and arrangement (the geometry) of the two conductors of the capacitor, and on the material between the conductors. Figures 25–1a to 25–1e illustrate different configurations of conductors that can act as capacitors.

CONCEPTUAL EXAMPLE 25–1 In the case of Fig. 25–1e, there is a charge of only one sign. Where is the charge of the other sign?

Answer An isolated conductor has a capacitance because when charge is placed on it, the charge must be brought in from infinity. Remember that charge must be conserved, so that the second piece of the capacitor is, in effect, at infinity. The potential difference in this case is the difference between the potential at infinity, normally taken to be zero, and the potential on the isolated conductor itself. Accordingly, an isolated conductor held high above Earth forms a capacitor. Even Earth itself is a capacitor—we'll calculate its capacitance below.

Why are capacitors important? Capacitors of differing capacitance allow us to hold different amounts of charge for a given potential difference or to maintain different potential differences for a given amount of charge. Capacitors thus control the storage and delivery of charge. Since it requires work to separate charges, capacitors store energy as well as charge. This energy is quite visible in certain circumstances, particularly through the often rapid release of the energy. A lightning strike is the spectacular discharge of a large capacitor formed by the system of a cloud and Earth. In its simplest form, a camera photoflash contains a capacitor that stores energy and then discharges it when the flash is fired. The slow but smooth delivery of energy when capacitors are coupled with other circuit elements represents another type of application. Emergency backup systems for computers and electrical power distribution systems use capacitors in this way. Lastly, because capacitors play a crucial role in controlling the time dependence of the electric charges that move through circuits, almost any device with an electronic circuit contains capacitors. We regularly use capacitors both large and small. Devices for the possible generation of fusion energy employ gigantic capacitors whose energy content rivals that of the largest lightning strike. In the realm of microscopic circuits—nanotechnology, or quantum engineering—capacitors are used that may consist of small collections of atoms.

Consider a capacitor formed by placing equal but opposite charges, $+Q$ and $-Q$, on two conductors. This is easily done by touching the two conductors with wire leads attached to the + and − terminals of a battery (Fig. 25–2). The battery will give the two pieces a potential difference, and the amount of charge that accumulates depends on the shape of the conductors and on their relative positions. We call Q the *charge on the capacitor* even though the *net* charge on the oppositely charged pair of elements is zero.

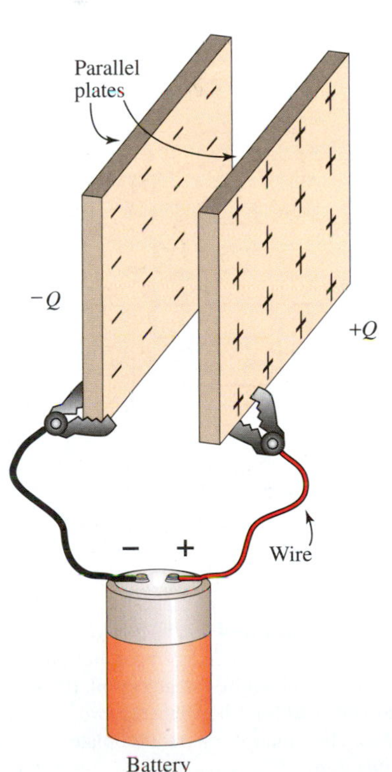

▶ **FIGURE 25–2** We can construct a capacitor by connecting two wires to two conductors such as the plates shown here, and then attaching one wire to the positive terminal of a battery and the other wire to the negative terminal of the battery. This will place charge $+Q$ on one conductor and charge $-Q$ on another conductor.

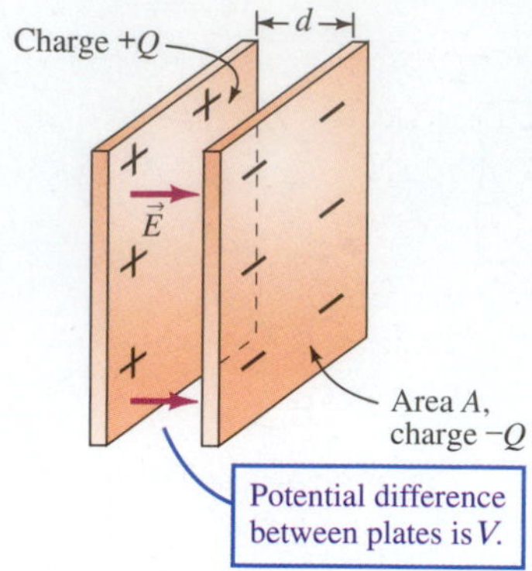

▲ **FIGURE 25–3** Two parallel plates with equal and opposite charges make up the most basic capacitor.

With the charge on our capacitor proportional to the potential difference, its capacitance, C, is defined by the relation

$$Q = CV. \tag{25–1}$$

In other words, the capacitance of the capacitor is defined as the ratio of the charge to the potential difference V that results when charges $\pm Q$ are placed on the two conductors:

$$C \equiv \frac{Q}{V}. \tag{25–2}$$

DEFINITION OF CAPACITANCE

When there is charge on a capacitor, we say it is *charged*. When a capacitor *discharges*—for example, when it is used to fire a flashbulb—it can deliver its stored energy rapidly as the charges on the plates flow off.

C is always taken to be positive; that is, Eq. (25–2) should contain magnitudes only. The unit of capacitance is coulombs per volt (C/V), but capacitance occurs so frequently that it has been given its own SI unit, the **farad** (F), in honor of Michael Faraday:

$$1\ \text{F} \equiv 1\ \text{C/V}. \tag{25–3}$$

In practice, the farad is inconveniently large, and for practical use units of μF, nF, and pF ("puffs") are more common.

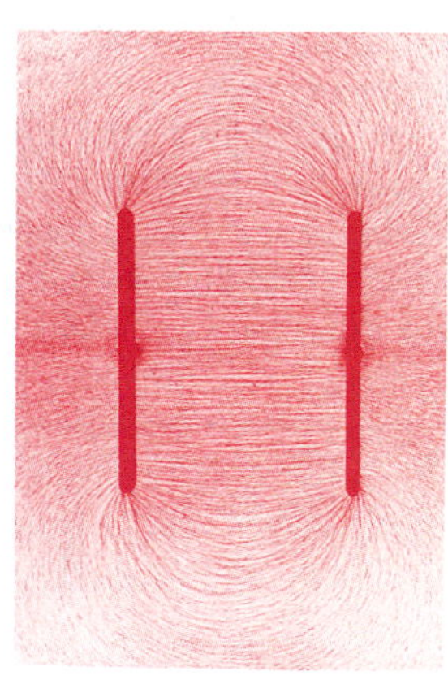

(a)

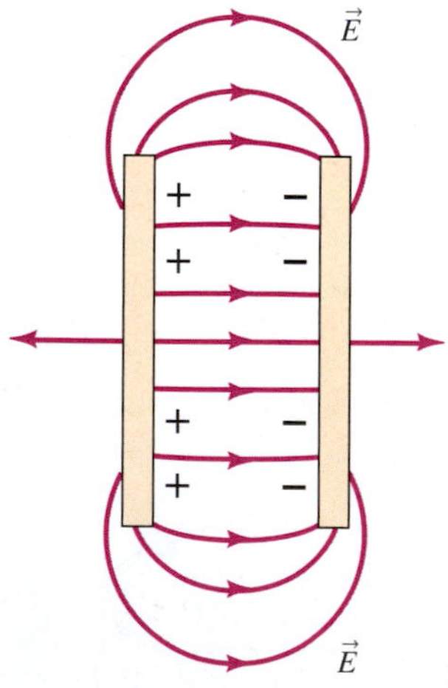

(b)

▲ **FIGURE 25–4** (a) The electric field lines due to a charged parallel-plate capacitor, shown by threads in oil. (b) The electric field lines between the two conducting plates of a parallel-plate capacitor.

Calculating Capacitance

The capacitance of a capacitor can be calculated easily if the geometry is simple. (Keep in mind that even if we can't calculate a capacitance, we can always measure it.) The most basic capacitor consists of two parallel conducting plates of area A, separated by a distance d, with charges $+Q$ and $-Q$, respectively, distributed uniformly over the plates (Fig. 25–3). If the dimensions of the plates are large compared with d, then the electric field between the plates is to a very good approximation constant. Neglecting (small) edge effects, we found previously that the field between the plates has magnitude $E = \sigma/\varepsilon_0$, where σ is the charge density Q/A and ε_0 is the permittivity of free space. The potential difference between the plates is $V = Ed$, and we can combine these results to find that

$$V = Ed = \frac{\sigma}{\varepsilon_0}d = \frac{Q}{A}\frac{d}{\varepsilon_0} = Q\frac{d}{\varepsilon_0 A}.$$

Thus $Q/V = \varepsilon_0 A/d$, and the capacitance $C \equiv Q/V$ of a parallel-plate capacitor is

$$C = \frac{Q}{V} = \frac{\varepsilon_0 A}{d}. \tag{25–4}$$

The fields near the edge, the *fringe fields*, are not uniform (Fig. 25–4). Their effect is small if the linear dimensions of the plates are much larger than the separation. Fringe fields do not affect the linear relationship between the charge and the potential, but do modify the simple form of Eq. (25–4). We will use the approximation of Eq. (25–4) throughout this chapter in reference to parallel-plate capacitors.

Equation (25–4) gives us a second commonly used unit for the permittivity of free space, ε_0, namely farads per meter (F/m):

$$\varepsilon_0 = 8.85 \times 10^{-12}\ \text{C}^2/\text{N}\cdot\text{m}^2 = 8.85 \times 10^{-12}\ \text{F/m} = 8.85\ \text{pF/m}.$$

Either unit is consistent with the SI.

EXAMPLE 25–2 (a) Calculate the capacitance C of parallel plates of area $A = 100\ \text{cm}^2$ separated by a distance $d = 1.0\ \text{cm}$. (b) Find the area of a parallel-plate capacitor with plate separation of 1.0 cm and a capacitance of 1.0 F.

Setting It Up The setup is similar to that of Fig. 25–3.

Strategy For part (a) we use Eq. (25–4) to determine the unknown capacitance. For part (b) we can again use this equation, but this time we solve it for A.

Working It Out (a) Substituting the known values into Eq. (25–4) gives

$$C = \frac{\varepsilon_0 A}{d} = \frac{(8.85\ \text{pF/m})(1.0 \times 10^{-2}\ \text{m}^2)}{1.0 \times 10^{-2}\ \text{m}} = 8.9\ \text{pF}.$$

The plate area is rather large, yet the capacitance is only 8.9 pF.

(b) We again use Eq. (25–4), this time solving for A:

$$A = \frac{dC}{\varepsilon_0} = \frac{(1.0 \times 10^{-2}\ \text{m})(1.0\ \text{F})}{8.85 \times 10^{-12}\ \text{F/m}} = 0.11 \times 10^{10}\ \text{m}^2.$$

This represents a square with sides of length $0.33 \times 10^5\ \text{m} = 33\ \text{km}$! Most common practical capacitors have capacitances much smaller than 1 F.

What Do You Think? Suppose the charge on each plate doubles. By how much does C change? *Answers to* ***What Do You Think?*** *questions are given in the back of the book.*

EXAMPLE 25–3 Consider a system made of a solid conducting cylinder of radius a surrounded by a thin coaxial conducting tube, or sheath, of radius b. Find the capacitance per unit length of this system, assuming that there is a vacuum between the central wire and the sheath. This is a model of a *coaxial cable*, a wire used for the high-speed transmission of signals. In a real coaxial cable, there would be some type of material, not a vacuum, between the central wire and the sheath, but for simplicity we assume here that there is a supporting material that has the electrical properties of a vacuum.

Setting It Up The system is sketched in Fig. 25–5. (Keep in mind that the material that mechanically supports the outer cylinder has the electrical properties of the vacuum.) We want to find the potential difference for a given charge. For an infinitely long system, however, we can specify only the charge *per unit length*, so we must similarly calculate the capacitance per unit length.

Strategy The capacitance is the ratio of the charge to the potential difference, and the capacitance per unit length is the ratio of the charge per unit length to the potential difference. Therefore we set a charge per unit length $+\lambda$ on the outer (and $-\lambda$ on the inner) conductor of the cable and calculate the resulting potential between the conductors. This calculation was already done in Example 24–11. There, we saw how Gauss' law gives the electric field, and hence the potential, outside a wire of finite radius. That example applies directly to the potential in the space *between* the conductors of our system.

Working It Out From Example 24–11, we have

$$V_b - V_a = \frac{\lambda}{2\pi\varepsilon_0} \ln\frac{b}{a}.$$

Dividing the charge per unit length λ by this potential difference gives the capacitance per unit length, a quantity we denote by c:

$$c = \frac{\lambda}{V_b - V_a} = \frac{2\pi\varepsilon_0}{\ln(b/a)} \tag{25–5}$$

Equivalently, the capacitance of a length L of the cable is L times this value.

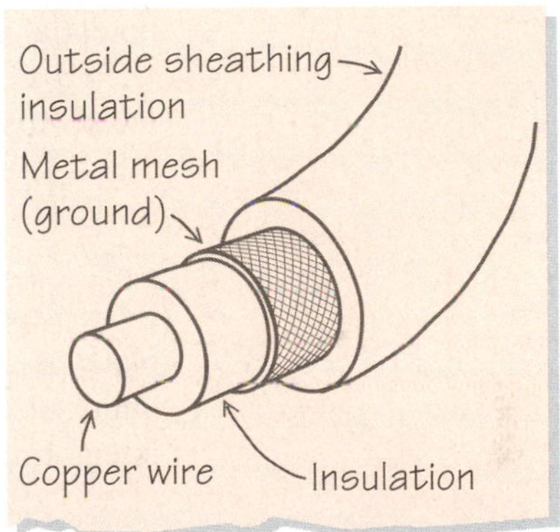

▲ **FIGURE 25–5** The coaxial cable has an inner solid wire of radius a and a cylindrical metal sheath of radius b. An insulator is placed between the two conductors.

EXAMPLE 25–4 *The Isolated Sphere.* What is the capacitance of an isolated conducting sphere of radius R? Calculate Earth's capacitance.

Strategy We place a charge Q on the sphere and find its potential V, assuming the potential at infinity is zero; because the "second piece" of the capacitor is at infinity, V is indeed the appropriate potential difference. The ratio Q/V is, by definition, the capacitance.

Working It Out We found in Example 24–12 that, for the potential at infinity set at zero, the potential of a conducting sphere is

$$V = \frac{Q}{4\pi\varepsilon_0 R}.$$

The capacitance of the isolated sphere is Q divided by V:

$$\text{for an isolated sphere: } C = \frac{Q}{V} = 4\pi\varepsilon_0 R. \tag{25–6}$$

Earth's capacitance is determined by setting R equal to Earth's radius, $R_E = 6.38 \times 10^6\ \text{m}$. Therefore

$$C = 4\pi\varepsilon_0 R = 4\pi(8.85 \times 10^{-12}\ \text{F/m})(6.38 \times 10^6\ \text{m}) = 7.10 \times 10^{-4}\ \text{F}.$$

It is interesting to compare this number to the value of the largest capacitors in use, of the order of 1 F.

What Do You Think? Perform a dimensional analysis to confirm the dependence of the capacitance on the radius.

The technique for determining the capacitance is always the same. We assume that the conductors have a charge $\pm Q$; then we find the potential difference V between the conductors due to this charge. The ratio Q/V gives the capacitance.

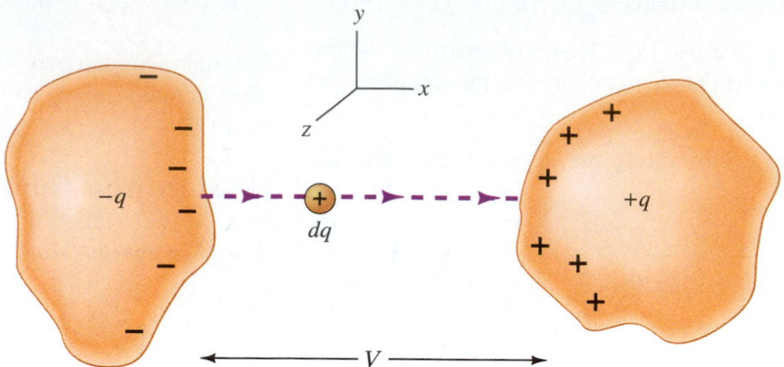

▶ **FIGURE 25–6** Work is done when an infinitesimal charge dq is moved between the two conductors of a capacitor already charged to a potential difference V.

25–2 Energy in Capacitors

There is an electric field between the two conductors of a charged capacitor, and this field can accelerate a test charge. Thus, a charged capacitor is capable of doing work and must contain energy. The energy contained in a charged capacitor is in fact the work required to charge it, and we can calculate this work to find the energy the capacitor contains. Start by taking a positive charge dq from one neutral conductor and moving it to the other neutral conductor. The second conductor then has charge $+dq$, and the first conductor has charge $-dq$. If we do this for a time, the conductors will be charged to a potential difference V with charge q (Fig. 25–6), and V is given in terms of q by $V = q/C$. To move an *additional* charge dq from the negatively charged conductor to the positively charged one, we must then do work

$$dW = V\,dq = \frac{q}{C}\,dq.$$

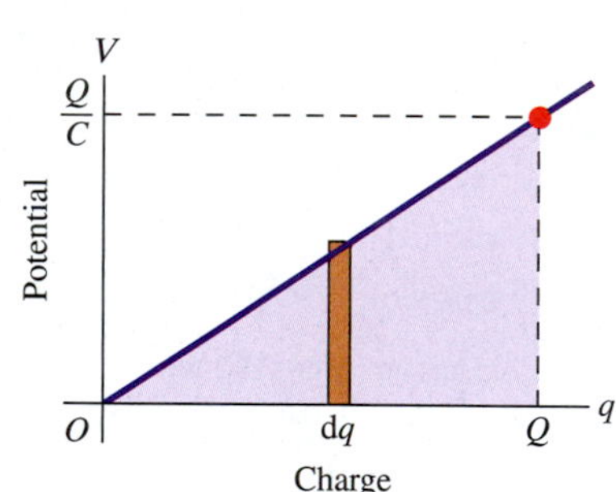

▲ **FIGURE 25–7** The amount of work $dW = V\,dq$ that is done to move a charge element dq from one part of a capacitor to the other increases as the charge on the conductors—and hence the potential V across them—builds up. The total work is found by adding the work elements together. This work is the area under a curve of potential difference versus charge on the plates: the integral of $V\,dq$.

As represented in Fig. 25–7, the infinitesimal work done to place a charge dq on the plates is zero when the plates are uncharged and increases as the charge on the plates builds up to $\pm Q$. The total work done as the charge increases from 0 to Q is obtained by integrating the above expression from $q = 0$ to $q = Q$:

$$W = \int dW = \int_0^Q \frac{q}{C}\,dq = \frac{1}{C}\int_0^Q q\,dq = \frac{Q^2}{2C}. \tag{25–7}$$

This is a general result for all capacitors.

Once charged, the stored energy in the capacitor can do work. We can think of the stored energy as a potential energy. This potential energy can be used to move a test charge placed between the conductors or to cause a flashbulb to flash.

To summarize, the potential energy of a charged capacitor is

$$U = \frac{Q^2}{2C}. \tag{25–8}$$

ENERGY IN A CAPACITOR

Because $Q = CV$, this result is equivalent to

$$U = \frac{CV^2}{2}. \tag{25–9}$$

ENERGY IN A CAPACITOR

The first form is used when the charge is known; the second when the potential is known. Another equivalent form that is useful when the charge and voltage are both known is

$$U = \frac{QV}{2}. \quad (25\text{–}10)$$

ENERGY IN A CAPACITOR

As we know from Chapter 24, the electric potential energy associated with the movement of a charge Q through a fixed potential V is $U = QV$. This expression differs by a factor of 2 from the capacitor energy $QV/2$, Eq. (25–10). The reason for this difference is that, as a capacitor is charged, the potential increases from 0 to V, and in effect the average potential as the charging takes place is $V/2$.

Batteries vs. Capacitors

A *battery* is a chemical device for the storage of energy. Whereas the potential of a capacitor decreases as the capacitor delivers its charge, a battery ideally maintains a fixed potential between two points (terminals) as it delivers charge. If we want to charge a capacitor to a certain potential, we can use a battery because it can hold the desired potential even as it delivers charge to the capacitor.

EXAMPLE 25–5 A 12-V car battery is used to charge a 100-μF capacitor. (a) How much energy is stored in the capacitor? (b) Compare this energy with the energy stored in the battery itself, if the battery is capable of delivering a total charge of $Q = 3.6 \times 10^5$ C at the given voltage. (This is the charge that can be delivered by a battery rated at 100 ampere-hours; the ampere-hour is a standard unit for charge.)

Setting It Up By a "12-V battery" we mean a battery with a 12-V potential difference across its leads. In the charging process the battery brings the potential difference across the capacitor plates to that same value.

Strategy (a) To find the energy in the capacitor given the capacitance and the potential difference across it, we apply Eq. (25–9). (b) The electric potential energy associated with the movement of a charge Q through a fixed potential V is $U = QV$. If the battery is capable of moving a given amount of charge through a given potential, as described here, then the battery must contain at least this amount of energy.

Working It Out (a) We have

$$U = \frac{CV^2}{2} = \frac{(100 \times 10^{-6}\,\text{F})(12\,\text{V})^2}{2} = 7.2 \times 10^{-3}\,\text{J}.$$

The answer is in joules—we have used SI units throughout.

(b) The potential energy in the battery is (at least)

$$U = QV = (3.6 \times 10^5\,\text{C})(12\,\text{V}) = 4.3 \times 10^6\,\text{J}.$$

The battery contains a factor of 6×10^8 more energy than is stored in the capacitor!

What Do You Think? Why is it possible to store so much more energy in the battery?

The car battery of Example 25–5 has a potential energy of about 10^6 J. This is far more energy than that of a typical practical capacitor of 100 μF charged to a moderate potential. A typical capacitor would require a potential of about 10^{10} V to store the amount of energy in a car battery. The largest available commercial capacitors have capacitance on the order of 1 F, and it would require a potential of about 1000 V for a 1-F capacitor to contain as much energy as that contained in the car battery. However, high voltages across capacitors present their own difficulties, and large capacitors can be charged to a potential of only a few volts. A battery's energy is stored in chemical bonds rather than in the macroscopic separation of charge. Batteries provide a practical way to store large amounts of energy for long periods, but they are not a practical way to quickly deliver electrical energy when it is needed. (By electrical energy, we essentially mean the movement of charge. Chemical energy can in other situations be delivered quickly, as in chemical explosives.) Conversely, whereas a capacitor cannot store as much energy as a battery, it can deliver electrical energy much more quickly. This is achieved if a charged capacitor is "shorted," meaning that the two plates are connected by a conducting wire, as in Fig. 25–8.

(a)

(b)

▶ **FIGURE 25–8** (a) Energy is stored in a capacitor by charging it. (b) Shorting the capacitor allows the charges on the capacitor to recombine and results in a rapid release of the stored energy, which makes a spectacular display of light and heat.

25–3 Energy in Electric Fields

The electric field is of fundamental physical significance. In this section, we develop the concept that the energy of a capacitor is contained in the electric field itself.

We start by relating the expression for the energy in a capacitor to the strength of the electric field in that capacitor. The parallel-plate capacitor is convenient for this purpose because both the capacitance and the field are known (Fig. 25–4). Equation (25–4) gives the capacitance for this case, $C = \varepsilon_0 A/d$, where A is the plate area and d is their separation. The field has constant strength E, and the potential difference between the plates is $V = Ed$. Thus Eq. (25–9) gives the energy

$$U = \frac{CV^2}{2} = \frac{\varepsilon_0 A}{2d}(Ed)^2 = \left[\frac{\varepsilon_0 E^2}{2}\right](Ad). \qquad (25\text{–}11)$$

We have written Eq. (25–11) so that the *volume* of the space between the plates, Ad, stands out. This is the volume that contains the electric field and, because the field is constant, the coefficient of the volume (the quantity in square brackets) in Eq. (25–11) is the **energy density**, u, or energy per unit volume:

$$u \equiv \frac{U}{\text{volume}} = \frac{\varepsilon_0 E^2}{2}. \qquad (25\text{–}12)$$

ENERGY DENSITY IN ELECTRIC FIELD

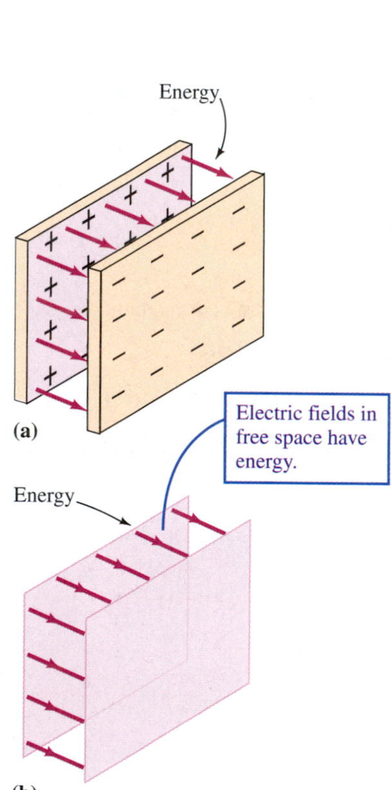

▲ **FIGURE 25–9** Electric fields have energy whether they are (a) inside a capacitor or (b) in free space.

This result suggests that the energy of a capacitor is located where the associated electric field exists, in the space between the plates (Fig. 25–9a). Now imagine that the plates aren't there but the field remains (Fig. 25–9b). Equation (25–12) is in fact *a general expression for the local energy density in free space, for a constant or variable electric field*. Wherever there is an electric field—even one that varies throughout space—the energy density, or energy per unit volume, at a particular location in space is found by squaring the electric field there and multiplying by $\varepsilon_0/2$. This makes sense if we use the reasoning that if a physical configuration can do work, it must contain energy—after all, if we place a test charge in a region of space where there is an electric field, it will accelerate, gaining energy in the process. In Section 25–5 we shall see that we must modify the coefficient $\varepsilon_0/2$ when dielectrics are present, but the energy density is always *proportional* to the square of the field.

CONCEPTUAL EXAMPLE 25–6 The plates of an isolated parallel-plate capacitor with a fixed charge on the plates are pulled slightly apart. Does the energy density change in the region between the plates? Does the overall energy in the field increase and, if so, why?

Answer We have argued that the energy density u depends on the electric field, but the field magnitude is σ/ε_0, and as long as the plate separation is small compared to their size, this doesn't depend on the distance between the plates. The energy density is constant. However, the volume between the plates has increased, so that the total energy in the field increases. This makes sense because the two oppositely charged plates attract, and it takes positive work to separate them further. We have added energy to the system by moving them apart, and this shows up in the greater total energy in the field.

EXAMPLE 25–7 An isolated spherical conductor of radius R carries a total charge Q. (a) Determine the energy density at each point in space as a function of the distance r from the sphere's center. (b) Use this energy density to compute the system's total energy. (c) Compare this total energy to the work done in charging the sphere.

Strategy (a) Equation (25–12) expresses the energy density in terms of the electric field. Equation (22–5) gives the electric field at a radius r outside the charged sphere. This field is radial and has magnitude $E = Q/(4\pi\varepsilon_0 r^2)$. Inside the conducting sphere, the field is zero.

(b) We must integrate the energy density over all space to find the total energy. As we have spherical symmetry, we can reduce the three-dimensional integral over all space to a one-dimensional integral over the radial distance by considering the volume as a series of concentric spherical shells. Because the energy density is zero inside the sphere, we need integrate only over radii larger than R.

(c) To calculate the work required to charge the sphere by bringing charge in from infinity, suppose that the sphere already has charge q and is at a potential of $V = q/(4\pi\varepsilon_0 R)$. The additional work to bring charge dq in is $dW = V\,dq$, and this must be integrated from $q = 0$ to $q = Q$.

Working It Out (a) Given our value for the electric field, the energy density at distance r is

$$u = \frac{\varepsilon_0 E^2}{2} = \frac{Q^2}{32\pi^2\varepsilon_0 r^4}$$

outside the sphere ($r > R$) and zero inside ($r < R$).

(b) The volume dV of a thin shell of thickness dr is the product of dr and the surface area of the shell, $4\pi r^2$. Then the volume element $dV = 4\pi r^2\,dr$, and the system's total energy is

$$U = \int_{\text{volume}} u\,dV = \frac{Q^2}{32\pi^2\varepsilon_0}\int_R^{\infty}\frac{4\pi r^2}{r^4}\,dr$$
$$= \frac{Q^2}{8\pi\varepsilon_0}\int_R^{\infty}\frac{dr}{r^2} = -\frac{Q^2}{8\pi\varepsilon_0}\frac{1}{r}\bigg|_R^{\infty} = \frac{Q^2}{8\pi\varepsilon_0 R}.$$

(c) We integrate Vdq from $q = 0$ to Q to determine the total work done. This gives

$$W = \int dW = \int V\,dq$$
$$= \int_0^Q \frac{q}{4\pi\varepsilon_0 R}\,dq = \frac{1}{4\pi\varepsilon_0 R}\int_0^Q q\,dq = \frac{Q^2}{8\pi\varepsilon_0 R}.$$

The work done to bring the charge in from infinity is indeed equal to the total energy of the system found in part (b). This is a good check on the validity of our calculations.

What Do You Think? Our results indicate that the energy density increases dramatically for smaller R. Does this make sense?

25–4 Capacitors in Parallel and in Series

Circuits

Here we begin our discussion of electric circuits, in which the physics of electric and magnetic fields have direct application to a myriad of electrical devices that we use each day. We have already mentioned two circuit elements: capacitors and batteries. We use the universal symbols shown in Fig. 25–10 for batteries and capacitors in circuit diagrams, which are schematic drawings that illustrate the connection of various circuit elements. The lines connecting circuit elements in these diagrams are assumed to be perfectly conducting wires, which means that they form equipotentials. We can use circuit diagrams to single out some interesting capacitor combinations. In a **parallel** combination, capacitors are connected as in Fig. 25–11a, with equal voltage drops across the capacitors; in a **series** combination, capacitors are connected as in Fig. 25–11b, with equal charges on the capacitors. We can simplify the analysis of circuits by finding the capacitance of the single *equivalent capacitor* that can replace these combinations without changing the potential across them.

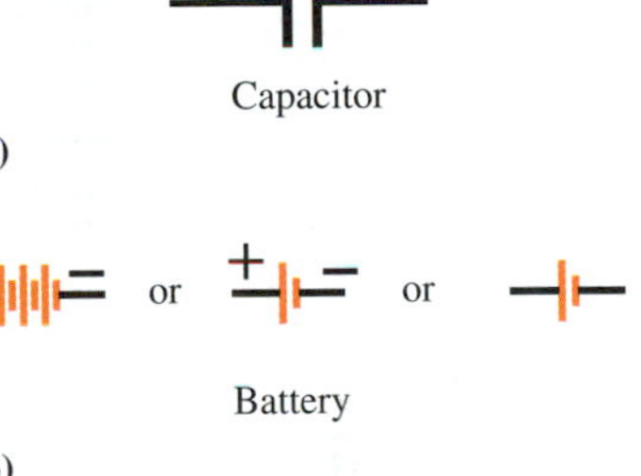

▲ **FIGURE 25–10** The symbols used to indicate (a) capacitors and (b) batteries in electric circuits.

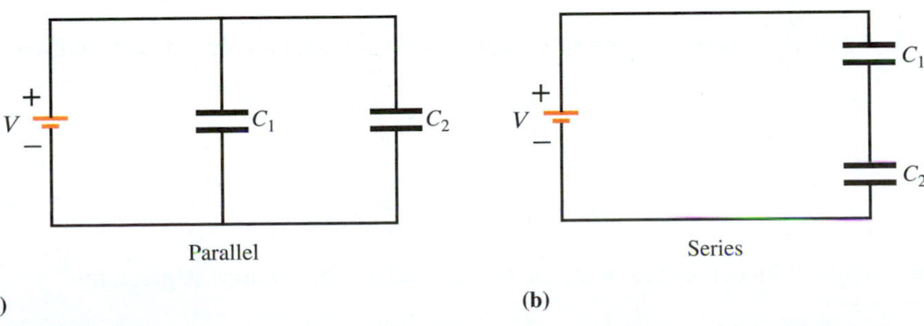

▲ **FIGURE 25–11** (a) Circuit with two capacitors connected in parallel. (b) Circuit with two capacitors connected in series.

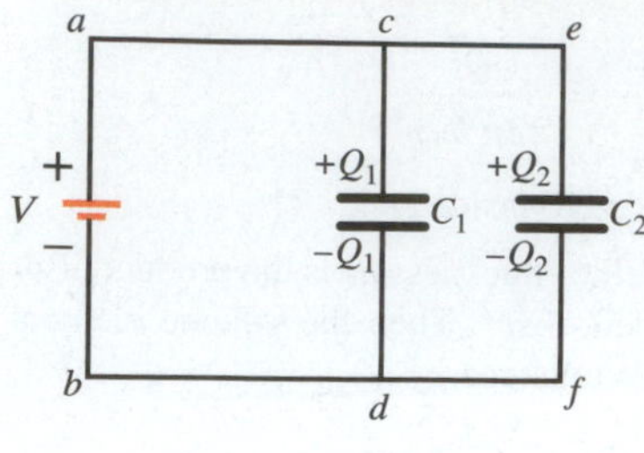

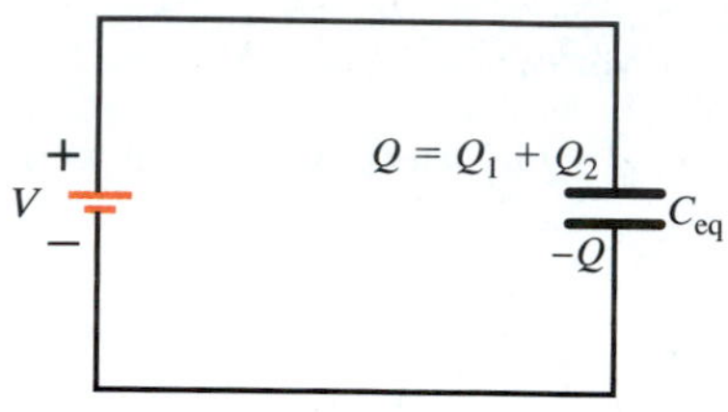

▲ **FIGURE 25–12** (a) A battery is used to place the same voltage V across the two capacitors connected in parallel. The charge on each capacitor depends on their individual capacitances. (b) The two capacitors can be replaced by an equivalent capacitor C_{eq}.

Parallel Connection

The battery in Fig. 25–12a maintains a potential V across the points a and b. Because the connecting wires are perfect conductors, the line ace is an equipotential, as is the line bdf. The potential difference from points c to d and from e to f must therefore also be V. *Capacitors connected in parallel have the same potential between their conductors.*

What is the equivalent capacitance, C_{eq}, of the single capacitor—defined by Fig. 25–12b—that could replace the parallel combination of capacitors C_1 and C_2 in such a way that the behavior of the circuit outside the parallel combination is unchanged? The equivalent capacitor will have the same potential difference V and total charge Q as the parallel combination of C_1 and C_2 and it is this condition that determines C_{eq}. The charges on C_1 and C_2 are related to the voltage across each capacitor:

$$Q_1 = C_1V \quad \text{and} \quad Q_2 = C_2V.$$

The charge for the equivalent capacitor is $Q = C_{eq}V$, and the total charge produced by the battery for the circuit in Fig. 25–12a is $Q_1 + Q_2$, so

$$Q = Q_1 + Q_2 = C_1V + C_2V = (C_1 + C_2)V = C_{eq}V.$$

The last equality shows that

$$C_{eq} = C_1 + C_2. \tag{25–13}$$

With n capacitors connected in parallel, we can similarly show that the equivalent capacitance is

$$C_{eq} = C_1 + C_2 + \cdots + C_n. \tag{25–14}$$

When capacitors are arranged in parallel, the total capacitance is larger than any of the individual capacitances.

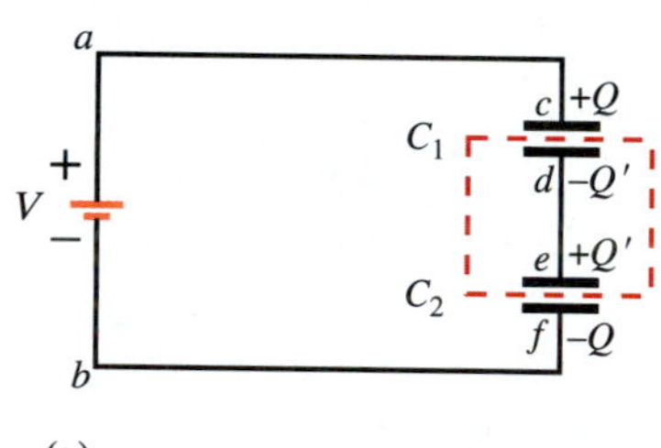

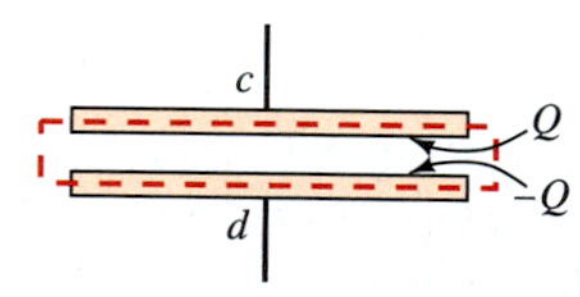

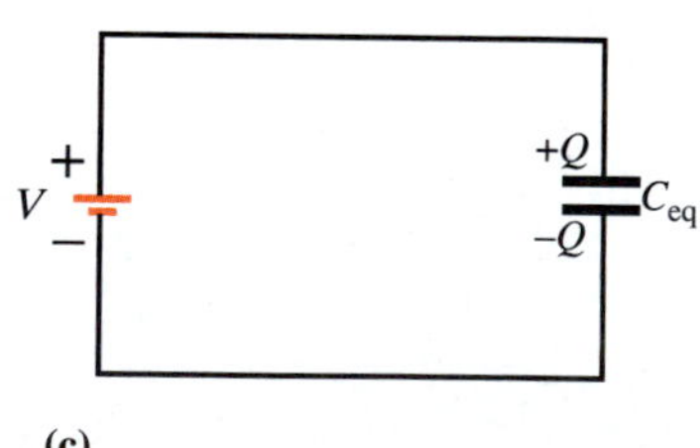

▲ **FIGURE 25–13** (a) Two capacitors connected in series must have identical charges $\pm Q$ but can have different voltages. The net charge within the dashed region is zero. (b) By drawing a Gaussian surface we establish that $Q' = Q$. (c) The two capacitors can be replaced by an equivalent capacitor C_{eq}.

Series Connection

We can also determine the single capacitor equivalent to capacitors connected in series (Fig. 25–13a). Because the battery maintains a fixed potential V, charge $+Q$ appears at c and charge $-Q$ at f. The positive charge at c induces a charge $-Q'$ at d, and because the isolated piece of metal enclosed by the dashed line in Fig. 25–13a is neutral, $+Q'$ also appears at e. We now show that $Q' = Q$ by drawing a Gaussian surface like that shown in Fig. 25–13b. Because there is no field within the metal plates and because there is no electric flux through the short side portions, the flux through the surface is zero, the net charge enclosed must be zero, and $Q' = Q$. Capacitors C_1 and C_2 thus have identical charges Q. *Capacitors connected in series have identical charges.*

The single equivalent capacitor C_{eq} in the circuit of Fig. 25–13c must carry the identical charges $+Q$ and $-Q$ and have the same potential difference V across it as exists between c and f in the circuit of Fig. 25–13a. Capacitor C_1 has potential $V_1 = Q/C_1$; similarly, capacitor C_2 has potential $V_2 = Q/C_2$. The total potential is $V = V_1 + V_2$, so we have

$$V = V_1 + V_2 = \frac{Q}{C_1} + \frac{Q}{C_2} = Q\left(\frac{1}{C_1} + \frac{1}{C_2}\right) = \frac{Q}{C_{eq}}.$$

The last equality shows that the value of the equivalent capacitance is determined by

$$\frac{1}{C_{eq}} = \frac{1}{C_1} + \frac{1}{C_2}; \tag{25–15}$$

$$C_{eq} = \frac{C_1C_2}{C_1 + C_2}. \tag{25–16}$$

With n capacitors connected in series, the equivalent capacitance is given by

$$\frac{1}{C_{eq}} = \frac{1}{C_1} + \frac{1}{C_2} + \cdots + \frac{1}{C_n}. \tag{25–17}$$

When capacitors are arranged in series, the total capacitance is given by the reciprocal of the expression in Eq. (25–17), and this capacitance is less than any of the individual capacitances.

CONCEPTUAL EXAMPLE 25–8 Suppose you add an additional capacitor in series to a group of capacitors already in series. Does the equivalent capacitance increase or decrease?

Answer According to Eq. (25–17), if we add another capacitor in series, the value of $1/C_{eq}$ increases, so that C_{eq} decreases.

EXAMPLE 25–9 Determine the equivalent capacitance for the capacitors in the circuit shown in Fig. 25–14a.

Strategy In this circuit, the capacitors are combined both in series and in parallel, and we want to find the single equivalent capacitance for the whole combination. Our first step is to combine the two capacitors in parallel into one capacitor of value C'_{eq} (Fig. 25–14b); we then combine the three remaining capacitors in series into one final equivalent capacitor (Fig. 25–14c).

Working It Out We use Eq. (25–14) to combine the parallel capacitors in Fig. 25–14a:

$$C'_{eq} = 10\ \mu\text{F} + 6\ \mu\text{F} = 16\ \mu\text{F}.$$

Having reduced the capacitor arrangement to the one shown in Fig. 25–14b, we combine the series capacitors by using Eq. (25–17), as in Fig. 25–14c:

$$\frac{1}{C_{eq}} = \frac{1}{5\ \mu\text{F}} + \frac{1}{16\ \mu\text{F}} + \frac{1}{2\ \mu\text{F}} = \frac{61}{80}(\mu\text{F})^{-1};$$

$$C_{eq} = 1.3\ \mu\text{F}.$$

What Do You Think? If the 2 μF capacitor is 1 μF instead, is the equivalent capacitance smaller, larger, or the same?

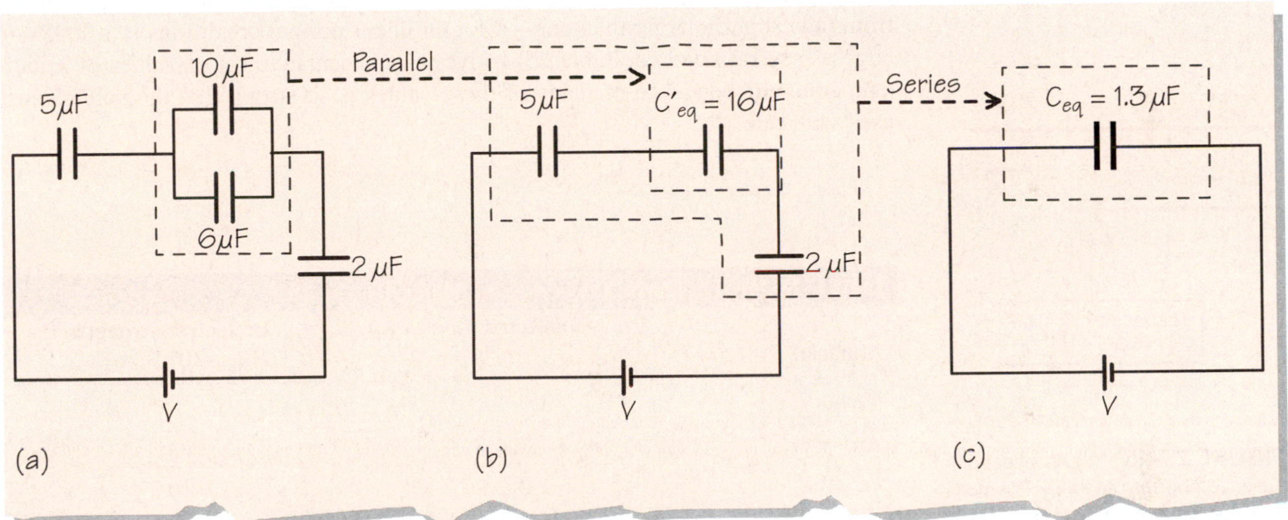

▲ **FIGURE 25–14** (a) Four capacitors of this electric circuit can be combined into one. (b) The two parallel capacitors are combined, giving three capacitors in series, (c) which are then combined into one *equivalent capacitor* C_{eq}.

25–5 Dielectrics

Many materials, including paper, plastics, and glass, do not conduct electricity easily; we referred to them earlier as *insulators*. Nevertheless, they modify the external electric fields in which they are placed. We call these materials **dielectrics**. We shall see that a dielectric placed in a capacitor allows more charge on the capacitor for a given voltage. A solid dielectric placed between a capacitor's two conductors also lends strength and mechanical stability to the capacitor. Finally, a dielectric can reduce the possibility of charge leakage and of sparking across the plates of a capacitor.

In this section, we take an experimental point of view of dielectrics, and in the next section we explore the microscopic origin of this behavior. Before we describe the basic experimental facts, however, it is worthwhile to take a peek ahead and describe in a conceptual way the underlying mechanism. Electrons are not free to move in dielectrics as they do in conductors, but that does not mean there is no response to an applied field. Dielectrics contain molecules that can behave as dipoles, either permanent ones or induced ones, and we can illustrate the possible effects with a collection of permanent

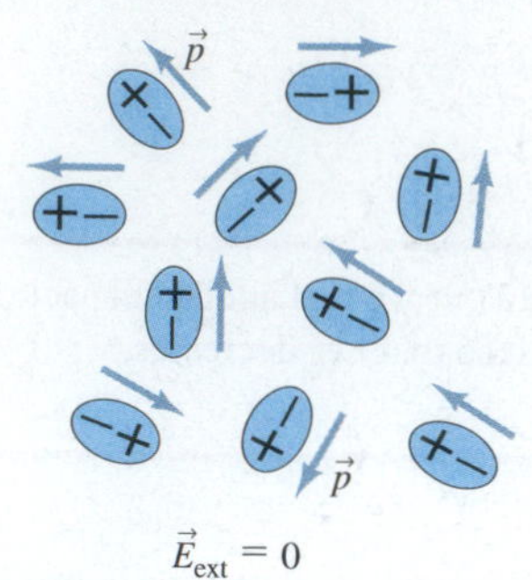

(a)

(b)

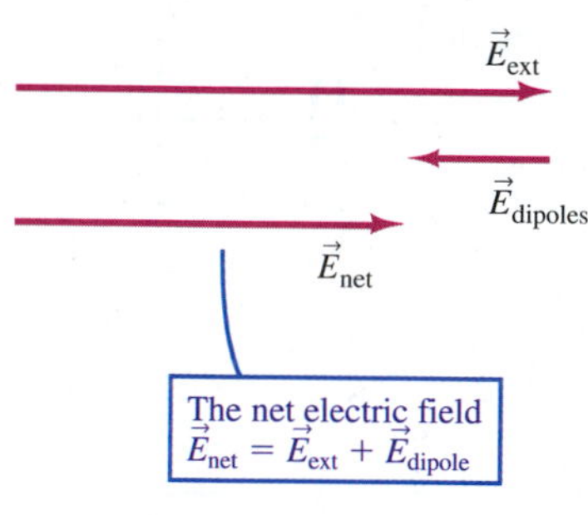

(c)

▲ **FIGURE 25–15** (a) A collection of permanent dipoles is randomly oriented. (b) In the presence of an external electric field, the dipoles tend to align themselves along the external field. (c) The aligned dipoles produce their own field $\vec{E}_{dipole}$ that tends to cancel the external field $\vec{E}_{ext}$ that aligned them, producing a net field $\vec{E}_{net}$.

dipoles. When such a collection starts completely unaligned (Fig. 25–15a) and is exposed to an external field, the dipoles align with the field as in Fig. 25–15b. But once there is a set of aligned dipoles, the set of dipoles can make its own electric field, which as in Fig. 25–15c opposes the external field. The net field within the material is thereby decreased. It is helpful to keep this qualitative picture in mind as we turn to the experimental facts.

Michael Faraday is generally given credit for having performed the first experiments, in the early nineteenth century, that showed that the capacitance increases when insulating materials are placed between the two conductors of a capacitor. If C_0 is the capacitance of a given capacitor in a vacuum (or in air), then the capacitance C of the same capacitor with dielectric between its conductors is larger than C_0 by a factor called the **dielectric constant**, κ. We have

$$C = \kappa C_0; \qquad (25\text{–}18)$$

CAPACITANCE MODIFICATION IN DIELECTRICS

$$\kappa \equiv \frac{C}{C_0}. \qquad (25\text{–}19)$$

The dielectric constant κ, which is larger than unity for all materials, depends on the material as well as on external conditions such as temperature. The value of κ can run from only slightly larger than one—κ for air under normal conditions is 1.0005—to as large as several hundred. Table 25–1 gives a representative set of values of κ, but the temperature dependence of many of these values is so strong that the values must be used with care.

TABLE 25–1 • Dielectric Properties of Materials†

Material	Dielectric Constant, κ	Dielectric Strength, E_{max} (10^6 V/m)
Vacuum	1.0	
Air	1.00054	3
Paraffin	2.0–2.5	10
Teflon	2.1	60
Polystyrene	2.5	24
Lucite	2.8	20
Mylar	3.1	
Plexiglas	3.4	40
Nylon	3.5	14
Paper	3.7	16
Fused quartz	3.75–4.1	
Pyrex	4–6	14
Bakelite	4.9	24
Neoprene rubber	6.7	12
Silicon	12	
Germanium	16	
Water	80	
Strontium titanate	332	8

†Values for some materials depend strongly on temperature and the frequency of oscillating fields.

You may have seen a device that makes use of the modifications due to dielectrics, an electric stud finder (Fig. 25–16). A unit with a metal plate acting as a capacitor in a circuit slides along a wall. Unless the plate is over a stud, there is air behind the wall, but when a stud is encountered, the capacitance changes because wood is a dielectric, and this is detectable in the circuit.

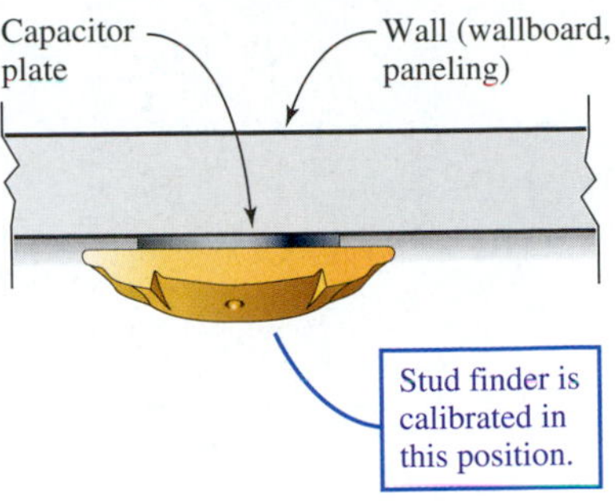

(a)

Stud

Light indicates stud is near.

(b)

▲ **FIGURE 25–16** A schematic diagram of a stud finder, as seen from above the wall looking down. In (a) the stud finder is calibrated with no dielectric material (as for example a building stud) behind the wall. (b) The capacitance changes when the stud finder approaches a stud.

Experimental Evidence for the Behavior of Dielectrics

To simulate the experiment that led Faraday to his conclusions about dielectrics, let's first use a battery to charge a parallel-plate capacitor in air to a potential V_0 and a charge $Q_0 = C_0V_0$ (Fig. 25–17a). (By "air" we mean an approximate vacuum—air itself has a dielectric constant, but it is very close to unity.) Here, the subscripts refer to the quantity in air—for example, C_0 is the capacitance when there is air between the plates. We disconnect the battery and measure the voltage, or potential difference (Fig. 25–17b). We then slide a dielectric, such as Plexiglas, between the plates (Fig. 25–17c). *The voltage is reduced.* The charge is unchanged—Plexiglas is an insulator—and the reduction factor is what we called above the dielectric constant:

$$V = \frac{V_0}{\kappa}.$$

To see that κ is indeed the dielectric constant defined above, note that the charge did not change, so we can interpret the reduction of the voltage as due to a change in the capacitance from the original value C_0 to a new value C:

$$C = \frac{Q_0}{V} = \frac{Q_0}{V_0/\kappa} = \kappa\frac{Q_0}{V_0} = \kappa C_0.$$

This verifies Eq. (25–18).

In another experiment we leave the battery connected to the capacitor after it is charged in air (Fig. 25–18a). After we insert the Plexiglas the potential remains the battery voltage V_0 (Fig. 25–18b). We observe, however, that *the charge on the conducting plates increases* by a factor of κ ($Q = \kappa Q_0$). Our experimental result remains in agreement with Eq. (25–18):

$$C = \frac{Q}{V_0} = \frac{\kappa Q_0}{V_0} = \kappa C_0.$$

Let's reinterpret these results in terms of permittivity. If we take a parallel-plate capacitor, for which $C_0 = \varepsilon_0 A/d$, we have

$$C = \kappa C_0 = \frac{\kappa\varepsilon_0 A}{d}. \qquad (25\text{–}20)$$

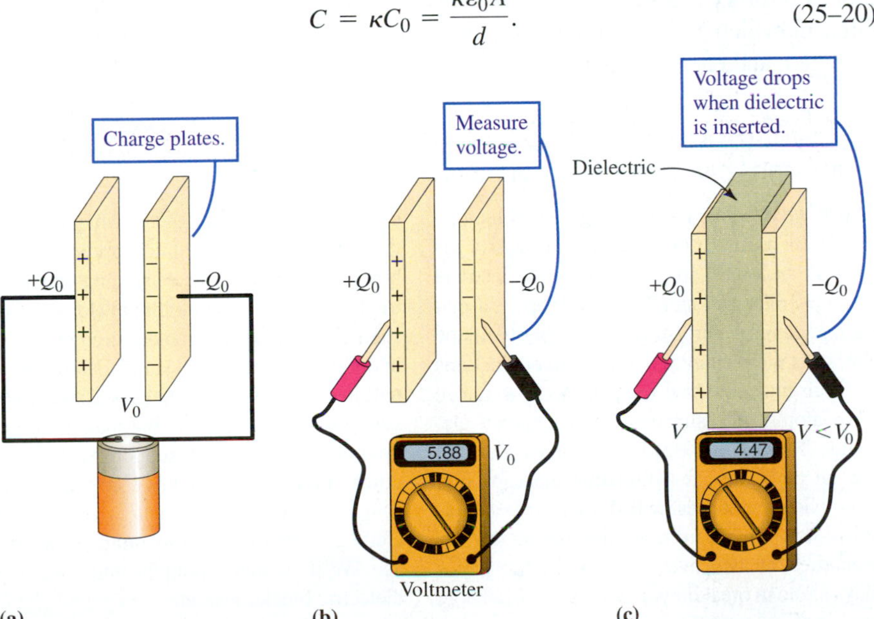

◀ **FIGURE 25–17** (a) A battery charges a capacitor to charge Q_0 and potential V_0. (b) If we take the battery away and measure the voltage with a voltmeter, we measure V_0 for the voltage. (c) If a dielectric is inserted into the capacitor, the voltage drops to $V < V_0$.

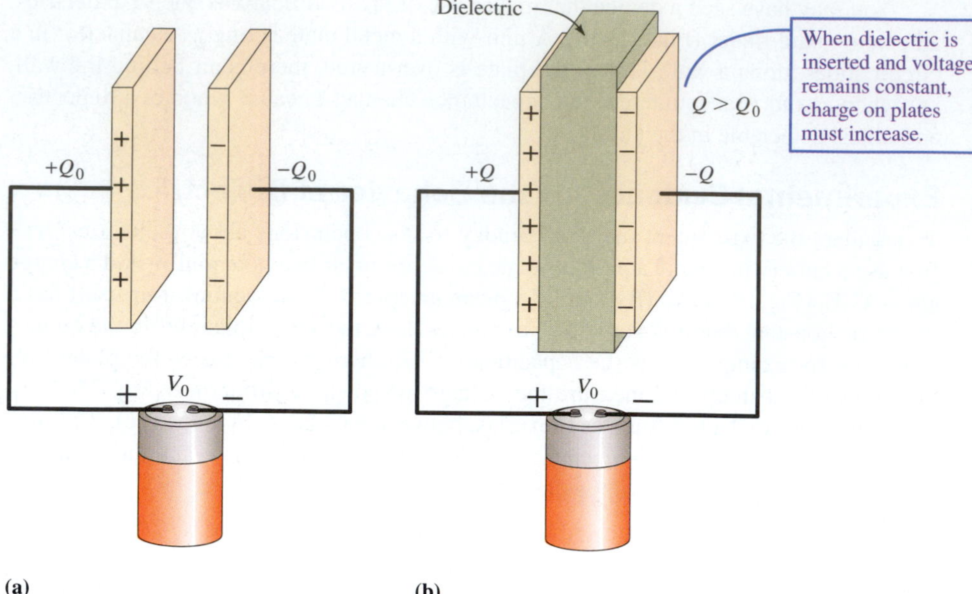

▶ **FIGURE 25–18** (a) Again, a battery charges the capacitor to charge Q_0 and potential V_0. (b) This time we leave the battery connected when we insert the dielectric. The potential must remain at V_0, but the new charge is $Q > Q_0$.

If the charge is held fixed, the new voltage is

$$V = \frac{V_0}{\kappa} = \frac{Q_0}{\kappa C_0} = \frac{\sigma_0 A}{\kappa C_0} = \frac{\sigma_0\, d}{\kappa \varepsilon_0}. \tag{25–21}$$

The new electric field between the plates is reduced in magnitude to

$$E = \frac{V}{d} = \frac{\sigma_0}{\kappa \varepsilon_0} = \frac{E_0}{\kappa}, \tag{25–22}$$

where $E < E_0$. Note that in each of Eqs. (25–20) through (25–22) we can *replace the permittivity of free space*, ε_0, *by a new permittivity*, ε, which depends on the dielectric used and on external conditions:

$$\varepsilon = \kappa \varepsilon_0. \tag{25–23}$$

PERMITTIVITY MODIFICATION IN DIELECTRICS

Although we have shown this simple rule only for the parallel-plate case, the substitution of ε for ε_0 when a dielectric is involved applies to all geometries and to all equations in which the permittivity appears, such as the expressions for field strength, potential, and energy density.

THINK ABOUT THIS. . .

HOW LARGE CAN THE VOLTAGE ACROSS A CAPACITOR BE?

Above we described a lightning bolt as the discharge of a capacitor. The voltage across the "plates" becomes so large that the electrical forces literally pull atoms in the air apart, accelerating charges across the space between the plates. As these charges collide with other atoms in the air they may have enough energy to ionize these atoms, releasing still more charged elements. There is a cascade of charges that will end on the plates and discharge the capacitor. This phenomenon is called *dielectric breakdown*. It is evident from our description that the properties of whatever is between the plates plays a role in breakdown; moreover, a discharge can carry enough energy to damage irrevocably this material and hence the capacitor itself. Thus how easily a given dielectric undergoes dielectric breakdown is an additional element, beyond such elements as mechanical strength, cost, etc., in deciding whether it is suitable to fill the space within capacitors. Each dielectric has a *dielectric strength*, E_{max}, which is the maximum electric field a dielectric will support without breakdown. Table 25–1 contains some representative values. Commercial capacitors accordingly include a limitation to a maximum allowable voltage. We'll say more about the interesting topic of dielectric breakdown later. ■

EXAMPLE 25–10 A parallel-plate capacitor has plate area 20.0 cm^2 and a plate separation 4.0 mm. (a) Find the capacitance in air and the maximum voltage and charge the capacitor can hold. (b) A Teflon sheet is slid between the plates, filling the entire volume. Find the new capacitance and maximum charge. (c) Before the insertion of the Teflon, the plates are set to a voltage of 24 V by a battery that is then disconnected. What are the energies in the capacitor before and after the Teflon is inserted?

Setting It Up We have a standard parallel-plate capacitor with known plate area A and separation d. We denote the unknown capacitance as C for air and C' for Teflon, according to what fills the space between the plates.

Strategy For parts (a) and (b), this is a matter of applying Eq. (25–4), but with ε_0 replaced by ε in part (b). We can look up the value of the dielectric constant of Teflon in Table 25–1 ($\kappa = 2.1$), and this gives us $\varepsilon = \kappa\varepsilon_0$; equivalently, $C' = \kappa C$. As for the maximum voltages, V_{max} and V'_{max} for air and Teflon respectively, that will be determined by the dielectric strengths of air and Teflon, also found in Table 25–1. Finally, the maximum charge that can be put on the plates is the capacitance multiplied by the maximum voltage.

For part (c), we fix the voltage across the capacitor in the two cases and use Eq. (25–9) to determine the energy given this voltage and the known capacitances.

Working It Out (a) For air we have

$$C = \frac{\varepsilon_0 A}{d} = \frac{(8.85 \times 10^{-12}\ \text{F/m})(2.00 \times 10^{-3}\ \text{m}^2)}{(4.0 \times 10^{-3}\ \text{m})}$$
$$= 4.4 \times 10^{-12}\ \text{F} = 4.4\ \text{pF},$$

while for Teflon,

$$C' = \kappa C = (2.1)(4.4\ \text{pF}) = 9.2\ \text{pF}.$$

(b) From Table 25–1, the dielectric strength of air and Teflon are 3×10^6 V/m and 6.0×10^7 V/m respectively, so for air

$$V_{max} = E_{max} d = (3 \times 10^6\ \text{V/m})(4.0 \times 10^{-3}\ \text{m})$$
$$\cong 10^4\ \text{V} = 10\ \text{kV},$$

and

$$Q_{max} = CV_{max} = (4.4 \times 10^{-12}\ \text{F})(10^4\ \text{V}) \cong 4 \times 10^{-8}\ \text{C}.$$

For Teflon

$$V'_{max} = E'_{max}\, d$$
$$= (6.0 \times 10^7\ \text{V/m})(4.0 \times 10^{-3}\ \text{m}) = 2.4 \times 10^5\ \text{V}$$

and

$$Q'_{max} = C'V'_{max} = (9.2 \times 10^{-12}\ \text{F})(2.4 \times 10^5\ \text{V}) = 2.2\ \mu\text{C}.$$

Both the maximum voltage and maximum charge are greatly increased after the Teflon is inserted.

(c) Before the Teflon is inserted, the energy is

$$U = \frac{CV^2}{2} = \frac{(4.4 \times 10^{-12}\ \text{F})(24\ \text{V})^2}{2} = 1.3 \times 10^{-9}\ \text{J}.$$

After the Teflon is inserted, C increases by the factor κ, whereas V decreases by the same factor. The product CV^2 thus *decreases* by a factor of κ:

$$U' = \frac{C'V'^2}{2} = \frac{(\kappa C)(V/\kappa)^2}{2}$$
$$= \frac{U}{\kappa} = \frac{1.3 \times 10^{-9}\ \text{J}}{2.1} = 6.2 \times 10^{-10}\ \text{J}.$$

What Do You Think? Did the capacitor do work as the Teflon was inserted? If so, what was its sign?

Example 25–10 shows that when the material in the space between the plates of a capacitor is replaced by a dielectric of higher dielectric constant, the energy decreases. As we implied in the "What do you think?" question for that example, the capacitor therefore does positive work as the new dielectric is inserted. In turn, this means that there must be a force that pulls the dielectric into the space between the plates, and with sensitive instruments, the tug on the dielectric can be measured.

EXAMPLE 25–11 Suppose that the Teflon sheet inserted between the capacitor plates in Example 25–10 is only 2.0 mm thick and fills just half the volume. The capacitor is isolated, and both before and after the Teflon is inserted, the capacitor carries a charge of $Q = 1.0$ nC. Find the electric field everywhere inside the capacitor, and find the new capacitance.

Setting It Up We show the new configuration in Fig. 25–19. Although the Teflon sheet is on the right in Fig. 25–19, this choice is immaterial.

Strategy We previously found the electric field between the plates in air to be $E_0 = \sigma/\varepsilon_0$. Where Teflon is present rather than air, ε_0 is replaced by $\varepsilon = \kappa\varepsilon_0$, or equivalently the electric field is $E_{Tef} = E_0/\kappa$ [Eq. (25–22)]. In this case, we will have two successive values of electric field, and by integrating the field all the way across the space, we find the voltage between the plates. Given the known charge, we can then determine the capacitance from Eq. (25–4).

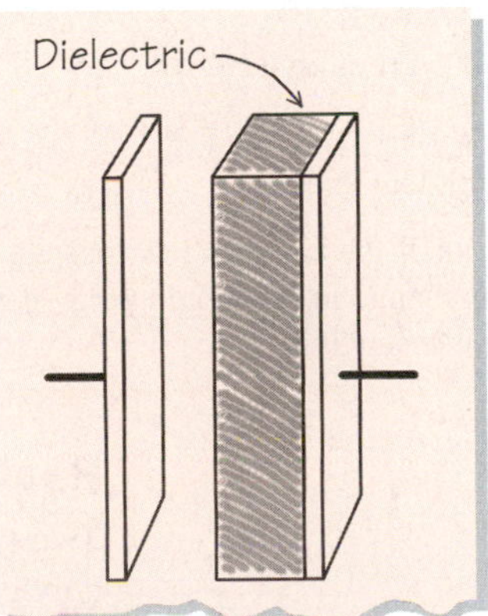

▲ **FIGURE 25–19** A dielectric inserted in a parallel-plate capacitor fills only half the volume.

(continues on next page)

Working It Out The electric field strengths in air and Teflon are

$$E_0 = \frac{\sigma}{\varepsilon_0} = \frac{Q}{\varepsilon_0 A} = \frac{1.0 \times 10^{-9}\ \text{C}}{(8.85 \times 10^{-12}\ \text{F/m})(2.0 \times 10^{-3}\ \text{m}^2)}$$
$$\cong 5.6 \times 10^4\ \text{V/m},$$
$$E_{\text{Tef}} = \frac{E_0}{\kappa} = \frac{5.6 \times 10^4\ \text{V/m}}{2.1} \cong 2.7 \times 10^4\ \text{V/m}.$$

The direction of these fields is from the positively charged plate to the negatively charged one. To find the total voltage drop across the plates (which we need to find the capacitance), we integrate the electric field over the distance between the plates:

$$V = \int_0^d E\,dx = \int_0^{2\ \text{mm}} E_0\,dx + \int_{2\ \text{mm}}^{4\ \text{mm}} E_{\text{Tef}}\,dx$$
$$= E_0 \int_0^{2\ \text{mm}} dx + E_{\text{Tef}} \int_{2\ \text{mm}}^{4\ \text{mm}} dx = E_0(2\ \text{mm}) + E_{\text{Tef}}(2\ \text{mm})$$
$$= (5.6 \times 10^4\ \text{V/m})(2 \times 10^{-3}\ \text{m}) + (2.7 \times 10^4\ \text{V/m})(2 \times 10^{-3}\ \text{m}) \cong 170\ \text{V}.$$

This calculation makes it clear that it does not matter whether the Teflon sheet is on one side or the other, or even in the middle.

Finally, the capacitance is, by definition, $C = Q/V$:

$$C = \frac{1.0 \times 10^{-9}\ \text{C}}{170\ \text{V}} = 5.9\ \text{pF}.$$

This value is intermediate to the capacitances of the system empty or filled with Teflon. This capacitor is equivalent to two capacitors in series—one empty and of width $d/2$; the other filled with Teflon and of width $d/2$.

What Do You Think? If we used Bakelite instead of Teflon, would the new capacitance be smaller, larger, or unchanged?

EXAMPLE 25–12 Consider the system shown in Example 25–3—what we'll refer to here as a coaxial cable. (We ignore the necessary mechanical presence of a support for the structure.) With equal but opposite charges per unit length λ on the two elements of the cable, the latter acts as a capacitor. A plug of insulating material of dielectric constant κ is inserted between the wire and the sheath to a depth x from the end. What is the change in potential energy of the charged cable? What electric force, if any, acts on the material as it is inserted? Assume that the charge density on the cable is unchanged as the material is inserted.

Setting It Up We show the coaxial cable in Fig. 25–20 with the radii of the central wire and outer tube a and b, respectively.

Strategy In the presence of the dielectric the charge on the elements is unchanged, while the field and hence the potential drop across the elements will change. In Example 25–3 we found the potential difference between the conductors in the coaxial cable to be $V_b - V_a = V = \frac{\lambda}{2\pi\varepsilon_0}\ln\frac{b}{a}$. By using a charge per unit length λ in place of a total charge Q, Eq. (25–10) will give us an energy per unit length, and we will want to see if this changes as the dielectric is inserted—we can anticipate from previous examples that it will. A change in the energy will imply a force, and its sign will in turn depend on the sign of the energy change: A decrease in the energy in the cable as the dielectric is inserted means a force pulling the material in. More precisely, if U is the energy in the cable as a function of the insertion distance x, the force will be $-dU/dx$.

Working It Out From Eq. (25–10),

$$\frac{U}{\text{unit length}} = \frac{1}{2}\lambda V = \frac{\lambda^2}{4\pi\varepsilon_0}\ln\frac{b}{a}.$$

In the region in which the dielectric is located, ε_0 is replaced by $\varepsilon = \kappa\varepsilon_0$. Thus the change in potential energy per unit length when the dielectric plug is inserted is

$$\frac{\Delta U}{\text{unit length}} = \frac{U_{\text{dielectric}}}{\text{unit length}} - \frac{U_{\text{air}}}{\text{unit length}} = \frac{\lambda^2}{4\pi}\left(\frac{1}{\varepsilon} - \frac{1}{\varepsilon_0}\right)\ln\frac{b}{a}.$$

Thus if the plug penetrates to a depth x, the total change in potential energy is

$$\Delta U = \frac{x\lambda^2}{4\pi}\left(\frac{1}{\varepsilon} - \frac{1}{\varepsilon_0}\right)\ln\frac{b}{a} = \left(\frac{1}{\kappa} - 1\right)\frac{x\lambda^2}{4\pi\varepsilon_0}\ln\frac{b}{a}.$$

Because $\kappa > 1$, $\Delta U < 0$. The energy decreases when the plug is inserted, so we expect the plug to be pulled into the space between the conductors. The force exerted on the dielectric as it moves into the cable is

$$F = -\frac{dU}{dx} = \left(1 - \frac{1}{\kappa}\right)\frac{\lambda^2}{4\pi\varepsilon_0}\ln\frac{b}{a}.$$

F is positive because $\kappa > 1$. The plug is indeed pulled into position.

What Do You Think? The force on the plug is independent of the depth to which it is inserted. Why?

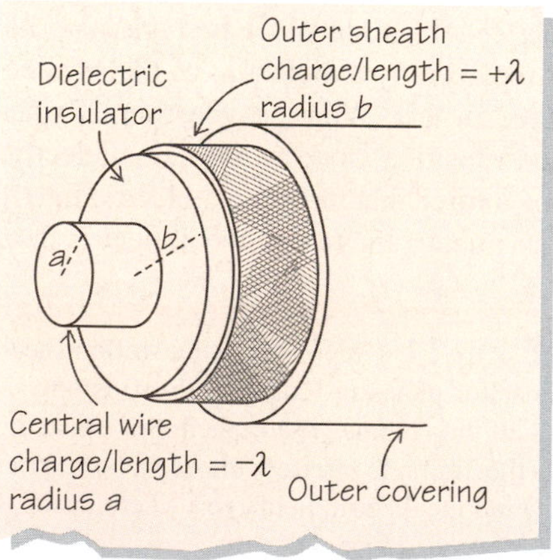

▲ **FIGURE 25–20** A coaxial cable.

Real Capacitors

Two types of capacitors comprise the bulk of modern capacitors. In *multilayer ceramic capacitors*, metal sheets separated by ceramic insulators with dielectric constants as high as 20,000 are folded into a compact form. The dielectric constant is so high that capacitances on the order of millifarads can be reached. *Electrolytic capacitors* can achieve capacitances of roughly the same size in even smaller volumes. In this case, the dielectric—a nonconducting metal oxide—is deposited in a thin layer on a sheet of metal. The second conductor is a conducting paste or liquid that adheres well to the metal oxide. The dielectric layer between the conductors can be made quite thin—as thin as 10^{-8} m. Moreover, by etching the metal before the dielectric layer is deposited, a series of sharp valleys is created in the metal, greatly increasing its surface area. If we recall that the capacitance of parallel plates is inversely proportional to the distance between the plates and proportional to the area of the plates, we see that electrolytic capacitors can achieve large capacitances.

THINK ABOUT THIS. . .

HOW ARE CAPACITORS WITH VERY LARGE CAPACITANCES CONSTRUCTED?

Capacitors with large capacitance are potentially important in many applications, and this has encouraged extensive work. The technology now exists for making capacitors with capacitances on the order of farads, but such capacitors cannot be built with a couple of flat plates! With a suitable dielectric, a 6-V potential difference across a gap of as little as 0.25 μm is possible without breakdown of the dielectric material. A 1-F capacitor for this situation must therefore have an area $A = (0.25 \times 10^{-6}\,\text{m})(1\,\text{F})/(8.85 \times 10^{-12}\,\text{F}\cdot\text{m}^{-1}) = 2.8 \times 10^4\,\text{m}^2$; this corresponds to a square more than 100 m on a side, twice as large as a football field. One way to make capacitors with capacitances this large is to make one of the "plates" from a conducting powder which is compressed only enough to join the grains but not so much as to close the pores. The result is a piece of material with a huge surface area. A dielectric is deposited to make a very thin layer on the compacted powder, and the entire piece is then immersed in a conducting liquid that forms the second "plate." While such capacitors can hold a lot of charge, the paths that current follows through them is long and, in fact, the capacitor cannot be discharged very rapidly.

25–6 The Microscopic Description of Dielectrics

We briefly argued at the start of the previous section that the presence of dipoles within dielectrics explains their macroscopic behavior in electric fields. In this section we shall expand on these comments. We refer to molecules with permanent electric dipole moments, such as H_2O, as *polar* molecules. In the absence of an external electric field, the directions of the dipole moments of polar molecules in a material are randomly distributed (Fig. 25–21a). However, when the material is placed in an external electric field, as

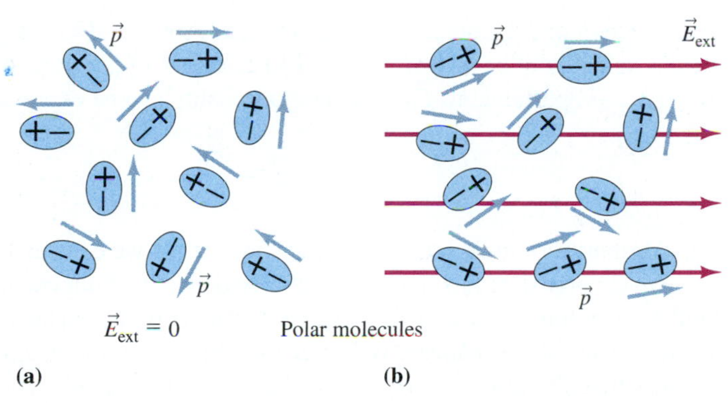

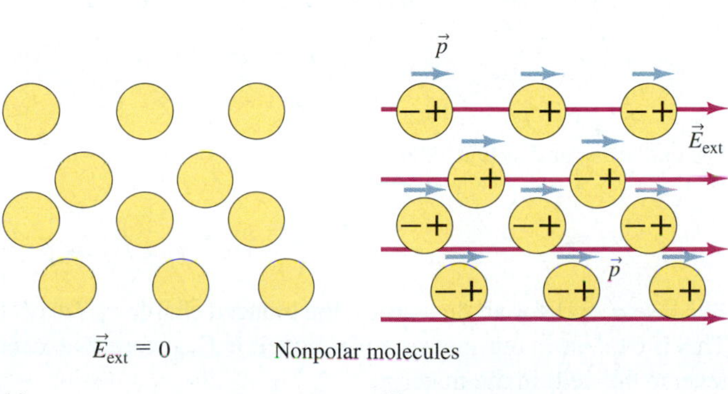

FIGURE 25–21 (a) In the absence of an external electric field, the dipole moments of polar molecules, here denoted $\vec{p}$, are randomly oriented. (b) In the presence of an external electric field, the dipole moments align themselves with the field. (c) In the absence of an external electric field, nonpolar molecules have no dipole moment. (d) In the presence of an external electric field, nonpolar molecules obtain an induced dipole moment aligned with the electric field.

in Fig. 25–21b, the dipoles are subject to a torque and tend to align themselves with the field. Thermal agitation disturbs the individual alignments, leaving only an average alignment which will be more pronounced for stronger electric fields and lower temperatures. The average degree of alignment grows *linearly* with the external electric field over a wide range of values of the field.

Nonpolar molecules are those without a permanent dipole moment. In the absence of an external electric field, their charge distributions are symmetric; that is, no particular direction is apparent (Fig. 25–21c). As we discussed in Section 22–5, when these molecules are placed in an external electric field, their charge components separate, and they acquire an induced dipole moment that is fully aligned with the field (Fig. 25–21d). The magnitude of this dipole moment increases as the external field increases, and again there is an important range of values of the external electric field for which the dipole moment grows *linearly* with the field.

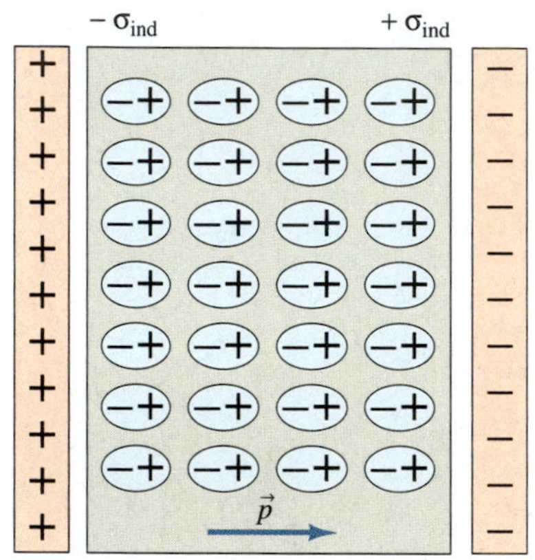

(a)

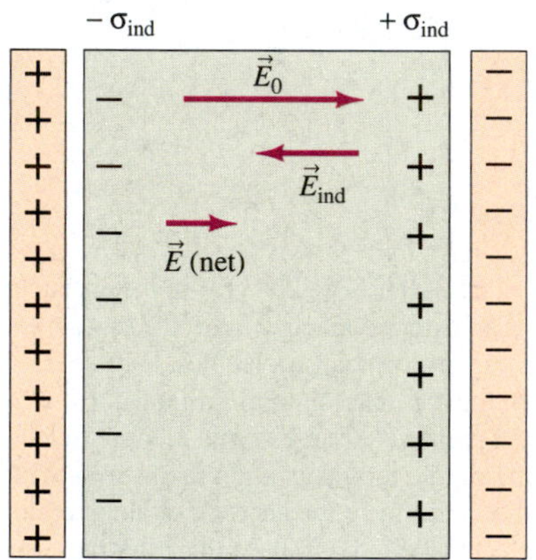

(b)

▲ **FIGURE 25–22** (a) When a dielectric is inserted in a capacitor, induced charges appear on the surface. (b) The induced charges cause an induced electric field $\vec{E}_{ind}$ opposite to the external electric field $\vec{E}_0$ caused by the free charge on the capacitor plates. The net effect is a reduced electric field $\vec{E} = \vec{E}_0 + \vec{E}_{ind}$ within the dielectric.

The effect of placing an insulating slab made of either polar or nonpolar molecules in a charged capacitor is shown in Fig. 25–22a. As opposed to a conductor, in an insulator there is no free charge to move through the material. However, the dipole moments, either permanent or induced, become aligned with the electric field. The inside of the dielectric remains electrically neutral, but the charge distribution is *polarized*, and *induced charge* appears on the two outside surfaces of the slab. We denote the induced surface charge density as σ_{ind}. The two dielectric surfaces have equal but opposite induced charge densities. The induced surface charge density is proportional either to the degree to which the permanent dipole moments of polar molecules are aligned, or, for a material consisting of nonpolar molecules, to the magnitude of the induced dipole moments. Both of these are linear in the external field over a large range of values of the external field, so there is an important range of values of the external field for which σ_{ind} *is proportional to the external field.*

By assuming that the induced charge is proportional to the external field, we can distinguish the electric fields that appear in dielectrics. There are three fields: The external field $\vec{E}_0$ is present whether the dielectric is present or not. An *induced electric field* $\vec{E}_{ind}$ is produced by the induced surface charge on the dielectric, and this is proportional to the external field. Because of the way the induced charge forms, $\vec{E}_{ind}$ is opposite to $\vec{E}_0$. Finally, the net electric field $\vec{E}$ inside the dielectric is

$$\vec{E} = \vec{E}_0 + \vec{E}_{ind}. \tag{25–24}$$

The direction of $\vec{E}$ is indicated in Fig. 25–22b. From Eq. (25–24), we can see that the resultant field $\vec{E}$ will be proportional to $\vec{E}_0$ because if $\vec{E}_{ind}$ (which we know is proportional to σ_{ind}) is proportional to $\vec{E}_0$. This proportionality can be expressed with the constant κ:

$$\vec{E} = \frac{\vec{E}_0}{\kappa}.$$

The constant κ is, in fact, the dielectric constant, as we can see from Eq. (25–22).

We can translate this in terms of charge densities. Suppose that the original electric field $\vec{E}_0$ is produced by a surface charge density σ on the capacitor plates. We refer to σ as the density of *free charge*. We know that the magnitude of $\vec{E}_0$ is $E_0 = \sigma/\varepsilon_0$ and that $\vec{E}_{ind}$ has magnitude $E_{ind} = \sigma_{ind}/\varepsilon_0$ and points to the left if $\vec{E}_0$ points to the right. From Eq. (25–22), we have $E = \sigma/\kappa\varepsilon_0$, so Eq. (25–24) becomes

$$\frac{\sigma}{\kappa\varepsilon_0} = \frac{\sigma}{\varepsilon_0} - \frac{\sigma_{ind}}{\varepsilon_0}. \tag{25–25}$$

We cancel ε_0 and solve for σ_{ind}:

$$\sigma_{ind} = \sigma\left(1 - \frac{1}{\kappa}\right). \tag{25–26}$$

Because $\kappa > 1$ for all dielectrics, the induced charge is always less than the free charge. This is evident in our microscopic model; if E_{ind} were to exceed E_0, we would actually reverse the field in the material.

Gauss' Law and Dielectrics

How does the fact that ε_0 is replaced by ε in a dielectric affect Gauss' law? For the Gaussian surface drawn in Fig. 25–23, Gauss' law gives

$$\int_{\text{closed surface}} \vec{E} \cdot d\vec{A} = \frac{Q_{\text{encl}}}{\varepsilon_0} = \frac{Q - Q_{\text{ind}}}{\varepsilon_0}, \tag{25–27}$$

where Q is the free charge enclosed, the charge that sets up the original field $\vec{E}_0$. The enclosed charge is the free charge minus the induced charge. But from Eq. (25–25),

$$\frac{Q}{\kappa} = Q - Q_{\text{ind}},$$

and this gives us an alternative form of Gauss' law when dielectrics are present, namely

$$\int_{\text{closed surface}} \vec{E} \cdot d\vec{A} = \frac{Q}{\kappa\varepsilon_0} = \frac{Q}{\varepsilon}. \tag{25–28}$$

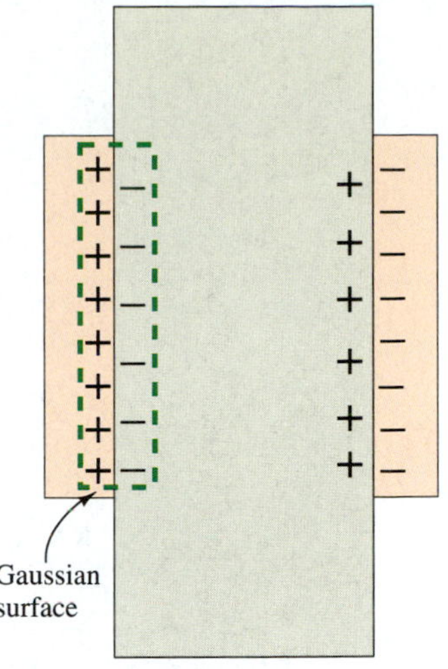

▲ **FIGURE 25–23** A dielectric fills the entire volume of a capacitor. A Gaussian surface surrounding the interface region between the dielectric and each plate surrounds both free and induced charges. The total charge enclosed by the Gaussian surface when Gauss' law is applied includes both charges.

The constant ε_0 is replaced by $\varepsilon = \kappa\varepsilon_0$ *provided Q is the free charge*. This is a general result even if Fig. 25–23 refers to a parallel-plate capacitor. If the dielectric is not uniform throughout, κ will not have the same value throughout. In this case, κ should be brought under the integral, and Gauss' law reads

$$\int_{\text{closed surface}} \kappa\vec{E} \cdot d\vec{A} = \frac{Q}{\varepsilon_0}. \tag{25–29}$$

Consequences of the Microscopic Model of Dielectrics

Our model of induced charges explains the experimental behavior of the dielectrics we have described to this point. This model is also the basis for a variety of testable experimental predictions:

1. There are two classes of dielectrics: those made of nonpolar molecules and those made of polar molecules. The dipole moments of induced dipoles are generally much smaller than those of permanent dipoles, so the value of κ for nonpolar dielectrics should be much closer to one than that for polar dielectrics. Water is an example of a polar dielectric, and you can see from Table 25–1 that its dielectric constant κ is large.
2. Thermal effects that disrupt the alignment are less strong, and polar dielectrics should line up more easily—have larger values of κ—at lower temperatures. Kinetic theory (see Chapter 19) shows that the dielectric constant takes the more precise form

   $$\kappa = 1 + \frac{\text{a constant}}{T}. \tag{25–30}$$

   This temperature dependence, which is *Curie's law*, holds rather well. The mechanism by which nonpolar dielectrics react to external fields suggests that they should not obey Curie's law, and this is indeed the case.
3. The polarization of crystalline solids with a permanent dipole moment can change if the planes of their lattice structure are stressed by being twisted or pressed. Under such stress, the internal electric fields change, and the changing fields produce an electrical signal. This phenomenon, known as *piezoelectricity*, is the principle behind the operation of some microphones and strain gauges.

Summary

Capacitors are devices for storing electric charge and energy and typically consist of two conductors with equal and opposite charges of magnitude Q and potential difference V. Capacitance is defined as

$$C \equiv \frac{Q}{V}. \tag{25–2}$$

The capacitance of a parallel-plate capacitor in air is given by

$$C = \frac{\varepsilon_0 A}{d}. \tag{25–4}$$

where A is the plate area and d is the plate separation. The SI unit of capacitance is the farad: $1\ \text{F} = 1\ \text{C/V}$.

The potential energy of a capacitor can be written as

$$U = \frac{Q^2}{2C} = \frac{CV^2}{2} = \frac{QV}{2}. \tag{25–8, 25–9, 25–10}$$

The energy density, or energy per unit volume, of an electric field is

$$u = \frac{\varepsilon_0 E^2}{2}. \tag{25–12}$$

Capacitors connected in parallel can be replaced by an equivalent capacitor with capacitance

$$C_{\text{eq}} = C_1 + C_2 + \cdots + C_n. \tag{25–14}$$

Capacitors connected in series can be replaced by an equivalent capacitor according to the relation

$$\frac{1}{C_{\text{eq}}} = \frac{1}{C_1} + \frac{1}{C_2} + \cdots + \frac{1}{C_n}. \tag{25–17}$$

Dielectrics are insulators with a characteristic property called the dielectric constant, κ, $\kappa > 1$. When a dielectric fills the space between the two conducting plates of a capacitor, the value of the capacitance is increased:

$$C = \kappa C_0, \tag{25–18}$$

where C_0 is the capacitance of the capacitor with a vacuum (or air) between its conductors. Our previous results can be modified for the presence of dielectrics by replacing the permittivity of free space, ε_0, by the permittivity ε given by

$$\varepsilon = \kappa\varepsilon_0. \tag{25–23}$$

Each insulator also has a characteristic property called the dielectric strength, which gives the approximate maximum electric field that the insulating material can withstand before it breaks down and ionizes.

The behavior of capacitors can be understood by considering the molecular structure of matter. Polar and nonpolar molecules of a dielectric become aligned with the external electric field, reducing the effects of that field. An alternative form of Gauss' law when dielectrics are present in a capacitor is

$$\int_{\text{closed surface}} \vec{E}\cdot d\vec{A} = \frac{Q}{\varepsilon}, \tag{25–28}$$

where Q is the free charge.

Understanding the Concepts

1. There are two common ways to write SI units of permittivity. Does the fact that there is more than one way present a problem?
2. You have two parallel plates, a battery, a voltmeter, and a piece of unknown plastic. Devise a method to determine the dielectric constant of the plastic.
3. What argument can you give to show that the electric field of a parallel-plate capacitor cannot drop abruptly to zero as we pass outside the region between the plates? Recall the fact that the voltage drop around any closed path must be zero.
4. What is the meaning of a capacitor with zero capacitance?
5. If the radius of the inner wire of the coaxial cable in Example 25–3 approaches zero, the capacitance per unit length of the coaxial cable also approaches zero. What is the physical significance of that?
6. It is not possible to break up every combination of capacitors into a sequence of parallel and series capacitors. Find an example of a combination that cannot be decomposed in this way.
7. We say that a capacitor has a charge Q. But is that what we really mean? Is the total charge on a capacitor actually Q?
8. From our discussion of the physical nature of dielectrics, can you imagine a physical system in which the dielectric constant is less than one?
9. The plates of a charged parallel-plate capacitor are disconnected from the charging battery and are pushed together. What happens to the potential difference, the capacitance, and the stored energy?
10. The plates of a parallel-plate capacitor, still connected to a battery with potential difference V, are pushed together. What happens to the charge on the plates, the capacitance, and the stored energy?
11. You are given a thin metal sheet of area A. You can make it into a spherical shell, roll it into two concentric cylinders, or cut it to make a parallel-plate capacitor. Which arrangement would give the largest possible capacitance?
12. What happens if you short out (connect with a conductor) the two plates of a large, charged capacitor? Could this be dangerous?
13. For finite parallel plates there is a fringe field (see Fig. 25–4). What effect would you expect this phenomenon to have on the capacitance of a parallel-plate capacitor?

14. Is it possible for a pair of nonconductors carrying equal but opposite charges to act as a capacitor? In what ways would such an arrangement differ from, or be similar to, the capacitors treated in this chapter?
15. Consider Example 25–5. Do you think you could charge more than one 12-V capacitor with a car battery? What about a thousand?
16. Why is it a good idea to short out (connect with a conductor) the plates of a large capacitor when the capacitor is not in use?
17. Would you expect the term "dielectric strength" to have meaning for a vacuum?
18. Air, particularly on humid days, can cause charge leakage. Why, then, can capacitors with air between their plates hold charge in a way that is useful for circuits?

Problems

25–1 Capacitance

1. (I) (a) What is the capacitance of two square metal plates, each 50 cm^2 in area, separated by 1 mm? (b) What is the radius of a conducting sphere with the same capacitance?
2. (I) A coaxial cable has an inner wire of radius 0.8 cm and an outer sheath of radius 1.2 cm. What is the capacitance of a kilometer of the cable?
3. (I) At different times, a 4-μF capacitor has a charge of (a) 4 μC, (b) 10 μC, and (c) 1 mC. What is the voltage across the capacitor in each case?
4. (I) How much charge can be stored on the plates of a 1-μF capacitor if the plates are attached to a battery that can give a potential difference of (a) 2 V? (b) 12 V?
5. (I) You must design a capacitor to store 2×10^{-6} C of charge, but you have only a 3-kV power supply and two metal plates of area 250 cm^2 each. What limits do you put on the separation between the plates?
6. (I) What is the capacitance of a piece of coaxial cable 180 cm long for which the radius of the inner conductor is 1.0 cm and the radius of the outer conducting sheath is 1.5 cm?
7. (II) Calculate the capacitance of two concentric spherical conductors of radii r and R, respectively. Discuss the limits of (a) finite r, $R \to \infty$; (b) $(R - r) \ll r$.
8. (II) Two concentric conducting spheres have radii of 3.0 cm and 15 cm, respectively, and an equal but opposite charge of 1.4×10^{-7} C. What is the potential difference between them? [*Hint*: Use the results of Problem 7.]
9. (II) A parallel-plate capacitor of area 0.040 m^2 carries a charge $q = 4.0 \times 10^{-8}$ C. The potential across the plates increases with time t according to the equation $V = 50.0\text{ mV} + (0.10\text{ mV/s})t$, as a result of a time-dependent increase of the separation between the plates. Find the function of time that describes the separation.
10. (II) A parallel-plate capacitor has square plates 6 cm on a side, separated by 0.3 mm. The capacitor is charged to 3 V, then disconnected from the charging power supply. What is the charge density on the plates? The total charge on each plate?
11. (II) The capacitance of a variable capacitor used in a radio varies from 0.2 μF to 0.01 μF. The capacitor is charged to a potential difference of 300 V at maximum capacitance and then isolated. At minimum capacitance, what is the voltage?

25–2 Energy in Capacitors

12. (I) A thundercloud has a charge of 900 C and a potential of 90 MV with respect to the ground 1 km below it. (a) What is the capacitance of the system? (b) How much energy is stored in the thundercloud system?
13. (I) A capacitor holds a charge of 0.068 C at a potential of 2900 V. How much energy was required to charge the capacitor?
14. (I) A capacitor in a computer with a capacitance of 0.7 pF has 2 V across it. How much energy is contained within the electric field of the capacitor?
15. (I) A fully charged flash attachment for a camera has electrical energy of 27 J. The potential across it is 300 V. What is the capacitance of the flash attachment?
16. (I) How much energy is stored on a metal sphere of radius 35 cm when a charge of 3.0×10^{-5} C is placed on it?
17. (II) A coaxial cable with an inner wire of diameter 3 mm and an outer sheath wire of diameter 8 mm has a potential of 1 kV between the wires. (a) What is the capacitance of 10 m of the cable? (b) How much energy is stored in the 10-m piece of cable? In a 1-km piece?
18. (II) Two concentric conducting spheres of radii 4.0 cm and 12.0 cm, respectively, are given equal but opposite charges of 6.0×10^{-8} C. How much energy is stored in the system?
19. (II) A capacitor consists of two parallel plates, each of area A. It is charged using a battery of potential V_0, which is then disconnected. (a) How much does the energy of the capacitor change if the separation of the plates is changed from d_0 to d_1? (b) How much work is done by the external force used to move the plates? (c) Suppose that the plates of the capacitor remain connected to the battery as they are moved. How much does the energy stored in the capacitor change under these conditions? (d) Is this change related to the work done by the force moving the plates?

25–3 Energy in Electric Fields

20. (I) The electric field in a large thunderstorm is 125,000 V/m. How much energy is contained in 1 m^3? In 1 km^3?
21. (I) The electric field due to an infinitely long, uniformly charged wire is calculated in Example 23–5. What is the electrical energy density in the space around the wire as a function of the distance from the wire?
22. (I) A Van de Graaff accelerator with a spherical dome of radius 0.75 m has a potential of 20,000 V in air. Assume that the accelerator is, in effect, a charged sphere. How much energy is stored in its electric field?
23. (I) Approximately how much energy is stored in a cube of sides 5 cm that is 1.0 m from a point charge of magnitude 5×10^{-4} C?
24. (I) The energy density in the space between the plates of a parallel-plate capacitor is 10^{-6} J/m^3. What is the voltage between the plates if the separation of the plates is 1 cm?
25. (II) An isolated metal sphere of radius 18 cm is at potential 8300 V. What is the charge on the sphere? What is the energy density of the electric field outside the sphere? Integrate this to obtain the total energy in the electric field.
26. (II) A metal sphere of diameter 25 cm carries a charge of 6.0×10^{-7} C. How much energy is contained in a spherical region of radius 50 cm that is concentric with the sphere?
27. (II) A Geiger–Muller tube is a device used to detect ionizing particles (radioactive products). It is a cylindrical capacitor with the outer metal cylinder at zero potential and the central wire at about 500 V (Fig. 25–24). (a) Calculate the capacitance of the tube if its length is 15 cm, the radius of the outer cylinder is

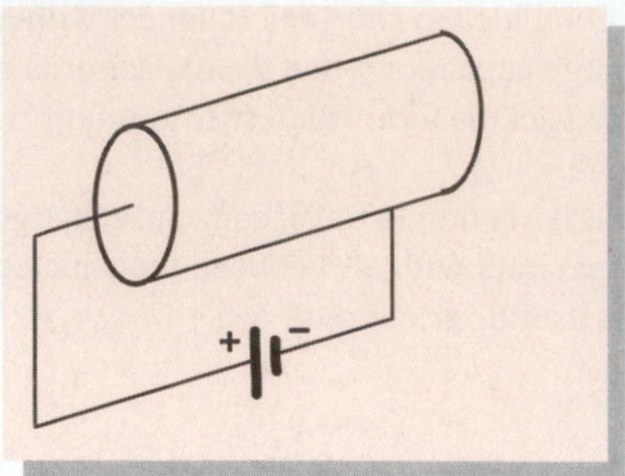

▲ FIGURE 25–24 Problem 27.

2 cm, and that of the central wire is 0.02 cm. (b) When an ionizing particle enters the detector, it creates free electrons and ions, the gas breaks down, and the capacitor discharges. How much energy is needed to recharge the Geiger–Muller tube?

28. (III) A nonconducting sphere of radius 0.070 m carries a uniformly distributed charge of 7.3×10^{-6} C. How much energy is contained in a spherical region of radius 25 cm that is concentric with the sphere?

29. (III) The plates of a parallel-plate capacitor are 400 cm^2 in area and 0.5 cm apart. The potential difference between the plates is 1500 V. (a) What is the field between the plates? (b) the charge on each plate? (c) the force exerted by the field on one of the plates? (d) Suppose that the plates are pulled apart so that the separation increases by 20 percent. What is the change in the stored energy? Is this consistent with the answer to part (c)? [If not, you probably answered (c) incorrectly.]

30. (III) Assume that an electron consists of a sphere of radius R with its charge distributed uniformly on the surface. (a) What is the electric field outside of the radius R? (b) What is the total electrostatic energy stored in the electric field? (c) Assume that all the energy of part (b) is solely responsible for the rest energy of the electron. (Rest energy is the energy associated with an object's mass, according to the theory of special relativity, even if the object is at rest. It takes the form mc^2, where in this case m is the electron's mass, 0.9×10^{-30} kg, and c is the speed of light, 3×10^8 m/s.) What must the radius R of the electron be?

25–4 Capacitors in Parallel and in Series

31. (I) When two capacitors are connected in parallel, the resulting combination has capacitance 6.5 μF. When the same two capacitors are connected in series, the resulting combination has capacitance 1.4 μF. What are the capacitances of the two capacitors?

32. (II) Find the capacitance of the parallel-plate system in Fig. 25–25. Can this system be represented by two pairs of parallel plates of half the total area connected in series or in parallel?

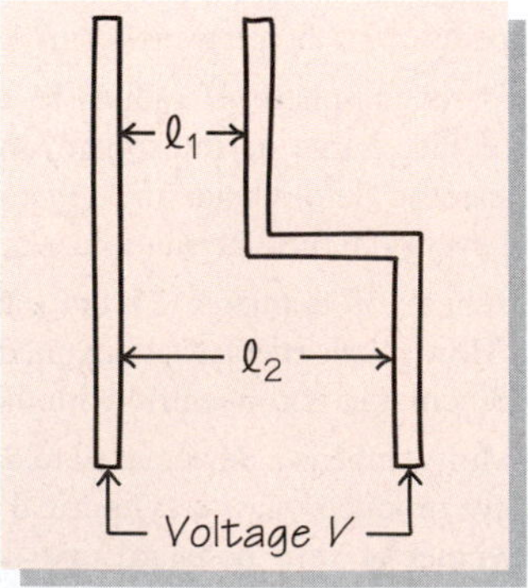

▲ FIGURE 25–25 Problem 32.

33. (II) Find the equivalent capacitance of the circuit shown in Fig. 25–26.

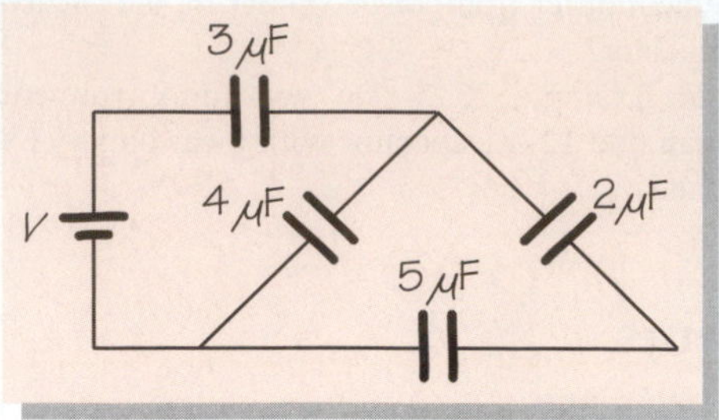

▲ FIGURE 25–26 Problem 33.

34. (II) Two large, thin metal plates of area A and thickness d, carrying charges Q and $-Q$, respectively, are placed a distance D apart (Fig. 25–27). Suppose that an uncharged, thin metal plate of the same area and thickness is placed between them, such that the distance between the uncharged plate and the positively charged plate is x. What is the capacitance of the combined system as a function of x?

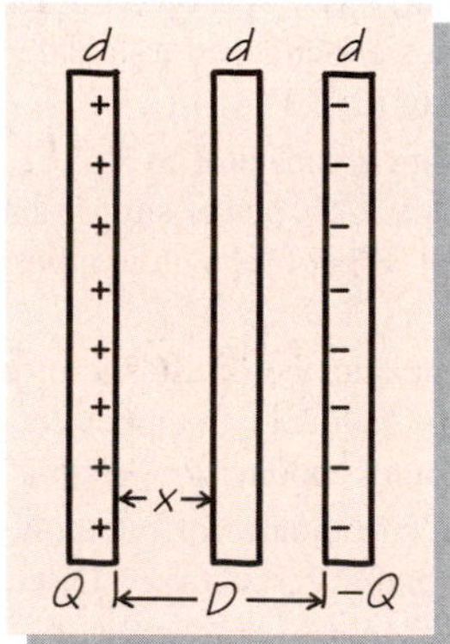

▲ FIGURE 25–27 Problem 34.

35. (II) What is the capacitance of the two concentric, spherical conductors of radii 3.0 mm and 12 mm, respectively, connected as shown in Fig. 25–28a? Suppose that the conductors are connected as shown in Fig. 25–28b. What is the capacitance now?

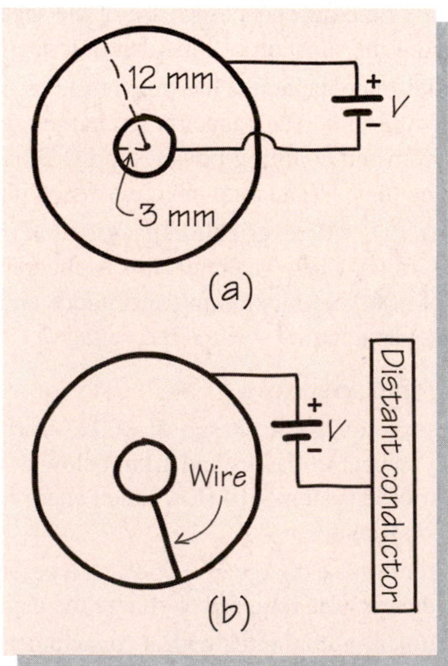

▲ FIGURE 25–28 Problem 35.

36. (II) Find the equivalent capacitance of the circuit shown in Fig. 25–29. The capacitance of each capacitor is 18 μF.

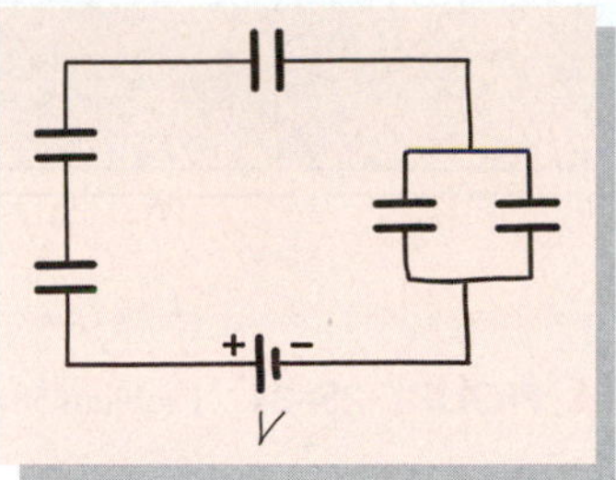

▲ **FIGURE 25–29** Problem 36.

37. (II) (a) Find the equivalent capacitance of the combination of capacitors shown in Fig. 25–30. (b) Assume that the potential difference between b and a is 300 V and find the charge on each of the capacitors in the figure.

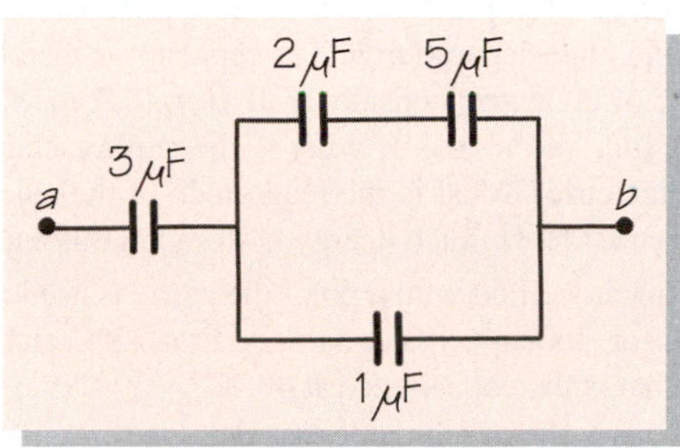

▲ **FIGURE 25–30** Problem 37.

38. (II) Figure 25–31 illustrates a set of five capacitors connected together across the points a and b. What is the value of a single capacitor that could replace this system and collect the same total charge for a given voltage drop V_{ab}?

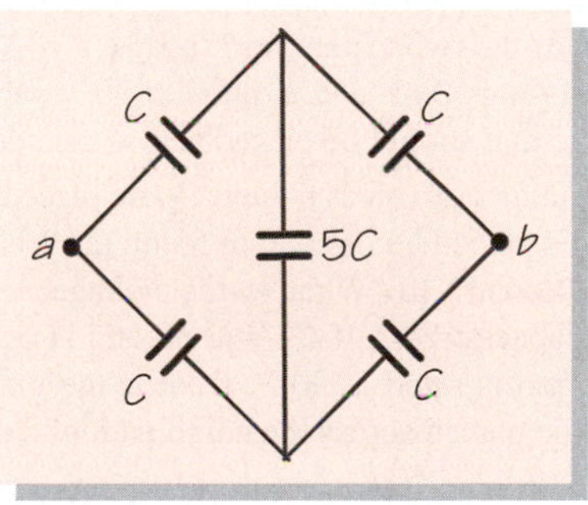

▲ **FIGURE 25–31** Problem 38.

39. (II) Figure 25–32 shows a network of identical capacitors. What is the equivalent capacitance between the points a and b? What is it between a and c? Between b and d?

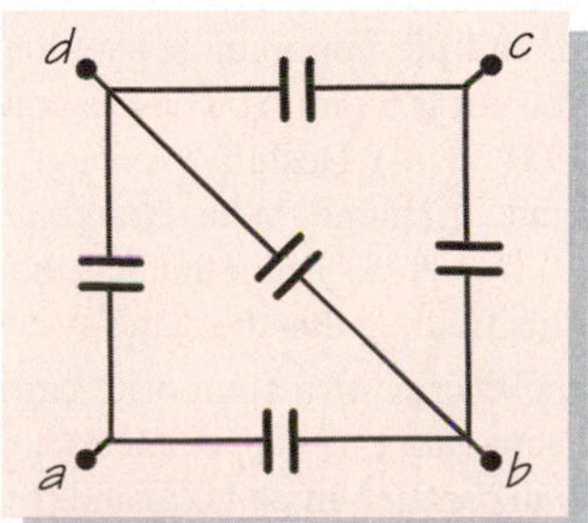

▲ **FIGURE 25–32** Problem 39.

40. (II) Capacitor C_1 has a capacitance of 175 μF; capacitor C_2 has a capacitance of 18 μF. A charge of $q = 4\ \mu$C is placed on C_1, whereas C_2 is brought to a potential difference between its plates of 3 V. (a) What is the total energy stored in the two capacitors? (b) The negatively charged plate of C_1 is connected to the positively charged plate of C_2. What will change in the system, if anything? Neglect the fringe fields at the ends of the capacitors.

41. (II) Consider the capacitors of Problem 40, with C_2 modified so that it holds a charge of 25 μC at a potential difference of 5 V between the plates. (a) How is the capacitance of C_2 modified? (b) What is the charge on the capacitor equivalent to the whole system when the negatively charged plate of C_1 is connected to the positively charged plate of C_2?

42. (II) You have four capacitors whose capacitances are 2 μF, 3 μF, 4 μF, and 5 μF, respectively. Describe a circuit with an equivalent capacitance smaller than the 5-μF capacitor by 0.032 μF.

25–5 Dielectrics

43. (I) Consider a parallel-plate capacitor in which the space between the plates is filled with Teflon. With the charge held fixed, the Teflon is replaced by Plexiglas. If the voltage across the capacitor was 600 V in the first case, what is it after the change?

44. (I) You have a piece of plastic whose dielectric constant you want to measure with two parallel plates, a 4-V battery, and a voltmeter. You charge the plates with the battery and then disconnect them. After you slide the plastic into the full volume between the plates, the voltmeter indicates a voltage drop from 4 V to 3.6 V. What is the dielectric constant?

45. (I) A 12-V automobile battery can store 4×10^6 J of energy. Find the area of a parallel-plate capacitor that can store the same amount of energy, if the separation between the plates is 1 mm and a dielectric with dielectric constant $\kappa = 3$ is between the plates.

46. (II) Repeat Problem 17 for polystyrene placed between the wires of the coaxial cable.

47. (II) A homemade capacitor is constructed out of two sheets of aluminum foil 20 cm by 15 cm in size. The foil is placed on either side of a single page of a telephone book. One hundred pages of the telephone book are 7.7 mm thick. (a) If the paper has $\kappa = 2.9$, what would the capacitance be? (b) The capacitance was measured to be 4.6 nF. What would you estimate for the actual value of the dielectric constant? (c) Make a plot of the capacitance for 1, 2, 4, 10, and 20 telephone pages between the foils. What does this curve look like? (Such low capacitances can be measured with a typical undergraduate physics laboratory multimeter. However, one does have to take account of the capacitance of the multimeter leads themselves.)

48. (II) Electrolytic capacitors involve a very thin coating of a dielectric between two conducting plates that are very close together. Let's imagine such a capacitor with plates of area 2 cm^2, but an insulating layer (and hence a separation) only 1 atom thick. If an atom is approximately 0.1 nm in diameter and has a dielectric constant of 10, what capacitance could we attain? Such a capacitor would have to operate at a relatively low voltage.

49. (II) Calculate the change in capacitance of an isolated sphere that becomes embedded in a dielectric with dielectric constant κ. If the capacitance change is due to a charge induced on the surface of the dielectric, what is the ratio of the induced charge density to that of the original surface charge density?

50. (II) Two large, parallel metal plates have a potential difference of 12 kV, and the electric field between them has a magnitude of 0.90×10^6 V/m. A material with a dielectric constant of 1.5 is inserted between the plates, with the plate separation adjusted so that the capacitance is unchanged. Calculate the new plate separation.

51. (II) A parallel-plate capacitor carrying charge q_0 is modified by the insertion of a dielectric with $\kappa = 1.8$ between the plates. As a consequence, the energy stored in the capacitor triples. What will the charge be after the dielectric is inserted?

52. (II) A coaxial cable has an inside wire of radius 3.5 mm and an outside metal sheath of radius 5.0 mm. The intermediate region is filled with a material of dielectric constant 2.2. (a) What is the capacitance of such a cable 100 m long? (b) If the potential difference between the inner and outer conductors is 500 V, what is the charge on the inner conductor, and how much energy is stored in 100 m of cable?

53. (II) A dielectric slab of thickness d and dielectric constant κ is inserted in the middle of a parallel-plate capacitor of plate separation D. What is the new capacitance of the capacitor, given that the area of each plate is A?

54. (II) A parallel-plate capacitor of dimensions 20 cm × 28 cm and separation distance 1.6 cm contains a dielectric slab of thickness 0.6 cm and dielectric constant 1.8. The potential difference between the plates is 600 V (Fig. 25–33). What are the electric fields in the empty space and inside the dielectric?

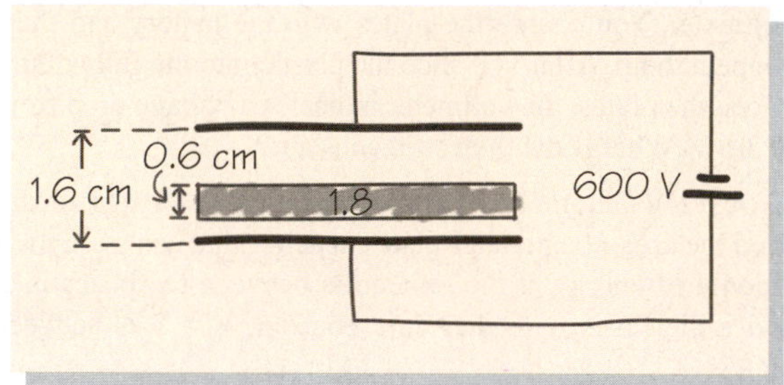

▲ **FIGURE 25–33** Problem 54.

55. (II) A parallel-plate capacitor of area 10 cm^2 and plate separation 5 mm holds how much free charge if the voltage between its plates is 300 V, and the following materials are inserted between its plates: air, paper, neoprene, Bakelite, and strontium titanate? (Use Table 25–1.)

56. (II) Two parallel plates of area 0.80 cm^2 with Plexiglas inserted between them break down when a voltage of 6 kV is applied to the plates. How much charge will the plates hold when the Plexiglas is removed? (Use Table 25–1.)

57. (II) A capacitor consists of two concentric spherical shells of radii r_1 and r_2, respectively. Calculate the capacitance if the space between the shells is filled with a dielectric of dielectric constant κ. If the capacitor starts out with air between the shells and carries a charge Q, and if the space is then filled with the dielectric, what is the change in energy?

58. (II) A parallel-plate capacitor has area $L \times L$ and separation $D \ll L$. One-half the space between the plates is filled with a dielectric for which $\kappa = \kappa_0$, and the other half with a dielectric for which $\kappa = \kappa_1$ (Fig. 25–34). Find the capacitance of this capacitor.

59. (II) A capacitor consists of 10 plates attached alternately to a positive and negative terminal. The plates are 6.0 cm × 8.0 cm in size and are 1.2 mm apart. What is the capacitance? Suppose that the region between the plates is filled with material of dielectric constant 2.8. What will the capacitance be?

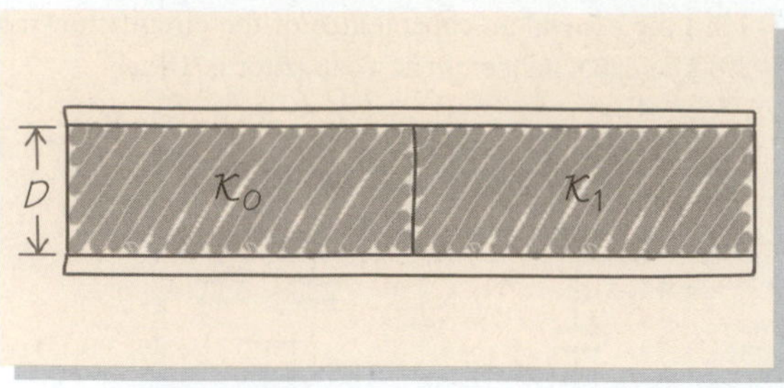

▲ **FIGURE 25–34** Problem 58.

25–6 The Microscopic Description of Dielectrics

60. (I) By measuring the capacitance and voltage of a capacitor containing a dielectric with dielectric constant 4.5, the free charge on the capacitor is measured to be 18 μC. What is the induced charge?

61. (II) Use Gauss's law and Eq. (25–24) to show, from Fig. 25–23, that $E_{\text{ind}} = \sigma_{\text{ind}}/\varepsilon_0$.

62. (II) A charge Q is placed on a parallel-plate capacitor of area $L \times L$ and plate separation d. The capacitor is then filled with a dielectric of dielectric constant κ. If $L = 0.22$ m, $d = 1.8$ mm, $Q = 0.3\ \mu$C, and $\kappa = 3.5$, what is the surface charge induced on the dielectric? What is the magnitude of the electric field in the dielectric? How much energy is stored in this capacitor?

63. (II) A capacitor filled with a polar dielectric is used as a temperature sensor. Its capacitance is 3.2 μF at 23°C and 2.65 μF at 87°C. What is the capacitance at 48° C?

General Problems

64. (I) An uncharged metal plate is inserted midway between the plates of a parallel-plate capacitor carrying charges Q and $-Q$ on the plates. Will the plate be sucked in or will it have to be pushed in? Give a simple explanation for your result.

65. (I) Two capacitors of values C_1 and C_2 are placed in series, and a voltage V is placed across the total. (a) What is the total voltage across each of the two capacitors? (b) If $C_1 = 2$ nF and is rated at 10 V, and $C_2 = 3$ nF and is rated at 30 V, what is the maximum voltage that should be placed across the combination?

66. (I) Two capacitors of values C_1 and C_2 are placed in parallel, and a voltage V is placed across one of them. (a) What is the voltage across the second? (b) What is the voltage across the single equivalent capacitor? (c) If $C_1 = 2$ nF and is rated at 10 V, and $C_2 = 3$ nF and is rated at 30 V, what is the maximum voltage that should be placed across the combination?

67. (II) *Estimate* how much charge you pick up when you walk across a carpet on a dry winter day. [*Hint*: View yourself as a good conductor, spherical in shape, and notice how close your hand has to come to a doorknob before the inevitable spark occurs. Use Table 25–1.]

68. (II) You have 300 cm^2 of aluminum plate (which you can cut into pieces) and a 500-cm^2 sheet of 5-mm-thick Bakelite (which you can also cut). Neither material can be sliced into thinner sheets or rolled, and the minimum separation between any aluminum plates you cut is 5 mm. You have a power supply of a single voltage, 1200 V. (a) Design a system that will hold the maximum amount of charge. What charge and energy can this system hold? (b) Design a system that has the maximum electric field, and find this field. Is this the same system as part (a)?

69. (II) Calculate the energy of a composite capacitor that consists of N identical capacitors of capacitance C_1 that are connected (a) in series; (b) in parallel. In parts (a) and (b), the total potential difference across the composite capacitor is V. (c) Assume that the total charge is Q, and repeat the calculation.

70. (II) A capacitor consists of two flat metal plates of area 0.28 m^2 and plate separation of $d = 1.5$ cm. A flat metal plate of the same area and of thickness 0.5 cm is inserted midway between the plates of the capacitor, leaving two spaces of thickness 0.5 cm each. (a) Find the new capacitance. (b) If the original capacitor has charge Q, what is the surface charge density induced on the intermediate plate? (c) Suppose that the original charge on the external plates remains the same. How does the energy of the new system compare to the energy of the system without the inserted plate? (d) Compare the capacitor with the metal inserted to the same capacitor with a dielectric of the same dimensions inserted.

71. (II) A parallel-plate capacitor has an area of $L \times L$ and a plate separation of $D \ll L$. It is filled with a nonuniform dielectric whose dielectric constant varies linearly across the capacitor (Fig. 25–35). At $x = 0$, $\kappa = \kappa_0$, and at $x = L$, $\kappa = \kappa_1$. We can express κ as a function of x: $\kappa = \kappa_0 + [(\kappa_1 - \kappa_0)x/L]$. Treat the capacitor plates as broken into a set of capacitors connected in parallel with plates that are strips of width dx, and calculate the capacitance.

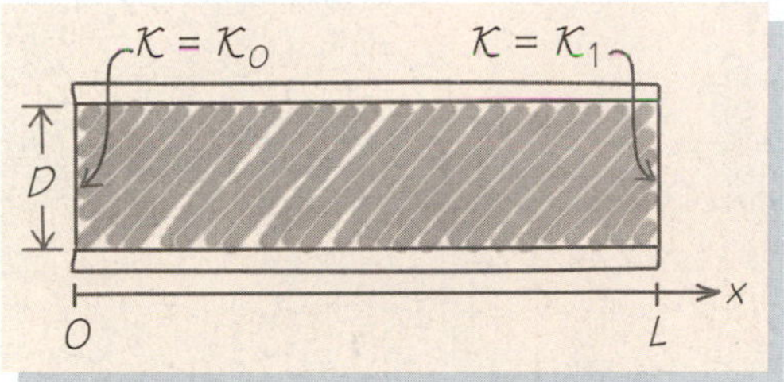

▲ **FIGURE 25–35** Problem 71.

72. (II) A thunderstorm is a fairly complicated phenomenon in terms of the distribution of charges, but we can estimate that there is a voltage drop of as much as 10^8 V between Earth and the bottom of a thundercloud, and the charges involved may run into the hundreds of coulombs. Estimate the capacitance of the Earth–cloud system and the energy contained in the space between the cloud and Earth.

73. (II) Consider the arrangement of the four initially uncharged capacitors shown in Fig. 25–36. Capacitors A, B, C, and D have capacitances 5.4 μF, 4.3 μF, 3.2 μF, and 2.1 μF, respectively. Suppose that a battery applies a potential difference of 3000 V across the circuit, which is then disconnected from the battery. What is the potential difference across each capacitor?

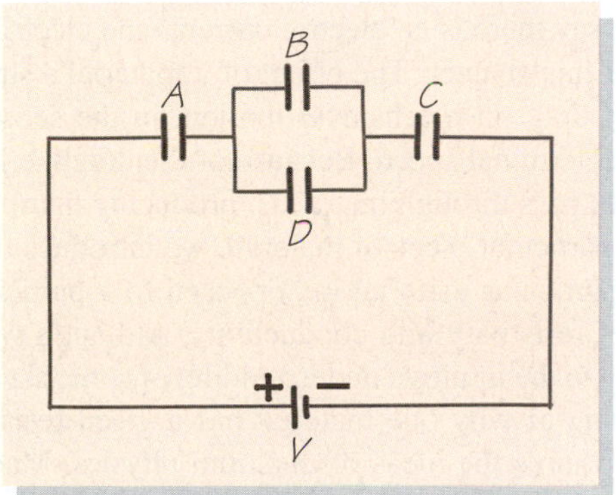

▲ **FIGURE 25–36** Problem 73.

74. (II) Consider a parallel-plate capacitor of plate area 0.40 m^2 and plate separation 3.0 mm. (a) Assume that the maximum electric field strength (before breakdown) in air is 2.7×10^6 V/m. What are the capacitance and the charge stored at the maximum voltage? (b) Suppose that the capacitor is immersed in oil of dielectric constant $\kappa = 6.0$, and the maximum charge that can be stored is a factor of 10 larger than that without the oil. What is the maximum field strength the oil can maintain?

75. (II) A parallel-plate capacitor has a capacitance of 3.0 μF. The plates are charged to 1500 V. What is the energy stored in the capacitor? How much work is required to insert a dielectric of $\kappa = 2.8$ between the plates? Assume that the capacitor is disconnected from the voltage source before the dielectric is inserted.

76. (II) A dielectric of dielectric constant κ is inserted a distance x into a parallel-plate capacitor with square plates of area A and plate separation d. What is the capacitance as a function of x? Calculate the amount of energy stored in the capacitor for a potential difference V.

77. (II) A parallel-plate capacitor has an area of $L \times L$ and a plate separation of $D \ll L$. It is filled with a nonuniform dielectric whose dielectric constant varies linearly from one plate to another (Fig. 25–37). At the bottom plate, the dielectric constant is κ_0; at the upper plate, it is κ_1. If y is the distance measured up from the bottom plate to the top plate, then $\kappa = \kappa_0 + [(\kappa_1 - \kappa_0)y/D]$. Treat the capacitor as a set of capacitors connected in series, and calculate the capacitance.

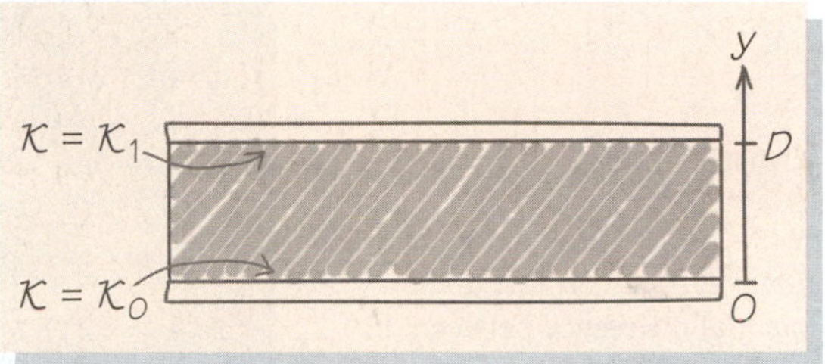

▲ **FIGURE 25–37** Problem 77.

78. (II) Three capacitors of strengths 2 μF, 4 μF, and 9 μF, respectively, can be connected in various ways between two points. What arrangement gives the smallest equivalent capacitance, and what arrangement gives the largest equivalent capacitance?

79. (II) Show that when capacitors are arranged in series, the total capacitance is less than any of the individual capacitances.

80. (III) Two identical capacitors of capacitance C are connected in series across a total potential V. A dielectric slab of dielectric constant κ can fill one of the two capacitors and is slowly inserted into that capacitor (Fig. 25–38). Compute the changes in the total electric energy of the two capacitors, in the charge on each capacitor, and in the potential drop across each capacitor. Account for any energy change by a corresponding change in energy in some other part of the system.

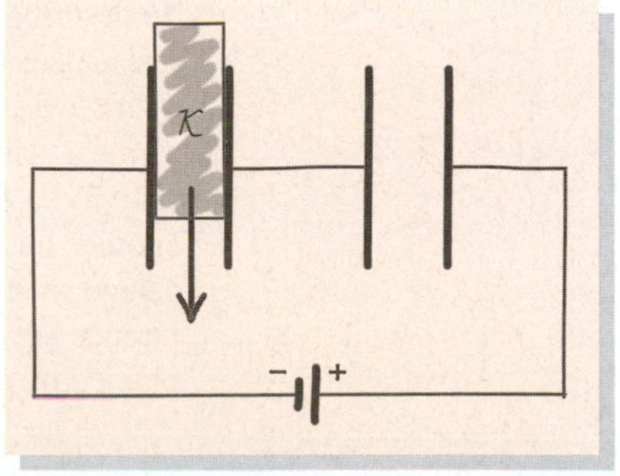

▲ **FIGURE 25–38** Problem 80.

Chapter 26

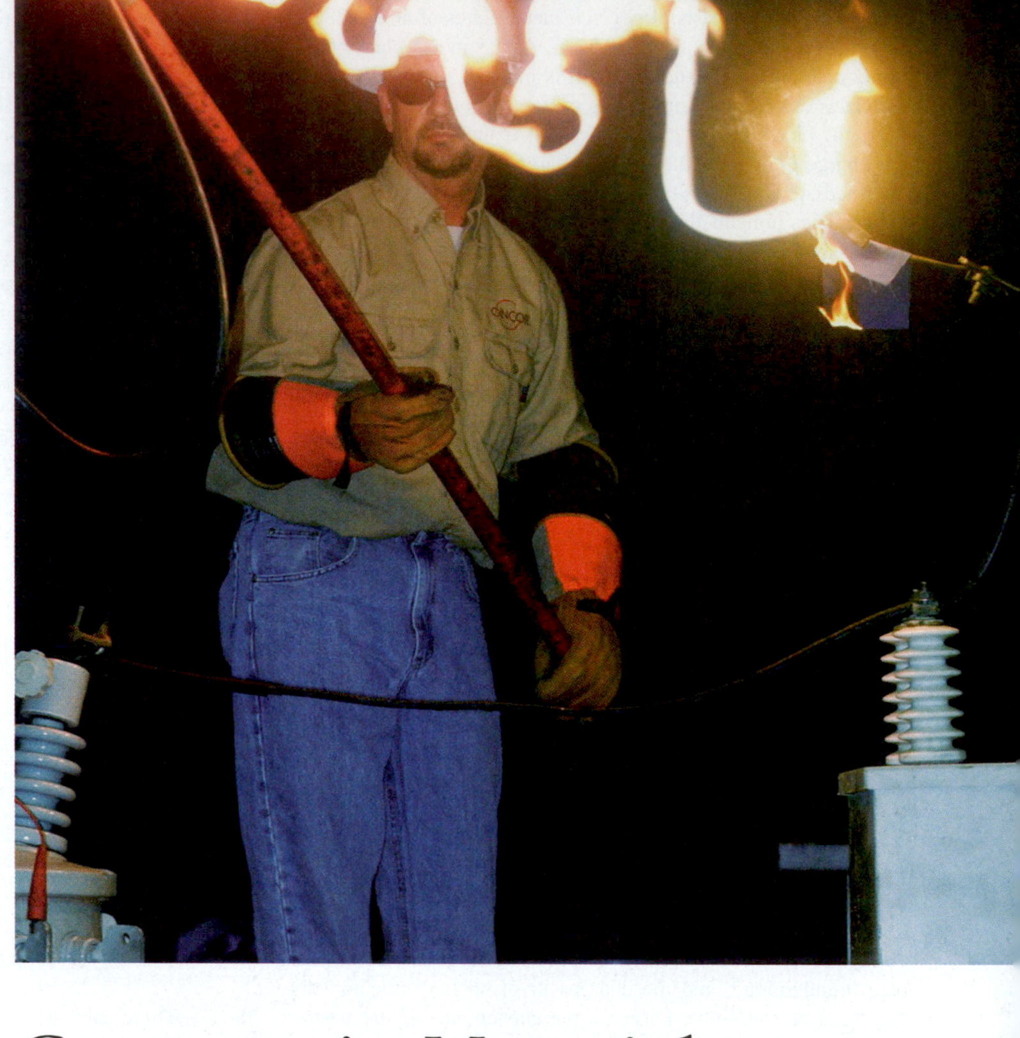

An electric arc longer than one meter jumps through air due to the large potential difference between the two electrodes. The current carried in such an arc can be quite large.

Currents in Materials

In the last few chapters we have been discussing electrostatics, but we also encountered some examples of charge movement, such as the charging and the rapid discharge of a capacitor in Chapter 25, and of the motion of electrons as an electron beam in Section 22–4. In fact the motion of charge within conducting materials—for example, within the wiring, appliances, and other circuits in your home—is everywhere around us. When charges move, we say there is an electric current, and electric currents within conductors form the topic of this chapter. The effect of a material's structure on electric current is like the effect of drag on mechanical motion, in the sense that the charges move on average at constant terminal speed. Because of the draglike forces, we must expend energy to make charges pass through materials, producing thermal energy. To describe these phenomena for a particular piece of material, we introduce the idea of resistance; to describe them for a particular material (as opposed to a particular piece of it), we introduce the concepts of resistivity and conductivity. Although we can describe a classical model that is an aid to the intuitive understanding of some aspects of resistivity, a fundamental understanding of why one material has a given resistivity and another has a different resistivity requires the ideas of quantum physics. Materials are categorized according to how easy it is for a current to move through them, and the difference between conductors, insulators, semiconductors, and superconductors also is explained by quantum physics.

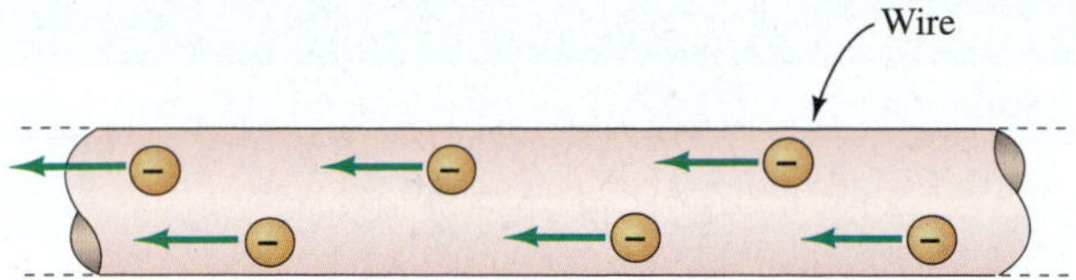

FIGURE 26–1 Charges move in a cross section of wire, meaning that a current is flowing.

26–1 Electric Current

Electric current (or just **current**) is defined as the total charge that passes through a given cross-sectional area per unit time. Current can be composed of moving negative charges such as electrons or positive charges such as protons; it may occur within an overall-neutral material such as a conductor, or it may occur as a charged beam such as the electron beam of a television tube. Here, we concentrate on the general notions of current—whether that current describes the motion of charges within free space or within conducting materials. In Fig. 26–1, we have drawn the charge that passes through a conducting wire. Recall that *charge is conserved* (see Chapter 21), meaning that unless charge accumulates within a region, as much charge enters that region as leaves it.

A simple picture may be helpful to you in thinking about electric current within the metals that conduct electric current easily. In this picture, electrons under the influence of the potential that makes them move are like the balls that pass through a pinball machine impelled by gravity. The parent ions that form the lattice structure of the material are like the pins of the pinball machine, and multiple collisions with these fixed ions introduce a random element in the motion of the electrons, slowing them down, giving them a terminal velocity on average. While this picture "explains" many features of electric current, don't take it too seriously; we'll see later in the chapter that it is quite incomplete!

If ΔQ is the amount of charge passing through the cross sectional area of a wire in a time interval Δt, then the *average current*, I_{av}, through the wire is defined as

$$I_{av} \equiv \frac{\Delta Q}{\Delta t}. \tag{26–1}$$

If the current changes with time, we define the *instantaneous* current, I, by taking the limit $\Delta t \to 0$, so that the current is the instantaneous rate at which charge passes through an area:

$$I \equiv \frac{dQ}{dt}. \tag{26–2}$$

DEFINITION OF ELECTRIC CURRENT

The unit of current is the coulomb per second; this unit is also called the *ampere* (A, or, frequently, "amp"), after André Marie Ampère, who performed pioneering work in electricity and magnetism early in the nineteenth century. While the ampere will be defined more precisely in Chapter 28, that definition is equivalent to the simple relation

$$1\ \text{A} \equiv 1\ \text{C/s}.$$

The ampere is a fairly large unit, and it is also convenient to express current in milliamps (mA, 10^{-3} A); microamps (μA, 10^{-6} A); or even nanoamps (nA, 10^{-9} A). You might note that since an electron has a charge magnitude 1.6×10^{-19} C, some 0.6×10^{19} electrons pass each second in a 1 A current.

Currents occur over a wide range of values (Table 26–1). When the current has no time dependence we refer to a direct current (DC). Currents that have a harmonic time dependence, alternating currents (AC), are commonplace; we shall study them in more detail in Chapter 33. The values in Table 26–1 that refer to AC represent the average magnitude of the oscillating current.

TABLE 26–1 • Values of Various Currents

Situation	Current (A)
Advanced-technology computer chips	10^{-12} to 10^{-6}
Electron beam of a TV set	10^{-3}
Minimum current that is dangerous when it passes through the human body (AC or DC)	10^{-2} to 10^{-1}
Proton beam of the Fermilab accelerator	0.3
Flashlight bulb	0.3
Household lightbulb (AC)	1
Automobile starter	200
Peak current in a lightning strike	10^4
Maximum current carried by a superconducting niobium wire of 1 cm^2 cross section	10^7

The Direction of Current

Current is a scalar quantity, as both charge and time are scalars, but it has a sign associated with it. It is useful to indicate the sign of the current by a directional arrow. Figure 26–2 depicts the direction of current flow within a conductor. By historical convention we associate the direction of current flow with the direction of "flow" of positive charges—even though it is actually the negative charges that move in conducting materials such as metals. In a conductor the positive charges—the atomic ions left behind by the electrons—are fixed in an ordered crystal lattice. In an ionized gas or a chemical solution, the charges that actually move and create the current may be positive or negative. This arbitrary convention for current direction causes no real problem, because a flow of positive charge to the right and a flow of the same amount of negative charge to the left represent the same current. By simply measuring the current, it is *not possible* to determine the sign of the charges that move (the *charge carriers*). By convention:

> **The direction of the current is the direction in which positive charge carriers would move, even if the actual charge carriers are negative charges moving in the opposite direction.**

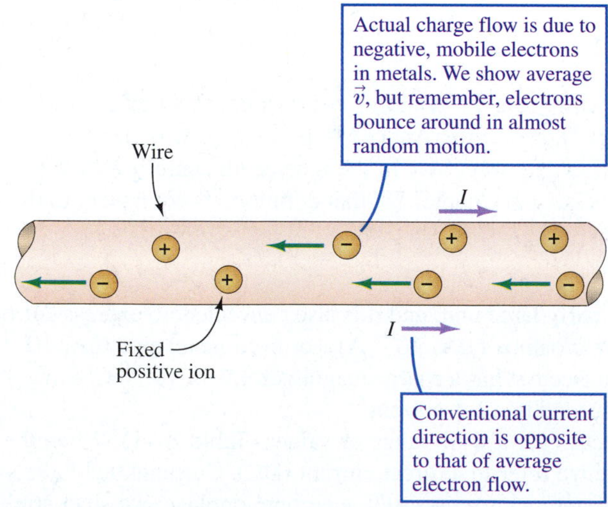

▲ **FIGURE 26–2** By convention, the current direction is the direction in which positive charge effectively moves, even though in conductors it is the negative charge, electrons, that actually moves.

EXAMPLE 26–1 An accelerator used for medical research emits protons at the rate of 2.0×10^{13} protons per second. What is the current carried by this beam of protons?

Setting It Up Let us call N the number of protons emitted per unit time. We will need to use the charge per proton, $e = 1.6 \times 10^{-19}$ C.

Strategy The current I is defined as the rate at which charge passes, i.e. the charge per unit time. Here that is simply given by $I = Ne$.

Working It Out

$$I = Ne = (2.0 \times 10^{13}\ \text{protons/s})(1.6 \times 10^{-19}\ \text{C/proton})$$
$$= 3.2 \times 10^{-6}\ \text{A}.$$

What Do You Think? In what way does this current differ from a current in a wire described above? *Answers to **What Do You Think?** questions are given in the back of the book.*

CONCEPTUAL EXAMPLE 26–2 A colliding beam experiment at the accelerator at the Fermi National Laboratory involves a beam of protons moving in the positive x-direction, and a beam of antiprotons moving in the opposite direction. Is there a net current here, and what is its sign? (For our purpose it is enough to know that antiprotons are particles with the same mass as protons, but with a charge of equal magnitude and opposite sign.)

Answer Suppose that the number of protons (charge e) that pass a given point each second is N, and the number of antiprotons (charge $-e$) per second is N'. Then the current associated with the protons is Ne in the positive x direction. The current of antiprotons is $N'(-e)$ in the *negative* x direction, and this is equivalent to a current of $N'e$ in the positive direction. Thus the total current is $(N + N')e$ in the positive x-direction. As far as current-carrying capacity is concerned, the antiprotons are equivalent to electrons.

What Do You Think? Consider also the fact that an antiproton and a proton can collide and annihilate each other, with the result that both disappear. Suppose that $N = N'$ and that all the antiprotons annihilate all the protons in some region where they can collide. Will this violate current conservation?

Current Density

Sometimes we must deal with the *details* of charge motion, not just an overall movement of charge, as for example within nonuniform materials, or even gases and plasmas. In that case we work with **current density**, $\vec{J}$, which is the rate of charge flow per unit area through an infinitesimal area. To define the current density, we must take into account the *local* magnitude and direction of the charge flow. Unlike current, which is a scalar, current density is a *vector*, with units of amperes per square meter. The direction of $\vec{J}$ is defined to be the direction of the net flow of positive charges at that particular infinitesimal element of area—since the area is tiny, it forms a plane and its direction is unambiguous.

What is the relation between current density and current? We determine this relation in a wire by dividing the finite area A through which the charge flows into infinitesimal areas $d\vec{A}$ (Fig. 26–3). This procedure is analogous to one we followed in treating fluid flow (in Chapter 16) or electric flux (in Chapter 23). The differential current dI flowing through $d\vec{A}$ is

$$dI = \vec{J} \cdot d\vec{A} = J\,dA\cos\theta, \qquad (26\text{–}3)$$

where θ is the angle between $\vec{J}$ and the area element $d\vec{A}$. From Eq. (26–3), we see that dI is a maximum when $\vec{J}$ and $d\vec{A}$ are parallel and dI is zero when $\vec{J}$ is perpendicular to $d\vec{A}$. The total current passing through the area A is a sum over the differential currents dI:

$$I = \int_{\text{surface } A} \vec{J} \cdot d\vec{A}. \qquad (26\text{–}4)$$

If the current density is uniform across a wire, the current through the wire is just the product of the wire's cross-sectional area with the magnitude of the current density.

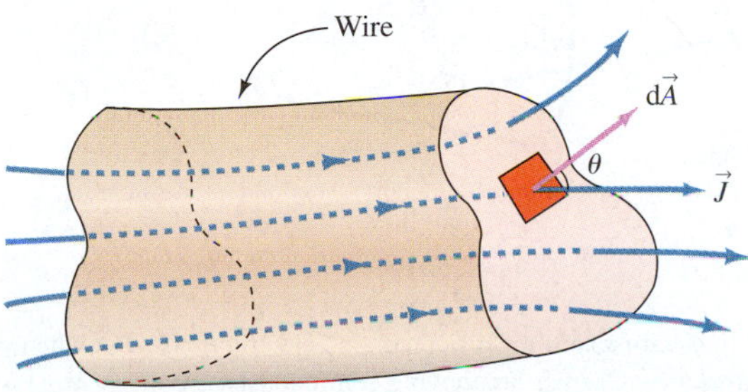

FIGURE 26–3 The area of a finite wire is divided up into differential areas $d\vec{A}$ with the current density, $\vec{J}$, whose direction is the direction of the local current, defined at every point. The direction of $d\vec{A}$ is normal to the differential area.

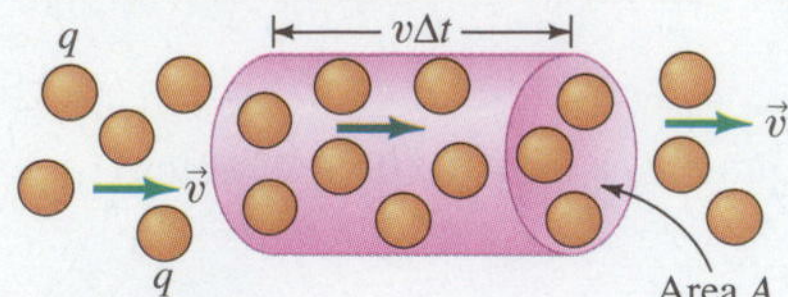

▲ **FIGURE 26–4** A collection of particles (each with charge q) with number density n_q all move to the right with velocity $\vec{v}$. The total charge passing through an area A in time Δt is $\Delta Q = n_q qvA\,\Delta t$.

Current Density of Moving Charges

Let us now connect current density to the microscopic motion of individual charges. Consider a collection of moving charges each of charge q, as in Fig. 26–4. In some small region, the number of charged particles per unit volume—the *number density*—is n_q. Suppose also that these particles all move with velocity $\vec{v}$. Then in a time interval Δt, the amount of charge passing through a given area A perpendicular to $\vec{v}$ is ΔQ, the charge contained in the volume $A(v\,\Delta t)$ swept out by the moving charges. We have

$$\Delta Q = \left(\frac{\text{charge}}{\text{volume}}\right)(\text{volume}) = (n_q q)(Av\,\Delta t) = n_q qvA\,\Delta t, \qquad (26\text{–}5)$$

where we have used the fact that the charge per unit volume is the number density of the charge carriers times the charge per particle. Thus the current is given by

$$I = \frac{\Delta Q}{\Delta t} = n_q qvA. \qquad (26\text{–}6)$$

Finally, the current density is I divided by A in the limit of *small* A, or $J = I/A$. The direction of $\vec{J}$ is specified by the direction of $\vec{v}$ with a sign determined by the sign of q:

$$\vec{J} = n_q q\vec{v}. \qquad (26\text{–}7)$$

26–2 Currents in Materials

We have referred to *conductors* as materials through which charge moves easily, *insulators* as materials through which charge does not move easily, *semiconductors* as materials intermediate to conductors and insulators, and *superconductors* as materials that under certain circumstances—in particular, at sufficiently low temperatures—carry current with no opposition whatsoever. How materials carry charge is of central interest for technology and of fundamental scientific importance. In Section 26–1 we mentioned a simple guiding picture (to be developed further in Section 26–5) in which electrons within conductors behave as though they were free to move, except for the presence of the stationary ions that form the crystal lattice of the metal, which form a set of obstacles for the electron motion. In this scenario electrons, constantly moving because the metal is at a finite temperature, undergo frequent collisions with the positive ions that form the crystal lattice of the metal whether the field is present or not (Fig. 26–5). When there is no field, the electrons do not, *on average*, move in any particular direction, and there is no macroscopic current. The motion of the electrons is random, like the motion of air molecules. In the presence of an electric field $\vec{E}$ each "free" electron experiences a force $\vec{F} = -e\vec{E}$ and thus accelerates between collisions in a direction opposite to the electric field.[†] In this case there is a *net* movement of electrons in

▶ **FIGURE 26–5** In one picture of current flow in a metal, an electron collides frequently with the ions and impurities and scatters randomly from the ions, here indicated with circles enclosing plus signs. When an electric field is present, the electron picks up a component of velocity opposite the field. The electron's path in the electric field is slightly parabolic. The extra component of velocity present due to the electric field is, on average, much less than the average speed of the electrons when no field is present.

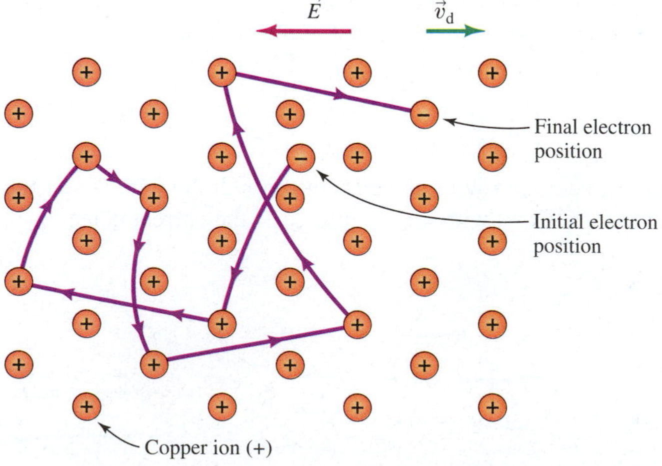

[†] In electrostatics, metals contain no electric fields. But when we deal with currents, we are no longer in electrostatics: Charges are moving continuously.

the direction of the electric force they experience. The collisions, in effect, give rise to a drag force on the flow of electrons. As with the fall of a parachute, drag acts to settle the motion to a steady flow in the direction of the force. The electrons move on average with a constant terminal velocity called the **drift velocity**, $\vec{v}_d$—this is the velocity indicated in Figs. 26–4 and 26–5. Equation (26–6) gives the relation between drift speed and current. Here the charge carriers are the electrons, so that n_q equals n_e, the density of free electrons in the metal, and Eq. (26–6) then gives

$$I = n_e q v_d A, \tag{26–8}$$

where A is the cross-sectional area of a metal wire.

We can solve Eq. (26–8) to find the drift speed in terms of the current:

$$v_d = \frac{I}{n_e q A}. \tag{26–9}$$

Equations (26–8) and (26–9) show the relations between current and drift speed. Remember that the direction of the electron's drift velocity is opposite to the defined direction of the current density because of the positive charge-carrier convention.

In Fig. 26–6, we show the relationships among the external electric field $\vec{E}$, the current I, the current density $\vec{J}$, and the electron's drift velocity $\vec{v}_d$. For the case of the wire, $J = I/A$, and from Eq. (26–9), we have

$$\vec{J} = n_e q \vec{v}_d. \tag{26–10}$$

Note the formal confirmation that $\vec{J}$ is opposite to the direction of $\vec{v}_d$: A negative sign is introduced when we recognize that $q = -e$.

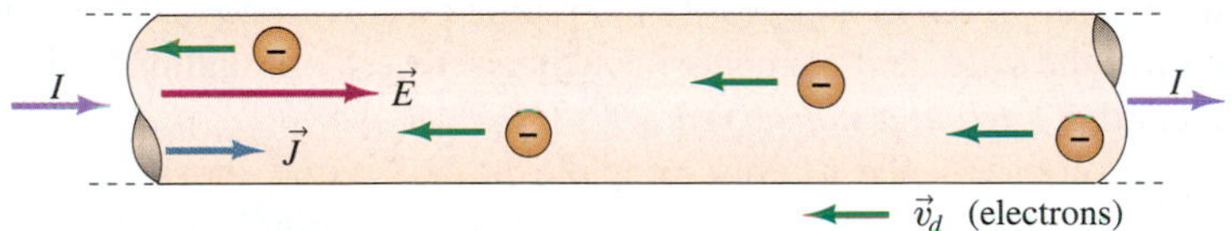

FIGURE 26–6 Electrons drift in the direction opposite that of the current, I, current density, $\vec{J}$, and electric field, $\vec{E}$.

EXAMPLE 26–3 Estimate the drift speed v_d for electrons in a copper wire of diameter $d = 1.0$ mm that carries a current $I = 100$ mA. Copper has about one free electron per atom available to carry charge and has a mass density of 8.92 g/cm^3 and a molecular weight of 63.5 g/mol.

Setting It Up We refer to Fig. 26–6 for a pictorial representation.

Strategy The drift speed can be calculated from Eq. (26–10) if the current density and electron density are known. The current density involves the total current and the area of the wire, both obtainable from the data given. We have for the magnitude of the current density $J = \dfrac{I}{A} = \dfrac{I}{\pi r^2} = \dfrac{I}{\pi(d/2)^2}$, where d is the diameter of the wire and I is the total current. To calculate the free electron density, we make use of the fact that there is one free electron per atom, so that the free electron density equals the density of copper atoms. Here we proceed as follows:

$$\begin{aligned}(\text{No. of atoms})/\text{m}^3 &= [(\text{No. of atoms})/\text{kg}][\text{mass density in kg/m}^3] \\ &= [(\text{No. of atoms/mole})/(\text{No. of kg/mole})][\text{mass density in kg/m}^3].\end{aligned}$$

In the last step, the number of atoms per mole is Avogadro's number N_A. To get the number of kg in 1 mole, we use the fact that the molecular weight of Cu is 63.5 g/mol = 63.5×10^{-3} kg/mol, and the mass density is 8.92 g/cm^3 = 8.92×10^3 kg/m^3. With all this information in hand we can then use Eq. (26–10) to find v_d.

Working It Out With the number of free electrons per atom denoted by n (here $n = 1$), we have

$$\begin{aligned} n_e &= nN_A \frac{1}{M}\rho_{\text{Cu}} \\ &= (1\ \text{electron/atom})(6.02 \times 10^{23}\ \text{atoms/mol}) \\ &\quad \times \frac{1}{63.5 \times 10^{-3}\ \text{kg/mol}}(8.92 \times 10^3\ \text{kg/m}^3) \\ &= 8.46 \times 10^{28}\ \text{electrons/m}^3. \end{aligned}$$

A calculation of the current density gives

$$J = \frac{I}{\pi(d/2)^2} = \frac{4(100 \times 10^{-3}\ \text{A})}{\pi(1.0 \times 10^{-3}\ \text{m})^2} = 1.3 \times 10^5\ \text{A/m}^2,$$

from which we calculate

$$\begin{aligned} v_d = \frac{J}{n_e q} &= \frac{1.3 \times 10^5\ \text{C/s}\cdot\text{m}^2}{(8.5 \times 10^{28}\ \text{electrons/m}^3)(1.6 \times 10^{-19}\ \text{C/electron})} \\ &= 9.6 \times 10^{-6}\ \text{m/s}. \end{aligned}$$

Does this seem like a small value to you? See the "Think about this . . ." box below.

What Do You Think? If the current is increased by a factor of 4 in this problem, by how much would the radius of the wire have to change to keep the current density constant?

THINK ABOUT THIS. . .

WHY DOES THE LIGHT APPEAR THE MOMENT YOU THROW THE LIGHT SWITCH?

In Example 26–3, we saw that the drift speed in a realistic wire carrying a realistic current is only a tiny fraction of a centimeter per second. Certainly you don't have to wait for hours for the electrons to drift to the lightbulb you want to put to use. In fact, household circuits carry alternating current (the subject of Chapter 33), in which there is no net flow of electrons in one direction or another! So what happens when we switch on a light in our house? When the switch is thrown, the electric field that influences the electrons to move in the wire is set up throughout the wire at speeds approaching the speed of light. The free electrons are spread throughout the wire, and they are each affected by the electric field almost simultaneously. They start moving at once in response to this field—those nearest the switch as well as those nearest the electrical appliance. A similar effect occurs in fluid flow, and we can compare an electric wire with free electrons to a garden hose full of water. If you want to move a sprinkler while you water the lawn, you turn the water off, move the sprinkler, and then turn the water back on. Because the hose is already full of water, the sprinkler starts immediately: The force of the water at the faucet end is quickly transmitted all along the hose, and the water at the sprinkler end of the hose flows from the sprinkler almost the moment the faucet is opened. Similarly, the electrons that move within the filament of the lightbulb are already present when the switch is thrown. When this happens the electric field and hence the current that it drives, are set up almost instantaneously.

Current and the Conservation of Charge

How does the conservation of charge affect currents in materials? The current in a wire describes the flow of charge along the wire. For steady flow (constant current), the rate at which charge enters a section of the current-carrying wire in a given time must equal the rate at which charge leaves the section: There is no accumulation of charge in any part of the wire, and net charge is neither created nor annihilated. This is just the statement that *current is conserved*.

A consequence of this principle is that if the cross-sectional area of the wire changes, the *current density* must change accordingly so that the product JA is unchanged (Fig. 26–7a). Equation (26–10) in turn implies that *for a fixed current*, the drift speed, like the current density J, is inversely proportional to the area of the wire (Fig. 26–7b). The conservation of current will continue to act as a constraint if the wire

FIGURE 26–7 (a) A steady current is the same in all parts of a wire, even if the area of the wire varies. This means that the current density and drift speed will vary with the area: Both are larger when the wire cross section is smaller. (b) More precisely, the drift speed and current density are inversely proportional to the wire area.

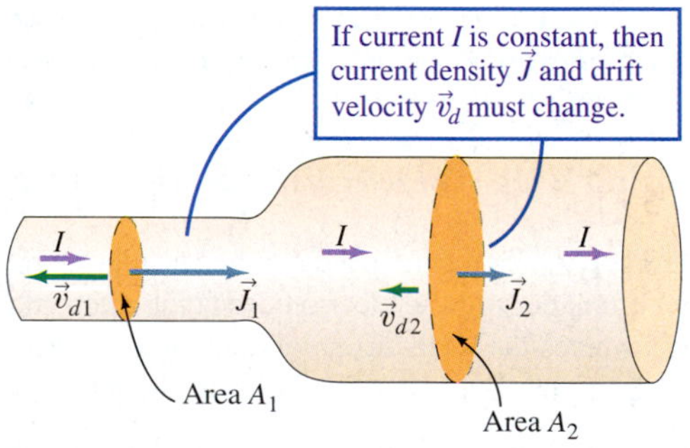

(a)

(b)

splits into two or more wires, even if, as we will learn in Section 26–4, the currents in the individual wires are determined by the properties of these wires. The situation is analogous to the flow of water along a pipe. As we learned in Chapter 16, for water in steady flow the conservation of mass implies the conservation of flux of the water. The rate at which water enters a pipe is equal to the rate at which water leaves. If the pipe narrows, the speed of the flow increases, while if the pipe widens, the speed of the flow decreases. If the pipe splits, the rate of flow in each section adjusts itself in a way that is consistent with conservation of the total flow rate.

26–3 Resistance

We have seen that a current flows when an electric field is applied to a conductor. We can equally consider the flow to be due to a potential difference V, with the current flowing from a higher to lower potential. The amount of current that flows through a material for a given potential difference across that material depends on the material's properties and its geometry.

The **electrical resistance** of a piece of material is a measure of how easily charge flows within that material. The resistance R is defined to be the ratio of the voltage (potential difference) across the material to the current that flows through it:

$$R \equiv \frac{V}{I}. \qquad (26\text{–}11)$$

DEFINITION OF RESISTANCE

If a material has a low resistance, then more current will flow for a given voltage than it will for a higher resistance. The units of resistance are volts per ampere, but a separate SI unit called the **ohm** (Ω) has been defined as the resistance through which a current of 1 A flows when a potential difference of 1 V is applied:

$$1\ \Omega \equiv 1\ \text{V/A}.$$

Units of kilohms (kΩ or $10^3\ \Omega$) or megohms (MΩ or $10^6\ \Omega$) are also commonly used.

Georg Simon Ohm was the first to study the resistance of different materials systematically. In 1826, he published his experimental result that, for many materials including most metals, *the resistance is constant over a wide range of potential differences*. This statement is called *Ohm's law*. It is not, properly speaking, a law, but rather an empirical statement about the behavior of materials. When the resistance of a material is constant over a range of potential differences, we say that the material is *ohmic*. We shall continue the traditional practice of referring to this linear relation between voltage and current for these materials as a "law" and writing it as

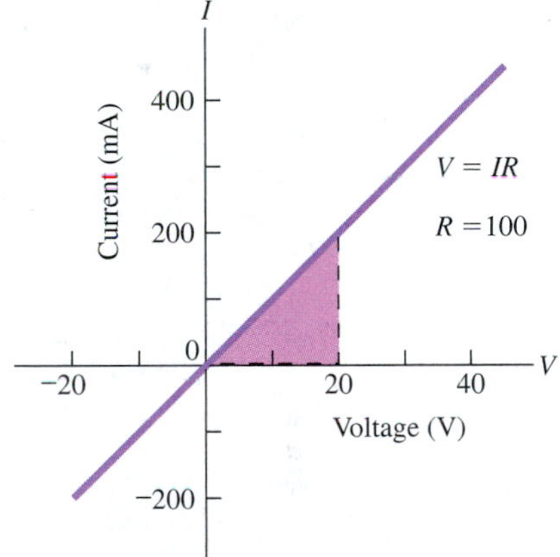

▲ **FIGURE 26–8** For ohmic materials, Ohm's law states that the ratio V/I is a constant.

$$V = IR, \qquad (26\text{–}12)$$

OHM'S LAW

where *R is independent of V*. Figure 26–8 illustrates the consequence of the independence of V and R.

Resistors

A **resistor** is a piece of ohmic material with a specific value of resistance. It forms a part of many circuits; indeed, since all circuit elements through which current flows have some resistance, it is ubiquitous in circuits. Common resistors have resistances that range from fractions of an ohm to millions of ohms (Fig. 26–9). A resistor of given resistance R with a given potential V between its terminals allows the flow of a current $I = V/R$. Resistors are represented in circuit diagrams by zigzag lines, -^^^-. They are connected to each other and to other elements such as capacitors or batteries by conducting wires that are generally assumed to have negligible resistance.

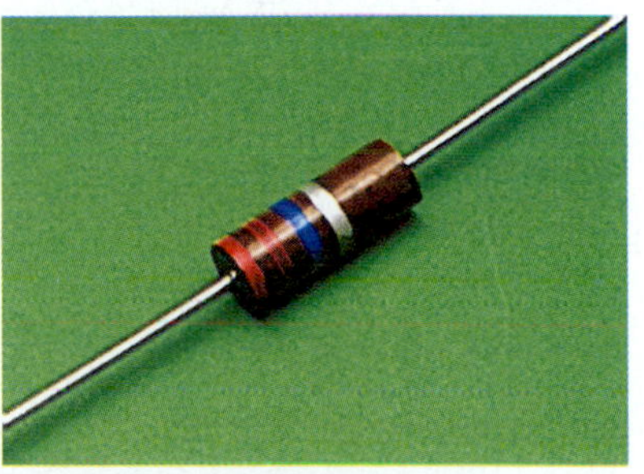

▲ **FIGURE 26–9** Resistors are color coded to indicate the value of their resistance.

THINK ABOUT THIS. . .

CAN NONOHMIC MATERIAL BE USEFUL IN CIRCUITS?

There are many *nonohmic materials*: materials for which the voltage and current do not obey the linear relation of Ohm's law. In fact, if you look closely enough, there are no perfectly ohmic materials, so this is partly just a question of degree. But some materials have conduction properties that differ spectacularly from Ohm's law (see Section 26–6). Moreover, one can tailor these conduction properties through precise control of the relative proportions of different atoms. Such materials and devices play an enormous role in computing and other electronic applications. Figure 26–10a shows current-versus-voltage curves (ideal and typical) for a **diode**. A diode is a device that transmits current easily when the voltage difference has one sign, but prevents charge flow (that is, it has a very high resistance) when the voltage difference has the other sign. The curve is very far from the linear curve demanded by Ohm's law. Diodes are used in many electric devices (Fig. 26–10b). They may be used to charge a battery while preventing it from discharging, or, more interestingly, they prevent the "backflow" in AC and therefore are important in devices that convert AC to DC.

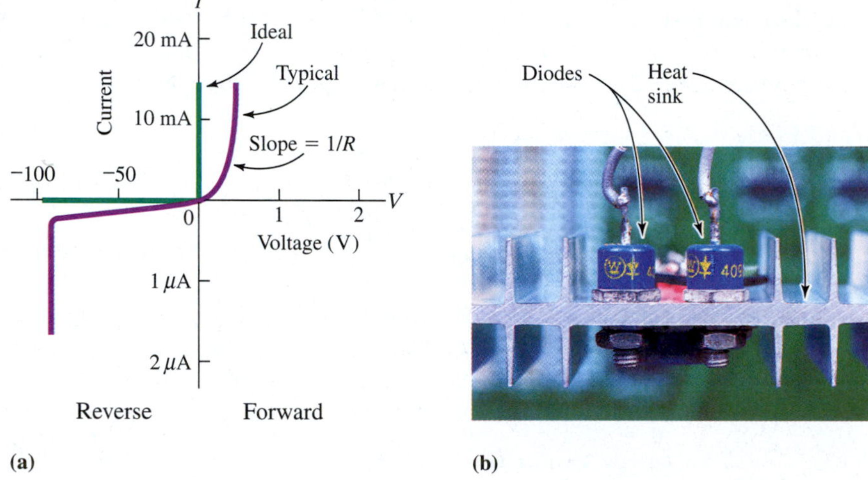

▲ **FIGURE 26–10** (a) Plot of current versus voltage for an ideal diode (green) and a typical real diode (purple). Note that, for the ideal diode, there is no current when the voltage is negative: The diode allows current to flow in only one direction. (b) Two (blue) diodes used to convert AC voltage to DC are bolted to an aluminum plate to dissipate thermal energy (heat sink).

Resistivity and Conductivity

The resistance of a conducting wire of a given material depends both on the shape (length and cross-sectional area) of the wire, and on the intrinsic properties of the material. Our guiding picture for resistance ascribes the difference between the acceleration of a charge in free space and the steady flow in a conductor to the effect of collisions with the ions within the material. These provide a retarding force that opposes the electric field E. This retarding force will clearly vary from material to material. Since the effect of the collisions is a local one—that is, it does not depend on the precise shape or size of the piece of material—we expect that it is $J = I/A$ that will depend on E and the properties of the material. If we define the intrinsic (geometry- and size-independent) **resistivity** ρ of a material by E/J, we see that this can be expressed in terms of I/A and the voltage V, which is related to the electric field by $E = V/L$, where L is the length of the wire. We write

$$\rho \equiv \frac{E}{J}. \qquad (26\text{–}13)$$

DEFINITION OF RESISTIVITY

Using $E = V/L$ and $J = I/A$, and recalling $R = V/I$, Eq. (26–13) reads $\rho = R(A/L)$, or

$$R = \rho \frac{L}{A}. \tag{26–14}$$

The reciprocal of the resistivity is the **conductivity**, σ.

$$\sigma \equiv \frac{1}{\rho}. \tag{26–15}$$

DEFINITION OF CONDUCTIVITY

Typical values of conductivity are listed in Table 26–2. The resistivities and conductivities of the materials shown in Table 26–2 vary over many orders of magnitude. The conductivity of a metal such as aluminum is a factor of 10^{21} higher than that of a good insulator such as Teflon. It is worthwhile noting here that while our guiding model—acceleration of free electrons with drag due to collisions—may provide us with a concrete picture to work with, it is not remotely adequate to explain the huge range of values for conductivity observed in nature. Only quantum mechanics can do that, and we shall discuss this a little in Section 26–6.

In the basic relation that defined the resistivity in Eq. (26–13), both the electric field and the current density are vectors, and a better version of this relation is

$$\vec{E} = \rho \vec{J}. \tag{26–16}$$

RELATION BETWEEN FIELD AND CURRENT DENSITY

TABLE 26–2 • Resistivities, Conductivities, and Temperature Coefficients (at 20°C)

Material	Resistivity, ρ $(\Omega \cdot m)$	Conductivity, σ $(\Omega \cdot m)^{-1}$	Temperature Coefficient, α $(°C)^{-1}$
Conductors			
Elements			
Aluminum	2.82×10^{-8}	3.55×10^{7}	0.0039
Silver	1.59×10^{-8}	6.29×10^{7}	0.0038
Copper	1.72×10^{-8}	5.81×10^{7}	0.0039
Iron	10.0×10^{-8}	1.0×10^{7}	0.0050
Tungsten	5.6×10^{-8}	1.8×10^{7}	0.0045
Platinum	10.6×10^{-8}	1.0×10^{7}	0.0039
Alloys			
Nichrome	100×10^{-8}	0.1×10^{7}	0.0004
Manganin	44×10^{-8}	0.23×10^{7}	0.00001
Brass	7×10^{-8}	1.4×10^{7}	0.002
Semiconductors			
Carbon (graphite)	3.5×10^{-5}	2.9×10^{4}	−0.0005
Germanium (pure)	0.46	2.2	−0.048
Silicon (pure)	640	1.6×10^{-3}	−0.075
Insulators			
Glass	10^{10} to 10^{14}	10^{-14} to 10^{-10}	
Neoprene rubber	10^{9}	10^{-9}	
Teflon	10^{14}	10^{-14}	

Equivalently, from the definition of conductivity, Eq. (26–15), we have

$$\vec{J} = \sigma\vec{E}. \tag{26–17}$$

The reason for writing out these equations is that Eqs. (26–16) and (26–17) are general results—not limited to ohmic materials. It is only for ohmic materials that ρ and σ do not vary with V or $\vec{E}$.

EXAMPLE 26–4 Determine the current density, resistance, and electric field for the copper wire of Example 26–3 if the wire is 10 m long.

Setting It Up In addition to the quantities given in the problem statement, we will want to use data for copper in Table 26–2.

Strategy Given the current and the wire's cross section, the current density is $J = I/A = I/(\pi r^2) = 4I/(\pi d^2)$. We are also given the length of the wire, so that the resistance can be calculated—$R = \rho L/A$—and we can look up $\rho = 1.72 \times 10^{-8}\ \Omega\cdot\text{m}$ for copper. Finally we find the desired field using $E = \rho J$.

Working It Out We calculate in order

$$J = \frac{4I}{\pi d^2} = \frac{4(100 \times 10^{-3}\ \text{A})}{\pi(1.0 \times 10^{-3}\ \text{m})^2} = 1.3 \times 10^5\ \text{A/m}^2.$$

Next we have

$$R = \rho\frac{L}{A} = \frac{(1.72 \times 10^{-8}\ \Omega\cdot\text{m})(10\ \text{m})}{\pi(0.50 \times 10^{-3}\ \text{m})^2} = 0.22\ \Omega.$$

Finally

$$E = \rho J = (1.72 \times 10^{-8}\ \Omega\cdot\text{m})(1.3 \times 10^5\ \text{A/m}^2)$$
$$= 2.2 \times 10^{-3}\ \text{V/m}.$$

What Do You Think? If the electric field we calculated above acts within an iron wire of the same size as the copper wire in this example, how will the current differ, if at all?

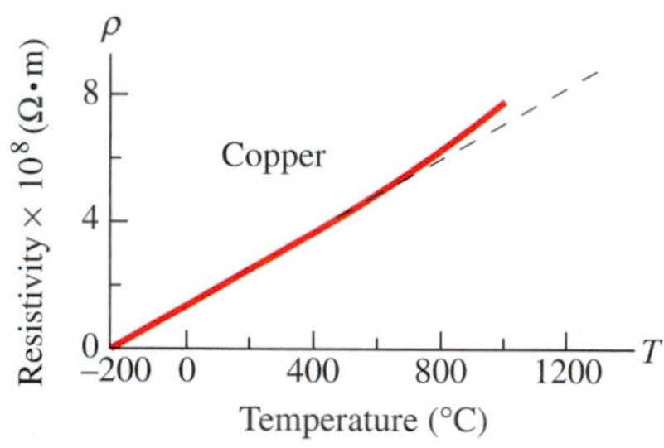

▲ **FIGURE 26–11** The resistivity of copper as a function of temperature.

The Temperature Dependence of Resistivity

Resistivities of some materials, copper for example, have a strong temperature dependence (Fig. 26–11). We can represent the temperature dependence with a linear approximation that is sufficiently accurate for most purposes:

$$\rho \cong \rho_0[1 + \alpha(T - T_0)]. \tag{26–18}$$

The parameter α is the *temperature coefficient of resistivity*, and ρ_0 is the resistivity at the reference temperature T_0, normally 20°C. Values of ρ, σ, and α are given in Table 26–2 for $T = 20$°C. Resistivities for most metals increase with temperature, as Fig. 26–11 shows for copper, and we'll discuss this further in the next section.

EXAMPLE 26–5 Find the resistance of a coil of platinum wire of diameter 0.5 mm and length 20 m at 20°C and at 1000°C.

Setting It Up The numbers for resistivity as well as for temperature dependence can be found in Table 26–2.

Strategy Calculation of resistance in terms of the dimensions of the wire and of the resistivity involves a simple application of Eq. (26–14). We are given the diameter d of the wire, and the area is then $A = \pi(d/2)^2$. The resistance is then given in terms of the resistivity. The latter has a well-defined temperature dependence, given by Eq. (26–18). If we take that into account, then

$$R = \rho\frac{L}{A} = \rho\frac{4L}{\pi d^2} = \rho_0[1 + \alpha(T - T_0)]\frac{4L}{\pi d^2}.$$

Here ρ_0 for platinum is $10.6 \times 10^{-8}\ \Omega\cdot\text{m}$, and the coefficient $\alpha = 3.9 \times 10^{-3}\ °\text{C}^{-1}$. All we need to do is to evaluate the above formula for two different values of T.

Working It Out For $T = T_0 = 20$°C,

$$R = \rho\frac{L}{A} = (10.6 \times 10^{-8}\ \Omega\cdot\text{m})\frac{4 \times 20\ \text{m}}{\pi(0.5 \times 10^{-3}\ \text{m})^2} = 11\ \Omega.$$

For $T = 1000$°C, we multiply the above result by the factor $[1 + \alpha(T - T_0)]$ to get

$$R = (11\ \Omega)[1 + (3.9 \times 10^{-3}\ °\text{C}^{-1})(1000°\text{C} - 20°\text{C})] = 53\ \Omega.$$

This large spread in values suggests that, depending on the materials used for resistors, the operation or design of a circuit must account for temperature dependence. This is particularly true since energy is dissipated within resistors in the form of thermal energy (see Section 26–7).

What Do You Think? The melting temperature of platinum is over 1700°C, while that of copper is a little over 1000°C. Consider their temperature coefficient of resistivity and decide which would be more useful for a lightbulb filament. (Real filaments are made of tungsten.)

26–4 Resistances in Series and in Parallel

In the previous chapter we studied capacitors, and we saw that a set of capacitors connected in a combination within an electric circuit could sometimes be replaced by a single "equivalent" capacitor, i.e. one that could replace the combination and leave the relevant circuit characteristics external to the combination unchanged. Resistors can be treated similarly, and in this section we develop the rules that tell us how to find equivalent resistors. As for capacitors, so long as any particular configuration of resistors can be reduced to a sequence of series and parallel connections, it is possible to find a single equivalent resistance. This can be helpful in any number of contexts, starting with the most direct: simplifying a circuit that has been "overdesigned."

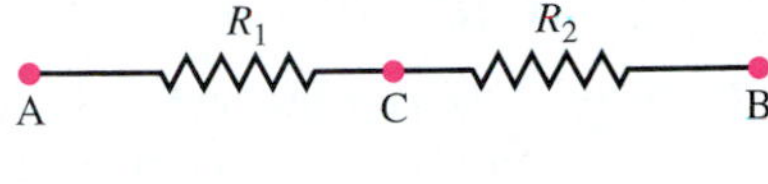

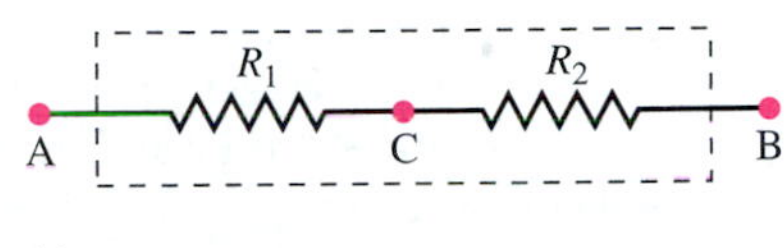

▲ **FIGURE 26–12** (a) A single resistor between points A and B on a wire, with potential difference V_{AB} between the points. (b) Two resistors in series between points A and B. (c) Two resistors in series act as a single equivalent resistor.

Resistors in Series

Current conservation allows us to calculate the effective resistance when a number of resistors are connected in series. Figure 26–12a shows a wire connecting two points A and B, with a resistor of resistance R in between. If the current flowing through the resistor R is I, then the potential difference between the points A and B is $V_{AB} = IR$. Suppose we now put two resistors along the same wire—we have connected them in series—and take a point C that lies on the wire between the two resistors (Fig. 26–12b). Current conservation demands that the current I is the same all along the wire. If the first resistor has resistance R_1, then the potential difference between A and C must be $V_{AC} = IR_1$. If the second resistor between C and B has resistance R_2, then the potential difference between these points is $V_{CB} = IR_2$. Let us now put a box around the two resistors in series, as in Fig. 26–12c. We see that the potential difference between the points A and B is given by $V_{AB} = V_{AC} + V_{CB}$. This sum, however, is $V_{AB} = IR_1 + IR_2 = I(R_1 + R_2)$. We may now ask for the resistance R_{eq} equivalent to the two resistors in series, by which we mean that $V_{AB} = IR_{eq}$. It follows that

$$R_{eq} = \frac{V_{AB}}{I} = R_1 + R_2. \qquad (26\text{–}19)$$

A repeat of the exercise with three resistors R_1, R_2, R_3 between the points A and B, with intermediate breaks at the points C and D, as in Fig. 26–13, shows that the equivalent resistance is

$$R_{eq} = \frac{V_{AB}}{I} = \frac{V_{AC} + V_{CD} + V_{DA}}{I} = R_1 + R_2 + R_3. \qquad (26\text{–}20)$$

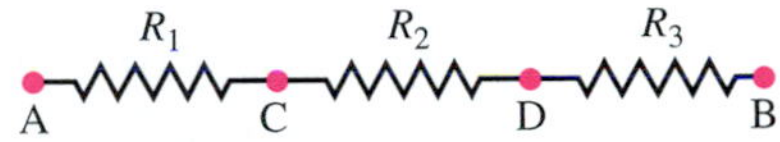

▲ **FIGURE 26–13** Three resistors in series, acting as a single equivalent resistor.

This is easily generalized to n resistors in series:

$$R_{eq} = R_1 + R_2 + R_3 + \cdots + R_n. \qquad (26\text{–}21)$$

Resistors in Parallel

Like capacitors, resistors can be placed in parallel, as in Fig. 26–14. A current I flows into a junction point A, from which n wires sprout, carrying resistors of resistance $R_1, R_2, \ldots R_n$. These wires come together again at the junction point B, out of which the original (*conserved*) current I flows. The potential difference between the points A and B is given as V_{AB}, and *this is the same as the potential difference across any of the resistors*. The current flowing out of the junction point A breaks up into parts that flow through the different resistors. How big these partial currents are is determined by the resistors that they flow through. Thus $I_1 = V_{AB}/R_1$, $I_2 = V_{AB}/R_2$, and so on. The equivalent resistance is given by rewriting the conserved current I as a sum of the partial currents, all of which is written in terms of the potential difference V_{AB} and the individual resistances. In other words, the equivalent resistance R_{eq} is defined by V_{AB}/I, so that $I = V_{AB}/R_{eq}$. Current conservation then implies that

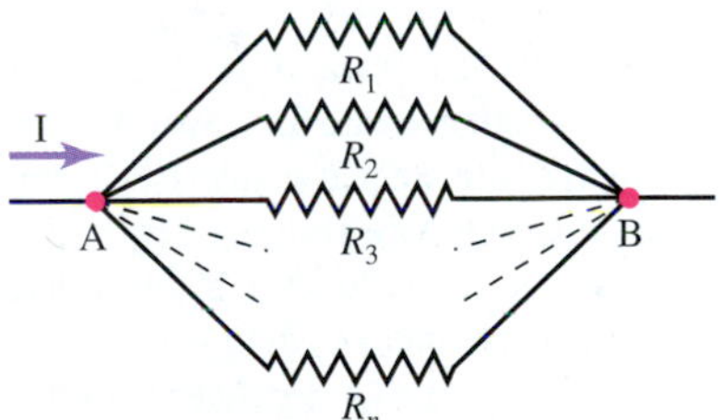

▲ **FIGURE 26–14** An arrangement of n resistors in parallel, carrying current from A to B, with potential difference V_{AB} from A to B.

$$I = \frac{V_{AB}}{R_{eq}} = I_1 + I_2 + \cdots + I_n = \frac{V_{AB}}{R_1} + \frac{V_{AB}}{R_2} + \cdots + \frac{V_{AB}}{R_n}$$
$$= V_{AB}\left(\frac{1}{R_1} + \frac{1}{R_2} + \cdots + \frac{1}{R_n}\right).$$

Dividing out V_{AB} yields the relation that holds for resistors **in parallel**,

$$\frac{1}{R_{eq}} = \frac{1}{R_1} + \frac{1}{R_2} + \cdots + \frac{1}{R_n}. \qquad (26\text{–}22)$$

EXAMPLE 26–6 Consider a current $I = 3.0$ A flowing as shown into the combination of resistors in Fig. 26–15. (a) Calculate the potential difference between the points A and B. (b) What is the potential difference across the 4 Ω resistor?

Setting It Up We remark here only that the desired potential differences are best labeled as V_{AB} for part (a), and for part (b), V_{CB} across the 4 Ω resistor.

Strategy The current entering at point A divides up into two branches, so that we have a situation with resistances in parallel. In each branch we have successive resistances, so there are also resistances in series. We can apply our rules to find equivalent resistances: first the equivalent resistance in each branch, then an overall equivalent resistance for the parallel combination. By finding the overall equivalent resistance R_{eq} we can find the potential difference between A and B, namely $V_{AB} = IR_{eq}$. This gives us the answer to (a). For part (b), we note that with the known V_{AB} and the equivalent resistance in each branch, we can find the current in each branch. The potential difference V_{CB} is then the current in the left branch times the resistance 4 Ω.

Working It Out The equivalent resistance in the left branch (the ACB path), which contains two resistors in series, is $R_{ACB} = 8\ \Omega + 4\ \Omega = 12\ \Omega$. The equivalent resistance in the right branch is similarly $R_{ADB} = 6\ \Omega + 2\ \Omega = 8\ \Omega$. Finally we want the overall equivalent resistance,

$$1/R_{eq} = 1/R_{ACB} + 1/R_{ADB} = 1/(12\ \Omega) + 1/(8\ \Omega) = 5/(24\ \Omega),$$

or $R_{eq} = (24\ \Omega)/5 = 4.8\ \Omega$. With these values we can proceed to the potential differences.

(a) We have immediately

$$V_{AB} = R_{eq}I = (4.8\ \Omega) \times (3\ \text{A}) = 14\ \text{V}.$$

(b) The current through the branch ACB is $I_{ACB} = V_{AB}/R_{ACB} = (14\ \text{V})/(12\ \Omega) = 1.2$ A. In turn,

$$V_{CB} = I_{ACB} \times (4\ \Omega) = (1.2\ \text{A}) \times (4\ \Omega) = 4.8\ \text{V}.$$

As a check we can calculate the potential difference between A and C, $V_{AC} = (1.2\ \text{A}) \times (8\ \Omega) = 9.6$ V, which when added to the 4.8 V yields the total potential difference of 14 V.

What Do You Think? In what way would the calculation be simpler if we were to interchange the 8 Ω and 6 Ω resistors?

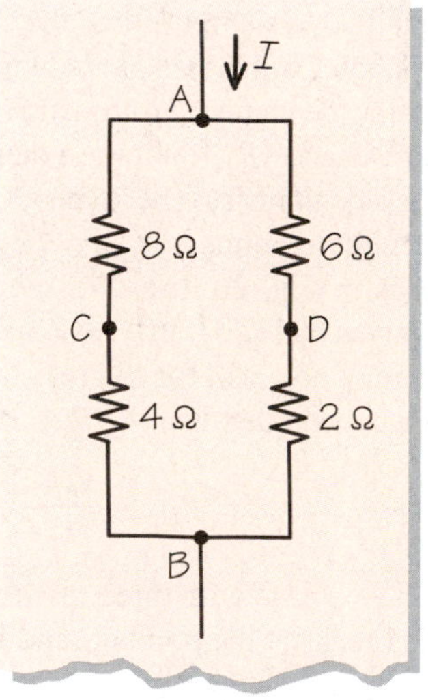

▲ **FIGURE 26–15**

CONCEPTUAL EXAMPLE 26–7 Equation (26–14) tells us that the resistance of a wire is proportional to its length L and inversely proportional to its cross-sectional area A. Is this consistent with the results obtained within this section?

Answer The linear dependence on the length is just an application of the equivalence rule for resistors in series. If we cut a uniform wire of length L into n equal segments, each of resistance r, then the rule for the calculation of the resistance equivalent to n identical resistors in series gives $R = nr$. Since r is the resistance per segment, and n segments give the total length of the wire, we find that R is proportional to the length L. The $1/A$ dependence of the resistance is a direct application of the equivalence rule for resistors in parallel. We may view a wire of total cross-section A as a total of N wires in parallel, each of cross-sectional area A/N. Suppose each of these N wires has a resistance r. Then the rule for calculating the equivalent resistances yields $1/R_{eq} = N(1/r)$, or $R_{eq} = r/N$. Now according to Eq. (26–14), $r = \rho\dfrac{L}{(A/N)} = N\rho\dfrac{L}{A}$. This gives us $R_{eq} = r/N = \rho L/A$, a consistent result. (Another way to see this is to note that R should be independent of N—of how many pieces you cut the original resistor into—and the only way that can happen is for r to be *inversely proportional* to the area A/N; this leaves R_{eq} inversely proportional to the area A.)

*26–5 Free-Electron Model of Resistivity

A more fundamental understanding of resistivity requires quantum mechanics. Nevertheless, there is a simple classical model of electrons and resistivity that is consistent with Ohm's law. It was first proposed in 1900 by Paul Drude and is known as the **free-electron model**, or the **Drude model**. Although the model has fundamental deficiencies, its study is worthwhile for two reasons: First, the model allows us to focus on the physics of resistivity. Second, the model illustrates how model-building in the physical sciences proceeds, and how we can judge the success or failure of a model.

We start with the idea that solids contain "free" electrons, which can move within the material and carry charge. The density of free electrons n_e depends on the material, and this is the factor responsible for the differences among conductors, insulators, and semiconductors (which we shall discuss in Section 26–6). In metals, the number of loosely attached electrons per atom (these are the electrons that behave as though they were free) lies on average in the range 1.0 to 1.3, although it can be as large as 3.5 (for aluminum).

The model postulates that free electrons form a "gas" of independent particles at temperature T. The electrons move erratically due to collisions with the atoms or ions that form the crystal lattice when there is no electric field, but when there is an applied field, they are accelerated by it in between the collisions. The result is an on-average movement that forms the current. The effect of the collisions, in other words, is to produce a drag force that results in a constant drift velocity, producing the steady current. As we saw in Chapter 5, one of the simplest drag forces is proportional to the electrons' speed. If we assume this form, then Newton's second law for the component of electron motion that is parallel to the applied field is

$$ma = -eE - (\text{a constant})v,$$

where m is the mass of an electron. The constant must have dimensions of mass/time, and we write it as m/τ, where τ is a quantity with dimensions of time. From the kinetic theory of Chapter 19 one can establish that τ is the *collision time*, the average time between successive collisions of an electron with the lattice ions. The acceleration drops to zero when the speed of the electrons reaches the drift speed, v_d, meaning that $ma = -eE - (m/\tau)v_d = 0$, or

$$v_d = -\frac{eE\tau}{m}. \tag{26–23}$$

The minus sign indicates that the direction of the drift velocity is opposite to that of the electric field, as must be the case with electrons in motion. When this expression is inserted into Eq. (26–10) for the current density, we find

$$J = n_e(-e)v_d = \frac{n_e e^2 \tau}{m}E. \tag{26–24}$$

Comparison with Eq. (26–17) yields

$$\sigma = \frac{n_e e^2 \tau}{m}, \quad \text{or} \quad \rho = \frac{1}{\sigma} = \frac{m}{n_e e^2 \tau} \tag{26–25}$$

for the conductivity σ and the resistivity ρ. The quantities e and m are independent of the type of material. The average time between collisions may be expressed in terms of the *mean free path* λ and the average speed v_{av} of the electrons in the free-electron "gas" by using Eq. (19–51), $\tau = \lambda/v_{av}$.

For electric fields not so large that the material itself is disrupted, none of the quantities in Eq. (26–25) depend on E, and thus the resistivity (or conductivity) is constant over a wide range of applied electric fields. This was the basis of Drude's (and independently Hendrik Lorentz's) claim to understand Ohm's law at an atomic level, dating from 1900.

EXAMPLE 26–8 What is the free-electron model's prediction for the collision time of current-carrying electrons in copper, given that the resistivity of copper is $1.7 \times 10^{-8}\ \Omega \cdot \text{m}$? You may use the parameters of Example 26–2.

Setting It Up The collision time τ is related to the resistivity ρ given above by Eq. (26–25). The other parameters in the equation are the known charge $e = 1.6 \times 10^{-19}$ C, the known electron mass $m = 0.91 \times 10^{-30}$ kg, and the electron density $n_e = 8.5 \times 10^{28}$ electrons/m^3, a number calculated in Example 26–3.

Strategy A rearrangement of Eq. (26–25) gives us the desired quantity, $\tau = \dfrac{m}{n_e e^2 \rho}$.

Working It Out We have

$$\tau = \frac{0.91 \times 10^{-30}\ \text{kg}}{(8.5 \times 10^{28}\ \text{electrons/m}^3)(1.6 \times 10^{-19}\ \text{C})^2(1.7 \times 10^{-8}\ \Omega \cdot \text{m})}$$

$$= 2.5 \times 10^{-14}\ \text{s}.$$

The Failure of the Free-Electron Model

The free-electron model assumes electrons moving among a lattice of ions. A closer look at the predictions of this model reveals significant discrepancies with experiment. In particular:

- The measured average speed of electrons in metals is *more than a factor of 10* higher than the model predicts for copper at room temperature.
- We know from Chapter 19 [see Eq. (19–33)] that the rms speed of the particles of a gas is proportional to $\sqrt{T}$, where T is the temperature. However, experimental values for the mean speed of conduction electrons are essentially independent of temperature.
- The mean free path—the average distance an electron travels between collisions [see Eq. (19–15)]—should be independent of temperature according to the model. Experimentally, this quantity is much larger than expected and has a $1/T$ dependence.

Although the free-electron model is *qualitatively* correct in many aspects, it cannot be taken too literally, because it is a classical model in what turns out to be the domain of quantum mechanics. A correct model of electrical conduction *requires* the use of quantum mechanics. Indeed, quantum physics comes in at several levels, the major ones as follows:

- Conduction electrons do not act as a classical gas of noninteracting electrons; rather, they obey a velocity-distribution law based on quantum physics.
- Quantum physics requires us to treat electrons as though they were waves scattering from the lattice structure of the material. In particular, this treatment leads to a surprise: Quantum physics predicts that *there would be no resistance to electron flow in a fixed, perfectly ordered crystal with no impurities*. Finite conductivities result from departures from the perfect lattice structure. These departures occur in two ways: impurities are present in real materials, and the positions of lattice atoms vibrate at finite temperatures due to thermal excitation. At high temperatures, resistivity to electron flow is caused primarily by thermal vibrations. At low temperatures, resistivity is due to electrons being scattered by impurities.

There is ample evidence that the quantum physics ideas are correct. Indeed, *all* the properties described are correctly explained with these ideas. We'll discuss them in more detail in the following optional section and then later in Chapter 43.

*26–6 Materials and Conductivity

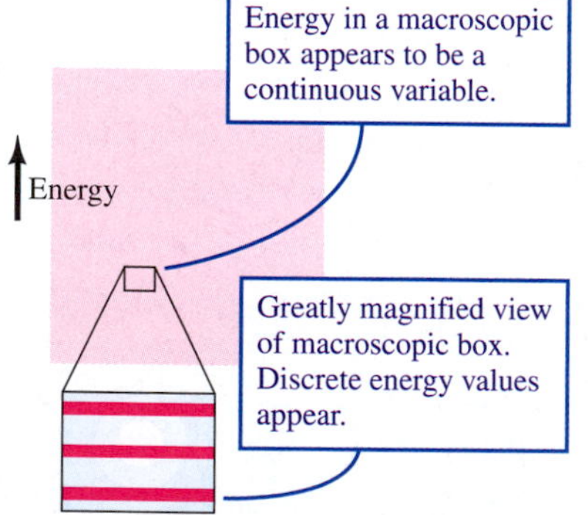

▲ **FIGURE 26–16** Energy diagram that shows the possible energy levels of an electron in a solid. (It takes no account of the crystalline structure formed by the parent ions of the electrons, whose presence has a strong effect on the possible energy levels—see Fig. 26–17). Classical physics predicts a continuum of possible energies, but quantum mechanics shows that the possible levels are actually discrete but so closely spaced that they are hard to distinguish.

Materials differ in their ability to conduct electricity over an enormous range. A good conductor might have a resistivity of $10^{-8}\ \Omega \cdot \text{m}$; a good insulator, about $10^{14}\ \Omega \cdot \text{m}$. The resistivity of semiconductors ranges from 10^{3} to $10^{-5}\ \Omega \cdot \text{m}$ and depends sensitively on temperature. Superconductors have no measurable resistance at all below a so-called critical temperature, a quantity that is specific to the material. We argued in the previous section that a proper quantitative explanation of the resistivity of all materials requires quantum physics. In this section, we employ a minimal amount of the ideas that explain quantum physics to describe the critical properties that distinguish conductors, insulators, semiconductors, and superconductors.

In classical physics, the energy of a "free" electron within a metal can take on any value; we say that the energy values form a *continuum*. In contrast, a quantum description of electrons confined to the interior of a metal but otherwise free shows that the possible energy values of such electrons are *quantized*; that is, the possible energies have discrete values. In other words, an electron cannot have *any* energy value, much as the frequencies of standing waves on a string cannot have any value, just a set of discrete values. In a sample of material whose size is large compared with atomic sizes (10^{-10} m), these energy values are so close together that they appear to be continuous, just as the separate dots in a newspaper photograph are not distinguishable from a large distance. Figure 26–16, an energy diagram, illustrates the allowed energy levels. It is important to keep in mind that this diagram illustrates only the *possible* energy levels.

We do not necessarily have electrons in each energy level. Some states are empty, while others have electrons in them and are said to be *occupied*.

When a set of atoms forming a regular background lattice is added to the picture, the possible energy values of the electrons are modified still further. The allowed energies of an electron are still discrete, but instead of a tiny separation between neighboring levels, there are *energy gaps*, which are large regions of energy forbidden to the electron. The regions where the energy levels are close together are called *allowed bands* of energy levels (Fig. 26–17). The gaps are quite sizable on the scale of atomic physics—of the magnitude of electron-volts.

According to quantum physics, *there are at most two electrons in any one energy level*. This property, proposed by Wolfgang Pauli in 1925 and called the *Pauli exclusion principle*, has no counterpart in classical physics, and it plays a crucial role in determining the properties of materials. Let's consider a solid with many "free" electrons. In an equilibrium state of that material, at least at low temperatures, these electrons fill the lowest energy levels available in the allowed bands—up to two in each level. When all the electrons are placed in the lowest possible energy states, we have two possible situations. In the first, the highest level to be filled is some intermediate level within a band; in the second, the electrons fill one or more bands completely.

Suppose that we now add some energy to the free electrons—by imposing an electric field, for example. The electrons in the lower energy levels cannot accept that energy, because they cannot move into a higher energy level that already has its quota of two electrons. The only electrons that can accept energy are those that lie in the top levels, and then only if there are nearby unoccupied levels into which they can move. This will be quite easily done with a small input of energy if the topmost occupied levels are in a partially filled band, but difficult if the topmost occupied levels fill out a band. Materials with a partially filled band are *conductors*. When the electrons in the highest occupied energy levels move freely into the empty energy levels immediately above, a current is produced. The electrons that jump from a lower level to a higher level are said to be *excited*. The energy-band structure for conductors is shown in Fig. 26–18a. *Conductors are characterized by having a highest-energy band with levels only partly occupied.*

If the highest-energy electrons of a material fill a band completely, then a small electric field will not give these electrons enough energy to jump the large energy gap to the bottom of the next (empty) band. We then have an *insulator* (Fig. 26–18b). An example of a good insulator is diamond (a form of carbon), whose energy gap is 6 eV.

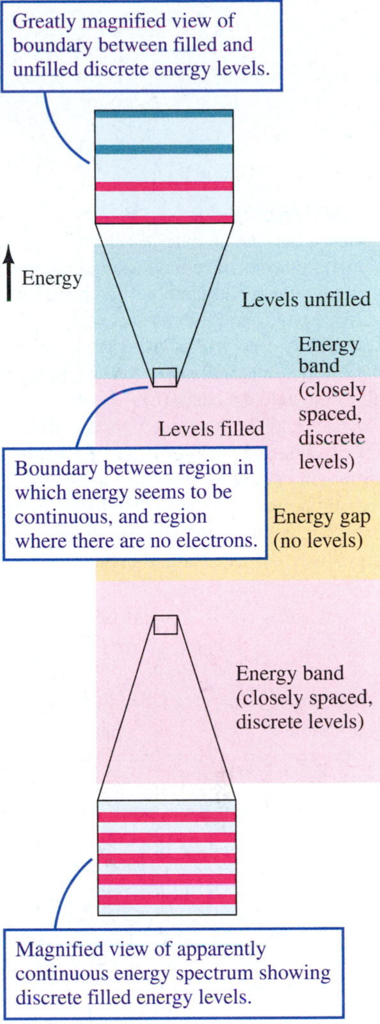

▲ **FIGURE 26–17** Energy diagram that shows the possible energy levels of an electron within a material made of a regular lattice of atoms. In contrast to the possibilities of Fig. 26–16, the electron energies are restricted to lie within allowed bands, and there is a large energy gap where no electrons are allowed. Even within the allowed bands, the possible electron energies are closely spaced discrete levels, as the magnified view shows. In the pink regions, the electron energy levels are filled; in the green regions, electron levels are present but are unfilled.

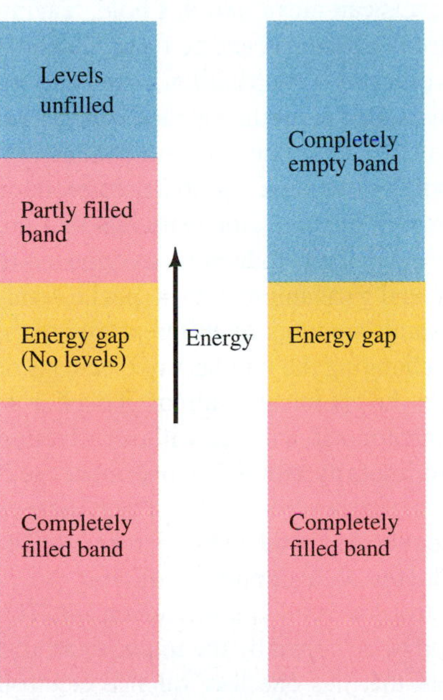

◀ **FIGURE 26–18** (a) Conductors have electrons in partly filled bands, whereas (b) insulators have an energy gap between a completely filled band and the next completely empty band. The pink and blue regions indicate where the allowed energy levels are filled and unfilled, respectively. Within each of the allowed bands, the possible energy levels form a set of closely spaced discrete levels.

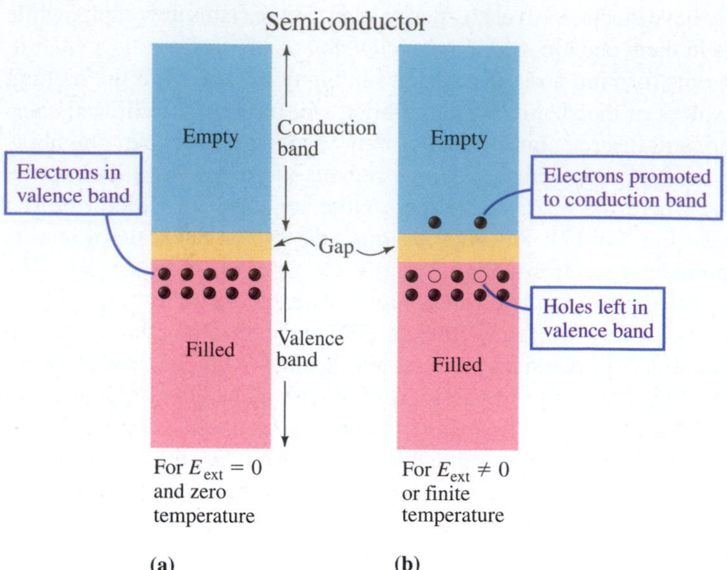

▶ **FIGURE 26–19** (a) For zero temperature and no external electric field, semiconductors have only a small energy gap between a completely filled band (the valence band) and the next highest, completely empty band (the conduction band). The dots indicate electrons, here all in the valence band. (b) A modest electric field $\vec{E}_{ext}$ or finite temperatures are enough to give some of the electrons sufficient energy to jump the energy gap, leaving holes (open circles) in the valence band and conduction electrons in the previously empty conduction band.

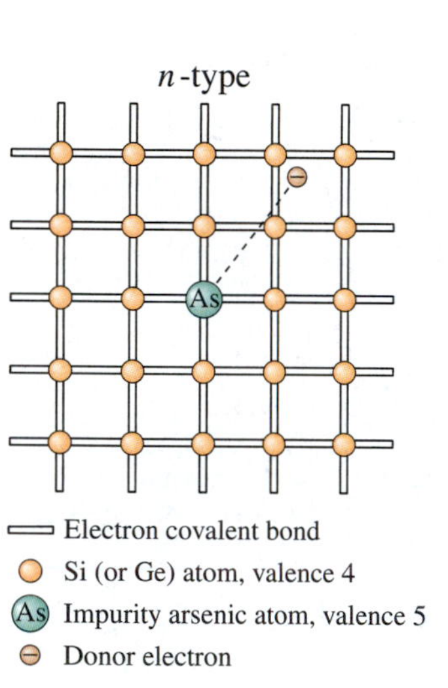

(a)

p-type

Ga

Impurity gallium atom, valence 3

⊕ Acceptor hole

(b)

▲ **FIGURE 26–20** An (a) n-type and a (b) p-type semiconductor are created by doping the original lattice with atoms that have, respectively, more and less valence electrons than the atoms of the original lattice have.

In semiconductors, the highest-energy electrons fill a band (the *valence band*) at $T = 0$, as in insulators. However, semiconductors have only a small energy gap between that band and the next, the *conduction band* (Fig. 26–19a). Silicon and germanium have energy gaps of 1.1 eV and 0.7 eV, respectively, and are semiconductors. Because the energy gap is so small, increasing the temperature or applying a modest electric field will allow some electrons to jump the gap, and electric current will flow (Fig. 26–19b). A semiconductor with an electric field too small to boost conducting electrons to the next energy band will act as an insulator; once the field increases above the strength necessary to allow electrons to "jump the gap," the material acts as a conductor. In other words, there is a minimum electric field under the influence of which a material changes from an insulator to a conductor. In addition, an increase in temperature will give a fraction of the electrons enough thermal energy to jump the gap in semiconductors and thus *lowers* the resistivity. For a conductor, a rise in temperature *increases* the resistivity because the atoms, which are obstacles to electron flow, vibrate more vigorously.

When an electron in the valence band of a semiconductor crosses the energy gap and conducts electricity, it leaves behind what is known as a **hole**. Other electrons in the valence band near the top of the stack of energy levels can move into this hole, leaving behind their own holes, into which still other electrons can move, and so forth. The hole behaves as a positive charge and, since it can move, forms a contribution to the current on its own as a positive charge carrier. An electron excited from the valence band to the conduction band is thus doubly effective at conducting electricity in semiconductors.

One of the major advances in materials technology has been our ability to produce new semiconductors, those with tailored properties. Semiconductor materials that are compounds, such as gallium arsenide, are called *hybrid* semiconductors, as opposed to *intrinsic* elemental semiconductors, such as silicon and germanium. Other special semiconductors are made by introducing impurities, small amounts of different elements, into the lattice. These impurities have the effect of changing the numbers of electrons or holes available to conduct. For example, an atom in the chemical group of phosphorus, arsenic, and antimony can replace one of the silicon atoms in a lattice without affecting the lattice itself too much. However, each of these impurity atoms has one more electron in its valence level than does a silicon atom; this extra electron, for which there is no room in the valence band, takes a place in the conduction band and therefore is available to carry current (Fig. 26–20a). A semiconductor with impurities of this sort is called an n-type semiconductor, and the extra electrons are called *donor* electrons. The semiconducting material, silicon in this case, is said to be *doped* by the impurity atoms.

We can alternatively add atoms of an element that has one less valence electron than silicon, such as boron, aluminum, or gallium. Figure 26–20b shows that in this case we are one electron short of what is needed to form the atomic-level bond that holds the lattice together. This electron must be provided by the electrons of the valence band of the lattice material, and so holes are created in this band. These holes act as positive charge carriers. The impurity atoms are called *acceptors*, and a semiconductor with impurities of this sort is called a *p*-type semiconductor.

Many electronic devices, such as the diode mentioned in Section 26–3, depend heavily on the properties of semiconductors. Probably the best known and most widely used of these devices are transistors, which can amplify electronic signals, and form the basic building blocks of logic devices.

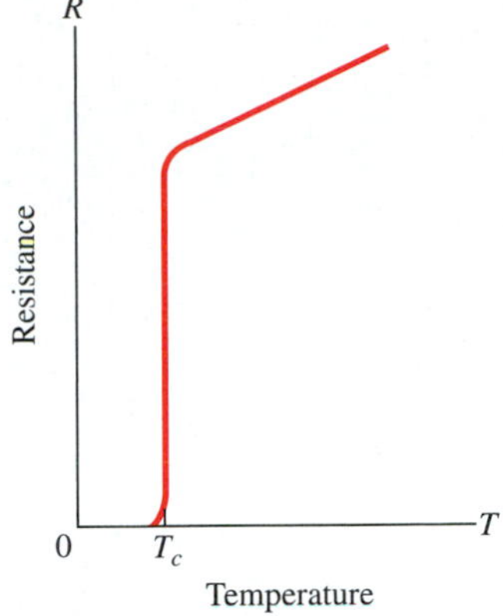

(a)

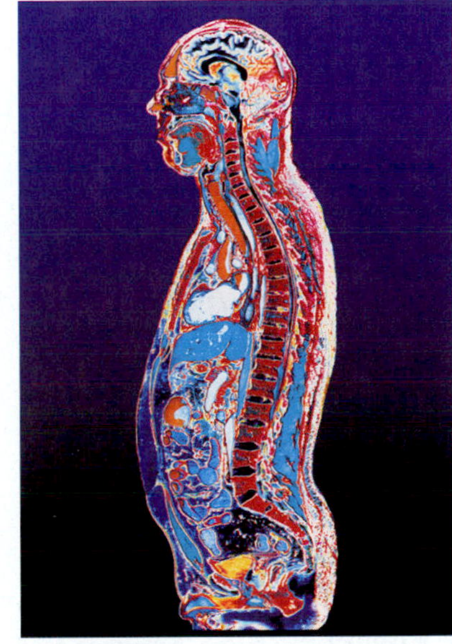

(b)

▲ **FIGURE 26–21** (a) For superconductors the resistance drops to zero at the critical temperature T_c. (b) An MRI image of a human made using a superconducting magnet.

Superconductors

In 1911, H. Kammerlingh Onnes found that mercury abruptly loses *all* of its resistance—at least as far as he could tell—at a *critical temperature* T_c of 4.1K (Fig. 26–21a). This state of zero resistance persists at temperatures below T_c. A material that exhibits zero resistance at some critical temperature is called a *superconductor*. An experiment on a superconducting ring in which a current had been induced showed that there was no observable decrease in the current after a full year. From the measurements in this experiment, it was possible to deduce that if there were any resistive decrease of the current, it had to occur over a period of at least 10^9 years!

The prospect of having an *electric current that lasts forever* is an enticing one. It implies, among other things, the cheap transmission of electricity. The phenomenon of superconductivity cannot be understood as an extension of ordinary conductivity. The abruptness with which resistance disappears completely suggests that an ordinary conductor makes a transition to a totally different state of matter at T_c, much as liquid water turns into a crystal (ice) at 273K. In 1957, John Bardeen, Leon Cooper, and Robert Schrieffer satisfactorily explained the *superconducting phase* with quantum physics in what is now known as the BCS theory.

Until 1986, the materials with the highest-known values of T_c became superconducting at 23K. Helium is liquid at such temperatures and is thus used for cooling superconductors. However, liquid helium is a relatively expensive medium, as is the equipment that keeps it cold. That limited the uses of superconductors to scientific applications, such as magnets for particle accelerators (see Chapter 28) or to important situations such as nuclear magnetic resonance imaging machines in hospitals (Fig. 26–21b). In 1986, however, K. Alex Muller and J. George Bednorz discovered a new class of materials for which T_c is much higher; superconductors are now known that have a T_c above 120K ($-153°C$). This discovery has great technological implications because such materials can be cooled relatively cheaply with nitrogen (which is liquid at 77K). Research in both the basic physics behind these materials and their technological applications is of active interest.

26–7 Electric Power

Electric energy is sent to our homes and workplaces and composes much of the energy used in our society. Efficient delivery of this energy is of paramount importance, and in this section, we shall look at the ways in which resistance affects the delivery of electric energy.

We have compared electrical resistance to mechanical drag. When there is drag in mechanical motion, mechanical energy is converted to thermal energy. The second law of thermodynamics (Chapter 20) shows that some of this thermal energy is irretrievably lost in the sense that it cannot all be converted to mechanical work. Similarly, just as mechanical friction generates heat, the passage of a current through a resistor generates heat, and in this way some electric energy is lost due to resistance. While sometimes we want to use the thermal energy, as in the heating element of an electric stove, it cannot all be converted to useful mechanical work, and in the transmission of electricity the thermal energy created is lost energy.

To calculate the energy lost per unit time (the power lost) when a charge moves in a material, consider a small charge dq that moves through a potential difference V. The change in the potential energy of the charge (dU) is equal to the work done (dW)

by the electric force due to the potential difference, and is given by $dU = V\,dq$. It follows that the power, the rate at which energy is expended by the force that pushes the charge, is

$$P = \frac{dW}{dt} = V\frac{dq}{dt}. \tag{26–26}$$

Because the current $I = dq/dt$, the electric power lost, which is the power that must be delivered to move I through the potential V, is

$$P = VI. \tag{26–27}$$

POWER LOST IN RESISTANCE

This result is a general one, independent of the type of material—in particular, whether the material is ohmic or nonohmic—and of the nature of the charge movement. Power has SI units of watts (W), with $1\ \text{W} = 1\ \text{J/s}$. By using Eq. (26–27), we have another unit for power:

$$1\ \text{W} = 1\ \text{V}\cdot\text{A}. \tag{26–28}$$

For ohmic materials, $V = IR$, where R is a constant. Thus the power expenditure for ohmic materials is

$$P = VI = V\left(\frac{V}{R}\right) = \frac{V^2}{R}. \tag{26–29}$$

Equivalently, we can use $V = IR$ in Eq. (26–29) to find that

$$P = I^2R. \tag{26–30}$$

Whether we use Eq. (26–29) or Eq. (26–30) depends on what is known in a particular application. The power lost (rate of energy loss) in a resistor appears in the form of thermal energy and is variously called *ohmic heating, Joule heating*, and I^2R loss.

EXAMPLE 26–9 Nichrome is an alloy of nickel, chromium, and iron often used as a heating element in electrical devices. A nichrome wire (1.0 m in length) is crisscrossed along the interior of a toaster (Fig. 26–22) that can carry a maximum current of 8 A when there is a 120-V potential difference from one end of this wire to the other†. If the resistivity of nichrome is $1.0 \times 10^{-6}\ \Omega\cdot\text{m}$, what is the radius of the wire? What power does the toaster use?

Setting It Up We call the given current I, the given voltage V, the given wire length L, and the given resistivity $\rho = 1.0 \times 10^{-6}\ \Omega\cdot\text{m}$. The desired wire radius will be labeled r. The current and resistance are known, so the power can be calculated.

▲ **FIGURE 26–22** The glowing nichrome wires of a toaster.

Strategy With the help of Eq. (26–14) we can calculate the cross-sectional area A once the resistance is determined: $A = \rho L/R$, and given A the wire radius follows from $A = \pi r^2$. Thus the first step is to calculate the resistance using $R = V/I$. The power can be calculated using $P = V^2/R$.

Working It Out We have

$$R = \frac{V}{I} = \frac{120\ \text{V}}{8\ \text{A}} = 15\ \Omega.$$

It follows from $A = \rho L/R$ that

$$r = \sqrt{\frac{A}{\pi}} = \sqrt{\frac{\rho L}{\pi R}} = \sqrt{\frac{(1.0 \times 10^{-6}\ \Omega\cdot\text{m})(1.0\ \text{m})}{\pi(15\ \Omega)}}$$
$$= 0.15 \times 10^{-3}\ \text{m}.$$

Finally, the power is

$$P = \frac{V^2}{R} = \frac{(120\ \text{V})^2}{15\ \Omega} = 960\ \text{W}.$$

What Do You Think? Suppose the toaster is accidentally connected into a 240-V outlet. With the voltage doubled, by how much will the power consumed change? (Assume that the toaster doesn't burn out, and don't try this!)

†Real household electricity involves an oscillating voltage difference and an oscillating (or alternating) current. Ignore these effects here and in Example 26–10.

We described the commercial units of electric energy in Section 6–5: The energy unit in the electric power industry is the kilowatt-hour (kWh), 3.6×10^6 J. When electric energy is delivered to a home, the energy delivered per unit time is also called the electric power.

EXAMPLE 26–10 A 100-W bulb is left on in an outdoor storage room to keep paint from freezing. The 100-W rating refers to the power dissipated in the bulb's filament, which is a resistor. If electricity costs 8 cents/kWh, about how much does it cost to burn the lightbulb for three months during winter?

Setting It Up The given power of 100 W expresses an energy consumed per second. We know the total time that the bulb burns, as well as the cost per kWh of energy.

Strategy The number of joules of energy used over three months is found by converting three months to seconds, then multiplying the power, 100 W, by the number of seconds the bulb is on. Since the cost rate is given in dollars per kWh, we'll want to convert the energy units to kWh.

Working It Out The number of seconds the bulb is lit is

$$(3\ \text{months})(30\ \text{days/month})(24\ \text{h/day})(3600\ \text{s/h}) = 7.8 \times 10^6\ \text{s}.$$

The energy used is

$$(100\ \text{W})(7.8 \times 10^6\ \text{s}) = 7.8 \times 10^8\ \text{J}$$
$$= (7.8 \times 10^8\ \text{J}) \times [1\ \text{kWh}/(3.6 \times 10^6\ \text{J})] = 220\ \text{kWh}.$$

From this we find the cost, $(8 \times 10^{-2}\ \$/\text{kWh}) \times (220\ \text{kWh}) =$ \$17.60.

Although the primary purpose of a lightbulb is to produce light, most of the electric energy it dissipates is converted into heat, not light.

What Do You Think? Assuming the voltage is fixed, will it be more or less expensive to have *two* of these bulbs in (a) series, (b) parallel?

EXAMPLE 26–11 As we described in Eq. (17–17), a piece of metal such as a wire heated to a temperature T (in units of degrees kelvin) radiates energy in the form of electromagnetic waves, with the power emitted per unit surface area given by Stefan's formula, $P_{\text{Stefan}} = \sigma T^4$, where the constant σ is roughly $5.7 \times 10^{-8}\ \text{W}/(\text{K}^4 \cdot \text{m}^2)$. A bus-stop shelter is heated by a 1-m-long metallic coil that is 1 mm in diameter. The temperature of the coil is 2000K. If the resistivity of the coil material at that temperature is given by $5 \times 10^{-7}\ \Omega \cdot \text{m}$, estimate the current that must flow through the wire to maintain its temperature. (We are assuming that at this temperature the fraction of the power dissipated due to convection of the air around the coils is small.)

Setting It Up In addition to knowing the temperature T, we are given the diameter d of the wire and its length L. These will suffice to find the power emitted in radiation, and since we are given the resistivity ρ, we can match this power emission to a resistive power loss in terms of the unknown current.

Strategy The first step is to use Stefan's formula to calculate the power P radiated by the wire, given that the temperature is $T = 2000\text{K}$. This will give for the total radiated power $P = P_{\text{Stefan}} \times A$, where A is the surface area of the wire. This power must come from resistive loss in the wire, because that is what heats the wire to high temperatures. Therefore we calculate the resistance R of the wire in terms of the resistivity of the material, using Eq. (26–14) and the wire dimensions, and match P to the resistive loss RI^2, a relation that allows us to solve for the desired current I.

Working It Out With the wire's surface area $A = \pi\, d \times L$, the emitted power is

$$P = \sigma T^4 \times (\pi\, d \times L) = [5.7 \times 10^{-8}\ \text{W}/(\text{K}^4 \cdot \text{m}^2)] \times (2 \times 10^3\ \text{K})^4 \times \pi \times (10^{-3}\ \text{m}) \times (1.0\ \text{m}) = 3 \times 10^3\ \text{W}.$$

The resistance of the wire is

$$R = \rho \frac{L}{A_{\text{cross section}}} = \rho \frac{L}{\pi (d/2)^2}$$
$$= (5 \times 10^{-7}\ \Omega \cdot \text{m}) \frac{1.0\ \text{m}}{\pi (0.5 \times 10^{-3}\ \text{m})^2} = 0.6\ \Omega.$$

Finally we use $P = RI^2$, with P the radiated power, so that

$$I = \sqrt{\frac{P}{R}} = \sqrt{\frac{3 \times 10^3\ \text{W}}{0.6\ \Omega}} = 70\ \text{A}.$$

What Do You Think? Suppose you would like to keep the material and the temperature of the radiating coil the same but at the same time reduce the current. Could this be done by changing the diameter d of the wire, and if so, how?

Resistors used in circuits are characterized not only by their resistance, but also by a power rating. This power rating states the maximum power that the resistor can dissipate without being damaged due to overheating. The power rating is measured in watts. According to Eq. (26–30), which states that the power dissipated in a resistor is $P = I^2 R$, we can deduce the maximum allowed current from the power rating. One class of relatively inexpensive resistors, so-called carbon film resistors, is limited to about 2 W; a second more expensive type known as wire-wound resistors have a power rating up to 50 W.

Summary

Electric current is the rate at which charge passes. The instantaneous current is given by

$$I \equiv \frac{dQ}{dt}. \tag{26–2}$$

The unit of current is the ampere (A), 1 C/s. Currents in wires are depicted as though the positive charges are moving, but it is actually electrons (negative charge) that are mobile.

The current density $\vec{J}$ is a vector quantity representing the current that passes through an area per unit time. The current is related to the current density by

$$I = \int_{\text{surface } A} \vec{J} \cdot d\vec{A}. \tag{26–4}$$

The free-electron model of conduction is useful as a qualitative description of current in a solid. The average, or drift, speed of the electrons that pass through the material is

$$v_d = \frac{I}{n_e q A}, \tag{26–9}$$

where n_e is the density of free electrons and q is the charge of an electron.

Electrical resistance R is the ratio of voltage to current

$$R \equiv \frac{V}{I}. \tag{26–11}$$

Many conducting metals show a linear relationship between voltage and current. The resistance is then constant over a wide range of voltages. This relation is called Ohm's law, $V = IR$.

Resistivity ρ is the quantity that distinguishes the part of the resistance that is intrinsic to each particular type of material. For wires of length L and cross-sectional area A, we have

$$R = \rho \frac{L}{A}. \tag{26–14}$$

The inverse of the resistivity is the conductivity σ which expresses how well a type of material conducts current:

$$\sigma \equiv \frac{1}{\rho}. \tag{26–15}$$

Both ρ and σ depend on temperature.

The electric field and current density are related by

$$\vec{E} = \rho \vec{J} \tag{26–16}$$

and by

$$\vec{J} = \sigma \vec{E}. \tag{26–17}$$

When a number of resistors of resistance $R_1, R_2, \ldots R_n$ are placed in series, then they act together as a single resistor of equivalent resistance

$$R_{eq} = R_1 + R_2 + R_3 + \cdots + R_n. \tag{26–21}$$

When a number of resistors of resistance $R_1, R_2, \ldots R_n$ are placed in parallel, then they act together as a single resistor of equivalent resistance given by

$$\frac{1}{R_{eq}} = \frac{1}{R_1} + \frac{1}{R_2} + \cdots + \frac{1}{R_n}. \tag{26–22}$$

The free-electron model, which provides a classical model for Ohm's law and the mechanism of electrical conduction, cannot fully account for the observed behavior of conductors, and the correct explanation of electrical conduction in metals is dependent on how electrons fill allowed energy bands and on the energy gap between these bands. Materials that conduct current easily have electrons in partially filled bands. Semiconducting materials, such as silicon and germanium, can be doped by impurity atoms to increase the density of charge carriers. The explanation of superconductivity requires both quantum mechanics and the presence of a new phase of matter in which electrons collectively transport electric current.

When a current moves through a potential difference, electric power P is dissipated (or produced), given by

$$P = VI. \tag{26–27}$$

For resistive materials, the power is also given by

$$P = \frac{V^2}{R} = I^2R. \qquad (26\text{–}29, 26\text{–}30)$$

Understanding the Concepts

1. Consider the electron beam in a cathode-ray tube. The velocity of the electrons in the beam changes as the electrons are accelerated. Is the current the same everywhere in the beam?
2. How does the free-electron model for electrical resistance account for power dissipation? Does our microscopic picture agree with the voltage/current result?
3. The same current passes through two similar wires of unequal areas. Which wire will get hotter, and why?
4. The same current passes through two wires of the same area. One of the wires is made of aluminum, whereas the other is made of brass. Which wire will get hotter, and why?
5. What factors determine the differences in drift velocity of electrons in wires if the dimensions and current are the same?
6. Knowing what you now know about parallel and series resistance, why does the potential drop linearly along the length of a resistor that is a uniform cylinder?
7. If the movement of charges in a wire is similar to the flow of water in a hose, why, when a new hose is hooked up to a faucet, do we have to wait for a while until the water comes out, but when we hook a new wire up to a circuit, we do not have to wait for charge to come out the other end when the switch is turned on?
8. According to the discussion of Section 26–5, the resistivity in the free-electron model should vary with the square root of the temperature and thus should be zero at $T = 0$. Is this reasonable? How would you interpret this result?
9. We know that the resistivity of a metal is temperature dependent, and so therefore is the resistance of a wire. In Chapter 17, we saw that the dimensions of a piece of metal—such as a wire—change when the wire is heated. Does this provide an additional reason to change the resistance of a wire as it undergoes Joule heating? Would you expect the effect to be large?
10. When you throw a switch and charge flows in a household wire, does the wire become charged?
11. Suppose that we orient a wire between the plates of a charged capacitor so that there is an electric field along the cross section of the wire. Will the resistance of the wire change because all the charge-carrying electrons crowd to one side of the wire, thus effectively reducing the wire's cross section?
12. Gauss' law states that free charge in a conductor moves to the surface of the conductor. Does this mean that the current flowing through a wire is actually on the wire's surface?
13. The resistivity of most metals is on the order of $10^{-8}\ \Omega \cdot \text{m}$. Discuss why this might be so in terms of the result given by Eq. (26–25).
14. Is it possible to break up any combination of resistors into a sequence of parallel and series resistances? If not, give an example of a combination that cannot be so decomposed.
15. What is likely to happen when a current is so large that the power dissipation in a resistor through which the charge flows exceeds the resistor's power rating? What mechanism is responsible for such a disaster scenario?
16. What considerations would you have to take into account in designing a lightbulb filament with trade-off between the length of filament and the diameter of the filament?
17. The resistance of a wire is proportional to the length of the wire. Think of a wire as consisting of a series of shorter wires placed together in series. Can you use this information to predict the effective resistance of two different resistors of resistance R_1 and R_2 when they are placed in series in a circuit?
18. In Fig. 26–7a the drift velocity is smaller in area A_2 than in A_1, yet the current is the same. How is that possible?
19. Consider three identical bulbs between points A and B as shown in Fig. 26–23. The potential difference between A and B is fixed. The switch between C and bulb 3 may be open or closed. Will bulb 2 be brighter when the switch is open, or if it is closed? Will bulb 1 be brighter when the switch is open or if it is closed?

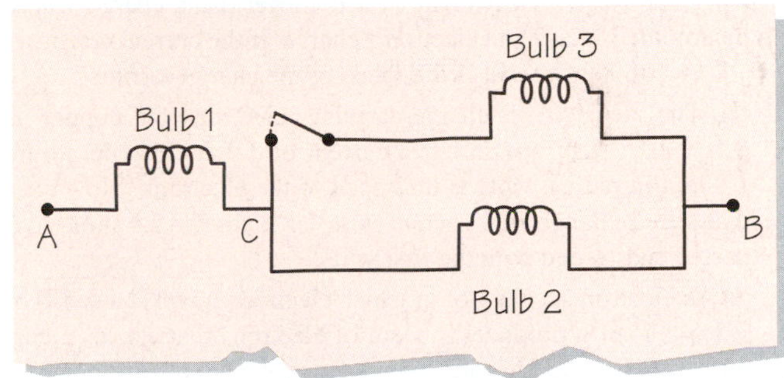

▲ **FIGURE 26–23** Question 19.

20. A copper rod of a given length and diameter has resistance R. Suppose the same rod is drawn into a wire with diameter 1/10 of what it was before. By what factor will the resistance change? Will it be larger or smaller?
21. Suppose all the resistors in Fig. 26–14 have the same resistance and the potential difference V_{AB} is held fixed. If the number of resistors is changed from n to $n + 1$, will the total current flowing between A and B increase or decrease?
22. Three resistors of resistance 1 Ω, 2 Ω and 4 Ω are placed in a network shown in Fig. 26–24. Can you decide in which places the resistors are to be put so as to minimize the overall resistance between A and B? What arrangement would maximize the overall resistance?

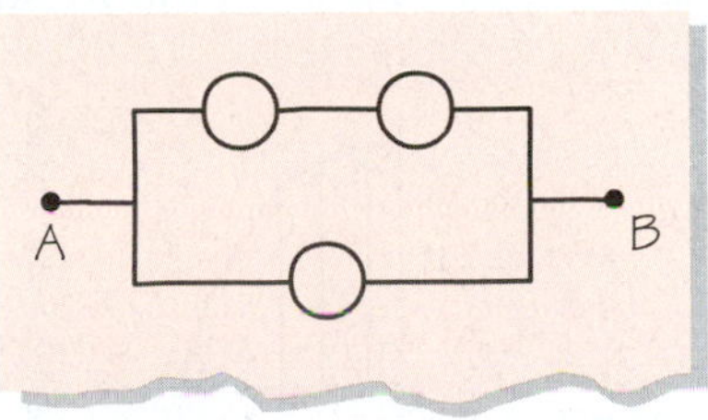

▲ **FIGURE 26–24** Question 22.

23. In high wattage lightbulbs, the filament is typically coiled. Why?

Problems

26–1 Electric Current

1. (I) A wire of diameter 2.2 mm carries a current of 0.46 A. What is the average current density? How much charge crosses a fixed point in the wire per second?

2. (I) There is a 1.2-A current in a wire of cross-sectional area $4.2 \times 10^{-5}\ \text{m}^2$. The drift speed of the electrons that carry the current is 0.32×10^{-5} m/s. Find the density of current-carrying electrons.

3. (I) A jumper cable used to start a car carries a current of 100 A and has a cross-sectional area of 36 mm^2 and a length of 2 m. The free-electron density in the cable is 8.5×10^{28} electrons/m^3. How long does it take a free electron to pass from one end of the cable to the other?

4. (I) Three straight wires of area 0.02 mm^2, 0.2 mm^2, and 2 mm^2, respectively, are aligned along the x-axis. They carry current densities along the x-axis of magnitude 3×10^5 A/m^2, 13×10^4 A/m^2, and 15×10^4 A/m^2, respectively. Find the current in each wire.

5. (I) A wire of radius 1.6 mm carries a current of 0.092 A. How many electrons cross a fixed point in the wire in 1 s?

6. (I) Charge carriers in a semiconductor have a number density $n_q = 3.5 \times 10^{24}$ carriers/m^3. Each carrier has a charge whose magnitude is that of an electron's charge. If the current density is 7.2×10^2 A/m^2, what is the speed of the charge carriers?

7. (I) The density of charge-carrying electrons in copper is 8.5×10^{28} electrons/m^3. If a current of 1.2 A flows in a wire 1.8 mm in radius, what is the speed of the electrons? How does that speed change in a second wire, of diameter 2.4 mm, connected end-to-end with the first wire?

8. (I) An electron accelerator in which electrons travel at a speed of 3.5×10^7 m/s produces a beam of electrons that carries a current of 6.1 mA. The effective area occupied by the beam is 0.50 cm^2. What is the density of electrons in the beam? Ignore all relativistic effects.

9. (II) In the National Synchrotron Light Source x-ray device at Brookhaven National Laboratory, there is an electron beam with an average current of 200 mA. The electrons have a kinetic energy of 2.5 GeV and a speed extremely close to the speed of light. How many electrons pass a given point in the accelerator per hour? How many electrons are contained in a 1-m length of the beam? Ignore all relativistic effects (a poor approximation, in this case).

10. (II) A cube of material is placed with one corner at the origin of a coordinate system; its sides, 1 cm long, are parallel to the three axes. The current density is $A\hat{i} + B\hat{j} + C\hat{k}$ throughout the cube. The units of A, B, and C are mA/cm^2. What are the currents along the x-axis, y-axis, and z-axis?

11. (II) The current density in a cylindrical wire of radius R is $J = J_0(1 - r^2/R^2)$, parallel to the axis of the wire (Fig. 26–25). Calculate the total current across a section perpendicular to the axis.

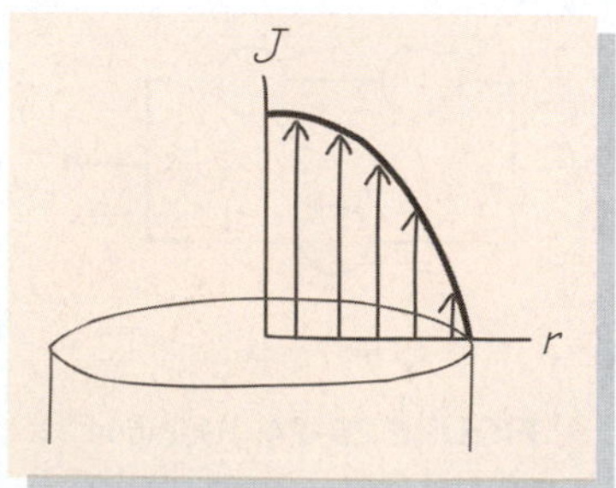

▲ **FIGURE 26–25** Problem 11.

12. (II) In a plasma containing equal densities n of electrons and (positive) ions, the ions move to the right. Their speed is a factor of 1.5×10^{-3} smaller than the speed with which the electrons move to the left. What is the (net) current density? Give its direction and magnitude.

13. (II) An aqueous solution contains 0.1 mol/L of NaCl. The NaCl is dissolved in the form of Na^+ and Cl^- ions. Calculate the velocities of the Na^+ and Cl^- ions, respectively, if there is a measured total current density of 40 A/m^2. Assume that the velocity of the Cl^- ions is about 50 percent greater than that of the Na^+ ions.

26–2 Currents in Materials

14. (I) Calculate the drift speed of electrons in the conduction cables of an automobile starter cable, which is made of copper and has a diameter of 4 mm, if you suppose that the cable carries 100 A. How would this speed change if the diameter of the wire were doubled? [*Hint*: Useful data are contained in Example 26–2.]

15. (I) A single charged elementary particle ($q = 1.6 \times 10^{-19}$ C) travels with a speed very close to that of light in a circular accelerator of diameter 5 km. What is the current represented by the particle, taking into account multiple traversals of the charge past a given point?

16. (I) How many particles like that described in Problem 15 must be present at a given time to give rise to a current of 42 mA?

17. (II) An aluminum wire of area 50 mm^2 placed along the x-axis passes 10,000 C in 1 h. Assume that there is one free electron for each aluminum atom. Determine the current, current density, and drift speed. The mass density of aluminum is 2.7 g/cm^3.

18. (II) Gold has one electron per atom available to carry charge. Given that the mass density of gold is 19.3×10^3 kg/m^3 and that its molecular weight is 197 g/mol, calculate the drift speed of the electrons in a gold wire that carries 0.3 A and has a circular cross section 0.5 mm in radius.

19. (II) Two parallel metal wires of diameter 0.2 cm and a charge-carrier density $n_e = 7 \times 10^{22}$ electrons/cm^3 carry a current of 3 A each. The wires join and then split into three identical but separate wires, each with a radius one-half that of the original wire (Fig. 26–26). All the wires are made of the same material. What are the drift speeds in both the larger and smaller wires? Can you explain the difference in speeds in terms of the speeds of water flow in pipes?

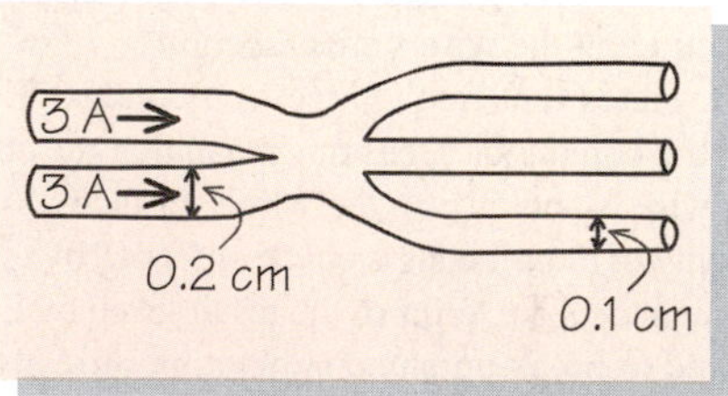

▲ **FIGURE 26–26** Problem 19.

20. (II) The charge carriers in a certain wire of circular cross section and radius R have a drift speed down the wire that is not constant across the wire. Instead, the drift speed rises linearly from zero at the circumference ($r = R$) to v_0 at the center ($r = 0$). Compare the total current carried by this wire with the current carried by a wire of the same radius, same density of charge carriers, and a constant drift speed of $v_0/2$.

21. (II) A thin copper wire carrying a current I is welded to the center of a circular copper plate capping a copper tube (Fig. 26–27). The radius of the tube is R, the thickness of its wall and the top plate is d, and $d \ll R$. What is the current density in the tube and in the top plate?

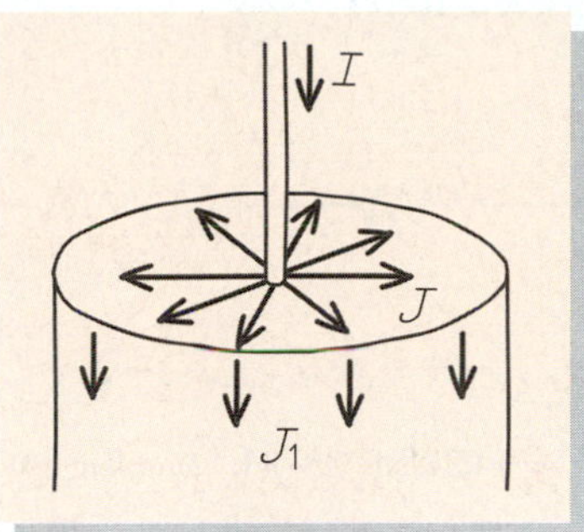

▲ **FIGURE 26–27** Problem 21.

22. (III) Charges q move longitudinally down a rod of circular cross section and radius R. The density of the charge carriers n decreases as a function of the radial distance r from the center of the rod according to $n = n_0 - n'r$. The speed v of the charge carriers varies with r according to $v = v_0 - v'r^2$, where n_0, n', v_0, and v' are constants. Calculate the current that passes through the rod.

26–3 Resistance

23. (I) You have two solid cylinders of the same material. Piece 2 has half the length and half the diameter of piece 1. What is the ratio of the resistances of the two pieces?

24. (I) The conductivity of silver is 1.5 times that of gold. What is the ratio of the diameter of a silver wire to that of a gold wire of the same length if both wires are designed to have the same resistance?

25. (I) An underground wire made of aluminum is 528 m long and has an area of 0.12 cm^2. (a) What is its resistance? (b) What is the radius of a copper wire of the same length and resistance?

26. (I) An old house is wired with AWG #18 copper wire, which has a diameter of 0.0403 in. (a) What is the wire's resistance per 100 ft? (b) One circuit consists of only one wire behind walls and has a resistance of 7.5 Ω. How long is this wire?

27. (I) The resistivity of copper is $1.72 \times 10^{-8}\ \Omega \cdot \text{m}$. What is the resistance of a section of gauge #10 wire (diameter 0.2588 cm) that is 10 m long?

28. (I) A carbon rod used in a welding machine is 5.0 mm in diameter and 20.0 cm in length. What is its resistance and how much current will pass through it if the welding machine puts a voltage of 380 V across it?

29. (I) Cables used to jump an automobile can get hot if used for more than a few seconds. Calculate the resistance at 20°C of a 2-m-long copper cable of cross-sectional area 36 mm^2. By how much does the resistance increase as the temperature rises from 20°C to 100°C?

30. (I) In the text, we refer to a power line with a total resistance of 10 Ω. Suppose that the power line is made of copper with a resistivity of $1.72 \times 10^{-8}\ \Omega \cdot \text{m}$, and is 175 km long. What is the radius of the wire?

31. (I) How long would a tungsten wire have to be if it is to be used in a toaster at 120 V and the current to be carried is 15 A? The cross-sectional area of the wire is 0.20×10^{-6} m^2.

32. (I) The electron current densities of copper and aluminum are 9×10^{28} m^{-3} and 18×10^{28} m^{-3}, respectively. A wire of copper is joined to a wire of aluminum of the same diameter. If the same current is passing through the wires, compare the electron drift velocities and the resistances per unit length.

33. (II) A current passes through a tungsten wire in an appliance. If you assume that there is a fixed potential drop from one end of the wire to the other, what is the fractional change in the power consumed as the temperature of the tungsten wire changes from 800°C to 1200°C? Ignore any effects due to the change in the wire's length by thermal expansion.

34. (II) An electrician tests for a short circuit by putting a potential difference of 1.5 V across two neighboring parallel wires that would be independent of each other if there were no short. A current of 0.14 A then flows in the wires. The wires consist of material with a resistivity of $1.7 \times 10^{-8}\ \Omega \cdot \text{m}$ and have a diameter of 0.24 mm (Fig. 26–28). Given that the short effectively makes the wires act like a single wire, how far away is the short?

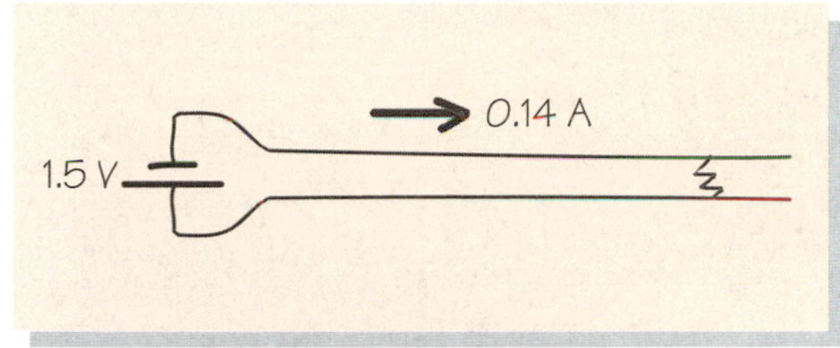

▲ **FIGURE 26–28** Problem 34.

35. (II) A nichrome wire of diameter 0.5 mm and length 50 cm is connected to a 50-V battery. What current passes through the wire at room temperature (25°C) and after the wire heats up to 400°C?

36. (II) The change in the resistance of a thin platinum wire can be used to measure temperature. Suppose that a constant 6.0 mA current passes through a platinum wire and that the potential drop measured at room temperature is 8.5 mV. What is the temperature of the wire when the potential drop is 8.7 mV?

37. (II) An aluminum wire of length L and a copper wire of length $5L$ have precisely the same resistance. Given that the resistivity of aluminum and copper are $2.8 \times 10^{-8}\ \Omega \cdot \text{m}$ and $1.7 \times 10^{-8}\ \Omega \cdot \text{m}$, respectively, what is the ratio of the radii of the two wires?

38. (II) You have a 100-m-long wire of area 0.5 mm^2 with a thin coating of insulation, but you cannot identify the type of material that makes up the wire. You have a 12.0-V battery and a device to measure current. When the battery is placed across the two ends of the wire, you measure a current of 1.07 A. What is the wire material? (Use Table 26–2.)

39. (II) A coil used to produce a magnetic field is made of copper wire of area 1.5 mm^2 wound many times around a spool of diameter 20 cm. The resistance of the wire is 1.35 Ω. We must know the number of turns of wire to know the magnetic field. How many turns of wire are there on the spool?

40. (II) You wish to double a current that flows through a wire of fixed length, but you can increase the voltage that drives the current by only a factor of 1.8. You have other wires made of the same material but of different radii. What is the smallest factor by which the radius of a replacement wire should differ from the radius of the original wire?

41. (II) Aluminum has a density of 2.7×10^3 kg/m^3. What is the resistance of an aluminum wire 0.12 cm in diameter and 80 m long? What is the mass of the wire? What is the mass of a copper wire, of density 8.9×10^3 kg/m^3, with the same length and same total resistance?

42. (II) What are the length and the radius of a copper wire (of circular cross section) whose resistance is 2 Ω and whose mass is 1.5 kg?

43. (II) How much silver (of density 10.5×10^3 kg/m^3) would be needed to make a wire 1 km long, with a resistance of 5 Ω?

44. (II) A copper pipe has an inside diameter of 2.75 cm and an outside diameter of 2.45 cm. What length of copper pipe will have a resistance of 3.5 Ω?

45. (II) A copper resistor has the shape of a cylindrical shell. What is the resistance of this resistor if its length is 1 m, its inner radius is 0.1 cm, and its outer radius is 0.2 cm? What is the radius of a solid wire of circular cross section with the same length and the same resistance? Compare the masses of the two resistors.

46. (II) A *zener diode*, named for Clarence Zener, has the *I–V* curve shown in Fig. 26–29. Sketch the resistance of the diode versus both current and voltage. What is special about the critical voltage V_c?

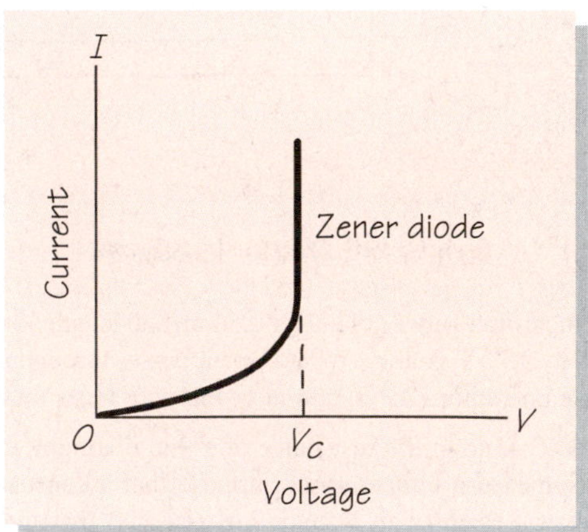

▲ **FIGURE 26–29** Problem 46.

26–4 Resistances in Series and in Parallel

47. (I) We have three identical lightbulbs available to us. Two of them are connected in series to a battery. What is the relative brightness of the two bulbs? If the third bulb is connected in parallel with the second of the above bulbs, what are the relative brightnesses of the three bulbs?

48. (I) Consider the circuits shown in Fig. 26–30. The lightbulbs in all circuits are identical and the batteries are the same in the two circuits. Before working with these circuits, you are asked to make some predictions: (a) What is the brightness of the bulbs in circuit II relative to each other and to the bulb in circuit I? (b) If one of the bulbs is removed in circuit II, how will the brightness of the other bulb be affected? Does it matter which bulb is removed?

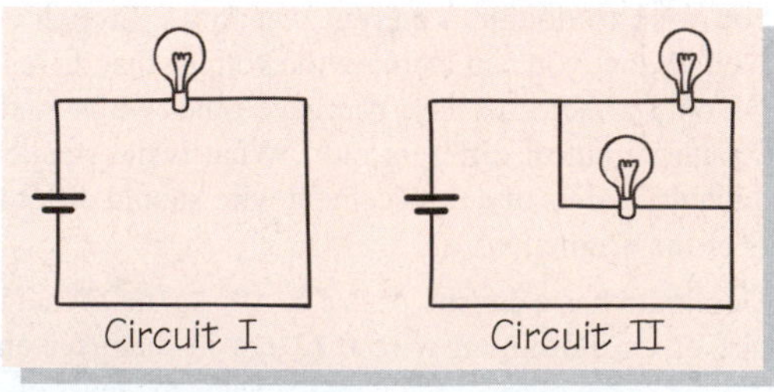

▲ **FIGURE 26–30** Problem 48.

49. (I) Consider the circuit in Fig. 26–31 in which there are two resistors in series. The resistance of one of them, x, is unknown. If the resistor R is 10 Ω, the voltage drop across x is 8 V. If R is 5 Ω, the voltage drop across x is 12 V. What is the resistance of x, and what is the total potential drop across the two resistors in the case that R is 10 Ω?

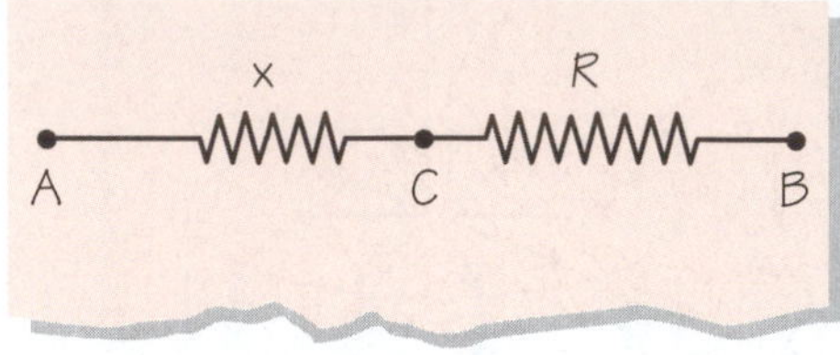

▲ **FIGURE 26–31** Problem 49.

50. (I) Suppose we have two resistors in parallel. Their values are 20 Ω and 12 Ω respectively. What resistance placed in parallel with these will make the total effective resistance equal to 4 Ω?

51. (I) Five resistors of 18 Ω each are connected in series. If the potential difference between the ends of this set of resistors is 16 V, what current flows through the resistors? What is the power expended in the circuit?

52. (I) Two 60-Ω resistors are placed in series across two terminals whose potential difference is 120 V. What is the total power dissipated?

53. (I) Two resistors are placed in parallel. One of the resistors has twice the resistance of the other; the lesser of the two resistances is 150 Ω. What is the resistance of the parallel combination?

54. (I) Find the equivalent resistance of the circuit shown in Fig. 26–32.

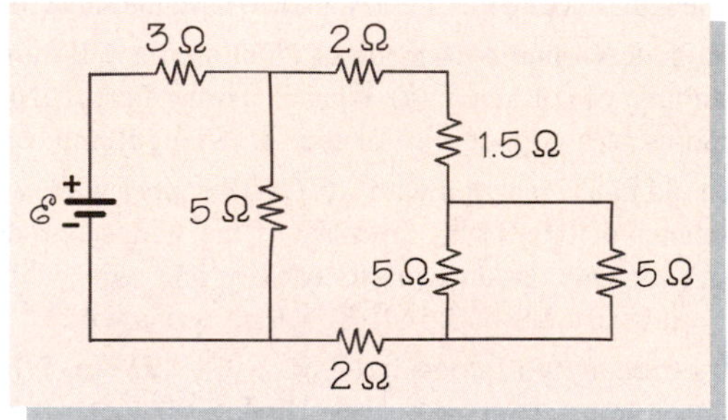

▲ **FIGURE 26–32** Problem 54.

55. (II) Consider the combination of resistors shown in Fig. 26–33. Calculate the current in each resistor, given that $V_{AB} = 16$ V.

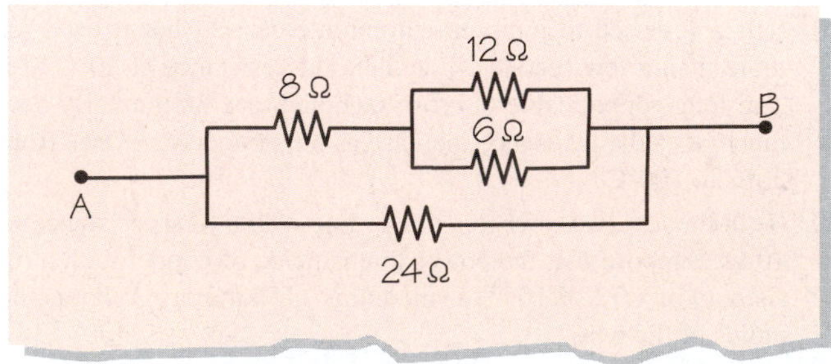

▲ **FIGURE 26–33** Problem 55.

56. (II) Take the resistor combination in Fig. 26–33. What is the power dissipated in each of the resistors?

57. (II) Consider the network of resistors shown in Fig. 26–34. If a total current of 20 A flows through the network from A to B, what is the current in each of the resistors, and what is the voltage across each of the resistors?

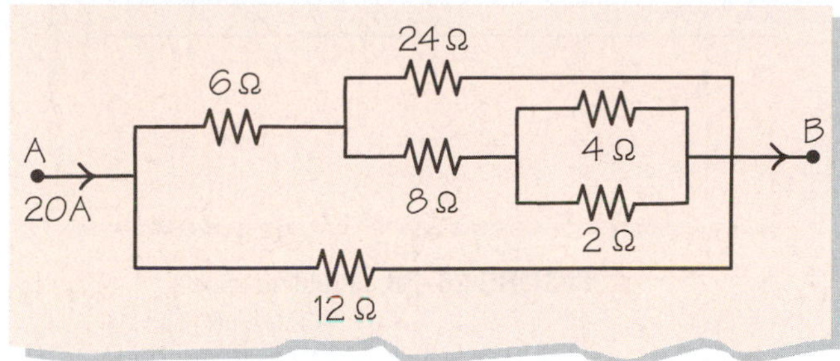

▲ **FIGURE 26–34** Problem 57.

58. (II) Consider the circuit shown in Fig. 26–35. What must the value of the resistance x of the unknown resistor be so that the total equivalent resistance of the network is also x?

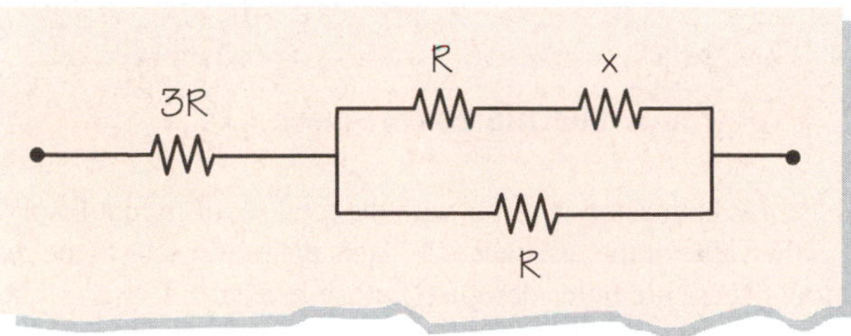

▲ **FIGURE 26–35** Problem 58.

59. (II) Points a and b are connected by the system of resistors shown in Fig. 26–36. A battery of 12 V and negligible internal resistance is connected across points a and b. (a) What is the equivalent resistance between points a and b? (b) The potential difference across the 75-Ω resistor? (c) The current flowing through the 33-Ω resistor?

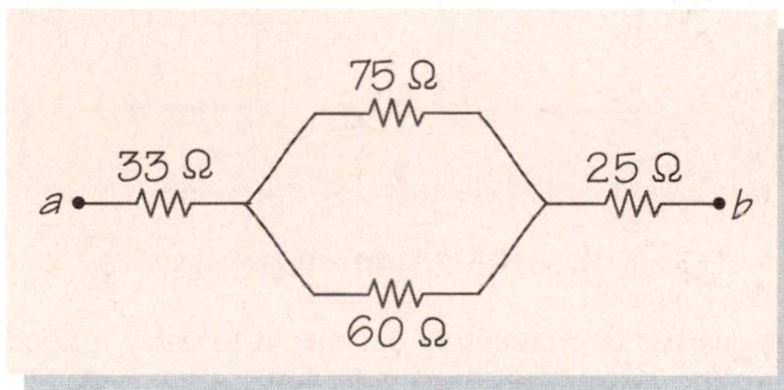

▲ **FIGURE 26–36** Problem 59.

*26–5 Free-Electron Model of Resistivity

60. (I) Using the average time between collisions as calculated in Example 26–5, determine the drift speed of charge carriers for a material in which the electric field is 2.0×10^{-3} V/m.

61. (I) Assuming the collision time from Example 26–8 and an average speed of 2.7×10^{6} m/s, estimate the mean free path for an electron in copper.

62. (I) Recall Eq. (19–46), which relates the collision cross section to the mean free path of a particle. Use that result together with the results of Problem 52 to estimate the collision cross section of an electron with an ion in a copper lattice.

63. (II) In Problem 20, we described a wire of radius R within which the drift speed of charge carriers varies with the distance from the center of the wire as $v_d = v_0[1 - (r/R)]$. Supposing that this wire is made of ohmic material, describe how the resistivity must vary with r to produce this drift-speed profile.

*26–6 Materials and Conductivity

64. (II) If you treat electrons as a gas of independent particles, at what temperatures would an average electron have sufficient energy to cross the energy gap for silicon (1.1 eV), germanium (0.7 ev), and carbon (6 eV)?

26–7 Electric Power†

65. (I) What is the resistance of a 65-W headlight used on a 12-V car battery?

66. (I) What is the maximum voltage that can be applied to a 1000-Ω resistor rated at 1.5 W?

67. (I) Your little sister leaves a 100-W lightbulb burning for an unnecessary hour. Assuming that electric power costs 10 cents per kilowatt-hour, what is the cost of her inaction?

68. (I) A graduate student in engineering has a collection of 100-Ω resistors with different power ratings of 1/8, 1/4, 1/2, 1, and 2 W. What is the maximum current that the student should use in each resistor?

69. (I) What is the maximum allowable current for (a) a 160-Ω, 5-W resistor? (b) A 2.5-kΩ, 3-W resistor?

70. (I) An electrostatic accelerator has a maximum attainable voltage of 8×10^{6} V. If a current of 100 μA is produced by accelerating charges in this potential difference, what is the nominal power required to operate the accelerator? Assume that this power can be converted to the accelerator voltage with 100 percent efficiency.

71. (I) Consider a resistor of resistance R. If the maximum allowed power dissipation is P, what is the maximum allowed operating voltage?

72. (I) An electric heater draws a current of 10 A from a 120-V circuit. What is the cost per hour of operating the heater if electrical energy costs 7 cents per kWh?

73. (I) A 2-m-long copper cable with cross-sectional area 36 mm^2 draws 100 A when it is used to jump start an automobile. It takes 20 s to start the engine. Given the resistance is 9×10^{-4} Ω, calculate how much energy is dissipated in the cable during this operation.

74. (II) A heater uses nichrome wiring ($\rho = 10^{-6}\ \Omega \cdot \text{m}$) and generates 1250 W when connected across a 110-V line. How long must the wire be if its cross-sectional area is 0.2×10^{-6} m^2?

75. (II) Consider the terminals of a 12-V battery connected by a copper wire. How long must the wire be if its cross-sectional area is 8×10^{-6} m^2 and if the power dissipated is 0.8 kW?

76. (II) Buildings have circuit breakers, devices that switch the current off when it exceeds a critical value, to protect the electrical system from damage. One circuit for a building's lights has a 15-A breaker. (a) What is the maximum power that can be delivered by a 110-V line to this circuit? (b) How many lightbulbs, each requiring 75 W, can this circuit handle?

† When we refer to household electricity applications, assume in each case that their currents and voltage differences are of the simple, constant type discussed in this chapter.

77. (II) A 6-V battery is connected to two metal wires dipped in a pot of water (Fig. 26–37). A current of 50 mA flows for 18 h. How much energy is taken out of the battery during that time?

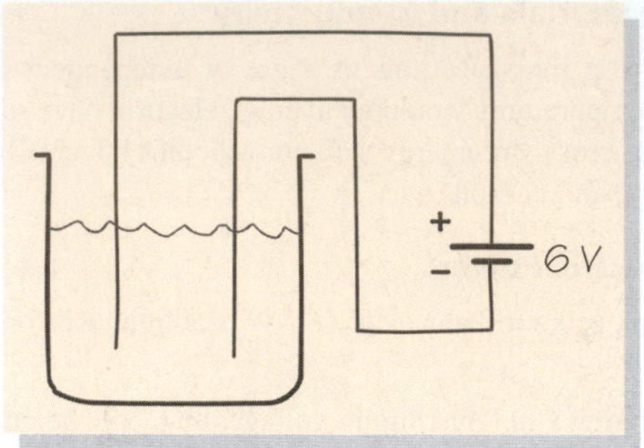

▲ **FIGURE 26–37** Problem 77.

78. (II) A 500-W electric heater is designed to operate on a line of 115 V. As the result of a brownout (a partial interruption of electrical power) the line voltage drops to 105 V. Assuming that the heating unit has a fixed resistance, what is the power of the heater now?

79. (II) Consider the bus stop in Example 26–11. If it were necessary to reduce the current in the heater by a given factor, one could proceed by decreasing the temperature T of the wire or the wire's diameter (see "What do you think?" for that example). What changes would be preferable and why? Ignore the change of the resistivity with temperature—this is typically a smaller effect than the others that you will deal with here.

General Problems

80. (I) The power dissipated in a resistor through which a current I passes is P_0. What is the power dissipated if the same current passes through three such resistors connected in series?

81. (II) An electric hot plate is used to boil water. The current drawn by the hot plate is 4 A. Calculate the voltage and resistance, based on a rough estimate of how long it takes to make a pot of tea.

82. (II) One month's electricity bill for an apartment is \$25.33, and the cost of electricity is 8 cents/kWh. All appliances used in this apartment work at 120 V. How many electrons passed through the apartment's electrical meter that month?

83. (II) A wire of resistance r is drawn—pulled like taffy—to double its length. Assuming a constant voltage and a fixed volume, by how much does the power dissipation change?

84. (II) A Van de Graaff accelerator delivers 4-MeV protons at a current of 5 μA through a target onto a piece of tungsten that serves to stop the proton beam. (a) How many protons stop in the tungsten in 1 h? (b) How much energy is delivered to the tungsten in 1 h? (c) What is the power of the proton beam?

85. (II) A piece of brass is machined into a long, tapering cylinder. Its radius is expressed by $r = r_0 + \alpha x$, where α is a constant and x is measured from the narrow end of the tapering cylinder and runs from 0 to L (Fig. 26–38). Find an expression for the resistance of this piece.

86. (II) The voltage at an electrical outlet is a constant 120 V. You have 10 identical lightbulbs whose maximum power consumption is 5 W (Fig. 26–39). (a) What is the resistance in each bulb if the power consumption is 50 W when the bulbs are connected in series? (b) If the 10 bulbs are connected in parallel, an additional resistor is needed so that the bulbs do not burn out. The resistor is connected in series with the total set of lightbulbs. What is the value of the resistance? What is the power loss in the resistor? (These are bulbs designed for use in a car.)

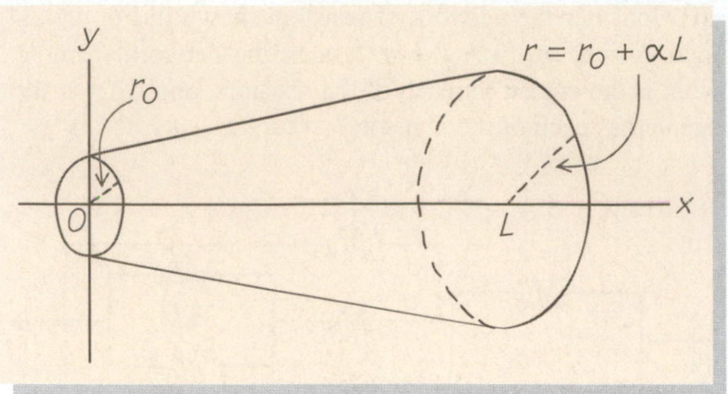

▲ **FIGURE 26–38** Problem 85.

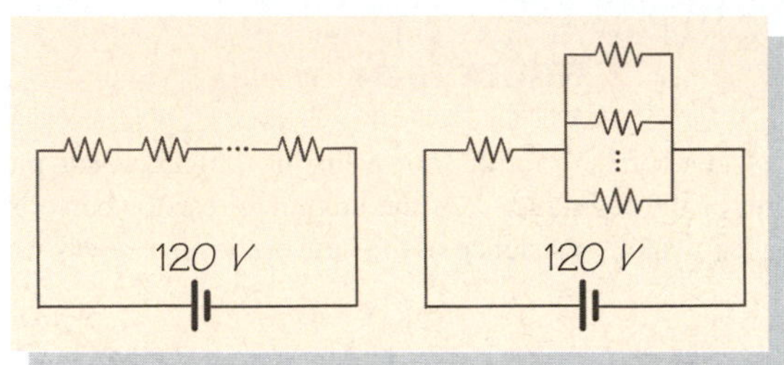

▲ **FIGURE 26–39** Problem 86.

87. (II) Consider the circuit shown in Fig. 26–40. Calculate the current and the power dissipated in the 4-Ω resistor as a function of the unknown resistance R_x.

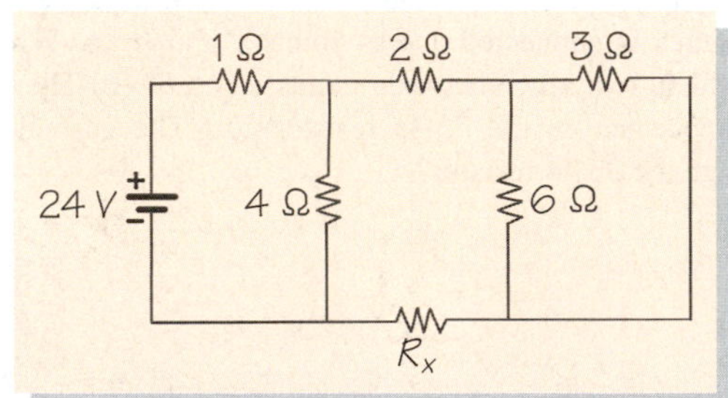

▲ **FIGURE 26–40** Problem 87.

88. (II) A bus bar (a conducting bar meant to carry a good deal of current) made of copper, of resistivity $1.72 \times 10^{-8}\ \Omega \cdot \text{m}$, is meant to carry 100 A over a distance of 0.25 m at a temperature of 300°C. What is the minimum cross-section of the bus bar if no more than 0.2 W of power is to be dissipated?

89. (II) A generator delivers 75 A at a voltage of 12 V. What power does the generator deliver? How long would it take to raise the temperature of $10^{-3}\ \text{m}^3$ of water by 7.5°C? How long would it take to boil away 0.5 L of water, starting at 25°C?

90. (II) Figure 26–10 shows the I–V curve of a typical semiconductor diode. Use the data from the figure to sketch the power dissipated in the diode as a function of the current. What is the power dissipated in the ideal diode, as shown in the figure?

91. (II) A potential is set up from one end of a copper wire to the other; as a result, current flows. The copper is thermally isolated to some extent. As the charge flows, the wire heats up, causing the resistivity to increase. Suppose that, during a short time

period, the temperature of the wire as a function of time t is given by $T = T_0 + kt^2$. (a) Describe the current in the wire during this period. (b) What is the power dissipated by the wire as a function of time? (c) From the change with time of the dissipated power, will the wire continue to heat up and, perhaps, melt?

92. (II) The density of charge-carrying electrons in copper is 8.5×10^{28} electrons/m^3, its resistivity is $1.7 \times 10^{-8}\ \Omega \cdot \mathrm{m}$, and the drift speed in a copper wire is 1.2×10^{-5} m/s. The wire has a diameter of 1 mm and a length of 3 m. At what rate must thermal energy be carried off by a cooling medium if the wire is to maintain its temperature?

93. (II) A single layer of 200 turns of closely spaced wire of radius $r_1 = 0.6$ mm is wound in a coil of diameter $D_1 = 5$ cm (Fig. 26–41). A second coil of the same length but of diameter $D_2 = 8$ cm is composed of a single layer of closely spaced wire of radius $r_2 = 0.4$ mm. The wires are made of the same material. Find the ratio of the resistances of the two coils.

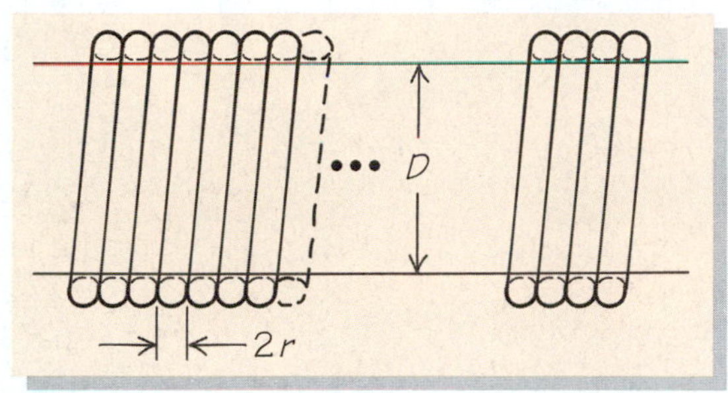

FIGURE 26–41 Problem 93.

94. (III) A thin wire of length L and cross-sectional area A oriented in the x-direction is made of an ohmic material whose resistivity varies along the wire according to the empirical law $\rho = \rho_0 e^{-x/L}$. (a) Describe how the field within the wire varies with position if the end at $x = 0$ is at a potential V_0 greater than the end at $x = L$. (b) How does the potential vary as you move along the wire? (c) What is the total resistance of the wire?

95. (III) If all the energy lost from Joule heating stays in a wire, and the temperature increases as a result, the resistivity will increase according to Eq. (26–18). The current will therefore change as a function of time, the Joule heating will change, and so forth. If the wire material has a constant heat capacity, the rate of energy loss in the wire will be proportional to the rate of change of temperature. Assuming that the potential stays constant, set up a differential equation that describes the rate of temperature change. If this equation is solved, how can the current be found as a function of time?

Chapter

27

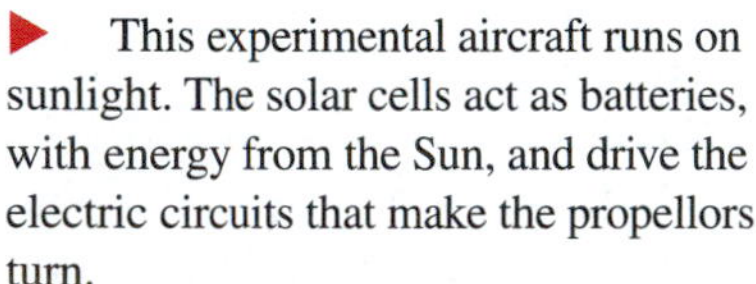

This experimental aircraft runs on sunlight. The solar cells act as batteries, with energy from the Sun, and drive the electric circuits that make the propellors turn.

Direct-Current Circuits

We have seen how charges move through potential differences under the influence of electric fields, and how resistors control current and capacitors store charge. When resistors, capacitors, and batteries are connected together by conducting wires, they form electric circuits. We can understand the flow of currents in circuits by applying just two simple physical principles: the conservation of current and the conservation of energy. In this chapter, we learn to apply these principles systematically to the analysis of circuits. Our earlier work on capacitors or resistors connected in parallel and in series will be useful in this analysis. We also discuss some of the instruments available to measure and monitor the current and voltage of electric circuits. The flow of energy to and from circuit elements is an important theme that leads us to the concept of time-varying currents and voltages.

27–1 EMF

The sources of energy that cause charges to move in electric circuits have historically been called sources of **electromotive force**. They are actually sources of energy, not of force, and because the word force is misleading here, we use the abbreviation **emf** instead. Briefly, a source of emf is a device that *maintains* a potential difference across a conductor, thereby allowing a current to run continuously through the conductor. (Contrast this to the potential difference supplied by, say, a capacitor. The charged capacitor does indeed have a potential difference between its plates, but the current running under its influence is composed essentially of the charge on the plates running off, and as it runs off, the potential difference quickly drops to zero.) When we think of sources of emf, we often think of batteries. A battery can be thought of as a device that expends chemical energy to pump charges, just as a water pump expends mechanical energy to pump water uphill to a tank with a higher gravitational potential energy. There is a wide variety of sources of electric energy beyond batteries. A battery converts chemical energy into an emf; a solar cell converts the energy of sunlight into an emf; a thermocouple produces an emf as a result of a difference in temperature; a large commercial electric power plant may burn coal, gas, or nuclear fuel, or use falling water, to drive a generator that produces an emf (Fig. 27–1).

In this chapter, we use the term "battery" to refer to any source of emf. We shall restrict ourselves to batteries for which the emf is *constant* with time. Up to Section 27–5, we focus on phenomena such as current flows or potential differences that are similarly constant in time. We refer to this as *equilibrium*, or *steady-state*, behavior, and use the terms **direct-current**, or **DC**, behavior as a label for it.

(a)

(b)

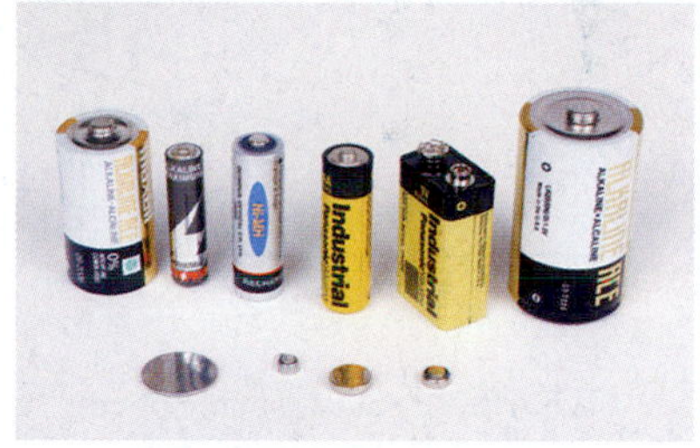

(c)

▲ **FIGURE 27–1** Various sources of emf that produce electrical energy include (a) solar panels (photovoltaic), (b) burning fuel, here nuclear, can heat water, produce steam, and turn turbines, (c) batteries, which produce their emf through chemical means.

Circuits

Sources of emf supply energy to circuits. But just what is a circuit? When batteries, resistors, capacitors, or other circuit elements (some of which will be introduced later) are connected by wires (ideally of negligible resistance), they form a **circuit**. For example, when a switch is closed—meaning that a gap in a wire is closed—and a battery establishes current through the filament of the lightbulb of a flashlight, a circuit has been formed. Figure 27–2 illustrates this simple circuit with the conventional symbols for resistors, ideal wires (wires with no resistance), and batteries; the lightbulb is a simple resistor.

Circuit analysis typically requires us to relate the currents and potential differences within the circuit. For example, we may want to know the potential drop across a capacitor or the current that passes through a resistor when there is a particular emf in the circuit.

The Meaning of Emf

The fact that the force involved in moving charges under the influence of an emf is a conservative one provides us with a powerful method for the analysis of circuits. Consider the circuit of Fig. 27–2, consisting of a battery and a single resistor known as the **load** resistance. The battery has a potential difference across its terminals called the **terminal voltage**. Current flows away from the battery terminal at the higher potential—the one marked positive. How much current flows depends on the load resistance (or, in more complicated circuits, on the characteristics of the components of the rest of the circuit). Because current is defined as moving in a direction opposite to that of electrons, it is helpful to imagine positive charges flowing to the negative terminal (the terminal at the lower potential), equivalent to electrons flowing to the positive terminal, which is in fact what occurs.

Inside a chemical battery, a chemical process carries the positive charges back to the positive terminal. It is this process—the internal pumping action of the battery—that gives a precise definition of the emf. Suppose that it takes work dW to move a charge dq from the negative to the positive terminal. Then the emf of the battery is defined to be

$$\mathcal{E} \equiv \frac{dW}{dq}. \qquad (27\text{–}1)$$

EMF DEFINED

The SI unit of emf is the volt, one joule per coulomb. The word *voltage* is sometimes used loosely to describe the emf $\mathcal{E}$, but voltage more properly refers to the potential difference or terminal voltage across the emf terminals, which as we shall see below may be different from $\mathcal{E}$.

When a battery is connected to a circuit such as that in Fig. 27–2, it sets charges into motion, driving a current from the positive (higher-potential) terminal around the circuit to the negative (lower-potential) terminal, and we say that the battery *discharges*. In discharging, the battery is expending its chemical energy. If a current is driven from the negative to the positive terminal, a process that can be accomplished in conjunction with a battery of larger emf, the battery of smaller emf is said to be *charging*; this process replaces its spent chemical energy.

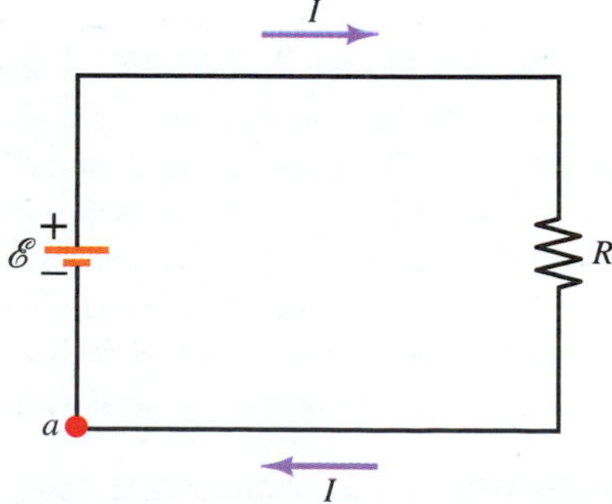

▲ **FIGURE 27–2** A simple circuit with a source of emf, $\mathcal{E}$, and a resistor, R.

Let us now turn to the analysis of our circuit. Start by assuming that the potential difference across the battery terminals in Fig. 27–2 is in fact the emf $\mathcal{E}$. (We'll refine this in the next subsection.) A current will flow around the circuit, and to find this current we can use the fact that the electric potential is associated with a conservative force. If the force is conservative, then the net work done by the force in sending a charge around a closed loop is zero. In turn, the total potential drop involved in any round trip that starts from any point on a closed loop must be zero. Make such a round trip that starts at point *a* of Fig. 27–2 and follow the current around the circuit. There is no change in potential as we pass through the ideal (resistanceless) wire. In crossing the battery from the negative to the positive terminal, the potential *increases* by $\mathcal{E}$. When we cross the ohmic resistance, the potential *decreases* by an amount IR [see Eq. (26–12)]. The potential drop implies a decrease in the potential energy of the charges. This potential energy is converted into thermal energy in the resistor. The net potential change as we travel once around the circuit is zero, so

$$\mathcal{E} - IR = 0. \tag{27–2}$$

This equation determines the current, I:

$$I = \frac{\mathcal{E}}{R}. \tag{27–3}$$

Internal Resistance

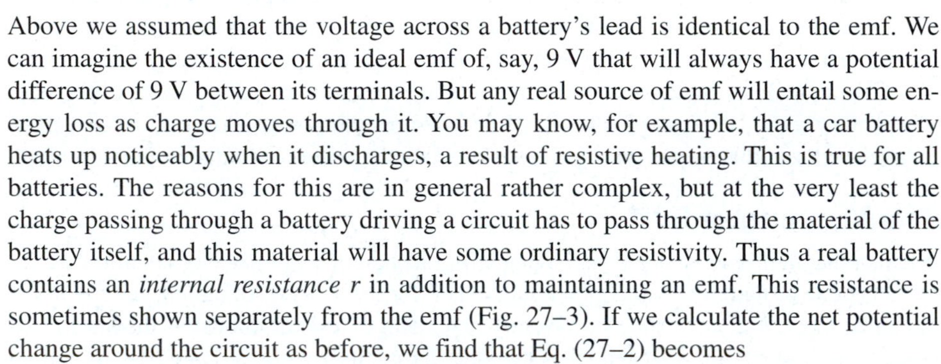

Above we assumed that the voltage across a battery's lead is identical to the emf. We can imagine the existence of an ideal emf of, say, 9 V that will always have a potential difference of 9 V between its terminals. But any real source of emf will entail some energy loss as charge moves through it. You may know, for example, that a car battery heats up noticeably when it discharges, a result of resistive heating. This is true for all batteries. The reasons for this are in general rather complex, but at the very least the charge passing through a battery driving a circuit has to pass through the material of the battery itself, and this material will have some ordinary resistivity. Thus a real battery contains an *internal resistance* r in addition to maintaining an emf. This resistance is sometimes shown separately from the emf (Fig. 27–3). If we calculate the net potential change around the circuit as before, we find that Eq. (27–2) becomes

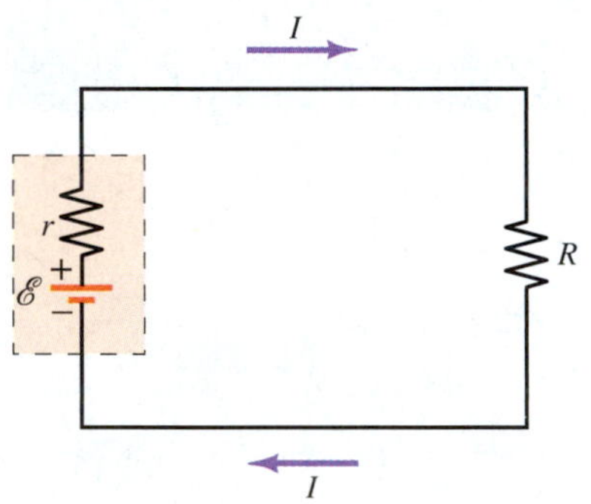

▲ **FIGURE 27–3** A source of emf also contains an internal resistance r, depicted in circuit diagrams as the emf symbol with a resistor in series close by.

$$\mathcal{E} - Ir - IR = 0. \tag{27–4}$$

Because of internal resistance, the potential difference across the battery terminals is no longer just $\mathcal{E}$; it is given instead by

$$\text{with internal resistance: } V = \mathcal{E} - Ir. \tag{27–5}$$

This potential difference is a function of the current. Depending on the direction of current flow, the voltage across the terminals of a battery can be greater or less than the battery's emf. A second modification that results from internal resistance is that the current depends on it. From Eq. (27–4), the current in our circuit is

$$I = \frac{\mathcal{E}}{r + R}. \tag{27–6}$$

Compare this to Eq. (27–3).

CONCEPTUAL EXAMPLE 27–1 An engineer has inadvertently placed a battery into a circuit (Fig. 27–3) that has a much larger internal resistance r than the resistance R outside the battery. As a result, which of the following, if any, is true: (a) The current through the circuit will be much less than the emf of the battery; (b) the voltage drop across the terminal will be much greater than it would have been if the internal resistance were less; (c) the current through the circuit doesn't depend very much on what R is.

Answer (a) is a nonsensical statement. The current and the emf may be related, but they are not comparable quantities. (b) is false—the voltage drop is in fact much less than it would have been if the internal resistance were small. In effect, the internal resistance "uses up" some of the voltage drop that the battery could otherwise supply for use within the circuit. (c) is true—the current is determined by the sum of the internal and external resistance, as in Eq. (27–6), and this is insensitive to R if $r \gg R$.

It is worth looking at (b) a little more closely, and in a quantitative way. The voltage drop across the terminals is given by Eq. (27–5), and if we insert the current from Eq. (27–6), we see the voltage drop is

$$\begin{aligned} V &= \mathcal{E} - Ir = \mathcal{E} - [\mathcal{E}/(r + R)]r \\ &= \mathcal{E}[r + R - r]/(r + R) = \mathcal{E}R/(r + R) \cong \mathcal{E}R/r, \end{aligned}$$

and this is a small fraction of the emf.

We can see from Eq. (27–5) that for small internal resistance the potential across the terminals is approximately the same as the emf. This is generally true if there is no current at all, so a reading of the potential across the terminals is also a reading of the emf when the external, or load, resistance in the circuit is very large. It is desirable that the internal resistance be "small," but we need to remember that we mean small in comparison with external resistances, which vary depending on the application. The typical internal resistance of a car battery is less than 0.01 Ω, but it may be as large as 0.1 Ω for a flashlight battery. In many situations, the internal resistance is so small relative to the resistances within the circuit that it can be ignored in electric-circuit analyses. Ordinary batteries run down with age not because their emf decreases, but because their internal resistance increases, which means that the current they can supply decreases.

EXAMPLE 27–2 One of two different resistors with respective resistances $R_1 = 5.00\ \Omega$ and $R_2 = 10.0\ \Omega$ can be placed into the circuit shown in Fig. 27–3. The emf $\mathscr{E}$ and internal resistance r of the battery are unknown. When only R_1 is inserted, the current is $I_1 = 0.291$ A; when only R_2 is inserted, the current is $I_2 = 0.147$ A. Find $\mathscr{E}$ and r.

Strategy The total resistance for the two cases under consideration are $R_1 + r$ and $R_2 + r$, and in both cases, when these resistances are multiplied by the appropriate currents I_1 and I_2, respectively, they yield the emf $\mathscr{E}$. We will then have two linear equations that we can solve for the two unknowns r and $\mathscr{E}$.

Working It Out The fact that the net potential drop around the circuit is zero for the two cases yields the two equations

$$\mathscr{E} - I_1 r - I_1 R_1 = 0,$$
$$\mathscr{E} - I_2 r - I_2 R_2 = 0.$$

Multiplying the first equation by I_2 and the second by I_1 and then subtracting to solve for $\mathscr{E}$ we find that

$$\mathscr{E} = \frac{I_1 I_2}{I_1 - I_2}(R_2 - R_1).$$

If we insert this into either of the first two equations, we can solve for r:

$$r = \frac{I_2 R_2 - I_1 R_1}{I_1 - I_2}.$$

Numerical evaluation gives

$$\mathscr{E} = \frac{(0.291\ \text{A})(0.147\ \text{A})}{0.291\ \text{A} - 0.147\ \text{A}}(10.0\ \Omega - 5.00\ \Omega) = 1.49\ \text{V}$$

and

$$r = \frac{(0.147\ \text{A})(10.0\ \Omega) - (0.291\ \text{A})(5.00\ \Omega)}{0.291\ \text{A} - 0.147\ \text{A}} = 0.104\ \Omega.$$

What Do You Think? Suppose R_2 had been $1.0 \times 10^5\ \Omega$, known to 2 significant figures. Could you have found r? *Answers to What Do You Think? questions are given in the back of the book.*

Electric Power and Batteries

A source of emf (or electric energy) is also a source of *electric power*. The power is the rate at which the source delivers energy. From Eq. (26–30), the power of the source is the potential drop across the source times the current that passes through it. For a source of emf $\mathscr{E}$ we have

$$P = I\mathscr{E}. \tag{27–7}$$

Let's see how this works for the circuit of Fig. 27–3. Equation (27–6) tells us that $\mathscr{E}$ is given by $I(r + R)$, where I is the current in the circuit and r and R are the internal and load resistances, respectively. If we use this relation for $\mathscr{E}$ in Eq. (27–7), we find that

$$P = I^2 R + I^2 r. \tag{27–8}$$

Energy conservation implies that we had to find this result. The electric power of the source of emf is balanced by the sum of the power dissipated in both the internal and load resistances.

EXAMPLE 27–3 (a) A battery of emf $\mathscr{E}$ and negligible internal resistance r acts in the circuit of Fig. 27–3 with a resistor R. What is the power delivered in terms of $\mathscr{E}$ and R? (b) Suppose r is not negligible. What is the power delivered in terms of $\mathscr{E}$, R and r?

Setting It Up We want the result for power not in terms of I and $\mathscr{E}$, or in terms of I and resistance, as in the text discussion above, but in terms of $\mathscr{E}$ and resistance.

Strategy By using the expression $P = \mathscr{E}I$ and expressing the current in terms of the emf and the resistances, we will find the power in terms of the emf and the resistances.

Working It Out For case (a) $I = \mathscr{E}/R$, while for case (b) $I = \mathscr{E}/(R + r)$. We substitute these into the expression for power $P = I\mathscr{E}$: For case (a)

$$P = I\mathscr{E} = \frac{\mathscr{E}}{R}\mathscr{E} = \frac{\mathscr{E}^2}{R},$$

while for case (b)

$$P = I\mathscr{E} = \frac{\mathscr{E}}{R + r}\mathscr{E} = \frac{\mathscr{E}^2}{R + r}.$$

What Do You Think? The value of r determines the maximum power P_{max} a battery of a given emf can deliver in this circuit. What is P_{max}?

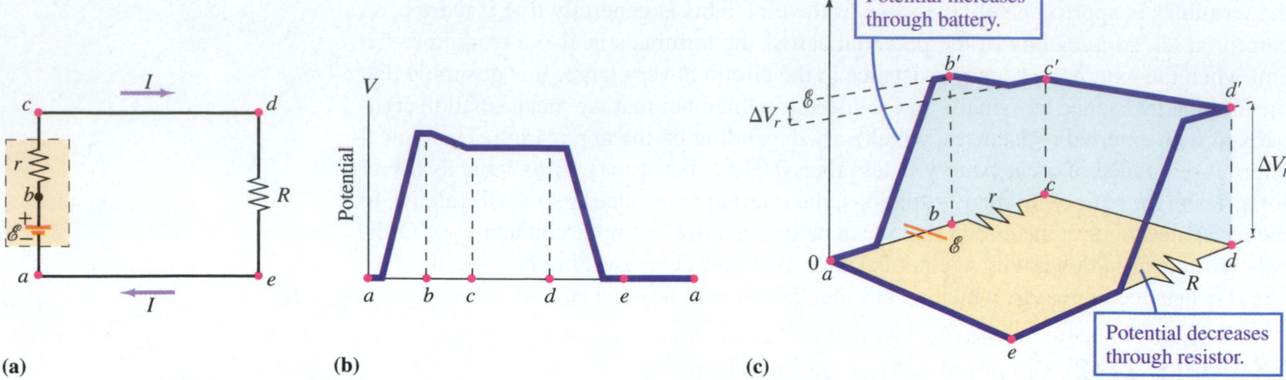

▲ **FIGURE 27–4** (a) A single-loop circuit showing the emf (ℰ), internal resistance (r), and load resistance (R). (b) The potential differences for the points labeled in part (a). (c) A three-dimensional view of part (b).

27–2 Kirchhoff's Loop Rule

A **single-loop circuit** is a circuit with a single path for the current. The simple circuit discussed in Section 27–1 (Fig. 27–3) is an example of such a circuit. Let's go around the loop and examine the potential change at every step. First we'll redraw the circuit in Fig. 27–4a and mark the points a through e. The potential is graphed in Fig. 27–4b, c, where we follow the circuit along the current direction, with zero potential chosen arbitrarily at point a. We draw the internal resistance, r, in Fig. 27–4a as though it were separate from the emf; in Fig. 27–4b, c, we draw the rise of the emf as gradual. There is no potential change in an ideal conducting wire. There is some very small resistance in a real wire, but it can usually be ignored, and we do so here.

As we observed earlier, the net potential change in traversing the complete circuit is zero. This is an expression of the conservation of energy: For this particular circuit the charge that moves round the loop gains energy from the emf and loses the same amount of energy in the resistors. If we had followed the circuit in the opposite direction—against the current—the changes would all be of the opposite sign, but the end result would remain: The potential change in a complete circuit is zero. In the context of circuits, this simple law is given a special name, after the nineteenth-century physicist Gustav Kirchhoff, **Kirchhoff's loop rule**:

The sum of the potential changes around a closed path is zero,

$$\sum_{\text{closed path}} \Delta V = 0. \tag{27–9}$$

KIRCHHOFF'S LOOP RULE

The loop rule is applicable to any closed path in any electric circuit. When a circuit is laid out in a diagram, many closed paths may be possible, and as we shall see in several of the examples that follow, the loop rule applied to these loops is an important tool for finding the desired circuit parameters, which is the aim of any circuit analysis. In Fig. 27–4a, there is only one closed loop.

In applying the loop rule, we must have knowledge of the potential differences across various parts of a circuit. It is therefore useful to summarize what we have learned here and in previous chapters about the potential changes ΔV across individual circuit elements—batteries, resistors, and capacitors—as a set of rules. Figure 27–5 is a summary figure for these rules. The quantity ΔV is the change in potential when moving from an initial point to a final point; for example, the potential change in going from point a to point b is $\Delta V = V_{ab} = V_b - V_a$.

Traversal direction

$$V_{ab} = V_b - V_a = +\mathscr{E}$$

$$V_{ba} = V_a - V_b = -\mathscr{E}$$

$$V_{ab} = -IR$$

$$V_{ba} = +IR$$

$$V_{ab} = +\frac{Q}{C}$$

$$V_{ba} = -\frac{Q}{C}$$

▲ **FIGURE 27–5** Rules for potential differences across various circuit elements.

1. In going from the negative to the positive terminal of a battery with emf $\mathscr{E}$, the potential change is positive, $\Delta V = +\mathscr{E}$. In going from the positive to the negative terminal, the potential change is negative, $\Delta V = -\mathscr{E}$. This rule ignores the presence of an internal resistance, which we treat as a separate resistance in series with the battery. The actual potential difference across the terminals is the voltage change $\mathscr{E}$ plus a term associated with the internal resistance that depends on the current (see point 2, below).
2. In moving across a resistance R *along* the direction of the current, I, the potential change is negative, $\Delta V = -IR$. The sign is opposite, $\Delta V = +IR$, in moving *against* the direction of the current.
3. In moving from the negatively to the positively charged plate of a capacitor of capacitance C and charge Q, the potential change is positive, $\Delta V = +Q/C$. The potential change is negative, $\Delta V = -Q/C$, when we move from the positively charged plate to the negatively charged plate. It is worth recalling here that no actual current flows across a capacitor.

The simplest application of the loop rule involves several resistors in series in a single loop. As an example, consider the loop shown in Fig. 27–6a. Here we have an emf and three resistors. (The internal resistance of the source of emf will be ignored.) We start at point a and move toward the battery in the (clockwise) direction of the assumed current. (We could just as well follow the circuit in the opposite direction; each term would change sign, but this is irrelevant if the sum of voltage changes is zero.) Kirchhoff's loop rule, Eq. (27–9), gives

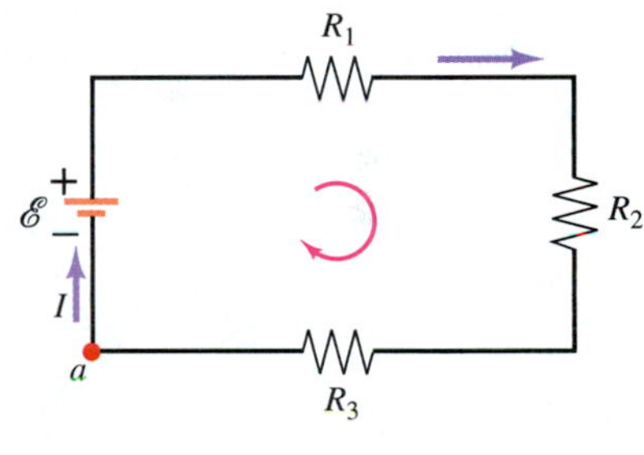

(a)

$$\sum \Delta V = \mathscr{E} - IR_1 - IR_2 - IR_3 = 0.$$

The solution for the current is

$$I = \frac{\mathscr{E}}{R_1 + R_2 + R_3}. \tag{27–10}$$

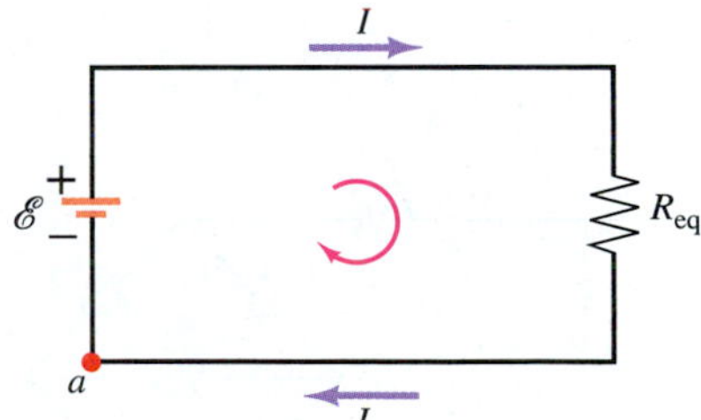

(b)

▲ **FIGURE 27–6** (a) The circuit with three resistors connected in series is equivalent to (b) the circuit with one resistor of value R_{eq}.

Of course, we know from Section 26–4 that we can replace resistors in series with an equivalent resistor R_{eq} given by the sum of the individual resistances, Eq. (26–21). This is illustrated in the circuit of Fig. 27–6b. There the current is $I = \mathscr{E}/R_{eq}$, and comparison of this result with Eq. (27–10) confirms Eq. (26–21).

In a single-loop circuit the internal resistance is simply included by adding it to the load resistance, since it always occurs *in series* with it. It is important in multi-loop circuits to place the battery's internal resistance adjacent to the battery.

EXAMPLE 27–4 Find the current for the two-battery circuit shown in Figs. 27–7a and b. The values of the emfs and resistances are $\mathscr{E}_1 = 12\text{ V}$, $r_1 = 0.4\ \Omega$, $R_3 = 3\ \Omega$, $\mathscr{E}_2 = 6\text{ V}$, $r_2 = 0.1\ \Omega$, and $R_4 = 10\ \Omega$.

Setting It Up We have indicated the direction of the current in the figure; this assumption is necessary because the voltage change across resistors depends on the choice of direction. The solution to the problem will give us the actual current direction; if I is positive, our assumed direction was correct, and if I is negative, the current flows the other way.

Strategy We can regard the loop rule as an equation for the current, and we know all the other quantities that appear in the loop rule, including the internal resistances associated with the two emfs. These emfs are oriented in opposite directions and so will appear in the loop rule with opposite signs.

Working It Out Proceeding counterclockwise from point a, the loop rule reads

$$\sum \Delta V = +Ir_1 - \mathscr{E}_1 + IR_4 + \mathscr{E}_2 + Ir_2 + IR_3 = 0,$$

and we can solve this equation for the current. The result is

$$I = \frac{\mathscr{E}_1 - \mathscr{E}_2}{r_1 + R_4 + r_2 + R_3}.$$

Numerically,

$$I = \frac{12\text{ V} - 6\text{ V}}{0.4\ \Omega + 10\ \Omega + 0.1\ \Omega + 3\ \Omega} = \frac{6\text{ V}}{13.5\ \Omega} \cong 0.4\text{ A}.$$

Note that the only time you might see two opposing emfs in a single-loop circuit is when one battery is charging another.

What Do You Think? What would the consequence of having drawn the current in the opposite direction have been?

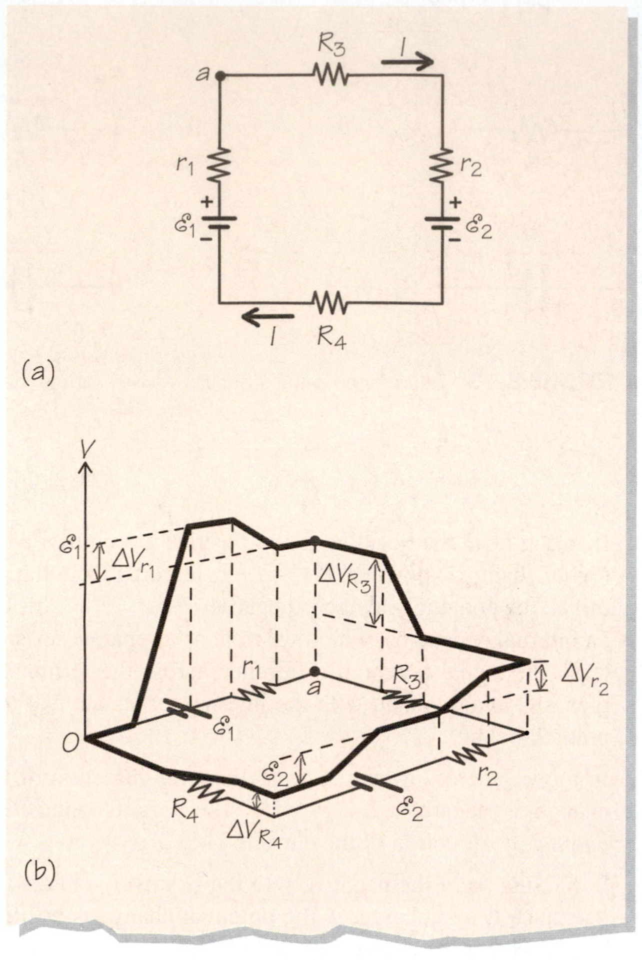

▲ **FIGURE 27–7**

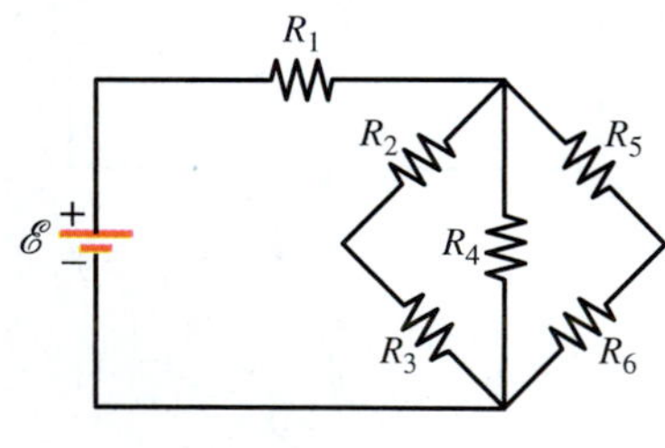

(a)

(b)

▲ **FIGURE 27–8** Two multi-loop circuits.

27–3 Kirchhoff's Junction Rule

Few circuits are as simple as the one-loop circuits discussed so far. Examples of more complex circuits—multi-loop circuits—are shown in Figs. 27–8a and 27–8b. These examples have four- and three-line junctions. *Current conservation*, an experimental fact whose meaning and consequences were discussed in some detail in Chapter 26, is an additional tool that will allow a complete analysis of multi-loop circuits. We can summarize current conservation with **Kirchhoff's junction rule**. The rule states that the sum of the currents that enter a junction equals the sum of the currents that leave the junction. If we view currents that leave a junction as *negative*, and currents that enter the junction as *positive*, then we may rephrase this as

The algebraic sum of the currents that enter a junction equals zero,

$$\sum I_{\text{in}} = 0. \tag{27–11}$$

KIRCHHOFF'S JUNCTION RULE

This equation applies for every junction. Application of Kirchhoff's two rules gives us further information for currents, voltage differences, etc., in multi-loop circuits.

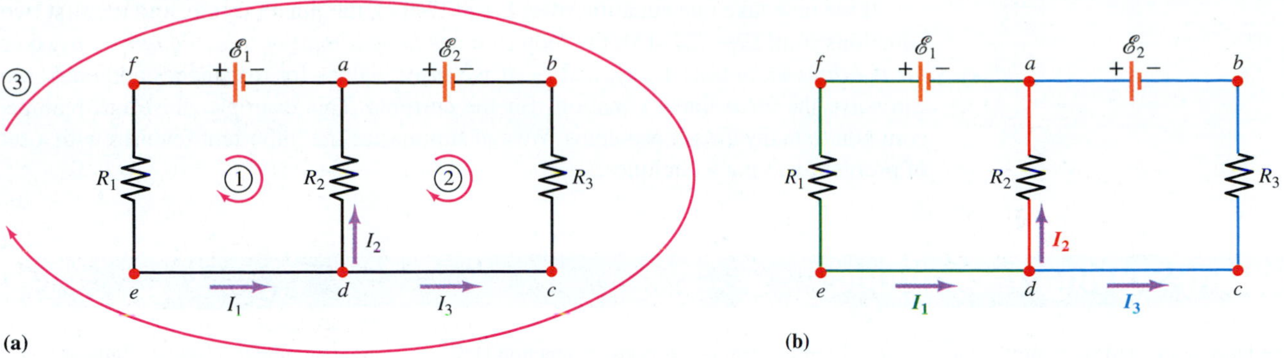

▲ **FIGURE 27–9** (a) A multi-loop circuit with two independent loops and two junctions. (b) The three legs of the circuit are shown in different colors.

Solving for the Behavior of Multi-Loop Circuits

Let's take a look at how the junction rule and loop rule help us solve for the unknown values in multi-loop circuits. Consider the circuit in Fig. 27–9a. Current flows between every junction. Thus we have currents I_1, I_2, and I_3 in the three separate *legs*, or *branches afed* (green), *da* (red), and *dcba* (blue), respectively (Fig. 27–9b). (A leg is any line of the circuit that starts and finishes on a junction but does not itself include a junction. For example, the three possible lines that connect *a* and *d* in Fig. 27–9b are all legs.) We can arbitrarily choose the direction of the three currents, and as we shall see, the algebraic equations for the currents will determine the actual direction for us: If these equations give a negative value for any of the currents, then that current actually travels in the opposite direction than the one we postulated. The junction rule can be applied at junctions *a* and *d* to obtain

$$\text{for junction } a\text{: } -I_1 + I_2 + I_3 = 0; \tag{27–12a}$$

$$\text{for junction } d\text{: } I_1 - I_2 - I_3 = 0. \tag{27–12b}$$

Notice that currents drawn as leaving the junction have minus signs, while currents that are drawn as arrows entering the junction have plus signs. Notice also that these two equations are not independent of each other; in fact, they are identical! There is a general rule—we won't prove it—that if there are N' junctions in a circuit, then there are $N' - 1$ *independent* junction rules—the meaning of an independent set of equations is that one equation cannot be derived as a combination of any others. But it is perhaps not so important to remember this rule; it is always possible to write an equation for each junction, then check whether or not they are independent.

How do we use the loop rule for this circuit? We must start by counting the number of *independent* loops. (While the counting is a bit more complicated than for the junctions, it is worthwhile looking at it in more detail here.) The rule for counting the number of such loops is as follows: For planar circuits (circuits that can be drawn in a plane with no two wires crossing over each other), the number of independent loops is just the number of enclosed areas through which you could poke a pencil. According to this idea, the circuit in Fig. 27–9a has two *independent* loops. But we can draw three possible loops, those indicated by the circled integers 1, 2, and 3 of Fig. 27–9a, where 3 is around the perimeter of the circuit. Only two of these three loops can be independent, and we can choose *any* two of the three possible loops as our independent loops. Before we choose, though, let's apply the loop rule for all three loops and verify that only two are independent. Point *a* is part of each of the three loops, so for convenience we'll begin at point *a* and apply the loop rule of Section 27–2. The internal resistances for the two emfs are neglected. We traverse each loop in the direction of the loop arrow:

$$\text{for loop 1: } +I_2R_2 + I_1R_1 - \mathscr{E}_1 = 0; \tag{27–13a}$$

$$\text{for loop 2: } -\mathscr{E}_2 + I_3R_3 - I_2R_2 = 0; \tag{27–13b}$$

$$\text{for loop 3: } -\mathscr{E}_2 + I_3R_3 + I_1R_1 - \mathscr{E}_1 = 0. \tag{27–13c}$$

Only two of these three loop-rule equations are independent; the sum of the first two produces the third.

If we now take one equation from Eqs. (27–12), the junction rule, and the first two equations from Eqs. (27–13), the loop rule, *we have a total of three equations to solve for the three unknowns* I_1, I_2, and I_3. If we insert values for resistances and emfs, we can solve the three linear equations for the currents. This example illustrates features common to many circuit problems. We can summarize the important features with a set of problem-solving techniques.

Problem-Solving Techniques

In problems associated with multi-loop circuits, we must find unknown circuit parameters (such as resistance or current) when other parameters are given. To solve these problems, the following procedure may be helpful.

1. Draw a diagram with sources of emf, resistors, capacitors, and so forth clearly labeled. List the known and unknown parameters.
2. Assign a separate current for each leg of the circuit, and indicate that current on the diagram. The direction of current flow may not be immediately obvious; any direction can be assumed for the current, and the final algebraic solution will determine the correct direction. If the solution for a current turns out to be negative, the actual direction of current is opposite to your initial guess.
3. Apply the junction rule for the currents at each junction. Currents that you have chosen to draw as incoming have plus signs; those you have chosen to draw as outgoing have minus signs. If the circuit has N' junctions, then $N' - 1$ of the equations relating currents at the junctions will be independent.
4. Identify the number of independent loops N by counting the number of different ways that a pencil can poke through the circuit—a simple procedure for planar circuits. Indicate N loops on the diagram (for example, the loops labeled 1 and 2 on Fig. 27–9a).
5. Apply the loop rule to each of these loops.
6. Check to see that the number of linear equations from steps 3 and 5 matches the number of unknowns.
7. Solve these equations for the unknowns—whether they are currents or other parameters of the circuit. It is usually best to solve these equations *algebraically* and substitute numerical values later. Any checks, in the form of special cases or simple limits, can easily be made this way.

EXAMPLE 27–5 Find the currents for the circuit of Fig. 27–10, given that $\mathscr{E}_1 = 6.00\ \text{V}$, $\mathscr{E}_2 = 12.0\ \text{V}$, $R_1 = 100\ \Omega$, $R_2 = 10.0\ \Omega$, and $R_3 = 80.0\ \Omega$.

Strategy We follow the Problem-Solving Techniques (p. 774). The circuit diagram is given, and all parameters are labeled on Fig. 27–10. In this case, the unknowns are I_1, I_2, and I_3. For the application of the junction rules, we note that there are only two junctions, and hence only one of them is independent. For the loop rules, we choose loop 1 and loop 2 as the two independent ones. In fact, we have already written the junction and loop equations for this circuit: Eqs. (27–12) and (27–13), respectively. Accordingly, our single independent junction equation will be Eq. (27–12b), and our two independent loop equations will be Eqs. (27–13a) and (27–13b). We have three equations to solve for the three unknowns.

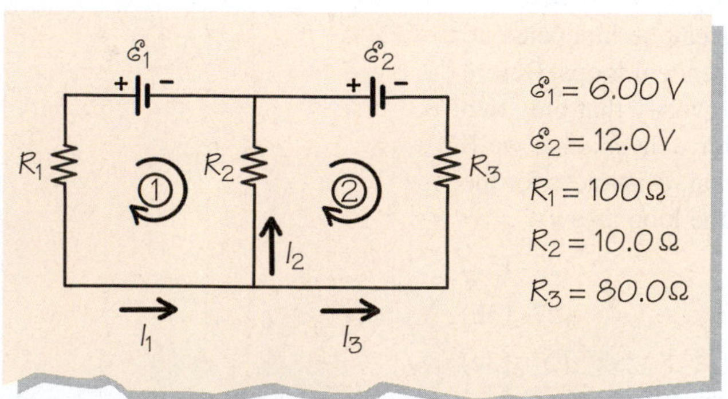

▲ **FIGURE 27–10**

Working It Out Equation (27–12b) gives $I_1 = I_2 + I_3$. We substitute this into the loop equations:

$$I_2R_2 + (I_2 + I_3)R_1 - \mathscr{E}_1 = 0; \tag{27–14a}$$

$$-\mathscr{E}_2 + I_3R_3 - I_2R_2 = 0. \tag{27–14b}$$

Solve Eq. (27–14b) for I_3:

$$I_3 = \frac{I_2R_2 + \mathscr{E}_2}{R_3}. \tag{27–15}$$

When this result is inserted into Eq. (27–14a), we find an equation for I_2, namely $I_2(R_1 + R_2) + (I_2R_2 + \mathscr{E}_2)\dfrac{R_1}{R_3} - \mathscr{E}_1 = 0$, and this has solution

$$I_2 = \frac{R_3\mathscr{E}_1 - R_1\mathscr{E}_2}{R_1R_2 + R_1R_3 + R_2R_3}. \tag{27–16}$$

If we substitute this result into Eq. (27–15), we get an expression for I_3:

$$I_3 = \frac{R_2(\mathscr{E}_1 + \mathscr{E}_2) + R_1\mathscr{E}_2}{R_1R_2 + R_1R_3 + R_2R_3}. \tag{27–17}$$

Finally, I_1 is the sum of I_2 and I_3:

$$I_1 = \frac{R_2(\mathscr{E}_1 + \mathscr{E}_2) + R_3\mathscr{E}_1}{R_1R_2 + R_1R_3 + R_2R_3}. \tag{27–18}$$

Equations (27–16), (27–17), and (27–18) constitute the desired algebraic solution. Numerically, we can most simply start with the denominator combination, $(100\ \Omega)(10.0\ \Omega) + (100\ \Omega)(80.0\ \Omega) + (10.0\ \Omega)(80.0\ \Omega) = 9.80 \times 10^3\ \Omega^2$. In turn,

$$I_1 = \frac{(10.0\ \Omega)(6.00\ \text{V} + 12.0\ \text{V}) + (80.0\ \Omega)(6.00\ \text{V})}{9.80 \times 10^3\ \Omega^2}$$

$$= 67.3\ \text{mA}$$

$$I_2 = \frac{(80.0\ \Omega)(6.00\ \text{V}) - (100\ \Omega)(12.0\ \text{V})}{9.80 \times 10^3\ \Omega^2} = -73.5\ \text{mA}$$

$$I_3 = I_1 - I_2 = 140.8\ \text{mA}.$$

Note that I_2 is negative, meaning that it flows in a direction opposite to the one we assumed in Fig. 27–10.

What Do You Think? What do you expect to happen if the resistance R_2 is much larger than any other resistance in the problem? Verify your expectation by using the expressions above for the currents.

EXAMPLE 27–6 Assume that the emf and resistors are known for the circuit in Fig. 27–8b. Express the linear equations that can be solved to find all currents.

Setting It Up Figure 27–11 identifies our choice of junctions and loops to which the Kirchhoff rules are to be applied.

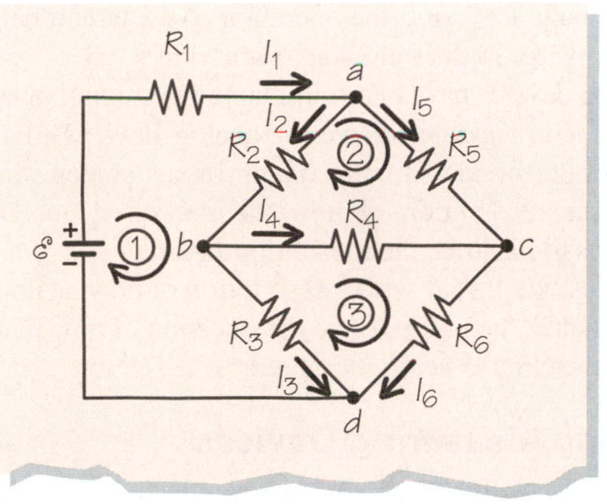

▲ **FIGURE 27–11** The circuit of Fig. 27–8b is labeled with loops and junctions to be solved by circuit analysis.

Strategy As we see from the figure, there are six currents, so we need six linear equations. Three independent loops in the circuit are drawn in Fig. 27–11. There are four junctions, hence three independent junction rule equations—one of the four junction rule equations will not be independent of the others. Thus we choose any three junctions to apply the junction rules, giving three equations. The three loop equations provide the necessary remaining three equations.

Working It Out The six equations are

for junction a:	$I_1 - I_2 - I_5 = 0;$
for junction b:	$I_2 - I_3 - I_4 = 0;$
for junction c:	$I_4 + I_5 - I_6 = 0;$
for loop 1:	$\mathscr{E} - I_1R_1 - I_2R_2 - I_3R_3 = 0;$
for loop 2:	$-I_5R_5 + I_4R_4 + I_2R_2 = 0;$
for loop 3:	$-I_6R_6 + I_3R_3 - I_4R_4 = 0.$

With sufficient patience, we can solve these six equations for the six unknown currents. This, however, is not required in this example.

What Do You Think? Suppose loop 1 were chosen so that it goes all the way around (battery–a–c–d) rather than (battery–a–b–d), as chosen in the example. Could you still solve for the unknowns?

EXAMPLE 27–7 Find the steady-state currents I_1 and I_2 in the circuit drawn in Fig. 27–12. Also find the resistance of resistor R_3 that will give a steady-state current $I_3 = 50$ mA. Finally, determine the potential drop across the capacitor. The values of the known elements are $\mathscr{E} = 6$ V, $R_1 = 100\ \Omega$, $R_2 = 80\ \Omega$, and $C = 2\ \mu$F.

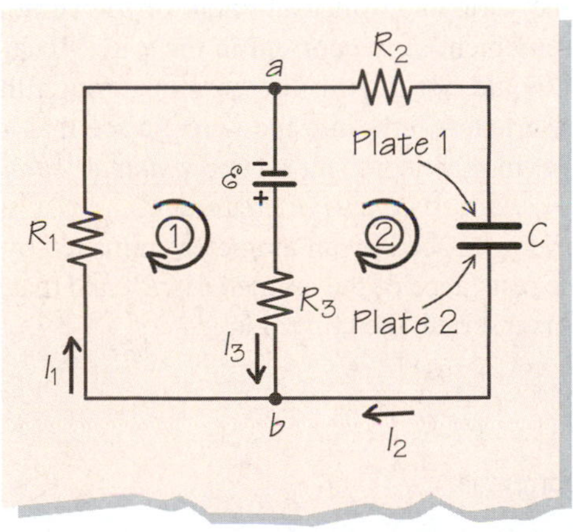

▲ **FIGURE 27–12** Notice that, in steady-state operation, the capacitor acts as an open switch despite the fact that it has voltage across its plates.

Strategy A capacitor acts as an open switch for steady-state current flow. Nevertheless, there is a voltage drop across the capacitor, because charge can build up on the capacitor plates under the impetus of the battery as the steady state is achieved. (Once steady state is reached, there is no further buildup of charge.) We must accordingly include the capacitor in application of the loop rule, and we cannot simply ignore, for example, the loop (a–capacitor–b–a). There will therefore be two loop equations and one junction equation—with two junctions, there is always only one independent equation. With three equations, we will be able to solve for the three unknowns I_1, R_3, and the potential drop across the capacitor.

Working It Out We first write down the loop equations, remembering that the currents are constant in time, so that no current passes through the leg containing the capacitor, and $I_2 = 0$. This simplifies the independent junction (b) equation which reads

$$I_1 = I_2 + I_3 = I_3.$$

For the loop equations: loop 1 gives

$$\mathscr{E} - I_3R_3 - I_1R_1 = 0,$$

from which we obtain the resistance

$$R_3 = \frac{\mathscr{E} - I_1R_1}{I_3} = \frac{\mathscr{E} - I_3R_1}{I_3} = \frac{\mathscr{E}}{I_3} - R_1.$$

(continues on next page)

Finally, the equation for loop 2 reads $-I_2R_2 + V_C + I_3R_3 - \mathscr{E} = 0$, where V_C is the potential difference across the capacitor, and with $I_2 = 0$ we get

$$V_C = \mathscr{E} - I_3R_3.$$

The numerical results are

$$R_3 = \frac{6\text{ V}}{50\text{ mA}} - 100\ \Omega = \frac{6\text{ V}}{50 \times 10^{-3}\text{ A}} - 100\ \Omega = 20\ \Omega$$

and

$$V_C = (6\text{ V}) - (50 \times 10^{-3}\text{ A})(20\ \Omega) = 5\text{ V}.$$

What Do You Think? (a) Could the equation for R_3 lead to a negative value for this resistance under certain circumstances? (b) Which of the two capacitor plates is at the higher potential?

27–4 Measuring Instruments

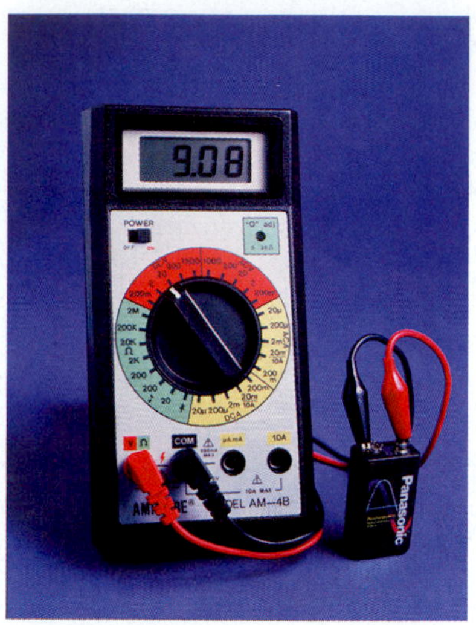

▲ **FIGURE 27–13** A multimeter can be used as a voltmeter, ohmmeter, or ammeter.

We have referred to the currents and voltages in various circuit elements, but we have not yet explained how these quantities are measured. A variety of instruments exists for this purpose: **Ammeters** measure current, **voltmeters** measure voltage, and **ohmmeters** measure resistance. These devices are often combined into one instrument called a **multimeter** (Fig. 27–13). We shall focus our attention on ammeters and voltmeters. Whatever the detailed operation of such instruments, a general principle must be respected: The measuring device should not distort the operation of the circuit being measured. We shall emphasize this principle in the following discussion.

For most applications, analog devices have been supplanted by digital ones, which are usually less expensive and more accurate. However, analog devices (dials)—the gasoline gauge of your automobile, or the moving hands of an analog watch, for example—remain superior for the recognition of how the measured quantities are changing. It will be simplest for us to illustrate the most important features of ammeters and voltmeters with analog devices. We'll start with a description of how analog ammeters and voltmeters can be constructed, then move on to the question of how they can be used in ways that do not disturb the circuits being measured.

How to Construct Analog Measuring Devices

Analog ammeters and voltmeters typically utilize a *galvanometer*, a device that relies on magnetic effects. The galvanometer, indicated by a circled *"G"* in circuit diagrams, consists of a coil of wire that rotates in the magnetic field produced whenever a current passes through the wire. (We'll explore this when we study the effects of magnetic fields, in Chapter 28.) A needle connected to the coil is deflected by an amount proportional to the current that passes through the coil. The position of the needle, properly calibrated, is the analog measurement.

Ammeters: An ammeter measures the current in a circuit wire by direct insertion into the wire—in other words, it is placed in series in the circuit segment through which the current to be measured passes. That means that the current flowing through the ammeter is the current flowing through the wire, and a measurement of the current passing through the ammeter itself is a measurement of the current in the wire. Recall now the general principle that an ammeter should not disturb the current through the circuit. From the fact that the ammeter is inserted directly into the wire we see that ideally the ammeter should have zero resistance. In practice, then, *a good ammeter should have a resistance that is small compared to other resistances in the circuit.*

Figure 27–14a shows a circuit being probed by an ammeter, indicated by a circled "A" in Fig. 27–14b. Suppose that the resistance of the ammeter is R_A and that the series resistance is R. Before the ammeter is inserted, the current is

$$I = \frac{\mathscr{E}}{R}.$$

After the ammeter is inserted, the current is

$$I = \frac{\mathscr{E}}{R + R_A}. \qquad (27\text{–}19)$$

The current will be the same with or without the ammeter attached only if $R_A \ll R$.

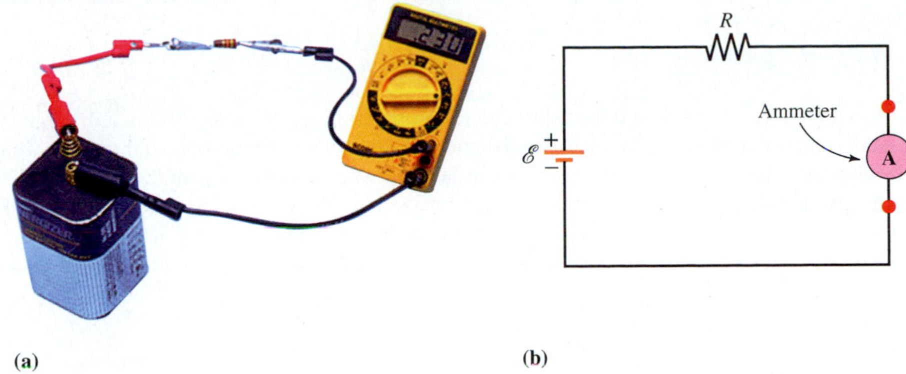

FIGURE 27–14 (a) An ammeter is placed in series where the current is to be measured. The resistance of an ammeter should be small so as not to change the current. (b) Schematic diagram of the circuit.

Voltmeters: Voltmeters measure potential differences across circuit elements. They do so by being placed in parallel with those elements, so that the potential drop across the voltmeter is the same as across the circuit element. Figure 27–15a shows a voltmeter—indicated by the circled "V" in Fig. 27–15b—used to measure the potential across a resistor. A voltmeter should not disturb the potential difference being measured, and so *a good voltmeter should have a large resistance*. Why is this? If the internal resistance of the voltmeter is R_V, the combination of voltmeter and resistance in Fig. 27–15b forms a parallel resistance circuit with equivalent resistance

$$\frac{1}{R_{eq}} = \frac{1}{R_V} + \frac{1}{R}.$$

When $R_V \gg R$, then $R_{eq} \cong R$, and none of the parameters of the original circuit are affected.

A galvanometer connected in series with a resistor of large resistance can serve as a voltmeter. If the current I passing through the galvanometer is measured (and that is just what the needle deflection measures), then the potential drop across the voltmeter is approximately IR_V. For example, take a galvanometer that can measure a maximum current of 100 μA. This is equivalent to a measurement of 10 V if the internal resistance is set at

$$R_V = \frac{10\ \text{V}}{100 \times 10^{-6}\ \text{A}} = 10^5\ \Omega.$$

An analog voltmeter is then a galvanometer with a series resistor of resistance R_V (Fig. 27–16). If there are potential differences within a circuit of 1000 V, then the 100 μA current would be exceeded and the resistor in our example would not be adequate. With a 10^5-Ω resistor the full-scale reading is 10 V. To measure a full-scale voltage of 1000 V with 100 μA, we must take $R_V = 10^7\ \Omega$.

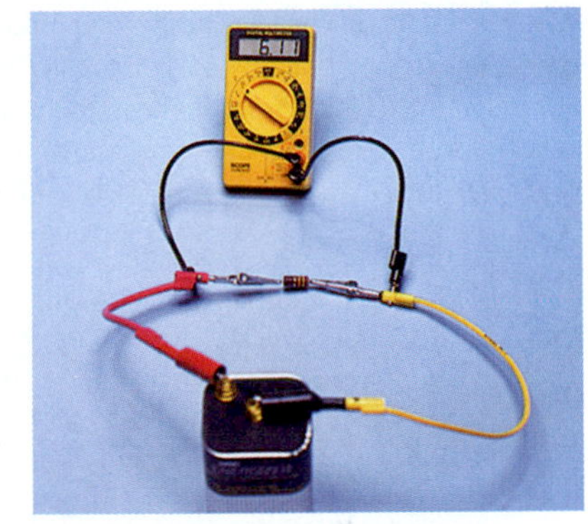

(a)

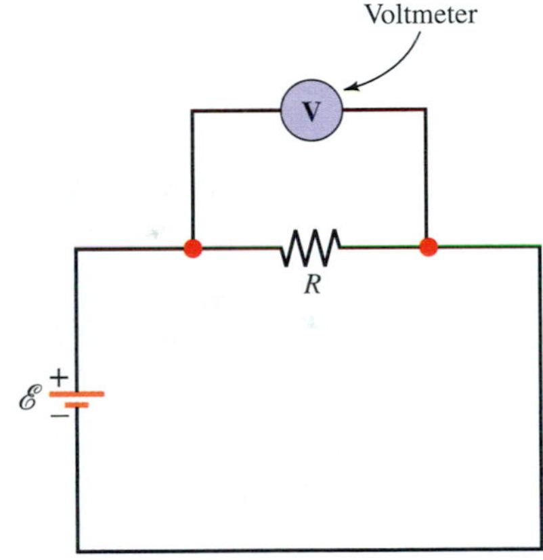

(b)

FIGURE 27–15 (a) A voltmeter is placed in parallel across the circuit element whose potential drop is to be measured. The resistance of a voltmeter should be large so as not to change the circuit. (b) Schematic diagram of the circuit.

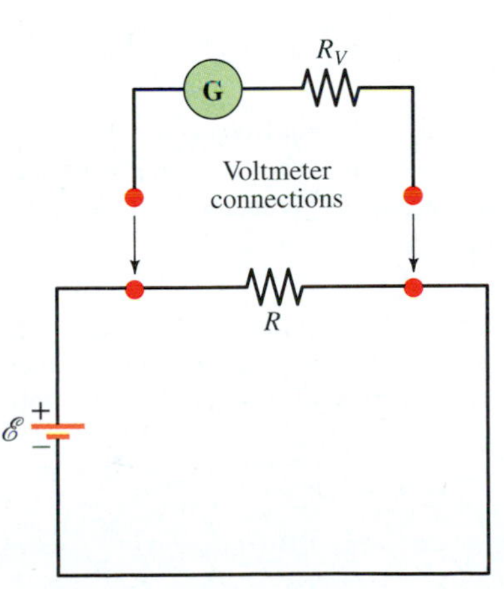

FIGURE 27–16 A galvanometer with a large series resistance can serve as a voltmeter.

CONCEPTUAL EXAMPLE 27–8 A voltmeter with a large resistance R_V is used as in Fig. 27–15. Will it be a good voltmeter for all circuits?

Answer The point of the apparatus is that the current that flows through the voltmeter be small compared to the current in the circuit. This means that R_V must be large compared to the parallel equivalent resistance. In this way, the total current, which is

$$I = \mathscr{E}\left(\frac{1}{R_V} + \frac{1}{R_{eq}}\right) = I_{voltmeter} + I_{circuit},$$

will be approximately equal to $I_{circuit} = \mathscr{E}/R_{eq}$. The word "large" in the question statement is not a very precise one, and the important point is whether it is large when compared with the corresponding quantity, here R_{eq}, for the circuit. When someone uses a word like "large," a suitable response is "large compared to what?"

EXAMPLE 27–9 A voltmeter with an internal resistance of $10^5\ \Omega$ is used to measure the voltage across resistor R_1 in the circuit of Fig. 27–17. To learn the error in measurement caused by the voltmeter itself compare the potential drop with and without the voltmeter for $\mathscr{E} = 6\ \text{V}$, $R_1 = 10\ \text{k}\Omega$, and $R_2 = 5\ \text{k}\Omega$.

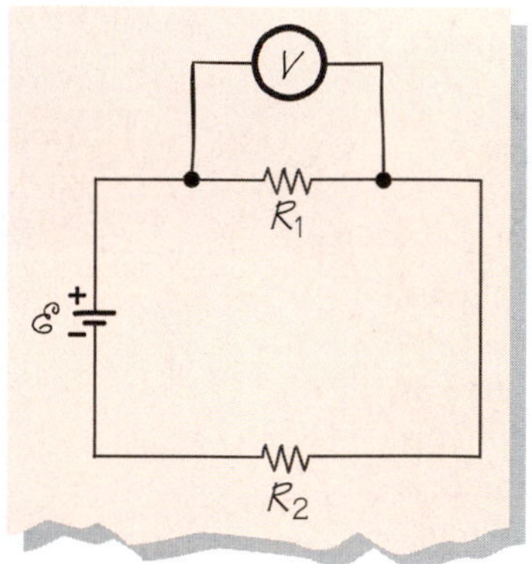

▲ FIGURE 27–17

Setting It Up As Fig. 27–17 makes clear, we are asked to compare the voltage across the resistance R_1 without and with the voltmeter placed in parallel with R_1.

Strategy In the case where the voltmeter is not present, we deal with R_1 and R_2 in series, in a single loop with a load resistance $R_1 + R_2$. In the case where the voltmeter is present, we can proceed either by applying the Kirchhoff rules to two loops or, more directly, by treating the voltmeter as a resistor parallel to R_1 and replacing the combination of these parallel resistors with a single equivalent resistance in a single-loop problem. The second alternative is simpler.

Working It Out With the voltmeter absent and a load resistance of $R_1 + R_2$, the current flowing is

$$I = \frac{\mathscr{E}}{R_1 + R_2}.$$

From this we calculate the voltage across the resistor R_1 to be

$$V_1 = IR_1 = \frac{\mathscr{E}R_1}{R_1 + R_2}.$$

When the voltmeter is connected, the resistance R_1 is replaced by the equivalent resistance R_1^* given by

$$\frac{1}{R_1^*} = \frac{1}{R_1} + \frac{1}{R_V} \quad \text{or} \quad R_1^* = \frac{R_1 R_V}{R_1 + R_V}.$$

The current is now

$$I^* = \frac{\mathscr{E}}{R_1^* + R_2}$$

and the voltage drop is just the voltage drop across the effective resistance R_1^*,

$$V_1^* = R_1^* I^* = \frac{\mathscr{E}R_1^*}{R_1^* + R_2}.$$

Numerically,

$$V_1 = IR_1 = \frac{6\ \text{V}}{10\ \text{k}\Omega + 5\ \text{k}\Omega}(10^4\ \Omega) = 4\ \text{V}.$$

With the voltmeter we have

$$R_1^* = \frac{R_1 R_V}{R_1 + R_V} = \frac{(10^4\ \Omega)(10^5\ \Omega)}{(10^5\ \Omega)(1 + 0.1)} = 0.9 \times 10^4\ \Omega$$

and therefore

$$V_1^* = R_1^* I^* = \frac{ER_1^*}{R_1^* + R_2} = \frac{(6\ \text{V})(0.9 \times 10^4\ \Omega)}{(0.9 \times 10^4\ \Omega) + (0.5 \times 10^4\ \Omega)} = 3.9\ \text{V}.$$

The difference between 4 V and 3.9 V is equivalent to a 3 percent error due to the presence of the voltmeter. The larger the voltmeter resistance, the smaller the error introduced. Whether this is a tolerable error depends on the requirements.

What Do You Think? A good voltmeter will have an internal resistance of 10 MΩ, a factor of 100 larger than the resistance used in this example. Would such a device give a significantly more accurate reading?

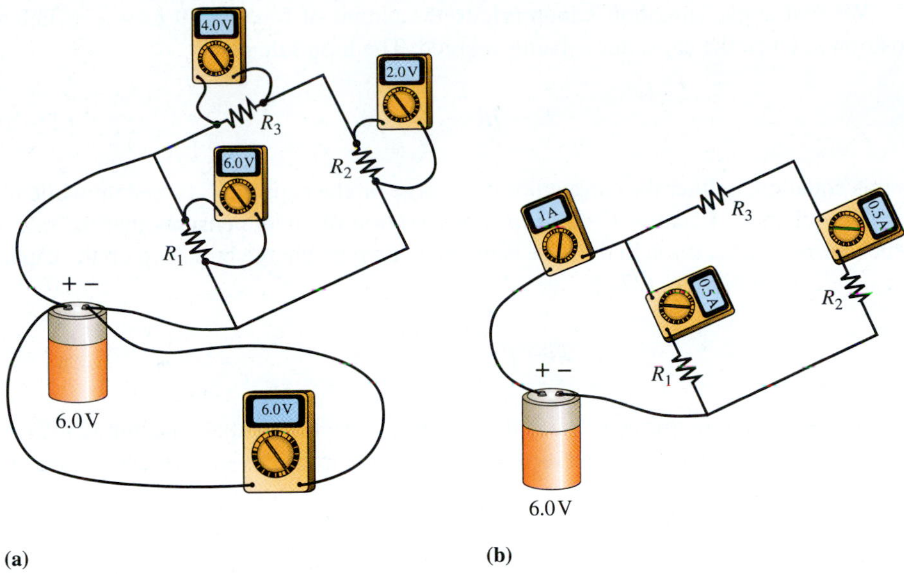

FIGURE 27–18 (a) Measuring voltage across circuit elements. (b) Measuring current through circuit elements.

In Fig. 27–18, we summarize the use of voltmeters and ammeters in a typical circuit.

27–5 *RC* Circuits

RC circuits are circuits that contain both resistors and capacitors. Unlike the circuits we have considered so far that have a steady current flowing through them, the currents and potentials of *RC* circuits can exhibit time-varying behavior when we introduce another element, a switch (Fig. 27–19). Even for circuits containing steady sources of emf, we introduce time dependence in a circuit every time we open or close the switch. *RC* circuits with switches have time-dependent effects that are useful for the control of motors, machinery, or computers.

We first observed the effect of a fully charged capacitor in an electric circuit in steady-state operation in Example 27–7. Now we want to examine the more complex, transient behavior that occurs when a capacitor is being charged and discharged. Consider the circuit shown in Fig. 27–19, with an initially uncharged capacitor. When the switch is closed (to position a) at $t = 0$, current begins to flow from the positive terminal of the battery, and positive charge begins to collect on plate 1 of the capacitor, while an equal amount of negative charge collects on plate 2. Current flows everywhere in the circuit *except* through the plates of the capacitor. Immediately after the switch is closed, the current has its maximum value, but the charge that builds up on the capacitor plates opposes further charge flow, and the current decreases. When the potential across the capacitor plates equals the emf and equilibrium is reached, the current is zero. This occurs when the charge on the capacitor plates, Q_0, is such that $\mathscr{E} = Q_0/C$.

After equilibrium has been reached and the current has become zero, we change the switch to position b, effectively taking the battery out of the circuit. The circuit now consists only of the charged capacitor and the resistor, and it is not in equilibrium. Current flows through the circuit from plate 1 of the capacitor to plate 2. The rate of flow is limited by the resistor. At first the current is high, but it decreases as the capacitor discharges through the resistor. Eventually the capacitor discharges completely, and the current again falls to zero when equilibrium is reached.

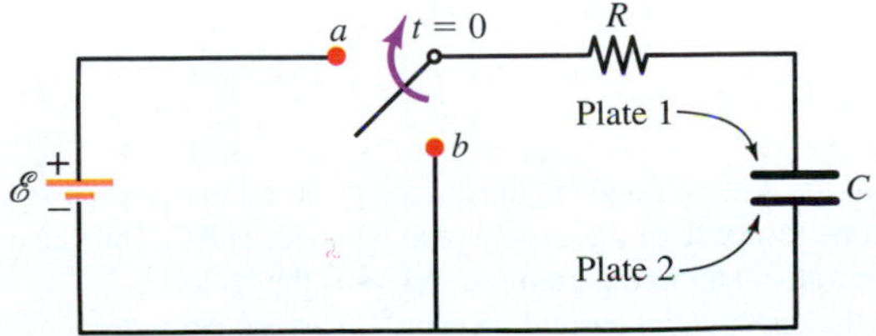

FIGURE 27–19 A circuit used to charge and discharge a capacitor through a resistor. When the switch is closed at a, the capacitor is charged by the source of emf, whereas the capacitor discharges through R when the switch is thrown to b.

We first apply Kirchhoff's loop rule to the circuit of Fig. 27–19 for the switch at position a, when the capacitor is being charged. The loop rule gives

$$\mathcal{E} - IR - \frac{Q}{C} = 0. \tag{27–20}$$

In this equation, neither the current nor the charge on the capacitor is constant while the capacitor charges. Because $I = dQ/dt$ (conservation of charge ensures that the rate at which current flows through the wire is the rate at which charge builds up on the capacitor), we can rewrite Eq. (27–20) as

$$\mathcal{E} - R\frac{dQ}{dt} - \frac{Q}{C} = 0. \tag{27–21}$$

The single variable in this equation is the charge Q. The differential equation (27–21) is straightforward to solve; let's omit the mathematical complexities and present its solution:

$$Q = C\mathcal{E}(1 - e^{-t/RC}). \tag{27–22}$$

By differentiating Eq. (27–22) with respect to time and substituting into Eq. (27–21), we can see that it satisfies Eq. (27–21) (see Problem 54). More important, does it agree physically with what we expect? According to Eq. (27–22), the charge on the capacitor is zero at $t = 0$ and builds smoothly to $C\mathcal{E}$ at large times, in agreement with our earlier discussion.

We can find the current in the circuit by differentiating Eq. (27–22) with respect to time:

$$I = \frac{dQ}{dt} = C\mathcal{E}\left(\frac{1}{RC}e^{-t/RC}\right) = \frac{\mathcal{E}}{R}e^{-t/RC}. \tag{27–23}$$

The sign of the current is positive, so we chose the correct current direction (clockwise). The maximum value of the current is $\mathcal{E}/R$, at $t = 0$, and the current is zero at $t = \infty$, which also agrees with our earlier discussion. Just after the switch is closed, the potential drop across the resistor is $\mathcal{E} = IR$, with no potential drop across the uncharged capacitor. As the capacitor charges, the current drops *exponentially* to zero.

Equations (27–22) and (27–23) show that the time dependence of both charge and current is determined by the product $\tau \equiv RC$; τ is called the **time constant**. It has units of time; with R and C in SI units, RC will be in seconds. The time constant determines how fast a capacitor charges and discharges. The smaller the value of RC, the more quickly the exponentials in the equations for Q and I fall; similarly, the larger the value of RC, the more slowly the exponentials change. Figure 27–20a and c show the current

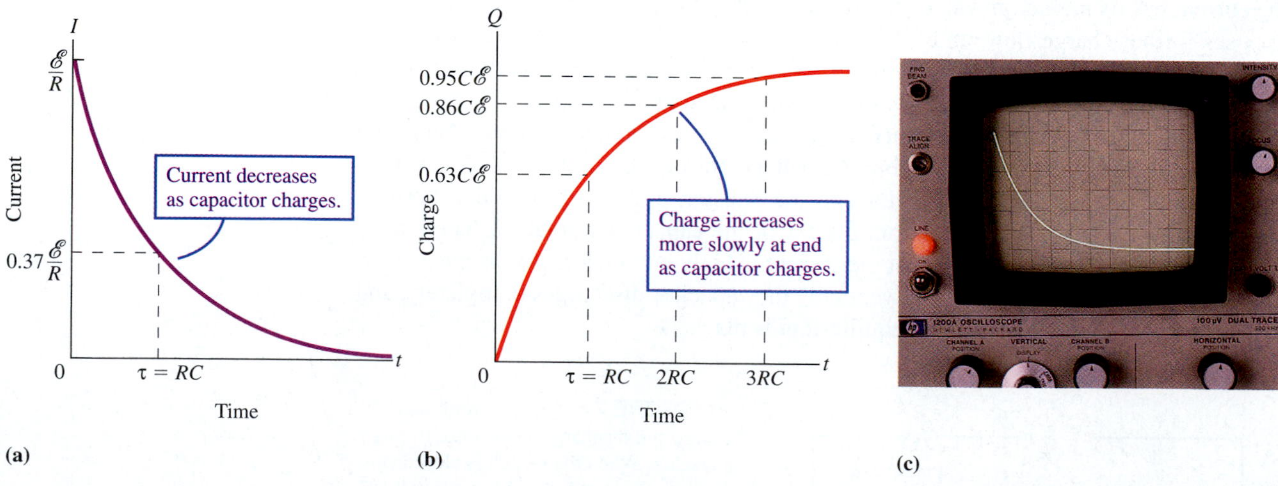

▲ **FIGURE 27–20** The time response of (a) the current I and (b) the charge Q stored in a capacitor as the capacitor is charged. The characteristic time response of the exponential behavior is RC. The value 0.37 in the graph of current is the factor e^{-1}; the value 0.63 in the graph of charge is the factor $(1 - e^{-1})$. (c) This oscilloscope screen shows the exponential current drop on a charging capacitor.

in the circuit, and Fig. 27–20b shows the charge on the capacitor as a function of time while the capacitor is being charged. After a time RC, the current has dropped to $e^{-1} \cong 0.37$ times its original value. After this same amount of time, the capacitor is $(1 - e^{-1}) \cong 63$ percent fully charged. It is 86 percent charged at time $2RC$ and 95 percent charged at time $3RC$.

Let's return to the circuit of Fig. 27–19. Suppose that the switch has been in position a for a long time, the capacitor is fully charged, and there is no current. At time $t = 0$, we throw the switch to position b. Only the discharging capacitor and the resistor are now in the circuit (Fig. 27–21). The positive charge is on plate 1, and we assume as before that the current is clockwise. The loop rule now gives

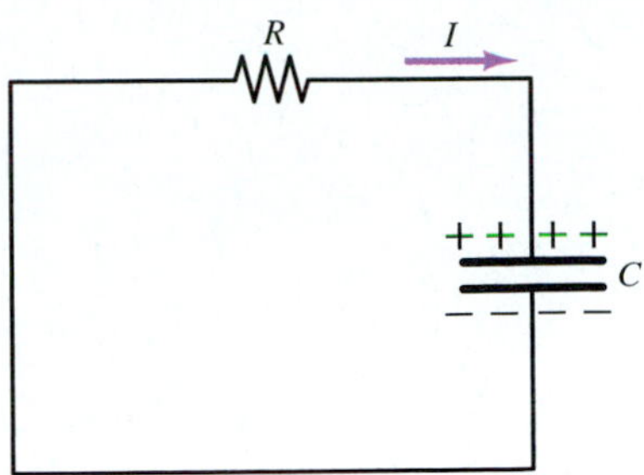

▲ **FIGURE 27–21** The circuit of Fig. 27–19 after the switch has been thrown to position b.

$$-IR - \frac{Q}{C} = 0. \tag{27–24}$$

Using $I = dQ/dt$, we have

$$R\frac{dQ}{dt} + \frac{Q}{C} = 0. \tag{27–25}$$

This differential equation is solved by the function

$$Q = Q_0 e^{-t/RC}, \tag{27–26}$$

where Q_0 is the initial charge on the capacitor when the switch is changed, $Q_0 = C\mathscr{E}$. Equation (27–26) may be substituted into Eq. (27–25) to verify that it is a solution (see Problem 55). The charge on the capacitor decreases exponentially with the time constant RC, and after a long time, there will be no charge on the capacitor.

We find the current by differentiating Eq. (27–26):

$$I = \frac{dQ}{dt} = -\frac{Q_0}{RC}e^{-t/RC}. \tag{27–27}$$

The current in this case is negative, indicating that the actual current is counterclockwise, opposite in direction to the current we assumed when we drew the diagram. It is again a maximum at $t = 0$, when the magnitude of the current is $Q_0/RC = \mathscr{E}/R$. After a long time, the current is again zero.

The behavior of the charge and current for the capacitor that discharges through a resistor is qualitatively what we expected from our earlier discussion. The magnitude of the current for this case is just as shown in Fig. 27–20a for the charging capacitor. The charge on the capacitor is plotted as a function of time in Fig. 27–22. Again, the factor 0.37 is e^{-1}.

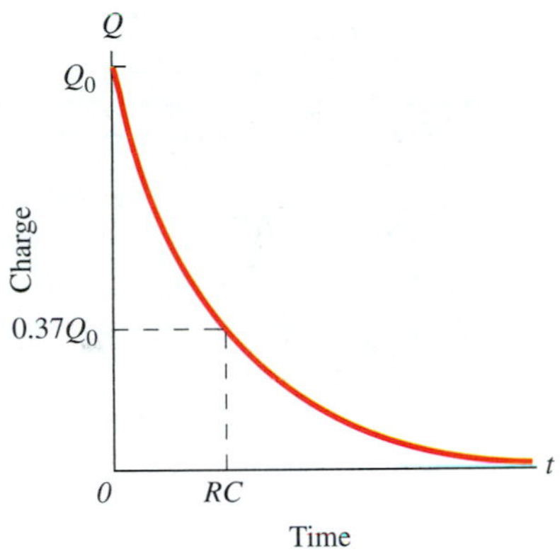

▲ **FIGURE 27–22** The capacitor charge as the capacitor of Fig. 27–21 discharges through the resistor as a function of time. The characteristic time response of the exponential behavior is again RC.

THINK ABOUT THIS. . .

CAN CAPACITORS REPLACE BATTERIES IN LAPTOP COMPUTERS?

Rechargeable batteries in laptop computers work only for a few hours, and recharging takes a long time. Could a capacitor replace the battery? The computer requires a steady current of around 2 μA but can work with a range of voltage from 6 V to a minimum voltage of 2.5 V (the lower limit is determined by the necessity to maintain memory function). If the charged capacitor is placed in series with a resistance, the current produced by the capacitor as it discharges will drop as the voltage drop V_C across the capacitor decreases. A constant current I calls for a non-ohmic device for which I is *independent* of the voltage drop V across it over some range of V. (The device that achieves this is a *zener diode*.) When such a device is placed in series with a capacitor, then the relation $Q = CV$ implies that the current obeys the equation $I = dQ/dt = C\,dV/dt$. Since the device produces constant I, V must drop linearly with time. If $I = 2\ \mu$A how long will it take before the potential difference across the capacitor changes from 6 V to 2.5 V? From the relation $I = C\,dV/dt$ with constant I we calculate $\Delta t = C \times (\text{change in } V)/I = C(3.5\text{ V})/(2 \times 10^{-6}\text{ A})$. With a capacitor of $C = 1$ F this time scale is $1.8 \times 10^6\text{ s} \cong 20$ days. Whether this is a practical option depends on the existence of capacitors with capacitance of farads—a very large capacitance—yet compact enough to be used in portable devices, and as we already discussed in Chapter 26 this technology is being actively pursued. ■

Energy in *RC* Circuits

Let's now examine the role of energy and its conservation during the charging of a capacitor. From the definition of potential, the amount of work done by the battery emf during the charging process is $\mathscr{E}$ times the total charge processed by the battery. This charge is the final charge $C\mathscr{E}$ on the capacitor plates after a long period of time. Thus, the work done by the battery, W_{bat}, is

$$W_{bat} = \mathscr{E}(C\mathscr{E}) = C\mathscr{E}^2. \tag{27–28}$$

How do we account for the energy that matches this work? In part, the energy is stored in the capacitor. We know from Eq. (25–9) that the total energy stored by a capacitor is $CV^2/2$. The voltage V in this case is $\mathscr{E}$, so the energy stored by the capacitor, E_{cap}, is

$$E_{cap} = \tfrac{1}{2}C\mathscr{E}^2. \tag{27–29}$$

Where has the other half of the work done by the battery gone? The only other circuit element is the resistor, and the other half of the work has gone into Joule heating of that resistor. From Eq. (26–27), we know that the power loss in the resistor is $P = I^2R$. We can integrate the power over time to find the energy loss in the resistor, E_{res}, using the current from Eq. (27–23)

$$\begin{aligned} E_{res} &= \int_0^\infty I^2R\,dt = \frac{\mathscr{E}^2}{R}\int_0^\infty e^{-2t/RC}\,dt \\ &= \frac{\mathscr{E}^2}{R}\left(-\frac{RC}{2}e^{-2t/RC}\right)\Bigg|_0^\infty = \frac{C\mathscr{E}^2}{2}. \end{aligned} \tag{27–30}$$

The thermal energy loss in the resistor accounts for the other half of the work done by the battery. This 50-percent split of energy between the resistor and the capacitor is *independent* of $\mathscr{E}$. For the case of the discharging capacitor, all the energy stored in the capacitor dissipates as heat in the resistor. A spectacular demonstration of the rapid release of the energy contained within a capacitor was shown in Fig. 25–8.

EXAMPLE 27–10 The charging circuit shown in Fig. 27–23 (with a switch thrown to position a at $t = 0$) has a 12 V emf, a 100 Ω resistance, and a 100 μF capacitance. (a) Find the time constant, the final charge on the capacitor, and the work done by the battery. (b) How long does it take for the capacitor to be charged to 99.9 percent of its final charge?

Setting It Up We label the given emf, resistance, and capacitance $\mathscr{E}$, R, and C, respectively.

Strategy (a) Knowing both R and C, the time constant is $\tau = RC$. The final charge on the capacitor is $Q_f = C\mathscr{E}$ (as we already know from Chapter 25), and the work done by the battery is $W_f = C\mathscr{E}^2$. As for part (b), this is a matter of solving the expression $Q = Q_f(1 - e^{-t/\tau})$ for t knowing that $Q = 0.999Q_f$.

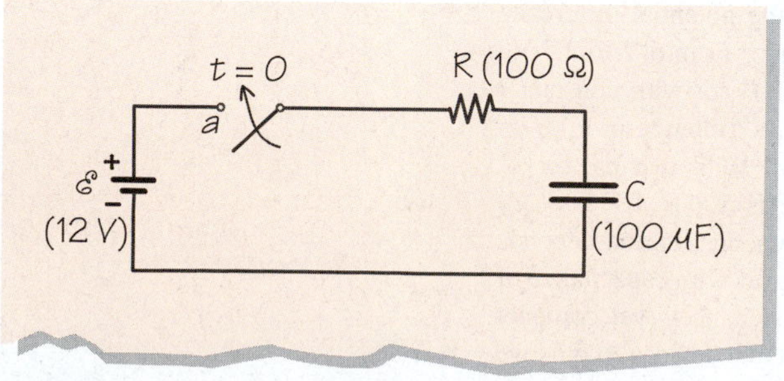

▲ FIGURE 27–23

Working It Out (a) We need only plug numerical values into the appropriate expressions:

$$\begin{aligned} \tau = RC &= (100\ \Omega)(10.0 \times 10^{-6}\ \text{F}) \\ &= 1.00 \times 10^{-3}\ \text{s} = 1.00\ \text{ms}, \\ Q_f &= C\mathscr{E} = (10.0 \times 10^{-6}\ \text{F})(12\ \text{V}) = 1.2 \times 10^{-4}\ \text{C}, \\ W_{bat} &= C\mathscr{E}^2 = (10.0 \times 10^{-6}\ \text{F})(12\ \text{V})^2 = 1.4 \times 10^{-3}\ \text{J}. \end{aligned}$$

(b) With $Q = 0.999Q_f$, we want to solve Eq. (27–22) for t. The factor Q_f cancels, leaving $0.999 = (1 - e^{-t/\tau})$, or

$$e^{-t/\tau} = 1 - 0.999 = 0.001.$$

Taking the natural logarithm of both sides gives

$$-\frac{t}{\tau} = -6.91.$$

The time to reach 99.9 percent of its final charge is thus

$$t = 6.91\tau = (6.91)(1.00\ \text{ms}) = 6.91\ \text{ms}.$$

What Do You Think? If it takes 6.91 ms to charge the capacitor to 99.9% of its final voltage, how long do you estimate it will take to reach 99.99%?

Summary

Sources of emf (electromotive force) $\mathcal{E}$, such as chemical batteries, are sources of electric energy. The emf is defined by the amount of work it can do to move charge:

$$\mathcal{E} \equiv \frac{dW}{dq}. \tag{27–1}$$

Batteries cause charges to move in circuits. The simplest circuits to analyze are direct-current circuits, in which no circuit parameters change with time.

Analysis of single- or multi-loop circuits is accomplished by the use of Kirchhoff's two rules. Kirchhoff's loop rule states that the sum of the potential changes around a closed path of a circuit is zero:

$$\sum_{\text{closed path}} \Delta V = 0. \tag{27–9}$$

We can specify the potential change across batteries, resistors, and capacitors. Any source of emf has an internal resistance r that may be large enough to require consideration.

Kirchhoff's junction rule is "circuit language" for the conservation of electric current. It states that the sum of the currents that enter a junction equals the sum of the currents that leave the junction. If we interpret an outgoing current as an ingoing current with a minus sign in front of it, then the junction rule takes the simple form:

$$\sum I_{\text{in}} = 0. \tag{27–11}$$

Ammeters, voltmeters, and ohmmeters measure current, voltage, and resistance, respectively. Ammeters must have a small internal resistance so that they do not affect the circuit leg in which the current is measured. Conversely, voltmeters need a large internal resistance because they are used in parallel with the circuit element being measured, and they too should not affect the circuit being measured.

A circuit exhibits time-varying behavior when a capacitor is being charged with a source of emf or when the capacitor is being discharged. For a simple circuit with an emf $\mathcal{E}$, a resistor R, and a capacitor C, an initially uncharged capacitor has charge

$$Q = C\mathcal{E}(1 - e^{-t/RC}) \tag{27–22}$$

and current

$$I = \frac{\mathcal{E}}{R}e^{-t/RC}. \tag{27–23}$$

When the source of emf is disconnected from the circuit and the capacitor is allowed to discharge through the resistor, the charge decreases exponentially:

$$Q = Q_0 e^{-t/RC}. \tag{27–26}$$

The time constant RC determines the time dependence of exponential increase or decrease of the charge and the current during the charging and discharging of a capacitor.

Understanding the Concepts

1. Why is it dangerous to be in a bathtub when an electrical appliance is standing on the edge of the tub?
2. One voltmeter measures the voltage of a flashlight battery as 0.9 V, whereas another measures 1.5 V. What might cause this difference?
3. Show that it is irrelevant whether, in drawing a circuit diagram, the internal resistance of a source of emf is placed before or after the emf itself.
4. What sense does it make to draw a circuit diagram with resistanceless wires when real wires always have some resistance?
5. By taking a special combination of batteries of constant emf, resistors, and capacitors, is it possible to construct a circuit in which the emf around a closed loop is not zero?
6. You measure a potential difference of 3 V across the terminals of a 3-V battery with some age to it and proclaim it to be just fine. Are you right?
7. A flash unit on a camera discharges by means of an RC circuit. Find out (say, from a local camera store) what the R and C values are, and obtain the RC time.
8. You have an RC circuit and three identical capacitors available to you. How should you arrange them to make the discharge time as short as possible?
9. It would appear superficially that, in Section 27–5, if we had chosen the current in the wrong direction, we would have ended up with the equation $IR - (Q/C) = 0$ rather than Eq. (27–7) to describe the current in an RC circuit. This would be disastrous

because the solution of our new equation would be a rising exponential, $e^{+t/RC}$, rather than a falling exponential, and this expression would grow without limit. What is wrong with this reasoning?

10. When you reverse the polarities of all batteries in a circuit, the magnitudes of all currents stay the same. Why?

11. Given two identical sources of emf and two identical light bulbs, how would you arrange these elements so that both bulbs glow with maximum brightness?

12. Suppose that you connect the terminals of two batteries of different emfs + to + and − to −. What do you expect to happen?

13. What happens if the emf of battery 2 in Example 27–4 is reversed?

14. It might appear that the only effect of the internal resistance of a battery in a circuit is to change the battery's emf from $\mathscr{E}$ to $\mathscr{E}' = \mathscr{E} - Ir$, where $\mathscr{E}'$ acts as though there were no internal resistance. Could this be true? If not, why not?

15. Show that the circuit in Fig. 27–8(a) can be reduced to a one-loop circuit. Is this also true of the circuit in Fig. 27–8(b)?

16. Two reckless teenagers hang by both hands on a wire that can be connected to a constant voltage source. One teenager hangs from a position in which there is a resistor between his hands; the other one does not. The resistance of the resistor is much larger than that of the wire, which is very small. When the wire is connected to the battery, how do the teenagers experience that fact?

17. What happens to the current I_1 of Example 27–7 if the resistance R_3 is increased?

18. To start a car in cold weather, one sometimes first turns the lights on for a while. How could this help?

Problems

27–1 EMF

1. (I) A 12-V car battery is rated at 80 A, meaning that it will send 80 A through a wire connected to its terminals. What is its internal resistance?

2. (I) The *Magellan* spacecraft that studied Venus in 1990 used two solar panels capable of producing 1200 W. If the solar array was capable of producing a total of 40 A, what was the terminal voltage of the device?

3. (I) A battery with an emf of 3.00 V sends a current of 1.99 A when it is connected in series with a 1.50-Ω resistor. What is the internal resistance of the battery?

4. (I) Nickel–cadmium batteries used in space flight are rated at 30 A · h (they can put out 30 A for 1 hr) and 30 V. How much energy do the batteries contain?

5. (I) A defibrillator used by emergency medical staff to restart an accident victim's heart has an internal resistance of 20 Ω, and a power supply that produces an average 5000 V over a short period of time. If the body resistance between the two electrodes is 230 Ω, how much current passes through the body to restart the heart?

6. (II) A flashlight battery with an internal resistance of 0.06 Ω produces a 170-mA current through a 15-Ω resistor. What is the emf of the battery? What is the terminal voltage of the battery in this usage?

7. (II) A certain automobile battery has an emf of 12 V. When it produces a current of 100 A, the terminal voltage reads 9.0 V. Calculate the internal resistance of the battery. What is the power dissipated in the battery when it produces this current?

8. (II) The resistance of the starter of a car, including the cables, is 0.11 Ω. On a frigid winter morning, cranking the engine for 10 s reduces the terminal voltage of the car's 12-V battery to 8 V. The car does not start. How much heat (in kilocalories) is produced in the battery during this time? Does it improve your chances of starting the car on a second try?

9. (II) The internal resistance of a battery whose emf is 12 V varies with the current according to the equation $r = (\alpha + \beta I)$, where $\alpha = 0.15\ \Omega$ and $\beta = 0.018\ \Omega/\text{A}$. Find the terminal voltages and the power dissipated in the battery when $I = 1.0$ A and $I = 10$ A.

27–2 Kirchhoff's Loop Rule

10. (I) Two bulbs in series are connected across an emf of 110 V. The normal operating ratings of the bulbs are given as (2.5 V, 0.5 A) for one bulb and as (110 V, 10 W) for the other. Calculate how much power will actually be developed in each lamp.

11. (I) For what value of R_2 in the circuit of Fig. 27–24 is the voltage across the points a and b zero? For what value is the current in the circuit zero?

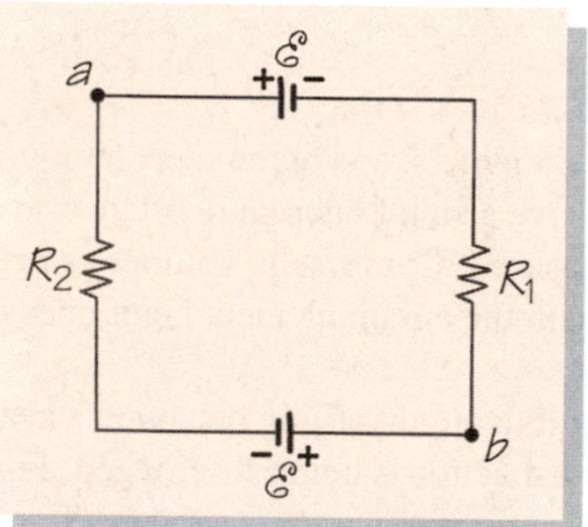

▲ **FIGURE 27–24** Problem 11.

12. (II) A generator (a "battery" that uses mechanical rather than chemical energy) of emf 110 V and internal resistance 0.50 Ω is used to charge a series of 20 batteries, each with emf 2.2 V and internal resistance 0.06 Ω. A series resistor is used to limit the charging current. (a) What is the terminal voltage of the generator? (b) the terminal voltage of the bank of batteries? (c) What series resistance must be included to allow a charging current of 15 A? (d) What is the power dissipated in all the resistors?

13. (II) A portion of a larger circuit is shown in Fig. 27–25. The potential drop between the pairs of points labeled in the figure is $V_{ba} = 2$ V, $V_{cb} = 3.5$ V, $V_{cd} = 2$ V, and $V_{df} = -0.5$ V. Find the potential differences V_{gf}, V_{ag}, and V_{ca}.

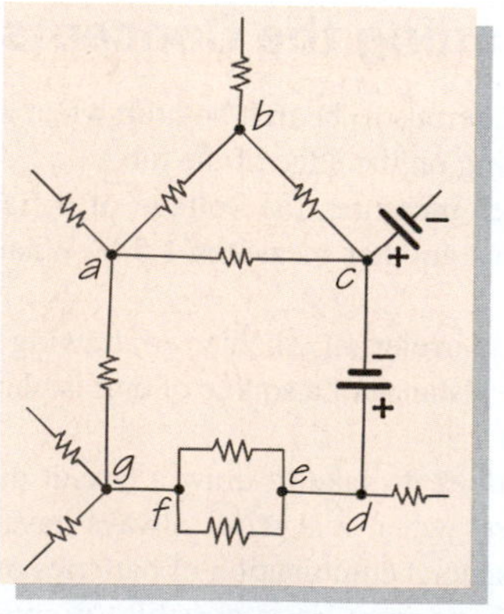

▲ **FIGURE 27–25** Problem 13.

14. (II) A flashlight consists of two 1.5-V batteries connected in series to a bulb with resistance 10 Ω. (a) What is the power delivered to the bulb? (b) Batteries run down when they acquire an (internal) resistance. How large is the additional resistance if the power delivered to the bulb has decreased by one-third of its initial value?

27–3 Kirchhoff's Junction Rule

15. (I) In Fig. 27–26, the currents $I_1 = 2$ A, $I_2 = 0.5$ A, $I_3 = -3$ A, $I_4 = -0.5I_6$, and $I_5 = -I_6$. Find the unknown currents I_4, I_5, and I_6.

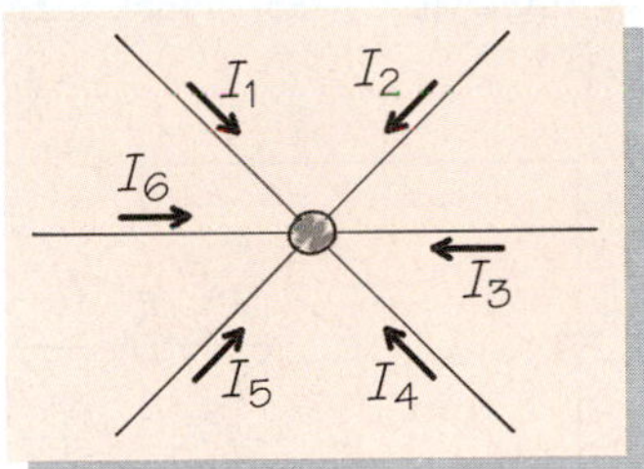

▲ **FIGURE 27–26** Problem 15.

16. (I) Five identical lightbulbs are placed in a circuit as shown in Fig. 27–27. What is the brightness of bulb #3 relative to bulb #1? Even though this problem can be worked without them, use the Kirchhoff rules.

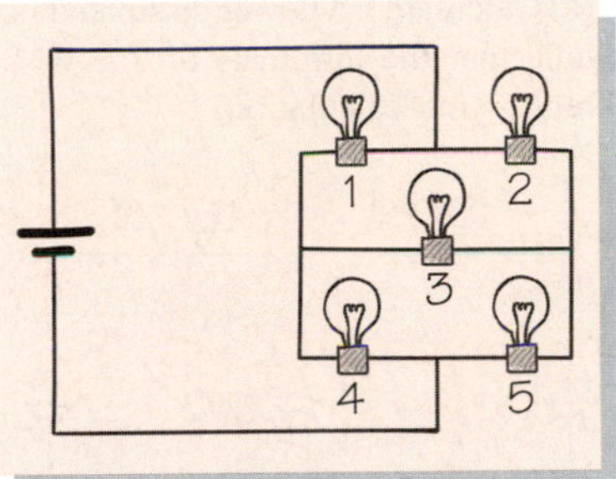

▲ **FIGURE 27–27** Problem 16.

17. (II) Consider the part of a circuit shown in Fig. 27–28. What is the total resistance if (a) the switch is open, and (b) the switch is closed?

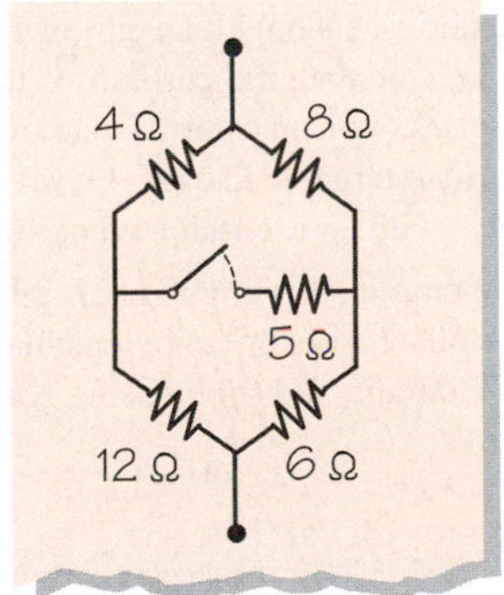

▲ **FIGURE 27–28** Problem 17.

18. (II) Consider the circuit in Fig. 27–29. Calculate the current through the 16 Ω resistor, and the voltage across *ab*.

19. (II) Find the current that passes through each of the resistors in the circuit shown in Fig. 27–30.

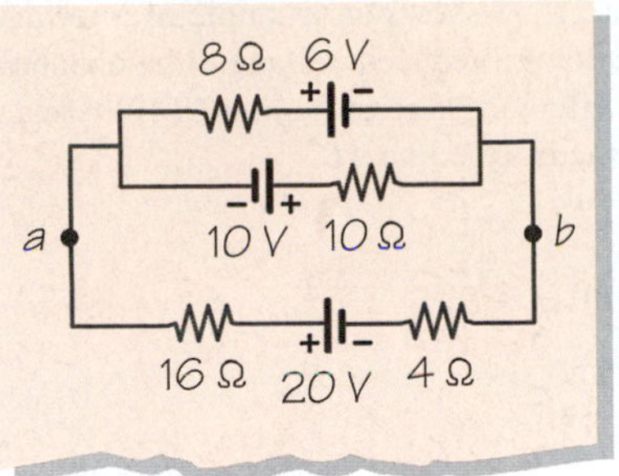

▲ **FIGURE 27–29** Problem 18.

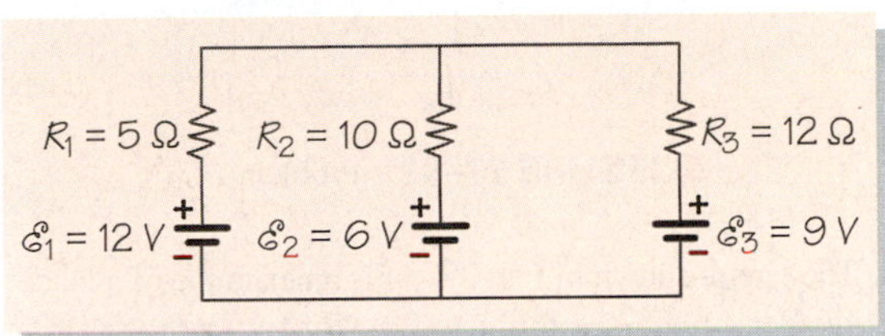

▲ **FIGURE 27–30** Problem 19.

20. (II) Find the current that passes through the 4-Ω resistor in the circuit shown in Fig. 27–31.

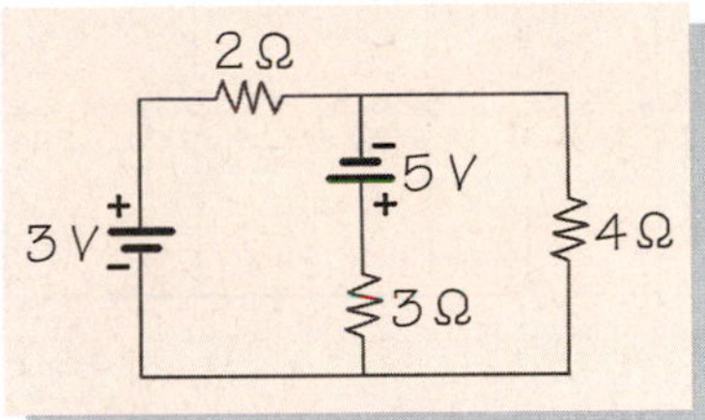

▲ **FIGURE 27–31** Problem 20.

21. (II) Three resistors connected in parallel have resistances of 250 Ω, 420 Ω, and 510 Ω, respectively. The total current passing through the set is 0.020 A. What is the potential difference across the set, and what are the currents in each of the resistors?

22. (II) Can the resistors of the circuit in Fig. 27–32 be reduced to a single equivalent circuit by application of the rules for circuits with connections in parallel and in series? Solve for the currents through the three resistors.

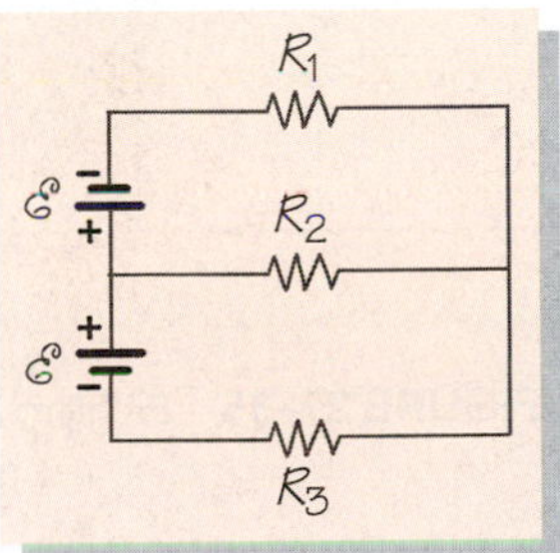

▲ **FIGURE 27–32** Problem 22.

23. (II) *N* identical batteries, with emf $\mathscr{E}$ and internal resistance *r*, are connected in parallel across a resistance *R*. Obtain the value for the current, and compare its value with that obtained if the batteries are connected in series.

24. (II) Figure 27–33 shows an example of a *voltage divider*, a device that allows a reduced voltage to be obtained. Calculate the potential difference across the line CD in terms of the potential difference across the line AB.

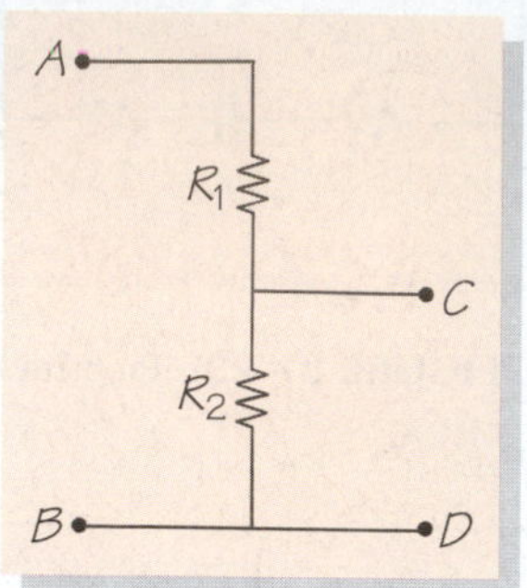

▲ **FIGURE 27–33** Problem 24.

25. (II) The circuit shown in Fig. 27–34 is an example of a loaded voltage-divider circuit (see Problem 24). By varying the values of R_1 and R_2, different values of V_L can be obtained; R_L represents the load. Let $R_1 = R_2 = 3.3\ \text{k}\Omega$ and $\mathscr{E} = 10\ \text{V}$. For load resistances of $20\ \text{k}\Omega$, $200\ \text{k}\Omega$, and $2\ \text{M}\Omega$, how much different is V_L than 5 V?

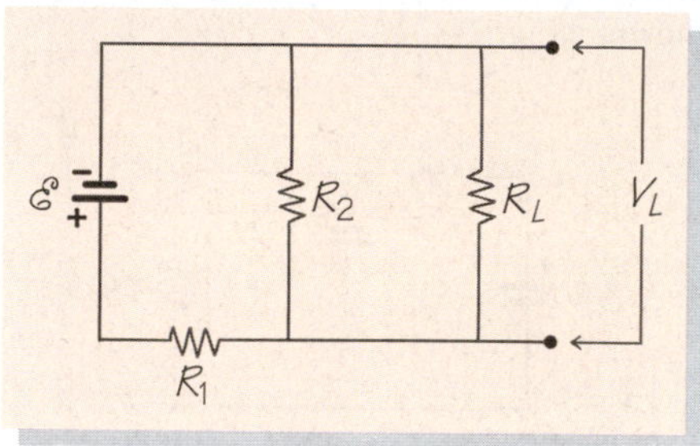

▲ **FIGURE 27–34** Problem 25.

26. (II) Consider the circuit shown in Fig. 27–35. If $\mathscr{E} = 2.8\ \text{V}$, and if all three resistors have identical resistances, find the resistance that ensures that the current I_3 will be 55 mA. Even though this problem can be worked without them, use the Kirchhoff rules.

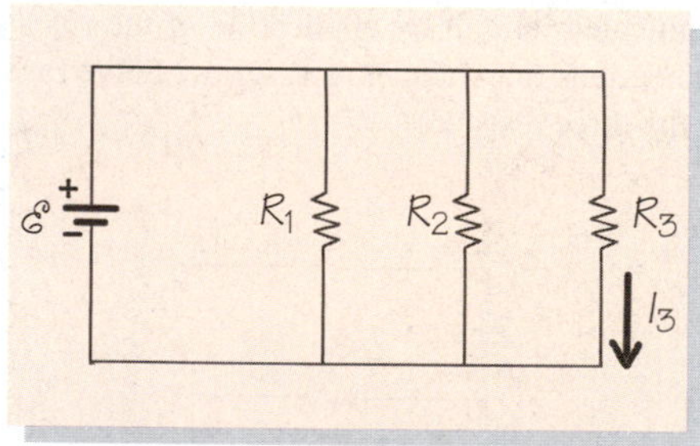

▲ **FIGURE 27–35** Problem 26.

27. (II) How many independent junctions are there in the circuit shown in Fig. 27–36? To verify your answer, solve for all currents.

28. (II) Replace the network of resistors in Fig. 27–37 by a single equivalent resistor. Can the combination of resistors be reduced to a single resistor by successive application of the rules for parallel and series resistors?

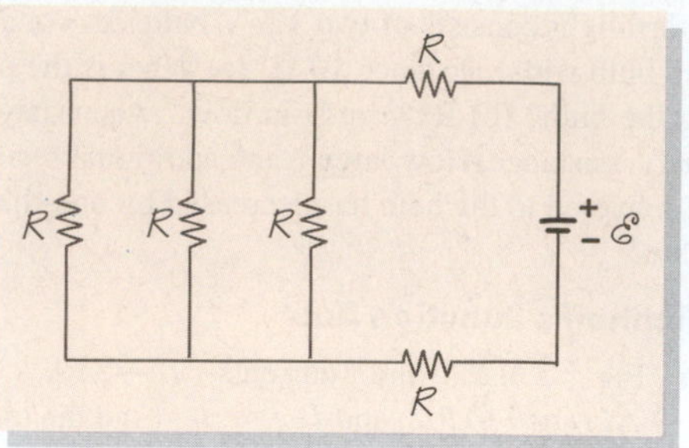

▲ **FIGURE 27–36** Problem 27.

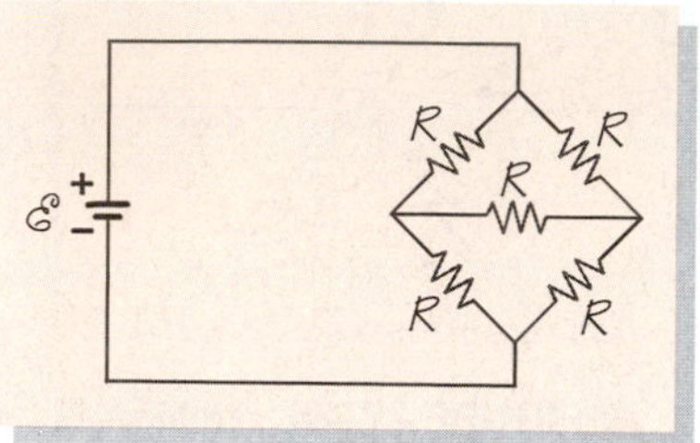

▲ **FIGURE 27–37** Problem 28.

29. (II) Two batteries with emf $\mathscr{E}_1 = 6\ \text{V}$ and $\mathscr{E}_2 = 9\ \text{V}$, respectively, are connected to resistors with the resistances as marked in Fig. 27–38. (a) Calculate the power dissipated in the 50-Ω resistor. (b) Assume that the terminals on the 6-V battery are reversed, and repeat your calculation.

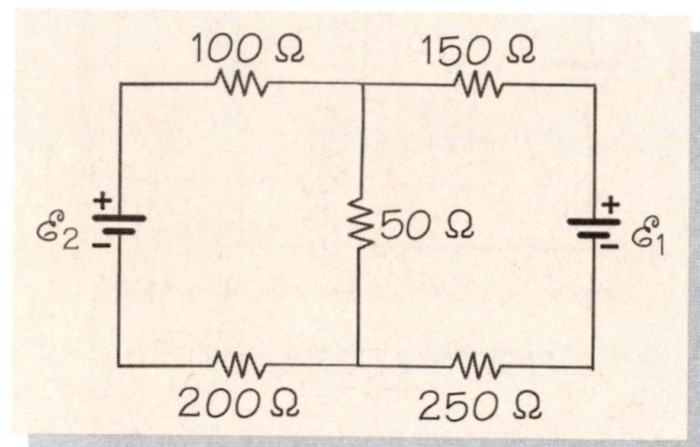

▲ **FIGURE 27–38** Problem 29.

30. (II) Given 20 batteries, each with an emf of 12 V and an internal resistance of 2 Ω, calculate the current for the following cases: (a) batteries in series, with an external load of 80 Ω; (b) batteries in parallel, with an external load of 20 Ω; (c) 4 parallel rows of 5 batteries in series, with an external load of 40 Ω.

31. (II) Consider the circuit shown in Fig. 27–39. Calculate the current through the 50-Ω resistor (a) by calculating the equivalent resistance for the circuit, and (b) by using Kirchhoff's rules.

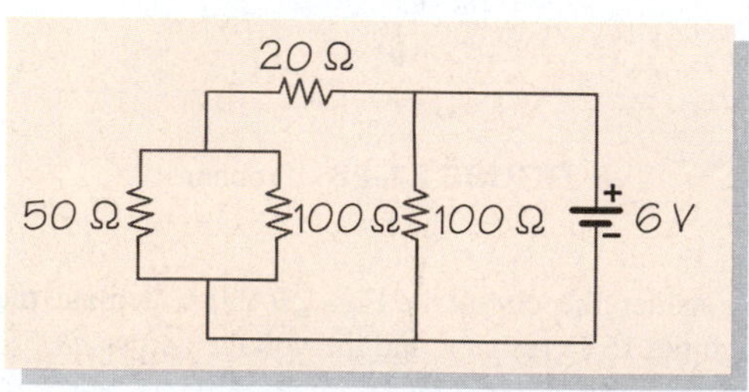

▲ **FIGURE 27–39** Problem 31.

32. (II) The known elements of the circuit in Fig. 27–40 are indicated. Find the value of R_3 that will give a current I_3 of 0.1 A with the indicated sign. Is there a value of R_3 that will give a current I_3 of the same magnitude but of opposite sign? If so, what is it?

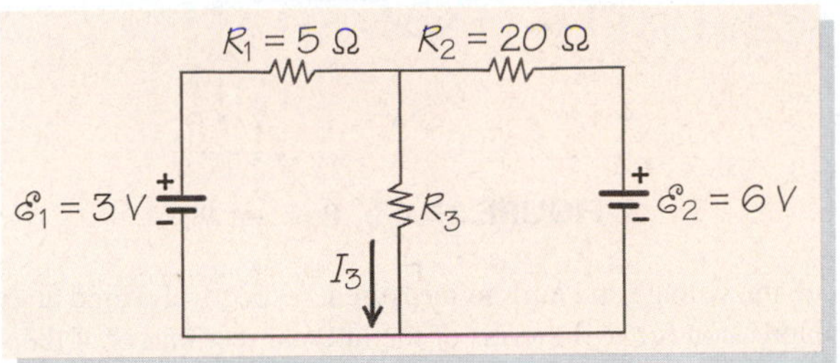

▲ **FIGURE 27–40** Problem 32.

33. (II) Two batteries are connected in parallel as in Fig. 27–41 and supply current to a load resistor of 5 Ω. One of the batteries is freshly charged, with an emf of $\mathscr{E}_1 = 12$ V and an internal resistance of $r_1 = 0.1\ \Omega$. The other one is almost dead, with an emf of $\mathscr{E}_2 = 10$ V and an internal resistance of $r_2 = 10\ \Omega$. What is the current through the load resistor? How much of this current is supplied by each of the batteries?

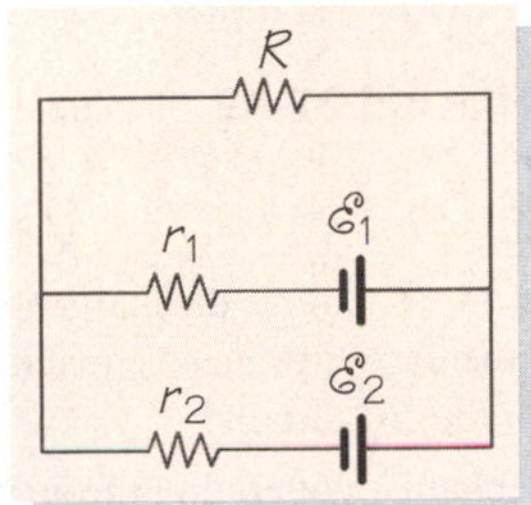

▲ **FIGURE 27–41** Problem 33.

34. (III) Consider a tetrahedron whose sides consist of identical wires, each with resistance 1 Ω (Fig. 27–42). Suppose that this arrangement is attached at two of its corners to a generator with potential 4 V. What is the power dissipated in each of the wires?

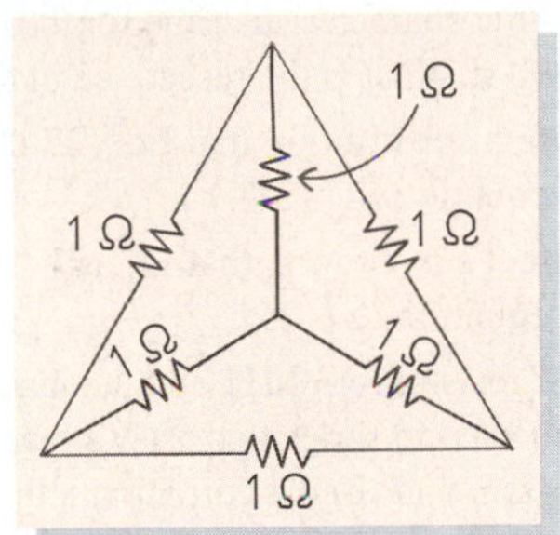

▲ **FIGURE 27–42** Problem 34.

35. (III) A cube consisting of identical wires, each of resistance R, is put across a line with voltage V (Fig. 27–43). What is the equivalent resistance of the cube? What is the current in each of the wires?

36. (III) Figure 27–44 shows a ladder of resistors with n rungs. (a) Find the equivalent resistance between points P_1 and P_2 for $n = 1$; (b) for $n = 2$; (c) for $n = 3$; (d) for the limit $n \to \infty$. [*Hint*: Write an expression for R_n (the equivalent resistance of a ladder of n rungs) in terms of R_{n-1} (the equivalent resistance of a ladder of $n - 1$ rungs) and R, and use that equation in the limit $n \to \infty$.]

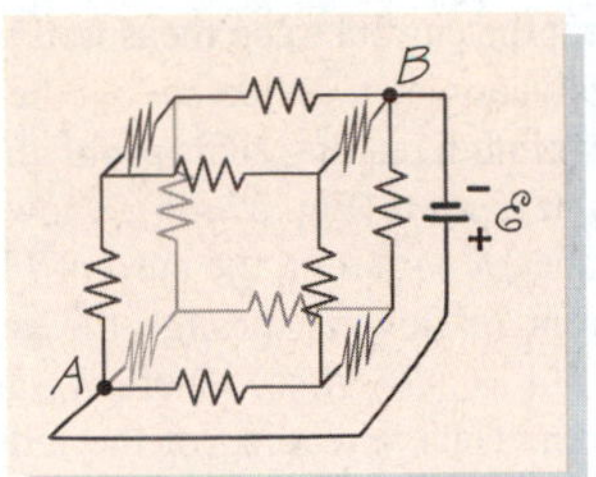

▲ **FIGURE 27–43** Problem 35.

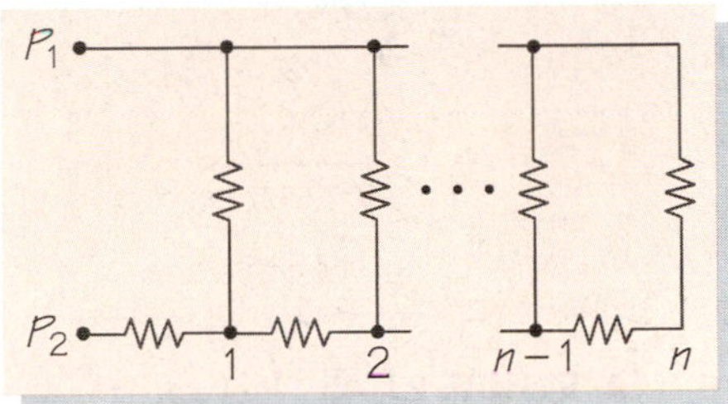

▲ **FIGURE 27–44** Problem 36.

27–4 Measuring Instruments

37. (I) Look at Example 27–9 again, with emf = 6 V, $R_1 = 1400\ \Omega$, and $R_2 = 10\ \text{k}\Omega$. Calculate the voltage across R_1 for the two cases of when the voltmeter's internal resistance is 200 kΩ and 10 MΩ.

38. (I) A voltmeter with an internal resistance of 60 kΩ measures the voltage of a D-cell flashlight battery (of nominal voltage 1.5 V) as 1.45 V. What is the internal resistance of the battery?

39. (I) Currents produced with a 12-V source of emf and a range of resistances from 10 Ω to 1000 Ω are to be measured to an accuracy of at worst 0.1 percent with an ammeter. How small must the resistance of the ammeter be?

40. (I) A voltmeter is to be used to measure the voltage across a range of resistances from 5 Ω to 5000 Ω. What is the minimum value of the internal resistance of the voltmeter such that a measurement can be carried out to 0.1 percent accuracy?

41. (I) A certain voltmeter has an internal resistance of $10^5\ \Omega$. The voltmeter is used to measure the potential drop across resistors of (a) 10 Ω, (b) $10^5\ \Omega$, and (c) 100 MΩ. In each case, what is the equivalent resistance of the voltmeter and the resistor across which it is placed? (The voltmeter is a suitable one if this equivalent resistance is as close as possible to the resistance of the resistor across which the potential drop is measured.)

42. (II) An ammeter that can measure a maximum current of 5 mA has an internal resistance of $1.8 \times 10^{-4}\ \Omega$. What series resistance will convert it to a 0-to-3-V voltmeter?

43. (II) A battery with emf $\mathscr{E}$ and internal resistance r is connected across a variable resistance. A voltmeter is also connected across the resistance. When the resistance R is set to 20 Ω then the voltmeter reads 23 V; if R is set to 5 Ω, the voltmeter reads 16 V. What will it read when R is set to 50 Ω?

44. (II) A galvanometer whose resistance is 20 Ω will go through its full deflection when a current of 2×10^{-4} A flows through it. What would you do to have its full deflection occur for a current of 2×10^{-3} A? What would you have to do to make it into a voltmeter so that the full deflection corresponds to 0.2 V?

45. (II) Suppose that the current to be measured by an ammeter is so large that a galvanometer deflected by the current would be pinned at its maximum reading. This problem can be resolved by the use of a *shunt resistor* (Fig. 27–45). Show that with the shunt resistor (resistance R_s) present, the current I is given in terms of a reduced current I_G flowing through the galvanometer by the formula $I = I_G[1 + (R_G/R_s)]$, where R_G is the resistance of the galvanometer. Thus, a reading of the reduced current I_G allows us to determine the current I.

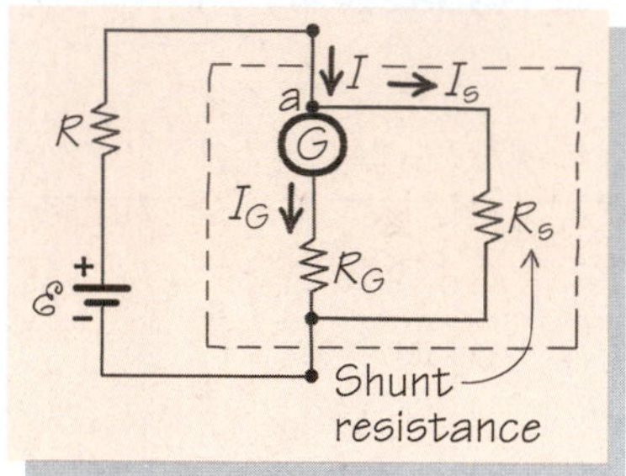

▲ **FIGURE 27–45** Problem 45.

46. (II) The output of the voltage-divider network shown in Fig. 27–46 is to be measured with two voltmeters of internal resistances 500 kΩ and 100 MΩ, respectively. What voltage will each indicate?

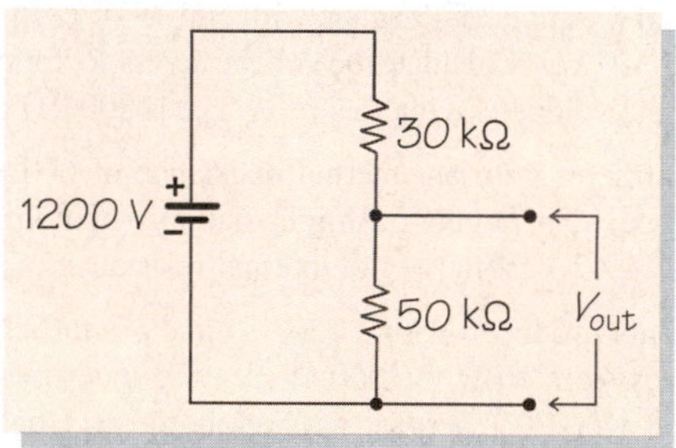

▲ **FIGURE 27–46** Problem 46.

47. (II) A *Wheatstone bridge* is a device that measures resistances. In the circuit shown in Fig. 27–47, R is an unknown resistance. The resistances R_1, R_2, and R_3 are variable. A galvanometer, G, can be used to determine when the potential difference between B and C is zero, given that the battery is connected between A and D. The variable resistances are varied until there is no current in the galvanometer when the circuit is closed at the switch S. Obtain an expression for R in terms of R_1, R_2, and R_3.

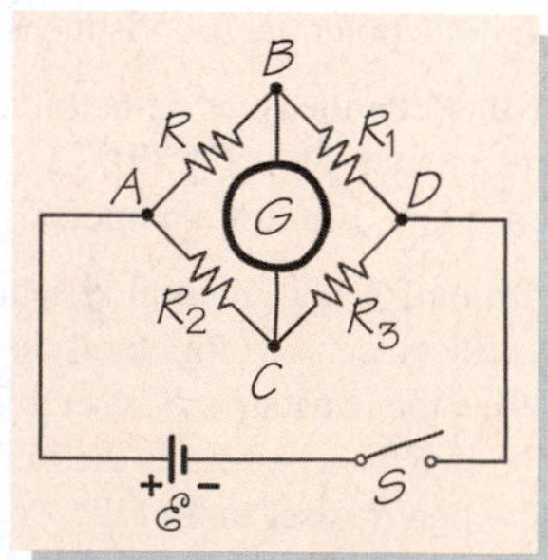

▲ **FIGURE 27–47** Problem 47.

48. (II) The circuit shown in Fig. 27–48 is used to measure the resistance R_x. Draw the circuit including internal resistances. V and I are the voltage and current measured, respectively. Find an exact expression for R_x in terms of the internal resistances of the voltmeter and ammeter. Under what conditions is $R_x = V/I$?

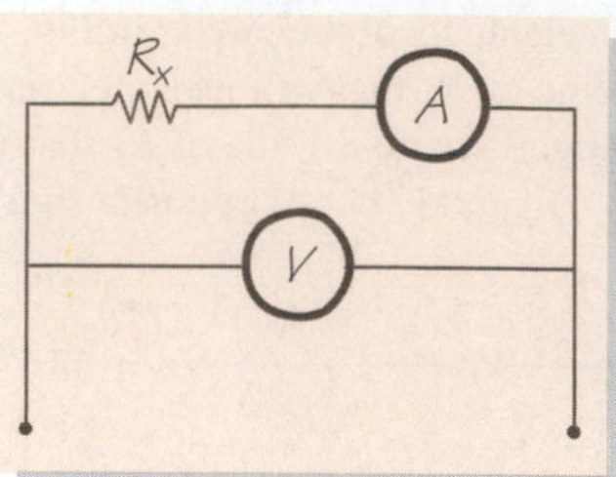

▲ **FIGURE 27–48** Problem 48.

49. (II) Repeat Problem 48 for the circuit shown in Fig. 27–49.

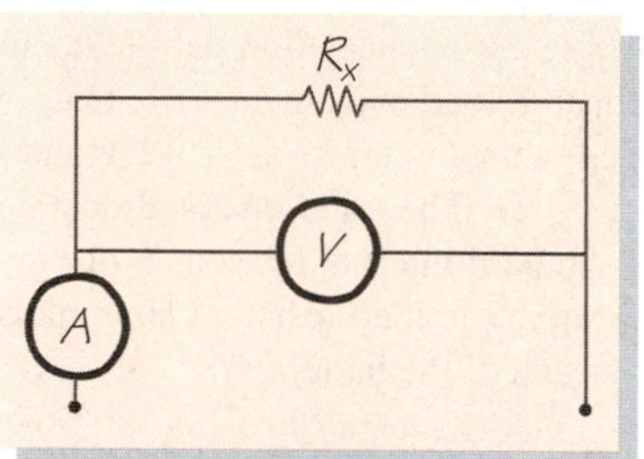

▲ **FIGURE 27–49** Problem 49.

27–5 *RC* Circuits

50. (I) A flashbulb in an *RC* circuit discharges with a time constant of 5×10^{-4} s. If the capacitor has a capacitance of 16 μF, what is the resistance in the *RC* circuit?

51. (I) A flashbulb mechanism operating through an *RC* circuit has a capacitor charged with a time constant of 2.0 s. If the resistance in the *RC* circuit is 10^5 Ω, what is the capacitance of the charging mechanism?

52. (I) Show that the product *RC* has units of seconds. Find the time constants for the following values of R and C: 5 MΩ, 30 μF; 8 kΩ, 3 μF; 20 Ω, 50 pF.

53. (I) The flash attachment of a camera has a capacitance of 600 μF. The flash time—the characteristic time for the discharge of the capacitor—is 1/500 s. What is the resistance in the *RC* circuit?

54. (II) Show by direct substitution that Eq. (27–22) is a solution for the differential equation (27–21).

55. (II) Show by direct substitution that Eq. (27–26) is a solution for the differential equation (27–25).

56. (II) A resistor of resistance 3 MΩ and a capacitor of capacitance 350 μF are connected in series to a 50-V power supply. Calculate (a) the time constant and (b) the current at a time when the charge on the capacitor has acquired 90 percent of its maximum value.

57. (II) Calculate the current in the battery as a function of time for the circuit shown in Fig. 27–50 if the switch S is closed at time $t = 0$.

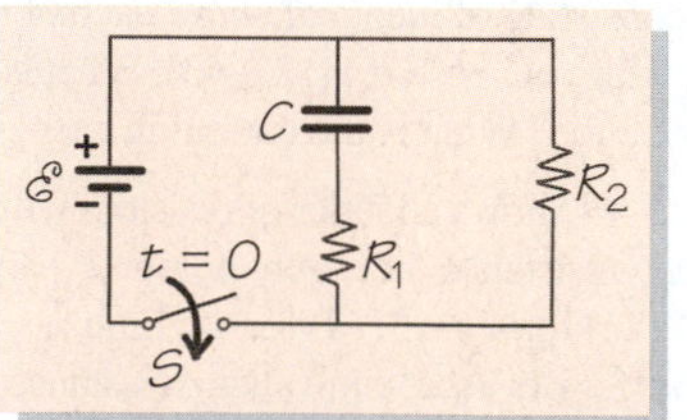

▲ **FIGURE 27–50** Problem 57.

58. (II) The circuit of Fig. 27–19 has $\mathscr{E} = 200$ V, $R = 350$ kΩ, and $C = 20\ \mu$F. What are the voltage across the resistor and the charge on the capacitor 4 s after the switch is closed to position a?

59. (II) You have two capacitors of capacitance 5 μF and three resistors, one of resistance 250 Ω and the remaining two of resistance 300 Ω. Find the connection between these elements that will make a circuit whose time constant is 1 ms.

60. (II) Show that the time constant of a parallel-plate capacitor filled with a dielectric with a finite resistivity is independent of the area and separation of the plates.

61. (II) Polycarbonate, a so-called polar polymer, is a material with a dielectric constant $k = 3.2$. It has a resistivity $\rho = 2 \times 10^{14}\ \Omega \cdot$m. Suppose that it is used to fill the space in a parallel-plate capacitor of area 0.03 m^2 and plate separation 0.50 mm. A charge of $Q = 2\ \mu$C is placed on the plates of the isolated capacitor. How long does it take for 70 percent of the charge to leak away?

62. (III) A capacitor C_1 is charged to Q_0 and connected to an uncharged capacitor C_2 via the resistor R (Fig. 27–51). Find the charge on each capacitor as a function of time, assuming that the switch is closed at $t = 0$.

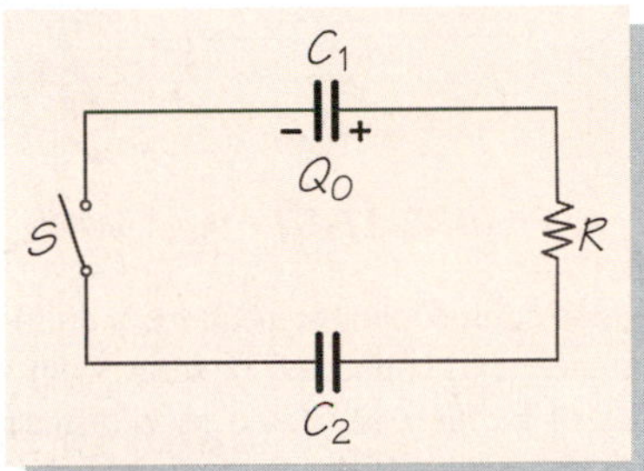

▲ **FIGURE 27–51** Problem 62.

General Problems

63. (I) Imagine that a household circuit uses direct current. Compare the current drawn from the main supply at 120 V if three household appliances of resistances 50 Ω, 60 Ω, and 20 Ω, respectively, are connected in parallel. (The appliances would not work in series, because each appliance requires a voltage drop of 120 V. Actual household circuits use alternating currents; see Chapter 33.)

64. (II) Consider the circuit shown in Fig. 27–52, in which a 12-V battery is used to charge a 6-V battery. The resistance in the circuit is 20 Ω. Calculate (a) the current in the circuit; (b) the rate at which the energy of the smaller battery increases; (c) the total rate of energy dissipation in the resistor.

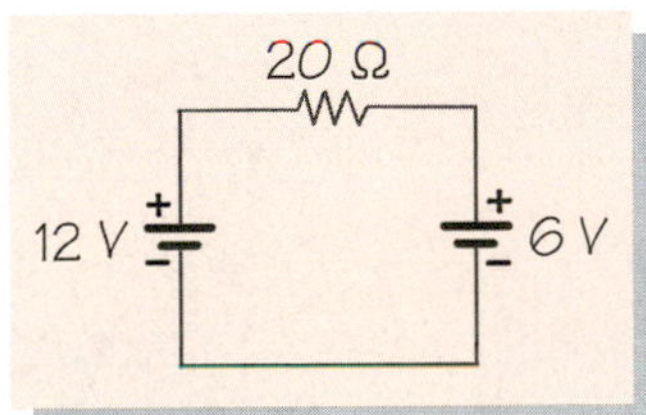

▲ **FIGURE 27–52** Problem 64.

65. (II) A student has a wide range of resistors, all rated for 5 W. How can a student combine identical resistors to obtain an effective resistance of 100 Ω rated for 30 W?

66. (II) If a battery of fixed emf and internal resistance r is connected to an external resistor of resistance R, show that the maximum power delivered to the external resistor occurs when $R = r$.

67. (II) Automobile batteries typically have 12.6 V output and an internal resistance of 0.05 Ω. They have a rating of ampere-hours, which demonstrates how long they can produce suitable currents for 20 hours. A good battery has a rating of 75 A $\cdot$ hr. The battery is charged at the rate of 2.5 A for 10 hours. (a) How much current is the battery capable of providing? (b) What is the voltage across the battery terminal during recharging? (c) What total electrical energy is stored in the battery after recharging?

68. (II) Suppose you have n identical batteries, each with emf $\mathscr{E}$ and internal resistance r. The cells may be placed in series, or in parallel across a load resistance R. Which arrangement should you use to maximize the current through the external resistor? Does your answer depend on how large r is?

69. (II) The battery considered in Problem 9 is connected to a load resistor R. Calculate and plot the current in the circuit as a function of R in the range $R = 0$ (short circuit) to $R = 5\ \Omega$. Plot the ratio of the power delivered to the load to that dissipated in the internal resistance of the battery.

70. (II) When separately connected across a line with voltage V, two resistors generate power P_1 and P_2, respectively. What is the power generated when the two resistors are connected in series? In parallel?

71. (II) A student picks up an electric heater at a yard sale. She discovers later that, even at the lowest setting, it delivers too much power to be used in her small room. She determines that the heater has two heating elements, one delivering 1 kW, the other 2 kW. The highest setting turns on both heaters simultaneously. How can she connect the elements to get a new setting lower than the lowest setting, and what is the power output for the new setting?

72. (II) A resistor R forms a single loop with an arrangement of two batteries of emf $\mathscr{E}$ and internal resistance r. The batteries are arranged (a) in series and (b) in parallel. Find the current through the resistance in both cases. Which arrangement gives the larger current for large R? for small R?

73. (II) An 800-W kitchen mixer, a 600-W vacuum cleaner, and a chandelier with 10 60-W bulbs are all plugged into the same outlet in a 120-V circuit. A *fuse* acts as a switch that opens if the current exceeds 15 A. How much current does each device draw? What is the minimum number of screwed-in bulbs that will blow the circuit? Do not worry about the oscillations of the current and voltage in real household circuits, but assume that all currents and voltages are DC.

74. (II) Consider three resistors of 30 Ω each. Each resistor can dissipate 2 W at most. What are the four possible distinct ways of arranging all three resistors, and calculate the maximum power that can be dissipated in each of the ways.

75. (II) Find the currents in each leg of the circuit shown in Fig. 27–53.

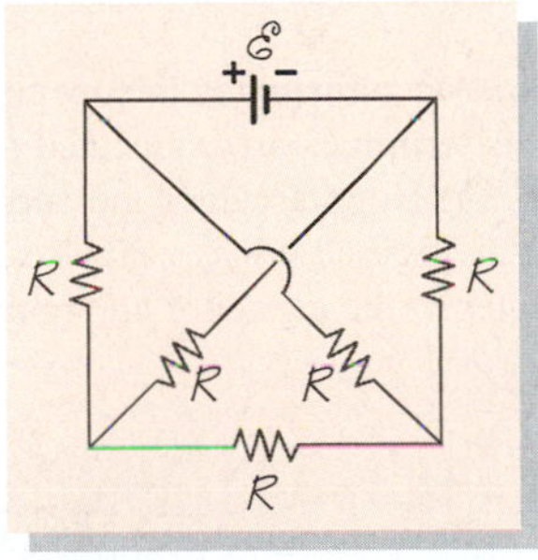

▲ **FIGURE 27–53** Problem 75.

76. (II) By using Table 26–2, compare the current density, electric field strength, and power loss in two cylindrical wires of the same length and same radius, one made of aluminum and the other made of copper. The wires are connected (a) in series and (b) in parallel. (c) If wires of a given length and radius were constructed from all the materials listed in Table 26–2, and if the same current were passed through each, which wire would have the largest current density, the weakest electric field, and the least power loss? Assume throughout that the temperature dependence is unimportant.

77. (II) A simple *potentiometer* circuit used to measure unknown voltages accurately is shown in Fig. 27–54. Here V_s is the known source voltage, V_x is the unknown voltage, and the resistor is a variable one from which the values R_1 and R_2 can be read from the position of the pointer. These resistances are varied until the current in the ammeter is zero. Show that the unknown voltage then has the value $V_x = V_s R_2/(R_1 + R_2)$.

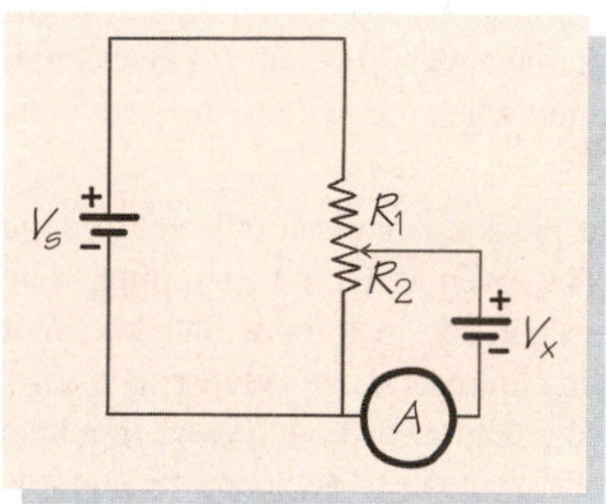

▲ **FIGURE 27–54** Problem 77.

78. (II) Two resistors and two capacitors are connected in series to a battery as shown in Fig. 27–55. Calculate the potential at B relative to that at A: (a) shortly after the closing of the switch and (b) a long time after the closing of the switch. (c) How fast does the circuit reach a steady state? (Give a time scale.)

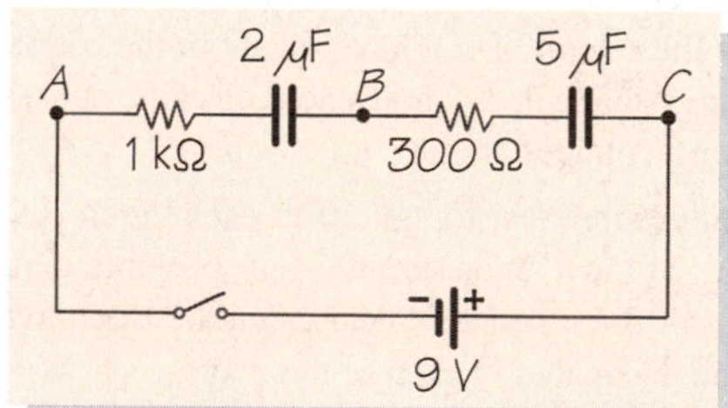

▲ **FIGURE 27–55** Problem 78.

79. (II) To avoid sparks accompanying the opening of a high-current circuit breaker, its terminals are connected to a large capacitor, as in Fig. 27–56. (a) How fast does the current decrease in the circuit shown in the figure? (Give a time scale.) (b) What is the charge on the plates of the capacitor a long time after the switch is opened?

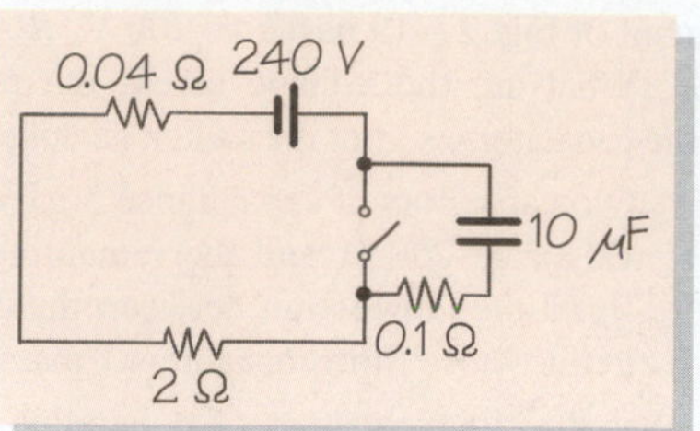

▲ **FIGURE 27–56** Problem 79.

80. (II) The circuit shown in Fig. 27–57 has been established for a long time. (a) What is the charge on the capacitor? Indicate which plate carries the positive charge and which one carries the negative charge. (b) Calculate the current flowing through the 35-Ω resistor.

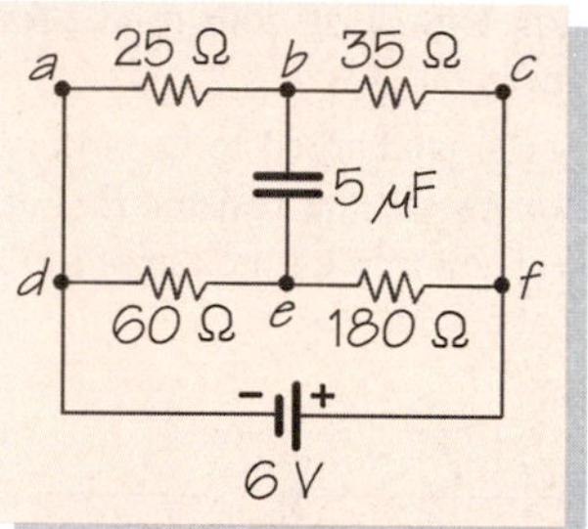

▲ **FIGURE 27–57** Problem 80.

81. (II) A single-loop circuit contains a battery of emf V_0 and negligible internal resistance and, connected in series with the battery, two circular plates of radius r separated by a distance $d \ll r$. The space between the plates is filled with a material of conductivity σ and dielectric constant 1. (a) What is the electric field between the two plates? (b) What is the current flowing in the circuit?

82. (III) Consider the infinite network of resistors shown in Fig. 27–58a. Calculate the resistance R^* of the network by noting that with an infinite set of resistors, adding one more rung to the ladder does not change the resistance. Thus the network may be broken up as shown in Fig. 27–58b.

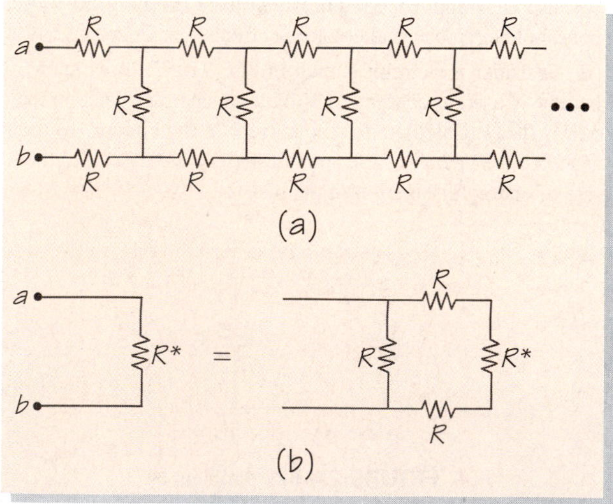

▲ **FIGURE 27–58** Problem 82.

Chapter 28

◀ The aurora is a phenomenon that depends on the existence of Earth's magnetic field. When the stream of charged particles known as the solar wind enters Earth's atmosphere under the influence of Earth's magnetic field, its interaction with the atoms of the upper atmosphere produce excitations that lead to the emission of light: the aurora.

The Effects of Magnetic Fields

We are all familiar with magnets. Sailors have used navigational compasses made from treated iron or lodestone—the natural mineral magnetite—for at least 800 years, and possibly far longer. The behavior of compass needles was systematically studied around the year 1600 by William Gilbert, an English physician. He correctly suggested that the action of a compass is a manifestation of magnetic forces. These forces arise because Earth is itself a giant lodestone, or magnet. Magnetism was not associated with electricity until 1820, when André Ampère used his own experiments and those of Hans Christian Oersted to show that magnetic effects arise when electric charges move. In the 1820s, Michael Faraday uncovered another connection between electricity and magnetism, but it was James Clerk Maxwell who, in the late 1860s, made the ultimate synthesis of electricity and magnetism. We now know that electrical and magnetic phenomena are both aspects of the interactions of electrically charged objects. The synthesis of electricity and magnetism is described entirely by Maxwell's equations, and our understanding of light and other electromagnetic waves rests on Maxwell's great achievement. Maxwell's equations, together with Newton's work, the ideas of thermodynamics, and Einstein's special theory of relativity, summarize virtually all of classical physics. In Chapters 28 to 34 we shall study magnetic phenomena, their connection to electrical phenomena, their practical applications, and other remarkable consequences of Maxwell's equations.

In this chapter we describe the laws of magnetic forces through experiments involving magnets and electric currents. We'll also learn how magnetic forces are associated with magnetic fields, just as electric forces are associated with electric fields. We will here concentrate on the effects of magnetic fields on test objects, leaving the description of how magnetic fields are generated for the next chapter.

28–1 Magnets and Magnetic Fields

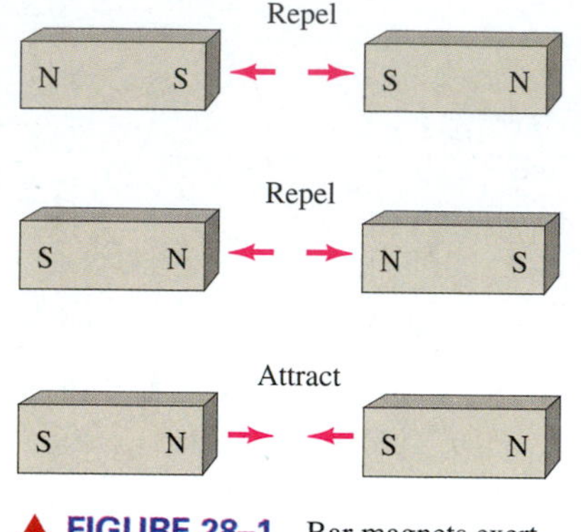

▲ **FIGURE 28–1** Bar magnets exert magnetic forces on each other.

When two bar **magnets** are brought close to each other, the forces between them—**magnetic forces**—become evident. These forces are of a type we have not yet encountered. In some positions they attract each other, while in other positions they repel (Fig. 28–1); in still other positions, they exert torques on each other. These forces suggest that bar magnets have an orientation, or an axis. We arbitrarily label the end of a magnet that is attracted to a point very near Earth's geographic South Pole as the *south pole*, S, while the other end of the magnet is called the *north pole*, N. If we experiment with two bar magnets labeled in this way, we find that the N end of one attracts the S end of the other, whereas the two N ends repel each other, as do the two S ends, as Fig. 28–1 shows. (The fact that the south pole of a bar magnet is attracted toward Earth's South Pole means that Earth's South Pole actually behaves like the north pole of a magnet! Similarly, Earth's North Pole behaves like the south pole of a magnet. These labelings are of course historical artifacts.) Based on our experience with charges, we might be tempted to conclude that a bar magnet contains "magnetic charges" (or *magnetic monopoles*) at each end and, further, that we could somehow extract them. A little bit of experimentation suggests that this is not possible; when you break a bar magnet in two, you end up with two bar magnets, not two separated magnetic charges. (Magnetic monopoles remain, nevertheless, the subject of modern searches.)

Iron and a few other materials (all known as *ferromagnetic* materials) have a particular property: If we place a piece of iron, for example, near a lodestone (a "natural" magnet), the piece of iron also becomes a magnet. Later we'll discuss the reasons why this occurs. For now we note that we can use this property to turn tiny shavings of iron (iron filings) into tiny magnets that can be used as test probes of magnetic forces.

Magnetic Fields

If we scatter iron filings on a sheet of plastic above a bar magnet, as in Fig. 28–2, the magnetic forces acting on them line up the iron filings in certain directions, which are dependent on their position in relation to the magnet. Moreover, the filings clump more densely in certain regions, such as near the poles. For different magnets, the scattered filings have different densities and alignments. Figure 28–2 shows the distribution of iron filings around a straight bar magnet, a horseshoe magnet, and—surprise!—a wire that carries current. This last example is a clue that magnetic forces are associated with moving charges as well as with magnets. The magnetic force acts at a distance, just like gravitational and electric forces. Just as an object with mass sets up a gravitational field and a charged object sets up an electric field, a magnet, a moving charge, or an electric current sets up a **magnetic field** throughout space. We denote this field with the symbol $\vec{B}$. The bar magnet, horseshoe magnet, and current-carrying wire each set up a characteristic magnetic field. Magnetic forces have a marked directional character, and the magnetic field, like the electric field, is a vector.

How do we find the magnetic field due to a magnet or a current? Just as an electric field can be mapped with a small test charge, the response of iron filings to the presence of a magnet or a current can be used to map out the magnetic field. The magnetic field $\vec{B}$ is oriented along the alignment direction of the filings, and the magnitude of $\vec{B}$ is proportional to the density of the filings. (Actually, the density-magnitude connection is only an approximation, which we will refine later.) We can picture the magnetic field by using magnetic field lines—continuous lines that run parallel to the direction of the field at every point and whose density (the number of lines per unit area) is proportional to the strength of the field, in analogy to electric field lines representing electric fields. As an example of this mapping process, take a look at Fig. 28–2a

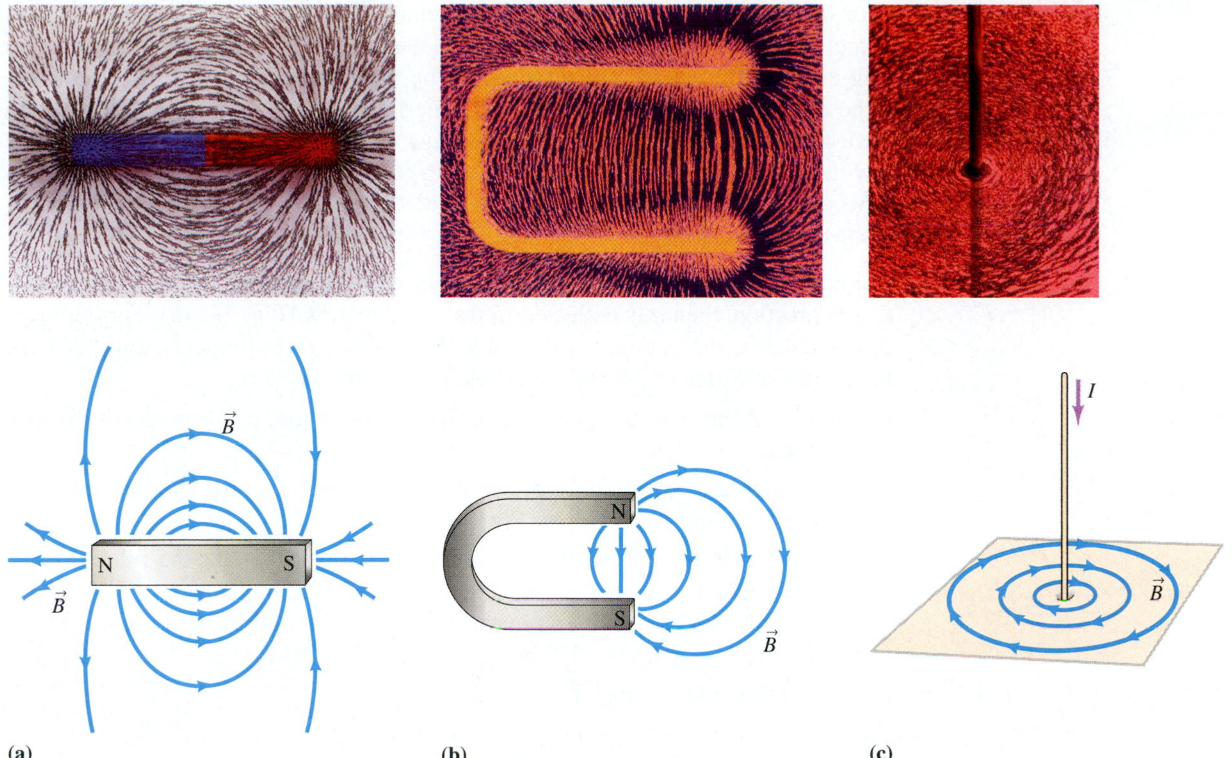

▲ **FIGURE 28–2** Iron filings map the magnetic field for (a) a straight bar magnet, (b) a horseshoe magnet, and (c) a current-carrying wire.

and b: The iron filings align themselves between the poles of magnets, and, therefore, the magnetic field lines associated with the magnet run from pole to pole. We take the direction of the magnetic field of a magnet to run *from the N pole to the S pole*, just as we assign the electric field to run from positive electric charges to negative charges. Notice, however, that the magnetic field around a current-carrying wire has no magnetic pole—it has no starting or ending point. Once we have mapped out a magnetic field in this way, or by the technique of Fig. 28–3, we can further investigate its effects and find the force laws associated with magnetism.

28–2 Magnetic Force on an Electric Charge

Experiments show that compass needles or iron filings are not the only objects to experience forces in the presence of magnetic fields. Moving charged particles also experience forces due to magnetic fields. The effect of a magnetic field on electric charge is most easily studied by using a bar magnet to deflect the electron beam of an oscilloscope. When a bar magnet is placed in different orientations near the beam, the beam deflects in various ways. The deflection allows us to measure the magnetic forces on the beam.

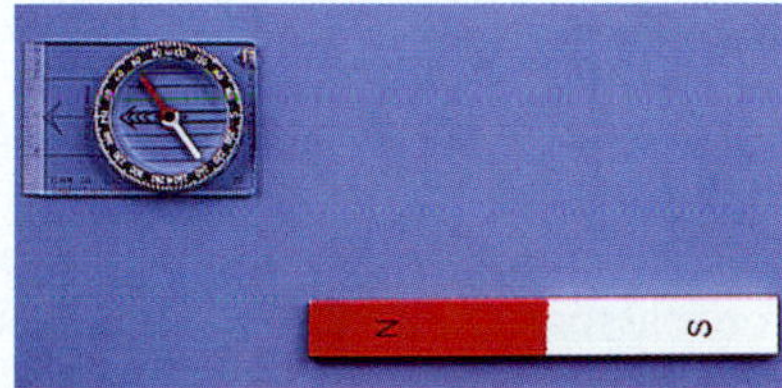

(a)

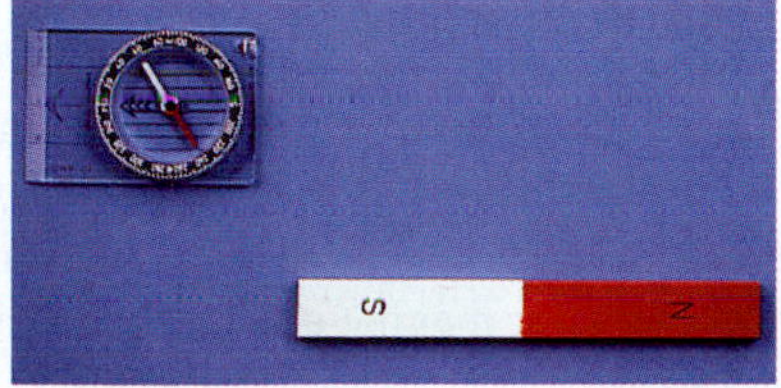

(b)

◀ **FIGURE 28–3** A compass needle is sensitive to a magnetic field and can be used to map the field due to a magnet.

Consider a bar magnet with its N pole oriented so that the magnetic field is in the $+y$-direction (Fig. 28–4a). The magnitude of the magnetic field at the position of a moving electric test charge q can be varied by altering the distance of the magnet from the charge. The results of a series of experiments are shown in Fig. 28–4, with the deflected particle chosen to have a *positive* charge. (If the actual experiment is done with an oscilloscope, keep in mind that the test charge, a moving electron, has a negative charge.) We observe the following (below the list, we summarize these observations in a different way):

1. If q moves at speed v in the $+z$-direction and the magnetic field points in the positive y-direction, then q is deflected in the $-x$-direction (Fig. 28–4b). Furthermore, the larger v is, the stronger is the force $\vec{F}$. Detailed measurements show that the magnitude of $\vec{F}$ due to the magnetic field is proportional to v.
2. If q moves in the $+x$-direction, $\vec{F}$ is in the $+z$-direction, again proportional to v (Fig. 28–4c).
3. If q moves in the y-direction ($+$ or $-$), there is no change in the charge's direction or speed; that is, there is no force (Fig. 28–4d).
4. If q moves at speed v in an arbitrary direction, $\vec{F}$ is proportional to the velocity component perpendicular to the magnetic field, $v_{\perp}$, and perpendicular to the directions of both $\vec{v}$ and $\vec{B}$. This result summarizes points 1 through 3. In particular, if the charge is at rest, so that $v = 0$, there is no force (Fig. 28–4e).
5. $\vec{F}$ is proportional to the magnitude of $\vec{B}$.
6. $\vec{F}$ is proportional to the sign and magnitude of q (Figs. 28–4f and 28–4g).

One important feature of this collection of results is the dependence on $\vec{v}$. In results 1 and 2, the initial velocity is purely in the z- or x-directions and is therefore perpendicular to $\vec{B}$. In result 4, there is no $\vec{v}$ and also no force. A stationary charge in a magnetic field experiences no force. Moreover, Figure 28–4 indicates that the force is always perpendicular to both $\vec{v}$ and $\vec{B}$.

To summarize:

The magnetic force $\vec{F}$ on a moving charge is proportional in magnitude to q, $v_{\perp}$, and B, where $\vec{v}_{\perp}$ is the velocity component perpendicular to the field, while the direction of $\vec{F}$ is perpendicular to both $\vec{B}$ and $\vec{v}$ and depends on the sign of q.

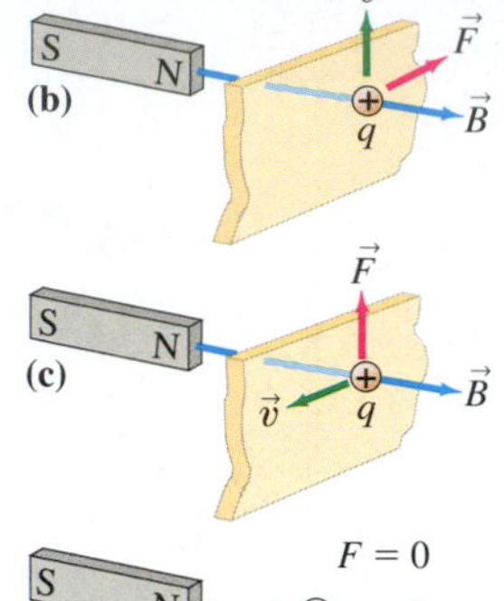

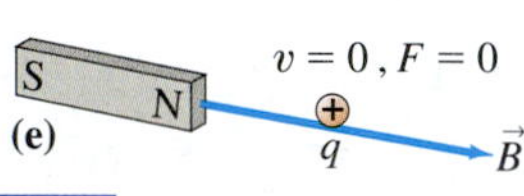

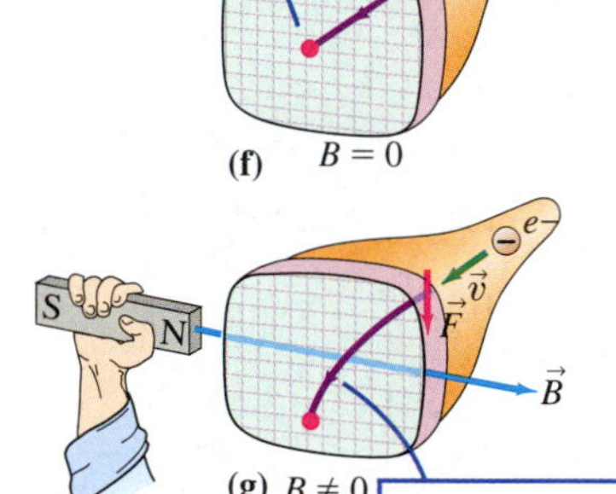

▲ **FIGURE 28–4** (a) The magnetic field of a bar magnet will be used to influence a moving charge. (b)–(e) Experiments on a positive charge moving in a magnetic field. (f) An oscilloscope can be used to measure the effects of a magnetic field on moving negative charges, here an electron in a cathode-ray tube. (g) In the orientation shown the electron is deflected down, whereas for the positive charge in part (c) the force was upward.

A direction perpendicular to both $\vec{v}$ and $\vec{B}$ can be represented by the *vector product*, or *cross product*, which was discussed extensively in Chapter 10 in connection with torques and rotational motion (see the problem-solving techniques box, p. 284). Recall that a vector $\vec{c} = \vec{a} \times \vec{b}$ (the vector product of $\vec{a}$ and $\vec{b}$) has magnitude $ab \sin\theta$, where θ is the angle between vectors $\vec{a}$ and $\vec{b}$ and is always taken to be less than 180° (Fig. 28–5). Vector $\vec{c}$ is perpendicular to both $\vec{a}$ and $\vec{b}$, in a direction determined by the right-hand rule. Thus our experiments have determined that the magnetic force on a test charge q moving with velocity $\vec{v}$ in a magnetic field $\vec{B}$ is given by

$$\vec{F} = q\vec{v} \times \vec{B}. \tag{28–1}$$

MAGNETIC FORCE LAW

This important result is the **magnetic force law**. If θ is the angle between vectors $\vec{v}$ and $\vec{B}$, the magnitude of $\vec{F}$ is given by

$$F = qvB \sin\theta = qv_{\perp}B. \tag{28–2}$$

Figure 28–6 shows how $\vec{F}$ is perpendicular to the plane formed by $\vec{v}$ and $\vec{B}$ according to a right-hand rule. Recall that the vector product of two *parallel* vectors is zero; this describes the fact that there is no magnetic force on a charge that moves along the axis of a bar magnet and no magnetic force associated with the component of $\vec{v}$ parallel to $\vec{B}$.

Equation (28–1) shows that the dimensions of the magnetic field are quite different from those of the electric field. The SI unit of magnetic field is called the *tesla* (T), in honor of Nikola Tesla, who made important contributions to the technology of electrical energy generation. In terms of previously defined SI units,

$$1\ \text{T} = 1\frac{\text{kg}}{\text{C}\cdot\text{s}}. \tag{28–3}$$

Another (non-SI) unit for the magnetic field that is in common use is the *Gauss* (G); 10^4 G = 1 T.

Table 28–1 contains some representative values of magnetic fields.

TABLE 28–1 • Some Magnetic Fields

Location or Source	Magnitude (T)
Interstellar space	10^{-10}
Near Earth's surface	5×10^{-5}
Refrigerator magnet for notes	10^{-2}
Bar magnet near poles	10^{-2}–10^{-1}
Near surface of Sun	10^{-2}
Large scientific magnets	2–4
Largest steady-state magnet	30
Largest pulsed field in laboratory	500–1000
Near surface of pulsar	10^8
Near surface of atomic nucleus	10^{12}

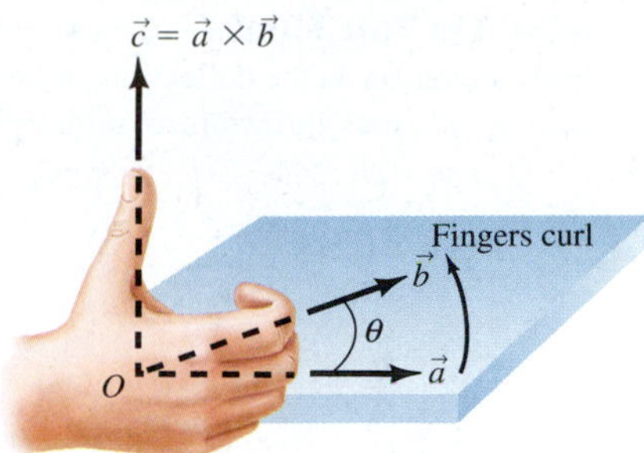

▲ **FIGURE 28–5** The vector product $\vec{c} = \vec{a} \times \vec{b}$ and the right-hand rule that determines the direction of the vector product. See Chapter 10, Fig. 10–5.

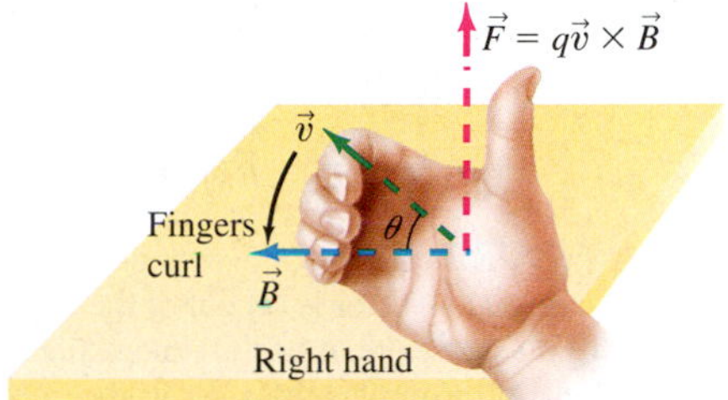

▲ **FIGURE 28–6** The right-hand rule for the magnetic force law $\vec{F} = q\vec{v} \times \vec{B}$.

CONCEPTUAL EXAMPLE 28–1 Consider the situation depicted in Fig. 28–4f, with the electron beam showing a spot on the oscilloscope screen. Suppose we now bring a bar magnet down from above to the top of the oscilloscope, with the S pole of the magnet closest to the beam. As we bring the magnet down slowly, how will the spot move on the screen?

Answer We see from Fig. 28–2a that our action effectively brings in an increasing B field that points up in the vertical direction. An application of the right-hand rule shows that the direction of the cross-product of the velocity and the field points in the left-side direction as seen when facing the screen. Note, however, that the electrons in the beam are *negatively* charged, so that the force points toward the right side of the screen, as seen from the front. The electrons are therefore accelerated to the right and will therefore strike the screen to the right of the original spot (always as seen from the front). As the magnet is slowly brought down from above, the field strength at the electron location increases, and the deflection will also increase, so the spot will be seen to move to the right.

EXAMPLE 28–2 The undisturbed electron beam of an oscilloscope moves along the x-direction. The north pole of a bar magnet approaches the cathode-ray tube from the right, as seen facing the screen, and deflects the beam. The magnitude of the magnetic field from the magnet is 0.050 T in the vicinity of the beam, and the speed of the electrons in the beam is 2.0×10^5 m/s. What is the magnitude of the magnetic force on the electrons? What is the direction of this force; that is, which way is the beam deflected?

Setting It Up We draw the situation in Fig. 28–7 and note that we are given both the magnitude of the velocity v and the magnitude of the magnetic field B, as well as their directions. The electron charge $-e$ is known.

Strategy We can obtain the direction of the force by using the right-hand rule, but to get some experience with the notion of vector products, we here set up a coordinate system and use unit vectors along the x-, y- and z-axes to work out the direction of the force. It is convenient to choose the direction of the beam as the positive x-axis. We also choose the horizontal line pointing to the right on the screen to be the y-axis, so that the conventional (right-hand) choice of z-axis is upward. We then simply apply the force law.

Working It Out We have $\vec{v} = v\hat{i}$ and $\vec{B} = -B\hat{j}$, and with the help of the right-hand rule for unit vectors, $\hat{i} \times \hat{j} = \hat{k}$, we find

$$\vec{F} = (-e)(v)(-B)\hat{k} = (-1.6 \times 10^{-19}\ \text{C})(2.0 \times 10^5\ \text{m/s})(-0.050\ \text{T})\hat{k}$$
$$= (1.6 \times 10^{-15}\ \text{N})\hat{k}.$$

The force is upward, and this is the direction in which the spot moves.

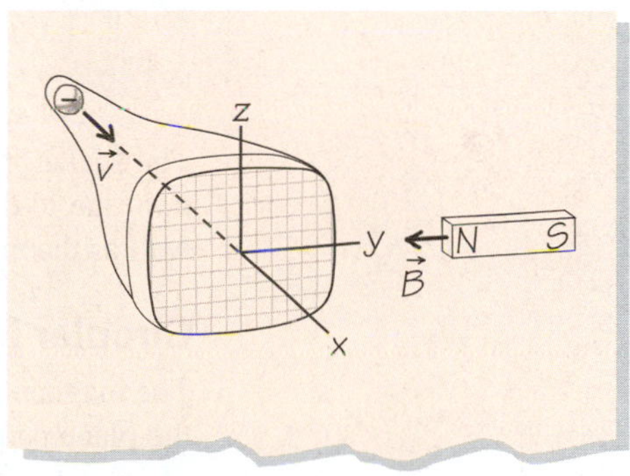

▲ **FIGURE 28–7** The magnetic field points away from the north pole of the bar magnet and is therefore oriented in the $-y$-direction. The spot at which the beam reaches the screen is deflected in the $+z$-direction.

(continues on next page)

What Do You Think? Given the force, we can easily calculate the acceleration. Is the deflection of the spot as simple to calculate as the drop of a rock thrown in an initially horizontal direction and subject to gravity? *Answers to* ***What Do You Think?*** *questions are given in the back of the book.*

The Lorentz Force

Further experimentation shows that charges react independently to electric and magnetic fields. Thus if an electric field is present in addition to a magnetic field, it produces an additional force $\vec{F} = q\vec{E}$ on a charge. The net force on a charged particle in an electric and a magnetic field is then

$$\vec{F} = q(\vec{E} + \vec{v} \times \vec{B}). \qquad (28\text{–}4)$$

LORENTZ FORCE LAW

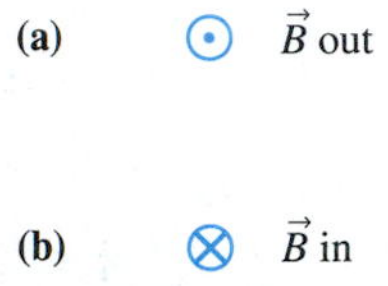

This equation is known as the **Lorentz force law**, named after the late-nineteenth-century physicist Hendrik A. Lorentz, who influenced the development of many areas of classical physics.

(a) ⊙ $\vec{B}$ out

(b) ⊗ $\vec{B}$ in

▲ FIGURE 28–8 Conventions to indicate that a vector is (a) out of the page or (b) into the page. A useful mnemonic is to think of the circle with a dot as the point of the arrow coming toward you, and the circle with an x as the feathered tail of the arrow moving away from you.

A Notation for Vectors Perpendicular to the Page

Because the three-dimensional aspect of magnetic forces is so important, it is useful to have a notation for vectors oriented perpendicular to the page. Figures 28–8a and 28–8b show a vector coming out of and going into the page, respectively. We shall often use this convention when we illustrate magnetic fields.

28–3 Consequences of the Magnetic Force on a Charge

Magnetic forces on charged particles have important implications that range from medical imaging devices to complicated phenomena in astrophysics and plasma physics. In this chapter, we'll keep things simple and assume that the magnetic fields are constant and do not change over time. When magnetic fields are independent of time, we are dealing with **magnetostatics**.

Energy of a Charged Particle in a Static Magnetic Field

Magnetic forces have a special property: *A charged particle moving in a static magnetic field has a constant kinetic energy*. One way to see this is to note that the force on a charge is perpendicular to its velocity, and hence to its infinitesimal displacement. The work done by the force is then zero, and by the work-energy theorem, the kinetic energy is unchanging. *A static magnetic field does no work on a charge*. Another way to see it is to calculate the rate of change of the kinetic energy. We have

$$\frac{d}{dt}\left(\frac{1}{2}m\vec{v}^2\right) = \frac{1}{2}m\frac{d}{dt}(v_x^2 + v_y^2 + v_z^2) = m\left(v_x\frac{dv_x}{dt} + v_y\frac{dv_y}{dt} + v_z\frac{dv_z}{dt}\right)$$

$$= m\vec{v}\cdot\frac{d\vec{v}}{dt} = m\vec{v}\cdot\vec{a}.$$

The scalar product on the right is zero, since magnetic forces, and hence the acceleration due to those forces, always point in a direction perpendicular to the velocity (as well as the magnetic field).

Circular Motion in a Constant Magnetic Field

The magnetic force law—Eq. (28–1)—states that only the component of the velocity in the plane perpendicular to $\vec{B}$ contributes to the expression for the force. *The component of the velocity of a charged particle parallel to the magnetic field is not affected by the field*, and it is therefore unchanging in the absence of any other forces. In addition, Eq. (28–1) states that *the force on the charge, and hence the charge's acceleration, is perpendicular to $\vec{B}$ and thus acts only in the plane perpendicular to $\vec{B}$.*

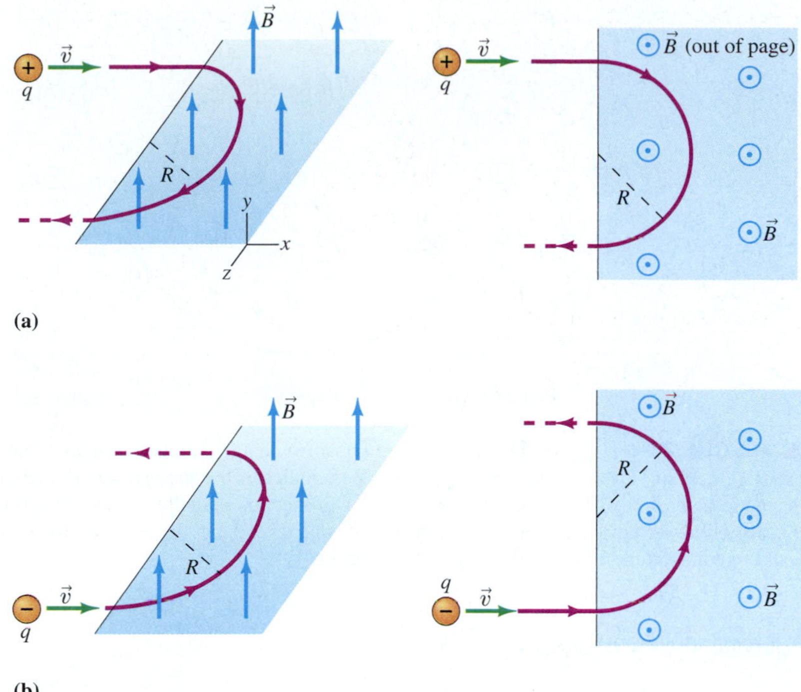

◀ **FIGURE 28–9** (a) A positively charged particle moves perpendicularly to a constant magnetic field, $\vec{B}$, shown in both an oblique perspective and a view from above. The charged particle traces a circular path in the plane perpendicular to $\vec{B}$, which is directed out of the plane of the page. (b) The direction of the curvature is opposite for a negatively charged particle.

To explore the consequences of these observations more closely, consider a magnetic field $\vec{B}$ that is uniform in some region of space and a test charge q that enters this region with a velocity $\vec{v}$ perpendicular to the field (Fig. 28–9). What is the consequent motion of the charge? According to Eq. (28–1), the force will be perpendicular to $\vec{v}$ and have magnitude $F = qvB$. We saw in Chapter 3 that when the acceleration (and hence the force) is constant in magnitude and perpendicular to the velocity, there is *circular motion at constant speed*. A charged particle moving perpendicularly to a constant, spatially uniform magnetic field will move in a circle (Fig. 28–9a). The magnitude of the acceleration for circular motion is $a = v^2/R$, where R is the radius of the particle's circular path; the direction of the acceleration is toward the center of the circular path. By Newton's second law, the force must have magnitude $ma = mv^2/R$. In our case, the force responsible for the acceleration has magnitude F, and Newton's second law ($F = ma$) becomes (Fig. 28–10)

$$F = qvB = \frac{mv^2}{R}.$$

We solve for R:

$$R = \frac{mv}{qB}. \tag{28–5}$$

R is proportional to the product of m and v—that is, to the momentum of the moving particle, $p = mv$—and is inversely proportional to the magnitudes of the charge q and of the field, $\vec{B}$. Equation (28–5) holds *only* when the velocity is perpendicular to $\vec{B}$. Whether the motion is clockwise or counterclockwise depends on the sign of the charge, according to the right-hand rule. Figure 28–9b depicts the motion for a test charge of opposite sign to the test charge in Fig. 28–9a. The larger B is, the larger is the magnetic force, and the "tighter" is the curved path, which corresponds to a smaller radius of curvature (the radius of the segment of a circle along which the charge moves at a given moment). The smaller the magnetic field, the smaller the force, and the larger is R. If the magnetic field varies in strength from place to place, then so will the radius of curvature of the path (Fig. 28–11).

The circular motion has a period $T = 2\pi R/v$, or, from Eq. (28–5),

$$T = \frac{2\pi}{v}\frac{mv}{qB} = \frac{2\pi m}{qB}. \tag{28–6}$$

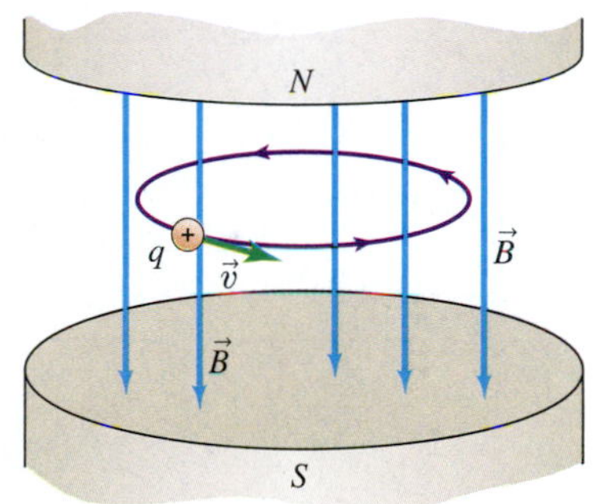

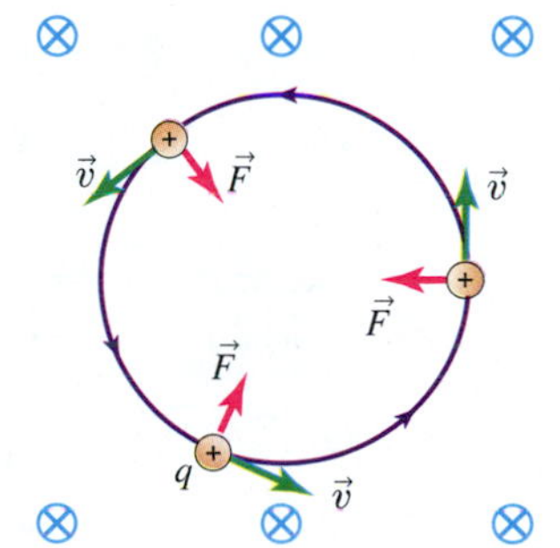

▲ **FIGURE 28–10** Forces on a charged particle that is moving perpendicularly to a uniform magnetic field.

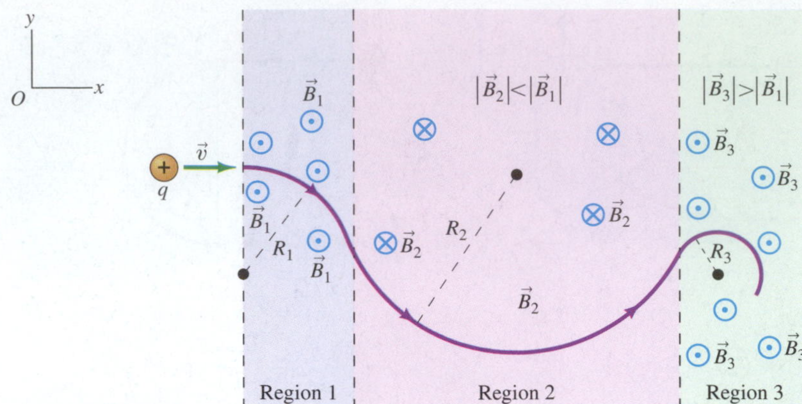

▲ **FIGURE 28–11** A particle of mass m and positive charge q moves with an initial velocity $\vec{v}$ in the x-direction. The curvature of the particle's path indicates the strength and direction of the magnetic field. In region 1, the magnetic field $\vec{B}_1$ is of medium magnitude and oriented in the $+z$-direction. In region 2, $\vec{B}_2$ is small in magnitude and oriented in the $-z$-direction. In region 3, $\vec{B}_3$ is of large magnitude and oriented in the $+z$-direction.

Equivalently, the frequency $f = 1/T$ is

$$f = \frac{qB}{2\pi m}. \tag{28–7}$$

CYCLOTRON FREQUENCY

This frequency is called the **cyclotron frequency**. Notice that the period and frequency are *independent* of the speed. A slow particle traces out a tight circle in the same time that a fast particle traces out a large circle. The constancy of the cyclotron frequency is a guiding principle of a device called the *cyclotron* (see Problem 31).

Equation (28–5), which specifies the radius of the circular path of a charged particle, has found application in many particle-detection devices—the *bubble chamber* illustrates the principle well. When charged particles produced in high-energy collisions speed through liquid hydrogen, they leave tracks that consist of very tiny bubbles, like a jet leaving a vapor trail in the atmosphere (Fig. 28–12). The momentum of these particles can be obtained by measuring the radius of curvature of their tracks when an external magnetic field is imposed. As we know from Chapter 8, information about momentum is helpful in deciphering collisions—in this case, the collisions of subatomic particles. (Although today bubble chambers are rarely used in such experiments, the more modern detectors that have replaced them rely on the same principles for the measurement of momentum.)

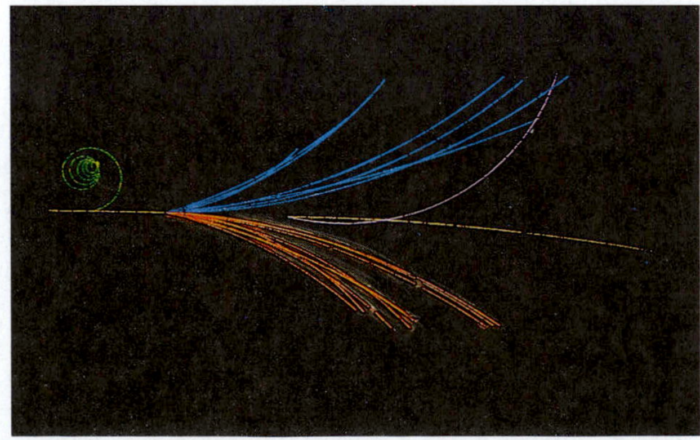

▲ **FIGURE 28–12** The tracks left by charged particles moving through a bubble chamber in a magnetic field. The colors are computer generated.

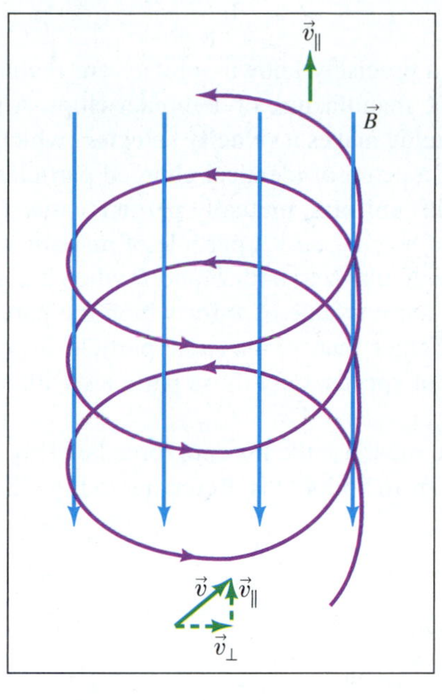

(a)

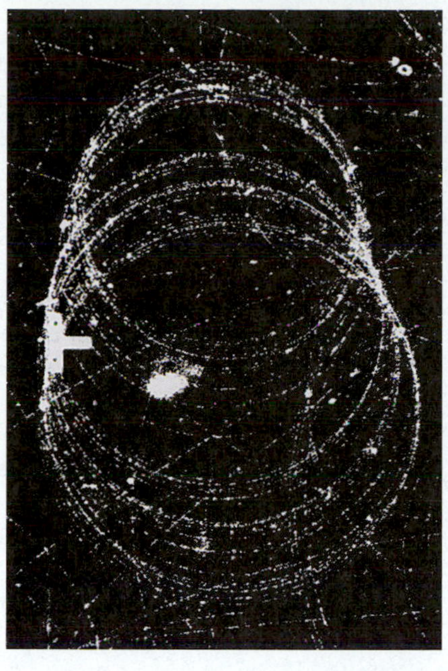

(b)

◀ **FIGURE 28–13** (a) A charged particle follows a helical path in a region where the magnetic field is constant. (b) An electron in a cloud chamber produced this 10-m-long spiral track. The electron's path begins at the bottom. The helix becomes more tightly wound about halfway up because the electron loses energy by radiation while moving in the helical path.

Finally, when there is a component $v_{\parallel}$ of velocity $\vec{v}$ that lies along $\vec{B}$, that component of the velocity does not change. The particle advances along $v_{\parallel}$ while it moves in a circle in the plane perpendicular to $v_{\parallel}$ (Fig. 28–13a). The resulting trajectory forms a *spiral* (or *helix*), with its axis along $\vec{B}$ (Fig. 28–13b). The circular motion in the plane perpendicular to $\vec{B}$ has a radius (the "radius of curvature") given by

$$R = \frac{mv_{\perp}}{qB}. \qquad (28\text{–}8)$$

EXAMPLE 28–3 A particle of unknown charge q and unknown mass m moves at speed $v = 4.8 \times 10^6\ \text{m/s}$ in the $+x$-direction into a region of constant magnetic field. The field has magnitude $B = 0.50\ \text{T}$ and is oriented in the $+y$-direction. The particle is deflected in the $+z$-direction and traces out a segment of a circle of radius $R = 0.10\ \text{m}$. What is the sign of the particle's charge, and what is the ratio q/m?

Setting It Up The situation described is drawn in Fig. 28–14.

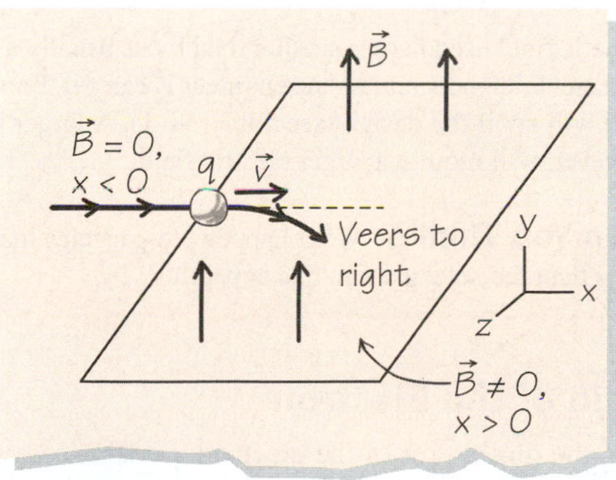

▲ **FIGURE 28–14** If the speed of a particle of unknown charge and mass that moves in a region of constant magnetic field is known, then a measurement of the radius of curvature of the particle's path gives the charge-to-mass ratio of the particle.

Strategy Given the velocity and the magnetic field, we can find the force. The acceleration will be in the same direction as the vector product $\vec{v} \times \vec{B}$ when the charge is positive, and opposite to the vector product if the charge is negative. The information on the deflection of the particle thereby tells us about the charge's sign. For the ratio q/m, we use the equation of motion, which links this ratio to the radius and the field magnitude if the field is constant, as in Eqs. (28–8) or (28–5).

Working It Out The standard rules for vector products give $\vec{v} \times \vec{B} = vB\hat{i} \times \hat{j} = vB\hat{k}$. For a positive charge this would lead to a deflection in the $+z$-direction, and since the observed deflection is in that direction, the *charge is positive*.

To calculate q/m, we use Eq. (28–5), which gives

$$\left|\frac{q}{m}\right| = \frac{v}{BR} = \frac{4.8 \times 10^6\ \text{m/s}}{(0.50\ \text{T})(0.10\ \text{m})} = 9.6 \times 10^7\ \text{C/kg}.$$

It is possible that this particle is a proton. We can see why if we assume that the unknown charge is that of a proton, $q = 1.6 \times 10^{-19}\ \text{C}$. Then $m = q/(9.6 \times 10^7\ \text{C/kg}) = (1.6 \times 10^{-19}\ \text{C})/(9.6 \times 10^7\ \text{C/kg}) = 1.7 \times 10^{-27}\ \text{kg}$, which is just the mass of a proton. Note, however, that the experiment described in this example can measure only the charge-to-mass ratio, not the charge or the mass alone.

What Do You Think? A deuteron is one type of hydrogen nucleus. It has the charge of a proton and (approximately) the mass of two protons. What fraction of the proton's speed would a deuteron have if it followed exactly the same path as the protons in the example above?

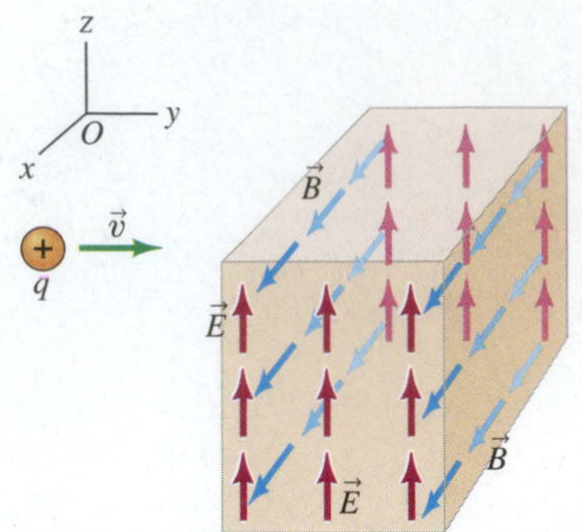

▲ **FIGURE 28–15** A charged particle enters a region of crossed electric and magnetic fields. If the speed of the particle is $v = E/B$, the particle will cross the region undeflected.

Velocity Selectors

Beams of charged particles moving with a precisely known velocity are required in a wide range of applications, including in the manufacture of computer chips. A particular arrangement of electric and magnetic fields makes a **velocity selector**, which passes only particles of a specific velocity out of a beam of identical charged particles with a variety of velocities. Consider a region with uniform, mutually perpendicular $\vec{E}$ and $\vec{B}$ fields (Fig. 28–15). (Such fields are said to be *crossed*.) A particle of mass m, charge q (positive), and velocity $\vec{v}$ is directed perpendicularly to both $\vec{E}$ and $\vec{B}$ when it enters this region. We will now show that there is a certain value of v for which the particle traverses the region undeflected. At a speed other than v, the same particle *is* deflected; thus, in a beam of particles with a variety of speeds, only those particles with a certain speed pass through undeflected.

Both $\vec{E}$ and $\vec{B}$ fields are present, so we must use the Lorentz force law [Eq. (28–4)] and compute the contributions to the force from both fields. Referring to Fig. 28–15, the electric force is

$$\vec{F}_E = qE\hat{k}.$$

By the right-hand rule, the magnetic force $q\vec{v} \times \vec{B}$ is

$$\vec{F}_B = -qvB\hat{k}.$$

The electric and magnetic forces will cancel if their magnitudes are equal, because they point in opposite directions; in this case, the particle will travel undeflected. This cancellation occurs for $qvB = qE$, so the speed of a charged particle that passes through the crossed fields undeflected is

$$v = \frac{E}{B}. \tag{28–9}$$

If we reverse the sign of charge q, the electric force points in the $-z$-direction while the magnetic force points in the $+z$-direction; the forces still cancel when v is given by Eq. (28–9).

EXAMPLE 28–4 A hole is cut in the wall of a container of a plasma (matter that is composed of positively charged ions and electrons acting as a kind of gas), creating a beam of particles of various charges. It is necessary to choose particles from the beam that have a speed of 3.2×10^6 m/s in order to carry out some materials testing. The engineer designing the speed selector uses crossed $\vec{E}$ and $\vec{B}$ fields, and the $\vec{B}$ field comes from a magnet with field strength $B = 6.5 \times 10^{-4}$ T. What must the engineer choose as the magnitude of the electric field? Will all the particles so chosen have the same momentum and energy?

Strategy With B and v given, there is enough information to calculate the magnitude of the electric field required, and we can use the relation $v = E/B$. Since that relation is independent of the charge q and the mass m of the particle, the particles selected by the apparatus are only characterized by the velocity v. The momentum and the energy of a particle depends not only on its velocity but on the mass of the particle, and this is not given, so we have no information on the momentum and energy. Thus, for example, electrons will be chosen in the same way as positive ions of a variety of charges. Both momentum and energy will take on a variety of values. This fact would make the particles so chosen not so useful to test materials, because their effect on materials depends more on their momentum and energy than on their speed.

Working It Out The relation in Eq. (28–9) leads to

$$E = vB = (3.2 \times 10^6 \text{ m/s})(6.5 \times 10^{-4} \text{ T}) = 2.1 \times 10^3 \text{ V/m}.$$

The magnetic field used here is smaller than those usually associated with permanent magnets; unless the engineer is careful, Earth's magnetic field will spoil the design (see Table 28–1). A larger magnetic field, however, will require a larger electric field.

What Do You Think? What happens to particles that move a little faster than the selected v in this apparatus?

The Charge-to-Mass Ratio of the Electron

In 1897 Sir Joseph John Thomson, the discoverer of the electron, performed a series of wide-ranging experiments whose results were crucial in the development of our understanding of the electrical nature of matter. A velocity selector was an important component of his experiment to measure the charge-to-mass ratio of the electron (Fig. 28–16). He first accelerated electrons in an electric field—not the electric field of the velocity selector—by passing them through an electric potential V. The work thereby done on

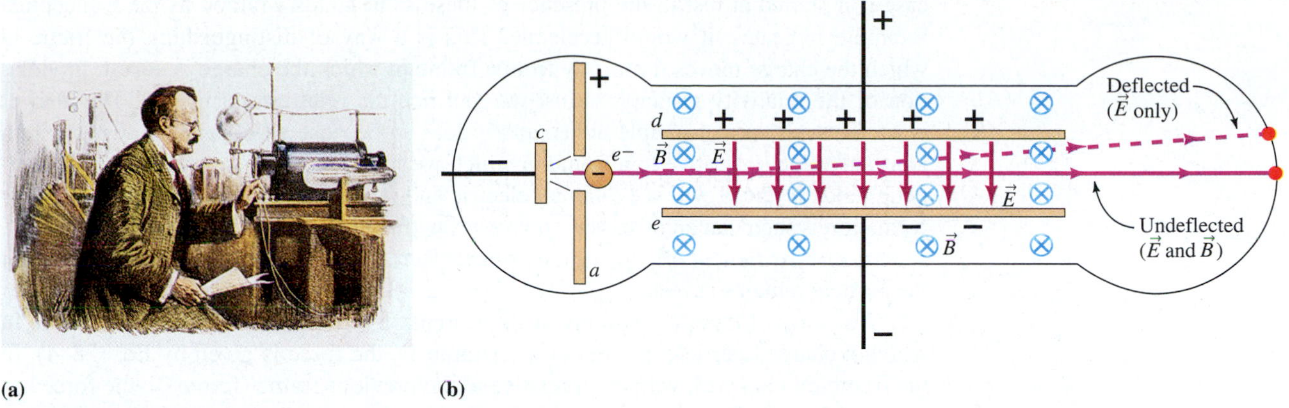

(a) (b)

▲ **FIGURE 28–16** (a) J. J. Thomson at work in his laboratory. (b) Schematic diagram of Thomson's apparatus for measuring the charge-to-mass ratio of the electron. The magnetic field is directed into the plane of the page.

the electrons is qV. Assuming they started from rest, the electrons gained a speed v determined by $mv^2/2 = qV$, so

$$v = \sqrt{\frac{2qV}{m}}. \qquad (28\text{–}10)$$

The electrons accelerated in this way continued into a region of crossed electric and magnetic fields. Thomson adjusted the magnitudes of these fields until the electrons passed through the apparatus undeflected. When we combine Eqs. (28–9) and (28–10), we find that

$$v = \frac{E}{B} = \sqrt{\frac{2qV}{m}}.$$

When both sides of this equation are squared, we can solve for q/m:

$$\frac{q}{m} = \frac{E^2}{2VB^2}. \qquad (28\text{–}11)$$

The electrons in Fig. 28–16b are accelerated between plates c and a. They travel on to the region between plates d and e where the magnetic field was adjusted until electrons were undeflected. The electric field and the required magnetic field may be used in Eq. (28–11) to obtain q/m. More refined experiments based on Thomson's scheme have led to a value for q/m of 1.759×10^{11} C/kg.

*The Mixing of Electric and Magnetic Fields

CONCEPTUAL EXAMPLE 28–5 Is there an arrangement of non-zero $\vec{E}$ and $\vec{B}$ fields that will leave a particle at rest if it starts at rest?

Answer This question has a simple answer: When a particle is at rest, so that $v = 0$, there is no magnetic force on the particle. In the presence of a non-zero electric field, there will be an electric force and therefore an acceleration. The particle cannot remain at rest unless the electric field is zero.

The Conceptual Example above is more interesting than it might appear. In Section 4–2 we described an overriding law of physics, what we referred to as the relativity principle: Observers in different inertial frames—frames that move at constant velocity with respect to one another—cannot decide by any experiment which of them is at rest and which is moving. In particular, observers in different inertial frames all see the same forces. But above we constructed a velocity selector, in which a combination of electric and magnetic fields produces no net force on a moving charge. The charge moves without acceleration in the fields of the velocity selector, yet this would not be the case if it started at rest in the presence of these same fields—rather, as the Conceptual Example indicates, it would accelerate. This is a way of distinguishing the frame in which the charge moves uniformly from a frame in which the charge is at rest, in violation of the relativity principle. Must we sacrifice the relativity principle? The answer is no. The relativity principle holds, and it does so because an observer moving with a constant velocity relative to a frame in which we have given constant electric and magnetic fields $\vec{E}$ and $\vec{B}$ will see *different* electric and magnetic fields $\vec{E}'$ and $\vec{B}'$. The transformation is such that an observer who moves with the charge, and thereby sees it at rest, sees *no electric field at all*, that is, the *primed* electric field vanishes. This guarantees that the particle remains at rest.

We can go farther by looking at the Lorentz force law, Eq. (28–4). In a frame in which a charged particle has velocity $\vec{v}$ (frame 1), the force is given by Eq. (28–4). In the frame moving with velocity $\vec{u}$ relative to the previous frame (frame 2), the force has the same form as Eq. (28–4), but with the changes $\vec{E} \rightarrow \vec{E}'$; $\vec{B} \rightarrow \vec{B}'$; $\vec{v} \rightarrow \vec{v} - \vec{u}$. Since according to the relativity principle the forces are the same in the two inertial frames, we must have

$$\vec{E} + \vec{v} \times \vec{B} = \vec{E}' + (\vec{v} - \vec{u}) \times \vec{B}'.$$

Since this holds whatever $\vec{v}$ is, we conclude that

$$\vec{B}' = \vec{B}, \quad \vec{E}' = \vec{E} + \vec{u} \times \vec{B}.$$

For our velocity selector, the combination $\vec{E} + \vec{v} \times \vec{B}$ is zero. We now see that if $\vec{u}$ is chosen to equal $\vec{v}$, so frame 2 is the frame in which the particle is at rest, then the electrical force on the object, $q\vec{E}'$, is also zero. This shows that the Lorentz force is not merely the sum of two forces, electrical and magnetic, that are independent of each other, but that it involves both the electric and the magnetic field intrinsically intertwined. The relativity principle inevitably leads to a close relation between electric and magnetic forces.

We conclude this brief discussion with a caveat that the results we have described are only approximate. When speeds are comparable to the speed of light, the discussion needs refinement. We'll discuss this further in Chapter 39.

Magnetic Fields in Outer Space

Magnetic fields exist in outer space. Throughout our galaxy, the magnetic field strength is in the range of 10^{-10} T (although the fields also have a great deal of structure). Charged particles (*cosmic rays*) are generated and accelerated by various stellar processes. If their momentum is less than a certain critical value p_c, they drift in gigantic circles within the galaxy due to the magnetic forces on them.[†] Cosmic rays with a momentum greater than p_c move on a circle with a radius of curvature greater than the galaxy's radius, and they therefore escape the galaxy. To estimate p_c for a cosmic ray whose charge has the magnitude e of the electron charge, we use the observation that the radius of the galaxy is about 5×10^{21} m. From Eq. (28–5), the critical momentum has magnitude

$$p_c = eBR = (1.6 \times 10^{-19}\,\text{C})(10^{-10}\,\text{T})(5 \times 10^{21}\,\text{m}) = 8 \times 10^{-8}\,\text{kg}\cdot\text{m/s}.$$

For a particle such as an electron or a proton, this momentum is enormously large. For comparison, an electron in the beam of a television picture tube typically has a momentum of 10^{-22} kg·m/s, while protons in the huge Fermilab proton accelerator attain momenta of 5×10^{-16} kg·m/s. Because cosmic rays with a momentum greater than p_c leave the galaxy, we should expect to detect more cosmic rays that strike Earth with momenta lower than p_c than with momenta greater than p_c. Experimental observations of particles arriving from outer space help us to estimate the value of the interstellar magnetic field.

[†] We use momentum rather than speed because a calculation of the speed here gives a critical speed greater than the speed of light. This indicates that special relativity is necessary, and special relativity shows that momentum should be used here.

THINK ABOUT THIS. . .

HOW DOES EARTH'S MAGNETIC FIELD CREATE THE NORTHERN LIGHTS?

Earth has a magnetic field like that of a huge bar magnet, directed from the geographic South Pole to the geographic North Pole. As Fig. 28–17 shows, the lines get denser near the poles. (The field extends down into the atmosphere, but ions within the atmosphere tend to be affected more by the presence of the atmosphere than by Earth's magnetic field.) The field interacts with charged particles, one evidence of which is the aurora, also known in the northern hemisphere as the northern lights. Charged elementary particles reach Earth from two sources. One is the *solar wind*, which consists of electrons and positive ions ejected from the sun. Most of these particles are of relatively low energy, and in the vicinity of Earth they are trapped by the magnetic field, moving in spiral paths along Earth's magnetic field lines. The helical orbits of particles around the field lines get flattened as the lines approach each other near the pole, and ultimately the particles are turned around. The mechanism for the reversal is known as a **magnetic mirror** (the process has to do with magnetic fields of varying strength; we'll look at the dynamics of this situation in Chapter 29.). In this way, the trapped charged particles, mainly electrons, bounce back and forth between the poles creating the *outer Van Allen radiation belt*. Particles also reach Earth as **cosmic rays**, consisting of very energetic particles coming from distant space. Some of these lose energy by collisions with atmospheric particles, and if they are charged, they get trapped along magnetic field lines closer to Earth than the outer Van Allen belt. Electrons and protons originating with cosmic rays have spiral paths wide enough that they touch the atmosphere and, because of collisions with atmospheric molecules, don't stay around very long. The neutral particles, *neutrons*, which are produced in the high-energy collisions of the cosmic rays with the atmosphere, will eventually decay into protons, electrons, and neutrinos. The protons that come from some of these neutron decays are captured by the magnetic field lines lying relatively close to Earth, and they also accumulate slowly to form the stable *inner Van Allen radiation belt*. The outer belt averages about 15,000 km from the surface of Earth, the inner one about 3000 km, although both approach Earth more closely in the polar regions.

An aurora occurs (see photo on page 791) when a sufficient number of charged particles from the belts (or directly from the solar wind) enter the atmosphere and excite the molecules in the air through collisions. When the excited atoms decay back to their normal configuration, they emit radiation with wavelength (i.e. color) that is characteristic of the state of excitation (see Chapter 41). The auroras are more noticeable near the poles, where the magnetic field lines, and therefore the Van Allen belts, dip toward Earth.

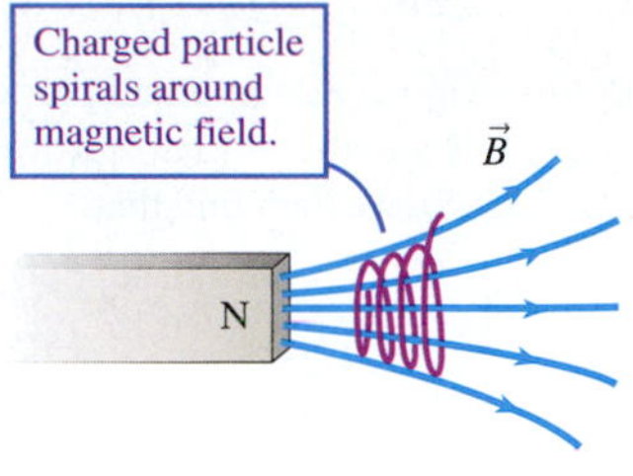

(a)

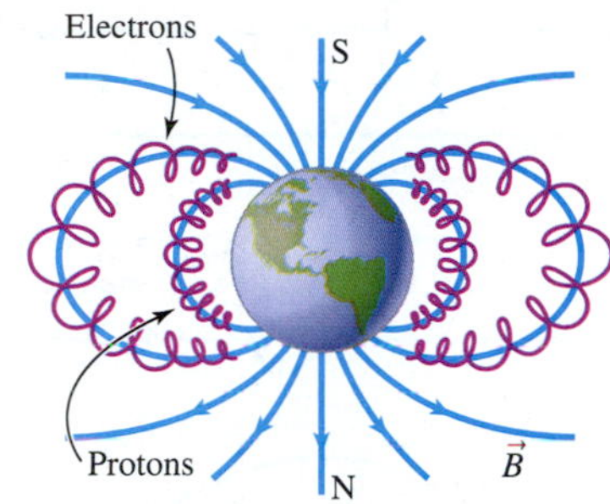

(b)

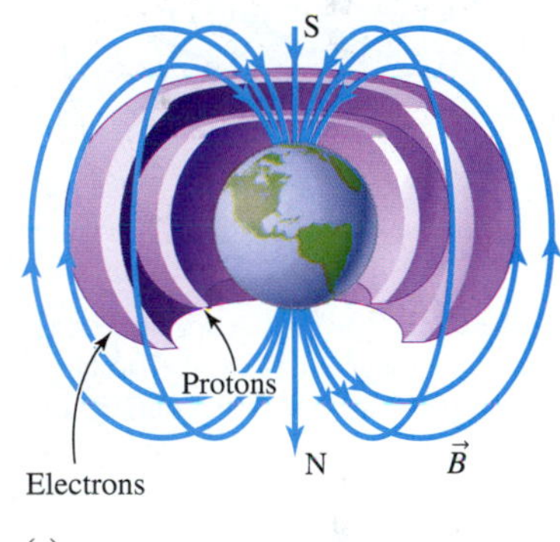

(c)

FIGURE 28–17 (a) Charged particles spiraling around the varying magnetic field near the pole of a magnet. The pitch, or inclination, of the helix decreases to zero as particles near the poles, where the field lines are denser, and the particles reverse the direction of their helical path. (b) When this is applied to particles spiraling around Earth's magnetic field, we see that the particles are trapped, bouncing back and forth between the polar regions. The trapped particles form the two Van Allen belts, one for electrons and one for protons. (c) A three-dimensional view of the magnetic field lines and of the belts.

EXAMPLE 28–6 Assume that a proton of speed 1.5×10^7 m/s approaches Earth at an angle of 40° to Earth's magnetic field lines and is captured in the lower Van Allen belt (at a mean altitude of 3000 km) without a change in speed. If the mean strength of the field at this altitude is 10^{-5} T, find the cyclotron frequency and the radius of curvature for the circular part of the proton's helical motion.

Setting It Up Figure 28–18 shows the path of the proton. The curvature is the radius of the circular part of the motion. Given the angle of approach of 40° and the speed, we know the speed perpendicular to the magnetic field. We also know the field strength B and therefore have all the data necessary for the calculation.

Strategy The cyclotron frequency, Eq. (28–7), depends only on B and on the proton mass $m_p = 1.67 \times 10^{-27}$ kg and charge $e = +1.6 \times 10^{-19}$ C. For the calculation of the curvature of the proton's motion, we note that the proton does not change speed when it starts its helical motion along the magnetic field line, whose B value is given. We can use Eq. (28–8). Given a speed v, the component of velocity that is perpendicular to the direction of B is $v \sin 40°$, and it is this quantity that appears in Eq. (28–8).

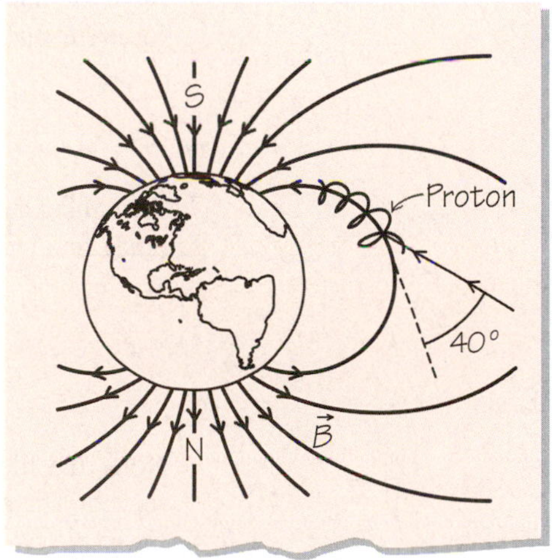

FIGURE 28–18 A proton approaches and passes into Earth's magnetic field.

(continues on next page)

Working It Out The cyclotron frequency is

$$f = \frac{qB}{2\pi m} = \frac{(1.6 \times 10^{-19}\,\text{C})(10^{-5}\,\text{T})}{2\pi(1.67 \times 10^{-27}\,\text{kg})} = 150\ \text{Hz}.$$

The proton's velocity component perpendicular to the magnetic field is $v_\perp = v \sin 40° = (1.5 \times 10^7\ \text{m/s})(0.64) = 9.6 \times 10^6\ \text{m/s}$. Using Eq. (28–8), we then find that

$$R = \frac{mv_\perp}{qB} = \frac{(1.67 \times 10^{-27}\,\text{kg})(9.6 \times 10^6\,\text{m/s})}{(1.60 \times 10^{-19}\,\text{C})(10^{-5}\,\text{T})} = 10^4\ \text{m} = 10\ \text{km}$$

The radius of curvature is much less than the altitude of the Van Allen belt. Since it is collisions with the atmosphere that tend to be most effective at removing particles from the belt, this proton is likely to stay within the belt for a while.

28–4 Magnetic Forces on Currents

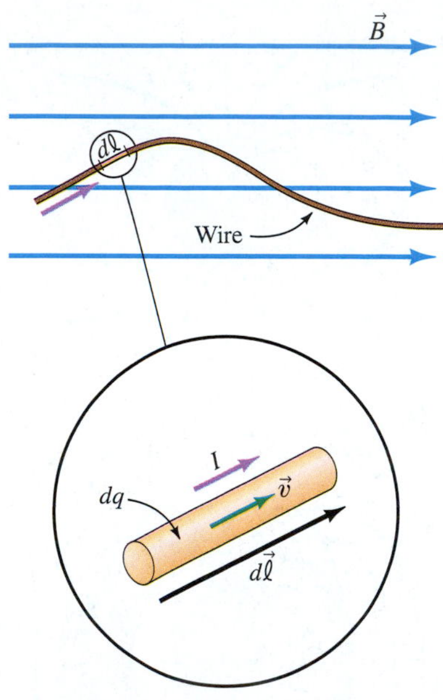

▲ **FIGURE 28–19** A wire carrying a current in a magnetic field. We isolate a segment $d\vec{\ell}$ of the wire that contains the moving charge dq.

In the previous sections, we have seen that moving charges in a magnetic field may experience a force due to the field. Because electric currents in wires consist of moving charges, we can expect that a magnetic field will exert a force on the charges in a current-carrying wire, and thus on the wire itself (Fig. 28–19), and experiment shows this is the case.

A wire contains moving charges throughout its length, and a magnetic field may vary significantly along its length. The total force on a current-carrying wire is the vector sum of the magnetic forces on all of the moving charges within it. To find the total force, we first determine the force on a small segment of a current-carrying wire, and subsequently sum (integrate) the infinitesimal force on each segment.

Magnetic Forces on Infinitesimal Wires with Currents

Let's denote the small segment of a thin current-carrying wire by $d\vec{\ell}$. It has both an infinitesimal magnitude $d\ell$ and a direction along the instantaneous current carried by the wire at the location of the segment. If the moving charge dq contained in the segment has velocity $\vec{v}$ along the wire (Fig. 28–19), its displacement $d\vec{\ell}$ in time dt is $d\vec{\ell} = \vec{v}\,dt$ so

$$\vec{v} = \frac{d\vec{\ell}}{dt}. \tag{28–12}$$

Because the current, I, is dq/dt by definition, the amount of moving charge within the segment is

$$dq = I\,dt. \tag{28–13}$$

Note that the magnetic field will be uniform over the length of the segment if the segment is small enough. With Eqs. (28–12) and (28–13) we can calculate the magnetic force $d\vec{F}$ that acts on our charge element dq and hence on the wire element:

$$d\vec{F} = dq(\vec{v} \times \vec{B}) = (I\,dt)\left(\frac{d\vec{\ell}}{dt} \times \vec{B}\right).$$

We cancel the factor dt to find the infinitesimal force on a wire element $d\vec{\ell}$ carrying current I in a magnetic field $\vec{B}$:

$$d\vec{F} = I\,d\vec{\ell} \times \vec{B}. \tag{28–14}$$

MAGNETIC FORCE ON WIRE SEGMENT

Note that the current is the same everywhere along the wire because current is conserved. The magnitude of the magnetic force $d\vec{F}$ is given by

$$dF = I(d\ell)B \sin\theta, \tag{28–15}$$

where θ is the angle between the direction of the wire segment (the current's direction) and the direction of the magnetic field.

CONCEPTUAL EXAMPLE 28–7 What is the direction of the force on a current-carrying wire (a) perpendicular to and (b) parallel to the direction of a constant magnetic field?

Answer We sketch the three different placements of a segment of current-carrying wire in a uniform magnetic field that points in the $+x$-direction in Fig. 28–20. In each case, the direction of the force on the wire segment is given by the right-hand rule and Eq. (28–14). In Fig. 28–20a, $d\vec{\ell}$ points in the $+y$-direction, so that the force points in the $-z$-direction. In Fig. 28–20b, $d\vec{\ell}$ points in the $+z$-direction, so the force points in the $+y$-direction. In Fig. 28–20c, $d\vec{\ell}$ points in the $+x$-direction, parallel to $\vec{B}$, and since the vector product of two parallel vectors is zero, there is no force.

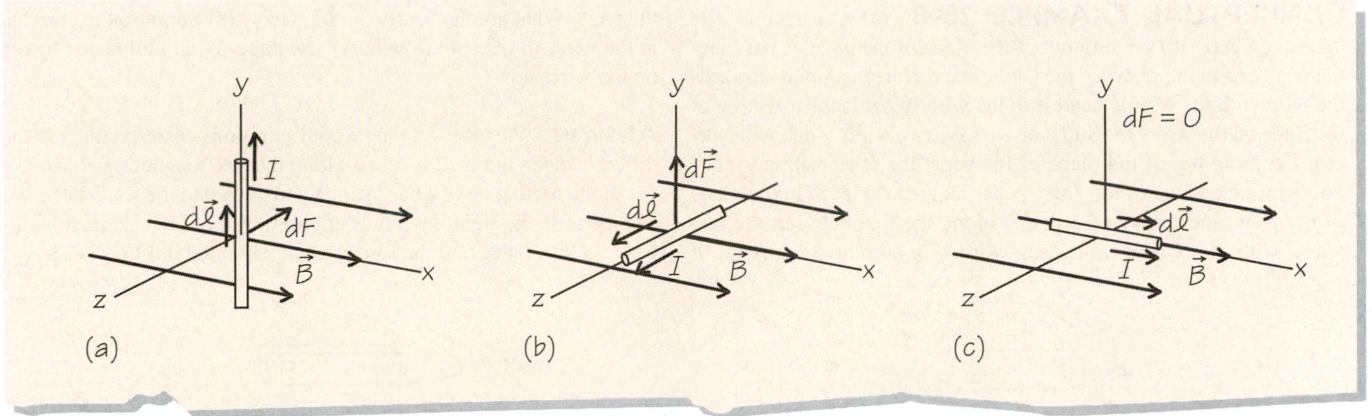

▲ **FIGURE 28–20** A wire segment, aligned with the (a) y-axis, (b) z-axis, and (c) x-axis, in a magnetic field aligned along the $+x$-direction, with the infinitesimal forces $d\vec{F}$ that act on the segment indicated.

Magnetic Forces on Finite Wires with Currents

The *net* force $\vec{F}$ on a *finite* section of wire is the vectorial sum of the forces on the various infinitesimal segments that make up the wire. We find the net force by integrating $d\vec{F}$ [Eq. (28–14)] over the total length of the wire. Because I is the same everywhere along the wire, the summation of Eq. (28–14) has the form

$$\vec{F}_B = I\int (d\vec{\ell} \times \vec{B}). \tag{28–16}$$

Whether the integral is easy to perform or not depends on the particular situation. A straight wire within a constant magnetic field represents an important special case. Let's suppose first that the wire, with length L and current I, is oriented perpendicular to the field (Fig. 28–21a). To perform the integration of Eq. (28–16), notice that each segment $d\vec{\ell}$ of the wire points in the y-direction, $d\vec{\ell} = d\ell\,\hat{j}$, and $\vec{B}$ is constant at each segment, $\vec{B} = B\hat{i}$. The infinitesimal force on each segment is therefore identical:

$$d\vec{F} = I[(d\ell\,\hat{j}) \times (B\hat{i})] = I(d\ell)B(\hat{j} \times \hat{i}) = I(d\ell)B(-\hat{k}),$$

which is a vector that points in the $-z$-direction. We can easily check the direction with the right-hand rule. The net force $\vec{F}$ is an integral over $d\vec{F}$:

$$\vec{F} = \int d\vec{F} = I\int d\ell B(-\hat{k}) = IB(-\hat{k})\int_0^L d\vec{\ell} = -IBL\hat{k}. \tag{28–17}$$

The net force points in the $-z$-direction (here, into the paper), and its magnitude is IBL.

Next, suppose that the wire makes an angle θ to the field (Fig. 28–21b). The only change we must make is in the expression for the magnitude of the infinitesimal force element: There is an additional factor $\sin\theta$ in dF [Eq. (28–15)]. The direction of the force remains in the $-z$-direction by the right-hand rule. Because θ is constant, it does not enter into the integration, which thus remains as in Eq. (28–17). The result is identical to Eq. (28–17) with an additional factor of $\sin\theta$:

$$\vec{F} = -IBL\sin\theta\,\hat{k}. \tag{28–18}$$

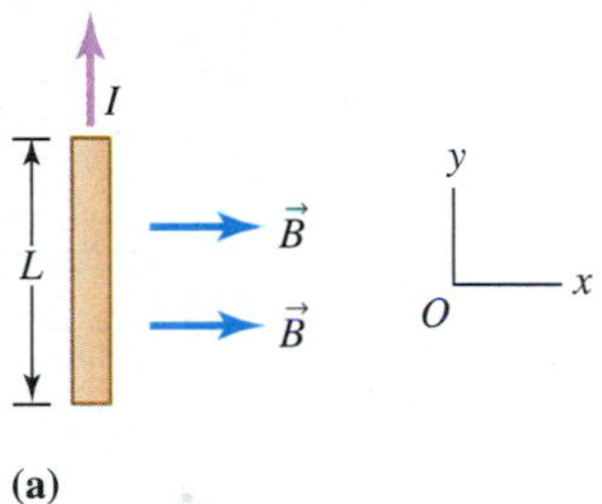

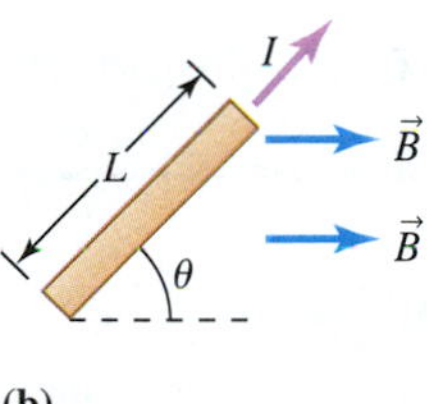

▲ **FIGURE 28–21** (a) A wire segment is oriented perpendicularly to a magnetic field. (b) The same wire segment at an angle to the field.

Equations (28–17) and (28–18) are useful and important results. They can be combined into one vectorial equation that gives the force on a thin, straight wire of length L in a magnetic field, namely,

$$\vec{F} = I\vec{L} \times \vec{B}, \qquad (28\text{–}19)$$

where the vector $\vec{L}$ is oriented along the wire in the direction of the current.

CONCEPTUAL EXAMPLE 28–8 (a) Consider a wire carrying a current I coming out of the plane of the page. A bar magnet is placed in the plane of the page. The bar magnet points towards the wire with the N pole nearest to the wire. What is the direction of the force on the wire? (b) Suppose we have two wires—one with current I coming out of the plane of the page, the other with current I going into the plane of the page. A bar magnet is placed in the plane of the page, located symmetrically along the line between the two wires, with the N pole nearest the wire with current coming out of the page. What are the forces on the wires? (c) Suppose the currents in the wires of (b) both flow out of the page. What will be the forces on the wires now?

Answer We have drawn the configurations corresponding to the different cases in Fig. 28–22, recalling that the magnetic field comes out of the north pole of a bar magnet and goes into the south pole. We can then use the right-hand rule imposed by Eq. (28–19) in straightforward fashion to find the forces, each indicated in Fig. 28–23.

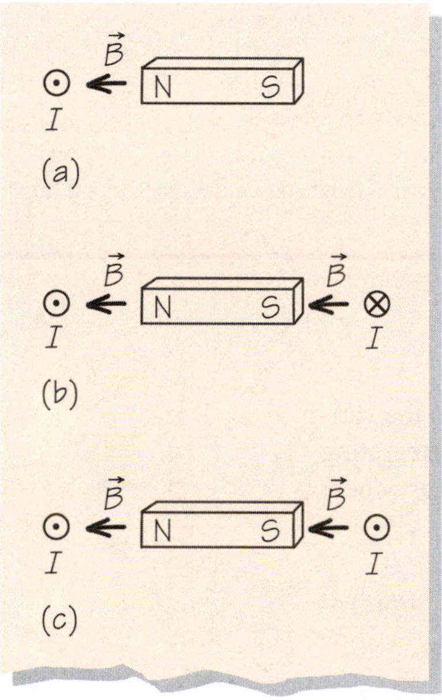

▲ **FIGURE 28–22** Magnet positions for each of the configurations described.

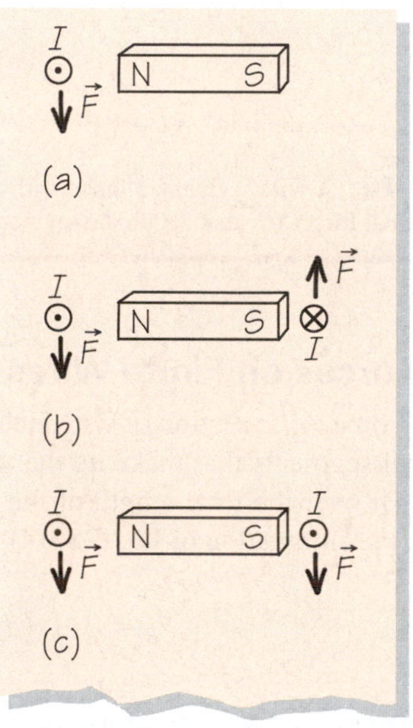

▲ **FIGURE 28–23** Forces for each of the configurations described.

EXAMPLE 28–9 A 12-cm-long straight segment of wire carrying a current of 7.2 A is maneuvered entirely within a region known to contain a constant magnetic field until the force on the wire has a maximum magnitude of 0.37 N. Find the magnitude of the magnetic field.

Strategy Figure 28–24 shows the orientation of a wire relative to the direction of the magnetic field in the plane which contains both the field and the direction of the wire. Since there is no force on the component of the wire parallel to the field, the maximum force occurs when the wire and the magnetic field are perpendicular to each other. In the relation $F = IBL \sin\theta$ [Eq. (28–18)] that expresses the magnitude of the force on the wire, the maximum occurs for $\theta = \pi/2$ (90°). The magnetic field can be calculated from this relation.

Working It Out For $\theta = \pi/2$, we have $F = ILB$, and hence

$$B = \frac{F}{IL} = \frac{0.37\ \text{N}}{(7.2\ \text{A})(12 \times 10^{-2}\ \text{m})} = 0.43\ \text{T}.$$

The measurement of forces on wires represents an important tool for the accurate measurement of magnetic fields.

What Do You Think? What is the direction of the magnetic force on the wire?

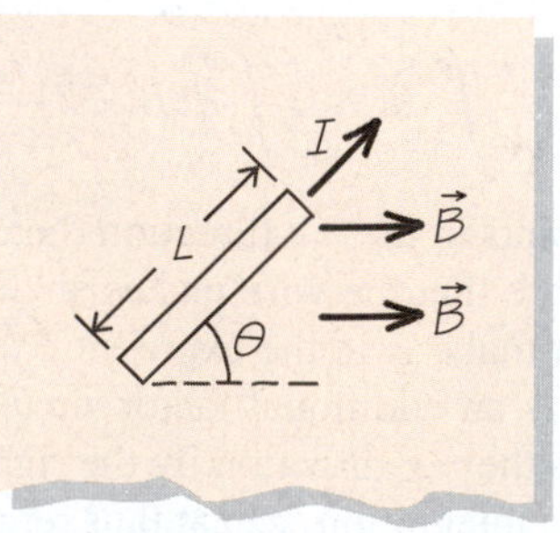

▲ **FIGURE 28–24**

The expression for the magnetic force on an isolated moving charge has led us directly to the expression for the magnetic force on a current-carrying wire. Historically, the order of discovery was just the reverse: Oersted, François Arago, and Ampère, who early in the nineteenth century performed the first quantitative experiments on magnetic forces, observed those forces on current-carrying wires. Their results then led to an understanding of magnetic forces on moving charges.

28–5 Magnetic Force on Current Loops

Magnetic fields exert forces on all kinds of current-carrying wires, including those of closed loops. As we shall see, a uniform magnetic field actually exerts only a torque on a current loop. This phenomenon provides the torque that runs direct-current electric motors and the galvanometer (the device cited in Chapter 27 for use in ammeters and voltmeters).

Figure 28–25a shows a stiff rectangular loop of wire carrying current I in the presence of a constant magnetic field along the $+x$-direction. The rectangular wire loop has sides of length a and b (denoted 1, 2, 3, and 4 in Fig. 28–25), and can be thought of as a series of straight wire segments. Figure 28–25b is a side view of the apparatus along the $+y$-direction. A perpendicular to the plane of the loop (the direction of the thumb when the fingers of the right hand follow the current direction) makes an angle ψ with the magnetic field. Fig. 28–25b also shows the angle θ between the direction of the magnetic field and the direction of the current for leg 1. We can calculate the force on each leg of the loop by using Eq. (28–19), and we find:

$$F_1 = IaB \sin\theta, \text{ in the } -y\text{-direction;} \quad (28\text{–}20a)$$

$$F_2 = IaB, \text{ in the } -z\text{-direction;} \quad (28\text{–}20b)$$

$$F_3 = IaB \sin\theta, \text{ in the } +y\text{-direction;} \quad (28\text{–}20c)$$

$$F_4 = IaB, \text{ in the } +z\text{-direction.} \quad (28\text{–}20d)$$

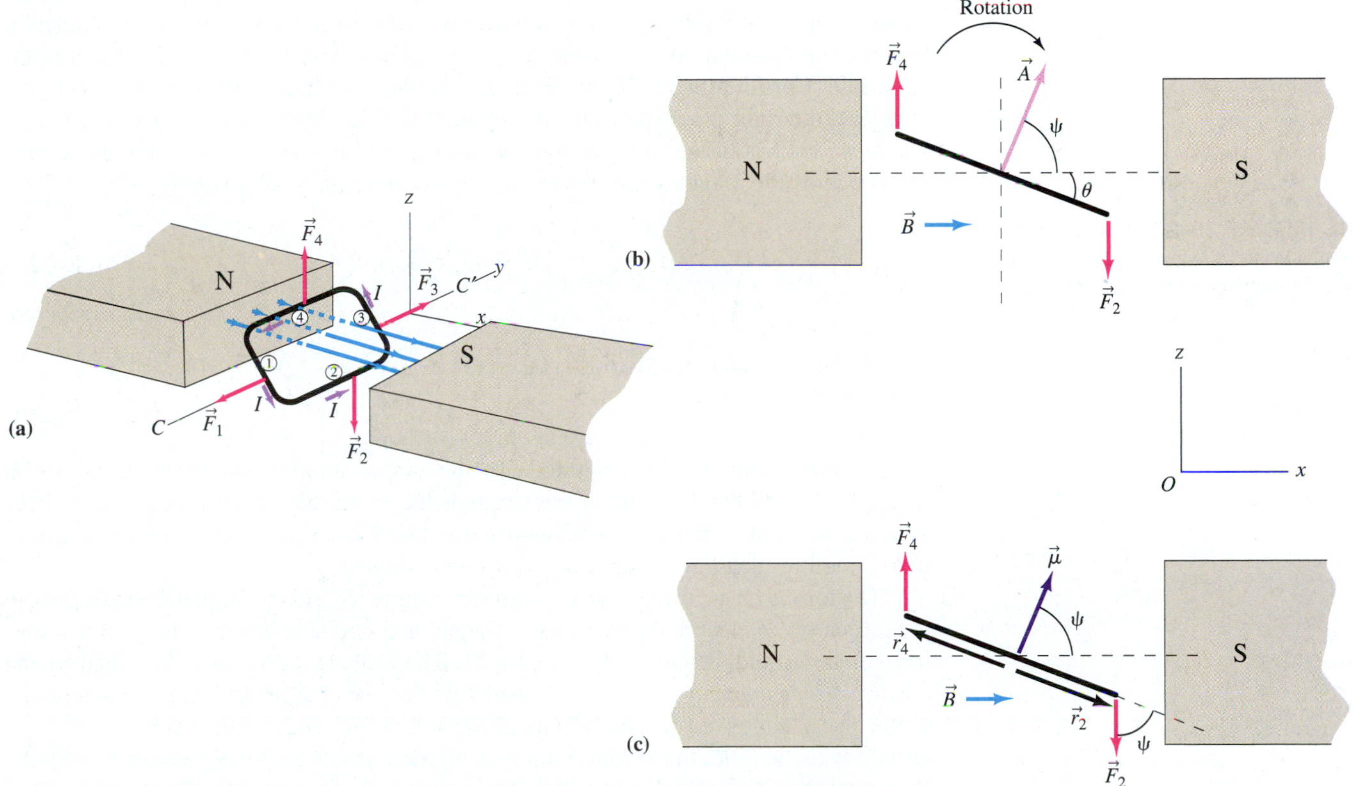

▲ **FIGURE 28–25** (a) A stiff, rectangular loop of wire is placed in a constant magnetic field. (b) A side view, looking along the y-axis, of the loop. (c) Geometry of the loop that allows us to calculate the torque on the loop. The torque tends to align the vector $\vec{\mu}$ with the magnetic field, $\vec{B}$.

These forces are indicated in Figs. 28–25a and 28–25b. Forces $\vec{F}_1$ and $\vec{F}_3$ are equal and opposite, as are forces $\vec{F}_2$ and $\vec{F}_4$, so *there is no net force on a current loop in a uniform magnetic field.* However, there is an important difference between these two sets of forces: $\vec{F}_1$ and $\vec{F}_3$ act along the same axis (CC' in Fig. 28–25a) and exert no torque on the loop. As Fig. 28–25b shows, $\vec{F}_2$ and $\vec{F}_4$ act along different axes and therefore produce a torque that causes the wire loop to rotate clockwise in the magnetic field. When the wire has rotated into the yz-plane (when $\theta = 90°$ in Figure 28–25b), $\vec{F}_2$ and $\vec{F}_4$ act along the same axis, and there is no torque. When the loop is in the xy-plane ($\theta = 0°$), the torque is a maximum. Finally, when θ changes sign, so does the torque, and the loop will tend to rotate counterclockwise.

We can find the torque about the central axis CC' in Fig. 28–25a by using the results of Chapter 10. From Eq. (10–6), the net torque about this axis is

$$\vec{\tau} = (\vec{r}_2 \times \vec{F}_2) + (\vec{r}_4 \times \vec{F}_4),$$

where $\vec{r}_2$ and $\vec{r}_4$ are the perpendicular vectors from axis CC' to legs 2 and 4, respectively (Fig. 28–25c). Both $\vec{r}_2$ and $\vec{r}_4$ have magnitude $a/2$. Figure 28–25c shows that ψ is the angle between $\vec{r}_2$ and $\vec{F}_2$ and between $\vec{r}_4$ and $\vec{F}_4$. The torque has magnitude

$$\begin{aligned}\tau = r_2F_2 \sin\psi + r_4F_4 \sin\psi &= (a/2)(IbB)\sin\psi + (a/2)(IbB)\sin\psi \\ &= IabB \sin\psi. \end{aligned} \tag{28–21}$$

Here, we have used the values for F_2 and F_4 given by Eq. (28–20). According to the right-hand rule, both terms in the equation for τ point in the $+y$-direction, so the net torque is in this direction.

The torque on a current loop in a magnetic field as given by Eq. (28–21) can be summarized and generalized. For a rectangular loop, the factor ab is the area A of the current loop. We can generalize to *any* planar loop of area A, whatever its shape, by the calculus technique of decomposing a planar loop of any shape into tiny rectangles and then treating each small rectangle in the way we have discussed above. We can neatly handle the vectorial nature of the torque by defining a vector $\vec{\mu}$ perpendicular to the plane of the loop. There are two possibilities for the direction of a vector perpendicular to any plane—which do we choose for $\vec{\mu}$? We choose the direction of $\vec{\mu}$ with a right-hand rule: Curl the fingers of the right hand in the direction of the current around the loop, and the right thumb gives the direction of $\vec{\mu}$ (Fig. 28–26). Try it for Fig. 28–25c, where we have indicated $\vec{\mu}$. The angle between $\vec{\mu}$ and $\vec{B}$ is ψ. We have thus shown that the direction and magnitude of the *torque on a current loop* are given by

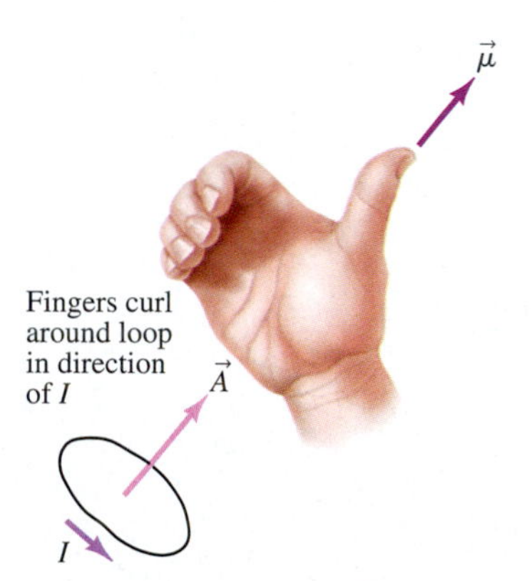

▲ **FIGURE 28–26** A right-hand rule indicates the direction of the magnetic dipole moment, $\vec{\mu}$, of a current loop.

$$\vec{\tau} = \vec{\mu} \times \vec{B}, \tag{28–22}$$

TORQUE ON A CURRENT LOOP

provided that the magnitude of $\vec{\mu}$ is taken to be

$$\mu = IA. \tag{28–23}$$

From these equations, the magnitude of the torque is $\tau = \mu B \sin\psi$, exactly as in Eq. (28–21), and the direction of the torque is the $+y$-direction. A torque in this direction acts to align $\vec{\mu}$ and $\vec{B}$. It is generally true that the torque tends to rotate a current loop or coil in such a way that $\vec{\mu}$ and $\vec{B}$ become aligned.

The torque on a current loop in a uniform magnetic field is completely analogous to the torque on an electric dipole (a pair of equal and opposite electric charges) in a uniform electric field, discussed in Chapter 21. The electric dipole was described by the *electric dipole moment* $\vec{p}$, a vector aligned with the two charges and equal in magnitude to the charge times the distance of charge separation. In terms of the response to an external magnetic field, the current loop is in all measurable respects a **magnetic dipole**. We therefore call $\vec{\mu}$, which plays a role analogous to $\vec{p}$, the **magnetic dipole moment** of the loop. One of the measurable properties—the defining property—of a dipole, magnetic or electric, is the characteristic field that the dipole itself *produces*. We deal with the calculation of the magnetic dipole field in Chapter 29.

CONCEPTUAL EXAMPLE 28–10 Suppose instead of one turn of wire we take N turns of wire, each surrounding the same plane area, thus creating a *coil*. What is the magnetic moment of the coil?

Answer Each turn of the coil experiences the forces we have described, and the torque is multiplied by N. This factor is included with the other factors intrinsic to the coil, so the magnetic dipole moment will now be multiplied by the factor N:

$$\mu = INA. \tag{28–24}$$

MAGNETIC DIPOLE MOMENT OF A COIL

We have seen another system that aligns itself with external magnetic fields: the iron filings we used to make our preliminary definition of magnetic fields. These iron filings behave like little bar magnets that are rotated by magnetic fields. Bar magnets react to fields just like current loops do. As we shall see in Chapter 29 when we find the magnetic fields *produced* by magnetic dipoles, bar magnets are themselves magnetic dipoles. We can explain this by looking at the microscopic level, where we find that metals contain the quantum-mechanical equivalent of circulating currents. We shall study this behavior in more detail in Chapter 31.

Galvanometers

In Chapter 27 we encountered the *galvanometer*—a device that measures currents. The principle behind a galvanometer is the fact that a magnetic field exerts a torque on a current loop. For example, we can attach a spring to a loop to balance the torque due to a known magnetic field, and the amount the spring stretches is a measure of the torque on the loop and hence of the current that passes through it.

THINK ABOUT THIS. . .

HOW DO ELECTRIC MOTORS WORK?

In one fundamental type of electric motor, we have a permanent magnet to supply a magnetic field and an emf to supply a current. The current is run through a loop that is placed in the magnetic field. With the apparatus shown in Fig. 28–25a, as soon as the wire loop rotates past the position in which it is aligned with the field due to the poles, the torque on it changes sign and becomes counterclockwise. In fact, the torque changes direction when ψ goes through 0° or 180°. This device oscillates and would not make a motor, which requires a continuous turning motion. However, if we can make the *current* switch directions every time the loop passes $\psi = 0$ or π (180°), then the torque will continuously produce a clockwise rotation. A device that does just this is known as a *split-ring commutator*, a device whose mechanism is visible in Fig. 28–27. With this device, the loop accelerates its rotation, always in the same direction, under the influence of a torque whose sign does not change, and an *electric motor* has been created. Once we have this arrangement, and a continuously turning loop, it is no great step to imagine tranferring this motion to, for example, a set of wheels. But that is another story more properly treated in a chapter on mechanics.

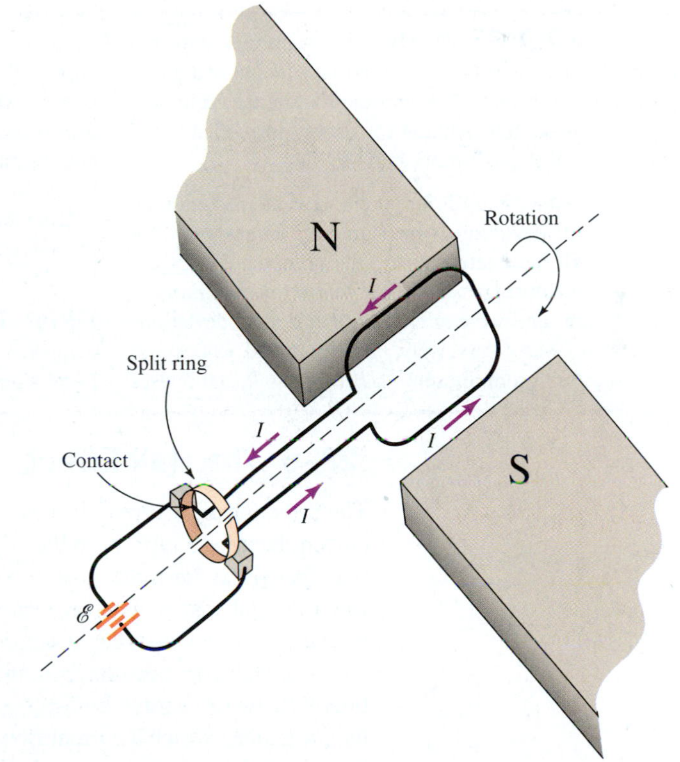

▲ **FIGURE 28–27** A current-carrying loop aligned in a magnetic field is fitted with a split-ring commutator, as shown in a schematic diagram. The torque on the loop serves to turn the loop and makes a motor.

Energy and the Torque on Loops

When a magnetic field rotates a current loop, the field does work. For a constant field, the only variable in the work is the angle of rotation, ψ. We know from Chapter 7 that we can use the concept of potential energy when the force (or the torque) depends only on position, as here. Accordingly we associate a potential energy $U(\psi)$ with a loop in a magnetic field, where ψ is the angle between $\vec{\mu}$ and $\vec{B}$. As always, only *changes* in the potential energy have physical consequences. The change in potential energy in rotating the coil from some initial angle ψ to a final angle of $\pi/2$ (90°) is given by the negative of the work done by the magnetic field in moving the coil through these angles:

$$U(\psi) - U(\pi/2) = -\int_{\psi}^{90°} \tau\, d\psi' = -\int_{\psi}^{90°} \mu B \sin\psi'\, d\psi'$$

$$= -\mu B \int_{\psi}^{90°} \sin\psi'\, d\psi' = \mu B \cos(\pi/2) - \mu B \cos\psi.$$

The cosine of $\pi/2$ is zero, so

$$U(\psi) - U(\pi/2) = -\mu B \cos\psi. \qquad (28\text{–}25)$$

We choose the location of $U = 0$ for convenience, and it is customary to choose U to be zero at $\psi = \pi/2$; that is, when $\vec{\mu}$ is perpendicular to $\vec{B}$. Setting the term $U(\pi/2)$ in Eq. (28–25) to zero gives *the potential energy of a current loop with a given magnetic dipole moment $\vec{\mu}$ in a constant magnetic field $\vec{B}$*:

$$U(\psi) = -\vec{\mu}\cdot\vec{B}. \qquad (28\text{–}26)$$

POTENTIAL ENERGY OF A MAGNETIC DIPOLE

The potential energy has a *minimum* when $\vec{\mu}$ is aligned along $\vec{B}$ (that is, when $\psi = 0$). Thus the orientation in which $\vec{\mu}$ is aligned with $\vec{B}$ is a stable equilibrium point. This agrees with our earlier result that the torque tends to rotate the loop to line up $\vec{\mu}$ and $\vec{B}$.

CONCEPTUAL EXAMPLE 28–11 A current loop in a constant magnetic field is initially aligned so that $\vec{\mu}$ points in a direction slightly different from that of $\vec{B}$, and the loop is then released. There is no mechanism such as friction for energy loss—that is, no damping. What is the subsequent motion of the loop?

Answer At the (stable) equilibrium point, $\vec{\mu}$ is aligned with the magnetic field. A small deviation is measured by an angle ψ. The subsequent motion will be determined by the potential energy expressed in terms of ψ for small ψ. This potential has a minimum around the equilibrium, and we can express it for small deviations from the minimum. More precisely, the potential energy is $U(\psi) = -\mu B \cos\psi$, with a minimum $-\mu B$ for $\psi = 0$, and it rises as ψ moves away from the minimum value, independent of the sign of ψ. Any such curve varies as $C\psi^2/2$ for ψ close enough to $\psi = 0$, where C is a constant that depends on the potential energy. This, however, is just the potential energy for a harmonic oscillator, and we may therefore expect that the motion will be simple harmonic, of the general form $\psi = \psi_m \sin(\omega t + \delta)$, where ω involves the coefficient C as well as the rotational inertia of the loop (see Problem 64). In Chapter 13 we saw that this is a general property of motion around stable equilibrium points.

What Do You Think? What is the behavior of an undamped compass needle slightly displaced from the direction of Earth's magnetic field?

28–6 The Hall Effect

The direction of a current does *not* itself determine the sign of the charge carriers in that current because a current to the right can be produced by the movement either of positive charges to the right or of negative charges to the left. The Hall effect allows us to find this sign, that is, whether a current is formed of moving electrons or of moving positive charges such as holes in semiconductors (see Section 26–6).

The Hall effect results from the fact that charges moving along a wire in a magnetic field experience a force whose sign depends on their charge. Consider a metal strip of length L along which a current flows. The strip is placed in a uniform magnetic field that is perpendicular to the strip, as in Fig. 28–28. As a result, there will be a potential difference between points a and b whose cause and qualitative nature we now describe. Equation (28–19) gives the total force on the strip, $\vec{F} = I\vec{L} \times \vec{B}$. This force, by the right-hand rule, is directed in the $-x$-direction—it acts to the *left* in Fig. 28–28. By using the equivalent force law $\vec{F} = q\vec{v} \times \vec{B}$, we can show that the force on the charge carriers acts to the left, *whatever the sign of the charge carriers*. If the moving charges are positive (ions) and the current flows in the $+y$-direction, then the velocity of the charges is also in the $+y$-direction. According to the right-hand rule, $\vec{F} = q\vec{v} \times \vec{B}$ is then directed toward point a in Fig. 28–28. If the moving charges are negative (electrons), however, then the velocity of the charges is in the $-y$-direction. The vector product $\vec{v} \times \vec{B}$ is directed to the right, but q is negative, and $q\vec{v} \times \vec{B}$ again points to the left, moving these negative charges toward a. Either way, there is a buildup of the charge carriers at the left side of the strip. This buildup cannot continue indefinitely: Once enough charge carriers have moved to the left, they will supply a repulsive Coulomb force against the movement of other charge carriers that is large enough to stop the process. An equilibrium is established in which an electric potential is set up between points a and b that prevents further leftward drift of charge carriers. Charges then move up the strip as they would if there were no magnetic field. (The charge separation has led to an electric field between points a and b. We have crossed $\vec{E}$ and $\vec{B}$ fields, with the value of the electric field such that the charges travel undeflected up the strip just as in the velocity selector discussed in Section 28–3.)

The **Hall effect** is that *the sign of the potential difference between points a and b determines the sign of the charge carriers.* If the charge carriers are negative, negative charges build up on the left side of the metal strip, and point a is at a lower potential than point b. Conversely, if the carriers are positive, positive charges build up on the left side of the strip, and point a is at a higher potential than point b. The first measurement of the sign of this *Hall potential* was performed by the American physicist Edwin H. Hall in 1879. His measurement proved that the carriers of current in metals are negatively charged.

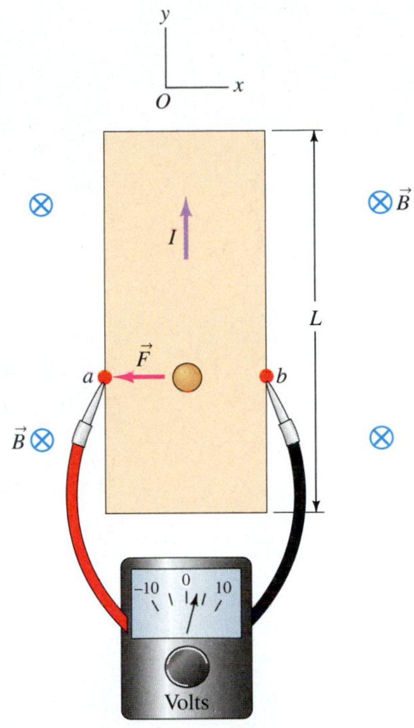

▲ **FIGURE 28–28** A conducting strip perpendicular to a constant magnetic field develops a potential called the Hall voltage between points a and b when the strip carries a current I.

THINK ABOUT THIS. . .

DOES THE HALL EFFECT HAVE TECHNOLOGICAL USE?

Our discussion of the Hall effect shows that the potential across the strip is proportional to the strength of the magnetic field. Potential changes are easy to measure accurately, and in a Hall strip they signal changes in the magnetic field. In particular, they can be used to sense the change in distance of a magnet from the Hall strip. This makes the Hall effect useful in the detection of the motion of magnets, even slow motion.

An example is in the computer keyboard. Constant and heavy use would wear out the keyboard rather rapidly if the use involved establishing direct contact between two solid pieces. The Hall effect is used to signal the application of force to a key on a keyboard without this type of contact. A small, permanent magnet is attached to the bottom of each key on the keyboard, and underneath it there is a Hall probe. Modern probes consist of a thin layer of a conductor deposited on a rigid substrate. Because of the spatial dependence of the magnetic field, the change in the position of the magnet relative to the conducting layer can change the magnetic field at the probe by an order of magnitude. Such a change leads to changes in a Hall potential that is then communicated to the computer, signaling that a given key has been pressed. In an electronic piano this application can be carried further. The rate of change in the Hall potential signals the speed at which a key is depressed, so that the keyboard can signal whether the key has been pressed rapidly or gently, and produce a loud or soft sound in response. ■

EXAMPLE 28–12 Figure 28–28 shows a Hall strip whose width is 1.0 cm in a magnetic field of 2.0 T with a voltage of magnitude 7.2 μV across the strip. What is the speed v of the charge carriers in the strip?

Strategy The velocity of the charge carriers is that of charges passing through a crossed electric and magnetic field. We are given the magnetic field B and can find the electric field in terms of the strip width d and the voltage V across the strip. The charge flow is that of a velocity selector, with $v = E/B$. The electric field is determined from the potential difference and the strip width to be $E = V/d$.

Working It Out Combining the equations above, the velocity is

$$v = \frac{V}{Bd} = \frac{7.2 \times 10^{-6}\ \text{V}}{(2.0\ \text{T})(1.0 \times 10^{-2}\ \text{m})} = 3.6 \times 10^{-4}\ \text{m/s}.$$

This is a measurement of an electron's speed in a current, what we earlier called the drift speed.

What Do You Think? The width of the strip is doubled. Which of the following is true: the voltage (a) is doubled, (b) is halved, (c) remains the same?

Summary

Magnets, moving electric charges, and electric currents all experience magnetic forces. These forces can be described in terms of a magnetic field, $\vec{B}$, whose spatial dependence can be mapped out with iron filings or by observing its effect on a moving electric test charge or a test current element. In terms of this field, the magnetic force on an electric charge q depends on the charge's velocity according to the magnetic force law,

$$\vec{F} = q\vec{v} \times \vec{B}. \qquad (28\text{–}1)$$

The SI unit of magnetic field is the tesla, T: $1\ \text{T} = 1\ \text{kg}/(\text{C} \cdot \text{s})$. When both magnetic and electric fields are present, the Lorentz force law holds:

$$\vec{F} = q(\vec{E} + \vec{v} \times \vec{B}). \tag{28–4}$$

In a static magnetic field, the component of a charged particle's velocity parallel to the field is unaffected by that field. The magnitude of the force on the particle due to the field is proportional to the component of the velocity perpendicular to the field, and the direction of the force is perpendicular to this component of the velocity and to the field itself. It follows that the kinetic energy of a charged particle in a magnetic field is unchanging. When the field is constant, a charged particle traveling perpendicular to the field moves in a circle of radius

$$R = \frac{mv}{qB}. \tag{28–5}$$

The frequency of the particle's circular motion is the cyclotron frequency,

$$f = \frac{qB}{2\pi m}, \tag{28–7}$$

which is independent of the particle's velocity. The general path followed by a moving charge is a spiral around the magnetic field lines.

When a charged particle has a particular velocity perpendicular to constant crossed electric and magnetic fields, the electric and magnetic forces cancel, and the particle passes through the fields undeflected. The magnitude of this special velocity is

$$v = \frac{E}{B}. \tag{28–9}$$

With the help of a velocity selector, an apparatus based in part on this phenomenon, the charge-to-mass ratio of the electron can be measured.

The infinitesimal magnetic force on an infinitesimal length of thin wire $d\ell$ that carries a current I in the presence of a constant magnetic field is

$$d\vec{F} = I\, d\vec{\ell} \times \vec{B}. \tag{28–14}$$

To find the net force on a wire of finite length in a magnetic field, Eq. (28–14) is integrated. For example, the force on a straight wire of length L in a uniform magnetic field is given by Eq. (28–-19), $\vec{F} = I\vec{L} \times \vec{B}$. Another important example is a wire that carries a current and is formed into a loop (or coil) of N turns; the area of the face of the loop is A. When it is placed in a constant magnetic field, such a loop experiences a torque

$$\vec{\tau} = \vec{\mu} \times \vec{B}. \tag{28–22}$$

The loop reacts as a magnetic dipole with magnetic dipole moment $\vec{\mu}$. For a coil of N turns, $\vec{\mu}$ has magnitude

$$\mu = INA \tag{28–24}$$

and direction perpendicular to the face of the coil, oriented by a right-hand rule on the current. The torque tends to rotate the loop so that $\vec{\mu}$ and $\vec{B}$ become aligned. The potential energy of the loop in a constant magnetic field can be expressed as

$$U(\psi) = -\vec{\mu} \cdot \vec{B}. \tag{28–26}$$

The Hall effect exploits the equivalence between the force on a moving charge and the force on a current-carrying wire. This effect proves that the current carriers in metals are negatively charged.

Understanding the Concepts

1. A wire carrying a current is electrically neutral, yet a magnetic field acts on it. Why?
2. Explain how you might define and measure a magnetic field if magnetic monopoles existed.
3. An electron beam in an oscilloscope is deflected to the right on the screen. Could this be caused by an electric field *or* by a magnetic field? Explain how you could distinguish these possibilities.
4. An electron beam makes a spot in the center of the screen of a cathode-ray tube. A bar magnet is brought in from the left side (as seen from the front of the tube), with the S pole nearest to the beam. Which way will the spot move? Suppose that the N end of the bar magnet is brought near the beam from above. Which way will the spot move?
5. Much of the description of magnetic forces depends on the use of a right-hand rule. Does the magnetic force depend fundamentally on the fact that we have chosen the right rather than the left hand?
6. Currents flow in two parallel wires. Wire 1 carries 2 A to the right, and wire 2 carries 1 A to the left. (a) What is the direction of the magnetic field near wire 2 due to wire 1? (b) What is the direction of the magnetic force on wire 2?
7. If you have just used a velocity selector for electrons and you wish to use it to choose positrons with the same speed, do you have to change any settings on the selector? Positrons are like electrons, but positively charged.

8. Induced charges give rise to electric forces even between electrically neutral objects. How do we know that the forces between bar magnets are not induced electric forces?
9. The aurora is a manifestation of Earth's magnetic field. Why do these lights occur frequently near Earth's poles and only rarely elsewhere?
10. Imagine that an electrically neutral wire carrying a current moves in the presence of an external magnetic field. Do you expect that there will be an additional force on the wire due to the movement?
11. You have a fixed length of wire and want to use it to make a magnetic dipole with the largest possible magnetic dipole moment. Into what shape should you wind it? Are you better off making a single loop or N loops?
12. A small bar magnet forms a magnetic dipole; a current-carrying wire in the shape of a small loop also forms a magnetic dipole. If that is the case, the current loop should give rise to a magnetic field. Use this analogy to sketch the magnetic field lines that would be generated by such a current loop.
13. Consider two small circular current loops. Suppose the two loops are placed flat on a table close to each other (but not touching) and the two currents both flow in a counterclockwise direction. Will the two loops attract or repel? What happens if the directions of the currents are opposite?
14. A proton coming from outer space reaches Earth and is now traveling across the magnetic north pole parallel to Earth's surface. Will it be deflected (a) up, (b) down, (c) to the left, (d) to the right, or (e) not at all?
15. Consider two small circular current loops. Suppose one loop is placed above the other (but not touching), with their areas oriented similarly. If the currents flowing in the two loops are both in the same direction, will the loops attract or repel?
16. Is it possible for an electron to move in a straight line through a magnetic field? If so, how?
17. Suppose that the coil of a direct-current electric motor consists of many turns rather than one turn of wire that carries a current I. Does the coil rotate faster than a single loop would? Does the split-ring commutator still work?
18. You have a large pail of water, a bar magnet with its N and S ends unmarked, a straight pin, and a cork. How could you make a compass? One of the things you need to know to construct this compass is how to distinguish north from south; you are allowed to watch the Sun to help with this part of the question.
19. Do magnetic north poles repel positive electric charges?
20. A classmate tells you that 1 T is 1 N/A·m. Is your classmate correct?
21. True or false? The magnetic force on an electron moving in Earth's magnetic field is zero only if the electron (a) is at the N and S poles, (b) is at the equator, (c) is moving slowly, (d) is moving along the magnetic field lines, or (e) is moving radially away from Earth.

Problems

28–1 Magnets and Magnetic Fields

1. (II) Sketch the magnetic fields for the arrangements of bar magnets shown in Fig. 28–29.

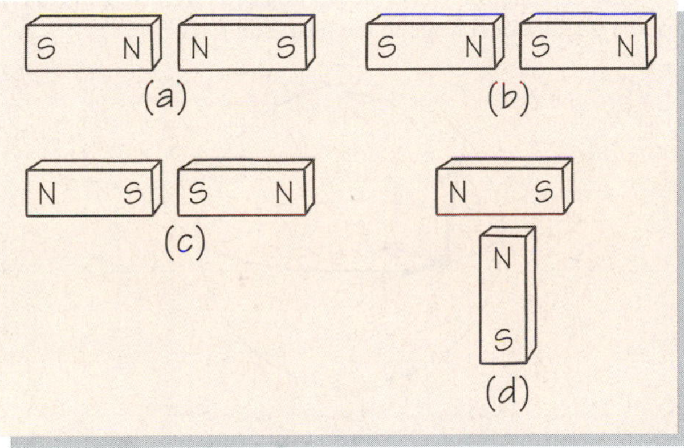

▲ **FIGURE 28–29** Problem 1.

2. (II) Consider the magnetic field generated by a current-carrying wire, as depicted in Fig. 28–2. Assuming that a reversal in the direction of the current also reverses the direction of the magnetic field, sketch the magnetic field due to two wires that are parallel to each other, and whose currents flow in the same direction. You may assume that the magnetic fields add vectorially, just like electric fields. Repeat the sketch for the situation in which the currents flow in opposite directions. Your sketches should show the field lines as seen by someone looking along the wires.

28–2 Magnetic Force on an Electric Charge

3. (I) An electron moving in the $-y$-direction enters a region of constant magnetic field and is observed to deflect to the $-x$-direction. What is the direction of the magnetic field?
4. (I) A proton with velocity $\vec{v} = (1.7 \times 10^6\ \text{m/s})\hat{\imath} + (0.8 \times 10^6\ \text{m/s})\hat{\jmath} - (4.5 \times 10^5\ \text{m/s})\hat{k}$ moves through a magnetic field $\vec{B} = (0.70\ \text{T})\hat{\imath} - (0.50\ \text{T})\hat{\jmath} + (0.10\ \text{T})\hat{k}$. Calculate the force on the proton.
5. (I) A proton of energy 100 keV moving in the $+x$-direction enters a region of uniform magnetic field perpendicular to the x-axis. Upon entry into that region, the proton experiences an acceleration of $3 \times 10^{12}\ \text{m/s}^2$ in the $+y$-direction. What are the magnitude and direction of the magnetic field?
6. (I) A pith ball charged to $+1\ \mu\text{C}$ falls vertically at the equator. At this location, the magnitude of Earth's magnetic field is 0.5 Gauss (0.5×10^{-4} T) and the field points to the north. When the ball reaches a speed of 5 cm/s, what is the magnetic force (magnitude and direction) on it?
7. (I) A proton moving with speed v enters a narrow (1.0 cm wide) region of magnetic field perpendicular to the direction of the proton's motion. As a result, the proton acquires a small component of speed perpendicular to its original direction of motion. This speed is much less than the original speed, and it is measured to be 3.3×10^5 m/s. What is the strength of the magnetic field?
8. (II) A cork ball carrying charge q has a mass of 0.4 g and is set in straight-line motion perpendicular to a uniform magnetic field of 0.007 T. What is the value of q if its direction of motion changes by 0.05° in 3.0 s?
9. (II) (a) A rapidly moving charged particle of charge e, mass m, and speed v passes through a region of magnetic field $\vec{B}$, which points in a direction perpendicular to the motion. The particle spends a time interval Δt in the region. Estimate the angle θ through which it will be deflected during Δt, assuming that θ is small. (b) The particle is a proton, with $m = 1.7 \times 10^{-27}$ kg, and $e = 1.6 \times 10^{-19}$ C, and the speed is 1.4×10^7 m/s. The size of the magnetized region is 0.1 m across. How large must B be to give rise to a deflection of 0.1 rad?

10. (II) In an oversimplified model of Earth's magnetic field, the field is parallel to the rotation axis, has a constant magnitude of 10^{-4} T up to a height of 100 km, and then quickly drops to zero. A cosmic-ray particle with charge 1.6×10^{-19} C and mass 9.5×10^{-26} kg moves at a speed of 10^8 m/s directly toward the equatorial region from above. (a) In what direction is the particle deflected? (b) Estimate how much it will be deflected from the point of impact it would have if it were uncharged. (In fact, this is not a realistic example. Cosmic-ray particles as massive as this have greater charge.)

11. (II) Electrons travel at a speed of 6.0×10^7 m/s in a television tube. The electrons are affected by Earth's magnetic field. The tube, which is 0.40 m long, is located at a region where the magnetic field has a vertical component of 18 μT and a horizontal component of 24 μT (Fig. 28–30). If the initial direction of the electron beam is in the same direction as the horizontal component of Earth's magnetic field, in which direction, and by how much, is the electron beam deflected?

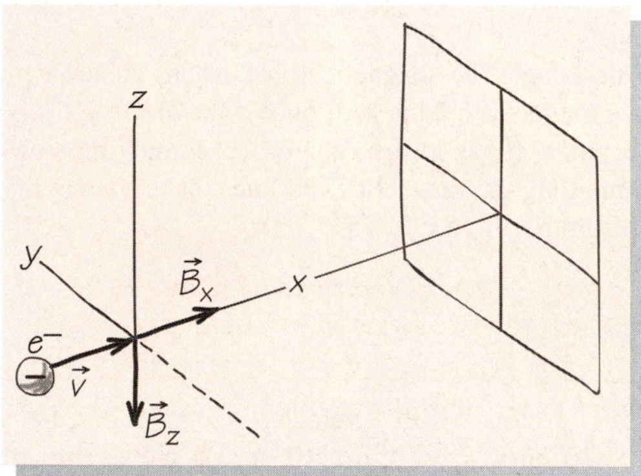

▲ **FIGURE 28–30** Problem 11.

28–3 Consequences of the Magnetic Force on a Charge

12. (I) A proton is sent into a region of constant magnetic field, oriented perpendicular to the proton's path. There the proton travels at a speed of 3×10^6 m/s in a circular path of radius 240 cm. What is the magnitude of the magnetic field?

13. (I) (a) Suppose that electrons from an electron gun with a voltage of 1600 V are injected into a region of constant magnetic field perpendicular to the electrons' velocity. What magnetic field will give the electrons a radius of curvature of 6 cm? (b) A magnetic field of what magnitude is necessary to give an alpha particle (charge $q = 2e$ and mass $m_\alpha = 7360$ times the mass of an electron) with a kinetic energy of 1200 eV a path with radius of curvature of 20 cm?

14. (I) Show that the radius of curvature of a proton moving at a velocity of 25 km/s in a magnetic field of 10^{-10} T is small compared to interplanetary distances. The protons therefore spiral around interplanetary magnetic field lines; we say that the protons are "tied to the magnetic field lines" in cosmic magnetic fields.

15. (I) The magnetic field at the surface of a neutron star has magnitude 3×10^7 T. What is the radius of the circular orbit of an electron that moves there at 0.1 percent of the speed of light? What is the magnitude of the magnetic force on the electron?

16. (I) With what frequency will deuterons, which have the same charge as protons but twice the mass, circulate in a cyclotron with a magnetic field of 0.65 T?

17. (I) If we want to triple the cyclotron frequency associated with a proton accelerator from an initial value of 6.1 MHz, what quantity must we change, and from what initial value to what final value?

18. (I) In a certain region, the average radius of curvature of the trajectory of electrons trapped in the Van Allen belt is 300 m and the average electron energy is 100 keV. What is the value of Earth's magnetic field in this region?

19. (I) Electrons of speed 10^6 m/s and protons of speed 10^4 m/s perpendicularly enter a region of constant magnetic field 10^{-5} T above Earth. What are the radii of their orbits? Why is the proton's radius greater? If the proton's speed were the same as that of the electron (10^6 m/s), what would be the radius of its orbit?

20. (II) An accelerator designer envisages circulating protons moving in a ring of radius 17 km by means of magnetic fields of magnitude 7.0 T. What is the magnitude of the momentum of a proton that moves in this way? For protons with this momentum, the energy is given to excellent accuracy by the formula $E = pc$, where c is the speed of light, about 3×10^8 m/s. Calculate the energy of the proton in megaelectron-volts; 1 MeV $= 1.6 \times 10^{-13}$ J.

21. (II) Assume that the electrons in a television picture tube have an energy of 10 keV and move perpendicularly to Earth's magnetic field (see Table 28–1). (a) Calculate the final velocity (vector) of an electron when it hits the screen if the horizontal distance the electron travels is 40 cm. (b) What is the deflection (distance) of the electron perpendicular to its original direction?

22. (II) Earth acts as a giant magnet whose field lines are like those of a bar magnet, running from the magnetic north pole to the magnetic south pole. The magnetic field at the equator is approximately constant, of magnitude 5×10^{-5} T, and runs from the geographic South Pole to the geographic North Pole (Fig. 28–31). If we ignore air resistance and the gravitational force, a charged object could orbit Earth at the equator as a result of the magnetic force if it has just the right velocity. Suppose that such an object has a charge of -1 mC and a mass of 1.0 g. (a) What would its velocity have to be for it to travel in such an orbit? (b) Suppose that the gravitational force acts on this object as well. What is the ratio of the gravitational force to the magnetic force?

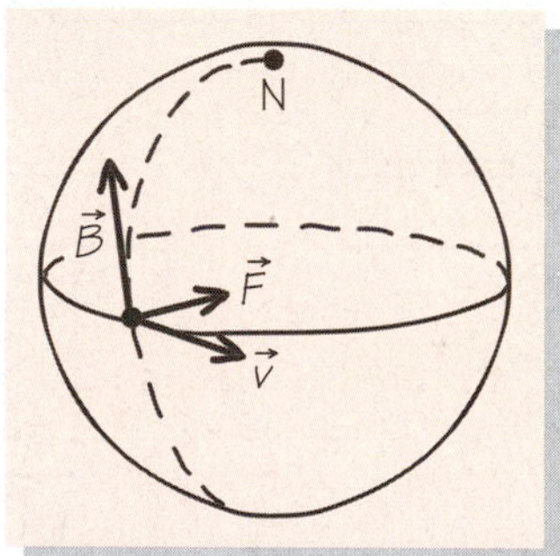

▲ **FIGURE 28–31** Problem 22.

23. (II) In Section 28–3, we calculated the critical momentum for an electron to stay within the galaxy. (a) Given that the energy of a high-energy particle is related to its momentum by $E = pc$ (see Problem 20), what is the energy of an electron with the critical momentum? What are the critical momentum and energy for (b) an alpha particle (charge $2e$ and mass four times the mass of a proton) and (c) an ion of uranium, with charge e and mass 240 times the proton mass?

24. (II) A proton moves horizontally perpendicular to a constant magnetic field oriented so as to deflect the proton instantaneously upward. The magnitude of the field is 0.7×10^{-2} T. What is the speed of the proton so that the magnetic force just cancels the gravitational force on the proton, leaving it in horizontal flight? This problem illustrates how very weak the gravitational force is compared to electromagnetic forces.

25. (II) An electron moves at a speed $v = 3.0 \times 10^5$ m/s in a region of constant magnetic field of magnitude 0.12 T. The direction of the electron when it enters this region is at 40° to the field, and the electron follows a helical path. When you look along the direction of the magnetic field, the path is a projected circle. How far has the electron traveled along the direction of $\vec{B}$ when one projected circle has been completed?

26. (II) An electron enters a bubble chamber that contains a constant magnetic field of strength 0.035 T and follows a helical path. The spacing between the turns of the path is 8.5 mm, as is the radius of the circular part of the path. Find the components of the velocity parallel and perpendicular to the field.

27. (II) A vacuum tube contains two axial cylinders (Fig. 28–32). The potential difference between these cylinders is 500 V. Some electrons are released from the inner cylinder and are accelerated by the electric field toward the outer cylinder; a small current thereby travels through the tube. Suppose that a uniform magnetic field is set up parallel to the axis of the tube. This curves the trajectories of the electrons, and at a critical magnetic field, the electrons will no longer reach the outer cylinder, and the current ceases to flow. What is the kinetic energy of the electrons that hit the outer cylinder at a magnetic field just below the critical value?

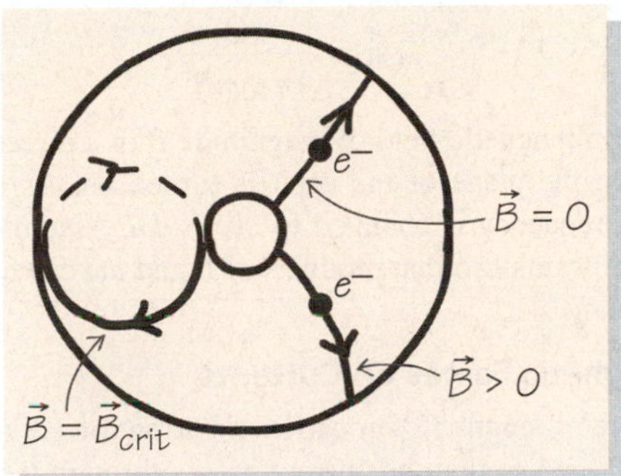

▲ **FIGURE 28–32** Problem 27.

28. (II) A proton and an alpha particle, which has twice the charge and four times the mass of the proton, are each accelerated through the same potential difference and enter a region of constant magnetic field perpendicular to their paths. (a) What is the ratio of the radii of their orbits? (b) What is the ratio of the frequencies of their orbits?

29. (II) An electron is injected at $t = 0$ s with velocity $\vec{v}_0 = (2 \times 10^6\ \text{m/s})\hat{i}$ into a region with parallel electric and magnetic fields $\vec{E} = (1500\ \text{V/m})\hat{j}$ and $\vec{B} = (-0.2\ \text{T})\hat{j}$, respectively. Calculate the subsequent motion.

30. (II) You want to be able to tune a velocity selector such that you have the capacity to select electrons that have been accelerated from rest by a potential that runs from 1500 V to 15,000 V. If the magnetic field, B, is fixed at 0.40 T, what range of electric field strengths must be available? If the electric field strength were fixed at 15 V/cm, what range of magnetic field strengths must be available?

31. (II) Figure 28–33 is a schematic diagram of a cyclotron. A charged particle starts out at the central point and, for a given magnetic field perpendicular to the plane of motion, follows a circular path. The cyclotron takes advantage of the fact that the time for the particle to execute a half-circle is independent of the particle's velocity. An alternating voltage is applied across the gap between the two "dees" (the semicircular regions), so that when the particle crosses the gap, the voltage acts to accelerate it. When the particle gets to the gap again after having completed a half-circle, the voltage has changed sign, and the particle is once again accelerated. The frequency of the oscillating voltage must match the cyclotron frequency. In this way, the particle is always accelerated, completing ever bigger circles in the same time, until the beam is extracted at the maximum radius. (a) If the magnetic field has strength 1.0 T and the circulating particle is a proton, $q = +e$ and $m = 1.7 \times 10^{-27}$ kg, what is the cyclotron frequency? (b) What is the maximum velocity of the proton for a maximum radius of 50 cm? (c) the corresponding maximum kinetic energy? (d) If the maximum voltage across the gap is 50 kV, how many full circles does the proton make before it reaches its maximum energy? (e) How much time does the proton spend in the accelerator?

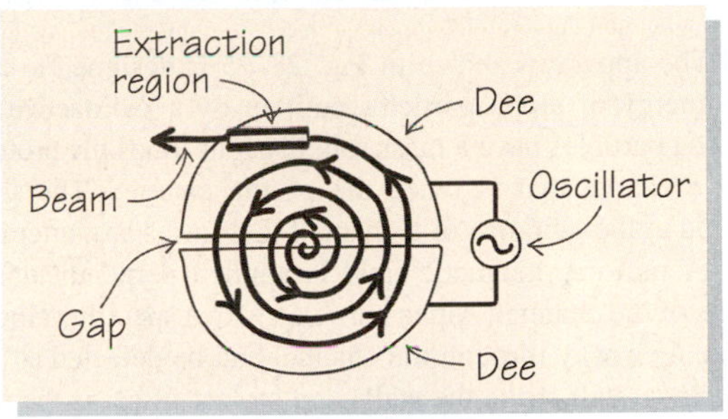

▲ **FIGURE 28–33** Problem 31.

32. (II) A cyclotron used for accelerating protons has a magnetic field of magnitude 1.7 T. The circular region in which the magnetic field exists has a radius of 40 cm. (a) What is the cyclotron frequency? (b) What is the largest kinetic energy that a proton accelerated in this machine can have? (c) Repeat parts (a) and (b) for a doubly ionized helium nucleus, $^4\text{He}^{++}$, with four times the mass of a proton and twice the charge.

33. (II) The particle accelerator at Fermilab, the Fermi National Accelerator Laboratory in Batavia, Illinois, can accelerate protons to relativistic speeds. The accelerator is circular and holds the protons in circular paths by increasing the strength of a magnetic field perpendicular to this path as the protons' momentum increases. (The momentum increases because the protons pass repeatedly through regions of electric potential.) The radius of the main Fermilab accelerator is 6.2 km, and the magnets are capable of maintaining magnetic field strengths between 1 T and 4.5 T. Given that the magnitude of a proton's electric charge is 1.6×10^{-19} C, what range of momenta can be accommodated in this accelerator? Because protons of such momenta are highly relativistic, their energies are given by the approximate relativistic formula $E = pc$. What range of energies can be reached at Fermilab? What would the speed of a baseball, mass 0.15 kg, be if it had the energy of the most energetic protons at Fermilab? (For the baseball, use the normal nonrelativistic formulas that relate energy and speed.)

34. (II) A proton, with charge $q_p = +e$ and mass m_p, is accelerated through an electric potential V. The proton then enters a region of constant magnetic field $\vec{B}$ oriented perpendicular to its path. In this region, the proton's path is circular with radius of curvature R_p. Another particle with the same charge as the proton but with mass m_x follows under the same conditions. Its radius of curvature in the magnetic field, R_x, is 1.4 times as large as R_p. What is the ratio of m_x to m_p?

The device we have described is a type of *mass spectrometer*, which can be used to identify a material by the masses of that material's constituent molecules (Fig. 28–34, see next page). Sometimes, instead of a simple electrostatic potential as in our example, a velocity selector of crossed $\vec{E}$ and $\vec{B}$ fields is used to select particles of a given speed.

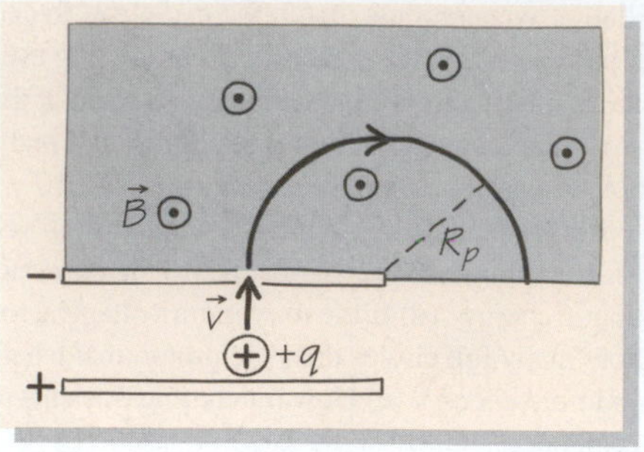

▲ **FIGURE 28–34** Problem 34.

35. (II) The apparatus shown in Fig. 28–35 is designed to measure the energy of alpha particles emitted by a radioactive source. (Alpha particles have a mass roughly four times the proton mass and a charge that is twice the proton charge.) The source is placed at the entrance of a channel that forms a quarter of a circle. A uniform magnetic field is applied perpendicular to the plane of the channel. Alpha particles with a specific velocity will make their way through the channel and be detected at the exit. All others will strike the walls and be lost. What is the range of values of B necessary to analyze alpha particles whose energies range up to 6 MeV?

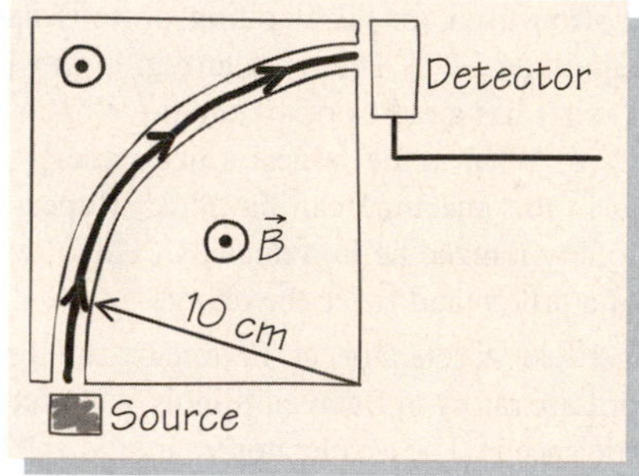

▲ **FIGURE 28–35** Problem 35.

36. (II) A 150-MeV proton, moving in the x-direction, enters a region in which there is a magnetic field. The proton experiences an acceleration of $7.0 \times 10^{12}\ \text{m/s}^2$ in the y-direction. What can you say about the magnetic field?

37. (II) A 2.3-keV electron is moving horizontally and passes perpendicular to the Earth's magnetic field at a location where the field magnitude is 0.52 Gauss. What is the magnetic force on the electron? What is the ratio of this force to the gravitational force on the electron? What energy electron would be needed to make the magnetic and gravitational forces equal?

38. (II) The 88-inch (diameter) cyclotron was put into operation at the Lawrence Berkeley Laboratory in 1962 and for 40 years produced significant scientific results. It was able to produce protons up to 55 MeV energy, and heavier ions of mass A up to 5 MeV/A. (a) What was the maximum magnetic field of the magnet? (b) If a ^{86}Kr ion with charge $+19e$ is accelerated in this cyclotron, what maximum energy can be expected?

39. (II) Constant electric and magnetic fields are perpendicular to each other, the electric field pointing from left to right on the page, the magnetic field from the bottom to the top of the page. A proton with charge q and speed v moves into the region of the fields in the direction out of the page. (a) What are the directions of the magnetic and electric forces? (b) What is the magnitude of the resulting force on the proton? (c) When could this force ever be zero?

40. (III) You have an apparatus that can form an electric field of 2000 N/C and a magnetic field of 0.3 T. You want to build a velocity selector to select electrons of speed 2×10^4 m/s. (a) Draw the orientation of your apparatus, showing $\vec{E}$, $\vec{B}$, and $\vec{v}$. (b) What are the minimum and maximum values of v that you can select? [*Hint*: Set the apparatus up so that $\vec{v}$ and $\vec{B}$ are not perpendicular to each other (Fig. 28–36).]

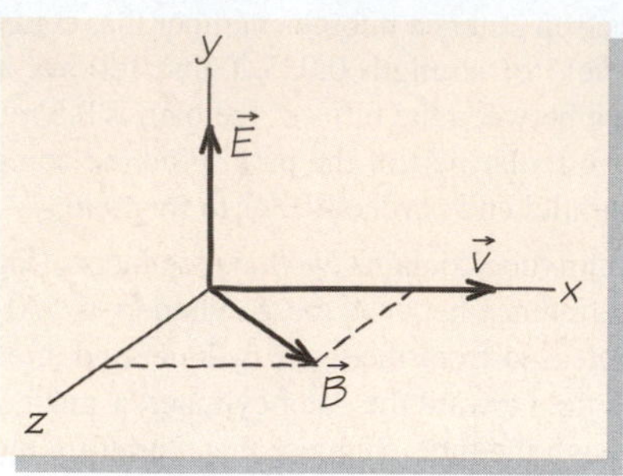

▲ **FIGURE 28–36** Problem 40.

41. (III) A particle of mass m and charge $-q$ moves in a circular orbit of radius R about a fixed charge Q. The angular frequency for the orbit is given by

$$\omega_0^2 = \frac{qQ}{4\pi\varepsilon_0 m R^3}.$$

A uniform magnetic field of magnitude B in a direction perpendicular to the plane of the orbit is turned on. As a result, the angular frequency is changed to $\omega_0 + d\omega$. Assuming that B is sufficiently small so that products of B and $d\omega$ can be neglected, calculate $d\omega$.

28–4 Magnetic Forces on Currents

42. (I) A wire of length 12 cm carries a current of 1.3 A. When the wire is placed so that the current passes through the wire along the x-direction, the force on the wire due to an external magnetic field B is measured to be $-(0.02\ \text{N})\hat{j}$. When the wire is moved so the current through the wire moves along the y-direction, the force is measured to be $(0.02\ \text{N})\hat{i}$. What is the magnetic field?

43. (I) A straight wire segment is placed in a region known to contain a constant magnetic field of unknown strength. A current of 6 A runs through the wire, which can be turned in various directions until the force per unit length acting on it takes on a maximum value of 0.18 N/m. What is the value of the magnetic field?

44. (I) A long wire carries a current of 6.5 A. A bar magnet is brought near the wire so that the charge carriers, of speed 1.7×10^{-4} cm/s, experience a magnetic field of 0.50 T perpendicular to their direction of motion. Calculate the force (a) on each moving charge carrier (electron) and (b) on a 1-m length of the wire.

45. (I) A thin, straight wire carries a current of 10 mA and makes an angle of 60° with a constant magnetic field of magnitude 10^{-6} T. The portion of the wire in this field has a length of 10 cm. Calculate the force, both direction and magnitude, on this segment of the wire.

46. (I) The length of a vertical lightning conductor from roof to ground is 20 m. A lightning stroke leads to a current of 10^4 A flowing through the conductor. Given that Earth's magnetic field is horizontal and of magnitude 0.5×10^{-4} T at the location of the building, what is the force on the conductor during the period the current flows?

47. (I) A straight wire is placed in a uniform magnetic field of magnitude 0.010 T. The direction of the field makes an angle of 30° with that of the wire, which carries a current of 10 A. What is the force on a 1.0-m segment of the wire?

48. (II) A current I flows through a circular wire loop of radius R that lies in the xy-plane (Fig. 28–37). Consider a constant magnetic field of magnitude B that points in the x-direction. Calculate the force on an element of the loop formed by an angle $d\theta$, located at an angle θ from the $+x$-axis.

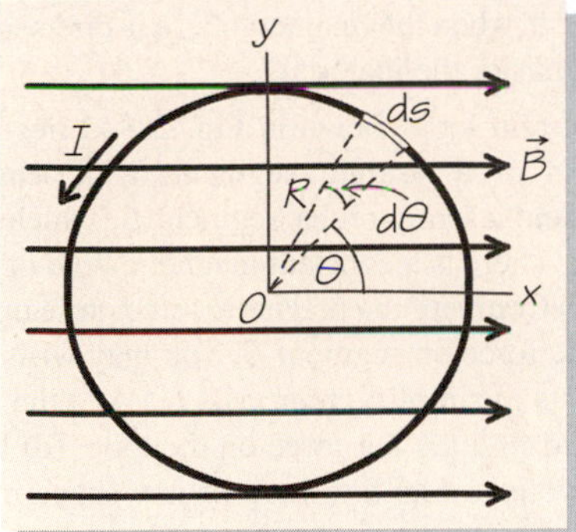

▲ **FIGURE 28–37** Problem 48.

49. (II) A wire of length L is suspended from two springs of spring constant k attached to a current source (Fig. 28–38). A magnetic field B, in a horizontal direction perpendicular to the wire (out of the page), is turned on; a current I then flows in the wire, which moves to a new equilibrium position. Which way will the wire move, and by how much?

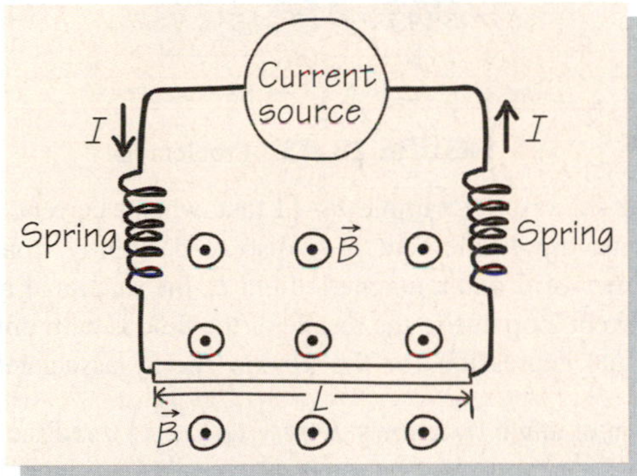

▲ **FIGURE 28–38** Problem 49.

50. (II) In a physics lecture demonstration, a thick copper wire of length 0.8 m and mass 70 g is attached to two thin wires and suspended so that it is horizontal (Fig. 28–39). A 0.03-T magnetic field pointing in the downward direction is turned on. What angle will the supporting wires make with the vertical if a current of 0.5 A flows through the wire?

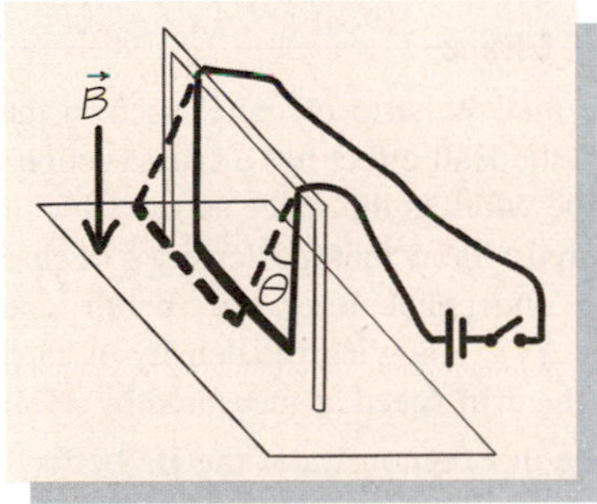

▲ **FIGURE 28–39** Problem 50.

51. (II) Figure 28–40 shows a possible device for measuring magnetic fields. A loop carrying a current I is dipped into a region of magnetic field. The loop is suspended from a spring of spring constant k that stretches if the magnetic field points in a certain direction. Here the loop has width $\ell = 1.2$ cm, $I = 100$ mA, the spring stretches 0.6 cm, $k = 5 \times 10^{-2}$ N/m, and the magnetic field is uniform. What is the magnitude of the field? How could such a device be used, or modified, to measure fields that are not uniform?

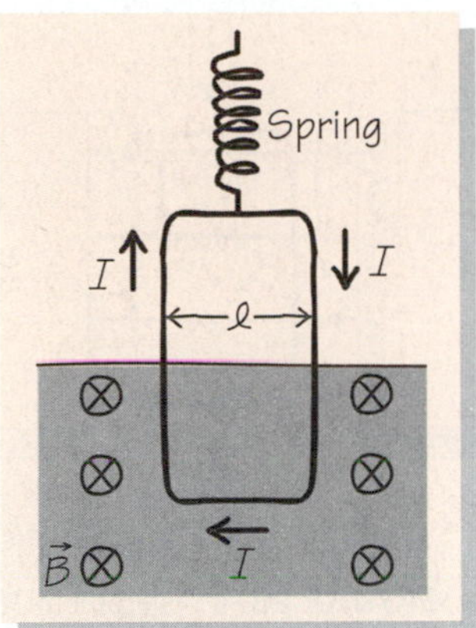

▲ **FIGURE 28–40** Problem 51.

28–5 Magnetic Force on Current Loops

52. (I) A wire coil of area 20 cm^2 with 180 turns experiences a maximum torque of 2×10^{-2} N · m when placed in a magnetic field of 0.3 T. What is the current through the coil?

53. (I) A rectangular wire loop of height 5 cm and width 3 cm consists of 60 turns and carries a current of 1.2 A. What are the magnitude and direction of the magnetic dipole moment? If a uniform magnetic field of 0.5 T is applied to the loop, and the field's direction makes an angle of 26° with the normal to the current loop, what is the torque (magnitude and direction) that acts on the loop?

54. (I) A circular coil of diameter 2.5 cm, consisting of 1500 turns of wire, carries a current of 50 mA. How much work must be done to flip the coil through 180° when it is placed in a uniform magnetic field of 0.75 T? The field makes an initial angle of 50° with the direction of the coil's dipole moment.

55. (I) A wire forms a circular coil of N turns and radius R and carries a current I. The coil's magnetic dipole moment is initially aligned with a fixed external magnetic field, $\vec{B}$. How much work must be done by an external torque to rotate the coil through an angle θ?

56. (I) A current loop of area 3.0 cm^2, carrying a current of 5.0 A, is placed in a uniform magnetic field of 0.25 T such that the normal to the loop is perpendicular to the direction of the magnetic field. There is a torque, and the loop changes direction. Because of friction in the bearings, it settles to the minimum energy orientation. How much energy was dissipated in the process?

57. (I) An atom can have a magnetic dipole moment of 10^{-23} J/T. Such an atom is placed in a magnetic field of 10 T. What is the range of potential energies involved?

58. (II) A wire carrying a current I splits into two channels of resistance R_1 and R_2, respectively, forming a circuit. The wire enters the space between the two poles of a magnet with a uniform magnetic field that runs from one pole piece to the other (Fig. 28–41, see next page). The circuit forms a loop; the field lies in the plane of the loop. What is the torque on the circuit about the wire axis, given that the wires are a distance d apart and that the length of the split is L?

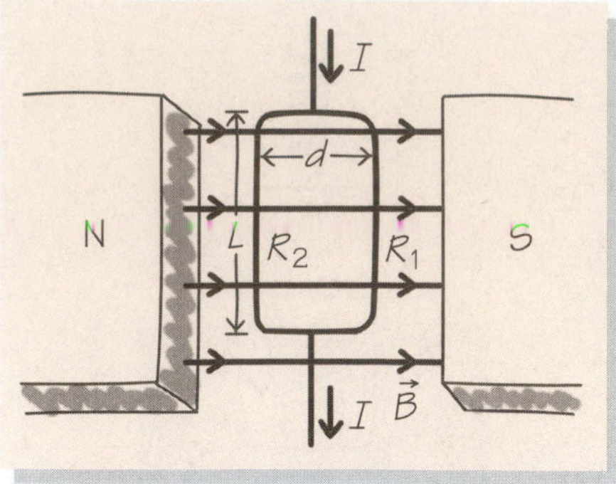

▲ **FIGURE 28–41** Problem 58.

59. (II) A circular wire coil of area 6 cm^2 has 50 turns. When the coil is placed in a magnetic field of 0.2 T, the maximum torque is 3×10^{-5} N·m. (a) What is the current in the coil? (b) What work is required to rotate the coil 180° in the magnetic field? Does the work depend on the initial angle?

60. (II) An electric motor consists of a current-carrying wire loop in a constant magnetic field $\vec{B}$ (Fig. 28–42). The field produces a torque that tends to rotate the loop so that the loop's magnetic dipole moment, $\vec{\mu}$, and $\vec{B}$ become aligned. When that happens, a split-ring commutator reverses the current direction, so that $\vec{\mu}$ changes its orientation by 180°, and the torque acts to continue the rotation. Suppose that $\vec{\mu}$ and $\vec{B}$ start out almost antiparallel. Plot the magnitude of the torque as a function of the angle between $\vec{\mu}$ and $\vec{B}$, as this angle runs from $-\pi$ to 0. At 0° the commutator reverses the current. Plot the torque through another half turn. What is the average value of the torque through a full turn if the current in the motor is 6.2 A, the magnitude of $\vec{B}$ is 0.45 T, and the area of the loop is 54 cm^2?

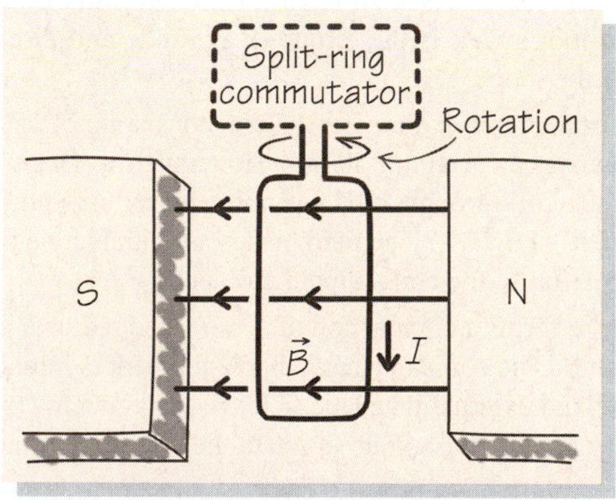

▲ **FIGURE 28–42** Problem 60.

61. (II) An electron, of charge $q = -1.6 \times 10^{-19}$ C, has a "size" of about 3×10^{-15} m, called its classical radius. The magnetic dipole moment of the electron is roughly 10^{-23} A·m^2. (a) Suppose that this magnetic moment were due to the entire charge q orbiting at the classical radius. What would the speed of the charge be to generate this magnetic moment? (b) Suppose that the electron's magnetic moment were perpendicular to a magnetic field of magnitude 1 T. What is the torque on the electron?

62. (II) (a) Calculate the magnetic dipole moment of a single atom, based on the following model: One electron travels at speed 2.2×10^6 m/s in a circular orbit of diameter 10^{-10} m. (b) The individual atomic magnetic dipoles of magnetic materials (such as iron) are preferentially lined up to point in the same direction. If a fraction f of the dipoles are so aligned along the long axis (with the rest oriented randomly so that their magnetic dipole moments add vectorially to zero), what is the net magnetic dipole moment of a piece of such material 1 cm^2 in area and 10 cm long? (The material may be viewed as an array of cubes, each of which contains one atom and is 10^{-10} m on a side.) (c) What is the torque experienced by the piece of material in part (b) in a field of 10^{-3} T when the magnetic field is directed at right angles to the long axis of the material?

63. (III) The current loop shown in Fig. 28–43 lies in the xy-plane and consists of a straight segment α of length $2R$ in the x-direction and a semicircular segment β, which has a radius of curvature R. There is a constant magnetic field of strength B into the page. (a) Compute the magnetic force on segment α. (b) Find the magnetic force on segment β. You may wish to use symmetry arguments to simplify your task. (c) Add the results of parts (a) and (b) to find the net force on the loop. (d) How could you generalize your results to a loop of any shape in the xy-plane?

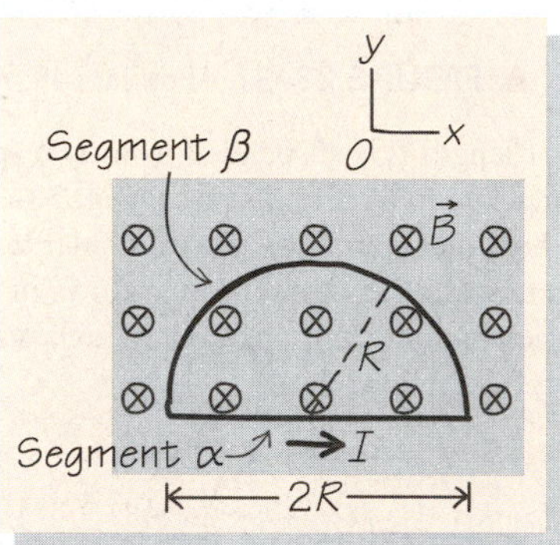

▲ **FIGURE 28–43** Problem 63.

64. (II) We showed in Example 28–11 that, when a current loop with magnetic dipole moment $\vec{\mu}$ is displaced slightly from perfect alignment of $\vec{\mu}$ and a magnetic field $\vec{B}$, the rotational motion of the current loop due to the torque of the field is harmonic. Given the usual expression for the kinetic energy associated with a rotation of angle ψ, namely $K = \frac{1}{2}I_M\left(\frac{d\psi}{dt}\right)^2$, and the approximate expression $\cos\psi \cong 1 - \psi^2/2$, calculate the angular frequency of the harmonic motion.

65. (III) A coil carrying current $I = 50$ mA has a moment of inertia $I_M = 7.5 \times 10^{-7}$ kg·m^2 about a rotational axis and an area of 6.0×10^{-4} m^2. The coil is placed in a magnetic field of magnitude 0.6 T, displaced 5° from alignment between its magnetic dipole moment, μ, and the field, and released from rest. Describe the subsequent motion. What is the maximum angular speed of the coil in that motion?

28–6 The Hall Effect

66. (II) Suppose that the strip of metal used in the apparatus that demonstrates the Hall effect has a cross section of width w and depth d_0. (The width is the space across which the Hall voltage ΔV is measured.) Show that the density n of charge carriers with charge e is independent of the width and is given by $n = IB/(d_0 e\,\Delta V)$. Knowing the density of carriers, find an expression for the drift speed as measured by a Hall apparatus.

67. (II) The probe that demonstrates the Hall effect is used to measure the density of charge carriers in an unknown sample of metal. A sample of the material 1.5 mm thick is placed in a magnetic field of 1.2 T. When a current of 1.8 A passes through the material, a Hall voltage of 6.2 μV is measured. What is the density of charge carriers?

68. (II) A Hall-effect probe can be used to measure the magnitude of a magnetic field. A researcher has lost the instruction booklet and forgotten the calibration procedure. However, when she places the Hall probe inside a known magnetic field of 7500 G, she measures a Hall voltage of 165 mV. What is the field of a magnet with a Hall voltage of 390 mV?

General Problems

69. (II) The wire coil of a galvanometer has an area of 2 cm^2 and 500 turns. The coil is placed in a magnetic field of magnitude 0.18 T and oriented so that its plane is initially parallel to the field. The restoring torque of the galvanometer spring is proportional to the angular deflection, with a proportionality constant of $10^{-8}\ \text{N}\cdot\text{m}/°$ (see Example 28–9). What current corresponds to a deflection of 70°?

70. (II) The masses of atomic ions of known charge can be precisely measured by finding the time an atom takes to complete a circular trajectory in a known magnetic field. With a magnetic field of magnitude 3.0 T and an apparatus capable of measuring times to an accuracy of 10^{-9} s, how accurately can the mass of an ion with charge $+e$ be measured in 1 rev? If the mass is to be measured to an accuracy of 5×10^{-31} kg, how many revolutions must be measured?

71. (II) When an electron orbits a proton, the smallest circular orbit is one with a radius of about 0.5×10^{-10} m, the Bohr radius. The proton's electric field must have what magnitude to make the electron follow this orbit? Compare the magnitude of the magnetic field that would be required to make an electron move in a circle of the same radius at the speed it would have if it were orbiting a single proton.

72. (II) A massive charge Q is fixed at the origin of a coordinate system. A magnetic field $\vec{B}$ points in the $+z$-direction. A light particle of charge q and mass m moves in a circular orbit of radius r about the origin. For what value of $\vec{B}$ (as a function of r) is such motion possible if Q and q have the same sign and if the angular momentum of the motion is a fixed constant L?

73. (II) Consider a parallel-plate capacitor with charge density $\pm 8.0 \times 10^{-7}\ \text{C/m}^2$ on the two plates and an electric field that points in the $+z$-direction. What magnetic field is necessary to provide a velocity selector for 60-keV deuterons that move in the $+y$-direction? A deuteron has a mass of 3.2×10^{-27} kg and a charge of 1.6×10^{-19} C; 1 keV $= 1.6 \times 10^{-16}$ J.

74. (II) A narrow beam of particles of mass m and charge q travels in free space at speeds between v_1 and v_2. It enters a region of length L with a constant magnetic field that points perpendicular to the beam direction and parallel to the boundary between the field-free region and the region with the field. There, it follows a circular path with radius of curvature R until it exits that region. Show that the beam widens when it emerges from the region with the field, and calculate the spread in terms of a range of angles.

75. (II) An electron moving in the xy-plane is subject to forces due to a constant magnetic field $\vec{B}$ that points in the $+z$-direction. Assuming that the electron loses 10 percent of its energy after 20 turns, as a consequence of frictional forces, what will the fractional change in the radius of the orbit be after 20 turns?

76. (II) For the motion described in Problem 75, (a) what will the fractional angular-momentum change be during the 20 turns? (b) What is the torque exerted by the frictional forces in terms of the initial kinetic energy?

77. (II) Electrons are injected into a region with a constant magnetic field $\vec{B}$ by an electron gun with known voltage V. The electrons move in a plane perpendicular to $\vec{B}$ and follow an arc of radius R. Determine the charge-to-mass ratio e/m for the electrons in terms of the given parameters.

78. (II) Particles with mass $M_A = A(1.6 \times 10^{-27}\ \text{kg})$ and charge $q = 1.6 \times 10^{-19}$ C are accelerated by a potential difference of 6.0×10^4 V and directed perpendicularly into a region of uniform magnetic field of strength 2.2 T. The region with the field is 35 cm deep. Calculate the angular deflection of the particles, θ, as a function of A.

79. (II) N electrons move at speed v in a circular orbit of radius R. (a) What is the angular momentum of the system of electrons? (b) The magnetic dipole moment associated with the current loop? (c) The ratio of the quantities in parts (a) and (b)?

80. (II) A rectangular wire loop of width a and height b is connected to a current source that, when turned on, gives rise to a current I in the wire. The loop is suspended in a uniform magnetic field $\vec{B}$ that points in a vertical direction (Fig. 28–44), and it would hang vertically if there were no current. We assume that the wire is massless, but two masses m are suspended at the lower corners. What is the angle θ at which the loop is in equilibrium? Calculate this in two ways: by using torques, and by expressing the potential energy as a function of θ and minimizing it. What happens if the direction of the current is reversed?

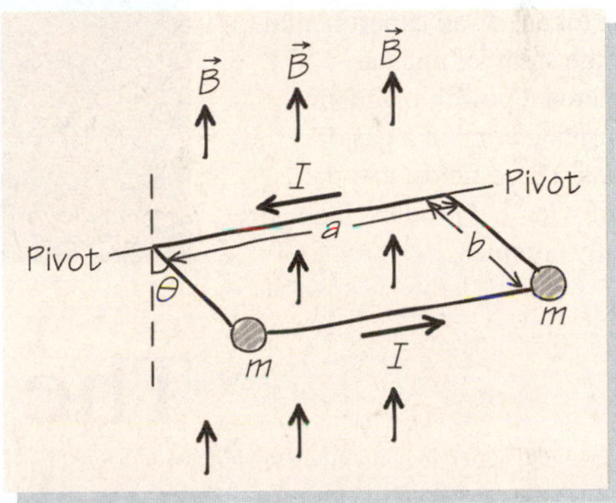

▲ **FIGURE 28–44** Problem 80.

81. (III) A particle of charge q and mass m is subject to gravity acting downward on the page, and a magnetic field of magnitude B acting in a direction perpendicular to the page, pointing into the page. We assume that the initial velocity of the particle is zero. (a) Show that the motion will only be in the plane of the page; (b) if there are many such particles, and their mutual Coulomb repulsion is negligible, what is the direction of the average current? (c) Check your conclusions by writing out the form of Newton's second law for the three components of the motion.

82. (III) Suppose that an experimental apparatus can have both electric and magnetic fields constant in magnitude and direction. In this apparatus a proton moving at a speed of 5.0×10^4 cm/s in the $+z$-direction does not accelerate, whereas a proton moving at a speed of 8.0×10^4 cm/s with no x-component at an angle of 42° with respect to the z-axis experiences an initial acceleration of magnitude $3.5 \times 10^8\ \text{m/s}^2$ in the $-x$-direction. A proton moving in the xy-plane has a circular orbit. Find the values of $\vec{E}$ and $\vec{B}$ in the apparatus.

Chapter 29

In the tokamak, an experimental device for the study of nuclear fusion–generated power, magnetic fields are used to contain a gas of positive ions. These fields, and the windings of wire that produce them, are topologically complex.

The Production and Properties of Magnetic Fields

We have seen how magnetic fields influence moving charges and current-carrying wires. But where do the magnetic fields themselves come from? We have seen that an electric charge creates an electric field, and that an electric field exerts a force on a charge. Similarly, a *moving* charge or a current *creates* a magnetic field, just as we know that a magnetic field exerts a force on a *moving* charge or a current. In other words, currents and moving charges exert magnetic forces on each other. In this chapter, we will describe and explore the ways in which magnetic fields are produced, and learn the properties of these fields, some quite different from those of electric fields. We shall also continue to explore the close relation between electric and magnetic fields.

29–1 Ampère's Law

Over the winter of 1819–1820, Hans Christian Oersted discovered that electric currents influence compass needles (Fig. 29–1). Until this discovery, there was only a suspicion of a connection between electricity and magnetism. Oersted, as well as André-Marie Ampère, soon showed that *current-carrying wires exert forces on each other*. Because such wires are everywhere electrically neutral, these forces are not electric. As we saw in the previous chapter, a current-carrying wire aligns iron filings on a plane perpendicular to the wire in a circular pattern (Figure 28–2c). This suggests that a current-carrying wire creates the magnetic field.

The Magnetic Field of a Straight Wire

Figure 29–2 shows a set of experiments that tell us much about the nature of the magnetic fields produced by a current-carrying wire. To start with, in Figs. 29–2a and 29–2b two parallel wires with current flowing in the same direction are attracted to each other, while in Figs. 29–2c and 29–2d we see that if the currents flow in opposite directions, the parallel wires repel. If we interpret the force between the two current-carrying wires in Fig. 29–2 as a magnetic force on one wire due to the magnetic field produced by the other and perform some measurements, we can determine the magnetic field of a straight wire. In Fig. 29–3a we redraw the situation with wire segments that are long, $L \gg d$, where L is the wire length and d their separation. In this limit all other segments of the wires are distant and can be ignored, because the forces between the wires weaken considerably as the separation between the wires increases. Only the relatively close long, parallel segments come into the picture. In Fig. 29–3a, the force on wire 2, which is directed to the left, is due to the magnetic field of wire 1. Equation (28–19) describes the force on a segment of current-carrying wire. Using this equation, the force on wire 2 is of the form

$$\vec{F}_2 = I_2\vec{L}_2 \times \vec{B}_1, \tag{29–1}$$

provided only that the field $\vec{B}_1$ due to wire 1 is the same all along wire 2, an assumption that is justified for arbitrarily long wires. Here the vector $\vec{L}_2$, of magnitude L, is oriented along the direction of I_2.

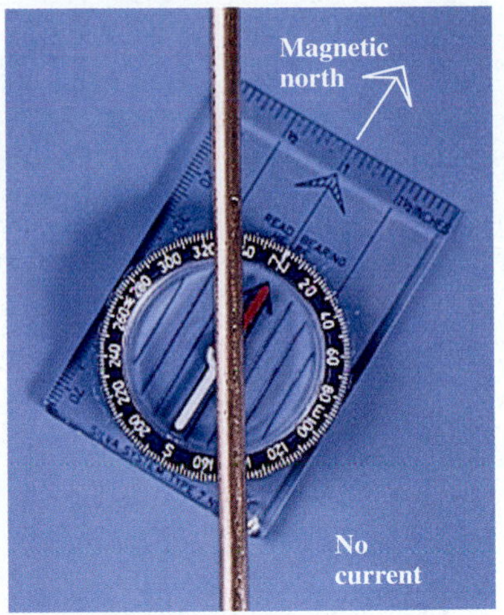

(a)

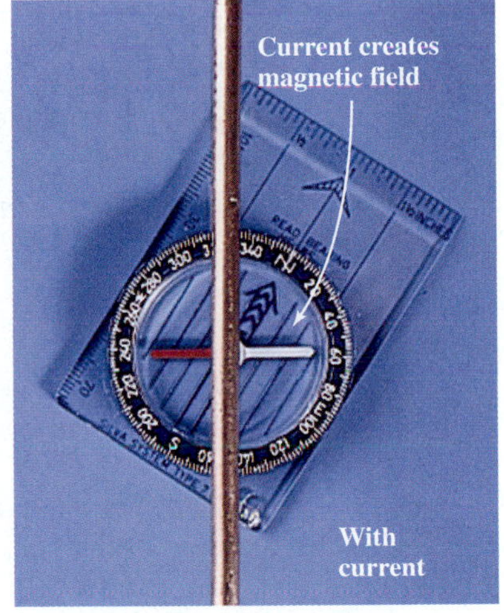

(b)

▲ **FIGURE 29–1** (a) The compass needle continues to point north when there is no current in the wire, but (b) when current flows through the wire from the bottom to the top of the photo, the needle reacts to the magnetic field produced by the current.

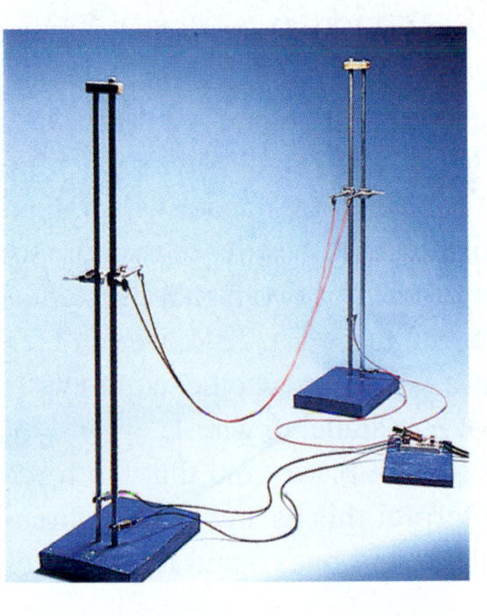

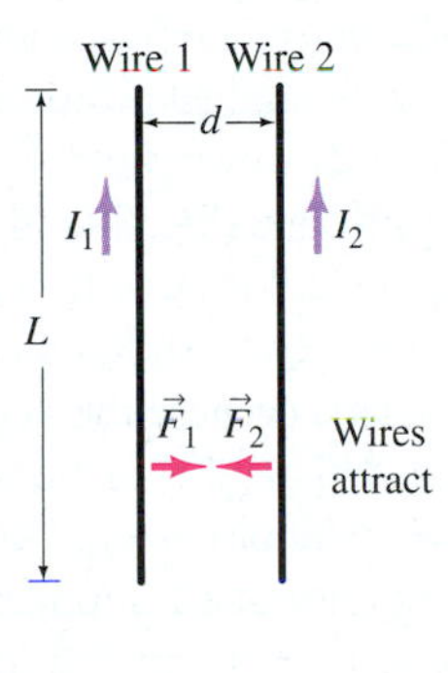

(a) (b)

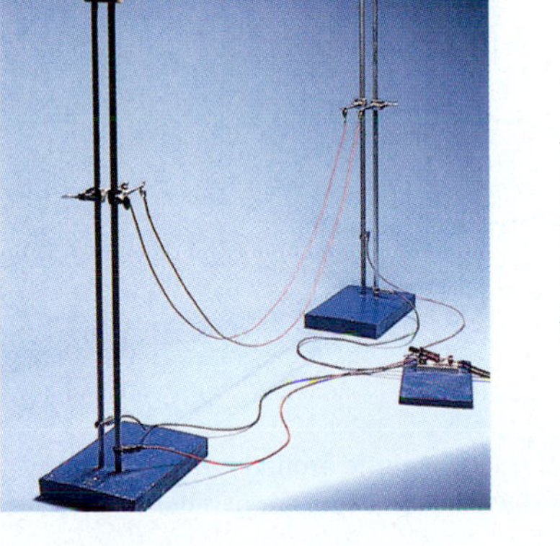

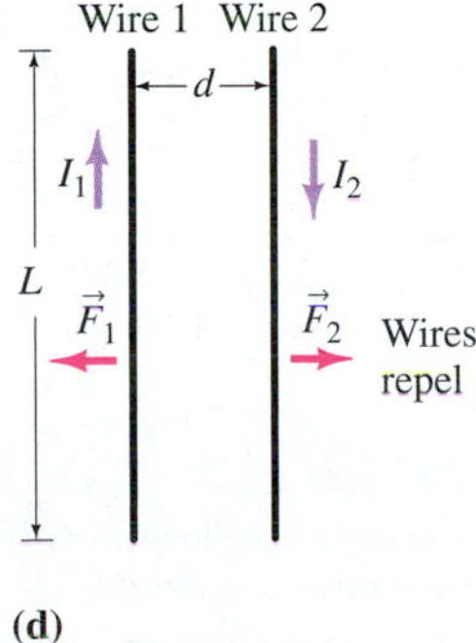

(c) (d)

◀ **FIGURE 29–2** Two parallel wires that carry currents exert forces on one another; these forces are larger when the wires are closer. (a) and (b) The currents are parallel, and the forces are attractive. (c) and (d) The currents are antiparallel, and the forces are repulsive.

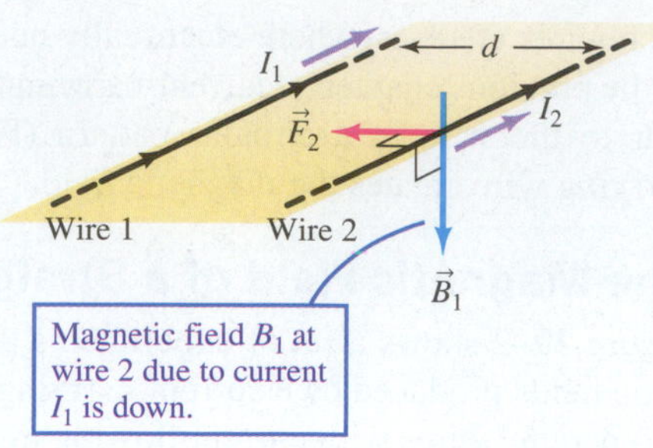

(a)

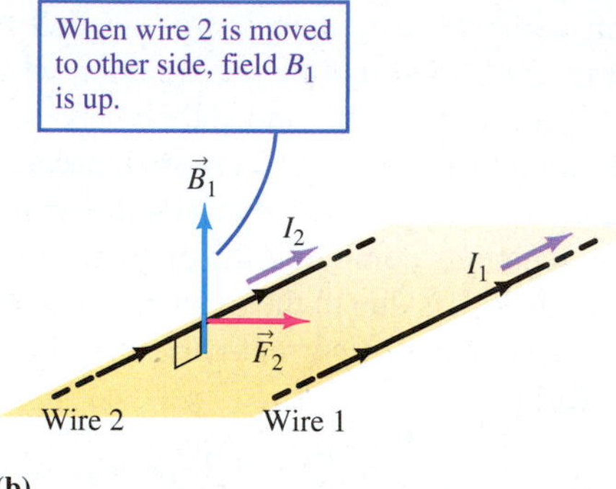

(b)

▶ **FIGURE 29–3** Determining the direction of the magnetic field due to wire 1. Currents I_1 and I_2 are parallel to each other. (a) According to the right-hand rule, $\vec{B}_1$ due to wire 1 is directed down when wire 2 is to the right of wire 1. (b) $\vec{B}_1$ due to wire 1 is directed up when wire 2 is to the left of wire 1.

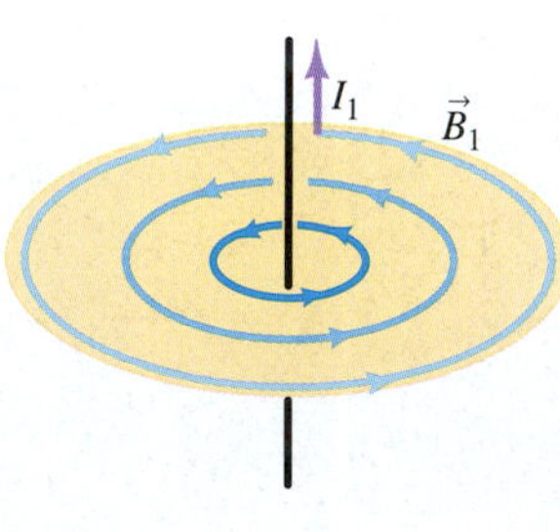

(a)

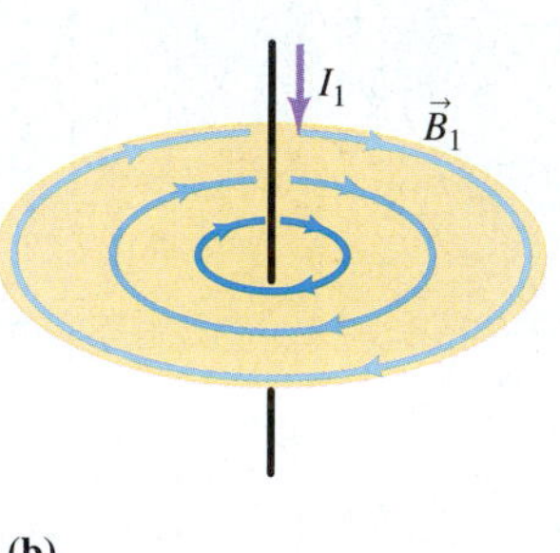

(b)

▲ **FIGURE 29–4** (a) The magnetic field due to wire 1, $\vec{B}_1$, traces out a circle around the wire in the direction shown. (b) If the current in wire 1 were reversed, the orientation of $\vec{B}_1$ would change.

If we move wire 2 around wire 1 in such a way that the wires remain parallel and at the same separation (moving from the position in Fig. 29–3a to that in Fig. 29–3b), we find that the force remains attractive and does not change in magnitude. This observation and Eq. (29–1) are consistent with the interpretation that *the magnetic field $\vec{B}_1$ due to wire 1 follows a circle around wire 1*. Application of a right-hand rule in Eq. (29–1) shows that $\vec{B}_1$ must be directed down when wire 2 is in its original position to the right of wire 1 (Fig. 29–3a). If wire 2 is moved to the left of wire 1, however, field $\vec{B}_1$ at wire 2 will be directed up because the two wires continue to attract each other (Fig. 29–3b). By using this argument for other positions, we find that the magnetic field lines due to wire 1 make circles about wire 1 (Fig. 29–4a; see also Fig. 28–2c). Equation (29–1) shows that force $\vec{F}_2$ is insensitive to any component of $\vec{B}_1$ that is *parallel* to the wires, because the vector product of two parallel vectors is zero—there may be a component of $\vec{B}_1$ parallel to the wire, but due to Eq. 29–1, we can't tell from the experiments that we have so far discussed. We have to resort to another argument—see Conceptual Example 29–1—to rule out a component of $\vec{B}_1$ parallel to wire 1.

If we reverse the current in wire 1 as we did in Fig. 29–2b, we find that the force on wire 2 is also reversed—the wires repel—and we interpret this as a change in direction of the magnetic field around wire 1 (Fig. 29–4b). The field lines again form circles around the wire, but in the opposite direction.

CONCEPTUAL EXAMPLE 29–1 In an experiment, a wire (wire 2) is wrapped in a circle around a straight wire (wire 1), as in Fig. 29–5. Measurements show that when constant currents run through them, there are no forces between them, and it is claimed that this shows that wire 1 produces no component of magnetic field parallel to itself. Is this correct?

Answer Yes. Symmetry tells us that any component of $\vec{B}_1$ oriented parallel to wire 1 will be the same all the way around wire 1, so that every segment of the circle formed by wire 2 will be in the presence of the same field $\vec{B}_1$. But then an application of the right-hand rule shows that a component of $\vec{B}_1$ parallel to wire 1 would cause a force to be exerted on wire 2. As Eq. (29–1) indicates, such forces would tend to expand or contract the circle traced by wire 2, and this effect would be measurable. Wire 1 does not produce a component of the magnetic field parallel to itself.

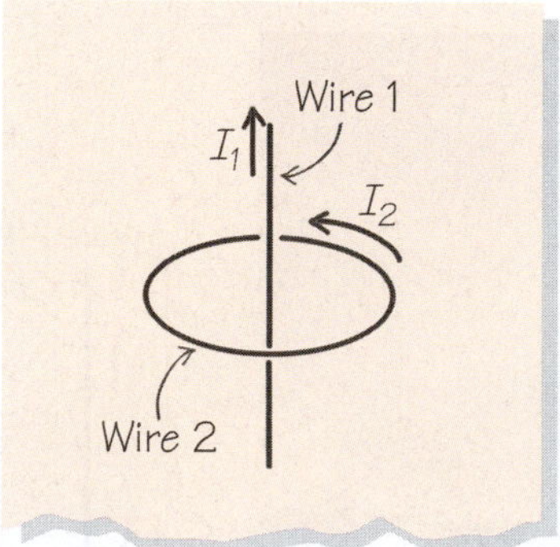

▲ **FIGURE 29–5** If wire 2 traced a circle around wire 1, it would react to any components of $\vec{B}_1$ due to wire 1 that are parallel to wire 1. No such forces are found.

We can summarize the results above by saying that the direction of the magnetic field produced by a straight wire is determined by a right-hand rule (Fig. 29–6):

If the thumb of the right hand is oriented along the direction of current flow in a wire, the fingers curl in the direction of the magnetic field.

We have found the direction of the magnetic field produced by the current. How do we find its magnitude? We do this by measuring the magnitude of the force between the wires, and such measurements show that the magnitude of the force between two parallel, straight segments of wire is

$$F = \frac{CI_1 I_2 L}{d},$$

where I_1 and I_2 are the currents in wires 1 and 2, respectively, d is the separation between the wire segments, and L is their length. The proportionality constant C depends on how we define the units of current. Conversely, if we use a *defined* proportionality constant, the force between two current-carrying wires determines the units of current. This latter alternative is the one used in the SI: C is defined according to

$$F = \frac{\mu_0 I_1 I_2 L}{2\pi d}, \tag{29–2}$$

where the constant μ_0, called the **permeability of free space**, is

$$\mu_0 \equiv 4\pi \times 10^{-7}\ \mathrm{T \cdot m/A}. \tag{29–3}$$

With this definition of C (that is, of μ_0) 1 A is defined as the current that travels in two long, parallel wires of length L that are 1 m apart, such that the attractive force between them is $(2 \times 10^{-7}\ N/m)\ L$. Is this result consistent with previous definitions of the current? In Chapter 21, we defined the coulomb as the charge on two pointlike objects such that there is a certain force between them, and we provisionally defined 1 A as 1 C/s in Chapter 26. The definition of the coulomb in terms of a force between charges depends on another constant, ε_0, in exactly the same way that the definition of the ampere depends on μ_0. Thus, for our relations to be consistent, ε_0 *must be a measured constant*. If both μ_0 and ε_0 are known, the same must be true for their product, which is given by

$$\mu_0 \varepsilon_0 \equiv c^{-2} = (2.99792458 \times 10^8\ \mathrm{m/s})^{-2}. \tag{29–4}$$

The constant c is precisely the speed of light! We shall see in Chapter 34 why this is so.

Comparison of Eqs. (29–1) and (29–2) shows that a long, straight wire that carries a current I gives rise to a magnetic field whose magnitude at a distance r from the wire is

$$B = \frac{\mu_0 I}{2\pi r}. \tag{29–5}$$

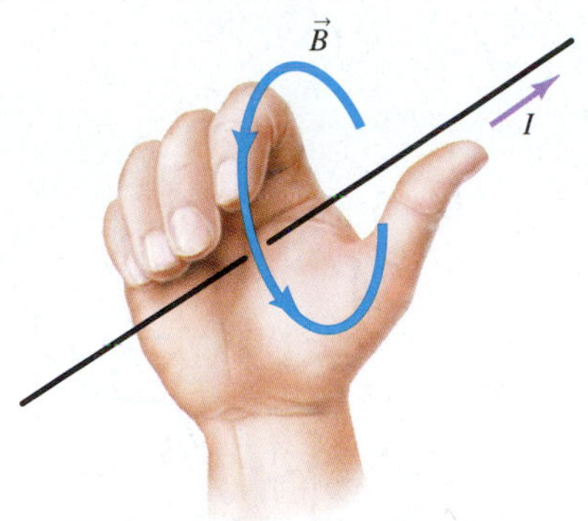

▲ **FIGURE 29–6** A right-hand rule determines the direction of the magnetic field around a current-carrying wire.

Ampère's Law

We can find a more universal form for the magnetic field produced by a current by expressing Eq. (29–5) in a different form, one that relates the magnetic field along a closed path to the electric current that the closed path encloses. This relation is what is known as Ampère's law. We'll see that we can use it to find the magnetic field in symmetric situations much as we use Gauss' law to find the electric field in symmetric situations.

To arrive at Ampère's law, imagine a line integral over the magnetic field $\vec{B}$ that follows a circular path of radius r all the way around a long wire, as in Fig. 29–7. The integration path, labeled C, thus follows the direction of $\vec{B}$. The path is broken into infinitesimal distance elements $d\vec{s}$ that are parallel to the magnetic field, so $\vec{B} \cdot d\vec{s} = B\,ds$. Finally, B is a constant when the distance r from the wire is constant and we can remove it from the integral:

$$\oint \vec{B} \cdot d\vec{s} = B \oint ds = B(2\pi r). \tag{29–6}$$

(The sign $\oint$ indicates that the path of the line integral is closed, going all the way around the circle.) The factor $2\pi r$ is the length of the path, the circumference of the circle of radius r. If we use Eq. (29–5), we find that

$$\oint \vec{B} \cdot d\vec{s} = \frac{\mu_0 I}{2\pi r} 2\pi r = \mu_0 I. \tag{29–7}$$

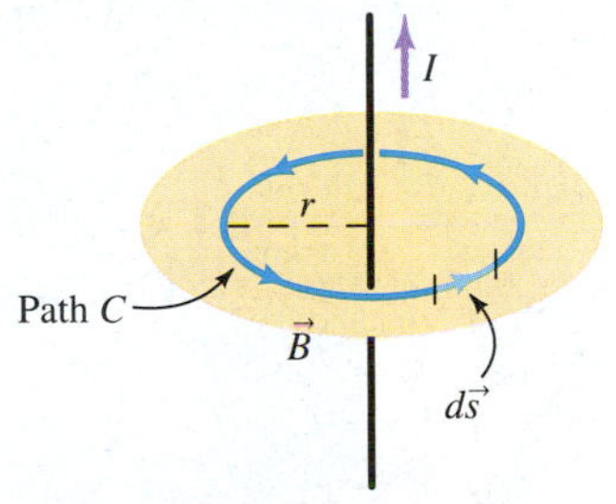

▲ **FIGURE 29–7** Path C circles a current-carrying wire at a constant distance r from the wire and follows the direction of the magnetic field, $\vec{B}$, around the wire.

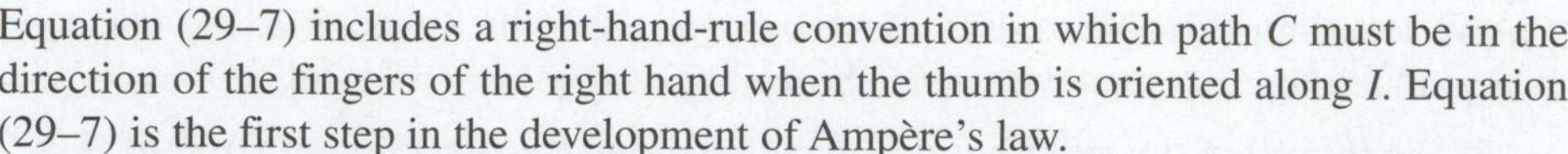

Equation (29–7) includes a right-hand-rule convention in which path C must be in the direction of the fingers of the right hand when the thumb is oriented along I. Equation (29–7) is the first step in the development of Ampère's law.

A current "pierces" the closed path C described above (or more properly pierces a surface whose edge is formed by the path C). Now let us consider a similar integral, but this time over a closed path *not* pierced by a current, for example path C' shown in Fig. 29–8. We thus wish to compute $\oint_{C'} \vec{B} \cdot d\vec{s}$. This time we shall see that the integral vanishes. We first break the path C' into the segment from a to b, the nearly full circle C_2, the segment from c to d, and the nearly full circle C_1. The total contribution of the two paths from a to b and c to d is zero, because $\vec{B}$ is perpendicular to the path there. Thus in the limit that the "nearly" full circles get arbitrarily close to closing,

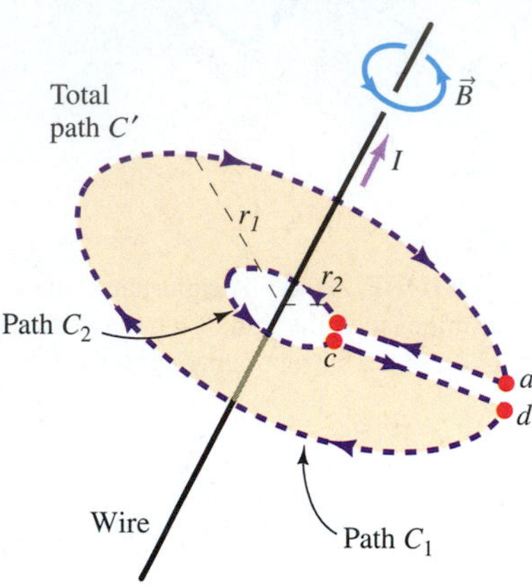

▲ **FIGURE 29–8** This path does not form a closed path around the wire. Path C' consists of a clockwise circle C_1 of radius r_1, a leg from a to b that moves inward to a distance r_2 from the wire, a counterclockwise circle C_2 of radius r_2, and a leg from c to d that moves outward to r_1.

$$\oint_{C'} \vec{B} \cdot d\vec{s} = \int_{C_1} \vec{B} \cdot d\vec{s} + \int_{C_2} \vec{B} \cdot d\vec{s} = -B_1(2\pi r_1) + B_2(2\pi r_2). \qquad (29\text{–}8)$$

Here B_1 is the magnitude of the magnetic field at a distance r_1 from the wire, and B_2 is the magnitude of the field at a distance r_2. The first term is negative because $\vec{B}$ is oriented opposite to the path direction on segment C_1. From Eq. (29–5) we see that the two terms on the right of Eq. (29–8) cancel:

$$\oint_{C'} \vec{B} \cdot d\vec{s} = -\frac{\mu_0 I}{2\pi r_1}(2\pi r_1) + \frac{\mu_0 I}{2\pi r_2}(2\pi r_2) = -\mu_0 I + \mu_0 I = 0. \qquad (29\text{–}9)$$

The difference between the integral over path C and that over path C' arises from the fact that path C encloses current I, whereas path C' encloses no current. This is the first step toward the following generalization. Let the quantity I_{enclosed} be the total current enclosed by *any closed path*. Then

$$\oint \vec{B} \cdot d\vec{s} = \mu_0 I_{\text{enclosed}}, \qquad (29\text{–}10)$$

AMPÈRE'S LAW

where the integral is taken around that closed path. Equation (29–10), which was formulated by Ampère in the 1820s during his extensive work on magnetism, is known as **Ampère's law**. The direction of the loop integral must be specified: If the fingers of the right hand curl in the same sense as the integral path, the thumb points in the direction a positive current takes in passing through the loop. The total current can include both positive and negative contributions, and *the path does not have to be circular, just closed*.

Ampère's law includes an experimental result that is worth pointing out: *The magnetic fields produced by different currents add, or superpose*, just as the electric fields of different charges add according to the superposition principle.

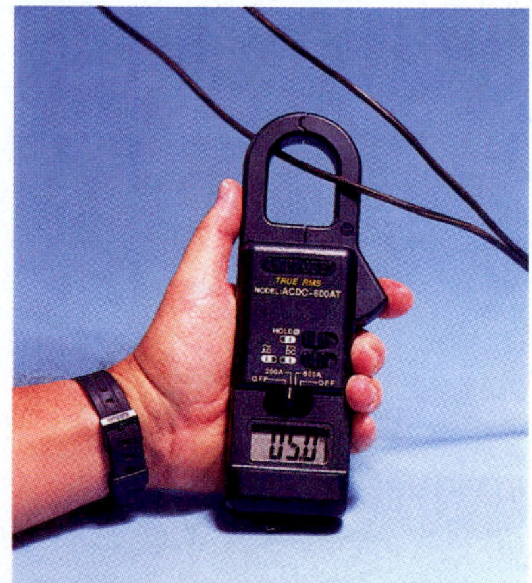

▲ **FIGURE 29–9** This clip-on ammeter uses Faraday's law to measure the current through the wire it surrounds, here 5.0 A. The jaws of the ammeter can be opened to allow its positioning.

THINK ABOUT THIS. . .

CAN YOU MEASURE A CURRENT WITHOUT ENTERING THE CIRCUIT?

It is not always easy to insert an ammeter directly into a circuit wire, for example a wire from your car's battery. If we want to measure the current carried by one or more wires in a bundle of many wires without physically entering the circuit, we can use a device known as a clip-on ammeter, shown in Fig. 29–9. It works by measuring the magnetic field that the current in the wire produces; Ampère's law then gives the current. In its simplest form the magnetic field is measured with a permanent dipole magnet pointer mounted on a center point bearing much like a magnetic compass. When a current is present in the wire, the magnetic field generated deflects the pointer. The deflection angle depends on the torque induced by the magnetic field—and hence the current. In more sophisticated versions of the clip-on ammeter, the magnetic field sensor is a Hall effect sensor (see Chapter 28). Clip-on ammeters can be found in automobile parts stores; mechanics use them to measure the current through a starter motor.

Using Ampère's Law to Find the Magnetic Field

If there is some symmetry that suggests that the integral over a particular path is simple, then Ampère's law [Eq. (29–10)] can be used to *find* the magnetic field, in analogy with the way we use Gauss' law to find electric fields. In the case of Gauss' law, the integral is taken over a closed surface, and $\vec{E}$ is related to the electric charge enclosed. In the case of Ampère's law, the integral taken is along a closed path, and $\vec{B}$ is related to the electric current enclosed by the path.

EXAMPLE 29–2 The current I within a wire that has a circular cross section of radius R is known to be distributed uniformly over that cross section. (Real currents in real wires would show some variance from this.) What is the magnetic field as a function of the distance r from the wire's axis outside the wire, and what is it within the wire?

Setting It Up We show current moving through the wire in Fig. 29–10. The wire has cylindrical symmetry—it looks the same as we move around it at a fixed distance.

Strategy The cylindrical symmetry means that any magnetic field will not vary with the angle around the wire; the field is a function only of the radial distance r from the central axis. We can apply Ampère's law—Eq. (29–10)—for a circular path of radius r centered on the middle of the wire, and use the fact that $\vec{B}$ will be the same all along this path. This will allow us to pull the magnitude of the magnetic field from the integration of Ampère's law. In turn, we can then use information about the current enclosed by the path to solve for the field as a function of r. The amount of current enclosed depends on whether the path lies outside or inside the wire. Figure 29–10a shows a path *outside* the wire that will determine the field outside, and Fig. 29–10b shows a path *inside* the wire that determines the field inside.

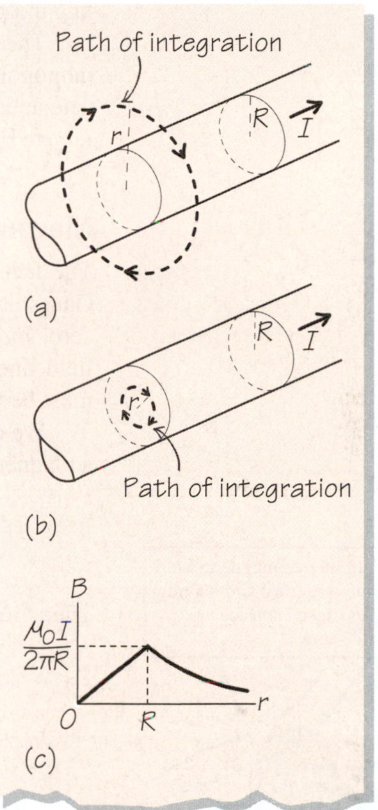

▲ **FIGURE 29–10** (a) A circular path of radius r is used to determine the magnetic field outside a wire that carries a current I. (b) A similar path inside the wire. (c) The magnitude of the magnetic field versus r.

Working It Out By the right-hand rule, $\vec{B}$ is oriented in the direction of the path, so $\vec{B}\cdot d\vec{s} = B\,ds$. The magnetic field magnitude is constant over the chosen path and thus comes out of the integral, leaving just the circumference of the path. If the circular path is outside the wire, the current enclosed is the *total* current carried by the wire. Thus Ampère's law becomes

$$\oint \vec{B}\cdot d\vec{s} = \oint B\,ds = B\oint ds = B(2\pi r) = \mu_0 I.$$

We can solve for B to find that

$$B = \frac{\mu_0 I}{2\pi r},$$

the same result we found for a thin wire. Note that the magnetic field outside the wire is independent of the size of the wire, just as the electric field outside a spherically symmetric charge distribution is independent of the size of the distribution.

We continue to use symmetry to find the field inside the wire, but this time we take our circular path *inside* the wire (Fig. 29–10b). The current enclosed by the path is I times the ratio of the area of the circle of radius r to the area of the wire:

$$I_{\text{enclosed}} = I\frac{\pi r^2}{\pi R^2}.$$

As before, Ampère's law gives

$$\oint \vec{B}\cdot d\vec{s} = B(2\pi r) = \mu_0 I\left(\frac{\pi r^2}{\pi R^2}\right).$$

If we solve for B, we find that

$$B = \frac{\mu_0 I}{2\pi R^2}r.$$

By analogy with Gauss' law for electricity, any current outside a circle of radius r makes no contribution to the net magnetic field at radius r. Inside the wire, the magnetic field decreases linearly to zero as r approaches zero. As a check, we see that the results for outside and inside the wire agree at $r = R$. Figure 29–10c is a graph of the magnitude of the magnetic field.

What Do You Think? Would this method give us the same magnitude and direction for the magnetic field if the current were reversed? *Answers to **What Do You Think?** questions are given in the back of the book.*

29–2 Gauss' Law for Magnetism

We have already studied Gauss' law for electricity. Is there a Gauss' law for magnetism similar to that for electricity? In this section, we shall see in what sense such a law holds.

There is an important difference between electricity and magnetism that determines the form of Gauss' law for magnetism: Despite much experimental effort, *magnetic charges* (**monopoles**) *have never been observed.* A magnetic monopole would be a source of magnetic field analogous to electric charges as sources of the electric field. Finding magnetic monopoles would be the equivalent of being able to isolate the N or S poles of a bar magnet. The bar magnet's magnetic field looks from the outside like the electric field of an electric dipole, and we can of course separate the plus and minus charges that form the electric dipole. But when we try to separate the N and S poles of a bar magnet by cutting the magnet in two, we end up with two smaller bar magnets.

The consequence of these observations for the magnetic field is that if magnetic monopoles analogous to electric charges existed, then magnetic field lines would originate and terminate on magnetic monopoles, just as electric field lines originate and terminate on electric charges. But since there are no magnetic monopoles, *magnetic field lines, unlike electric field lines, must form closed curves.*

Magnetic Flux and Gauss' Law for Magnetism

The fact that there are no magnetic charges means that a relation similar in form to Gauss' law for electricity holds for magnetism, but with the electric charge replaced by zero. In other words, Gauss' law for magnetism is equivalent to the statement that if field lines are continuous, then the number of field lines entering any closed surface must be the same as the number leaving.

We come to the explicit statement of the law by defining the **magnetic flux** Φ_B for a magnetic field $\vec{B}$ over a surface S, open or closed, by

$$\Phi_B(S) \equiv \int_{\text{surface } S} \vec{B} \cdot d\vec{A}. \tag{29–11}$$

Then **Gauss' law for magnetism** is

$$\text{for a closed surface: } \Phi_B = \int_{\text{closed surface}} \vec{B} \cdot d\vec{A} = 0. \tag{29–12}$$

GAUSS' LAW FOR MAGNETISM

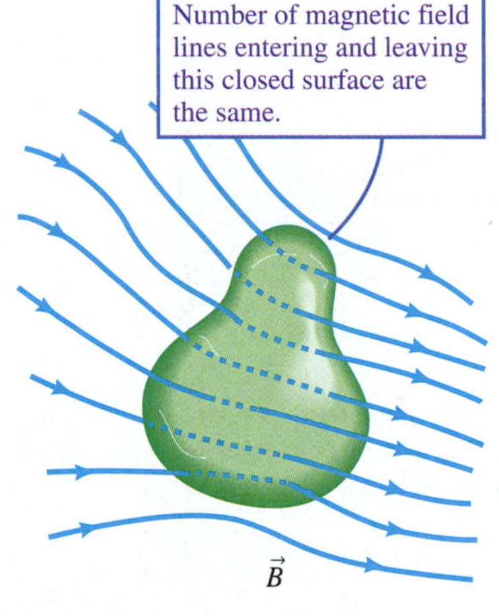

▲ **FIGURE 29–11** Magnetic field lines are everywhere parallel to the magnetic field; their density measures the field's strength. There are no magnetic charges, so magnetic field lines do not end, as there is nothing for them to originate or end upon, and the magnetic flux through a closed surface is zero. This is Gauss' law for magnetism.

As for the electric flux, infinitesimal surface elements $d\vec{A}$ are perpendicular to the surface and, for a closed surface, are oriented outward. We can easily see why this holds by thinking in terms of field lines and noting that the field lines for a magnetic field do not terminate. The magnetic flux is a measure of the number of field lines that go out of a surface minus the number of lines that go in. But since magnetic field lines are continuous, the number of magnetic field lines that enter a closed surface minus the number that leave the surface is zero (Fig. 29–11). Any magnetic field line entering a closed surface must leave it somewhere because there are no magnetic monopoles inside on which magnetic field lines can begin or end.

The SI unit for magnetic flux is the unit of magnetic field times area, that is, tesla times square meters $(\mathrm{T \cdot m^2})$. This unit occurs often enough to be given its own name in SI, the **weber** (Wb), after Wilhelm Eduard Weber:

$$1\ \mathrm{Wb} \equiv 1\ \mathrm{T \cdot m^2}. \tag{29–13}$$

The Field Lines of a Bar Magnet

When we drew the magnetic field lines for a bar magnet in Chapter 28, it may have seemed natural to think of the field lines as starting on the north pole and ending on the south pole. In light of what we have just learned, we now realize that magnetic field lines never start or stop—they are *continuous*. Therefore, we must reconsider our view of the

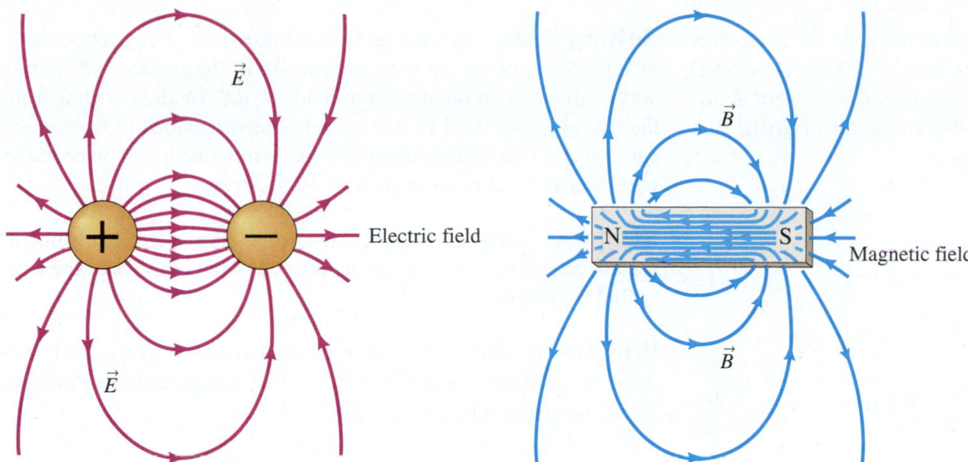

▲ **FIGURE 29–12** Although electric field lines outside an electric dipole resemble magnetic field lines outside a bar magnet, the fields within are quite different. The electric field lines begin and end on the electric charges, whereas the magnetic field lines are continuous.

magnetic field lines for a bar magnet. In fact, the field lines do not start or stop at the poles *but pass through the bar magnet*. This remark underlines a critical difference between the electric field of an electric dipole (equal and opposite charges) and the magnetic field of a bar magnet, both shown in Fig. 29–12. "Outside" the ends, the field lines have the same form. However, while the field lines for the electric dipole begin and end on the charges, the field lines that run from the north pole to the south pole outside the magnet *return within the magnet to form closed loops*. This view is consistent with Gauss' law for magnetism, Eq. (29–12), because for any closed surface that can be drawn in and around a bar magnet, the same number of field lines enter the surface as leave it.

Using Gauss' Law to Find Magnetic Fields

Gauss' law for magnetism is useful for limiting the forms a magnetic field may take. As an example, let's use Gauss' law for magnetism to show that the magnetic field around a straight current-carrying wire can have no radial component, as we concluded in Section 29–1. We need a suitable closed (imaginary) surface to construct about the wire in order to exploit any symmetries, and for a straight wire cylindrical symmetry is appropriate. Our closed surface will therefore be a cylinder of radius R and length L whose central axis lies on the wire (Fig. 29–13). The wire looks the same from any point on the surface of the cylinder, so the magnetic field cannot depend on the angle around the axis of the cylinder. Thus, if there were a radial component B_r at some fixed radial distance, it would have to be the same all around the wire.

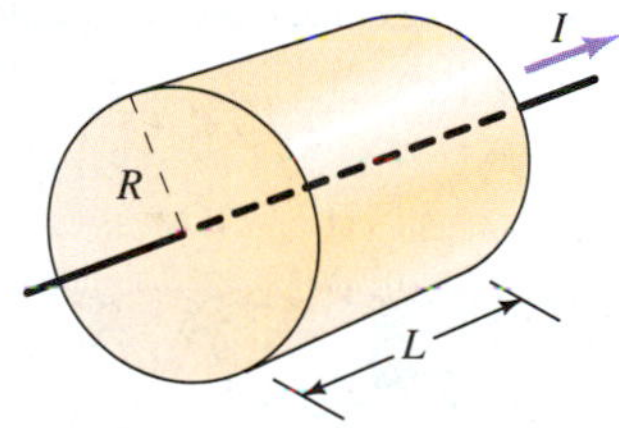

▲ **FIGURE 29–13** A Gaussian surface that exploits the symmetry of a long, straight wire.

To find the net magnetic flux through the closed cylinder, Φ_B, we must consider contributions from its ends and sides. Only a longitudinal component of the field (along the wire) contributes to the flux at the ends. But the contribution from one end must cancel the contribution from the other end—if the longitudinal component enters the surface at one end, it must leave the surface at the other end. The net flux through the ends is therefore zero.

The contribution to the magnetic flux from the sides is due to the radial component of the field, B_r. We have for the net flux

$$\Phi_B = \int_{\text{end}} \vec{B}\cdot d\vec{A} + \int_{\text{side}} \vec{B}\cdot d\vec{A} = \int_{\text{side}} \vec{B}\cdot d\vec{A} = B_r(2\pi RL),$$

where $2\pi RL$ is the area of the cylinder's sides. This must equal zero by Gauss' law and, because the area of the cylinder sides is not zero, B_r must be zero. We have shown by Gauss' law that there can be no radial component of the magnetic field.

We shall see in later chapters that the magnetic flux plays an important role in other fundamental laws of electromagnetism, and it is important to know how to calculate the magnetic flux for both closed and open surfaces. Example 29–3 is an exercise of this type.

EXAMPLE 29–3 The region between the poles of an electromagnet† contains a constant magnetic field, $B = 0.0030$ T, oriented in the $+x$-direction. A square wire loop of sides $L = 1.0$ cm is oriented at a 30° angle to the field (Fig. 29–14a). Find the magnetic flux through the loop.

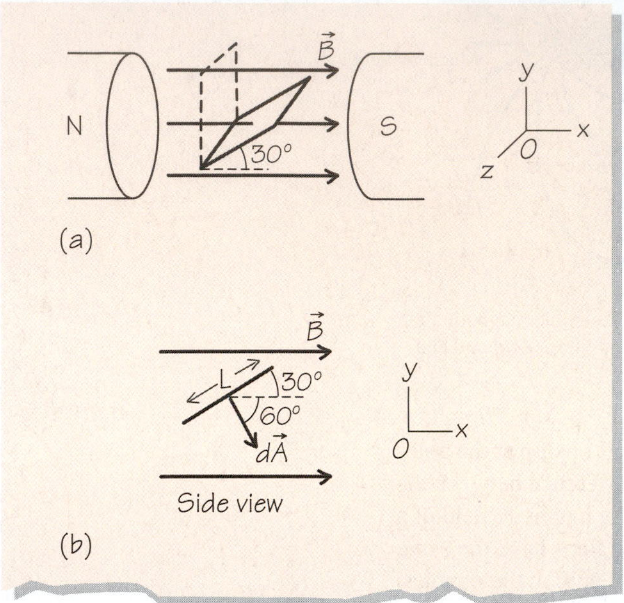

▲ **FIGURE 29–14** (a) A tabletop electromagnet for which $\vec{B}$ is oriented in the $+x$-direction. (b) The surface element $d\vec{A}$ of the square wire loop is oriented perpendicular to the surface.

Setting It Up In order to find the flux through a given surface, we must first specify an area element $d\vec{A}$ of the surface and find its scalar product with the magnetic field, $\vec{B} \cdot d\vec{A}$. We draw a view from the side in Figure 29–14b that includes the orientation of the surface element $d\vec{A}$. This surface element is perpendicular to the plane of the wire loop and makes an angle $\theta = 60°$ with $\vec{B}$.

Strategy Having established the orientation of the area element, we calculate $\vec{B} \cdot d\vec{A}$, then integrate over the whole surface of the loop to find the magnetic flux.

Working It Out The scalar product $\vec{B} \cdot d\vec{A} = B \cos\theta \, dA$. Because B and θ are constants, they can be removed from the integral for the magnetic flux [Eq. (29–11)]:

$$\Phi_B(S) = \int_{\text{surface } S} \vec{B} \cdot d\vec{A} = \int_{\text{surface } S} B \cos\theta \, dA$$

$$= B \cos\theta \int_{\text{surface } S} dA = BL^2 \cos\theta.$$

The numerical value is

$$\Phi_B(S) = (0.0030 \text{ T})(1.0 \times 10^{-2} \text{ m})^2 (\cos 60°)$$
$$= 1.5 \times 10^{-7} \text{ Wb}.$$

What Do You Think? For what orientation of the loop will the flux be a maximum?

†An electromagnet is a magnet whose field is produced by appropriately circulating electric currents. It is typically set up so that there are two parallel flat faces separated by a space, with a roughly constant magnetic field running from one face (the N pole) to the other (the S pole)—in the next section, we'll describe how this might be done. It thus resembles a horseshoe magnet whose tips have been formed into the two flat faces.

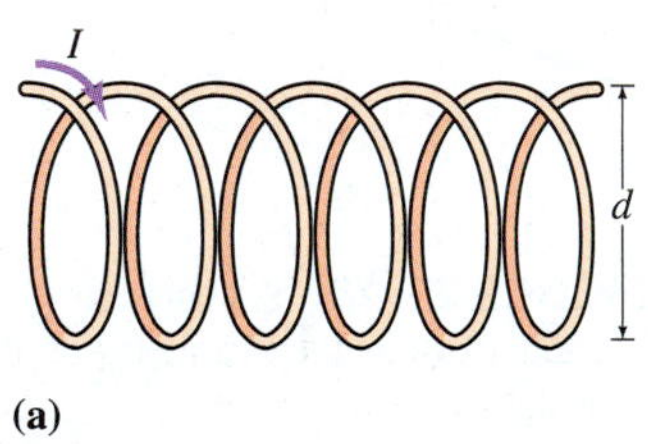

(a)

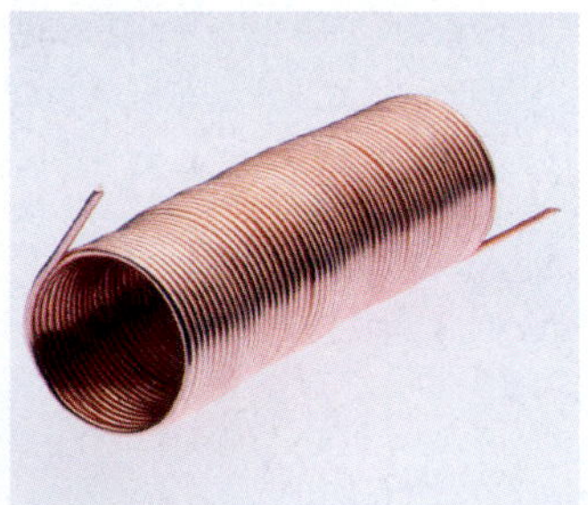

(b)

▲ **FIGURE 29–15** (a) An ideal solenoid is an infinitely long cylinder made from a uniformly wound coil carrying current I (view shown is exaggerated). (b) A real solenoid.

29–3 Solenoids

A parallel-plate capacitor produces a uniform, constant electric field between its plates. The equivalent device for magnetic fields, a **solenoid**, is a length of wire coiled uniformly into a long cylinder, ideally infinitely long (Fig. 29–15). A solenoid generates a constant magnetic field in the interior of the cylinder it forms just as a parallel-plate capacitor sets up a constant electric field in the space between its plates. In Chapter 32 we'll see that, like capacitors, solenoids play an important role in circuits. We mention one use as a mechanical switch: An iron piece near an end of a solenoid moves in response to a magnetic force that appears when a current starts to run through the solenoid.

Let's sketch what the magnetic field of a solenoid might look like. Consider a solenoid whose cylinder diameter is d carrying current I, with the wires wound so that there are n turns per unit length, where the length is measured along the axis of the solenoid. In Fig. 29–15a we have exaggerated the spacing between the wires, which normally are tightly wound. Figure 29–16a is a cross-sectional view of several loops of the solenoid, which are again spaced more widely than in reality. Very near the wire, the magnetic field lines form circles around the wire because the field approximates that of a single straight wire. Figure 29–16a shows that *between* adjacent turns of the wire where two segments of wire are near one another, the fields tend to cancel. *Within* the solenoid, the fields from adjacent turns add together to form a large component that points to the right along the axis of the solenoid. *Outside* the solenoid the fields from the bottom of the wire loop tend to cancel the fields from the top of the wire loop, so that as the number of loops becomes large, the field outside is small (Fig. 29–16b).

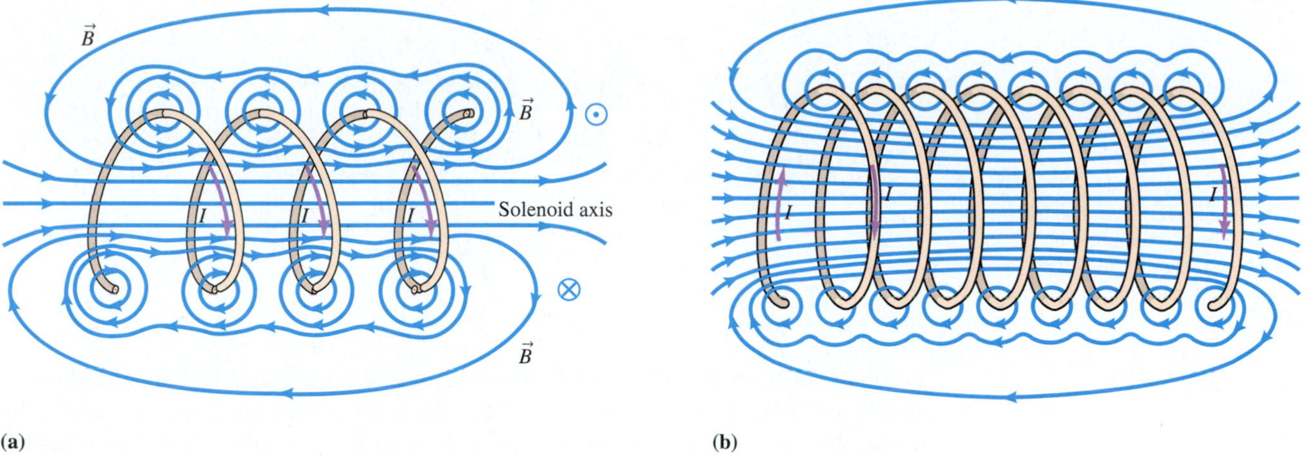

▲ **FIGURE 29–16** (a) One three-turn section of a solenoid, showing the superposed magnetic fields (view shown is exaggerated). (b) As the winding density increases, the magnetic field takes on a simpler form.

To summarize, the fields from the different loops of the coil reinforce inside the cylinder to create a net magnetic field that is parallel to the cylinder axis (Fig. 29–17a) and whose direction is determined by a right-hand rule: If the fingers curl around the solenoid in the direction of the current, the thumb shows the direction of the magnetic field (Fig. 29–17b). Outside the cylinder, the field points primarily in the opposite direction and is much weaker. A quantitative analysis shows that *even though the magnetic field is not exactly zero outside a real solenoid, to a good approximation the field there is insignificant.*

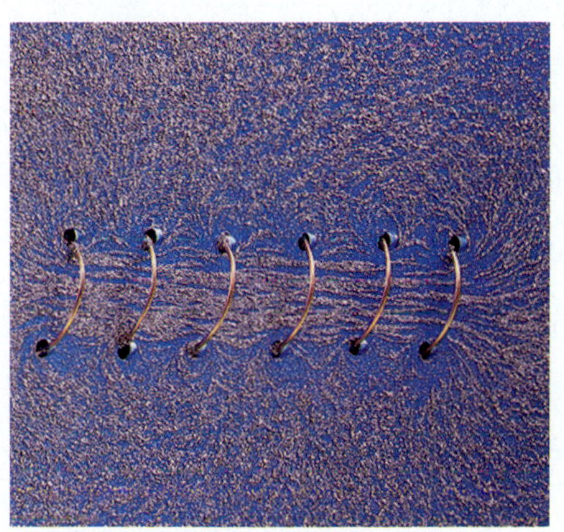
(a)

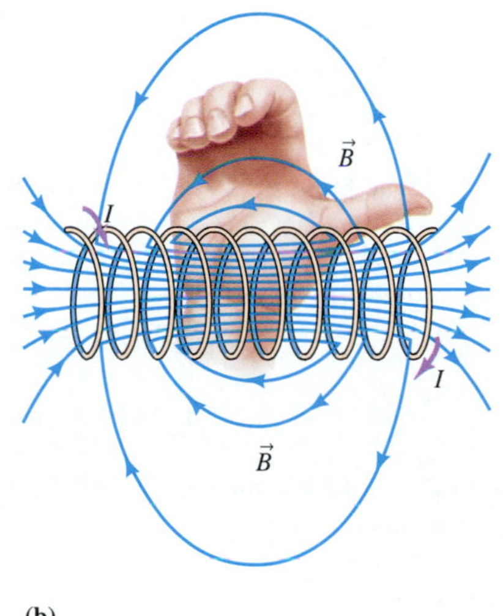

(b)

◀ **FIGURE 29–17** (a) Magnetic field lines of a solenoid, as shown by iron filings that align with the field. (b) A right-hand rule gives the direction of the magnetic field within a solenoid.

Using Ampère's Law to Find the Magnetic Field in a Solenoid

Now that we understand qualitatively that a long solenoid has a large magnetic field inside—parallel to the solenoid axis—and a weak field outside, we can apply Ampère's law to calculate quantitatively the magnetic field inside an ideal solenoid—one so long that the field outside can be taken to be zero. Figure 29–18 shows a solenoid that carries a current I, and an *imaginary* closed area, bounded by a loop consisting of four legs in a rectangle of length ℓ and height w, on which to apply Ampère's law. The wire of the solenoid passes N times from above through the imaginary loop. The path about the imaginary loop is taken to be clockwise, so the net current into the imaginary loop, NI, is positive by the right-hand rule. We now calculate the line integral on the left-hand side of Eq. (29–10). There is only a very small contribution from leg 2 (point b to point c), because the field outside is insignificant. There is no contribution from leg 1 (point a

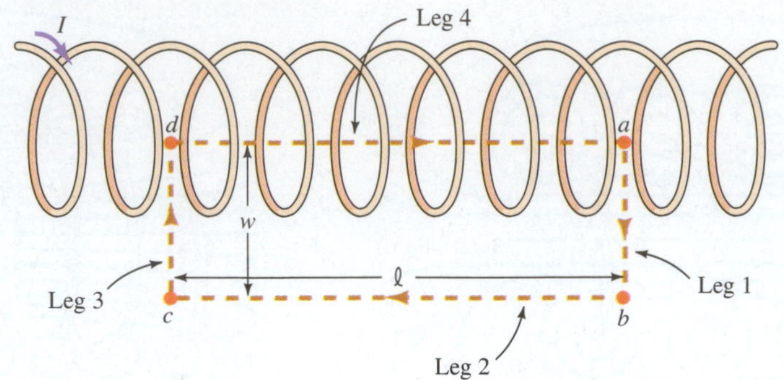

▶ **FIGURE 29–18** An imaginary rectangular loop is drawn half inside and half outside a solenoid. This loop provides a path for the application of Ampère's law.

to point *b*) or from leg 3 (point *c* to point *d*) for two reasons. First, the field outside is insignificant, and the field inside is parallel to the cylinder axis and hence perpendicular to the path. Second, any contributions from these two legs would cancel each other because they are in opposite directions. From point *d* to point *a* (leg 4), the field is parallel to the path. Along this portion of the path, the field has a constant unknown value *B*. The contribution to the integral is $B\ell$, and thus Ampère's law gives

$$\oint \vec{B} \cdot d\vec{s} = B\ell = \mu_0 I_{\text{enclosed}} = \mu_0 NI. \tag{29–14}$$

We can eliminate the explicit dependence on ℓ in Eq. (29–14) by noting the total number of turns $N = n\ell$, where *n* is the *turn density* of the solenoid, the number of turns per unit length. We have

$$B\ell = \mu_0 n\ell I,$$

and the *interior magnetic field of a long solenoid* has magnitude

$$B = \mu_0 nI. \tag{29–15}$$

MAGNETIC FIELD WITHIN A SOLENOID

Note that Eq. (29–15) contains no reference to the distance from the axis on the inside of the loop. Our derivation is completely independent of how close the imaginary path in Fig. 29–18 comes to the solenoid axis, and any choice of this distance would give the same field. *The magnetic field inside a long solenoid, not too close to the ends, is uniform.* The magnetic field depends linearly on the current.

EXAMPLE 29–4 A solenoid consists of wires—each of diameter $d = 0.6$ mm—that can carry a maximum current of $I = 0.03$ A; the wires are tightly wound in a single layer. What is the maximum magnitude of the field inside the solenoid?

Strategy Equation (29–15) gives us the magnitude of the unknown magnetic field inside the solenoid in terms of the current *I* and the turn density, *n*. We are given *I*, but we must calculate *n*. If the wire has diameter *d*, and the wires are tightly wound in a single layer, then we have one turn every length *d*, and $n = 1/d$.

Working It Out We have

$$B = \mu_0 nI = \frac{\mu_0 I}{d} = \frac{(4\pi \times 10^{-7}\ \text{T}\cdot\text{m/A})(0.03\ \text{A})}{0.6 \times 10^{-3}\ \text{m}}$$
$$= 0.6 \times 10^{-4}\ \text{T}.$$

This is comparable in magnitude to Earth's field.

What Do You Think? What would happen to the magnetic field if we double the number of turns of wire by wrapping another row on top of the existing one?

CONCEPTUAL EXAMPLE 29–5 Assume that the maximum current a wire can carry in the previous example is proportional to the area of the wire, and that the wire's diameter is the only variable under consideration. The winding will continue to be in a single layer. How should we change the diameter of the wire to double the magnetic field inside?

Answer If *B* is to double, then the product *nI* must double according to Eq. (29–15). Let's see how these factors depend on the diameter *d*. Because the solenoid is tightly wound, the turn density $n \propto 1/d$. The maximum current is proportional to the wire area, which in turn is proportional to the wire diameter squared. Thus $I \propto d^2$. Combining our results, $nI \propto (1/d)d^2 \propto d$. If the field is to be doubled, the diameter of the wire used in the solenoid should also be doubled.

We have used Ampère's law to calculate the magnetic field inside a long, cylindrical solenoid, and we can also apply the law in the same way to noncylindrical geometry—say, a solenoid with square cross section—with exactly the same results. Equation (29–15) holds *even if the winding does not form a cylindrical tube*. We require only that the solenoid be long and that the cross-sectional area be constant.

Our results for the ideal solenoid hold rather well for a solenoid of finite length. Figure 29–19 shows the magnetic field lines, numerically calculated, of a solenoid in a plane that cuts through the center of the solenoid, whose length is four times its diameter.

The exterior field of the solenoid of finite length illustrated in Fig. 29–19 looks just like the magnetic field of a bar magnet, Fig. 28–2a. Does this mean that the field of a bar magnet has the same physical origin as that of a solenoid? As Ampère himself suggested, the answer is yes. A bar magnet is made of the equivalent of aligned current loops of atomic size. In fact, the interior field of a bar magnet is also the same as the interior field of a solenoid. The origin of magnetism in matter is discussed further in Chapter 31.

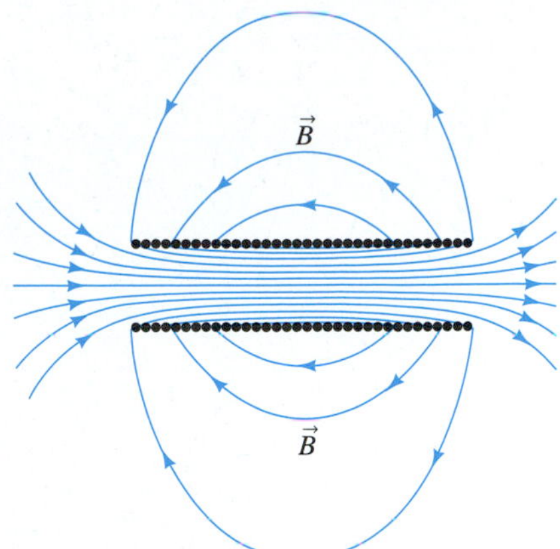

▲ **FIGURE 29–19** The magnetic field of a solenoid of finite length. (After E. M. Purcell, *Berkeley Physics Course: Electricity and Magnetism*, McGraw-Hill, 1990, p. 229.)

A Toroidal Solenoid

A real solenoid has finite length, and therefore its magnetic field departs from its constant value as one approaches either end. These end effects can be eliminated by making the solenoid into a doughnut shape, or *torus* (Fig. 29–20). This shape does, however, introduce some variation in the magnetic field within the solenoid. To look at this variation, consider a toroidal solenoid whose coil radius is r_0 and whose overall radius—the distance from the center to the circular axial line—is R. Symmetry implies that the magnetic field within the coil must be parallel to the cylinder walls. The same arguments that we gave for the straight solenoid imply that this field is oriented in the direction of the thumb when the fingers of the right hand are curled in the direction of the current. Ampère's law can then be used to find the magnitude of $\vec{B}$. Take a circular path for Ampère's law that lies within the coil a distance R' from the center (Fig. 29–20). The magnitude of the field is the same all along the path, so

$$\oint \vec{B} \cdot d\vec{s} = B(2\pi R') = \mu_0 NI.$$

Here, N is the total number of loops in the coil. We solve to find the field magnitude at a distance R',

$$B = \frac{\mu_0 NI}{2\pi R'}. \qquad (29\text{–}16)$$

Although there are no end effects, the field does depend on R'—it is not constant across the cross-sectional area of the torus. But if the coil's radius r_0 is much less than the overall radius R of the torus, the possible values of R', from $R - r_0$ to $R + r_0$, do not vary much, and the magnetic field within the torus will not vary very much either.

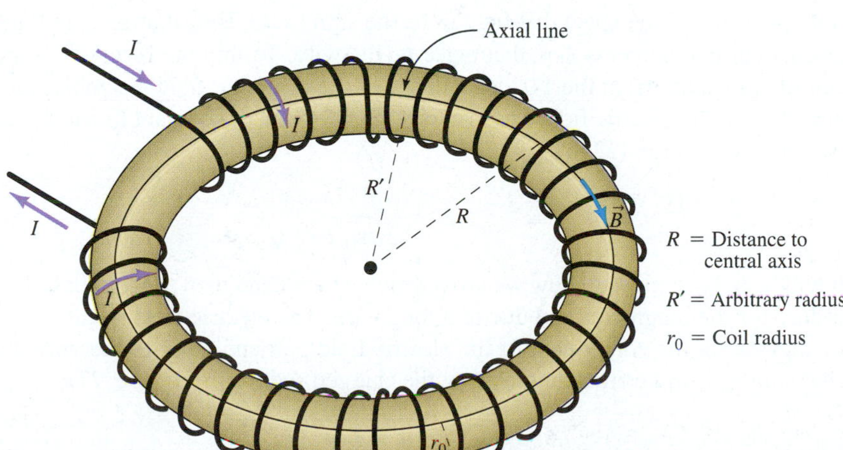

◀ **FIGURE 29–20** A torus wrapped with a wire that carries a current I has a magnetic field inside, which we can calculate by using Ampère's law. The overall radius of the torus is R, whereas the radius of the coil is r_0. The distance R' is not equal to R.

THINK ABOUT THIS. . .

WHAT CAUSES EARTH'S MAGNETIC FIELD?

In the regions outside Earth itself, Earth's field is that of a magnetic dipole, which is the form of the magnetic field due to a bar magnet or to a current loop or coil. The field's origin is in Earth's high-temperature core, which is mainly composed of iron, with a solid inner core and a fluid outer core. In fact, Earth's core cannot be a gigantic bar magnet, which is formed by the alignment of microscopic magnetic dipoles in substances like solid iron (Section 31–3), because when a bar magnet is too hot, the alignment is spoiled by the high temperature, and Earth's core is far too hot to maintain such an alignment. It is rather associated with the convective motion of the fluid iron, as well as with the rotation of the solid core at a slightly greater speed than Earth's surface. These motions effectively carry current that result in magnetic field lines with the same form as that of a bar magnet or current loop. The behavior of this system is too complex to yield an analytic solution, but it has been successfully modeled. The best mathematical model, which requires large supercomputers to run, has some surprising features. This model confirms the speed of rotation of Earth's solid core. It also explains the hitherto puzzling fact that Earth's magnetic field reverses rather suddenly every hundred thousand years or so (although the reversal is not entirely regular). These reversals are obvious in the geological rock record, because molten rock solidifies with a record of Earth's magnetic field frozen within it. (The Sun reverses its magnetic field every 11 years, the difference with Earth being that the Sun does not have a solid core.) ■

29–4 The Biot–Savart Law

Ampère's law is a general one, but its usefulness as a tool for calculating magnetic fields depends on the symmetry of the system of currents that create the magnetic field. Here we find a direct expression known as the *Biot–Savart law* for the magnetic field produced by a current, one that can be applied even when there is no symmetry. There is a simple analogue to this procedure in electrostatics. When there is symmetry in a charge distribution, Gauss' law provides a powerful tool for finding the electric field. When there is no symmetry, we can always find the net electric field by using the superposition of the electric fields of point charges (as determined by Coulomb's law). Similarly, the Biot–Savart law gives us the magnetic field due to an infinitesimal distribution of current segments. We then use the superposition principle to determine the magnetic field of a finite arrangement of currents.

Let's start with a result we already know. The magnitude of the magnetic field at a radial distance r from a long, straight wire that carries a current I is

$$B = \frac{\mu_0 I}{2\pi r},$$

Eq. (29–5). The field lines form circles around the wire with the direction given by the right-hand rule. We expect this field to be the sum of the contributions of all the infinitesimal current elements $I\,d\ell$ that make up the wire. To find the form of the individual contributions, note that the $1/r$ dependence of the magnetic field resembles the $1/r$ dependence of the electric field due to a long, charged rod of constant linear charge density λ, as given in Eq. (22–31):

$$E = \frac{1}{2\pi\varepsilon_0}\frac{\lambda}{r}.$$

It will be helpful to recall how we arrived at this result and then to look for a similar procedure for the magnetic field due to a long wire. The electric field result was obtained by integrating the component of the electric field perpendicular to the wire due to the charge in an infinitesimal length $d\ell$ of the charged rod, from Eq. (22–27):

$$dE_\perp = \frac{1}{4\pi\varepsilon_0}\frac{\lambda\,d\ell}{r^2}\cos\phi.$$

Here, the particular element of the charged rod is a distance $r = \sqrt{L^2 + d^2}$ from the point where the field is measured, and ϕ is as shown in Fig. 29–21a. The factor $\dfrac{\lambda\, d\ell}{4\pi\varepsilon_0 r^2}$ is just the electric field strength for a point charge $dq = \lambda\, d\ell$. The second factor, $\cos\phi$, is present only because we are looking at the perpendicular component. If we compare Eqs. (22–31) and (29–5), we see that they have the same form, but with $1/\varepsilon_0$ replaced by μ_0 and λ replaced by I. This suggests that the individual contribution to the magnetic field of a length $d\ell$ of a wire that carries a current I can similarly be found by taking the contribution to the electric field from an element of the charged wire, Eq. (22–27), and replacing $1/\varepsilon_0$ with μ_0 and λ with I:

$$dB = \frac{\mu_0}{4\pi}\frac{I\, d\ell}{r^2}\cos\phi. \tag{29–17}$$

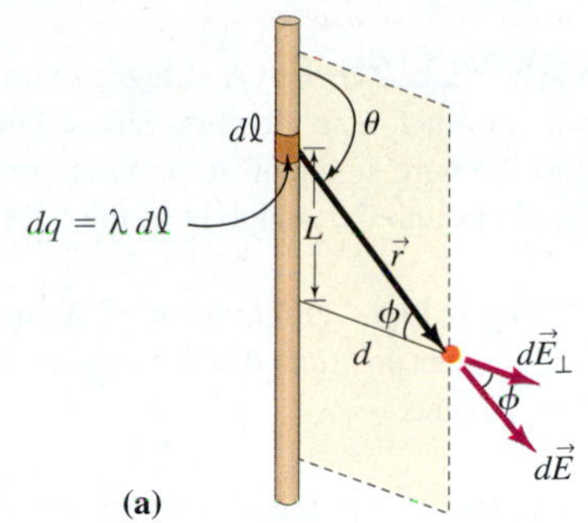

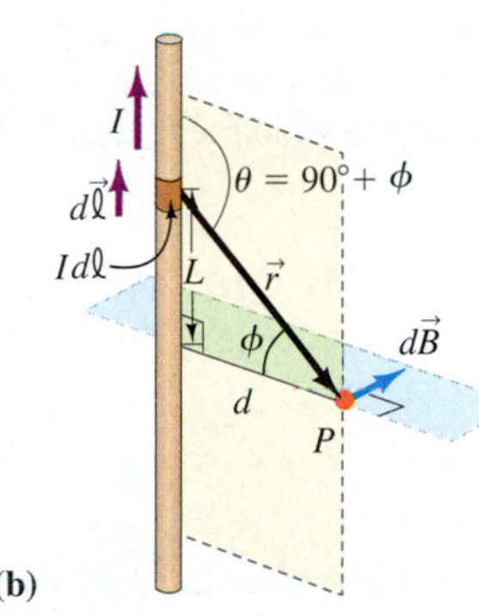

▲ **FIGURE 29–21** (a) A rod that carries a net charge density λ can be broken up into segments that contribute an electric field at any point. Integration over the contributions from the segments gives the net electric field. (b) A wire carrying current I similarly has a magnetic field that can be calculated by integrating the contributions of segments of the wire. For a segment of length $d\vec{\ell}$, the vector $d\vec{\ell}$ is oriented along the segment in the direction of I.

There is, however, a difference in the vectorial aspect of the electric and magnetic fields. The electric field is directed radially away from the charge element, whereas the magnetic field lines form circles. The direction of $\vec{B}$ is shown in Fig. 29–21b for point P and follows the right-hand rule. The direction of the current is indicated by making the infinitesimal length $d\ell$ a vector $d\vec{\ell}$ whose direction is along I. We see that $\vec{B}$ is perpendicular to both $d\vec{\ell}$ and $\vec{r}$. You should recall that there is a direct way to produce a vector that is perpendicular to two other vectors, and that is by using the vector product (see Chapter 10). Using trigonometry, we can see that if $\theta = \phi + 90°$ (as in Fig. 29–21b), then $\cos\phi = \cos(\theta - 90°) = \cos\theta\cos 90° + \sin\theta\sin 90° = \sin\theta$. Equation (29–17) becomes

$$dB = \frac{\mu_0}{4\pi}\frac{I\, d\ell\sin\theta}{r^2} = \frac{\mu_0}{4\pi}I\frac{r\sin\theta}{r^3}\,d\ell \tag{29–18}$$

in the direction perpendicular to $d\vec{\ell}$ and $\vec{r}$. *The magnetic field $d\vec{B}$ produced by a segment of wire $d\ell$ that carries a current I at a displacement $\vec{r}$ from the segment is:*

$$d\vec{B} = \frac{\mu_0}{4\pi}\frac{I\, d\vec{\ell}\times\vec{r}}{r^3}. \tag{29–19}$$

THE BIOT–SAVART LAW

The vector product $d\vec{\ell}\times\vec{r}$ has a magnitude $d\ell\, r\sin\theta$, and for the situation in Fig. 29–21 the direction is as drawn there. Note that there is no ambiguity as to the vector $d\vec{\ell}$; if the segment is short enough, it may be treated as a straight line, and that has a definite direction. Equation (29–19) is the **Biot–Savart law**, named after the two physicists who first formulated it, Jean-Baptiste Biot and Félix Savart. This law is analogous to Coulomb's law in electricity. It even has the same overall distance dependence of $1/r^2$—note that the magnitude r appears in the numerator of Eq. (29–19). The angular factors are quite different, however.

Using the Biot–Savart Law

We can use the Biot–Savart law to find the magnetic field due to nonsymmetric current distributions, in just the same way that Coulomb's law is used to find the electric field. The net magnetic field is found by integrating over $d\vec{B}$:

$$\vec{B} = \int d\vec{B} = \frac{\mu_0}{4\pi}\int\frac{I\, d\vec{\ell}\times\vec{r}}{r^3}. \tag{29–20}$$

Evaluation of this expression depends on the details of the currents. For most cases Eq. (29–20) is too complicated to give an analytic expression. In that case it is always possible to make a numerical evaluation of the integral. Sophisticated computer programs exist that allow one to design the form of the windings of wire needed to produce a desired field. Such programs superpose infinitesimal field contributions $d\vec{B}$, each given by the Biot–Savart law. Finally it is of course always possible to measure the magnetic fields due to nonsymmetric current distributions experimentally.

EXAMPLE 29–6 A straight segment of wire of length L carries a current I. Use the Biot–Savart law to find the magnetic field due to the wire segment in the plane perpendicular to the wire and passing through the midpoint of the wire segment.

Setting It Up Orient the wire along the x-axis, as in Fig. 29–22, with its center positioned at the origin. We then wish to find the field in the yz-plane.

Strategy Since the wire looks the same from anywhere on any circle in the yz-plane that is centered on the wire, we can pick any point on such a circle. Here we choose to find the field at a distance D from the wire along the y-axis with $z = 0$. The right-hand rule shows that the quantity $d\vec{\ell} \times \vec{r}$ points out of the plane where $d\vec{\ell}$ and $\vec{r}$ are as shown in the figure. The field forms circles around the wire, as before.

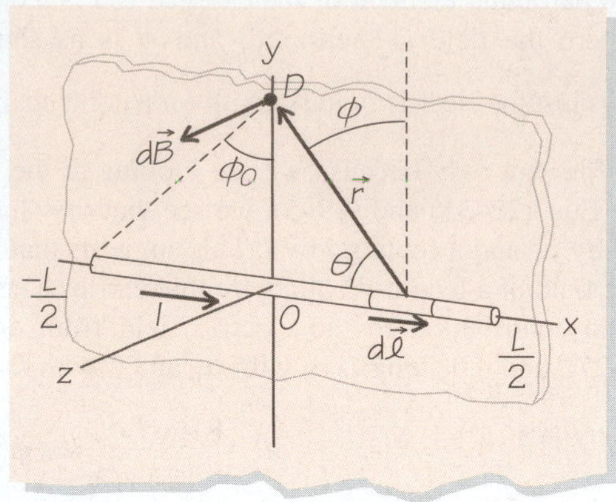

▲ **FIGURE 29–22** A straight segment of wire of length L, carrying a current I, is oriented on the x-axis and centered at the origin.

Working It Out From Eq. (29–18), the magnetic field at the chosen point due to the segment $d\vec{\ell}$ has magnitude

$$dB = \frac{\mu_0 I}{4\pi} dx \frac{\sin(\pi - \theta)}{r^2} = \frac{\mu_0 I}{4\pi} dx \frac{\sin\theta}{r^2} = \frac{\mu_0 I}{4\pi} dx \frac{\cos\phi}{r^2},$$

where we have replaced $d\ell$ by dx and used the angle ϕ defined in Fig. 29–22. To find the net magnetic field, we sum over the contributions of segments from $x = -L/2$ to $x = L/2$:

$$B = \frac{\mu_0 I}{4\pi} \int_{-L/2}^{L/2} \frac{\cos\phi}{r^2} dx.$$

Both ϕ and r depend on x. The integral is computed most simply if we use trigonometric variables; we therefore change variables from x to ϕ. We have by simple geometry $x/D = \tan\phi$. Hence

$$dx = D\,d(\tan\phi) = D(\sec^2\phi)\,d\phi = \frac{D}{\cos^2\phi} d\phi.$$

In addition, $r = D/\cos\phi$, so the combination that appears in the integral becomes

$$\frac{\cos\phi}{r^2} dx = \cos\phi \frac{1}{(D/\cos\phi)^2} \frac{D}{\cos^2\phi} d\phi = \frac{1}{D}\cos\phi\,d\phi.$$

Thus

$$B = \frac{\mu_0 I}{4\pi D} \int_{-\phi_0}^{+\phi_0} \cos\phi\,d\phi,$$

where $\pm\phi_0$ is the limit of integration—the largest values taken on by ϕ. The integral of the cosine is the sine, and so

$$B = \frac{\mu_0 I}{4\pi D}[\sin\phi_0 - \sin(-\phi_0)] = \frac{\mu_0 I}{2\pi D}\sin\phi_0.$$

We can reexpress this result in terms of L and D by using the geometrical relation

$$\sin\phi_0 = \frac{L/2}{\sqrt{(L/2)^2 + D^2}}$$

to find that

$$B = \frac{\mu_0 I}{4\pi} \frac{L}{D\sqrt{(L/2)^2 + D^2}}. \tag{29–21}$$

This form shows that the field depends not only on the distance D from the wire but also on the relative magnitudes of D and L.

What Do You Think? Find the magnetic field when $L \gg D$. Compare your result with Eq. (29–5), which is the field of an infinitely long wire.

What is the numerical significance of the fact that the wire of Example 29–6 has finite length? Suppose that the current is 1 A and that the wire segment is 10 cm long. Then, 1 cm from the wire, Eq. (29–21) gives a field of 1.96×10^{-5} T, whereas the magnetic field of the infinitely long wire carrying the same current [Eq. (29–5)] is 2.00×10^{-5} T. In this case, we are a distance from the wire of one-tenth the length of the wire and the field is only 2 percent less than the infinite length result. But let's now move away from the wire so that we are one wire's length away: $L = 0.1$ m, and $D = 0.1$ m. In this case, the field from the wire has a value some 2.2 times smaller than the field of the wire of infinite length.

EXAMPLE 29–7 A wire forms a circular loop of radius $R = 12$ cm. A current $I = 8.0$ A flows counterclockwise in the wire. Find the magnetic field at the center.

Setting It Up We show the wire loop and an element of the wire $d\vec{\ell}$ in Fig. 29–23 along with an appropriate coordinate system.

Strategy There is no path along which the magnetic field is constant, so we cannot use Ampère's law. We must use the Biot–Savart law and integrate over the contributions $d\vec{B}$ of the different elements of the wire to find the unknown total field $\vec{B}$. Vector $\vec{r}$ runs from the current element $d\vec{\ell}$ to the center of the circle, the point where we want to find $\vec{B}$. The quantity $d\vec{\ell} \times \vec{r}$ for the current element shown in Fig. 29–23 is directed along the x-axis, and this will be true for all the current elements that make up the loop. The net magnetic field at the center is thus directed along the x-axis.

Working It Out The magnitude of the field due to the element shown is [Eq. (29–18)]

$$dB = \frac{\mu_0}{4\pi}\frac{I\,d\ell}{R^2}.$$

There is no sine factor here because $d\vec{\ell}$ is perpendicular to $\vec{r}$. The integral of $d\vec{\ell}$ around the circle is the circumference $2\pi R$, so the net field at the center has magnitude

$$B = \int dB = \int \frac{\mu_0}{4\pi}\frac{I\,d\ell}{R^2} = \frac{\mu_0 I}{4\pi}\frac{1}{R^2}\int d\ell = \frac{\mu_0 I}{4\pi}\frac{2\pi R}{R^2} = \frac{\mu_0 I}{2R}. \quad (29\text{–}22)$$

The numerical value is

$$B = \frac{(4\pi \times 10^{-7}\ \mathrm{T\cdot m/A})(8.0\ \mathrm{A})}{2(0.12\ \mathrm{m})} = 4.2 \times 10^{-5}\ \mathrm{T}.$$

This is roughly the size of Earth's magnetic field at Earth's surface.

What Do You Think? Is the orientation of the magnetic field we have calculated consistent with a right-hand rule?

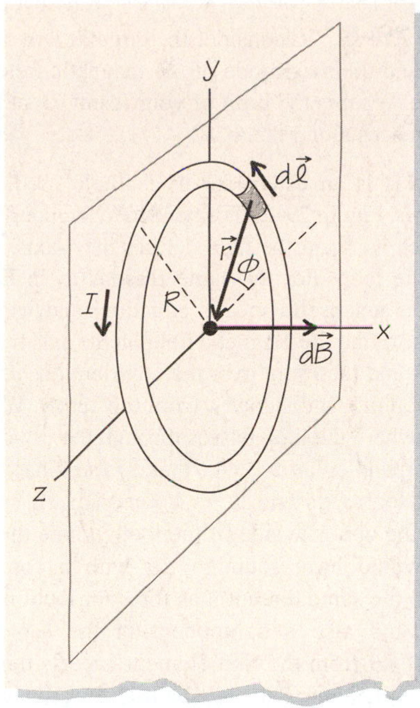

▲ **FIGURE 29–23** An integration is required to find the net magnetic field at the center of the loop.

Problem-Solving Techniques

In the examples of this chapter and Chapter 28, we have studied two aspects of problems on static magnetic fields: We may need to find the magnetic fields produced by a given time-independent set of currents, or we may need to find the magnetic forces on currents or on moving charges. Two sets of key laws contain all that is generally necessary to approach such problems, and you should understand the symbols in these formulas and what the laws mean. First, we have the laws that determine the magnetic field due to currents, which can be written in the two forms

Ampère's law: $$\oint \vec{B}\cdot d\vec{s} = \mu_0 I_{\text{enclosed}}; \quad (29\text{–}10)$$

Biot–Savart law: $$d\vec{B} = \frac{\mu_0}{4\pi}\frac{I\,d\vec{\ell}\times\vec{r}}{r^3}. \quad (29\text{–}19)$$

Second, we have the laws that express the force on a moving charge or a current due to a given magnetic field, namely,

$$\vec{F}_B = q\vec{v}\times\vec{B}, \quad (28\text{–}1)$$

$$d\vec{F} = I\,d\vec{\ell}\times\vec{B}. \quad (28\text{–}14)$$

Each set of laws involves a right-hand rule. For Ampère's law, if the thumb of the right hand follows the current, the fingers curl in the direction of the integration path. For the force laws and the Biot–Savart law, the right-hand rule for a vector product applies.

Based on these laws, we can suggest a list of habits to develop when solving static magnetic field problems. Many of these are the very same habits that are useful for solving *any* problem in physics.

1. Draw a figure that indicates the physical situation with the quantities known; include directions if appropriate.
2. Write down what is known and what is to be determined. Are you dealing with moving charges or with currents?
3. What physical principles connect the unknown quantities to the known ones?
4. If the problem concerns a force, do you have sufficient information to determine the force directly from the force laws? If not, you may need to compute a magnetic field or integrate an infinitesimal force.
5. In a situation with enough symmetry (for example, for long, straight wires), we can use Ampère's law to calculate a magnetic field. When it is applicable, Ampère's law will usually give the answer more easily than the Biot–Savart law.
6. If the system is not sufficiently symmetric to use Ampère's law, the Biot–Savart law is always available. In using it, be sure that the infinitesimal element $d\vec{\ell}$ and the position vector $\vec{r}$ are identified properly. A partial symmetry may rule out one or more directions for the magnetic field. If only one direction is indicated by symmetry, then the other components will cancel in the calculation of the integral over $d\vec{\ell}$, and you need only integrate for the component desired.
7. Superposition can be a useful tool; the field due to a complex system of currents can sometimes be found by adding the fields due to pieces of the current. For example, the field due to a wire that has an angle bend in it can be found by adding the fields due to the two straight segments.
8. It is always useful to check dimensions and units.
9. Substitute numbers only at the last stage. Any checks you can find of limits or special cases are always helpful.

EXAMPLE 29–8 Reconsider the circular wire loop of Example 29–7. (a) Find the magnitude of the magnetic field all along the axis of the loop. What is the limit of your result (b) at the center and (c) at large distances along the axis?

Strategy This is an extension of Example 29–7 to include all points on the axis. Figure 29–24 is extended to include such points; the axis along which we want the field defines the x-axis, with $x = 0$ at the center of the loop. For the same reasons as in Example 29–7, Ampère's law is not useful in this situation, and we must use the Biot–Savart law to find the magnetic field on the axis from a particular current element and then sum over the contributions of the elements.

Consider a point P a distance x from the center. We have chosen an element $d\vec{\ell}$ where the loop passes through the $+y$-axis. Its contribution $d\vec{B}$ is perpendicular to both $d\vec{\ell}$ and $\vec{r}$, so it has *both* a component along the loop axis and a $+y$-component. If we had chosen an element on the opposite side of the loop, where the loop cuts the $-y$-axis, we would have found a $d\vec{B}$ with a component along the loop axis in the same direction as the contribution from the first loop element, and also a component in the $-y$-direction. The y-component of $d\vec{B}$ from the first element cancels the y-component from the second element. This will be true for all pairs of elements around the loop, so we must calculate only the component of $d\vec{B}$ along the loop axis, which is the x-component. This understanding of the role played by symmetry is important, and a moment or two spent looking for such symmetries is time well spent.

From Fig. 29–24, we see that the vectors $d\vec{\ell}$ and $\vec{r}$ are perpendicular, so the cross product $d\vec{\ell} \times \vec{r} = (d\ell)(r)$. Then

$$dB_x = \frac{\mu_0}{4\pi}\frac{I\,d\ell}{r^2}\cos\gamma.$$

We sum these elements to find the net field. Notice that the x-dependence of the field is contained both in the distance r and in the angle γ.

Parts (b) and (c) simply require substitutions and limits of the more general result.

Working It Out

(a) From Fig. 29–24 we can use trigonometry for the following simplifications: $d\ell = R\,d\phi$, $\cos\gamma = \dfrac{R}{\sqrt{R^2 + x^2}}$, and $r = \sqrt{R^2 + x^2}$, so that

$$dB_x = \frac{\mu_0}{4\pi}\frac{IR^2\,d\phi}{(R^2 + x^2)^{3/2}}.$$

The net field is then

$$B_x = \int dB_x = \frac{\mu_0 IR^2}{4\pi(R^2 + x^2)^{3/2}}\int_0^{2\pi} d\phi = \frac{\mu_0 IR^2}{2(R^2 + x^2)^{3/2}}. \qquad (29\text{–}23)$$

(b) At the center of the loop, $x = 0$ in Eq. (29–23). The result of substituting $x = 0$ correctly gives Eq. (29–22).

(c) At large distances, $x \gg R$, the axial magnetic field in Eq. (29–23) reduces to

$$B = \frac{\mu_0 I}{2}\frac{R^2}{x^3} = \frac{\mu_0}{2\pi}\frac{I\pi R^2}{x^3}. \qquad (29\text{–}24)$$

What Do You Think? Does the $1/r^3$ (i.e. $1/x^3$ in this case) behavior of the magnetic field remind you of the electric field of (a) a point charge or (b) an electric dipole?

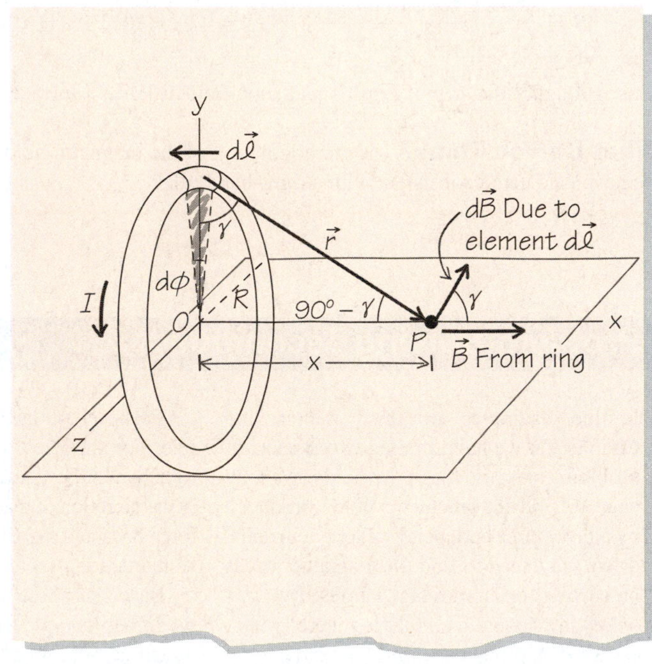

▲ **FIGURE 29–24** The net magnetic field along the loop axis due to a current loop carrying a current I is oriented along the axis, according to a right-hand rule. The contribution from an infinitesimal element $d\vec{\ell}$ has components along its axis and in other directions as well.

Magnetic Dipoles

Calculation of the magnetic field over all space due to the current-carrying loop of Examples 29–7 and 29–8 shows that the magnetic field lines shown in Fig. 29–25 are just like the electric field lines of the electric dipole (Fig. 22–12), apart from the region between the two electric charges of the electric dipole. In fact, it is the form of the field, not how that field is produced, that labels the magnetic field of the current loop as a *magnetic dipole field*. The loop forms a magnetic dipole whose strength is characterized by a magnetic dipole moment, μ. The magnetic dipole moment μ plays the same role for the magnetic dipole that the electric dipole moment p plays for the electric dipole. (Don't confuse μ with the permeability of free space, μ_0!) By looking at Eq. (29–24), we see that a sensible definition is

$$\text{for a circular current loop: } \mu \equiv I\pi R^2. \qquad (29\text{–}25)$$

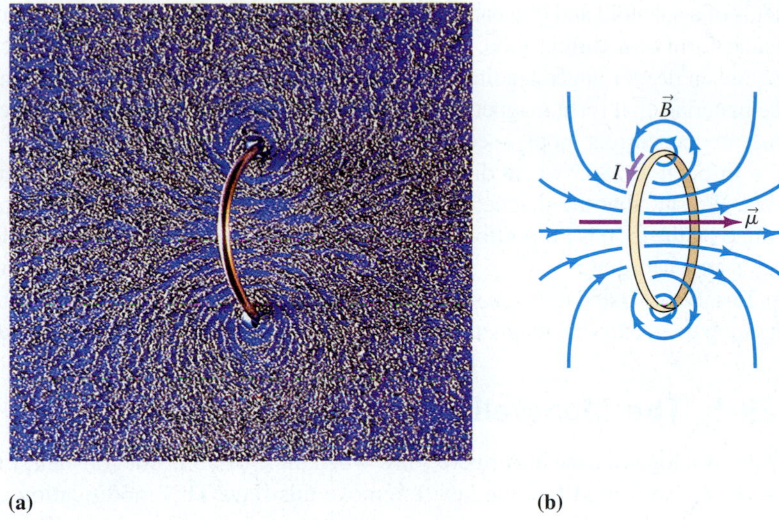

▶ **FIGURE 29–25** (a) Magnetic field lines for a circular loop of current, as shown by iron filings. (b) The field for such a loop is a magnetic dipole field.

From Eq. (29–24) we know that at a distance x far from the loop the magnitude of the magnetic field along the axis is

$$\text{along the axis: } B = \frac{\mu_0}{2\pi}\frac{\mu}{x^3}. \tag{29–26}$$

In fact, for *any* closed loop of area A that carries a current I, there is a similar result: The magnetic field decreases as $1/r^3$ far from the loop. The strength of the field is proportional to a dipole moment μ that is equal to the product of the current in the loop and the area of the loop:

$$\mu = IA. \tag{29–27}$$

The magnetic dipole moment, like the electric dipole moment, forms a vector $\vec{\mu}$: For any plane loop, $\vec{\mu}$ is perpendicular to the plane of the loop according to a right-hand rule—curl the fingers of the right hand around the direction of current flow in the loop, and the thumb points in the direction of $\vec{\mu}$. The vectorial aspect of the magnetic dipole moment ensures that the magnetic field of the current loop has an appropriate directionality.

One feature of this discussion is quite striking. The electric dipoles that we know about are formed by a pair of equal and opposite electric charges, while the magnetic dipole above is formed by a current loop. Magnetic charges do not exist; the magnetic dipole *cannot* be formed from a pair of equal but opposite magnetic charges, rather it is formed from a closed loop of current. The magnetic field of a current loop (Fig. 29–26a) has the same form as the

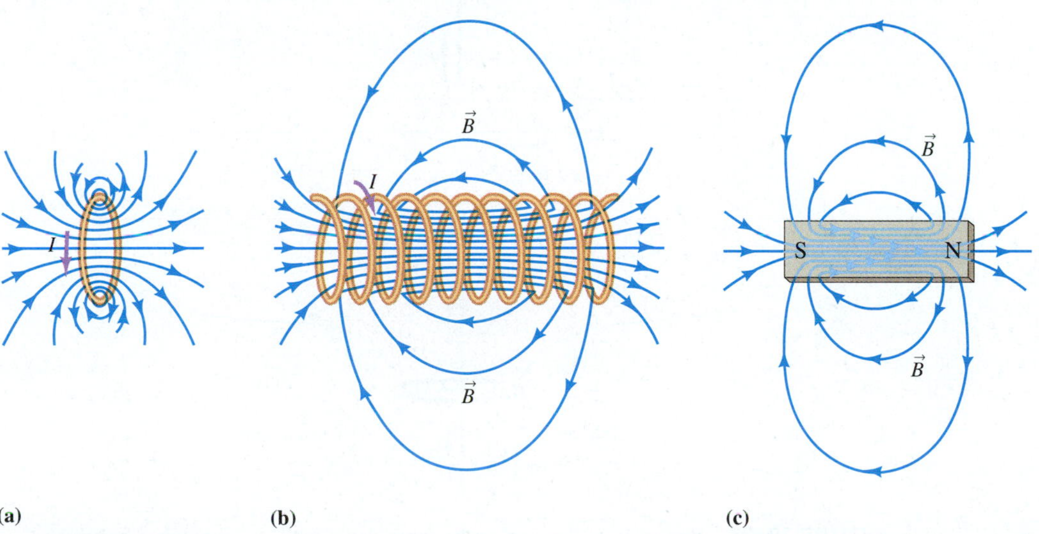

▲ **FIGURE 29–26** (a) The magnetic field lines of a circular loop of current have the same form as the magnetic field lines of (b) a solenoid and of (c) a bar magnet.

fields of a solenoid and of a bar magnet (Figs. 29–26b, c). A solenoid is of course just a particular form of a current loop. The bar magnet is another story, and to understand it fully requires a deeper understanding of materials at the atomic level. But we can say here that the materials that form magnets behave at a microscopic level as if they were composed of microscopic current loops—see Chapter 31 for more on this subject.

In Section 28–5 we studied the *response* of a current loop to an external magnetic field. We saw that the magnetic dipole responds to an external field by rotating, and the torque on the loop is proportional to the magnetic dipole moment. In more detail, the field $\vec{B}_{ext}$ exerts a torque $\vec{\tau} = \vec{\mu} \times \vec{B}_{ext}$ on the loop that tends to line up the vector $\vec{\mu}$ with $\vec{B}_{ext}$. As Eqs. (29–27) and (28–22) show, both the magnetic field of the loop and the reaction of the loop to an external magnetic field are given in terms of the dipole moment.

29–5 The Maxwell Displacement Current

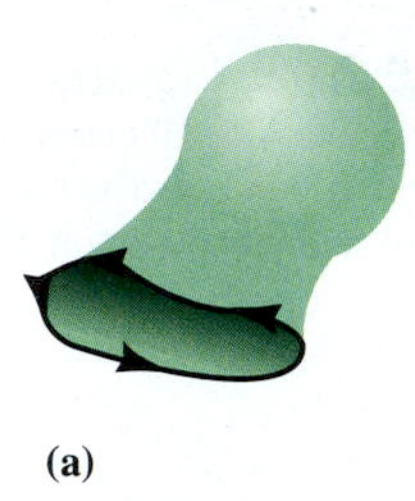

(a)

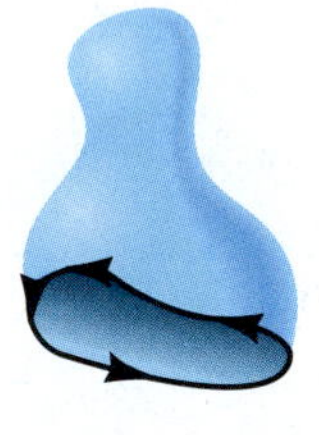

(b)

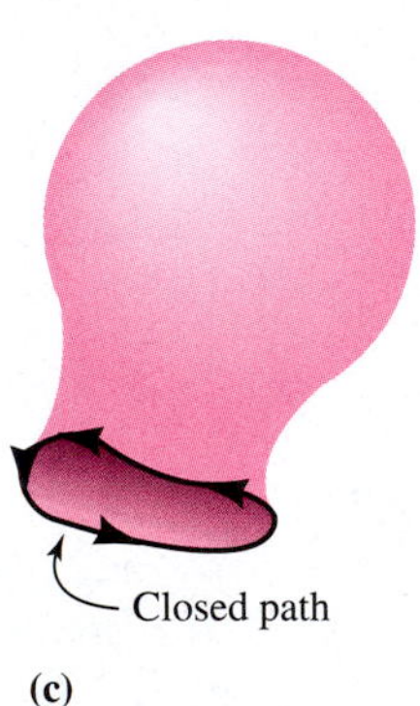

(c)

▲ **FIGURE 29–27** A closed path defines an infinite number of surfaces.

There is a logical flaw in Ampère's law when the current is not constant. In 1865, James Clerk Maxwell modified the law to remove this flaw. This modification was crucial to the completely unified theory of electricity and magnetism that will be discussed in Chapter 34.

Ampère's law is applied with an integration over some closed path. The right-hand side of Ampère's law, Eq. (29–10), contains what we called the current enclosed by a path. By "the current enclosed by a path," we mean the rate of charge flow through a surface whose boundary is the closed path. Such a surface can be chosen in many different ways (Fig. 29–27), but *when the current is continuous, the current that crosses any one of these surfaces must be the same as the current that crosses any other*. For constant current, the freedom to choose a surface therefore presents no problem. There is a situation, however, in which the freedom to choose the surface presents a difficulty, and this situation arises, for example, when the current deposits charge on the plates of a capacitor. Figure 29–28 shows two surfaces with the same loop as their boundary. A current I crosses surface 1 in the positive sense, while no current crosses surface 2, as the charge collects and remains at the capacitor plate. Ampère's law is ambiguous in this situation, because either surface is allowed in its expression.

Maxwell noted that even if no current passes through surface 2, there is a distinguishing feature for this surface: *There is a changing electric flux through it*. As the charge builds up on the plates of the capacitor, the resulting electric field between the plates also builds up, and hence so does the electric flux in the space between the plates.

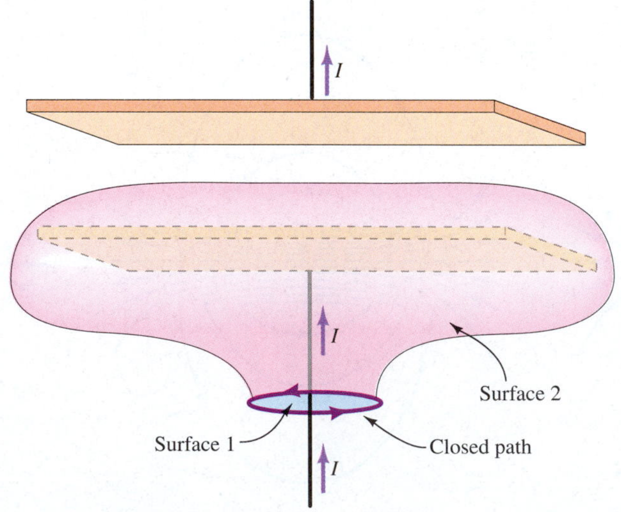

▲ **FIGURE 29–28** Two surfaces bounded by the same closed path. A current passes through surface 1 but not through surface 2.

Suppose that we have a capacitor with large, planar plates of area A (Fig. 29–29). As the current I flows, it deposits a charge $+q$ on one of the plates. We can then use the results of Chapters 22 and 25: The field $\vec{E}$ of such a capacitor is uniform between the plates and small outside the region between the plates. The field points from the plate with positive charge (where the charge accumulates) to the plate with negative charge. The electric flux associated with a surface that passes between the plates, Φ_E, is then just EA. The magnitude of the electric field is given by $E = (1/\varepsilon_0)(q/A)$. Because $\Phi_E = EA$ in this case, this result is equivalent to

$$\varepsilon_0 \Phi_E = q. \tag{29–28}$$

If we take a time derivative, we find the relation

$$\varepsilon_0 \frac{d\Phi_E}{dt} = \frac{dq}{dt} = I. \tag{29–29}$$

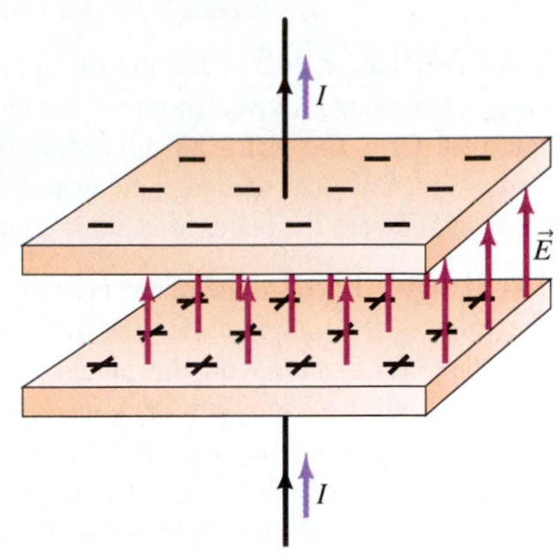

▲ **FIGURE 29–29** The electric field between two parallel plates that carry charges $+q$ and $-q$, respectively, is uniform inside and small outside the region between the plates.

Equation (29–29) implies that whatever the value of the current that passes through the wire that leads to the capacitor, that current equals the quantity $\varepsilon_0\, d\Phi_E/dt$ between the plates. Therefore, if we replace I in Ampère's law by the *sum* of the two terms in Eq. (29–29),

$$I + \varepsilon_0 \frac{d\Phi_E}{dt}.$$

Ampère's law would be satisfied for *any* surface we could draw for the path of Fig. 29–28. For surface 1, only the term I in this sum applies; for surface 2, only the changing flux term applies. The second term, $\varepsilon_0\, d\Phi_E/dt$, is written in a way that does not refer explicitly to the plane geometry. Indeed, Maxwell was able to show that if the sum of these two terms is used, any surface gives the same answer in Ampère's law. Maxwell called the changing flux term the **displacement current**, I_d:

$$I_d \equiv \varepsilon_0 \frac{d\Phi_E}{dt}. \tag{29–30}$$

DISPLACEMENT CURRENT

The term "current" is used here because the displacement current has the same dimensions as, and appears in the place of, the current composed of moving charge. Note that the displacement current is present only when there are changing electric fields.

Maxwell's insight that changing electric flux gives rise to a magnetic field, and his quantification of the magnetic field that results, allows Ampere's law to be extended to cover all situations, including those in which there is varying current. Even though the current is not continuous when capacitors are present, *the sum of the ordinary current and the displacement current is continuous.*

Maxwell's generalized form of Ampère's law is accordingly

$$\oint \vec{B} \cdot d\vec{s} = \mu_0(I + I_d) = \mu_0 I + \mu_0\varepsilon_0 \frac{d\Phi_E}{dt}. \tag{29–31}$$

GENERALIZED AMPÈRE'S LAW

Here, the sum $I + I_d$ is calculated with reference to *any* surface that spans the closed path defining the line integral of the magnetic field. We have concluded from this generalization that changing electric flux produces magnetic field just as moving charges do.

That a changing electric flux produces a magnetic field has great importance for electromagnetic waves, as we shall see in Chapter 34. It may also be of importance when a very large capacitor discharges very quickly, but as Example 29–9 shows, it may otherwise have very little practical effect.

EXAMPLE 29–9 The planar circular plates of a capacitor are being charged. At a given moment, the charge is being built up at the rate of 1 C/s. The plates have a radius $R = 0.1$ m and a separation $d = 1$ cm. Calculate the magnetic field due to the displacement current midway between the plates at a radius equal to half the plate radius.

Strategy Because the plates are circular, symmetry requires that the value of the magnetic field be the same everywhere on path C, a circular path centered on the plates' axis and of radius $R/2$ (Fig. 29–30). Then we can apply the generalized Ampère's law using this path, extracting the magnetic field magnitude B from the integration. The problem statement gives us enough information to find the rate of change of the electric field across the plates and hence the rate of change of electric flux through our path. We can then solve for B.

Working It Out The line integral in Ampère's law is taken in the sense drawn. Because $\vec{B}$ has constant magnitude on this path and points along the path, we can remove it from the integral. The remaining integral is the circumference of the path, $2\pi(R/2)$:

$$\oint \vec{B}\cdot d\vec{s} = B\oint ds = B\left(2\pi\frac{R}{2}\right).$$

To calculate the displacement current, we note that, in terms of the charge q on the plates, the electric field, which is uniform across the region between the plates, has magnitude

$$E = \frac{1}{\varepsilon_0}\frac{q}{\pi R^2}.$$

We must now calculate the electric flux through the area bounded by path C (and not the *total* electric flux in the capacitor). The flux through path C is E times the area $\pi(R/2)^2$:

$$\Phi_E = \frac{1}{\varepsilon_0}\frac{q}{\pi R^2}\pi\left(\frac{R}{2}\right)^2 = \frac{1}{4\varepsilon_0}q.$$

Thus the displacement current is

$$I_d = \varepsilon_0\frac{d\Phi_E}{dt} = \varepsilon_0\left(\frac{1}{4\varepsilon_0}\frac{dq}{dt}\right) = \frac{1}{4}\frac{dq}{dt}.$$

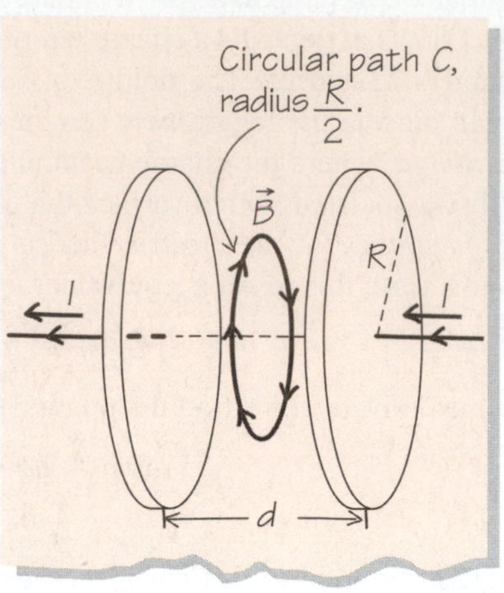

▲ **FIGURE 29–30** Symmetry requires that the value of the magnetic field be the same everywhere on path C.

We can find the magnitude of the magnetic field $\vec{B}$ by using Eq. (29–31):

$$B\left(2\pi\frac{R}{2}\right) = \mu_0 I_d = \frac{\mu_0}{4}\frac{dq}{dt};$$

then

$$B = \frac{\mu_0}{4\pi}\frac{1}{R}\frac{dq}{dt}.$$

Numerically, $dq/dt = 1\text{ C/s} = 1\text{ A}$ and $R = 0.1$ m, so

$$B = \frac{4\pi\times 10^{-7}\text{ N/A}^2}{4\pi}\frac{1}{0.1\text{ m}}(1\text{ A}) = 10^{-6}\text{ T}.$$

This is a small field; for comparison, recall that Earth's magnetic field at Earth's surface is around 10^{-4} T.

What Do You Think? What happens if the distance d between the plates is halved?

Summary

Magnetic fields are produced by electric currents. The magnetic field lines about a long, straight wire that carries a constant current form circles around the wire in the plane perpendicular to the wire. The direction of the field lines in these circles is determined when the thumb of the right hand points along the direction of the current flow: The fingers then curl in the direction of the magnetic field. The magnitude of the field at a radial distance r from the wire is

$$B = \frac{\mu_0 I}{2\pi r}. \qquad (29\text{–}5)$$

The defined constant $\mu_0 = 4\pi\times 10^{-7}\text{ T}\cdot\text{m/A}$ is the permeability of free space.

The magnetic fields produced by unchanging currents obey Ampère's law:

$$\oint \vec{B}\cdot d\vec{s} = \mu_0 I_{\text{enclosed}}. \qquad (29\text{–}10)$$

Here, the line integral follows any closed path through which the current I_{enclosed} passes. A second law obeyed by the magnetic field results from the absence of magnetic equivalents to the electric charge. Because there are no magnetic charges on which magnetic field lines begin or end, magnetic field lines must close on themselves. This fact is expressed by Gauss' law for magnetism:

$$\text{for a closed surface: } \Phi_B = \int_{\text{closed surface}} \vec{B}\cdot d\vec{A} = 0. \qquad (29\text{–}12)$$

This law states that the magnetic flux, Φ_B, through any closed surface is zero; equivalently, the number of magnetic field lines that enter a closed surface is the same as the number of lines that leave the surface.

Ampère's law is an important practical tool for determining magnetic fields when there is enough symmetry to allow a path choice in which the integral simplifies, as in the determination of the interior field of a long solenoid. A solenoid is a wire wound uniformly into a coil to form a tube. When current flows, a magnetic field is produced within the tube that has a constant magnitude and is aligned with the tube axis. The magnitude of the interior field is

$$B = \mu_0 n I, \tag{29–15}$$

where n is the number of windings of wire per unit length of the solenoid. Because its interior magnetic field is constant, a solenoid is to magnetism what a capacitor is to electricity.

When there is not enough symmetry to allow Ampère's law to be used to determine the magnetic field produced by a given configuration of currents, the Biot–Savart law can be used instead. According to this law, the magnetic field $d\vec{B}$ produced by a segment of wire $d\vec{\ell}$ that carries a current I at a displacement $\vec{r}$ from the segment is given by

$$d\vec{B} = \frac{\mu_0}{4\pi}\frac{I\, d\vec{\ell} \times \vec{r}}{r^3}. \tag{29–19}$$

The magnetic field from an infinitesimal segment can be integrated to find the net magnetic field due to a finite segment of wire.

Application of the Biot–Savart law shows that the magnetic field due to a ring of current, or the exterior field of a solenoid of finite length, is a magnetic dipole field. The form of this magnetic dipole field is the same as that of a bar magnet or, equivalently, has the same form as the exterior electric field produced by an electric dipole. The current loop forms a magnetic dipole, characterized by a magnetic dipole moment μ which is aligned perpendicular to the surface of the loop according to a right-hand rule. Its magnitude is

$$\mu = IA, \tag{29–27}$$

where A is the loop area.

If currents are not constant in time, as when wires are interrupted by the presence of charging capacitor plates, then one surface that spans a closed path might not cross the wire that another surface might cross, and the concept of the current enclosed by a path becomes ambiguous. This ambiguity is remedied by Maxwell's modification of Ampère's law to

$$\oint \vec{B}\cdot d\vec{s} = \mu_0(I + I_d) = \mu_0 I + \mu_0\varepsilon_0\frac{d\Phi_E}{dt}. \tag{29–31}$$

The quantity I_d, proportional to the rate of change of electric flux, is known as the Maxwell displacement current. The surface through which the sum of I and I_d passes is any surface that spans the closed integration path. We conclude from this modification that a changing electric flux creates a magnetic field.

Understanding the Concepts

1. Suppose you move a compass needle near a straight wire that carries a current. Describe how the compass needle reacts when the compass is moved slowly in a circle centered on the wire and perpendicular to it.
2. In the definition of the ampere, does the length of the two parallel wires matter?
3. A wire connected to a battery is placed in the yoke of a tabletop electromagnet when a switch is open. When the switch is closed, the wire may take a big jump upward or downward, according to which side of the battery terminals the wires are attached. Why?
4. Suppose the torus of Figure 29–20 is replaced by a tube that has an almost rectangular cross section and whose shape is irregular (e.g., an extended ellipse). Current-carrying wire is wrapped closely around the tube. What can you say about the magnetic field?
5. In the definition of the ampere, must we worry about the Coulomb forces between the charges in the two wires?
6. Does the statement of the generalized Ampère's law, including displacement current, imply that without the displacement current charge would *not* be conserved?
7. Why is the Biot–Savart law written in differential form? Explain why it cannot be written as in Eq. (29–19) but without the differential signs.
8. Consider the solenoid, length L, of Example 29–4. You increase the number of windings, always using the same wire, so that the magnetic field is doubled. How will the resistance of the coil change?
9. Why is it preferable to define current in terms of the force between two long, parallel wires rather than in terms of the rate at which charge passes a point?

10. Is it possible to arrange a set of electric currents and produce a magnetic field that, at large distances from the apparatus responsible, is everywhere directed radially away from the apparatus? Feel free to choose your apparatus, and give either a proof that it is impossible or a description of the apparatus.
11. Suppose that the space between the plates of the capacitor discussed in Section 29–5 is not empty but is filled with a dielectric. How would the treatment in that section, and the determination of the displacement current, change?
12. Suppose that magnetic charges were discovered. What are some practical consequences?
13. In Example 29–8 we looked at the field along the axis of a loop. Suppose you had two identical loops parallel to one another and aligned along the same axis. Consider the point P that is on the axis and midway between the loops. What, qualitatively, is the field at P if the current travels in the same direction in both loops? If the current travels in opposite directions?
14. What are the SI units of the ratio E/B, where E is an electric field and B is a magnetic field?
15. When two bar magnets are placed side by side, they will (a) attract or (b) repel if the adjacent poles are (a) opposite or (b) the same. If you draw magnetic field lines for the combination of two magnets in both cases, the net magnetic field between the magnets will tend to (a) cancel or (b) be doubled. What conclusions can you draw?
16. A current passes through a helical coil. Which of the following statements is true? (a) The helical coil will tend to get shorter; (b) the helical coil will tend to get longer; (c) the coil will get charged; (d) no magnetic field will get produced.

Problems

29–1 Ampère's Law

1. (I) Sketch the magnetic field lines due to two current-carrying wires that are parallel to each other for the cases that (a) the currents move parallel to each other, and (b) the currents move in opposite directions.
2. (I) Consider an array of parallel, current-carrying wires arranged so that they all lie in a plane and are separated by equal distances. Sketch the magnetic field due to this array. Assume that all the currents point in the same direction.
3. (I) What is the force per meter between two long, parallel wires, each carrying 50,000 A but in opposite directions, if the two wires are 30 cm apart? Such currents are normal in the electrolytic production of aluminum.
4. (I) A long, straight wire carries a current of 17 A. What is the magnetic field at a distance of 35 cm from the wire?
5. (I) A lightning conductor carries a current of 5×10^4 A for a short period. During that time, what is the magnitude of the magnetic force per unit length exerted on a parallel wire 4 m away in which a current of 100 μA flows?
6. (I) Two 12-cm-long parallel wires in a handheld calculator are 1.2 cm apart. The currents are parallel, running in the same direction, and have values 35 μA and 8 μA, respectively. What is the force between the wires due to the currents?
7. (I) Find the dimensions of μ_0 and ε_0, and use your expressions to show that the product $\mu_0\varepsilon_0$ has dimensions of $(1/\text{speed})^2$. Find the value of that speed in SI units.
8. (I) A coaxial cable consists of a central wire that carries current I to the right and a tube centered on the central wire that carries the same current to the left. Find the magnetic field outside the cable.
9. (II) (a) In a thick, straight wire carrying a current that is uniform through its cross section, where is the magnetic field the greatest? (b) If the radius of the wire is R and the current is I, what is the value of the maximum magnetic field? (c) What is the minimum magnetic field, and where does this occur? Consider regions both inside and outside the wire. (d) Plot the magnetic field as a function of the distance from the center of the wire.
10. (II) The current-carrying capacity of superconducting wires is limited by the fact that superconductivity breaks down if a large magnetic field is present. Estimate the largest current that can be transported via a NbTi wire, 0.8 mm in diameter, if the critical (breakdown) magnetic field is 10 T. [*Hint*: What is the wire's own magnetic field?]
11. (II) Plot the curves of constant magnetic field in the xy-plane for values B_0, $2B_0$, $3B_0$, and $4B_0$ of the field about a straight wire that carries current along the z-axis. B_0 is some field value that you can choose. These curves are the intersections with the xy-plane of the surfaces of constant field.
12. (II) A very thin, infinitely long metal sheet lies in the xy-plane, between $x = -w$ and $x = w$. A current of density h A/m flows in the $+y$-direction (Fig. 29–31). What are the magnitude and direction of the magnetic field at a distance $z \ll w$ above and below the sheet? Neglect end effects.

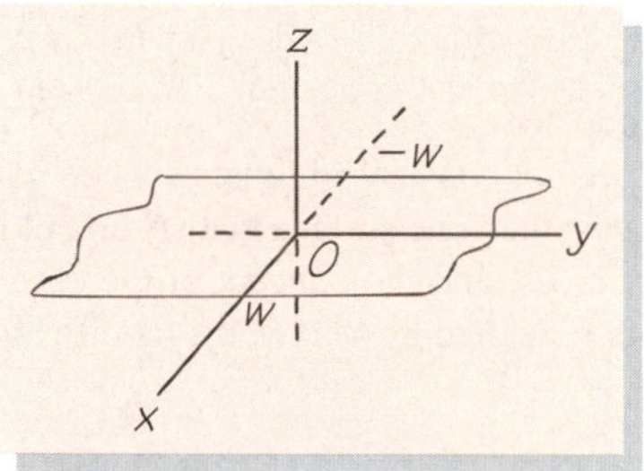

▲ **FIGURE 29–31** Problem 12.

13. (II) Current is carried from a battery to a device by a copper "ribbon" 1 in wide and 1/32 in thick. What is the magnetic field over the surface of the ribbon, if the current it carries is 120 A?
14. (II) Consider two parallel metal sheets, such as the sheet of Problem 12, with currents flowing in opposite directions. What are the magnetic fields between and outside the sheets? What is the situation when the currents are parallel rather than anti-parallel?
15. (II) Consider a wire that passes through the origin along the z-axis and carries a current I (Fig. 29–32a). (a) Calculate the x- and

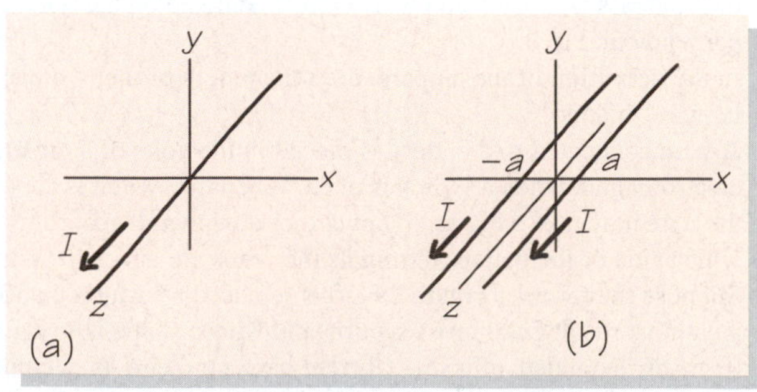

▲ **FIGURE 29–32** Problem 15.

y-components of the magnetic field at a point whose coordinates are $(x, y, 0)$. (b) Use this result to obtain the magnetic field due to two wires that are parallel to the z-axis, cross the xy-plane at $(a, 0)$ and $(-a, 0)$, and carry current I in the $+z$-direction (Fig. 29-32b). (c) What are the fields when the currents are in opposite directions?

16. (II) A uniform current with current density J A/m flows parallel to the z-axis on a cylindrical metal sheath, where the radius of the cylinder is R. What is the magnetic field outside the sheath? inside the sheath?

17. (II) Current flows up the inner cylinder of a coaxial cable and returns on the outside cylinder. The radius of the inner cylinder is 0.1 cm, and the radius of the thin outer cylindrical shell is 0.5 cm. Calculate the magnetic field on the cylindrical surface midway between the inner and outer surfaces, given that the current is 10 A. Ignore end effects.

18. (II) Two long, parallel wires carrying a current I in the same direction each have a mass density λ. The wires are initially a distance D apart and are then released. Write a differential equation for the distance between the wires that describes the relative motion of the wires. Ignore all forces other than the magnetic force.

19. (III) An electron beam contains electrons that move along the $+x$-axis at $0.020c$. The beam enters a region of length 1.0 m and runs parallel to a wire that carries a 0.20-A current in the $+x$-direction. The beam is 10.0 cm from the wire. (a) Specify the direction in which the beam is deflected, if at all. (b) Find the deflection of the beam as it passes through the 1-m region by calculating the impulse it receives during its brief passage through that region. (c) After it has passed the wire, does the beam have the same energy it had when it entered the region that contains the wire?

29–2 Gauss' Law for Magnetism

20. (I) Figure 29–33 shows the magnetic field lines that emerge from one pole of a bar magnet; these lines resemble the electric field lines that emerge from one end of an electric dipole. Sketch the magnetic field lines in the region of the pole inside the magnet. For comparison, also sketch the electric field lines in the central region of an electric dipole. What would the magnetic lines look like if the N and S poles of a magnet represented magnetic monopoles, which would be point sources of magnetic field?

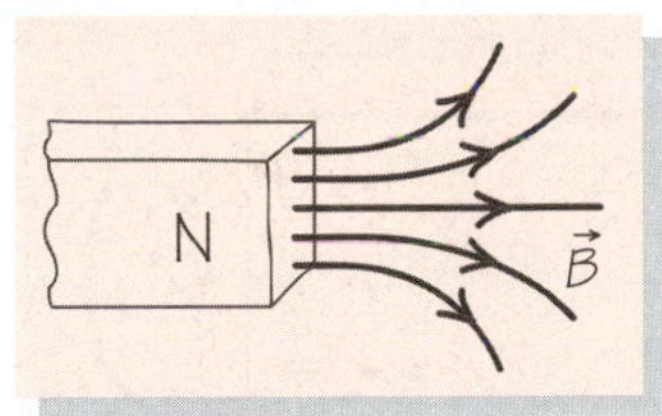

▲ **FIGURE 29–33** Problem 20.

21. (II) A long, current-carrying wire is oriented vertically; next to it is drawn a square whose area lies in the same plane as the wire (Fig. 29–34). Using the distances indicated, find the magnetic flux through the square.

22. (II) Using Gauss' law for magnetism, show that a magnetic field with only an x-component must be constant as x varies.

23. (II) Show that Gauss' law is satisfied for the magnetic field due to a straight wire that carries a current I in the $+z$-direction for a volume that represents a portion of a cylindrical shell of height h, extending from a radius r to a radius R and formed by an angle θ (Fig. 29–35).

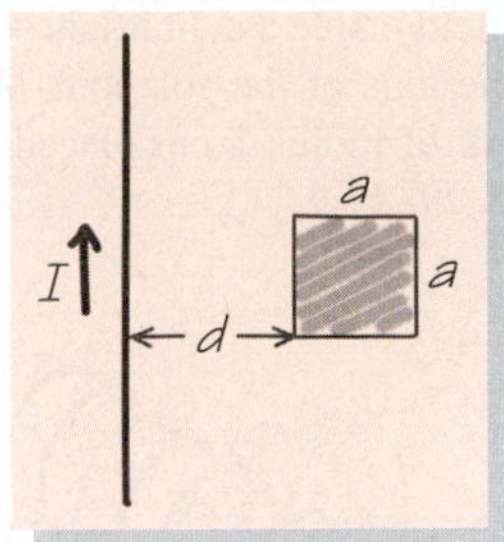

▲ **FIGURE 29–34** Problem 21.

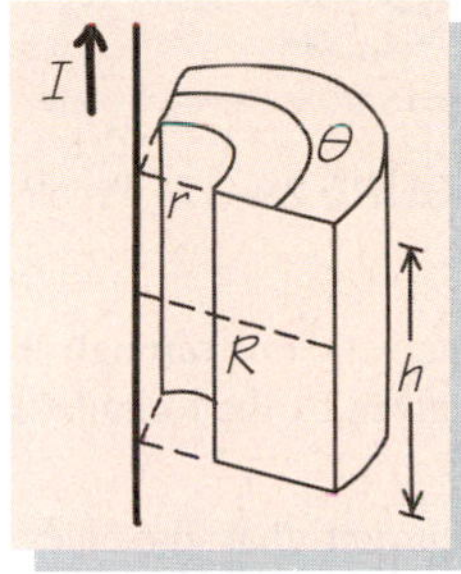

▲ **FIGURE 29–35** Problem 23.

24. (III) Apply Gauss' law for magnetism to a parallelepiped of dimensions a, b, and c, one of whose corners is located at point (x, y, z), as in Fig. 29–36. Assume that the dimensions in the x-, y-, and z-directions $(a, b, \text{and } c)$ are small enough so that $B(x + a, y, z) = B(x, y, z) + a\,\partial B/\partial x$, and so on. Show that Gauss' law leads to the condition $(\partial B_x/\partial x) + (\partial B_y/\partial y) + (\partial B_z/\partial z) = 0$ in this limit.

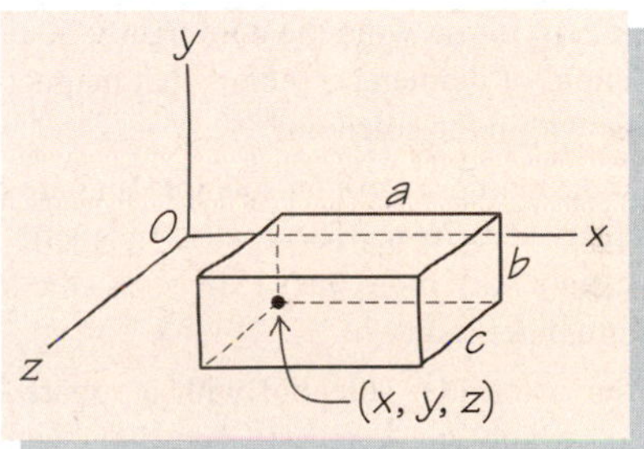

▲ **FIGURE 29–36** Problem 24.

29–3 Solenoids

25. (I) A solenoid of diameter 5 cm has a length of 25 cm and 320 turns of wire. What is the magnetic field at the center of the solenoid when the current in the coil is 3 A?

26. (I) A long, superconducting solenoid is wound with fine niobium–tin wire so that there are 16×10^4 turns/m. If a power supply produces 12 A, what is the magnetic field inside the solenoid?

27. (I) You are told that a toroidal solenoid, carrying a current of 0.36 A through 2500 turns, produces a magnetic field of 5.1×10^{-4} T at its central axis. What can you conclude about the radius of the circle made by that axis?

28. (I) A wire is wound around a torus with outer radius of 32 cm and inner radius of 30 cm. There are 6400 turns in all and the wire carries a current of 2.2 A. What is the range of the magnetic field inside the torus? What percentage change is there from the center to the outside?

29. (I) The magnetic field inside a cylindrical solenoid of area 4 cm^2 is 0.15 T along the axis of the solenoid. What is the magnetic flux through a disk of radius 3 cm placed perpendicular to the solenoid axis (Fig. 29–37)?

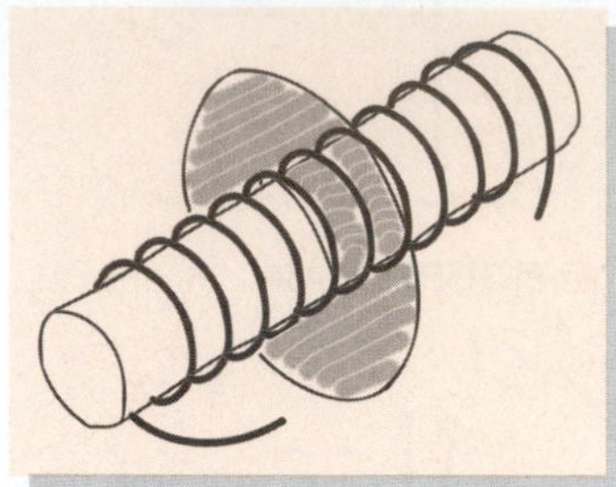

▲ **FIGURE 29–37** Problem 29.

30. (I) Show that the magnetic flux through an ideal cylindrical solenoid of radius R is given by the formula $\Phi_B = \mu_0 n I \pi R^2$, where n is the turn density.

31. (I) What is the current that produces a magnetic field of 0.40×10^{-4} T in the middle of a solenoid that is 1.6 m long and that has 800 turns?

32. (II) A toroidal solenoid consists of a cylinder bent into a circle. The axis of the cylinder is 60 cm from the center of the torus. The cross-sectional area of the cylinder is 70 cm^2, and the coil has 1200 turns. Given that the current in the coil is 25 A, what is the magnetic field inside the solenoid, along the axis of the cylinder?

33. (II) You have 20 m of #16 AWG copper wire (0.051 inch diameter) that can carry a maximum current of 22 A. (a) If you form the wire into a large circle and pass the maximum current through it, what magnetic field would you produce at the center of the circle? (b) If you wind the wire tightly in a single layer to form a solenoid of diameter 3.0 cm, what magnetic field would you produce within the solenoid?

34. (II) You are designing a toroidal solenoid and are constrained to an inner radius of 0.76 m. The central magnetic field must be 3.0 T and cannot vary more than 15% across the toroid. What is the maximum outer radius?

35. (II) Consider a toroidal solenoid with a square cross section, each side of which has length L. The inner wall of the torus forms a cylinder of radius R. The torus is wound evenly with N loops of wire, and a current I flows through the wire. What is the total magnetic flux through the torus?

36. (II) A toroidal solenoid—similar to the one considered in Problem 35—has a square cross section of side length 1.1 cm and an inner radius of $R = 15$ cm. It is wound with 500 turns of 0.25-mm-diameter copper wire. The wire is connected to a 1.5-V battery with negligible internal resistance. (a) Calculate the largest and smallest magnetic field across the cross section of the toroid. (b) Calculate the magnetic flux through the torus. (c) Do you need to cool the solenoid?

29–4 The Biot–Savart Law

37. (I) Two long wires are placed along the y- and z-axes, respectively. They carry the same current I in the positive directions. Calculate the magnetic field along the x-axis.

38. (I) A single loop of wire forms a rectangle whose sides have lengths 17 mm and 150 mm. The wire carries a current of 36 mA. What is the magnetic dipole moment of the loop?

39. (II) An infinitely long L-shaped wire is placed so that a current I flows in along the y-axis toward the origin, then out from the origin along the x-axis. What is the magnetic field at a point on the z-axis at a height H above the origin?

40. (II) Consider a straight segment of wire of length L that carries a current I. Use the Biot–Savart law to find the magnetic field along the axis of the wire, beyond the wire itself, due to this segment.

41. (II) A differential length $d\vec{\ell}$ of wire carrying a current of 2 A is positioned at the origin of a coordinate system and points in the $+x$-direction. Find the magnetic field due to this wire segment at the following (x, y, z) positions, given in centimeters: (a) $(0, 0, 3)$, (b) $(0, 6, 0)$, (c) $(3, 0, 0)$, and (d) $(6, 0, 6)$. Give both the magnitude and direction of the magnetic field.

42. (II) Consider a thin dielectric ring 7 cm in diameter that rotates around a stem perpendicular to the plane of the ring and through its center at the rate of 120 rev/s. Assume that the ring is charged uniformly and carries a total charge of 8×10^{-7} C. What is the magnetic field produced at the center of the ring by the rotating charge?

43. (II) Repeat the calculation of Problem 42 for a solid disk 5 cm in diameter, with the same total charge.

44. (II) A wire carries a current of 10 A, starting from $x = -\infty$. The wire is laid along the negative x-axis to the point $x = -15$ cm. The wire then follows in the positive y direction for 10 cm, then continues parallel to the x-axis to the point $(x = 15$ cm, $y = 10$ cm$)$, then drops back in a straight line to the x-axis, and continues along the positive axis. What is the magnetic field at the origin?

45. (II) Calculate the magnetic field at the center of a rectangular wire, with sides a and b respectively, if the current flowing through the wire is I. What happens when $a = b$? What happens when $b \gg a$?

46. (II) A current loop consists of a square with sides of length L. A current I circulates counterclockwise around the loop. Find the direction and magnitude of the magnetic field at the center of the square. Compare this to the field at the center of a circular loop of diameter L that carries the same current.

47. (II) Consider the wire shown in Fig. 29–38. Calculate the magnetic field at point P, the center of the half-circle of radius R around which the wire turns, as a function of R and the current I carried by the wire.

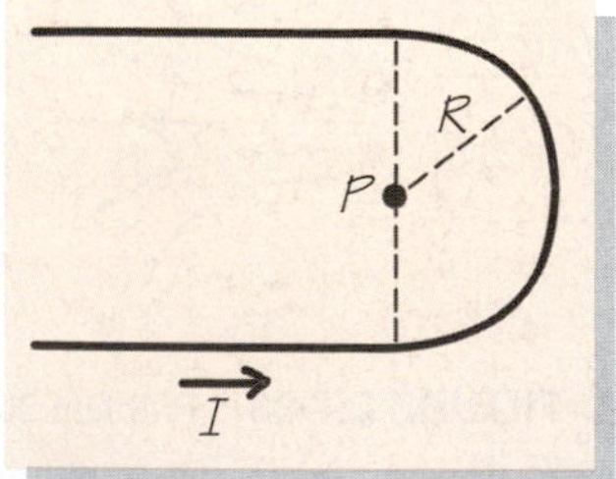

▲ **FIGURE 29–38** Problem 47.

48. (II) A very long wire is aligned along the $+x$- and $+y$-axes, making a right angle at the origin. A current I travels in the $-y$-direction and continues in the $+x$-direction. What is the magnetic field at the point (x, y), where both x and y are positive?

49. (II) Consider the wire shown in Fig. 29–39, see next page, with the inner and outer radii of the semicircle given as 5 cm and 8 cm, respectively. Given that the current in the wire is 12 A, what is the magnetic field at point P, the center of the semicircles?

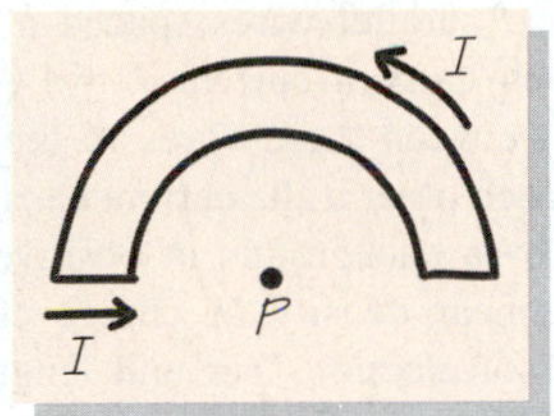

▲ **FIGURE 29–39** Problem 49.

50. (II) Calculate the magnetic field at the center of a wire square that consists of 80 loops and has sides of length 5 cm and carries a current of 0.7 A.

51. (II) A charge q moves at instantaneous speed v when it crosses the axis of a ring of current with a magnetic dipole moment μ. At that instant, q is located a distance d from the center of the ring in the direction of the dipole moment vector and is moving perpendicular to the axis (Fig. 29–40). What is the resulting instantaneous motion of the charge? Find the instantaneous radius of curvature of its motion.

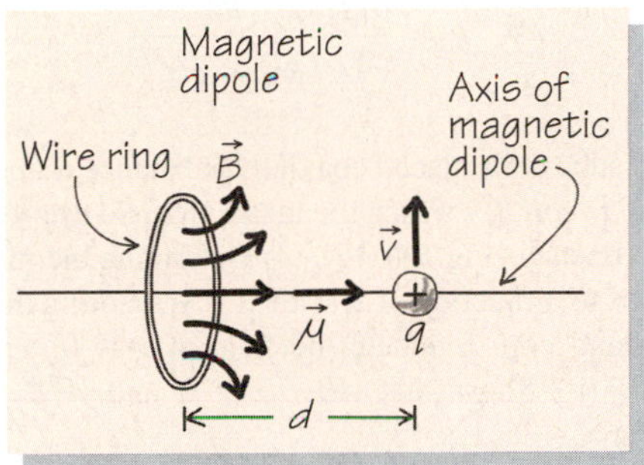

▲ **FIGURE 29–40** Problem 51.

52. (II) A circular current loop of radius R produces a magnetic field. At what distance along the axis of the loop does the field have magnitude 0.5 times the magnitude at the center of the loop? At what distance is the magnitude of the field reduced to $1/100$ the value at the center? Give your answer in units of R.

53. (II) Find the magnetic field at point P in Fig. 29–41 if a current of 8 A flows in the infinitely long wire; the radius R of the semicircle is 1.2 cm.

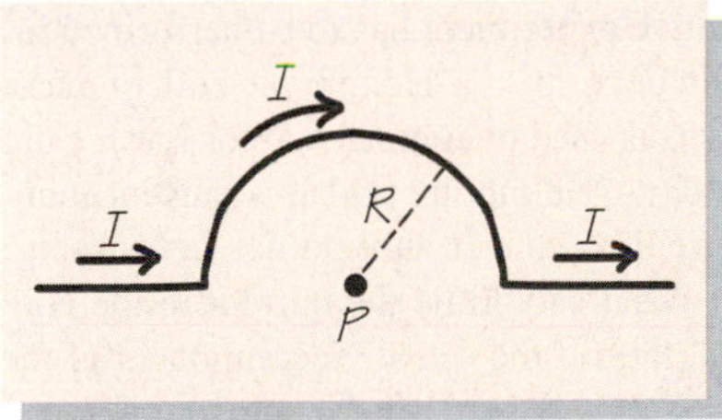

▲ **FIGURE 29–41** Problem 53.

54. (III) A segment of wire forms a straight line of length L and carries a current I. Find the magnetic field due to the wire segment in the plane perpendicular to it and passing through one end.

55. (III) By integration, find the magnetic dipole moment of a spherical shell of radius R that carries a total charge Q, distributed uniformly, if the shell rotates with angular velocity ω oriented along the z-axis.

56. (III) Consider a long, thin-walled metal pipe that carries a total current I distributed evenly along the walls of the pipe. A simple application of Ampère's law indicates that the magnetic field inside the pipe is zero. Show by a simple geometric argument that the same result follows from the Biot–Savart law.

29–5 The Maxwell Displacement Current

57. (I) Consider the RC circuit shown in Fig. 29–42. Switch S is closed at time $t = 0$. Calculate the displacement current in the capacitor as a function of time.

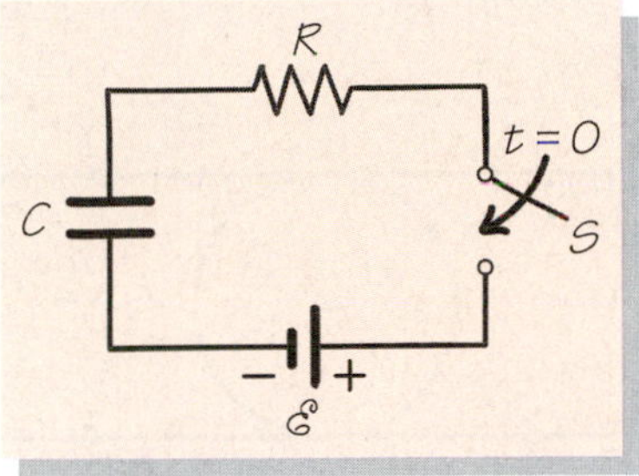

▲ **FIGURE 29–42** Problems 57, 58.

58. (I) Consider the RC circuit shown in Fig. 29–42; this time the switch is closed. At some time, the switch is opened. What is the displacement current?

59. (I) A parallel-plate capacitor is being charged at a rate of $I = 0.2$ A. The plates have an area of 0.25 m^2 and are separated by 1.0 cm (Fig. 29–43). What is the value of $\int \vec{B} \cdot d\vec{\ell}$ for a closed path midway between the plates and covering an area of $5.0 \times 10^{-2}\ m^2$?

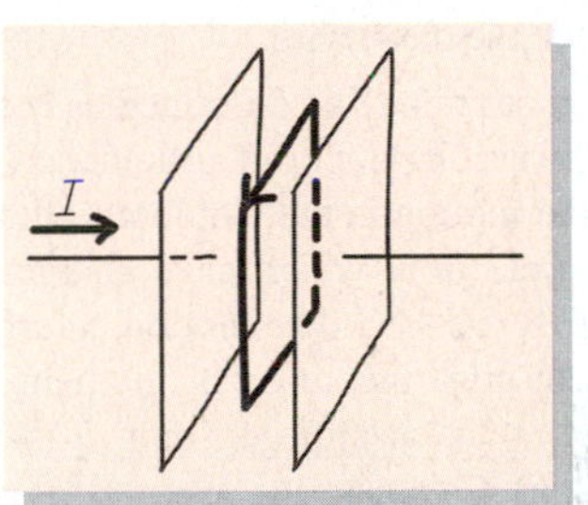

▲ **FIGURE 29–43** Problem 59.

60. (II) A 15-μA current starts flowing in a circuit with a 3.5 μF capacitor of area 40 cm^2 at $t = 0$ s. (a) How fast is the voltage across the capacitor plates changing at $t = 0$ s? (b) Use the result of (a) to calculate *explicitly* $d\Phi_E/dt$ and the displacement current at $t = 0$ s.

61. (II) An alternating voltage of the form $V = V_0 \cos(\omega t)$ is connected across a capacitor C. What is the displacement current in the capacitor?

62. (II) A voltage of the form $V = V_0 \cos(\omega t)$, with $\omega = 2 \times 10^4$ rad/s and $V_0 = 0.1$ V, is applied across the plates of a 5-nF capacitor; the plates are 1.5 cm apart. (a) What is the maximum rate of change in electric field between the plates? (b) The maximum value of current leading to the capacitor?

63. (II) A conducting sphere of radius R initially has a uniform surface-charge density σ_0. Beginning at $t = 0$, this charge is drained off over a period t_0 such that $\sigma = \sigma_0[1 - (t/t_0)]$. Find the displacement current at the surface of the sphere as a function of time. Compare the displacement current to the current carried off by the wire.

General Problems

64. (I) Three wires lie in a plane, placed parallel to one another and equally spaced by 0.20 m. If the wires are oriented up and down on the page, the right-hand wire carries 100 A to the top, the middle wire 300 A to the bottom, and the left-hand wire carries 200 A to the top. What are the forces per unit length on each of the wires?

65. (II) Two wires shown in Fig. 29–44 have identical currents flowing. Use the superposition principle and symmetry to obtain as much information as you can about the magnetic field at point P.

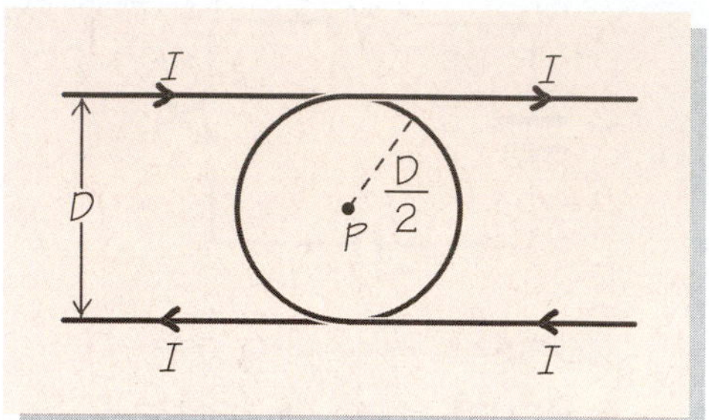

▲ **FIGURE 29–44** Problem 65.

66. (II) Calculate the force per unit area between two metal sheets that carry identical currents in the same direction. The sheets carry a current of linear density h A/m as in Problem 12.

67. (II) Equal but opposite currents I travel in the inner and outer wires of a coaxial cable. As a function of the distance from the central axis, find the magnetic field (a) inside the inner wire; (b) in the region between the wires; (c) in the outer (tubular) wire; (d) outside the outer wire.

68. (II) A hydrogen atom may be described as consisting of an electron that moves in a circular orbit around a proton. The force that gives rise to the motion is the Coulomb attraction between the proton and the electron, which have charges $\pm e$, respectively, where $e = 1.6 \times 10^{-19}$ C. The motion is further constrained by the requirement that the angular momentum has the value $nh/2\pi$, where n is an integer and $h = 6.63 \times 10^{-34}$ J·s, Planck's constant. Calculate the magnitude and direction of the magnetic field at the location of the proton. What is the magnetic moment of the current loop?

69. (II) Find the force between the long, straight wire and the rectangular wire loop shown in Fig. 29–45 for currents $I_1 = 10$ A and $I_2 = 5$ A.

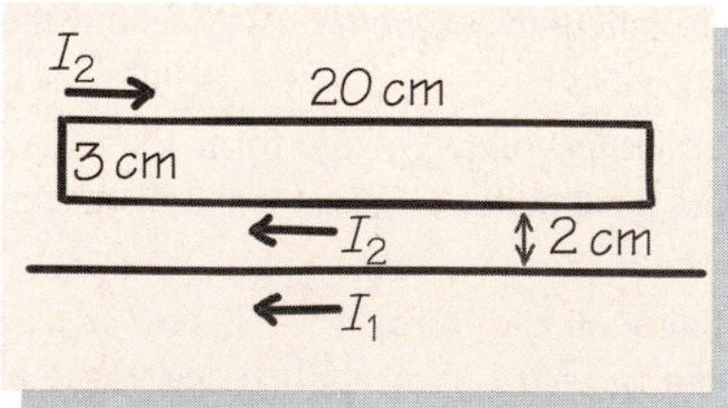

▲ **FIGURE 29–45** Problem 69.

70. (II) The mechanical integrity of solenoids may present a problem that has to be anticipated in technical design. To illustrate the forces that can be present, calculate the force between two neighboring turns of a superconducting solenoid. The radius of the solenoid is 3 cm, the diameter of the wire is 0.8 mm, and the current in the solenoid is 150 A.

71. (II) Consider two parallel wires spaced a distance $d = 1$ cm apart, which each carry a current $I = 1$ A. (a) Compare the magnetic force between these wires to the electric force they would exert on each other if the current carriers (electrons) were not neutralized by a background of positive charges. Use 10^{21} per cm as the linear density of charge carriers in the wire. (b) What excess of electrons per unit length over the positive background would make the electric force equal the magnetic force between the wires? (c) What fraction of the total number of charge carriers is the excess calculated in part (b)?

72. (II) A long wire carries a current I_1. A segment of a second wire, which carries a current I_2, is oriented radially away from the first wire. The segment has length L, and its closest end is a distance d from the first wire. Calculate the torque, direction and magnitude, on the wire segment about the axis defined by the long wire.

73. (II) In Example 29–8, we found the magnetic field due to a circular wire of radius R, carrying a current I, at a point a distance x away from the center of the ring but along the axis to be

$$B = \frac{\mu_0 I}{2} \frac{R^2}{(R^2 + x^2)^{3/2}}.$$

A pair of such coils placed coaxially a distance R apart makes up a *Helmholtz coil*, for which the magnetic field everywhere inside is fairly constant (Fig. 29–46). (a) Determine the magnetic field on the axis as a function of x, with $x = 0$ marking the location of the left-hand coil. Evaluate the field at $x = 0$, $x = R/4$, and $x = R/2$. (b) Show that $dB_x/dx = 0$ and $d^2B_x/dx^2 = 0$ at $x = R/2$.

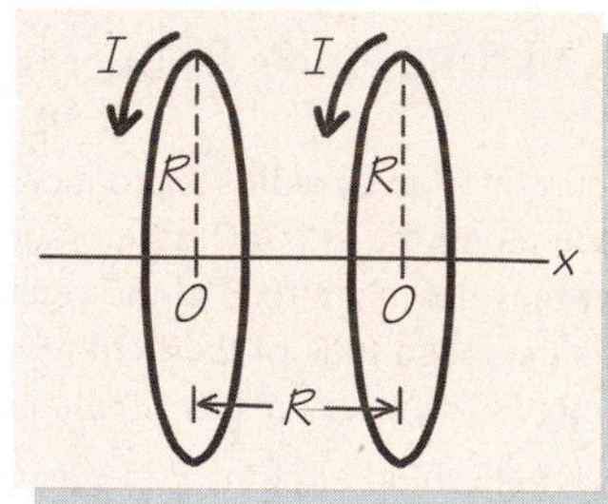

▲ **FIGURE 29–46** Problem 73.

74. (II) A sensitive experiment has to be performed in zero magnetic field. To achieve this, a Helmholtz coil of radius $R = 50$ cm with 50 turns is used to compensate for Earth's magnetic field of 5×10^{-5} T. (A Helmholtz coil is a current loop that sets up a magnetic field to cancel an external field, such as that due to Earth.) (a) What should be the current in the coil? (b) *Estimate* the residual field if the sensitive component of the equipment is confined to a thin cylindrical volume 10 cm long.

75. (III) A demonstration apparatus consists of a large glass bulb containing a small electron gun. The bulb is filled with rarified inert gas, which makes the trajectory of the electron visible. By placing the equipment in the magnetic field of a Helmholtz coil, the experiment described in Example 28–3 can be performed, and the ratio e/m for an electron can be determined. Design the equipment and select appropriate parameters for this demonstration.

76. (III) A certain electric current distribution produces a magnetic field of the form $\vec{B} = \beta(y\hat{\imath} - x\hat{\jmath})$ near the origin of a coordinate system. Find the current distribution responsible.

Michael Faraday delivering one of his famous public lectures in 1856. These lectures earned him great popular success.

Faraday's Law

Our treatment of the magnetic field, its sources, and its effects on moving charges and currents brought in elements of the intimate connection between electricity and magnetism. In this chapter, a new feature of this connection is introduced in the form of a new physical law: Faraday's law. This law describes how changes in magnetic fields produce electric fields. Faraday's law has far-reaching technological applications. It lies behind our entire system of electrical power generation and plays a role in most of the electronic devices we use.

Before we look further at Faraday's law, we should acknowledge Faraday the man. Apprenticed as a bookbinder at age 13, he was inspired at age 22 by a series of lectures at the Royal Institution in London to become a scientist. In spite of the fact that he had little mathematical training, in 1813 he became an assistant to Sir Humphry Davy, an already famous scientist. Davy soon recognized Faraday's qualities, and Faraday gradually became more independent in his experimental inquiries. After Oersted discovered in 1819 that an electrical current in a wire deflected a magnetic compass, considerable excitement developed around the possible connections between electrical currents and magnetism. In particular, the inverse question presented itself: Could magnetic fields produce an electric current? Faraday pursued this sort of question for several years—an entry dated 1822 in his notebooks sets the goal "Convert magnetism into electricity." Faraday exhibited all the qualities of a great scientist in the discovery of what we now know as Faraday's law, which states that *changing magnetic fields generate electric fields*. We will study the details of his discovery and some of its ramifications in this chapter.

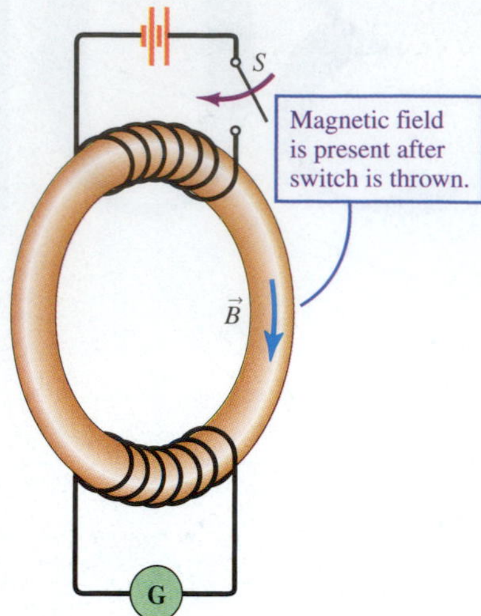

▲ **FIGURE 30–1** Faraday's ring. A changing magnetic flux in the iron ring induces a current in the galvanometer coil at the bottom; the changing flux is due to the opening or closing of a switch connected to the battery of the coil on the top.

30–1 Faraday's Discovery and the Law of Induction

A great experimentalist such as Michael Faraday recognizes the significance of an odd or unexpected measurement. He or she realizes that a small effect is not always experimental error, and pursues the effect systematically, checks its reality, and considers its ramifications from as many points of view as possible. In 1831, Faraday carried out the experiment shown schematically in Fig. 30–1. The battery sends a current through the coil on the top side of an iron ring, which acts as a solenoid. The galvanometer is used to indicate any current in the coil on the bottom side of the ring. The only unfamiliar element is the iron ring, which does two things: It carries the magnetic field set up by the upper coil within the torus, and hence through the bottom coil (recall that magnetic field lines are closed), and—as we shall discuss in Chapter 31—it *magnifies* the size of the field set up in the upper coil. We say that the iron ring *links* the two coils.

To Faraday's disappointment, he observed no effect on the galvanometer when a *steady* current passed through the upper coil, but Faraday's intuition served him well when he noticed a very small twitch of the galvanometer *when the switch that controlled the flow of current in the upper coil was opened or closed*. Within several days of his observations of this small effect, he completed a series of experiments that revealed essentially all the aspects of magnetic induction.

One of the first things Faraday did was to eliminate the possibility that the battery was itself important to the effect. In a second experiment, illustrated in Fig. 30–2a, two bar magnets make a V shape. A large magnetic flux passes through an iron rod when it touches the ends of the two magnets as shown. This rod is surrounded by a coil attached to a galvanometer. The galvanometer deflects—indicating a current in the coil—both when the rod is brought into contact with the ends of the two bar magnets and when the rod is pulled away. A more modern version of this experiment is illustrated in Fig. 30–2b, c; when a bar magnet remains stationary in the presence of a coil, there is no effect (Fig. 30–2b), but when the magnet moves toward the coil, a current is observed (Fig. 30–2c). The critical observation is that it is the *change* in the magnetic flux through the coil that leads to the creation of the current. The general result, which we will describe in detail below, is that the *change* of the magnetic flux through any surface bounded by a closed line causes an emf around that line. If this closed line is a circuit, then this emf can induce a current in the circuit. Faraday referred to the emf as an **induced emf** and called the current produced by a changing magnetic flux an **induced current**; he called the general phenomenon **magnetic induction**.

▼ **FIGURE 30–2** (a) A figure from one of Faraday's lectures. The movement described in the text creates a changing magnetic flux and induces a current even when no battery is present. Adapted from Faraday's book *Experimental Researches in Electricity* in 1839; the labeling is our own. (b), (c) A more modern version of Faraday's experiment. There is a current produced in the coil only when the bar magnet moves toward (or away from) the coil.

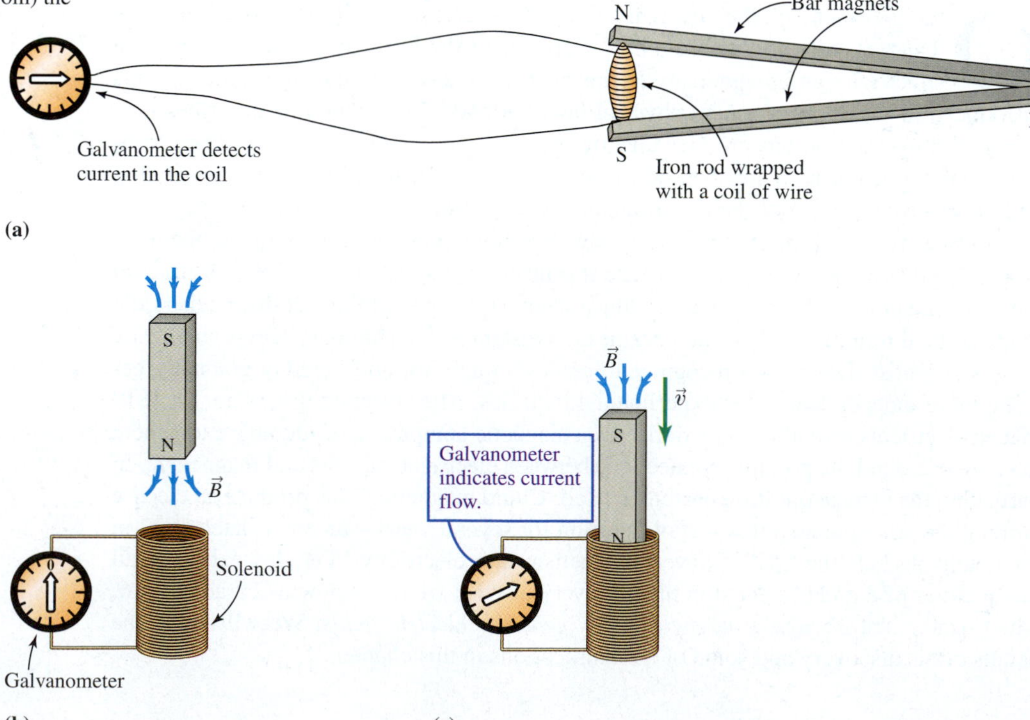

Faraday's discovery was greeted enthusiastically. With it, the possibility of converting mechanical energy to electrical energy became a reality, and electricity generation worldwide is based on Faraday's results. Faraday's discovery could be said to have had a greater effect on the material welfare of humans than any other discovery before or since.

Faraday's Law of Magnetic Induction

We can clarify and add detail to Faraday's law of magnetic induction through a series of experiments, shown in Fig. 30–3.

1. Figure 30–3a shows a bar magnet in the vicinity of a wire loop. If we move the bar magnet toward or away from the wire loop, we observe a current in the wire—an induced current. In Section 29–2 we discussed the magnetic flux, which is associated with the number of magnetic field lines passing through a surface. We have such a flux here, and when the magnet moves as described, the flux through the loop made by the wire changes. Current flows *only* when the magnetic flux through the loop changes. A faster movement—a more rapidly changing flux—results in a larger current than does a slower movement. If the magnet's motion is reversed, then the current reverses.
2. Figure 30–3b shows that as a switch closes in the circuit of a first loop, an induced current momentarily appears in the second loop. Current flow in the first loop produces a magnetic field. When the switch closes, this magnetic field changes as the current in the first loop builds up from zero to its steady value. In turn, this results in a changing magnetic flux through the second loop, and a changing flux through the second loop creates an emf and hence an induced current within it.
3. In Figure 30–3c, there is a constant current in the first (upper) loop, and therefore the magnetic field it generates is constant. But if we move the second loop up or down, the flux through that loop due to the field from the first loop changes. We find that once again an induced current flows in the second loop while it is in motion within the magnetic field of the first loop, and the size of the induced current depends on how fast we move the second loop up or down.
4. Figure 30–3d shows an experiment similar to the third, except that in this case the magnetic flux through the second loop due to the magnetic field of the first loop changes because we change the orientation of the two loops. Again a current flows in the second loop.
5. A last experiment is revealing. We can take the lower coil in Fig. 30–3c and simply squeeze it, thereby changing its area. The flux through it accordingly changes, and once again we observe an induced current, and the size of the current depends on how fast we change the area of the second loop.

▼ **FIGURE 30–3** Ways to make a magnetic flux through a loop change and thereby induce a current I_{ind}. (a) The distance between a wire loop and a bar magnet changes, and a current is induced in the loop. (b) A switch is closed to start a current in one loop, and a current is induced in a second, nearby loop. (c) The distance between a current-carrying loop and a second loop changes, and a current is induced in the second loop. (d) A second loop rotates in the presence of the current-carrying loop, and a current is induced in the second loop.

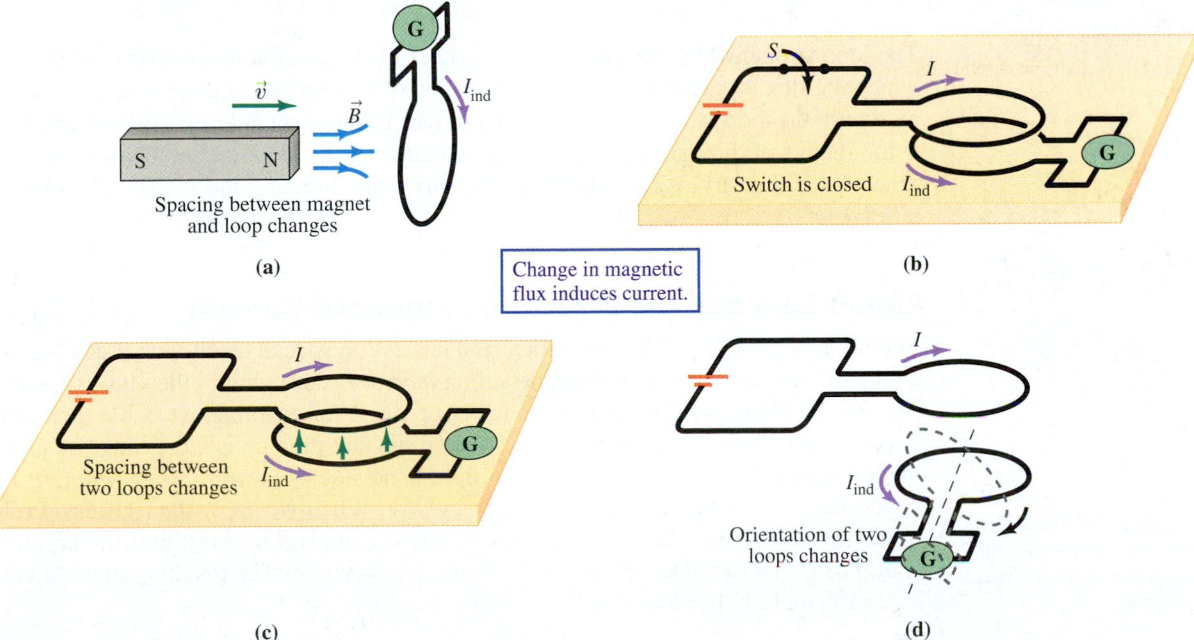

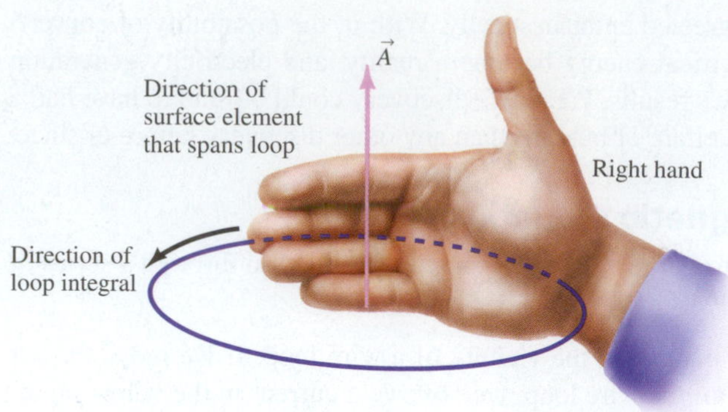

▶ **FIGURE 30–4** When the direction around a loop is given, the orientation of the surface that spans the loop is specified by a right-hand rule. Here we show a single vector $\vec{A}$ for the entire surface, which is flat. For a curved surface, the directions of infinitesimal areas $d\vec{A}$ vary from point to point.

The common feature to all these experiments is that a changing magnetic flux through a loop induces an emf around that loop—if the loop is formed by a wire, then the induced emf leads to an induced current. How fast the flux through the loop changes determines how large the emf is. Faraday's law summarizes these observations, and also the sign: The negative of the *time rate of change* of the magnetic flux through a surface, Φ_B, equals an emf around the closed loop that bounds the surface. The negative sign dictates the direction of the induced emf, as we will discuss in the next subsection. We know from Chapter 28 that an emf $\mathscr{E}$ is the line integral of an electric field. In this case we are interested in the line integral around a *closed* loop:

$$\mathscr{E} = \oint \vec{E} \cdot d\vec{s}. \tag{30–1}$$

The precise statement of **Faraday's law of induction** is

$$\mathscr{E} = \oint \vec{E} \cdot d\vec{s} = -\frac{d\Phi_B}{dt}. \tag{30–2}$$

FARADAY'S LAW OF MAGNETIC INDUCTION

Here Φ_B is the magnetic flux through the surface S that spans the loop, a quantity defined in Chapter 29:

$$\Phi_B = \int_{\text{surface } S} \vec{B} \cdot d\vec{A}. \tag{29–11}$$

The loop around which the emf is defined, Eq. (30–1), must bound the surface through which the flux is calculated, and the orientation of that surface is determined by the direction of the loop integral and a right-hand rule. This right-hand rule works as follows: If the fingers of the right hand curl in the direction of the loop, the thumb indicates the direction of the surface for calculating the flux—the direction the surface element $d\vec{A}$ takes (Fig. 30–4).

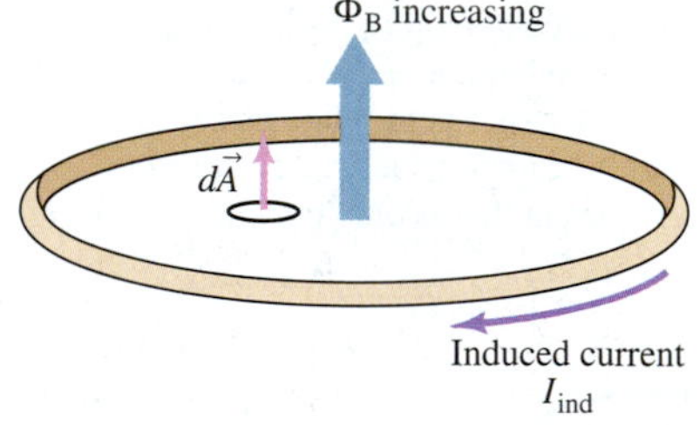

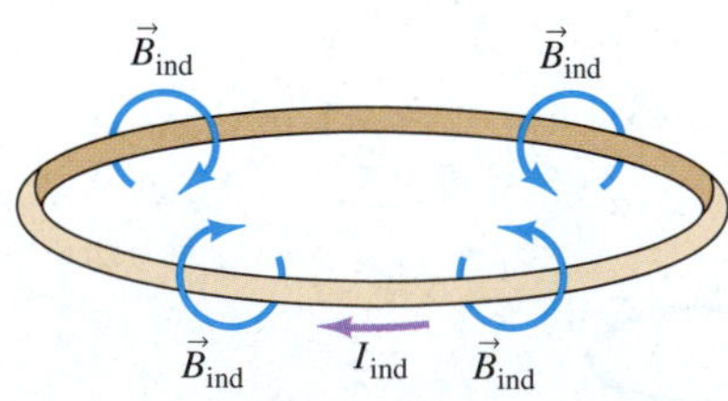

▲ **FIGURE 30–5** (a) A magnetic flux change induces a current. (b) The induced current produces its own magnetic field; that field tends to oppose the flux change that induced the current.

Lenz's Law and the Direction of Induced Current

The minus sign in Eq. (30–2) is critical and deserves a special discussion. Let's look at a loop of wire and suppose the magnetic flux *increases* through it in the sense shown in Fig. 30–5a. (This could happen in a number of ways; for example, we could thrust the N pole of a magnet toward the loop.) When we say that the flux is increasing, we mean that its time derivative is positive, so the right-hand side of Faraday's law [Eq. (30–2)] is negative. The induced emf is therefore *negative*. When we apply the right-hand rule to Fig. 30–5a, we see that the positive direction is counterclockwise; thus the negative sign means that the induced emf is *clockwise* in that figure. The resulting induced current will similarly be clockwise.

We know that currents produce magnetic fields, and the induced current is no exception. By using the right-hand rule, we can see that the magnetic field produced by the induced current is directed down through the loop (Fig. 30–5b). The direction of this field is such that it tends to *decrease* the magnetic flux through the loop. Because the original flux change that induced the current in the first place was positive, we conclude that the induced current has acted to oppose the flux change that caused it. Further analysis shows that the induced current always opposes the *change* in flux and therefore tends to keep the flux from changing. This way of thinking about Faraday's law is due to Heinrich Emil Lenz, and it is called **Lenz's law**:

> **Induced currents produce magnetic fields that tend to oppose the flux changes that induce those currents.**

Lenz's law is useful in determining the *direction* of an induced current.

CONCEPTUAL EXAMPLE 30–1 The north pole of a bar magnet is thrust toward the face of a fixed metal ring (Fig. 30–6). Use Lenz's law to determine the direction of any induced current in the ring.

Answer As the north pole of the magnet approaches the ring, the magnetic field lines near the ring become denser. Hence the magnetic flux through the surface of the ring, which is perpendicular to the magnet, *increases*. Lenz's law states that the induced current will *oppose* the change of magnetic flux that passes through the ring. The induced current must therefore produce a field that serves to decrease the magnetic flux (and magnetic field) through the ring. This induced magnetic field will be directed to the left. If we use the right-hand rule, the current in the ring that will produce this field is oriented clockwise as seen by an observer looking toward the north pole of the magnet.

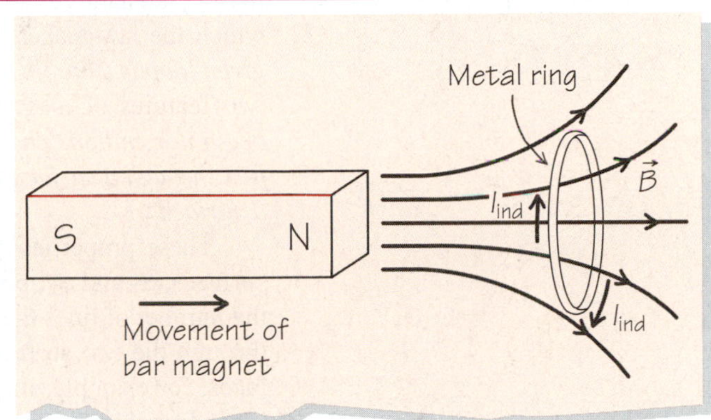

▲ **FIGURE 30–6** When the bar magnet approaches the ring, a current is induced in the ring.

EXAMPLE 30–2 A constant magnetic field has only a y-component B_0 in a large region for which $x < 0$, and is zero for $x > 0$ (Fig. 30–7). A square metal loop with sides of length L is oriented in the xz-plane and pulled through the field in the $+x$-direction with steady velocity $\vec{v} = v\hat{i}$. The total resistance of the loop is R. Find the magnitude and direction of any induced current in the wire as a function of time, assuming that the front edge of the square crosses the line $x = 0$ at $t = 0$. Evaluate your result for $B_0 = 1.0\text{ T}$, $L = 0.10\text{ m}$, $R = 0.065\ \Omega$, and $v = 10.0\text{ cm/s}$.

Strategy Starting at $t = 0$, the wire loop passes out of the region where there is magnetic field ($x < 0$). As it does so, the magnetic flux upward through the loop decreases, and an emf is induced in the loop. We find the magnetic flux through the planar surface of the loop and in particular its time dependence. We can then use Faraday's law, Eq. (30–2), to determine the emf and subsequently the induced current in the wire loop. The loop integral as seen from above is arbitrarily chosen to be counterclockwise. With this orientation for the surface element, the infinitesimal surface elements $d\vec{A}$ that make up the integral are oriented upward, in the same direction as $\vec{B}$, and $\vec{B}\cdot d\vec{A} = B_0\,dA$. The constant B_0 can then be removed from the integral.

Working It Out For $t < 0$, the flux through the loop has a constant value,

$$\Phi_B = \int_{\text{surface } S} \vec{B}\cdot d\vec{A} = B_0 \int_{\text{surface } S} dA = B_0 L^2,$$

so there is no induced emf and no current.

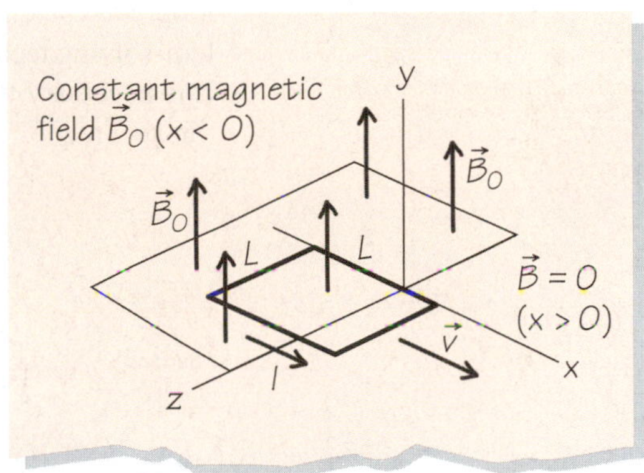

▲ **FIGURE 30–7**

In the time period $t = 0$ to $t = L/v$, the loop is in the process of leaving the region of the magnetic field, and the magnetic flux through it changes. The portion of the loop in the field runs from $x = 0$ to the back of the loop, which is at $x = -(L - vt)$. In other words, only an area $(L - vt)L$ remains in the field, and

$$\Phi_B = B_0(L - vt)L.$$

This flux is not constant, and for this time period

$$\frac{d\Phi_B}{dt} = -B_0 vL.$$

(continues on next page)

The emf counterclockwise around the loop equals the *negative* of this value: $\mathcal{E} = +B_0vL$. A counterclockwise current is induced in the loop during the time interval $t = 0$ to $t = L/v$:

$$I = \frac{\mathcal{E}}{R} = \frac{B_0vL}{R}. \tag{30–3}$$

The numerical value of this induced current is

$$I = \frac{(1.0\ \text{T})(10.0 \times 10^{-2}\ \text{m/s})(0.10\ \text{m})}{0.065\ \Omega} = 0.15\ \text{A}.$$

Finally, for $t > L/v$, the loop has moved out of the region of constant field, so the flux takes on a constant value (zero), and there is neither an induced emf nor a current.

What Do You Think? What happens if the loop is pushed back into the region of magnetic field? (a) nothing, (b) the induced current is in the same direction as before, (c) the induced current is in the opposite direction as before. *Answers to* ***What Do You Think?*** *questions are given in the back of the book.*

What Sets the Surface Used in Faraday's Law?

In our statement of Faraday's law, we did not specify the surface formed by the loop to which the law makes reference. We shall now show that *any surface bounded by the given loop is suitable because the flux is the same through any such surface*. Let's recall two features of magnetic flux. *All magnetic field lines are continuous; they neither begin nor end on "charges"*; indeed, there are no magnetic charges. Also, *the magnetic flux through a surface is proportional to the net number of field lines that pass through a surface*.

These properties of magnetic field lines have some consequences: Consider two surfaces, S_1 and S_2, bounded by a loop (Fig. 30–8a). Because the lines are continuous, the number of lines that pass through the two surfaces must be the same; hence the flux through the two surfaces must be the same. Some lines may not pass through all surfaces, for example, line 1 in Fig. 30–8b does not pass through S_1, but if a line passes into some third surface S_3 and not into surface S_1, then the line must also pass out of surface S_3 and therefore does not contribute to the net flux through S_3.

From this analysis we can conclude that *the magnetic flux through one surface bounded by a closed loop is the same as the magnetic flux through any other surface bounded by the same loop. Both surfaces must be oriented by the right-hand rule*. This is a very helpful result because it shows that, within the constraint of a given bounding loop, we can choose any surface we like to calculate the magnetic flux. A good problem-solving technique is to find a surface over which the flux is easily calculated when it is necessary to compute the flux, as it is in Faraday's law. Example 30–3 illustrates this technique.

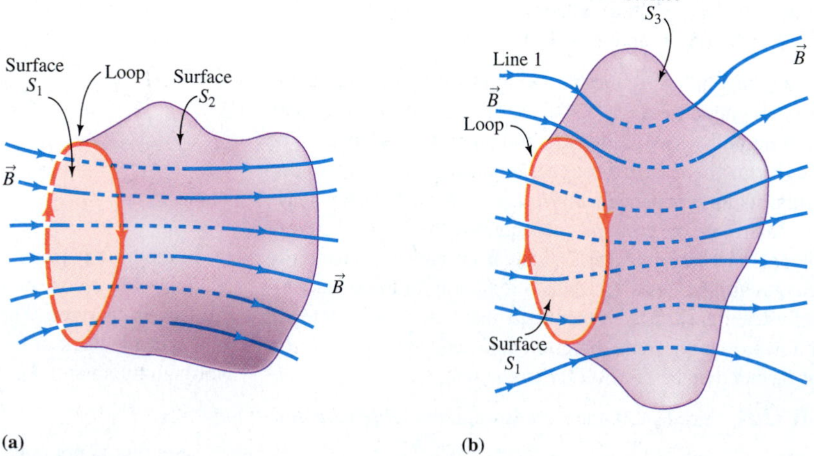

▲ **FIGURE 30–8** (a) The same net number of magnetic field lines pass through any two surfaces S_1 and S_2 bounded by a closed loop. (b) Surfaces S_1 and S_3 are both bounded by a loop. If magnetic field line 1 passes through surface S_3 but not through surface S_1, it must pass through surface S_3 again and hence does not contribute to the net magnetic flux through surface S_3.

EXAMPLE 30–3 Suppose that a certain region has a constant magnetic field of magnitude B_0. Find the magnetic flux upward through a hemisphere of radius R whose base is perpendicular to the field.

Setting It Up In Fig. 30–9 we sketch the situation, including making a choice of axes: We choose the field in the y-direction, $\vec{B} = B_0\hat{j}$, so that the base of the hemisphere forms a circle on the xz-plane.

Strategy We can find the flux through the hemisphere by finding the flux through any other surface bounded by the hemisphere's bounding perimeter, namely the circle of radius R in the xz-plane. The simplest surface that spans this perimeter is the planar disk in the xz-plane. For the disk, $\vec{B}$ is parallel to $d\vec{A}$.

Working It Out Mathematically, the equality of the flux through the hemisphere and through the planar disk can be written as

$$\Phi_{B\text{ hemisphere}} = \int_{\text{planar disk}} \vec{B}\cdot d\vec{A} = B_0 \int_{\text{planar disk}} dA = B_0\pi R^2.$$

Although the direct calculation of the flux through the hemisphere would be complicated, the calculation becomes trivial when the flat surface is used.

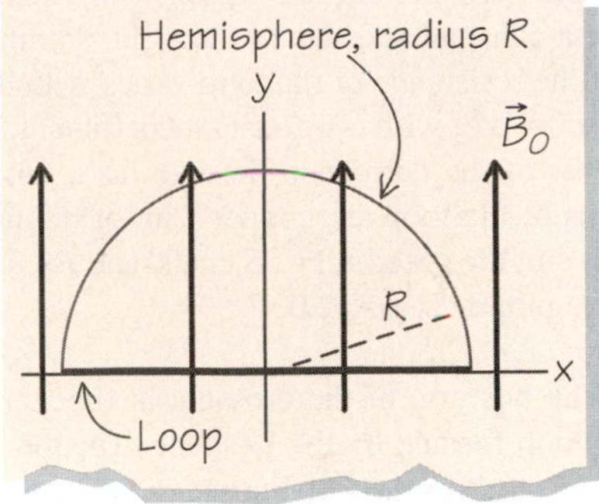

▲ **FIGURE 30–9**

CONCEPTUAL EXAMPLE 30–4 A loop of wire and a straight wire each lie on a tabletop (Fig. 30–10a). The straight wire, which is connected to a battery and carries a current in the direction shown, is moved toward the loop. Is a current induced in the loop? If so, what is its direction?

Answer First, does the movement of the straight wire lead to an induced emf in the loop? In Fig. 30–10b, we have drawn some of the magnetic field lines for the constant current in the straight wire. This field is stronger near the wire, so as the wire moves toward the loop, there is an increase in the flux down through the loop into the table. Because there is a changing flux through the loop, there is indeed an induced emf around it. What is its direction (which will also be the direction of the induced current)? We can use Lenz's law to answer a question about the sign of the induced emf. If we look down on the table, the current is induced in the counterclockwise direction, and the magnetic field associated with this current will come up through the center of the loop, creating a flux that opposes the flux change from the movement of the straight wire (Fig. 30–10c). Thus the direction of the induced current is counterclockwise.

What Do You Think? Will there still be a current induced in the loop if instead of the straight wire with the battery moving toward the loop, the loop moved toward the straight wire?

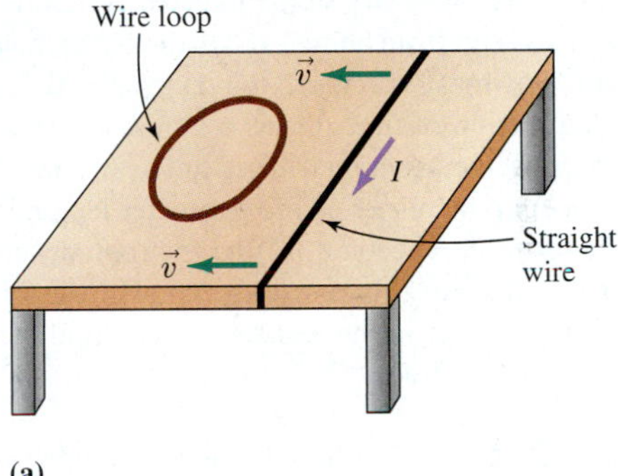

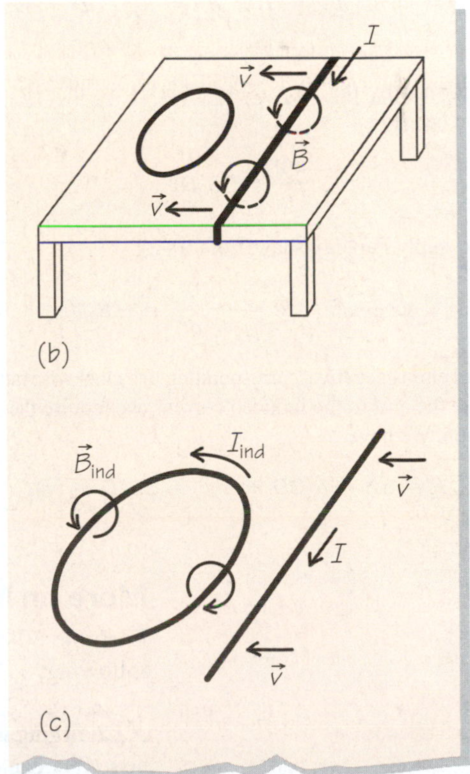

▶ **FIGURE 30–10** (a), (b) The field associated with the current in the wire. (c) The current induced in the loop produces a magnetic field of its own.

EXAMPLE 30–5 A closed loop is constructed of a fixed wire shaped as a square-ended U (Fig. 30–11a) and a conducting crossbar that is free to move in the x-direction and makes electrical contact with the U-shaped section. The loop lies in the xz-plane with the square base of the U-shaped section at $x = 0$. A constant magnetic field is oriented in the y-direction, $\vec{B} = B_0\hat{j}$. (a) The movable crossbar is pulled at a constant speed v to the right, starting from $x = 0$ when $t = 0$. If the resistance of the loop varies with the total length L according to $R = \alpha L$, with α a constant coefficient, what is the direction and value of the current in the loop as a function of time? (b) All segments of the loop are copper wire of radius 0.25 cm, the length $D = 7.0$ cm, the speed $v = 28$ cm/s, and $B_0 = 0.18$ T. What is the current magnitude at $t = 2.0$ s?

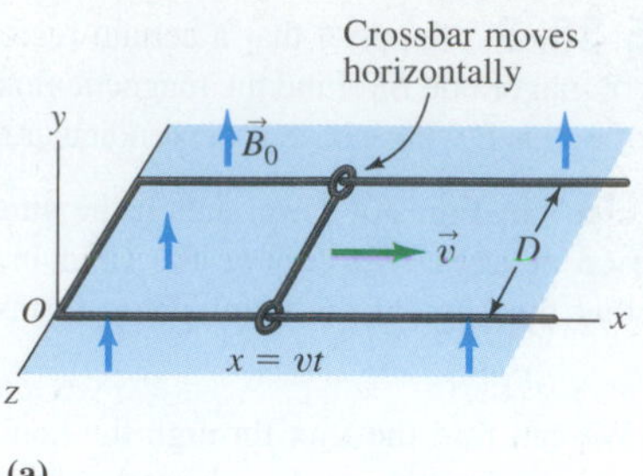

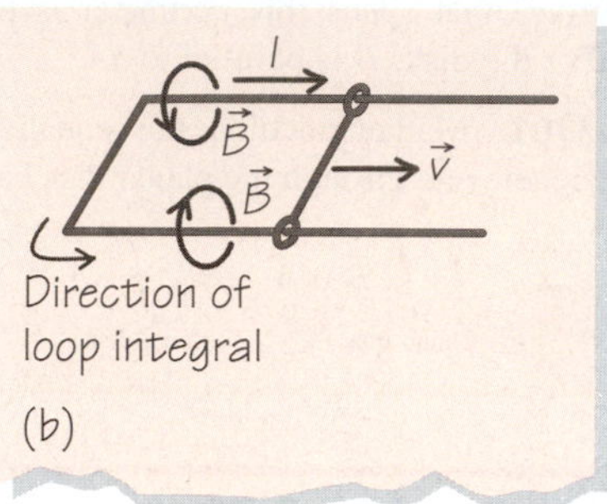

▲ **FIGURE 30–11** (a), (b) Lenz's law shows that the current induced in the circuit that contains the moving crossbar will be in the direction shown.

Strategy The position of the crossbar at time t is $x = vt$, and the area of the loop formed by the U-shaped section and the crossbar increases. In the presence of the magnetic field, and with the area of the loop oriented upward, parallel to the magnetic field, that means that the flux through the loop also increases, and Faraday's law of induction applies. Lenz's law states that the magnetic field due to the induced current must be directed down in order to decrease the flux (Fig. 30–11b), so the induced current is in the clockwise direction as seen from above. To calculate the magnitude of the induced emf, we must calculate the magnetic flux Φ_B through the loop as a function of time. Once we know the induced emf, we can find the induced current by using Ohm's law. In this case the resistance of the loop increases as the perimeter length increases, and that must be taken into account to find the time dependence of the current. Finally, for part (b) we use the known resistivity of copper to find the coefficient α in the resistance, then find the value of the current at a particular time.

Working It Out (a) The flux through the loop is

$$\Phi_B = \int_{\text{surface}} \vec{B}\cdot d\vec{A} = \int_{\text{surface}} B_0\, dA = B_0 \int_{\text{surface}} dA = B_0 A.$$

The area formed by the loop is $A = (vt)D$, so that $\Phi_B = B_0 A = B_0 Dvt$, and hence

$$\frac{d\Phi_B}{dt} = B_0 Dv.$$

We now apply Faraday's law, Eq. (30–2):

$$\oint \vec{E}\cdot d\vec{s} = \mathscr{E} = -\frac{d\Phi_B}{dt} = -B_0 Dv.$$

The induced emf is negative, corresponding to a clockwise induced emf.

To relate the emf to the induced current, we require the resistance R of the loop. We have

$$R = \alpha L = \alpha(2D + 2vt) = 2\alpha(D + vt).$$

The induced current is therefore

$$I_{\text{ind}} = \frac{\mathscr{E}}{R} = -\frac{B_0 Dv}{2\alpha(D + vt)}.$$

Try checking the dimensions of this current.

(b) To find α, note that the resistivity of copper is $\rho = 1.72 \times 10^{-8}\ \Omega\cdot\text{m}$ and that, according to Eq. (27–14), the resistance of a wire of length L and cross-section A is $R = \rho(L/A)$. By comparing the form $R = \alpha L$, we see that

$$\alpha = \frac{\rho}{A} = \frac{1.72 \times 10^{-8}\ \Omega\cdot\text{m}}{\pi(0.25 \times 10^{-2}\ \text{m})^2} = 8.8 \times 10^{-4}\ \Omega/\text{m}.$$

At this point, all quantities in the equation for the current are known and, at $t = 2$ s, the magnitude of the induced current is

$$\begin{aligned} I_{\text{ind}} &= \frac{B_0 Dv}{2\alpha(D + vt)} \\ &= \frac{(0.18\ \text{T})(0.070\ \text{m})(0.28\ \text{m/s})}{2(8.8 \times 10^{-4}\ \Omega/\text{m})[(0.070\ \text{m}) + (0.28\ \text{m/s})(2.0\ \text{s})]} \\ &= 3.2\ \text{A}. \end{aligned}$$

What Do You Think? Copper has a low resistivity. Nichrome has a much higher one. If nichrome wire is used throughout, the induced current will (a) increase, (b) decrease, (c) stay the same, (d) go to zero.

More on Magnetic Induction

Example 30–5 illustrates some important features of Faraday's law, among them the following:

A Changing Flux Does Not Necessarily Mean a Changing Magnetic Field: The magnetic flux can change not only because the magnetic field changes with time but also because the area of the loop through which the flux is calculated may change with time.

In Example 30–5, the magnetic field is constant in some region, yet there is nevertheless an induced emf. In the next section we see some immediate applications of this fact.

Induced Electric Fields Are Nonconservative: We note that induced fields differ fundamentally from the electric fields we have previously encountered. In our earlier work, electric fields were always associated with *conservative* forces. The work done by those fields in moving a charge around a closed loop is always zero:

$$\text{conservative: } \mathscr{E} = \oint \vec{E} \cdot d\vec{s} = 0,$$

and this fact is what allows us to describe such electric forces with a potential. This is precisely what is *not* true for the fields that result from Faraday's law; the emf about a closed loop is specified by the changing flux:

$$\text{nonconservative: } \mathscr{E} = \oint \vec{E} \cdot d\vec{s} = -\frac{d\Phi_B}{dt}.$$

The induced electric field cannot be described by a potential that is a function of space.

30–2 Motional EMF

We have now developed the idea that an emf can be induced in a conductor that moves in a magnetic field, as in Example 30–5. We call this emf a **motional emf**. Figure 30–12a illustrates this in the case of a conducting rod (length L) moving with constant speed v in a constant magnetic field of magnitude B; in Fig. 30–12a, the axes are chosen so that the position of the rod along the x-axis is $x = vt$. The rod is not part of a circuit, and there is no current flowing through it, but there is an effect on the charge carriers within the rod: There is an accumulation of positive charges at the near end of the rod and negative charges at the far end. The charge accumulation leads to an emf that cancels the motional emf and allows an equilibrium to be established. This effect is described alternatively by Faraday's law or by the Lorentz force law, and here we want to see how that occurs.

Let us first look at Faraday's law applied to this situation. The rod itself does not form a loop, so you may wonder how Faraday's law applies here. But Faraday's law applies to *any* loop; we can think of the rod as forming one leg of a loop that is part real conductor and part imaginary, as indicated by the rod and the fixed dashed lines that complete the loop in Fig. 30–12a. The loop formed by the imaginary line and the rod is situated in the xz-plane with its area elements $d\vec{A}$ oriented in the $+y$-direction. Its area

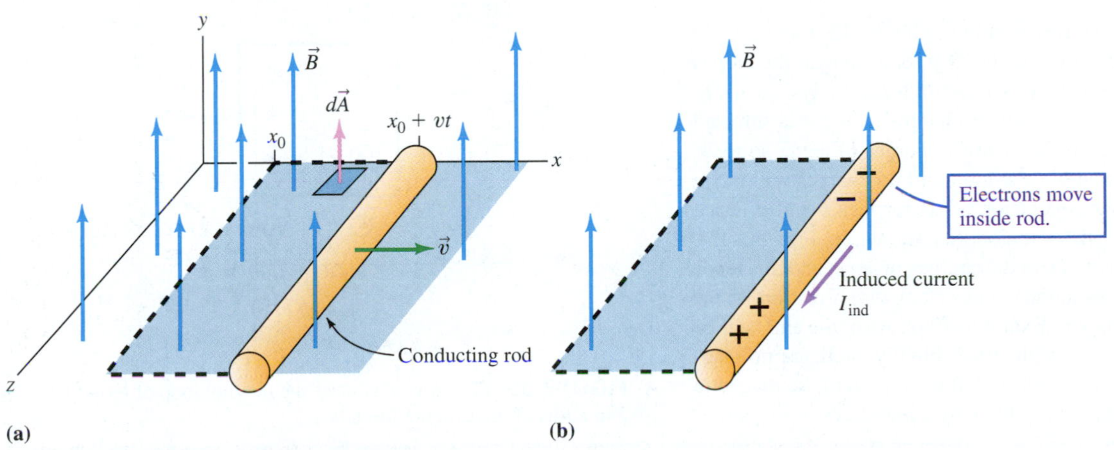

▲ **FIGURE 30–12** (a) A conducting rod of length L moves with constant speed v in the $+x$-direction through a constant magnetic field directed in the y-direction. The rod is oriented in the z-direction. The dashed lines represent the imaginary closure of a loop, to which we apply Faraday's law. (b) The induced emf in the rod can be computed by Faraday's law.

is $L(vt - x_0)$. The flux through the closed loop is therefore

$$\Phi_B = BL(vt - x_0),$$

and the rate of change of this flux is

$$\frac{d\Phi_B}{dt} = BLv. \tag{30–4}$$

The emf taken in the counterclockwise direction looking down—the direction that must be taken if we orient the loop area in the $+y$-direction—is then

$$\mathscr{E} = -BLv. \tag{30–5}$$

$\mathscr{E}$ is the *motional emf.* If the dashed line were conducting wire, this emf would drive a current clockwise (as seen from above) around the circuit. Because there is no actual closed circuit, there is movement of charge, and therefore current flow, only until sufficient positive charges accumulate toward the near end of the rod to set up an emf that cancels the motional emf (Fig. 30–12b). At that point, charge carriers no longer move within the rod. Note that the charges move in such a way that the potential will be higher at the near end of the bar.

The effect of the motional emf—the fact that positive charge moves to accumulate at the near end—is simple to understand in terms of the Lorentz force law. Each charge carrier in the rod is moving in a magnetic field and therefore feels a force that equals $q\vec{v} \times \vec{B}$. This force acts in the $+z$-direction for positive charges, just as the motional emf argument in the paragraph above indicates. The force per unit charge, or the electric field that produces this force, has magnitude vB. Because this electric field is constant along the entire length of the rod, the potential difference from one end of the rod to the other is $\Delta V = EL = BvL$. This potential is just the emf that cancels the motional emf, Eq. (30–5). *We conclude that we can view the effect of motional emf as due either to the Lorentz force law or to Faraday's law of induction.*

CONCEPTUAL EXAMPLE 30–6 In Example 30–2, we looked at how Faraday's law describes induced currents in a square loop of wire as it is pulled through a region of constant magnetic field into a region of no magnetic field. Will a Lorentz force analysis lead to the same results? Assume at $t = 0$ that the loop of wire is entirely within the magnetic field region and is about to start to leave it.

Answer We redraw the situation in Figure 30–13. According to Faraday's law, there is no current in the loop for as long as the loop is entirely within the constant field. A Lorentz force analysis leads to the same result: The force on positive charges in leg a is directed downward, as is the force on the positive charges in leg c, so these forces oppose each other within the loop and no charges will move around the loop. A similar conclusion holds for legs b and d. Once leg a passes out of the field region (as shown in Fig. 30–13), however, there is a net force due to the charges in leg c, and this tends to push the positive charges in the counterclockwise direction, as we deduced from Faraday's law in Example 30–2. Also, the emf will be B_0vL, just as we found in Example 30–2. Finally, there are no magnetic forces once the loop has entirely left the region of the magnetic field, and hence no current, just as Faraday's law states.

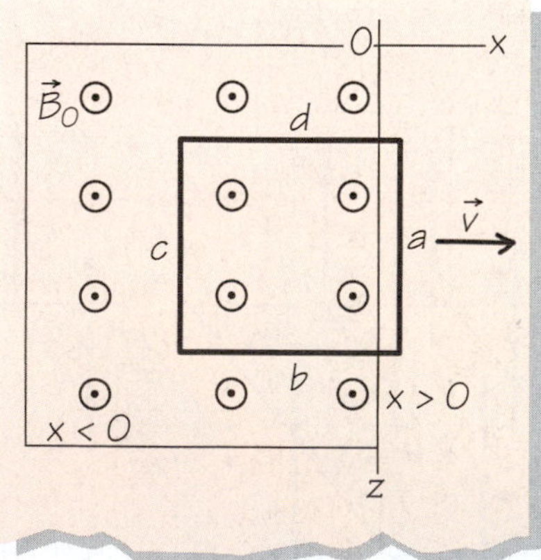

▲ **FIGURE 30–13** An analysis of the moving loop of Example 30–2 in terms of the Lorentz force law.

EXAMPLE 30–7 A rod of length L rotates counterclockwise with constant angular velocity ω about one end in a constant magnetic field of magnitude B_0 oriented perpendicular to the rod's plane of rotation. Find the motional emf in the rod by applying Faraday's law of induction.

Setting It Up We illustrate this situation in Fig. 30–14a, which also includes specification of a coordinate system.

Strategy This is a case similar to the one described at the start of this section. The rod itself does not make a loop, but we construct one through the dashed line shown in Fig. 30–14b. We can then apply Faraday's law of induction by calculating the time dependence of the flux through the loop. It is simplest to orient the area elements of the loop parallel to the magnetic field. Then the flux through the loop is just the constant field times the area of the loop. We shall first find the area of our imaginary loop. This will allow us to find the magnetic flux and subsequently to find its change with time that will give us the emf.

Working It Out With the angle $\theta = \omega t$ specified in the figure, the loop area is

$$\frac{1}{2}\theta L^2 = \frac{1}{2}\omega t L^2.$$

Since the area is oriented upward, the loop integral is followed counterclockwise as seen from above. The magnetic flux through the loop is B_0 times the area; the rate of change of the flux is

$$\frac{d\Phi_B}{dt} = \frac{d}{dt}\left(\frac{1}{2}B_0\omega t L^2\right) = \frac{1}{2}B_0\omega L^2.$$

The motional emf is then

$$\mathscr{E} = -\frac{1}{2}B_0\omega L^2.$$

The sign corresponds to a clockwise emf—that is, the emf drives positive charges radially out from the origin.

What Do You Think? Does the application of the Lorentz force law imply the same direction of charge flow?

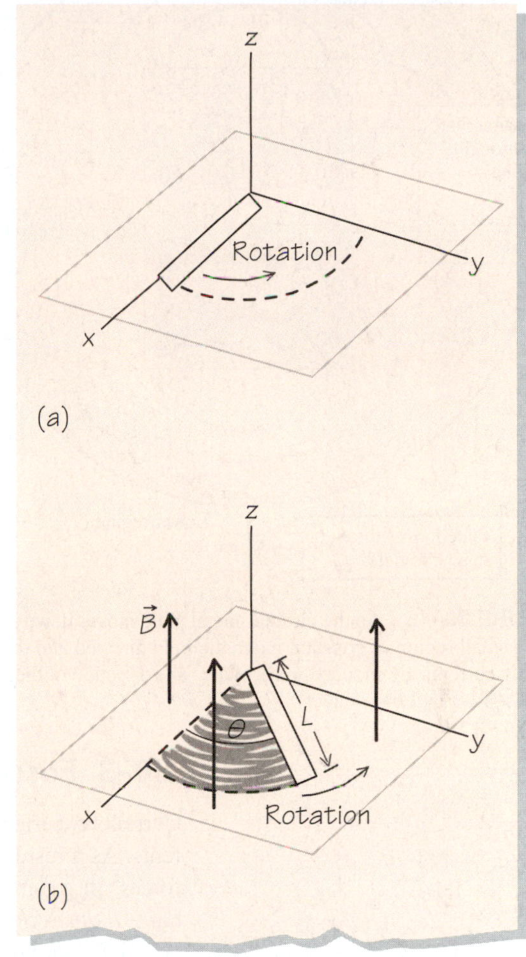

▲ **FIGURE 30–14** (b) The dashed lines represent the imaginary closure of a loop, to which we apply Faraday's law.

Eddy Currents

We have talked about induced emfs in the motion of wires and rods. But in many applications large pieces of metal move in a magnetic field, and there the effects of induced emfs can be significant, either desirable or undesirable. In large pieces the induced currents are spread through the material, and we refer to them as **eddy currents**. Figure 30–15 shows the eddy currents set up in a flat, vertically oriented plate moving downward through a region with a horizontal magnetic field.

Eddy currents are dissipated in Joule heating through the resistivity of the metal plate. The eddy currents or the related Joule heating can be a significant advantage in certain applications; eddy currents in a piece of metal that moves through a magnetic field act as brakes. Brakes of this type have practical applications that range from large electric motors to sensitive measuring devices within which oscillations or other motion would be a disadvantage. When eddy currents are undesirable, they can be reduced by eliminating paths for the current flow. This is done either by cutting slots in the metal plate (Fig. 30–16) or by laminating the metal with an insulator.

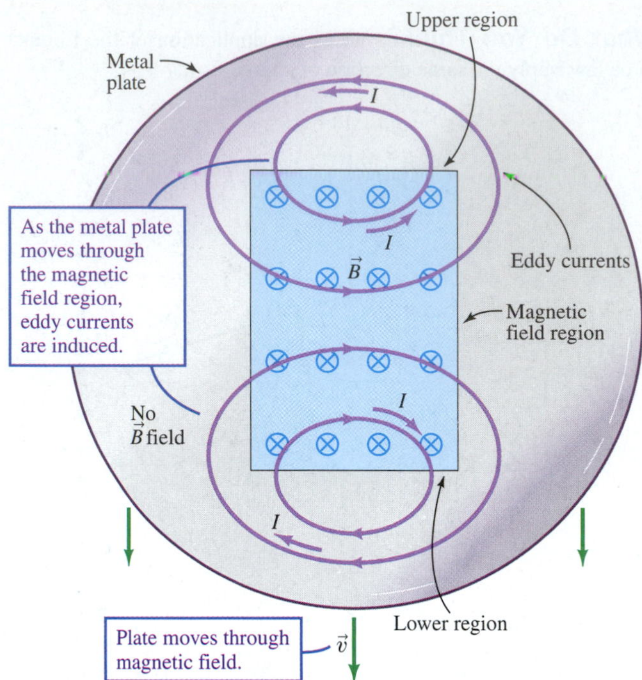

▲ **FIGURE 30–15** As the circular metal plate moves down through a small region of constant magnetic field directed into the plate, eddy currents are induced in the plate. The direction of these currents is given by Lenz's law.

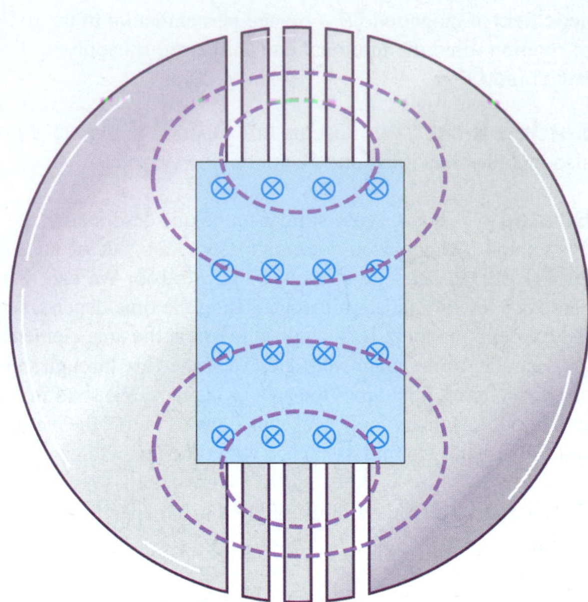

▲ **FIGURE 30–16** To inhibit the development of eddy currents in the moving metal plate, slots can be cut in the plate.

30–3 Forces and Energy in Motional EMF

Currents experience forces in magnetic fields, and this is also the case for any induced currents. As a result, wires or other materials in which currents are induced will experience forces. In general, *the magnetic force on an induced current always inhibits the motion that produces the motional emf.* As we'll now argue, this is a consequence of Lenz's law.

In the moving loop of Example 30–2 (Fig. 30–17), an induced current I appears as the loop leaves the region of magnetic field. The force on the wire is given by Eq. (28–16),

$$\vec{F}_B = I \int d\vec{\ell} \times \vec{B}, \tag{30–6}$$

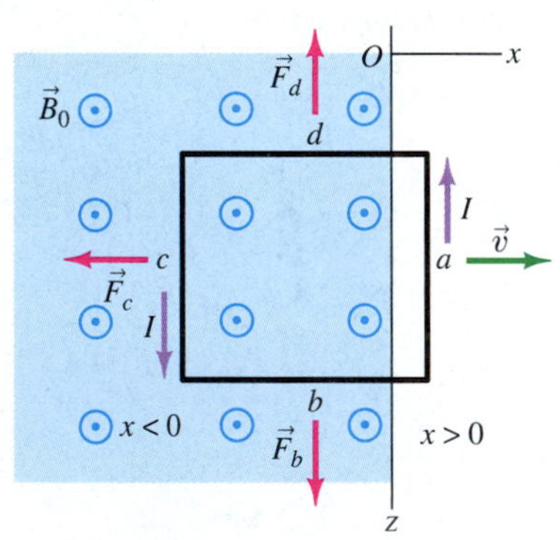

▲ **FIGURE 30–17** The magnetic forces on the loop of Example 30–2 act to slow down the loop.

where $d\vec{\ell}$ describes an element of the wire and $\vec{B}$ is the magnetic field at that wire element. Let's once again draw the loop—this time including the forces on each leg (Fig. 30–17). There is no force on those legs or portions of legs that are out of the magnetic field. Using the right-hand rule, the force on the portion of leg b that is located in the field is directed down, and it is canceled by the force on the portion of leg d located in the field. The only contribution to the net force comes from leg c. Here, application of Eq. (30–6) gives a force

$$F_c = ILB_0 \text{ to the left.} \tag{30–7}$$

Recall from Example 30–2, Eq. (30–3), that the magnitude of the current is $I = B_0 vL/R$, where R is the total resistance of the loop and v is the speed with which the loop moves. The force on the loop is thus

$$F_c = \frac{vL^2B_0^2}{R}. \tag{30–8}$$

This force acts to slow down the loop—it is the typical drag force of a viscous medium.

Because the magnetic force tends to slow the motion, the loop can move at constant velocity only if there is an external force that cancels the magnetic drag. We can imagine that we pull the loop in the direction of the velocity in Fig. 30–17. In this case, we are the external agent supplying a force $\vec{F}$ to the right. The magnitude of this force is F_c, Eq. (30–8).

By applying the external force, we are doing work, or equivalently we are supplying energy at a certain rate. In other words, there is an expenditure of power given by

$$P = \vec{F} \cdot \vec{v} = F_c v = \frac{v^2 L^2 B_0^2}{R}. \qquad (30\text{–}9)$$

CONCEPTUAL EXAMPLE 30–8 The expenditure of power by an external force in a system whose kinetic energy does not change means that there is an energy dissipation in the system. Where is the energy lost in the moving loop of Example 30–2?

Answer There is one place where we lose energy in electrical phenomena, and that is in current flow through a resistance: Joule heating. And we do indeed generate a current in this situation, the induced current which is running through a system with a resistance—the wire. Therefore we have a mechanism to account for the power supplied by the external force. An analytic calculation of the rate of Joule heating should match Eq. (30–9), and we show this below.

Let us check that the power lost in the resistive heating matches the rate at which energy must be supplied to the loop to maintain its motion at a constant speed. The power loss (or energy loss per unit time) in the resistance of the wire is

$$P = I^2 R = \left(\frac{B_0 v L}{R}\right)^2 R = \frac{v^2 L^2 B_0^2}{R}. \qquad (30\text{–}10)$$

Equations (30–9) and (30–10) are the same. *The power loss due to the current flow through the resistor is matched by the power required to keep the loop moving*. The principle of conservation of energy suggests that this result was a foregone conclusion.

The magnetic force on the loop [Eq. (30–8)] is proportional to the speed at which the loop moves and hence is a typical viscous drag (see Chapter 16). This is generally true for the forces on induced currents due to motional emf because their origin is the Lorentz force, which is also proportional to the speed.

EXAMPLE 30–9 A square loop of wire has sides of length 5.0 cm. This loop falls at speed v under the influence of gravity through a region with a constant magnetic field of magnitude 15 T (only magnets built for scientific purposes can attain such a high field), into a region with no magnetic field (Fig. 30–18). The loop is constrained to remain vertically oriented, and the field is horizontal—perpendicular to the loop and into the page. The total resistance of the loop is 1.0 Ω, and its mass is 150 g. (a) Find the terminal speed of the loop as it passes the boundary between the two fields. (b) Calculate the total energy lost to Joule heating in the loop during this period. As an approximation, assume that the loop moves at its terminal speed when it enters the magnetic field region and that this speed remains constant during the loop's passage.

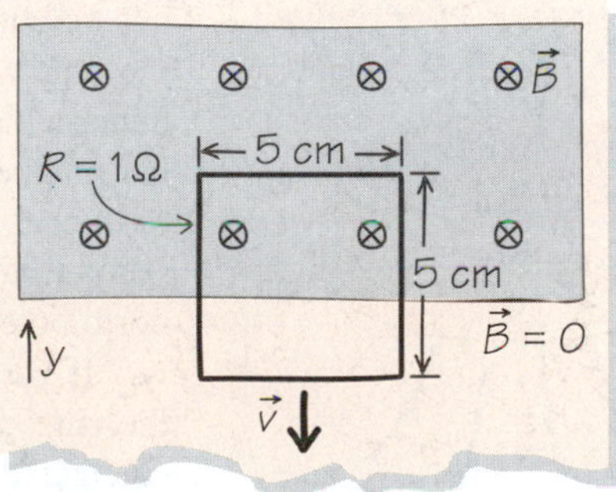

▲ **FIGURE 30–18**

Strategy (a) There will be a drag force for the situation shown in Fig. 30–18 only when the loop of wire is either partway into or partway out of the region with magnetic field. If this is the case, we can use Eq. (30–8) for the magnetic force on the loop, which will be upward. The magnetic force, $F = vL^2B_0^2/R$, will be equal in magnitude to the force of gravity mg when the loop acceleration is zero and the loop reaches its terminal speed v_t.

Working It Out We set $v_t L^2 B_0^2/R = mg$ and solve to obtain the terminal speed v_t,

$$v_t = \frac{mgR}{L^2 B_0^2}. \qquad (30\text{–}11)$$

Numerically,

$$v_t = \frac{(0.15\ \text{kg})(9.8\ \text{m/s}^2)(1.0\ \Omega)}{(5.0 \times 10^{-2}\ \text{m})^2 (15\ \text{T})^2} = 2.6\ \text{m/s}.$$

Strategy (b) As long as the loop moves at its terminal speed, the total energy lost to Joule heating will be a constant given by the product of Joule power and the time the loop spends in the transition region. Given that the loop is moving at v_t, the time spent in the transition region is $t = L/v_t$, where $L = 0.050$ m is the length of a side.

Working It Out The Joule power is, according to Eq. (30–10),

$$P = I^2 R = \frac{v_t^2 L^2 B_0^2}{R}.$$

Thus the energy loss is

$$\Delta E = Pt = \frac{v_t^2 L^2 B_0^2}{R} \frac{L}{v_t} = \frac{v_t L^3 B_0^2}{R}.$$

(continues on next page)

Substituting for v_t from Eq. (30–11), we find that

$$\Delta E = \frac{mgR}{L^2B_0^2}\frac{L^3B_0^2}{R} = mgL.$$

This is just the change in the loop's gravitational potential energy. Numerically

$$\Delta E = (0.15\text{ kg})(9.8\text{ m/s}^2)(0.050\text{ m}) = 0.074\text{ J}.$$

What Do You Think? The calculated energy change is just the change in the gravitational potential energy of the loop as it falls through the transition region. Why?

(a)

THINK ABOUT THIS. . .

HOW DOES MAGNETIC LEVITATION WORK FOR SUPERCONDUCTORS?

As we have remarked above, magnetic drag can act to slow the fall of a metal plate in a magnetic field. We can carry this to an interesting extreme by supposing that the metal plate is a superconductor, a material with no resistance and with the property that the magnetic field does not penetrate its interior. (We'll discuss superconductors and their magnetic properties further in Chapter 31.) If we drop such a piece of material into the field region, the piece is more than slowed down by its entry into the field; it is repelled and bounces back up. This bounce is perfectly elastic because there is no Joule energy loss in a material with no resistance. In this extreme thinking about drag, we are led to a system without energy losses; we have an ideal form of magnetic levitation (Fig. 30–19).

A related application is in magnetically levitated trains, with the levitation used both to hold the train off the surface and guide it along its track (Fig. 30–20). This is useful because there is no rolling friction in the operation of such a train.

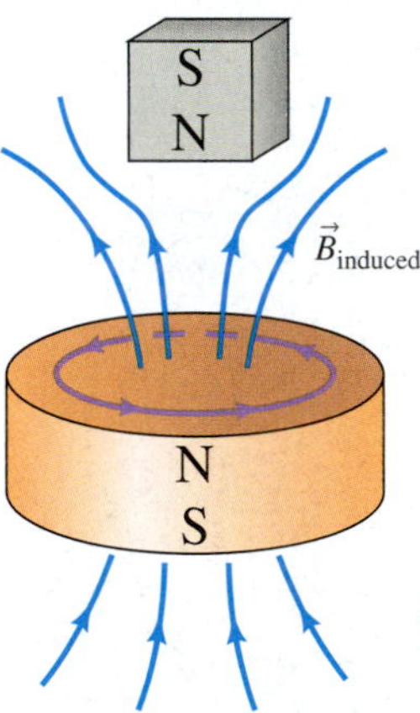

(b)

▲ **FIGURE 30–19** (a) A levitated magnet. A bar magnet moves toward the superconducting material, inducing persistent currents in the superconductor. (b) The magnetic forces between the superconductor and the magnet are repulsive and sufficiently strong to support the magnet's weight.

Forces and Lenz's Law

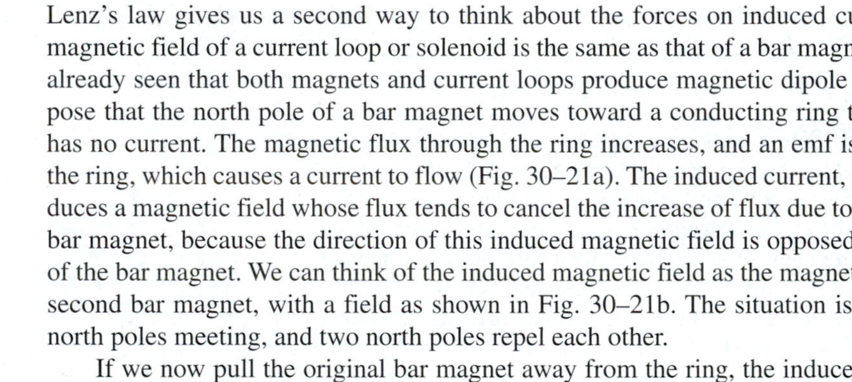

Lenz's law gives us a second way to think about the forces on induced currents. The magnetic field of a current loop or solenoid is the same as that of a bar magnet. We have already seen that both magnets and current loops produce magnetic dipole fields. Suppose that the north pole of a bar magnet moves toward a conducting ring that initially has no current. The magnetic flux through the ring increases, and an emf is induced in the ring, which causes a current to flow (Fig. 30–21a). The induced current, in turn, produces a magnetic field whose flux tends to cancel the increase of flux due to the moving bar magnet, because the direction of this induced magnetic field is opposed to the field of the bar magnet. We can think of the induced magnetic field as the magnetic field of a second bar magnet, with a field as shown in Fig. 30–21b. The situation is one of two north poles meeting, and two north poles repel each other.

If we now pull the original bar magnet away from the ring, the induced current in the ring points in the opposite direction (Fig. 30–21c). The induced magnetic field changes direction, and the ring's magnetic field is like that of a bar magnet, with its south pole adjacent to the north pole of the original bar magnet (Fig. 30–21d). We know that opposite poles attract each other, so the magnet pulls the ring along with it.

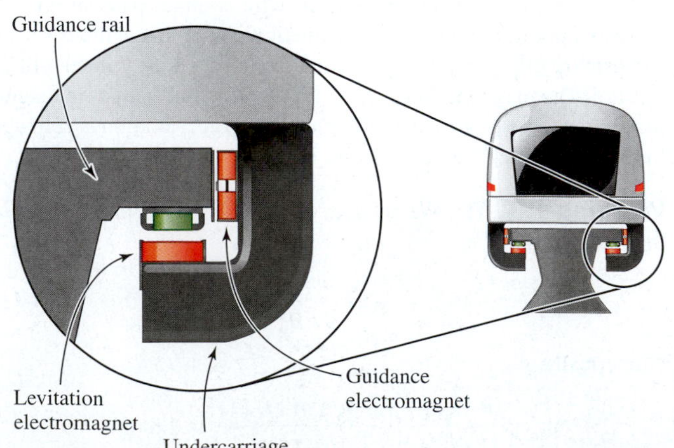

▶ **FIGURE 30–20** Magnetically levitated trains use magnetic induction to hold the train above the travel track as well as to guide it within the track. Even the propulsion system is contained in the time-dependent fields of the magnets. Friction is reduced to a minimum.

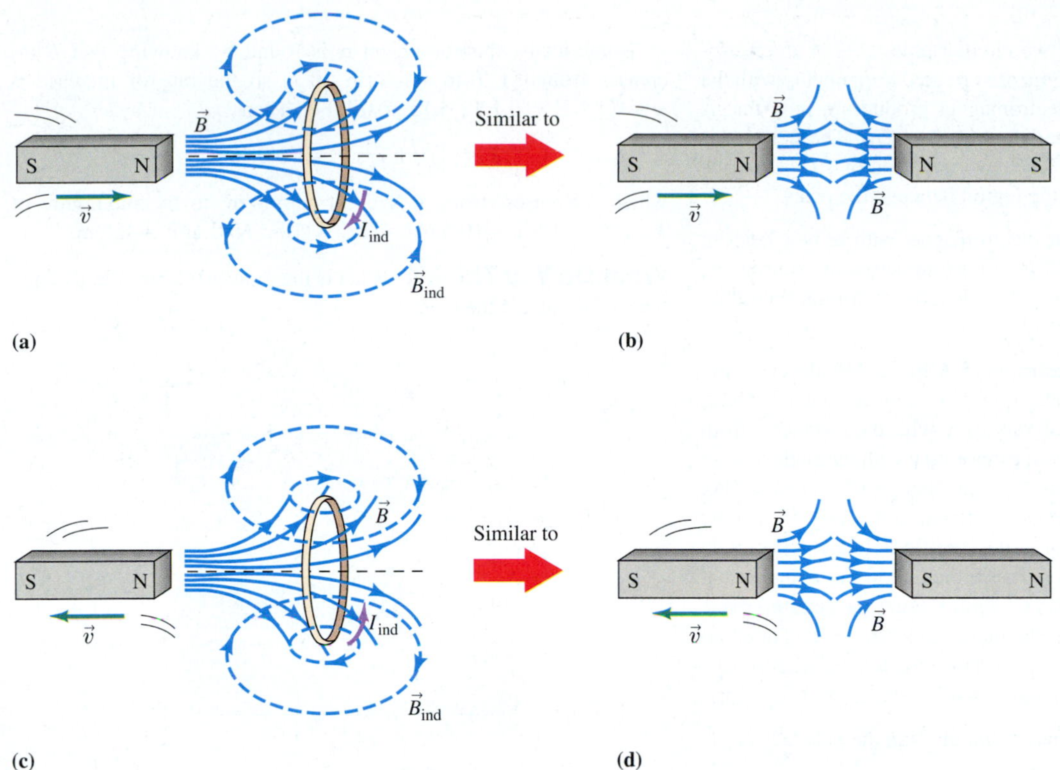

▲ **FIGURE 30–21** (a) The magnetic flux through a conducting ring is increasing because the north pole of a bar magnet moves toward it. The ring is repelled. (b) The direction of the induced current in the ring gives the ring the field of a bar magnet with its north pole to the left, and two north poles repel. (c) The bar magnet is pulled away. In this case, a current is induced in the ring in the opposite direction. (d) The magnetic field of the ring is then that of a bar magnet with its south pole to the left, and it is attracted to the receding bar magnet.

30–4 Time-Varying Magnetic Fields

The magnetic flux through a loop can change in a variety of ways.

1. The loop can move or rotate in the presence of a magnetic field that is not changing with time.
2. The source of the magnetic field can move, as when a bar magnet moves.
3. The source of the magnetic field and hence the field itself can have explicit time dependence, such as when the current through a solenoid is made to change.

We have seen that for case (1) it is possible to analyze the situation using the Lorentz force law. This is not true for cases (2) and (3). Yet Faraday's law doesn't care whether it is the loop that has moved or the field that has changed. *Motional* emf can be interpreted in terms of the Lorentz force law, but a time-varying magnetic field and the emf induced by it is a truly new aspect of Faraday's law. In this section, we look more closely at this situation.

A New Way to Make an Electric Field

When a magnetic field changes with time, then an electric field is induced in space that satisfies Eq. (30–2). *If* there is sufficient symmetry in a situation, then it is possible to calculate the induced electric field in a way similar to the way in which we used Ampère's law to determine a magnetic field. This is most easily illustrated with an example.

EXAMPLE 30–10 The two circular pole faces of an electromagnet, both of radius $R = 0.5$ m, are oriented horizontally with the north pole underneath. The electromagnet produces a field that is uniform throughout the volume between the faces. The field is increased linearly from 0.1 T to 1.1 T over a period of 10 s. Describe the electric field that results in the region between the poles.

Setting It Up We show the electromagnet with its pole faces in Fig. 30–22a. The magnetic field $\vec{B}$ is uniform across the entire face, oriented upward. Its magnitude varies linearly with time according to $B = B_0 + \alpha t$, with both B_0 and α given.

Strategy The unknown electric field is induced by the changing magnetic flux. There is cylindrical symmetry between the pole faces, so the induced electric field can vary only with the distance r from the central axis of the pole faces; it cannot vary with the angle around this axis. By Faraday's law, the rate of change of magnetic flux through a horizontal disk of radius r centered on the axis of the pole faces will determine the integral of the electric field along the circle forming the edge of the disk. Symmetry then allows us to argue that this integral is the (constant) field strength E times the circumference $2\pi r$, and that will determine E. The induced electric field is oriented along this circumference—Faraday's law will tell us whether the field lines run clockwise or counterclockwise.

Working It Out The magnetic flux through the pole face is

$$\Phi_B = \pi r^2 B.$$

Using the given time dependence of B, we have

$$\frac{d\Phi_B}{dt} = \pi r^2 \frac{dB}{dt} = \pi r^2 \frac{d}{dt}(B_0 + \alpha t) = \pi r^2 \alpha.$$

Faraday's law, Eq. (30–2), then gives

$$\oint \vec{E} \cdot d\vec{s} = -\frac{d\Phi_B}{dt} = -\pi r^2 \alpha. \qquad (30\text{–}12)$$

The direction of the loop integration is counterclockwise as we look down the north pole because the flux is oriented upward (Fig. 30–22b). With the minus sign in Eq. (30–12), the induced electric field lines run clockwise. Application of the symmetry argument gives

$$\oint \vec{E} \cdot d\vec{s} = E 2\pi r = -\pi r^2 \alpha$$

or

$$E = \frac{\pi r^2 \alpha}{2\pi r} = \frac{1}{2}\alpha r.$$

The magnitude of the induced electric field increases as we go out from the center.

Numerically, the coefficient α is found by knowing that B increases from 0.1 T to 1.1 T in 10 s, so the rate of increase is $\alpha = (1.1\text{ T} - 0.1\text{ T})/10\text{ s} = 0.1\text{ T/s}$. Then

$$E = \tfrac{1}{2}(0.1\text{ T/s})r = (0.05\text{ T/s})r,$$

which increases from 0 N/C at $r = 0$ m to a maximum of $E_{max} = (0.05\text{ T/s})(0.5\text{ m}) = 2.5 \times 10^{-2}$ N/C at $r = 0.5$ m.

What Do You Think? What is the induced electric field at distances far outside the poles?

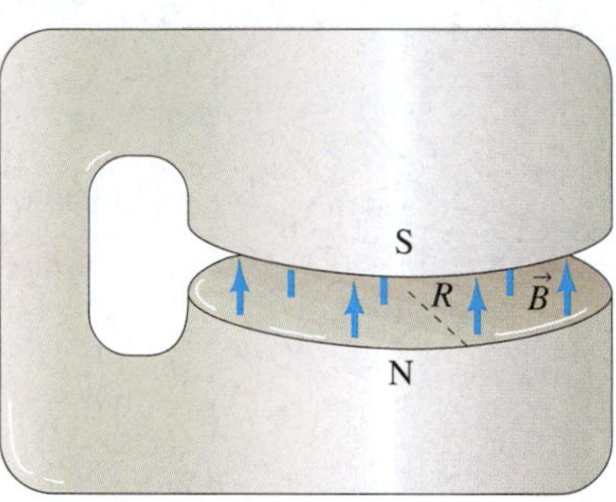

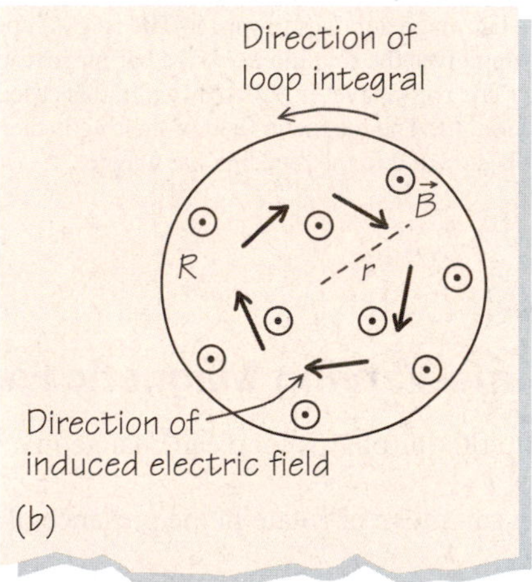

▲ **FIGURE 30–22** (a) The current windings that produce the field are not shown. (b) A view straight down at the north pole. A changing magnetic field between the pole faces of an electromagnet induces an electric field. Symmetry allows us to specify the electric field, not just its integral around an arbitrary loop.

Is a Magnetic Field Present Where a Current Is Induced?

A changing magnetic flux induces electric fields even where the magnetic field itself is very small. This aspect of Faraday's law is not very intuitive and deserves special attention. Suppose that charged particles such as protons are moving, but they are so far away from the region of the pole faces of an electromagnet that the magnetic field in their location is very small. Would the protons accelerate when the magnetic field is changed? Is an electric field induced even in regions with no magnetic field whatsoever, as in the space outside a toroidal solenoid, or as in the situation described in Example 30–10, "What Do You Think?" Experiment confirms that the answers to our questions are affirmative. According to Faraday's law, all that counts is the change of the flux through the loop in question, regardless of the loop's location.

30–5 Generators

The generation of electric energy in our society is based largely on the Faraday induction law. The mechanical energy of the rotating blades of a steam turbine turned by steam or by rushing water is converted to the energy of moving charges—electric current—with the **alternating-current (AC) generator**.

Imagine a coil of N turns of wire that makes a circle of area A. The coil is placed in a constant magnetic field, $\vec{B}$, and rotated at angular speed ω around an axis perpendicular to the field (Fig. 30–23a). The ends of the wire that make up the coil are brought to the exterior through some sort of sliding contact with a fixed wire. As the coil rotates in the magnetic field, the magnetic flux through it changes, and an emf is induced. Figure 30–23b, which is a side view of the coil, shows that the magnetic flux through the loop is $\Phi_B = \vec{B} \cdot \vec{A} = BA \cos\theta$. If we imagine starting the rotation at $t = 0$, so that $\theta = \omega t$, the time derivative of the magnetic flux is

$$\frac{d\Phi_B}{dt} = BA\frac{d}{dt}\cos\omega t = -BA\omega\sin\omega t. \tag{30–13}$$

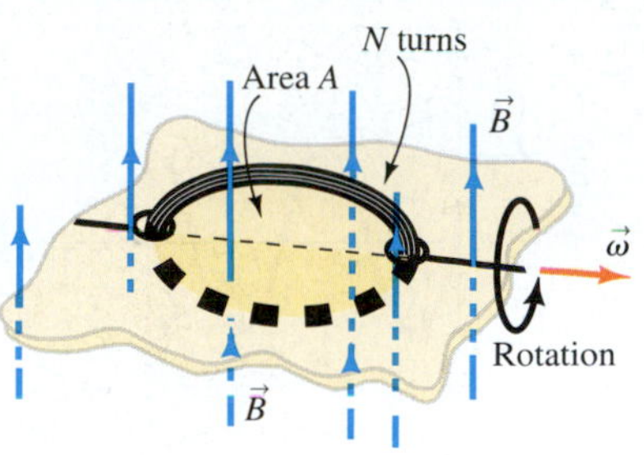

▲ **FIGURE 30–23** (a) An external force rotates a coil with angular velocity $\vec{\omega}$ in a magnetic field. An emf with sinusoidal time dependence of angular frequency ω is induced in the coil. (b) Side view of the process, looking at $\vec{\omega}$.

There are N turns of wire, and the total emf induced across the two ends of the coil is

$$\mathcal{E} = -N\frac{d\Phi_B}{dt} = NBA\omega\sin\omega t. \tag{30–14}$$

This arrangement makes up our generator, denoted in circuits by a circle enclosing a wavy line (Fig. 30–24).

If the wire of the generator coil is connected as a series element of a circuit with a resistance R, then a current is generated in the circuit:

$$I = \frac{\mathcal{E}}{R} = \frac{NAB\omega}{R}\sin\omega t. \tag{30–15}$$

This *alternating current* oscillates in sign and has a maximum magnitude of $NAB\omega/R$.

The power P delivered to this circuit is the product of the emf and the current:

$$P = \mathcal{E}I = INAB\omega\sin\omega t. \tag{30–16}$$

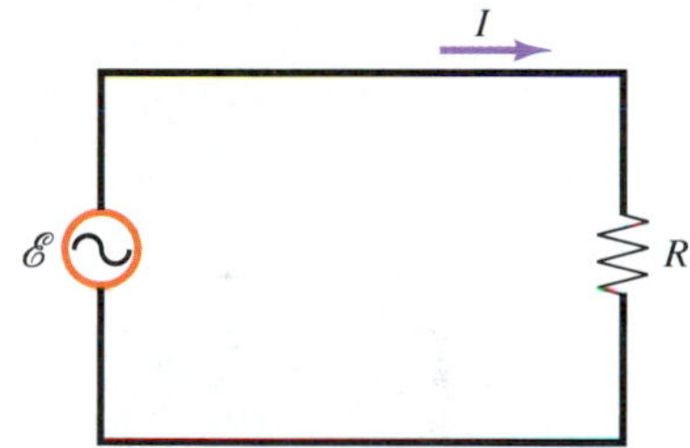

▲ **FIGURE 30–24** When the sinusoidally varying emf that results from rotating a coil in a constant magnetic field (as in Fig. 30–22a) is part of a circuit, a sinusoidal current with the same angular frequency ω results.

The mechanical force that rotates the loop is the source of this power. We know that a loop that carries a current forms a magnetic dipole; we also know that a magnetic dipole experiences a torque that tends to align it with the direction of the magnetic field (Section 28–5). Thus the force that rotates the coil must do work against this torque. Let's compute the rate at which this work is done. The torque on a dipole of magnetic dipole moment $\vec{\mu}$ in a field $\vec{B}$ has magnitude

$$\tau = |\vec{\mu} \times \vec{B}| = \mu B\sin\theta,$$

where we refer to Fig. 30–23b for θ and recall that the magnetic dipole moment is perpendicular to the current loop. The mechanical power P_{mech}, or work per unit time, that must be expended by the force that rotates the loop against this torque is

$$P_{mech} = \tau\omega = \mu B\omega\sin\theta.$$

The magnetic dipole moment of a current loop with N turns is INA [Eq. (28–24)]. Thus

$$P_{mech} = INAB\omega\sin\theta. \tag{30–17}$$

This result is the same as that given by Eq. (30–16). As expected, the electric power is accounted for entirely by the mechanical power expended.

The explicit time dependence of the power is found by taking the product of the current, Eq. (30–15), and the emf, Eq. (30–14) or, more simply, by evaluating the product V^2/R. We then find

$$P = \frac{(NAB\omega)^2}{R}\sin^2\omega t. \tag{30–18}$$

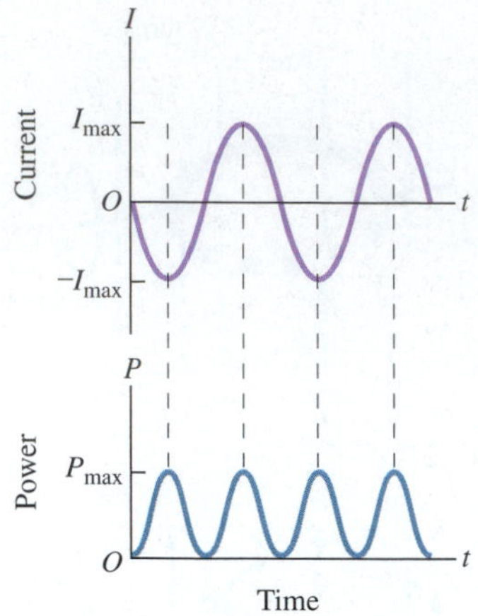

▲ **FIGURE 30–25** The power dissipated in the circuit of Fig. 30–24 is always positive, unlike the current, which alternates in sign. The maximum current is $I_{max} = NAB\omega/R$, while the maximum power is $P_{max} = I_{max}^2 R$.

This quantity is always positive, as opposed to the emf or the current, both of which alternate in sign. The distinction is illustrated in Fig. 30–25, which shows a plot of the current and the power in the circuit of Fig. 30–24 as a function of time.

In practical situations it is important to consider the *time average* of the power. To do so, we note that the sine-squared function, which oscillates between 0 and 1, averages over one period to 1/2. The average power dissipated in this circuit is then

$$P_{av} = \frac{1}{2}\frac{(NAB\omega)^2}{R} = \frac{1}{2}\frac{V_{max}^2}{R}. \tag{30–19}$$

Electric power is generated by using mechanical energy to produce electric current; the current is transmitted as an alternating current (AC)—see Chapter 33. If we run a generator in reverse, we can convert electric energy into mechanical energy, and this is the basis of an *electric motor*. As we described in Chapter 28, the proper operation of a motor requires a split-ring commutator or something similar to periodically reverse the torque on a loop such that the torque is always in the same direction as the loop turns.

*30–6 The Frame Dependence of Fields

In Section 28–3 we discussed the fact that in order to maintain the relativity principle, electric and magnetic fields had to "mix," meaning that observers in different inertial frames see different combinations of electric and magnetic fields. You should recall that according to the relativity principle, there should be no way to tell which of two inertial frames—frames moving at constant velocity with respect to one another—is moving and which is not. The laws of motion should be the same in all inertial frames, so they do not provide a way to tell which frame is moving. Thus an observer O and an observer O′, who is moving with constant velocity $\vec{u}$ relative to observer O, should express the laws of motion in exactly the same way. In particular, Newton's second law,

$$\frac{d\vec{p}}{dt} = \vec{F},$$

with $\vec{p} = m\vec{v}$, must look the same to observer O, who sees a particle move with velocity $\vec{v}$, and to observer O′, who sees the same particle move with velocity $\vec{v}' = \vec{v} - \vec{u}$. This will be the case if $\vec{u}$ is constant (as it must be for inertial frames) and if the force on the particle is the same to both observers.

If the particle is charged and subject to a Lorentz force, there is an apparent difficulty because the Lorentz force depends on the velocity of the particle. The difficulty is that since the force describes the motion of the object on which the force acts, observation of the motion would represent a way to decide which of the observers is moving, and this is a violation of the principle of relativity. The problem would only be an apparent one if both observers saw the same Lorentz force. We applied this argument in Chapter 28 with the velocity selector and suggested that the problem is resolved if the fields seen by our two observers take the form

$$\vec{B}' = \vec{B}, \quad \vec{E}' = \vec{E} + \vec{u} \times \vec{B}. \tag{30–20}$$

Equation (30–20) is the correct solution to the problem of reconciling the forces when speeds are not large compared to the speed of light. *Under the transformation from one inertial frame to another, electric and magnetic fields get mixed up* (*transform among themselves*) in a very special way, as specified in Eq. (30–20).

It is interesting to think about how stationary and moving observers view the source and effect of motional emf. Suppose that an observer O is in a reference frame where there is a magnetic field but no electric field. If a conducting rod moves through this field, there is a motional emf. Its source, according to observer O, is the magnetic force on the conducting electrons in the rod (Section 30–3). Observer O′, moving with the rod, sees the rod at rest and also sees the same constant magnetic field seen by observer O, from Eq. (30–20). Therefore observer O′ sees no magnetic force. However, the same transformation equation shows that observer O′ also sees an electric field $\vec{E}'$, and this electric field has a magnitude that makes the conducting electrons in the rod accelerate in just the way that observer O sees them accelerate. Each observer attributes the observed effects to different combinations of fields.

The question of the consistency of physical law for observers moving with respect to one another is a very important one; it led Albert Einstein to formulate the theory of special relativity, the subject of Chapter 39.

Summary

When the magnetic flux Φ_B through an open surface changes with time, an emf $\mathcal{E}$ is induced around the line that bounds the surface. Faraday's law states the relation:

$$\mathcal{E} = \oint \vec{E} \cdot d\vec{s} = -\frac{d\Phi_B}{dt}. \qquad (30\text{–}2)$$

Here, the line integral is over the closed bounding line that forms the edge of the surface. When the loop is a physical object capable of carrying current, such as a loop of wire, then the induced emf results in an induced current. In this case, the minus sign in Faraday's law can be interpreted in more physical terms as Lenz's law: Induced currents produce magnetic fields that tend to cancel the flux changes that induce them.

The flux change to which Faraday's law refers can occur either because the magnetic field changes with time or because the area or orientation of the surface through which the flux is calculated changes with time. In the latter case, the induced emf is called a motional emf, and it can be derived directly from application of the Lorentz force law. Application of Lenz's law to motional emfs shows that the induced current must lead to forces that inhibit the motion of the object in which the emf is induced. The power loss due to resistive flow of induced currents is matched by the power required to keep the conductor moving. When an induced emf occurs because magnetic fields change with time, Faraday's law predicts a new type of electric field, one that cannot be described with a conventional potential and that is therefore nonconservative.

When a changing magnetic flux passes through a conducting solid, Faraday's law manifests itself by the induction of eddy currents in the material.

The AC generator, the foundation of electrical power generation, is an application of Faraday's law. When a coil rotates in a magnetic field, an emf is induced in the coil. The mechanical energy of the rotation is thus transformed into electric energy in the form of a current in circuits connected to the coil.

Understanding the Concepts

1. A spherical surface is placed in a changing magnetic field. Will there be an induced electric field along the equator?
2. Must there be a real conducting loop in a region with a changing magnetic flux in order for an electric field to be induced?
3. When the end of a magnet is brought to a stationary loop of wire, which of the following are true: (a) the magnetic flux through the loop decreases, (b) Lenz's law does not apply, (c) a voltage but no current are induced in the wire, (d) the induced voltage depends on the resistance in the wire, (e) the induced current depends on the resistance in the wire.
4. Electric leads (the wires that run from one part of the apparatus used for experiment to another) for sensitive experiments are almost never separated but are close together or even twisted around one another. Explain why this might be done.
5. Can the magnetic field change over some region without a change in the magnetic flux through a surface in the region? If so, give as many examples as you can.
6. A bar magnet is dropped vertically through a loop of wire, with the north pole crossing the loop first. Looking down at the loop from above, the induced current will be (a) clockwise and then counterclockwise, (b) counterclockwise and then clockwise, (c) zero, (d) only clockwise, or (e) only counterclockwise.
7. Each part of Fig. 30–26 shows a current being induced in a conducting loop by a changing magnetic flux through the loop. In each case, is the direction of the induced current correct as shown? (a) A magnet approaches a loop; (b) a current-carrying conducting loop approaches a loop at rest; (c) a switch is closed in the first loop, causing a current to flow; (d) a switch is closed in the straight wire, causing a current to flow.

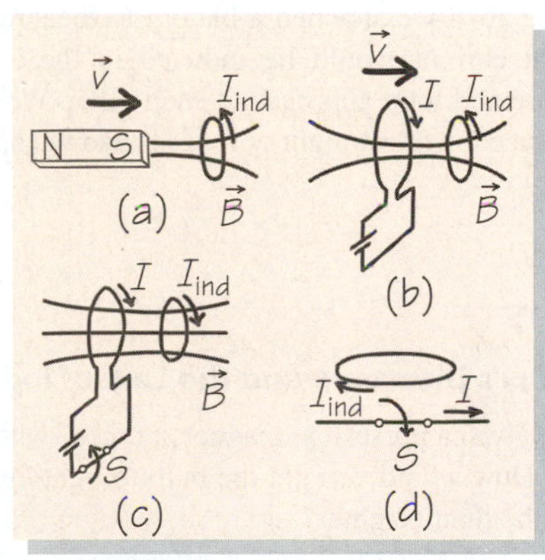

▲ **FIGURE 30–26** Question 7.

8. A conducting hoop is rolled in a straight line at a constant speed in an east–west direction in the northern hemisphere through Earth's magnetic field. Will a current flow in the hoop? If so, in what direction will it circulate?

9. A rectangular loop is moving across a uniform magnetic field such that the induced emf is zero. Can you tell how the loop is oriented?
10. In Example 30–2 the induced current jumped from zero to a constant value and then dropped to zero again. Sketch, roughly, what the time-dependence of the induced current would look like if the loop were circular.
11. A sheet of metal is placed between the pole pieces of a permanent magnet, perpendicular to the direction of the field lines. Does it take positive work to pull the sheet of metal out? If so, why?
12. What would happen if you dropped a strong cylindrical magnet down inside a long vertical copper tube? Down an aluminum tube? Down a plastic tube?
13. When a bar magnet is moved toward a current loop, a current is induced in the loop. How, if at all, will physically measurable quantities change if the loop is moved toward the magnet rather than vice versa?
14. If a flat metal plate hung by a cord and oriented parallel to the pole faces of a magnet—the faces are in a vertical plane, parallel to each other—moves as a pendulum bob through the pole faces, it slows down. If the magnet is sufficiently strong, the plate comes to rest. Why? How could this phenomenon be prevented?
15. The Tennessee Valley Authority constructed several dams on the Tennessee River to produce electricity. How is water used to produce electricity?
16. When a gas is heated sufficiently, its atoms ionize into electrons and positive ions. This type of material is called a plasma. Plasma flow may be viewed as a superposition of equal and opposite currents. If the plasma is forced to flow in a channel perpendicular to a magnetic field, an electric potential builds up across the channel. The device based on this phenomenon is the *magnetohydrodynamic* (*MHD*) *generator*. Given the magnetic field strength and the potential, we can calculate the velocity of the plasma. The charge and density of the charge carriers in the plasma do not enter into the result. Why not? [*Hint*: Review the Hall effect.]
17. What happens when a bar magnet is dropped down a long, vertical copper tube?
18. In Example 30–4 we attached a battery to the straight wire and asked what current would be induced in the loop when the straight wire and loop approached each other. Would there be a current induced in the straight wire if instead the battery were on the loop?
19. In a demonstration, an aluminum ring is placed around a projection of an iron core wound with a wire and connected to a battery (Fig. 30–27). The ring jumps when the circuit is closed. Why? What happens if a gap is cut in the ring?

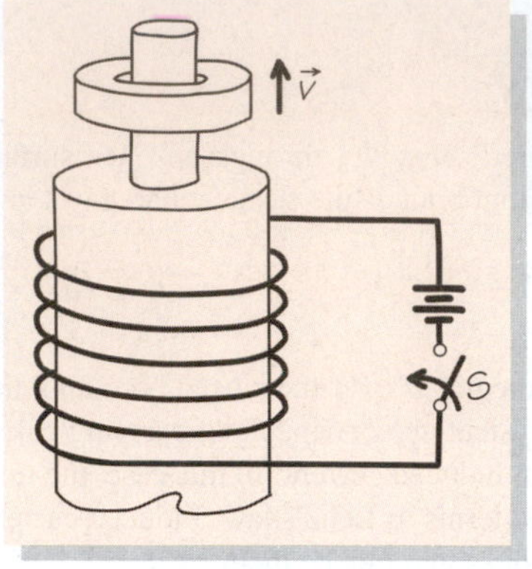

▲ **FIGURE 30–27** Question 19.

20. A cylindrical piece of iron is inserted inside a solenoid to increase the magnetic field. A voltage varying harmonically with time is placed across the solenoid leads. A copper ring is slipped down over the solenoid, so that the solenoid passes through the ring, and is held there. (a) Explain why the copper ring becomes hot even though nothing touches it. (b) What is the source of the thermal energy? (c) Explain how energy is conserved in this case.
21. In Question 20, the solenoid axis is vertical. It is possible to find a particular ring of copper that, when slipped over the solenoid and placed horizontally, remains suspended in space around the solenoid. (a) Why does this work? (b) What are the criteria for selecting the particular piece of copper?
22. Discuss how a bicycle light generator utilizing the Faraday effect might work.
23. At many traffic lights cars drive over a wire embedded in the pavement, and the traffic light soon changes from red to green. The wire forms a loop of about 1.5 m on a side. How might such a system work?
24. Describe some of the interesting things you might see if the sign of the flux change term in Faraday's law were positive.

Problems

30–1 Faraday's Discovery and the Law of Induction

1. (I) You are given a horseshoe magnet, a coil of wire, and a flashlight bulb. How would you get the bulb to light up? How would you make the light brighter?
2. (I) Explain in words why the induced currents are as shown in each case in Figure 30–3. Describe the induced magnetic field.
3. (I) A loop of wire of area 12 cm^2 is placed between the pole pieces of an electromagnet, at right angles to the direction of the magnetic field lines. What is the emf generated around the loop if the magnetic field is changed at a uniform rate from 1.5 T to 2.0 T in 5.7 s? Assume that the magnetic field is uniform across the area of the loop.
4. (I) Suppose that the wire in Problem 3 has a resistance of 7.7 Ω. How much power will be lost to ohmic heating while the magnetic field increases?
5. (I) A magnetic field that changes with time but is uniform in space is directed along the x-axis. A conducting ring of diameter 7 cm and resistance 1.5×10^{-3} Ω is placed in the yz-plane. If the current in the ring is 2 A, how fast is the magnetic field changing?
6. (II) The magnetic field in a region is uniform. It varies with time as shown in Fig. 30–28 (see next page). Plot the current through a ring that has an area of 14 cm^2 and a resistance of 0.02 Ω, and whose plane is perpendicular to the magnetic field.
7. (II) A square wire loop of dimensions $L \times L$ oriented in the xz-plane enters a region where the magnetic field is first oriented in the $+y$- and then in the $-y$-direction (Fig. 30–29, see next page). The width of each region is L. The loop moves at speed v in the $+x$-direction. Find the emf, sign and magnitude, induced in the loop as it enters and passes through the regions with the magnetic field.

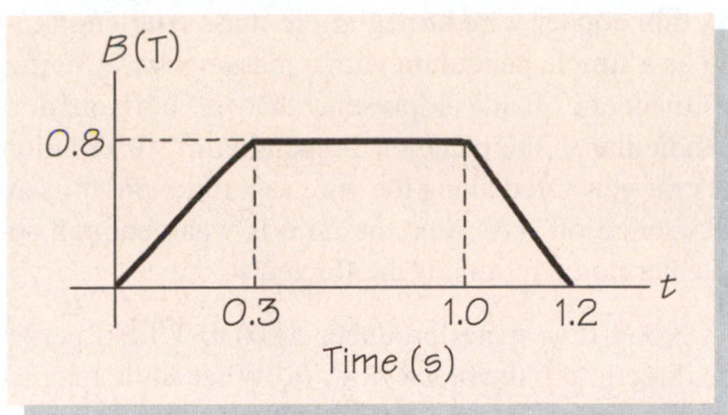

▲ **FIGURE 30–28** Problem 6.

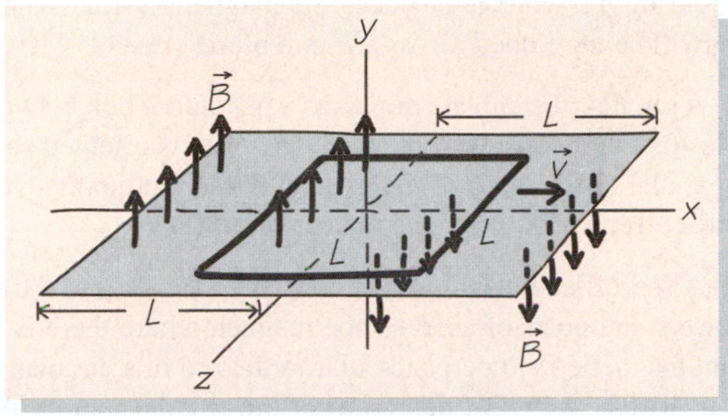

▲ **FIGURE 30–29** Problem 7.

8. (II) A long, straight wire oriented in the z-direction carries a current of 120 mA. A square loop with sides of length 2.5 cm is in the xz-plane with its nearest edge 15 cm from the wire (Fig. 30–30). In a time of 0.05 s, the square loop moves uniformly 1.0 cm closer to the wire. What is the emf induced in the loop while it is moving? Ignore the variation in the wire's magnetic field *across* the loop.

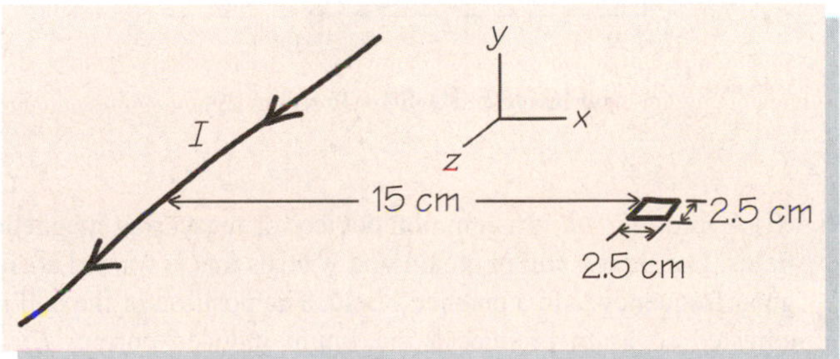

▲ **FIGURE 30–30** Problem 8.

9. (II) What is the peak emf produced by a 100-turn square coil 8.0 cm on each side, rotating on a diagonal axis with a frequency of 15 Hz in a magnetic field of 0.30 T perpendicular to the axis?

10. (II) A coil with 450 turns, a radius of 2.5 cm, and a resistance of 12 Ω is rotating about a diameter in a uniform magnetic field of 0.35 T. How fast must it rotate to produce a maximum current of 3.0 A in the coil?

11. (II) There is a constant magnetic field $\vec{B} = B_0(\hat{i} + \hat{j} + \hat{k})$ in the region $x > 0$, $y > 0$, $z > 0$. A square loop of dimensions $L \times L$ whose sides are parallel to the x- and y-axes moves with constant velocity $\vec{v} = v_0(\hat{i} + \hat{j})$ in the xy-plane such that its center moves along the line $x = y$. Calculate the emf induced in the loop, given that its leading corner passes the origin at the time $t = 0$.

12. (II) A vertical loop rotates with angular velocity $\vec{\omega}$ as shown in Fig. 30–31. At time $t = 0$, it is aligned perpendicular to a constant magnetic field oriented in the x-direction. Use Lenz's law to find the direction of the emf induced in the loop at $t = 0$, $t = T/4$, $t = T/2$, and $t = 3T/4$, where T is the rotation period of the loop.

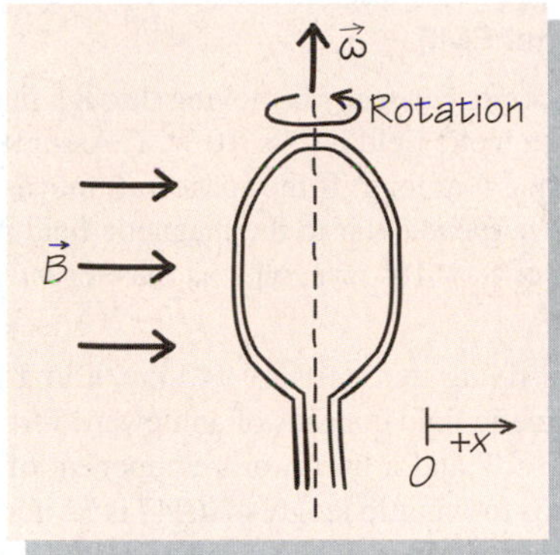

▲ **FIGURE 30–31** Problem 12.

13. (II) A closed loop is constructed of a fixed wire shaped as a squared-off U and a crossbar free to move in the x-direction, all in the xz-plane. The square base of the U-shaped segment is at $x = 0$. A magnetic field oriented in the y-direction varies with x according to $\vec{B} = Cx\hat{j}$; it is zero at $x = 0$. The situation and the relevant dimensions are as in Fig. 30–32. Suppose that the movable crossbar is pulled at a constant speed v to the right, starting at $x = 0$ when $t = 0$. Its position at any time is $x = vt$. If the resistance of the loop varies with the total length L according to $R = \alpha L$, what is the current in the loop as a function of time? Compare your answer with Example 30–5, and explain any differences.

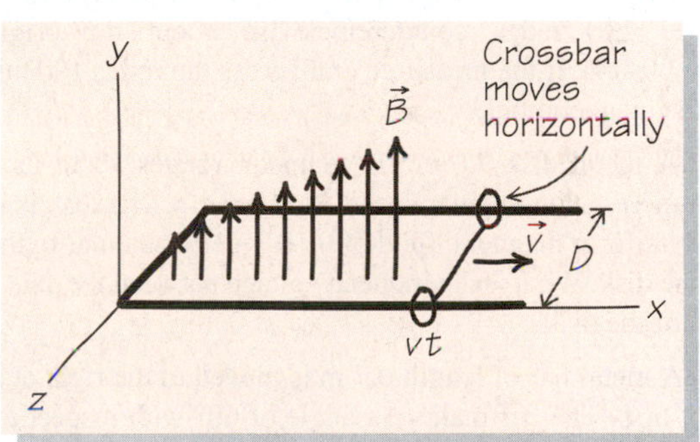

▲ **FIGURE 30–32** Problem 13.

14. (II) Suppose the magnetic field in Example 30–5 is a constant field oriented in the z-direction, $\vec{B} = B_0\hat{k}$. Find the induced current as a function of time.

15. (II) Suppose the magnetic field in Example 30–5 varies linearly with z and is oriented in the y-direction, $\vec{B} = Cz\hat{j}$. Find the induced current as a function of time.

16. (II) A metal ring is constructed so as to expand or contract freely. In a region with a constant magnetic field $\vec{B}_0$ oriented perpendicular to it, the ring expands, with its radius growing linearly with time as $r = r_0(1 + \alpha t)$. As the ring expands and grows thinner, its resistance *per unit length* changes according to the empirical rule $R = R_0(1 + \beta t)$. Find the current induced in the ring as a function of time. Specify the direction as well as the magnitude of the current.

17. (II) A circular loop of area A rotates with angular frequency ω about its vertical diameter. The rotating loop is placed in a horizontal constant magnetic field, B. What is the emf induced in the loop?

18. (II) Work Example 30–3 by direct computation of the magnetic flux through the hemispherical surface.

30–2 Motional EMF

19. (I)The spacecraft *Voyager I* is moving through interstellar space, where the magnetic field is 2×10^{-10} T. Assume that *Voyager I* has an antenna 5 m long. If the spacecraft moves so that the antenna rod is perpendicular to the magnetic field when *Voyager I* has a speed of 8×10^3 m/s, what is the emf induced across the antenna?

20. (I) A 747 is flying due north at 940 km/h in a location where Earth's magnetic field consists of an upward vertical component of 2.8×10^{-5} T and a northward component of 2.5×10^{-5} T. If the wingtip-to-wingtip length of a 747 is 35 m, find the emf induced across the wings. If the airplane were flying due east instead of due north, how would your answer change?

21. (I) A metal rod is pulled through a magnetic field perpendicular to it with a velocity perpendicular to both the rod and the magnetic field as in Fig. 30–12a. The rod has length 0.25 m, its speed is 1.7 cm/s, and the magnetic field has magnitude 0.069 T. What is the magnitude of the potential difference, if any, from one end of the rod to the other?

22. (I) A metal rod 6.0 cm long falls to the ground from a height of 6.0 m. It stays horizontal and oriented in an E–W direction throughout the fall. If we assume that, in the region where the metal falls, Earth's magnetic field is 1.5×10^{-4} T and points in the N–S direction, then at what rate does the potential difference between the ends of the rods increase?

23. (I) A rod 10 cm long lying in the *xy*-plane pivots with angular speed 100 rad/s counterclockwise about the origin (see Fig. 30–14). If the measured emf across the rod is 100 mV, what is the magnetic field?

24. (II) A metal disk 7.0 cm in diameter rotates about its axis of symmetry at an angular speed of 150 rad/s. The disk is situated in a uniform magnetic field of 0.10 T perpendicular to the plane of the disk. What is the induced voltage between the axis and the rim of the disk?

25. (II) A metal bar of length 0.7 m is moved to the right at a speed of 5 m/s. The bar makes an angle of 60° with respect to its direction of motion. It is passing through a region of uniform magnetic field of magnitude 5×10^{-3} T oriented perpendicular to the plane swept out by the bar (out of the page in Fig. 30–33). What is the potential difference between the two ends of the bar as it moves through the magnetic field?

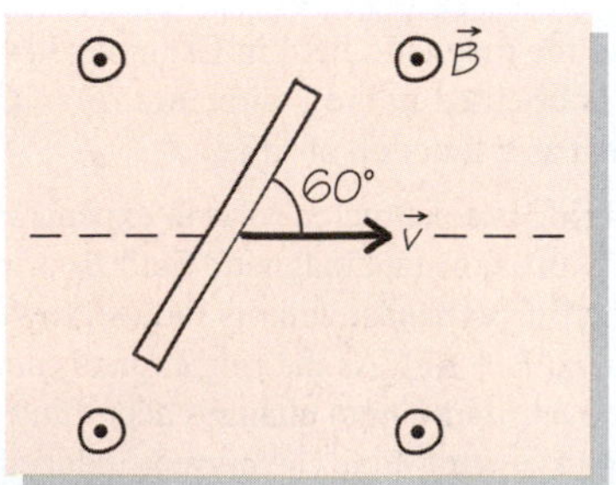

▲ **FIGURE 30–33** Problem 25.

26. (II) A thin copper wire of negligible mass and length L is oscillating as a simple pendulum with a mass bob m. A uniform, constant magnetic field is present that is horizontal, oriented perpendicular to the plane of the pendulum's oscillations. What is the emf generated along the wire as a function of its angle θ of small oscillation? (Assume the field is weak enough so that the simple harmonic motion is unaffected.)

27. (II) A blood-flow meter produces a 0.030 T field perpendicular to the direction of the blood flow. (a) What is the magnetic force on an ion of charge e flowing with the blood at 0.25 m/s? (b) Positive and negative ions in the blood are forced in opposite directions, producing an electric field and a voltage across the blood vessel. The ion separation stops when the electric force caused by this voltage balances the magnetic force. What voltage will be developed across a 3-mm blood vessel?

28. (II) A circular metal plate moves as a pendulum bob between the poles of a tabletop electromagnet. The plate is oriented so that it is parallel to the faces of the magnet. Describe qualitatively the eddy currents induced in the plate as it moves.

29. (II) A rod of length L moves at constant speed v into the region between the poles of a horseshoe magnet, where there is a constant magnetic field perpendicular to the rod in a circular region (Fig. 30–34). $L = 2R$, the radius of the circular region. What is the emf induced in the rod as a function of time?

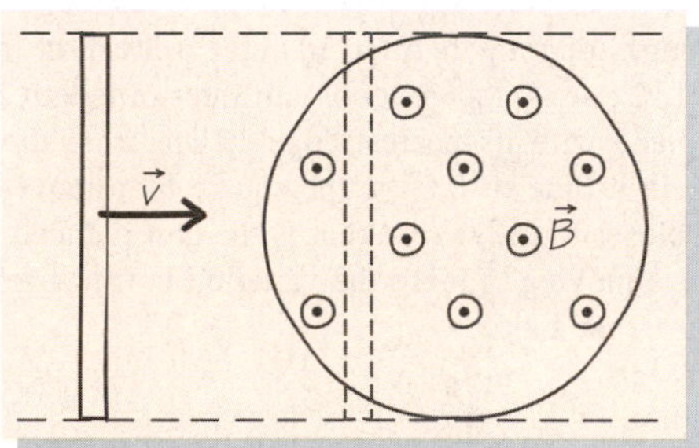

▲ **FIGURE 30–34** Problem 29.

30. (II) A *rotating coil* is a common device for measuring magnetic fields. Consider a coil of area A and N turns that is rotated at angular frequency ω in a magnetic field. The position of the coil is adjusted so as to produce a maximum induced current I_{max}, which can be measured by using an appropriate ammeter. R is the total resistance of the coil circuit. Find the relationship between the unknown magnetic field and I_{max}.

31. (II) If the rotating coil of Problem 30 is used with the split-ring commutator described in Chapter 29, DC current can be measured with a sensitive galvanometer. (a) Sketch the current as a function of time for several periods, where the period T is given by $2\pi/\omega$. (b) Calculate the average of this rectified current. (c) Find the relationship between the unknown magnetic field and the average measured DC current. [*Hint:* The average of an oscillating function with period T is $(1/T)\int_0^T f(t)\,dt$.]

32. (II) A long, straight wire carries a current of 650 mA. A thin metal rod 45 cm long is oriented perpendicular to the wire and moves with a speed of 2.2 m/s in a direction parallel to the wire. What are the size and direction of the emf induced in the rod if the nearest point of the rod is 3.5 cm away from the wire, and if the rod moves in a direction opposite to the current?

30–3 Forces and Energy in Motional EMF

33. (I) A loop of metal, total resistance $R = 25\ \Omega$, moves with speed 35 m/s through a region of magnetic field such that at time $t = 0$ the rate of change of magnetic flux through the loop is given by $17\ \text{T}\cdot\text{m}^2/\text{s}$. The magnetic field in the vicinity of the loop has instantaneous value 0.16 T at $t = 0$. Assume the shape of the loop is such that the resulting net force on the loop is due entirely to a straight section of the loop, 12 cm in length, that is perpendicular both to the magnetic field and to the direction of the loop's motion. What is the drag force on the loop at $t = 0$, and what is the instantaneous power at $t = 0$ expended by the force that must be used to keep the loop moving with constant velocity?

34. (II) A square wire loop of dimensions $L \times L$ lies in a plane perpendicular to a constant magnetic field. The field exists only in a certain region, with a sharp boundary (Fig. 30–35). The sides of the loop make a 45° angle with this boundary, and an external force moves the loop at a speed v out of the region of constant field. How much power must be supplied by the external force as a function of time?

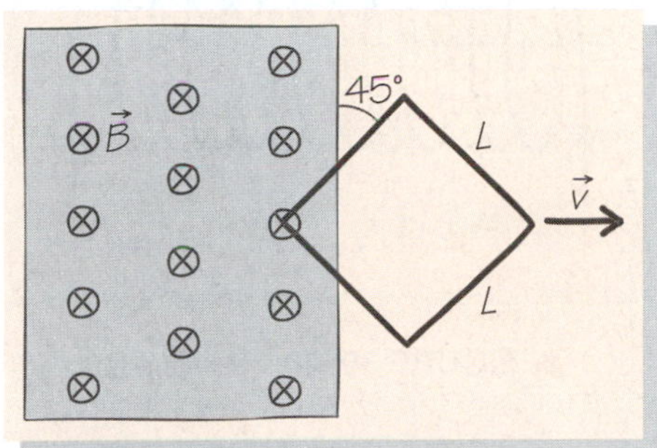

▲ **FIGURE 30–35** Problem 34.

35. (II) A conducting bar slides frictionlessly on two parallel horizontal rails 30 cm apart. The bar and rails form a closed circuit with a resistor of resistance 0.050 Ω, assumed to be constant throughout the motion. The circuit is placed in a uniform vertical magnetic field of 0.28 T perpendicular to the circuit's plane. The bar is pulled at a constant speed of 60 cm/s along the rails. (a) What is the magnitude of the force required to pull the bar? (b) What is the rate of Joule heating in the resistor?

36. (II) A circuit with a moveable cross wire is indicated in Fig. 30–36. The resistance is 2.5 Ω, the battery voltage is 5 V, the moveable wire length is 12 cm, and there is a uniform, constant magnetic field into the plane of the circuit (the page) with magnitude 0.3 T. (a) What is the current if the moveable wire is locked into position? (b) What is the force on the moveable wire at the moment it is released from the locked position? (c) If there is an opposing force of 0.02 N on the wire, how fast will it move as a function of the time after release?

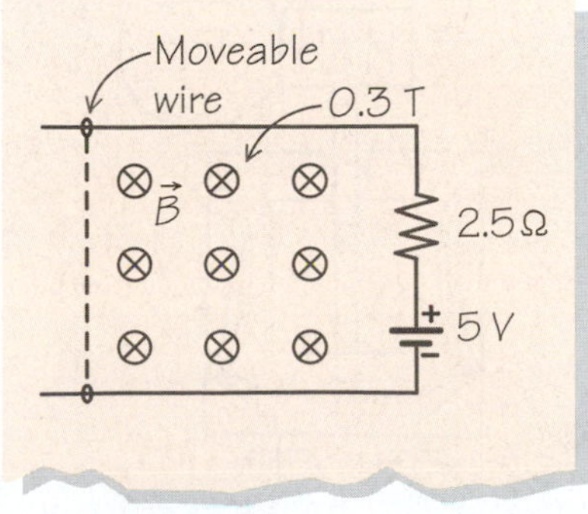

▲ **FIGURE 30–36** Problem 36.

37. (II) A horizontal wire of mass m is free to slide on vertical rails a distance L apart. These are connected at the bottom with a resistance R between them. There is a constant, uniform magnetic field B in a horizontal direction, normal to the area enclosed by the rails and sliding wire. Use the fact that the power generated by gravity is equal to the power dissipated by ohmic heating, to show that the wire ultimately moves with a constant velocity, and obtain an expression for this velocity in terms of mg, B, R and L.

38. (II) A long, straight wire carries a constant current I_0. A square loop with sides of length L and two sides parallel to the wire is pulled away at uniform speed v in a direction perpendicular to the wire. The nearest side of the loop is initially a distance D from the wire; the resistance of the loop is R. (a) Calculate the force necessary to pull the loop. (b) At what rate is work being done by the force? (c) How does your answer to part (b) compare with the Joule heating in the loop?

39. (II) What happens if the initial speed of the loop in Example 30–9 is (a) less than v_t, and (b) greater than v_t?

40. (II) When eddy currents are induced in a piece of metal moving through a magnetic field, drag forces that are proportional to the velocity of the metal act on it. Consider a thin metal disk rotating in a plane between the poles of a magnet. Show that the equilibrium angular velocity of the disk is proportional to the torque on the disk. Note that in household electricity meters, this arrangement allows us to correlate the number of turns of the disk with the power consumption.

30–4 Time-Varying Magnetic Fields

41. (I) Consider a length of wire looped back on itself in a magnetic field B. The shape of the loop is not given, but it lies in a plane. The wire has negligible resistance, but there is a resistor R at one end. A cross bar, which rests on opposite legs of the loop, is pulled along so that the flux enclosed by the wire and crossbar varies with time as $\Phi_B(t)$. What is the instantaneous power needed to pull the crossbar?

42. (I) A long, straight wire carries a current $I = I_0 \cos(220\pi t)$, where t is time. Two sides of a fixed rectangular loop are 9.0 cm long and are parallel to the wire; the other sides are 0.80 cm long. The nearest long side is 2.0 cm from the wire. What is I_0 if the maximum emf induced in the loop is 1.3 μV? (Ignore the small variation of the magnetic field across the loop.)

43. (I) A loop of wire is placed between the poles of a large electromagnet. The loop is oriented so that the vector that characterizes the orientation of its planar surface runs from one pole to the other. When the magnet is turned on, its magnetic field builds up according to the formula $B = B_0(1 - e^{-at})$. We may assume that the magnetic field has the same value all across the surface of its poles and runs from one pole to another. (This formula gives $B = 0$ at $t = 0$ and $B = B_0$ at $t = \infty$.) What is the magnitude of the emf around the loop?

44. (I) A metal ring of diameter 3.0 cm is left between the pole tips of an electromagnet, such that the plane of the loop is perpendicular to the magnetic field. The electromagnet is turned on and reaches its full magnetic field of 2.7 T in 50 ms at a linear rate. If the ring has resistance 0.15 Ω, how much current passes through the ring during the 50 ms?

45. (II) A solenoid of radius r wound with n turns per unit length carries a current given by $I = I_0 \cos(\omega t)$, where t is the time. What are the magnitude and direction of the induced electric field just outside the solenoid?

46. (II) A wire of 5.5 Ω resistance and 20 m length is wound into a coil of radius 4.3 cm. A magnetic field of 2.4 T is perpendicular to the plane of the coil. (a) If the current that forms the magnetic field is switched to the opposite direction, what is the total charge that passes through the wire coil? Assume the current switching takes place in 85 ms. (b) What is the average current and emf induced in the coil over that period?

47. (II) A single-turn circular loop of radius 0.08 m and resistance 6 Ω is coaxial with a solenoid of length 90 cm and a radius of 0.04 m, with 1800 turns. A variable resistor, as indicated in Fig. 30–37, is varied in such a way that the solenoid current falls linearly from 5 A to 1.2 A in 0.3 sec. (a) What is the induced current in the loop? (b) In what direction does the induced current flow?

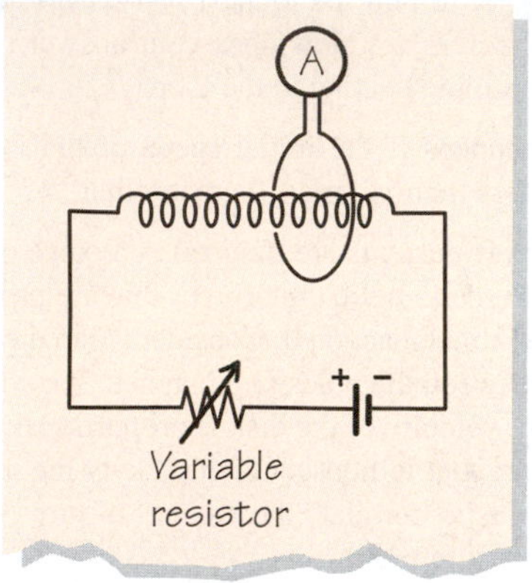

▲ **FIGURE 30–37** Problem 47.

48. (II) A very long cylindrical solenoid of radius r made from n turns of wire per unit length carries a current with the time dependence $I = I_0 e^{-t/t_0}$. Coaxial with and surrounding the solenoid are two turns of wire that make a circular loop slightly larger than the circular cross section of the solenoid (Fig. 30–38). The loop with two turns is far from the ends of the solenoid and has a resistance R. Find the current in the loop with two turns, I', as a function of time.

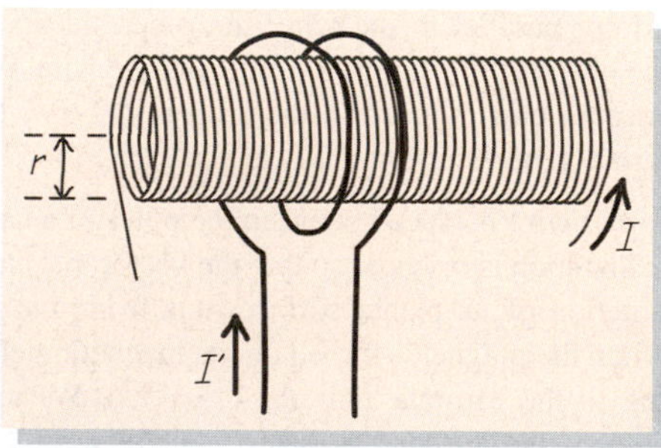

▲ **FIGURE 30–38** Problem 48.

49. (II) The uniform magnetic field of the electromagnet of Example 30–10, with circular pole faces of radius $R_0 = 0.08$ m, decreases linearly from 1.5 T to 0.7 T in 25 ms. What is the emf induced around the path drawn in Fig. 30–39 that consists of quarter arcs at radial distances $R_0/4$ and $R_0/2$, connected by radial lines? The path is clockwise.

50. (II) A long solenoid of radius R and n turns per unit length carries an alternating current $I = I_0 \sin(\omega t)$ (Fig. 30–40). What are the electric fields induced within the solenoid at a distance $R/2$ and outside the solenoid at a distance $2R$? [*Hint*: Apply Faraday's law to the two paths shown, and use symmetry.]

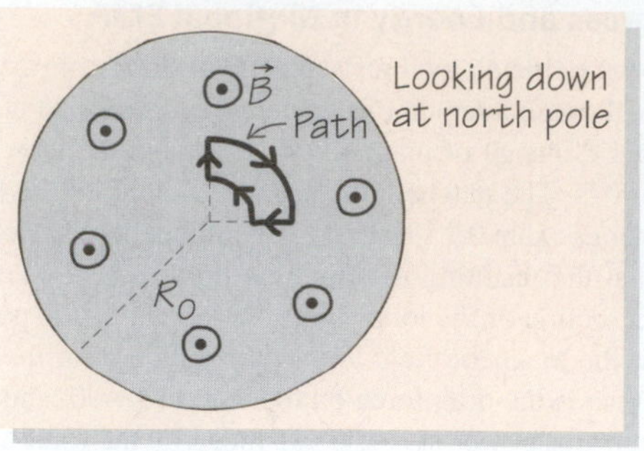

▲ **FIGURE 30–39** Problem 49.

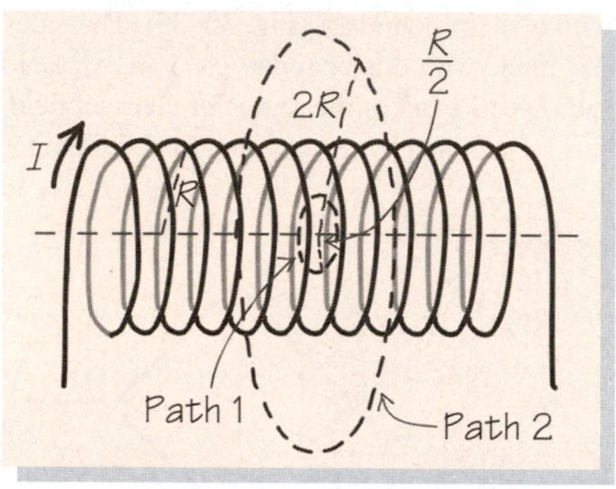

▲ **FIGURE 30–40** Problem 50.

30–5 Generators

51. (I) A coil of area 6.0 cm^2 with 180 turns of wire is connected to a resistor of resistance 3 Ω. It is rotated by hand at a frequency of 0.6 rev/s in a magnetic field of 0.40 T. (a) What is the maximum amount of current produced? (b) the average power produced?

52. (II) You have 18 m of wire, a constant magnetic field of 0.45 T, and a device that can rotate a coil at a fixed frequency of 300 Hz. What size circular coil will produce an AC emf of maximum voltage 120 V?

53. (II) The headlight of a bicycle is powered by a small generator that is driven by a wheel of the bicycle. The generator contains two coils fixed at the sides of the generator and connected in series with appropriate polarity (Fig. 30–41). Each coil consists of 70 turns and has an area of 8 cm^2. A small permanent magnet is rotated in front of the coils, so that the magnitude of the magnetic field in the coils varies between 0.1 T and zero. At what speed of the bicycle will the maximum emf be 6.4 V, given that the radius of the friction wheel is 1 cm?

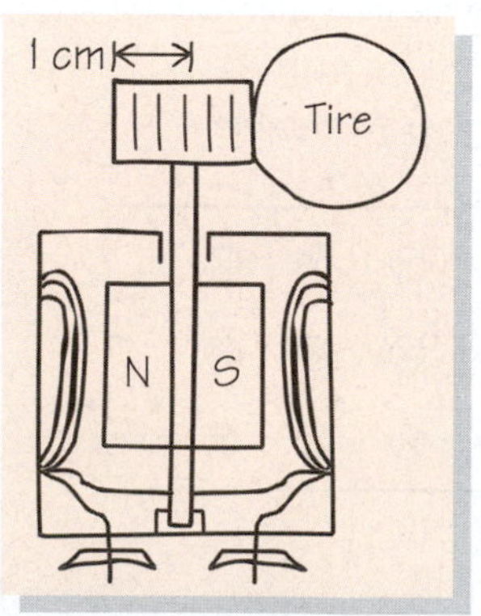

▲ **FIGURE 30–41** Problem 53.

54. (II) A bicycle wheel of radius $R = 33$ cm rotates at angular speed 53 rad/s in a plane perpendicular to a constant magnetic field of magnitude 0.55 T. What is the emf generated between the center of the wheel and its rim? When one end of a wire is attached to the center and the other end to a circular track in contact with the rim, a direct current is generated in the wire. Such a device is called a *homopolar generator*.

*30–6 The Frame Dependence of Fields

55. (II) Suppose that observer O sees an electric field $\vec{E} = E\hat{i}$ and a magnetic field $\vec{B} = B\hat{k}$. In what direction and at what (constant) speed u should a second observer move so as to see no electric field whatsoever? Use the nonrelativistic relation Eq. (30–20). If $E = 10^3$ V/m, for what range of values of B is the nonrelativistic approximation appropriate?

General Problems

56. (II) A 120-cm-long wire of square cross section with a mass of 65 g and a resistance of 2.0 Ω slides without friction down parallel conducting rails of negligible resistance (Fig. 30–42). The rails are connected to each other at the bottom by a resistanceless rail parallel to the wire so that the wire and rail form a closed rectangular conducting loop. The plane of the rails makes an angle of 15° with the horizontal, and a uniform vertical magnetic field of 0.68 T, pointing upward, exists throughout the region. What is the steady speed of the wire?

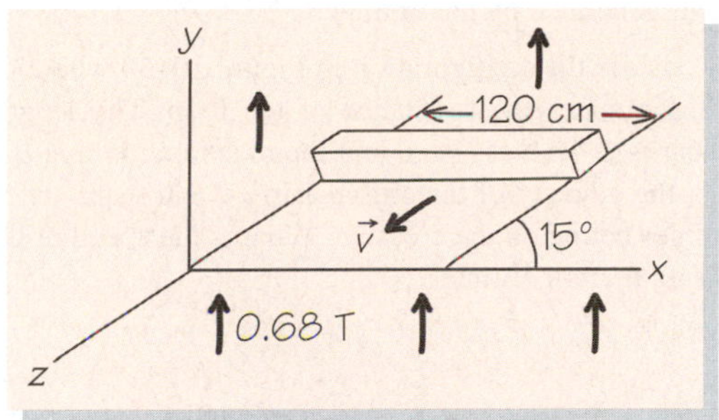

▲ **FIGURE 30–42** Problem 56.

57. (II) A straight wire carries a current $I = 150$ A near a rod that moves across two conducting wires (Fig. 30–43). The resistor has $R = 0.20$ Ω, and the rod moves at speed 45 cm/s. (a) What is the emf induced in the rod? (b) What is the current in the circuit? (c) How much work is done to move the rod 100 cm to the right? What force does this work?

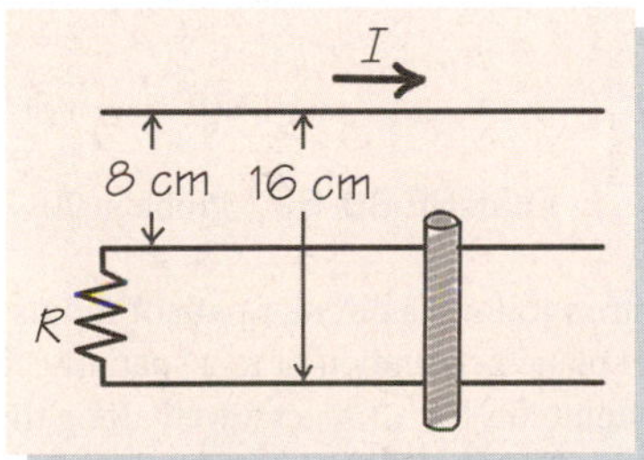

▲ **FIGURE 30–43** Problem 57.

58. (II) A conducting crossbar bracketing two vertical conducting wires slides down the wires. The wires are connected with a resistor R to form a closed circuit (Fig. 30–44a). (a) If there is a horizontal magnetic field B perpendicular to the plane of the loop, how fast does the bar fall after the initial accelerating period? (b) A battery is added to the circuit (Fig. 30–44b). What polarity and emf of the battery are needed to lift the bar with the same velocity?

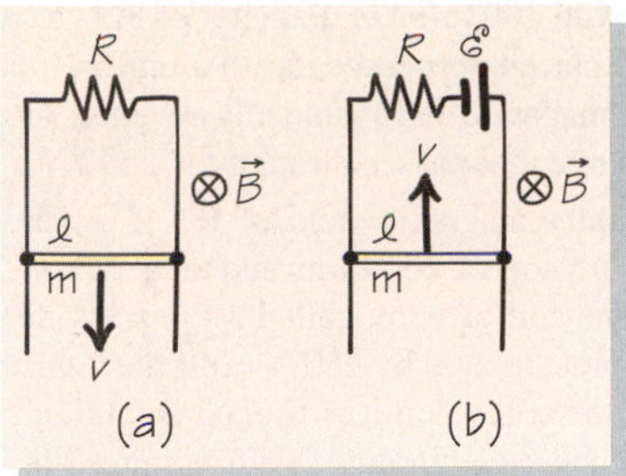

▲ **FIGURE 30–44** Problem 58.

59. (II) A large, circular coil of N turns and radius R carries a steady current I and is rotated at a constant angular speed ω about a horizontal diameter. At the center of this coil is a small, fixed, horizontal circular ring of radius r. (a) What is the emf induced in the small ring? (b) What is the angle between the plane of the coil and that of the ring when this emf is a maximum?

60. (II) Consider a 9-V battery attached to two conducting, frictionless rails 0.1 m apart. There is a magnetic field $\vec{B}$ of magnitude 0.5 T perpendicular to the rails, and a conducting bar can slide over the rails perpendicular to them as well as to the field (Fig. 30–45). The bar is placed on the rails, starts from rest, and accelerates. (a) What is the direction of its motion? (b) the direction of the emf induced? (c) Given that the total resistance of the closed circuit is 3 Ω, calculate the current in the bar when its speed is 3 m/s.

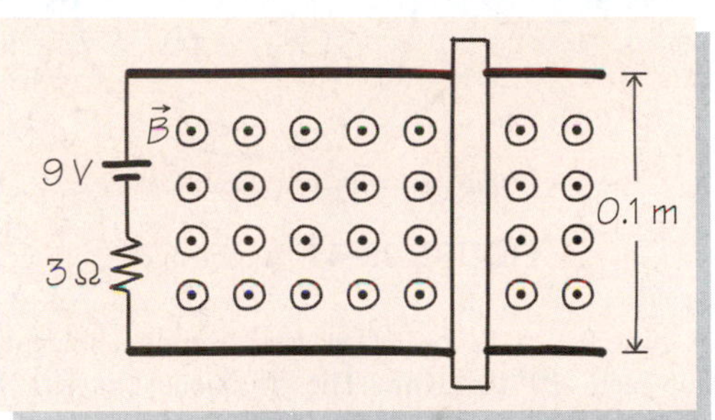

▲ **FIGURE 30–45** Problem 60.

61. (II) A wire carrying a current I is oriented in a horizontal direction. To its side, a wire loop is oriented so that it and the straight wire lie in the same horizontal plane. The straight wire is moved toward the loop. If a current is induced in the loop, what is its direction, and what is the direction of the force on the loop?

62. (II) If the plasma in a magnetohydrodynamic generator (see Question 16) is forced to flow in a channel perpendicular to a magnetic field, an electric potential builds up between points a and b, which are 1 m apart (Fig. 30–46). If the magnetic field has a strength 2.5 T, what must the speed of the plasma be in order that the potential be 1000 V?

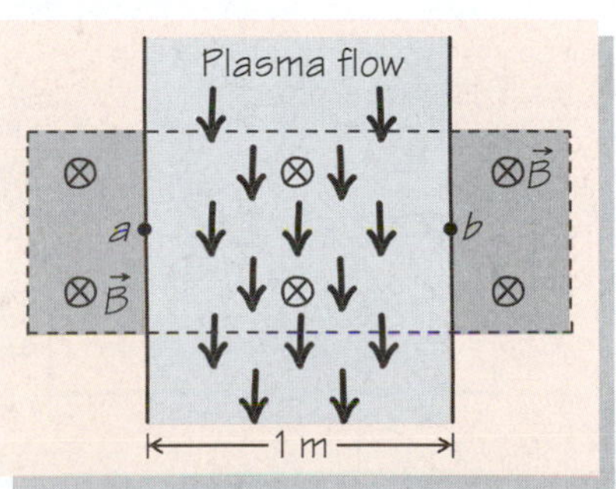

▲ **FIGURE 30–46** Problem 62.

63. (II) A coil with 200 turns, a diameter of 8.0 cm, and a resistance of 5.6 Ω is placed perpendicular to a uniform magnetic field of 1.4 T. The magnetic field suddenly reverses direction. What is the total charge that passes through the coil?

64. (II) A constant magnetic field of 0.5 T is directed along the x-axis. A wire coil of 200 turns and area 12 cm^2 is placed in the yz-plane. The coil of wire, called a *flip coil*, is then turned over (in other words, rotated by 180°). (a) If the total charge that passes through the coil when it is flipped is 0.007 C, what is the resistance of the coil circuit? (b) The same flip coil is used to measure an unknown magnetic field. The coil is flipped in several directions until it attains its maximum charge of 0.02 C, when the coil is flipped with its face in the xy-plane. What is the magnitude of the magnetic field? (c) What is the direction of the magnetic field in part (b)?

65. (II) A circular ring of area 100 cm^2 is connected to a 15-μF capacitor. The circuit has a resistance of 2 Ω. A uniform time-dependent magnetic field of magnitude $B = (0.03\ \text{T/s})t$ is perpendicular to the ring (Fig. 30–47). Calculate the current in the ring and the charge on the capacitor. Give the direction of the current and the polarity of the charge.

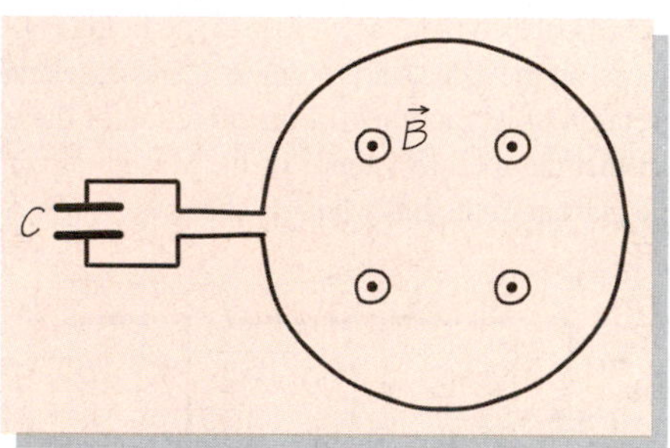

▲ **FIGURE 30–47** Problem 65.

66. (II) A current $I = I_0 \cos(\omega t)$ passes through a solenoid of area 10 cm^2 and 10^5 turns/m. The frequency is 60 Hz, and $I_0 = 10$ A. A small coil—a sense coil—is used to sense the changing flux. This sense coil has an area of 20 cm^2 with 10 turns and is placed across the solenoid so that the face of the coil is perpendicular to the solenoid axis; the two coils are concentric. (a) What is the emf induced in the sense coil? (b) If the resistance of the sense coil circuit is 5 Ω, what is the current?

67. (II) A wire that is bent into a semicircle is rotated with angular velocity ω about its diameter, as shown in Fig. 30–48. The bent wire and its supports are placed in a uniform magnetic field perpendicular to the plane of the supports. What is the emf induced in the circuit shown? If the resistance of the closed loop is R, what is the average power dissipated?

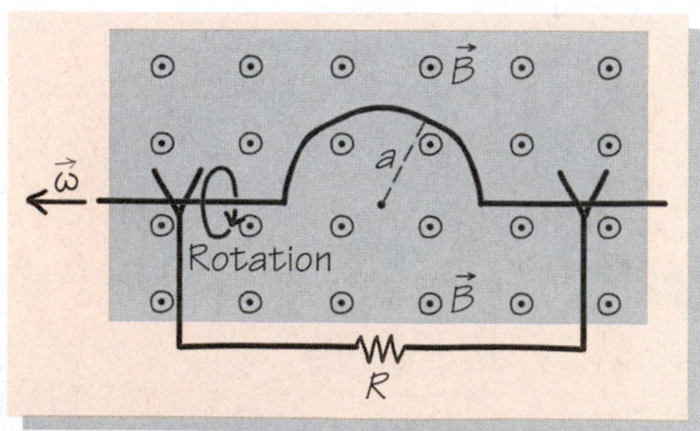

▲ **FIGURE 30–48** Problem 67.

68. (II) A pendulum consists of a metal bar suspended from two thin wires attached to a fixed conducting bar (Fig. 30–49). The resistance of this closed circuit is 0.030 Ω. The pendulum is placed in a vertical magnetic field of magnitude 0.12 T. The pendulum is displaced by a small angle from the equilibrium position and allowed to oscillate. What is the ratio of the power dissipated to the energy of the oscillator?

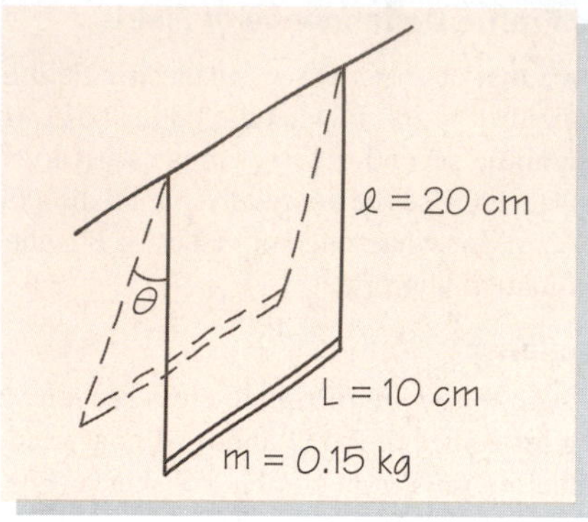

▲ **FIGURE 30–49** Problem 68.

69. (II) A train travels in the northerly direction at 60 mph on tracks that are 1.5 m apart. In the region, Earth's magnetic field is 0.7×10^{-4} T making an angle of 30° with the vertical. What is the emf generated by the train?

70. (II) Consider the configuration in Figure 30–50 where the resistor R dominates the resistance of the loop. The length of the crossbar is L, its mass is m, and the magnetic field is B. At time $t = 0$, the velocity of the crossbar is v_0, and there are no external forces acting on the crossbar. What is the speed of the crossbar as a function of time?

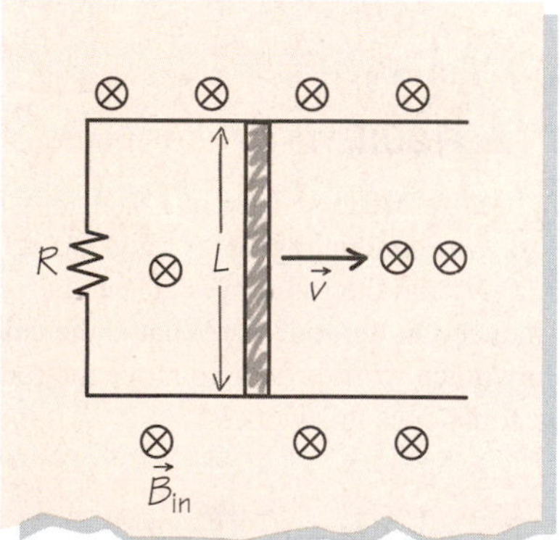

▲ **FIGURE 30–50** Problem 70.

71. (III) An electron follows a circular path of radius $R = 1$ m while traveling in a plane perpendicular to a spatially constant magnetic field of magnitude 10^{-6} T. As viewed along the magnetic field lines, the electron follows a counterclockwise path. (a) What is the speed of the electron? (b) Assuming that the motion of the electron is nonrelativistic, what is the energy E of the electron? (c) The magnitude of the magnetic field is reduced smoothly by a certain percentage during an interval Δt. Show that the fractional energy change of the electron, $\Delta E/E$, is independent of the radius of the electron's orbit as well as of the electron's speed. (This effect is the basis for low-energy operation of a particle accelerator called the *betatron*.) (d) If the magnetic field is reduced in time $\Delta t = 5$ s by 10 percent, estimate $\Delta E/E$.

Chapter 31

As part of the manufacturing process for high-temperature superconductor wire, technicians at American Superconductor operate capstans that pull the wire through a metal die, reducing its diameter and increasing its length. The successful production of high-capacity wire made of a high-temperature superconductor would revolutionize many aspects of our power distribution system.

Magnetism and Matter

While we have made frequent use of bar magnets in the previous chapters, we have not asked questions as to why the materials that make up such magnets behave the way they do. Is there something in such materials that is like the electric currents that produce magnetic fields according to the laws we have developed? How do we explain the fact that magnets pick up needles but not pieces of paper? Why is it possible to make a slab of iron act like a bar magnet at some times but not at other times? Why can't the same be said for a piece of aluminum? What explains the magnetic properties of superconductors? In other words, what lies behind the magnetic properties of matter? The answers to these types of questions lie in the microscopic structure of matter, all the way down to the subatomic level. We utilize the magnetic properties of material when we construct computer disk drives, electric motors, generators, transformers, particle accelerators, and medical scanners, so that understanding the magnetic properties at a fundamental level is essential. When we considered the electric properties of materials, we saw that the dielectric properties of materials depend on the polarizability of atoms and molecules. In the same way, the magnetic properties of materials depend on the magnetic properties of atoms and molecules and their constituents. This chapter is concerned with the origins of these properties.

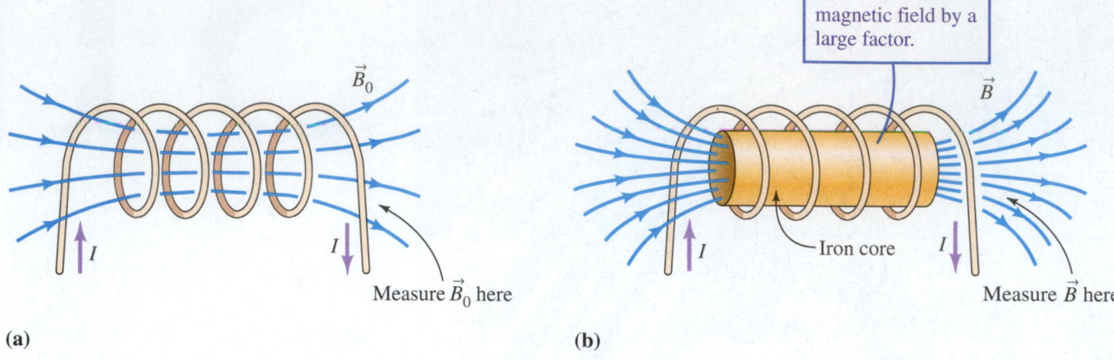

▲ **FIGURE 31–1** (a) A solenoid that carries a current has a magnetic field. (b) The cylindrical volume of the solenoid is filled with an iron core.

31–1 The Magnetic Properties of Bulk Matter

We begin with a description of the different kinds of behavior that are observed on an everyday scale. We can explore the magnetic properties of various materials using a solenoid. We measure the magnetic field $\vec{B}_0$ produced by, and near one end of, a current-carrying solenoid (Fig. 31–1a) and ask how the presence of various materials affect the magnetic field. (Throughout this chapter $\vec{B}_0$ represents any field, not necessarily constant, that is present before we introduce materials.) If we insert a wooden or copper core into the solenoid and repeat the measurement, it would take a very sensitive instrument to reveal that the field has been changed by one part in 10^6. But if we place an iron core in the solenoid (Fig. 31–1b), the measured field is increased by a factor many times that of the original field of the hollow solenoid. If we now remove it, the iron core will act like a bar magnet, even if it showed no such properties before insertion. The solenoid somehow magnetizes the iron core; the wooden or copper cores show no such effect. Magnetic properties depend on the material!

The magnetic behavior of bulk materials—gases, liquids, and solids—is characterized by the **magnetization**, $\vec{M}$, which we define as the *magnetic dipole moment per unit volume* of material. The field outside a material with a net magnetization is that of a magnetic dipole, the same as that of a bar magnet. A magnetic moment has dimensions of current times area [see, for example, Eq. (29–27)], so magnetization has dimensions of current times area divided by volume, or current per length. The SI units of $\vec{M}$ are amperes per meter (A/m). Our aim here is to find a connection between an applied field $\vec{B}_0$ and the magnetization.

The net magnetic field of a solenoid S_1 with a core of some material in it is the vector sum of contributions from the original magnetic field of the solenoid, $\vec{B}_0$, and the contribution of the magnetization of the core. The core can be viewed as a second solenoid S_2 as any magnetic dipole has a magnetic field with the same general form as that of a solenoid. Suppose that S_2 can be treated as a solenoid of area A, length L, and N loops that carry current I. From Eq. (29–15), the magnetic field of S_2 is

$$B_2 = \frac{\mu_0 NI}{L} = \frac{\mu_0 NIA}{LA} = \mu_0 \frac{m_2}{V},$$

where† m_2 is the magnetic moment of S_2, $m_2 = NIA$, and μ_0 is the permeability of free space. The factor m_2/V is the magnetic moment per unit volume—the magnetization size $|\vec{M}|$. Thus *the contribution of the core to the total magnetic field is* $\mu_0 \vec{M}$, and the total field of the solenoid with the core present is

$$\vec{B} = \vec{B}_0 + \mu_0 \vec{M}. \tag{31–1}$$

†To avoid confusion with the magnetic permeability μ to be introduced shortly, we use the notation m for the magnetic dipole moment throughout this chapter. Don't confuse it with a mass!

Equation (31–1) applies whatever the source of the original field $\vec{B}_0$ in which the bulk material is placed, and it need not be the field due to a solenoid. This situation is similar to that of dielectrics, where an electric field within a material has contributions from an original electric field due to free charges and from an internal, induced distribution of charge. We saw in Chapter 25 that we can separate the effect of these types of charges through the introduction of a dielectric constant and a generalized permittivity. Similarly, we want to separate the effects of the internal magnetization, which has contributions from what must be the equivalent of internal currents, from the magnetic fields due to ordinary currents. We refer to the latter as *free*, or *real*, currents. The effects of free currents are isolated by defining the **magnetic intensity** $\vec{H}$, a quantity for which the effect of the magnetization of the material is subtracted out:

$$\vec{H} \equiv \frac{\vec{B}}{\mu_0} - \vec{M}. \tag{31–2}$$

Notice that the dimensions of $\vec{H}$ are those of $\vec{M}$, *not* of $\vec{B}$. By replacing $\vec{B}$ in Eq. (31–2) with $\vec{B}_0 + \mu_0\vec{M}$ according to Eq. (31–1), we find

$$\vec{B}_0 = \mu_0\vec{H}. \tag{31–3}$$

Equation (31–3) shows that *the magnetic intensity measures the magnetic field due to free currents*. Another form for the relation among $\vec{B}$, $\vec{H}$, and $\vec{M}$ is found by combining Eq. (31–1) and (31–3):

$$\vec{B} = \mu_0\vec{H} + \mu_0\vec{M}. \tag{31–4}$$

Most materials have no magnetization unless they are in the presence of an external magnetic field $\vec{B}_0$ that induces magnetization. The important exceptions to this are iron and some other materials that we will refer to below as ferromagnetic—these are the materials that make permanent magnets. We know by experiment that over a large range of conditions, *the size of the magnetization of nonferromagnetic materials varies linearly with the external magnetic field*. For these materials, $\vec{M}$ depends linearly on $\vec{B}_0$, and hence on $\vec{H}$. The **magnetic susceptibility** χ_m is defined as the coefficient of the linear relation between the magnetization and the magnetic intensity:

$$\vec{M} \equiv \chi_m\vec{H}. \tag{31–5}$$

The direction of the magnetization is more complicated: The original field $\vec{B}_0$ and the field $\mu_0\vec{M}$ due to the magnetization are parallel for one class of materials but antiparallel for a second class. If the susceptibility of a material is positive, its magnetization is aligned along the external field; if the susceptibility is negative, the magnetization is aligned opposite to the external field. Because $\vec{M}$ and $\vec{H}$ have the same dimensions, χ_m is dimensionless. Table 31–1 gives a range of susceptibilities found in nature. Examination of the table reveals the three general classes of materials: those with small negative susceptibility (**diamagnetic materials**), those with small positive susceptibility (**paramagnetic materials**), and iron and its cousins (**ferromagnetic materials**), which have a large positive susceptibility and can have a magnetization even with no external field present. The nature and properties of these classes of materials is the major topic of this chapter.

With our definition of magnetic susceptibility χ_m, we can express the relation between the magnetic field in a material and the magnetic intensity. From Eq. (31–1),

$$\vec{B} = \vec{B}_0 + \mu_0\vec{M} = \mu_0\vec{H} + \mu_0\chi_m\vec{H} = \mu_0(1 + \chi_m)\vec{H}. \tag{31–6}$$

We define the coefficient of $\vec{H}$ in this equation as the **permeability**, μ, of the material:

$$\mu = \mu_0(1 + \chi_m). \tag{31–7}$$

TABLE 31–1 • Some Magnetic Susceptibilities (at 20°C unless indicated otherwise)

Material	Susceptibility, χ_m
Diamagnetic	
Water	-13×10^{-6}
Copper	-5.5×10^{-6}
Silver	-2.0×10^{-5}
Carbon (diamond form)	-5.9×10^{-6}
Bismuth	-2.4×10^{-4}
Paramagnetic	
Sodium	1.6×10^{-6}
Cupric oxide (CuO)	2.6×10^{-4}
Aluminum	1.7×10^{-5}
Liquid oxygen (90 K)	8.7×10^{-3}
Oxygen gas	3.5×10^{-3}
Ferromagnetic	
Iron (annealed)	5.5×10^{3}
Permalloy (55% Fe, 45% Ni)	2.5×10^{4}
Mu-metal (77% Ni, 16% Fe, 5% Cu, 2% Cr)	1×10^{5}

The relation between the total magnetic field in a material and the magnetic intensity, which is a measure of the effect of free currents, is then

$$\vec{B} = \mu\vec{H}. \tag{31–8}$$

Just as when materials are present, the electric permittivity ε replaces the permittivity of free space ε_0 in expressions for electric fields that contain only the free charge, so too when materials are present, μ replaces μ_0 in expressions for magnetic fields that contain only the free current. Equation (31–3) shows that when there is a magnetic field in a vacuum, that field is related to the intensity by a relation like that of Eq. (31–8), but with μ_0 appearing in the place of μ; thus μ_0 is the permeability of the vacuum. From Table 31–1 we can see that μ is very close to μ_0 for nonferromagnetic materials.

The various quantities we have defined are all useful in characterizing the bulk magnetic behavior of materials. In Table 31–2, we summarize these quantities and their relations.

TABLE 31–2 • Magnetic Bulk Properties and Their Relations

Symbol	Property
$\vec{B}_0$	Applied magnetic field, produced independently of type of material by a nearby magnet or currents
$\vec{H}$	Magnetic intensity, proportional to the applied magnetic field
$\vec{M}$	Magnetization, the magnetic dipole moment per unit volume of a material
$\vec{B}$	Net magnetic field, the sum of the applied magnetic field and a term proportional to the magnetization
μ_0	Permeability of free space
χ_m	Magnetic susceptibility of a material
μ	Permeability of a material, $\mu = \mu_0(1 + \chi_m)$.

Some Relations

	$\vec{B}_0$	$\vec{H}$	$\vec{M}$	$\vec{B}$
$\vec{H} =$	$\vec{B}_0/\mu_0$	—	$\vec{M}/\chi_m$	$\vec{B}/\mu$
$\vec{M} =$	$\chi_m\vec{B}_0/\mu_0$	$\chi_m\vec{H}$	—	Not used
$\vec{B} =$	$(1 + \chi_m)\vec{B}_0$	$\mu_0(1 + \chi_m)\vec{H}$	$\dfrac{\mu_0(1 + \chi_m)}{\chi_m}\vec{M}$	—

EXAMPLE 31–1 A straight solenoid of diameter 5 cm and length 25 cm is wrapped with 200 turns of wire that carries a current of 5 A. The solenoid is filled with iron, magnetic susceptibility $\chi_m = 5.5 \times 10^3$. Find (a) the magnetic intensity within the solenoid and (b) the magnetic field within the solenoid. (c) By what factor is the magnetic field changed due to the presence of the material?

Setting It Up We assign labels of the known and unknown quantities: The current is I, the number of turns of wire per unit length is n, the magnetic intensity has magnitude H, and the magnetic field has magnitude B.

Strategy The first step is to calculate the magnetic intensity $H = B_0/\mu_0$, where B_0 is the magnetic field associated with the free currents in the solenoid. This is given by $B_0 = \mu_0 nI$, so that $H = nI$. We can then calculate $B = \mu H$, using, for example, Eqs. (31–8) and (31–7), and hence the factor by which the field changes due to the material in the solenoid. The diameter of the solenoid does not enter here, except to confirm that the solenoid is long compared to its width, so that our formulas for ideal solenoids apply.

Working It Out (a) With $n = (200 \text{ turns})/(25 \text{ cm}) = 800 \text{ turns/m}$, and $I = 5$ A,

$$H = nI = (800 \text{ turns/m})(5 \text{ A}) = 4000 \text{ A/m}.$$

(b) The magnetic field includes the effect of the field due to the iron that fills the solenoid. We have

$$\begin{aligned} B &= \mu H = \mu_0(1 + \chi_m)H \cong \mu_0\chi_m H \\ &= (4\pi \times 10^{-7}\,\text{T}\cdot\text{m/A})(5.5 \times 10^3)(4000 \text{ A/m}) = 28 \text{ T}. \end{aligned}$$

(c) The factor by which the field changes is

$$\frac{B}{B_0} = 1 + \chi_m \cong \chi_m.$$

For a ferromagnetic material such as iron, this is a large factor, and very substantial enhancements of magnetic fields occur using these kinds of materials.

What Do You Think? Do you expect B inside to be (a) larger, (b) smaller, or (c) unchanged if copper were used in the core? *Answers to **What Do You Think?** questions are given in the back of the book.*

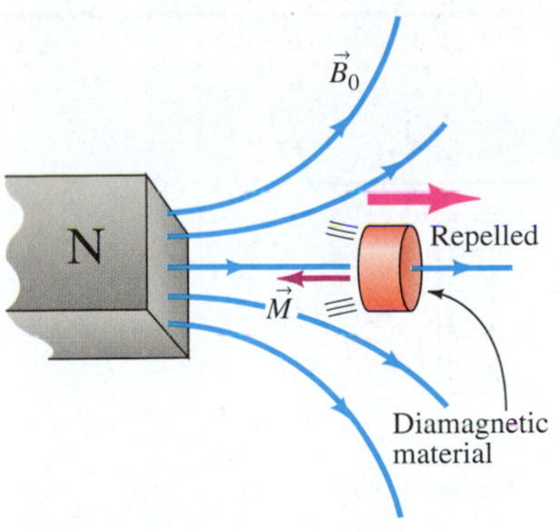

▲ **FIGURE 31–2** Diamagnetic substances are repelled by one pole of a nearby bar magnet.

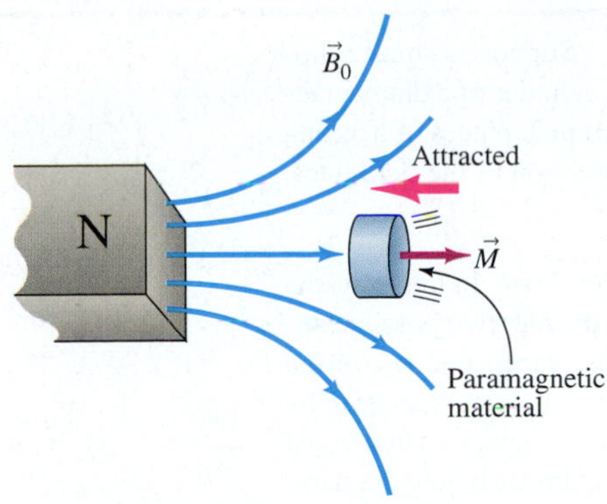

▲ **FIGURE 31–3** Paramagnetic substances are attracted to one pole of a nearby bar magnet.

Let us summarize the properties of the three broad classes of materials that Table 31–1 reveals. The class of materials that are most often used for their magnetic properties is composed of *ferromagnetic* materials, those with large positive susceptibilities. These substances can have magnetization without the presence of an external field, forming permanent magnets.

Diamagnetic substances are those with very small negative susceptibilities. In such materials, the magnetization direction is *opposite* to the direction of the inducing field. The magnetic field inside such materials is *reduced* a little from its value outside the material. If a diamagnetic material is placed near the north pole of a magnet, the magnetization produces a field that points toward that pole (Fig. 31–2). The diamagnetic material acts as though it has a north pole adjacent to the external north pole: The diamagnetic material is *repelled* by the magnet. The behavior of diamagnetic substances is similar to that of dielectrics, for which polarization effects tend to cancel the electric field associated with free charges.

Paramagnetic substances are those with small positive susceptibilities. The magnetization points in the same direction as the field of an external magnet (Fig. 31–3), as though the paramagnetic substance has a south pole oriented toward the magnet's north pole: The piece of paramagnetic material is *attracted* to the magnet.

Table 31–3 gives the forces acting on some samples near a sizable (3.0 T) electromagnet. Only the relative scale of the numbers matters in the table, since we are providing no details about the geometry of the magnet. The force is expressed in units of the weight of each sample. The minus sign reflects repulsion (diamagnetism); the plus sign reflects attraction (paramagnetism and ferromagnetism). Note the differences in the sizes of these forces for the different classes of materials.

We'll discuss the underlying atomic origins of ferromagnetism, diamagnetism, and paramagnetism in Sections 31–3 and 31–4.

TABLE 31–3 • Magnetic Forces on Materials near a Large Electromagnet

Material	Material Class	Force (in units of sample weight)
Copper (pure)	Diamagnetic	-1.3×10^{-3}
Lead	Diamagnetic	-19×10^{-3}
Graphite	Diamagnetic	-56×10^{-3}
Sodium	Paramagnetic	$+10.2 \times 10^{-3}$
Copper chloride	Paramagnetic	$+143 \times 10^{-3}$
Iron	Ferromagnetic	$+20$
Magnetite	Ferromagnetic	$+61$

CONCEPTUAL EXAMPLE 31–2 Suppose a small cylinder of a paramagnetic substance and a small cylinder of a diamagnetic substance are placed between the (parallel) pole pieces of a strong magnet, so that they are aligned along the direction of the field lines. Will they attract or repel each other?

Answer We visualize the situation in Fig. 31–4. In this way, the magnetic field lines point upward. We now consider two possible situations, one in which the two cylinders are placed one above the other (Fig. 31–4a) and another in which the two cylinders are side by side (Fig. 31–4b). For case a, we place the paramagnetic cylinder just below the S pole piece and the diamagnetic cylinder below the paramagnetic cylinder. (You can verify that the order does not matter.) The external magnetic field induces magnetism in the paramagnetic substance so that the cylinder acts like a very weak magnet lined up in such a way that the field points up. Thus the top of the paramagnetic cylinder acts like a N pole of a magnet, and the bottom like the S pole of a magnet. The situation with a diamagnetic cylinder is just the opposite. The top of the diamagnetic cylinder acts like the S pole, and the bottom like a N pole. Thus the two S poles of the materials are adjacent, and they repel each other. Figure 31–4b shows that when the two cylinders are placed side by side, they attract.

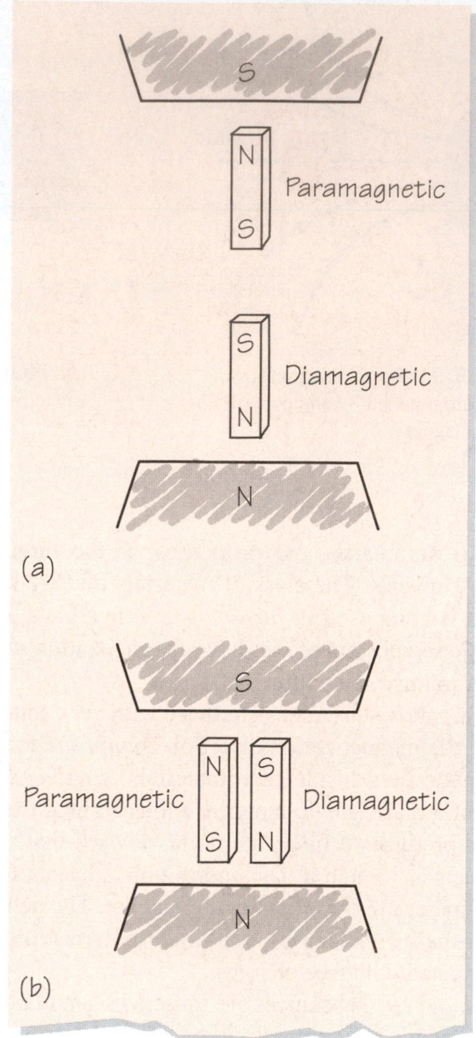

▶ **FIGURE 31–4** In (a) the samples are one above the other; in (b) they are side by side.

31–2 Atoms as Magnets

The magnetic field of a solenoid is the same as that of a bar magnet. How can two such apparently dissimilar systems have the same magnetic field? The answer lies in the magnetic properties of atoms. To understand these properties qualitatively, we start with a classical planetary model for atoms, in which electrons orbit a nucleus, and add some necessary quantum mechanical features. The orbiting electrons form ring currents like those of a solenoid, and the atoms then act as magnetic dipoles, complete with magnetic dipole moments. In most situations, the magnetic moments of a large assembly of atoms point in random directions, so that the net magnetic moment of a macroscopic material adds to zero. The exception occurs in ferromagnetic materials, in which forces between atoms act to line up atomic magnetic moments with their neighbors, giving rise to a net magnetization.

The magnetic properties of individual atoms are affected by a (quantum-mechanical) tendency of their electrons to pair off such that the magnetic moment due to an individual electron's current points in the opposite direction from that of its partner. The net result is that a pair of electrons has no magnetic moment. In a classical picture of the atom as a collection of electrons orbiting about a nucleus, this effect would be interpreted as the cancellation of moments due to clockwise-moving electrons with the moments due to counterclockwise-moving electrons. This suggests that in many cases, atoms with an even number of electrons will have no magnetic dipole moment, whereas only a last unpaired electron makes a difference in atoms with an odd number of electrons.

The Magnetic Dipole Moment of Atoms

Consider the simplest possible *classical* planetary model of an atom, in which a single electron of charge $-e$ moves at speed v in a circular orbit of radius r around a heavy nucleus—as in Fig. 31–5a. The orbital period is $T = 2\pi r/v$. Because electric current is charge per unit time, the current around the nearly stationary nucleus is

$$I = \frac{-e}{T} = -\frac{ev}{2\pi r}. \tag{31–9}$$

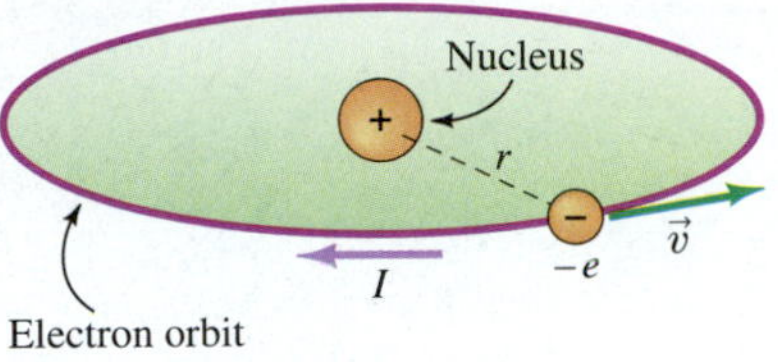

(a)

The minus sign indicates that the current is in a direction opposite to that of the motion of the electron. The current loop has an area of πr^2; following Eq. (29–27), the magnitude of the *orbital magnetic dipole moment* is therefore given by

$$m_{\text{orbital}} = |I|\pi r^2 = \frac{ev}{2\pi r}\pi r^2 = \frac{1}{2}evr. \tag{31–10}$$

The direction of the magnetic moment vector $\vec{m}$ is determined by a right-hand rule (Fig. 31–5b).

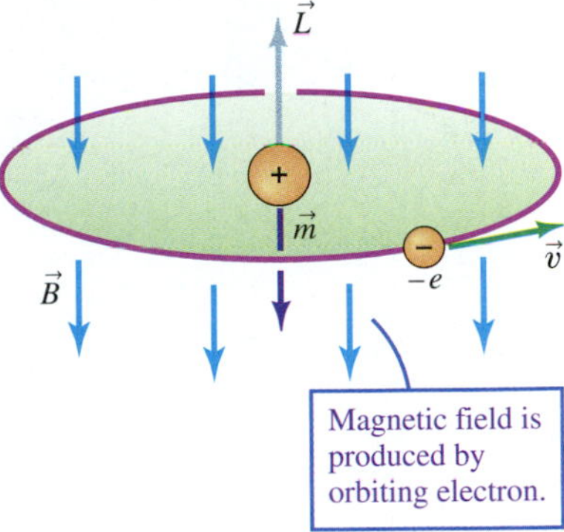

(b)

▶ **FIGURE 31–5** (a) An electron in a circular orbit around a nucleus. Over numerous orbits, the effect is the same as a continuous current ring about the nucleus. (b) The circulating electron forms a magnetic dipole with magnetic moment $\vec{m}$ ($\vec{m}_{\text{orbital}}$) oriented downward, opposite to the angular momentum $\vec{L}$.

EXAMPLE 31–3 Estimate an atomic orbital magnetic moment by taking the radius of the electron orbit to be of roughly atomic size, 10^{-10} m, and the kinetic energy to be a typical atomic energy of 1 eV. Compare this to the magnetic moment of a macroscopic loop of area 1 cm^2 that carries a current of 1 mA.

Strategy To use Eq. (31–10) for the magnetic moment, we must first calculate the speed. With the kinetic energy $K = m_e v^2/2$, where m_e is the electron mass, we get $v = \sqrt{2K/m_e}$. Thus

$$m_{\text{orbital}} = \frac{1}{2}er\sqrt{\frac{2K}{m_e}}.$$

The moment of the specified current loop is IA, where A is the area of the loop.

Working It Out If we convert to SI units, our energy estimate of 1 eV is 1.6×10^{-19} J. Thus

$$v = \sqrt{\frac{2(1.6 \times 10^{-19}\text{ J})}{m_e}} \cong \sqrt{\frac{3.2 \times 10^{-19}\text{ J}}{10^{-30}\text{ kg}}} \cong 6 \times 10^5\text{ m/s},$$

where we have used (to the same accuracy as the orbit radius) $m_e \cong 10^{-30}$ kg. Equation (31–10) then gives the magnetic moment

$$m_{\text{orbital}} = \tfrac{1}{2}(1.6 \times 10^{-19}\text{ C})(6 \times 10^5\text{ m/s})(10^{-10}\text{ m}) \cong 5 \times 10^{-24}\text{ A}\cdot\text{m}^2.$$

For comparison, the magnetic moment of the macroscopic loop is

$$m = (10^{-3}\text{ A})(10^{-4}\text{ m}^2) = 10^{-7}\text{ A}\cdot\text{m}^2,$$

which is some 2×10^{16} times larger than the magnetic moment of the single atom.

What Do You Think? We learned in mechanics that in planetary motion, both the light particle (the electron here) and the heavy particle (here, the positively charged nucleus) rotate about the center of mass of the system. Do we need to worry about this complication?

In Example 31–3, we can think of a magnetic dipole moment as due to a circulating charge, with the magnetic moment proportional to the angular momentum of the circulating charge. It is useful to express magnetic moments in terms of angular momentum because angular momentum is a fundamental physical quantity. For an electron in an atom, the orbital magnetic dipole moment [Eq. (31–10)] has the magnitude

$$m_{\text{orbital}} = \frac{1}{2}evr = \frac{e}{2m_e}m_e vr = \frac{e}{2m_e}L, \tag{31–11}$$

where m_e is the electron's mass and $L = m_e vr$ is the angular momentum of the electron in its circular orbit. If we include the vectorial properties of both the angular momentum and the magnetic moment, then Eq. (31–11) becomes

$$\vec{m}_{\text{orbital}} \equiv g_L\vec{L}. \tag{31–12}$$

The coefficient g_L connecting the magnetic moment and the angular momentum is known as the **gyromagnetic ratio**. For the orbital motion, we have just seen that

$$g_L = -\frac{e}{2m_e}. \tag{31–13}$$

The minus sign is present because $\vec{m}_{\text{orbital}}$ and $\vec{L}$ point in opposite directions (Fig. 31–5b).

At this point, we make a necessary leap to a nonclassical atom. According to classical mechanics, the magnitude of the orbital angular momentum depends on the size of the orbit and the velocity of the electron, but in fact atoms cannot be described in this way. Quantum mechanics shows that orbital angular momentum cannot take on a continuous range of values; rather, the magnitude of the orbital angular momentum is limited to the values $\ell\hbar$, where $\hbar = 1.05 \times 10^{-34}\ \text{J}\cdot\text{s}$ is a constant that appears in all aspects of atomic physics and ℓ is a positive integer (or zero).[†] When we use this restriction in Eq. (31–12), we find that this also places restrictions on the possible values of the magnetic moment:

$$m_{\text{orbital}} = \left(\frac{e\hbar}{2m_e}\right)\ell \equiv m_B\ell. \tag{31–14}$$

The quantity m_B is the **Bohr magneton**, after Niels Bohr, one of the founders of quantum mechanics. Its numerical value is

$$m_B = \frac{e}{2m_e}\hbar = 9.27 \times 10^{-24}\ \text{A}\cdot\text{m}^2. \tag{31–15}$$

This is a value that is characteristic of all magnetic moments on the atomic scale.

In addition to a magnetic moment associated with orbital motion, experiments show that electrons also carry an *internal* magnetic moment that cannot be identified with any real current. *The electron itself behaves as a tiny magnet!* This is another quantum phenomenon, one with no analog in classical physics. As a consequence, we must add the contribution of the electron's *intrinsic magnetic moment*, $m_{\text{intrinsic}}$, to the orbital magnetic moment. The value of $m_{\text{intrinsic}}$ turns out to be approximately m_B to an accuracy of 0.1%.

Bulk Effects Are Due to the Alignment of Atomic Magnetic Dipoles

The magnetic field of an individual atom is tiny compared to the magnetic field of a bar magnet. But there are so many atoms that if all the atomic magnetic moments were perfectly aligned in a material, we would have a truly huge effect. As Example 31–4 shows, the alignment of the atomic magnetic moments need be only very slight to produce noticeable bulk effects.

EXAMPLE 31–4 Consider 1 mol of atoms with individual magnetic moments $m_0 = 10^{-23}\ \text{A}\cdot\text{m}^2$. Assume that the magnetic moments can point only in the $+z$- and $-z$-directions, with a fraction f pointing "up" and a fraction $1 - f$ pointing "down." What value of f gives the same magnetic moment as a 1-cm² wire loop that carries a current of 10 mA?

Strategy The total magnetic moment is the vectorial sum of individual magnetic moments $m_0 = 10^{-23}\ \text{A}\cdot\text{m}^2$. In this case the vectors that represent the individual moments take only the two directions *up* and *down*. Thus with altogether N_A atoms in a mole, where N_A is Avogadro's number 6×10^{23}, the total magnetic moment is the algebraic sum of fN_A moments with a plus sign (pointing up) and $(1 - f)N_A$ identical moments with a minus sign (pointing down). This must match the magnetic moment IA of the wire loop, where A is the area of the loop and I is the current running through it. We can then solve this equation for f.

Working It Out The relation above reads

$$m_{\text{tot}} = N_A f m_0 - N_A(1 - f)m_0 = N_A m_0(2f - 1) = IA.$$

Solving this relation for f, we find

$$f = \frac{1}{2}\left(1 + \frac{IA}{N_A m_0}\right).$$

Numerical evaluation gives

$$f = \frac{1}{2} + \frac{(10^{-2}\ \text{A})(10^{-4}\ \text{m}^2)}{2(6 \times 10^{23}\ \text{atoms})(10^{-23}\ \text{A}\cdot\text{m}^2/\text{atom})}$$
$$= \frac{1}{2} + (8 \times 10^{-8}).$$

In other words, a departure from complete randomness ($f = \frac{1}{2}$) of only one part in 10 million leads to macroscopic effects.

What Do You Think? Suppose we place the atoms between the poles of a magnet. The magnetic field will to some degree line up the magnetic moments so that they tend to point in the direction of the magnetic field. Why don't we expect f to be 1 in this case?

[†] We would have to refine this further to be in accord with the full predictions of quantum mechanics; for now this is a sufficiently good description of the situation.

CONCEPTUAL EXAMPLE 31–5 Take a bar magnet 30 cm long and cut it into two 15- cm pieces. On the basis of our understanding of the magnetism of materials, would you predict that the magnetization of the two pieces is (a) larger, (b) smaller, or (c) comparable to the magnetization of the original magnet?

Answer If we use the simple model of a magnet treated in Example 31–4, we see that net magnetism is a consequence of a small imbalance between atomic magnetic moments pointing *up* and *down*. There is no reason to believe that this imbalance is in any way distributed inhomogeneously throughout the sample, so that the fraction f discussed in Example 31–4 should be the same across different parts of the sample. Since the magnetization is the magnetic moment per unit volume, the size of any pieces is not relevant—magnetization is an intensive quantity. We thus expect the magnetization of the two pieces to be the same as that of the single larger piece from which they came.

THINK ABOUT THIS. . .

WHY AREN'T MOST MATERIALS MAGNETIC?

We saw in Example 31–4 that only a small deviation from randomness leads to significant bulk effects provided that individual atoms or molecules have *some* magnetic moment (as indeed they do). Some random fluctuation of this size might then produce macroscopic effects. But in fact the "small" deviation in the example is not so small. In statistics, $\sqrt{N}$ is a typical fluctuation from the mean when N objects or events are involved. Thus in a sample of 10^{24} atoms, statistical fluctuations away from an average magnetic moment of zero will lead, on average, to an excess of only 10^{12} atoms pointing, say, up. This translates to a fraction $f = 10^{12}/10^{24} = 10^{-12}$ of atoms pointing up, and this corresponds to a net magnetic moment much smaller than the one we saw in Example 31–4. It is only when there are forces between atoms that cause neighboring atoms to line up with each other that we form permanent magnets.

The Connection Between Microscopic and Macroscopic Quantities

We have seen that a piece of material will have significant magnetization if the directions of the magnetic dipole moments of its many component atoms or molecules are not completely random. In that case, the vector sum of the atomic magnetic moments will not be zero. If we divide the vector sum of the magnetic moments by the number of atoms, we get an *average* magnetic moment $\vec{m}_0$ per constituent (atom or molecule). The magnetization is then

$$\vec{M} = n\vec{m}_0, \tag{31–16}$$

where n is the number of constituents per unit volume. Once we have determined $\vec{M}$, we can determine the other bulk magnetic properties from the discussion in Section 31–1.

31–3 Ferromagnetism

Ferromagnetic materials, which include the elements iron, cobalt, nickel, gadolinium, and dysprosium, together with their alloys, can have large permanent magnetizations, and the direction and size of the magnetization can be set by an external magnetic field. These materials and their properties come into play every time you use computer memory or a credit card, as well as throughout the electrical transmission network.

In ferromagnetic materials, the intrinsic magnetic dipole moments of the electrons in atoms align themselves to some degree and lead to large magnetic effects. The explanation for the alignment of the intrinsic magnetic moments, which was suggested in 1928 by Werner Heisenberg, is purely quantum mechanical—there is no classical mechanism that results in a sufficiently strong alignment. As a consequence of the quantum-mechanical exclusion principle (see Chapter 26), electrons with parallel *intrinsic* magnetic moments arrange themselves in ferromagnetic materials in orbits that tend to maximize the distance between them. This reduces the potential energy of Coulomb repulsion between them and makes a state with parallel magnetic moments a state of lower energy. As a result, there is a preference based on energy for the intrinsic magnetic moments of electrons to line up parallel with one another. This argument shows that the forces that tend to keep the magnetic moments parallel are electrostatic forces rather than the much weaker magnetic forces, and this explains why ferromagnetic materials must be heated to hundreds of degrees Kelvin before they become demagnetized.

▶ **FIGURE 31–6** Photomicrograph of magnetic domains in a sample of iron with 3 percent silicon. A strong net magnetic field is associated with each domain. Domains with different orientations appear in different colors.

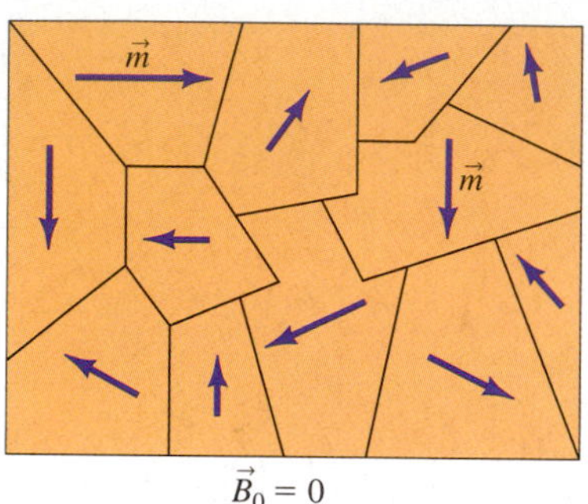

(a)

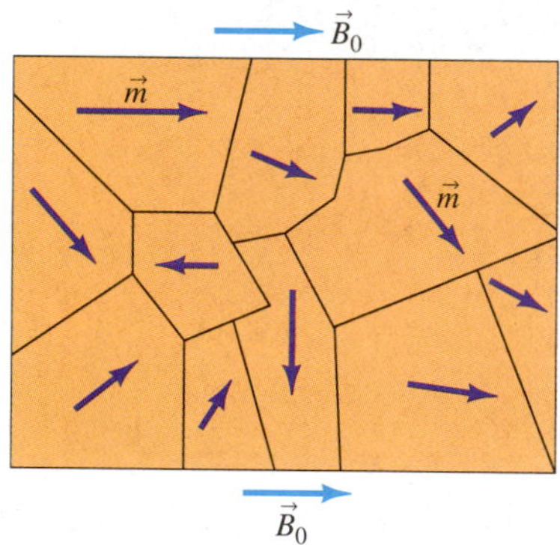

(b)

▲ **FIGURE 31–7** (a) Domain formation in ferromagnetic materials in the absence of an external magnetic field. The arrows indicate the magnetic moments of individual domains. (b) The presence of an external magnetic field influences the domains, making some larger and realigning others.

The intrinsic magnetic moments of the unpaired electrons of different atoms—for example, each iron atom has two such electrons—do not ordinarily become aligned *throughout* a piece of ferromagnetic material. Rather, the alignment takes place between adjacent atoms in regions called *magnetic domains*, which may contain 10^{17} to 10^{21} atoms and occupy a volume on the order of 10^{-12} to 10^{-8} m^3 (volumes from 0.1 mm to 1 mm on a side). The magnetic field within these domains is quite large, but a piece of ferromagnetic material may be made of thousands of such domains, and because each domain will have its magnetization aligned differently, the magnetization of the entire material will average to zero. Figure 31–6, a photograph of iron, clearly shows the boundaries (*domain walls*) between domains. Figure 31–7a is a schematic diagram of the domains showing their individual magnetic moments.

Given the description above, how does a piece of iron gain and maintain a significant magnetization? An external magnetic field $\vec{B}_0$ provides the mechanism that can align the magnetizations of different domains. This field acts in two ways. First, the size of domains with their magnetic moments already aligned with $\vec{B}_0$ may enlarge at the expense of neighboring domains. Second, the magnetic moments of some of the domains may rotate to the direction of $\vec{B}_0$ through an overall realignment of their constituents (Fig. 31–7b). Both of these mechanisms work because a state with a magnetic moment aligned along $\vec{B}_0$ is a state of lower energy.

The process we have described is a little like that of a marching band whose members face in random directions. The band leader orders them to face the same direction but fails to say *which* direction. Influenced by immediate neighbors, small groups of the band may align themselves together, but these groups may not be aligned with each other. If the band leader adds an instruction as to a precise direction, the groups will realign along that direction and there will be uniformity across the entire band.

When a magnet made of ferromagnetic material—we'll refer to this more simply as a ferromagnet—is heated, the increased movement of the atoms leads to a randomization of their orientation and thus to a decrease in the alignment. At the *Curie temperature*, T_c (after Pierre Curie), the randomization is complete, and the material loses its magnetization—it is no longer a ferromagnet. The value of T_c varies from material to material; in iron, $T_c = 1043\text{K}$, in gadolinium, $T_c = 292\text{K}$. Now we drop the temperature again. As T drops below T_c, ferromagnetism appears through a phase change, just as water forms the ordered lattice we know as ice below 273K. The ordering occurs in domains, as described by our marching-band analogy. Thus when the material cools below T_c, it will not act as a permanent magnet, because the domains are randomly oriented. Water supplies another physical example of this behavior. When a lake surface freezes, the freezing takes place in domains rather than as one huge ice crystal.

EXAMPLE 31–6 Estimate the maximum possible magnetization in a piece of iron.

Setting It Up To solve this problem, we require some knowledge about iron: We may know the intrinsic magnetic moment of a single electron, but we also need the knowledge that there are *two* unpaired electrons per iron atom, and both will contribute to the total magnetic dipole moment. In addition, to find the number of electrons per unit volume we also need the number of iron atoms per cubic meter; this is information that can come from the mass density of iron and the atomic weight of iron.

Strategy The maximum possible magnetization corresponds to having *all* the magnetic moments of the contributing electrons lined up—only one domain, with perfect alignment within the domain. The magnetization is then the number of contributing electron magnetic moments in a unit volume, times the individual electron magnetic moment $m_B = 9.3 \times 10^{-24}\ \text{A}\cdot\text{m}^2$ [Eq. (31–15)]. The number of iron atoms in a cubic meter of iron is

$$n_{Fe} = \frac{N_A \text{ atoms}}{1 \text{ mol}} \frac{1 \text{ mol}}{\text{mass of iron in 1 mol}} \times \frac{\text{mass of iron}}{\text{per unit volume}};$$

the number of unpaired electrons per atom is twice this value.

Working It Out The mass of iron in 1 mol is 56 g, and the mass density of iron—the third factor on the right in the expression for n_{Fe}—is 7.8 g per cm^3 = 7.8×10^6 g per m^3. Using $N_A = 6.02 \times 10^{23}$, we then find $n_{Fe} = 0.84 \times 10^{29}$ atoms per m^3, hence n = the number of unpaired electrons per m^3 = 1.7×10^{29}. This has to be multiplied by the magnetic moment per unpaired electron to get the maximum magnetization:

$$M_{max} = nm_B = (1.7 \times 10^{29} \text{ electrons/m}^3)(9.3 \times 10^{-24}\ \text{A}\cdot\text{m}^2 \text{ per electron}) = 1.6 \times 10^6 \text{ A/m}.$$

Compare this value to the magnetic dipole moment of the macroscopic loop in Example 31–3—the value found here is quite large by comparison.

What Do You Think? An experimental value of M_{max} for iron in a particular device is 1.7×10^4 A/m. What does this say about our calculation or about the piece of iron?

Hysteresis

The relation between the magnetic field B and the magnetic intensity H is more complicated in ferromagnets than in other materials. In order to measure the relation between B and H in a ferromagnetic material, that material is first demagnetized by heating. It is then cooled, shaped into a ring, and wound with a wire that carries a current I. This experimental arrangement is called a *Rowland ring* (Fig. 31–8). Without the ferromagnetic material, the magnetic field inside the ring, or toroidal solenoid, has the nearly constant value

$$B_0 = \mu_0 H = \mu_0 nI, \tag{31–17}$$

provided that the torus is "thin," like a bicycle tire. Here n is the number of windings per unit length. As we know, when the ferromagnetic material is inserted into the torus, the magnetic field increases tremendously to a new value B. We measure B by using a sense coil outside the torus (Fig. 31–8). The sense coil measures an induced emf proportional to the time rate of change of the magnetic field. As we raise the current in the toroidal coil at a given rate, we know the magnetic intensity $H = nI$, and the sense coil measures B. Figure 31–9 shows one example of a measured relation between H and B; a plot of H versus B is called a *magnetization curve*. Knowing H and B, we can determine the magnetization from the relation $B = \mu_0 H + \mu_0 M$.

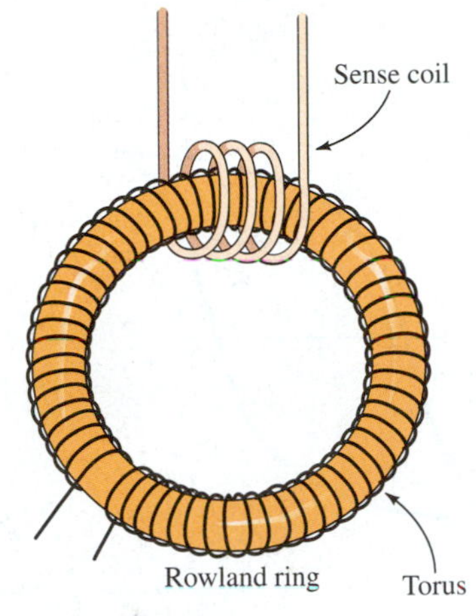

▲ **FIGURE 31–8** A Rowland ring is a wound core (solenoid) of material that may be used to measure the relation between $\vec{B}$ and $\vec{H}$ of that material. A sense coil measures changes in $\vec{B}$.

For ferromagnetic materials, we observe a magnetization curve like that shown in Fig. 31–9; this curve is known as a **hysteresis loop**, and it indicates the phenomenon of **hysteresis**. The presence of hysteresis demonstrates an irreversibility in the magnetization process. When the current I in the solenoid is slightly changed and then changed back again, the original magnetization is generally not recovered. For example, if we start on curve c in Fig. 31–9 at a value of 10^{-4} T for $\mu_0 H$, B in the ferromagnetic material is negative. If $\mu_0 H$ is increased to 3×10^{-4} T and then brought back to 10^{-4} T, B is now positive, following curve b. Hysteresis results from the fact that the magnetic domains do not return to their original zero-external-field status when the current decreases. They "remember" the rise in field and do not automatically revert to their original alignments.

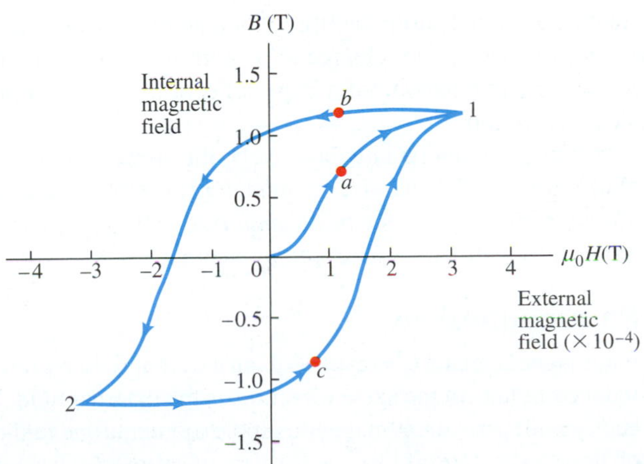

◀ **FIGURE 31–9** A magnetization curve illustrates the phenomenon of hysteresis in ferromagnetic materials. The material starts at the origin with zero magnetization. When a magnetic intensity H is applied, the material responds by becoming magnetic and is magnetic even when H is again zero.

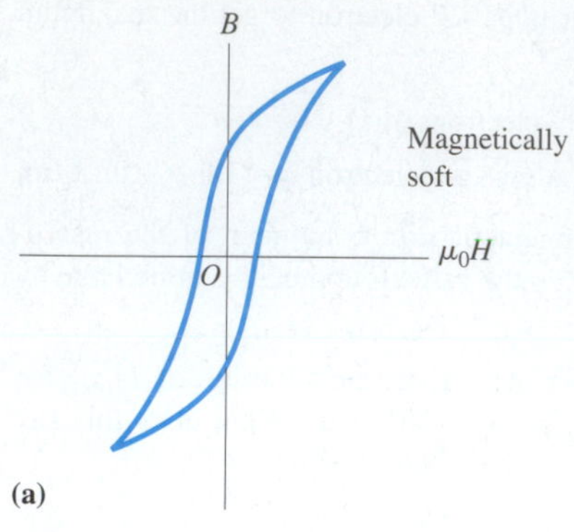

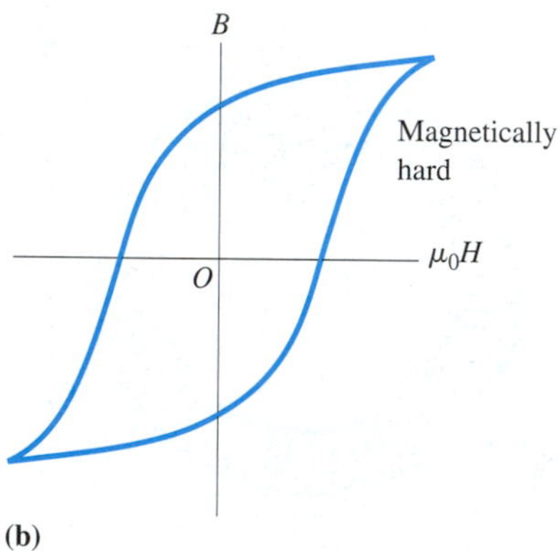

▲ **FIGURE 31–10** Hysteresis loops for materials that are (a) magnetically soft and (b) magnetically hard.

Some materials have narrow hysteresis loops, meaning that the alignment of the domains follows the external field rather closely (Fig. 31–10a). This is the type of curve that materials considered *magnetically soft*, such as properly prepared iron, will follow. Such materials are often used in transformer cores. (Transformers are devices that transform AC currents—or voltages—from one value to another.) Other materials have broad hysteresis loops, meaning that their domains respond only to large external fields (Fig. 31–10b). Such materials, including alloys of iron with carbon or tungsten, are said to be *magnetically hard*. They are difficult to magnetize but, once magnetized, they retain much of the magnetization and make good permanent magnets because they are equally difficult to demagnetize. Magnetically hard materials are especially important for making magnetic tapes or memory disks because such materials are stable against changes due to nearby magnetic fields (Fig. 31–11). Which materials are magnetically hard and which are magnetically soft is a complex question whose understanding demands a detailed understanding of atomic and solid state structure.

*31–4 Diamagnetism and Paramagnetism

In this section we'll examine the origin of the phenomena of *diamagnetism* and *paramagnetism*, both generally small but nevertheless significant manifestations of the way materials respond to external magnetic fields.

Diamagnetism

Diamagnetic materials respond to an external magnetic field $\vec{B}_0$ by reducing it slightly; the induced field is *opposite* to $\vec{B}_0$. The resulting net magnetic field is *less* than $\vec{B}_0$. Diamagnetism, as distinguished from ferromagnetism and paramagnetism, occurs as a primary effect in materials whose atoms have no permanent magnetic dipole moments, either orbital or intrinsic, and is also a secondary effect in any material. A classical model helps us to understand the phenomenon qualitatively. Let's consider two electrons with identical orbits, except that the motion is counterclockwise in one and clockwise in the other (Fig. 31–12a). With no external magnetic field, the orbital magnetic moments of the two electrons cancel ($\vec{m}_1 + \vec{m}_2 = 0$), and there is no magnetization.

Now suppose that an applied magnetic field $\vec{B}_0$ perpendicular to the orbits of the electrons is turned on (Fig. 31–12b). For the electron on the left of the figure, the flux through its orbit increases as the external field increases; by Lenz's law, the electron responds to counter the increasing flux. Accordingly, the negatively charged electron speeds up, increasing its angular momentum as well as the magnitude of its orbital magnetic moment $\vec{m}_1$, which points downward. When the external field levels off, angular momentum conservation ensures that the new value of $\vec{m}_1$ persists. Similarly, the electron on the right slows down to oppose the increase in flux through its orbit, so the magnitude of its magnetic moment $\vec{m}_2$, which points upward, is reduced. The result is that $\vec{m}_1 + \vec{m}_2$ now has a net value that points downward, and a magnetic field is produced that opposes the increasing external field. This is the origin of the negative magnetic susceptibility. The classical model must be revised for a proper quantum mechanical treatment of the atom because, in quantum mechanics, the angular momentum is *quantized*, and cannot change only slightly when the field changes slightly. Nevertheless, a correct treatment of a large collection of atoms reproduces the effect of the classical discussion.

When it is applied quantitatively, the simple model just described leads to reasonable estimates for the size of diamagnetic effects. The model correctly implies that *diamagnetism is present in all materials*, although it is masked for materials whose atoms have permanent magnetic moments.

▲ **FIGURE 31–11** Closeup view of the magnetic structure of recording media, here a hard disk. These consist of a series of magnetized regions laid out along grooves. Here the grooves are some 3 nm wide, and the storage density on this still-experimental medium would be 1.7 Gbit/in^2.

Paramagnetism

Paramagnetic materials respond to an external field by reinforcing the field slightly; the induced field is in the same direction as the external field. These materials contain molecules with permanent magnetic dipole moments due to the intrinsic magnetic moments of unpaired electrons. In the absence of an external magnetic field, these dipoles are

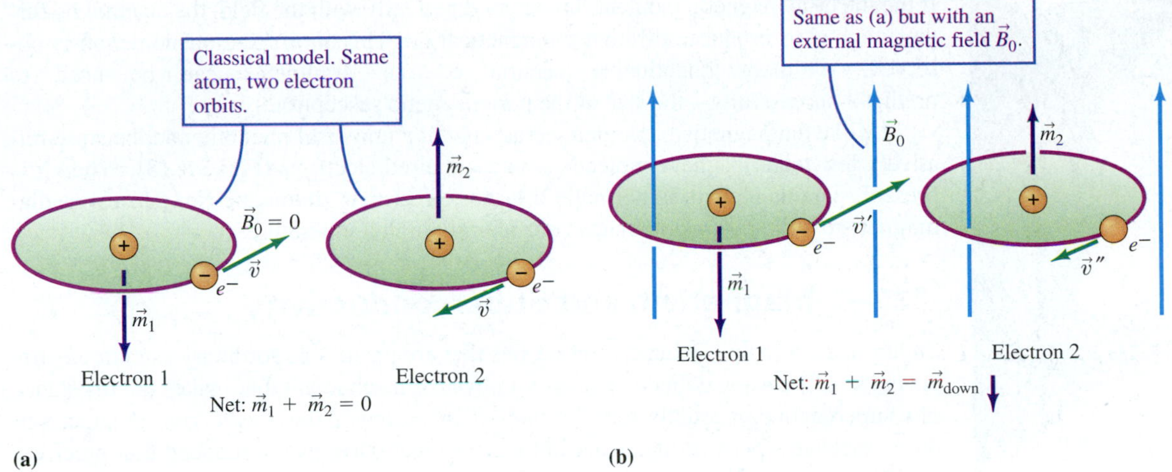

▲ **FIGURE 31–12** (a) A two-electron atom with a net orbital magnetic moment of zero in the absence of an applied magnetic field. (b) In the presence of an applied magnetic field, the orbital magnetic moment associated with each electron is changed, and there is a net magnetic moment that points down, opposite to the external field.

randomly oriented due to thermal motion, and the net magnetization of the materials is zero. Recall from Chapter 28 that the energy of a magnetic dipole moment $\vec{m}$ in a magnetic field $\vec{B}$ is, by Eq. (28–26), $U = -\vec{m} \cdot \vec{B}$. The lowest energy occurs when $\vec{m}$ and $\vec{B}$ are *parallel*. Thus an external magnetic field $\vec{B}$ tends to align the atomic magnetic moments along $\vec{B}$ and produces a positive magnetic susceptibility.

Two effects determine the extent to which the permanent magnetic dipoles become aligned. The first is the external field, which by the energy argument above encourages alignment, and the second is the thermal motion, which randomizes the alignment. The relative importance of these two factors is measured by the relative size of the magnetic energy factor mB and the thermal energy factor kT, where T is temperature. If T is so large that $kT \gg mB$, the average alignment over a large number of electrons will be weak. Conversely, if T is so low that $kT \ll mB$, the average alignment will be strong. For intermediate temperatures and fields, the average alignment is proportional to the ratio of these energies, i.e. to $(mB)/(kT)$. At room temperature, the intrinsic magnetic moments of most paramagnetic materials are only very slightly aligned, but as we already saw in Example 31–4, large bulk effects can come from very small alignments. In 1895, Pierre Curie observed the linear relation that we now call *Curie's law*:

$$\vec{M} = C\frac{\vec{B}}{T}, \tag{31–18}$$

where C is *Curie's constant*, a quantity that varies from one material to another. This law, which holds best for small values of mB/kT, is often expressed in terms of magnetic susceptibility, defined according to Eq. (31–5) as $\vec{M} = \chi_m \vec{H}$. If we anticipate that the susceptibility will be small, as it is for paramagnetic materials, then we can replace $\vec{B}$ in Eq. (31–18) by $\mu_0 \vec{H}$:

$$\vec{M} = C\frac{\mu_0 \vec{H}}{T},$$

or

$$\chi_m = \frac{\mu_0 C}{T}. \tag{31–19}$$

The value of C depends on the material. The susceptibility is positive, which is characteristic of paramagnetism.

The temperature dependence in Eq. (31–18) is the same as that of the analogous phenomenon for dielectrics [see Eq. (25–30)]. There, too, this dependence is called *Curie's law*. We expect the law to fail at sufficiently low temperatures and/or large fields.

▲ **FIGURE 31–13** Oxygen is paramagnetic and is therefore attracted by the poles of a magnet. Here, liquid oxygen poured between two poles is held in place by the forces between it and the permanent magnet.

If the intrinsic magnetic moments are aligned perfectly with the field, they cannot be further aligned to produce still higher magnetization. This *saturation* phenomenon is observed. A more quantitative version of our arguments can be used to predict—successfully—the size of the paramagnetic susceptibility.

Unlike diamagnetism, paramagnetism is not a universal phenomenon, because relatively few materials have molecules with unpaired electrons (Fig. 31–13). When it is present, paramagnetism is normally a larger effect than diamagnetism. However, diamagnetism will always dominate at sufficiently high temperatures.

*31–5 Magnetism and Superconductivity

Superconductors have magnetic properties that are just as extraordinary as their electric properties. The same collective quantum physical mechanism that makes the resistance of a superconductor exactly zero *also makes the magnetic field inside zero*. A superconductor acts as a perfect diamagnet in the sense that currents are induced that precisely cancel any magnetic field inside. Alternatively, we say that the magnetic field lines are *expelled* from the superconductor—a phenomenon known as the *Meissner effect*. In fact, superconductors come in two types according to just how the field is expelled: In Type I superconductors, the field is expelled entirely (Fig. 31–14). In Type II superconductors, the field is isolated in nonsuperconducting filamentary structures within the material (Fig. 31–15). Currents circulate on the surfaces of these filaments, shielding the rest of the material from the magnetic field within the filaments. Quantum physics sets a minimum amount of magnetic flux within each filament, and the flux found within any given filament is an integer multiple of this minimum amount. This is yet another example of how physical quantities are quantized in a quantum-mechanical world.

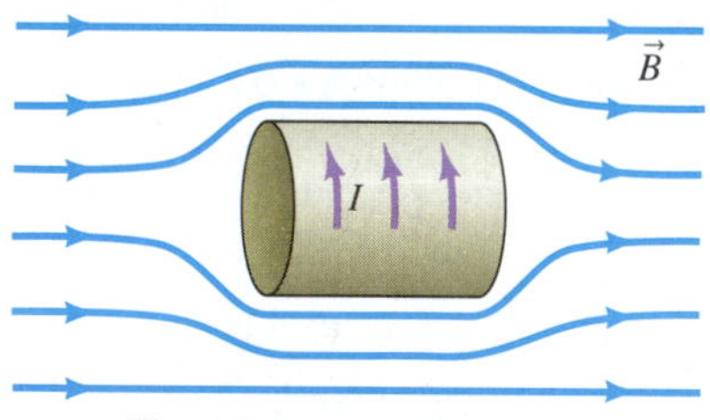

Type I superconductor

▲ **FIGURE 31–14** A Type I superconductor expels magnetic field from its interior by acting as a perfect diamagnet: Surface currents that just cancel the applied field inside are established.

The expulsion of magnetic fields from the interior of a superconductor, whether Type I or II, translates into a statement about its magnetic susceptibility. Equation (31–6) expresses the internal field in terms of the magnetic intensity $\vec{H}$, and this internal field must be zero:

$$\vec{B} = \mu_0(1 + \chi_m)\vec{H} = 0.$$

The magnetic intensity is due to the free currents and is not zero. Thus $1 + \chi_m = 0$, or

$$\chi_m = -1.$$

If there is no magnetic field inside a superconductor, then there can be no currents inside. To see this, imagine that some internal region of a superconductor carries current. Then we can draw a loop around this region and apply Ampère's law. If there were a current through the loop, there would be a magnetic field, but there is no such field. We conclude that *all current carried by superconductors must be carried on their surfaces* (any boundary between superconducting and nonsuperconducting phases). For example, current can be carried on the walls of the filaments in Fig. 31–15.

In the presence of a large enough magnetic field (a *critical field*), a material in a superconducting phase jumps back to the normal (nonsuperconducting) phase even at a fixed temperature, a phenomenon that can cause serious practical problems. For example, many large electromagnets are made from superconductors because they do not undergo Joule heating in spite of the large currents in them. However, the large magnetic field may itself destroy the superconductivity, and one is left with a lot of current in nonsuperconducting wires and a potential meltdown from Joule heating. To overcome this problem, Type II superconductors are generally a better choice for superconducting wires in the electromagnets, because they channel the magnetic field into filaments, providing a way to make superconductors with much higher critical fields.

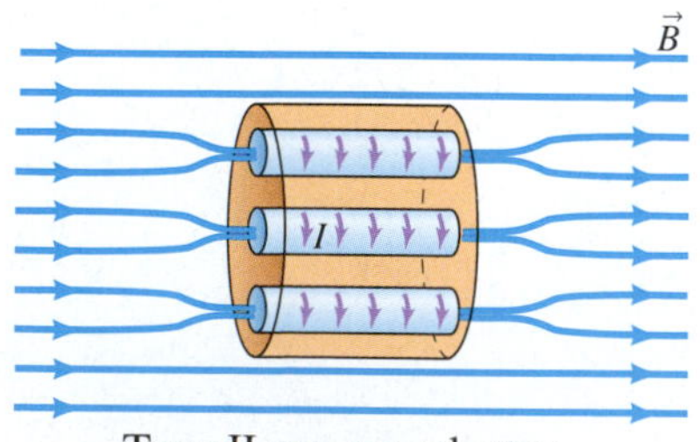

Type II superconductor

▲ **FIGURE 31–15** In Type II superconductors, the magnetic field is confined to filamentary structures. Inside the filaments, the material is not in its superconducting phase.

*31–6 Nuclear Magnetic Resonance

Atomic nuclei consist of protons and neutrons, far more massive than electrons. Because it is so massive, the orbital motion of the nucleus itself about the atomic center of mass is negligible, and the small size of the nucleus allows us to ignore orbital motions of the protons and neutrons within the nucleus. However, protons and neutrons, like electrons, have intrinsic magnetic dipole moments. These magnetic moments are some 2000 times smaller than that of the electron because protons and neutrons are some 2000 times more massive than electrons [see Eq. (31–15)]. They arise from an intrinsic angular momentum called the *spin* and labeled $\vec{S}$. Unlike the orbital angular momentum, which according to quantum mechanics can occur only in integer multiples of $\hbar$ (Planck's constant, h, divided by 2π), the electron or proton spin has the value $\frac{1}{2}\hbar$. If the proton magnetic moment is labeled $\vec{m}_p$, then as for the orbital angular momentum [see Eq. (31–12)], there is a gyromagnetic ratio g_p that gives the relation between the spin and the magnetic moment,

$$\vec{m}_p \equiv g_p\vec{S}. \tag{31–20}$$

What is the effect of an external magnetic field $\vec{B}$ on the proton? Consider the torque on it due to such a field, given by $\vec{\tau} = \vec{m}_p \times \vec{B}$. Now, the torque on the proton is the rate of change of its internal angular momentum, $d\vec{S}/dt$, so with $\vec{S} = \vec{m}_p/g_p$ we have

$$\frac{1}{g_p}\frac{d\vec{m}_p}{dt} = \vec{m}_p \times \vec{B}. \tag{31–21}$$

This is an equation that gives the rate of change of the magnetic moment of a proton in a magnetic field. As the analysis of this equation in Chapter 10 showed, the magnitude of $\vec{m}_p$ cannot change, but its direction can, and Eq. (31–21) describes the *precessional* motion of $\vec{m}_p$ about the direction of $\vec{B}$ (Fig. 31–16). This precession, called *Larmor precession*, is analogous to the precession of a spinning top under the influence of gravity (see Section 10–8). In a classical picture of Larmor precession, the direction of $\vec{m}_p$ traces a cone around the direction of $\vec{B}$, as in Fig. 31–16. However, quantum mechanics tells us that the analog of only two such cones are allowed: one with spin "up" and one with spin "down." In Problem 41, we find the angular speed of classical precession,

$$\omega_0 = g_pB. \tag{31–22}$$

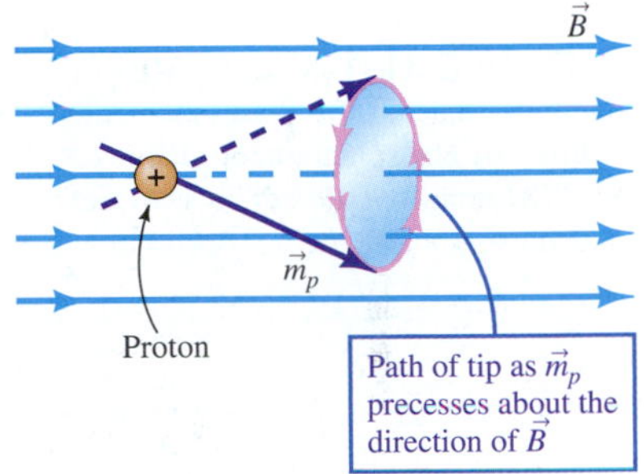

▲ **FIGURE 31–16** The magnetic moment of a proton, $\vec{m}_p$, precesses about the direction of an external magnetic field with an angular speed ω_0.

Finally, let's recall one other feature of this motion: There is a potential energy associated with a magnetic moment $\vec{m}$ in an external field, given by $U = -\vec{m}\cdot\vec{B}$, and the motion with the "up" cone has lower energy than does the motion with the "down" cone.

Now consider exposing the atom to an oscillating magnetic field. (Such a field occurs in electromagnetic waves; see Chapter 34.) This field would normally not have much effect, but when the angular frequency ω of the oscillating field *exactly matches* the angular frequency ω_0 of the precession (a condition known as *resonance* and described in Section 13–8), the direction of the magnetic moment can flip. In this case, the oscillating magnetic field either supplies *just the precise amount of energy*—$2\vec{m}_p\cdot\vec{B}$—necessary to flip the spin of the proton from up to down, or absorbs exactly this amount of energy to flip the spin from down to up.

This effect is called **nuclear magnetic resonance** (**NMR**). When the spin flips due to a transfer of energy between the oscillating field and the proton, there is a detectable signal. Thus, by tuning the frequency of the oscillating magnetic field, we can measure the frequency $\omega_0 = g_pB$ with very high precision.

If the external magnetic field is known, the NMR method may be used to measure the gyromagnetic ratio, g_p. It is in this way that the gyromagnetic ratios for protons and neutrons are known to contain the coefficients 2.79 and -1.91, respectively, multiplying an expression like Eq. (31–13) (but with the proton or neutron mass in place of the electron mass). These coefficients suggest that the internal structures of protons and neutrons are more complicated than that of electrons. NMR measurements can also be extended to nuclei, where they are used to study nuclear structure and the forces that give rise to it. For example, the deuterium nucleus—which may be described as a bound state of a proton and neutron structured in such a way that the intrinsic spins, and therefore the intrinsic magnetic moments, are parallel to each other—is expected to have a magnetic moment that is the sum of the moments of the proton and the neutron. NMR measurements give a slightly smaller result than this sum. It is possible to conclude from this that the deuteron is slightly cigar-shaped, and this gives us information about the forces that bind it.

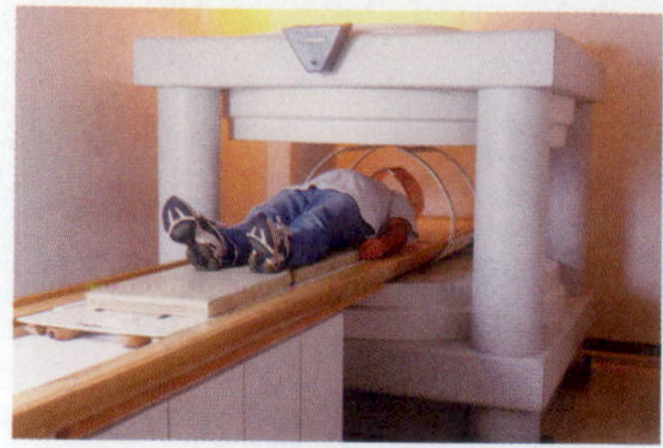

(a)

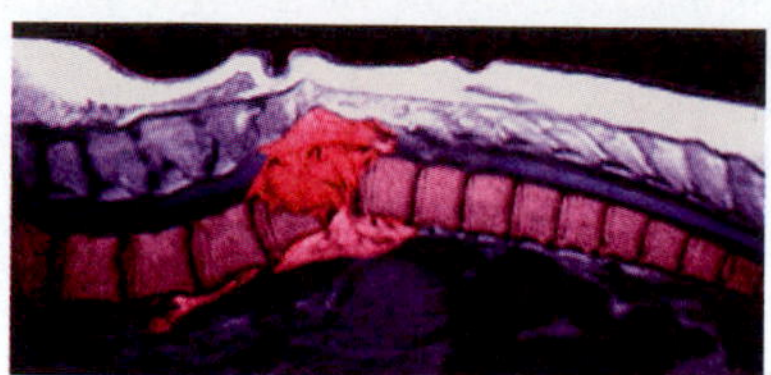

(b)

▲ **FIGURE 31–17** (a) An MRI diagnostic machine, as used in a medical facility. (b) The spinal tumor visible on this MRI image would not be as evident in a conventional X-ray.

THINK ABOUT THIS. . .

WHAT IS MAGNETIC RESONANCE IMAGING?

NMR has important applications in the study of materials and in medical diagnostics, where the procedure is called *magnetic resonance imaging* (MRI). (The word "nuclear" was dropped due to patients' fears that nuclear radiation was being used. In fact, whatever it is called, it is an especially safe procedure.) One knows the value g_p associated with the nucleus of hydrogen, and by placing a complex combination of materials such as a human body within a field whose values vary with position, one can tell by setting up the conditions for spin flip and hence absorption of energy from the field just where there are large numbers of hydrogen atoms—only at locations where the *B*-field has just the right value will the spin flip occur. In this way MRI locates concentrations of hydrogen atoms in patients. Fat, which has a high concentration of hydrogen, can be distinguished from muscle, which has a much lower hydrogen concentration, and tumors can be distinguished from nerve tissue. Bones have little hydrogen and are hardly seen at all (Fig. 31–17). Before MRI it was difficult to image soft tissues, which don't show in X-rays.

Summary

The magnetic properties of bulk matter are summarized in the magnetization, $\vec{M}$, the magnetic dipole moment per unit volume. In the presence of an external **magnetic field** $\vec{B}_0$, there is a total magnetic field in a material given by

$$\vec{B} = \vec{B}_0 + \mu_0\vec{M}. \tag{31–1}$$

The effect of free (real) currents (as opposed to the induced atomic effects) is contained in the magnetic intensity, $\vec{H} = \vec{B}_0/\mu_0$:

$$\vec{H} \equiv \frac{\vec{B}}{\mu_0} - \vec{M}. \tag{31–2}$$

The magnetic susceptibility χ_m describes the response of a material to a magnetic field of external origin:

$$\vec{M} \equiv \chi_m\vec{H}. \tag{31–5}$$

The net magnetic field is given in terms of χ_m by

$$\vec{B} = \mu_0(1 + \chi_m)\vec{H} = \mu\vec{H}. \tag{31–6, 31–8}$$

Here, μ is the permeability of the material:

$$\mu = \mu_0(1 + \chi_m). \tag{31–7}$$

Magnetism in matter is due ultimately to the magnetism of its atomic constituents and particularly to the unpaired electrons of atoms. An orbiting electron produces an atomic orbital magnetic moment

$$\vec{m}_{\text{orbital}} = g_L\vec{L}, \tag{31–12}$$

where g_L is the gyromagnetic ratio. Quantum mechanics implies that these magnetic moments take the value

$$m_{\text{orbital}} = \left(\frac{e\hbar}{2m_e}\right)\ell \equiv m_B\ell, \tag{31–14}$$

where the factor m_B is the Bohr magneton and ℓ is an integer. In addition, electrons have intrinsic magnetic moments equal in magnitude to m_B. Even a very slight alignment of atomic magnetic moments leads to large magnetic effects in bulk matter.

Ferromagnetic materials have large permeabilities. The atomic dipole moments are lined up in small regions called domains due to forces of quantum mechanical origin. The imposition of an external field leads to the dipole moments of the domains lining up together and produces permanent magnets. The fact that a ferromagnetic material "remembers" the orientation of the external field that magnetizes it leads to the phenomenon of hysteresis, in which the magnetization curve depends on how the magnetization was produced.

Diamagnetic materials have small negative susceptibilities that are ultimately due to Faraday's law. Diamagnetism is always present but may be masked by paramagnetic or ferromagnetic effects. Paramagnetic materials have small positive susceptibilities due to the intrinsic magnetic moments of unpaired electrons, which find it energetically favorable to line up with an external field. Paramagnetism is strongly temperature dependent. As well as having no measurable resistance, superconductors also have some remarkable magnetic properties; namely, they completely expel magnetic field from their superconducting regions.

In nuclear magnetic resonance (NMR), the intrinsic magnetic moments of nuclei and nuclear constituents precess about an applied magnetic field. This precession is detected by the response of a material to the imposition of an electromagnetic wave of just the right frequency—a frequency that is characteristic of the material.

Understanding the Concepts

1. When we calculated the magnetic dipole moment associated with orbital motion, why was it reasonable to think of the nucleus as stationary and the electron as circulating around it?
2. When an electron orbits the nucleus in a planetary model, the system forms an electric dipole. Why does this electric dipole not produce a measurable electric dipole field around the atom?
3. Devise an experiment that would determine whether a certain material was diamagnetic or paramagnetic.
4. Under what circumstances will Gauss' law for the magnetic field also hold for the magnetic intensity?
5. In a Rowland ring measurement of the magnetic field inside a piece of magnetic material, is it helpful to wrap the sense coil around the material many times?
6. Why would speaker coils and computer disks use magnetically hard material, while transformers and computer read/write heads use magnetically soft material?
7. Does iron exhibit diamagnetic properties? How could you determine them?
8. Aluminum is separated in junk yards by using large magnets. How is this possible?
9. Should the magnetic latch on a refrigerator door be made from magnetically hard or soft material?
10. Why should computer floppy disks not be made from magnetically soft material?
11. Why does diamagnetism dominate over paramagnetism in most materials at sufficiently high temperatures?
12. Explain how a permanent bar magnet attracts an unmagnetized iron needle.
13. Will a paramagnetic sample be attracted or repelled from a region of increasing magnetic field?
14. You are given two identical iron rods—one magnetized, the other not. How can you determine which is the magnet, without using a third magnet (for example, Earth)?
15. Suppose that an electron in a circular orbit around a nucleus is placed in an external magnetic field. Will the angular momentum of the electron change if the field is aligned perpendicular to the plane of motion? Parallel to the plane of motion?
16. Is it possible to arrange for a classical current loop to have a magnetic moment but no angular momentum? Assume first that you have both positive and negative charge carriers to work with, and then that you have only negative ones.
17. What is the value of $\vec{H}$ in an isolated permanent magnet?
18. It takes an external field to establish a macroscopic magnetization inside a permanent magnet cooled below its Curie temperature. What could have done this for lodestones, which are permanent magnets found in nature?
19. In a uniform magnetic field, a magnetic dipole experiences no net force, only a torque. How do two bar magnets repel or attract each other?
20. Suppose you are given the magnetic moment due to a single charge orbiting in a circle about a center. Describe qualitatively how you would calculate the magnetic moment of (a) a uniformly charged ring rotating with angular momentum perpendicular to its plane; (b) a uniformly charged disk rotating with angular momentum perpendicular to its plane; and (c) a uniformly charged sphere rotating about an axis through its center.

Problems

31–1 The Magnetic Properties of Bulk Matter

1. (I) A cylindrical rod of palladium (magnetic susceptibility $\chi_m = 8 \times 10^{-4}$), of radius 1 cm and length 5 cm, is placed in and aligned with a uniform magnetic field of 1.0 T. What is the magnetic dipole moment of the rod?

2. (I) A thin, toroidal coil of total length 34 cm is wound with 1600 turns of wire. A current of 0.62 A flows through the wire. What is the magnitude of $\vec{B}$ inside the torus if the core consists of a ferromagnetic material of magnetic susceptibility $\chi_m = 2.8 \times 10^3$? What is the magnitude of $\vec{H}$?

3. (I) The coil of a solenoid wound with a turn density of 3400 turns/m is filled with a material of unknown magnetic susceptibility χ_m. When the wire carries 0.450 A, the magnetic field within is 1.907×10^{-4} T. What is χ_m?

4. (I) A solenoid magnet wound with a turn density of 1300 turns/m, with permalloy inserted inside the windings, has a magnetic field of 3.5 T inside. How much current flows in the windings?

5. (I) A permalloy magnet is 5 cm in diameter, 30 cm long, and has magnetic intensity $\vec{H} = 30$ A/m at its pole. How many turns/m must an empty solenoid of the same dimensions have to give rise to the same intensity if it carries a current of 6 A?

6. (I) What is the magnetic moment of the equivalent solenoid in Problem 5?

7. (I) Earth's magnetic moment is about 10^{23} A$\cdot$m^2. If the core, which is responsible for the magnetic moment, is about 20% of Earth's volume, what would be the core's magnetization? Assume the core is spherical.

8. (I) The core of a solenoid wound with 400 turns/m is filled with some material. When the current is $I = 3.1$ A, the magnetic field inside the core is measured to be 0.20 T. What is the magnetic susceptibility of the material placed in the core?

9. (I) A toroidal solenoid has a mean outer radius of 8.7 cm with 487 turns of wire. How much current is required to produce a magnetic field of 0.40 T if the solenoid is filled with permalloy?

10. (I) A toroid of core radius of 1.5 cm and toroidal radius of 20 cm is wound with silver wire. There are 1200 turns of wire and a current of 8 A flows through the wire. What is the magnetization per turn of wire?

11. (II) In a vacuum, a solenoid with a current I has a magnetic field B_0. (a) If silver is placed inside the solenoid, what is the change in the magnetic field? (b) What happens if cupric oxide is placed inside the solenoid?

12. (II) A 0.80-cm^3 cube of copper is placed between the poles of a magnet with a magnetic field of 0.48 T. What is the induced magnetization in the copper?

13. (II) A long solenoid filled with ferromagnetic material of permeability $\mu = 1320\mu_0$ is wound with wire so that there are 15 turns per cm. What current must flow through the wire to produce a magnetic field of 0.16 T within the solenoid?

31–2 Atoms as Magnets

14. (I) Suppose that 1 mol of atoms in a material have individual magnetic moments of $1.8 \times 10^{-23}\ \mathrm{A \cdot m^2}$. In the absence of any alignment, the magnetic moments form an *average* angle of 90° with some external axis. By how much does the average angle differ from 90° if the material has the same magnetic moment as a 1-cm^2 loop of wire that carries a current of 0.3 A? (The magnetic moment of the loop is aligned with the external axis.) Assume that the components of the atomic magnetic moments add algebraically.

15. (I) The atomic number of iron is 26, its atomic weight is 55.8, and its density is 7.87 g/cm^3. (a) How many electrons are there in 10 cm^3 of iron? (b) Suppose that each electron has the magnetic moment estimated in Example 31–3 ($5 \times 10^{-24}\ \mathrm{A \cdot m^2}$), and that the magnetic moments "up" and "down" make up the fractions $\frac{1}{2}(1 + 2 \times 10^{-7})$ and $\frac{1}{2}(1 - 2 \times 10^{-7})$, respectively. What is the magnetization of the iron?

16. (II) Consider an electron in a circular orbit around a single proton (a hydrogen nucleus) whose total energy is −13.5 eV. Find the value of the orbital magnetic moment.

17. (II) Consider a ring of radius R rotating about the axis through its center and perpendicular to the plane of the ring with an angular velocity ω. It carries a uniform charge, with charge density λ C/m. What is the magnetic moment of the ring?

18. (II) A disk of radius R rotates with angular velocity ω about an axis through its center, perpendicular to the surface of the disk. The disk carries a uniform charge, with charge density σ C/m^2. What is the magnetic moment of the disk?

19. (II) The electron has a *classical radius* given by $r_0 \equiv e^2/4\pi\varepsilon_0 m_e c^2 = 2.8 \times 10^{-15}$ m. This quantity is suggested by dimensional analysis: The particular combination of classical quantities is the only one that can be formed with dimensions of length. Use Eq. (31–11), with m_{orbital} equal to the Bohr magneton, m_B, to show that any charge at the distance of the classical radius will be moving faster than the speed of light. (Assume that all the charge is concentrated at a belt of radius r_0.) Treatment of the magnetic moment of an electron as a classical quantity leads to trouble!

20. (II) The current I in a circular loop of radius R is due to the flow of free electrons in the loop. Show that the gyromagnetic ratio of the loop is independent of I, R, and the density of atoms.

31–3 Ferromagnetism

21. (I) A torus is wound with 800 turns/m of wire. A current of 5 A runs through the wire. If the core of the torus is iron, the internal magnetic field in the core is 1.8 T. What is the magnetization? What is the value of μ/μ_0 for the iron core?

22. (I) An electromagnet with a ferromagnetic core, $\chi_m = 15{,}000$, produces a maximum magnetic field of 0.36 T. What is the maximum current carried by the coil if the turn density of the coils is 14 turns/cm?

23. (I) A long, tightly wound solenoid contains a magnetic field of magnitude $B = 2.4 \times 10^{-3}$ T. An iron core, with susceptibility χ_m, is inserted so that it fills the space inside. What is the new value of B?

24. (II) We may view the iron core in the previous problem as equivalent to another solenoid that is concentric with the outer one and essentially coincident with it. How could you view the situation if the iron core were quite a bit shorter than the length of the solenoid? If the radius of the iron core were smaller than that of the solenoid? Draw magnetic field lines for these cases.

25. (II) A current of 0.5 A flows through a solenoid with 400 turns/m. An iron bar, with $\mu/\mu_0 = 640$, is placed along the solenoid axis (Fig. 31–18). (a) What is the magnetic field inside the iron bar? (b) Outside the iron bar, but still within the solenoid?

▲ **FIGURE 31–18** Problem 25.

26. (II) A disk-shaped permanent magnet has a thickness of 3 mm and a diameter of 1.5 cm. It is magnetized perpendicular to the plane of the disk, with the magnetic field on the axis near its north pole of magnitude 0.05 T. What is the current carried by a 80-turn coil of the same dimensions that gives this same value of the magnetic field on the axis (Fig. 31–19)?

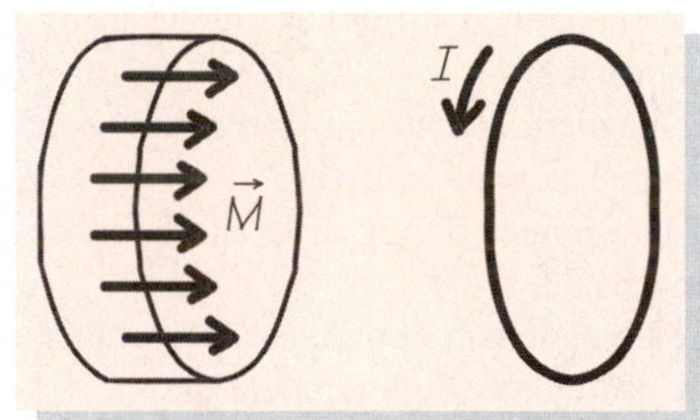

▲ **FIGURE 31–19** Problem 26.

27. (II) A current of 0.16 A is carried by a 50-turn coil 5.5 cm in diameter and 1.0 cm in length. Suppose a piece of iron with susceptibility 4.8×10^3 is placed inside the coil. What is the magnetic field inside the iron? What is the magnetic intensity there?

28. (II) A torus with a central radius of 25 cm and a tube radius of 2.0 cm is filled with iron of permeability $2800\mu_0$ (Fig. 31–20). There are 1200 turns around the torus. How much current must flow in the winding coil to produce a magnetic field of 1.5 T inside the torus? Treat the torus as having a constant magnetic field equal to the field at the central radius of the torus.

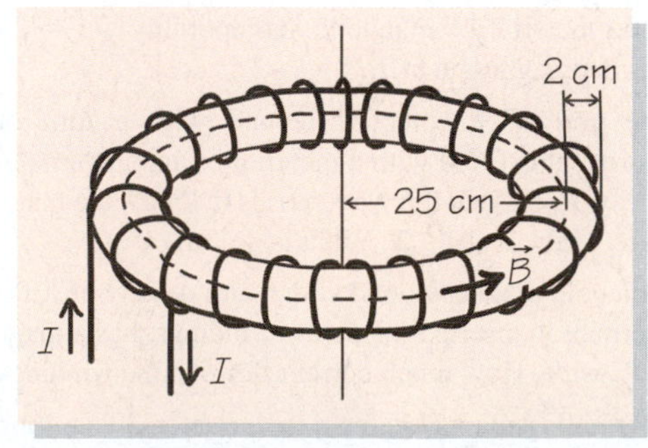

▲ **FIGURE 31–20** Problem 28.

29. (II) Iron (susceptibility = 6000) is used as the core of a transformer (see Chapter 33 for a discussion of this device). If large distortions of current are to be avoided, the transformer must be designed in such a way that the proportionality between $\vec{B}$ and $\vec{H}$ applies even for the strongest field. Estimate the maximum allowable value of H, knowing that the magnetic moment of individual iron atoms is 2.2 m_B.

30. (II) A Rowland ring measures the charge Q that passes through a sense coil by integrating the current in the sense coil over time (see Fig. 31–8). Both the sense coil and primary coil are wrapped tightly around a material. These coils have an area A. The number of turns in the sense coil is N. The emf induced in the sense coil by a changing magnetic flux in the material (due to a switch that passes current through the primary coil when the switch closes) is $\mathscr{E}$, and the sense coil has resistance R. Obtain a relation between the change in magnetic field, ΔB, and the charge Q. [*Hint*: Remember that $I = dQ/dt$.]

31. (II) A sense coil with a resistance of 0.1 Ω is wrapped tightly in 40 turns around a magnetic material of area 0.02 m^2. When a switch is closed in the primary coil, a charge of 5 mC flows through the sense coil (Fig. 31–21). If the magnetic field was initially zero, what is the new magnetic field in the material? (See Problem 30.)

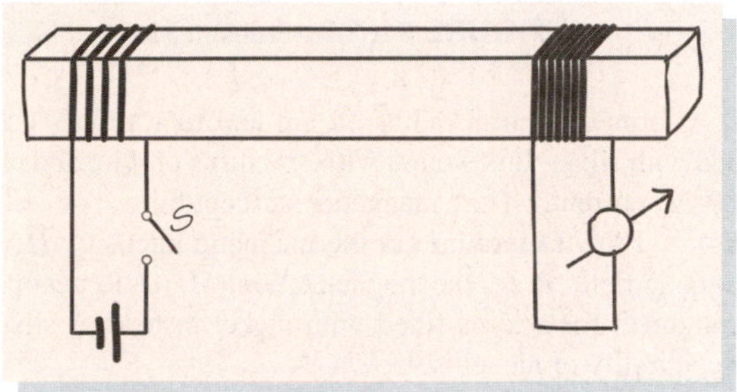

▲ **FIGURE 31–21** Problem 31.

*31–4 Diamagnetism and Paramagnetism

32. (III) An electron under the influence of some central force moves at speed v_i in a counterclockwise circular orbit of radius R. A uniform magnetic field $\vec{B}$ perpendicular to the plane of the orbit is turned on (Fig. 31–22). Suppose that the magnitude of the field changes at a given rate dB/dt. (a) What are the magnitude and direction of the electric field induced at the radius of the electron orbit? (b) The tangential force on the electron due to the induced electric field increases the electron's speed. Find the value of dv/dt. (c) Assuming that the initial orbital speed was v_i, find the final speed v_f as the magnitude of the magnetic field steadily increases from zero to a final value B_f by integrating dv/dt with respect to time. (d) Using your result for the change in speed, find the change in orbital angular momentum. (e) Use Eq. (31–11) to relate a change in the orbital magnetic moment to the change in the angular momentum.

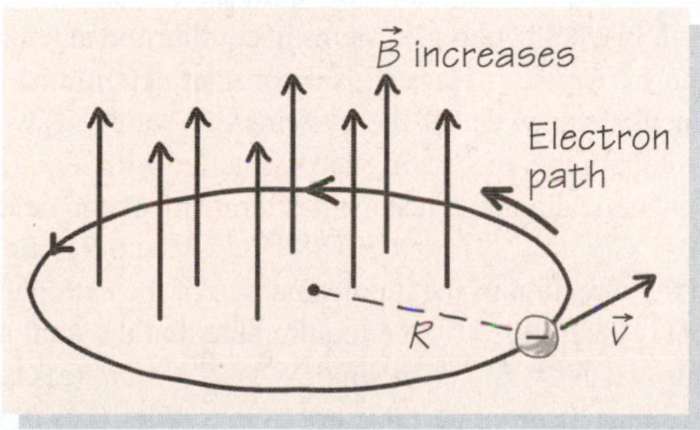

▲ **FIGURE 31–22** Problems 32 and 33.

33. (III) Refer to Fig. 31–22, but this time assume that the electron circulates clockwise rather than counterclockwise at speed v_i. By applying the same sequence of steps, show that the change in the magnetic moment of the electron's orbit is opposite the direction of change in the external field, just as in the case in which the electron circulates counterclockwise.

34. (III) Refer to Problems 32 and 33. Suppose that there are now two electrons moving at speed v_i in circular orbits of radius R, one clockwise and one counterclockwise. (a) What is the net orbital magnetic moment when the external field is zero? (b) After the external field has reached $\vec{B}_f$? (c) Show that the magnetic susceptibility for this system is $\chi_m = -(\mu_0 e^2 R^2/4m_e)\rho_e$, where ρ_e is the electron density.

35. (III) Using the techniques of Problems 32 through 34, estimate the magnetic susceptibility of copper, which has 29 electrons per atom. Assume that all the electrons move in orbits of the same radius, and that 14 move clockwise while 15 move counterclockwise. You will need to calculate the number density of electrons in copper.

36. (II) The temperature of a sample of $FeCl_3$ (the ferric ions have an intrinsic magnetic moment) inside a magnetic field is held constant as the field is increased. Sketch the induced magnetic moment as the magnetic field is increased.

37. (II) A long, straight conducting wire is embedded within an insulating paramagnetic material of magnetic susceptibility 2.6×10^{-4} at 300K and carries a current of 10 mA. Find the value of the magnetic intensity as a function of the distance from the wire, as well as the magnetic field. What is the change in the magnetic field when the temperature is lowered to 86K?

*31–6 Nuclear Magnetic Resonance

38. (I) Find the magnetic moment of the neutron, given that its gyromagnetic ratio is $-3.82e/2m_n$.

39. (II) Assume that it is possible to align perfectly the magnetic moments of protons in 1 mol of hydrogen gas at standard temperature and pressure. What are the magnetization and magnetic field inside the gas?

40. (II) In ^{17}O (oxygen with 17 nucleons in its nucleus) the nuclear magnetic moment is -9.54×10^{-27} A·m^2. The atomic electrons are lined up in such a way that they make no contribution to the magnetic moment of the atom. Suppose it were possible to align the oxygen atoms such that 50.05 percent pointed in one direction and 49.95 percent pointed in the opposite direction. What would be the magnetization of 1 mol of ^{17}O gas under those conditions at standard temperature and pressure?

41. (II) Express the equation $d\vec{m}/dt = g_p\vec{m} \times \vec{B}$ relevant to NMR in component form for the case that $\vec{B} = B\hat{k}$ and $\vec{m} = m_x\hat{i} + m_y\hat{j} + m_z\hat{k}$. (a) Show that m_z is a constant; (b) that $m_x^2 + m_y^2 + m_z^2$ is a constant; and (c) that $m_x = m_1\cos(\omega t)$ and $m_y = -m_1\sin(\omega t)$ satisfy the equation of motion, where ω is *the angular frequency of precession.*

42. (II) Archaeological objects are often located by detecting their minute influence on Earth's magnetic field. The equipment used for such measurements is the *proton magnetometer*, which measures the intensity of the magnetic field by measuring the angular frequency of protons in that field. Determine the angular frequency of protons in Earth's field at a typical location ($B = 80\ \mu$T) and the change in frequency caused by a change $\Delta B = 12$ nT.

43. (II) Calculate the frequency of precession (see Problem 42) for a proton's magnetic moment in a field of 10^{-1} T. This frequency is in the so-called rf (radio-frequency) range.

44. (II) Gauss' law for magnetism,

$$\int_{\text{closed surface}} \vec{B} \cdot d\vec{A} = 0,$$

remains unchanged when materials are present because materials do not give rise to magnetic monopoles. Use this law to show that the magnetic field $\vec{B}$ does not change at the interface of two materials if the interface is perpendicular to the direction of the field. [*Hint*: Recall the method used to derive the electric field due to a surface distribution of charges.]

45. (II) If magnetic matter is present, Ampere's law changes to

$$\oint (\vec{B}/\mu) \cdot d\vec{s} = \oint \vec{H} \cdot d\vec{s} = I_{\text{enclosed}}.$$

Use this equation to find the magnetic field in a narrow gap cut through the core of a toroidal coil (Fig. 31–23). (The purpose of such an arrangement is to allow samples to be placed within the gap.) The average radius of the torus is 30 cm, $\mu = 1200\mu_0$ for the core material, and the coil consists of 500 turns carrying a current of 8 A. Calculate B for gap widths of 1 mm and 3 cm.

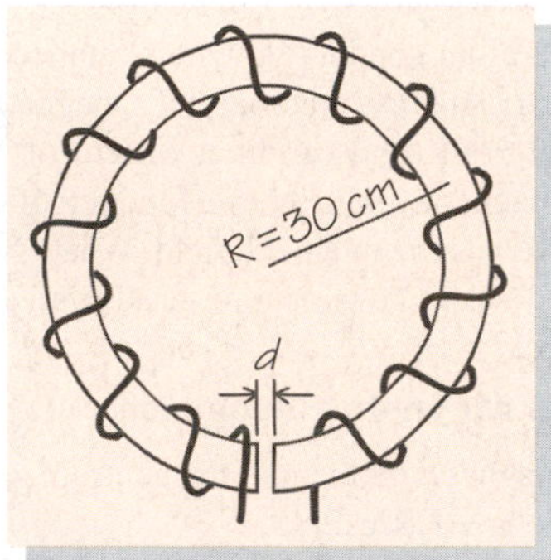

▲ **FIGURE 31–23** Problem 45.

General Problems

46. (II) Estimate the diamagnetic susceptibility of the diamond form of carbon, using the formula derived in Problem 34. Take the density of diamond to be 3.5 g/cm^3; the atomic weight, 12; the atomic number, 6; and the atomic radius, 0.75×10^{-10} m. Assume that all the electrons circulate at this radius. Your estimate should be rather good. Compare this result to that found in Table 31–1.

47. (II) Large magnets typically consist of wound toruses of ferromagnetic material. There is a gap in the torus that forms a space between pole faces. Show that the magnetic field across the pole faces is the same as the magnetic field inside the ferromagnet by applying Gauss' law for magnetism to a closed surface partly in and partly out of one of the pole faces. (Gauss' law for magnetism holds independent of the presence of materials. Its validity rests on the fact that there are no magnetic monopoles, and materials introduce no such animals.)

48. (II) Two parallel conducting strips are each 3.0×10^{-3} m thick and 2.5 cm wide and are separated by a distance of 1.0 cm. The space between the strips is filled with a ferromagnetic material whose permeability is $650\mu_0$. Each strip carries a uniform current of 2.0 A, in opposite directions. Find the value of the magnetic field and magnetic intensity in the space between the strips.

49. (II) Cobalt has atomic mass number $A = 59$, and mass density of 8.7×10^3 kg/m^3. If each atom has a magnetic dipole moment of the order of 1.7 Bohr magnetons, what is the maximum magnetic moment of a cubic centimeter of Co?

50. (II) The magnetic moment of a uniformly charged sphere of radius a and total charge Q rotating with angular velocity ω is $m = Q\omega a^2/5$. A neutral spherical object has a net magnetic moment 0.45μ. If we were to model this by a sphere of charge $-Q/2$ of radius r, surrounded by a spherical shell of charge $+Q/2$ and thickness $a - r$, what is the ratio r/a?

51. (II) Earth's magnetic field is close to that of a dipole, and the strength of the field at the magnetic north pole is about 0.6×10^{-4} T. Calculate Earth's magnetic moment. If this magnetic moment is due to a magnetized iron core whose radius is half Earth's radius, what is the magnetization of the core (Fig. 31–24)? If the magnetic moment were due to a circulating belt of current at the radius of the core, what would be the magnitude of this current (Fig. 31–24)?

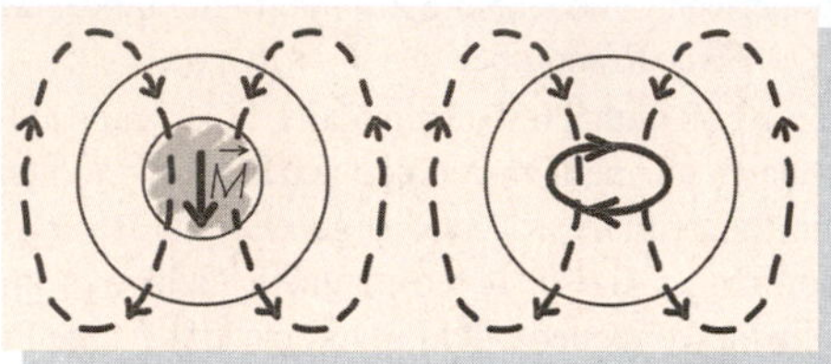

▲ **FIGURE 31–24** Problem 51.

52. (II) A torus of central radius 12 cm and tube radius 1.5 cm is filled with silver. It is wound with 450 turns of wire and carries a 0.80-A current. The magnetic susceptibility of silver is -2.4×10^{-5}. Determine (a) the magnetic intensity, $\vec{H}$; (b) the magnetic field, $\vec{B}$; (c) the magnetization, $\vec{M}$. (d) Repeat parts (a) through (c) for a torus filled with nickel instead of silver. The susceptibility of nickel is 95.

53. (II) In one of two simple classical models of the electron spin, the charge circulates at the classical electron radius r_0 (see Problem 19). In the other, the total electron charge is spread uniformly over a disk whose radius is the classical radius. Calculate the ratio of magnetic moments for the case in which the overall charge occurs entirely at the classical radius versus the case in which the charge is spread over the disk.

54. (II) The neutron has an internal spin $\vec{S}$ of magnitude $\hbar/2$ and a magnetic moment related to the spin by the gyromagnetic ratio g_p, as in Eq. (31–20). The gyromagnetic ratio is $g_p = -3.82(e/2m_n)$. Suppose that a neutron consists of a heavy, positively charged particle of mass M and magnetic moment $e\hbar/2M$, with a lighter, negatively charged particle of mass m but no intrinsic magnetic moment orbiting the heavier particle with orbital angular momentum $\hbar$. What would mass m have to be to explain the observed magnetic moment of the neutron? (For simplicity, ignore the motion of the heavier particle about the center of mass.)

55. (III) In our discussion of kinetic theory (Chapter 19), we noted that, according to Boltzmann, the number of systems with a given energy E in a collection of systems in equilibrium at temperature T is given by $Ce^{-E/kT}$. Here, C is a constant determined by the requirement that, when all the systems are summed, we find the same total number of systems that we started with. For a collection of N magnetic dipoles at rest in an external magnetic field, we have $N(T) = Ce^{-(-\vec{m}\cdot\vec{B})/kT} = Ce^{(mB\cos\theta)/kT}$, where θ is the angle between the direction of the dipole and that of the external magnetic field. C is determined by the requirement that the total number of systems N is $N = C\int_0^{\pi} 2\pi \sin\theta e^{(mB\cos\theta)/kT}\, d\theta$. (a) Calculate C. (b) Calculate the average value of $\cos\theta$. (c) Plot $\langle\cos\theta\rangle$ as a function of mB/kT.

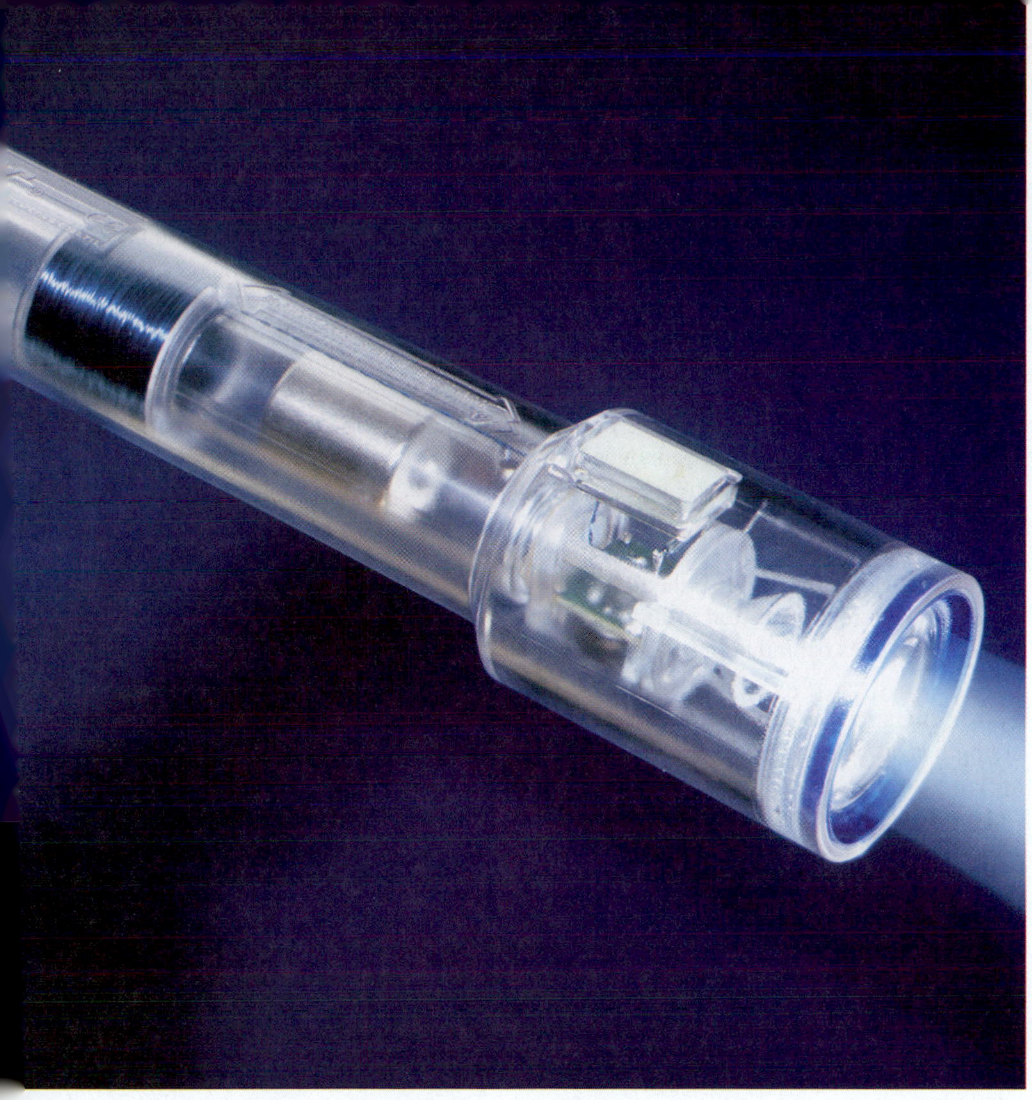

Chapter

Mechanical motion charges this flashlight. When it is shaken the plug of magnetized material visible at the center moves through the coil and through Faraday's law generates an electric current. With the addition of a capacitor or small chargable battery this is a flashlight that will never run out on you.

Inductance and Circuit Oscillations

We have already seen that we can store energy in the electric field of a capacitor. Energy can also be stored in a magnetic field, using *inductors* as the circuit elements. Their operation is based on Faraday's law, which describes the phenomena that occur when magnetic fields (or more generally magnetic fluxes) change. These phenomena occur frequently in circuits because so many applications of circuits, from computers to televisions to the functioning of the power grid, involve time dependence in currents and hence in magnetic fields. For this reason, inductors are crucial to the control of time dependence in circuits. We will see that electric circuits containing inductors, capacitors, and resistors are analogous to damped harmonic oscillators, and all the features of such mechanical systems are also seen in these circuits.

32–1 Inductance and Inductors

When you are at home vacuuming the rug, you may see a spark at the wall socket if you accidentally pull the plug from it. Why? This is just Faraday's law at work, holding up the currents that were already flowing. When a circuit contains a *changing* electric current, the magnetic field associated with that current also changes, changing the magnetic flux through the circuit. According to Faraday's law, that means that the circuit's *own* changing current will induce an additional emf in the circuit. Lenz' law tells us that this additional emf will tend to maintain the existing current.

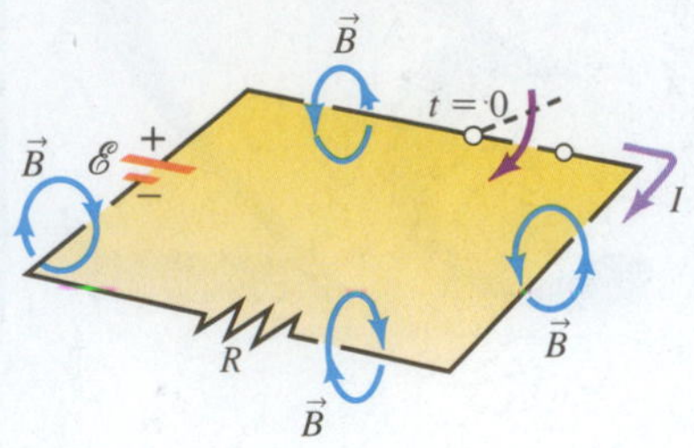

(a)

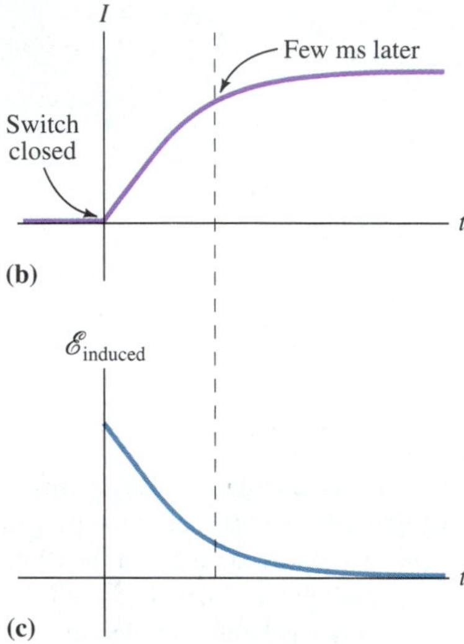

▲ **FIGURE 32–1** (a) When the current in a circuit changes, the flux through the circuit also changes. (b) The current in the circuit increases as a function of time. (c) The induced emf is proportional to the derivative of the current according to Faraday's law.

For example, Fig. 32–1a shows a circuit with a switch that closes at $t = 0$. When as a result the current increases from zero (Fig. 32–1b), the magnetic field around the wire also increases. As the magnetic field grows, the magnetic flux through the area enclosed by the loop increases—in the figure, the flux is directed downward. According to Faraday's law, as formulated by Lenz, an emf is induced in the loop and opposes this increase in flux (Fig. 32–1c). The induced emf *opposes* the emf of the battery and slows down the flow of current. The principle illustrated here is a simple one: *Changing currents in circuits lead to effects that act to reduce the rate of change of those currents.*

If a second circuit loop is in the general vicinity of the first, there may be a changing magnetic flux through the second circuit due to the magnetic field produced by the first circuit, and a current will be induced in the second circuit. This in turn produces a changing flux that can affect the first circuit, and so forth. In this case, the two circuits are said to be **linked**.

When the first loop induces an emf in itself, we say that there is a **self-inductance**, or **inductance** for short. When the first loop induces a current or emf in a second loop, we say that there is a **mutual inductance** between the two loops. Faraday's law is the principle that lies behind both self-inductance and mutual inductance.

Self-Inductance: When a wire carries a current I, a magnetic field is set up whose strength is proportional to I. In turn, the magnetic flux appearing through a loop of that wire is also proportional to the current. The proportionality constant is defined to be the *inductance L*. L depends on the particular geometry of the loop around which the emf is induced as well as on the number of turns of the loop, and is defined by

$$\Phi_B = LI. \tag{32–1}$$

If the current and therefore the flux changes, then according to Faraday's law, the emf induced in this loop, $\mathscr{E}$, is the rate of change of the flux through the loop:

$$\mathscr{E} = -\frac{d\Phi_B}{dt} = -L\frac{dI}{dt}. \tag{32–2}$$

EMF BY SELF-INDUCTANCE

The minus sign in this equation is the manifestation of Lenz' law. We will see that the presence of this induced emf can have a marked effect on how charges flow through the circuit.

Mutual Inductance: Let's now consider the two adjacent circuits shown in Fig. 32–2. If a current I_1 flows in loop 1 and a current I_2 flows in loop 2, there is a magnetic flux $\Phi_B(1)$ through the area of loop 1 given by

$$\Phi_B(1) = L_1 I_1 + M_{12} I_2. \tag{32–3}$$

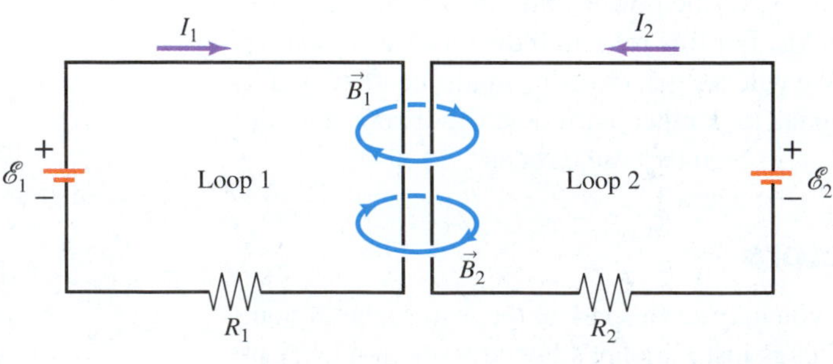

(a)

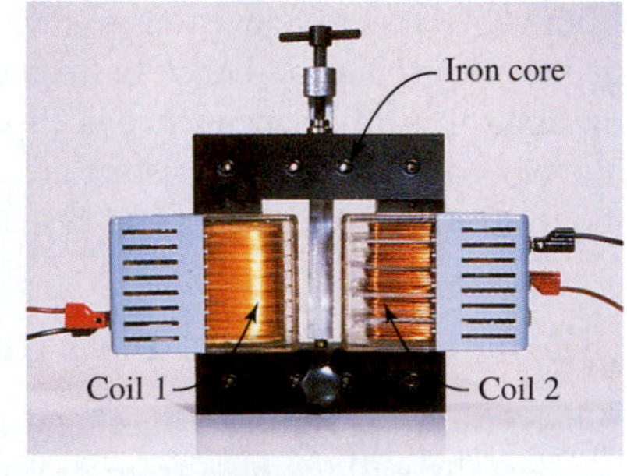

(b)

▲ **FIGURE 32–2** (a) The flux through a circuit or circuit element may be due to its own current or to the current carried by an adjacent circuit or circuit element. (b) The two coils mounted on the iron core demonstrate mutual induction.

The first term is due to the current flowing in loop 1, and the constant of proportionality L_1 is the self-inductance of loop 1. The second term is due to the current flowing in loop 2 (Fig. 32–3a), and the constant M_{12} is what we define as the *mutual inductance* of loop 1 due to loop 2. Both L_1 and M_{12} are positive by definition. They depend *only* on the geometry of the loop and on the materials in its vicinity, but not on the currents themselves.

The term *mutual* implies a degree of symmetry between the two loops. The magnetic flux through loop 2 has a term proportional to its own current and also a term proportional to the current in loop 1 (Fig. 32–3b):

$$\Phi_B(2) = L_2 I_2 + M_{21} I_1. \tag{32–4}$$

The second term introduces what might appear to be a new constant, M_{21}, the mutual inductance of loop 2 due to loop 1. We will not provide the proof, which is not simple, but *the mutual inductances are equal*: $M_{12} = M_{21}$. It is customary to drop the subscripts and write $M = M_{12} = M_{21}$, the mutual inductance of two loops.

Faraday's law gives the emf induced in loop 2 due to the change in current of loop 1:

$$\mathcal{E}_{21} = -M\frac{dI_1}{dt}. \tag{32–5}$$

EMF BY MUTUAL INDUCTANCE

A similar expression gives the emf induced in loop 1 due to the current in loop 2.

The inductances L and M have SI units of magnetic flux divided by current, or webers per ampere (Wb/A). Inductance is given its own unit in the SI, the **henry** (H), named after Joseph Henry (Fig. 32–4):

$$1\ \text{H} = 1\ \text{Wb/A} = 1\ \text{T}\cdot\text{m}^2/\text{A}. \tag{32–6}$$

To give you an idea of how large a henry is, a cylindrical solenoid of area 10 cm^2, length 20 cm, and a winding density of 10 turns/cm has an inductance of 0.25 mH. An inductance of 1 H is large but not unrealizable; typical values of self- and mutual inductance range from μH to tens of mH. Self-inductance is usually the more important effect in circuits; the role of mutual inductance in linked circuits is most important in *transformers*. These are devices used to change the magnitude of time-varying voltages and are crucial elements of the electric-power distribution system. We shall encounter transformers again in Chapter 33.

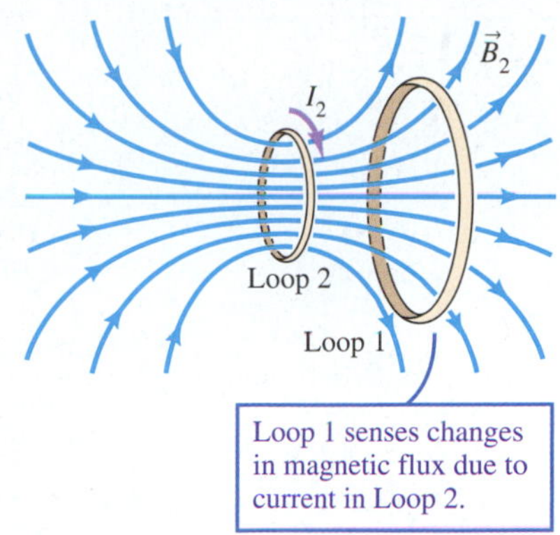

(a)

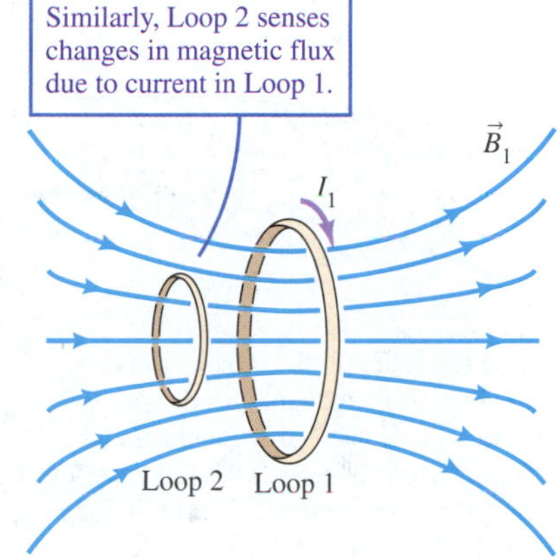

(b)

▲ **FIGURE 32–3** Mutual inductance. (a) There is a magnetic flux through loop 1 due to the magnetic field from the current I_2 in loop 2. (b) There is a magnetic flux through loop 2 due to the current I_1 in loop 1.

◀ **FIGURE 32–4** Joseph Henry, as depicted in a stained-glass window in the First Presbyterian Church of Albany, New York, the site of Henry's baptism. Henry investigated many effects of induction at about the same time as did Faraday.

Elements within circuits with a significant self-inductance—we refer to them as **inductors**—provide sources of emf, which must be included when Kirchhoff's loop rule is used for potential changes around a circuit. Inductors, usually in the form of solenoids, are useful devices that are placed in circuits to control time dependence and to store energy. They join capacitors, resistors, and batteries as the basic elements of circuits. They are represented in circuit diagrams by the symbol ⅏. We add the following to our list of rules for application of Kirchhoff's loop rule to circuits:

> **In moving across an inductor of inductance L *along* (or *against*) the presumed direction of the current I, the potential change is $\Delta V = -L\,dI/dt$ (or $+L\,dI/dt$, respectively).**

The actual sign of the potential drop depends on the sign of the rate of change of the current and is determined when the circuit equations are solved.

Finding the Inductance

To use the loop rule with inductors present in a circuit, we must know the values of the self-inductance or mutual inductance. As for capacitance, there are only a few simple, but important, geometries for which inductance can easily be calculated. The most important of these is the ideal solenoid. Consider the ideal solenoid of Fig. 32–5, which has length ℓ and radius R. For $\ell \gg R$, the magnetic field within the solenoid is longitudinal and constant and is given by Eq. (29–15):

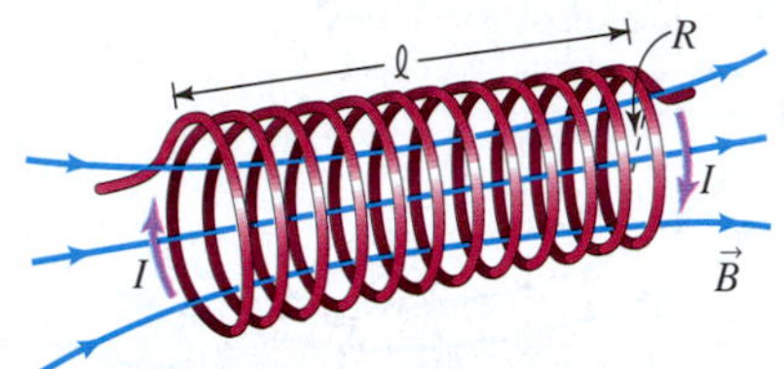

▲ **FIGURE 32–5** A solenoid of length ℓ, radius R, and turn density n carries a current I. Only the magnetic field $\vec{B}$ inside is shown; the field outside the solenoid is zero in the limit that the solenoid is infinitely long.

$$B = \mu_0 n I,$$

where μ_0 is the permeability of free space, n is the number of turns per unit length of solenoid, and I is the current the solenoid carries. The magnetic flux through one turn of the solenoid is the field B times the cross-sectional area A, $\Phi_B = BA = \mu_0 A n I$. The total magnetic flux is this value times the *total* number of turns $N = n\ell$:

$$\Phi_B = \mu_0 A n^2 \ell I. \tag{32–7}$$

By comparison with Eq. (32–1), the self-inductance is the coefficient of the current:

$$\text{for an ideal solenoid: } L = \mu_0 A \ell n^2. \tag{32–8}$$

EXAMPLE 32–1 During a short time period, the current in a cylindrical coil of length 10 cm, radius 0.5 cm, and 1000 turns of wire in a single layer is increased at the steady rate of 10^3 A/s. Find the emf induced during this period.

Strategy We can treat the cylindrical coil as an ideal solenoid, as it is much longer than its coil width. The induced emf is given by Eq. (32–2); the inductance that appears in that equation is in turn given for a coil by Eq. (32–8). Hence this problem is simply a matter of inserting numbers into these equations.

Working It Out We start with the inductance. The turn density is $n = (1000\text{ turns})/(0.1\text{ m}) = 10^4$ turns/m. The area of the solenoid is given by $A = \pi r^2$, and

$$\begin{aligned} L &= \mu_0 A \ell n^2 \\ &= (4\pi \times 10^{-7}\,\text{T}\cdot\text{m/A})[\pi(0.005\text{ m})^2](0.1\text{ m})(10^4\text{ m}^{-1})^2 \\ &= 10^{-3}\text{ H}. \end{aligned}$$

As for the induced emf, the current change rate is $dI/dt = 10^3$ A/s, so if we follow the circuit in the direction of the current, we have

$$\mathscr{E} = -L\frac{dI}{dt} = -(10^{-3}\text{ H})(10^3\text{ A/s}) = -1\text{ V}.$$

The induced emf is negative.

What Do You Think? Which of the following will increase the induced emf the most? In each case the change mentioned is the *only* change. (a) Doubling the length of the solenoid, (b) doubling the rate at which the current changes, (c) doubling the area of the solenoid, (d) doubling the number of turns in the solenoid. *Answers to **What Do You Think?** questions are given in the back of the book.*

CONCEPTUAL EXAMPLE 32–2 A friend claims that he can increase the inductance L of a solenoid with given area and length and that must be constructed with a single layer of wire by using finer wire of the same material. Is he correct? What are the implications of his method of increasing L on the resistance R of the solenoid?

Answer The inductance is proportional to the square of the turn density n^2, and for a single layer $n \propto 1/d$, where d is the diameter of the wire. Thus

$$L \propto 1/d^2,$$

and decreasing d will certainly increase L. To see the implication of the use of finer wire for the resistance, we can recall from Chapter 26 that the resistance of a wire of length ℓ and area A made of a material of resistivity ρ is $R = (\ell/A)\rho$. In our case, with fixed solenoid length and area, each turn uses the same wire length, and the total number of turns N is proportional to the turn density $n \propto 1/d$. Thus the length of wire needed is proportional to $1/d$. The wire cross-section $A \propto d^2$, so $\ell/A \propto d^{-1}/d^2$, i.e.

$$R \propto 1/d^3.$$

R increases faster than L does as the wire diameter decreases. Whether this has important consequences depends on the circumstances, but the changed value of R would almost certainly have to be taken into account in determining the properties of the circuit.

As an example of a calculable mutual inductance, consider a solenoid (of length ℓ_1, radius R_1, winding density n_1, and current I_1) that contains within it a single loop of radius R_2 whose area $A_2 = \pi R_2^2$ is oriented perpendicular to the axis of the solenoid (Fig. 32–6a). The magnetic field of the solenoid is given by Eq. (30–15), $B = \mu_0 n_1 I_1$. The magnetic flux that passes through the single loop is

$$\Phi_B = BA_2 = \mu_0 A_2 n_1 I_1.$$

By definition, the mutual inductance, M, is the coefficient of I_1:

$$M = \mu_0 A_2 n_1. \qquad (32\text{–}9)$$

If the single loop is replaced with a second solenoid (Fig. 32–6b) with a total number of turns N_2, then the total flux that links that second solenoid contains a factor N_2, and M must be increased by this same factor:

$$M = \mu_0 A_2 n_1 N_2. \qquad (32\text{–}10)$$

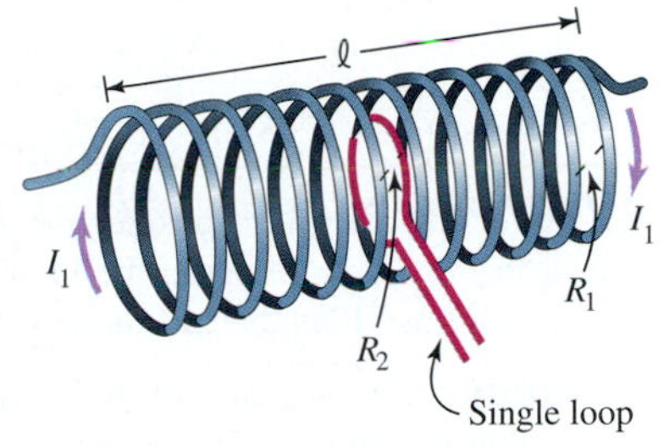

(a)

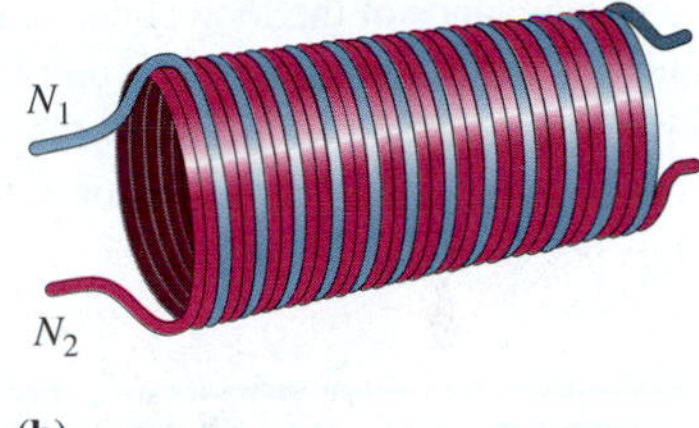

(b)

▶ **FIGURE 32–6** (a) The area of the single, small loop is oriented perpendicular to the axis of the solenoid. (b) Similar qualitative effects are associated with the replacement of the single loop (a solenoid with but a single turn) by a solenoid with multiple turns.

EXAMPLE 32–3 Consider the two solenoids in Fig. 32–6b and take their radii, and hence the areas, to have the same value. Compare the ratio of mutual inductance M to self-inductance L_1 of solenoid 1.

Setting It Up We denote the common area as A. We must assume values for the turn density n_1 and the length ℓ_1 of solenoid 1, as well as the total number of turns N_2 in solenoid 2.

Strategy This is a straightforward use of Eqs. (32–8) and (32–10).

Working It Out We have from Eqs. (32–8) and (32–10)

$$\frac{M}{L_1} = \frac{\mu_0 A n_1 N_2}{\mu_0 A n_1^2 \ell_1} = \frac{\mu_0 A n_1 N_2}{\mu_0 A n_1 N_1}.$$

In the last step, we have used the fact that the factor $n_1 \ell_1$ is the total number of turns N_1 in solenoid 1. Both the area A and the turn density n_1 cancel, and we are left with

$$\frac{M}{L_1} = \frac{N_2}{N_1}.$$

If solenoid 2 has many fewer windings than solenoid 1, then the effect of the mutual inductance on solenoid 1 is small compared to that of the self-inductance.

What Do You Think? (a) If the radius R_1 is doubled, what happens to L_1 and M? (b) If instead, the radius R_2 is doubled, what happens to L_1 and M?

The Effects of Magnetic Materials on Inductance

In Chapter 31, we considered modifications to a magnetic field due to a current in the presence of materials with magnetic properties. We saw that the presence of materials can modify, and in the case of ferromagnetic materials greatly magnify, the magnetic field present and hence the flux. This is made explicit in expressions that include the free (or real) current by replacing the permeability of free space μ_0 with the permeability of the material μ. The permeability is given by Eq. (31–7),

$$\mu = \mu_0(1 + \chi_m).$$

Here, χ_m is the magnetic susceptibility of the material, which is negative and small for diamagnets, positive and small for paramagnets, and positive and large for ferromagnets. If a solenoid is filled with a magnetic material, its self-inductance would change, with the replacement of μ_0 by μ in Eq. (32–8). For ferromagnetic materials, self-inductances can thereby be increased manyfold (Fig. 32–7). In Fig. 32–8, we illustrate the application of inductors that use the magnetic properties of materials in a doorbell.

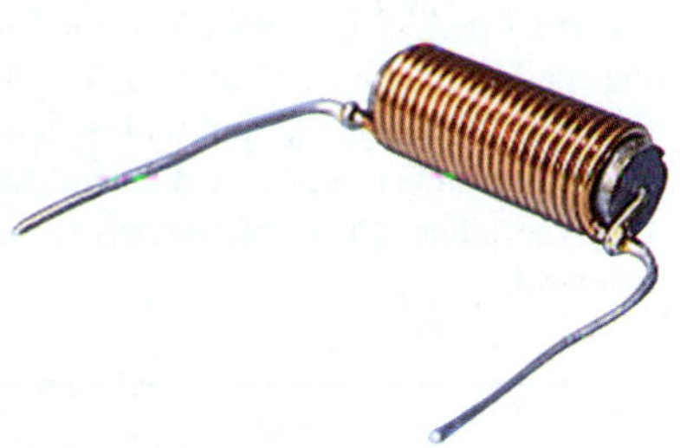

▲ **FIGURE 32–7** The inclusion of an iron core in this solenoid (an actual inductor) increases the magnetic field produced by a given coil current.

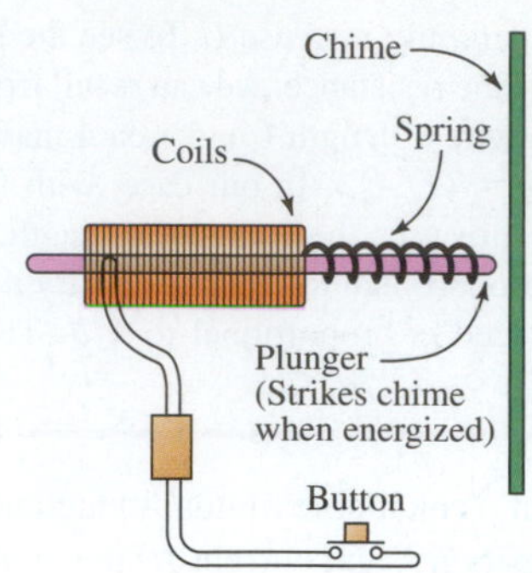

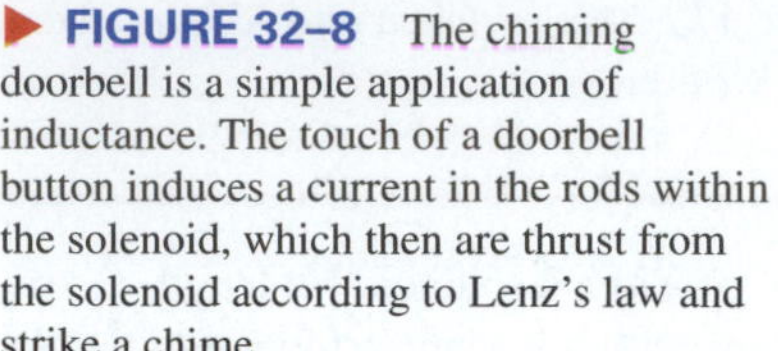
▶ **FIGURE 32–8** The chiming doorbell is a simple application of inductance. The touch of a doorbell button induces a current in the rods within the solenoid, which then are thrust from the solenoid according to Lenz's law and strike a chime.

EXAMPLE 32–4 Consider the single loop and the solenoid shown in Fig. 32–6. Suppose the loop carries a current I_2 that is a function of time. Find the emf in the solenoid induced by current I_2.

Strategy This is a case of mutual inductance—we want the effect within solenoid 1 of a changing current in loop 2. The single loop of radius R_2 carries the current I_2 in this case, which changes as dI_2/dt. To apply Faraday's law as in Eq. (32–5), we must find the magnetic flux Φ_B due to loop 2 through solenoid 1. This flux is the mutual inductance of the loop and solenoid times current I_2. A direct calculation of the flux might be complicated here, but we can fortunately use Eq. (32–9) for the inductance together with the "mutuality" of the inductance—the equation gives M_{21}, while what we want is M_{12}, but the two are equal.

Working It Out Using Eq. (32–9) for M gives

$$\Phi_B = MI_2 = \mu_0 \pi R_2^2 n_1 I_2.$$

The emf in the solenoid is then the negative time derivative of this flux:

$$\mathscr{E} = -\frac{d\Phi_B}{dt} = -\mu_0 \pi R_2^2 n_1 \frac{dI_2}{dt}.$$

What Do You Think? If a ferromagnetic material is placed inside the solenoidal coil (without the single loop), the self-inductance of the coil will (a) increase, (b) decrease, (c) stay the same.

CONCEPTUAL EXAMPLE 32–5 A hollow solenoid is carrying a steady current I when a cylindrical piece of iron is sent through the cylindrical space and out the other side. The cylinder of iron fits the interior of the solenoid and is a little shorter than the solenoid. Describe the emf induced in the circuit containing the solenoid and include a sketch. What is the effect on the current?

Answer As the iron passes into the solenoid (Fig. 32–9a), the magnetic field (and magnetic flux) inside the solenoid will suddenly increase, because the space is increasingly filled with a material with a permeability much larger than the permeability of the air, which is to a very good approximation that of the vacuum. According to Lenz's law, an emf will be generated within the solenoid that opposes this increase in flux. The sign of the emf corresponds to this "opposition," and we will call it negative, as shown in Fig. 32–9b. If the speed of the cylinder entering the space is held constant, then the rate of increase in flux will be constant, and the induced emf will be constant. The emf will induce a current I_{induced} in the coil that will be opposite to the original current. Once the iron cylinder is entirely within the solenoid, there is no change in flux and no further emf; for this brief period the induced emf is zero. Finally, the cylinder exits the space, this time leading to a decreasing flux and an induced emf that tries to hold up the flux. This positive emf is also indicated in the figure; it results in a positive current. Just as quickly, the induced emf (and the induced current) returns to zero once the iron has left the solenoid.

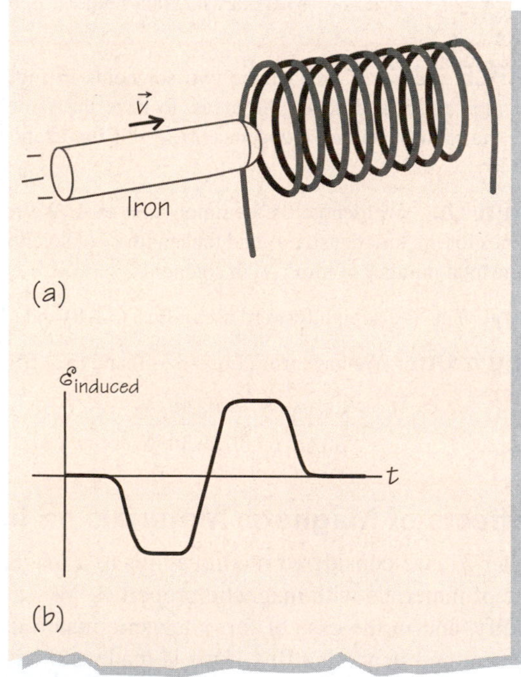

▲ **FIGURE 32–9** (a) A ferromagnetic cylinder moves into and through a coil. (b) The emf induced in the coil as a result of steady movement of the cylinder.

32–2 Energy in Inductors

Just as a capacitor is a device for storing energy in an electric field, an inductor is a device for storing energy in a magnetic field. Because any emf induced in the inductor opposes the change in current, work must be done by an external source, such as a battery, to cause a current to pass through an inductor. By the work–energy theorem, just how much work is done is a measure of the energy stored in the inductor. To calculate this energy, we proceed as we did for the capacitor in Chapter 25, and calculate the work that must be done by some external emf to pass a current through the inductor.

We derive the general expression for the rate dW/dt (the power) at which an external emf $\mathscr{E}_{\text{ext}}$ does work when a current I flows from Eqs. (26–26) and (26–27):

$$\frac{dW}{dt} = I\mathscr{E}_{\text{ext}}.$$

If we have only the external emf and an inductor, the external emf must be equal but opposite to the induced emf in the inductor, given by Eq. (32–2). Thus

$$\frac{dW}{dt} = +LI\frac{dI}{dt}. \qquad (32\text{–}11)$$

If the current is increasing, the power is positive, meaning that the external source must do positive work in supplying energy to the inductor; the internal energy U_L in the inductor is increasing. If the current is decreasing, the power is negative, meaning that the external source takes energy from the inductor; the inductor's internal energy is decreasing. The net change ΔU_L in the total magnetic energy of the inductor as the current changes from a value I_1 to a value I_2 between the times t_1 and t_2 can be found by integrating the work done by the external source as the current changes. We integrate Eq. (32–11) for dW/dt from an initial time t_1 to a later time t_2:

$$\begin{aligned}\Delta U_L &= \int_{t_1}^{t_2}\frac{dW}{dt}\,dt = \int_{t_1}^{t_2} LI\frac{dI}{dt}\,dt = L\int_{I_1}^{I_2} I\,dI \\ &= \frac{1}{2}LI_2^2 - \frac{1}{2}LI_1^2.\end{aligned} \qquad (32\text{–}12)$$

In particular, if the inductor carries a current I, then the increase in energy as the current increases from zero up to I, which we refer to simply as the energy of the inductor, is

$$U_L = \frac{1}{2}LI^2. \qquad (32\text{–}13)$$

ENERGY IN AN INDUCTOR

In applications to circuits (see Section 32–4 as well as later chapters), we'll see that it is not only the storage of energy that is interesting but the time dependence of that process.

Equation (32–13) should be compared to the expression for the energy U_C contained in a capacitor of capacitance C that carries charge Q from Eq. (25–8):

$$U_C = \frac{1}{2}\frac{Q^2}{C}.$$

EXAMPLE 32–6 A solenoid is designed to store $U_L = 0.10\ \text{J}$ of energy when it carries a current I of 450 mA. The solenoid has a cross-sectional area A of $5.0\ \text{cm}^2$ and a length ℓ of 0.20 m. How many turns of wire must the solenoid have?

Strategy We know the energy, which is a function of the inductance L and the given current; L is in turn a function of N, through Eq. (32–8). The only modification necessary in Eq. (32–8) is to write it in terms of the total number of turns N rather than the turn density n by using $N = n\ell$, where ℓ is the length of the solenoid. We can then solve for N in terms of L and hence in terms of the energy.

Working It Out With the change to Eq. (32–8), we have

$$L = \frac{\mu_0 AN^2}{\ell}.$$

The expression for the energy, Eq. (32–13), then reads

$$U_L = \frac{1}{2}\frac{\mu_0 AN^2}{\ell}I^2.$$

We solve this for N and insert numbers:

$$\begin{aligned}N &= \frac{1}{I}\sqrt{\frac{2U_L\ell}{\mu_0 A}} \\ &= \frac{1}{4.5\times 10^{-1}\ \text{A}}\sqrt{\frac{2(0.10\ \text{J})(0.20\ \text{m})}{(4\pi\times 10^{-7}\ \text{N/A}^2)(5.0\times 10^{-4}\ \text{m}^2)}} \\ &= 1.8\times 10^4\ \text{turns}.\end{aligned}$$

What Do You Think? What is the easiest way to increase the energy stored in a solenoid if its size is fixed and the wire (of which you have an abundant supply) can carry no additional current?

An inductor has an energy given by Eq. (32–13) even if the current is steady. We have, however, argued that the origin of the effects of inductance is Faraday's law, which involves changes in current. How can we reconcile these two facts? As we are going to describe in more detail in the next section, the energy of an inductor is in the magnetic field within it, just as the energy of a capacitor lies within the electric field between the plates. The energy of an inductor carrying steady current must arise from the original buildup of current—even if it occurred in the distant past. It is when the current changes that the magnetic field changes and, along with it, the energy in the magnetic field.

32–3 Energy in Magnetic Fields

In Chapter 25, we demonstrated that the electric energy associated with a capacitor is located in the electric field within the capacitor. Similarly, the energy of an inductor is located in its magnetic field. Just as the ideal parallel-plate capacitor was a means to learn about energy in the electric field, the ideal solenoid presents us with a tool to explore the energy density in a magnetic field.

The inductance of an ideal solenoid of area A and length ℓ is given by Eq. (32–8), so from the expression for the total energy of an inductor [Eq. (32–13)], we find

$$U_L = \frac{1}{2}LI^2 = \frac{1}{2}\mu_0 A\ell n^2 I^2. \tag{32–14}$$

We also know that the magnetic field in the solenoid is proportional to the current. Equation (29–15) gives the precise connection, $B = \mu_0 nI$. If we substitute for I in terms of B in Eq. (32–14), we obtain

$$U_L = \frac{1}{2}\frac{B^2}{\mu_0}A\ell. \tag{32–15}$$

The volume enclosed within the solenoid is $A\ell$. Because the magnetic field is uniform within the solenoid, we can identify the **energy density** u_B, the energy per unit volume of the magnetic field, as

$$u_B = \frac{1}{2}\frac{B^2}{\mu_0}. \tag{32–16}$$

ENERGY DENSITY IN A MAGNETIC FIELD

This result generalizes to the case of a nonuniform magnetic field, no matter how it is produced. It should be compared to our expression for the energy density of an electric field, Eq. (25–12):

$$u_E = \frac{1}{2}\varepsilon_0 E^2,$$

a result derived in a similar way. It is important to realize that *energy is located within the electric and magnetic fields themselves.*

When both magnetic and electric fields are present, the energy density is the sum of both magnetic and electric energy densities:

$$u = u_B + u_E = \frac{1}{2}\left(\frac{B^2}{\mu_0} + \varepsilon_0 E^2\right). \tag{32–17}$$

EXAMPLE 32–7 A large electromagnet produces a magnetic field of 1 T. Compare the energy density associated with this field to that of the largest electric field in air, about 10^6 V/m (beyond this value, there is breakdown).

Strategy Equations (32–16) and (25–12) express the energy density in magnetic and electric fields, respectively, and we can directly obtain the ratio of the magnetic and electric energy densities from them.

Working It Out We have

$$\frac{u_B}{u_E} = \frac{\frac{1}{2}\frac{B^2}{\mu_0}}{\frac{1}{2}\varepsilon_0 E^2} = \frac{1}{\mu_0\varepsilon_0}\frac{B^2}{E^2}.$$

In our case, the magnitudes of both the magnetic and electric fields are given, and

$$\frac{u_B}{u_E} = \frac{1}{(4\pi \times 10^{-7}\ \text{N/A})(8.85 \times 10^{-12}\ \text{F/m})} \frac{(1\ \text{T})^2}{(10^6\ \text{V/m})^2}$$
$$= 9 \times 10^4.$$

Depending on the medium, electric breakdown properties may limit the size of electric fields that can be easily maintained and make magnetic fields more suitable for storing energy at high density. How rapidly that energy can be delivered back for other uses is another story, however, and capacitors remain the best way to deliver a lot of energy as fast as possible.

What Do You Think? What puts a limitation on the production of large magnetic fields?

32–4 Time Dependence in *RL* Circuits

The fact that the potential drop across an inductor depends on how rapidly the current passing through it changes leads to new time-dependent behavior in circuits with inductors. For example, if we attempt to stop a current already flowing—say, by opening a switch—then an emf is induced that attempts to keep the current flowing, possibly creating an arc across the switch. The simplest illustration of time-dependent behavior involves a circuit with a source of emf $\mathscr{E}$, a resistor of resistance R, and an inductor of inductance L (Fig. 32–10). We call such a circuit an ***RL* circuit**. A switch allows us to control the initial conditions. When we close the switch, the inductor acts to oppose the changing current, and as a result, the current cannot jump suddenly but must build up over time. If we apply the loop rule to the circuit in the direction of the pink arrow in Fig. 32–10, we can see this quantitatively. We have

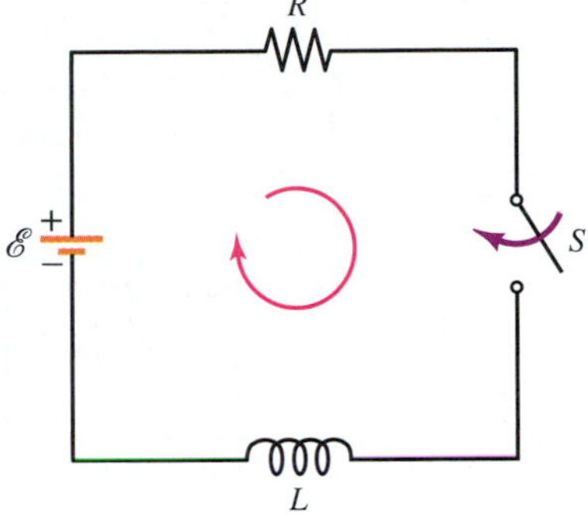

▲ **FIGURE 32–10** An *RL* circuit, including a source of constant emf.

$$\mathscr{E} - IR - L\frac{dI}{dt} = 0. \tag{32–18}$$

To find the solution to this differential equation for I, compare Eq. (32–18) with Eq. (27–21), which comes from applying the loop rule to the *RC* circuit shown in Fig. 32–11:

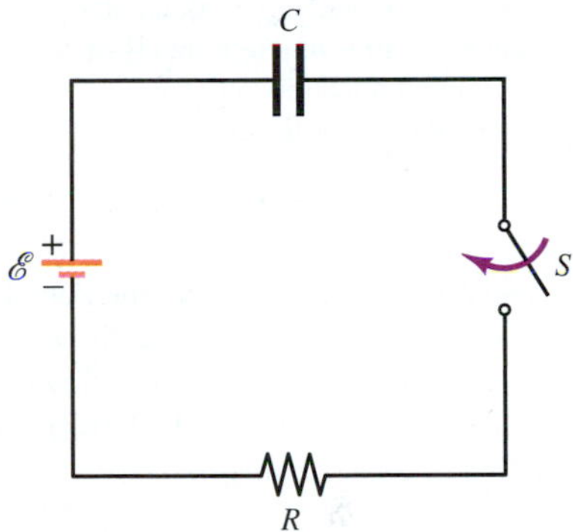

▲ **FIGURE 32–11** An *RC* circuit is analogous to the *RL* circuit of Fig. 32–10.

$$\mathscr{E} - \frac{Q}{C} - R\frac{dQ}{dt} = 0. \tag{27–21}$$

Here, Q is the charge on the capacitor. The first differential equation determines the current in *RL* circuits and has exactly the same *form* as the equation for the charge in *RC* circuits. The similarities between the *RC* and *RL* circuits are summarized in Table 32–1, and the solution of Eq. (32–18) for the current in the circuit is the same as the solution of Eq. (27–21) for charge if we make the substitutions indicated in the table. In particular, if we replace RC with L/R, the current in the *RL* circuit will have the time dependence $\exp[-t/(L/R)]$. The *RL* circuit is said to have a *time constant* L/R. Note that, as for the *RC* circuit, the time dependence is transient. Large values of the time constant (large L and/or small R) mean that the transient behavior is slow to disappear, so that long times are required for the current to build up or decay; small values of the time constant (small L and/or large R) mean that the transient behavior quickly disappears, and current builds up or decays rapidly. Of course, the "large" and "small" times are relative terms.

For complete solutions of Eq. (32–18) for the current, including the initial conditions, we can directly apply the results of Section 27–5 for the solution of the differential equation for the *RC* circuit. As in Section 27–5, it is important to understand

TABLE 32–1 • Analogy Between *RC* and *RL* Circuits

	***RC* Circuit Parameter**	***RL* Circuit Parameter**
Variable	Q	I
Coefficient of variable	$1/C$	R
Coefficient of $\frac{d}{dt}$ (variable)	R	L
Time constant	RC	L/R

physically how different initial conditions will affect the solution. As a guiding principle, keep in mind the following:

The current in an inductor never changes instantaneously, but after the current settles down to a constant value, the inductor plays no role in the circuit.

Finally, we should mention that real (as opposed to ideal) inductors may contain relatively long lengths of wire and always have some resistance. Depending on the circuit, this resistance can be significant.

Let's look again at time dependence in an *RL* circuit, this time being explicit about some boundary conditions. Suppose that the switch in Fig. 32–10 has been open for a very long time and is suddenly closed at $t = 0$. What happens? Since the inductor acts to prevent any sudden change in current, the current must ramp up smoothly from zero following the switch closing. We also know that after a very long time, the system will have stabilized, with any transient effects having disappeared. The inductor will play no role at these late times, and the circuit effectively consists of only an emf $\mathscr{E}$ and a resistance R (including any resistance in the inductor). The constant current will be $I = \mathscr{E}/R$. In between these times the current has gone from 0 at $t = 0$ to $\mathscr{E}/R$ at large times with a time constant L/R—in other words, the current must reach its asymptotic value exponentially over this time scale.

Without going through the formal steps (see Problem 47 to verify the solution to the differential equation), we can say that the function that fits these criteria is

$$I = \frac{\mathscr{E}}{R}\left[1 - e^{-t/(L/R)}\right] = \frac{\mathscr{E}}{R}\left(1 - e^{-Rt/L}\right). \tag{32–19}$$

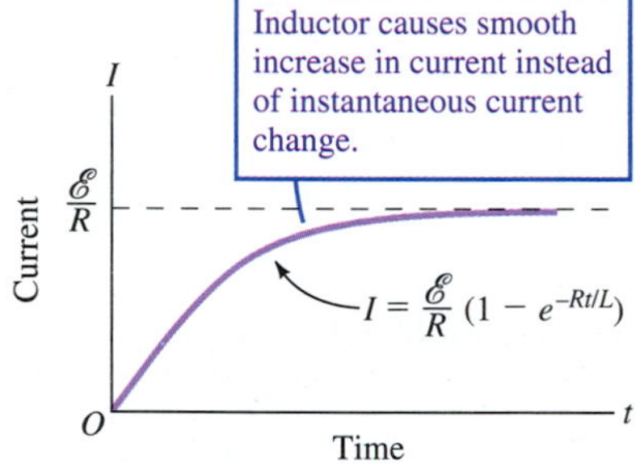

▲ **FIGURE 32–12** The switch of Fig. 32–10 is closed at $t = 0$, and the current rises from zero to a steady-state value only after a period of time determined by the ratio L/R.

Figure 32–12 is a graph of current versus time for this case. At $t = 0$, the exponential term is unity and $I = 0$, consistent with the principle that the current cannot change instantaneously. As $t \rightarrow \infty$, the transient exponential term drops out and I approaches $\mathscr{E}/R$—as though the inductor were not present at all.

EXAMPLE 32–8 Find the potential magnitude across the resistance and the inductor in the circuit of Fig. 32–10 as a function of time, assuming that the switch is closed at $t = 0$. Draw graphs of the potential across the resistor and the potential across the inductor as a function of time, starting at $t = 0$.

Strategy We have already found, in Eq. (32–19), the current as a function of time in this situation. This is sufficient to find the potentials across these elements, which in the case of the resistor is IR and in the case of the inductor is $L\,dI/dt$.

Working It Out The potential across the resistor is just Eq. (32–19) multiplied by R; we won't write that out again. For the magnitude of the potential across the inductor, we must evaluate

$$V_L = L\frac{dI}{dt} = L\frac{\mathscr{E}}{R}\left[-\frac{-R}{L}e^{-Rt/L}\right] = \mathscr{E}e^{-Rt/L}.$$

This starts at $t = 0$ with a maximum value of $\mathscr{E}$, then drops off exponentially to zero at long times; there is a potential across an ideal inductor only when the current through it is changing. In this case, it is changing fastest at $t = 0$ and not changing at all at large values of time.

We have plotted V_R and V_L in Figs. 32–13a and b.

What Do You Think? If L is arbitrarily large, the argument of the exponential in the expression for V_L is zero and the potential across the inductor never drops to zero. Is that correct?

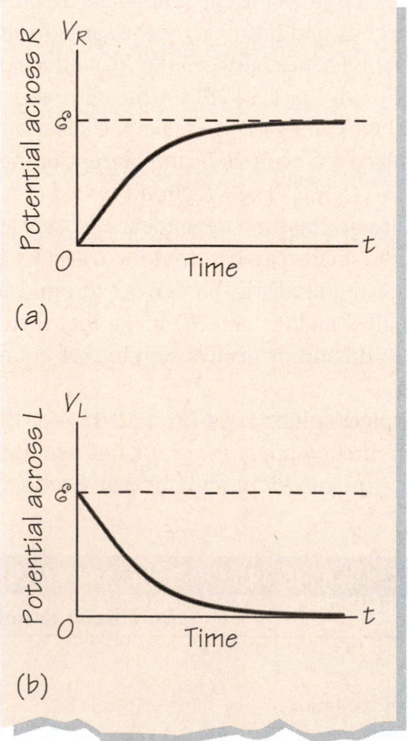

▲ **FIGURE 32–13** (a) The potential across the resistor in the *RL* circuit. (b) The potential across the inductor.

THINK ABOUT THIS. . .

HOW WOULD YOU MEASURE INDUCTANCE?

We have described the calculation of inductance for simple geometries—for example, a long coil—but in many cases the geometry is complicated, and it is not possible to calculate L. The time dependence we have described in this section and the section to come provides us with very simple tools to measure L directly. Here we have seen that the presence of an inductor in a circuit that additionally contains a resistor of known resistance R exhibits exponential time dependence with a time constant L/R. Measurement of a curve of rise time or of fall time allows us to extract L. In the next section we'll see that circuits with an inductor and a capacitor but no (or negligible) resistance exhibit oscillatory behavior whose frequency depends on L and C. Since it is possible to measure the time for N oscillations of such a circuit and to count N with precision, one can learn the value L of an unknown inductance very accurately this way. Finally, in the next chapter we shall learn and discuss still another way in which an unknown L can be determined, using a resonance phenomenon.

32–5 Oscillations in *LC* Circuits

A single-loop circuit with inductance and capacitance but negligible resistance exhibits a behavior we have not yet encountered in electric circuits but which played a central role in our study of mechanics—oscillatory behavior. Such a circuit is called an ***LC* circuit**, and it is analogous to a mass on the end of a spring—a familiar mechanical system. If we look at Fig. 32–14, we can see what the oscillatory behavior is and why it is plausible. Imagine that the capacitor is fully charged at an initial time. Then current will flow off the positively charged plate of the capacitor through the wire toward the negatively charged plate; without the inductor, the discharge through the wire would be rapid and direct. With the inductor present, the current is held back initially and grows smoothly from zero. Moreover, once the current reaches some maximum value, the inductor will induce an emf that tends to maintain this maximum current. This allows the discharge to "overshoot," with positive charge building up on the initially negatively charged plate. Conditions are then right for the process to repeat in the other direction.

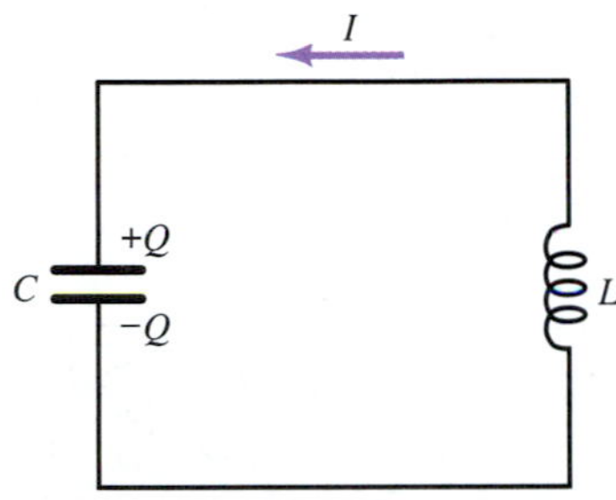

▲ **FIGURE 32–14** An *LC* circuit, consisting simply of an inductor and a capacitor in series.

At this point we need to establish not simply oscillatory behavior but harmonic behavior, and this must be done in a quantitative fashion. We do this by establishing a direct correspondence between the LC circuit and the mass on an ideal spring. We start with the loop rule for our LC circuit (Fig. 32–14). Following the clockwise direction, we have

$$\frac{Q}{C} + L\frac{dI}{dt} = 0, \tag{32–20}$$

where Q is whatever charge is on the capacitor and I is the current in the circuit with direction indicated. The current I is dQ/dt, so this expression is actually a *second-order* differential equation for the charge:

$$\frac{Q}{C} + L\frac{d^2Q}{dt^2} = 0. \tag{32–21}$$

At this point recall the equation of motion for a mass m on the end of an ideal spring of spring constant k, namely $-kx = ma$. The acceleration is the second derivative of displacement, so that this equation of motion is

$$kx + m\frac{d^2x}{dt^2} = 0. \tag{32–22}$$

This equation for x has the same form as the equation for Q, and so the solutions for these variables have the same form. But we already know that the mechanical motion is harmonic—$x = A\sin\omega t + B\cos\omega t$—and so the same is true for the charge on the capacitor. We must change the variable names using the correspondences summarized in Table 32–2. In particular, the angular frequency of the mechanical system is $\omega = \sqrt{k/m}$, so from the table of equivalences the angular frequency of the LC circuit is

$$\omega = \frac{1}{\sqrt{LC}}, \tag{32–23}$$

TABLE 32-2 • Analogy Between *LC* Circuit and Mass on a Spring

	Mass on Spring Parameter	*LC* Circuit Parameter
Variable	x	Q
Coefficient of variable	k	$(1/C)$
Coefficient of $\frac{d^2}{dt^2}$ (variable)	m	L
Natural frequency	$\sqrt{k/m}$	$1/\sqrt{LC}$

and the charge on the capacitor varies as

$$Q = Q_1 \sin(\omega t) + Q_2 \cos(\omega t). \tag{32–24}$$

Equivalently, this can be written as

$$Q = Q_0 \cos(\omega t + \phi). \tag{32–25}$$

The constants, either Q_1 and Q_2 or Q_0 and ϕ, are determined by the initial conditions. For example, if we know that the capacitor has a given total charge at time $t = 0$ and that the current at $t = 0$ is also zero (because a switch has been open until that time), then we can determine the two unknown constants.

Once we know the charge on the capacitor, we also know the current flowing through the inductor; we need only apply the relation $I = dQ/dt$. For example, Eq. (32–25) gives

$$I = \frac{d}{dt}[Q_0 \cos(\omega t + \phi)] = -Q_0\omega \sin(\omega t + \phi). \tag{32–26}$$

Thus the current, like the charge, is oscillatory, and with the same angular frequency $\omega = 1/\sqrt{LC}$. Here we have rather directly used the equivalent of the equations of motion, but we shall see in Section 32–7 that it is also possible to understand the oscillations in terms of energy flow.

EXAMPLE 32–9 The capacitor ($C = 0.80\ \mu\text{F}$) in the circuit of Fig. 32–14 carries a charge in the form $Q = Q_0 \cos(\omega t)$. Find the voltage drop across the inductor ($L = 12\ \mu\text{H}$) and the period of oscillation of that voltage.

Strategy The current through the circuit is the time rate of change of the charge on the capacitor, so the voltage across the inductor is

$$V_L = -L\frac{dI}{dt} = -L\frac{d^2Q}{dt^2}.$$

The time dependence of Q is given, and the derivative can be carried out to find V_L. Since two derivatives of a cosine give the cosine back again, the time dependence of V_L, like that of the charge, will be $\cos \omega t$. The angular frequency ω is given for our circuit by Eq. (32–23), $\omega = 1/\sqrt{LC}$, and from this the period $T = 1/f = 2\pi/\omega = 2\pi\sqrt{LC}$ can also be evaluated.

Working It Out We have

$$\frac{d^2Q}{dt^2} = Q_0\frac{d}{dt}\left(\frac{d}{dt}\cos \omega t\right)$$
$$= Q_0\frac{d}{dt}[-\omega \sin \omega t] = -\omega^2 Q_0 \cos \omega t = -\omega^2 Q.$$

Thus

$$V_L = \omega^2 LQ = \omega^2 LQ_0 \cos \omega t = (Q_0/C) \cos \omega t,$$

where in the last step we used $\omega^2 = 1/LC$. The oscillation period of this voltage drop is given by

$$T = \frac{2\pi}{\omega} = 2\pi\sqrt{LC} = 2\pi\sqrt{(12 \times 10^{-6}\ \text{H})(0.80 \times 10^{-6}\ \text{F})}$$
$$= 1.9 \times 10^{-5}\ \text{s}.$$

Circuits such as this are useful where internal timing is necessary—here we have a clock with a 20 μs "tick," or equivalently a 0.05 MHz frequency. When you realize that FM radio is in the range of 100 MHz, then you can begin to see why circuits such as this have direct relevance to FM radio broadcasting; for example, decreasing the inductance by a very reasonable factor will lead to oscillations in the FM realm.

What Do You Think? Does the maximum voltage across the inductor depend more strongly on the inductance L or the capacitance C in this example?

THINK ABOUT THIS. . .

ARE THERE TECHNOLOGICAL APPLICATIONS FOR CURRENT OSCILLATIONS?

Every time you use your cell phone you are making use of the phenomenon. In what has become a very important application, wireless communication relies on circuits with natural frequencies of what are effectively *LC* circuits produced on chips with a so-called CMOS (Complementary Metal Oxide Semiconductor) technology. These circuits, which are microscopic and involve very little energy, are used for both the generation and reception of signals—the reception occurs through what we will learn are resonance phenomena analogous to those seen in mechanical systems.

Another class of application is in sound synthesizers. We have seen in Chapter 15 that the sounds we hear are a complex combination of different frequencies, and successful simulation of, say, a trumpet or a violin is possible only when the entire set of frequencies and their relative amplitudes approximates the spectrum of a real instrument. For this purpose *LC* circuits with enough of a variety of inductances and capacitances can generate fairly complex spectra.

32–6 Damped Oscillations in *RLC* Circuits

When resistance, inductance, and capacitance are all present in a single-loop circuit, as in Fig. 32–15, we have an ***RLC* circuit**. The addition of a resistance to the *LC* circuit gives us an element for which there is energy loss—energy dissipated in Joule heating—and as we shall see next, the resistance term in Kirchhoff's loop rule is analogous to a drag force, proportional to the speed of the mass, in a mechanical harmonic oscillator. Thus we would expect behavior characteristic of damped mechanical oscillations in *RLC* circuits. It is worth noting that since any inductor has some resistance, all *LC* circuits are really *RLC* circuits.

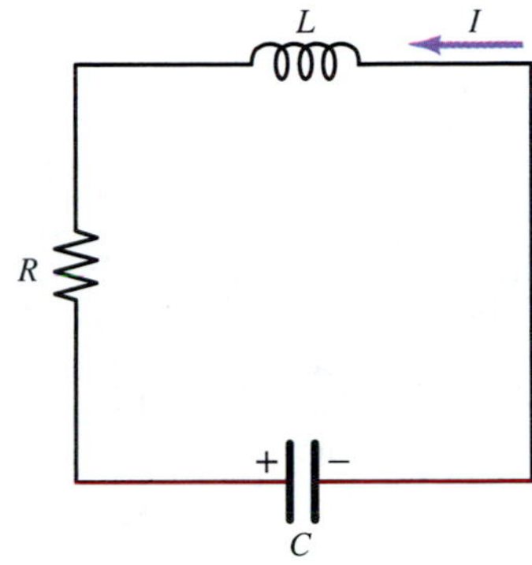

▲ **FIGURE 32–15** A basic *RLC* circuit.

To analyze this case, we assume that a current is present in the circuit and apply the loop rule by following the direction of the current:

$$-L\frac{dI}{dt} - IR - \frac{Q}{C} = 0. \qquad (32\text{–}27)$$

Here Q is the charge on the capacitor. Because $I = dQ/dt$, Eq. (32–27) can also be written as

$$L\frac{d^2Q}{dt^2} + R\frac{dQ}{dt} + \frac{Q}{C} = 0. \qquad (32\text{–}28)$$

This equation is a differential equation for the charge Q; with the appropriate initial conditions, it determines the charge on the capacitor, including its time dependence. The current in the circuit is then found by differentiation of the charge.

Equation (32–28) has an analogue in mechanics problems that involve masses on springs in the presence of drag: To Newton's second law for a mass on an ideal spring [Eq. (32–22)], add a term for a drag force. The drag force $-bv$ is proportional to the velocity or first derivative of x. With this term Newton's second law for the damped harmonic oscillator becomes

$$m\frac{d^2x}{dt^2} + b\frac{dx}{dt} + kx = 0. \qquad (32\text{–}29)$$

This equation [which is Eq. (13–45)] describes the motion of a mass at the end of a spring immersed in a fluid that gives rise to a drag force, and it is a physical system about which we have prior knowledge—it was the subject of Section 13–7. Intuition about this system and the mathematical equivalence of Eqs. (32–28) and (32–29), described in detail in Table 32–3, is immensely helpful in understanding the *RLC* circuit. We note in particular that the mechanical drag term $bv = b\,dx/dt$ corresponds to the circuit resistance term $RI = R\,dQ/dt$.

As you can review in Section 13–7, damping modulates the harmonic behavior of a mass on an ideal spring by changing the period slightly and imposing an envelope on the harmonic motion in the form of a falling exponential. The harmonic motion occurs with an exponentially decreasing amplitude. In the same way, the charge on the capacitor of an *RLC* circuit will display harmonic oscillatory time dependence within an envelope that

TABLE 32–3 • Analogy Between *RLC* Circuits and Damped Harmonic Motion

	Damped Harmonic Motion Parameter	*RLC* Circuit Parameter
Variable	x	Q
Coefficient of variable	k	$1/C$
Coefficient of $\frac{d}{dt}$ (variable)	b	R
Coefficient of $\frac{d^2}{dt^2}$ (variable)	m	L

falls exponentially with time. Current is the time derivative of charge, and because the derivatives of sines, cosines, and exponentials are cosines, sines, and exponentials, the current will also be harmonic in time within an exponentially decaying envelope.

We now flesh this out in a quantitative way. We can follow the techniques of Section 13–7 to find a solution to the loop rule equation for the *RLC* circuit, Eqs. (32–27) or (32–28). Using our solution to the damped oscillator, Eq. (13–46), along with the equivalences in Table 32–3, we find

$$Q = Q_0 e^{-\alpha t}\cos(\omega' t + \phi). \tag{32–30}$$

The constants α and ω' are determined either by substitution back into the original loop rule equation, Eq. (32–28), or simply by using the correspondences in Table 32–3 together with the results of Section 13–7. They are

$$\alpha = \frac{R}{2L} \tag{32–31}$$

and

$$\omega'^2 = \frac{1}{LC} - \frac{R^2}{4L^2} = \frac{1}{LC} - \alpha^2 = \omega^2 - \alpha^2. \tag{32–32}$$

The quantity α determines the rate of exponential damping and has dimensions of inverse time. The constants Q_0 and ϕ are determined from the initial conditions.

The exponential damping constant α depends only on L and R, as in *RL* circuits. This factor shows that the resistive element is the crucial element in damping; when there is no resistance, there is no damping. (Recall that pure *LC* circuits oscillate without damping.) The damping factor $e^{-\alpha t}$ forms a decreasing envelope for the harmonic behavior within the envelope. Note how the angular frequency ω' differs from the angular frequency $\omega = 1/\sqrt{LC}$ of the undamped circuit (an *LC* circuit). If the damping constant α is *small* compared to ω, then ω' is only slightly less than ω. In Fig. 32–16, we plot the behavior of the capacitor charge for a circuit in which $L = 1$ H, $C = 1$ F, and $R = 0.3\ \Omega$, and compare it to the previous case of $R = 0$. The period in the damped case is only slightly larger than the period for the undamped case, and this difference is not very apparent in the figure.

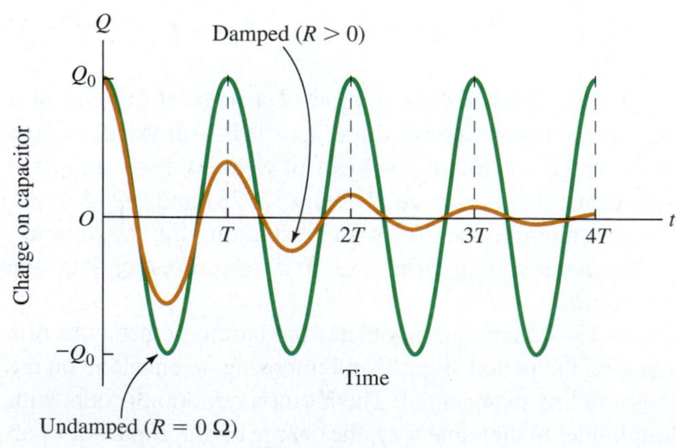

FIGURE 32–16 Comparison of an *RLC* circuit with and without damping: the charge on the capacitor versus time. The damped case has a very slightly larger period, so close to that of the undamped case that it cannot be seen in the figure. T is the period of the undamped oscillator.

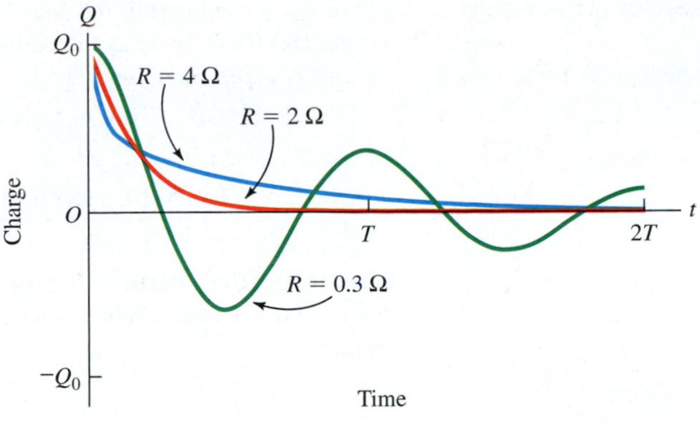

FIGURE 32–17 The *RLC* circuit of Fig. 32–15 for various values of *R*. The critical value R_c is $R = 2\ \Omega$. There is no oscillatory behavior, only damping, when $R > R_c$.

Figure 32–17 shows what happens to the charge if *R* is increased to $2\ \Omega$ and to $4\ \Omega$. What is the explanation for this behavior? Equation (32–32) shows that when *R* is increased to a critical value R_c, ω'^2 decreases to zero. For $R = R_c$, we have

$$0 = \frac{1}{LC} - \frac{R_c^2}{4L^2},$$

which has the solution

$$R_c = 2\sqrt{\frac{L}{C}}. \qquad (32\text{–}33)$$

When ω'^2 is zero, there is no more oscillation; as in Chapter 13, we refer to this case as *critical damping*. Judicious use of a resistance that will lead to critical damping will *eliminate* current oscillations in situations where it is not wanted. The value $R = 2\ \Omega$ in Fig. 32–17 represents this case: When $L = 1$ H and $C = 1$ F, we have $R_c = 2\ \Omega$. For values of *R* that are larger than the critical value, there is *overdamping*. This kind of motion has its mechanical analogue in the motion of a mass at the end of a spring when the mass is moving in a jar of thick molasses. There is no oscillation, just a slow movement directly to the equilibrium position. An example of this behavior is illustrated in Fig. 32–17, where we have chosen $R = 4\ \Omega$ (a value greater than R_c).

EXAMPLE 32–10 A switch is used to introduce a resistor of $0.052\ \Omega$ as a series element into an *LC* circuit that is undergoing oscillatory behavior (Fig. 32–18). The values of the inductance and the capacitance are 75 mH and 16 μF, respectively. How much time passes between $t = 0$, when the switch is thrown, and the moment when the amplitude of the oscillations has decreased to one half of its value prior to $t = 0$? How many oscillations does the circuit undergo during this time?

Strategy This problem is a straightforward application of our expression for the capacitor charge *Q*, Eq. (32–30). The envelope factor is $\exp(-\alpha t)$, and at the time $t_{1/2}$ when this factor is one half, we say by definition that the amplitude of the oscillations will have been cut in half. We know all the factors that go into α, so the relation $\exp(-\alpha t_{1/2}) = \frac{1}{2}$ is a relation that we can solve for $t_{1/2}$. We then see how many periods $T = 2\pi/\omega'$ will fit into $t_{1/2}$, and this is the answer to the second part of the question.

It is worth noting that if $\omega \gg \alpha$, the envelope that characterizes the decay of the oscillations changes slowly compared to the oscillation time. Many oscillations take place during the time that the envelope decreases only a little, and it is meaningful to speak of the decay of the amplitude itself, according to

$$\text{amplitude} = A_0 \exp(-\alpha t).$$

We can verify numerically whether this situation holds or not.

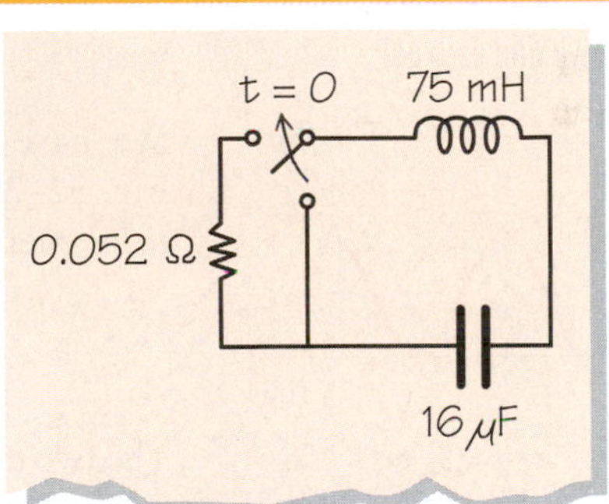

FIGURE 32–18

Working It Out Equation (32–31) is used to determine the damping factor α:

$$\alpha = \frac{R}{2L} = \frac{(0.052\ \Omega)}{2(75 \times 10^{-3}\ \text{H})} = 0.35\ \text{s}^{-1}.$$

By comparison, the angular frequency of the oscillations of the (undamped) *LC* circuit is

$$\omega = \frac{1}{\sqrt{LC}} = \frac{1}{\sqrt{(75 \times 10^{-3}\ \text{H})(16 \times 10^{-6}\ \text{F})}}$$
$$= 9.1 \times 10^2\ \text{rad/s}.$$

(continues on next page)

Indeed $\omega \gg \alpha$, and the concept of the decay of the amplitude is reasonable.

For $t_{1/2}$, we have $\exp(-\alpha t_{1/2}) = \frac{1}{2}$, or taking the natural logarithm of both sides,

$$-\alpha t_{1/2} = \ln(1/2) = -\ln 2 = -0.69.$$

Thus

$$t_{1/2} = \frac{0.69}{\alpha} = \frac{0.69}{0.35\ \text{s}^{-1}} = 2.0\ \text{s}.$$

For the second part of the question, with $\omega \gg \alpha$, $\omega \cong \omega'$, and we can take $T = 2\pi/\omega$ as the period of the oscillator. Then during the time $t_{1/2}$, the circuit undergoes

$$\frac{t_{1/2}}{T} = \frac{t_{1/2}}{2\pi/\omega} = \frac{\omega t_{1/2}}{2\pi}$$
$$= \frac{(9.1 \times 10^2\ \text{rad/s})(2.0\ \text{s})}{2\pi} \cong 290\ \text{oscillations.}$$

What Do You Think? The original resistor is replaced by one that is 10 times larger. What effect does this have on the number of oscillations?

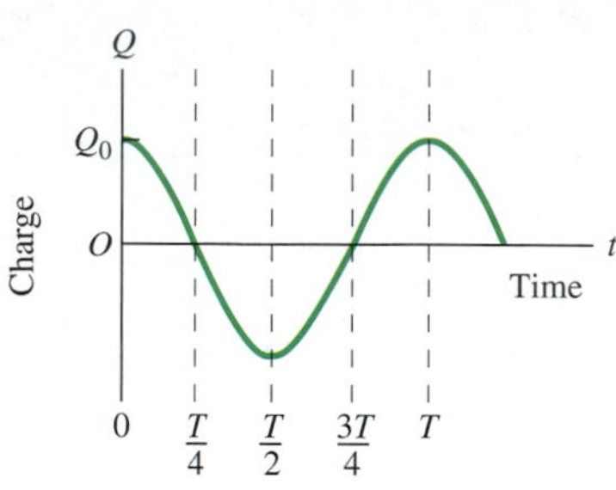

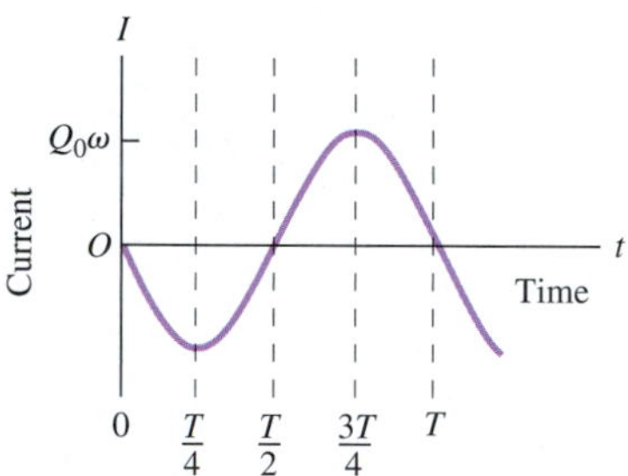

▲ **FIGURE 32–19** The capacitor charge and the current in a circuit that contains only inductance and capacitance are harmonic functions of time. Note the relative positions of the maxima and minima for current and charge.

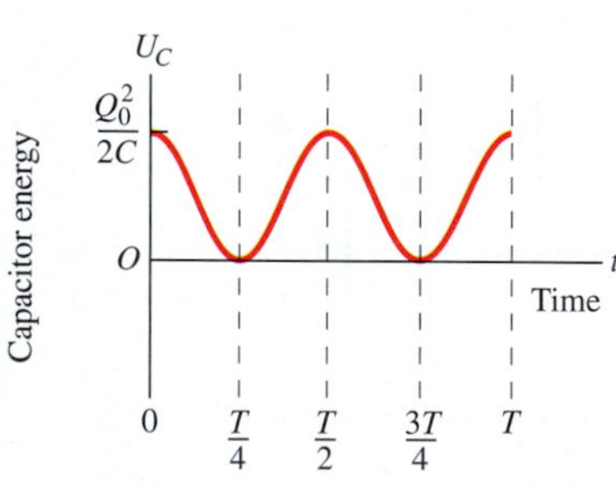

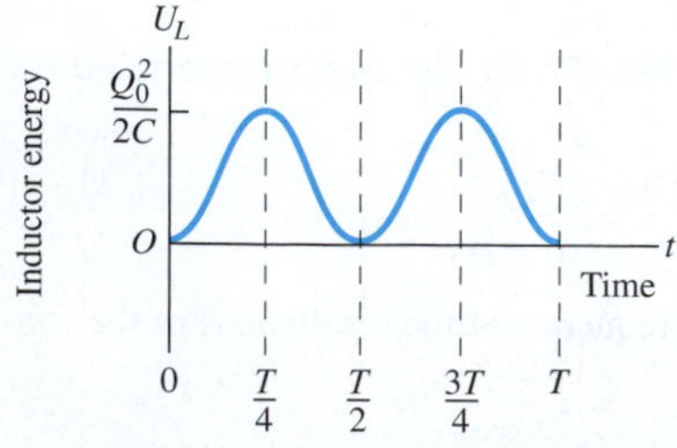

▲ **FIGURE 32–20** The energies of capacitor and inductor for charge and current of the circuit in Fig. 32–15, but with $R = 0$. These oscillate out of phase.

32–7 Energy in *LC* and *RLC* Circuits

Energy is a useful concept in *RLC* circuits, just as it is for the harmonic oscillator. Let's reconsider instances without and with damping.

No Resistance: We set $R = 0$, i.e. there is no damping. We take initial conditions such that the charge on the capacitor is

$$Q = Q_0 \cos \omega t. \tag{32–34}$$

The current in the circuit is then

$$I = \frac{dQ}{dt} = -\omega Q_0 \sin \omega t. \tag{32–35}$$

One full period of both of these functions is plotted in Fig. 32–19.

The energy contained in a capacitor is given by Eq. (26–8), $U_C = Q^2/2C$, or

$$U_C = \frac{Q_0^2}{2C}\cos^2(\omega t). \tag{32–36}$$

Equation (32–14) gives the magnetic energy in an inductor, $U_L = \frac{1}{2}LI^2$:

$$U_L = \frac{1}{2}L\omega^2 Q_0^2 \sin^2(\omega t) = \frac{Q_0^2}{2C}\sin^2(\omega t). \tag{32–37}$$

We have used $\omega = 1/\sqrt{LC}$. The (positive) functions U_C and U_L are plotted in Fig. 32–20. The energy of one rises to a maximum as the other falls to zero. But the *total* energy in the inductor and capacitor is *constant*:

$$U = U_C + U_L = \frac{Q_0^2}{2C}[\cos^2(\omega t) + \sin^2(\omega t)] = \frac{Q_0^2}{2C}. \tag{32–38}$$

The two circuit elements swap the constant total energy back and forth harmonically, just as in the mechanical oscillator, for which the constant total energy is made up of a back-and-forth exchange of the potential energy of the spring and the kinetic energy of the attached mass. In both the mechanical and circuit case, this is an expression of the conservation of energy. In Figs. 32–21a to 32–21e, we have drawn the progression in a series of snapshots over a full period T, starting with no current and a fully charged capacitor at $t = 0$. Without a resistor to dissipate energy, the oscillation will continue forever.

Resistance Is Introduced: We can easily understand the role of resistance in terms of energy. The power P_R dissipated in the resistor is voltage times current:

$$P_R = IV_R = I^2R, \tag{32–39}$$

where we have used Ohm's law, $V_R = IR$. This power is proportional to the current squared and is always positive. Energy is *always* lost to Joule heating in a resistor, regardless of the sign of the current. This is the origin of the exponential damping in *RLC* circuits. The power loss in resistors should be contrasted to the equivalent expressions for inductors ($P_L = IV_L$) or capacitors ($P_C = IV_C$). In each case, the rate of energy expenditure can be positive or negative, according to the situation [see Eq. (32–11)]. Unlike the resistor, these elements sometimes take energy from the other circuit elements and sometimes give it back.

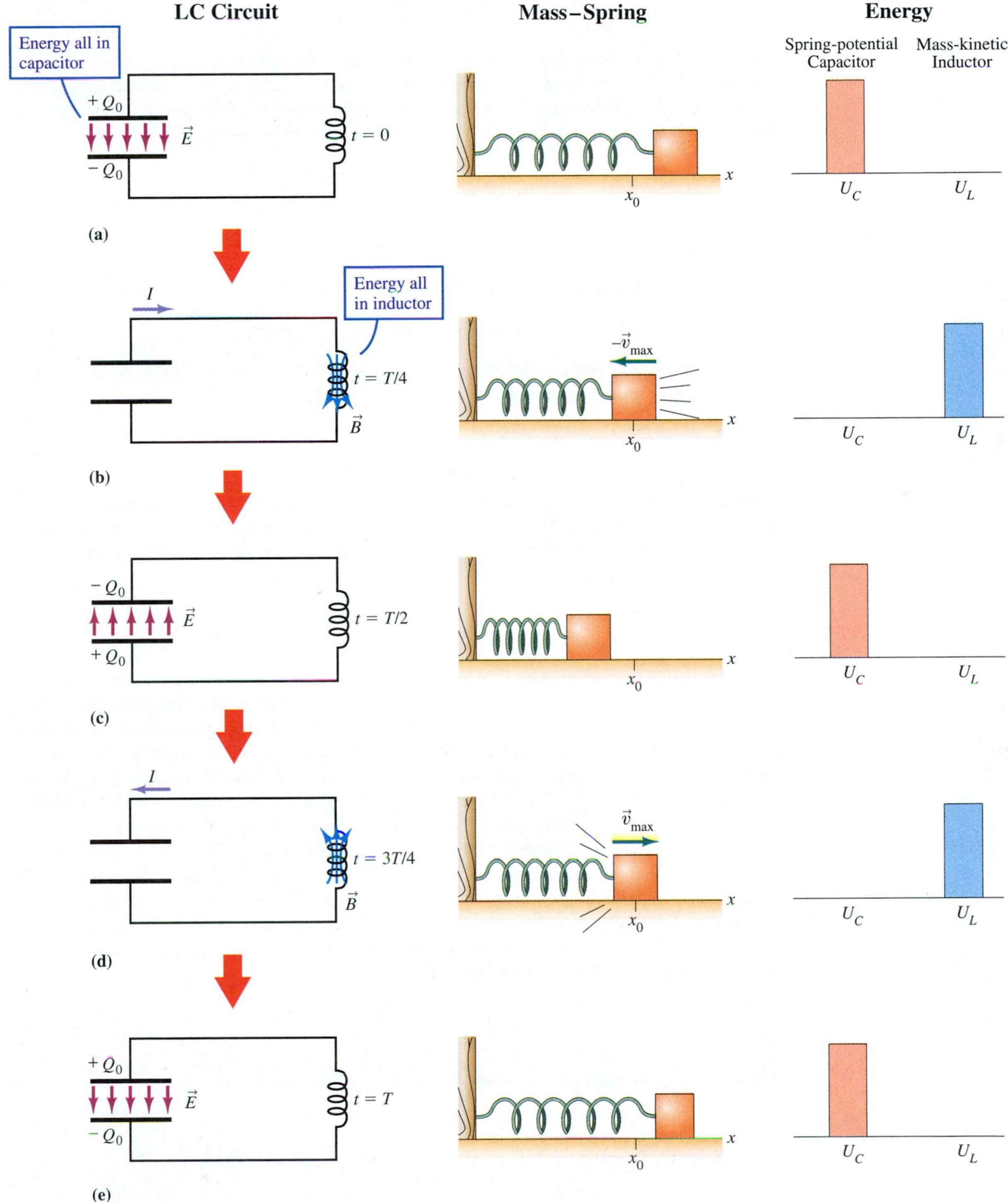

▲ **FIGURE 32–21** (a)–(e) A time sequence showing how the electric and magnetic fields, and therefore the energies, of Fig. 32–20 are realized within the capacitor and inductor. We also show the analogous mechanical system at corresponding times, as well as the balance of energy in the capacitor and inductor for those times.

Summary

An inductor is a circuit element that behaves as a current-carrying loop, or solenoid. It has an inductance L defined by the ratio between the magnetic flux and the current that passes through it:

$$\Phi_B = LI. \tag{32–1}$$

By Faraday's law, the emf induced in this circuit element is

$$\mathscr{E} = -\frac{d\Phi_B}{dt} = -L\frac{dI}{dt}. \tag{32–2}$$

Faraday's law also shows that when there are adjacent loops in a circuit (or pair of circuits), the changing current in one loop induces an emf in the adjacent loop. In this case, the geometrical factor is the mutual inductance M, which measures both the emf induced in loop 1 due to the current in loop 2 and the emf induced in loop 2 due to the change in current in loop 1,

$$\mathscr{E}_{21} = -M\frac{dI_1}{dt}. \tag{32–5}$$

Inductance is measured in henries (H) in the SI. The emf in an inductor is one more term to add to the loop rule.

A simple, calculable inductance is that of an ideal solenoid:

$$\text{for an ideal solenoid: } L = \mu_0 A\ell n^2. \tag{32–8}$$

Here, A is the area, ℓ is the length of the solenoid, and n is the number of turns per unit length.

The energy carried in an inductor is given by

$$U_L = \frac{1}{2}LI^2. \tag{32–13}$$

Just as the energy of a capacitor is carried by the electric field in the capacitor, the energy of an inductor is in the magnetic field. The energy density, or energy per unit volume, carried by a magnetic field is found by comparing the known field within a solenoid with the energy carried by the solenoid, and is given by

$$u_B = \frac{1}{2}\frac{B^2}{\mu_0}. \tag{32–16}$$

The combination of inductance, capacitance, and resistance in circuits with and without batteries leads to interesting time dependence for currents and charges. The current in the inductor cannot change instantaneously. When an inductor is placed in a circuit with a battery and a resistor, we have an *RL* circuit, and the loop rule produces an equation for the current characterized by transient exponential behavior, with time constant L/R. When a capacitor is added to the circuit, the loop rule produces an equation for the charge on the capacitor whose solution is a capacitor charge that varies harmonically when there is no resistance (we have an *LC* circuit) and whose capacitor charge behaves as damped harmonic motion when there is resistance (an *RLC* circuit). The current in these two cases also exhibits harmonic oscillations or damped harmonic oscillations, respectively. These phenomena can also be viewed in terms of energy, with the harmonic oscillations of the *LC* circuit characterized by a continual exchange of energy between the capacitor and the inductor, and the resistance providing a mechanism of energy loss. The angular frequency of the free oscillations in an *LC* circuit is

$$\omega = \frac{1}{\sqrt{LC}}. \tag{32–23}$$

When resistance is added, the capacitor charge, for example, has the time dependence

$$Q = Q_0 e^{-\alpha t}\cos(\omega' t + \phi). \tag{32–30}$$

The exponential damping factor is governed by the size of the resistance,

$$\alpha = \frac{R}{2L}, \tag{32–31}$$

while the frequency is shifted to the value

$$\omega'^2 = \omega^2 - \alpha^2. \tag{32–32}$$

Understanding the Concepts

1. Two electric circuits are placed near one another. Each circuit has self-inductance. Must there be a mutual inductance?
2. Does it take more work to cause current to flow through a coil of wire than through the same wire when it is straight?
3. Consider two circular coils that are in a variety of configurations (Fig. 32–22). Assuming that the separation between the coils is roughly the same for the various configurations, can you order the mutual inductances from largest to smallest?

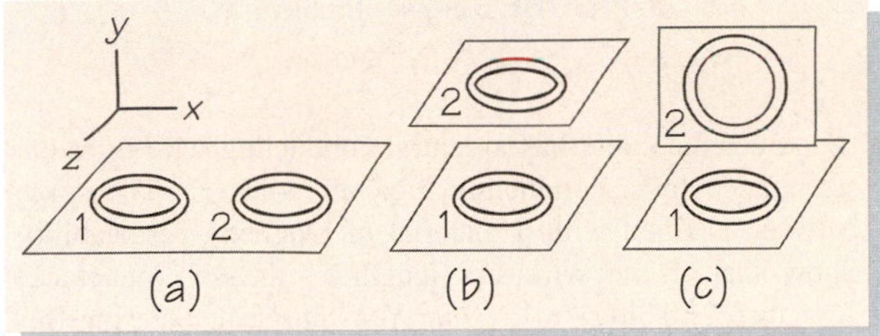

▲ **FIGURE 32–22** Question 3.

4. Why might Faraday's law cause the lights in a house to dim when an electrical appliance that uses a lot of energy, such as an electric clothes dryer, is turned on?
5. Is it possible to calculate the mutual inductance between a straight wire and a wire loop?
6. If the energy in a given region is equally split between energy in the electric and the magnetic fields, what does this tell you about the relative magnitude of the magnetic and electric fields?
7. The time for the current amplitude to drop to $\frac{1}{2}$ its initial value in Example 32–10 is (a) larger than, (b) the same as, or (c) smaller than the time for the charge amplitude to be one half its initial value.
8. Why do you sometimes see a spark at a light switch when the switch is turned off? Is there a spark when the switch is turned on? Why or why not?
9. Describe how you could measure inductance with a battery of known emf, a known resistor, a voltmeter, and a timer.
10. A lightbulb is placed in series with a resistor and in parallel with a coil of large inductance and negligible resistance. When a switch that connects a battery to this circuit is closed, the lightbulb flashes before glowing dimly. When the switch is opened, the bulb flashes again before going out. Explain.
11. In the oscillations of an LC circuit, the energy is transferred from the electric field in the capacitor to the magnetic field in the inductor. How does the energy get from one place to the other?
12. Given your knowledge of the largest and smallest practical sizes of capacitors and inductors, what would you estimate is the electronic oscillator with the smallest frequency possible? The largest?
13. Consider Example 32–9. What determines the value of the maximum charge on the capacitor in that example?
14. Given a certain length of wire, you are asked to wind it in the form of a solenoid in a way that maximizes the self-inductance. You may change the radius of the core and the number of turns per unit length, but must take into account that the wire has a certain fixed thickness. What should you do?
15. A solenoid has magnetic flux outside as well as inside, because magnetic field lines must close on themselves. Does this mean that there is magnetic energy outside the solenoid as well as inside it?
16. In Section 32–3, the magnetic energy calculation for a solenoid used the inductance of a portion of the solenoid, and that was translated into the energy density expression $B^2/2\,\mu_0$. The magnetic field must come around the outside of the solenoid because all magnetic field lines are closed. Why then does the above calculation give the correct answer for the energy density?
17. When $R = 0$, the time constant for the RL circuit is infinite. Physically, why?
18. If a diamagnetic material is placed inside the solenoidal coil, the self-inductance of the coil will (a) increase, (b) decrease, (c) stay the same.
19. We have made an analogy between a damped harmonic oscillator and an RLC circuit. What mechanical quantities are analogous to the energies $LI^2/2$ and $Q^2/2C$ of the RLC circuit?
20. How would you go about finding a generalization of Eq. (32–13) when two circuits with different currents are placed in such close proximity that their mutual inductance plays a role?
21. The magnetic energy density $B^2/2\,\mu_0$ has the dimensions of pressure and may be viewed as a magnetic pressure. Use this interpretation to justify the attraction/repulsion of two parallel wires that carry currents in the same/opposite directions.
22. If you have two parallel wires carrying currents, then the sign of the force (attraction or repulsion) depends on whether the currents travel in the same or opposite directions. Can you interpret this in terms of the magnetic energy density in the region between the wires?

Problems

32–1 Inductance and Inductors

1. (I) A wire loop has an inductance of 2 mH when a current of 30 mA passes through the circuit. What is the value of the magnetic flux that passes through the loop?
2. (I) What is the self-inductance per unit volume of a solenoid?
3. (I) Calculate the mutual inductance of a solenoid 25 cm long of radius 1.8 cm with 600 turns, and a single loop of radius 3.0 cm centered on the solenoid, with its area perpendicular to the axis of the solenoid.
4. (I) A current that changes at the rate of 1.25 A/s passes through a solenoid; an induced emf of 520 mV is the result. The length and diameter of the solenoid are 12 cm and 0.50 cm, respectively. What is the number of turns?
5. (I) The emf induced in an isolated circuit when the current in the circuit is changing by 10 A/s is 0.3 V. What is the self-inductance?
6. (II) An electrical engineer needs an inductor capable of producing an emf of 150 mV. The current source available produces current of the form $I = I_0 \cos(\omega t)$, with $I_0 = 0.60$ A and $\omega = 2.7 \times 10^2$ rad/s. What size inductor should be used?
7. (II) A current of 1 A flows through a circuit placed in isolation. A magnetic flux of 0.010 T·m^2 passes through the circuit area. When this circuit is placed near another circuit with a current flow of 2 A, the magnetic flux through the first circuit increases to 0.012 T·m^2. (a) What is the mutual inductance of the two circuits? (b) How much magnetic flux passes through the second circuit, whose self-inductance is 1 mH?

8. (II) What is the self-inductance of the single inductor that is equivalent to two inductors of values L_1 and L_2, respectively, placed in series? Neglect the mutual inductance.

9. (II) A solenoid of length L consists of two coils tightly placed on top of each other. One coil has N_1 turns, the other N_2, and the area of the coils is A. Calculate the self-inductances of the coils (a) if only one of the two coils is used; (b) if the two coils are connected in series with their windings going in the same direction; (c) if the two coils are connected in series with their windings going in the opposite direction. (d) Calculate the mutual inductance of the two coils.

10. (II) Consider two inductors with inductances L_1 and L_2, respectively, connected in parallel. What is the value of the single equivalent inductance that could replace the two inductances, assuming that the mutual inductance can be neglected?

11. (II) A solenoid of length L and area A contains two windings—one tightly placed on top of the other—with N_1 and N_2 turns, respectively. What happens if the two windings are connected in parallel and the composite coil is included in a circuit with a variable current?

12. (II) Equation (32–8) was derived for an ideal cylindrical solenoid. Show that this result holds also for a solenoid of any shape cross section, provided that the length is large compared to any cross-sectional measure.

13. (II) Consider the cylindrical solenoid and ring illustrated in Fig. 32–23. The solenoid, of diameter $d_1 = 2.0$ cm, has length $\ell = 20$ cm and 120 turns of wire. The ring of wire inside, with diameter $d_2 = 1.5$ cm and area perpendicular to the solenoid's axis, is connected by two wires to a single resistor of resistance $R = 33\ \Omega$. The current I_1 is in the form of a pulse that starts to rise linearly at $t = 0$ s. The current reaches a maximum of 30 A at $t = 0.30$ s, then starts to descend linearly; when the current reaches 0 A at $t = 0.60$ s, it ceases to flow. Find the current I_2 induced in the ring as a function of time.

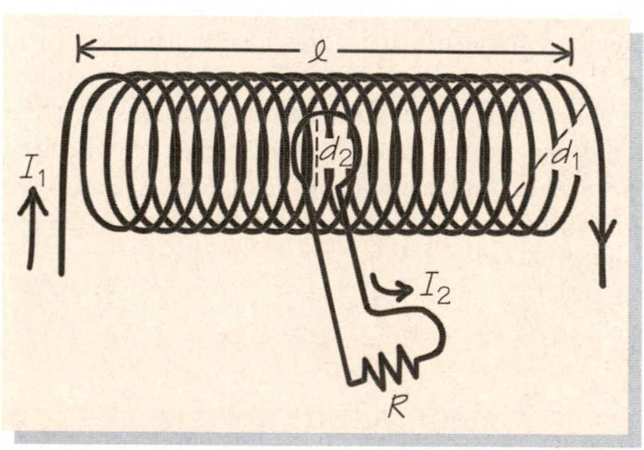

▲ FIGURE 32–23 Problem 13.

14. (II) The ring contained within the cylindrical solenoid in Fig. 32–6a is replaced by a second cylindrical solenoid, of length ℓ_2, radius R_2, and turn density n_2. Calculate the mutual inductance of this system.

15. (II) The current in an inductor has the periodic triangular form plotted in Fig. 32–24, with an amplitude of 0.50 A and a period of $T = 0.45$ s. What is the voltage across the inductor as a function of time if $L = 2.3 \times 10^{-3}$ H? Express your answer algebraically or plot it.

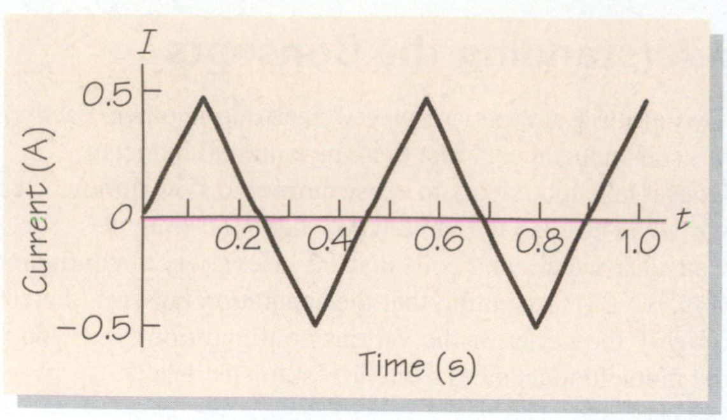

▲ FIGURE 32–24 Problem 15.

16. (II) A coaxial cable has a central conducting wire of radius r_0 surrounded by a conducting tube of radius r_1. The space in between is filled with a material of magnetic permeability μ. Show that if the wire has length ℓ, the self-inductance is $L = (\mu\ell/2\pi)\ln(r_1/r_0)$. [*Hint*: You must calculate the flux in the region between the cylinders.]

17. (II) Consider two identical solenoids placed end to end, with the windings going in the same direction. Prove that the total self-inductance of the combined system is $2(L + M)$, where L is the self-inductance of either solenoid and M is the mutual inductance.

18. (II) A torus of rectangular cross section with width w, height h, and inner radius R is wound with N turns of wire (Fig. 32–25). What is the self-inductance of the torus? Use the approximation $\ln(1 + x) = x$, valid for $x \ll 1$, to discuss the case where $R \gg w$ and its relation to the self-inductance of a solenoid.

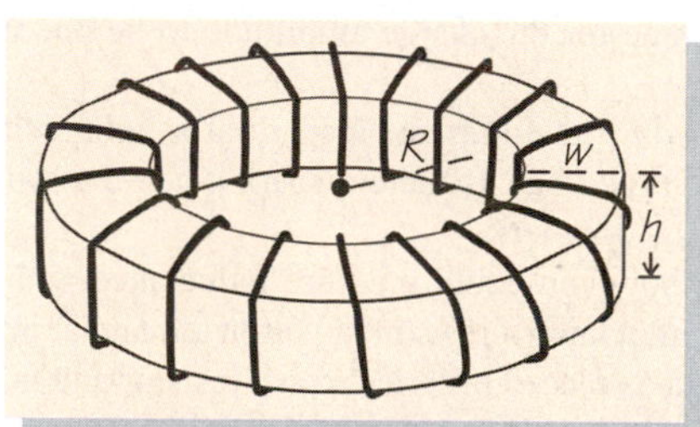

▲ FIGURE 32–25 Problem 18.

19. (II) Consider a torus of square cross section. The radius of the torus (distance from the symmetry axis to the center of the square) is 35.0 cm; the sides of the square are 5.00 cm. The torus is wound with 1650 turns of wire. (a) What is the self-inductance of the torus? (b) What is the self-inductance if the core of the torus is made of soft iron, with $\mu = 4200\mu_0$?

20. (II) Consider a toroidal coil wound around an empty core whose self-inductance is 15 mH. The current in the coil changes uniformly by 120 mA in 0.50 s. (a) What is the induced emf? (b) If the hollow center of the torus is filled with an iron core, with $\mu = 3400\mu_0$, what is the induced emf?

21. (II) Calculate the inductance of an elongated rectangular circuit, such as a length of a two-wire ribbon cable (Fig. 32–26). The length of the circuit is L and its width is $a \ll L$. If the radius of the wire is much less than a, can it (the radius) be neglected?

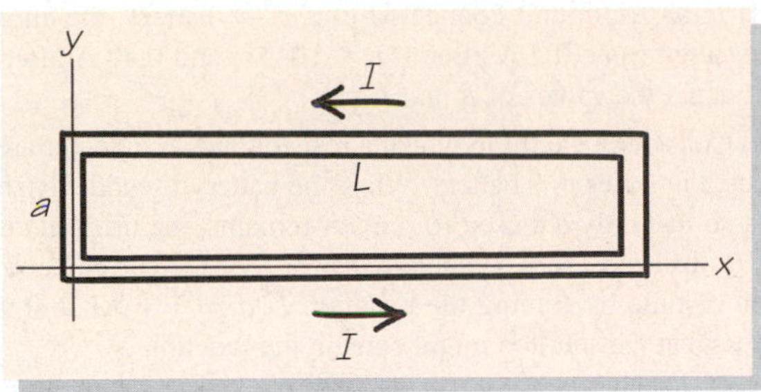

▲ **FIGURE 32–26** Problem 21.

22. (II) Figure 32–27 shows a straight wire that carries a current I and a square loop of wire with one side oriented parallel to the straight wire a distance d away. The square has sides of length a. Calculate the mutual inductance of this system. [*Hint*: The magnetic field due to the straight wire through a slice of the square of width dx parallel to the wire is constant, so the flux through this slice is easily calculable. Integrate to find the total flux through the square.]

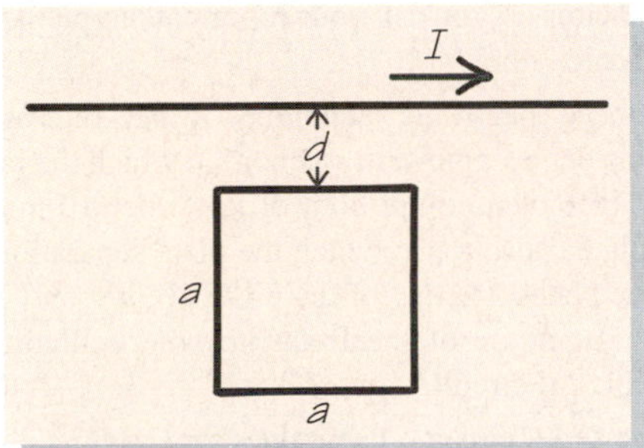

▲ **FIGURE 32–27** Problem 22.

23. (III) Calculate the mutual inductance of the two elongated rectangular circuits shown in Fig. 32–28. Assume that $L \gg a, b$.

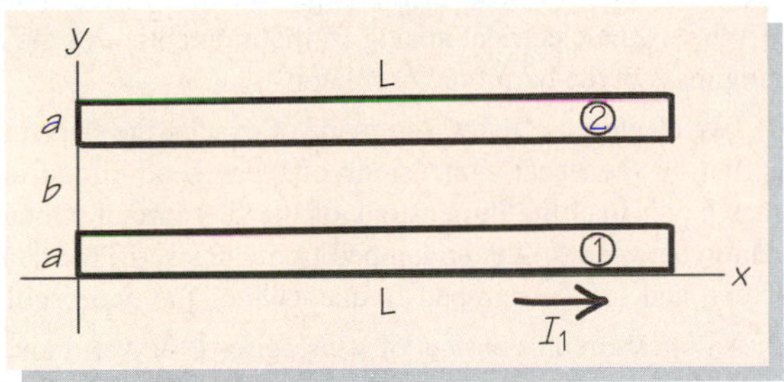

▲ **FIGURE 32–28** Problem 23.

32–2 Energy in Inductors

24. (I) A cylindrical solenoid of radius 0.75 cm is wound with a turn density of 140 turns/m. It carries a current of 0.45 A. How much energy is stored per meter length of the solenoid?

25. (I) Consider an inductor with $L = 16$ H and an internal resistance of 0.10 Ω. We wish to use this inductor to store 0.10 MJ of energy. What is the rate at which energy is lost to Joule heating in this system? It is not practical to store large amounts of energy in large inductors unless the wire is superconducting.

26. (I) An inductor with $L = 1.6$ mH has an internal resistance small enough to be ignored. How much work would a battery have to do to increase the current through the inductor from 88 mA to 98 mA?

27. (I) A capacitor with $C = 0.020\ \mu$F has a charge of 15 μC. What is the equivalent steady current that should be carried by an inductor of $L = 20\ \mu$H if the inductor is to store the same amount of energy?

28. (I) A doorbell circuit contains a solenoid with 600 turns of wire, a cross section of 6.0 cm^2, and a length of 12 cm. When the doorbell button is pushed, 100 mA passes through the circuit. How much energy is contained within the solenoid at this time?

29. (II) A current with time dependence $I = I_0 e^{-\alpha t}$ passes through an inductor with $L = 2$ mH; $I_0 = 4.0$ A and $\alpha = 0.02\ \text{s}^{-1}$. Compute the power expended in the inductor as a function of time.

30. (II) The voltage across an inductor with $L = 3.0$ mH is fixed at 6.0 V. The current is increased (a) from 0.00 A to 0.25 A, (b) from 0.25 A to 0.35 A, and (c) from 0.35 A to 0.40 A. What average power must be supplied from an external source in each step?

31. (II) Consider an inductor with $L = 1$ H and a capacitor with $C = 1$ F. (a) Compare the energy contained in the inductor when a current of 10 A flows through it with the energy in the capacitor if the charge is the amount of charge contained in the 10-A current, flowing for 1 s. (b) Repeat part (a) for a current of 1 mA.

32. (II) An electrical engineer constructs a cylindrical solenoid of area 8 cm^2 and length 25 cm from 150 m of thin wire. The wire will handle a maximum current of 50 mA. (a) What is the inductance of the solenoid? (b) How much energy can the inductor store?

33. (II) The inductance of a small superconducting solenoid is 8 H. The current is gradually increased from 0 A to 40 A. (a) How much energy is stored in the solenoid? (b) When the current reaches 40 A, the solenoid "quenches"; i.e. the wire of the solenoid loses its superconductive property because of the large magnetic field. The current decreases to zero rapidly, and the magnetic energy is dissipated in the liquid helium coolant. Given that the latent heat of vaporization of helium is 2.7×10^3 J/L, how much of the liquid helium coolant is evaporated during the quench?

32–3 Energy in Magnetic Fields

34. (I) The magnetic field in interstellar space has an approximate magnitude of 10^{-10} T. How much magnetic field energy does this contribute to the spherical region around the Sun of radius matching the mean radius of Neptune's orbit?

35. (I) The two circular pole pieces of a magnet are 63 cm in diameter and 21 cm apart. The magnetic field between them is 0.10 T. What is the magnetic energy stored in the field?

36. (I) An ideal cylindrical solenoid carrying a current of 115 mA has a winding density of 15 turns/cm. If the core is filled with iron, $\chi_m = 5500$, what is the energy density contained in the magnetic field within?

37. (II) A straight wire carries a current $I = 20$ A. Find the energy density in the surrounding magnetic field as a function of the distance r from the wire. At what distance from the wire does the energy density equal that of a parallel-plate capacitor with a charge of 10^{-7} C and a capacitance of 6.3×10^{-9} F, if the separation between the plates is 1.5 mm?

38. (II) (a) What is the energy density of the magnetic field outside a straight wire of radius a that carries a current I? (b) What is the total energy per unit length, due to that magnetic field, that is contained in a cylinder of radius R ($R > a$) centered about the wire?

39. (II) Consider a torus of radius R, wound with n turns per unit length of a wire that carries a current I. The cross section of the torus forms a square with sides of length b; $b \ll R$. We know that the magnetic field inside the torus has the nearly constant value $B = \mu_0 nI$. Use this result and the two expressions related to the magnetic energy [Eqs. (32–13) and (32–15)] to show that, for this torus, $L = 2\pi\mu_0 n^2 Rb^2$.

40. (II) Consider an iron ring of cross-sectional area A. The upper half of it has a certain number of turns of wire wrapped around it. The wire is connected to a battery so that a current I flows through the wire. The ring is cut so that there is a small gap of width y. Within the gap, the magnetic field has magnitude B. (a) What is the force exerted by the upper half of the ring on the lower half? (b) Suppose the gap is filled with some material of large (positive) susceptibility. Will the force increase or decrease?

41. (II) Consider a hollow solenoid inside of which there is a magnetic field B. The solenoid is vertical, and a small cylinder of iron that just fits into the solenoid without touching the sides is dropped into the empty space. Will there be a force on the small cylinder? If so, in what direction will it point? How large will it be?

42. (II) (a) What is the magnetic field energy density inside a straight wire of radius a that carries current I uniformly over its area? (b) What is the total magnetic field energy per unit length inside the wire?

43. (II) A coaxial cable consists of a wire 0.15 cm in diameter with a return path for the current in the shape of a very thin cylindrical conductor of diameter 0.80 cm. A current of 0.25 A flows through the cable. Calculate the magnetic energy per unit length of cable within the inner wire. [*Hint*: It will help to do Problem 42 first.]

32–4 Time Dependence in *RL* Circuits

44. (I) You wish to make a circuit in which a resistor and an inductor are connected in series to a battery such that when the switch is closed, the current builds up to within 18% of its steady-state value in 5×10^{-4} s. You have a series of inductors with inductances ranging from 0.010 H to 0.10 H. You must acquire a set of resistors with what range of resistance?

45. (I) Show that the time constant L/R that characterizes RL circuits has dimensions of time.

46. (II) Consider the RL circuit of Fig. 32–10; the switch is closed at time $t = 0$. For the circuit elements, $\mathcal{E} = 6$ V, $R = 3.3$ kΩ, and $L = 2.5$ mH. Using Eq. (32–19), find how much charge flows in the circuit during the first (a) 1 μs, (b) 1 ms, (c) 1 s.

47. (II) Show by direct substitution that Eq. (32–19) is a solution of Eq. (32–18).

48. (II) Consider the RL circuit shown in Fig. 32–29. The switch is opened at time $t = 0$ after it had been closed for a long time. How long will it take for the current in the inductance to drop to 25% of its initial value? How long will it take for the energy in the inductance to drop to 25% of its initial value?

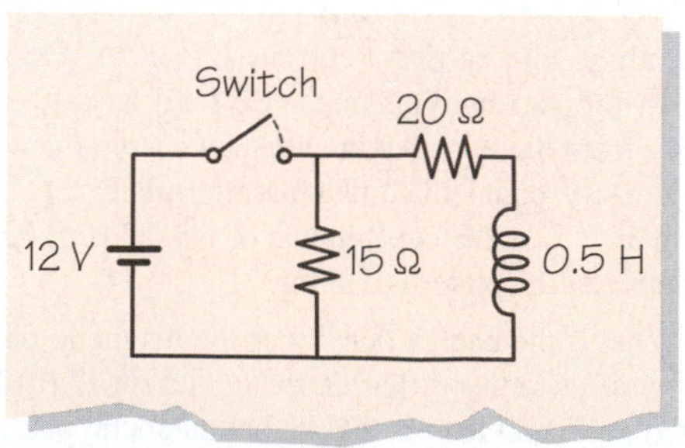

▲ **FIGURE 32–29** Problem 48.

49. (II) In an RL circuit connected to a 12-V battery, the current is measured to be 0.2 A after 1.2×10^{-4} s, and 0.48 A after 10 s. What are the values of R and L?

50. (II) Consider a circuit in which a resistor and an inductor are connected in series to a battery. When the battery is suddenly shorted out so that only a closed RL circuit remains, the original current, V/R, decays to zero. Calculate the form of the current as a function of time by solving the equation $L(dI/dt) + RI = 0$ with a guess that the solution might contain the function $e^{-\alpha t}$.

51. (II) Consider the situation discussed in Problem 50. Calculate the total energy dissipated in the resistor from the time when the switch that shorts out the battery is thrown to the time $t = \infty$. Show that this is the energy stored in the inductor just before the switch is closed.

32–5 Oscillations in *LC* Circuits

52. (I) An electric oscillator consists of a parallel-plate capacitor and a long, cylindrical solenoid. If the resonant frequency of the oscillator is ω_0, what is the frequency of a similar oscillator in which both the capacitance and inductance are reduced by a factor of 12?

53. (I) You have an inductor with an inductance of 40 mH. Using it, you want to make a circuit with oscillations of frequency 20 Hz. What capacitor do you ask your roommate to pick up at the corner electronics store?

54. (II) Advanced electronic techniques utilize microscopic structures. Consider a single-turn solenoid in which the radius and the length of the solenoid are both of the order 10 microns, and a parallel plate capacitor in which the plate separation and the radius of the plates are also of the order 10 microns. *Estimate* the order of magnitude of the frequency of oscillation of such a microscopic LC circuit.

55. (II) Design an LC circuit—give values for C and L—that has an angular frequency of 4.32×10^4 rad/s and a stored energy of 0.30 mJ. The maximum voltage drop across the capacitor must be 20.0 V.

56. (II) Suppose that at time $t = 0$ the current in an ideal LC circuit is zero but the charge on the capacitor is $Q(0)$. (a) What is the energy in terms of $Q(0)$, L and C? (b) If the charge on the capacitor is zero at $t = 0$, and the current has the value $I(0)$, what is the energy? (c) Starting at $t = 0$, when is the first time that the energy is equally stored in the capacitor and in the inductor? (d) What is the electrical analog of the momentum of the oscillating mass in the harmonic oscillator?

57. (II) Two electric oscillators are made of exactly the same materials, but all the linear dimensions of the second circuit are ten times larger than the dimensions of the first circuit. Obtain the relation between (a) the undamped frequencies, (b) the damping factors, and (c) the damped frequencies of the two oscillators.

58. (II) An open circuit consists of a capacitor C and an inductor L connected in series. A charge q is placed on the capacitor, and the circuit is closed at time $t = 0$ by means of a switch. Find the maximum value of the current, as well as the times for which this maximum value occurs.

32–6 Damped Oscillations in *RLC* Circuits

59. (I) An RLC circuit is composed of a resistor $R = 0.883$ Ω, an inductor $L = 1.75$ H, and a capacitor $C = 133$ pF, all arranged in series. What is the angular frequency of current oscillations in this circuit?

60. (I) An RLC circuit has $R = 85$ mΩ, $L = 0.60$ mH, and $C = 55$ μF. (a) Find the damping factor and angular frequency. (b) If the resistance is variable, what value of R will give critical damping?

61. (I) Consider a series RLC circuit for which the initial capacitor charge is Q_0. If R is chosen such that there is critical damping, what is the instantaneous power consumption in the resistor? [*Hint*: Try the formula in Eq. (32–30).]

62. (II) Consider an RLC circuit at critical damping, with $L = 68$ mH. What is the value of R if the current decays by 15 percent in 8.0 ms?

63. (II) Show that Eqs. (32–30) through (32–32) solve Eq. (32–27).

64. (II) Suppose that the values of R, L, and C in a series RLC circuit are such that $\omega'^2 < 0$. Assuming that the solution for the charge on the capacitor takes the form $Q = Q_1 \exp(-\alpha_1 t) + Q_2 \exp(-\alpha_2 t)$, find the values of α_1 and α_2.

65. (II) Consider the basic RLC circuit. By making an appropriate approximation of Eq. (32–32), show that when α is small compared to $\omega = 1/\sqrt{LC}$ the modified angular frequency ω' of the damped RLC circuit is $\omega' \simeq \omega - R^2\sqrt{C/L}/8L$. Find a similar relation for the periods of the undamped and slightly damped cases.

32–7 Energy in *LC* and *RLC* Circuits

66. (II) Calculate the energy in an LC circuit, assuming that the initial conditions are such that the charge on the capacitor is $Q = Q_0 \cos(\omega t + \delta)$. Show that the energy is constant.

67. (II) A circuit consists of a capacitor of capacitance $C = 20$ nF connected in series with an inductor of inductance $L = 2 \times 10^{-5}$ H. If a charge of 30 nC is put on the capacitor, there is an oscillation in the circuit. (a) What is the maximum current that moves through this circuit? (b) Find the maximum energy within the inductor. (c) What is the ratio of the maximum energy in the inductor to the maximum energy in the capacitor?

68. (II) An LC circuit consists of a 15-mH inductor and a 120-μF capacitor. If the maximum energy stored in the circuit is 3.0×10^{-4} J, what are the maximum charge on the capacitor and the maximum current in the circuit? What are the minimum values?

69. (III) The 3-mF capacitor of an RLC circuit is initially charged to 30 μC. The 1.5-mH inductor has a very small resistance. At a particular instant, after 100 oscillations, the current through the inductor is zero while the capacitor is still charged to 5 μC. (a) What is the resistance of the circuit? (b) What are the energies of the circuit before and after the 100 oscillations? (c) Why are the two values of the energy in part (b) different? Where has the energy gone?

70. (III) Consider an RLC circuit. The energy is given by $E = LI^2/2 + Q^2/(2C)$. Show that the rate of change of this energy is equal to the power loss in the resistor (the ohmic heating power).

General Problems

71. (II) By considering the definition of inductance, show that if the voltage V across an inductor changes with time, the total current passing through the inductor in that time is given by $I = \dfrac{1}{L}\displaystyle\int V\,dt$.

72. (II) Suppose that a square wave of voltage, as plotted in Fig. 32–30, is applied across an inductor with $L = 0.005$ H. Use the result of Problem 71 to plot the current as a function of time.

73. (II) The switch in the circuit shown in Fig. 32–31 has been closed for a long time. (a) What is the current in each leg of the circuit? (b) When the switch is opened, the current in the inductor drops by a factor of 2 in 8 μs. What is the value of the inductance? (c) What is the current passing in each leg at 12 μs?

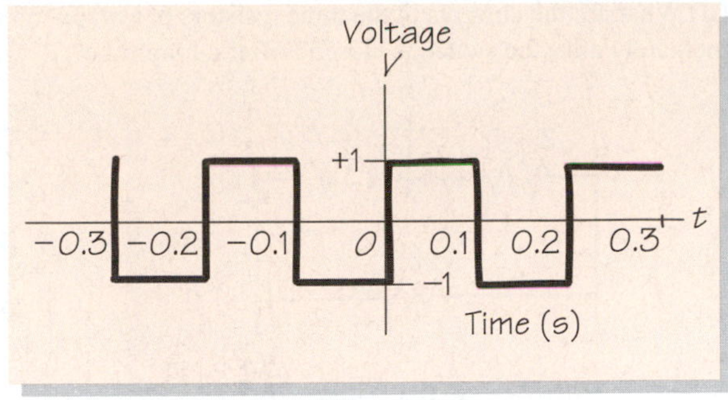

▲ **FIGURE 32–30** Problem 72.

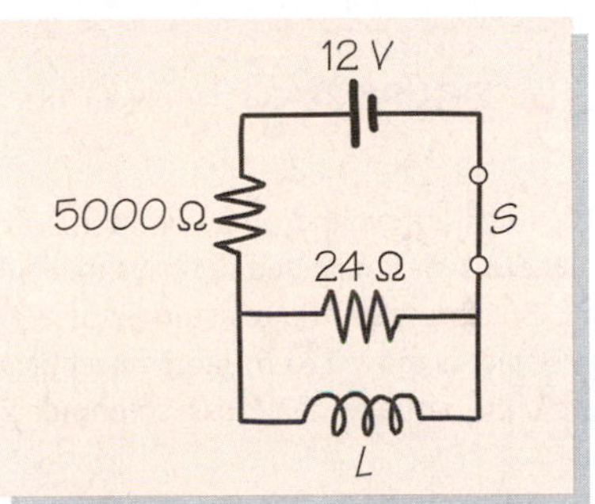

▲ **FIGURE 32–31** Problem 73.

74. (II) As a way of preventing arc formation between the terminals of a switch, a capacitor is connected to the two terminals (Fig. 32–32). What is the minimum capacitance of the capacitor if no voltage larger than 200 V is to be allowed in the circuit? [*Hint*: Assume that very little power can be dissipated during the time that the capacitor charges.]

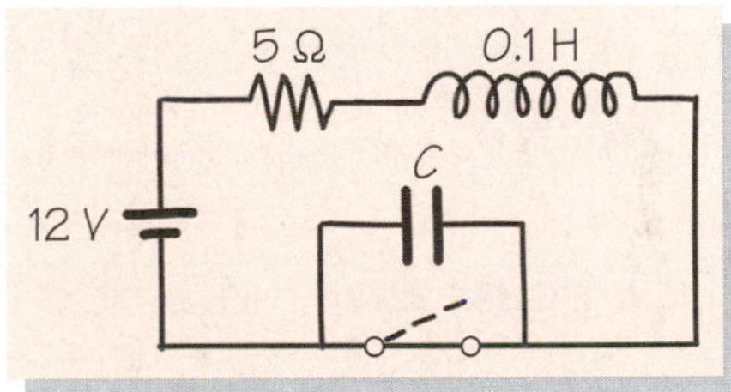

▲ **FIGURE 32–32** Problem 74.

75. (II) A coaxial cable has an inner, solid wire of radius r_1 and an outer, hollow wire of radius r_2. A current I flows through the inner wire and returns through the outer wire. Assuming that the cable is infinitely long, find the magnetic field energy per unit length. Include any field energy inside the inner wire and outside the outer wire.

76. (II) Molybdenum is paramagnetic, with a magnetic susceptibility of 1.2×10^{-4} at 300K, which is about one half its value at 20K. Suppose that the self-inductance of a solenoid filled with molybdenum is $L = 0.35$ mH at 300K. What is the fractional change in self-inductance between 300K and 20K?

77. (II) Consider a cavity uniformly filled with oscillating electric and magnetic fields. (a) Show that the ratio of the amplitude of these fields, E_0 and B_0, respectively, has the dimensions of $[\text{velocity}]^{-1}$. (b) For what value of this ratio is the magnetic energy density equal to the electric energy density?

78. (II) What are the currents in the three resistors of Fig. 32–33 immediately after the switch is closed? After a long time?

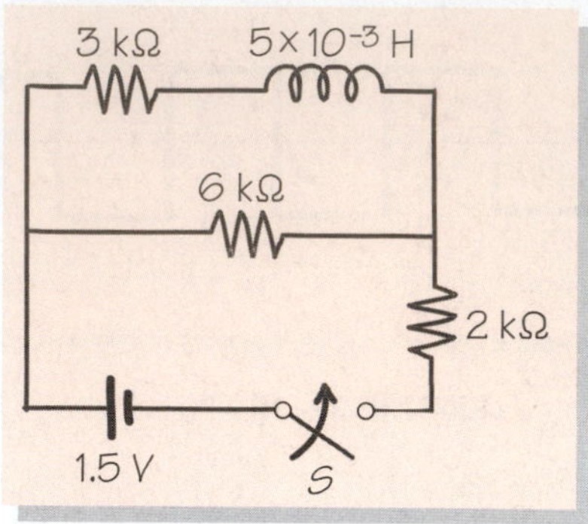

▲ **FIGURE 32–33** Problem 78.

79. (II) Two solenoids are wound on a common soft iron core (Fig. 32–34). Solenoid S_1 is connected in series to a battery and a variable resistor. Starting with the resistor set at A (low resistance), the sliding contact is moved to B (large resistance) and back to A again. Sketch the voltage V across solenoid S_2 while this is happening.

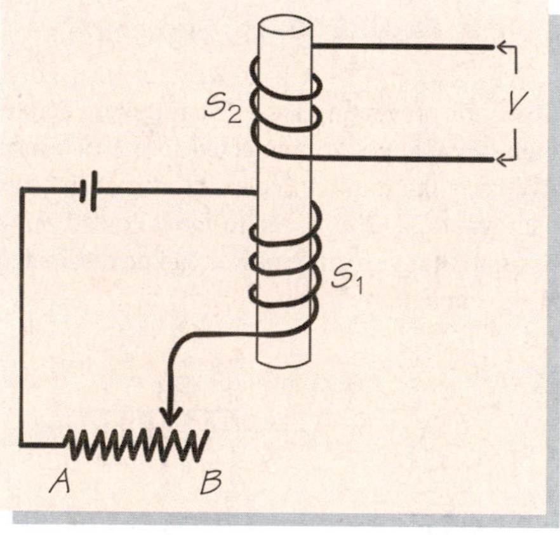

▲ **FIGURE 32–34** Problem 79.

80. (II) The two identical coils in the circuit of Fig. 32–35 are placed close to each other, and their mutual inductance is 0.7 mH. Suppose that the switch has been closed for a long time and is then opened at $t = 0$. Calculate the current in the circuit at $t = 18$ ms.

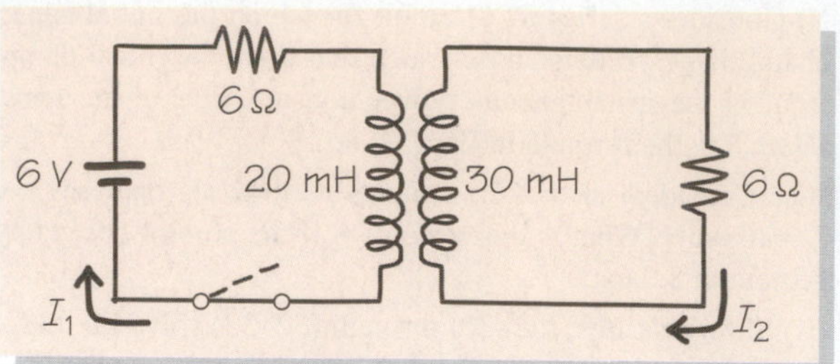

▲ **FIGURE 32–35** Problem 80.

81. (II) A ferromagnetic torus is part of a device to be used in a region where the magnetic permeability has the constant value $\mu = 2500\mu_0$. The torus has a circular cross section of 4 cm^2. Over its total length of approximately 35 cm, the torus is wrapped with 220 turns of wire. Immediately surrounding this winding is a secondary winding of 40 turns of (insulated) wire. What is the mutual inductance of the two windings? What is the role of the iron core, if any, in determining this mutual inductance?

82. (II) A torus of inner radius r_i and outer radius r_0 has a square cross section (Fig. 32–36). It is wound with N turns of wire that carries a current I. (a) Use Ampère's law to find the magnetic field inside the torus. (b) Calculate the magnetic energy density within the torus. (c) Integrate the magnetic energy density to find the total magnetic energy within the torus. (d) Use the formula $U_L = \frac{1}{2}LI^2$ to compute the self-inductance of this torus.

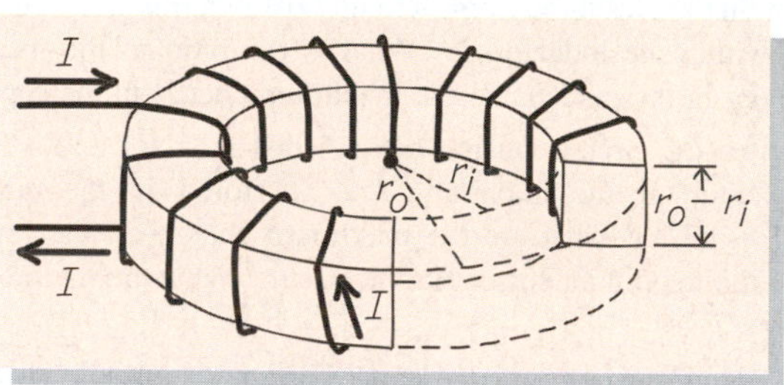

▲ **FIGURE 32–36** Problem 82.

83. (II) An LC circuit oscillates with an angular frequency of 1.2×10^6 rad/s. When a second capacitor is inserted in series with the original one, the angular frequency becomes 1.6×10^6 rad/s. If the capacitors are replaced by a resistor of 0.02 Ω, the current drops to 1/2 of its initial value in 3.5 ms. What are the values of the two capacitors and of the inductance L?

84. (III) Consider two adjacent circuits as shown in Fig. 32–2. Show that the total energy is given by $U = \frac{1}{2}L_1I_1^2 + \frac{1}{2}L_2I_2^2 + MI_1I_2$ and that $M \leq (L_1L_2)^{1/2}$.

Chapter

This bank of transformers at an electricity substation forms a crucial part of the electricity distribution system. Transformers use Faraday's law to raise or lower the amplitude of time-varying voltages, allowing electrical energy to be transported efficiently at high voltages and consumed at safer low voltages.

Alternating Currents

In Chapter 30 we learned how a changing magnetic flux induces an emf—Faraday's law. In particular, when a coil rotates in the presence of a magnet, an emf is induced in the coil that varies sinusoidally with time. The induced emf produces an alternating current (AC), which is a source of AC power. AC generators use induction to convert the mechanical energy of falling water or the pressure of hot steam into electric currents that vary with time. Such generators are the starting point for the delivery of electric power to home and industry. Alternating current circuits are at the heart of most household equipment, and alternating current flows every time you switch on a light. The ability to vary the maximum voltage of the harmonically oscillating emf that AC power provides is an important element in the delivery of electric power, and learning how to do this is one of our aims here. AC sources of emf in circuits that include resistors, inductors, and capacitors provide currents and voltages with new types of time-dependent behavior. In particular, such circuits exhibit the same kinds of resonance phenomena that we saw in mechanics and are the basis for devices like radio tuners.

33–1 Transformers

An alternating current is characterized by harmonic (sine and cosine) time dependence, as are the other variables of the circuit, such as voltages[†]. The possibility of being able to vary the maximum AC voltage (the *voltage amplitude*) is of interest because high or

[†]"AC" stands for any kind of current or voltage that varies harmonically in time.

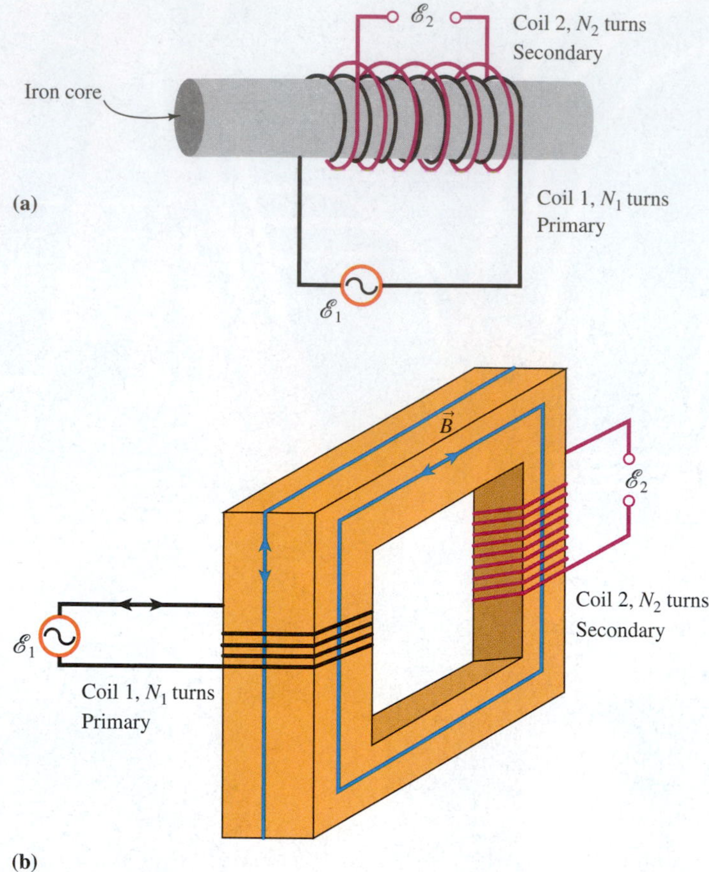

▲ **FIGURE 33–1** Two methods for creating fully linked coils: (a) one coil tightly wound over the other; (b) two coils wrapped around a common core of ferromagnetic material, which has the property of keeping the magnetic field lines within it.

low voltages are useful in differing circumstances. For example, it is more economical to transport electric energy at high voltage (see the "Think About This" box on p. 920). High voltages, however, are dangerous, as well as inefficient in small appliances.

Let's suppose that an AC generator produces an emf of the general form

$$\mathcal{E} = V_0 \sin \omega t \tag{33–1}$$

[see Eq. (30–14)]. The factor V_0 is the voltage amplitude of the source of emf, and this is the quantity we want to vary. In this section we describe a device that can take an AC emf as input and produce another AC emf with a *different* voltage amplitude. This device is called a **transformer**, and it is constructed using the principle of mutual inductance.

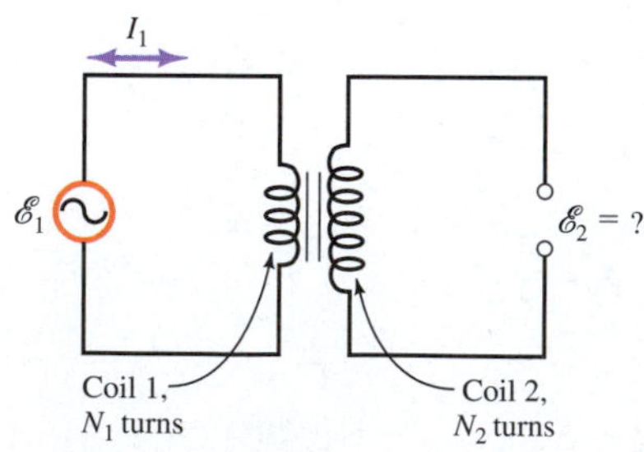

▲ **FIGURE 33–2** Schematic circuit-diagram symbol for two fully linked coils. When one coil is in the same circuit with an AC source of emf, this combination acts as a transformer.

Consider two fully linked ideal solenoids. We can arrange this in either of the two ways shown in Fig. 33–1. What we mean by "fully linked" is that the magnetic flux through one turn is equal to the magnetic flux through one turn in the other. The solenoids have a different total number of turns: N_1 and N_2, respectively. Across the first coil—the *primary coil*—there is an AC emf $\mathcal{E}$ with an amplitude V_0, as in Eq. (33–1):

$$\mathcal{E} = V_0 \sin \omega t. \tag{33–2}$$

This emf is changing with time, and you can see from Fig. 33–2 that the voltage drop $V_1(t) = V_{10} \sin \omega_1 t$ across the full primary coil has exactly the same form as Eq. (33–2). The source of this potential is Faraday's law, and this law tells us that the voltage drop across *each turn* is proportional to the rate of change of the flux through the primary coil. It is irrelevant for our purposes that this proportionality has something to do with the self-inductance of the coil; the only thing that matters is that the voltage per turn, V_1/N_1, across the primary coil is proportional to this change in flux. Now, since the coils are fully linked, we see that the voltage across *each turn* of the secondary coil is proportional in exactly the same way to the rate of change of the magnetic flux.

The voltage drop across *each turn* of the secondary coil is thus identical to the voltage drop across each turn of the primary. We learn two things. First, the time dependence of the voltage across the secondary coil is exactly the same as the time dependence of the voltage across the primary coil. Thus the voltage across the secondary coil is $V_2(t) = V_{20} \sin \omega_2 t$, and the time dependence of the voltage across both coils is identical:

$$\omega_1 = \omega_2 = \omega. \tag{33–3}$$

The second thing we learn is a relation between the voltage *amplitudes* in the two coils. We have argued that

$$V_1(t)/N_1 = V_2(t)/N_2, \tag{33–4}$$

RATIO OF VOLTAGES IN A TRANSFORMER

and when we cancel the identical time-dependent factors $\sin \omega t$, we are left with $V_{10}/N_1 = V_{20}/N_2$ or, rearranging,

$$\frac{V_{20}}{V_{10}} = \frac{N_2}{N_1}. \tag{33–5}$$

The two-solenoid device we have described is a transformer, a tool for manipulating voltage amplitudes. Transformers find use throughout the power distribution network, and you have no doubt frequently used them for your electronic tools, which may require voltage amplitudes a good deal less than the 120-V or 220-V voltages supplied by a wall plug.

We see from Eq. (33–5) that when the number of turns on the secondary coil is greater than the number of turns on the primary coil $(N_2 > N_1)$, the transformer is a *step-up transformer*, and the voltage amplitude in the secondary coil is greater than that in the primary coil. When $N_2 < N_1$, the transformer is a *step-down transformer*; the voltage amplitude in the secondary coil is smaller than that in the primary coil. Note that the terms primary and secondary do not imply any fundamental distinction between the two coils, and either coil could be primary or secondary according to the application.

CONCEPTUAL EXAMPLE 33–1 Assuming that resistance has been reduced to a minimum, how could you use the power associated with an inductor to find an expression for the ratio of currents in the primary and secondary coils of a transformer?

Answer The meaning of the resistance being reduced to a minimum is that there is minimal energy loss to Joule heating—we can ideally assume this to be zero—and any energy flow into or out of the primary coil will be matched by an energy flow out of the secondary coil. The power, or rate of energy flow, in an inductor is given by the product IV. Thus the equality $I_1V_1 = I_2V_2$ can be used to find the ratio of currents in terms of the ratio of emfs, and Eq. (33–4) gives us the latter quantity in terms of the number of turns in each solenoid when the solenoids are fully linked.

Conceptual Example 33–1 shows us how we can find the ratio of currents in the two coils in terms of their numbers of turns: Assuming insignificant losses to Joule heating, conservation of energy gives us $I_1V_1 = I_2V_2$. Rearrangement gives $I_1/I_2 = V_2/V_1$, and then use of Eq. (33–4) implies

$$\frac{I_1}{I_2} = \frac{N_2}{N_1}. \tag{33–6}$$

RATIO OF CURRENTS IN A TRANSFORMER

In other words, the current amplitude in the secondary coil of a step-up transformer $(N_2 > N_1)$ is decreased, while it is increased in the secondary coil of a step-down transformer $(N_2 < N_1)$. We should add here that when there is resistance in the secondary coil or in the circuit of which the secondary coil is a part—we say that there is a *load* on the secondary coil—then things are more complicated, with the inductances playing an explicit role.

EXAMPLE 33–2 A step-down transformer has 5,000 turns in the primary coil, which handles an AC current with voltage amplitude $V_{10} = 20{,}000\text{V}$, and 220 turns in the secondary coil. If the current amplitude desired in the secondary coil is 100 A, what is the maximum power that must be delivered by the primary coil to the secondary coil? Assume negligible resistance.

Setting It Up Imagine a transformer similar to those shown in Fig. 33–1. We know the number of turns of wire, N_1 and N_2 respectively, in both the primary and secondary coils.

Strategy We must find the power P_2 delivered to the secondary coil. Maximum power is related to the unknown voltage amplitude V_{20} and to the known current amplitude I_{20} by the relation $P_{2,\,\max} = I_{20}V_{20}$. (This is actually true only when the load is zero, i.e. when there is no resistance. We'll assume this to be the case here.) We can find V_{20} from the known V_{10} and the known numbers of turns using Eq. (33–5).

Working It Out From Eq. (33–5),

$$V_{20} = \frac{N_2}{N_1}V_{10} = \frac{220}{5{,}000}(20{,}000\ \text{V}) = 880\ \text{V}.$$

With the maximum current I_{20} carried by the secondary coil at 100 A, the maximum power that it carries is $P_{2,\,\max} = (100\ \text{A})(880\ \text{V}) = 88\ \text{kW}$. With negligible resistance, this is equal to the maximum power carried in the primary coil.

What Do You Think? Must the currents in the primary and secondary coils be wound in the same direction for a transformer to work? *Answers to **What Do You Think?** questions are given in the back of the book.*

THINK ABOUT THIS. . .

WHY IS ELECTRIC POWER TRANSMITTED ALONG HIGH-VOLTAGE LINES?

The long-distance transmission lines that bring power from generating plants to users carry AC with a voltage amplitude as high as $V = 1.0$ MV. (Compare this to the 160-V amplitude—rms 110 V—for typical appliance use.) Why such high voltages? If we label the rate of energy delivery as P, then the current that runs through these lines can be characterized according to $P = IV$. (We are ignoring the "alternating" aspect here—this doesn't have much effect on our argument.) On the way, some of this power will be dissipated as Joule heating. This depends on the resistance, R, of the transmission line—the power dissipated, or lost, in the line will be $P_{\text{lost}} = I^2R = P^2R/V^2$. Now a measure of the efficiency of transmission is the ratio of the power delivered to the power lost in Joule heating—one would want this ratio to be as large as possible. Using our expressions for P and P_{lost},

$$\frac{P}{P_{\text{lost}}} = \frac{V^2}{PR}. \qquad (33\text{–}7)$$

This ratio increases rapidly as V increases, and that is why it is best to deliver the power at as high a voltage as possible. For example, a transmission line delivering power $P = 1.0$ MW may have a total resistance of 10 Ω. If the power is delivered at 110 V, the ratio in Eq. (33–7) would be intolerably low: $(110\ \text{V})^2/(1.0\ \text{MW})(10\ \Omega) = 1.2 \times 10^{-3}$; 99.9 percent of the electrical power that set off from the generating plant would be lost through Joule heating in the wires! If the power is delivered at 500,000 V, the ratio is $(500{,}000\ \text{V})^2/(1.0\ \text{MW})(10\ \Omega) = 2.5 \times 10^4$, and nearly all of the electric energy produced is delivered to the end of the line. Then transformers can reduce the voltage amplitude and make the energy safe to use.

In addition to their ubiquitous use in power transmission and distribution, and their role in allowing electronic instruments to take their power from a wall socket, transformers are crucial to the operation of every automobile. The spark that is fired by a spark plug and ignites gasoline in an automobile cylinder requires a very high potential difference between the elements of a spark plug. However, the voltage differences supplied by the automobile battery or alternator are much more modest—the battery maintains 12 V DC, for example. The spark coil is the device that makes the large potential. The coil contains a secondary coil with a large number of windings. When the current in a primary coil is interrupted suddenly, by either mechanical or electronic means, the large change in flux induces a potential difference across the secondary coil that is large enough to cause a spark to form. Timing devices ensure that the potential is set up at the right time for the engine operation.

33–2 Single Elements in AC Circuits

We now turn to the question of how AC is used in circuits. We start with an examination of the effects of placing an AC source of emf, for which

$$\mathcal{E} = V_0 \sin \omega t,$$

in a circuit containing only a single resistor, capacitor, or inductor.

Resistive Circuit

We begin with the resistive circuit shown in Fig. 33–3a. The loop rule for the potential change around the circuit is

$$V_0 \sin \omega t - IR = 0. \tag{33–8}$$

The voltage across the resistor is then $V_R = IR = V_0 \sin(\omega t)$, and the current I through the resistor is

$$I = \frac{V_R}{R} = \frac{V_0 \sin(\omega t)}{R}. \tag{33–9}$$

The current through the resistor and the voltage across the resistor have the same sinusoidal time dependence (Fig. 33–3b). With peaks and valleys that occur at the same time, we say that the current and the voltage are *in phase*. The current amplitude is $I_{max} = V_0/R$.

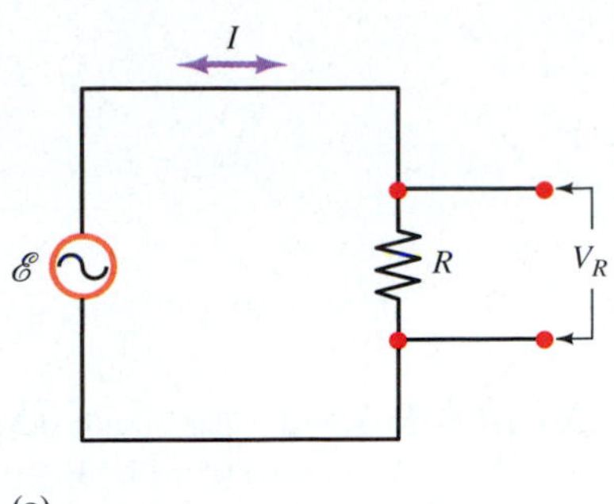

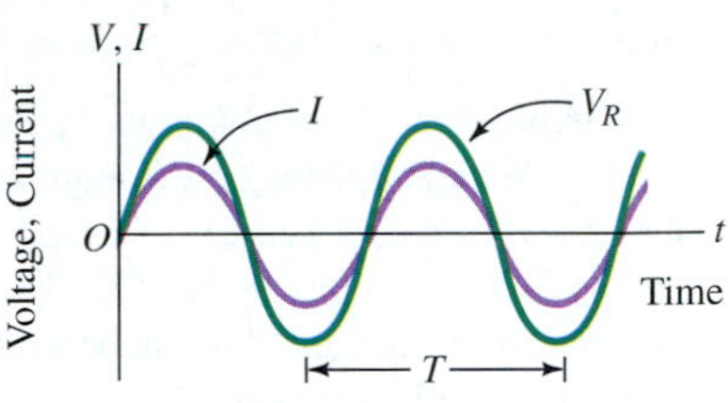

▲ **FIGURE 33–3** (a) A resistor connected in series with an AC source of emf. (b) The voltage across the resistor and the current through it are in phase. T is the period of the oscillation.

Capacitive Circuit

We now place the AC emf across a pure capacitance in the circuit, so that the voltage across the capacitor is $V_C = V_0 \sin(\omega t)$ (Fig. 33–4a). To find the current in the circuit, we apply the loop rule:

$$V_0 \sin(\omega t) - \frac{Q}{C} = 0. \tag{33–10}$$

This expression gives us the charge Q on the capacitor,

$$Q = CV_0 \sin \omega t. \tag{33–11}$$

The current is then the time derivative of the charge:

$$I = \frac{dQ}{dt} = \omega C V_0 \cos(\omega t). \tag{33–12}$$

With the aid of the identity $\sin[\theta + (\pi/2)] = \cos\theta$, Eq. (33–12) becomes

$$I = \omega C V_0 \sin\left(\omega t + \frac{\pi}{2}\right). \tag{33–13}$$

The current amplitude in this circuit is

$$I_{max} = \omega C V_0. \tag{33–14}$$

We can compare this equation to a similar equation for the resistive circuit, for which the corresponding relation was $I_{max} = V_0/R$. The *effective resistance* for a capacitive circuit is called the **capacitive reactance**, X_C, defined by

$$X_C \equiv \frac{1}{\omega C}. \tag{33–15}$$

Equation (33–14) now takes the form

$$I_{max} = \frac{V_0}{X_C}. \tag{33–16}$$

The capacitive reactance has units of ohms.

For this circuit, the effective resistance to current flow *decreases* at higher frequencies. This is physically reasonable because a high frequency means that the voltage and current are changing rapidly, and for very rapid oscillations the current must be large enough to allow the charge on the capacitor to build rapidly and thereby oppose the driving emf. In other words, the current grows with frequency. On the other hand, a low frequency means an almost constant current, and no constant current can pass through the wires leading to a capacitor. Once enough charge has been deposited on the plates, there is a large enough potential difference to stop a *steady* current. To repeat this in another way, keep in mind that while current never actually flows across the capacitor, current can flow in the wires leading to and from the capacitor plates as the charge builds up or

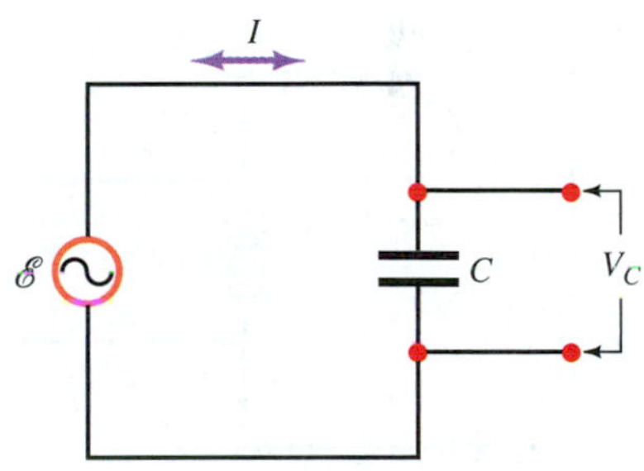

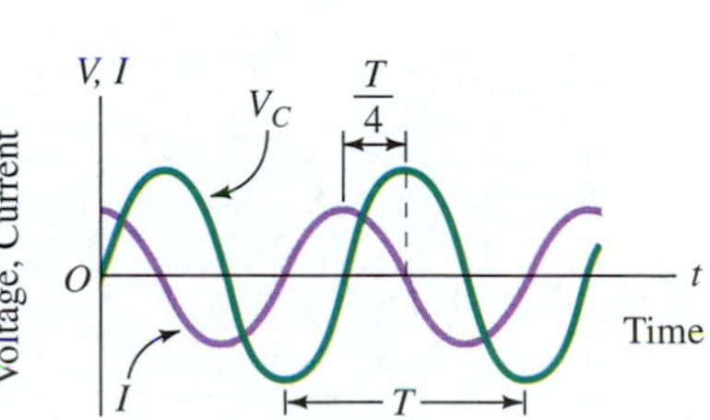

▲ **FIGURE 33–4** (a) A capacitor connected in series with an AC source of emf. (b) The current in the circuit leads the voltage across the capacitor by 90°.

flows off the plates. A steady current in the wires would mean an unlimited buildup of charge on the plates—an impossibility.

In Fig. 33–4b, we plot the current I and the voltage V_C versus time. The voltage across the capacitor is zero at time $t = 0$, but the current in the circuit is at a maximum. The phase of the current differs by a quarter cycle, $\pi/2$ rad $(90°)$, from that of the voltage, and we say that the current in the circuit *leads* the voltage across the capacitor by a phase $\pi/2$.

EXAMPLE 33–3 The circuit shown in Fig. 33–4a has an emf given by $\mathscr{E} = V_0 \sin \omega t$, with $V_0 = 6.0$ V, and a capacitance $C = 1.0\ \mu$F. (a) What are the peak currents for frequencies of 60.0 Hz and 6.00 MHz? (b) What are the currents I and voltages V_C at time 2.0 ms for the 60-Hz frequency?

Strategy (a) The unknown peak current is found from Eq. (33–14) and depends on the angular frequency. The angular frequencies ω for the two cases are found from $\omega = 2\pi f$. (b) Once we have the current amplitude, Eq. (33–13) gives us the full time dependence of the current, and this can be evaluated at any specific time, as can the given voltage, $V_0 \sin \omega t$.

Working It Out We begin with the evaluation of the angular frequencies, $\omega = 2\pi(60.0\text{ Hz}) = 2\pi(60.0\text{ s}^{-1}) = 377\text{ rad/s}$ and $2\pi(6.00 \times 10^6\text{ s}^{-1}) = 3.77 \times 10^7$ rad/s, respectively.

(a) From Eq. (33–14),

for 60 Hz: $I_{max} = (377\text{ rad/s})(1.0 \times 10^{-6}\text{ F})(6\text{ V}) = 2.3\text{ mA};$

for 6 MHz: $I_{max} = (3.77 \times 10^7\text{ rad/s})(1.0 \times 10^{-6}\text{ F})(6\text{ V})$
$= 230\text{ A}.$

The maximum current is proportional to the frequency.

(b) We evaluate Eqs. (33–13) and the input voltage $V_0 \sin \omega t$ at 2.0 ms. For $f = 60$ Hz, we have $I_0 = 2.3$ mA, $\phi = \pi/2$ rad, $V_0 = 6.0$ V, and $\omega = 377$ rad/s. Thus

$$I(t = 2.0\text{ s}) = (2.3\text{ mA}) \sin\left[(377\text{ rad/s})(0.0020\text{ s}) + \frac{\pi}{2}\right]$$
$$= 1.7\text{ mA},$$

and

$$V_C(t = 2.0\text{ s}) = (6.0\text{ V}) \sin[(377\text{ rad/s})(0.0020\text{ s})]$$
$$= 4.1\text{ V}.$$

By $t = 2.0$ ms, the current is coming down from its peak toward zero, while the voltage is rising toward its peak. At $t = 2.0$ ms, they are both about 70 percent of their peak values.

What Do You Think? Why does the higher frequency make such a big difference in the maximum current?

Inductive Circuit

Repeating the previous procedure for the circuit in Fig. 33–5a, which contains a single inductor, we first note that the voltage drop across the inductor is the emf $V_L = V_0 \sin \omega t$. For the current, we apply the loop rule to the potentials around the circuit:

$$V_0 \sin \omega t - L\frac{dI}{dt} = 0. \tag{33–17}$$

This gives us dI/dt in terms of known quantities; to find I itself, we integrate over time:

$$I(t) = \int^t \frac{dI}{dt'}\, dt' = \frac{V_0}{L}\int^t \sin(\omega t')\, dt' = -\frac{V_0}{\omega L}\cos \omega t + \text{a constant}.$$

The constant must equal zero, because there is no constant emf to drive a constant current term. We now use the trigonometric identity $\cos\theta = -\sin[\theta - (\pi/2)]$ (Appendix IV–4) to rewrite our equation as

$$I = \frac{V_0}{\omega L}\sin\left(\omega t - \frac{\pi}{2}\right). \tag{33–18}$$

The maximum current through the inductor, or current amplitude, is

$$I_{max} = \frac{V_0}{\omega L}. \tag{33–19}$$

If we compare our expressions for I_{max} with the similar one from the purely resistive circuit, $I_{max} = V_0/R$, we see that the effective resistance for an inductive circuit is ωL. We call this the **inductive reactance**, defined by

$$X_L \equiv \omega L. \tag{33–20}$$

The inductive reactance has units of ohms.

I $\mathscr{E}$ L V_L

(a)

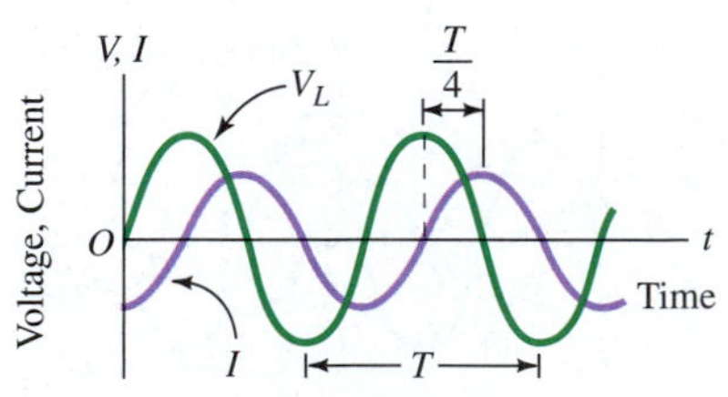

(b)

▲ **FIGURE 33–5** (a) An inductor connected in series with an AC source of emf. (b) The current in the circuit lags the voltage across the inductor by 90°.

For this circuit the effective resistance to current flow *increases* at higher frequencies. This is physically reasonable because inductors react to oppose any *change* in the current flow through them. The more rapidly the current is changing, the larger the opposing induced emf. On the other hand, the inductor is transparent to the unchanging current that corresponds to the limit $\omega = 0$.

We plot the current and voltage of the inductor versus time in Fig. 33–5b. As for the capacitive circuit, one sinusoidal curve is displaced from the other by a quarter cycle, although the role of the current and voltage curves is reversed in the two cases. This time the current in the circuit *lags* the voltage across the inductor.

EXAMPLE 33–4 Use the parameters of Example 33–3, but replace the capacitor with an inductor of inductance $L = 1.00$ mH. Calculate the inductive reactances.

Strategy We simply evaluate Eq. (33–20) to determine the inductive reactances. We have evaluated the angular frequencies in Example 33–3.

Working It Out The inductive reactances are

for 60.0 Hz:

$$X_L = \omega L = (377\ \text{rad/s})(1.00 \times 10^{-3}\ \text{H}) = 0.377\ \Omega;$$

for 6.00 MHz:

$$X_L = \omega L = (3.77 \times 10^{7}\ \text{rad/s})(1.00 \times 10^{-3}\ \text{H}) = 3.77 \times 10^{4}\ \Omega.$$

What Do You Think? A capacitor acts as an open switch when there is no time dependence in the current. Under what conditions does the inductor act as an open switch?

*Some Mathematical Devices

Two techniques simplify the treatment of circuits with time dependence. We treat them each here in abbreviated and optional form—in particular, we refer to time dependence with a single frequency. Further study in electrical engineering will involve one or the other of these techniques in more detail than we can supply here. The first involves **phasors**, which make it easier to follow phases by graphically summarizing phase and amplitude relations in AC circuits. The second, **complex analysis**, is a powerful tool that working engineers use a great deal. It simplifies all aspects of circuit analysis problems with time dependence.

Phasors: Any quantity in an AC circuit (or in any other problem) that has harmonic time dependence can be associated with a rotating vector known as a phasor. Thus for the function

$$f(t) = f_0 \sin(\omega t + \phi), \tag{33–21}$$

we define the phasor as follows: The phasor lies in the xy-plane with its tail fixed at the origin, and its length is the function amplitude, f_0. The time dependence is described by a counterclockwise rotation of the phasor with angular speed ω such that the function $f(t)$ itself is the instantaneous projection of the phasor on the y-axis (Fig. 33–6). For an alternating current, the length of the phasor represents the amplitude of the voltage or current, with its angular speed given by the angular frequency of the alternating current. For example, the function $V(t) = V_0 \sin \omega t$ has a phasor that starts at $t = 0$ aligned with the positive x-axis. As time increases and the phasor rotates counterclockwise, the y-component of the phasor increases until it reaches a maximum when $\omega t = \pi/2$. The y-component then decreases as the phasor moves into the second quadrant.

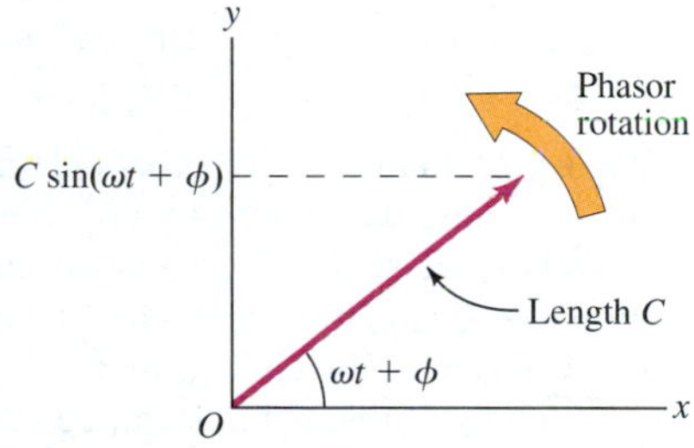

▲ **FIGURE 33–6** The projection of the phasor on the y-axis gives the value of the associated harmonic function.

By comparing the phasors of different harmonic functions that appear in an AC problem, we learn about the relative phases of these functions; that is, which function leads or which function lags. To see how this works, let's apply it to the purely inductive AC circuit of Fig. 33–5. The input voltage—and hence the voltage V_L across the inductor—has the form $V_0 \sin \omega t$, whereas the current I_L takes the form $I_{max} \sin(\omega t - \pi/2)$ according to Eq. (33–18). The phasors for these two quantities (Fig. 33–7a) rotate in the counterclockwise sense as time advances; the phasor for the voltage is always ahead of the phasor for the current. In this diagram, the idea that the voltage *leads* the current for the inductive circuit is easy to visualize and understand. We can easily perform the same exercise for the purely capacitive and purely resistive circuits, whose phasor diagrams are shown in Fig. 33–7b and 33–7c, respectively. Again, we see the phase relationship

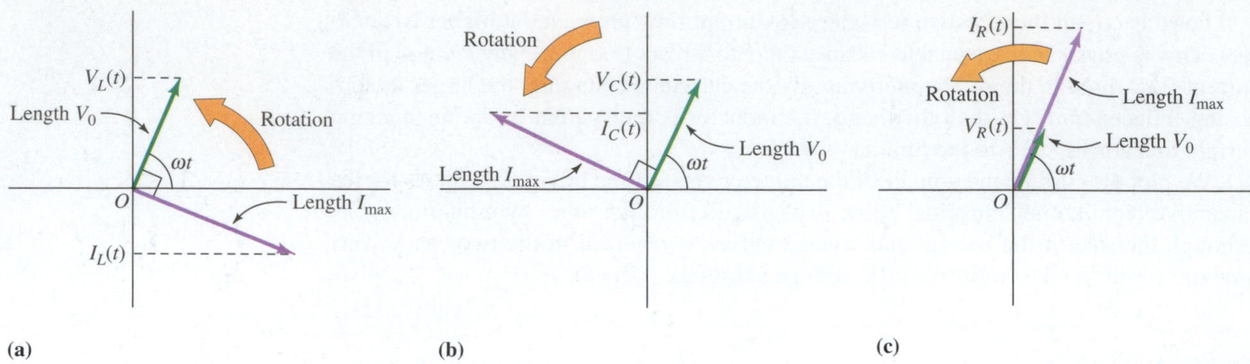

▲ **FIGURE 33–7** Phasors associated with the voltage across and current through (a) the inductor; (b) the capacitor; and (c) the resistor of an *RLC* circuit.

clearly in these phasor diagrams: The voltage across the capacitor in the purely capacitive AC circuit lags the current, and the corresponding current and voltage are in phase in the purely resistive circuit.

When elements are combined in certain AC circuits, the phasors combine as vectors, and this can simplify the analysis of these circuits. This is one way of treating the series *RLC* circuit to be discussed in Section 33–3. Here we content ourselves with a brief example.

EXAMPLE 33–5 Consider an AC circuit with a driving emf $V_0 \sin \omega t$, an inductor of inductance L, and a resistor of resistance R in series. Does the driving voltage lead or lag the current, and by what amount?

Setting It Up Our circuit diagram (Fig. 33–8a) is accompanied by a phasor diagram (Fig. 33–8b) that specifies the voltages across the elements. In accordance with the discussion above, the projection on the y-axis of the respective phasors $\vec{V}_L$, $\vec{V}_R$, and $\vec{I}$ represents the real values of the corresponding physical quantities. (Note that here the overarrow refers to the fact that the phasor is a rotating vector, not to any new "vector" properties of potential or current.) We also note here what is being asked: When we say that the driving voltage leads or lags the current, we want to know whether its phasor is ahead of or behind the current phasor in its counterclockwise rotation, and when we ask "by how much," we are asking for the value of the angle between these two phasors.

Strategy Figure 33–8b gives only the phasors $\vec{V}_L$, $\vec{V}_R$, and $\vec{I}$, with $\vec{V}_L$ leading the current by 90°. Since $\vec{V}_R$ and $\vec{I}$ are in phase (Ohm's law), these phasors are drawn together. To find the phasor $\vec{\mathscr{E}}$ for the driving emf, we note that at any given time its projection on the y-axis must match the sum of the projections of the phasors for the voltage across the inductor and the resistor. This is guaranteed for vectors if we take $\vec{\mathscr{E}} = \vec{V}_L + \vec{V}_R$. Thus graphical vector addition will give us the phasor for the driving emf; simple geometry will give us its angular location relative to the current.

Working It Out Figure 33–8c shows the required addition of the phasors. We see from the figure that the driving emf leads the current and that the angle between them is determined by

$$\tan \phi = \frac{V_L}{V_R} = \frac{IX_L}{IR} = \frac{X_L}{R}.$$

The angle ϕ is less than 90°, as we can see most easily from Fig. 33–8c.

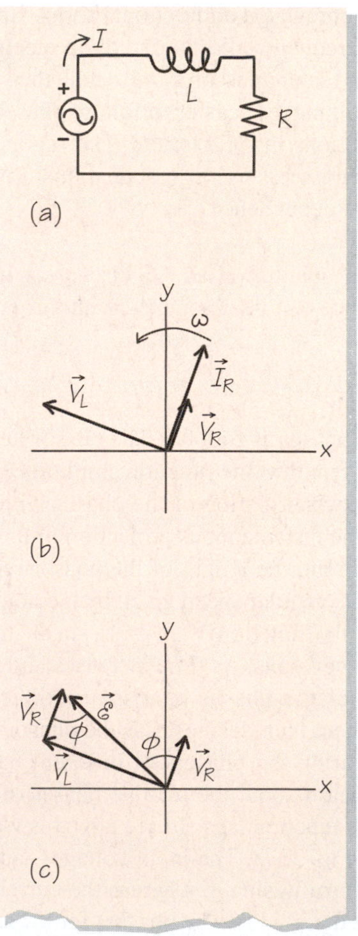

▲ **FIGURE 33–8** (a) A series *RL* circuit. (b) Phasor diagram of the current, voltage across the resistor, and voltage across the inductor. (c) The phasor for the driving emf is the sum of the phasors for voltage across the elements R and L.

Complex Analysis: Complex analysis is a powerful technique that is the working engineer's basic tool for the analysis of circuits with time dependence. Complex analysis depends on two ideas. The first is that a *complex variable* z can represent two physical quantities through the two variables x and y. This is accomplished through the use of $i \equiv \sqrt{-1}$; z is then given the form $z = x + iy$. x and y are extracted from z by taking the real and imaginary parts of z—just the coefficients of i^0 and i^1, respectively. We say that $x = \text{Re}(z)$ and $y = \text{Im}(z)$. (Higher powers of i, which may appear through algebraic manipulation, reduce to i^0 or i^1; for example, $i^2 = -1 = -i^0$.) In Fig. 33–9, we locate z on the xy-plane. You can see that if we think of z as a vector, then it is just the phasor that we spoke of above. But the general use of complex variables is more algebraic than pictorial.

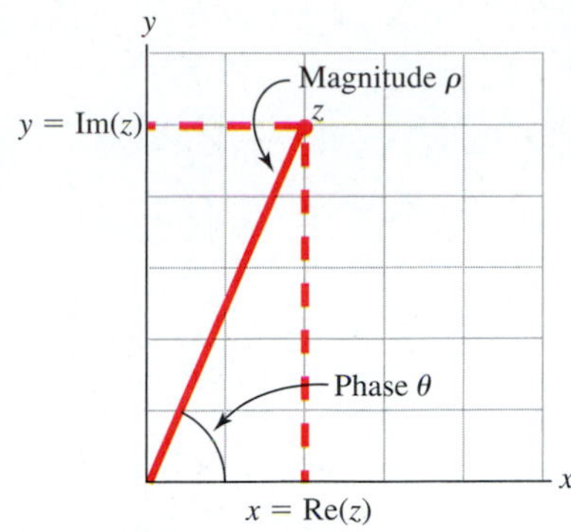

▲ **FIGURE 33–9** A complex number z is characterized either by its real and imaginary parts, x and y, respectively, or by a magnitude and phase, ρ and θ, respectively.

The second idea of complex analysis is one of the most remarkable relations in mathematics:

$$e^{i\theta} = \cos\theta + i\sin\theta.$$

This result follows from the series expansions of the sine, the cosine, and the exponential. Using this identity, we can write an input voltage like that of Eq. (33–1) as

$$V_0 \sin\omega t = \text{Im}(V_0\, e^{i\omega t}),$$

where $\text{Im}(z)$ is the imaginary part of z, that is, the coefficient of i in the expression for z. We then represent all oscillating functions—currents, or potentials across individual elements—as exponentials rather than as sines or cosines, including possible phases. At the end, we take the imaginary part of the complex quantity that we have calculated. Using the relation $e^{i\theta} = \cos\theta + i\sin\theta$, one additional representation of z is useful in the procedure, namely

$$z = x + iy = \rho\, e^{i\theta},$$

where $\rho = \sqrt{x^2 + y^2}$ and $\tan\theta = y/x$. We refer to ρ as the magnitude of z and to θ as the phase of z. Figure 33–9 labels these alternative variables and makes clear the relation between (x, y) and (ρ, θ).

What is the advantage of this procedure? The answer is simple: Upon differentiation, exponentials remain exponentials. Therefore, the differential equations that describe the circuit behavior contain overall powers of exponentials that ultimately cancel. Accordingly, these differential equations reduce easily to algebraic equations, which are simpler to deal with. Problems 91 to 95 illustrate complex analysis. We present a simple example here.

EXAMPLE 33–6 Consider the AC circuit treated in Example 33–5: a driving emf $\mathscr{E} = V_0 \sin\omega t$ and an inductor and resistor in series. Find the current amplitude in the circuit as a function of V_0, R, L, and ω.

Strategy Let us use the subscript "C" to indicate complex quantities, and employ the rule that any physical quantity A is found from the imaginary part of its complex counterpart A_C: $A = \text{Im}(A_C)$. We therefore write the driving emf as $\mathscr{E}_C = V_0 \exp(i\omega t)$, as well as $I_C = I_0 \exp(i\omega t)$ for the current. We will want to use these expressions in the loop rule. We should find that after derivatives are taken, the loop rule becomes a very simple algebraic expression for I_0, the desired current amplitude.

Working It Out Referring to Fig. 33–8a, the loop rule reads $\mathscr{E}_C - L(dI_C/dt) - I_C R = 0$. With the explicit complex forms for $\mathscr{E}_C$ and I_C above, and with $\frac{d}{dt}e^{i\omega t} = i\omega e^{i\omega t}$, we have

$$V_0 \exp(i\omega t) - i\omega L I_0 \exp(i\omega t) - I_0 \exp(i\omega t) = 0.$$

The factor $\exp(i\omega t)$ cancels out, and we are left with an algebraic equation for I_0 whose solution is

$$I_0 = \frac{V_0}{R + i\omega L}, \quad \text{or} \quad I_C = \frac{V_0}{R + i\omega L} e^{i\omega t}.$$

We are asked to find the amplitude of the current; that is, a real quantity, the *magnitude* of I_C. But what is this magnitude? For that we write the combination $R + i\omega L$ that appears in the denominator of our expression for I_C as a magnitude and a phase:

$$R + i\omega L = \sqrt{R^2 + (\omega L)^2}\exp(i\phi),$$

so that

$$I_C = \frac{V_0 \exp(i\omega t - i\phi)}{\sqrt{R^2 + (\omega L)^2}}.$$

The magnitude of this quantity is the coefficient of the exponential, and that is the required relation. We have not written the phase explicitly, because that is not called for in the question. But we could have, and it would have given us the phase that was found in Example 33–5. The square root that appears in our answer is called the impedance—it plays the role of a resistance and will be discussed in more detail below.

The main thing to note here is the simplicity of the algebraic steps.

33–3 AC in Series *RLC* Circuits

In mechanics, a *driven harmonic oscillator* is a device in which a harmonic external force acts on, for example, a mass fixed to a spring. If the driving force has been acting for some time, the mass will move with the *driving frequency*—the angular frequency ω of the force—even though the mass is attached to a spring and undergoes some damping. As we saw in Section 13–8, this system illustrates the important physical phenomenon of *resonance*, characterized by a large amplitude motion when the driving frequency is near the natural frequency[†] ω_0, the frequency with which the mass would move if there were no driving force.

In Chapter 32, we noted the similarities between the damped mass-spring system and series *RLC* circuits without a driving term. If we add an AC source of emf to a series *RLC* circuit, the analogy is extended to a damped mass-spring system with a harmonic driving force. Using this analogy, we can expect such a circuit to exhibit *resonant behavior*.

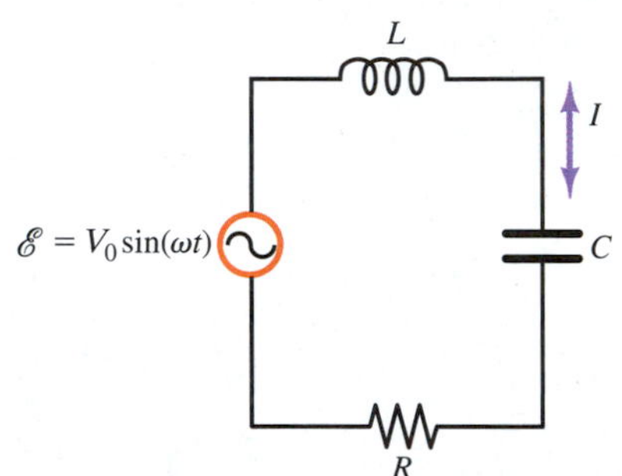

▲ **FIGURE 33–10** An *RLC* circuit is driven by an AC emf.

Let us treat our circuit—the *RLC* circuit with an AC source of emf—more explicitly (Fig. 33–10). Applying the loop rule for the potential changes around the circuit gives

$$V_0 \sin \omega t - L\frac{dI}{dt} - \frac{Q}{C} - IR = 0. \qquad (33\text{–}22)$$

Because $I = dQ/dt$, we can reexpress this result in terms of the single variable Q, the charge on the capacitor:

$$V_0 \sin \omega t - L\frac{d^2Q}{dt^2} - R\frac{dQ}{dt} - \frac{Q}{C} = 0. \qquad (33\text{–}23)$$

The unknown quantity in this differential equation is Q, and the solution will give us Q as a function of time. Once we solve for Q, differentiation with respect to time will give the current. We can then find the voltage drops across the various circuit elements.

The analogy to the mechanical system is evident in the comparison between Eq. (33–23) and Eq. (13–52), which expresses Newton's second law for a driven harmonic oscillator with damping (Fig. 33–11). The equation for the mechanical system is rewritten here in the form

$$F_0 \sin \omega t - m\frac{d^2x}{dt^2} - b\frac{dx}{dt} - kx = 0, \qquad (33\text{–}24)$$

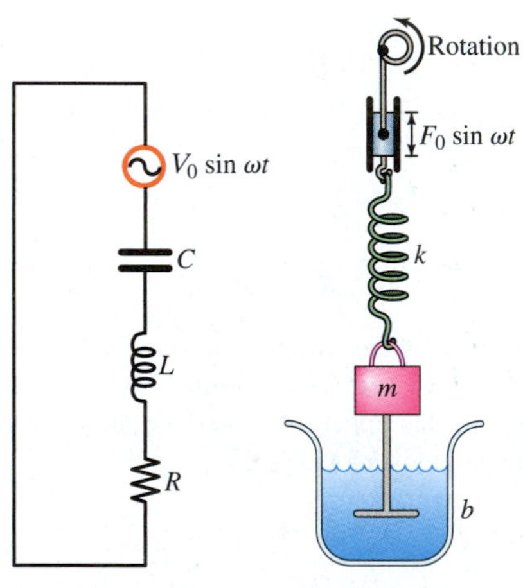

▲ **FIGURE 33–11** The *RLC* circuit and a damped mass-spring system are analogous. The analogies between the different elements of the two systems are understood in the vertical ordering of the elements.

where the first term is the driving force, with amplitude F_0. The structure of Eqs. (33–24) and (33–23) are identical, and once we recall the solution for the mechanical system, we can immediately make the substitutions summarized in Table 33–1 to find the solution for the driven *RLC* circuit. Now, as we described in Section 13–8, Eq. (33–24) for the

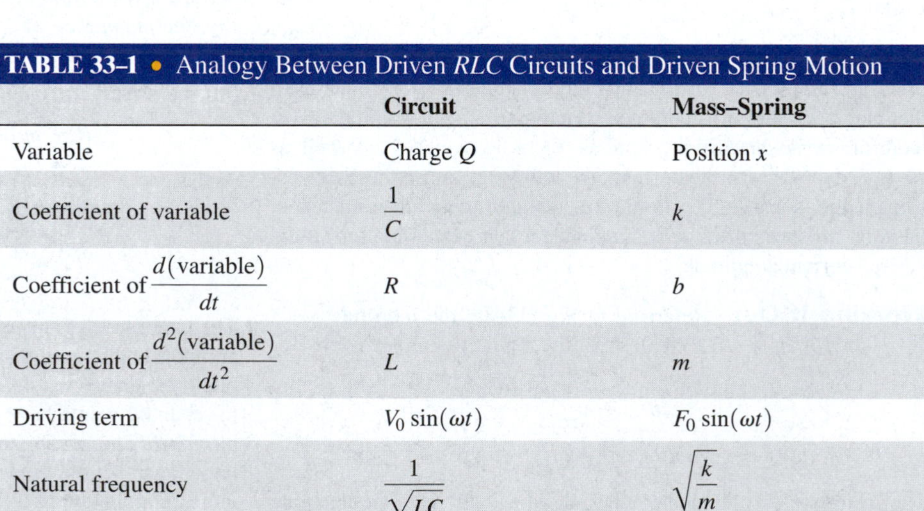

TABLE 33–1 • Analogy Between Driven *RLC* Circuits and Driven Spring Motion

	Circuit	Mass–Spring
Variable	Charge Q	Position x
Coefficient of variable	$\frac{1}{C}$	k
Coefficient of $\frac{d(\text{variable})}{dt}$	R	b
Coefficient of $\frac{d^2(\text{variable})}{dt^2}$	L	m
Driving term	$V_0 \sin(\omega t)$	$F_0 \sin(\omega t)$
Natural frequency	$\frac{1}{\sqrt{LC}}$	$\sqrt{\frac{k}{m}}$

[†]When no confusion is possible, we use the term "frequency" rather than "angular frequency" for ω.

position $x(t)$ has a solution in which the position of the mass oscillates with the angular frequency of the driving force. This solution is given by Eq. (13–53); it is convenient to shift the phase of that solution and write it as

$$x = -A\cos(\omega t - \phi). \tag{33–25}$$

Not only is the frequency determined, but *the amplitude A and phase ϕ are also determined*. We find their values by direct substitution into the differential equation, with the results

$$A = \frac{F_0}{\sqrt{m^2(\omega^2 - \omega_0^2)^2 + b^2\omega^2}} \tag{33–26}$$

and

$$\tan\phi = \frac{1}{b}\left(\omega m - \frac{k}{\omega}\right). \tag{33–27}$$

Here, ω_0 is the natural frequency of the oscillator, given by $\omega_0 = \sqrt{k/m}$.

We also recall that for the mechanical oscillator, the force and the position are both harmonic with the same frequency but out of phase. For example, the function $\sin\omega t$—proportional to the force—rises from zero at $t = 0$, whereas the function $\cos(\omega t - \phi)$—proportional to the position—rises from zero when the argument of the cosine is $-90° = -\pi/2$ rad, that is, when $t = (\phi - \pi/2)/\omega$.

Now we can move on to the *RLC* circuit with an AC source of emf. The solution for the charge on the capacitor in the loop equation for this circuit [Eq. (33–23)] is found by making the formal substitutions of Table 33–1, which relates the parameters of the harmonic oscillator to the circuit parameters. The solution for the charge is

$$Q = -Q_{max}\cos(\omega t - \phi) = -\frac{V_0}{\sqrt{L^2(\omega^2 - \omega_0^2)^2 + R^2\omega^2}}\cos(\omega t - \phi), \tag{33–28}$$

where Q_{max} is defined in comparing the second and third terms, and

$$\tan\phi = \frac{1}{R}\left(\omega L - \frac{1}{\omega C}\right). \tag{33–29}$$

The current is found from evaluation of dQ/dt:

$$I = \frac{dQ}{dt} = I_{max}\sin(\omega t - \phi) = \frac{\omega V_0}{\sqrt{L^2(\omega^2 - \omega_0^2)^2 + R^2\omega^2}}\sin(\omega t - \phi). \tag{33–30}$$

Again I_{max} is directly defined here. Note that $I_{max} = Q_{max}\omega$.

As we would expect, all the results given here reduce to the cases we treated in Section 33–2, in which only one circuit element—the resistor, the capacitor, or the inductor—is present at a time. We need only replace the values of *L*, *R*, or $1/C$ by zero, as appropriate (keeping in mind that $\omega_0 = 1/\sqrt{LC}$).

When we dealt in Chapter 32 with *RLC* circuits without AC sources of emf, neither the amplitude nor the phase were determined; rather, they were a matter of initial conditions. What is different here? A clue is provided by the equation satisfied by the charge, Eq. (33–23). If the AC source is not present, then the equation is homogeneous in *Q*, meaning that if there is a solution for some charge, then a multiple of that charge is also a solution. The overall multiplicative constant would cancel from the equation. But once we add the AC emf, Eq. (33–23) is no longer homogeneous in *Q*, and the overall scale of *Q* is determined. The same reasoning applies to the phase, where without the AC source, any phase is possible. With the AC emf present, that emf "sets the clock" with its definite $\sin\omega t$ form that vanishes at $t = 0$, and the phase of the solution is no longer arbitrary. These remarks about amplitude and phase actually apply to long-time solutions—transient solutions do have properties set by initial conditions, but these solutions die away with time.

Impedance

We have already defined the reactances $X_L = \omega L$ and $X_C = 1/\omega C$ [Eqs. (33–15) and (33–20)]. They enter immediately into the phase [Eq. (33–29)] that appears in the driven *RLC* circuit according to

$$\tan\phi = \frac{1}{R}(X_L - X_C). \tag{33–31}$$

The reactances play the role of an effective resistance, dependent on frequency, for the single element circuits. As we shall see, the effective resistance of the *RLC* circuit is the **impedance** Z defined by

$$Z \equiv \sqrt{\left(\omega L - \frac{1}{\omega C}\right)^2 + R^2} = \sqrt{(X_L - X_C)^2 + R^2}. \tag{33–32}$$

The impedance has units of ohms.

In contrast to resistance, impedance depends on the frequency; we can understand this on physical grounds. Inductance opposes a change in current, and larger values of angular frequency mean more rapid changes in the current. However, an inductor is transparent to a static potential, corresponding to $\omega \rightarrow 0$. These properties are reflected in the frequency dependence of $X_L = \omega L$. A capacitor has just the opposite properties: No constant current can pass through a capacitor, but the capacitor has little effect when the current changes so rapidly that little charge can accumulate. These properties are reflected in the frequency dependence of $X_C = 1/\omega C$. The combination of the effects of both the inductance and the capacitance implies that the impedance is high in the limit of both large and small values of the frequency ω.

To see in more concrete terms how the impedance plays the role of a resistance, let's express the current in terms of it. It is a matter of a little algebra (see Problem 38) to show that, in terms of these quantities, the amplitude in Eq. (33–28) becomes $Q_{max} = V_0/\omega Z$ and hence, with $I_{max} = \omega Q_{max}$, we have $I_{max} = V_0/Z$. In other words,

$$I = I_{max}\sin(\omega t - \phi) = \frac{V_0\sin(\omega t - \phi)}{Z}. \tag{33–33}$$

The current takes the form of an AC emf divided by the impedance. This equation is analogous to the DC equation $I = V/R$. *Impedance thus plays the role of resistance in an AC circuit.* Note that $I_{max} = V_0/Z$, but because of the phase, $I(t)$ is not generally equal to $V(t)/Z$.

EXAMPLE 33–7 The series *RLC* circuit in Fig. 33–10 is driven with an AC source of emf of the form $\mathscr{E} = V_0 \sin \omega t$, where V_0 is exactly 110 V and the frequency f is exactly 60 Hz. If $R = 20.0\ \Omega$, $L = 5.00 \times 10^{-2}$ H, and $C = 50.0\ \mu$F, find the potential drops across the inductor at $t = 0$ and at time t_1, the first time after $t = 0$ that $\mathscr{E}$ reaches a maximum.

Strategy We use $V_L = -L\,dI/dt$ to find the potential drop across the inductor, and utilize Eq. (33–33) for the current I in this circuit:

$$V_L = -L\frac{d}{dt}\left[\frac{V_0}{Z}\sin(\omega t - \phi)\right] = -\frac{LV_0\omega}{Z}\cos(\omega t - \phi).$$

This can easily be evaluated at $t = 0$. For the next part of the problem, we note that the first time after $t = 0$ that the source emf reaches a maximum is time t_1, determined by $\omega t_1 = \pi/2$, or

$$t_1 = \frac{\pi}{2\omega} = \frac{\pi}{4\pi f} = \frac{1}{4f}.$$

Again, V_L can be evaluated at this time.

Working It Out At $t = 0$, V_L takes the form

$$V_L = -\frac{LV_0\omega}{Z}\cos(-\phi) = -\frac{LV_0\omega}{Z}\cos\phi,$$

while at $t = t_1$ we have

$$V_L = -\frac{LV_0\omega}{Z}\cos\left(\frac{\omega}{4f} - \phi\right) = -\frac{LV_0\omega}{Z}\cos\left(\frac{\pi}{2} - \phi\right)$$
$$= -\frac{LV_0\omega}{Z}\sin\phi.$$

To evaluate these results numerically, note that $\omega = 2\pi(60\ \text{Hz}) = 377$ rad/s. We insert the values of ω, R, L, and C to obtain $X_L = \omega L = 18.9\ \Omega$, $X_C = 1/\omega C = 53\ \Omega$, $Z = \sqrt{(X_L - X_C)^2 + R^2} = 39.6\ \Omega$, and $\tan\phi = -1.71$ [or $\phi = -1.04$ rad$(-59.7°)$]. Thus at $t = 0$,

$$V_L = -\frac{LV_0\omega}{Z}\cos\phi = -26.4\ \text{V}.$$

At $t = t_1$,

$$V_L = -\frac{LV_0\omega}{Z}\sin\phi = 45.2\ \text{V}.$$

What Do You Think? At what time after t_1 will the voltage across the inductor first reach a negative maximum?

Resonance in Driven *RLC* Circuits

The current amplitude contained in Eq. (33–30) exhibits the phenomenon of resonance—as do the amplitude of the charge on a capacitor and the voltage amplitudes across any of the elements of a driven *RLC* circuit. The amplitudes are inversely proportional to the impedance. When Z is a minimum, these amplitudes reach a maximum—we say they are *peaked*, or *resonate*. This occurs when the driving frequency ω is at the undamped natural frequency ω_0, which as we know from Section 32–5 is given by $\omega_0^2 = 1/LC$ (Fig. 33–12). (This can be seen by setting the derivative of I_{max} with respect to ω equal to zero.) The amount and sharpness of the peaking depends on the damping factor, which is the resistance R. Without R, the amplitudes would be infinite when the driving frequency ω equals ω_0, something you can easily see from the form of the impedance, Eq. (33–32). However, there is always some resistance in real circuits. Figure 13–25 illustrates the resonance peak for several values of the damping parameter b of a mechanical oscillator. Similar plots apply to *RLC* circuits. The smaller the damping—the smaller the resistance—the larger and sharper the resonance peaks. The maximum in this peak occurs when $X_L = X_C$,

$$\omega L = \frac{1}{\omega C}.$$

This condition implies that

$$\omega^2 = \omega_0^2 = \frac{1}{LC}. \tag{33–34}$$

At this value of the angular frequency, $Z = R$. The current amplitude at this value of the driving frequency is given simply by

$$\text{at resonance: } I_{max} = \frac{V_0}{R}. \tag{33–35}$$

At resonance both *the maximum value of current amplitude and the sharpness of the resonance peak* increase as R decreases. It is useful to note one last feature: At resonance, the tangent of the phase [Eq. (33–29)], and hence the phase itself, is zero.

Resonance phenomena have many applications in circuits. The most familiar concern radio and television reception (Fig. 33–13). A receiver can be tuned by changing either an inductance or a capacitance, thus changing the resonant frequency. The receiver then preferentially picks up broadcast signals at that frequency.

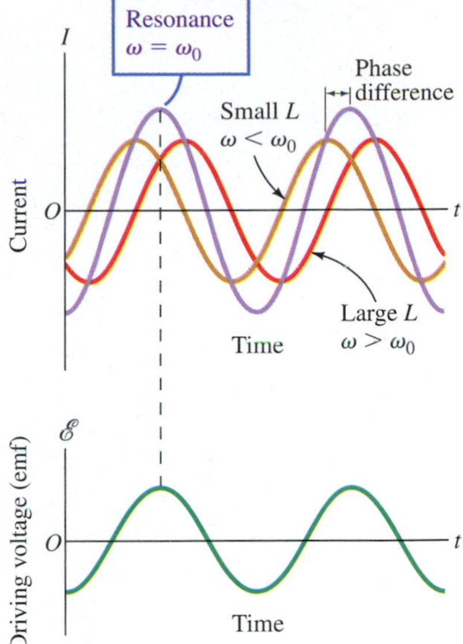

▲ **FIGURE 33–12** Given the driving voltage shown, the current through an *RLC* circuit varies harmonically with the frequency of the driving voltage but differs from it in amplitude and phase in a way that depends on whether or not the driving frequency ω equals the undamped natural frequency ω_0. At resonance ($\omega = \omega_0$), the current amplitude is a maximum and the phase is that of the driving voltage. In the plot of current versus time, we have controlled the natural frequency by varying L.

(a)

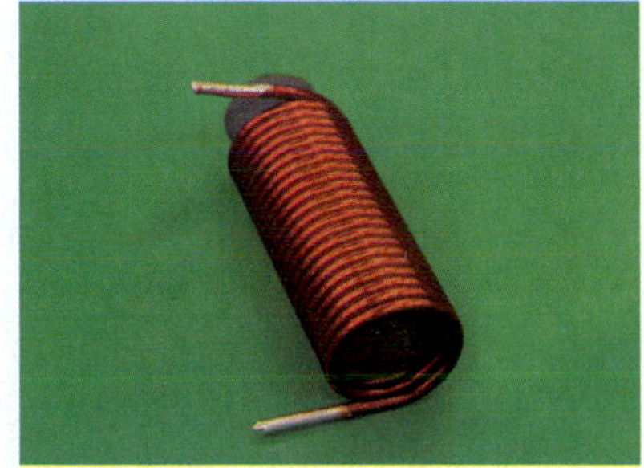

(b)

▲ **FIGURE 33–13** Receiver tuners can be constructed by (a) rotating the overlapping areas of the plates to change the capacitance or by (b) moving an iron core in and out of a solenoid to vary the inductance.

33–4 Power in AC Circuits

Energy flows in an *RLC* circuit. The capacitor and inductor store this energy for some parts of the cycle and release it for others, but no energy is lost (dissipated) in these elements. Energy is, however, lost in the resistance of the circuit. The power dissipated in the circuit is accordingly

$$P = I^2R = \frac{[V_0 \sin(\omega t - \phi)]^2}{Z^2}R. \tag{33–36}$$

This power is always positive, oscillating between zero and a maximum of V_0^2R/Z^2. For engineering purposes, it is often sufficient to know the *average* power dissipated over time. Let's indicate time-averaged quantities with angle brackets, $\langle\ \rangle$. The average of a sine (or cosine) squared over one cycle is one-half:

$$\langle \sin^2(\omega t - \phi) \rangle = \frac{1}{2},$$

so

$$\langle P \rangle = \frac{V_0^2R}{Z^2}\langle \sin^2(\omega t - \phi) \rangle = \frac{1}{2}\frac{V_0^2R}{Z^2}. \tag{33–37}$$

If we substitute Eq. (33–32) for the impedance, we find an explicit form for the average power:

$$\langle P\rangle = \frac{1}{2}\frac{V_0^2 R}{[\omega L - (1/\omega C)]^2 + R^2} = \frac{1}{2}\frac{V_0^2 R\omega^2}{L^2(\omega^2 - \omega_0^2)^2 + \omega^2 R^2}. \tag{33–38}$$

The resonant behavior of AC circuits is evident in this result for the power dissipated—the denominator is smallest, and so the average dissipated power is largest, when the driving angular frequency ω is near the "natural" angular frequency $\omega_0 = \sqrt{1/LC}$. As the driving frequency ω increases through ω_0, the power dissipated has the typical peaked behavior of resonance. (We'll discuss this in more detail below around Fig. 33–14.) At resonance, the expression for power simplifies to

$$\langle P\rangle_{\text{res}} = \frac{1}{2}\frac{V_0^2}{R}. \tag{33–39}$$

The power displays the same resonant behavior as does the current.

Equation (33–33) shows that the current oscillates with time. Just as it is useful to characterize the power with average values, we can also work with an *rms (root mean square)* value for the current (and other harmonically varying quantities in AC). The rms value, x_{rms}, of any quantity x is defined as the square root of the time average of the square of that quantity:

$$x_{\text{rms}} \equiv \sqrt{\langle x^2\rangle}.$$

In particular, if x varies harmonically—if $x = x_0\cos(\omega t - \phi)$—we can use the fact that the time average of the cosine squared is one-half to show that

$$x_{\text{rms}} = \frac{x_0}{\sqrt{2}}. \tag{33–40}$$

When we apply this concept to the AC current, we see from Eq. (33–33) that

$$I_{\text{rms}} = \frac{V_0}{(\sqrt{2})Z} = \sqrt{\frac{V_0^2\omega^2/2}{L^2(\omega^2 - \omega_0^2)^2 + \omega^2 R^2}}. \tag{33–41}$$

From Eq. (33–38), we can see that I_{rms} and the average dissipated power, $\langle P\rangle$, obey the same power–current relation that DC quantities obey, namely

$$\langle P\rangle = I_{\text{rms}}^2 R. \tag{33–42}$$

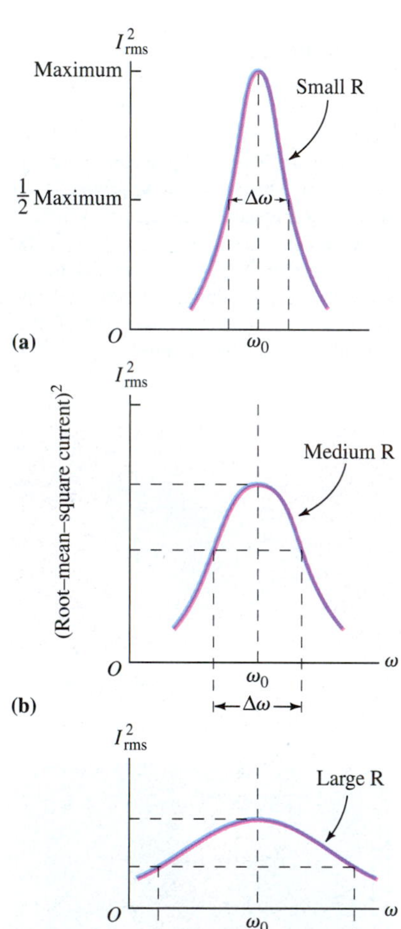

▲ **FIGURE 33–14** The rms current squared in a series *RLC* circuit with an AC source of emf such as the circuit of Fig. 33–10. (a) There is a resonance phenomenon when the driving frequency ω matches the natural frequency ω_0 of the circuit. The bandwidth $\Delta\omega$ measures the width of the rms current squared. (b, c) As *R* increases, the distribution of the current as a function of ω broadens.

Figure 33–14 plots I_{rms}^2 as a function of the driving angular frequency ω for three values of resistance. The sharpness in the peak of the average power (or of I_{rms}^2) versus ω is characterized by the *width* of the peak or, more precisely, the *total width at half-maximum* $\Delta\omega$. This is commonly called the **bandwidth** in the context of AC. To calculate the bandwidth, we find the angular frequencies at which the power drops to half the peak value and take the difference between these angular frequencies. This calculation shows that, for small values of resistance, the bandwidth is given by

$$\Delta\omega = \frac{R}{L}. \tag{33–43}$$

The smaller the resistance and the larger the inductance, the smaller the bandwidth. We can understand the importance of a small bandwidth by thinking about a radio or television receiver whose tuning circuit depends on the resonance phenomenon. If the resonance is sharp, the receiver will more effectively pick out only the desired frequency over others nearby (Fig. 33–14a). Conversely, if the resonance is broad, the circuit will respond to frequencies in the AC signal far from the desired frequency (Fig. 33–14b, c).

Another measure of the sharpness is the ratio $\omega_0/\Delta\omega$. This quantity is defined as the *quality factor*, or *Q-factor*,

$$Q \equiv \frac{\omega_0}{\Delta\omega} = \frac{\omega_0 L}{R}. \tag{33–44}$$

This factor is often used by electrical engineers to represent the sharpness of a resonant circuit.

EXAMPLE 33–8 Two FM radio stations broadcast at the same strength from the same nearby distance, one at a frequency of 91.3 MHz and the other at 91.1 MHz. You prefer the former broadcast over the latter, and you want to construct a simple series *RLC* circuit to act as a receiver that is unique to your favorite station. Given an inductor with an inductance L of exactly 1 μH and adjustable resistance and capacitance, what values should you choose for R and C in order to limit the power received from the unwanted station to 1 percent of the power received from the desired station?

Strategy There are two requirements for the R and C values in our circuit. First, it must be resonant at $\omega_0 = 2\pi f = 2\pi(91.3\text{ MHz}) = 5.74 \times 10^8\text{ s}^{-1}$, and second, the resonant peak must be sharp enough to limit the power from the station broadcasting at $\omega_1 = 2\pi(91.1\text{ MHz}) = 5.72 \times 10^8\text{ s}^{-1}$. The resonant frequency is determined from L and C alone; since the value of L is known, we use this to find C. The sharpness requirement determines R. The power delivered by the signal at resonance is given by Eq. (33–39), whereas the power delivered off resonance is given by Eq. (33–38).

Working It Out We use $\omega_0^2 = 1/LC$ to find C:

$$C = \frac{1}{\omega_0^2 L} = \frac{1}{(5.74 \times 10^8\text{ s}^{-1})^2(1\ \mu\text{H})} = 3.04 \times 10^{-12}\text{ F}.$$

For R, we look at the power delivered by the two stations. With signals arriving with the same strength, it is appropriate to use the same value of V_0. Thus

$$\frac{\langle P\rangle_{\omega_1}}{\langle P\rangle_{\text{res}}} = 0.01 = \left[\frac{1}{2}\left(\frac{V_0^2}{R}\right)\right]^{-1}\frac{1}{2}\frac{V_0^2 R\omega_1^2}{L^2(\omega_1^2 - \omega_0^2)^2 + \omega_1^2 R^2}$$
$$= \frac{R^2\omega_1^2}{L^2(\omega_1^2 - \omega_0^2)^2 + \omega_1^2 R^2}.$$

This gives

$$[L^2(\omega_1^2 - \omega_0^2)^2 + \omega_1^2 R^2](0.01) = R^2\omega_1^2.$$

To a good approximation, we can ignore $0.01\omega_1^2 R^2$ on the left compared to $\omega_1^2 R^2$ on the right. We then take the square root of both sides:

$$L(\omega_0^2 - \omega_1^2)(0.1) = L(\omega_0 - \omega_1)(\omega_0 + \omega_1)(0.1) = R\omega_1.$$

The factor $(\omega_0 + \omega_1)$ is, to a good approximation, equal to $2\omega_1$, so we have

$$2\omega_1 L(\omega_0 - \omega_1)(0.1) \cong R\omega_1.$$

Thus, canceling the factor ω_1,

$$R = 2L(\omega_0 - \omega_1)(0.1)$$
$$= 2(10^{-6}\text{ H})[(5.74 \times 10^8\text{ s}^{-1}) - (5.72 \times 10^8\text{ s}^{-1})](0.1)$$
$$= 0.4\ \Omega.$$

You can check with this value of R that our approximations were justified.

What Do You Think? Suppose instead of a factor 0.01 (1%), you wanted a "discrimination requirement" 100 times better, i.e. the power from the unwanted station is 0.0001 of that of the desired station. What value would you now choose for R?

The Power Factor

The power in AC circuits is commonly given in a form other than that given in Eq. (33–38). We find this form with the aid of the trigonometric identity

$$\cos^2\phi = \frac{1}{\tan^2\phi + 1}.$$

If we now use Eq. (33–31) for $\tan\phi$, we find that

$$\cos^2\phi = \frac{1}{[(1/R)(X_L - X_C)]^2 + 1} = \frac{R^2}{(X_L - X_C)^2 + R^2} = \frac{R^2}{Z^2};$$
$$\cos\phi = \frac{R}{Z}. \qquad (33\text{–}45)$$

Then, using Eqs. (33–41) and (33–43), we see that Eq. (33–42) becomes

$$\langle P\rangle = I_{\text{rms}}^2 R = I_{\text{rms}}^2 Z\cos\phi. \qquad (33\text{–}46)$$

The term $\cos\phi$ in Eq. (33–46) is called the *power factor*. For a circuit without resistance, it is zero, whereas for a pure resistance it is a maximum, with a value of one.

33–5 Some Applications

Most electronic circuits in use today involve elements beyond those we have studied here. For example, transistors perform amplifying functions, while diodes have resistance that depends on the direction of current flow. Modern circuits are typically constructed in integrated form with many thousands of elements included together from the start, and perform many complicated functions. Nevertheless, we can understand a good deal more with a small addition to the elements we have in place.

Diodes and Rectifiers

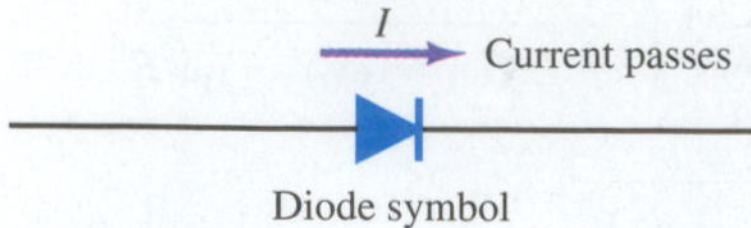

▲ **FIGURE 33–15** The symbol for a diode in a circuit diagram. Current can flow only in the direction shown.

While many sources of electric power supply AC voltage, many applications of that power require DC voltage. For example, rechargeable batteries require a DC current for charging but are charged with a device that plugs into household AC. A process called *rectification* transforms AC voltage to DC voltage using a diode. A **diode** is a semiconductor device with a high resistance to current that flows in one direction, but a low resistance to current that flows in the other direction—the direction of the arrow in the diode symbol (Fig. 33–15). In effect, the diode allows current flow only in one direction.

The diode can be used to construct a **rectifier**, a circuit element that changes AC into DC. Let's consider the circuit shown in Fig. 33–16a. The voltage across the load resistor can be negative or positive. When a diode is placed in the circuit, however, the negative voltages are blocked, leaving only positive voltages across the load resistor (Fig. 33–16b). Such a circuit is called a *half-wave rectifier*. This circuit can suffice as an approximation to a source of DC voltage although the voltage between points a and b, V_{ab}, is neither smooth nor constant.

We can improve the situation considerably with the circuit shown in Fig. 33–16c, called a *full-wave rectifier*. The diodes in this circuit are arranged such that *the voltage V_{ab} is always positive* even though the input voltage oscillates from positive to negative. When the emf produces positive voltage, positive current flows clockwise and passes through the path *cabd* in the direction of the rectifier arrows. The voltage V_{ab} is positive. When the emf produces negative voltage, positive current flows counterclockwise and the path of the current is *dabc*. In this case also, the voltage V_{ab} is positive. Note that now the voltage V_{ab} is positive for all half-cycles, and the rms voltage is higher than it is for the half-wave rectifier. The use of *filters* (see below) allows the voltage peaks to be smoothed, producing a voltage that is much closer to being constant.

THINK ABOUT THIS. . .

HOW DOES A CAR PRODUCE THE ELECTRICITY IT NEEDS?

Although a car's battery is essential for starting, once the automobile engine is running, the engine itself produces the emf necessary to keep the engine running, to power the various "extras" such as lights or radio, and to recharge the battery. The mechanism involves an engine-driven belt that leads to the automobile's *alternator*. The belt runs over a pulley on whose shaft there is a wire coil rotating inside a magnetic field. We learned in Section 30–5 that such a device produces alternating current. But because AC current is not always suitable for the car's electrical system, and certainly not suitable for charging the battery, a rectifier circuit similar in principle to that in Fig. 33–16c is present to change the AC current to DC current.

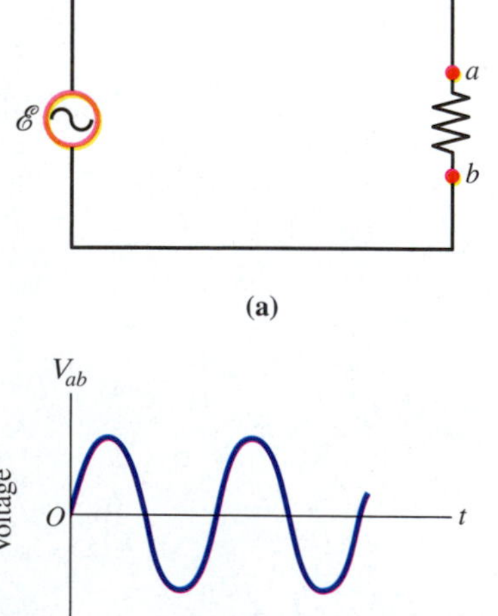

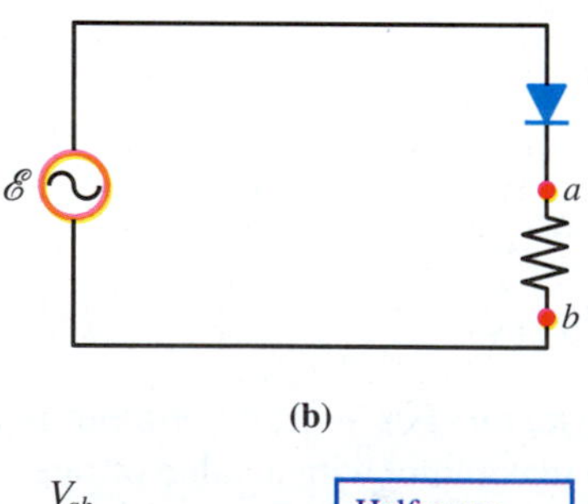

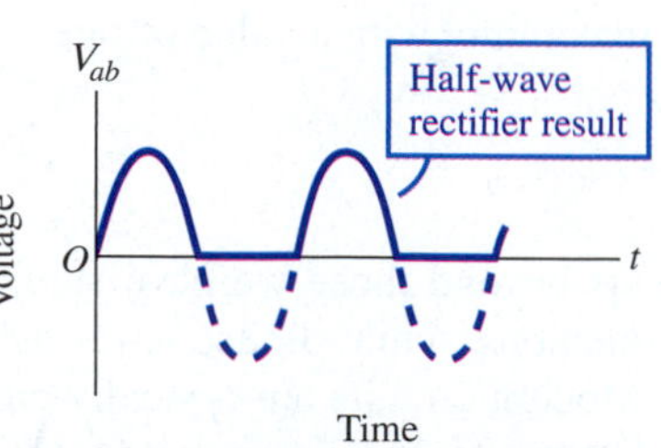

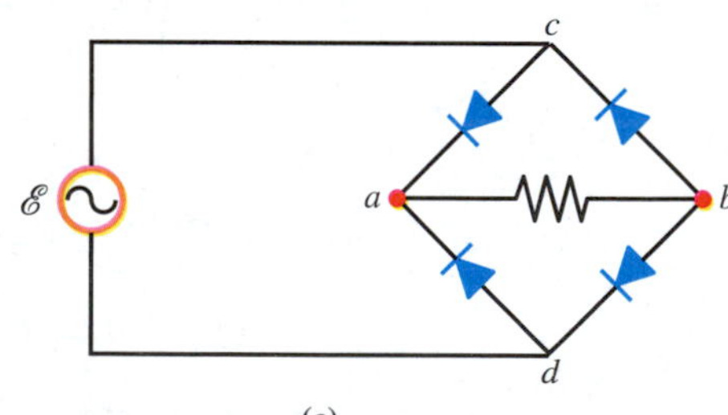

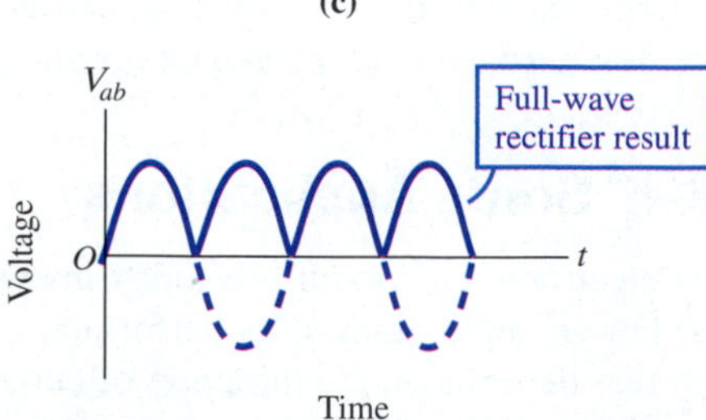

▲ **FIGURE 33–16** (a) The AC voltage across a resistor. (b) A half-wave rectifier is applied. (c) A full-wave rectifier is applied.

Filters

A **filter** is a device that takes an input signal from one part of a circuit that may be a mixture of AC and DC and passes only the AC or only the DC signal to a different part of the circuit. More generally, a filter will change the balance of high and low frequency components in a signal that may contain many different frequencies. Our discussion of the capacitive and inductive reactances shows that either a capacitor or an inductor can act as such a filter. Consider Fig. 33–17, in which the current I from the input is a mixture of DC and AC,

$$I = I_0 + I_1 \sin \omega t.$$

Here I_0 and I_1 are constants. A constant current cannot pass across a capacitor, whereas the impedance of the capacitor goes to zero if ω becomes large. Thus for the capacitor in Fig. 33–17a, only the AC part of the current passes to the other side. For the capacitor in Fig. 33–17b, AC passes through the capacitor to ground, and the DC passes through to the output side of the circuit. An inductor works in just the opposite way: DC passes through without impedance ($X_L \to 0$ as $\omega \to 0$, and $\omega = 0$ corresponds to DC), whereas the impedance is large for AC with large ω. Thus for the inductor in Fig. 33–17c, DC passes through to the output; for the inductor in Fig. 33–17d, AC passes through to the output and DC passes through the inductor to ground.

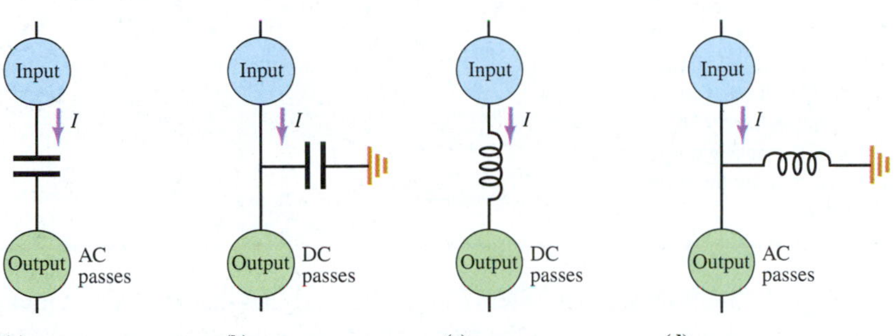

FIGURE 33–17 AC and DC filters formed from capacitors or inductors. With a capacitor: (a) AC current passes through to the output side. (b) DC current passes to the output side. With an inductor: (c) DC passes through to the output side. (d) AC passes through to the output side.

EXAMPLE 33–9 The generator in the circuit shown in Fig. 33–18 produces a combination of DC and AC in the form $\mathcal{E} = V_0 + V_1 \sin \omega t$, where $V_0 = 0.10$ V, and $V_1 = 0.25$ V. The values of the capacitance and resistance in the circuit have the values $C = 1.0\ \mu$F and $R = 0.20\ \Omega$. What is the value of ω for which the voltage amplitude across the resistor is 50 percent of the value of the maximum voltage of the generator?

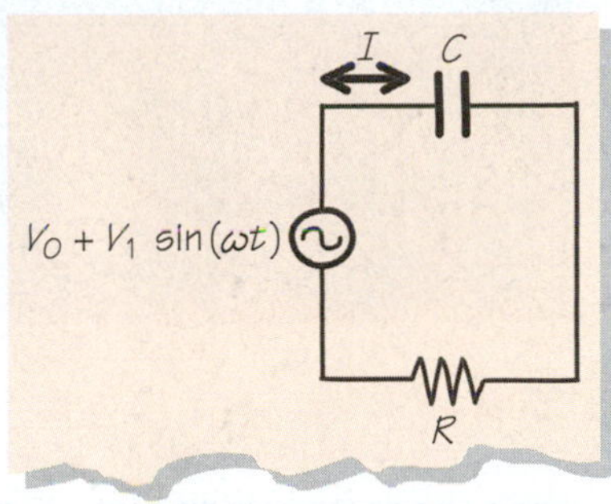

▲ **FIGURE 33–18**

Strategy We can apply the superposition principle by applying the loop rule that corresponds to the DC and AC terms separately, and then add the voltage drops. For the AC term, the current in the circuit is, from Eq. (33–33),

$$I_{AC} = \frac{V_1 \sin(\omega t - \phi)}{Z} = \frac{V_1}{\sqrt{(1/\omega C)^2 + R^2}} \sin(\omega t - \phi).$$

The voltage drop across the resistor from the AC is $I_{AC}R$ and therefore has the amplitude

$$\frac{V_1 R}{\sqrt{(1/\omega C)^2 + R^2}}.$$

For the DC term the capacitor acts as a perfect filter for the constant term in the input voltage because no constant current can pass. There is thus *no* voltage drop across the resistor associated with the V_0 term.

The maximum value of the input voltage is $V_0 + V_1$. We calculate the ratio of the voltage amplitude across the resistor to the maximum input voltage and set it equal to $\frac{1}{2}$. This equation determines ω.

Working It Out The equation for the ratio of potentials described above is

$$\frac{1}{V_0 + V_1} \frac{V_1 R}{\sqrt{(1/\omega C)^2 + R^2}} = 0.5.$$

We solve for ω:

$$\omega = \frac{1}{RC\sqrt{\dfrac{V_1^2}{(V_0 + V_1)^2(0.50)^2} - 1}}$$

$$= \frac{1}{(0.20\ \Omega)(1.0 \times 10^{-6}\ \text{F})\sqrt{\dfrac{(0.25\ \text{V})^2}{(0.10\ \text{V} + 0.25\ \text{V})^2}\dfrac{1}{(0.50)^2} - 1}}$$

$$= 4.9 \times 10^6\ \text{s}^{-1}.$$

What Do You Think? At very high frequencies what becomes of the ratio of the voltage amplitude across the resistor to the maximum input voltage?

CONCEPTUAL EXAMPLE 33–10 Consider the *RC* circuit shown in Fig. 33–19a. The diode allows only the positive voltage to be passed through the circuit. Assume the values of *R* and *C* produce an *RC* time constant that is much longer than the period of the emf source voltage. What does the output voltage V_{output} across the resistor look like?

Answer During the rising voltage period of the source emf ($V = V_{max} \sin \omega t$), the capacitor will be charged to a full voltage V_{max}. As soon as the source voltage passes its peak and begins to decrease, the fully charged capacitor will start to discharge. Because the time constant is much greater than the period of the source voltage, the resulting voltage across the resistor will hardly decrease from its peak voltage V_{max} due to the capacitor discharging before the source voltage comes back to a peak again and recharges the capacitor. In effect, this circuit has rectified the AC voltage to one of full DC. We sketch the result in Fig. 33–19b. Application of a filter to the rectified voltage shown in Fig. 33–16b, for example, will further smooth out the peaks and valleys in the curve of voltage versus time.

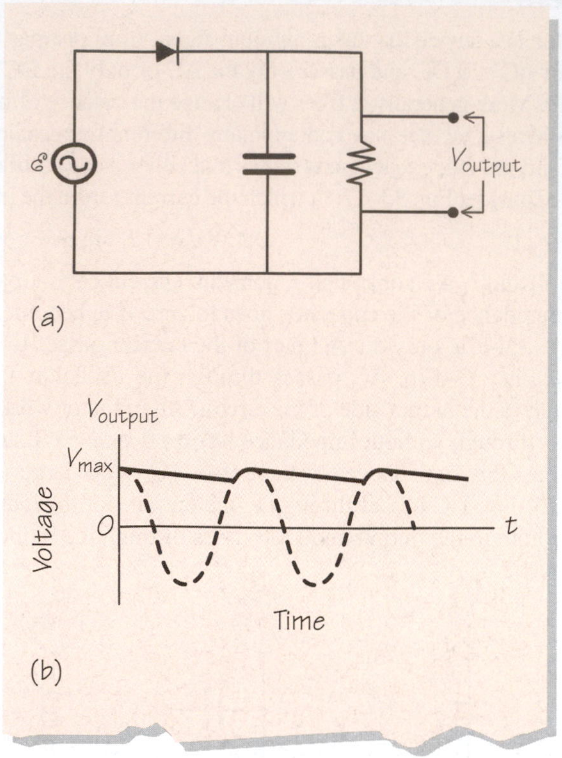

▶ **FIGURE 33–19** (a) An *RC* circuit acts as a filter for rectified AC voltage. Such a filter can produce a voltage that is nearly DC. (b) The slow decrease of the nearly constant-voltage segments is governed by the time constant of the *RC* circuit.

Figure 33–20 shows how filters can modify a signal that is a mix of many different frequencies. When you change the treble and bass balance on a music system you are applying filters. We can also add that when you sequentially apply a high- and low-pass filter, you are making a band-pass filter, one that allows a given band of frequencies to pass.

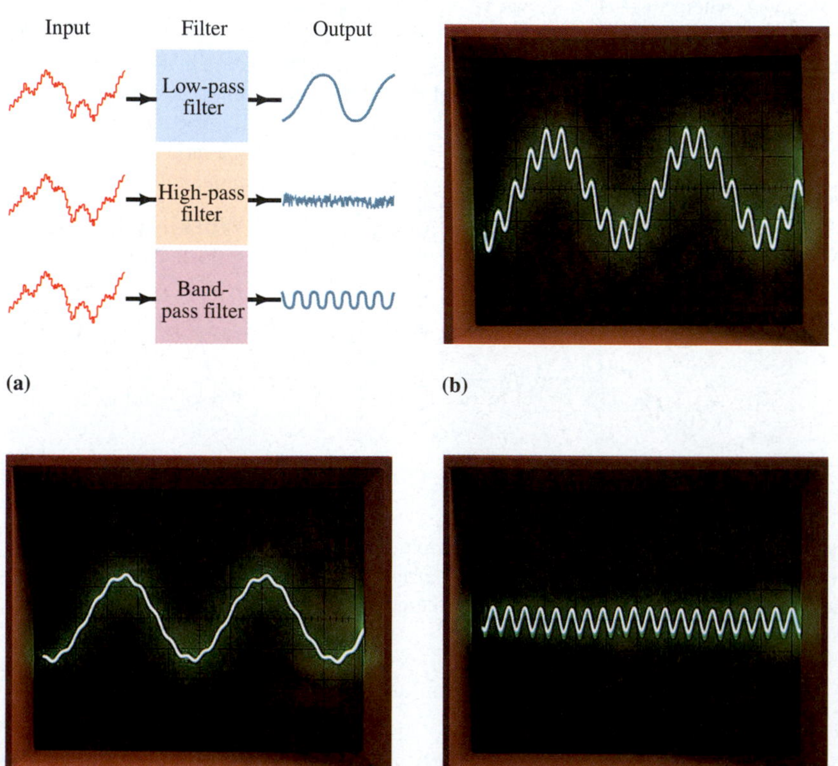

▶ **FIGURE 33–20** (a) Filters allow the low frequencies contained in an input signal to pass. (b) An input voltage containing both low and high frequencies. (c) The input voltage in part (b) has been sent through a low-pass filter, which eliminates the high frequencies. (d) The input voltage in part (b) has been sent through a high-pass filter, which eliminates the low frequencies.

*Impedance Matching

Another aspect of AC of practical importance concerns **impedance matching**, which refers, as in our discussion of filters, to the *connection* between different parts of a circuit. Figure 33–21a shows such a situation, in which some combination of circuit elements makes up circuit 1, connected at points a and b to circuit 2. The two circuits have impedances Z_1 and Z_2, respectively. We are not concerned here with the origin of currents in these circuits as much as we are with our ability to deliver power from circuit 1 to circuit 2. We therefore assume that the origin of these currents is within circuit 1 and break that circuit down as in Fig. 33–21b. The primary question is, if Z_1 is fixed, what are the requirements for Z_2 so that the power delivered to circuit 2 is a maximum? If, for example, a stereo amplifier is connected to a loudspeaker, what should the loudspeaker's impedance be in order that maximum power is delivered to it?

The answer is found by computing the average power $\langle P \rangle$ to circuit 2, which, from Eq. (33–42), is $I_{\text{rms}}^2 R_2$. The current in the loop of Fig. 33–21b is given by

$$I_{\text{rms}} = \frac{\mathcal{E}_{\text{rms}}}{Z_{\text{total}}}. \tag{33–47}$$

Here, $\mathcal{E}_{\text{rms}}$ is the rms value of the generator, whose maximum voltage, or amplitude, is V_0. If the generator produces a sinusoidal emf of the form of Eq. (33–1), then Eq. (33–40) shows that $\mathcal{E}_{\text{rms}} = V_0/\sqrt{2}$. The total impedance Z_{total} is found by separately adding the capacitive reactances, inductive reactances, and resistances, a result that follows from our knowledge of how series combinations of C, L, and R add (see Problem 44):

$$Z_{\text{total}} = \sqrt{[(X_{L_1} + X_{L_2}) - (X_{C_1} + X_{C_2})]^2 + (R_1 + R_2)^2}. \tag{33–48}$$

Thus the average power delivered to circuit 2 is

$$\langle P \rangle = \frac{\mathcal{E}_{\text{rms}}^2 R_2}{Z_{\text{total}}^2} = \frac{\mathcal{E}_{\text{rms}}^2 R_2}{[(X_{L_1} + X_{L_2}) - (X_{C_1} + X_{C_2})]^2 + (R_1 + R_2)^2}. \tag{33–49}$$

That there is a value of the parameters of Z_2 that maximizes this power is clear: If Z_2 is too small, the factor R_2 will also be small and $\langle P \rangle$ will be small; if Z_2 is too large, its parameters will dominate the denominator of Eq. (33–49), and $\langle P \rangle$ will again be small. An intermediate value of the parameters of Z_2 will give a maximum value of $\langle P \rangle$. Two independent parameters are involved here: the resistance R_2 and the total reactance term for circuit 2, $X_{L_2} - X_{C_2}$. Formally, we find the value of the parameters that maximize $\langle P \rangle$ by taking the derivative of $\langle P \rangle$ with respect to these quantities and setting it equal to zero. From this exercise the power is maximized when

$$R_2 = R_1 \quad \text{and} \quad X_{L_2} - X_{C_2} = -(X_{L_1} - X_{C_1}). \tag{33–50}$$

The second condition—that the reactance term of Z_2 is equal but opposite to that of Z_1—follows because it means that the reactance terms in the denominator of Eq. (33–49) cancel, thus maximizing $\langle P \rangle$ whatever the value of the resistances. The first condition—that the resistances be equal—is perhaps less intuitive but nevertheless follows directly from the requirement that the derivative of $\langle P \rangle$ is zero (see Problem 72). When the conditions of Eq. (33–50) are met, the impedances are said to be matched.

Impedance matching is desirable when you wish to deliver maximum power to one part of a circuit. It is worth noting that we do not always wish to deliver maximum power. An AC voltmeter, for example, should have an impedance *mismatch* because we want it to draw as little current as possible.

The subject of circuit analysis is highly developed. We have been able to do no more than describe its principles, and this chapter will not have taught you to fix, much less design, TVs or computers. But the principles we have described here apply to all AC circuits.

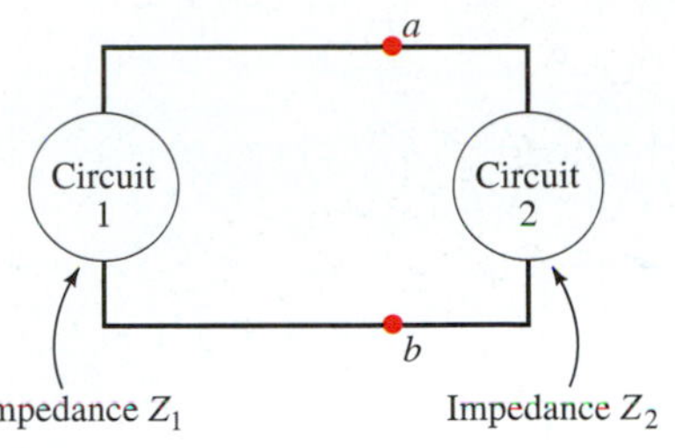

(a)

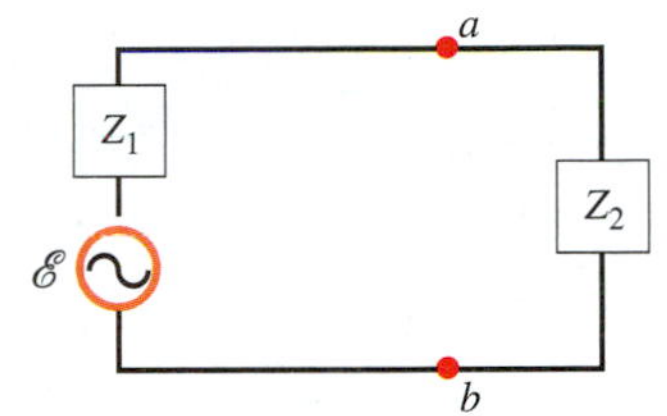

(b)

▲ **FIGURE 33–21** (a) Circuit diagram to illustrate impedance matching. (b) Circuit 1 of part (a) is broken down into a source of emf $\mathcal{E}$ and an impedance Z_1. Circuit 2 is assumed to include only an impedance Z_2.

Summary

The presence of AC sources of emf in circuits with resistors, inductors, and capacitors introduces a variety of new possibilities of phenomena and applications. Transformers allow us to vary the voltage amplitude of AC emfs, of importance in power transmission on large and small scales.

The relation between the voltages and the numbers of turns of the primary and secondary coils of a transformer is

$$\frac{V_2}{V_1} = \frac{N_2}{N_1}. \qquad (33\text{–}4)$$

Conservation of energy implies that the currents carried by the respective coils are related inversely:

$$\frac{I_1}{I_2} = \frac{N_2}{N_1}. \qquad (33\text{–}6)$$

A series *RLC* circuit with an AC source of emf of frequency ω behaves like a damped harmonic oscillator driven by a harmonically varying force. Solutions for currents, voltages, and charges in such circuits can be found by using the solutions already developed for the driven harmonic oscillator. For such circuits the impedance Z is a quantity that plays the role of a resistance. The impedance is frequency dependent:

$$Z \equiv \sqrt{\left(\omega L - \frac{1}{\omega C}\right)^2 + R^2} = \sqrt{(X_L - X_C)^2 + R^2}. \qquad (33\text{–}32)$$

where X_C is the capacitive reactance and X_L is the inductive reactance. The current in the driven circuit is then

$$I = I_{\text{max}} \sin(\omega t - \phi) = \frac{V_0 \sin(\omega t - \phi)}{Z}, \qquad (33\text{–}33)$$

where $V_0 \sin \omega t$ is the driving emf. The phase ϕ is given by

$$\tan \phi = \frac{1}{R}\left(\omega L - \frac{1}{\omega C}\right). \qquad (33\text{–}29)$$

Such circuits exhibit resonant behavior when the driving frequency is near the natural frequency $\omega_0 = \sqrt{1/LC}$. This type of behavior is most clearly seen in the power dissipated in driven *RLC* circuits. Averaged over time, the power lost is

$$\langle P \rangle = I_{\text{rms}}^2 R = I_{\text{rms}}^2 Z \cos \phi, \qquad (33\text{–}46)$$

where $I_{\text{rms}} Z = V_0/\sqrt{2}$ gives the rms current. Near resonance, the power dissipated is a maximum, and the width of the peak of average power versus driving frequency has a width at half-maximum of

$$\Delta\omega = \frac{R}{L}, \qquad (33\text{–}43)$$

a result that holds as long as the resistance is not too large.

If we add diodes—devices that allow current to pass in one direction only—to our arsenal of circuit elements, we can construct a variety of electronic devices, including rectifiers, which produce a positive (or negative) emf from an AC source, and filters, which take a mixed AC and DC signal and pass predominantly either the constant (DC) part or the variable (AC) part of a given frequency. The term impedance matching refers to constraints that describe how different parts of a circuit can be connected together with minimal power loss.

Understanding the Concepts

1. Why is the material used to make the core of transformers so important?
2. What are some applications for step-up and step-down transformers?
3. Why does a capacitance act as a short circuit at high frequencies?
4. Without R, the current amplitude in a series *RLC* circuit would be infinite when the driving frequency ω equaled ω_0, but this possibility could never happen, because there is always some resistance in real circuits. How do you reconcile this statement with the existence of superconductors?
5. To find the rms current [Eq. (33–41)], we square the current, then take the time average, and then take the square root. Why do we not simply take the time average of the current?
6. Consider a lightbulb connected in series with an inductance L, a capacitor C, and an AC emf. The frequency ω of the power supply can be tuned. For what value of ω will the bulb be brightest?
7. In discussing AC current, we found it useful to describe its root mean square; why didn't we bother with this in our discussion of power?
8. In Example 33–9, we took the input emf to be a mixture of DC and AC, then treated the effect of the AC and DC parts separately and added the two parts. How is this procedure justified?
9. What is the distinction between which side of a transformer is the primary or secondary coil? Could it easily be switched around and operated the opposite way? What would change?
10. A capacitor and a lamp are connected in series with an AC generator of constant voltage but variable frequency (Fig. 33–22, see next page). Which of the following three statements is true? The lamp will (a) not light, because the capacitor is connected in series with the lamp; (b) burn brightest when the frequency is high; (c) burn with the same brightness for all frequencies.

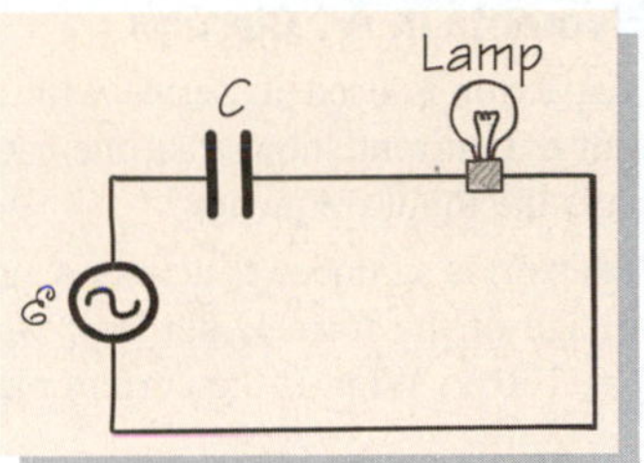

▲ **FIGURE 33–22** Question 10.

11. A capacitor, lamp, and resistor are connected to an AC generator of constant voltage but variable frequency (Fig. 33–23). Which of the following statements is true? The lamp will (a) not burn, because the capacitor shorts out the lamp; (b) burn brightest when the frequency is low; (c) burn brightest when the frequency is high; (d) burn with the same brightness for all frequencies.

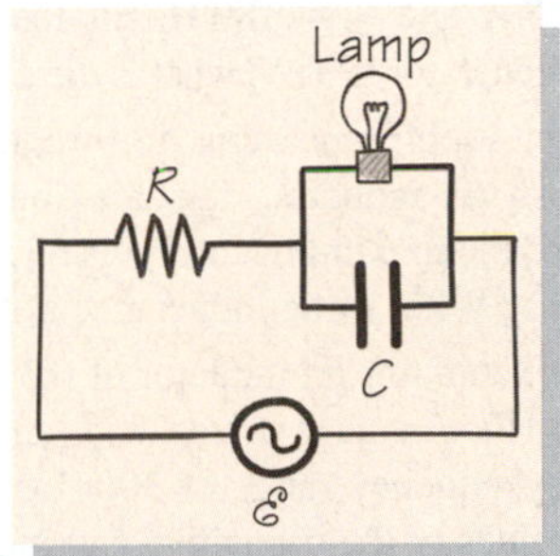

▲ **FIGURE 33–23** Question 11.

12. A particular appliance or household circuit is rated for a maximum current. Why is that, and why must currents not exceed that maximum?
13. The purpose of a ballast in a fluorescent light is to start and control the lamp. A voltage at least three times as large as the steady state voltage is needed to start the lighting process, and current spikes must be controlled. Which of the three circuit elements (R, L, C) are you likely to find in a ballast? Explain what it might do.
14. The primary purpose of an electric heater is to produce heat. Why would such a heater require a 220-V socket rather than a 120-V socket?
15. In the complex analysis approach, two combinations of circuit elements have complex impedances Z_1 and Z_2, respectively. When these combinations are placed in series with one another, their (complex) total impedance is given by the addition $Z = Z_1 + Z_2$. For application to physical circuits, the actual impedance is the magnitude of the complex impedance. Will the magnitudes of the complex quantities Z_1 and Z_2 add to give the magnitude of the complex quantity Z?
16. The reactance X_C is infinite when the input voltage is DC. Does that mean that the impedance is not defined for this situation?
17. If a capacitor has a large impedance for DC and an inductor has a large impedance for AC, how can a series LC circuit pass any current?
18. If the frequency of an AC voltage across an inductance is doubled, the inductive reactance of the inductor is (a) increased by a factor of 4; (b) decreased by a factor of 4; (c) doubled, (d) none of these.
19. Television antenna wires normally have negligible resistance and an impedance of 75 Ω. Why is it important to use antenna wires with the same impedance throughout?
20. Some appliances that operate off a 120-V line and draw in excess of 20 A of current have different plugs than regular 120-V household devices have. Why are these plugs different? What might happen if they were not different?
21. Three lightbulbs, a capacitor, and an inductor are connected to an AC power supply as shown in Fig. 33–24. Under what circumstances do you expect bulbs 1 and 2 to be very bright, while 3 is quite dim?

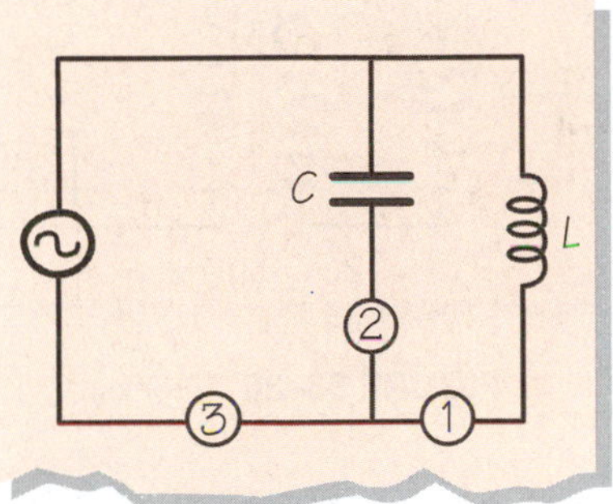

▲ **FIGURE 33–24** Question 21.

22. Do all CRT-based television sets have transformers? How might transformers be used in a television set?
23. If electricity is transported in power transmission lines at 200 kV to 500 kV, does the power have to be generated at these voltages? Why or why not?
24. In a capacitor, there is a gap between the two plates. Why then is there any current at all flowing in a circuit with a capacitor? Is this current carried by sparks?

Problems

33–1 Transformers

1. (I) A transformer has 100 turns in the primary coil and 1,500 turns in the secondary coil. If the amplitude of the AC voltage in the primary coil is 600 V, what is the voltage amplitude in the secondary coil?
2. (I) Suppose that electric power costs 18 cents/kWH. Consider a transmission line that delivers 500 kW of power and has a resistance of 150 Ω. Calculate the dollars lost annually due to the transmission line if the power is delivered at (a) 750,000 V and (b) 1,440 V.
3. (I) Many electrical devices, such as doorbells or buzzers, operate on 12 V AC. A small transformer used to produce this voltage has a primary coil of 550 turns and takes an input of 110 V AC. How many turns must the secondary coil have?
4. (I) The primary coil of a step-down transformer is connected to house current, 115 V at 60 Hz. If the secondary coil of the transformer delivers a current with an amplitude of 2.0 A at 24 V, what is the current drawn by the primary coil? Ignore losses in the transformer.
5. (I) A transformer whose output voltage can be varied is used to obtain AC power from a 120-V, 10-A supply. The secondary coil consists of 1200 turns of wire. The variable transformer works by connecting different numbers of turns of wire on the secondary coil. The secondary voltage can thereby be regulated. When all 1200 turns act as the secondary coil, the output voltage has an amplitude of 120 V. How many turns of wire should be used to obtain 45 V (Fig. 33–25, see next page)? How much current will flow in this case?

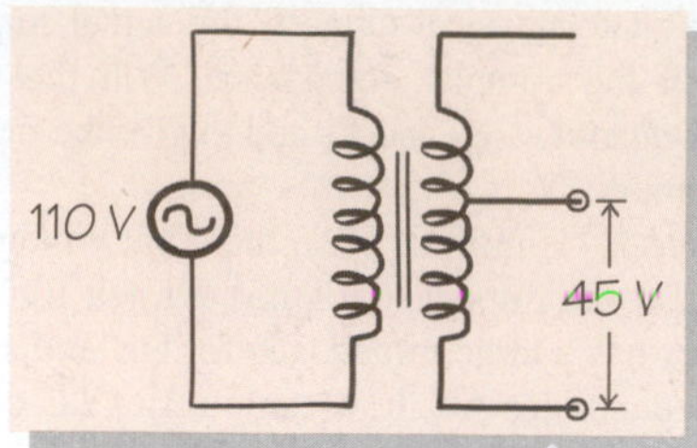

▲ **FIGURE 33–25** Problem 5.

6. (I) A transformer has one coil with an inductance of 74 mH, area 35 cm^2, and length 20 cm. It is fully linked with another coil having the same area and length, but not the same number of turns. Their mutual inductance is 43.5 mH. How many turns does each coil have?

7. (II) Figure 33–26 shows an ideal transformer with 220 V on the primary coil supplying power to a resistor of resistance R. If the resistor dissipates 88 W, what is the current in the primary coil?

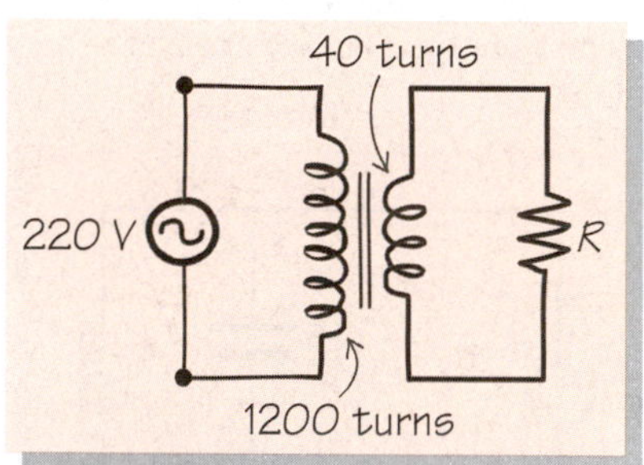

▲ **FIGURE 33–26** Problem 7.

8. (II) A step-down transformer has a turn ratio (N_1/N_2) of 15:1. (a) If the primary coil is connected across a 220-V oscillating-voltage generator, what voltage appears across the secondary coil? (b) Assuming that there are no power losses in the transformer, what current would have to flow through the primary coil so that a 15-Ω resistor placed across the secondary coil draws all the power of the circuit? (c) What resistance connected across the 220-V voltage generator would draw the same total power?

9. (II) The transformer shown in Fig. 33–27 has two secondary windings; one supplies 220 V, the other, 11 V. The input voltage at the primary coil is 110 V. If the 220-V secondary coil has 1000 turns, how many turns does the 11-V secondary coil have?

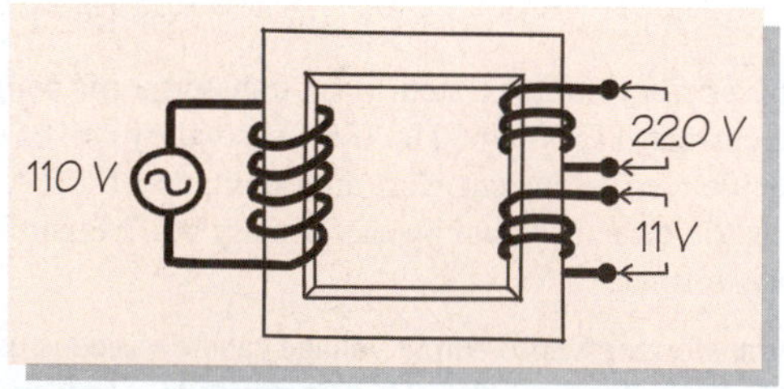

▲ **FIGURE 33–27** Problem 9.

10. (II) Suppose that a transformer consists of two separate windings of wire on the same core. The core material has a magnetic permeability μ. How does the ratio of the emfs in the two coils depend on μ?

33–2 Single Elements in AC Circuits

11. (I) A 12-μF capacitor is used in series with an AC generator. Measurement of the current shows that the capacitive reactance is 1.0 Ω. What is the input frequency?

12. (I) A 25-Ω resistor is connected across a power supply that produces a voltage of the form $V_0 \sin(\omega t)$, where $f = \omega/2\pi =$ 60 Hz and $V_0 = 130$ V. What is the current passing through the resistor?

13. (I) An alternating current of maximum value 2 A in a solenoid of self-inductance $L = 15$ mH induces an emf of maximum value 330 V. What is the angular frequency of the alternating current?

14. (I) An AC power supply with frequency 60 Hz is connected to a capacitor of capacitance $C = 40\ \mu$F. The maximum instantaneous current that passes through the circuit is 2.26 A. What is the maximum voltage?

15. (I) A current flowing through a circuit that contains only a capacitor and an AC power supply has the form $I_0 \cos[2\pi ft - (\pi/6)]$, where $I_0 = 2.45$ A and $f = 180$ Hz. If the maximum voltage supplied by the generator is 95 V, what is the capacitance?

16. (I) An AC power supply operating at a frequency of 220 Hz is connected across an inductor. The maximum voltage of the source is 4 V, and the maximum current is 65 mA. What is the inductive reactance? What is the inductance of the circuit?

17. (I) An AC circuit contains an inductor of 0.3 H and capacitor of 2 μC in series. The circuit is driven with an AC source of emf with an angular frequency range of 300–1,000 rad/s. What are the maximum values of the capacitive and inductive reactances?

18. (II) A current $I = I_0 \sin(\omega t - \pi/3)$ flows in a circuit for which $I_0 = 2.3$ A and $\omega = 2\pi(60\text{ Hz})$. (a) At what times does the peak current flow? (b) If the current flows through an inductance of 0.25 H, what is the peak voltage on the inductor? At what times does this peak voltage occur?

19. (II) The average of the square of the voltage in an inductive circuit (a circuit with no capacitors and no resistors) driven by an AC emf is $(30\text{ V})^2$, and the average of the square of the current is $(2\text{ A})^2$. What is the inductive reactance? If the inductance is 25 mH, what is the frequency of the alternating current?

20. (II) The voltage across an inductor takes the form $V(t) = (0.3\text{ V}) \sin[(400\text{ s}^{-1})t] + (0.3\text{ V}) \sin[(2{,}700\text{ s}^{-1})t]$. Determine the current through the inductor if $L = 40\ \mu$H.

21. (III) (a) Draw the phasor for the function $D \cos(\omega t + \phi)$ on the graph that contains the phasor for the function $C \sin(\omega t + \phi)$. Which phasor is more advanced in phase—that is, points in a direction corresponding to a larger angle, as measured from the $+x$-direction? (b) What is the angle between the two phasors on the plot you drew for part (a)? Which phasor leads? (c) Repeat the exercise for the function $f(t) = A \cos(\omega t) + B \sin(\omega t)$.

33–3 AC in Series *RLC* Circuits

22. (I) Consider an LC circuit driven by an AC source of emf. It differs from the one shown in Fig. 33–10 in that the inductor and the capacitor are in parallel, while there is no resistor. Given that the input voltage is $V_0 \sin \omega t$, determine the form of the current through the inductor without using Kirchhoff's rules.

23. (I) Consider a radio circuit with a fixed inductance of 14 μH. What is the value of the tunable capacitance for the reception of a 42-m radio wave?

24. (I) What is the range needed for a variable capacitor to be combined with a 12-mH coil so that a tuned circuit could be formed to cover the range of broadcast-band frequencies from 540 kHz to 1,600 kHz?

25. (I) The driving frequency in a driven *RLC* circuit is at the resonant frequency. The maximum current carried by the circuit is found to be insufficient for the desired application. By what factor should the resistance be changed to double the maximum current?

26. (I) An AC generator with a voltage amplitude of 50 V and a frequency of 750 Hz is built to drive a circuit meant to be resonant. The resistance of the circuit is 0.5 Ω, and the inductance is 5 mH. What must the value of the capacitance be?

27. (I) A series *RLC* circuit of frequency 60 Hz has a maximum current of 100 mA. What is the maximum charge on the capacitor? If the impedance is 40 Ω, what is the emf?

28. (I) FM radio stations operate at frequencies spaced at 0.2 MHz intervals from 88.1 to 107.9 MHz, but many radio tuners operate down to 87.9 MHz. However, TV channel 6 operates at 88.0 MHz, with the audio signal of TV channel 6 operating at 87.75 MHz. You have a *RLC* series tuning circuit with a 20 pF capacitor. What inductance do you need to receive the Channel 6 audio circuit?

29. (II) An inductance is connected in series with a resistor of 150 Ω. When these are placed in series across an rms 110 V AC line with $\omega = 60$ radians/s, the voltage across the resistance is 40 V. Calculate the value of L and its resistance.

30. (II) A series *RLC* circuit has parameters $R = 18\ \Omega$, $L = 30.0$ mH, and $C = 15.0\ \mu$F. Find the capacitive reactance, inductive reactance, and impedance for the frequencies (a) 60 Hz, (b) 500 Hz, and (c) 20,000 Hz.

31. (II) A series *RLC* circuit consists of a 1200.0-Hz AC emf with $V_0 = 80$ V; $R = 500\ \Omega$, $L = 92$ mH, and $C = 2\ \mu$F. Find X_C, X_L, Z, Q_{max}, ϕ, and I_{max}.

32. (II) You want to build an AC series circuit with the smallest possible impedance. You have a fixed frequency generator with an angular frequency of 1000 rad/s, and the following circuit elements available: two capacitors, of 1 μF and 100 μF; two inductors, of 10 mH and 25 mH, and two resistors, of $10^3\ \Omega$ and $3 \times 10^3\ \Omega$. You may use only one of each type element. What is the lowest circuit impedance and which values of R, L, and C would you choose to make this circuit?

33. (II) Find the voltages across the capacitor and inductor in the AC circuit of Problem 31 at $t = 0.10000$ s if the emf is switched on at $t = 0$ s. All circuit elements initially have no charge or current.

34. (II) An unknown impedance Z is investigated with an oscilloscope. It is connected in series with a 12-Ω resistor, and connected to a 60-Hz AC power supply. The horizontal deflection plates of the oscilloscope are connected to the known resistor so that the horizontal deflection of the electron beam is proportional to the potential drop on the resistor. The potential drop on the impedance Z is measured by the vertical displacement of the electron beam (Fig. 33–28). (a) Sketch the shape of the figure on the oscilloscope's screen if Z consists of a coil with inductance $L = 0.95$ mH and resistance $r = 650$ mΩ. (b) Repeat the calculation if Z is a 0.5-mF capacitor in series with a resistance $r = 650$ mΩ. (c) How can you tell if Z is capacitive or inductive?

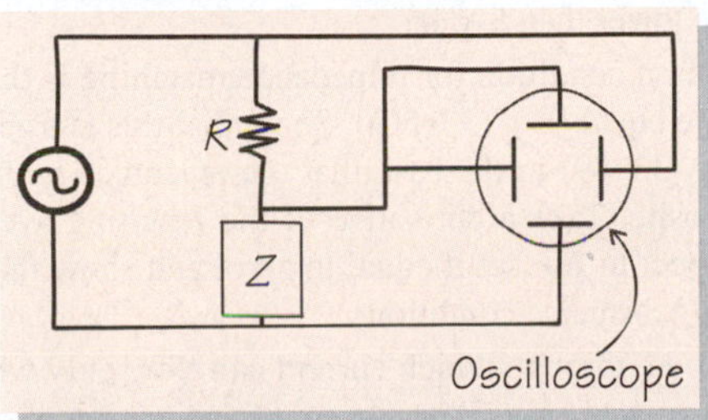

▲ **FIGURE 33–28** Problem 34.

35. (II) Given that the maximum voltage in the circuit shown in Fig. 33–29 is 110 V and the frequency of oscillation is 60 Hz, calculate the maximum current and the maximum potential drops across the resistor, capacitor, and inductor.

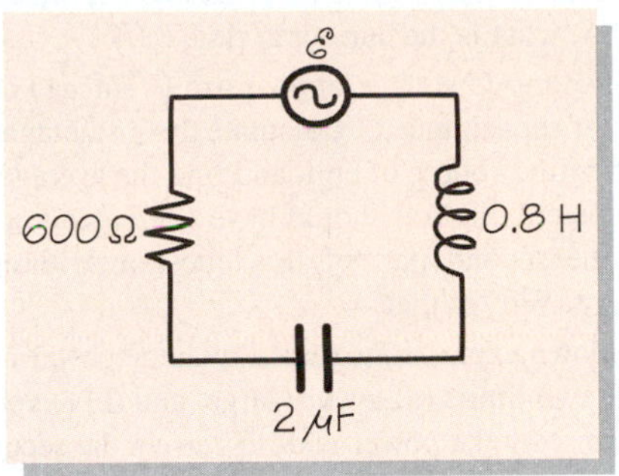

▲ **FIGURE 33–29** Problems 35 and 36.

36. (II) What is the resonant angular frequency ω_0 of the circuit in Problem 35? Suppose that the voltage generator has a variable angular frequency ω. For what values of ω will the current have half the value it has at resonance?

37. (II) An AC circuit consists of a parallel-plate capacitor and a long, cylindrical solenoid. Suppose that all the dimensions of the apparatus, including the wire sizes, are scaled down by a factor of 2. (Note that the turn density doubles.) How would the resonant frequency of the circuit change? Assume that there are changes in resistance.

38. (II) Show that $Q(t)$ in Eq. (33–28) satisfies Eq. (33–23) by direct substitution. Determine the maximum charge Q_{max} on the capacitor in terms of the impedance.

39. (II) Sketch the current and voltage for the following AC series circuits: (a) a pure capacitive circuit, (b) a pure inductive circuit, (c) an *RL* circuit, (d) an *RC* circuit, and (e) an *LC* circuit.

40. (II) A resistor draws 3 A when connected to a 80-V, 60-Hz line. A capacitor of what capacitance, when connected in series with the resistor, will drop the current to 1.5 A? What are the voltage drops across the capacitor and the resistor?

41. (II) A 16-μF capacitor is connected in series with a coil whose resistance is 30 Ω and whose inductance can be varied. The circuit is connected across a 12-V, 60-Hz generator. What is the potential difference across the capacitor and across the inductor–resistor combination when the frequency is the resonant frequency?

42. (II) Suppose that the maximum voltages across the resistor, capacitor, and inductor of a series *RLC* circuit driven by an AC generator of frequency f are identical. If the resistor has a resistance R, find the values of C and L in terms of R and f.

43. (II) A series *RLC* circuit contains a 70-nF capacitor and a 0.2-Ω resistor. If the circuit is resonant at a frequency of 180,000 Hz, what is the inductance?

44. (II) Consider an *RLC* circuit in which two resistors, R_1 and R_2, are connected in series, as are two capacitors, C_1 and C_2, and two inductors, L_1 and L_2. Show that the resulting total impedance is of the form

$$Z_{total} = \sqrt{[(X_{L_1} + X_{L_2}) - (X_{C_1} + X_{C_2})]^2 + (R_1 + R_2)^2}.$$

45. (III) There is an AC source of emf in a single-loop circuit that produces a potential drop in the form $V(t) = V_0 \sin(\omega t)$, while the current in the circuit takes the form $I(t) = I_0 \sin(\omega t - \phi)$. Make a phasor diagram for the current and potential drop across each element if the circuit contains (a) a resistor and a capacitor, and (b) a resistor and an inductor.

33–4 Power in AC Circuits

46. (I) What is the average power dissipated in the resistor for the circuit in Problem 12?

47. (I) An AC power supply with a frequency of 75 Hz dissipates energy at a rate of 150 W in a 12-Ω resistor. If the current at time 0 s is 5.0 A, what is the current at time 0.04 s?

48. (I) Consider an AC voltage of the form $V_0 \sin(\omega t)$ connected to a capacitor of capacitance C. Calculate the instantaneous power VI delivered by the source of emf, and find the average power dissipated in the circuit. You should have been able to obtain the answer to the second part of this question without doing any calculations. Why is that?

49. (I) Write down expressions for the average power in an RLC circuit in the two limits (a) ω very large, and (b) ω very small. Can you explain why the power goes to zero in the second case?

50. (I) A portable electric heater operating on AC voltage of amplitude 110 V is rated at a power of 800 W. (a) What is the resistance of the heater? (b) Find the rms current. (c) Find the maximum current.

51. (I) What are the power factors for (a) pure capacitive circuits, (b) pure inductive circuits, and (c) pure resistive circuits?

52. (II) Show that, on average, no power is dissipated in a purely inductive circuit (a circuit with neither capacitors nor resistors).

53. (II) What are the power factors for (a) RL circuits, (b) RC circuits, and (c) LC circuits?

54. (II) A coil draws a current with a peak amplitude of 0.4 A from a peak 170 V source operating at a frequency $f = 60$ Hz. The power consumed is 18 W. What are the impedance, the resistance, and the inductance?

55. (II) Consider an RLC circuit, with $L = 170\ \mu$F and $C = 24$ mH. (a) Calculate the resonant frequency. (b) How large should the resistance R be if a 1% shift in the frequency is to reduce the power received by a factor of 50?

56. (II) An AC source of emf operating at a frequency of 60 Hz produces an rms voltage of 78 V. Find the voltage amplitude. The source of emf is connected in series with an impedance of $Z = 20\ \Omega$. Find the rms current and the current amplitude.

57. (II) When a coil draws 200 W from a $V_{rms} = 110$-V, 60-Hz line, the power factor is 0.6. If the same coil with a capacitor added in series is to draw the same power from a $V_{rms} = 220$-V, 60-Hz line, what must the capacitance be? If the aim were to maintain the same power factor rather than the same rms power, how would your answer change?

58. (II) A machine shop uses 120 A from a 220-V, 60-Hz line. However, due to the primarily inductive load—motors—the voltage and current are out of phase by 40°, wasting a lot of heat in the cables. A large capacitor connected parallel to the machines can solve this problem (Fig. 33–30). (a) How large should the capacitance be? (b) What will be the current in the main cable with the capacitor attached? (c) What is the total power of the machines in the shop?

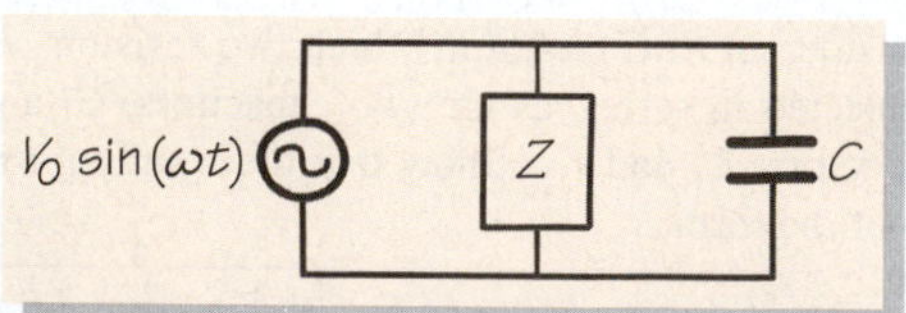

▲ **FIGURE 33–30** Problem 58.

59. (II) An electric motor consumes 5 kW of power at 220 V (voltage amplitude) with a power factor of 0.80. This motor is to be run at the end of a power transmission line with a total resistance of 2.5 Ω. What voltage and power must be supplied at the input end of the transmission line?

60. (II) An AC transmission line transfers energy to a device with a power factor of 0.85 at the rate of $\langle P \rangle = 4$ kW and a voltage of 220 V. If the transmission line has a resistance of 15 Ω, how much energy is lost to Joule heating in the transmission line?

61. (II) A 220-V generator has a current-carrying capacity of 80 A. What is the maximum rate at which energy can be taken from this generator by an impedance with a power factor of 0.55? for a power factor of 0.95?

62. (II) A 20-Ω resistor and a 4.5-μF capacitor are connected in series to a 110-V, 60-Hz power supply. What are the rms current, power, and power factor? How will these numbers change if an inductance of 0.035 H is connected in series with this circuit?

63. (II) An RLC circuit draws a peak 3.00 A from an rms 220 V AC power supply. The power consumed is 200 W. When the capacitor is short-circuited, the current falls to 2.20 A. Find the values of R, the capacitive reactance, and the inductive reactance, assuming that $X_L < X_C$.

64. (II) A series RLC circuit is known to have a resonance at 2.2 MHz. The half-power point is 3.0 KHz away from resonance. Given that the capacitance is 6.6 pF, what are L and R?

65. (II) House current, which has an rms voltage of 110 V and a frequency of 60 Hz, drives a resistor of a variable resistance set at $R = 50\ \Omega$, a capacitor of fixed capacitance $C = 20\ \mu$F, and an inductor of variable inductance, connected in series. (a) What is the power absorbed by the circuit if $L = 10$ mH? (b) What would the power drawn be if the resistance were halved without changing the setting of the inductance? (c) What is the maximum power drawn in part (b)?

66. (II) For a driven series RLC circuit, show that

$$\frac{R}{Z} = \frac{1}{\sqrt{1 + Q^2\left(\dfrac{\omega}{\omega_0} - \dfrac{\omega_0}{\omega}\right)^2}}.$$

67. (II) Plot R/Z in Problem 66 for values of ω/ω_0 from 0.4 to 2.5 and values of Q of 1, 10, and 100. Use a computer program and graphics output, if available.

68. (II) For a driven series RLC circuit, show that Q is related to $\Delta\omega$ by the relation

$$\frac{\Delta\omega}{\omega} = \frac{1}{Q}\frac{\omega_0}{\omega} \cong \frac{1}{Q}.$$

33–5 Some Applications

69. (II) Consider the circuit treated in Example 33–9 and drawn in Fig. 33–18. Take $C = 5$ nF and $R = 120\ \Omega$, but assume now that the input emf has the purely sinusoidal form $V_1 \sin(\omega t)$, where $V_1 = 0.20$ V. Calculate the potential across the capacitor for (a) $f = 100$ Hz, (b) $f = 10^5$ Hz, and (c) $f = 10$ MHz.

70. (II) Design a high-pass RC filter that will remove voltages with frequencies lower than 8 kHz.

71. (II) Design a high-pass RL filter for filtering out signals with frequencies lower than 8 kHz.

72. (II) The first condition for impedance matching is that the resistances are equal [Eq. (33–50)]. Show that this is true by starting with Eq. (33–49) in the case that the reactance terms are equal and opposite. Take a derivative of the resulting average power with respect to R_2, set it equal to zero, and show that this gives the equal resistance condition.

73. (II) A diode, through which current can flow only when the emf is positive, acts as a filter for an AC generator of angular frequency ω. The current has maximum magnitude I_0. Find its average and rms values.

74. (III) An RC filter circuit like that shown in Fig. 33–18 is called a *high-pass* filter circuit when the voltage output is taken across the resistor. Plot the ratio V_{out}/V_{in} as a function of frequency. Why does such a circuit block signals of low frequency but allow high-frequency signals to pass?

75. (III) An RC filter circuit like that shown in Fig. 33–18 is called a *low-pass* filter circuit when the voltage output is taken across the capacitor. Plot the ratio V_{out}/V_{in} as a function of frequency. Why does such a circuit block high-frequency signals but allow low-frequency signals to pass?

76. (III) Consider the LC filter of Fig. 33–31 with the emf $V_0 \sin(\omega t)$. Assume that $X_L \gg X_C$ (or $\omega \gg \omega_0$). (a) Show that $V_{out} = (X_C/X_L)V_0$. (b) Show that the circuit of Fig. 33–31 is generally effective in reducing the AC components, but not the DC components, of emf.

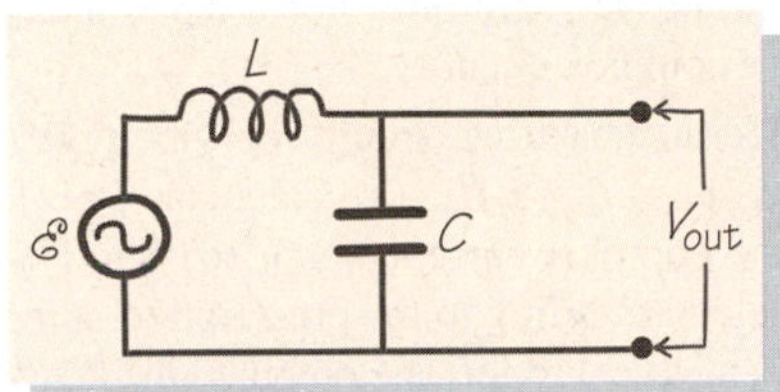

▲ **FIGURE 33–31** Problem 76.

77. (III) Given that the driving voltage of the RLC circuit shown in Fig. 33–32 is $V = V_0 \cos(\omega t)$, calculate the currents in the three elements. Is there a resonant frequency? [*Hint*: Write down the circuit equations, and substitute the trial solution $I = I_0 \cos(\omega t + \phi)$].

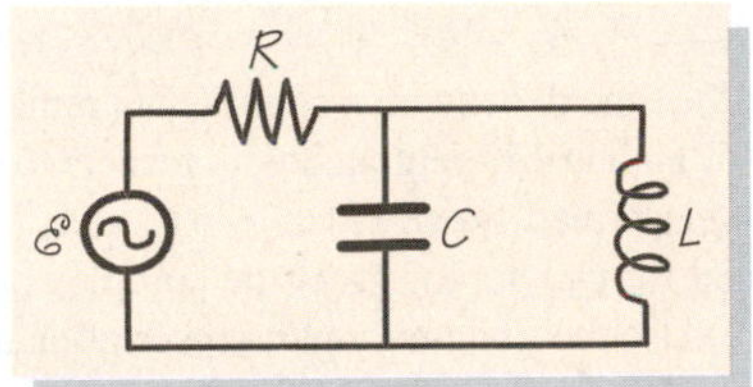

▲ **FIGURE 33–32** Problem 77.

78. (III) What is the relation between the resistances if the average power dissipated in them is the same at $\omega = 1/\sqrt{LC}$ (see Fig. 33–33)?

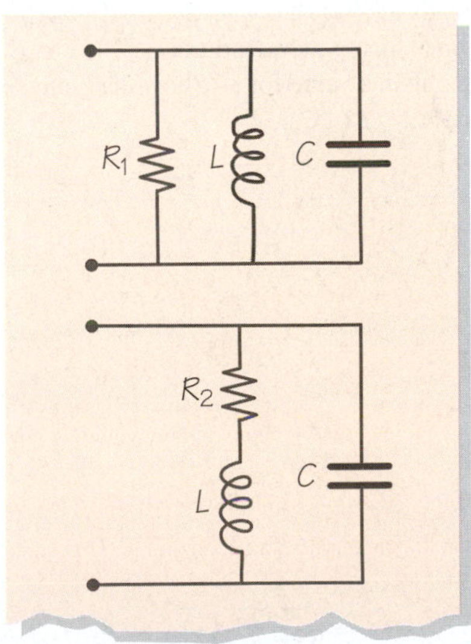

▲ **FIGURE 33–33** Problem 78.

General Problems

79. (II) Calculate (a) the maximum instantaneous voltage across each capacitor; and (b) the maximum instantaneous voltage across the inductor for the circuit shown in Fig. 33–34. Use the parameters specified in Problem 80.

80. (II) Consider the circuit shown in Fig. 33–34. The emf has an amplitude of $V_0 = 4$ V and a frequency of 600 Hz; $L = 70\ \mu$H, $C_1 = 4\ \mu$F, and $C_2 = 9\ \mu$F. Find (a) the maximum current; (b) the resonant frequency.

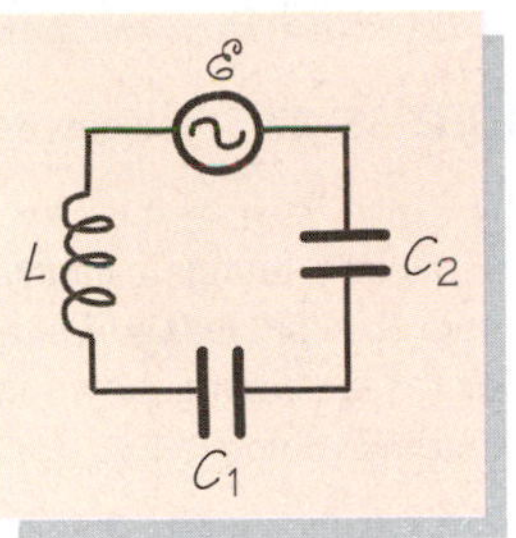

▲ **FIGURE 33–34** Problem 80.

81. (II) Show that Eq. (33–31) for the phase angle can be determined from

$$\tan\phi = \frac{(\omega^2 - \omega_0^2)L}{\omega R}.$$

82. (II) An amplifier with an equivalent impedance of 3,000 Ω is to be connected to an 8-Ω speaker through a transformer (Fig. 33–35). What should the turn ratio of the transformer be?

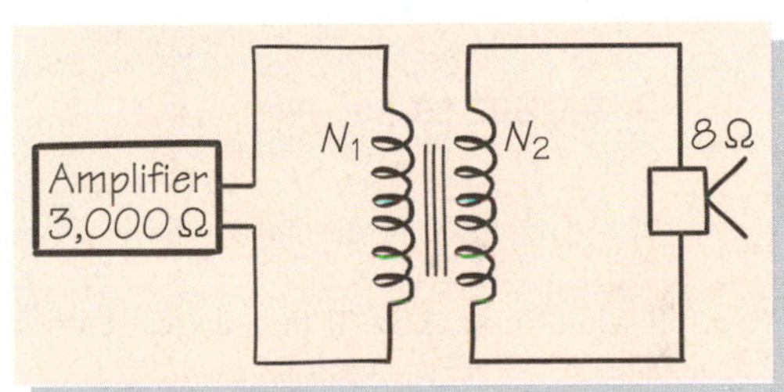

▲ **FIGURE 33–35** Problem 82.

83. (II) The impedance Z_1 in Fig. 33–36 can be regarded as a pure resistance $R_1 = 15\ \Omega$, whereas the impedance Z_2 is associated with a series resistance $R_2 = 8\ \Omega$ and a capacitance $C = 2\ \mu$F. If $f = 3000$ Hz and $V_0 = 3$ V, what is the power dissipated in Z_2?

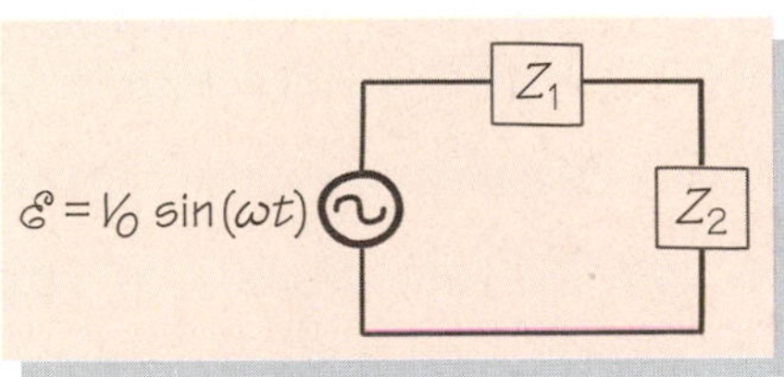

▲ **FIGURE 33–36** Problem 83.

84. (II) Consider the circuit shown in Fig. 33–37. The emf has an amplitude of $V_0 = 12$ V and a frequency of 400 Hz; $L = 10$ mH, $C_1 = 20\ \mu$F, and $C_2 = 30\ \mu$F. Find (a) the maximum current in each leg, and (b) the resonant frequency.

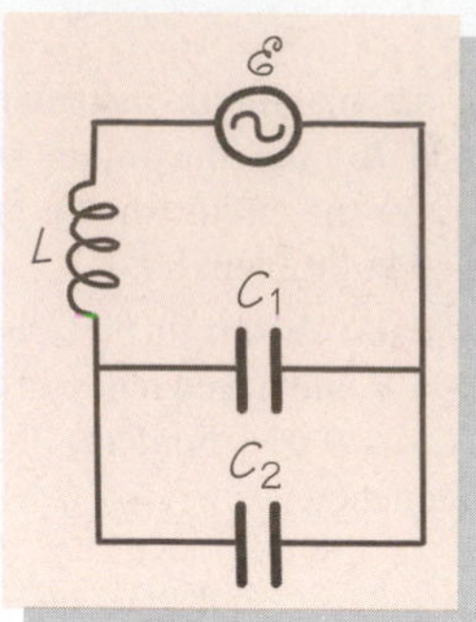

▲ **FIGURE 33–37** Problems 84 and 85.

85. (II) Calculate (a) the maximum instantaneous voltage across each capacitor; and (b) the maximum instantaneous voltage across the inductor for the circuit shown in Fig. 33–37. Use the parameters specified in Problem 84.

86. (II) Write down the two equations that specify the currents I_1 and I_2 in the two loops of the circuit shown in Fig. 33–38.

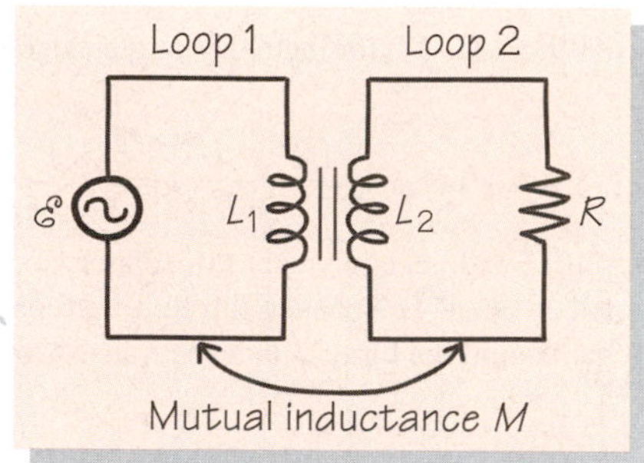

▲ **FIGURE 33–38** Problem 86.

87. (II) A series *RLC* circuit is to be designed to have a resonant frequency of 18 MHz, and the curve of power versus frequency f is to have a full width of 4.0 kHz. If the only capacitor available has a capacitance of 33 pF, what must R and L be?

88. (II) A resistor with $R = 62\ \Omega$ draws a current from a wall plug; a capacitor is connected in parallel with this resistor. The current source has an amplitude of 110 V and a frequency of 60 Hz, and the reactance of the capacitor is 3 Ω at this frequency. What is the current drawn by the parallel combination?

89. (II) A 15-μF capacitor connected in series with a resistor of variable resistance R is connected to a $V_{rms} = 110$-V, 60-Hz AC supply. Plot the variation of the rms current with R, and calculate the value of R for which the power delivered is maximum.

90. (II) An AC circuit supplies $V_{rms} = 220$ V at 60 Hz to a 10-Ω resistor, a 35-μF capacitor, and an inductor of variable self-inductance in the 30 mH to 300 mH range, all in series. The capacitor is rated to stand a maximum voltage of 1200 V. (a) What is the largest current possible that does no damage to the capacitor? (b) To what value can the self-inductance be safely set?

Problems with Complex Variables

In the following set of problems, we suppose there is an AC source of emf that produces a potential drop of the form $V_0 \sin(\omega t)$. In the complex variable technique, we work instead with the complex form $V_C(t) = V_0 \exp(i\omega t)$, with the instruction that the original potential drop is given by the imaginary part of this form, $\mathrm{Im}[V_C(t)]$. All quantities with oscillating time dependence are given this treatment. Note: The subscript "C" here does not refer to a capacitor, but instead indicates a complex quantity.

91. (III) Consider a capacitive circuit, for which the loop rule takes the form $V_C(t) = Q_C(t)/C$. (a) Calculate $Q_C(t)$, and use it to calculate the complex current $I_C(t)$. (b) Show that the current $I(t)$ calculated according to the prescription at the head of these problems is identical to the one obtained in Eq. (33–12).

92. (III) Consider an inductive circuit, for which the loop rule takes the form $V_C(t) = L\, dI_C(t)/dt$. (a) Calculate $I_C(t)$ in two ways: (i) by direct integration of the equation; (ii) by noting that no matter how often the function $\exp(i\omega t)$ is differentiated with respect to time, the time dependence remains $\exp(i\omega t)$. (b) Show that the current $I(t)$ calculated according to the prescription at the head of these problems is the same as the one obtained in Eq. (33–18).

93. (III) In complex analysis, the loop rule for the series *RLC* circuit is $V_C(t) = L(dI_C(t)/dt) + Q_C(t)/C + RI_C(t)$. (a) Make use of the fact that the time dependence of $\exp(i\omega t)$ remains the same no matter how often it is differentiated with respect to time in order to calculate Q_{0C}, defined by $Q_C(t) = -iQ_{0C}e^{i\omega t}$. (b) Use the formula obtained for Q_{0C} to construct the complex quantities $Q_C(t)$ and $I_C(t)$. (c) Use the imaginary part prescription described at the head of these problems to calculate $Q(t)$ and $I(t)$.

94. (III) Consider the *RLC* circuit of Problem 93. Define the complex impedance Z_C to be $V_C(t)/I_C(t)$, and use the results of Problem 93 to show that Z_C is independent of time, with an absolute magnitude given by Eq. (33–32).

95. (III) Consider an *RLC* circuit without a driving emf, so that the loop rule reads $L(dI_C(t)/dt) + Q_C(t)/C + RI_C(t) = 0$. Solve this equation by using the substitution $Q_C(t) = Q_{0C}e^{i\omega t}$. Use the equation to obtain a value for ω. Show that your solution leads to the results of Eq. (32–32).

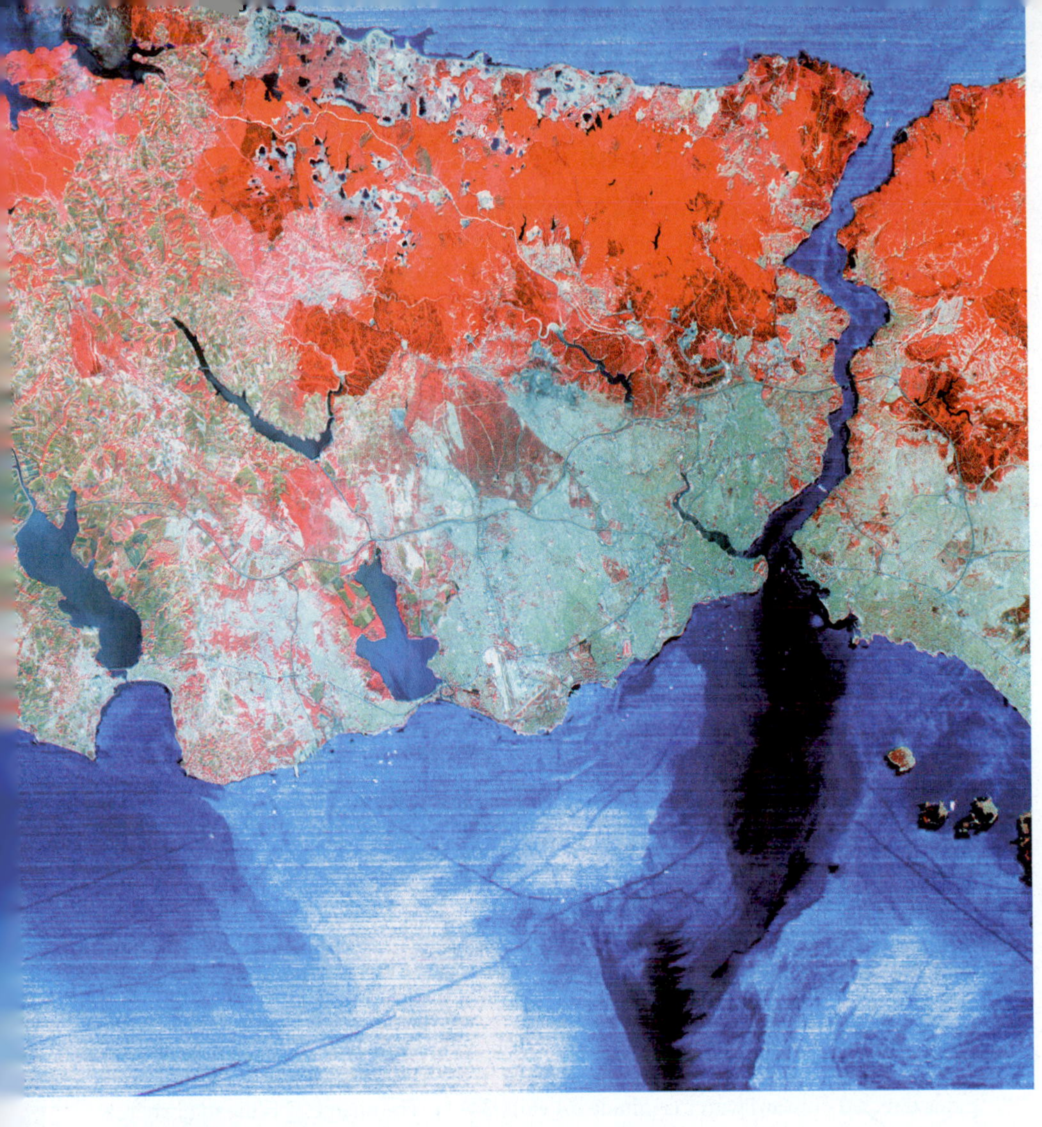

Terra satellite image of the straits of Bosporus. This image is a combination of visible and infrared light images. Vegetation appears red while urban areas appear blue-green; the different water colors are measures of the water temperature.

Chapter 34

Maxwell's Equations and Electromagnetic Waves

Faraday's law shows that electricity and magnetism are fundamentally connected; James Clerk Maxwell's introduction in 1864 of the displacement current enhances this connection. We have a complete, consistent set of fundamental laws for electricity and magnetism, collectively known as Maxwell's equations. The individual experiments that led to the discovery and completion of each of these equations—Gauss' law for electric fields, Gauss' law for magnetic fields, Ampère's law, and Faraday's law—never hinted at the rich implications of Maxwell's equations taken *together*. The most dramatic prediction of the full set of Maxwell's equations is the existence of electromagnetic waves that propagate through empty space at a predictable speed—the speed of light. Light itself is such a wave. It is the intimate coupling between electric and magnetic fields that leads to these waves, with, in effect, one field generating the other in a self-sustaining way. In this chapter, we discuss the nature and properties of these *electromagnetic waves*: the orientation and relationship of the electric and magnetic fields contained in them, the energy and momentum they carry, and polarization. While these waves are described in terms of the mechanical waves we met in Chapters 14 and 15, they differ in a profound way in that they do not require a medium in which to propagate. This observation contains the seeds of Einstein's special relativity. We'll also see here how quantum physical phenomena intrude into the properties of classical electromagnetic waves.

34–1 Maxwell's Equations

Let's first recall and then comment on **Maxwell's equations**, which fully describe the electric and magnetic fields in the presence of electric charges and currents. In this discussion we'll largely ignore the effects of matter, which although very well understood, would be a diversion from the important points to be described in this chapter.

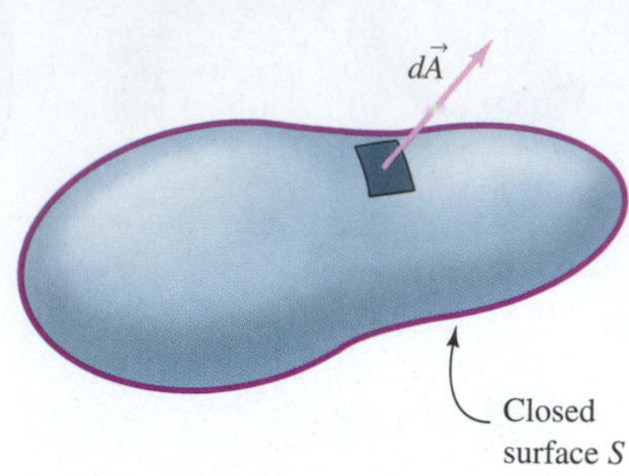

▲ **FIGURE 34–1** An infinitesimal surface element $d\vec{A}$ on the closed surface S.

I. Gauss' law for electric fields

$$\int_{\text{closed surface}} \vec{E} \cdot d\vec{A} = \frac{Q}{\varepsilon_0}. \tag{34–1}$$

II. Gauss' law for magnetic fields

$$\int_{\text{closed surface}} \vec{B} \cdot d\vec{A} = 0. \tag{34–2}$$

III. Generalized Ampère's law

$$\oint \vec{B} \cdot d\vec{s} = \mu_0 I + \mu_0 \varepsilon_0 \frac{d}{dt} \int_{\text{surface}} \vec{E} \cdot d\vec{A}. \tag{34–3}$$

IV. Faraday's law

$$\oint \vec{E} \cdot d\vec{s} = -\frac{d}{dt} \int_{\text{surface}} \vec{B} \cdot d\vec{A}. \tag{34–4}$$

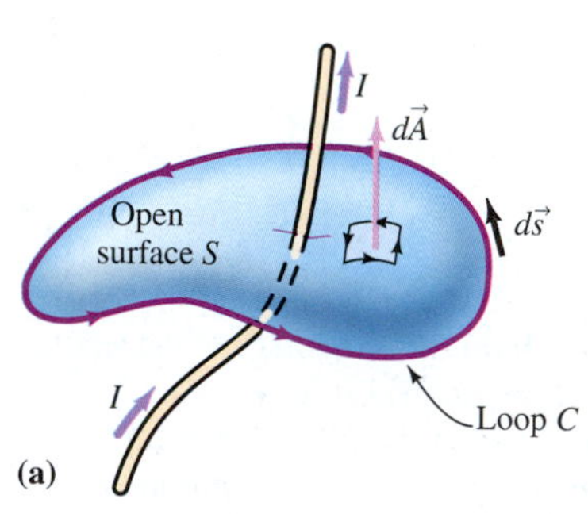

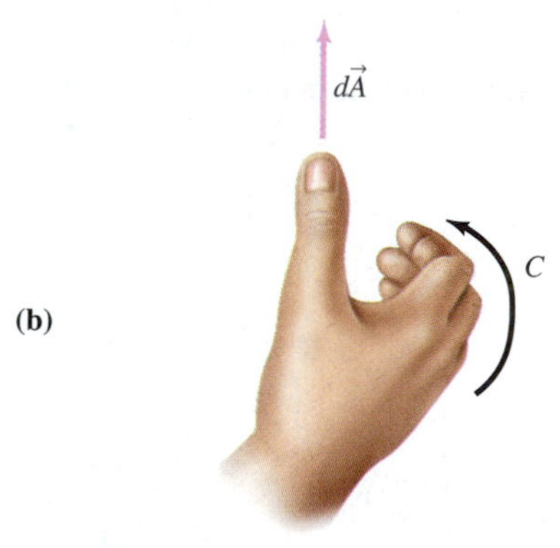

▲ **FIGURE 34–2** (a) A surface S bounded by the closed loop C. A current I passes through the surface. If the integration along C proceeds counterclockwise, (b) then the direction of the surface elements $d\vec{A}$ that make up S is given by a right-hand rule.

I. Gauss' law is equivalent to Coulomb's law in static situations and hence shows that the electric field due to a static charge varies inversely with the square of distance from the charge. It relates the electric flux through a closed surface (the surface can be imaginary) to the charge enclosed [see, for example, Eq. (23–7)] and states that charges are the sources of electric field. The surface element $d\vec{A}$ in Eq. (34–1) is normal to the surface S and is directed outward with magnitude dA (Fig. 34–1). The charge Q is the total charge contained within the closed surface. The factor ε_0 (the permittivity of free space) is associated with our choice of units. Gauss' law holds even for time-dependent electric fields.

II. Magnetic monopoles—which would be the magnetic analogues of electric charge—have never been discovered. Their nonexistence leads to Gauss' law for magnetic fields [Eq. (29–12)], which states that the flux of the magnetic field through a closed surface is zero, implying that magnetic field lines have no beginning or ending. This equation also holds even for time-dependent magnetic fields.

III. Ampère's law states that magnetic fields arise from currents or changing electric fluxes. The left-hand side of this equation is the expression for the integral of the magnetic field's tangential component along an arbitrary closed loop C (Fig. 34–2). The right-hand side has two contributions: One is the total current flowing through any surface S bounded by the closed loop C; the other is the rate of change of the electric field flux through such a surface, the displacement current contribution. As we described in Chapter 29 [see Eq. (29–31)], Maxwell was responsible for introducing the displacement current. As you can review in Chapter 29, he did this to fill a gap in the logic of Ampère's law. The presence of the parameter μ_0 (the permeability of free space) is a consequence of SI units.

IV. Faraday's law describes the induced electric field loop generated by a changing magnetic flux [see Eq. (30–2)] and shows that the electric field encircles any area in which the magnetic flux is changing. The left-hand side is the integral of the tangential component of the induced electric field around an arbitrary closed loop C. The right-hand side measures the rate of change of the magnetic flux through any surface S bounded by C, just as in Fig. 34–2. Equation (34–4), as well as Eq. (34–3), implies a sign convention given by a right-hand rule. The minus sign is very important: It represents the fact that the induced electric field, were it to act on charges, would give rise to an induced current that opposes the change in the magnetic flux (Lenz's law).

The four Maxwell equations display a degree of symmetry between electric and magnetic fields. This symmetry is not perfect, because magnetic monopoles apparently do not exist. Faraday's law contains no term like the $\mu_0 I$ term in Ampère's law, because there is no free magnetic charge to form a magnetic current. In a vacuum there are neither currents nor charges, and then the symmetry between $\vec{E}$ and $\vec{B}$ is complete. This is a situation in which the propagation of electromagnetic waves occurs.

In the presence of matter, Maxwell's equations are modified. Assuming that we don't have to deal with boundaries between different materials, then for most types of materials we can simply replace ε_0 by $\varepsilon = \kappa\varepsilon_0$, where κ is the dielectric constant. Except for ferromagnetic materials, μ is very close to μ_0, so that the additional rule that the permeability of the vacuum (μ_0) is to be replaced by the material's permeability (μ) does not much affect things.

34–2 Electromagnetic Waves

Equations (34–3) and (34–4) clearly show that time-dependent electric and magnetic fields influence each other; they are said to be *coupled*. As a consequence of this coupling, electric and magnetic fields can transport energy (and momentum) over much larger distances than might be suggested by the $1/r^2$ falloff of the electric field in Coulomb's law or the magnetic field in the Biot–Savart law. As we shall see, the coupled fields transport energy through traveling waves called **electromagnetic waves**. These waves are all around us in the form of radio, television, and cell phone signals, microwaves, visible light, and X rays, for example. In fact, electromagnetic waves can be produced with essentially an unlimited range of wavelength or frequency, and the examples listed above just correspond to different wavelength or frequency ranges.

The following argument conveys the physical mechanism by which the electric and magnetic fields couple and electromagnetic waves propagate. Consider a straight wire that is aligned with the x-axis and carries a time-varying current I (Fig. 34–3a). A magnetic field encircles the wire; if the current changes with time, so does the magnetic field. As the current increases, so does the magnetic field. As the magnetic field increases, so does the magnetic flux through an area A_1 in the xz-plane. According to Faraday's law, Eq. (34–4), a changing magnetic flux induces an emf around the boundary of this area. This emf is associated with the induced electric field shown in Fig. 34–3b. Lenz's law determines the orientation of the field. (Note that while we have drawn electric fields in this figure, all that Faraday's law really tells is that an induced electric field is present whose integral along the boundary is given. The drawn electric fields are simply present to help you follow this argument more closely. Later we'll see how to tighten up the argument.)

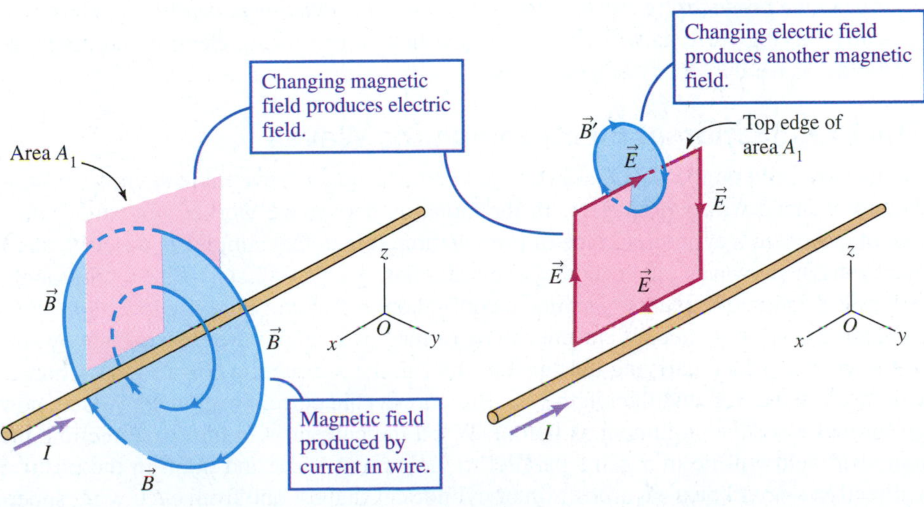

FIGURE 34–3 (a) As we know from Ampère's law, a current-carrying wire aligned in the x-direction has a magnetic field that forms circles in the yz-plane. (b) If the current in the wire changes with time, the magnetic field it produces changes with time, inducing a changing electric field, which in turn induces a changing secondary magnetic field, and so forth.

Let's consider now the top edge of area A_1. Along that edge, the electric field has been induced in the $-x$-direction. This induced electric field is time varying because it is due to a changing magnetic field; at the instant shown in our example, the electric field is increasing. Now, according to the generalized Ampère's law, Eq. (34–3), we do not need flowing charges to induce a magnetic field. *A changing electric field also produces a magnetic field by giving rise to a displacement current* (Section 29–5). The displacement current in this case is along the direction of the changing electric field, which is the direction of the original current. The displacement current, however, is spread through space rather than localized at the wire like the original current. In particular, there is a displacement current away from the location of the original current. At this point, we can begin to see how the propagation works: The displacement current produces a secondary magnetic field $\vec{B}'$ at still larger values of z (Fig. 34–3b). The field $\vec{B}'$ is perpendicular to the xz-plane. Because the displacement current varies with time, $\vec{B}'$ is also changing. Therefore this secondary magnetic field produces an induced emf aligned in the x-direction at still larger values of z, and the process repeats itself to larger and larger z values.

We could have taken our plane area A_1 to intersect the cylinder around the wire at any angle relative to the z-axis, since our picture is symmetric about the x-axis. The same sequence of events results from the coupling however we rotate our figure about the x-axis. This means that when we use the words "larger value of z," we really mean "larger value of the perpendicular distance from the wire." As the current changes with time, so do the secondary fields $\vec{B}'$. We therefore have electric and magnetic fields that *propagate* on the surface of an ever growing cylinder whose axis lies along the wire.

CONCEPTUAL EXAMPLE 34–1 Suppose the current oscillates for a while and is then turned off. What happens to the fields?

Answer If you look at the discussion above, you can see that to learn about a field farther from the wire, we discuss the fields closer to the wire, not the current itself. The induced electric field, the secondary magnetic field whose source is the displacement current due to this induced electric field, and all their successors have acquired a life of their own. While they were initially created by the current in the wire, the fields have detached themselves from their original source and are propagating "alone." If the current is turned off at some time $t = T$, then from that moment on, no new fields are produced, but the fields produced earlier continue their cylindrical propagation outward.

What qualitative conclusions can we can draw from this discussion?

1. With the changing current *restricted to a line*, the fields propagate with cylindrical symmetry outward from the current line. The electric field is aligned parallel to the current, and the magnetic field is aligned perpendicular to both the electric field and to the direction of propagation. *These are general features of electromagnetic waves.*
2. The current must change in time if it is to give rise to propagating fields, as a steady current merely produces a static magnetic field. We can translate this into a statement about the charges whose flow gives rise to the current: *The charges that give rise to the propagating electric and magnetic fields must be accelerating.* Harmonically varying currents will give rise to harmonically varying electric and magnetic fields, as we will verify below.

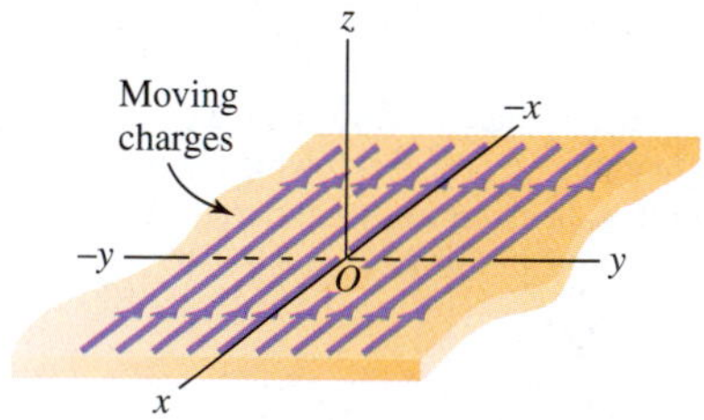

▲ **FIGURE 34–4** Current flows in a sheet along the $-x$-direction. It can be approximated by aligning wires side by side in the x-direction. If the current is oscillatory, charges move first in the $-x$-direction, then in the $+x$-direction.

The Propagation of Electromagnetic Waves

Let us now build on the points above to gain a more quantitative understanding of how electromagnetic waves propagate. In the situation above, we worked outward from a line of current to a cylindrical type of propagation. It is in fact simpler to describe plane electromagnetic waves rather than cylindrical ones. So we will start with a brief qualitative description of a *sheet of current*, a configuration that will give rise to a *plane electromagnetic wave*. A sheet of current can be formed by a set of wires placed side by side in the xy-plane, each carrying the same current in the x-direction (Fig. 34–4). Charges will move in the $+x$- and then in the $-x$-direction as the current oscillates. We can now go through the same argument as before. When the current is in the $-x$-direction, the magnetic field will lie in a plane parallel to the current sheet and point in the positive y-direction—it will be a superposition of cylindrical shapes, one from each wire, superimposed side by side, as in Fig. 34–5. In the limit of wires very close to each other, the

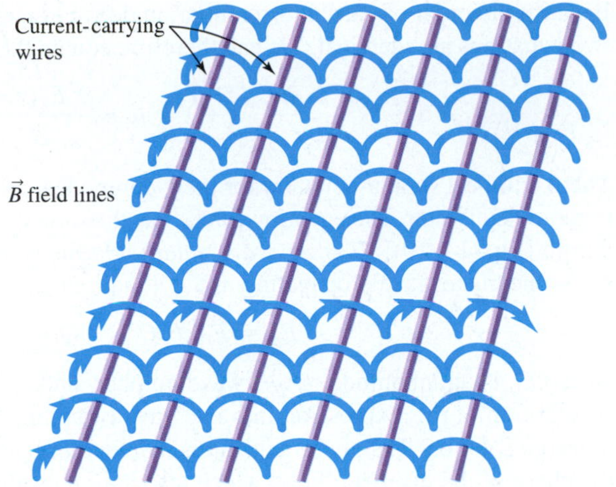

FIGURE 34–5 In the limit that the wires carrying the current are close to one another, the magnetic field of a series of closely spaced and parallel currents forms planes parallel to the plane of the currents.

field lines of the net magnetic field form a corrugated surface; taking the wires still closer to one another, the field forms two planes, one above the current sheet and the other the same distance away below the current sheet. The electric field will also lie in a plane parallel to the current sheet, and point in the direction of the current. These fields in turn lead to other fields, and if you follow the same argument that we made for the single wire, you will find that these fields will propagate in the z-direction (both + and −), perpendicular to the sheet. These fields will depend on time in a way that mirrors the time dependence of the current. The electromagnetic fields form *plane waves*, which we recall from Chapters 14 and 15 refers to waves that advance along planar wave fronts—in this case, planes parallel to the xy-plane.

To understand quantitatively how we can get fields that behave like this, we assert that Maxwell's equations—in particular, Faraday's law and the generalized Ampère's law—are equivalent to a set of differential equations for the components of these fields. Not surprisingly, these are *coupled* differential equations—how the magnetic field changes determines how the electric field changes and vice versa. We derive these equations in the Appendix to this chapter on p. 967. Here, in order to get directly to plane waves propagating in the z-direction, we write them down for the special case of fields that are independent of x and y, but which do depend on z and, of course, on the time t. The components B_y and E_x in particular satisfy

$$-\frac{\partial B_y(z,t)}{\partial z} = \mu_0 \varepsilon_0 \frac{\partial E_x(z,t)}{\partial t} \tag{34–5}$$

and

$$-\frac{\partial B_y(z,t)}{\partial t} = \frac{\partial E_x(z,t)}{\partial z}. \tag{34–6}$$

Although a changing current was necessary to produce the initial changing fields, these equations are not directly dependent on that current—this is what we meant when we said that the fields take on "a life of their own." The field components B_y and E_x both depend on the value of z and on the time t. (Recall that partial derivatives appear whenever quantities such as fields depend on two or more variables. In taking a partial derivative with respect to one variable, the other variables are held fixed.)

Equations (34–5) and (34–6) lead directly to wave equations for the field components. These two equations couple the two fields. We can combine and simplify them by a straightforward procedure. The partial derivative of Eq. (34–5) with respect to time gives

$$-\frac{\partial^2 B_y(z,t)}{\partial t\, \partial z} = \mu_0 \varepsilon_0 \frac{\partial^2 E_x(z,t)}{\partial t^2}.$$

Similarly, the partial derivative of Eq. (34–6) with respect to z gives

$$-\frac{\partial^2 B_y(z,t)}{\partial z\, \partial t} = \frac{\partial^2 E_x(z,t)}{\partial z^2}.$$

Because the order of partial differentiation does not matter, the left-hand sides of these two equations are identical. We can therefore equate the right-hand sides:

$$\frac{\partial^2 E_x(z,t)}{\partial z^2} = \mu_0 \varepsilon_0 \frac{\partial^2 E_x(z,t)}{\partial t^2}. \tag{34–7}$$

There will be a similar equation for the y-component of the magnetic field, as we'll see in more detail later. The equation for E_x has the same form as an equation we have seen before [Eq. (14–10)]: *It is the wave equation!* One solution of this wave equation is a harmonic plane wave propagating in the $+z$-direction:

$$E_x = E_0 \cos(kz - \omega t + \phi), \tag{34–8}$$

where E_0 is an amplitude, k is a wave number, and ω is an angular frequency. Direct substitution of this expression into the wave equation, Eq. (34–7), will verify that it is indeed a solution. The phase angle ϕ is included because we shall want to look at how the phase of the magnetic field, which also has an oscillating solution, is related to that of the electric field. It is important to understand what this equation means: the electric field has an x-component (always aligned along the x-axis), and that component changes as one looks at different values of z and t.

The Propagation Speed of Electromagnetic Waves

We recall from our discussion in Chapter 14 of wave motion that Eq. (34–8) represents a wave of wavelength $\lambda = 2\pi/k$ and frequency $f = \omega/2\pi$. The propagation speed is $v = \lambda f = \omega/k$. This speed is found immediately from the wave equation itself, as comparison with the original form of the wave equation, Eq. (14–10), shows. We have

$$v^2 = \frac{1}{\mu_0 \varepsilon_0}. \tag{34–9}$$

When we use the numerical values for μ_0 and ε_0, we find

$$v^2 = \frac{1}{(1.257 \times 10^{-6}\,\mathrm{T \cdot m/A})(8.854 \times 10^{-12}\,\mathrm{C^2/N \cdot m^2})} = 8.999 \times 10^{16}\,\mathrm{m^2/s^2}$$
$$= (3.00 \times 10^8\,\mathrm{m/s})^2.$$

Maxwell recognized that the magnitude of v is the *speed of light c!* One of the most commonplace of all the physical phenomena around us, light, was now explained in terms of the laws of electricity and magnetism, or more properly, electromagnetism. This surely represents one of the greatest discoveries of science, equal in every way to Newton's understanding of gravitation. (And there was more, much more, to come from Maxwell's equations, including special relativity and a great portion of today's technology.) Equation (34–9) tells us that the speed of light is determined by constants of the equations of electromagnetism:

$$c = \frac{1}{\sqrt{\mu_0 \varepsilon_0}}. \tag{34–10}$$

ELECTROMAGNETIC WAVE SPEED IN EMPTY SPACE

What distinguishes *visible* light? The wavelengths of electromagnetic waves are not restricted to any particular value or set of values. The kinds of experiments with optics that we'll be discussing later in the book show that what we call visible light corresponds to a limited range of wavelengths for which, through evolutionary adaptation, our eyes have become particularly good detectors. This range of wavelength is one in which the Sun emits radiation strongly and for which the waves pass easily through the atmosphere. Within this visible spectrum, we interpret different wavelengths as colors. The shortest wavelengths of the visible spectrum are violet; the longest wavelengths are red. The speed which Maxwell confirmed was the speed of visible light, because that was the phenomenon that to that point had been, quite literally, visible and whose speed had been measured. But Maxwell's discovery opens a door onto the full spectrum of electromagnetic radiation, a subject we'll discuss more fully later in this section.

Keep in mind the important relation $v = \lambda f$ between speed, wavelength, and frequency. If the speed of propagation is fixed, as it is in empty space, then wavelength and frequency have an inverse relation, with large wavelengths corresponding to small frequencies and vice versa.

The Relation Between *E* and *B* in an Electromagnetic Wave

To see how E and B for an electromagnetic wave are related, we can start with Eqs. (34–5) and (34–6) and show that B_y also obeys a wave equation similar to that of E_x; namely,

$$\frac{\partial^2 B_y(z,t)}{\partial z^2} = \mu_0\varepsilon_0 \frac{\partial^2 B_y(z,t)}{\partial t^2}.$$

Like the x-component of the electric field, the y-component of the magnetic field forms a wave that propagates at speed c in the z-direction. However, because Eqs. (34–5) and (34–6) couple the fields, the waves of B_y do not propagate independently from those of E_x. If we have a wave solution for E_x, Eq. (34–8), then from Eq. (34–5),

$$\begin{aligned}\frac{\partial B_y}{\partial z} &= \mu_0\varepsilon_0 \frac{\partial E_x}{\partial t} = -\mu_0\varepsilon_0 \frac{\partial}{\partial t}[E_0 \cos(kz - \omega t + \phi)] \\ &= -\mu_0\varepsilon_0\omega E_0 \sin(kz - \omega t + \phi).\end{aligned} \qquad (34\text{–}11)$$

Equation (34–6) becomes

$$\frac{\partial B_y}{\partial t} = -\frac{\partial E_x}{\partial z} = -\frac{\partial}{\partial z}[E_0 \cos(kz - \omega t + \phi)] = kE_0 \sin(kz - \omega t + \phi). \qquad (34\text{–}12)$$

From these two expressions for the derivatives of B_y, it is easy to check that the following equation has the correct spatial and time dependence:

$$B_y = B_0 \cos(kz - \omega t + \phi). \qquad (34\text{–}13)$$

Relations Between Amplitudes: The amplitude B_0 of the magnetic field wave is not independent of the amplitude E_0 of the electric field wave, as Example 34–2 shows.

EXAMPLE 34–2 Consider the electromagnetic traveling wave for which the electric and magnetic fields are given by Eqs. (34–8) and (34–13). Use the derivative relations we have found to show that the amplitudes are related by $E_0 = cB_0$.

Setting It Up Here we are proving a relation between the two coupled quantities E and B, so we must use the equations that couple them.

Strategy Equation (34–11) relates a derivative of B_y to a derivative of E_x. With E_x given by Eq. (34–8), and B_y given by Eq. (34–13), we can compute the partial derivative of B_y with respect to z using Eq. (34–11), and we can compute it directly. Comparing these two results will give us information about the magnitude of the amplitudes as well as the phases.

Working It Out We have from Eq. (34–11) with E_x given by Eq. (34–8)

$$\frac{\partial B_y}{\partial z} = -\mu_0\varepsilon_0\omega E_0 \sin(kz - \omega t + \phi).$$

With B_y given by Eq. (34–13), we can compute the partial derivative directly:

$$\frac{\partial B_y}{\partial z} = \frac{\partial}{\partial z}[B_0 \cos(kz - \omega t + \phi)] = -kB_0 \sin(kz - \omega t + \phi).$$

We equate these two results:

$$-kB_0 \sin(kz - \omega t + \phi) = -\mu_0\varepsilon_0\omega E_0 \sin(kz - \omega t + \phi).$$

The sine factor cancels, and we are left with

$$B_0 = \frac{\mu_0\varepsilon_0\omega}{k} E_0.$$

The factor $\omega/k = c$, whereas $\mu_0\varepsilon_0 = 1/c^2$, so we are left with $B_0 = E_0/c$, the relation we needed to show.

What Do You Think? Suppose the wave in this example impinges on a metal sheet in which there are free electrons. Will it be the electric field or the magnetic field that predominantly determines the motion of the electrons? [*Hint:* Think about the size of B_0.] *Answers to **What Do You Think?** questions are given in the back of the book.*

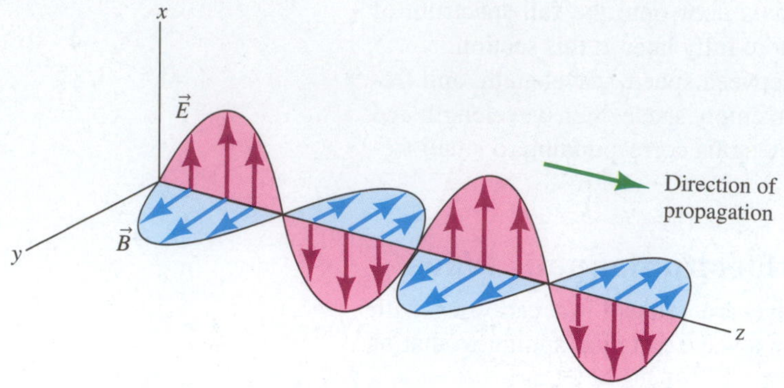

▲ **FIGURE 34–6** (a) A view at one particular time of the transverse electric and magnetic fields that propagate along the z-axis. (b) A view downward from the $+z$-direction of the electric and magnetic fields of an electromagnetic wave in an xy-plane at one particular z-value over time.

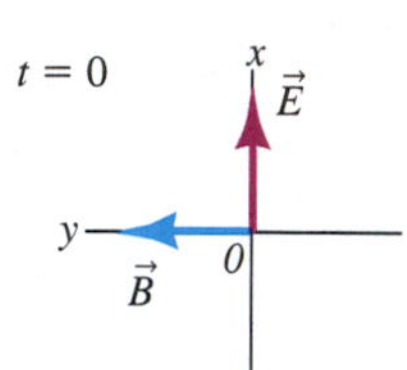

$t = \frac{T}{8}$

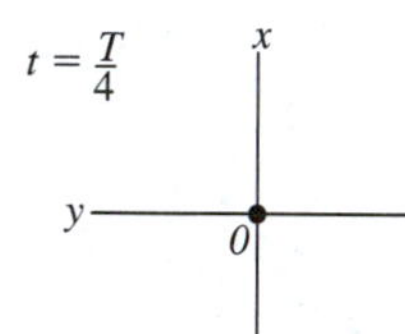

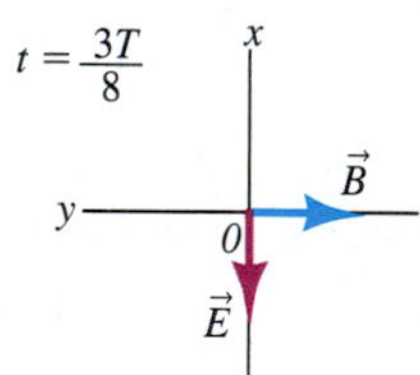

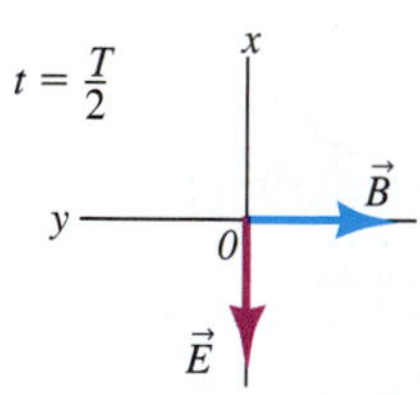

In the example above, we did not mention the currents that set up the original wave, and in fact the relation between the electric and magnetic field amplitudes in an electromagnetic wave is independent of those currents. We have, in general,

$$E = cB, \tag{34–14}$$

RELATION OF FIELD AMPLITUDES

where E and B are the amplitudes of the fields in an electromagnetic wave.

The Transversality of Electromagnetic Waves: The fields described by Eqs. (34–8) and (34–13) and pictured in Fig. 34–6 form traveling waves that propagate in the z-direction. Even though the fields are oriented in the x- and y-directions, they do not depend on x or y—we are describing plane waves, as we stated earlier. The currents that we used to set up the waves in the first place guarantee that the waves will be plane waves, and by changing the originating current, we can form other configurations of waves. In particular, there is nothing special about the x- and y-directions. If we had set up our currents to run in the y- rather than the x-direction, we would have found another set of solutions, with $\vec{E}$ in the y-direction and $\vec{B}$ in the x-direction. The wave propagation would still have been in the z-direction. It is generally true that *the electric field and the magnetic field in an electromagnetic wave are perpendicular to each other,*

$$\vec{E} \cdot \vec{B} = 0. \tag{34–15}$$

ORTHOGONALITY OF FIELDS

Moreover, an electromagnetic wave is *transverse* because the direction of the fields involved is perpendicular to the direction of wave propagation. *Neither the electric field nor the magnetic field has a component in the direction of propagation of the wave* (Fig. 34–6). We saw that both of these properties hold for the cylindrical wave emanating from the wire at the beginning of this discussion. We will limit ourselves to the plane and cylindrical wave discussions, but it is not difficult to see that these properties hold generally.

The Electric Field and Magnetic Field Are in Phase: The phases that appear in the harmonic expressions for B_y and E_x in Eqs. (34–13) and (34–8), respectively, are exactly the same. When the electric field is a maximum, the magnetic field is also a maximum; when one is zero, the other is zero, and so forth. The fields oscillate together as shown in Fig. 34–6. The fields are *in phase*.

Figure 34–6 illustrates each of the features of electromagnetic waves described above: The fields are in phase, transverse (perpendicular to the direction of propagation), and perpendicular to each other.

Electromagnetic Waves Are Real

When Maxwell introduced the displacement current and predicted electromagnetic waves in 1864, a number of the leading physicists of his time found these notions difficult to accept, and it was more than 20 years before experiment made all such resistance collapse. Experimental confirmation of the existence of electromagnetic waves was not possible when Maxwell proposed them because there was no technology to create AC currents of sufficiently high frequency and amplitude to provide detectable radiation. Heinrich Hertz devised the first direct test of Maxwell's waves in 1887. Hertz used the sparks that form when there is a large potential difference between the two points of a "spark gap" (Fig. 34–7a). The sparks have a rhythm associated with a back-and-forth motion of charge in the gap. To confirm that this oscillatory motion of charges produces electromagnetic waves, or *radiation*, Hertz took a wire bent into a circle with a (second) gap and placed it near the original spark gap (Fig. 34–7b). The electromagnetic wave that propagated in the space between the spark gap and the circular wire loop gave rise to sparks in the secondary gap, which thereby acted as a detecting antenna. Hertz also reflected waves from metallic surfaces, focused them with a concave metallic mirror, and found that they generally shared many of the properties of light that we shall study in Chapters 35 and 36.

The frequencies of the electromagnetic waves studied by Hertz are quite different from the frequencies of the waves that form visible light. The wave equation for electromagnetic waves admits solutions for *any* frequency, and the collection of all frequencies is known as the **electromagnetic spectrum**. In the century since Hertz's

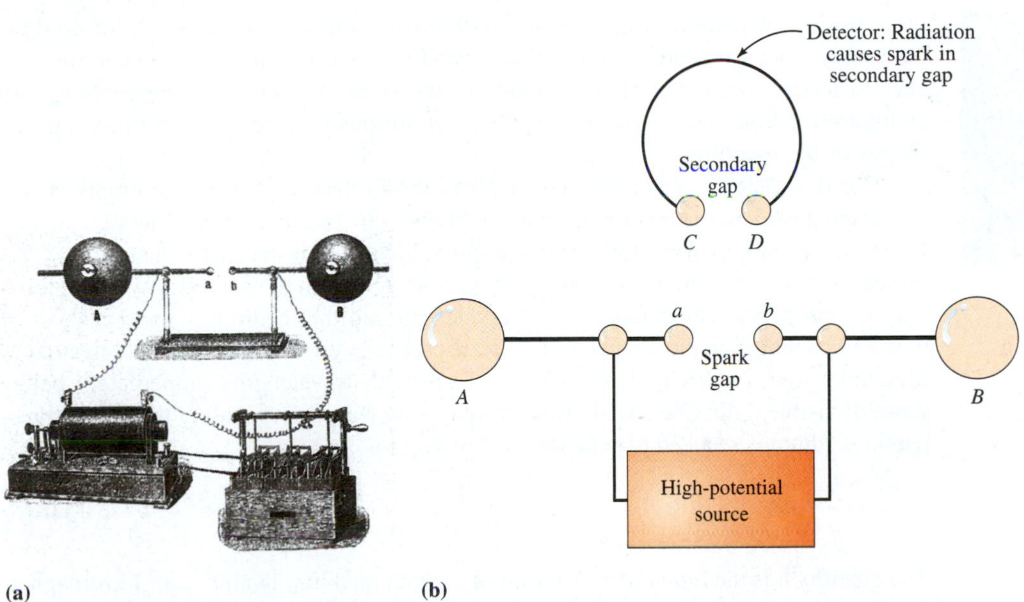

▲ **FIGURE 34–7** (a) Hertz's apparatus for the detection of electromagnetic radiation. (b) Schematic diagram of Hertz's apparatus. The radiation propagates from the region between the oscillating spark *ab* to the gap *CD*, which detects the radiation produced at gap *ab* by forming its own sparks.

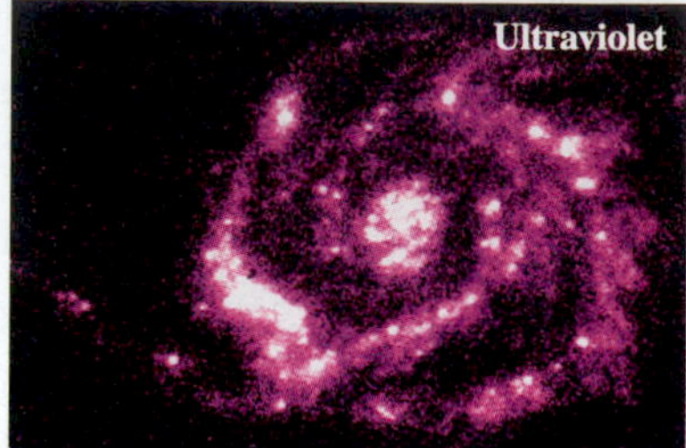

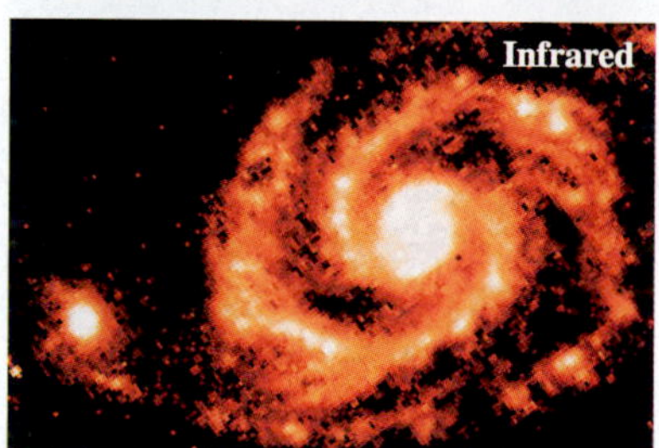

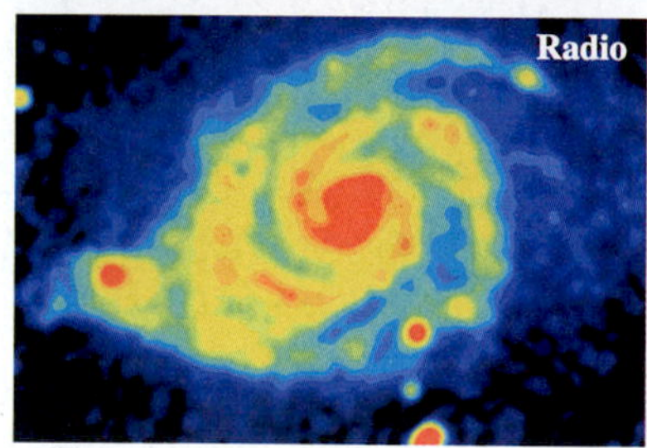

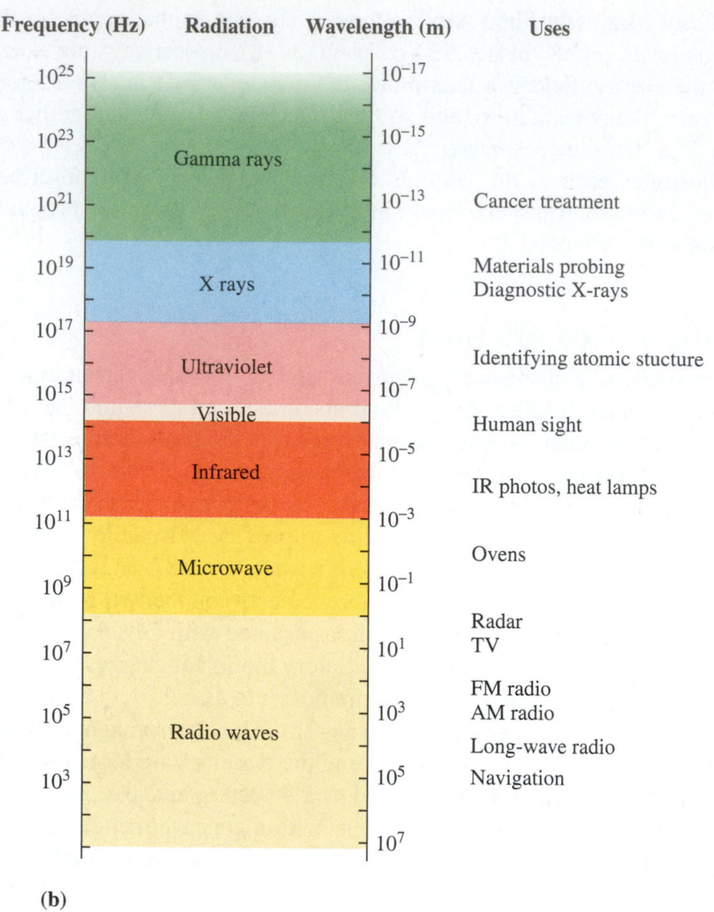

▲ **FIGURE 34–8** (a) These spectacular photos taken of the Whirlpool galaxy reveal different details, because they record radiation in different frequency ranges. (b) Different frequency regions of the full spectrum of electromagnetic radiation have specific names, as well as uses.

work, the electromagnetic spectrum has been explored across an enormous range of frequencies (Fig. 34–8). The radiation corresponding to a given part of the spectrum often has its own name—this includes visible light, ultraviolet radiation, infrared radiation, microwaves, radio waves, X rays, and gamma rays. The entire spectrum occurs in nature, but many of these names have been introduced in connection with technologies. You can say that we have been enormously successful at harnessing vast ranges of the spectrum.

Our discussion to this point has involved the formation of waves in empty space. Electromagnetic waves also propagate in matter. In transparent nonmetallic media, Maxwell's equations are modified only slightly. We mean by the term "transparent" that waves pass through—in other words, they propagate within the medium. Waves may or may not propagate depending on the wavelength and the medium. Ozone, a layer of which occurs high in the atmosphere, is transparent to visible light but much less so to ultraviolet light, which is composed of waves with wavelengths immediately below those of visible light. The speed of electromagnetic waves propagating through a transparent medium is reduced by a factor n according to

$$v = \sqrt{\frac{1}{\mu\varepsilon}} \equiv \frac{c}{n}. \qquad (34\text{–}16)$$

The quantity n is the **index of refraction** of a given medium. In all except ferromagnetic materials (introduced in Chapter 31), μ is very close to μ_0 and $\varepsilon = \kappa\varepsilon_0$, where κ is the dielectric constant of the medium. Thus

$$v = \sqrt{\frac{1}{\mu_0\varepsilon_0\kappa}} = \frac{c}{\sqrt{\kappa}}; \qquad (34\text{–}17)$$

in other words, the index of refraction $n = \sqrt{\kappa}$. It should be noted that the dielectric constant, and hence the wave speed, can depend on the frequency of the electromagnetic wave. When the speed of the wave depends on the frequency, the medium is said to be *dispersive*. The rainbow is an example of a phenomenon associated with dispersion.

THINK ABOUT THIS. . .

HOW DO RADIO WAVES TRAVEL AROUND THE GLOBE?

Driving at night in northern Minnesota, you may well hear a clear AM radio broadcast emanating from Florida. But the horizon seen from the antennas that produce these radio waves, which are electromagnetic radiation with wavelengths measured in tenths of kilometers, is substantially less than 100 km. The AM radio waves that you receive from so far away must travel through the atmosphere in what is effectively a curved path around Earth. The mechanism for this process lies in the *ionosphere*. This is a complex region of the atmosphere that extends from about 50 km to about 300 km from the ground. Short-wavelength radiation from the Sun and cosmic-ray particles that strike the upper atmosphere are energetic enough to separate some electrons from the atoms in this region, and this process of *ionization* leads to the presence of some *free electrons* in the ionosphere, forming a gas of electrons and ions known as a *plasma*. The structure of the ionosphere is that of a series of layers, with the most persistent ionization in the upper layers.

Radio waves that arrive at the ionosphere cause the free electrons to oscillate, and we have already pointed out that an accelerating charge produces radiation. This reradiation makes waves whose frequency matches the frequency of the oscillating charge. It sends back radio waves like those that have impinged on the plasma—we can refer to this as reflection. Through a process that involves the group behavior of large numbers of free electrons in the plasma, strong reflection occurs over a limited range of frequencies: Only radiation with a frequency below what is called the *plasma frequency* is reflected, or, in other words, radiation with wavelengths above a certain limit. The plasma is transparent to waves above that frequency. Now the plasma frequency varies with the density of free electrons n_e as $\sqrt{n_e}$, and for typical numbers in the ionosphere the plasma frequency is of the order of 10^7 Hz. This frequency is in the range of FM radio transmission but a factor of 100 above the frequency range of AM transmission. Thus AM radio signals are well reflected by the ionosphere, whereas FM signals pass through it. This allows AM signals to "bounce" their way to other locations through successive reflections from the bottom of the ionosphere and Earth's surface (Fig. 34–9).

Where, then, does the day–night difference come in? During the day, the Sun's radiation creates ionization. At night, when there is no ionizing radiation from the Sun, electrons recombine with their parent ions, and the ionosphere's composition changes. In particular, the lowest layer of the ionosphere, the layer from 50 km to 100 km high, essentially disappears as a plasma. The layer from which the radio waves reflect is now at a much higher altitude than during the day, and many fewer bounces are needed to reach a distant location. Since some energy is lost from the wave with each bounce, the distance the wave will travel at night is greatly increased over the distance it travels during the day.

The entire subject of the relation between the ionosphere and different communication bands (ranges of wavelength) is a rich one, and today's technologies make extensive use of knowledge about these effects.

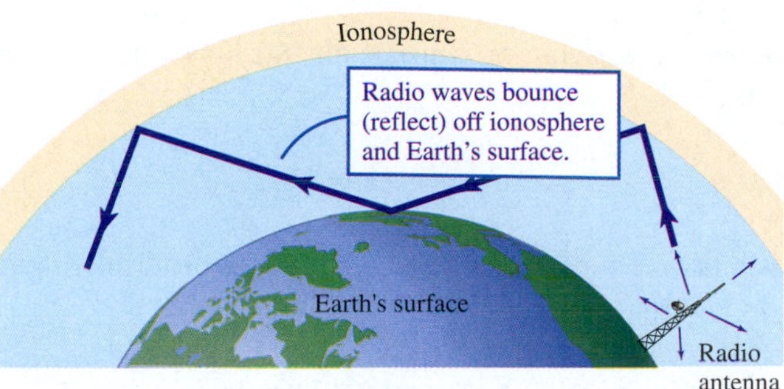

▲ **FIGURE 34–9** For a certain range of frequencies the waves emitted by an antenna will be reflected back toward Earth by the ionosphere. That range includes AM radio waves.

34–3 Energy and Momentum Flow

The Energy of Electromagnetic Waves

We can get sunburned because electromagnetic waves carry energy. We can calculate the energy in electromagnetic waves from our earlier results, in particular Eq. (32–17), on the energy density in electric and magnetic fields:

$$u = \frac{1}{2}\left(\frac{B^2}{\mu_0} + \varepsilon_0 E^2\right) = \frac{\varepsilon_0}{2}\left(\frac{B^2}{\mu_0\varepsilon_0} + E^2\right) = \frac{\varepsilon_0}{2}(c^2B^2 + E^2). \qquad (34\text{–}18)$$

Let's apply this result to an electromagnetic wave traveling in the z-direction. The fields are given by Eqs. (34–8) and (34–13):

$$E_y = E_0 \cos(kz - \omega t + \phi) \quad \text{and} \quad B_x = -B_0 \cos(kz - \omega t + \phi),$$

where $E_0 = cB_0$. For this wave, the energy density is

$$u = \frac{\varepsilon_0}{2}(c^2B_0^2 + E_0^2)\cos^2(kz - \omega t + \phi). \qquad (34\text{–}19)$$

In this expression, the two terms are the contributions of the magnetic and electric parts of the wave, respectively. Because $E_0 = cB_0$, *the energy contained in an electromagnetic wave is shared equally between the magnetic field and the electric field.* Equivalently, we could take the contribution of either the electric or the magnetic terms and multiply by 2 to find the total energy density in an electromagnetic wave:

$$u = \varepsilon_0 E^2 = \frac{1}{\mu_0}B^2. \qquad (34\text{–}20)$$

For practical purposes, the oscillations in electromagnetic waves are so rapid that we can simply consider the average of the energy density over one period, which we write as $\langle u \rangle$. The average of the cosine-squared factor in Eq. (34–19) over one period is one-half, so that

$$\langle u \rangle = \frac{\varepsilon_0}{2}E_0^2 = \frac{1}{2\,\mu_0}B_0^2. \qquad (34\text{–}21)$$

The Transport of Energy

The $\cos^2(kz - \omega t + \phi)$ time and space dependence of the energy density in Eq. (34–19) shows that the energy in an electromagnetic wave is itself *transported* as a wave; it travels at speed $v = \omega/k = c$ in the z-direction. The amount of energy dU_t transported across a surface of area A perpendicular to the transport direction in a time interval dt is the energy contained in the volume of area A times the distance $c\,dt$ (Fig. 34–10); that is, the energy density u times this volume,

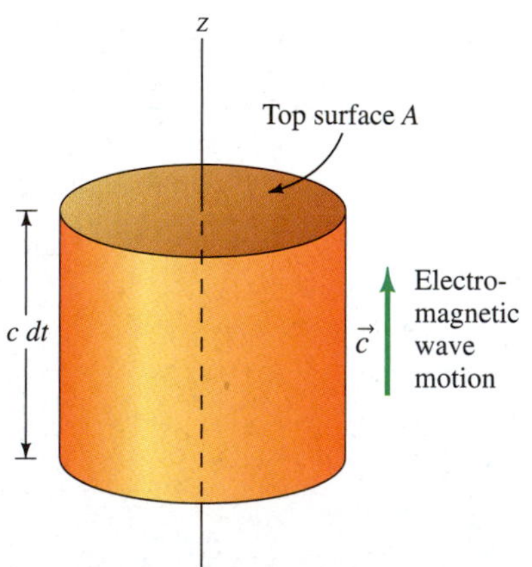

▲ **FIGURE 34–10** Electromagnetic energy contained in a volume $Ac\,dt$ is delivered in time dt to the area A.

$$dU_t = u(Ac\,dt).$$

Thus the rate of energy transport, or, equivalently, the power delivered by the electromagnetic wave, is

$$\frac{dU_t}{dt} = cuA.$$

Finally, the power delivered per unit area to a surface perpendicular to the direction of propagation—the *energy flux*—is given by

$$S = \frac{1}{A}\frac{dU_t}{dt} = cu. \qquad (34\text{–}22)$$

This flux has a direction associated with it and is more properly described as a vector. The vector $\vec{S}$ that describes the energy flux is the **Poynting vector**, given by

$$\vec{S} = \frac{1}{\mu_0}\vec{E} \times \vec{B}. \tag{34–23}$$

THE POYNTING VECTOR

Let's check that the magnitude and direction of the vector $\vec{S}$ are indeed correct. Because the fields $\vec{E}$ and $\vec{B}$ are at right angles to each other, we notice that the magnitude of $\vec{S}$ is just EB/μ_0. This can be rewritten in terms of the electromagnetic energy density, $u = \varepsilon_0 E^2$, as

$$S = \frac{1}{\mu_0}EB = \frac{\varepsilon_0}{\varepsilon_0\mu_0}E\left(\frac{E}{c}\right) = \varepsilon_0 c^2\frac{E^2}{c} = c\varepsilon_0 E^2 = cu. \tag{34–24}$$

This is the same magnitude we found in Eq. (34–22). As for the direction, recall that the vector product of two vectors is perpendicular to both of them. From Fig. 34–6, we see that the direction of the vector product $\vec{E} \times \vec{B}$ is the $+z$-direction, the direction of wave propagation. More generally, the transversality of the electromagnetic wave will always lead to a Poynting vector that lies along the direction of propagation.

The energy density and the magnitude of the Poynting vector each vary with time. The value of the magnitude of $\vec{S}$ time-averaged over one cycle of the electromagnetic wave is called the **intensity** I of the radiation. Equation (34–24) allows us to relate the intensity to the amplitude E_0 of the electric field in the wave:

$$I = \langle S \rangle = c\varepsilon_0\langle E^2 \rangle = \frac{1}{2}c\varepsilon_0 E_0^2. \tag{34–25}$$

Note from Eq. (34–21) that the intensity is also related to the average energy density in the wave, $I = c\langle u \rangle$.

EXAMPLE 34–3 A characteristic number for the rate per unit area at which solar energy is delivered to a spot on Earth's surface is $1000\ \text{W/m}^2$. (This energy consists mainly of electromagnetic radiation in the visible range of wavelengths.) Use this number to estimate the amplitude of the electric and magnetic fields in the waves that deliver this energy.

Strategy We use the fact that the power (energy per unit time) delivered by an electromagnetic wave per unit area is S, whose average value is simply related to the square of the amplitude of the electric field. This result is independent of the wavelength of the radiation. More particularly, the equation $I = \langle S \rangle = \frac{1}{2}c\,\varepsilon_0 {E_0}^2$ gives the electric field amplitude in terms of the power delivered per unit area—we can solve this for E_0. Once this is calculated, we can easily obtain B_0 using $B_0 = E_0/c$.

Working It Out The relation between I and E_0 gives

$$E_0 = \sqrt{\frac{2I}{c\varepsilon_0}} = \sqrt{\frac{2(1000\ \text{W/m}^2)}{(3 \times 10^8\ \text{m/s})(9 \times 10^{-12}\ \text{C}^2/\text{N}\cdot\text{m}^2)}}$$
$$= 0.9 \times 10^3\ \text{V/m}.$$

In turn,

$$B_0 = \frac{E_0}{c} = \frac{0.9 \times 10^3\ \text{V/m}}{3 \times 10^8\ \text{m/s}} = 0.3 \times 10^{-5}\ \text{T}.$$

What Do You Think? An inventor seeking financing claims that he can block just the electric field in a light wave, leaving all the power in the magnetic field. Should you invest?

CONCEPTUAL EXAMPLE 34–4 Compare the electric field in sunlight to the electric field in the beam of a 0.1-W laser that covers an area of $1\ \text{cm}^2$. (The laser rating is the power in the beam.)

Answer The field is proportional to $\sqrt{I}$. For sunlight I is known to be $1000\ \text{W/m}^2$, and for the laser it is $(0.1\ \text{W})/(1\ \text{cm}^2) = 10^3\ \text{W/m}^2$. The fields will be comparable.

Lasers with much higher power than the one in the example above can generate fields strong enough to rip atoms apart! On a less dramatic level, an interesting device makes direct use of the electric fields in light for manipulation of large molecules and nanostructures. In the *optical tweezers*, a laser beam is focused to a very sharp point; the rms electric field is largest at that point and falls off away from it. When the focal point is near a tiny piece of dielectric material that may be chemically attached to a large molecule such as DNA, a dipole moment is induced in the material and in the same way that bits of paper with induced dipole moments are attracted to larger values of nonuniform

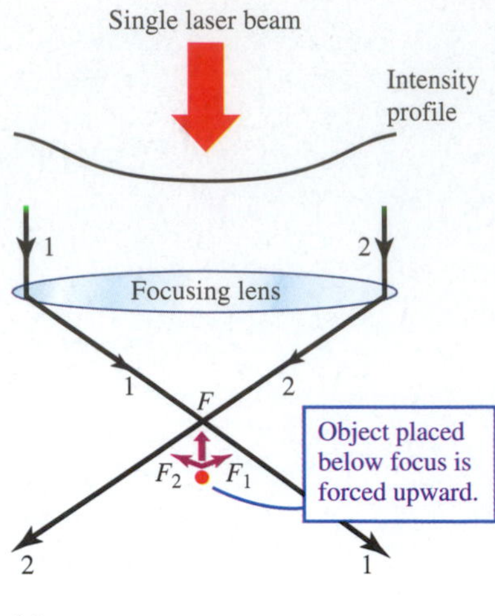

(a)

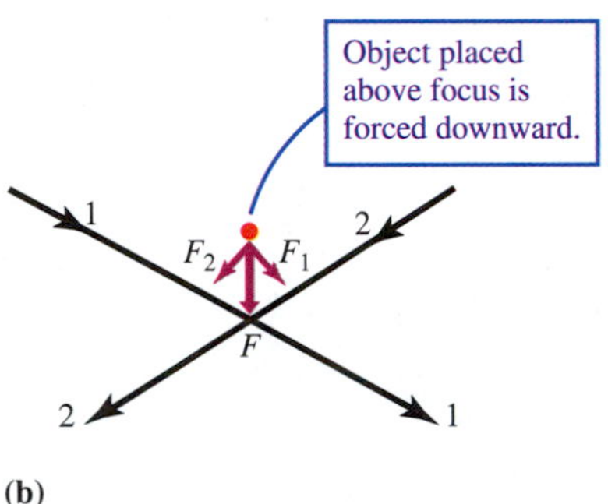

(b)

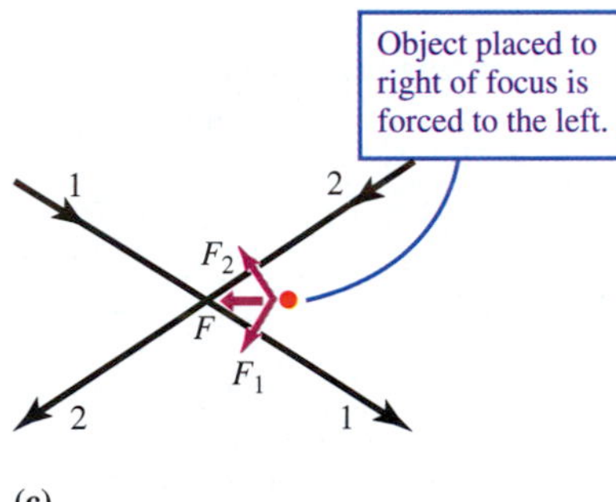

(c)

▲ **FIGURE 34–11** The focal point of an intense beam of light, a location where the beam is concentrated, acts to attract objects within which the electric field associated with the beam induces a dipole moment. When a laser provides the beam, the focal point can be so "tight" that the forces it exerts make a kind of optical tweezer that can manipulate nanoscale objects such as DNA molecules.

fields, it is attracted to the focal point of the nonuniform rms electric field of the laser beam. Figure 34–11 describes the action of the tweezers in schematic form. Once the system has been "gripped," the laser beam can be moved, moving the system with it.

Momentum in Electromagnetic Waves

An electromagnetic wave carries momentum as well as energy. To see this qualitatively, let's reconsider a plane wave that travels in the z-direction, with the electric and magnetic fields along the x- and y-directions, respectively. When such a wave impinges on a particle (you can think of an electron if you like, although we'll assume for simplicity that the charge is positive) with charge $+q$, the fields exert forces on the particle. Suppose that, at a given time, the oscillating electric field of the wave points in the $+x$-direction so there is a force qE in the $+x$-direction. The charge accelerates and moves with some velocity $\vec{v}$ in the $+x$-direction. If $\vec{E}$ points in the $+x$-direction, then $\vec{B}$ (which oscillates in phase with $\vec{E}$) must point in the $+y$-direction. The magnetic force $q\,\vec{v} \times \vec{B}$ on the charge acts in the $+z$-direction and pushes the charge in that direction. When the electric field later reverses sign, the electric force on the charge acts in the $-x$-direction, and the velocity then has a component in the $-x$-direction. The magnetic field has also reversed sign, but the magnetic force *continues to act in the* $+z$*-direction*. All the forces in the x- and y-directions average to zero, but the force in the z-direction is always positive, and there is a net force in the $+z$-direction. The charge has an increased momentum in the $+z$-direction; by momentum conservation, this momentum had to have been supplied by the electromagnetic wave.

By evaluating the amount of momentum the charged particle above picks up in a period of time dt, and recalling that the wave advances with speed c, so that a length $c\,dt$ of wave impinges on the particle in that time, we can show that the **momentum density** of an electromagnetic wave—the amount of momentum carried by the wave per unit volume—is $\vec{S}/c^2$. The magnitude of the momentum density is given by

$$\frac{S}{c^2} = \frac{u}{c}, \tag{34–26}$$

and the direction is that of $\vec{S}$, along the direction of wave propagation.

Radiation Pressure: When electromagnetic waves are absorbed or reflected as they enter into matter, they transfer momentum to the material on which they impinge. The particle upon which the wave fell above is an example. The rate at which momentum is transferred per unit area is a force exerted per unit area; that is, a pressure: **radiation pressure**. When an electromagnetic wave is absorbed, which happens when light falls on a black surface, all the momentum carried by the wave is transferred to the surface. The amount of radiation-produced momentum that falls perpendicularly on a surface A in a time interval dt is given by the momentum density multiplied by the volume $A(c\,dt)$. Thus the momentum dp transferred is

$$dp = \left(\frac{S}{c^2}\right)(Ac\,dt) = \frac{S}{c}A\,dt.$$

The force per unit area (radiation pressure) is given by

$$\frac{F}{A} = \frac{1}{A}\frac{dp}{dt} = \frac{1}{A}\frac{S}{c}A = \frac{S}{c} = u, \tag{34–27}$$

where we have used Eq. (34–22). This expresses the radiation pressure when radiation is totally absorbed. When the electromagnetic wave is reflected, which happens when it falls on a shiny, metallic surface, then the momentum of the wave is reversed upon reflection. Thus the momentum density transferred to the metallic surface is double the previous result, $2u/c$, and the radiation pressure is $2u$. (As we remarked above, a microscopic view of reflection is that the wave imparts momentum to a charged constituent, which then in turn reradiates and recoils in the process. This accounts for the factor of two.)

CONCEPTUAL EXAMPLE 34–5 Above we assumed that when light falls on a mirror and is reflected, the energy reflected is equal to the incident energy. Is this reasonable?

Answer Reflection is actually the reradiation of the wave by free electrons in the material. Metallic reflectors contain some free electrons, and that is why they make mirrors. (As for the shiny surface, that is another story, to be discussed in Chapter 35.) But the motion of these electrons is a local current, and there will be some small resistance, ohmic heating, and associated energy absorption. This means that the reflected energy will be slightly lower than the incident energy. This is typically a small effect.

EXAMPLE 34–6 Consider a 10^4-W searchlight that projects a cylindrical beam 0.6 m in diameter. What is the radiation pressure on a metallic mirror placed at right angles to the beam? Ignore the spreading of the beam.

Strategy The power delivered by the electromagnetic wave to a surface at right angles to the beam is given by

$$P = (\text{energy flux})(\text{area}) = SA = cuA,$$

where u is the energy density in the beam at the surface and A is the area of the beam. Given the area of the beam $A = \pi r^2$, we can calculate the energy density u and therefore the radiation pressure.

Working It Out The area is $A = \pi r^2 = \pi(0.3\ \text{m})^2 \cong 0.3\ \text{m}^2$. Thus

$$u = \frac{P}{Ac} = \frac{10^4\ \text{J/s}}{(0.3\ \text{m}^2)(3 \times 10^8\ \text{m/s})} \cong 10^{-4}\ \text{J/m}^3.$$

In turn, the radiation pressure is

$$\frac{F}{A} = 2u \cong 2 \times 10^{-4}\ \text{N/m}^2.$$

What Do You Think? Just how small is this pressure? What thickness of a water layer on the ground would exert this kind of pressure?

EXAMPLE 34–7 The intensity (average energy flux) of solar radiation that falls on Earth is $1.4 \times 10^3\ \text{W/m}^2$. Compare the force exerted by solar radiation on a totally absorbing dust particle of diameter 10^{-6} m and mass density $3 \times 10^3\ \text{kg/m}^3$ with the gravitational force on the particle due to the Sun. The particle is located at a distance from the Sun equal to the Earth–Sun distance, $R = 1.5 \times 10^{11}$ m. The mass of the Sun is $M_{\text{Sun}} = 2 \times 10^{30}$ kg.

Strategy We are given the intensity $I = uc$, so that we have the data to find the radiation pressure, just u for an absorbing object like the dust particle. (We assume time averages throughout.) Since we can work out the area A of the dust particle, the radiation force F on the particle is pressure × area $= uA = IA/c$. Knowing the size of the particle and its density, as well as the mass of the Sun, and the distance from the Sun, Newton's law of gravitation determines the gravitational force on it, magnitude F_g.

Working It Out The area presented by the dust particle is $A = \pi(d/2)^2 = \pi(0.5 \times 10^{-6}\ \text{m})^2 = 0.8 \times 10^{-12}\ \text{m}^2$. This gives for the radiation force

$$F = uA = \frac{IA}{c} = \frac{(1.4 \times 10^3\ \text{W/m}^2)(0.8 \times 10^{-12}\ \text{m}^2)}{3 \times 10^8\ \text{m/s}}$$
$$= 0.4 \times 10^{-17}\ \text{N}.$$

As for the gravitational force on the particle, the mass of the dust particle is $m = \rho V$, and the force of gravity is $F_g = \dfrac{GmM_{\text{sun}}}{R^2}$. Numerically, the mass m of the dust particle is

$$m = \rho V = \frac{4}{3}\pi\left(\frac{d}{2}\right)^3 \rho = \frac{4}{3}\pi(0.5 \times 10^{-6}\ \text{m})^3(3 \times 10^3\ \text{kg/m}^3)$$
$$= 1.6 \times 10^{-15}\ \text{kg},$$

hence the gravitational force on it is

$$F_g = \frac{GmM_{\text{sun}}}{R^2}$$
$$= \frac{(6.67 \times 10^{-11}\ \text{N}\cdot\text{m}^2/\text{kg}^2)(1.6 \times 10^{-15}\ \text{kg})(2 \times 10^{30}\ \text{kg})}{(1.5 \times 10^{11}\ \text{m})^2}$$
$$= 0.9 \times 10^{-17}\ \text{N}.$$

We see that the two forces are comparable, so the radiation pressure can keep the dust particle at its current location. It is therefore no coincidence that this kind of dust grain is typical of those found in interplanetary space.

34–4 Dipole Radiation

Accelerating charges produce electromagnetic waves. For waves with wavelengths much longer than light (radio, TV, cell phones, etc.), we refer to the systems in which accelerating charges initiate electromagnetic waves as *broadcasting antennas*; we refer to the systems in which we detect the response of charges to the fields of an electromagnetic wave as *receiving antennas*. Here, we shall describe one of the simplest systems that can act as an antenna, the *dipole antenna*. Radiation emitted with the characteristic pattern of this antenna is called **dipole radiation**.

A dipole antenna is formed by charges that move back and forth in harmonic motion along a line. The charge configuration within the antenna is that of a set of electric dipoles, with one or both of the two charges making up each dipole oscillating (Fig. 34–12). Such an antenna is easy to construct using an AC generator. When the dimensions of the antenna are small compared with the wavelength of the radiation, the current

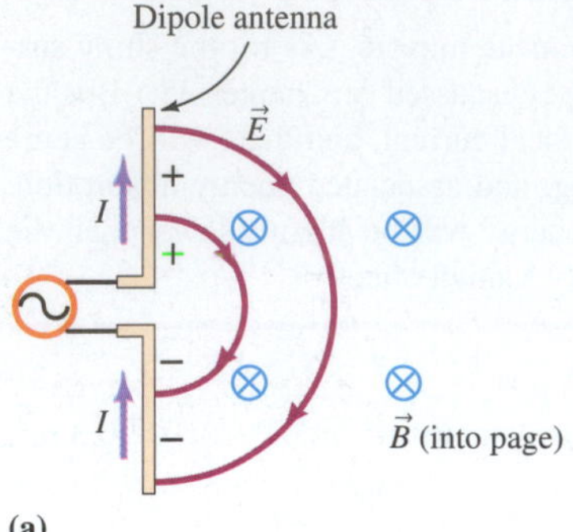

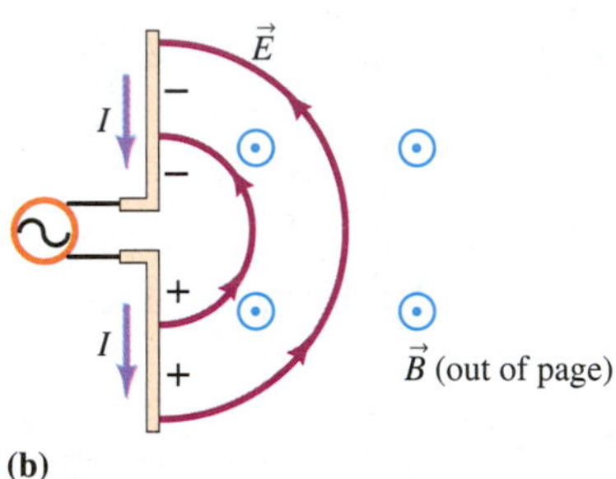

▲ **FIGURE 34–12** Pairs of equal but opposite charges move in simple harmonic motion along a line (vertical, here). These pairs of charges form a dipole antenna. In (a) the current is moving in one direction and in (b) it is moving in the opposite direction. As a result in (a) and (b) the two sides of the antenna are oppositely charged, and the field directions are reversed.

throughout the antenna is in phase, and the resulting electric and magnetic fields are oriented as shown in Fig. 34–12. They form an outgoing electromagnetic wave whose frequency is that of the oscillating charges.

How the Intensity of Radiation from an Antenna Decreases with Distance

Because electromagnetic waves are so often used for communication, it is clear that one of the most important characteristics of the electromagnetic waves radiated by an antenna is the rate at which the intensity decreases with increasing distance from the antenna. To understand this feature, we can consider any radiator—it is not important that the antenna be a dipole antenna. Let us consider an antenna that radiates electromagnetic waves symmetrically in all directions, such as the Sun or a lightbulb. From a distance, the electromagnetic waves emitted by the Sun appear to come from a point source, and we can use this fact to study the magnitudes of the electric and magnetic field strengths. As we learned in Section 34–3, the energy flux (the rate of flow of energy per unit area) is given by $S = cu = c\varepsilon_0 E^2$. The total energy flow per unit time (the power) across any surface is

$$P = \int_{\text{surface}} \vec{S} \cdot d\vec{A}.$$

If the magnitude of the electric field is independent of direction, as we would expect for a point source, then the total rate of energy flow across a sphere of radius R centered on the source is

$$P = c\varepsilon_0 E^2 (4\pi R^2). \tag{34–28}$$

But all the radiation emitted must eventually pass through any sphere that surrounds the source, whatever its radius, so P does not depend on R. From our expression for P, we see that this is possible only if the electric field decreases as $1/R$. The magnetic field must similarly fall off as $1/R$, because the magnetic and electric fields only differ by a factor of c in an electromagnetic wave. Contrast this result with the typical $1/R^2$ behavior of static electric fields (Chapter 22).

We can express the result of Eq. (34–28) in terms of intensity. The quantity $c\varepsilon_0 E^2$ is the magnitude of the Poynting vector, and its average value, which is defined as the intensity I, is one-half this value [Eq. (34–25)]. Thus

$$P = 2I(4\pi R^2). \tag{34–29}$$

Because P is independent of R, the intensity of the electromagnetic wave from a point source decreases as $1/R^2$. The next two examples illustrate this important property.

CONCEPTUAL EXAMPLE 34–8 In Example 34–7 we studied the balance of radiation pressure from sunlight with the gravitational force from the Sun for a dust particle as distant from the Sun as is Earth. We found that these forces balanced, suggesting that this is a mechanism for holding these particles at their current location. How would the argument change if the dust particle were as far away from the Sun as Jupiter?

Answer There is no change; the forces would still balance. Both the intensity of the Sun's radiation and the gravitational force due to the Sun have inverse square dependence on the distance from the Sun. If they balance at Earth's orbit, they will balance at Jupiter's.

EXAMPLE 34–9 A 100-W lightbulb emits electromagnetic radiation equally in all directions. Assume that 10 percent of the 100 W is converted into radiation in the visible spectrum. What is the intensity of the visible radiation 1.5 m from the bulb?

Setting It Up We are given the total power P and the power P_0 that appears as visible light, as well as the radius R of the sphere over whose surface the power is distributed. The desired intensity is denoted as I.

Strategy We may then use the relation $P = 2I(4\pi R^2)$ to calculate I as a function of R.

Working It Out We have $P_0 = 10\%$ of 100 W, so that $P_0 = 10$ W. This gives

$$I = \frac{P_0}{8\pi R^2} = \frac{10\text{ W}}{8\pi(1.5\text{ m})^2} = 0.2\text{ W/m}^2.$$

Compare this value to the 1400 W/m^2 in sunlight incident at the top of Earth's atmosphere, or to the 1000 W/m^2 of solar energy that reaches Earth's surface. About half this solar energy is in light in the visible part of the spectrum, whereas the lightbulb emits most of its energy in the infrared region of the spectrum.

What Do You Think? Does the fact that a lightbulb radiates equally in all directions mean that it is not an antenna?

In our discussion of static electric fields due to a point source, we used symmetry considerations to argue that the *electric field vector points in a radial direction*. This cannot be the case for electromagnetic waves emitted from a point source, however, because we showed that the fields for such waves are transverse to the direction in which they travel. When the waves come from a point source, they travel outward radially. So given the symmetry, how does the electric (or magnetic) field "know" in what direction to point for a point source? The resolution to what appears to be a paradox is simple: *There are no truly pointlike sources of electromagnetic radiation*. The Sun is not a point; it radiates because charges within it move and accelerate. As far as this discussion is concerned, we can think of the Sun as a large collection of dipole antennas with random orientation.

The Angular Pattern of Dipole Radiation

The variation of the intensity of electromagnetic radiation with the angle of observation is an important property of radiation from an antenna. Such patterns are an important element of the design of real antennas. No broadcaster wants to use expensive electric power to send a signal where nobody lives. In our simple dipole antenna (Fig. 34–12), charges execute simple harmonic motion along the antenna direction (we shall call this the z-axis). The motion of the charge determines a preferred direction—along the z-axis. An observer looking along the z-axis would see no motion. An observer looking along a line perpendicular to the z-axis would see the full range of motion of the charges. An observer at an angle θ to the z-axis would see the charges move harmonically with an amplitude reduced from the full amplitude by a factor $\sin\theta$. The electric field that the observer sees is thus proportional to $\sin\theta$. Because the intensity is proportional to the square of the electric field in the wave, the intensity of the radiation emitted by a dipole antenna along the direction of θ is proportional to $\sin^2\theta$:

$$S \propto \frac{\sin^2\theta}{R^2}. \tag{34–30}$$

Here we have also included the $1/R^2$ factor that describes how the intensity varies with the distance R from the antenna. This intensity pattern describes the **angular distribution** of the power emitted by charges oscillating along a line (Fig. 34–13): no signal along the direction of the antenna and a maximum signal perpendicular to it.

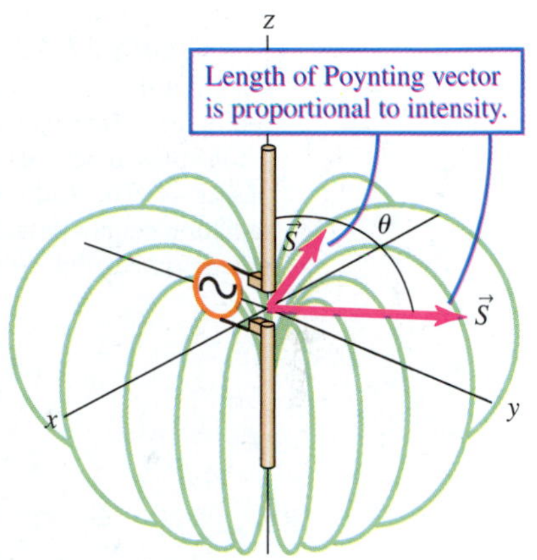

▲ **FIGURE 34–13** The intensity distribution S for a radiating dipole antenna. The curve illustrates the relative amount of power emitted as a function of the angle θ. The Poynting vectors for two different positions are drawn.

34–5 Polarization

A little experimentation with *polarizing* sunglasses at the seashore shows that a change in the orientation of the glasses' axis results in a change of the intensity of the light transmitted. This occurs because the sunglasses are made of a material that is sensitive to the direction of the electric field. As we shall see, light reflected from water or sand is **polarized**, meaning that its electric field is oriented in a particular way; the glasses "detect" the polarization of the electromagnetic wave (light). These sunglasses also have the effect of passing light whose electric field is aligned in a certain direction, and if two pairs of glasses are aligned in a "crosswise" fashion, no light passes the pair (Fig. 34–14).

▲ **FIGURE 34–14** The extent to which polarizing materials pass light or other electromagnetic waves depends on how an internal "axis" that they possess is oriented relative to the orientation of the electric field vector of the light. Little light passes through the region where the glasses' axes are crossed—light passing through one pair cannot pass through the second pair. But when the axes are aligned, the same amount of light passes as for a single lens.

If polarized light contains its electric field oriented in a particular direction, what is unpolarized light? Unpolarized light consists of a mixture of light waves with their electric fields aligned in different directions—always perpendicular to the propagation direction. The alignment must be such that *no* direction of $\vec{E}$ is preferred. In sampling a beam of unpolarized light, one would be just as likely to find the electric field aligned in one transverse direction as in another.

Let us reconsider a charge that oscillates along the z-axis, as in Fig. 34–13. We found that, if we look along the x-direction, we would detect an electromagnetic plane wave that propagates along the x-direction, with an electric field aligned along the z-direction: $\vec{E} = E_z\hat{k}$, with $E_z = E_0\cos(kx - \omega t)$. (Setting the phase $\phi = 0$ won't

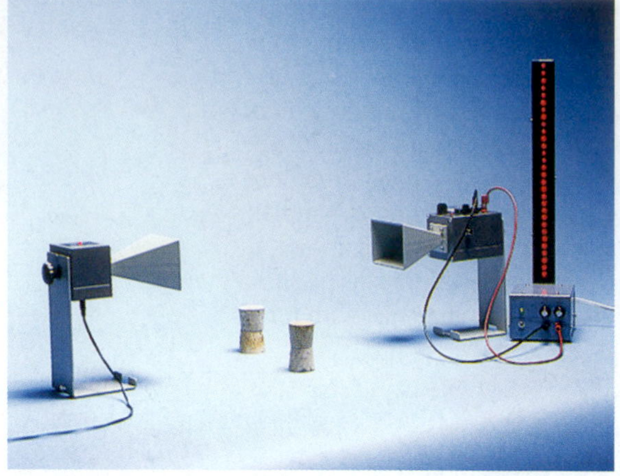
(a)

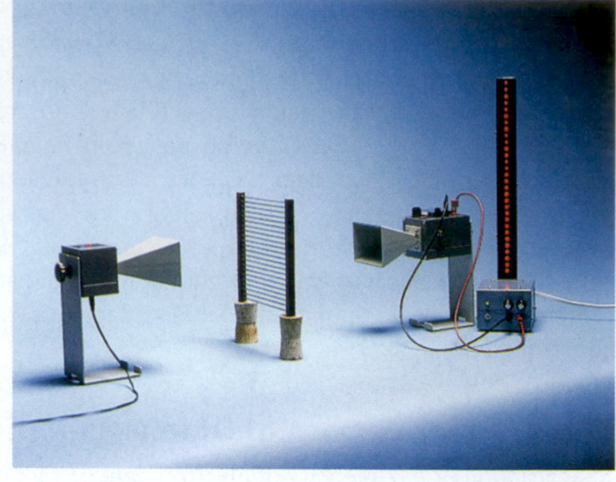
(b)

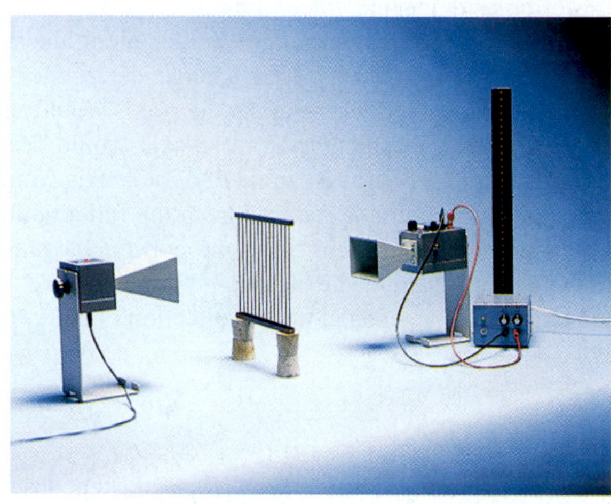
(c)

FIGURE 34–15 (a) A receiver and detector for determining the polarization of microwave radiation. The red lights indicate the presence of a signal. (b) A horizontal grid is placed between them, oriented so that the radiation passes. (c) The grid is now oriented vertically so that the radiation cannot pass. The grid's orientation reveals the polarization of the radiation, which is vertical.

change anything here.) We say that the light, or indeed electromagnetic radiation of any wavelength, is **linearly polarized** along the direction of the electric field vector—in this case, the light propagating along the x-direction is polarized along the z-direction. Suppose that a dipole antenna emits radiation with a wavelength in the centimeter range. The polarization can be detected as follows: A current is induced in a receiving antenna, and the rms current detected can be measured (Fig. 34–15). Place a metal grid (such as an oven rack) between the transmitter and the receiver. The diameter of the wires in the grid should be much less than the wavelength of 1 cm, and the grid spacing should be on the order of 1 cm or less. Then, the intensity of the radiation at the receiver depends on the orientation of the metal grid, and we say that the grid acts as an **analyzer** for the polarization.

Here is why the grid acts as an analyzer. The electrons in the grid wires are accelerated by the electric field of the wave along the field direction. When the wires in the grid are parallel to the electric field, the electrons in the grid wires can move in response to the field (Fig. 34–16a). Because they are set into motion, they *absorb large amounts*

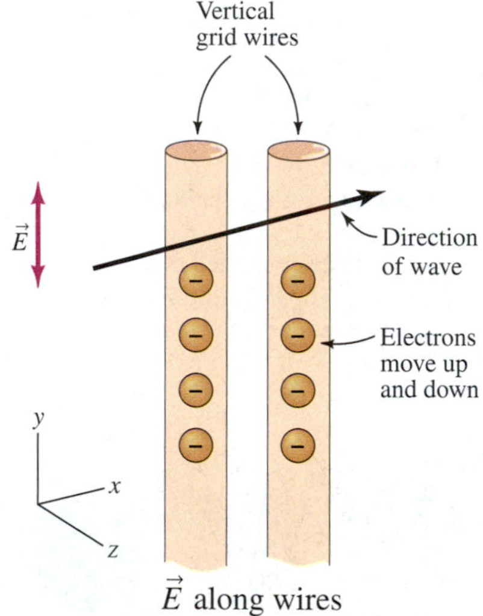

FIGURE 34–16 (a) When the grid wires are oriented in the direction of the electric field of an incoming wave, the electrons of the grid wires can respond and absorb energy from the wave. The transmitted wave is reduced in amplitude. (b) When the wires are perpendicular to the electric field of the wave, the electrons of the grid wires are constrained and cannot respond. Little energy is absorbed, and the wave passes through with little attenuation.

of energy from the field. This energy is lost in ohmic heating. The electric field of the radiation that passes through is reduced in magnitude because energy has been removed from the incident wave. In effect, the grid is opaque to the polarized radiation when it is oriented along the electric field vector. When the wires in the grid are perpendicular to the z-direction (Fig. 34–16b), the electrons in the metal are accelerated across the diameter of the wire. But, because the diameter is small, the electrons in the grid wires cannot respond fully and cannot absorb large amounts of energy from the incident wave. The energy remains in the transmitted wave. The grid acts as if it were transparent when it is oriented perpendicular to the polarization direction of the wave.

Certain materials, such as Polaroid, are analyzers for visible light. They are made of long molecules aligned parallel to each other. Electrons can easily move along the molecules but not across them and, because the molecular spacings are appropriate to the wavelengths of visible light, these materials behave like the microwave grid does.

A microwave grid or a piece of Polaroid is not simply an analyzer; it is also a **polarizer**: The microwave radiation that passes through the grid becomes polarized perpendicular to the grid wires. This is easily understood. Suppose that unpolarized microwave radiation approaches the grid. *Unpolarized radiation* is radiation that consists of a mixture of waves whose electric field vectors are as likely to point in any one direction as in another, as long as the direction is perpendicular to the direction of wave propagation. As we have seen, only those waves with the electric field oriented perpendicular to the grid can pass through, whereas the waves with the electric field parallel to the grid are absorbed. Thus the radiation that passes through the grid has become polarized perpendicular to the direction of the grid.

Malus's Law

When unpolarized radiation moving in the z-direction falls on a polarizer whose polarizing axis (the axis perpendicular to the "grid wires" within the polarizing material) makes an angle θ with the x-axis, for example, then only the component of any electric field along the polarizing axis will pass through. What emerges is radiation that is linearly polarized along a line that makes an angle θ with the x-axis. We take the magnitude of the electric field that has passed through the polarizer to be E_0. The corresponding intensity is then

$$I_0 = \langle S \rangle = (\text{a constant})E_0^2. \tag{34–31}$$

Let's now place a second polarizer so that its axis lies along the x-axis (Fig. 34–17). The amplitude for the electric field in the wave incident on the polarizer is

$$\vec{E}_0 = (E_0 \cos\theta)\hat{i} + (E_0 \sin\theta)\hat{j}. \tag{34–32}$$

Only the component that is parallel to the axis of the second polarizer—the x-axis—passes through. Thus the field behind the second polarizer (which acts here as an analyzer) is given by $E_0 \cos\theta\, \hat{i}$. The intensity of the transmitted light is therefore

$$I = (\text{a constant})(E_0 \cos\theta)^2, \tag{34–33}$$

and the intensity of the light is reduced:

$$I = I_0 \cos^2\theta. \tag{34–34}$$

Equation (34–34) is known as **Malus's law**. In particular, when the axes of the polarizer and analyzer are perpendicular to each other ($\theta = \pi/2$), radiation is not transmitted. This case is illustrated well for the sunglasses of Fig. 34–14.

One of the important consequences of Malus's law is that when unpolarized light passes through a polarizer, it has *half* its original intensity (see Example 34–10).

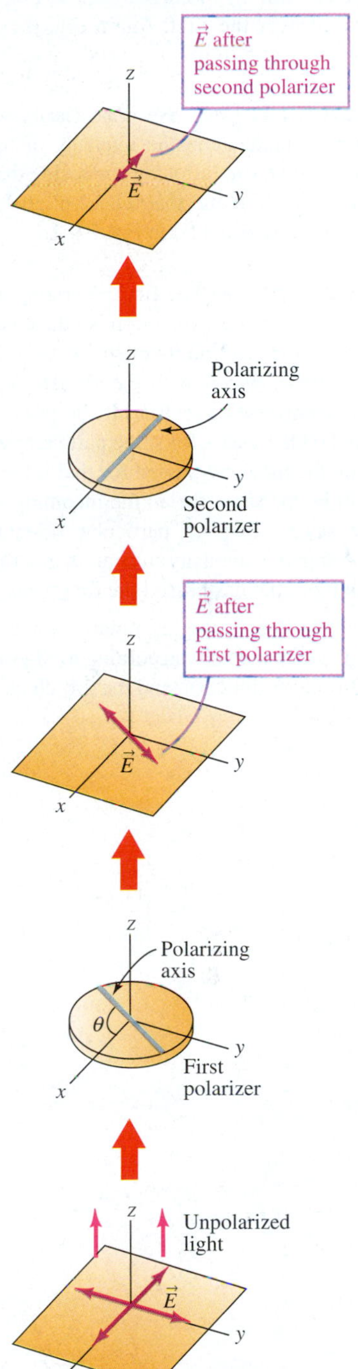

▲ **FIGURE 34–17** An unpolarized beam passes first through a polarizer whose axis makes an angle θ with the x-axis, and the beam is then polarized linearly in this direction. A second polarizer aligned with the x-axis allows only the component of the electric field aligned along the x-axis to pass.

EXAMPLE 34–10 Light passes through the glass plate of a transparency projector and emerges unpolarized with intensity I_0. (a) A Polaroid sheet is placed on the glass plate with its polarizing axis aligned with the 12-o'clock position. What are the polarization and intensity of the emerging light? (b) A second Polaroid sheet, with its polarizing axis along the 2-o'clock position, is placed over the first. Again find the polarization and intensity of the emerging light.

Setting It Up We must clarify what is meant by *unpolarized* light to understand the intensity of the emerging light in part (a). We set up the solution by supposing that the light wave propagates in the z-direction and that the 12-o'clock position is aligned along the $+y$-axis, as sketched in Fig. 34–18.

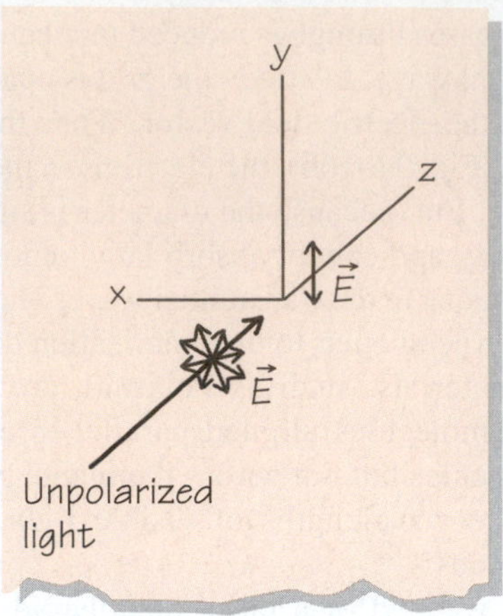

▲ **FIGURE 34–18**

Strategy (a) The first polarizing sheet passes light with its polarization in the y-direction, so the emerging light is polarized in the y-direction. To find its intensity, we recall that unpolarized light is a mixture of waves with the electric field equally likely to lie along any (tranverse) direction. If the projection of the incoming electric field on the y-axis for some particular wave in the mixture is $E \cos \theta$, then the intensity passed for that wave is $I = I_0 \cos^2 \theta$. Note that I_0 will be the same for all the incoming waves—that is what we mean by saying that no particular orientation is preferred. We must *average* this intensity over all θ, and that will give us the intensity of the light that has passed the first polarizer.

(b) The light emerging polarized along the 12-o'clock axis will be reduced in intensity according to Malus's law. The angle between the 12-o'clock direction and the 2-o'clock direction is $\pi/3$ radians.

Working It Out The average value of $\cos^2 \theta$ is $1/2$, giving the following answers for part (a): The polarizer is aligned with the x-direction, so that the light will be polarized in the x-direction, and the intensity of the light will be $I = I_0\langle \cos^2 \theta \rangle = \frac{1}{2} I_0$.

(b) From Malus's law we get the intensity

$$I_1 = I \cos^2(\pi/3) = \frac{I_0}{2}\frac{1}{4} = \frac{I_0}{8}.$$

What Do You Think? Both Polaroid sheets will experience some heating due to absorption of light. In what ratio do you expect the two sheets to acquire thermal energy?

How to Produce Polarized Radiation

We have already described two ways to produce polarized radiation: by accelerating charges in an oriented dipole antenna and by passing unpolarized radiation through a polarizer. Two more ways are important in many situations.

Polarization by Scattering: If you look at the beam of an automobile headlight from the side in a rainstorm or a snowstorm, it is quite visible, because there is a pronounced scattering of the light. The light beam is much less visible from the side on a dry night. Nevertheless, even in the absence of water droplets or dust particles, light and other forms of electromagnetic radiation are scattered by air molecules. The scattering mechanism is quite different from the mechanism that operates when the light scatters from droplets, because the size of the droplets is much larger than the wavelength of the light. It can be described as follows: The oscillating electric field $\vec{E}$ of the incoming radiation sets in motion the electrons in the air molecules. The electrons act like oscillators subject to an external harmonic force and oscillate with the frequency of the incoming field. The electrons move in a plane perpendicular to the incident radiation and, if the incident wave is unpolarized, then there is no preferred direction to the electron motion as long as it occurs in the plane. An observer looking at an electron from a direction close to that of the incident radiation will see a radiated field that is unpolarized because there is no preferred direction. In contrast, an observer looking at the electron from a direction perpendicular to the direction of the incident radiation will see the electron moving in just one direction (and will not see the component of the motion toward or away from him or her). This observer thus sees 100 percent linearly polarized light (Fig. 34–19). The polarization is partial for angles between these directions. If you live where the atmosphere is clear, you can easily observe this by holding a piece of Polaroid and looking 90° away from (*but not at*) the Sun. The light intensity will change when the Polaroid is rotated, showing that the light scattered by the air molecules is polarized.

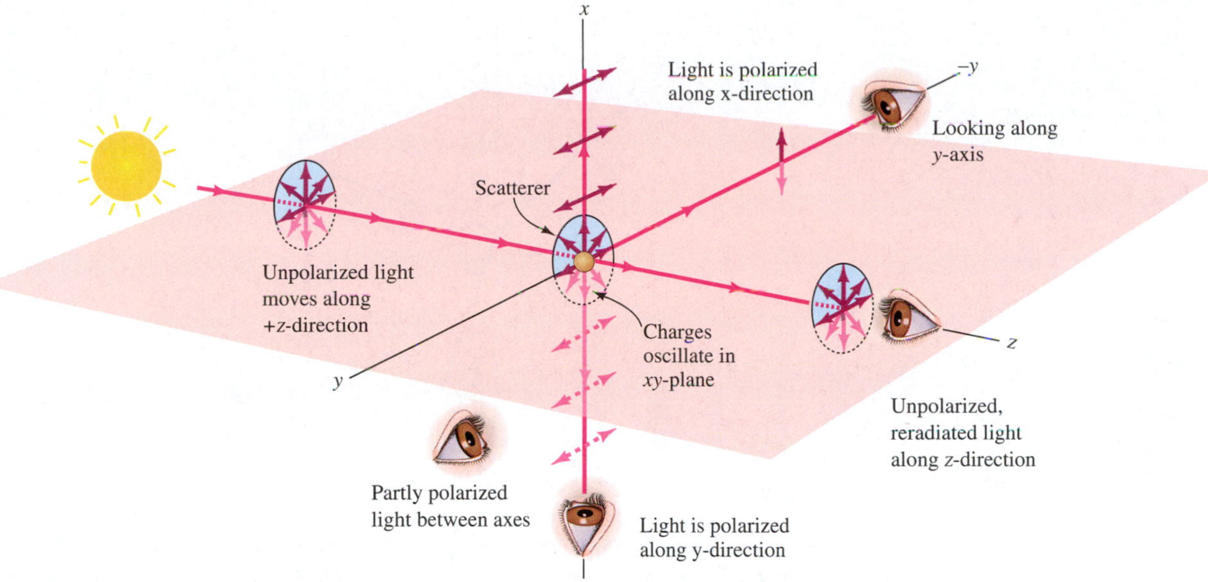

▲ **FIGURE 34–19** The polarization of radiation by scattering. An electromagnetic wave can propagate through a material because the wave's electric and magnetic fields cause electrons in the material to oscillate at the radiation frequency; these electrons in turn radiate new waves of the same frequency. The electric field of these new waves is aligned with the electrons' motion. Here, unpolarized radiation is perpendicularly incident on the *xy*-plane in a gas; the electric fields of the radiation lie in that plane but are otherwise unrestricted. An observer along the *z*-axis sees the full range of motion of the electrons in that plane and hence sees unpolarized light. An observer at 90° to the original wave direction can see a side view of the plane from which the light is radiated and hence sees light fully polarized; the polarization direction is parallel to the plane's edge. At intermediate angles, the reradiated light is partly polarized.

Polarization by Reflection: When unpolarized radiation is reflected from a surface such as glass, the reflected light is partly polarized (Fig. 34–20a). When the angle of incidence is just right, the reflected light is fully polarized (Fig. 34–20b). This is for much the same reason that scattered light is polarized (Fig. 34–19). Unpolarized light incident at an angle θ_i (*the angle of incidence*) impinges on a surface. In general, we may decompose the electric field of the incident wave into two components—each perpendicular to the direction of propagation. As Fig. 34–21 illustrates, one of these directions, the *z*-direction, is perpendicular to the surface of the page and parallel to the reflecting surface; we label the other the *a*-direction. When the wave arrives at the surface, its electric field accelerates electrons. These accelerated charges reradiate and give rise to both the transmitted and the reflected wave.

(a)

(b)

◄ **FIGURE 34–20** Radiation is polarized by reflection. (a) Here we see a shop window with oblique reflections, which are partially polarized. (b) The same scene, but with the camera lens fitted with a polarizing filter. The reflected light passing through the filter is greatly reduced.

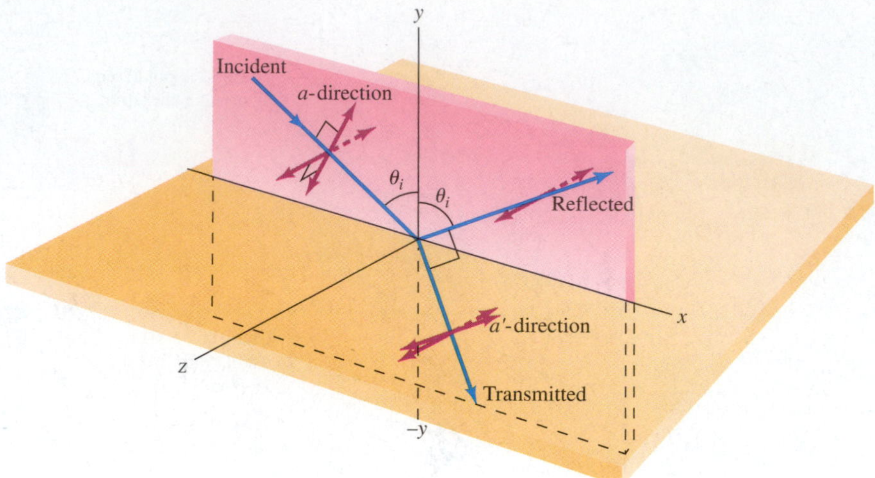

▶ **FIGURE 34–21** There is an angle of incidence θ_i for which the reflected wave is fully polarized. The electric field components are perpendicular to the rays, both in the marked plane and out of it.

Let's first discuss the radiation caused by the component of $\vec{E}$ in the z-direction, which is perpendicular to the plane of the paper. The electrons accelerated by that component of the incoming electric field move at right angles to the direction of the reflected wave. An observer looking back along the line of the reflected wave sees the full motion of these electrons. Thus there is strong reflection of this part of the incident wave. Next, let's consider the radiation induced by the component of the electric field of the incident wave in the a-direction. The electrons that absorb this incident radiation move parallel to the a-direction. An observer who looks along the line of the reflected wave sees a foreshortened motion of the electrons and thus only a limited amount of reflected radiation. Thus there is a preferential polarization direction for the reflected light. In the special case that the direction of the reflected wave is along the a'-direction (perpendicular to the direction of the transmitted radiation), there is no reflected radiation polarized along the a'-direction because the motion of the absorbing and reradiating electrons along the a'-direction cannot be seen. *The reflected radiation is plane-polarized with an electric field in the z-direction, parallel to the plane of the reflecting surface.* For this special angle, the reflected and transmitted waves must be at 90° to one another.

The angle of incidence for which the reflected and transmitted (or refracted) rays are perpendicular to one another are easily found once the rules for these rays (Snell's law) are developed, which we will do in Chapter 35. We give the result here: When the *angle of incidence* is the angle θ_B, known as **Brewster's angle**, the reflected ray is linearly polarized. This angle, which is the incident angle in Fig. 34–21, is given by

$$\tan \theta_B = \sqrt{\frac{\varepsilon}{\varepsilon_0}} = n. \qquad (34\text{–}35)$$

As we have already noted, the effect is present but less dramatic for other angles. An analyzer whose polarizing axis is oriented in a direction perpendicular to the z-direction (the direction of polarization of the reflected wave) will absorb most of the reflected radiation. Thus Polaroid sunglasses, worn to cut down the glare of reflected light from roads, beaches, car hoods, and other horizontal surfaces, must have their polarizing axis aligned in a vertical direction.

*34–6 Electromagnetic Radiation as Particles

In one of the most astonishing discoveries of the early part of the twentieth century, we learned that *electromagnetic radiation consists of particles*. The research of Max Planck, of Albert Einstein, and of Arthur Compton established that what we call an electromagnetic wave consists of a large number of individual particles called **photons**. These particles are indivisible: It is not possible to have 0.3 photons, for example. For radiation characterized by a frequency $f = \omega/2\pi$, the energy carried by a single photon is

$$E = hf, \qquad (34\text{–}36)$$

where $h = 6.63 \times 10^{-34}\ \text{J} \cdot \text{s}$ is Planck's constant. A photon also carries momentum, given by

$$p = \frac{E}{c} = \frac{hf}{c} = \frac{h}{\lambda}. \tag{34–37}$$

The particle nature of electromagnetic radiation was established through Compton's experiments on the scattering of radiation by free electrons in carbon. Photons scattered through a given angle have energy that can be calculated by treating each photon as a relativistic billiard ball that collides elastically with an electron at rest. The momentum of the outgoing photon depends on the collision angle. Equation (34–37) then implies that the frequency of the scattered radiation also depends on the collision angle in a way that can easily be calculated.

That h is small explains why we think of light as a continuous phenomenon rather than a series of individual photons. Someone standing under Niagara Falls wouldn't feel as if he or she is being bombarded by droplets of water! We now have instruments that can routinely detect individual photons. Had the evolutionary history of the human eye been somewhat different—so that the eye could easily respond to a single photon—the notion of radiation as consisting of particles would have been obvious to everyone.

EXAMPLE 34–11 At what rate does a 60-W lightbulb emit photons? For simplicity, assume that the light is emitted with a single wavelength of 590 nm.

Strategy We are given the wavelength, and therefore the frequency of the light, from which we can calculate the energy per photon, namely $E = hf = h(c/\lambda)$. The given wattage P of the bulb is the energy emitted per second, and P is the number of photons per second N times the energy per photon, $P = NE$. We can then solve for N.

Working It Out We have

$$E = h\frac{c}{\lambda} = \frac{(6.6 \times 10^{-34}\ \text{J} \cdot \text{s})(3 \times 10^{8}\ \text{m/s})}{(590 \times 10^{-9}\ \text{m})} = 3.4 \times 10^{-19}\ \text{J}.$$

From this we learn the number of photons per second,

$$N = \frac{P}{E} = \frac{60\ \text{W}}{3.4 \times 10^{-19}\ \text{J}} \cong 1.8 \times 10^{20}\ \text{photons/s}.$$

This is a very large number, one that would not permit you to sense the presence of individual photons.

What Do You Think? Do you expect photons to carry momentum? What do you think is the momentum of a photon with frequency f?

THINK ABOUT THIS. . .

WHEN PHOTONS GET REFLECTED BY A MIRROR, DO THEY DROP IN FREQUENCY?

We argued earlier that when electromagnetic waves are reflected by a mirror, a little of the energy that sets the electrons in motion is absorbed by ohmic heating, so that a little of the incident energy is absorbed, and the energy reflected is a little smaller than the incident energy. In terms of photons, which for monochromatic incident waves each have the same energy hf, a loss of energy would imply a drop in the frequency, and therefore an increase in the associated wavelength ($f = c/\lambda$). Since color of radiation is associated with frequency, does this mean that the reflected light is somewhat redder than the incident light? This is certainly not what is observed. A charge that is set into oscillation by an external field will reradiate with the same frequency, which is what we *do* observe. The loss in energy is not due to a loss of energy for individual photons, but rather in the reflection of *fewer* photons. Some of the photons emitted by individual oscillators are absorbed in the medium and their energy goes into kinetic energy of the atoms that make up the medium—causing ohmic heating.

Summary

Maxwell's equations—which comprise Gauss' laws for electric and magnetic fields, the generalized Ampère's law, and Faraday's law [Eqs. (34–1) to (34–4)]—imply that it is possible to have propagating electric and magnetic fields even in the absence of currents and charges. In the absence of free charges, the electric and magnetic fields obey the wave equation, which has the generic form, here written for the x-component of the electric field,

$$\frac{\partial^2 E_x(z, t)}{\partial z^2} = \mu_0 \varepsilon_0 \frac{\partial^2 E_x(z, t)}{\partial t^2}. \tag{34–7}$$

In the case that $E_z = 0$ and $B_z = 0$, the waves propagate along the z-direction. Whatever the direction, the speed of propagation is given by

$$v^2 = \frac{1}{\mu_0 \varepsilon_0}. \tag{34–9}$$

This speed is the speed of light, $v = c \cong 3 \times 10^8$ m/s. In material media characterized by the dielectric constant κ, the speed of propagation is $c/\sqrt{\kappa} = c/n$, where $n = \sqrt{\kappa}$ is the index of refraction. There are solutions of the wave equation (electromagnetic waves) in which the fields have the harmonic form

$$E_x = E_0 \cos(kz - \omega t + \phi), \tag{34–8}$$

and

$$B_y = B_0 \cos(kz - \omega t + \phi). \tag{34–13}$$

These waves propagate in the z-direction. More generally, electric and magnetic fields of waves that propagate in a given direction are transverse to that direction. The electric and magnetic field amplitudes are related by

$$E = cB, \tag{34–14}$$

and the fields are perpendicular to each other:

$$\vec{E} \cdot \vec{B} = 0. \tag{34–15}$$

Electromagnetic waves carry energy with energy density

$$u = \frac{\varepsilon_0}{2}(c^2 B^2 + E^2). \tag{34–18}$$

This energy is carried in equal amounts by the electric and magnetic fields. Electromagnetic waves also carry momentum, with momentum density $\vec{S}/c^2$, where $\vec{S}$ is the Poynting vector, given by

$$\vec{S} = \frac{1}{\mu_0} \vec{E} \times \vec{B}. \tag{34–23}$$

Thus radiation can transfer momentum; when a material absorbs radiation, there is a radiation pressure on the material, given by

$$\frac{S}{c} = u. \tag{34–27}$$

Charged particles radiate when they are accelerated. For a charge q undergoing an acceleration along the z-direction, the energy flux is proportional to

$$S \propto \frac{\sin^2 \theta}{R^2}, \tag{34–30}$$

where θ is the angle with the z-axis and R is the distance from the charge. Radiation with a $\sin^2 \theta$ angular dependence is called dipole radiation.

The polarization of an electromagnetic wave is the direction of the transverse electric field vector. It can be measured because polarizers transmit electromagnetic waves only along a particular polarization axis. Polarizers may be used to detect as well as to polarize electromagnetic waves. If a second polarizer is placed with its axis making an angle θ with the first one, then the electric field E of the transmitted wave is reduced in magnitude from the electric field E_0 of the incident wave according to $E = E_0 \cos \theta$. Thus the intensity I (the average of the energy flux) of the transmitted light is reduced from the incident intensity I_0 according to Malus's law:

$$I = I_0 \cos^2 \theta. \tag{34–34}$$

Waves can be polarized by reflection. If light falls on a medium of dielectric constant κ at an angle θ_B (Brewster's angle), for which

$$\tan \theta_B = n, \tag{34–35}$$

(n is the material-dependent index of refraction), then the reflected light is polarized in a direction perpendicular to both the incoming direction and the reflected direction of the wave. Light can also be polarized by scattering.

Appendix Getting Maxwell's Equations in Differential Form

Starting from a set of accelerating charges and Maxwell's equations, let's derive the equations that lead us directly to electromagnetic waves. The particular set of charges that we use form currents in the xy-plane, oscillating back and forth in the x-direction, as in Fig. 34–4. As in the qualitative discussion of Section 34–2, we know that the moving charges will give rise to changing electric and magnetic fields. We concentrate on time-dependent fields that vary with z but not with x and y. This implies that for a given z, the fields are the same out to infinity in the x- and y-directions. This cannot strictly be true in a physical situation; thus, we shall keep in the back of our minds that somewhere, for large enough values of x and y, the fields actually taper off to zero.

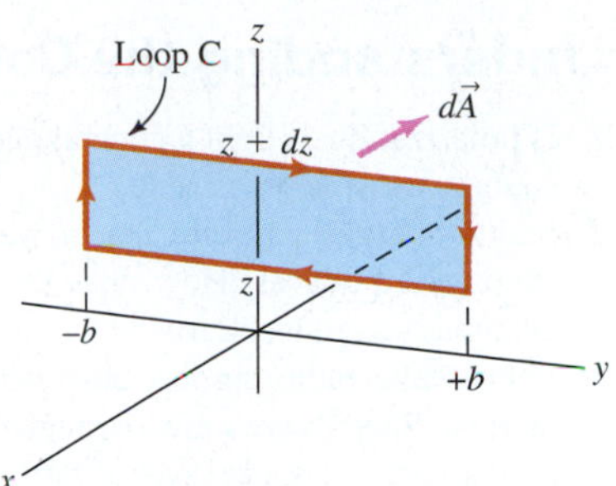

▲ **FIGURE 34A–1** A loop used to derive a relation between $\partial B_y/\partial z$ and $\partial E_x/\partial t$ using Ampère's law.

Let's draw an imaginary loop C in the yz-plane (at $x = 0$) that goes from $y = b$ to $y = -b$ at some value of z and returns from $y = -b$ to $y = b$ at $z + dz$ (Fig. 34A–1). We are going to apply the generalized Ampère's law to the loop. Sides at $y = \pm b$, going from z to $z + dz$, are very short. We shall ignore the contribution from the short sides because we can make these sides infinitesimally short. Moreover, our qualitative argument in Section 34–2 gives us no reason to believe that there is a field B_z. (This can be verified with the help of Gauss' law.) Application of the generalized Ampère's law, Eq. (34–3), now becomes easy. All we need to calculate for the line integral in Ampère's law are the contributions from the long (horizontal) sides of the loop. We have

$$B_y(z + dz, t)(2b) - B_y(z, t)(2b) = \mu_0\varepsilon_0 \frac{d}{dt} \int_{\text{loop area}} \vec{E} \cdot d\vec{A}. \tag{34–A1}$$

From the definition of a derivative, the difference $B_y(z + dz, t) - B_y(z, t)$ is the rate of change of B_y with respect to z times dz, so

$$2b[B_y(z + dz, t) - B_y(z, t)] = 2b\left(\frac{\partial B_y}{\partial z} dz\right). \tag{34–A2}$$

The partial derivative appears because we keep t constant in $B_y(z, t)$.

Now let's consider the right-hand side of Eq. (34–A1). In using Ampère's law, a right-hand rule dictates that for loop C in the direction shown in Fig. 34A–1, the surface element $d\vec{A}$ is oriented in the $-x$-direction, so $\vec{E} \cdot d\vec{A} = -E_x\, dA$. In addition, the area $A = 2b\, dz$ is infinitesimally small, so we can assume that E_x does not vary over the surface and we can remove it from the integral. Finally, the time derivative on the right-hand side of Eq. (34–18) acts only on E_x, because the surface is itself fixed. Thus

$$\mu_0\varepsilon_0 \frac{d}{dt} \int_{\text{loop area}} \vec{E} \cdot d\vec{A} = -\mu_0\varepsilon_0 \frac{\partial}{\partial t} E_x \int_{\text{loop area}} dA = -\mu_0\varepsilon_0 \frac{\partial E_x}{\partial t} A$$

$$= -\mu_0\varepsilon_0 \frac{\partial E_x}{\partial t}(2b)\, dz. \tag{34–A3}$$

We have used a partial derivative because z is a second variable that is held fixed. We now equate the two right-hand sides of Eqs. (34–A2) and (34–A3):

$$2b\left(\frac{\partial B_y}{\partial z} dz\right) = -\mu_0\varepsilon_0 \frac{\partial E_x}{\partial t} 2b\, dz;$$

that is,

$$\frac{\partial B_y}{\partial z} = -\mu_0\varepsilon_0 \frac{\partial E_x}{\partial t},$$

which is Eq. (34–5).

We next make use of Faraday's law, Eq. (34–4), the fourth of Maxwell's equations. We apply it to a loop C' that goes from $x = a$ to $x = -a$ at some value of z and returns from $x = -a$ to $x = a$ at $z + dz$ (Fig. 34A–2). Then, a nearly identical derivation to the one that led us to Eq. (34–5) leads us to Eq. (34–6),

$$\frac{\partial B_y}{\partial t} = -\frac{\partial E_x}{\partial z}.$$

Equations (34–5) and (34–6) are the ones we use in Section 34–2 to find the wave equation for electromagnetic waves.

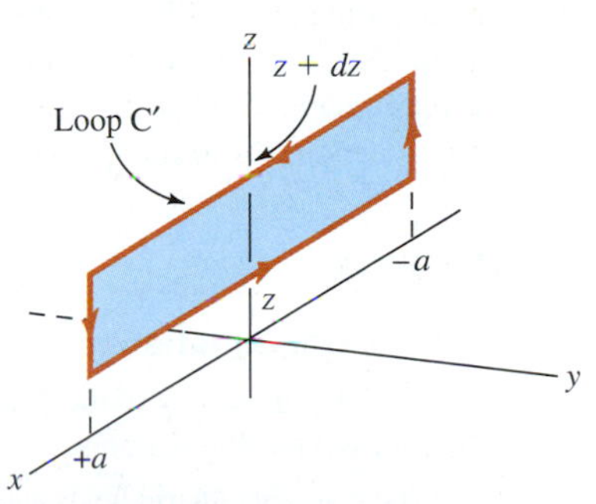

▲ **FIGURE 34A–2** A loop used to derive a relation between $\partial B_y/\partial t$ and $\partial E_x/\partial z$ using Faraday's law.

Understanding the Concepts

1. Is polarization of light a general phenomenon for all waves? Can sound waves be polarized?
2. Stable charged particles that move uniformly produce no electromagnetic waves. How does the conservation of energy suggest that this must be true?
3. Short-wave radio signals have wavelengths of several tens of meters. Such waves are particularly well reflected by Earth's ionosphere (an upper layer of the atmosphere that contains many free charges). Why would Earth's ionosphere reflect, rather than absorb, these waves?
4. Two identical wires with identical lightbulbs connected to the middle of each wire are placed in an electromagnetic plane wave moving out of the plane of the page. The two wires make an angle of 30 degrees with each other. Bulb B1 is totally dark, bulb B2 is not. Can bulb B1 be made to shine brightly, and if so, how?
5. The production and detection of polarization depends on the electric field vector. Can there be, in principle, a polarization associated with the magnetic field vector?
6. Can there be standing electromagnetic waves as well as traveling ones? Recall that mechanical standing waves on a string are possible when certain boundary conditions are satisfied, such as the ends of the string being fixed. How can we control the values of electric or magnetic fields on fixed boundaries?
7. You have a series of simple polarizers that you can orient as you choose in succession along a beam. Can you arrange them so that if the beam were polarized to start with it would be unpolarized when it has passed through your succession of polarizers?
8. Would the presence of magnetic monopoles analogous to electric charges change the nature of electromagnetic waves in free space?
9. A *solar sail* is a large surface on which the radiation pressure of the Sun's radiation can act and thereby push the sail along. Solar sails have been proposed for spaceships to travel throughout our solar system. What properties must such a sail have, and what difficulties do you see in the proposal?
10. Rockets are propelled forward when mass is ejected backward from them. Could a source of light (or other electromagnetic radiation) be used in place of the mass?
11. How can you tell whether or not light is linearly polarized?
12. Incident light is linearly polarized along the x-axis. We would like to rotate the direction of polarization so that it lies along the y-axis. Can this be done with one polarizer? Can it be done with two? What is the minimum reduction in intensity when two polarizers are used? Can there be even less intensity reduction with three polarizers?
13. Consider a metal rod with a lightbulb connecting the two halves (Fig. 34–22). The lightbulb glows when the rod is in the vicinity of a radio station. What happens when the rod is rotated in a plane perpendicular to the line between the rod and the station?

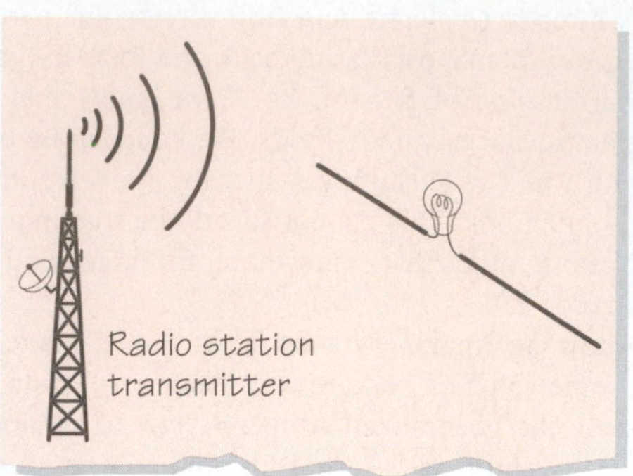

▲ **FIGURE 34–22** Question 13.

14. In Hertz's test of the existence of electromagnetic waves, sparks appear in a secondary gap as the result of an AC current in a primary circuit. Hertz interpreted these sparks as due to the effect of electromagnetic waves and not to the effects of Faraday induction. What sort of checks did Hertz need to make in order to rule out Faraday induction?
15. A 100-W lightbulb hangs from the ceiling of a windowless and thermally isolated room with the door closed. How much of the 100 W goes into heating the room?
16. Can you use the example and the arguments given in Section 34–2 to prove in a more general way than was done there that electromagnetic waves are transverse? Are there any pitfalls?
17. We showed that electromagnetic radiation carries momentum by thinking about its effect on a free charge. Consider its effect on an electric dipole to see whether it might carry angular momentum as well. Start by orienting the dipole with its axis along the direction of the electric field vector of the electromagnetic wave.
18. Consider the electromagnetic wave shown in Fig. 34–6(a). Suppose the green arrow showing the direction of the propagation of the wave were missing. Could you figure out how to replace it?
19. Consider the solar sail described in Question 9. Is it better to make a solar sail reflective (shiny) or absorbing (black)?
20. In the subsection on momentum in electromagnetic waves, we mentioned that an electromagnetic wave will accelerate a charged particle, giving it momentum at the same time. Does this mean the wave loses energy and momentum? In other words, if we sit behind a receiving antenna, do we pick up less radiation because of the presence of the antenna?
21. When electromagnetic radiation interacts with matter—for example, when light propagates in a crystal—it is always the electric and not the magnetic field that determines the behavior. Can you explain this, recalling the role of the two fields in the Lorentz equation?
22. How is it possible to clearly hear a 50,000-watt radio station that is 375 miles away, but not possible to clearly hear a local 5000-watt radio station that is 5 miles away?

Problems

34–1 Maxwell's Equations

1. (I) Verify the consistency of the dimensions of both sides of each of the four Maxwell equations.
2. (II) Gauss' laws for electric fields and for magnetic fields differ due to the lack of magnetic charges. Assume that magnetic monopoles (magnetic charges) exist; denote them by the symbol M. Rewrite Gauss' law for magnetic fields, and give the SI units of M.
3. (II) Ampère's and Faraday's laws differ due to the lack of a currentlike term in Faraday's law. Assume that magnetic monopoles exist (call them M), and rewrite Faraday's law. Discuss the physical significance of any new terms added.
4. (II) A region is bounded by an imaginary closed surface. Cut the region into two subregions with an arbitrary surface. Show that if Maxwell's first and second equations (which involve the surface

integrals) are valid for both subregions, they are also valid for the whole region. Show that the validity of the third and fourth equations (which involve line integrals) for the two parts of the surface of the regions created by the cut implies the validity of these equations for the full region.

34–2 Electromagnetic Waves

5. (I) If the electric field for a plane wave is given by $E_x = 0$, $E_y = E_0 \cos(kz + \omega t)$, what are $\vec{B}$ and the direction of propagation of the wave?

6. (I) Use dimensional analysis to show that $1/\sqrt{\mu_0 \varepsilon_0}$ has the dimensions of speed, $[LT^{-1}]$.

7. (I) An FM radio station announcer identifies the station as "Q94;" the number 94 stands for the frequency in some units. What is the wavelength and frequency of the waves emitted by the radio station?

8. (I) What is the relation between the amplitudes of the electric and magnetic fields in an electromagnetic wave propagating in a medium whose dielectric constant is κ? Assume the magnetic permeability of the medium is that of the vacuum.

9. (I) A superposition of electromagnetic waves traveling in the $+z$-direction and electromagnetic waves traveling in the $-z$-direction gives rise to standing waves. Check that a standing wave whose x-component of electric field has the form $E_0 \sin(kz)\cos(\omega t)$ satisfies the wave equation [Eq. (34–7)].

10. (II) Find the approximate wavelength, wave number, frequency, and angular frequency for electromagnetic waves associated with (a) your favorite AM station; (b) your favorite FM station; (c) a microwave oven; (d) yellow light; (e) X rays.

11. (II) Use Gauss' law to show that electromagnetic waves must be transverse. [*Hint*: Choose as your Gaussian surface a pill-box, with one of the plane surfaces chosen such that $\vec{E}$ or $\vec{B}$ vanishes on it.]

12. (II) Starting from Eqs. (34–5) and (34–6), derive a wave equation for the y-component of the magnetic field. What is the speed of the resulting wave?

13. (II) Write the counterparts of Eqs. (34–5) and (34–6) for electromagnetic fields B_y and E_z that lie in the yz-plane and propagate in the x-direction. [*Hint*: Start with Figs. 34A–1 and 34A–2. Then relabel the axes according to $x \rightarrow y \rightarrow z \rightarrow x$.]

14. (II) A plane harmonic wave of electromagnetic radiation with wavelength λ is propagating in the $-x$-direction. The z-component of the electric field has magnitude E_0, and there is no y-component. (a) Write an expression for the electric field. (b) Use this expression and the result of Problem 13 to calculate the magnetic field. What vector components will this field have?

15. (II) A plane wave propagates along the direction in the xy-plane that makes an angle θ with the x-axis. Show that the electric field is given by $\vec{E}_0 \cos(kx\cos\theta + ky\sin\theta - \omega t + \phi)$. What directions can $\vec{E}_0$ have?

16. (II) A plane wave of wavelength 17 m propagates in the z-direction. The electric field points in the y-direction and has an amplitude of 0.16 V/m. Write an expression for the magnetic field, including its amplitude in SI units. Assume that the electric field is at its maximum at $z = 0$, $t = 0$.

17. (II) An electromagnetic wave of wavelength 600 nm propagates in the z-direction. The magnetic field points in the y-direction, and has a magnitude of 10^{-8} T. Write an expression for the electric field, including numerical values and units. Assume that the magnetic field is maximum at $z = 0$ m, $t = 0$ s.

18. (II) An electromagnetic traveling wave is generated at the left-hand end of a tube oriented in the z-direction; the wave travels in the $+z$-direction. At the ends of the tube, $z = 0$ and $z = L$, are highly reflective mirrors. The electric field of the incident wave is $\vec{E} = E_1\cos(kz - \omega t)\hat{\imath}$, and the electric field of the wave reflected at $z = L$ is $\vec{E} = E_1\cos(kz + \omega t + \phi)\hat{\imath}$. Show that the net electric field forms a standing wave and, by computing B_y, that the associated magnetic field B_y also has the form of a standing wave.

19. (II) Consider the standing electromagnetic wave in Problem 9. If the standing wave is confined to a region lying between $z = 0$ and $z = L$ by two metallic plates, what is the relation between the allowed wavelengths of the radiation and L? (Recall from Chapter 22 that the electric field along a conducting surface must vanish on that surface.)

20. (III) A pulse of electromagnetic radiation travels in the $-z$-direction. The electric field is oriented in the x-direction and is given by $\vec{E} = E_0\exp(-(z + ct)^2/a^2)\hat{\imath}$. What is the orientation of the magnetic field? Make a guess of the space–time dependence of the magnetic pulse, and use Eqs. (34–5) and (34–6) to find a form for $\vec{B}$ that satisfies Maxwell's equations.

34–3 Energy and Momentum Flow

21. (I) The intensity of an electromagnetic wave is $6 \times 10^6\ \mathrm{W/m^2}$. What is the amplitude of the magnetic field in this wave?

22. (I) A radio station emits a signal with a power of 18 kW. What are the values of the electric field and magnetic field at distances of 3.5 km and 10.5 km? Assume that the signal far from the antenna is transmitted with equal intensity in all directions. (Real radio stations cannot afford to transmit their energy in this way, and their antennas distribute energy with a high degree of directionality.)

23. (I) The electric field for a given electromagnetic wave has a peak value of 140 mV/m. What is the intensity of the wave?

24. (I) A laser emits a beam with an intensity of $0.40 \times 10^{13}\ \mathrm{W/m^2}$ across an area of $1.5\ \mathrm{mm^2}$. What force would the laser beam exert on a black (perfectly absorbing) object?

25. (I) A harmonic plane wave of wavelength $0.45\ \mu$m and an electric field amplitude of 3 V/m impinges on a totally reflecting surface of area $200\ \mathrm{cm^2}$. What is the radiation pressure exerted by the wave?

26. (I) A plane electromagnetic wave with maximum electric field amplitude of 120 V/m is incident on a perfectly absorbing surface perpendicular to the direction of propagation. What is the rate of energy absorption per unit area of the surface?

27. (I) The rate at which the Sun emits energy in the form of radiation is 3.8×10^{26} W. (a) Calculate the magnitude of the Poynting vector at a distance of 1.5×10^{11} m from the Sun. (b) What is the radiation pressure exerted on a totally absorbing surface perpendicular to the direction of the radiation?

28. (II) (a) Sketch on the same graph $\sin x$ and $\cos x$ as a function of x. (b) On a separate graph sketch $\sin^2 x$ and $\cos^2 x$. Observe that the periodic functions $\sin^2 x$ and $\cos^2 x$ are identical to one another, except that one is displaced from the other by an interval $\pi/2$. (c) Use your sketch from part (b) to show that the area under the $\sin^2 x$ curve in the interval $0 \le x \le 2\pi$ is the same as the area under the $\cos^2 x$ curve in the same interval. (d) Given the fact that $\sin^2 x + \cos^2 x = 1$, use the results obtained in parts (a)–(c) to show that the averages $\langle \sin^2 x \rangle$ and $\langle \cos^2 x \rangle$ are equal to each other and thus equal to 1/2.

29. (II) The magnetic field for a given electromagnetic wave has an rms value of 7×10^{-9} T. What is the intensity of the wave? How much energy is transported per minute through a 0.1-m^2 area?

30. (II) A typical lecture-demonstration laser of power 0.75 mW has a beam of diameter 0.90 mm. (a) What are the peak values of the electric and magnetic fields? (b) Suppose—as is in fact possible—that the beam is focused to a circular area with diameter of one wavelength. What is the peak value of the electric field, given that $\lambda = 650$ nm?

31. (II) Assume that a 75-W lightbulb emits light equally in all directions. What are the peak and rms values of the electric and magnetic fields at a distance of 0.50 m?

32. (II) A 75-W lightbulb radiates uniformly in all directions, and 9 percent of this energy is emitted as electromagnetic radiation in the visible light range. What is the electromagnetic energy density of visible light at a distance of 130 cm from the bulb? What are the rms values of the corresponding electric and magnetic fields there?

33. (II) The total electromagnetic power emitted by the Sun is 3.8×10^{26} W. What is the radiation pressure exerted on a totally reflecting surface a distance $r = 1.0 \times 10^{10}$ m from the Sun?

34. (II) What are the dimensions and SI units for the Poynting vector? Reduce your answer to the dimensions and units of mass, length, and time, then reexpress it in terms of watts and meters.

35. (II) Solar energy delivered to a horizontal surface in Washington, D.C., averaged over a full year is 160 W/m^2. Assuming that this radiation is fully absorbed on a particular square meter of ground, what is the approximate total momentum delivered to this area in 1 y? Compare this number to an estimate of the momentum absorbed by a baseball catcher in catching a single pitch.

36. (II) The radiation pressure of a beam of electromagnetic radiation is equal to atmospheric pressure. Calculate the intensity, energy density, and rms electric and magnetic fields of this beam. Assume that the beam is totally absorbed.

37. (II) Sunlight exerts an average radiation pressure of 5×10^{-6} N/m^2. Consider a rectangular mirror of dimensions 20 cm × 500 cm attached to a vertical wire, so that it is oriented perpendicular to the sunlight. How should you attach it to get the maximum torque exerted by the sunlight? What is the torque?

38. (II) What is the radiation pressure on the walls of a microwave oven in which the rms electric field is 500 V/m? Assume that the waves are traveling waves of the sort we have studied in this chapter, and if it is necessary to know the frequency, estimate the appropriate value.

39. (II) Suppose that you want to use the radiation pressure from a beam of light to suspend a piece of paper in a horizontal position; the paper has an area of 50 cm^2 and a mass of 0.20 g (Fig. 34–23). Assume that there is no problem with balance, that the paper is dark and absorbs the beam fully, and that the entire beam can be used to hold the paper against the pull of gravity. How many watts must the light produce? Given your answer, what do you suppose would happen to the paper?

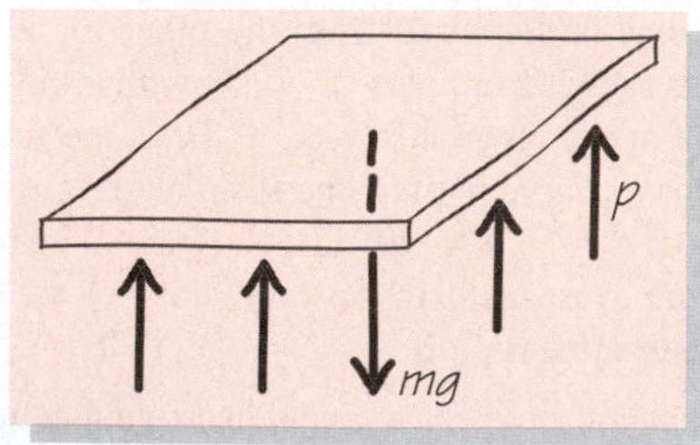

▲ **FIGURE 34–23** Problem 39.

40. (II) Tiny flakes of mica are kept aloft by a beam of light projected vertically upward. If the mass of a typical flake is 5.4×10^{-9} kg, and if on the average the area presented to the beam by a flake is 0.06 mm^2, what is the intensity of the beam? Assume that all of the light is reflected.

41. (II) A light beam with a given Poynting vector falls on a flat, fully reflecting surface at an angle of incidence θ (with respect to the vertical) (Fig. 34–24). What is the momentum transferred to the surface per unit area?

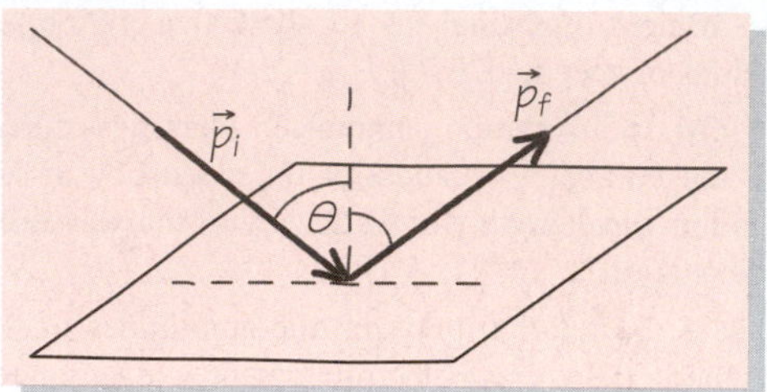

▲ **FIGURE 34–24** Problem 41.

42. (II) A laser delivers 1.8×10^2 J of energy in a pulse that lasts 5×10^{-9} s. What are the peak electric and magnetic fields for a laser beam of diameter 0.5 mm?

43. (II) Consider an electromagnetic wave propagating in the negative x-direction with frequency of 6×10^9 Hz. The wave exerts a pressure of 10^{-4} N/m^2. Write down an expression for the electric and magnetic fields. You may choose your axes such that the electric field is polarized along one of the axes. Specify your choice.

44. (II) Consider a standing electromagnetic wave for which the electric field is $\vec{E} = \hat{j} E_0 \sin kx \cos \omega t$. (a) What is the wavelength of the electromagnetic wave? (b) What is the magnetic field? (c) What is the value of the Poynting vector?

45. (III) The short side of a thin, stiff rectangle 3.0 cm × 1.0 cm is attached to a vertical axis. Half of each side is painted black and is fully light absorbent; the other half is a shiny, reflecting metal (Fig. 34–25). The back of each half is different from the front. There is no friction at the axis. The apparatus is bathed in a well-collimated (nonspreading) beam of light whose Poynting vector has magnitude 0.5 kg/s^3 and travels perpendicular to the vertical axis. Is there a net torque on the rectangle's surface? If so, what is its average value due to the light over a full, uniform rotation of the rectangle about the axis?

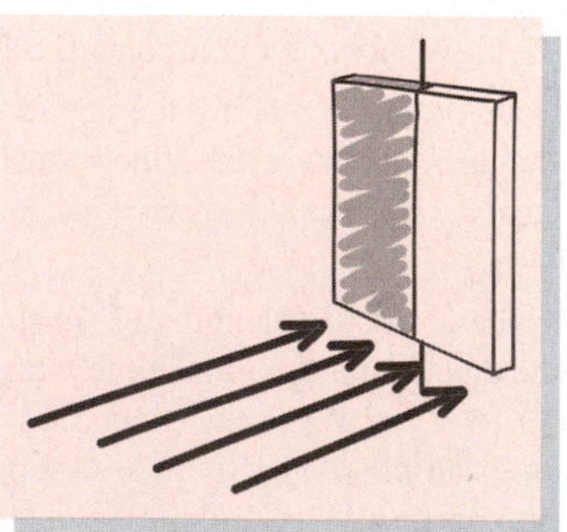

▲ **FIGURE 34–25** Problem 45.

46. (III) The total power of a broadcasting dipole antenna is 20 MW. Calculate the intensity of its radiation at a distance of 1000 m, in the direction of the intensity maximum. Compare this to the intensity that would have been obtained if the intensity were distributed uniformly in every direction.

34–4 Dipole Radiation

47. (I) Suppose that a vertical tower 120 m tall acts as a dipole antenna, with currents running back and forth along the tower to generate electromagnetic waves in a dipole pattern. If the wavelength of each electromagnetic wave is the height of the tower, what is the period of the current oscillation in the tower?

48. (I) A charge moves harmonically along an 8-m length in the z-direction, emitting dipole radiation. Two observers detect this radiation. Observer A is at a position that is 10 km from the charge and at an angle of 25° with respect to the z-axis, while observer B is at a position that is also 10 km from the charge, but at an angle of 58° with respect to the z-axis. What is the ratio of the intensity detected by the two observers?

49. (II) A broadcasting dipole antenna is oriented along the y-axis. For the geometry shown in Fig. 34–26, give the following information for a point P far away along the z-axis: (a) the direction of the electric field; (b) the direction of the magnetic field; (c) the direction of the Poynting vector. (d) Repeat parts (a)–(c) for the electromagnetic wave one-half cycle later.

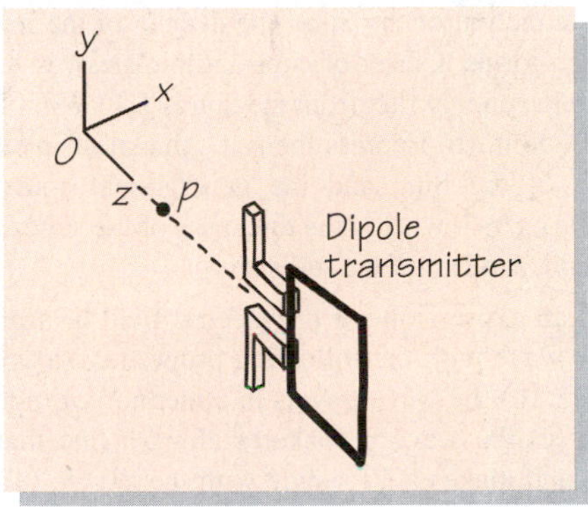

▲ **FIGURE 34–26** Problem 49.

50. (III) A cross-shaped antenna lies in the xy-plane, centered at the origin (Fig. 34–27). The charges oscillate with the same frequency within each arm of the cross. Find the Poynting vector along the z-axis as a function of z if charges moving in the $+x$-direction in the x-arm pass the origin at the same moment that (a) charges moving in the $+y$-direction in the y-arm pass the origin; (b) charges moving in the $-y$-direction in the y-arm pass the origin.

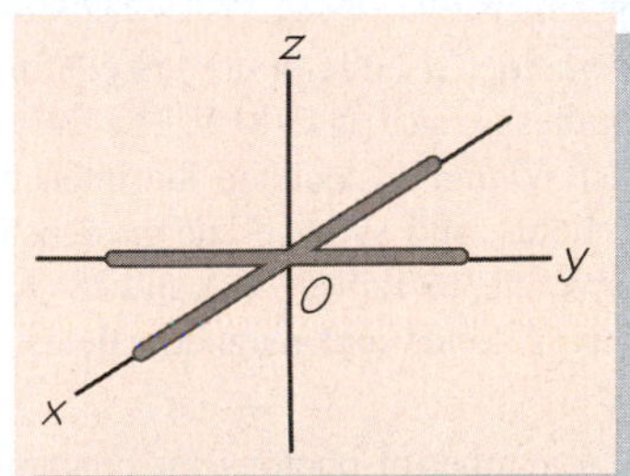

▲ **FIGURE 34–27** Problem 50.

34–5 Polarization

51. (I) At what angle should the axes of two ideal Polaroid sheets be placed to reduce the intensity of a given source of unpolarized light to (a) 7/10; (b) 3/10; (c) 3/20; (d) 1/20?

52. (I) The axes of four ideal Polaroid sheets are stacked, each at 28° with respect to the previous one. What fraction of initially unpolarized light passes through all four sheets?

53. (I) The Moon reflects light off a still pond at night. At what angle above the horizon is the polarization a maximum? The index of refraction of water is 1.33.

54. (I) Polarized light of intensity $1.0 \times 10^6\ \text{W/m}^2$ is incident on a Polaroid sheet placed perpendicular to the light beam with the polarizing axis of the sheet at an angle of 40° to the polarization vector of the light. What is the intensity of the beam after passing through the polarized sheet?

55. (I) The beam of Problem 54, after passing through the Polaroid sheet described in that problem, then passes through another Polaroid sheet, this one with its polarizing axis at an angle of 80° to the original polarization vector. What is the final intensity of the beam?

56. (II) A beam of light propagating in the z-direction is polarized in the y-direction. Two superposed Polaroid sheets are placed perpendicular to the beam. The polarization axis of one makes a 33° angle with respect to the y-direction, and the axis of the other makes a 51° angle with respect to the axis of the first sheet (Fig. 34–28). What is the intensity of the transmitted beam?

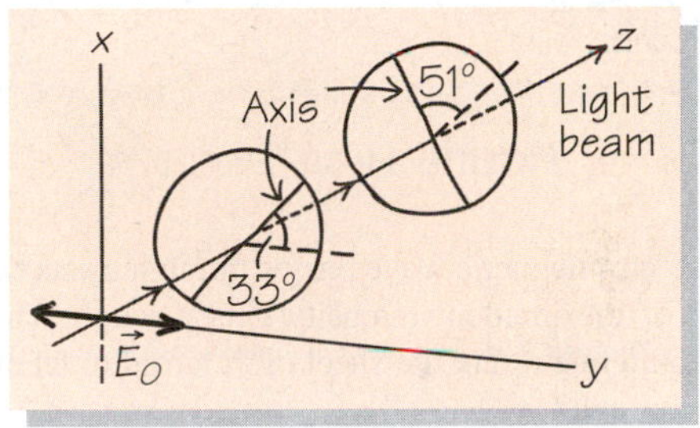

▲ **FIGURE 34–28** Problem 56.

57. (II) What fraction of initially unpolarized light passes through two Polaroid sheets placed at right angles to each other? What happens if a third sheet is placed between the two sheets, with its axis at an angle of 45° to the two?

58. (II) If light of intensity I_0 moving in the z-direction is polarized linearly in the x-direction, it will not pass through a piece of Polaroid that passes light polarized in the y-direction. Figure 34–29 shows a way in which this light can pass the y-direction analyzer if a second analyzer is used. If the lower analyzer makes an angle of θ with respect to the x-direction, what is the intensity of the light that passes the upper analyzer?

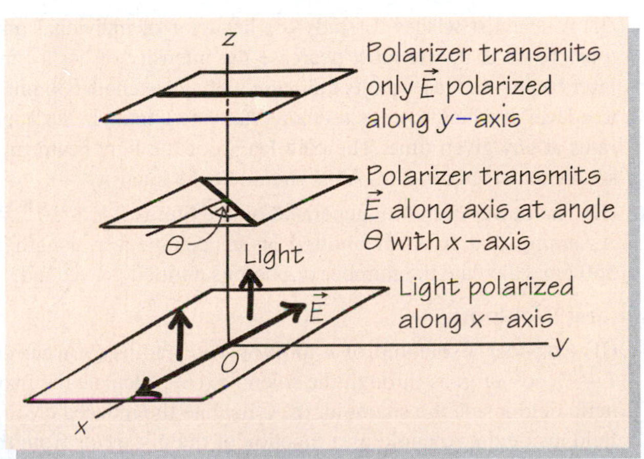

▲ **FIGURE 34–29** Problem 58.

59. (II) Unpolarized light of intensity I_0 passes through two pieces of Polaroid successively. (a) What is the intensity of the light after it passes through the first piece of Polaroid? (b) The second piece is rotated so that the intensity of the transmitted light goes to zero. What angle does the polarizing axis of the second piece make with that of the first piece? (c) A third piece of Polaroid is inserted between the two pieces in place. Calculate the intensity as a function of the angle θ that the axis of the third piece makes with the axis of the first piece (Fig. 34–30). (d) Show that the intensity of the transmitted light is no longer zero unless the axis of the third piece is parallel to that of either of the other pieces.

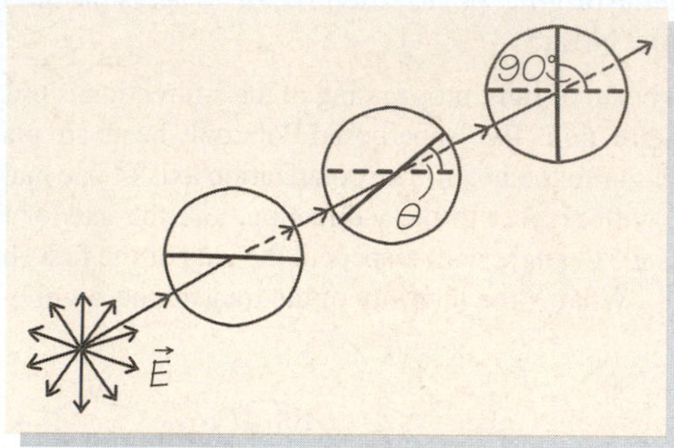

▲ **FIGURE 34–30** Problem 59.

60. (II) An electromagnetic wave passes through a sheet of Polaroid, and 3/4 of the incident intensity gets through. Through what angle should one rotate the sheet of Polaroid to let all the radiation get through?

61. (III) We have *circular polarization* when the electric field of a wave propagating in the z-direction takes the form of the superposition of a wave linearly polarized with the electric field along the x-direction and another wave linearly polarized with the electric field along the y-direction. The phases of these waves must differ by $\pi/2$, and this gives a net electric field of the form $\vec{E} = [E_0 \cos(kz - \omega t)]\hat{i} + [E_0 \sin(kz - \omega t)]\hat{j}$. Show that this electric field at any given z-value forms a vector that rotates uniformly in the xy-plane. Why is this superposition of solutions a physically acceptable wave?

*34–6 Electromagnetic Radiation as Particles

62. (II) Calculate the number of photons emitted by an FM radio station that broadcasts at a frequency of 4.5×10^8 Hz that are required to equal the energy contained in one photon of visible light at a wavelength of 450 nm.

63. (II) A scientist wishes to study the behavior of individual photons. To do that, she must decrease the intensity of her 1-mm^2 laser beam—the laser emits radiation with wavelength 630 nm—to a level at which there is no more than one photon in her apparatus at any given time. The path length of the light beam from source to detector is 2 m. What should be the intensity?

64. (II) The power currently generated by the Sun is 3.8×10^{26} W. Assuming that it is all emitted at an average wavelength of 550 nm, calculate the number of photons emitted per second.

General Problems

65. (II) Consider a solenoid of n turns/m with radius R. A current $I = I_0 \cos \omega t$ goes through the solenoid. (a) Calculate the magnetic field inside the solenoid. (b) Calculate the induced electric field inside the solenoid as a function of the distance r from the axis. (c) Calculate the Poynting vector, $\vec{S}$. In particular, find its direction at different times during one cycle.

66. (II) The electric and magnetic fields of an electromagnetic wave act on a charge q. With what speed must the charge move so that the magnetic force on the charge is, at most, 30 percent of the electric force? If the electromagnetic wave is traveling in the z-direction and the electric field has only an x-component, what is the direction (or directions) of motion of q so that the magnitude of the magnetic force is greatest?

67. (II) Consider the solar sail described in Question 9. A solar sail can be aligned with its area perpendicular to a radial line from the Sun so that the sail is pushed straight outward. Show that in this configuration the force on the sail always has the same sign and is proportional to $1/r^2$, where r is the distance from the sail to the Sun. (Assume that only the radiation pressure and the gravitational force due to the Sun act on the sail.) This economical method of propulsion has been proposed for travel to the far reaches of the solar system when transit time is not an important factor.

68. (II) A solar sail (see Question 9) is to be designed such that, when it is aligned perpendicular to the Sun's rays and is 1.5×10^{11} m from the Sun, the radiation pressure on it, P, just cancels the gravitational attraction of the Sun. The density of the material of the sail, which forms a sheet of constant thickness, is ρ. (a) Find P, given that the energy flux from the Sun is 1.4 kW/m^2 at the radius of Earth's orbit. (b) Express the sail's thickness in terms of ρ, P, the mass of the Sun, and the gravitational constant. If ρ is 2.0×10^3 kg/m^3, what is the thickness of the sail material? Your result is independent of the sail's area.

69. (II) Find an expression for the electric field of a plane electromagnetic wave with the following properties: (a) the frequency is 10^{14} Hz; (b) the wave travels in a medium of index of refraction 1.4; (c) the wave propagates along a line that lies in the xy-plane and makes a 30°-angle with the x-axis; (d) the wave is polarized along the z-axis; (e) the average value of the Poynting vector is 500 W/m^2.

70. (II) A swimming pool has underwater lights. What is Brewster's angle for reflection off the upper surface of water? The index of refraction of water is 1.33.

71. (II) The amount of solar energy reaching your body when you sunbathe on an ocean beach in summer is about 800 W/m^2. Assume that your body absorbs 40 percent of this incident radiation and that your exposed body area is 0.5 m^2. How much solar energy do you absorb in 1 h? Estimate how much perspiration must evaporate to dissipate this energy (see Chapter 17).

72. (II) A high-powered, pulsed laser used to confine plasma for nuclear fusion studies is rated at 15 MW. The laser beam is focused on an area of 0.60 mm^2. Calculate the intensity, peak electric and magnetic fields, and average energy density in this beam. Compare your results to Tables 22–1 and 28–1, which list some values for electric fields and magnetic fields, respectively, in other contexts.

73. (II) What is the number of photons/m^3 contained in a beam of electromagnetic radiation in a plane wave with a wavelength of 2 cm and an electric field amplitude 10 V/m?

74. (II) The solar energy flux at a distance $R_0 = 1.5 \times 10^{11}$ m from the Sun (the radius of Earth's orbit) is 1400 W/m^2. (a) What is the total energy flow from the Sun in watts? (b) Use your result to calculate the rate at which photons are emitted. Assume an average wavelength of 600 nm. (c) Using the result of part (b), find the number of photons/s that strike a 1 mm × 1 mm surface at a distance R_0. The surface is oriented perpendicular to the Sun.

75. (II) A laser emits N photons of frequency f. The beam strikes a mirror that is moving with speed v in the direction of propagation of the laser beam. Assuming that the kinetic energy of the mirror is much larger than that of the beam, use energy conservation and momentum conservation to find the frequency of the reflected beam. Treat the photon as a particle of energy hf and momentum hf/c.

76. (II) A wire is bent into a loop. The two ends are attached to the terminals of a battery and a current flows through the wire. Sketch the direction of the Poynting vector field near the wire.

77. (III) Many people believe solar sails are a useful source of propagation in space. What acceleration would a 150-kg space probe have if it had a 100 m^2 sail perpendicular to the line to the Sun and starts at Jupiter's orbit? The intensity of sunlight at Earth is 1400 W/m^2; use this fact to find the corresponding quantity at the radius of Jupiter's orbit, all the way to Saturn's orbit. How long would it take this probe to reach the orbit of Saturn?

78. (III) Consider a current I that flows through a cylindrical wire of length L, radius b, and resistance R (Fig. 34–31). The current flows uniformly across the cross section of the wire. Calculate the electric fields inside and on the surface of the wire. The current in the wire gives rise to a magnetic field, which you can calculate. Use these fields to find the direction and magnitude of the Poynting vector on the surface of the wire. Show that the rate of energy flow into the wire through its surface is IR^2, the power dissipated in ohmic heating.

79. (III) Consider Eq. (34–3), in which the current I passes through the surface dA, so that $I = \int_{\text{surface}} \vec{J} \cdot d\vec{A}$. If the wave propagates in a medium in which there is some conductivity, we may write the right hand side of Eq. (34–3) as

$$\mu_0\varepsilon_0 \frac{d}{dt}\int \vec{E}\cdot d\vec{A} + \frac{\mu_0}{\rho}\int \vec{E}\cdot d\vec{A}.$$

This means that in Eq. (34–5) we should make the replacement $\frac{\partial}{\partial t}E_x \rightarrow \frac{\partial}{\partial t}E_x + \frac{1}{\varepsilon_0\rho}E_x$. How does this affect the wave equation for E_x?

80. (III) Consider a plane electromagnetic wave of frequency f that propagates in the z-direction in a cubic box whose sides are length L, with L much larger than the wavelength. The electric field of the radiation has the form $\vec{E} = E_0 \sin(kz - \omega t)\,\hat{i}$. Alternatively, we can say that the radiation consists of N photons, each propagating in the z-direction with energy hf, where h is Planck's constant. Use two alternative expressions for the energy of the radiation to express E_0 in terms of h, f, N, and L.

81. (III) Consider a capacitor that consists of two circular metal plates of radius R a distance d apart (Fig. 34–32). R is so much larger than d that all fringe fields can be neglected. If the charge on the plates, Q, changes with time, then according to Ampère's law a magnetic field will be induced in the region between the plates. (a) What is the induced magnetic field? (b) Using the induced magnetic field and a calculation of the electric field between the plates, find the Poynting vector. (c) Show that with this Poynting vector, the net energy flow into the capacitor is the rate of change of the capacitor energy $Q^2/2C$.

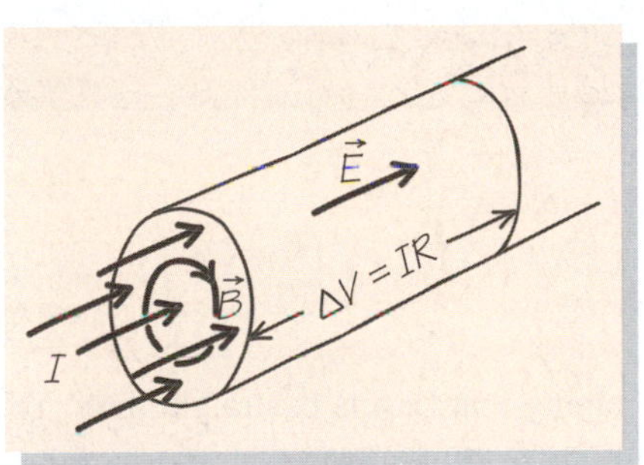

▲ **FIGURE 34–31** Problem 78.

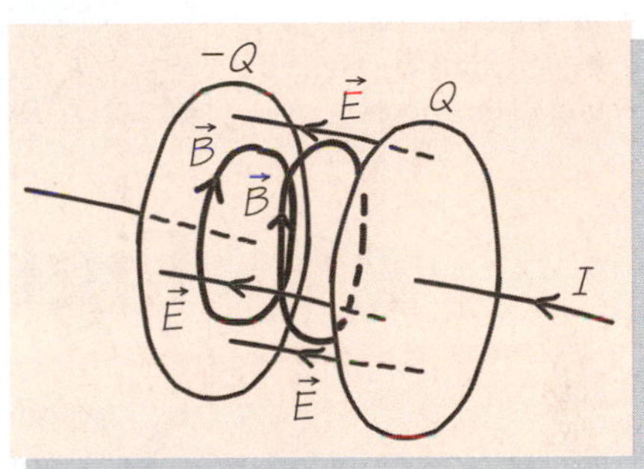

▲ **FIGURE 34–32** Problem 81.

Chapter 35

This spectacular shuttle launch produced a very special and memorable effect. The launch took place just after sunset, and the lower part of the plume was in shadow while the upper part still received the rays of the Sun. The launch also took place very close to a full Moon, so that the Sun, Moon and Earth were nearly aligned. The upper part of the plume cast a shadow, visible in the form of an "anticrepuscular ray," a shadow that is approximately aligned along the direction to the Moon.

Light

You may think of light as something that travels in straight lines. You may have come to this conclusion from observing the rays that appear when light penetrates a forest on an early morning. This property of light strongly suggests that light is composed of particles emitted by a source, and Isaac Newton, whose earliest work was on optics, supported that view. The phenomena he considered in coming to a conclusion include many with which we have some everyday experience: the reflection of light from mirrors, refraction as light passes through glass lenses or water, and observations of rainbows and the prismatic separation of colors. The particle model provided such a good explanation of these observations that it is surprising that the idea that light consists of waves could have taken root in Newton's day. Yet Robert Hooke's idea that light is some type of oscillatory activity in an unidentified medium led Christian Huygens to propose a wave theory of light in 1687. In this chapter, we shall show that the wave theory of light can explain almost everything that the particle theory can, as well as the interference and diffraction phenomena that cannot be explained by a particle picture.

By the early nineteenth century, it had become apparent that certain observations could not be explained by the particle theory; an explanation of these observations demanded that light behave like a wave. For example, when we look very closely, light does not cast *sharp* shadows, and so to some extent light bends around corners. (Newton did not have the equipment to make this observation, and, in fact, he argued against the wave theory on the basis that light does *not* appear to bend around corners!) Under controlled conditions, we can also see that beams of light interfere with each other in just the same way as the waves discussed in Chapter 15 interfere. Definitive experiments by

Thomas Young in 1801 on the wave aspects of light eventually established the preeminence of the wave theory (Fig. 35–1). The phenomena associated with the wave aspects of light are the subject of Chapters 37 and 38. The prediction from Maxwell's equations that light is an electromagnetic wave would seem to have settled the question of whether light is a particle or a wave once and for all. In the twentieth century, however, we had to revise our view once more, as new experimental evidence suggested that some aspects of light can be explained only if light sometimes behaves as particles. Today we are not forced to choose between a particle picture and a wave picture of light; a quantum mechanical explanation encompasses them both.

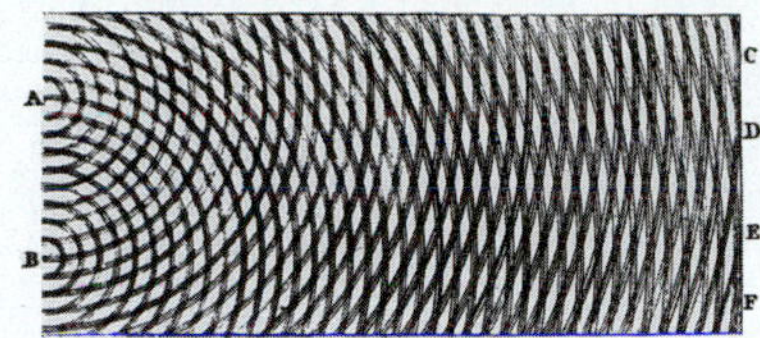

▲ **FIGURE 35–1** Young's view of the wave nature of light. In this sketch published in 1807 from his lectures, points A and B represent pinholes; points where the waves reinforce each other on the screen are marked.

35–1 The Speed of Light

Light travels so rapidly that it requires a good deal of ingenuity to show that its transmission is not instantaneous. Although Galileo had thought that the speed of light might be finite, he failed to find a time delay in the passage of light from one mountaintop to another. Ole Roemer observed the eclipses of the moons of Jupiter in 1675 and found that the timing of these phenomena could be explained if light traveled with a large, but finite, speed. The solar system data available to Roemer at the time gave a value of 2×10^8 m/s for the speed of light—certainly a result of the correct order of magnitude. The first terrestrial measurements were made in 1849 by Hippolyte Fizeau, who used the device shown in Fig. 35–2. A light source is placed behind a toothed wheel that can be rotated at high speeds. The light passes through an inclined glass plate and then between two teeth of the rotating wheel. It then travels to a mirror and is reflected straight back. If the speed of light were infinite, light would be reflected back through the gap before the wheel had moved at all (Fig. 35–2b). Light traveling at a finite speed would also pass

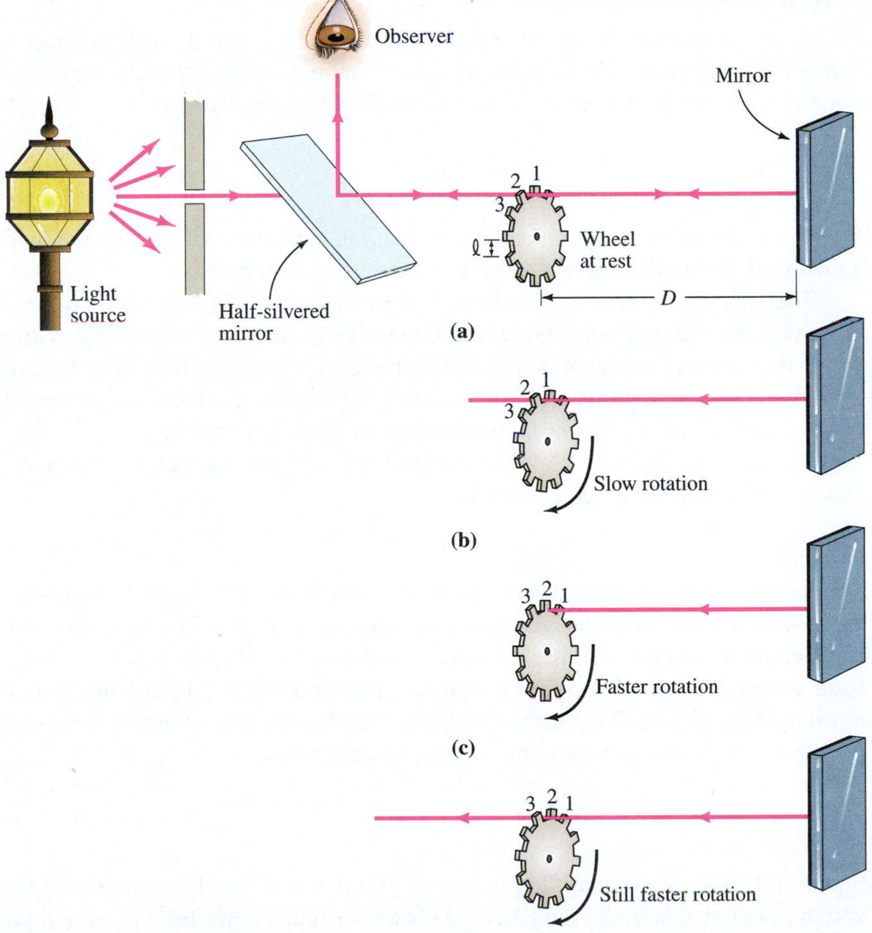

◀ **FIGURE 35–2** Fizeau's method for measuring the speed of light: (a) a sketch of the apparatus he used. (b) Incident and reflected light pass through the same gap when the wheel rotates slowly. (c) As the rotation of the wheel increases, it will become fast enough so that the reflected light fails to pass through the original gap, and instead strikes the cog between. (d) When the rotation speed is still higher, the reflected light passes through the next gap by traveling the distance $2D$ in the time the wheel rotates to the next gap.

through the same gap if the rotational speed of the wheel were small. But if the wheel is rotating fast enough and the speed of light has a finite value c, then, at a certain value of angular velocity, the wheel moves enough during the time the light travels to the mirror and is reflected back that the light will strike cog 2 (Fig. 35–2c), and no light reaches the observer in this case. As the wheel rotates even faster, light once again reaches the observer, but this time it passes through the *next* gap in the wheel, the one between cogs 2 and 3 (Fig. 35–2d). If the edge of the wheel is moving at speed v when Fig. 35–2d first applies, then we can equate the time $2D/c$ for light to make a round trip from the wheel to the mirror (D is the distance from wheel to mirror) and back to the time ℓ/v (ℓ is the cog spacing) for the wheel to move a distance of one gap:

$$\frac{\ell}{v} = \frac{2D}{c}.$$

If D is much larger than ℓ, then a value of v much smaller than c would suffice to measure c accurately. In Fizeau's 1849 experiment, the light traveled almost 20 km, while a wheel with more than 700 cogs rotated tens of times per second. Fizeau reported a result of about 3.1×10^8 m/s, a few percent different from the correct value.

Better measurements improved the value of c. Ultimately, because a measurement of c is the ratio of a distance to a time, the best measurement of c would be limited by our ability to measure time (today, to one part in 10^{13}) and distance (today, to four parts in 10^9). Because distance is harder to measure accurately than time, today we pass this difficulty by *defining* the speed of light in a vacuum to be $c = 299{,}792{,}458$ m/s and use time along with the definition of c to measure distances. In this way of doing things, the meter is no longer defined but is rather measured: One meter is 1/299,792,458 times the distance traveled by light in 1 s (see Section 1–2). For practical purposes, you can use $c = 3.00 \times 10^8$ m/s.

The Index of Refraction

TABLE 35–1 • Indices of Refraction for Various Substances (λ = 600 nm)

Material	Index of Refraction, n
Air (1 atm, 0°C)	1.00029
Carbon dioxide (1 atm, 0°C)	1.00045
Ice	1.31
Water (20°C)	1.33
Ethyl alcohol	1.36
Castor oil	1.48
Benzene	1.50
Fused quartz	1.46
Glass (crown)	1.52
Glass (flint)	1.66
Diamond	2.42

Fizeau also found that the speed of light in transparent materials such as water or glass is *less* than the speed of light in empty space. We reserve the symbol c for the speed of light in empty space, and express the speed of light in a material as

$$v_m = \frac{c}{n}, \tag{35–1}$$

where n is the *index of refraction* of the material, a quantity introduced in Chapter 34. Table 35–1 lists indices of refraction for a variety of materials.

The speed of light in materials also generally varies with wavelength, or, put another way, *the index of refraction is a function of wavelength*. For example, violet light, which has a shorter wavelength than red light, travels more slowly in glass than does red light. We shall see that this property explains the separation of white light into the colors of a rainbow by a prism or by water droplets in the atmosphere.

We saw in Chapter 34 [Eqs. (34–16) and (34–17)] that the index of refraction for a material with a dielectric constant κ is

$$n = \sqrt{\kappa}. \tag{35–2}$$

(We have assumed that the relative magnetic permeability $\mu_m/\mu_0 \cong 1$, which is a good approximation for substances that are transparent to light.) The variation of n with wavelength occurs because the dielectric constant generally varies with wavelength to some extent. We must therefore use κ at the appropriate wavelength rather than its static value in Eq. (35–2). Frequency f and wavelength λ are related by $f\lambda = v$, so we find from Eq. (35–1) that in a medium of index of refraction n,

$$f\lambda = \frac{c}{n}. \tag{35–3}$$

Equation (35–3) shows that the product of f and λ is inversely proportional to n, and you might think that both f and λ could change as light passes from one medium to another. But when the medium changes *the frequency does not change*. This is easy to understand: Consider two observers on either side of an air–glass interface. Each wave front that passes one observer must pass the other—otherwise wave fronts would pile up or disappear, neither of which happens. As a consequence, *it is the wavelength of light that changes with the index of refraction* in such a way that $c/f = n\lambda$ is constant [see Eq. (35–3)]. Thus, when light passes between media 1 and 2,

$$n_1\lambda_1 = n_2\lambda_2. \tag{35–4}$$

35–2 When Can Light Waves Be Treated as Rays?

A particle model of light accounts for many of the apparent features of light propagation, including the facts that light seems to travel in a straight line and casts a sharp shadow. But a wave model can also explain these features, as well as how they break down when we take a closer look. These issues are of some importance because one of the goals of this chapter is to establish the validity of a wave picture of light.

In Chapter 34, we discussed electromagnetic waves that propagate along the $+z$-axis. It will be helpful to review the properties of those waves here. The space dependence and time dependence of the electric or magnetic fields are described by a function such as $\cos(kz - \omega t)$. This function has a series of crests and troughs, with, for example, crests occurring at $kz - \omega t = 0$, or

$$z = \frac{\omega t}{k} = ct.$$

Thus the crest propagates at speed c. We referred to this as a plane wave because all points in the xy-plane defined by a fixed value of z have the same fields, whatever the x- or y-value (Fig. 35–3a). This allows us to make the idea of the wave fronts in electromagnetic waves more explicit: The planes for which the argument $kz - \omega t$ is constant represent the fronts. Figure 35–3b shows a sequence of wave fronts transverse to the direction of the electromagnetic (light) wave. It is useful to think of these fronts as representing a particular set of field values along the wave. For example, the wave fronts could represent the planes on which the electric field is a maximum or the points where the field is zero. (Figure 35–3b arbitrarily sets the wave fronts at the points where the electric field is maximal; since the magnetic and electric fields are in phase, these are also the points where the magnetic field is maximal.) In empty space the sequence of wave fronts moves at speed c along the original direction of propagation. Light is observed to move in a straight line, and the description above is in fact in accord with the observation. Christian Huygens, who was one of the earliest proponents of the wave theory of light, was able to strengthen this agreement between the wave picture and observation with a special approach to the process.

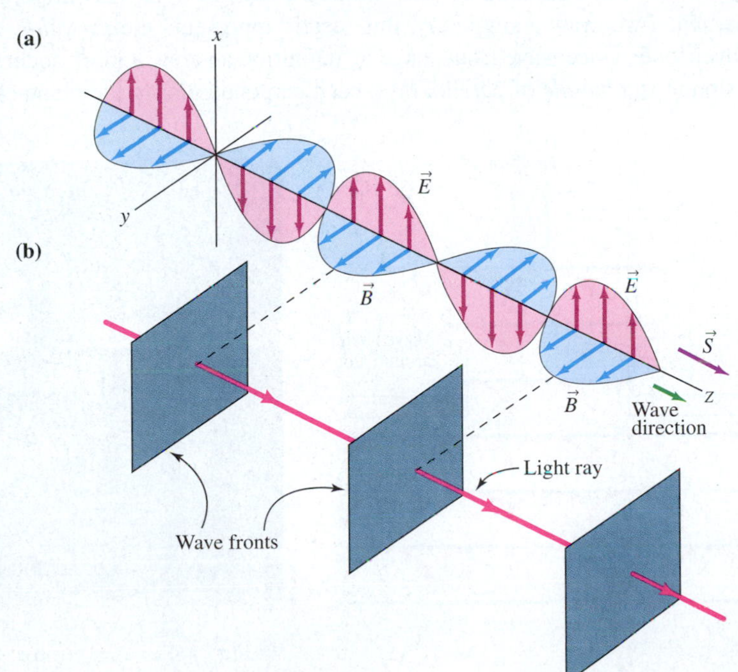

FIGURE 35–3 (a) The electric and magnetic fields of an electromagnetic wave propagating in the direction of the Poynting vector $\vec{S}$ (see Chapter 34). (b) Wave fronts are chosen arbitrarily at points where the electric and magnetic fields are maximal; the fronts could be chosen at the fields' zero points instead.

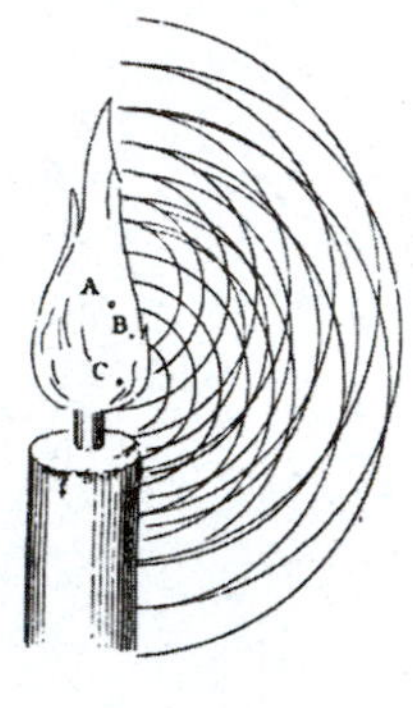

▲ **FIGURE 35–4** (a) Huygens' construction of wave fronts. Wavelets emitted at each point along the wave front add to a new wave front and produce plane waves. (b) Huygens' illustration of wavelets, here for a candle flame, from his book *Traité de la Lumière* (1885).

Huygens' Principle

Consider, as Huygens did in a publication dating from 1690, a wave front perpendicular to the $+z$-direction (Fig. 35–4a). Huygens located the wave front after a time interval Δt by viewing *every* point on the original wave front as a source of light emitting a spherical pulse (or *wavelet*) of radiation (Fig. 35–4b). The radius of the wavelet sphere in empty space is $c\,\Delta t$, the distance the light travels in time Δt. (In a medium in which the speed of light is c/n, the radius of the sphere is reduced by a factor of n to $(c/n)\,\Delta t$.) In the limit that the separation between all the emission points is small, the envelope of all these tiny spheres, taken in the direction of propagation of the initial wave front, is the new wave front. As long as no matter boundaries are encountered, the wave fronts generated in this way remain planes parallel to the xy-plane, and the straight-line propagation of wave fronts is assured. This treatment of waves, in which each point on an advancing wave front emits wavelets, and the wavelets add to the future configuration of the entire wave, is called **Huygens' principle**. Huygens' principle can be justified from a detailed study of the behavior of waves in Maxwell's equations, although we shall not do so here.

We know from our study of mechanical waves in Chapter 15 that when a wave front such as that formed by water waves approaches a slot, or opening, in a barrier, the wave spreads as it passes the edge of the opening, a phenomenon known as *diffraction*. Huygens' principle can be applied to this situation (Fig. 35–5). When the wave front arrives at the wall, only the part of the wave at the opening can continue to propagate beyond the wall. This part of the wave front generates waves that travel past the slot, with the additional feature that the spherical wavelets emitted near the edges of the slot have no neighboring wavelets to maintain a parallel wave front, and *a wave that spreads away from the slot edges is generated past the slot*. In effect, the light bends around the corner of the opening. Huygens' principle suggests that the spreading is significant (in terms of the fraction of energy in the bent waves) only if the wavelength is about the same as, or larger than, the size of the slot. If the slot width is much larger than the wavelength, only a small fraction of the energy goes into the bent waves, and it is adequate to view the entire slot as a source of a plane wave front. Light has wavelengths around 5×10^{-7} m; therefore the slot must not be too much larger than this size for the effect to be significant. Diffraction had in fact been observed in Newton's day, and it presented a difficulty, although for some reason not a mortal one, for the particle picture. We'll study this phenomenon as it applies to light in Chapter 38.

If the slot is large compared with the wavelength of light, then diffraction can be ignored. The wall casts a shadow, but to a good approximation that is its only effect. We have a *beam* of light, and we can talk about **rays** of light. It is important to remember that when we draw a single ray, this merely represents the *direction* of the motion of wave fronts; since wave fronts have by definition an area, a more accurate graphical description is a *bundle* of parallel rays, each perpendicular to the plane of the wave front.

▶ **FIGURE 35–5** (a) Huygens' construction of wave fronts that approach and pass through an open slot in a wall. Past the slot, the wave fronts bend around the slot edges. (b) A concrete realization of the effect, in a ripple tank, with an opening small enough to act as a single point source.

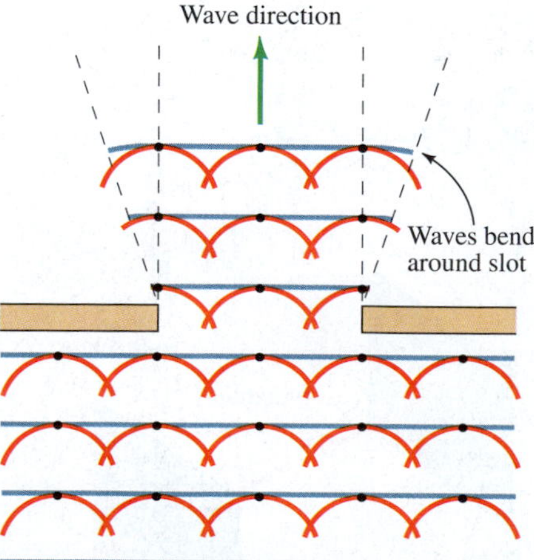

(a)

(b)

Light entering a darkened room through a pinhole or the beam of a searchlight or laser provides vivid images of the propagation of light in the form of rays. We normally do not *see* rays, but they can be made visible by, for example, putting dust particles in the path of the ray (Fig. 35–6). A small part of the light is scattered to the side and is visible from there. The description of light based on the straight-line propagation of rays is called **geometric optics**. This description is perhaps more intuitively referred to as particle-like, because straight-line propagation in free space is a particle characteristic.

In the next section we introduce the two principal laws that allow a complete treatment of geometric optics, laws that describe the reflection and refraction of light rays. In Section 35–4 we describe how both of these laws can be derived directly from Huygens' principle. Even though Huygens' principle is based firmly on the idea that light is a wave phenomenon, to the extent that it explains the behavior of geometric optics, it describes something that more clearly resembles a particle phenomenon. And, as we have seen in our discussion of the passage of light through slots, Huygens' principle will also account for the characteristically wavelike features of light. As we proceed, keep in mind that our treatment of wave phenomena is reserved for Chapters 37 and 38. Here we concentrate on geometric optics.

▲ **FIGURE 35–6** A beam of light generated by a laser. The beam is visible because particles in the air have scattered the light. Lasers have many uses—this beam is used to make a reference for astronomical instruments that can adjust for atmospheric turbulence. It supplies a view of the instantaneous turbulence, which can then be corrected for.

35–3 Reflection and Refraction

Reflection

A light ray **reflects**—it "bounces back"—when it strikes a smooth surface such as that formed by a mirror. (This, of course, oversimplifies much of what happens to reflected light. Virtually everything we see comes from light that has reflected from objects. But these reflections are for the most part quite complicated, because of both the nature of the material that is doing the reflecting and the surface structure of that material. Here you should be thinking of a mirror, something whose surface is essentially invisible but that reflects light rays in the ways we'll describe now.) The *incident ray* makes an angle θ with a line normal to the surface at the point of reflection (Fig. 35–7). The *reflected ray* lies in the plane formed by the incident ray and the normal. The angle θ' that the reflected ray makes with the normal obeys the equation known as the **law of reflection**:

$$\theta' = \theta. \tag{35–5}$$

LAW OF REFLECTION

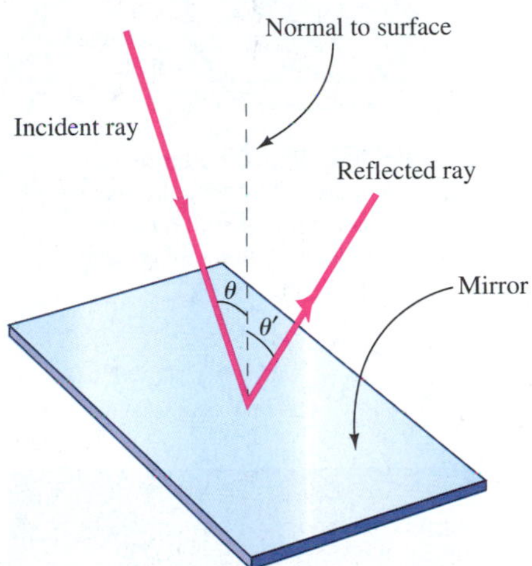

▲ **FIGURE 35–7** The angle of incidence θ equals the angle of reflection θ'.

The consequences of this law are shown in Fig. 35–8a for the reflection of a set of parallel incident rays (a *bundle* of rays) from a flat mirror and in Fig. 35–8b for that from a smooth, curved surface. For the curved surface, the angles of incidence and reflection are indeed equal, but the direction of the normal to the surface varies from point to point, and the reflected rays radiate in various directions.

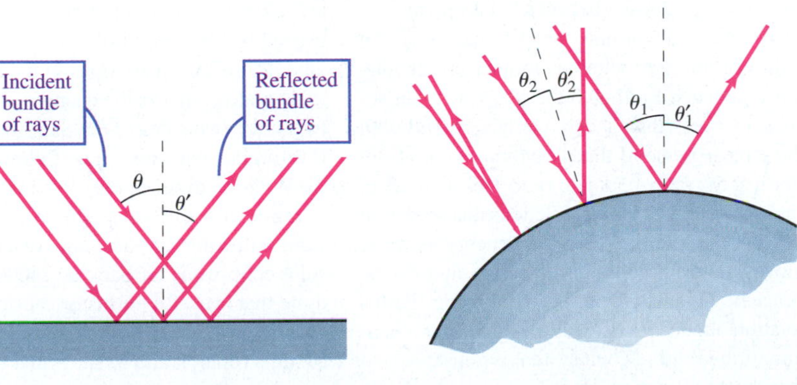

◀ **FIGURE 35–8** Reflection of rays from (a) flat and (b) curved surfaces.

CONCEPTUAL EXAMPLE 35–1 Show that two flat mirrors placed at 90° with respect to each other form a perfect reflector if the incident ray is in the plane perpendicular to the two mirrors. By a perfect reflector we mean that after two successive reflections the outgoing ray will travel back parallel to the line of incidence for an incoming ray at any incident angle.

Answer By drawing the two mirrors perpendicular to the page, as in the sketch of Fig. 35–9, the incident ray is in the plane of the page. Then we need only note that angle ACB—the angle between two normals to surfaces perpendicular to each other—is 90°. We can then establish that the sum of the remaining two interior angles of triangle ACB add to $180° - 90° = 90°$, that is, $\theta + \phi = 90°$. That means the angle of the outgoing ray with respect to the incoming ray, which the diagram shows is $2\theta + 2\phi$, is 180°. The incoming and outgoing rays are parallel.

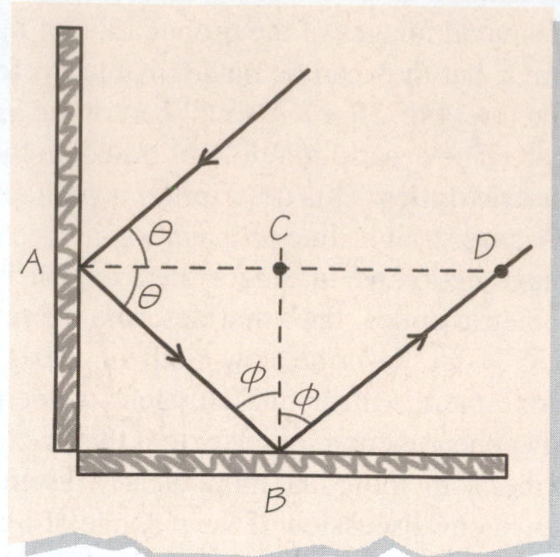

▲ **FIGURE 35–9** The incoming ray that strikes point A eventually emerges along BD, parallel to the incoming ray.

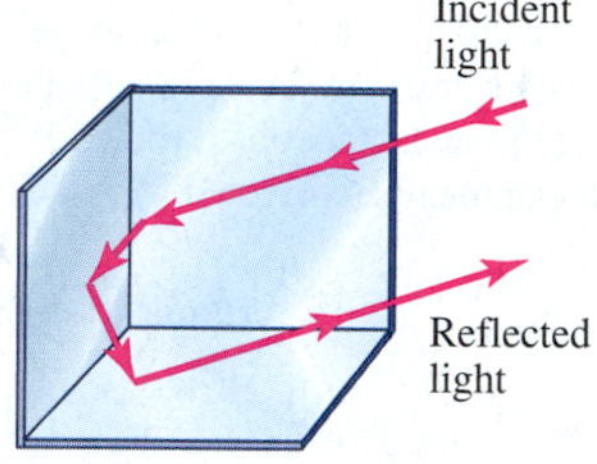

▲ **FIGURE 35–10** A corner reflector. A laser beam that has been sent into the device produces a reflected beam that is parallel to the incoming beam.

Corner Reflectors: Example 35–1 illustrates the principle behind the construction of *corner reflectors* (Fig. 35–10). In a corner reflector, *three* mirrors are placed together at mutual 90° angles, like the walls of an interior corner of a room, or an array of such corners is placed adjacent to one another. Geometry of the kind used in Example 35–1 shows that a ray incident at *any* angle reflects out from a corner reflector along a ray parallel to the incident ray. Corner reflectors are used on some highway signs so that the reflection of light from a vehicle's headlight automatically heads back to the vehicle, making the presence of the reflector very evident to the driver. Other applications stem from the fact that it is possible to measure time delays with great accuracy; a pulse from a laser directed at a corner reflector comes back, and the time delay gives a measurement of distance. In this way, for example, slight movements across fault lines on Earth's surface can be accurately surveyed.

▲ **FIGURE 35–11** This array of corner reflectors was left on the Moon by the *Apollo 14* astronauts. By shining a laser beam from Earth at the reflector and looking for the return light on Earth, the distance to the Moon can be accurately measured. The red lines indicate the directions of the incoming and outgoing laser beams.

THINK ABOUT THIS. . .

WHAT HAVE WE LEARNED FROM CORNER REFLECTORS ON THE MOON?

Starting with the *Apollo 11* lunar landing in 1969, several corner reflectors have been placed on the Moon (Fig. 35–11). Flashing laser beams sent through terrestrial telescopes and aimed at the reflectors allow us to capture a fraction (actually only one in about 50 of the very brief flashes ever sends back as much as a single photon—the minimum possible amount—to the detectors on Earth) of the sent beam and measure the time elapsed between departure and arrival. Precision atomic clocks that time the return of the light allow us to measure the distance between Earth and the Moon with an accuracy of less than 1 cm! We have found with these measurements that the Moon recedes from Earth at the rate of 3.8 cm/year. This increase in separation agrees with predictions that start with the fact that the tides dissipate energy as water moves along the bottom of the sea and on the beaches. This has the effect of slowing Earth's rotation about its axis. But the angular momentum of the Earth–Moon system is conserved, so that the angular momentum associated with the Moon's orbiting the Earth must increase to compensate, and for this to happen, the radius of the Moon's orbit increases—in other words, the Earth–Moon separation increases. The measurement of the rate gives us confidence that we understand the process. We have learned other things from the Earth-Moon distance measurements: (1) There is a constant change in Earth's shape, resulting from the fact that land masses are slowly recovering from the compression caused by the weight of glaciers in the last Ice Age. More local variations in the height of tectonic plates, such as the changing difference between the two sides of the San Andreas Fault in California, can also be followed in time. (2) With the distance data in hand, it is possible to retrodict the Moon's orbit to very high precision, allowing an accurate determination of solar eclipses as far back as 1400 B.C.E., something that is useful for historical studies. (3) The data allows us to look in more detail at the Moon's rotation and to show that the Moon has a liquid core. ■

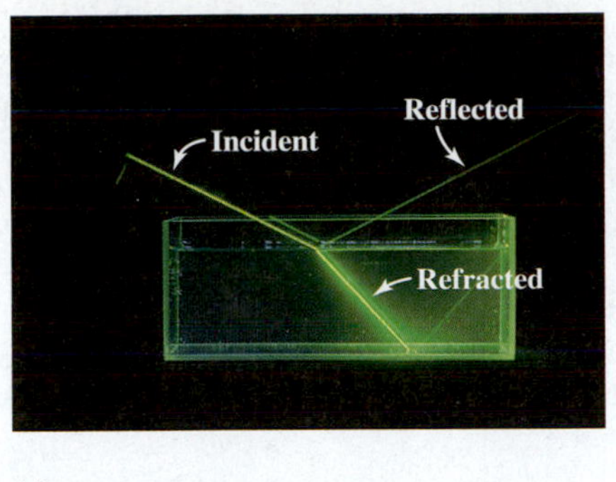

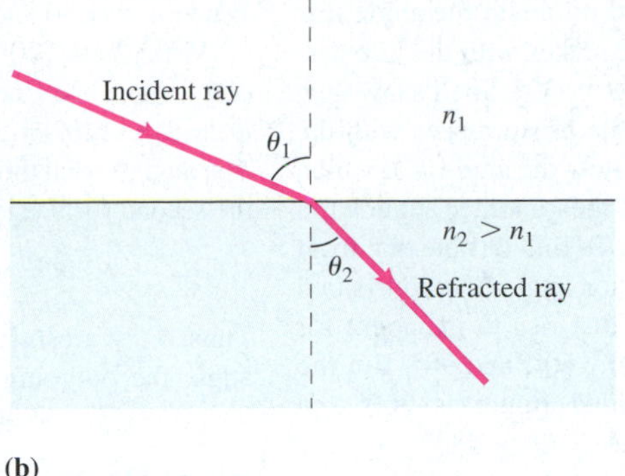

(a) (b)

▲ **FIGURE 35–12** (a) A beam of light is refracted as it enters a tank of water. (b) Refraction from a medium with index of refraction n_1 into a medium with index of refraction n_2. In this case $n_2 > n_1$, and the refracted ray is bent toward the normal to the boundary surface. If n_2 had been less than n_1, the refracted ray would have bent away from the normal.

Refraction

We are all aware of the fact that some media are transparent and that light can pass from one transparent medium into another; air and water form one familiar pair, as do air and glass. We have referred to this process as *transmission*. When the light forming a ray undergoes transmission, the incident ray does not continue along a single straight line but instead changes direction at the boundary between the media. The ray is said to undergo **refraction** (Fig. 35–12a).

Let the index of refraction of the medium with the incident ray be n_1 and that of the medium with the *refracted ray* be n_2. The angles that the incident and refracted rays make with the line normal to the boundary between the media are θ_1 and θ_2, respectively (Fig. 35–12b). Then one finds that

$$n_1 \sin \theta_1 = n_2 \sin \theta_2. \tag{35–6}$$

SNELL'S LAW

This is **Snell's law**, discovered by Willebrord Snell in 1621. The index of refraction of air is very close to unity, so the angle of the refracted ray θ_2 at the interface for light that passes from air into a medium with index of refraction n is given by

$$\sin \theta_1 = n \sin \theta_2. \tag{35–7}$$

Because n is generally larger than one, it follows that $\theta_2 < \theta_1$; that is, *the light is bent toward the normal to the boundary surface*. Equation (35–6) also shows that when light enters a medium with a lower index of refraction, such as when a ray of light travels from water to air, the ray is bent farther away from the line normal to the boundary. This makes a ray coming from an underwater object move closer to the horizontal upon reaching the air, so that an eye receiving this ray and mentally tracing it back will "place" the object at a shallower position (Fig. 35–13).

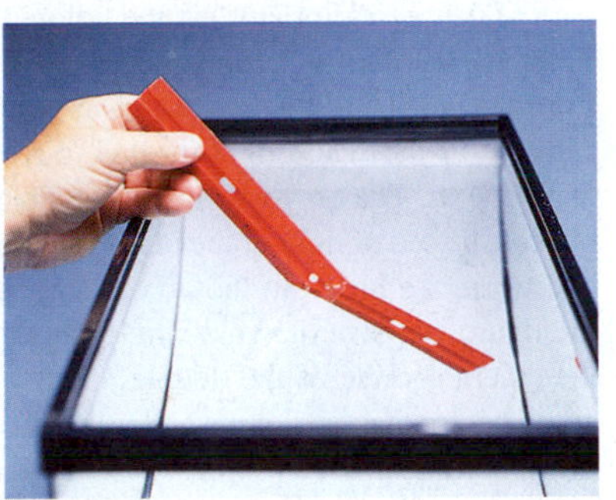

▲ **FIGURE 35–13** Water has a higher index of refraction than air, so the immersed part of this ruler appears in a position that is shallower than its actual position.

EXAMPLE 35–2 Consider a horizontal ray of light approaching a prism, a piece of glass shaped as a rod whose cross section makes a triangle. In this example, the triangle is an equilateral triangle, and the ray is perpendicular to the axis of the rod. The ray is refracted once as it enters the glass and then a second time as it exits to the air. What is the total deflection of the ray, given that the index of refraction of the prism glass is 1.50?

Setting It Up The essential part of this problem is the sketch (Fig. 35–14), because the tools necessary for this problem are those of simple geometry. The sketch defines a series of angles that must be calculated to find the angle of the final ray.

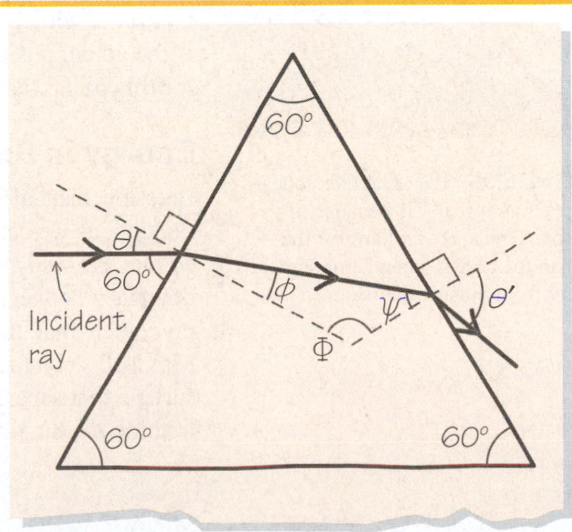

▶ **FIGURE 35–14** Geometric construction of the path followed by a ray incident on a prism.

(continues on next page)

Strategy Simple geometry can be used to obtain the angle that the incident ray, drawn parallel to the base, makes with the line perpendicular to the face of the prism, namely $\theta = 30°$. Snell's law then gives us the angle ϕ that the beam inside the prism makes with the perpendicular to the first face. We need to know the angle ψ at which the interior ray meets the second face, and then another application of Snell's law will give us the final angle θ'. To find ψ, note that from the figure, the interior angle Φ is an interior angle of a four-sided figure (the two dashed lines and the lines that rise to the apex) for which the other three interior angles are 90°, 60°, and 90°. But the sum of the interior angles of a four-sided figure is 360°, so $\Phi = 120°$. Then we find ψ by using $\phi + \psi + \Phi = 180°$.

Working It Out To find ϕ, we use Eq. (35–7):

$$\sin\theta = n\sin\phi.$$

When $n = 1.50$ and $\theta = 30°$, this expression gives $\phi = 19.5°$.

With $\Phi = 120°$, the angle ψ that our once-refracted ray makes with the line normal to the second surface is given by $\phi + \psi + 120° = 180°$, or $\psi = 60° - \phi = 60° - 19.5° = 40.5°$. The angle θ' that the second refracted ray makes with the line normal to the second surface is then given by a second application of Snell's law,

$$\sin\theta' = n\sin\psi = 1.50\sin(40.5°) = 0.97.$$

Thus, $\theta' = \arcsin(0.97) = 77°$, and we see from Fig. 35–14 that the angle the outgoing light ray makes with the base of the prism is $\theta' - \theta = 77° - 30° = 47°$.

What Do You Think? If we want to make the exit angle θ' smaller, how do we change the angle θ? ***Answers to What Do You Think? questions are given in the back of the book.***

CONCEPTUAL EXAMPLE 35–3 The study of refraction originated with the discovery by Isaac Newton that white light falling on a prism is split into the colors of the rainbow, and his interpretation of this as white light being composed of all the colors at once. Some people argued that another interpretation was possible: that sunlight, falling on the prism, generated the emission of a variety of colors. Can you propose one or two experiments, using a second, identical prism, to argue that it is indeed white light that is composed of the rainbow colors?

Answer (a) Put the second prism upside down so that the face of the first prism from which the spread of colors emerges is parallel to the face of the second prism that the colored rays will enter (Fig. 35–15). This reverses the refraction process of the first prism. What emerges from the far side of the two prisms is white light, illustrating the reconstitution of the white light through direct combination of its colorful components. The only way this is possible is that the white light is simply constituted of light of the rainbow colors, and that in adding the second prism we have put those colors back together again. Note that in the figure we have drawn a single ray entering the prism, whereas on the emerging side of the double prism we have parallel rays. But we noted before that rays represent only a direction, with the actual incoming light more properly represented as a bundle of rays. The correct representation of our situation is that there is a bundle of incoming white light rays and a bundle of outgoing white light rays. (b) Again insert the second prism, but this time place a mask on the surface of the second prism so that only light of one color enters (say green). The result of this is that only green will emerge on the far side of the prism. This can be done with each of the colors. Again, the constituent colors of white light are nothing more than that—constituents—and the only role of the first prism is to separate them out.

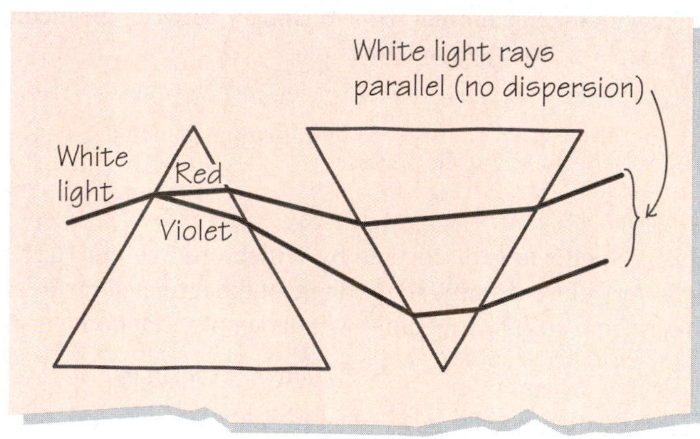

▲ **FIGURE 35–15** Arrangement of two prisms that shows that the colors are the components of white light.

▲ **FIGURE 35–16** The fish appears less deep than it really is because of refraction. This makes it hard for the fisherman to spear it when he or she aims the spear from outside the water.

Refraction is responsible for some curious optical effects, as Fig 35–13 shows. You may be familiar with the problem of spearing a fish (Fig. 35–16). The source of the light ray from the fish is seen to originate from a point that lies on a straight line along the direction at which the ray enters the observer's eye. Thus the observer thinks the image I is the actual position of the fish, and if the observer is a spear fisherman who is aiming for this point, the fish will not be touched.

Energy in Reflection and Refraction

Like mechanical waves, transmission (with refraction) is generally accompanied by reflection, as we can see in Fig. 35–12a where a reflected ray is visible as well as a refracted one. As we saw in Chapter 34, the incident ray carries electromagnetic energy. At the boundary between media, this energy is apportioned among the reflected and refracted rays such that the total energy is conserved. Here we quote a result without proof: Maxwell's equations can be used to show that when light is perpendicularly incident on a surface that separates a medium of index of refraction n_1 from a medium of index of refraction n_2, the intensity of the reflected light, I_r, is related to the incident intensity, I_0, by

$$\frac{I_r}{I_0} = \frac{(n_2 - n_1)^2}{(n_2 + n_1)^2}. \tag{35–8}$$

For light perpendicularly incident—the angle of incidence is zero—from air ($n = 1.0$) into glass ($n = 1.5$), only 4 percent of the incident light is reflected. Note the squares in this equation. The ratio is the same for light coming from air and moving into glass as it is for light coming from glass and passing into air. More generally the intensity of the reflected (and refracted) light varies with the angle of incidence, with the intensity of the reflected ray a minimum at perpendicular incidence.

CONCEPTUAL EXAMPLE 35–4 What is the intensity of the transmitted light that falls in a perpendicular direction on a thin sheet of glass ($n = 1.5$) and then passes out of the far side of the sheet?

Answer The index of refraction of air is approximately unity, so from Eq. (35–8) the fraction of energy reflected at the first surface is $f_r = (1.5 - 1.0)^2/(1.5 + 1.0)^2 = 0.04$. By energy conservation, the fraction that penetrates is $1 - 0.04 = 0.96$. This intensity of light falls on the second interface, and here again 4 percent is reflected, and therefore the total fraction of light transmitted is $(0.96)^2 = 0.92$. Some of the light reflected at the second interface will be re-reflected at the first interface and head back out the far side, but there we are dealing with fractions like 4 percent of 4 percent, which we can neglect. We should add that when monochromatic light is involved, wavelike phenomena can make the fraction reflected vary dramatically with wavelength or glass thickness (see Chapters 37 and 38).

Total Internal Reflection

For some incident angles, *all* the incident energy is contained in the reflected ray. This situation, known as **total internal reflection**, can occur only when light travels from a medium with a larger index of refraction toward a medium with a smaller index of refraction, such as when light passes from water toward air. Simple geometry explains this phenomenon.

Let's consider a light-ray incident from a medium with an index of refraction n_1 to a medium with an index of refraction n_2; this time $n_1 > n_2$. Snell's law, Eq. (35–6), may be written in the form $\sin\theta_2 = (n_1/n_2)\sin\theta_1$. As θ_1 increases, θ_2 reaches 90° before θ_1 does, because the factor n_1/n_2 is larger than unity. Figure 35–17 shows the reflected and refracted rays for various values of θ_1. When $\theta_2 = 90°$, the ray in medium 2 skims along the interface of the two media. This occurs when θ_1 reaches a critical angle θ_c such that $(n_1/n_2)\sin\theta_c = \sin 90° = 1$, or

$$\sin\theta_c = \frac{n_2}{n_1}. \qquad (35\text{–}9)$$

When θ_1 exceeds θ_c, there is no angle θ_2 that can satisfy Snell's law. The electromagnetic energy carried by the incident ray must go somewhere, and the ray is reflected. There is no diminution of the intensity of the reflected ray, and the reflection is total.

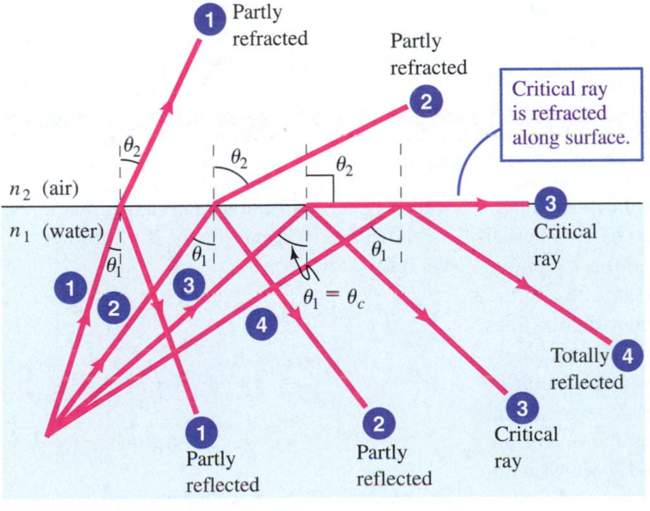

(a)

(b)

▲ **FIGURE 35–17** (a) Various rays traveling from a medium with a larger index of refraction (water) to a medium with a smaller index of refraction (air). When the incident angle is θ_c, there is total internal reflection. (b) Refraction and total internal reflection off the air–water interface in a water tank.

CONCEPTUAL EXAMPLE 35–5 Suppose a ray of white light originates inside the tank in Fig. 35–17 and that the ray is at an angle larger than the critical angle for all the colors in white light. The ray is then rotated so that the incident angle decreases. What color light emerges first into the air?

Answer The observation of the decomposition of white light by a glass prism shows that the index of refraction for blue light is larger than that for red light: $n(\mathrm{B}) > n(\mathrm{R})$. Taking the index of refraction of air as unity, Eq. (35–9) reads $\sin\theta_c = 1/n$ and thus $\theta_c(\mathrm{B}) < \theta_c(\mathrm{R})$. This means that *red* light emerges first as the white-light source rotates upward to make a smaller and smaller angle of incidence in Fig. 35–17.

EXAMPLE 35–6 A fisherman knows of a fishing spot whose horizontal distance from a river bank is a distance $R = 1.5$ m from the bank. Thinking to sneak up on a fish which might be at any water depth, he stealthily approaches the bank with his eyes at ground (and water) level. Describe what he might see.

Setting It Up We draw in Fig. 35–18 an illustration. We have drawn a particular ray, with angle of incidence θ, that meets the surface very close to where the eye is located.

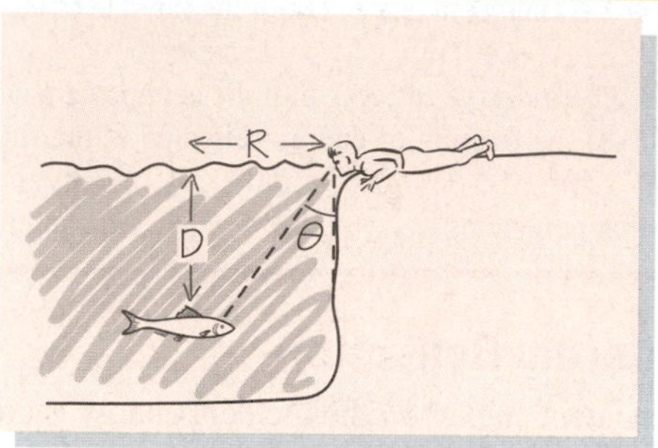

▲ **FIGURE 35–18**

Strategy and Working It Out If we model the fish as a point, then for a given depth D, there is only a single ray from the fish that reaches the fisherman's eyes. When D is large (the fish is deep) and the angle of incidence is smaller than the critical angle, then when the ray leaves the water it continues upward, although more parallel to the surface than the incident ray. The only such ray able to reach the eye at the edge must then strike the surface very close to the edge. The fisherman extrapolates this ray back and mentally places the fish closer to the surface, just as for the spear fisherman of Fig. 35–16.

We can now imagine that D decreases, until at a critical depth D_c the incident angle in the figure is the critical angle θ_c. The refracted ray comes off parallel to the surface, so the fisherman sees the fish right at the surface. The corresponding ray that comes from any fish closer to the surface than D_c does not make it out of the water, because the incident angle is greater than the critical angle. But a ray from that fish that is incident at the critical angle, and which intersects the surface at a point in between the edge and the horizontal location of the fish, will propagate in air parallel to the surface and ends at the eye. The fisherman sees *all* the fish from just below the surface to a depth of D_c at once, squashed into a flat pancake at the surface and hardly recognizable. Hanging in the region from $D = 0$ to $D = D_c$ is a good way for the fish to hide from this particular fisherman!

What is D_c? We can easily find it from geometry. The figure shows that generally $\sin\theta = R/\sqrt{D^2 + R^2}$. With $\sin\theta_c = 1/n_{\text{water}}$, we then have

$$\frac{R}{\sqrt{D_c^2 + R^2}} = \frac{1}{n_{\text{water}}} \quad \text{or} \quad \frac{\sqrt{D_c^2 + R^2}}{R} = n_{\text{water}}.$$

We solve this for D_c:

$$D_c^2 = R^2(n_{\text{water}}^2 - 1) = (1.5\ \text{m})^2(1.33^2 - 1) = 1.7\ \text{m}^2;$$
$$D_c = 1.3\ \text{m}.$$

What Do You Think? Can a fish see the fisherman sneaking up to the bank?

THINK ABOUT THIS. . .

HOW DOES LIGHT PROPAGATE WITHIN OPTICAL FIBERS?

Fiber optics represents one of the most important technological applications of total internal reflection. The principle behind this technique of conducting light from one place to another is straightforward: A transparent quartz fiber (typically of diameter 50 μm—the thickness of a human hair) will serve as a conductor of light if any ray inside the fiber undergoes total internal reflection upon striking the side of the fiber (Fig. 35–19a). Figure 35–19b shows a ray in air ($n = 1$) entering a cylinder of diameter D at an angle θ_i with the axis of the cylinder. If n_f is the index of refraction of the fiber, then the angle that the ray makes with the axis inside the fiber is θ_f, where $\sin\theta_f = \sin\theta_i/n_f$. This ray will strike the wall of the cylinder at an angle $(90° - \theta_f)$ with the normal to the wall. There will be total internal reflection if $n_f \sin(90° - \theta_f) > 1$; that is, if $n_f \cos\theta_f > 1$. We have

$$n_f \cos\theta_f = n_f\sqrt{1 - \sin^2\theta_f}$$
$$= n_f\sqrt{1 - \frac{\sin^2\theta_i}{n_f^2}}$$
$$= \sqrt{n_f^2 - \sin^2\theta_i} > 1.$$

Because $\sin^2\theta_i \le 1$, we have

$$\sqrt{n_f^2 - \sin^2\theta_i} \ge \sqrt{n_f^2 - 1}.$$

Thus we automatically satisfy the condition for total internal reflection, $n_f \cos\theta_f > 1$, if

$$\sqrt{n_f^2 - 1} > 1. \qquad (35\text{–}10)$$

Because the largest value of $\sin\theta_i$ is 1 (the light first enters the cylinder from the end), Eq. (35–10) is a condition for internal reflection for *all* of the light that enters the fiber. Equation (35–10) is satisfied for any material with $n_f > \sqrt{2}$. A typical fiber has an index of refraction of 1.62, which is larger than the critical value. Note that once a ray is in the fiber, it remains inside *even if the fiber curves*, at least if the bend is not too sharp. Information is carried when the light is pulsed in an order that contains the information, and in this way the light carries the information of a telephone call or an exchange between computers. There is a considerable advantage to optical fibers: Compared to the density of information that can be carried in an old-fashioned coaxial cable, a fiber can carry a much greater density of information. That translates into a capacity to carry many telephone calls at once, or to carry the enormous amounts of information exchanged by modern computers. This is in addition to the fact that for the same weight or diameter of an ordinary cable, one can put together a great number of fibers to carry information in parallel.

The situation outlined above is only an ideal. The internal reflection is somewhat less than total if there are impurities such as moisture, dust, or oil on the surface, because electromagnetic energy can leak across the thin "barrier" formed by the air layer between the fiber and the impurity. In long-distance transmission, light may be reflected many times, and it is therefore important to have no leakage of light. This problem is controlled largely by *cladding*—coating each fiber with a transparent covering whose index of refraction is lower than that of the fiber. In addition, the light intensity generally decreases as the ray propagates in a medium because the medium is not perfectly transparent. This effect is reduced by making the fiber from fused quartz, a highly transparent material, and purifying it to remove all traces of water. For the trans-Atlantic cable TAT-8, which can carry 40,000 conversations over two pairs of glass fibers simultaneously, it is necessary to boost the signal only every 50 km with a repeater station. This is much less expensive than systems of metal wire, which require boosting every kilometer.

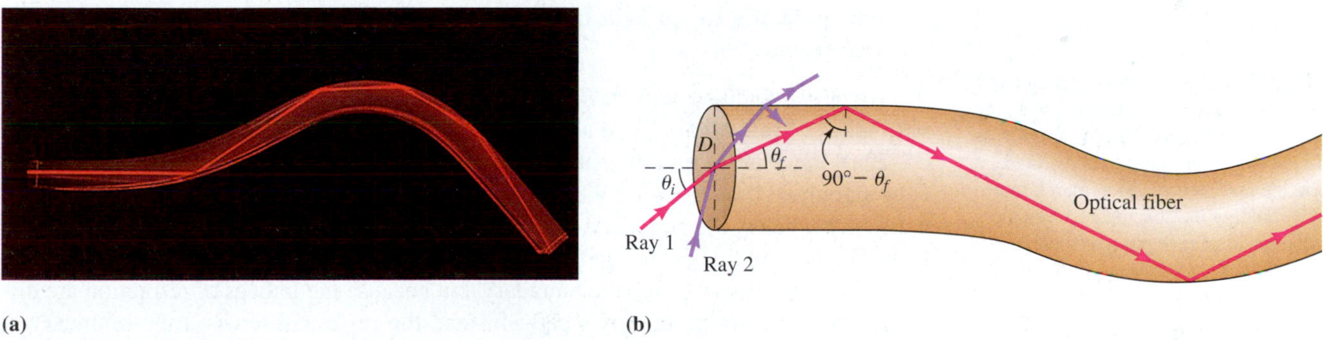

▲ **FIGURE 35–19** (a) Total internal reflection in an optical fiber. (b) Detailed construction of ray angles in a curved fiber.

*35–4 Fermat's Principle

Fermat's principle provides us with an interesting way to understand reflection and refraction. In order to set the stage for Fermat's principle, let us first look at how Huygens' principle leads to the law of reflection [Eq. (35–5)]. Figure 35–20a shows a sequence of wave fronts as they approach a mirror. In Fig. 35–20b, point C_2 is the center of a reflected spherical wave, one of many along the mirror. An outgoing (reflected) wave front—here, the line tangential to point D_2 of the semicircle centered on point C_2—forms. The distance the wave travels in time Δt is the same for incoming and outgoing waves, so a simple geometrical argument yields the result described by Eq. (35–5); namely, that the angle of reflection equals the angle of incidence. Figure 35–20c shows a later part of the sequence.

Snell's law may also be obtained by an application of Huygens' principle. The bending of the wave front is associated with the slowing down of the light waves in the medium. The bending can be visualized by analogy with the direction change of a band whose members march at an angle toward a sidewalk. Each row is like a wave front. The band members are given orders that each one must walk more slowly, without changing the distance between marchers in each row, as soon as that marcher steps on the sidewalk.

Incident wave fronts

Mirror

(a)

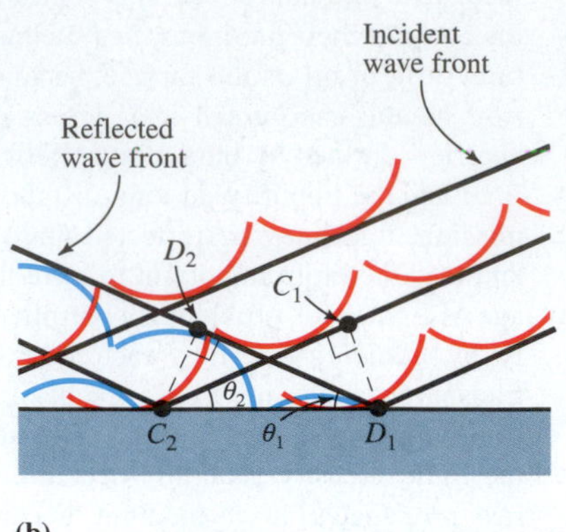

(b)

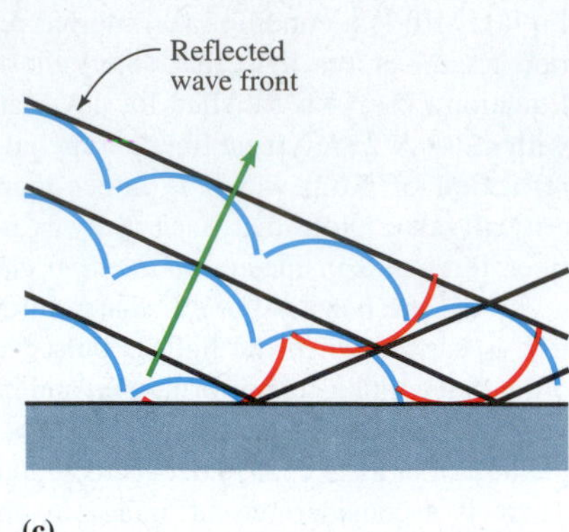

(c)

▲ **FIGURE 35–20** (a) Incident wave fronts approaching a plane mirror. (b) The wave fronts reflect. (Note that for clarity, fewer wavelets are shown.) The reflected wave fronts are generated by Huygens' construction. The relation $C_2D_2 = C_1D_1$ leads by geometrical reasoning to equal angles of incidence and reflection. (c) Most of the wave fronts have been reflected.

$\vec{v}_1$

$\vec{v}_2$

▲ **FIGURE 35–21** A marching band forms rows that change direction when the speed of the marchers change across a row. Here the rows bend sharply at a boundary where the marchers begin to march more slowly, holding the spacing between the marchers within each row constant.

The result is that the front formed by the row of band members bends (Fig. 35–21). Rather than going through the derivation of Snell's law from the Huygens construction, which leads to a picture like that of the marching band, let's demonstrate it from the principle enunciated by Pierre de Fermat in 1657. **Fermat's principle** states that

The path of a ray of light between two points is the path that minimizes the travel time.

To derive Snell's law from Fermat's principle, let's consider a point A in medium 1 with index of refraction n_1 and a point B in medium 2 with index of refraction n_2 (Fig. 35–22a). We want to find the path between points A and B that takes a ray of light the least amount of time to travel. We choose A to be a distance d above the boundary and B a distance d below the boundary; we choose the horizontal distance between A and B to be b. The straight line connecting A to B crosses the boundary at a distance $b/2$ from the normal dropped from A onto the boundary, but because the indices of refraction are different, this will not be the ray's path; instead the ray's path crosses the boundary at some point P that we want to find. Figure 35–22a shows that the distance from A to the intersection point P is $\sqrt{d^2 + x^2}$, and the distance from the intersection point P to B is $\sqrt{d^2 + (b - x)^2}$. The time for the ray to travel a distance D in a medium of index of refraction n is given by $t = D/v = D/(c/n) = nD/c$. Thus the total travel time is

$$t_{AB} = t_{AP} + t_{PB} = \frac{n_1\sqrt{d^2 + x^2} + n_2\sqrt{d^2 + (b - x)^2}}{c}. \tag{35–11}$$

Figure 35–22b is a graph of t_{AB} as a function of x. The minimum travel time is obtained by finding the place at which the slope of t_{AB} as a function of x is flat; that is, the value of x at which

$$\frac{dt_{AB}}{dx} = 0.$$

This condition implies that

$$\frac{dt_{AB}}{dx} = \left(\frac{1}{c}\right)\left[\frac{n_1 x}{\sqrt{d^2 + x^2}} - \frac{n_2(b - x)}{\sqrt{d^2 + (b - x)^2}}\right] = 0. \tag{35–12}$$

Now observe from Fig. 35–22a that

$$\frac{x}{\sqrt{d^2 + x^2}} = \sin\theta_1 \tag{35–13a}$$

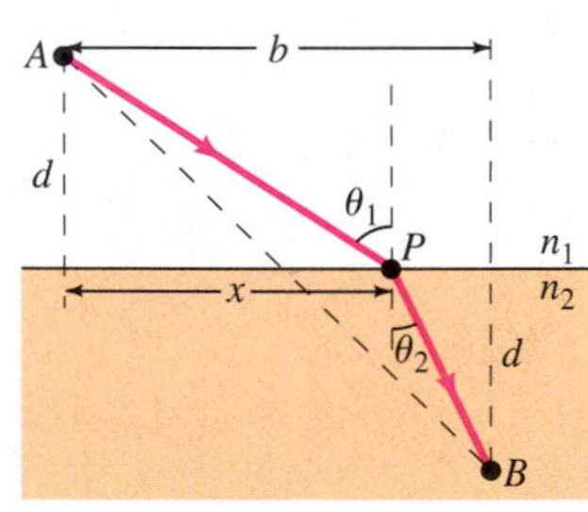

(a)

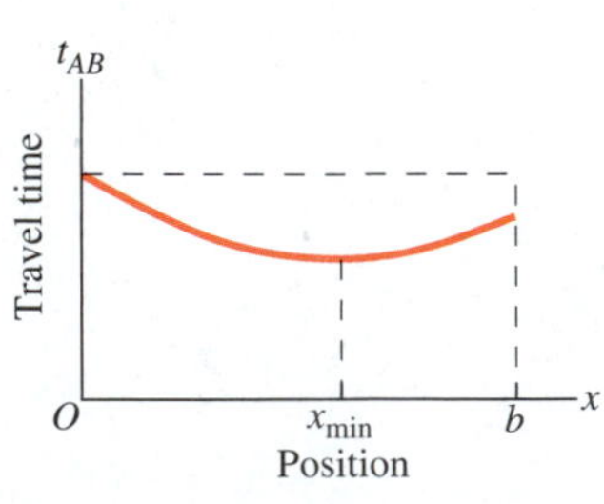

(b)

▲ **FIGURE 35–22** (a) Geometry for proving Snell's law by Fermat's principle. (b) The travel time t_{AB} for the ray as a function of x.

and that

$$\frac{b - x}{\sqrt{d^2 + (b - x)^2}} = \sin\theta_2, \qquad (35\text{–}13b)$$

where θ_1 and θ_2 are the angles the two rays make with respect to the normals in their respective media. Thus Eq. (35–12) may be rewritten as

$$n_1 \sin\theta_1 = n_2 \sin\theta_2, \qquad (35\text{–}14)$$

which is just Snell's law.

Both the straight-line propagation of light in a single medium and the law of reflection can also be derived from Fermat's principle. Fermat's principle in fact follows directly from Maxwell's equations, although we do not perform the derivation here. As Example 35–7 shows, principles such as Fermat's principle (more generally termed minimum principles) can apply in surprising circumstances.

EXAMPLE 35–7 A girl located at point *B* in Fig. 35–23a spots a ball at point *A*. Point *A* is in tall grass, where the girl can run at 1.1 m/s, and point *B* is in short grass, where the girl can run at 2.2 m/s. The whole area is flat. At what point should she cross the boundary between the grasses so that she retrieves the ball as quickly as possible?

Strategy Fig. 35–23a shows that the problem is conceptually identical to the problem of deriving Snell's law from Fermat's principle. We want an expression for the total travel time, consisting of the time in the short grass where the speed is 2.2 m/s and the time in the tall grass, where the speed is 1.1 m/s. The distances in the two regions can be calculated in terms of the unknown crossing point, whose position is at *x*.

For a given (to be found) value of *x*, the distance to be covered in the short grass is given by $L_{short} = \sqrt{d^2 + (d - x)^2}$. If the speed of travel in the short grass is v_{short}, then the time spent in the short grass is $t_{short} = \dfrac{\sqrt{d^2 + (d - x)^2}}{v_{short}}$. The distance traveled in the tall grass is $L_{tall} = \sqrt{d^2 + x^2}$, hence the time in the tall grass is $t_{tall} = \dfrac{\sqrt{d^2 + x^2}}{v_{tall}}$. Our task then is to consider the total time as a function of *x* and find the value of *x* for which this sum is a minimum.

Working It Out The expression for the total time involves $d = 5.0$ m, $v_{short} = 2.2$ m/s, and $v_{tall} = 1.1$ m/s. We thus want to minimize

$$t(x) = \frac{\sqrt{25 + x^2}}{1.1} + \frac{\sqrt{50 - 10x + x^2}}{2.2}.$$

We plot $t(x)$ as a function of *x* in Fig. 35–23b. There is indeed a minimum, at about $x = 1.5$ m. Compare this to the 2.5-m value *x* would take if the child were to run in a straight line. It is best to run a little farther in the short grass.

Rather than plotting, we could also find the desired value of *x* by solving for *x* in the algebraic equation $\dfrac{dt(x)}{dx} = 0$. But this is a rather messy equation and needs more numerical work than the plot of Fig. 35–23b.

The path followed looks like the path a light ray would take in traveling from a medium with a larger value for the speed of light to a medium with a smaller value for the speed of light.

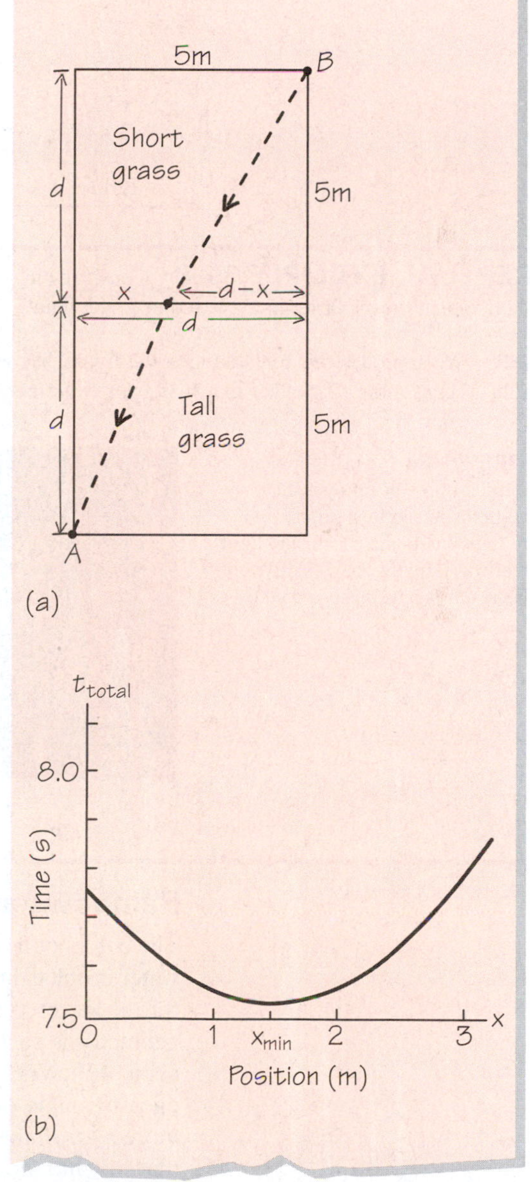

▲ **FIGURE 35–23** (a) Bird's-eye view of girl's path. (b) The travel time as a function of where the child crosses the border between the tall and short grass.

35–5 Dispersion

In this section, we explore another property of the index of refraction, a property with some truly spectacular consequences: In general, the index of refraction depends on the wavelength (or color) of the light being transmitted. Table 35–2 shows how n varies with wavelength for glass, for wavelengths near and including the wavelengths of visible light. Different wavelengths are refracted to different degrees. In this way, white light (a mixture of different wavelengths) can be separated into its constituent colors of the rainbow (Fig. 35–24). As we mentioned in Section 34–2, the dependence of refraction on the wavelength of light is called *dispersion*.

TABLE 35–2 • Index of Refraction of Glass as a Function of Wavelength

Wavelength in Air (nm)	$\omega = (2\pi c/\lambda)$ rad/s in units of 10^{15}	n	Color
361	5.22	1.539	Near ultraviolet
434	4.35	1.528	Blue
486	3.87	1.523	Blue-green
589	3.19	1.517	Yellow
656	2.86	1.514	Orange
768	2.45	1.511	Near infrared
1200	1.58	1.505	Infrared

CONCEPTUAL EXAMPLE 35–8 Consider the setup in Fig. 35–24. Which moves faster in glass, red or violet light?

Answer We approach this by looking at the figure. We see that violet light is bent more. (The fact that it is bent twice because it passes through two surfaces does not change this conclusion.) By Snell's law this implies that n for violet light is larger than for red light, and since the speed of light is c/n, this implies violet light moves more slowly than red light. We confirm this by looking at Table 35–2, which shows that n is indeed larger for violet light.

(a)

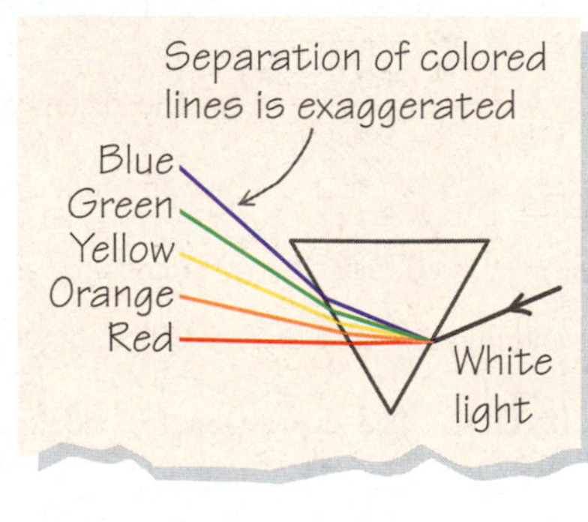

(b)

FIGURE 35–24 (a) Bright white light is formed into a beam by a slit before being dispersed by a prism. (b) White light enters the prism, and light of different wavelengths follows different paths. The result is a beam separated by color.

Rainbows and the Blue Sky

The colors of a rainbow result from dispersion in the reflection of light from individual water droplets in the air. When sunlight shines on a raindrop, light is reflected once before it leaves the drop. Many paths are possible; two are shown in Fig. 35–25a. The geometry is such that no ray can emerge after one reflection at an angle *steeper* than about 42°. When the Sun is behind the viewer, only drops that lie within a cone with an opening angle of about 42° reflect sunlight back to the observer's eye (Fig. 35–25b); moreover, *all* the drops in this cone reflect light to the observer. (We shall refer to a disk that fits into the cone because the depth of the cone is irrelevant.) One other feature of the disk is that light is reflected most strongly from raindrops at the edge, around 42°.

Dispersion has played no role in our discussion so far. The effect of dispersion is to make the angle of the outer radius of the disk slightly different for different colors. As Fig. 35–25c indicates, the disk for red light is larger than the disk for blue light. Because

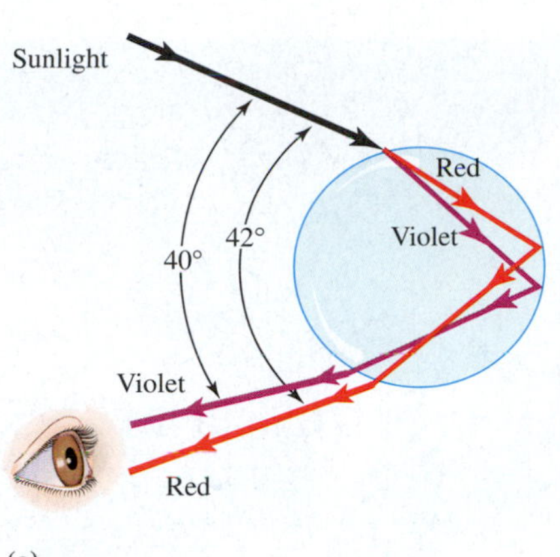

(a)

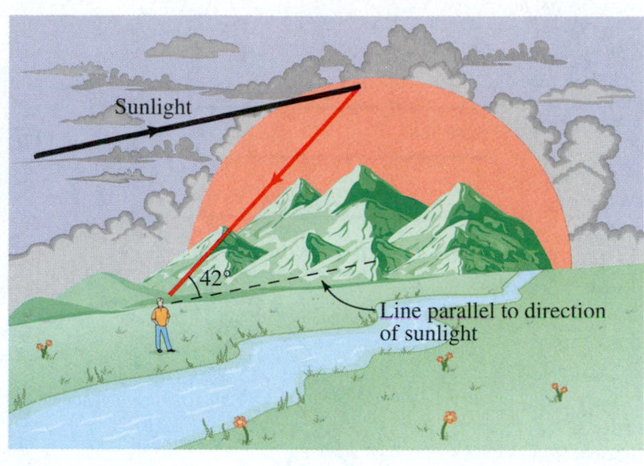

(b)

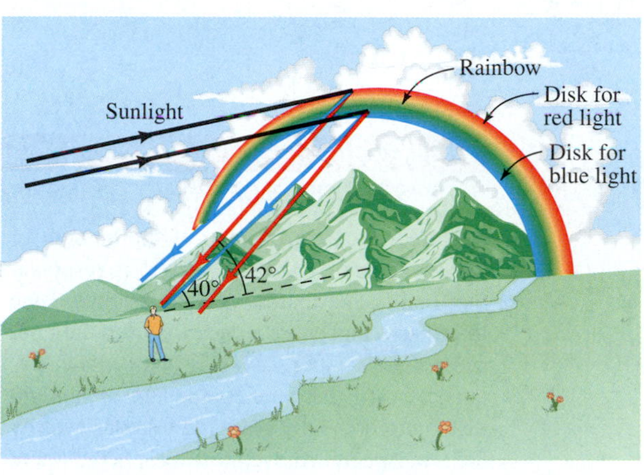

(c)

▲ **FIGURE 35–25** (a) When sunlight enters a raindrop and exits after one reflection, no light exits at an angle steeper than about 42° from the incident ray. (b) As a result, light comes back to an observer with the Sun behind him from all the raindrops that lie within a cone with an angle of about 42°, as seen by the observer. (c) Sunlight is a variety of colors. Due to dispersion, the disks that fit into the cones are of slightly different sizes for different colors, with red forming the largest cone and violet the smallest. This figure exaggerates the effect.

the intensity of the light in the disk is strongest at the edges, we see a red ring outside a blue ring (with other colors placed accordingly). All the disks overlap inside the rainbow, giving white light. A *secondary rainbow* can be produced when there are two internal reflections within the raindrops (Fig. 35–26). The order of the colored disks produced by the raindrops will now be reversed, with red light at the bottom and blue light at the top of the secondary rainbow. Figure 35–27a illustrates how an observer sees rainbows and how the pattern of dispersion leads to the color inversion of a secondary rainbow compared with a primary rainbow. The light is brightest inside (below) the primary rainbow and outside (above) the secondary rainbow because the disks of each of the colors overlap in these regions; it is relatively darker between the two rainbows (Fig. 35–27b).

Dispersion in the more general sense of phenomena that depend on frequency (or wavelength) is quite common; for example, the scattering of light by matter has such dependence. In 1872 Lord Rayleigh showed that the fraction of incident light scattered by air molecules varies as f^4 for light in the visible range. This explains the color of the setting Sun. As the Sun sets, the rays of light pass through more and more atmosphere to reach our eyes, and an increasing number of the high-frequency (low-wavelength) components are scattered away, leaving the lower frequency in the direct rays from the Sun. The Sun's color changes from white to yellow to orange and finally to red, as it moves closer to the horizon, as the higher frequencies are scattered away from the observer.

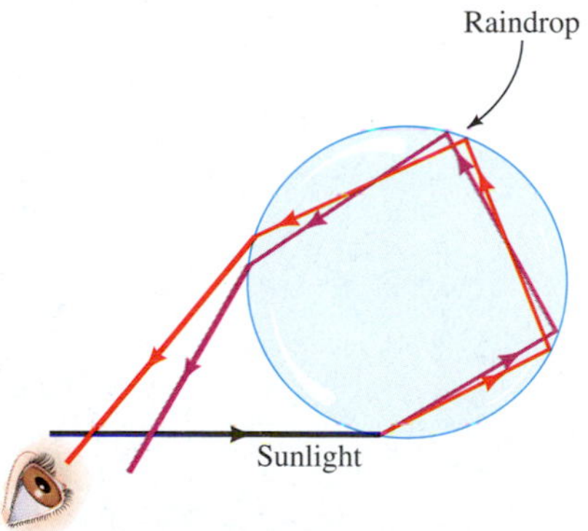

▲ **FIGURE 35–26** The light that reaches the eye from a secondary rainbow has undergone two internal reflections in a set of raindrops. Light of shorter wavelengths (violet light) emerges at a steeper angle than light of longer wavelengths (red light) does, in contrast to the light that undergoes only one internal reflection, which forms the primary rainbow, and the colors of the secondary rainbow are inverted.

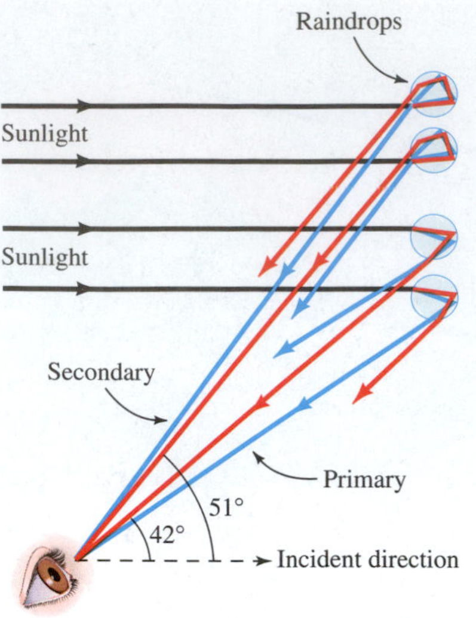

(a) (b)

▶ **FIGURE 35–27** (a) When the eye sees a region of sky that contains raindrops illuminated by sunlight, the light reflected from individual drops forms a primary rainbow (one internal reflection within each drop) and a secondary rainbow (two internal reflections). (b) The brighter, primary rainbow is on the bottom. The order of colors is reversed in the two rainbows due to the extra reflection, which produces the fainter secondary rainbow. The disks overlap, so it is relatively brighter below the primary rainbow and above the secondary rainbow, but darker between them.

THINK ABOUT THIS. . .

WHY IS THE SKY BLUE WHILE CLOUDS ARE WHITE?

The frequency dependence of the amount of light scattered by the atmosphere is responsible for the fact that the sky looks blue. Blue light has a higher frequency than red, and because there is more scattering in the higher frequencies, it is the blue component of sunlight that is preferentially scattered into our eyes by the atmosphere. Above the atmosphere, where there are no molecules to scatter the light, astronauts see a black sky.

If the sky is blue, why are clouds white? The answer to this question has to do with the size of the objects doing the scattering. The f^4 law (equivalently what we could call the $1/\lambda^4$ law) applies only to the scattering of light by objects much smaller than the wavelength of the light. Thus it applies to the scattering of light by individual molecules in the air. However, the water droplets that make up a cloud are larger than the wavelength of light impinging on them. For such objects, a set of molecules lying within an area whose linear size is λ will act *together* as a "single reflector" to reflect the light. The amplitude of the reflected wave is proportional to the effective area of the "single reflector," and this is proportional to λ^2. The intensity of the reflected light is in turn proportional to the square of the amplitude, that is, to λ^4. This just cancels the $1/\lambda^4$ dependence of the individual contributions of the molecules, and consequently there is no preferential wavelength in scattering by droplets. All wavelengths are scattered equally, so that white light scattered by a cloud stays white, and that is how we see the cloud. ■

The Atomic Theory of Dispersion

Dispersion occurs because of the atomic structure of dielectric media. The atoms that make up the medium contain electrons bound to their respective nuclei (in opposition to conductors, in which some of the electrons are in effect free to roam). The size of the bound system has linear dimensions of about 0.1 nm. Let's think of each atom as a single electron (of mass m) that oscillates about a positive ion as though the electron were bound to that ion by a spring. If the spring constant is k, then the angular frequency of oscillation (the natural frequency) is ω_0, given by $\omega_0^2 = k/m$. If no other forces act, the motion of the electron along the z-axis is of the form

$$z = A \cos \omega_0 t, \tag{35–15}$$

where A is the amplitude of motion. While this may not immediately remind you of an atom, you can see that it is relevant if you recall that an object in a circular orbit has a motion whose projection along an axis in the orbit plane is simple harmonic motion (see Chapter 13).

Now suppose that a plane electromagnetic wave oscillating with angular frequency ω is incident on an atom, with the electric field oriented in the z-direction. The electric force on an electron in the atom is then oscillatory with frequency ω. The situation is that of a driven harmonic oscillator. The motion of the electron is oscillatory with the driving frequency ω. The amplitude exhibits resonance, becoming large when $\omega_0 \cong \omega$. Thus, based on our knowledge of the driven harmonic oscillator, the motion is of the form

$$z \propto \frac{1}{\omega_0^2 - \omega^2} \cos \omega t. \tag{35–16}$$

We assumed here that ω_0 is not exactly at ω; in fact, for materials such as water and glass, ω_0 is on the order of 5 to 6 times larger than the characteristic angular frequencies of visible light. As you may recall from our work on the driven harmonic oscillator, there is no actual singularity when $\omega = \omega_0$, because there is an additional additive term in the denominator of Eq. (35–16) associated with damping. This term can be ignored unless ω_0 and ω are very close.

An accelerating charge (the electron) *radiates* electromagnetic energy, and the intensity of the radiation is proportional to the average of the acceleration squared. (While we are not going to show this in detail, it is implicit in our original discussion of electromagnetic waves in Chapter 34.) From Eq. (35–16), the average acceleration is proportional to

$$\frac{d^2z}{dt^2} \propto \frac{\omega^2}{\omega_0^2 - \omega^2} \cos \omega t \cong \frac{\omega^2}{\omega_0^2} \cos \omega t.$$

In the last step, we have used the fact that $\omega_0^2 \gg \omega^2$ for visible light in these materials. The intensity I of the radiation is proportional to the acceleration squared. Thus I varies as ω^4 or, equivalently, as f^4. (The average of $\cos^2 \omega t$ over many periods is just $1/2$.) The wavelength λ is related to ω by $\lambda = c/f = 2\pi c/\omega$, so *the intensity of the radiation emitted by a charge set in oscillation by an external electric field is proportional to* $1/\lambda^4$, *where* λ *is the wavelength of the oscillating field.* This fact, first obtained by Lord Rayleigh, explains the blue sky: blue light, which is a high-frequency component of visible light, is scattered preferentially from atmospheric molecules—they reradiate this light equally in all directions (see the "Think About This" box on p. 990).

This discussion shows how the original electric field is modified by the addition of a radiated field as a result of the electron's motion; in fact, it is such a modification that is described by a dielectric constant κ, and hence an index of refraction n given by Eq. (35–2), $\kappa = n^2$. We accordingly expect n to depend on the frequency; that is, to exhibit dispersion. A more advanced treatment of how the external field is modified in the presence of atomic electrons shows that

$$\frac{1}{n^2} = 1 - \frac{C}{\omega_0^2 - \omega^2}, \tag{35–17}$$

where C is a constant that is proportional to the density of atoms. This equation describes how the index of refraction varies with frequency. Because $\omega_0 \gg \omega$ for visible light, *the index of refraction increases as the frequency of the light increases*. This is certainly in accord with Table 35–2. In fact, atoms and molecules have many resonant frequencies, so a more accurate version of Eq. (35–17) must contain several terms of the form $C_k/(\omega_{0k}^2 - \omega^2)$ added together.

Summary

The wave theory of light allows us to understand almost all the classical properties of light, in particular straight-line propagation in open space, reflection, and refraction. Light waves propagate with speed $c = 3.00 \times 10^8$ m/s in a vacuum; in transparent media, the speed of propagation is c/n, where n is the index of refraction of the medium. In general, the index of refraction depends on the wavelength of the light.

The propagation of light can be described either in terms of wave fronts, which form an envelope of spherical wavelets built upon earlier wavelets (Huygens' principle), or in terms of rays, which are lines perpendicular to the wave fronts. Light rays travel in straight lines unless they meet boundaries. Upon reflection from a surface, the angle θ that the incident ray makes with the normal to the surface is equal to the angle θ' that the reflected ray makes with the surface (the law of reflection):

$$\theta' = \theta. \tag{35–5}$$

In the passage from a medium of index of refraction n_1 to a medium of index of refraction n_2, the incident angle θ_1 and the refracted angle θ_2 are related by Snell's law of refraction:

$$n_1 \sin\theta_1 = n_2 \sin\theta_2. \tag{35–6}$$

These results can be established using the geometry of wave fronts. They can also be derived with the help of Fermat's principle, which states that the path taken by a light ray between two points is the path that takes the shortest time. One consequence of Snell's law is that total internal reflection occurs when light moving in a medium with index of refraction n_1 strikes a boundary of a medium with index of refraction n_2, where $n_1 > n_2$. This holds true provided that the angle of incidence is larger than a critical angle θ_c, given by

$$\sin\theta_c = \frac{n_2}{n_1}. \tag{35–9}$$

The dependence of the index of refraction on wavelength is called dispersion. Dispersion causes the different wavelengths in a beam of white light to refract through different angles. The colors of the rainbow and the blue sky are naturally occurring dispersion phenomena. Dispersion can be understood in terms of the atomic theory of matter.

Understanding the Concepts

1. If light travels only in straight lines, how does a light burning in one room give light in another room?
2. When light reflects from a surface, there is a change in (a) frequency, (b) wavelength, (c) speed of light, (d) all of these, (e) none of these.
3. How difficult would it be to reflect light back to Earth from the Moon by using two perpendicular plane mirrors? Why does it help if there are three mutually perpendicular mirrors?
4. If fish could think, they might realize that the relative indices of refraction of water and air allow them to outwit fishermen. Why?
5. A person swimming underwater sees a lifeguard who is standing in the shallow part of the pool; the water comes up to the lifeguard's waist. In what way does the swimmer see the lifeguard's upper body distorted?
6. If you lie at the bottom of a pool and look up, it appears that you are lying at the bottom of a conical hole. Why is that? Estimate the angle that apparent walls make with the vertical.
7. A fisherman standing up to his waist in a lake appears, to an observer outside the lake, to have shorter-than-normal legs. How will a fish in a horizontal position near the bottom of the lake appear to the observer?
8. A plane wave of radiation has an electric field of the form $\vec{E}_0 \cos(kz - \omega t)$ when it propagates in empty space. How do k and ω change when the plane wave enters a medium with index of refraction n?
9. When a beam of light of frequency f passes from air into glass (whose index of refraction is 1.5), which of the following happens: (a) f increases by 1.5; (b) f decreases by 1.5; (c) f becomes zero; (d) f is unchanged; (e) f decreases very slightly.
10. As the Sun sets, its color changes from white to yellow to orange and finally to red. As the lowest part of the Sun sinks below the horizon, the Sun appears squashed, more egg-shaped than circular. Why?
11. A coin lies at the bottom of a pool of water. Starting from a point immediately above the coin, you observe the coin from the level of the surface. You then move your head horizontally away from the coin across the surface of the water. Is there a horizontal distance at which the coin is no longer visible?
12. Light from the sky refracts near the surface of hot sand, giving the impression that there is a bright surface that could be interpreted as water: a mirage (Fig. 35–28). The air near the surface of hot sand is hotter than the surrounding air. Does light travel faster or slower in hot air than in cold air?

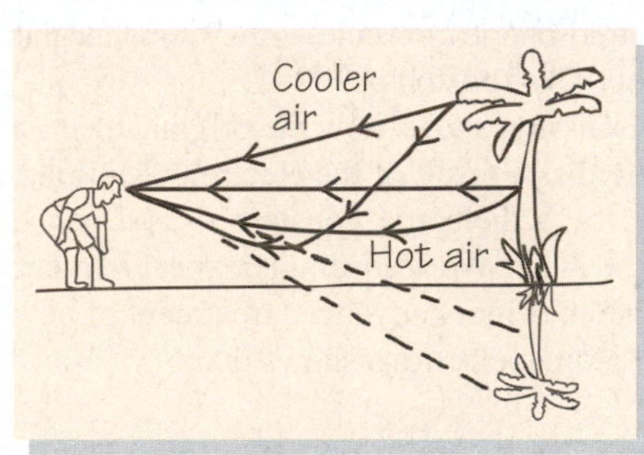

▲ **FIGURE 35–28** Question 12.

13. Mirages can occur when a layer of cold air lies closer to the surface. How would such an air layer affect the appearance of distant houses?
14. Consider Fizeau's experiment modified so that the light between the rotating wheel and the reflecting mirror goes through a pipe filled with water. Should the wheel be speeded up or slowed down in the repetition of the experiment described in the text? What would the result be if light traveled faster in water than in air?
15. Why does the sky look black rather than blue, as it does from Earth, to astronauts in orbit?
16. What is the index of refraction of a vacuum?
17. For a moment, you are lying in the middle of a circular swimming pool—at the bottom of the pool—which is filled to a depth of 1 m with water and is surrounded by trees. A 2-m-tall lifeguard is standing in the water about 3 m from you. What do things look like as you scan in all directions?

18. Laser light directed into the end of a glass rod comes out the other end with almost the original intensity. If another glass rod touches the side of the first rod, making a 30° angle with the lengthwise direction of the first rod, nothing happens. But if the point of contact is lubricated with glycerin, some of the original light beam is "stolen" by the second rod. Explain what happens.

19. White light is incident onto a pane of glass. Is there a dispersion of colors in the reflected light?

20. Stick a pin into the underside of a cylinder of cork, then float the cork in water. Even if you do not stick the pin in very far, you may not be able to see it from outside the water. Why not?

21. Why isn't the Moon red when it sets?

Problems

35–1 The Speed of Light

1. (I) What are the speeds of light in ice, ethyl alcohol, benzene, and diamond?

2. (I) The nearest star to our solar system (aside from the Sun) is Alpha Centauri, some 4.2 ly from Earth. How far is this in meters?

3. (I) A light wave of red light ($\lambda = 650$ nm) passes from air into water, where the index of refraction is 1.32. What are the wavelength and frequency of the light in water?

4. (I) Light of frequency 5.6×10^{14} Hz impinges on glass, $n = 1.45$. What are the wavelengths of this light in a vacuum and in glass? What is the index of refraction of a material within which the wavelength of yellow light is one-half its value in a vacuum?

5. (II) Suppose that you have a version of Fizeau's apparatus in which the round-trip distance for the light beam is $2D = 1{,}000$ m. The width of the opening between the teeth on the cogged wheel is 0.70 mm, and the center-to-center distance between these gaps is 1.5 mm. The wheel has a radius of 15.0 cm. What would the minimum rotational speed be, in revolutions per minute, so that light entering through the center of one gap would come out through the center of the next gap? Is such an apparatus realizable?

6. (II) Figure 35–29 shows an exaggerated view of the eclipsing of Io, the innermost moon of Jupiter, as seen from two different points on Earth's orbit around the Sun. If Earth were stationary at a point nearest Jupiter, N, a particular eclipse would begin at a precise time. When Earth is at point F, the eclipse starts somewhat later than expected because the light has to travel the additional distance of a diameter of the Earth–Sun orbit. The mean distance from Earth to the Sun is 1.50×10^{11} m. How much later will the eclipse be seen at point F compared with point N?

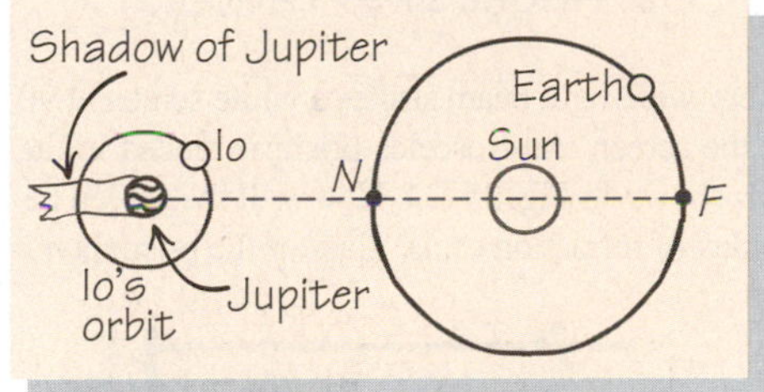

▲ **FIGURE 35–29** Problem 6.

7. (II) Personal computers can perform as many as 5×10^7 steps every second. This means that some leads connecting different parts of the computer may carry this many pulses per second. If you assume that the pulses travel at the speed of light, what is the distance between pulses? Does this result have implications for the design of these machines?

8. (II) Telephone connections between Europe and North America can be carried by cable or by the use of a geosynchronous communication satellite. Estimate the time it takes for a signal to travel 10,000 km via cable, assuming the speed is close to the speed of light. How does this compare to the time required for the same signal to travel via satellite, 40,600 km from the center of Earth?

9. (II) The speed of light in a vacuum is defined to be 299,792,458 m/s. A lunar-ranging experiment measures the time for a light pulse to reach the Moon and reflect back to Earth. Such experiments allow us to determine the distance between the Moon and Earth, which is approximately 3.84×10^8 m, to an accuracy of 15 cm. What is the smallest time interval that can be measured by the clock used to determine the time it takes for light to go to the lunar reflector and back?

10. (II) Galileo attempted to measure the speed of light with the help of lights and a clock on two adjacent mountains. In essence, a shutter over a light was opened on the first mountain, an observer on the second mountain saw that signal and returned a second signal, and the experimenter on the first mountain looked for a delay between the time the shutter was opened and the time the signal was returned. Use your knowledge of human reaction time to estimate the time measured by the first experimenter for the total round trip. How long would it actually take for light to travel back and forth between two mountaintops separated by 4 km? Your answers explain why Galileo's attempt did not work.

11. (II) Imagine experiment similar to Fizeau's, with a cogged wheel of diameter 20 cm. A laser beam shines through one opening, travels 1,500 m, and is reflected back. Given that the fastest rotation rate of the wheel is 1.2×10^5 rev/min, what should be the separation between adjacent cogs on the rim of the wheel?

35–3 Reflection and Refraction

12. (I) A fixed projector emits a narrow beam of light onto a plane mirror. At what angle with respect to the beam should you place the mirror in order to turn the beam by 75°?

13. (I) The critical angle for a particular material (used in air) is observed to be 38°. What is the material's index of refraction?

14. (I) A horizontal beam of light is reflected from a plane mirror that revolves about a vertical axis at a rate of 30 rev/min. The reflected beam sweeps across a screen that, at the point nearest the mirror, is 20 m away (Fig. 35–30). With what speed does the spot of light move across the screen at the point nearest the mirror?

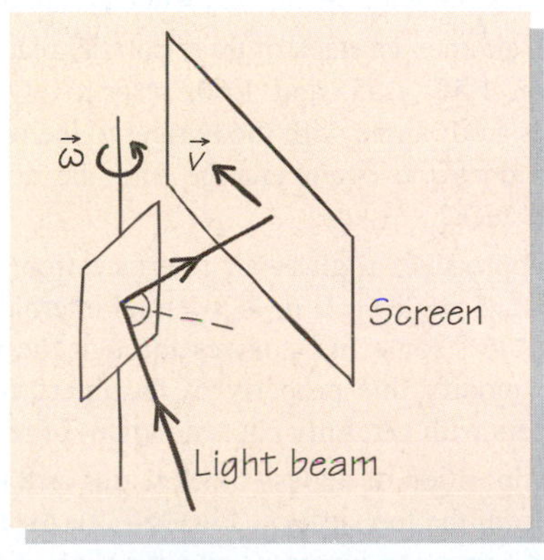

▲ **FIGURE 35–30** Problem 14.

15. (I) An intense light beam is incident at 45° to the surface of a clear lake. If the lake is 500 m deep and has a flat bottom, how far does the light beam travel before it hits bottom?

16. (I) A beam of light is sent from medium 1, index of refraction n_1, into a medium 2, index of refraction n_2; here, $n_2 > n_1$. We know that a refracted ray is bent toward the perpendicular to the boundary. Are there any incident angles for which the angle of the refracted ray is 90°? If not, what is the largest possible angle of refraction? Give a numerical value to your answer in the case of water-to-glass, $n_1 = 1.33$ and $n_2 = 1.50$.

17. (I) A burglar stands in front of a department store window and directs his flashlight into the store. What fraction of the light is reflected at the window's surface, assuming that the index of refraction of the glass is 1.43? Ignore all reflections except for that at the outside interface between the glass and air.

18. (I) A swimmer is at the bottom of a large, shallow swimming pool. Through what angle must she move her eyes so that her direct gaze swings across the whole sky? Water's index of refraction is 1.33.

19. (I) The index of refraction of air is $1 + (2.93 \times 10^{-4})$. Assume that the atmosphere may be treated as a uniform medium of thickness 8.3 km, which covers Earth's surface; further, suppose a ray of light hits the top of the atmosphere parallel to the top of the atmosphere—grazing incidence. What is the angle that the refracted ray makes with the *horizontal*?

20. (II) What is the critical angle for total internal reflection in a glass (used in air) for which $n = 1.46$? Is it possible to use a 45° – 45° – 90° triangular prism of crown glass (see Table 35–1) to make a perfect reflector of light (Fig. 35–31)?

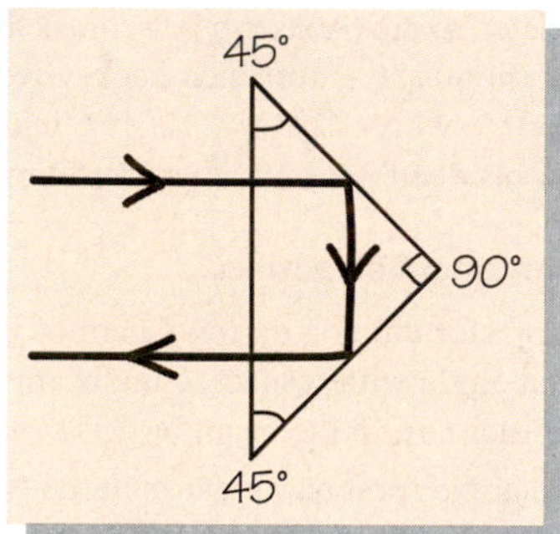

▲ **FIGURE 35–31** Problem 20.

21. (II) A thick glass plate ($n = 1.53$) lies on the bottom of a tank of water ($n = 1.33$). A light ray enters the water from air, making an angle of 72° with the normal to the surface. What angle does the ray make with the normal when the ray is in the water? What angle does it make with the normal when it is in the glass?

22. (II) Light in air enters a stack of three parallel plates with indices of refraction 1.50, 1.55, and 1.60, respectively. The incident beam makes a 60° angle with the normal to the plate surface. At what angle does the beam emerge into the air after passing through the stack?

23. (II) Light approaches a glass–air interface from the glass side ($n = 1.6$) at an angle θ_i. If $\theta_i > \theta_c$, total internal reflection occurs; if $\theta_i < \theta_c$, some light passes through the interface. Is it possible to modify this property of the interface by adding a stack of layers with carefully chosen indices of refraction?

24. (II) The composition of a glass block varies as a function of the distance x from the top surface (Fig. 35–32). As a consequence, the index of refraction increases as a function of x according to $n(x) = 1.54 - (0.18\ \text{cm}^2)/(x + 1\ \text{cm})^2$, with x in centimeters. A beam of light strikes the surface at an angle of incidence of 55° from the vertical. What will be the direction of the beam deep inside the block?

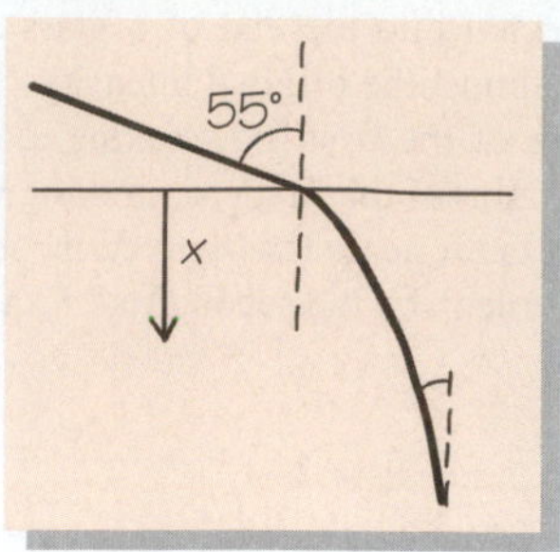

▲ **FIGURE 35–32** Problem 24.

25. (II) A glass sphere ($n = 1.6$) is centered at the origin of a coordinate system, with its equatorial plane defining the xy-plane. A beam of light enters the glass sphere at a latitude of 40°, parallel to the x-axis in the xz-plane. Make a careful drawing to determine the angle at which the beam will strike the back of the sphere. Will there be total internal reflection?

26. (II) White light is refracted by the triangular prism shown in Fig. 35–33. A beam of light enters the prism along a path parallel to the prism base. The light is observed on a screen that is located 10 m from the prism and is perpendicular to the emerging rays. How far apart on the screen are the spots of blue light ($n = 1.528$) and red light ($n = 1.514$)?

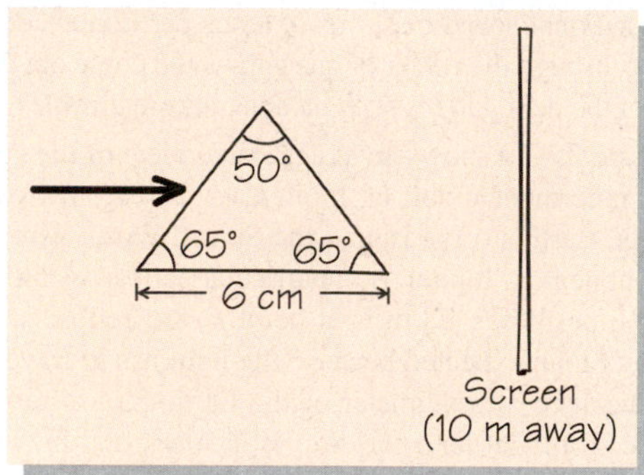

▲ **FIGURE 35–33** Problem 26.

27. (II) A very wide light beam strikes a white screen at 90° to the surface of the screen. An isosceles prism is placed in the way of the beam, as shown in Fig. 35–34. How will the screen be illuminated if the index of refraction of the glass of the prism is $n = 1.5$?

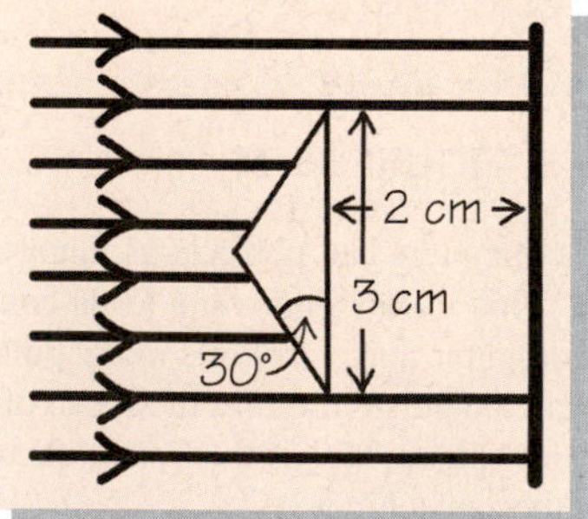

▲ **FIGURE 35–34** Problem 27.

28. (II) A lifeguard whose eyes are 1.78 m from his feet stands in water 90 cm deep. From a vantage point at the bottom of the pool, a swimmer sees the lifeguard's head to be along a line at a 46° angle to the vertical. How far is the swimmer's eye from the lifeguard's feet? (For water, $n = 1.33$.)

29. (II) Suppose that you look at an aquarium with your eyes at the level of the water surface (Fig. 35–35). A duck swims on the surface of the water. When you look at the duck from the front, everything seems normal. However, when you look at the duck at an angle to the glass surface, the duck seems to be split in half, with the feet paddling ahead of the upper body. Explain this phenomenon. Suppose that both the duck and your eyes are at a distance of 1 m from the glass, and the line connecting them forms a 30° angle with the glass. Calculate the difference between the directions of the line of sight of the upper and lower halves of the duck.

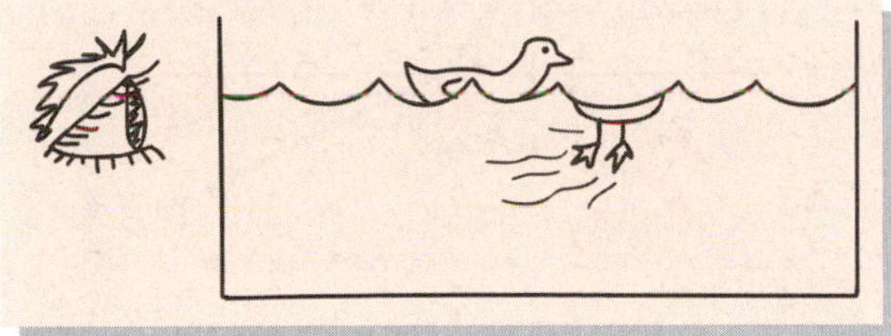

▲ **FIGURE 35–35** Problem 29.

30. (II) A transparent tank of water, of index of refraction n_1 and thickness t_1, is placed on top of a slab of glass of index of refraction n_2 and thickness t_2. A laser beam strikes the upper surface of the tank at an angle θ. At what horizontal distance from the point of entry will the beam emerge from the bottom surface of the glass? What is the answer if the tank and the slab of glass are interchanged?

31. (II) A beam of light from a flashlight is reflected by a mirror which is placed under a sheet of glass (index of refraction 1.6) that is 5 cm thick. The beam makes an angle of 60° with the vertical. If the light source is 15 cm above the surface of the glass, how far below the surface of the glass will you see a source of light matching the original? You are locating the *image* of the light source. [*Hint*: In this exercise in trigonometry, use $\tan\theta = \sin\theta/\cos\theta$; $\cos\theta = \sqrt{1-\sin^2\theta}$.]

32. (II) A narrow beam of light is incident at a 30° angle from the normal onto a glass pane 6 mm thick. Describe the position of the exit beam of light. What is its direction? Is it displaced from the incident beam? If so, by how much? (For the glass, $n = 1.60$.)

33. (II) Light is incident on an equilateral triangular prism ($n = 1.55$) at a 35° angle from the normal to one of the faces (Fig. 35–36). What is the exit angle?

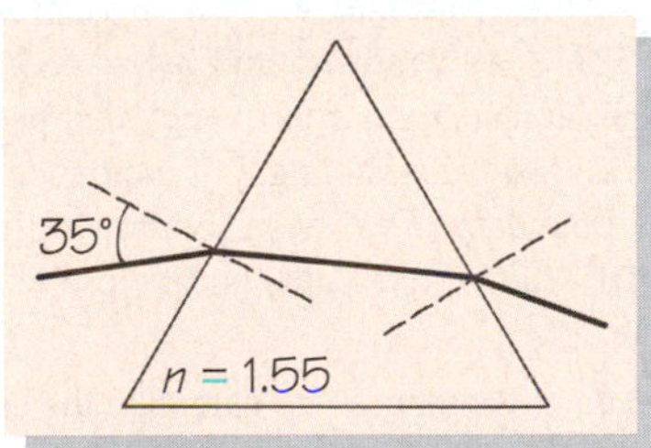

▲ **FIGURE 35–36** Problem 33.

34. (II) Consider a solid glass rod of length 75 cm and diameter 1.5 cm, with index of refraction 1.46. The ends of the rod are perpendicular to the lengthwise direction. (a) Light enters the center of the end of the rod from air. What is the maximum angle of incidence for which the light is totally reflected inside the rod? (b) Repeat part (a) for a similar rod totally immersed in water ($n = 1.33$).

35. (II) A ray of light impinges at a 60° angle of incidence on a glass pane of thickness 5 mm and index of refraction 1.54. The light is reflected by a mirror that touches the back of the pane (Fig. 35–37). By how much is the beam displaced compared with the return path it would have if the pane were absent?

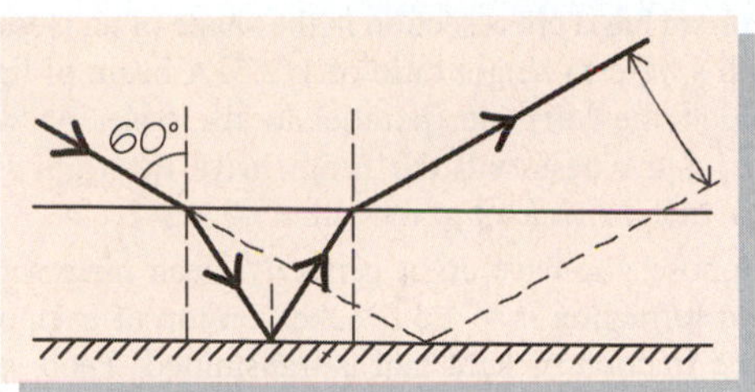

▲ **FIGURE 35–37** Problem 35.

36. (II) When a light beam is reflected by a conventional mirror, part of the light is reflected by the front surface of the glass pane, and part by the silvered back surface. What is the distance between the two reflected beams if the mirror is 2.0 mm thick, if it is made of glass with $n = 1.45$, and if the angle of incidence is 70°? (To avoid this double reflection, many optical instruments use mirrors with their front surfaces silvered.)

37. (II) At noon, a 2.0-m-long vertical stick casts a shadow 1.0 m long. If the same stick is placed in a flat-bottomed pool of water half the height of the stick (still at noon), how long is the shadow on the floor of the pool? (For water, $n = 1.33$.)

38. (II) You have three transparent liquids labeled 1, 2, and 3 that do not mix. When light is sent from liquid i to liquid j, there is an angle of incidence θ_i and an angle of refraction θ_j. Two separate experiments show the following: $1 \rightarrow 2$, $\theta_i = 22°$ and $\theta_j = 29°$; $2 \rightarrow 3$, $\theta_i = 35°$ and $\theta_j = 53°$. Find the ratios of the indices of refraction for each pair of liquids.

39. (II) Consider light that is perpendicularly incident on a triangular prism of the kind shown in Fig. 35–38. The index of refraction of the prism material is $n_1 = 1.814$. Suppose that the two reflecting sides are coated with a thin, uniform layer of a dielectric with index of refraction $n_2 = 1.380$. Will the glass–dielectric interface be totally reflecting? How large can n_2 be so that the interface is still totally reflecting?

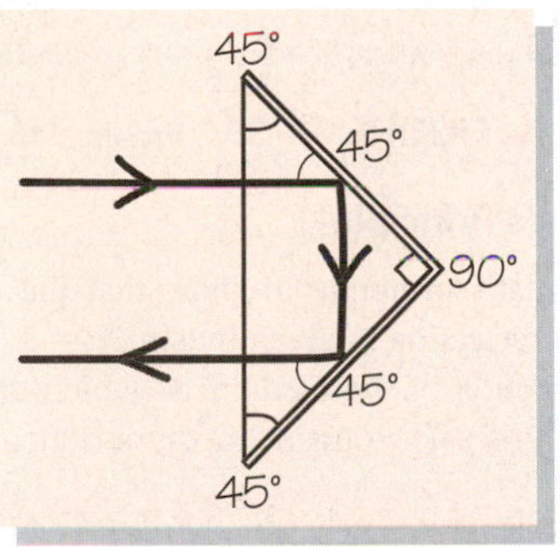

▲ **FIGURE 35–38** Problem 39.

40. (II) A sodium lamp emitting light with $\lambda = 589$ nm is placed at the bottom of a pool 4 m deep. (a) Seen from the edge of the pool, where will the light appear? (b) The lamp is taken out of the pool and the light now shines into the pool. What will be the frequency, wavelength, and speed of the light as it appears to a swimmer under the surface of the pool? ($n_{water} = 1.33$)

41. (II) A prism made of glass with index of refraction n and whose cross section is an equilateral triangle deflects light of wavelength 550 nm. As the beam is moved to make different angles with respect to the prism, the minimum angle through which the beam is deflected is 35°. What is the value of n?

42. (II) A prism has a cross section in the shape of an isosceles triangle with a base-to-height ratio of $1/2.5$. A beam of light is incident upon the left side, parallel to the base. At what angle relative to the base will the beam leave the right side of the prism, which is made of glass with $n = 1.58$?

43. (II) Suppose you have eight perfectly clean microscope slides, index of refraction $n = 1.5$, stacked on top of each other. Estimate the fraction of light that is transmitted. There are tiny air gaps between the slides; ignore multiple reflections.

44. (II) Zeno and his friends are discussing light incident perpendicularly on a glass pane in air. The pane has two surfaces at which reflection can take place—we'll call them A and B, with A being the first surface encountered by the light. The discussion centers around the fact that some light reflects back from A directly, but more light comes back to the A side of the pane after having undergone reflection at surface B, then from double reflection at surface B, having bounced once from A, and so on. They reason that since each bounce contributes something to the A side, an infinite amount will end up there! But of course this is just Zeno's paradox, and you know that this is really a matter of a limit. What is the total intensity of the light reflected back to the A side, including an arbitrary number of bounces at the boundaries? Assume an incident intensity I_0. Take the index of refraction of air to be 1, that of the glass n, and assume that there is no absorption within the glass.

45. (III) Use Huygens' construction to prove Snell's law by working out the geometrical details in Fig. 35–39.

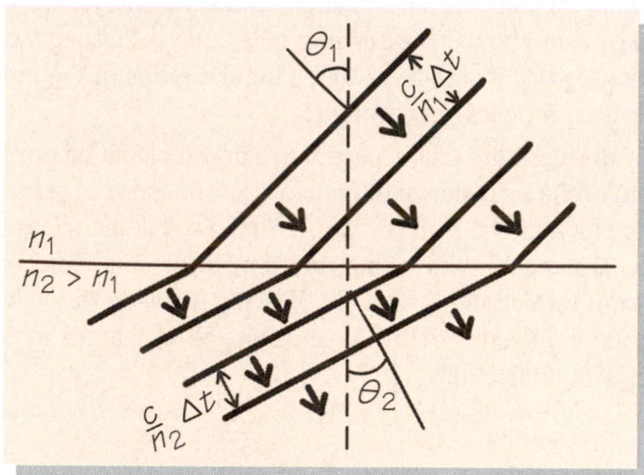

▲ **FIGURE 35–39** Problem 45.

*35–4 Fermat's Principle

46. (II) Use Fermat's principle to show that the critical angle for total internal reflection is given by $\sin\theta_c = 1/n$, where n is the index of refraction of the medium in which the light ray originates (Fig. 35–40). The outside medium is air.

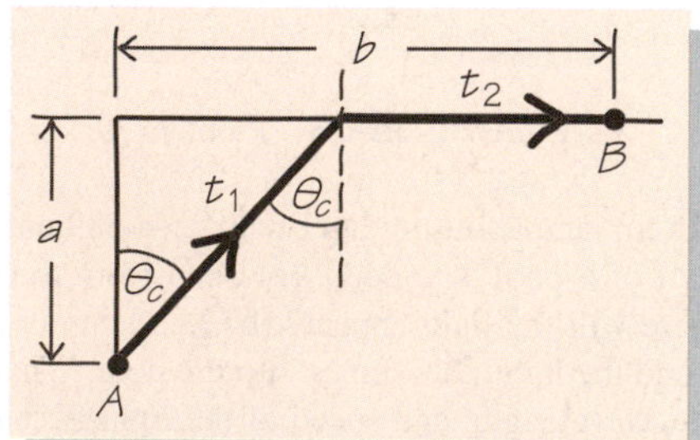

▲ **FIGURE 35–40** Problem 46.

47. (II) Show that the law of reflection follows from Fermat's principle.

48. (II) By using Fermat's principle, show that if two media have exactly the same index of refraction, then a beam of light travels in a straight line when it crosses the boundary between them.

49. (II) Use Fermat's principle to show that a beam of light that enters a plate of glass of uniform thickness emerges parallel to its initial direction (Fig. 35–41).

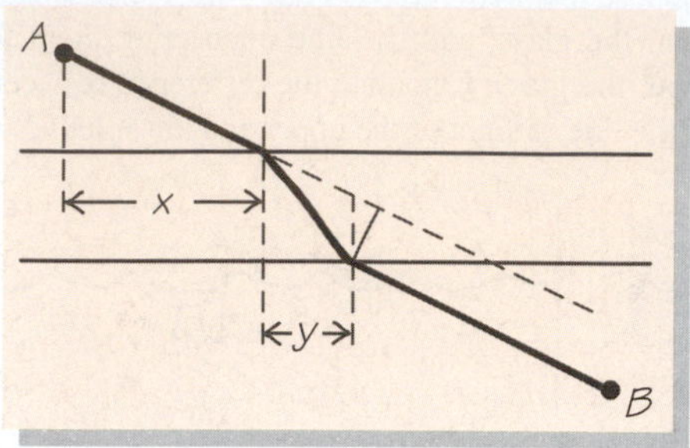

▲ **FIGURE 35–41** Problem 49.

50. (III) Calculate the parallel displacement of a beam of light that strikes a vertical slab of glass, with index of refraction n and thickness D, at an angle ϕ with the horizontal (use Fermat's principle). (In Problem 49, we showed that a ray of light that passes through a slab of glass emerges parallel to its initial direction. The ray is, however, displaced from its original line, and that displacement is what we want here.)

35–5 Dispersion

51. (I) At 0°C, the index of refraction in water of light of wavelength 397 nm (violet) is 1.3444, whereas it is 1.3319 for a wavelength of 656 nm (red). What is the difference in angles of refraction for rays refracting from water near the freezing point into air for these two wavelengths? The angle of incidence is exactly 30° in each case. Take $n = 1$ for air and ignore dispersion in this medium.

52. (I) By what percent does the speed of red light in a type of glass ($\lambda = 656$ nm, $n = 1.522$) exceed that of blue light in the same glass ($\lambda = 486$ nm, $n = 1.545$)?

53. (II) A beam of white light, whose frequencies are mixed with equal intensity, passes within a piece of glass and impinges on a boundary to the air at an angle of incidence θ. The index of refraction of the glass increases with increasing angular frequency according to the formula $n^2 = 1 + [C/(\omega_0^2 - \omega^2 - C)]$, where $C = 529 \times 10^{30}\ \text{rad}^2/\text{s}^2$ and $\omega_0^2 = 685 \times 10^{30}\ \text{rad}^2/\text{s}^2$. (a) What is the largest angular frequency that passes through the glass into the air? (b) At what angle of incidence should the light approach the boundary if we wish to allow only frequencies of $\omega = 3.2 \times 10^{15}$ rad/s (red light) and below to pass through to the air?

54. (II) Use the data in Problem 53 to calculate the critical angles for total internal reflection for five values of wavelengths in the range 430 nm to 770 nm. Plot your results.

55. (III) We wish to select a glass to construct a prism that can separate the yellow ($\lambda = 590$ nm) component of light from the blue-green ($\lambda = 490$ nm) component. The prism is to be a bar with the cross section of an equilateral triangle. If a ray of white light arrives parallel to the base of the prism, it must leave the prism with the two colors separated by at least 2°. What must the difference in indices of refraction be for the two colors? [*Hint*: Because the difference of angles is small, so is the difference of indices of refraction. Keep only leading terms in differences of angle and of index of refraction.]

General Problems

56. (I) Light of wavelength 450 nm enters a piece of glass with index of refraction 1.50. What are the wavelength, frequency, and speed of that light in the glass?

57. (II) A pin is partly inserted perpendicularly into the flat surface of a cork with a 1.5-cm radius (Fig. 35–42). The cork, with the pin on the underside, is set afloat in a pool. A length of 1.2 cm of cork is under the water surface. Because of the effects of refraction, much of the pin is hidden from view from above the surface. What length of pin can be hidden in this way?

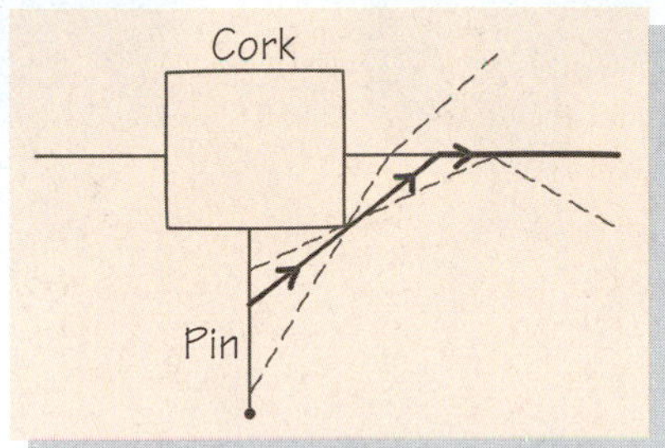

▲ **FIGURE 35–42** Problem 57.

58. (II) A beam of light is incident at an angle of 30° to the vertical on a horizontal glass plate of thickness 2.0 cm. The index of refraction of the glass is $n = 1.52$. The beam emerges on the other side. What is the perpendicular distance between the straight-line extrapolation of the incident ray and the ray refracted by the glass plate?

59. (III) Show that if an incident ray of white light that is parallel to the base of a prism in the shape of an isosceles triangle (apex angle 2ϕ) is separated into two components that exit the prism with an angular separation $\Delta\theta \ll 1$, then the difference in the indices of refraction for the two colors is proportional to $\Delta\theta$. Find the equation that expresses the relation between the differences in the indices of refraction and in $\Delta\theta$. [*Hint*: Consider the angle of emergence for a given n, and then find Δn as a function of $\Delta\theta$.]

60. (III) Sound can refract like light. Suppose that a submarine lies flat 180 m below the water surface, and that there are three thermal layers of water (each 60 m deep) of different temperatures (Fig. 35–43). The speed of sound in water depends on temperature. In the bottom layer, the speed is 1.16 times that in the top layer; in the middle layer, the speed is 1.05 times that in the top layer. A detection device at surface level determines that sound from the submarine arrives at the surface at a 36° angle with the horizontal. What is the horizontal distance between the detector and the submarine?

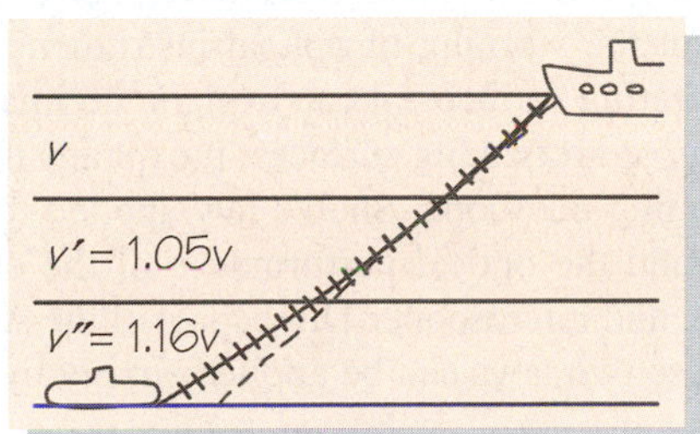

▲ **FIGURE 35–43** Problem 60.

61. (III) A ray of light is incident at an angle of incidence θ_i on one surface of a prism whose cross section is an isosceles triangle (apex angle 2ϕ). The light exits the prism at a total deflection angle θ (Fig. 35–44). The prism has index of refraction n and is in a vacuum, which has an index of refraction of exactly 1. For what angle θ_i is the angle of deflection θ a minimum?

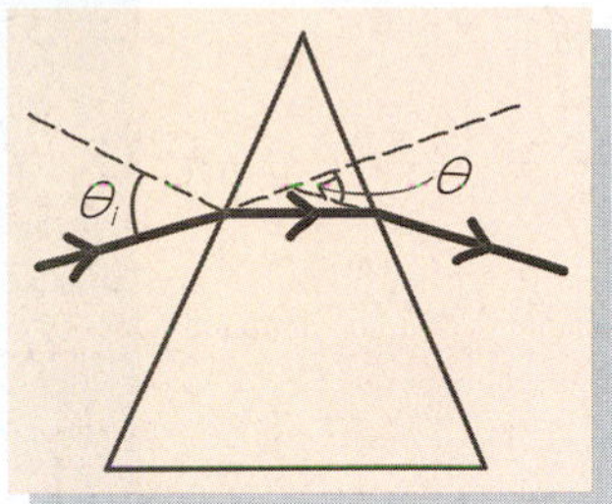

▲ **FIGURE 35–44** Problem 61.

62. (III) A ray of light incident from air onto a glass pane is partly reflected and partly refracted at the two surfaces of the pane (Fig. 35–45). The glass has an index of refraction n and a thickness d. Express in terms of n, d, and θ_i the displacement d' of the ray drawn, which enters the glass, reflects off the back surface, and exits.

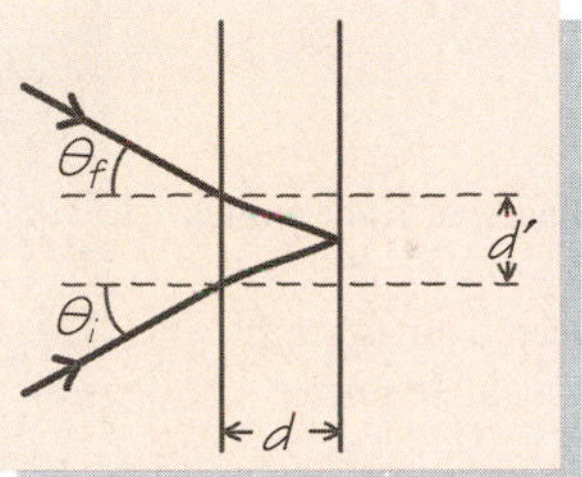

▲ **FIGURE 35–45** Problem 62.

63. (III) The first successful measurement of the speed of light, made by Ole Roemer in 1675, was based on the following method. The mean orbital period of Io, a moon of Jupiter, is 42.5 h; however, that period is measured to be about 15 s less than this value when Earth in its orbit is approaching Jupiter, and about 15 s more when Earth is receding from Jupiter. (a) Given that Earth's orbital speed around the Sun is about 30 km/s, and that Earth is on a part of its orbit when it is moving toward Jupiter, how much closer will Earth have moved toward Jupiter during one orbit of Io? (b) Use the information given to estimate the speed of light.

Chapter 36

Two mirrors placed at right angles gives us three images of a candle. Are all three images reversed left-to-right, as a single image in a single mirror would be?

Mirrors and Lenses and Their Uses

Instruments that can explore previously inaccessible domains often open new doors to understanding nature. For example, astronomy owes its progress to the invention of the telescope, and modern biology could not have been created without the microscope. In this chapter, we shall discuss the ideas that govern the construction of optical instruments such as these. The law of reflection and Snell's law, both introduced in Chapter 35, provide the foundation for the working of optical instruments. When the law of reflection is applied to flat reflecting surfaces, it can explain the images that we see in mirrors, and when applied to curved reflecting surfaces, it explains the functioning of rearview mirrors and the reflecting telescope. Snell's law applied to curved refracting surfaces can help us understand the optical performance of the eye, camera, magnifying glass, refracting telescope, and microscope. This aspect of the study of light is called geometric optics because these two laws can be applied simply by tracing the geometrical paths of light rays.

36–1 Images and Mirrors

The simplest reflecting surface is a flat (or plane) mirror. When you look into a mirror, you see an image of yourself. What is an image, and how is it formed? Let's begin with rays going directly from a point source *S*—think of a small lightbulb—to a person's eye

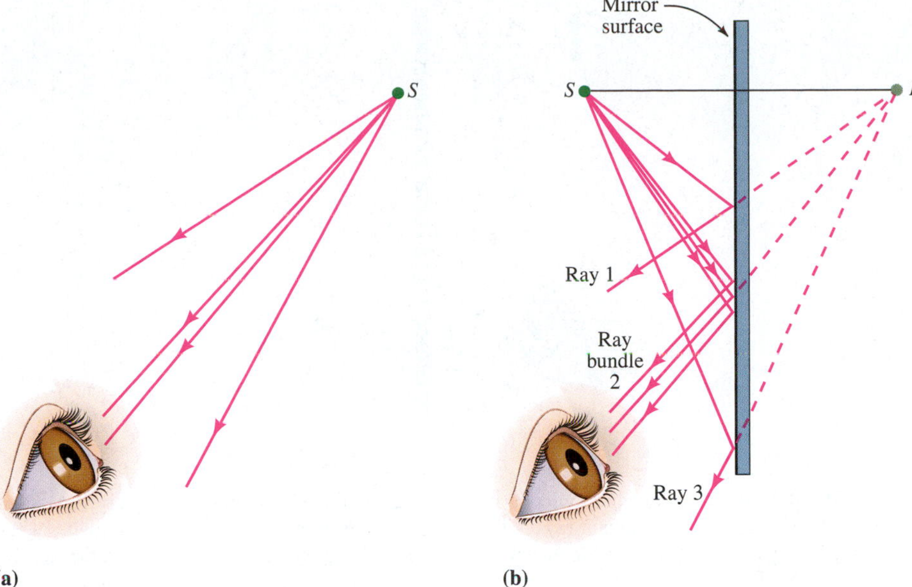

▲ **FIGURE 36–1** Rays leaving source point *S* (a) go to an eye and (b) reflect from a plane mirror before going to the eye. A bundle of such rays enters the eye, apparently from point *I*.

(Fig. 36–1a), together with the light from the source reflecting from a plane mirror according to the law that the angles of incidence and reflection are equal (Fig. 36–1b). The rays we draw follow the direction of motion of the wave front that emanates from the light source, and we could, in fact, draw an infinite number of such rays as close to one another as we like. As we described in Chapter 35, rays that are near one another form *bundles*, visible in Fig. 36–1b.

When we look toward the mirror, we see an *image* of the light source. What exactly do we mean by this? The simple geometry in Fig. 36–1b allows us to see that *all the reflected rays from S trace back to the same point I*. To see this, look at rays 1 and 3 in Fig. 36–2, which shows in more detail the situation in Fig. 36–1. We have indicated the equal angles of incidence and reflection θ_1 and θ_3 for these rays, respectively, as well as the angles α_1 and α_3. The angle formed by BP_1I is then equal to α_1. If point *B* is formed by dropping a perpendicular line to the mirror from point *S* and if point *I* lies along the continuation of this line, triangles BIP_1 and BSP_1 are similar triangles. By the same method, so are triangles BIP_3 and BSP_3. Because both rays 1 and 3 emanate from the same point *S*, the distance *BS* forms the base of both triangles to the left of the mirror (the *object side*), and the distance *BI* forms the base of both triangles to the right of the mirror (the *image side*). The (imaginary) continuations of rays 1 and 3 to the image side meet at point *I*, as would the continuation of *any* reflected ray.

We have calculated the location of point *I*. Because BIP_1 and BSP_1 are similar triangles, the distances *BS* and *BI* are equal. How does the eye/brain "know" where to put *I*? Two eyes (or one eye that moves a little) sense a bundle of rays rather than a single ray. The eye/brain can measure their degree of divergence and is capable of extrapolating this diverging bundle back to point *I*.

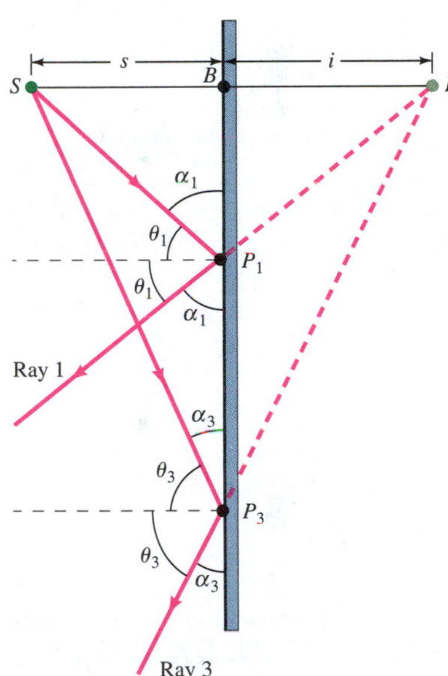

▲ **FIGURE 36–2** All the reflected rays from point *S* trace back to point *I*. The geometry implies that the perpendicular distance *s* from the mirror to point *S* equals the distance *i* from the mirror to point *I*.

The Image of an Extended Object

Suppose now that our light source is extended rather than being a point source—think this time of a candle flame or, perhaps better yet, think of an object that is reflecting light and acting as a source, such as a person. Two different points on the source form two different image points in the mirror. Moreover, the second image point is as close to the first image point as the second source point is to the first source point. Indeed, a set of nearby source points forms nearby image points (Fig. 36–3). The entire **object**, or *source* (we use the terms interchangeably), forms a set of matching image points, which together constitute an **image**. *An image is a set of contiguous points to which reflected rays lead when the rays are extrapolated back in straight lines.* Figure 36–3 illustrates

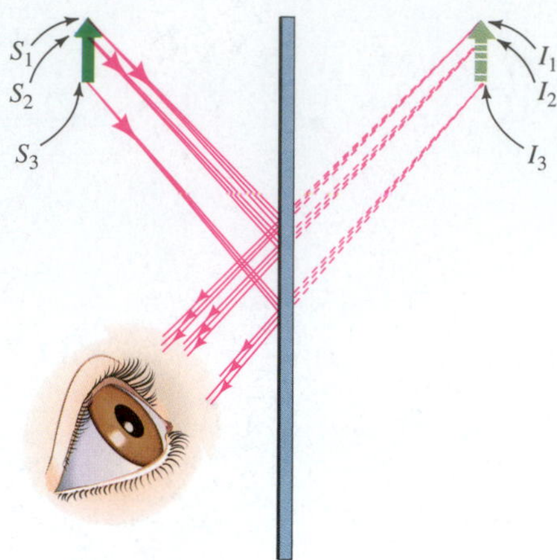

▲ **FIGURE 36–3** When the source is an extended one, there is an image point for every source point. This means that a bundle of rays will enter your eye from every image point no matter what your position before the mirror. Geometry can be used to locate the image $I_1I_2I_3$ of source $S_1S_2S_3$.

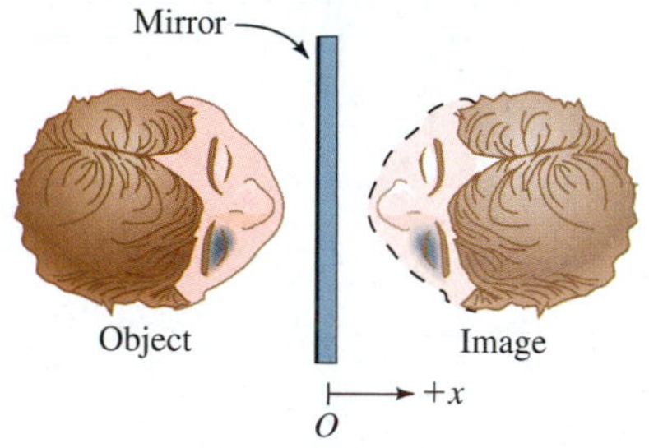

▲ **FIGURE 36–4** An image is reversed front to back. This means that if your right eye is black, your image has a black left eye.

that the source and the image formed by a plane mirror have the same size. The idea of the size of the image will be of some importance throughout this chapter. We can also see that the image is as far behind the mirror as the object is in front.

No light rays actually emanate from the image formed by a plane mirror; thus, we call it a **virtual** image. Light rays actually do pass through a **real** image on their way to your eye, and we shall see how this can occur when we discuss mirrors with curved surfaces. We could summarize the difference between real and virtual images by saying that if you were to place a screen at the location of a virtual image, no light would strike the screen, whereas a screen placed at the location of a real image would reveal the presence of the image.

There is one striking peculiarity of plane mirrors. If your right eye is blackened, your image, viewed as if you were meeting yourself on the street, has a left eye that is blackened. From Fig. 36–4, however, we can see that the *actual* reversal is a front-to-back reversal (the nose of the object points in the $+x$ direction in the figure, whereas the nose of the image points in the $-x$ direction), and this is what lies behind the left-to-right reversal of the image.

CONCEPTUAL EXAMPLE 36–1 How are the multiple reflections shown in Fig. 36–5 formed?

Answer The image of some source that is made by a mirror comes from the reflections of the rays emitted by the source. These reflected rays form a set of diverging rays, which result in the image. But the reflected rays can approach and reflect from a second mirror, and we have no way to tell whether the diverging set of rays incident on the second mirror come from an actual source or from the reflection from the first mirror of the source. In other words, *an image can act as a source for a second image*. It makes no difference whether an image is virtual or real—the only thing that matters is that the rays are diverging as they approach and then reflect from the second mirror. In the case of Fig. 36–5, the successive images must be due to a second mirror in front of the chess pieces, not visible in the photo. You might note that while the image produced by one reflecting surface is reversed left-to-right, the image of the original source produced by successive reflection from two (or more generally an even number of) mirrors is not.

▲ **FIGURE 36–5** Multiple reflections can be obtained with two plane mirrors. Where is the second mirror in this case?

EXAMPLE 36–2 A horizontal ray of light is incident at angle θ on a vertically suspended plane mirror. If the mirror is rotated about a vertical axis through an angle α, by what angle ϕ is the reflected ray rotated?

Setting It Up Figure 36–6 is an overhead view of the situation, including the initial and final normals to the mirror, N and N', respectively, as well as the reflected ray, labeled I.

Strategy A rotation of the mirror through an angle α means that the new angle of incidence is $\theta + \alpha$, which is also the new angle of reflection from N'. It is then a matter of geometry to find the required angle ϕ.

Working It Out From the figure, the angle between the incident ray and the new reflected ray I' increases from 2θ to $2(\theta + \alpha)$. Because the incident ray has not moved, the reflected ray is rotated by $\phi = 2\alpha$.

What Do You Think? Suppose you want to measure by how much an object has rotated and that the rotation angle is very small. How might you go about it? *Answers to* ***What Do You Think?*** *questions are given in the back of the book.*

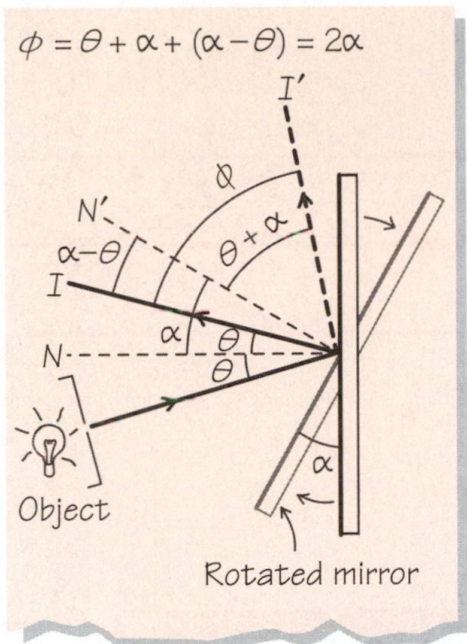

▲ **FIGURE 36–6** Overhead view of a horizontal ray of light incident on a mirror that rotates about a vertical axis.

36–2 Spherical Mirrors

Figure 36–3 shows that plane mirrors produce images that are the same size as the object. We can construct mirrors that produce images of altered sizes by using curved surfaces. In this section, we shall study mirrors whose surfaces form a segment of a sphere. We look at both *concave* (Fig. 36–7a) and *convex* (Fig. 36–7b) mirrors: For concave mirrors the source (object) is on the same side of the surface as the sphere's center, while for convex mirrors the source is on the other side. There are some particularly simple and significant rays to follow in these situations, and the ray-tracing techniques of Section 36–1 will be useful here. We simplify things by studying only objects on or near the *axis*, the line perpendicular to the center point of the mirror, as shown in the figure. The tips of the arrows shown in Fig. 36–7 are a distance h from the axis, and if h is small compared to the radii of curvature of the mirrors, we say that the object is *near the axis*. We consider only rays that are so close to being parallel to the axis that we can use small-angle approximations in studying their reflections. Such rays are said to be **paraxial**.

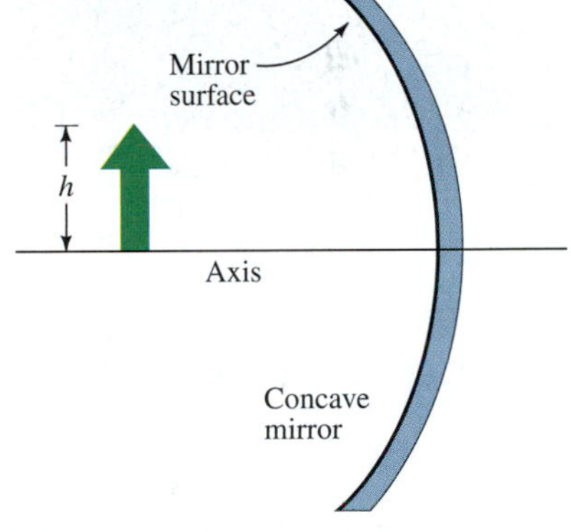

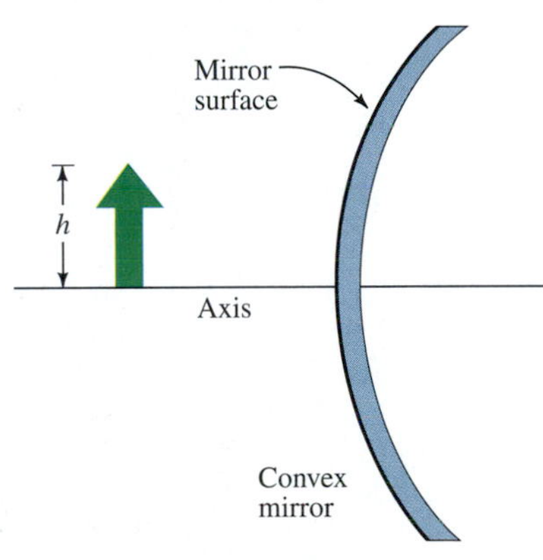

▲ **FIGURE 36–7** (a) Concave and (b) convex spherical mirrors. The object shown, an arrow, acts as an extended source.

The Concave Mirror

We will first consider rays from a very distant point source (to the left of the figure) on the axis CB of the concave mirror in Fig. 36–8a. When a source is so far away that *all the rays from it arrive practically parallel to each other*, we say that the source is *at infinity*. Point C indicates the position of the center of the sphere (of radius R) of which the mirror is a segment. The position of C (called the *center of curvature*) is therefore a distance R from the mirror surface, and all lines from point C to the mirror are perpendicular to the mirror.

Location of the Focal Point: Look at ray 1, which is reflected at point A in the direction AF in Fig. 36–8a. Angle θ is the angle of incidence and the angle of reflection (line CA is perpendicular to the mirror). Note that triangle ACF is isosceles with a base of length R. Thus, by dropping a perpendicular from point F to the base of the triangle AC, we see that the distance CF is $CF = (R/2)/\cos\theta$. For small θ, $\cos\theta \cong 1$; hence $CF = R/2$, or $BF = R - CF = R/2$, *independent* of θ.[†] *All* the parallel rays near the axis reflect through point F, a distance $R/2$ from the mirror (Fig. 36–8b, c). This is the position at which we will see an image point of a very distant source point. Is the image we see at point F a real or virtual image? Unlike the image points produced by a plane mirror, point F is a real image point because rays actually cross there.

[†]This result is accurate to 1 percent for angles θ less than about 10°.

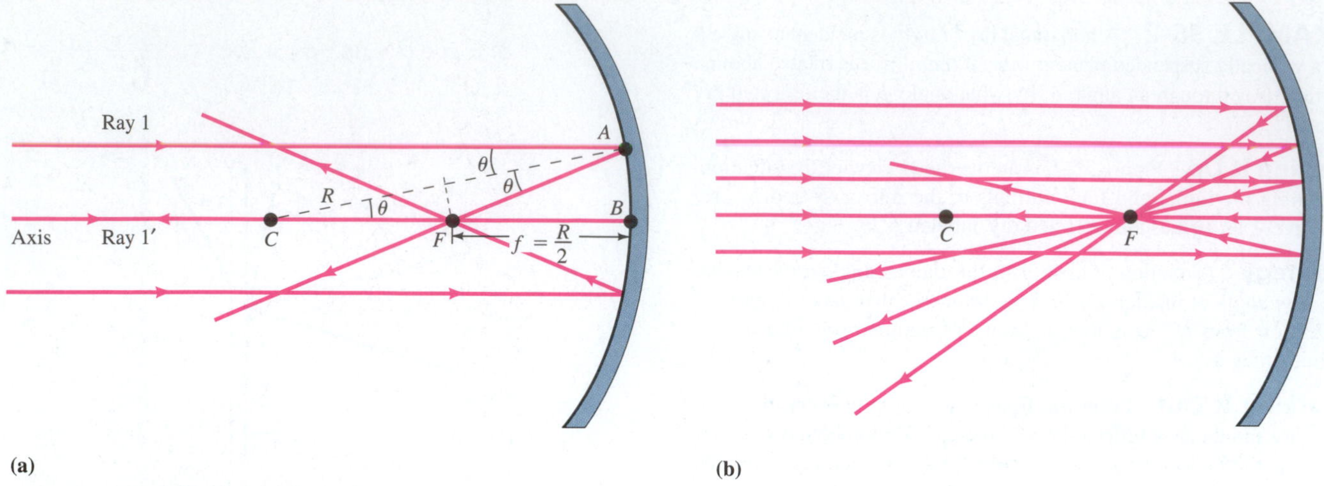

▲ **FIGURE 36–8** (a) Rays emitted by an object at infinity are all parallel to the axis. Ray 1 is reflected by the concave mirror surface and passes through the focal point, F. (b) To a good approximation, F is independent of θ; that is, any incoming ray parallel to the axis is reflected through it. Thus all rays from infinity cross the axis at F, which is therefore an image point. (c) A demonstration of the construction of part (b).

The paraxial rays from a source point at infinity are brought together to form an image at the point F; we say they are *focused* at point F, which is known as the **focal point**, or *focus*, and we say that its distance f from the mirror is the **focal length**. These terms can be applied to any optical system that produces images—including plane mirrors, whose focal length is infinite. We have shown that for concave mirrors

$$f = R/2. \tag{36–1}$$

CONCEPTUAL EXAMPLE 36–3 How could you use a concave mirror to make a flashlight?

Answer The purpose of a flashlight is to send out a parallel beam—a beam that in effect goes to a point far away. Since the path of any ray can be followed in reverse, this can be done by reversing the picture in Fig. 36–8a. We place a source of rays (a lightbulb) at the focal point, and the result will be a set of parallel rays leaving the mirror.

The Image of an Extended Object: Let's take an extended object—as, for example, the arrow in Fig. 36–7—that is small compared to the radius of curvature of the mirror and close enough to the axis so that the rays are paraxial. In Fig. 36–9a, we label two points on the object, which is *upright*, with the letters S and S'. If we follow bundles of light rays coming from a given spot on the object, we see that after reflection the rays pass through a corresponding spot in space, thus forming image points I and I' of the two points, and indeed an image of the entire object. We want to determine the position and size of the image. To do so, we use the principal-ray technique described in detail in the Problem-Solving Techniques, p.1004. In Fig. 36–9b, we draw the principal rays to find point I of the image that corresponds to point S of the source, and in Fig. 36–9c, we draw the principal rays to find the image point I'' that corresponds to source point S''. These sets of rays are

- ray 1, which approaches the mirror parallel to the axis and is reflected through the focus F;
- ray 2, which passes through the focus and reflects off the mirror parallel to the axis;
- ray 3, which passes through the center of curvature of the mirror and is reflected back in the direction from which it came;
- ray 4, which goes to the center of the mirror surface and whose reflection makes the same angle with the axis as the incident angle.

The rays leaving source point S do indeed cross at the image point I, and those from S'' cross at I''. All other points of the source have image points that can be constructed in

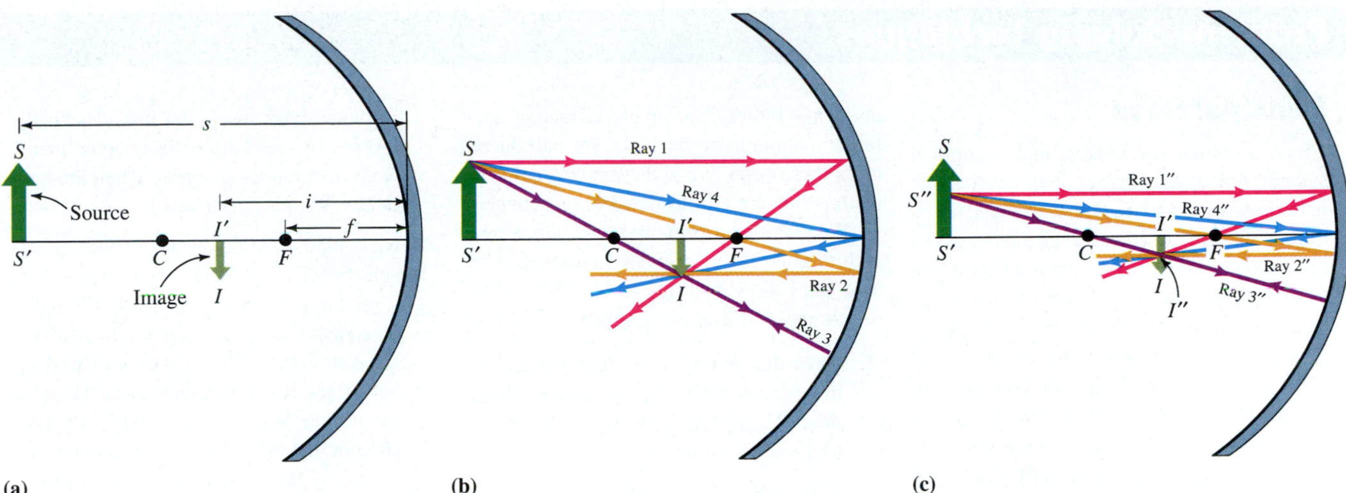

FIGURE 36–9 (a) An extended object a distance s from a mirror forms an image at a distance i from the mirror. The object is outside the focal length of the mirror. (b) Ray tracing for the concave mirror, with the principal rays for source point S. (c) Principal rays for source point S''. By repeating this exercise, we can build up the entire image. The principal rays are a guide; *any* ray from S that reflects will cross the image point I. (d) The object has produced an inverted and reduced image in a concave mirror.

the same way, and the image is thus constructed as in Figs. 36–9b and 36–9c. Note that both image point I' and source point S' lie along the optical axis. Our construction shows (and it is generally true) that a vertical source gives a vertical image, so that we can compute the location of just one image point rather than many. For example, the entire image can be constructed if we find only image point I of the top of the source (that is, of source point S), and, in fact, any two of the four principal rays are sufficient to determine point I. For example, rays 1 and 2 are sufficient for locating point I in Fig. 36–9b.

In this discussion, the source is *outside* the focal point, i.e., further away from the mirror than F. The image in Fig. 36–9a, located by the procedure in Figs. 36–9b and 36–9c, is *real* (real light rays pass through points I and I' and those in between), in contrast to the virtual image produced by plane mirrors. The image of the object in Fig. 36–9d is *inverted* (*upside down*) and *reduced in size*.

Let's now consider the situation depicted in Fig. 36–10, with the source *inside* the focal point of the mirror. In this case rays 2 and 3 only *behave* as though they pass through points F and C, respectively. The reflected rays from source point S do not actually cross but are aligned as though they come from behind the mirror at image point I. In other words, the image is virtual. It is also *upright* and enlarged. These are useful features of the concave mirrors used for shaving or applying makeup, or in dentistry.

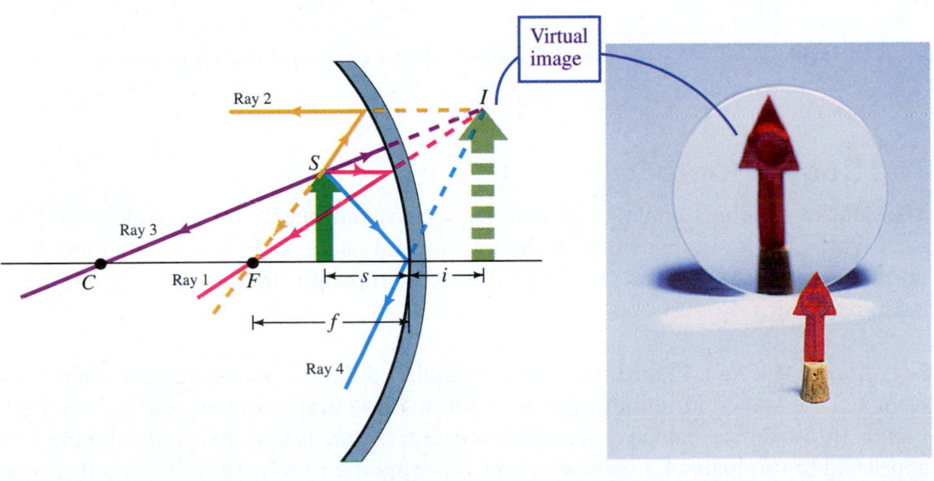

FIGURE 36–10 (a) Ray tracing with the principal rays for a concave mirror, for a source closer to the mirror than that of Figure 36–9b. The image becomes virtual when the source moves inside focal point F. (b) An object is placed within the focal length of a concave mirror. This time the image it produces is upright and magnified.

Problem-Solving Techniques

Principal Rays

Given a source (or object) and an optical system made of mirrors and lenses, one generally wants to find the size and location of the image. Because all rays cross at the image point of a source point (or behave as though they do in the case of virtual images), we need only find the crossing points of a few rays from any point on the object to be able to find both the position and size of the image. We refer to these rays as *principal rays*. Even if the optical system is such that a ray does not actually exist—for example, there may be a hole at the center of a mirror—we can pretend that it does and draw a ray through it. That is because the image is actually formed by rays coming from all parts of the mirror, and if part of the mirror is missing, the image is still formed in the same way (although it may be less intense). In other words, we simply use the principal rays as a tool to learn where the rest of the rays go.

In this box, we describe the principal rays for a convex mirror (Fig. 36B–1) and a converging lens (Fig. 36B–2; see Section 36–4), but the method applies to concave mirrors, diverging lenses, and single refracting surfaces. Although we have not yet introduced all of these cases, we will eventually come to them—they are all illustrated in this chapter.

We count four principal rays, numbered 1 through 4, from a given source point S. You can follow each ray in Fig. 36B–1 for the example of a convex mirror:

1. Rays that enter the system parallel to the optical axis. By definition, these paraxial rays are reflected or refracted to the focal point F.
2. Rays that pass through (or for a virtual image are aligned so that if they are extrapolated, they would pass through) the focal point as they enter the system. These rays are just reversed versions of type 1 rays and thus after reflection or refraction leave the system parallel to the axis.
3. Rays that pass through (or are aligned as though they pass through) the center of curvature C of the sphere from which a mirror or refracting surface is formed. These rays are perpendicularly incident on the surface and will be reflected or refracted back along the line of arrival. (As we will see later in the chapter, there is no useful analogous ray when the optical system is a thin lens.)
4. Rays that strike the center of the mirror surface. The reflected rays make the same angle with the axis as do the incident rays (except for sign). (As we will see later in the chapter for the case of thin lenses, the ray drawn directly to the center of the lens passes through it in a straight line. Also, there is no useful analogous ray for a single refracting surface.)

By drawing these principal rays from any given point S on a source, we find where the reflected or refracted rays cross (or for a virtual image appear to cross) and learn the location of the image point I of source point S. When an optical system has more than one reflecting or refracting surface (an "element"), we can apply the simple rule that the image formed by one element serves as a new object for the next element. In that case, *the principal rays must be redrawn for the new object* as they apply to the next optical element to locate the next image.

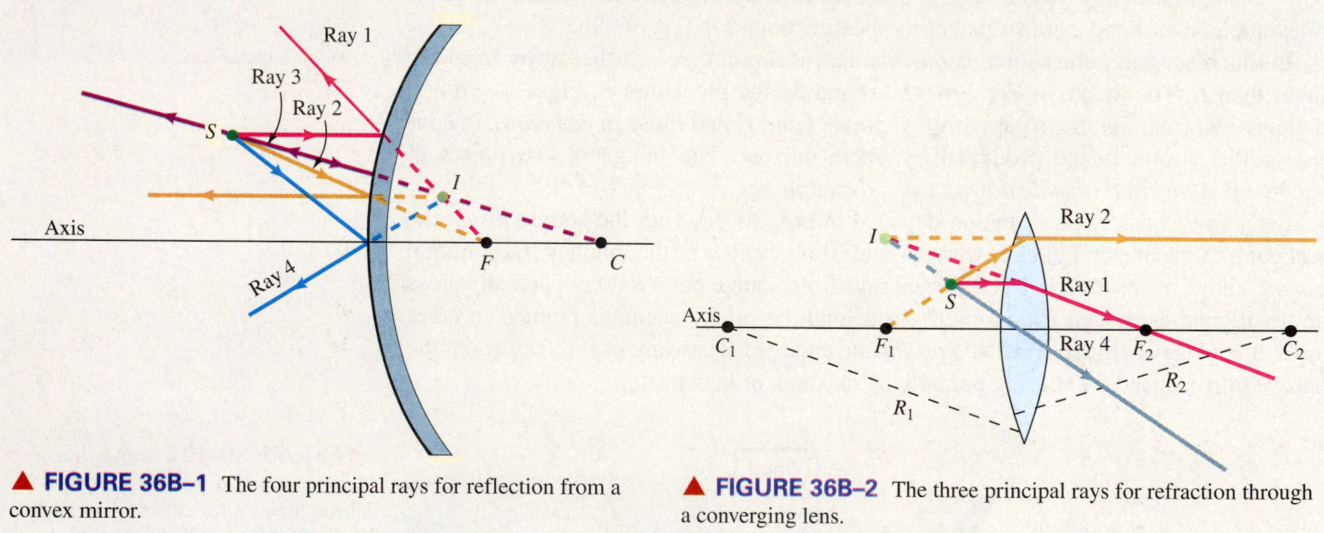

▲ **FIGURE 36B–1** The four principal rays for reflection from a convex mirror.

▲ **FIGURE 36B–2** The three principal rays for refraction through a converging lens.

The Convex Mirror

The same ray-tracing techniques we used for concave mirrors allow us to understand convex mirrors. Point C in Fig. 36–11 is the center of curvature of the sphere (of radius R) of which the convex mirror is a segment. All lines from point C to the mirror are perpendicular to the mirror.

Location of the Focal Point: We start by finding the focal point, the spot where rays from a point source at infinity (that is, a set of rays parallel to the axis) are focused. Figure 36–11 shows that the reflected rays diverge, so *the image is virtual*, with the rays appearing to originate at a common point F behind the mirror. By following the same trigonometric reasoning we used for the concave mirror and by using Fig. 36–11, we

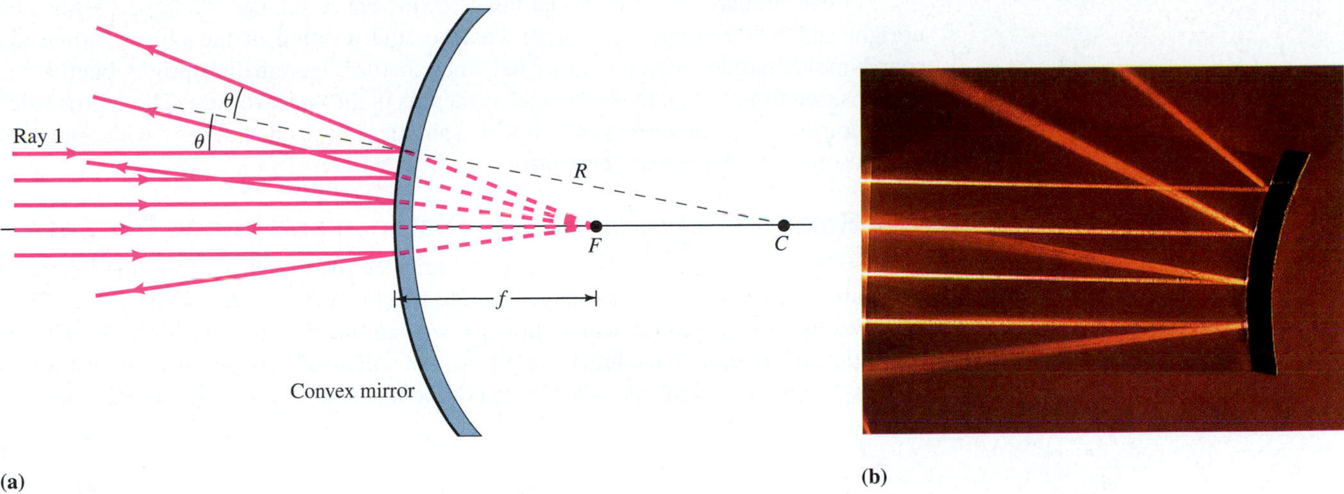

▲ **FIGURE 36–11** (a) When a spherical mirror is convex, the focal point lies behind the mirror, as ray tracing shows. The reflected rays diverge, and their extensions all lead back to the focal point. (b) Parallel rays of light reflected by a convex mirror.

can see that the distance f is again given by Eq. (36–1), $f = R/2$ (see Problem 14). Note that the focal point of a convex mirror is on the side opposite the object, unlike the case of the concave mirror, for which the focal point is on the same side as the object.

The Image of an Extended Object: We trace the four principal rays from source point S of the extended object shown in Fig. 36–12: ray 1—parallel to the optical axis, and whose reflection extends back along the line from the mirror to point F; ray 2—drawn as though it would pass through F, and whose reflection is parallel to the axis; ray 3—drawn as though it would pass through C, and whose reflection returns along the line of incidence; and ray 4—striking the center of the mirror surface, and whose reflection makes the same angle with the axis as the incident angle. A careful drawing shows that the reflected rays diverge from each other, but all four (indeed, *all* rays from S) would originate at point I if they were traced back through the mirror. *Point I is the virtual image of point S.*

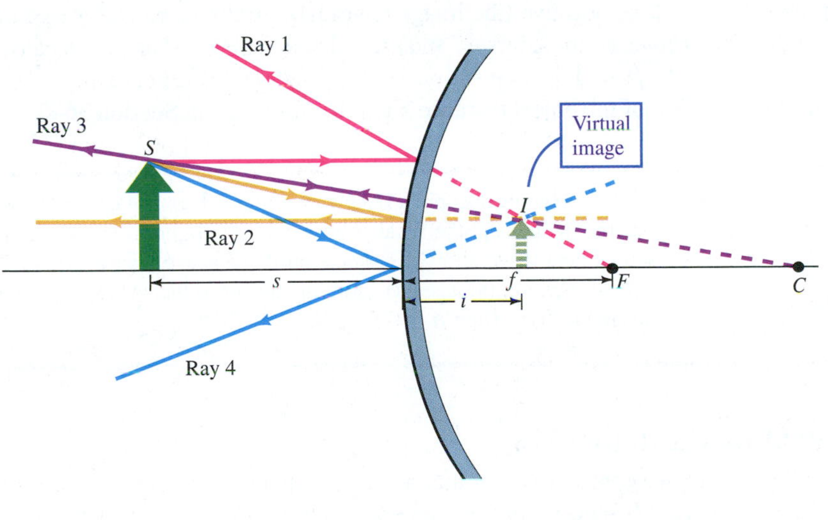

▲ **FIGURE 36–12** (a) Ray tracing describes the formation of a virtual image by a convex spherical mirror. (b) The image produced by a convex mirror is upright, reduced, and behind the mirror.

We can similarly find the virtual image of the entire source, which will always be upright and smaller than the source, whatever the position of the object. When the source moves farther away, the image becomes smaller and remains upright, but there is no transition from virtual to real image, as there is in the concave case. These properties make convex spherical mirrors, which with a smaller image encompass a wider range of view, useful for vehicle rearview mirrors.

The Relation Between Source Distance and Image Distance

In Figs. 36–9a, 36–10, and 36–12 we have indicated the distance s from the mirror to the source, the distance i to the image, and the focal length $f = R/2$ (with R the radius of curvature). Using geometrical arguments, we can find a relation between these three quantities. In addition to its direct usefulness, we will also be able to use the relation to help us find image height and whether or not the image is inverted. The relation is

$$\frac{1}{s} + \frac{1}{i} = \frac{1}{f}. \qquad (36\text{–}2)$$

SOURCE–IMAGE–FOCAL-LENGTH RELATION

(The details of the argument for a concave spherical mirror are contained in the optional subsection that follows.) With an appropriate set of conventions about signs, *the same relation holds for the convex mirror*. Equation (36–2) is immediately understood in two limits. When the object is far away ($s \to \infty$), then $1/s \to 0$ and $i = f$ (which is the definition of f). When the object is at the focus, $s = f$, then $1/i = 0$: The image is very far away.

Equation (36–2) contains information on the sign of the image point i. We'll discuss this information here, but point out that later the information on the sign will be summarized in a table. To start, note that if the object is between the concave mirror and the focus, as in Fig. 36–10, then s is smaller than f, and Eq. (36–2) implies that *i must be negative*. We associate a negative i with the image on the far side of the mirror—that is, with a virtual image. Likewise, i is positive when the image is real. Equation (36–2) may be applied to a convex mirror if we follow the convention that *the focal length f is negative when the focus is on the "virtual image" side of the mirror*. This is equivalent to saying that if the mirror's center of curvature is on the back (nonreflecting) side of the mirror, f is negative. In the application of Eq. (36–2) to a convex mirror, s is always positive and f is always negative. If s is positive (as it is when our optical system consists only of the convex mirror), then i will be negative (the image is virtual). Furthermore, the image in this case must be between the mirror and the focal point. That is because $1/s = 1/f - 1/i = -1/|f| + 1/|i|$ is positive, so $1/|i| > 1/|f|$, and hence $|i| < |f|$.

The rules for the sign of the object distance s will be discussed in Section 36–3.

CONCEPTUAL EXAMPLE 36–4 You are given a concave mirror and cannot measure its radius of curvature directly. How could you determine this radius with optical techniques?

Answer If we can determine the focal point of the mirror, then its radius of curvature is $R = 2f$, according to Eq. (36–1). As Eq. (36–2) confirms, this is done with a source at infinity, $s \to \infty$, so that the image is at the focal point, $i = f$. Sunlight provides such a source, and if we allow it to shine on the concave mirror, we need only to measure the distance from the mirror to the "hot spot"—the image location. Then $R = 2f = 2i$.

*How to Obtain Eq. (36–2)

To arrive at Eq. (36–2), we consider two points on an optical axis—the light source (or object), S, and its image, I—and a concave spherical surface. We can see from Fig. 36–13 and from the fact that the sum of the internal angles (in radians) of a triangle is π, that the following relationships hold:

$$\gamma = \beta + \alpha; \qquad (36\text{–}3)$$

$$\delta = \gamma + \alpha = \gamma + (\gamma - \beta) = 2\gamma - \beta. \qquad (36\text{–}4)$$

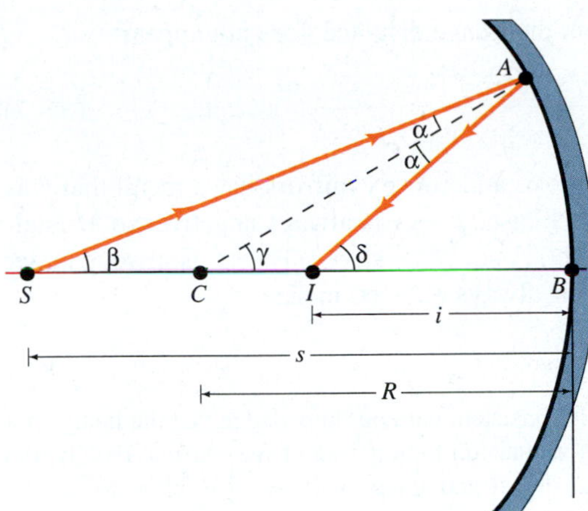

◀ **FIGURE 36–13** Geometric construction for deriving Eq. (36–2) for a spherical mirror.

In arriving at Eq. (36–4), we have used Eq. (36–3) as an intermediate step to eliminate α. The distances of Fig. 36–13 are related to the angles by the exact relation $AB = R\gamma$ and by the *approximate* (small-angle) equations $AB = i\delta = s\beta$. These relations allow Eq. (36–4) to be rewritten as

$$\frac{AB}{i} = \frac{2AB}{R} - \frac{AB}{s}.$$

We divide out the common factor AB and use the focal length $f = R/2$ for a spherical surface. We immediately obtain Eq. (36–2).

Magnification

Our geometric constructions show that an image may not be the same size as its source. Consider the convex mirror in Fig. 36–14. Ray 4 to the center point A of the mirror is useful because all the angles marked θ are the same, so triangles $AS'S$ and $AI'I$ are similar triangles. Thus, the magnitude of the **magnification** M defined as the ratio of the heights of the source and image, is

$$|M| = \frac{|II'|}{|SS'|} = \frac{|i|}{|s|}. \qquad (36\text{–}5)$$

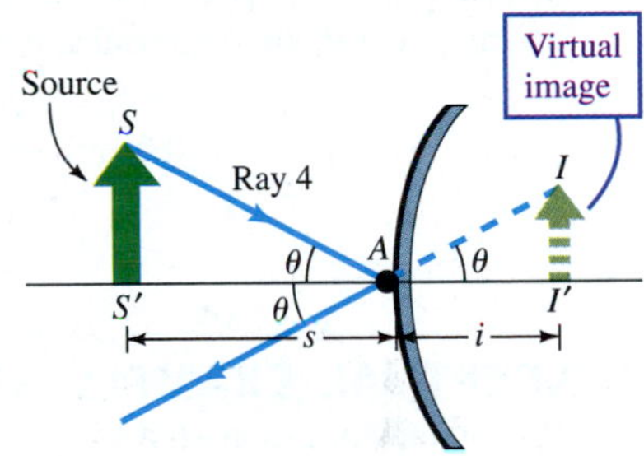

▲ **FIGURE 36–14** Geometry for the calculation of magnification.

We also specify whether the image is upright or inverted by writing

$$M = -\frac{i}{s}. \qquad (36\text{–}6)$$

MAGNIFICATION

If M is negative, the image is inverted; if M is positive, the image is upright. We can verify that this form works in the following explicit cases:

1. When the mirror is concave and the source is outside the focal point, the image is real (i is positive). By Eq. (36–6), M is then negative and the image should be inverted, as it is in Fig. 36–9a.
2. When the mirror is concave and the source is inside the focal point, the image is virtual (i is negative). By Eq. (36–6), M is then positive and the image should be upright, as it is in Fig. 36–10.
3. When the mirror is convex, the image is virtual (i is negative). By Eq. (36–6), M is then positive and the image should be upright, as it is in Fig. 36–12.

Equation (36–2) can be rewritten as

$$\frac{1}{i} = \frac{1}{f} - \frac{1}{s} = \frac{s - f}{fs}.$$

We can thereby find M in a form in which the image distance does not appear:

$$M = -\frac{i}{s} = -\frac{fs/(s-f)}{s} = \frac{f}{f-s}. \qquad (36\text{–}7)$$

This equation can be applied to both concave and convex mirrors if we recall that f is negative for convex mirrors. In the convex case, $f - s$ is always negative, so M is always positive; the image is always upright. Also, $|f - s| = |f| + |s|$ is always larger than $|f|$ for convex mirrors, so the image is always reduced in size.

EXAMPLE 36–5 A convex spherical mirror of radius of curvature R of magnitude 20.0 cm produces an upright image precisely one-quarter the size of the candle that is the object. What is the separation distance between the object and its image?

Setting It Up We draw the mirror in Fig. 36–15, indicating the center of curvature C and the focal point F. We know the position of neither source nor image.

Strategy We find f from Eq. (36–1), $f = R/2$. We then rearrange Eq. (36–7) to find s in terms of the known quantities f and M. We can then use Eq. (36–6) to find i.

Working It Out First we have numerically $f = R/2 = -10.0$ cm. (The negative sign indicates that the mirror is convex.) The appropriately rearranged Eq. (36–7) is

$$s = f\left(1 - \frac{1}{M}\right) = (-10.0\text{ cm})\left(1 - \frac{1}{\frac{1}{4}}\right)$$
$$= (-10.0\text{ cm})(-3) = 30.0\text{ cm}.$$

(We have taken $M = \frac{1}{4}$, positive because the image is upright.) Next we rearrange Eq. (36–6) to determine i:

$$i = -sM = -(30.0\text{ cm})\left(\frac{1}{4}\right) = -7.50\text{ cm}.$$

The minus sign is consistent with our knowledge that the image of a convex mirror is virtual (on the far side of the mirror). Finally, the distance between object and image will be $|s| + |i| = 30.0$ cm + 7.50 cm = 37.5 cm.

Alternative Solution The ray-tracing construction in Fig. 36–15 confirms these conclusions qualitatively; to confirm the actual numbers, the ray-tracing technique requires a very accurate drawing.

What Do You Think? What happens to the image when the candle is slowly moved farther away from the mirror?

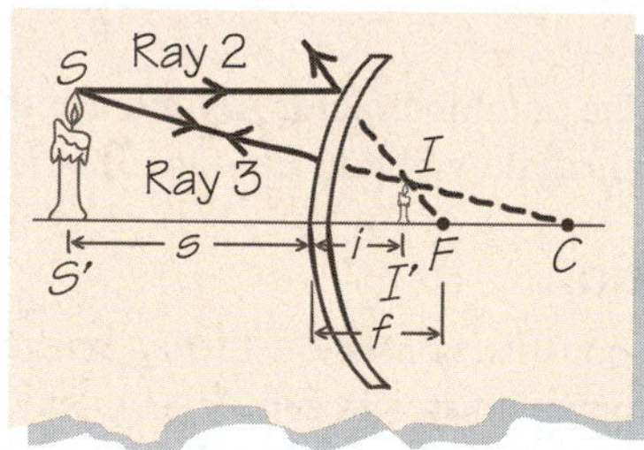

▲ **FIGURE 36–15** Ray tracing to find the image.

CONCEPTUAL EXAMPLE 36–6 What are the focal length f and magnification M of a plane mirror?

Answer A plane mirror is a special case of a concave or convex mirror in which $R \to \infty$. Since $f = R/2$, if $R \to \infty$, then $f \to \infty$ as well. We have seen that by using the relation between s, i, and f, we can determine the magnification in terms of any two of these. Here we use Eq. (36–7), which gives M in terms of f and s:

$$M = \frac{f}{f-s} \to \frac{f}{f} = 1.$$

We have used the fact that $f \gg s$, and the answer is no magnification (image size = object size) for any value of s. Note that M is positive, indicating an upright image—every morning, when you look in the bathroom mirror, you verify this!

THINK ABOUT THIS. . .

ARE RAY-TRACING TECHNIQUES REALLY USEFUL?

In this chapter our primary tools for analyzing optical systems are algebraic equations, not ray-tracing techniques. But ray tracing is the basis for the design of real optical systems, especially the most sophisticated. We may want an optical system to produce a very sharp image over a very limited range of source distances; for example, the lenses used in orbital satellites to image Earth's surface will never have to make an image of a very close source. Or, as in many cameras, we may want to sacrifice a sharp image in order for an optical system to operate in dim light. Real systems may have nonspherical mirrors, or thick, multielement lenses in which the elements move relative to one another, as in zoom lenses. To attain the desired optical properties, designers of such systems use computer programs capable of tracing large numbers of rays in a system design, which allows them to preview the quality and placement of the image and to test modifications in the design. ■

The relation between source distance, image distance, and focal length [Eq. (36–2)], the expression for magnification [Eq. (36–6)], and the ray-tracing techniques are applicable to lenses as well as to mirrors. For the lenses that we will study in this chapter, these three elements provide all the information we need.

We conclude this section with a comment on signs. Table 36–1 gives the signs of all the quantities necessary for mirrors, refracting surfaces, and lenses. However, in our opinion it is not necessary to keep track of the signs of the various quantities we have discussed. Develop your ray-tracing techniques, and you will be able to rederive the signs on your own. In this way a ray-tracing diagram is perhaps as useful as a free-body diagram.

TABLE 36–1 • Sign Conventions for Mirrors, Refracting Surfaces, and Lenses

In applying the information in this table, we must distinguish two "sides" to a reflecting or refracting surface:

Side *A*, the side from which light originates, and

Side *B*, the side to which light passes.

For mirrors, side *B* is identical to side *A*; for refracting surfaces and lenses, the two sides are opposite. Only the sign of the source position is determined by side *A*. All other quantities are determined by reference to side *B*.

Determined by Side *A*	
Source distance s	Positive if object is on side *A* (real object)
	Negative if object is on side opposite to side *A* (virtual object)
Determined by Side *B*	
Image distance i	Positive if image is on side *B* (real image)
	Negative if image is on side opposite to side *B* (virtual image)
Curvature R	Positive if center of curvature is on side *B*
	Negative if center of curvature is on side opposite to side *B*
Focal point	Positive if on side *B*
	Negative if on side opposite to side *B*

36–3 Refraction at Spherical Surfaces

Mirrors change the direction of rays of light and create real or virtual images of objects. Lenses do the same, using pieces of transparent material to refract light. With Snell's law of refraction, we can use rays to determine the behavior of lenses in the same way that we handled mirrors using the law of reflection. The most basic type of lens typically has two curved refracting surfaces, and it is best to approach this case by first thinking about refraction through a single curved surface, which is the subject of this section. By repeatedly applying the rules we develop for a single boundary, we shall be able to understand the passage of light through lenses. We study, in particular, surfaces that are spherical sections without too much curvature, and we consider only paraxial rays. This simplifies the calculations and leads to the right qualitative conclusions.

Consider, then, light that crosses the boundary between one medium with index of refraction n_1 and another medium with index of refraction n_2 (Fig. 35–9). The angles of incidence and refraction satisfy Snell's law, Eq. (35–6):

$$n_1 \sin \theta_1 = n_2 \sin \theta_2.$$

Here we apply this law to a boundary that is not flat but rather forms a segment of a sphere of radius of curvature R. Let's take a convex surface, one whose center of curvature—point C in Fig. 36–16—is in the region to which light passes. Although we choose $n_1 < n_2$, so that the light that passes from medium 1 to medium 2 bends toward the perpendicular to the surface, the results will be more general.

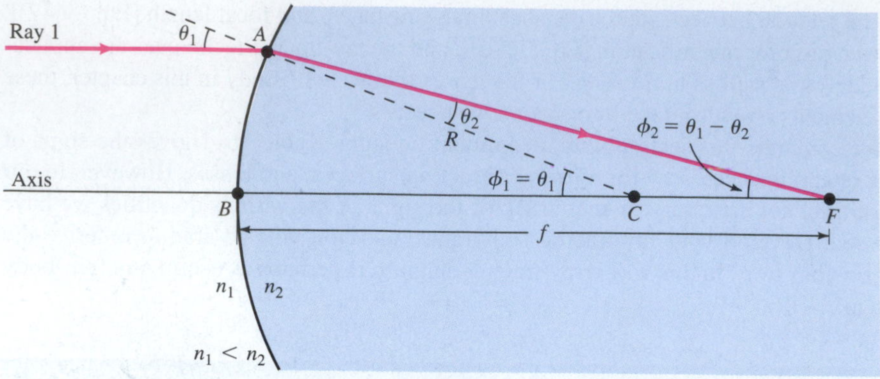

▶ **FIGURE 36–16** Ray tracing of a ray that enters a medium whose index of refraction is different from than that of the medium from which the ray came requires us to use Snell's law of refraction. Here we see refraction at a convex spherical surface.

The Focal Point of a Single Refracting Surface

As for a spherical mirror, a single refracting surface has a focal point F that we find by tracing rays that come from a very distant source, parallel to the axis. For the convex surface in Fig. 36–16, ray 1 bends toward the axis and crosses it at a point F. This point will be a focal point if all the incident rays that are parallel to the axis cross at F. Here we show that this is the case for paraxial rays, where the angle of incidence θ_1 and that of refraction θ_2 are both small, so the relation $\sin\theta \cong \theta$ is a good approximation. In that case Snell's law becomes

$$n_1\theta_1 \cong n_2\theta_2. \tag{36–8}$$

Simple geometry shows that in Fig. 36–16 $\phi_1 = \theta_1$, and therefore $\phi_2 = \theta_1 - \theta_2$. For small angles, the relation between BF and the arc length AB is given by

$$BF(\theta_1 - \theta_2) \cong AB.$$

Because $AB = R\theta_1$, this result, along with Eq. (36–8), implies that

$$BF \cong \frac{R\theta_1}{\theta_1 - \theta_2} \cong \frac{Rn_2}{n_2 - n_1}.$$

This distance is independent of θ_1 for small angles, so *all* parallel rays near the axis pass through point F, and F is the image of a point source at infinity. The focal length f is the distance BF:

$$f = \left(\frac{n_2}{n_2 - n_1}\right)R. \tag{36–9}$$

The focal point for a single refracting surface is farther from the surface than the center of curvature, as in Fig. 36–16 or by noting that in Eq. (36–9), $f > R$ if $n_2 > n_1$. Although we have derived Eq. (36–9) for a convex surface, we can derive it for a concave surface just as easily (Fig. 36–17). The center of curvature C of the concave surface is on the side from which the light is incident. We find exactly the same formula—except that the focal point is to the left of the surface, on the same side as C. We see in Fig. 36–17 that an image at the focal point F for such a surface is virtual.

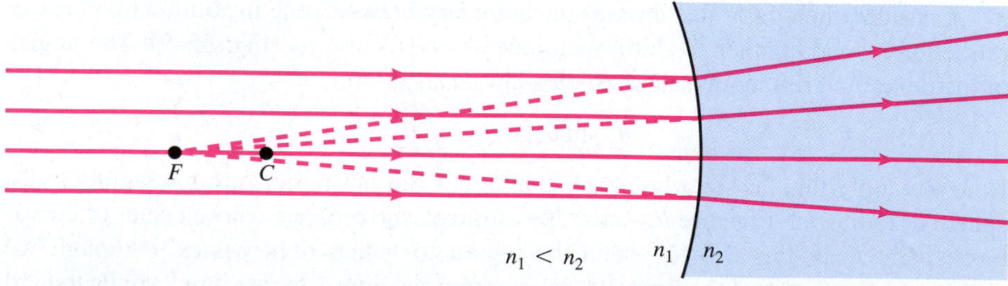

▲ **FIGURE 36–17** Refraction at a concave spherical surface.

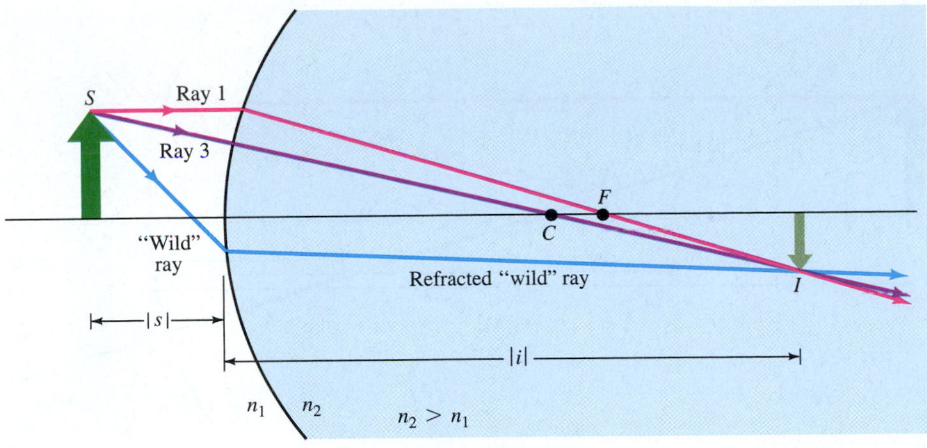

FIGURE 36–18 Ray tracing shows how a real image is formed by a convex spherical refracting surface.

The Image of an Extended Object

Convex Surface: Consider next a vertical object that stands erect on the optical axis. We already know enough about the principal rays to proceed with ray tracing. For a single refracting surface, only two of the four principal rays from any source point, such as point S, are useful. In Fig. 36–18, ray 1 is incident parallel to the axis and refracts such that it crosses the axis at F; ray 3 forms the straight line through C. Ray 3 is perpendicular to the surface and so passes into the medium without deflection. The two refracted rays meet at point I. Again, these rays are only two of an unlimited number of rays that leave point S and pass through I. For example, we have drawn a "wild" ray in Fig. 36–18. We will not carry out the detailed geometry that shows that the wild ray passes through I, but this is the case. By drawing the principal rays for any point on the object, we can reconstruct the entire image, which, given the distance s, is *real* and *inverted.*

Concave Surface: In Fig. 36–19, we offer three possibilities that depend on whether the source distance s is larger than the focal length of the lens f, smaller than the radius of curvature R, or smaller than f but larger than R. In each case, we use the same two principal rays from point S that we did for the convex case to locate the image of S at point I. In Figs. 36–19b and 36–19c, ray 3 may not actually pass through the curved surface. This is not a problem, because the principal rays are just tools for determining where all the rays that do pass through the surface cross. In each case, the image is upright and virtual (the diverging rays appear to come from point I).

The Relation Between Source Distance and Image Distance

The relationship between the positions of a source and an image for a single (concave) refracting surface is analogous to Eq. (36–2), which holds for a mirror. It takes the form

$$\text{for a refracting surface: } \frac{n_1}{s} + \frac{n_2}{i} = \frac{n_2 - n_1}{R}. \tag{36–10}$$

The derivation of this equation is like that of Eq. (36–2). We will not derive Eq. (36–10) but rather leave the derivation to Problem 28. In the derivation it is easiest to assume that the surface is convex; that is, that the center of curvature of the surface is on the side of the surface to which the light passes. Let's suppose that this corresponds to a positive value of R, as for the concave mirror. In addition, s is positive from the start. If we then find i from Eq. (36–10), it could be either positive or negative. When i is positive, it is on the side of the surface to which light passes and the image is real, meaning that light passes through it. When i is negative, it is on the side from which light is emitted and the image is virtual, meaning that light only *appears* to radiate from it when it is observed from medium 2. At the start of the discussion, we assumed that $n_2 > n_1$, but the result does not depend on this (see Problem 29), although a geometrical drawing does.

We can also repeat this exercise with a concave spherical surface between the two media (see Problem 30). In this case, the center of the spherical surface is on the side of the light source, and the important result is that *Eq. (36–10) continues to hold, but with a*

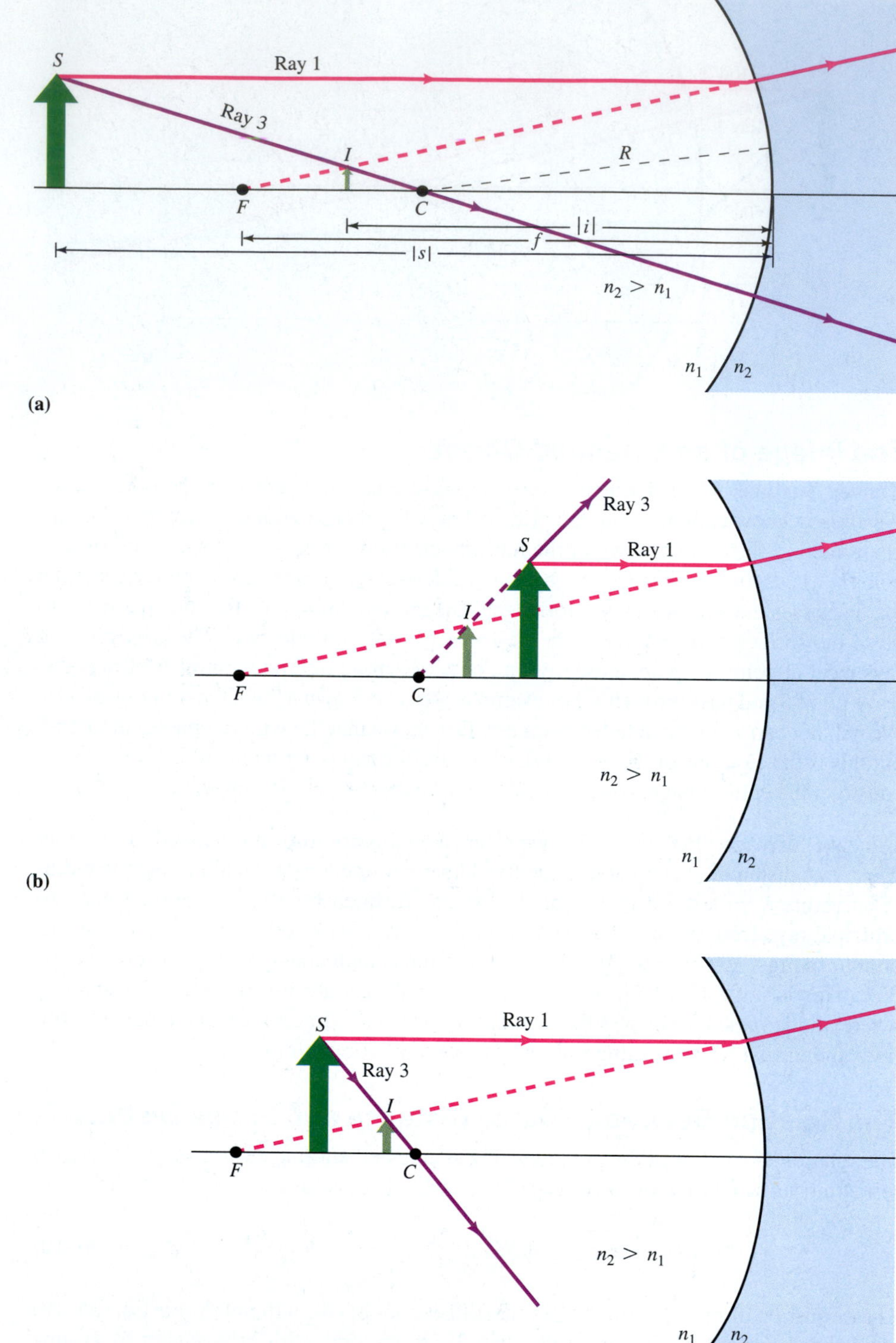

▶ **FIGURE 36–19** Ray tracing with principal rays for image formation for a concave spherical refracting surface: situations in which (a) $s > f$; (b) $R > s$; (c) $f > s > R$.

negative value of R. Summarizing, when the image is on the side to which light goes, i is positive and the image is real, while R is positive when it is on that side. When the image is on the side from which light radiates, i is negative and the image is virtual, while R is negative when it is on that side. Table 36–1 contains a summary of the various signs.

The Sign of the Object Distance

In order to be able to apply the results of this section to material with two surfaces—lenses—it will be necessary to understand one more aspect of signs. We have established that a negative image distance and radius of curvature have meaning. Is it possible for s

to take on a negative value? A positive s corresponds to a real object on the side from which the light radiates. For positive s, the light rays diverge from the object as they approach the boundary surface. Negative values of s correspond to the rays that *converge* as they approach the boundary, so their extrapolation would be on the side of the boundary to which the light passes. This cannot occur for a real object. But it is possible if the image produced by one surface—boundary 1, say—acts as the source object for a second surface: boundary 2 (Fig. 36–20a). We can break up the problem and find first the image point I_1 produced by boundary 1 (Fig. 36–20b). In actuality, the light never forms the image point I_1 because boundary 2 intervenes. However, the image I_1 becomes the virtual object S_2 for the light refracted at boundary 2 (Fig. 36–20c). According to our convention, the source distance s_2 is negative because the rays converge toward boundary 2. The light comes from the left side of boundary 2, but the virtual object is on the right side. Equation (36–10) holds for the refraction at boundary 2, with a negative object distance s_2. The actual paths of the light rays through both boundaries are shown in Fig. 36–20d.

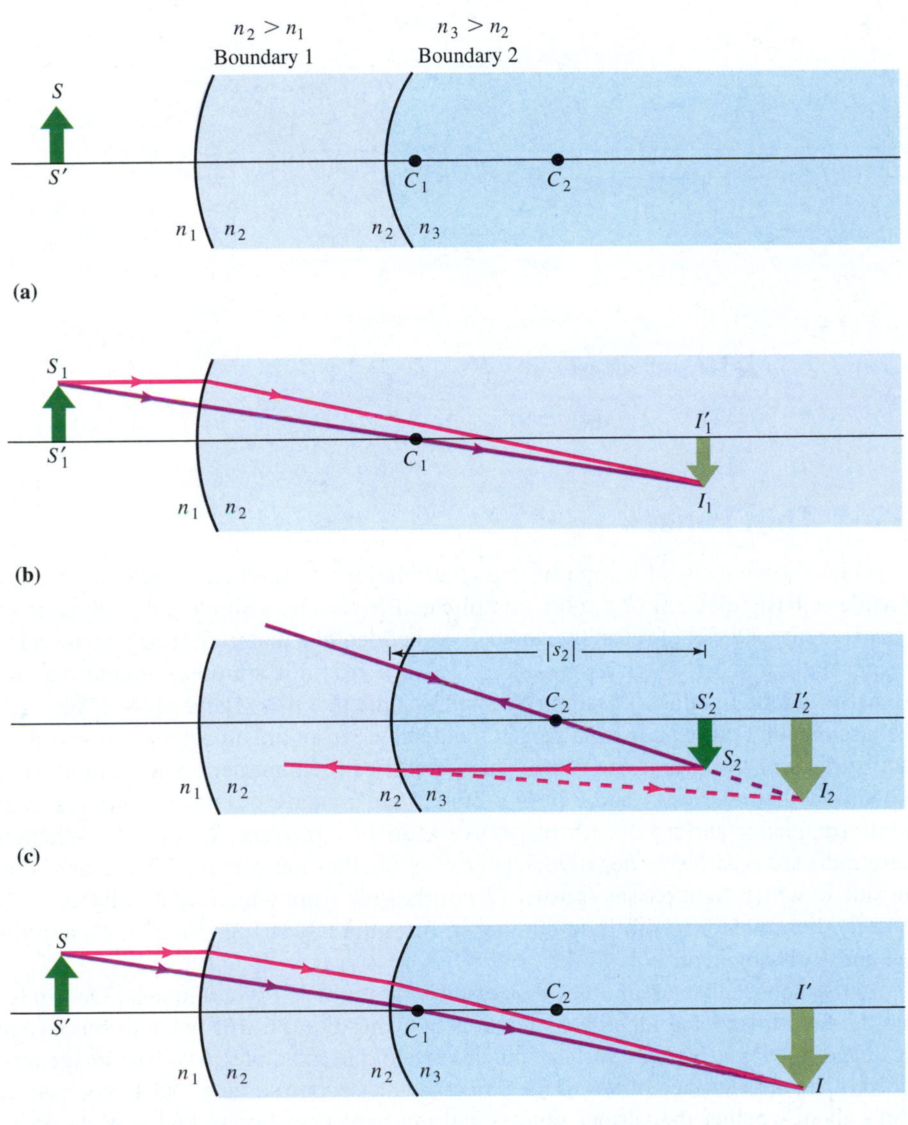

▲ **FIGURE 36–20** (a) Image construction when two refracting surfaces are involved. To simplify, (b) we can split the problem up by first finding the image point I_1 from boundary 1. (c) The resulting image serves as a virtual object; we use virtual source point S_2 for the interaction with boundary 2 to find the final image, at image point I. (d) The actual ray path.

EXAMPLE 36–7 Consider a cylinder of glass 50 cm long, with $n = 1.6$, in air (Fig. 36–21). Surface 1 has a radius of curvature $R_1 = 0.20$ m; surface 2 has a radius of curvature $R_2 = 0.40$ m. Both surfaces "bulge out." A small object (a leaf) is placed perpendicular to the optical axis at a distance of 120 cm from surface 1. (a) Find the location of the object's image due to refraction at surface 1. (b) Let this image be the source object for surface 2, and find the location of *its* image as light passes through surface 2.

Strategy Equation (36–10) applied successively to the two surfaces leads to the location of the final images. We first find the image formed by surface 1, and this image then becomes the object for surface 2.

Working It Out (a) We calculate the distance i_1 of the image point I_1 from surface 1. We have $n_2 = 1.6$ and $n_1 = 1.0$ (air). The center of curvature is on the side to which light passes, so R_1 is positive. Finally, $s = +1.20$ m. We then have

$$\frac{n_1}{s} + \frac{n_2}{i} = \frac{n_2 - n_1}{R}, \quad \text{or} \quad \frac{1.0}{1.20\text{ m}} + \frac{1.6}{i_1} = \frac{1.6 - 1.0}{0.20\text{ m}};$$

$$i_1 = +0.74\text{ m}.$$

The image is real and located (not shown) 74 cm to the right of surface 1.

(b) Because the surfaces are separated by 50 cm, this new object S_2 (which is the image I_1 for surface 1) is 24 cm to the right of surface 2. The object is on the side to which light passes, so its distance from surface 2 is negative: $s_2 = -0.24$ m. For this second step, $n_1 = 1.6$, $n_2 = 1.0$, and $R_2 = -0.4$ m (surface 2 is concave, so its center of curvature is on the side from which light comes). Thus Eq. (36–10) now gives

$$\frac{1.6}{-0.24\text{ m}} + \frac{1}{i_2} = \frac{1.0 - 1.6}{-0.4\text{ m}};$$

$$i_2 = +0.12\text{ m}.$$

This is positive, so the second image is real, or to the right of surface 2.

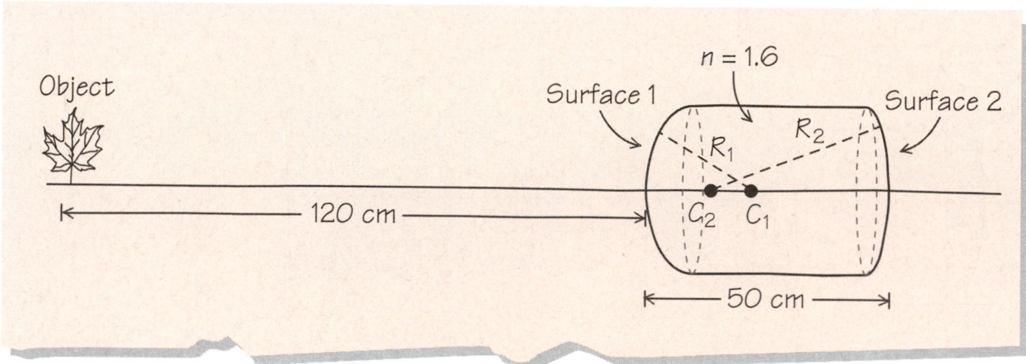

▲ **FIGURE 36–21**

36–4 Thin Lenses

(a)

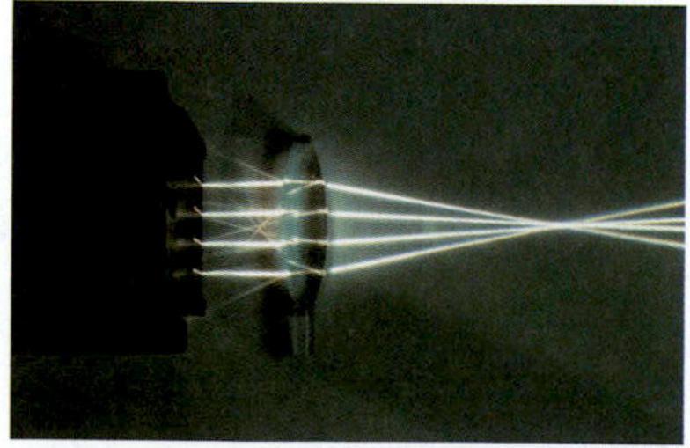

(b)

▲ **FIGURE 36–22** (a) Light passing through a diverging lens and (b) light through a converging lens.

A real lens can consist of a complicated combination of individual elements. We can consider a basic element of a more complicated lens to be a single element made of a transparent material of refractive index n embedded in a material of refractive index n_1, normally air, for which $n_1 = 1$ (Fig. 36–22). It is this simpler element that we refer to as a **lens** in this discussion. We shall assume that $n > 1$ and $n_1 = 1$. We shall also assume that our lens is thin, so that the distance from the object and the image to each surface of the lens is the same. This simplifies the treatment considerably. The two surface boundaries (1 and 2 in Fig. 36–23) are concave or convex spherical segments (or planar surfaces) with respective radii of curvature R_1 and R_2. Whether these radii are positive or negative depends on whether the center of curvature is on the side to which light passes (positive R) or the side from which light radiates (negative R). For example, with light coming in from the left in Fig. 36–23a, R_1 is positive and R_2 is negative.

Let's suppose that a real object is a distance s_1 to the left of a thin lens. We can locate the final image and identify its features by using Eq. (36–10) twice in succession for image-making at a single surface, much as we did in Example 36–7. The image produced by the first surface serves as the object for the second surface. We do not have to worry about whether the various objects and images are real or virtual, upright or inverted, because the equation will automatically handle these questions. At surface 1, we have

$$\frac{1}{s_1} + \frac{n}{i_1} = \frac{n - 1}{R_1}, \tag{36–11a}$$

which we rewrite as

$$\frac{1}{i_1} = \frac{n-1}{nR_1} - \frac{1}{ns_1}. \tag{36–11b}$$

Now, the image point I_1 produced by surface 1 serves as an object point S_2 for surface 2, producing a final image point at I_2. What is the sign of i_1? If i_1 is positive, the image is on the right of surface 1 and hence on the right of surface 2. This corresponds to an object distance s_2 for surface 2 that is negative. Similarly, if i_1 is negative, the image is to the left of both surfaces, corresponding to a positive object distance s_2 for surface 2. We must then reverse the sign of i_1 when we use it as the source distance s_2 for surface 2. Finally, note that in applying Eq. (36–10) a second time, $n_1 = n$ and $n_2 = 1$. Thus

$$\frac{n}{s_2} + \frac{1}{i_2} = \frac{1-n}{R_2};$$

$$-\frac{n}{i_1} + \frac{1}{i_2} = \frac{1-n}{R_2}.$$

When we substitute Eq. (36–11b), we find that

$$-n\left(\frac{n-1}{nR_1} - \frac{1}{ns_1}\right) + \frac{1}{i_2} = \frac{1-n}{R_2}.$$

If we now write $s_1 = s$ for the original object and $i_2 = i$ for the final image, we find (upon rearrangement)

$$\text{for a thin lens in air: } \frac{1}{s} + \frac{1}{i} = (n-1)\left(\frac{1}{R_1} - \frac{1}{R_2}\right). \tag{36–12}$$

LENS-MAKER'S EQUATION

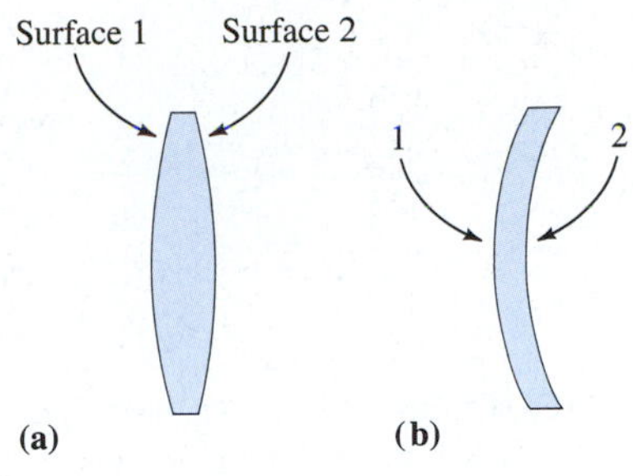

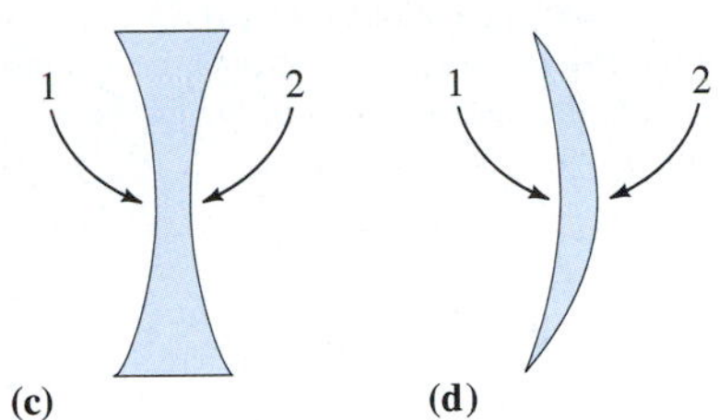

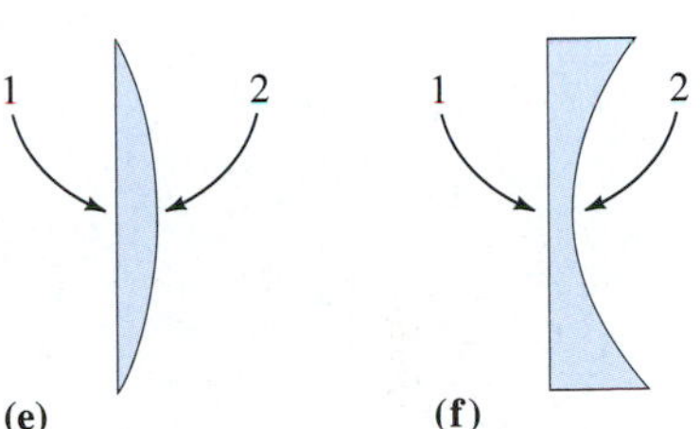

▲ **FIGURE 36–23** Six types of simple thin lenses with surfaces of different radii of curvature: (a) $R_1 > 0$, $R_2 < 0$; (b) $R_1 > 0$, $R_2 > 0$; (c) $R_1 < 0$, $R_2 > 0$; (d) $R_1 < 0$, $R_2 < 0$; (e) $R_1 = \infty$, $R_2 < 0$; (f) $R_1 = \infty$, $R_2 > 0$.

Equation (36–12), which applies *only* to thin lenses in air, is the *lens-maker's equation*. By Eq. (36–12), the image can be positive or negative; that is, real or virtual. The signs are summarized in Table 36–1, and ray tracing will alternatively allow you to understand the image—large or small, upright or inverted, real or virtual.

Equation (36–12) can be used to find the focal point of a lens. By definition, the image is at the focal point when $s \to \infty$. Therefore in this limit

$$\frac{1}{f} = (n-1)\left(\frac{1}{R_1} - \frac{1}{R_2}\right). \tag{36–13}$$

If we substitute this result into Eq. (36–12), we get Eq. (36–2), which we originally derived for mirrors—Eq. (36–2) is a general one that holds for most of the optical systems we study. The sign of f is determined by the signs of the radii of curvature, but we can say that f is positive if the image of a point source at infinity is on the side to which light passes (real image); f is negative if the image of the source at infinity is on the side from which light radiates (virtual image).

We now turn to some of the rays that help us understand the image. Ray 1, which comes in parallel to the optical axis of the lens, crosses (or behaves as though it crosses) the axis at f. This ray is drawn in Fig. 36–24a. Note that there is a symmetry in Eq. (36–12). When light arrives from the right of the lens rather than the left, R_1 and R_2 reverse their signs, and light from infinity coming from the right is focused the same distance from the lens as the first focal point, but on the opposite side. In turn, if light radiates (or behaves as though it does) from one of the two symmetric focal points of the lens (ray 2), the light emerges as a set of parallel rays.

Ray 4, which is drawn in Fig. 36–24a to the center of the lens and behaves as though it passes straight through, is a last useful ray. (Remember, principal ray 3 is not applicable to lenses.) This ray is shown in more detail at the enlarged section (Fig. 36–24b).

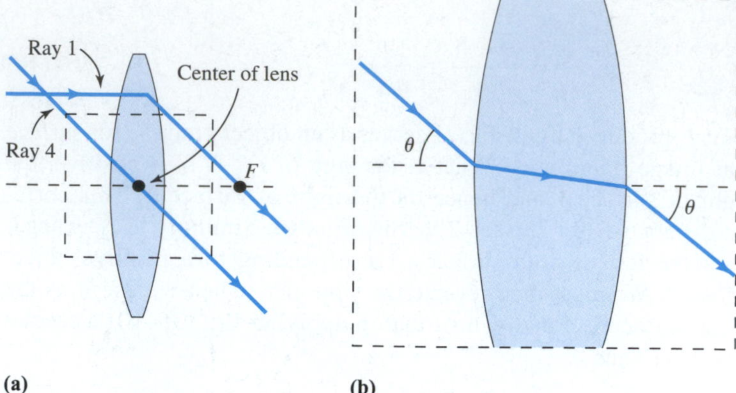

FIGURE 36–24 (a) The ray to the center of a lens passes through without changing angle, because at its axis the lens is like a pane of glass. (b) An enlarged view of the same lens. If the lens is thin, the displacement of the ray is small.

Since the two lens surfaces are in the middle of the lens, the ray behaves to a good approximation like a ray that passes through a thin pane of glass. There is a *small* displacement of the ray, but it drops to zero as the lens becomes thinner.

These three principal rays can be used to find an image. For example, Fig. 36–25 shows a lens that collects light from an object, and with the principal rays, we can easily find the image point I of object point S. The construction works for any point on the object. The image is real and inverted in this case. In general, if a lens causes rays that pass through it to come together, it is called a *converging lens*, and if it causes rays that pass through it to spread out, it is a *diverging lens*. Converging lenses have positive focal lengths, whereas diverging lenses have negative focal lengths. Some simple ray tracing will show that a lens like that of Fig. 36–23a is a converging lens, and one like that of Fig. 36–23c is a diverging lens.

Magnification

A thin lens produces a perfect image, to the extent that the small-angle approximation is valid. Thus we can find the magnification by direct use of similar triangles. In Fig. 36–25, the magnification of the image has magnitude

$$M = \frac{II'}{SS'}.$$

From the geometry of the similar triangles $SS'P$ and $II'P$, we see that the magnitude of the magnification is $|M| = |i|/|s|$. Just as for mirrors, a systematic look at signs shows that we can decide with a single sign whether the image is upright or inverted:

$$M = -\frac{i}{s}.$$

This is Eq. (36–6)—the same form we found for mirrors. If M is positive, the image is upright; if it is negative, the image is inverted. From Eq. (36–2), we have the alternate form

$$M = \frac{f}{f - s},$$

which is Eq. (36–7), also applicable to mirrors.

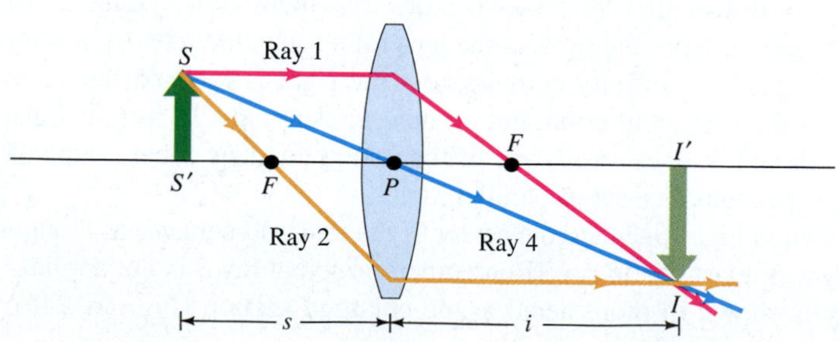

FIGURE 36–25 Ray tracing shows how a real image is formed with one type of thin lens. Point P marks the center of the lens.

EXAMPLE 36–8 A converging lens like that shown in Fig. 36–23a has surfaces with radii of curvature $R_1 = 80$ cm and $R_2 = 36$ cm. An emerald that is 2.0 cm tall is placed 15 cm to the left of the lens, for which $n = 1.63$. Where will the image be located, and what will its size be?

Setting It Up We have sketched Fig. 36–26 to include three rays. Such a diagram is useful if one wants to use ray tracing to verify results.

Strategy We first calculate the focal length from Eq. (36–13). Then we use Eq. (36–2) to determine the image distance i. At that point we will have both s and i and can find M.

Working It Out The radius of curvature of the first surface is positive, $R_1 = 80$ cm, whereas the second surface has negative curvature, $R_2 = -36$ cm. Thus Eq. (36–13) gives

$$\frac{1}{f} = (n-1)\left(\frac{1}{R_1} - \frac{1}{R_2}\right)$$

$$= (1.63 - 1)\left(\frac{1}{80\text{ cm}} - \frac{1}{-36\text{ cm}}\right) = 0.025\text{ cm}^{-1}.$$

The object distance is positive, $s = 15$ cm, so Eq. (36–2) gives

$$\frac{1}{i} = \frac{1}{f} - \frac{1}{s} = 0.025\text{ cm}^{-1} - \frac{1}{15\text{ cm}} = -0.041\text{ cm}^{-1}.$$

Thus $i = -24$ cm. The minus sign indicates that the image is virtual and on the same side as the light source. The magnification is given by

$$M = -\frac{i}{s} = -\frac{-24\text{ cm}}{15\text{ cm}} = 1.6.$$

The positive value indicates that the image is upright.

The rays in Fig. 36–26 confirm the qualitative aspects of our results.

What Do You Think? Suppose the lens were replaced by one with a smaller value of n. How would the magnification change?

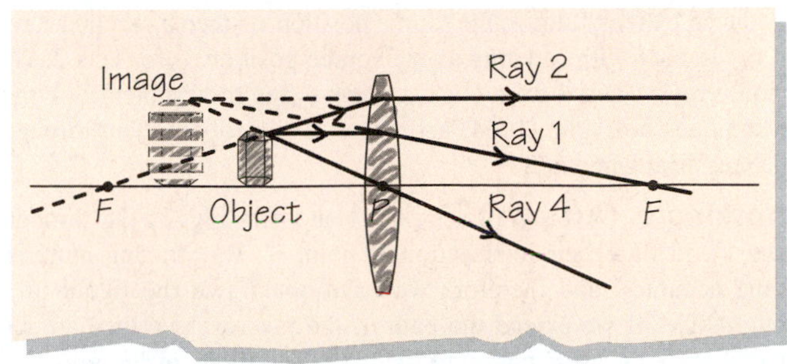

▲ **FIGURE 36–26** The horizontal and vertical scales are different here. When the object lies inside the focal point of the lens, ray tracing shows that the image formed is virtual.

We saw in Section 36–3 how the image produced by refraction at one surface acts as an object for the second surface. This principle extends to combinations of two or more lenses and lies at the heart of the design of more complicated lenses or optical instruments. Figure 36–27 gives an image construction for two thin converging lenses. The object SS' lies inside the focal length of lens 1 and thus gives rise to a virtual, enlarged image $I_1I'_1$. That image serves as an object $S_2S'_2$ for lens 2. Ray tracing uses the parallel ray $I_1A_2F_2I$ and I_1P_2I to determine the position of the real image, but the particular rays chosen really follow paths like SA_1F_1AI and SP_1BI. This example shows that it is possible to obtain a magnified *real* image with two converging lenses in conditions where it is not possible with one lens.

In Example 36–9, the object for the second lens is a negative distance from the lens.

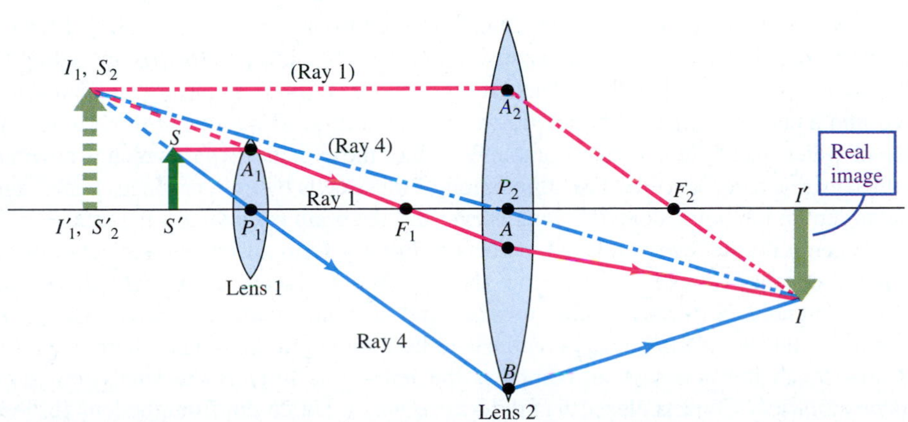

▲ **FIGURE 36–27** Ray tracing shows how two converging lenses produce a real magnified image.

EXAMPLE 36–9 Consider the two-lens arrangement shown in Fig. 36–28, with the source just outside the focal point of lens 1. The focal length of each lens is known. You place your eye at the position shown and look for the image. Use ray-tracing techniques to describe qualitatively that image.

Setting It Up If these were not already included in the figure, the starting point would be to draw in the focal points F_1 and F_2, the source distance s_1, and the lens separation distance L. The figure should be to scale. Note that with the eye placed as shown, the "image" in question is the one that is produced by lens 2 using the image produced by lens 1 as a source.

Strategy We must use ray tracing to find the image position i_1 produced by lens 1 due to the source position s_1, then a second round of ray tracing with i_1 acting as the source position s_2 for lens 2. We choose rays 1 and 4 from our list of rays (see the Problem-Solving Techniques box on p. 1004) and successively follow them through the lens arrangement.

Working It Out We trace rays 1 and 4 in Fig. 36–28 through lens 1 and find their intersection at point i_1. Ray tracing requires some accuracy, and therefore we have not drawn the figure in a rough style. If we extend the path of the rays, we see they are refracted by lens 2 and pass into the eye. Extensions of the rays that enter the eye give an image i_2 that is behind the lens and very far away. That is because we have placed the relative position of the two lenses so that the image I_1 formed by lens 1 is very close to the focal point F_2 of lens 2, so that the image I_2 formed by lens 2 is very distant. The image in this case will be very large. (Note that to keep the exercise simple, we have not actually traced the principle rays for lens 2 formed by I_1.)

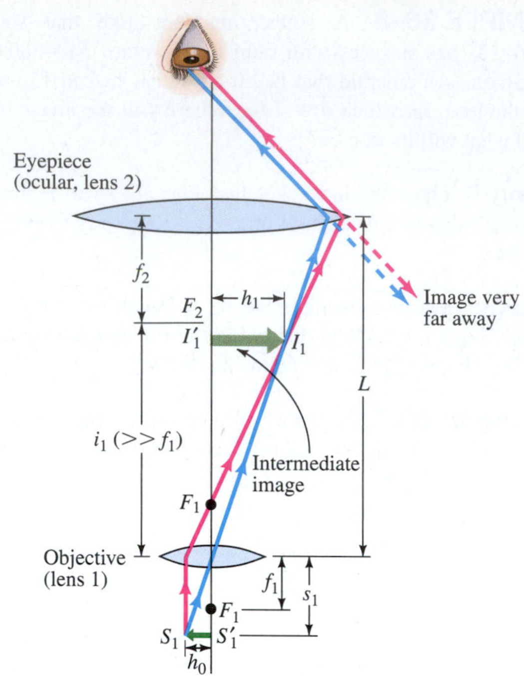

▲ **FIGURE 36–28** Ray tracing for the arrangement of the example.

This arrangement describes a **compound microscope**. The lens near the original source is called the *objective* lens, and the lens near the eye is called the *ocular* or *eyepiece*.

What Do You Think? Should lens 2 be moved up or down to obtain a taller final image?

36–5 Optical Instruments

The example of the compound microscope above shows how an arrangement of thin lenses can work together to produce some desired optical goal. As we implied, real camera lenses are in fact combinations, often rather complicated ones, of thin (or thick) simple lenses. There are many other examples of combinations that fulfill precise needs, and we look at some in this section. We start with the instrument that takes the light reflecting off the ink on this page and transforms it into a form that your brain can use to reconstruct the image.

The Eye

The typical vertebrate eye—the basic structure of which is shown in Fig. 36–29—is a remarkable optical instrument. Light enters the eye through the *pupil*, the size of which can be changed by contraction or expansion of a membrane called the *iris* according to the intensity of the incident light. The light then passes through a convergent *crystalline lens* into a chamber filled with the *vitreous humor*, a fluid with index of refraction near that of water. The light is focused onto the back of the eye, the *retina*, which is covered with sensitive receptor cells. The stimulation of these cells by light produces a message that is sent to the brain along the *optic nerve*, and the brain reconstructs the image.

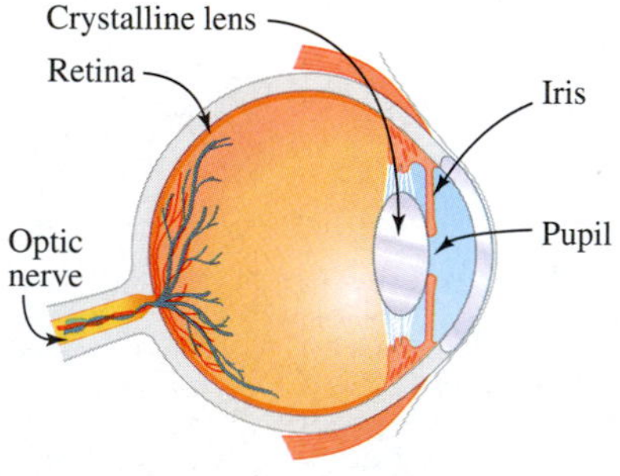

▲ **FIGURE 36–29** Schematic diagram of the human eye and some of its important features.

When a normal eye is relaxed, objects at infinity form an image precisely on the retina, a distance of about 1.7 cm from the lens. When objects are brought closer, the lens is compressed by surrounding muscles and becomes more convergent. The focal length is reduced, and the image continues to be focused on the retina. There is a limit to how much the muscles can compress the lens—this limit is the lens's power of *accommodation*. Objects closer than the *near point*, about 25 cm from the lens (or less for younger people), appear blurred. The near point tends to increase with age because the lens becomes unable to compress as far as it once did, and the image of a near object

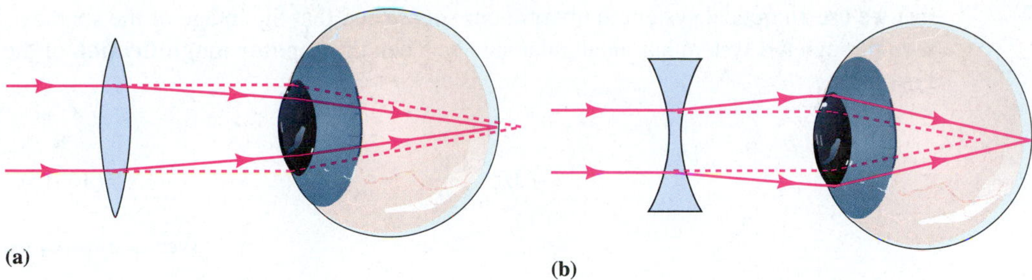

▲ **FIGURE 36–30** The dashed lines indicate the paths rays would take if no correcting lens were present. The solid lines mark the path of rays when a correcting lens is included. (a) A converging lens causes rays from an object, in this case at infinity, to focus closer to the lens of the eye. Such a lens corrects farsightedness by allowing the near point to be moved closer to the eye. (b) A diverging lens causes rays from an object, in this case at infinity, to focus farther from the lens of the eye. Such a lens corrects nearsightedness.

is beyond the location of the retina. Converging lenses correct this problem (Fig. 36–30a). In cases of nearsightedness, the image of an object at infinity is in front of the retina. A diverging lens will provide the necessary correction (Fig. 36–30b).

The Camera

With one important difference, the camera is optically equivalent to an eye. There is a converging lens in front, and the *film* (or receiving surface in a digital camera), which plays the role of the retina, is in back. There is an *aperture*, an opening equivalent to the pupil, and a *shutter*, which provides an approximation to an instantaneous image to avoid blurring of the picture due to motion. The difference between the simple camera and the eye is that the focal length of the lens changes in the eye, whereas the focal length is fixed in a simple camera. Instead, the camera lens moves in and out (changing the image distance) to enable objects of different source distances to produce a focused image on the film.

Angular Magnification

For optical instruments used for observing the world closely, *angular magnification* is a critical concept, and we shall discuss it before we cover some other instruments.

From Eq. (36–7), we see that the magnification of a lens or mirror is infinite when $s = f$. This is less important than it might appear to be because the image distance i also becomes infinite in that case. More important than the actual size of the image is the angle the image takes up in our field of vision. Given the limits of our own vision, *it is this angular coverage that determines how much detail we can see in an observed source.*

Imagine that you are a distance d from some object of height h (Fig. 36–31). For a source that does not cover an enormous part of your vision, the angular size θ_s of the source is

$$\theta_s \cong \frac{h}{d}. \qquad (36\text{–}14)$$

For normal, unaided vision, this angular size can be maximized when the object is brought to the near point of vision, around $d = d_{\text{min}} = 25\text{ cm}$, and it is $\theta_s \cong h/(25\text{ cm})$ that is used as a reference for the angular magnification. Suppose now

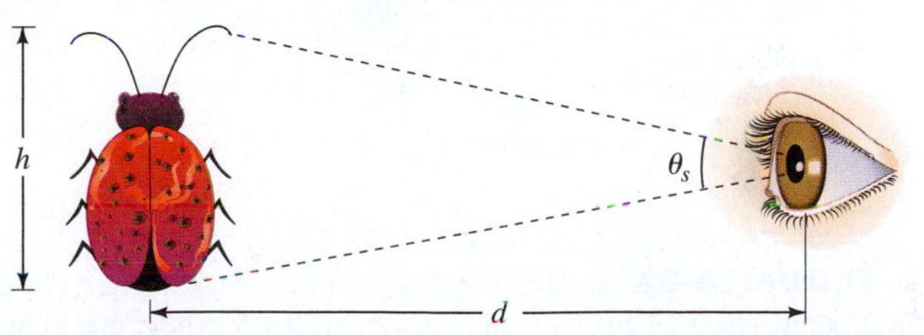

◀ **FIGURE 36–31** The angular size of an object, θ_s, is the relevant quantity for our ability to see detail in the object.

that we use an optical system to observe our source and that the image of the source as seen through the system has an angular size θ_i. Then the **angular magnification** of the system is

$$M_\theta \equiv \frac{\theta_i}{\theta_s}. \tag{36–15}$$

ANGULAR MAGNIFICATION

We do not bother with signs here and keep track only of the magnitudes of the angular sizes. If we know the angular magnification of two elements that are superposed in an optical system, then the net angular magnification is the product of the angular magnifications of each element.

The Simple Magnifier

A converging lens has a positive focal length. By Eq. (36–2), $\frac{1}{i} = \frac{1}{f} - \frac{1}{s}$. For a real object, i passes from positive (real image) to negative (virtual image) as the object moves toward the lens through the point $s = f$. At this point, i shifts to $-\infty$. A *simple magnifier* is a converging lens with the object placed near $s = f$ (Fig. 36–32). If the object size is h, the image size is, by definition, $h_i = Mh$, where $M = i/s$ is the magnitude of the magnification. The image size is infinite if i is infinite, but the *angular size of the image is finite*. When $s = f$, we have for the angular size

$$\theta_i = \frac{h_i}{i} = \frac{Mh}{i} = \frac{h}{s}\bigg|_{s=f} = \frac{h}{f}. \tag{36–16}$$

Note that we have no trouble seeing an image at infinity. At the near point, $d_{min} = 25$ cm, the angular size of our object is $\theta_{object} = h/d_{min}$. Thus the angular magnification of the magnifier is

$$M_\theta = \frac{\theta_i}{\theta_{object}} = \frac{h/f}{h/d_{min}} = \frac{d_{min}}{f}. \tag{36–17}$$

If we choose a converging lens with a focal length of 2 cm, we get an angular magnification of $(25\text{ cm})/(2\text{ cm}) = 12.5$.

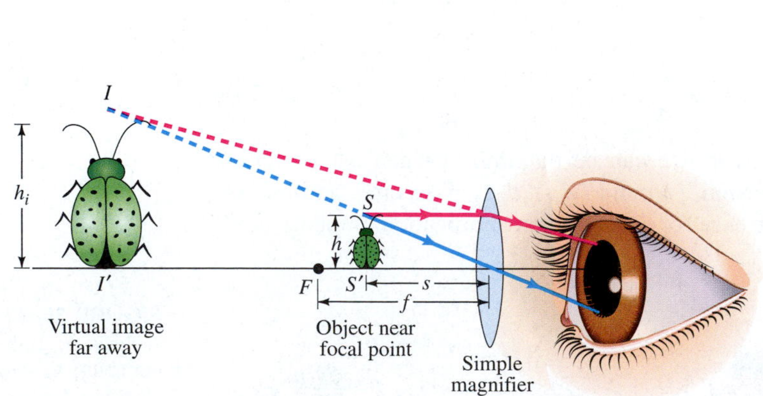

▲ **FIGURE 36–32** (a) The simple magnifier is a converging lens with an object placed near the focal point. The image is virtual and far away. (b) A simple magnifier in use.

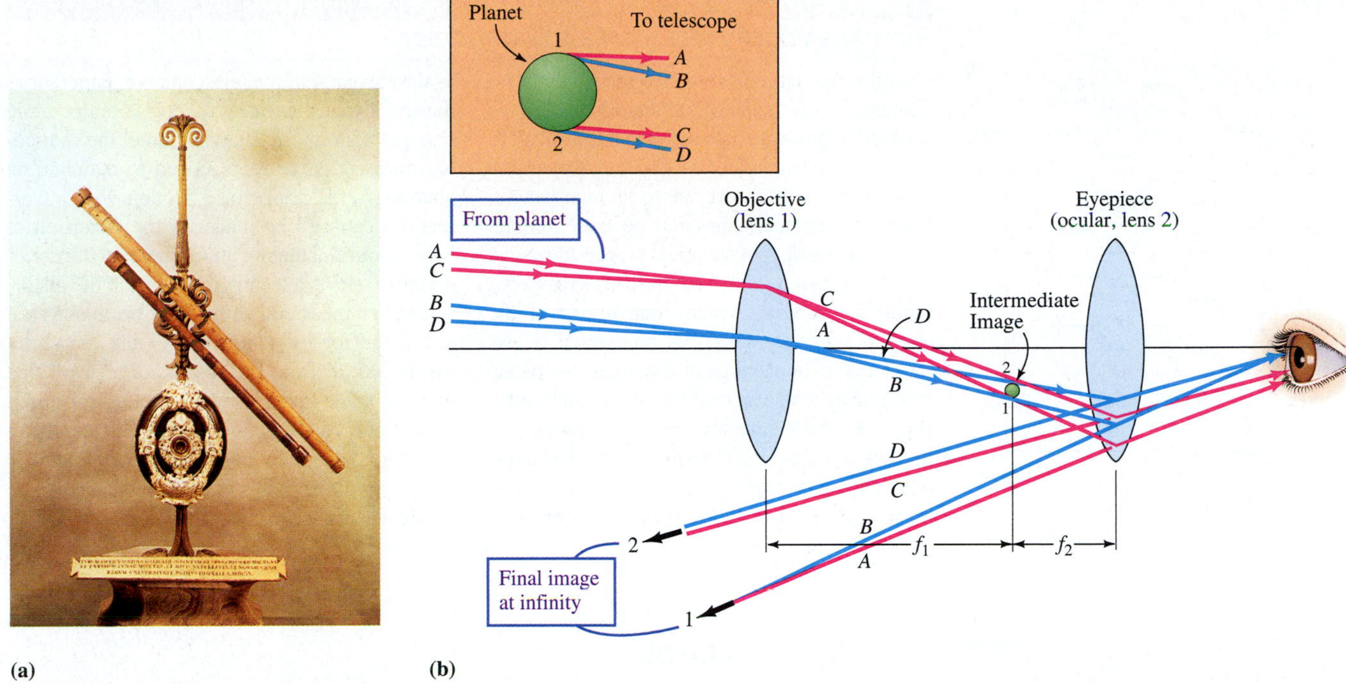

▲ **FIGURE 36–33** (a) Galileo's refracting telescope, used for viewing distant objects. (b) Schematic diagram of a refracting telescope.

The Telescope

The **telescope** magnifies very distant objects. It was invented in Holland at the beginning of the seventeenth century and made an impact on astronomy soon thereafter. Galileo built his own telescope in 1609 (Fig. 36–33a).

The *refracting telescope* (a telescope with only refracting elements) is a system designed around the fact that the original object is very distant—in effect at infinity (Fig. 36–33b). The first lens, the *objective*, creates an intermediate image very close to the focal point of that lens. If that point coincides with the focal point of the eyepiece, then the eyepiece acts as a simple magnifier. The final image is magnified. Let's calculate the angular magnification for an object that has angular size θ_s. (The Moon, for example, has angular size of $1/2°$. With the naked eye, we can distinguish stars separated by about $1'$ of arc [$1/60$ of $1°$].) If the original object has size h_0, the objective produces an image of size $h_1 = Mh_0 = ih_0/s = i\theta_s = f_1\theta_s$. The final image then has an angular size given by Eq. (36–16) with $h \to h_1$ and $f \to f_2$, namely, $\theta_i = h_1/f_2 = \theta_s f_1/f_2$. In turn, the angular magnification is

$$M_\theta = \frac{\theta_i}{\theta_s} = \frac{f_1}{f_2}.$$

This quantity will be large for large f_1: The objective lens of a telescope should have as long a focal length f_1 as is practical, and that is why a refracting telescope is long.

The study of distant galaxies depends on an examination of the spectrum of the light they emit and of their energy output. The incident light from very distant objects is rather low in intensity, and more light is needed at the eyepiece in order to study spectra. To be most efficient at collecting light, the diameter of the optical system must be large. Large lenses are more difficult to construct than large mirrors, so most large telescopes are *reflecting telescopes* rather than refracting telescopes. In a reflecting telescope, a mirror replaces the objective for the purpose of creating an intermediate image, which is then magnified by the eyepiece (Fig. 36–34). Another advantage of such a telescope is that it has no chromatic aberration (see Section 36–6).

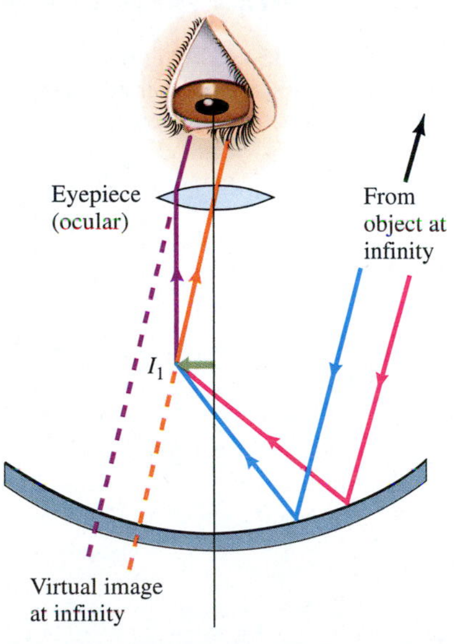

▲ **FIGURE 36–34** Schematic diagram of a reflecting telescope.

THINK ABOUT THIS. . .

HOW ARE MODERN TELESCOPES CONSTRUCTED?

Useful single large mirrors like that of the 5-m-diameter mirror in the Mount Palomar telescope are hard to produce, both because it is difficult to cast a single large piece of glass and because a large mirror will distort due to its own weight. Large, modern telescopes may be made from an array—a mosaic—of spherical mirrors, arranged on a frame that positions the mirrors across a paraboloidal surface. As any one mirror of the array is relatively small, its production is less problematic. In one type of design, the elements of the array are hexagons that fit together neatly (Fig. 36–35). Computer-controlled positioning devices are then used to mount the hexagons onto the main support frame. Each hexagonal mirror can be oriented by computer command; as the elevation of the telescope is changed or as thermal expansion due to ambient temperature changes distorts the support frame, the individual mirrors can be repositioned to maintain an aberration-free image (see Section 36–6). Corrections can also be made for the distortion of light by our fluctuating atmosphere by the use of *adaptive optics*, a technique by which the distortion is first measured with a laser beam and then accounted for by adjustment of the array elements. Recall from our discussion of rays that one can make holes in a lens or a mirror and it would still work to gather light and focus it at a detector. Indeed, in other versions of today's telescopes the elements of the array that make up the device may be widely separated.

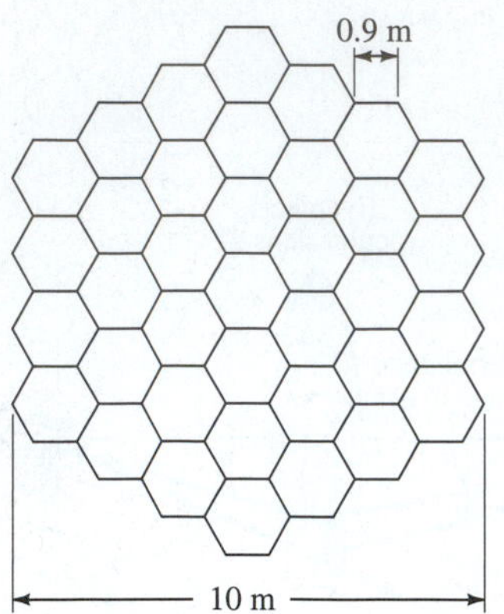
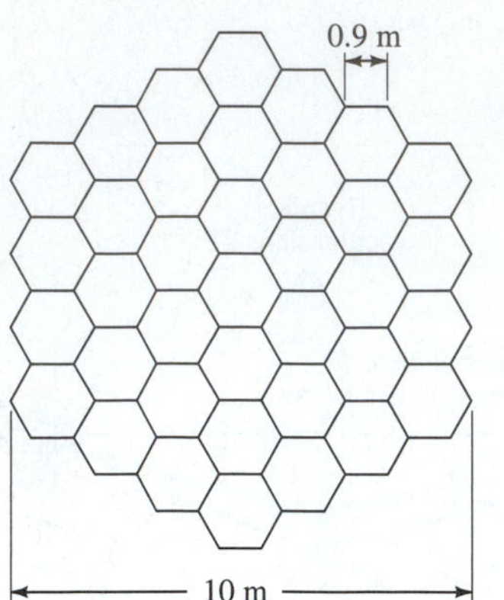

FIGURE 36–35 A large collecting mirror is made by assembling hexagonal elements.

*36–6 Aberration

FIGURE 36–36 This poor image is the result of spherical aberration, a type of monochromatic aberration.

An accurate calculation would show that all rays that arrive at a spherical mirror or refracting surface from infinity cross in a small but finite region rather than at a single point. This is but one example of **aberration** (Fig. 36–36). Aberration should be distinguished from *distortion*, in which an image is not identical in form to the object, as in a fun-house mirror. For scientific purposes, the fun-house image is not necessarily a bad one, because every ray from the object is correctly positioned at a precise location in the image. Aberration concerns what we might call the quality of an image, not its geometric form.

We can distinguish two important types of aberrations in geometric optics. *Monochromatic aberrations* describe the fact that, in real optical systems, the rays from a given point on an object are not focused on a single image point (Fig. 36–37a). The correction for this type of aberration depends on the application. An optical system that collects images only from distant objects will have no aberration when a parabolic surface is used (Fig. 36–37b). Although such surfaces are difficult to construct from glass, a pool of mercury spinning about a vertical axis forms a parabolic surface, and such surfaces are employed in some modern telescopes. These telescopes only point up! Mosaic telescopes can also avoid monochromatic aberration. Alternatively, this type of aberration is minimized when the spherical section of the lens surface or mirror is small, although then the system collects less light.

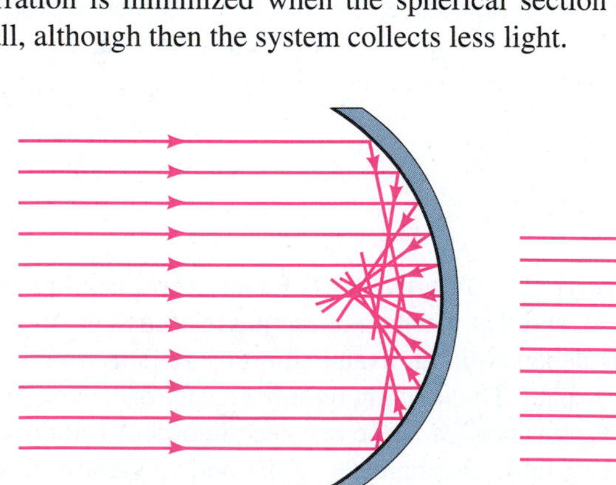

FIGURE 36–37 (a) In monochromatic aberrations, all the rays from infinity do not pass through the same point for a spherical mirror, so the focus is not sharp. (b) This type of aberration is eliminated by use of a parabolic mirror.

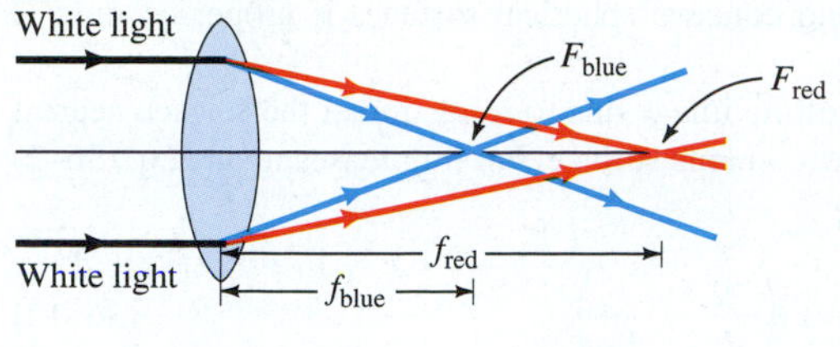

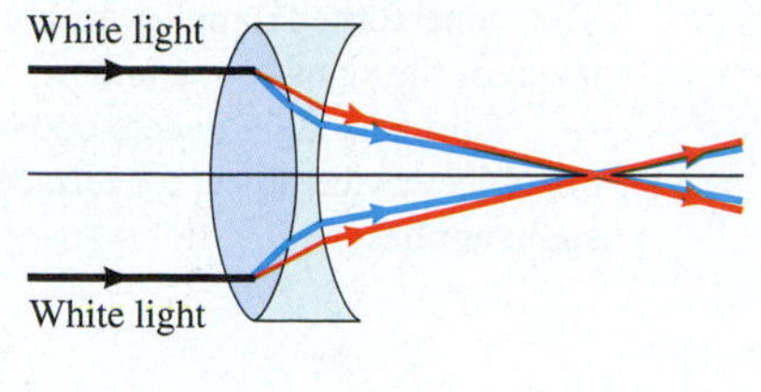

▲ **FIGURE 36–38** (a) In chromatic aberrations, the focal point of a converging lens may be different for different wavelengths. Here we show rays only for red light and blue light. (b) This type of aberration is eliminated by combining the lens with another lens with a different dispersion.

Chromatic aberrations appear in refracting systems but not in mirrors. We treated dispersion in Chapter 35, in which the index of refraction of a material depends on the wavelength of the light. The optical path of a ray at one wavelength will differ from that of a ray at another wavelength (Fig. 36–38a). If a given point on an object is the source of a mixture of wavelengths (as is true of white light), then the image of this point is spread out according to the wavelength. A simple correction for chromatic aberration is to use filters that allow only a narrow band of wavelengths to pass. More commonly, several lenses are superposed (Fig. 36–38b). Different elements are designed to have canceling dispersion to minimize the net dispersion.

A good camera lens may consist of a dozen elements of different types of glass, with complicated geometric relations, which, in the case of zoom lenses, are also variable. Unfortunately, even with all the corrections such lenses provide, the wave nature of light provides a fundamental and unavoidable limitation on the ability of optical systems to produce sharp images (see Chapters 37 and 38).

Summary

Geometric optics is based on two basic laws of the behavior of light rays: In reflection, the angle of incidence on a reflecting surface is equal to the angle of reflection. In refraction, Snell's law, $n_1 \sin\theta_1 = n_2 \sin\theta_2$, holds, where θ_1 and θ_2 are the angles of incidence and refraction, respectively, for a ray incident from medium 1 on medium 2. Ray tracing is a technique that allows us to locate the image of a given source.

A spherical reflecting or refracting surface forms an image of an object that is a distance s from the surface. Bundles of light rays pass through an image point (for a real image) or are directed as though they all come from such a point (for a virtual image). The image point is a distance i from the surface. One limit of such a surface is the plane mirror, for which the distance of the virtual image from the mirror is given by $i = -s$.

Parallel rays falling on a spherical reflecting or refracting surface approximately converge to the focal point, a distance f from the element. For a mirror, $f = R/2$ [Eq. (36–1)], where R is the radius of curvature of the spherical section. The distance of the object and the distance of the image from the surface, and the focal length for a spherical mirror are related by

$$\frac{1}{s} + \frac{1}{i} = \frac{1}{f}. \tag{36–2}$$

Equation (36–2) applies to both convex and concave mirrors if proper account is taken of the signs of s, i, and R. The image size is magnified by a factor M, the magnification, times the object size, where

$$M = -\frac{i}{s} = \frac{f}{f - s}. \tag{36–6, 36–7}$$

For a positive M, the image is upright; for a negative M, the image is inverted.

For a spherical boundary between a medium of refractive index n_1 and a medium of refractive index n_2, with light incident from medium 1, Eq. (36–2) is replaced by

$$\frac{n_1}{s} + \frac{n_2}{i} = \frac{n_2 - n_1}{R}. \tag{36–10}$$

The same formula applies to both convex and concave spherical surfaces if proper account is taken of the signs of s, i, and R.

Thin lenses are understood by thinking of the image due to refraction at the surface nearest the object as the object for refraction at the second lens surface. For thin lenses in air, Eq. (36–2) again applies, with

$$\frac{1}{f} = (n - 1)\left(\frac{1}{R_1} - \frac{1}{R_2}\right). \tag{36–13}$$

Moreover, Eqs. (36–6) and (36–7) apply also to thin lenses.

Thin lenses can be used singly or in combination to make up optical instruments, including magnifiers, eyeglasses, cameras, microscopes, and telescopes. Angular magnification, which measures the ratio of the angular size of an object as seen through the instrument to the object's angular size as the naked eye sees it, is a fundamental consideration for those instruments whose explicit purpose is to magnify, as is the quality of the image they produce.

Understanding the Concepts

1. Why is "ƎƆИA⅃UᗺMA" written on the front of an ambulance?
2. Consider a large room with walls covered with mirrors; at the center of the room is a candelabra with burning candles. Is the room brighter than a comparable room with black drapes in place of the mirrors?
3. The image of a distant candle is projected by a converging lens on a screen placed at the focal length of the lens. A piece of paper is taped over the lower half of the lens. Will only half of the image be seen?
4. Draw a right-handed coordinate system and its image in a plane mirror. Is the image a right-handed or a left-handed coordinate system? (In a right-handed system, the vector product $\hat{i} \times \hat{j}$ points along $\hat{k}$.)
5. A physicist stands in front of a mirror. The floor is the xy-plane at $z = 0$; the mirror is the xz-plane. He has learned that reflections involve the change $(x, y, z) \rightarrow (-x, -y, -z)$. Will he be surprised when he looks in the mirror? What is the rule that gives him the correct description of what he sees?
6. Would a dental mirror, the small mirror a dentist uses to examine your teeth, be concave, convex, plane, or sometimes one or another?
7. The sideview mirrors of some cars are labeled "Objects seen in this mirror may be closer than they appear." Is the mirror plane, convex, or concave?
8. For each of the simple lenses shown in Fig. 36–23, Eq. (36–6)—for magnification—shows that the size of the image of a ball placed at the focal point is infinite. Can you see by ray tracing why this must be?
9. Figure 36–39 shows the reflection made by a spherical surface. Parts of all four walls of the room are visible in the image. Why?
10. To form a taller image in Example 36–5, the object in Fig. 36–15 should be moved (a) to the left, (b) to the right, (c) can't be done.
11. Does the focal length of a lens change when the lens is in water?
12. When a magnifying glass is lined up perpendicular to the line between it and the Sun, a hot spot forms on the side of the lens away from the Sun. What is the relation between the distance of this hot spot from the lens and the focal length of the glass? Why does the spot become hot?
13. A camera works by forming a real image on a film plate. Can a camera take a picture of a virtual image?
14. In William Golding's novel *The Lord of the Flies* (1954), some boys rediscover fire with the aid of the Sun shining through the eyeglasses of Piggy, a nearsighted boy. Has Golding made a mistake?

▲ **FIGURE 36–39** Question 9.

15. Are any principal rays useful for a point *on* the axis of an optical system?
16. You recently noticed that someone who used to wear thick glasses now wears glasses that are much thinner. Assuming that their eyesight has not improved, why might this be?
17. When you have an eye exam, even one as simple as reading an eye chart, the examiner may dilate (open) the pupil by putting drops in your eye. Why is that useful?
18. The n-dependence of the bending of light allows for the making of a flat eyeglass lens with material in which the index of refraction varies with position. Sketch the profile of the index of refraction for a converging lens constructed in this way.
19. Legend has it that Archimedes, acting as an advisor to the ruler of Syracuse, devised an optical system made of shields that could concentrate sunlight sufficiently well to set enemy boats on fire from a distance. How plausible is this legend?
20. Can the image ever be smaller than the object for a converging lens?

Problems

36–1 Images and Mirrors

1. (I) Consider two mirrors at right angles to each other (Fig. 36–40). How many virtual images will a pointlike light source have?

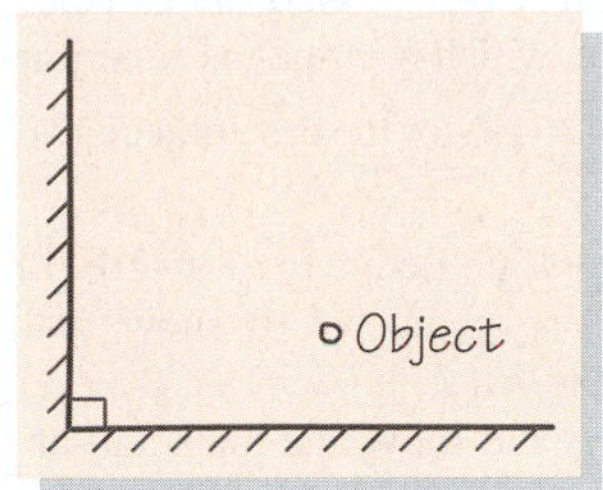

▲ **FIGURE 36–40** Problem 1.

2. (I) Consider two parallel mirrors that face each other, placed along the x-axis at $x = a$ and $x = -a$. Assume that a point source of light is placed at $x = x_0$ between the mirrors. What are the locations of the four images of the point source with the smallest values of image distance i?

3. (I) Two mirrors, each 2.0 m wide, are placed facing each other and parallel to each other and are separated by 10 cm. A ray of light enters the gap between them, grazes the edge of one mirror and strikes the other mirror at an angle of 30° with respect to the normal to the mirrors. At each reflection, the intensity of the light beam is attenuated by 5 percent. By how much is the beam attenuated when it finally leaves the space between the mirrors?

4. (II) A mirror is exactly half your height, and the top of the mirror is aligned with the top of your head. (a) If your eyes were at the top of your head, how close would you need to be to the mirror in order to be able to see your feet? (b) If your height is 158 cm and your eyes are 12 cm below the top of your head, what would have to be done with the mirror so that you could see both the top of your head and your feet?

5. (III) A kaleidoscope contains three plane mirrors forming a prism with an equilateral triangular base of side a (Fig. 36–41). Consider a small object placed on the axis of the kaleidoscope. Construct the position of the images formed by single as well as double reflection. How far are these images from the axis?

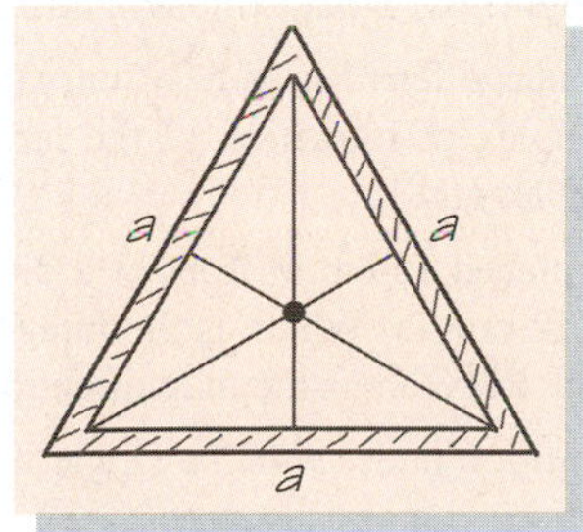

▲ **FIGURE 36–41** Problem 5.

6. (III) Suppose that two plane mirrors meet at an angle of 60° (Fig. 36–42). An object is placed between the mirrors on the line that bisects this angle. Use graphical methods or trigonometric methods to locate all the images.

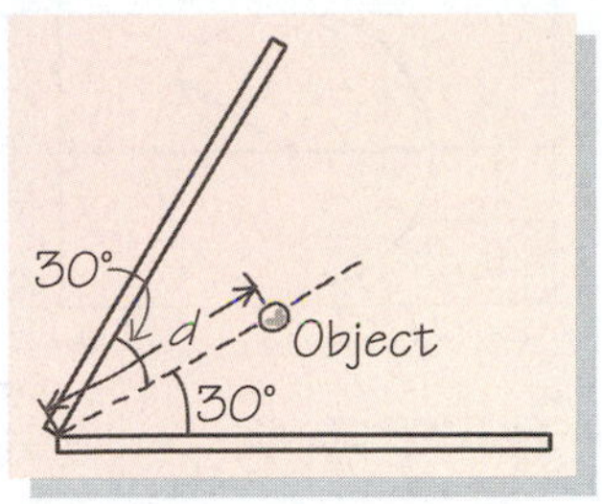

▲ **FIGURE 36–42** Problem 6.

36–2 Spherical Mirrors

7. (I) A dime 60 cm away from and on the optical axis of a concave spherical mirror produces an image 20 cm away from the mirror. If the dime is moved on the axis to 35 cm from the mirror, where will the image move? How large is the radius of the sphere of which the mirror is a section? Draw the system for the second case described.

8. (I) A paper clip is placed on the axis 28 cm away from a convex mirror, part of a sphere of radius 44 cm. Where will the image be located, and what is the magnification? Make a sketch, including rays.

9. (I) An object of height 2.0 cm is placed 20 cm from a concave mirror. The real image is found to be 8.0 cm from the mirror. On which side of the mirror is the image and how tall is it? Is it inverted?

10. (I) A concave mirror has a radius of curvature of 176 cm. What is the size of the image of an object 6.00 cm tall that is placed 133 cm from the mirror?

11. (I) A concave mirror is cut from a spherical surface of radius of curvature 2.0 m. A pencil 10 cm long is placed perpendicular to the axis of the mirror at a distance of 80 cm from the mirror. Where is the image and how large is it?

12. (II) Consider an object in front of a convex mirror. A plane mirror is placed along the axis between the object and the convex mirror, and moved along till the images of the object in the two mirrors coincide. If the distance of the object to the plane mirror is x and the distance of the plane mirror to the convex mirror is y, what is the focal length of the convex mirror?

13. (II) A bird flies from far away toward a concave mirror, with constant speed v. What is the velocity of the image as a function of the distance of the bird from the mirror? When does the bird meet its image? [*Hint*: Use the relation between source distance and image distance and differentiate with respect to time.]

14. (II) By using the same reasoning that we used in the text for the case of the concave mirror, show that the reflection of ray 1 in Fig. 36–11 appears to originate at point F, independent of the angle θ. Your argument shows that, at point F, there is an image of a source point at infinity.

15. (II) Use ray tracing for parallel rays far from, as well as near to, the optical axis to show that parabolic mirrors more accurately focus parallel rays than do spherical mirrors.

36–3 Refraction at Spherical Surfaces

16. (I) A sphere of glass ($n = 1.50$) of radius 3.0 cm is immersed in water ($n = 1.33$). A small flower (at point B) is 1.5 cm outside the sphere (Fig. 36–43, see next page). What are the location and nature (real or virtual) of the flower's image made by refraction at the first surface?

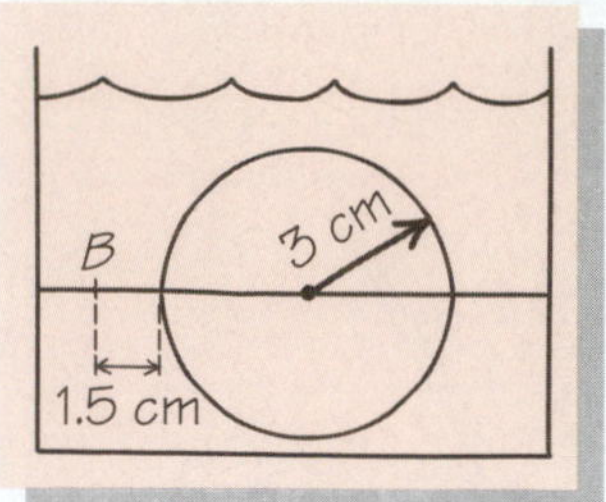

▲ **FIGURE 36–43** Problem 16.

17. (I) An object is placed 15 cm in air from the convex surface (radius of curvature 10 cm) of a very thick piece of glass ($n = 1.5$). Where is the image?

18. (I) The single refracting surface of a piece of glass in air has a radius of curvature $R = 8.5$ cm. A ray parallel to the axis of the curved piece of glass is bent toward the axis inside the piece of glass and crosses that axis at a point 13 cm into the glass. What is the index of refraction of the glass?

19. (I) A fish is located at a distance of 40 cm from the glass pane of an aquarium. How far from the glass does the fish appear to be located to an observer looking from the outside? (Use $n_{water} = 1.33$.)

20. (I) A small fish is cast into the center of a glass sphere of radius $R = 5$ cm and $n = 1.5$. Where will an observer see the fish? Where will the observer see a decorative background pattern painted on the back side of the sphere?

21. (II) A glass rod of refractive index $n = 1.6$ and diameter 1.6 cm has a hemispherical cap (Fig. 36–44). There is a fault in the glass 2.3 cm from the end. Can you see this fault if you look at the rod through the spherical cap? From about how far away should you look?

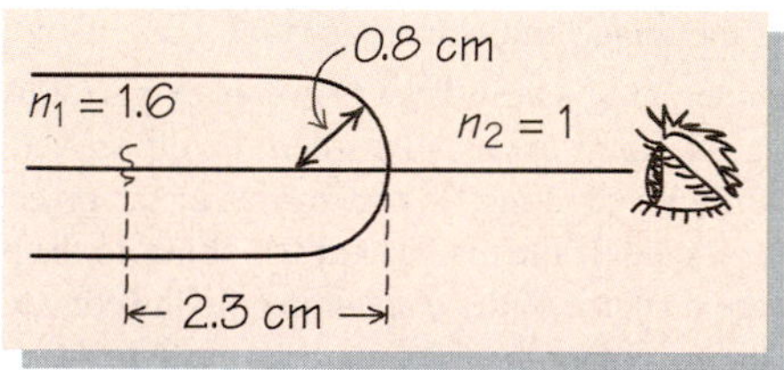

▲ **FIGURE 36–44** Problem 21.

22. (II) By applying Eq. (36–10), show that, if light is incident on a convex refracting surface with $n_2 > n_1$ (see Fig. 36–18), there is a critical distance s_c such that the image of an object closer than s_c will be virtual. Find s_c, and show by ray tracing that the virtual image when $s < s_c$ is upright and magnified.

23. (II) Consider the situation described in Problem 22. Use ray tracing to find the image when $s = s_c$, and when $s = 3s_c$.

24. (II) Consider a convex spherical boundary between two media with an upright object whose extreme point is at S, as in Fig. 36–18. Suppose that $n_2 < n_1$ rather than $n_2 > n_1$. Find the nature of the image (inverted or upright, virtual or real, reduced or magnified) by tracing rays from S. Is there a critical distance at which the nature of the image changes, as in Problem 22?

25. (II) A convex spherical boundary produces an image whose distance from the boundary surface is governed by Eq. (36–10). Suppose that $n_2 > n_1$. (a) Show that when an object is very far from the surface, the image is a distance $i = n_2R/(n_2 - n_1)$ from the surface, and that the image is inverted, reduced, and real. (b) What is the distance s at which the image distance becomes infinite? (c) What is the position of the image for s just less than the critical value found in part (b)? Is it real? (d) As s continues to decrease, what happens to the position of the image?

26. (II) Consider a concave surface of radius of curvature R that separates two media with indices of refraction n_1 and n_2, where $n_2 > n_1$ (see Fig. 36–19). Find the distance s of an object for which the image, which is virtual, is superimposed on the object.

27. (II) Derive Eq. (36–9) for the case of a concave surface shown in Fig. 36–17.

28. (II) Use Fig. 36–45 to derive Eq. (36–10). Let $n_1 < n_2$ and use small angles. Use geometry to show that $\theta_2 = \beta - \alpha$ and $\theta_1 = \beta + \gamma$, and $n_1(\beta + \gamma) = n_2(\beta - \alpha)$. Use the exact relation $AB = R\beta$ and show that the small-angle approximations give $AB = s\gamma = i\alpha$ to finally obtain Eq. (36–10).

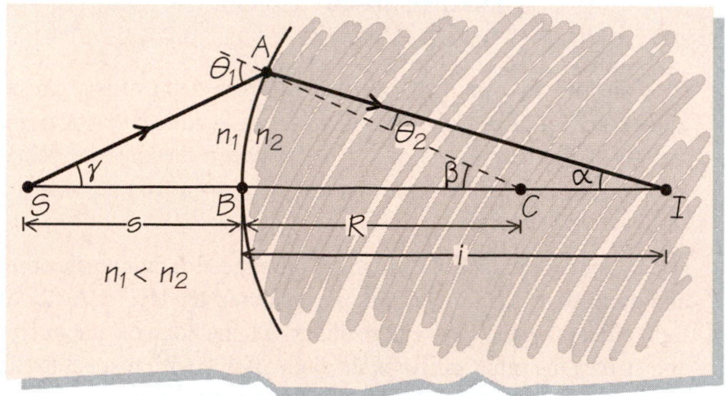

▲ **FIGURE 36–45** Problem 28.

29. (III) In deriving Eq. (36–10) in the previous problem, we took $n_1 < n_2$ in Fig. 36–45. Make a new drawing appropriate to $n_2 < n_1$ for a convex surface. Apply the same kind of reasoning, using small angles to show that the same algebraic formula applies whatever the relative sizes of n_2 and n_1.

30. (III) Show that Eq. (36–10) holds for a concave refracting surface with $n_2 > n_1$ by drawing a figure analogous to Fig. 36–45 (Problem 28) and by making small-angle assumptions.

36–4 Thin Lenses

31. (I) The image of an object placed 24 cm away from a thin lens forms at a distance of 51 cm on the other side of the lens. (a) What is the focal length? (b) What type of lens is it? (c) Is the image real? upright? (d) What is the magnification?

32. (I) A double concave lens has radii of curvature of 6.0 cm and 7.5 cm. If the index of refraction of the lens material is 1.56, what is the focal length?

33. (I) An apple is placed 15 cm in front of a diverging lens with a focal length of 22 cm. (a) Where is the image? (b) Is the image real? (c) upright? (d) What is the magnification?

34. (I) Find the condition under which a single thin lens produces a real image, starting with a real source.

35. (II) We want to form an image of an insect magnified twofold by using a converging lens with a focal length of 25 cm. (a) Where should the object be placed for the image to be real? (b) Repeat part (a) for a virtual image.

36. (II) An object 4.5 cm high is placed on one side of a thin converging lens of focal length 43 cm. What are the location, size, and orientation of the image when the object is (a) 86 cm from

the lens, (b) 50 cm from the lens, (c) 40 cm from the lens, (d) 15 cm from the lens?

37. (II) The two surfaces of a thin lens have radii of the same sign and magnitude. Show by ray tracing that the focal length of this lens is infinite. Is the image produced by this lens real or virtual?

38. (II) A thin converging lens forms an image of a distant mountain at a distance of 38 cm from the lens. (a) What is the focal length of the lens? (b) A pine cone is placed 75 cm from the lens. Describe the resulting image: its magnification and distance from the lens, and whether it is real or virtual, upright or inverted. (c) The lens glass has an index of refraction of 1.55. The lens is immersed in water, index of refraction 1.33. What is its focal length in water?

39. (II) Consider the thin lenses shown in Figs. 36–23a through 36–23d. Suppose that in each case the magnitudes of the radii of curvature are $R_1 = 25$ cm and $R_2 = 60$ cm, and that $n = 1.55$. (a) Find the focal lengths for each of the four lenses, and use the sign of the focal lengths to obtain the locations of the image of a source 10 m from each lens. (b) In each case, is the image upright or inverted, real or virtual? (c) Calculate the magnification M from Eq. (36–6) and check that it is consistent with your results in part (b).

40. (II) An object is placed 25 cm to the right of each of the lenses of Problem 39. For each case, locate the image, state whether it is upright or inverted and real or virtual, and give the magnification.

41. (II) Repeat Problem 40 for an object placed 65 cm to the right of each of the lenses.

42. (II) Consider an object on the left side of a thin lens with $R_1 > 0$ and $R_2 < 0$. The image will be on the right of the lens. Let X_0 be the distance of the object from the focal point on the left and X_1 be the distance of the image from the focal point on the right. Show that $X_0X_1 = f^2$. This was Newton's original formulation of the lens equation, given in his *Opticks* in 1704.

43. (II) Show that the thin lens equation [Eq. (36–12)] follows from Fermat's Principle, according to which all light rays leaving an object and traveling to the image do so in minimum times. [*Hint*: You may need the following geometric result. Consider a line that is drawn perpendicular to a particular radius at a distance $R - x$ from the center. Then the distance from the radius to the point where the line touches the circle, y, is given by the relation $y^2 = x(2R - x)$ which for small x reduces to $y^2 = 2Rx$.]

44. (II) Two thin lenses of focal length f_1 and f_2, respectively, are aligned along the same axis and placed very close together. Show that the focal length f of the combination is given by

$$\frac{1}{f} = \frac{1}{f_1} + \frac{1}{f_2}.$$

36–5 Optical Instruments

45. (I) The eyes of an elderly person have near points of 70 cm. What must the focal length of corrective lenses be in order for this person to read a book at a distance of 30 cm?

46. (I) A nearsighted person has near and far points of 12 cm and 41 cm, respectively. (The *far point* is the farthest point at which a person can see clearly.) (a) Determine the lens required for this person to be able to see clearly at infinity. (b) What does the lens correction of part (a) do to the near point? Can the person still easily read a book?

47. (I) You have a thin lens with $f = 9$ cm. If you want to see an insect magnified by a factor of 3, how close should you hold the glass to the insect? (Let the image be virtual.)

48. (I) What is the magnification of a telescope that has an objective lens with a focal length of 80 cm and an eyepiece with a focal length of 1.7 cm?

49. (I) You are trying to construct a compound microscope given two lenses with focal lengths $f_1 = 1.0$ cm and $f_2 = 4.0$ cm. How far apart should you place the lenses in order to obtain an angular magnification of 60?

50. (II) The two lenses of a telescope with magnification of 120× are separated by 70 cm. What are the focal lengths of the lenses?

51. (II) Calculate the angular magnification of the reflecting telescope shown in Fig. 36–33b.

52. (II) Galileo's original telescope had a convex objective and a concave eyepiece. The focal points of the two lenses coincided, as shown in Fig. 36–46. What is the angular magnification for a distant (but not infinitely far) object, and what is it for an infinitely far object?

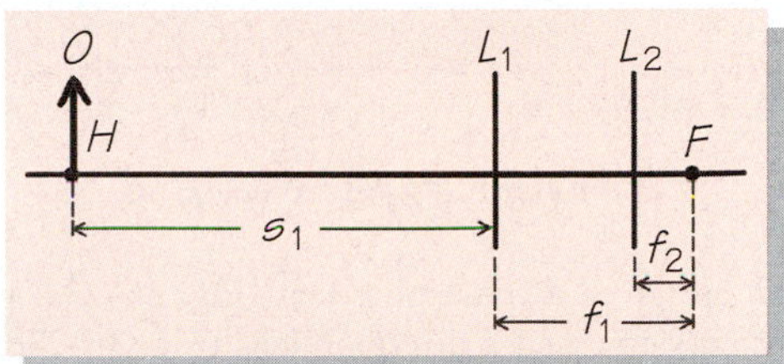

▲ **FIGURE 36–46** Problem 52.

*36–6 Aberration

53. (II) Consider a spherical mirror without making the paraxial approximation (Fig. 36–47). Show that when a ray parallel to the axis makes an angle θ with the radius R at the point of contact, then f, here the distance at which the ray crosses the axis, is given by

$$f = R\left(1 - \frac{1}{2\cos\theta}\right).$$

Show that for small angles, this formula reduces to $f = R/2$. Note that F is not the focal point here (there is no sharp focus), but only the point at which some particular ray crosses the axis.

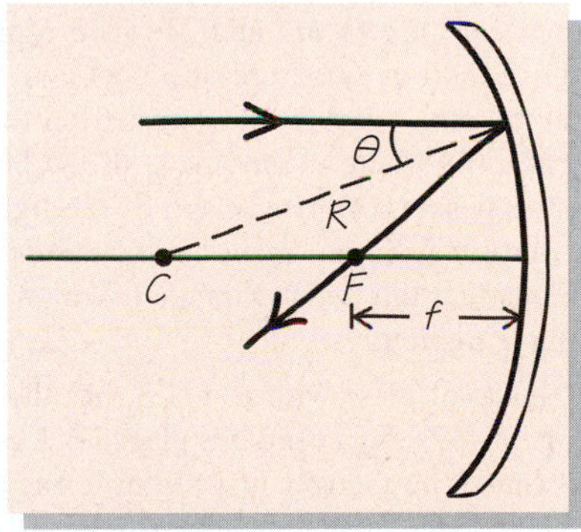

▲ **FIGURE 36–47** Problem 53.

54. (II) Use the result of Problem 53 to calculate the spread in values of f for a spherical mirror of radius 0.18 m and arc length 46 cm.

55. (II) The index of refraction of optical glass used for a thin lens with $R_1 = +20.00$ cm and $R_2 = +28.75$ cm is $n = 1.48523$ for light of wavelength $\lambda = 587.6$ nm and $n = 1.48135$ for light of wavelength $\lambda = 768.2$ nm. What is the difference in the focal length for these two wavelengths?

General Problems

56. (II) Consider a circular concave mirror of focal length f and diameter d, where $f \gg d$. This mirror's optical axis is aligned with the Sun. What is the area of the spot that contains the reflected rays as a function of the distance L from the mirror if $L < f$ (Fig. 36–48)? Sunlight has an intensity I as it arrives at the mirror. Find the intensity of the reflected rays as a function of L. Treat the Sun as a point source.

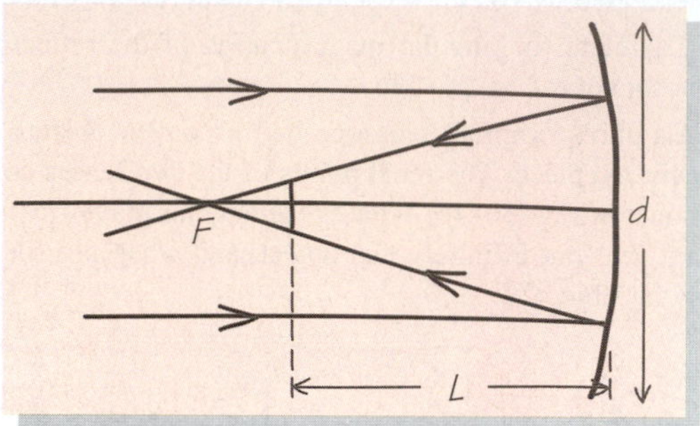

▲ **FIGURE 36–48** Problem 56.

57. (II) You are given a converging lens (Fig. 36–23a) with equal radii of curvature and a diverging lens (Fig. 36–23c) with the same radii of curvature as those of the converging lens. The lenses are made of material with $n = 1.50$, and the radii of curvature are all 35 cm. They are placed at opposite ends of a tube 15 cm long, and the nearer lens is 10 cm from an object. What is the location of the image that results from the two refractions? Does it make a difference whether the converging or the diverging lens is closer to the object?

58. (II) Consider a 45-cm-long cylinder of glass in air, with $n = 1.6$, like the cylinder shown in Fig. 36–21. The two ends are shaped into sections of spheres; each has radius 18 cm. A small object is placed perpendicular to the optical axis at a distance of 12 cm from one of the spherical surfaces. (a) Find the location of the object's image due to refraction at surface 1. (b) Let this image be the object for surface 2, and find the location of *its* image as light passes through surface 2. (c) Use ray tracing to determine if the final image is upright or inverted.

59. (II) Two concave mirrors M_1 and M_2 face each other. They have respective radii of curvature of 32 cm and 14 cm and are separated by 50 cm. A lightbulb is placed on the optical axis 7 cm from M_1. (a) Where is the image of the bulb formed by M_1? Draw the system. (b) The image of the lightbulb formed by M_1 can in turn form an image as the result of reflections from M_2. Construct this second image by ray tracing, starting from the source lightbulb.

60. (II) A lens, made of glass with $n = 1.5$, has the configuration shown in Fig. 36–49, and a candle is placed 62 cm from surface 1. The lens cannot be thought of as thin; it has a thickness of 4 cm. (a) Where is the image made by surface 1? Is it inverted? What is the magnification? (b) By using the image made by surface 1 as an object for surface 2, find the final image's location relative to the candle as well as the magnification of the image. Is it inverted or upright?

61. (II) A thin lens with focal length f_1 is placed a distance d in front of a concave mirror with focal length f_2. What is the focal length of the combination?

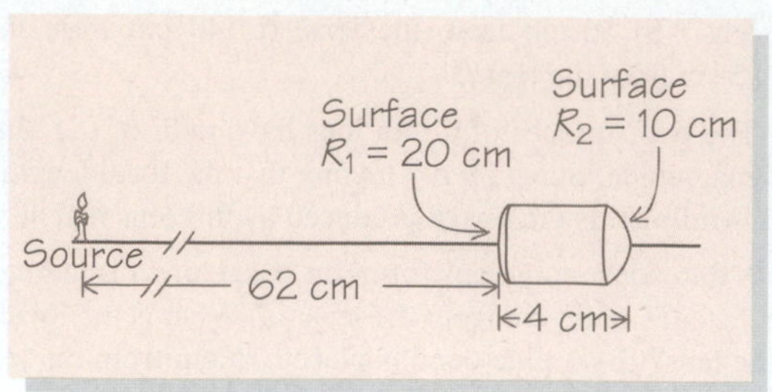

▲ **FIGURE 36–49** Problem 60.

62. (II) Consider the sphere of glass with $n = 2$ in Fig. 36–50. Any incoming ray is parallel to an axis through the middle of the sphere and will be refracted, striking the rear surface of the sphere at the axis. Demonstrate that this holds true for paraxial rays. If the back surface is painted with a reflecting material, symmetry shows that the ray will come back out in the opposite direction. Tiny spheres of this type are used for highway reflectors.

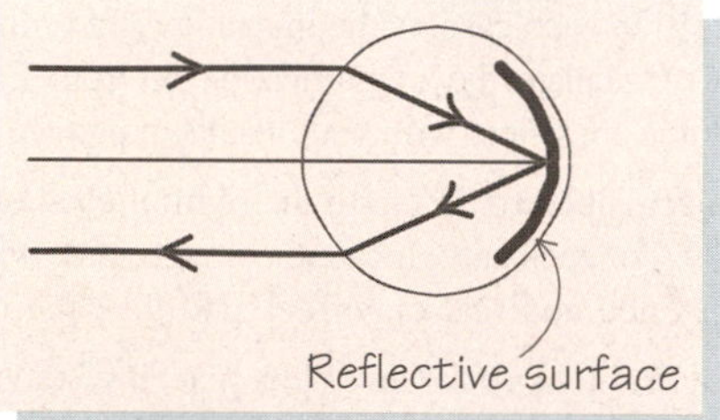

▲ **FIGURE 36–50** Problem 62.

63. (III) Rays of light strike a spherical glass surface parallel to the optical axis (Fig. 36–51). The incoming ray makes an angle θ with the normal to the surface. Show that the rays will cross the optical axis at a distance $d = R/\left(\sqrt{n^2 - \sin^2\theta} - \cos\theta\right)$ beyond the center of the sphere, where n is the index of refraction of the glass. To what does this expression reduce for small angles?

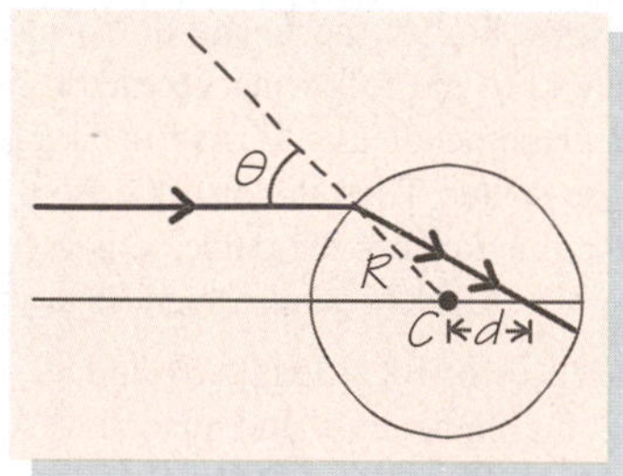

▲ **FIGURE 36–51** Problem 63.

64. (III) The index of refraction of a particular type of glass varies from 1.615 (for blue light) to 1.596 (for red light). Use the result of Problem 63 to calculate the color spread on the axis for light that strikes a hemispherical cap at the end of a glass rod, at an angle $\theta = 0.8$ rad. Take the radius of curvature of the sphere to be $R = 1.50$ cm. What is the spread for paraxial rays?

65. (III) An optical system contains a thin lens, $n = 1.4$, with positive curvature of radius $R_1 = 25$ cm for surface 1 and negative curvature of radius $R_2 = -25$ cm for surface 2. This lens collects light from the right side. Where to the left of the lens should a flat plate of thickness t of the same glass be placed, and how thick should it be, if you want the light that radiates from a distant object to be focused on a screen 35 cm to the left of the lens?

Chapter

37

◀ The brilliant colors of the peacock's feathers are due not to pigmentation, but to interference of the light reflected from them.

Interference

In Chapters 35 and 36, we emphasized the geometrical properties of light. We discussed reflection and refraction by treating light in terms of rays, but did not address the fact that light is a wave phenomenon. However, if we look more carefully at the behavior of light when obstacles or holes have dimensions comparable to the wavelength of the light, geometric (or ray) optics is inadequate, and the wave nature of light becomes important. Geometric optics cannot explain the colors observed in oil slicks or soap bubbles, and if we look closely at shadows, we find that they are not completely sharp, in contradiction to the predictions of geometric optics. These phenomena are due to interference and diffraction, the subjects of this chapter and Chapter 38, subjects already discussed in terms of mechanical waves in Chapter 15. *Physical optics*, which takes into account the wave nature of light, explains a wider range of observations than does geometric optics.

37–1 Young's Double-Slit Experiment

When two or more harmonic waves superpose, they interfere—whether they are water waves, waves on a string, sound waves, or light waves. We saw how this worked for waves in mechanical systems in Chapter 15. Here we shall see how interference between two light waves occurs because *the electric (or magnetic) fields of the two*

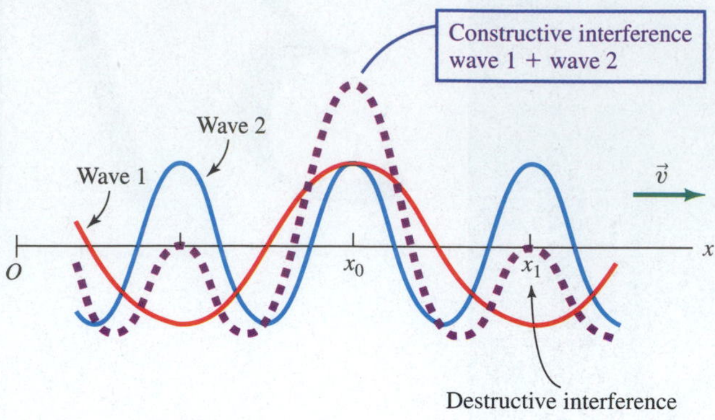

▲ **FIGURE 37–1** Constructive interference between two waves occurs at point x_0 when the peaks coincide. Destructive interference between two waves occurs at point x_1 when their amplitudes cancel.

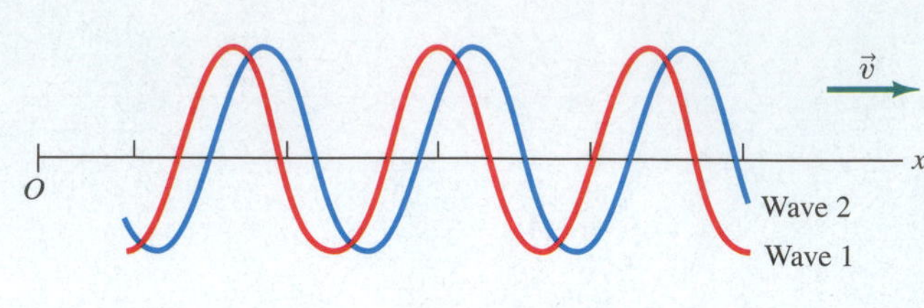

▲ **FIGURE 37–2** Two coherent waves have a constant phase difference.

waves add vectorially. Consider the superposition of two light waves from different sources at particular points in space that are propagating along the x-axis at a given time (Fig. 37–1). Where the two waves add to produce a wave with a larger amplitude (we refer here to, say, the electric field amplitude), we say that the waves interfere constructively, as, for example, at point x_0 in Fig. 37–1. The two waves interfere destructively where they cancel each other, as they do, for example, at point x_1. Any degree of interference between maximally constructive or maximally destructive is possible.

Light waves from two sources can produce an interference pattern in space when there is a definite relation between the respective wavelengths and phases *at their respective sources*. In other words, the waves must be *coherent*. The two light waves shown in Fig. 37–2 each have the same wavelength and a constant phase difference. We can also call monochromatic waves from a single source coherent when those waves form a long train of purely harmonic form. The "coherent" single source becomes two coherent sources that can interfere when the light is split; the interference occurs when the optics are arranged so that the two parts recombine later. A laser emits very long wave trains like these, and it is easy to demonstrate the interference pattern produced by laser light in the classroom (see Chapter 38). If the waves emitted at one or both sources consist of a mixture of waves of different wavelengths and phases, then there is no interference pattern. For example, an incandescent lightbulb produces light from many independent atomic sources at different times and places within the filament—this light is *incoherent*.

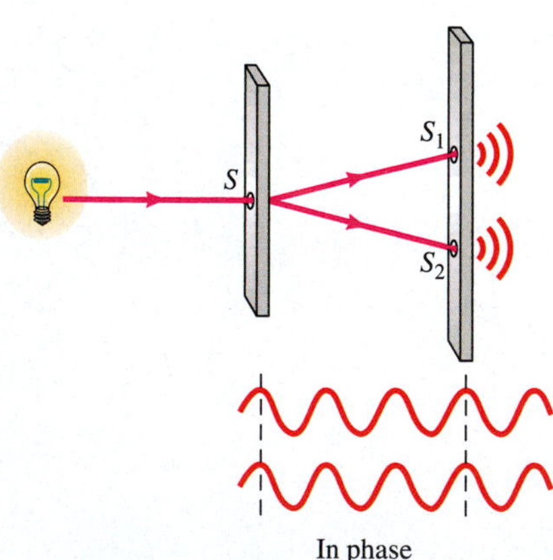

▲ **FIGURE 37–3** The light waves that pass through slits S_1 and S_2 are coherent. These waves are in phase if the light travels the same distance from S to S_1 as the distance from S to S_2. Even when the path lengths are different, there is a definite constant phase difference.

Thomas Young observed interference phenomena between two sources of light at the very beginning of the 19th century—he was the first to see such effects in light. One can produce coherent light at two sources by illuminating a single aperture S (a slit or a hole) with a source of monochromatic light. If the source of this light is a lightbulb, the light comes in series of bursts from individual atomic sources, a quantum phenomenon. (The monochromaticity can be taken care of with a prism.) These bursts last for a time on the order of $\tau \cong 10^{-8}$ s, and the length of the resulting individual wave trains is therefore $c\tau$, or several meters. The aperture S must be so small that only one wave train enters at a time. The single wave train of light that passes through illuminates two other apertures, S_1 and S_2 (Fig. 37–3). We can suppose that the size of the two apertures is the same, so that the amplitudes of the waves that come from them are identical. *These two apertures are two sources of coherent light*. If S_1 and S_2 are equidistant from S, the light from S travels the same distance to reach S_1 and S_2, and the light is in phase as it passes through the two apertures (Fig. 37–3). If S_1 and S_2 are not equidistant from S, the light waves that pass through them are still coherent because they have a definite, time-independent phase difference.

CONCEPTUAL EXAMPLE 37–1 Will there be interference between the (monochromatic) light passing through the slits of S_1 and S_2 if the light exiting the two slits is exactly 180° out of phase?

Answer We rearrange the slits of Fig. 37–3 as shown in Fig. 37–4 to make the light exiting the two slits to be π rad out of phase. The two sources of light have a strict coherence. At any point equidistant from the two slits (point A in Fig. 37–4, for example), the sources of light will remain out of phase and will interfere destructively. On the other hand, the two waves will be in phase at a point such as B placed so that its distance from the two slits differs by $\lambda(n + 1/2)$, where $n = 0, 1, 2, \ldots$ is an integer. At these points there will be constructive interference (Fig. 37–4). Similar remarks can be made if the phase difference for the light passing through the two slits has *any* value, as long as it is a definite value and not something that is constantly changing.

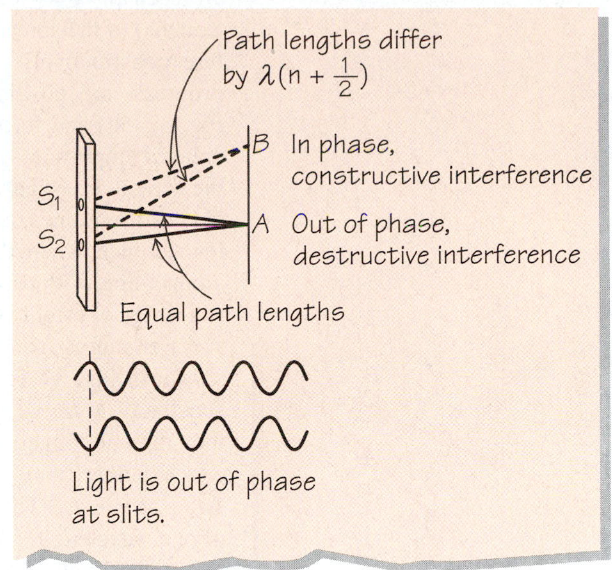

FIGURE 37–4 The light sources at S_1 and S_2 are out of phase. Whether they are out of phase or in phase at points A and B depends on the difference in path lengths to these points.

The Two-Source Interference Pattern

Let's review the spatial interference pattern produced when light from two sources of coherent waves interfere, a subject we developed in Chapter 15. We suppose that we have two sources, S_1 and S_2 (vertical slits), that emit monochromatic light in phase (Fig. 37–5a). Figure 37–5b is a view from above. Waves of the same frequency and phase emanate from S_1 and S_2 in the form of spreading cylinders from the sources, and these

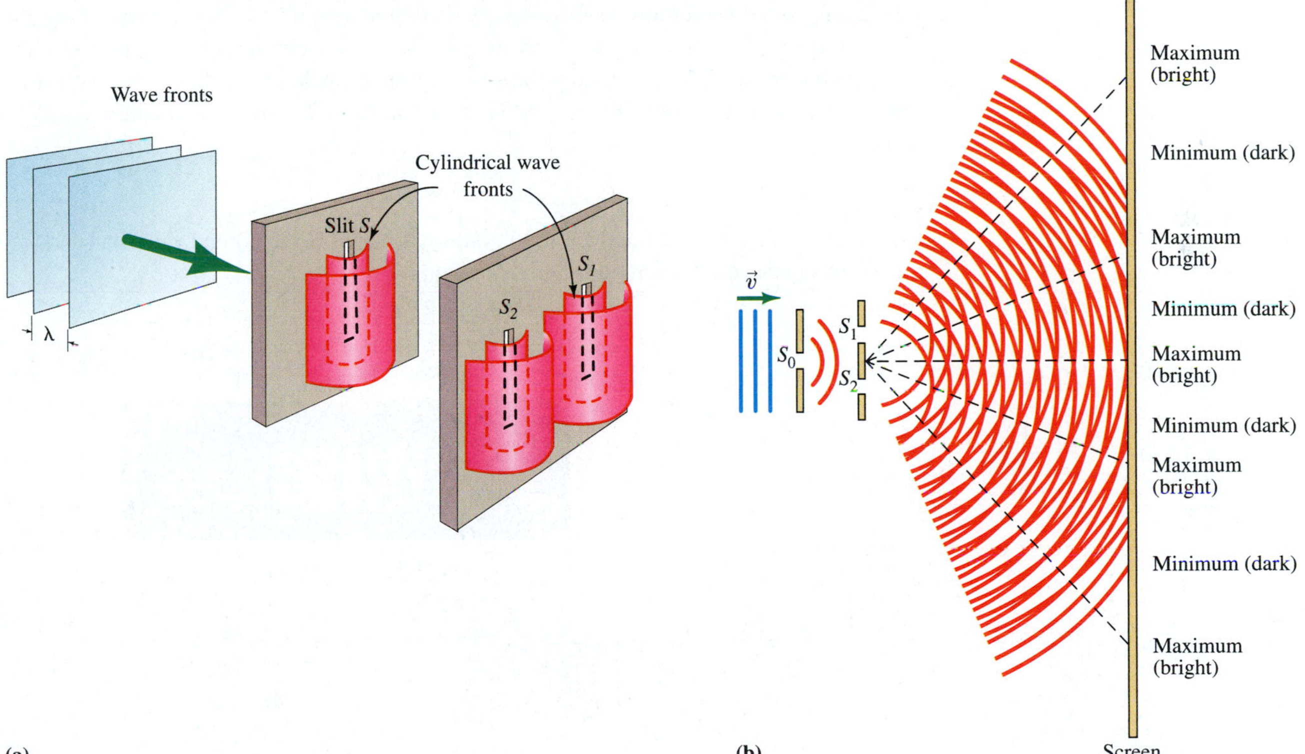

FIGURE 37–5 (a) Light coming from a single source S is split when it passes through the two slits S_1 and S_2, which act as two coherent sources, producing coherent cylindrical waves. (b) The view from above. Constructive interference occurs everywhere along the directions where the concentric circles, representing the crests of the spreading waves, overlap, because the waves are in phase along these directions. Alternating bright and dark places will be observed on a distant screen.

are seen in cross-section as circles in the figure. The circles represent the crests (or troughs) of the spreading waves. Where the crests (or troughs) overlap, the waves interfere constructively. The pattern of these overlap points is apparent: As the waves progress, the positions of these points advance (indicated by the dashed lines in Fig. 37–5b) and form lines. There is constructive interference—wave motion with increased amplitude—*all along* these lines; therefore, the places where the lines intersect the screen are bright. There is partial constructive interference in the regions on either side of these lines, and destructive interference in between the regions of constructive interference, where the screen is dark. (Again, the interference is maximally destructive along a line, with partial destructive interference in a region around that line.) The result is a series of bright and dark areas on the screen.

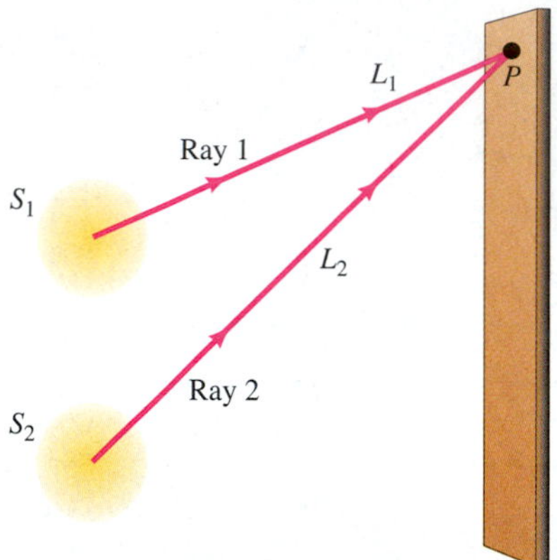

▲ **FIGURE 37–6** Light waves, indicated by rays 1 and 2, may not be in phase at point P despite being in phase at their sources, S_1 and S_2.

Let's investigate this double-slit configuration more closely. Consider the geometry shown in Fig. 37–6. Along ray 1 and ray 2, the waves travel distances L_1 and L_2, respectively, to arrive at point P on the screen. Because the rays travel different distances, they may no longer be in phase at P, although they were in phase at the sources S_1 and S_2. Whether they are in phase or not depends on the *path-length difference* $\Delta L = L_2 - L_1$. The waves arrive in phase if ΔL is zero or if ΔL is an integral multiple of one wavelength $(\Delta L = \lambda n)$, but will be 180° out of phase if this difference is a half-integer multiple of one wavelength $[\Delta L = \lambda(n + 1/2)]$. The interference is constructive where the waves are in phase and destructive when the waves are 180° out of phase:

$$\text{for constructive interference:} \quad \Delta L = n\lambda, \quad n = 0, \pm 1, \pm 2, \ldots; \tag{37–1a}$$

$$\text{for destructive interference:} \quad \Delta L = \left(n + \frac{1}{2}\right)\lambda, \quad n = 0, \pm 1, \pm 2, \ldots. \tag{37–1b}$$

The resulting series of bright and dark lines on the screen is indicated in Fig. 37–5 and Fig. 37–7.

The geometry shown in Fig. 37–8 determines conditions for constructive and destructive interference. We assume that the distance R to the screen is much greater than the distance d between the two slits. A very distant screen means that the rays from the slits to a given point on the screen become parallel, so we obtain a good approximation to the path difference by using only the single angle θ made by the line from the center point between the slits to P. From Fig. 37–8, we see that the angle formed by S_2S_1K is also θ. Thus

$$\Delta L = d \sin\theta. \tag{37–2}$$

According to Eqs. (37–1a) and (37–1b), maxima (bright regions) and minima (dark regions) thus occur on the screen for angles given by

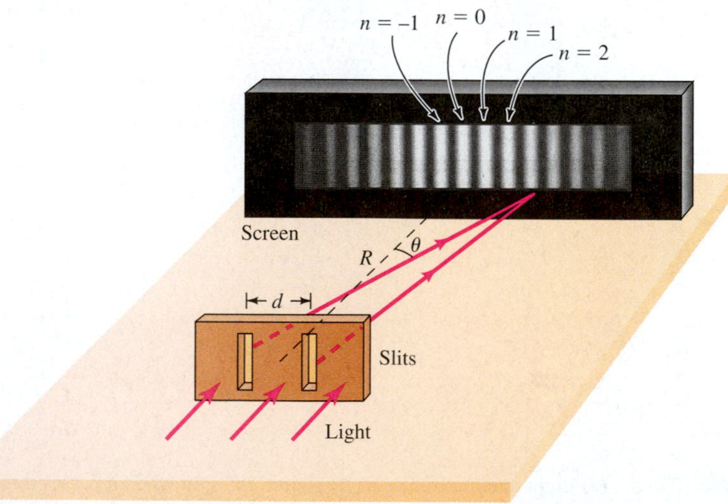

▲ **FIGURE 37–7** The interference pattern produced by double vertical slits is a series of alternating bright and dark vertical lines on a screen. The falloff of the intensity toward the edge of the figure is a single-slit effect (see Chapter 38).

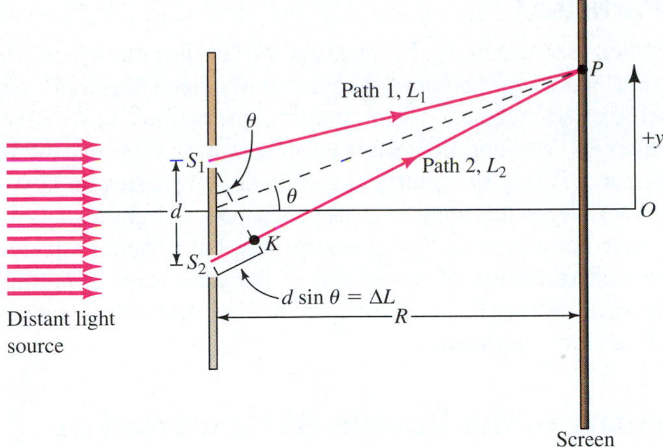

▲ **FIGURE 37–8** The geometry used to find the interference-pattern conditions for the light that reaches point P. The path-length difference $\Delta L = d \sin\theta$.

$$\text{for constructive interference: } \sin\theta = n\frac{\lambda}{d}, \quad n = 0, \pm 1, \pm 2, \ldots; \qquad (37\text{–}3a)$$

$$\text{for destructive interference: } \sin\theta = \left(n + \frac{1}{2}\right)\frac{\lambda}{d}, \quad n = 0, \pm 1, \pm 2, \ldots. \qquad (37\text{–}3b)$$

INTERFERENCE CONDITIONS FOR TWO SOURCES

The point that is aligned with the sources ($\theta = 0$) is a maximum ($n = 0$). Alternating minima and maxima lie on either side of this centerline. The value of n that labels the maxima is known as the *order*. The central maximum is the zeroth order, and the maxima on either side of the central maximum are first-order maxima ($n = \pm 1$). If θ is small, so that $\sin\theta \cong \theta$, the maxima and minima are equally spaced in θ. This experiment shows conclusively that light behaves as a wave. Geometric optics cannot explain the result shown in Fig. 37–7.

EXAMPLE 37–2 In a double-slit experiment, y is the distance along the screen from the center maximum. Find the positions of the maxima as a function of y. If the source-to-screen distance $R = 3.0$ m, the source separation $d = 0.20$ mm, and light comes from a helium– neon laser ($\lambda = 633$ nm) far away, determine y for the ninth-order maximum.

Setting It Up Fig. 37–8 illustrates the experimental schematic and shows the path difference $\Delta L = L_2 - L_1$.

Strategy Geometry tells us that $\Delta L = d \sin\theta$, and then the values of θ for which maxima occur is given by Eq. (37–3a). After determination of the angles, we use $y = R \tan\theta$ to find the desired y-values. If $R \gg y$, $\tan\theta \cong \sin\theta$, and we can directly insert the value of $\sin\theta$ from Eq. (37–3a) for the maxima.

Working It Out The maxima are located at

$$y = R\tan\theta \cong R\sin\theta = R\frac{n\lambda}{d}. \qquad (37\text{–}4)$$

For the ninth order, we have $n = 9$, and $y = 9\lambda R/d$. If we substitute the given values for λ, R, and d, we get for the ninth-order maximum

$$y = \frac{9(633 \times 10^{-9}\,\text{m})(3.0\,\text{m})}{0.20 \times 10^{-3}\,\text{m}} = 8.5\,\text{cm}.$$

The distance between each of the maxima is therefore about 1 cm.

What Do You Think? Could such measurements be used to determine the wavelength λ? *Answers to **What Do You Think?** questions are given in the back of the book.*

CONCEPTUAL EXAMPLE 37–3 Show that the observed pattern of lines on the screen in the double-slit experiment spreads out as the wavelength increases and/or the separation between the sources decreases.

Answer Equation (37–4) shows that the distance between adjacent maxima on the screen equals $\lambda R/d$. With R a constant, the separation Δy is proportional to λ/d. The pattern is most evident—it spreads over large angles—when the wavelength is comparable to the source separation. Thus to see the interference pattern in a water ripple tank where the wavelength is measured in centimeters, the slit separations should be on the order of centimeters. This is a feature characteristic of all wave phenomena.

Waves or Particles?

Young's observations seemed to resolve once and for all the controversy about whether light was a particle or a wave phenomenon, particularly since Maxwell's equations successfully predicted electromagnetic waves and their properties. As we have already described in Chapter 34, the question was reopened toward the end of the 19th and through the first part of the 20th century. In certain circumstances far beyond the physical realms that we study in this chapter, a particle aspect of light reveals itself. Light is neither a pure wave nor a pure particle phenomenon, but is rather a phenomenon that displays behavior characteristic of waves and of particles, depending on the circumstances. The fact that light has this double nature is completely nonclassical, explicable only in terms of quantum mechanics.

37–2 Intensity in the Double-Slit Experiment

The previous discussion relied on geometrical arguments to determine the angles for which maxima and minima can be obtained. We now turn our attention to the *intensity* of the light that reaches the screen. In this section, we are going to assume that the two sources are both slits that are very narrow. The case where even a single slit has a finite width is by itself interesting, and we'll reserve that case for Chapter 38.

The intensity (or brightness, for light) measures the energy delivered by a wave per unit time per unit area. The energy in a given mechanical wave or superposition of waves is proportional to the displacement squared. For light, the quantity that plays the role of displacement is the electric (or magnetic) field. The intensity of a light wave (the energy delivered by the wave per unit time per unit area) is the time average of the Poynting vector (see Chapter 34), which is proportional to the product of electric and magnetic field vectors in the wave. Because the magnetic field is itself proportional to the electric field in an electromagnetic wave, the intensity is proportional to the electric field squared ($I \propto E^2$). In order to find the intensity of a collection of waves, we add the electric fields of all the waves and square the sum of the net field. For example, with two sources of equal intensity I_0, the maximum electric field is twice the electric field E_0 from each source, so

$$\frac{I_{\max}}{I_0} = \frac{(E_0 + E_0)^2}{E_0^2} = 4,$$

where $I_{\max}$ is the maximum intensity. The maximum intensity $I_{\max} = 4I_0$. Similarly, the minimum intensity occurs when the electric fields exactly cancel, and $I_{\min} = 0$.

The simple argument just given, which is based on energy, is so useful that we shall develop it further. The intensity at any point P on the screen in Fig. 37–8 is proportional to the net Poynting vector, which is in turn proportional to the square of the net electric field. The net instantaneous electric field $\vec{E}_{\text{net}}$ at P is the sum of the instantaneous electric fields of the light waves emitted at the two sources: $\vec{E}_{\text{net}} = \vec{E}_1 + \vec{E}_2$. The net Poynting vector therefore has magnitude $S = (\vec{E}_1 + \vec{E}_2)^2 = E_1^2 + E_2^2 + 2\vec{E}_1 \cdot \vec{E}_2$. But light waves oscillate rapidly, and thus it is the *time average* of the Poynting vector (that is, the intensity, I) at P that is of interest.

It is the coherence of the two light waves that is important for the time-averaged intensity value. If we denote time averages with triangle brackets, then

$$I_{\text{net}} \propto \langle E_1^2 \rangle + 2\langle \vec{E}_1 \cdot \vec{E}_2 \rangle + \langle E_2^2 \rangle. \tag{37–5}$$

For incoherent light, there is no correlation—no definite phase relation—between the electric fields from the two sources. One moment the sources have one relative phase, the next moment the relative phase is different, and *the term* $\langle \vec{E}_1 \cdot \vec{E}_2 \rangle$ *is zero*. Thus

$$I_{\text{incoh}} = I_1 + I_2. \tag{37–6}$$

For coherent waves, $\langle \vec{E}_1 \cdot \vec{E}_2 \rangle$ in Eq. (37–5) is not zero. If, at a given time, there is constructive interference at point P, where $\vec{E}_1 = \vec{E}_2$, the constructive interference will persist because the waves are coherent. Similarly, if there is destructive interference at a given time, where $\vec{E}_1 = -\vec{E}_2$, it also persists through later times. For destructive interference, $\langle \vec{E}_1 \cdot \vec{E}_2 \rangle \propto -I_1$, and Eq. (37–5) gives $I_{net} = I_1 - 2I_1 + I_1 = 0$.

Suppose then that the electric fields of the light waves from our coherent sources S_1 and S_2 at a single point P in space are identically oriented and have magnitudes

$$E_1 = E_0 \sin \omega t, \tag{37–7a}$$

$$E_2 = E_0 \sin(\omega t + \phi). \tag{37–7b}$$

The phase difference ϕ for E_2 results from the path-length difference between the waves. If $\phi = 2\pi n$, where n is an integer, the fields are identical, and there is constructive interference. This phase difference of $2\pi n$ corresponds to a path-length difference of $\Delta L = n\lambda$. The ratio of ϕ to $2\pi n$ is the same as ΔL to $n\lambda$, so we have

$$\frac{\phi}{2\pi n} = \frac{\Delta L}{n\lambda};$$

$$\frac{\phi}{2\pi} = \frac{\Delta L}{\lambda}. \tag{37–8}$$

For the distant-screen geometry of Fig. 37–8, we can use $\Delta L = d \sin \theta$ [Eq. (37–2)] to transform Eq. (37–8) to

$$\phi = 2\pi\frac{\Delta L}{\lambda} = \frac{2\pi}{\lambda} d \sin \theta. \tag{37–9}$$

Now the net electric field at P has magnitude

$$E_{net} = E_1 + E_2 = E_0[\sin \omega t + \sin(\omega t + \phi)].$$

If we apply the equation $\sin \theta_1 + \sin \theta_2 = 2 \cos[(\theta_1 - \theta_2)/2] \sin[(\theta_1 + \theta_2)/2]$ (see Appendix IV–4), with $\theta_1 = \omega t$ and $\theta_2 = \omega t + \phi$, we find

$$E_{net} = 2E_0 \cos\left(\frac{\phi}{2}\right) \sin\left[\omega t + \left(\frac{\phi}{2}\right)\right]. \tag{37–10}$$

The Poynting vectors $\vec{S}_1$ and $\vec{S}_2$ of the light from the individual sources have magnitudes

$$S_1 \propto E_1^2 = E_0^2 \sin^2(\omega t) \quad \text{and} \quad S_2 \propto E_2^2 = E_0^2 \sin^2(\omega t + \phi), \tag{37–11}$$

respectively, whereas the net Poynting vector at P has magnitude

$$S_{net} \propto E_{net}^2 = 4E_0^2 \cos^2(\phi/2) \sin^2\left[\omega t + \left(\frac{\phi}{2}\right)\right]. \tag{37–12}$$

To find the intensities (the time averages of the Poynting vectors), we need know only that the time average of $\sin^2(at + b) = \frac{1}{2}$. If we write the individual intensities as $I_0 \propto E_0^2/2$, then the net intensity from the two sources in terms of I_0 is

$$I_{net} = 4I_0 \cos^2\left(\frac{\phi}{2}\right). \tag{37–13}$$

When the phase ϕ in Eq. (37–7b) is related to the path-length difference by Eq. (37–8), then Eq. (37–13) for the intensity on the distant screen becomes

$$I_{net} = 4I_0 \cos^2\left(\frac{\pi d}{\lambda} \sin \theta\right). \tag{37–14}$$

This is the expression for the intensity in Young's classic double-slit experiment.

The maxima and minima occur at the angles specified by Eq. (37–3). Figure 37–9 is a plot of the intensity at the screen as a function of $\sin \theta$. This figure also serves as a plot of the intensity as a function of the distance y from the center maximum along the screen: For small θ, $y \cong R \sin \theta$, where R is the distance from the screen. If the apertures are narrow vertical slits, then the bright maxima on the screen are vertical lines called *fringes*.

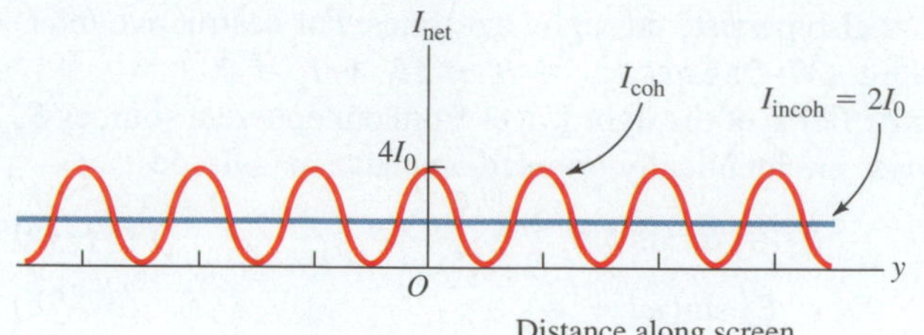

FIGURE 37–9 The net intensity of light from the double slit as a function of the distance from the center point on the screen ($y \cong \sin\theta$). Compare the results for coherence and incoherence. The same amount of light energy reaches the screen in both cases, but in the coherent case, it occurs in peaks and valleys. We have assumed that each slit is infinitely narrow; that is why the falloff visible at the edges of Fig. 37–7 is not visible here.

We have also plotted in Fig. 37–9 the intensity $2I_0$ that would be present on the screen if the light sources were incoherent, represented by the blue line in the plot. The result for incoherence is constant and shows no interfering maxima and minima. However, *averaged over the entire screen*, the energy reaching the screen is exactly the same in the two cases, as required by the conservation of energy. To average the energy that reaches the screen in the case for coherence, we need only use the fact that the average of the cosine-squared factor in Eq. (37–14) is $\frac{1}{2}$, and $4I_0 \times \frac{1}{2} = 2I_0$. The energy emitted at each source is the same whether the light from these sources is coherent or incoherent, and the total energy arriving at the screen must also be the same as that emitted. The energy is spread evenly over the screen when the sources are incoherent, whereas it is distributed in peaks and valleys when the sources are coherent.

EXAMPLE 37–4 Suppose you live at point H, 20 km from a vertical radio dipole antenna that broadcasts at a frequency of 1100 kHz from point B. How well your radio picks up the signal is a direct function of the intensity of the signal. A second antenna is constructed at point A, located $d = 100$ m from the first, so that AB makes an angle of 15° (0.26 rad) with AH. The two antennas are fed by the same source, and the two signals are in phase at the sources—they have the same wavelength and are coherent. Find the new intensity at your radio in terms of the old intensity. Is your signal improved?

Setting It Up The diagram in Fig. 37–10, which is a vertical view, helps us determine the path difference to the house from the two sources.

Strategy The situation is like the double-slit situation for light because there are two sources of coherent radiation, with the difference that in this case the relevant wavelengths are much longer. Because the distance from the transmitting antennas is much greater than their separation distance, the geometrical approximations we used in discussing the double-slit experiment apply. These approximations tell us that the difference in distances between the house and the two antennas is $R_A - R_B = d\cos\theta$ (Fig. 37–10). To find the net electric field at the house, we add the fields from each source, using the same methods that we did for light. Note that the electric fields, which are parallel to the antennas, are vertical (perpendicular to the page).

Working It Out Suppose that the first antenna (at point B) broadcasts a signal whose z-component at point H is $E_B = E_0\cos(\omega t)$ and that the original intensity is $I_0 = CE_0^2$, where C is some constant. The second antenna's electric field takes the same form except that there is a phase difference due to the fact that the antenna is a distance $R_A - R_B$ farther away:

$$E_A = E_0\cos(\omega t + \phi),$$

where $\phi = 2\pi(R_A - R_B)/\lambda = (2\pi d\cos\theta)/\lambda$. The net field at point H is then

$$E_{net} = E_A + E_B = E_0[\cos\omega t + \cos(\omega t + \phi)].$$

This sum is similar to the one we needed in order to find Eq. (37–10), $E_{net} = 2E_0\cos(\phi/2) \times \sin[\omega t + (\phi/2)]$. The net intensity is given similarly in Eq. (37–13),

$$I_{net} = CE_{net}^2 = 4CE_0^2\cos^2\left(\frac{\phi}{2}\right) = 4I_0\cos^2\left(\frac{\phi}{2}\right).$$

It remains to calculate the factor $\cos^2(\phi/2)$ and see by what factor the intensity is changed. We require that

$$\cos\left(\frac{\phi}{2}\right) = \cos\left(\frac{2\pi d\cos\theta}{2\lambda}\right) = \cos\left(\frac{\pi d\cos\theta}{\lambda}\right).$$

We have $d = 100$ m and $\theta = 0.26$ rad. The wavelength λ comes from the frequency $f = 1100\text{ kHz} = 1.1 \times 10^6$ Hz. We get $\lambda = c/f = (3.0 \times 10^8\text{ m/s})/(1.1 \times 10^6\text{ s}^{-1}) = 270$ m. Thus

$$\cos\left(\frac{\phi}{2}\right) = \cos\left[\frac{\pi(100\text{ m})(\cos 0.26\text{ rad})}{270\text{ m}}\right] = \cos(1.12\text{ rad}) = 0.43.$$

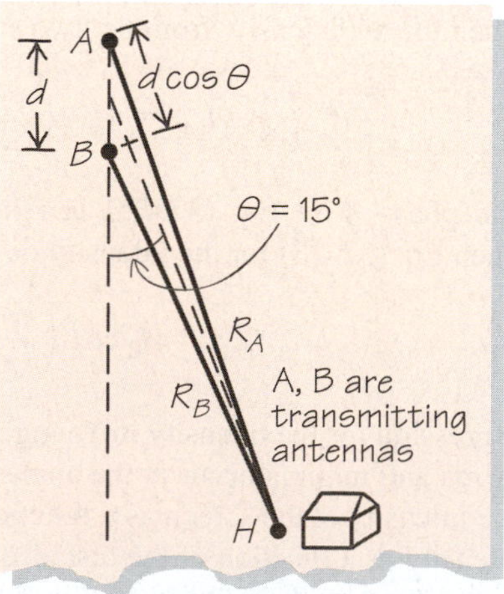

FIGURE 37–10 R_A and R_B are the distances between your home and two antennas at points A and B, respectively.

The net intensity $I_{net} = 4I_0\cos^2(\phi/2)$ is a factor $4[\cos^2(\phi/2)] = 4(0.44)^2 = 0.77$ times the original intensity I_0. The signal you receive at your radio at point H has actually become weaker because there is partial destructive interference between the signals of the two antennas at that point.

What Do You Think? Suppose that the length of the cables from the radio station to the two transmitting antennas were each changed by different amounts. Will the signals still interfere at your house?

37–3 Interference from Reflection

A ray in air, say, may be partly reflected from a surface and partly transmitted through the surface into a medium such as glass. The transmitted ray may then be subsequently reflected from a second surface and reemerge into the air. If the original wave train is long compared with the distance traversed in the glass, the two reflected rays are coherent and can interfere.

Interference Fringes from the Space Between Two Glass Plates

Let's look at two glass plates, each with one very flat surface. They have been placed together with a spacer at one side, so that the flat surfaces touch along one side and are separated by a small distance d at the other (Fig. 37–11). When these plates are illuminated from above with monochromatic light, a series of alternating light and dark bands is seen, starting with a black band along the side where the plates touch.

To study this phenomenon further, let the plates have indices of refraction n, the space between the plates be air ($n = 1$), and the wavelength of the light be λ. We can think of the light as consisting of rays perpendicularly incident on the plates. These rays are *perpendicular to the wave fronts*; when we speak of rays interfering, we are referring to the interference of the waves that form those fronts. In this situation, the rays can reflect and refract to form new rays in various ways, but one possibility leads to the observed interference here: An incident ray in Fig. 37–11 passes almost vertically through the top piece of glass and is partly reflected (ray 1) and partly transmitted (ray 2) at point P_1 on the bottom surface of the top plate. Ray 2 then continues through the air gap and is reflected at point P_2 at the top of the bottom plate. To a good approximation, ray 2 travels a distance $2P_1P_2$ in air farther than ray 1, because the incident ray is almost vertical. Rays 1 and 2, exiting at the top, will interfere, resulting in the series of bright and dark bands corresponding to different values of $2P_1P_2$.

Now, if the points P_1 and P_2 move systematically closer to the side where the plates touch (point C), the path-length difference distance $2P_1P_2$ decreases, and we expect to see a series of bright and dark regions corresponding to constructive and destructive interference. What happens when the plates touch at point C? Observation reveals that there is a dark line at the edge where the plates touch, indicating destructive interference. But at point C each of the rays travel exactly the same distance, so the destructive interference cannot come from any path-length difference. The only explanation is that *destructive interference occurs because one of the rays undergoes a 180° phase change during reflection.* The light undergoes a phase change of 180° when it is reflected at P_2, whereas the light undergoes no such change when it is reflected at P_1. Let us consider the origin of this fact.

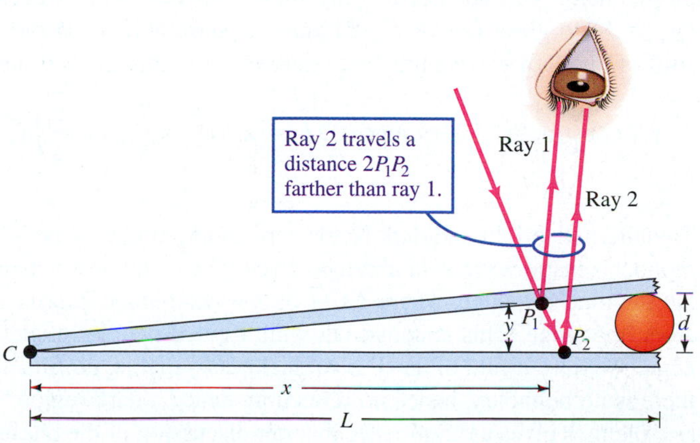

FIGURE 37–11 Two pieces of glass are placed together with a spacer (exaggerated in size) at one edge. The adjacent surfaces are flat on the scale of the wavelength of light.

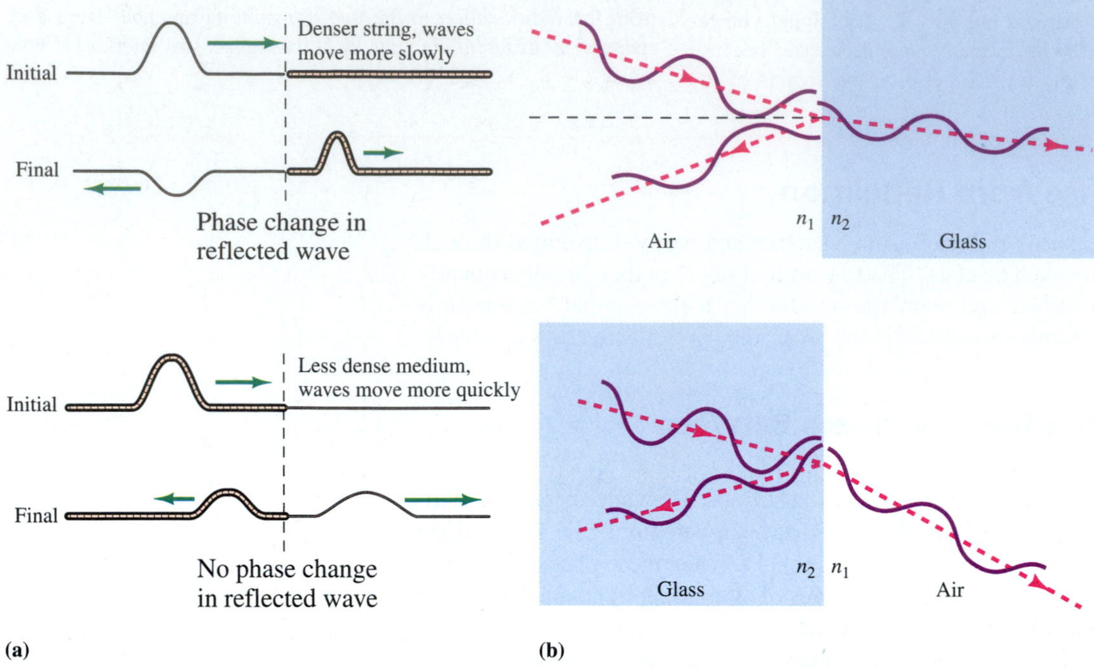

▲ **FIGURE 37–12** (a) The phase change of light upon reflection is similar to the inversion that occurs when a pulse that moves along a string meets a denser string. No phase change occurs when the string encounters a less dense string. (b) For light, a phase change of 180° occurs when the second medium has a higher index of refraction. When the second medium at the reflection boundary has a lower index of refraction, there is no phase change in the reflection.

In Section 15–5, we discussed a corresponding phenomenon in the reflection of one-dimensional waves on strings. When two strings of different densities are connected (Fig. 37–12), the connection point forms a boundary at which there may be reflection and transmission. We showed that if the density on the side from which the wave comes is less than the density on the far side of the boundary, then the reflected wave is inverted, corresponding to a phase shift of 180° if the wave is harmonic. Similarly, if the pulse comes from a string of larger density than the string on the far side of the connection point, the pulse is not inverted on reflection.

Electromagnetic waves behave similarly: Maxwell's equations determine the form of the electric and magnetic fields of an electromagnetic wave at a boundary between two dielectric media. The electric field changes sign or does not change sign according to whether the wave speed $(v = c/n)$ in the medium on the far side of the boundary is greater than or less than the wave speed in the medium from which the wave comes. (There is no such change of sign for the magnetic field of the wave.) When the electric field changes sign, the result is a 180° phase change in the reflected electromagnetic wave. The result for the phase change can therefore be stated as:

The phase of an electromagnetic wave that moves from a medium of index of refraction n_1 toward a medium with index of refraction n_2 will change by 180° upon reflection when $n_2 > n_1$ and will not change when $n_2 < n_1$.

This information allows us to understand the interference pattern for the plates. The phase change of 180° occurs only for the reflection at P_2 because ray 2 goes from air $(n_1 = 1)$ to glass $(n_2 > 1)$. Because a phase shift of π rad (180°) corresponds to a shift of one-half wavelength, the condition for constructive interference becomes

$$\text{for constructive interference: } \Delta L = 2P_1P_2 = \left(m + \frac{1}{2}\right)\lambda, \quad m = 0, \pm 1, \pm 2, \ldots. \tag{37–15}$$

The alternate bright and dark bands correspond to plate separation distances for which a monochromatic wave is in alternate constructive and destructive interference. In particular, the touching edge, where $\Delta L = 0$, has destructive interference; there is a dark band along that edge. This is consistent with a much simpler model of what happens at that edge: We have a film of air of zero thickness—that is, no film at all—and it is as though there is no boundary, hence no reflection, hence a dark region. The consistency of these two pictures gives us confidence that our discussion of the phase change is correct.

EXAMPLE 37–5 Two flat glass plates of length $L = 10\,\text{cm}$ touch at one end but are separated by a wire of diameter $d = 0.01$ mm at the other end (Fig. 37–11). Light shines almost perpendicularly on the glass and is reflected into the eye as shown. What is the distance x between the observed maxima if the incident (violet) light has $\lambda = 420$ nm?

Strategy We must here take the conditions for constructive interference that we previously described and translate those conditions into a band separation. We utilize Eq. (37–15) for the constructive interference condition, and by noting that $2y \cong 2P_1P_2$, where y is the spacing between the plates, we have

$$2y = \left(m + \frac{1}{2}\right)\lambda, \quad m = 0, \pm1, \pm2, \ldots .$$

To translate this condition into a condition for the x-position of the bands we use a geometrical relation between similar triangles in Fig. 37–11: $x/y = L/d$ or $x = yL/d$.

Working It Out We insert the y-values for constructive interference into our equation for x and find the x-positions where there is constructive interference:

$$x = \frac{L}{d}y = \frac{L}{d}\frac{1}{2}\lambda\left(m + \frac{1}{2}\right).$$

The difference in x from one maximum to another corresponds to a shift in m of 1, so

$$\Delta x = \frac{L}{d}\frac{\lambda}{2}\left\{\left(m + \frac{1}{2}\right) - \left[(m-1) + \frac{1}{2}\right]\right\} = \frac{L}{d}\frac{\lambda}{2}$$

$$= \frac{(10 \times 10^{-2}\,\text{m})(420 \times 10^{-9}\,\text{m})}{(0.01 \times 10^{-3}\,\text{m})2} \cong 2\,\text{mm}.$$

For glass plates 10 cm long, there will be about 50 bands of constructive interference.

What Do You Think? Would the number of bands of constructive interference increase or decrease if red light is used instead of blue light?

Newton's Rings

Newton's rings form a variant on Example 37–5; they are named after Isaac Newton, who studied them in the 1700s along with his contemporaries Robert Hooke and Robert Boyle. Here, a curved piece of glass is placed on a flat piece of glass and illuminated from above with white light (Fig. 37–13a). Observation from above reveals rings of color (Fig. 37–13b). If monochromatic rather than white light shines down on the glass, then a series of bright and dark concentric rings appear (Fig. 37–13c). Most significantly, the region around point C—the place where the glass pieces touch—is dark (Fig. 37–13c). From our earlier discussion, we know why this occurs: The region around C can be thought of as a region without a reflecting boundary at all or the result of a phase change where ray 2 reflects from the second glass surface with no phase shift due to different path lengths.

How do we explain the colors we observe when white light such as sunlight, which consists of all wavelengths, is incident on the glass? Refer to Fig. 37–14. For a particular radial distance from point C, there may be only one wavelength in the visible range, say, for the color blue, for which there is *destructive* interference. At that radius, we see the color of sunlight with blue subtracted. If we look slightly farther away from C, where the distance P_1P_2 is greater, there will be destructive interference for a slightly larger wavelength, say, for green, and we see sunlight with green subtracted. These colors are not as vivid as those of the rainbow because they remain mixtures of different frequencies with some frequencies subtracted.

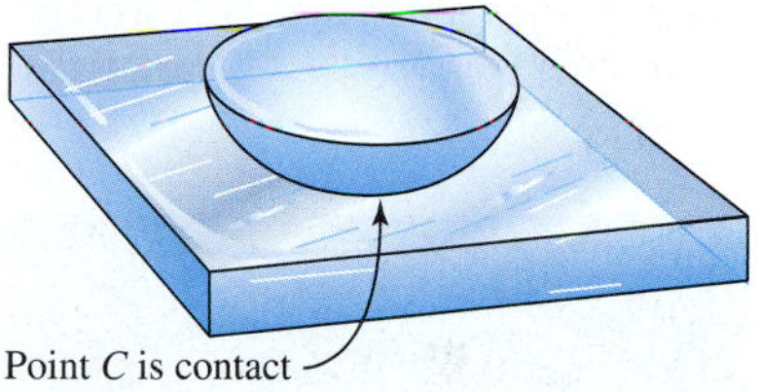

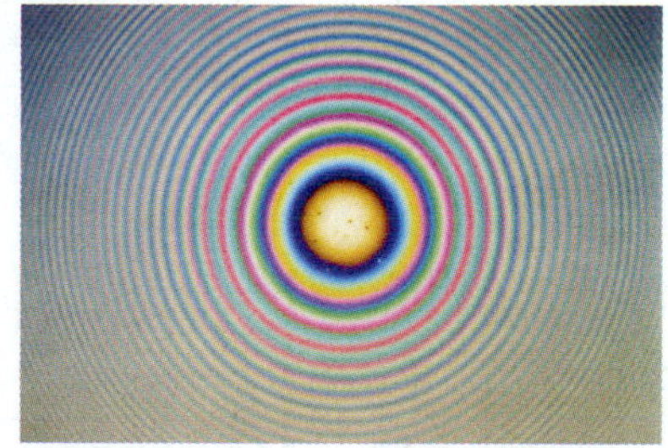

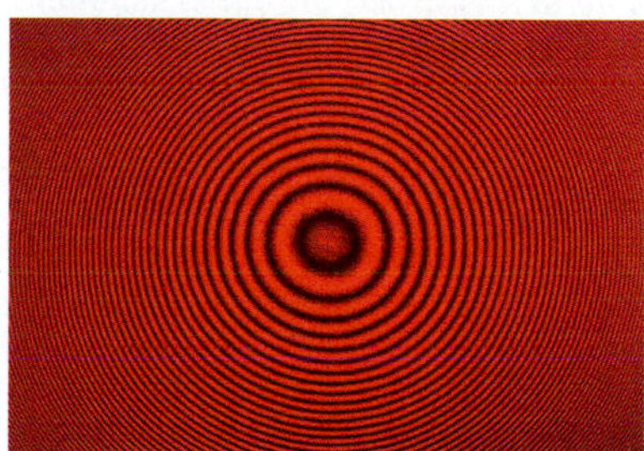

▲ **FIGURE 37–13** (a) A piece of glass whose bottom surface is curved rests on a flat glass surface. (b) The system is illuminated with white light from above, and a vertical view reveals colored rings called Newton's rings. (c) Concentric rings of alternating bright and dark appear when Newton's rings form from monochromatic light.

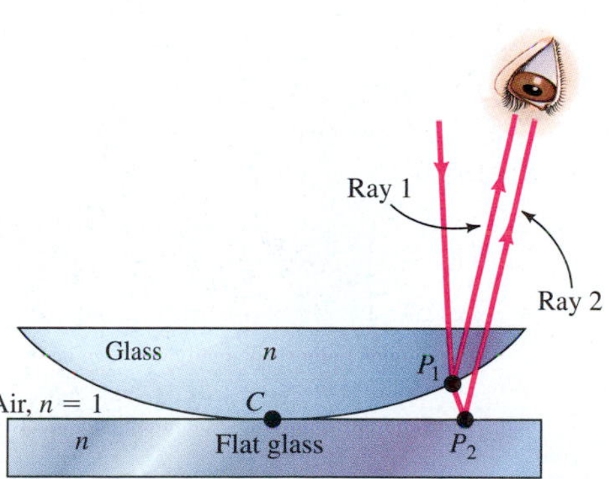

▲ **FIGURE 37–14** The geometry used to obtain the conditions for constructive interference in Newton's rings. The two rays moving toward the eye interfere after they reflect from different surfaces.

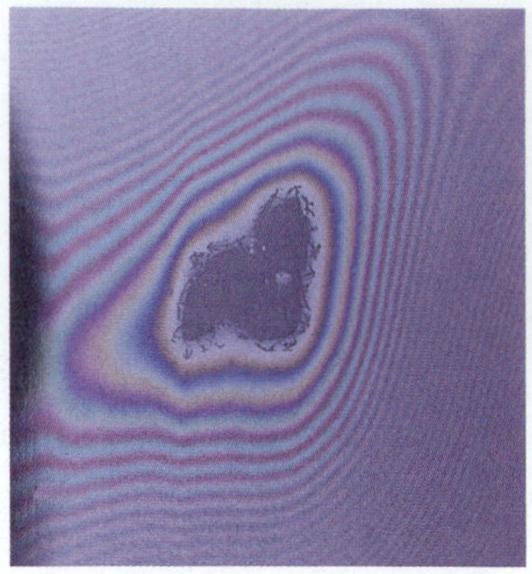

▲ **FIGURE 37–15** Fringes that occur when two nonflat pieces of glass are placed on top of one another. If each surface were perfectly flat and parallel, there would be no interference pattern. If the surfaces were flat but not parallel, the fringes would form straight lines. This is a good test of flatness for mirrors.

THINK ABOUT THIS. . .

HOW WOULD YOU CHECK THE FLATNESS OF A GLASS SURFACE?

The interference phenomena we have described in this section give us a practical method for determining how flat a given glass surface is (Fig. 37–15). If the surface to be tested is placed on top of an *optical flat* (a glass with a surface known to be flat to a fraction of a wavelength of visible light), then if the tested surface is indeed flat, no regions of constructive interference will appear. In practice, a tiny spacer may be placed between the piece to be measured and the optical flat to open a wedge-shaped space between them. If the two pieces are so nearly flat and parallel that, without the spacer, *no* interference fringes are observed, the pieces may bond together and be virtually impossible to pull apart! In effect, the two pieces become one. (Weeks of grinding and much money will have been lost if this happens).

Mirrors are commercially available that are flat to better than 5 percent of one wavelength, or about 25 nm. They are used in lasers as well as in *interferometers*—devices that use the interference of light to measure distance down to a fraction of one wavelength (see Section 37–4). ■

▲ **FIGURE 37–16** The colors of films of soapy water are due to thin-film interference of light.

Thin-Film Interference

The colors seen in soap bubbles and oil slicks are a manifestation of *thin-film interference*, which is another example of interference from reflection (Fig. 37–16). The interference occurs between the light reflected from the two surfaces of the thin film that forms the bubble. Consider light ray 1 that is incident on the thin film in Fig. 37–17. Part of the light is reflected at boundary I and forms ray 2. Part of ray 1 is refracted at boundary I and then reflected at boundary II. This light wave is partly refracted again at boundary I before forming ray 3.

Because rays 2 and 3 both originate from ray 1 at point P_1, the conditions for constructive or destructive interference depend on the path-length difference $\Delta L = P_1P_2P_3 - P_1P_4$ as well as on any phase changes that may occur during the reflection. The rule discussed earlier for phase changes upon reflection indicates that ray 1 undergoes a phase change upon reflection at surface I but not at II. Thin films such as soap bubbles have varying thicknesses, so different wavelengths will interfere destructively on different parts of the bubble, and the colors that appear in the reflected light represent the original light minus the wavelength that interferes destructively. There is also an enhancement of those colors for which there is constructive interference. For an oil film floating on water, the oil may have an index of refraction between that of air and water, in which case there is a 180° phase change at *both* the air-oil surface and the oil-water surface. The phase changes due to reflection then cancel each other in the interference from light reflected from these two surfaces, and any phase difference is due only to a difference in the path length.

A feature in thin-film interference that does not arise in the case of Newton's rings is that the additional path length lies *within* the thin film of material. The phase change is then computed according to Eq. (37–9), but *the wavelength that appears is the wavelength within the material*, $\lambda_{\text{film}} = \lambda/n$, where n is the index of refraction of the material.

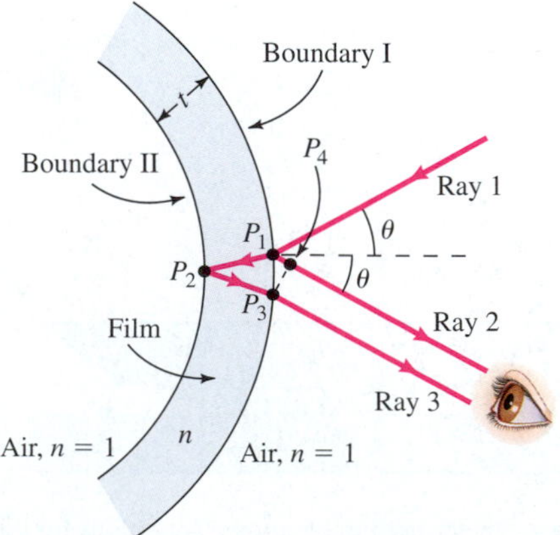

▲ **FIGURE 37–17** Geometry for thin-film interference. A light ray reflects from both the front and back boundaries of the film, and the reflected waves interfere.

EXAMPLE 37–6 The soap bubble shown in cross section in Fig. 37–17 has thickness t and index of refraction n. Light of wavelength λ in air falls vertically on the bubble and is reflected back. (a) Express the condition for constructive interference for the reflected light. (b) If $t = 400$ nm and $n = 1.3$, what color or colors will interfere constructively in the reflected light?

Strategy (a) There is constructive interference when $\phi = 2\pi m$, where m is an integer. To find the value of ϕ, note that when the incident light in Fig. 37–17 is vertical, distance $P_1P_4 = 0$. Light reflects at P_1 and P_2, undergoing a 180° phase change only at P_1, where the index of refraction of the medium on the far side—the soap film—is greater than that of air. The path-length difference is $\Delta L = 2t$, and the phase difference $\phi_{\Delta L}$ between the two reflected waves due to ΔL is found from Eq. (37–8) in terms of the wavelength λ_n in the soap film:

$$\frac{\phi_{\Delta L}}{2\pi} = \frac{2t}{\lambda_n} = \frac{2tn}{\lambda}.$$

The overall phase difference between the two reflected waves is then $\phi = \pi + \phi_{\Delta L}$, where π appears as an additive factor because of the 180° phase change at P_1.

Working It Out We have

$$\phi = \pi + \phi_{\Delta L} = \pi + \frac{4\pi tn}{\lambda} = 2\pi m.$$

Thus

$$\text{for constructive interference: } 4nt = (2m - 1)\lambda,\ m = 1, 2, 3, \ldots. \qquad (37\text{–}16)$$

Strategy (b) We need only solve the condition for constructive interference [Eq. (37–16)] for λ.

Working It Out Given $t = 400$ nm and $n = 1.3$, we have

$$\lambda = \frac{4nt}{2m - 1} = \frac{4(1.3)(400\ \text{nm})}{2m - 1} = \frac{2100\ \text{nm}}{2m - 1}.$$

For $m = 1$ through $m = 4$, these values of λ are

$$\lambda \cong 2100\ \text{nm}, 700\ \text{nm}, 420\ \text{nm}, 300\ \text{nm}.$$

Only the wavelengths 700 nm (red) and 420 nm (violet) are in the visible spectrum and these are the colors that interfere constructively in the reflected light.

What Do You Think? Estimate the wavelength of the color, if any, that is missing between the maxima at 700 nm and 420 nm.

EXAMPLE 37–7 A thin film of water, $n_2 = 1.33$, floats on cinnamon oil, which is denser than water and has index of refraction $n_3 = 1.65$. White light reflected at 45° has a maximum intensity for a wavelength around 600 nm. What is the minimum possible thickness of the film?

Setting It Up Figure 37–18 illustrates the paths of the two reflected rays that arrive in parallel at a distant point.

Strategy *Both* reflected rays undergo a phase shift of 180°, so that no net phase difference is associated with reflection. There is still a phase shift associated with different path lengths for the two rays—the phase shift ϕ_1 from the additional path P_1P_4 of ray 1,

$$\frac{\phi_1}{2\pi} = \frac{P_1P_4}{\lambda};$$

and the phase shift ϕ_2 for the additional path $P_1P_2 + P_2P_3$ of ray 2,

$$\frac{\phi_2}{2\pi} = \frac{P_1P_2 + P_2P_3}{\lambda_{H_2O}} = \frac{(P_1P_2 + P_2P_3)n_2}{\lambda}.$$

Note that the relevant wavelength for ray 1 is λ, the wavelength in air $(n_1 = 1)$, whereas the relevant wavelength for ray 2 is $\lambda_{H_2O} = \lambda/n_2$. It is the difference $\phi_2 - \phi_1$ that enters into the net phase difference between the two waves. The geometry necessary to calculate the path lengths is shown in Fig. 37–18. Once we know the phase difference, the usual conditions relate it to constructive and destructive interference; we will want to solve the condition for a first maximum to find the thickness t.

Working It Out The net phase difference is

$$\phi = \phi_2 - \phi_1 = (P_1P_2 + P_2P_3)\frac{2\pi n_2}{\lambda} - (P_1P_4)\frac{2\pi}{\lambda}.$$

The incident ray enters at an angle θ perpendicular to the surface and is either reflected at angle θ or refracted into the water at angle θ'. We note that $P_1P_2 = P_2P_3 = t/(\cos\theta')$. Also, $P_1P_4 = P_1P_3\cos(90° - \theta) = (2t\tan\theta')\cos(90° - \theta)$. The phase difference is then

$$\phi = \left(\frac{2t}{\cos\theta'}\right)\frac{2\pi n_2}{\lambda} - (2t\tan\theta')\cos(90° - \theta)\frac{2\pi}{\lambda}.$$

We solve the equation for t, the film thickness:

$$t = \frac{\lambda\phi}{4\pi}\frac{1}{(n_2/\cos\theta') - \tan\theta'\cos(90° - \theta)}.$$

From Snell's law, $\sin\theta' = (\sin\theta)/n_2$, and with $\theta = 45°$, $\theta' = 32°$. The problem states that there is a maximum in the intensity (that is, constructive interference) for $\lambda = 600$ nm. For constructive interference, $\phi = m(2\pi)$, $m = 1, 2, 3, \ldots$. (Is $m = 0$ allowed here?) Thus

$$t = \frac{\lambda m(2\pi)}{4\pi}\frac{1}{(n_2/\cos\theta') - \tan\theta'\cos(90° - \theta)}$$

$$= \frac{600\ \text{nm}}{2}m\frac{1}{(1.33/\cos 32°) - \tan 32°\cos(90° - 45°)}$$

$$= (266\ \text{nm})m.$$

The minimum film thickness, for $m = 1$, is 266 nm.

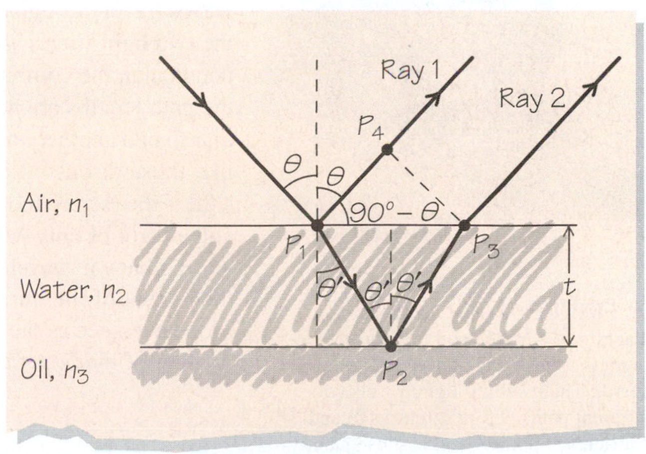

▲ **FIGURE 37–18**

▲ **FIGURE 37–19** Thin coatings of materials can serve to reduce reflection by taking advantage of destructive interference. MgF_2 is often used as a coating. Here the incident and reflected rays are essentially vertical.

THINK ABOUT THIS. . .

WHY DO SOME EYEGLASSES REFLECT LIGHT WHILE OTHERS DON'T?

The lenses of many eyeglasses, cameras, and other optical devices are coated with a thin layer of material whose purpose is to reduce the intensity of the reflected light through destructive interference. Many optical devices have multiple lenses, and if each lens typically reflects 4 percent of the energy of incident light, there may be an intolerable loss of light that is meant to be transmitted.

If light is reflected directly back from both surfaces in Fig. 37–19, and if the coating material has an index of refraction between that of air and the glass beneath it, there is a phase change in each reflection. Destructive interference will occur when the difference $2t$ in optical path lengths is a half-integer multiple of the wavelength λ_n within the coating and t is the thickness of the coating; in other words,

$$2t = \left(m + \frac{1}{2}\right)\lambda_n = \left(m + \frac{1}{2}\right)\frac{\lambda}{n}. \qquad \text{(AB–1)}$$

The coatings are applied as thinly as possible (that is, for $m = 0$). For $m = 0$ the coating thickness must be

$$t = \frac{\lambda}{4n}, \qquad \text{(AB–2)}$$

what is called a *quarter-wave* thickness. Of course, there can be destructive interference for only one wavelength in the visible range of light. A 100-nm thickness of MgF_2, a material commonly used to coat glass lenses, reduces reflectivity at 550 nm, in the middle of the visible range. The reflected light consists of red and blue light, which explains why such lenses have a purple cast to them. Antireflective coatings also increase the intensity of the transmitted light, and for this reason they are used to coat solar cells (components of solar batteries).

Sometimes we want to *increase* the intensity of reflected light. Lasers require mirrors that reflect light strongly at their operating wavelengths. For this purpose, a single coating must have thickness $t = \lambda/2n$, so that there will be constructive interference between the two reflected light waves. Multiple coatings are even more effective in increasing the reflected intensity. It is possible in this way to reflect more than 99 percent of the energy.

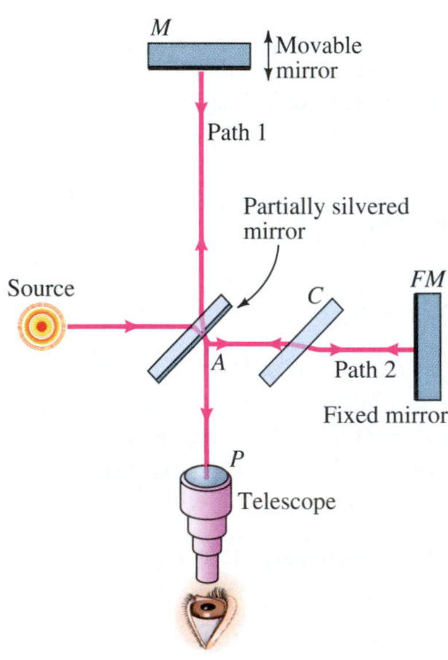

(a)

(b)

▲ **FIGURE 37–20** (a) Schematic diagram of a Michelson interferometer. Light is split by the partially silvered mirror. The resulting light travels two different paths before it returns to and interferes at point A and is subsequently observed through the telescope. (b) A modern Michelson interferometer.

*37–4 Interferometers

Optical interferometers are devices that utilize the interference between light waves to measure quantities such as wavelength, small path-length differences, wave speeds, and indices of refraction, all to high precision. Figure 37–20a is a schematic diagram of one type of optical interferometer called the **Michelson interferometer** (Fig. 37–20b). In this device, developed by Albert Michelson in the 1880s, a light source is split by a beam splitter — for example, a partially silvered mirror—into two coherent waves that may travel different distances or through different media before they rejoin and interfere.

Monochromatic light from the source in Fig. 37–20a is split at the mirror at point A. The two beams then travel along paths 1 and 2 before they rejoin at A. The recombined beam is formed from the superposition (and therefore interference) of the two beams that arrive at A. The element C (a *compensator*) is added to make sure that the two light waves travel through the same amount of glass. If the path lengths are exactly the same, the two light waves will constructively interfere. With mirrors M and FM precisely perpendicular, the combined beam undergoes constructive interference and is bright. But if the path lengths are not precisely the same because the mirrors are not quite perpendicular to one another, the interference will produce alternating dark and bright lines, much like those discussed in Example 37–3, between two flat glass plates. The fringes will shift if the (screw-mounted) movable mirror is moved slightly. A movement of the movable mirror of only $\lambda/2$ will cause a shift from one fringe maximum to the adjacent one.

Unknown wavelengths of light can be determined by accurately measuring the movement of the mirror and counting the number of maxima that pass across the telescope eyepiece as the mirror moves. If ΔL is the mirror movement distance and N is the number of maxima that pass across the eyepiece for this movement, then

$$N = \frac{\Delta L}{\lambda/2} = \frac{2\Delta L}{\lambda},$$

or solving for the unknown wavelength,

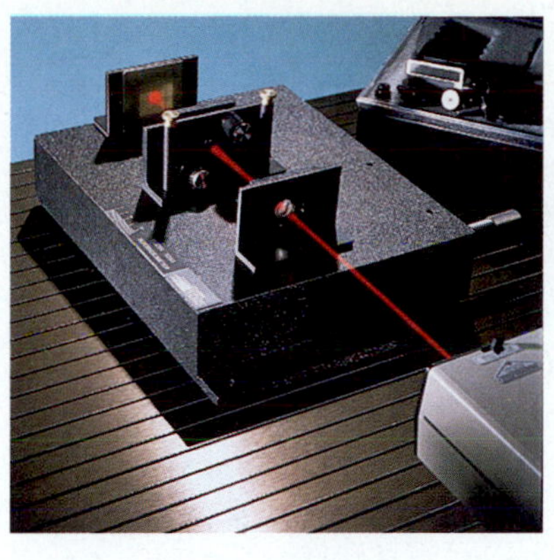

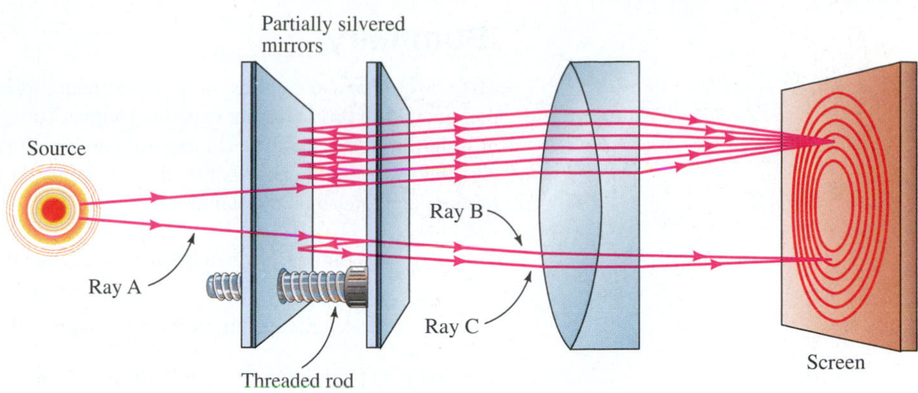

(a) (b)

▲ **FIGURE 37–21** (a) A Fabry–Perot interferometer. (b) Schematic diagram of a Fabry–Perot interferometer. Some of the light is transmitted and some reflected at the partially silvered mirrors. Rays B and C may constructively or destructively interfere, depending on the extra path length of ray C. Multiple reflection is an important part of the operation of the Fabry–Perot interferometer and serves to make the maxima sharper and more easily locatable.

$$\lambda = \frac{2\Delta L}{N}. \tag{37–17}$$

If the wavelength is known, the same technique can be used to measure very small distances, in this case, ΔL.

The Fabry–Perot Interferometer

The most widely used interferometer is the *Fabry–Perot interferometer* (Fig. 37–21a), invented by Charles Fabry and Alfred Perot and illustrated schematically in Fig. 37–21b. It contains two end plates (partially silvered mirrors) that are precisely parallel and flat and are connected by a rod that allows the distance between the plates to be changed smoothly by a screw thread. An incident laser beam (ray A) is partly transmitted (ray B) and partly reflected; the reflected ray is in turn partly reflected and partly transmitted to make ray C, which interferes with ray B. The major improvement incorporated into the Fabry–Perot interferometer is that the plates are silvered such that multiple reflections are possible, and the interference is between the multiple rays formed by these multiple reflections. The multiple reflections reinforce the regions of constructive interference, making the maxima stronger. The maxima become easier to locate and the pattern becomes more distinct. It then becomes easier to tell when these maxima have been shifted, so distances can be measured with more precision. If we compare Figs. 37–22a and 37–22b—interference patterns produced by a Michelson and by a Fabry–Perot interferometer, respectively—the greater sharpness of the Fabry–Perot pattern is evident.

To measure distances with a Fabry–Perot interferometer, we begin with the plates at known positions and then count the number of interference maxima changes (fringes) as the plate separation varies by the desired distance. Because the location of maxima can be determined to great accuracy, the distance change can be measured to within an error of only a fraction of a wavelength of the laser light. The fringe counting is done automatically by electronic sensors. The number of fringes involved in distances of about 1 m is on the order of the number of wavelengths of visible light contained in 1 m, around 50 million.

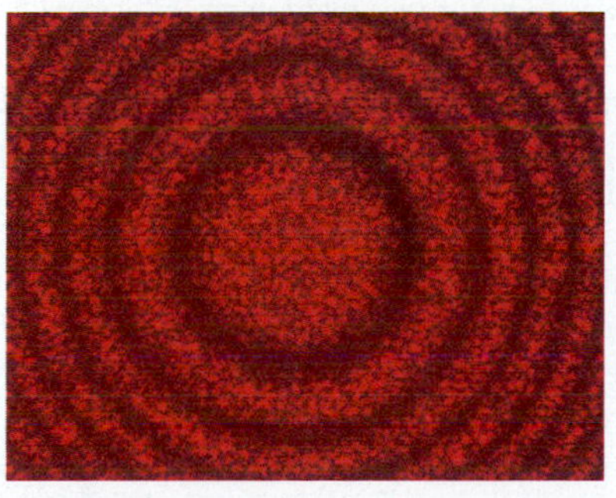

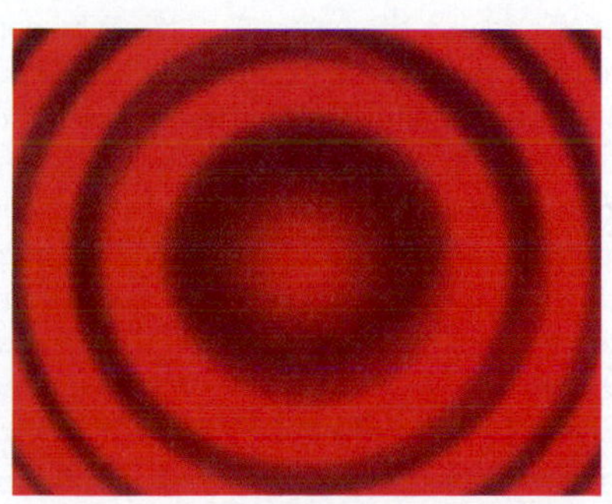

(a) (b)

◀ **FIGURE 37–22** The interference patterns produced by (a) a Michelson interferometer and (b) a Fabry–Perot interferometer. The Fabry–Perot fringes are noticeably sharper.

Summary

The wave theory of light explains experimental phenomena such as interference and diffraction that geometric optics cannot explain. Thomas Young was the first to substantiate the wave nature of light through his double-slit experiment, which produces interference maxima and minima. If λ is the wavelength and d is the distance between the (narrow) slits, there will be maxima and minima at angles θ on a distant screen given by

$$\text{for constructive interference:} \quad \sin\theta = n\frac{\lambda}{d}, \quad n = 0, \pm1, \pm2, \ldots; \tag{37–3a}$$

$$\text{for destructive interference:} \quad \sin\theta = \left(n + \frac{1}{2}\right)\frac{\lambda}{d}, \quad n = 0, \pm1, \pm2, \ldots. \tag{37–3b}$$

The intensity pattern for the double slit is

$$I_{\text{net}} = 4I_0 \cos^2\left(\frac{\pi d}{\lambda}\sin\theta\right). \tag{37–14}$$

where I_0 is the maximum intensity for a single slit and I_{net} is the total observed intensity.

We have looked at a variety of common situations where the geometry and other properties of reflecting surfaces results in characteristic interference patterns. Newton's rings appear when light shines vertically down upon a curved glass that rests on a flat piece of glass. The light reflecting from two surfaces interferes. For monochromatic light, alternating rings of bright and dark are observed. For sunlight, colored rings are observed. In both cases, the center is dark because the phase of an electromagnetic wave will change by 180° upon reflection when the wave moves from one medium to another of higher index of refraction. The condition for constructive interference in Newton's rings is

$$\Delta L = \left(m + \frac{1}{2}\right)\lambda, \quad m = 0, \pm1, \pm2, \ldots, \tag{37–15}$$

where ΔL is the difference in path length of the interfering rays.

Thin-film interference is responsible for the colors observed in bubbles and oil slicks. For light falling perpendicularly from air onto the surface of the film of thickness t, the condition for constructive interference is

$$4nt = (2m - 1)\lambda, \quad m = 1, 2, 3, \ldots, \tag{37–16}$$

where n is the index of refraction of the film. Light reflected from the two surfaces of the thin film interferes destructively for some wavelengths but not for others. If the incident light is white, the wavelength undergoing destructive interference is subtracted from the reflected light. This phenomenon is used in the application of thin coatings to lenses and other optical surfaces in order to reduce the intensity of reflected light.

Optical interferometers utilize the interference of light waves to measure distances with great precision. Two interferometers with practical applications are the Michelson interferometer and the Fabry–Perot interferometer.

Understanding the Concepts

1. Why is a laser a more practical source of coherent light than a bright lightbulb and a series of slits?
2. In Fig. 37–5b, the intersections represent places where constructive interference occurs on the crests of the advancing wave fronts. The troughs also reinforce each other between the circles. Do such reinforced troughs also represent a region of constructive interference? In the regions between the crests and troughs, where each wave is zero, the waves add to zero. Is this a region of constructive interference?
3. In Fig. 37–5b, the intersections represent sites where constructive interference occurs. Will these bright spots be seen in air, or is a screen required for us to see the alternating bright and dark spots?
4. We refer to lines, or fringes, when we use two elongated slits for interference. If we used two holes rather than two slits, would we still observe lines or would we observe something else?
5. Why does a single thin coating produce destructive interference in the reflection of light for only one wavelength in the visible region of light?
6. When we look up, stars twinkle and planets do not, while astronauts in orbit do not see any twinkle at all. Can you explain this?
7. In discussing interference from thin films, we have mentioned viewing the reflected light from afar. Why? Is there no interference if the light is viewed from close to the film?
8. Why don't we study (or see) interference effects with thick films?
9. There is no light intensity and therefore no energy at an interference minimum from two sources of coherent light. Yet each of the two interfering sources alone would produce energy at that point. Why does this situation not violate the conservation of energy?
10. When there is a minimum in the intensity of light reflected from thin films, there is less energy in the reflected beam. Does this mean that energy conservation is violated? If not, what happens to the energy that would otherwise have been in the reflected light? It may be helpful to think about what the light that passes all the way through a Newton's rings apparatus must look like.
11. Is it possible to use two beams of light that travel in the z-direction—one polarized with its electric field vector aligned

along the x-direction, and the other polarized with its electric field vector aligned along the y-direction—to make an interference pattern?

12. We estimated in Example 37–5 that almost 50 bands of constructive interference can occur along the glass plates. Can all these bands be observed? Will the color still appear blue?
13. The discussion of antireflective coatings suggested that the index of refraction of the coating material should be in between that of air and that of the glass that it coats. Why? (See Problem 52.)
14. In our discussion of Newton's rings, we did not consider light that reflects from the top surface of the curved piece of glass, nor that from the bottom surface of the bottom plate. Why not?
15. In Example 37–4, a second antenna has been constructed whose signal is as strong as that of the first, yet the signal you receive weakens. How is this consistent with the conservation of energy?
16. For small angles, the maxima for the double slit interference pattern (see Fig. 37–7) are equally spaced. Is this true for larger angles, ones for which the small-angle approximations are no longer useful?
17. How is it possible to use interferometry to measure a small fraction of a wavelength when the distance between maxima represents a full-wavelength difference?
18. When you look into a good-quality camera lens, you will see a color tint (generally purple). What is the origin of the observed color?
19. It might appear that a given antireflective coating will work for any surface to which it is applied, provided that the thickness of the coating is given by Eq. (AB–2) on p. 1042; however, this is not true. Why not? You can go back to the discussion of the energy in reflection (in Chapter 36) to find out why the index of refraction of the surface to be coated must be considered.
20. Why is it important that each slit be as narrow as possible in the double-slit experiment?
21. When you look out of a window with mini-blinds, you don't see interference effects. Why?
22. Antireflective coatings are always applied to the front surface—the side from which light comes—of an optical element, never to the back surface. Why?

Problems

37–1 Young's Double-Slit Experiment

1. (I) A coherent source of monochromatic light of unknown wavelength shines on double slits separated by 0.20 mm. Bright spots separated by 0.70 cm appear on a screen 3.0 m away. What is the wavelength of the light?
2. (I) Red light ($\lambda = 630$ nm) shines on a double slit with slit separation $d = 0.25$ mm. How far away from the central axis will the first minimum be on a screen 2.0 m from the double slit?
3. (I) A double-slit interference experiment is done in a ripple tank. The slits are 3.5 cm apart, and a viewing screen is 0.8 m from the slits. The wave speed of the ripples in water is 0.12 m/s, and the frequency of the vibrator producing the ripples is 12 Hz. How far from the centerline of the screen will the first maximum be found?
4. (I) The source for a double-slit experiment has a wavelength of 525 nm. The slits are a distance of 120 μm apart, and a screen is 40 cm away from the wall that contains the slits. How far from the center will the third maximum occur?
5. (I) Light of wavelength 590 nm falls on a wall with two slits 0.12 mm apart. A photographic plate is placed at a distance R from the wall. The $n = 3$ maximum appears 18 cm from the central maximum on the photographic plate. How far is the plate from the wall?
6. (I) Two small speakers are 65 cm apart. They broadcast a signal of frequency $f = 380$ Hz; the signals are in phase. A sensitive microphone is placed 2.5 m from the midpoint of the two speakers, on the line perpendicular to the line joining the two speakers. Is the intensity of sound a minimum or a maximum there? How far would the microphone have to be moved along an arc of radius 2.5 m centered on the midpoint in order to pick up a signal of maximum intensity?
7. (II) A double-slit experiment produces fringes on a distant screen. How does the linear separation between the bright maxima on the screen change when (a) the wavelength of the light doubles? (b) The separation between the slits doubles? (c) The distance between the slits and the screen doubles? (d) The intensity of the light doubles?
8. (II) A laser emitting light with $\lambda = 595$ nm shines on a double slit with a separation of 0.15 mm and produces interference fringes. If the maxima are separated by 0.50 cm, how far away is the screen on which the fringes are observed?
9. (II) In a double-slit experiment to determine an unknown wavelength of light, the measured total distance between 16 maxima (8 on each side of the central maximum) is 16.8 cm. The screen is located 3.45 m from the double slits, whose centers are 0.21 mm apart. What is the wavelength?
10. (II) Suppose that a double slit illuminates a distant screen. The light from sources S_1 and S_2 has come from a single monochromatic source S that is one-half wavelength closer to S_1 than to S_2. Use the geometry of Fig. 37–8 for the relation between the screen and the double slit to express the locations of maxima and minima. Is the point at $\theta = 0$ a maximum, a minimum, or neither?
11. (II) A double slit with variable separation d is superimposed on a single slit at right angles to the double slit, leading to a two-point source (Fig. 37–23). Then the double slit is rotated relative to its original direction by an angle ϕ. If light of wavelength λ shines through the system, determine the position of the interference maxima on a screen a distance R away. Do the fringes move inward or outward as ϕ increases?

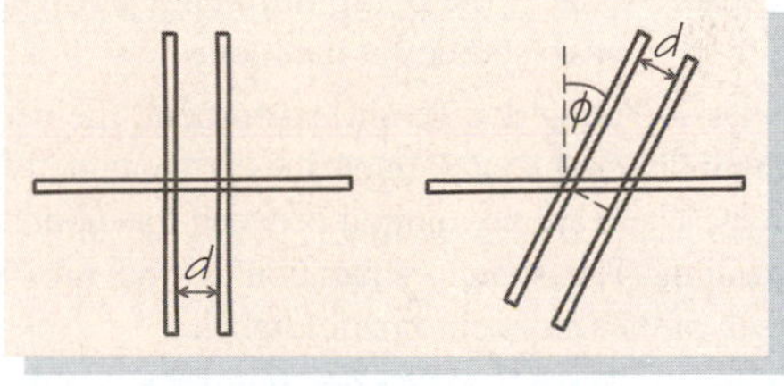

▲ **FIGURE 37–23** Problem 11.

12. (II) Two microwave sources are 18 cm apart. They radiate coherently with a frequency of 3.7×10^{10} Hz but with a phase difference α between the two sources. A microwave detector is moved along a line 1.6 m away from the sources. How far from the center ($\theta = 0$) will the first maximum occur, as α varies from 0 to 2π?
13. (II) Two sources radiate at almost identical frequencies f and $f + \Delta f$. How fast will the interference fringes on a screen a distance R away move, assuming that the sources are a distance d apart and that the velocity of wave propagation is v? Evaluate the result for two sources in a ripple tank ($f = 10$ Hz, $\Delta f = 10^{-6} f$, $R = 1$ m, $d = 5$ cm, $v = 0.15$ m/s). Do the same for an optical double-slit experiment ($f = 4.7 \times 10^{14}$ Hz, $\Delta f = 10^{-6} f$, $R = 1$ m, $d = 0.25$ mm, $v = 3.0 \times 10^8$ m/s). What can you conclude about the degree of coherence in the two cases?

14. (II) Light of two different wavelengths, λ_1 and λ_2, is incident on a double slit. On a distant screen, the twentieth maximum of λ_1 overlies the nineteenth minimum of λ_2. Show that the relative difference $(\lambda_1 - \lambda_2)/\lambda_1$ is small, and find a numerical value for this ratio.

15. (II) Consider two narrow slits illuminated from behind; the light impinges on a screen that is close rather than far (Fig. 37–24). Assume that the wavelength of the light λ is comparable to the separation d between the slits and to the screen distance R. The angle θ is measured from the point midway between the slits. (a) Show that the point corresponding to $\theta = 0$ continues to be a maximum of the pattern of light that reaches the screen. (b) At what angle θ is there a first maximum? (c) Show that your result for part (b) reduces to the distant-screen result for $R \gg d$ (and $R \gg \lambda$).

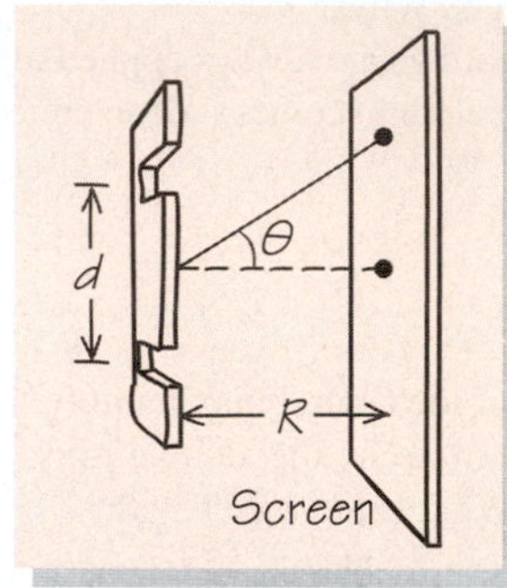

▲ **FIGURE 37–24** Problem 15.

37–2 Intensity in the Double-Slit Experiment

16. (I) Light of wavelength 500 nm shines on two slits separated by 0.3 mm. Find the intensity ratio I/I_0 at positions 0.6 mm and −0.5 mm from the central maximum on a screen 1 m from the slits.

17. (I) Two coherent light sources of the same wavelength, each with intensity of $1.0 \times 10^3\ \mathrm{W/m^2}$, interfere at a point at which the phase difference is 60°. What is the net intensity of light at this point?

18. (I) The net intensity from two coherent sources of equal strength is exactly twice the intensity due to either source at a certain point in space. What is the phase difference between the waves arriving from the two sources at that point?

19. (I) In a classic Young double-slit experiment, the net intensity is 1/4 of the individual source intensities at an angle of 27° off the central axis. There are no minima between this point and the central maximum. The source separation is 485 nm. What is the wavelength of the (monochromatic) light?

20. (I) The intensity at the central maximum of a double-slit diffraction pattern is I_{max}. If the wavelength of the light is 490 nm and the nearest location of maximum intensity is 1.7×10^{-3} rad from the central axis, what is the separation of the slits?

21. (II) Use the result of Young's classic double-slit experiment to find the average intensity on the screen by integrating the intensity over the surface of the screen. This result should be twice the average intensity from one slit alone and shows that energy conservation holds even when there is interference.

22. (II) Consider the double-slit arrangement shown in Fig. 37–25. The center of the screen C is a point of constructive interference. A container of thickness w, holding a liquid of refractive index n, is placed in the path of the ray from slit S_2 to C. Plot, qualitatively, the intensity of light at C as a function of w, assuming that the separation between the slits is d, and that the screen is a distance L from each slit. Ignore any intensity loss due to absorption in the fluid or at the fluid boundaries.

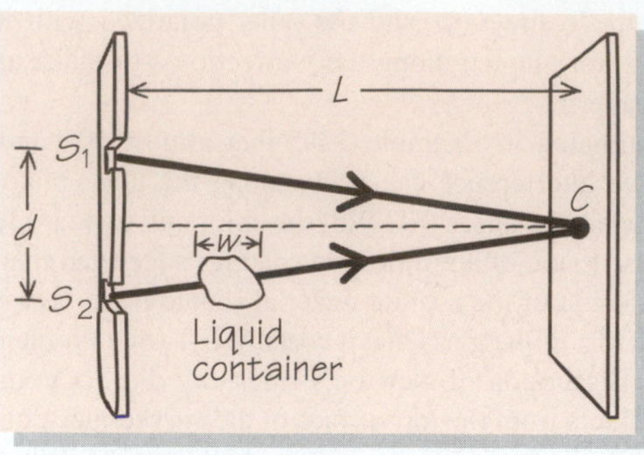

▲ **FIGURE 37–25** Problem 22.

23. (II) The *angular width* of a maximum of the intensity pattern due to double-slit interference is defined to be the angular separation $\Delta\theta$ of the points where the intensity is half its maximum value. Express the width of the central maximum in terms of the wavelength and the slit separation. Are the widths of all the maxima the same?

24. (II) A He–Ne laser, $\lambda = 633$ nm, shines on double slits separated by 0.35 mm. At what minimum angle θ is the intensity 50 percent of the maximum? If the screen is located 1.8 m away, what is the distance between the two angles on either side of the maximum for which this intensity occurs? This distance is the *full width at half maximum* of the central peak.

25. (II) Two point sources of radio waves, 12 m apart, radiate in phase with a frequency of 3.8×10^7 Hz. (a) If the average intensity of each single source is $5.0 \times 10^{-4}\ \mathrm{W/m^2}$ at a certain distance, what is the direction in which the combined intensity is maximized? (b) What is the magnitude of the maximum intensity? (c) At what angle will the intensity have fallen to half its maximum value?

26. (II) Suppose that the two slits in a double-slit experiment are not exactly the same size, so the electric field from one of the slits at a particular point P on the screen is $E_1 \sin(\omega t)$, whereas the other one is $E_2 \sin(\omega t + \phi)$, where the phase ϕ is due to the path-length difference [compare Eq. (37–7a, b)]. Show that the intensity at P is given by

$$\langle I_{net}\rangle = \langle I_1\rangle + \langle I_2\rangle + 2\sqrt{\langle I_1\rangle\langle I_2\rangle}\cos\phi,$$

where $\langle I_1\rangle$ and $\langle I_2\rangle$ are the intensities due to the light from the individual slits. [*Hint*: You will need $\langle \sin^2(\omega t)\rangle = \frac{1}{2}$; $\langle \sin(\omega t)\cos(\omega t)\rangle = 0$.]

27. (II) Point sources S and S' radiate with the same intensity and the same frequency, corresponding to a wavelength of 0.020 m (Fig. 37–26). They are 45° out of phase and 2.5 m apart. Plot the intensity as a function of distance along the x-axis for values of x much larger than the source separation.

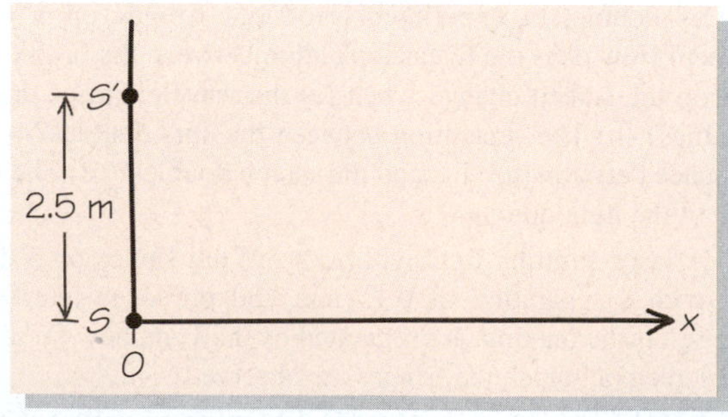

▲ **FIGURE 37–26** Problem 27.

28. (III) Find an expression for net intensity for the situation in Problem 27 in terms of x, the wavelength (λ), and the source separation (d). Take into account that the electric field of the electromagnetic wave decreases as the inverse power of the distance between the source and the receiver, and that the distance between source S and any point on the axis is different from the distance between source S' and that point. [*Hint*: Use the results of Problem 26.]

37–3 Interference from Reflection

29. (I) Consider the two glass plates of Example 37–5 and the configuration of Fig. 37–11. The plates are 25 cm long. When light of wavelength 656 nm from hydrogen shines down perpendicularly to the glass, 102 interference fringes appear. How thick is the wire that separates the two glass plates at one end?

30. (I) Two rectangular pieces of glass are laid on top of one another on a plane surface. A thin strip of paper is inserted between them at one end, so that a wedge of air is formed. The plates are illuminated by perpendicularly incident light of wavelength 615 nm, and 17 interference fringes per centimeter-length of wedge appear. What is the angle of the wedge?

31. (I) For what thicknesses of a soap bubble $(n = 1.3)$ that are less than 500 nm thick will blue light $(\lambda = 420\text{ nm})$ interfere constructively?

32. (I) A thin, uniform film of oil is spread on a glass plate. The oil has an index of refraction between that of air and of the glass. Write the condition for constructive interference for light of wavelength λ in air perpendicularly incident from air and reflecting back into the air from the air–oil–glass interface.

33. (II) You are an inventor who has never heard of tinting, and you want to use a plastic film of index of refraction 1.32 to place on the outside of glass to increase reflection, thereby keeping the intensity of the transmitted light down. What is the minimum thickness of the film that will accomplish this for light of average wavelength 550 nm? As this might be too thin to work with, suggest two other thicknesses that would work.

34. (II) A gardener accidentally spills oil (index of refraction 1.38) into a swimming pool. The oil spreads over the surface in a uniform way. (a) As you look directly down at the surface, you see light whose color suggests that the wavelength 520 nm is not being reflected. What is the minimum thickness of the oil film? (b) What will be the primary transmitted light frequency observed by someone swimming under water in the pool?

35. (II) A curved piece of glass in the form of the cap of a sphere of radius R is placed on a plane surface of glass. What will be the radius of the first dark Newton ring for light of wavelength λ? [*Hint*: It follows from plane geometry that, for a spherical cap, the radius of the circle bounding the cap, r, is related to the radius of the sphere R and h, the maximum thickness of the cap, by the relation $r^2 = h(2R - h)$, which can be approximated by $2Rh$ for $h \ll R$.]

36. (II) When two flat glass plates are placed on top of one another and a slip of paper is inserted between them at one edge, a thin wedge filled with air is produced between them. Interference bands form in reflection when monochromatic light falls vertically on the plates. Is the first band near the edge where the plates are in contact light or dark? Why?

37. (II) A wedge of air is formed between two glass plates that are 8.0 cm long, with one edge touching, and the other separated by a wire that is 1.0 mm in diameter. How far apart are the maxima for the reflected light with $\lambda = 655$ nm?

38. (II) Two plane glass plates touch at one end and are separated at the other by a wire, so that they form a wedge of air. When light with wavelength $\lambda = 590$ nm is reflected by the combination, 110 dark stripes appear. What can you say about the diameter of the wire?

39. (II) In a standard Newton's rings experiment, there is a dark spot where the convex surface touches the flat plate. Light of wavelength 500 nm is perpendicularly incident on the system. The convex lens is pulled slowly away from the flat plate until the minimum of the convex lens is 0.25 mm from the flat plate. A series of maxima and minima will appear at the center as the lens moves. How many maxima pass? Do the rings appear to move in to the center or away from the center?

40. (II) Consider the Newton's rings apparatus of Problem 39, with the minimum of the convex lens 0.85 mm from the flat plate. Water is poured into the space between the plates. Do the rings appear to move in to the center or away from the center, and how many maxima pass?

41. (II) A lens whose curved surface is part of a sphere of radius 5.0 m is placed over a flat glass plate and Newton's rings are observed. Determine the diameter of the fourth and seventh dark fringe for a wavelength of 520 nm.

42. (II) Light with a wavelength of 560 nm gives rise to a system of Newton's rings formed with a convex lens resting on a plane surface. The twentieth bright ring is at a radial distance of 0.98 cm. What is the thickness of the air film there, and what is the radius of curvature of the lens surface?

43. (II) The radius of curvature of a convex surface used for a Newton's rings apparatus is R (Fig. 37–27). Find the position x, measured from the point where the convex surface touches the flat surface, of the nth dark ring for light of wavelength λ perpendicularly incident from above.

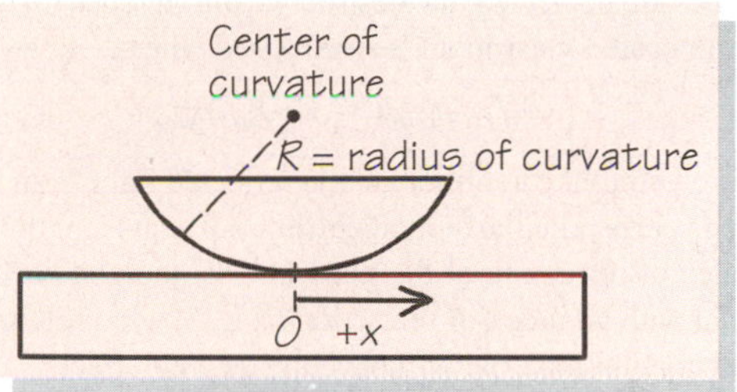

▲ **FIGURE 37–27** Problem 43.

44. (II) Constructive interference occurs when a soap bubble reflects light of wavelength 460 nm. What is the minimum thickness of the bubble if its index of refraction is 1.35?

45. (II) What minimum thickness of antireflective coating of MgF_2 is required to minimize the reflection of red light at 650 nm (the wavelength in air)? For MgF_2, $n = 1.38$.

46. (II) For light emitted by a He–Ne laser, $\lambda = 633$ nm, what *nonzero* minimum thickness of MgF_2 coating allows *maximum* reflectivity?

47. (II) The reflected light from an oil film floating on water shows constructive interference for light of wavelengths 434 nm and 682 nm incident along the normal. The index of refraction of the oil is $n = 1.51$. What is the film's minimum possible thickness?

48. (II) The reflected light from an oil film floating on water shows destructive interference for light of wavelengths 550 nm, 610 nm, and 685 nm incident along the normal. The index of refraction of the oil is $n = 1.40$. What is the film's minimum possible thickness?

49. (II) White light reflected at perpendicular incidence from a uniform soap film has an interference maximum at 666 nm and a minimum at 555 nm with no minima between 666 nm and 555 nm. If $n = 1.34$ for the film, what is the film thickness?

50. (II) A thin layer of CaF_2 ($n = 1.41$) is deposited onto glass, with $n = 1.52$. The layer is viewed in reflected light at 45° using a white light source (Fig. 37–28). For what layer thicknesses will light of 640 nm show constructive interference? Are there possible thicknesses for which light of 480 nm interferes constructively as well?

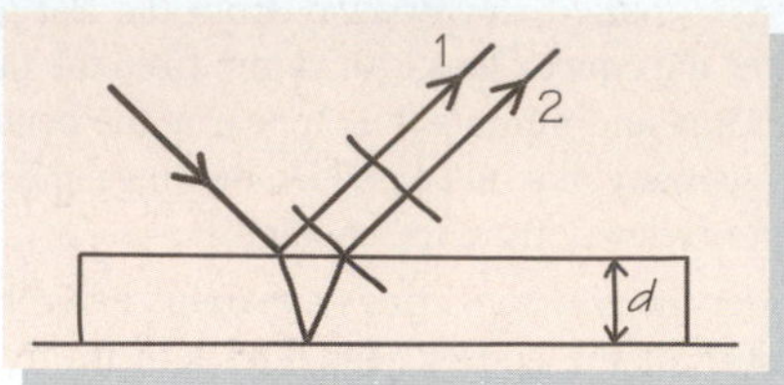

▲ **FIGURE 37–28** Problem 50.

51. (II) Light is perpendicularly incident on an oil film with $n = 1.2$, suspended in air. (a) If green light ($\lambda = 550$ nm) is reflected back most strongly, what is the minimum thickness of the film? (b) If n were increased, would the maximally reflected light have a longer or shorter λ? (c) If the film were suspended on the interface between water ($n = 1.33$) and air, what would be seen?

52. (II) We mentioned in Chapter 36 that when light is perpendicularly incident from a medium of index of refraction n_1 and refracts into a medium of index of refraction n_2, then the intensity of the reflected light, I_r, is related to the incident intensity, I_0, by $I_r/I_0 = (n_2 - n_1)^2/(n_2 + n_1)^2$. In order for a coating to eliminate reflections, it is not enough that the light reflecting from the two surfaces differs in phase by 180°; the interference is totally destructive only if the amplitudes are equal. Show that, when multiple reflections at interfaces are neglected, the destructive interference in light reflecting from coated glass in air is maximized when

$$(n_{air}/n_{coat}) = (n_{coat}/n_{glass}).$$

53. (II) You would like to eliminate the reflected light from a flat glass pane for perpendicularly incident light of wavelength 600 nm. If the index of refraction of the glass is 1.55 and you have a coating material with an index of refraction of 1.25, what minimum thickness of coating material will have the desired effect?

54. (II) Blue light, $\lambda = 485$ nm, is perpendicularly incident on a vertical soap film held in a plane. A sequence of bright horizontal bands appears in the reflected light (Fig. 37–16). Note the film is illuminated with white, not monochromatic, light. What is the rate of change with height of the thickness of the soap film if the horizontal bright bands are 0.6 cm apart? For the soap solution, $n = 1.36$.

55. (II) The material to be used for an antireflective coating has index of refraction of 1.25. How thick should the coating be to give the best result for $\lambda = 550$ nm and an angle of incidence of 30° with the normal?

*37–4 Interferometers

56. (I) Laser light with $\lambda = 633$ nm enters a Michelson interferometer. How many fringes will pass through the field of view if one of the mirrors is moved 0.10 mm?

57. (I) If the mirror of one arm of a Michelson interferometer is moved along the arm by 0.31 mm, 980 fringes traverse the field of view. What is the wavelength of the light used?

58. (II) A very sharp wedge of glass of index of refraction 1.55 is introduced perpendicularly in the path of one of the interfering beams of a Michelson interferometer illuminated by a narrow beam of light of wavelength 426 nm. This causes 750 dark fringes to sweep across the field of view. Calculate the thickness of the glass wedge at the point where the beam passes through it.

59. (II) Identical glass tubes, each 10 cm long, are placed in the two optical paths of a Michelson interferometer. Both of them are evacuated, then one is slowly filled with a gas until it reaches atmospheric pressure. During the filling time the pattern has moved by 90 fringes. What is the index of refraction of the gas?

60. (II) A thin glass plate of index of refraction 1.58 is introduced into one of the beams of a Michelson interferometer. This causes a displacement of 25 fringes for light with $\lambda = 589$ nm. What is the thickness of the plate?

61. (II) A scientist wants to measure the wavelength of yellow light ($\lambda \cong 590$ nm) to a precision of 0.1 percent. The minimum motion of the movable mirror is 0.03 mm. What is the minimum number of fringe shifts that must be counted?

62. (II) A laser with light of wavelength 582.5 nm is used to calibrate a Fabry–Perot interferometer. As the screw controlling the position of one end plate rotates exactly 100 turns, 714 fringes are counted. Calculate the wavelength of another light source that shifts only 593 fringes for 100 turns of the screw.

General Problems

63. (II) Two ordinary lightbulbs S_1 and S_2 are 1 m apart, each emitting light waves with intensity I, mainly at a wavelength of 550 nm. What is the pattern of intensity on a screen 100 m away?

64. (II) AM radio waves with a wavelength of 480 m travel 13 km to your home. Halfway between the transmitting tower and your home, but off to the side, is a building that reflects radio waves; the reflected wave has no phase shift due to the reflection. How far off the direct line is the building if destructive interference occurs between the direct waves and the reflected waves (Fig. 37–29)?

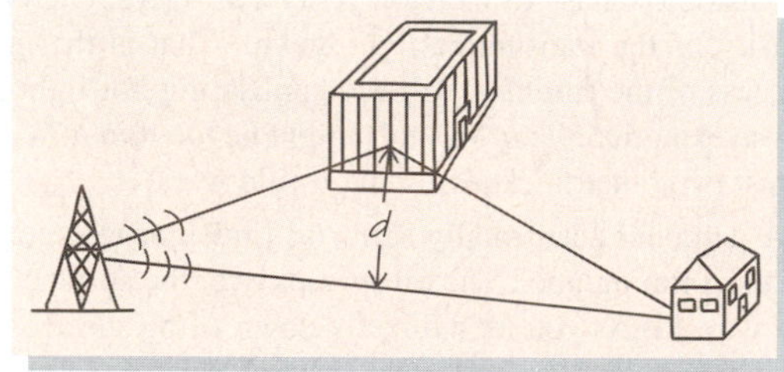

▲ **FIGURE 37–29** Problem 64.

65. (II) Two point sources of identical strength radiate in phase with the same frequency f. They are separated by a distance L. What is the energy density (discussed in Section 34–3) as a function of distance from one of the sources along the line that connects the sources in the central region, where the variation of the amplitudes with distance can be neglected (Fig. 37–30)?

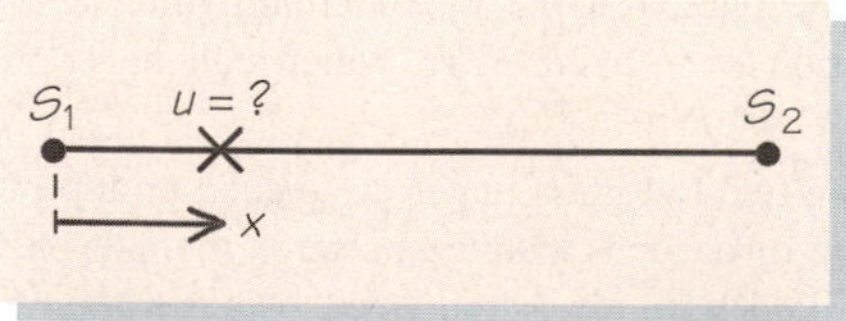

▲ **FIGURE 37–30** Problem 65.

66. (II) You want to hear a radio station that broadcasts at 97.9 MHz. You live on a direct line between the antenna and a large building that acts as a mirror for the radio waves broadcast by the station; you are exactly 100 m from the building (Fig. 37–31, see next page). Calculate the intensity of the signal you receive in terms of the intensity you would receive if the building were not present. Assume that the reflection from the building is total, with no phase shift, and ignore any decrease of the signal with distance.

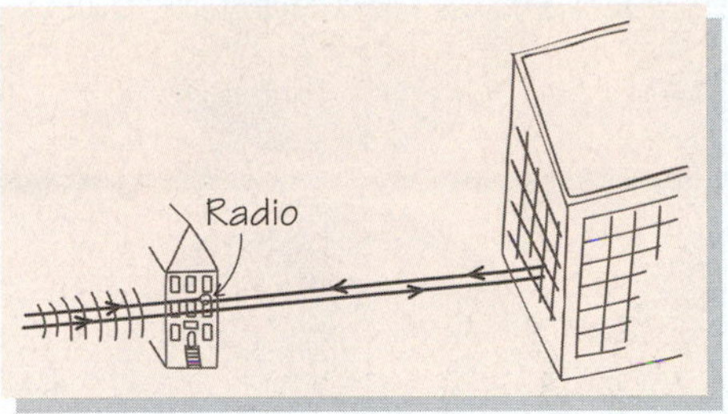

▲ **FIGURE 37–31** Problem 66.

67. (II) Coherent microwave radiation reflects from two identical obstacles (Fig. 37–32). Each obstacle, of size a, is much smaller than the wavelength of the radiation, λ, and much smaller than the obstacle separation d. Find an expression for the angles θ', defined in the figure, for which there are maxima on the distant screen.

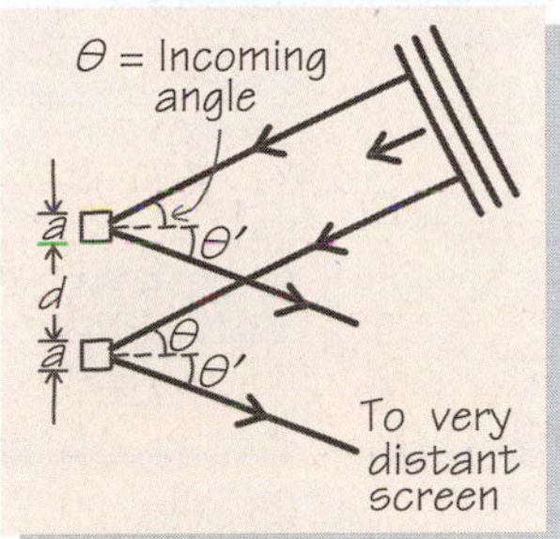

▲ **FIGURE 37–32** Problem 67.

68. (II) A radio wave undergoes a phase shift of 180° when it reflects from the calm surface of the ocean. A ship in a calm port nearing a shore station receives a 230-MHz signal from the station's antenna. This antenna is located 15 m above the sea surface, as is the ship's receiving antenna. The direct and reflected signals interfere, and a succession of maxima and minima are heard in the interfering signal at the ship (Fig. 37–33). How far is the ship from the station the first time the signal passes through a minimum? How fast is the ship moving if the time between this first minimum and the next one is 170 s?

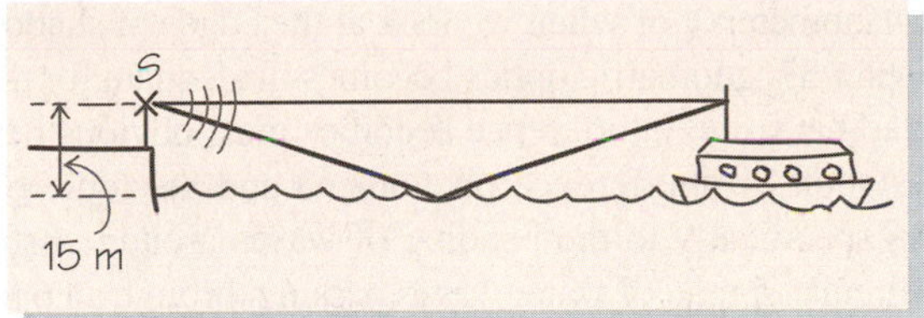

▲ **FIGURE 37–33** Problem 68.

69. (II) Sources S_1 and S_2 illuminate a distant screen; the distance to the screen is much larger than the separation between the sources. Each source emits light rays that are in phase at the sources, but the intensity I_1 of the light from S_1 is twice the intensity I_2 of the light from S_2. Give the ratio of the maximum to the minimum intensity of the light observed on the screen. (See Problem 26.)

70. (II) Figure 37–34 is an overhead view of two dipole radio antennas—vertical towers separated along the x-axis by a distance $\lambda/2$. This arrangement allows the radio signal to be beamed with greater intensities in some directions than in others, whereas either antenna alone would radiate its signal with the same intensity I_0 for any angle θ. (a) Find the intensity radiated by the antenna pair very far from the antennas as a function of θ, assuming that the signals of the two antennas are in phase. Describe the signal for all values of θ, from 0° to 360°. (b) The signal in the antennas is now 180° out of phase. How, if at all, does the distant intensity pattern change?

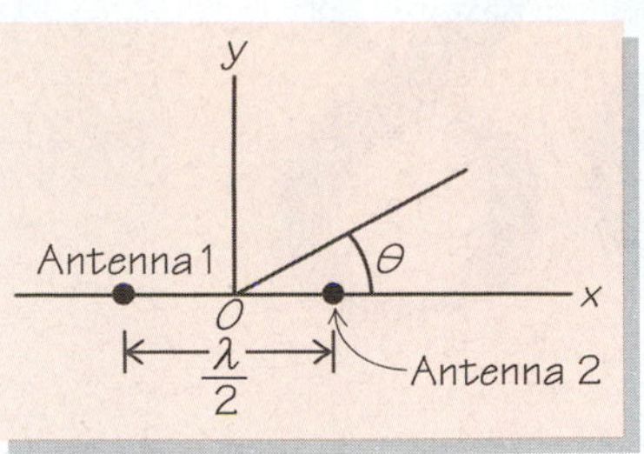

▲ **FIGURE 37–34** Problem 70.

71. (II) Consider the pair of dipole antennas in Problem 70. Suppose that the antennas are separated by one-quarter wavelength and that the signal in antenna 1 lags the signal in antenna 2 by 90° (one-quarter cycle). Show that the signal has a maximum in one direction, not two.

72. (II) An interference experiment uses a triplet slit, with slit separations d. What are the positions of the maxima on a screen a distance R away, if light of wavelength λ shines through the slits. What are the intensities of the maxima in terms of the intensity that each slit would produce alone?

73. (II) Four identical loudspeakers are placed at the corners of a square with sides of length $\lambda/\sqrt{2}$. The loudspeakers emit sound coherently with wavelength λ. A listener is situated very far from the square along one of the diagonals. If the intensity with just one loudspeaker on is I_0, what are the intensities when two, three, and four speakers are on? In a table, list all combinations and the resultant intensities.

74. (II) Light of wavelength λ is incident at a normal angle θ_1 to a double slit with slit separation d. Show that if you observe the exit light at a normal angle of θ_2, interference peaks will be seen when $m\lambda/d = \sin\theta_1 + \sin\theta_2$, where m is an integer.

75. (III) Light of wavelength λ is incident on a slab of glass of thickness h and index of refraction n resting on a mirror. The ray makes an angle θ with the vertical. The light reflected at the top of the glass surface and that reflected at the mirror will interfere. For what angles will the interference be totally constructive? [For angles even slightly away from grazing incidence (see Problem 76), light undergoes no phase shift upon reflection by a conductor such as a mirror.]

76. (III) *Lloyd's mirror* is a mirror that reflects light at large angles of incidence from a point source to a screen (Fig. 37–35). As the angle of incidence nears 90° (a grazing angle), the source and image become close. (a) Will the direct light and reflected light interfere constructively or destructively in this limit? (b) Suppose that the (monochromatic) source is 20 cm from the center of the mirror and 0.50 mm above the plane of the mirror and that the screen is distant. What will the angular separation be between successive maxima of the interference pattern, where the angle is measured from the source point to the screen?

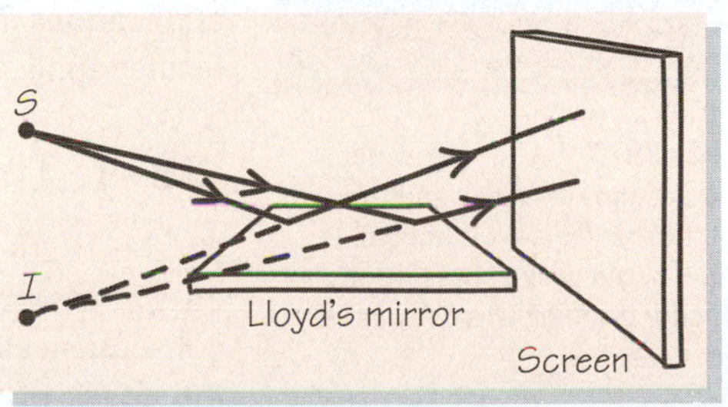

▲ **FIGURE 37–35** Problem 76.

Chapter 38

▶ The regular grooves in these compact disks act as diffraction gratings, which here exhibit maxima in reflection at angles that are different for light waves of different colors.

Diffraction

We can most easily discern the wave nature of light when it is incident on obstacles or apertures with sizes comparable to the wavelength of visible light (roughly 400 to 700 nanometers), or when we look at the edges of shadows at a similar scale. As we saw in Chapter 37, geometric optics becomes inadequate for the description of the behavior of light at this scale. Interference becomes most obvious here, as does *diffraction*, another manifestation of interference phenomena and the topic of this chapter. Diffraction usually refers specifically to the bending of waves around obstacles. We are already familiar with the circular spread of water waves passing through an opening narrower than their wavelength from Chapter 15, and we shall see that for light a single, narrow slit behaves similarly, producing characteristic patterns. This phenomenon is a diffractive effect. So is the pattern of maxima and minima that spreads across a screen in Young's double-slit experiment. The term diffraction also refers to interference between waves that emanate from a large number, or even a continuous set, of sources. Diffraction gratings consist of many slits or sources of coherent light, and such gratings have important applications in the study of atomic systems and crystalline materials. Holography is a spectacular application of the diffraction phenomenon with some interesting consequences.

▲ **FIGURE 38–1** The way water waves bend around obstacles and through openings in obstacles is well known. Here, parallel water waves pass through a hole, producing circular wave fronts on the far side.

38–1 The Diffraction of Light

By the 1820s, Young's double-slit experiment had reinforced the wavelike interference effects associated with light, and serious attempts were under way to understand the consequences of these light waves. The bending of light waves around obstacles or at the edges of apertures (Fig. 38–1) is not part of our everyday experience. This is because *interference effects require coherent wave sources* and because *diffraction effects are typically most significant when the sizes of the apertures or obstacles involved are comparable to the wavelength.* For light, the coherence condition is not realized in most situations, and wavelengths—several hundred nanometers—are tiny compared to the sizes of familiar objects. Nevertheless, these effects can be very important for the kind of high-precision instruments that today's technology requires.

We can observe diffraction effects with a source of coherent light, an intermediate object in the form of an obstacle or a wall with holes, and a viewing screen (Fig. 38–2). For example, the shadow of a razor blade in Fig. 38–3 shows a diffraction pattern at its edges. Around 1810 the French physicist Augustin Fresnel systematically investigated the interference patterns from apertures, edges, and small obstacles, and these diffraction patterns form what is now known as *Fresnel diffraction*. Fresnel treated various interference phenomena using the Huygens construction (see Section 35–2). He showed that even a single aperture creates its own diffraction pattern, because waves passing through different parts of the aperture interfere with each other. Similarly, even a single obstacle creates a diffraction pattern because parts of the original plane wave have been blocked by the obstacle and no longer participate in the Huygens regeneration of the wave. By 1821, Fresnel's research had progressed to the point where he was able to use a primitive version of an interferometer to make the first quantitative measurement of the wavelength of light.

If both the source and the screen are far from the intermediate aperture or object that forms the pattern, the mathematics is considerably simplified (Fig. 38–4). This special case is known as *Fraunhofer diffraction*, after the Bavarian physicist Joseph von Fraunhofer, a contemporary of Fresnel. The Fraunhofer limit is easy to treat because the lines that originate at each position of the intermediate object and that reach a given point on the screen are nearly parallel; approximating them as parallel simplifies the calculation of path-length differences and phase differences. We used the geometry appropriate to the Fraunhofer case in Chapter 37 when we studied double-slit interference patterns. Starting with Section 38–2, we shall restrict ourselves to this case.

The types of things seen by Fresnel and Fraunhofer are visible to the naked eye with a little ingenuity. For example, try looking at a distant mercury-vapor street lamp through the narrowest possible slit you can make between your fingers. As you look through the slit, what do you see?

Before we proceed to more detail, let's look at one rather spectacular demonstration of diffraction effects observed by Arago in 1818. Suppose that we place a perfectly round obstacle in the path of a point source of coherent light, as in Fig. 38–5a. If every point on the rim of the disk is equidistant from the source, the light falling on the rim is perfectly in phase. According to the Huygens construction, we can think of each of these points as a new source, and they are all in phase. All the rim points are equidistant from the point P on a screen that lies on the symmetry axis between the disk and the source. Because the light reemitted in the Huygens construction from all points on the rim arrives at P in phase, there is constructive interference at P, and hence a bright spot appears on the screen at the center

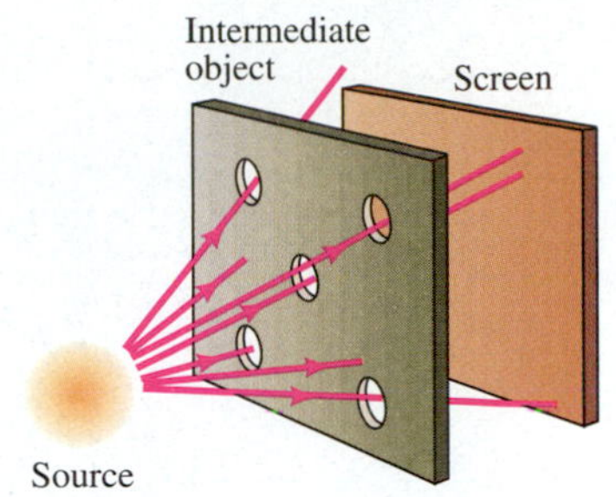

▲ **FIGURE 38–2** Diffraction effects can be observed by placing an intermediate object in the path of light that passes from a source to a viewing screen.

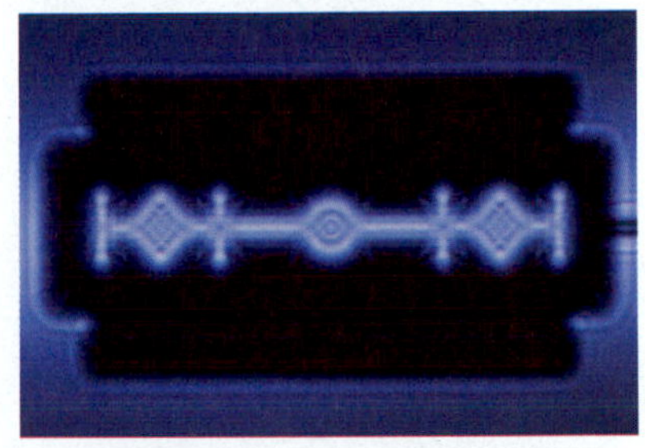

▲ **FIGURE 38–3** The diffraction of light around a razor blade.

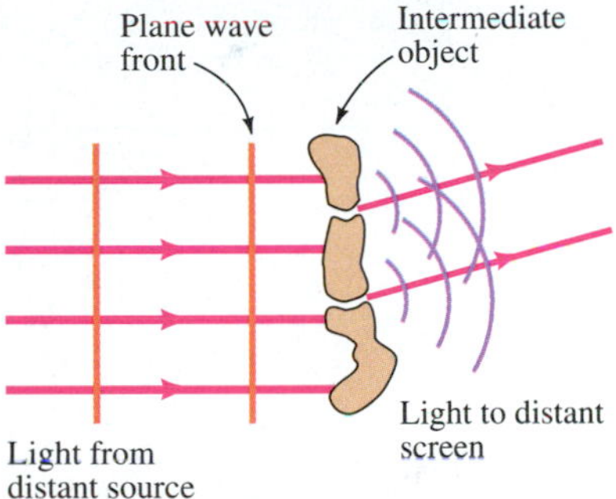

▲ **FIGURE 38–4** Parallel rays from a distant source are diffracted by an intermediate object and then viewed on a distant screen. When the screen is distant, we can treat the outgoing rays as parallel. This is the case of Fraunhofer diffraction.

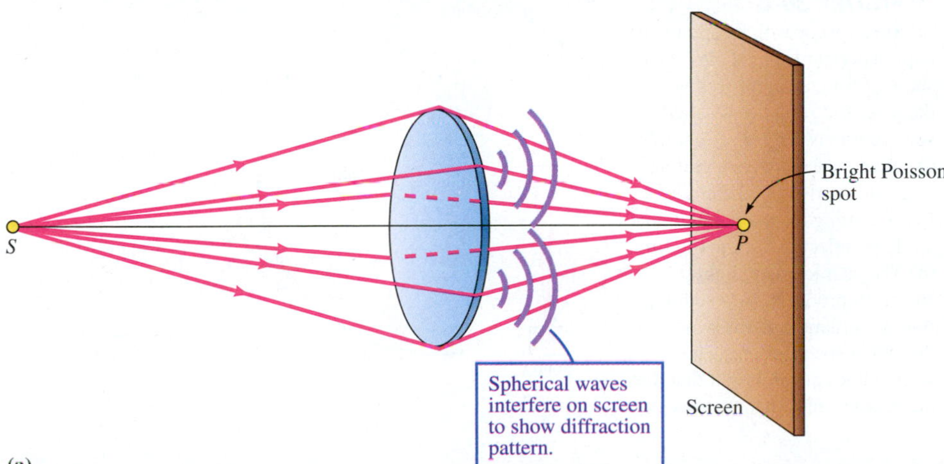

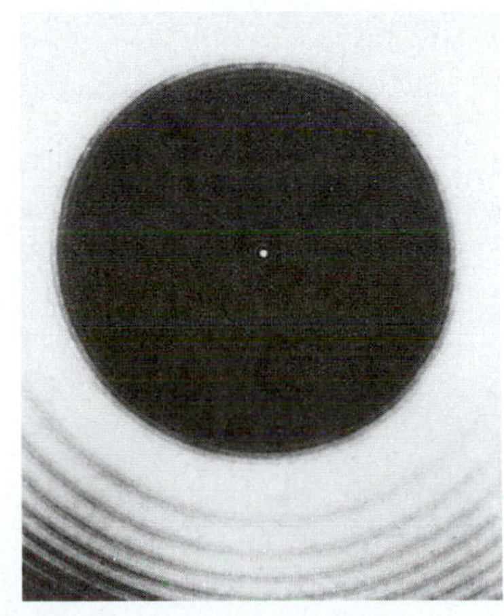

▲ **FIGURE 38–5** (a) A round, opaque object in the path of a point source of coherent light. Diffraction causes a bright spot (the Poisson spot) to be seen along the optical axis at point P. (b) The Poisson spot of a penny.

of the shadow, a central maximum known as the *Poisson spot*. This diffractive effect is certainly inconsistent with geometric optics! Even an ordinary penny can be used as the object; the Poisson spot at the center of its shadow in Fig. 38–5b is clearly visible.

38–2 Diffraction Gratings

A simple way to generalize the double-slit interference experiment that we studied in Chapter 37 is to increase the number of narrow slits. *If the slits are regularly spaced*, then a characteristic interference pattern is obtained; a screen with such an arrangement is called a **diffraction grating**. Diffraction gratings are important for two reasons: First, the multiple slits allow more light through than do two slits, thus increasing the intensity; second, the interference maxima are much sharper than they are for two slits, allowing the wavelength of the light to be measured more precisely.

A diffraction grating can take many forms. We simply require an array of obstacles to serve as pointlike sources for the reradiation of spherical wavelets. For example, when light passes through a glass plate scratched with rulings, the scratches act as sources for the regeneration of spherical wavelets. This type of grating is called a *transmission grating*. When the scratches are made on metal plates, they act as regular point sources of reflected rather than transmitted light; this grating is called a *reflection grating*. Even the marks on an ordinary ruler can be a reflection grating for laser light. What counts is that the light be scattered from regularly spaced centers.

THINK ABOUT THIS. . .

WHAT IS THE SIGNIFICANCE OF DIFFRACTION GRATINGS?

We stated above, and we'll show below in ample detail, that light that passes through a diffraction grating exhibits very sharp interference maxima and that this enables the measurement of light wavelengths to a very high precision. The ability to produce such gratings in a reliable way in the last half of the nineteenth century led to the discovery that light is emitted by excited atoms in discrete frequencies, and this discovery laid a good part of the foundations for quantum mechanics and, eventually, to an understanding of the structure of the atom. The unique set of wavelengths (or frequencies) produced by excited states of each type of atom, ion, or molecule is called its *spectrum*. *Spectroscopes*—optical devices used to measure so-called spectral lines, which are the intensity maxima corresponding to a particular wavelength—are typically based on diffraction gratings. Figure 38–6a is a schematic diagram of a spectroscope, and Fig. 38–6b shows the result of using a spectroscope to observe the light from a distant sodium-type street lamp. Diffraction gratings are still widely used in science and technology. Today we have a good understanding of the spectra of atoms and molecules. Knowing these wavelengths, which are completely characteristic of atoms and molecules of particular species, we can use a spectroscope as a very sensitive tool to recognize the presence of particular atoms or molecules in a distant star, or in a complex material. Observation of a given set of spectral lines is conclusive evidence of the presence of the emitting atom or molecule.

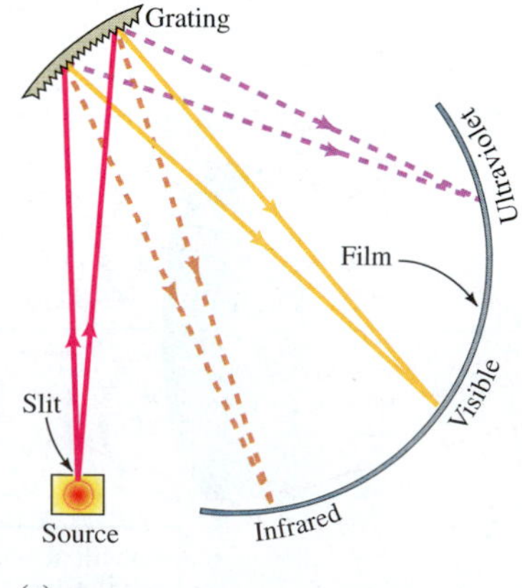

(b)

FIGURE 38–6 (a) Spectroscopes that capture an entire spectrum spread over a large sheet of film, as in this schematic diagram, are sometimes called spectrographs. The spectrograph depicted here employs a grating that is located along the circle. Light coming from a point on that circle is reflected back to the film. (b) The characteristic light spectrum produced by a sodium-type street lamp, as observed through a spectroscope. The yellow lines are so bright that they mask other lines that are present.

Energy Conservation and Intensity

Let's turn now to the analysis of the pattern made by a diffraction grating on a distant screen—the Fraunhofer diffraction pattern. Figure 38–7 shows a few of the slits of a grating with N slits, separated from each other by a distance d; a monochromatic plane wave of wavelength λ approaches from a distant source. The plane wave arrives with the same phase at each slit, and spherical waves are emitted in phase at each slit. Consider wave propagation along the lines labeled W, with lines $W1$, $W2$, $W3$, and so on oriented at an angle θ to the original wave-propagation direction. We allow the screen to be so distant that these lines are approximately parallel even though they all point toward a particular spot on the screen (or a lens focuses them on that spot).

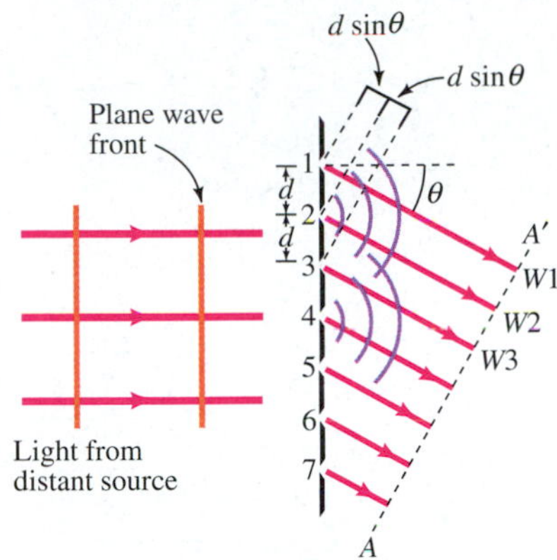

▲ **FIGURE 38–7** The geometry of a diffraction grating. Light passing through individual slits spreads in all directions. The interference of the light at a distant screen, along the lines indicated by $W1$, $W2$, $W3$, and so on, depends on the path-length differences $d \sin\theta$ between adjacent slits. The distant screen is not shown.

If the wavelets are in phase along the front AA', defined by θ, then the light that eventually reaches the distant screen at this angle will also be in phase, and there will be constructive interference—a maximum. The condition under which the waves along AA' are in phase is that each path length differs from any other by integral multiples of the wavelength λ. The path-length difference between adjacent waves is $d\sin\theta$ (Fig. 38–7). Thus there are *principal maxima*, where the light from all the slits interferes constructively, at angles such that

$$d\sin\theta = m\lambda, \qquad \text{where } m = 0, \pm1, \pm2, \ldots. \tag{38–1}$$

PRINCIPAL MAXIMA FOR A GRATING

As is the case for the two-source pattern, the integer m specifies the *order* of the principal maxima. In this result, we see a universal diffraction phenomenon: *The pattern spreads in angle as the ratio λ/d increases.* Equation (38–1) is the same as Eq. (37–3a), which determines the maxima of the double-slit pattern. The intensity pattern on the screen, however, takes quite a different form.

Before performing the mathematical analysis for the intensity pattern, let's first do a qualitative analysis of intensity. If there were no interference whatsoever, the average intensity over the entire screen due to N slits would be NI_0, where I_0 is the average intensity for just one slit. Energy must be conserved whether or not there is interference, so the average intensity over the entire screen must be NI_0 even with interference. If the light intensity is zero in regions of destructive interference, there must be a higher light intensity in regions of constructive interference. What is the intensity at the principal maxima, where there is constructive interference? At such points, the electric or magnetic fields of the waves from all the slits add to N times the field from one slit. Because the intensity is proportional to the *square* of the fields, the intensity at any maximum is N^2 times the intensity I_0 due to one slit:

$$I_{\max} = N^2 I_0. \tag{38–2}$$

This result agrees with the calculation of the double-slit pattern, where we found that the heights of the maxima are $4I_0 = (2^2)I_0$. If the maximum intensity increases so markedly for N slits, then, in order to conserve energy, the space over which the maximum occurs must be much smaller. Suppose that the principal maxima have a *width* of $\Delta\theta$. (Qualitatively, you can think of the width as some reasonable measure of the spread in angle for which there is some substantial intensity. We make this quantitative by defining the width as the angular spread of the peak where the intensity is half its maximum height of $I_{\max}$.) To find $\Delta\theta$, we equate the intensity in a single maximum with the averaged intensity between maxima. Thus

$$\begin{aligned} I_{\max}\Delta\theta &\cong NI_0 \times (\text{angular separation of successive maxima}) \\ &\cong NI_0(\lambda/d). \end{aligned}$$

We solve this equation for the width, and when we use Eq. (38–2), we find

$$\Delta\theta = \frac{NI_0(\lambda/d)}{I_{\max}} = \frac{NI_0(\lambda/d)}{N^2 I_0} = \frac{1}{N}\frac{\lambda}{d}. \tag{38–3}$$

To summarize, we have found that *as the number of slits N increases, the height of the principal maxima increases as N^2. In addition, the width decreases—the principal maxima become sharper—as $1/N$.* It is this feature that makes diffraction gratings so useful. A tall and narrow peak implies that the peak is easily seen even if the source is relatively weak, and that the peak can be located with great precision. Thus the wavelengths for which the principal maxima occur can be determined with great precision from Eq. (38–1).

Intensity Pattern

Let us turn now to a more precise treatment of the pattern. We can sum the electric fields at the screen associated with each of the slits in the N-slit pattern and by squaring the time average of the *net* field we find the intensity. Although we shall not go through the details here, the result for the intensity is

$$\text{for multiple slits: } I = I_0\left[\frac{\sin(N\beta)}{\sin\beta}\right]^2. \tag{38–4a}$$

The quantity 2β is the slit–to–slit phase difference. It is given by the product of the path length difference $d \sin\theta$ (see Fig. 38–7) and the wave number $2\pi/\lambda$:

$$2\beta = \frac{2\pi d \sin\theta}{\lambda}. \tag{38–4b}$$

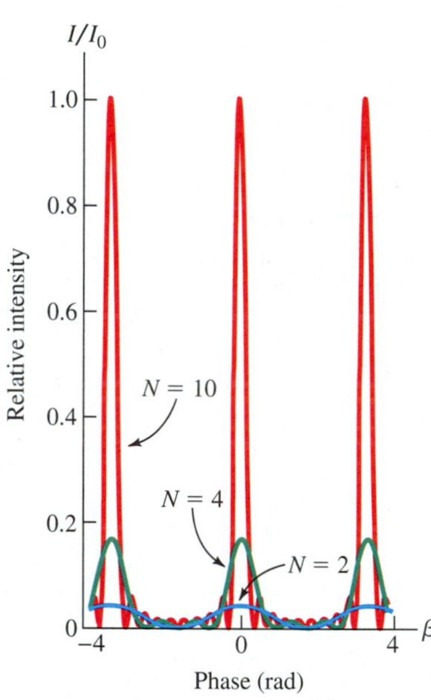

▲ **FIGURE 38–8** The ratio I/I_0 plotted against the phases $\beta = (\pi d \sin\theta)/\lambda$ for $N = 2$, 4, and 10 slits. The intensity pattern changes dramatically with N. The widths of the principal maxima decrease as $1/N$.

We can show that this result agrees with the earlier qualitative argument based on energy. Using Eq. (38–1), we see that $\sin\beta$ is zero at principal maxima. As $\sin\beta \to 0$, l'Hôpital's rule tells us that the ratio $\sin(N\beta)/\sin\beta$ approaches N. In this limit, Eq. (38–4a) agrees with Eq. (38–2).

Equation (38–4a) is most easily understood by plotting it, as in Fig. 38–8 for $N = 2$, 4, and 10: The ratio I/I_0 is plotted against β. The figure shows the principal maximum at $\beta = 0$ (corresponding to $\theta = 0$), the central spot on the screen, as well as the first principal maxima to the sides, corresponding to $m = \pm 1$ in Eq. (38–1). The widths of the principal maxima do indeed decrease as $1/N$. (Note that there are also $N - 2$ small secondary maxima between each pair of principal maxima. They have intensities on the order of I_0 itself. For diffraction gratings in ordinary use, N is in the thousands, so the secondary maxima can safely be ignored. They occur because there is the possibility that the light from two or more of the non-neighboring slits may be in phase at a given angle, even though the light from *all* the slits is not in phase.)

We can check our general result for N slits by setting $N = 2$ and comparing with our result for the double slit. If we let $N = 2$ in Eq. (38–4) with $\sin(2\beta) = 2\sin\beta\cos\beta$, we find

$$I = I_0\left[\frac{\sin(2\beta)}{\sin\beta}\right]^2 = I_0\frac{(2\sin\beta\cos\beta)^2}{(\sin\beta)^2} = 4I_0\cos^2\beta.$$

This is indeed the result for two slits, given in Eq. (37–13).

Resolution of Diffraction Gratings

Angular Dispersion: Diffraction gratings were formerly used to identify the characteristic wavelengths of elements. Now that these wavelengths are known, gratings are used primarily to *identify* elements, ions, and compounds through the characteristic light they emit or as a tool for understanding the structure of molecules. Because d, the distance between slits, is usually known, the angular location of a principal maximum ($m \neq 0$) gives λ according to Eq. (38–1). In this case, an important limitation is the ability to separate the spectral lines of nearly equal wavelengths λ_1 and λ_2. Two quantities determine the effectiveness of spectroscopic instruments. One is the **angular dispersion**, defined as $\Delta\theta/\Delta\lambda$, which measures the difference $\Delta\theta$ in the angles of the principal maxima of a given order due to two nearly equal wavelengths that differ by $\Delta\lambda$. Larger values of angular dispersion are desirable when one wants to distinguish two lines. We find the angular dispersion by considering two nearly identical wavelengths, λ and $\lambda + \Delta\lambda$, with maxima at the angles θ and $\theta + \Delta\theta$, respectively. Equation (38–1) gives us the difference in the angular position of their maxima

$$\sin(\theta + \Delta\theta) - \sin\theta = \frac{m}{d}[(\lambda + \Delta\lambda) - \lambda] = \frac{m(\Delta\lambda)}{d}.$$

If $\Delta\lambda$ is small, then so is $\Delta\theta$ and the difference $\sin(\theta + \Delta\theta) - \sin\theta$. We can approximate $\sin(\theta + \Delta\theta) = \sin\theta\cos(\Delta\theta) + \sin(\Delta\theta)\cos\theta \cong \sin\theta + \Delta\theta\cos\theta$, and hence $\sin(\theta + \Delta\theta) - \sin\theta \cong \Delta\theta\cos\theta$. The relation above can then be rearranged to read

$$\frac{\Delta\theta}{\Delta\lambda} = \frac{m}{d\cos\theta}. \qquad (38\text{–}5)$$

The angular dispersion increases for higher orders, is inversely proportional to the distance between slits, and increases away from the central maximum.

Resolution: The angular dispersion alone does not tell us whether we can visually separate two similar wavelengths. This aspect of the effectiveness of a grating is characterized by the **resolving power** of the grating, defined by

$$R \equiv \frac{\lambda}{\Delta\lambda}. \qquad (38\text{–}6)$$

Here, $\Delta\lambda$ is the smallest wavelength difference that can be observed with the grating. (The maxima of two wavelengths that are too close together lie so close to one another that they cannot be distinguished.) The larger the value of R, the better the grating can distinguish the relative wavelength difference of two closely spaced lines. Two closely spaced lines can be separated if the peaks are sharp. Detailed analysis of the peak widths of an N-slit system allows us to see that the resolving power of a grating is given by

$$R = mN. \qquad (38\text{–}7)$$

The resolving power improves as the number of slits N increases and is better for larger orders; that is, for larger integers m. The improvement in the resolving power as N gets large is quite visible in Fig. 38–8.

EXAMPLE 38–1 Heated sodium is an easily available source of light. It emits light of a characteristic yellow–orange color with two intense wavelengths of 589.00 nm and 589.60 nm, called a *doublet*. (a) How many slits are required in a grating that resolves the doublet at the first-order maxima? (b) If the screen is 4.000 m from a grating with exactly 2000 slits/cm, what are the screen positions of the two principal maxima of first order? Assume that the screen is far enough away so that the conditions of Fraunhofer diffraction apply.

Strategy The solution to this problem lies in our discussion of the resolving power and the position of principal maxima. For part (a), the minimum resolving power needed to resolve two lines whose wavelengths differ by $\Delta\lambda$ is given by Eq. (38–6), $R = \lambda/\Delta\lambda$. Given R, Eq. (38–7) then allows us to find N. For part (b), the angular positions of the first-order principal maxima are given by Eq. (38–1) with $m = 1$.

Working It Out (a) We have $\Delta\lambda = 589.60\text{ nm} - 589.00\text{ nm} = 0.60\text{ nm}$, and hence the resolving power R is

$$R = \frac{\lambda}{\Delta\lambda} = \frac{589.0\text{ nm}}{0.60\text{ nm}} = 9.8 \times 10^2.$$

Equation (38–7), with $m = 1$, gives $N = R = 9.8 \times 10^2$. About 1000 slits are required.

(b) To use Eq. (38–1), we need the slit separation $d = 1/(2000\text{ slits/cm}) = 5.000 \times 10^{-6}$ m. Then Eq. (38–1), with $m = 1$, gives

$$\text{for } \lambda_1\text{: } \sin\theta_1 = \frac{\lambda_1}{d} = \frac{589.0\text{ nm}}{5.000 \times 10^{-6}\text{ m}} = 0.1178;$$

$$\text{for } \lambda_2\text{: } \sin\theta_2 = \frac{\lambda_2}{d} = \frac{589.6\text{ nm}}{5.000 \times 10^{-6}\text{ m}} = 0.1179.$$

The respective angles are $\theta_1 = 0.1181$ rad and $\theta_2 = 0.1182$ rad, and the distance from the centerline of the screen is $y = L\tan\theta$. These respective distances are

$$y_1 = (4.000\text{ m})\tan(0.1181\text{ rad}) = 0.4745\text{ m};$$
$$y_2 = (4.000\text{ m})\tan(0.1182\text{ rad}) = 0.4750\text{ m}.$$

The images are separated by only 0.5 mm, but this resolution is sufficient to distinguish the two spectral lines.

What Do You Think? If the order $m = 1$ is not sufficient to separate the two sodium lines, would it help to look at higher orders? *Answers to* ***What Do You Think?*** *questions are given in the back of the book.*

CONCEPTUAL EXAMPLE 38–2 Take a compact disc (CD) and hold it below a lightbulb in a horizontal position such that when you look down on the CD, the reflection of the lightbulb overlaps the hole in the middle. As you tilt the CD away from the horizontal, you will see a swath of brilliant colors coming in from the rim toward the center. Why is this? Which colors come first? Will the colors repeat? Finally, could you estimate the spacing between the grooves on the CD from your experiment?

Answer The CD, which contains very finely spaced grooves, acts as a reflection grating. You start by looking straight down on the disc, that is, with an angle of reflection θ of zero. As you tilt it, you increase the reflection angle, and with it the groove-to-groove phase difference β [see Eq. (38–4b)]. This phase difference depends inversely on the wavelength λ so that it is larger for blue light than for red light, and the blue light will be seen first, with a sequence of the colors of the rainbow all the way to red light. Further tilt will bring in higher orders (secondary maxima), so there will be a repeat of the sweep of colors. Finally, β also depends on the groove spacing d, so that if you know the wavelength of, say, blue light and can estimate the tilt angle, you can measure d. You can certainly try this out for yourself (Fig. 38–9).

What Do You Think? What would you expect to see if you did the same experiment in the light of a helium-neon laser ($\lambda = 633$ nm) in an otherwise dark room?

▲ **FIGURE 38–9** The grooves on a CD can act as a diffraction grating, here separating the colors of a light bulb.

38–3 Single-Slit Diffraction

Coherent light passing through even a *single* slit produces a diffraction pattern. In the language of Huygens' principle, this pattern forms because wavelets that have regenerated at different places across the single slit interfere with each other. If the width a of the single slit is comparable to or smaller than the wavelength of coherent light that passes through it, the diffraction pattern is quite evident. We can explain it by treating the single slit as an infinitely large number of infinitesimal sources of wavelets.

Let's consider a plane light wave of wavelength λ that moves toward a wall with a narrow rectangular slit of width a ($a > \lambda$). A coherent wave arrives at the slit and regenerates spherical waves at each point across it. In Fig. 38–10a, we examine the light from the slit that continues along the initial direction toward the center point P of a distant screen. If the screen is far enough, we can assume that to a good approximation the waves from all parts of the slit travel the same distance, so that the regenerated wavelets are all in phase in this direction, and the central point of the screen is bright.

Along the direction that leads to point P' on the screen (Fig. 38–10b), with the angle given by $\sin\theta = \lambda/a$, there will be destructive interference under certain conditions. The path lengths from the tilted line s_1s_2 to point P' are all the same because the screen is distant, so we must consider only the phase relations of the light waves at line s_1s_2. The wave emitted at the top of the slit has traveled a distance λ to point s_2, and the wave emitted from the midpoint of the slit has traveled a distance $\lambda/2$ to point s_3. Thus, along the line s_1s_2, the wave emitted at the top is *out of phase* with the wave emitted at the center of the slit, and the waves will have this same phase relation at point P'. Similarly, the wave emitted just below the top of the slit is out of phase with the wave emitted from just below the center. We can follow the points *in pairs* along the slit. For every point in the top half of the slit, there is a point in the bottom half, and the waves from the two points are precisely out of phase with each other. The result is destructive interference—a minimum (or dark spot)—on the screen at P' at the angle given by $\sin\theta = \lambda/a$.

Plane wave front
Slit, width a
a
P
Light from distant source
(a)

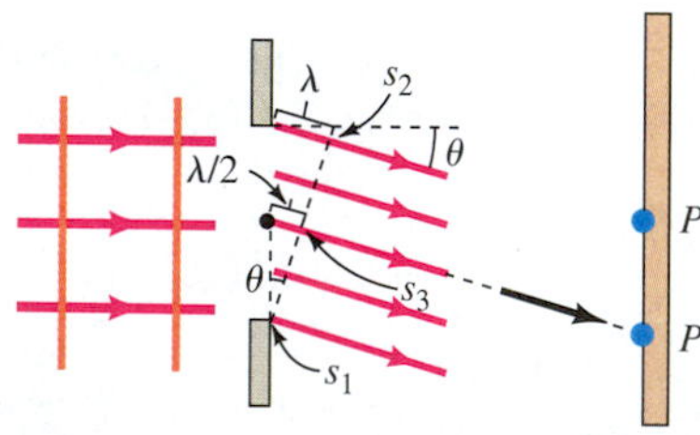

(b)

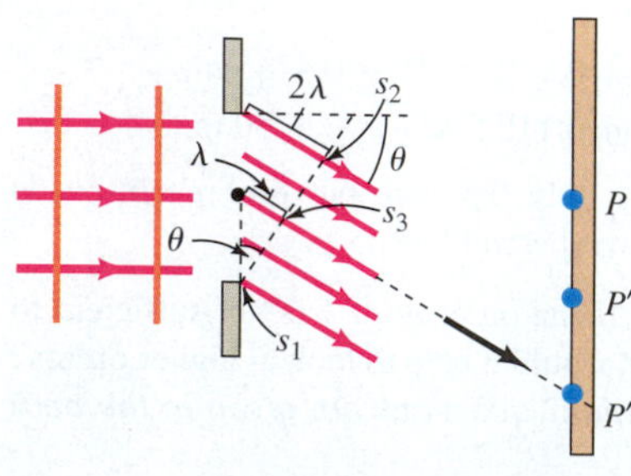

(c)

◀ **FIGURE 38–10** Monochromatic plane waves of light enter a narrow slit of width a (exaggerated here). We show three positions (P, P', and P'') on a distant screen. (a) We look along the incident direction at P. (b) We look along angle θ at P'. If the light from positions in the top half of the slit are out of phase with corresponding positions in the bottom half, destructive interference occurs at P'. (c) We look along a different angle θ at P''. Destructive interference can occur also at P'', as demonstrated by breaking the slit into halves and treating each half as we did for the direction toward point P'.

Along the direction given by $\sin\theta = 2\lambda/a$ (Fig. 38–10c), the wavelet emitted from the top of the slit travels a distance 2λ farther than the wavelet emitted from the bottom, and a distance λ farther than the wavelet emitted from the center point. We can think of the slit of width a as separated into two slits of width $a/2$. Along the direction chosen, we are in a situation similar to that in Fig. 38–10b. There is destructive interference at point P'' due to net destructive interference from both the top half and the bottom half of the slit. For example, for every wavelet emitted from a particular point in the top half of the slit, we can find a second wavelet emitted from another point in the top half that destructively interferes with the first wavelet, because their path-length difference is $\lambda/2$. The same happens for the bottom half. We will have another intensity minimum (dark spot) on the screen for angle $\sin\theta = 2\lambda/a$.

If we continue our analysis, we find that every time there is an additional path-length difference of λ between the top and bottom of the slit, destructive interference and a screen minimum result. Thus we have

$$\sin\theta = \frac{m\lambda}{a}, \qquad \text{where } m = \pm1, \pm2, \pm3, \ldots. \qquad (38\text{–}8)$$

MINIMA FOR A SINGLE SLIT

The value $m = 0$ is not part of this sequence of minima: For $m = 0$, $\sin\theta = m\lambda/a = 0$, and we have seen that this central point P must always be a maximum.

The interference pattern (Fig. 38–11) has the typical behavior of diffractive phenomena: Larger values of a/λ give a smaller angular spread in the interference pattern, and smaller values of a/λ give a larger angular spread. In the limit that $a \gg \lambda$, the spread decreases so much that there is only a bright central spot on the screen. Do not confuse the angular spread on the screen with a projection of a slit width in geometric optics. The screen is very distant, and if geometric optics were to hold, the projection of the slit on the screen due to direct incident light would be a line with precisely the width of the slit. *But geometric optics does not hold; the smaller the slit width compared to the wavelength, the larger the angular spread of the interference pattern.*

Approximately halfway between the successive minima, we have conditions for constructive interference and an intensity maximum (a bright region). In other words, in addition to the constructive interference at $\theta = 0$ (the central maximum), we also have

$$\text{for constructive interference: } \sin\theta \cong \frac{\left(m + \frac{1}{2}\right)\lambda}{a}, \qquad \text{where } m = \pm1, \pm2, \ldots. \qquad (38\text{–}9)$$

We'll work out more details of the intensity pattern in the two subsections to follow.

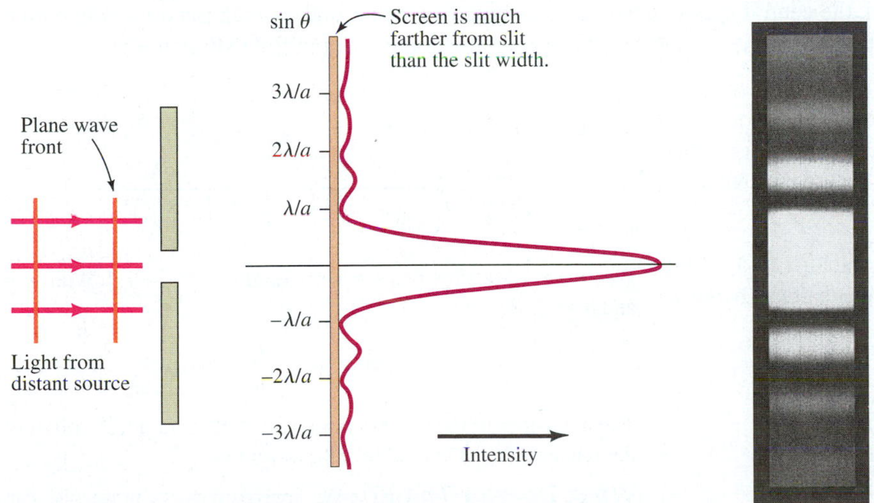

▲ **FIGURE 38–11** The interference pattern of single-slit diffraction and the relative intensities of such a pattern. Most of the light energy is in the bright central peak. The central peak is twice as wide as the secondary peaks.

EXAMPLE 38–3 Helium–neon laser light of wavelength 633 nm passes through a single slit of width 0.10 mm. The diffraction pattern is observed on a screen 3 m away. What is the distance between the two minima on either side of the central maximum?

Strategy The conditions for the single-slit pattern apply here; in particular, the slit width a is some 160 times larger than the wavelength. We can use Eq. (38–8) to find the angular positions of the minima that correspond to $m = 1$ and -1, and with the angular positions known geometry gives the positions on the screen.

Working It Out The minima for $m = \pm 1$ occur at angles

$$\sin\theta = \frac{m\lambda}{a} = (\pm 1)\frac{633 \times 10^{-9}\ \text{m}}{0.10 \times 10^{-3}\ \text{m}} = \pm 0.0063;$$

$$\theta = \pm 0.36°.$$

At a distance of 3 m away, these minima will be (3 m) tan 0.36° = 1.9 cm from the central line ($\theta = 0$). The total distance between them is thus 2(1.9 cm) = 3.8 cm.

What Do You Think? If coherent blue light is used instead of the red light of the helium-neon laser, would the distance between minima increase or decrease?

The Intensity Pattern of Single-Slit Diffraction

As we shall show in the next subsection, the intensity pattern on a distant screen due to a single slit of width a takes the form

$$\text{for single slits: } I = I_{max}\frac{\sin^2\alpha}{\alpha^2}, \tag{38–10}$$

where

$$\alpha = \frac{\pi a \sin\theta}{\lambda}. \tag{38–11}$$

As we learned at the start of this section, the angle α, which is measured in radians, is simply the phase difference between the top and the middle of the slit.

As is confirmed by Fig. 38–11, there is a central maximum at $\theta = 0$: At $\theta = 0$, Eq. (38–11) shows that $\alpha = 0$. Because the limit of $(\sin\alpha)/\alpha \to 1$ as $\alpha \to 0$, the intensity at this point on the screen is just I_{max}. The intensity drops off rapidly as α increases. The intensity is zero for angles given by Eq. (38–8), and this occurs when α is an integral multiple of π.

$$\text{for minima: } \alpha = n\pi = \frac{\pi a \sin\theta}{\lambda}, \qquad \text{where } n = \pm 1, \pm 2, \pm 3, \ldots. \tag{38–12}$$

The intensities at the secondary maxima are estimated in Example 38–4.

By looking at Fig. 38–12, for which we have chosen $a = 4\lambda$, we can identify two important features. First, the secondary maxima are not very strong—most of the light is contained in the wide central maximum at $\theta = 0$. This contrasts with the pattern made by a grating, for which there is no difference in the strength of the successive maxima. Second, the central maximum is *twice* as broad as the secondary maxima are—a feature that distinguishes the single-slit pattern from the double- or multiple-slit patterns.

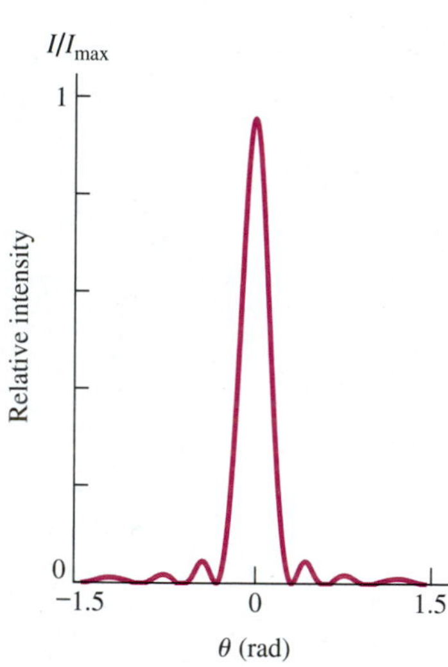

▲ **FIGURE 38–12** The intensity ratio I/I_{max} as a function of angle θ [from Eqs. (38–10) and (38–11)] for the slit width $a = 4\lambda$.

EXAMPLE 38–4 Estimate the ratios of the intensities of the first and second maxima to the intensity of the central maximum for a single slit.

Strategy We can use our approximate relation, Eq. (38–9), to locate the maxima. This relation places the maxima halfway between the minima, and in terms of α this means

$$\text{for maxima: } \alpha \cong (n + \frac{1}{2})\pi, \qquad n = \pm 1, \pm 2, \ldots.$$

The intensity need only be evaluated at these locations.

Working It Out Denote the intensity at the secondary maxima by I_n. Then Eq. (38–10) gives

$$\frac{I_n}{I_{max}} = \left\{\frac{\sin\left[\left(n + \frac{1}{2}\right)\pi\right]}{\left(n + \frac{1}{2}\right)\pi}\right\}^2 = \frac{1}{\left(n + \frac{1}{2}\right)^2\pi^2}.$$

[We have used $\sin^2(\pi/2) = \sin^2(3\pi/2) = \ldots = 1$.] With $n = 1$ and $n = 2$, we get

$$\frac{I_1}{I_0} = 0.045 \quad \text{and} \quad \frac{I_2}{I_0} = 0.016.$$

The intensities of the secondary maxima fall off rapidly, with most of the intensity in the central bright maximum.

What Do You Think? We approximated a midway location for the maxima; how can we find an exact expression for the location of the maxima in single-slit diffraction?

*Deriving Single-Slit Intensity Values

Let's imagine breaking up a single slit of width a into N strips of width d (Fig. 38–13), so that $Nd = a$. Each strip acts as a separate slit. In the limit that the number of strips $N \to \infty$, d must approach zero in order to keep a constant. Because each strip is infinitely narrow, we can treat these strips as the thin slits of a grating. We can then use the diffraction grating result of Eq. (38–4) directly to find that the intensity at the screen at angle θ is

$$I = \lim_{N \to \infty} I_0 \left[\frac{\sin(N\beta)}{\sin\beta} \right]^2.$$

This expression includes a slit width $d = a/N$ that approaches zero as $N \to \infty$ through the quantity β. According to Eq. (38–4), $\beta = [\pi d \sin\theta]/\lambda = [\pi(a/N)\sin\theta]/\lambda = \alpha/N$, where the definition of α is contained in Eq. (38–11). Thus the intensity takes the form

$$I = \lim_{N \to \infty} I_0 \left[\frac{\sin\alpha}{\sin(\alpha/N)} \right]^2.$$

In the limit of large N, the factor $\sin(\alpha/N) \cong \alpha/N$, and

$$I = N^2 I_0 \frac{\sin^2\alpha}{\alpha^2}.$$

The factor I_0 is the intensity due to one of the subslits of width d. We need only interpret the factor $N^2 I_0$ as the maximum possible intensity I_{max} of the single slit of width a to get Eq. (38–10).

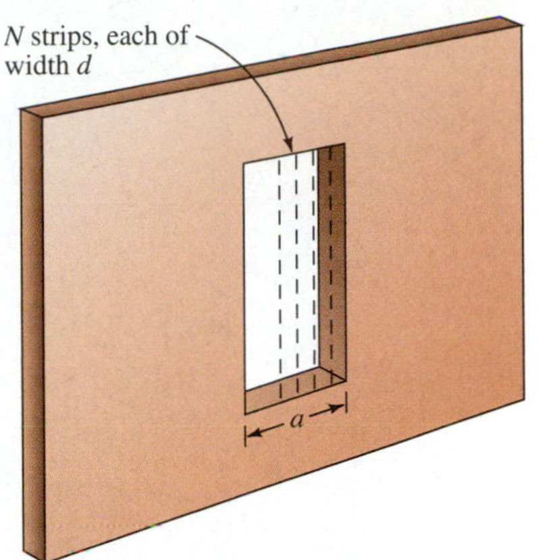

▲ **FIGURE 38–13** We can understand the intensity pattern of a single slit by supposing that the slit is composed of a large number of strips N and then using the result we derived for diffraction gratings.

38–4 Resolution of Optical Instruments

We have already seen that the ability of a grating to resolve closely spaced lines is limited by the width of the principal maxima. Similarly, the fact that there is some spreading of light in a single aperture due to diffraction *intrinsically* limits the capacity of optical instruments to resolve objects. The resolution of instruments such as telescopes or microscopes is also limited by lens aberration (see Chapter 36), but better lens design decreases aberration, and there is no theoretical limit to such improvement. The limitation due to diffraction, however, is set by the aperture of the instrument and the wavelength of light, and this limit is intrinsic to the wave nature of light.

Most optical instruments rely on circular lenses or mirrors, so we shall concentrate on circular apertures. Sir George Airy worked out the diffraction pattern from a distant point source that passes through a circular aperture (Fig. 38–14) in the 1830s. The bright central area—containing some 85 percent of the light intensity—is called an *Airy disk*. The rings outside the central area are the minima and secondary maxima of the diffraction pattern. The position of the first minimum occurs at an angle from the central axis given by

$$\theta_{min} = 1.22\frac{\lambda}{D}, \tag{38–13}$$

where D is the diameter of the aperture. This result can be compared to Eq. (38–8), which gives the angle of the first minimum for a slit of width $a \gg \lambda$ as $\theta_{min} = \lambda/a$. (Recall that $\sin\theta \cong \theta$ for small θ.) The factor 1.22 arises because the "width" of a circular aperture varies (in effect, $a \cong D/1.22$). But it is enough to use a factor 1 instead of the precise factor 1.22 to understand the effects of diffraction. Thus we give an approximate θ_{min} as

$$\theta_{min} \cong \frac{\lambda}{D}. \tag{38–14}$$

MINIMUM OBSERVABLE ANGULAR SEPARATION

How does the presence of a diffraction pattern limit our ability to make images? In many applications high resolution is a necessity; for example, we may want to observe two closely spaced objects, such as a double star, or we may need to see detail in an X

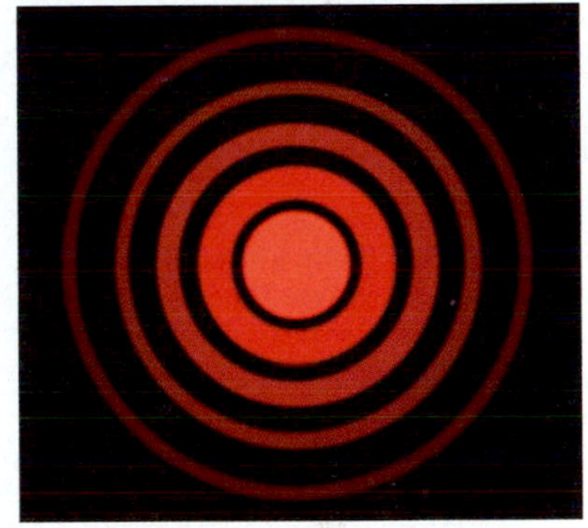

(a)

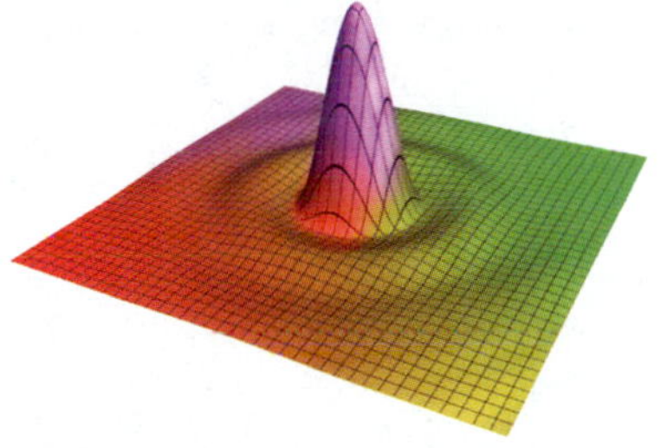

(b)

▲ **FIGURE 38–14** (a) Diffraction pattern of light from a distant source after the light passes through a circular aperture. Some 85 percent of the light intensity is contained in the bright central maximum, called the Airy disk. (b) A three-dimensional representation of the light intensity on a distant screen from a circular aperture.

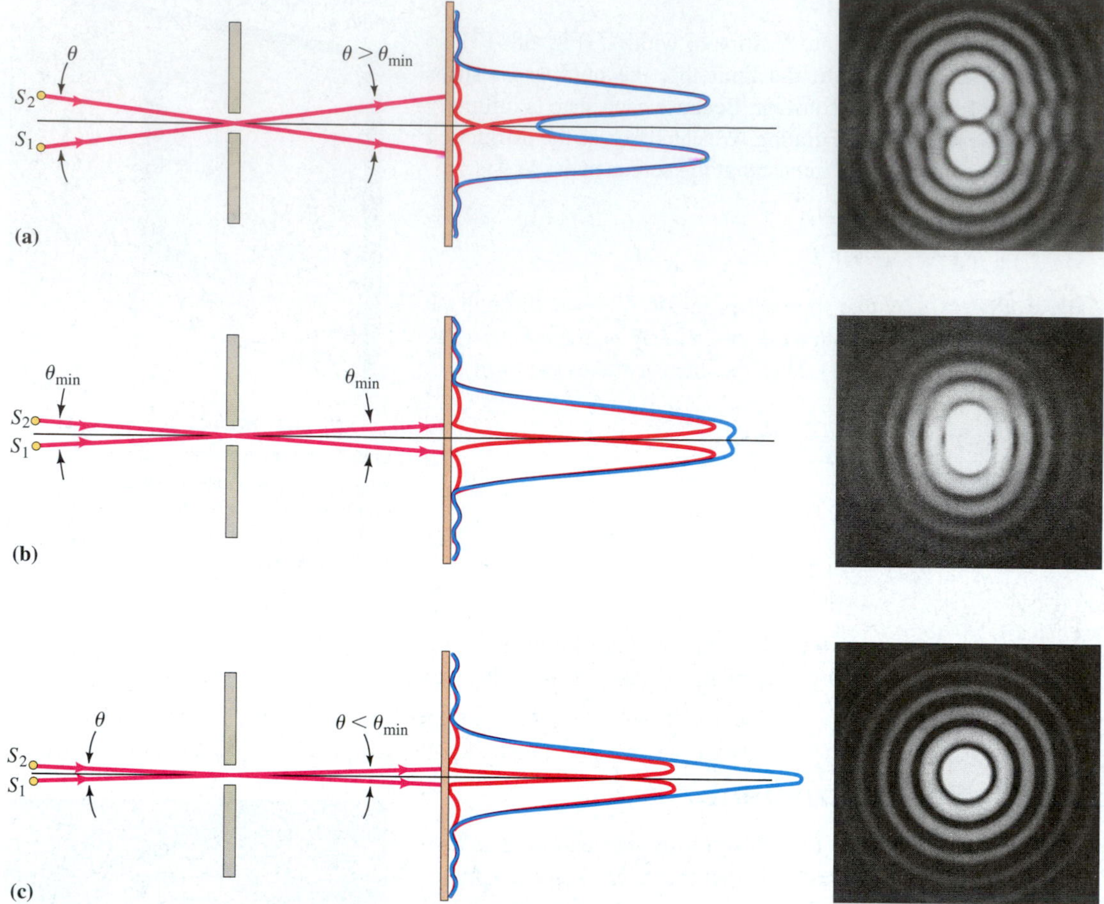

▲ **FIGURE 38–15** Light from two sources passes through an aperture. (a) The objects are easily resolved. (b) The objects are just barely resolved. The angular separation is θ_{min}. (c) The objects are not resolved. The blue lines in each represent the sum of the two intensities.

ray taken of suspected stress fractures in piping for a power plant. The presence of the Airy disk means that even a very distant star does not produce a pointlike image but rather a disklike image with an angular spread described by Eq. (38–14). The image in a telescope of two stars that are so close together that their Airy disks overlap cannot be recognized as an image of two stars. In Fig. 38–15a, two objects have a sufficient angular separation θ that they are easily resolved. In Fig. 38–15b, the objects are just barely resolved. When they are even closer, as in Fig. 38–15c, the central diffraction peaks overlap too much to be resolved. It has become customary to describe the limiting case—shown in Fig. 38–15b—as the *Rayleigh criterion*:

Two point sources are just resolved if the peak of the diffraction image of the first source overlies the first minimum of the diffraction image of the second source.

The Rayleigh criterion is satisfied when the angular separation of the objects is just θ_{min}, defined by Eq. (38–13). Equation (38–14) provides an approximate alternative. Remember, the quality of the optical instruments is not the issue here; the limitation on the resolution of images is due to light's intrinsic wave character and can be improved only by making the instrument aperture larger.

The angular separation of an image formed by an optical system is also the angular separation of the objects. We can show this using the methods we used when we studied optical instruments in Chapter 36. These angular separations are the same because the principal rays through the center of a lens are undeviated. These rays are shown in Fig. 38–16 for two objects at P_1 and P_2 respectively, with the objects a distance L from a lens of diameter D. When this angle is less than that specified by the Rayleigh criterion, the objects cannot be separated. This translates into a minimum spatial separation S_{min} of the two objects through the relation $S_{\text{min}} = L\theta_{\text{min}}$:

$$S_{\text{min}} = L\theta_{\text{min}} \cong \frac{L\lambda}{D}. \tag{38–15}$$

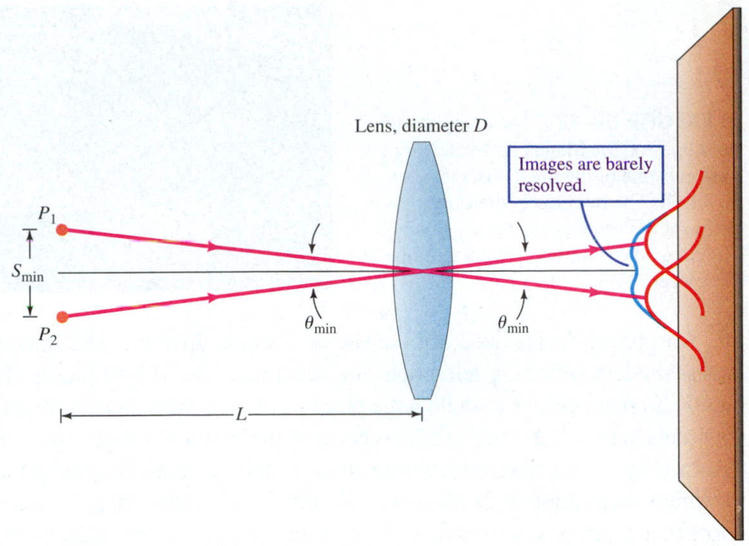

◀ **FIGURE 38–16** The minimum separation distance that is resolvable for a lens depends on θ_{min}.

EXAMPLE 38–5 Estimate the minimum separation between two objects such that the human eye can still perceive them as separate (the *minimum visible object separation*) if the objects are (a) at the near-point distance (25 cm) and (b) at a distance of 5 m. Take the pupil diameter to be 2.5 mm.

Setting It Up We can apply Eq. (38–15) if we can estimate the eye's aperture as well as a wavelength. You can look in a mirror and estimate the pupil diameter to be 2.5 mm, and for the wavelength, we choose a median range of visible wavelengths, about 550 nm.

Strategy This is straightforward application of Eq. (38–15).

Working It Out (a) For $L = 25$ cm (the near point), we have

$$S_{min} = \frac{L\lambda}{D} = \frac{(0.25\ \text{m})(550 \times 10^{-9}\ \text{m})}{0.0025\ \text{m}} = 0.055\ \text{mm}.$$

This is about the diameter of a thin thread or a human hair. It is also, roughly, the separation between the cells of the retina. In other words, the cells that receive light and send that message to the brain are no closer than the minimum separation that could ever be resolved, an admirable example of the economy of biological systems.

(b) At 5 m, the minimum separation distance becomes

$$S_{min} = \frac{L\lambda}{D} = \frac{(5.0\ \text{m})(550 \times 10^{-9}\ \text{m})}{0.0025\ \text{m}} = 1.1\ \text{mm}.$$

Thus, from a few meters away, we have an *intrinsic* inability to distinguish the millimeter markings on a meter stick.

What Do You Think? What is the separation distance that can be resolved when the two separated objects are at infinity?

The Resolution of Telescopes

Depending on the case, diffraction effects can limit the effectiveness of telescopes. For the visible-light region of the electromagnetic spectrum, one of the world's largest telescopes is the 200-in-diameter Hale telescope on Mt. Palomar in southern California. Its diffraction limit alone implies an angular resolution of about 0.03 seconds of arc. Its real resolution is some 300 times worse than this due to the effects of atmospheric turbulence and aberration. The New Technology Telescope in Chile, another visible-light telescope, has resolved 0.36 seconds of arc, much closer to the diffraction limit. However, as the next example shows, the Hubble Space Telescope achieves a resolution that is very close to the diffraction limit.

EXAMPLE 38–6 The primary mirror on the Optical Telescope Assembly on the Hubble Space Telescope (HST), which orbits 600 km above Earth, has a diameter of 2.4 m. (a) Calculate the minimum angular separation that it might resolve for visible light (about 550 nm). (b) Assume that the telescope is viewing Earth's surface. What is the separation of the most closely spaced objects that it might resolve? Ignore all atmospheric effects.

Strategy (a) This problem is a direct application of the resolution angle of Eq. (38–14). For (b) we simply use $S_{min} = L\theta_{min}$, where $L = 600$ km and S_{min} is the separation distance.

Working It Out (a) Equation (38–14) gives

$$\theta_{min} \cong \frac{\lambda}{D} = \frac{(550 \times 10^{-9}\ \text{m})}{2.4\ \text{m}} = 2.3 \times 10^{-7}\ \text{rad} = 0.047''\ \text{of arc}.$$

The actual resolution of the instrument is 0.1 seconds of arc, to a good approximation the diffraction limit. The HST can do so much better than comparable Earth-based telescopes because the HST avoids all atmospheric turbulence.

(b) $S_{min} = L\theta_{min} = (600\ \text{km})(2.3 \times 10^{-7}\ \text{rad}) = 0.14\ \text{m} = 14\ \text{cm}.$

The HST would not avoid atmospheric turbulence if it looked down, so 14 cm is better than the actual resolution that would be obtainable with this instrument if we were to point it toward the surface. Earth-observation satellites such as Landsat and Spot have resolution capabilities of a few meters, and classified spy satellites are reported to have resolutions of less than 1 m.

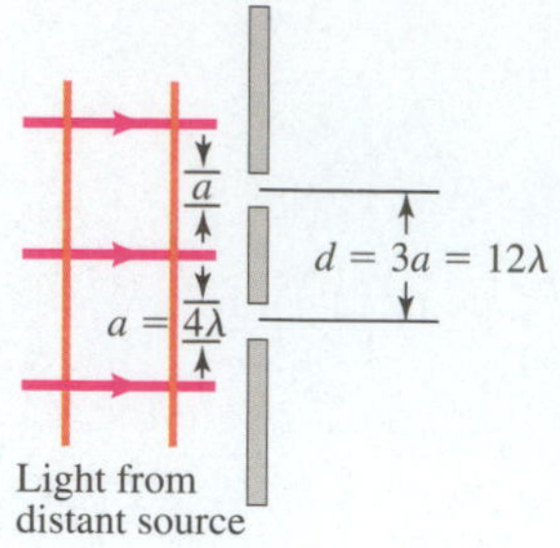

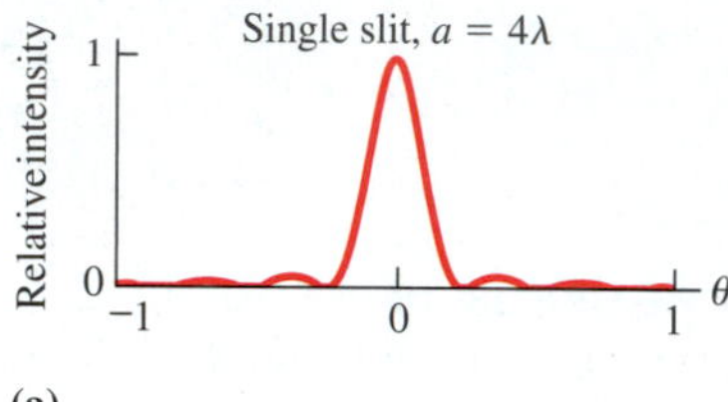

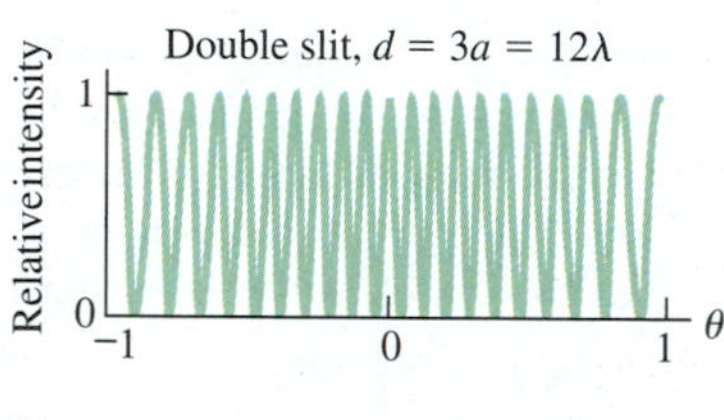

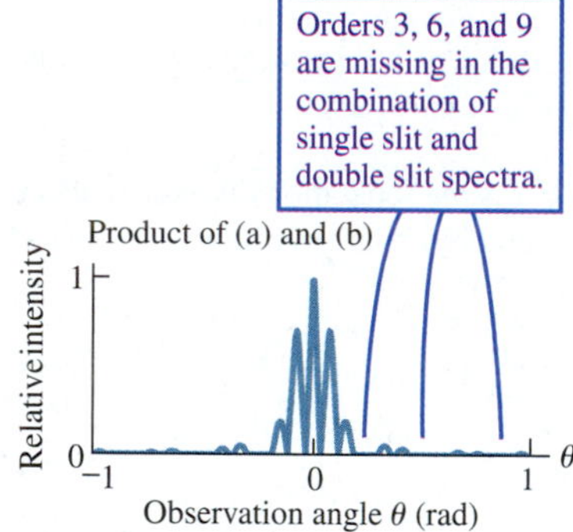

▲ **FIGURE 38–18** Intensity patterns as a function of observation angle θ for diffraction from multiple slits must include the effects of single-slit diffraction. For $a = 4\lambda$, (a) the intensity pattern for a single slit, (b) a double slit $(d = 12\lambda)$, and (c) their product, which is the observed pattern. Missing orders occur; in this case, $d = 3a$, and the 3rd, 6th, 9th, ... orders are missing from the overall pattern.

▲ **FIGURE 38–19** Diffraction pattern for multiple slits where $d = 10a$. Note the missing orders.

▶ **FIGURE 38–17** The complex of telescopes on the Mauna Kea volcano in Hawaii is one of the great observatories of the world. The twin Keck telescopes, visible at the center, can act as a single telescope whose mirror is separated into two pieces.

An optical device need not consist of a single aperture. The mirrors of many of the largest modern reflecting telescopes are separated into several pieces (Fig. 38–17), like a mosaic, in part because a single large piece of glass is difficult to form and handle. Such instruments diminish diffractive effects because the minimum angle that can be resolved is determined by the interference from the most widely separated pieces of the apparatus. If this maximum separation is D, then the minimum resolvable angle between objects is λ/D. (Note that D refers to a *transverse* separation; the distance between an eyepiece and an objective does not enter into diffractive effects.) An array of small electronically connected optical telescopes on the Moon spread over a region with a diameter of 10 km would have a resolution 100,000 times better than that of the best telescope on Earth. Were it not for Earth's atmosphere, such an instrument could pick up a newspaper headline on Earth!

38–5 Slit Width and Grating Patterns

Our discussions of double- and multiple-slit diffraction patterns have treated each slit as a point source of light that emits a single spherical wave. The angular spread of the diffraction pattern depends on the parameter d/λ, where d is the slit separation. We have now seen that single slits of finite width have their own diffraction pattern. The angular spread of this pattern depends on a/λ, where a is the slit width. What effect does a finite slit width have on the multiple-slit pattern? For Fraunhofer diffraction, the overall intensity distribution is the product of the two intensity patterns. The pattern I_{mult}, corresponding to the double or multiple slit, is multiplied by the pattern I_{single}, corresponding to the single slit. The multiple-slit intensity is given by Eq. (38–4), whereas the single-slit intensity is given by Eq. (38–10). Thus their product is

$$I = I_{multi} I_{single} = I_{max}\left[\frac{\sin(N\beta)}{\sin\beta}\right]^2\left(\frac{\sin\alpha}{\alpha}\right)^2, \tag{38–16}$$

where we recall that

$$\beta = \frac{\pi d \sin\theta}{\lambda} \quad \text{and} \quad \alpha = \frac{\pi a \sin\theta}{\lambda}.$$

In the equation for the intensity, we have combined the maximum intensity factors into a single maximum intensity I_{max}.

The fact that the intensity patterns are multiplied means that the broader pattern (usually due to the single slit) acts as an envelope for the narrower pattern. For example, suppose that $d = 3a$ for a double slit $(N = 2)$. In this case, the individual slit pattern is much broader than the multiple-slit pattern. At the same time, let $a = 4\lambda$, so that the single-slit pattern is easily distinguishable. Figure 38–18 shows the single-slit pattern, double-slit pattern, and combined pattern. Note that certain maxima of the double-slit pattern are absent from the combined pattern because they fall where the minima of the single-slit diffraction pattern occur. These missing maxima are called *missing orders*. The locations of missing orders are independent of λ, as Problem 43 illustrates. A measurement of the pattern described here, but with $d = 10a$, is shown in Fig. 38–19.

*38–6 X-Ray Diffraction

We have been emphasizing the use of gratings as a tool for the exploration of diffracted light. Light is just one form of electromagnetic radiation, and other electromagnetic waves can also be diffracted. Let's take a look at the diffraction of X rays, an important tool in the understanding of crystalline solids. Diffraction gratings work because the apertures or obstacles serve as rescatterers. A powerful constructive interference occurs among rescattered light from *all* the apertures or obstacles because the sources of wavelets are in a *regular* pattern. A crystalline solid contains a regular array of obstacles and therefore forms a natural diffraction grating. In a crystalline solid the array is spread over three dimensions rather than two, but the effect works in just the same way. If the electromagnetic waves can penetrate the material, each atom in the array can serve as a rescatterer, and the diffraction pattern will be characteristic of the particular way in which the rescatterers are arranged.

In 1895, Wilhelm Roentgen discovered that radiation was produced when he bombarded metal with high-energy *cathode rays* (now called electrons). This radiation was unlike any seen previously, and Roentgen called it *X rays*. Shortly thereafter, he produced the first X-ray picture, a human hand (Fig. 38–20). We know now that X rays are just electromagnetic radiation with wavelengths in the range of about 0.01 nm to 10 nm. This radiation is produced when atomic electrons change states within atoms, or when electrons are accelerated (or decelerated).

In the early 1900s, it was suspected that X rays might be some form of electromagnetic radiation. A diffraction experiment reported in 1899 vaguely suggested that X rays might have wavelengths of about 0.1 nm, much smaller than those of visible light. At the same time, some scientists suspected that solids might be made of atoms arranged in regular arrays. In 1912, Max von Laue had the idea of scattering X rays from solids. If X rays had about the same wavelength as the distance between the arrays of atoms (about 0.1 nm), then diffraction effects would be significant. Von Laue was interested in both a tool for the precise measurement of the wavelengths of X rays and a tool for the exploration of crystals. He convinced two of his colleagues, Friedrich and Knipping, to perform an experiment, and the observation of X-ray diffraction soon followed. Von Laue's idea was a crucial step in the measurement of X-ray spectra and led to a revolution in our ability to study the nature of solids and the molecules that compose them. The precise knowledge that table salt, NaCl, has the three-dimensional structure shown in Fig. 38–21 is a consequence of X-ray diffraction experiments; virtually all of our knowledge of crystalline structure comes from such experiments. It was the use of X-ray diffraction on a crystallized form of DNA that led to the discovery of that molecule's double-helical structure (Fig. 38–22).

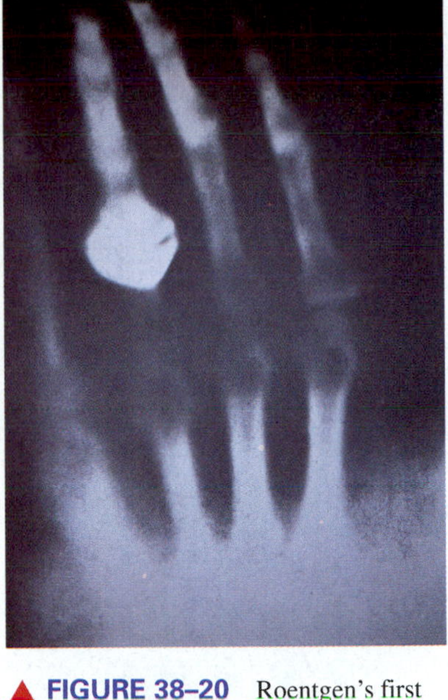

▲ **FIGURE 38–20** Roentgen's first X-ray photograph of a human—that of his wife's hand. X rays have the well-known property of penetrating matter.

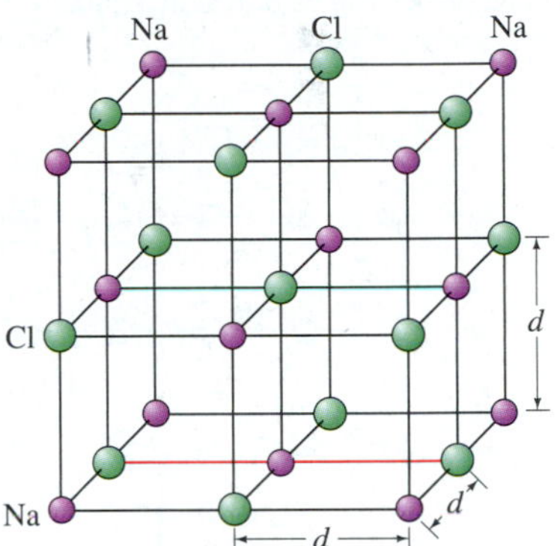

▲ **FIGURE 38–21** Crystals have three-dimensional structure with their atoms in regular arrays. This diagram shows one of the simplest, NaCl (table salt), which has a cubic structure.

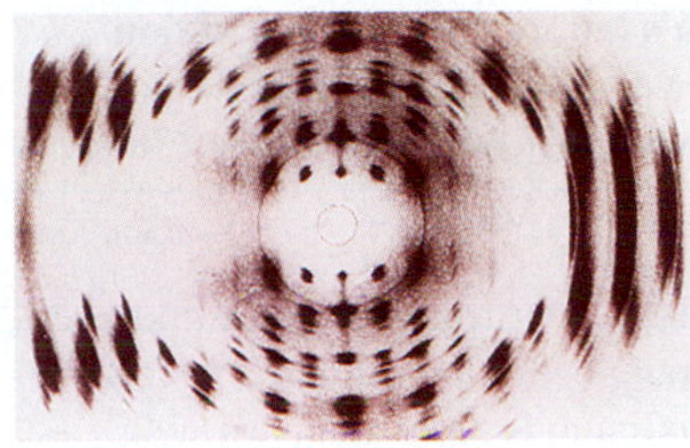

(a)

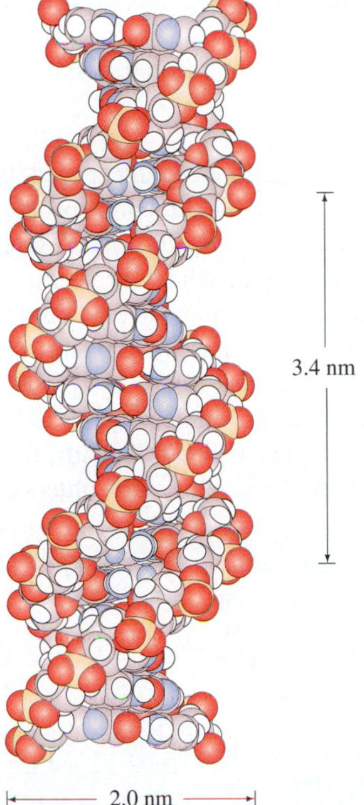

(b)

▲ **FIGURE 38–22** (a) Analysis of thousands of diffraction patterns produced by crystals of the large biological molecule deoxyribonucleic acid (DNA) showed that (b) the molecule has the shape of a double helix.

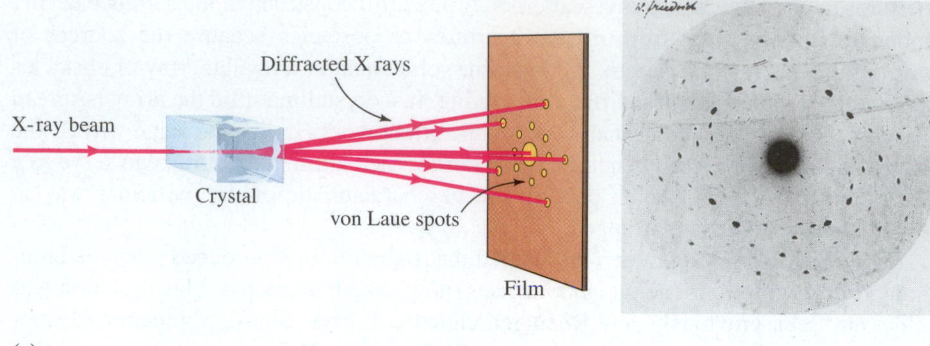

▶ **FIGURE 38–23** (a) Schematic diagram of the von Laue experiment for the diffraction of X rays. (b) Von Laue spots in one of the first X-ray diffraction patterns. The large spot is undiffracted radiation.

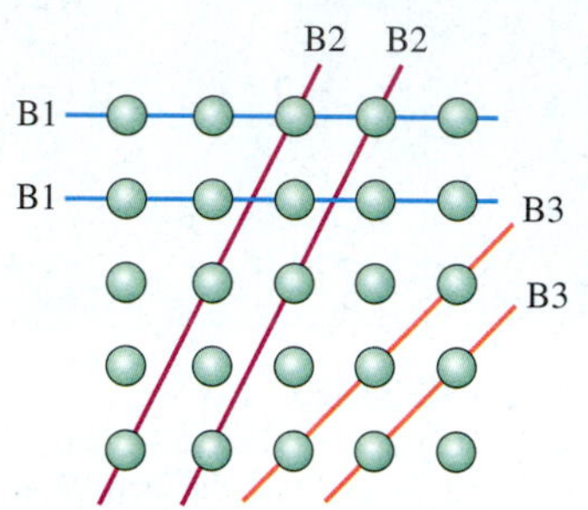

▲ **FIGURE 38–24** Many parallel Bragg planes can be drawn in a three-dimensional crystal.

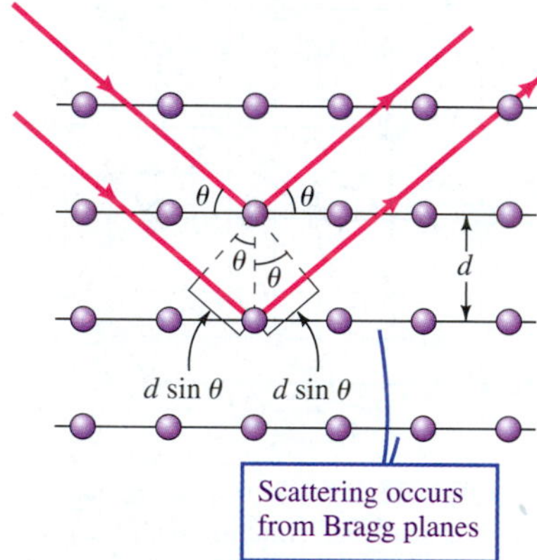

▲ **FIGURE 38–25** The geometry of X-ray diffraction between adjacent Bragg planes.

Because the rescattering centers (the atoms of a solid) are pointlike and three-dimensional rather than slitlike, the diffraction pattern of a crystalline solid consists of a regular array of spots rather than lines. The von Laue experiment on a crystalline solid, shown schematically in Fig. 38–23a, leads to a set of spots like those shown in Fig. 38–23b. Von Laue's idea was clarified almost immediately by W. L. Bragg in 1912, who proposed a simple and systematic way of showing just how the positions of the spots would be determined by the solid's crystalline structure. Bragg pointed out that in any crystal, many sets of parallel planes (called *Bragg planes*) can be drawn that pass through the positions of the atoms, and that the planes of a set are separated by characteristic distances (*Bragg spacings*). Figure 38–24 shows where some of these planes cut a two-dimensional cross section of a cubic lattice similar to that formed by NaCl. We can think of each family of parallel planes as a slit-type diffraction grating for the X rays. Figure 38–25 shows two rays scattered from two parallel planes within a crystal. These rays scatter from a given plane for which the reflection angle equals the incident angle because, at that reflection angle, the wavelets re-emitted by each atom within that plane add constructively. Now consider the interference between the scattered waves of *different* planes, which occurs because the X rays penetrate the crystal. If the separation between the planes is d, then, from the geometry of Fig. 38–25, the difference in path lengths for the two lines is $2d \sin \theta$. Note that angle θ is measured from the plane surface rather than from the normal to the plane. Constructive interference for scattering from these two adjacent planes occurs when this path-length difference is an integer multiple of the wavelength. This relation is known as **Bragg's law**, or the **Bragg condition**:

$$2d \sin \theta = n\lambda, \qquad \text{where } n = 1, 2, 3, \ldots. \qquad (38\text{–}17)$$

BRAGG'S LAW

Because the planes are equally spaced, the waves that scatter from the atoms in the entire set of planes in the direction specified by the Bragg condition *all* add constructively and, as in the case of a diffraction grating, the maximum is large and narrow.

This discussion, while laying out the principles behind the X-ray spectrometer (Fig. 38–26), is inadequate to explain the *intensities* of the spots. We may state generally that, if a particular family of planes contains more atoms than another does, the maxima those planes give are more intense, and intensity information is very important in determining the crystalline structure. More advanced mathematical methods are necessary for a complete model of the diffraction pattern, which is quite complicated because of the many possible planes and orders of scattering.

EXAMPLE 38–7 Figure 38–26 shows an X-ray tube, which produces a continuous distribution of wavelengths in the X-ray range. If these wavelengths are scattered from a particular set of parallel planes of rock salt (NaCl) with a spacing $d = 0.282$ nm, what wavelengths will appear in the first and second orders at 25°?

Strategy To use Bragg's law, either the wavelength or the atomic-plane spacing must be known. In this case, the plane spacing is known, and Bragg's law, Eq. (38–17), can be used to identify unknown wavelengths.

Working It Out We use Eq. (38–17) to determine the wavelengths:

$$\lambda = \frac{2d \sin \theta}{n} = \frac{2(0.282\ \text{nm})(\sin 25^\circ)}{n} = \frac{0.238\ \text{nm}}{n}.$$

The wavelengths at 25° are 0.238 nm and 0.119 nm for the first ($n = 1$) and second ($n = 2$) orders, respectively.

What Do You Think? If $\theta = 25^\circ$, what is the actual overall deflection from the original beam?

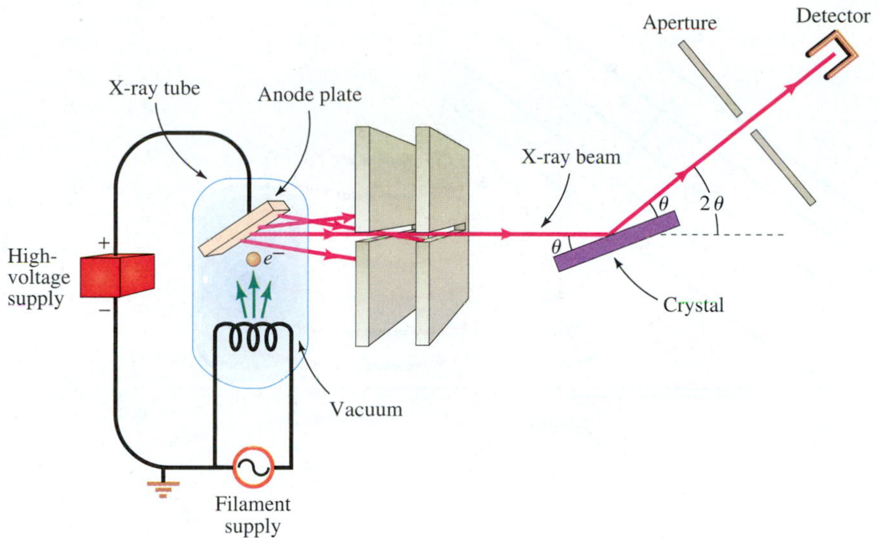

FIGURE 38–26 Schematic diagram of an X-ray spectrometer used to study properties of crystals. Electrons bombard the anode plate, producing X rays that are collimated before being scattered by the crystal. A movable detector records X-ray intensity as a function of θ to determine where constructive interference occurs.

*38–7 Holography

In 1947, Dennis Gabor proposed that interference effects between light emitted by an object (a source) and a second coherent beam can be recorded on film, which effectively becomes a diffraction grating. When light is passed through this diffraction grating, it is diffracted and forms a fully three-dimensional image of the object, an image that can be viewed from different positions and angles, just like the original object. This process is **holography**, and the film on which the interference pattern is stored is a *hologram*.

In order to understand the principles, let's start with a distant point source that sends plane waves directly toward a piece of film (Fig. 38–27). At the same time, we send a second beam—*coherent with the light from the source*—toward the film from an angle θ_r. This second beam is known as the *reference beam*. The reference beam interferes with the light from the source. Suppose that the wave in the reference beam interferes constructively with the source wave at point P_1. There will also be constructive interference at P_2, a distance d from P_1, if the wave path along line ℓ_2 differs from the path along line ℓ_1 by λ (or an integer m times λ; we consider only the case $m = 1$). When there is constructive interference at both P_1 and P_2, the relation between θ_r, the wavelength of the light, λ, and the separation d is

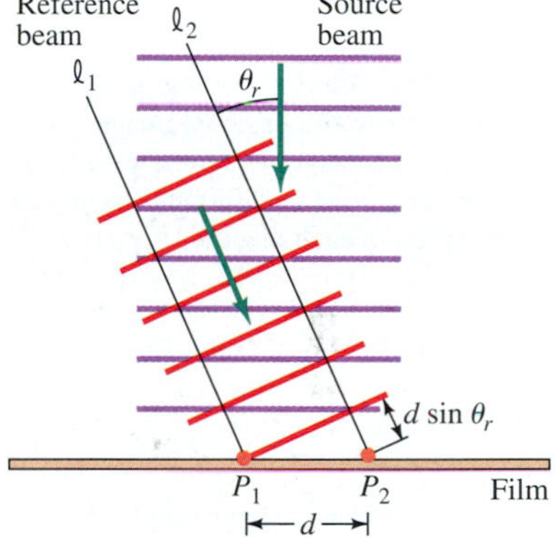

FIGURE 38–27 Coherent light from a distant source beam and from a reference beam interfere on film, producing a hologram.

$$d \sin\theta_r = \lambda. \qquad (38\text{–}18)$$

On the slice of film shown in Fig. 38–27, constructive interference will occur at a series of equally spaced points, which will be recorded as dark spots on the film. (Recall that the film makes a negative.) These points are parts of continuous lines into or out of the page. The full interference pattern on the film thus consists of a set of curving lines that represent all the places where there is constructive interference. The film can record with shadings of gray places where the interference is not totally destructive.

Let's now turn to the question of how the image is viewed (or *reconstructed*). Suppose that we project a beam just like the reference beam, and at the same angle, onto the back of the film (Fig. 38–28). The dark areas on the film act as obstacles that rescatter the light. The direction indicated, *that of the original light from the source*, is a direction for which the diffracted light is a maximum as the geometry in Fig. 38–28 shows. Thus a viewer placed at point E will see light as though it comes from a distant point I, which we may think of as an image of the original source. Note that there is no requirement that the beam that produces the image be identical to the original reference beam, as long as it is coherent across the film. If the angle of the new beam is different, the only effect is to shift the angle of the viewed image.

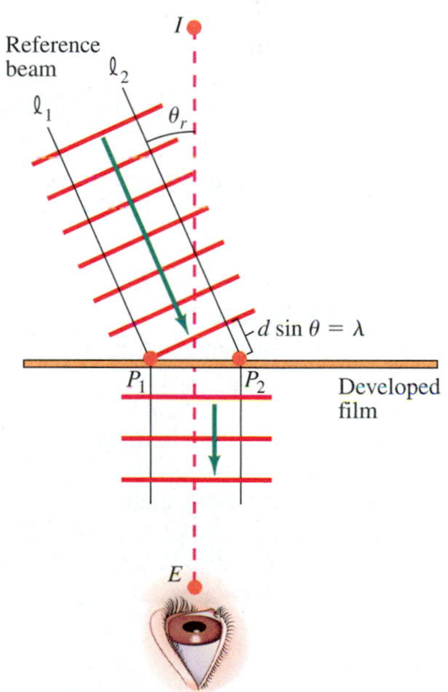

FIGURE 38–28 When a reference beam shines on holographic film, the image of the original object is reconstructed.

Suppose now that the point source is closer to the film when the image is made. In this case, the spots of constructive interference will not be spaced equally across the screen (Fig. 38–29). At region A_1, the situation we have described is reproduced, but at

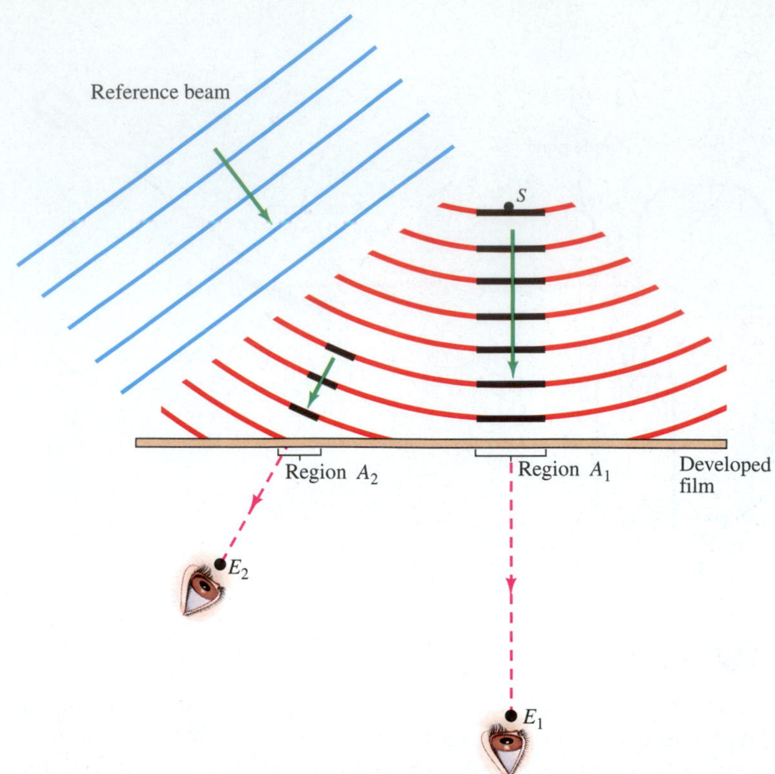

▶ **FIGURE 38–29** Holograms produce a three-dimensional image that allows the original object to be seen from different positions. A different image will be observed in each place.

region A_2, the points where there is constructive interference between the beams are different. When the exposed film is illuminated by a reference beam, a viewer at E_1 will see a plane wave along the direction from E_1 back to region A_1; that is, the observer will be looking back at the source from one angle. However, when the viewer is at E_2, the maximum of the diffraction pattern will indicate an image back along the direction from E_2 to A_2. *The viewer will be looking at the source from another angle*. There is a true three-dimensional image, which can be viewed from different angles.

When the object is more complicated than a point source, light arrives at any given point on the film from many points of the object. The interference pattern that this light makes with the reference beam is far more complicated and irregular, but it is nevertheless unique to the object. Once this pattern is recorded, it serves as a diffraction grating for light from a reference beam to make a unique pattern that reproduces the light emitted by the original object, and from many angles. Although we have treated the film as a transmission grating, if the interference pattern can be recorded as scratches on a shiny surface, then it will act as a reflection grating, such as that in a hologram on a credit card.

How is the reference beam made coherent with the light from the source? We can take a laser beam and split it in two. One part illuminates the object while the other is routed to serve as a reference (Fig. 38–30). Figure 38–31 shows one of the many uses of holograms.

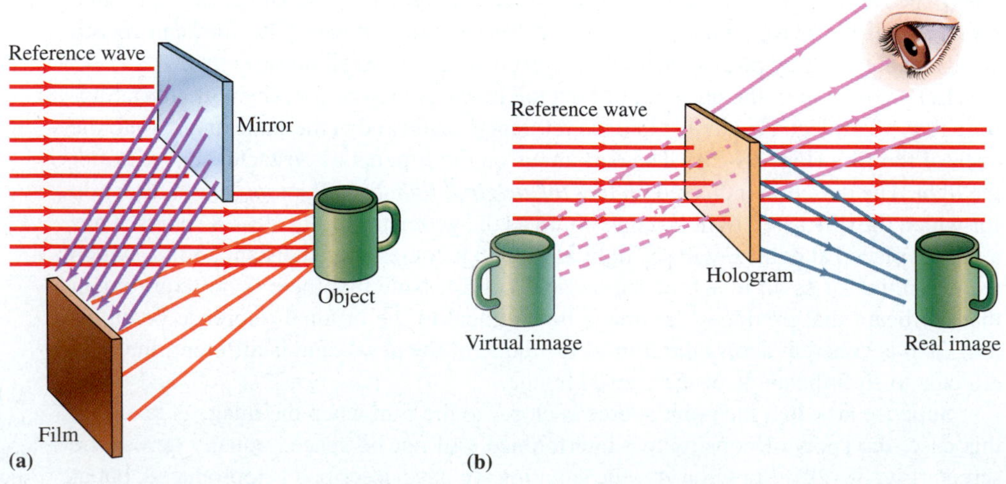

▶ **FIGURE 38–30** (a) Schematic diagram of the formation of a hologram. The reference beam and the light reflected from the object must be coherent so that they can interfere to make the hologram. (b) The holographic image is formed when the hologram acts as a diffraction grating.

FIGURE 38–31 You can observe a hologram like those visible on these visa documents by looking at a credit card. Holograms such as these are difficult to counterfeit.

THINK ABOUT THIS. . .

WHAT ARE THE USES OF HOLOGRAPHY?

The uses of holograms go beyond the simple beauty of the image. Holography has the potential to provide an extremely compact system of information storage. Because the light from every point on a printed page reaches every point of a hologram, every region of the film larger than several wavelengths across can reproduce the entire page, albeit with less detail. Moreover, successive holograms of successive pages can be made within thick photographic emulsions. If the exposure of each page is made with a reference beam oriented at a slightly different angle, then, by illuminating the resulting hologram with a light beam of that particular angle, only the corresponding page appears from a particular vantage point. For example, all the paintings in a museum could be so recorded with great accuracy in a very small space indeed.

Another important use of holography involves making two holograms of the same object on the same film at successive times. If the object has moved slightly between the moments when the holograms are made, then the two images interfere with each other like the light from two surfaces of a soap film does. Figure 38–32 shows an interference hologram of a violin in motion under the influence of a vibrating string. This interference pattern, not to be confused with the interference that makes the hologram itself, reveals detail about the motion not otherwise visible. Similarly, density variations in air are visible as the interference between two successive images of the air made on the same hologram. In this way, the mechanisms by which a candle heats the air above it or an airplane produces shock waves can be studied.

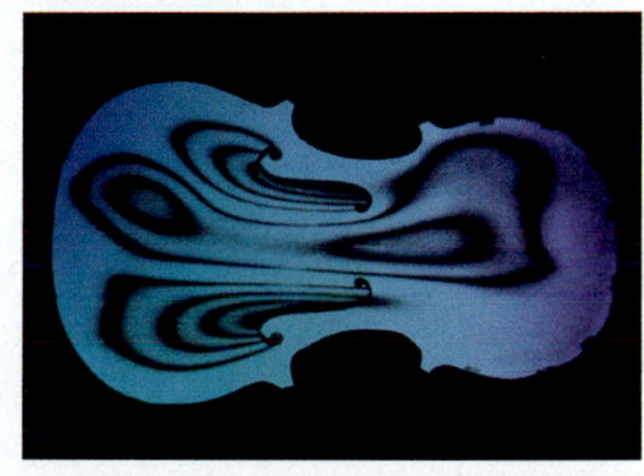

FIGURE 38–32 Interference and diffraction were crucial in producing this image of a violin. The lines indicate the motion of the vibrating instrument and are associated with the interference of two holographic images.

Summary

Diffraction is a manifestation of interference among waves. Examples include the pattern produced by screens with evenly spaced multiple slits (diffraction gratings) and the patterns made by light that passes through single apertures or around obstacles.

If d is the distance between slits in a diffraction grating and θ is the angle of observation from the direction of incident light, principal maxima are observed for the condition

$$d \sin\theta = m\lambda, \qquad \text{where } m = 0, \pm 1, \pm 2, \ldots. \tag{38–1}$$

Here, m is the order of the principal maxima. If the average intensity reaching the screen from any one slit is I_0, then the intensity of the light from the grating at the principal maxima is $N^2 I_0$, where N is the number of slits. In addition, the width of the principal maxima depends on N as $1/N$, so the diffraction peaks become sharper as N increases. This dependence on the parameters of the grating is contained in the expression for the intensity pattern:

$$I = I_0\left[\frac{\sin(N\beta)}{\sin\beta}\right]^2, \tag{38–4a}$$

where

$$\beta = \frac{\pi d \sin\theta}{\lambda}. \tag{38–4b}$$

Angular dispersion represents the change in observation angle θ as a function of a change in wavelength and is given by

$$\frac{\Delta\theta}{\Delta\lambda} = \frac{m}{d\cos\theta}. \tag{38–5}$$

The resolving power, R, is the ability of a grating to separate closely spaced lines:

$$R = \frac{\lambda}{\Delta\lambda} = mN. \tag{38–6, 38–7}$$

A single slit produces a diffraction pattern that can be derived by considering the slit to be composed of a large number of very thin slits. The criterion for destructive interference is

$$\sin\theta = \frac{m\lambda}{a}, \qquad \text{where } m = \pm 1, \pm 2, \pm 3, \ldots \tag{38–8}$$

and a is the width of the slit. The intensity pattern of a single slit is

$$I = I_{\max}\frac{\sin^2\alpha}{\alpha^2}, \tag{38–10}$$

where the angle α is given by

$$\alpha = \frac{\pi a \sin\theta}{\lambda}. \tag{38–12}$$

The minima are given in terms of α by

$$\alpha = n\pi, \qquad \text{where } n = \pm 1, \pm 2, \pm 3, \ldots. \tag{38–12}$$

Most of the light from the single slit is contained in the central peak; the secondary peaks are much less intense. The narrower the slit, the broader the diffraction pattern.

The Rayleigh criterion specifies that two point sources are just resolved if the peak of the diffraction image of the first source falls on the first minimum of the diffraction image of the second source. The minimum separation angle of two closely spaced sources obtained by a circular aperture of diameter D is approximated by

$$\theta_{\min} \cong \frac{\lambda}{D}. \tag{38–14}$$

The minimum separation $S_{\min}$ of two closely spaced objects a distance L from a lens of diameter D is given by

$$S_{\min} \cong \frac{L\lambda}{D}. \tag{38–15}$$

The practical limitation of Earth-based telescopes is due to air turbulence and not to diffraction limits.

X rays are diffracted by the atom centers of regularly spaced Bragg planes—planes formed by the regular array of atoms in a crystal. The technique is important for studies of crystalline structure. Bragg's law gives the observation angles θ (as measured from a plane surface in a crystal lattice) for which constructive interference is obtained from planes of spacing d:

$$2d\sin\theta = n\lambda, \qquad \text{where } n = 1, 2, 3, \ldots. \tag{38–17}$$

Holography represents a special process by which three-dimensional images are captured. A hologram is a kind of diffraction grating formed on film by scattering coherent light from an object; the three-dimensional image of the object can be reconstructed from the film when an appropriate light beam passes through it.

Understanding the Concepts

1. Describe what happens to the pattern you observe for single-slit diffraction as the width of the slit is slowly reduced.
2. Would the diffraction of water waves around the timbers of a pier be reduced by decreasing the diameter of the support poles? By increasing their diameter?
3. There are tentative plans to build telescopes for waves of various wavelengths, including visible light, on the Moon. What would the advantages of such facilities be?
4. Discuss how a Poisson spot might be obtained from a bowling ball. Would you want the source and screen to be close to or far away from the bowling ball? Explain.
5. A CD is "burned" with a laser—information is inscribed with the beam on the disk. If you had to maximize the information the CD contained, should you use a short or a long wavelength laser?
6. Is it possible to obtain better resolution with a microscope with blue light than with red light? Why or why not?
7. Two waves are linearly polarized. The electric field of one wave is aligned with the x-axis and the other is aligned with the y-axis. In the absence of matter that might change the polarization, can these waves interfere with each other?
8. In a demonstration of diffraction peaks that involves the reflection of laser light from an ordinary ruler, does the light have to be at a glancing angle?
9. The spreading of light due to diffraction in an optical instrument is greater when the instrument uses red as opposed to blue light. Why?
10. A hologram contains information about an entire object, even in just a small portion of the film. Would you expect the image made by a small portion of the hologram to be as sharp as the image made by the entire hologram?
11. For diffraction by two slits whose centers are a distance a apart, the condition for maximum intensity is $n\lambda = a\sin\theta$. When the center between the slits is removed, so that we now have a single slit of width b, the same formula with a replaced by b describes minima. What is the role of the centerpiece in producing opposite effects in the apparently identical formulae?
12. What are the differences between the interference patterns formed on a distant screen by coherent light that passes through a diffraction grating with thousands of rulings at a particular spacing and a double slit separated by the same spacing?

13. A lightbulb emits light with a spectrum characteristic of blackbody radiation (see Chapter 17). What pattern will this light produce when it is observed through a grating?
14. You are standing in the ocean, and a wave passes around you. Is this an example of diffraction?
15. How do the X rays used in X-ray diffraction "know" that there is a given set of planes of atoms for which a diffraction pattern appears?
16. Does the fact that light bends around corners mean that, with a sensitive camera, you could read a newspaper from around a corner? (This is a serious question: Try to estimate the amount of bending that the smallest obstacle would give for light, and how much information that light could contain.)
17. In the so-called 3-D movies introduced in the 1950s, a three-dimensional effect is achieved when different images are sent to each of your two eyes. How could you tell that they are not holographic images?
18. When monochromatic light falls with normal incidence on a screen with a circular hole, and we look at the transmitted light falling on a second parallel screen, there is a diffraction pattern on the second screen. At a given distance between the two screens the intensity at the center of the diffraction pattern is zero. As the second screen is moved away, the center of the pattern brightens, darkens again and then brightens, remaining the brightest part of the pattern at all greater distances. Explain these observations.

Problems

38–2 Diffraction Gratings

1. (I) Laser light is diffracted from a grating with 400 lines/cm. The central peak and the fourth peak are 10.34 cm apart on a screen 1.44 m away. The screen is perpendicular to the ray that makes the central peak. What is the wavelength of the light?
2. (I) A grating has a line density of 1200/cm, and a screen perpendicular to the ray that makes the central peak of the diffraction pattern is 1.5 m from the grating. If light of two wavelengths, 620 nm and 635 nm, passes through the grating, what is the separation on the screen between the second-order maxima for the two wavelengths?
3. (I) A student finds a diffraction grating but does not know the spacing of the ruled lines. She shines light from a laser with $\lambda = 680$ nm through the grating and examines the maxima on a screen 265 cm away. If the distance between the tenth maxima on either side of the central peak is 14.3 cm, what is the rule spacing of the grating?
4. (I) A spectrum is formed with a diffraction grating, which has 4500 lines per cm. Sodium light has two nearby lines with $\lambda = 589.0$ nm and $\lambda = 589.6$ nm. What is the resolution in the second-order spectrum for this grating? What is the angular separation for the sodium lines in the second-order spectrum?
5. (II) Mercury has visible optical spectral lines of wavelengths 422.729 and 422.787 nm. You have an apparatus that allows the light to fall on only 0.800 cm of a transmission diffraction grating. How many lines per centimeter does your diffraction grating need in order to resolve the third-order maxima of the two mercury lines?
6. (II) A grating with 2.0×10^4 rulings spaced uniformly over 3.0 cm is illuminated at normal incidence by light of wavelength 530 nm. (a) What is the dispersion of the grating in the second order? (b) What is the smallest wavelength interval that can be resolved in the second order near $\lambda = 530$ nm?
7. (II) What is the resolving power of a 3.0-cm-wide diffraction grating with 5000 lines/cm, for the first three orders? If light consisting of a series of discrete wavelengths around 420 nm is incident on the grating, what is the minimum wavelength separation that can be resolved in these three orders?
8. (II) Estimate the line spacing between two closely spaced lines near 530 nm if they are barely resolved in the third order by a grating with $N = 11{,}000$.
9. (II) A grating is to be inscribed on a 4.00-cm-wide glass plate so as to resolve two spectral lines with wavelengths 618.32 nm and 618.34 nm, respectively, in the first order. What is the minimum number of lines that must be ruled on the plate? What is the dispersion of the grating with this number of lines?
10. (II) Argon has wavelengths of 454.6 and 457.9 nm. (a) What resolving power does a transmission diffraction grating need to resolve the two wavelengths? (b) If the diffraction grating has 450 lines/mm, how wide does the grating have to be in order to resolve the second-order maxima?
11. (II) Parallel light falls at an angle of 30° from the vertical on a horizontal reflection grating with 4000 lines per cm. At what angle will you see the first-order spectrum of sodium light ($\lambda = 589$ nm) reflected from the grating?
12. (II) A grating is made of five similar, uniformly spaced, narrow slits. For light of wavelength $\lambda = 633$ nm perpendicularly incident on the slits, the angular position of the first principal order is 0.18° to the normal. What is the slit separation? What is the angular position of the first principal order when the first and fifth slits are covered? When the second and fourth slits are covered?
13. (II) The resolving power of a certain grating for the first-order spectrum is 10^4. If the grating is 2 cm long, what angle separates the first- and second-order images for light with $\lambda = 580$ nm at normal incidence?
14. (II) White light shines on a diffraction grating with 3600 lines/cm. The diffracted light is observed on a screen 0.90 m away. Find the second- and third-order positions for blue light (440 nm), green light (560 nm), and red light (720 nm). Sketch a view of the screen.
15. (II) Visible light extends from wavelengths of 430 nm to 680 nm. If blackbody radiation, which contains all these wavelengths, is incident on a 5.0-cm-wide grating with 2500 slits/cm, what range of angles is covered for these wavelengths in the first-order maximum? In the second-order maximum?
16. (II) An atomic source emits two strong spectral lines, a red one of wavelength 615 nm and a blue one of wavelength 475 nm. The light falls on a diffraction grating with 5000 lines/cm that is 1.2 cm across, and passes to a screen 2.00 m away. On the screen, how far from the central maximum are the second-order maxima ($m = 2$) of the spectral lines? What is the width of these maxima?
17. (II) Light of wavelength λ is incident at an angle α to the normal of a transmission grating with spacing d between each slit (Fig. 38–33). At what angles β to the normal will diffraction maxima be located?

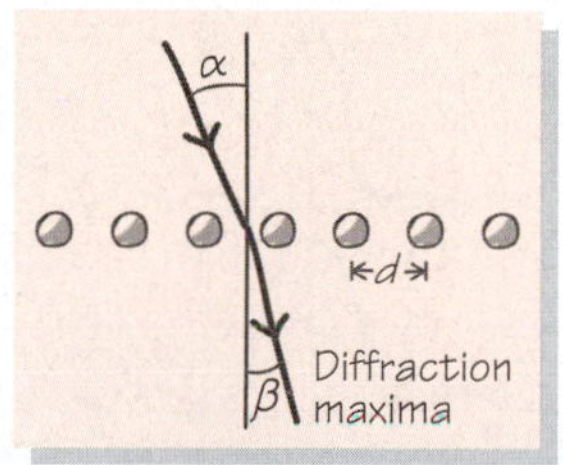

▲ **FIGURE 38–33** Problem 17.

38–3 Single-Slit Diffraction

18. (I) Light of wavelength $\lambda = 535.0$ nm falls on a slit of width $a = 0.35$ mm. At what angle θ from the normal to the wall in which the slit is cut does the second dark fringe occur?

19. (I) A single slit of width 2.8×10^{-5} m diffracts light of wavelength 495 nm to a screen. The distance between the minima on either side of the central maximum is 1.8 cm. How far away is the screen?

20. (I) A single slit diffracts laser light of wavelength 635 nm onto a screen 2.5 m away. The distance between the first-order maxima on either side of the central peak is 6.0 mm. How wide is the slit?

21. (I) Microwaves of wavelength 2.6 cm are perpendicularly incident on a single slit of width 4.2 cm, and then onto a screen. (a) At what angles are maxima and minima found on the screen? (b) If the screen is 0.8 m away from the slit, what is the distance of the secondary maximum from the central maximum?

22. (II) Plane waves of light of wavelength 610 nm are incident on a single slit of width 75 μm. A lens focuses the plane waves on a screen 120 cm away (Fig. 38–34). (a) What is the width of the central maximum on the screen? (b) What is the intensity ratio between the central maximum and the first-order maximum?

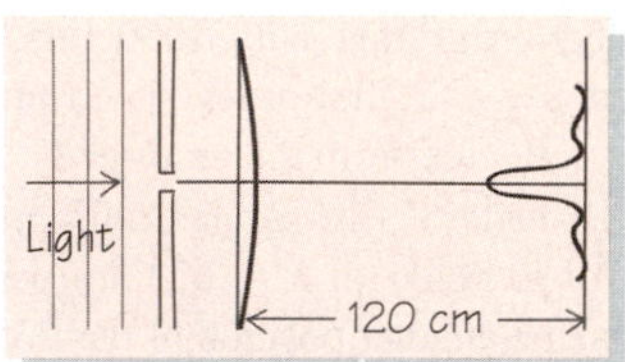

▲ **FIGURE 38–34** Problem 22.

23. (II) Blue light ($\lambda = 470$ nm) passes through a slit 10 μm wide. What is the ratio between the maximum intensity of the central peak and the maximum intensity of the next adjacent peak? At what angle θ from the horizontal will the intensity of the central peak be half its maximum value? Would θ increase or decrease if red light ($\lambda = 670$ nm) were used instead?

24. (II) A diffraction pattern is formed by an adjustable slit. If the width of the slit is doubled, how do the following quantities change? (a) The distance of the first minima on the two sides of the central maximum; (b) the intensity at the central maximum; (c) the total power reaching the screen.

25. (II) A single slit produces a diffraction pattern on a distant screen. Show that the separation distance between the two minima on either side of the central maximum is twice as large as the separation distance between all the other neighboring minima. Compare your result to the corresponding case for a double-slit pattern with very narrow slits.

26. (II) When light of wavelength 540 nm passes through a single slit of unknown width, the diffraction pattern displays a second maximum where the first minimum of light of an unknown wavelength had been observed to fall (Fig. 38–35). What is the unknown wavelength?

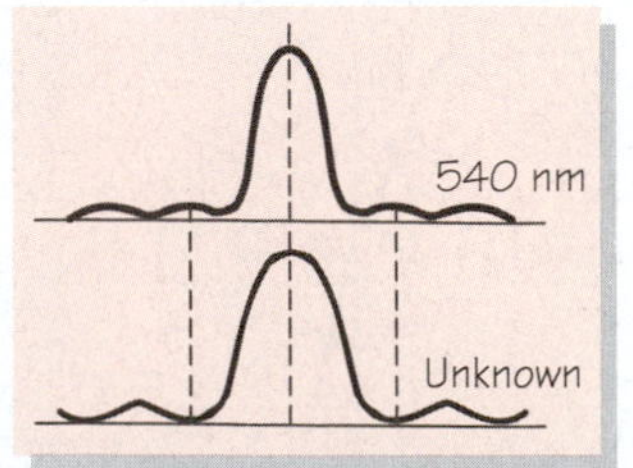

▲ **FIGURE 38–35** Problem 26.

27. (II) The width of the central peak of a single-slit diffraction pattern can be characterized by the distance of the first-order minima on both sides of the maximum or the full width at half maximum, the latter defined by the points where the intensity decreases to 50 percent (Fig. 38–36). Compare the values of these widths.

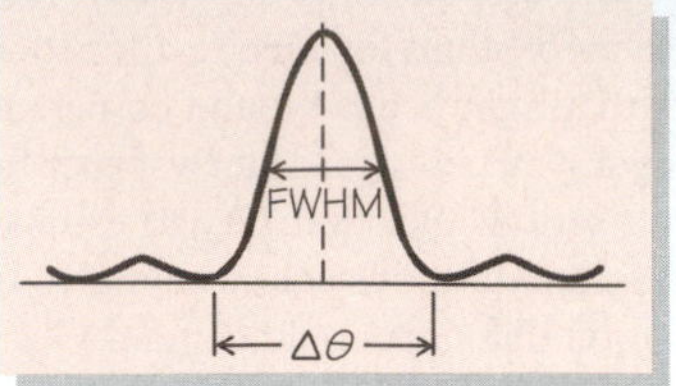

▲ **FIGURE 38–36** Problem 27.

28. (II) Monochromatic light falls with normal incidence on a screen with a circular hole of diameter 2.2 mm. When a second screen is placed 0.60 m away, the intensity at the center of the diffraction pattern is zero (first minimum). As this screen is moved away, the center of the pattern brightens, darkens again and then brightens, remaining the brightest part of the pattern at all greater distances. What is the wavelength of the light? What was the distance between the screens at the second minimum?

29. (II) Light of wavelength λ arrives at a single slit of width a; the plane wave fronts arrive at the slit at an angle θ_i (Fig. 38–37). Find the angles θ for which minima appear on a very distant screen. Is there a "central maximum" in the direction defined by the incoming wave; that is, at $\theta = \theta_i$?

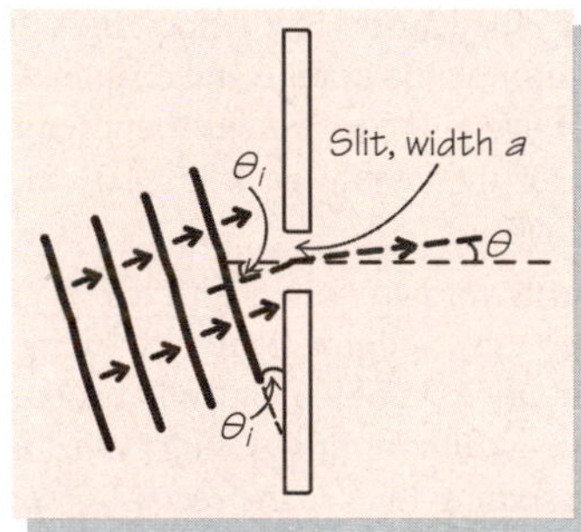

▲ **FIGURE 38–37** Problem 29.

30. (II) Suppose that light falls on a single slit at an angle ϕ with the normal to the wall that contains the slit (Fig. 38–38). Show that Eq. (38–10) still holds, but $\sin\theta$ must be replaced by $(\sin\theta + \sin\phi)$ in the expression for α [Eq. (38–11)].

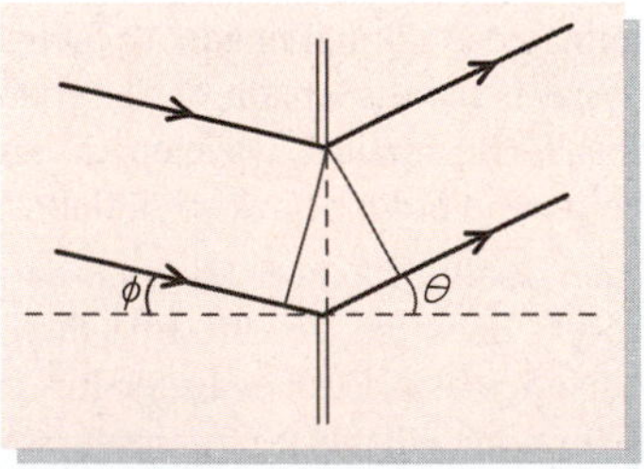

▲ **FIGURE 38–38** Problem 30.

31. (III) When we determined the position of the minima of the Fraunhofer diffraction pattern for a single slit, we argued that the maxima are located midway between the minima. To look at the accuracy of this assumption, (a) show that the maxima of the intensity pattern $(\sin^2\alpha)/\alpha^2$ are determined by the solutions of the transcendental equation $\alpha = \tan\alpha$. (b) Compare a numerical solution of this equation for the first and second maxima with

angles that are midway between the first and second, and second and third, minima (you will need a calculator for the trigonometric values). (c) By plotting the intersection points of $y = \tan\alpha$ and $y = \alpha$, show that the approximation improves as the order of the maximum increases.

38–4 Resolution of Optical Instruments

32. (I) A plane wave of microwave radiation, $\lambda = 2.3$ cm, passes through a circular aperture of diameter 4.7 cm. What is the angular position of the first minimum of the resulting Fraunhofer diffraction pattern?

33. (I) Astronauts leave two lunar rovers 5.00 km apart on the Moon. An Earth-based telescope of what minimum diameter is required to resolve laser beams ($\lambda = 650$ nm) emitted by the rovers toward the telescope? The rovers are 3.0 m long. A telescope of what diameter is required for the rovers themselves to be detected? Ignore air turbulence. (The Earth–Moon distance is 3.83×10^8 m.)

34. (I) An amateur astronomer uses a reflecting telescope of diameter 20 cm and focal length 200 cm to observe light of $\lambda \cong 600$ nm from a star. (a) What minimum angular resolution can the astronomer obtain? (b) What is the diameter of the Airy disk? (c) What is the minimum separation distance of two objects on the Moon that the telescope can resolve?

35. (II) An astronaut in a satellite can barely resolve two point sources on Earth 220 km below. What is the separation distance between the sources, assuming ideal conditions, $\lambda = 480$ nm, and a pupil diameter of 1.5 mm?

36. (II) The two stars of a binary star system are just resolvable when observed by a telescope with a resolution of 18 seconds of arc and are 125 ly from Earth. Estimate their separation.

37. (II) A spy satellite is announced to be capable of distinguishing detail 10 inches across. If the satellite orbits at a height of 220 mi, what must be the minimum size of the lens aperture, assuming maximum sensitivity at $\lambda = 525$ nm? Would it be better if the film (or other sensor) were sensitive to shorter or to longer wavelengths?

38. (II) What must be the lateral separation between two objects located 1.0 km from a camera that must resolve them? The camera lens's aperture is 5.0 mm in diameter, and the film is sensitive to light of wavelength 550 nm.

39. (II) The lens of a 35-mm camera is set in such a way that the image of a very distant object is ideally sharp on the film. The focal length of the objective is 50 mm. At what setting of the camera's aperture will the sharpest possible image of an object 5.0 m away form? [*Hint*: The image will be blurred both because it is off focus and because of diffraction. Find the aperture at which these two sources give equal contributions.]

40. (II) Use the Rayleigh criterion and make assumptions to estimate the distance at which the human eye should be able to resolve the headlines in a newspaper. Carry out an experiment to see how good your estimate is!

41. (II) The SR-71 Blackbird reconnaissance airplane could fly at over 70,000 ft. If the pilot's pupil has a diameter of 1.5 mm on a bright day, what is the distance between two objects on Earth that the pilot could just resolve from 60,000 ft? Take the wavelength of light to be 520 nm.

42. (II) The headlights of a car are 1.8 m apart. At night, the pupils of an oncoming driver have expanded to 4.2 mm. How close must the two cars approach before the headlights can be resolved? Take the wavelength of the light to be 510 nm.

38–5 Slit Width and Grating Patterns

43. (II) Calculate the lowest missing order of a double-slit interference pattern if the separation of the two slits is three times their individual widths, $d = 3a$.

44. (II) The separation distance between two narrow slits is ten times the width of either slit. What is the intensity of the tenth interference maximum, taking the center as the first, when monochromatic light passes through the two slits and falls on a distant screen?

45. (II) Light of wavelength 690 nm from a ruby laser impinges on two slits 1.1 mm apart. Each slit is 0.20 mm wide. Find the intensity ratio I/I_0 on a screen 3.0 m away at the following distances from the central maximum: 0.050 mm, 0.50 mm, 1.5 mm, and 3.0 mm.

46. (II) Figure 38–39 shows the intensity as a function of diffraction angle (in radians) for a double slit with light of wavelength 550 nm. Estimate the separation of the slits as well as their widths.

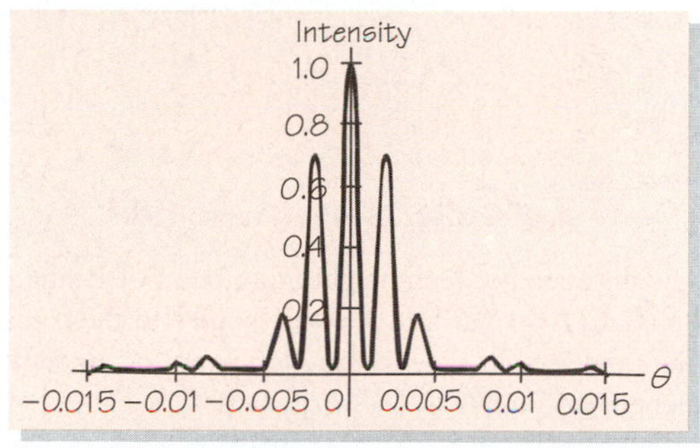

▲ **FIGURE 38–39** Problem 46.

47. (II) Figure 38–40 shows the intensity as a function of diffraction angle (in radians) for a multiple slit with light of wavelength 600 nm. Compare the pattern with the double-slit pattern shown in Fig. 38–39 of the previous problem and determine the number of slits, their separation, and their width. [*Hint*: Look at Fig. 38–8.]

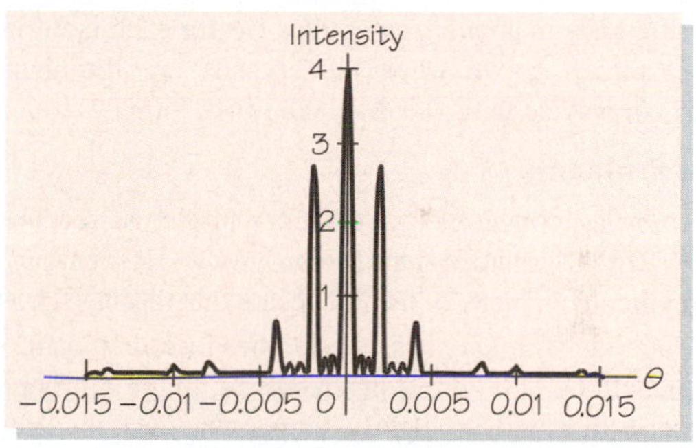

▲ **FIGURE 38–40** Problem 47.

48. (II) Light of wavelength 600 nm is perpendicularly incident on a diffraction grating. Two adjacent maxima occur at $\sin\theta = 0.30$ and $\sin\theta = 0.36$, respectively. The fourth order is missing. (a) What is the separation distance between adjacent slits? (b) What is the smallest possible individual slit width? (c) Name all orders that appear on the screen, consistent with the answers to parts (a) and (b).

49. (II) The centers of a double slit are separated by 1.2 mm; each slit is 0.4 mm wide. Are there missing orders? If so, at what angles are they missing on a distant screen if $\lambda = 589$ nm?

50. (II) The slit widths of a grating with 900 slits/cm are one-third the slit spacing. What is the ratio of the intensities of the second-order and first-order principal maxima of the grating?

51. (II) Light of wavelength 625 nm is perpendicularly incident on a screen in which double slits of width $a = 0.25$ mm have been cut. The slits are a distance $d = 0.30$ mm apart. Find the first angle away from the central axis for which the intensity on a distant screen is exactly one-half the maximum intensity.

52. (II) A grating consists of slits of width a whose centers are separated by a distance d. Sketch the diffraction pattern for (a) $d \gg a$ and (b) $d - a \ll a$ (the slits are wide compared to the strips between them).

*38–6 X-Ray Diffraction

53. (I) X rays of wavelength 0.14 nm are aimed at an unknown crystal in a diffractometer. A first-order peak occurs at 38.2° (Fig. 38–41). What is the corresponding Bragg-plane spacing for the crystal?

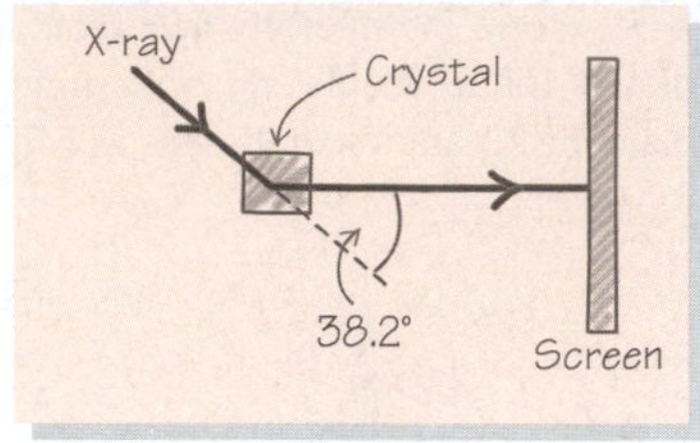

▲ **FIGURE 38–41** Problem 53.

54. (II) The distance between neighboring pairs of Bragg planes in calcite ($CaCO_3$) is 0.3 nm. At what angles to these planes will the first- and second-order diffraction peaks occur for X rays of wavelength 0.55 nm?

55. (II) Consider a crystal consisting of identical cubes with atoms at the vertices. The spacing between adjacent atoms is 0.28 nm. X rays of wavelength 0.14 nm scatter elastically from a set of planes parallel to the face of the cubes. At what angles will first-order Bragg diffraction be observed?

56. (II) Mica has a set of Bragg planes with spacing of 1.0 nm, whereas a set of planes in rock salt has a spacing of 0.28 nm. For an X ray of wavelength 0.10 nm, which material produces a diffraction pattern with the greater angular separation? What is the difference in angular separation $\Delta\theta$ for each material for the Bragg planes above when the crystals are illuminated with X rays of wavelengths 0.096 nm and 0.104 nm?

General Problems

57. (I) A ruby laser of wavelength 690 nm with a cross-sectional area of 1.0×10^{-3} m^2 is aimed at the Moon, 3.84×10^8 m away. Estimate the minimum diameter of the light beam that reaches the Moon.

58. (I) A grating 4 cm long has 16,000 lines inscribed on it. A line of wavelength 623.000 nm is just resolved, in the third order, from a second line with a slightly longer wavelength. What is the wavelength of the second line?

59. (II) Radar is used to study the shapes of airplanes from as far away as 100 km. (a) Assuming that the distance scale determining a plane's shape (the size of the curves that distinguish one plane from another) is 1 m, what angular resolution is needed in the radar system? (b) Estimate the wavelength of the radar waves if the reflected radar signals are gathered in a dish of diameter 2.5 m.

60. (II) Deep-ocean waves move in linear fronts directly toward a harbor opening of width 50 m (Fig. 38–42). For what wavelength will there be a minimum within the harbor at an angle of 50° from the axial line of the opening?

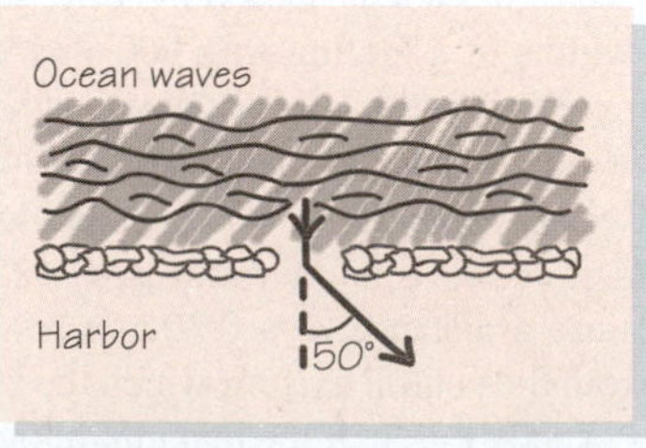

▲ **FIGURE 38–42** Problem 60.

61. (II) By varying the spacing between two vertical dipole antennas as well as the phase of the signal generated by each antenna, the antennas can give signals that are stronger in some directions than in others (see Chapter 37). Suppose that N antennas are lined up along the x-axis (Fig. 38–43). The total distance between the first and last antennas is λ, so the spacing between the antennas is $\lambda/(N - 1)$. Any one antenna would radiate its signal with the same intensity I_0 for any angle θ. (a) Find the intensity radiated very far from the array by the system of antennas as a function of angle θ in terms of I_0, assuming that the signals of all the antennas are in phase. (b) Describe the signal for all values of θ from 0° to 360°.

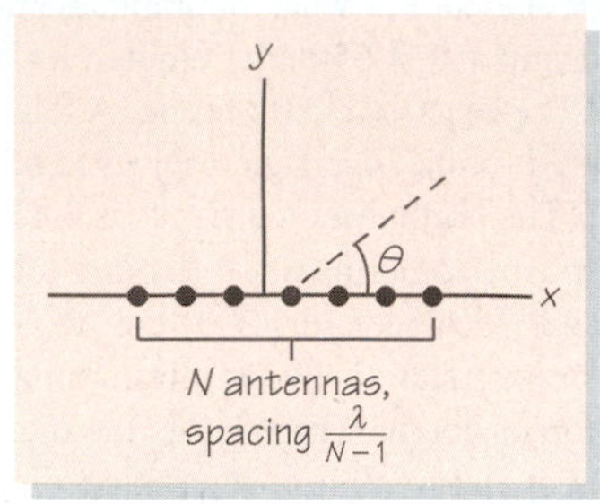

▲ **FIGURE 38–43** Problem 61.

62. (II) A grating with alternate perfectly clear and perfectly opaque spaces gives a spectrum in which all the even orders are missing. What is the ratio of the width of the clear and opaque spaces?

63. (II) A researcher embeds a single slit of width 5×10^{-4} m in a piece of glass of index of refraction n. He then shines laser light of wavelength 560 nm through one surface of the glass, which is diffracted through the slit and passes through another surface of the glass and on to a distant screen. The screen in which the slit is cut, the second surface of the glass, and the distant screen are all parallel to one another and perpendicular to the incoming laser beam. He hopes to measure the index of refraction n of the glass by seeing how the position of the first diffraction minimum differs from the position of the same minimum when the slit is in air. Can the researcher measure n this way?

64. (II) In Chapter 40, we shall see that an electron behaves like a wave whose wavelength, λ, is related to its momentum p by $\lambda = h/p$. Here, h is Planck's constant, $h = 6.63 \times 10^{-34}$ J·s. Electrons used in an electron microscope can be diffracted, and electron microscopes have a diffraction limit. If the energy of the electrons used in an electron microscope is 8.0 keV and if the aperture through which the electrons are channeled has a diameter of 0.06 mm, what, approximately, is the smallest angular separation the microscope can distinguish in an object?

65. (III) *Babinet's principle* is useful for the treatment of the diffraction of light by obstacles. It states that if light is incident on an opaque screen in which a hole (of any shape) is cut, then the diffraction pattern produced is the same (except at $\theta = 0$) as that obtained if the screen were removed and the hole were replaced by an obstacle. Use Babinet's principle to estimate the size of an opaque obstruction on a glass slide if a narrow laser beam (with $\lambda = 633$ nm) perpendicularly incident on the slide spreads to a spot of diameter 0.70 cm on a screen 2.5 m from the slide.

66. (III) What diffraction pattern is produced on a distant screen when light of wavelength λ is perpendicularly incident on a plane that contains N very thin hairs, each spaced a distance d apart from the next hair. [*Hint*: See Problem 65.]

67. (III) Electromagnetic radiation of frequency 1.25×10^{23} Hz is scattered by a nucleus of radius 3.2×10^{-15} m. The nucleus is totally radiation-absorbent and thus is a perfect obstacle. At what angle will the first diffraction minimum lie? [*Hint*: Use Babinet's principle (Problem 65).]

The fact that the speed of light is the same in all frames has deep implications for how we see the world when it involves objects moving with speeds comparable to that of light. The trolley car in the top image is at rest with respect to an observer standing in the street. In the bottom image, the same trolley car is moving to the left at a speed that is a significant fraction of the speed of light, here 0.866 of the speed of light, and the same observer sees it in a very different way.

Special Relativity

There appears to be an important difference between electromagnetism and mechanics, at least superficially. The laws of mechanics look the same in all inertial frames—all reference frames that move with uniform velocity with respect to some standard inertial frame (for example, a frame at rest relative to distant stars). Electromagnetism, however, appears to violate this general law. According to Maxwell's equations, electromagnetic waves propagate at the speed c, with no restrictions on the state of motion of the source or detector. This suggests the existence of an absolute frame for electromagnetism.

The special theory of relativity, proposed by Albert Einstein in 1905, extended to electromagnetism the principle that *the fundamental laws of physics look the same in all inertial frames*. This was accomplished by modifying certain assumptions about our notions of space and time, assumptions that had gone unquestioned until 1905. Here we explore these ideas of space and time as well as the physical consequences of Einstein's theory.

39–1 Is an Ether Necessary?

In the years following Maxwell's mid-nineteenth-century discovery of the nature of light, the absolute value of the speed of propagation of electromagnetic waves caused little concern. In an age of mechanical models, it was believed that electromagnetic waves need a medium to support them (just as sound waves need air). This presumed medium was thought to fill the universe and was called the **ether**. The ether was assumed to be at rest relative to the fixed stars. Maxwell's theory was assumed to give c

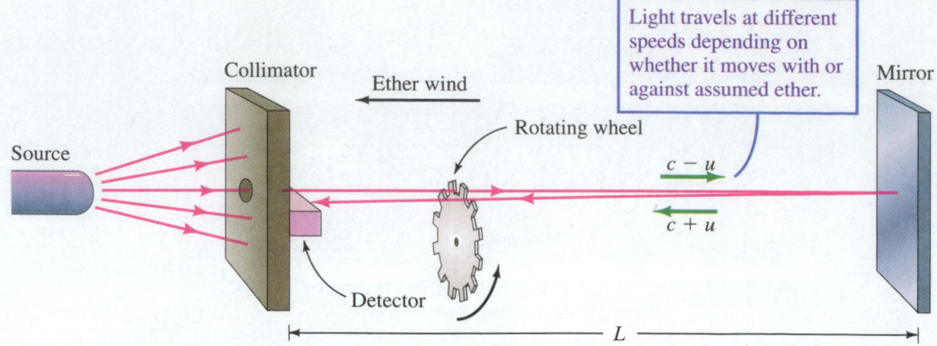

▶ **FIGURE 39–1** A "standard" measurement of the speed of light, which measures the time for a light beam to travel in a round trip over a known distance. While such methods can be used to measure the speed, they are by no means sufficiently precise to find the effect of any motion through the ether.

for the speed of propagation of electromagnetic waves *relative to the ether's rest frame*, just as the speed of sound is given as 330 m/s relative to stationary air. This assumption was accepted even though the derivation of the speed of light in Maxwell's equations contains no reference to a frame and no reference to a medium of any kind.

The existence of the ether means that in a reference frame moving at speed u relative to the ether, the speed of light emitted by a source at rest relative to the ether would be $c + u$ if the frame were moving toward the source, and $c - u$ if the frame were moving away from the source. Earth represents such a moving frame because it travels at a speed of approximately 30 km/s relative to the fixed stars in its motion around the Sun. From the point of view of a frame fixed to Earth, the ether moves past at a speed of 30 km/s. Detecting the *ether wind*, though, was likely to be very difficult.

We can see why detection of the ether wind might be hard by considering the following. A standard speed-of-light measurement is set up with the light beam propagating along an axis that lies in the direction of the supposed ether wind. If the distance from the source (and detector) to a mirror is L, then in the absence of ether wind, the time for a single traversal of light from source to mirror and back is $t_0 = 2L/c$ (Fig. 39–1). If the ether wind is blowing against the source, the speed of light that travels toward the mirror is $c - u$, and the speed of light that returns is $c + u$. Thus the time for a single traversal is

$$t_1 = \frac{L}{c-u} + \frac{L}{c+u} = \frac{2Lc}{c^2 - u^2} = \frac{2L/c}{1 - (u^2/c^2)}. \qquad (39\text{–}1)$$

For $u = 30$ km/s and $c = 3 \times 10^5$ km/s, the factor $u^2/c^2 = 10^{-8}$. Thus from the denominator alone, a time sensitivity of better than one part in 10^8 is necessary. The problem is compounded by the fact that the time factor L/c is also very small—if, for example, $L = 1$ m, then $2L/c$ is of the order of 10^{-8} s. The detection of the ether wind by this method would require a sensitivity of one part in 100 million in a measurement of one one-hundredth of a microsecond!

The Michelson–Morley Experiment

In 1887 Albert A. Michelson and Edward W. Morley carried out a high-precision experiment to measure the possible effect of an ether wind. They used an interferometer designed by Michelson (see Section 37–4 for a general discussion of such devices). Let's suppose that the ether wind were aligned in the direction shown in Fig. 39–2a. The distance from the half-silvered mirror to mirror M_1 in the direction aligned with the ether wind is L. The time that it takes for light to travel to mirror M_1 and back is t_1, given by Eq. (39–1). Let's next find t_2, the time it takes for the light to travel to mirror M_2 and back in the direction perpendicular to the presumed ether wind. Mirror M_2 is also a distance L away from the half-silvered mirror. (The assumption that the arms of the interferometer are equal in length is not essential, as we shall see later.) Because the second beam is perpendicular to the presumed ether wind, the light would have to travel a distance larger than $2L$. The beam would be blown "off course" by the ether wind, like the boat crossing the river in Section 3–6. As Fig. 39–2b shows, it travels a distance given by the hypotenuse of a triangle in which one leg has length L and the

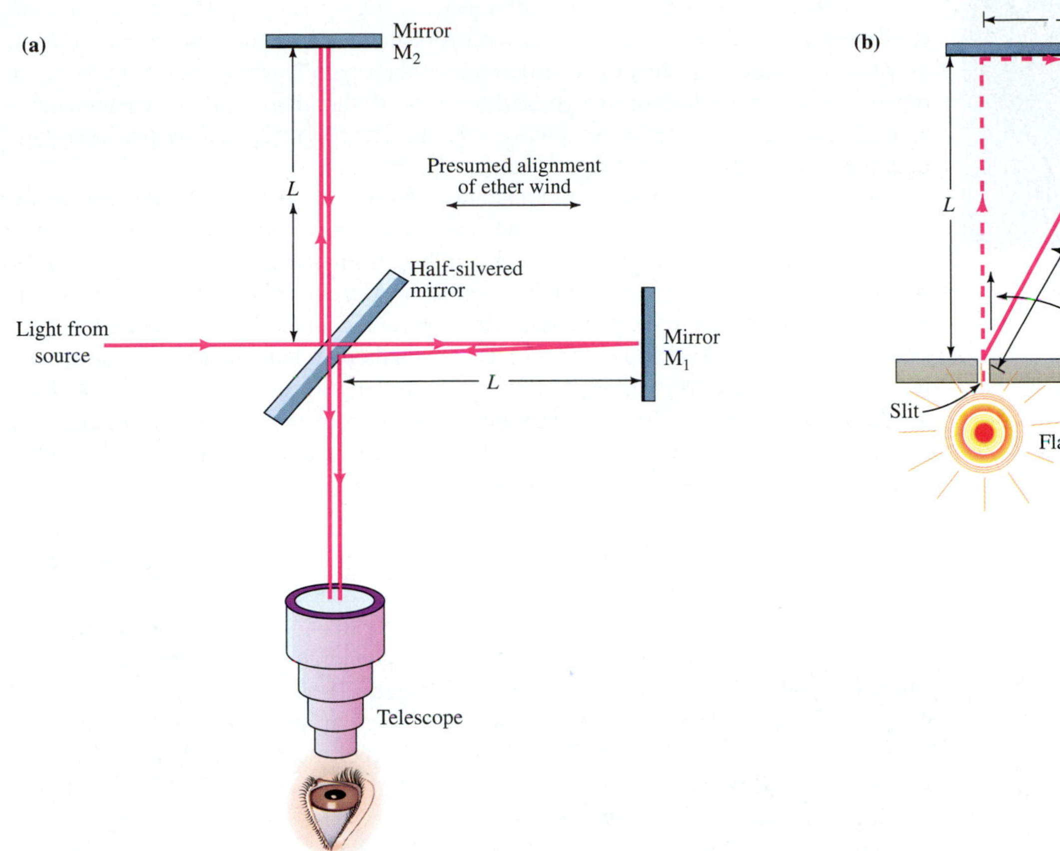

▲ **FIGURE 39–2** (a) Schematic diagram of the Michelson–Morley experiment. Light is split into two beams by a half-silvered mirror. The beams reflected by mirrors M_1 and M_2 recombine before they enter the telescope, where interference fringes are produced. (b) The beam that is perpendicular to the supposed ether wind is carried off course. Because the distance covered is larger than the direct line L, the time for a round trip to the mirror and back is longer than it would be if the distance were $2L$.

other leg is the transverse distance the beam is blown in time $t_2/2$; that is, $ut_2/2$. The distance is $\sqrt{L^2 + (ut_2/2)^2}$. Because the speed of light is c, we find

$$\sqrt{L^2 + \left(\frac{ut_2}{2}\right)^2} = \frac{ct_2}{2}.$$

We square both sides to find

$$L^2 + \left(\frac{u^2}{4}\right)t_2^2 = \left(\frac{c^2}{4}\right)t_2^2.$$

From this equation, it follows that

$$t_2^2 = \frac{4L^2}{c^2 - u^2};$$

$$t_2 = \frac{2L/c}{\sqrt{1 - (u^2/c^2)}}.$$

We now use the fact that for small x, $(1 - x)^{-1} \cong 1 + x$ and $(1 - x)^{-1/2} \cong 1 + \frac{1}{2}x$ and apply these formulas for $x = u^2/c^2$. We find that the time difference between the arrivals of two parts of a wave pulse is

$$\Delta t \equiv t_1 - t_2 = \frac{2L}{c}\frac{u^2}{2c^2}. \qquad (39\text{–}2)$$

▲ **FIGURE 39–3** Interference fringes observed in a Michelson interferometer.

This corresponds to a path-length difference of $c\Delta t = Lu^2/c^2$. The two beams are combined upon their return from the two mirrors and, because they started out in phase, they will interfere according to the difference of their path lengths—in other words, the two beams will have developed a phase difference. Rather than a direct measurement of a small time difference, it is this phase difference that is measured by observation of maxima or minima in the combined beams.

It is impossible to construct an apparatus in which the paths to the mirrors are exactly the same. In addition, because the mirrors are not exactly perpendicular to the beams, the path-length difference will vary slightly from one side of a mirror to another, and a view through the telescope yields a set of interference fringes (Fig. 39–3). Fortunately, any effects due to the apparatus itself can be accounted for by rotating the apparatus through 90°. The apparatus effects are unaltered, but the rotation effectively interchanges M_1 and M_2 and thus changes the path length to $-Lu^2/c^2$ (for unequal arm lengths, L is replaced by the average length). If there were an effect due to the ether, the fringe pattern would *shift* accordingly when the apparatus is rotated. The total path-length difference for the two orientations is $\Delta L = 2Lu^2/c^2$.

Result of the Michelson–Morley Experiment: A change in path length of $\Delta L = 2Lu^2/c^2$ leads to a shift of interference fringes of magnitude $\Delta L/\lambda = 2(L/\lambda)(u/c)^2$. Although the ratio u/c is very small, the shift in interference fringes is not an impossibly small number, because L is so much larger than λ. The apparatus was capable of detecting a shift of as few as 0.04 fringes. If $(u/c)^2$ were as small as 10^{-8} (the value that would follow from Earth's movement around the Sun), the apparatus would give a shift of 0.4 fringes. The result of the experiment was that no shift was observed; that is, if there were any shift at all, it had to be less than 0.04 fringes. In other words, *there was no experimental evidence for the existence of an ether wind.*

More recent experiments performed with lasers show that the shift is less than 10^{-3} of the result that would be "expected" for Earth's movement through the ether. The Michelson–Morley experiment sharpened the difference between mechanics and electromagnetism. Maxwell's equations predict a definite speed of light, and all previous experience with mechanical waves had suggested that any wave speed refers to a definite reference frame. This frame would have been the "preferred" frame of electromagnetism, yet the Michelson–Morley experiment showed that motion through this preferred frame cannot be detected.

39–2 The Einstein Postulates

Albert Einstein was unaware of the Michelson–Morley experiment when he formulated the laws that explain its result. He conjectured that the laws of electricity and magnetism (electrodynamics), like those of mechanics, are the same in all inertial reference frames—frames that move with constant velocity with respect to one another, and in which Newton's laws held. The **special theory of relativity** is the result of this conjecture. Einstein himself referred to the following two postulates:

1. The laws of physics are the same in all inertial reference frames. In other words, no experiment can be performed to decide who in a set of inertial frames is moving and who, in any absolute sense, is standing still.
2. The speed of light in empty space is the same in all inertial frames.

Strictly speaking, the second postulate is part of the first because Maxwell's equations do not specify a frame when they predict the speed of light. But as we have seen in Section 39–1, the second postulate might appear to be incompatible with the first. The combination of these apparently irreconcilable assertions led to the revolutionary insights into space and time that underlie the special theory of relativity.

The first postulate as applied to mechanics was introduced in Chapter 4, where we noted, for example, that an observer in a steadily moving train could not tell by observing the motion of a ball that he tosses that he is moving in any absolute sense. We also saw in Chapter 4 that one way of characterizing this fact is to say that the laws of mechanics are unchanged under the transformation

$$\vec{r}' = \vec{r} - \vec{u}t. \tag{39–3}$$

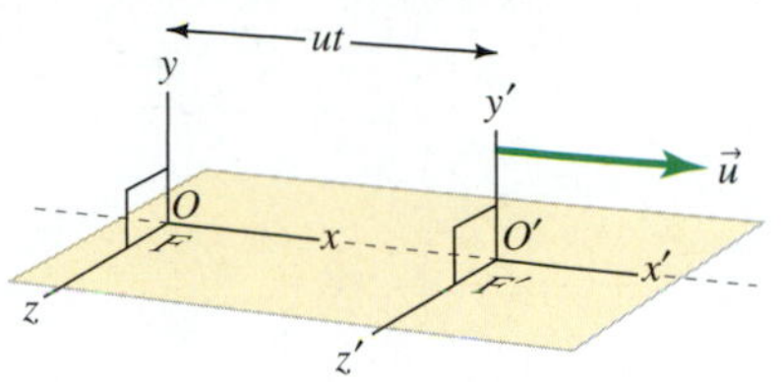

▲ **FIGURE 39–4** Reference frame F′ moves at velocity $\vec{u}$ with respect to frame F. An observer is in a particular frame when he or she measures the position and time of events in terms of the coordinate system and clocks of that frame.

Figure 39–4 shows the two frames to which these equations apply. The primed variables are the coordinates in a coordinate system (or reference frame) F′. Unprimed variables are the coordinates in frame F. Frames F and F′ move relative to one another.

The transformation in Eq. (39–3) has no reference to time in the primed frame. Newton realized that this implied that time is the same in all frames. This may seem quite obvious to you; it certainly is in accord with our everyday experience. Formally this statement reads

$$t' = t. \tag{39–4}$$

This transformation law is a second one to be added to Eq. (39–3). Together these equations form the *Galilean transformations*.

At this point we can see that *the Galilean transformation laws, Eqs. (39–3) and (39–4), are incompatible with Einstein's second postulate*. To see why, suppose that we have a light source at rest at the origin of a frame F that emits a flash of light at time $t = 0$. The flash spreads in the form of a spherical wave that expands at the speed of light; the front's location is given by (see Fig. 39–5)

$$x^2 + y^2 + z^2 = \vec{r}^2 = c^2t^2. \tag{39–5}$$

Now let's consider how the light behaves according to an observer in a second frame, F′, that moves with respect to F, as in Fig. 39–4. Suppose that origin O' of F′ coincides with origin O of F at $t = t' = 0$, when the light flashes. The second postulate implies that *in* F′, *the pulse also forms a spherical wave front* because the speed of light is the same in F′ as it is in F, whatever the direction. Thus

$$x'^2 + y'^2 + z'^2 = \vec{r}'^2 = c^2t'^2. \tag{39–6}$$

If we set $t' = t$, Eq. (39–6) would imply that $\vec{r}'^2 = \vec{r}^2$. But this cannot be; suppose, for example, that $\vec{u}$ points in the x direction. Then we would require

$$(x - ut)^2 + y^2 + z^2 = x^2 + y^2 + z^2,$$

a relation that is patently wrong. There appears to be an irreconcilable conflict between the invariance under the transformation in Eq. (39–3), the statement in Eq. (39–4), and the frame-independence of the speed of light. In a beautiful paper in 1905, Einstein set out to investigate this conflict. He found it necessary to examine what we *really* mean when we talk about measurements of space and time. Einstein did not give up either on invariance or on Maxwell's equations, but he did find that *both* transformation equations, including the frame-independence of time measurement, needed modification. His work led to a revolution in our thinking about space and time, to a series of testable predictions, and to a set of totally unexpected new phenomena.

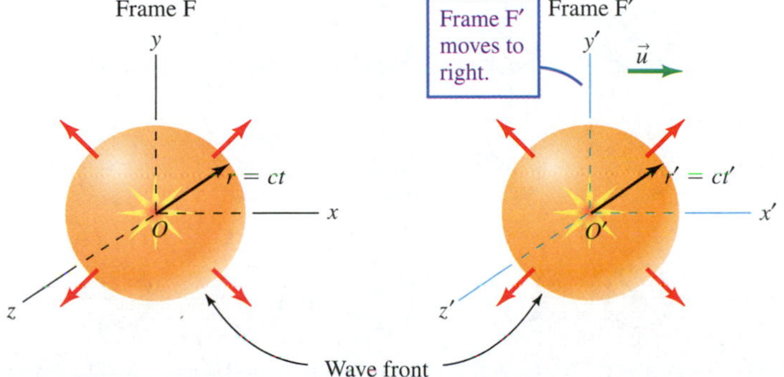

◀ **FIGURE 39–5** A spherical burst of light is emitted at the origin of frame F. To an observer in that frame, the light makes a spherical wave front centered at origin O. If origin O' of frame F′, moving with respect to frame F, is coincident with O when the light burst occurs, then an observer at O' moving with frame F′ will claim that the light makes a spherical wave front in frame F′ centered at O'.

39–3 Space, Time, and Simultaneity

What do we mean by a measurement of time or a measurement of space? Let us start with the idea of time and hence with the concept of a clock. Our basic requirement is that our clock be periodic: Time intervals between "ticks" are the same. We can put our clock and a light source at the origin of a coordinate system and then shine light along the x-axis. We place a mirror at some point x_1 and measure the time that it takes for the light to reach x_1 and return after reflection. If that time interval is two "ticks," we say that x_1 is one unit of length away from $x = 0$. Specifically, if the length of a tick is τ, then the distance to x_1 is $c\tau$. We now move the mirror farther away until the time for the light to reach there and back is four ticks. That point will be two length units $(2c\tau)$ away from $x = 0$ along the x-axis. Proceeding in this way, we can in principle assign a coordinate (x, y, z) to every point in space.

To be able to discuss time at each point in the coordinate system, we put a clock at every point for which x, y, and z are integer multiples of the unit of distance $c\tau$. We can synchronize all these clocks—set them all to the same time—as follows: At the origin at noon, a light signal is sent out to the point $x = 1, y = z = 0$. When the light ray arrives there, the clock operator at that point sets the clock to "one tick after noon," which corresponds to the point $x = 1, y = z = 0$. The operator at $x = 2, y = z = 0$ will set the clock there to "two ticks after noon" when the wave front reaches that point, and so on. In this way, all the times are synchronized and we have a reference frame in which space and time are well defined (Fig. 39–6). We have transmitted our signals in an unambiguous way because the speed of light is, by postulate (and experiment!), independent of any motion. The determination of the coordinate system and of the notion of synchronization depends on the use of light.

Although time intervals and distances within our given frame F (or another frame F′) are well defined by our setup, we must be careful about how we specify times and distances as seen from a moving frame. Let's reconsider frames F and F′ of Fig. 39–4. An observer at O' in frame F′ who wishes to see how fast a clock ticks in frame F must receive signals from the clock attached to O every time the clock in frame F ticks. Because O' is moving away from O in Fig. 39–4, the signals will be delayed. We'll see that this affects what we mean by simultaneity.

Simultaneity

We define two events at different points in a given reference frame to be *simultaneous* when they occur at the same time in that frame. *This is what we mean when we say that two events are simultaneous in a given inertial frame*. The concept of simultaneity is crucial because it enters subtly into many kinds of measurements. As an example, suppose that an observer in the F′ frame wants to measure the length of a train at rest in the F frame (the observer sees the train move). To measure the length, the observer in F′

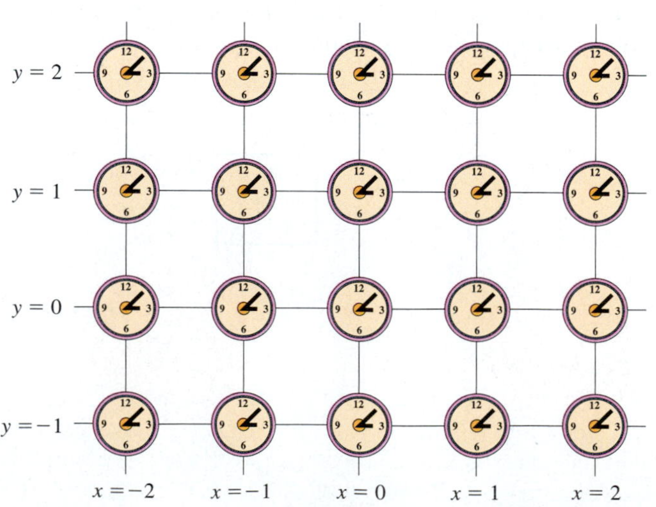

FIGURE 39–6 Clocks attached to lattice points that represent space coordinates separated by fixed distances. The location and the clock reading define the space–time coordinates of an event. Note that all the clocks show the same time as they have been synchronized by means of light signals.

must take care that the locations x'_1 and x'_2, corresponding to the front and rear ends of the train, are marked off *at the same time*—simultaneously. Marking the position of the rear end of a moving train at midnight and the front end at 2 min past midnight and taking the difference between these two positions will not give a correct reading of the length of the train. Thus the notion of simultaneity enters into the length measurement. But as Einstein pointed out, our ordinary notion of simultaneity is strongly affected by the existence of a maximum speed for signals, the speed of light.

Let's take a more careful look at the problem of measuring a train's length. Figure 39–7a shows a train of length L initially at rest with respect to a platform. L is measured while the train is at rest, so there is no difficulty in making the measurement; we merely lay meter sticks down and count them. The rear and front of the train are labeled B and C, respectively; two persons, A and A', are stationed at the exact midpoint of the train. Person A is *inside* the train, and person A' is *outside* on the platform. For the moment, everything is measured with respect to a single inertial reference frame F. As we have already described, there is a set of synchronized clocks in this frame. We label all times in frame F as t. If A sends out a spherical light pulse at $t = 0$, light reaches B and C simultaneously, at $t = L/2c$.

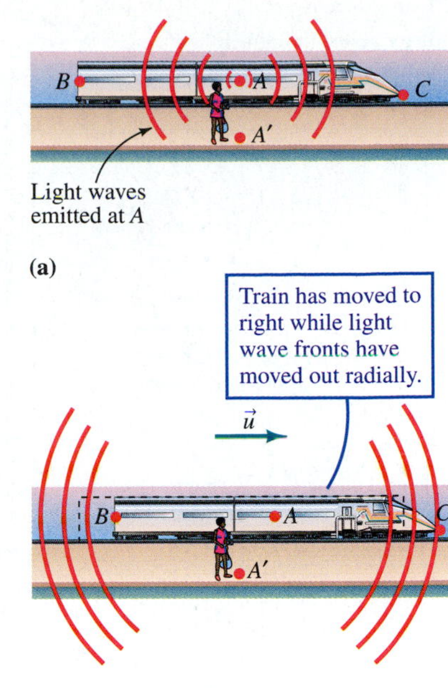

▲ **FIGURE 39–7** (a) Light emitted at point A, located midway between points B and C, reaches points B and C at the same time. (b) When B is moving toward A' and C away from A', the light reaches B before it reaches C.

Now suppose that the train is moving at uniform speed u (Fig. 39–7b). The frame at rest with respect to the train is F, whereas F′ is a new frame at rest with respect to the platform. Frame F′ has its own system of clocks along the railroad tracks, and times in that frame are labeled t'. At the moment person A is adjacent to person A', person A fires a light pulse, and we can set the clocks to $t = 0 = t'$. From the point of view of frame F, all is as it was in the original situation: The light pulse reaches both points B and C at $t = L/2c$. *But this cannot be true from the point of view of frame* F′ *if the speed of light is the same for both frames*. Person A' sees point B approach even as the light pulse moves toward B at the (fixed) speed of light, and the light pulse reaches B at a time $t' = t'_B$ somewhat earlier. Similarly, the light pulse arrives at point C at a time $t' = t'_C$ somewhat later. In fact, if the train moves a distance ut'_B during the time point B moves toward the pulse, then the distance the beam moving toward B covers is $(L/2) - ut'_B$; according to Einstein's second postulate, this distance is ct'_B. Thus $(L/2) - ut'_B = ct'_B$, an equation that we can solve for t'_B:

$$t'_B = \frac{L/2}{c + u}. \tag{39–7}$$

Similarly, according to an observer in F′, the light beam must travel an additional distance ut'_C to reach C. By the same argument,

$$t'_C = \frac{L/2}{c - u}. \tag{39–8}$$

Time t'_B is different from time t'_C so *events that are simultaneous in* F *are not simultaneous in* F′! This rather counterintuitive idea that the concept of simultaneity is not absolute will play an important role in much of what is to follow. Note that the time difference between t'_B and t'_C is very small for $u \ll c$, and we do not see these time differences in everyday experience; this fact explains the origin of our nonrelativistic intuition. If we lived in a world where ordinary speeds were comparable to the speed of light, we would have developed a different intuition.

39–4 Time Dilation and Length Contraction

The Einstein postulates, or the fact that events that are simultaneous in one reference frame are not simultaneous in a frame that is moving with respect to the first, have two dramatic consequences. These are the slowing down of moving clocks—time dilation—and the shortening of moving rods aligned with the direction of motion—length contraction. Let's examine both consequences.

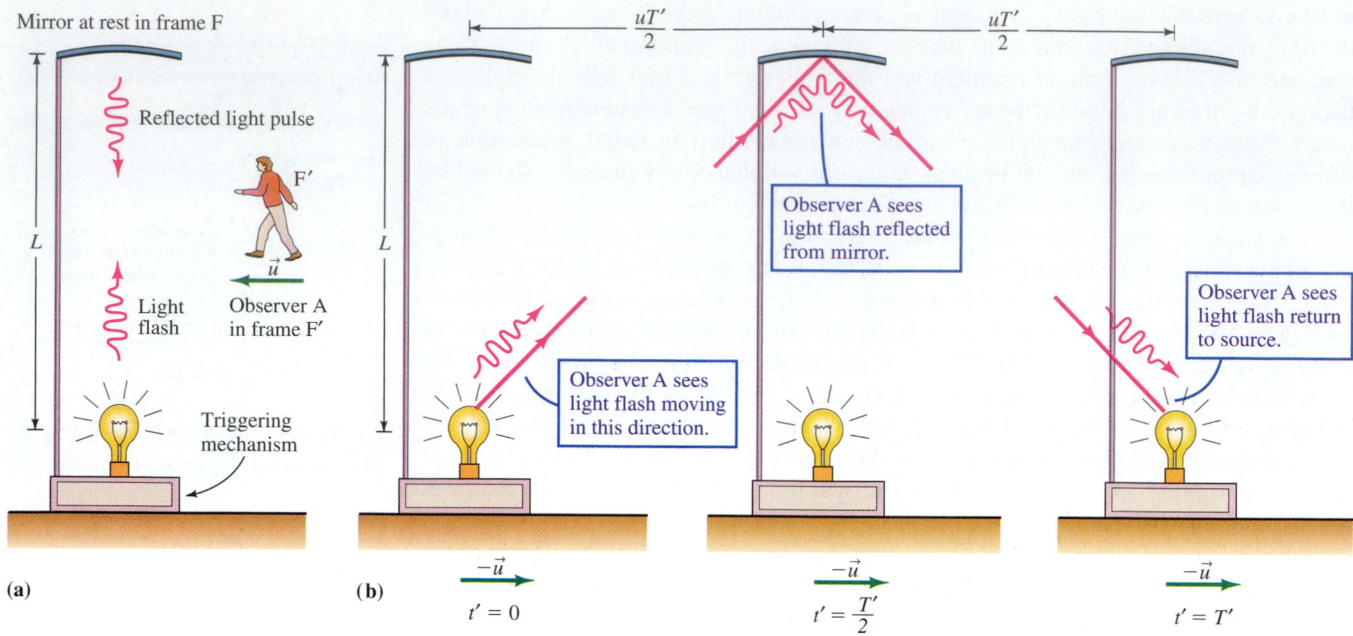

▲ **FIGURE 39–8** (a) Schematic diagram of a clock. The lightbulb at one end of the rod flashes whenever it receives a light flash reflected off the mirror a distance L away from the bulb. The time interval between flashes in the frame of the clock ("ticks") is $2L/c$. (b) The path that the light ray must take according to an observer for whom the clock is moving.

Time Dilation

To clarify our discussion,[†] consider a very simple clock (Fig. 39–8a). It consists of a rod with a lightbulb at one end and a mirror at the other end, a distance L apart. A mechanism is attached to the lightbulb that makes the bulb flash instantaneously whenever a previous flash returns after reflecting off the mirror. According to an observer at rest relative to the clock, the bulb flashes with a period $T = 2L/c$.

A moving observer sees this clock behaving rather differently. Suppose that this observer is in an inertial frame F′ that moves at speed u to the left (Fig. 39–8a). An observer in F′ will see the clock receding to the right (Fig. 39–8b). In F′, the light still travels at speed c to the mirror, but it now has farther to go: The mirror moves during the time that the light travels to it from the lightbulb. As in our discussion of the transverse light beam in the Michelson–Morley experiment, the time that it takes for the light to travel to the mirror and back is such that

$$\frac{cT'}{2} = \sqrt{L^2 + \left(\frac{uT'}{2}\right)^2};$$

that is, $T' = (2L/c)/\sqrt{1 - (u^2/c^2)}$, or

$$T' = \frac{T}{\sqrt{1 - (u^2/c^2)}}. \qquad (39\text{–}9)$$

TIME DILATION

[†]Here we follow N. D. Mermin in *Space and Time in Relativity*, New York: McGraw-Hill, 1968.

Time T' is greater than time T by a factor of $1/\sqrt{1-(u^2/c^2)}$. The observer in frame F′ sees longer "ticks" for the clock; in other words, *the moving clock is slower by a factor of* $\sqrt{1-(u^2/c^2)}$. This effect is known as **time dilation**: *Moving clocks are observed to run more slowly than clocks at rest do.* It is not that the clocks are physically altered; rather, time intervals that are observed in different inertial frames differ.

CONCEPTUAL EXAMPLE 39–1 Suppose there were a clock at rest in frame F' identical to the clock at rest in F. Would the observer in F see the clock in F' run faster or slower, and by what amount?

Answer We can go through a calculation similar to the one we just carried out, or we can appeal to the first postulate, which can be rephrased as follows: There is no experiment that tells us exactly which inertial frame is moving or which is not. The postulate itself demands that the observer in F sees the clock in F' run slower, and by exactly the same amount as the observer in F' sees the clock in F run slow.

There are two important features of the time-dilation effect. First, it is a symmetric effect, as Conceptual Example 39–1 makes clear. Second, although the particular clock we have considered (the "light clock") is an odd one, it is universal in the sense that *every* clock imaginable must behave like it. An additional clock in F of any type can be synchronized with the light clock so that its ticks are directly tied to the light clock's ticks. The ticks of a clock can take diverse forms from the periodic vibrations of an atomic system to the frequency of a light wave to the beating of a heart. *All* these clocks run slow according to an observer who is moving with respect to them.

THINK ABOUT THIS. . .

ARE EXPERIMENTAL TESTS OF TIME DILATION POSSIBLE?

The time-dilation effect is real. We can produce experimental evidence with measurements of the half-lives of radioactive nuclei or unstable particles in motion. We shall learn in subsequent chapters that there exist radioactive nuclei and unstable particles; starting with a collection of such objects, a well-defined and accurately measurable fraction of them decay every "tick" of the clock. This fraction is characteristic of the unstable species. Note this doesn't mean that every last one of such particles lasts exactly the same amount of time. It means, rather, that half of them (say) decay in a fixed time interval. In particular, the decays of a specific species are described by a parameter, the *lifetime*, which is the time after which the fraction remaining undecayed is $1/e = 0.3679$; i.e., about 63 percent of the original number will have decayed in one lifetime. For example, a collection of the unstable fundamental particle called the *muon* has a lifetime of 2.197×10^{-6} s when the muons are at rest. We can regard the length of time it takes for 63 percent of a given large sample of at-rest muons to decay as the tick of a clock. These decays are thus a visible measure of the tick of the "clock."

With this "clock" we can test time dilation. Muons can be produced in a particle accelerator and will then travel at a speed u determined by the characteristics of the accelerator and which is a sizable fraction of the speed of light. Then in the laboratory frame of the accelerator, one finds that 63 percent of the moving muons decay after a time $t = \left[2.197/\sqrt{1-(u^2/c^2)}\right]\,\mu\text{s}$, *not* after a time 2.197 μs. In modern high-energy accelerators, it is possible to accelerate the muons to such a high velocity that the time-dilation factor $1/\sqrt{1-(u^2/c^2)}$ can be as large as 10^6, and 63 percent of these muons decay in a period $10^6 \times (2.197\,\mu\text{s}) = 2.197$ s.

The time-dilation effect was also checked in a much more pedestrian way in 1972 (Fig. 39–9). A very accurate cesium clock was flown in a commercial airplane around the world and used to confirm time dilation to an accuracy of about 10 percent. This test contains some extra elements due to accelerations associated with Earth's rotation and the presence of gravity, and these lead to some additional corrections. To eliminate these effects the experiment was carried out twice, once with the clock traveling in the easterly direction and once with the clock traveling in the westerly direction.

The time-dilation effect has been confirmed so often that there is no doubt as to its reality nor of the accuracy of Eq. (39–9). Indeed, it is now an essential element of technology; for example, the Global Positioning System must include the effect to operate properly. ■

▲ **FIGURE 39–9** The clock used to test time dilation on an around-the-world flight took up a good number of seats.

EXAMPLE 39–2 Consider a clock taken on an airplane that travels at 1.0×10^3 km/h around the world along the equator. If the clock is synchronized with a stationary clock on departure, by how much will the two clocks differ after one round trip? (Take into account only the time-dilation effect of special relativity.)

Setting It Up We will need to calculate the number of seconds it takes to go around the world, and for that we need Earth's circumference, approximately 4.0×10^4 km.

Strategy The stationary observer sees the traveling clock tick slower by a factor of $\sqrt{1-(u/c)^2}$, where u is the airplane's speed relative to the surface. This factor must be applied to the total time T for which the motion occurs, which is $T =$ (distance traveled in km)/(speed in km/s).

Working It Out The speed of the plane in km/s is $u = (1.0 \times 10^3\ \text{km/h})/(3600\ \text{s/h}) = 0.28$ km/s. Thus

$$T = \frac{40 \times 10^3\ \text{km}}{0.28\ \text{km}} = 1.4 \times 10^5\ \text{s}.$$

Also, $\dfrac{u}{c} = \dfrac{0.28\ \text{km/s}}{3.0 \times 10^8\ \text{m/s}} \times \dfrac{10^3\ \text{m}}{\text{km}} = 9.3 \times 10^{-7}$, or $(u/c)^2 = 8.6 \times 10^{-13}$. This is quite small, and we can use the small-x approximation $\sqrt{1-x} \approx 1 - x/2$ for our dilation factor. The number of seconds lost is the difference between T and the dilated value of T, i.e.,

$$T - T\sqrt{1-(u/c)^2} \cong T - T\left(1 - \frac{u^2}{2c^2}\right) = T\frac{u^2}{2c^2}.$$

The numerical value of this loss is quite small, $(1.4 \times 10^5\ \text{s}) \times (0.5 \times 8.6 \times 10^{-13}) = 6.0 \times 10^{-8}$ s.

The Twin Paradox

Time dilation applies to clocks that move with uniform velocities. If you don't pay enough attention to this restriction, you may be led to the *twin paradox*. Consider a set of identical twins. One of them takes off on a long journey. After traveling outward for a long time at a high speed v, that twin turns around and retraces her steps back home. When the traveling twin finally arrives at home after years of travel, the stay-at-home twin will observe that the traveling twin is now the younger of the two—no wrinkles, gray hair, or other signs of age. The twins are no longer identical. The stay-at-home twin reasons that this is to be expected because, relative to herself, the traveling twin was moving with uniform velocity: The traveling twin's clock slowed by a factor of $\sqrt{1-(v^2/c^2)}$, and it is not just the mechanical clock but all types of clocks, including the twin's metabolism, heart rate, and so on that are slowed down. She decides that the deceleration and acceleration at the turning point occur in such a short time that they do not affect this conclusion.

The paradox appears if the traveling twin considers herself to be at rest while the stay-at-home twin was traveling at speed v in the opposite direction. The traveling twin would expect the stay-at-home twin to be younger. Surely both cannot be right! But there is no paradox: The traveling twin is not always in an inertial frame. She moves at uniform speed most of the time, but she does experience a deceleration and then an acceleration for the return. These are very real to her, and she knows that she has not been in the same inertial frame all the time. Thus she cannot make the same statements about the slowing down of clocks as her sister can. From the point of view of special relativity, only the stay-at-home twin, who is always in an inertial frame, can apply the theory to herself. In fact, a careful use of special relativity can show just how much older the stay-at-home twin is at the end of the journey.

The paradox may already have occurred to you in our discussion of the clock that traveled around the world in an airplane. An observer on Earth sees that clock running more slowly and arriving having lost time. A naïve symmetry argument would say there was a paradox, because an observer on the airplane might claim that he is not moving and it should be an Earth-based clock that runs more slowly. But the airplane is in a noninertial frame—it travels in a circle. There is no symmetry here, and therefore no reason to invoke a paradox. The clock on the airplane was indeed the moving clock.

Length Contraction

The slowing down of moving clocks is accompanied by the contraction in length of moving objects along their direction of motion. We can begin by giving an argument that uses time dilation. Consider again the muon, an unstable particle with a lifetime of $\tau \cong 2\ \mu\text{s}$. Experiments with balloons at the top of the atmosphere measure the number of muons that arrive at that level as well as their speeds. Given those numbers and the value of τ, one can calculate that without time dilation, the number of muons arriving at sea level would be completely negligible, even if they were traveling at (nearly) the speed of light. But in fact, the number of muons that make it all the way to the surface is quite consistent with a muon lifetime increased by the time-dilation factor. We may state this quantitatively: If the muons all have speed u, the lifetime is $\tau/\sqrt{1-(u/c)^2}$, meaning that 37 percent of the initially present muons will travel a distance

$$L = \frac{u\tau}{\sqrt{1-(u/c)^2}} \tag{39–10}$$

without decaying. This length is much greater than it would be if there were no time-dilation effect, because the square-root factor approaches zero as u approaches c. The length could be larger than the height of the atmosphere; in this way, muons could reach the ground before they decay. We'll suppose that the speed of the muons is such that the "average" muon just reaches the ground before decaying, so L in Eq. (39–10) is the height of the atmosphere.

At this point we'll shift our attention to how this looks to an observer who is moving with the muon—an observer to whom the muon is at rest. He will measure the muon's lifetime to be its original value τ. This observer, however, will also detect that the muon reaches the ground if the Earth-based observer does; the collision with the ground is an event no observer could dispute. That means that for him it takes a time τ for the whole atmosphere to move past him at speed u. The only way this can happen is for our observer to measure the atmosphere to have a height L', where $L' = u\tau$, or, from Eq. (39–10),

$$L' = L\sqrt{1-\frac{u^2}{c^2}}. \tag{39–11}$$

LENGTH CONTRACTION

The observer moving with the muon measures the atmosphere to be thinner than an Earth-based observer does. To the moving observer, the atmospheric height, or any length in the direction of his motion, has undergone a **length contraction** by a factor of $\sqrt{1-(u^2/c^2)}$.

Another way of seeing that there must be a length contraction along the direction of motion is to modify the clock constructed at the beginning of this section. We add an identical rod at right angles (transverse) to the original rod (Fig. 39–10). The mechanism is modified so that the bulb relights only when both reflected beams reach the lightbulb at the same time. This can be achieved by making the length of the rods identical. Each flash of the bulb is an event, and these events are observed from any inertial frame. Suppose now that the clock moves at speed u in the direction of the added rod with respect to an observer in frame F′. According to the observer in frame F′, the round-trip time for the light on the transverse rod is

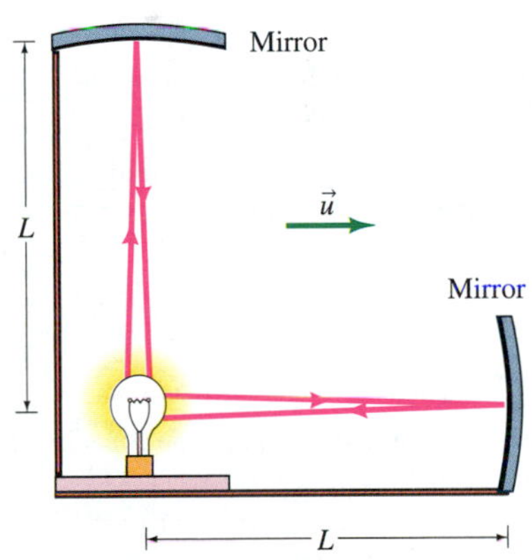

▲ **FIGURE 39–10** Schematic diagram of a two-armed clock used to exhibit length contraction. Note the similarity to the Michelson–Morley apparatus.

$$T' = \frac{2L}{c}\frac{1}{\sqrt{1-(u^2/c^2)}}. \tag{39–12}$$

The round-trip time for the light that travels along the horizontal rod is the time t_1' it takes to get to the mirror added to the time t_2' it takes to return. The mirror is moving to the right, so the light on its outward trip has an extra distance ut_1' to travel. If, according to the observer in frame F′, the length of the rod is L' (the quantity we want to find), then

$$ct_1' = L' + ut_1'.$$

For the return trip, the bulb approaches the mirror at speed u, so the time t_2' to return is determined by

$$ct_2' = L' - ut_2'.$$

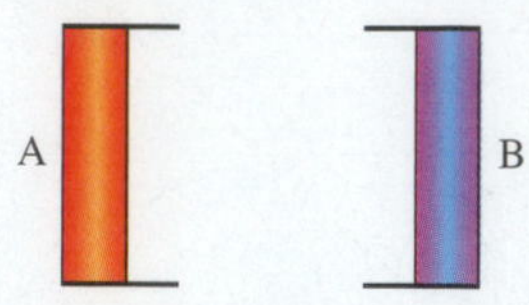

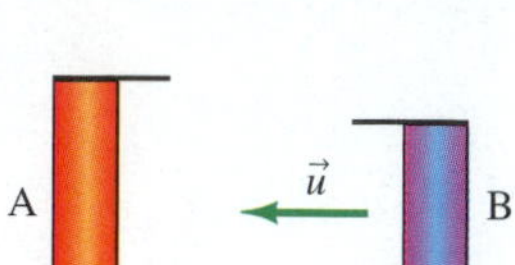

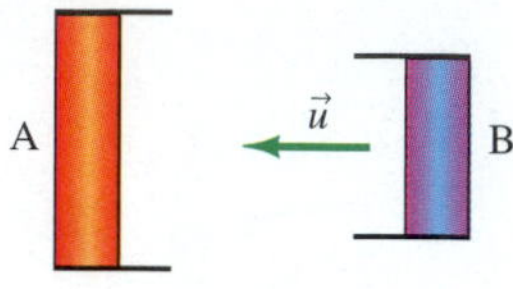

We solve these two equations for t_1' and t_2', respectively, and add:

$$t_1' + t_2' = \frac{L'}{c - u} + \frac{L'}{c + u} = \frac{2L'/c}{1 - (u^2/c^2)}. \tag{39–13}$$

This, however, must equal T'—remember, that is the only way the bulb reflashes—and a comparison of Eqs. (39–12) and (39–13) gives $L' = L\sqrt{1 - (u^2/c^2)}$, which is the length contraction of Eq. (39–11).

Note that length contraction occurs only *along* the direction of motion. We can see that there is no change in directions transverse to the motion by the argument in Fig. 39–11. We made the unspoken assumption that there is no transverse length contraction in our derivation of time dilation; that assumption is justified by the argument in Fig. 39–11.

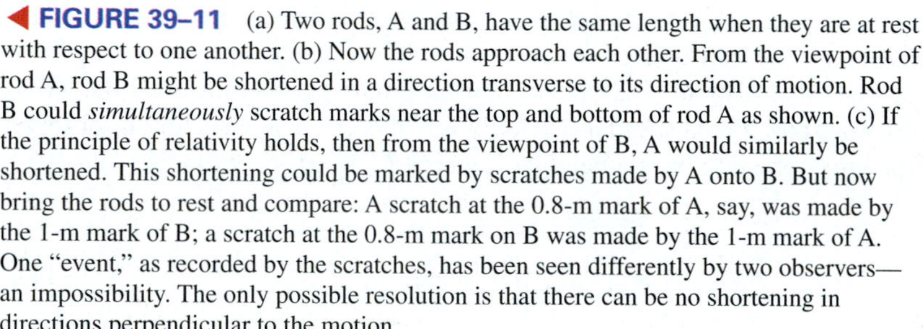

FIGURE 39–11 (a) Two rods, A and B, have the same length when they are at rest with respect to one another. (b) Now the rods approach each other. From the viewpoint of rod A, rod B might be shortened in a direction transverse to its direction of motion. Rod B could *simultaneously* scratch marks near the top and bottom of rod A as shown. (c) If the principle of relativity holds, then from the viewpoint of B, A would similarly be shortened. This shortening could be marked by scratches made by A onto B. But now bring the rods to rest and compare: A scratch at the 0.8-m mark of A, say, was made by the 1-m mark of B; a scratch at the 0.8-m mark on B was made by the 1-m mark of A. One "event," as recorded by the scratches, has been seen differently by two observers—an impossibility. The only possible resolution is that there can be no shortening in directions perpendicular to the motion.

EXAMPLE 39–3 The radius of our galaxy is 3.0×10^{20} m. (a) How fast would a spaceship have to travel to cross the entire galaxy in 300 yr as measured from within the spaceship? (b) How much time would elapse on Earth during the traversal?

Strategy (a) An observer at rest relative to the spaceship sees the galaxy passing with some speed v. The galaxy is contracted from its original length L to $L' = L\sqrt{1 - (v/c)^2}$. The whole galaxy traveling with this speed v would need to pass the observer in time T, so that $v = L'/T$. Thus

$$v = \frac{L'}{T} = \frac{L\sqrt{1 - (v/c)^2}}{T}.$$

We may solve this expression for v in terms of the given quantities L and T.

Working It Out We rewrite the last formula by squaring it and cross-multiplying, so that

$$v^2T^2 = L^2\left(1 - \frac{v^2}{c^2}\right) = L^2 - \frac{L^2}{c^2}v^2.$$

Rearranging, $v^2\left(T^2 + \frac{L^2}{c^2}\right) = L^2$, and hence

$$\frac{v^2}{c^2} = \frac{L^2}{L^2 + (cT)^2} = \frac{1}{1 + (cT/L)^2}.$$

Numerically $L = 2 \times (3.0 \times 10^{20}\ \text{m})$ and $cT = (3.0 \times 10^8\ \text{m/s}) \times (300\ \text{y}) \times (3.15 \times 10^7\ \text{s/y}) = 2.8 \times 10^{18}\ \text{m}$, so that $(cT/L)^2 = \left(\frac{2.8 \times 10^{18}}{6 \times 10^{20}}\right)^2 = 2.2 \times 10^{-5}$. Thus $(cT/L)^2$ is a very small number, and we may write

$$\frac{v}{c} = \sqrt{\frac{1}{1 + (cT/L)^2}} \cong 1 - \frac{1}{2}\left(\frac{cT}{L}\right)^2 = 1 - 1.1 \times 10^{-5}.$$

The spaceship must travel almost (but not quite) with the speed of light to cross the entire galaxy in 300 years.

Strategy (b) and Working it out The observer on Earth does not see the galaxy contracted, and sees the spaceship move with speed v, so that the time is $t = L/v$. To a very good approximation $v = c$, so that

$$t \cong \frac{L}{c} = \frac{6.0 \times 10^{20}\ \text{m}}{3.0 \times 10^8\ \text{m/s}} = 2.0 \times 10^{12}\ \text{s}$$

$$t = (2.0 \times 10^{12}\ \text{s}) \times \frac{1}{3.15 \times 10^7\ \text{s/y}} = 6.3 \times 10^4\ \text{y}.$$

What Do You Think? If the Earth-based observer could read the spaceship clock, what would he read on the face of this clock when the passage is complete? *Answers to **What Do You Think?** questions are given in the back of the book.*

39–5 The Relativistic Doppler Shift

The Doppler shift for sound describes the changes in pitch of a train whistle as the train approaches, passes, and recedes from an observer. When a moving source that emits sound waves with frequency f travels toward an observer at rest relative to the air, the observed frequency f' is shifted from the source frequency according to Eq. (14–49).

$$f' = \frac{f}{1 - (u/c)} \tag{39–49}$$

We have changed the notation slightly, representing the source speed by u (instead of v_s) and the speed of sound by c (instead of v). If the source is at rest relative to the air and the observer is moving toward the source, then the frequency heard by the observer is Eq. (14–53), with a similar change of notation,

$$f' = f\left(1 + \frac{u}{c}\right) \tag{39–53}$$

The frequencies f' are not the same in the two cases, so it is possible—by an accurate measurement of the frequency shift and a knowledge of the relative speed—to determine whether it is the source or the receiver that is moving relative to the medium (the air). The reason for the difference between the two shifts is that, for sound, there *is* a preferred frame, namely, the frame at rest relative to the air. The Doppler shift for electromagnetic radiation (including light) cannot distinguish between the two situations and therefore has a different, more symmetric, form.

To find the Doppler shift for light, let's consider a periodically flashing light that moves at speed u toward an observer. The source is placed at the origin of frame F′. Suppose that one pulse of light is emitted for every time interval τ_0, so the frequency of emission seen by someone moving with the source is $f_0 = 1/\tau_0$. The stationary observer sees frame F′ moving toward him at speed u. Let's calculate the frequency of the flashes that he observes. First, the flashing is time dilated; the moving clock runs slow according to the stationary observer, with period

$$\frac{\tau_0}{\sqrt{1 - (u^2/c^2)}}.$$

But there is a second effect. The time between two successive pulses is reduced because the flashing clock has moved toward the stationary observer between a first and a second pulse. If he—the stationary observer—says that the light is a distance L from him when the first pulse is sent, then he measures the light to be at a reduced distance

$$L - \frac{u\tau_0}{\sqrt{1 - (u/c^2)}}$$

when the second pulse is sent. This reduces the time between pulses because the second pulse has a shorter distance to travel. If the first pulse arrives at a time $t_1 = L/c$, then the second pulse arrives at a time t_2 given by $\tau_0/\sqrt{1 - (u/c)^2}$ plus the reduced distance divided by c:

$$t_2 = \frac{\tau_0}{\sqrt{1 - (u/c)^2}} + \frac{1}{c}\left[L - \frac{u\tau_0}{\sqrt{1 - (u/c)^2}}\right].$$

The stationary observer measures the period of the pulses to be $\tau = t_2 - t_1$; that is,

$$\tau = t_2 - t_1 = \frac{\tau_0}{\sqrt{1 - (u/c)^2}} + \frac{1}{c}\left[L - \frac{u\tau_0}{\sqrt{1 - (u/c)^2}}\right] - \frac{L}{c} = \frac{\tau_0}{\sqrt{1 - (u/c)^2}}\left(1 - \frac{u}{c}\right).$$

If we use $1 - x^2 = (1 - x)(1 + x)$, we can write this result as

$$\tau = \tau_0 \sqrt{\frac{1 - u/c}{1 + u/c}}. \qquad (39\text{–}14)$$

Recognizing that the frequency is the inverse of the period, we have

$$f_1 = f_0 \sqrt{\frac{1 + (u/c)}{1 - (u/c)}}, \qquad (39\text{–}15a)$$

DOPPLER SHIFT FOR LIGHT

where f_0 is the frequency of the source in its rest frame and f_1 is the frequency observed from a frame that moves at speed u relative to the source. The two frames are moving toward each other at relative speed u, and the frequency of flashes is increased. Instead of the frequency, we can use $\lambda = c/f$ to express the wavelength λ_1 seen by the observer in terms of the wavelength λ_0 at the source:

$$\lambda_1 = \lambda_0 \sqrt{\frac{1 - (u/c)}{1 + (u/c)}}. \qquad (39\text{–}15b)$$

We call this situation—in which the source moves toward the observer and the observed wavelength decreases—a *blueshift*. The speed u can be interpreted either as the speed of the source toward a stationary receiver or as the speed of the receiver toward a stationary source. According to the principles of special relativity, these two possibilities are not distinguishable. If the source is moving away from the observer, then we must change the sign of u in our results, and

$$f_1 = f_0 \sqrt{\frac{1 - (u/c)}{1 + (u/c)}}. \qquad (39\text{–}16a)$$

Equivalently, the observed wavelength is

$$\lambda_1 = \lambda_0 \sqrt{\frac{1 + (u/c)}{1 - (u/c)}}. \qquad (39\text{–}16b)$$

Thus the frequency decreases (the wavelength increases) in this case. The visible spectrum is shifted toward the red colors, and Eqs. (39–16) are said to describe a relativistic *redshift*.

CONCEPTUAL EXAMPLE 39–4 A very bright source of radiation is in a (presumed) circular gravitational orbit about a large mass. The plane of the orbit is along the line of vision. Can we determine the size of the orbit even if its angular size seen from Earth is too small to be observed directly?

Answer When the source emerges from behind the central mass, it comes toward us, and the spectrum will be blueshifted. When it is moving away from us on the other side of the orbit, its spectrum is redshifted. The amount of the shift gives us the speed of the source. The time between the blueshifted and redshifted stages is half the period of the orbit, and given both the speed and the orbital period, the size of the orbit is easily calculated.

The Doppler Shift for Light and Cosmology

Measurements of the Doppler shift of starlight have proved crucial in the evolution of modern astrophysics and cosmology. In one application, the Doppler shift is used to establish the velocities of stars or other radiating bodies. As we know from our discussion of spectral lines in Chapter 38, radiation emitted by atoms and molecules is characterized by *spectral lines*, discrete or very narrow frequency bands of especially intense radiation. These spectral lines provide a unique signature for elements and compounds. If an entire sequence of spectral lines in starlight is observed to correspond to a

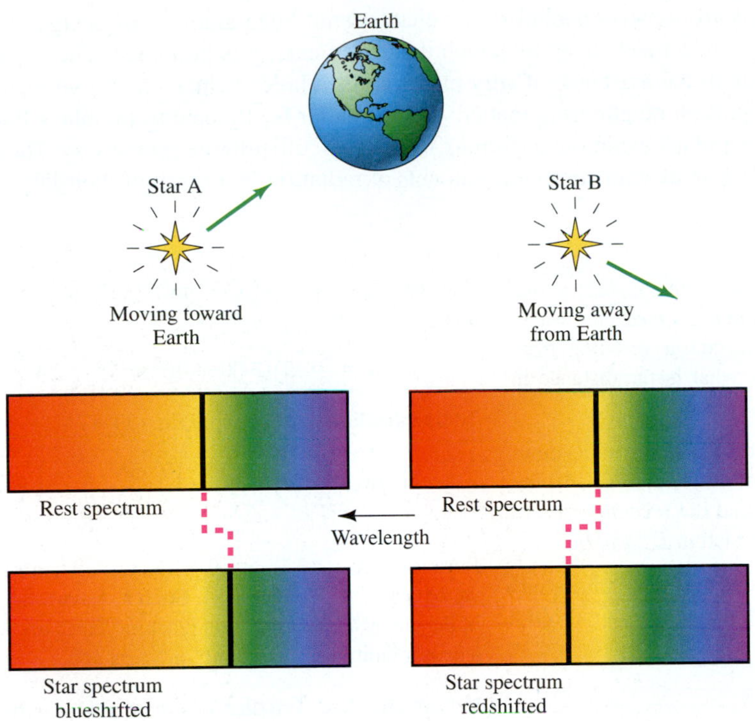

◀ **FIGURE 39–12** The spectral lines that can be attributed to a specific element are shifted if the source moves relative to the observer. The spectral lines emitted by Star A, moving toward Earth, are shifted to the blue, while those of Star B, moving away from Earth, are shifted to the red.

sequence of laboratory-observed spectral lines all shifted by the same factor, then we know that the source of the starlight is the same as the laboratory source but moves at a velocity that can be calculated from equations such as Eqs. (39–16) (Fig. 39–12).

One of the most interesting uses of the Doppler shift was made by the astronomer Edwin Hubble. In the 1920s, 1930s, and 1940s he studied the spectral lines of a large number of stars in distant galaxies, estimating their distance from Earth using known characteristic brightness of these stars. Hubble discovered that the spectra of most of these distant stars are redshifted, which means that their galaxies are receding from us. He found that *the recession velocity of the galaxies relative to our galaxy is proportional to their distance from Earth*. This result is known as **Hubble's law**, and it takes the mathematical form

$$D = u/H. \tag{39–17}$$

Here, D is the distance to a galaxy, u is the recession speed relative to us, and H is the so-called *Hubble parameter*, measured today with 1 percent accuracy to be $H \cong 2.32 \times 10^{-18}\ \mathrm{s}^{-1}$. The fact that the distance and speed are measured relative to Earth appears superficially to give Earth a central position, but some thought shows that conclusion is wrong. If all stars and galaxies are moving away from each other according to Hubble's law, then an observer located at *any one of them* would report the same effect. This is easily visualized if we consider a simple model of dots painted uniformly on a balloon (Fig. 39–13a). As the balloon is inflated, all the dots move farther away from each other, and each dot "sees" the others moving away from it (Fig. 39–13b). In fact, a more detailed analysis shows that a picture in which the recession speed is linearly dependent on the distance, just as in Eq. (39–17), is the *unique* picture that does not privilege any one position. This picture, in which galaxies move away from each other according to Hubble's law, is part of the cosmological theory of the *Big Bang*, which proposes that the universe started from a tiny region with all of space undergoing a subsequent expansion, an expansion that still continues. In this theory, the age of the universe is on the order of H^{-1}, roughly 13 billion years.

Measurements of the Doppler shift, together with Hubble's law, allow us to calculate distances to galaxies. In the 1960s, astronomers working with radio telescopes discovered very powerful pointlike sources of radiation in which very large redshifts ($f_1/f_0 \cong 0.3$) were observed. These sources are *quasars*, or *quasistellar objects*.

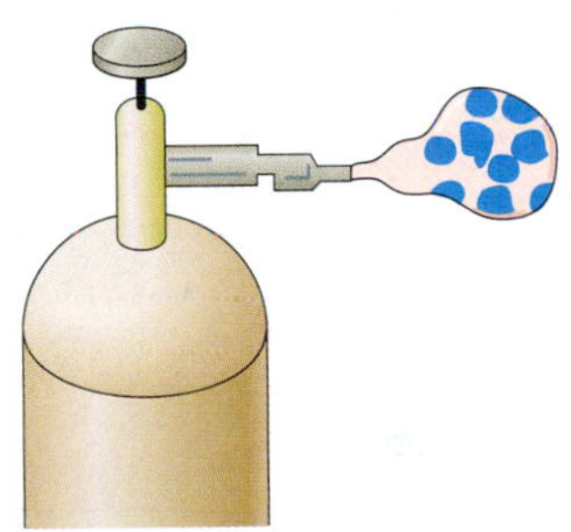

(a)

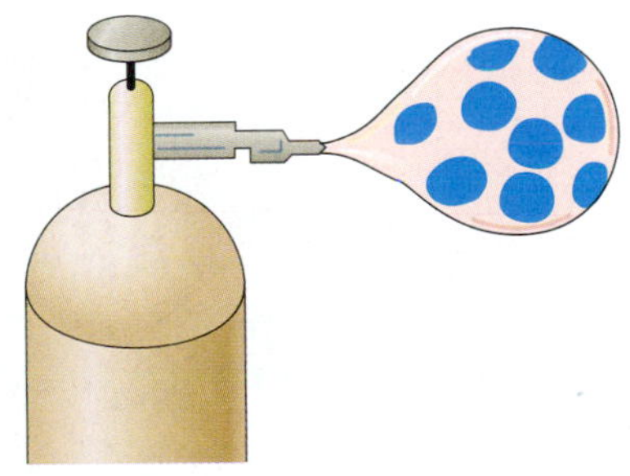

(b)

▲ **FIGURE 39–13** (a) Dots painted on the surface of a balloon represent an analogy to Hubble's expanding universe. (b) As the balloon expands, the dots move away from one another at a speed that depends on the distance between them.

Astronomers concluded that quasars emit huge amounts of energy, which left a puzzle as to a mechanism by which that much energy is produced. The explanation has to do with the existence of very massive black holes, which exert a very strong gravitational pull on neighboring matter. As that matter is attracted to the black hole, it undergoes a rapid acceleration, becoming ionized by collisions along the way. The resulting charged particles emit enormous amounts of radiation as a result of their large acceleration.

EXAMPLE 39–5 Studies of a quasar show that a spectral line whose wavelength in the laboratory is 121 nm has a measured wavelength of 358 nm. With what speed is the quasar receding from Earth? Assuming that Hubble's law holds, what is the distance in light-years of the quasar from Earth?

Strategy Given the wavelength in the rest frame and in the moving frame, λ_0 and λ_1, respectively, we can invert Eq. (39–16), the Doppler shift formula for wavelengths, to find the recession speed. The distance can then be obtained from the relation $D = u/H$.

Working It Out Equation (39–16b) is the Doppler-shift formula for wavelengths for source and receiver moving apart (a redshift). We can solve it for the ratio u/c:

$$\frac{u}{c} = \frac{(\lambda_1/\lambda_0)^2 - 1}{(\lambda_1/\lambda_0)^2 + 1}.$$

The data give $\lambda_1/\lambda_0 = (358\text{ nm})/(121\text{ nm}) = 2.96$, so $u/c = 0.79$, and

$$u = (0.79)(3.00 \times 10^8\text{ m/s}) = 2.4 \times 10^8\text{ m/s}.$$

To find the distance, we apply Eq. (39–17):

$$D = \frac{u}{H} = \left(\frac{2.4 \times 10^8\text{ m/s}}{2.32 \times 10^{-18}\text{ s}^{-1}}\right) = 1.0 \times 10^{26}\text{ m}.$$

Because $1\text{ ly} = (3.15 \times 10^7\text{ s})(3.00 \times 10^8\text{ m/s}) = 0.95 \times 10^{16}\text{ m}$, we obtain $D = 1.1 \times 10^{10}$ ly. Such a distance is nearly at "the edge of the universe," and the light reaching Earth gives information about a quasar emitting light 10 billion years ago.

What Do You Think? Is there a limit to how much the wavelength of a given spectral line can be redshifted?

The Relativistic Addition of Velocities

Suppose that observer A measures the velocity of an object as $\vec{v}_1$; in turn, observer B measures observer A to move with velocity $\vec{u}$ with respect to him. According to the Galilean law of velocity addition, observer B would measure the object to move with velocity $\vec{v}_2 = \vec{v}_1 + \vec{u}$. It should be no surprise that this simple result cannot be consistent with special relativity; for example, for $v_1 = v_2 = 0.8c$, $v_1 + v_2$ is $1.6c$, and our expressions for time dilation and length contraction [Eqs. (39–9) and (39–11)] do not allow speeds greater than c.

We can use the Doppler shift for light to find the important relation that describes how velocities add—that is, what observer B would actually measure for the velocity of our object. Suppose that a source emits light with the rest-frame frequency f_0 (Fig. 39–14). This source is our object. An observer A moving away from the source at speed v_1 along the x-axis receives a redshifted frequency f_1. If that observer immediately

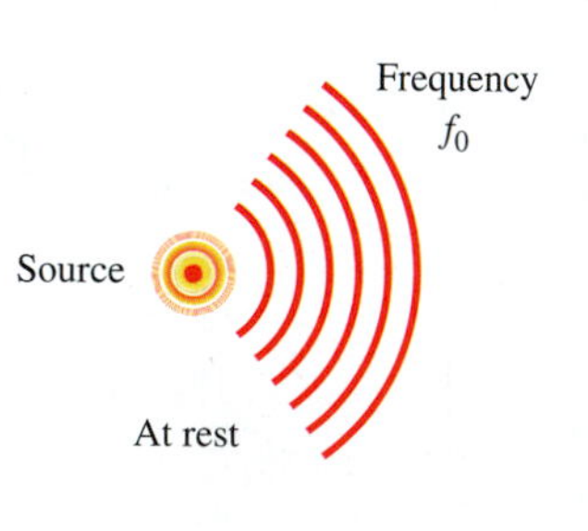

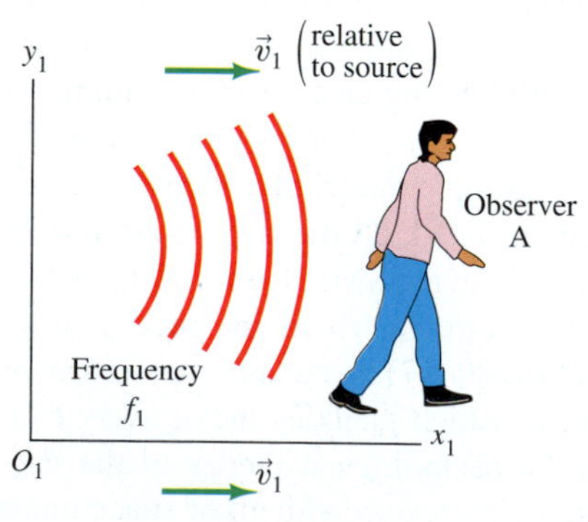

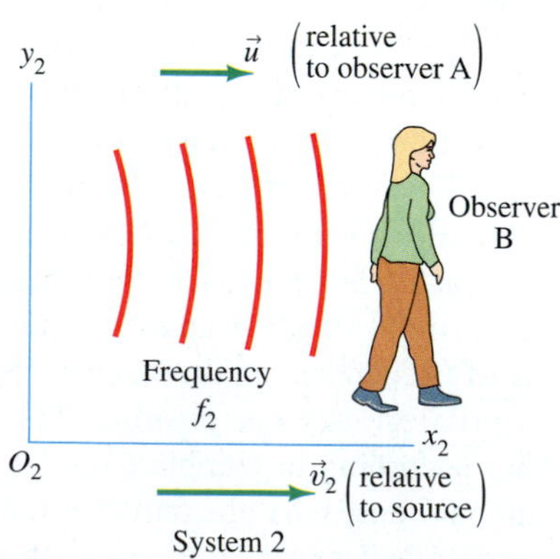

▲ **FIGURE 39–14** A light source emits a frequency f_0, seen as frequency f_1 by an observer A who moves at speed v_1 to the right. Observer A re-emits the light with frequency f_1, which is seen by an observer B who moves to the right at speed u with respect to observer A (Observer B moves at speed v_2 with respect to the original source.) Observer B measures a light frequency f_2, consistent with the law of addition of velocities.

reradiates with frequency f_1 to observer B, who is moving away in the same direction at speed u with respect to A, we would expect that the relationship between the received frequency f_2 and the original frequency f_0 would be the frequency f_0 shifted by the speed v_2 of observer B relative to the source. We can thus extract the correct relativistic expression for v_2 in terms of v_1 and u by finding f_2 in terms of f_1 and f_0.

We do this by starting with the frequency f_1 measured by observer A: According to Eq. (39–16a),

$$f_1 = f_0\sqrt{\frac{1 - (v_1/c)}{1 + (v_1/c)}}.$$

If observer A immediately reradiates the light toward observer B, moving away at speed u relative to observer A, then the frequency seen by observer B is

$$f_2 = f_1\sqrt{\frac{1 - (u/c)}{1 + (u/c)}}.$$

We now eliminate f_1 by expressing f_2 in terms of f_0 and the (as yet unknown) velocity v_2 of observer B relative to the source. We have

$$f_2 = f_0\sqrt{\frac{1 - (v_1/c)}{1 + (v_1/c)}}\sqrt{\frac{1 - (u/c)}{1 + (u/c)}},$$

which, if v_2 is the speed with which observer B sees the original object move, must also take the form

$$f_2 = f_0\sqrt{\frac{1 - (v_2/c)}{1 + (v_2/c)}}.$$

When we square (f_2/f_0) in its two forms and equate the two, we obtain

$$\frac{1 - (v_2/c)}{1 + (v_2/c)} = \frac{[1 - (v_1/c)][1 - (u/c)]}{[1 + (v_1/c)][1 + (u/c)]}.$$

If we solve this equation for v_2, the result is:

$$v_2 = \frac{v_1 + u}{1 + (v_1 u/c^2)}.$$

In order to remove the emphasis from what is an observer and what is an object and make the result appear more symmetric, it is best at this point to change the notation a bit. The symbol v_1 will continue to represent the velocity of an object 1 with respect to some origin O, while rather than using the symbol u to represent the velocity of another object with respect to a second origin centered on object 1, we use the symbol v_2. Then V will represent the velocity of the second object with respect to origin O, rather than v_2 as in the equation above. We are supposing that the motion is along a single direction, so that the vectorial aspect is handled with a sign. Our result for the *law of addition of velocities* is therefore

$$V = \frac{v_1 + v_2}{1 + (v_1 v_2/c^2)}. \qquad (39\text{–}18)$$

ADDITION OF VELOCITIES

We see that in the nonrelativistic limit where v_1/c and v_2/c are both small, Eq. (39–18) reduces to the Galilean law of addition of velocities: $V = v_1 + v_2$. But if, for example, $v_1 = v_2 = 0.8c$, then Eq. (39–18) gives $V = 0.98c$, not $1.6c$. If $v_1 = c$, we get $V = c$, *independent of* v_2. The speed of light is in fact the maximum speed allowed, and it is never exceeded. Equation (39–18) is a beautiful illustration of how relativistic kinematics smoothly adjoins the nonrelativistic (Galilean) form while providing a very different general result.

EXAMPLE 39–6 A light source flashes with a frequency of 1.0×10^{15} Hz. The radiation is reflected by a mirror that moves at a speed of 100 km/s away from the source. How much does the frequency of the reflected radiation received back at the source differ from the original emitted frequency?

Strategy We can proceed as we did to work out the relativistic addition of velocities. The source has frequency f_0 at the source, while the mirror moves with speed u away from a source at the origin. It therefore "sees" (absorbs) the frequency redshifted down to a value f_1. It then reemits at f_1, and the detector at the source receives a frequency f_2 shifted down from f_1 because the reemitter (the mirror) is moving away with speed u. By superposing the two shifts, we find f_2 in terms of the original frequency f_0.

Working It Out The frequency seen at the mirror is

$$f_1 = f_0\sqrt{\frac{1 - u/c}{1 + u/c}}.$$

The mirror reemits light of this frequency, and because the mirror is moving away from the observer at the source, the observer sees f_1 shifted down (redshifted) by the same factor:

$$f_2 = f_1\sqrt{\frac{1 - u/c}{1 + u/c}} = f_0\frac{1 - u/c}{1 + u/c}.$$

The difference between f_2 and f_0 is

$$f_2 - f_0 = f_0\left(\frac{1 - u/c}{1 + u/c} - 1\right) = -f_0\frac{2u/c}{1 + u/c}.$$

The minus sign properly indicates that $f_2 < f_0$. The factor u/c is small [order of 10^{-4}], and we can drop it compared to 1 in the denominator. We are left with

$$f_2 - f_0 \cong -f_0(2u/c) \cong -7 \times 10^{11}\ \text{Hz},$$

only 7×10^{-4} times the original frequency.

This example illustrates the operation of a police speed radar. There are two Doppler shifts as the car whose speed is measured reflects a beam of a given emitted frequency. The returned beam is very nearly the same frequency as the emitted one; the very small shift, which as above is approximately proportional to the vehicle's speed, is detected by means of a "beat" (see Chapter 15).

Alternative Solution We can also solve this problem by thinking of the image of the source in the mirror as radiating directly back to the source. The mirror sees the original source receding from it at speed u, and the mirror equally sees the image receding from it at the same speed. Therefore, an observer at the source sees the image receding from it at a speed V given by Eq. (39–18) with $v_1 = v_2 = u$. The Doppler shift with V as a recession speed will then give f_2.

What Do You Think? What does the color of the image of the light source reflected in the mirror look like to an observer back at the source?

39–6 The Lorentz Transformations

In this section we find the transformation laws between measured values of time and space in different inertial frames. We are looking for a set of relations like the *Galilean transformations*, Eqs. (39–3) and (39–4). We have already argued that the Galilean transformations are inconsistent with the Einstein condition that the speed of light is the same in all inertial frames. But then what are the correct transformation laws, the ones that must hold in a world in which the speed of light is the same in all inertial frames?

To proceed, we focus on an **event**, something that happens at a particular place and at a particular time. Examples are the collision of two particles, the tick of a clock, or an explosion. In an inertial frame F, the event will be described by the *space-time* coordinates (x, y, z, t). What are the coordinates (x', y', z', t') of the event in another inertial frame F′, a frame that moves with velocity u along the x-axis relative to frame F (Fig. 39–15)? (The vectorial aspect of u is entirely described by its sign.) We'll assume that the origins of the two frames coincide at times $t = t' = 0$. The expression for the coordinates (x', y', z', t') of our event in terms of the event's coordinates (x, y, z, t) in F are our desired transformation laws.

As we'll confirm below, the correct transformation laws—known as the **Lorentz transformations** for their discoverer, who formulated the laws in 1890[†]—are given by

$$x' = \gamma(x - ut) \qquad (39\text{–}19)$$

$$t' = \gamma\left(t - \frac{ux}{c^2}\right). \qquad (39\text{–}20)$$

THE LORENTZ TRANSFORMATIONS

[†]This date is well before 1905, when Einstein published his work. Einstein's key role was perhaps less in discovering new formulas than in formulating a conceptual whole.

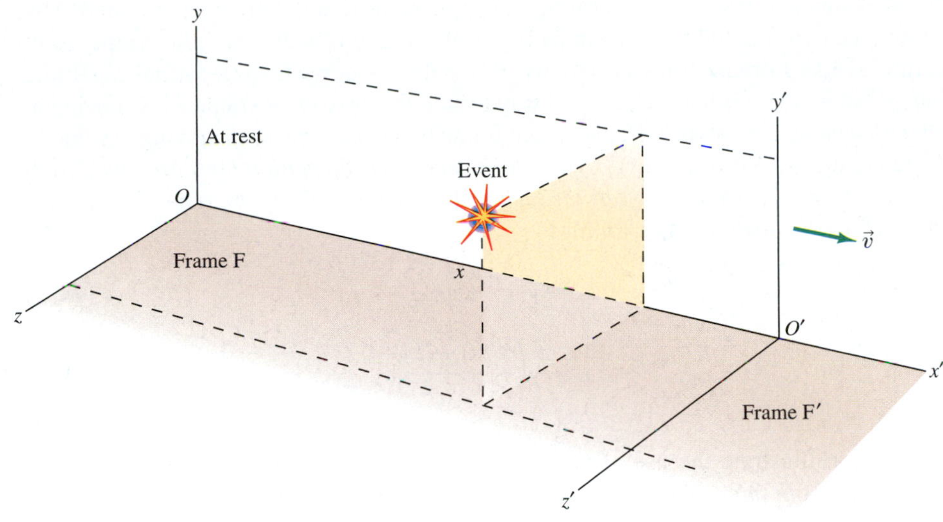

◀ **FIGURE 39–15** The origins of the coordinate systems of frames F and F′ coincide at time $t = t' = 0$. An event occurs some time later. An observer in frame F would describe the space–time coordinates of the event differently than an observer in frame F′.

Here γ is defined by

$$\gamma \equiv \frac{1}{\sqrt{1 - (u^2/c^2)}}. \tag{39–21}$$

To these equations, we should add two others:

$$y' = y; \tag{39–22}$$

$$z' = z. \tag{39–23}$$

Equations (39–22) and (39–23) reflect the fact that with the relative motion of the frames along the x-direction, the description of the y- and z-coordinates of an event is the same in frames F and F′.

Note carefully the nature of these transformations. Not only does the event location x' in the frame F′ depend on both the position x and time t of the event in the frame F, *but so does the time t' of the event*. That x' depends on x and t is not surprising—it is that way in the intuitive Galilean transformation as well. But the idea that the time of the event in frame F′ also depends on both the time and position of the event in the frame F is both new and unintuitive.

As a first step to confirming the Lorentz transformation, we observe that when the relative speed of the two frames is small compared to c $(u/c \ll 1)$, as would be the case for most everyday situations we experience, then $\gamma \cong 1$ and the Lorentz transformations reduce to the Galilean transformations, Eqs. (39–3) and (39–4). This is why the necessity for the Lorentz transformations is not evident from everyday experience. It is also easy to show that the location of a wave front for a light pulse emitted from the joint origin of frames F and F′ at $t = t' = 0$ is given by $x^2 + y^2 + z^2 - c^2t^2 = 0 = x'^2 + y'^2 + z'^2 - c^2t'^2$ (see Problem 44) or, in other words, that the light pulse spreads at the speed of light along a spherical front in both frames. This result reflects the necessity that the speed of light be the same in both frames and is the crucial step that distinguishes the Lorentz transformations. In fact, the Lorentz transformations show more: *In general*, you can easily verify that

$$x'^2 + y'^2 + z'^2 - c^2t'^2 = x^2 + y^2 + z^2 - c^2t^2. \tag{39–24}$$

Equation (39–24) states that the quantity $x^2 + y^2 + z^2 - c^2t^2$ is *invariant*, meaning that its value is the same in all inertial frames.

We can invert the Lorentz transformation and obtain x and t in terms of x' and t' by solving the two simultaneous algebraic Eqs. (39–19) and (39–20). We refer to the result as the inverse Lorentz transformations. But rather than perform a formal algebraic manipulation, there is a more physical way to find the answer: If frame F′ is moving at velocity u along the x-axis of frame F, then frame F may be viewed as moving at velocity $-u$ along the x'-axis of frame F′. Thus *the inverse transformation laws are found from the original transformation laws by the exchange of the primed and unprimed coordinates and the simple substitution of $-u$ for u*:

$$x = \gamma(x' + ut'); \tag{39–25}$$

$$t = \gamma\left(t' + \frac{ux'}{c^2}\right). \tag{39–26}$$

We can illustrate the use of Lorentz transformations by working out the length contraction from them. This is a useful exercise because it obliges us to think carefully about what is meant when we talk about a length measurement. Consider a rod at rest in frame F. The measurement of its length L involves "events" that consist of each end being at a particular place at a particular time. The length of an object in its own rest frame is called its **proper length**. In frame F the coordinates of the ends of the rod are $x_1 = 0$ and $x_2 = L$. Since the rod is at rest in frame F, there is no confusion about when the measurements are made. But what is the length of the rod in frame F′? As we emphasized earlier, when an object is moving in a certain reference frame, a length measurement makes sense only when the coordinates of the two ends are located simultaneously. Let us denote by t'_1 the time in frame F′ at which we measure the position x'_1 of one end and by t'_2 the time in frame F′ at which we measure the position x'_2 of the other end. Then to measure the length we require that $t'_1 = t'_2$. We can choose this common time to be zero; that is, we make the length measurement at time $t'_1 = t'_2 = 0$. Let's see what these times are in frame F. We have from the Lorentz transformations

$$t'_1 = \gamma\left(t_1 - \frac{ux_1}{c^2}\right).$$

Because $x_1 = 0$ and $t'_1 = 0$, we obtain

$$t_1 = 0.$$

We also have

$$t'_2 = \gamma\left(t_2 - \frac{ux_2}{c^2}\right).$$

With $t'_2 = 0$ and $x_2 = L$, we find that

$$t_2 = \frac{uL}{c^2}.$$

We note that $t_1 \neq t_2$: The length measurements are made at different times in frame F. Events that are simultaneous in one frame are not simultaneous in another. With t_1 and t_2 given, we may now use the Lorentz transformations to find the coordinates of the two ends in frame F′:

$$x'_1 = \gamma(x_1 = ut_1) = 0;$$

$$x'_2 = \gamma(x_2 - ut_2) = \gamma\left(L - \frac{u^2L}{c^2}\right) = L\gamma\left(1 - \frac{u^2}{c^2}\right) = \frac{L}{\gamma}. \tag{39–27}$$

The observer in frame F′ measures a length $L' = x'_2 - x'_1 = L/\gamma$, just the length contraction of Eq. (39–11).

CONCEPTUAL EXAMPLE 39–7 Suppose you wanted to use the Lorentz transformation to confirm the time dilation factor. What "events" would you want to measure and compare in two inertial frames, and how would you use the Lorentz transformations to make this comparison?

Answer The question concerns the time interval between ticks of a clock, and these successive ticks are the events. Our two frames are F, within which the clock is at rest, and F′ moving with velocity u along the x-direction. (With this arrangement, we need only keep track of the x-coordinate in space.) Our clock's ticks occur at times t_1 and t_2 in frame F and are the time coordinates of the events in F. We can always set $t_1 = 0$, and we shall do so here. We are also free to place the spatial origin in F at the clock's location in F—in other words, the two events are at $x_1 = x_2 = 0$. We want now to give the time coordinates t_1' and t_2' of the two events in the frame F′; the time interval $t_2' - t_1'$ in terms of $t_2 - t_1$ give the time dilation. But times t_1' and t_2' are given directly by the Lorentz transformations, in particular Eq. (39–20). (The Lorentz transformations assume that the origins of F and F′ coincide at the times $t = 0 = t'$. This is just a question of setting the origins of time and space and has no consequence for physical manifestations such as time dilation.)

The use described above of the Lorentz transformation, Eq. (39–20), is so simple that we can carry it through before leaving the example:

$$t_1' = \gamma\left(t_1 - \frac{ux_1}{c^2}\right) = 0; \qquad t_2' = \gamma\left(t_2 - \frac{ux_2}{c^2}\right) = \gamma t_2.$$

Thus $t_2' - t_1' = \gamma(t_2 - t_1)$. A tick in the moving frame is a factor γ longer than T, and we have confirmed the time-dilation factor.

What Do You Think? Suppose you have a clock that consists of a mass attached to a spring oriented along the x-axis and oscillating with some frequency. An observer moving with uniform speed along the x-axis will see the spring contracted. Does this explain the time-dilation effect?

CONCEPTUAL EXAMPLE 39–8 A perfectly smooth horizontal road has in it an open manhole 1 m in diameter. A stick of proper length 2 m is aligned with the road (Fig. 39–16a, which shows the stick and the manhole at rest with respect to one another). The stick slides along the road with a constant speed u such that its length in the rest frame of the manhole is 0.5 m. Someone working in the manhole therefore says that when the stick slides over the manhole, it will fall in (possibility 1, as in Fig. 39–16b). However, someone in the rest frame of the stick sees the manhole approaching with the same speed u and thus sees the length of the hole along the direction of motion to be only 25 cm. That observer says the stick won't fall in (possibility 2, as in Fig. 39–16c). How can these two points of view be reconciled? (Ignore the fact that the stick doesn't have time to fall very far if the acceleration of gravity were that of Earth's—you can always imagine that the acceleration of gravity is arbitrarily large.)

Answer The first point to observe is that the notion of a rigid object is not compatible with special relativity. If there were such a thing as a rigid stick, one could push one end of it and move something with the other end *simultaneously*, in violation of the requirement that no signal can move faster than the speed of light. The resolution of the two points of view follows from a careful examination of the simultaneity or nonsimultaneity of events. In the rest frame of the manhole, the front and rear ends of the stick are unsupported *at the same time* and the stick will indeed fit in. In the rest frame of the stick, the front end and rear ends of the stick are never simultaneously over the hole. In this frame the stick "snakes" into the hole, with the front entering first, and the rear following later, when the front end is already under the road. This distortion is possible because of our comments about rigidity made above.

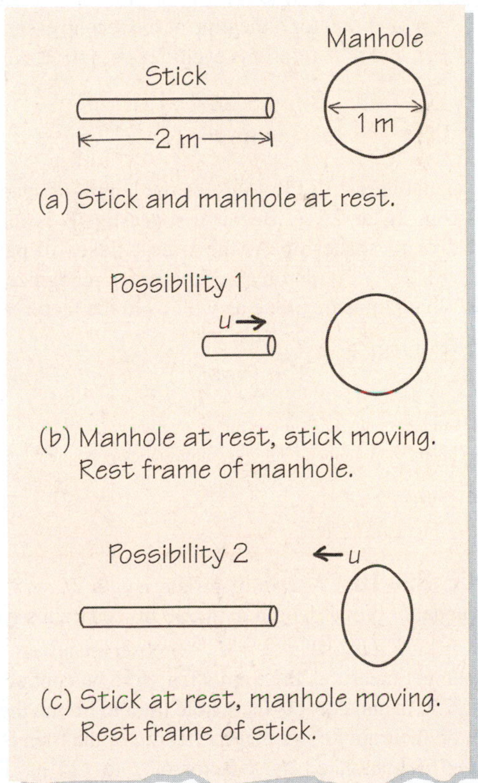

▲ **FIGURE 39–16** A stick sliding over a manhole cover presents us with a possible conflict typical of special relativity.

EXAMPLE 39–9 As seen from Earth, spaceship A of proper length L is traveling east at speed v_1 and spaceship B of proper length $2L$ is traveling west at speed v_2. The pilot of spaceship A sets his clock to zero when the front of spaceship B passes him. (The spaceship pilots sit in the nose cones.) Use the Lorentz transformations to calculate the time at which, according to the pilot of spaceship A, the tail of spaceship B passes him.

Setting It Up This involves the use of two frames and two events that must be precisely specified. We will take frame F to be the rest frame of spaceship A, with the origin at the pilot and coordinates (x, t). The frame in which spaceship B is at rest is frame F′, with (x', t') describing events and a spatial origin at the pilot.

(continues on next page)

This means that the front of spaceship A is at $x = 0$, and the back is at $x = -L$, as the sketch in Fig. 39–17 shows, while the front of spaceship B is at $x' = 0$ and its back is at $x' = +2L$. Finally we set the respective clocks so that when the nose cones of the two ships pass each other, $t = 0 = t'$.

Strategy We must first specify the events whose coordinates will be related by the Lorentz transformations. The first event is when the two nose cones pass each other, which occurs at $(x, t) = (0, 0)$ and $(x', t') = (0, 0)$. The second event is the coincidence of the back of spaceship B with the front of spaceship A. The coordinates of this event are $(x, t) = (0, t)$ and $(x', t') = (2L, t')$. The Lorentz transformation now allows us to find the time t at which the second event is seen in the frame A. In particular, we can use the transformation $x' = \gamma(x + ut)$ and solve it for t in terms of the known quantities x' and x. The plus sign appears here because spaceship B is moving west (leftward, or toward negative x) with respect to spaceship A. One last issue: The Lorentz transformation involves the *relative* velocity of the two frames, so that the transformations between the two frames contain a relative velocity u given by the relativistic addition of the two velocities v_1 and v_2.

Working It Out As described above, we solve the Lorentz transformation $x' = \gamma(x + ut)$ for t, the time of the second event in frame A, using $x' = 2L$ and $x = 0$ for this event: $2L = \gamma(0 + ut)$, or

$$t = \frac{1}{u}\frac{2L}{\gamma}.$$

This is a reasonable result: Pilot A sees the length of spaceship B contracted from $2L$ to $2L/\gamma$. Because spaceship B is moving at speed u relative to spaceship A, the time it takes to pass is its observed length $2L/\gamma$ divided by u. Finally we recognize that we must find u, which appears alone as well as in the factor γ. This is given by Eq. (39–18), $u = \dfrac{v_1 + v_2}{1 + (v_1 v_2/c^2)}$.

What Do You Think? What time does the pilot of B record for the second event? Is the answer what you expected?

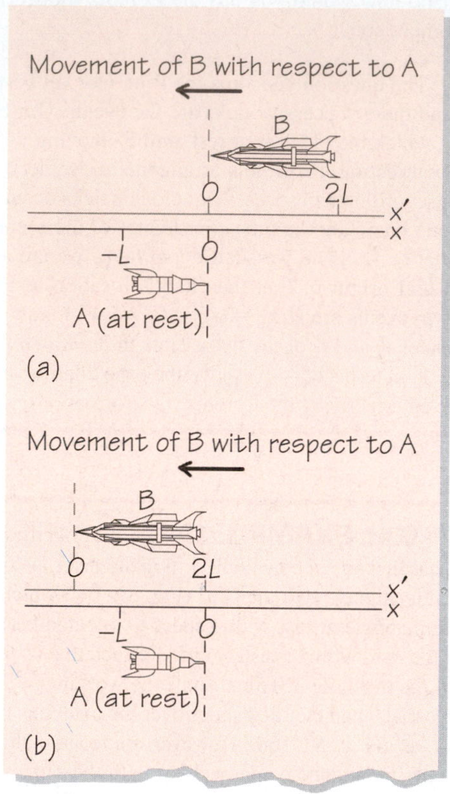

▲ **FIGURE 39–17** (a) The event specified by the passing of the fronts of spaceships A and B. (b) The event specified by the passing of the rear of spaceship B by the front of spaceship A. Both cases are from the viewpoint of an observer in spaceship A, so spaceship B is shortened.

EXAMPLE 39–10 A train of proper length $2L = 500$ m approaches a tunnel of proper length $L = 250$ m. The train's speed v is such that $\gamma = [1 - (v^2/c^2)]^{-1/2} = 2$. An observer at rest with respect to the tunnel measures the train's length to be contracted by a factor of 2 to 250 m and expects the whole train to fit into the tunnel. An observer on the train knows that the length of the train is 500 m, and the tunnel is contracted by a factor of 2 to 125 m. Thus the observer on the train argues that the train will not fit into the tunnel. Who is right?

Setting It Up We label two frames: frame F, the rest frame of the tunnel, and frame F′, the rest frame of the train, with the usual unprimed and primed coordinates. We set $x = 0$ at the front of the tunnel and $x' = 0$ at the front of the train. We start both clocks at 0 when the points $x = 0$ and $x' = 0$ coincide (i.e., when the front of the train enters the tunnel, Fig. 39–18a). The right side of the tunnel—the exit—is $x = L$ in frame F (Fig. 39–18b), while the rear of the train is $x' = -2L$ in frame F′ (Fig. 39–18c).

Strategy Conceptually this problem is similar to that of a rapidly sliding stick falling into a manhole. We recognize events, in this case the coincidence of the front of the train with the exit of the tunnel (event 1) and of the back of the train with the entrance of the tunnel (event 2). These two events are marked by arrows in Fig. 39–18b and c respectively, with Fig. 39–18a marking the entry of the front of the train. The event coordinates in space and time are:

event 1: $(x, t) = (L, t_1 = L/v)$ and $(x', t') = (0, t_1')$;

event 2: $(x, t) = (0, t_2)$ and $(x', t') = (-2L, t_2' = 2L/v)$.

We are able to find the times t_1 and t_2' without thinking about frame dependence. Time t_1 is the time in the tunnel frame for the train to reach the exit: The length of the tunnel is L in the tunnel frame, and the front of the train moves at speed v. Time t_2' is the time in the train frame for the back of the train to reach the entrance: The train length is $2L$ in the train frame, and the tunnel rushes by at speed v. We are left with two unknowns here,—t_1', the time for the front of the train to reach the entrance according to the train-based observer, and t_2, the time that the rear of the train reaches the entrance according to the tunnel-based observer. We can use the Lorentz transformations to find out these times. In doing so we will check that the train does indeed fit into the tunnel as measured in frame F, while it does not as measured in frame F′. The manner in which this happens will resolve the apparent paradox.

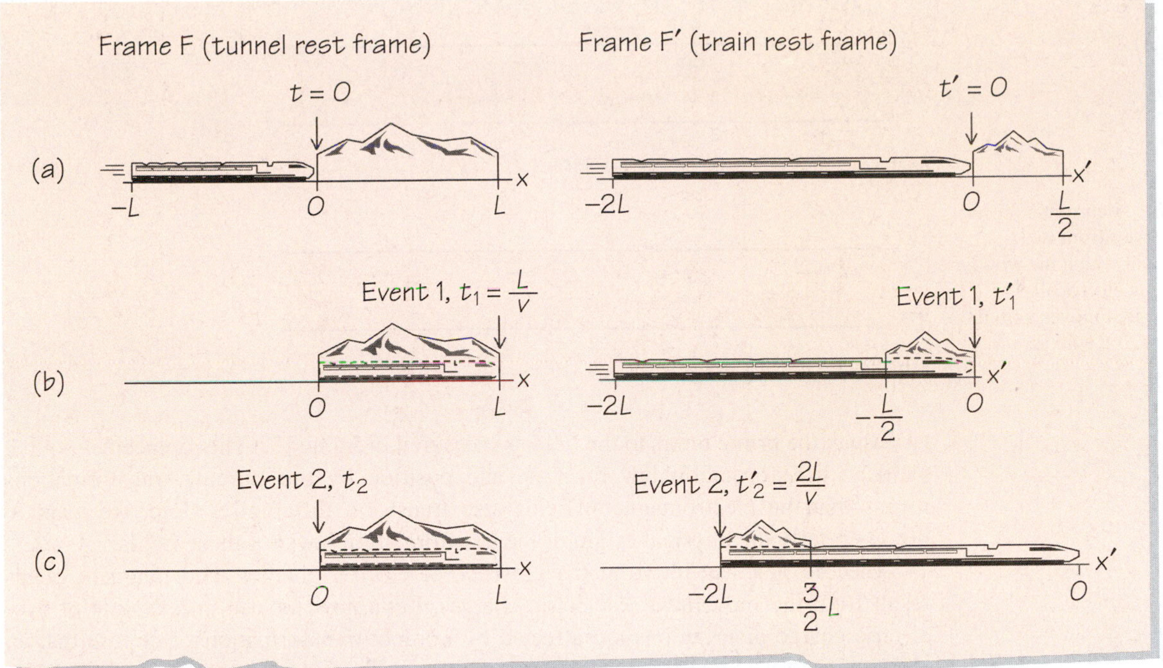

▲ **FIGURE 39–18** (a) The unnamed event specified by the coincidence of the front of the train with the tunnel entry—this event sets origins of space and time. (b), (c) Two events associated with the passage of a very fast train through a tunnel: event 1, the coincidence of the front of the train first with the right end (exit) of the tunnel, and event 2, the coincidence of the back of the train first with the left end (entry) of the tunnel. Both parts show the events according to two observers, one in the rest frame of the tunnel and another at rest in the train.

Working It Out Let's first find time t_2 corresponding to event 2. It follows from the inverse transformation Eq. (39–26) that

$$t_2 = \gamma\left(t_2' + \frac{vx_2'}{c^2}\right) = \gamma\left(\frac{2L}{v} + \frac{v(-2L)}{c^2}\right) = \gamma\frac{2L}{v}\left(1 - \frac{v^2}{c^2}\right) = \frac{2L}{v\gamma},$$

where in the last step we have recognized that $(1 - v^2/c^2) = 1/\gamma^2$. But in this problem $\gamma = 2$, so the time $t_2 = L/v$: In the tunnel frame the back of the train is at the tunnel entrance *at exactly the same time that the front of the train is at the tunnel exit*. The front and back are at the two ends of the tunnel *simultaneously* in the tunnel frame, and this is what we mean by the train fitting the tunnel (Figs. 39–18b and c).

We find the second unknown, t_1', the time in the train frame that the front of the train is at the exit (event 1), from the Lorentz transformation Eq. (39–20),

$$t_1' = \gamma\left(t_1 - \frac{vx_1}{c^2}\right) = \gamma\left(\frac{L}{v} - \frac{vL}{c^2}\right) = \gamma\frac{L}{v}\left(1 - \frac{v^2}{c^2}\right) = \frac{L}{v\gamma}.$$

But this time is not t_2', the time in the train frame when the back of the train is at the tunnel entrance—for $\gamma = 2$, it is exactly one quarter that value (Fig. 39–18b). In other words, for the observer in the train, only one quarter of the train will be in the tunnel when the train's front end is at the exit, and it will be some time after the front of the train has left the tunnel exit before the rear of the train reaches the tunnel entrance—the tunnel is shorter than the train to the train-based observer.

There is no contradiction. The differences between the interpretations of the two observers stem from their different notions of simultaneity. These differences allow both observers to be correct in their claims!

What Do You Think? Is all this consistent with our nonrelativistic world?

Lorentz Transformations of Electric and Magnetic Fields

Maxwell's equations predict the speed of light. Therefore, electric and magnetic fields play a distinct role in special relativity. In Section 30–6, we saw that in order to maintain Galilean invariance (invariance under the transformation laws $\vec{r}' = \vec{r} - \vec{u}t$ and $t' = t$) of the Lorentz force equation, the electric and magnetic fields must mix among themselves when we observe them from different reference frames. In Eqs. (30–20) we derived Galilean transformation laws for these fields:

$$\vec{E}' = \vec{E} + (\vec{u} \times \vec{B})$$

$$\vec{B}' = \vec{B}.$$

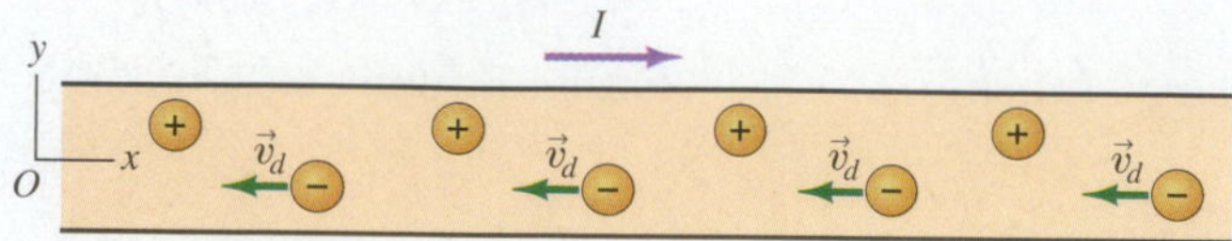

(a) Frame F: Positive charges (wire) at rest

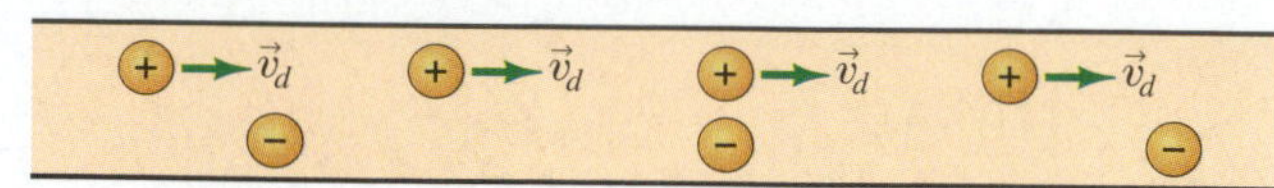

(b) Frame F′: Negative charges at rest

FIGURE 39–19 A current-carrying wire as observed from two frames. (a) The frame in which the wire is at rest. Here, the wire is electrically neutral. (b) The frame that moves with the negative charges. Here, the wire acquires a net charge density.

(As usual, the prime refers to the field as measured in frame F′.) The replacement of the Galilean transformation law for time and position by the Lorentz transformations means that the electromagnetic fields also transform differently. Here, we want to discuss briefly the physical origin of the transformation laws for these fields.

Electric fields result from the presence of electric charges, and magnetic fields result from the movement of electric charges. Let's now look at an example of how electric charge distributions are affected by Lorentz transformations. Suppose that an electrically neutral wire carries a current in the $+x$-direction in some frame F. The current consists of negative charges that move at some drift speed v_d in the $-x$-direction against a background of stationary positive charges with the same spacing (Fig. 39–19a). This wire produces a magnetic field but no electric field.

Now let's consider the same wire as observed from a frame F′ that moves at the drift velocity $\vec{v}_d$ (Fig. 39–19b). An observer in frame F′ would see the electrons in the wire at rest and the positive charges (the ions) move in the $+x$-direction. And there is a surprise: Special relativity implies that *the wire is not electrically neutral in frame* F′. Because of the Lorentz contraction, an observer in frame F would measure the electrons to have less space between them than the F′ observer does. In other words, the F′ observer measures *more* space between the electrons than does the F observer. Moreover, the F′ observer measures *less* space between the positive ions than does the F observer: To the F′ observer, the positive ions are moving. Thus if the F observer sees the same spacing between electrons as between positive ions—a neutral wire—the F′ observer will not, and the wire will no longer be electrically neutral to the F′ observer. The F′ observer sees an electric field.

We have thus seen that the space and time aspects of special relativity imply that the presence of a magnetic field alone in one frame introduces an electric field in another. In other words, an observer in one frame sees a different combination of electric and magnetic fields than an observer in another frame. It is in this way that transformation laws between fields come about. A careful quantitative analysis of situations such as the one just described gives the full set of Lorentz transformations between electric and magnetic fields. These transformation laws, together with the Lorentz transformations for space and time, leave the physical consequences of Maxwell's equations invariant. Because the speed of light is one of the physical consequences of Maxwell's equations, that speed is the same in all frames. We have come full circle with a consistent description.

39–7 Momentum and Energy in Special Relativity

Momentum

The need to modify our notions of space and time suggests that the definitions of other kinematical quantities that are based on measurements in space and time also require modification. In nonrelativistic mechanics, the *momentum* of a particle that moves with velocity $\vec{v}$ is

$$\vec{p} = m\vec{v}.$$

We sometimes call the coefficient of $\vec{v}$ in this expression the *rest mass m*. In the absence of external forces, the sum of the momenta of interacting particles is constant; that is, the total momentum is conserved:

$$\sum \vec{p}_i = \vec{P} = \text{a constant.}$$

Momentum conservation has its origin in Newton's third law and is a principle that holds both nonrelativistically and relativistically. However, as we shall now see, the momentum takes a different form when relativity is taken into account.

Momentum in relativity is a quantity ascribed to moving particles and has the following properties: (a) In the absence of external forces, the sum of momenta of interacting particles is conserved, and (b) in the limit that $\vec{v} \to 0$, $\vec{p} \to m\vec{v}$. On purely dimensional grounds, we expect that

$$\vec{p} = mf(v)\vec{v},$$

where the function $f(v)$ must be 1 for $v = 0$ and $f(v)$ is dimensionless. The function $f(v)$ depends only on the magnitude of $\vec{v}$, so $f(v)$ must be a function of v^2. Because f is dimensionless, it must be a function of v^2/c^2 or, equivalently, of the now familiar combination $\gamma = 1/\sqrt{1 - (v^2/c^2)}$.

An analysis of collisions between equal-mass particles—an analysis too long to present here—leads to the result that $f(v) = \gamma$, giving us the relativistic momentum

$$\vec{p} = m\gamma\vec{v} = \frac{m\vec{v}}{\sqrt{1 - (v^2/c^2)}}. \tag{39–28}$$

RELATIVISTIC MOMENTUM

For $v/c \ll 1$, this reduces to the familiar low-velocity result $\vec{p} = m\vec{v}$. Newton's second law now reads

$$\vec{F} = \frac{d\vec{p}}{dt} = m\frac{d}{dt}(\gamma\vec{v}). \tag{39–29}$$

One consequence of the relativistic modification of the expression for momentum is that $\vec{F}$ and $d\vec{v}/dt$ no longer have to point in the same direction (see Problem 65).

Kinetic Energy

The work-energy theorem together with Eq. (39–29) leads to the relativistic expression for the kinetic energy K: The work–energy theorem tells us that the work done to bring a particle of mass m from rest to speed v will be the kinetic energy of the particle. The result of this calculation is that

$$K = mc^2(\gamma - 1) = mc^2\left(\frac{1}{\sqrt{1 - (v^2/c^2)}} - 1\right). \tag{39–30}$$

As we already pointed out in Section 6–6, $[1 - (v^2/c^2)]^{-1/2} \cong 1 + (v^2/2c^2)$ for small v^2/c^2, so that in the low-velocity limit K reduces to the familiar $mv^2/2$.

To derive the result above from the work–energy theorem, we must evaluate

$$K = \int F\,dx = \int \frac{d(mv\gamma)}{dt}\frac{dx}{dt}\,dt = \int \frac{d(mv\gamma)}{dt}v\,dt.$$

The evaluation of the integral is long but straightforward and yields Eq. (39–30).

Energy Associated with Mass

Einstein first showed that energy and inertial mass are closely related. A thought experiment demonstrates this fact. Consider a railroad car of mass M and length L standing on rails. Imagine a flashbulb attached to the left interior wall of the car (Fig. 39–20a). At a particular time, the bulb emits a burst of light toward the right wall of the railroad car. As we learned in Chapter 34, if the energy of the light pulse is E, then there is momentum of magnitude E/c associated with the pulse. Momentum conservation implies that the railroad car must move with an equal and opposite momentum toward the left (Fig. 39–20b). Because the mass of the car is M, the car will move with a velocity such that

$$Mv = E/c.$$

The time t that the pulse spends between the walls is given by

$$ct = L - vt.$$

The right side is less than L because the right wall is now moving toward the light pulse. Solving for t, we find

$$t = \frac{L}{c + v} = \frac{L}{c + (E/Mc)}.$$

The distance traveled by the car in that time is

$$D = vt = \left(\frac{E}{Mc}\right)\frac{L}{c + (E/Mc)} = \frac{EL}{E + Mc^2}. \tag{39–31}$$

Finally the car comes to a stop after it has moved a distance D when the light along with its momentum is absorbed by the right wall of the car.

If momentum is to be conserved, the center of mass of the railroad car with the flashing bulb must not move. Yet the car has moved to the left. From this, we must infer that the energy in the light is equivalent to a mass μ carried by the light flash and that this mass moves to the right when the car moves to the left such that the center of mass remains stationary. We can find μ by equating the position X of the overall center of mass before and after the flash event. Place the initial position of the left-hand side of

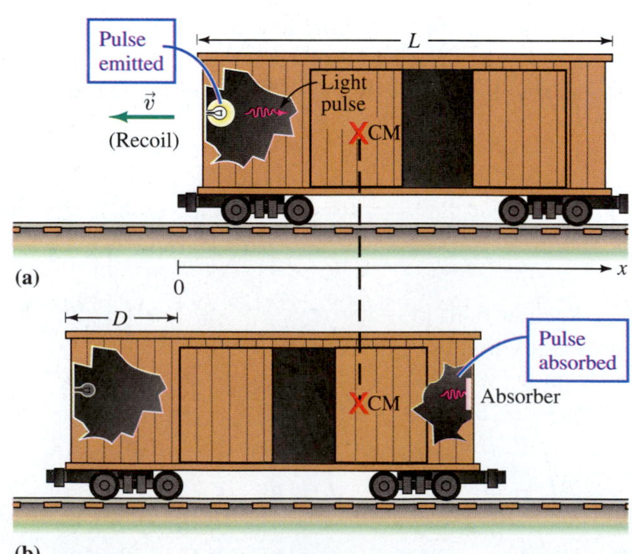

FIGURE 39–20 (a) A light source in a railroad car emits an electromagnetic pulse that carries energy E. The recoil momentum $p = E/c$ sets the car in motion. (b) This motion ceases when the pulse is absorbed at the other end of the car. (The displacement of the car is greatly exaggerated here.)

the car at $x = 0$. The car can be treated as a point mass M, initially at $x = L/2$. Then, before the light is emitted,

$$X = \frac{(\mu)(0) + (M)(L/2)}{\mu + M} = \frac{M}{\mu + M}\frac{L}{2}.$$

After the light has been absorbed at the right-hand side, at position $x = L - D$,

$$X = \frac{\mu(L - D) + M[(L/2) - D]}{\mu + M}.$$

We equate these and solve for the mass μ:

$$\mu = \frac{MD}{L - D}.$$

From Eq. (39–31), $L - D = LMc^2/(E + Mc^2)$; when we substitute for $L - D$ as well as for D in the equation for μ, we get

$$\mu = \frac{MEL/(E + Mc^2)}{LMc^2/(E + Mc^2)} = \frac{E}{c^2}.$$

E is therefore equivalent to a mass μ given by

$$E = \mu c^2. \quad (39\text{–}32)$$

MASS-ENERGY EQUIVALENCE

This very important result shows that mass and energy are interchangeable concepts. We can reverse the reasoning here to say that if an object has a mass m, then it has an energy that we refer to as the **rest energy** $E = mc^2$. Light has no mass in the usual sense, but any energy it has is equivalent to a mass E/c^2. The total energy E of any object is now the sum of the kinetic energy, the rest energy (or mass energy, E_{mass}), and the potential energy. For a particle on which no forces act, there is no potential energy, and

$$E = E_{\text{mass}} + K = \frac{mc^2}{\sqrt{1 - (v^2/c^2)}} = \gamma mc^2. \quad (39\text{–}33)$$

It follows from this result and from Eq. (39–28) that

$$\vec{v} = \frac{c^2\vec{p}}{E}. \quad (39\text{–}34)$$

RELATIVISTIC EXPRESSION FOR VELOCITY

The mass-energy equivalence has immediate experimental consequences, and it has been tested innumerable times in a large variety of nuclear reactions (see Example 39–11). Perhaps the most dramatic confirmation of this law came with the discovery of *antimatter*. Quantum mechanics and relativity together show that, for each particle, there is a corresponding *antiparticle*. Particles and antiparticles each have the same mass, while the antiparticle has a charge opposite that of the particle; an antiparticle and a particle can annihilate each other to produce electromagnetic radiation, all their mass going into the energy of the radiation (Fig. 39–21). Similarly, a particle-antiparticle pair, complete with their masses, can be created out of radiation energy alone. The conversion of energy (in the form of radiation) into mass is thereby exhibited unambiguously. We'll revisit this issue in our study of nuclei and elementary particles in Chapters 44 and 45.

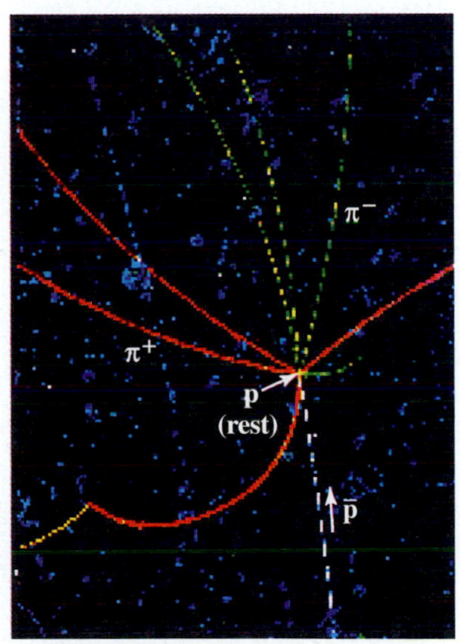

▲ **FIGURE 39–21** In this color-enhanced bubble-chamber photo, an incoming antiproton (light blue) strikes and annihilates with a proton at rest, producing 4 positive pions (red) and 4 negative pions (green), which are antiparticles of the positive pions.

EXAMPLE 39–11 The nucleus ^{8}Be is an unstable isotope of beryllium. It decays into two alpha particles (helium nuclei), $^8\text{Be} \to {}^4\text{He} + {}^4\text{He}$. The masses of these nuclei in atomic mass units u are $M(^8\text{Be}) = 8.005305$ u and $M(^4\text{He}) = 4.002603$ u. Assume that a nucleus ^{8}Be decays while it is at rest. Find the kinetic energy of the helium nuclei (the alpha particles) in MeV.

Strategy The basic principles that apply here are energy and momentum conservation. The energy must include the mass energy—you can see from the numbers above $2M(^4\text{He}) \neq M(^8\text{Be})$, so that the mass energies do not cancel from the energy conservation equation.

We can start with the conservation of momentum, whose consequence is very simple here: The initial momentum is zero, so that the two (identical) helium nuclei must go off back-to-back, with the same momentum magnitude and hence the same kinetic energy K. This observation tells us that the conservation of energy equation can be used to find the single remaining unknown K.

We now turn to energy conservation. The initial energy is that of the beryllium nucleus, and since it is at rest, this energy is $E_{\text{initial}} = M(^8\text{Be})c^2$. The final energy is the rest energy of the two helium nuclei together with the total kinetic energy in the final state, so that $E_{\text{final}} = 2M(^4\text{He})c^2 + 2K$. From the equality of the initial and final energies we can get K.

Working It Out $2K = [M(^8\text{Be}) - 2M(^4\text{He})] = [8.005305\text{ u} - 2(4.002603\text{ u})]c^2 = (0.000099\text{ u})c^2$.

The conversion $1\text{ u} = 931.5\text{ MeV}/c^2$ shows that each helium nucleus carries kinetic energy

$$K = \frac{1}{2}(0.000099\text{ u})\left(\frac{931.5\text{ MeV}/c^2}{\text{u}}\right)c^2 = 0.046\text{ MeV}.$$

The fact that the masses of the two alpha particles are less than the mass of the parent ^{8}Be nucleus allows the alpha particles to have nonzero kinetic energy. The overall "mass defect" is the total kinetic energy in the final state, $2K$, also referred to by specialists in nuclear reactions as the *Q-value* of the reaction.

What Do You Think? Suppose the two decay products were some other nuclei whose masses were not equal. Could one still calculate the separate kinetic energies?

EXAMPLE 39–12 A 1.0-kg meteorite of antimatter strikes Earth. (This is just a thought experiment; there is no evidence at all of antimatter in such large lumps.) How much energy is liberated in the annihilation process in which all the antimatter and an equal amount of matter are converted to radiant energy? Neglect the kinetic energy of the meteorite; $(v/c) \ll 1$.

Strategy The calculation is a simple matter of conversion. The rest energy of mass M is Mc^2. The antimatter interacts with an equal amount of matter to annihilate, so the amount of energy liberated is $2Mc^2$.

Working It Out

$$E = 2Mc^2 = 2(1.0\text{ kg})(3 \times 10^8\text{ m/s})^2 = 1.8 \times 10^{17}\text{ J}.$$

To gain some idea of the significance of this number, the amount of (chemical) energy in 1 ton of TNT is 4.2×10^9 J. The meteorite explosion would generate the equivalent of 4×10^7 tons of TNT or about 40 hydrogen bombs. Antimatter–matter annihilation has been suggested as a source of fuel for manned planetary journeys because it is the most efficient fuel possible.

What Do You Think? Can you think of reasons why this proposal for interplanetary travel is at this time technologically impossible?

The Relation Between the Momentum and Energy of a Particle

In nonrelativistic physics there is a direct relation between the momentum and energy of a moving particle, namely $E = p^2/2m$. This expression does not include the rest energy. The correct relativistic analogue, which must reduce to the nonrelativistic case for speeds much less than that of light, is of some importance. From Eq. (39–33) we have

$$E^2 = \frac{m^2c^4}{1 - (v^2/c^2)}.$$

Moreover, Eq. (39–28) gives us an expression for p^2c^2, a quantity with the same dimension as E^2:

$$p^2c^2 = \frac{m^2v^2c^2}{1 - (v^2/c^2)}.$$

The speed does not appear in the difference between these results,

$$E^2 - p^2c^2 = \frac{m^2c^4}{1 - (v^2/c^2)}\left(1 - \frac{v^2}{c^2}\right) = m^2c^4. \tag{39–35}$$

This relation, or its square root, gives us the desired relation between E and p:

$$E = \sqrt{p^2c^2 + m^2c^4}. \tag{39–36}$$

RELATIVISTIC ENERGY-MOMENTUM RELATION

This relation helps us understand the velocity of particles. From Eq. (39–34) we have

$$v = \frac{c^2 p}{E} = c\frac{\sqrt{E^2 - (mc^2)^2}}{E} = c\sqrt{1 - \left(\frac{mc^2}{E}\right)^2}. \tag{39–37}$$

For light, $m = 0$ and therefore $v = c$. The lightest particles we know about are *neutrinos*, symbol ν, particles whose mass is of the order of one millionth of that of an electron, $m_\nu c^2 < 2$ eV. (While we do know that the neutrino has a mass, we do not yet know the actual mass, only that it must be less than this value. See Chapter 45 for more.) Thus in any nuclear reaction, which always involve energies at least of the order of kilovolts, the square root factor in Eq. (39–37) can be replaced by unity. Neutrinos always move at very nearly the speed of light.

The energy momentum relation, Eq. (39–36), also shows that the combination $E^2 - p^2c^2$ is *invariant*. Invariant quantities have the same value in every inertial frame, just as $c^2t^2 - x^2$ does [see Eq. (39–24)]. This is a valuable tool in the analysis of relativistic collision phenomena.

THINK ABOUT THIS. . .

IS THE ELECTRON CHARGE AN INVARIANT?

The electron mass is an invariant. We also know that electric charge is an invariant—by this we mean that the total electric charge is measured to have the same value in any inertial frame. What is the evidence? An electrically neutral piece of metal contains positive ions and some nearly free electrons in addition to a lot of atoms in which the electrons are closely attached to the ions. Suppose we heat up that piece of metal. We know from equipartition that the constituents of the metal increase their kinetic energy. Electrons are thousands of times less massive than ions, so that for a given increase in kinetic energy their velocities increase much more than those of the ions. If the electric charge were to increase, for example, as a function of velocity, then a piece of metal that was neutral at one temperature would be charged at another. This is an effect that could be easily measured, and there is no evidence of a departure from neutrality as temperature changes. Finally, you may wonder if this is consistent with our remarks about how a current seen from a moving frame appears to have charge—we used this argument to motivate the mixing of electric and magnetic fields in Section 39–6. But there we spoke about the density of charge, not a total charge, and it is not surprising that charge density, a charge per unit volume, is not an invariant, because the volume is not an invariant.

*39–8 Beyond Special Relativity

The Equivalence Principle

Special relativity expresses the physical equivalence of all inertial reference frames. In noninertial, or accelerating, frames, another physical equivalence holds. It is expressed by the *equivalence principle*, formulated by Einstein in 1911:

Provided that the observations take place in a small region of space and time, it is not possible by experiment to distinguish between an accelerating frame and an inertial frame in a suitably chosen gravitational potential.

This is the principle behind *general relativity*, also known as Einstein's theory of gravitation. We described a number of consequences of the equivalence principle in Section 12–8. We list these again:

A. The Equality of Gravitational Mass and Inertial Mass: The *gravitational mass* m_g, the attribute of an object that appears in Newton's expression for the gravitational force, of magnitude

$$F = -\frac{Gm_{1g}m_{2g}}{r^2}$$

and the *inertial mass* m_i, the attribute of an object that appears in the expression for the proportionality of force and acceleration,

$$F = m_i a,$$

must be equal. As a consequence, all objects fall at equal rates in a given gravitational field: If an object of mass m is subject to a gravitational force due to an object of mass M, the relation $F = ma$ reads

$$\frac{Gm_g M}{r^2} = m_i a;$$

if $m_i = m_g$, the acceleration does not depend on the mass of the object. The equality of the inertial mass and gravitational mass has been verified experimentally to an accuracy of one part in 10^{11}.

CONCEPTUAL EXAMPLE 39–13 An automobile at a stop light, with all its windows closed, contains a balloon filled with helium floating over the back seat. The light changes and the automobile accelerates. Will the balloon move relative to the back seat, and if so, in which direction?

Answer This can be answered by a simple application of the equivalence principle. The forward acceleration of the automobile is experienced by the driver as a pressure on her back. Just as the effect of an upwardly accelerating elevator is indistinguishable from the gravitational force due to a mass below one's feet, the effect on everything and everyone within the automobile is indistinguishable from the gravitational force due to a mass behind the car. Effectively the acceleration is equivalent to the sudden placement of a planet behind the car. Just as near Earth's surface helium balloons rise as a result of their buoyancy, here the balloon moves forward with respect to the back seat!

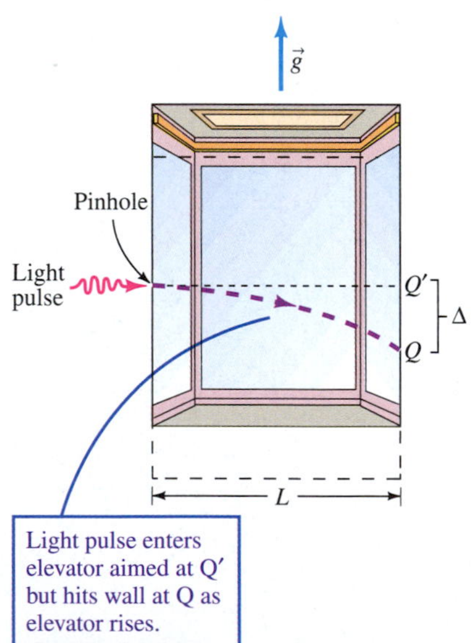

▲ **FIGURE 39–22** A pulse of light is directed into an elevator. If the elevator is at rest, the pulse would eventually arrive at point Q'. If the elevator accelerates upward with magnitude g as the light pulse crosses, an observer within the elevator would see the light pulse follow the parabolic path shown, eventually arriving at point Q on the opposite wall, a distance Δ below Q'.

B. The Gravitational Deflection of Light: If a horizontal beam of light enters a pinhole in an elevator that is accelerating upward with an acceleration g, then, in the time t that it takes for the light to cross the elevator, the horizontal level of the pinhole will have moved upward by a distance $gt^2/2$. If the width of the elevator is L, the light will hit a spot Q that is a distance

$$\Delta = \frac{1}{2}g\left(\frac{L}{c}\right)^2$$

below the spot Q', which is horizontally across from the pinhole; indeed, if you followed the light path you would see a parabola (Fig. 39–22). In other words, an observer within the elevator sees the light *falling*. According to the equivalence principle, no experiment can determine whether the elevator has accelerated upward or has remained at rest within the influence of a gravitational force. Therefore light must fall under the influence of the gravitational force due to a mass. Note that Δ is the same distance that any massive body would fall. The implications are twofold:

1. Everything that has energy falls downward with the same acceleration as a result of the influence of a gravitational force.
2. Light is deflected toward a star as it passes it, a phenomenon whose consequences are discussed in Chapter 12 (Fig. 39–23).

C. The Gravitational Redshift: When light "falls," it undergoes a frequency shift. Consider a source at a height x above the ground in the presence of local gravity,

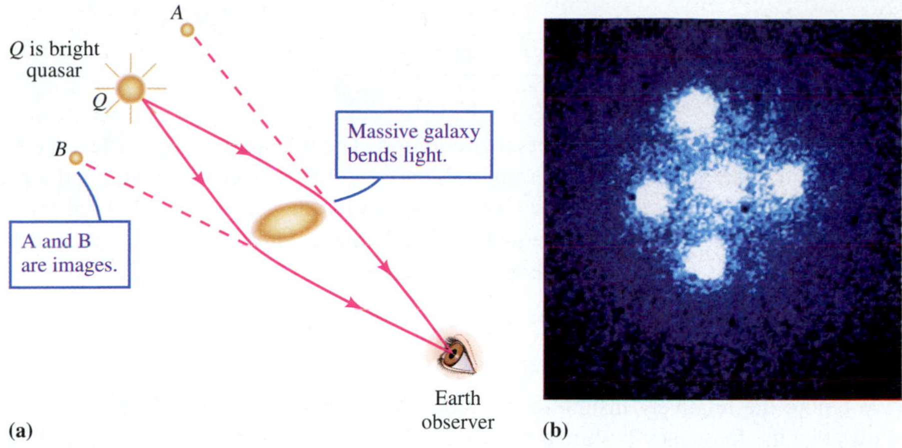

FIGURE 39–23 (a) The formation of an image by a gravitational lens. If a light ray "bends" toward mass, we should see the effect as additional images in the sky. (b) A mass (a relatively nearby galaxy) has bent the light from a quasar behind it such that four distinct images of the quasar can be observed from Earth. The fuzzy spot in the center is the galaxy's core.

emitting radiation with frequency f (Fig. 39–24a). According to the equivalence principle, the physics should be described equally well by an observer who sees the system being accelerated upward with acceleration g in empty space. If both the source and the detector are at rest at the time of emission, then, according to the observer who sees the entire apparatus accelerating upward, in the time that the radiation has reached the detector ($t = x/c$), the detector will have acquired an upward velocity of magnitude $v = gt = gx/c$ (Fig. 39–24b). Thus the detector sees the radiation with a frequency f', Doppler shifted upward from the emitting frequency f. These are related by

$$\frac{f'}{f} = \sqrt{\frac{1 + (v/c)}{1 - (v/c)}} \cong 1 + \frac{v}{c} = 1 + \frac{gx}{c^2}.$$

The factor gx may be regarded as a *gravitational potential*, ϕ: We express the potential energy of a mass m in the vicinity of Earth as $mgx = m\phi$. In this way, the effect of Earth is isolated in the factor $\phi = gx$. Our result for the frequency ratio then takes the form

$$\frac{f'}{f} = 1 + \frac{\phi}{c^2}, \tag{39–38}$$

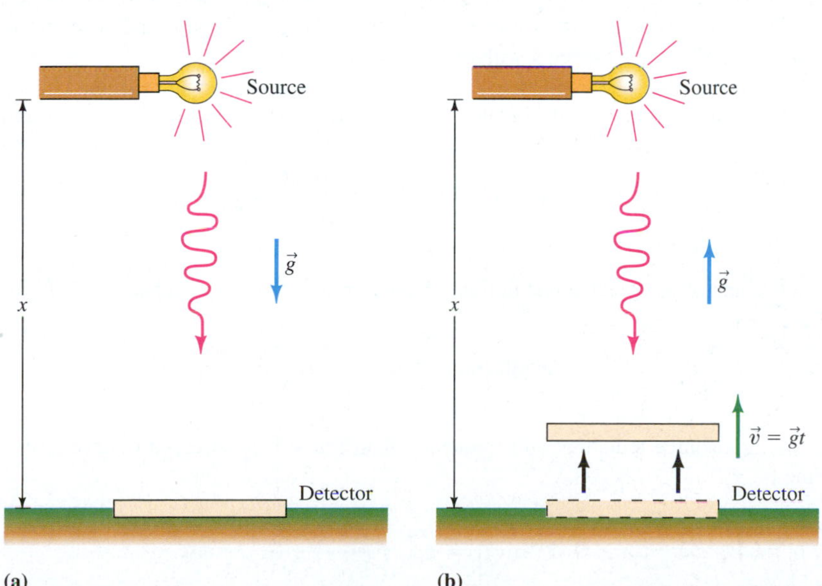

FIGURE 39–24 (a) An apparatus to measure the frequency of emitted light on Earth, in Earth's gravitational field; $\vec{g}$ is the (downward) acceleration due to gravity. (b) The measurement of a Doppler shift due to the acceleration of the detector (as part of the whole system) is indistinguishable from what is observed when the apparatus is in a gravitational potential; $\vec{g}$ is the (upward) acceleration of the system.

so

$$\frac{\Delta f}{f} \equiv \frac{f' - f}{f} = \frac{\phi}{c^2}. \tag{39–39}$$

Let's now apply our result to universal gravitation. The potential energy difference for a mass m and a star (of mass M and radius R) between the surface of the star and a point far from the star is $-GmM/R$. Thus we can assign to the star a corresponding gravitational potential of $-GM/R$. The frequency of light that is emitted from the star's surface and arrives at an Earth-based telescope is shifted by

$$\frac{\Delta f}{f} = -\frac{GM}{Rc^2}.$$

(We ignore the relatively insignificant "fall" down to Earth's surface.) The minus sign shows that the frequency is shifted downward. Thus the wavelength shifts upward, indicating a *gravitational redshift*. For the Sun, $M = 2 \times 10^{30}$ kg and $R = 7 \times 10^8$ m, so $\Delta f/f = 2 \times 10^{-6}$. Measurements of some characteristic spectral lines of sodium in the solar spectrum have confirmed this to an accuracy of 5 percent.

In a terrestrial measurement of the gravitational redshift carried out in 1960 by Robert Pound and Glen Rebka, light was "dropped" from a tower at Harvard University. This experiment measured a fractional shift of 3.3×10^{-15} to an accuracy of 1 percent.

When the gravitational potential is very large, the frequency shift $\Delta f/f$ becomes large. In the extreme case that $\Delta f/f = 1$, the frequency of the light is shifted to zero; that is, no light can be seen. A *black hole* has been formed. This occurs when $(GM/Rc^2) > 1$. When Einstein's full theory of gravitation is taken into account, it is for the modified condition $GM/Rc^2 > \frac{1}{2}$ that a black hole is formed. Although when they were first proposed, black holes may have seemed at most an abstract possibility, they turn out to be ubiquitous in the universe. The center of our galaxy, the Milky Way, for example, almost surely contains a very large one.

Summary

The special theory of relativity is based on the postulates that

1. The laws of physics are the same in all inertial reference frames.
2. The speed of light in empty space is the same in all inertial frames.

The first postulate generalizes the notion of Galilean invariance that we applied to the laws of mechanics. The second postulate is justified by the experimental result of Michelson and Morley.

These postulates require a rethinking of the concepts of space and time. If two inertial frames F and F′ move at speed u with respect to each other, then:

Two events that are simultaneous in frame F are not simultaneous in frame F′.

A time interval measured as T on a "clock" at rest in frame F is given by T' in frame F′:

$$\text{time dilation: } T' = \frac{T}{\sqrt{1 - (u^2/c^2)}}. \tag{39–9}$$

A length of an object at rest in frame F measured as L in frame F has length L' in frame F′:

$$\text{length contraction: } L' = L\sqrt{1 - \frac{u^2}{c^2}}. \tag{39–11}$$

A light source radiating with frequency f_0 in frame F is observed to have frequency f_1 in frame F′:

$$f_1 = f_0\sqrt{\frac{1 - (u/c)}{1 + (u/c)}}. \tag{39–16a}$$

If two objects are moving at velocities v_1 and $-v_2$, respectively, with respect to an observer, then the velocity of one object as seen in the rest frame of the other is

$$\text{law of addition of velocities: } V = \frac{v_1 + v_2}{1 + (v_1 v_2/c^2)}. \tag{39–18}$$

The relativistic Doppler shift for light has important astronomical implications. The measurement of the redshift of the light from a galaxy, given by Eq. (39–16), leads to a value for the speed u with which that galaxy recedes from Earth. That speed is in turn related to the distance D of the galaxy by Hubble's law:

$$D = u/H; \tag{39–17}$$

here H is the Hubble parameter, a number characteristic of the age of the universe, around 13 billion years.

The space–time coordinates of an event are described in reference frames F and F′ by (x, t) and (x', t'), respectively. These coordinates are related by the Lorentz transformations

$$x' = \gamma(x - ut) \tag{39–19}$$

and

$$t' = \gamma\left(t - \frac{ux}{c^2}\right), \tag{39–20}$$

where $\gamma \equiv 1/\sqrt{1 - (u^2/c^2)}$. An event can be specified by any two of (x, t, x', t'), and the other two coordinates can then be found from the Lorentz transformation laws, which must be supplemented by $y' = y$ and $z' = z$ when the relative motion is along the x-axis.

The relativistic momentum is defined by

$$\vec{p} = \frac{m\vec{v}}{\sqrt{1 - (v^2/c^2)}}. \tag{39–28}$$

The general relation between energy and inertial mass μ is $E = \mu c^2$, Eq. (39–32). The total energy of a particle is

$$E = \frac{mc^2}{\sqrt{1 - (v^2/c^2)}}, \tag{39–33}$$

and consists of a rest energy associated with the mass of the particle, $E_{\text{mass}} = mc^2$, plus the relativistic kinetic energy,

$$K = mc^2\left[\frac{1}{\sqrt{1 - (v^2/c^2)}} - 1\right]. \tag{39–30}$$

The velocity of a particle is given in terms of the momentum and energy according to

$$\vec{v} = \frac{c^2\vec{p}}{E}.$$

The energy and momentum are related by

$$E = \sqrt{p^2c^2 + m^2c^4}. \tag{39–36}$$

Einstein extended his theory beyond inertial reference frames with the equivalence principle, according to which it is not possible by experiment to distinguish between an accelerated frame and an inertial frame in a suitably chosen gravitational potential, provided that the observations take place in a small region of space and time. This principle has some important consequences, including:

Inertial and gravitational masses are equal, a result known to be accurate to one part in 10^{11}.

Light falls in a gravitational field.

A source radiating with frequency f will be observed in a gravitational potential ϕ at a frequency f' such that

$$\frac{f'}{f} = 1 + \frac{\phi}{c^2}. \tag{39–38}$$

Understanding the Concepts

1. Can there be such a thing as a perfectly rigid object?
2. Does the statement "moving clocks run slow" depend on the direction in which a clock is moving?
3. The rest mass of a proton is given as 937 MeV/c^2. How can a mass involve energy units?
4. Describe the constituents of an antiatom of ^{4}He.
5. Two events occur simultaneously at the same position in space in a given inertial frame of reference. Is it possible that the events will not occur simultaneously in another inertial frame of reference?
6. As measured from Earth, what is the shortest possible travel time between Earth and Alpha Centauri, the second nearest star system to us, which is located 4.3 ly away? Why can it not be made any shorter?
7. Suppose that the Michelson–Morley experiment were carried out over one arbitrarily short time period, much less than 1 day, and showed no sign of movement through an ether. Is this result enough to rule out the presence of an ether?
8. According to one solution proposed to make the presence of an ether consistent with the results of the Michelson–Morley experiment, there is an *ether drag*: For some reason, Earth carries a bubble of the ether with it as it moves through space. Can you think of experimental consequences that could be used to rule out such an idea?
9. If a mirror recedes at speed $0.75c$ from a light source, does the image recede at speed $1.5c$ from the source?
10. By looking at a very distant quasar some years ago, astronomers found that the separation distances of certain peaks in brightness increase at a rate of 0.2 ms of arc/yr. The quasar is so far away that the separation speed translates into $v \cong 8c$! Is this the death knell of the special theory of relativity?
11. Suppose you are traveling relative to Earth, with speed u such that $1 - u/c = 10^{-5}$. If you are holding a mirror and looking at your reflection, will there be any distortion in shape or in the color of the image?
12. In Fig. 39–2, light makes a single round trip between the half-silvered mirror and the two remaining mirrors. In a real version of the Michelson–Morley experiment, the light bounces back and forth many times. Why?
13. Suppose an experimenter found that some particle, such as a neutrino, travels in a vacuum just a bit faster than the speed of light in vacuum. Would we have to give up the special theory of relativity?
14. Folklore has it that, as a teenager, Einstein worried about what would happen if somebody were looking in a mirror while accelerating to a speed faster than that of light. What could Einstein have been worried about?
15. Suppose that current in a wire is carried by little green men who pass negative charges along a chain from person to person while standing on positive charges, such that the wire is electrically neutral. Would the modified mechanism for the current flow change the observed behavior of a charge q that lies outside the wire? How would this look to an observer who is moving with some velocity along the wire?
16. A closed box of little mass sits on a horizontal frictionless surface. The inside walls are perfect mirrors and reflect back all radiation. A laser is inside on the left-hand wall and projects a very short burst of light directly at the right-hand wall. What is observed from the outside, and why?
17. According to Hubble's law, a single spectral line characteristic of a single atom is redshifted by an amount proportional to the distance of the star from Earth. How do we know that the radiation of a particular color seen to come from a distant star is the redshifted radiation of a particular known spectral line?
18. The length-contraction experiment seems to imply that a meter stick accelerated to the speed of light would shrink to a point, and all the calibration markings on the meter stick would be lost. Is there something wrong with this reasoning?
19. Light falls in an accelerating elevator. But if an elevator moves upward at a constant velocity, a horizontal light beam would hit a spot *below* the horizontal projection on the opposite wall of the elevator. Does this mean that light falls in an elevator that moves at constant velocity?
20. When the supernova 1987a occurred, bursts of neutrinos—particles that have mass that is too small to be detected directly in accelerator-based experiments—arrived at detectors at various places on Earth's surface. These neutrinos are thought to have been emitted by the supernova all at once. How could differences in the arrival times of the neutrinos be used to test whether or not neutrinos have mass?
21. A very long fence has a sinusoidal curve painted on it. Will an observer moving rapidly along the fence see the curve as Doppler-shifted?

Problems

39–1 Is an Ether Necessary?

1. (I) An airplane flies at an air speed of 600 mi/h. It travels east from town A to town B and returns to A without stopping. In the absence of wind, the journey takes exactly 4 h. A town C is the same distance away from A and is located due north of A. Suppose that a wind with ground speed 60 mi/h is blowing east to west. Calculate the times it takes for the plane to make journeys ABA and ACA.

2. (II) In one version of the Michelson–Morley experiment, light of wavelength 633 nm emitted by a He–Ne laser travels through a total path length of 4.7 m in each arm of the interferometer. To the accuracy of the apparatus, a shift of 1/15 of a fringe, no shift was seen. Estimate the greatest value possible for the speed of Earth through the ether.

39–2 The Einstein Postulates

3. (II) A small, powerful laser is placed on a turntable that rotates at 900 rev/s. The laser, whose beam makes a 25° angle with the horizontal, shines on clouds 70 km away. Calculate the speed with which the light spot on the clouds moves. Does this speed violate the limitation of the speed of light? Explain.

39–4 Time Dilation and Length Contraction

4. (I) A muon (a subatomic particle) moves at a speed of $0.40c$. How much slower does its "clock" tick than if it were at rest?
5. (I) Two twins wave good-bye to each other. One twin, an astronaut, travels to Mars. The trip takes 1.0 yr in each direction, and the average speed with respect to Earth is 20,000 km/h. What will the approximate time difference in the twins' clocks be when they are together again on Earth?

6. (I) Proxima Centauri, the star nearest our own, is some 4.2 ly away. (a) If a spaceship could travel at a speed of $0.24c$, how long would it take to reach the star according to the spaceship's pilot? (b) What would someone in the frame that moves along with the spaceship measure as the distance to Proxima Centauri?

7. (I) According to a passenger, how long will it take a spaceship moving at $0.999c$ to cross a galaxy with a diameter, as measured in the galaxy's rest frame, of 2.5×10^{19} m?

8. (I) The space shuttle orbits Earth at 16,300 mi/h in 111 min. How much time will an astronaut's atomic clock have lost during a total trip that takes 7 d?

9. (I) The diameter of our galaxy is about 10^5 ly, or 10^{21} m. Suppose that a proton moves at a speed such that $\sqrt{1 - (v^2/c^2)} \cong 10^{-7}$. (Such speeds correspond to the most energetic cosmic rays known). How long does it take the proton to cross the galaxy in (a) the galaxy's rest frame? (b) The proton's rest frame?

10. (I) Spaceship A passes the control tower of a spaceport at a speed of $0.76c$. Automatic instruments in the control tower measure the length of the moving spaceship to be 143 m. What is the length of the spaceship once it lands (assuming you could wait long enough to make the measurement!)

11. (I) A researcher has a device that can measure lengths to an accuracy of one part in 10^{12}. What is the minimum speed for which he could measure a Lorentz contraction?

12. (II) A meter stick is tilted to make an angle of 30° with the x-axis. How will an observer, at rest in a frame F′ that moves at velocity $v = 0.85c$ in the $+x$-direction relative to the meter stick, describe the stick?

13. (II) A student must complete a test in 1 h in the teacher's frame of reference F. The student puts on his rocket skates and soon is moving at a constant speed of $0.75c$ relative to the teacher. When 1 h has passed on the teacher's clock, how much time has passed on a clock that moves with the student, as measured by the teacher?

14. (II) As measured by an observer in an inertial frame, a small clock moving at a constant speed of $0.75c$ traverses a distance of 60 km. The moving clock records 10,000 ticks during the passage. How many ticks pass on an identical clock at rest relative to the observer?

15. (II) A spaceship of length 30 m travels at $0.60c$ past a satellite. Clocks in frame S′ of the spaceship and S of the satellite are synchronized within their respective frames of reference and are set to zero so that $t' = t = 0$ at the instant the front of the spaceship F passes point A on the satellite, located at $x' = x = 0$ (Fig. 39–25). At this time, a light flashes at F. (a) What is the length of the ship as measured by an observer on the satellite? (b) What time does the observer on the satellite read from her clock when the trailing edge B of the spaceship passes her? (c) When the light flash reaches B at the rear of the spaceship, what is the reading t'_1 of a clock at B? (d) What is the reading t_1 on the clock on the satellite when, according to the observer on the satellite, the flash reaches B?

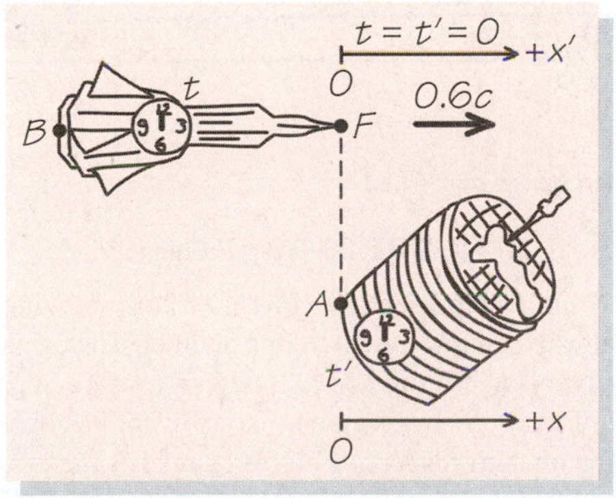

▲ **FIGURE 39–25** Problem 15.

16. (II) One can construct a helpful graph called a *Minkowski diagram* that represents the motion of a particle as a function of time. The x-axis represents the distance traveled (we deal here with straight-line motion only), and the y-axis represents the time, or more accurately ct. Draw (a) the "path" of a particle that is at rest; (b) the line representing the motion of a wave front of light; (c) a line representing the motion of a particle moving with speed $c/2$. (d) Does the line making an angle of 30° with the x-axis represent something physical?

17. (II) Consider the graphical representation of particle motion in the previous problem. On the same graph plot the motion of the two twins in the twin paradox. Suppose the non-traveling twin sends out regular light signals to her sibling. Will these be received regularly? Is there a difference between the frequency of the pulses received on the way out and that of the pulses received on the return journey?

18. (III) Jessica embarks on a cosmic journey at a speed of $(12/13)c$ relative to Earth. Before leaving, she tells her twin brother Tom, who stays on Earth, that she will travel outward for 26 yr of Earth time, then back for another 26 yr of Earth time. Tom will thus be 52 yr older when she returns. She promises to send a radio message on each of her birthdays (Fig. 39–26). According to an Earth-based clock, when will these messages reach Tom, and how much older than the age at which she leaves will Jessica be when she returns to Earth?

▲ **FIGURE 39–26** Problem 18.

19. (III) A relativistic sprinter running at speed v, near the speed of light, passes beneath a victory arch a height h above his eyes. Show that he will continue to see the arch, even though his eyes face forward, until he has run a distance $hv/\left[c\sqrt{1 - (v^2/c^2)}\right] = \gamma hv/c$ *beyond* the arch. [*Hint*: Work in the rest frame of the sprinter, and think of the top of the arch as emitting pulses of light, the last of which can be seen when it travels vertically downward toward the sprinter.]

39–5 The Relativistic Doppler Shift

20. (I) The sodium doublet refers to light waves emitted by sodium in a closely spaced pair of frequencies. The wavelengths of this doublet are at 589.0 nm and 589.6 nm. Suppose that the lower-wavelength member of this doublet is Doppler redshifted to a wavelength of 593.2 nm in the light emitted by a certain star. What happens to the wavelength of the second member of the doublet?

21. (I) A spaceship accelerates at a rate of 0.1 m/s^2 away from Earth. How long will it take (as measured in Earth's reference system) before a yellow beacon on Earth ($\lambda = 600$ nm) looks green ($\lambda = 500$ nm) to the crew of the spaceship?

22. (I) The wavelength of a spectral line in the laboratory is measured to be 108 nm. The same line is observed in light coming from a distant galaxy; in this observation, the wavelength is found to be 124 nm. What is the speed of motion of the galaxy relative to Earth?

23. (I) A particular spectral line measured in the emission of light by the star Alpha Centauri has wavelength $\lambda = 512.311$ nm. That same line measured in the laboratory has wavelength $\lambda = 512.350$ nm. Determine the radial velocity of Alpha Centauri relative to Earth.

24. (I) Spaceship A, moving at a speed of $0.92c$ relative to a certain star, is pursued by spaceship B, which is moving in the same direction at speed $0.98c$ relative to the same star. (a) With what speed does the pilot of spaceship A see spaceship B approaching? (b) Spaceship C passes both moving in the opposite direction with speed $0.75c$ relative to the star. At what speeds does the pilot of C see spaceships A and B passing?

25. (II) An observer in a large spaceship traveling east to west with speed $0.6c$ relative to Earth measures the distance between two points on Earth aligned in the east-west direction to have separation 500 m. A flea in the spaceship has developed high intelligence and is itself traveling within a tiny spaceship within the first spaceship, in the same direction with respect to Earth as the large spaceship and at a speed $0.2c$ with respect to the large spaceship. What does the flea measure for the separation of the two points on Earth?

26. (II) A driver was caught running a red light. His defense is that he saw the light as green, as a result of the Doppler shift. He is arrested. What for? Estimate the seriousness of his transgression.

27. (II) Yellow light at 587.6 nm, characteristic of helium, is found to be redshifted as it is observed in a certain star; the wavelength is measured to be 611.7 nm. (a) How fast is the star receding from Earth? (b) Use Hubble's law to estimate the distance of the star from Earth.

28. (II) For a particular quasar, $(\lambda - \lambda_0)/\lambda_0 = 2.45$, where λ_0 is the wavelength of the radiation emitted as measured in the quasar's rest frame. What is the speed of the quasar relative to Earth, assuming that it is traveling in a radial direction away from Earth? How far away is the quasar according to Hubble's law?

29. (II) A source radiates light with a frequency of 2×10^{15} Hz. The signal is reflected by a mirror that is moving at speed 1 km/s away from the source. What is the shift of the frequency of the reflected radiation, as observed at the source?

30. (II) The equation $\lambda/\lambda_0 = \sqrt{(1+\beta)/(1-\beta)}$, where $\beta = v/c$ and v is the speed of a source that is moving away from an observer or of an observer who is moving away from the source, takes a simple form if v is small compared to c. Show that if $\lambda = \lambda_0(1 + x)$, then for small β, $x \cong \beta$.

31. (III) During the journey described in Problem 18, Tom sends a radio message to Jessica on each of his birthdays adding to a total of 52 messages. With what interval in her rest frame does Jessica receive these messages during the outward part of the journey? During the return trip? Use this information to calculate how much Jessica ages during her trip according to an Earth-based clock.

32. (III) A source emits pulses with a frequency f_0. A spaceship moving at speed v_1 away from the source will receive a redshifted frequency f_1. Suppose that the spaceship immediately reemits the signals with the frequency f_1. A second spaceship, moving at speed v_2 relative to the first spaceship and in the same direction, will receive the signals with a redshifted frequency f_2. (a) Calculate f_1 and f_2. (b) If we were to eliminate the first spaceship, we could view f_2 as the redshifted frequency received by the second spaceship, which moves at some speed v relative to the source. Show that if both $v_1 \ll c$ and $v_2 \ll c$, then $v = v_1 + v_2$, as expected from the ordinary rules that govern relative motion. (c) Calculate v for arbitrary values of v_1 and v_2. This result is the relativistic law of addition of velocities, which differs from $v = v_1 + v_2$ when v_1 and v_2 are not very small compared with c.

39–6 The Lorentz Transformations

33. (I) Measurements of distant galaxies show that all galaxies are receding from one another at a speed proportional to their intergalactic distances. Suppose that we see galaxy 1 move away from us at a speed of $0.4c$ along the South Pole, and galaxy 2, equally far away, move away from us at the same speed along the North Pole (Fig. 39–27). What would an observer in galaxy 1 measure for the speed with which galaxy 2 moves away from him?

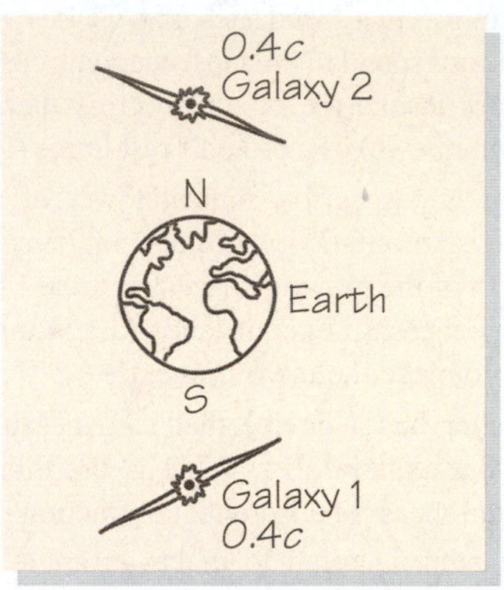

▲ **FIGURE 39–27** Problem 33.

34. (I) Events A and B are simultaneous in frame F and are 18 km apart on a line that defines the x-axis. A series of spaceships all pass at the same speed in the $+x$-direction, and they have synchronized their clocks so that together they make up a moving frame F′. They time events A and B to be separated by 0.80 μs. What is the speed of the spaceships? How far apart in space do they measure the two events to be?

35. (II) Two friends decide to demonstrate the Lorentz contraction of a train. They sit at the two ends of the 100-m-long train, with their watches properly synchronized. At $t = 0$, each drops a small bag out of the window; these bags act as markers (Fig. 39–28). Later, they go back and measure the distance between the bags. (a) Does this distance represent the length of the train relative to a coordinate system fixed to the ground? (b) What is the distance between the bags if the speed of the train is $0.7c$? (Neglect the time it takes for the bag to reach the ground!)

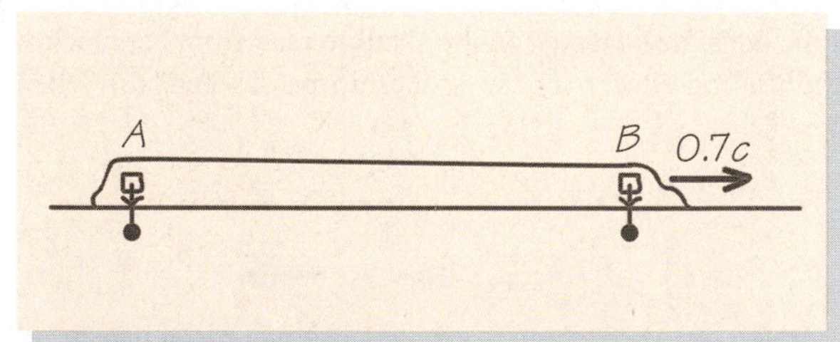

▲ **FIGURE 39–28** Problem 35.

36. (I) Let F and F′ represent two inertial frames moving at speed u along the x-axis with respect to one another. The origins and axes of these two frames coincide at time $t = t' = 0$. Suppose an observer in F and one in F′ agree that an event has occurred at the same time on their respective clocks, namely time $t = t' = t_0$. Is it possible that the event also occurred at the same place—that is, the same space coordinate—in each frame?

37. (I) An unmanned spaceship launched with speed $0.8c$ explodes after 24 s as measured in its own rest frame. (a) Assuming that the launching coordinates are $x = t = 0$ in the spaceship coordinate system, what are the coordinates of the explosion point in the coordinate system of the spaceship? (b) If the launching coordinates in the rest frame of the launching pad are $x' = t' = 0$, what are the coordinates of the explosion point in the frame of the launching pad?

38. (II) In a given reference frame, event 1 occurs at time $t_1 = 0$ s and position $x_1 = 0$ m, while event 2 occurs at $t_2 = 3.6 \times 10^{-4}$ s and $x_2 = 0.60 \times 10^5$ m. Is there a second frame in which these events could be at the same position but different times? If so, specify its motion with respect to the first frame. If not, what is the frame in which the events have the least possible separation in distance? [*Hint*: Use invariants.]

39. (II) A new Klingon battleship races at a top speed of $0.20c$ away from the planet XG4T. The starship *Enterprise* follows at a speed of $0.25c$ relative to the Klingon ship. With what speed does the *Enterprise* appear to catch up with the Klingon ship according to an observer on the planet?

40. (II) Spaceship A moves with velocity $0.50c$ in the positive x-direction of a reference frame S. Spaceship B, moving in the same direction with a speed of $0.60c$, is 3.0×10^9 m behind (Fig. 39–29). At what times, in reference frame S and in the reference frame of spaceship A, will B catch up with A?

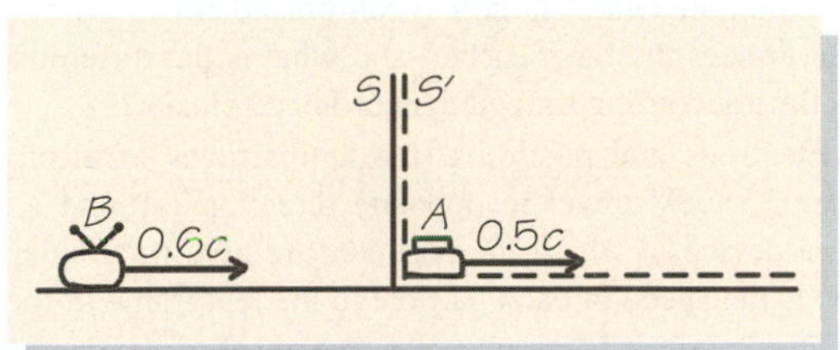

▲ **FIGURE 39–29** Problem 40.

41. (II) An observer in frame S measures two events to occur at the same point in space and separated by a time interval Δt. Show that in every other inertial frame, these events are separated by a larger time interval.

42. (II) A particle in frame F has velocity $v_x\hat{i} + v_y\hat{j}$. What is the velocity seen by an observer at rest in frame F′, a frame that moves at velocity $u\hat{i}$ relative to frame F?

43. (II) You shine light that moves at speed c/n through a medium of index of refraction n (Fig. 39–30). Suppose that the medium moves at speed u relative to you, parallel to the direction of the light. What is the speed of light in the medium as seen by you? [The result for $(u/c) \ll 1$ gives a result that is different from $(c/n) + u$, and it was first obtained by Augustin Fresnel in 1818. The measurements confirming the result were made by Hippolyte Fizeau in 1851.]

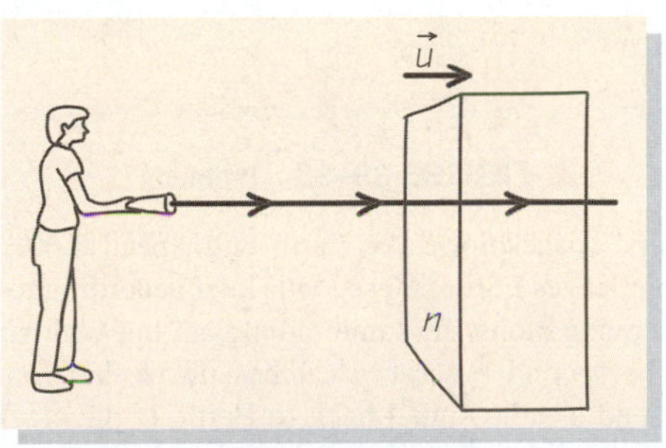

▲ **FIGURE 39–30** Problem 43.

44. (II) Use the Lorentz transformations to show that, as in Eq. (39–5), the location of a wave front for a plane wave emitted at $t = t' = 0$ from the joint origin of reference frames F and F′ is given by $x^2 - c^2t^2 = 0$ and by $x'^2 - c^2t'^2 = 0$.

45. (III) The electric field obeys the wave equation

$$\frac{\partial^2 E}{\partial t^2} - c^2\frac{\partial^2 E}{\partial x^2} = 0$$

when electromagnetic waves propagate along the x-axis. What is the form of the equation for a wave seen in a Lorentz-transformed frame that moves at speed u along the x-axis? [*Hint*: Use the Lorentz transformation laws to obtain expressions for $\partial/\partial t$ and $\partial/\partial x$ in terms of $\partial/\partial t'$ and $\partial/\partial x'$.]

39–7 Momentum and Energy in Special Relativity

46. (I) A spaceship of mass 7.5×10^3 kg has a kinetic energy that is 0.12 times its rest energy. What is its total energy?

47. (I) Assume here that the neutrino has a rest mass of zero. What is the momentum, in SI units, of a neutrino with energy 4 MeV? (The neutrino mass is actually small but nonzero.)

48. (I) *Estimate* the mass lost when 1 million tons of TNT explodes. Assume that each chemical reaction between individual molecules involves 10 eV of energy.

49. (I) Humans generate energy at a rate of some 10^{13} W worldwide. (The United States, with less than 10 percent of the world's population, uses about 25 percent of the energy.) At what rate is mass being lost due to relativistic effects?

50. (I) A particle, the π^0, has a rest mass of 135 MeV/c^2. It decays at rest into two identical massless particles. What is the momentum of each of the two decay products of the π^0?

51. (II) Energy from the Sun reaches Earth (above the atmosphere) at a rate of about 1400 W/m^2. How fast is the sun losing mass due to energy radiation?

52. (II) What value of v/c must a particle of rest mass m have in order for its momentum to have magnitude $p = 8.0mv$?

53. (II) How much work has to be done on a proton to accelerate it (a) from rest to $0.01c$; (b) from $0.8c$ to $0.81c$; (c) from $0.9c$ to $0.91c$; (d) from $0.99c$ to c?

54. (II) An electron that is accelerated in the Stanford Linear Accelerator in California has a total energy of 56 GeV. How much of this is kinetic energy? What is the momentum of the electron? What is its speed?

55. (II) A proton accelerated at Fermi National Laboratory in Illinois has a momentum of 746 GeV/c. (a) What is the proton's velocity? (b) The proton's kinetic energy?

56. (II) An electron and its antiparticle of identical mass, the positron, annihilate each other and produce two photons (Fig. 39–31). Both the electron and the positron were initially at rest. What are the energy and momentum of each photon?

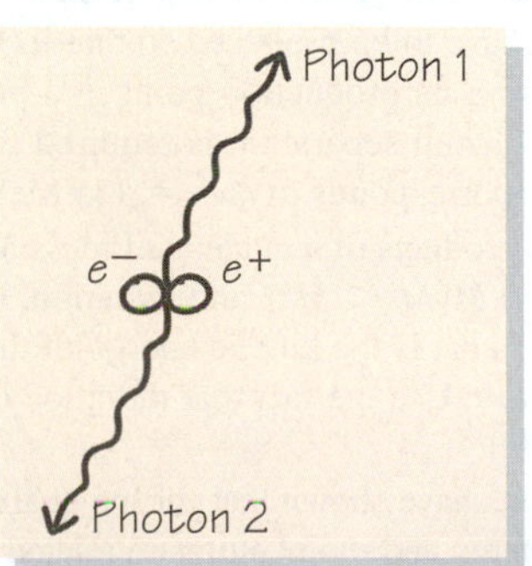

▲ **FIGURE 39–31** Problem 56.

57. (II) A photon is the quantum unit of light. It has an energy $E = hf$, where h is Planck's constant and f is the frequency of the light. Show that when a photon is absorbed by a free electron, without anything else occurring, energy and momentum cannot be conserved simultaneously.

58. (II) Show that energy and momentum conservation do not allow a high-energy photon to turn into a positron–electron pair with no other result. This process, called pair production, *can* occur if another object participates in the reaction (Fig. 39–32). What is the minimum energy required for the photon to give rise to an electron–positron pair, if the third object is very massive compared with the electron? [*Hint*: For the electron and positron $mc^2 = 511$ keV, and the third object, if it is a lead nucleus, has mc^2 about 4×10^5 times larger.]

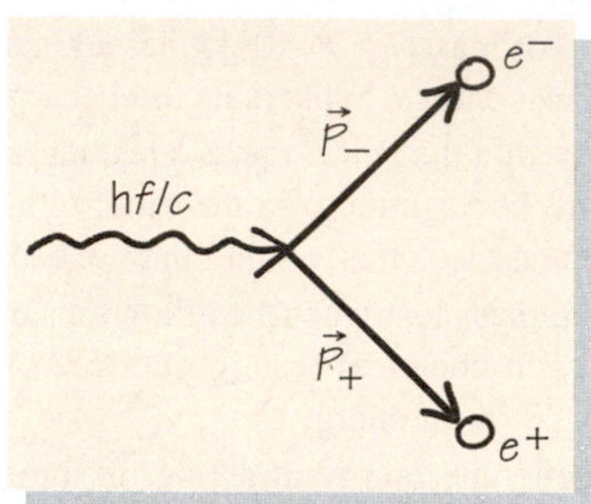

▲ **FIGURE 39–32** Problem 58.

59. (II) You analyze the track of a particle in a photographic plate placed in a magnetic field, and find that the total energy of the particle is 1130 MeV. The bending in the magnetic field gives information about momentum, and you learn that the particle's momentum is $p = 830$ MeV/c. What is the mass of the particle?

60. (II) In a generalization of Example 39–11, a ^{8}Be nucleus is moving in the x-direction with 3.5 MeV of kinetic energy when it decays into two alpha particles. Both alpha particles move off along the x-axis. What are their kinetic energies? [*Hint*: Note that $Q \ll M(^4\text{He})c^2$.]

61. (II) In the nuclear reaction $^{241}\text{Am} \rightarrow {}^4\text{He} + {}^{237}\text{Np}$, there is an energy release of 6 MeV. In the approximation in which the value of mc^2 for the Am nucleus is (241)(938 MeV), that of the He nucleus is 4(938 MeV) and that of the Np nucleus is (237)(938 MeV), calculate the value of pc (p = momentum) of the ^{4}He nucleus emitted from a ^{241}Am nucleus at rest (a) using nonrelativistic kinematics and ignoring the recoil of the ^{237}Np nucleus; (b) using nonrelativistic kinematics and taking the recoil motion of the ^{237}Np nucleus into account; and (c) using relativistic kinematics and taking the recoil motion of the ^{237}Np nucleus into account.

62. (II) The lifetime of a particle called the neutral pion, π^0, is 0.9×10^{-16} s in the particle's rest frame. With what energy would one π^0 have to be produced so that its decay point is distinguishable from its production point in a photographic plate? Assume that a 1-mm separation is required for a measurement. The pion mass corresponds to $mc^2 = 135$ MeV.

63. (II) The decay products of a nucleus of mass M^* include another nucleus of mass $M(M < M^*)$ and radiation. If the decaying nucleus is at rest, what is the kinetic energy of the remnant nucleus of mass M? [*Hint*: Use the fact that radiation of energy E carries momentum E/c].

64. (III) Experiments have shown that for the quantum of radiation (a photon), the energy and momentum are related by $E = pc$, corresponding to a particle with mass $m = 0$. Suppose that in the observation of a supernova 170,000 ly away, the first bursts of photons with an energy range of E = 10 eV to 10^4 eV arrive within 10^{-8} s of each other. What limits does this set on the mass of a photon? [*Hint*: Use the fact that mc^2 is small, so $E = pc + (m^2c^3/2p)$ is a good approximation.]

65. (III) Calculate an expression for the force as defined by $m\dfrac{d(\gamma\vec{u})}{dt}$, and show that the force and the acceleration $d\vec{u}/dt$ do not necessarily point in the same direction.

*39–8 Beyond Special Relativity

66. (II) A neutron star has a mass of 2.4×10^{30} kg and a radius of 8.5 km. What is the gravitational redshift of radiation emitted with a frequency of 2×10^{19} Hz from the star's surface?

67. (III) A clock on a disk rotating with angular speed ω, when placed at a distance R from the center of the disk, experiences an acceleration toward the center of the disk. What gravitational potential will an observer at rest relative to the clock assume that he or she is in? (Use the equivalence principle.) Will the clock be slow or fast relative to a clock at the center of the disk?

General Problems

68. (I) A photographer takes a flash-illuminated photo of a train car as the car passes moving to the left at 12 m/s. The flash goes off when the midpoint of the car is adjacent to the camera. Will two observers, one at each end of the 18-m-long train car, receive the flash simultaneously on their synchronized clocks? If not, which observer sees the flash earlier, and what is the difference in arrival time according to their synchronized clocks?

69. (II) Electrons and positrons (the antiparticles of electrons) of energy 35 GeV travel in opposite directions around a storage ring, a device in which the particles are held in circular orbits. What is the speed of each particle in the rest frame of the other?

70. (II) The Stanford Linear Accelerator accelerates electrons to a total energy of 50 GeV. How long does a meter stick at rest appear to a hypothetical observer at rest with respect to one such electron?

71. (II) An astronomer on Mars measures the optical spectrum of Earth. Averaged over a long period, she will see the spectral lines broadened by the Doppler effect due to Earth's rotation (Fig. 39–33). Calculate the width of the spectral line at 650 nm; i.e., the difference between the longest and shortest wavelength of a 650-nm line as seen by the astronomer, assuming that she is positioned in Earth's equatorial plane. Ignore the motion of Mars.

▲ **FIGURE 39–33** Problem 71.

72. (II) A first spaceship leaves Earth with speed 0.60c. A second spaceship leaves Earth 100 seconds later according to a clock on Earth, moving along the same radial path but with speed 0.75c. When the second spaceship catches up to the first one, they jointly send a radio signal back to Earth. If the clock on Earth had been set to 0 at the moment the first spaceship leaves, what will that clock read when the signal is received on Earth?

73. (II) Consider the situation described in the previous problem. How long, according to a clock in the lead spaceship, will it be till the second one catches up? How long will it be, according to a clock in the second spaceship? Each clock starts at the departure time of the first spaceship.

74. (II) In 1990, an SR-71 Blackbird reconnaissance airplane on its way to retirement at the Smithsonian Air and Space Museum set several speed records. The plane averaged 2153 mi/h during the 2300-mi trip from Los Angeles to Washington D.C., and 2242 mi/h during the 311-mi trip from St. Louis to Cincinnati. What would have been the difference in time elapsed for the two record-setting segments between an atomic clock placed in the airplane and another atomic clock on the ground?

75. (II) A cosmic ray is approaching Earth from outer space. A hypothetical observer in a frame that moves with the cosmic ray measures Earth as a flattened ball whose thickness is 3/7 of its diameter (Fig. 39–34). (a) With what speed is the cosmic ray approaching Earth? (b) The cosmic ray is identified as a proton, with mass m and mass energy given by $mc^2 = 1$ GeV. What is the energy of the approaching proton, as seen from Earth?

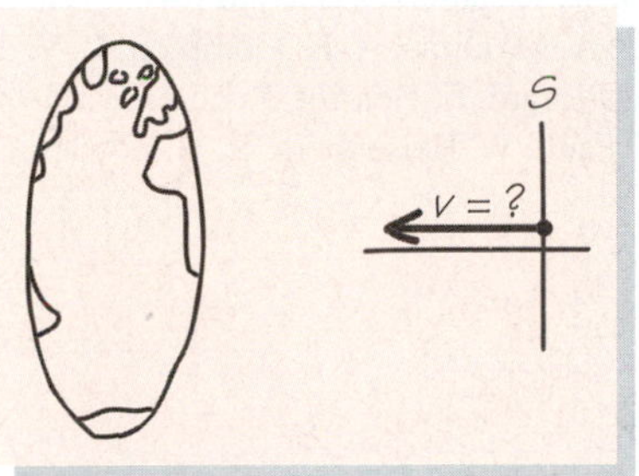

▲ **FIGURE 39–34** Problem 75.

76. (II) A new Klingon battleship has a proper length of 780 m and travels at speeds of $0.26c$ with respect to its home planet. The Klingons prepare to battle the *Enterprise*, which is moving at the same speed with respect to the same planet (Fig. 39–35). If the Klingons are heading straight at the *Enterprise*, what is the length of the Klingon ship as measured by Captain Kirk?

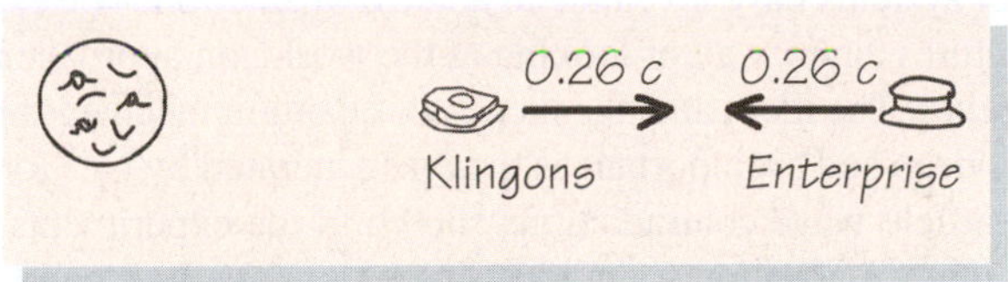

▲ **FIGURE 39–35** Problem 76.

77. (II) A particle of mass M is at rest. It decays into two identical particles, each of mass m, with $2m < M$. (a) If one of the two decay particles moves north with a momentum of magnitude p, what is the momentum of the other particle? (b) Use energy conservation to find p.

78. (II) In a quantum mechanical model, a proton and an antiproton annihilate each other and produce a pair of photons, light quanta whose frequency is related to their energy by the relation $E = hf$, where h is Planck's constant ($h \cong 6.63 \times 10^{-34}\ \text{J}\cdot\text{s}$). The proton and the antiproton are nearly at rest when they annihilate. Find the frequencies of the emitted photons (Fig. 39–36). What are these frequencies if the proton and the antiproton are approaching each other in a head-on collision in which each particle has a kinetic energy of 1600 MeV? For both protons and antiprotons, $mc^2 \cong 938$ MeV.

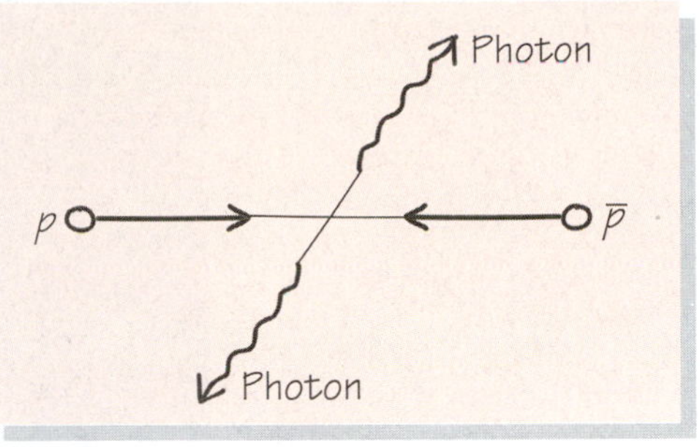

▲ **FIGURE 39–36** Problem 78.

79. (II) According to Chapter 19, the distribution function for the z-component of the velocity of a gas at temperature T is

$$G(v_z) \propto e^{-(mv_z^2/2kT)},$$

where k is Boltzmann's constant and m is the mass of one gas molecule. If a molecule at rest emits a spectral line, light at a characteristic frequency of f_0, then what is the distribution of frequencies of the light given off by the gas of such molecules when heated to temperature T? Assume that you are looking along the z-direction and take into account only motion in this direction. The effect described here is known as *Doppler broadening* of a spectral line. It is a tool for determining the temperature of stars and interstellar gases.

80. (II) Particles with energies as high as 10^{18} eV have been observed. Suppose that one of those particles collides with a photon of cosmic background radiation of wavelength $\lambda = 10^{-3}$ m. After a head-on collision, what will the final wavelength of the photon be? [*Hint:* These energies are so high that the particles can be treated as massless.]

81. (II) A charged pi meson (a particle of mass about 140 MeV/c^2) is ejected from a nuclear collision with a kinetic energy of 250 MeV. Pi mesons have a half-life (the time over which half of a given collection will decay radioactively) of about 1.5×10^{-8} s. Calculate (a) the pi meson's speed and (b) momentum. (c) How far will a collection of pi mesons travel before half of them decay?

82. (II) The *rapidity* V of a moving body is defined by $\tanh(V/c) = v/c$, where v is the body's relativistic speed. An observer in frame S′, moving at speed u in the $+x$-direction with respect to a frame S, measures a body to have speed v along the x-axis. Show that an observer in frame S measures the body to have rapidity W, given by $W = U + V$. Here, U is the rapidity of frame S′ with respect to frame S. The rapidity thus adds like a Galilean velocity.

83. (III) A proton moves with a momentum of magnitude p in the $+x$-direction. It strikes a second proton, which is at rest. Three protons and one antiproton result from the collision. The four particles in the final state remain together; that is, they have no motion relative to one another. Use energy and momentum conservations to find p. For both protons and antiprotons, $mc^2 \cong 938$ MeV.

Chapter 40

Conferees in the 1927 Solvay Conference, including many of the founders of quantum mechanics. First row, left to right: I. Langmuir, M. Planck, M. Curie, H. A. Lorentz, A. Einstein, P. Langevin, C. E. Guye, C. T. R. Wilson, O. W. Richardson. Second row, left to right: P. Debye, M. Knudsen, W. L. Bragg, H. A. Kramers, P. A. M. Dirac, A. H. Compton, L. V. de Broglie, M. Born, N. Bohr. Standing, left to right: A. Piccard, E. Henriot, P. Ehrenfest, E. Herzen, T. De Donder, E. Schrödinger, E. Verschaffelt, W. Pauli, W. Heisenberg, R. H. Fowler, L. Brillouin.

Quantum Physics

The results of a vast array of experiments that explore systems on both the macroscopic and microscopic scale cannot be explained by classical physics; indeed, they are often in direct contradiction with it. These experiments show that radiation *also* exhibits particle-like behavior, that matter *also* exhibits wavelike behavior, and some aspects of the behavior of systems have no classical analog whatsoever. The explanation of these phenomena requires a new way of looking at the world, an approach that is provided by *quantum mechanics*. The ideas and techniques of quantum mechanics represent a scientific revolution as deep and as important as the one generated by Newtonian mechanics. It reconciles the particle-wave contradictions shown in the experiments, although in doing so it requires that we abandon some ideas about how well we can measure quantities such as position and velocity. Most important of all, the theory abandons the traditional causal view of physical phenomena: It makes no predictions about individual events, but rather predicts the *probabilities* that these events occur.

Quantum mechanics does not have an evident role to play in the everyday phenomena around us. But it has been enormously successful in the quantitative description of atoms, nuclei, and more fundamental particles and their interactions, as well as of the behavior of materials. In recent decades it has also had a big impact on technology, through its role in *mesoscopic* systems, those that lie in scale between the atomic and "human" scale. For example, many properties of the circuits that act in computers rely to some degree on quantum phenomena. Some familiarity with quantum mechanics is an essential part of a scientific education. In this chapter we describe the particle nature of radiation, the wave nature of particles, and the limitations on the concepts of position and momentum, all basic ideas of quantum mechanics. In a succeeding part of this book we explore a large range of phenomena that can only be understood with the help of quantum physics.

▲ **FIGURE 40–1** Max Planck.

40–1 The Particle Nature of Radiation

Blackbody Radiation

The discovery of quantum mechanics did not come through a study of atoms or other microscopic systems. The first hints of it came through the work of Max Planck (Fig. 40–1), who was trying to understand the experimental data on *blackbody radiation*, the name given to a spectrum of electromagnetic radiation in thermodynamic equilibrium at a temperature T, a situation most easily arranged within a cavity (Chapter 17). The radiation inside a cavity can be studied by making a tiny hole in the cavity walls and looking at what emerges. One finds that blackbody radiation has energy distributed across a continuum of frequencies. The distribution is described by an energy density $u(f, T)$, whose meaning is that *the electromagnetic energy of the radiation in the frequency range from f to $f + df$ within a cavity of unit volume in equilibrium at temperature T is $u(f, T)\,df$*. We considered this function in Section 17–5, and you may want to review that discussion.

The allowed frequencies for blackbody radiation in a cavity do not in fact form a continuum. Standing waves in a cavity, such as those formed by the electromagnetic field in a microwave oven, resemble the standing waves in a guitar string fixed at the two ends, and the possible frequencies are discrete, depending on the shape and size of the cavity. There is a fundamental frequency as well as a first, second, . . . nth harmonic. Each allowed frequency represents a degree of freedom for the radiation as far as the equipartition of energy is concerned (see Chapter 19). Using this classical principle, one expects the energy density for a given frequency to be proportional to the temperature T. In 1900 Lord Rayleigh and James Jeans accordingly predicted that for radiation in a cavity, the energy density should have the form

$$u(f, T) = \frac{8\pi f^2}{c^3} kT. \tag{17–18}$$

This result agrees with experiment for low frequencies but strongly disagrees with observations for high frequencies. The same year the Rayleigh–Jeans result was published, Planck arrived at a remarkable formula, Eq. (17–16), that fits the observed energy density over the full range of frequencies:

$$u(f, T) = \frac{8\pi h}{c^3} \frac{f^3}{e^{hf/kT} - 1}. \tag{17–16}$$

This formula involves a new fundamental constant h, now called Planck's constant, with the value $h = 6.63 \times 10^{-34}\ \text{J}\cdot\text{s}$. Figure 17–19 shows the precise agreement between measured values of $u(f, T)$, and the Planck formula for a temperature of 2.7K.

CONCEPTUAL EXAMPLE 40–1 You have no doubt seen pictures of steel mills, with molten iron at 2000°C visible through the doors of the furnaces. Why does this fact show that the Rayleigh–Jeans formula [Eq. (17–18)] cannot be correct?

Answer We must start with the assumption that looking through a small hole into a cavity such as the door to a furnace allows us to "see" blackbody radiation. There is then a striking reason to doubt the Rayleigh–Jeans formula. This formula does not correlate the color of the radiation with the temperature—the formula picks out no prominent frequency, or color, that changes with the temperature. Yet a cavity at room temperature certainly has a different color than one at the temperature of molten iron. More generally, as the temperature increases, the color of a fire changes systematically, from dull red through a bright orange to a brilliant yellow. The full Planck formula does indeed contain such a correlation, and we discussed the fact that it has a maximum as a function of frequency that changes with temperature in Section 17–5.

Planck was able to derive his formula only by making a strange assumption: The walls of the cavity constantly absorb and reemit the radiation, and Planck assumed that the absorption and emission of radiation was limited to "bundles" of energy. Each bundle had an energy proportional to its frequency, according to the formula

$$E = hf. \tag{40–1}$$

This assumption, which has no grounding in classical physics, was translated by Albert Einstein in 1905 into a more general assertion: *Electromagnetic radiation consists of* ***quanta****, or identical, indivisible units, each carrying energy hf, where f is the frequency of the radiation.* In other words, the walls are observed to absorb and emit in bundles of radiation because the radiation only comes in such bundles.

We shall soon discuss other experiments that confirm these quanta of radiation and show how they behave as particles. These particle-like quanta of radiation are called **photons**. Because the momentum and energy of any particle that travels at the speed of light are related by $p = E/c$, a photon of energy $E = hf$ carries momentum of magnitude

$$p = \frac{E}{c} = \frac{hf}{c}. \tag{40–2}$$

For radiation,

$$\lambda = \frac{c}{f} = \frac{hc}{hf} = \frac{hc}{E} = \frac{h}{p}. \tag{40–3}$$

This formula, connecting a wavelength and a momentum through Planck's new constant, was later adopted for matter by de Broglie in his daring conjecture concerning the wave properties of matter (see Section 40–2).

EXAMPLE 40–2 How many photons in the visible range (around 550 nm) are emitted every second by a 60-W lightbulb? Only about 10 percent of the total power emitted as radiation by an incandescent lightbulb is in the visible range.

Strategy If the total power emitted by the bulb is P, then the power emitted in the visible range, which we assume is concentrated at $\lambda = 550$ nm, is $(0.1)P$. Since the energy carried by a single photon is $hf = hc/\lambda$, the number of photons of visible light emitted per second is $n =$ (total energy emitted as visible light/sec)/(energy per visible-light photon) $= (0.1P)/(hf)$.

Working It Out Numerical evaluation gives

$$n = \frac{(0.1)P}{(hc/\lambda)} = \frac{(0.1)P\lambda}{hc} = \frac{(0.1)(60\ \text{J/s})(550 \times 10^{-9}\ \text{m})}{(6.6 \times 10^{-34}\ \text{J}\cdot\text{s})(3 \times 10^{8}\ \text{m/s})}$$
$$= 1.7 \times 10^{19}\ \text{photons/s}.$$

It is obvious from this number that direct detection of individual photons from such a bulb would be quite difficult.

What Do You Think? What happens to the other 90 percent of the power? *Answers to **What Do You Think?** questions are given in the back of the book.*

EXAMPLE 40–3 A bright star, easily visible to the naked eye, emits radiation such that the intensity (power per unit area) at Earth's surface is $I = 1.6 \times 10^{-9}\ \text{W/m}^2$ at a wavelength of 560 nm. Estimate the rate at which photons enter the night-adapted eye from such a star.

Setting It Up We have the power per unit area, and to find the power or the number of photons per unit time that enters the eye, we need to assume an area for the pupil. We shall take the pupil in night-time conditions to be a circle with a diameter of 0.5 cm.

Strategy The frequency of the radiation needs to be calculated using the relation $f = c/\lambda$. The energy per photon is $hf = hc/\lambda$, where λ is the wavelength of the light. The intensity I can be converted into the number of photons per m^2 per second by dividing I by the energy per photon. The number of photons entering the eye is this result multiplied by the area $\pi d^2/4$ of the pupil.

Working It Out We write the intensity as NE, where N is the number of photons striking Earth per square meter per second. Thus

$$N = \frac{I}{E} = \frac{I\lambda}{hc} = \frac{(1.6 \times 10^{-9}\ \text{J/m}^2\cdot\text{s})(560 \times 10^{-9}\ \text{m})}{(6.63 \times 10^{-34}\ \text{J}\cdot\text{s})(3.0 \times 10^{8}\ \text{m/s})}$$
$$= 0.45 \times 10^{10}\ \text{photons/m}^2\cdot\text{s}.$$

The area of the pupil is estimated to be $A = \pi(d/2)^2 = \pi(2.5 \times 10^{-3}\ \text{m})^2 = 2.0 \times 10^{-5}\ \text{m}^2$, so the number of photons that enter the eye in 1 s is

$$NA = (0.45 \times 10^{10}\ \text{photons/m}^2\cdot\text{s})(2.0 \times 10^{-5}\ \text{m}^2)$$
$$= 0.90 \times 10^{5}\ \text{photons/s}.$$

The human eye can register as few as several photons per second (see Problem 17).

What Do You Think? Assume that the human eye responds to a rate of 15 photons/s. At this wavelength, how much farther away could this star be and still be seen?

The Photoelectric Effect

Support of a different kind for the quantum nature of radiation came from the work of Albert Einstein, who used the quantum nature of radiation to explain the *photoelectric effect* in 1905. This effect was discovered by Heinrich Hertz in 1887 in experiments shown schematically in Fig. 40–2.

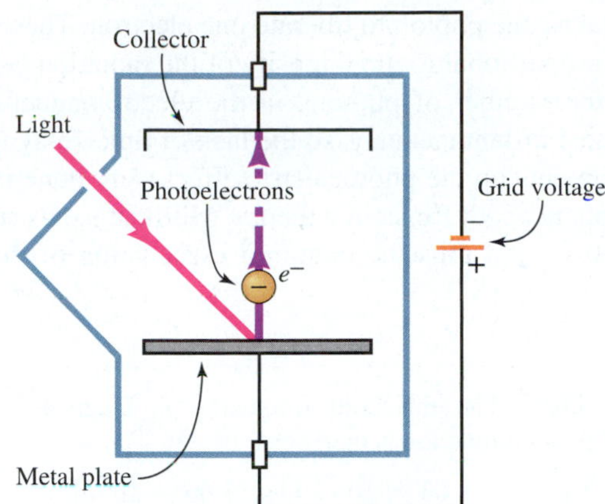

◀ **FIGURE 40–2** Schematic diagram of an experimental setup for measuring the photoelectric effect. Light strikes a metal plate in an evacuated chamber. The electron current is measured by a collector, and the kinetic energy is determined by the grid voltage needed to stop the electrons from reaching the collector.

The simplest statement of the photoelectric effect is that when a polished metal plate is exposed to electromagnetic radiation, it may emit electrons. These electrons are sometimes termed *photoelectrons*. The mere fact that electrons are emitted from metals subjected to electromagnetic radiation can be understood without invoking quantum ideas. Metals contain free electrons, and because electrons do not leak out of a metal freely, it is reasonable to expect that a minimum of energy must be deposited in the metal to liberate electrons. Moreover, in classical electromagnetic theory the energy delivered by radiation to the metal is proportional to the square of the electric field, $\vec{E}$; that is, to the intensity of the incoming radiation. We would therefore expect the energy carried off by the electrons to be proportional to the intensity. For example, a doubling of the intensity would double the number of electrons emitted with a given kinetic energy. And indeed, if the frequency is held constant, the magnitude of the emitted current of electrons is proportional to the intensity of the light source.

However, when one looks more closely at the characteristics of the photoelectric effect, one finds that there are many features that don't make much sense in the classical picture. For example, in the classical picture, electrons should not be emitted at all at very low intensities. Yet they are, and the maximum kinetic energy of the emitted electrons is independent of the intensity of the light source. In the classical picture, the maximum kinetic energy of the emitted electrons should not depend on the frequency of the radiation, only on its intensity, yet the maximum kinetic energy of the emitted electrons in fact varies linearly with the frequency of the incident light (Fig. 40–3). In the classical picture, energy should be delivered for all radiation frequencies, yet in fact no matter how large the intensity, electrons are emitted only if the frequency of the incident light exceeds a threshold value—that is, $f > f_0$. (The value of the *threshold frequency* f_0 may vary with the particular metal.) In the classical picture, we would expect that with low-intensity radiation, the energy required to liberate a certain number of electrons would have to be collected over some time, and that there would be a time delay (increasing with decreasing intensity) before the electrons would appear. Yet in fact experiments show that to an accuracy of 10^{-9} s, there is no measurable time delay between the arrival of the radiation and the appearance of the electron current.

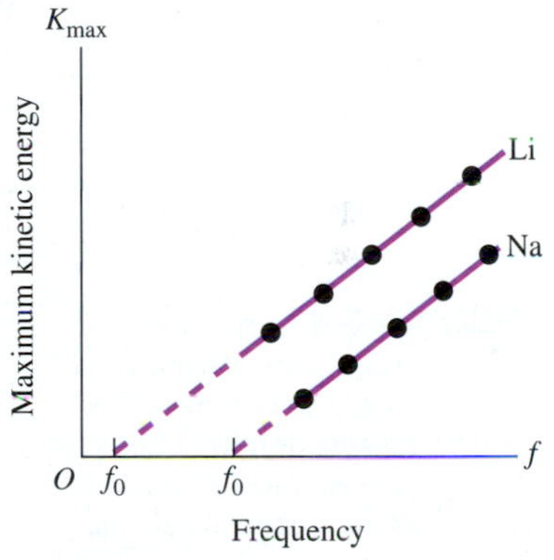

▲ **FIGURE 40–3** Data for the photoelectric effect, showing the maximum kinetic energy of the emitted electrons (of lithium, Li, and sodium, Na) as a function of light frequency. Note the linear relationship and the presence of a minimum frequency f_0.

Einstein explained these phenomena by postulating that electrons are emitted because individual electrons absorb individual photons. The photons that correspond to radiation with frequency f carry energy $E = hf$. If there is a minimum energy W required to liberate an electron, then no electrons will be emitted when hf is less than W. When hf exceeds W, the excess energy can go into kinetic energy of the emitted electrons:

$$\frac{1}{2}mv^2 = hf - W. \tag{40–4}$$

The quantity W is a kind of potential energy that must be acquired before the electron can be liberated; it is called the *work function*. The work function is a characteristic of the particular metal that emits electrons, typically taking on a value in the range of several eV. (Recall 1 eV $= 1.6 \times 10^{-19}$ J; using the fact that kT is an energy, one finds that the equivalent of an eV in temperature is around 1.2×10^4 K. The electron-volt is discussed in detail in Section 24–2.) It takes one photon to liberate one electron. Therefore, the current of emitted electrons is proportional to the intensity of the radiation because the intensity is proportional to the number of photons in the electromagnetic wave. An electron absorbs a photon almost instantaneously, so the lack of time delay is also explained. The first accurate experiments on the photoelectric effect were done in 1916 by Robert Millikan, who was skeptical about Einstein's theory. Millikan's experiments unequivocally confirmed Eq. (40–4) and thus the quantum explanation of the photoelectric effect.

EXAMPLE 40–4 The longest wavelength of light that will induce a photoelectric effect in potassium is 564 nm. Calculate the work function for potassium in electron-volts.

Strategy We must recognize that the longest wavelength of light corresponds to the lowest or threshold frequency f_0 necessary to induce the photoelectric effect, with the photoelectrons barely moving as they leave the metal—they have minimal kinetic energy. We then have an application of Eq. (40–4) for $v = 0$, and the work function is $W = hf_0$.

Working It Out The threshold frequency f_0 is given by $f_0 = c/\lambda_{\max}$. The work function is then given by

$$W = hf_0 = \frac{hc}{\lambda_{\max}} = \frac{(6.63 \times 10^{-34}\,\text{J}\cdot\text{s})(3.00 \times 10^8\,\text{m/s})}{5.64 \times 10^{-7}\,\text{m}}$$

$$= 3.53 \times 10^{-19}\,\text{J} = (3.53 \times 10^{-19}\,\text{J})\frac{1\,\text{eV}}{1.60 \times 10^{-19}\,\text{J}}$$

$$= 2.20\,\text{eV}.$$

What Do You Think? What do you expect to happen if radiation of a shorter wavelength is incident on potassium? What if the radiation has a longer wavelength?

The Compton Effect

More compelling evidence for the particle properties of photons came from experiments of Arthur Compton in 1922 (Fig. 40–4a). Compton sent X rays through thin metallic foils (Fig. 40–4b) and discovered that the scattered X rays emerge with one of two wavelengths. One component emerges with the same wavelength λ_0 as the incident radiation. The other component emerges with a longer wavelength λ'. This result is in contrast to the prediction of classical radiation theory, in which the electrons absorb radiation and reradiate it as dipole radiation without any change in wavelength. The experiments show that λ' varies with the scattering angle θ of the X ray (Fig. 40–5), satisfying the relation

$$\lambda' - \lambda_0 = \frac{h}{mc}(1 - \cos\theta), \tag{40–5}$$

where m is the electron mass. The presence of h indicates that quantum mechanical effects are involved, and because the result does not depend on the metal used in the foils and the electron mass is present in the formula, Compton concluded that the effect had nothing to do with the metal's crystal structure.

Compton was able to derive Eq. (40–5) (see Problem 16) by treating the photon as a particle of energy hf and, as in Eq. (40–2), momentum hf/c colliding elastically with a

▼ **FIGURE 40–4** (a) Arthur Compton. (b) Schematic diagram of a setup for Compton's experiment. The scattered X rays are diffracted by a crystal, with the angle α used to determine the wavelength of the scattered radiation.

(a)

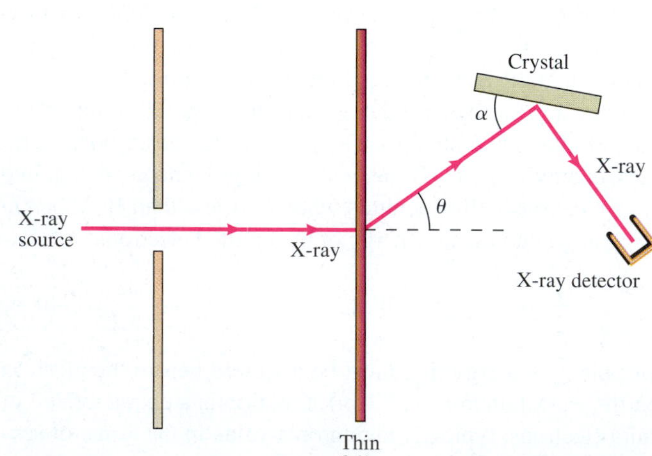

(b)

target electron at rest—in other words, a standard two-body collision, complete with momentum and energy conservation (Fig. 40–6). The treatment of the collision follows the methods for elastic collisions in space (see Chapter 8), with the difference that it is necessary to take the electron's final energy in the relativistic form $E_e = \sqrt{p^2c^2 + m^2c^4}$, where p is the final electron momentum. The energy of the scattered photon can be calculated, and it differs from that of the incident photon, just as would be true in a collision of billiard balls. And because the energy of the scattered photon is different from that of the incident photon, so is its wavelength. Equation (40–5) correctly emerges. The quantity h/mc, called the *Compton wavelength* of the electron, has the dimensions of length and magnitude 2.4×10^{-12} m. Why are there two peaks in Fig. 40–5? The peak at the original wavelength λ_0 corresponds to the collision of a photon with an ion, which is thousands of times more massive than the electron. This is like the elastic scattering of a classical particle from a wall; the photon loses no energy, and the shift of the photon wavelength is very nearly zero.

Experiments such as the scattering of X rays from matter, designed to study the particle nature of radiation, do in fact see radiation exhibiting particle behavior. At the same time, experiments that probe the wave character of radiation, such as interference experiments, confirm the wave character of radiation, even at extremely high frequencies—Bragg's law, for example, concerns the diffraction of X rays from regular structures. *Light has a dual wave-particle nature.*

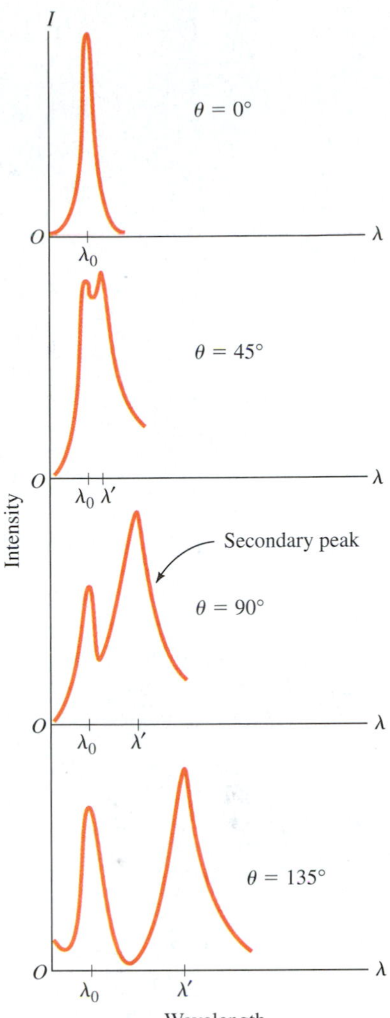

▲ **FIGURE 40–5** Experimental data for Compton's experiment. The secondary peak, due to X-ray scattering by free electrons, becomes more pronounced as the scattering angle increases.

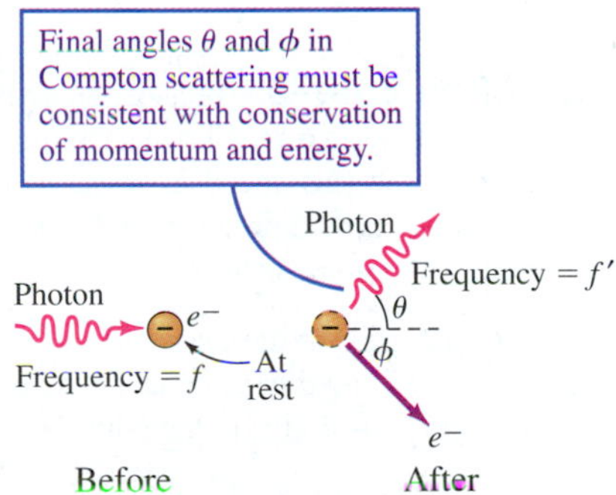

▲ **FIGURE 40–6** Light of frequency f scatters from an electron as though the light were a particle (photon). As in the collision of any two particles, the photon's energy changes when the photon scatters, and by the laws of quantum mechanics, its frequency changes to f'. The phenomenon is known as the Compton effect.

THINK ABOUT THIS. . .

CAN ONE ACTUALLY "SEE" INDIVIDUAL PHOTONS?

The human eye is an instrument of incredible efficiency, but not quite good enough to respond to a single photon. However, single photons can be detected by using a device called a *photomultiplier*. This device amplifies the effect of a single photon in the following way. The incident photon strikes a material and produces a photoelectron. The material is said to form a *photocathode*. The photoelectron is projected into an evacuated region in which an electric potential is set up. This potential accelerates the electron, and when it strikes another surface, the *dynode*, it has enough energy to release several more electrons through collisions with the material. These electrons are then accelerated by another potential until each electron is energetic enough to again release a small number of electrons in collisions with another surface. By the time many such stages are completed, the single incoming photon has produced a cascade of up to 10^9 electrons, enough to generate a measurable electric current. The magnitude of the pulse created by the current can be translated into a measurement of the energy deposited by the photon. This cascading mechanism is utilized in the *image-intensifier tube*, which is at the heart of one type of night-vision viewer (Fig. 40–7). The device works by collecting individual photons, then using the photoelectric effect to amplify their presence. A lens system sends any collected light to a glass plate coated on the back side with a photoelectric material. The photoelectrons are accelerated through a potential difference of several hundred volts to a "channel plate" containing many holes, typically 10 microns in diameter. There are many such holes, and they are designed to "hold the image together" as the cascade proceeds. When a photoelectron strikes the sides of a hole, it releases several electrons. These collide further with the sides of the hole and produce still other electrons. The electrons may be passed on to other channel plates in a series of repeating steps, ending with a large number of electrons aligned with the holes, and since few photons will produce few electrons, the numbers of electrons in various locations will reproduce the image. A final acceleration stage to a fluorescent screen produces photons and hence a visible image.

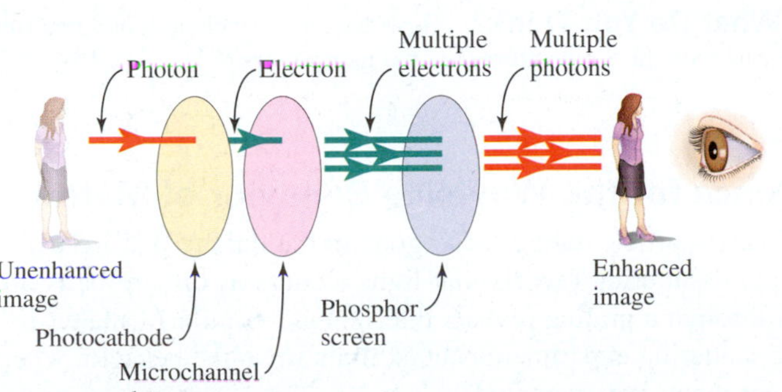

◀ **FIGURE 40–7** An *image-intensifier tube* produces a visible image from only a few photons—even one photon produces a cascade of electrons that can in turn form a visible signal.

40–2 The Wave Nature of Matter

The presence of both wavelike and particle-like aspects in light stimulated Louis de Broglie to inquire whether particles such as electrons would also exhibit wavelike behavior. In his 1924 doctoral thesis, he proposed that when particles are subject to experiments that test wavelike properties—basically interference experiments—they do indeed exhibit wavelike properties. More precisely, particles with momentum p act like waves with a de **Broglie wavelength** λ given by

$$\lambda = h/p. \tag{40–6}$$

This expression involves *the same Planck's constant* as appears in the photon energy-frequency relation. In fact, it is *exactly* the relation between momentum and wavelength that holds for photons [Eq. (40–3)].

De Broglie's wavelength-momentum relation may also be written in terms of a wave number $k = 2\pi/\lambda$:

$$k = \frac{2\pi}{\lambda} = \frac{2\pi p}{h} = \frac{p}{\hbar}. \tag{40–7}$$

We have here written the commonly occurring combination $h/2\pi$ as $\hbar$:

$$\hbar \equiv h/2\pi \cong 1.05 \times 10^{-34}\ \mathrm{J \cdot s}. \tag{40–8}$$

The small size of h makes the wave character of matter evident only on a very small scale, where the relevant momenta are extremely small and the wavelengths large enough to have observable effects. For example, a dust particle of mass 10^{-6} g traveling at a speed of 1 m/s has a wavelength of

$$\lambda = \frac{h}{p} = \frac{h}{mv} = \frac{(6.63 \times 10^{-34}\ \mathrm{J \cdot s})}{(10^{-9}\ \mathrm{kg})(1\ \mathrm{m/s})} \cong 7 \times 10^{-25}\ \mathrm{m}.$$

This wavelength is far too small to detect: Dust particles, baseballs, and airplanes do not reveal their wavelike aspects. On the atomic scale, however, things are quite different. Electrons ($m_e = 9.1 \times 10^{-31}$ kg) moving at a speed of 10^6 m/s, typical of electron speeds in atoms, have a wavelength

$$\lambda = \frac{(6.63 \times 10^{-34}\ \mathrm{J \cdot s})}{(9.1 \times 10^{-31}\ \mathrm{kg})(10^6\ \mathrm{m/s})} \cong 0.7\ \mathrm{nm}.$$

This wavelength is of the same magnitude as interatomic spacing in matter and so can be tested by diffraction experiments, such as those described in Chapter 38 for X rays.

EXAMPLE 40–5 What is the de Broglie wavelength of a neutron (mass $m = 1.6 \times 10^{-27}$ kg) with a speed $v = 1500$ m/s? (If v is taken as the rms speed of a gas of neutrons, the corresponding equilibrium temperature is around 35K. These neutrons are often referred to as "cold neutrons.")

Strategy Given the mass and velocity of a particle, its momentum $p = mv$ is readily calculated, and by the de Broglie formula, the wavelength $\lambda = h/p$ can immediately be obtained.

Working It Out The wavelength is

$$\lambda = \frac{h}{mv} = \frac{(6.63 \times 10^{-34}\ \mathrm{J \cdot s})}{(1.6 \times 10^{-27}\ \mathrm{kg})(1.5 \times 10^3\ \mathrm{m/s})} = 0.28\ \mathrm{nm}.$$

This value is comparable to the typical spacing between atoms in a crystal, so that we would expect diffraction to be visible in the scattering of such neutrons from a crystal.

What Do You Think? How does the wavelength of a neutron scale with the temperature T of the neutron gas?

Experimental Evidence for the Wavelike Behavior of Matter

Every particle has a dual wave–particle nature, analogous to the duality that light displays. For example, Compton scattering reveals that light scatters as billiard balls do, while the passage of light through a grating reveals interference. Similarly, matter behaves classically in simple scattering experiments but exhibits wavelike behavior when the proper interference experiments are attempted.

The first experiments to confirm de Broglie's conjecture about the wavelike aspects of matter were carried out in 1927 by Clinton J. Davisson and Lester H. Germer, and independently by George Paget Thomson. They found that certain scattering directions are preferred when electrons are scattered by a crystal, a behavior typical of the scattering of a wave from a regular array such as a crystal. We found the interference condition for this situation, *Bragg's law*, in Chapter 38, in our discussion of the scattering of electromagnetic radiation, which we expect classically to show interference. Bragg's law [Eq. (38–17)] states that when waves of wavelength λ are reflected from a succession of crystal planes separated by a distance d, there will be constructive interference for angles θ that satisfy (Fig. 40–8)

$$k(2d)\sin\theta = 2\pi n, \tag{40–9}$$

where n is an integer and $k = 2\pi/\lambda$ is the wave number. From Eq. (40–9), we have constructive interference at angles θ that satisfy

$$\sin\theta = \frac{n\lambda}{2d}. \tag{40–10}$$

In the Davisson–Germer experiment, it was not radiation but rather a beam of electrons that was sent through a crystal (Fig. 40–8). The spacing between the scattering planes in the crystal had been determined by X-ray diffraction experiments to be $d = 0.091$ nm. Davisson and Germer then accelerated electrons through a potential of 54 V, giving them an energy of 54 eV, or 86.4×10^{-19} J. They scattered these electrons and observed a diffraction maximum, and therefore *wavelike behavior* at 65°. The kinetic energy of the electron corresponds to a momentum p:

$$p = \sqrt{2m_eE} = \sqrt{2(9.1 \times 10^{-31}\text{ kg})(86.4 \times 10^{-19}\text{ J})} = 39.7 \times 10^{-25}\text{ kg}\cdot\text{m/s}.$$

Knowing the momentum, we can find the angle of constructive interference from Eq. (40–10),

$$\sin\theta = \frac{n\lambda}{2d} = \frac{n}{2d}\frac{h}{p} = \frac{n}{2(9.1 \times 10^{-11}\text{ m})}\frac{(6.63 \times 10^{-34}\text{ J}\cdot\text{s})}{(39.7 \times 10^{-25}\text{ kg}\cdot\text{m/s})} = 0.92n.$$

For $n = 1$, this yields $\theta = 67°$, in agreement with the measured value to the accuracy of their experiment.

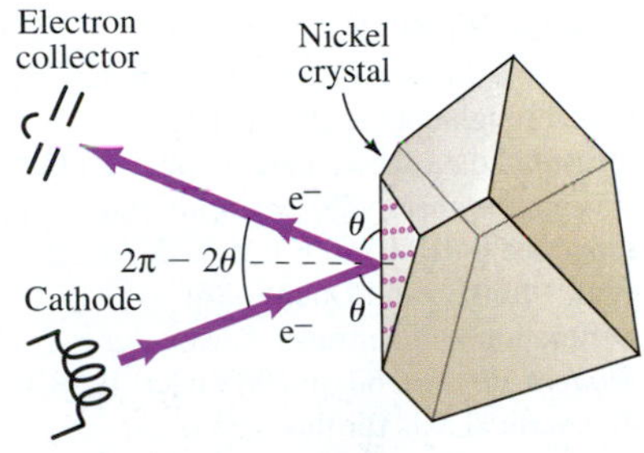

▲ **FIGURE 40–8** Experimental setup for the experiment of Davisson and Germer. Electrons from a cathode strike a surface of a nickel crystal and are scattered to an electron collector. The dots represent the regular array of nickel atoms that make up the crystal.

EXAMPLE 40–6 At what angles do diffraction peaks occur for electrons of kinetic energy 120 eV incident on a crystal whose scattering planes are 0.12 nm apart?

Strategy We must find the conditions for maxima as a function of the wavelength of the electrons, and this is given by Eq. (40–10). The wavelength itself is determined in terms of the momentum according to $\lambda = h/p$. The momentum is in turn given in terms of the energy by the nonrelativistic kinetic energy–momentum relation $p = \sqrt{2m_eE}$. We must make sure that all the units are consistent and convert where necessary to SI.

Working It Out The energy of the electrons is 120 eV $= (120\text{ eV})(1.6 \times 10^{-19}\text{ J/eV}) = 1.9 \times 10^{-17}$ J. Thus the wavelength is

$$\lambda = \frac{h}{p} = \frac{h}{\sqrt{2m_eE}} = \frac{6.6 \times 10^{-34}\text{ J}\cdot\text{s}}{\sqrt{2(9.1 \times 10^{-31}\text{ kg})(1.9 \times 10^{-17}\text{ J})}}$$
$$= 1.1 \times 10^{-10}\text{ m} = 0.11\text{ nm}.$$

With a crystal plane separation of $d = 0.12$ nm, the angles for constructive interference are

$$\sin\theta = \frac{n\lambda}{2d} = \frac{n(0.11\text{ nm})}{2(0.12\text{ nm})} = 0.47n.$$

There will be diffraction peaks at $\theta = 28°$ ($n = 1$) and at 70° ($n = 2$).

What Do You Think? Would there be more or fewer diffraction peaks with neutrons at the same kinetic energy?

Diffraction experiments have been performed with a variety of particles. In Example 40–5, we calculated the wavelength of neutrons that move at a certain speed; the wavelengths of these neutrons satisfy the conditions for substantial diffractive effects in scattering from crystals. The diffraction of neutrons by crystal surfaces is of practical importance in the study of those surfaces. Neutrons are ideal for such experiments because they can be slowed by collisions in hydrogenous materials such as paraffin, and slower neutrons have

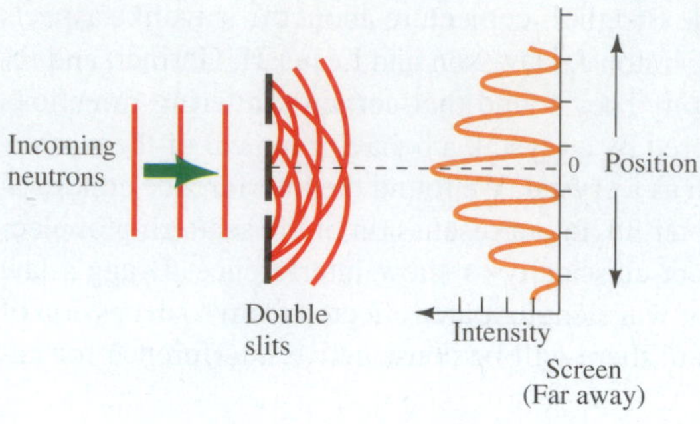

(a)

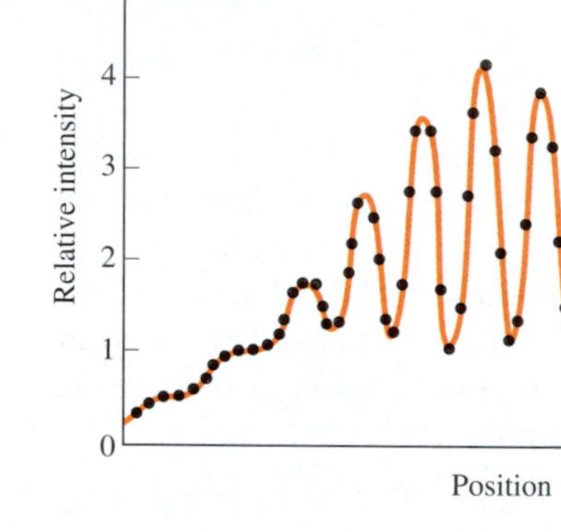

(b)

▶ **FIGURE 40–9** (a) Neutrons produce a diffraction pattern when they pass through double slits. (b) The measured data shown here involve double slits approximately 20 μm wide and separated by 104 μm (after A. Zeilinger et al., "Single- and Double-Slit Diffraction of Neutrons," *Reviews of Modern Physics,* 60, p. 1067 [Oct. 1988]). The vertical axis for this data is proportional to the rate at which neutrons arrive at a screen.

longer wavelengths. Figure 40–9 shows the results of a 1988 experiment in which neutrons of de Broglie wavelength 2 nm were incident on a screen with two slits approximately 100 μm apart, a classical wave phenomenon of the most basic type.

The simplicity of the effects described here should not mask their extraordinary nature. In many respects, electrons and neutrons are like classical particles. Electrons were discovered by subjecting them to electric and magnetic fields and studying their trajectories through observation of their impact on the screen of a cathode-ray tube. These trajectories are those of a particle of mass m_e and charge $-e$ subject to the force law $\vec{F} = -e(\vec{E} + \vec{v} \times \vec{B})$. Similarly, neutrons produced in nuclear collisions move in paths that are described by Newton's laws. In contrast, an ordinary water wave does not move as a classical particle. Yet electrons and neutrons deflected by regular structures produce interference patterns, just as a water wave would!

The mystery—How can something be simultaneously a particle and a wave?—is a repetition of the mystery of the particle properties of photons. On one hand, photons participate in collisions as if they were point masses with energy and momentum; on the other hand, we know that light is a wave phenomenon. Both of these mysteries are reconciled by quantum mechanics, as we shall discuss in Section 40–3.

CONCEPTUAL EXAMPLE 40–7 You want to explore the shape of a certain molecule by scattering electrons of momentum p from a gas of the molecules and studying the deflection of the electrons. You will be able to see finer details in the molecules by (a) increasing p; (b) decreasing p; (c) not worrying what p is.

Answer The process of observation with a beam of electrons is analogous to the study of an object with light. The limits to resolution in both cases are based on diffraction phenomena, with the resolution improving as the wavelength decreases. You can achieve better resolution with electrons of shorter wavelength and thus, by the de Broglie relation, with a larger value of momentum.

Tunneling

The phenomenon known as *tunneling*, or *barrier penetration*, is a more indirect manifestation of the wavelike properties of matter. Experiments and other natural phenomena such as nuclear fission show that particles, electrons or alpha particles (^{4}He nuclei) for example, are able to pass through a potential energy barrier from one region in space to another. Consider a classical particle with energy E_1 that is confined to a region $r_1 < r < r_2$ by some potential energy function in the energy diagram of Fig. 40–10. As long as the energy of that particle is unchanged, it will always remain in that region. In the region $r_2 < r < r_4$ of Fig. 40–10 the particle's total energy would be less than its potential energy, so that its kinetic energy would be negative, an impossibility given its form $p^2/2m$. We say that the potential energy function forms a barrier. However, quantum mechanics allows a real particle that starts in the region $r_1 < r < r_2$ to appear in the region $r > r_4$, on the other side of the barrier.

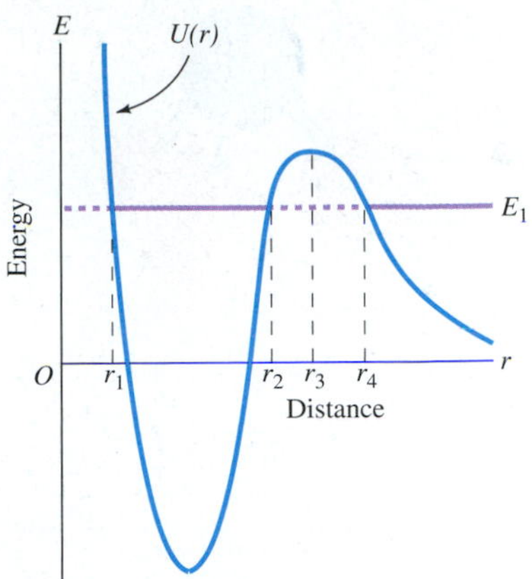

▲ **FIGURE 40–10** Energy diagram in which the potential energy forms a barrier for a classical particle with less energy than the potential energy maximum at distance r_3.

The idea of particle tunneling is completely unintuitive, inexplicable in terms of the particle aspects of matter. Still, the effect is real and is necessary to explain a large number of phenomena, including nuclear fission (Chapter 44). Tunneling today plays a large role in technology. One application you may have heard of is the tunneling microscope (Chapter 43), and there are applications in many microcircuit devices as well.

Tunneling can be explained in terms of the wavelike properties of matter. There is in fact such a thing as tunneling in optics, although it does not fall under that name. We can approach it by thinking about the phenomenon of total internal reflection in optics. When light traveling through glass reaches a glass–air surface at an angle that exceeds a critical angle, the light is completely reflected. We saw this as a consequence of Snell's law in Chapter 35, but it is really a consequence of Maxwell's equations together with the behavior of electric and magnetic fields at the interface between two different media. Solutions of Maxwell's equations reveal that the fields just outside the glass drop exponentially, not abruptly, to zero (Fig. 40–11a). If the air outside the glass forms a thin layer—no larger than a few wavelengths of the light—before another piece of glass is encountered, an exponentially reduced field remains at the second interface. Starting from there, the fields can again propagate as sinusoidal waves in the second piece of glass, although their amplitudes will be reduced by the amount of the exponential falloff that occurred in the air layer (Fig. 40–11b). *The light waves have tunneled through the air gap* with a calculable intensity. A laboratory experiment demonstrating this effect is shown in Fig. 40–12.

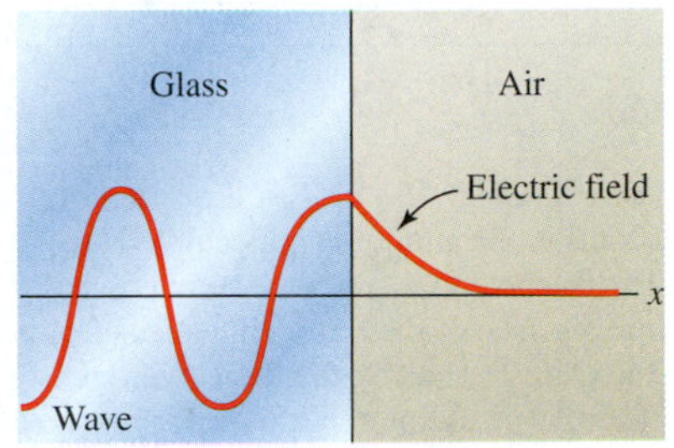

(a)

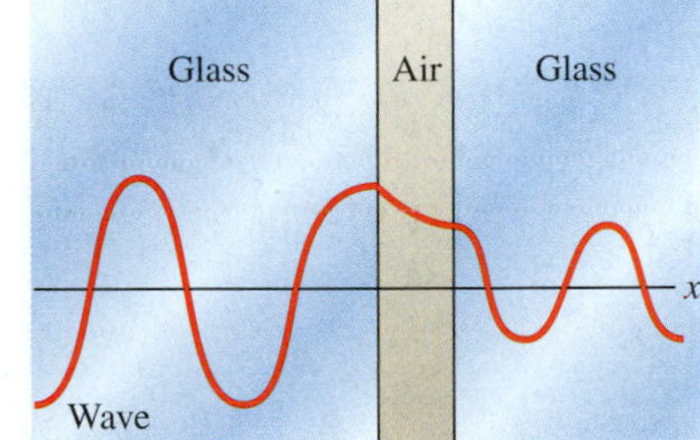

(b)

▲ **FIGURE 40–11** (a) Electric field at a glass–air interface for the case of total internal reflection. (b) Electric field at a narrow air gap between two pieces of glass, which shows the tunneling of an electromagnetic field through the air gap.

The mathematics of quantum mechanical tunneling is very similar to what we just described for classical electric and magnetic fields, although the interpretation of the quantities involved is very different. In quantum mechanics one calculates the reduction of the intensity of quantum mechanical "matter waves" that tunnel through a potential barrier. This turns out to be a calculation of *the fraction of the number of particles* that tunnel through the barrier. This fraction is very small under ordinary circumstances, which is why the phenomenon is not intuitively familiar to us. The fraction is given approximately by

$$\text{Fraction of particles getting through} \cong \exp\left[-(2a/\hbar)\sqrt{2m(\langle U\rangle - E)}\right]. \tag{40–11}$$

Here a is the barrier's width and $\langle U\rangle$ is its average potential energy. For any macroscopic barrier the fact that $\hbar$ is so small means that this is the exponential of a huge negative quantity—the fraction is extremely small. Still, there are regimes in which a is small and the barrier height is small, and in those regimes tunneling is an observable fact.

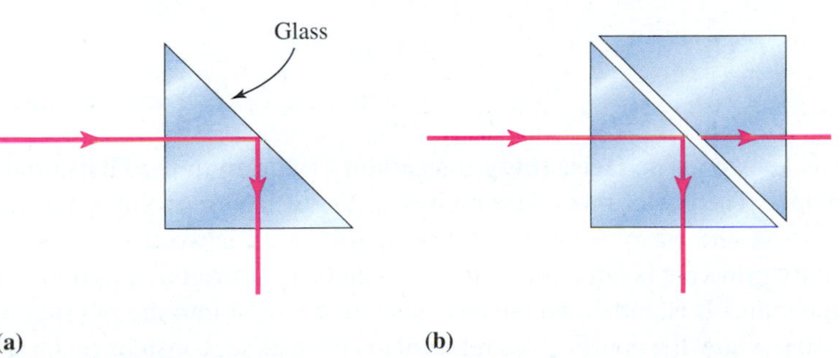

▲ **FIGURE 40–12** (a) Light incident from glass onto a glass–air interface that meets the proper geometric conditions for total internal reflection. (b) If a second piece of glass is very close to the first piece, some light will cross the air gap.

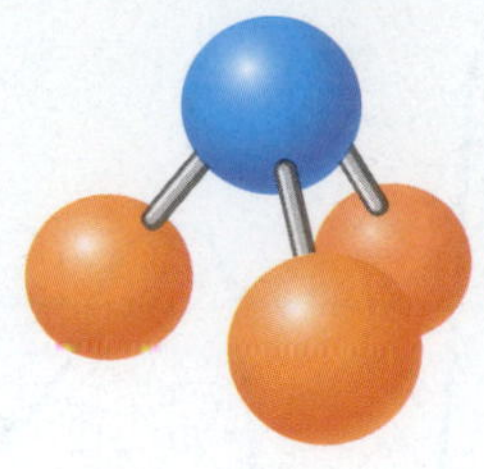

(a)

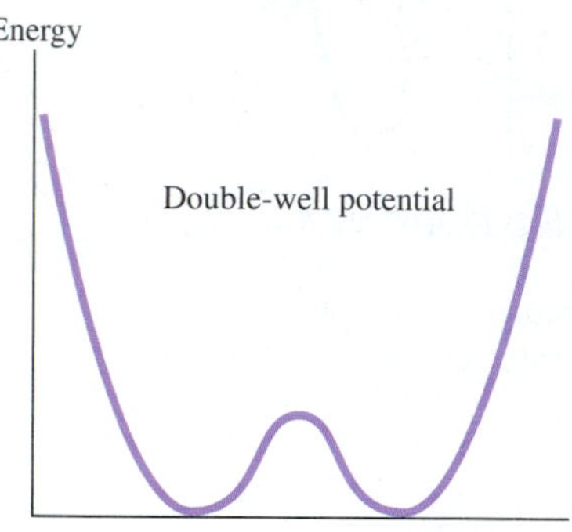

(b)

▲ **FIGURE 40–13** (a) Location of atoms in the ammonia molecule NH_3. (b) The potential energy as a function of the distance of the nitrogen nucleus from the plane formed by the three hydrogens. There is a minimum in this energy located above the plane and a second minimum located equally far below the plane.

THINK ABOUT THIS. . .

HOW IS TUNNELING USED IN ATOMIC CLOCKS?

A nice example of tunneling that makes use of the structure of the ammonia molecule led to one of the earliest atomic clocks. The ammonia molecule NH_3 has a tetrahedral structure (Fig. 40–13a). The nuclear "skeleton" of the molecule is formed by three hydrogen nuclei at the corners of an equilateral triangle lying in the xy-plane with the nitrogen nucleus lying on the z axis, which goes through the center of the triangle. The nitrogen nucleus lies above the plane of the triangle, its precise location determined by the location of a minimum of the electric potential energy shown in Fig. 40–13b created by the electrons and the hydrogen nuclei—moving the nitrogen nucleus farther away from, or closer to, the plane costs energy. There is, however, a symmetry: The potential energy *below* the triangle is identical to the one above, and the potential energy has *two* minima symmetrically placed above and below the plane formed by the hydrogen nuclei, with a potential barrier in between (Fig. 40–13b). Thus there are two equally possible equilibrium positions for the nitrogen nucleus, and if the nitrogen starts out in one, it has a certain probability, per unit time, of tunneling through to the other position. In other words, the interpretation in which a certain fraction of a large number of particles can tunnel through a barrier translates in this case into a rate at which a single particle—here the nitrogen nucleus—tunnels from one potential energy minimum to another. This tunneling rate is determined by quantum mechanics, and the regular back-and-forth tunneling of the nucleus in turn translates into an *oscillation frequency* between one minimum and another. This frequency is 23.87×10^9 Hz, and it provides a high-precision and highly reproducible clock, whose "readout" is performed by electronic circuits tuned to be in resonance with this frequency. This clock was one of the first atomic clocks because its frequency is in the so-called *radio-frequency* range, and technology using frequencies around this value was highly advanced due to the World War II development of radar. ■

40–3 The Heisenberg Uncertainty Relations

Although the electron behaves as a wave in an experiment that explictly tests for wavelike behavior, and as a particle in an experiment that tests for particle-like behavior, there are many experiments—we'll look at examples below—in which we can look at the behavior in both ways, and as we shall see, in these types of experiments there is an apparent conflict. Similar problems arise for electromagnetic radiation. We should emphasize that the difficulties are primarily conceptual; quantum mechanics gives unambiguous and correct predictions for all such experiments. Werner Heisenberg (Fig. 40–14) provided a framework in which these conflicts are resolved—or rather sidestepped. He found that within quantum mechanics one cannot assign to either matter or radiation both a well-defined position and a well-defined momentum (or velocity). Either one of them can be determined to an arbitrary degree of accuracy, but not both of them. If an experiment determines the position of a particle moving in the x-direction with an accuracy Δx, and the momentum in that same direction with an accuracy Δp_x, then there is an intrinsic limitation on these quantities given by the inequality

▲ **FIGURE 40–14** Werner Heisenberg.

$$\Delta x \, \Delta p_x > \hbar. \qquad (40\text{–}12)$$

HEISENBERG UNCERTAINTY PRINCIPLE FOR POSITION AND MOMENTUM

This is known as the **Heisenberg uncertainty relation** or the **Heisenberg uncertainty principle**. The use of this relation gives us the means of avoiding the conflict between the particle and wave views of a given experiment. It is worth emphasizing that the uncertainty principle is not a limitation that can be improved by improving an experimental apparatus. It is, rather, an intrinsic uncertainty built into the physical world.

To see how the conflicts we referred to above arise, consider first a simple example: a beam of light with frequency f strikes a screen. There is a slit of width a in the screen. We know from optical observations that on passing through the slit, the beam will spread out and appear on a screen to have an angular spread of the order of $\theta \cong \lambda/a$.

But we now know that this beam actually consists of photons, and it is this beam of photons that spreads. How is this possible? After all, a single photon can only go to a single place. One possible explanation is that when there are lots of photons, they act collectively, the way a lot of individual water molecules act in an ocean wave. This explanation won't work, however, because it is possible to perform this experiment with a beam of such low intensity that each photon was many centimeters away from the one before and the one after it, and the results remain the same. (In one realization of this experiment, it took three months' worth of radiation to build up the diffraction pattern.)

The uncertainty relation tells us how to resolve the difficulty correctly. The answer is the following: We don't know the y-position of the photon when it passes through the slit, where y measures the distance *across* the slit, which means the uncertainty in its position is of order of a. In other words, $\Delta y \cong a$, and from Eq. (40–12) its transverse momentum (its momentum in the direction across the slit) is only known to an accuracy satisfying

$$\Delta p_y > \frac{\hbar}{\Delta y} \cong \frac{\hbar}{a}.$$

But an uncertainty in its transverse momentum means an uncertainty in its angle of deflection θ from the central line of the beam:

$$\theta \cong \frac{\Delta p_y}{p} \gg \frac{\hbar}{ap} = \frac{h}{2\pi ap} = \frac{1}{2\pi}\frac{\lambda}{a}.$$

This uncertainty in the direction holds for any single photon, and this accounts for the observed spread. A collection of photons, each one of which has an angular uncertainty given as above, will produce a pattern spread over the screen in a way that is identical to the pure wave treatment of the light.

Below we'll discuss this and other examples of how conflicts are resolved in more detail. Before we do this, we want to add a couple of remarks. First, there is another relation that emerges from quantum mechanics similar to Eq. (40–12), and it reads

$$\Delta E\, \Delta t > \hbar. \tag{40–13}$$

HEISENBERG UNCERTAINTY PRINCIPLE FOR TIME AND ENERGY

The meaning of this statement is that it takes time to determine the energy of a system to a certain degree of accuracy. Finding the energy with an uncertainty ΔE requires a time of *at least* $\Delta t \cong \hbar/\Delta E$. As we shall see later (specifically in Chapter 45), the consequences of this relation include the possibility that energy conservation can be violated if it is done over a short enough period of time.

We also emphasize that the uncertainty principle recognizes an *intrinsic* uncertainty in measurement, and not one that can be fixed by making better and better instruments. The value of Planck's constant, $h \cong 7 \times 10^{-34}\ \text{J}\cdot\text{s}$, is so small that we are guaranteed that both forms of the uncertainty principle are important only on an atomic scale. For example, if we know the location of a dust particle to an accuracy of 1 micron, 10^{-6} m, then the uncertainty principle constrains our simultaneous knowledge of its momentum to an accuracy of $10^{-28}\ \text{kg}\cdot\text{m/s}$. But this momentum uncertainty is so tiny that it is overwhelmed by other more mundane experimental uncertainties, the ones due to imperfect instruments. Thus, the uncertainty principle has no practical role in the world of cars or dust particles.

When we deal with electrons in an atom, however, the situation is quite different. The mass of an electron is about 10^{-30} kg, and its speed in an atom is in the range of 10^{6} m/s. The momentum of an electron in an atom is then about $10^{-24}\ \text{kg}\cdot\text{m/s}$. The diameter of an atom is on the order of 10^{-10} m. If we try to pin down the location of an atomic electron to within 10 percent of the atom's size ($\Delta x \cong 10^{-11}$ m), then the momentum becomes uncertain to about $10^{-23}\ \text{kg}\cdot\text{m/s}$, *10 times the value of the electron's momentum in its classical atomic orbit*. The momentum becomes so uncertain that we are not even sure that the electron will stay within the atom! The uncertainty relation is so important for atoms and nuclei that Newtonian momentum is a concept that must be used with care when dealing with matter at this scale.

A closely related uncertainty relation arose in our discussion of pulses in Section 15–7. We pointed out there that a wave pulse is made of a superposition of waves with different wavelengths (or equivalently different wave numbers $k = 2\pi/\lambda$). Even though it is constructed from waves, a pulse, if it is short, resembles nothing more than a particle, and we are justified in thinking about this situation as another example of duality. We noted in Eq. (15–27) that if the pulse has a spatial width of Δx, then the spread Δk of wave numbers that go into making it up is constrained by the relation

$$\Delta x \, \Delta k \cong 1.$$

If we recall that the de Broglie relation between k and p is $\hbar k = p$, we see that this result is equivalent to the Heisenberg relation $\Delta x \Delta p \cong \hbar$.

In the remainder of this section, we shall explore in more detail how the uncertainty principle resolves conflicts inherent in a dual wave–particle model. We shall see that the uncertainty relations are also useful for making numerical estimates when quantum effects are important.

The Double-Slit Dilemma and Its Resolution

The double-slit experiment distills the conceptual difficulties of quantum mechanics to their starkest form. Diffraction experiments carried out with electrons, protons, neutrons, and a variety of molecular beams exhibit the same kinds of interference patterns seen for other waves and verify that these particles possess the wave properties predicted by quantum theory. Consider, then, a source of electrons emitted at some rate, so many per second, and impinging on a screen after they have passed through two slits in a wall (Fig. 40–15a). If we think of electrons as particles, we expect that each electron will go through

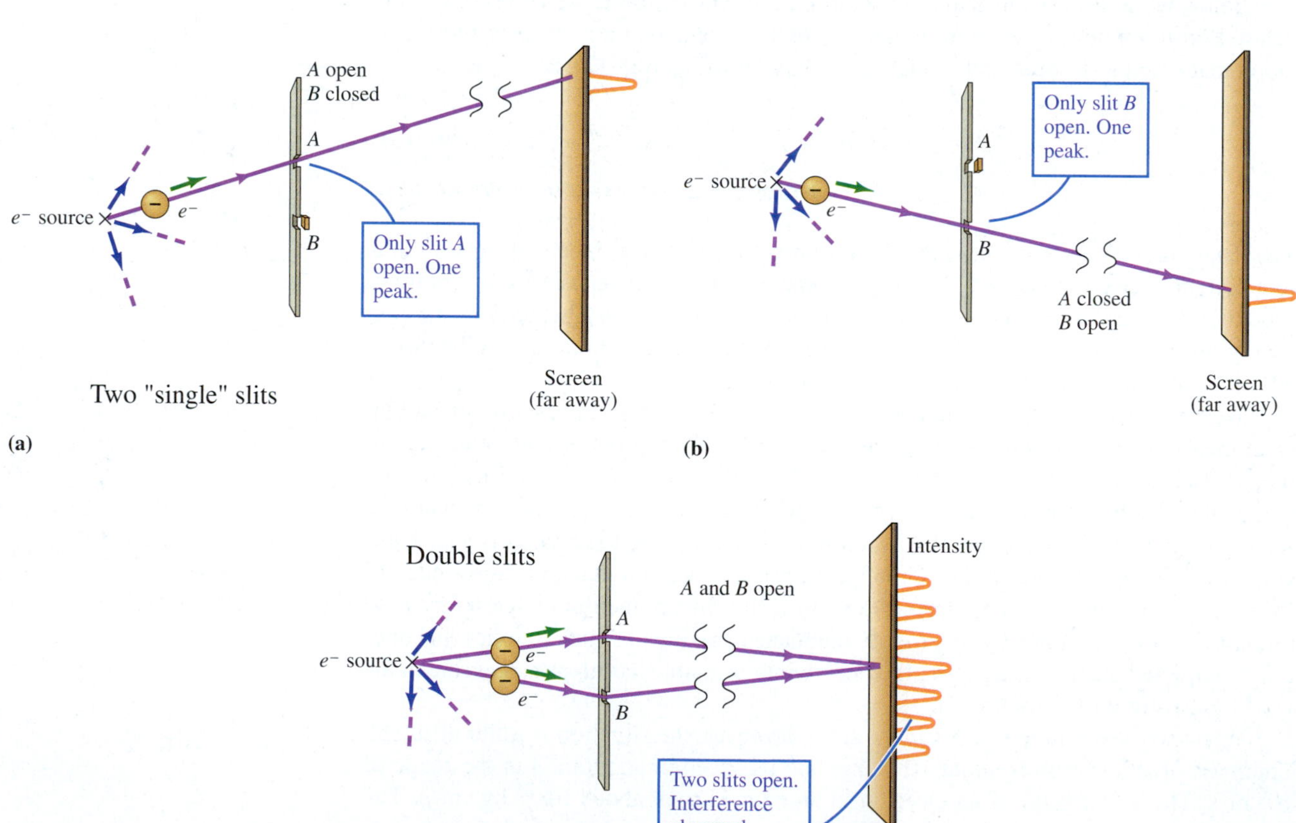

▲ **FIGURE 40–15** Schematic diagrams of the rate at which electrons arrive at a screen in a double-slit experiment: (a, b) the arrival-rate distribution, single peaks, when one slit is open at a time; (c) the arrival-rate distribution, an interference pattern, when both slits are open.

one slit or the other. In fact, if slit A is open for 5 min while slit B is closed, and then B is opened while A is closed for 5 min, then the electrons will arrive at the screen in two well-defined locations, with no interference pattern visible (Fig. 40–15b). But if the slits are open simultaneously for a total of 10 min, an interference pattern very much like that in Fig. 37–7 for light forms on the screen (Fig. 40–15c). There are maxima and minima, and we know from our earlier work on waves that for classical waves, these maxima and minima are built up through the superposition and interference of that part of the wave that passes through one slit and that part that passes through the other.

Here is the problem: The electrons leave their source as particles. The beam may be deflected and guided by crossed electric and magnetic fields until it impinges on the wall with slits. At the end, the electrons again act as particles. If they come with a low enough intensity, one can see the buildup of darkened spots on the photographic plate *one by one*. Nevertheless, the diffraction pattern, whose existence requires "part" of the electron to pass through one slit and "part" to pass through the other, emerges. Couldn't we just watch and see through which slit each electron passes? In fact, we could do better: We could watch closely enough to decide toward which slit the electron was heading and then close the other one. If we do, how can the diffraction pattern appear?

You might find the answer to this last set of questions surprising. The system of slits, together with the alert observer/monitor, does indeed give the same result as one finds by opening one slit at a time. *Somehow, observing the electrons closely enough to know through which of the two slits each electron passes destroys the interference pattern. But if we don't watch the electrons closely enough to know through which slit they pass, the interference pattern appears.* Quantum mechanics accounts for this very peculiar behavior through the position–momentum uncertainty relation.

In order to be able to tell which slit a given electron passes through in our double-slit electron experiment, an observer must use a monitor of some kind. But as we shall now see, any monitor that enables us to determine which slit each electron passes through will destroy the interference pattern. Any such monitor must be able to locate the electron's y-coordinate close to the slits to an accuracy $\Delta y < d/2$, where d is the separation distance between the slits and y is taken to be the direction across the slits (Fig. 40–16). The monitor must interact with the electrons to "see" where they are going. For example, the monitor may consist of a beam of light that reflects off the electron. The light carries momentum, and some of this momentum is transferred to the electron in a direction parallel to the screen (the y-direction in Fig. 40–16). If this momentum transfer is Δp_y, then the uncertainty relation states that

$$\Delta p_y > \frac{\hbar}{\Delta y} > \frac{2\hbar}{d}.$$

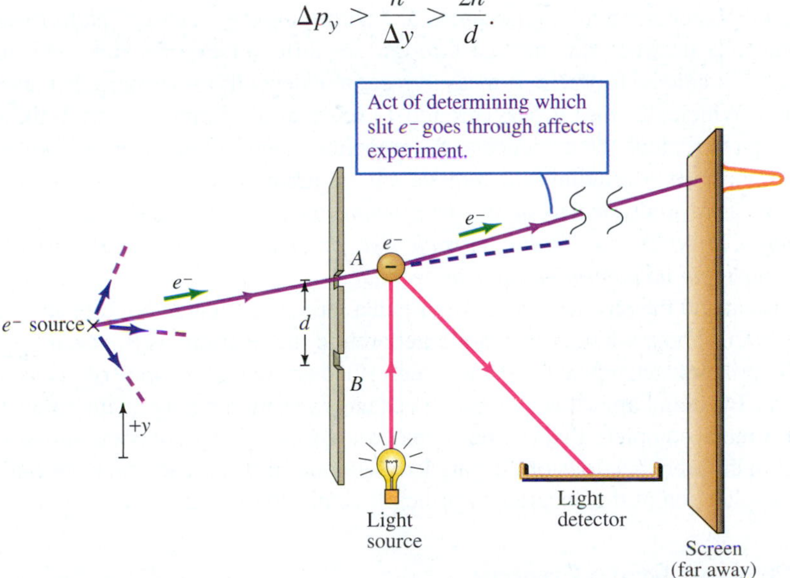

▲ **FIGURE 40–16** Schematic diagram of a monitor for a double-slit experiment designed to detect which slit an electron passes through. One mechanism: Shine light toward the slits and signal the passage of the electron by the pattern of reflected light.

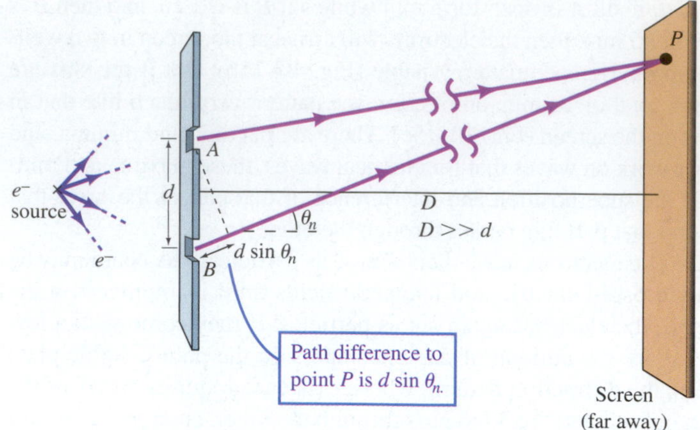

▶ **FIGURE 40–17** Geometry of path-length differences for a double-slit experiment.

This much momentum imparted to the electron is sufficient to erase the interference pattern. For a slit separation d, the angles for constructive interference—interference maxima—are given by

$$d \sin \theta_n = n\lambda,$$

where θ_n is the angle that the line leading to the nth maximum makes with the central axis (Fig. 40–17). The distance between adjacent maxima on that screen is

$$D \sin \theta_{n+1} - D \sin \theta_n = \frac{(n+1)D\lambda}{d} - \frac{nD\lambda}{d} = \frac{D\lambda}{d},$$

where D is the distance to the viewing screen. But if the monitor gives the electron a sideways (y-direction) "kick," giving it an additional momentum Δp_y in the y-direction, the angle of deflection is changed by $\Delta p_y/p$ and thus its end-point at the screen by $D(\Delta p_y/p)$. In other words, using the uncertainty relation, the displacement at the screen has a new uncertainty

$$D\frac{\Delta p_y}{p} > \frac{D(2\hbar/d)}{p} = \frac{2D/d}{k} = \frac{D\lambda}{\pi d}.$$

This displacement uncertainty is comparable to the separation between the maxima. Thus the interference pattern is wiped out.

We have shown that there is no paradox. A pure double-slit experiment and an experiment that includes a monitor capable of determining the electrons' paths to an accuracy that tells us what slit they go through are different experiments, and different patterns are predicted for them. A measurement that depends on the particle nature of an electron ("Which slit does it pass through?") must, at a minimum, disturb the system just enough to remove the evidence of the wavelike nature of the electron. Quite generally, the uncertainty relation will remove any contradiction between the particle and wave aspects of a physical system. *Any attempt to determine whether an electron (or other physical system) is "really" particle-like, or "really" wavelike, disturbs the system so much that no determination can be made.*

Einstein and the physicist Niels Bohr had a famous exchange of correspondence in the late 1920s about whether one could get around the restrictions imposed by the uncertainty principle and make the paradox described above real. In spite of his many contributions, Einstein hoped that quantum mechanics was nothing but an intermediate step toward a more complete theory and invented a series of experimental situations that brought in different versions of the paradox; without fail Bohr each time showed how a correct application of the uncertainty principle removed the difficulty.

The Ground State Energy

Quantum mechanics tells us that the minimum possible energy of a particle is not the classical minimum in which a particle sits at rest at the very bottom of a potential well, but something a little greater. The minimum energy is called the **ground state energy**, and the values of these energies for atoms and molecules is of great importance to chemistry and physics, because those states are the stable ones toward which an atom or molecule will normally gravitate. In Chapter 41 we'll see how quantum mechanics determines these energies precisely. Here we use the uncertainty relations to estimate the smallest possible energy of a particle under the influence of a given force.

Consider a particle with a potential energy $U(x)$. We choose a coordinate system such that the minimum of the potential energy is located at $x = 0$, and we set the zero of the potential energy so that $U(0) = 0$. Because the total energy of the particle is given by $E = (p^2/2m) + U(x)$, the energy is lowest, $E = 0$, when both the kinetic energy and the potential energy are lowest; classically, this occurs for $p = 0$ and $x = 0$. However, quantum mechanics does not permit a perfect localization in both p and x. If we suppose that the particle is at $x = 0$ with an uncertainty Δx, then we impose an uncertainty in the momentum p of magnitude larger than $\hbar/\Delta x$. This means that p^2 can only be known to an accuracy $(\Delta p)^2 > (\hbar/\Delta x)^2$. Thus the energy is a function of Δx. We can find the value of Δx for which the energy has its lowest value, *but that minimum energy value cannot be zero*.

To understand how this works, take the example of a particle of mass m subject to the influence of a spring of spring constant k. The potential energy is $U(x) = m\omega^2x^2/2$, where $\omega = \sqrt{k/m}$ is the angular frequency of the classical motion. The particle's energy is

$$E = \frac{p^2}{2m} + \frac{m\omega^2x^2}{2}. \tag{40–14}$$

If the particle's position is known only to an accuracy $b(\Delta x = b)$, then the uncertainty in the momentum is $\Delta p > \hbar/b$. Thus the lowest value of the energy must obey the inequality

$$E > \frac{(\hbar/b)^2}{2m} + \frac{m\omega^2b^2}{2}. \tag{40–15}$$

The right-hand side of Eq. (40–15), which we call $f(b)$, is plotted in Fig. 40–18. We see that it has a minimum as a function of b. We find this minimum from the condition that the slope of $f(b)$ is zero at the minimum, $df/db = 0$:

$$\frac{df}{db} = \frac{-\hbar^2}{mb^3} + m\omega^2b = 0.$$

When we solve this equation for b^2, we find $b^2 = \hbar/m\omega$. Substituting this value into the expression for $f(b)$, we obtain the minimum value of $f(b)$. According to Eq. (40–15), this is the minimum value of E, or ground state energy:

$$E_{\text{min}} = \frac{\hbar^2}{2m(\hbar/m\omega)} + \frac{1}{2}m\omega^2\frac{\hbar}{m\omega} = \hbar\omega. \tag{40–16}$$

This estimate can be compared to a full quantum mechanical calculation, $E_{\text{min}} = \hbar\omega/2$. The minimum energy is never zero; the particle on the end of the spring can never be brought completely to rest. This is certainly a nonclassical result!

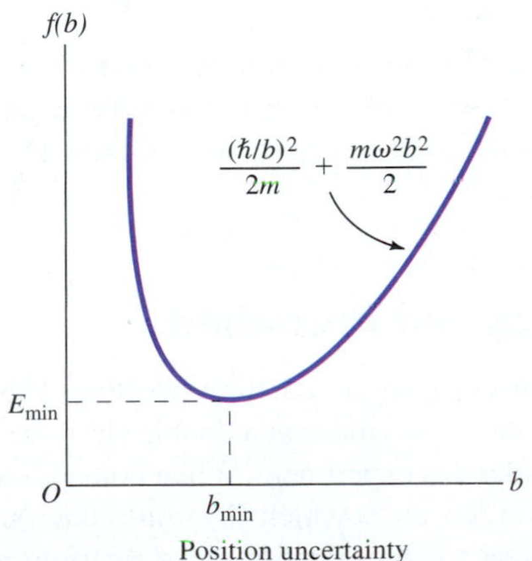

FIGURE 40–18 The right-hand side of Eq. (40–14), labeled $f(b)$, as a function of the position uncertainty b.

EXAMPLE 40–8 How would you use the position–momentum uncertainty relation to estimate the lowest energy of a particle of mass m in a one-dimensional box of width L?

Setting It Up What do we mean by saying the particle is in a box? We mean that there is a potential that confines the particle to the region from, say $x = 0$ to $x = L$, and leaves the particle free within that region. In effect, the potential is infinitely large outside the region $0 \leq x \leq L$.

Strategy This is another application of the technique described for the mass on the end of the spring. The energy consists of the purely kinetic term $p^2/2m$. This is as small as possible if the momentum itself as well as its uncertainty is as small as possible, making $E_{\rm min}$ take the form $\Delta p^2/2m$. But Δp is as small as possible according to uncertainty if Δx is as large as possible and this largest possible value is $\Delta x = L$—all we know about the particle is that it is *somewhere* in the box. In this case $\Delta p > \hbar/L$.

Working It Out $E_{\rm min} = (\Delta p)^2/2m$, or

$$E_{\rm min} = \hbar^2/2mL^2.$$

Note that this is an *estimate*. Had we used an expression for the uncertainty principle $\Delta x \Delta p_x > 2\hbar$, which is also acceptable, then $E_{\rm min}$ would have come out a factor 4 larger. The important result is that the energy is not zero. For comparison, the exact methods of quantum mechanics give for the ground state energy $E_{\rm min} = \pi^2\hbar^2/2mL^2$. Note also that as the box becomes smaller, the particle is "squeezed" into a smaller and smaller space, and the minimum energy increases because in effect its minimum momentum is larger.

What Do You Think? Couldn't we have found this result by dimensional analysis?

If we accept the introduction of $\hbar$ into dimensional analysis of quantum mechanical energies, as in the "What do you think?" question for Example 40–8, then we can use dimensional analysis as another tool for estimating ground state (and other) energies. Consider as an example a hydrogen atom, an electron orbiting a proton. Classically, this system is like that of a planet that rotates around the Sun. The lowest classical energy of this system corresponds to an orbital radius of zero; the planet is directly on top of the Sun. The position–momentum uncertainty relation suggests that the configuration of minimum energy has a finite radius. The fact that there is a minimum energy is of paramount importance—it accounts for the stability of all atoms. Without this, electrons orbiting nuclei would radiate away energy and spiral into the nucleus.

To make a dimensional analysis of the hydrogen atom, we note that the parameters that enter into the problem are the electron mass m, the combination $e^2/4\pi\varepsilon_0$ (from Coulomb's law), and $\hbar$. (You might guess that the speed of light could also enter, but later analysis shows that the motion is basically nonrelativistic.) The dimensions of our three quantities are obtained as follows: Mass is dimension M. Since $e^2/(4\pi\varepsilon_0 R)$ is a potential energy, with dimensions ML^2T^{-2}, the dimensions of $e^2/4\pi\varepsilon_0$ are ML^3T^{-2}. Finally, the dimensions of $\hbar$ are energy/frequency; that is, ML^2T^{-1}. We now write the energy in the form

$$E = m^a(e^2/4\pi\varepsilon_0)^b\hbar^c = M^a(ML^3T^{-2})^b(ML^2T^{-1})^c$$
$$= M^{a+b+c}L^{3b+2c}T^{-2b-c}.$$

Since this should have the dimensions ML^2T^{-2}, we require that $a + b + c = 1$, $3b + 2c = 2$, $-2b - c = -2$, and this gives $a = 1$, $b = 2$, $c = -2$, and dimensional analysis then gives

$$E = \text{constant} \times m\left(\frac{e^2}{4\pi\varepsilon_0\hbar}\right)^2.$$

The fact that the constant is negative could be obtained from the argument that since the electron is bound by the Coulomb potential, the total energy has to be negative. But dimensional analysis gives us no further clues to the value of the constant. As we shall see in Chapter 41, the constant is in fact $-\frac{1}{2}$.

40–4 Quantum Mechanics and Probability

The phenomena we have discussed in this chapter are far from intuitive. This is perhaps most pronounced for an electron beam that passes through a double slit—the experiment discussed in Section 40–3. Let's reconsider this experiment. When both slits are open, an interference pattern is produced, whatever the rate at which electrons go through the double-slit system. Suppose that we significantly reduce the density of electrons in the beam.

Then the interference pattern is built up by a one-by-one accumulation of electrons in regions of constructive interference, while no electrons accumulate in the regions of destructive interference. The arrival of any one electron is causally disconnected from the arrival of any other, so *each* electron must somehow carry information about the final interference pattern. Similarly, when we discussed using the uncertainty relation to reconcile particle and wave properties of photons going through a single slit, we said that the pattern would build up even if the photons came through spaced widely apart.

Quantum mechanics accounts for the double-slit experiment by assigning a **wave function** to the system. The wave function, which is a function of position, contains all the information about the diffraction pattern. The wave function (or strictly speaking, its square) at any given location on the screen is the *probability distribution function* for finding the electron at that location. The probability is largest where the magnitude of the wave function is largest. The probability that an electron will land on the screen where the waves from the two slits interfere destructively is very small; the probability for it landing where the interference is constructive is large. This implies that the outcome of the journey of any one electron is not determined; only the *probability* for a set of outcomes can be known.

An analogous situation arises when polarized light passes through an analyzer (see Chapter 35). If the polarizer makes a 45° angle with the polarization vector, then the intensity of the passed light is half that of the incident light. Classically, this is easy to understand—one component of a classical wave is blocked. But the classical description no longer works if light is formed from photons. Imagine an incident light intensity that is so low that the photons arrive one at a time. How will a particular photon "decide" whether to pass the analyzer or not? Remember that photons cannot be split into pieces that separately do different things. The resolution of this problem is that in quantum mechanics, a wave function describes the photon, the polarizer, and the analyzer, and with this wave function we can predict *only* that a given photon has a certain probability of passing the analyzer. We cannot predict whether or not a given photon will pass.

CONCEPTUAL EXAMPLE 40–9 Classical electromagnetic radiation can be polarized in directions perpendicular to the direction of propagation. Photons too can carry such polarization information. Consider photons propagating in the z-direction and passing through a polarizer arranged so that its axis is in the x-direction. As a result, all the photons that pass through the polarizer have polarization given by $\vec{e} = \hat{i}$. Let a second polarization screen—an analyzer—be placed so that its axis makes an angle θ with the x-axis. What will the polarization of one of our photons be after it passes the analyzer?

Answer As we discussed above, the probability of a single photon passing through the analyzer is $\cos^2\theta$. Just as a classical wave will have a polarization imposed by the analyzer, once the photon has passed through, it *must* have a polarization along the new axis. This polarization is $\vec{e} = \hat{i}\cos\theta + \hat{j}\sin\theta$. This is in fact confirmed by experiment. The passage through the analyzer has changed the *state* of the photon, another example of how a measurement affects the system being measured.

Example 40–9 leads to an interesting observation. Suppose the analyzer were at 90° to the axis of the polarizer. Then *no* photons would get through. If, however, we put an intermediate analyzer making an angle of 45° between the polarizer and the first analyzer, then the intermediate analyzer would pass 50 percent of the photons, and *then* the original analyzer would let 50 percent of those photons through. At the end, the photons that made it through (all 25 percent of them) will be polarized, along the y axis!

Saying that an electron has a certain probability to land at a certain spot on a screen, or that a photon has a certain probability to pass a polarizer, doesn't sound very much like Newtonian physics. Quantum mechanics is different from all other theories that we have studied so far in that it *does not make predictions about the outcome of single events. It makes predictions only about the probabilities of different outcomes.*

A particularly germane example of this type arises in nuclear radioactivity. Radioactivity occurs when a nucleus decays with the emission of, for example, an electron, an alpha particle (^{4}He nucleus), or a photon. Quantum theory predicts (and experiment confirms) that if we start with a certain number of radioactive nuclei N_0, then after a time t the number of nuclei left will be

$$N = N_0 e^{-t/\tau}. \qquad (40\text{–}17)$$

The parameter τ has the dimensions of time and is called the *lifetime*, or *mean life*, of the radioactive nucleus. (If one knows enough about the decay mechanism, the lifetime can be calculated using quantum mechanical methods.) We see from Eq. (40–17) that after a time $t = \tau$, the number of nuclei that remains is N_0e^{-1}, or about 0.37 N_0. Equivalently, the probability that a given nucleus will have decayed after a time $t = \tau$ is $1 - 0.37 \cong 0.67$. And here is what may surprise you: The value of τ is the same for a given undecayed nucleus once the particular nucleus is observed, whether the nucleus has just been artificially produced in the laboratory or has been sitting in a rock for a billion years. How does a given nucleus "know" when to decay? There is no evidence that the nucleus contains an internal clock that tells it when to decay. Because the quantum mechanical description of a single nucleus cannot contain information about what other nuclei are going to do, the only interpretation possible is that the *probability* that a nucleus will last a certain time t is part of the wave function of that nucleus, whereas a determination of precisely when that particular nucleus will decay is not.

It might appear superficially that this is no different from the problem of life expectancy in a population. There is some probability that 100-yr-old people exist in a population, but actuarial tables make no predictions for an individual. There is, however, a profound difference: People do have internal clocks, and an examination of the habits and jobs of individuals can give us more information about their life expectancy. With sufficiently detailed medical information, we could at least *in principle* make a prediction about a given lifetime. But nuclei do not age! If we start with a large initial sample of N_0 radioactive nuclei with lifetime τ, then after a time τ, there will be 0.37 N_0 nuclei left. This sample of surviving nuclei acts as a whole new sample of fresh nuclei, smaller in number, but in no way different from the ones in the initial sample. It will take an additional time τ for this new sample to be reduced to 0.37 times its initial number. This is very different from a human population. Once 100 years have passed with a certain number of people still alive, these people are not in effect reborn, and we can be quite sure that there will be none of them left after an additional 100 years have passed! This effect plays an important role in the applications of radioactivity to radiometric dating.

CONCEPTUAL EXAMPLE 40–10 You are given two samples of pure ^{235}U—one from an ancient cave, the other from a centrifuge used for separating different isotopes of uranium. There are equal numbers of nuclei in the two samples. The two samples get mixed up. Can you use radioactive decay to separate the "old" uranium from the "new" uranium?

Answer Both samples of uranium have the same decay rates. The decay rates are characteristic of the nucleus and its structure and do not depend on the history of the sample. If we currently have one milligram of each sample, the "old" one originally consisted of many more milligrams, but that is the only difference, and we cannot test for this by studying the current decay rate of the sample. One can easily imagine a sample embedded in ore from an ancient deposit—in this case an analysis of the decay products of ^{235}U that have been trapped within the ore sample would give us an idea of how old the sample is.

These concepts are far removed from what we would call "common sense." Remember, however, that common sense about the physical world is developed through observation, and there is no reason why the microscopic world should conform to what observation of the macroscopic world has established for us as sensible.

Summary

The phenomena that comprise quantum physics are most important in microscopic systems such as atoms, molecules, and nuclei. These phenomena are rather unintuitive, because our experience is based on classical physics. We have learned that what we think of as particles behave in some respects like waves, and what we think of as waves (electromagnetic radiation, for example) behave in some respects like particles. Quantum mechanics provides us with a unified explanation of all these phenomena.

When we look closely, we find that electromagnetic "waves" have particle properties. Electromagnetic radiation of frequency f behaves as though it consists of particles (photons) with energy

$$E = hf, \tag{40–1}$$

and momentum of magnitude

$$p = hf/c, \tag{40–2}$$

where $h = 6.63 \times 10^{-34}$ J·s is Planck's constant. Radiation exhibits its particle-like properties in the spectrum of blackbody radiation, in the photoelectric effect, in which electrons absorb incident photons and are ejected with specific energies from metals, and in the Compton effect, in which the wavelengths of photons scattered by electrons change in a manner dictated by the classical collisions of particles.

Just as electromagnetic radiation exhibits particle-like properties, a classical particle of momentum p has properties of a wave of de Broglie wavelength

$$\lambda = h/p. \tag{40–6}$$

The frequency associated with a particle can similarly be related to the particle's energy. The wavelike properties of particles include interference, which has been observed for particles such as electrons and neutrons in diffraction experiments, as well as in the phenomenon of tunneling. The wave properties of matter are not evident on a macroscopic scale because h is so small.

Quantum mechanics sets limits on the use of classical variables such as position and momentum. The Heisenberg uncertainty relations state that any attempt to specify the x-position with a precision Δx implies a limit with which the x-component of momentum can be measured,

$$\Delta x \, \Delta p_x > \hbar, \tag{40–12}$$

where $\hbar = h/2\pi = 1.05 \times 10^{-34}$ J·s. Similarly, there are limits on the precision of measurements of energy and time. An energy measurement is limited to a precision ΔE by the duration of time Δt that the measurement takes:

$$\Delta E \, \Delta t > \hbar. \tag{40–13}$$

These limitations are intrinsic to the quantum mechanical world, not improvable through the improvement of an experiment. They resolve potential inconsistencies between a simultaneous particle and wave description; for example, interference phenomena intrinsic to waves require a wave to be spread out in space and time, while particle-like properties demand that a particle have a definite, well-defined location. They are also useful in making estimates of ground state energies. In particular, a particle cannot be at rest at the minimum level of potential energy, so the lowest energy of a quantum system (the ground state energy) is always larger than what is expected by classical reasoning.

The difference between classical physics and quantum mechanics is most pronounced in the different types of predictions that can be made from each viewpoint. Classical physics makes direct cause-and-effect predictions of the consequences that follow from forces acting in different situations. Quantum mechanics, the paradigm that supercedes the classical picture, makes only predictions about the probabilities of different possible outcomes of a given dynamical situation.

Understanding the Concepts

1. The shorter the wavelength of a photon, the more the photon behaves like a particle. Why?
2. The uncertainty relations provide a reason why the temperature $T = 0$ cannot be reached. What is that reason?
3. A lit cigarette can be seen at a distance of 500 m on a dark night. Outline how you would estimate the rate at which photons from the cigarette hit the retina of a night-adapted eye.
4. To probe very tiny regions of space (such as the inside of a proton) with electron beams, you need electron beams of very high energy. Why? Can you estimate the kind of energy needed to study a region of diameter d?
5. Given that electrons behave like waves, how is a Doppler shift described in terms of momentum?
6. Why do we use ^{14}C for radiometric dating—why not something more plentiful, such as ^{16}O or ^{208}Pb?
7. Suppose the charge on the electron were twice its value of 1.602×10^{-19} C. Would you expect the minimum frequency for the photoelectric effect to change? If so, why?
8. Before the Planck formula was discovered Rayleigh and Jeans had obtained the expression $u(f, T) = (8\pi f^2/c^3)kT$. How could we tell that something is wrong with this expression, even with no experimental data on the subject of blackbody radiation?
9. Suppose that half of a sample of radioactive nuclei has decayed in a given time T. How long will it be before half the remaining nuclei will have decayed?
10. You start with a large sample of radioactive nuclei; how much time will elapse before you can be sure that all the radioactivity has died away to zero?
11. The lifetime τ that measures the decay rate of a sample of radioactive particles is affected by the considerations of special relativity; that is, moving radioactive particles decay more slowly than stationary ones. How do the particles "know" that they are moving and that they should decay more slowly?
12. One electron is sent through a double-slit apparatus. In what sense, if any, can we say that there is an interference pattern on the screen?
13. Consider the one-dimensional box of length L treated in Example 40–8. Is there a connection between the ground state energy and the number of wavelengths of the particle that fit the box?
14. You don't have an X-ray machine in your lab, and you want to do an experiment on the Compton effect. Is there a difficulty with using ordinary light ($\lambda = 450$ nm, say) for your purposes?
15. Does the fact that all particles, however large, have wavelike properties mean that there is some probability that a baseball can tunnel through a catcher's mitt?

16. In discussing blackbody radiation, we spoke of a cavity. What does the cavity provide? Do we mean a real cavity in bulk material?
17. Do the uncertainty relations taken together imply that there are restrictions on the simultaneous measurement of position and time?
18. An electron microscope operates by the reflection of electrons, rather than by the reflection of light, from an object. Does the use of particles such as electrons eliminate the problems associated with diffraction through the viewing aperture of the microscope?
19. Some people claim that the conservation of energy principle can be violated. How might this happen in light of the Heisenberg uncertainty principle of Eq. (40–13)? Is it really a violation of the conservation of energy?
20. In the search for a site for the disposal of radioactive waste, a stringent requirement has been imposed, which is that the waste should not have a chance to leak out for at least 10,000 years. What do you think sets this age scale?
21. Does the fact that the speed of light is a definite, predictable quantity conflict with the uncertainty relations?
22. What increases the probability of tunneling the most: halving the energy difference between U and E or halving the barrier width?

Problems

40–1 The Particle Nature of Radiation

1. (I) For what kinetic energy is the de Broglie wavelength of an electron equal to its Compton wavelength? Express your answer in units of m_ec^2 in doing the calculation, and then use $m_ec^2 =$ 0.5 MeV.
2. (I) What are the energy and momentum of a photon in He-Ne laser light of wavelength 633 nm?
3. (I) Find the energy of a photon for each of the following cases: (a) a microwave of wavelength 1.5 cm: (b) red light of wavelength 660 nm; (c) a radio wave of frequency 96 MHz; (d) an X ray of wavelength 0.17 nm.
4. (I) Use the fact that the human eye can detect as few as 15 photons per second of red light to *estimate* the minimum power needed to trigger the optic nerve.
5. (I) The work function of cesium is 2.14 eV and that of iron is 4.31 eV. What is the largest wavelength of light that will produce photoelectrons from both of these materials? If you wanted to produce a photocell that operated with visible light, would either of these materials be useful?
6. (I) A light source at $\lambda = 600$ nm emits radiation at a rate of 0.01 W. How many photons per second are emitted by that source?
7. (I) Light of frequency 0.85×10^{15} Hz falls on a metal surface. If the maximum kinetic energy of the photoelectrons is 1.7 eV, what is the work function of the metal?
8. (I) The work function of sodium is 2.75 eV. What is the energy of electrons emitted in the photoelectric effect, if any, when light of wavelength 668 nm bombards sodium?
9. (I) Calculate the maximum wavelength for the initiation of a photoelectric current in the following elements: aluminum (work function $W = 4.28$ eV), cesium (2.14 eV), nickel (5.15 eV), and lead (4.25 eV).
10. (I) A metal has a work function of 1.6 eV. What is the maximum kinetic energy of a photoelectron if radiation of 450 nm wavelength falls on the surface?
11. (II) Estimate the energy of a photon of each of the following radiation types: (a) visible light; (b) X rays; (c) microwaves; (d) television signals; (e) AM radio.
12. (II) Calculate the number of photons from an FM radio station broadcasting at a frequency of 98 MHz that is required to equal the energy contained in one photon of visible light at a wavelength of 510 nm.
13. (II) The power generated by the Sun is 4.0×10^{26} W. Assuming that it is emitted entirely at an average wavelength of 500 nm, calculate the number of photons emitted per second.
14. (II) The energy density of electromagnetic radiation in some region of space is 3.2×10^{-8} J/m^3. Assume that the radiation has a wavelength 610 nm. What is the photon density?
15. (II) Show that the Planck distribution reduces to the classical Rayleigh–Jeans distribution for high temperatures.
16. (II) Derive the Compton effect formula (Eq. 40–5) by following the steps below:
   (a) Show that if the incident photon momentum is $\vec{p}$, the final photon momentum is $\vec{p}\,'$ and the electron momentum is $\vec{p}_e$, then momentum conservation implies that the magnitudes are related to the angular deflection of the outgoing photon θ by $p^2 - 2pp' \cos\theta + p'^2 = p_e^2$.
   (b) Show that energy conservation, $pc + m_ec^2 = p'c + \sqrt{m_e^2c^4 + p_e^2c^2}$, combined with the result of (a) leads to $2pp'(1 - \cos\theta) = 2mc(p - p')$. Then use the relation between the momentum and the wavelength.
17. (II) Use the fact that the human eye can pick up as few as 5 photons/s in the visible range to estimate the intensity of the dimmest star that can be detected by a night-adapted eye. What is the ratio of this intensity to the intensity of noon sunlight, some 1400 W/m^2? This large intensity range means that the eye is indeed a very adaptable instrument.
18. (II) Show that the total energy density in a cavity filled with blackbody radiation at temperature T in degrees kelvin—that is, of energy density $u(f, T)$, given by the Planck radiation law, Eq. (17–16)—is $U(T) = aT^4$. This result is the *Stefan–Boltzmann law*. Calculate the value of the constant a, given the integral
$$\int_0^\infty \frac{x^3}{e^x - 1}\,dx = \frac{\pi^4}{15}.$$
19. (II) Use the Stefan–Boltzmann law [see Eq. (17–14)] to calculate the temperature of the Sun, given that the Sun has a radius of 0.70×10^9 m, it is 1.5×10^{11} m from Earth, and the rate at which the total solar radiation reaches Earth is 1.36×10^3 J/m·s.
20. (II) The maximum energy of photoelectrons from aluminum is 2.3 eV for radiation of wavelength 200 nm and 0.90 eV for radiation of 216 nm. Use these data to calculate Planck's constant and the work function of aluminum.
21. (II) The threshold wavelength for the photoelectric effect in tungsten is 270 nm. Calculate the work function of tungsten, and calculate the maximum kinetic energy that a photoelectron can have when radiation of 120 nm falls on tungsten.
22. (II) A photon of energy 580×10^3 eV (580 keV) collides with an electron at rest. The photon is scattered through 70° (Fig. 40–19, see next page). What is its energy after the collision? What is the kinetic energy of the electron after the collision?

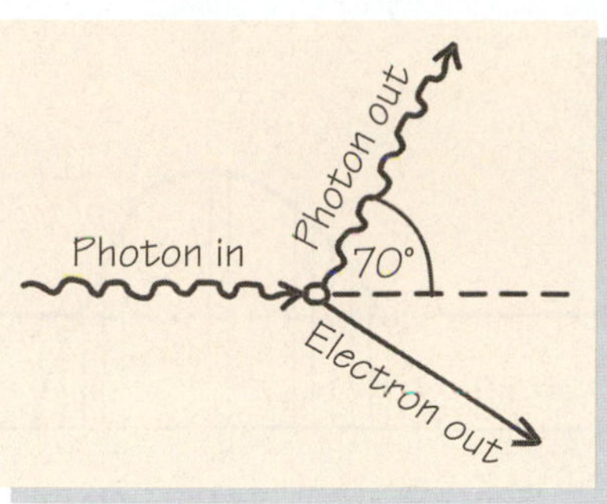

▲ **FIGURE 40–19** Problem 22.

23. (II) The wavelength of the incoming X rays in a Compton scattering experiment is 7.078×10^{-2} nm, and the wavelength of the outgoing X rays is 7.314×10^{-2} nm. At what angle was the scattered radiation measured?

24. (II) Consider a case of Compton scattering in which a photon collides with a free electron and scatters backward while it gives up half its energy to the electron (Fig. 40–20). (a) What are the frequency and energy of the incident photon? (b) What is the electron's velocity after the collision?

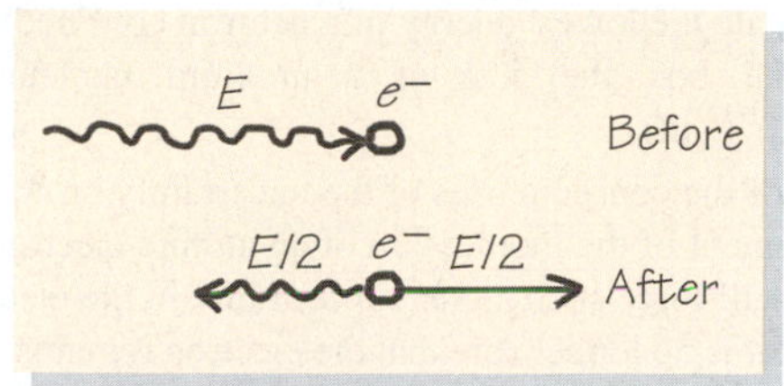

▲ **FIGURE 40–20** Problem 24.

25. (III) Consider a cavity that contains blackbody radiation at 6000°C. Calculate the energy density for radiation in the wavelength range 690 nm to 710 nm, and compare it with the energy density for radiation in the range 440 nm to 460 nm (Fig. 40–21). [*Hint*: To calculate energy density as a function of wavelength, use $u(f,T)\ df = u(f,T)\ d\lambda \times df/d\lambda$ and calculate the factor $df/d\lambda$. You must substitute $\lambda f = c$ in $u(f,T)$.]

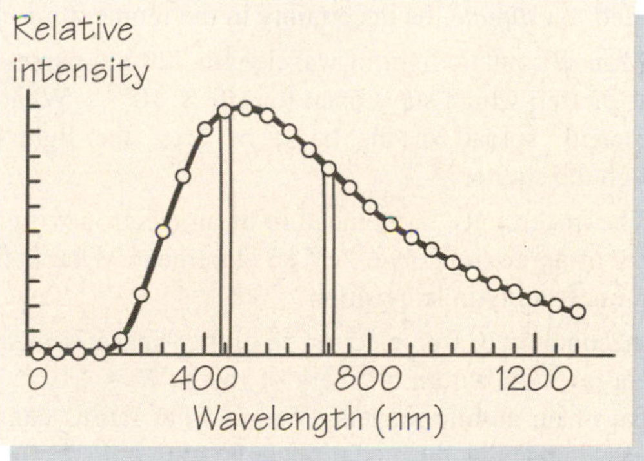

▲ **FIGURE 40–21** Problem 25.

26. (III) The Sun's radiation peaks at a wavelength of 500 nm. How much less is the radiation intensity at 450 nm and at 550 nm? Use the approach of Problem 25.

27. (III) A consequence of the Stefan–Boltzmann law (see Problem 18) is that a blackbody at a temperature T emits radiation at a rate $P(T) = \sigma T^4$, where $\sigma = 5.68 \times 10^{-8}\ \mathrm{W/m^2 \cdot K^4}$. Suppose we have two large plates, one maintained at a temperature of 300K and the other at a temperature of 6000K, and that we place a small black plate in between these two. The fact that it is black implies that it absorbs all the radiant energy falling on its surface. Assuming that at equilibrium this plate emits as much energy as it absorbs, what temperature will it acquire on reaching equilibrium?

40–2 The Wave Nature of Matter

28. (I) What is the kinetic energy of an electron whose de Broglie wavelength is that of visible red light, 620 nm?

29. (I) What is the de Broglie wavelength of an electron whose kinetic energy is (a) 5.0 eV? (b) 50 eV? (c) 50,000 eV? (d) 5.0×10^8 eV? (e) What size targets would you need in order to observe diffraction of electrons of each of these wavelengths?

30. (I) What is the de Broglie wavelength of a proton (mass 1.7×10^{-27} kg) with kinetic energy (a) 1.0 MeV; (b) 10 MeV; (c) 300 MeV? Neglect relativistic effects.

31. (I) Ultracold neutrons can have speeds as low as 100 m/s. What is the de Broglie wavelength of such neutrons?

32. (I) The spacing between scattering planes in a crystal is 0.45 nm. What is the scattering angle from such a crystal with electrons of energy 22 eV for which a first maximum is observed?

33. (I) Consider a crystal with a planar spacing of 0.074 nm. (a) What energies would electrons need for you to be able to observe up to four interference maxima? (b) Repeat the problem for neutrons.

34. (I) In a neutron two-slit diffraction experiment, the slits are 100 μm apart. If the third diffraction maximum is located at an angle of 2.0×10^{-7} rad, what is the kinetic energy of the neutrons?

35. (I) Express the de Broglie wavelength of a particle of mass m and kinetic energy E in terms of m and E. Suppose that the particle moves at a speed close to the speed of light. How is that relationship modified?

36. (I) You are thinking of testing de Broglie's conjecture by shooting raisins (m about 1 g) at a set of slits that are suitable for microwave interference experiments (λ about 1 cm). Accordingly you adjust the momenta of the raisins so that their wavelength is about 1 cm. Estimate the time it takes for a raisin to get from the raisin gun to the screen with the slits 1 m away from the gun.

37. (II) Suppose that you want to carry out diffraction experiments with the protons of Problem 30. What spacing of scatterers would you need in each of the three cases of energy 1.0 MeV, 10 MeV, and 300 MeV?

38. (II) What is the de Broglie wavelength of a neutron whose kinetic energy is equal to the average kinetic energy of a gas of neutrons of temperature $T = 85$K?

39. (II) Crystals are studied by means of electron and neutron diffraction as well as by X-ray diffraction. Recall that the typical interatomic distance in a crystal is 10^{-8} cm. Estimate the energy an electron must have to be useful for diffraction experiments on crystals. Repeat the exercise for neutrons.

40. (II) Although the working of an electron microscope does not depend on the wave nature of matter, the waves associated with electrons do set a limit on the resolving power of such instruments. (a) If the electrons in an electron microscope have a kinetic energy of 2.5×10^4 eV and the aperture of the microscope is 3.5×10^{-4} m, estimate the smallest angle that can be resolved.

(b) How much energy would electrons need so that two objects separated by 5 nm could be resolved? Give your answer in electron-volts.

41. (II) A beam of electrons with energy 1.0 eV approaches a potential barrier with $\langle U \rangle = 2.0$ eV, whose width is 0.10 nm (Fig. 40–22). Estimate the fraction of electrons that tunnel through the barrier.

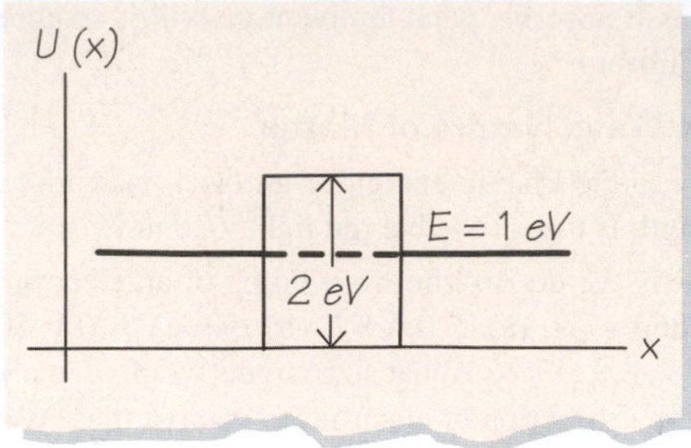

▲ **FIGURE 40–22** Problem 41.

42. (II) Suppose that the barrier in Problem 41 is 0.25 nm thick. What fraction of electrons tunnel through the barrier?

43. (II) A truck of mass 3000 kg rolls, with its transmission in neutral, at 2.0 m/s and approaches a smooth bump whose average height is 25 cm and whose average width is 10 m (Fig. 40–23). Estimate the tunneling factor for the truck from Eq. (40–11).

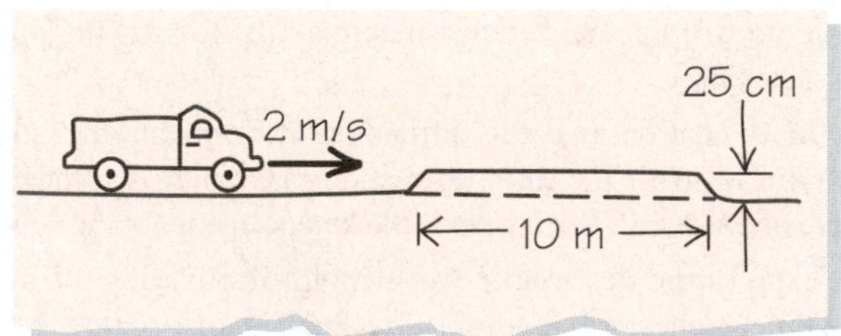

▲ **FIGURE 40–23** Problem 43.

44. (II) An electron in a semiconductor device is contained by a potential barrier of width 2.5 nm. The difference between the average potential energy and the energy of the electron is 1.4 eV. By what factor is the probability of tunneling changed if (a) the barrier width is halved? (b) The energy of the electron is increased by 1.2 eV? (c) The barrier width is halved and E is increased by 1.2 eV?

45. (III) The fraction F_n of particles of mass m and energy E that tunnel through a barrier whose potential energy is $U(x_n)$ and (minimal) thickness is Δx_n is given by Eq. (40–11).

$$F_n = e^{(-2/\hbar)\Delta x_n \sqrt{2m[U(x_n)-E]}}.$$

Suppose that a sequence of adjacent barriers at positions x_1, $x_2, \ldots$ have potential energy $U(x_1)$, $U(x_2), \ldots$ and thickness Δx_1, $\Delta x_2, \ldots$ respectively (Fig. 40–24). The fraction of particles that penetrate these barriers is given by the product of probabilities $F(x_1)$, $F(x_2), \ldots$ that the particle can penetrate the individual barriers independently, namely $F = F(x_1)F(x_2), \ldots$ Show that in the limit that each barrier is infinitesimally thin, we get

$$e^{(-2/\hbar)\int dx \sqrt{2m[U(x_n)-E]}}$$

for the fraction of particles that penetrate the barrier. Note that Eq. (40–11) is an approximation to this result.

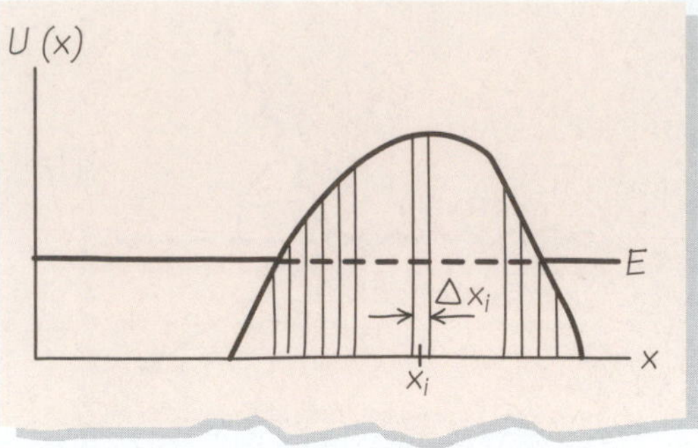

▲ **FIGURE 40–24** Problem 45.

40–3 The Heisenberg Uncertainty Relations

46. (I) An electron is localized inside a cubic region of sides 1.5 nm. What is the uncertainty in its kinetic energy?

47. You are given a detector that can time the arrival of a particle to within an accuracy of 10^{-15} s. What is the accuracy of the energy determination possible with such a detector?

48. (I) The speed of an electron emitted from an atom is measured to a precision of ± 0.08 cm/s. What is the smallest uncertainty possible in the electron's position?

49. (I) Estimate the lowest energy of a neutron confined in a one-dimensional box the size of a uranium nucleus (diameter 1.5×10^{-14} m).

50. (I) One of the consequences of the uncertainty principle is that a measurement of the momentum of an atomic electron that is too precise will force such a great uncertainty in the electron's position that it is no longer sure that the electron remains in the atom. In other words, the measurement has knocked the electron out of the atom. Estimate that precision. Compare your estimate of Δp with the typical momentum of an atomic electron, which corresponds to a kinetic energy on the order of 27 eV.

51. (I) A proton in a tin nucleus is known to lie within a sphere whose diameter is about 1.2×10^{-14} m. What are the uncertainties in the momentum and kinetic energy of the proton?

52. (I) One measure of the uncertainty in the position of a particle in a simple harmonic oscillator is to set that uncertainty equal to the amplitude of the motion. In turn, we find the amplitude by writing the energy in terms of the classical amplitude $E = m^2\omega^2 A^2/2$ and setting E equal to $n\hbar\omega$. Use the amplitude you find to *estimate* the uncertainty in the momentum.

53. (II) Monochromatic light of wavelength 720 nm passes through a fast shutter, which stays open for 1.0×10^{-9} s. What will the wavelength spread of the beam be after the light emerges through the shutter?

54. (II) The uncertainty in momentum of an electron with a kinetic energy of approximately 6 keV is 12 percent. What is the minimum uncertainty in its position?

55. (II) A completely free electron in empty space is measured to have a position within a sphere of radius $R = 1.0 \times 10^{-14}$ m, typical of an atomic nucleus. Within what radius can you say with assurance that the electron will be found after 1.0 s? Repeat the problem for an electron initially measured to lie within a sphere of radius $R = 1.0 \times 10^{-10}$ m, the radius of an atom.

56. (II) A radar gun measures the speed of a projectile, mass 28 g, to be 450 mi/h, to an accuracy of 0.5 percent. How well can the position of the projectile be determined in principle? Such accuracy cannot be achieved in practice.

57. (II) A wide beam of electrons of momentum p impinges on a slit of width a. Classically, the width of the beam of electrons passing through the slit and impinging on a screen that is a distance D from the slit would be a (Fig. 40–25). Even if we continue to treat electrons as particles, quantum effects operating through the uncertainty principle will modify this result. Using the uncertainty relations, find the spread of the beam on the screen. For what value of a will the width of the beam be a minimum?

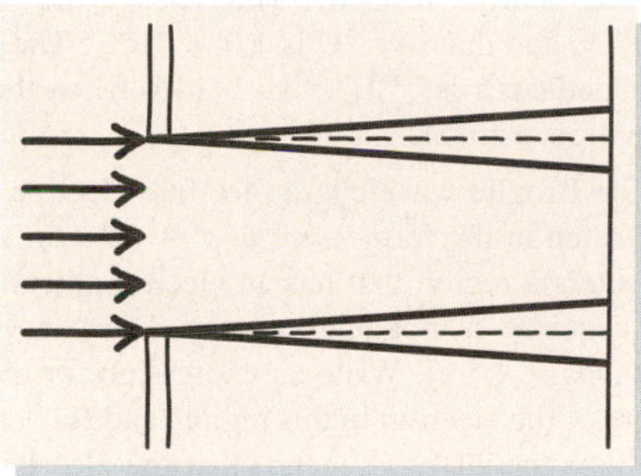

▲ **FIGURE 40–25** Problem 57.

58. (II) Consider a particle of mass m with potential energy of the form $U = -U_0(x/a)$ for $x < 0$ and $U = +U_0(x/a)$ for $x > 0$ (Fig. 40–26). Use the uncertainty relations to estimate the lowest energy the particle can have.

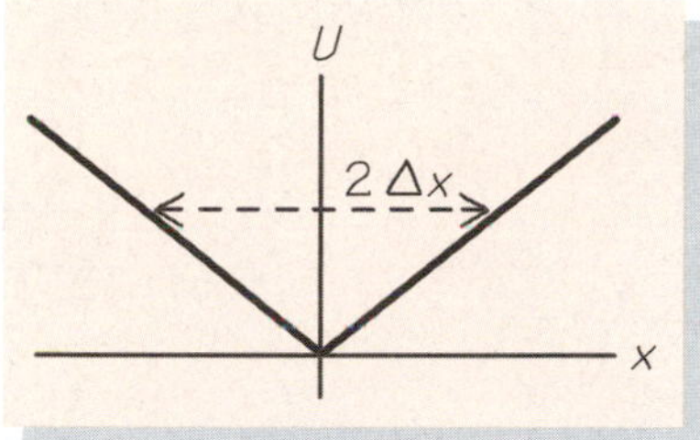

▲ **FIGURE 40–26** Problem 58.

59. (II) A neutron, mass $m = 1.6 \times 10^{-27}$ kg, is localized inside a carbon nucleus, of radius 6.0 fm ($1 \text{ fm} \equiv 10^{-15}$ m). Use the uncertainty relations to calculate a minimum (negative) potential energy for neutrons inside heavy nuclei.

60. (II) A hydrogen molecule can be modeled as a dumbbell in which the two masses—the hydrogen atoms—are connected by a spring whose unextended length is 0.074 nm (Fig. 40–27). The hydrogen molecule can vibrate along its axis with frequency 1.32×10^{14} Hz. (a) Calculate the minimum energy of this vibration. (b) A deuterium molecule D_2 differs from the picture we describe only in the masses of the atoms, which are twice that of hydrogen. Estimate the vibration frequency and zero point energy of the deuterium molecule.

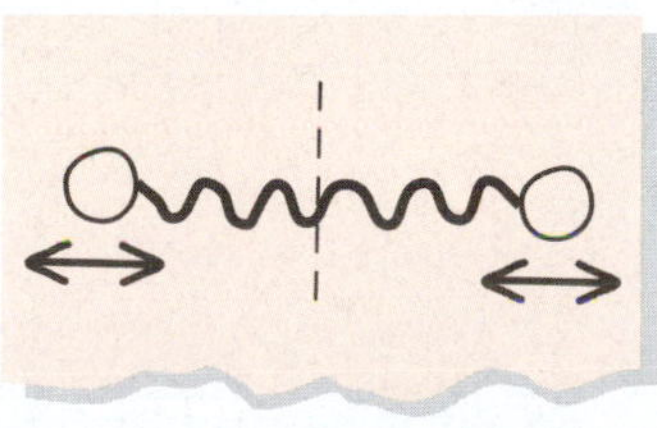

▲ **FIGURE 40–27** Problem 60.

40–4 Quantum Mechanics and Probability

61. (I) A beam of light is sent along the $+z$-axis through a polarizer so that it is polarized in the direction of the $+x$-axis. This beam impinges on a second polarizer that makes a 55° angle with the x-axis. What is the probability that a photon will get through the second polarizer?

62. (I) The number of atoms in an excited state whose mean lifetime for single-photon decay is τ is given by $N(t) = N(0)e^{-t/\tau}$. What is the number of photons emitted per second?

63. (II) The half-life of ^{14}C is about 5730 yr (see Problem 64). Organisms accumulate this isotope from the atmosphere while they live but cease doing so upon dying. The skeleton of a mammoth is found to have a concentration of ^{14}C that is 15 percent of the atmospheric value (Fig. 40–28). When did the mammoth live? Assume that the concentration of atmospheric ^{14}C does not change.

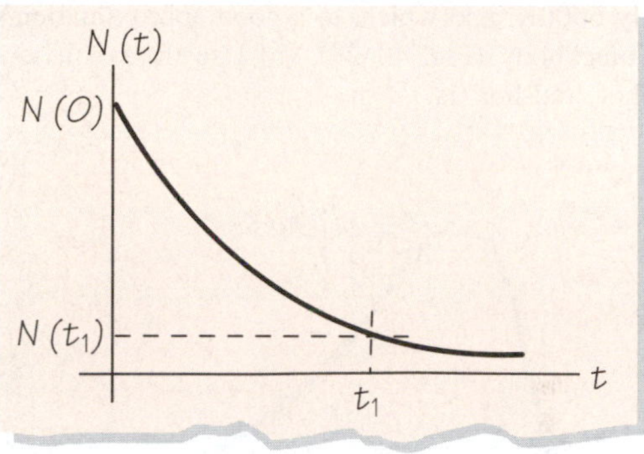

▲ **FIGURE 40–28** Problem 63.

64. (II) The *half-life* of a set of radioactive nuclei is the time in which half of the nuclei decay. Express the half-life in terms of the lifetime, τ, that appears in Eq. (40–17).

65. (II) Equation (40–17) may be written in the form $dN = -(N/\tau)\, dt$. Interpret this equation, including the sign. Does your interpretation support the assertion that the decay of any one radioactive nucleus is unaffected by the presence of others?

66. (II) Volcanic eruptions can be dated by analyzing the potassium and argon contents of rocks from the eruption. Most rocks contain potassium (K), 0.012 percent of which is ^{40}K, a radioactive isotope that decays to ^{40}Ar (argon) with a half-life of 1.3×10^9 years. Since argon is an inert gas, it is likely that all of the argon in the rock originates from the decay of ^{40}K. Suppose that a test of a rock shows that the ratio of the number of non-decayed ^{40}K and ^{40}Ar isotopes is found to be 24 : 1. When did the eruption happen?

67. (II) What is the activity of 1 g of ^{232}Th (thorium), whose lifetime is 1.4×10^{10} y? (See Problem 68.)

68. (II) A sample of radioactive material undergoing 3.7×10^{10} disintegrations/s is said to have an activity of 1 *curie* (Ci), where the activity is the rate at which decays occur. The activity of 1 g of Ra (radium) is 1 Ci. Given that $A = 226$ for radium, estimate the lifetime of ^{226}Ra.

General Problems

69. (I) What is the fractional energy loss of a photon in the Compton effect for a scattering angle of 60°, given an initial photon energy 2×10^3 eV?

70. (II) When high resolution is needed in biological microscopy, electron microscopes rather than light microscopes are used. Compare the resolution limit of an electron microscope that uses electrons of energy 12 keV with that of a light microscope.

71. (II) Find the average de Broglie wavelength of a gas of nitrogen molecules (each of mass 4.7×10^{-26} kg) at a temperature of 300K. Assume that the gas is ideal and is at 1 atm of pressure. Calculate the interatomic spacing. What would the temperature have to be for the same amount of nitrogen gas in the same volume to have a de Broglie wavelength equal to the interatomic spacing?

72. (II) *Estimate* the rate at which photons in the visible range are emitted by a flashlight. Assume that 50 percent of the energy is emitted in the visible range and that the average wavelength in this range is 550 nm.

73. (II) It is a consequence of the Planck radiation law that the energy density expressed as a function of wavelength and temperature, $u(\lambda, T)$, peaks at $\lambda = \lambda_{max}$, where $\lambda_{max} = 2.9 \times 10^{-3}\,(\text{m}\cdot\text{K})/T$, and T is the absolute temperature (see Problem 25). (a) Use this result to calculate λ_{max} for the radiation emitted by the Sun, whose surface temperature is approximately 6000K, and which, to a good approximation, radiates as a blackbody (Fig. 40–29). (b) Use the result to estimate how hot "red-hot" is.

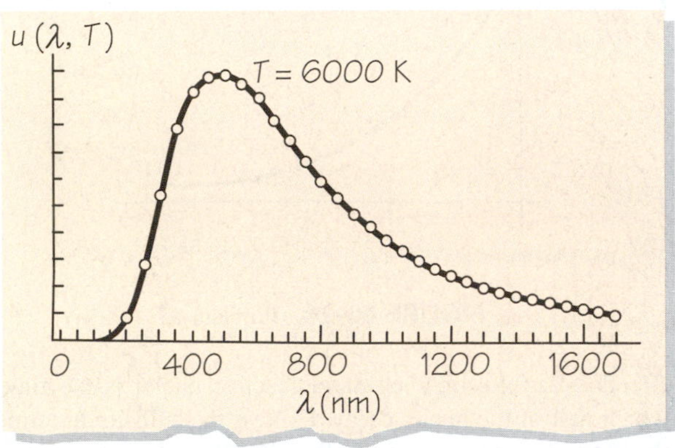

▲ **FIGURE 40–29** Problem 73.

74. (II) Show that, in the simple process of photon absorption by a free electron, without anything else occurring, energy and momentum cannot be conserved simultaneously.

75. (III) A tungsten block contains a microcrack of width 1 nm. What is the probability that an electron will tunnel from one side of the crack to the other, if the work function of tungsten is 4.6 eV?

76. (III) The abundances of ^{238}U and ^{235}U are assumed to have been equal when Earth formed. The mean life of ^{238}U is 0.6×10^{10} y, and that of ^{235}U is 1.0×10^{9} y. The present ratio of their abundances is $^{238}\text{U}/^{235}\text{U} = 140$. From these data and the assumption, estimate Earth's age.

77. (III) The de Broglie wavelength of a free electron of energy E may be written in the form $\lambda_0 = h/p = h/\sqrt{2m_e E}$. When the electron enters a region that has an electric potential $V(x)$, the electron acquires a potential energy $U(x) = -eV(x)$, and $E = (p^2/2m) + U(x)$. Write an expression for the de Broglie wavelength of the electron in this region, and show that the presence of the potential $V(x)$ means that the electron behaves as though it were moving in a region of refractive index $n(x) = \sqrt{[E - U(x)]/E}$.

78. (III) A copper crystal emits 8.3×10^{10} photoelectrons/m$^2\cdot$s. Given that the atomic weight of copper is $A = 64$ and the mass density of the crystal is 8900 kg/m^3, and assuming that the photoelectrons are supplied by the five layers of copper atoms nearest the surface of the crystal, how many atoms, on average, furnish one photoelectron/s? Assume that the atoms form a cubic lattice.

The light emitted by the various sources in this picture have colors that are associated with electromagnetic radiation of various wavelengths. These wavelengths are emitted by different species of atoms in characteristic patterns that depend among other things on the energy levels of those atoms.

Atomic and Molecular Structure

In classical physics, any particle or system of particles has a continuous range of possible energies—whether it is free or confined to orbits. Experiments carried out on atoms and molecules have shown, however, that energy can take on only discrete values. This surprising fact can be explained by quantum mechanics and has to do with the wavelike behavior of matter. The discreteness (or quantization) of energy values is mathematically similar to the discreteness of standing-wave modes of a string fixed at two ends. Another unexpected phenomenon is the quantization of angular momentum. The quantization of both energy and angular momentum plays an important role in the description of the structure of atoms and molecules. When we couple this quantization with the so-called exclusion principle, we can understand many otherwise inexplicable features of microscopic systems, including the structure and stability of atoms, of molecules, and as we will see in Chapter 44, of nuclei.

41–1 The Quantization of Energy

We have stressed throughout this book that matter consists of atoms. Chemical reactions, which could be understood in terms of exchanges of units such as H (hydrogen), C (carbon), N (nitrogen), and so on, provide the first evidence for the atomic nature of matter. The discovery of the electron as a component of atoms made the understanding of the atomic structure a high-priority research goal. Ernest Rutherford's early-twentieth-century experiments revealed an atom made of negatively charged electrons orbiting a positively charged nucleus (the nucleus approximately 10^5 times smaller than the atom), like a miniature solar system; elliptical or circular electron orbits of any size would be allowed. Such variability is in sharp conflict with experiment. One hydrogen atom (^1H) is, *literally*, indistinguishable from another. We see a strictly limited number of forms of hydrogen, of helium, and of iron—there would indeed be only one form of each, if not for the possibility of differing but highly limited numbers of neutrons in the nucleus. In contrast, one would never find an exact copy of our solar system in astronomical observations. Quantum mechanics was needed to explain the atomic observations, and it did this as follows: In contrast to classical systems, *quantum systems can have only certain—quantized—values of energy*. Possible energy values are always separated by gaps. The allowed energy values have come to be called *energy levels*. In atoms, these correspond to the possible energy values of the bound electron-nucleus system. To change the configuration of an atom requires an input of energy that equals or exceeds the gap between an *occupied* energy level—there is an electron actually in it—and an unoccupied level with a larger energy, so that under ordinary circumstances, only the (unique) configuration corresponding to this *ground-level energy* is seen in the world around us. For example, virtually all the atoms in a bottle of helium at room temperature will be in their lowest energy level, the *ground state*. Ordinary thermal collisions between these helium atoms cannot supply enough energy to change very many helium atoms from the ground state to the next allowed state (an *excited state*). At sufficiently high temperatures, some non-negligible fraction of the atoms may be "kicked" up into the next, higher state; these atoms are said to be *excited*. Excited atoms can "jump" back to the lowest energy level and emit light of *discrete frequencies*. The light carries an energy E that is the difference in energies between the excited state and the ground state. In this way, energy is conserved. The emitted light consists of photons with a frequency f determined by $E = hf$, the frequency–energy relation for photons. Thus atomic energy levels can be studied by looking at the discrete radiation frequencies emitted by atoms at high temperatures (Fig. 41–1), an experimental field known as *spectroscopy*. This work was essential to the understanding of the underlying simplicity of atomic structure.

We should emphasize that the energies of all microscopic systems, including molecules, nuclei, and—as far as we understand their structure—elementary particles, are known to be quantized.

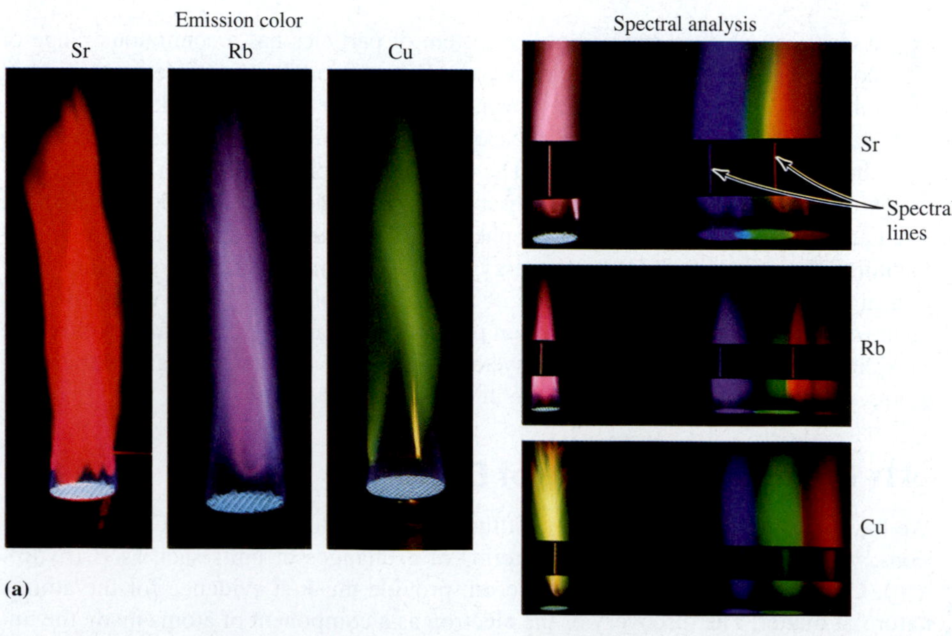

FIGURE 41–1 (a) The color emitted by atoms after they have been excited by heat is characteristic of the particular element they comprise: strontium, rubidium, and copper. (b) When the color of each element in part (a) is analyzed spectrally, it is seen to contain discrete frequencies (lines).

The Wave Nature of Matter and Energy Quantization

The wave properties of matter suggest how energy quantization might come about. Let's consider a particle confined to a one-dimensional box of length b. If the particle inside the box behaves like a standing wave on a string of length b that is fixed at its ends, then the wavelengths are constrained by the condition

$$\lambda = \frac{2b}{n}$$

[Eq. (14–8)], where $n = 1, 2, 3, \ldots$ (Fig. 41–2). The wavelength is related to the momentum p by the de Broglie relation, $\lambda = h/p$ (Eq. 40–6), so the particle's momentum has the form

$$p = \frac{nh}{2b}.$$

The energy of the particle in the box then takes on only the discrete values

$$E = \frac{p^2}{2m} = \frac{n^2h^2}{8mb^2}. \tag{41–1}$$

This expression gives us both the ground-state energy $(n = 1)$ and the gaps between the allowed energies. Our more precise calculation improves the estimate we made in Chapter 40 with the help of the position–momentum uncertainty relation (see Example 40–8).

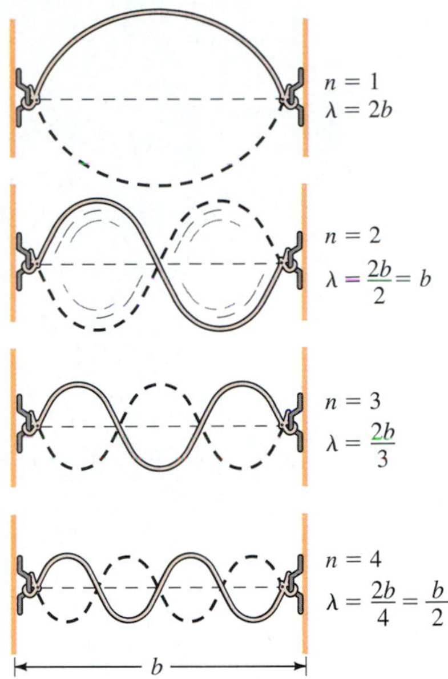

▲ **FIGURE 41–2** The wavelengths of standing waves on a string fixed at both ends are given by the length of the string divided by half-integers.

The Bohr Model of Hydrogen

In the description above, the key step is the de Broglie relation, which links intrinsically wavelike properties, including the quantization of standing waves, to the behavior of particles. But in 1913, more than a decade before de Broglie's deep insight, Niels Bohr (Fig. 41–3) daringly proposed a set of postulates to cure the difficulties of the Rutherford model of the atom. Although these postulates violated the laws of classical physics, they proved very successful in explaining many details of atomic structure.

We start with a classical description of the hydrogen atom, consisting of one electron of mass m_e and charge $-e$, and a nucleus that usually consists of a single proton,[†] of mass $m_p \gg m_e$ and charge $+e$. The classical configuration is that of the electron in circular (or elliptical) orbits with the proton at the center (or a focus), in analogy to planetary orbits (Chapter 12). For a circular orbit of radius r, Newton's second law, $F = ma$, becomes

$$\frac{e^2}{4\pi\varepsilon_0 r^2} = m_e a = \frac{m_e v^2}{r}. \tag{41–2}$$

The left side of this equation is the Coulomb force. The total energy is

$$E = K + U = \frac{p^2}{2m_e} - \frac{e^2}{4\pi\varepsilon_0 r}, \tag{41–3}$$

where zero potential energy has been chosen to be at $r = \infty$. Newton's second law, Eq. (41–2), implies that

$$\frac{p^2}{2m_e} = \frac{1}{2}m_e v^2 = \frac{e^2}{8\pi\varepsilon_0 r}.$$

When this result is inserted into Eq. (41–3), we find

$$E = \frac{e^2}{8\pi\varepsilon_0 r} - \frac{e^2}{4\pi\varepsilon_0 r} = -\frac{e^2}{8\pi\varepsilon_0 r}. \tag{41–4}$$

The energy is negative, as we would expect.

▲ **FIGURE 41–3** Niels Bohr.

[†]Deuterium is hydrogen-like in that it only has one electron, but the nucleus consists of a proton and a neutron; tritium has a nucleus consisting of one proton and two neutrons.

So far all of this is classical, and because of that the model as it stands has a fatal flaw. A charge moving in a circular orbit has a constant acceleration, and we saw in Chapter 35 that an accelerating charge radiates energy. Consequently, an orbiting electron would steadily lose energy by radiation; as it did so, the radius would decrease [see Eq. (41–4)] until the proton swallowed up the electron. Detailed estimates (see Problem 18) show that this would happen in only 10^{-10} s!

To fix the flaw, Bohr incorporated an energy quantization condition. Here are the assumptions Bohr made:

1. The classical orbits, whose parameters can take on any values, are replaced by *stationary states*, so called because the energies of those states are fixed, and hence electrons in these states cannot radiate classically. The energies of these stationary states take on only discrete values.
2. The energy values of stationary states are determined by a well-defined set of rules. For circular orbits, the rules reduce to the condition that the angular momentum, $L = mvr$, is quantized in integer units of $\hbar$:

$$L = n\hbar, \quad \text{where } n = 1,2,3,\ldots. \tag{41–5}$$

BOHR QUANTIZATION CONDITION

The integer n is the *principal quantum number* for the orbit. Equation (41–5) is called the *Bohr quantization condition*.

3. Atomic electrons can make transitions (jumps) from one allowed level with an initial energy E_i to another allowed level with a final energy E_f. When $E_i > E_f$, energy is released. The released energy can manifest itself in the appearance of a photon that carries off the excess energy $E_i - E_f$. Because a photon of energy E has frequency f given by $E = hf$, the frequency of the emitted photon is determined by

$$hf = E_i - E_f. \tag{41–6}$$

POSSIBLE ATOMIC EMISSION FREQUENCIES

An electron in an excited state, a state with energy above the ground-state energy, will undergo a *spontaneous* transition to a lower state. An electron can also make a transition to a higher energy state, but to do so it must absorb energy. This may be provided by collisions or by the absorption of a photon. In the latter case the transition is said to be *induced*. The Bohr theory made no predictions about when the transition would take place, and it explicitly ruled out the possibility of catching an electron in the middle of a transition.

EXAMPLE 41–1 A certain laser radiates with a wavelength of 3390 nm. What is the energy difference, in electron volts, between the two energy levels involved in producing this light?

Setting It Up Although we don't know the actual energy values of the levels involved in the transition, Bohr's third postulate, Eq. (41–6), brings in only the energy difference of these levels, and that is directly related to the frequency, and hence to the wavelength of the radiation emitted.

Strategy The use of Eq. (41–6) can be rewritten in terms of the wavelength by using $f = c/\lambda$, and this expression then gives us the the energy difference of the levels.

Working It Out We have immediately

$$\Delta E = hf = \frac{hc}{\lambda} = \frac{(6.63 \times 10^{-34}\,\text{J}\cdot\text{s})(3.00 \times 10^{8}\,\text{m/s})}{(3390 \times 10^{-9}\,\text{m})(1.60 \times 10^{-19}\,\text{J/eV})}$$

$$= 0.367\ \text{eV}.$$

What Do You Think? Suppose light is emitted in two steps: first a transition from E_{initial} to $E_{\text{intermediate}}$ and then a transition from $E_{\text{intermediate}}$ to E_{final}, with $E_{\text{intermediate}}$ lying between E_{initial} and E_{final}. Are the wavelengths of the light emitted greater than, less than, or equal to 3390 nm? *Answers to* ***What Do You Think?*** *questions are given in the back of the book.*

To see how the Bohr quantization condition determines the energy values, we start with Newton's second law $\vec{F} = m\vec{a}$, which for an electron in a circular orbit of radius r reads $\dfrac{e^2}{4\pi\varepsilon_0 r^2} = m\dfrac{v^2}{r}$. (A unit vector pointing to the center has been dropped here.)

When we multiply this equation by $m_e r^3$, we find

$$m_e(e^2/4\pi\varepsilon_0)r = m_e^2 v^2 r^2 = L^2. \tag{41–7}$$

We have recognized that the right side of this equation is the square of the electron's angular momentum, so that *for circular orbits* the Bohr condition yields $m_e(e^2/4\pi\varepsilon_0)r = n^2\hbar^2$. We can solve this for r to find the allowed values of the orbit radius:

$$r_n = \frac{n^2\hbar^2}{m_e e^2/4\pi\varepsilon_0} = n^2 a_0, \tag{41–8}$$

where the *Bohr radius* a_0 is the radius corresponding to $n = 1$:

$$a_0 = \frac{\hbar^2}{m_e e^2/4\pi\varepsilon_0} = 0.53 \times 10^{-10}\ \text{m}. \tag{41–9}$$

The allowed radii are discrete, with an index n corresponding to the quantum number n for L.

Once the radii can take on only certain discrete values, Eq. (41–4) shows that the energy also has certain discrete values. When the allowed values of r from Eq. (41–8) are inserted into Eq. (41–4) for the energy, we obtain the allowed energy values of a hydrogen atom in the Bohr model:

$$E_n = -\frac{e^2}{8\pi\varepsilon_0 r_n} = -\frac{e^2}{8\pi\varepsilon_0}\frac{1}{[(n^2\hbar^2)/(m_e e^2/4\pi\varepsilon_0)]}$$
$$= -\frac{m_e}{2n^2}\left(\frac{e^2}{4\pi\varepsilon_0\hbar}\right)^2 = -\frac{21.8 \times 10^{-19}\ \text{J}}{n^2} = -\frac{13.6\ \text{eV}}{n^2}. \tag{41–10}$$

The ground-state energy ($n = 1$) is $E_1 = -13.6$ eV.

Figure 41–4 illustrates the energy values predicted by Eq. (41–10). The excited-state energies get closer and closer together as n increases, and the energy approaches zero as $n \to \infty$. Positive energies correspond to the situation in which the electron is not bound to the proton. There the energy can take on continuous values because the electron can have any kinetic energy, even though it still feels an attractive force due to the positive nucleus. The minimum energy required to remove an electron in the ground state from the atom is called the *ionization energy*. For hydrogen, this energy is (0 eV) − (−13.6 eV) = 13.6 eV.

We can show by analogy with our discussion of the energy quantization in a one-dimensional box that the Bohr condition emerges from the fitting of de Broglie waves into circular orbits. The circumference of a circular orbit is $2\pi r$, and we would require that n de Broglie wavelengths fit into this length. This is a condition that sets allowed discrete radii: $2\pi r_n = n\lambda$, where λ is the de Broglie wavelength, or

$$\frac{2\pi r_n}{\lambda} = n. \tag{41–11}$$

But the wavelength itself should have an index n, because it is given generally by $\lambda = h/p$, and for a given orbit radius r_n the momentum is determined, $p = p_n$. Thus Eq. (41–11) takes the form

$$\frac{2\pi r_n}{h/p_n} = \frac{r_n p_n}{h/2\pi} = \frac{L}{\hbar} = n,$$

which is identical to Eq. (41–5).

This result was noted by de Broglie in his original work. We should stress, however, that the Bohr approach is much more closely related to what quantum mechanics ultimately says about atoms, as we shall see in Section 41–2.

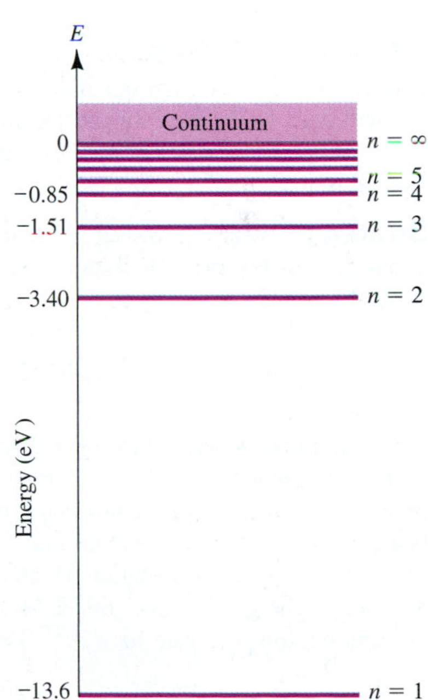

▲ **FIGURE 41–4** The energy levels in a hydrogen atom for circular orbits in the Bohr model, obtained from Eq. (41–9). The energy levels (here, not to scale) bunch up as the quantum number n increases. Above the ionization point $E = 0$, the electron and the proton are no longer bound together.

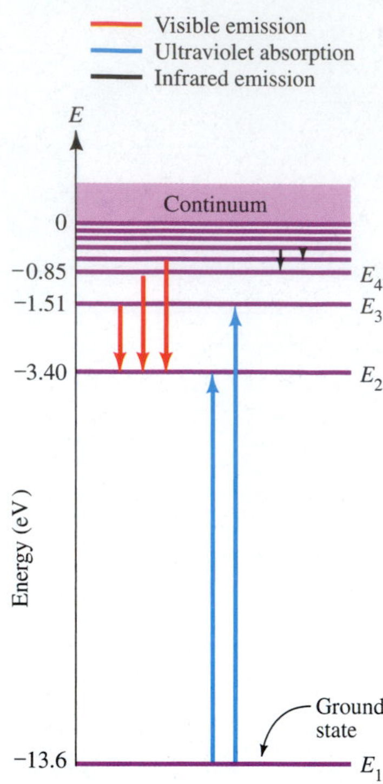

▲ **FIGURE 41–5** An atomic electron may jump down from one level to another with the emission of a photon or jump up with the absorption of a photon. The frequency of the photon is determined by the energy change.

Emission and Absorption of Photons from Atoms

Bohr's approach gave a concrete rule for the frequencies of the light, or more generally the electromagnetic radiation, emitted by atoms, namely Eq. (41–6). Because of how they appear in a spectrographic separation, one refers to these discrete frequencies as *lines*. In the special case of hydrogen, we have a formula for the possible values of the electron energies, and thus the predictions of the theory are tested experimentally by the observation of the light frequencies that hydrogen atoms emit after they have been excited (see Fig. 41–5).

For hydrogen, the possible energies E_i and E_f are given by Eq. (41–10). We may rewrite Eq. (41–11) in terms of the wavelength λ rather than the frequency by using the relation $f = c/\lambda$. Then the wavelengths of the photons emitted when an electron jumps down from an initial excited state to a lower-energy final state $(E_i > E_f)$ are restricted to the values

$$\frac{1}{\lambda} = \frac{E_i - E_f}{hc} = -R_\infty\left(\frac{1}{n_i^2} - \frac{1}{n_f^2}\right), \tag{41–12}$$

where n_i and n_f are the quantum numbers of the initial and final energies, respectively, and $R_\infty \equiv (m_e/2hc)(e^2/4\pi\varepsilon_0\hbar)^2 = 1.0974 \times 10^7\ \mathrm{m}^{-1}$ is the *Rydberg constant*, named for the spectroscopist Johannes Rydberg. Note the sign: With $n_i > n_f$ this expression is properly positive. The predicted wavelengths are in good agreement with the measured wavelengths of the spectral lines of hydrogen. In fact, by 1885 Johannes Balmer had already made a purely empirical fit to the four frequencies of the emission spectrum of hydrogen known at the time, using a formula like that of Eq. (41–12). In Bohr's time, frequencies corresponding to many values of n_f and $n_i (n_f > n_i)$ could be closely approximated by this formula, and new frequencies were correctly predicted on its basis. Bohr was on the right track.

EXAMPLE 41–2 Find the magnitude of the energy difference between the lowest energy of a hydrogen atom $(n = 1)$ and the first excited state $(n = 2)$. Estimate the lowest temperature at which about 20 percent of a gas of hydrogen atoms would occupy the first excited state.

Strategy What is involved in the first part is a straightforward application of Eq. (41–10), with the energy difference given by the general formula

$$E_{n_2} - E_{n_1} = (-13.6\ \mathrm{eV})\left(\frac{1}{n_2^2} - \frac{1}{n_1^2}\right).$$

We'll want to evaluate this for $n_2 = 2$ and $n_1 = 1$.

For the second part of the question, we require that the relative probabilities of hydrogen atoms being in the first excited state and being in the ground state are of the same order of magnitude. If we recall that for a given temperature the probability for finding atoms in the state of energy E is given by the Maxwell-Boltzmann probability distribution, proportional to $e^{-E/kT}$ [see Eq. (19–35)], then our requirement is that the ratio $e^{-E_2/kT}/e^{-E_1/kT} = e^{-(E_2-E_1)/kT} = e^{-\Delta E/kT}$ should be of the order of (20 percent)/(80 percent) = 0.25. In making this estimate we ignore excitations to higher levels. For conversion purposes we note that 1 eV is equivalent to roughly 1.2×10^4 K of temperature.

Working It Out For the first part we evaluate

$$E_2 - E_1 = E_1\left[\frac{1}{(2)^2} - \frac{1}{(1)^2}\right] = (-13.6\ \mathrm{eV})\left(\frac{1}{4} - 1\right)$$
$$= 10.2\ \mathrm{eV}.$$

For the second part we want $\exp(-\Delta E/kT) = 0.25$. Taking the natural log of both sides and using $\ln(0.25) = -1.4$, we solve for kT:

$$kT = (10.2\ \mathrm{eV})/(1.4) \cong 7.3\ \mathrm{eV}, \quad \text{or} \quad T \cong 8.8 \times 10^4\ \mathrm{K}.$$

What Do You Think? How would the numbers change if instead of dealing with hydrogen you were dealing with twice-ionized lithium, i.e. a $Z = 3$ nucleus with only one electron present?

The estimate in Example 41–2 dramatically confirms our earlier assertion that, under normal circumstances, matter will be in its ground state. We can nevertheless study the spectra of elements in the laboratory because a small fraction of atoms is excited, even at low temperatures.

The principles on which the Bohr model is based can be applied to any classical system. Example 41–3 presents another use of Bohr-model techniques.

THINK ABOUT THIS. . .

ARE THERE OTHER WAYS TO VERIFY THAT ATOMIC ENERGY LEVELS ARE DISCRETE?

In 1914, shortly after Bohr proposed his theory of the atom, James Franck and Gustav Hertz performed an important experiment (Fig. 41–6a). They constructed a tube containing a filament that, on heating, emitted electrons. These electrons were attracted to a grid held at a higher potential. On passing through the grid they went on to a collector plate, which measured the current as a function of the potential. When mercury vapor was introduced into the tube, the plot of current as a function of potential showed sharp breaks at well-defined intervals of about 4.9 V (Fig. 41–6b). The interpretation of these results required the basic picture of the Bohr atom. As the potential increases, electrons accelerate and their kinetic energy increases, leading to an increase in the current. When the free electron has enough energy to excite an atomic electron from its ground level to an excited state in an inelastic collision with a mercury atom, the free electron loses energy, and the measured current drops. The free electron reaccelerates until it again reaches the threshold for the excitation of an atomic electron from its ground state to its first excited state. Franck and Hertz also observed that above the 4.9-V threshold energy, light of a discrete wavelength was emitted from the mercury vapor. This wavelength, 253.6 nm, corresponds to the 4.9 eV energy difference between the lowest and first excited state of the mercury atom, reemitted when the atom decayed from the excited state back to the ground state. The Franck–Hertz experiment provided an immediate and dramatic confirmation of the Bohr picture.

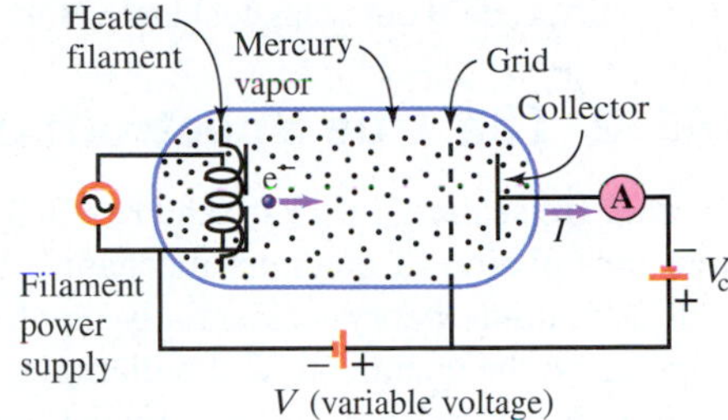

(a)

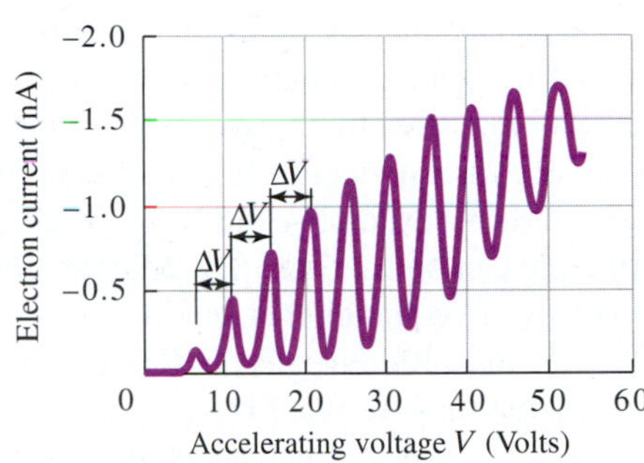

(b)

▲ **FIGURE 41–6** (a) Schematic sketch of apparatus for the Franck–Hertz experiment. (b) Current as a function of voltage in a Franck–Hertz experiment.

EXAMPLE 41–3 Consider a particle of mass m moving in a circular orbit about a fixed center under the influence of the radial force $\vec{F} = (-kr)\hat{r}$, a specific orbit for a two-dimensional harmonic oscillator. Calculate the energy spectrum of this system.

Strategy We start with the classical expression for the energy of a system, composed of a kinetic and a potential energy term. The latter is of the form $\frac{1}{2}kr^2$, and with the notation $k = m\omega_0^2$, the energy takes the form

$$E = \frac{1}{2}mv^2 + \frac{1}{2}kr^2 = \frac{m}{2}(v^2 + \omega_0^2 r^2).$$

We note here that the angular momentum, magnitude $L = mvr$, is conserved, because the force is central.

Our expression for the energy is a function of both v and r, whereas the quantized energy analogous to Eq. (41–10) contains neither. We have two additional tools to eliminate v and r. One is Newton's second law, the dynamical equation. The other is the essential quantum ingredient, the quantization of angular momentum. Once we know v and r, we can use these in the expression for the energy to find its quantized values.

Working It Out We start with Newton's second law. Both the force and the acceleration are centripetal, and the magnitude of the law reads $\frac{mv^2}{r} = kr = m\omega_0^2 r$. We solve for the speed:

$$v = \omega_0 r.$$

As the next step, we apply the angular momentum quantization condition, $L = mvr = n\hbar$, where $n = 0, 1, 2\ldots$ When we insert $v = \omega_0 r$ into this expression, we have $m\omega_0 r^2 = n\hbar$, or

$$r^2 = \frac{n\hbar}{m\omega_0} \quad \text{and} \quad v^2 = \frac{n^2\hbar^2}{m^2r^2} = \frac{n\hbar\omega_0}{m}.$$

At this point we have both r and v, and we can insert these into the expression for the possible values of energy:

$$E_n = \frac{m}{2}(v^2 + \omega_0^2 r^2) = \frac{m}{2}\left(\frac{n\hbar\omega_0}{m}\right) + \frac{m}{2}\omega_0^2\left(\frac{n\hbar}{m\omega_0}\right) = n\hbar\omega_0,$$

where $n = 1, 2, 3, \ldots$ (41–13)

In contrast to the hydrogen atom, in this case the possible energies are equally spaced.

What Do You Think? Suppose we had a collection of harmonic oscillators in equilibrium at a given high temperature, what we might call a gas of harmonic oscillators. Assuming that the only transitions allowed are those that occur between neighboring levels, what would a spectroscopist see when examining the spectrum of such a gas?

Despite the successes of Bohr's quantization rules, which work well for single-electron atoms, the rules cannot be successfully applied to multi-electron atoms. Among additional problems that the Bohr rules raised was that there was no understanding of when an electron would decide to "jump" from one orbit to another, nor of where the electron was during its jumps. Moreover, the spectrum displays some features that are not accounted for in Bohr's predictions. Perhaps most unsatisfying of all was the fact that the rules were rather artificially grafted onto classical laws. Bohr's efforts were only provisional.

41–2 The True Spectrum of Hydrogen

Werner Heisenberg (in 1925) and Erwin Schrodinger (in 1926) made the leap to the correct formulation of quantum mechanics by generalizing their extensive studies of the "old" quantum theory—essentially the theory proposed by Bohr, together with knowledge about the quantization of radiation and de Broglie's assignment of wave properties to particles. The details are beyond the scope of this book; we simply quote some of the results of quantum mechanics without attempting to derive them. The complete hydrogen spectrum is one of the topics that must be treated in this way.

Heisenberg used the **Bohr correspondence principle** as a guide for his construction of quantum mechanics. This principle, which grew out of work on the original Bohr atom, states that quantum mechanical results should reduce to those that follow from a classical treatment when quantum numbers—the principal atomic quantum number n, for example—are large. (Problems 19, 20 illustrate how the correspondence principle can be applied.) Schrodinger arrived at his formulation by attempting to construct a wave equation that would give rise to de Broglie waves. Both approaches turned out to be just different ways of formulating the same quantum mechanics, and both gave exactly the same results for the hydrogen atom, as well as for all other problems.

A full quantum mechanical treatment applied to the hydrogen atom contains several noteworthy features. We cite in particular the discarding of classical concepts in favor of a probabilistic description, a more elaborate angular momentum quantization than the Bohr quantization rule, and a more complex set of allowed energy levels. Let us discuss these features in sequence.

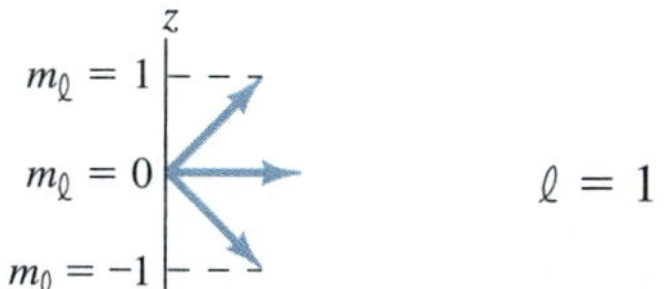

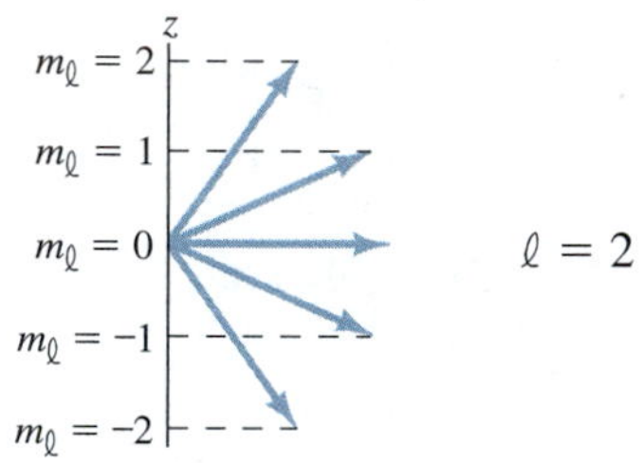

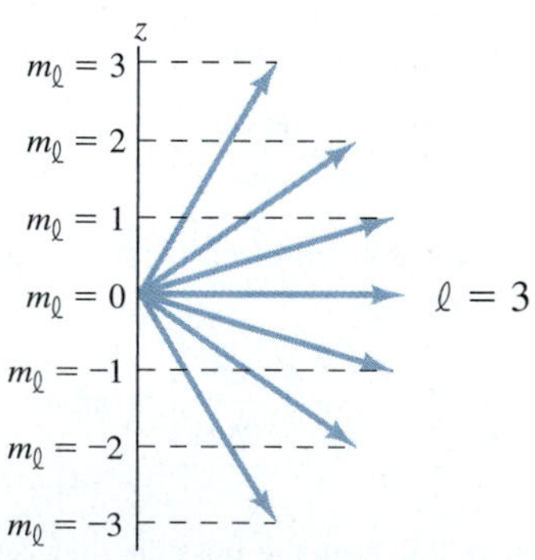

▲ **FIGURE 41–7** The directions of angular momentum vectors for $\ell = 1, 2,$ and 3 allowed by the vector angular-momentum quantization conditions. The attempt to depict a quantum phenomenon with a classical model of restricted motion is successful only in that it indicates that not all spatial orientations are allowed.

The Wave Function

In quantum mechanics the concept of orbits disappears completely. When an electron is bound to a nucleus, it is described by a *wave function* (see Chapter 40). The wave function is a solution of the **Schrodinger equation**, whose form depends on the potential energy of the electron in the presence of the nucleus. The square of the wave function at a given point is the probability distribution function—in the sense described in Chapter 19—of finding the electron at that point. The electron can exist in one of a number of *states*, each of which is described by a solution of the Schrodinger equation. These solutions, like the solutions of a wave equation for a rope with both ends fixed, are characterized by the quantization of certain physical quantities. For hydrogen, which is a three-dimensional system, these are the energy, the total angular momentum, and the projection of angular momentum along some axis. All of these variables are quantized through their dependence on a set of *quantum numbers*, numbers that take on integer values. Once the quantum numbers are specified, the wave function (and hence all the properties of the atom) is determined.

Angular Momentum

The electron's total angular momentum in the atom is quantized, but not in the way that Bohr's hypothesis specifies. In quantum mechanics it is the *square* of the angular momentum that takes on discrete values, with L^2 taking on the values $\ell(\ell + 1)\hbar^2$, where $\ell = 0,1,2,3,\ldots$ In addition, only one component of angular momentum, along a definite but arbitrary axis (which we shall call the z-axis), can be specified, and for a given value of ℓ it can only take on the values $m\hbar$, with $m = \ell, \ell - 1, \ell - 2, \ldots 1, 0, -1, \ldots -(\ell - 1), -\ell$. In other words, the z-component L_z of the angular momentum vector is itself quantized. Figure 41–7 shows the angular momentum ℓ and its allowed projections for $\ell = 1$, $\ell = 2$, and $\ell = 3$. This behavior is totally at variance with classical mechanics, although for very large L (such that $\ell\hbar$ is some macroscopic number such as $1\ \text{g}\cdot\text{cm}^2/\text{s}$), the deviations from the classical description are negligibly small.

It is important to note that the specification of the integer ℓ does not give us all of the angular momentum information, since the z-component can take on a total of $(2\ell + 1)$ different m-values, corresponding to $m = \ell, \ell - 1, \ldots, -\ell$.

In 1921, Otto Stern and Walter Gerlach observed the quantization of angular momentum directly by seeing how atoms react in a magnetic field. Their experiment depends on the fact that the magnetic dipole moment of an atom is proportional to the angular momentum. Equation (31–12) gives the classical result, and there is a quantum-mechanical equivalent in which the z-component of the magnetic moment $\vec{\mu}$ is proportional to the z-component of the angular momentum, so that $\mu_z = \text{constant} \times m$. Figure 41–8 uses a classical analogy to illustrate that the atoms will experience a force if the magnetic field is not uniform. The Stern–Gerlach experiment makes direct use of this fact. In this experiment (Fig. 41–9), a beam of atoms is passed through a region of thickness D in which there is a magnetic field $\vec{B}$ oriented in the z-direction.The atoms are characterized by a magnetic dipole moment $\vec{\mu}$, and the potential energy of the atoms in the magnetic field is given by Eq. (28–26):

$$U_{\text{mag}} = -\vec{\mu} \cdot \vec{B} = -\mu_z B.$$

If $\vec{B}$ varies with z, then there is a force on the atom proportional to its magnetic dipole moment, in the z-direction:

$$F_z = -\frac{dU_{\text{mag}}}{dz} = \mu_z \frac{dB}{dz}. \tag{41–14}$$

Classically, μ_z, which is proportional to the angular momentum component of the atom in the z-direction [see Eq. (31–12)], is not quantized. From the classical point of view, therefore, the atoms would reach the far side of the apparatus distributed in the z-direction with a uniform spread. The maximum deflection angle θ characterizing this spread would be given approximately by

$$\theta \cong \frac{\Delta p}{p} = \frac{(\text{force})(\text{time in region containing the field})}{mv}$$

$$= \mu_z \frac{dB}{dz} \frac{1}{mv} \frac{D}{v} = \frac{D}{mv^2} \mu_z \frac{dB}{dz}, \tag{41–15}$$

where $p = mv$ is the momentum of an atom of the beam.

Rather than seeing a uniform spread, Stern and Gerlach saw that their beams were split into components at specific angles. The number of components varied with the species of atom. For helium, there was only one (undeflected) component; a beam of oxygen atoms split into five components; and a beam of silver atoms split into two components. In all cases, the components were symmetrically distributed about the point of no deflection. These observations imply that the magnetic moments must be quantized, which in turn implies that the angular momentum is quantized. The initial beam of a species of atom with angular momentum quantum number ℓ consists of a mixture of atoms in each of the allowed $2\ell + 1$ states that correspond to quantization of its z-component. Each of these $2\ell + 1$ states has a different magnetic dipole moment

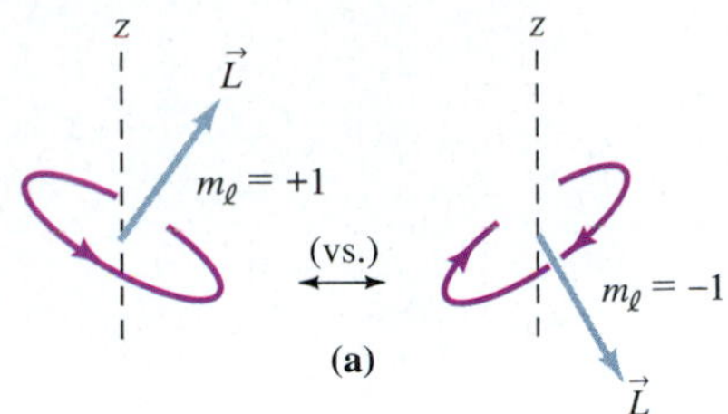

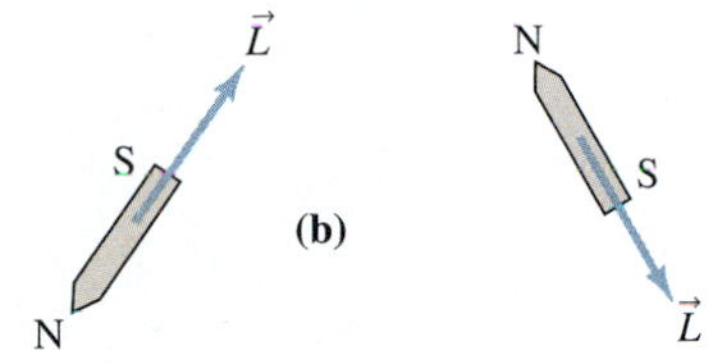

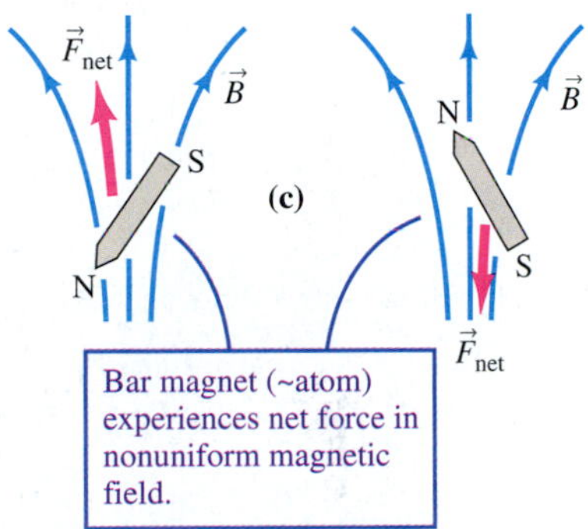

▲ **FIGURE 41–8** (a) The atom with oriented angular momentum is analogous to (b) an oriented bar magnet. (c) The bar magnet will experience a net force whose direction depends on the bar magnet's orientation if it is placed in a nonuniform magnetic field. There is a net force because the force on one end of the magnet is different from the force on the other end; in a uniform field, there is a torque but no net force.

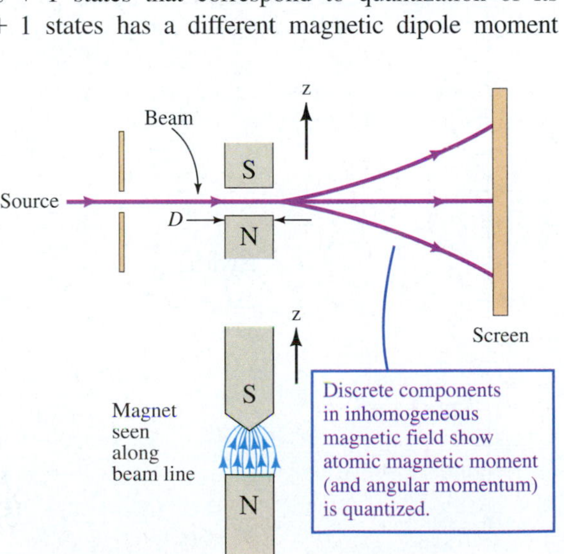

◀ **FIGURE 41–9** A Stern–Gerlach experiment. The magnetic field is not uniform. In this case, the beam is split into three components.

z-projection, proportional to $\ell, \ell - 1, \ldots, -\ell$. Thus the single component of helium corresponds to $\ell = 0$, and the five components of oxygen correspond to $\ell = 2$. We shall discuss the case of silver, for which two components are observed, in Section 41–3.

The Spectrum of Hydrogen

In quantum mechanics, the energy values of the hydrogen atom are quantized according to

$$E_n = -\frac{13.6\text{ eV}}{(n_r + \ell + 1)^2} \equiv -\frac{13.6\text{ eV}}{n^2}. \tag{41–16}$$

This is close to the prediction of the Bohr model, but in a generalized form. The original integer n of Bohr's model is replaced by $n_r + \ell + 1$, where $n_r = 0, 1, 2, 3, \ldots$ and ℓ is restricted to the integer values $0, 1, 2, 3, \ldots$ Note, however, that because n_r takes only non-negative values, $n \geq \ell + 1$; in other words, $\ell \leq n - 1$ for a given n. In addition, a state with a given ℓ really refers to the collection of $2\ell + 1$ different states—all of which have the same value of angular momentum squared.

Equation (41–16) leads to a series of energy levels that are more complex than the naïve Bohr model in that each energy level contains several states. The lowest state—the state with the lowest energy value—has $n_r = 0$ and $\ell = 0$ (or $n = 1$), and it corresponds to a single state. The next level, corresponding to $n = 2$, consists of one state for which $n_r = 1$ and $\ell = 0$, and $2\ell + 1 = 3$ states with $n_r = 0$ and $\ell = 1$; that is, there are four states for $n = 2$. The next level, corresponding to $n = 3$, consists of 1 state for which $n_r = 2$ and $\ell = 0$, 3 states with $n_r = 1$ and $\ell = 1$, and 5 states with $n_r = 0$ and $\ell = 2$—a total of 9 states. These results generalize in a straightforward way, with the total number of states labeled by n equal to n^2.

There is one more aspect of the emission spectrum of hydrogen that requires comment: The spectrum is limited by the constraint of angular momentum conservation. We have already noted that the conservation of energy and linear momentum continue to hold in quantum mechanics. The same is true of angular momentum conservation in the absence of external torques. This affects the state in which a given electron will end up when it makes a jump with the emission of a photon. Electromagnetic radiation carries angular momentum, and the angular momentum of a photon is for our purposes $L_{\text{photon}} = \hbar$. As a result, the initial and final angular momenta of the electron states must differ by one unit of $\hbar$. The set of restrictions imposed by angular momentum conservation are known as *selection rules*. Some of the possible transitions are sketched in Fig. 41–10. All the results described here are in extremely good agreement with experiment.

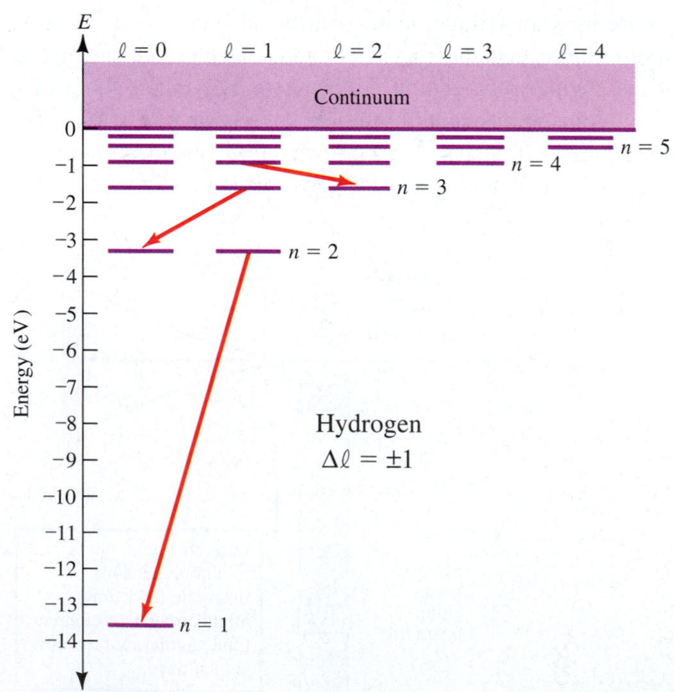

FIGURE 41–10 Some possible transitions of electrons in the quantum mechanical view of a hydrogen atom.

EXAMPLE 41–4 The *Balmer series* is a series of spectral lines that correspond to atomic transitions in hydrogen that end with a principal quantum number $n = 2$. Sketch the allowed emission transitions that lead to this series, and compute the longest and shortest wavelengths in this series.

Setting It Up An emission transition is one in which the final atomic state is at a lower energy than that of the initial state. The sketch in Fig. 41–11 shows the transitions from states with energies above the level with $n = 2$ that end up at the $n = 2$ level, each transition originating at levels characterized by $n = 3, 4, 5, \ldots$ In each case ℓ will change by 1.

Strategy We can find the wavelengths of the light emitted in the transitions from the Rydberg formula, Eq. (41–12), with the final index set to $n_f = 2$ and the initial index $n_i = 3, 4, 5, \ldots$ Thus the set of wavelengths are given by

$$\frac{1}{\lambda_{n_i,2}} = \frac{E_i - E_f}{hc} = -R_\infty\left(\frac{1}{n_i^2} - \frac{1}{2^2}\right).$$

Working It Out We have

$$\lambda_{n_i,2} = \frac{1}{R_\infty}\frac{4n_i^2}{n_i^2 - 4} = 365 \times 10^{-9}\left(\frac{n_i^2}{n_i^2 - 4}\right)\text{ m}.$$

The shortest wavelength, corresponding to $n_i = \infty$, is 365 nm; the longest wavelength, corresponding to $n_i = 3$, is 656 nm.

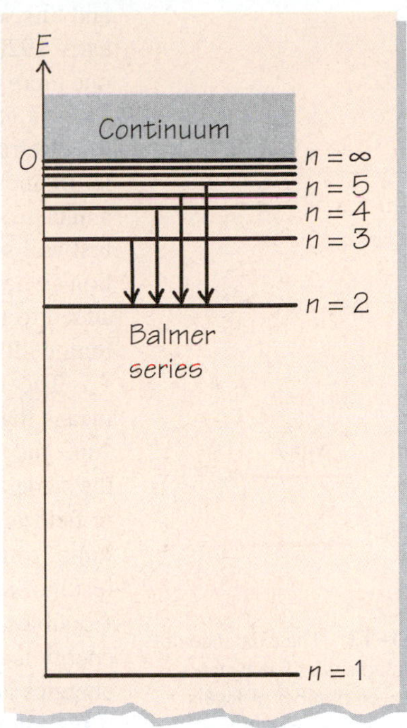

▲ **FIGURE 41–11** This series of transitions is known as the Balmer series.

CONCEPTUAL EXAMPLE 41–5 In principle we can imagine hydrogen atoms in states with very large quantum numbers, for example $n = 30$. Under what circumstances could a gas of such atoms exist?

Answer In a gas of such atoms, it would not take much energy to ionize the atoms, since the ionization energy is just $(13.6\text{ eV})\frac{1}{30^2} =$ 0.015 eV. To maintain such a gas, it is necessary that any collisions with other atoms should be gentle, and this can be accomplished only if the temperature of the gas is low enough. We could estimate the temperature T be such that $kT \ll 0.015$ eV or equivalently that $T \ll 180$ K. There is moreover a second requirement and more restrictive condition. According to Eq. (41–8), the radius of an atom with $n = 30$ is $(30)^2 = 900$ times larger than the Bohr radius, and hence the volume taken up by the atom is $(900)^2 \cong 800{,}000$ times larger than the atom in its bound state. These atoms will always be in each other's way unless the gas is very dilute. You can use the techniques of Chapter 19 to make a calculation as to just how dilute the gas must be—for example, you may want the mean separation to be a factor of 10 or 20 times larger than the atomic radius, as is typical for a normal ideal gas at STP.

41–3 The Exclusion Principle and Atomic Structure

The Spin of the Electron

When an atom with a magnetic dipole moment is placed in an external magnetic field, it has a potential energy $U_{mag} = -\vec{\mu}\cdot\vec{B}$ [Eq. (28–26)]. As we already saw in our discussion of the Stern–Gerlach experiment, each one of the $2\ell + 1$ orientations of an atom with angular momentum quantum number ℓ has a magnetic dipole moment with a different z-component. When such an atom is placed in a magnetic field, the $2\ell + 1$ orientations no longer have the same energy. If hydrogen is placed in an external magnetic field, the energies of the first excited state with $\ell = 1$ are therefore slightly *split* (Fig. 41–12). The frequencies of the radiation emitted (or absorbed) in a transition to or from one of the three (now split) levels are accordingly not quite the same, and this is detectable by experiment.

States of a given ℓ that have a common energy in the absence of a magnetic field break up into $(2\ell + 1)$-member *multiplets* with slightly different energies in the presence of a magnetic field. Because $\ell = 0, 1, 2, 3, \ldots$, only odd-valued multiplets were expected. This was not, however, what was seen: For some atoms—silver, for example—*doublets* appear; that is, there are two components, as we mentioned in our discussion

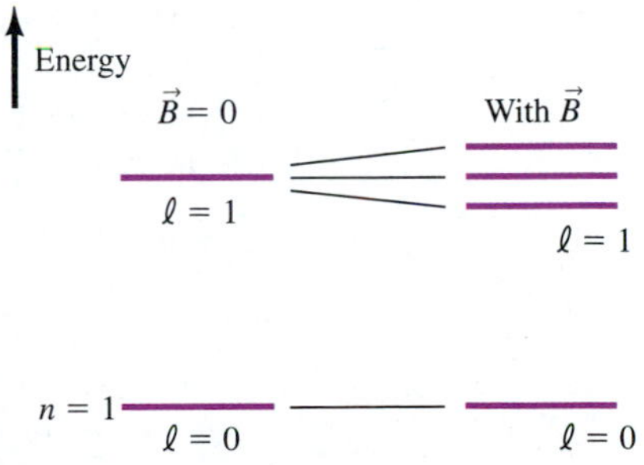

▲ **FIGURE 41–12** When a hydrogen atom is subject to an external magnetic field, the three states corresponding to $\ell = 1$, which all have the same energy in no magnetic field, have slightly different energies—the levels are split. The amount of splitting depends on the strength of the magnetic field. Note that the $\ell = 0$ state consists of only one level and is therefore not split.

of the Stern–Gerlach experiment. For a doublet, $2\ell + 1 = 2$, so $\ell = \frac{1}{2}$ for these atoms, and this was forbidden by the rules of quantum theory as they were understood in the early 1920s. In 1924, Wolfgang Pauli decided that the electron had to be described by one more quantum number, which could take only two values. A year later, George Uhlenbeck and Samuel Goudsmit explained the Pauli quantum number by proposing that the electron has an *intrinsic angular momentum*, or *spin*, $\hbar/2 \equiv s\hbar$. Whereas the angular momentum L discussed thus far is associated with the motion of an electron around a nucleus, the spin is an *internal* property of the electron. This means that an electron at rest will still have its intrinsic angular momentum. Even though there is no orbital motion associated with the spin, we know that it is an angular momentum because it can be added to the orbital angular momentum of an electron to give a total angular momentum, with its various characteristics.

Energy

$\vec{B} = 0$ With $\vec{B}$

$m_s = +\frac{1}{2}$

$m_s = \pm\frac{1}{2}$

$\ell = 0$

$m_s = -\frac{1}{2}$

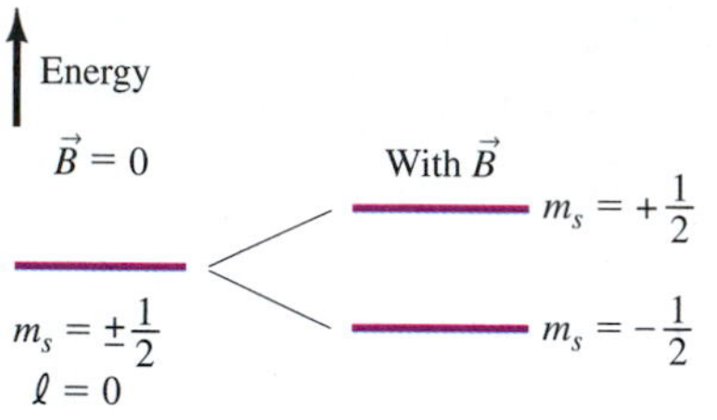

▲ **FIGURE 41–13** The existence of electron spin explains why some states are split into an even number of levels when a magnetic field is applied.

The fact that electrons have an intrinsic angular momentum $s\hbar$, with $s = {}^1\!/_2$, means that $2s + 1 = 2$, and an electron can appear in two states. We call these states "up" and "down." In the absence of a magnetic field, the energy of an "up" electron is the same as that of a "down" electron in the hydrogen atom. However, when a magnetic field is present, the energies of these two states differ slightly. When the electrons jump from these two states, the frequencies of the photons emitted differ slightly. As a result of electron spin, the number of possible electron states that correspond to a given ℓ doubles from $2\ell + 1$ to $2(2\ell + 1)$. When $\ell = 0$, the number of states with the same energy is 2 (Fig. 41–13). For $\ell = 1$, there are $2 \times 3 = 6$ states, and so on. When the energies are observed closely in the presence of a magnetic field, the $\ell = 0$ state is always a doublet, the $\ell = 1$ state contains 6 levels, and so forth.

EXAMPLE 41–6 Given that the magnetic dipole moment of an electron is $\vec{\mu} = -\frac{e}{m_e}\vec{S}$, where $\vec{S}$ is the spin angular momentum of the electron, what is the energy difference between the "up" and the "down" states of an electron in a magnetic field of 1.0 T?

Strategy We start with the energy of a magnetic dipole in a magnetic field. We define the z-axis by the direction of $\vec{B}$ and then make use of the fact that the two possible values of S_z are $\pm\hbar/2$. Therefore

$$U = -\vec{\mu}\cdot\vec{B} = -\mu_z B = \frac{e}{m_e}S_z B = \frac{e\hbar B}{m_e}\left(\pm\frac{1}{2}\right).$$

The desired energy difference ΔE is the difference between these two values.

Working It Out We have

$$\Delta E = \frac{eB\hbar}{m_e}.$$

Numerically,

$$\Delta E = \frac{(1.6 \times 10^{-19}\,\text{C})(1.0\,\text{T})(1.05 \times 10^{-34}\,\text{J}\cdot\text{s})}{(0.91 \times 10^{-30}\,\text{kg})(1.6 \times 10^{-19}\,\text{J/eV})}$$
$$= 1.2 \times 10^{-4}\,\text{eV}.$$

This number is small but not negligible on the atomic scale, and the effect of the energy modifications is quite visible in spectroscopic studies.

What Do You Think? It is believed that magnetic fields on the surface of a neutron star can be as large as 10^8 T. What spin state would electrons occupy in that environment?

Multi-Electron Atoms and the Exclusion Principle

Here is a naïve picture of multi-electron atoms. We have a central charge Ze and Z electrons, where Z is the atomic number. The nucleus is several thousand times more massive than the electrons, and we can treat it as fixed. We are therefore concerned only with a description of the electrons. Each electron moves in the attractive Coulomb potential of the nucleus plus a repulsive potential due to the presence of the other $(Z - 1)$ electrons. Treating the electrons as single particles in a potential is a good approximation. The electrons close to the nucleus experience the full charge Ze of the nucleus. The ones that are far away experience a smaller charge, because the inner electrons superimpose a negative charge on the positive nuclear charge. We can also assume that the modified force acting on the outer electrons remains central, so that angular momentum is conserved. The energy-level structure is accordingly the same as that of the hydrogen atom in many respects. There will still be an n-quantum number, which labels the total energy, and an ℓ-quantum number, which labels the angular momentum of the electron in that energy level. But it would no longer be true that, for a given n, the $\ell = 0$,

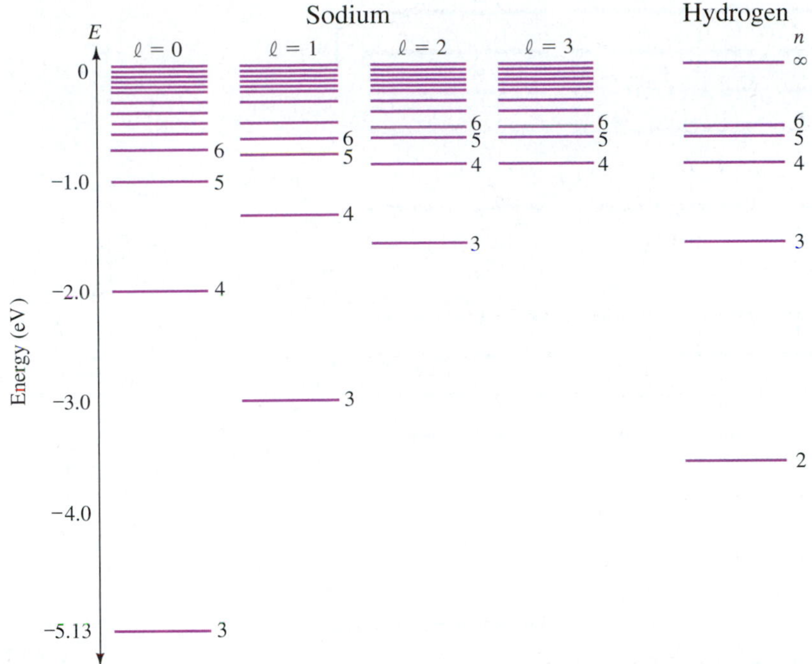

▲ **FIGURE 41–14** The schematic diagram of energy levels for an atom (here, sodium) is qualitatively similar to the hydrogen atom spectrum. The multiplets all have very nearly the same energy. For atoms with $Z \geq 2$, the energies for a given n value, but for varying values of ℓ, are no longer equal.

$\ell = 1, \ldots$ levels have the same energy. Instead, we expect that the electrons in large angular momentum states will be, on average, farther from the nucleus. These electrons are therefore more effectively screened by the inner electrons, and we expect that for fixed n, the electrons with larger ℓ will have higher energy values (Fig. 41–14).

The trouble with the description above is that its predictions are essentially all wrong. For example, it predicts that in the ground state of any atom, all the electrons would be in the lowest energy level. The radiation spectrum from such an atom would be qualitatively the same as that of hydrogen, although the numbers would be different because the central charge is Ze and the chemistry of the elements would all resemble that of hydrogen to some extent. Such a spectrum, and such chemistry, would bear little resemblance to what we know.

Pauli (Fig. 41–15) pointed out that an entirely new ingredient was needed to understand the structure of multi-electron atoms $(Z \geq 2)$. He proposed the **exclusion principle**, according to which *each quantum state can accommodate only two electrons, one in the "up" state and one in the "down" state.*

◀ **FIGURE 41–15** Wolfgang Pauli.

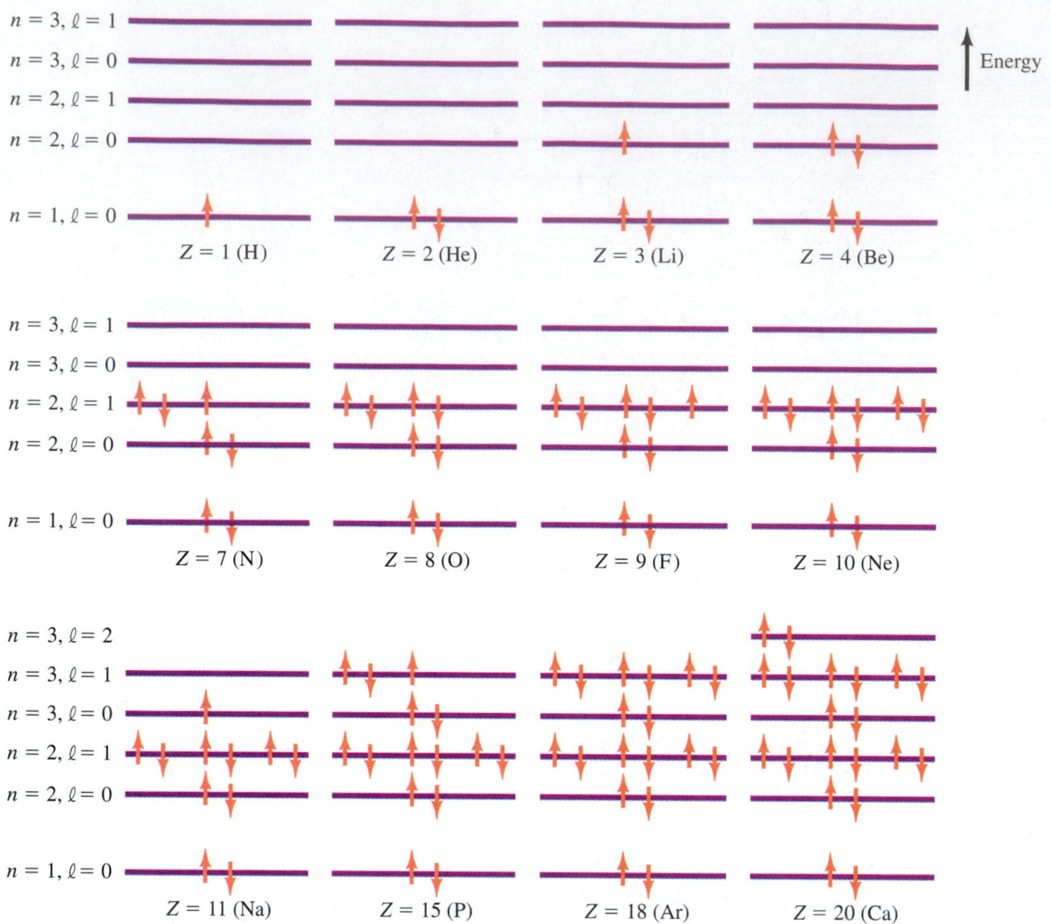

▲ **FIGURE 41–16** Pattern of electron energy-level occupation for elements from $Z = 1$ to $Z = 20$. The level splittings are not to scale. Each arrow represents an electron with its spin orientation.

Let's examine what emerges when we start filling energy levels according to the exclusion principle, a process that we can follow on Fig. 41–16. Helium, $Z = 2$, has two electrons; both can fit into the $n = 1$, $\ell = 0$ state. There is no room for another electron in the lowest state. Helium is said to form a *closed shell*. Next let's consider lithium, $Z = 3$. Two electrons fit into the $n = 1$, $\ell = 0$ state, and the third electron has to go into the next lowest energy state, which is the $n = 2$, $\ell = 0$ state. The third electron is farther from the nucleus than the other two electrons are (remember that $r \propto n^2$), and the positive charge $+3e$ of that nucleus is partly screened by the negative charge of the two electrons in the $n = 1$ orbit. As a result, the third electron is less tightly bound to the nucleus and can therefore be pulled more easily from the atom and into the orbit of another atom. Thus a lithium atom can bind with another atom to form a molecule and, like other atoms that have one electron outside a closed shell, lithium atoms are chemically very active.

For beryllium, $Z = 4$, we again fill a shell, the $n = 2$, $\ell = 0$ shell, and we expect beryllium to lose its electron less easily and be less chemically active than lithium. This is indeed the case. For $Z = 5$ through $Z = 10$, the $n = 2$, $\ell = 1$ levels are successively filled. The element $Z = 10$ is neon, which corresponds to another major closed shell, the $n = 2$ shell; it is an inert gas, one of a group of elements noted for their chemical inactivity. Fluorine, $Z = 9$, is one electron short of having a filled shell. Elements such as fluorine, which have a *hole* in a shell, react particularly strongly with atoms such as lithium, which have one electron outside a filled shell. Fluorine has the lowest Z value of the *halogens* (atoms with a single hole in a shell), just as lithium has the lowest Z value of the *alkali metals* (atoms with one electron outside a filled shell).

All the details of the periodic table (see Appendix V) can be understood both qualitatively and quantitatively in a quantum mechanical description. Note that the existence of discrete energy levels, spin, and the exclusion principle are purely quantum mechanical phenomena. There is no classical hint of their existence.

EXAMPLE 41–7 An atom has $Z = 37$ electrons. What are the values of n and ℓ for the electron that is least tightly bound?

Strategy The problem is one of systematically counting states and, starting with the states of lowest energy, occupying each of them with a maximum of 2 electrons. The states, again starting with the lowest ones, have n values 1, 2, 3, ... For a given n value, there are states corresponding to the ℓ-values $0, 1, 2, \ldots, n - 1$, and each ℓ state is itself subdivided into $2\ell + 1$ different states corresponding to the angular momentum projections.

Working It Out The simplest way to proceed is to make a table in which the states are enumerated, counting two electrons for each state. The first entry corresponds to the lowest level.

n	ℓ	Number of Electrons	Cumulative Total of Electrons
1	0	2	2
2	0	2	4
2	1	6	10
3	0	2	12
3	1	6	18
3	2	10	28
4	0	2	30
4	1	6	36
4	2	10	46

Thus the thirty-seventh electron is expected to lie in the $n = 4$, $\ell = 2$ shell.

What Do You Think? Would you call the atom described above inert, halogen, or alkali?

In Example 41–7, we applied simple counting rules to determine the quantum numbers of the outermost electron in an atom with a certain number of electrons. As Z increases, these simple rules fail, because new effects enter. For example, the dynamics of the orbital motion leads to a tendency for an electron to align its spin with or against the orbital angular momentum. There is then a term in the energy that depends on the alignment of the spin and the orbital angular momentum, a so-called spin–orbit term. These effects may lead to slight differences in the order in which the levels are filled; deviations from our simple rules start at $Z = 19$.

Do All Particles Obey the Exclusion Principle?

Electrons have spin $\hbar/2$. Nuclear physics experiments have shown that protons and neutrons also have spin $\hbar/2$, and according to a very general theorem, *all particles with spins* $\hbar/2$, $3\hbar/2$, $5\hbar/2$, ... *obey the exclusion principle*. This only means that no two *identical* particles can be in the same state, including the same spin state. The exclusion principle does not apply to a neutron and a proton, for example, which can perfectly well be in the same state. We call particles with half-integral spins **fermions**, named after Enrico Fermi. This has an important bearing on the structure of nuclei, which are made up of protons and neutrons. Nuclei have a shell structure analogous to that of atomic electrons. In nuclei, the average potential energy is the result of the mutual attraction of all the protons and neutrons by a *nuclear force*, with some effects from the coulomb repulsion of the protons.

There are particles that do not obey the exclusion principle. Those particles with intrinsic spin of the form $s\hbar$, where $s = 0, 1, 2, \ldots$, behave differently from particles with spin $\hbar/2$. Unlike fermions, which are unable to share the same quantum state with one another, such particles in some sense "prefer" to be in the same state. Particles in this class, whose spin is an integer multiple of $\hbar$, are known as **bosons**, named after Satyendra Nath Bose. The photon is an example of a boson; it has intrinsic angular momentum $\hbar$. *Photons, like all particles with integer spin, show a preference for congregating in the same quantum state*. We shall see some physical consequences of this congregation effect when we study lasers and liquid ^{4}He.

41–4 Molecular Structure

The Formation of Molecules

Molecules are bound combinations of two or more atoms. Even the lightest nucleus is some 2000 times more massive than an electron, so the nuclei in a molecule move much more slowly than their electrons. It is thus a good approximation to take the nuclei to be fixed, and this fact leads to considerable simplification in the description of molecules. The electrons move rapidly around the nuclei and effectively create a smear of negative charge in which the nuclei are embedded.

Many of the important properties of simple molecules can be understood by focusing on a particular example. Let's consider the molecule H_2, which consists of two hydrogen atoms. We'll study the energy as a function of R, the separation between the two hydrogen nuclei. There are several contributions to the energy. First, there is the electrostatic repulsion between the positively charged nuclei. For the H_2 molecule, this contribution to the potential energy is $U(R) = e^2/(4\pi\varepsilon_0 R)$.

We must also include the attraction of each of the two electrons to each nucleus and the electron–electron repulsion. Although a precise calculation requires quantum mechanics, we can estimate these contributions to the energy (the "electron contribution"). When R is large, the energy is lowest when one electron is close to one of the H nuclei and the other is close to the other H nucleus. In other words, for large nuclear separation, the system of two hydrogens is just two independent hydrogen atoms. On the other hand, when the separation of the nuclei is small, so that the two H nuclei are directly on top of one another, the electrons see a nucleus of charge $2e$, which, for atomic purposes, is like a helium nucleus. Thus, the system looks like a helium atom for small nuclear separation.

We take these facts into account by pulling together some numbers. The energy of a single hydrogen atom in its ground state is -13.6 eV, so for two widely separated atoms, the total electron contribution for molecular hydrogen is twice this value, -27.2 eV. At small separation, the electron contribution is that of a helium atom, and this energy can be evaluated by noting that it has two components. First, there is the electron interaction with the nuclei: Each electron sees a nucleus of charge Ze, with $Z = 2$. The electron–nucleus energy is then $Z^2(-13.6\text{ eV}) = -54.4$ eV for each electron, a total of -108.8 eV. Second, there is the mutual repulsion of the two electrons, which are separated by roughly half a Bohr radius. In that case the energy associated with the electron–electron repulsion is about double 13.6 eV and positive. When we put these two pieces together, the electron contribution at small nuclear separation is $(-108.8\text{ eV}) + (27.2\text{ eV}) \cong -80$ eV. This is quite a good estimate; a correct quantum mechanical calculation of the binding energy of helium, and experiment, gives the binding energy of helium as -78.5 eV.

We now have estimated the electronic contribution to the energy for large and small nuclear separation R. To estimate it for other values of R, we draw a smooth line between the small-R result (-78.5 eV) and the large-R result (-27.2 eV). Figure 41–17 shows the internuclear repulsion term, $e^2/(4\pi\varepsilon_0 R)$; the electron contribution; and the sum of these two terms. *This sum has a minimum* at a nuclear separation of approximately 0.07 nm, and this is a stable equilibrium point. The experimental value of the separation of the H atoms in the H_2 molecule is very close to this value. This pattern

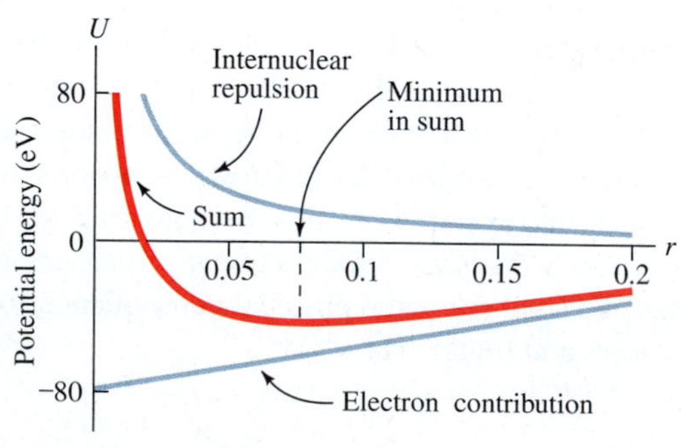

▲ **FIGURE 41–17** Energy diagram for the formation of the H_2 molecule. The upper curve is the potential energy associated with internuclear repulsion; the lower curve is an estimate of the potential energy associated with the electrons; the middle curve is the sum of these two terms. The sum has a minimum at $r \cong 0.07$ nm, very close to the observed nuclear separation in the H_2 molecule.

holds for other stable diatomic molecules. Most simply stated, *the forces of attraction between the electrons and nuclei cancel the forces of repulsion between the electrons and between the nuclei at the position of stable equilibrium.*

Only certain combinations of atoms can form molecules. Qualitatively, molecules can form under the following circumstances:

1. The two atoms have *paired* electrons outside closed shells, and the orbits of those electrons overlap. Here, "paired" means that one electron has spin-up while the other has spin-down. Because the electrons have opposite spin orientations, the Pauli exclusion principle does not prevent them from moving close to one another, into the region between the nuclei. In that region, the attraction of each electron to the "opposite" nucleus more than compensates for their mutual repulsion, and there is a net attraction. Each pair of electrons of this type form a *bond*: Their interaction binds the two atoms together. The larger the number of bonds, the stronger the binding of the atoms.
2. Only electrons that are not in closed shells can form bonds. An electron in a closed shell already has a second electron acting as its "partner" within its own shell. Thus only electrons in outer shells determine chemical properties.
3. An electron may not be in a closed shell but may nevertheless already be paired within its own, partly filled shell. Such electrons cannot pair up with an electron from another atom. More precisely, whether electrons pair up with partners from their own or from another atom is a matter of which configuration has the lower energy. This is not easily calculated, but chemists have developed a set of useful empirical rules for this question.

The H_2 or CH_4 (methane) molecules are examples of bonding between atoms that have unfilled shells. We say that such molecules are formed by *valence bonding*. *Ionic bonding* occurs when an atom with an electron outside a closed shell combines with an atom in which the outermost shell has one vacancy (a "hole"). As an example, consider sodium fluoride (NaF). Sodium has one electron outside a closed shell, and fluorine has a hole in its outer shell. Energy must be expended for the outermost electron of a sodium atom to be freed. This energy is the ionization energy of sodium. Its value is 5.1 eV, which is compensated by energy released if the free electron occupies the hole in a fluorine atom's outer shell. This released energy is the negative of the energy it would take to remove that electron from a now filled shell of fluorine, -3.5 eV. The net energy cost to move the electron from the sodium atom to the fluorine atom is $5.1\text{ eV} - 3.5\text{ eV} = 1.6\text{ eV}$. However, we now have a positive (Na^+) ion and a negative (F^-) ion. These ions attract one another, and the energy associated with this attraction more than makes up for the 1.6 eV, so NaF has a lower energy than the separated Na and F atoms.

Van der Waals Forces

The conditions that we listed for the formation of molecules are not hard and fast. Under special circumstances, the inert gases—whose atoms have no electrons outside of closed shells—*do* form molecules. Ar_2 (diatomic argon) is an example. These molecules form because there is a residual electromagnetic force between (neutral) atoms, the *van der Waals force*. This force falls off much more rapidly than does the Coulomb force with distance. Despite the fact that the van der Waals force is rather weak, it is always attractive and thus can bind atoms. The van der Waals force arises for the same reason that a comb attracts uncharged bits of paper. When one atom approaches another, its charge distribution affects the other by giving rise to a small displacement of charges, such that an electric dipole structure is created. This electric dipole has an electric field, which interacts with the electric dipole of the first atom. The two dipoles attract with a force proportional to $1/r^7$ (see Problem 45).

Although very weak, van der Waals forces are responsible for a number of additional phenomena, such as the adhesion between a liquid and the sides of its container, and the departure of a gas from ideal gas behavior (see Chapter 17).

▲ **FIGURE 41–18** Comparison of the spectra of (a) molecular hydrogen and (b) atomic hydrogen. The far richer structure of the molecular spectrum is evident even in this case, where the molecule is as simple as can be imagined.

Molecular Spectra

In molecules, nuclei separated by roughly 0.1 nm form nearly fixed centers of attraction for electrons. We may expect from the general ideas about quantum mechanics in Section 41–1 that molecules will have a series of electron energy levels separated by the kinds of energies that characterize atomic levels; that is, gaps on the order of 1 eV to 10 eV. Experiments, however, reveal a far richer spectrum—even for simple diatomic molecules (Fig. 41–18). The additional structure is due to quantized motions not seen in the hydrogen atom: vibrational and rotational motion.

Vibrational Motion: For diatomic molecules, we saw that the potential energy curve has a minimum (Fig. 41–17). The minimum point is a point of stable equilibrium. Moreover, as we learned in Chapter 7, the potential energy curve can normally be approximated near the minimum by a parabolic curve; that is, by a harmonic oscillator potential. As a consequence, there will be energy levels associated with the vibrational motion of the nuclei in this harmonic oscillator potential. We saw in Example 41–3 by using Bohr's techniques that these energy levels are given by $E_{\text{vib}} = n\hbar\omega$, where n is an integer and ω is a characteristic frequency given by $\sqrt{k/M}$. Here, k is the "spring constant" and M is the *reduced mass* of the two nuclei in vibrational motion, given by $M = M_1M_2/(M_1 + M_2)$, where M_1 and M_2 are the two nuclear masses.

We have seen that the minimum in the potential energy (the "spring") is created by the electron cloud in which the two nuclei are embedded, and we might therefore expect that it is parameters such as the size of the atomic orbits that determine the value of k. We can *estimate* the value of k by dimensional analysis. The dimensions of any spring constant such as k are $[k] = [EL^{-2}]$, where $[E]$ is the dimension of energy, $[ML^2T^{-2}]$. To estimate k, we use this relation with typical atomic energies and distances, $E \cong e^2/(8\pi\varepsilon_0 a_0)$ and $L \cong a_0$, where a_0 is the Bohr radius. Then

$$k \cong \frac{e^2}{8\pi\varepsilon_0 a_0^3} = \frac{e^2}{8\pi\varepsilon_0}\left(\frac{e^2 m_e}{4\pi\varepsilon_0\hbar^2}\right)^3 = \frac{1}{2}\frac{(m_e)^3}{\hbar^2}\left(\frac{e^2}{4\pi\varepsilon_0\hbar}\right)^4. \tag{41–17}$$

Dimensional analysis cannot specify any additional numerical factors and we must add a little additional physical reasoning to improve our estimate. In particular, because molecules are somewhat larger than atoms, we could replace a_0 by a somewhat larger radius—take $2a_0$ just for illustration. This leads to the estimate

$$k \cong \frac{1}{16}\frac{(m_e)^3}{\hbar^2}\left(\frac{e^2}{4\pi\varepsilon_0\hbar}\right)^4. \tag{41–18}$$

Once we know the effective spring constant, we can find the allowed energies of oscillation from Eq. (41–13):

$$E_{\text{vib}} = n\hbar\omega = n\hbar\sqrt{\frac{k}{M}} = \frac{n}{2}m_e\left(\frac{e^2}{4\pi\varepsilon_0\hbar}\right)^2\sqrt{Z_1Z_2\frac{m_e}{M}}, \tag{41–19}$$

where n is an integer and M is the reduced mass of the two nuclei.

We have written Eq. (41–19) in this way because it exposes the factor $(m_e/2)(e^2/4\pi\varepsilon_0\hbar)^2$, which we recall from Eq. (41–9) is the magnitude of the ground-state energy of hydrogen, 13.6 eV. Let's call this factor E_0. Then

$$E_{\text{vib}} = nE_0\sqrt{Z_1 Z_2 \frac{m_e}{M}}. \tag{41–20}$$

Numerically, the factor by which nE_0 is multiplied is on the order of 10^{-2}; that is, we estimate that the vibrational energies are on the order of 10^{-1} eV.

Rotational Motion: Independent of vibrations, the nuclei of a diatomic molecule can rotate like a dumbbell of length r_0. Here r_0 is the separation for which the potential energy has a minimum. Our molecule will then have energy levels associated with these rotations. The classical energy E is given in terms of the angular momentum L by $E = L^2/2I$, and we find the quantized energies when we insist that L be quantized.

We must evaluate the rotational inertia. We are treating a diatomic molecule such as OH as two masses, M_1 and M_2, connected by a massless rigid rod of length r_0. The molecule rotates about an axis perpendicular to the rod and passing through the center of mass of the system. For this system the rotational inertia I is given by (see Chapter 9)

$$I = Mr_0^2,$$

where M is the reduced mass of the system. We now quantize L, $L^2 = \ell(\ell + 1)\hbar^2$. The rotational energy levels are then

$$E_{\text{rot}} = \frac{L^2}{2I} = \frac{\ell(\ell+1)\hbar^2}{2I} \cong \frac{\ell(\ell+1)\hbar^2}{2Mr_0^2} = \ell(\ell+1)\frac{m_e}{2}\left(\frac{e^2}{4\pi\varepsilon_0\hbar}\right)^2\frac{m_e}{4M}.$$

We have substituted $r_0 = 2a_0 = 8\pi\varepsilon_0\hbar^2/e^2 m_e$, as we did in the discussion of the vibrational energy levels. We again recognize the factor E_0, so

$$E_{\text{rot}} = \frac{\ell(\ell+1)E_0 m_e}{4M}. \tag{41–21}$$

The energy levels are suppressed by an additional factor $\sqrt{m_e/M}$ from the factors of the vibrational levels and are on the order of 10^{-3} eV.

Let's summarize our results on molecular spectra. There are electronic levels on the order of electron volts apart. Associated with each level is a series of vibrational levels, separated by roughly 10^{-2} of the electronic levels; associated with each of *these* levels is a series of rotational levels, with separations about 10^{-4} of the electronic levels (Fig. 41–19). The latter two sets of levels are described as vibrational bands and rotational bands. The study of rotational bands is important to chemistry because these are most easily excited. Molecular spectra involving transitions between the rotational levels consist of wavelengths on the order of 10^4 times atomic wavelengths and thus involve infrared rather than optical spectroscopy.

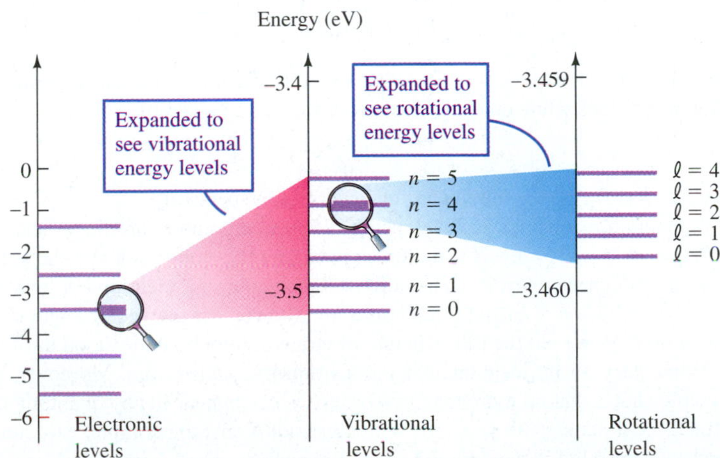

FIGURE 41–19 Molecular levels. The striking feature of these spectra is that the energy levels in each part are separated by energies 100 times finer than the energy-level separations of the previous part. The three parts represent electronic, vibrational, and rotational motions.

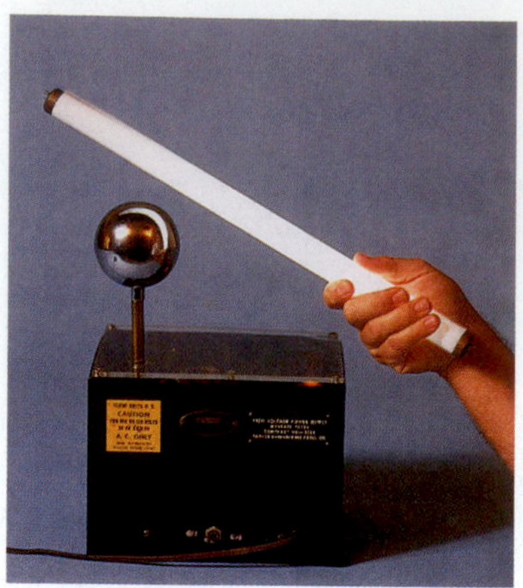

▲ **FIGURE 41–20** The high electric field produced by the power supply is enough to ionize the molecules inside the nearby fluorescent tube, which then produces light.

The complex structure of molecular energy levels is responsible for many of the subtleties of organic chemistry and of the chemical reactions that occur in biological systems. A technological application illustrates one consequence of the complexity of molecular structure. The high electric potentials that occur within transformers can cause the electrical breakdown of pockets of air in the transformer. In breakdown, high voltage ionizes air molecules. [A fluorescent tube can be lit simply by holding it near a high-voltage supply (Fig. 41–20).] The electrons liberated are accelerated by the large electric fields. In colliding with successive air molecules, they liberate more electrons, which are accelerated. An avalanche of electrons results, leading to currents that short out the transformer. An excellent method to prevent this type of breakdown is to introduce a gas that allows the liberated electrons to lose energy in ways that do not lead to further ionization. Sulfur hexafluoride (SF_6), which is one gas used in this way, has a very rich molecular structure, and there are many non-ionizing ways in which it can be excited by an electron colliding with it.

THINK ABOUT THIS. . .

WHERE DOES THE STEPLIKE STRUCTURE OF MOLECULAR SPECIFIC HEATS COME FROM?

In Chapter 19 we discussed equipartition, that is, the equal distribution of stored energy in a molecule among all of its degrees of freedom. This led to a prediction that the molar heat capacity for H_2 would be $7R/2$, where R is the universal gas constant. Figure 19–11 shows that this value is finally reached at rather high temperatures, with the molar heat capacity rising in steps from a value of $3R/2$ through $5R/2$ to $7R/2$. Our discussion of molecular excitations readily explains this.

At low temperatures we may think of molecules as inert objects like billiard balls. Thermal energy will be stored in the kinetic energy of these objects, and there the equipartition theorem predicts a heat capacity of $3R/2$. As the temperature increases, molecules can store energy in rotational degrees of freedom. This can only occur when the thermal energy is large enough to excite rotational levels. We argued in the text that these begin at energies of the order of 10^{-3}–10^{-2} eV, which corresponds to a temperature of $T = 10\text{K}–100\text{K}$, with significant contributions occurring at the higher end. There are, at this stage, no contributions from the vibrational degrees of freedom, since the threshold for their onset is at an estimated 0.1 eV, that is, 10^3K. It is above these temperatures that the vibrational excitations begin to share in the storage of energy. These estimates are in good agreement with the data shown in Fig. 19–11.

Summary

One of the important consequences of quantum mechanics is that energies of bound systems can take on only discrete values. In the simplest version of the Bohr model, the quantization of atomic energy levels follows from the restriction of angular momenta to integral multiples of $\hbar$:

$$L = n\hbar, \quad \text{where } n = 1, 2, 3\ldots \tag{41–5}$$

With this condition the allowed energy values are

$$E_n = -\frac{m_e}{2n^2}\left(\frac{e^2}{4\pi\varepsilon_0\hbar}\right)^2, \quad \text{where } n = 1, 2, 3, \ldots \tag{41–10}$$

in agreement with experiment. Energy conservation allows electrons to jump between levels with different n values while emitting photons of frequency given by

$$hf = E_i - E_f, \tag{41–6}$$

where E_i and E_f are the initial and final energies, respectively.

Quantum mechanics as developed by Heisenberg and Schrodinger shows that the structure of the possible energy levels is more complex than the Bohr model predicts. For each value of n, there are n^2 energy levels characterized by angular momentum $\ell\hbar$, where $\ell = 0, 1, 2, \ldots, (n-1)$, and $2\ell + 1$ spatial orientations are allowed for the vector angular momentum characterized by ℓ. However, the full structure of atoms cannot be understood until we add the fact that electrons carry an intrinsic angular momentum $\hbar/2$ called spin. Moreover, the Pauli exclusion principle shows that no more than two electrons can appear in any quantum state (corresponding to the $2s + 1$ states with $s = {}^1\!/_2$). With these additions, the complex structure of multi-electron atoms—as revealed in the periodic table of elements—can be explained.

Molecules form when the forces of attraction between the electrons and nuclei cancel the forces of repulsion between the electrons and between the nuclei at the position of stable equilibrium. This can occur if the atoms involved have electrons outside closed shells. Molecular spectra reveal energy levels of a molecule that are associated with vibrations of the atoms and rotations of the entire molecule. The electronic, vibrational, and rotational modes form a hierarchy.

Understanding the Concepts

1. Why is the ground state of a particle confined to a box (Section 41–1) characterized by $n = 1$ rather than $n = 0$?
2. Can we determine the atomic composition of distant objects by studying the wavelengths of their emitted photons?
3. What determines the shortest and longest wavelengths that a hydrogen atom can emit?
4. On the one hand, we say that electrons in atoms have discrete energies; on the other hand, we say that there is inherent uncertainty in our ability to measure energies. Is there a conflict here?
5. We know that metals contain free electrons. Using our illustration of energy quantization in a box, would you expect to be able to measure spectroscopically discrete energy levels for the electrons in a piece of metal of volume 1 cm^3?
6. In the Stern–Gerlach experiment, atoms of silver were observed to have only two components of angular momentum. Does that mean that all the electrons of a silver atom have angular-momentum quantum number ℓ equal to zero?
7. Is it true that all hydrogen atoms are indistinguishable from each other? How is it possible to distinguish between hydrogen with a nucleus consisting of a single proton (^{1}H), hydrogen with a nucleus consisting of a single proton and a single neutron (^{2}H, deuterium), and hydrogen with a nucleus consisting of a single proton and two neutrons (^{3}H, tritium)?
8. Why does an atom with a magnetic dipole moment need a magnetic field that varies spatially in order to deflect an atom? Would a constant magnetic field suffice?
9. Consider a hydrogen atom on Earth and another one on the Moon. Each one of them is in the ground state, and therefore the electrons have the same quantum numbers ($n = 1, \ell = 0$). Doesn't the exclusion principle (rather absurdly) tell us that this is impossible?
10. Van de Graaff accelerators, which have terminals at very high voltages inside a pressurized tank, are used to accelerate nuclear particles to high energy for nuclear reactions. Would air or sulfur hexafluoride be better as an insulating gas in a van de Graaff accelerator?
11. Why does it take about twice as much energy to excite an electron from the $n = 1, \ell = 0$ state to the $n = 2, \ell = 1$ state for He^+ as it does for He (neutral helium)?
12. Would you expect the orbital radius of the lowest orbit in a helium atom to be less than, equal to, or greater than that in a hydrogen atom? Why?
13. In discussing the formation of molecules, we stated that the minimum of the net potential energy in the interaction of two hydrogen atoms is the position where the attractive forces between the electrons and the nuclei cancel the repulsive forces between each electron and between each nucleus at the position of stable equilibrium. How do we translate a statement about the potential energy minimum to a statement about the force?
14. Why do the arguments we made about the formation of the diatomic hydrogen molecule, H_2, not apply to the formation of a diatomic helium molecule, He_2, from two helium atoms?
15. In a H_2 molecule there are two protons, and these have spin $\frac{1}{2}\hbar$, that is, they are fermions. If we just look at the two protons, would you expect their spins to be parallel or antiparallel in the ground state of the H_2 molecule?
16. Suppose that we add an electron to hydrogen. The second electron could be in the same orbit as the first (the spins would then have to point in opposite directions). What might prevent the existence of such a negatively charged atom? Would the existence of an atom consisting of one proton and three electrons be as likely, or unlikely?
17. In the levels of sodium that are described in Fig. 41–14 we have displayed the level with quantum numbers $n = 4, \ell = 0$ below the level $n = 3, \ell = 2$. Is this necessarily a misprint?
18. The molecules N_2 and CO have the same number of electrons but different binding energies. Why might this be? [*Hint*: Think about the energy levels of the valence electrons.]
19. In Example 41–2, we estimated the temperature at which 20% of the atoms were in their first excited state. If you had used the same method to find the temeprature at which 50% of the atoms were in the first excited state, what temperature would you have found? Why this peculiar result?
20. We stated that electrons under the influence of the nuclear Coulomb force form states whose energy is lowest when the orbital angular momentum is zero. Is this classically possible?

Problems

41–1 The Quantization of Energy

1. (I) What wavelength of radiation is necessary to ionize hydrogen? [*Hint*: Recall that to ionize an atom, it is necessary to raise the energy of the electron to be emitted from its ground state to at least $E = 0$.]

2. (I) Consider double ionized lithium (lithium in which two electrons have been removed). How much energy is needed to remove the third electron?

3. (I) What are the energy and wavelength of the photon emitted when a hydrogen atom jumps from its second excited state ($n = 3$) to its ground state ($n = 1$)? From $n = 5$ to $n = 3$ (Fig. 41–21)?

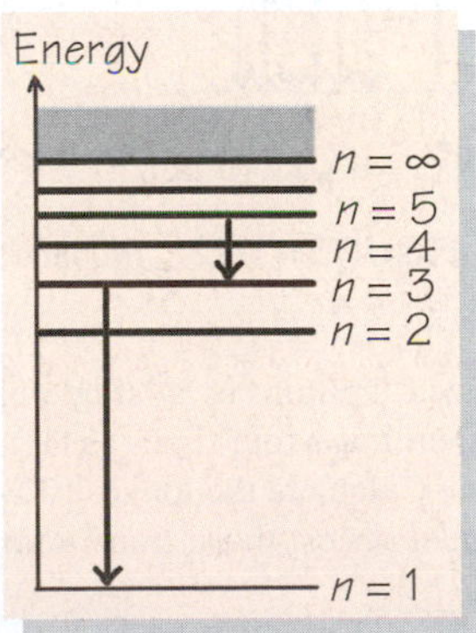

▲ **FIGURE 41–21** Problem 3.

4. (I) Bohr suggested an addition to the model named after him; specifically, that emitted radiation is very faint unless n changes by one unit. For what range of values of n will the radiation from singly ionized helium lie in the visible range; that is, with wavelengths in the range 400 nm to 700 nm (Fig. 41–22)?

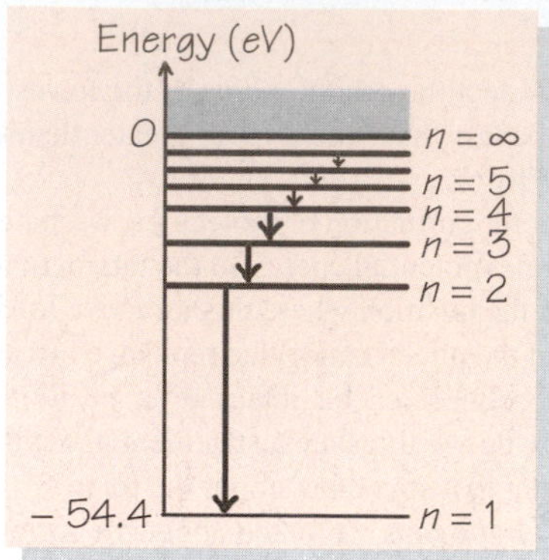

▲ **FIGURE 41–22** Problem 4.

5. (I) The negative muon (symbol μ^-) is a heavy version of the electron; it has a mass $m_\mu = 210m_e$. What is the ground state of the atom formed by a μ^- and a proton atom, assuming the Bohr rules apply?

6. (I) Consider the μ^--proton atom described in Problem 5. What is the energy difference between the first excited state and the ground state?

7. (II) The *Lyman series* is a series of spectral lines for hydrogen whose wavelengths correspond to Eq. (41–12) when $n_1 = 1$ (Fig. 41–23). (a) What are the quantum numbers of the states involved in the three transitions of the Lyman series with the longest wavelengths? (b) Calculate the wavelengths for the transitions of part (a). Are these wavelengths in the visible or ultraviolet regions?

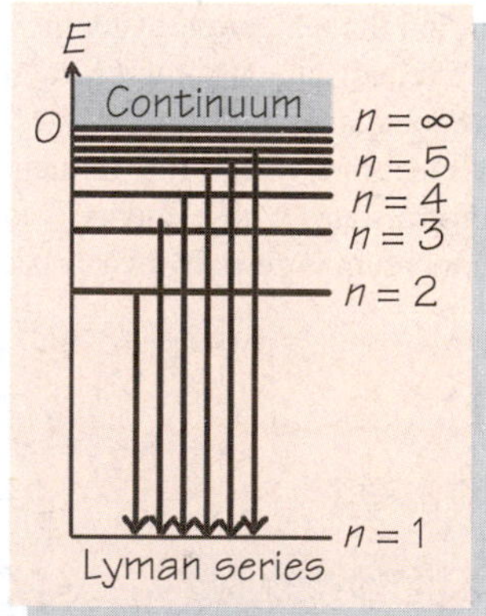

▲ **FIGURE 41–23** Problem 7.

8. (II) Singly ionized helium is a single-electron atom with $Z = 2$ and a nuclear mass four times as large as the mass of the hydrogen nucleus. Calculate the longest wavelength in the analogue of the Lyman series, those transitions that end in $n = 1$ (see Problem 7).

9. (II) What is the wavelength of the first line of the Lyman series (see Problem 7) in triply ionized beryllium ($Z = 4$)?

10. (II) The *Paschen series* is a series of spectral lines for hydrogen whose wavelengths correspond to Eq. (41–12) when $n_1 = 3$ (Fig. 41–24). (a) What are the quantum numbers of the states involved in the three transitions of the Paschen series with the longest wavelengths? (b) Calculate the wavelengths for the transitions of part (a). Are these wavelengths in the visible or infrared regions?

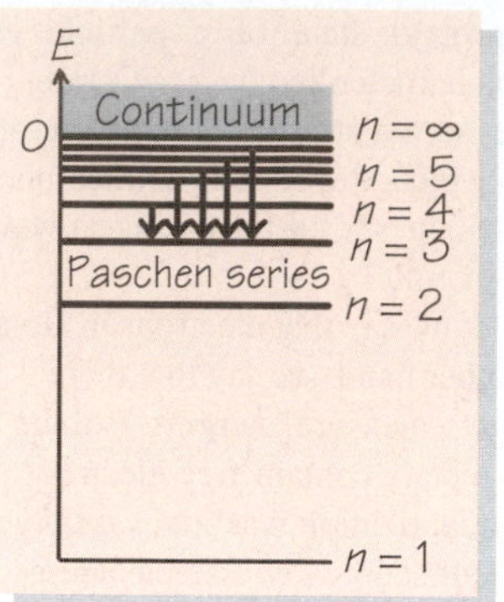

▲ **FIGURE 41–24** Problem 10.

11. (II) Calculate the wavelengths of the *Paschen* ($n_1 = 3$), *Brackett* ($n_1 = 4$), and *Pfund* ($n_1 = 5$) series transitions. (These series all refer to hydrogen.) Sketch the spectrum of atomic hydrogen for the 1000-nm to 5000-nm range of wavelengths. Calculate only wavelengths for values of $n_2 \leq 8$ in the n_2 to n_1 transitions.

12. (II) Derive a formula for the wavelengths of the transitions in single-electron ions whose nuclei have charge $+Ze$. For what range of Z values will radiation for $n = 2$ to $n = 1$ transitions be in the 20-nm range? In the 0.2-nm range?

13. (II) It is useful to introduce the *fine-structure constant* $\alpha \equiv e^2/4\pi\varepsilon_0\hbar c$ in problems that involve atoms. (a) What are the dimensions and value of α? What is the value of $1/\alpha$, to the nearest integer? (This is a useful number to remember.) (b) Express in terms of α the energy of the nth level of the hydrogen atom, E_n. (c) Calculate in terms of α the speed of the electron in the lowest Bohr orbit of hydrogen.

14. (II) What are the orbital radius, speed, momentum, and energy of an electron in the $n = 3$ state of hydrogen? Assume a classical model to calculate the momentum and speed of the electron.

15. (II) The most strongly bound $n = 1$ electrons of a large atom are closer to the nucleus than the other electrons. As a consequence, their energy is primarily determined by their interaction with the nucleus, without much influence from the other electrons. Use this idea to *estimate* the energy required to remove an electron from the lowest energy state of copper ($Z = 29$) and tungsten ($Z = 74$) atoms. (Experiments by Moseley showed that the nucleus is somewhat shielded by the other electrons, an effect that can be taken into account for $n = 1$ electrons by replacing Z of the nucleus by $Z - 1$.)

16. (II) Consider the Rutherford planetary model of circular orbits. (a) Use the expression for the radius given in Eq. (41–7) to calculate the speed of an electron in terms of $e^2/4\pi\varepsilon_0$, the angular momentum L and m_e. (b) Use the result of part (a) to calculate the electron's acceleration. (c) Calculate the period of the electron's orbital motion.

17. (II) According to classical electromagnetism, the power radiated by a particle of charge e that undergoes an acceleration a is

$$P = \frac{2}{3}\frac{e^2}{4\pi\varepsilon_0 c^3}a^2.$$

Use this formula and the results of Problem 16 to calculate the energy radiated by an electron per unit time for a circular orbit of angular momentum L.

18. (II) (a) Use the results of Problem 17 to show that the fraction of energy radiated by an electron in a single period T is given by

$$\frac{\Delta E}{E} = \frac{8\pi}{3}\left(\frac{e^2}{4\pi\varepsilon_0 Lc}\right)^3.$$

(b) Use your result to find the number of periods over which the electron would lose all its energy, and the time over which this would take place. Evaluate this result numerically for L when the orbital radius is 0.05 nm. (Your result should be on the order of 10^{-10} s.)

19. (II) Calculate the frequency of the radiation emitted by an electron in a hydrogen atom when it jumps from a level of quantum number $n + k$ to a level of quantum number n. Find a simple expression for it when $n \gg k$. Express your result in terms of $L = n\hbar$, and compare it with the classical frequency $f_{cl} = 1/T$, where T is the period calculated in Problem 16. Show that the two will be equal only for $k = 1$, so the requirement that the results of quantum mechanics coincide with classical results for very large values of n (the Bohr correspondence principle) means that only transitions with changes in n of value $\Delta n = 1$ are allowed.

20. (III) (a) Express the power $(P = \Delta E/\Delta t)$ radiated by an accelerating electron (see Problem 17) in terms of $L = n\hbar$. The energy will be radiated in the form of a photon of energy hf, where f is the frequency that corresponds to a transition $\Delta n = 1$. (b) Find $\Delta t = \Delta E/P$ in terms of n. Use this formula to calculate a numerical value for Δt when $n = 1$. This time interval may be viewed as the time constant for the $n = 2$ state of the hydrogen atom to jump to the $n = 1$ (ground) state. (c) Compare your time constant with the period of an electron in the $n = 2$ orbit. From your comparison, could you argue that the $n = 2$ orbit is almost stable? (d) Roughly, Δt is the longest time you have available for measuring the energy of the $n = 2$ state. Use the time–energy uncertainty relation to estimate the spread in energy of the $n = 2$ state, and compare this spread with the transition energy $E_2 - E_1$.

21. (III) *Estimate* the deflection of a beam of electrons in a Stern–Gerlach experiment. Assume that the kinetic energy of the electrons are 100 eV each, and that they travel through a region of length 0.20 m, in which there is a magnetic field gradient of 1.0 T/cm.

41–2 The True Spectrum of Hydrogen

22. (I) What are the possible values for the quantum numbers ℓ and its corresponding m-value for an atomic electron in a state labeled with $n = 3$?

23. (II) A correction should be supplied to the formulas for the energy levels of hydrogen-like atoms: The electron orbits about the center of mass of the electron–proton system rather than about the proton itself (Fig. 41–25). This correction, known as the *reduced-mass effect*, is small because the proton, of mass m_p, is much more massive than the electron, of mass m_e. The result is that the energy levels of hydrogen should be corrected to

$$E_n = -\frac{m_e}{1 + (m_e/m_P)}\frac{1}{2n^2}\frac{e^2}{(4\pi\varepsilon_0\hbar)^2}.$$

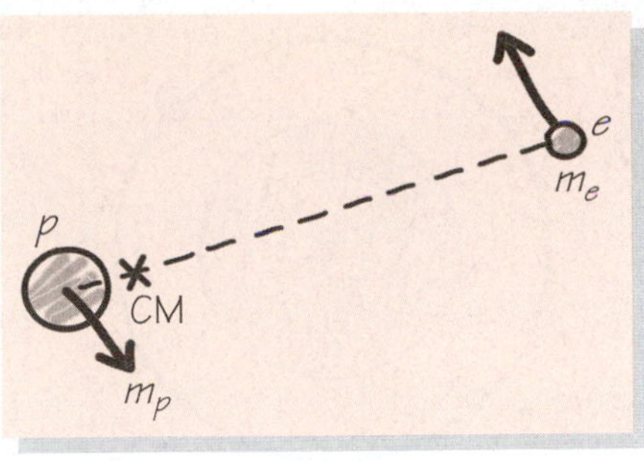

▲ **FIGURE 41–25** Problem 23.

The energy levels for deuterium, an atom with a nucleus whose charge is that of the hydrogen nucleus but whose mass is about twice that of hydrogen, obey the same formula. Find the difference in the wavelengths of radiation emitted in the transition between the $n = 2$ and $n = 1$ states for the two atoms. It was the observation of this difference that led to the discovery of deuterium by Harold Urey in 1931.

24. (II) A heavy version of the electron, called the *muon*, μ, differs from the electron only in that its mass is $m_\mu \cong 207m_e$. In the muonic hydrogen atom (a proton–muon atom), the reduced-mass effect of Problem 23 is much more important than in ordinary hydrogen. (a) What is the radius of the muonic hydrogen atom in its ground state? (b) Calculate the wavelength of radiation emitted in the transition of a muonic hydrogen atom from the $n = 2$ state to the $n = 1$ state.

25. (II) When an electron and a positron get close together at very low energies, they can form a bound state. This $e^- - e^+$ "positronium atom" lasts for a relatively long time before the electron and the positron annihilate. What are the energy levels of the positronium atom, assuming that the potential energy results from a pure Coulomb attraction? Which transitions are in the ultraviolet, the visible, and the infrared ranges?

41–3 The Exclusion Principle and Atomic Structure

26. (I) For what value of Z is the $n = 3$ level filled?

27. (I) Give the n and ℓ quantum numbers of the levels that are filled in the ground state of an atom of phosphorus, $Z = 15$.

28. (I) An atom has $Z = 34$ electrons. What are the n and ℓ values for the electron that is least tightly bound?

29. (I) What is the lightest element with a single electron in an $n = 3$ level?

30. (I) Sketch the pattern of electron energy-level occupation for the elements $Z = 13$ and 16 (see Fig. 41–16).

31. (II) In multi-electron atoms, the ordering of levels does not coincide with hydrogen-like atoms. In a particularly stable multi-electron atom, the following states are fully occupied: $n = 1, \ell = 0$; $n = 2, \ell = 0, 1$; $n = 3, \ell = 0, 1, 2$; $n = 4, \ell = 0, 1, 2$; $n = 5, \ell = 0, 1$. What is the Z value of this atom?

32. (II) An atom with $Z = 10$ has a closed shell; an atom with $Z = 11$ (sodium) may be viewed as a "nucleus" with a net charge of $+e$ and one electron on the outside (Fig. 41–26). In terms of this simplistic depiction, what would you expect sodium's ionization energy to be? [*Hint*: What levels are filled in the "nucleus"?]

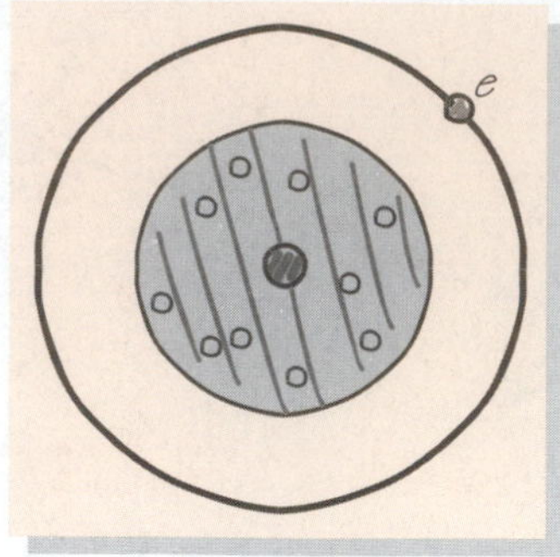

▲ **FIGURE 41–26** Problem 32.

33. (II) When an electron in an atomic state characterized by the quantum numbers (n, ℓ) is placed in a magnetic field, then the $(2\ell + 1)$ possible states no longer have the same energy $E_{n,\ell}$. The magnetic field *splits the degenerate levels* so the energy levels have the values $E_{n,\ell} + \kappa\ell B$, $E_{n,\ell} + \kappa(\ell - 1)B, \ldots,$ $E_{n,\ell} - \kappa(\ell - 1)B$, $E_{n,\ell} - \kappa\ell B$, where κ is a constant. (a) In the presence of a magnetic field, how many different spectral lines are there in the transition $(n = 2, \ell = 1) \rightarrow (n = 1, \ell = 0)$? (b) in the transition $(n = 3, \ell = 2) \rightarrow (n = 2, \ell = 1)$? [*Hint*: Different transitions in which the energy change ΔE is the same give a single spectral line of frequency $f = \Delta E/h$.]

34. (II) The potential energy U of a magnetic dipole with magnetic dipole moment μ in a magnetic field $\vec{B}$ is $U = -\vec{\mu} \cdot \vec{B}$. The magnetic dipole moment of an electron has magnitude $\mu = (e/m_e)S$. How much more (or less) energy does an electron with spin up have than an electron with spin down in the presence of an external magnetic field with magnitude $B = 0.3$ T, assuming that the field is parallel to the "up" direction?

35. (II) The halogens are a set of elements whose outermost shells of a given n are filled, except for one electron. Find the Z values of all the halogens for which $Z < 100$. [The approximation of successive filling of levels in the order $n = 1, 2, 3, \ldots,$ $\ell = 0, 1, 2, \ldots, (n - 1)$ is incorrect for Z greater than about 20. Thus your numbers will not agree with the periodic table at the high end.]

36. (III) The binding energy of the $n = 1$ electron in an atom for which the nuclear electric charge is $+Ze$ is obtained by taking the hydrogen atom results and replacing e^2 by $(Ze)(e) = Ze^2$. The ionization energies of the least-tightly bound electrons for some atoms with Z values from 1 to 29 (that is, through the $n = 3$ levels) are as follows:

Z:	2	10	11	18	19	28	29
Ionization energy (eV):	24.6	21.6	5.1	15.8	4.3	7.6	7.7

Compare these values with those that you would obtain for the removal of an $n = 1$ electron. To what do you ascribe the huge discrepancy in most cases?

41–4 Molecular Structure

37. (I) What is the minimum (nonzero) rotational energy of the hydrogen molecule?

38. (I) Use the Boltzmann factor $e^{-E/kT}$ to estimate the temperature at which 15 percent of a gas of harmonic oscillators with angular frequency 10^{15} Hz will be excited from the ground state to the first excited state.

39. (I) Use the Boltzmann factor $e^{-E/kT}$ to estimate the temperature at which 25 percent of a gas of diatomic molecules in the ground state $(\ell = 0)$ will be excited to the $\ell = 1$ rotational state, given that the rotational inertia of the molecule is $3 \times 10^{-47}\ \text{kg} \cdot \text{m}^2$.

40. (II) What are the energies of the three lowest levels in the vibrational spectrum of the HCl molecule (containing ^{35}Cl)? Assuming that the allowed transitions correspond to $\Delta n = 1$, find the wavelengths of the allowed transitions between these levels.

41. (II) The wavelength of the $n = 1 \rightarrow n = 0$ transition in the vibrational spectrum of NaCl is 2.63×10^{-5} m. Use this to estimate the spring constant k in Eq. (41–18). Approximate the mass of the sodium and chlorine nuclei as 23 and 35 times the mass of a hydrogen nucleus, respectively, to compare your result with a calculation from Eq. (41–18). The discrepancy that you will find between the two numbers suggests that Eq. (41–18) is a very crude approximation to the spring constant.

42. (II) The $\ell = 1 \rightarrow \ell = 0$ rotational transition in the lowest electronic state of the KCl molecule has a wavelength of 43.4 mm. Estimate the rotational inertia of the KCl molecule and the equilibrium separation of the atoms. Approximate the mass of the K and Cl nuclei as 39 and 35 times the proton mass, respectively.

43. (II) The energy difference between the lowest state in the CN molecule and the first excited electronic state would give rise to a single spectral line at a wavelength near 387.4 nm if there were no rotations (or vibrations). However, the lowest and first excited states actually consist of a series of rotational states with superimposed energies $\ell(\ell + 1)\hbar^2/2I_0$ and $\ell(\ell + 1)\hbar^2/2I_1$, respectively (Fig. 41–27). Calculate I_0 and I_1 from the following data: Transition $(n = 1, \ell = 1) \leftrightarrow (n = 0, \ell = 0)$ gives $\lambda = 387.4608$ nm; transition $(n = 1, \ell = 2) \leftrightarrow (n = 0, \ell = 1)$ gives $\lambda = 387.3998$ nm; transition $(n = 1, \ell = 0) \leftrightarrow (n = 0, \ell = 1)$ gives $\lambda = 387.5763$ nm. Why, physically, should I_0 and I_1 be different?

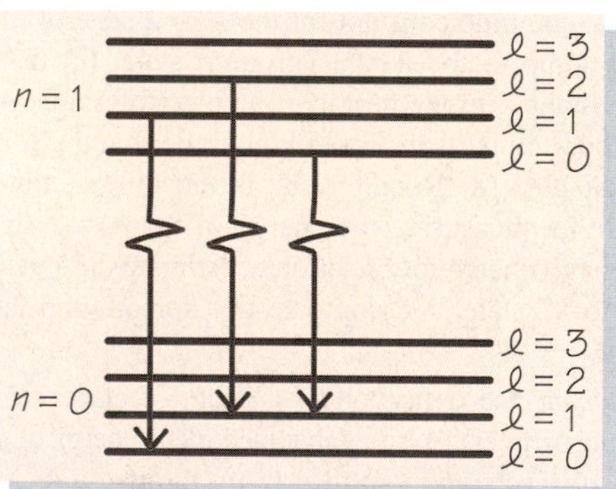

▲ **FIGURE 41–27** Problem 43.

44. (II) (a) Use the data from Problem 43 to find the internuclear separation between the C $(A = 12)$ and N $(A = 14)$ nuclei for the two electronic states $n = 1$ and $n = 0$. (b) Calculate the wavelengths for the transitions $(n = 1, \ell = 3) \rightarrow (n = 0, \ell = 2)$. Compare your results with the measured value, $\lambda = 387.3369$ nm.

45. (II) Suppose that when one atom is separated from another by a distance r, its charge distribution affects the other by giving rise to a small displacement of charges, with separation d less than the atomic radius, thus creating an electric dipole. We know from Eq. (22–14) that the electric field due to such a dipole has magnitude $E \cong ed/4\pi\varepsilon_0 r^3$. (We have ignored angular factors such as sines or cosines.) The charge separation created in the second atom induces a dipole moment of magnitude αE. The coefficient α has the dimensions $4\pi\varepsilon_0(\text{length})^3$, as can be seen from $E \cong ed/4\pi\varepsilon_0 r^3$, and this length turns out to be the charge separation d. Now the induced dipole interacts with the electric field E. Show that the resulting interaction leads to a potential energy in the interaction of the atoms of the general form

$$U(r) = -(\text{a constant})\frac{e^2}{4\pi\varepsilon_0}\frac{d^5}{r^6}.$$

This potential energy is what leads to the van der Waals force between the atoms.

46. (III) The rotational motion of a diatomic molecule affects the equilibrium position of the nuclei. If R_0 is the separation for zero angular momentum and R is the average separation when there is rotation, then the rotational energy is $\ell(\ell + 1)\hbar^2/2MR^2$, where M is the reduced mass of the nuclei. In addition, for a given vibrational frequency ω, the vibrational potential energy in the presence of rotation is $\frac{1}{2}M\omega^2(R - R_0)^2$. Calculate the new equilibrium separation R by minimizing $E(R)$, the sum of the two new terms. Treat $R - R_0$ as small. How is the rotational inertia changed, and what effect does this have on the rotational spectrum of the molecule?

General Problems

47. (I) For what value of the angular momentum quantum number ℓ does the quantum mechanical angular momentum have the value $1.00\ \text{kg}\cdot\text{m}^2/\text{s}$?

48. (II) Suppose that two electrons are in orbit around one proton (an H^- ion), both in an $n = 1$ level. By listing all the potential energy contributions, make a crude guess of how much energy it would take to ionize one of the electrons.

49. (II) An electron in an atom jumps from the first excited state to the ground state. The mean duration for the transition is 4.5×10^{-11} s. What is the uncertainty in the energy value of the first excited state? Give your answer in electron volts and as a fraction of the energy of the state, which is 2.6 eV.

50. (II) By using the assumptions of the Bohr model, calculate the radius of the ground-state electron orbit for doubly ionized lithium, Li^{2+}.

51. (II) A marble of mass $m = 20$ g moves in a circular orbit near the bottom of a circular bowl (Fig. 41–28). The height of the sides of the bowl is given by $h = \alpha r^2$, where r is the radial distance from the bottom of the bowl. Given that $\alpha = 0.25\ \text{cm}^{-1}$, find the separation between the successive allowed energies of the marble in the bowl. It is not surprising that we have no intuitive feel for quantum mechanical phenomena.

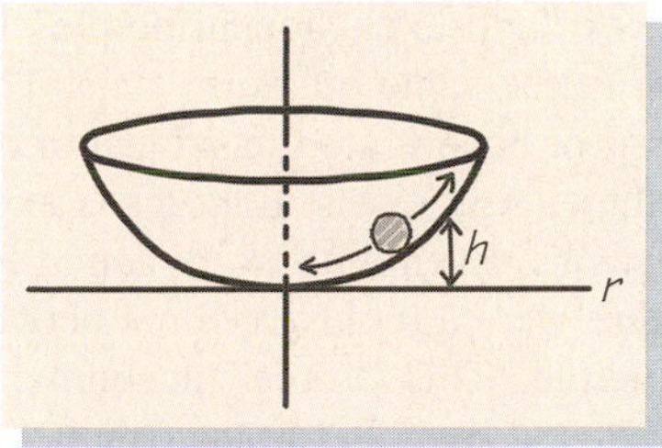

▲ **FIGURE 41–28** Problem 51.

52. (II) All integer values of the principal quantum number n, even very large ones, are allowed in atoms. In practice, it is very hard to excite orbits that correspond to large n values in an atom unless the atom is totally isolated. Estimate the largest value of n that would be possible if you could make a gas of *atomic* hydrogen of density $r = 6.0 \times 10^{-10}\ \text{g/cm}^3$. For practical purposes, we regard an interatomic spacing of at least three times the diameter of the large-n atom as total isolation.

53. (II) By finding the momentum of an electron in a circular orbit with orbital angular momentum L and by using your knowledge of the magnetic dipole moment due to a current loop (Fig. 41–29), show that this orbiting electron has a magnetic dipole moment

$$M = \frac{e\hbar}{2m_e}\frac{L}{\hbar}.$$

The quantity $e\hbar/2m_e$, known as the *Bohr magneton*, is the minimum quantized value for the possible magnetic dipole moments of electrons in their orbits.

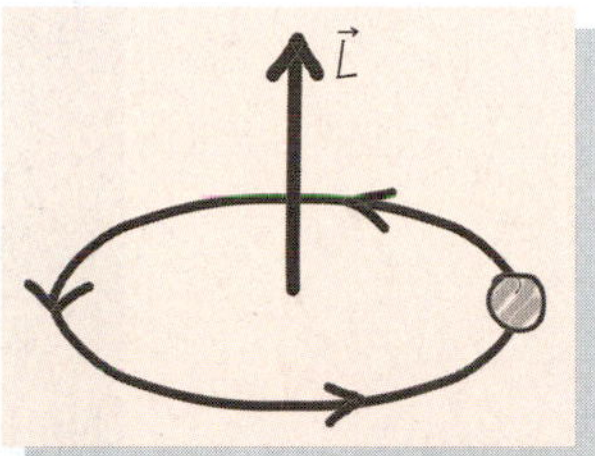

▲ **FIGURE 41–29** Problem 53.

54. (III) Repeat your calculation of the energy levels of the Bohr atom, but now assume that the potential energy is given by $U(r) = -(e^2/4\pi\varepsilon_0 r) + (\sigma/r^2)$, where σ is a constant. Are your energy values larger or smaller than the corresponding values for the pure Coulomb case? Is your result physically reasonable? Is the effect of the added term more important for large n or for small n, and is that result plausible?

55. (III) According to the Heisenberg uncertainty relations, it is impossible to measure to high precision the position of an atomic electron without making its momentum highly uncertain. By finding the magnitudes of the momentum for the ground state and the first excited state of an electron in a hydrogen atom, calculate just how well the position of an electron in the ground state of hydrogen can be located before you can no longer be sure whether it is in the ground state or the first excited state. It is for the reasons outlined here that it makes little sense to think of an electron as following a classical orbit like those of the planets.

56. (III) Use the position–momentum uncertainty relation to prove that it is not possible to detect orbits in hydrogen, by the following argument: (i) a measurement of the nth orbit must be such that $\Delta x \ll r_{n+1} - r_n$. Calculate Δx. (ii) This gives rise to an uncertainty in the momentum of the electron in the orbit, and hence in the energy. Calculate the uncertainty in the energy. (iii) Show that the energy is larger than the energy difference between adjacent orbits, so an orbit cannot be "photographed."

57. (III) Consider a particle of mass m that moves in a circular orbit of radius r around a center of attraction that exerts a force on the particle. The particle's potential energy is $U = U_0(r/a)^4$, where U_0 is a constant with dimensions of energy, and a is a constant with dimensions of length. (a) Calculate the total energy (including kinetic energy) in terms of r by using the relation $mv^2/r = |F| = |dU/dr|$. (b) Use the Bohr quantization rules to obtain an expression for the quantized energy values of the particle.

Chapter 42

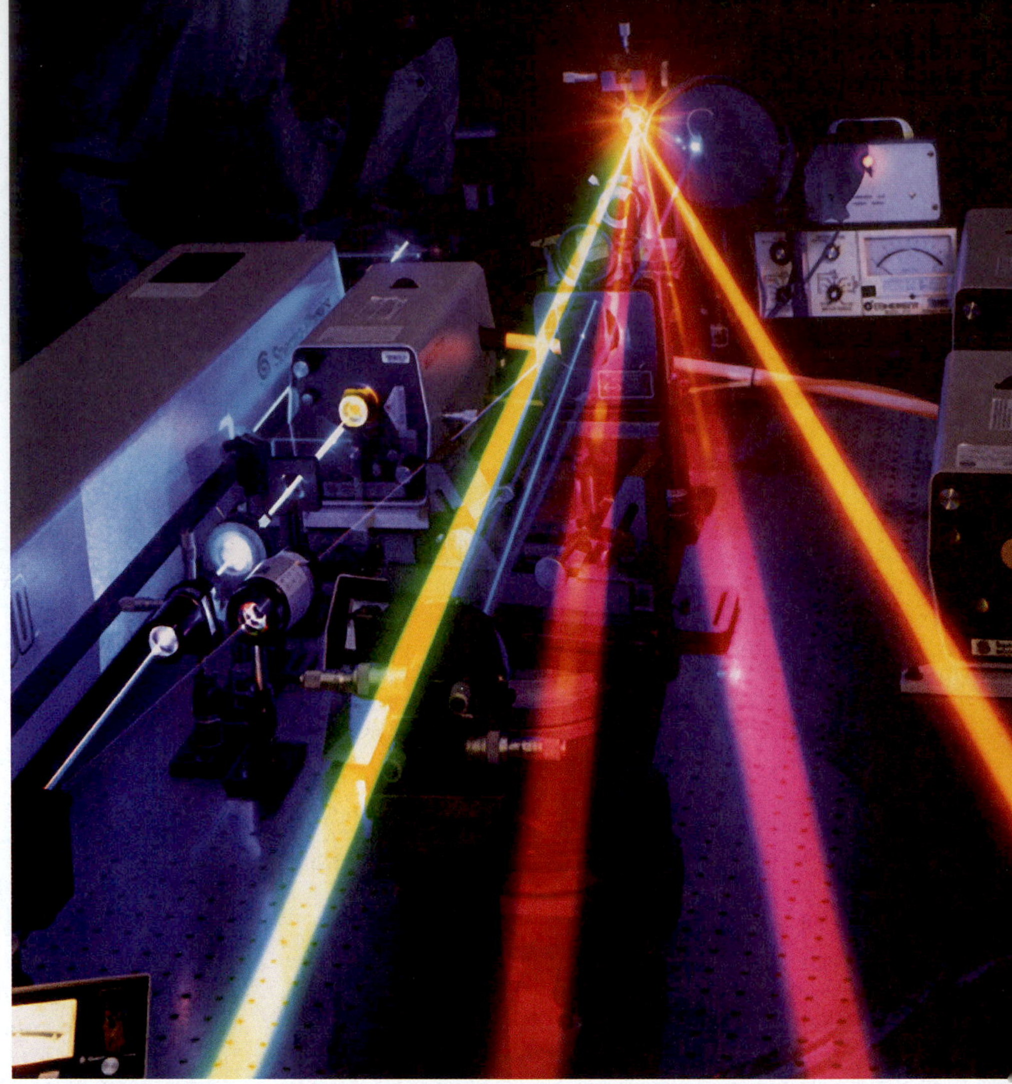

Smoke in the air makes these laser beams visible. Lasers, whose effects are visible on a macroscopic scale, produce intense beams of coherent light through quantum-mechanical effects.

Quantum Effects in Large Systems of Fermions and Bosons

The particles that compose matter are of two types: Fermions have angular momentum $\left(\ell + \frac{1}{2}\right)\hbar$, where $\ell = 0, 1, 2, \ldots$, and bosons have angular momentum $\ell\hbar$. Identical fermions obey the Pauli exclusion principle, which states that at most one fermion can be in a given quantum state (given by n, ℓ, ℓ_z, s, s_z). In contrast, identical bosons obey a kind of "anti-exclusion principle" that translates into a tendency for them to congregate in the same quantum state. The consequent behaviors of systems with many fermions or bosons are the subject of this chapter. In Chapter 41 we saw how the filling of the energy levels in atoms is explained by the exclusion principle as it applies to electrons. Even with no dynamic forces present, identical fermions behave as though there were a repulsive force between them, just because they obey the exclusion principle. This "force" accounts, among other things, for the incompressibility of solids and liquids. Large-scale behavior can also be influenced by the behavior of bosons. The congregating effects of identical bosons explain the laser, Bose-Einstein condensates, and superfluidity. Under the right circumstances, identical fermions may in effect congregate in pairs, such that a large number of them behaves like a large system of bosons. This phenomenon lies behind the behavior of superconductors and leads to such properties as flux quantization and the Josephson effect—properties that go beyond the mere absence of resistivity in superconductors.

42–1 The Exclusion Principle in Bulk Matter

Electrons in Metals and the Fermi Energy

The classical treatment of the electrical conductivity of metals (Section 26–5) starts with free electrons in metals. These electrons move under the influence of an externally imposed electric field. Resistance to current flow is due to the collisions between the electrons and ions, which leads to a retarding force and an average terminal speed for the electrons called the *drift speed*. The quantum mechanical description of conductivity is also based on the premise of free electrons in metals. The description of the motion of those electrons must be quantum mechanical, however, and this leads to some major differences with the classical estimates.

Picture the electrons as being confined to a one-dimensional box of macroscopic length L—several centimeters, for example. From Eq. (41–1), we get the energy levels for a single electron confined to a one-dimensional box:

$$E = \frac{\pi^2\hbar^2 n^2}{2m_e L^2}.$$

In a three-dimensional box, there are three such contributions, corresponding to motion in the x-, y-, and z-directions. The result is that

$$E = \frac{\pi^2\hbar^2(n_1^2 + n_2^2 + n_3^2)}{2m_e L^2}, \tag{42–1}$$

with each of the integers n_1, n_2, and n_3 allowed to take the values 1, 2, 3, ... These integers label a state, and with this in mind it is evident that many states have the same energy. The lowest energy level is the one for which $n_1 = n_2 = n_3 = 1$, and it is composed of a single unique state. The next energy level consists of three states: $(n_1, n_2, n_3) = (2, 1, 1)$, $(1, 2, 1)$, and $(1, 1, 2)$. Different states that have the same energy are said to be *degenerate*; as the n value increases, the degree of degeneracy becomes high.

Suppose that we now start filling the levels described by Eq. (42–1) with electrons—we are now viewing a piece of metal as a three-dimensional box that contains the valence (free) electrons of the metal. According to the Pauli exclusion principle, each state can accommodate a maximum of two electrons, one for each of the two spin states of an electron, "up" and "down." With one or more valence electrons per atom, the "box" is filled with many electrons. *The lowest possible electron energy for a metal "box" that contains N_e electrons corresponds to the states filled from the bottom—the lowest energy level—with two electrons per state.* The values of $n_1^2 + n_2^2 + n_3^2$ under consideration are very large. Suppose that an electron in the metal has energy $1\text{ eV} = 1.6 \times 10^{-19}\text{ J}$—a value typical of the energy of electrons in conductors. Let's also take $L = 1\text{ cm}$, so that we are dealing with a small but definitely macroscopic piece of metal. It follows from Eq. (42–1) that

$$(n_1^2 + n_2^2 + n_3^2) = 2m_e\frac{EL^2}{\pi^2\hbar^2} \cong \frac{2(0.9 \times 10^{-30}\text{ kg})(1.6 \times 10^{-19}\text{ J})(0.01\text{ m})^2}{\pi^2(1.05 \times 10^{-34}\text{ J}\cdot\text{s})^2}$$
$$\cong 2.6 \times 10^{14}.$$

Thus we are dealing with values of n_1, n_2, and n_3 that are very large. If, for example, we let $n_1 = n_2 = n_3 = 10^7$, then the energy *difference* ΔE between the corresponding level and an adjacent level, with n_1 increased by 1 (but n_2 and n_3 unchanged), is determined by

$$\frac{\Delta E}{E} = \frac{(n_1 + 1)^2 - n_1^2}{n_1^2 + n_2^2 + n_3^2} \cong \frac{2n_1}{n_1^2 + n_2^2 + n_3^2} \cong 10^{-7}.$$

The energy levels are so close together that we say they *almost* form a continuum.

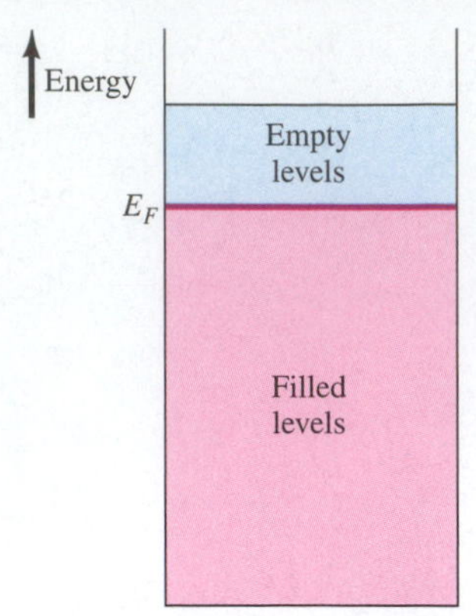

▲ **FIGURE 42–1** The energy levels of "free" electrons in a metal at $T = 0$. All the levels are filled to the Fermi energy, E_F, of the material, leaving unoccupied levels above.

When an electron accelerates under the influence of an external field, its energy increases smoothly according to the classical view. In quantum mechanics, the electron energy must jump by a discrete amount. Here, the exclusion principle plays an important role: An electron cannot jump to a state of higher energy if that higher-energy state is fully occupied by other electrons. This means that only electrons at the top of the filled levels—the levels that have energy above this are not fully occupied—can be accelerated by the electric field (Fig. 42–1).

The energy of the *highest* filled energy level is called the **Fermi energy**, E_F, after Enrico Fermi (Fig. 42–2), who with Paul Dirac first clarified the importance of the exclusion principle in many-fermion systems. In general, the de Broglie wavelength of a particle roughly corresponds to the space it occupies. Thus the closest that two electrons with the same energy and angular momentum can get to each other is about half a de Broglie wavelength. Any closer distance would effectively superimpose the electrons, a situation forbidden by the exclusion principle. While this estimate is not precise, it is a useful guide to the qualitative calculation of E_F that follows.

To estimate the Fermi energy, we suppose that the closest possible distance between two electrons is one-half the de Broglie wavelength that corresponds to the *Fermi momentum* $p_F = \sqrt{2m_eE_F}$. If we denote this closest distance by d, then the total number of electrons N_e in a cubical box of sides L is

$$N_e = \left(\frac{L}{d}\right)^3; \tag{42–2}$$

$$d = \left(\frac{N_e}{L^3}\right)^{-1/3} = n_e^{-1/3}. \tag{42–3}$$

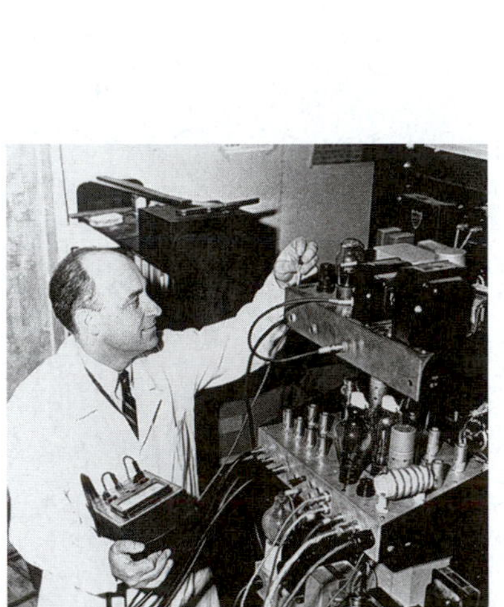

▲ **FIGURE 42–2** Enrico Fermi excelled as an experimentalist and as a theorist. Among his other accomplishments, he led the successful effort to build the first controlled nuclear fission reactor, completed in December 1942 in Chicago. His theory of beta decay was a crucial step along the way to a deeper understanding of fundamental interactions.

Here, n_e is the number density of free electrons in the metal. When we equate this closest distance to half the de Broglie wavelength at the Fermi energy, λ_F, we find that

$$d = \frac{\lambda_F}{2} = \frac{h}{2p_F} = \frac{h}{2\sqrt{2m_eE_F}} = \frac{\hbar\pi}{\sqrt{2m_eE_F}}. \tag{42–4}$$

We can combine Eqs. (42–3) and (42–4) and solve for E_F: $E_F = (\hbar^2/2m_e)(\pi^3 n_e)^{2/3}$. A more precise calculation of the number of electrons that can be accommodated with energy $E < E_F$, a calculation that takes into account the presence of degeneracy, leads to the replacement of the factor π^3 by $3\pi^2$, a very minor change:

$$E_F = \frac{\hbar^2}{2m_e}(3\pi^2 n_e)^{2/3}. \tag{42–5}$$

The magnitude of the Fermi energy depends on the density of free electrons. Copper, for example, has one free electron per atom. For copper, then, with an atomic weight of 63.5 g/mol and a mass density of 8.95 g/cm^3,

$$n_e = \left(\frac{1\ \text{electron}}{\text{atom}}\right)\left(\frac{6.02\times10^{23}\ \text{atoms}}{\text{mol}}\right)\left(\frac{1\ \text{mol}}{63.5\ \text{g}}\right)\left(\frac{8.95\ \text{g}}{\text{cm}^3}\right)\left(\frac{10^6\ \text{cm}^3}{\text{m}^3}\right)$$
$$= 8.48\times10^{28}\ \text{electrons/m}^3.$$

Substituting this result into Eq. (42–5), we find $E_F = 1.13\times10^{-18}\ \text{J} = 7.1\ \text{eV}$.

EXAMPLE 42–1 Calculate the Fermi energy for N *neutrinos*, particles of spin $\hbar/2$, symbol ν, confined to a cubic volume with sides of length L. The neutrino mass is so small that the relation between energy and momentum is the relativistic relation $E = pc$ rather than the nonrelativistic one $E = p^2/2m$.

Strategy With spin $\hbar/2$, neutrinos are fermions, and the exclusion principle still applies; that is, $\lambda_F = 2d$, where d is the spacing between neutrinos. The only difference between the treatment of neutrinos and our earlier treatment of electrons is that for neutrinos, we use the massless-particle relation $E_F = p_F c$.

Working It Out We combine the relation $d = \lambda_F/2$ with the energy-momentum relation to get

$$d = \frac{\lambda_F}{2} = \frac{h}{2p_F}.$$

With $E_F = p_F c$, this reads $d = hc/2E_F$, or

$$E_F = \frac{hc}{2d} = \pi\hbar c n_\nu^{1/3} = \pi\hbar c\left(\frac{N}{L^3}\right)^{1/3}.$$

What Do You Think? How would the calculation be modified if we asked for the Fermi energy of neutrinos in a two-dimensional box? *Answers to* ***What Do You Think?*** *questions are given in the back of the book.*

With the picture we have sketched in mind, we can ask what happens when a weak electric field is applied to a metal. Only the electrons at the top of the filled levels can accelerate and move to states of higher energies because, as we have already mentioned, only they find empty levels to occupy (see Fig. 42–1). In the expression for electrical conductivity, Eq. (26–25), with the collision time given by Eq. (19–45), the appropriate quantity to use is the Fermi speed v_F, defined by $E_F = \frac{1}{2}m_e v_F^2$, rather than v_{rms} given by the kinetic theory of gases. Recall that the use of v_{rms} gave an incorrect $\sqrt{T}$ temperature dependence for the resistivity (see Section 26–5). The effect of temperature changes on the behavior of electrons with a given Fermi energy is negligible. A 100K change in temperature changes the energy of an electron by $kT = (1.4 \times 10^{-23}\ \text{J/K})(100\text{K}) = (1.4 \times 10^{-21}\ \text{J})/(1.6 \times 10^{-19}\ \text{J/eV}) \cong 10^{-2}\ \text{eV}$. Compared with the Fermi energy of copper—7 eV—the factor kT is negligible. Thus the resistivity will be temperature independent for reasonable temperature changes.

The Incompressibility of Matter

The exclusion principle plays a crucial role in explaining the incompressibility of matter. A measure of this incompressibility is given by the bulk modulus, B, defined by Eq. (11–20),

$$B = -\frac{\Delta p}{\Delta V/V}.$$

Here, Δp is a pressure change that brings about a fractional change $\Delta V/V$ in the volume of some sample of matter. Because an infinitesimal volume change dV is brought about by an infinitesimal pressure change, we rewrite this definition as

$$B = -V\frac{dp}{dV}. \tag{42–6}$$

Suppose now that the sample of material forms a cylinder of cross-sectional area A and that the pressure is applied to the ends. The work done in compressing the material along the cylinder's axis by an amount dL is $dW = -F\,dL$. Work dW is done, so energy $dE = dW$ is added to the sample. We have

$$dE = dW = -F\,dL = -\left(\frac{F}{A}\right)(A\,dL) = -p\,dV,$$

so the pressure p is the negative of the ratio of the energy change to the volume change:

$$p = -\frac{dE}{dV}. \tag{42–7}$$

When the volume of a metal changes, the number density changes, and so does the total energy. Let's calculate the bulk modulus under the assumption that the *only* resistance of a metal to compression is due to this energy change. The total energy of the free electrons in the material is equal to the number of electrons N_e multiplied by an average energy, which is a value somewhere between 0 and E_F. A calculation that we omit because of its length yields the result that the average energy factor is $\frac{3}{5}E_F$ (see Problem 51):

$$E \cong \frac{3}{5}E_F N_e = \frac{3}{5}\frac{\hbar^2}{2m_e}\left(3\pi^2\frac{N_e}{V}\right)^{2/3} N_e. \tag{42–8}$$

Thus, from Eq. (42–7), the pressure, which is called the *degeneracy pressure* of the electrons, is given by

$$p_e^{\text{degen}} = -\frac{dE}{dV} = \frac{2}{5}\frac{\hbar^2}{2m_e}(3\pi^2)^{2/3}\left(\frac{N_e}{V}\right)^{5/3}. \tag{42–9}$$

We emphasize that the exclusion principle, not an explicit repulsive force between the electrons, provides this countering pressure. According to Eq. (42–6), one further derivative of this pressure is necessary to find the corresponding bulk modulus:

$$B = -V\frac{dp_e^{\text{degen}}}{dV} = -V\frac{d}{dV}\left[\frac{2}{5}\frac{\hbar}{2m_e}(3\pi^2)^{2/3}\left(\frac{N_e}{V}\right)^{5/3}\right]$$

$$= -V\left[\frac{2}{5}\frac{\hbar^2}{2m_e}(3\pi^2)^{2/3}(N_e)^{5/3}\right]\left(-\frac{5}{3}\right)\frac{1}{V^{8/3}}.$$

$$= \frac{2}{3}\frac{\hbar^2}{2m_e}(3\pi^2)^{2/3}\left(\frac{N_e}{V}\right)^{5/3} = \frac{2\pi^{4/3}}{3^{1/3}}\frac{\hbar^2}{2m_e}(n_e)^{5/3}. \tag{42–10}$$

For copper, $n_e = 8.5 \times 10^{28}\ \text{m}^{-3}$, so $B = 6.4 \times 10^{10}\ \text{N/m}^2$. The experimental value of B for copper is $13.4 \times 10^{10}\ \text{N/m}^2$. Given the uncertainties of our estimates, our neglect of interaction between electrons, and the Coulomb repulsion between the ions, the fact that our rough approximation is within a factor of 2 of the experimental result is impressive. The effective repulsion between electrons that is supplied by the exclusion principle plays a major role in the high degree of incompressibility of matter.

White Dwarfs and Neutron Stars

The exclusion principle also plays a critical part in the evolution of stars: It is responsible for preventing a star's collapse under the mutual gravitational attraction of its mass. Stars are generally formed from the clumping together of large clouds of gas through mutual gravitational attraction (mainly hydrogen, which forms most of the raw material of the universe). When the hydrogen atoms fall together, gravitational potential energy is converted into kinetic energy. The density and temperature increase. The hydrogen atoms are ionized, and as further compression occurs, the protons that form the hydrogen nuclei come close enough to undergo a variety of reactions. For example, through a sequence of reactions to be described in Chapter 44, four protons can combine to form a helium nucleus and two positrons (the antimatter of electrons). A great deal of energy is released in the process, which is responsible for the luminosity of many stars and represents a kind of thermonuclear burning. While the reactions occur, the temperature of the star remains high, and an equilibrium is established. Once the hydrogen fuel has burned to the point where the process can no longer occur, there is a substantial fraction of helium nuclei in the star. The gravitational pressure shrinks the size of the star further until the temperature is high enough that the helium nuclei begin to undergo reactions that produce still more energy. In a sequence of processes, heavier and heavier elements, up to iron, are produced. Iron does not take part in further thermonuclear reactions, and the reactions stop. However, the material continues its gravitational contraction.

To estimate the gravitational pressure that acts to compress the matter, we proceed as follows. Suppose that the star contains N nucleons (protons and neutrons), each of mass M. (We can neglect the electron mass here.) A characteristic gravitational force at the star's surface has magnitude $G(NM)^2/R^2$, and thus a characteristic gravitational pressure (force per unit area) at the surface is

$$p_g = \frac{G(NM)^2/R^2}{4\pi R^2} = \frac{G(NM)^2}{4\pi\left(\frac{3}{4}V/\pi\right)^{4/3}} = 0.54\frac{G(NM)^2}{V^{4/3}},$$

where the volume $V = 4/3\,\pi R^3$. Note that the gravitational pressure is directed *inward*. It actually varies through the star; a more rigorous calculation shows that the correct value for our constant is 3/5 of 0.54 and hence

$$p_g = 0.32\frac{G(NM)^2}{V^{4/3}}. \tag{42–11}$$

As V decreases, the pressure grows and, without some countering pressure, the star would collapse. The degeneracy pressure p_e^{degen} of the electrons [Eq. (42–9)], which points outward, provides this counterbalance. (Note that the degeneracy pressure is largest when the mass of the fermion involved is smallest, which is why we use the degeneracy pressure of electrons rather than that of, say, protons.) When the electron degeneracy pressure matches the gravitational pressure, equilibrium is reached. For less massive stars, the end product is a *white dwarf* (Fig. 42–3). Given that N_e = number of protons $\cong N/2$, we can calculate the radius of the resulting star. We get a radius of some 7000 km for a star the mass of the Sun, 2×10^{30} kg (see Problem 54). This corresponds to a density of 7.5 metric tons/cm^3! Note that the radius of our Sun is about 1.4 million km, and a quick comparison of the densities of the Sun and our white dwarf shows that the difference between ordinary matter and degenerate matter is enormous.

For a more massive star, p_g is larger and, to maintain equilibrium, the factor N_e/V in the degeneracy pressure [Eq. (42–9)] must increase. This means, however, that the Fermi energy of the electrons grows. When the Fermi energy of an electron is comparable to the electron's rest mass, we can no longer use nonrelativistic formulas but instead must treat the electron relativistically. In the extreme relativistic case, the result of Example 42–1 must be used. The total energy of the electron is again N_e multiplied by the average energy; this time, though, p_e^{degen} is proportional to $(N_e/V)^{4/3}$. This implies that the factor V cancels from the balancing equation $p_e^{\text{degen}} = p_g$, which then can no longer be satisfied by a judicious choice of V. The gravitational pressure always exceeds the electron degeneracy pressure when a star is more massive than about 1.4 solar masses (the *Chandrasekhar mass*) and without the possibility of balance, gravitational collapse continues.

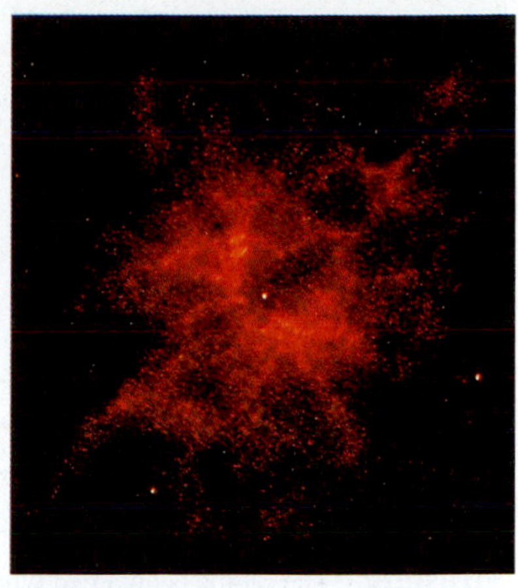

▲ **FIGURE 42–3** A white dwarf star is at the center of the gas cloud.

EXAMPLE 42–2 Show that if the electron's energy E in a star is so high that its rest mass can be ignored (see Example 42–1), then the gravitational pressure is on the order of magnitude of the degeneracy pressure for a star of one solar mass. (A star is electrically neutral, and to a good approximation the number of neutrons equals the number of protons when the star is formed.)

Strategy The gravitational pressure is given by Eq. (42–11). We want to compare this to the electron degeneracy pressure, which can be calculated from the relation $p_e^{\text{degen}} = -dE/dV$, Eq. (42–7). We again use $E \cong 3/5\, E_F N_e$, but now we need the expression for the Fermi energy applicable to massless (or highly relativistic) particles, $E_F = \pi\hbar c(N_e)^{1/3}V^{-1/3}$ (see Example 42–1). We must also relate the number of electrons to the mass of the star. We do this as follows: Electrical neutrality gives us $N_e = N_p$. The total mass of the star, $M_\odot$ (one solar mass), is the total number N of neutrons and protons times the mass M of a proton (or neutron), $M_\odot = NM$. Finally, as we stated above, the number of protons equals the number of neutrons, so $N_e \cong N/2 = (NM)/2M = M_\odot/2M$.

Working It Out The degeneracy pressure is

$$p_e^{\text{degen}} = -\frac{dE}{dV} = -\frac{d}{dV}\left[\frac{3}{5}E_F N_e\right] = -0.6N_e^{4/3}\pi\hbar c\frac{d(V^{-1/3})}{dV}$$
$$= +0.2\pi\hbar c N_e^{4/3}V^{-4/3}.$$

We now express the electron number in terms of the solar mass, giving

$$p_e^{\text{degen}} = +0.2\pi\hbar c\left(\frac{M_\odot}{2M}\right)^{4/3}V^{-4/3}.$$

Thus, the ratio of the degeneracy pressure to the gravitational pressure is

$$\frac{p_e^{\text{degen}}}{p_g} = \frac{0.2\pi\hbar c(0.5M_\odot/M)^{4/3}V^{-4/3}}{0.32G(NM)^2/V^{4/3}}$$
$$= \frac{0.2\pi\hbar c(0.5M_\odot/M)^{4/3}}{0.32GM_\odot^2} \cong 1.2.$$

This crude calculation does give the right order of magnitude. Chandrasekhar showed the more general result that if a star has a mass $sM_\odot$, then $p_e^{\text{degen}}/p_g = 1.2s^{-2/3}$.

What Do You Think? Would this calculation apply to the Sun, whose mass is 2×10^{30} kg and radius is 7×10^8 m?

What happens when a star exceeds the Chandrasekhar mass and collapses? The process can crudely be described by the statement that the consequent squeezing forces the electrons in the star to combine with protons into neutrons and neutrinos via the reaction

$$e^- + p \rightarrow n + \nu.$$

As we remarked in Example 42–1, neutrinos are very nearly massless, and more important, they interact so weakly with matter that they immediately escape from the star. What is left is a star made of N neutrons only: a *neutron star* (Fig. 42–4).

Neutrons are fermions and hence also obey the exclusion principle. We have a new equilibrium condition, $p_g = p_n^{\text{degen}}$, where p_n^{degen}, the neutrons' degeneracy pressure, has the same form as that of p_e^{degen} but replaces N_e with N and m_e with M. The balancing relation is then

$$p_g = 0.32\frac{G(NM)^2}{V^{4/3}} = p_n^{\text{degen}} = \frac{2}{5}\frac{\hbar}{2M}(3\pi^2)^{2/3}\left(\frac{N}{V}\right)^{5/3}. \qquad (42\text{–}12)$$

▲ **FIGURE 42–4** At the center of the crab nebula, shown here, is a neutron star, identifiable through its properties as a pulsar.

This equation can be solved for V and thus for R. For a star of a few solar masses, numerical calculation yields R on the order of 10 km! The neutron star is as dense as a nucleus—that is roughly a billion times denser than the white dwarf we talked about above—and the compactness of neutron stars is responsible for the many properties that continue to intrigue astrophysicists. The existence of neutron stars was predicted in 1934 by Walter Baade and Fritz Zwicky. The *pulsars* first discovered by Anthony S. Hewish and Jocelyn Bell Burnell in 1967 were identified by Tom Gold in 1968 as rapidly rotating neutron stars.

EXAMPLE 42–3 Calculate the radius of a neutron star whose mass is three solar masses.

Strategy The radius is obtained when the gravitational pressure is equated to the degeneracy pressure. Since we are dealing with a neutron star, the degeneracy pressure that applies corresponds to a degenerate set of neutrons, p_n^{degen}, as described above.

Working It Out The equation $p_g = p_n^{\text{degen}}$ yields

$$0.32\frac{G(3M_\odot)^2}{V^{4/3}} = \frac{2}{5}\frac{\hbar^2}{2M}(3\pi^2)^{2/3}\left(\frac{3M_\odot}{MV}\right)^{5/3},$$

where we have replaced N, the number of neutrons, by $3M_\odot/M$; $M_\odot$ is the solar mass, and M is the neutron mass. From this expression, we find

$$V^{1/3} = \left(\frac{4\pi}{3}\right)^{1/3} R = \frac{\frac{2}{5}(\hbar^2/2M)(3\pi^2)^{2/3}(3M_\odot/M)^{5/3}}{0.32G(3M_\odot)^2},$$

where R is the radius of the neutron star. Thus,

$$R = \left(\frac{3}{4\pi}\right)^{1/3}\frac{\frac{2}{5}(\hbar^2/2M)(3\pi^2)^{2/3}(3M_\odot/M)^{5/3}}{0.32G(3M_\odot)^2} = 1.0 \times 10^4\ \text{m}.$$

What Do You Think? If the mass of the star is n solar masses, what is the variation of the radius with n?

If the mass of a star is so large that even the pressure due to the Pauli exclusion principle applied to the neutrons proves inadequate to resist collapse, continued gravitational contraction occurs, and this time there is no mechanism to stop it. A black hole forms. It is clear that the exclusion principle is a crucial ingredient in the understanding of matter, from its ordinary terrestrial form to how stellar cores behave.

THINK ABOUT THIS. . .

WHAT IS THE CONNECTION BETWEEN PULSARS AND NEUTRON STARS?

We briefly described above how gravitational pressure in massive stars (two to three solar masses) forces electrons and protons to combine into neutrons and neutrinos, thus forming a neutron star. The star begins its collapse when its primary fuel supply has been exhausted, creating a supernova. As part of the process, a fraction of the mass of the star is blown off, leaving an expanding shell of material surrounding a neutron star at the core. Most of the neutron star is indeed tightly packed neutrons, but on the surface (about 1 km thick) there is an iron crust containing protons and electrons as well. Typically the original star has a radius of the order of 10^6 km and a surface magnetic field of 10^{-2} T. It also has some angular momentum—the Sun, for example, rotates over a nearly 25-day span, $\omega \cong 3 \times 10^{-6}$ radians/s. When a collapsing star contracts in radius by a factor of 10^5, the magnetic flux is conserved, and thus the magnetic field increases to $10^{-2} \times (10^5)^2 = 10^8$ T, while the spin rate grows to a typical value of 10–100 revolutions/s as a result of angular momentum conservation. The rapid motion of magnetic field lines generates an emf that accelerates some of the electrons on the surface. This acceleration gives rise to strong beams of radiation. On Earth a light beam is seen that repeats with the rotation period of the star, like that of some vast and distant lighthouse. The period can be measured to very high precision—the first such example to be discovered had a period of 1.33730113 seconds! The regularity of the emission from these objects led them to be dubbed as pulsars.

An isolated neutron star will very slowly decrease the rate of its rotation due to the effects of general relativity. But the star's shape is correlated with the speed of rotation, and its iron crust is rigid, so that as the star slows, it may be stressed and suddenly develop cracks. The consequent rapid change, what may properly be called a "star-quake," gives rise to sudden changes in the otherwise nearly steady period of rotation. These sudden tiny changes in the period have indeed been observed.

42–2 Lasers and the Behavior of Bosons

The tendency of identical bosons to cluster into the same quantum mechanical state has its most direct application in the functioning of lasers. To understand lasers, we need some further detail about photon-producing transitions in atoms and molecules.

Transitions Between Energy Levels

The quantum of electromagnetic radiation—the photon—has a wavelength λ or a frequency f related to λ by $f = c/\lambda$. It also has a momentum vector $\vec{p}$ oriented in the direction of propagation of the photon. The magnitude of this momentum is given by Eq. (40–2):

$$p = \frac{hf}{c}.$$

Photons are emitted (or radiated) when the electrons of atoms (or of molecules) undergo a transition ("jump") from a higher energy level to a lower energy level, as in Fig. 42–5a. Such photons are said to be *spontaneously emitted*. We have already mentioned that transitions in which photons are emitted are subject to the law of conservation of angular momentum; moreover, because photons act as though they have angular momentum quantum number $\ell = 1$ (with small mixtures of $\ell = 2$ and higher values), the most frequent transitions occur between atomic or molecular levels for which $|\ell_i - \ell_f| = 1$.

In order for an atom to radiate, it must first be excited into a higher energy level. To excite the atom, an atomic electron must be given the proper energy and angular momentum. One method of doing this is to bombard the atom with photons whose energy is the energy required for the transition. In this case, an atomic electron *absorbs* a photon and jumps into a higher energy state (Fig. 42–5b). If electrons in the ground state have $\ell = 0$, then (to high accuracy) the only states that can be reached in this way are those in which the electrons have $\ell = 1$. It is possible to reach other states—in this case, $\ell = 2, 3, \ldots$—by other mechanisms, such as collisions with other atoms. In fact, the discovery of states with $\ell > 1$ was possible only through these other excitation methods. How does an electron in a state with $\ell = 2$ fall back to the ground state? If an $\ell = 1$ energy level lies between the $\ell = 2$ state and the $\ell = 0$ ground state, which is always the case for low-Z atoms, the electron can jump first to the intermediate $\ell = 1$ state by emitting a photon, then jump to the $\ell = 0$ state by emitting a second photon (Fig. 42–6). What happens, though, if the atomic level structure is complicated enough that there is no intermediate $\ell = 1$ state through which the electron can cascade downward? A jump accompanied by photon emission is still possible, but it is on the order of 10^4 times less probable. If the change in ℓ is 3 rather than 2, then a jump accompanied by photon emission is 10^8 times less probable. These probabilities can be illustrated by thinking of a given excited energy level as a tub filled with water. A large drain that empties it in a short time corresponds to the possibility of transitions in which the angular momentum changes by one unit of $\hbar$. There are two tiny holes that are 10^4 and 10^8 smaller in area than the drain hole. If the drain is open, we can ignore any leakage through the small holes, but if the drain is blocked (if there are no possible $\Delta\ell = 1$ transitions), leakage occurs through the next smallest hole. It takes much longer to empty the tub through the tiny hole. If there are no possible $\Delta\ell = 1$ transitions, then the energy level is called *metastable*, or nearly stable. Metastable states play a critical role in the operation of the laser.

The discussion so far concerns what are known as *spontaneous transitions* in atoms. In 1917, Albert Einstein used thermodynamics together with the rudimentary quantum theory in existence at the time to predict the possibility of *stimulated transitions*. Consider an electron in a metastable state. It will stay in that state for a long time (long on the atomic scale, typically 10^{-8} s) before it decays with the emission of a single photon into the ground state. The energy of the photon will be $hf = E_i - E_f$, where E_i and E_f are the initial and final electron energies. Einstein discovered, however, that if photons of frequency f are present in the vicinity of the atom, then the very presence of these photons will make the transition occur more rapidly, and the rapidity

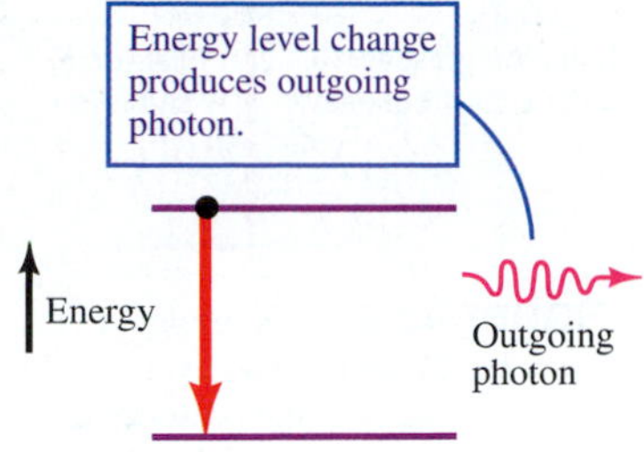

(a) Spontaneous emission

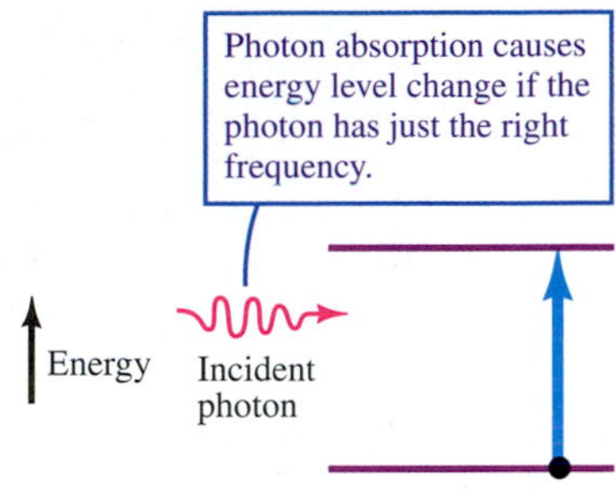

(b) Absorption

▲ **FIGURE 42–5** (a) In spontaneous emission, a photon is produced as an electron drops from an excited atomic state to a lower-lying level. (b) An atom can absorb a photon if the photon's frequency matches the energy difference between two atomic levels of the proper ℓ values.

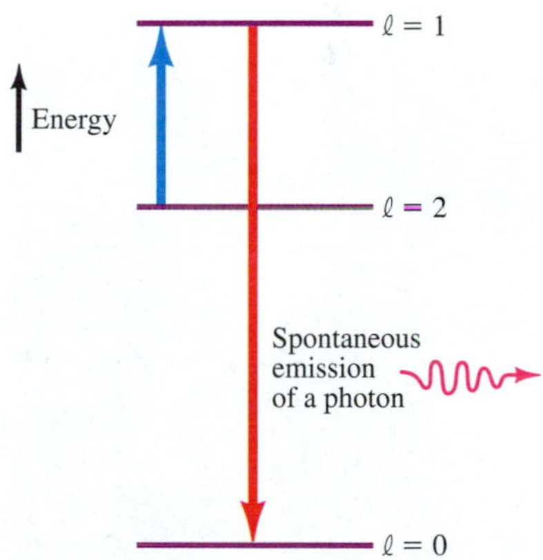

▲ **FIGURE 42–6** The de-excitation of metastable $\ell = 2$ level is by the allowed excitation to the $\ell = 1$ level, followed by the allowed transition to the $\ell = 0$ state. The metastable level is populated by the inverse process: allowed excitation to the $\ell = 1$ state followed by the allowed decay to the $(\ell = 2)$ state. Direct excitation from the $\ell = 0$ level to the $\ell = 2$ level is also possible in collisions.

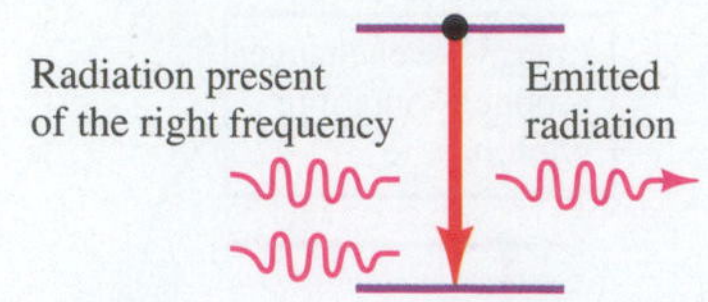

▲ **FIGURE 42–7** In stimulated emission, the decay from an excited state occurs more readily in the presence of photons in the same state as the photon to be emitted.

increases as the number of photons grows. We say that there is *stimulated emission* (Fig. 42–7). The more external photons, the more rapidly the stimulated emission occurs. There is another very important effect: *Because they are bosons, the photons emitted by stimulated emission will preferentially be in the same quantum state as the stimulating photons*; that is, they will have the same momentum, frequency, and phase. A *coherent* state of many photons is formed, a state in which the electromagnetic fields associated with the photons reinforce. This coherent state describes a single, intense, monochromatic plane wave, or beam. Moreover, because the photons have the same momentum (both magnitude and direction), the beam is extremely well collimated. The functioning of lasers relies on the existence of such stimulated transitions. The term *laser* is in fact an acronym for "light amplification by stimulated emission of radiation."

Lasers

Suppose that we have a collection of atoms in a cavity, each of which is in a metastable state. The presence of many photons of energy hf will stimulate a very rapid transition to the ground state, where f is the frequency corresponding to the transition to the ground state. Without the presence of these photons, the atoms will remain in the metastable state for a long time. We may then picture the following sequence: After a long time, one atom decays to the ground state and emits a photon of energy hf. The photon (as well as many others to follow) is constrained to stay in the cavity by means of mirrors positioned to reflect the photon back into the cavity (Fig. 42–8a). Other atoms in the metastable state, stimulated by the first photon, undergo a slightly faster transition to the ground state and produce more photons in the same state. The ever increasing number of the "right" photons—those that induce stimulated emission—produces an avalanche of decays. These photons can form a well-collimated, coherent *laser beam* (Fig. 42–8b).

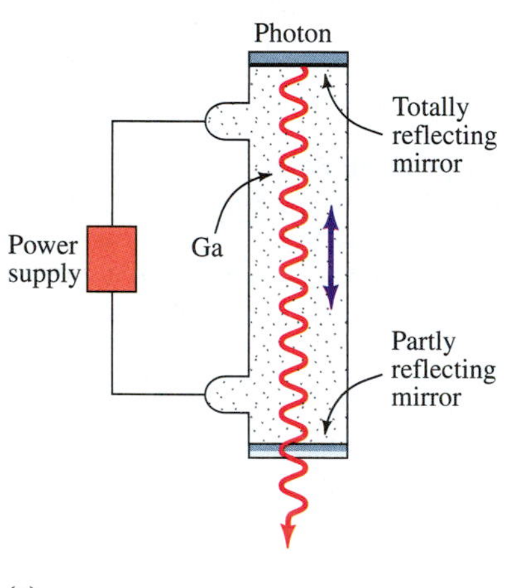

(a)

The fraction of the radiation reflected back into the tube is over 99 percent, allowing the cascade of stimulated emission to occur rapidly. To form a coherent beam, many of the photons bouncing back and forth between the reflectors must be extracted, so one of the mirrors at the ends of the tube is not a perfect reflector. Photons that emerge from the tube do so as a coherent, monochromatic, and well-collimated beam through the partly reflecting mirror. The coherent state is so intense that even though only 1 percent of the photons emerge, the beam is intense enough to be useful.

In summary, the construction of a laser requires:

1. A collection of atoms that contain a metastable energy level into which many atoms can be excited.
2. A mechanism for holding emitted photons in a cavity containing excited atoms so that there is a massive stimulated deexcitation of the excited atoms (a laser transition).
3. A mechanism for repopulating the excited level after the laser transition has taken place, basically a repeat of step 1.
4. A way for the coherent laser beam to be extracted sufficiently well to be useful.

(b)

▲ **FIGURE 42–8** (a) Schematic diagram of a laser. The mirrors are present to contain photons and provide the conditions necessary for stimulated emission. (b) A krypton laser under test.

Some Uses of Lasers

We described in Chapter 35 how the ability to measure time intervals accurately allows us to use lasers to perform *ranging*, or a measurement of distance (see the discussion on corner reflectors on p. 980). The coherence of the laser beam is crucial in *holography*, as we described in Chapter 38. Lasers are an important part of optical readers, found in CD players and checkout counters. They hold promise as a way to manipulate systems as small as individual atoms,[†] and the ability to do this will be of increasing importance in quantum engineering (see Chapter 43).

In Chapter 19 we discussed isotope separation and its significance. An application of lasers occurs in a commercially important method of isotope separation that works as follows: The spectra of different isotopes of an element differ slightly because of tiny differences in their nuclear masses and magnetic dipole moments. Both properties affect the

[†]See S. Chu, "Laser Trapping of Neutral Particles," *Scientific American*, Feb. 1992, p. 70.

atomic energy levels. If atoms consisting of a mixture of isotopes are irradiated with a laser beam tuned very precisely to a transition frequency of only one of the isotopes, then the atoms of only those isotopes will be excited. For example, a suitably tuned laser beam can excite atoms of ^{235}U but not atoms of ^{238}U. A second laser beam with enough energy to ionize the already excited ^{235}U atoms but not ^{238}U atoms is then applied. The charged ions of ^{235}U can now be separated by electric fields from the un-ionized atoms of ^{238}U.

Lasers also have important applications in medicine. One of their first uses was to reattach detached retinas. The lens of the eye focuses a laser beam onto a small area of the retina, which fuses to the tissue from which it has become detached. Because the energy is delivered in a short time, there is no need to immobilize the eye. Infections are less likely to occur than with surgical procedures. Lasers can also be used to cauterize internal wounds and stop bleeding. We can mention also the potential application of lasers for controlled thermonuclear fusion reactions (to be discussed further in Chapter 44). Laser beams focused on lightweight nuclei may be able to provide the energy necessary for them to fuse and produce additional energy. Figure 42–9 illustrates an apparatus designed for this job.

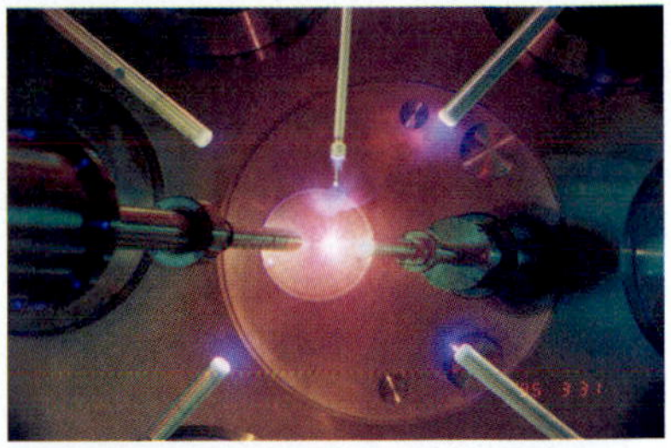

▲ **FIGURE 42–9** The intense laser beams of the OMEGA system at the University of Rochester focus simultaneously on a pellet in order to produce a state hot enough and dense enough to cause thermonuclear fusion.

Lasers play an important role in the scientifically interesting process of the slowing down, or *cooling*, of atoms. From our knowledge of the Doppler shift, we can say that an atom moving with speed v toward the source of a laser beam of a certain frequency f sees the frequency shifted upward to $f(1 + v/c)$. If we choose f to be just under the frequency at which a photon is readily absorbed, then if the atom moves toward the source, it absorbs the photon and its momentum and is therefore slowed down. The photon is, of course, reemitted, but the direction in which this happens is not related to the direction of the initial motion. The net effect is that on average the atom is slowed down. If there are two beams, one coming along the $+x$-axis and one along the $-x$-axis, then any atom moving along the x-axis will be slowed down, whether it moves to the right or to the left. A set up of six laser beams, all tuned to a frequency just below the absorption frequency, will slow the atoms down no matter what their direction of motion. The electromagnetic field in the environment of our atom has been described as *optical molasses*. The creation of optical molasses requires a tunable laser with an extremely accurate control of its frequency. This method of cooling atoms was essential to the creation of the first Bose-Einstein condensate.

Bose-Einstein Condensation

As the temperature is lowered in a gas of bosons, a phase transition can occur at a critical temperature T_c. In this transition, a large number of bosons accumulate in a single lowest-energy state composed of large numbers of these bosons. In his original work on this subject, Einstein was able to find an expression for T_c.

Classical kinetic theory (see Chapter 19) shows that for a gas of atoms, the root mean square speed v_{rms} and hence the root mean square momentum is proportional to $T^{1/2}$. The de Broglie wavelength is h/p and therefore increases as the temperature is lowered. Thus it is only at low temperatures, when the de Broglie wavelength becomes comparable to the interparticle spacing, that collective quantum effects can become important. At the same time, we want to work where the density of the atoms is very small in order to avoid the effects of interatomic forces. Most of the time, the atoms condense to a liquid and then to a solid well before the quantum domain is reached. When the gas of atoms is very dilute, however, so that the interparticle forces are almost negligible, then quantum effects become important. In particular, a gas of atoms of integer angular momentum is a gas of bosons, and at a critical temperature the atoms cease their independent existence. As the temperature is lowered, more and more join together to form a single quantum state.

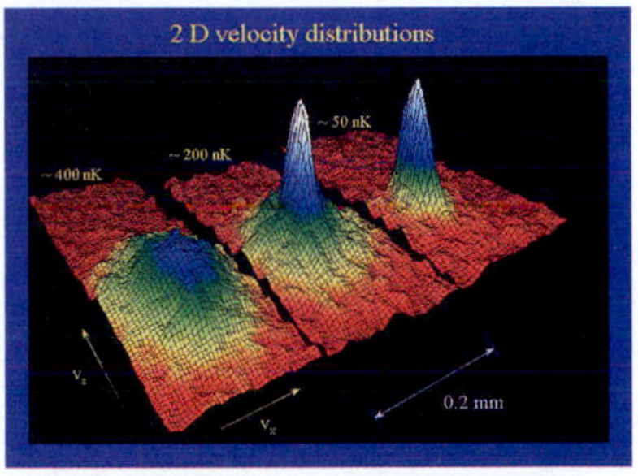

▲ **FIGURE 42–10** A computer-generated image of the velocity distribution in the first Bose-Einstein condensate created at the University of Colorado. As the temperature is lowered from 400nK to 200nK (corresponding to the leftmost and center image, respectively), the atoms condense to form a single quantum system with zero velocity, the effect of which here is the sharp peak. The peak is stable and is even more prominent in the rightmost image, which is at a temperature of 50nK.

In 1995 Eric Cornell and Carl Wieman cooled a gas of ^{87}Rb atoms to a temperature of only 20×10^{-9} K. This was achieved by a combination of laser cooling (which brought the temperature down to 10^{-4}K) and a technique known as evaporative cooling, in which the more energetic atoms are allowed to leave the collection of atoms, leaving behind only the coldest (least energetic) ones. The condensate, which is held in a kind of magnetic trap, manifests itself as a cloud. When the trap is suddenly switched off, the central part of the cloud expands very slowly, showing that it consists of a coherent object at effectively zero speed. Figure 42–10 shows how the lowering of temperature from 400nK to 200nK produces a peak around zero in the speed plot. Individual atoms will not give such a peak; only a large quantum state consisting of millions of atoms can do it.

The Bose-Einstein condensate produced in this way is a new state of matter—so new, in fact, that its technological usefulness has not yet been explored. If the field is anything like the field of lasers, then given what has happened with lasers since the first one was made in the early 1960s, the potential for Bose-Einstein condensates is enormous.

EXAMPLE 42–4 The expression found by Einstein for the critical temperature T_c at which a Bose-Einstein condensate forms is

$$n = \frac{2.31}{4\pi^2}\left(\frac{2mkT_c}{\hbar^2}\right)^{3/2},$$

where n is the number density and m is the mass of the bosonic atoms. What is the density of ^{87}Rb atoms for a critical temperature of 2×10^{-7} K?

Strategy The calculation is a direct application of the expression above. The mass of an ^{87}Rb atom is, to a good approximation, 87 times a nucleon mass, 1.67×10^{-27} kg.

Working It Out We have

$$n = \frac{2.31}{4\pi^2}\left[\frac{2(87 \times 1.67 \times 10^{-27}\text{ kg})(1.38 \times 10^{-23}\text{ J/K})(2 \times 10^{-7}\text{ K})}{(1.055 \times 10^{-34}\text{ J}\cdot\text{s})^2}\right]^{3/2}$$

$$= 3.6 \times 10^{19}\text{ atoms/m}^3.$$

This is a very dilute gas indeed; the density of air at STP is about 3×10^{25} molecules/m^3.

What Do You Think? What is the de Broglie wavelength at T_c, and how does it support our qualitative argument about the onset of quantum phenomena in gases?

CONCEPTUAL EXAMPLE 42–5 Hydrogen atoms are bosons that are composites of fermions: two particles of half-integer spin, an electron and a nucleus composed of a single proton. Does this mean that we can squeeze two hydrogen atoms together until they overlap strongly?

Answer As soon as the electrons or the protons get to within a de Broglie wavelength of each other, the effective repulsion due to the exclusion principle for fermions becomes significant. We know from the previous chapter that the de Broglie wavelength of an electron in hydrogen is of the order of an atomic orbit, so squeezing will become hard due to fermionic "repulsion" once the atoms are immediately adjacent, and this is part of what determines the size of a hydrogen molecule.

42–3 Superconductivity

In Section 26–6 we described the phenomenon of superconductivity, with a further discussion of the remarkable magnetic properties of superconductors in Section 31–5. Superconducting materials have no resistance at temperatures below a critical temperature, in which case we say they are in the superconducting state. In addition, they either expel magnetic field completely or confine it to filaments when they are in the superconducting state. The phenomenon has technological importance in a number of domains, including the large magnets used in medical resonance imaging—the current flows necessary to produce these magnetic fields would produce an unmanageable amount of thermal energy were they not carried in superconducting wires. We are at this point equipped to understand a little of the physics that lies behind the phenomenon of superconductivity.

Cooper Pairs and the BCS Theory

In our discussion of fermions and the Fermi energy in metals, we ignored all dynamic interactions among the fermions. How can this make sense, given that electrons repel each other because of their electric charges? The exclusion principle explains why electron–electron interactions are not very important in ordinary metals. Consider two electrons in different states. Their interaction would manifest itself in some change of state—that is the only way we would see that something had happened. However, the vast majority of electrons cannot change states because the states above and below their own state are already fully occupied, and the exclusion principle forbids further occupation. Interactions do lead to some modification of the predictions of the free-electron theory, and these modifications are well understood.

The phenomenon of superconductivity also involves interactions, but between the electrons and the vibrating lattices of ions in a metal. For complicated reasons that were first understood by Leon Cooper, the electron–ion interaction has as a secondary effect a weak *attraction* between electrons with energies that are close to the Fermi energy. This attraction leads to the formation of weakly bound pairs of electrons known as **Cooper pairs**. Their binding is so weak that they can be quite far apart from one another, as much as hundreds of lattice spacings—in quantum-mechanical language this is the statement that the wave function of the bound state is very spread out. (Indeed, this is the only way the pair can exist in the presence of a Coulomb repulsion—the Coulomb force between the electrons is screened by the presence of the lattice when the electrons are far apart.) The spread-out wave function means that many Cooper pairs may overlap each other, and we cannot think of Cooper pairs as "atoms" consisting of two electrons, because such a term implies that the bound system is localized, with each pair well separated. Although different pairs overlap spatially, there is a sense in which two electrons that make up a Cooper pair are tightly correlated, acting as a true pair: The electrons have equal and opposite momenta, and opposite spins. Because the pairs have lower energies than free electrons—it takes some energy to separate the electrons that form a pair—an energy gap of width $2\Delta \cong 10^{-3}$ eV develops about the Fermi energy, meaning that there are no electron states with energies that lie between $E_F - \Delta$ and $E_F + \Delta$. The gap width is small compared to the Fermi energy itself.

John Bardeen, Leon Cooper, and Robert Schrieffer constructed a theory in 1957—the so-called BCS theory—that explained satisfactorily all of the observations about superconductors known at the time. For most superconducting materials a Cooper pair has a total angular momentum of 0. Although, strictly speaking, a pair cannot be treated as a boson such as a helium atom, a pair shares some crucial properties with bosons. In particular, the BCS theory proposed that the Cooper pairs all condense into the same state when the material's temperature drops below a critical temperature T_c that is approximately given by the gap width divided by the Boltzmann constant. This condensation signals an abrupt change in the properties of the metal, a change somewhat analogous to the phase change from liquid to solid.

The BCS Theory and the Electromagnetic Properties of Superconductors

The most striking property of superconductors is that their resistivity is zero below T_c. This phenomenon is explained by the coherence of the superconducting state. In a normal metal, when a current starts and there is no potential difference to maintain it, the electrons scatter from the lattice and give up energy. The energy appears as ohmic (I^2R) heating and the current rapidly decays. In a superconductor, however, a large number of electron pairs move together in a coordinated way. For the superconductor to lose energy, *the entire coherent state would have to be broken up*, and this would require a large amount of energy to be supplied at once. The mechanisms of ordinary electrical resistance cannot work if a macroscopic number of electron–electron pairs (say, 10^{22}) must be slowed down *all at once*. An analogy is the difference between pushing your hand through liquid water (analogous to a "normal" metal) and pushing your hand against an icicle (analagous to a "superconductor"). In the first case, there are easy ways to lose just a little energy: You can form arbitrarily small waves and eddies. In the second case, you can lose a large amount of energy only all at once, when you break the icicle.

The Meissner effect, which as we saw in Chapter 31 states that no magnetic field can penetrate a superconductor and that a magnetic field already present is expelled when a metal is cooled below T_c (Fig. 42–11), is also described by the BCS theory. The inability to establish a magnetic field within a superconductor is once again a consequence of the coherence of the superconducting state. If you were to try to increase a magnetic field inside a superconductor, you would induce an emf within the material by Faraday's law. A current would be induced to oppose the change in magnetic flux. But because all the Cooper pairs act together, even the tiniest change in magnetic flux can generate a current sufficiently large to cancel entirely the inducing field within the superconductor.

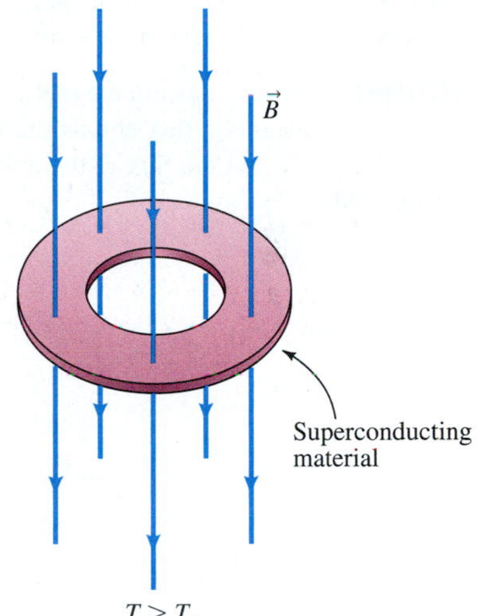

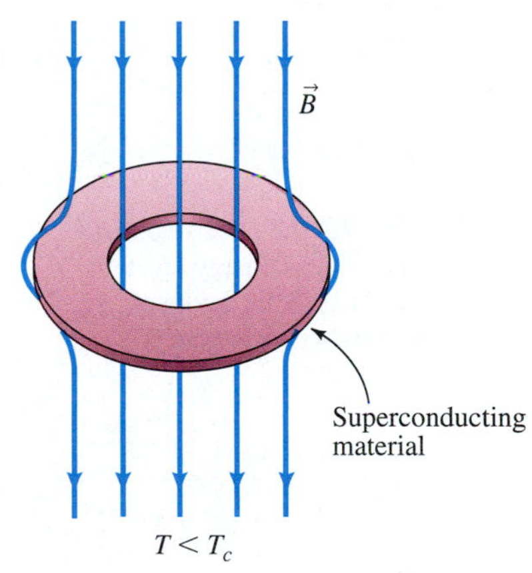

▲ **FIGURE 42–11** Magnetic field lines expelled by a ring of material that makes a transition to the superconducting state when the temperature drops below the critical temperature T_c. Flux is trapped by the hole.

Flux Quantization

Suppose that we place into a magnetic field a ring of superconducting material that is above its critical temperature. The ring lies in a plane perpendicular to the field direction. We now lower the temperature of the material below T_c so that the material becomes superconducting. The field is expelled, and all the field lines will be either outside the ring or confined within the ring (Fig. 42–11).

The BCS theory can be used to show that the coherence of the superconducting state implies that *the magnetic flux through the ring is quantized in units of* $h/2e$:

$$\Phi_B = \frac{nh}{2e} \equiv n\Phi_0, \qquad \text{where } n = 1, 2, \ldots. \tag{42–13}$$

The quantity Φ_0 is the *magnetic flux quantum*, with value 2.07×10^{-15} Wb = $2.07 \times 10^{-15}\ \text{T} \cdot \text{m}^2$. The charge $2e$ that appears here is the charge of a Cooper pair. The prediction that flux is quantized has been successfully tested by experiment.

EXAMPLE 42–6 A single unit of the quantized flux Φ_0 is confined within a cylinder of copper wire of diameter 1.3×10^{-5} m. What is the magnitude of the magnetic field in the wire, assuming that it is uniform throughout the wire?

Strategy With a uniform magnetic field across a cross section of the wire, the magnetic flux equals the magnitude of the field times the wire area. We set the flux to the unit of quantized flux and solve for the field.

Working It Out The unit of flux is $\Phi_0 = 2.07 \times 10^{-15}\ \text{T} \cdot \text{m}^2$. With the magnetic field magnitude written as B and the wire radius R, the flux is given by the surface integral of the component of the magnetic field perpendicular to the area elements over the total area, $\Phi_0 = \int \vec{B} \cdot d\vec{A} = B \times \pi R^2$. We can solve for B:

$$B = \frac{\Phi_0}{\pi R^2} = \frac{2.07 \times 10^{-15}\ \text{T} \cdot \text{m}^2}{\pi (0.65 \times 10^{-5}\ \text{m})^2} = 1.6 \times 10^{-5}\ \text{T}.$$

This is only about a factor of 6 less than Earth's magnetic field and is quite measurable.

Tunneling of Pairs and the Josephson Effects

We saw in our discussion in Sections 7–2 and 40–2 that quantum-mechanical tunneling, in which particles such as electrons can cross potential barriers—a classically forbidden process—has important technological implications. Tunneling can occur across junctions between normal metals, between normal metals and superconductors, and between superconductors. Here, "junction" refers to a very thin insulating strip between two pieces of metal. The insulating strip is formed from a material that can be evaporated and then deposited in thin, uniform layers. If the strip is thin enough (10 nm to 20 nm), an electron can tunnel across it from one metal to the other. Let's consider two normal metals (materials 1 and 2), with Fermi energies E_{F1} and E_{F2}, respectively. If $E_{F2} > E_{F1}$ (Fig. 42–12a), then, since they find lower energy levels to occupy, electrons can tunnel through the barrier formed by the insulating strip, moving from metal 2 to metal 1 (Fig. 42–12b).

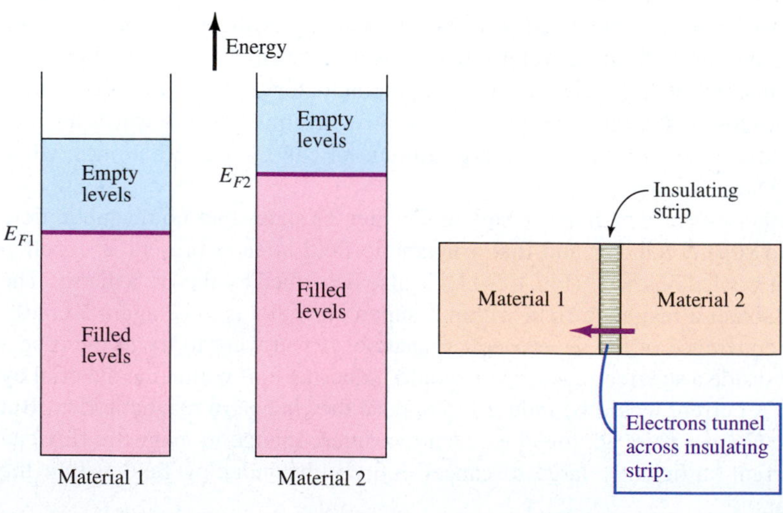

FIGURE 42–12 (a) Two metals with different Fermi energy levels. (b) The metals are joined with an insulating strip between them.

If the two metals are identical, there will be no tunneling—at least at very low temperatures—because there are no empty states to occupy. However, if an external potential V_{ext} is imposed, the levels on one side of the junction will be lowered relative to those on the other side of the junction, and tunneling is again possible (Fig. 42–13).

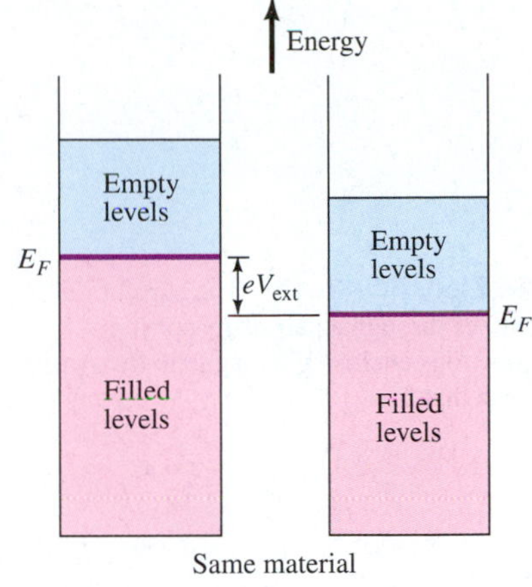

▲ **FIGURE 42–13** When an electric potential that lowers the levels on the right-hand side compared to those of the left-hand side is applied, a tunneling current flows from left to right across the insulating strip. In this case, the metals on both sides of the strip are the same.

Now suppose that we have a junction between a metal in its normal state and a metal in a superconducting state (Fig. 42–14a) and that each piece has the same Fermi energy. With no external potential, no electrons flow. Suppose that an external potential V_{ext} that lowers the Fermi energy of the superconductor is applied. A current will flow between normal metals, but because of the energy gap in the superconductor, the external potential must exceed a minimum value before empty levels become available for tunneling into (Fig. 42–14b). This minimum value is given by $eV_{min} = \Delta$ and, by varying the external potential and observing the onset of tunneling, it is possible to measure the gap size. Figure 42–14c illustrates the relationship between current and potential in this case.

Let's now consider two superconductors separated by a thin insulating strip, an arrangement known as a *Josephson junction*. At first there might appear to be little current across the barrier. If the probability of a single electron tunneling through is very small, then, because the probability of two independent electrons tunneling through is the product of the tunneling probabilities of the individual electrons, there would be very little tunneling for an electron pair. In 1962, Brian Josephson noted that if the barrier is less than 1 nm thick, the two superconductors form a single, coherent quantum system. As a result, what tunnels through the barrier is not two individual electrons, but a pair of electrons as a unit. This has two consequences (we shall prove neither here): First, *even in the absence of a potential difference between two identical superconductors, a tunneling current will flow across a Josephson junction.* This is called the *DC Josephson effect*. Second, *when there is a constant potential difference V across the barrier between two superconductors, the current that flows between them oscillates with angular frequency*

$$\omega = \frac{2e}{\hbar}V. \tag{42–14}$$

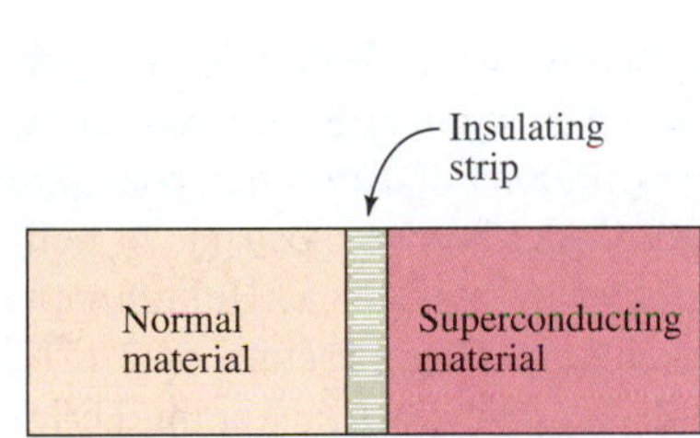

(a)

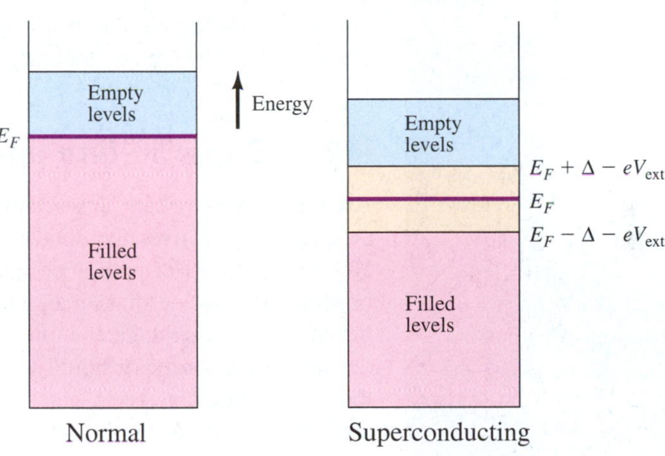

(b)

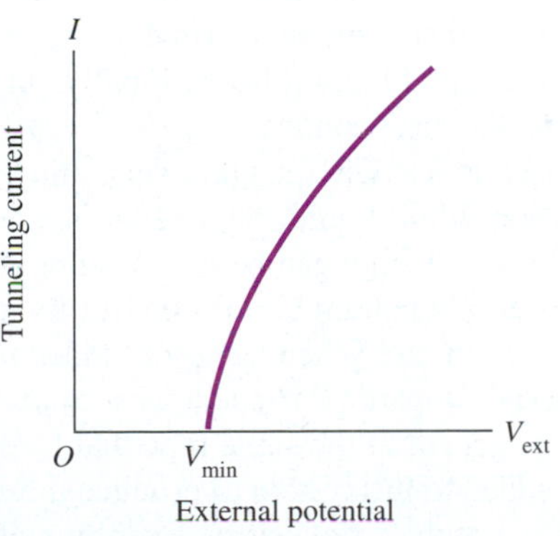

(c)

▲ **FIGURE 42–14** (a) A metal in its normal state separated by a thin insulating strip from the same metal in its superconducting state. (b) An energy diagram of this situation, including an external potential, V_{ext}, applied so that the energy levels in the superconducting metal are lowered with respect to those in the normal metal. Because an energy gap is present in the superconductor, the potential must be a certain minimum size before electrons from the normal metal can tunnel into the empty levels of the superconductor. This minimum value of V_{ext} is given by $eV_{min} = \Delta$. (c) The flow of tunneling current in the junction described in Fig. 42–14a as a function of V_{ext}. V_{ext} must have the minimum value V_{min} before current can flow.

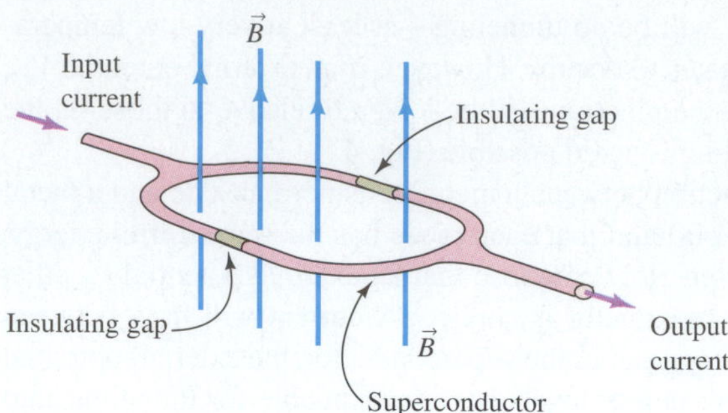

▶ **FIGURE 42–15** The current that passes through a pair of Josephson junctions enclosing a magnetic flux varies with that flux.

This phenomenon is called the *AC Josephson effect.* Because frequencies can be measured with great precision, the AC Josephson effect provides physicists and engineers with the most accurate way known of standardizing voltage measurements. Equivalently, it allows the most precise measurement possible of the fundamental ratio $e/\hbar$.

Consider two Josephson junctions connected in parallel (Fig. 42–15). Like the interference pattern produced by an electron passing through a double slit, an interference pattern appears in the net current. The interference pattern is affected by a magnetic flux enclosed by the two arms of the pair of junctions. The flux plays the same role in the pair of Josephson junctions that the slit separation plays in a double-slit experiment. The dependence of the current on the enclosed magnetic flux Φ_B is given by

$$J = J_{\text{max}} \cos\left(\frac{e\Phi_B}{\hbar}\right). \tag{42–15}$$

This phenomenon allows us to measure magnetic flux precisely. The SQUID (Superconducting QUantum Interference Device), which is based on the phenomenon described, is widely used for high-precision measurements of very tiny magnetic fluxes. Magnetic fields as small as 10^{-13} T have been measured. With SQUIDs, an esoteric quantum mechanical phenomenon has become a major technological tool in medical diagnostics, where tiny variations in the electric currents generated by the heart or the brain can be measured by the magnetic fields they produce.

▲ **FIGURE 42–16** The fountain effect of superfluid helium, which flows without viscosity and exhibits a number of phenomena not seen in ordinary fluids—here we see it flowing against gravity through an orifice under the influence of interparticle forces and continuing through to form a type of fountain.

42–4 Superfluidity and Liquid Helium

Another technologically important low-temperature phenomenon, distinct from superconductivity, involves the condensation of bosons into a macroscopic coherent state. Helium atoms, ^{4}He, with a nucleus consisting of two protons and two neutrons, are bosons; their nucleus has an angular momentum of zero, and the two electrons of helium atoms have an angular momentum that is an integral multiple of $\hbar$. Helium condenses into an ordinary liquid at 4.2K; more remarkably, at a temperature of 2.17K, there is another change of state. As the temperature decreases below this point, an increasing fraction of the liquid helium flows with no internal friction (viscosity). This property is called **superfluidity**. Superfluid helium can flow freely through the narrowest of channels. It creeps up the sides of a beaker containing it and flows to the outside, even escaping through a hole to make a fountain (Fig. 42–16). It can be used to make the most nearly perfect (frictionless) heat engine—one that converts thermal energy to kinetic energy with no moving parts save that of the liquid helium itself. Finally, and most important technologically, it makes a nearly perfect heat conductor.

The fact that atoms of ^{4}He are bosons condensed in a single quantum state allows us to explain these properties. In an ordinary, nonsuperfluid liquid, the excitations do not involve the entire fluid. Arbitrarily small amounts of energy can be lost, leading to viscosity at any speed. In superfluid helium, however, the helium atoms congregate in the same (coherent) ground state, a Bose-Einstein condensate. When energy is added to the system, excitations of the entire fluid are produced. In particular, the lowest excitations are compressional, or sound, waves. (These waves are of the same type that form sound waves in solids.) Very little energy must be added to the system to produce these excitations. However, the energy of the excitation is not the whole story. Even though sound waves are the excitations of lowest energy, it is difficult to excite them mechanically, and this class of excitations is not useful for carrying off mechanical energy; that is, they produce no internal friction. To understand the significance of this remark, let's consider the process by which friction is produced.

When we drag a sphere of mass M through liquid helium, we will create an excitation only if the sphere gives up some energy and momentum to the liquid. Such excitations are discrete quantum mechanical states that correspond to motion of the entire system of liquid helium, and the excitations have a definite momentum $\vec{p}$ and energy E. Momentum conservation states that

$$M\vec{v}_i = \vec{p} + M\vec{v}_f,$$

where $\vec{v}_i$ is the initial velocity of the sphere and $\vec{v}_f$ is its velocity after it has given up some energy. Then $\vec{v}_f = \vec{v}_i - \vec{p}/M$. Similarly, energy conservation gives

$$\frac{1}{2}Mv_i^2 = E + \frac{1}{2}Mv_f^2.$$

We substitute $\vec{v}_f$ into the energy conservation equation:

$$\frac{1}{2}Mv_i^2 = E + \frac{1}{2}\left(Mv_i^2 - 2\vec{v}_i \cdot \vec{p} + \frac{p^2}{M}\right) \cong E + \frac{1}{2}Mv_i^2 - \vec{v}_i \cdot \vec{p}.$$

We dropped the last term on the right because M is large (macroscopic). Solving for E,

$$E = \vec{v}_i \cdot \vec{p}. \tag{42–16}$$

As long as this relation can be satisfied, it will be possible for the sphere to lose kinetic energy by causing excitations in the liquid, and there will therefore be resistance to the motion of the sphere. But if it cannot be satisfied, the sphere has no way to lose energy.

Let's see whether it is possible for sound waves to be excited in this way. Now sound waves themselves obey the rules of quantum mechanics, meaning that if one looks at sound of very low energy, one will see that it comes in quantized packets. The relation between energy and momentum in these packets of sound is given by

$$E = v_s p,$$

where v_s is a constant, the velocity of a sound wave. This is analogous to the relation between energy and momentum for photons. In fact, the quanta of the acoustic field that manifests itself as sound waves are called *phonons*. We see that for $v_i < v_s$, Eq. (42–16) cannot be satisfied, and *there is no way for the sphere to lose energy*. It therefore travels with *no viscosity*—superfluidity.

THINK ABOUT THIS. . .

OF WHAT USE IS SUPERFLUID HELIUM?

The technological importance of the properties of superfluid helium is closely linked to the importance of superconductivity. Superconducting systems such as the large superconducting magnets used in particle accelerators or MRI apparatus must be kept at temperatures below their critical temperature. The failure of superconductivity in such a system can be very destructive, with the large currents suddenly producing large amounts of thermal energy. The only practical superconductors that can be used for such magnets are those for which liquid helium is sufficiently cold to form a coolant. Here, the fact that the liquid ^{4}He is in a single quantum state plays an important role. First, it is a nearly perfect heat conductor—its coefficient of thermal conductivity is hundreds of thousands of times greater than that of copper. That is because a less-than-perfect thermal conductivity is associated with collisions in which whatever carries off the thermal energy (in the case of liquid helium it is carried off by vibrational excitations) can lose small amounts of that energy in collisions, and there is no way to lose only small amounts of thermal energy in a collision with a single quantum state. Moreover, and for similar reasons, boiling cannot occur at any one location, and this offers crucial insurance against any of the magnet coils heating beyond its critical temperature. In other words, no "hot spots" can form. The Large Hadron Collider accelerator, under construction at CERN, headquartered in Geneva, Switzerland, will operate with a large array of magnets cooled by a connected system of superfluid helium.

The creeping of liquid helium up the sides of a beaker depends on the difficulty of dissipating arbitrarily small amounts of energy. In this situation the superfluid can respond vigorously to very small forces such as the van der Waals forces between neighboring helium atoms and between the atoms of helium and of the beaker walls. While the details of the mechanisms that produce these strange properties are more complicated than we can get into here, the phenomena themselves provide another demonstration of quantum effects on a large scale.

Summary

In large systems for which quantum mechanics governs the dynamics, the Pauli exclusion principle plays a crucial role if the system is composed of fermions. This principle states that only one fermion—for our purposes, a particle with spin $\hbar/2$—of a given type can occupy a particular state such as an energy level. In metals, this means that very closely spaced energy levels are filled with electrons up to the Fermi energy, E_F. The Fermi energy of electrons is

$$E_F = \frac{\hbar^2}{2m_e}(3\pi^2 n_e)^{2/3}, \tag{42–5}$$

where n_e is the electron number density in the metal. Only electrons near the top of the filled levels take part in dynamic processes such as current flow.

The exclusion principle is responsible for the incompressibility of matter. In ordinary matter the incompressibility can be estimated to within a factor of 2 by assuming that the electrons in the matter are free. The exclusion principle prevents stars that are not too massive from collapsing under the mutual gravitational attraction of their constituents.

A system composed of bosons—particles whose spins are integral multiples of $\hbar$ such as photons and ^{4}He atoms—manifests quantum mechanical effects quite different from those of fermionic systems. Identical bosons have an enhanced probability of aggregating in the same quantum state. As a consequence, there can be stimulated emission of radiation (photons) into states already occupied by other photons. The operation of the laser is based on stimulated emission and on the bosonic properties of photons. To construct a laser, we need a way to populate an excited energy level in a collection of atoms; a way to build up many photons of the "proper" energy to provoke a massive stimulated transition from the excited level; and a way to extract the coherent laser beam formed by the stimulated emission.

Fermions can pair up to form bosons, and those correlated pairs tend to congregate in the same quantum state. This is the mechanism behind superconductivity. Superconductors form when a collection of electron–electron pairs known as Cooper pairs occupy a single, coherent quantum state at low temperatures. Cooper pairs can tunnel, and this phenomenon has advanced our ability to observe very small magnetic fields. Superfluidity, characterized by flow without viscosity in superfluid ^{4}He, is another phenomenon that follows from the tendency of bosons to condense in the same quantum state at low temperatures.

Understanding the Concepts

1. How would the behavior of electrons in materials change if the Pauli exclusion principle were not applicable?
2. Most heavenly bodies rotate and have some angular momentum. Astrophysicists believe that pulsars are rapidly rotating neutron stars, some with periods as short as 0.005 s. Explain how such a small, massive object could rotate so fast.
3. Where does the fact that photons tend to congregate in the same quantum state play a role in our discussion of lasers?
4. You have perhaps heard that neutrinos interact only very slightly with matter and therefore could not be confined in the way that is described here. Does this make the exercise in Example 42–1 totally useless? In thinking about the answer to this question, ask yourself whether there might be highly relativistic systems of fermions in nature in which the constituents are effectively confined.
5. The *Fermi temperature*, T_F, is defined by $E_F \equiv kT_F$. What does it mean if T_F in a given material is much larger or much smaller than the material's temperature?
6. Describe the steps in the evolution of a star that becomes a black hole.
7. Is the neutron star described in Example 42–3 relativistic? That is, do its constituents move relativistically? What is the general criterion for deciding when such a system is relativistic?
8. An electron on the Moon and an electron on Earth both have spin up and are in the ground state of hydrogen. Given that both electrons are in the same state, how is the Pauli exclusion principle bypassed?
9. Suppose that a crystal could be made from the atoms of the inert gases (by going to low temperatures, for example). Would such a crystal be likely to act as a metal?
10. The spins of the two electrons in the ground state of helium can be either parallel or antiparallel. The Pauli exclusion principle does not apply to antiparallel spins, but electrons with parallel spins must avoid each other. Would you expect the two possible spin arrangements to have the same energy? If not, which energy would be lower?

11. How could you tell if electrons in very distant galaxies obey the Pauli exclusion principle?
12. Given our discussion of stellar evolution, why is Jupiter not a star? (Assume that Jupiter is an aggregate of hydrogen atoms held together by gravity.)
13. An electron traveling through a collection of protons in a plasma is slowed down by collisions and captured into an orbit with a large principal quantum number, far from the proton it orbits. Given that radiative transitions are largely those for which $\Delta \ell = \pm 1$, would the electron end up in the ground state through one transition or through a sequence of transitions?
14. The filament of a lightbulb has many atoms that become thermally excited into the same excited states. Why doesn't the bulb act as a laser?
15. What difficulties might you encounter in an attempt to make an X-ray laser?
16. Use the argument as to why the flow of superconducting helium around an obstacle is nonviscous when the flow occurs at a speed less than that of sound to explain why electron waves propagate without energy loss through a rigid crystal lattice at $T = 0$. Why will the electrons experience resistance at finite temperatures?
17. In our treatment of the incompressibility of ordinary matter, we spoke about metals rather than a more general class of solids. Does our discussion apply to insulators? (See Chapter 26 for the difference between conductors and insulators.)

Problems

42–1 The Exclusion Principle in Bulk Matter

1. (I) What is the value of the Fermi energy of calcium, which has two valence electrons per atom? The gram-atomic weight of calcium is 40.1 g/mol, and its mass density is 2.30 g/cm^3.
2. (I) Find the Fermi temperature (see Question 5) of gold, for which the Fermi energy is 5.51 eV.
3. (I) Calculate the Fermi energies of sodium ($n_e = 2.65 \times 10^{28}\ \mathrm{m}^{-3}$), potassium ($n_e = 1.40 \times 10^{28}\ \mathrm{m}^{-3}$), and aluminum ($n_e = 18.1 \times 10^{28}\ \mathrm{m}^{-3}$). All these metals have a single valence electron per atom.
4. (I) Calculate the bulk moduli, B, of sodium, potassium, and aluminum, using the data given in Problem 3. (The experimental values of B for these materials are 0.64, 0.28, and 7.6, respectively, in units of $10^{10}\ \mathrm{N/m^2}$.)
5. (I) Zinc, a metal with two free electrons per atom, has a gram-atomic weight of 65.4 g/mol and a density of 7.13 g/cm^3. What is the density of electrons in zinc?
6. (I) The density of electrons in silver is 5.85×10^{28} electrons/m^3. Calculate the bulk modulus of silver. (The experimental value is $10.1 \times 10^{10}\ \mathrm{N/m^2}$.)
7. (I) What values of $n_1 (= n_2 = n_3)$ give an energy close to 1.0 eV for an electron in a cubical box whose sides are precisely 100 nm long? What is the actual energy of this state? Of the state in which n_1 and n_2 are unchanged but n_3 is increased by 1?
8. (I) Consider a copper "nanoparticle," a cube of $10 \times 10 \times 10$ atoms (Fig. 42–17). Given that $n_e = 8.47 \times 10^{28}$ electrons/m^3, what is the energy difference between adjacent energy states in the neighborhood of the Fermi energy? Take the distance between neighboring atoms to be 0.25 nm.

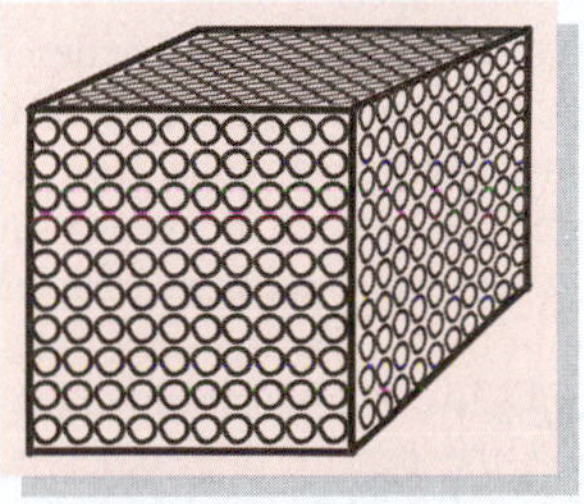

▲ **FIGURE 42–17** Problem 8.

9. (I) The Fermi energy of electrons in a metal, E_F, may be used to define the Fermi momentum, p_F, defined by $E_F \equiv p_F^2/(2m_e)$. Express the Fermi momentum in terms of n_e. Use your results to calculate the Fermi speed, v_F, the speed of an electron with the Fermi energy.
10. (I) What is the speed of an electron with the Fermi energy in magnesium, given that $n_e = 8.60 \times 10^{28}$ electrons/m^3?
11. (I) Estimate the gravitational pressure at the surface of the Sun.
12. (II) When the Fermi speed is, say, $0.1c$, where c is the speed of light, it is more appropriate to use a relativistic connection between E_F and p_F rather than the nonrelativistic connection that we used in Problem 9. Estimate the value of n_e for which the relativistic connection is necessary.
13. (II) The conduction electrons of a metal have to be treated relativistically when the Fermi speed is about $0.1c$ or more. Because they are confined to a small volume, the inner-shell electrons may also reach high speeds when Z is large. (a) Derive the condition on the density of conduction electrons that allows a nonrelativistic treatment. (b) Derive the condition on Z for which the innermost electrons can be treated nonrelativistically.
14. (II) Seven electrons are in an infinitely deep one-dimensional potential energy well of width L. (a) What is the lowest possible energy value for this system? (b) What is the next highest energy value? In both parts, ignore Coulomb interactions between the electrons.
15. (II) Consider N identical, noninteracting fermions of mass m in an infinitely deep one-dimensional potential energy square well—a box—of width L. Assume that $N \gg 1$. Find (a) the Fermi energy, and (b) the total energy of this system of particles in its lowest energy state. (c) What is the smallest amount of energy that can be absorbed by the system; that is, what is the minimum excitation energy of the system from its ground state?
16. (II) Assume that the nucleus is composed of free neutrons and protons. What is the equation analogous to Eq. (42–1) for the case of a nucleus? Assume that the nucleus is a cubic box with sides L. L is such that the box volume is the same as a sphere of radius $R = r_0 A^{1/3}$, where A is the total number of neutrons and protons and $r_0 \cong 1.2 \times 10^{-15}$ m.
17. (II) Calculate the Fermi energy for the neutrons confined to the nucleus ^{96}Mo, which roughly forms a sphere of radius 5.5×10^{-15} m. (There are 54 neutrons in the ^{96}Mo nucleus.)

18. (II) The general definition of the pressure in a gas with internal energy U is $p = -(\partial U/\partial V)$, where V is the volume of gas. (This follows from the first law of thermodynamics and from the fact that the infinitesimal work done when a system changes volume by dV is $dW = p\,dV$.) Given the fact that for a gas of identical fermions at $T = 0$, $U = 3/5\ NE_F$, show that at $T = 0$, $p = (2/3)\ (U/V)$ for the fermion gas. This result is the *same* as our result in Chapter 19, where we studied the atomic origin of pressure. (That result did not depend on the precise distribution, so it is not surprising that we recover it here.)

19. (II) Using the result of Problem 18, (a) calculate the pressure of an ideal gas of N identical fermions at $T = 0$ in terms of the volume, and (b) evaluate the pressure of the degenerate electron gas at $T = 0$ in a sample of sodium. The density of free electrons in sodium is 2.65×10^{22} electrons/cm^3.

20. (II) Calculate the ratio of the degeneracy pressure of electrons to that of protons for a star such as our Sun. Assume that the star is electrically neutral.

21. (II) Suppose that electrons are confined to a plane at a density of n_e electrons per unit area. Follow the steps that lead to Eq. (42–5) to calculate the Fermi energy for such a system.

22. (II) An iron nucleus consists of 26 protons and 30 neutrons in a sphere of radius 6.0×10^{-15} m. Assume that none of the particles interact with each other. Calculate the Fermi energies of the protons and the neutrons.

23. (II) (a) Calculate the radius of a neutron star as a function of its mass. (b) Find the Fermi energy and Fermi momentum of a neutron star in terms of its radius and its mass. (c) Estimate the mass for which one should start to use the relativistic expression in part (b). Express your results in solar masses.

24. (III) A proper calculation of the Fermi energy in a cubical box with sides L amounts to counting the number of electrons that fill all the states up to an energy E_F. There are two states for each set of positive integers $\{n_1, n_2, n_3\}$ that satisfy the condition $n_1^2 + n_2^2 + n_3^2 = 2(m_e E_F L^2/\pi^2\hbar^2) \equiv R^2$. In a "continuum approximation" to the problem of counting the number of states, appropriate to large values of n_1, n_2, and n_3, the number of states is twice the volume of an octant (one-eighth) of a sphere of radius R; that is, $2(\frac{1}{8})(4\pi/3)R^3$ (Fig. 42–18 is a two-dimensional version). Compare the answer in terms of the density with the approximate result that we gave in the text, $E_F \cong (\hbar^2/2m_e)(\pi^3 n_e)^{2/3}$.

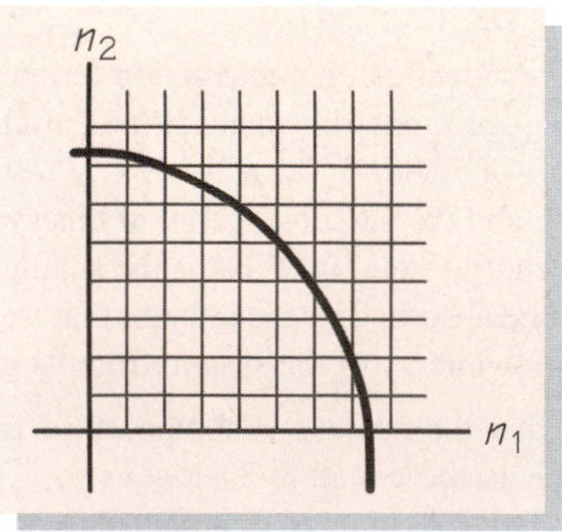

▲ **FIGURE 42–18** Problems 24, 25.

25. (III) By counting the number of states (that is, by using the techniques described in Problem 24), calculate the Fermi energy for a two-dimensional gas of free, noninteracting electrons (Fig. 42–18). Show that $d = \lambda_F/2$, an approximation we made in our text discussion, is not as good as it is for a three-dimensional gas.

42–2 Lasers and the Behavior of Bosons

26. (I) A laser emits radiation of wavelength 590 nm and delivers power of 65 mW. How many photons are emitted per second?

27. (I) The characteristic red light of a He–Ne gas laser is due to stimulated emission between neon levels at 20.66 eV and 18.70 eV. Calculate the wavelength and frequency of the photon emission.

28. (I) A He–Ne laser emits radiation with a power of 42 mW at a wavelength of 632.8 nm. How many photons are emitted per second?

29. (I) A laser emits radiation at a wavelength of 555 nm. Photons are emitted at a rate of $8.5 \times 10^{18}\ \text{s}^{-1}$. What is the power of the laser?

30. (II) Free electrons in a magnetic field could form the basis of a tunable laser, in which the two levels involved in the transition are the "up" and "down" spin states of the electron in a magnetic field. How large a magnetic field is needed to operate the laser at 500 nm?

31. (II) Consider N identical bosons of mass m in an infinitely deep one-dimensional potential energy well of width L (Fig. 42–19). (a) What is the total energy of this system of particles in its lowest energy state? (b) What is the smallest amount of energy that can be absorbed by the system; that is, what is the minimum excitation energy of the system from its ground state?

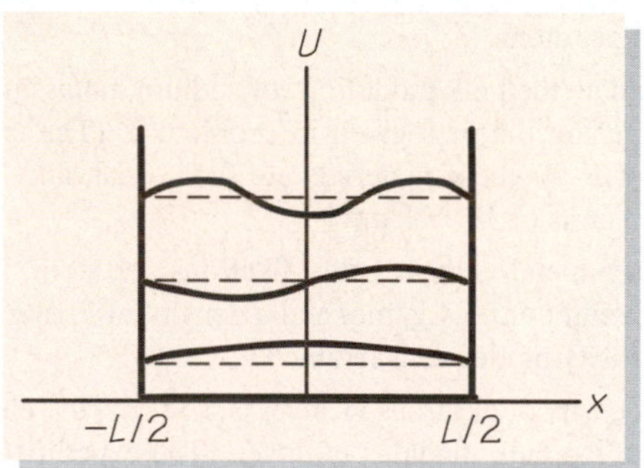

▲ **FIGURE 42–19** Problem 31.

32. (II) The collimation of a laser beam is limited by diffraction; that is, the angular beam spread $\Delta\theta$ is at least as large as $\Delta\theta \cong \lambda/D$, where D is the diameter of the laser's aperture. What is the diameter of the beam spot projected on the Moon, 3.8×10^8 m from a laser of aperture diameter 18 mm emitting light of wavelength 550 nm?

33. (II) A laser beam projects light of wavelength 663 nm at the rate of 3.0×10^{17} photons/s. (a) What is the power of the laser? (b) What is the radiation pressure exerted by the beam on a shiny surface if the laser beam projects a circle 3.5 mm in diameter?

34. (III) In 1917, Einstein wrote a paper that describes the stimulated-emission process. We can summarize his results as follows: Suppose that an atom has two states. State 1 is the ground state; state 2 is an excited state. The atom is in an enclosure filled with blackbody radiation, which we recall from Chapter 17 has the energy-density function $u(f, T)$. The transition rate for excitation from state 1 to state 2, $\Gamma_{2\to1}$, is proportional to the number of atoms in state 1, N_1, times the density of photons present that have the proper frequency (such that the photons have the transition energy): $\Gamma_{2\to1} = N_1 u(f, T)B_{12}$, where B_{12} is a proportionality constant. The transition rate for deexcitation from state 2 to state 1, $\Gamma_{2\to1}$, has two terms: a stimulated-emission term proportional to the number of atoms in state 2 times the density of photons present with the proper frequency, $N_2 u(f, T)B_{21}$, and a spontaneous-emission term proportional to the number of atoms in state 2. Thus $\Gamma_{2\to1} = N_2[u(f, T)B_{21} + A]$, where A is a constant. (a) In thermodynamic equilibrium, the transition rate in the two directions must be equal. Use this fact, together with the Boltzmann relation for the populations, $(N_1/N_2) = e^{-E_1/kT}/e^{-E_2/kT}$, to show that

$$u(f, T) = \frac{A}{B_{12}e^{hf/kT} - B_{21}}.$$

(b) In the classical limit, $u(f, T) = 8\pi f^2 kT/c^3$. Show that the quantum mechanical expression for u can match this form for large T only if $B_{12} = B_{21}$, and find the value of the ratio A/B_{12} from the matching requirement.

35. (III) From Problem 34, the rate per atom at which atoms in equilibrium with radiation at temperature T emit photons of frequency f has the form

$$\frac{\Gamma_{2\to1}}{N_2} = C\left(1 + \frac{1}{e^{hf/kT} - 1}\right),$$

where C is a constant. (a) Use the hint at the end of this problem to show that the average number of photons of frequency f per unit volume, $\langle n \rangle$, is given by

$$\langle n \rangle = \frac{1}{e^{-hf/kT} - 1}.$$

(b) Use this result to show that the deexcitation transition rate per atom is given by

$$\frac{\Gamma_{2\to1}}{N_2} = A(1 + \langle n \rangle).$$

This establishes the result that the transition rate for deexcitation increases linearly with the number of photons of the proper frequency already present, as is required for laser operation. [*Hint*: The distribution function to find n photons of energy hf is given according to Planck's work on blackbody radiation as $P_n = e^{-nhf/kT}(1 - e^{-hf/kT})$. The sum representing the average number is $\sum_n P_n n$, and this sum can be performed by noticing that $\sum_n x^n$ is a geometric series, and that $\sum_n nx^n = x(d/dx)\sum_n x^n$.]

42–3 Superconductivity

36. (I) The critical temperature of a superconductor, T_c, varies with the isotopic mass of the element making up the superconductor, M, according to the relation $T_c\sqrt{M}$ = a constant. In lead, $T_c = 7.184\text{K}$ for the isotopic molar mass 207.7 g. What is the critical temperature for the isotopic molar mass 206.3 g?

37. (I) An unknown constant potential difference V is placed across a Josephson junction, and an alternating current of frequency 23.3 GHz is produced. What is the value of V?

38. (I) What is the frequency of the AC component of the current when a DC voltage of 1.75 V is placed across a Josephson junction? A DC voltage of 0.80 μV?

39. (II) Verify that the units of magnetic flux are consistent with Eq. (42–15).

40. (II) The current-carrying ability of superconductors is limited by the fact that superconductivity breaks down in a large magnetic field. How large a current can be carried by a niobium-tin wire of diameter 0.60 mm if the critical magnetic field is 1.0 T?

41. (II) When a current passes through a Josephson junction, the resulting AC current can be regarded as corresponding to photons whose frequency is that of the AC current. For a particular Josephson junction, if these photons have an energy of 5.0×10^{-6} eV, what is the DC potential difference across the junction?

42. (II) The loop of a SQUID surrounds an area of 0.40 cm^2. If the phase of the interference current can be determined to an accuracy of $\pi/2$, to what accuracy can a magnetic field be determined? What current produces a field like this 12 cm away?

42–4 Superfluidity and Liquid Helium

43. (II) According to one approach that takes into account the quantum mechanical "clustering" behavior of the bosons that make up a superfluid, the temperature at which the superfluid forms is given by $T_c = [N/(2.612)(0.886)C]^{2/3}$, where $C = 2\pi V(2mk/h^2)^{3/2}$, V is the volume in which the N bosons are confined, and m is the mass of the boson. Apply this result to liquid ^{4}He. Take the density of the liquid to be 0.147 g/cm^3, and calculate T_c. (The experimental value of T_c for this system is 2.2K.)

General Problems

44. (II) (a) The energy-distribution function of electrons at $T = 0$ is given in Fig. 42–20. Use this to calculate the speed-distribution function. (b) Calculate $\langle v^4 \rangle$ by using the result of part (a).

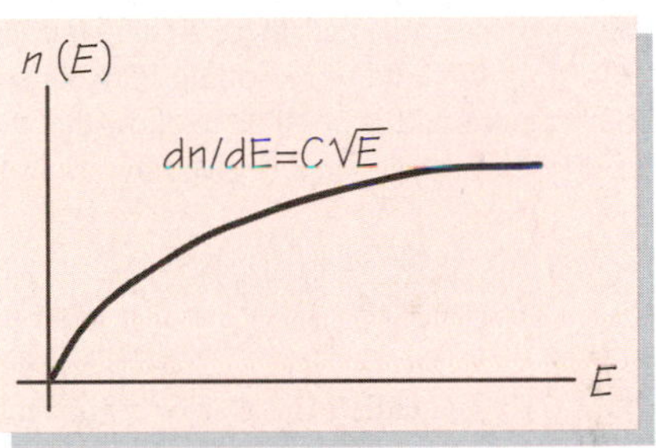

▲ **FIGURE 42–20** Problem 44.

45. (II) What is the radius of a star with a mass of 10^{-3} solar masses, assuming that the star has the density of the Sun? Compare your answer with the radius of Jupiter.

46. (II) Assume that the Fermi energy depends only on $\hbar$, the electron density n_e, and the mass of the electron. Use dimensional analysis to find that dependence.

47. (II) Assume that the Fermi energy depends only on $\hbar$, the electron density n_e, and the speed of light c. Use dimensional analysis to find that dependence.

48. (II) You want to verify the isotopic effect described in Problem 36 with a set of data you collected. Show that if an isotope equation of the type $T_c M^\gamma$ = a constant were correct, the data on a plot of $\ln(T_c)$ versus $\ln(M)$ would be a straight line whose slope determines γ. What is the curve if $\gamma = \frac{1}{2}$?

49. (II) Show that the isotopic-effect equation $T_c\sqrt{M}$ described in Problem 36 implies that T_c is proportional to v, where v is the speed of sound in a material with isotopic mass M.

50. (II) (a) Using the methods in the text, show that in a gas of free electrons in a box with sides of length L at $T = 0\text{K}$, the total number of electrons with energy less than E is

$$N = \left(E\frac{2m_e}{\hbar^2\pi^2}L^2\right)^{3/2}\frac{\pi}{3},$$

where $E < E_F$. (b) Use the results of part (a) to show that the number of electrons with energies between E and $E + dE$ is given by

$$n(E)\,dE = \frac{V}{2\pi^2}\left(\frac{2m}{\hbar^2}\right)^{3/2}\sqrt{E}\,dE,$$

where $n(E)$ is the energy-distribution function of the electrons. [*Hint*: The expression you want is given by $n(E)\,dE \equiv N(E + dE) - N(E)$.]

51. (II) Use the energy-distribution function of Problem 50 to show that the total energy of a gas of electrons in a box with sides of length L at $T = 0\text{K}$ is given by $E_{\text{tot}} = \frac{3}{5}NE_F$.

52. (III) The behavior of the atoms of a crystal at temperature T under the influence of the springlike forces between the atoms is analogous to the behavior of light in a cavity at temperature T (blackbody radiation). The behavior of the lattice can be described by *phonons*, collective oscillations of the lattice that can be treated as particles analogous to the photons in a hot cavity. The phonons are identical bosons that obey the quantum rules for such particles. There is one important difference between the lattice vibrations and blackbody radiation: The number of possible modes of a lattice made up of N atoms is limited to $3N$ (corresponding to the three-dimensional motion), and there is a maximum frequency f_{max}. By following the kind of reasoning Planck used in deriving his blackbody spectrum, it is possible to show that the total lattice vibration energy, U, of the crystal is given by

$$U = \frac{9NkT}{\Lambda^3}\int_0^{\Lambda}\frac{x^3}{e^x - 1}\,dx,$$

where $\Lambda \equiv T_D/T$; T_D, called the *Debye temperature*, is given by $T_D = hf_{\text{max}}/k$. (a) Show that in the limit of $T \gg T_D$, the energy of the solid is the classical expression $3NkT$. (b) Show that in the limit $T \ll T_D$, $U \propto T^4$. (c) What is the temperature dependence of the specific heat at low temperatures? The curve formed by the expression for U matches well the experimental data over a large temperature range in many solids (Fig. 42–21; recall $C_V = dU/dT$). (After G. Burns, *Solid State Physics*, Academic Press, Inc., 1985, p. 369.)

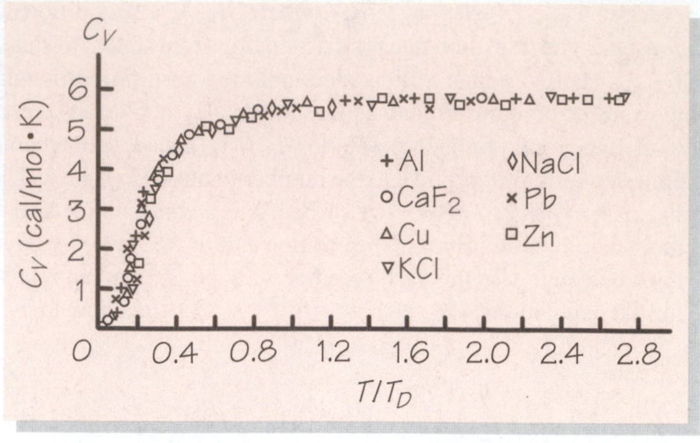

▲ **FIGURE 42–21** Problem 52.

53. (III) For stars whose mass is roughly one solar mass (2×10^{30} kg) and whose thermonuclear reactions have stopped because the fuel has been used up, the repulsive effects due to the Pauli exclusion principle *for electrons* are strong enough to keep the star from collapsing gravitationally. In that case, white dwarfs are formed. (a) Calculate the Fermi energy of the electrons in a white dwarf of mass M. (b) Calculate the radius of a white dwarf of mass M. If $M = M_\odot$ is one solar mass, what is the ratio of your result to the present radius of the Sun? [*Hint*: Follow our procedure for a neutron star. The gravitational pressure is the same as before, because only the protons and neutrons account for the mass. The number of electrons is the same as the number of protons, which is roughly half the total of protons plus neutrons. The radius of the Sun is 6.96×10^8 m.]

54. (III) When all the hydrogen in a star has burned, helium nuclei and electrons remain. Suppose that there are N electrons, and that the star is a sphere of radius R and mass $M \cong (N/2)m_{\text{He}}$. (a) Use Eq. (42–9) to estimate the degeneracy pressure. (b) The gravitational force of the helium nuclei exerts an inward pressure, which is approximately $p_{\text{grav}} = E_{\text{grav}}/V \cong (GM^2/R)/V$. Find p_{grav} in terms of R. (c) The star will collapse until the two pressures cancel, when it has a radius R_f. Calculate R_f in terms of m_e, m_{He}, and N. (d) Evaluate R_f for a star of one solar mass. (The result will be surprisingly small.)

55. (III) Use the methods of Problem 24 to calculate the Fermi energy of a gas of electrons, assuming that the mass of electrons can be neglected, so that $E = pc$. Under what conditions will the star of Problem 54 *not* collapse?

56. (III) What is the bulk modulus of a nucleus that consists of N neutrons and Z protons and has a radius $R = r_0A^{1/3}$, where $r_0 \cong 1.2 \times 10^{-15}$ m?

Chapter 43

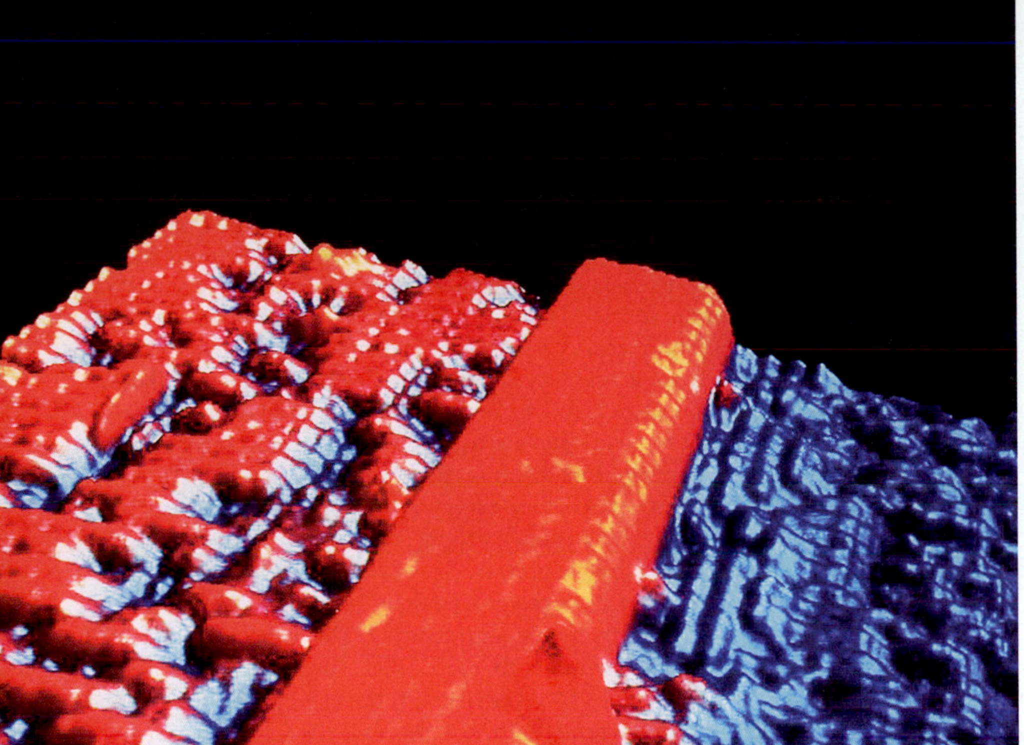

This scanning tunneling micrograph of a nanowire is just 10 atoms wide. The wire (red color) is a compound of a rare earth metal (lanthanide) and silicon and is laid down on a silicon substrate. Such wires are projected to be used in miniaturization of circuits.

Quantum Engineering

The close interplay of science and technology enriches both. At times, technology has led science, as in the case of thermal physics, which was stimulated by the need to understand the engines that propelled the Industrial Revolution of the early nineteenth century. At other times, science has led technology, as with the development of the electric power industry out of the discoveries of Faraday and Maxwell. Such processes occurred frequently throughout the twentieth century. The discovery of quantum mechanics has led to an ever better understanding of the detailed properties of materials; this understanding has been accompanied by important developments in the fabrication of materials required to satisfy the needs of technology. Semiconductors provide an important example. In Section 26–6, we described how the conduction properties of matter are determined by the presence of energy bands—ranges of energies within which possible levels are very closely spaced, spaced by gaps within which there are no allowed levels—and whether or not those bands are fully occupied by electrons. In this chapter, we begin with an explanation of why these bands are formed and how the exclusion principle, as it applies to electrons, explains how they are filled. We then move on to the study of semiconductors and their applications. This is followed by a qualitative tour of more recent technological developments that rely on quantum physics. The chapter ends with a brief discussion of some of the devices whose potential for making major contributions to future technology has already been demonstrated.

43–1 Energy Bands

Band structure in crystalline solids plays an important role in determining many of the properties of those solids, including, as we saw in Section 26–6, electrical properties such as conductivity. Crystalline solids have regions of energy with no allowed energy levels. No electrons within a given solid can have energies in the forbidden regions for the solid; the electron energies are restricted to certain finite ranges of energy—the bands of allowed energies. The presence of energy bands and the numbers of electrons that occupy the allowed energy levels determine why some materials are conductors, whereas others are insulators or semiconductors (Section 26–6). We are now in a better position to elaborate on why bands form. The band structure has such a powerful influence on the behavior of conductors and semiconductors that it is appropriate to discuss it here.

Let's start by recalling our discussion of the energy levels for two hydrogen atoms, Section 41–4. We found that we can think about a helium atom as something that results from moving two hydrogen atoms together. However, our naïve expectation for the binding energy of helium—that the value of the ground-state energy of helium is a simple multiple of the binding energy of hydrogen—is incorrect. This is because the electrons of the two hydrogen atoms interact with each other. In other words, a new interaction added to an existing system will generally shift the energy levels of the existing system. Just how are the energy levels shifted in this case? When the two hydrogen atoms are far apart, there are two electrons in a pair of identical levels of energy -13.6 eV (the ground state of an isolated hydrogen atom). We say that the two electrons are in *degenerate* energy levels. (Actually, there are four degenerate levels, because each electron can be in a spin "up" or spin "down" state, and these states have the same energy values.) As we saw earlier, when the atoms are brought so close together that helium forms, there is at least one level with energy less than -13.6 eV, namely, the true ground state of helium. The remaining three levels are pushed *upward* and correspond to excited states of helium. This is a general characteristic of energy levels:

> **When there are degenerate energy levels and an additional interaction comes into play, the levels are split—some moving upward and some moving downward relative to their original positions.**

This observation plays an important role in band theory. Let's study N atoms of sodium, which have filled shells plus a single electron (the valence electron) in an $n = 3$, $\ell = 0$ state. There are $2N$ degenerate (equal-energy) states available to the valence electron when the atoms are far apart and there is no interaction between the atoms. (As before, the factor of 2 is present because the electron can be in a spin-up or spin-down state, which each have the same energy.) Now suppose that the N atoms form a crystal lattice with closely spaced lattice sites. Then the originally degenerate valence-electron states are split due to interactions between the atoms; N energy levels are pushed down, and N are pushed up, each by a slightly different amount, and there is a finite energy gap between the two sets of levels. The original $2N$ degenerate energy levels form two separate bands, each composed of a set of N very narrowly spaced levels.

Since electrons are fermions, there can be at most two—spin-up and spin-down—electrons in each of the levels, which fill in this way, starting at the level with the lowest energy and working upward. The N valence electrons of the sodium atoms fill only half of the band with the lower energy, which is made up of the N lowest-lying energy levels (Fig. 43–1). As we discussed in Section 26–6, an electric field can easily lift some

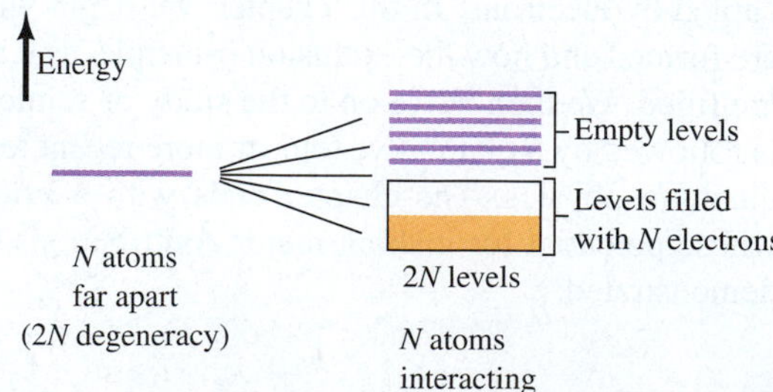

FIGURE 43–1 When many atoms are brought together to form a crystalline solid, the degenerate levels of N atoms split to form band structures, in which there are ranges of allowed electron energies and ranges (gaps) of forbidden energies. In sodium, the valence electrons fill one-half of a band.

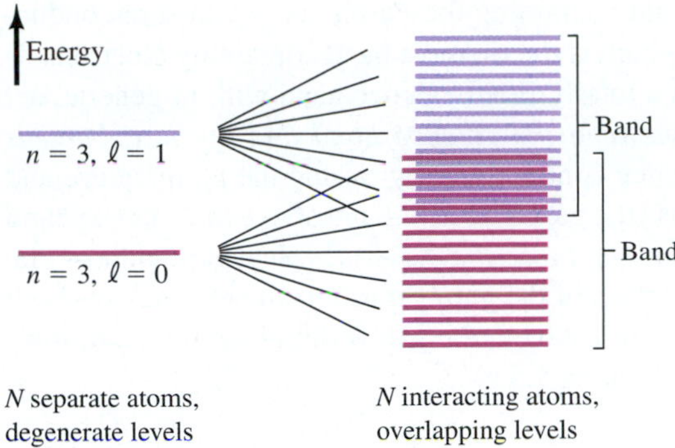

◀ **FIGURE 43–2** The origin of the band structure of N atoms of magnesium in a crystalline solid. The splitting of the degenerate levels with quantum number $n = 3$, $\ell = 0$ leads to a band that is completely filled with electrons. The degenerate levels $n = 3$, $\ell = 1$ also split, and the allowed band they form overlaps with the band formed from the $n = 3$, $\ell = 0$ levels.

electrons at the top of the filled levels to the empty levels, where they can move without hindrance. Metallic sodium is thus a good conductor.

For contrast, we can consider N atoms of magnesium. The Z value of magnesium is one higher than that of sodium. Magnesium has the same atomic structure as sodium, except that there are *two* electrons in the $n = 3$, $\ell = 0$ states. The band formed from the originally degenerate states when the magnesium atoms combine into a lattice is therefore completely filled. Magnesium could be an insulator. However, empty levels that originally came from the $n = 3$, $\ell = 1$ states, which were above the $n = 3$, $\ell = 0$ levels when the two sets of levels were each degenerate, fan out into a second band. In magnesium the two bands overlap (Fig. 43–2). Thus there is room for electrons, under the influence of an electric field, to move into empty levels, and magnesium is a good conductor. There are other cases in which bands of this type do not overlap. Indeed, all the possibilities discussed in Chapter 26 can be realized.

Let us now turn back to the question of the "new" interactions that occur when the atoms are assembled into a regular lattice, which lead to the splitting of degenerate electron energy levels into a band structure. These are the interactions of electrons with ions other than the ones to which they were originally attached. In particular, if the ions were not present, the electrons would move freely in the "box" formed by the dimensions of the material. The electrons' wavelike aspect makes them behave analogously to waves on a string, namely as standing waves whose wavelengths must be integral multiples of the appropriate dimensions of the box. These waves are the waves formed by the wave function, and the electron is likely to be found where those waves have crests (or troughs). The crests and troughs are regularly spaced. At this point, we can imagine adding the ions at regularly spaced sites. Two modifications to our original depiction result: First, only those electrons whose wavelengths correspond to the ionic spacing can be supported in the material for reasons closely related to the Bragg scattering arguments we gave in Chapter 40. Second, half of the allowed waves of the electrons have peaks (or troughs) where the ions are found, and half have peaks (or troughs) exactly in between the ions. The energies of the half whose peaks are where the ions are found are lowered because these electrons are more likely to be near the ions to which they are attracted. The energies of the half whose peaks are in between the ion locations are raised because these electrons do not feel the force of attraction as strongly as they did when they were in an isolated atom.

43–2 Semiconductors

Two major features govern the electrical conduction properties of solids: the Pauli exclusion principle, which is obeyed by all electrons (see Chapter 42), and the presence of a band structure, with gaps. The Pauli exclusion principle implies that there are at most two electrons in each energy level, and as we have already reminded you, at low temperatures these levels are filled in order, starting from the lowest. The energy of the highest filled level is the *Fermi energy*, E_F. The band structure expresses the presence of energy gaps between allowed bands; there are no levels within such gaps, so no electrons can have energies there. In conductors, levels are filled to the Fermi energy, but

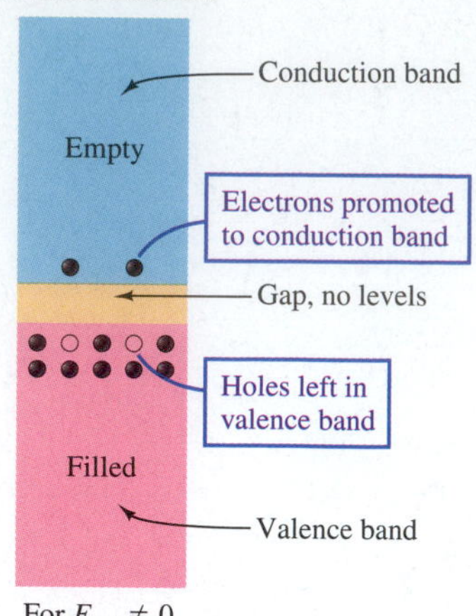

▲ **FIGURE 43–3** Semiconductors are characterized by a filled band (the valence band) with a relatively small energy gap to the next band (the conduction band), which is empty at zero temperature. Finite temperatures suffice to boost some electrons to the conduction band.

TABLE 43–1 • Gap Widths of Some Semiconductors†

Material	Gap Width (in eV)
Tin (gray form)	0.08
Tellurium	0.35
Lead sulfide	0.37
Germanium	0.67
Silicon	1.12
Gallium arsenide	1.43
Boron	1.5
Selenium	1.8

†At room temperature.

there are empty energy levels in the band containing the Fermi energy. In semiconductors and insulators, the *valence band*, which is the topmost band containing electrons, is filled right up to a gap. Materials with a totally filled valence band will, in general, act as insulators *unless the energy gap between the valence band and the next band is small*. In that case, it is possible to induce conductivity by raising the temperature and thermally exciting some electrons across the gap. The next band above the valence band is therefore referred to as the *conduction band* (see Fig. 43–3). Typical gap widths—the energy difference from the top to the bottom of the gap—at room temperature are listed in Table 43–1 for some semiconductors. In contrast, the gap width of the diamond form of carbon, which is regarded as an insulator, is 5.5 eV.

The Effects of Temperature

The behavior of semiconductors depends crucially on temperature. To study the effects of finite temperatures, it is simplest to look first at finite temperature effects in conductors. At $T = 0\text{K}$ (what we'll refer to as $T = 0$), we are certain to find an electron in every state up to an energy of E_F. If we suppose that the possible levels are so closely spaced as to represent a continuum, then at $T = 0$, the probability $f(E)$ that the state with energy E is occupied by an electron (actually two electrons, due to spin) is

$$\begin{aligned}&\text{for } E < E_F\text{: } f(E) = 1,\\ &\text{for } E > E_F\text{: } f(E) = 0.\end{aligned} \tag{43–1}$$

When $T > 0$, some of the electrons will be thermally excited into previously unoccupied levels, leaving vacant energy levels behind them. For temperatures that are not too large (see the more detailed discussion below), this can happen only for electrons with energies near E_F because electrons with energies far below E_F are unlikely to get a sufficiently large thermal "kick." If electrons behaved according to classical rules, they would obey the Maxwell–Boltzmann distribution (see Chapter 19). But because electrons obey the Pauli exclusion principle, their energy distribution is instead described by the **Fermi–Dirac distribution**:

$$f(E) = \frac{1}{1 + e^{(E-E_F)/kT}}. \tag{43–2}$$

THE FERMI–DIRAC DISTRIBUTION

You can easily check that we recover Eq. (43–1) from this distribution in the limit $T \to 0$. Figure 43–4 shows the distribution for the increasing temperatures $T = 0$, T_1, and T_2. Notice that for $(E - E_F) \gg kT$, the exponential factor dominates in the denominator, and we get the approximation

$$f(E) \cong e^{-(E-E_F)/kT}. \tag{43–3}$$

Equation (43–3) has the form of the Maxwell–Boltzmann distribution, so the distribution is classical when the energy difference $E - E_F$ is large compared to kT. What is a typical value of kT? A useful number to remember is that room temperature, $T = 300\text{K}$, corresponds† to 1/40 eV.

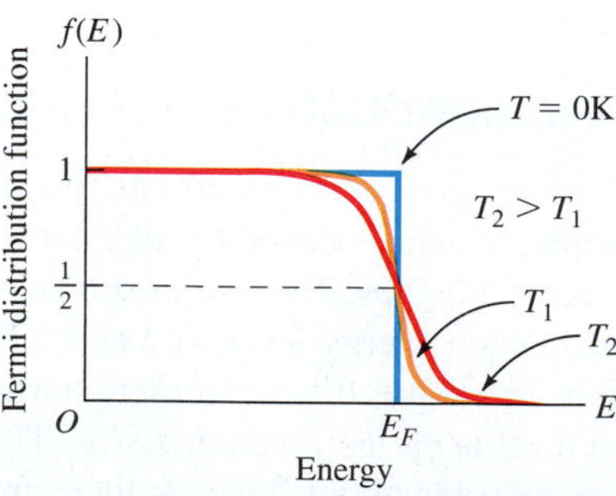

▶ **FIGURE 43–4** The Fermi–Dirac distribution, $f(E)$, which describes the probability of finding one of a set of identical fermions with an energy E. The distribution is sketched for three successively higher temperatures, $T = 0$, T_1, and T_2.

†For $T = 300\text{K}$, $kT = (1.38 \times 10^{-23}\ \text{J/K})(300\text{K})/(1.6 \times 10^{-19}\ \text{J/eV}) = 2.6 \times 10^{-2}\ \text{eV} = 1/40\ \text{eV}$.

EXAMPLE 43–1 Assume that electrons in copper are distributed according to the Fermi–Dirac distribution. The Fermi energy in copper is $E_F = 7.04$ eV. (a) The *Fermi velocity* has magnitude v_F, the speed of an electron that moves with a kinetic energy equal to E_F. What is v_F for copper? (b) From the Fermi–Dirac distribution, $f(E_F) = 1/2$. What is the value of $f(E)$ at a temperature of 300K for an electron with a speed 1 percent higher than v_F? (c) Estimate the temperature at which the probability $f(E)$ is 10^{-10} if E is 1.2 eV above E_F.

Strategy (a) Assuming that the electrons are nonrelativistic, we can use $E_F = \frac{1}{2}m_e v_F^2$, and solve for v_F. The assumption about being nonrelativistic would be justified if v_F turns out to be a small fraction of c.

(b) This is a straightforward evaluation of the distribution function. It is helpful to note here that when the speed of an electron increases by 1 percent, the energy—proportional to the speed squared—increases by approximately 2 percent, and we can then make our estimate by taking E to be 2 percent larger than E_F.

(c) We set $f(E)$ equal to 10^{-10} and solve for kT, given the value of $E - E_F$. When $f(E) = 10^{-10}$, the exponential term in the denominator of $f(E)$ in Eq. (43–2) is much larger than 1, and we can use the approximation $f(E) \cong e^{-(E-E_F)/kT} = 10^{-10}$.

Working It Out (a) We have

$$v_F = \sqrt{\frac{2E_F}{m_e}} = \sqrt{\frac{2(7.04\text{ eV})(1.60 \times 10^{-19}\text{ J/eV})}{0.911 \times 10^{-30}\text{ kg}}}$$
$$= 1.57 \times 10^6\text{ m/s}.$$

This is only about half a percent of the speed of light.

(b) We take E to be 2 percent larger than E_F, $E = (1.02)(7.04\text{ eV}) = 7.18$ eV. With $kT = 2.6 \times 10^{-2}$ eV at $T = 300$K, $(E - E_F)/kT = 5.4$, and hence

$$f(E) = \frac{1}{1 + e^{5.4}} \cong 4.5 \times 10^{-3}.$$

(c) According to the strategy, we have

$$f(E) \cong e^{-(E-E_F)/kT} = 10^{-10},$$

with $E - E_F = 1.2$ eV. Taking the natural log of both sides, $-(1.2\text{ eV})/kT = \ln(10^{-10}) = -10\ln(10) = -23$. Thus, $kT = (1.2\text{ eV})/23 = 0.052$ eV. This is twice the value of kT at $T = 300$K [see part (b)], so the desired estimate is 600K.

What Do You Think? We made a post-calculation check as to whether the electrons were nonrelativistic. Is there any way we could have argued that had to be so without a numerical evaluation of v_F? *Answers to **What Do You Think?** questions are given in the back of the book.*

Semiconductors, Electrons, and Holes

To apply our ideas about finite temperatures to semiconductors, we must reconsider the role of the energy gap. Suppose that the band structure is that of a semiconductor; that is, at $T = 0$, the valence band is filled, the conduction band is empty, and the gap energy $E_g \equiv E_c - E_v$ is relatively small. Here, E_v is the maximum energy of the valence band (the *edge of the valence band*), and E_c is the lowest energy of the conduction band (the *edge of the conduction band*) (Fig. 43–5).

But just what is the Fermi energy for this situation? If we think of the Fermi energy as the energy above which there are no electrons (at zero temperature), then *any* energy value within the energy gap would serve, and that is not an answer to our question. Our procedure is as follows: We use Eq. (43–2) but replace the Fermi energy with an unknown quantity μ:

$$f(E) = \frac{1}{1 + e^{(E-\mu)/kT}}. \tag{43–4}$$

The quantity μ is a thermodynamic variable—the *chemical potential*—but the detailed properties of this variable need not concern us further. For the moment, μ is unknown, but we shall determine its value from a condition of the conservation of charge. In the following discussion, we shall conform to established practice for semiconductor engineering by writing E_F rather than μ throughout. When you see formulas for semiconductors with the parameter E_F, keep in mind that this is not the Fermi energy in the sense in which we first defined it.

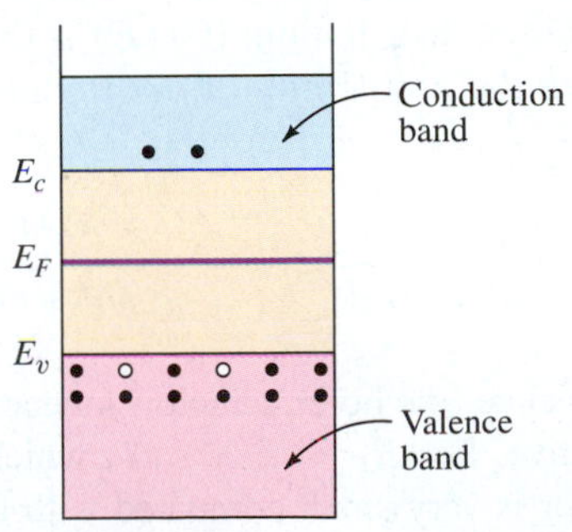

FIGURE 43–5 The band structure of an intrinsic semiconductor is that of a filled valence band, a relatively narrow band gap, and, at $T = 0$, an empty conduction band. The solid circles represent electrons, and the open circles represent holes. At $T = 0$, there are no electrons in the conduction band and no holes in the valence band.

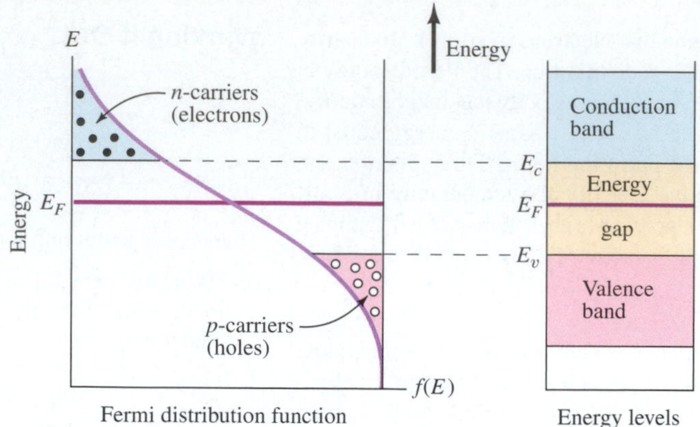

▶ **FIGURE 43–6** Electrons near the top of the valence band can jump the gap in a semiconductor at finite temperature. The Fermi–Dirac distribution describes the probability that an electron will have enough energy to be promoted to the conduction band. For every electron that can be promoted, a hole is left behind in the valence band.

To determine E_F, we must look more closely into the question of what happens when electrons move into the conduction band. When $T \neq 0$, some electrons will acquire enough energy to go into the conduction band, and the valence band will be correspondingly depleted of electrons (Fig. 43–6). At room temperature ($kT \cong 1/40\ \text{eV}$), $E_g \gg kT$. We can then estimate that at room temperature the relative number of electrons able to jump the gap to the conduction band in, say, silicon, for which $E_g \cong 1\ \text{eV}$, is $e^{-E_g/kT} \cong e^{-40} \cong 10^{-18}$, only a very small fraction of the electrons. This same reasoning would hold if, instead of E_g, we used the smaller quantity $E_c - E_F$, as long as $E_c - E_F$ were of the same order of magnitude as E_g (as it typically is). We would thus have a situation in which an energy distribution of the same form as Eq. (43–3) holds (although we have yet to determine E_F). Because the function in Eq. (43–3) decreases so rapidly as energy increases, the number of electrons per unit volume (the number density n) in the conduction band is well approximated by replacing E in Eq. (43–3) with the lowest energy in the conduction band, E_c:

$$n = N_c e^{-(E_c - E_F)/kT}. \qquad (43\text{–}5)$$

The proportionality constant N_c will be treated in detail later, but we can say here that it represents a normalization that describes the total number of electrons that have been promoted to the conduction band. The electrons in the conduction band are free to move when an electric field is applied and form a current. These electrons are called *n-carriers* (n for negative).

Electrons that are thermally "promoted" into the conduction band leave vacancies in the valence band. Only the electrons with energies near E_v are promoted, so the vacancies are located at those energies. Such a vacancy is called a **hole**. Holes can move: When an electron moves to fill a hole, that leaves a hole in the electron's original position, which in turn is filled by a second electron, and so on. The hole movement traces the positions of the electrons that have filled the hole's previous position. If there is an electric field pointing in the $+x$-direction, then the succession of hole-filling electrons will have moved in the $-x$-direction, and thus the hole will flow in the $+x$-direction. (A helpful analogy might be an air bubble floating upward in a tank of water. The "vacancy" in the water, the air bubble, is constantly replaced by water falling down under the force of gravity; this water leaves its own vacancy, which thus continually moves upward.) The hole thus acts as a positively charged particle; we call such a hole a *p-carrier* (p for positive). The probability of finding a hole with energy E can be determined by the Fermi–Dirac distribution. If $f(E)$ is the probability of finding an electron with energy E, then the probability of *not* finding an electron with that energy is

$$1 - f(E) = 1 - \frac{1}{1 + e^{(E-E_F)/kT}} = \frac{e^{(E-E_F)/kT}}{1 + e^{(E-E_F)/kT}} = \frac{e^{-(E_F-E)/kT}}{1 + e^{-(E_F-E)/kT}}. \qquad (43\text{–}6)$$

Here, E is the energy of the empty electron state (the hole), which is an energy in the valence band, so the factor $E_F - E$ is positive. For $E_F - E \gg kT$, which is typically the case, the exponential in the denominator is very small compared with unity, and the right-hand side of Eq. (43–6) reduces to $e^{-(E_F-E)/kT}$. This exponential factor decreases so rapidly as the factor $E_F - E$ increases that, to a good approximation, we can replace $E_F - E$ by $E_F - E_v$. Thus the number density of holes is accurately given by

$$p = N_v e^{-(E_F-E_v)/kT}. \qquad (43\text{–}7)$$

We shall give the proportionality constant N_v later. Note that the product of the n-carrier concentration and the p-carrier concentration is given by

$$np = N_c e^{-(E_c-E_F)/kT} N_v e^{-(E_F-E_v)/kT} = N_c N_v e^{-(E_c-E_v)/kT} = N_c N_v e^{-E_g/kT}. \qquad (43\text{–}8)$$

CONCEPTUAL EXAMPLE 43–2 Suppose $E_F - E \gg kT$. Will the probability of finding a hole with energy E be large or small? Does your answer make sense physically?

Answer The probability of finding a hole with energy E is proportional to $1 - f(E)$, Eq. (43–6). This expression takes the approximate form $e^{-(E_F-E)/kT}$ and is not only small in our limit but has the classical Maxwell–Boltzmann form. The reason that this probability is small is that for large values of $E_F - E$, the hole is created by the promotion of an electron to the conduction band from deep within the valence band, rather than near the top of the valence band, which is a relatively small jump. This is a large jump in energy and is unlikely, exactly as thermodynamics tells us.

When each hole in the valence band is created by the promotion of an electron into the conduction band, the number of p-carriers equals the number of n-carriers. Semiconductors for which this is the case are called *intrinsic semiconductors*. We can now use this condition to determine E_F. With a subscript i to indicate that we are dealing with an intrinsic semiconductor, Eq. (43–8) implies that

$$n_i = p_i = \sqrt{n_i p_i} = \sqrt{N_c N_v} e^{-E_g/2kT}. \qquad (43\text{–}9)$$

Comparison of the exponential factor in this result with the exponential factors of Eqs. (43–5) and (43–7) is valid as long as N_c and N_v do not vary too rapidly with temperature, and shows that

$$E_c - E_F = E_F - E_v = E_g/2. \qquad (43\text{–}10)$$

In other words, E_F *(or more properly, the chemical potential) is at the halfway point of the gap*. This result justifies our expectation that $E_c - E_F$ and $E_F - E_v$ are much larger than kT if E_g is.

An additional effect is important in semiconductors. The energy of a free electron with momentum p is $E = p^2/2m_e$, as in Fig. 43–7a. Figure 43–7b shows the energy–momentum relation that holds for the electrons in the conduction band if those electrons are free. Electron interaction with the lattice, however, changes this relationship. For small momentum, the principal effect is to change the steepness of the parabola, and this is equivalent to changing the electron mass to an *effective electron mass*, denoted by m^*. For semiconductors, m^* is typically an order of magnitude *smaller* than the true electron mass; for gallium arsenide (GaAs), $m^*/m_e \cong 0.067$, and for indium antimonide (InSb), this ratio is only 0.015. Furthermore, the effective mass of n-carriers need not be the same as that of p-carriers; generally m^* is much

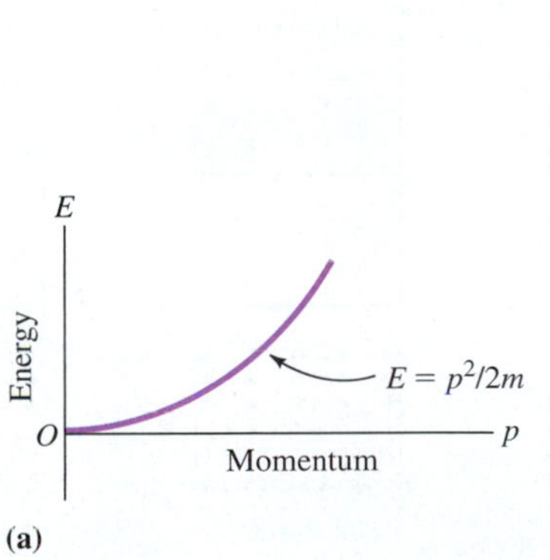

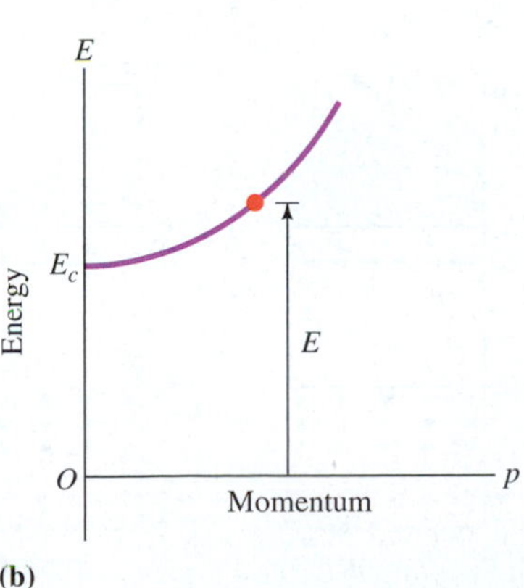

FIGURE 43–7 (a) Energy versus momentum for a free electron. (b) Energy versus momentum for an electron in the conduction band. The lowest possible energy for the electron is E_c.

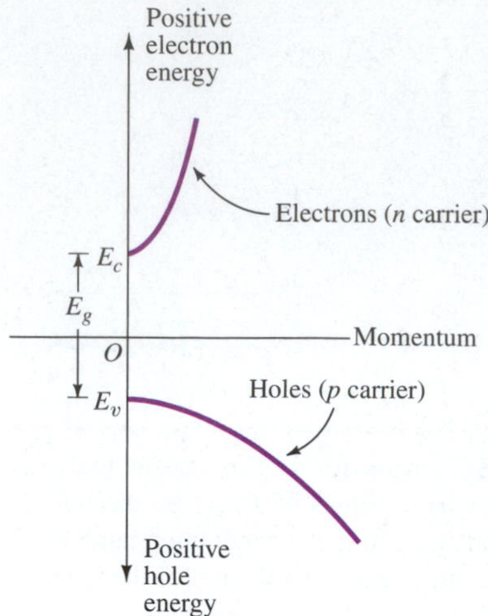

▲ **FIGURE 43–8** The energy–momentum relation for an electron in the conduction band (Fig. 43–7b) is modified because the electron behaves as though it has a modified mass m^*. Also shown is the energy–momentum relation for a hole, which in general has a different effective mass than does an electron.

smaller for n-carriers (Fig. 43–8). An understanding of why effective masses take on their particular values is beyond the scope of our discussion. Roughly speaking, the ratio m^*/m_e is small if E_g is small compared to the width of the conduction band.

Doping

In contrast to intrinsic semiconductors, in *extrinsic*, or *impurity*, *semiconductors*, additional n-carriers or p-carriers are supplied by the addition of atoms of another element—impurities—to the crystal lattice. The addition of impurities (whether artificially or naturally) is known as *doping*. It works as follows: Suppose that we add some arsenic atoms to germanium. The crystal structure will remain unchanged (each arsenic ion replaces one germanium ion in the lattice), but the electronic structure will change. Germanium has four valence electrons, whereas arsenic has five; the arsenic atom makes one more electron available for conduction and is said to be a *donor impurity*. When n-carriers are added in this way, the doped material is said to be an *n-type semiconductor*.

If the additional electron were not bound at all to the arsenic ion, it could fill an energy level only in the conduction band and therefore could easily carry current. Such an electron would have energy on the order of E_c. The additional electron is in fact bound to the arsenic ion, but only very weakly. If the arsenic were not part of the semiconductor lattice, we might expect the electron to be bound to the arsenic ion with an energy like the Bohr energy [Eq. (41–10) with $n = 1$]:

$$E = -\frac{e^4 m_e}{2(4\pi\varepsilon_0\hbar)^2} \cong -13.6 \text{ eV}.$$

Within the lattice of the semiconductor, this binding energy is reduced in magnitude for several reasons. First, the mass of the electron is effectively reduced—typically by more than a factor of 10—from m_e to m^*. Second, the semiconductor medium has a dielectric constant κ(defined by $\varepsilon = \kappa\varepsilon_0$) that is typically more than 10 times larger than the value of 1 for a vacuum. These two factors change the binding energy to

$$E = -\frac{e^4 m^*}{2(4\pi\varepsilon\hbar)^2}. \tag{43–11}$$

Our numerical values for m^* and ε show that the magnitude of this energy is less than $\frac{1}{1000}$ that of the typical Bohr energy. Moreover, with the change to an effective mass the radius of the electron's Bohr orbit [see Eq. (41–8)] becomes so large that the electron is circulating within the larger environment of the semiconductor lattice and screening by other electrons occurs, reducing the magnitude of the binding energy still further. The small binding energy, on the order of 0.01 eV or less, manifests itself in the presence of additional energy levels very close to the bottom of the conduction band (Fig. 43–9a). The amount by which these energies differ from E_c is on the order of kT or less at room temperature. Thus the additional electron, which is bound at $T = 0$, is very easily excited to the conduction band at room temperature. In this circumstance, the donor electrons are far more important for conduction than the intrinsic electrons.

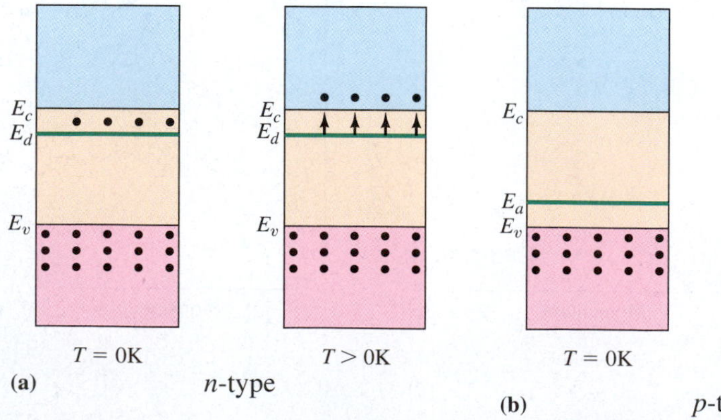

▶ **FIGURE 43–9** Energy-band structure for doped semiconductors. (a) In n-type semiconductors, E_F (labeled E_d, label d for donor) lies close to E_c. The levels of electrons (solid circles) from donor impurities lie above E_F and are easily excited at $T > 0$ into the conduction band. (b) In p-type semiconductors, E_F (labeled E_a, label a for acceptor) lies close to E_v. There are levels associated with the acceptor impurities that are quite close to E_v, so at finite temperatures, electrons from the valence band are easily promoted to these levels, leaving holes (open circles) in the valence band.

EXAMPLE 43–3 Calculate the ionization energy for a donor electron in doped indium antimonide, a semiconductor in which $\varepsilon/\varepsilon_0 = 17.9$ and $m^*/m_e = 0.015$.

Strategy The ionization energy E is the negative of the binding energy of the ground state of the donor atom [Eq. (43–11)]. We need only insert the new values of dielectric constant and mass.

Working It Out

$$E = \frac{e^4 m^*}{2(4\pi\varepsilon\hbar)^2} = (13.6\text{ eV})\left(\frac{m^*}{m_e}\right)\left(\frac{\varepsilon_0}{\varepsilon}\right)^2$$
$$= (13.6\text{ eV})(0.015)\left(\frac{1}{17.9}\right)^2 = 6.4 \times 10^{-4}\text{ eV}.$$

This much energy corresponds to a temperature of $T = E/k = 0.025\text{K}$.

If an impurity has a deficit of valence electrons, a surplus of p-carriers is created because holes form when electrons from the intrinsic semiconductor atoms become strongly attached to the impurity atoms. Such impurities are known as *acceptor impurities*. Boron, for example, has three valence electrons and is an acceptor impurity for germanium. Semiconductors doped with acceptor impurities are known as *p-type semiconductors*. By the same sort of reasoning that we used for n-type semiconductors, p-type semiconductors have the level structure shown in Fig. 43–9b, with some additional levels close to E_v. The n-carriers from the valence band are easily promoted to the acceptor levels, leaving holes in the valence band that are very effective at carrying charge.

As in intrinsic semiconductors, the effective Fermi energy of doped semiconductors determines the probability of promoting electrons and holes to levels where they can conduct. The Fermi energy E_F is found by correctly relating the densities of n-carriers and p-carriers. Since the carrier densities depend on the difference between the energy and E_F, as in Eq. (43–3), we might expect that the Fermi energy is close to E_c (i.e., it increases) in n-type semiconductors and is close to E_v (i.e., it decreases) in p-type semiconductors, and this is indeed the case. These results can be formally developed by starting with Eq. (43–8), which holds whether or not the semiconductor is doped. The constants N_c and N_v in Eq. (43–8) can be obtained by summing the number of electrons in the conduction band and the number of holes in the valence band, respectively:

$$N_c = 2\left(\frac{m_n^* kT}{2\pi\hbar^2}\right)^{3/2} \quad \text{and} \quad N_v = 2\left(\frac{m_p^* kT}{2\pi\hbar^2}\right)^{3/2}, \tag{43–12}$$

where m_n^* and m_p^* are the effective masses of the n-carriers and p-carriers, respectively. Then Eq. (43–8) reads

$$np = n_i^2 = N_c N_v e^{-(E_c - E_v)/kT} = N_c N_v e^{-E_g/kT}. \tag{43–13}$$

We have rewritten Eq. (43–8) as Eq. (43–13) because we have given it a greatly extended reach. Equation (43–13) relates the n-carrier and p-carrier densities in a *doped* semiconductor to the intrinsic carrier densities, whereas in its original form [Eq. (43–8)] it related the carrier densities in the undoped semiconductor.

Equation (43–13) has important consequences. For example, if we dope an intrinsic semiconductor with acceptor impurities to increase the density of holes, the density of electrons must decrease so that Eq. (43–13) continues to be satisfied. It is typical that the n-carriers dominate in an n-type extrinsic semiconductor, whereas p-carriers dominate in a p-type extrinsic semiconductor.

EXAMPLE 43–4 The effective masses in the semiconductor gallium arsenide are $m_n^*/m_e = 0.067$ and $m_p^*/m_e = 0.48$; the band gap is $E_g = 1.43$ eV. Find the intrinsic concentration of n-carriers and p-carriers in gallium arsenide at room temperature, where $kT = 0.026$ eV.

Strategy The concentration of intrinsic carriers in a semiconductor (doped or not) is given by Eq. (43–9). To evaluate this expression, we must include the normalization factors N_c and N_v.

Working It Out With the help of Eq. (43–12), we obtain

$$n_i = \sqrt{N_c N_v}\, e^{-E_g/2kT} = 2\left(\frac{m_e kT}{2\pi\hbar^2}\right)^{3/2}\left(\frac{m_n^* m_p^*}{m_e^2}\right)^{3/4} e^{-E_g/2kT}$$
$$= 2\left[\frac{(0.91 \times 10^{-30}\text{ kg})(1.38 \times 10^{-23}\text{ J/K})(300\text{K})}{2\pi(1.05 \times 10^{-34}\text{ J}\cdot\text{s})^2}\right]^{3/2} [(0.067)(0.48)]^{3/4} e^{-(1.43\text{ eV})/2(0.026\text{ eV})}$$
$$= 2.2 \times 10^{12}\text{ carriers/m}^3.$$

What Do You Think? We asked for the number of p-carriers as well as the number of n-carriers. Why didn't we carry out a separate calculation?

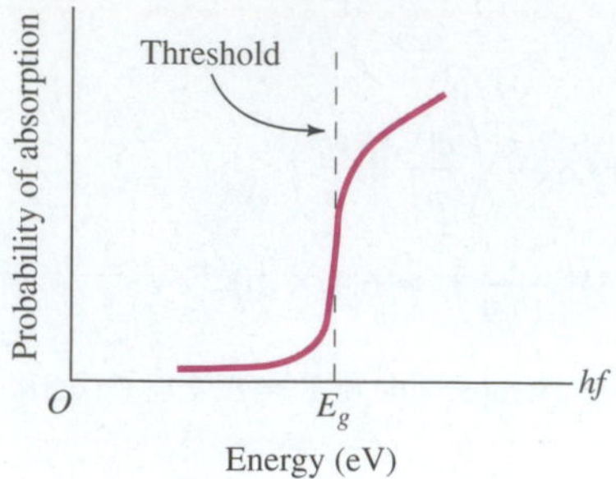

▲ **FIGURE 43–10** The probability of photon absorption in a semiconductor has a threshold when the photon energy equals the gap energy, because no energy levels are available for the absorbing electron at lower energies. Because a photon's energy is proportional to its frequency, the energy threshold can be expressed as a threshold in frequency.

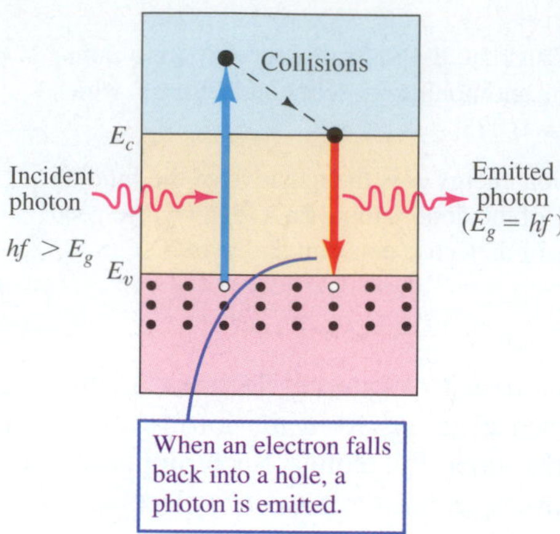

▲ **FIGURE 43–11** An electron excited to the conduction band by photon absorption loses energy by collision until its energy reaches E_c; at that point it recombines with a hole in the valence band, with photon emission.

Optical Effects in Semiconductors

When a photon—a quantum of light—strikes a semiconductor, there is a probability that it will be absorbed by the electron system. This process has a threshold, given by a function of the photon's frequency f (Fig. 43–10). An electron in the material can absorb the photon only if the energy of the photon, hf, is greater than the gap energy; only at such energies are there empty levels available for electron occupation in the conduction band. Thus, some materials are transparent (a photon passes through without being absorbed) in certain frequency ranges but cease to be so for higher frequencies. An absorbing electron that is kicked up into the conduction band will lose energy by colliding with lattice atoms until its energy falls to near E_c; here, further energy loss is inhibited by the exclusion principle and/or the absence of energy levels below E_c. At that stage, the only way in which the electron can lose energy is to fall back into a hole in the valence band, a process called *recombination*. The electron–hole recombination can produce radiation called *photoluminescence* whose frequency is such that $hf_g = E_g$ (Fig. 43–11). The electrons can alternatively be initially excited into the conduction band by the bombardment of the material by high-energy electrons (*cathodoluminescence*) or by the action of an electric current in the material (*electroluminescence*).

When an electron–hole pair recombines rapidly, on the order of 10^{-8} s, the process is called *fluorescence*. A slower transition may arise when there are impurity levels in the gap, and the excited electron first undergoes a transition to those levels. If the impurity level has quantum numbers such that selection rules forbid the direct transition of an electron from it to the hole, the electron is said to be *trapped*. It will ultimately undergo a transition to a vacancy in the valence band (electron–hole recombination), but this may take as much as seconds or even minutes to occur. This process is termed *phosphorescence*, and the materials in which this occurs are called *phosphors*. The energy of the emitted photon is determined by the energy difference between the intermediate impurity level and the top of the valence band. We can thus choose the impurities such that radiation will be emitted in a variety of colors. In a color television tube, the screen is coated with phosphor dots that emit radiation in the primary colors (Fig. 43–12). The phosphors are excited by three incoming electron beams (cathodoluminescence), which are deflected to the proper spot on the screen by electric and magnetic fields determined by the incoming signal to the television. Each of the beams excites a different color.

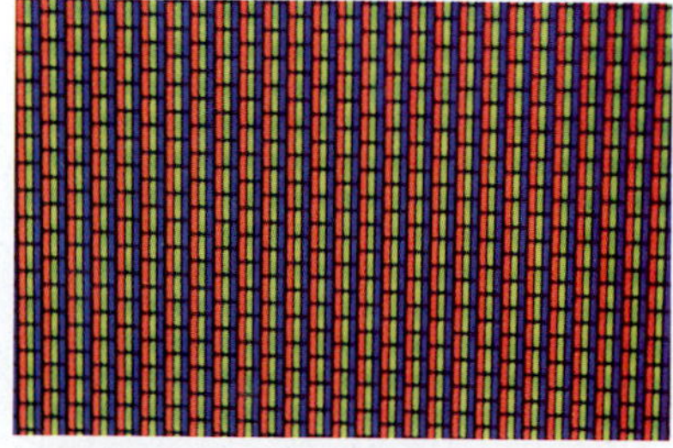

▲ **FIGURE 43–12** Excited red, green, and blue phosphors are visible in this close-up view of a color television screen.

A semiconductor exposed to radiation for which $hf > E_g$ has an increased number of *n*-carriers and *p*-carriers and thus an increased conductivity. Devices whose conductivity changes when they are exposed to light—*photoconductive* devices—are used for lights that automatically turn on at dusk and off at dawn, or for measuring light intensity, as in exposure meters for cameras.

43–3 Semiconductor Structures

In applying our knowledge of energy distributions in single semiconductors to semiconductor structures (combinations of semiconductors in contact so that charge carriers can pass between them), one principle determines essentially all the important features: Like temperature, *the chemical potential*—which for semiconductors we call the Fermi energy —*has the same value throughout a system in equilibrium*. This follows from the fact that in the absence of external potentials and in thermodynamic equilibrium, there is no net flow of charge or of energy across the boundary between two materials. The materials may be the same type of semiconductor doped differently, two different intrinsic semiconductors, or even a semiconductor and a metal. You can prove this important result in Problem 54.

The *p-n* Junction

When a semiconductor doped to n-type and the same semiconductor doped to p-type are brought into contact, a *p–n junction* is formed. (We can also work with junctions between different semiconductors, known as *heterojunctions*.) Such a junction is of interest because it acts as a diode (see Section 33–5). We consider p–n junctions in which both semiconductors are made by doping the same intrinsic semiconductor. In Fig. 43–13a, we have used p_p and p_n to denote the hole densities and n_p and n_n to denote the electron densities in p-type and n-type semiconductors, respectively. The sketches are highly distorted because the ratio of the dominant carriers to the carriers of opposite sign on each side of the junction is typically on the order of 10^{11}; drawn to scale, the n_p and p_n levels would be indistinguishable from the horizontal axis. We shall henceforth refer to the regions of n-type and p-type semiconductors as the n-side and p-side, respectively.

The p-side has an excess of mobile holes, and the n-side has an excess of mobile electrons, though not necessarily with the same density. Electrical neutrality is preserved on the p-side by the (immobile) negative acceptor ions in the crystal lattice and on the n-side by the (immobile) positive donor ions. When the two materials are put in contact, the mobile holes on the p-side tend to diffuse into the n-side, and the mobile electrons on the n-side tend to diffuse into the p-side. This mixing continues until the Fermi energies in the two materials reach the same level. This occurs by the following mechanism: Charges flow until the net positive charge remaining on the n-side (the right) and the net negative charge remaining on the p-side (the left) set up an electric field that stops the diffusion (Fig. 43–13b). The n-side will be at a higher potential, V_0, than the p-side. But electrons have a *negative* charge, so the energy of electrons on the right is *reduced* by eV_0.

Let's look at the p–n junction with the aid of energy diagrams. Figure 43–14a shows the energy levels of the semiconductors before they are brought together. The gap energies are the same because both semiconductors are made by doping the same intrinsic semiconductor. When the junction is formed, the conduction and valence bands will be distorted (Fig. 43–14b). The internal potential V_0 is what brings the Fermi

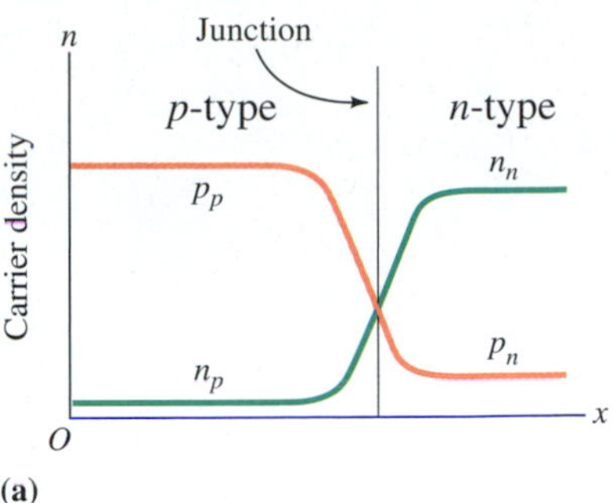

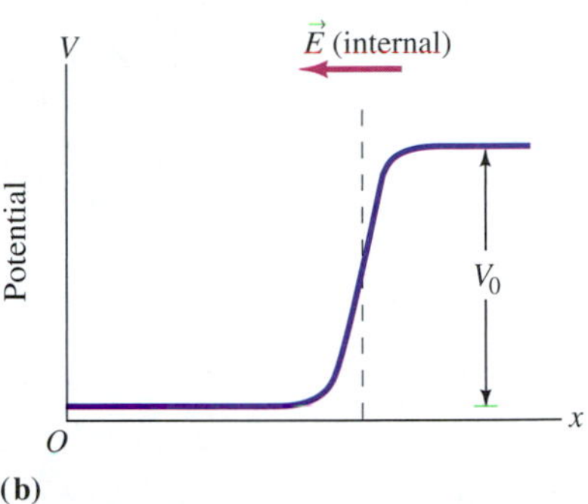

▲ **FIGURE 43–13** A p–n junction is formed by placing an n-type and a p-type semiconductor in contact. The variable x is the linear distance across the boundary. (a) The number densities of p-carriers (p_p and p_n, respectively), and the number densities of n-carriers (n_p and n_n, respectively) in p-type and n-type semiconductors. These densities are not drawn to scale. (b) As described in the text, the diffusion of carriers causes a potential difference to build up across the boundary. The direction of the resulting electric field is shown.

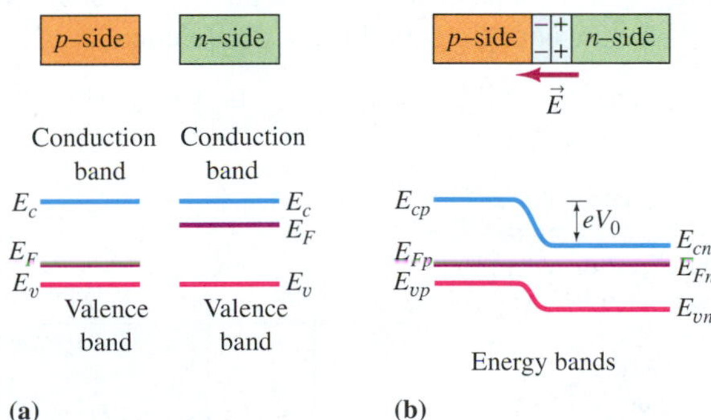

▲ **FIGURE 43–14** (a) The energy-level structure of n-type and p-type semiconductors before they are joined in a junction. (b) After the junction is formed and equilibrium is established, the energy levels adjust in such a way that the Fermi levels of the two materials equalize. This lowers the electron energy levels on the n-side and raises the electron energy levels on the p-side. (After B. G. Streetman, *Solid State Electronic Devices*, Prentice Hall, 1980, p. 141.)

energies to the same level. From Figs. 43–13 and 43–14, we see this statement is equivalent to the relation

$$E^0_{Fn} - E^0_{Fp} = eV_0,$$

where E^0_{Fn} and E^0_{Fp} are the Fermi energies in the n-type and p-type semiconductors, respectively, *before* the materials are brought into contact. *After* the materials are brought into contact,

$$E_{cp} - E_{cn} = eV_0, \tag{43–14}$$

where E_{cp} and E_{cn} are the edges of the conduction band in the p-type and n-type semiconductors, respectively. This relation follows directly from Fig. 43–14b. A similar expression holds for the edges of the valence bands.

Note that V_0, which is known as the *contact potential*, is *not* an externally imposed potential. It is a property of the junction itself.

Let's suppose that the potential jump across the boundary is abrupt and study the consequences. (In real materials, the potential change occurs across a finite width called the *transition region*, or *depletion region*, as in Fig. 43–15. However, we shall ignore effects due to the transition region.) The densities of n-carriers on the two sides are, according to Eq. (43–5),

$$n_n = N_c e^{-(E_{cn}-E_F)/kT} \quad \text{and} \quad n_p = N_c e^{-(E_{cp}-E_F)/kT}.$$

Thus

$$\frac{n_p}{n_n} = \frac{N_c e^{-(E_{cp}-E_F)/kT}}{N_c e^{-(E_{cn}-E_F)/kT}} = e^{-(E_{cp}-E_{cn})/kT} = e^{-eV_0/kT}, \tag{43–15}$$

where we have used Eq. (43–14) in the last step. Similarly, the ratio of the equilibrium hole densities on the two sides is

$$\frac{p_n}{p_p} = e^{-eV_0/kT}. \tag{43–16}$$

The ratios in Eqs. (43–15) and (43–16) help us to understand just how the currents in the two directions balance. Let's consider only the electrons—the flow of holes works in the same way. From Eq. (43–15), the concentration of electrons on the p-side is less than the concentration on the n-side by the factor $e^{-eV_0/kT}$, and these electrons have no potential barrier to climb to get to the n-side. There is a large density of electrons on the n-side, but to get to the p-side they need enough energy to climb a potential barrier of height eV_0, and as we know from our study of statistical mechanics, only a number reduced by the Maxwell–Boltzmann factor $e^{-eV_0/kT}$ will have that much energy. In this way, the flow of electrons in the two directions matches in equilibrium.

It follows from Eq. (43–14) that each of the products $p_n n_n$ and $n_p p_p$, which, from Eqs. (43–15) and (43–16) are equal in equilibrium, is also equal to n_i^2. Thus

$$p_n n_p = N_c N_v e^{-(E_g+eV_0)/kT} \quad \text{and} \quad p_p n_n = N_c N_v e^{-(E_g-eV_0)/kT}. \tag{43–17}$$

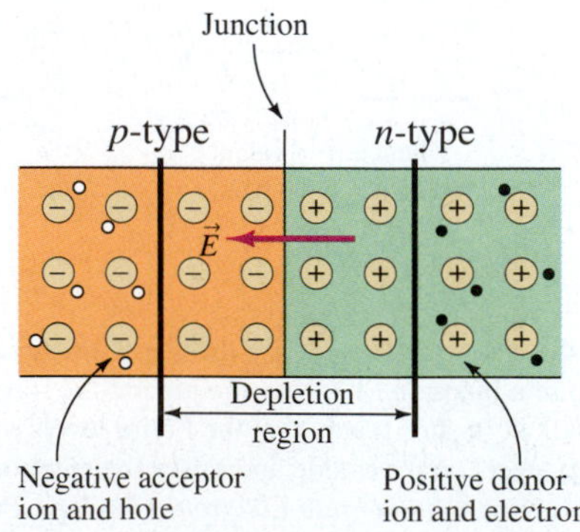

▶ **FIGURE 43–15** The potential in a p–n junction changes value over a region of finite width called the depletion region. There is an internally generated electric field across the depletion region. (After Narciso Garcia and Arthur C. Damask, *Physics for Computer Science Students*, John Wiley & Sons, 1986, p. 457.)

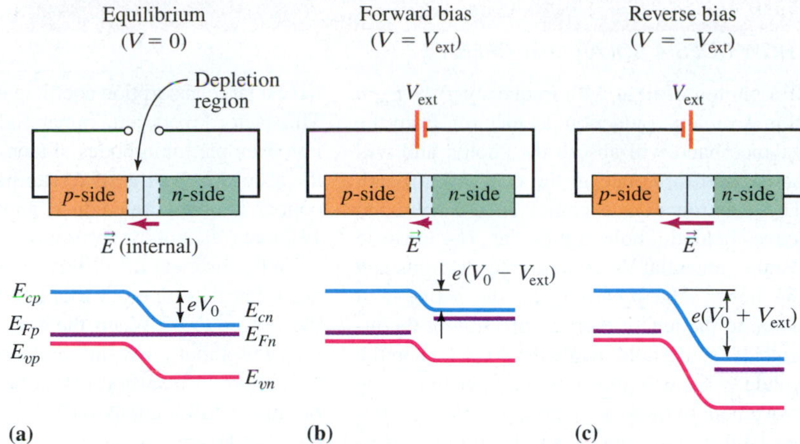

FIGURE 43–16 (a) In the unbiased *p–n* junction, the intrinsic contact potential leads to equilibrium. (b) In forward biasing of a *p–n* junction, the energy levels on the *n*-side are raised relative to those on the *p*-side. (c) In reverse biasing of the same junction, the energy levels on the *n*-side are lowered relative to those on the *p*-side. (After B. G. Streetman, *Solid State Electronic Devices*, Prentice Hall, 1980, p. 151.)

Biasing

Let's now apply an external potential, V_{ext}, across a *p–n* junction, which is then said to be *biased*. In this case, there can be a net flow of charge and a net flow of energy, in contrast to the unbiased case of Fig. 43–16a. Let's first assume that the external potential is such that the energies on the *p*-side are brought closer to the energies on the *n*-side (Fig. 43–16b). This external potential, by which the magnitude of the potential difference between the *n*-side and the *p*-side is *reduced* from $|V_0|$ to $|V_0| - |V_{\text{ext}}|$, is described as a *forward bias* voltage. The electrons that diffuse from the *n*-side to the *p*-side have less of a potential barrier to overcome. The same is true for the holes that diffuse from the *p*-side to the *n*-side (the potential difference is the same, but the charges of the carriers have opposite signs, so both have the same "hill" to climb). There is a net current from the *p*-side to the *n*-side, which rapidly increases as $|V_0| - |V_{\text{ext}}|$ decreases in magnitude. Note that there is also a small fixed current from the *n*-side to the *p*-side, associated with movement of electrons and holes without a potential barrier to overcome.

If the external potential is applied in such a way that the potential difference between the *n*-side and the *p*-side is *increased* from $|V_0|$ to $|V_0| + |V_{\text{ext}}|$, we have a *reverse bias* voltage (Fig. 43–16c). Both electrons from the *n*-side and holes from the *p*-side have an even harder time overcoming the additional barrier, and there is little current due to those charge carriers. However, the small fixed current from the *n*-side to the *p*-side is unaffected.

To make our argument more quantitative, let's suppose that a positive V_{ext} is a forward bias and a negative V_{ext} is a reverse bias. Consider first the current due to the motion of the holes, with positive current corresponding to positive charges moving to the right. The potential barrier $e(V_0 - V_{\text{ext}})$ inhibits positive carriers from the left (of density p_p) from flowing, and only the fraction $p_p e^{-e(V_0 - V_{\text{ext}})/kT}$ gets through. Positive carriers from the right do not encounter such a barrier, but their density is smaller by a factor $e^{-eV_0/kT}$, independent of V_{ext}. (The same is true separately for the *n*-carriers.) Take I_0, which is proportional to $e^{-eV_0/kT}$, to be the current of *p*-carriers from the right and *n*-carriers from the left. Without the external potential, I_0 would be canceled by the current of *p*-carriers from the left and *n*-carriers from the right, but those currents have now been modified by the factor $e^{-eV_{\text{ext}}/kT}$. Thus the *net* current flowing to the right is given by

$$I_{\text{net}} = I_0(e^{-eV_{\text{ext}}/kT} - 1). \tag{43–18}$$

This current–voltage relation, called the *I–V characteristic* of the junction, is shown in Fig. 43–17.

The *I–V* characteristic shows that the *p–n* junction behaves like a *diode*, a device that allows current to pass in one direction but not in another. Such diodes can be used to construct rectifiers (see Chapter 33). The *p–n* junction can also act as a *solar cell* or a *light-emitting diode*.

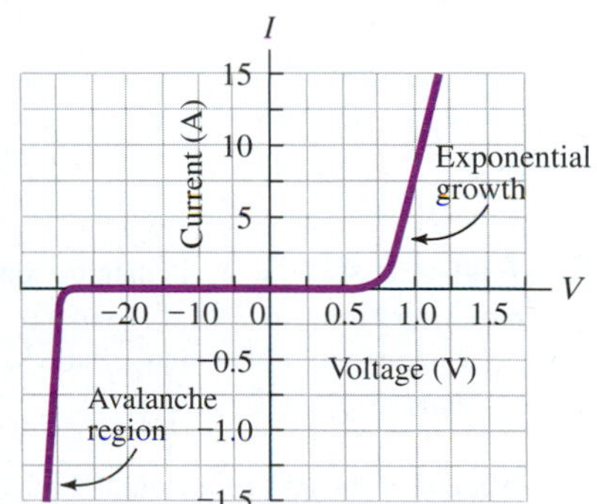

FIGURE 43–17 *I–V* characteristic of a *p–n* junction. The junction behaves as a diode under an external potential V_{ext}. Note the change in scale for negative voltage.

THINK ABOUT THIS. . .

HOW DOES A SOLAR CELL OPERATE?

If a photon of light with frequency $f > E_g/h$ shines on a p–n junction, an electron from the valence band will absorb the photon and will be excited to a state in the conduction band. The electron now becomes an n-carrier and leaves behind a hole, a p-carrier. The intrinsic contact potential V_0 separates the electrons and the holes: The p-carriers on the p-side will move to the junction in the direction of the intrinsic electric field, while the n-carriers on the n-side will also move to the junction. The movement of these new carriers produces a potential difference opposite to V_0. It thus acts as a forward bias voltage across the junction, even in the absence of V_{ext}. A current is produced by this *photovoltaic effect*, and power can be delivered to an external circuit. In silicon, for example, photons of wavelength shorter than 1200 nm will give rise to a photovoltaic effect, meaning visible light will produce current in the junction.

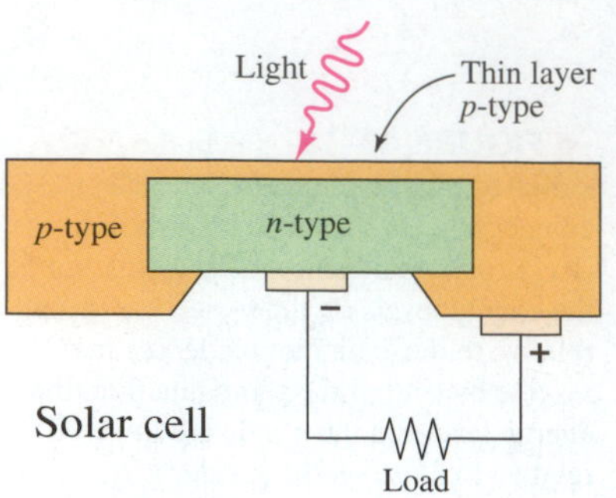

▲ **FIGURE 43–18** Schematic diagram of a solar cell, which generates current across the load (resistor) when light shines on the p-side of a p–n junction.

▲ **FIGURE 43–19** This artist's concept of the Mars Rover that successfully explored the surface of Mars has solar cells on almost its entire top surface to supply the electrical power for the various tools, cameras, and communication equipment.

A solar cell, a device that produces electricity from sunlight, is a p–n junction constructed in such a way that the solar photons can be absorbed in the area closest to the built-in contact potential—near the junction (Fig. 43–18). Sunlight is allowed to fall onto the top semiconductor, which is covered with an antireflective coating. The light penetrates the semiconductor and is absorbed in the junction region below. The materials are chosen with a band gap small enough to absorb even the long-wavelength component of solar light. It is also important to have a large absorption coefficient for the light. This factor favors certain materials over others. For example, amorphous silicon has 50 times the absorptive power of crystalline silicon, and copper indium diselenide is another factor of 10 more efficient for absorption.

Why, in view of all this, don't we simply use solar cells on every roof and dispense with the power grid? When the Sun is shining, it provides about 1 kW/m^2, and a typical household might use several kW at peak times. But because of limitations on light collection and on photon absorption, there is a theoretical maximum of 28 percent for conversion of solar energy into electric energy in silicon, and real silicon solar cells are even less efficient, on the order of 10 percent. Thus a large area of solar cells would be needed. Of course the Sun does not shine all the time, so one must collect enough energy for later use and store it as well. This all adds up to a prohibitive cost. The capital cost of an installation that could power a small house in ideal conditions can be counted in tens of thousands of dollars. Power generation by solar cells is cost-effective only in regions so remote from the grid that the cost of transmission from the grid exceeds the cost of the power cells, as for example on distant farms, or power for remote traffic control, buoys, or planetary exploration (Fig. 43–19). At a more manageable scale, you may be using solar cells or their equivalent on a daily basis in a watch or a calculator.

LEDs

A light-emitting diode (or LED), which is a p–n junction acting as a signal light, is essentially a solar cell that operates in reverse. When a forward bias is applied across the junction, electrons flow from the n-side to the p-side and holes from the p-side to the n-side. When the electrons arrive at the p-side, they recombine with available holes and emit light in the process. The holes that arrive at the n-side recombine with available electrons and also emit light. LEDs are widely used as electronic signaling lights because they are long-lasting, use little energy, can be switched on and off rapidly, and are compact.

EXAMPLE 43–5 Calculate the ratio of the currents through a p–n junction diode for positive voltages to those for negative voltages, for voltages whose magnitude is given by $eV_{ext} = 0.1$ eV, 0.2 eV, and 0.3 eV at $T = 300$K.

Strategy The current through a junction diode biased by a voltage is given by Eq. (43–18). Thus the ratio of the currents for positive to negative bias, R, is given in terms of $K \equiv eV_{ext}/kT$ by

$$R = \left|\frac{e^K - 1}{e^{-K} - 1}\right| = \frac{e^K - 1}{1 - e^{-K}}.$$

Working It Out At $T = 300$K, $kT = 2.6 \times 10^{-2}$ eV, so the parameter K has the values 3.9, 7.7, and 12 for $eV_{ext} = 0.1$ eV, 0.2 eV, and 0.3 eV, respectively. The respective values of R are then 48, 2.2×10^3, and 1.0×10^5. The rapid growth of these values as eV_{ext} increases illustrates the dramatic biasing power of relatively small potential differences.

Transistors: The *transistor* is a three-terminal device that can be used for sensitive control of currents flowing through circuits, particularly for amplifying or switching currents. One of the first transistors, the *bipolar junction transistor* (BJT), started a revolution in electronics that is still going on. That transistor and the theory of its operation were developed in 1948 and 1949 by John Bardeen, Walter Brattain, and William Shockley (Fig. 43–20). While we won't go into any of the details here, it is worth mentioning that a transistor is constructed by judiciously combining two oppositely doped semiconductors in a structure that places two junctions close by. It is possible in this arrangement to apply biasing voltages that allow a very small change in the current in one part of the transistor to lead to a very large change in the current of another part. In this way the transistor is a kind of amplifier. Depending on how the amplified current is used, the transistor can act as a very sensitive switch. Devices like these are today combined into building blocks for today's electronics. By the techniques to be described in Section 43–4, thousands or even millions of tiny semiconductor structures making up complicated circuits are fabricated as one unit; we call these circuits *integrated circuits* (Fig. 43–21). These circuits have enabled us to produce powerful desktop computers that take up very little space. The layers of material that make up the devices in integrated circuits are measured in fractions of nm, and the currents that flow in them may be on the order of mA.

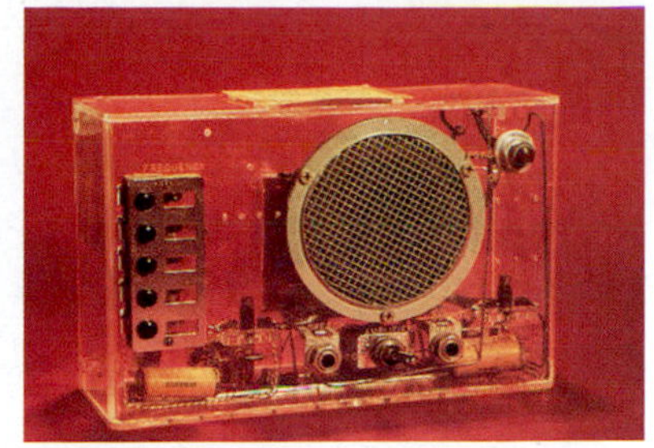

▲ **FIGURE 43–20** "Bardeen's box," constructed in 1949, contains a circuit with transistors as elements. The instant turn-on of the circuit impressed audiences who had been used to waiting a long time for circuits containing tubes to warm up.

▲ **FIGURE 43–21** Photomicrograph of the surface of an integrated circuit.

43–4 Band-Gap Engineering

Technological advances in the fabrication of semiconductor structures have created a new tool chest for the electronics engineer and computer designer. One advance involves the ability to place thin layers of one or several materials on a single underlying layer (a *substrate*) of another material with very similar crystal structure. (Such techniques are the basis of integrated-circuit construction.) This procedure, known as *epitaxy*, has reached its highest refinement with *molecular-beam epitaxy* (Fig. 43–22). Examples are provided in structures in which single crystals of gallium arsenide form a substrate and compounds of aluminum and gallium arsenide ($Ga_{1-x}Al_xAs$, which are alloys of variable content, where x represents a fraction) form junctions with the substrate. Gallium, aluminum, and arsenic are evaporated in a vacuum chamber and projected as beams onto the gallium arsenide substrate. The beams form layers at a rate of about one per second on a substrate whose thickness can be controlled. The components of the deposited layers can be varied by exposing the substrate to different beams in different proportions. Impurities are easily added in desired proportions.

Lithography is another indispensable technology for integrated circuits. With this process, we can create detailed patterns on a crystal surface. In scanning-electron-beam lithography, an opaque film on a transparent base is removed in a precise pattern by a moving electron beam. The resulting transparent features are as small as several nm. Light passes through this "negative" onto a semiconductor blank that has been coated with a polymer that disintegrates when light hits it. The coated semiconductor can then be etched by chemicals such as chlorine gas that act where the polymer has disintegrated. Alternatively, other materials can be deposited on the areas where the polymer has disintegrated, or doping ions can be introduced in those areas. A process such as this can be repeated to produce elaborate structures with many layers.

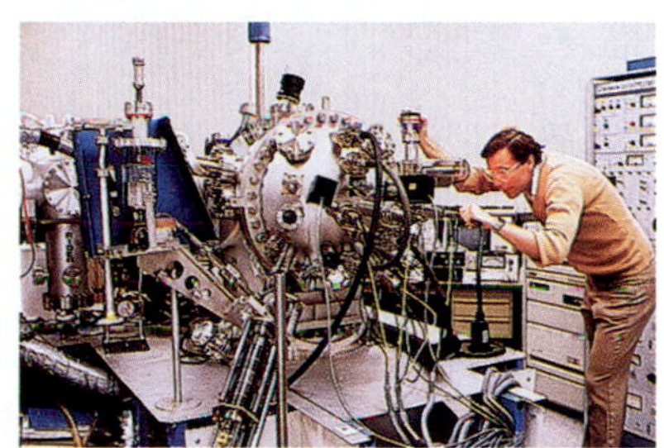

▲ **FIGURE 43–22** The molecular-beam epitaxy machine shown here is used to construct microscopic structures molecular layer by molecular layer.

Quantum Wells, Quantum Wires, and Quantum Dots

When heterojunctions are combined in various ways, the condition that the Fermi energy is constant across junctions has an effect on the energies of the edges of bands, and these edges may form regions in which electrons have "walls" on two sides. Figure 43–23a shows a thin GaAs layer sandwiched between two AlGaAs layers, surrounded by GaAs that contains the donor ions; Fig. 43–23b is an energy diagram for this structure. The electrons in the GaAs layer move in a *quantum well*. Such potential wells are characterized by quantized energy levels. We can find the values of these energies by thinking of standing waves set up in the direction of the (small) distance a between the "walls" that confine the electron, just as atomic energies in the Bohr model follow from

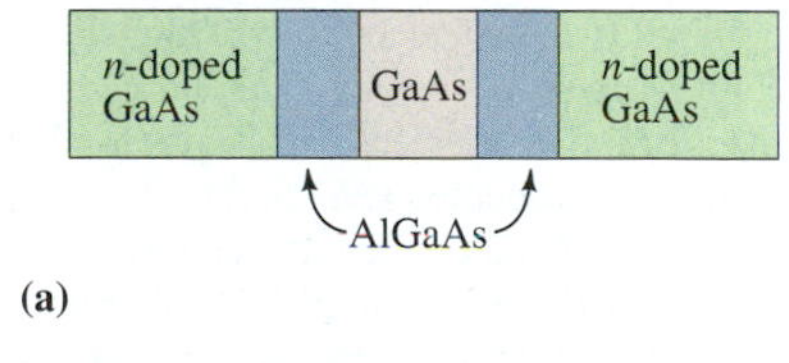

(a)

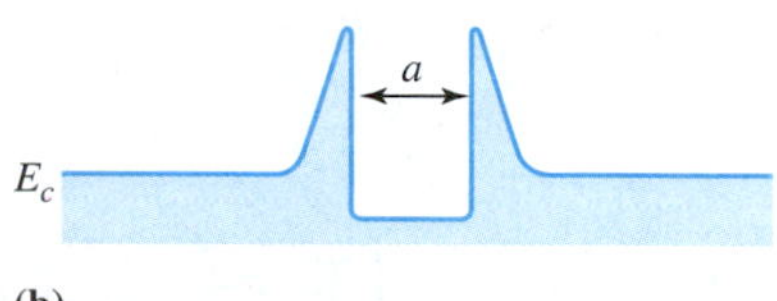

(b)

▲ **FIGURE 43–23** (a) A heterostructure within which (b) a potential energy well of width a forms and confines electrons.

thinking about fitting standing waves into atomic orbits. The allowed energies of the quantum well are then given by

$$E_n = \frac{n^2\hbar^2\pi^2}{2m^*a^2}, \tag{43–19}$$

where n is any integer. The confining distance a is typically 10 to 100 nm. Although a is large compared with atomic dimensions, the fact that m^* is small partly compensates for this effect. Only the energy of lateral motion is quantized: The motion in the direction perpendicular to the plane of the page for Fig. 43–23a is not quantized.

EXAMPLE 43–6 Calculate the energy difference between the ground state and the first excited state in a quantum well of width 20 nm, given that the effective mass is $m^* = 0.070m_e$. What is the wavelength of a photon emitted when an electron undergoes a transition between the two states?

Strategy We use Eq. (43–19) to calculate the energy difference ΔE between levels characterized by $n = 2$ (the first excited state) and $n = 1$ (the ground state). Once we have this energy difference, the frequency of a photon emitted in a transition between these two levels is determined by $\Delta E = hf$, and the corresponding wavelength is $\lambda = c/f$.

Working It Out We have

$$\Delta E = E_2 - E_1 = \frac{(2^2 - 1^2)\hbar^2\pi^2}{2m^*a^2} = \frac{3\hbar^2\pi^2}{2m^*a^2}$$
$$= \frac{3(1.05 \times 10^{-34}\,\text{J}\cdot\text{s})^2\pi^2}{2(0.070)(0.91 \times 10^{-30}\,\text{kg})(2.0 \times 10^{-8}\,\text{m})^2}$$
$$= 6.4 \times 10^{-21}\,\text{J} = \frac{6.2 \times 10^{-21}\,\text{J}}{1.6 \times 10^{-19}\,\text{J/eV}} = 4.0 \times 10^{-2}\,\text{eV}.$$

Therefore the wavelength of a photon emitted when there is a transition between these levels is

$$\lambda = \frac{c}{f} = \frac{hc}{\Delta E} = \frac{(6.6 \times 10^{-34}\,\text{J}\cdot\text{s})(3.0 \times 10^8\,\text{m/s})}{6.2 \times 10^{-21}\,\text{J}}$$
$$= 3.2 \times 10^{-5}\,\text{m}.$$

A quantum well can act as a very sensitive switch. Figure 43–24 shows how to arrive at the I–V characteristic for a device based on it. With zero voltage between points A and B (Fig. 43–24a), the quantum well presents a barrier to the n-carriers. There is no tunneling because the energy levels on both sides of the quantum well are occupied equally. When a voltage is applied, the n-carriers on the left are at a higher energy, and they can tunnel into empty levels (Fig. 43–24b). Thus, the current rises with V, as shown in Fig. 43–24e. When the bottom of the conduction band on the left is lifted high enough to match one of the low-lying quantized levels inside the well (Fig. 43–24c), there is a drastic increase in the flow of the current (Fig. 43–24e). A further voltage rise decreases the current because the bottom of the conduction band on the left no longer matches a well level (Fig. 43–24d). As the voltage increases further, the current decreases and then increases again as the second energy level is approached, as we see in Fig. 43–24e. The ability to pick the applied voltage that produces substantial tunneling makes for a sensitive switch.

Quantum wires and *quantum dots* represent further steps in the fabrication of systems in which quantum behavior is dominant. They correspond to electron-confining "pipes" and "boxes," respectively. Electrons in a quantum dot, the equivalent of a tiny three-dimensional box, have widely separated energy levels (see Problem 34). Such structures are promising as very precise switching devices.

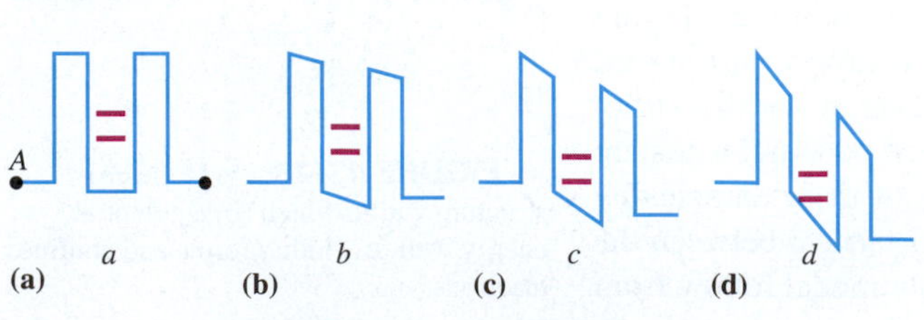

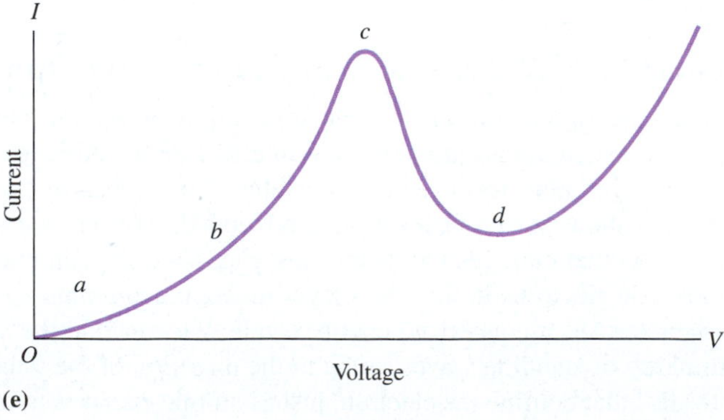

▼ **FIGURE 43–24** The quantum well of Fig. 43–23 subjected to an external potential. (a) through (d) As the potential is increased, there is an enhanced probability that an electron will tunnel from one side of the well to another when the location of allowed levels in the well matches an occupied level in the conduction band on the left. (e) I–V characteristic for the situation, with points marked according to the biasing in parts (a)–(d).

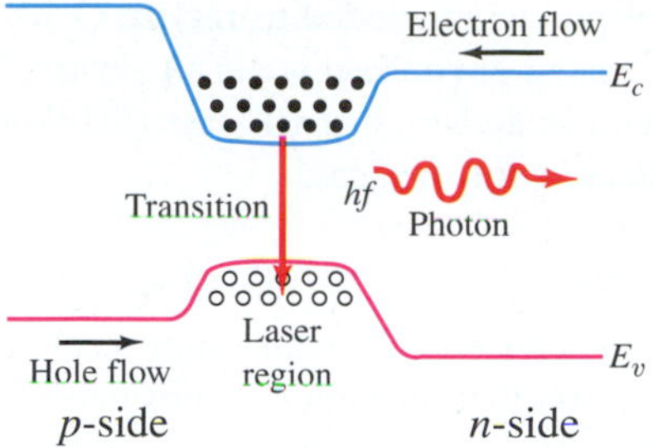

FIGURE 43–25 Energy diagram for a semiconductor laser. The laser light is produced in the recombination of n-carriers and p-carriers at the central region.

FIGURE 43–26 Scanning tunneling microscope (STM) image of a microlaser. The multilayer structure is evident.

Semiconductor Lasers

Semiconducting materials are well suited to the construction of microscopic lasers. Such lasers have already found uses in compact-disc players, amplifiers in satellite receivers, and communications with fiber optics. An energy diagram for a heterostructure illustrates how these lasers may be constructed. The semiconductor within which the laser light will be produced (for example, GaAs) is sandwiched between two layers of appropriately doped AlGaAs. The right side of the energy diagram in Fig. 43–25 is doped to n-type, and the left side is doped to p-type. A potential barrier keeps the electrons from flowing to the left near the bottom of the conduction band, and a counterpart barrier keeps the holes from flowing to the right near the top of the valence band. If a voltage is applied that causes more electrons to flow into the GaAs region and more holes to flow into that region, and the barrier prevents both n-carriers and p-carriers from flowing back, a population inversion is built up. Note that, unlike the lasers described in Chapter 42, the "pumping" is done automatically in these materials. The electrons and holes can recombine by the emission of photons, and these photons stimulate further, rapid recombination. This describes the process we treated in Chapter 42 for the production of laser light.

Figure 43–26 is an image of a set of *microlasers* that consist of a thin InGaAs quantum well, with stacks of AlAs, GaAs, and AlGaAs layers both above and below it. The laser action takes place in the quantum well and, because the well is narrow, the power required to make the laser work is very small. Many reflections of the light are required to get suitable amplification, and the large number of layers provide a certain amount of reflection at each interface. Even though the reflectivity at each interface is less than 1 percent, the large number of layers gives a total reflectivity back to the lasing material of 99 percent. Many lasers—millions of them—each with cylindrical geometry, can be formed on a single substrate by molecular-beam epitaxy and subsequent etching. Such laser arrays are useful for optical communications in fiber optics (Fig. 43–27).

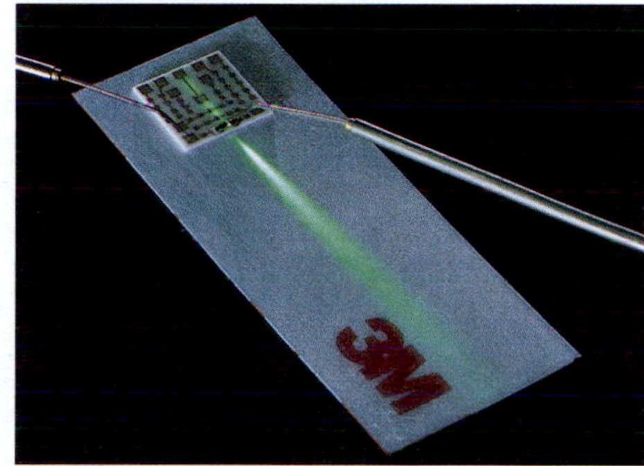

FIGURE 43–27 Blue-green light is emitted by this semiconductor laser.

43–5 Scanning Microscopy

Our ability to see an object is diffractively limited by the wavelength of the illuminating radiation. Thus, to resolve atoms, whose size is roughly 0.1 nm, the radiation must have a wavelength on the order of 0.1 nm or less. For electromagnetic radiation, this corresponds to a frequency of $f = c/\lambda = (3 \times 10^8\,\text{m/s})/(10^{-10}\,\text{m}) = 3 \times 10^{18}\,\text{Hz}$. A photon of this frequency has an energy $hf = (6.62 \times 10^{-34}\,\text{J}\cdot\text{s})(3 \times 10^{18}\,\text{s}^{-1}) = 2.0 \times 10^{-15}\,\text{J} = (2.0 \times 10^{-15}\,\text{J})/(1.6 \times 10^{-19}\,\text{J/eV}) = 12\,\text{keV}$. This energy is 1000 times larger than an atomic ionization energy, and such photons are not readily available. The situation is improved somewhat when electrons are used, but the required electron energy is about 150 eV, which is still very large. J. A. O'Keefe pointed out in 1956 that it would be possible to bypass the diffraction limits with a new type of microscope. The basic idea is simple: Light passes through a tiny hole in a screen and illuminates the object to be viewed, which is on the other side of the hole. The transmitted light (or reflected light) is recorded as the object is moved back and forth across the hole. *The size of the hole determines the resolution rather than the wavelength of the light.* The use of *piezoelectric* materials has provided the technology for controlling

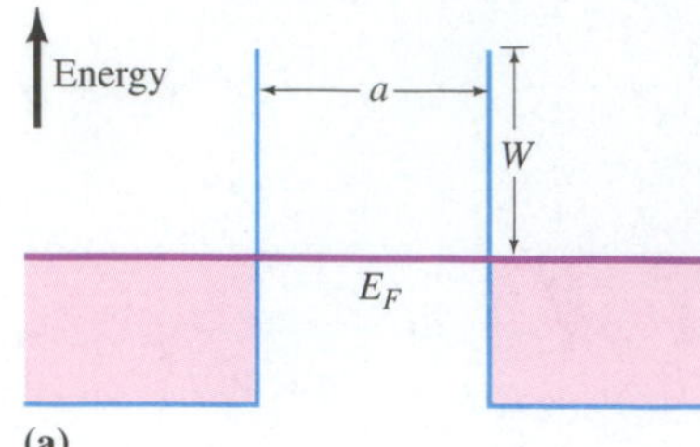

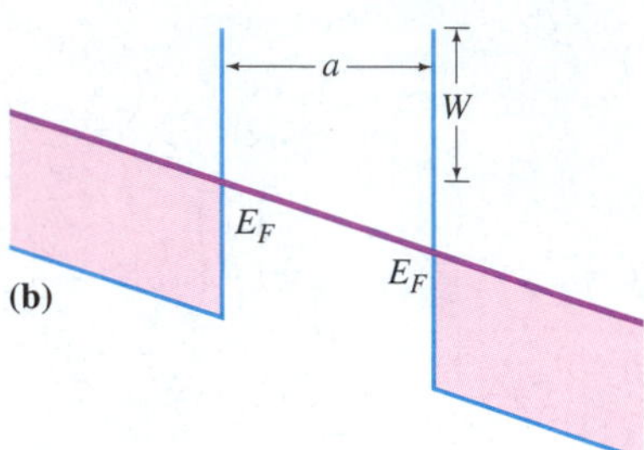

▲ **FIGURE 43–28** (a) When two metals are separated by a gap, there is no tunneling in equilibrium. (b) If an external potential is applied, the Fermi energy within one of the metals is lowered, leaving empty levels into which electrons from the side with the higher electron energies can tunnel. The height of the barrier to overcome is the metal's work function W.

the position of an object with the kind of precision needed to realize O'Keefe's idea. Piezoelectrics are ceramic materials that expand or contract when an electric field is applied to them. The ability to move them with an accuracy of 10^{-5} nm (!) has made scanning microscopy into a superb tool for the study of surfaces.

Scanning Tunneling Microscopy

The scanning tunneling microscope (STM) is based on the quantum mechanical result that electrons can tunnel from one region to another through domains that are classically inaccessible. Figure 43–28a shows the energy levels in two metallic samples separated by a vacuum of width a. Electrons at the Fermi energy can cross over from one sample to the other in significant numbers only if (1) there are empty energy levels of the same energy, so that the exclusion principle does not prevent tunneling, and (2) the tunneling probability is sufficiently large. If an electric field is applied, thereby lowering the Fermi energy on one side as in Fig. 43–28b, then condition (1) is satisfied and tunneling can occur. An application of Eq. (40–11) shows that the rate at which electrons cross such a barrier is

$$f = e^{-(2/\hbar)a\sqrt{2mW}}, \tag{43–20}$$

where m is the electron mass and W is the work function of the metal (the height of the potential barrier between the metals). The important feature of Eq. (43–20) is the extreme sensitivity of the tunneling fraction (that is, the current) to the magnitude of the separation.

Figure 43–29 (see also Fig. 24–33) shows how an STM is built. A tungsten tip acts as a "hole in the screen." When such a tip is poised above a surface at a potential different from that of the tip, then, according to Eq. (43–20), the current passing through the tip is a sensitive indicator of its distance to the surface. The resolution depends on the tip size. Tips that end in a *single* atom are made by heating a crude tungsten tip and applying a strong electric field to it. The field pulls atoms away layer by layer, leaving a single-atom tip. With such tips it is possible to resolve features 0.1 nm across: *We can "see" single atoms on a surface.*

There are two ways of scanning a surface. In one method, piezoelectric supports that hold the tip are arranged to sense the tunneling current and move the tip up or down

▶ **FIGURE 43–29** Schematic diagram of a scanning tunneling microscope.

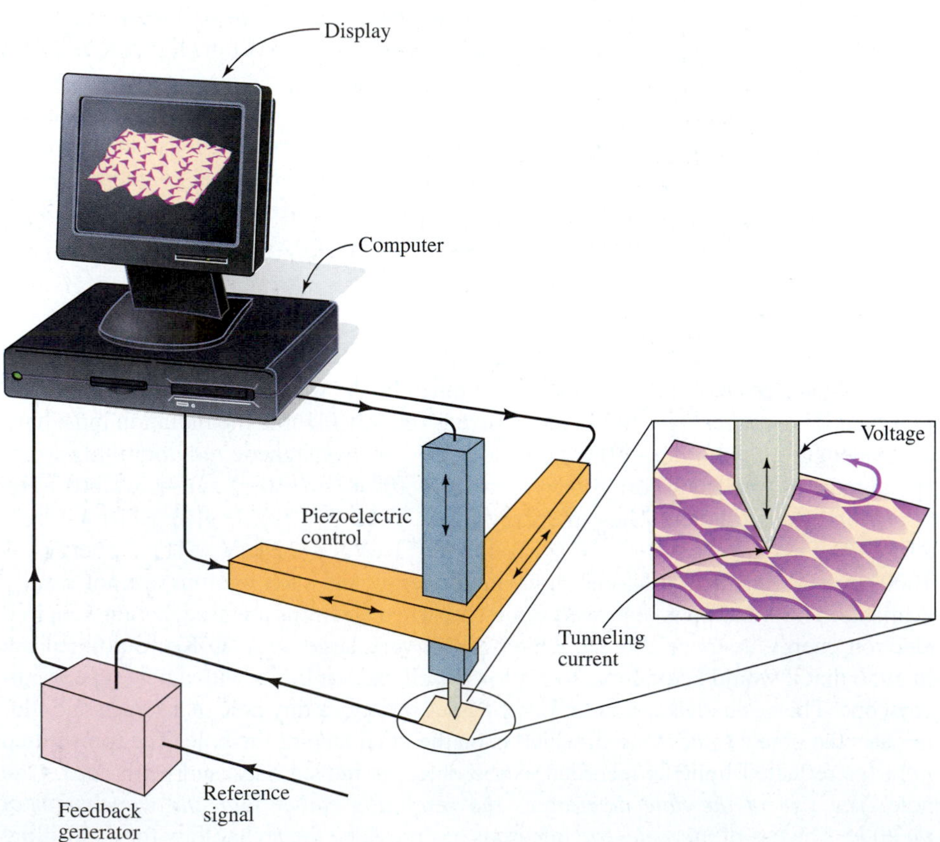

so as to maintain a constant current or, equivalently, a constant distance from the surface. The voltages applied to the piezoelectric supports during this process provide a record of the surface topography. A second way of scanning the surface is to move the tip horizontally across the surface and see how the tunneling current varies for a fixed applied voltage. STMs are especially effective because miniaturization of the apparatus helps in the reduction of "noise": The smaller the apparatus, the higher the frequency of the random thermal vibrations, and this noise is relatively easy to filter out. In addition, STMs work better in air than in a vacuum because the random bombardment by air molecules also tends to average out noise.

STMs are used in a variety of surface studies. Electron charge distributions at surfaces can be measured, as can the way the distributions are affected by the deposition of very thin films on the surfaces. Another application involves the use of the STM tip to manipulate individual atoms or groups of atoms and move them around on the surface of the sample (Fig. 43–30; see also Fig. 24–35). There is potential for the creation of tailor-made molecular structures for specific tasks in microelectronic devices.

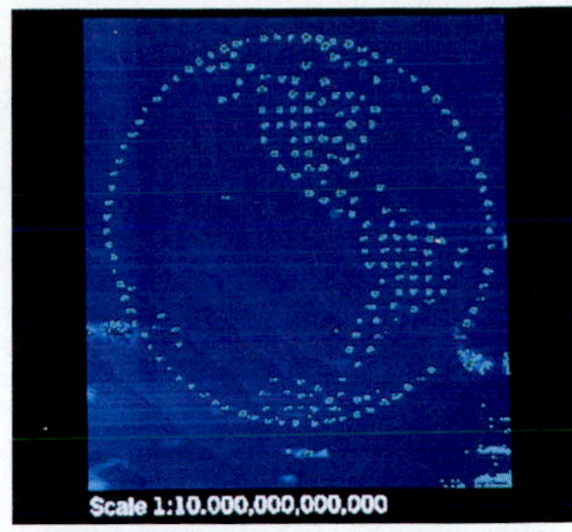

▲ **FIGURE 43–30** A map made of tiny dots that consist of just a few thousand atoms of gold. The map, about 1 micron across, was made by manipulating the gold dots with the tip of a scanning tunneling microscope.

Atomic Force Microscopy

Another imaging device, the *atomic force microscope* (AFM), images a surface by measuring the effect of the force it exerts on a small cantilever (like a diving board) rather than by probing it with a current (Fig. 43–31). This force is an atomic force, and the idea that it should be directly measurable sounds absurd. Nevertheless, if we calculate an interatomic spring constant $k = m\omega^2$, with $\omega = 10^{13}$ Hz (a typical atomic frequency) and $m = 10^{-25}$ kg, we find that $k = 10$ N/m. For comparison, the spring constant of a piece of aluminum foil 4 mm long and 1 mm wide acting as a cantilever is about 1 N/m. Because the atomic spring constant and the mechanical cantilever spring constant are comparable, atomic forces can be measured by sensing small displacements (0.1 nm) of such a cantilever.

In some AFMs, the cantilever is a piece of silicon or silicon oxide about 100 μm long and 1 μm thick. Such cantilevers have a spring constant in the range of 0.1 to 1 N/m. The sensor that detects the motion of the cantilever must detect displacements of less than 0.1 μm. One way of detecting this motion is to shine laser light on it and record the position of the reflected beam with a light detector. The light detector activates the piezoelectric base on which the sample is placed, and a current controls the base such that the distance

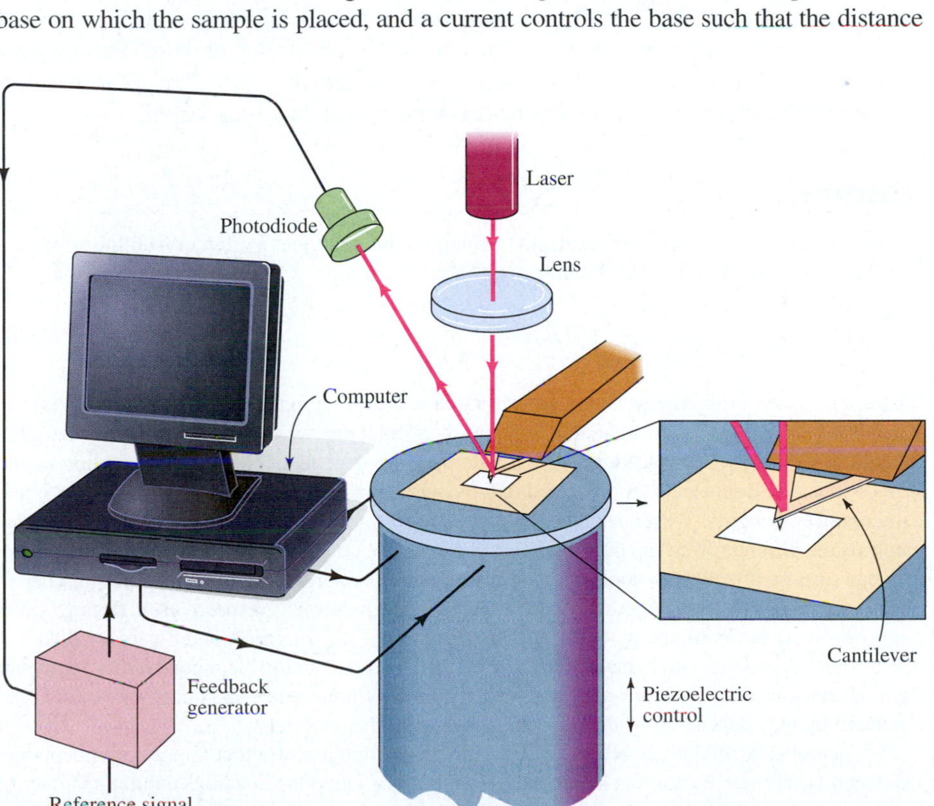

◀ **FIGURE 43–31** Schematic diagram of an atomic force microscope (AFM).

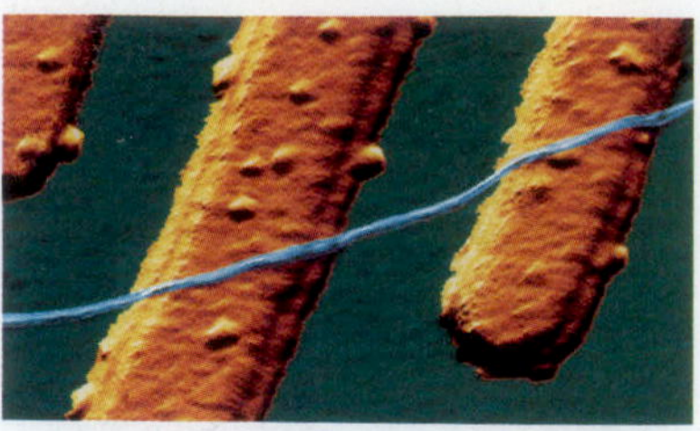

▲ **FIGURE 43–32** This atomic force micrograph (AFM) shows a blue carbon nanotube wire of only 10 atoms width (1.5 nm) lying across platinum electrodes (yellow).

between the sample and tip is held constant. The movement of the sample is translated into an image of the surface. Figure 43–32 illustrates the kind of image possible with an AFM.

The effect of the external environment on the AFM is quite negligible. Typical building vibration frequencies are about 10 to 20 Hz, whereas the resonant frequencies of the cantilever are about 10 to 100 kHz. Amplitude distortions due to external vibrations of frequency f are on the order of $(f/f_{res})^2$. Building vibrations thus contribute a correction of less than 10^{-3} nm to the displacement.

THINK ABOUT THIS. . .

DOES THE ATOMIC FORCE MICROSCOPE HAVE ADVANTAGES OVER THE SCANNING TUNNELING MICROSCOPE?

In the STM we make pictures of conducting materials; it works, after all, by measuring a current that passes between its tip and the material's surface. Some of the uses of the STM are described above. With an AFM, we can make images of the surface of *any* type of solid material, including insulators. Moreover, AFMs can be operated in a variety of environments: in air, in water, at low temperatures, or in a vacuum. The materials being probed also require little preparation, whereas the conducting properties of materials probed with STMs will change if, say, there is a layer of oxide on the surface. The AFM has also been used to study atomic-scale structure on the surfaces of graphite, mica, sodium chloride, and other nonorganic materials of this type. They have also been applied to the imaging of biological molecules such as amino acids, DNA, and proteins and have even imaged certain chemical processes *as they occur*. While the resolution of STMs takes us down to single atoms, the best AFMs cannot see quite as sharply, having resolutions on the range of 1 to 10 nm.

The Ultimate in Quantum Engineering

In this chapter, we have emphasized the interplay of physics and modern technology. We have touched on only a few topics—concentrating on those for which quantum physics is important. The ultimate aim of the kind of work we have described here is the production of devices in which single atoms or single electrons play the major role. For example, a computer memory that stores information by the presence or absence of an individual atom in a given surface site could conceivably store all the knowledge of the human race on a surface of area 1 cm^2! And transistors that operate by allowing single electrons to tunnel could compose the smallest possible circuit. The realization of these ideas may not be impossibly distant. The link between physics and technology in the area of materials is so strong that the distinction between the two fields has all but disappeared.

Summary

Because electrons obey the Pauli exclusion principle, their energies within crystalline solids are distributed according to the Fermi–Dirac distribution,

$$f(E) = \frac{1}{1 + e^{(E-E_F)/kT}}, \tag{43–2}$$

where E_F is the Fermi energy. Semiconductors have a band structure for which the (filled) valence band is separated from a conduction band by a small energy gap (with the Fermi energy in the middle of the gap). This leads to strongly temperature-dependent electrical conduction within semiconductors. Both electrons promoted to the conduction band and the holes they leave behind can conduct electricity. Within semiconductors, electrons (and holes) have effective masses that can be quite different from the mass of free electrons. The conduction properties of semiconductors can also be affected by doping, the introduction of impurities that increase the number of negatively or positively charged carriers (n-carriers and p-carriers, respectively). Doping shifts the position of the Fermi energy but leaves the densities of the n-carriers and p-carriers linked in a way that depends on the behavior of the undoped semiconductor. Semiconductors may emit light when n-carriers and p-carriers recombine, leading to fluorescence and phosphorescence, and conductivity may depend on whether or not light strikes the material.

A pair of semiconductors placed in contact with one another, whether the same semiconductor doped differently or different semiconductors, forms a junction. Junctions exhibit a variety of behaviors and are the basis for much of the electronic technology in use today. The conservation of energy and of charge leads to the principle that the Fermi energy across a junction is constant.

This principle dictates that an internally generated electric potential, the contact potential, is generated across the boundary of a *p–n* junction, a semiconductor structure consisting of a semiconductor doped with *p*-carriers in contact with the same semiconductor doped with *n*-carriers. The contact potential explains why *p–n* junctions act as diodes, devices for which current I_{net} in effect passes in one direction only when an external potential V_{ext} is applied:

$$I_{net} = I_0(e^{-eV_{ext}/kT} - 1), \tag{43–18}$$

where I_0 is a current associated with the contact potential. A *p–n* junction can also act as a light-emitting diode or as a solar cell.

The transistor is a three-terminal semiconductor device in which small changes in potential across one pair of terminals can lead to large amplification of the current that passes through another pair of terminals. Such devices are therefore useful for the sensitive control of currents and of voltages in circuits.

Modern techniques of material manipulation permit the elaboration of heterostructures, single-crystal structures built in a precisely controlled way from different materials. The electrical behavior of heterostructures depends intimately on quantum phenomena, even beyond the collective quantum effects that lead to any band structure. Microscopic semiconductor lasers and quantum wells are manifestations of this new technology whose limits are not yet known.

Scanning tunneling microscopes employ quantum tunneling to explore conductors, and atomic force microscopes allow us to explore nonconductors. Both instruments avoid resolution difficulties associated with diffraction and permit the study of matter at close to atomic scales.

Understanding the Concepts

1. You have just calculated the number density of *n*-carriers in undoped silicon, a semiconductor, at room temperature. Is it necessary to make a second calculation to find the number density of holes?
2. When a hole moves from the *p*-side of a junction to the *n*-side, why does its energy increase?
3. Does the Fermi energy still lie in the middle of the energy gap when the effective masses of the *n*-carriers and *p*-carriers are different?
4. Is the mechanism that makes metals opaque to visible light but transparent to X rays similar to the mechanism that makes semiconductors opaque to visible light but transparent to infrared light?
5. There are several methods by which electrons can be removed from a solid's surface. How can holes be removed?
6. What happens to the Fermi energy at low temperatures when a semiconductor is so heavily doped that the energy levels of the donor states overlap the conduction band?
7. How can the energy of a hole be raised as it travels from the *p*-side of a *p–n* junction to the *n*-side?
8. Why, physically, is it true that the smaller the effective mass of an *n*-carrier, the larger the drift velocity of the carrier?
9. We described why new states for electrons are available near the bottom of the conduction band in an *n*-type extrinsic semiconductor. Why are new states for holes available just above the top of the valence band in a *p*-type extrinsic semiconductor?
10. Is there a type of viscosity that acts on holes in a semiconductor and gives them a terminal velocity in an electric field, or do holes just accelerate under the influence of such a field? If there is a terminal velocity, is it the same as that of the electrons?
11. We referred to a current amplification in our discussion of transistors. Can this mean that the power is amplified? If so, where does the necessary energy come from?
12. The gap widths of the semiconductors in Table 43–1 are given at room temperature. How could a gap width depend on temperature?
13. Experiment shows that many insulators are transparent to visible light, many semiconductors are opaque to visible light and transparent to infrared light, and metals are opaque to all forms of visible light. Why might this be true?

Problems

43–2 Semiconductors

1. (I) In silicon, $E_g = 1.12$ eV, and the effective mass of the *n*-carriers is $m^* = 0.31m_e$, where m_e is the electron mass. Find the number densities of *n*-carriers at 100K and at 300K. Compare your result with the number density of "free" electrons in copper, 8.5×10^{22} electrons/cm^3.
2. (I) What is the density of electrons in the conduction band of silicon, for which $m_n^* = 0.31m_e$, at temperatures of (a) 90K, (b) 293K, and (c) 450K? [*Hint*: See problem 1.]
3. (I) For sodium ($E_F = 3.23$ eV) at room temperature (300K), calculate the occupation probability of an electron state of energy (a) 2 eV, (b) 3 eV, (c) 3.2 eV, (d) 3.26 eV, and (e) 4 eV.
4. (I) At what temperature is the probability that an energy state at 7 eV will be populated equal to 40 percent for copper ($E_F = 6.95$ eV)?
5. (I) Find the energy gap of an intrinsic semiconductor in which 1.02×10^{-30} of the available energy levels near the bottom of the conduction band are occupied at $T = 126$K.
6. (I) The value of the Fermi–Dirac distribution function $f(E)$ for a particular energy value within a collection for electrons at a temperature $T = 65$K is found to be equal to 0.4. What is the value of that function for the same energy value when $T = 85$K?
7. (I) What is the probability of finding an electron at the bottom of the conduction band in GaAs, for which $E_g = 1.43$ eV, at $T = 300$K?
8. (I) The Fermi temperature T_F is defined by E_F/k, where k is Boltzmann's constant. What is T_F for (a) silver ($E_F = 5.1$ eV), (b) copper ($E_F = 6.95$ eV), and (c) a white dwarf star ($E_F \cong 2 \times 10^5$ eV)?
9. (I) What is the maximum wavelength of light that will excite an electron from the valence band to the conduction band in (a) InSb ($E_g = 0.18$ eV), (b) GaSb ($E_g = 0.68$ eV), and (c) GaAs ($E_g = 1.43$ eV)?

10. (I) The following table gives the energy above the edge of the valence band, ΔE, for the energy levels of various acceptor impurities in silicon:

Impurity:	Boron	Gallium	Indium	Nickel	Zinc
ΔE (in eV):	0.045	0.065	0.16	0.22	0.31

Find the maximum wavelength of the radiation required to excite electrons from the top of the valence band to these acceptor levels.

11. (I) Calculate the intrinsic carrier concentration in silicon at 400K, given that $m_n^* = 0.31m_e$ and $m_p^* = 0.56m_e$.

12. (I) At room temperature (300K), undoped GaSb has an energy gap of 0.68 eV. (a) What is the probability that an electron occupies an energy state at the bottom of the conduction band? (b) What is the probability that there is a hole state at the top of the valence band?

13. (II) The effective masses of n- and p-carriers in germanium are given by $m_n^* = 0.55m_e$ and $m_p^* = 0.37m_e$, respectively. The concentration n_i of n-carriers at 300K is $2.5 \times 10^{19}\ \text{m}^{-3}$. Use these data to calculate the gap width of germanium.

14. (II) The effective masses of n- and p-carriers in GaAs are given by $m_n^* = 0.067m_e$ and $m_p^* = 0.48m_e$, respectively. The gap width of GaAs is 1.43 eV. Use this information to find the intrinsic carrier density at 380K.

15. (II) A sample of silicon is doped so that the donor density is $n = 2.0 \times 10^{23}\ \text{m}^{-3}$. Find $E_c - E_F$ at $T = 300\text{K}$, given that the effective n-carrier mass in silicon is $m_n^* = 0.31m_e$. Compare your result with E_g for silicon.

16. (II) A sample of silicon is doped so that the hole density is $p = 2 \times 10^{22}\ \text{m}^{-3}$. Find $E_F - E_v$ at $T = 330\text{K}$, given that the effective p-carrier mass in silicon is $m_p^* = 0.56m_e$.

17. (II) The intrinsic semiconductor InSb has a gap energy of 0.18 eV, and the effective masses of the electrons and holes are $0.015m_e$ and $0.39m_e$, respectively. Find the carrier densities at $T = 100\text{K}$ and 300K.

18. (II) The *mobility*, μ, of a charge carrier in a material is defined as the drift speed of the carrier, v_d, divided by the magnitude of the electric field driving the current, $\mu \equiv v_d/E$. For germanium at 300K, the mobilities of electrons and holes are about $0.45\ \text{m}^2/\text{V}\cdot\text{s}$ and $0.35\ \text{m}^2/\text{V}\cdot\text{s}$, respectively. (a) Use the definition of μ and the relation between v_d and the current to calculate the free-electron density, n, if the conductivity is $8.3 \times 10^{-5}(\Omega\cdot\text{m})^{-1}$ and $n \gg p$. (b) Repeat part (a) for the hole density, p, assuming doping such that $p \gg n$.

19. (II) Calculate the electrical conductivity at room temperature of germanium doped to n-type if the density of n-carriers and p-carriers are $2.5 \times 10^{18}\ \text{m}^{-3}$ and $7.0 \times 10^{13}\ \text{m}^{-3}$, respectively. [*Hint*: Use the results of Problem 18.]

20. (II) The Hall Effect (see Chapter 28) is probed on an n-type semiconductor with carrier density n; the density of holes is negligible. Determine the magnitude and polarity of the Hall voltage if there is a current to the right with density j (Fig. 43–33). Repeat the derivation for a p-type semiconductor. Notice the difference in polarity.

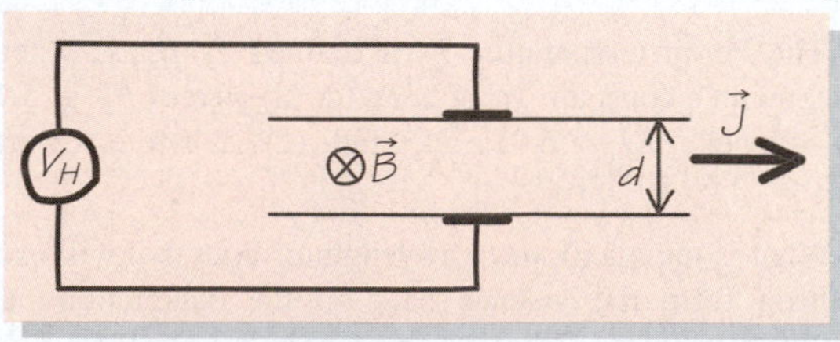

▲ **FIGURE 43–33** Problem 20.

21. (II) Use the energy distribution for a collection of identical fermions at finite temperature T to (a) show that the distribution of speeds (the probability that a fermion has a speed between v and $v + dv$) is given by

$$n(v)\,dv = \frac{8\pi m^3}{h^2}\frac{v^2\,dv}{1 + e^{[(mv^2/2)-E_F]/kT}}.$$

(b) What is the distribution for $kT \ll E_F$? [*Hint*: See Problem 50.]

22. (II) The Fermi energy can be defined as the energy at which the probability of finding an electron is $\frac{1}{2}$. (a) Show that this is the case at any temperature. (b) Refine this idea by using the fact that if the probability that an electron has an energy E *above* E_F is P, then the probability that an electron has an energy E *below* E_F is $1 - P$.

23. (III) Using a computer, plot the Fermi distribution as a function of energy for a material with a Fermi energy of 11 eV at temperatures 20K, 100K, 300K, and 1000K.

43–3 Semiconductor Structures

24. (I) A silicon solar cell produces electric current when the frequency of the light that falls on it is above 2.68×10^{14} Hz. What is the width of the band gap?

25. (I) Plot the factor $e^{eV_{\text{ext}}/kT} - 1$ [from Eq. (43–18)] as a function of temperature for a sample in which $eV_{\text{ext}} = 0.02$ eV over the range $100\text{K} < T < 350\text{K}$. Plot the same factor as a function of eV_{ext} (positive and negative) for $T = 300\text{K}$.

26. (I) The ratio of carrier densities p_n/p_p for a p–n junction at 350K is determined to be 0.86. What is the difference between Fermi energies for the n-type and p-type semiconductors?

27. (II) The Fermi energy of a heavily doped n-type semiconductor is close to E_c, whereas the Fermi energy of a heavily doped p-type semiconductor is close to E_v. (a) By using an energy diagram, show that when a p–n junction is formed from these semiconductors, a sufficient amount of reverse biasing will allow electrons to tunnel from the p-side to the n-side—a phenomenon known as the *Zener effect*. (b) The tunneling current is a sensitive function of the width of the barrier across which the electrons must tunnel. Show by means of energy diagrams that this width systematically decreases when the reverse bias increases, leading to a quickly building current. [*Hint*: Recall that the transition region has a finite width.] The Zener effect can be used in a *Zener diode*, in which a current flows when a reverse bias potential reaches a certain value.

28. (II) Consider a p–n junction formed from two samples of doped silicon. The two samples are as described in Problems 15 and 16 (Fig. 43–34). What is the value of eV_0 for this junction?

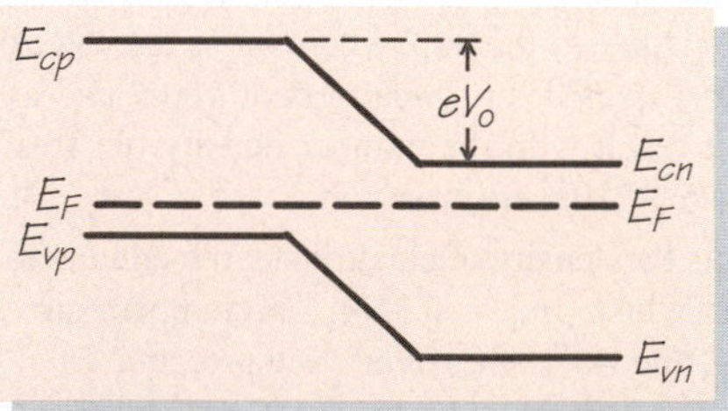

▲ **FIGURE 43–34** Problem 28.

29. (II) A germanium p–n junction has $n = 3.5 \times 10^{22}\ \text{m}^{-3}$ on the n-side and $p = 2.8 \times 10^{23}\ \text{m}^{-3}$ on the p-side. For germanium, $E_g = 0.67$ eV, $m_n^* = 0.55m_e$, and $m_p^* = 0.37m_e$. Find the locations of the Fermi energies before the junction is formed. Use your results to find eV_0 at (a) $T = 80\text{K}$ and (b) $T = 300\text{K}$.

30. (II) Many LEDs give off red light. Given this fact, estimate the width of the band gap for the materials from which such LEDs are constructed.

31. (II) A current of 2.5 mA flows through an LED with a band-gap energy of 1.4 eV. Assume that each current-carrying electron drops into a hole, and a single photon is thereby emitted. (a) What is the power emitted in the light? (b) How many photons/s are emitted?

32. (II) Consider a heterostructure in which an intrinsic semiconductor with a narrow band gap is sandwiched between an intrinsic semiconductor with a larger energy gap. (a) Draw an energy diagram for this system. (b) What happens to any n-carriers in this system in the absence of external potentials?

43–4 Band-Gap Engineering

33. (II) A quantum well with a width of 9 nm is produced, and spectroscopic measurements show that the energy difference between the ground state and the first excited state within this well is 2.28×10^{-2} eV. What is the effective mass of electrons within the well?

34. (II) For a quantum dot, in which an electron is confined within three dimensions, the allowed energy values have the form

$$E = \frac{\hbar^2\pi^2 n_1^2}{2m^* a_1^2} + \frac{\hbar^2\pi^2 n_2^2}{2m^* a_2^2} + \frac{\hbar^2\pi^2 n_3^2}{2m^* a_3^2},$$

where m^* is the effective mass of an electron; a_1, a_2, and a_3 are the three confining dimensions of the box, and the n_i values are integers. Suppose that a particular quantum dot has dimensions 8 nm × 12 nm × 12 nm and that $m^* = 0.55m_e$. (a) Calculate the energy gap between the first and second excited states. (b) What is the wavelength of the radiation emitted in the transition between these two states?

43–5 Scanning Microscopy

35. (I) The current measured in a scanning tunneling microscope whose tip height is held fixed increases by a factor of 20 when the tip moves from point A to point B on the surface. How much closer to the tip is point B than point A? The work function of the sample is 2.3 eV.

36. (I) An atomic force microscope has a spring constant of 0.3 N/m. If a displacement as small as 0.8 nm must be detected, how much work must be done to displace the cantilever?

37. (II) The cantilever of an atomic force microscope has a spring constant of 0.6 N/m, and a laser sensor detects that it has been deflected by 0.2 nm (Fig. 43–35). What is the ratio of the force acting on the cantilever to that of the force of gravity on a fly of mass 0.1 g?

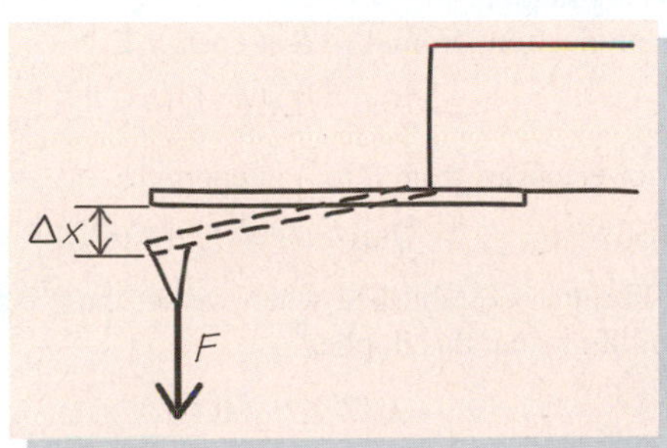

▲ **FIGURE 43–35** Problem 37.

General Problems

38. (I) An electric field of magnitude 10^6 V/m is applied in the $+x$-direction in a piece of intrinsic Si, for which the mobilities of electrons and holes at room temperature are 0.14 m²/V · s and 0.050 m²/V · s, respectively (see Problem 18). What are the drift velocities (direction and magnitudes) of electrons and holes at room temperature?

39. (I) A germanium sample is doped with 1.1×10^{23} atoms/m³ of arsenic. Given that the intrinsic n-carrier density is $n_i = 7.0 \times 10^{16}$ carriers/m³, what is the hole concentration at equilibrium (Fig. 43–36)?

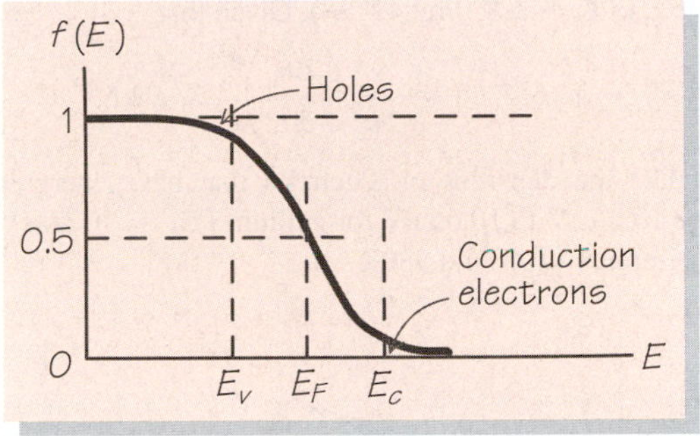

▲ **FIGURE 43–36** Problem 39.

40. (II) The effective electron mass in silicon is given by $m_n^* = 0.55m_e$. Calculate the value of N_c [from Eq. (43–12)] at 300K, and use your result to find the location of E_F relative to the edge of the conduction band for the doped silicon sample of Problem 39.

41. (II) A donor electron moves in doped indium antimonide, for which $\varepsilon/\varepsilon_0 = 17.9$ and $m^* = 0.015m_e$. Find the radius of a circular atomic orbit of such an electron in terms of the Bohr radius a_0 (Fig. 43–37). The effective nuclear charge for such a loosely bound electron is $Z = 1$.

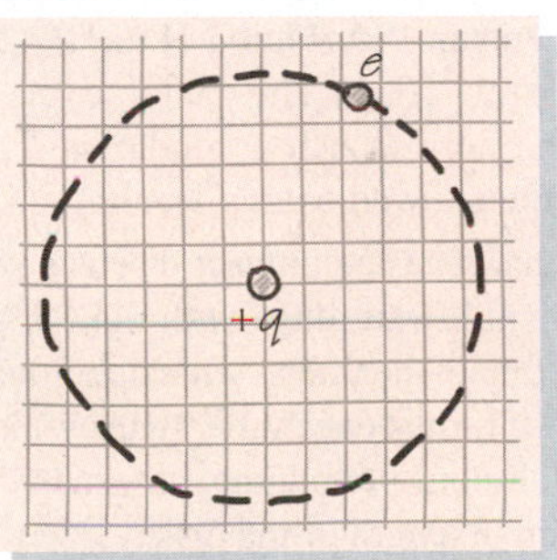

▲ **FIGURE 43–37** Problem 41.

42. (II) An electron in germanium moves about a positive center of charge $+e$ to form a hydrogen-like atom in its ground state. Calculate the wavelength of the photon needed to raise the electron to the first excited state. What is the change in the radius of the orbit? ($\kappa = 15.8$ and $m_n^* = 0.55m_e$ in germanium.)

43. (II) The impurity concentration in semiconductors must be low in order to keep the interaction between individual electrons or hole charge carriers negligible. Estimate the maximum concentration of As impurities in germanium that avoids an overlap between the ground state wave functions about adjacent impurities. Express your result in terms of a volume density (impurities/cm³) as well as in terms of the ratio impurities/atoms. (For germanium, $\kappa = 15.8$ and $m_n^* = 0.55m_e$, $\rho = 5.32$ g/cm³ and $A = 72.6$.)

44. (II) Repeat the calculation of Problem 43 for InSb with Sb impurities. (For InSb, $\kappa = 17.9$ and $m_n^* = 0.015m_e$, $\rho = 5.9$ g/cm³ and $A_{In} = 114.8$, and $A_{Sb} = 121.8$.)

45. (II) A magnetic field of 1.80 T is applied to doped indium antimonide. What is the angular frequency of the donor electrons? [*Hint*: See Problem 41 for the parameters of this semiconductor, and Chapter 28 for a discussion of the cyclotron frequency.]

46. (II) The number of electrons with energy in the range E to $E + \Delta E$ contained in a volume V is given by $f(E)\Delta N(E)$, where $f(E)$ is the Fermi–Dirac distribution function and $\Delta N(E)$ is the number of states available to the electrons in the range E to $E + \Delta E$ (Fig. 43–38). Given that

$$\Delta N(E) = \frac{V}{2\pi^2}\left(\frac{2m_e}{\hbar^2}\right)^{3/2}\sqrt{E}\Delta E,$$

calculate the densities of electrons that have energies in the range 10.63 eV to 10.65 eV for gallium ($E_F = 10.40$ eV) at the temperatures 110K and 380K.

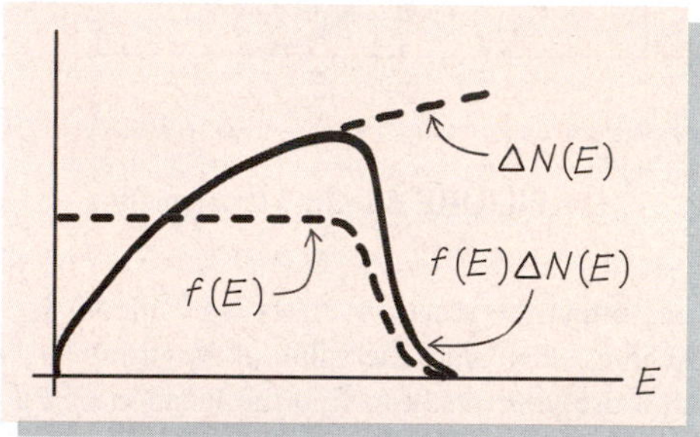

▲ **FIGURE 43–38** Problem 46.

47. (II) Repeat the calculation of Problem 46 for the energy interval between 12.65 eV and 12.67 eV.

48. (II) For a set of identical fermions at high temperature, the average occupation number of any given energy level $f(E) \ll 1$. Show that the Fermi–Dirac distribution approaches the Maxwell–Boltzmann distribution [Eq. (19–28)] in the limit of high temperatures. [*Hint*: The factor $\exp(-E_F/kT)$ is determined from the normalization condition $\Sigma_i n_i = N$ where the index i labels the allowed energy levels.]

49. (II) The laser that reads a compact disc is a semiconductor laser of power in the milliwatt range with light of wavelength around 800 nm. (a) Estimate the rate at which photons are emitted. (b) If the compact disc turns beneath the (narrow) laser beam at several hundred revolutions per minute, how many photons will strike a region 0.1 mm in length of the disk?

50. (II) Consider a group of N atoms that obey Maxwell–Boltzmann statistics. The atoms have only two energy states, a ground state at energy E_0 and an excited state at energy E_1. The energy difference $\Delta E = E_1 - E_0 \gg kT$. Show that the average energy of the atoms at temperature T is $E_0 + \Delta E e^{-\Delta E/kT}$.

51. (II) Monochromatic light shines on a thin film of silicon ($E_g = 1.12$ eV), and the electrical resistance is measured across the film. (a) What happens to the resistance as the wavelength of the light increases, starting with a wavelength in the ultraviolet range? (b) At what critical wavelength does a change in resistance occur? (c) Does the resistance increase or decrease at the critical wavelength?

52. (III) Use the expression for $\Delta N(E)$ given in Problem 46 to express (a) the total number of electrons in a volume V and (b) the average energy of an electron in a volume V at temperature T. (c) Find the limits of your results for parts (a) and (b) for $T \to 0$.

53. (III) A degenerate semiconductor is a semiconductor that is so heavily doped that the doping ions participate in the formation of their own band structure. The Fermi energy may lie within the original conduction band in a degenerate n-type semiconductor, and within the original valence band in a degenerate p-type semiconductor. When a p–n junction is formed from a semiconductor that is degenerate on both the n- and p-sides, a *tunnel diode* is formed. Under these circumstances tunneling is possible when there is an external voltage. Draw an energy diagram (a) for the equilibrium state (no external voltage across the junction); (b) for a reverse bias. How will the tunneling current vary as the reverse bias increases from zero? (c) Draw an energy diagram for a *small* forward bias. How will the tunneling current vary as the small forward bias increases from zero? (d) Show that when the forward bias increases beyond a critical value, the tunneling current begins to *decrease*. This phenomenon is called *negative resistance*.

54. (III) In this problem, we show that the Fermi energy is constant across a boundary. Consider two materials, A and B, in contact so that charge carriers can pass between them; the materials are in thermal equilibrium. At thermal equilibrium, there is no net energy flow and no net charge flow across the boundary. Let the number of states in a unit volume per unit energy interval around the energy E that electrons can occupy in each material be $n_A(E)$ and $n_B(E)$, respectively. The conditions of no energy flow and no charge flow mean that the flow of electrons from A to B at any E value equals the flow of electrons from B to A at that same E value. The flow of electrons from A to B at energy E is proportional to the number of electrons present in A times the number of empty states present in B; that is, to the density of occupied states in A times the density of empty states in B. Now the density of occupied states is

density of occupied states =

$$\text{(density of available states)} \times \text{(probability of occupation)} = n(E)f(E),$$

where $f(E)$ is the probability of occupation given by the Fermi–Dirac distribution, Eq. (43–2).

(a) Show that
density of unoccupied states =

$$\text{(density of available states)} \times \text{(probability of vacancy)} = n(E)[1 - f(E)].$$

(b) For our situation,
flow of electrons from A to B at energy $E \propto$

$$\{n_A(E)f_A(E)\}\,\{n_B(E)[1 - f_B(E)]\};$$

flow of electrons from B to A at energy $E \propto$

$$\{n_B(E)f_B(E)\}\,\{n_A(E)[1 - f_A(E)]\}.$$

Equilibrium is established when we set these equal to each other. Show that this implies

$$f_A(E) = f_B(E).$$

(c) By comparing the result of part (b) with Eq. (43–2), show that the parameters E_F in the two materials are equal.

Chapter 44

This marvelous image was made in neolithic times on the wall of a cave, where it has remined protected from the elements for 30,000 years. The age was established by analyzing the quantities of carbon atoms containing different nuclear types within the pigments. Radiometric dating of this kind depends on knowledge of the lifetimes of the various nuclear species.

Nuclear Physics

The nucleus, which lies at the center of the atom, contains all the positive charge and almost all the atomic mass. Understanding the properties of the nucleus has had enormous consequences for technology and for society in general. We shall describe the nuclear constituents, the static properties of nuclei, and the radioactive decays of certain nuclei. We shall also see how nuclei react with one another. The practical applications of nuclear physics are widespread and include power generation, radiometric dating, investigation of materials through imaging, food conservation, and cancer treatment.

44–1 Static Properties of Nuclei

Nuclear Constituents

The *nucleus* contains more than 99.9 percent of the atom's total mass but has a radius some 10^5 times smaller than the radius of the atom as a whole. The effects of the nucleus were seen in scattering experiments guided by Ernest Rutherford in 1911. Even before Rutherford made his discovery, Joseph J. Thomson noted the presence of electrons in atoms. The number of electrons in a neutral atom is equal to the *atomic number Z*, a quantity directly tied to chemical properties. Another number used for atoms is atomic weight, which is proportional to the atom's mass. If you take the atomic weight as a multiple A of the atomic weight of hydrogen, for which $Z = 1$, you could ask whether A matches Z as Z increases. In fact, one finds that as Z increases, A eventually becomes more than twice as large as Z.

(a)

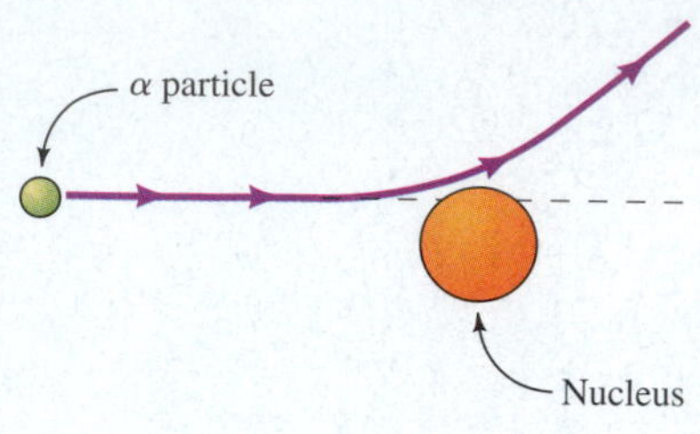

(b)

▲ **FIGURE 44–1** (a) Ernest Rutherford (here, on the right with Hans Geiger) directed the scattering experiments that revealed the atom's nuclear structure. (b) Alpha particle scattering from a nucleus.

In addition, Thomson had measured the charge-to-mass ratio of the electron, and coupled with H. A. Wilson's and Robert Millikan's measurements of the electron charge, the electron's mass was deduced to be much less than the atomic mass. Because the atom is neutral, it must contain a positive charge $+Ze$ in addition to the electron charge $-Ze$. What was *not* known was how the positive and negative charges were distributed within the atom or what carried the positive charge.

The atomic model favored at the turn of the twentieth century—that proposed by Thomson and known as the Thomson model (or the *plum pudding* model)—had point-like electrons embedded within a uniform distribution of positive charge. We described in Section 8–8 how we could learn of the existence of a steel pellet in a volume of cotton candy by firing BBs at the candy. Similar reasoning applies to the exploration of atoms by scattering. Rutherford undertook his study of atoms by scattering *alpha* (α) *particles* from atoms in a thin gold foil (Fig. 44–1). Alpha particles are produced when certain heavy elements decay, as we shall see in Section 44–3. Rutherford had earlier analyzed the charge and mass of α particles and knew that they consisted of doubly ionized helium atoms (charge $+2e$).

CONCEPTUAL EXAMPLE 44–1 What happens when an energetic alpha particle approaches an atom that is constructed according to the Thomson model?

Answer An α particle is some 8000 times more massive than an electron and would scarcely be deflected by a collision with an electron. According to the Thompson model, the positive charge is spread over the full volume of the atom, and when an α particle enters the positive charge distribution, it is only deflected by the imbalance of charge between what is "above" and "below" the α trajectory. In fact, the largest effect occurs when the α particle passes the edge of the charge distribution, and that is an effect that is demonstrably small. (See material leading to Eq. (44–1).)

Rutherford's experiment showed that far too many α particles scattered through large angles—a factor of 10^{10} for angles greater than 90°—than could be explained by the Thomson model (see Conceptual Example 44–1). However, if both the positive charge and the mass of the atom were concentrated in a central structure, the scattering angle could be large. We can estimate the typical scattering angle θ of an incident α particle deflected by its electrical interaction with a positively charged sphere of radius R as follows: $\theta \cong \Delta p/p$, where $p = m_\alpha v$ is the momentum magnitude of the α particle, v is its speed, and Δp is the sideways momentum transfer that the α particle receives as it passes the spherical charge distribution. The typical deflection occurs when the α particle passes close to the edge of the positive charge distribution (Fig. 44–1b). In this case, we can approximate Δp by $F\Delta t$, where $F = (Ze)(2e)/(4\pi\varepsilon_0 R^2)$ is the magnitude of the Coulomb force between the charge distribution and the α particle that reaches the edge of the distribution, and Δt is a characteristic time interval over which the force acts to deflect the α particle. We approximate this time interval by a characteristic size R of the target divided by the speed v of the projectile, $\Delta t \cong R/v$. Thus

$$\theta \cong \frac{F\Delta t}{p} = \frac{(Ze)(2e)}{4\pi\varepsilon_0 R^2}\frac{R/v}{m_\alpha v} = \frac{2Ze^2}{4\pi\varepsilon_0 m_\alpha v^2 R}. \tag{44–1}$$

The factor $1/R$ in this result indicates that the smaller the radius of the nucleus, the larger the typical scattering angle will be. For R equal to the atomic radius (about 0.1 nm), the deflection is negligible. Rutherford's results could be explained with R on the order of 10^{-5} of the atomic radius.

EXAMPLE 44–2 An α particle with kinetic energy of 5.0 MeV moves directly toward a gold nucleus ($Z = 79$) at rest. The gold nucleus is much heavier than the α particle. How close does the α particle come to the gold nucleus before it turns around?

Setting It Up We illustrate this situation in Fig. 44–2.

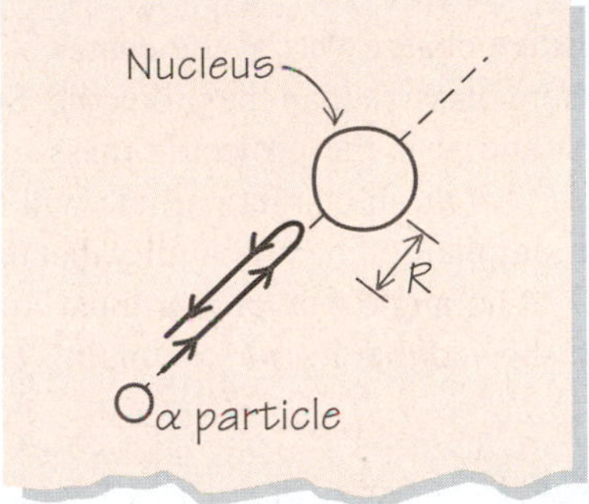

▲ **FIGURE 44–2** A head-on collision in which an α particle is deflected straight back from a nucleus.

Strategy It is the Coulomb repulsion between the gold nucleus and the α particle that causes the α particle to slow down, stop, and then turn around. To analyze this situation more quantitatively, assume that the gold nucleus is so heavy that it remains fixed. This makes it simple to use the conservation of energy to solve the problem. The initial total energy of the system is $E = K + U$. We take the Coulomb potential energy U to be zero when the colliding particles are far apart. We have for the Coulomb potential energy of two objects with charges Z_1e and Z_2e separated by a distance r

$$U(r) = \frac{Z_1Z_2e^2}{4\pi\varepsilon_0}\frac{1}{r}.$$

By evaluating the total energy E when the target and projectile are far apart, we see that E is equal to K, or 5 MeV, and it remains that value throughout. When the α particle reaches the turnaround point, its kinetic energy is zero, and the energy is entirely potential energy. At this point, the separation is R (Fig. 44–2).

Working It Out We apply the conservation of energy by equating the energies of the α particle-gold nucleus system at $r = \infty$ and $r = R$, setting $Z_1 = 2$ (Z_1e is the α particle charge) and $Z_2 = 79$ (Z_2e is the gold nucleus charge):

$$K\Big|_{r=\infty} + U\Big|_{r=\infty} = K\Big|_{r=R} + U\Big|_{r=R};$$

$$K + 0 = 0 + \frac{(2)(79)e^2}{4\pi\varepsilon_0}\frac{1}{R}.$$

We solve for R:

$$R = \frac{(2)(79)e^2}{4\pi\varepsilon_0}\frac{1}{K} = (2)(79)(1.6 \times 10^{-19}\,\mathrm{C})^2$$
$$\times\ (9 \times 10^9\,\mathrm{N\cdot m^2\cdot C^{-2}}) \times \frac{1}{5.0 \times 10^6\,\mathrm{eV}}\frac{1\,\mathrm{eV}}{1.6 \times 10^{-19}\,\mathrm{J}}$$
$$= 4.6 \times 10^{-14}\,\mathrm{m}.$$

This distance is well inside even the innermost Bohr radius, but outside the nucleus itself (see below for more on the nuclear size). You might recall from discussion on fundamental forces in Section 5–5 that nuclear forces are short range and the nuclear force will be negligible at this distance (see Section 44–2). The technique we used here to estimate R is the one Rutherford himself used as a first estimate.

What Do You Think? The de Broglie wavelength of a 5-MeV α particle is about 6×10^{-15} m. Discuss whether quantum mechanical effects might be evident. *Answers to **What Do You Think?** questions are given in the back of the book.*

Scattering Distributions

Rutherford was able to make a more sophisticated calculation of what happens when a collimated beam of α particles is fired at a set of atoms constructed on the basis of the nuclear model. Some of the α particles follow a line directly toward a nucleus, and some follow a line off to the side (Fig. 44–3). For a given line, the force anywhere along that line is the Coulomb force, and the trajectory of the α particle can be calculated. If the beam of α particles is spread uniformly across a region with a collection of target nuclei, then an α particle that arrives at a particular transverse distance b (Fig. 44–3) from a head-on collision with the nucleus will scatter at a particular angle. Rutherford was able to predict a precise *distribution* of the number of α particles deflected at any given angle, assuming always that the α particle never actually penetrates the nucleus.†

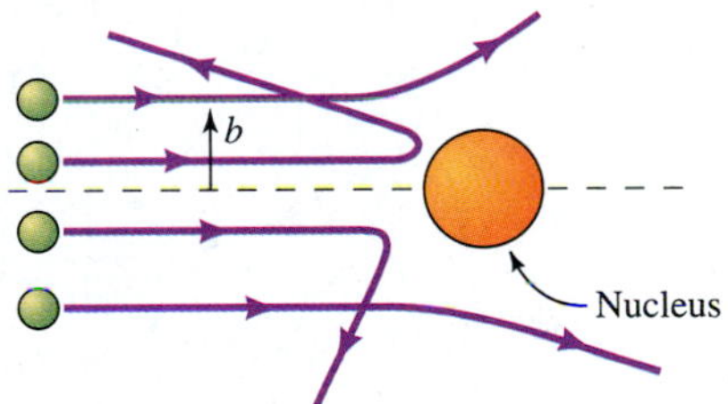

▲ **FIGURE 44–3** A beam of α particles sent into a sample of material will scatter from a nucleus at many different angles.

†Rutherford could make only a classical calculation, even though we now know that quantum physics dominates at these scales. By an extraordinary stroke of luck, the inverse-square law obeyed by the Coulomb force is the *only* force law for which a quantum mechanical calculation gives a distribution identical to that given by a classical calculation.

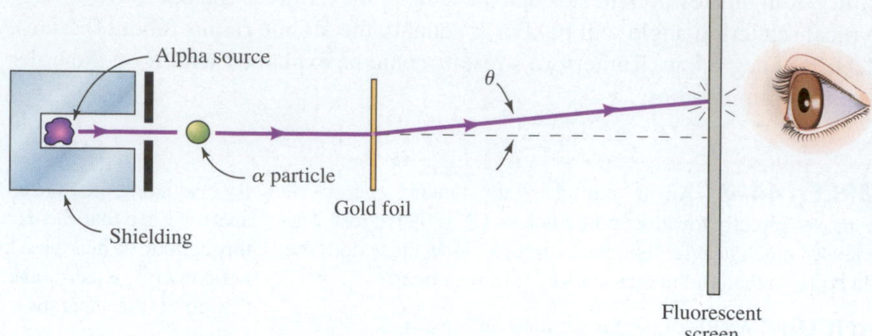

▶ **FIGURE 44–4** Schematic diagram of the experiment in which Geiger and Marsden observed the scattering of α particles from gold atoms in gold foil.

Rutherford assumed that *all* the positive charge was at the center of the atom and was contained in a concentration much more massive than the projectile. Suppose that M is the target mass (of charge Z_2e) at rest and m is the projectile mass (of charge Z_1e) moving with kinetic energy K. Then if $M \gg m$, the target particle will not recoil. We will consider only this case because of its simplicity. The probability that the particle will be scattered into a small solid angle $d\Omega$ at an angle θ is proportional to $d\Omega$ times the *collision cross section*, $\sigma(\theta)$. Rutherford showed that for $M \gg m$, this cross section is given by

$$\sigma(\theta) = \frac{(Z_1 Z_2 e^2)^2}{16(4\pi\varepsilon_0)^2} \frac{1}{K^2} \frac{1}{\sin^4(\theta/2)}. \tag{44–2}$$

The critical features are the inverse-square dependence on K and the strong dependence on θ.

Rutherford's assistants Hans Geiger and Ernest Marsden painstakingly counted the scattering of α particles from a thin gold foil by observing the particles that arrived on a fluorescent screen (Fig. 44–4). These experiments verified Eq. (44–2) and hence also the existence of the nucleus and Rutherford's concept of the atom. Rutherford also noticed a deviation from the prediction of Eq. (44–2). When an α particle is on a path directly toward a nucleus, it backscatters at 180° (see Example 44–2). If the Z value of the target is small enough and the energy of the incident α particle is high enough (greater than 5 MeV in Rutherford's experiment), the α particle has enough energy to overcome the Coulomb force and penetrate into the nucleus. The force that the α particle experiences is thereby modified, and Eq. (44–2) is no longer correct. The penetrating α particle experiences a nuclear force, and Rutherford's observation was the first time the nuclear force was seen in scattering.

Neutrons: The nucleus cannot be composed of protons alone: Almost all the mass of an atom is in its nucleus, and the atomic weight is roughly twice the atomic number. Rutherford speculated in 1920 that a nucleus could consist of Z protons and $N = A - Z$ particles of about the same mass as the proton but with no electric charge—particles later called **neutrons**. Here, A is the *mass number*, or the number of protons plus neutrons in the nucleus. But massive neutral particles are hard to detect. When charged particles pass through matter, Coulomb forces leave a path of ionized atoms that is easy to observe by various detection techniques. When neutral particles such as neutrons pass through matter, they do not ionize the matter very easily.

How, then, were neutrons observed? When atoms of boron or beryllium are bombarded with α particles, observation of the recoiling nuclei shows that neutral particles are produced. These neutral particles are not necessarily neutrons; they could, for example, be photons. In 1932, James Chadwick showed these particles to be neutrons by allowing them to pass into material that contains a good deal of hydrogen, such as paraffin. He observed that a rather energetic proton was occasionally produced (Fig. 44–5). He interpreted this as the result of a collision between the neutral component and a hydrogen nucleus, which consists of just a single proton. By measuring the momentum of the observed proton and by using momentum conservation, Chadwick verified

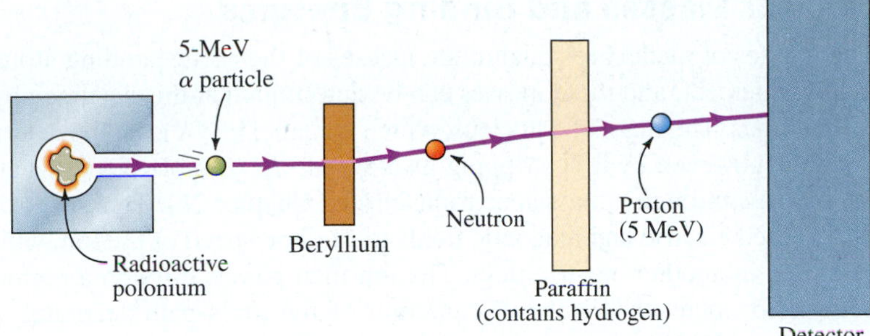

◀ **FIGURE 44–5** Schematic diagram of Chadwick's experiment to discover neutrons. Paraffin contains a high percentage of hydrogen and hence of nuclei consisting of single protons that act as targets for any incoming projectiles. A kinematic analysis shows that only neutrons scattering from hydrogen nuclei could have knocked forward the observed protons.

that the neutral component could not be a photon. Instead, he concluded that the particle, which is now called the *neutron*, must have about the same mass as the proton. Thus, some 20 years after Rutherford's discovery of the nucleus, the nature of the nuclear components was firmly established.

The proton and the neutron collectively are known as **nucleons** because they are both found in the nucleus and they share many of the same properties. An important common property is that they have an intrinsic angular momentum $\hbar/2$—we say that they have "spin $\frac{1}{2}$," like the spin of the electron, and like the electron the nucleons are fermions. We'll have more to say about the spin of the nucleons and of the nucleus later; see also Conceptual Example 44–3. An important difference between protons and neutrons is that unlike protons, free neutrons are not stable but decay with a lifetime $\tau \cong 890$ s. Neutrons do *not*, however, decay within stable nuclei.

The properties of the nucleus depend on the force that holds it together, called the *nuclear*, or *strong*, force. We know that such a force must exist because, without it, the repulsive Coulomb force between protons would cause the nucleus to fly apart.

CONCEPTUAL EXAMPLE 44–3 Before the neutron was discovered, the only possible constitution of a nucleus with a given (A, Z) was A protons and $(A - Z)$ electrons, leaving a net charge of $+Ze$. Early indications were that the nitrogen nucleus $(A = 14, Z = 7)$ had a total angular momentum that was an integral multiple of $\hbar$. Is this consistent with a nucleus made of A protons and $A - Z$ electrons? How does the inclusion of neutrons change things?

Answer In Chapter 41 we discussed angular momenta and how they add, so we know that orbital angular momentum is always an integral multiple of $\hbar$ ("integer angular momentum"), and the only way the total angular momentum can also be an integer is for the addition of all the spins to add to an integer. Now 14 protons (each with spin $\frac{1}{2}$) have a total spin that is integer-valued, while the total spin angular momentum of the $14 - 7 = 7$ electrons has a half-integer value. Thus the total angular momentum in this case would be a half-integer multiple of $\hbar$, and this contradicts the data for nitrogen. Introducing neutrons, the nucleus is made of Z protons and $A - Z$ neutrons; for $Z = 7$ and $A - Z = 7$, we have altogether an even number of spin $\frac{1}{2}$ particles, with a total spin angular momentum that has to be integer-valued in accordance with observation. This is yet another argument against the nucleus being composed of electrons and protons.

Some Terminology

Recall that the number of nucleons in a single nucleus is the mass number A. This quantity very nearly determines the mass of the nucleus; the number of protons is the atomic number Z; and the number of neutrons is $N = A - Z$. A **nuclide** is a nucleus of an element with a given Z and A and is written as ${}^{A}_{Z}\text{X}$, where X is the chemical symbol of the element. For example, the nuclide ${}^{4}_{2}\text{He}$ is a helium nucleus $(Z = 2)$, with 2 protons and 2 neutrons. The nuclide ${}^{3}_{2}\text{He}$ also occurs naturally; it has only a single neutron. This notation is redundant because the chemical symbol alone specifies Z, so we frequently write the simpler form ${}^{4}\text{He}$. The proton is a hydrogen nucleus, ${}^{1}_{1}\text{H}$, but we shall more frequently use the standard symbol p for the hydrogen nucleus. As we have just seen for helium, the nucleus of a given chemical element has a particular Z value, but a range of N values are possible. For example, neon has three stable nuclides (${}^{20}\text{Ne}$, ${}^{21}\text{Ne}$, and ${}^{22}\text{Ne}$) and several unstable ones. Nuclides with the same Z value are known as **isotopes** of the given element (see Chapter 19). The gram-atomic weight of a chemical element—the mass of 1 mol of that element—is actually an average over the different naturally occurring isotopes of that element weighted by the relative abundance of those isotopes.

TABLE 44–1 • Some Atomic Masses

Element	Atomic Mass (u)
$^{1}_{1}$H	1.007825
$^{2}_{1}$H	2.014102
$^{3}_{2}$H	3.016029
$^{4}_{2}$He	4.002603
$^{6}_{3}$Li	6.015121
$^{7}_{3}$Li	7.016003
$^{10}_{5}$B	10.012937
$^{12}_{6}$C	12.000000
$^{14}_{6}$C	14.003242
$^{14}_{7}$N	14.003074
$^{15}_{7}$N	15.000109
$^{16}_{8}$O	15.994915
$^{23}_{11}$Na	22.989768
$^{24}_{12}$Mg	23.985042
$^{27}_{13}$Al	26.981539
$^{35}_{17}$Cl	34.968853
$^{40}_{20}$Ca	39.962591
$^{55}_{25}$Mn	54.938047
$^{56}_{26}$Fe	55.934939
$^{60}_{28}$Ni	59.930788
$^{60}_{27}$Co	59.933820
$^{74}_{32}$Ge	73.921177
$^{90}_{40}$Zr	89.904703
$^{138}_{56}$Ba	137.905232
$^{181}_{73}$Ta	180.947992
$^{208}_{82}$Pb	207.976627
$^{209}_{82}$Pb	208.981065
$^{210}_{84}$Po	209.982848
$^{232}_{90}$Ge	232.038051
$^{235}_{92}$U	235.043924
$^{238}_{92}$U	238.050785

Nuclear Masses and Binding Energies

The masses of nuclei are equal to the masses of the corresponding atoms to within less than 0.5 percent, and these masses can be determined at this accuracy by purely chemical methods. The mass of a nucleus, which we label $m(^{A}_{Z}X)$, can be determined more accurately, however, by first stripping its atom of one or more electrons and then sending the nucleus through a mass spectrometer (see Chapter 28). This spectrometer may first use crossed electric and magnetic fields to find the speed of the ion, which has been accelerated in another electric field. The ion then passes through a region of fixed magnetic field; there, the radius of curvature of the ion's path accurately determines the momentum. If we know the speed and the momentum, we can determine the mass. With such devices, the masses of ions can be determined to an accuracy of better than 1 part in 10^8. The fact that atomic masses and not nuclear masses are obtained is not a handicap, because the electron masses normally cancel in calculations with nuclei.

Note that nuclear abundances, the relative amounts of isotopes that occur in nature, can also be established from mass spectrometry. Ionized atoms from a naturally occurring sample of the element pass through the spectrometer. Because their masses are different, each isotope follows a different path, and the relative number that follow each path is a measure of their abundance.

A useful mass unit for nuclear and atomic physics is the *atomic mass unit* (u), which is defined as one-twelfth the mass of a ^{12}C atom. Such a mass standard is useful because the *relative* masses of atoms (and nuclei) can be measured with high precision, even when we can express the atomic mass unit in terms of the kilogram with much less precision:

$$1\ \text{u} = 1.66057 \times 10^{-27}\ \text{kg}. \qquad (44\text{–}3)$$

We can also express masses in eV/c^2 or MeV/c^2 by using the relativistic relation between mass and energy. We find

$$1\ \text{u} = 931.494\ \text{MeV}/c^2. \qquad (44\text{–}4)$$

In atomic mass units, the nucleon masses are

$$m_p = 1.00728\ \text{u} = 938.272\ \text{MeV}/c^2 = 1.67262 \times 10^{-27}\ \text{kg} \qquad (44\text{–}5a)$$

and

$$m_n = 1.00860\ \text{u} = 939.566\ \text{MeV}/c^2 = 1.67493 \times 10^{-27}\ \text{kg}, \qquad (44\text{–}5b)$$

whereas the mass of the electron is

$$m_e = 5.48 \times 10^{-4}\ \text{u} = 0.510995\ \text{MeV}/c^2 = 0.910939 \times 10^{-30}\ \text{kg} \qquad (44\text{–}5c)$$

and the mass of the hydrogen atom is

$$M(^{1}\text{H}) = 1.00783\ \text{u} = 938.283\ \text{MeV}/c^2 = 1.67356 \times 10^{-27}\ \text{kg}. \qquad (44\text{–}5d)$$

Note in Eqs. (44–5) the standard notation of M for atomic mass and m for nuclear and particle masses. Table 44–1 gives the values of some atomic masses in u. The corresponding nuclear mass can be found by subtracting Zm_e from the atomic mass.

Binding Energy: From Eqs. (44–5), the sum of the masses of 6 protons, 6 neutrons, and 12 electrons is 12.0989 u—about 1 percent greater than the mass of a ^{12}C atom, exactly 12 u. This discrepancy is due entirely to the nuclear mass: The mass of a carbon nucleus is less than the sum of the masses of 6 protons and 6 nucleons by this same amount. *In general, the mass of a stable nucleus is less than $Zm_p + Nm_n$ due to the energies of binding the nucleons into the nucleus and to special relativity.* Separating a nucleon (or collection of nucleons) from a stable nucleus takes energy.

The *binding energy E_b* of a nucleus is the energy released when the nucleus is formed from A independent constituent nucleons. To understand the relation between binding energy and nuclear mass, we must use mass–energy equivalence (see Chapter 39). Because energy must be added to a nucleus to transform it to free nucleons, the energy of the nucleus is less than the sum of the energies of free nucleons; moreover, by the equivalence of mass and energy, the mass of the nucleus is less than the sum of the masses of the

TABLE 44–2 • Some Nuclear Binding Energies

Nucleus	Total Binding Energy (MeV)	Binding Energy per Nucleon (MeV/nucleon)
$^{2}_{1}H$	2.23	1.1
$^{4}_{2}He$	28.3	7.1
$^{12}_{6}C$	92.1	7.7
$^{16}_{8}O$	127.5	8.0
$^{56}_{23}Fe$	492.3	8.8
$^{63}_{29}Cu$	552.1	8.8
$^{238}_{92}U$	1803	7.6

constituents. These considerations hold for *any* bound system, but this effect is a small one for atoms and the other bound systems we have considered previously. For the nucleus, the attractive forces between nucleons are so strong that the binding energy is a significant fraction of the total mass. Thus, by the relativistic relation between mass and energy, *the mass of a nucleus is on the order of 1 percent less than the sum of its nucleon masses*. More precisely, the binding energy is c^2 times the difference between the atomic mass and the sum of the masses of Z hydrogen atoms and N neutrons:

$$E_b \equiv [ZM(^1\text{H}) + Nm_n - M(^A\text{X})]c^2. \tag{44–6}$$

This quantity is positive for stable nuclei. Table 44–2 gives the total binding energies of some nuclei. In the case of ^{12}C, we obtain $E_b = (12.0989\text{ u} - 12.0\text{ u})c^2(931.494\text{ MeV/u}c^2) = 92.12\text{ MeV}$, from Eq. (44–6).

Table 44–2 also gives the *binding energy per nucleon*, E_b/A, as well as the total binding energy. Once A is 12 or larger, the binding energy per nucleon is fairly constant, between 7 and 9 MeV/nucleon. We plot this quantity as a function of the mass number, A, in Fig. 44–6. Note the position on this curve of ^{4}He, which has an exceptionally large E_b/A value and is therefore strongly bound, and the region around ^{56}Fe, for which the curve exhibits a broad maximum. The fact that the binding energy per nucleon increases from zero for small nuclei and decreases as A increases for $A > 60$ tells us a lot about nuclear stability. We see that the total mass–energy per nucleon *decreases* (the total binding energy increases) when two small nuclei combine into an intermediate one (for instance, E_b/A for ^{2}H is 1.1 MeV, whereas E_b/A for ^{4}He is 7.1 MeV). Similarly, the total mass–energy per nucleon decreases when one large nucleus, such as ^{235}U, for which $E_b/A \cong 7.5$ MeV, forms two intermediate nuclei, for which E_b/A may be around 8.5 MeV, or when a large nucleus forms a remnant nucleus plus a ^{4}He nucleus (an α particle). Thus, decays of large nuclei into smaller nuclei are to be expected. Whether these decays actually occur or not, and the rate at which they occur, depend on details of the structure of the decaying nucleus. For example, a decay may occur only when there is tunneling through a potential barrier (see Chapter 40).

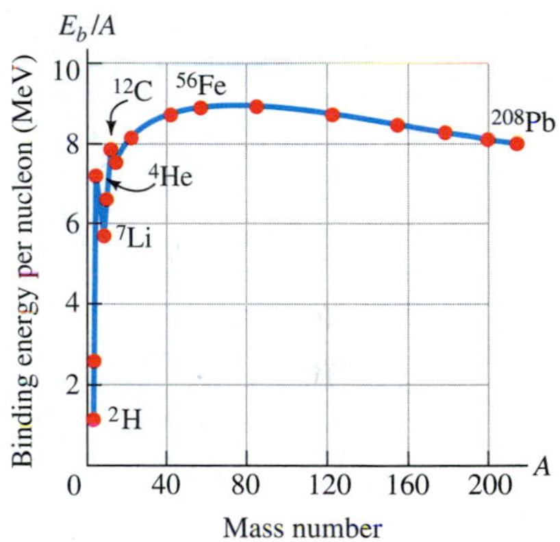

▲ **FIGURE 44–6** The binding energy per nucleon, E_b/A, as a function of the mass number A.

Other Properties

Size and Internal Distribution of Mass and Charge: When the energy of a particle bombarding an atom is increased to the point at which the particle can penetrate the nucleus, we can study the *internal* structure of the nucleus by analyzing the distribution of the scattering. This is a refinement of Rutherford's experiments. If the nucleus is bombarded with electrons, which interact *only* through the Coulomb force with the nucleus, we learn about the distribution of charge within the nucleus. If the nucleus is bombarded with α particles, protons, or neutrons, we learn instead about the distribution of nucleons within the nucleus. Both types of experiments give roughly the same picture of both the radius and the internal makeup of the nucleus; in other words, protons and neutrons are distributed in much the same way within the nucleus. In particular, both the mass and the charge of a nucleus of mass number A are distributed *uniformly* out to a radius R_A given by

$$R_A = r_0 A^{1/3}, \tag{44–7}$$

▲ **FIGURE 44–7** Nucleons are closely packed within nuclei. We have used red to indicate neutrons and blue to indicate protons.

where $r_0 \cong 1.2 \times 10^{-15}$ m. The mass (and charge) density within this volume is very nearly the same for all nuclei. Beyond R_A, the mass and charge densities drop rapidly to zero. The fact that R_A is proportional to $A^{1/3}$ means that the volume, approximately equal to $\frac{4}{3}\pi R_A^3$, is proportional to A: volume $= \frac{4}{3}\pi R_A^3 = \frac{4}{3}\pi(r_0 A^{1/3})^3 = \frac{4}{3}\pi r_0^3 A$. In the classical model, the nucleus resembles a set of closely packed marbles (Fig. 44–7).

We can also mention here that nucleons themselves have an internal structure. One way to learn about that is with further use of the Rutherford type of experiment. The substructure of nucleons is one of the subjects of Chapter 45.

EXAMPLE 44–4 We discussed neutron stars in Section 42–1. Compare the density of a nucleus to that of a neutron star.

Strategy In Example 42–3 we determined the radius of a neutron star of mass three times that of our star to be 1.0×10^4 m. From this we can determine the density of a neutron star by a straightforward calculation of M/V. To determine the density of a nucleus we use the mass of a nucleon ($m_N = 1.67 \times 10^{-27}$ kg) and the nuclear radius ($r = r_0 A^{1/3}$) already discussed in this section.

Working It Out We find the two densities to be

$$\rho(\text{neutron star}) = \frac{3M_{\text{Sun}}}{(4\pi/3)r_{\text{Neutron star}}^3} = \frac{3(2.0 \times 10^{30}\text{ kg})}{(4\pi/3)(1.0 \times 10^4\text{ m})^3} = 14 \times 10^{17}\text{ kg/m}^3;$$

$$\rho(\text{nucleus}) = \frac{M}{V} = \frac{m_N A}{4\pi r_0^3 A/3} = \frac{3 \times 1.67 \times 10^{-27}\text{ kg}}{4\pi(1.2 \times 10^{-15}\text{ m})^3} = 2.3 \times 10^{17}\text{ kg/m}^3.$$

Thus a neutron star is some $14/2.3 = 6$ times more dense than a nucleus.

What Do You Think? Can you give some reasons why the two densities are so close?

Stability: Although nuclides with Z values up to $Z = 92$ (uranium) occur naturally, not all of these nuclides are stable. The nuclide $^{209}_{83}\text{Bi}$ is the heaviest stable nucleus.† Even though uranium is not stable, however, its longest-lived isotope, ^{238}U, has a half-life (see Chapter 40) of some 4 billion years. At low Z values, nuclei tend to have the same number of neutrons and protons ($N \cong Z$), but as Z increases, N tends to exceed Z. Figure 44–8 shows that the stable nuclides lie in a region called the *line of stability*. This curve has the rough form $N \cong Z$ for small Z and $N \cong 1.6\,Z$ for large Z.

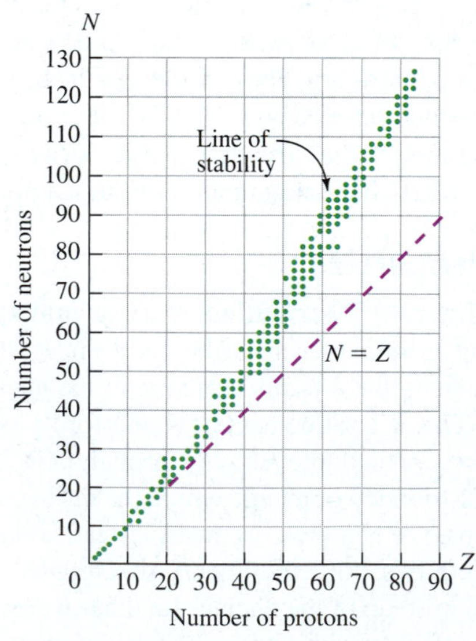

▶ **FIGURE 44–8** The stable nuclides plotted on a graph of neutron number, N, versus proton number, Z. Note that for heavier nuclides, N is larger relative to Z. The stable nuclides group along a curve called the line of stability.

†Actually, it is not perfectly stable, but its lifetime is so large—on the order of 10^{19} years—that its possible decay has no practical consequence.

CONCEPTUAL EXAMPLE 44–5 Why do heavier nuclei have more neutrons than protons? [*Hint*: Think about the different kinds of interactions that exist between these constituents.]

Answer While the nuclear force is short range, no more than a few $\times\ 10^{-15}$ m, the Coulomb force is long range, with r-dependence $1/r^2$. Nucleons, both neutrons and protons, interact via the nuclear force mainly with their nearest neighbors in a way that holds the nucleus together, while protons repel each other via the Coulomb force. If we consider nuclei with more nucleons, the attractive force on each nucleon stays the same, because the force is short range and the number of nearest neighbors no longer changes. But because it is a long-range force, the Coulomb repulsion on any additional protons increases, with all the protons present acting on that additional proton. The repulsive Coulomb interaction becomes more and more important, and it is less energetically favorable to have as many protons as neutrons. The nuclide ^{40}Ca is the heaviest stable nuclide for which $N = Z$. (See the discussion on p. 1218)

There are approximately 250 stable isotopes on the line of stability. When the number of stable isotopes is expressed in terms of whether Z and N are even or odd, a striking pattern emerges. About 60 percent of the stable isotopes have both Z and N even; about 20 percent have Z even, N odd; and 20 percent have Z odd, N even. Only 5 stable isotopes have both Z and N odd. The strong preference for nuclei to have both Z and N even can be understood by a pairing force of the kind that leads to Cooper pairs in superconductivity (see Chapter 42). In addition to the stable nuclides shown in Fig. 44–8, there are several thousand unstable nuclides with varying N and Z that surround the line of stability. We'll describe the unstable nuclides further in Section 44–4.

EXAMPLE 44–6 (a) From data on atomic masses, determine the total binding energy and binding energy per nucleon of ^{56}Fe. (b) How much energy does it take to remove a proton from ^{56}Fe?

Strategy (a) Equation (44–6) allows us to find the total binding energy and subsequently the binding energy per nucleon for any nucleus. Table 44–1 and Eqs. (44–5) contain the masses necessary for us to apply Eq. (44–6) to the case of iron.

(b) The same principle that led us to the binding energy leads to the answer to this question. We find the difference between the mass of a parent nucleus (here ^{56}Fe) and the sum of the masses of the atom with a single proton as its nucleus (here ^{1}H) and the atom that is left behind when the single proton is removed from the parent (here ^{55}Mn). The energy that corresponds to this mass difference is called the *separation energy*, and here we are finding the proton separation energy. Generally, the separation energy for a parent nucleus $^{A}_{Z}X$ is

$$S_p(^{A}_{Z}X) = [M(^{1}\text{H}) + M(^{A-1}_{Z-1}X') - M(^{A}_{Z}X)]c^2. \qquad (44\text{–}8)$$

Working It Out

(a) When Eq. (44–6) is applied to ^{56}Fe, we find

$$\begin{aligned}E_b(^{56}\text{Fe}) &= [Z_{^{56}\text{Fe}}M(^{1}\text{H}) + N_{^{56}\text{Fe}}m_n - M(^{56}\text{Fe})]c^2\\ &= [26M(^{1}\text{H}) + 30m_n - M(^{56}\text{Fe})]c^2\\ &= [26(1.007825\ \text{u}) + 30(1.008665\ \text{u}) - 55.934939\ \text{u}]c^2\\ &= (0.528461\ \text{u}c^2)(931.494\ \text{MeV/u}c^2) = 492.258\ \text{MeV}.\end{aligned}$$

The binding energy per nucleon is (492.3 MeV)/(56 nucleons), about (8.8 MeV)/nucleon.

(b) From the discussion above, Eq. (44–8), the proton separation energy for ^{56}Fe is

$$\begin{aligned}S_p(^{56}\text{Fe}) &= [M(^{1}\text{H}) + M(^{55}\text{Mn}) - M(^{56}\text{Fe})]c^2\\ &= [1.007825\ \text{u} + 54.938046\ \text{u} - 55.934939\ \text{u}]c^2\\ &= (0.0109\ \text{u}c^2)(931.494\ \text{MeV/u}c^2) = 10.18\ \text{MeV}.\end{aligned}$$

The proton separation energy for ^{56}Fe is particularly large—10.2 MeV—because ^{56}Fe is a particularly strongly bound nucleus.

What Do You Think? Write down the equation for the neutron separation energy of ^{56}Fe. Do you think the neutron separation energy will be somewhat larger, somewhat smaller, or equal to the proton separation energy?

Spins and Magnetic Dipole Moments: Nuclei have angular momentum, just as atoms do. This angular momentum is usually called the *spin* of the nucleus because, without knowing the internal structure of the nucleus, we can regard it as intrinsic. Nucleons have a spin of $\hbar/2$. As is true of electrons in atoms, the nucleons that compose the nucleus can have an orbital angular momentum. The total angular momentum of a nucleus results from adding the orbital angular momentum and the spins according to the rules of quantum mechanics, supplemented by the exclusion principle. (The exclusion principle states, for example, that two protons cannot move "in the same orbit" and both have spin-up.)

Associated with the angular momentum of a nucleus is a magnetic dipole moment, and this magnetic moment provides the means of detecting the angular momentum of the nucleus. We described *nuclear magnetic resonance* in Section 31–6, in which the nuclear magnetic dipole moment is detected because of a resonance phenomenon.

Another means of detecting the nuclear magnetic dipole moment is to note that a magnetic dipole produces a magnetic field and, because atomic electrons have both spin and orbital angular momenta, the energy of these electrons depends on this nuclear magnetic field. The small resulting shift in electron energy levels is referred to as *hyperfine splitting* of the atomic spectra, a splitting some thousand times smaller than the fine-structure splitting discussed in Chapter 41.

The orders of magnitude of nuclear magnetic moments are those of the magnetic dipole moments of the proton and neutron themselves. We would estimate these magnetic dipole moments in turn to be on the order of $e\hbar/2m_pc$ (a quantity known as the *nuclear magneton*) for the proton and zero for the (neutral) neutron, similar to the magnetic dipole moment of $e\hbar/2m_ec$ for the electron. (The composite nature of protons and neutrons—see Chapter 45—modifies this expectation somewhat.) Note that the nuclear magnetic moment is some 2000 times smaller than the electron magnetic moment because of the masses involved. That is why the hyperfine splitting of atomic spectra is some 1000 times smaller than the fine-structure splitting.

44–2 Nuclear Forces and Nuclear Models

One of the major tasks of nuclear physics is to discover the nature of the nuclear force. It must be strong enough to overcome the electrical repulsion between protons. An energy of 5 to 10 MeV is required to free a nucleon from a nucleus. This value can be compared with the 13.6 eV of energy needed to ionize hydrogen or the 3 to 5 eV needed to free an electron from a metal (the work function). By this measure, the nuclear forces are about a million times stronger than the electric forces that bind atoms.

The fact that the nucleon separation energy is on the order of a few MeV even for nuclei as heavy as lead ($A \cong 208$) implies that the number of "bonds" that attach a nucleon to a nucleus does not grow as the nucleus grows; we say that the nuclear force becomes *saturated* when a nucleon is surrounded by other nucleons. *The nuclear force has a short range comparable to the spacing between nucleons in a nucleus.* By the range, we mean the distance beyond which the nuclear force rapidly decreases. From our formula for the nuclear radius, Eq. (44–7), we can deduce that the spacing between nucleons is on the order of 1 **fermi** (fm), or 10^{-15} m (also known as a *femtometer*). The conclusion that the nuclear force has a short range is supported by scattering experiments, which also allow us to measure the range quantitatively.

We can summarize the quantitative behavior of nuclear interactions as follows. Except at distances much less than 1 fm, a reasonable representation of the nucleon–nucleon potential energy is

$$U(r) = -g^2\frac{e^{-r/R}}{r}, \tag{44–9}$$

where r is the distance between the two nucleons, the parameter R is called the range, approximately 2×10^{-15} m, and g^2 is a measure of the strength of the potential energy. Figure 44–9 illustrates this potential energy as well as a potential energy of the same strength but with a Coulomb form, $U_c = -g^2/r$, for comparison; these two potential energies have the same behavior for small r. The potential energy of Eq. (44–9) is known as the *Yukawa potential*, named for its originator, Hideki Yukawa. It can be justified in the context of quantum mechanics (as we shall see in Chapter 45).

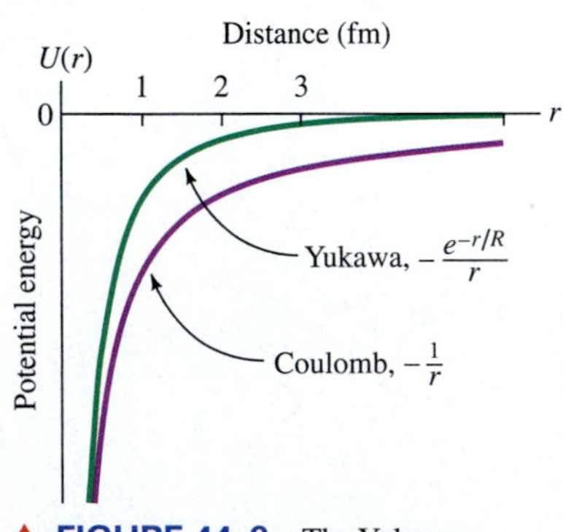

▲ **FIGURE 44–9** The Yukawa potential $U(r) \propto -e^{-r/R}/r$, and the Coulomb form $-1/r$ for comparison. We have chosen $R = 2.0$ fm.

In our discussion, we have referred to the nucleon–nucleon force, not the proton–proton force, the proton–neutron force, or the neutron–neutron force. Numerous scattering experiments, as well as the study of nuclear binding energies for nuclei with the same A value but different Z values and N values, show that, basically, the only differences in the interactions of neutrons and protons are due to electromagnetism.

A full description of nuclear potentials takes us far beyond the Yukawa potential. First, there is a repulsion (a *core repulsion*) between the nucleons at separation distances less than about 1 fm. Second, nuclear forces have a substantial dependence on the orientation of the spin of the nucleons. Third, and perhaps most important, nuclear forces normally involve more than just two interacting particles. In atomic physics this does not present much of a problem, because the forces are relatively weak, and mathematical tools exist by which we can handle such forces when many particles are involved. In nuclear physics, the forces are *strong*, as measured by the parameter g^2 in Eq. (44–9). Analytic mathematical techniques for the systematic treatment of such forces do not exist. The problems posed by nuclear forces are too difficult to be solved from a fundamental theory. Despite this difficulty, scientists know much about nuclei. As in the case of bulk materials, models and approximations that explain different, limited features of nuclei have been developed for describing the nucleus. We shall discuss two of these models—the shell model and the liquid-drop model.

The Shell Model

In Chapter 41, we learned that the energy levels of electrons in an atom form a shell structure under the influence of the Coulomb attraction of the nucleus. Despite the fact that nuclear forces are quite different from Coulomb forces, individual nucleons in a nucleus also appear to have well-defined energy levels. The description of the nucleus based on this idea is known as the **shell model** (or as an *independent-particle model*).

The *experimental* values of Z and N for which the binding energy is particularly large are 2, 8, 20, 28, 50, 82, 126, and these values are called *magic numbers*. A calculation assuming a three-dimensional spring force gives values of 2, 8, 20, 40, 70, ... These predicted numbers do not represent a perfect match to the experimental magic numbers and indicate that the order of filling levels is different in reality than a simple harmonic oscillator potential would suggest. In 1949, Hans Jensen and Maria Goeppert-Mayer independently pointed out that the energy should contain terms not included earlier that involve both the spin of a single nucleon and its orbital angular momentum. This *spin–orbit coupling* is the same type of coupling that causes splitting of atomic levels (see Section 41–3). By including this type of coupling, splitting of levels occurs and the ordering of levels is changed. The observed magic numbers are thereby more correctly predicted, as is the complicated pattern of level splittings seen in nuclei.

The Liquid-Drop Model

The short range of nuclear forces and the small amount of space between the constituent nucleons suggested to Niels Bohr that a nucleus should behave in some ways like a continuous fluid, such as water. The model based on this idea, the **liquid-drop model**, contains no reference to the behavior of individual nucleons and in this sense is complementary to the shell model. We note two similarities between nuclear data and the behavior of incompressible fluids. First, the density of all nuclear matter is roughly constant, just as for an incompressible fluid. Second, the binding energy per nucleon, E_b/A, is roughly constant from nucleus to nucleus.

Following this analogy, we can write an *empirical* formula for the mass of a nucleus, $M(^{A}X)$, based on the energy of a liquid drop, as follows:

1. A zeroth-order term is the sum of the rest masses of the nucleons:

$$ZM_H + Nm_n = ZM_H + (A - Z)m_n. \tag{44–10}$$

2. A term proportional to the volume of the nucleus (or $\propto A$, as we have shown) takes into account the nearly constant binding energy per nucleon:

$$-a_V A, \tag{44–11}$$

   where a_V is positive because a binding energy reduces the mass of the nucleus.

3. A term proportional to $A^{2/3}$ or, equivalently, to the surface area of the nucleus,

$$+a_S A^{2/3}, \tag{44–12}$$

   accounts for the fact that a nucleon near the surface is not bound as strongly as one in the interior, so a_S takes into account a reduced binding energy and is positive. This term is the analogue of a surface-tension term in a liquid drop.

The remaining three terms take into account some simple observations specific to the nucleus:

4. A term that accounts for the Coulomb repulsion between all the protons in the nucleus. This term tends to increase the mass of the nucleus. If we suppose that the protons are distributed uniformly throughout a sphere of radius $R_A = r_0 A^{1/3}$ [Eq. (44–7)], then the energy required to assemble that charge is[†]

$$\frac{3}{5}\frac{Z^2 e^2}{4\pi\varepsilon_0 r_0 A^{1/3}} = 0.72\frac{Z^2}{A^{1/3}}\text{MeV} \tag{44–13}$$

(see Problem 24–91). The contribution of the energy to the mass is this quantity divided by c^2.

5. A term that has a minimum for $N = Z$ accounts for the tendency for the number of neutrons to equal the number of protons. If terms 1 through 4 were the only terms present, then the energy could be lowered by taking Z to zero; that is, nuclei would consist exclusively of neutrons. In fact, there are no nuclei with many more neutrons than protons, and nuclei are particularly stable when $N \cong Z$. This term is written as

$$+a_A\frac{(A - 2Z)^2}{A}, \tag{44–14}$$

where a_A is positive.

6. Several terms describe the tendency for nuclei with even numbers of protons and/or neutrons to be more deeply bound than nuclei with odd numbers of protons and/or neutrons. This tendency occurs because the spins of two nucleons in each "shell" are antiparallel. Such terms have the empirical form

$$\text{for } Z, N \text{ even:} \quad \Delta = -\frac{a_p}{\sqrt{A}}. \tag{44–15a}$$

$$\text{For } Z \text{ even, } N \text{ odd; or } Z \text{ odd, } N \text{ even:} \quad \Delta = 0. \tag{44–15b}$$

$$\text{For } Z, N \text{ odd:} \quad \Delta = +\frac{a_p}{\sqrt{A}}. \tag{44–15c}$$

The sum of all these terms is the **semiempirical mass formula**:

$$M_A = ZM_H + (A - Z)m_n - a_V A + a_S A^{2/3} + \frac{a_C Z^2}{A^{1/3}} + a_A\frac{(A - 2Z)^2}{A} + \Delta. \tag{44–16}$$

The parameters are determined empirically by fitting this relation with many different nuclear masses. One set of parameters is given in Table 44–3. For these values, the curve in Fig. 44–6 for $A = 20$ and above is in good agreement with Eq. (44–16). For very light nuclei, the notion of a nucleus that looks like a droplet with a volume and a surface loses meaning, and the mass formula does not work very well, but otherwise the fit is excellent.

TABLE 44–3 • Parameters of the Semiempirical Mass Formula

Parameter	Value (u)
a_V	1.7×10^{-2}
a_S	1.8×10^{-2}
a_C	7.5×10^{-4}
a_A	2.5×10^{-2}
a_p	1.3×10^{-2}

EXAMPLE 44–7 (a) Use the semiempirical mass formula to calculate the atomic masses of ^{208}Pb and ^{209}Pb. (b) Calculate the neutron separation energy for ^{209}Pb.

Strategy (a) This is simply a matter of evaluating Eq. (44–16) with the parameters of Table 44–3. (b) We discussed the neutron separation energy in Example 44–6, and its value for ^{209}Pb is

$$S_n(^{209}\text{Pb}) = [m_n + M(^{208}\text{Pb}) - M(^{209}\text{Pb})]c^2.$$

The masses here were evaluated in part (a).

Working It Out

(a) Equation (44–16) gives

$$\begin{aligned} M(^{208}\text{Pb}) &= (82)(1.007825\ \text{u}) \\ &\quad + (208 - 82)(1.008665\ \text{u}) - (1.7 \times 10^{-2}\ \text{u})(208) \\ &\quad + (1.8 \times 10^{-2}\ \text{u})(208)^{2/3} + (7.5 \times 10^{-4}\ \text{u})\left[\frac{(82)^2}{(208)^{1/3}}\right] \\ &\quad + (2.5 \times 10^{-2}\ \text{u})\left\{\frac{[208 - 2(82)]^2}{208}\right\} - \left(\frac{1.3 \times 10^{-2}\ \text{u}}{\sqrt{208}}\right) \\ &= 207.921\ \text{u}; \end{aligned}$$

[†]In some instances, the calculated coefficient 0.72 of the factor $Z^2/A^{1/3}$ in Eq. (44–13) is replaced by the empirical parameter $a_c \cong 0.60$. This is a very small difference in the contribution of the energy to the masses of nuclei. In Table 44–3, we use the calculated form.

$$M(^{209}\text{Pb}) = (82)(1.007825\text{ u}) + (209 - 82)(1.008665\text{ u}) - (1.7 \times 10^{-2}\text{ u})(209) + (1.8 \times 10^{-2}\text{ u})(209)^{2/3} + (7.5 \times 10^{-4}\text{ u})\left[\frac{(82)^2}{(209)^{1/3}}\right] + (2.5 \times 10^{-2}\text{ u})\left\{\frac{[209 - 2(82)]^2}{209}\right\} + 0 = 208.915\text{ u}.$$

(b) The neutron separation energy of ^{209}Pb is

$$S_n(^{209}\text{Pb}) = [m_n + M(^{208}\text{Pb}) - M(^{209}\text{Pb})]c^2 = [1.008665\text{ u} + 207.912\text{ u} - 208.915\text{ u}]c^2 = (0.0057\text{ u}c^2)(931.494\text{ MeV/u}c^2) = 5.3\text{ MeV}.$$

The experimental values of the masses are $M(^{208}\text{Pb}) = 207.97663$ u and $M(^{209}\text{Pb}) = 208.98107$ u, and the experimental value of the neutron separation energy is 3.9 MeV. The 5.3 MeV separation energy that we calculated from the semiempirical mass equation does not take into account the shell-model magic numbers, for which nuclei are particularly strongly bound. On average, the nucleon binding energy is around 8 MeV.

What Do You Think? The nuclide ^{208}Pb is an example of a "doubly magic" nucleus ($Z = 82$, $N = 126$), so it should be *very* strongly bound. Why might this mean that the neutron separation energy for ^{209}Pb is relatively low?

A classical droplet, even one made of an incompressible fluid, can deform in shape, and the same is true for nuclei in the liquid-drop model. This changes the energy, because the surface area changes. Surface forces tend to return the droplet to a spherical shape, but the Coulomb energy term favors a larger deformation. When we compare these two effects for nuclei, we find that the liquid drop becomes unstable for more massive nuclei, and these nuclei may break up into two smaller droplets. The liquid-drop model thus provides an explanation of the fission process (see Section 44–5).

44–3 Energetics of Nuclear Reactions

By nuclear reactions, we generally mean the processes that occur when nuclei interact through collisions. All the possibilities for different types of collisions outlined in Chapter 8 occur. Two nuclei may have an elastic collision or various types of inelastic collisions. A nucleus may absorb energy and then decay by one of several modes, which we shall discuss in Section 44–4. Mass may be transferred from one nucleus to another by the exchange of nucleons. Colliding nuclei may coalesce—a process called **nuclear fusion**. The decay of a single unstable nucleus may be regarded as a nuclear reaction. Examples are α decay, in which a nucleus decays to a helium nucleus and another nucleus; and **nuclear fission**, in which the decay products include two nuclei of more or less equal size. In all these interactions, the conservation laws we have developed throughout this book apply, including the conservations of energy, momentum, angular momentum, and charge.

At the simplest level, collisions between nuclei allow us to measure kinematic quantities such as masses or angular momenta. A terminology special to nuclear reactions has been developed. Consider a nuclear collision in which both the initial and final states consist of two nuclei:

$$A + B \rightarrow C + D.$$

Energy conservation applies, but we must use relativistic relations. For any one nucleus (or particle) of mass M, this energy takes the form

$$E = K + Mc^2,$$

where K is the kinetic energy of the nucleus. The value of K depends on the frame of reference, but the conservation of energy, $E_A + E_B = E_C + E_D$, applies in any reference frame. If we suppose that nucleus B is the target nucleus, at rest in the laboratory frame of reference, then the conservation of energy takes the form

$$(K_A + M_Ac^2) + M_Bc^2 = (K_C + M_Cc^2) + (K_D + M_Dc^2) \qquad (44\text{–}17)$$

in that reference frame. Depending on the masses of the nuclei involved, the reaction can be *exothermic* (having more kinetic energy in the final state than in the initial state) or *endothermic* (having less kinetic energy in the final state). A kinematic parameter that describes such properties is the *Q value* of the reaction:

$$Q \equiv K_C + K_D - K_A - K_B. \qquad (44\text{–}18)$$

By including the possibility that nucleus B has kinetic energy, we have written a general relation. A positive Q value corresponds to an exothermic reaction, and a negative Q value corresponds to an endothermic reaction. By moving all the kinetic energy terms in Eq. (44–17) to one side of the equation and all the mass terms to the other, we can express the Q value in terms of atomic masses:

$$Q = (M_A + M_B - M_C - M_D)c^2. \tag{44–19}$$

By measuring kinetic energies in a reaction and comparing Eqs. (44–18) and (44–19), we can use the Q value for mass measurements that cannot otherwise be made. Note that although we have used laboratory frame quantities in Eq. (44–18), our original definition of Q, Eq. (44–19), shows that Q is actually an invariant.

The variety of measurements of different nuclear reactions—including their dependence on different variables of the collision process, such as the projectile energy and angle—gives us much of our knowledge of nuclear structure and nuclear forces. The most useful concept for collisions is the *collision cross section*, which, as we described in Chapter 19, measures the effective area taken up by the colliding nuclei. The cross section is a measure of the probability of interaction; the larger the cross section, the more probable the collision and hence the reaction. We can thus define a cross section *for each possible reaction* that measures the relative probability of each of these processes.

One of the important features of the cross section is the presence of *resonances*, which we define here as bumps in the curve of cross section versus collision energy of the same form as the resonance curve of Fig. 13–23. A bump in the cross section centered around E_0 of width ΔE implies that the nucleus is strongly affected by the incident projectile when the total energy is in the range $E_0 - \Delta E/2$ to $E_0 + \Delta E/2$. Here, the nucleus–projectile system behaves like a nucleus with an excited state around E_0, but spread over an energy range ΔE. In other words, *the presence of a resonance peak in a cross section signals the presence of an excited state at that energy*. Indeed, these excited states are often referred to as resonances.

The role of resonances in this process is most clearly understood by thinking of the reaction as $A + B \rightarrow$ resonance $\rightarrow C + D$, where A and B are nuclei that interact to form the resonance, and C and D are the nuclei, among several other possibilities, that result from the decay of the resonance. The resonance, which is also referred to as a *compound nucleus*, may live for a relatively long time—time enough to allow the compound nucleus to "forget" how it was formed. There are a number of predictive consequences based on the idea that how the decay of the resonance occurs is independent of how the resonance was formed.

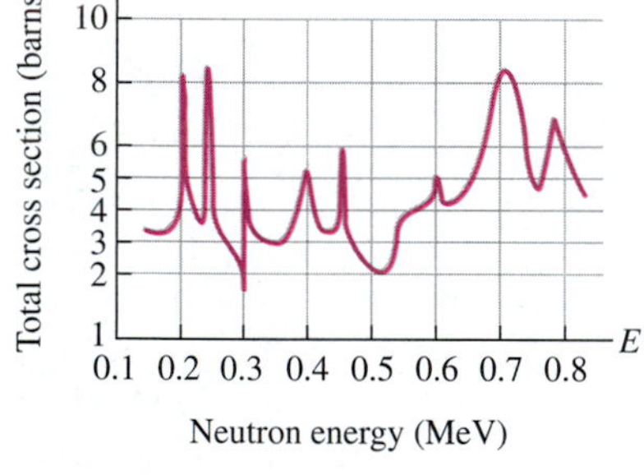

▲ **FIGURE 44–10** The total cross section for the absorption of neutrons by ^{23}Na. Energy is measured in the laboratory frame of reference. The peaks, also known as resonances, correspond to excited states of ^{24}Na. (After S. F. Mughabghab et al., *Neutron Cross Sections*, Academic Press, 1984.)

The lifetime τ of this excited state is related to the width of the bump in the curve of the cross section by the Heisenberg uncertainty principle,

$$\tau \Delta E \cong \hbar. \tag{44–20}$$

Thus *the width in energy of the resonance is $\hbar$ divided by the lifetime of the resonance*: the shorter-lived the resonance, the broader the resonance peak. In nuclear physics, resonances may have widths ranging from about 0.1 eV, corresponding to a lifetime of 10^{-14} s, to as much as 1 MeV, corresponding to a lifetime of 10^{-21} s. In Fig. 44–10, we see resonance shapes in the probability (total cross section) for neutron absorption by ^{23}Na. Each peak corresponds to an excited state of the nuclide ^{24}Na, which then decays. In some interactions of nucleons and other particles that participate in the nuclear force, resonances occur with widths as large as 100 MeV. This corresponds to a lifetime of 10^{-23} s, about the time it takes for light to cross the nucleus!

44–4 Radioactivity

Henri Becquerel discovered radioactivity in 1896, even before the existence of the nucleus had been established. **Radioactivity**—the name was coined by Marie Curie (Fig. 44–11)—is the phenomenon of nuclear decay. Nuclei can decay in a variety of *modes* or *channels*—for example, by emitting photons or α particles. Different physical laws may govern different decay modes, so each mode may have a different probability of occurring. The lifetime τ of a collection of radioactive nuclei is a constant with dimensions of time that determines the rate at which a set of such nuclei decay: If N_0 is the number of nuclei present at $t = 0$, then $N(t)$, the number of nuclei that remain after a time t, is given by Eq. (40–17), $N(t) = N_0 e^{-t/\tau}$. An alternate way to write this equation is to define the **decay constant** λ as $\lambda \equiv 1/\tau$, so that

$$N(t) = N_0 e^{-\lambda t}. \tag{44–21}$$

The rate at which nuclei decay, the *decay rate*, is $-dN/dt$. From Eq. (44–21),

$$\frac{dN}{dt} = -\lambda N_0 e^{-\lambda t} = -\lambda N. \tag{44–22}$$

The decay rate λN, also called the *activity*, is high when the number of unstable nuclei is large and when the decay constant is large (the lifetime is small). We also employ the *half-life*, $t_{1/2}$, the time for half of a given sample of unstable nuclei to decay. The half-life is related to the lifetime by $t_{1/2} = (\ln 2)\tau = 0.693\,\tau$ (see Problem 40–64). The SI unit of activity is the *becquerel* (Bq): 1 Bq = 1 decay/s.

If a nucleus can decay in two or more ways, then the total decay rate is a sum of decay rates for the various decay modes. Think of several holes in a leaky can: The total rate of flow from the can is the sum of the rates from each of the holes. The decay rate from the mode labeled by the subscript j (for example, j may label the α-decay mode) is proportional to the decay constant λ_j for that mode, so the total decay rate is proportional to the sum of all the decay constants. In other words, there is a total decay constant λ given by

$$\lambda = \sum_j \lambda_j. \tag{44–23}$$

If the decay in one mode, say, mode 1, is much more rapid than that in all the other modes, then λ_1 is much larger than the other decay constants, and the λ_1 term dominates the sum in Eq. (44–23). In that case, $\lambda \cong \lambda_1$. It is as though one hole in our leaky can is much larger than the others. We say that the nucleus *decays predominantly through the channel* labeled by λ_1.

With the exception of fission reactions (to be discussed in Section 44–5), a typical radioactive decay involves a nucleus (the *parent nucleus*) that emits a particle and is thereby converted into a *daughter* nucleus. Three different types of decays can occur, labeled α, β, or γ according to the particles emitted.

▲ **FIGURE 44–11** Marie Curie in her laboratory at the Sorbonne, Paris, circa 1908. Curie and her husband, Pierre Curie, performed important early work on radioactivity for which she earned two Nobel prizes.

EXAMPLE 44–8 A sample of uranium ore emits α radiation characteristic of ^{235}U at a rate of 9.3×10^5 decays/s. What mass of chemical uranium is present in the ore? The half-life of ^{235}U is 7.04×10^8 yr, and the abundance of naturally occurring ^{235}U in a sample of uranium is 0.72 percent.

Strategy We are given the decay activity and half-life for ^{235}U. From the half-life, we can find the decay constant λ; with λ and the activity, we can find the number of ^{235}U nuclei. Then we can use the abundance of ^{235}U to find the total quantity of uranium.

Working It Out We have $\lambda = 1/\tau = 0.693/t_{1/2}$. From the relation for the activity, Eq. (44–22), we find the number of ^{235}U atoms to be

$$N = -\frac{1}{\lambda}\frac{dN}{dt} = -\frac{t_{1/2}}{0.693}\frac{dN}{dt}$$
$$= -\frac{(7.04 \times 10^8\ \text{yr})(3.16 \times 10^7\ \text{s/yr})}{0.693}(-9.3 \times 10^5\ \text{decays/s})$$
$$= 3.0 \times 10^{22}.$$

Given this number of ^{235}U atoms, we can use Avogadro's number to find the mass m of ^{235}U:

$$m = (3.0 \times 10^{22}\ \text{atoms})\left(\frac{1\ \text{mol}}{6.02 \times 10^{23}\ \text{atoms}}\right)(235\ \text{g/mol}) \cong 12\ \text{g}.$$

Because ^{235}U makes up 0.72 percent of naturally occurring uranium, the total amount of uranium is

$$(1/0.0072)(12\ \text{g}) = 140(12\ \text{g}) = 1.7\ \text{kg}.$$

What Do You Think? The density of uranium is about 19 times that of water. Estimate the size of the piece of uranium referred to in this example. Could anyone lift a basketball made of solid uranium unaided?

Alpha Decay

Decay in the **α-decay mode** is written as

$$^{A}_{Z}X \rightarrow ^{A-4}_{Z-2}X' + ^{4}_{2}\text{He}, \tag{44–24}$$

where X and X' are the chemical symbols of the parent and daughter nuclei, respectively. Here, we have recognized that the emitted α particle is actually a ^{4}He nucleus. An example of this type of decay is $^{238}_{92}\text{U} \rightarrow ^{234}_{90}\text{Th} + ^{4}_{2}\text{He}$.

Alpha decay involves the quantum mechanical tunneling of an α particle through a potential barrier. For α decay to occur, the mass of the parent nucleus must be greater than the sum of the masses of an α particle and the daughter nucleus. The systematics of nuclear masses implies that α decay is increasingly more likely to occur as the size of a nucleus increases. Many nuclei above $Z = 83$ are unstable in this mode.

Nuclear forces are understood well enough to allow us to predict the probabilities of α decay. Because the rate at which tunneling occurs is extremely sensitive to the width and height of the potential barrier, the range of lifetimes of nuclei in the α-decay mode is huge, from 3×10^{-7} s to 1.4×10^{17} s!

EXAMPLE 44–9 ^{241}Am (an isotope of americium) decays in the α-decay mode: $^{241}\text{Am} \rightarrow \alpha + ^{237}\text{Np}$. Calculate the kinetic energy of the α particle if the ^{241}Am nucleus decays at rest and the atomic masses are $M(^{241}\text{Am}) = 241.05682$ u, $M(^4\text{He}) = 4.002603$ u, and $M(^{237}\text{Np}) = 237.04817$ u.

Strategy Both conservation of energy and conservation of momentum must be applied here. From the conservation of energy, we have $M(^{241}\text{Am}) = M(^4\text{He}) + M(^{237}\text{Np}) + K_f/c^2$, and this can be solved for K_f, the kinetic energy of the two-body final state. As long as this is much less than the rest mass of either final nucleus, we can treat the motion nonrelativistically, and use $K_f = K(^4\text{He}) + K(^{237}\text{Np})$, where $K = p^2/2M$ for the respective nuclei. Finally, the conservation of momentum applied to the final state gives $0 = p(^4\text{He}) + p(^{237}\text{Np})$, or $p(^{237}\text{Np}) = -p(^4\text{He})$, leaving only the single quantity $p(^4\text{He})$ in our expressions and allowing us to solve for it and hence for $K(^4\text{He})$.

Working It Out We solve the equation above for K_f:

$$\begin{aligned} K_f &= [M(^{241}\text{Am}) - M(^4\text{He}) - M(^{237}\text{Np})]c^2 \\ &= (241.05682\text{ u} - 4.002603\text{ u} - 237.04817\text{ u})c^2 \\ &= 0.006047\text{ u}c^2 = (0.006047\text{ u}c^2)(931.474\text{ MeV/u}c^2) \\ &= 5.633\text{ MeV}. \end{aligned}$$

This kinetic energy is indeed small enough to allow us to use nonrelativistic relations. We use the equation for K_f to solve for $K(^4\text{He})$, writing $p \equiv p(^4\text{He}) = -p(^{237}\text{Np})$:

$$\begin{aligned} K(^4\text{He}) &= K_f - K(^{237}\text{Np}) = K_f - \frac{p^2}{2M(^{237}\text{Np})} \\ &= K_f - \frac{p^2}{2M(^4\text{He})}\frac{M(^4\text{He})}{M(^{237}\text{Np})} = K_f - K(^4\text{He})\frac{M(^4\text{He})}{M(^{237}\text{Np})}. \end{aligned}$$

We can solve this relation for $K(^4\text{He})$:

$$K(^4\text{He}) = \frac{K_f}{1 + \left[\dfrac{M(^4\text{He})}{M(^{237}\text{Np})}\right]} = \frac{5.633\text{ MeV}}{1 + (4.00\text{ u}/237\text{ u})} = 5.539\text{ MeV}.$$

What Do You Think? Is the kinetic energy of the ^{237}Np nucleus (a) much larger than, (b) much smaller than, or (c) about equal to the kinetic energy of the α particle?

Beta Decay

In the **β-decay mode** of a nucleus, an electron and an *antineutrino* ($\bar{\nu}$) (or a positron and a neutrino) are emitted, leaving a daughter nucleus. The antineutrino is the antiparticle of the neutrino (ν), just as the positron is the antiparticle of the electron. The properties of the antineutrino and the neutrino are, for our purposes, identical.

Wolfgang Pauli postulated the existence of the neutrino in 1930 on the basis of experimental data on β decay. An important feature of the neutrino, which has zero charge and spin $\hbar/2$, is that it interacts so weakly with matter that any one neutrino will traverse *light-years* of ordinary matter before the probability of interacting with the matter is significant. It is therefore not surprising that the neutrino was not *independently* observed to initiate a collision process until 1956, 60 years after β decay was first observed. In β decay we observe the electron and (sometimes) the daughter nucleus but, as a result of the neutrino's weak interaction, never the neutrino. Finally, the neutrino has one other important property: Its mass is very small—at most a few eV$/c^2$, less than 10^{-5} times the electron's mass—and this value is negligible in all the situations we deal with here.

When an electron is emitted in nuclear β decay, the daughter nucleus has a charge $+e$ more than that of the parent nucleus; that is, *Z increases by 1 in this β-decay process*. As we did in Eq. (44–24), we write this β-decay process symbolically as

$$^{A}_{Z}X \rightarrow {}_{Z+1}^{\;\;A}X' + e^- + \bar{\nu}. \tag{44–25}$$

An example of this type of decay is $^3H \rightarrow {}^3He + e^- + \bar{\nu}$. In this form of β decay, a neutron has in effect been converted to a proton, an electron, and an antineutrino within the nucleus. Indeed, the most primitive β decay is that of the neutron itself:

$$n \rightarrow p + e^- + \bar{\nu}. \tag{44–26}$$

For *free* neutrons, the lifetime for this process is approximately 890 s. Thus neutrons by themselves are unstable. In contrast, neutrons within nuclei that do not undergo β decay are stable. The primary β-decay mode, Eq. (44–25), is possible only for the mass condition

$$M_P > M_D \tag{44–27}$$

(see Problem 32), where M_P and M_D are the atomic masses of the parent and daughter nuclei, respectively. For nuclei that undergo β decay, the kinetic energy of the electrons ranges up to 10 MeV, with 1 MeV being typical.

If the neutrinos in β decay cannot be observed directly, how was it possible for Pauli to predict their existence and properties as early as 1930? The existence of the neutrino was inferred on the basis of several types of conservation laws. First, conservation of momentum and of energy have rather different consequences for three-body decay (one in which the parent nucleus decays to three particles; Fig. 44–12a) than for two-body decay (Fig. 44–12b). In two-body decay of a nucleus at rest, the energies of each of the two decay products are fixed: The conservation of momentum and of energy *uniquely* determine the magnitudes of the momentum and the energy of the decay products. This is not true in three-body decay, in which different configurations of the three bodies allow for a range of energies for all three bodies. Various configurations allow the electron's kinetic energy to range from zero up to a maximum value, as shown in Fig. 44–13. A similar plot for a two-body decay would show a single energy value at the energy maximum K_{max}.

Pauli surmised from the broad curve observed in Fig. 44–13 that three particles—including the otherwise invisible neutrino—must be present. Angular momentum conservation leads to the same conclusion. If the decay of the neutron were the two-body decay $n \rightarrow p + e^-$ rather than a three-body decay, we would have a discrepancy. Neutrons, like protons and electrons, have a spin of $\hbar/2$, but the rules of angular momentum addition do not allow two $\hbar/2$ spins to add to a total spin of $\hbar/2$. The addition of a third particle with a spin of $\hbar/2$ can resolve the discrepancy; in this way, Pauli was able to predict that the invisible third particle, the neutrino, was a fermion. Although the form of the curve depends in principle on the mass of the neutrino, in fact to the accuracy available to us the curve is the one that would be obtained by assuming that neutrinos are massless, confirming our remark that any mass it has is negligible.

Lifetimes of β decays vary from about 1 s to more than 10^{20} s. The energy dependence and low rate of β decay of some nuclei suggest that the β-decay process is not a quantum mechanical tunneling process but rather a primary manifestation of the *weak force* (one aspect of the electroweak force; see Chapter 5).

We can mention two other forms of β decay here—both of them consequences of the weak force. The first is *positron emission*, which takes the form

$${}_{Z+1}^{A}X \rightarrow {}_{Z}^{A}X' + e^+ + \nu. \tag{44–28}$$

This process is allowed under the mass condition

$$M_X > M_{X'} + 2m_e. \tag{44–29}$$

The second variant is *electron capture*, in which an electron from an atomic orbit is absorbed by the nucleus:

$${}_{Z+1}^{A}X + e^- \rightarrow {}_{Z}^{A}X' + \nu. \tag{44–30}$$

Electron capture is possible under the mass condition

$$M_X > M_{X'}. \tag{44–31}$$

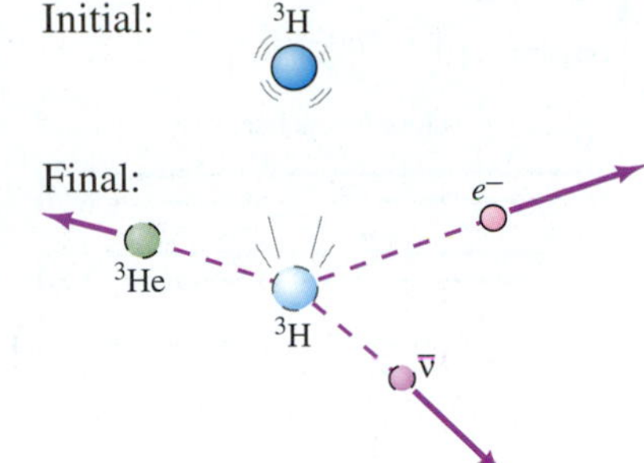

(a)

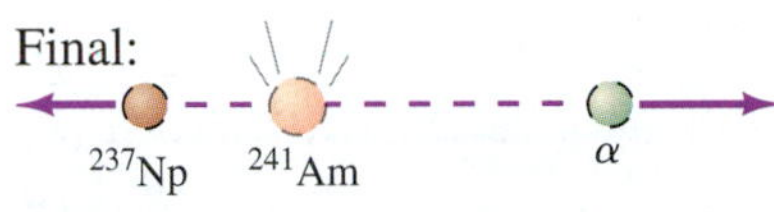

(b)

▲ **FIGURE 44–12** (a) Beta decay is a three-body decay, as opposed to (b) α decay, a two-body decay. In each case the parent nucleus is at rest. In the three-body decay, a variety of directions and energies are possible for the decay products, subject to overall energy and momentum conservations. In the two-body decay, the decay products emerge back-to-back and with fixed energies.

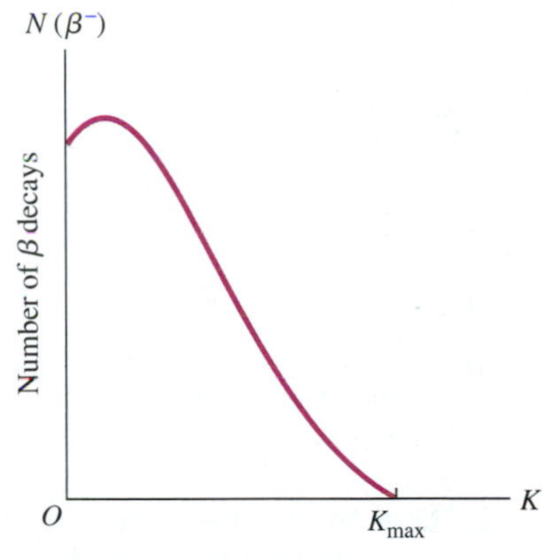

▲ **FIGURE 44–13** The energy of the electron for a large sample of β decays. The energy is not fixed, but rather spread, a consequence of the kinematics of three-body decay. The maximum possible electron energy, K_{max}, is the energy the electron would have if there were no neutrino in the decay.

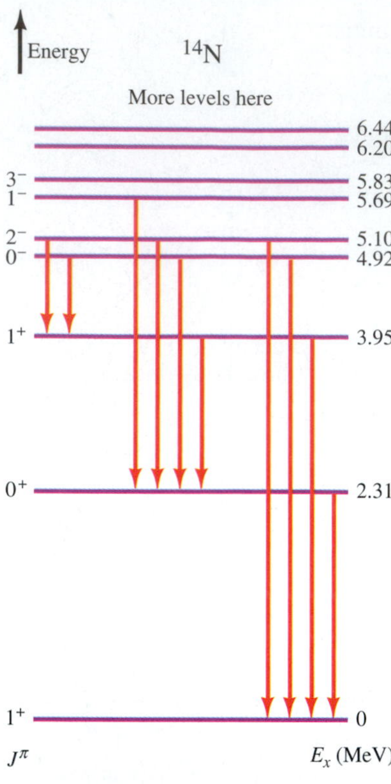

▲ **FIGURE 44–14** Some γ-ray decay transitions in ^{14}N. The notation J^π describes the respective quantum states labeled by quantum numbers not discussed here. E_x labels the energy of the states relative to the ground state. Not all possible transitions are shown. Many more levels exist above 6.44 MeV in ^{14}N.

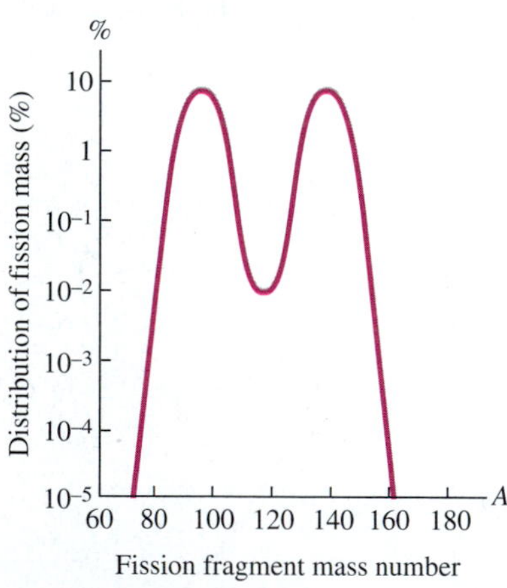

▲ **FIGURE 44–15** The relative probability of producing a fragment of atomic number A when a ^{236}U nucleus undergoes fission. When slow neutrons induce fission, a variety of daughter nuclides (fragments) are produced. Note that it is *less* likely that fragments with exactly half the size of the parent, in this case ^{236}U, are produced than that the fragments are of unequal size. (After W. E. Burcham, *Nuclear Physics*, McGraw-Hill, 1963.)

Note that the mass conditions for the primary decay mode and electron capture are identical. However, β decay tends to occur for neutron-rich nuclei, and electron capture for neutron-poor nuclei. Practically all the kinetic energy in electron capture goes to the neutrino because of the conservation of energy and linear momentum (see Example 44–9).

Gamma Decay

A nucleus in an excited state can decay in the **γ-decay mode**: It emits photons as it decays to its ground state or to lower-lying excited states (Fig. 44–14), just as an atom in an excited electronic state can. Because the typical energy differences between nuclear excited states and the ground state are some 10^6 times greater than the corresponding energy differences in atoms, lighting a match is not enough to excite the nucleus, even if it is enough to excite an atom or molecule. However, daughter nuclei are often produced in their excited states in α, β, or even γ decay, and these residual states can then decay by the γ-decay mode. Photons with energies as high as a few MeV were not at first recognized as electromagnetic radiation and were therefore given a special name, γ-rays; we speak of the γ-decay mode of the nucleus. The lifetimes of γ decay range from 10^{-17} s to as much as 10^8 s.

Radioactivity and Life

Radioactivity has significant effects on biological systems. Some effects are desirable—medical diagnosis and treatment, for example—and some are just the opposite. Undesirable effects occur when one of the forms of radiation deposits energy in a normal living cell. That energy is typically absorbed when ionization occurs. Chemical bonds are broken, destroying needed substances in the cell. New ions may induce abnormal chemical reactions. We can distinguish three levels of damage: The chemical functioning of cellular processes may be so badly interrupted that the cell dies or is unable to reproduce; internal controls on the cell's behavior may be lost, so that the cell becomes cancerous; or the gene structure of the cell may be changed, resulting in abnormal offspring. On the positive side, we know that cancerous cells as well as healthy ones can be destroyed with radiation. In many cases lives are saved or extended for many years because of radiation treatment that targets its effects on cancerous cells. The danger of a given radioactive substance is associated with the activity, the energy of the radioactive decay products, and the amount of that energy deposited in the body. One measure of possible damage is the energy of the radiation absorbed per mass of the organism. The corresponding SI unit is the *gray* (Gy): $1\ \text{Gy} \equiv 1\ \text{J/kg}$. A dose of several Gy over the entire human body may be lethal, and 10 Gy or more is lethal within a short period. Because damage depends on just how the energy is deposited, a more precise measure takes into account the type of radioactive decay products that produce a given dose.

We are constantly exposed to cosmic rays from outer space and from natural radioactivity of materials around us on Earth, receiving a dose of some 10^{-3} Gy/yr from such sources. Exposure to radioactivity from human-made sources must be compared to this natural background. For example, an inhabitant of the United States receives on average less than half the natural dose in diagnostic X rays, and the background radioactivity associated with properly controlled nuclear-power generation is only about 3×10^{-5} of this value.

44–5 Fission and Fusion

Fission

We have mentioned that α decay is a tunneling process. It is a type of **fission**, a process in which a nucleus breaks into two pieces of more or less equal size. Fission occurs in large nuclei because the energetics favor it. The liquid-drop model helps us to understand why fission occurs. When a large nucleus breaks into two fragments, the sum of the Coulomb repulsion terms from Eq. (44–13) for the two pieces is considerably less than that of the corresponding single term of the parent nucleus. The Coulomb term alone implies that the most energetically favorable fission process is one in which the parent nucleus breaks into two equal pieces. In real fission processes, the two fragments are typically of different sizes (Fig. 44–15).

THINK ABOUT THIS. . .

WHAT IS THE SIGNIFICANCE OF RADIOACTIVE SERIES?

It is often true that the daughter nucleus produced in a radioactive decay is itself unstable. The result is a *radioactive series*, in which a succession of unstable nuclei decay in a cascade to some stable nucleus or nuclei. The existence of these series helps us to understand a great deal about the kinds of nuclei that are found in nature.

Along the steps of the cascade that makes up a series, some nuclei may have two competitive decay modes, such as an α-decay mode and a β-decay mode, in which case we say that there is *branching*. Figure 44–16 shows the uranium series, which starts with ^{238}U. Beta decays pull the branching line closer to the line of stability. Of the three possible decay modes, only α decay changes A. Therefore, if a series starts with a parent whose A value has the form $A = 4n$, where n is an integer, then all the nuclides with $A = 4n - 4$, $4n - 8$, and so forth will be encountered until a stable nucleus is obtained. Similarly, if the starting nucleus has $A = 4n + 1$, then nuclei with $A = 4n - 3$, $4n - 7, \ldots$ are reached. A third series with $A = 4n + 2$ and a fourth with $A = 4n + 3$ are also possible, but the "fifth" series, with $A = 4n + 4$, is identical to the $A = 4n$ series. Thus *there are only four different radioactive series*. The series shown in Fig. 44–16 for $^{238}_{92}$U is the $A = 4n + 2$ series; a series starting with $^{232}_{90}$Th is the $A = 4n$ series.

The lifetimes of the original parent nuclei are 6.4×10^9 yr for $^{238}_{92}$U and 2.0×10^{10} yr for $^{232}_{90}$Th. The fact that these numbers are comparable to the age of the solar system explains why we can still find these nuclides in nature: If the lifetimes of the parent nuclei were shorter, all the parents would have decayed by now. Moreover, the fact that the daughter nuclei are continuously being replenished through decays of the parent nuclei explains why we find extremely short-lived nuclei in nature. For example, in the decay series starting with $^{238}_{92}$U, the sixth step produces $^{222}_{86}$Rn, and this nucleus undergoes α decay to $^{218}_{84}$Po with a half-life of 3.8 d, certainly much less than the age of the solar system. Yet $^{222}_{86}$Rn can be found by chemical means in uranium ore.

The end product of a radioactive series is a stable nucleus. The series starting with $^{238}_{92}$U includes 14 steps with 4 different branchings. In the series starting with $^{232}_{90}$Th, there are 10 steps with 1 branching. Despite the branchings, there is a unique end product in each case: the stable nuclides $^{206}_{82}$Pb and $^{208}_{82}$Pb, respectively.

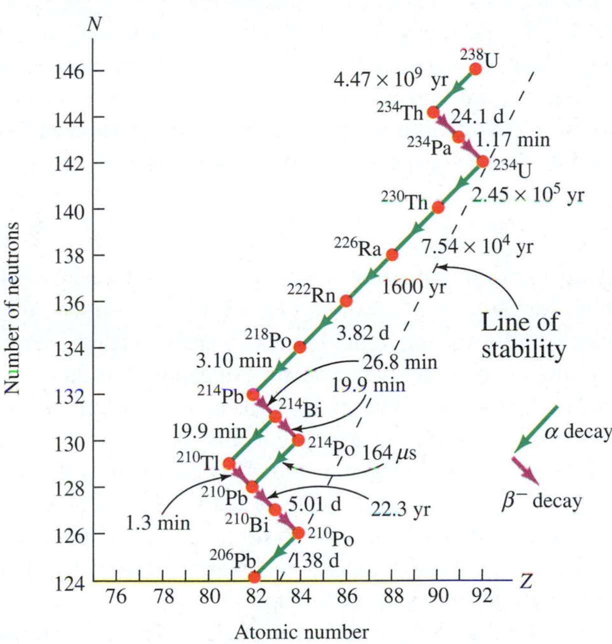

FIGURE 44–16 The decay series of ^{238}U plotted on a graph of N versus Z. Half-lives of each nuclide are also shown. Only one of the four different branchings of this series is shown.

EXAMPLE 44–10 Find the difference in energies due to the Coulomb term alone between a $^{238}_{92}$U nucleus and two $^{119}_{46}$Pd (palladium) nuclei.

Strategy We find the energy due to the Coulomb energy term, Eq. (44–13), for a parent nucleus with Z protons and A nucleons, and then we do the same for each of the two daughter nuclei with $Z/2$ protons and $A/2$ nucleons. We want to find the difference ΔE between the term for the parent and the sum of the terms for each (identical) daughter.

(continues on next page)

Working It Out The steps above lead in a direct way to

$$\Delta E = \frac{3}{5}\frac{Z^2e^2}{4\pi\varepsilon_0 r_0 A^{1/3}} - 2\left[\frac{3}{5}\frac{(Z/2)^2e^2}{4\pi\varepsilon_0 r_0 (A/2)^{1/3}}\right]$$
$$= (1 - 2^{-2/3})\frac{3}{5}\frac{Z^2e^2}{4\pi\varepsilon_0 r_0 A^{1/3}}.$$

We are interested in the case $Z = 92$ and $A = 238$. Numerical evaluation gives

$$\Delta E \cong (1 - 2^{-2/3})(9.8 \times 10^8 \text{ eV}) = 3.6 \times 10^8 \text{ eV} = 360 \text{ MeV}.$$

This is a substantial amount of energy even for a nuclear reaction. But it is not the whole story, because other terms contribute to energy differences of this type. It does, however, give a reasonable order-of-magnitude estimate of the kinds of energies that are released when nuclear fission occurs.

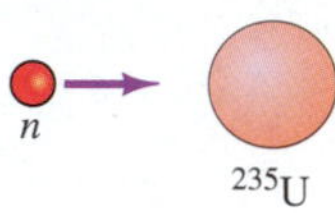

(a)

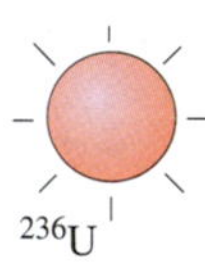

(b)

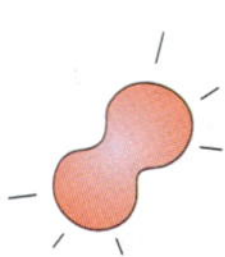

(c)

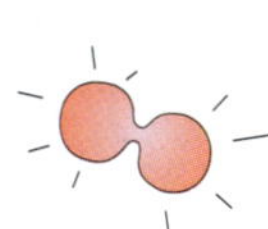

(d)

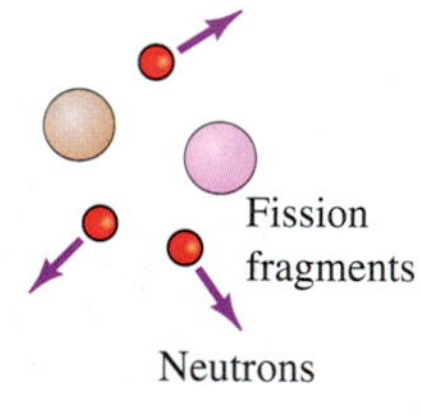

(e)

▲ **FIGURE 44–17** The fission process. (a), (b) The ^{236}U nucleus produced by neutron bombardment of ^{235}U is in unstable equilibrium, and (c)–(e) breaks into two daughter nuclei. Several free neutrons are also produced.

If the Coulomb energy destabilizes large nuclei, why do such nuclei even exist? The answer is that the presence of an energy proportional to the surface of a liquid drop provides a barrier against breakup. The surface term in the energy is positive and is minimized when the nucleus is spherical [Eq. (44–13)]. In Fig. 44–17, we show a sequence in which a large liquid-drop nucleus formed by neutron absorption in ^{235}U breaks into two fragments. As the fragments start to separate, the surface—and hence the surface energy—*increases* without the Coulomb energy changing very much. As the fragments separate further, the surface-energy term makes it energetically favorable for the separating fragments to "neck off." It is only when the nuclei have separated into two spheres that the surface energy no longer changes. The fact that the sum of surface and Coulomb energies initially increases as the fragments start to separate means that there is a potential barrier. This barrier is on the order of 5 MeV for nuclei such as uranium.

Because of the potential barrier, fission of the type we described can only proceed through quantum mechanical tunneling. This process is called *spontaneous fission*; it occurs at a slow rate. Another process, called *induced fission*, occurs when free neutrons are captured by heavy nuclei such as uranium. The resulting nucleus is an excited state, with the added neutron having a binding energy of about 5 MeV. This is about the height of the potential barrier against fission, and the new nucleus thus has enough energy to undergo fission easily. For ^{233}U and ^{235}U, the energy is carried *above* the potential barrier even for very slow neutrons, so fission is guaranteed. For ^{238}U, the neutron must also supply at least 1 MeV of kinetic energy; this will not happen if the neutrons are not sufficiently energetic. Induced fission is important in the possibility of using sustained fission for power production (we shall explore this possibility in Section 44–6).

Fusion

Heavy elements release energy during fission because the curve of binding energy per nucleon decreases as A increases for large A values (see Fig. 44–6). Conversely, this curve increases as A increases for *small* A values; light elements thereby release energy during **fusion**, the combining of small nuclei into larger nuclei. An example is the combining of free nucleons into nuclei: 2.23 MeV of energy is released when a free proton and a free neutron combine and form a deuteron, ^{2}H. More typically, the combining nuclei are each charged, and because the Coulomb force has long range whereas the nuclear force does not, each of the combining nuclei must have considerable energy in order to surmount the potential barrier due to the Coulomb force. Once the potential barrier is passed and the nuclei come close enough to fuse, the energy released by their fusion is much more than the total kinetic energy of the nuclei.

Fusion reactions are central to the "burning" processes of stars. These reactions occur in cycles of reactions or decays, the primary example of which is the *proton cycle*:

$$p + p \rightarrow {}^2\text{H} + e^+ + \nu, \text{ with 0.4 MeV of energy released;} \quad (44\text{–}32a)$$

$$^2\text{H} + p \rightarrow {}^3\text{He} + \gamma, \text{ with 5.5 MeV of energy released;} \quad (44\text{–}32b)$$

$$^3\text{He} + {}^3\text{He} \rightarrow {}^4\text{He} + 2p + \gamma, \text{ with 13.0 MeV of energy released.} \quad (44\text{–}32c)$$

The internal temperature of stars is high enough to give some colliding nuclei sufficient energy to overcome the Coulomb repulsion involved in all three of these reactions. The *net* effect of this cycle is to convert four protons into an α particle along with the emission of energy in the form of photons and two neutrinos. Although the energy given to a neutrino is lost (in the sense that the neutrino does not interact further in a way that deposits its energy), the photons continue to heat the interior. This is the source of the energy that eventually arrives on Earth from our own star, the Sun.

EXAMPLE 44–11 Calculate the total energy released each time a ^{4}He nucleus is produced in the proton cycle. To the energies described in Eqs. (44–32), add the energy released when each of the 2 positrons produced annihilates with 2 electrons (already present in the star), producing 4 photons and a kinetic energy equivalent to four times the electron mass—the positron and electron have the same mass, 0.51 MeV/c^2—namely, 2.0 MeV: $2e^+ + 2e^- \rightarrow 4\gamma + 2.0\text{ MeV}$.

Strategy We must write out the full set of reactions in the proton cycle, measuring the energy that is released at each step. For the third reaction of the cycle, Eq. (44–32c), two ^{3}He nuclei are present, so the reaction of Eq. (44–32b) must occur twice. Similarly, the reaction of Eq. (44–32a) must also occur twice.

Working It Out The chain of reactions that produces one ^{4}He nucleus is

$$\begin{aligned}
&2(p + p) + p + p + 2e^- \\
&\rightarrow 2(^2\text{H} + e^+ + \nu + 0.4\text{ MeV}) + p + p + 2e^- \\
&\rightarrow 2(^3\text{He} + \gamma + 5.5\text{ MeV}) + (2e^+ + 2\nu + 0.8\text{ MeV}) + 2e^- \\
&\rightarrow (^4\text{He} + 2p + \gamma + 13.0\text{ MeV}) + (2\gamma + 11.0\text{ MeV}) \\
&\qquad + (2e^+ + 2\nu + 0.8\text{ MeV}) + 2e^- \\
&\rightarrow (^4\text{He} + 2p + \gamma + 13.0\text{ MeV}) + (2\gamma + 11.0\text{ MeV}) \\
&\qquad + (2\nu + 0.8\text{ MeV}) + (4\gamma + 2.0\text{ MeV}) \\
&= {}^4\text{He} + 2p + 7\gamma + 2\nu + 26.8\text{ MeV}.
\end{aligned}$$

What Do You Think? We have produced a ^{4}He nucleus at the end. What happens to the remaining particles of the proton cycle?

EXAMPLE 44–12 The total power output of the Sun, its *luminosity*, is about 3.9×10^{26} W. Assume that the Sun is made entirely of protons, that its luminosity remains constant, and that it "burns" protons via the proton cycle until all the protons have been converted into α particles. How long can the Sun burn in this way? You may use the results of Example 44–11.

Strategy From the results of Example 44–11, four protons are required to produce one α particle, so the energy released *per proton* is $26.8\text{ MeV}/4 = 6.7$ MeV. We can use this to determine how many protons/second are used to produce the Sun's power output. We can calculate the total number of protons from the total mass of the Sun, and comparison between the total number of protons present and the number used per second for the helium cycle allows us to determine how long the Sun will burn this way.

Working It Out The number of protons in the Sun is $N_p = (\text{mass of Sun})/m_p = (2.0 \times 10^{30}\text{ kg})/(1.67 \times 10^{-27}\text{ kg/proton}) = 1.2 \times 10^{57}$ protons. The number of protons converted per second N from the Sun's total power is

$$N = \frac{\text{Sun's power}}{\text{energy/proton}} = \frac{3.9 \times 10^{26}\text{ J/s}}{(6.7\text{ MeV/proton})(1.602 \times 10^{-13}\text{ J/MeV})} = 3.6 \times 10^{38}\text{ protons/s}.$$

The time T it would take for all of the Sun's protons to be burned is then

$$T = \frac{\text{Number of protons}}{\text{Rate protons react}} = \frac{1.2 \times 10^{57}\text{ protons}}{3.6 \times 10^{38}\text{ protons/s}} = 3.3 \times 10^{18}\text{ s}.$$

This amounts to about 10^{11} yr, or 100 billion yr. The actual burning time for the Sun is more like 10 billion yr (half of which has passed) because only the innermost 10 percent of the Sun's mass is hot enough to burn through the proton cycle.

What Do You Think? Before physicists became aware of nuclear energy generation, they made some predictions of how long the Sun would burn. Make an estimate of the Sun's lifetime assuming that the reactions were chemical rather than nuclear. [*Hint*: Compare typical energies released in chemical and nuclear reactions.]

44–6 Applications of Nuclear Physics

Radiometric Dating

The radioactive nucleus ^{14}C can be used to date biological materials. The lifetime involved is 8268 yr, and that is the rough order of magnitude of ages that can be measured with this method. Other radioactive decays, with longer lifetimes, can be used to date longer geologic times, in the range 10^6 yr to 10^{10} yr (Fig. 44–18). For example, rocks form when liquid magma solidifies; after that point, chemically induced separation of elements is more difficult. As long as there is no lead in a sample of rock when that rock forms, any ^{206}Pb found in the rock must be the product of ^{238}U decays (for which $\tau = 4.5$ billion yr). Therefore the ratio of ^{238}U to ^{206}Pb in the rock allows us to find the rock's age. Dating techniques of this type typically use radioactive processes whose lifetimes are comparable to the time scale being probed.

▲ **FIGURE 44–18** The Drakensberg Mountains in South Africa were formed out of hardened lava outflows and are believed to be 190 milion years old based on radioactivity determinations.

Radioisotopes

Radioisotopes, unstable isotopes produced in nuclear reactions, have characteristic lifetimes for decay. Atoms with these nuclides behave chemically like stable isotopes of the same atom. Observation of the location of the decay of a radioisotope helps us to understand where a chemical or physical process has taken the radioisotope. The applications of radioisotopes include the use of ^{153}Gd, ^{67}Ga, ^{201}Pb, and ^{123}I in medical diagnosis—a cancerous tumor, for example, may concentrate one of these elements, and the observed location of α decays indicates the tumor site. Moreover, a nuclide can be placed in a tumor; upon decaying, it deposits large amounts of energy that kills cells there. There are also applications in structural engineering. For example, ^{85}Kr is used in leak detection—krypton can penetrate very small cracks, and observation of decays in certain locations indicate that there are cracks there.

Nuclear Power Generation

Here, we discuss two ways in which nuclei can be used for electric power generation. The first uses the fission process, and it is a proven, well-established technology. The second uses the fusion process, and it is not yet commercially feasible. In both cases, the energy released in the respective nuclear reactions is converted through collisions to thermal energy, which is then used to turn turbines and generate electric power.

Controlled Fission: It is possible to generate energy from fission because the proportion of mass in neutrons is larger for heavy nuclei, such as those of uranium, than for light nuclei. (In Example 44–6, we treated the decay of ^{238}U into two ^{119}Pd nuclei, but in fact the heaviest *stable* isotope of palladium has $A = 110$.) This fact, a consequence of the Coulomb repulsion of protons, implies that when a parent nucleus undergoes induced fission into two fragments, the daughter nuclei have relatively too many neutrons to be stable. They may undergo β decay, or more typically, a few neutrons may be produced along with the two daughter fragments in fission (see Fig. 44–17e).

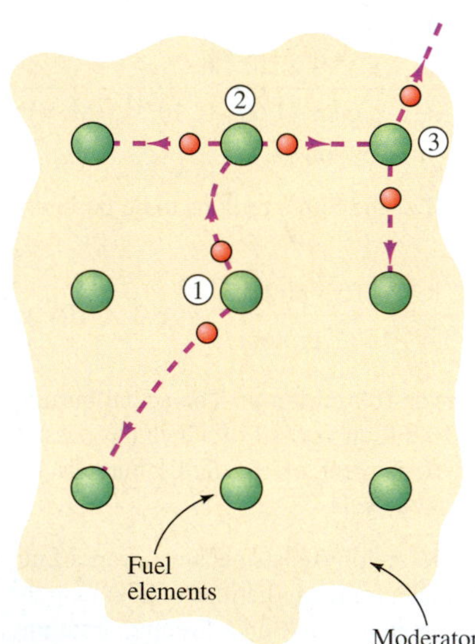

▲ **FIGURE 44–19** A chain reaction, in which the free neutrons produced in a fission process induce the fission of still other nuclei, indicated in the sequence ①, ②, ③. The moderator slows down the released neutrons and increases the likelihood that they will produce another fission.

The induced fission process tends to occur more readily when the bombarding neutrons are slow. Although the neutrons that are *produced* in the fission process generally have too much kinetic energy to induce efficiently the fission of other uranium nuclei, their kinetic energy can be reduced by allowing them to collide within a *moderator*, a material whose atoms have light nuclei: The recoiling neutrons lose a substantial fraction of their kinetic energy in such collisions. Once this has happened, these neutrons can induce other fission processes. Of course, some neutrons do not do so, but in each fission reaction as many as 2 to 3 neutrons are produced, on average. If enough fissionable nuclei (^{233}U or ^{235}U, for example) are present, so that at least one released neutron induces another fission, the process can be sustained. We thereby have a *chain reaction* (Fig. 44–19), which takes place within a *reactor*.

By increasing or decreasing the relative amounts of either fissionable nuclei or moderating material in the form of "control rods," the reaction process can be either speeded up or slowed down and thus satisfactorily controlled. The problem of disposing of the waste products has proven to be more difficult to resolve. These waste products, which include the fission fragments themselves, are typically radioactive and have lifetimes that are long compared to a human lifespan. Because radioactivity can cause genetic damage and cancer, we must dispose of such waste material in such a way that it is isolated from humans and other living things for many generations.

Controlled Nuclear Fusion: Controlling the processes that power the Sun for direct power generation on Earth is an old idea. Deuterium, a hydrogen atom with a ^{2}H nucleus, which is abundant in seawater, could supply fuel for the fusion reaction, and the radioactive waste problem for nuclear fusion appears to be less acute than that for fission. But the technology of controlling fusion reactions has proved to be a difficult one to master. The essence of the problem is not simply to make nuclei collide with enough energy to overcome the Coulomb repulsion barrier and fuse, but to make such collisions occur in large numbers. A collection of nuclei could in principle be brought to energies high enough to fuse by heating them, but the temperatures required to do so correspond to energies on the nuclear scale rather than the atomic scale. We can estimate this energy to be $E \cong e^2/4\pi\varepsilon_0 R$, the Coulomb energy between two singly charged nuclei separated by a nuclear radius R, where R is on the order of 10^{-14} m. Upon substituting, we find that $E \cong 150$ keV. This energy corresponds to a temperature of about 10^9K, far higher than any ordinary container could withstand. Indeed, at such temperatures, atoms lose their electrons, and matter breaks down to a fully ionized gas of electrons and nuclei, a *plasma*. Although in the Maxwell distribution there are enough particles with high speeds so that a temperature of some 10^8K is still sufficient to produce reactions, this temperature is still much too high for conventional treatment.

Before we discuss current ideas of how to overcome the problem of handling hot plasma, let's discuss the reactions themselves. The most promising of these reactions is the so-called D–T reaction. The deuteron, D, is the ^{2}H nuclide, and the triton, T, is the ^{3}H nuclide. (The atom with a triton nucleus is tritium.) The D–T fusion reaction is

$$D + T \rightarrow {}^4He + n.$$

The Q value of this reaction is 17.6 MeV. A second reaction of possible interest is the D–D reaction,

$$D + D \rightarrow {}^3He + n,$$

in which a total of 4 MeV of kinetic energy is produced. If the D–D reaction could be mastered, in particular, a virtually unlimited supply of energy would be available, because deuterium occurs so abundantly in water.

Two rather different approaches have been taken to address the problem of confinement of the hot plasma. The first is *inertial confinement*, a scheme in which tiny pellets of material that contain nuclei suitable for fusion are compressed by powerful laser beams or ion beams until the necessary temperatures and densities are reached (Fig. 44–20). In this scheme, there is no effort to hold the plasma confined for long periods; the beams are fired in pulses, and the pellets are replaced with each pulse. The principal difficulties of this approach are associated with understanding just how the pellets will behave when the energy-supplying beams strike them, with the construction and handling of the pellets, and with the guidance of the beams.

In the second scheme, *magnetic confinement*, the plasma, which is given additional thermal energy by the absorption of electromagnetic waves, is held in place by magnetic forces. The confinement can last for long periods. A difficulty faced in this approach is that magnetic fields do not affect the component of a charged particle's motion that is along the direction of the magnetic field lines, so the plasma tends to "squirt out" in one direction or another. To avoid this problem, complicated configurations of magnetic fields must be used. The *tokamak* is a device that attempts to hold the plasma with magnetic fields contained within a finite volume (Fig. 44–21). Large tokamaks can come close to achieving self-sustaining reactions. Although both inertial confinement and magnetic confinement schemes show promise, we are still very far from making fusion a commercial technology.

THINK ABOUT THIS. . .

DID HUMANS MAKE THE FIRST NUCLEAR REACTOR?

Sufficiently rich deposits of uranium occurred on Earth many billions of years ago to have sustained a "natural" chain reaction. Such ore deposits must have had not only a large concentration of uranium but also deuterium—which is contained in water—to act as a moderator. The ^{235}U-to-^{238}U ratio that exists today is not large enough for a natural reactor to operate. But because ^{235}U decays with a shorter lifetime than ^{238}U does, we know that ^{235}U was relatively more abundant in the past, and ore bodies could have sustained a chain reaction at earlier times. The remnants of a possible *natural reactor* that shut itself down after a certain amount of ^{235}U was burned have been discovered in Africa. The evidence includes a depleted supply of ^{235}U and a characteristic distribution of daughter nuclei descended from fission products.

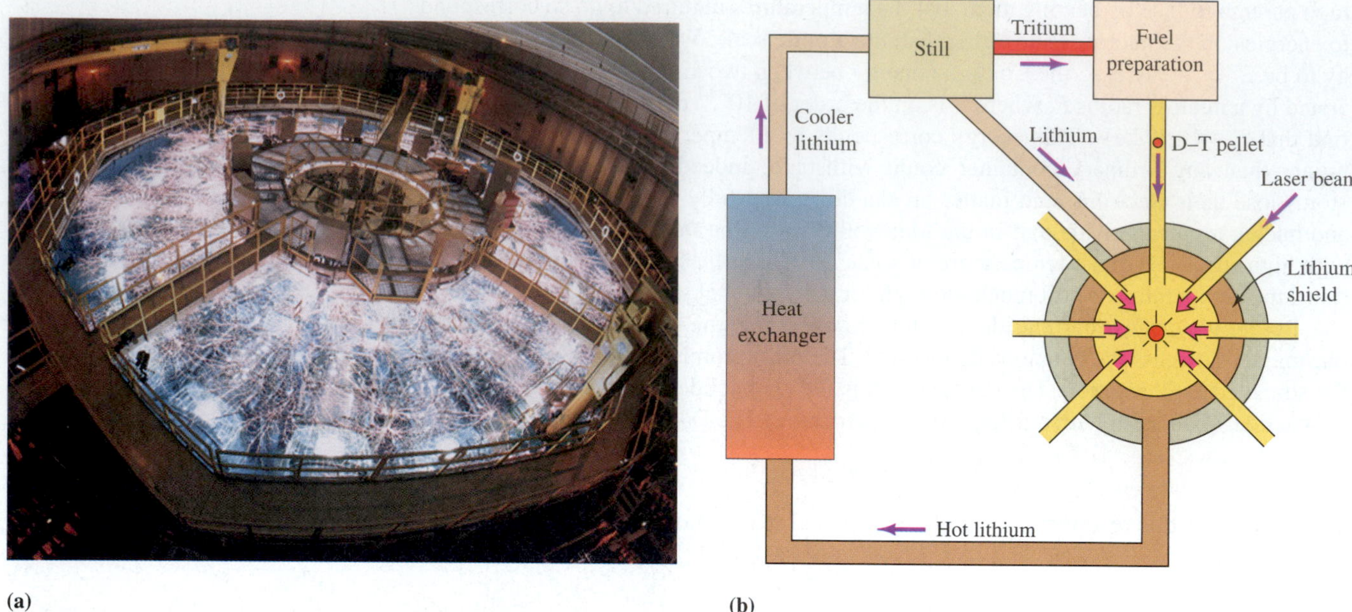

▲ **FIGURE 44–20** (a) The Particle-Beam Fusion Accelerator II at Sandia National Laboratories in Albuquerque, New Mexico, used for inertial-confinement experiments. (b) Schematic diagram of fusion by inertial confinement with laser beams or particle beams. D–T pellets are sent to a site where momentum from many laser or particle beams implodes them, bringing them to high enough temperatures and densities for significant amounts of fusion to occur. The neutrons from the fusion process produce tritium in the lithium shield, which also becomes hot. The tritium can be distilled out of the hot lithium and used to produce more fuel pellets. Deuterons are abundant in water. Electricity could be made from the thermal energy removed at the heat exchanger.

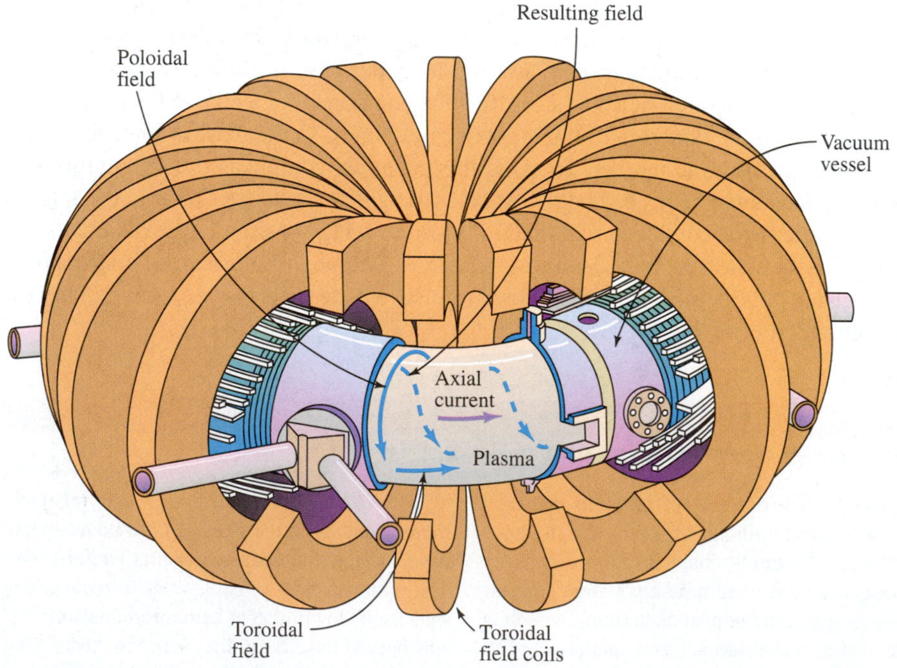

▲ **FIGURE 44–21** Schematic diagram of fusion by magnetic confinement, containment in the magnetic fields of a tokamak. The so-called poloidal field forms circles around the toroidal field. [After W. M. Stacey, Jr., "Fusion Reactor Development: A Review," *Advanced Nuclear Science and Technology*, **15**: 131 (1983).]

Summary

The nucleus was discovered by means of scattering experiments that showed significant numbers of incident α particles scattered through large angles, which follows if the atom has a small, massive, positive core structure. The nucleus, held together by the nuclear force, is composed of Z protons and $N = A - Z$ neutrons, where A is the mass number. Nuclei with the same Z value but different N values are isotopes of a given atomic species. The masses of nuclei and of atoms are conveniently measured in atomic mass units, u, defined so that the mass of a ^{12}C atom is exactly 12 u. The binding energy, the energy released when a nucleus is formed from its A constituents (nucleons), is significant; a binding energy per nucleon of 7 to 9 MeV is typical of larger nuclei. This binding energy is so large that due to the relativistic equivalence of mass and energy the mass of a nucleus is on the order of 1 percent less than the sum of its constituent masses. To a good approximation, the constituents of the nucleus are closely packed into a sphere of radius

$$R_A = r_0 A^{1/3}, \tag{44–7}$$

with $r_0 \cong 1.2$ fm. The lifetimes of unstable nuclei, the spectrum of states, spins, and magnetic dipole moments all provide useful ways to understand nuclear behavior.

For distances greater than about 1 fm, the nuclear force between nucleons is well described in terms of the Yukawa potential,

$$U(r) = -g^2 \frac{e^{-r/R}}{r}, \tag{44–9}$$

where g^2 measures the strength of the force and R is the range, approximately 2 fm. A useful set of models has been developed to explain nuclear behavior. The shell model is an approach in which individual nucleons move in a central nuclear potential. The resulting spectrum, including the presence of magic numbers—Z and N values for which a nucleus is particularly strongly bound—can be explained by this model. The liquid-drop model treats the nucleus as a continuous medium and is able to explain behavior such as fission, the splitting of large nuclei. The semiempirical mass formula, which is inspired by the liquid-drop model, successfully fits the masses of nuclei.

The strong binding of nuclei has important implications for whether nuclear reactions, including scattering and decay, can occur. Short-lived resonances, excited states of nuclei, are visible in cross sections for scattering, and their widths are inversely proportional to their lifetimes, according to the Heisenberg uncertainty principle.

Nuclei can decay through three principal processes: β decay, α decay, and γ decay, depending on whether the decaying nucleus emits an electron, an α particle, or a γ-ray (a highly energetic photon), respectively. Collectively, these processes comprise the phenomenon of radioactivity. The rate of a single decay process is described by the equation

$$N(t) = N_0 e^{-\lambda t}. \tag{44–21}$$

Tunneling is an essential feature of α decay. Natural radioactivity is associated with decay chains, processes in which the products of radioactive decay are themselves radioactive and decay. The presence of such processes supplies us with useful radiometric dating techniques. Certain artificially produced radioactive nuclides have a variety of uses, including radioactive and medical applications.

The systematics of nuclear masses shows that fission, the breakup of large nuclei into smaller ones, and fusion, the combining of small nuclei into larger ones, are energetically favorable. When these processes occur, energy is released. However, both can occur only if a potential energy barrier is overcome. Fusion processes are the means by which the stars obtain energy. Fission induced by neutrons can occur in controlled chain reactions, in which a fissioning nucleus releases neutrons that can initiate other fission reactions. Such chain reactions are a commercially important source of energy.

Understanding the Concepts

1. Alpha particles have a charge of $+2e$ and a mass of roughly four nucleon masses. Describe experiments that would allow you to measure these quantities.
2. Neutron stars are stars that have collapsed under the influence of the gravitational force into what are essentially enormous nuclei composed mostly of neutrons. Why do these neutrons not decay?
3. In our discussion of why some nuclei undergo fission, we did not mention the effect of the energy term proportional to A in the liquid-drop model. Why not?
4. Protons and electrons each have intrinsic spins $\hbar/2$. Use the fact that the spin of ^{6}Li is $\hbar$ to show that the ^{6}Li nucleus cannot consist exclusively of protons and electrons.
5. Why is it easier to produce ions with a positive charge rather than with a negative charge from neutral atoms?
6. How can a mass spectrometer be used to determine the isotopic abundances of nuclides of an element such as oxygen?
7. Experiments to measure the inner structures of the nucleus typically use electrons as incident projectiles. Why might electrons be more useful than α particles for such experiments?
8. Of the calcium isotopes ^{39}Ca, ^{40}Ca, ^{41}Ca, and ^{42}Ca, which would you expect to have the smallest neutron separation energy, and why?
9. Of the nuclides ^{15}N, ^{16}O, and ^{17}F, which would you expect to have the largest proton separation energy, and why?

10. The mass conditions for β decay and electron capture are the same. Under what conditions is each one more likely to occur for a given nuclide?
11. Certain chemical elements tend to concentrate in bone marrow. Why should you avoid any such elements that are radioactive?
12. Why are light nuclei more efficient moderators for controlled nuclear fission than are heavy nuclei?
13. Neutrons that have been slowed down by a moderator are unable to induce rapid fission in ^{238}U nuclei. Why?
14. When we gave the (final) kinetic energies of the ^{4}He and neutron products in the D–T fusion reaction, we assumed that the collision takes place at rest. Yet we know there is a Coulomb energy potential barrier to overcome. Why was it possible for us to ignore the initial kinetic energy?
15. Most of the daughter nuclei that result from fission are radioactive. Why?
16. Why might you expect a neutron to penetrate farther into a sample of matter than would an α particle of the same kinetic energy?

Problems

44–1 Static Properties of Nuclei

1. (I) How many neutrons and how many protons do the following nuclides have: ^{9}Be, ^{13}C, ^{22}Na, ^{31}P, ^{57}Fe, ^{72}Ge, ^{107}Ag, ^{131}Cs, ^{208}Pb, ^{241}Am?

2. (I) What is the total binding energy of ^{6}Li? Its binding energy per nucleon?

3. (I) What is the Rutherford collision cross section of a 8.0-MeV α particle $(Z = 2)$ on a gold nucleus $(Z = 79)$ at a scattering angle of 60° (Fig. 44–22)? Express your answer in units of *barns* $(1 \text{ barn} \equiv 10^{-24} \text{ cm}^2)$.

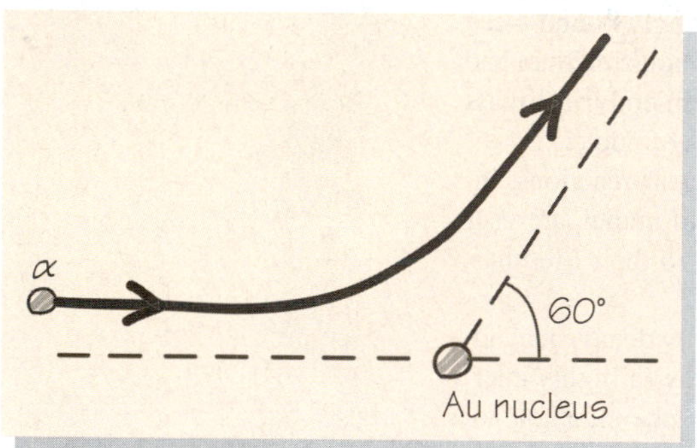

▲ **FIGURE 44–22** Problem 3.

4. (I) Consider the deflection of an α particle of kinetic energy 8 MeV by a tungsten atom, for which $Z = 74$. Estimate the typical angular deflection (a) for $R_1 = 0.01$ nm; (b) for $R_2 = 10$ fm (Fig. 44–23).

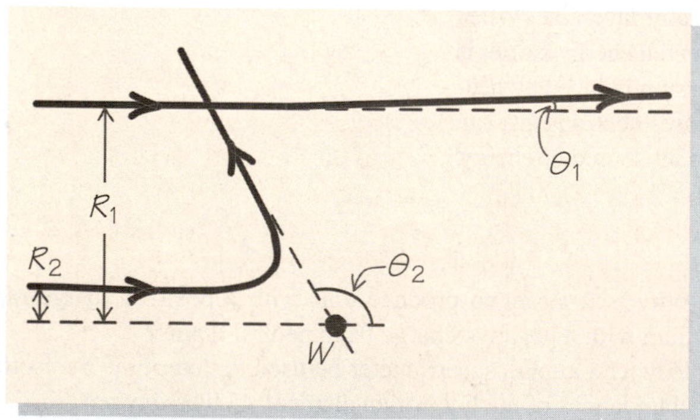

▲ **FIGURE 44–23** Problem 4.

5. (II) An α particle with (nonrelativistic) speed v is approaching the center of a nucleus of radius R and charge $+Ze$. Calculate the distance of closest approach. Assume that the nucleus is much heavier than the α particle. For what value of v will the distance of closest approach equal R?

6. (II) Calculate the distance of closest approach that a 5.5-MeV α particle makes when it scatters head-on with tungsten nucleus (see Fig. 44–2). Neglect recoil of the tungsten nucleus. Compare this with the sum of the nuclear radii for helium and tungsten.

7. (II) A proton with 6.0 MeV of energy scatters at a 60° angle from a gold atom. (a) How close did the proton come to the center of the nucleus? (b) To the surface of the gold nucleus? (Assume that the proton is pointlike.)

8. (II) The masses of the stable isotopes of calcium are 41.95863 u for ^{42}Ca, 42.95878 u for ^{43}Ca, and 43.95549 u for ^{44}Ca. Calculate the binding energies of each nuclide.

9. (II) Express the Rutherford collision cross section in terms of the momentum transfer, $\Delta\vec{p}$, of a scattered particle, which has an initial momentum $mv\hat{i}$ and a final momentum $(mv\cos\theta)\hat{i} + (mv\sin\theta)\hat{j}$ (Fig. 44–24).

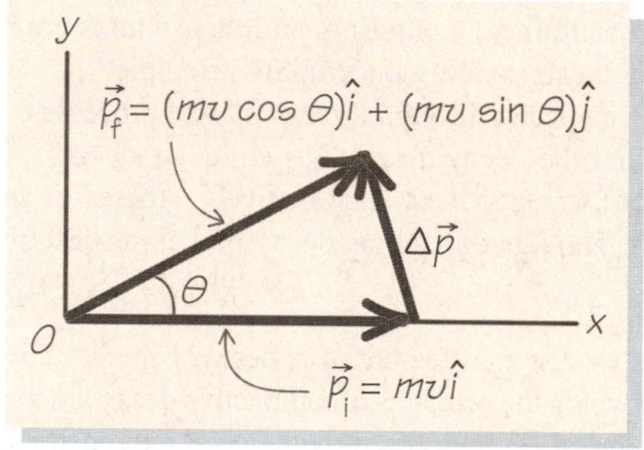

▲ **FIGURE 44–24** Problem 9.

10. (II) If we want to use an electron to probe distances to the size of a uranium nucleus, what energy should the electron have? What if we want to probe distances to the size of a nucleon, about 1 fm?

11. (II) Use the work–energy theorem to solve Example 44–2. Integrate the work over the distance from infinity to the turnaround distance R, the point at which the Coulomb force has done work equal to the initial kinetic energy of the α particle.

12. (III) *Mirror nuclei* are pairs of nuclei for which Z and N of each nucleus are opposites: For example, $^{23}_{11}$Na and $^{23}_{12}$Mg are mirror nuclei with $Z = 11$, $N = 12$ and $Z = 12$, $N = 11$, respectively. The idea that the *only* difference in the energies of these nuclei is due to a difference in electrostatic energy gives us another way to measure nuclear radii and in particular the constant r_0 in Eq. (44–7). If you solved Problem 24–91, you showed that when a charge Q is distributed uniformly throughout a spherical volume $(4/3)\pi R^3$, the electrostatic energy is $\frac{3}{5}Q^2/4\pi\varepsilon_0 R$. Use this result to express the *difference* ΔE in the electrostatic energies of the pair of mirror nuclei above, and use that expression to find r_0, as defined by $R = r_0 A^{1/3}$, in terms of ΔE.

13. (III) Evaluate r_0 by comparing the binding energy and electrostatic energy of the following pairs of mirror nuclei (see Problem 44–12): ^{7}Li(7.016005) and ^{7}Be(7.016930); ^{9}B(9.01333) and ^{9}Be(9.012183); ^{11}B(11.009305) and ^{11}C(11.011433); ^{21}Ne(20.99385) and ^{21}Na(20.99765); ^{23}Na(22.9898) and ^{23}Mg(22.9941). The quantity in parenthesis is the mass of the particular nucleus in units of u.

44–2 Nuclear Forces and Nuclear Models

14. (I) (a) The energy of the first excited state of the nucleus ^{70}Ge is 1.04 MeV. Calculate the frequency and wavelength of a photon needed to excite this state from germanium atoms at rest. (b) Do the same for the nuclide ^{181}Ta, whose first excited state has an energy of 0.0062 MeV.

15. (I) According to the shell model, closed shells of protons occur for $Z = 2$, $Z = 8$, $Z = 20$, $Z = 40$, $Z = 70$, and similarly for neutrons. Use the shell model to determine whether the following nuclides have a closed shell (magic number) for either neutrons or protons: ^{15}N, ^{39}K, ^{40}Ca, ^{56}Fe, ^{140}Ce. Does your answer correlate to whether the respective nucleus is tightly bound?

16. (I) Use the semiempirical mass formula to find the ratio of the surface contribution to the Coulomb contribution for the nucleus ^{6}Li.

17. (II) Assume that the two protons inside a ^{4}He nucleus are approximately 4 fm apart. (a) Calculate the Coulomb force between them. (b) Use the binding energy of ^{4}He to *estimate* the nuclear force. Compare the two forces.

18. (II) Find the force that corresponds to the Yukawa potential energy, Eq. (44–9). Sketch it as a function of the separation distance; on the same plot, sketch an inverse-square force, assuming that that force corresponds to the same potential energy as Eq. (44–9) for small r.

19. (II) The energy required to assemble Z protons in a uniform distribution throughout the volume of a sphere of radius $r_0A^{1/3}$ is $\frac{3}{5}Z^2e^2/(4\pi\varepsilon_0 r_0 A^{1/3})$. The parameter r_0 is roughly 1.2×10^{-15} m. What fraction of the mass of a carbon nucleus does this energy represent?

20. (II) Use the semiempirical mass formula to determine the neutron, proton, and α-particle separation energies for the nuclide ^{56}Fe.

21. (II) Use the semiempirical mass formula to find the Z value that minimizes the total mass $M(A, Z)$ for fixed A by solving $\partial M/\partial Z = 0$.

22. (III) Use the semiempirical mass formula to determine the relation between A and Z such that spontaneous fission into two identical fragments occurs; that is, for which $M(A, Z) - 2M(A/2, Z/2) > 0$.

44–3 Energetics of Nuclear Reactions

23. (II) The first nuclear reaction ever observed, by Ernest Rutherford, was the reaction $\alpha + {}^{14}\text{N} \rightarrow p + {}^{17}\text{O}$. If 5 MeV α particles are used to bombard ^{14}N at rest, determine the sum of the kinetic energies of the two outgoing particles from the atomic masses in Table 44–2 $[M(^{17}\text{O}) = 16.999131\text{ u}]$.

24. (II) The α particle Rutherford used to initiate the first observed nuclear reaction (see Problem 23) had to overcome the Coulomb barrier between the protons in ^{4}He and those in ^{14}N. Assume that the α particle must come within 1 fm of a proton in the target nucleus for a nuclear reaction to occur. Calculate the minimum energy the α particle must have to initiate the reaction. Compare your result with the 5.26 MeV energy for the α particle in Rutherford's experiment. Was the reaction likely?

25. (II) One way to produce a beam of energetic neutrons is to bombard ^{3}H with accelerated ^{2}H. The reaction $^3\text{H} + {}^2\text{H} \rightarrow {}^4\text{He} + n$ releases 17.6 MeV. What is the energy of a neutron produced by a 2 MeV ^{2}H beam, assuming that the ^{4}He nucleus moves in the direction of the ^{2}H beam (Fig. 44–25)?

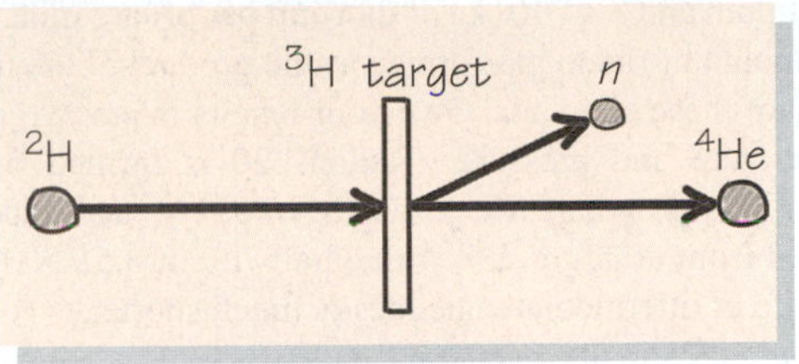

▲ **FIGURE 44–25** Problem 25.

26. (II) Nucleons are not static, but move within nuclei. The approximate energy of a nucleon inside a nucleus is about 20 MeV. Use this kinetic energy and the uncertainty principle to *estimate* the dimensions of a sphere that might contain the nucleon. Is this size approximately equal to that of a nucleus? Interpret your result.

27. (II) The nuclear reaction James Chadwick used in his experiment to identify the neutron was $\alpha + {}^{11}\text{B} \rightarrow n + {}^{14}\text{N}$. Assume that the α particle had energy 5.3 MeV and that the ^{11}B was at rest; $M(^{11}\text{B}) = 11.009305$ u. If the kinetic energy of the ^{14}N is 0.8 MeV, what energy would the neutron have?

28. (II) Is the nuclear reaction ${}^{16}_{8}\text{O} + {}^{24}_{20}\text{Mg} \rightarrow {}^{40}_{20}\text{Ca}$ possible if the nuclei in the initial state are at rest? If so, how much energy is released? If not, how much extra energy is needed in the form of kinetic energy?

29. (II) The most frequently used laboratory neutron source works as follows: A source of ^{226}Ra emits 4.78 MeV α particles that are used to bombard beryllium. Neutrons are produced in the reaction ${}^{9}_{4}\text{Be} + {}^{4}_{2}\text{He} \rightarrow {}^{12}_{6}\text{C} + n$. The atomic mass of ${}^{9}_{4}$Be is 9.012183 u. Find the other masses from Table 44–1 and use these to estimate the maximum kinetic energy of the neutrons.

30. (II) Which of the following reactions are not possible, and why not: (a) $\alpha + {}^{57}\text{Fe} \rightarrow n + {}^{60}\text{Ni}$; (b) $p + {}^{58}\text{Ni} \rightarrow {}^{3}\text{He} + {}^{57}\text{Co}$; (c) $n + {}^{157}\text{Gd} \rightarrow d + {}^{156}\text{Eu}$; (d) ${}^{241}\text{Am} \rightarrow \alpha + {}^{237}\text{Pu}$?

31. (III) A foil made of 20 mg of ^{58}Co (metallic cobalt) is exposed for 1 h to a beam of neutrons with a flux of 5×10^{13} neutrons/cm^2 s. If the *capture cross section* (the reaction ${}^{58}\text{Co} + n \rightarrow {}^{59}\text{Co}$) for such neutrons is 1700 barns (see Problem 3), how many ^{59}Co nuclei are formed? [*Hint*: Find the effective area taken up by the target cobalt atoms.]

44–4 Radioactivity

32. (I) Show that the primary β-decay process, Eq. (44–25), is possible only when the atomic mass of the parent nucleus is greater than the atomic mass of the daughter nucleus.

33. (II) Show that positron emission, Eq. (44–28), is possible only when $M_P > M_D + 2m_e$, where M_P is the atomic mass of the parent nucleus and M_D is the atomic mass of the daughter nucleus.

34. (I) Show that electron capture is possible only when the atomic mass of the parent nucleus is greater than the atomic mass of the daughter nucleus.

35. (I) The number of unstable nuclei $N(t)$ present at time t diminishes with time according to the differential equation

$$\frac{dN(t)}{dt} = -\frac{1}{\tau}N(t),$$

where τ is the lifetime. Show by direct substitution that the solution to this equation is indeed $N(t) = N_0 e^{-t/\tau}$.

36. (I) What is the activity of a 0.03-kg sample of ^{235}U? The half-life of ^{235}U is 7.04×10^8 yr.

37. (II) The upper 0.3 m of soil contains on the average about 10^3 kg of uranium and 3×10^3 kg of thorium per square mile. Calculate the amount of radon gas formed as the product of the radioactive decay of these elements. Give your results in cm^3/yr at standard temperature and pressure. (Radon-220 is formed from thorium-232 with a half-life of 1.39×10^{10} yr, and radon-222 is formed from uranium-238 with a half-life of 4.51×10^9 yr. The half-life of intermediate nuclides is much shorter.)

38. (II) The theory of α decay as a tunneling process allows us to predict an experimental relation between the half-life $t_{1/2}$ and the kinetic energy K of α particles emitted during radioactivity. An approximate version of this prediction is

$$\ln t_{1/2} = \frac{Z(3.97\ \text{MeV}^{1/2})}{\sqrt{K}} - 123,$$

where K is in MeV. (a) Fill in the missing values in the following table:

Nucleus	K (MeV)	$t_{1/2}$
^{215}Ra	?	1.6 ms
^{223}Th	7.7	?
^{244}Pu	?	8×10^7 yr
^{248}Cm	5.16	?

(b) In order to perform an experiment, we need a source of α particles with a half-life of at least 5 days. What is the highest-energy source we are likely to find?

39. (II) By looking up the masses of the neutron, proton, and electron in Appendix II, find the maximum energy, in MeV, of an electron produced in the decay of a neutron.

40. (II) In a sample of $^{211}_{86}Rn$ nuclei, 26 percent decay by α decay to $^{207}_{84}Po$ and the rest decay by electron capture to $^{211}_{85}At$. What is the ratio of the lifetime for α decay, τ_α, to the lifetime for electron capture, τ_β?

41. (II) The half-life of ^{14}C is 5730 yr, and the tissues of organisms accumulate this isotope from the atmosphere while the organisms are living. The skeleton of a mammoth is found to have 15 percent as much ^{14}C as the atmosphere has. When did the mammoth live? Assume that the concentration of atmospheric ^{14}C does not change.

42. (II) The age of marine sediments can be determined by their thorium-230 content. ^{230}Th is the decay product of ^{234}U and it decays to ^{226}Ra with a half-life of 75,200 yr. Measurements indicate that the ^{230}Th content of a sediment is 6 times larger at the surface than at a depth of 40 m. What is the deposition rate at that location in mm/g?

43. (II) ^{80}Br is one of a small number of nuclides that is able to decay by all three β-decay processes. Show that this is possible. $M(^{80}Br) = 79.918528$ u, $M(^{80}Kr) = 79.916376$ u, $M(^{80}Se) = 79.916521$ u.

44. (II) Suppose that the lifetime for γ decay of a nucleus at rest is so long that, for all practical purposes, the natural width of the spectral line can be ignored (that is, the γ energy takes on a single, fixed value). When the nucleus is part of a collection of atoms at finite temperature, it moves randomly, sometimes toward and sometimes away from a γ-ray detector, and the spectral line width is determined by the Doppler shift of the emitted γ-rays. This phenomenon is known as *Doppler broadening*. Estimate the width due to Doppler broadening for these atoms at a temperature $T = 600$K.

45. (III) A radioactive nucleus a decays with a lifetime τ_1 into nucleus b, which decays with a lifetime τ_2 into nucleus c. The change in the number of nuclei a obeys the relation $dN_a = -(N_a/\tau_1)\,dt$ (see Problem 40–65). The number of nuclei b does not obey a simple equation of this type, because those nuclei are continuously being depleted through decays into nuclei c while they are replenished through decays of nuclei a. Show that the correct expression for the change in the number of nuclei b is $dN_b = -(N_b/\tau_2)\,dt + dN_a$. How will this expression generalize to a chain of decays $a \to b \to c \to d \ldots$ with lifetimes $\tau_1, \tau_2, \tau_3, \tau_4, \ldots$, respectively (Fig. 44–26)?

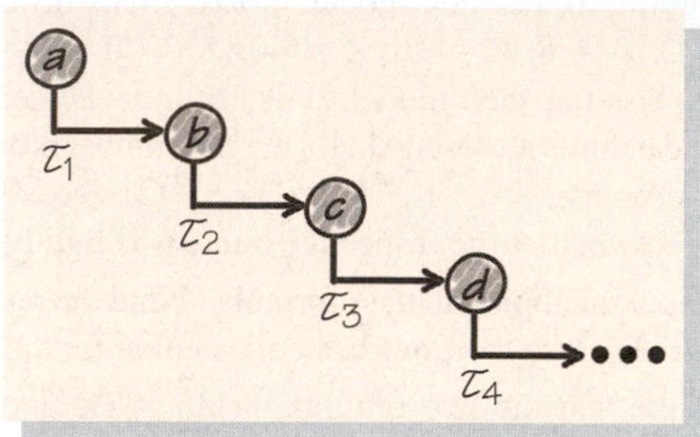

▲ **FIGURE 44–26** Problem 45.

46. (III) We have a chain of decays in a three-level system of nuclei a, b, c with $M_a > M_b > M_c$. Level a decays both to level b (with lifetime τ_{ab}) and level c (with lifetime τ_{ac}), while level b decays to level c (with lifetime τ_{bc}). Thus the number of nuclei at level a at time t obeys the differential equation

$$\frac{dN_a(t)}{dt} = -\frac{1}{\tau_{ab}}N_a(t) - \frac{1}{\tau_{ac}}N_a(t) \equiv -\frac{1}{\tau_a}N_a(t).$$

(a) Show that the solution to this equation is $N_a(t) = N_0 e^{-t/\tau_a}$. (b) Show that the number of nuclei of level b at time t obeys the differential equation

$$\frac{dN_b(t)}{dt} = \frac{1}{\tau_{ab}}N_a(t) - \frac{1}{\tau_{bc}}N_b(t).$$

(c) Show by substitution that the number of level b nuclei is

$$N_b(t) = N_{0a}\frac{\tau_a\tau_{bc}/\tau_{ab}}{\tau_{bc} - \tau_a}\left(e^{-t/\tau_{bc}} - e^{-t/\tau_a}\right).$$

(d) Show that there is no need to write an equation for N_c once N_a and N_b are known as functions of time, because $N_c(t) = N_{0a} - N_a(t) - N_b(t)$ if we start with only a collection of N_{0a} nuclei at time $t = 0$.

47. (III) Show that when the rate of decay from nucleus a in Problem 46 is much smaller than the rate of decay from nucleus b to nucleus c, then after a time that is long compared to τ_{bc}, $N_b(t) \cong (\tau_{bc}/\tau_{ab})\,N_a(t)$. (You may use the results derived in Problem 46.)

44–5 Fission and Fusion

48. (I) The energy of nuclear weapons is expressed in terms of megatons of TNT, where 1 megaton of TNT $= 4.3 \times 10^{15}$ J. What minimum mass of ^{235}U is required to produce a nuclear weapon of 200 kilotons? [*Hint*: Each fission of ^{235}U produces about 200 MeV of energy.]

49. (II) The temperature of the interior of stars is about 10^7 K. (a) Estimate the typical kinetic energy of the colliding protons that participate in the proton cycle. (b) Compare your answer to part

(a) with an estimate of the kinetic energy that protons initially distant from one another must have to come within 1 fm of each other, a distance at which the strong forces can lead to fusion.

50. (II) Calculate the difference between the Coulomb energy for a parent nucleus (Z, A) and the sum of the Coulomb energies of two daughter nuclei (fZ, fA) and $[(1-f)Z, (1-f)A]$, respectively, where f is some fraction between 0 and 1. Show that this energy difference is maximized when $f = \frac{1}{2}$; that is, from the Coulomb energy terms alone it is energetically favorable for fission to produce two equal-sized fragments.

44–6 Applications of Nuclear Physics

51. (II) A 440-MW (electrical output) nuclear power plant uses ^{235}U as its fuel. The uranium fuel contains 6 percent ^{235}U, and each fission reaction produces 200 MeV of energy. Assume that the power plant is 25 percent efficient in producing electric energy. (a) Calculate the amount of uranium fuel used in 1 yr. (b) Determine the amount of thermal energy released to the environment in that time. (c) How many fission events occur per second?

52. (II) One of the reactions in the proton cycle is Eq. (44–32b), $^2H + p \rightarrow {}^3He + \gamma$. Calculate the energy released in this reaction. Use the data given in Table 44–1, but take into account that atoms inside burning stars are completely stripped of their electrons.

General Problems

53. (I) The *carbon cycle*, a stellar burning process secondary to the proton cycle, consists of the following sequential series of nuclear reactions and decays:

$^{12}C + p \rightarrow {}^{13}N + \gamma$, with 1.9 MeV of energy released;
$^{13}N \rightarrow {}^{13}C + e^+ + \nu$, with 1.2 MeV of energy released;
$^{13}C + p \rightarrow {}^{14}N + \gamma$, with 7.6 MeV of energy released;
$^{14}N + p \rightarrow {}^{15}O + \gamma$, with 7.4 MeV of energy released;
$^{15}O \rightarrow {}^{15}N + e^+ + \nu$, with 1.7 MeV of energy released;
$^{15}N + p \rightarrow {}^{12}C + {}^4He$, with 5.0 MeV of energy released.

(a) What is the net effect of one cycle, and how much energy is produced per cycle? (b) Consider a star of mass 3×10^{30} kg, consisting mainly of hydrogen, with 0.1 percent of its mass in ^{12}C. The characteristic duration of a carbon cycle in the star is 5×10^6 yr. Estimate the energy produced each year by the carbon cycle, assuming that each carbon nucleus in the star acts as a catalyst for the cycle.

54. (II) (a) Use the semiempirical mass formula to determine an expression for the binding energy of a nucleus. (b) Apply your result to calculate the binding energy of ^{12}C.

55. (II) A fissioning ^{235}U nucleus produces two fission fragments, ^{139}Cs and ^{93}Rb (Fig. 44–27). Assume that the two fragments are spherical in shape and that the charge is uniformly distributed in both of them. What is the Coulomb potential energy of the system if the two fragments are just touching?

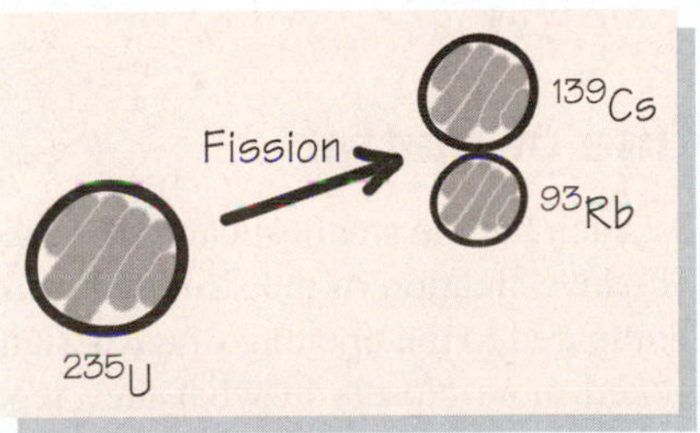

▲ **FIGURE 44–27** Problem 55.

56. (II) Use the semiempirical mass formula to find the neutron separation energy for the nuclides ^{16}O and ^{17}O. Explain the difference.

57. (II) Two protons within a star, each with energy 12 keV, collide head-on. What is the distance of their closest approach?

58. (II) A bone containing 20 g of carbon emits β particles from ^{14}C at the rate of 8.5 per second. At the time the ^{14}C is taken from the air and incorporated into bone, the activity of natural carbon due to its ^{14}C content is 15 Bq/mol. The lifetime of ^{14}C is 8268 yr. Estimate the bone's age.

59. (II) Two identical and neutral helium atoms approach each other. Because there is no Coulomb repulsion between them, they should be able to get very close together, close enough for the nuclei to touch. (a) What is wrong with this argument? (b) Given that the nuclear radius of helium is about 2 fm and that the center of mass of two approaching helium atoms is at rest, what energy must the helium atoms have for the nuclei to touch?

60. (II) A beam of neutrons of kinetic energy 3.7 GeV is produced through the bombardment of nuclei by protons. What distance does this beam travel before 5 percent of the neutrons have decayed (Fig. 44–28)? [*Hint*: The neutrons are relativistic, and time-dilation effects are important.]

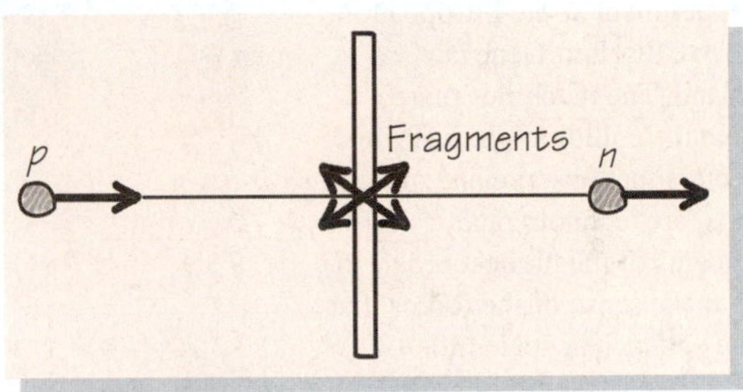

▲ **FIGURE 44–28** Problem 60.

61. (II) One of the early suggestions about the composition of the nucleus was that it contains A protons and A–Z electrons. Use the uncertainty principle to estimate that if an electron is confined to a sphere with a radius of about 10^{-14} m, the size of a nucleus, then its momentum must be larger than about 10^2 MeV/c. Given that β rays emitted by the nucleus have energy on the order of 1 MeV, this estimate rules out the possibility that there are electrons in the nucleus.

62. (II) Calculate the overall kinetic energy produced in the carbon cycle (see Problem 53) after the two positrons produced in the cycle have annihilated against electrons.

63. (II) The energy of a collection of noninteracting fermions of mass m with number density n in a volume V is given by $E = V(\hbar^2\pi^3/10m)(3n/\pi)^{5/3}$. Use this expression to calculate the energy of the protons, E_p, in a nucleus in terms of A, Z, and r_0, where r_0 is the nuclear radius scale in Eq. (44–7). Repeat your calculation for the energy of the neutrons, E_n.

64. (III) Starting with the results of Problem 63, sum the proton and neutron energy terms, using the variables A and $\tau \equiv N - Z$. You will want to make the substitutions $Z = (A - \tau)/2$ and $N = (A + \tau)/2$. Approximate your result for $\tau \ll A$ by using $(1 + x)^p \cong 1 + px + \frac{1}{2}p(p-1)x^2$, an approximation good for $x \ll 1$. Show that this result has one term proportional to A and another term proportional to τ^2/A. Compare the coefficients of these two terms with the corresponding terms in the semiempirical mass formula. What does this tell you about the model of a nucleus as a collection of noninteracting protons and neutrons in a box?

Chapter 45

The results of the spectacular collision of two lead nuclei moving at very high energy, seen in the so-called NA49 experiment at the European laboratory CERN, in Geneva, Switzerland. The machines that accelerate the colliding particles, and the detectors that observe and analyze the debris, are technological achievements of the highest order. In order to make sense of the forces that act at the subnuclear scale, it is necessary to make sense of collisions such as these.

Particles and Cosmology

The "inner-outer connection," by which we mean the coming together of the realms of the very small and the very large, is a new, fertile field of study, barely three decades old. To understand how the fundamental structure of matter and the evolution of the universe affect each other requires some background knowledge of the two areas. The story of these fields and of the connections between them involve the interplay of brilliant experiments and great leaps of imagination in the development of the theories that describe them. The progress that has been made has depended on ever improving observational tools: A series of powerful accelerators have provided knowledge of the structure of matter at the smallest scales, and a series of powerful telescopes have been essential to our understanding of the structure of the universe. In this chapter, we outline how these instruments have led to a deep understanding of the realms of particles and cosmology and of how these realms are described in many ways by the same physics.

45–1 Probing the Structure of Matter

A "Russian doll" is a set of nested dolls; each but the smallest can be opened to reveal a smaller version of itself (Fig. 45–1). The investigation of the structure of matter and of the forces that hold it together can be compared to the opening of a Russian doll. As the investigators of any given era have looked more closely at what they regarded as the fundamental constituents and interactions of matter, they have found at smaller scales within that matter other constituents interacting through other forces. The constituents and interactions at larger scales can be explained in terms of the physics at smaller scales. The first real step in "opening the Russian doll" was made by John Dalton in 1808. He knew that when chemical elements combine chemically, they do so in precise proportions, and he proposed that the simplest explanation for this phenomenon is that a variety of different types of atoms exist, one for each chemical element. The atoms of the chemical elements supply the building blocks for the construction of molecules. The interatomic forces that supply the "glue" for this construction are electrical in origin.

But atoms are not truly indivisible. We saw in Chapter 41 how an atom is constructed from electrons and a nucleus. The structure of the nucleus and of electrons represents the Russian doll within an atom. The atom is held together by electromagnetic forces between the electrons and the nuclei.

We have seen that nuclei have structure. They consist of nucleons—protons and neutrons. The nucleus is held together by the nuclear, or strong, force. This force differs from the electromagnetic force in that it has finite range. And compared with the electron rest energy, the energy required to remove a neutron or proton from a given nucleus with the help of electrons is so large that relativistic effects—those that relate energy to mass—are significant for the first time in our brief overview of the structure of the matter.

▲ **FIGURE 45–1** Each Russian doll hides inside its larger companion. This toy is a useful analog to the structure of the physical world.

Do electrons and nucleons have structure? If so, what forces hold those structures together? As far as we know today, there is no evidence to suppose that electrons are composed of more fundamental constituents. This is not the case for nucleons and other strongly interacting particles, known collectively as **hadrons**. Scattering experiments carried out with particle accelerators have provided us with powerful evidence regarding their structure and the nature of the interactions involved.

Seeing Subatomic Systems

How do we determine the shape and size of an object in everyday life? We look at it, by which we mean that we observe light (photons) bouncing off the object. Our eyes serve as a detector for observing a pattern of reflected light.

What prevents us from seeing atoms in this way? Light diffraction limits our ability to discern details of small objects, as we learned in Chapter 38. Resolution of details in an object of size d requires light of wavelength $\lambda \leq d$. Thus to study atoms in this way, we would need light of short wavelength. Because of the relation $\lambda = h/p$ [Eq. (40–3)], our light consists of photons with large momentum. The fact that matter has wavelike properties extends this limitation $(p > h/d)$ to the case in which we scatter massive projectiles—electrons, for example.

In Chapter 8, we discussed the connection between the momentum (or energy) of a projectile and the size of the target that can be studied. Let's review that subject, using our knowledge of diffraction patterns (Chapter 38). The characteristic angular size of the diffraction pattern made by a projectile of wavelength λ scattering from an object of size d is $\sin\theta \cong \lambda/d$. The size of the momentum transfer in a scattering experiment is characteristically $\Delta p \cong p\sin\theta$, as we learned in Chapter 8. Therefore,

$$\Delta p \cong p\sin\theta \cong p\frac{\lambda}{d} = \frac{h}{d}, \tag{45–1}$$

where we have used the de Broglie momentum–wavelength relation, $p = h/\lambda$, in the last step. Thus an increasing momentum is needed to study ever smaller objects. In the laboratory frame (a moving projectile and a fixed target particle), much of the energy of the system is contained in the motion of the center of mass, which, because of momentum conservation, continues unaltered after the collision. This energy is unavailable for probing the target. The correct statement is that *Eq.* (45–1) *actually applies to the center-of-mass momentum and momentum transfer*.

The diffraction limit expressed in Eq. (45–1) is essentially contained in the Heisenberg uncertainty relations. Recall that the momentum–position uncertainty relation is $\Delta x\,\Delta p > \hbar$, Eq. (40–12), where $\hbar = h/2\pi$. If we interpret d as the position uncertainty Δx, then the connection is evident.

Let's look at some numerical aspects of Eq. (45–1). An atomic radius is approximately 10^{-10} m. The nucleus is roughly 10^5 times smaller, and we can take the size of a single proton to be characteristically 10^{-15} m. If we want to see the structure within a region of radius 1 percent of the size of the proton, we want $d \cong 10^{-17}$ m. According to Eq. (45–1), this requires us to have a momentum transfer of magnitude $\Delta p \cong 6 \times 10^{-17}\ \text{kg}\cdot\text{m/s}$ in our collider. This corresponds to a particle energy that can only be produced in a large accelerator. The highest-energy accelerator currently operating is that of Fermilab, near Chicago, and it circulates protons of momenta ten times this amount. In the Large Hadron Collider (LHC), now under construction near Geneva, Switzerland, the colliding protons will have thirty times this momentum. From Eq. (45–1), we see that the distance to be probed at the LHC is on the order of $d \cong 10^{-19}$ m.

EXAMPLE 45–1 Alpha (α) particles with kinetic energy of 2 MeV are scattered by atoms of a heavy element. The experiment shows that a fraction of the α particles are scattered through 90° in the laboratory frame. What can one deduce from this about the size of the target responsible for the scattering?

Setting It Up We shall use Eq. (45–1) to estimate the dimensions of the scatterer, with $\sin\theta$ of order 1. We must therefore calculate the momentum of a 2-MeV α particle, which is a particle with mass very close to 4 times a nucleon mass. We need not worry about relativistic effects, since the kinetic energy is very small compared with the rest mass energy of the α particle, and since the target is an atom of a heavy element, we need not worry about the distinction between the center of mass frame and the laboratory frame. In other words, the scattering may be viewed as from an infinitely heavy atom.

Strategy The relation $p = h/(d \sin\theta)$ allows us to estimate the size of the target, $d = h/(p \sin\theta)$. This equation shows that a large target leads to small-angle scattering for a given p, whereas a small target leads to large-angle scattering for the same p. In terms of the (nonrelativistic) kinetic energy $K = p^2/2M$, we obtain

$$d = \frac{h}{\sqrt{2MK}\sin\theta}.$$

The problem statement states that the scattering can take place at 90°, and we find the smallest value of d—the target size—by replacing $\sin\theta$ by unity.

Working It Out With $\sin\theta = 1$, and the α mass equal to four nucleon masses,

$$\begin{aligned} d &= \frac{h}{\sqrt{2MK}\sin\theta} \\ &\cong \frac{6.6 \times 10^{-34}\ \text{J}\cdot\text{s}}{\sqrt{(2 \times 4 \times 1.67 \times 10^{-27}\ \text{kg})(2\ \text{MeV})(1.6 \times 10^{-13}\ \text{J/MeV})}} \\ &\cong 10^{-14}\ \text{m}. \end{aligned}$$

This number is a factor of about 10^4 smaller than an atomic radius. An experiment of this type was carried out by Rutherford at the beginning of the 20th century and led Rutherford to the conclusion that the particle that was responsible for the scattering of the α particle had to be very much smaller than the atom.

What Do You Think? Could the electrons in the atom have been responsible for the large-angle scattering of the α particles? *Answers to **What Do You Think?** questions are given in the back of the book.*

EXAMPLE 45–2 The proton has a radius of approximately 1.2 fm. You want to use electrons as projectiles to study the structure of protons on a scale of 0.5 to 1.2 fm. Assuming that the maximum observable electron-scattering angle in the center-of-mass system is 30°, what center-of-mass momentum must the incident electrons have? Is it necessary to use relativistic kinematics?

Strategy The uncertainty relation combined with the de Broglie relation can tell us the relation between the projectile momentum p and the distance scale d of the target observable with a scattering experiment:

$$p \cong \frac{h}{d\sin\theta}.$$

To decide whether this involves relativistic kinematics, we can compare the quantity (momentum × speed of light) with the rest energy of the electron. If $pc > m_e c^2$, the motion is indeed relativistic.

Working It Out Given that $p > h/(d\sin\theta)$, the largest value of p is determined by the smallest value of d, namely, $d = 0.5$ fm:

$$\begin{aligned} p &\cong \frac{6.63 \times 10^{-34}\ \text{J}\cdot\text{s}}{(0.5\ \text{fm})(\sin 30°)} = \left(\frac{6.63 \times 10^{-34}\ \text{J}\cdot\text{s}}{(0.5\ \text{fm})(0.5)}\right)\left(\frac{1\ \text{fm}}{10^{-15}\ \text{m}}\right) \\ &= 2.7 \times 10^{-18}\ \text{kg}\cdot\text{m/s}. \end{aligned}$$

Thus

$$\begin{aligned} pc &= (2.7 \times 10^{-18}\ \text{kg}\cdot\text{m/s})(3.0 \times 10^8\ \text{m/s}) \\ &= (8.0 \times 10^{-10}\ \text{J})\frac{1\ \text{eV}}{1.6 \times 10^{-19}\ \text{J}} = 5.0 \times 10^9\ \text{eV}, \end{aligned}$$

or 5.0 GeV. Because $m_e c^2 \cong 0.5$ MeV for an electron, the value of pc required is some 10,000 times the rest mass, and the motion is highly relativistic.

Note that an electron with 5 GeV of energy in the center-of-mass reference frame is an electron with much higher energy in the laboratory reference frame, namely about 53 GeV, or 53,000 MeV (see Problem 8).

What Do You Think? The target proton is about 2000 times more massive than the electron. Is its motion also relativistic in the center-of-mass reference frame?

Elastic and Inelastic Scattering

We have confined our discussion to elastic scattering to this point, and elastic scattering has indeed been a fruitful process to study. In Chapter 40, for example, we discussed Compton scattering, in which the scattering of light by electrons provides strong evidence for photons, the particles of light. We also know about Rutherford's experiments on the elastic scattering of α particles by atoms and how they demonstrate the presence and size of an atomic nucleus. The elastic scattering of electrons by protons has helped show that protons are not pointlike and explore their charge distribution.

Inelastic scattering is another type of process that can supply information about the structure of scattering particles. The oldest example arises in spectroscopy: A beam of electrons in a cathode ray tube collides with atoms of a gas that fills the tube, neon atoms for example. The electrons excite the atoms from their ground state to an excited state that then decays into a photon and neon in the ground state. This is an inelastic collision. Depending on which atomic level the electron in the atom is excited to, the photon (symbol γ) in the decay process will have the energy that is predicted by energy conservation. Consider, for example, this process in a neon gas, namely

$$e^- + \text{Ne} \rightarrow e^- + \text{Ne}^* \rightarrow e^- + \text{Ne} + \gamma.$$

We can use the characteristics of the process to study the excited states of neon, denoted by Ne*. The observed energy of the photon tells us the energies of the various Ne* states and how strongly these processes occur helps us discover features such as the angular momentum of these states.

To take a different example, consider the scattering of protons by protons. At low energies we can study the shape of the proton–proton potential energy function with the elastic process

$$p + p \rightarrow p + p.$$

Beyond a certain threshold, in the vicinity of projectile energy 290 MeV with a stationary target, there is a large inelastic component, with a large cross section[†] for the process

$$p + p \rightarrow p + n + \pi^+.$$

On the right, we have a new particle, the π^+ (positive pion), first seen in this way in the late 1940s, when accelerators powerful enough to produce protons in the 300 MeV range became available. Its discovery was not unexpected, having been predicted by Hideki Yukawa in 1935 (see Section 45–3 for more).

The pions (π^+ as well as its neutral and negatively charged partners π^0 and π^-, respectively) are produced copiously in reactions like those cited above. While not stable, the pion has a lifetime long enough to allow one to form beams of pions by deflecting them and then focusing them with strong magnetic fields. One can then observe how pions scatter from protons, as for example in elastic scattering,

$$\pi^+ + p \rightarrow \pi^+ + p. \qquad (45\text{–}2)$$

The cross section for the process in Eq. (45–2) was found to have a marked peak as a function of the energy in the initial configuration (Fig. 45–2a). This peak corresponds to the production and subsequent decay of the analog of an atomic excited state, or resonance—a state of rest mass 1232 MeV/c^2 and angular momentum 3/2 (in units of $\hbar$)—that we label with the symbol Δ (Fig. 45–2b). The fact that the cross section has a peak of finite width means that the mass of the Δ resonance (and other states like it) has a width that corresponds to a spread in the value of the mass. This is consistent with the energy-time uncertainty principle, in which the lifetime represents the spread in time, and the width of the resonance in energy represents the energy uncertainty (see Section 40–3 and Conceptual Example 45–3).

The sequence of events that produce Eq. (45–2) can be written as

$$\pi^+ + p \rightarrow \Delta; \quad \text{then } \Delta \rightarrow \pi^+ + p. \qquad (45\text{–}3)$$

Just as atoms have many excited states of different quantum numbers, studies of inelastic collisions reveal a very large number of such excited states of the proton analogous to the Δ, but with different spins. We term these states *resonances*, given the shape of the cross section as a function of energy. As for the Δ resonance, the width of any of these resonances is associated with its lifetime.

[†]See Section 19–7 for a discussion of the cross section.

(a)

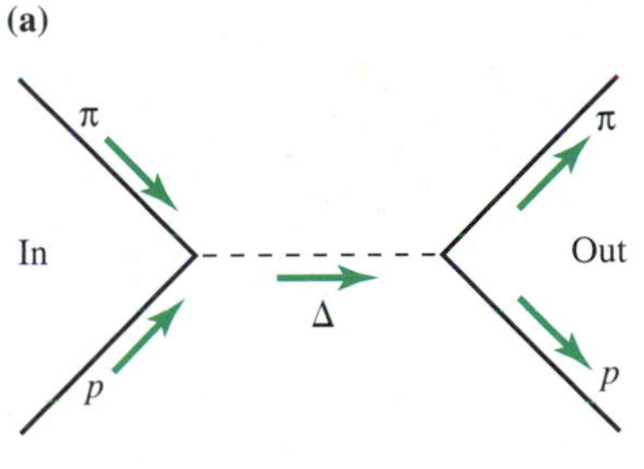

(b)

▲ **FIGURE 45–2** (a) The cross-section for the elastic scattering of positive pions from protons exhibits a peak at a center-of-mass energy of about 1230 MeV. This peak has a substantial width. (b) The interpretation of the peak is that the pion–proton system has a resonant state, or an unstable particle, the Δ. The width of the peak is interpreted according to the uncertainty principle in terms of the lifetime of the Δ.

CONCEPTUAL EXAMPLE 45–3 The rest mass of the Δ resonance, including its width, is given by $Mc^2 = (1232 \pm 60)$ MeV. A more recently discovered particle, the J/ψ, has a mass given by 3096.87 MeV ± 90 keV. Are both particles unstable? If so, which has the longer mean life?

Answer The very fact that the masses of these "particles" have an uncertainty is an indication of a finite lifetime and hence of instability. The term "resonance" is in fact just another term for "unstable particle." The time-energy uncertainty relation $\Delta E \Delta t \cong h$ shows that the mean life of an unstable system is inversely proportional to the energy spread, which translates to a spread in rest mass. In this case the Δ has a much larger (by a factor of 600) uncertainty in its rest mass, and therefore the Δ has a mean life that is shorter by that same factor.

As we have already argued, pion beams allow us to study processes such as the ones that we used to look at the Δ:

$$\pi^+ + p \rightarrow X^{++}, \quad \text{or} \quad \pi^- + p \rightarrow X^0.$$

The quantity labeled by X can consist of an unstable particle (or as we shall term it, a "state") like the Δ, but with different quantum numbers (see below). The collection of all these states is known as the **hadrons**. The origin of this name is in the Greek word for "strong," appropriate because the particles involved in the reactions are all associated with nuclear forces. Today the hadrons are understood as being composed of a small group of more fundamental building blocks of matter called **quarks**—an example of the "Russian doll" picture. We'll discuss hadrons and quarks further in Sec. 45–3.

In Fig. 45–3 we outline the constituent structure of matter, analogous to the Russian doll. A figure of this type gives us an oversimplified view of the larger picture, and we shall want to look more deeply into that picture.

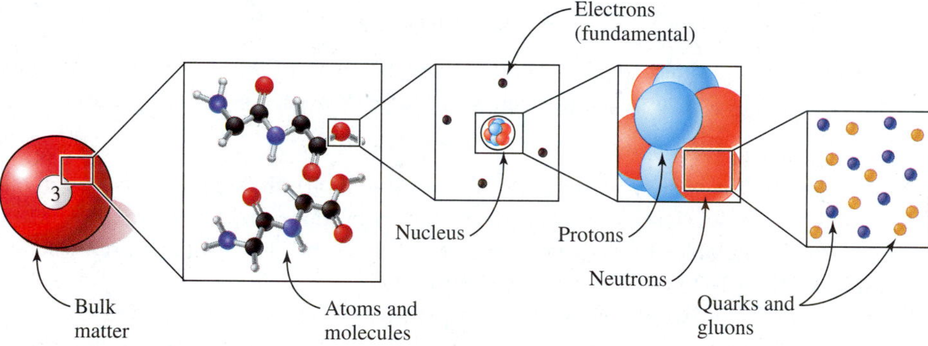

▶ **FIGURE 45–3** A schematic picture of the hierarchical structure of matter.

45–2 New Quantum Numbers

The Discovery of Antimatter

Relativity, which has thus far entered our discussion only in connection with kinematics, turns out to play a fundamental role in the physics of elementary particles. In 1928 Paul Dirac discovered an equation describing electrons that combines quantum mechanics and relativity. One of the predictions of this equation is that the electron should have a partner which could be called an anti-electron. This new particle, the **positron**, had to have exactly the same mass as the electron, the same spin $\frac{1}{2}$, but a positive charge (hence its name). When a positron makes contact with an electron, the two particles can annihilate each other, with the consequent production of neutral particles. At low energies, the annihilation produces photons, the most common reaction being

$$e^+ + e^- \rightarrow 2\gamma. \tag{45–4}$$

The products of such an annihilation reaction must have total charge zero, since charge conservation continues to hold in quantum mechanics. (Charge conservation is believed to be exact, by which we mean that it is *never* violated.) In fact, in relativistic quantum mechanics *all* particles have their corresponding antiparticles: the proton has its antiproton, the neutron its antineutron, and so on. The photon is truly neutral and is said to be its own antiparticle. (Why then is the neutron not its own antiparticle? We'll discover the answer to this in the next subsection.)

The positron was observed just three years after its prediction, and although it does not persist in nature—we live in a electron-rich world and a positron will sooner or later meet an electron and annihilate—it is easily produced in inelastic scattering processes. It can be slowed down and become bound to an electron for short periods, giving rise to a hydrogen-like neutral "atom" called *positronium* whose presence is signaled by a characteristic spectrum.

It is worth noting that in the description of processes, the "cancellation" of matter and antimatter implies that in a reaction one can transfer a particle on one side of a reaction to the other by replacing it with its antiparticle. (The underlying process is still subject to energy and momentum conservation.) Thus, assuming that there is sufficient energy available in the initial state, a reaction such as

$$e^- + p \to p + e^- + \gamma \tag{45–5}$$

will be closely related to the following processes:

$$\begin{aligned} e^- + e^+ &\to \bar{p} + p + \gamma \\ p + \bar{p} &\to e^- + e^+ + \gamma \\ \gamma + p &\to p + e^- + e^+. \end{aligned} \tag{45–6}$$

Baryon Number

There are some reactions, or decays, that are never seen. When this happens it may be because of the conservation of quantum numbers. To take an example, we never see the energetically possible decay $p \to e^- + \gamma$, and you could quickly say that that decay is forbidden because at the least it violates the law of charge conservation. Other examples conserve charge and are energetically possible, yet still do not occur in nature; for example,

$$p \to e^+ + \gamma, \quad \text{or} \quad p \to e^+ + e^+ + e^-. \tag{45–7}$$

Similarly, the annihilation process

$$e^- + p \to \text{photons} \tag{45–8}$$

does not occur—if it did, there would be no atoms in the universe, only photons. Finally, the electrically neutral neutron is *not* its own antiparticle. If it were, then nuclei containing more than one neutron would be blown apart by the energy generated in the annihilation of a neutron with another neutron (= anti-neutron). To explain all this, Eugene Wigner introduced a new quantum number analogous to the electric charge, the **baryon number** B. By assigning appropriate values of B and insisting on *baryon number conservation*, it is possible to explain the absence of these reactions. The following assignments will accomplish the task of ruling out unobserved reactions:

$B = 1$: proton and neutron;

$B = -1$: antiproton and antineutron;

$B = 0$: electrons and positrons, photons, pions.

For all practical purposes, baryon number conservation acts exactly like charge conservation in restricting the kind of final particles that can emerge from a given two-particle collision.

CONCEPTUAL EXAMPLE 45–4 Which of the following reactions are allowed and which are forbidden, and why? [*Hint*: Do not forget to pay attention to the conservation of angular momentum.]

(a) $e^+ + \gamma \to p + \gamma$;

(b) $e^+ + e^- \to p + \bar{p}$;

(c) $p + \bar{n} \to e^+ + \gamma$;

(d) $p + \bar{n} \to \pi^+ + \gamma$.

Answer (a) Forbidden. Baryon number conservation is violated. (b) Allowed. Both charge and baryon number are conserved; furthermore, there are two spin-$\frac{1}{2}$ particles on each side, so that angular momentum conservation is not manifestly violated. In an experiment, the initial energy must, of course, be large enough to create the proton–antiproton pair. (c) Forbidden. Baryon number and charge are conserved. However, the total angular momentum on the left side will always be an integer, while on the right side, which has only one spin-$\frac{1}{2}$ particle, the total angular momentum will always be half-integral, and angular momentum *cannot* be conserved. (d) Allowed. Baryon number and charge are conserved, and there are no manifest angular momentum conservation problems.

THINK ABOUT THIS. . .

HOW WOULD WE KNOW IF THE LAW OF BARYON CONSERVATION WERE VIOLATED?

We should start by saying that the question is badly posed. We can never make an absolute statement about a conservation law such as this; we can only set limits on how small an eventual violation has to be. And based on the remarkable stability of matter, we have very good limits on the conservation of baryon number. Consider the baryon-number violating decay

$$p \rightarrow e^{+} + \pi^{0}. \quad (45\text{–}9)$$

This decay is forbidden only by baryon number conservation. If it took place on a time scale much less than the age of the universe, it would make our universe atom-free. But might the reaction occur very rarely? This question is one that can only be decided by experiment, and a series of extraordinary experiments designed to measure the reaction show that *if the proton lifetime is not infinite, it is nevertheless greater than* 10^{30} *years*. How can we say this if the universe has existed for only some 10^{10} yr? We can measure such a lifetime by assembling 10^{30} protons and watching them carefully; if the lifetime were 10^{30} yr, we would expect one decay per year. In one experiment, a tank filled with 8000 tons of water (about 10^{32} protons) was surrounded by suitable detectors; no decays like those in Eq. (45–9) occurred over a period of several years (Fig. 45–4). The tank was placed in a salt mine deep underground to avoid interference from high-energy particles coming from outer space.

Although baryon number is conserved like electric charge, baryon number differs from electric charge in that there are no Coulomb-like forces associated with baryon number. Another way in which baryon number differs from electric charge is that according to certain ideas and under certain circumstances, the conservation of baryon number may be very slightly violated. Conditions in the early stages of the universe could have provided these circumstances.

▲ **FIGURE 45–4** The very pure water in this tank, in an underground laboratory in Japan, is surrounded by phototubes, detectors sensitive to the tiniest trace of light. Here the apparatus is being checked by a scientist with scuba equipment. The phototubes are activated after the lights are turned out; they are capable of capturing the tiny flash of light associated with a proton decay somewhere in the tank, if such a decay were to occur.

EXAMPLE 45–5 The discovery of the antiproton was made in collisions of a proton beam with a stationary proton target. What is the minimum energy required for the incoming proton so that an antiproton can be produced?

Strategy First of all, we need to see what reaction requires the minimum energy. In a collision $p + p$, the initial charge is $Q = 2$ and baryon number is $B = 2$. Thus the final state must have $Q = 2$ and $B = 2$. Since the antiproton has $Q = -1$ and $B = -1$, the remaining particles that form part of the final state must collectively have $Q = 3$ and $B = 3$. The smallest such aggregate is three protons, so that we are talking about the reaction

$$p + p \rightarrow p + p + p + \overline{p}.$$

To find the minimum energy necessary to make this process occur, we can first visualize what happens in the center-of-mass frame. There the two initial protons collide head-on, and the total momentum is zero. The lowest-energy state occurs when all the four particles in the final state just sit there, at rest. Any other state would require more energy, the kinetic energy of their relative motion. The picture of four particles all at rest relative to each other looks the same in any reference frame, except that the foursome all move with a fixed momentum. This then means that in the laboratory reference frame, the initial particles have momentum q and 0, respectively, and their energies are $\sqrt{(qc)^2 + (Mc^2)^2}$ and Mc^2, respectively. In the final state, we must again have total momentum q, but there is no internal momentum: The four particles act as if there were only a single particle of mass $4M$. Thus energy conservation implies that

$$\sqrt{(qc)^2 + (Mc^2)^2} + Mc^2 = \sqrt{(qc)^2 + (4Mc^2)^2}.$$

To answer the question posed, we can solve this expression for q, and hence for the incoming proton's total energy, $E_q = \sqrt{(qc)^2 + (Mc^2)^2}$. The kinetic energy is the total energy minus the rest energy.

Working It Out Squaring both sides of the energy conservation equation gives

$$(qc)^2 + (Mc^2)^2 + 2E_qMc^2 + (Mc^2)^2 = (4Mc^2)^2 + (qc)^2.$$

The factor $(qc)^2$ cancels, and we find

$$E_q = 7Mc^2.$$

The minimum proton kinetic energy required to produce an antiproton is then $6Mc^2$, or about 5630 MeV.

What Do You Think? Won't the antiproton sitting on top of three protons annihilate immediately, presenting a serious problem to your hopes of discovering the antiproton?

Leptons and Their Quantum Numbers

Both the electron and the neutrino ν are fundamental constituents of matter. These particles do *not* participate in the strong (nuclear) force and are known as **leptons**. We recall from Chapter 44 that the neutrino has no charge and little or no mass. But along with the electron, it does carry a conserved **lepton number** analogous to the baryon number. Both the electron and the neutrino, each with spin $\frac{1}{2}$, have the lepton number $+1$; their antiparticles e^+ and $\bar{\nu}$ have the lepton number -1. This law explains, for example, why the fundamental beta decay process $n \rightarrow p + e^- + \bar{\nu}$ can occur, while the process $\pi^+ + p \rightarrow e^+ + e^+ + n$ cannot.

The story of the leptons does not, however, end here. Less than a decade after the discovery of the positron, a new particle called the muon (symbol μ^-) was discovered. It has spin $\frac{1}{2}$, it has its own antiparticle (μ^+), and except for its mass (about $207m_e$), it appears to be identical to an electron. However, a certain number of reactions involving the muon are not observed even though they are allowed by charge, energy, and simple lepton number conservation. The most easily observed would be

$$\mu^- \rightarrow e^- + \gamma.$$

To account for the nonobservation of this reaction and of others, one introduces an electron lepton number *and* a muon lepton number, each of which is separately conserved. Neutrinos are associated with muonic processes in the same way that they appear in some electronic ones, and this means that we must put a subscript on neutrinos, with the muon having its own associated neutrino ν_μ, distinct from the electron neutrino ν_e. Each of these neutrinos carries its respective lepton number. The lepton quantum number assignment gives (e^-, ν_e) and (μ^-, ν_μ) electron and muon lepton number $+1$, respectively, with their antiparticles having lepton number -1. Thus for example the reaction $\mu^- + p \rightarrow n + \nu_\mu$ is allowed, while the process $\mu^- + p \rightarrow n + \nu_e$ is not.

Neutrinos interact so weakly that these ideas are hard to test, and it requires careful experiments to confirm the hypothesis. For example, it is possible to make a neutrino beam by using the decay

$$\pi^+ \rightarrow \mu^+ + \nu_\mu.$$

When the beam is directed at a target of nuclei, one learns that negative muons are produced in the (rare) collisions, never electrons. The appearance of an electron would imply the violation of separate muon lepton number and electron lepton number conservation.

CONCEPTUAL EXAMPLE 45–6 A group in a laboratory claims to have discovered a new, stable particle with positive charge. The group states that its mass is approximately 10 times that of the electron, and that when it collides with a proton, it leads to the annihilation of the proton with the production of three positrons and one electron. What can you say about the lepton and baryon number of this particle? Can it form hydrogen-like atoms with electrons? Why can you say with some confidence that such a particle cannot really exist? [*Hint*: All particles have antiparticles.]

(continues on next page)

Answer Let us label the particle by X. It is said to annihilate with a proton according to

$$X + p \to 3e^+ + e^-.$$

Baryon conservation implies that X has $B = -1$. Furthermore, charge conservation implies that X has unit positive charge. Finally, as far as lepton number is concerned, the lepton number on the right side is $N_e = -2$ and therefore the electron lepton number of the X is -2. Consider now a possible bound state of the X with an electron. While the X and the electron attract each other, they cannot annihilate, because the annihilation product would have to include an antiproton, and this is energetically impossible. Thus a hydrogen-like "atom" could indeed exist.

What is the argument that might lead one to doubt the existence of this state at a mass ten times that of the electron? If X exists, so does its antiparticle $\overline{X}$, which would have $B = 1$ and charge -1. But then the proton could decay through a process that is energetically allowed and charge- and B-conserving, namely

$$p \to \overline{X} + 3e^+ + e^-.$$

This proton instability would be quite visible.

45–3 The Fundamental Constituents of Matter: Leptons and Quarks

Decades of experiments with electrons and muons, at ever increasing energies, have revealed no evidence that these particles, the leptons, have a substructure. As far as we know, the pairs (e^-, ν_e), (μ^-, ν_μ) and their antiparticles are "pointlike." Pointlike is a good way to describe a fundamental particle; it says that the particle is not "made" of something more basic, in the way that atoms are made of electrons and nucleons. We might say that these particles are the smallest Russian doll, keeping in mind that other particles will also have this description. There is one complication—there is yet a third pair of leptons, denoted (τ^-, ν_τ), with their own conserved lepton number. But there is good experimental evidence that there are *only* three lepton pairs, so that the set of six leptons and their antiparticles make up all the leptons. The neutrinos of each of the three pairs are all much less massive than their charged partners, and indeed than the electron. The charged partners, the e^-, μ^-, and τ^-, have a large range in mass, with the τ some 3500 times more massive than the electron.

The Quark Model

The baryons and the mesons, or in other words the ensemble of hadrons, are certainly not pointlike, and there is evidence of a substructure. This substructure is described in terms of a set of fundamental constituents known as **quarks**.

The large number of hadrons discovered in scattering experiments is reminiscent of atoms and their excited states. The Bohr model of hydrogen allows us to explain all the states of hydrogen in terms of an electron, a proton, and the forces between them. Similarly, the *quark model* economically describes baryons and mesons in terms of a few constituents and their interactions. This model was proposed independently by Murray Gell-Mann and George Zweig in 1964.

In the quark model, neutrons, protons, and pions are composed of quarks, each with spin $\hbar/2$: the *u-quark* ("up" quark) and the *d-quark* ("down" quark). Both types of quarks have an antiparticle, $\overline{u}$ and $\overline{d}$, respectively. The quantum numbers of these (and other) quarks are listed in Table 45–1. The electric charge of the quarks is a fraction of the electron charge. This does not mean that charge is no longer quantized, but that the

TABLE 45–1 • Properties of Quarks

Quark	Charge in units of e	Baryon no.	Spin in units of $\hbar$	Rest energy*
$u, \overline{u}$	$\frac{2}{3}, -\frac{2}{3}$	$\frac{1}{3}, -\frac{1}{3}$	$\frac{1}{2}$	1.5–4.5 MeV
$d, \overline{d}$	$-\frac{1}{3}, \frac{1}{3}$	$\frac{1}{3}, -\frac{1}{3}$	$\frac{1}{2}$	5–8.5 MeV
$c, \overline{c}$	$\frac{2}{3}, -\frac{2}{3}$	$\frac{1}{3}, -\frac{1}{3}$	$\frac{1}{2}$	1.0–1.4 GeV
$s, \overline{s}$	$-\frac{1}{3}, \frac{1}{3}$	$\frac{1}{3}, -\frac{1}{3}$	$\frac{1}{2}$	80–155 MeV
$t, \overline{t}$	$\frac{2}{3}, -\frac{2}{3}$	$\frac{1}{3}, -\frac{1}{3}$	$\frac{1}{2}$	174 ± 5 GeV
$b, \overline{b}$	$-\frac{1}{3}, \frac{1}{3}$	$\frac{1}{3}, -\frac{1}{3}$	$\frac{1}{2}$	4.0–4.5 GeV

*Since individual quarks have never been observed, the values are inferred indirectly, using theoretical input as well as experimental information.

true quantization is in units of $\frac{1}{3}e$. The fact that quarks and antiquarks have spin $\hbar/2$ means that the spin of an even number of quarks plus antiquarks is an integer times $\hbar$ whereas the spin of an odd number of quarks plus antiquarks is a half-integer times $\hbar$.

The proton is a composite of three of these quarks, *uud*; the neutron has the structure *udd*. The antiproton is made of the combination $\bar{u}\bar{u}\bar{d}$, whereas the antineutron is $\bar{u}\bar{d}\bar{d}$. A combination of three quarks can have a variety of angular momenta, and these different combinations successfully explain all the baryons thus far discovered. For example, among the observed baryons is a quartet of states, partners of the previously discussed Δ^{++}. All four, Δ^{++}, Δ^{+}, Δ^{0}, and Δ^{-}, have spin $3\hbar/2$ and baryon number $B = 1$. These are formed from the combinations *uuu*, *uud*, *udd*, and *ddd*, respectively. The quark model also accounts for mesons. Mesons have $B = 0$ and are bosons (their spin is an integral multiple of $\hbar$), and they are properly described as pairs of quarks and antiquarks. In particular, $\pi^{+} = (u\bar{d})$ and $\pi^{-} = (d\bar{u})$, whereas π^{0} is a combination of $u\bar{u}$ and $d\bar{d}$.

Complementary evidence for the existence of quarks is provided by scattering experiments. When electrons with high energy were first used to probe protons in a hydrogen target in the 1950s, the results quite early revealed a distribution of charge within the proton (that is, the proton was found not to be pointlike). Figure 45–5 shows the ratio of the observed distribution to the distribution a pointlike proton would give. If the proton were pointlike, this ratio graph would be a flat line rather than the observed curve. There was widespread expectation that further experiments would continue to verify that the proton's charge is distributed smoothly within it, much as the champions of the "plum pudding" model of the atom would have expected for α particles scattering from the atom. If the proton charge were distributed uniformly, electrons passing through the proton (and possibly breaking it apart) would be deflected very little. To the surprise of the physics community, a series of experiments of this type in 1968 at previously unexplored energies revealed a very different behavior: Electrons were scattered in large numbers at large angles and in a pattern characteristic of a proton constructed of three pointlike particles, each carrying a fraction of the proton charge. These particles are the very quarks that we used to describe the pattern of observed hadrons. The existence of a subset of hadrons related to the proton in a more complicated way was explained by the existence of yet another quark, the *strange* quark *s* (and its antiparticle). These three quarks are called the light quarks. In addition to the three so-called light quarks (u, d, s), there are three so-called heavy quarks, denoted by the letters (c, b, t). These were anticipated by theorists who found that certain interactions required three families of fundamental particles, and they are elements of still other series of hadrons.† We will learn more about what lies behind the quark model in the next section, in our discussion of the fundamental forces.

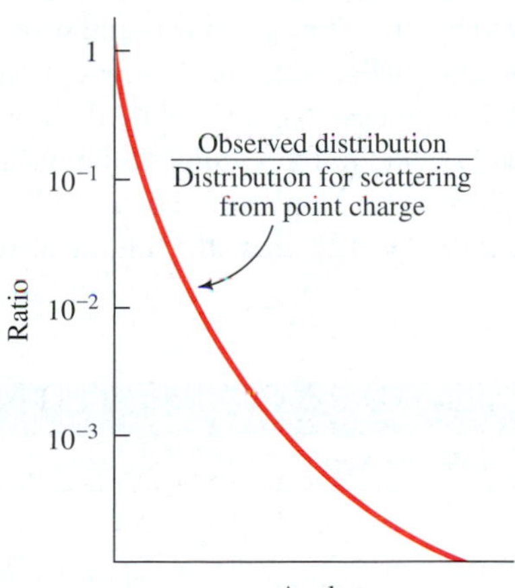

FIGURE 45–5 Elastic electron–proton scattering experiments show that the proton is not pointlike. The horizontal axis is the scattering angle; we have already seen that this quantity measures how closely the target is probed. The vertical axis is the ratio of the measured cross section to the cross section that would follow if the proton were pointlike. This curve would be flat, not decreasing, if the proton were pointlike. We know from other scattering experiments that the corresponding curve for quarks is flat, to the accuracy with which we can measure it.

†The $(udcstb)$ labels are often called *flavor* labels. They represent quantum numbers that are conserved in some aspects of the fundamental interactions but not in others.

EXAMPLE 45–7 What is the charge, baryon number, and lowest possible value of the total angular momentum of the following combination of quarks: $(u\bar{s})$, (tsd), $(c\bar{c})$?

Strategy We make use of the quantum numbers listed in Table 45–1. The charge and baryon numbers are additive, and working these out for the three combinations given is straightforward. In the case of angular momentum we note that all the quarks and antiquarks have spin $\frac{1}{2}$. Thus a two-quark system will have a total angular momentum of $\frac{1}{2} \pm \frac{1}{2}$ with possible orbital angular momentum added in as well. This means that for the two-quark system the lowest value of angular momentum will be zero: The spins point in opposite directions, and there is no orbital angular momentum. For a three-quark system, two of the spins can point in opposite directions, but a spin $\frac{1}{2}$ will always be left over. With no orbital angular momentum, that spin determines the lowest value of total angular momentum.

Working It Out

$$(u\bar{s}): Q = \tfrac{2}{3} + \tfrac{1}{3} = 1; B = \tfrac{1}{3} - \tfrac{1}{3} = 0, \quad \text{total angular momentum} = 0;$$

$$(tsd): Q = \tfrac{2}{3} - \tfrac{1}{3} - \tfrac{1}{3} = 0; B = \tfrac{1}{3} + \tfrac{1}{3} + \tfrac{1}{3} = 1; \quad \text{total angular momentum} = \tfrac{1}{2};$$

$$(c\bar{c}): Q = \tfrac{1}{3} - \tfrac{1}{3} = 0; B = \tfrac{1}{3} - \tfrac{1}{3} = 0; \quad \text{total angular momentum} = 0.$$

What Do You Think? Which of these particles can be its own antiparticle?

In spite of the great success of the quark model in classifying all of the observed hadrons, all attempts to find free quarks in collision experiments have thus far failed. The mystery of why single quarks have not been seen lies in the way in which they interact with each other to form nucleons, mesons, and all the hadrons. This is one of the themes of the next section.

45–4 The Fundamental Forces and Their Carriers

We first talked about fundamental forces in Chapter 5. Now we are in a position to add to that discussion. We can identify three fundamental forces acting in nature (Table 45–2): gravitation; the electroweak force, which is a unification of the electromagnetic force and the weak force originally associated with beta decay in nuclei; and the strong, or nuclear, force. There is very little we can say about gravitation beyond what we said in Chapter 12, because we do not know of a way to bring gravitation and quantum mechanics together. There is, however, a great deal more we can say about the other forces, and in order to set the stage for that discussion, let us consider how it is that forces are transmitted in a quantum-mechanical and relativistic world.

The Carriers of Forces: The Yukawa Theory

Our discussion begins with electromagnetism. Classically, charged particles—a pair of electrons, say—interact through the electromagnetic fields they set up. How does this result carry over into the quantum domain? In this domain the electromagnetic field is described by photons, and the interaction between the electrons occurs when one electron emits a *virtual photon* that is then absorbed by the other. To understand what we mean by a virtual photon, it is necessary to understand that an electron cannot emit a real photon and just remain an electron, because such a process would violate the conservation of energy and momentum. This is immediately evident if you think of the initial electron at rest; the final state would consist of an electron and a photon moving off back-to-back, and that configuration necessarily has more energy than the initial at-rest electron.

TABLE 45–2 • Fundamental Forces

Force	Particles on Which Force Acts	Particles that Transmit Force
Gravitation	All particles	Gravitons[†]
Electroweak force	Particles with electric charge or "weak" charge	Photons, weak bosons
Strong force	Quarks, gluons	Gluons

[†]Gravitons have never been observed in any sense, nor has the law of gravitation (general relativity) been satisfactorily integrated with quantum ideas.

But thanks to the uncertainty principle, quantum mechanics allows for our process to occur. The uncertainty relation, $\Delta E \Delta t \geq \hbar$, allows for deviations from energy conservation, but only for times Δt determined by the degree of energy uncertainty. During the period Δt the photon, whose production sidesteps energy conservation, is said to be *virtual*, and as long as the photon is reabsorbed (by the second electron) quickly enough, there is no measurable violation of energy conservation. A quantum-mechanical calculation of the effect of this virtual photon exchange shows that it gives rise to the usual Coulomb force of repulsion between two electrons and attraction between the oppositely charged electron and positron.

This way of thinking about forces in quantum mechanics came out of studies of the nuclear forces by Hideki Yukawa in 1934–35. Yukawa proposed that in analogy with the electromagnetic field responsible for the electron–electron interaction, there existed a new field, the *meson* field, whose quanta would play the role of photons. The characteristics of the nuclear force, and in particular the fact that it is a short-range force (Chapter 44), lead to an important difference between photons and the new quanta: The new quanta had to have a rest mass m. Their *virtual* existence now requires an energy deficit of at least mc^2. This is sustainable, as we argued above, for a time of the order of $\Delta t \cong \hbar/\Delta E < \hbar/mc^2$. The virtual particle could propagate a distance no larger than $c\Delta t$ in that time. Thus its effectiveness as a carrier of force is limited to *the range* of the force,

$$R < c\Delta t = \hbar/mc. \tag{45–10}$$

Yukawa's meson, the carrier particle of the nuclear force, is the *pion* mentioned earlier in this chapter. Using Eq. (45–10), we can estimate its mass, m_π, if we take $R \cong 1.4$ fm, an experimental value for the range of nuclear forces:

$$m_\pi = \frac{\hbar}{Rc} = \frac{1.05 \times 10^{-34}\ \mathrm{J \cdot s}}{(1.4 \times 10^{-15}\ \mathrm{m})(3.0 \times 10^{8}\ \mathrm{m/s})} = 2.5 \times 10^{-28}\ \mathrm{kg} = 0.15 m_p, \tag{45–11}$$

where m_p is the proton mass. Thus, $m_\pi c^2 \cong (0.15)(938\ \mathrm{MeV}) = 140\ \mathrm{MeV}$. Yukawa also calculated the potential energy as a function of particle separation r for the force associated with meson exchange:

$$U(r) = g^2 \frac{e^{-m_\pi rc/\hbar}}{r}. \tag{45–12}$$

Here g^2 is a measure of the strength of the force between the interacting particles, analogous to $e^2/4\pi\varepsilon_0$ for the Coulomb force. In fact, the Coulomb force fits Yukawa's framework perfectly well if it is regarded as the result of the exchange of a massless particle—the photon.

A couple of comments about Yukawa's idea are worth making:

1. The meson must be a boson; that is, its intrinsic angular momentum (spin) must be an integral multiple of $\hbar$. That is because the particles that emit and absorb the meson—the nucleons—are fermions, and if angular momentum is to be conserved, they cannot themselves emit or absorb a particle with half-integral spin and remain fermions.
2. Nuclear forces occur with nearly equal range between neutrons and neutrons, protons and protons, and neutrons and protons. This is a consequence of the fact that Yukawa's mesons occur in the three charge states π^0, π^+, and π^-, with the three particles having (nearly) the same mass.

The discovery of pions with all the predicted characteristics had a lasting influence on how we think about the fundamental interactions, even if we no longer think of the exchange of pions as representing a fundamental interaction.

The Electroweak Force

Let us turn back to a discussion of what today we regard as the fundamental forces. Just as electromagnetism is the result of a unification between two forces that were initially thought to be independent, electric and magnetic forces, so the electroweak force is the result of a unification of electromagnetism and the weak force. It will be useful to look first at these pieces of the electroweak force.

Quantum electrodynamics is a quantum version of classical electromagnetism. Quantum electrodynamics describes precisely how charged particles interact through the exchange of photons, massless particles with spin 1 (in units of $\hbar$). By itself, this description is the most accurate theory that we know of. It predicts certain effects of electromagnetism on the magnetic moments of electrons and muons to an accuracy of about one part in a billion, and so far a set of extremely accurate experiments have confirmed every unambiguous prediction.

The **weak interactions** take their name from the fact that their effects are extremely small compared to the effects of electromagnetism or the strong force. They are only visible because they lead to reactions that cannot occur through the other interactions. Among others, the weak interaction governs the muon decay reaction $\mu^+ \to e^+ + \bar{\nu}_\mu + \nu_e$ as well as beta decay, $n \to p + e^- + \bar{\nu}_e$, as well as some scattering processes, $\mu^- + p \to \nu_\mu + n$ for example. We can note that the latter two processes don't involve what we think of as fundamental particles, and indeed these two processes actually occur because the weak interactions operate at the quark level. Thus beta decay of the neutron involves the transformation of a d-quark in a neutron to a u-quark through the weak process $d \to u + e^- + \bar{\nu}_e$. You can verify for yourself that this is consistent with the quark model description of the neutron and proton.

We have spoken of the weak and electromagnetic interactions as separate things. A more fundamental understanding of the weak interaction came about in the late 1960s, when it became clear that the weak interactions were on some level very similar to the electromagnetic interactions. At first, the differences look overwhelming: In addition to differences in strength—manifest in the rate of reactions—the weak force has a very short range. These differences were reconciled by S. Weinberg, A. Salam, and S. Glashow, who recognized that there was a single more fundamental interaction, the **electroweak interaction**. They proposed that at *high* energies, above some critical energy, the weak interactions should be almost identical to the electromagnetic interactions, with a triplet of massless spin-1 particles acting as carriers of the force, much like the photon, with two of them charged and one neutral. These particles are the *weak bosons*, W^+, W^-, and Z^0. Processes involving the weak force can now be thought of in terms of exchange of these particles, subject to the conservation of electric charge, of baryon number, and of lepton number. Thus the underlying basis of β decay, namely $d \to u + e^- + \bar{\nu}$, can be visualized as a two-step process in which a d-quark within a neutron is transformed to a u-quark with emission of a W^- while the W^- then transforms into ("couples to") two leptons:

$$d \to u + W^-; W^- \to e^- + \bar{\nu}_e.$$

At energies below the critical value, the weak bosons W^+, W^-, and Z^0 acquire large masses, some 86 and 97 times the mass of a proton for the W (W^+ and W^- both) and Z^0, respectively. (By saying that a massless particle "acquires" a mass, we mean that the particle behaves as though it had a mass due to its own interactions with other elements of the theory.) The large mass explains the very short range of the weak interaction. The large mass also has the effect of making low energy processes in which exchange is involved occur at a much slower rate, thus explaining the "weakness" of the force. The unification is manifest only at very high energies, beyond our current ability to test directly, but even below these highest energies we can observe effects of the unification coming into play, thus giving experimental confirmation that these ideas are correct. For example, the production and decay characteristics of the W^+, W^-, and Z^0 are predicted, and these have all been well tested. Even more directly, the dramatically increased rate at which reactions such as $\bar{\nu}_e + p \to n + e^+$ occur at high energies is a consequence of the fact that the "weak" force becomes as strong as the electromagnetic force at high energies. The unification of the electromagnetic and weak forces is an important step in our understanding of the fundamental forces, and to this point, all the predictions of the electroweak unification that have been tested have confirmed that this unification is correct.

The Strong Force

Yukawa's original ideas concerning the hadrons must be modified to take account of what we now know are the hadronic constituents, the quarks. Thus our ideas about the

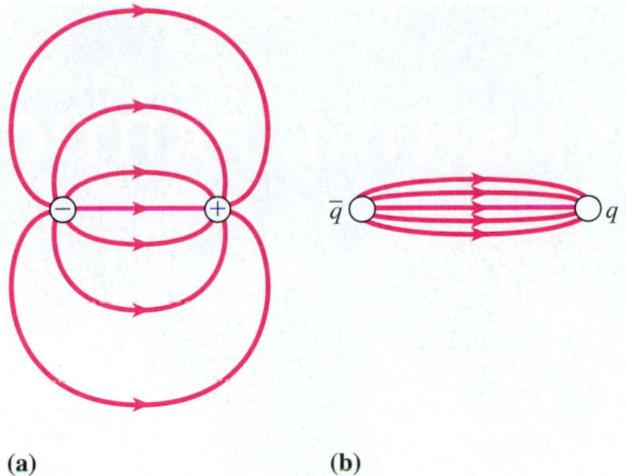

FIGURE 45–6 (a) Electromagnetic field lines between equal and opposite charges spread widely. (b) Gluon field lines between quark and antiquark remain within a "tube" connecting the quark–antiquark pair.

strong force are based on the theory of **quantum chromodynamics**, a theory that describes interactions between quarks and a set of particles called **gluons**, with spin $\hbar$, that act as the carrier particles. The quarks, whether u, d, s, c, t, or b, carry an additional (non-electrical) charge, called the color charge. It is analogous to the electric charge in that, like electric charge, it is both quantized and conserved. The quantum number should not be confused with the colors of the visible-light spectrum; it is merely a label to help us keep track of the charge, much as the label + or − keeps track of electric charge. The label is a bit more complicated than $\pm$, because while electric charge comes in two different types, the color quantum number occurs in three different types. Gluons are emitted and absorbed by particles with a color quantum number (quarks), just as photons are emitted and absorbed by particles with electric charge.

But there is a crucial difference between electromagnetism and the strong force: Photons do not carry electric charge, *but gluons carry the color quantum number and thus can themselves interact via the strong force through the exchange of other gluons*. This difference is crucial, and makes the forces between quarks and gluons quite different from other forces we have studied. One effect of this interaction is that single quarks cannot be isolated. When the interaction between a quark and an antiquark is worked out, it turns out that the gluon field lines (analogous to the electromagnetic field lines in the interaction between opposite charges) do not spread as for an electric dipole (Fig. 45–6a) but rather stay in tubelike regions, connecting quarks with antiquarks (Fig. 45–6b). Instead of being spread out all over space, the field energy is confined to what is roughly a cylindrical volume, and separating the quark from the antiquark would require more and more energy as the separation increases. In fact, a picture of what happens when the quark and antiquark are pulled apart is that the cylinder breaks and a new $q\bar{q}$ pair is produced (Fig. 45–7). Thus if one tries to separate a π-meson, say, into its constituent quarks one instead ends up breaking it up into two mesons. This picture bears some resemblance to what happens when you try to break a bar magnet into two pieces. Instead of ending up with an isolated N-pole and an isolated S-pole, one ends up with two bar magnets.

Even though the theory of quantum chromodynamics is well defined, we do not know how to calculate all the processes that are governed by it, largely because of a lack of mathematical tools. The technical details simplify at high energies, while at low energies some processes are easier to calculate than others. In spite of the calculational difficulties, there seems to be little doubt that quantum chromodynamics is the correct theory of the strong interactions.

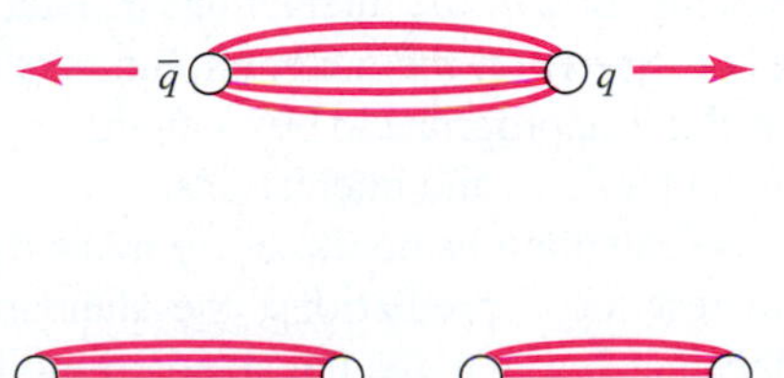

FIGURE 45–7 When an attempt is made to separate a quark from an antiquark by pulling them apart, it is energetically favorable to create a new quark–antiquark pair. This corresponds to the creation of two mesons from one.

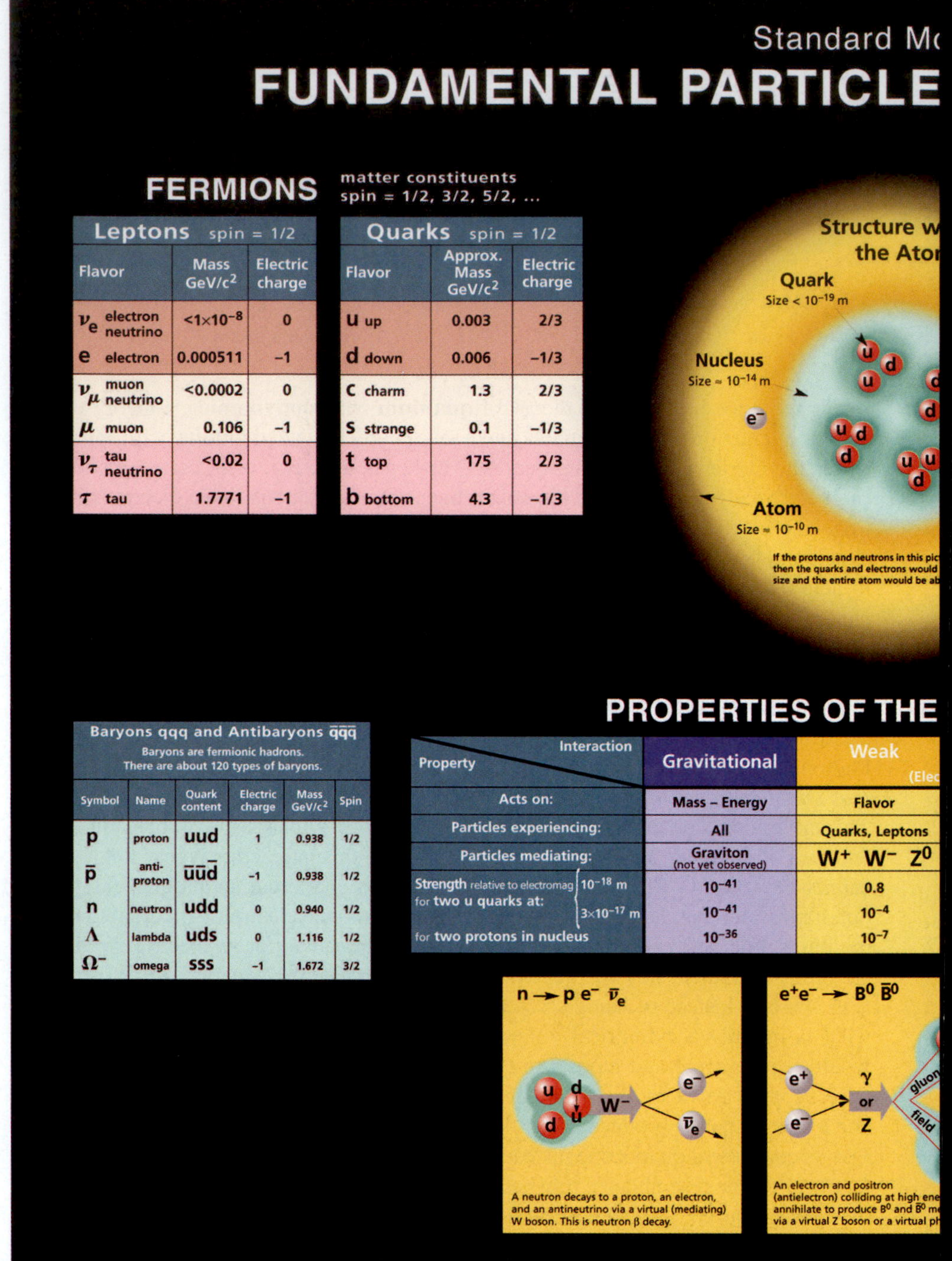

Leptons spin = 1/2		
Flavor	Mass GeV/c²	Electric charge
ν_e electron neutrino	<1×10⁻⁸	0
e electron	0.000511	−1
ν_μ muon neutrino	<0.0002	0
μ muon	0.106	−1
ν_τ tau neutrino	<0.02	0
τ tau	1.7771	−1

Quarks spin = 1/2		
Flavor	Approx. Mass GeV/c²	Electric charge
u up	0.003	2/3
d down	0.006	−1/3
c charm	1.3	2/3
s strange	0.1	−1/3
t top	175	2/3
b bottom	4.3	−1/3

Baryons qqq and Antibaryons $\bar{q}\bar{q}\bar{q}$
Baryons are fermionic hadrons. There are about 120 types of baryons.

Symbol	Name	Quark content	Electric charge	Mass GeV/c²	Spin
p	proton	uud	1	0.938	1/2
$\bar{p}$	anti-proton	$\bar{u}\bar{u}\bar{d}$	−1	0.938	1/2
n	neutron	udd	0	0.940	1/2
Λ	lambda	uds	0	1.116	1/2
Ω⁻	omega	sss	−1	1.672	3/2

Property \ Interaction	Gravitational	Weak
Acts on:	Mass – Energy	Flavor
Particles experiencing:	All	Quarks, Leptons
Particles mediating:	Graviton (not yet observed)	W⁺ W⁻ Z⁰
Strength relative to electromag for two u quarks at: 10^{-18} m	10^{-41}	0.8
3×10^{-17} m	10^{-41}	10^{-4}
for two protons in nucleus	10^{-36}	10^{-7}

▶ **FIGURE 45–8** A summary view of the fundamental particles and their interactions. Electric charges of the particles listed here are given in units of the proton's charge, and spins are given in units of $\hbar$. There are many more fermionic hadrons and bosonic hadrons than are described here, all of which can be accounted for by quantum chromodynamics. More complete copies of this chart can be obtained from Science Kit & Boreal Laboratories, 777 East Park Drive, Tonawanda, NY 14150. (Courtesy of Contemporary Physics Education Project.)

The Standard Model

Above we treated the electroweak and strong forces separately. But the same particles participate in these forces in ways that link reactions and decays. Therefore it is necessary to think about these forces, and the particles that interact through them, together. Additionally, there are self-consistency requirements in the form of the interactions that act between leptons and quarks. All this places strong constraints on the quantum numbers of the fundamental particles and the force carriers, and it is appropriate to bring all this together in what is called a **standard model** of fundamental particles and interactions.

To the extent that we can perform the calculations necessary to make a prediction or perform the experiments necessary to test those predictions, the standard model has passed every test. Figure 45–8 is a chart that summarizes the standard model and hence much of the current state of knowledge about fundamental particles and forces. There are still holes to fill—for example, we do not yet know the size nor the nature of neutrino

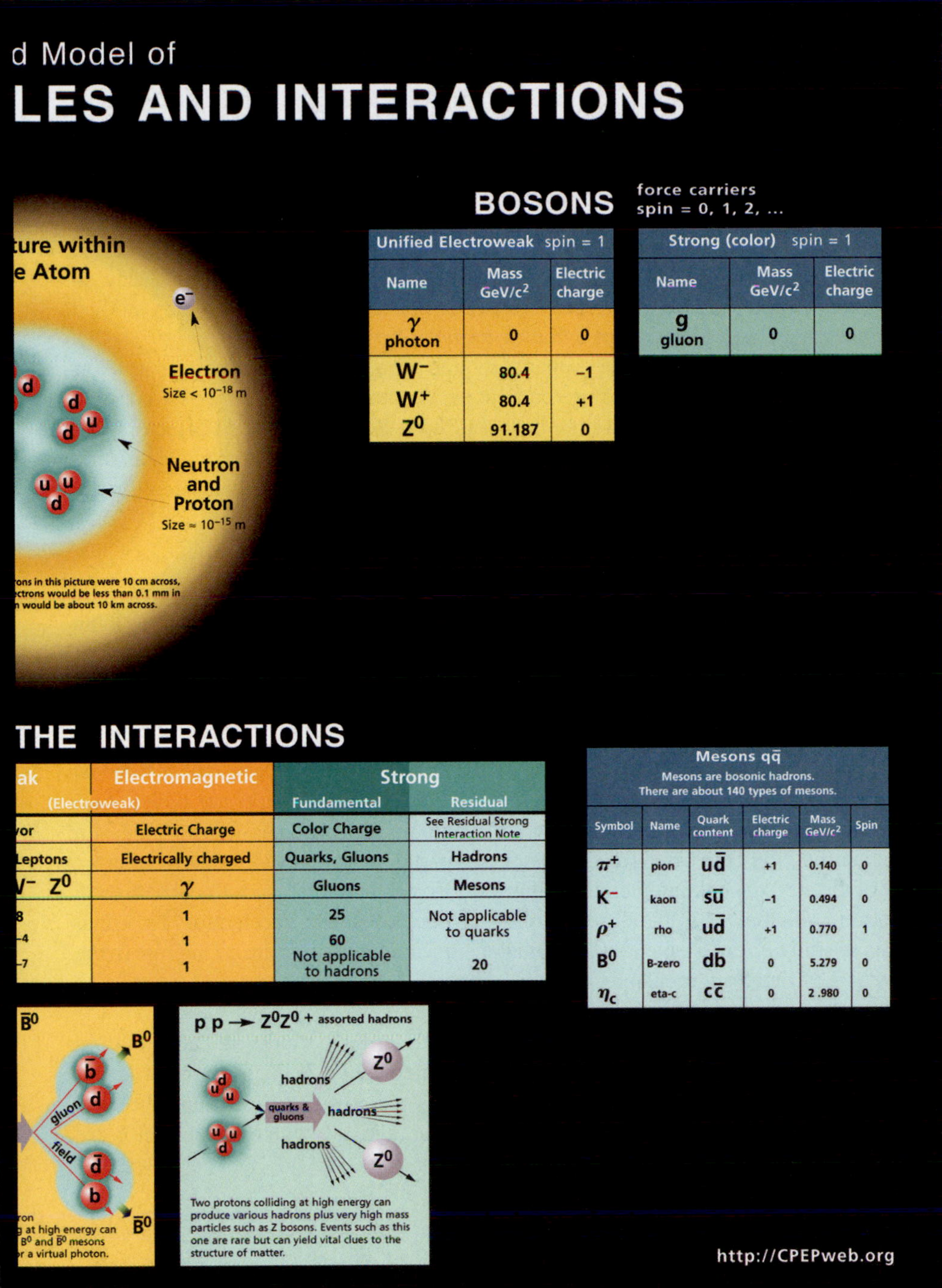

masses. And the standard model contains many parameters whose origins and values one would like to understand more deeply. In this way, the standard model sets out a firm platform from which searches for "new physics" can be launched.

The standard model is not really a single theory of both electroweak and strong forces as much as it is a way to bring these two theories together under a single umbrella. One important aspect of the standard model is that the theories of electroweak and strong forces are constructed in identical ways. There are of course differences; these lead to massive carriers for the weak interactions, massless ones for the electromagnetic and strong ones, and to confined gluons and quarks. Yet in many ways there are more similarities than differences, and these have led to attempts to construct a single unified theory. Still more ambitious attempts include gravitation, but although there are many interesting ideas around, there has been little testable progress in this area. Certainly a grand unification is one of the major goals of fundamental research. If we are lucky, progress will be made along these lines during the lifetime of this book!

45–5 The Tools of Particle Physics

Richard Feynman compared the process of probing the behavior of elementary particles to seeing how a watch works by throwing two watches at each other and studying the array of gears and springs that come out. Actually, the situation is somewhat worse. We require high energies to see well at short distances, but high energies bring with them relativistic effects such as pair production (pairs of "watches and antiwatches" are produced). In addition, because of confinement, the gears and wheels cannot themselves be final products—just more watches, perhaps of a different design but made from those same gears and wheels. Many of the particles that result are unstable and decay so quickly that we can see only their decay products. Finally, the collisions being studied are relatively rare, so a high flux of colliding particles is necessary.

The problem of designing experiments to study the behavior of particles at short distances is twofold: First, we require particle accelerators that have the best possible combination of high energies and high beam currents. Second, we need detectors that enable us to analyze the results of collisions. In constructing accelerators and detectors, physicists have pushed technologies to their limits. The results have been scientifically and technologically rewarding.

Accelerators

The principle of modern particle accelerators is the repeated acceleration of a stable charged particle by an electric field. A magnetic field controls the path that the beam of accelerated particles follows. Beyond these basic ideas, a variety of techniques are used in different accelerators.

Electrons or Protons: Electrons have revealed no substructure: They appear to be as elementary as quarks. In addition, it appears that electrons and quarks are *both* necessary in quantum mechanical models with calculable predictions. These facts suggest a possible unification between strongly interacting particles and particles that interact by means of the electroweak interactions. Thus there is as much interest in the observation of electron collisions as in the observation of proton collisions; accelerators exist that allow us to study pp, $p\bar{p}$, ep, and e^+e^- collisions. Proton collisions are of interest in the sense that they represent collisions involving quarks.

Fixed-Target Machines or Colliders: Until the 1960s, virtually all accelerators were *fixed-target machines*, in which an accelerated beam is incident on a target (Fig. 45–9). In a fixed-target machine, the beam is typically extracted and directed to a fixed target such as liquid hydrogen. The advantages of such an accelerator are

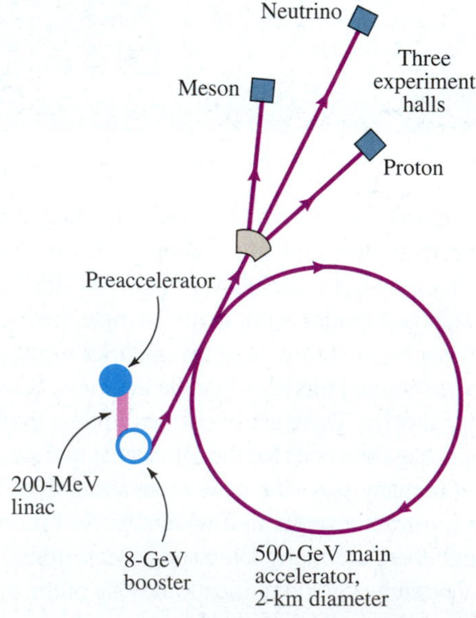

FIGURE 45–9 Schematic diagram of a fixed-target machine, in this case an early version of the accelerator at the Fermi National Accelerator Laboratory. Note the several stages of acceleration (beginning with the preaccelerator and a linear accelerator section, or linac) and the extensive target areas (experiment halls), where beams extracted from the ring of the main accelerator collide with fixed targets.

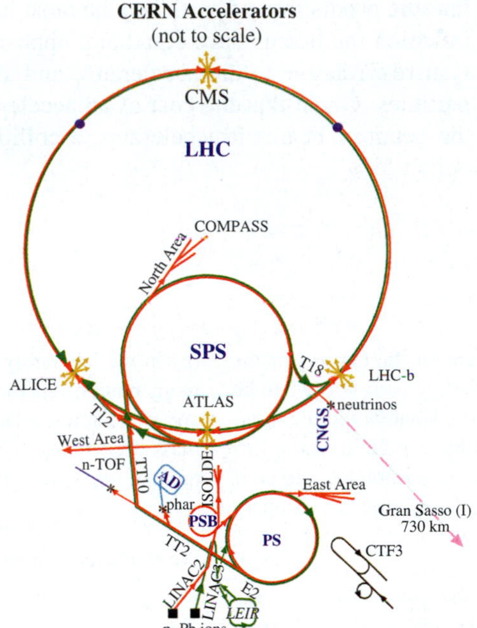

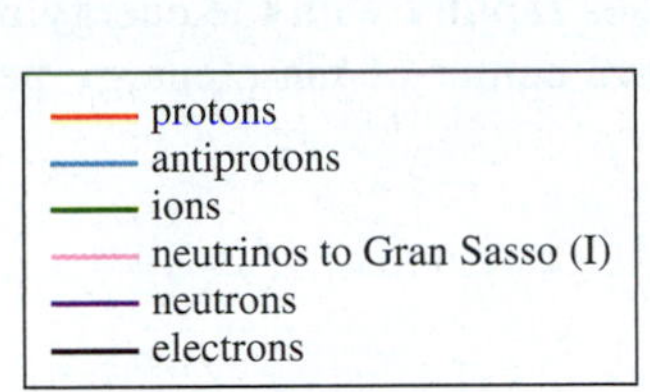

LHC: Large Hadron Collider
LHC-b: Detector probing matter-antimatter differences
SPS: Super Proton Synchrotron
AD: Antiproton Decelerator
ISOLDE: Isotope Separator OnLine DEvice
PSB: Proton Synchrotron Booster
PS: Proton Synchrotron
LINAC: LINear ACcelerator
LEIR: Low Energy Ion Ring
CNGS: Cern Neutrinos to Gran Sasso
ATLAS: General purpose detector
CMS: General purpose detector

▲ **FIGURE 45–10** Schematic diagram of the various CERN accelerators including the large hadron collider (LHC) currently under construction near Geneva, Switzerland. The entire LHC ring is not to scale—the large ring has a circumference of some 27 km. Protons and antiprotons are injected into the LHC accelerator and travel in opposite directions before colliding.

that a huge number of target particles can be spread throughout any volume and that there is only one beam to manage. These machines have systematically been replaced by *colliders*, in most of which two moving beams of particles with equal but opposite momenta interact in head-on collisions with one another (Fig. 45–10). Table 45–3 lists the characteristics of some advanced colliders. We know that the momentum of the center of mass of a system of interacting particles is constant. When a moving particle collides with a stationary target, the center of mass has nonzero momentum, which is maintained after the collision. Thus a large fraction of the incident energy remains in the motion of the center of mass rather than going

TABLE 45–3 • Characteristics of Some Colliders

Name	Laboratory, Location	Particles	Maximum Beam Energy (GeV)
TEVATRON	Fermi National Accelerator Laboratory (FNAL), Batavia, Illinois	$p\bar{p}$	1000
HERA	Deutsches Elektronen–Synchrotron (DESY), Hamburg, Germany	ep	e: 26; p: 820
Stanford Linear Collider (SLC)	Stanford Linear Accelerator Center (SLAC), Stanford, California	$e\bar{e}$	50
RHIC	Relativistic Heavy Ion Collider, Brookhaven National Laboratory, Brookhaven USA	pp	250
		Au-Au	100/u
LHC (under construction) Large Hadron Collider	European Laboratory for Particle Physics, CERN, Geneva, Switzerland	pp	7,000
		Pb-Pb	2760/u

into the production of particles, the most interesting part of the reaction. In colliders in which the beams have equal and opposite momenta, the system's center of mass is at rest relative to the accelerator, and all the energy goes into the production of particles. Given that the cost of an accelerator increases rapidly with the energy of the beam or beams it accelerates, a collider buys more center-of-mass energy per dollar.

EXAMPLE 45–8 The W^+ particle, one of the carriers of the electroweak force, has a mass of about 80 GeV/c^2. (a) For W^+ to be produced in the reaction $p + \bar{p} \rightarrow W^+ + \pi^-$, what would the minimum energy of the proton (and antiproton) have to be in a collider? (b) What would the minimum energy of the antiproton have to be if the same reaction were to occur in a collision with a stationary proton?

Setting It Up Figure 45–11 shows the $p\bar{p}$ collision in the frames of reference appropriate to the collider and to the fixed-target machine, with the momenta and energies appropriately labeled.

Strategy In both cases, we write down the energy and momentum conservation equations. In both cases the minimum energy required is one in which the W^+ and the π^- have no relative momentum, that is, they act as a single particle of mass $M_W + m_\pi$. Since the pion mass is a factor of about 500 smaller than the W mass, we can ignore it in the expressions for energy as a function of momentum.

Working It Out (a) We first express energy conservation in the center-of-mass reference frame, in which the initial particles have the same energy E. This yields $2E = E_W + E_\pi$. The minimum energy, $E = E_{min}$, corresponds to the W^+ and π^- both produced at rest, so $E_W = M_W c^2$ and $E_\pi = m_\pi c^2$. With $m_\pi \ll M_W$, we have $E_\pi \ll E_W$,

$$E_{min} \cong \frac{1}{2}M_W c^2 \cong \frac{1}{2}(80 \text{ GeV}) = 40 \text{ GeV}.$$

(b) In the laboratory reference frame, the total momentum is k, the momentum of the initial projectile $\bar{p}$. At the minimum energy, the W and π are at rest relative to each other and act like a single particle of mass $M_W + m_\pi \cong M_W$. Momentum conservation ensures that the momentum of the final W–π pair is k. Energy conservation then gives

$$E_{\text{lab, min}} + m_p c^2 = \sqrt{k^2c^2 + M_W^2 c^4}.$$

We square both sides of this equation to get

$$E_{\text{lab, min}}^2 + 2E_{\text{lab, min}} m_p c^2 + m_p^2 c^4 = k^2c^2 + M_W^2 c^4.$$

Finally, we use the fact that in the laboratory reference frame, the energy of the initial $\bar{p}$ is related to its momentum by $E_{\text{lab, min}} = \sqrt{k^2c^2 + m_p^2c^4}$. We square this relation and substitute for the term $E_{\text{lab, min}}^2$ in the preceding equation:

$$k^2c^2 + m_p^2c^4 + 2E_{\text{lab, min}} m_p c^2 + m_p^2 c^4 = k^2c^2 + m_W^2 c^4,$$

an equation we can solve for $E_{\text{lab, min}}$. Since $m_p \ll M_W$, we get

$$E_{\text{lab, min}} = \frac{M_W^2 c^4}{2m_p c^2} \cong \frac{(80 \text{ GeV})^2}{2(0.94 \text{ GeV})} = 3400 \text{ GeV}.$$

The huge difference between the answers to parts (a) and (b) illustrates the advantage of colliders. It is far cheaper to accelerate two beams to 40 GeV each than to accelerate one beam to 3400 GeV.

What Do You Think? The top quark has mass such that $M_t c^2 = 175$ GeV. Would you expect to need about $175/2 = 88$ GeV per particle to produce a top quark in a $p\bar{p}$ collider like that described above?

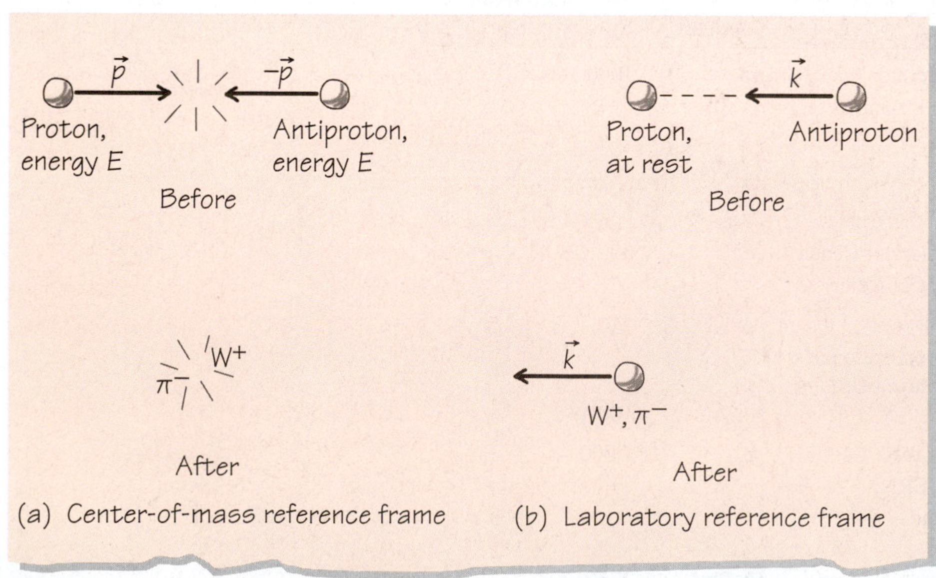

FIGURE 45–11 (a) The reaction as seen in the center-of-mass reference frame. (b) The reaction as seen in the laboratory reference frame.

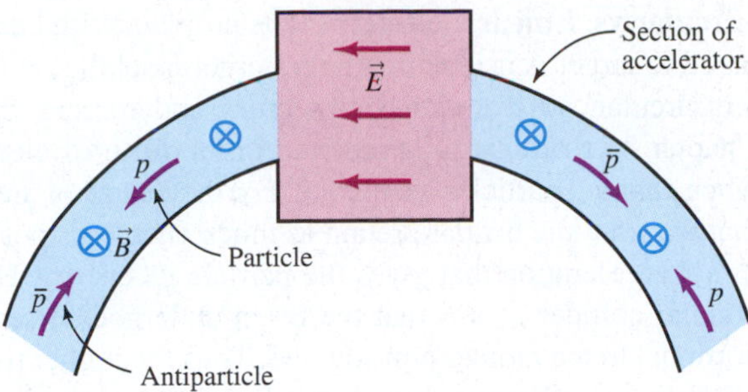

FIGURE 45–12 In a collider, the same magnetic fields guide both particle p and antiparticle $\bar{p}$, and the same electric fields accelerate both p and $\bar{p}$.

The difficulty in controlling the two beams of a collider is not as great as you might imagine. In the large ring of Fig. 45–10, within which both particles and antiparticles circulate in opposite directions, the same electric field that accelerates particles in one direction accelerates their antiparticles in the other direction, and the same magnetic field that forces charged particles to follow a counterclockwise path also forces their antiparticles to follow a clockwise path (Fig. 45–12). Thus the same fields can be used for a beam of both electrons and positrons, or protons and antiprotons. The major complication is that the two beams must be concentrated and steered so that they only cross at isolated points where the collisions occur; at those points, elaborate detectors, whose locations are marked on Fig. 45–10, monitor the results of the collisions.

We have shown that colliders are energetically advantageous. A second advantage of a collider concerns the distribution of the many particles that are produced in a collision. In the center-of-mass reference frame, the particles produced are spread over all angles (Fig. 45–13), and this is the way the reaction appears in a collider. The fact that the center of mass moves in a fixed-target machine means that the particles produced are carried forward with it and hence concentrated in a narrow cone in the forward direction. As Fig. 45–14 shows, this concentration makes it hard to separate the products.

Each beam of a collider provides a target for the other beam that contains far fewer target particles than does a fixed-target machine. Thus the rate at which collisions occur is much smaller in a collider than in a fixed-target machine. This is not always a disadvantage because the reactions we observe at very high energies produce so much information that even the best computers cannot keep up with data that arrive too rapidly. But when rare reactions are to be studied, a fixed-target machine has advantages.

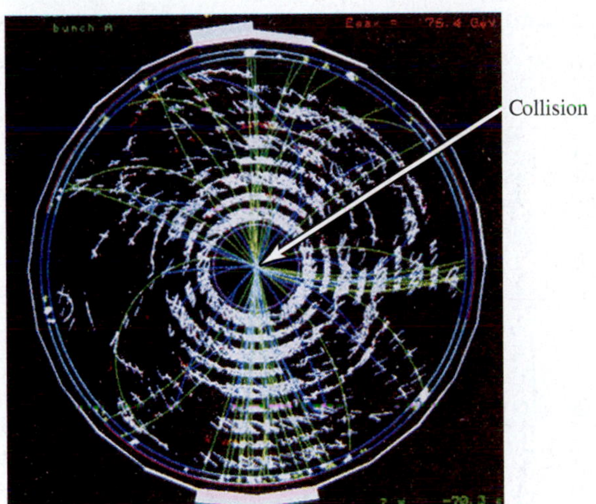

FIGURE 45–13 A collision at the TEVATRON collider at the Fermi National Accelerator Laboratory, as recorded in the Collider Detector Facility. The white and green lines are produced in the concentrically oriented cylindrical detectors and indicate the passage of particles. The view is an end-on one—that is, along the direction of the colliding beams. A substantial number of particles are produced at right angles to the beams.

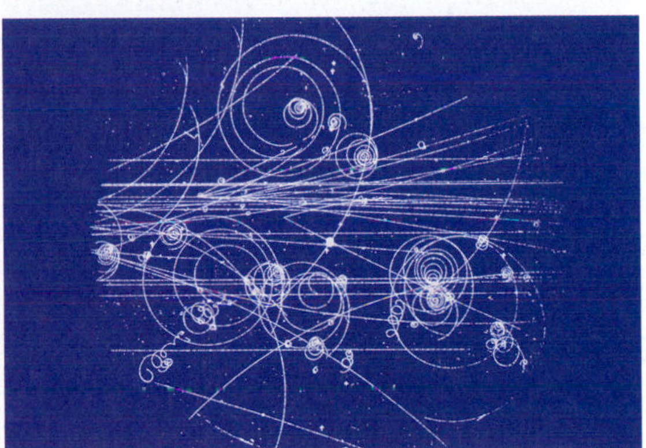

FIGURE 45–14 This image shows the tracks left in a bubble chamber by the passage of charged particles in a bath of liquid hydrogen colliding with the hydrogen in the chamber. By adjusting the pressure within the chamber, bubbles condense around the ions that are left along the trail of an energetic particle. The chamber is contained within a magnetic field, and the curvature of the tracks due to the presence of the field is evident.

Circular vs. Linear Colliders: The problem with beams that are not very dense (compared to target densities in a fixed-target machine) is in part overcome in colliders that are circular, so that the beams cross and recross frequently. However, recall from Chapter 34 that energy is radiated in the form of electromagnetic waves (or photons) when charged particles accelerate. For the beams of an accelerator, this is undesirable in that we want the beam to retain as much energy as possible. This is unavoidable for the initial acceleration that gives the particle its energy. However, the circular motion in a circular collider means that the beam undergoes a centripetal acceleration that is proportional to the momentum squared. Thus the higher the energy of the beam, the greater the radiation of energy. Moreover, the particles' acceleration is inversely proportional to their mass, so lighter particles, such as electrons, radiate energy at a higher rate than do heavier particles, such as protons. LEP, a now-defunct circular machine that accelerated electrons and positrons to 60 GeV each in a ring of radius 5 km was at about the limit for an electron collider in the form of a ring—at higher energies, the electron beam loses too much of its energy with each turn. The alternative is to construct a linear accelerator in which a new electron beam is repeatedly generated along a straight beam tube without following a circular trajectory. A linear collider—two linear machines with their beams meeting head-on—provides the best solution. The major disadvantage of such a machine, which has indeed been constructed on a demonstration basis (SLC in Table 45–3), is that the beams cross only once, so there is relatively little chance for collision. To overcome this difficulty, the beams are concentrated to only a few microns in size, a tiny fraction of the size of the beams of other accelerators; these beams must then cross with a remarkable precision. Such machines appear to be the only way to construct electron colliders with energies much greater than those of LEP.

Detectors

The devices that observe and analyze collisions in particle accelerators are called *detectors*. Detectors have undergone continual evolution since Joseph J. Thomson first detected the electron in 1897. Subnuclear particles can be detected because they leave a trace of energy as they pass through or decay within a detector. For example, a charged particle that passes through a *photographic emulsion* (a suspension used to coat photographic plates, film, and paper) leaves the same kind of trace that light does. Stacks of photographic emulsion formed the earliest detectors (Fig. 45–15)—most of their use occurred up to the mid-1950s—and allowed scientists to learn much about the charged particles that enter Earth's atmosphere (*cosmic rays*). The resolution of emulsions—their ability to pinpoint a particle track—is less than a micron (10^{-6} m). A similar detection technique uses the fact that a liquid near its boiling point forms bubbles of vapor along

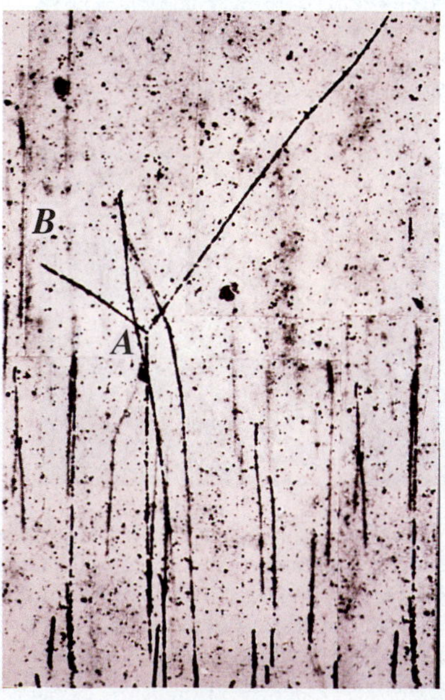

▶ **FIGURE 45–15** This image has been formed within a photographic emulsion. Where charged particles pass, a black line forms. At point A a reaction has occurred. One of the reaction products has either come to rest or decayed to neutral products at point B. Detailed track analysis can tell which.

FIGURE 45–16 A large modern detector facility such as the L3 detector in an accelerator at the European laboratory CERN, in Geneva, Switzerland is a huge and complex operation. Collisions of highly energetic beams take place deep within the detector. The multiple layers of detecting devices that surround the collision point analyze the results of the collision. The scale of this detector is impressive—note the catwalks in the photo.

the path of a passing charged particle. In use until the 1980s, *bubble chambers* filled with liquid hydrogen utilized this technique (Fig. 45–14). More modern types of detectors register the passage of a charged particle or of neutral particles, such as photons, by their characteristic effects on the electronic properties of the detector.

A detector may be sensitive to the total amount of energy a particle deposits as it is brought to a stop within the detector (a *calorimeter*), or the rate at which it loses energy as it passes through. In a supplementary technique, the particle's momentum is determined by measuring the amount of curvature in the path of a passing particle in a magnetic field of large, known magnitude within a detection region. Knowing both the momentum and energy, we can calculate the mass, so the particle can be identified. Particle identification is important and not so simple at high energies.

Photographic emulsions record everything that passes through them. Techniques that use electronic *triggers* avoid this difficulty. For example, suppose that you are interested in viewing only those events in which a muon (see Section 45–2) of high energy is produced. Muons pass through matter more easily than do protons or other strongly interacting particles. Then it is possible to surround the interaction region with a great deal of matter—such as many tons of steel—and outside this matter to place a detector that registers electronically the passage of charged particles. This detector emits a signal that the particle is likely to be a muon and thereby acts as a trigger to register the rest of the event. Modern detectors are huge and complex (Fig. 45–16). The electronic registration of the passage of dozens or even hundreds of particle tracks requires vast information storage facilities and powerful computers to extract the most useful information. The computers and analysis techniques developed in these experiments have filtered down to more general use. The World Wide Web is a direct by-product of the need of particle physicists to exchange large amounts of information.

45–6 Cosmology and the Expanding Universe

Giordano Bruno was burned at the stake in 1600 in large part for claiming that the universe is infinite. Although passions have calmed somewhat, people have always had a lively interest in the nature of the cosmos as a whole (*cosmology*). Powerful telescopes have allowed us to see far back into the past, and as a consequence, we know a great deal today about the very early stages of our universe.

Once Newton discovered the universal law of gravitation, it became possible to ask whether the universe was finite or infinite. Newton pointed out that in a finite universe, the seemingly uniform distribution of stars was unstable. Suppose two stars start moving closer to each other. They get further away from their other neighboring stars, so that any attraction impelling them back to their stable positions is decreased. As they get close together, they create a larger mass, which is then a stronger point of attraction for neighboring stars. You can easily see that there would be a tendency for ever more gravitational clumping of stars. Newton came to the conclusion that the only way to avoid this problem was for the universe to be infinite. This point of view, however, presents its own problems in the form of what is known as *Olbers' paradox*.

THINK ABOUT THIS. . .

WHY ISN'T THE SKY BLAZING WITH LIGHT?

Olbers' paradox was stated in 1823, when telescopic observations had made it clear that the universe contains many, many more stars than are visible with the naked eye. Heinrich Olbers argued that, if the universe were infinite and unchanging, then, regardless of the direction in which we looked, we would see a star and star light would cover the entire sky. We cannot dispute this argument by arguing that distance makes stars too faint to see. Suppose that N stars are contained in a thin spherical shell a distance R away from us. Then, because the area of a sphere increases as the radius-squared increases, there would be $4N$ stars in a shell a distance $2R$ away from us. The intensity of the light decreases as the distance-squared increases, so each star in the more distant shell is only one-fourth as intense as those in the nearer shell, but the total intensity is *independent* of the distance of the shell. The net intensity of an infinite number of concentric shells would be infinite. Yet—and this is what constitutes Olbers' paradox—the night sky is not bright. If the universe were infinite and had been in existence forever, as in Newton's picture, the paradox would be a real problem. But the Big Bang model (see below), in which the universe has existed for only a finite time and is moreover of finite size, resolves Olbers' paradox. If the universe has existed for only a finite time, then any one star has been radiating for a finite time, and there has not been enough time for the radiation from all stars—even if there were an infinite number of them—to have filled the universe and created a bright night sky.

The *science* of cosmology really started after Einstein discovered the general theory of relativity and with it the connection between gravity and the geometry of space. Einstein's starting point was based on the observations that the distant objects in the sky are (a) distributed isotropically—things look more or less the same whatever direction we look in, and (b) are distributed homogeneously—on the average there are the same number of objects in any large volume of a given size that one looks at. These basic assumptions form the **Cosmological Principle**. Einstein's original equations were found by Alexander Friedmann to lead to a nonstatic universe, which meant that the universe would expand, then possibly undergo a subsequent collapse. The Friedmann solutions soon found support in the work of the astronomer Edwin Hubble.

Hubble's Law

Before 1923, nebulae were thought to be glowing clouds of gas within our own galaxy. (It was only a few years before this that Harlow Shapley had determined that what we now know to be our galaxy has a definite shape and that our Sun is far from the galaxy's center.) In 1923, Edwin Hubble identified the nebulae as galaxies like our own and thereby took a giant step toward the understanding of the universe. Hubble used the Mount Wilson telescope in southern California, the first telescope sufficiently powerful to distinguish individual stars in the nebulae. Hubble also measured the spectral lines of stars in the galaxies and showed that, on average, those lines are redshifted, indicating a movement away from us. Hubble's law (1929) relates the average speed u with which another galaxy retreats from our own to a distance D away from us [Eq. (39–17)]:

$$D = \frac{u}{H}. \tag{45–13}$$

H is the *Hubble parameter*, measured to be $H \cong 2.5 \times 10^{-18}\ \mathrm{s}^{-1}$. The redshift gives the speed of recession directly, but how did Hubble know the distance of a given galaxy from our own? He made observations of the cepheids, a class of powerful stars whose systematic waxing and waning was found by Henrietta Leavitt in 1907 to be strongly correlated with their luminosity. The luminosity, or intensity of emitted radiation, of cepheids observed in distant galaxies could be compared with similar cepheids within our own galaxy, and using the inverse square law for the decrease in intensity, their distance could be determined.

There is also *independent* evidence that objects with large redshifts, such as quasars, are very distant: We see some of them through gravitational lenses (Chapter 39). This is possible only if these quasars are much farther than the galaxies that produce their images through the bending of the quasar light. The distances calculated from the lensing effect are consistent with those provided by Hubble's law.

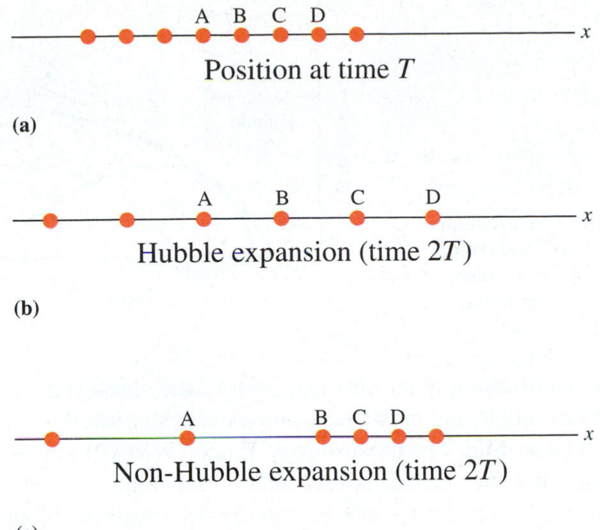

▲ **FIGURE 45–17** (a), (b) Two successive times in a one-dimensional version of Hubble's law. The points (galaxy positions, for example) remain equally spaced along the line. (a), (c) The same two successive times in a (non-Hubble) expansion law in which all the points recede from point A at the same speed, independent of their distance from A. Point A is a special point in the non-Hubble expansion law.

Hubble's law states that *the universe is expanding*. More accurately, we say that *space is expanding*. You might think that Hubble's law places Earth (or our galaxy) in a privileged position in the universe. In fact, it means just the opposite, as the following one-dimensional analogue shows. Figures 45–17a and 45–17b show a line of galaxies at times T and $2T$, respectively, after the moment that the expansion started, which we place at $t = 0$. At time T the galaxies are evenly spaced, and there is no preferred point in this arrangement. We have identified our galaxy (galaxy A) and supposed that Hubble's law holds. Therefore, galaxy C, which is twice as far from us as galaxy B, recedes from us at twice the speed with which galaxy B recedes. We see that at time $2T$, galaxy C is still twice as far from us as galaxy B. The galaxies continue to be uniformly spaced, and no point along the line is preferred. Hubble's law can be contrasted with a law in which other galaxies retreat from ours at constant velocity. While the galaxies may be evenly spaced at time T, this is no longer the case at time $2T$, and galaxy A occupies a special point (Fig. 45–17c). In fact, Hubble's law is the unique law for galactic speeds by which no one galaxy occupies a special point within the cosmos (see Problem 38). A two-dimensional analogue of the situation described by Hubble's law is provided by dots painted on a spherical balloon (Fig. 39–13). When the balloon is blown up, *all* the dots recede from each other, with no dot occupying any special position, and a law like Hubble's law holds. If we lived on the surface of such a balloon, any voyage we took would eventually take us back home, and we would conclude that there is no "outside."

It has taken centuries of discussion for humankind to understand that it is not at the universe's physical center. There *is* no center. It is not as though we are now living in a space in which an explosion once occurred; such a description would imply a privileged point, the position of the explosion itself. Rather, all of space began as a small volume *with no outside*.

Hubble's law with a constant value of H can be used to find the time required for galaxies to separate by given amounts (see Problem 42). Unfortunately, if the law is used as it stands to find the time at which all matter was together, that time is infinite (see Problem 43)! Thus if there was an initial moment when the universe started to expand, Hubble's law with a constant value of H cannot hold all the way back to that time. We can, however, use the law to make an estimate: If we assume that the recession speed is constant, Eq. (45–13) states that the expansion time (the *Hubble time*) is just H^{-1}, some 13 billion years. Other information (for example, theoretical knowledge of stellar evolution

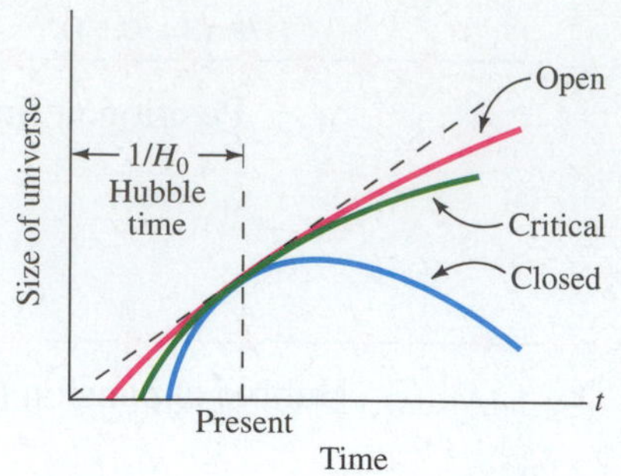

FIGURE 45–18 The evolution with time of open, closed, and critical universes. The vertical axis is some measure of the size of the universe.

and experimental identification of the characteristics of the oldest stars) confirms this estimate to within a factor of 2, and in fact measurements from a satellite-based experiment known as the Wilkinson Microwave Anisotropy Probe (WMAP) allows us to say with one percent accuracy that the universe is 13.7 billion years old.

Will the universe expand forever? Figure 45–18 illustrates some possibilities. Which of these outcomes holds depends, roughly speaking, on whether the gravitational pull of the energy contained in the universe is strong enough to slow down, and perhaps reverse, the expansion. A serviceable analogy is that of the escape velocity from a massive body. If there is too little mass, then an object hurled out with a certain energy will escape from the body; if the mass of the body is large, then the object will fall back. There is a critical density of energy for which the expansion will go on forever, but will slow down and come asymptotically to a stop at infinite time. This critical energy density is given by the formula

$$\rho_{\text{crit}} = \frac{3H^2}{8\pi G}, \tag{45–14}$$

where H is Hubble's parameter and G is Newton's constant. Less density than this and the expansion goes on forever; more density and the universe recollapses. This information is often stated in terms of the ratio Ω of the density to the critical density. If $\Omega > 1$ the universe will recollapse.

What is the value of Ω? If we look at luminous matter, mainly consisting of the 10^{11} stars in each of 10^{11} galaxies, we find $\Omega \cong 0.02–0.05$ or so. That, however, is not the whole story, since there is much evidence that *dark matter* exists (see Chapter 12), by which we mean matter that does not shine. Estimates of the amount of dark matter can be made in various ways, including the observed rotation profiles of galaxies and the frequency of gravitational lensing, and the conclusion to such estimates is that there is even more dark matter than luminous matter. Nevertheless, inclusion of dark matter boosts Ω only to 0.2–0.3.

A value for Ω of, say, 0.3 is a rather unsatisfactory number; in fact, any number other than $\Omega = 1$ presents problems. We can define a unit of time appropriate to cosmology which involves Planck's constant $\hbar$, c, and Newton's gravitational constant G. This unit is known as the *Planck time* τ_P:

$$\tau_P = \sqrt{\frac{\hbar G}{c^5}} \cong 5.4 \times 10^{-44}\ \text{s}. \tag{45–15}$$

In terms of this unit the universe is 10^{62} units old. This is *very old*, and any random value of $\Omega > 1$ in the early universe would have the universe collapse very quickly, or, if $\Omega < 1$, expand much too rapidly to allow for galactic evolution. In fact, calculations show that the only way to have a value of Ω in the vicinity of $\Omega = 1$ today is to have an early-universe value of Ω equal to one with a precision of one part in 10^{57}! The kind of fine tuning that would be necessary to lead to some value of Ω like 0.3 right now is very unnatural, and it is this kind of thinking that favors Ω *exactly* equal to one both in the early universe and therefore also today.

Current thinking about Ω relies on new observations that seem to suggest that the expansion of the universe, instead of slowing down, is actually accelerating. This can be interpreted as being due to something called *dark energy* which acts as a form of antigravity. These are new issues, and there are no compelling arguments about what the nature of this dark energy is. With the inclusion of dark energy, it is possible to arrive at $\Omega = 1$. All this is more than theoretical speculation: Measurements from the aforementioned WMAP experiment have led us to the conclusion that to within experimental error, $\Omega = 1$.

The Big Bang Model

The physicist George Gamow was the first to conclude that if other galaxies recede from ours at a speed proportional to their distance from us, then *at some time in the distant past, all galaxies must have been very close together*. This fact defines what is conventionally called the Big Bang model of the history of the universe.

Gamow was interested in how the elements were formed, and he went through the exercise of "running the film of the universe expansion backwards." One would see matter getting denser, and as we know from kinetic theory this would mean a rise in temperature. At some point the temperature would be such that electrons would be stripped from nuclei. At still higher temperatures—earlier times—the nuclei would break up into protons and neutrons, and in due course these would break up into quarks. (Of course, Gamow could not have known about this last step at the time he thought about this.) Let us look more closely at various stages in the lifetime of the universe.

We are currently living in what is called a *matter-dominated* era. The luminous matter in the universe is confined to galaxies and there is much empty space between galaxies. We may speak of an *average density* of matter, and it is small.

EXAMPLE 45–9 Given that at present there are 10^{11} stars per galaxy and on the order of 10^{11} galaxies, *estimate* the density of visible matter across the universe. You may assume that a typical star is like the Sun and that the size of the universe is of the order of 14×10^9 light-years.

Strategy The total mass M of visible matter is estimated by multiplying the number of stars, 10^{22}, by the mass of the Sun, 10^{30} kg. Then to estimate the density, we divide this mass by the volume,

$$\rho = \frac{M}{4\pi R^3/3},$$

with the radius $R = 1.4 \times 10^{10}$ light-years.

Working It Out Note that 1 light-year is the distance light travels in one year, or 3.15×10^7 s. Thus

$$\begin{aligned}\rho &= \frac{3M}{4\pi R^3}\\ &= \frac{3 \times 10^{22} \times 10^{30}\text{ kg}}{4\pi[(1.4 \times 10^{10}\text{ ly}) \times (3.15 \times 10^7\text{ s/y}) \times (3 \times 10^8\text{ m/s})]^3}\\ &\cong 10^{-27}\text{ kg/m}^3.\end{aligned}$$

Compare this tiny number with the density of liquid water, $10^3\ \text{kg/m}^3$!

What Do You Think? How would you express your answer in terms of hydrogen atoms per cubic meter?

As we go back in time, the density of matter increases, and with less gravitational potential energy there is more kinetic energy. The temperature thus rises. At temperatures of the order of 1.5×10^5K, atoms ionize, and we enter a period in which matter is in the form of a *plasma* of charged electrons and ions. Charged particles radiate, but the radiation is re-absorbed by the charged particles, and equilibrium between radiation and matter is established. As the temperature rises further and particle energies increase, electron–positron pairs are created. Moving backward in time, nuclei dissociate into protons and neutrons when kT approaches the 8–10 MeV range. Earlier still, the protons and neutrons fall apart into their constituent quarks. We are dealing with a kind of quark–gluon plasma, a state of matter not yet seen in the laboratory. At higher temperatures, quark–antiquark pairs are produced, and all of the particles, antiparticles, and gluons are in equilibrium. We can trace the evolution all the way back to 10^{-11} s after the Big Bang using physics that we understand!

It is instructive to go forward in time from a period when quarks have reassembled into neutrons and protons. This is about *one thousandth of a second* into the existence of the universe, when the temperature is about 10^{11}K. There are protons, neutrons, electrons, positrons, neutrinos and antineutrinos present and, of course, photons. This is a *radiation dominated* era, and the physics that governs the behavior of the fundamental particles at this temperature is well understood. The number of protons and neutrons is equal, because they can convert into one another through the weak interactions (the leptons must also play a role here). As the expansion proceeds, it takes about a second for the density to drop by a factor of 10^4. Neutrino interactions, which are important at high energies (high temperatures), become much weaker as the temperature falls, and they pretty well stop playing a role in the interconversion of protons and neutrons. It is still too hot for nuclei to form. About 14 seconds later, the temperature has dropped to three billion degrees. There is now not enough energy in collisions to create electron–positron pairs, and when the electrons and positrons that are already present annihilate, they are not recreated. They leave behind photons, and an excess of electrons of one part in 10^9.

It takes a full three minutes from our starting point of one-thousandth of a second after the Big Bang before the temperature has dropped to 10^9K, a number that is about 70 times hotter than that at the Sun's center. This is a significant temperature because below it an initial step in the formation of helium and tritium takes place. The formation of these nuclei must occur in stages in which the reaction

$$n + p \rightarrow \text{deuteron} + \gamma$$

is an essential step. The deuteron is a rather weakly bound state of neutron and proton, and it takes another minute or so before the reactions

$$\text{deuteron} + \text{deuteron} \rightarrow {}^3\text{He} + n$$

and

$$\text{deuteron} + {}^3\text{He} \rightarrow p + {}^4\text{He}$$

can occur. By now, pretty much all of the neutrons have either beta-decayed or been bound in the helium nucleus, and *no more nuclear reactions* take place. Detailed calculations show that the nuclei that are formed are 75 percent protons (hydrogen nuclei) and about 25 percent helium.

THINK ABOUT THIS. . .

WHY IS THERE MORE MATTER THAN ANTIMATTER IN THE UNIVERSE?

Looking around us, we observe that *there is an apparent domination of matter over antimatter in the universe*. Although we have no direct evidence that distant galaxies are made of matter rather than antimatter (antimatter would form antiatoms with exactly the same emission spectra as atoms), there is much indirect evidence that the amount of antimatter in the universe is severely limited. For example, even though Earth is continually bombarded from distant space by cosmic rays, these particles consist of matter, not antimatter. The collision and annihilation of antimatter and matter would produce spectacular emissions of radiant energy. Yet if a hot early universe had equal amounts of matter and antimatter, which is a fair assumption given our description, how is it that today there is so little visible antimatter?

Experimental evidence from the weak decays of certain particles indicates that physical processes are not quite symmetric under the reversal of time. This is not a statistical time's arrow, like that discussed in Chapter 20, but rather a real, if tiny in its effect, directionality of physical processes. To use the language of Chapter 20, a film of fundamental physical processes that is run in the forward direction would look slightly different than the same film run backward. Although we do not yet fully understand the origin of this effect, its implications have been explored. One of its consequences is that to a very slight degree matter is preferentially created over antimatter in the ensemble of possible reactions that could have occurred up to time $t = 10^{-12}$ s. For every 10^9 antiquarks created, roughly $10^9 + 1$ quarks are created, and the same ratio has to hold for the electrons. Later, when the universe had cooled to the point where the average collision energy in the cosmic soup was insufficient to make particle–antiparticle pairs, as in the case of electrons and positrons discussed in the text, then production of those pairs no longer took place. The existing particles and antiparticles annihilate each other and are not re-created in collisions. At this point, the slight predominance of matter over antimatter becomes all-important, with the excess matter persisting, even an excess as small as one part in 10^9.

At three minutes it is still far too hot for atoms to form. During the subsequent 700,000 years, any electrons briefly bound to nuclei are immediately knocked away from them, mainly through the absorption of photons—these photons, like all the constituents of the plasma, are in thermal equilibrium at whatever the temperature is at any time, and so they are on average highly energetic. Once the temperature drops enough so that the electrons remain attached to nuclei, atoms can form. The photons are no longer absorbed and can travel easily through what is now a collection of atoms—the universe becomes *transparent to photons*. We say that there is a *decoupling* between matter and radiation at some critical temperature T_C. There is another effect: The photon distribution (the Planck blackbody distribution at $T = T_c$) no longer changes, *except* that the further expansion of the universe leads to a predictable cooling of the photon "gas" as a function of time—Gamow and collaborators predicted that blackbody radiation at a temperature of about 5K would be found today. The distribution of such radiation has a peak in the microwave region. Nobody paid much attention to this prediction until the early 1960s, when Arno Penzias and Robert Wilson of the Bell Laboratories detected microwave radiation of just this type, now known as the *cosmic background radiation*. While the measurement did not find the temperature that Gamow and his collaborators had predicted, more accurate calculations by a group led by Robert Dicke at Princeton University, who at the time were on the verge of measuring this radiation, predicted that the temperature would be closer to 3K. More recent measurements give a value of this temperature of 2.725K $\pm$ 0.001K, an accuracy of 1 part in 10^4. You can read more about this fascinating story in Section 17–5.

The correct prediction of the cosmic blackbody background radiation and of the measured helium abundance in the universe were the "smoking guns" that have led to the acceptance of the Big Bang as the correct description of the evolution of the universe. Nuclei heavier than helium (except for a trace of light nuclei like lithium) are formed in stars. Very detailed calculations using our best knowledge of nuclear physics reproduce the observed abundance of all the elements in the universe.

The Uniformity of the Universe

Matter distribution in the universe is granular, or clumped. Galaxies (clumps of stars) are grouped into clusters, and there are large regions with no visible matter (Fig. 45–19). The granular structure is thought to have developed from small, random fluctuations in a uniform energy density, forming small clumps of higher density. Such clumps act as seeds, gathering matter into them under the influence of the gravitational force. The process is much like the formation of clouds, whose droplets require "nuclei" of some kind. The initial density fluctuations that started the process must have begun at

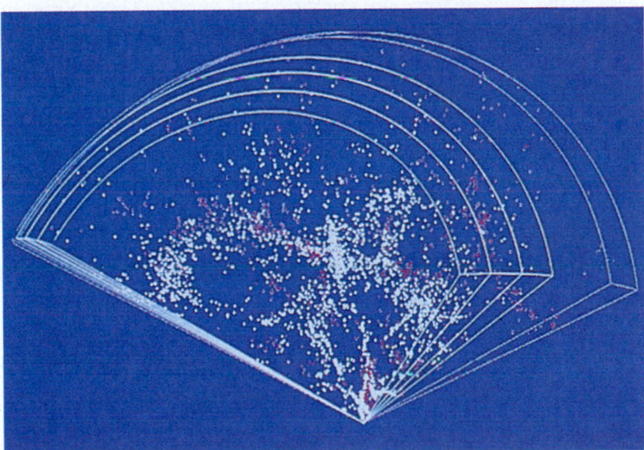

▲ **FIGURE 45–19** This view of many galaxies reveals a well-developed structure of clumping and voids. One of these features, the roughly horizontal band of galaxies about half-way up from the lower point, has received a name—the "great wall."

a time not too late in the evolution of the universe. The 3K background radiation should contain some "memory" of these fluctuations. High-precision studies of this cosmic background radiation, and in particular of the pattern of tiny fluctuations in it across the sky, have provided evidence in favor of the conjecture that during a very early period known as the *inflation period*, at times of the order of 10^{-35} seconds, the expansion of the universe went through a brief phase of very rapid, *exponential* expansion. This conjecture can be shown to lead to a density of energy given by $\Omega = 1$, which as we described above is very much the value favored by workers in this field.

45–7 A Few Last Words

In the introduction to this text, we spoke of macroscopic and microscopic realms. For centuries, technologies developed through the manipulation of matter in the macroscopic realm on the basis of empirical knowledge. The fact that matter consisted of atoms, and that the behavior of atoms was governed by quantum mechanics, initially had very little impact on the way engineers worked. Scientists, by contrast, have always been interested in the fundamental questions of their own time, many of which seemed to be rather remote from practical application. Today this has all changed. Technology develops to a larger and larger extent on the basis of scientific knowledge. The bulk of today's technology is built on quantum principles and on an understanding of the quantum nature of matter. It is unimaginable that our society could operate without the quantum physics–based circuit elements that make computers or electronic communication possible, or that a chemical engineer could construct new molecules without knowledge of their microscopic structure. While engineers continue to build bridges, those bridges are built on a solid understanding of the microscopic behavior of the materials that comprise the bridges and with the aid of computers that depend on the microscopic realm.

We are still learning how to apply quantum mechanical ideas to bulk materials. We are still in the process of evaluating how to make the best use of our scientific understanding of atoms and nuclei. And we have not yet seen how we can directly apply the knowledge we have gained in the areas of particle physics and cosmology. But we can say with confidence that if our society continues in anything like its present form, we shall certainly use our new knowledge.

Summary

At the microscopic scale, matter exhibits a series of levels, one within the other. Bulk matter is made of atoms and molecules (themselves made of atoms); atoms are built from electrons and nuclei; nuclei are constructed from nucleons (protons and neutrons); and nucleons and other particles that participate in the nuclear force are comprised of quarks. These different levels, which correspond to smaller and smaller distances, are probed by microscopes that must, because of the Heisenberg uncertainty relation between momentum and distance, use beams of higher and higher momenta. This fact implies that special relativity becomes important when we want to probe distances of nuclear size and less.

At higher energies, the conversion of energy into matter–antimatter combinations becomes important. At this stage it is no longer possible to speak of a simple potential energy. Forces themselves are the consequence of the emission and absorption of particles, the force carriers. The fundamental forces are characterized by carriers that share many characteristics. The electroweak force, a fusion of the weak force and electromagnetism, is carried by a triplet of massive spin-1 particles ($W^{\pm}$ and Z^0) and by photons. The strong force that acts between quarks is carried by a set of spin-1 gluons, which also interact among themselves. A consequence of the self-interaction of gluons is that quarks and gluons are *confined*, that is, free quarks and gluons are never seen as isolated particles. The third fundamental force, gravitation, has not yet been successfully reconciled with the ideas of quantum mechanics.

Various conservation laws apply to the fundamental forces and manifest themselves in certain rules for how reactions can proceed. The conservation of electric charge, of baryon number, and of lepton number applies in strong and electroweak interactions.

Figure 45–8 summarizes much of what we have studied about the nature of the fundamental constituents and interactions.

The reactions that occur in high-energy collisions form the most important pieces of evidence for the deduction of the fundamental laws. Colliders, accelerators in which beams of electrons or protons collide head-on, allow us to produce reactions with the highest energies.

Experiments on high-energy reactions share a common ground with astronomical observations. Hubble's law describes an expanding universe that started from a tiny initial volume, at the so-called Big Bang. When the expansion started, the universe was hot and dense, conditions appropriate to the high-energy reactions where the fundamental forces act. As the universe expanded, the temperature dropped, and a succession of energy scales resulted. By applying their knowledge of the behavior of matter at these different energy scales, physicists have successfully accounted for the background blackbody radiation measured to be present as well as the observed amounts of hydrogen and helium in today's universe. The very early evolution of the universe must involve gravitation, but as yet we do not understand how to include gravitation in a way that is consistent with quantum mechanics. Accounting for these earliest times is a formidable challenge for scientists.

Understanding the Concepts

1. Whether a surface has curvature can be determined from certain geometrical properties of that surface. The sum of the angles of a triangle drawn in a flat plane is 180°. Show that this is not the case for Earth's surface by summing the angles of a triangle that links the North Pole with any two points on the equator (Fig. 45–20).

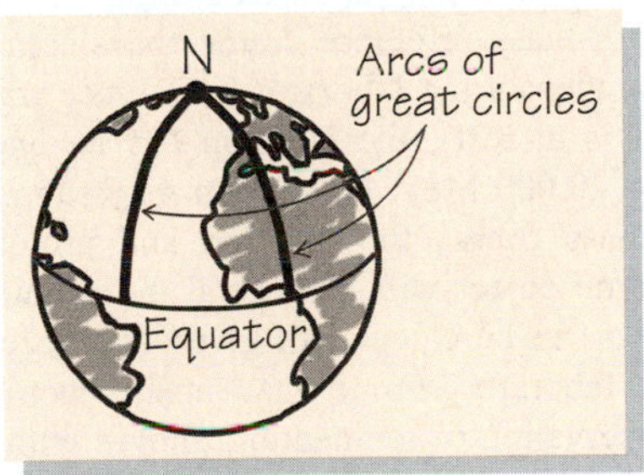

▲ **FIGURE 45–20** Question 1.

2. There are at least three known "families" of leptons. For example, in addition to the electron, there is a particle known as the muon (μ), which behaves in all respects like the electron except that it is more massive. By virtue of the extra mass, the muon is unstable. We also believe that the muon has a muon–lepton number, which, like the lepton number associated with the electron, is conserved. What experiment might tell you that the lepton numbers of the muon and the electron are conserved separately?
3. The lifetime of the proton is at least 10^{30} s, much greater than the age of the universe. Yet it is possible to create and destroy protons in particle reactions. How?
4. Why are electrons rather than protons used to study the charge distribution in the nucleus?
5. What fundamental forces most nearly account for the following phenomena: (a) walking; (b) the structure of gasoline molecules; (c) moonrise; (d) the source of sunlight; (e) the passage of sunlight from the Sun to Earth; (f) neutron stars (stars made mainly of neutrons packed to a density on the order of that of the nucleus)?
6. Can a particle be a lepton and hadron simultaneously? Meson and hadron? Hadron and baryon?
7. If a neutral meson could be made from a $(t\bar{u})$ combination, could such a meson decay *strongly* into a pair of pions? If not, would this mean that the meson is stable?
8. Suppose that electrons could not be removed from atoms, isolated, and studied independently. Would we still have experimental evidence of their existence?
9. Cosmic rays, consisting of protons and nuclei, arrive from outer space with energies that sometimes surpass those of the largest particle accelerators. What are the advantages and disadvantages of using such particles for beams for the study of high-energy reactions?
10. Earth orbits the Sun, the Sun moves within our galaxy, and our galaxy moves within its own galactic group. With all this movement, how could Hubble have discovered Hubble's law?
11. By energy and momentum conservations, an isolated electron cannot emit a photon and remain an electron. Why, then, can this happen in a virtual process?
12. Suppose that an accelerator can accelerate either electrons or protons to a given energy and that these projectiles are to collide with protons at rest. Which type of projectile should be chosen to obtain the greatest center-of-mass energy in the collision?
13. If the proton lifetime is expected to be of the order of 10^{35} s, what size apparatus would you need to see direct evidence of the proton's decay?
14. We know that the early universe was denser than it is now. What are some pieces of evidence that indicate that the early universe was hotter?
15. Recently astrophysicists have looked at supernovae with very large redshifts. Supernovae are very well understood, and the visible energy emitted is very predictable. Given a redshift (and thus, by Hubble's law, their distance from our galaxy), we can predict the luminosity as seen on Earth. Observations show that the luminosity is smaller than expected. What are some possible explanations for this fact?
16. In Example 1, we used large-angle scattering to estimate the size of a nucleus. Does that mean there is no small-angle scattering of α particles from nuclei, and if there is, what is the interpretation of d?
17. Is the W a fermion or a boson? Could the W^+ decay into (a) proton + antiproton + photon? (b) Positron + photon?

Problems

45–1 Probing the Structure of Matter

1. (I) Robert Hofstadter received a 1961 Nobel Prize for experiments in which 500-MeV electrons scattered from stationary nuclei. He found that the nucleus has a fairly constant density but, over a thickness of about 2.4 fm at its surface, the density drops to zero. With what spatial resolution could Hofstadter probe the nucleus with 500-MeV electrons?

2. (I) The Stanford Linear Accelerator accelerates electrons up to an energy of 52 GeV. These electrons can collide with stationary targets. (a) Determine the spatial accuracy with which such a probe can discern details of the nucleus. (b) In the 1960s, experiments performed there with 20-GeV electrons showed that the proton contains pointlike particles, the quarks. What is the best spatial resolution obtainable with such an electron beam?

3. (II) We want to use electrons as projectiles to study the structure of protons by means of the scattering reaction $e + p \rightarrow e + p$. The target proton is initially at rest in the laboratory; we measure those electrons that recoil directly backward. Suppose that all the motion is ultrarelativistic (you can ignore all rest masses except that of the particle at rest). Calculate the momentum transferred to the proton target (Fig. 45–21). If the energy of the incident electron is doubled, how much does the momentum transferred to the proton increase?

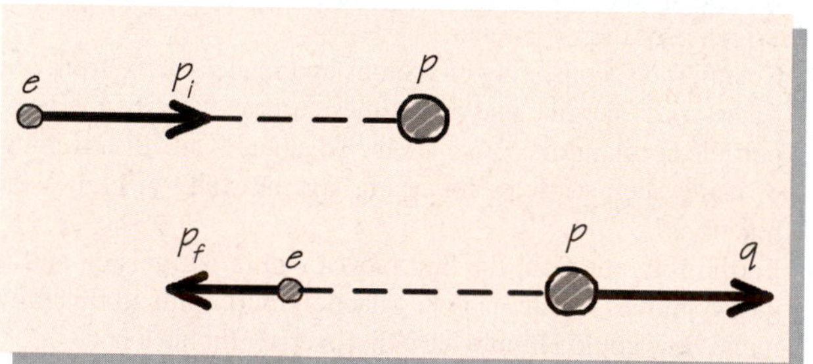

▲ **FIGURE 45–21** Problem 3.

4. (II) Problem 3 describes the ultrarelativistic collision of an electron with a stationary proton. Compare the final momenta obtained there with the momenta obtained in a similar situation with nonrelativistic kinematics.

5. (II) Consider the ultrarelativistic $e + p$ collision described in Problem 3, but now consider those events in which the electron is scattered at 90° to its initial direction (Fig. 45–22). What is the direction and magnitude of the proton's momentum after the collision?

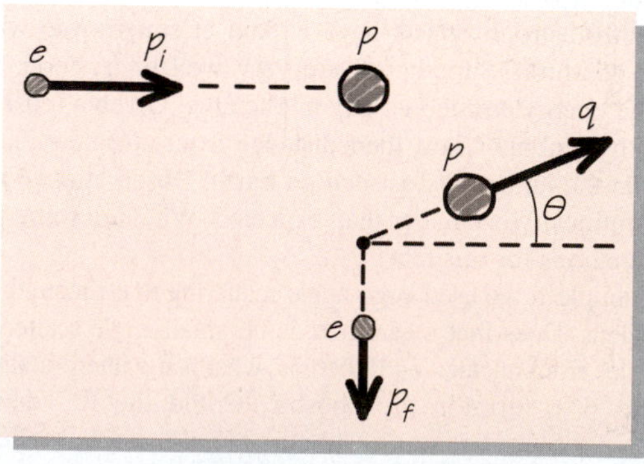

▲ **FIGURE 45–22** Problem 5.

6. (II) (a) Repeat Problem 3, but keep the rest mass of the proton, so that $E = \sqrt{(m_p c^2)^2 + (qc)^2}$ for a proton of momentum q. Continue to treat the ratio of the mass of the electron to that of the proton as zero. (b) Is there an initial electron energy for which the electron motion must be treated relativistically while the motion of the proton can be treated nonrelativistically?

7. (II) Equation (44–2) gives the collision cross section of a *pointlike* projectile of charge $Z_1 e$, mass m, and kinetic energy K scattering from a *pointlike* target particle of charge $Z_2 e$ and mass $M (M \gg m)$ at rest, as a function of the scattering angle θ:

$$\sigma(\theta) = Z_1^2 Z_2^2 \left(\frac{e^2}{4\pi\varepsilon_0}\right)^2 \frac{1}{16} \frac{1}{K^2 \sin^4(\theta/2)}.$$

Suppose that the initial and final momenta of the projectile are $\vec{p}_i$ and $\vec{p}_f$, respectively, with $|\vec{p}_i| = |\vec{p}_f| = p$. Then for large M the magnitude of the momentum transfer $\vec{\Delta} \equiv \vec{p}_f - \vec{p}_i$ in the collision is given by $\Delta^2 = 2p^2(1 - \cos\theta)$. (a) Express the cross section in terms of Δ^2 and plot it as a function of Δ^2. (b) Suppose that the target is not pointlike but spread out over a sphere of radius r_0. Give some qualitative arguments based on the position–momentum uncertainty relation of how the dependence of the cross section on Δ^2 would change.

8. (III) Consider a colliding electron and proton. Show that if in the center-of-mass reference frame the electron has energy 3000 MeV, then in the laboratory reference frame (in which the proton is at rest), the electron has an energy of about 20 GeV, or 20,000 MeV. [*Hint*: You may suppose that in the center-of-mass frame, the electron and proton collide and produce some object with zero total momentum, and with a mass M that can be calculated, given the electron's momentum. In the laboratory frame, that same object is created and, by the conservation of momentum, moves with a momentum p_{lab} equal to that of the incoming electron. The energy of the object in the laboratory frame is then given by $E_{object} = \sqrt{(Mc^2)^2 + (p_{lab}c)^2}$. You may then use energy conservation, $p_{lab}c + m_p c^2 = E_{object}$, and your calculation of the mass M to find $p_{lab}c$.]

9. (III) An electron collides with a second electron at rest. What is the minimum energy the incident electron must have to produce a proton–antiproton pair? The reaction takes the form $e + e \rightarrow e + e + (p\bar{p})$. Ignore the electron mass. (See the hint to Problem 8. Figure 45–23 shows the reaction in the center-of-mass frame.)

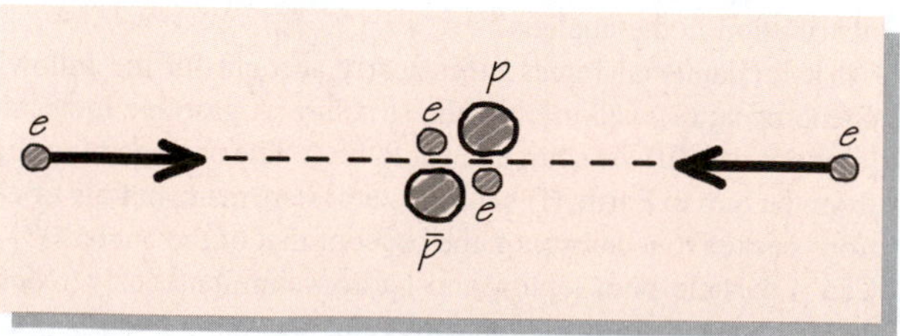

▲ **FIGURE 45–23** Problem 9.

45–2 New Quantum Numbers

10. (I) Use Table 45–1 to find the electric charge, baryon number, and spin of a π^-, which consists of a $d\bar{u}$ quark combination.

11. (I) The K^+ meson (the *kaon*) decays by the process $K^+ \to \pi^+ + \pi^0$. Assume that in this decay the conservation laws of charge, baryon number, and lepton number are obeyed. What are the values of these quantum numbers for the K^+ meson?

12. (I) Show that the conservation laws of charge, baryon number, and lepton number are satisfied in neutron decay, $n \to p + e^- + \bar{\nu}$.

13. (I) The reaction $p + e^- \to \bar{p} + e^+$ does not occur in sensitive experiments designed to observe it. Why not?

14. (II) The π^- meson decays as $\pi^- \to \mu^- + \bar{\nu}$. What is the baryon number of the μ^-? What is its lepton number? Is it a boson or a fermion? What is its electric charge?

45–3 The Fundamental Constituents of Matter: Leptons and Quarks

15. (I) Muonium is an atom consisting of a muon and a proton bound to each other. Is it a stable atom?

16. (I) Describe the quark content of a π^+-antiproton system.

17. (II) The pion is the exchange boson for nuclear forces: It is exchanged between two protons or two neutrons and leads to the nuclear force. Given that the transition $n + \pi^0 \to n$ is possible (virtually), where π^0 is the uncharged pion, describe the minimum quark content for this particle.

18. (II) Positive pions (π^+) can be produced in the reaction $p + p \to \pi^+ + D$, where D stands for the deuteron, the ^{2}H nucleus. Use this information and the known quark content of protons (*uud*) and neutrons (*udd*) to get the quark content of these pions.

45–4 The Fundamental Forces and Their Carriers

19. (I) The W^+ meson has a mass of 81 GeV$/c^2$. Calculate the range of the force transmitted by the exchange of the W^+ meson.

20. (II) You have 1.0 kg each of hydrogen (gram-atomic weight 1.0 g/mol) and uranium 235 (gram-atomic weight 235 g/mol). What are the baryon numbers of the two samples of material?

21. (II) There exist mesons like the pion that can be emitted and absorbed by nucleons, but with mass some 5.5 times that of the pion. The exchange of these particles gives rise to a repulsive nucleon–nucleon force. (a) Estimate the range of this repulsive force. (b) Estimate at what energy, in the center-of-mass reference frame of a proton–proton collision, the repulsive force will begin to affect the scattering.

22. (II) Estimate the repulsive force between protons in a uranium nucleus. Compare that force to the attraction between the uranium nucleus and the innermost electron in a uranium atom.

23. (II) The quark contents of some strongly interacting particles are (*uud*) for protons, (*udd*) for neutrons, $(u\bar{d})$ for π^+, $(d\bar{u})$ for π^-, and $(u\bar{u} - d\bar{d})/\sqrt{2}$ for π^0. Figure 45–24 is a sketch of the reaction $p + \pi^- \to n$ that uses this information. Here all lines represent particles moving to the right, with right-pointing arrows for quarks and left-pointing arrows for antiquarks. Moreover, quark lines never end in such a graph. Use this technique to draw graphs for the reactions (a) $p + \pi^- \to n + \pi^0$, (b) $p + \bar{p} \to \pi^+ + \pi^- + \pi^0$, and (c) $p + \pi^- \to p + n + \bar{p}$.

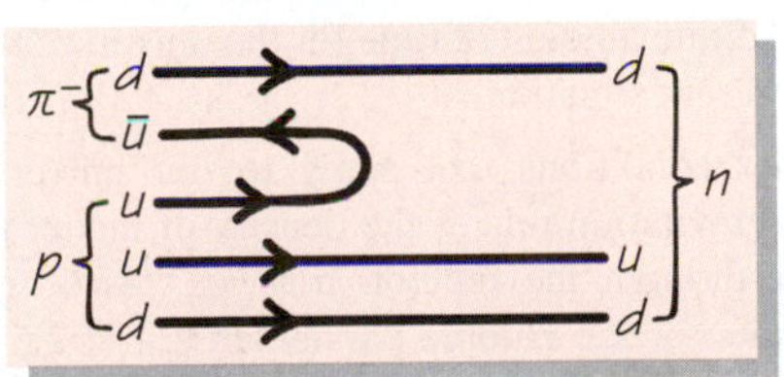

▲ **FIGURE 45–24** Problem 23.

24. (II) Combine the quark-line rules of Problem 23 with the additional graphical rules for emission of a W^- boson by a *d*-quark, emission of a W^+ boson by a *u*-quark, and $W^- \to e^- + \bar{\nu}$, as drawn in Fig. 45–25. Use these rules to give the graphical representation of the processes (a) $n \to p + e^- + \bar{\nu}$, (b) $\pi^+ \to \pi^0 + e^+ + \nu$, and (c) $\pi^- \to e^- + \bar{\nu}$, all of which proceed with a *W* boson present at an intermediate stage.

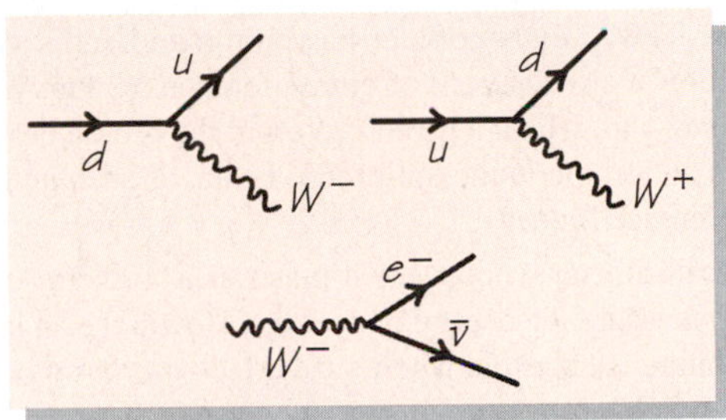

▲ **FIGURE 45–25** Problem 24.

45–5 Tools of Particle Physics

25. (I) According to some theories, a proton at rest can decay (with very long lifetime!) in the mode $p \to e^+ + \pi^0$. Use $m_p \cong 940$ MeV$/c^2$ and $m_\pi \cong 135$ MeV$/c^2$ to calculate the momentum, q, of the π^0 in the decay; neglect the electron mass (Fig. 45–26). Express your answer in terms of the combination qc, in units of MeV. Detectors looking for the decay would use this characteristic momentum as a signal.

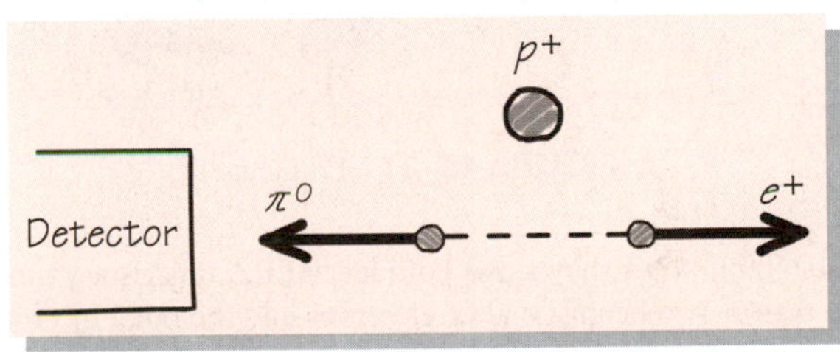

▲ **FIGURE 45–26** Problem 25.

26. (II) A proton–antiproton collider is actually a quark–antiquark collider, if we think of each quark in a proton (or each antiquark in an antiproton) as carrying one-third the momentum of the proton (or antiproton). What is the proton (and antiproton) energy in a symmetric $p\bar{p}$ collider that produces quark–antiquark collisions with the same total energy as the collisions produced by an electron–positron collider in which each electron carries a momentum of 11 GeV$/c$? Ignore the masses of the electron, positron, quark, and antiquark but not those of the proton and antiproton.

27. (II) Calculate the momentum in the center of mass of each incident proton for an accelerator in which 300-GeV protons are incident on stationary proton targets. (See the hint for Problem 8.)

28. (II) Antiprotons of energy 1000 GeV collide with protons of the same energy but opposite momentum in the TEVATRON collider (see Table 45–3). What is the momentum of antiprotons incident on stationary protons if the center-of-mass energy of the $p\bar{p}$ system is the same as that of the TEVATRON?

29. (II) An experimenter wants to study the reaction $e^+ + e^- \rightarrow Z^0$, where the Z^0, of mass 91.1 GeV/c^2, is the neutral intermediate boson of the electroweak interactions. The experimenter has a fixed-target machine that accelerates a beam of positrons that can be directed at atoms, which provide a stationary target of electrons. What is the minimum energy of the positron beam for which the reaction can occur?

30. (II) The *luminosity* of a colliding beam machine is the rate at which particles pass in one beam times the number of collisions per unit area of a second beam. The luminosity of the TEVATRON is $10^{37}/\text{m}^2 \cdot \text{s}$. If the total collision cross section for proton–proton interactions is 100 millibarns ($1 \text{ barn} = 10^{-28} \text{ m}^2$), how many interactions per second are there at the TEVATRON?

31. (II) The TEVATRON collider has proton and antiproton beams that follow a circular path of circumference 6.3 km. What is the frequency with which a given proton in the proton beam crosses a given antiproton in the antiproton beam, as seen in the laboratory reference frame?

32. (II) A calorimeter whose active material is 100 cm^3 of water is used to measure the energy of protons. How large an increase of temperature is expected when 1.0×10^{12} protons of kinetic energy 1.0 GeV are stopped in the calorimeter?

33. (II) A proton–antiproton pair can be produced in the reaction $\gamma + p \rightarrow p + \bar{p} + p$ (Fig. 45–27). If the target proton is at rest, what is the minimum energy that the photon must have to produce this reaction?

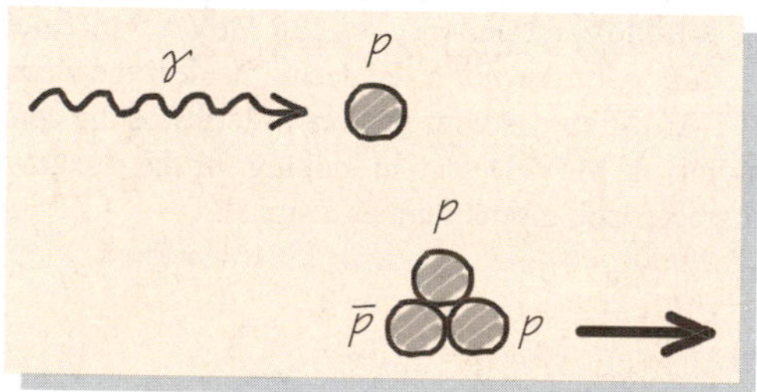

▲ **FIGURE 45–27** Problem 33.

34. (III) Table 45–3 shows one collider, HERA, that is asymmetric in that its two beams, one of electrons and the other of protons, are of different momenta. This collider is thus a kind of hybrid between symmetric colliders and fixed-target machines. In the following, neglect the baryon-number and lepton-number conservation laws that must be obeyed in proton–electron collisions. (a) What is the maximum number of neutral pions that can be produced in HERA collisions? (b) What momentum would beams of protons and electrons of equal momentum have in order to be able to produce the number of pions in part (a)? (c) What momentum would a beam of electrons incident on protons at rest have to produce the number of pions in part (a)?

35. (III) Electron accelerators that produce copious quantities of X rays for research on atomic and condensed-matter physics employ electron beams of several GeV. The X rays are produced when the electrons scatter from photons of lesser energy. (For example, the National Synchrotron Light Source at Brookhaven National Laboratory accelerates electrons to an energy of 2.5 GeV and produces photons with energies up to 310 MeV.) Consider laser photons (with wavelength of, say, 590 nm) that scatter head-on from 8-GeV electrons. Calculate the maximum energy of the recoil photon produced in such a (Compton scattering) reaction (Fig. 45–28). At what angle would this photon be produced? [*Hint*: You must keep the electron mass in the relativistic expression for the electron energy, E, even though $E \gg m_e c^2$. You may, however, make the approximation $\sqrt{(pc)^2 + (m_e c^2)^2} \cong pc + \dfrac{(m_e c^2)^2}{(2pc)}$ in this case.]

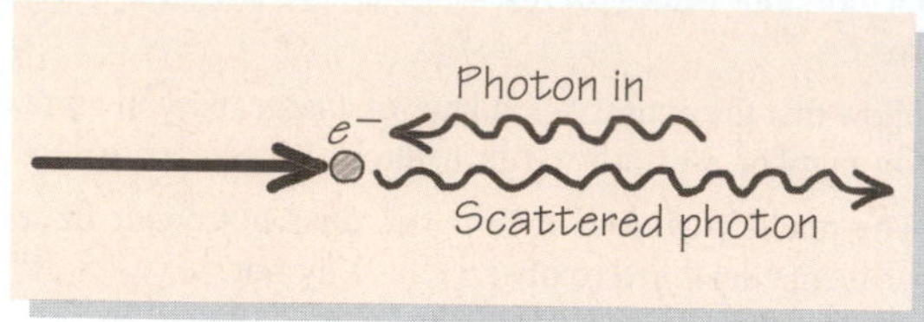

▲ **FIGURE 45–28** Problem 35.

45–6 Cosmology and the Expanding Universe

36. (I) In addition to the Planck time, there is a distance referred to as the *Planck length*. Use dimensional analysis to form a length from the constants $\hbar$, G, and c.

37. (I) In addition to the Planck time, there is a mass referred to as the *Planck mass*. Use dimensional analysis to form a time from the constants $\hbar$, G, and c.

38. (II) Consider the one-dimensional line of galaxies in Figs. 45–17a and 45–17b. Show that an inhabitant of galaxy B would agree that Hubble's law holds, but with the distance being the distance from galaxy B. In this way you confirm that Hubble's law is the unique law consistent with no special central point.

39. (I) Show that the Planck mass, 2.2×10^{-8} kg, is equivalent to an energy of 1.2×10^{19} GeV and to a temperature of 1.4×10^{32} K.

40. (II) The universe is expanding, but the expansion may be slowing down. (a) If R varies with time according to $R = kt^n$, where k and n are constants, find the acceleration of R. (b) The acceleration parameter, q, is defined by $q \equiv -R(d^2R/dt^2)/(dR/dt)^2$. Calculate the value of q.

41. (II) Assume that the size of the universe is determined by a distance scale factor R. This factor is increasing with time t at the rate $R = kt^n$, where k and n are constants. Hubble's law then becomes $dR/dt = HR$. Show that the Hubble parameter H is not constant, and find its time dependence.

42. (II) Use Hubble's law to find the time it would take for the distance from Earth to some galaxy to increase by 10 percent over what it is today. Assume that for the period over which you make your calculation, H is constant. [*Hint*: First find the time interval for the distance to change from x to $x + dx$, assuming that dx is an infinitesimal quantity. Then integrate your result.]

43. (II) By using the techniques of Problem 42, show that if Hubble's law with a constant value of H has held ever since the time when there was no separation between galaxies, it would have taken an infinite amount of time for the universe to have expanded to its present situation.

44. (III) Under conditions that apply to our universe, Einstein's theory of gravitation relates the density of matter to the Hubble parameter through the relation $\rho = 3H^2/8\pi G$. (a) Given the present value of the Hubble parameter, $H_0 = 2.5 \times 10^{-18} \text{ s}^{-1}$, calculate the present value of the density of matter, ρ_0. Assuming

that matter consists (almost) entirely of hydrogen, with $M_H = 1.7 \times 10^{-27}$ kg, find the number density of hydrogen. (b) Show that the conservation of baryon number implies that the time dependence of the density is given by $\rho(t) = \rho_0[R_0/R(t)]^3$, where R_0 is the current distance scale of the universe and $R(t)$ is the distance scale at a general time t. (c) Combine the result of part (b) with the general relation between ρ and $H \equiv (1/R)\, dR/dt$ to show that $R(t) = kt^{2/3}$, and find k. [*Hint*: Substitute the general form $R(t) = kt^n$ into your equations.]

45. (III) Suppose that at the time when stable atoms formed, neutrons and protons were present in the ratio of 1 : 4. (a) Show that if hydrogen and helium formed with these nucleons (plus an appropriate number of electrons), then the number of moles of hydrogen formed was 6 times the number of moles of helium formed. (b) Show that the total mass of hydrogen formed is roughly 1.5 times that of helium formed then. (The actual evolution was such that the neutron-to-proton ratio was 1 : 7 when stable atoms formed.)

General Problems

46. (II) What is the maximum number of antiprotons that can be created in a scattering reaction with 70-GeV protons incident on a stationary hydrogen target?

47. (II) Below what temperature will stable carbon atoms form? He^+ ions? The ionization energies of H and C are 13.6 eV and 11.3 eV, respectively.

48. (II) *Cerenkov counters* are detectors that measure the presence and speed of charged particles by observing the angle of the "wake" of light (Cerenkov radiation; see Fig. 14–36) produced by those particles when their speed through a medium is greater than the speed of light in that medium. What is the minimum energy of electrons that produce Cerenkov radiation in toluene ($n = 1.494$)?

49. (II) Consider a fixed-target machine with protons of momentum 3.2×10^5 GeV/c. If the largest magnetic field we could use to guide the beam in a circular path has magnitude 9.5 T, what is the radius of such a machine?

50. (II) Find a relationship between the de Broglie wavelength and the kinetic energy K of ultrarelativistic particles (particles for which $K \gg mc^2$).

51. (II) In a reaction in which 40-GeV electrons hit 40-GeV positrons head-on (as has been done at the Stanford Linear Accelerator Center), what is the maximum number of π mesons (and antimesons) that can be created for each colliding pair? The mass of a π meson is about 140 MeV/c^2.

52. (II) A pi meson (of mass approximately 140 MeV/c^2) is ejected from a nuclear collision with a kinetic energy of 350 MeV. π mesons at rest have a lifetime of about 2.6×10^{-8} s. Calculate (a) the speed of the pi meson, (b) its momentum, and (c) how far, on average, such a pi meson will travel before it decays.

53. (II) The muon, symbol μ, is for all practical purposes just like the electron, except that it is some 207 times more massive. The muon can be destroyed by a proton in the reaction $\mu^- + p \to n + \nu$. A muon sent into hydrogen gas slows and is finally captured in a hydrogen-like Bohr orbit. Once it is in the lowest orbit (radius R_0), it will be destroyed by the reaction above if it comes within a distance $L = 0.2$ fm of the proton (Fig. 45–29). The probability that a muon in the lowest orbit will come that close to the proton is estimated from quantum mechanics to be the ratio of the volume of a sphere of radius L to that of a sphere of radius R_0. Suppose that the lifetime for destruction from the lowest Bohr orbit in hydrogen is τ_c. Show that the lifetime for destruction from the ground state of a Bohrlike atom in which the muon orbits a nucleus with Z protons is τ_c/Z^4.

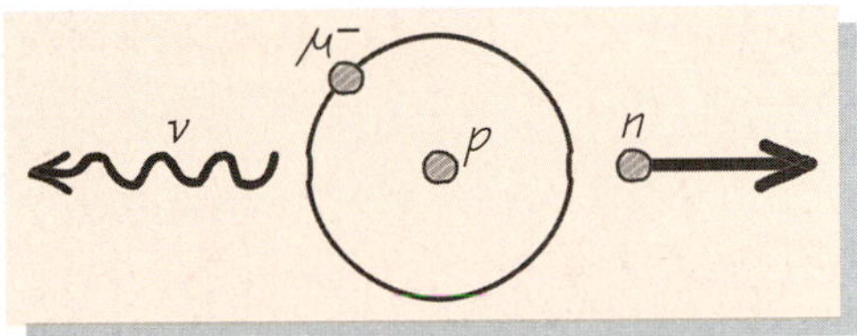

▲ **FIGURE 45–29** Problem 53.

54. (II) Consider muon destruction from the lowest state in a hydrogen-like Bohr atom (see Problem 53). What is the energy of the neutron that emerges from the reaction? If the process occurs when the muon is in orbit around a heavy nucleus, will the reaction look the same as in the hydrogen-like atom?

55. (II) Using energy conservation and momentum conservation, prove that an isolated electron cannot emit a photon and remain an electron (see Question 11).

56. (II) Free neutrons have a lifetime of approximately 890 s. Given that the ratio of neutrons to protons was 1 : 5 when the universe was 1 s old, *estimate* the ratio at 1000 s. Take into account only the loss of neutrons through their decay.

57. (III) When an antiproton slows down by interacting with matter and approaches a proton, it can be captured and form a hydrogen-like atom called *protonium* through its Coulomb attraction to the proton. Find (a) the binding energy of the atom in electron volts and (b) the radius of the atom in its ground state. (Do not forget the reduced-mass effect.)

58. (III) The probability that a proton and an antiproton in the ground state of protonium (see Problem 57) will come within a distance R_1 of each other is estimated from quantum mechanics to be the ratio of the volume of a sphere of radius R_1 to that of a sphere of radius $2R_0$, where R_0 is the ground-state radius of protonium. If the lifetime for annihilation of a proton and an antiproton is 10^{-22} s when they are within 0.2 fm of each other and is infinite beyond this distance, estimate how long the protonium atom lasts in its ground state.

APPENDIX I

The Système Internationale (SI) of Units

I–1 SOME SI BASE UNITS

Physical Quantity	Name of Unit	Symbol
length	meter	m
mass	kilogram	kg
time	second	s
electric current	ampere	A
thermodynamic temperature	kelvin	K
amount of substance	mole	mol

I–2 SOME SI DERIVED UNITS

Physical Quantity	Name of Unit	Symbol	SI Unit
frequency	hertz	Hz	s^{-1}
energy	joule	J	$kg \cdot m^2/s^2$
force	newton	N	$kg \cdot m/s^2$
pressure	pascal	Pa	$kg/m \cdot s^2$
power	watt	W	$kg \cdot m^2/s^3$
electric charge	coulomb	C	$A \cdot s$
electric potential	volt	V	$kg \cdot m^2/A \cdot s^3$
electric resistance	ohm	Ω	$kg \cdot m^2/A^2 \cdot s^3$
capacitance	farad	F	$A^2 \cdot s^4/kg \cdot m^2$
inductance	henry	H	$kg \cdot m^2/A^2 \cdot s^2$
magnetic flux	weber	Wb	$kg \cdot m^2/A \cdot s^2$
magnetic flux density	tesla	T	$kg/A \cdot s^2$

I–3 SI UNITS OF SOME OTHER PHYSICAL QUANTITIES

Physical Quantity	SI Unit
speed	m/s
acceleration	m/s^2
angular speed	rad/s
angular acceleration	rad/s^2
torque	$kg \cdot m^2/s^2$, or $N \cdot m$
heat flow	J, or $kg \cdot m^2/s^2$, or $N \cdot m$
entropy	J/K, or $kg \cdot m^2/K \cdot s^2$, or $N \cdot m/K$
thermal conductivity	$W/m \cdot K$

I–4 SOME CONVERSIONS OF NON-SI UNITS TO SI UNITS

Energy:
1 electron-volt (eV) $= 1.6022 \times 10^{-19}$ J
1 erg $= 10^{-7}$ J
1 British thermal unit (BTU) = 1055 J
1 calorie (cal) = 4.185 J
1 kilowatt-hour (kWh) $= 3.6 \times 10^6$ J

Mass:
1 gram (g) $= 10^{-3}$ kg
1 atomic mass unit (u) $= 931.5\ \text{MeV}/c^2 = 1.661 \times 10^{-27}$ kg
1 $\text{MeV}/c^2 = 1.783 \times 10^{-30}$ kg

Force:
1 dyne $= 10^{-5}$ N
1 pound (lb or #) = 4.448 N

Length:
1 centimeter (cm) $= 10^{-2}$ m
1 kilometer (km) $= 10^3$ m
1 fermi $= 10^{-15}$ m
1 Angstrom (Å) $= 10^{-10}$ m
1 inch (in or ″) = 0.0254 m
1 foot (ft) = 0.3048 m
1 mile (mi) = 1609.3 m
1 astronomical unit (AU) $= 1.496 \times 10^{11}$ m
1 light-year (ly) $= 9.46 \times 10^{15}$ m
1 parsec (ps) $= 3.09 \times 10^{16}$ m

Angle:
1 degree (°) $= 1.745 \times 10^{-2}$ rad
1 min (′) $= 2.909 \times 10^{-4}$ rad
1 second (″) $= 4.848 \times 10^{-6}$ rad

Volume:
1 liter (L) $= 10^{-3}\ m^3$

Power:
1 kilowatt (kW) $= 10^3$ W
1 horsepower (hp) = 745.7 W

Pressure:
1 bar $= 10^5$ Pa
1 atmosphere (atm) $= 1.013 \times 10^5$ Pa
1 pound per square inch (lb/in^2) $= 6.895 \times 10^3$ Pa

Time:
1 year (yr) $= 3.156 \times 10^7$ s
1 day (d) $= 8.640 \times 10^4$ s
1 hour (h) = 3600 s
1 minute (min) = 60 s

Speed:
1 mile per hour (mi/h) = 0.447 m/s

Magnetic field:
1 gauss $= 10^{-4}$ T

APPENDIX II

Some Fundamental Physical Constants†

Constant	Symbol	Value	Error
speed of light in a vacuum	c	2.99792458×10^8 m/s	exact
gravitational constant	G	6.67259×10^{-11} m^3/kg · s^2	128
Avogadro's number	N_A	6.02214×10^{23} mol^{-1}	0.1
universal gas constant	R	8.31447 J/mol · K	8.4
Boltzmann's constant	k	1.38065×10^{-23} J/K	1.7
elementary charge	e	1.60218×10^{-19} C	0.004
permittivity of free space	ε_0	$8.85418781762 \times 10^{-12}$ C^2/N · m^2	exact
	$1/4\pi\varepsilon_0$	8.987552×10^9 kg · m^3 · s^{-2} · C^{-2}	
permeability of free space	μ_0	$4\pi \times 10^{-7}$ T · m/A	exact
electron mass	m_e	9.10939×10^{-31} kg	0.1
proton mass	m_p	1.67262×10^{-27} kg	0.1
neutron mass	m_n	1.67493×10^{-27} kg	0.1
Planck's constant	h	6.62607×10^{-34} J · s	0.1
$h/2\pi$	$\hbar$	1.05457×10^{-34} J · s	0.1
		$= 6.58212 \times 10^{-22}$ MeV · s	0.1
	$\hbar c$	197.327 MeV · fm	0.3
electron charge-to-mass ratio	$-e/m_e$	-1.75882×10^{11} C/kg	0.1
proton-electron mass ratio	m_p/m_e	1836.15	0.15
molar volume of ideal gas at STP		22414.0 cm^3/mol	1.7
Bohr magneton	μ_B	9.27401×10^{-24} J/T	0.1
magnetic flux quantum	$\Phi_0 = h/2e$	2.06783×10^{-15} Wb	0.1
Bohr radius	a_0	0.529177×10^{-10} m	0.005
Rydberg constant	R_∞	1.09737×10^7 m^{-1}	0.00001

†P. J. Mohr and B. N. Taylor, "The 1998 CODATA Recommended Values of the Fundamental Physical Constants, Web Version 3.1," available at physics.nist.gov/constants (National Institute of Standards and Technology, Gaithersburg, MD 20899, 3 December 1999).

We have given values of the measured constants to six significant figures, even though they may be known to greater accuracy. The error, which expresses the uncertainty in the values of these constants, is in parts per million. Defined constants have no error, and we give their full definition; they are indicated by the notation "exact" in the error column.

APPENDIX III

Other Physical Quantities

III–1.1 SOME ASTRONOMICAL CONSTANTS

Constant	Symbol	Value
standard gravity at Earth's surface	g	$9.80665\ \text{m/s}^2$
equatorial radius of Earth	R_e	$6.378 \times 10^6\ \text{m}$
mass of Earth	M_e	$5.976 \times 10^{24}\ \text{kg}$
mass of Moon		$7.350 \times 10^{22}\ \text{kg}$ $= 0.0123\ M_e$
mean radius of Moon's orbit around Earth		$3.844 \times 10^8\ \text{m}$
mass of Sun	$M_\odot$	$1.989 \times 10^{30}\ \text{kg}$
radius of Sun	$R_\odot$	$6.96 \times 10^8\ \text{m}$
mean radius of Earth's orbit around Sun	AU	$1.496 \times 10^{11}\ \text{m}$
period of Earth's orbit around Sun	yr	$3.156 \times 10^7\ \text{s}$
diameter of our galaxy		$7.5 \times 10^{20}\ \text{m}$
mass of our galaxy		$2.7 \times 10^{41}\ \text{kg}$ $= (1.4 \times 10^{11})\ M_\odot$
Hubble parameter	H	$2.5 \times 10^{-18}\ \text{s}^{-1}$

III–1.2 PLANETARY DATA

Planet	Diameter (in km)	Diameter Relative†	Relative Mass†	Average Density (in g/cm³)	Period of Rotation	Surface Gravity† (in g)	Escape Speed (in km/s)	Semimajor Axis (AU)	Period of Solar Orbit	Average Orbital Speed (in km/s)
Mercury	4,800	0.38	0.05	5.4	58 d 15 h	0.38	4.3	0.387	87.96 d	47.8
Venus	12,100	0.95	0.82	5.2	243 d 4 h	0.90	10.3	0.723	224.7 d	35.0
Earth	12,750	1.00	1.00	5.5	23 h 56 min	1.00	11.2	1.000	365.26 d	29.8
Mars	6,800	0.53	0.11	3.9	24 h 37 min	0.38	5.0	1.524	687.0 d or 1.88 yr	24.1
Jupiter	142,800	11.21	317.8	1.3	9 h 50 min	2.53	59.5	5.20	11.86 yr	13.1
Saturn	120,660	9.45	95.2	0.7	10 h 39 min	1.07	35.5	9.58	29.46 yr	9.7
Uranus	51,000	4.00	14.5	1.3	17 h	0.91	21.3	19.20	84.01 yr	6.8
Neptune	49,500	3.88	17.1	1.6	16 h	1.14	23.5	30.05	164.79 yr	5.4
Pluto	2,390	0.18	0.002	0.32	6 d 9 h 17 min	0.05	1.1	39.24	247.68 yr	4.7

†Relative to Earth.

III–2 ENERGY SUPPLY AND DEMAND†

†From the *Physics Vade Mecum,* Ed. Herbert L. Anderson, American Institute of Physics (New York, 1981); and U.S. Congress, Office of Technology Assessment, *Changing by Degrees: Steps to Reduce Greenhouse Gases,* OTA-O-482 (Washington, D.C.: U.S. Government Printing Office, February 1991).

III–2.1 FUEL RESOURCES (1980, ESTIMATED)

Resource	U.S. Resources	World Resources
coal (recoverable)	5×10^{21} J	2×10^{22} J
oil (not including oil shales)	10^{21} J	10^{22} J
natural gas	2×10^{21} J	10^{22} J
hydroelectric	10^{22} J/yr (North America)	6×10^{22} J/yr

III–2.2 ANNUAL USAGE OF RESOURCE (2001, PERCENTAGE OF TOTAL) *Source:* www.energy.gov

Resource	U.S. Usage (total = 1×10^{20} J)	World Usage (total = 4×10^{20} J)
coal	23	24
oil	40	39
natural gas	24	23
nuclear	8	6
hydroelectric	2	7
biomass	3	1

III–2.3 ENERGY CONTENT OF FUELS

Fuel	Energy Content (in J/kg)
bread	10×10^6
glucose ($C_6H_{12}O_6$)	16×10^6
white pine wood	20×10^6
methyl alcohol (CH_4O)	23×10^6
anthracite coal	31×10^6
domestic heating oil	45×10^6
propane (C_3H_8)	50×10^6
natural gas (96% CH_4)	51×10^6
fission of U^{235}	5.8×10^{11}
perfect mass-energy conversion	9×10^{16}

III–2.4 SOLAR ENERGY OUTPUT

total radiated power from the Sun	4×10^{26} W
power per unit area at the top of Earth's atmosphere	1.4 kW/m^2
average power per unit area delivered to an average horizontal surface in the United States in 1 yr	0.2 W/m^2

III–2.5 ENERGY CONSUMPTION IN TRANSPORTATION

Mode	Energy Consumption (J/passenger · km)
bicycle	5×10^4
foot travel	1.5×10^5
automobile	1.9×10^5
intercity bus	6×10^5
intercity train	9×10^5
747 jet airplane	2.3×10^6
snowmobile	6×10^6

III–2.6 ENERGY CONSUMPTION OF ELECTRICAL APPLIANCE (*See* http://www.ianr.unl.edu/pubs/consumered/heg94.htm)

Appliance	Power (in W)	Energy Use per Year (in kWh)
window air conditioner	3750	3750
clock	2	17
dishwasher	1200	363
window fan	200	170
hair dryer	750	38
iron	1000	144
microwave oven	1450	190
radio	71	86
refrigerator-freezer	615	1830
stove	12,200	1175
color television	200	440
vacuum cleaner	630	46
washing machine	512	107

APPENDIX IV

Mathematics

IV–1 SOME MATHEMATICAL CONSTANTS†

Constant	Value
π	3.14159
e (Euler's constant)	2.71828
$\sqrt{2}$	1.41421
$1/\sqrt{2}$	0.707107
ln(10)	2.30259
ln(2)	0.693147
1 rad	57.2958°
1°	0.0174533 rad

†To six significant figures.

IV–2 SOLUTION OF QUADRATIC EQUATIONS

Quadratic equation:

$$ax^2 + bx + c = 0$$

Two solutions:

$$x = \frac{-b \pm \sqrt{b^2 - 4ac}}{2a}$$

IV–3 BINOMIAL THEOREM

$$(x + y)^n = \sum_{k=0}^{n} \binom{n}{k} x^{n-k} y^k$$

where

$$\binom{n}{k} = \frac{n!}{(n-k)!k!}$$

The factorial $m! \equiv 1 \cdot 2 \cdot 3 \ldots \cdot m$; $0! \equiv 1$. Some particular cases of the binomial theorem:

(1) $(x \pm y)^2 = x^2 \pm 2xy + y^2$;
(2) $(x \pm y)^3 = x^3 \pm 3x^2y + 3xy^2 \pm y^3$;
(3) $(x \pm y)^4 = x^4 \pm 4x^3y + 6x^2y^2 \pm 4xy^3 + y^4$.

IV–4 TRIGONOMETRY

1. For a right triangle with sides a, b, and c (the hypotenuse), where the angle opposite side a is θ_a (Figure A–1),

$$\text{sine of } \theta_a = \sin\theta_a \equiv \frac{a}{c};$$

$$\text{cosine of } \theta_a = \cos\theta_a = \frac{b}{c};$$

$$\text{tangent of } \theta_a = \tan\theta_a = \frac{a}{b}.$$

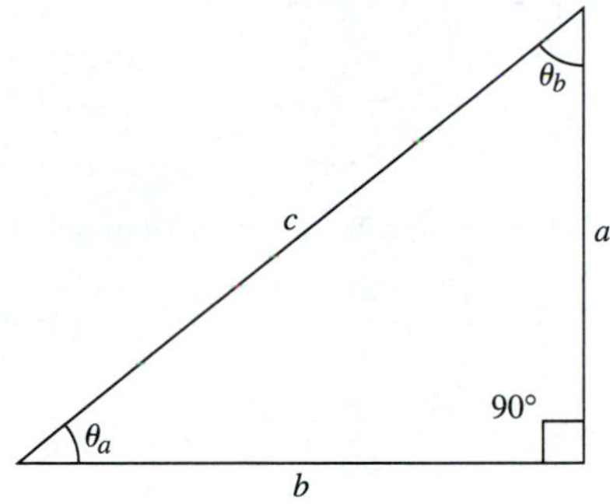

▲ FIGURE A–1

2. The cosine function is even, $\cos(-x) = \cos x$; the sine function is odd, $\sin(-x) = -\sin x$.

3. (1) $\tan\theta = \dfrac{\sin\theta}{\cos\theta}$

   (2) $\sec\theta = \dfrac{1}{\cos\theta}$

   (3) $\operatorname{cosec}\theta = \dfrac{1}{\sin\theta}$

   (4) $\cot\theta = \dfrac{1}{\tan\theta}$

4. (1) $\sin^2\theta + \cos^2\theta = 1$
   (2) $\sec^2\theta - \tan^2\theta = 1$
   (3) $\operatorname{cosec}^2\theta - \cot^2\theta = 1$

5. (1) $\sin(\theta_1 \pm \theta_2) = \sin\theta_1\cos\theta_2 \pm \cos\theta_1\sin\theta_2$
   (2) $\cos(\theta_1 \pm \theta_2) = \cos\theta_1\cos\theta_2 \mp \sin\theta_1\sin\theta_2$
   (3) $\sin\theta_1 \pm \sin\theta_2 = 2\sin\left(\dfrac{\theta_1 \pm \theta_2}{2}\right)\cos\left(\dfrac{\theta_1 \mp \theta_2}{2}\right)$
   (4) $\cos\theta_1 + \cos\theta_2 = 2\cos\left(\dfrac{\theta_1 + \theta_2}{2}\right)\cos\left(\dfrac{\theta_1 - \theta_2}{2}\right)$
   (5) $\cos\theta_1 - \cos\theta_2 = -2\sin\left(\dfrac{\theta_1 + \theta_2}{2}\right)\sin\left(\dfrac{\theta_1 - \theta_2}{2}\right)$
   (6) $\tan(\theta_1 + \theta_2) = \dfrac{\tan\theta_1 + \tan\theta_2}{1 - (\tan\theta_1)(\tan\theta_2)}$
   (7) $\cos\left(\theta \pm \dfrac{\pi}{2}\right) = \mp\sin\theta$
   (8) $\sin\left(\theta \pm \dfrac{\pi}{2}\right) = \pm\cos\theta$
   (9) $\sin\theta_1\sin\theta_2 = \dfrac{1}{2}[\cos(\theta_1 - \theta_2) - \cos(\theta_1 + \theta_2)]$
   (10) $\cos\theta_1\cos\theta_2 = \dfrac{1}{2}[\cos(\theta_1 - \theta_2) + \cos(\theta_1 + \theta_2)]$
   (11) $\sin\theta_1\cos\theta_2 = \dfrac{1}{2}[\sin(\theta_1 - \theta_2) + \sin(\theta_1 + \theta_2)]$

6. (1) $\sin(2\theta) = 2\sin\theta\cos\theta = \dfrac{2\tan\theta}{1+\tan^2\theta}$

(2) $\cos(2\theta) = \cos^2\theta - \sin^2\theta = 2\cos^2\theta - 1 = 1 - 2\sin^2\theta$

(3) $\tan(2\theta) = \dfrac{2\tan\theta}{1-\tan^2\theta}$

(4) $\sin\left(\dfrac{\theta}{2}\right) = \pm\sqrt{\dfrac{1-\cos\theta}{2}}$

(5) $\cos\left(\dfrac{\theta}{2}\right) = \pm\sqrt{\dfrac{1+\cos\theta}{2}}$

7. Expansions of trigonometric functions (θ in rad):

(1) $\sin\theta = \theta - \dfrac{\theta^3}{3!} + \dfrac{\theta^5}{5!} - \dfrac{\theta^7}{7!} + \cdots \qquad (\theta^2 < 1)$

(2) $\cos\theta = 1 - \dfrac{\theta^2}{2!} + \dfrac{\theta^4}{4!} - \dfrac{\theta^6}{6!} + \cdots \qquad (\theta^2 < 1)$

(3) $\tan\theta = \theta + \dfrac{1}{3}\theta^3 + \dfrac{2}{15}\theta^5 + \dfrac{17}{315}\theta^7 + \cdots \qquad \left(\theta^2 < \dfrac{\pi^2}{4}\right)$

IV–5 GEOMETRICAL FORMULAS

1. (circumference of a circle of radius r) $= 2\pi r$
2. (area of a circle of radius r) $= \pi r^2$
3. (area of a sphere of radius r) $= 4\pi r^2$
4. (volume of a sphere of radius r) $= \frac{4}{3}\pi r^3$
5. (area of a rectangle with sides of lengths L_1 and L_2) $= L_1 L_2$
6. For a right triangle with sides a, b, and c and angles θ_a and θ_b opposite the sides a and b, respectively (Fig. A–1):

   (1) $a^2 + b^2 = c^2$ (the Pythagorean theorem)

   (2) area $= \frac{1}{2}(\text{base})(\text{height}) = \frac{1}{2}ab$

7. For a triangle with sides a, b, and c opposite the angles θ_a, θ_b, and θ_c, respectively (Figure A–2):

   (1) $\theta_a + \theta_b + \theta_c = 180° = \pi$ rad

   (2) $a^2 = b^2 + c^2 - 2bc\cos\theta_a$

   (3) $\dfrac{a}{\sin\theta_a} = \dfrac{b}{\sin\theta_b} = \dfrac{c}{\sin\theta_c}$

   (4) $a = b\cos\theta_c + c\cos\theta_b$

   (5) area $= \dfrac{1}{2}(\text{base})(\text{height}) = \dfrac{1}{2}ab\sin\theta_c = \dfrac{1}{2}a^2\dfrac{\sin\theta_b\sin\theta_c}{\sin\theta_a}$

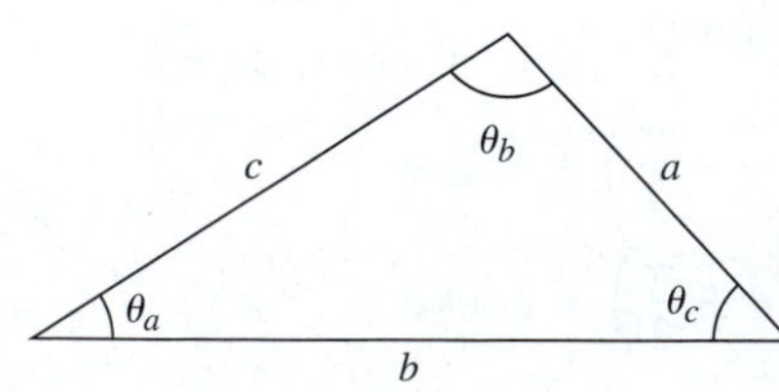

▲ FIGURE A–2

8. volume of a right cylinder of height h and radius $r = \pi r^2 h$

IV–6 SOME PROPERTIES OF ALGEBRAIC FUNCTIONS

1. General properties

   (1) $a^x a^y = a^{x+y}$

   (2) $a^0 = 1$

   (3) $(ab)^x = a^x b^x$

2. Properties of exponential of x, $\exp(x)$ or e^x:

   (1) $\exp(\ln x) = x$

   (2) $\exp(x_1)\exp(x_2) = \exp(x_1 + x_2)$

   (3) $\exp(0) = 1$

   (4) expansion: $e^x = 1 + x + \dfrac{x^2}{2!} + \dfrac{x^3}{3!} + \cdots$

3. Properties of the natural logarithm of x, $\ln(x)$:

   (1) $\ln(e^x) = x$

   (2) $\ln(x_1 x_2) = \ln(x_1) + \ln(x_2)$

   (3) $\ln(x_1/x_2) = \ln(x_1) - \ln(x_2)$

   (4) $\ln(1) = 0$

   (5) expansion:

   $$\ln(1+x) = x - \frac{x^2}{2} + \frac{x^3}{3} - \frac{x^4}{4} + \cdots \qquad (x^2 < 1)$$

IV–7 DERIVATIVES

In the following, b and p are constants, and u and v are functions of x:

1. $\dfrac{db}{dx} = 0$

2. $\dfrac{d}{dx}(bu) = b\dfrac{du}{dx}$

3. $\dfrac{d}{dx}(u+v) = \dfrac{du}{dx} + \dfrac{dv}{dx}$

4. $\dfrac{d}{dx}(uv) = v\dfrac{du}{dx} + u\dfrac{dv}{dx}$

5. $\dfrac{dx^p}{dx} = px^{p-1}$

6. Chain rule: If u is a function of y, and y is in turn a function of x, then $\dfrac{du}{dx} = \dfrac{du}{dy}\dfrac{dy}{dx}$.

7. $\dfrac{d}{dx}(\sin x) = \cos x$

8. $\dfrac{d}{dx}(\cos x) = -\sin x$

9. $\dfrac{d}{dx}(\tan x) = \dfrac{1}{\cos^2 x}$

10. $\dfrac{d}{dx}(e^{bx}) = be^{bx}$

11. $\dfrac{d}{dx}\ln(x) = \dfrac{1}{x}$

IV–8 TAYLOR EXPANSION

If $f(x)$ is well behaved near point $x = x_0$,

$$f(x) = f(x_0) + \left.\frac{df}{dx}\right|_{x=x_0}(x - x_0) + \frac{1}{2!}\left.\frac{d^2f}{dx^2}\right|_{x=x_0}(x - x_0)^2 + \cdots$$

IV–9 INTEGRALS

In the following, b and p are constants, and u and v are functions of x:

1. $\int \frac{du}{dx}\,dx = u$

2. $\int_{x_1}^{x_2} \frac{du}{dx}\,dx = u(x_2) - u(x_1)$

3. $\int bu(x)\,dx = b\int u(x)\,dx$

4. $\int (u + v)\,dx = \int u\,dx + \int v\,dx$

5. $\int u\frac{dv}{dx}\,dx = uv - \int v\frac{du}{dx}\,dx$ (integration by parts)

6. If u is a function of y and y is in turn a function of x, then

$$\int u\,dy = \int u\frac{dy}{dx}\,dx$$

7. $\int x^p\,dx = \frac{x^{p+1}}{p+1} \qquad (p \neq -1)$

8. $\int \frac{dx}{x} = \ln x$

9. $\int (\sin x)\,dx = -\cos x$

10. $\int (\cos x)\,dx = \sin x$

11. $\int e^{bx}\,dx = \frac{1}{b}e^{bx}$

12. $\int xe^{bx}\,dx = e^{bx}\left(\frac{x}{b} - \frac{1}{b^2}\right)$

13. Some definite integrals:

(1) $\int_0^\infty x^n e^{-x}\,dx = n!$

(2) $\int_0^\pi (\sin^2 x)\,dx = \int_0^\pi (\cos^2 x)\,dx = \frac{\pi}{2}$

(3) $\int_0^\infty e^{-b^2x^2}\,dx = \frac{\sqrt{\pi}}{2b} \qquad (b > 0)$

(4) $\int_0^\infty xe^{-x^2}\,dx = \frac{1}{2}$

(5) $\int_0^\infty x^2e^{-x^2}\,dx = \frac{\sqrt{\pi}}{4}$

(6) $\int_0^\infty \frac{b}{b^2 + x^2}\,dx = \begin{cases} \frac{\pi}{2} & (b > 0) \\ 0 & (b = 0) \\ -\frac{\pi}{2} & (b < 0) \end{cases}$

IV–10 SOME EXPANSIONS APPROPRIATE FOR $x^2 < 1$

1. The following expression is good for any n, positive or negative, integer or noninteger:

$$(1 + x)^n = 1 + nx + \frac{n(n-1)}{2!}x^2 + \frac{n(n-1)(n-2)}{3!}x^3 + \cdots$$

2. $\sin x = x - \frac{x^3}{3!} + \frac{x^5}{5!} + \cdots$

3. $\cos x = 1 - \frac{x^2}{2!} + \frac{x^4}{4!} + \cdots$

4. $\tan x = x + \frac{x^3}{3} + \frac{2}{15}x^5 + \cdots$

5. $e^{ax} = 1 + ax + \frac{(ax)^2}{2!} + \frac{ax^3}{3!} + \cdots$

IV–11 SOME MATHEMATICAL NOTATION

1.	$=$	is equal to
2.	$\cong$	is approximately equal to
3.	$\propto$	is proportional to
4.	$\equiv$	is defined to be
5.	$\neq$	is unequal to
6.	$>$	is greater than
7.	$\geq$	is greater than or equal to
8.	$<$	is less than
9.	$\leq$	is less than or equal to
10.	Δx	the change in x
11.	$\lvert x\rvert$	the absolute value of x
12.	$O(N)$	on the order of the magnitude of N
13.	$\pm$	plus or minus
14.	$\mp$	minus or plus
15.	$\langle x\rangle$	average of x
16.	$\sum_{i=i_1}^{i_2} f_i$	the sum of all f_i over the integers i from a smallest integer i_1 to a largest integer i_2
17.	$\ln(x)$	natural logarithm of x
18.	$\log_{10}(x)$	logarithm to the base 10 of x
19.	$\int$	integral
20.	$\oint$	line integral around a loop

APPENDIX V

Periodic Table of the Elements

PERIODIC TABLE

Atomic Properties of the Elements

Adapted from:
U.S. DEPARTMENT OF COMMERCE
Technology Administration
National Institute of Standards and Technology
http://physics.nist.gov/PhysRefData/IonEnergy/periodic-table.pdf

For a description of the atomic data, visit physics.nist.gov/atomic

Period	Group IA	IIA	IIIA	IVA	VA	VIA	VIIA	VIIIA	VIIIA	VIIIA	IB	IIB	IIIB	IVB	VB	VIB	VIIB	VIII
1	1 **H** Hydrogen 1.00794																	2 **He** Helium 4.00260
2	3 **Li** Lithium 6.941	4 **Be** Beryllium 9.01218											5 **B** Boron 10.811	6 **C** Carbon 12.0107	7 **N** Nitrogen 14.00674	8 **O** Oxygen 15.9994	9 **F** Fluorine 18.99840	10 **Ne** Neon 20.1797
3	11 **Na** Sodium 22.98977	12 **Mg** Magnesium 24.3050											13 **Al** Aluminum 26.98154	14 **Si** Silicon 28.0855	15 **P** Phosphorus 30.97376	16 **S** Sulfur 32.066	17 **Cl** Chlorine 35.4527	18 **Ar** Argon 39.948
4	19 **K** Potassium 39.0983	20 **Ca** Calcium 40.078	21 **Sc** Scandium 44.95591	22 **Ti** Titanium 47.867	23 **V** Vanadium 50.9415	24 **Cr** Chromium 51.9961	25 **Mn** Manganese 54.93805	26 **Fe** Iron 55.845	27 **Co** Cobalt 58.93320	28 **Ni** Nickel 58.6934	29 **Cu** Copper 63.546	30 **Zn** Zinc 65.39	31 **Ga** Gallium 69.723	32 **Ge** Germanium 72.61	33 **As** Arsenic 74.92160	34 **Se** Selenium 78.96	35 **Br** Bromine 79.904	36 **Kr** Krypton 83.80
5	37 **Rb** Rubidium 85.4678	38 **Sr** Strontium 87.62	39 **Y** Yttrium 88.90585	40 **Zr** Zicronium 91.224	41 **Nb** Niobium 92.90638	42 **Mo** Molybdenum 95.94	43 **Tc** Technetium (98)	44 **Ru** Ruthenium 101.07	45 **Rh** Rhodium 102.90550	46 **Pd** Palladium 106.42	47 **Ag** Silver 107.8682	48 **Cd** Cadmium 112.411	49 **In** Indium 114.818	50 **Sn** Tin 118.710	51 **Sb** Antimony 121.760	52 **Te** Tellurium 127.60	53 **I** Iodine 126.90447	54 **Xe** Xenon 131.29
6	55 **Cs** Cesium 132.90545	56 **Ba** Barium 137.327		72 **Hf** Hafnium 178.49	73 **Ta** Tantalum 180.9479	74 **W** Tungsten 183.84	75 **Re** Rhenium 186.207	76 **Os** Osmium 190.23	77 **Ir** Iridium 192.217	78 **Pt** Platinum 195.078	79 **Au** Gold 196.96655	80 **Hg** Mercury 200.59	81 **Tl** Thallium 204.3833	82 **Pb** Lead 207.2	83 **Bi** Bismuth 208.98038	84 **Po** Polonium (209)	85 **At** Astatine (210)	86 **Rn** Radon (222)
7	87 **Fr** Francium (223)	88 **Ra** Radium (226)		104 **Rf** Rutherfordium (261)	105 **Db** Dubnium (262)	106 **Sg** Seaborgium (263)	107 **Bh** Bohrium (264)	108 **Hs** Hassium (265)	109 **Mt** Meitnerium (268)	110 **Uun** Ununnilium (269)	111 **Uuu** Unununium (272)	112 **Uub** Ununbium						

Metals ← | → Nonmetals

□ Solids
□ Liquids
□ Gases
□ Artificially Prepared

57 **La** Lanthanum 138.9055	58 **Ce** Cerium 140.116	59 **Pr** Praseodymium 140.90765	60 **Nd** Neodymium 144.24	61 **Pm** Promethium (145)	62 **Sm** Samarium 150.36	63 **Eu** Europium 151.964	64 **Gd** Gadolinium 157.25	65 **Tb** Terbium 158.92534	66 **Dy** Dysprosium 162.50	67 **Ho** Holmium 164.93032	68 **Er** Erbium 167.26	69 **Tm** Thulium 168.93421	70 **Yb** Ytterbium 173.04	71 **Lu** Lutetium 174.967
89 **Ac** Actinium (227)	90 **Th** Thorium 232.0381	91 **Pa** Protactinium 231.03588	92 **U** Uranium 238.0289	93 **Np** Neptunium (237)	94 **Pu** Plutonium (244)	95 **Am** Americium (243)	96 **Cm** Curium (247)	97 **Bk** Berkelium (247)	98 **Cf** Californium (251)	99 **Es** Einsteinium (252)	100 **Fm** Fermium (257)	101 **Md** Mendelevium (258)	102 **No** Nobelium (259)	103 **Lr** Lawrencium (262)

Key: Atomic Number — 58; Symbol — **Ce**; Name — Cerium; Atomic Weight† — 140.116

†Based upon ^{12}C. ( ) indicates the mass number of the most stable isotope. For a description and the most accurate values and uncertainties, see J. Phys. Chem. Ref. Data, **26** (5), 1239 (1997).

March 1999

APPENDIX VI

Significant Dates in the Development of Physics

History can rarely be stated as a simple series of dates, and the history of science is no exception. Throughout the text we have alluded to important discoveries in physics. The list below is a personal choice and should be thought of as a guide. It oversimplifies some of the history, including stories that are covered more thoroughly in the text. Some of the dates are to be taken with a grain of salt, because discoveries are rarely made in a single identifiable moment. Our list includes some names (and discoveries) not mentioned in the text. Far more numerous are the names not listed, the names of those who built the experimental foundations, those who explored the false paths and cleared the way for those whose names we remember today, or those who verified the speculations that are now called laws.

1583	Galileo	Pendulum motion
1600	Gilbert	Study of magnets
1602	Galileo	Early statement of Newton's first law
1602	Galileo	Laws of falling bodies
1609	Kepler	First two laws of planetary motion
1619	Kepler	Third law of planetary motion
1620	Snell	Law of refraction
1648	Pascal	Atmospheric pressure
1650	Grimaldi	Diffraction of light
1661	Boyle	Chemical elements
1669	Newton	Light dispersion in prisms
1678	Huygens	Wave propagation
1687	Newton	Laws of motion; universal gravitation
1760	Black	Calorimetry
1785	Coulomb	Coulomb's law
1789	Lavoisier	Conservation of mass
1798	Cavendish	Measurement of G
1800	Volta	Electric battery
1801	Young	Interference of light
1801	Dalton	Laws of chemical combination
1802	Charles; Gay-Lussac	Ideal gases
1807	Dalton	Atomic theory
1812	Fourier	Decomposition of waves
1815	Fraunhofer	Discrete spectral lines
1819	Fresnel	Wave picture of light
1820	Oersted	Magnetic fields from currents
1820	Biot; Savart	Law of magnetic field produced by current
1824	Carnot	Second law of thermodynamics
1827	Ohm	Ohm's law
1827	Ampère	Ampère's law
1831	Faraday; Henry	Magnetic induction
1842	Joule	Mechanical equivalent of heat
1847	Helmholtz	Conservation of energy
1849	Fizeau	Direct measurement of the speed of light
1865	Maxwell	The laws of electricity and magnetism; light waves
1877	Boltzmann; Gibbs	Statistical mechanics
1879	Stefan	Blackbody radiation
1885	Osmond	Crystalline structure of metals
1887	Hertz	Electromagnetic waves
1887	Michelson and Morley	Constancy of the speed of light
1896	Becquerel	Radioactivity
1897	Thomson	Charge-to-mass ratio of the electron

1900	Planck	Quanta in blackbody radiation
1903	Rutherford; Soddy	Isotopes
1905	Einstein	Special relativity; quanta in photoelectric effect
1908	Kammerlingh Onnes	Superfluidity
1911	Kammerlingh Onnes	Superconductivity
1911	Rutherford	Nuclear structure of atom
1911	Millikan	Quantization of charge
1912	von Laue	X-ray diffraction in crystals
1912	Bragg and Bragg	Analysis of crystal structure with x-rays
1913	Bohr	Atomic structure
1916	Einstein	General relativity
1922	Compton	Scattering of x-rays
1923	Hubble	Discovery of galaxies
1924	de Broglie	Wave nature of particles
1925	Pauli	Exclusion principle
1925	Heisenberg	Formulation of quantum mechanics
1925	Goudsmit and Uhlenbeck	Electron spin
1926	Davisson and Germer; Thomson	Diffraction of electrons by crystals
1926	Schrödinger	Alternate formulation of quantum mechanics
1926	Born	Probabilistic interpretation of quantum theory
1927	Lemaitre	Big bang universe introduced
1927	Heisenberg	Uncertainty relations
1929	Hubble	Hubble's law
1930	Dirac	Antiparticles
1932	Anderson	The positron
1932	Lawrence and Livingston	The cyclotron
1932	Chadwick	The neutron
1934	Yukawa	Nuclear forces and the pi meson
1942	Fermi	Nuclear chain reaction
1948	Feynman; Schwinger; Tomonaga	Electromagnetism as a quantum theory
1954	Townes	The maser
1956	Reines and Cowan	Neutrinos observed
1957	Lee and Yang	Nonconservation of parity
1957	Bardeen, Cooper, and Schrieffer	Theory of superconductivity
1962	Josephson	Josephson junction
1964	Gell-Mann; Zweig	Quarks
1964	Penzias and Wilson	Background radiation of the universe
1967	Bell and Hewish	Neutron stars discovered (pulsars)
1967–1970	Glashow; Salam; Weinberg	Unification of electromagnetic and weak forces
1981	Binnig and Rohrer	Scanning tunneling electron microscope
1983	Taylor and Hulse	Gravitational radiation
1983	CERN	$W^{\pm}$ and Z^0 particles discovered
1986	Bednorz and Müller	High-temperature superconductors
1990	Davis and Koshiba	Neutrino oscillations and inference of neutrino mass

Answers to "What Do You Think?" Questions

CHAPTER 1

1. Nothing is special about a year. One year refers to an astronomical accident in a sense, since if Earth traveled a slightly different orbit, the year would be different. The second is just as arbitrary. The importance of these or any units is that they establish a common standard with which we can all agree. A light-year is the same to anyone anywhere in the world.

2. Just because this answer is an exact one does not mean that you are obliged to use all the figures in application. It would in fact be an unusual situation for you to want to use the exact answer. You can always round off if fewer digits are called for.

3. The conversion between the millimeter and meter involves a definition, so no new uncertainty is introduced in the conversion.

4. If your calculator does not have enough digits, you might have to use a larger computer or you might have to be clever and use mathematical approximations. In fact, we could have used the approximation $\sin\theta \cong \theta - (\theta^3/3!) + \cdots$, correct if θ is small, to arrive at our numerical answer. But don't throw your calculator away; for almost all applications it will suffice.

6. No, you have a choice of any units you like with the right dimensions. For example, you could take the mass in slugs, the time in minutes, and the length in furlongs, in which case the units of G would be cubic furlongs per slug per minutes squared. But we would more normally use mks units.

7. If the radius of a sphere increases, so does the area of the sphere's surface. Thus there would be more area available for the same number of people, and the area available per person would increase.

8. The estimate l for the size in the example is in fact the cube root of the space taken up by each molecule. The only difference is that with the conditions stated the linear size of a molecule—the length of the side of a cube of the molecular volume—is $\frac{1}{10}l$. In other words, answer (a) is the correct one.

9. It would be unchanged. Displacements don't depend on the choice of origin. You can very easily check this explicitly in this case.

10. How the displacement is broken into components *can* differ according to the coordinate system. But the vector itself, that is its magnitude and direction in space, is independent of its description in a particular coordinate system.

CHAPTER 2

2. See Fig. 2–38.

3. You could certainly take the data in the table and plot it. The plot would contain no more and no less information than the table. When you draw a smooth line connecting the points of the plot, then you are interpolating (guessing, with an assumption of smoothness) between the points. You could use the interpolated curve to find the velocity: The instantaneous velocity at a given time (or position) is given by the slope of the curve of position versus time at the given time. You would draw the straight line tangent to the curve at the given time, then just measure the slope of that straight line. Since the line is straight, you can measure its slope by examining any two points on it, just as we did in our discussion of Fig. 2–5. However, you should keep in mind that when you are taking the slope of a plot that is just a guess (an interpolation between points), you cannot *guarantee* that you have a precise value.

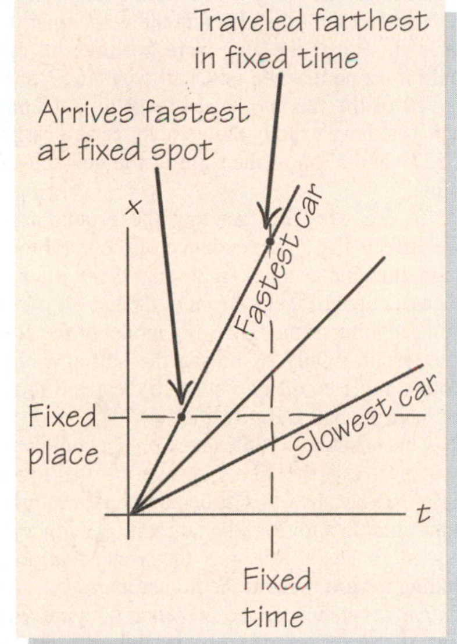

▲ **FIGURE 2–38** Discussion, What Do You Think? Example 2–2.

5. (1d) The ball first moves to the right, then to the left, and so on. When does any acceleration occur? Is it large? (2b) The two runners have different speeds, and the handoff occurs in an idealized fashion, with runners at their top speed. When is the acceleration of the baton? (3a) When the ball is first tossed, it has a large positive velocity; it slows down, passing through $v = 0$ at the top of its toss, at which point it falls with an increasing negative velocity. Since the acceleration is the slope of this curve and the slope is constant (and negative), the acceleration here is constant (and negative). (4c) The car is stopped, accelerates to a constant velocity, then has to stop again for another red light, and so on.

7. The last term in dv/dt is proportional to t^2 and therefore dominates the formula at large values of t in comparison to either the constant term or the term linear in t. Since its coefficient is negative, the expression for the acceleration will eventually become negative. This will in turn eventually lead to a negative velocity, so that the rocket would eventually be moving faster and faster in the backward direction. This is a very peculiar situation, certainly not what the rocket designer had in mind; for that matter, with only a finite amount of fuel it is not possible for the acceleration to continue forever. You could surmise that the term in the acceleration proportional to t^2 is an approximate one meant to be used only for a brief period of time at the beginning, when it is not dominant.

8. Friction—here rolling friction—between the ball and the floor may be small, but it is not zero, and there is also air resistance. Often when we look at physical situations we simplify them by saying that we can neglect air resistance and friction, but these factors are important in many real situations involving motion. Here we have in fact exaggerated their effects. In a real bowling alley the degree to which the alley is level is what most influences the ball's motion.

10. See Fig. 2–39.

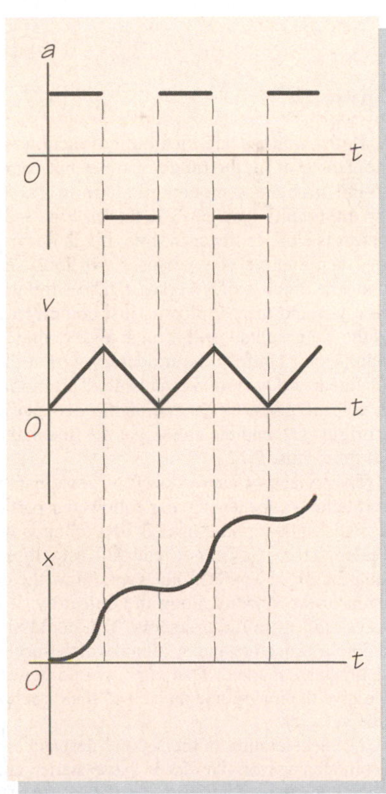

▲ **FIGURE 2–39** Discussion, What Do You Think? Example 2–10.

11. (b) The acceleration due to gravity is constant and downward with magnitude g, regardless of initial conditions, which here happens to be a positive (upward) initial velocity.
13. Under constant acceleration, the ball would fall four times as far in the first two time intervals as it did in the first time interval. (We are assuming that all the time intervals are the same.) This is a measurement that could easily be made on the photo in Fig. 2–23 and we encourage you to try it.
14. Refer to the sketches in Fig. 2–40. The acceleration is a constant, the same at the beginning of the motion, the end of the motion, and anywhere in between, including at the top of the trajectory (see Fig. 2–19).

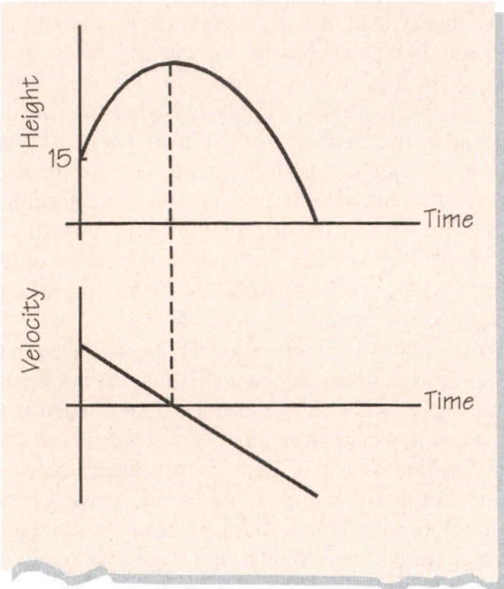

▲ **FIGURE 2–40** Discussion, What Do You Think? Example 2–14.

CHAPTER 3

1. Both x and y are quadratic functions of time. However, in the range of time that comes into this problem, the quadratic term in x is not very important—you can see this in Fig. 3–3b, where x is close to linear in time. If x is linear in time, then a curve of y versus x will look very much like a curve of y versus t. The similarity of the y–x and the y–t plots is just coincidence, and the x–t graph doesn't look at all like the x–y graph! Don't confuse the trajectory with a plot of a linear dimension versus time. The motion can also be given by means of the distance to the origin $r(t)$, and the angle that the line makes with the x-axis, $\theta(t)$.
2. The trajectory curve does not contain time information—whether the car followed a particular path while zipping or crawling along is not revealed. Thus you cannot read full velocity information off it. You can, however, know the direction of the velocity along the trajectory: It is always tangent to the trajectory. You could also imagine a trajectory curve along which successive times are marked. That curve would in principle give the magnitude as well as the direction of the velocity.
3. The acceleration is the second derivative of the position vector. We could have started this with the position vector as a function of time, as taking a double derivative is the same as taking two single derivatives, so we would have "passed through" the calculation of Example 3–2 in any case.
5. The motion in the x- and y-directions are independent of each other. The horizontal travel distance is determined by a combination of things: how fast it moves horizontally (a constant given by the initial condition) and how much time it has to travel (also determined by initial conditions, but this time by the initial vertical component of velocity, zero in this problem). Of course, if the cliff were higher, the ball would be in flight for a longer time and would fall further out in the ocean.
6. The parabola describes the motion for a given acceleration and with the condition that the tee is at the top of the cliff and the velocity at the tee is $\vec{v}_0$. But if the cliff were not present, one could imagine that the golf ball was hit 57 m to the *left* of the tee and at sea level in a manner such as to have exactly the velocity $\vec{v}_0$ at a height of 52 m and 57 m to the right of the new tee-off point.
7. To see why the fact that the ground isn't level affects the θ_0-dependence of the maximum range, imagine as an example that there is a 6-ft fence a couple of feet in front of the tee-off point. (You can think of the fence as a model of the possible terrain.) Only by hitting the ball at a high enough angle could one hope to clear the fence and have the ball travel further.
9. The answer to this question is somewhat subjective, but the likely answer is the lower angle 36° because this trajectory has a shorter flight time. In addition, the target might conceivably more easily observe the water balloons heading toward them at the higher angle.
10. Air resistance or the presence of wind will affect the trajectory of both the rock and the coconut, as we'll discuss in more detail in Chapter 5.
11. Definitely not. If the automobile were rounding a curve with a larger radius of curvature, the acceleration would be less; as for the ramps onto an interstate highway, the car can negotiate a circular arc with a larger radius without danger. The extreme case is straight-line motion, for which 54 mph would pose no problem whatsoever. We'll see in Chapter 5 why too much acceleration can be dangerous, but you know from experience that it has something to do with the car skidding out.
12. The expression $a = \omega^2 r$ shows that the outermost parts of the top move with the largest acceleration. The angular speed is the same for all parts, so $a \propto r$.
13. (b) The rower would say that there is less current to fight, that he or she wouldn't have to row partly upstream just to stay even with the point across the river from the dock. With no current, the frames are identical; $u = 0$, so that $\cos\theta = u/v' = 0$, and the boat should be directed straight across. It would move across with speed v', rather than $v'\cos\theta < v'$. The crossing time is $(0.15\text{ km})/(4\text{ km/h}) = 0.04\text{ h} \cong 2\text{ min}$.
14. It is the solution to Example 3–13 that supplies the answer to this one. In effect, the flight path of the airplane must be longer (the hypotenuse of a right triangle) through the air in order to hold the north–south direction relative to the ground (a side leg of the right triangle). Thus the jet stream adds to the flight time even if is not directly opposed to the motion. You may want to think about the relative angles of flight direction and jet stream direction for which the jet stream can actually help.

CHAPTER 4

1. Not always. Following our rules for addition of vectors (tail to tip, as discussed in Chapter 1), the sum of the five forces for the fixed rods will end up as a single vector pointing somewhere in the three-dimensional space. We can get a zero net force if the two adjustable force vectors—the ropes—can add up to give a vector equal and opposite to this single vector. This is equivalent to asking whether three straight lines of fixed length can form a triangle, and this is only possible if the sum of the lengths of the adjustable force vectors is equal to or larger than the vector representing the sum of the fixed forces.
3. For a constant force, and therefore a constant acceleration, the final velocity is equal to $a\,\Delta t = F\,\Delta t/m$ and is therefore inversely proportional to the mass. If M is the mass of the box and m that of the bowling ball, the ratio of the final speed with one ball to the final speed with n balls is $(M + nm)/(M + m)$. The ratio of the speeds could still be used to count balls, but you would have to know the ratio, not necessarily an integer, of M to m. Notice that if $M \gg nm$, then the ratio is very nearly unity.
4. During the brief contact between the hand and the spaceship the astronaut exerts a force on the spaceship, and the spaceship exerts an equal and opposite force on the astronaut. They are both accelerated in opposite directions (inversely proportional to their masses) and move away from each other until a tether stops their separation by exerting forces on both of them. As the astronaut has less mass than the spaceship, she will have greater acceleration.
5. We can easily imagine that the passenger accelerates forward along with the sled. If he accelerates, there is a force on him. This force must be a contact force coming from the sled. But then, by the third law, the passenger must be exerting a force on the sled, and it must be backward. Why doesn't the sled accelerate backward as a result? That is because there is an additional force on the sled, one that comes from the person who is doing the pushing. The net force on the sled is forward.
7. Professor A would see her own tennis ball bounce on the floor and rise up vertically to her hand. She would conclude that the ball suffers no horizontal acceleration and that there is no horizontal force acting on the ball. Professor B would see that his ball does *not* bounce back into his hand. The reason is that the ball as it leaves the hand has the same instantaneous velocity as he does. While it falls, it maintains that velocity, whereas the professor and the train are accelerating. Thus, by the time the tennis ball comes back up, professor B has moved a little away from the ball. Professor B thinks that the same mysterious force—it has no visible source—that makes professor A accelerate backward also acts on the tennis ball.
8. The glue here ensures a sufficient amount of friction between the blocks. As long as these friction forces are large enough, nothing changes. If, however, the forces between the two blocks along the direction of motion are significantly reduced—for example, the blocks sliding past one another in the absence of friction—then there would be no way to transmit the pull from block 2 to block 1, and block 1 would be not be affected by the force acting on block 2—it would be left behind. As we shall see in the next chapter, friction has an "adjustable" quality up to a certain point.

11. During the contact time, the ball is indeed accelerated both horizontally and upward and acquires its initial velocity for its subsequent motion, during which only the force of gravity acts.
12. Look at Eq. (4–23): The x-position changes in time in a way that depends not only on the acceleration but also on v_{0x}. So we cannot make a rule about how far the nugget goes based on acceleration alone.

CHAPTER 5

2. Either fishermen are eternally optimistic about the size of fish they will catch or, more likely, fish can use their muscles and exert a force on the line much larger than their own weight, and the fishermen don't want the fish to be able to break the line.
3. Absolutely nothing would be different. The acceleration, Eq. (5–12), is independent of the mass of the sliding object. Everything placed on the ramp would slide down in the same way. Without friction neither the golf ball nor the car would roll. We are assuming the ramp will not collapse under the weight of a car.
4. If the ropes are not massless, we must include the gravitational force on them in the force equation. We would have to lift them! If the pulleys are not frictionless, then the tension throughout all the ropes is not the same, and it is no longer true that $T_1 = T_2 = T_3 = T_4$. The entire problem would become more complicated. All these effects reduce the ease with which the person could lift an object with the block and tackle, and for real machinery we wish to minimize these effects so that lifting is easier.
5. The masses would have to be equal. If $m_1 > m_2$, then m_1 will rest on the scale, and the scale would still have a nonzero reading.
6. The critical angle is determined by the *ratio* of the masses [Eq. (5–20)]. In part (a), the masses have the same ratio as in the example, and the system will remain motionless. (b) The mass m_1 will fall down pulling m_2 up the ramp. (c) The mass m_2 slides down the ramp and pulls m_1 up.
8. A force F_{hand} directed into the surface of the ramp would be added to the free-body diagram for the box. To balance this, we would have to increase the normal force from the ramp to the value $F_N = mg\cos\theta + F_{\text{hand}}$, and if the normal force increases, so does the maximum value of static friction and hence the angle at which slipping will start. This makes good sense; we know we can keep an object from sliding down a ramp by pushing down on it.
9. No, the y-component of the force will have a term mg down, which will complicate the last two equations in the example enough to lead to an angle that is not independent of F_{prof}. In effect, we have included a new dimensional quantity that has the same dimensions as F_{prof} and thus allows F_{prof} to enter through the dimensionless ratio mg/F_{prof}.
10. There is more weight on the front tires because the engine is in front. With more normal force pressing on the snow, the maximum static force is larger for the front tires than for the rear tires. This makes a significant difference in traction when μ_s is small and means that if the drive power is coming from the front, it can be larger than if it were at the back without the drive wheels slipping.
12. Gas efficiency for a given velocity would improve if the drag coefficient of the car is reduced, and for racing cars speed can be increased for a given engine power in this way. Drag coefficients for the boxy cars of years ago were as high as 0.7, whereas modern streamlined cars have C_D near 0.3. Gas mileage increases by 5 percent for every 10 percent improvement in drag coefficient. (Of course, the popularity of SUVs is evidence that gas efficiency may not be an important factor for some drivers.) Airplane wings have drag coefficients as low as 0.05.
13. Oil becomes "thicker," or more correctly more viscous (a word that describes the size of the coefficients in drag force—the higher the viscosity, the larger these coefficients), at low temperatures, and the marble in this example would move more slowly. Oil is a much better lubricant when it flows more easily, that is, at higher temperatures. But with today's multi-viscosity oils, car owners don't worry so much about warming up the engine before driving off.
14. From experience we know that the book will move toward the outside of the merry-go-round and fall off. It would move in a combination of radial and tangential directions because once static friction is replaced by kinetic friction (typically smaller than the largest value of static friction), the book will tend to continue motion in its current direction, which is initially tangential but as it moves out also attains a radial component.
15. To the inhabitants of the station, "up" is radially inward, opposite to the "force" that keeps them on the outer wall. Walking on the innermost wall would be like your walking on the ceiling, and walking on the side walls would be like walking on the side walls of your room! In both cases there would be no force to keep them accelerating centripetally toward the center of the wheel, and they couldn't walk on any surface except the outermost wall.
16. We can say that because static friction acts within a range, there will be some leeway for error that is not present when there is no static friction. There will be a range of speeds within which the car can negotiate the curve. See Problem 51, which treats this topic in some detail.
18. A skater will slow down when he or she hits the rough patch. The coefficient of friction of 0.10 in this example is small. A more realistic value, perhaps three or four times larger, would cause the speed to drop quickly enough to present a real danger: The skater might not be able to slow his or her upper body as quickly as his or her legs are slowed by the sand, and the skater may topple over! We are touching here on effects due to the finite size of the body that go beyond the motion of pointlike or perfectly rigid objects that are the main topics of our attention in these early chapters. Later in the book we'll come back to this point in much more detail.

CHAPTER 6

2. As we have not yet dealt with energy other than kinetic energy, we can only answer this question with our intuitive knowledge. We have the work–energy theorem, so we know that a change in kinetic energy implies that there is work done. The furniture mover has the ability to do work because his muscles have the ability to do work. The muscles are able to do work as they have an energy source they can call on—chemical energy from the metabolism of nutrition. The depletion of the chemical energy in the mover's body is measured by the kinetic energy acquired by the sofa.
3. For (i), the answer is (a): The work done is proportional to the force and the force of gravity is proportional to the mass, so that the work done by gravity is doubled. For (ii), the answer is (c): The relation between velocity, acceleration, and displacement is a kinematic one, which does not involve the mass of the object, so that the velocity is the same.
4. Initially the elevator (with its contents) is at rest. The downward force of gravity is exactly canceled by the upward force of the friction of the brakes and the tension of the cable. When the brakes are released and the tension of the cable is increased, there is a net force upward. The net work done is positive, giving rise to the kinetic energy of the elevator and contents. When the brakes grab and the tension decreases, the net work done is large and negative, canceling the kinetic energy possessed by the elevator before it stops.
5. The "pushing" force is unchanged, but the friction force that acts against it slowly increases. Since work done on the box depends on the *net* force, contributions to it become increasingly negative so that, by the theorem, the kinetic energy decreases until it reaches zero. The box comes to rest. At this point the static friction is just large enough to balance the external force and the box does not move.
6. We assume that when the box hits a downward slope, the push is still parallel to the floor. The force of friction is reduced because it is proportional to the normal force, which on the slope only involves the component of weight perpendicular to the slope. There is also an accelerating force due to the component of the force of gravity parallel to the floor. The result is that the net force is no longer zero, and the kinetic energy (and therefore the speed) of the crate increases as it goes down the slope.
7. The component of the force perpendicular to the ramp increases, so that the force of friction increases. The component of the force along the ramp decreases, so that the *net* force along the ramp decreases. There will be an angle at which the net force will go through zero and the crate will start slipping down.
9. Friction is certainly not a spring force. But in this case the friction force varies linearly with distance, so that the integration necessary to find the work done by it takes the same form as the integration necessary to find the work done by a spring force.
10. The circular motion itself implies that a centripetal force must also be acting. It is supplied by a tension in the rod that holds the ball. It does no work because the force is perpendicular to the motion; hence the scalar product that appears in the work is zero. Because it does no work, it cannot increase the speed; only a force with a tangential component can do that. Gravity also acts, but it is canceled by the contact force at the pivot that keeps the rod moving in the horizontal plane.
12. If the work done is independent of the path, then $W(A, B) = W(A, C) + W(C, B)$. This is equivalent to $\int_A^B F(x)dx = \int_A^C F(x)dx + \int_C^B F(x)\,dx$. This, however, is a general property of one-dimensional integrals (see the box Integration), so it is true for the spring force with the variable spring constant.
13. When an object slides on a rough surface, the motion is impeded by irregularities in the surfaces in contact. When these push past each other, energy is dissipated in bending them or in

breaking them off (abrasion). In addition, bonds may be formed between atoms coming from the two different materials, and these bonds break when the object moves (see Section 5–2). In molasses, the object, in order to move, must push the fluid out of the way. Molasses is "hard to move" (liquids such as molasses are said to be viscous, and there are strong electrical bonds between neighboring molecules), while water, say, takes less energy to move it.

14. The horsepower is a rate, an energy per unit time. This can change from instant to instant, and it is therefore an "instantaneous" quantity. If the rate changes slowly, then an average rate over some interval of time may give a good approximate value for the instantaneous rate of delivery of energy.

CHAPTER 7

2. The shift $y \rightarrow y + y_0$ is just a shift in the coordinate system, nothing more. After you make this shift, you are measuring height y from the top of your building: $y = 1$ m in the new system is $y = 101$ m in the old system. If the expression for U in the old system is Mgy, its expression in the new system is $Mgy + mgy_0$. In the new system $U = 0$ at $y = -100$ m. This is a perfectly sensible result. You have not changed the *location* of the zero of U; that remains at sea level. But in the new coordinate system sea level is located 100 m below the building top, at $y = -100$ m.

3. One way to look at this is to note that the reverse process could be represented by a movie of the initial process *run backward*. Time does not enter into energy conservation equations, and for the square of the velocity (i.e., kinetic energy), it does not matter whether the movie runs backward or forward. We can conclude from this that the speed of the brick at roof level would have to be 15 m/s.

4. We have moved from one equilibrium situation to another, with no motion and hence no kinetic energy involved. The potential energy has, however, changed: the spring is compressed so that it has acquired a positive potential energy, while on the other hand the kilogram mass is now *lower*, so that the gravitational potential energy is lower. The total energy has changed because displacing the mass and compressing the spring involved work.

5. When the system is initially at rest, the spring has been extended to its equilibrium point, whose location is determined by the spring constant and *the sum of the masses attached to it*. This is not changed by the motion of the masses and so that motion will be the same. In terms of free-body diagrams we may enclose the pulley and the masses in a box, and then the relative positions of the masses and their motion are a matter that is internal to the box.

6. A sketch of the energy diagram shows that the $E = 0$ line just touches the top of the potential energy hill at $x = 0$. Our object may start at rest at the leftmost turning point, but in sliding down into the first valley it speeds up, reaching its maximum kinetic energy at $x = -2$. In climbing back up the other side of the first valley it loses kinetic energy, that is, it slows down, until at the top of the hill at $x = 0$ it has zero velocity. Will it go on to the right-side valley or will it slide back to the left-side valley? Actually it gets to the top and stays there since there are no perturbing forces that move it to one side or the other.

If you are inclined to worry how a solution of a perfectly well-defined $F = ma$ equation can have a solution that describes motion (the falling into the valley floor) and then at some later time describes a mass at rest, the answer turns out to be that it takes an infinitely long time for the mass to creep up to the very top of the hill. Remember, time information is something you cannot get from energy considerations; for that you need the full apparatus of Newton's second law.

7. No, we cannot tell. The speed would be the same in either case.

8. We see that as r becomes very large (approaches infinity), $U(r)$ goes to zero. The masses are very far apart, and since the force falls off with distance, it is reasonable that there be no potential energy where the force itself is zero. The total energy at large r ("at infinity") is just the kinetic energy.

9. Since the energy of the pendulum systematically decreases, its speed decreases. For example, at the bottom of the swing the energy is equal to the kinetic energy there, so the speed at the bottom of the swing will systematically decrease. The period of the clock, which is what marks the time, is the time for the pendulum to get back to its maximum elevation. After each swing both the speed and the distance that has to be traveled by the mass decrease, so it is conceivable that the period may be unchanged. This is indeed exactly what happens as long as the swing angle is small. The fact that the period is independent of the angle (for small starting angles) is what makes the pendulum a good clock. For more details, see Chapter 13.

CHAPTER 8

2. This is a matter of counting unknowns. The final state is formed of two balls each moving in the same plane. That means that the final state is specified with four pieces of information, the two velocity components of each ball. Originally we knew two pieces of information about the final state (the two components of the final velocity of the cue ball), and that was just enough to allow us to find the two components of the final velocity of the struck ball. Now we are replacing two pieces of information with one, the conservation of kinetic energy. This is no longer enough to allow us to find the final velocity of the struck ball.

3. We can suppose that the force applied to a ball has to do with your stroke and therefore is the same for either the strung rackuet or the solid bat. The strung racket is the choice to make because the ball remains in contact with it for a longer period as the strings "give." The impulse is then larger from the strung racket because of the larger Δt.

4. The more inelastic the better. In an elastic collision relatively more energy goes into the recoil of the car (as opposed to its distortion in collapse), the momentum change of the car is larger, and hence the impulsive force acting is greater. In other words, the acceleration of the occupants is greater in an elastic collision.

5. (b) Energy is lost as the boater and dog "give" in order to bring the dog to rest with respect to the boat—as they change their shape there is heating in the muscles and so forth. The collision was inelastic—the dog and boat have "coalesced" by means of friction between them. The answer cannot be (a) because all the motion is horizontal; it cannot be (c) because there is no friction between the boat and the ice.

6. We can often simplify a situation by using an approximation. In many situations a large quantity plus a small quantity can be closely approximated by the large quantity alone. Thus in the equation for v we can replace the denominator by m_{block}. But care must be taken—we certainly cannot ignore the mass of the bullet in the numerator; we would find $v = 0$ in that case!

7. With the initial object at rest, the momentum conservation equation reads $0 = m_{\text{light}}v_{\text{light}} + m_{\text{heavy}}v_{\text{heavy}}$, with solution $v_{\text{light}} = (m_{\text{heavy}}/m_{\text{light}})v_{\text{heavy}}$; the coefficient of v_{heavy} is greater than 1, so the answer is (b). None of this depends on details of the explosion.

8. With equal masses, the target ball moves off with the speed of the ball that struck it, while the projectile comes to rest in the place of the target ball. The struck ball then can rise to exactly the height from which the projectile started. The motion is symmetric and repetitive.

11. (b) The size of each of the orbs is taken care of by first recognizing that their center of mass lies at their respective centers, by symmetry. The center of mass of the entire system is then that of two *point* masses making a dumbbell. The center of mass is between the two ends of the dumbell, relatively closer to the more massive end. There is no reason that point could not lie within Earth's radius.

12. Certainly. The calculation of a center of mass is additive, and you can do addition in any combination you like. Try the suggested combination to show that this is the case.

13. Allowed angles in the laboratory frame are determined by the possible angles made by the vectors $\vec{p}\,' + \vec{p}/2$ and $\vec{p}\,' - \vec{p}/2$. To take a concrete example, suppose that nearly the maximum amount of energy were lost. This corresponds to the final two objects moving off arbitrarily slowly (i.e., very nearly sticking together—see Section 8–3). Then $\vec{p}\,'$ is very nearly zero, and the direction of $\vec{p}\,' \pm \vec{p}/2$ is that of $\vec{p}$ itself. In the laboratory frame the objects continue to move, nearly together, along the x-axis. They will not make a large angle with the direction of the projectile.

14. There are many objects that have an external center of mass. Some simple ones are a jackknife folded at a 90° angle—the center of mass will be somewhere within that angle—or a ring donut, which has the center of mass in the middle of the hole.

15. Symmetry places the center of mass at the center of the sheet, that is, $\left(\frac{1}{2}\text{ m}, \frac{1}{2}\text{ m}\right)$. If the mass density had not been constant, the calculation would have been more complex.

16. As we will learn in more detail in Chapter 12, the acceleration of gravity at the surface of the Moon is much smaller than its value at Earth's surface. It is possible to reach higher speeds far away from the Moon (and Earth) with a launch from the Moon, and some have proposed a Moon base just for this purpose.

CHAPTER 9

1. Not at all. The definition of the *average* acceleration depends only on the initial and final angular speeds, not on how they were reached.

2. This is a rotating rigid body: The angular acceleration is the same for all the points within it. We are of course repeating here what was used in the answer to the original question.

7. No. All the integrals would look the same. Of course, because the mass per unit area is just the total mass divided by the area, the correct total mass M would have to appear in the answer.
8. The rotational inertia "weights" points in an object according to their distance squared from the reference axis. Points in our object can be farther from the reference axis when the axis is at a corner than when it is at the center of mass.
10. True, as the equation for ω (the next-to-last equation of the example) shows; it contains the factor $F \times t$. This echoes the analogous result for linear velocity.
11. (c) Since the direction of both quantities is the same, the kinematic equation (9–34) becomes an equation for the magnitude of the angular velocity in terms of the magnitude of the angular acceleration. One doesn't know the time dependence of the change until the time dependence of α is known. In this case Eq. (9–33) shows that α is constant, and this will lead ω to change linearly with time.
13. Friction acts, and it is a nonconservative force, so energy is not conserved. This collision is analogous to the perfectly inelastic linear collision of Chapter 8, in which two objects collide and stick. That is in fact the most inelastic linear collision possible, and a similar remark holds for this "rotational" collision.
14. All points on the wheel have the same angular speed ω about the contact point. Because the top of the wheel is at a distance $2R$ from the point of contact, $v_{top} = 2R\omega = 2v$.
15. The ratio of the energy in rotation to the energy in linear motion is of the form $I\omega^2/(Mv^2)$. But for a fixed R the factor ω/v is fixed in rolling. Thus the percentage of the energy in rotation is greater for larger I given the same mass and radius.
16. The C factor for a thick cylinder lies in between its value for the solid case $\left(C = \frac{1}{2}\right)$ and the thin-walled case $(C = 1)$ From Eq. (9–52), the thick-walled cylinder will finish the race ahead of the thin-walled cylinder but behind the solid cylinder. The same equation reminds us that the results are independent of the radius.

CHAPTER 10

2. Crucially, there would no longer be a radial component to the angular momentum. This can be seen in our equation for L_{radial}: The angle θ would be $90°$, $d = R \sin\theta = R$, and hence $L_{radial} = 0$. When the rod lies in the plane of motion, its tension, which maintains the circular motion, is directed to the attachment point and hence exerts no torque.
3. It is unchanged. The instantaneous motion of the ball is the same, even if later the ball will go off as a projectile moving under only gravity. Since its motion looks the same at the moment the rope breaks, its angular momentum is the same.
4. As we had remarked in the "What do you think?" question for Example 10–2, the angular momentum is in this case purely vertical, and unchanging. One would then expect the torque to vanish, and it does because the radial component of tension (which is still present to maintain the circular motion) lies parallel to the radial vector; hence the cross product is zero.
5. (d) Answers (b) and (c) are closely related; as the moment arm is independent of distance fallen, it translates into a moment arm independent of time. Note that this problem involves no rotations at all, yet there is a torque and hence there is an angular momentum. As we argued right at the beginning of Section 10–1, even a linearly moving point mass will generally have an angular momentum about some/any origin.
7. (c) The linear momentum transfer to the center of mass—the linear impulse—is independent of where the impulse is applied. This is not the case for the angular impulse.
9. (b) The angular momentum depends only on the mass of the comet, the speed at closest approach, and the distance of closest approach. The shape of the orbit is not a direct indication of the angular momentum.
11. The stool will stop rotating. Angular momentum is conserved, so two flips, bringing the wheel back to its original orientation, gets us back to the original situation, with the student and stool at rest.
12. Spool B has relatively more of its mass at larger radius and thus will have a larger rotational inertia than spool A. (The general form of I for a spool of mass M and radius R is CMR^2, where C is a geometrical factor, larger for spool B than for spool A.) Thus v is larger for spool A than for spool B.
13. The infinitesimal work done by the impulse is zero, because the scalar product of the angular impulse and the change in angle is zero. Thus the magnitude of the angular momentum is unchanged, although its direction is not.

CHAPTER 11

3. The mass density of the book must not be constant, because the book's center of mass must be somewhere directly over the table in equilibrium. This could be possible if, for example, the spine side were on the average denser than the side that opens.
4. The same fact that gravity acts on the center of mass tells us that if the map is placed with its center of mass on the point it will be in equilibrium; if not, it will rotate off the point.
5. (a) The coefficents of mg in the expressions for both F_A and F_B determine the answer to this question. If L decreases, then both F_B and F_A decrease.
6. If the crane is not fixed to the ground, the crane could pivot about the front tires and tip over! The torque that could make this happen would come from the suspended mass.
7. (b) As the window washer moves up the ladder the lever arm of the washer's weight with respect to the contact point with the ground increases. The torque due to the washer's weight about the contact point with the ground increases, until the torque about this point due to the normal force with the wall can no longer counteract the effect of the torque due to the weight. Note that (d) is in fact a consequence of (b).
8. Doing pull ups requires the lower arm to be rotated about the elbow to bring the body upwards. Thus the attachment distance x may be smaller for the person who cannot do pull ups as easily. Smaller x leads to a smaller torque about the elbow (xF_B).
9. The mass density is constant, so the volume V does not change. Therefore, $V = AL = A'(L + \Delta L)$ where A' is the new cross section. We see $A' = AL/(L + \Delta L)$; the new area will be smaller by 1.2×10^{-3} or 0.1%. This is indeed a small correction, and the assumption that there was no change in the cross section was valid.

CHAPTER 12

2. (a) Look at the equation for the sun's mass M in this example. If, on the right hand side of the equation, we have a larger r, we must also have a larger T. This is independent of Mars' mass.
3. No. Once again a sketch is helpful. The centripetal force on the satellite always points to Earth's center, so that its orbit must be a circle in a plane that contains that point. The plane of a "great circle" passing though Washington makes an angle with the axis of Earth's rotation. This means that if at a given time the satellite is above Washington, a little later Washington will have moved along a circle at a fixed latitude, while the satellite will have moved to a different latitude and longitude.
4. The definition of a stable point can help us devise a test for this problem. We could imagine that $x = x_0 + \delta$, where $\delta \ll x_0$, and see whether the resulting (small) net force points back to x_0. In doing so one would want to use approximations based on the fact that $\delta/x_0 \ll 1$. This would be a test for stability along the line between Earth and the Moon. One could think about stability for motion perpendicular to this line in a similar way, with the vector aspect of the forces playing a more important role. The zero-point is in fact unstable: if, for example, the satellite is a little to the right of the zero-point, it tends to continue moving that way. See Problem 12–70.

You can also respond to the question by thinking about the potential energy as a function of the position on the line between Earth and the Moon. A graph of the potential energy would form a hill, and the zero-point is the top of the hill. The fact that you have a hill rather than a valley means that the equilibrium at the top of the hill is unstable; if the satellite goes a little off the top of the hill, it goes all the way off.
5. From Eq. (12–10), the escape speed depends on the ratio of M/R, which is a factor of 22 less for the Moon than for the Earth, so the escape speed from the Moon is significantly less than from Earth. This is why some people have proposed a moon base for initiation of longer voyages in space.
6. (d) The total energy is negative for a bound system, as in this example, and positive for an unbound system.
7. You cannot answer this without some algebraic thinking. Kepler's relation gives $R^3/T^2 =$ a constant. The orbital speed $v = 2\pi R/T$, and we can invert this to give $T = (2\pi R)/v$. This in turn implies that $R^3 \times v^2/(2\pi R)^2 =$ a constant, or $Rv^2 =$ a constant, or $v^2 \propto 1/R$, therefore as R increases the orbital speed *decreases*.
8. Keep in mind in thinking about orbital motions that all the conic sections described at the beginning of Section 12–2, and drawn in Fig. 12–4b, are possible. We can see from that figure that orbits can be elliptic $(E < 0)$, parabolic $(E = 0)$, or hyperbolic $(E > 0)$. Information that a comet's orbit grazes the Sun is not enough to distinguish these.
9. (d) You can answer this one by thinking of the extreme case: If all the mass were concentrated right at the center, then the force would have the $1/r^2$ dependence characteristic of a point mass.

CHAPTER 13

2. The acceleration is proportional to the displacement; it has maximum magnitude where the displacement has maximum magnitude, at $z = z_{min}$ and z_{max}. By the same reasoning the acceleration has minimum magnitude, namely 0, where $z = 0$.

4. (b) The time is still a quarter period, and the period does not depend on the speed at $t = 0$.

5. Simple harmonic motion is repetitive. The mass moves back out to $x = -X$, then returns to the origin, and the same maximum speed, a half period later. For this system, with an ideal spring, the pattern repeats forever. The times when $x = 0$, and hence the speed is a maximum, are at $t = T/4, 3T/4, 5T/4$, etc.

6. One would still expect simple harmonic motion. Although the springs may not be in a relaxed state in the stable equilibrium position, the restoring forces will always bring the mass toward this position as long as the spring is not stretched to the point that it no longer acts like a spring. As we saw earlier, any small motion about a stable equilibrium is harmonic, barring special circumstances.

7. The validity of an approximation such as the one we use for pendula depends on whether the first term of a mathematical expansion comes close enough to the exact answer. How close that is is a numerical question which can only be answered by first specifying a criterion for "close enough." In the case of the pendulum, the expansion is Eq. (13–32), and hence whether $\theta_{max}^3/3! \ll \theta_{max}$, or more usually whether the ratio $(\theta_{max}^3/3!)/\theta_{max} \ll 1$. For example, you may insist that $(\theta_{max}^3/3!)/\theta_{max} < 0.05$. In this example, $\theta_{max} \cong (5.0\text{ cm})/(2\text{ m}) = 0.025 \cong 1.5°$, so that your criterion is satisfied.

9. (a) The monkey makes three things change: The total mass increases, the length increases, and the distribution of the mass over the length has shifted to the monkey's end. The period is independent of the mass, so the first change is irrelevant. But the period increases with increased distance to the center of mass as well as with increased mass distribution toward the end (so that rotational inertia is increased.). Both the second and third changes increase the period.

10. At first sight we might say we can tell nothing whatsoever; the frequency of oscillation (and hence the period) is modified by the damping parameter b, but it depends on more than b alone. A little thought shows that there is something interesting we can say. You may know that sound is an oscillatory phenomenon, and the fact that one can continue to hear the gong even as its loudness decreases means that this system is not critically damped or overdamped. This establishes what you may already know: Sound involves oscillations that are much more rapid than the time scale associated with the damping of this gong.

CHAPTER 14

1. No. One of the inputs necessary to satisfy the wave equation is that the displacements can't be too large. If the string is infinitely long, then no matter how small a is, there will be x-values (positive and negative) for which the distortion is large. Of course, the function we gave could describe only a portion of a propagating distortion. No matter; this function displays the feature we are interested in here: It moves with speed v.

2. Greater. The relation $v = \lambda f$ shows this. If the system is really the same—for example, if water depth remains the same—v is unchanged. Then if λ decreases, f must increase to compensate.

4. The velocity will approach zero as either $T \to 0$, or $\mu \to \infty$, or both. The first condition of no tension implies that there is no force acting on any element of the wire. This implies no acceleration in the vertical direction and therefore no displacement of the wire from its equilibrium position. The second condition implies that the wire is more and more massive, and in order to get a transverse acceleration, the force (tension) would also have to get larger and larger. For a fixed T we get no acceleration in the limit.

6. Gravity causes the rope to sag, and the hand and attachment point will both have to supply a vertical component of force to allow for a vertical component of tension. We know from experience that if the rope is not too long, then the sag will be small, so that locally the rope will be straight and the conditions that went into our derivation of the wave equation will continue to be satisfied: The tension magnitude will be roughly the same, and the waves will be unchanged to first approximation.

8. We now have two separate situations. On the left we have a string whose length is 0.2 m, so that our expression for the wavelength changes because $L \to L/5$. Thus the wavelengths on the left side take the values $\lambda = \frac{2}{n_L}\frac{L}{5}$, that is, each allowed wavelength is shortened by a factor of 5. On the right side the length is changed from L to $4L/5$. Thus the wavelengths take the value $\lambda = \frac{2}{n_R}\frac{4L}{5}$—that is, they are also shortened, but only by a factor of 5/4. Now we can certainly find values of n_L and n_R such that the wavelength has a common value. For example, $n_R = 4$, $n_L = 1$ have the same wavelength as the original string for $n = 5$ (which has a node at the newly clamped point). Nevertheless, these modes will not necessarily match the appearance of an unclamped string, because the amplitudes and the phases on either side of the clamping point have to match.

9. Our expression for the relation between f and T shows that f is proportional to $T^{1/2}$. Thus an increase of 1% in f means that $T^{1/2}$ must increase by 1%, and therefore T must increase by 2%. (To see this, use the fact that $(1 + x)^{1/2} \cong 1 + \frac{1}{2}x$ for x small.) Piano strings differ from one another in mass density as well as tension. T is proportional to f^2, so if strings of different mass density were not used, the range of tensions in a piano would be enormous.

10. This is most easily answered by considering a long pipe with a rectangular cross section. Then it is clear that there can be standing waves across the height, as well as across the width of the pipe, since there is a wave equation for disturbances in those directions, and since the boundary conditions are the same as for the ends of the closed pipe. If the pipe is much longer than it is wide or high, then the wavelengths of the standing waves across the pipe are much smaller, and the frequencies much higher than for the waves along the pipe axis. Therefore they do not enter into the answer to this question. If the cross section is circular, then the standing waves in the directions perpendicular to the axis of the pipe look like Fig. 14–6. Again the distance between crests is much smaller than the wavelengths of quite a large number of harmonics for waves along the pipe axis.

11. As we shall see in Chapter 16, blowing air across the opening of the bottle causes a change of pressure across the top of the bottle, which can set up a standing wave. This wave is analogous to the one-end-open organ pipe wave, so that the fundamental wavelength is roughly four times the length of the bottle. This is a fairly long wavelength, and therefore a fairly low frequency, or deeply pitched, wave. For a tube open at both ends, we have the situation shown in Fig. 14–27. These boundary conditions produce their own harmonic series.

13. Since the pitch does not change on reflection, we may view the process as the arrival of a sound wave at the cliff, and the subsequent emission of a sound wave with the same pitch by the cliff. The pitch as received by the cliff is raised from f_0 because the source is moving toward the receptor. As far as the reflected wave is concerned, we now have a moving observer dealing with a *raised* frequency of emission (but a stationary emitter). Thus the overall effect will again be a raised frequency. For a numerical evaluation

$$f_{final} = \left(1 + \frac{v_{car}}{v}\right)f_{refl}$$
$$= \left(1 + \frac{v_{car}}{v}\right)\frac{f_0}{1 - v_{car}/v} = \frac{v + v_{car}}{v - v_{car}}f_0.$$

CHAPTER 15

1. We have set the wall separation so that the (fundamental) wavelength is fixed. The relation to this wavelength and the frequency with which the string vibrates is determined by the speed of waves on the string, i.e., by the tension and mass density of the string. There is no reason why this speed is the same as the wave speed in the medium. Thus there is no reason why the frequencies of the wave and the medium must match.

2. If we think of the vibrations of the prongs of a tuning fork as a harmonic oscillator, we recall that $\omega = \sqrt{k/m}$, where k is the spring constant, characteristic of the structure. Thus an increase in mass (such as the additional mass of the tape) reduces the frequency, and that is what happens when the tape is added.

3. In the first case, the different starting time would just reflect a different phase. *Nothing* is changed if we define a new time $t' = t - 5$ s and write the same equations that we wrote down with t replaced by t'. All we are doing is resetting the clock. In the second case we note that generically $\cos(kx - \omega t) = \cos\omega(t - x/v)$. Since the traveling waves move with the same velocity, a shift in the location where we observe the mutual maximum merely changes the origin of the t variable, and nothing really changes.

4. The motion is in a radial direction so that the angle θ does not change. The listener will still be at the first maximum. There are other effects: There is a Doppler shift, but this is very tiny since the shift is of magnitude $v/v_{sound} = 3/330 \cong 0.01$. The main effect is the loudness, since that falls off as $1/R^2$. This means a reduction of the sound intensity by a factor of 4.

5. Assuming constructive interference, the first maximum occurs on the center line with $\theta = 0$. The largest value of θ for an experimental setup as described is 90°, and this will be the limiting case of a "completely" spread-out pattern if that value of θ corresponds to $n = 1$. This occurs when d decreases to the size $d = \lambda$.

6. This wave is a standing wave in that at a given value of x the displacement varies between limits with a time dependence given by $\sin \omega t$. Since the wall was chosen to be at $x = 0$, the standing wave pattern holds for all values of $x < 0$. This pattern is identical to one in which a string is fixed at $x = 0$ and at $x = -2\pi/k$, for example, but its existence does not depend on having *two* ends fixed.

7. If we think in terms of a string attached to a second string, then the *continuity* of the strings—that is, the absence of a break at the connecting point (e.g. $x = 0$) provides a condition relating the incident wave, the reflected wave, and the transmitted wave at that point. Since this holds for all values of t, we will get another equation relating the three amplitudes. Unlike the energy condition we used here, which is quadratic in the amplitudes, this will be linear. We'll have enough information to solve for the amplitudes of the reflected and transmitted waves in terms of that of the incident wave.

CHAPTER 16

1. (a). The radius doubles, so the volume increases by a factor of 8. For density, which is mass per unit volume, the numerator (mass) doubles while the denominator (volume) goes up by a factor of 8; the density therefore drops by a factor of 4.

3. As in the example here, when the valve area is small it takes much less force to obtain a significant pressure across it, because $p = F/A$. This pressure is then transmitted everywhere within the fluid—in this case within the tire tube.

4. It is the difference in pressure between inside and outside the tube that gives rise to the force that pushes the mercury up inside the evacuated tube. If the top of the tube were open to the outside, the mercury would not rise at all. If there were, say, a third of an atmosphere above, then the net force up would be 2/3 of an atmosphere, and the height would be 2/3 of 760 mm.

5. The same reasoning that explains why the column of water (less dense) is higher than the column of amyl bromide also explains than the column of heptyl ether would be higher than the water column. A quick repeat of the calculation with heptyl ether would give $h_{he}/h_w = 1.23$.

7. (a) The density of salt water is greater than that of fresh water, and the final answer for the density would therefore be larger by a few percent according to the last equation in the example.

8. In contrast to the completely submerged sphere of Example 16–7, a floating object has a buoyant force that cancels the weight completely. The scale would read zero.

9. To reach a further distance, the water should come out of the nozzle at a higher speed—a projectile fired with a larger initial speed will go further. Thus a smaller nozzle diameter is required.

10. If the area changes, then the conservation of flux vA shows that the speed must change as well, increasing as the artery narrows. The full conservation law, Eq. (16–22), applies. A look at that equation shows that if both v and h increase as we go to the head, then p at the head will decrease even more than it did when only h increased. In turn, the pressure at the heart would have to be still higher to compensate. Narrowed arteries are dangerous, as blood flow to the body is reduced and the heart has to exert more effort.

11. An algebraic calculation as in the example would allow you to find the stream speed as a function of hole height. This could then be converted into an algebraic calculation of the horizontal distance carried by the stream as a function of hole height. One could then maximize this by setting the derivative of the horizontal distance with respect to height to zero. You can see that neither a hole near the top nor one near the bottom would give the farthest distance from the tank. For a hole near the top, the water will exit with a low speed, because the pressure is low. For a hole near the bottom, the interior pressure and hence the initial speed will be large, but the stream hits the ground very soon. Fig. 16–25b shows the trajectory of fluid for a variety of hole positions.

CHAPTER 17

2. Temperatures cannot be measured with this type of thermometer when the temperature and pressure are extremely low, because gases liquefy or freeze in this region of extreme cold. There are, however, other kinds of thermometers that we can use in this region, as we'll discuss later.

5. At the higher temperature the bridge would have expanded to fill up the 50-cm gap. At the lower temperature the bridge would have a 50-cm gap, enough to swallow small children if there were only a single joint. With many joints this is less of a problem! In any case, the joints tend to be closed in summer when it will be more pleasant to cross.

6. The inequality of the β-values states that the gasoline will shrink more than the container when the temperature goes down. You can safely fill to the brim on the hottest summer day, and when you look inside the container in winter, the container will not be full.

9. Air consists primarily of oxygen (molecular weight 32.0), nitrogen (molecular weight 28.0), and argon (molecular weight 39.9). The number 29.0 is a weighted average of the constituents' molecular weight according to their percentages in air: oxygen (21%), nitrogen (78%), and argon (1%). We have $M(\text{air}) = 0.21(32.0) + 0.78(28.0) + 0.01(39.9) = 29$.

11. None of them. The constant temperature line is shown on Fig. 17–13, while constant p is a horizontal line and constant V is a vertical line.

13. As it works its way to the surface, the radiation emitted from the Sun's interior is absorbed and reradiated at wavelengths characteristic of the successive temperatures of successive distances from the center.

CHAPTER 18

1. Because the temperature is constant, the transformation would be represented as a straight line parallel to the p axis as the pressure changes. The line would be at the fixed value of T.

3. One of the most common misunderstandings about thermodynamics is the incorrect assumption that heat and temperature are practically the same thing. Just because the temperature is unchanged does not mean there is no heat flow! In fact, there will generally be heat flow at constant temperature if other thermodynamic variables change.

4. (a) higher. The specific heat of aluminum is higher than that of iron (see Table 18–1). That means that for a given heat flow the temperature of the aluminum would change less, and that corresponds to a larger system temperature.

6. If the latent heat of fusion were much larger, then the freezing of lakes would be much rarer, since the air would have to extract a great deal more thermal energy from the lakes. Once a lake froze, melting would correspondingly be much more difficult. Ice cubes would not melt in drinks. (But they would still make them cold!) On the other hand, if the latent heat of fusion were very very small, the lakes would essentially have the same temperature as air, freezing quickly when air temperature fell below the freezing point, with the delay coming only from the time it takes water to conduct heat. Ice cubes would still have a cooling role, although a smaller one, and they would quickly melt and dilute your drink.

7. The rate of heat loss is proportional to the surface area, and a sphere has the smallest area for a given volume.

8. An examination of Eq. (18–12) shows that any term with a small R-value has a big effect. See the discussion following this example.

9. Yes. Kinetic energy will be converted to thermal energy when the containers crash on the road below. It does not matter how the containers receive the kinetic energy.

10. You are not reversing the entire process; in fact you are adding energy, whose source is your muscles, raising the weight and heating the water still further. We'll see in Chapter 20 that the impossibility of reversing the results of chaotic processes such as the churning of water is the basis of the second law of thermodynamics.

11. (b) You can see from the figure that the area enclosed is more than doubled. If you prefer an algebraic solution, the ratio of the new work to the old is $(2p_2 - p_1)/(p_2 - p_1) = (2p_2 - 2p_1)/(p_2 - p_1) + p_1/(p_2 - p_1) = 2 + p_1/(p_2 - p_1)$, a result larger than 2.

12. Because the cycle is closed, the internal energy over one cycle is unchanged, and the net heat flow in will always equal the work done. The heat flow in is positive if the path is clockwise and negative if it is counterclockwise.

13. The heat flow has to come from the thermal reservoir. The reservoir is, by definition, so large that you can take as much heat flow from it as you like without changing its temperature.

14. Since the temperature is constant, the internal energy does not change. Since the gas does work, thermal energy had to flow in, the same amount as the work done, so that positive thermal energy comes out of the reservoir.

CHAPTER 19

1. The internal energy enters in the expression for the collision rate in the form $(U)^{1/2}$; doubling the internal energy increases the collision rate by a factor of $2^{1/2}$.

3. The argon atoms are very few in number, so as soon as equilibrium sets in, they will be characterized by the same temperature as the helium atoms. Thus they have the same average kinetic energy as the helium atoms. Since the kinetic energy is of the form $m\langle v^2\rangle/2$, and since argon atoms are much more massive than helium atoms, their rms speed will be smaller.

4. The b values measure the space occupied by a mole of molecules, and as the calculation above shows, the molecular radii of these substances must therefore be within a factor of $2^{1/3} = 1.3$ of each other. Positive a values decrease the pressure that the gas exerts on the surface of the containing vessel. This comes about because there is a force that slows down the molecules as they approach the container walls, so to speak, and therefore reduces the momentum transfer per wall collision and thus the pressure. This suggests that molecules exert an attractive force on each other.

6. The answer is "obviously" 1.65 m. If we were asked to justify this, our answer would be based on the symmetry of the U shape. An algebraic demonstration of this would be tedious but quite straightforward.

8. The dependence of the mean free path on the number density, and in particular $\lambda \propto 1/n = V/N$, is the relevant point here. The compression reduces the volume (the number of molecules does not change) and therefore the mean free path is reduced. Just how much the volume is reduced depends on whether the compression is adiabatic or not, but that is not our concern here. It makes sense that the mean free path is reduced: It is harder to get through a dense crowd than a dispersed one.

CHAPTER 20

1. We can count without writing out each case by saying that there are three ways to assign the first number, then two ways to assign the second, and only one choice for third number, making a total of 6. The traditional way to approach this is to think of three boxes and three integers, and count the ways in which they can be filled. For the first box: there are three integers, and any one of them can fill that box. When that is done there are two boxes left and two integers. There are two ways of filling the second box, and finally the third integer goes uniquely into the third box, so that the number of ways is $3 \times 2 \times 1 = 6$. This way of doing things would prove superior if you had to do this problem for four dice.

2. To increase the efficiency, decrease the ratio T_c/T_h. For the second part, probably the most sensible thing would be to scale the heat flow for both reservoirs. There are practical limits to how high we can make T_h, and to get to T_c below the temperature of available coolants, such as river water for power plants, is costly.

6. It is clear from Eq. (20–12) that for a small temperature difference the coefficient of performance of the refrigerator is large, and this means, according to the definition in Eq. (20–10), that little work needs to be done. In the limiting case, when your kitchen is at 40° F, you can just unplug the refrigerator!

7. There is no need for a complicated calculation. We need only recall that the entropy is a state function, and therefore the change in entropy from an initial state to a final one is *independent of the path taken*. The result is unchanged as long as the initial and final states are the same.

10. The process is irreversible, since the gases, once mixed, will not spontaneously unmix. This means that $\Delta S > 0$.

11. A look at the expression for the entropy suggests that since the new volume, 3.0 L, is smaller than the original one, the entropy change would be negative, forbidden by the second law. But that assumes that all of each gas goes into the small container. In fact, such a process would not take place spontaneously. Some of the argon and some of the neon would pass into the common 3.0 L container, but much of the gas would stay where it was. The argon would increase its volume from 5.0 L to 8.0 L, and this would increase the entropy. The neon would be increasing its volume from 12.5 L to 15.5 L, and its entropy would also increase.

CHAPTER 21

1. The missing mass will be (number of electrons) $\times$ (mass of an electron) $= (6.9 \times 10^{11}$ electrons$)(9.11 \times 10^{-31}$ kg/electron$) \cong 10^{-18}$ kg. There is no instrument we can use to measure such a small mass, let alone a mass difference of that size.

2. (a) The charge left behind would have the same magnitude as the total electron charge, 1.1×10^6 C, because the coin was initially neutral. (b) If one of the 79 electrons in each gold atom were removed, the total charge would be only 1/79 of the charge just found, or 1.4×10^4 C, which is a huge electric charge.

4. No, just the opposite: Charge opposite to that of the smaller piece would move closer to that smaller piece, because it is attracted, while charge the same as that of the smaller piece would move farther because it would be repelled. The net effect would be an even stronger attraction. This phenomenon, charge induction, is what lies behind the separation of the electroscope leaves in Conceptual Example 21–3.

5. Most objects have little or no net electrical charge, and there is no Coulomb law force between such objects. The Earth is practically neutral, so that there is little or no Coulomb force between us and Earth, as we are also electrically neutral. But Earth is extremely massive, and the gravitational force between us and Earth is adequate to hold us at the surface.

6. The electric force between the two cork balls would be unchanged. The lower ball's mass is irrelevant—it sits on a surface. The force of gravity on the upper ball would now be twice as large as above and would match precisely the Coulomb force acting on it. It would now be in equilibrium and would not move farther.

7. Charges q_1 and q_2 are identical. If we place q_3 precisely on the x-axis between the other two charges, q_3 will be equally attracted to the other two charges. If q_3 is displaced slightly toward either q_1 or q_2, its attraction to the nearer charge will be larger than its attraction to the farther one, because the Coulomb force increases as $1/r^2$ when r decreases. Thus q_3 will not return to equilibrium—the equilibrium is unstable.

9. When $L = 0$, the charge is at the middle of the ring, and, by symmetry, the net force should be zero, as in Conceptual Example 21–9. This is indeed a property of our result, and a second check of the calculation.

10. The right side of the rod is positively charged, so we expect on physical grounds that the repulsion on the point charge will become very strong. This is verified with a numerical evaluation of the force. We can also see it in the analytic expression: Our limit is $R \to L/2$, and the dominant term in the force in this limit is the term $1/(R - L/2)$, which is indeed positive. The force on the point charge is large and to the right.

CHAPTER 22

2. There would be no change in the electric field due to q. However, the force on a q' of opposite sign would be a force with the same magnitude but reversed direction. The force would be repulsive, not attractive.

3. Remembering that the electric field produced by a charge points away from the charge if the charge is positive and toward the charge if the charge is negative, we see that the fields from all three charges are to the right in the region between q_2 and q_3. Thus the net field also points to the right.

4. Very close to one of the charges the net field is to a good approximation the field of the close charge, and this does not depend on L in any way. It is also possible to show that the field depends on p alone at distances $r \gg L$, although this requires a detailed analysis.

6. Although the lines will initially bend toward $-q$, very far away they will be pointing radially outward from a net charge $+q$ $(+2q - q = +q)$; at great distances we see a net charge of $+q$, and the field will be indistinguishable from the field of a point charge $+q$. The fact that 12 lines remain is consistent with our original choice of 24 lines for the line density (Fig. 22–13c).

7. To a point charge very close to the rod, the rod appears to have infinite length, and there is an infinite amount of charge in an infinite rod with a finite charge density. The summation over all the fields from charges in the rod, including the ones that are very distant from the point at which the field is measured, builds up a net field that decreases more slowly than the field of a finite charge distribution.

8. (b)

10. (c). The surface charge density would be larger because the area of the Moon is smaller than that of Earth, so the Moon's surface electric field would be correspondingly larger. If you calculate the (repulsive) Coulomb force between Earth and the Moon you would find about 10^{-5} N, a completely negligible effect.

11. The electron would not move, because it would be attracted to the positively charged plane.

12. (b) Our expression for y is proportional to E. This is the feature that makes the tube we have described here so useful.

CHAPTER 23

3. Gauss' law is now applied to a total charge of $q + 2q - 7q + 4q = 0$. There will be no *net* flux through the surface. Again we stress *net* flux, because there will certainly be flux going out of, and coming into, the surface, as could be seen by putting a cube around the two equal and opposite charges in Fig. 23–12b.

4. We would now use the principle of superposition of electric fields. Each face would get no contribution from the four charges at its edges, but would get an equal contribution from each of the four charges around the opposite face. Thus the flux through each face is $4 \times (q/24\varepsilon_0) = q/6\varepsilon_0$.

5. The symmetry is the same as in the example. If we take a Gaussian surface that is cylindrical with the same axis as the charged cylinder with a radius larger than that of the charged cylinder, exactly the same argument holds, and we obtain the same result. On the other hand, if we draw our Gaussian cylinder *inside* the charged cylinder, then the symmetries are exactly the same, but the enclosed charge is zero. This is what we would expect, as the electric field inside the cylinder is zero.

6. The Gaussian surfaces will again be concentric spheres. In each case we end up with the result $E = \dfrac{Q_{\text{enclosed}}}{4\pi\varepsilon_0 r^2}$. When r is larger than the radius of the outer shell, then the total enclosed charge is zero, so that $E = 0$. When r lies between the shells, then the total charge enclosed is Q and that is to be inserted in the above equation. When r is inside the inner shell, then no charge is enclosed and $E = 0$ again. There is a non-zero field only between the shells.

7. The motion is determined by the force, and the force on the charge is directed toward the center, with magnitude

$$F = -qE = -\frac{qQ}{4\pi\varepsilon_0}\frac{r}{R^3}$$

This is a "restoring" force proportional to the displacement from the origin. The motion is therefore simple harmonic motion, and the charge will oscillate about that point.

8. We know that at a great distance from the charge distribution the electric field lines from any localized charge, such as found on a finite plane, are spread in a spherically symmetric pattern, as if the charge were a point charge. This means that the field lines will ultimately bend and not point straight away from the charged surface. In a region close to the charged surface, where the distance to the surface is small compared with the distance to the nearest edge, we may make the approximation that the field lines are perpendicular to the charged surface. If we take for a Gaussian surface a very shallow squat cylinder half into the surface, the argument that led to Eq. (23–11) goes through as before, and the result applies here too.

CHAPTER 24

3. False. The sign of this result makes sense. The electric potential at point b is negative. The new charge is positive and will be attracted to the negative potential. Franklin would have to do positive work to bring the same charge back out to infinity. It is better to understand what is happening physically than to rely on your ability to avoid algebraic error.

4. The energy of the at-rest electron-proton system once the electron has been moved far away is zero. Therefore you must add positive energy to the system to make the separation.

5. If we could freeze the charges in position, then there would indeed be a dipole field. However, the two charges form a kind of planetary system, revolving around each other; from a given point P the angle θ that appears in Eq. (24–24) varies uniformly with time. Thus the average dipole potential involves the average of $\cos\theta$. But $\cos\theta$ varies between -1 and 1, and its average value is zero.

8. Yes. The electric field points from the positively charged plate to the negatively charged plate (see for example Fig. 22–19). We also know that the electric field points from the higher to the lower potential surface. This then implies that it is the lower potential surface—the left side plate—that carries the negative charge.

9. The solution is obtained by setting $x = 0$. It is $V = \dfrac{Q}{4\pi\varepsilon_0 R}$. The result follows very simply from superposition. The fact that all the charges whose potentials add up lie on a circle is irrelevant, because the potential is a *scalar* quantity that does not involve directions. The only thing that matters here is that the center is equidistant from all the elements of charge.

10. (d). This is evident from the fact that at a great distance the disk is indistinguishable from a point charge. If we want to take a mathematical limit, we cannot just drop the R^2 term in the square root, because we then get zero. We instead need to work out the term in parentheses by making use of the fact that

$$\frac{x}{\sqrt{x^2+R^2}} - 1 = \frac{x}{x\sqrt{1-(R^2/x^2)}} - 1 = \frac{1}{\sqrt{1-R^2/x^2}} - 1,$$

then making an expansion of 1 divided by the square root for small R/x.

11. Writing the potential as $V = \displaystyle\int_{-\infty}^{\infty}\frac{\lambda\,dz}{4\pi\varepsilon_0\sqrt{R^2+z^2}}$ is based on choosing the potential due to an element of charge as vanishing at infinity. As argued in the solution, this is not legitimate here. In fact, the integral written above is independent of R (easily seen by changing variables to $u = z/R$), and therefore gives a zero electric field, an incorrect result.

12. The relevant fact is that the potential is *constant*. This means that its derivatives are zero, consistent with zero electric field. The particular constant value of the potential ensures that the potential is *continuous*. If the potential were not continuous at $r = R$, then the motion of a point charge in that potential would not be physical; a jump in the potential energy would require a discontinuous jump in the kinetic energy, that is, in the velocity of a point charge.

CHAPTER 25

2. The potential difference doubles when Q doubles, but C is unchanged. C depends only on factors such as geometry, and is independent of the amount of charge on the capacitor.

4. The permittivity, ε_0, has units F/m and so must be multiplied by a length to give an acceptable capacitance. The only length in this problem is the radius of the sphere, so $C \propto \varepsilon_0 R$.

5. The energy stored in a battery is local and is held chemically, molecule by molecule. In a capacitor, each time we transfer another charge, it requires more work, because there is an increasing amount of charge already placed on the capacitor. We are fighting an uphill battle.

7. In regions outside the sphere itself, the sphere behaves as if all its charge is concentrated at the center. Thus it requires more work to bring charge from infinity and place it on the smaller conducting sphere because in effect you are closer to the center when you are charging a smaller sphere. Therefore the energy of the system also becomes large.

9. Consider the extreme case where we make this capacitance very small. From Eq. (25–17) we see that a small capacitor has a large effect when it is placed in series with other capacitors; indeed a capacitor with a very small capacitance C_{small} will dominate all the terms, giving as a good approximation $C_{eq} = C_{small}$. Thus decreasing the capacitance of the 2 μF capacitor will decrease the equivalent capacitance.

10. Yes, work was done because the electrical energy changed. The sign of the work done by the capacitor is positive, because the capacitor's energy decreases as the Teflon was inserted.

11. Bakelite has a larger value of the dielectric constant κ, so that for a given charge, the field within the Bakelite would be smaller, the voltage drop across the entire space would be smaller, and hence the capacitance larger.

12. The *change* in energy, which is what determines the force, depends only on how much *additional* length of plug is inserted. For example, the energy change is the same whether the plug is first inserted 1 mm in as it is when the plug is already 1 cm in and is then inserted an additional 1 mm.

CHAPTER 26

1. In a proton beam the charge carriers are positively charged, whereas in a wire the charge carriers are electrons, which have negative charge. Moreover, a wire is electrically neutral, because there is a background of positive ions that match the current-carrying electrons, whereas a proton beam in space is not neutral.

2. The question is, Is the rate at which charge goes into the annihilation region the same as the rate of charge that goes out? If we go to the right of the annihilation region, we only have a flux of antiprotons, and the entering rate of charge from the right is $N'e = Ne$. On the left of the annihilation region we have only a flux of incoming protons, and the rate at which charge enters is $N(-e)$. When we add these two terms together, we see that the net rate at which charge enters is zero, the same as the rate at which it leaves. Current conservation is a necessary consequence of charge conservation.

3. For a fixed current density, a quadrupling of the current implies a quadrupling of the area, and since the area is proportional to the radius squared, the radius of the wire will have to be doubled.

4. Table 26–2 shows that the resistivity of iron is about six times larger than that of copper. This means that for fixed E, the current density will be about six times smaller, and for a wire of the same cross section the current will also be a factor of six times smaller. Copper is used widely in the wires of circuits, and appliances, as it is a good conductor and has good mechanical properties.

5. A crucial requirement for a filament is that it should not melt! The other is that it is best to have a higher resistance, since it is the resistance that is responsible for the dissipation of energy, and this dissipation manifests itself in the heating of the coil, and ultimately in the radiation of the coil. Our result above tells us that at a temperature a bit below the copper melting point, the resistance per unit length of our platinum wire is $53/20 = 2.65\ \Omega/\text{m}$. The coefficient α for platinum is the same, but ρ_0 is a factor of 6.2 larger, so that even at the same temperature, the resistance of the platinum filament is that much larger. In addition, the platinum filament can be heated to a higher temperature, increasing the resistance still further. For these reasons the platinum filament is to be preferred.

6. With this exchange, the resistances on both sides would be equal (10 Ω each). With two equal resistances in parallel, the currents would be divided equally. This can be calculated but is also evident from symmetry—two identical paths will carry the same current. This means that the current in each branch is 1.5 A, and the potential difference overall is the same on both sides, i.e. $(1.5\text{ A}) \times (10\ \Omega) = 15$ V. The potential difference across the 4 Ω resistor is $(4\ \Omega) \times (1.5\text{ A}) = 6$ V.

9. With $P = V^2/R$, we can immediately tell that a doubling of the voltage implies an increase of the power by a factor of 4.

10. For a fixed voltage, the power is given by V^2/R. Whatever the resistance R of the bulb is, the power will be halved in case (a), since the resistance is doubled. The power will be doubled in case (b), since the reciprocal of the resistance is doubled. If the purpose of the lighting is to generate heat to keep the paint from freezing, the first alternative may be inadequate and the second too expensive.

11. The current is given by $(P/R)^{1/2}$. The radiated power is proportional to the overall area of the (cylindrical) wire, hence proportional to d; the resistance is inversely proportional to the cross section of the wire, hence proportional to d^{-2}. Thus, for a given $I \propto (d/d^{-2})^{1/2} = d^{3/2}$. You should therefore reduce the diameter of the wire to decrease the current. That means that the power will be reduced, so the wire will not be quite as effective at heating its surroundings.

CHAPTER 27

2. If we start from the expressions for $\mathscr{E}$ and r, we see that in the formula for $\mathscr{E}$, R_1 must be taken to be zero to the accuracy that the resistances are measured. Similarly, in the numerator of the expression for r, the second term I_1R_1 must also be taken to be zero to the accuracy under consideration. This then implies that $r = \mathscr{E}/I_1$. This can be calculated to two significant figures.

3. For a given battery, with characteristic emf and internal resistance, the expression $P = \mathscr{E}I$ shows that P is largest when I is largest. This is achieved in our circuit by reducing the external resistance as much as possible. Setting $R = 0$ gives $P_{max} = \mathscr{E}^2/r$.

4. As stated in "Setting It Up," the solution to the problem decides the current direction. If we had initially chosen the opposite current direction our algebraic solution would have been positive. Because the emf $\mathscr{E}_2$ is larger than the emf $\mathscr{E}_1$, we could have anticipated that the current will run counterclockwise.

5. If R_2 is large, there is so much resistance in the leg that contains it that the current I_2 should drop to zero. In effect, the segment containing R_2 would be eliminated from the circuit. In this limit our expressions give I_1, $I_2 \rightarrow (\mathscr{E}_1 + \mathscr{E}_2)/(R_1 + R_3)$, and $I_2 \rightarrow 0$, just as we expect.

6. As long as the loop equations are independent, the Kirchhoff rules will always give the correct answers. In this case it is particularly easy to see that this applies: All you have to do is flip the segment (d–battery–a) to the other side of the diamond. In that case the new loop is just the mirror image of the old loop.

7. (a) The resistance can never be negative, whatever the signs of the emf or the current direction. (b) The sign of the potential difference between the two plates shows that the potential increases when we go from plate 1 to plate 2, meaning that plate 2 is at a higher potential than plate 1. This is because positive charges have accumulated on plate 2.

9. Qualitatively the corrections to V are of the general form R/R_V, where R is the generic resistance in the circuit (e.g., combinations of R_1 and R_2 in this example). When the denominator is a factor 100 times larger, the error, instead of being in the 3% range, is in the 0.03% range.

10. A look at Fig. 27–20b shows that we are dealing with the very flat part of the curve on the extreme right. To answer this quantitatively we need to know something about exponential functions, or equivalently about natural logs. As we saw in the example, the time is determined by $e^{-t/\tau} = 1 - p$, where p is the desired percentage. In the flat part of the curve we are dealing with p very close to 1 (0.999, and now 0.9999), so that if we write $p = 1 - x$ (with x very small), we get $-t/\tau = \ln x$ or $t/\tau = \ln(1/x)$. If we look up $\ln(10) = 2.30$, we see that each factor of 10 in the percentage accuracy increases the time (in units of RC) by an additional 2.30 units. In this case we would increase 6.91 ms to 9.21 ms.

CHAPTER 28

2. When a rock is thrown, the gravity force perpendicular to its initial motion is constant, and it always points in the same direction. In the case of a magnetic field, the force changes direction as the velocity changes direction. Furthermore the velocity is constant in magnitude. The path of the electron is therefore quite different from the familiar motion of a thrown rock.

3. For a given charge (and the deuteron and the proton have equal charges), the radius of curvature is proportional to the momentum, the product mv. The deuteron mass is double the proton mass, so it would have the same momentum as the proton if the speed were halved.

4. Let us assume that the charged particle comes out of the opening hole traveling in the $+z$-direction. If the magnetic field is used to define the $+x$-direction, then the magnetic force points in the $+y$-direction, and the electric field must therefore point in the $-y$-direction. Suppose the particle is positive and has an excess velocity that is originally in the $+z$-direction. The magnetic force will now initially be in the $+y$-direction. The particle will acquire a small component of velocity in that direction and will start a circular motion (with angular frequency qB/m) in a clockwise direction as seen by somebody looking along the x-axis. The path will be similar to that of a point on the rim of a wheel whose center is moving with a uniform speed while the wheel is rotating in a direction opposite to what its rolling rotation would be like. A negative particle would have motion like a spot on the rim of a rolling wheel.

9. The wire and field form a plane, and the force is perpendicular to the plane. Whether it goes into or out of the plane depends on using the right-hand rule in accordance with Eq. (28–19).

11. We have already mentioned that bar magnets are influenced by magnetic fields in just the same way as are current loops. This suggests, correctly, that the compass needle, which is a small bar magnet, behaves much like the loop in this example, oscillating about the stable equilibrium position in the absence of damping.

12. The drift velocity does not depend on the width of the strip. The electric field will be of magnitude vB, so that it does not depend on the width of the strip. Therefore the potential $V = Ed$ will double if the width of the strip doubles.

CHAPTER 29

2. The magnitude would be unchanged. We didn't specify the direction of the current in the first place, so there is no real information on direction. But if the current is reversed, the direction of the magnetic field would be reversed. The direction of the field is determined by the right-hand rule: If the thumb of the right hand is in the direction of current, the fingers indicate the direction of the magnetic field.

3. The flux will be a maximum when $\cos\theta = 1$, so that $\theta = 0$. This corresponds to the loop perpendicular to the field.

4. With another row of wire, the number of turns per unit length n would double, and the magnetic field would also double.

6. We work directly from Eq. (29–21), which for $L \gg D$ becomes $B \cong \dfrac{\mu_0 I}{4\pi}\dfrac{L}{D(L/2)} = \dfrac{\mu_0 I}{2\pi D}$, which is the same result as given by Ampère's law for the infinitely long wire.

7. Yes. We curl the fingers of the right hand along the current, and the thumb indicates the direction of the field. This orientation rule is the same one that we found for the magnetic field of a solenoid; indeed, the loop is nothing more than a compressed solenoid.

8. An electric dipole; see Eq. (22–14).

9. Remember that the electric field between the plates depends only on the charge density and is independent of d as long as $R \gg d$. The displacement current and magnetic field between the plates will not change.

CHAPTER 30

2. (c) The induced current will always be in the direction to oppose the change in magnetic field through the loop. As the loop goes back into the region of field, the induced current will be opposite to what it was as the loop left the field. The magnetic field that the induced current creates will oppose what is now an increasing magnetic flux.

4. The motion of the loop toward the straight wire would produce the same change in magnetic flux through the loop as the motion of the straight wire toward the loop. The effect would be exactly the same.

5. (b) The induced emf is the same, but the resistance increases; hence the induced current will be less.

7. The Lorentz force is $q(\vec{v} \times \vec{B})$, and its direction on a positive charge is indeed radially outward, in agreement with the result we found using Faraday's law.

9. Because the kinetic energy doesn't change, the change in total energy (dissipated in Joule heating) comes from the change in gravitational potential energy.

10. We can imagine a large circle, radius R, well outside the pole faces but centered at the center of the faces. The magnetic flux through this circle is to a good approximation $\Phi_B = B\pi r_{\text{face}}^2$, where r_{face} is the radius of the magnet. The same symmetry argument gives the electric field lines forming circles centered at the center of the pole face. As in the example, the magnitude of the field is now determined by $E \times 2\pi R = \pi r_{\text{face}}^2 \alpha$, or $E = r_{\text{face}}^2\alpha/(2R)$; the field falls off as $1/R$. If R is larger than the whole magnet then $E = 0$.

CHAPTER 31

1. (b) Far smaller. The susceptibility of copper is much less than 1, so the expression for B includes $(1 + \chi_m) \cong 1$ instead of the very large value of the χ_m for iron. This means that B inside will be practically unchanged from its vacuum value.

3. The expression for the magnetic moment is proportional to vr which can be rewritten in the form ωr^2. When the nucleus and the electron rotate about their center of mass, the angular velocity ω is the same for both of them. However, the center of mass is very close to the massive nucleus, so that to good approximation $r_N = (m_e/M_N)r_e$. Thus the contribution of the nucleus to the orbital magnetic moment is negligible.

4. The terrestrial environment, in particular terrestrial temperatures, are such that the thermal energy of the tiny magnets (of order kT) dominates the energy of their interaction with any magnetic field B. In particular, the difference in the energies between alignment and anti-alignment ($2m_BB$) is tiny compared with kT. Thermal effects destroy order, and in this case, with $kT \gg m_BB$, they dominate the ordering that the field attempts to impose. Only at ultra-low temperatures, say below 10K for realistic fields, would you expect the alignment to be almost perfect.

6. Our calculation was based on perfect alignment of all unpaired electrons. But the assumption of a single domain is not realistic for everyday situations. This does not mean that near-perfect alignment is impossible: A nearly perfect alignment in ferromagnetic materials occurs with the application of external fields of several Tesla.

CHAPTER 32

1. (d) Doubling the number of turns without changing anything else doubles the turn density, which appears squared in the expression for the inductance of a solenoid. This will increase the induced emf by the greatest amount.

3. (a) The area of solenoid 1 increases by a factor of 4 and so does the value of L_1. There is no change in M, which does not involve A_1. (b) There is no change in either L_1 or M. Even though R_2 increases, the overlap with R_1 does not if, as here, the areas started out with the same value, so that M does not change.

4. a. The permeability is larger, indeed much larger as the material in question is ferromagnetic.

6. We must change something in the inductor, and since the geometry is fixed, that leaves the number of turns, which increases the stored energy as N^2.

7. It is limits on the current that can be carried by wires that has the most effect on our ability to make large magnetic fields. The appropriate use of ferromagnetic cores does enable these fields to attain many Tesla.

8. This is incorrect and is a matter of understanding the limits. No matter how large L is, if it is fixed there will always be a time much greater than R/L for which V_L will drop to zero.

9. The voltage doesn't depend on the value of the inductor at all. It is solely determined by the original voltage across the capacitor that in turn depends on Q_0 and C. Even though V_L in the example has an L dependence, it will cancel out if we insert the value $\omega^2 = 1/LC$.

10. The decay constant α is ten times larger, while it remains much less than ω. The reasoning that allows us to approximate the period is the same, and since $t_{1/2}$ is ten times smaller, there are ten times fewer oscillations in that time span.

CHAPTER 33

2. No; the only conceivable effect would be on the direction in which the current is induced, and in AC that is obviously unimportant.

3. For the higher frequency, 6 MHz, the capacitive reactance is small and the circuit has a small resistance to current flow. This is a manifestation of a capacitor's transparence to rapidly changing currents.

4. The resistance to current flow in the inductive circuit increases dramatically for higher frequencies—the opposite behavior from that of the capacitive circuit. The inductor acts as an open switch in the limit of very high frequencies.

7. The time dependence of V_L is proportional to $-\cos(\omega t + \phi)$, and this takes its most negative value when $\cos(\omega t + \phi) = +1$, or $\omega t + \phi = 0, 0, 2\pi, \ldots$, or $\omega t = -\phi, -\phi + 2\pi, \ldots$. If we recall that $\omega t_1 = \pi/2$, we see for $\omega t = -\phi \cong +60°$, $t < t_1$. Thus we take the next occurrence, $\omega t = -\phi + 2\pi$.

8. In the calculation of the example, R is proportional to the square root of the discrimination requirement, i.e. $(0.01)^{1/2} = 0.1$. Thus by choosing R a factor of 10 smaller we can improve the discrimination requirement by a factor of 100.

9. Examination of the equation for the ratio of potentials indicates that for $\omega \to \infty$, the left-hand side reduces to $V_1/(V_0 + V_1)$. For high frequencies, the circuit acts as though the capacitor is not present, and voltage V_1 is dropped across the resistor.

CHAPTER 34

2. Recall that the motion of the electron in the presence of fields is determined by the equation $\vec{F} = -e(\vec{E} + \vec{v} \times \vec{B})$. With the magnitude of B given by E/c, we see that the second term is of order v/c compared with the first one. Electrons in metals move with nonrelativistic velocities so that the first term is by far the more important one.

3. The electric and magnetic fields in a wave are intrinsically coupled through Maxwell's equations, and as Eq. (34–6), for example, indicates, the coupling is *local*, that is, at every point in space. This means that elimination of the electric field will also eliminate the magnetic field. Don't invest.

6. The pressure is the weight of a film of water of height h acting on an area of 1 m^2, namely $(10^3\ \text{kg/m}^3)hg$. If we set this equal to the radiation pressure and solve for h we find $h \cong 2.4 \times 10^{-8}$ m, a height only about 50 atoms deep! Radiation pressure is in most circumstances a small effect.

9. The light emitted by a lightbulb consists of the (incoherent) superposition of radiation by a huge number of oscillating charges. Each such oscillator emits dipole radiation, but because all these oscillators act incoherently, all the angular dependences average out and the distribution is spherically symmetric. Is this an "antenna"? It is as long as we agree that by this word we mean any system that radiates. If we insist upon *coherent* radiation, a lightbulb is not an antenna.

10. With the assumption that all of the power lost in the Polaroid sheets goes into heating, the ratio of acquisition of thermal energy is the ratio of intensity *lost* in the two sheets. In the first sheet, half the intensity is lost. In the second sheet, the intensity lost is the fraction $1 - \cos^2\theta = 3/4$. Thus the ratio is $(1/2)/(3/4) = 2/3$.

11. The classical answer (i.e. when there are *many* photons) is that the ratio of momentum to energy is $1/c$. This must also be valid for individual photons, and we therefore expect that the momentum of a single photon of frequency f will be hf/c.

CHAPTER 35

2. Figure 35–14 shows that if θ' is to be decreased, then ψ must decrease (Snell's law). This, however, means that ϕ will increase, and this can only happen if θ increases.

6. In this case we are talking about rays that go from air to water. Every such ray is bent to the vertical, meaning every such ray can enter the water and there is no analog to total internal reflection. The fish sees the entire upper hemisphere concentrated within a cone whose opening angle is the critical angle. That includes the fisherman's eyes at water's edge. Given that this is the case, the fisherman might be better off standing up—by being sufficiently above the surface, the fish can be visible to the fisherman at every depth.

CHAPTER 36

2. Attach a plane mirror to the object that rotates and then reflect a ray from the mirror onto a screen a distance L away during the rotation. The image on the screen will move across an arc length $2L\alpha$, and if L is large, this will be measurable. In effect, the distance magnifies the rotation.

5. It is easiest to see from rays 2 and 3 in Fig. 36–15 that the image will become smaller and move away from the mirror toward the focal point. Only ray 3 changes.

8. The magnification is less, as you can most easily see from the rays in Fig. 36–26 (the rays are refracted less).

9. Down. This is most easily seen if you used rays 2 and 3; ray 2 wouldn't change, but ray 3 would.

CHAPTER 37

2. Yes, if we measure y, R, and d, we can solve Eq. (37–4) for λ.

4. Changing the length of the feeding cables by different amounts will introduce a phase difference for the signals at the antennas, corresponding to the new path-length difference. However, the signals remain coherent, and there will be interference at your house. Whether this is destructive or constructive is a matter of the numerical value of the new phase difference.

5. We see from the solution that Δx is proportional to λ. Red light has a larger wavelength than blue, so that Δx will be larger for red light, and there will be fewer bands for red light than for blue.

6. While our list of wavelengths in the solution shows that the spacing in wavelengths between successive values for constructive interference is not equal, it will certainly be a good approximation to put a minimum (destructive interference) equidistant between 700 and 420 nm, namely $\lambda \cong 560$ nm. How would you find it exactly?

CHAPTER 38

1. Higher orders do indeed help, because as Eq. (38–7) shows the resolution increases with order, and it would be easier to resolve two closely spaced peaks.

2. The starting point will be the laser beam shining perpendicular to the CD. You could then either tilt the CD or move to the side so that you are observing the CD at an angle, and at an angle corresponding to the first- or higher-order maxima for the laser light, you will see a bright reflection.
3. The blue light has shorter wavelength, making the angles of the minima smaller, and the spacing between them on the screen is less than for red light.
4. In order to find the exact positions of the maxima, we must take the derivative of the intensity with respect to α (or θ) and set that equal to zero (see Problem 38–31).
5. If the two objects are arbitrarily far away—think of two stars—they may be arbitrarily far apart and still make an angle as small as that specified by the Rayleigh criterion. For very distant objects, it is in fact only the angular resolution—the minimum angular separation that can be resolved—that matters, and separation distance is not a quantity that is relevant here.
7. The overall deflection is $2\theta = 50°$.

CHAPTER 39

3. The Earth-based observer if situated at the end of the galaxy, would read 300 yrs on the clock, just like the spaceship pilot.
5. No, the equation in the solution shows that $u/c < 1$ no matter how large $(\lambda_1/\lambda_0)^2$ is.
6. The image is moving away from the light source, and therefore its light will be redshifted. The Alternate Solution describes this redshift, which corresponds to a shift with a velocity V that is given by adding the two velocities u according to the relativistic velocity addition formula.
7. The detailed mechanism of the clock is irrelevant to the time dilation. The effect is a consequence of the relation between space and time. In any case, the amplitude of a harmonic oscillator does not enter into the frequency. Only the mass and the spring constant do, and their transformation properties under Lorentz transformations must be such that the result of our calculation is intact.
9. The Lorentz transformation gives $t' = \gamma t$, and in view of our answer for t, we get $t' = 2L/u$. This is certainly a reasonable result. All the clocks read alike in the rest frame of spaceship B, by definition. The nose cone of A travels with speed u and traverses the length of B (in its rest frame), so that the time elapsed is $2L/u$.
10. Yes, since our nonrelativistic world deals with velocities $v \ll c$. In the limit that we may treat c as infinite, $\gamma = 1$. The events are simultaneous, the train fits into the tunnel for both observers, and all is well.
11. Once we know the connection between energy and momentum, the answer is yes. Momentum conservation still means that the two particles must have equal and opposite momenta. Then with a relativistic connection between energy and momentum [Eq. (39–36)], the conservation of energy becomes an equation for a single unknown, which can always be solved.
12. There are two huge technological difficulties. (1) There is no antimatter lying around. If there had ever been any, it would long since have annihilated with matter. Antimatter can be made in accelerators, but that costs a lot of energy, and having a large laboratory on a spaceship is at present hard to imagine. Perhaps there will be a future technology that will make accelerators very compact. (2) Antimatter is hard to keep around. It will annihilate with the matter of the walls of any container, and again give rise to huge explosions. There may one day be a technology that uses electromagnetism to hold antimatter away from container walls that are made of matter.

CHAPTER 40

2. All of the power is emitted in the form of radiation, but most of it is in the form of waves whose wavelengths do not lie in the visible range. A lightbulb generates a great deal of radiation with wavelengths in the infrared range, what we perceive as heat. In fact, incandescent bulbs are highly inefficient ways of delivering light, and in our society a great deal of money is spent to cool the spaces that they heat.
3. The reduction in the number of photons is by a factor of $(0.9 \times 10^5)/15 = 6 \times 10^4$. Since the intensity falls as the inverse of the square of the distance, the star could be seen if it were a factor $\sqrt{6 \times 10^4} \cong 250$ times farther away.
4. With a shorter wavelength, and therefore higher frequency, the photoelectrons would be emitted with a non-zero kinetic energy. With a longer wavelength, no photoelectrons would be emitted. Radiation could still be absorbed and/or reflected.
5. We know from kinetic theory that T is proportional to the root mean square velocity v in the gas, so that v scales as $\sqrt{T}$. This means that the momentum scales in the same way, while the wavelength scales as $1/\sqrt{T}$.
6. The wavelength scales with the mass m of the (nonrelativistic) particle as $1/\sqrt{m}$, so that it is much smaller for neutrons than it is for electrons. The maximum value of $\sin\theta$ is 1, and this means that the value of n can be much larger for a smaller λ, that is, many more diffraction peaks will fit in the 90° range of possible θ-values.
8. Yes. In dimensional analysis you ask what constants of the problem can enter and then combine them to make a quantity with the desired dimensions, which here is energy. That m and L could play a role as parameters is clear; what is new is that Planck's constant can also appear. The numerical factor is not defined, just as in the case of the use of the uncertainty relation.

CHAPTER 41

1. The energy of the photon is proportional to the frequency and therefore inversely proportional to the wavelength. In the two-step loss of energy, the energy is smaller in each step, and therefore the wavelength is larger in both steps.
2. Twice-ionized lithium is like hydrogen, with $Z = 3$ instead of 1. This means that the energies, which are proportional to $(Ze^2)^2$ [a factor of e is the electron charge and a factor of Ze is the nuclear charge], are 9 times larger than those of hydrogen. This changes the ionization energy to 122 eV and the estimated temperature to 4×10^5 K.
3. Because the level spacings are equal, the transitions between neighboring levels all have the same frequency and therefore the same wavelength. There would thus be only one spectral line corresponding to a change of energy $\hbar\omega_0$.
6. With a field of magnitude 10^8 T, the energy difference between the two spin states is of the order of 10^4 eV, huge on the atomic scale. All of the spins would be pointing "down," because spin-down has a lower energy than spin-up.
7. As you can see from the chart, 36 electrons make a closed shell. The thirty-seventh electron starts a new shell, and the atom with $Z = 37$ is therefore an alkali.

CHAPTER 42

1. The change is visible in Eq. (42–2). We now have a square two-dimensional "box," so that this equation will read $N_{ie} = (L/d)^2$. As a consequence, E_F is proportional to $n_v^{1/2}$ rather than $n_v^{1/3}$.
2. The calculation would not apply to the Sun, whose mass is the same but whose radius is 1.4 million km rather than 7000 km. The electron energy will be nonrelativistic. At this stage of its existence the Sun maintains its equilibrium not by internal degeneracy pressure, but by the pressure associated with the energy released in nuclear reactions.
3. We see from the expression for R that $R \propto n^{5/3-2} = (1/n)^{1/3}$. Taken alone, this result suggests that the pressure balance could be maintained at smaller and smaller values of R as n increased. But as noted above, the energy-momentum relation for the fermions changes when n increases significantly, and the balance of pressures is not sustainable.
4. The de Broglie wavelength is

$$\lambda = \frac{2\pi\hbar}{p} = \frac{2\pi\hbar}{mv_{rms}} = \sqrt{\frac{4\pi^2\hbar^2}{3mkT}}$$

so that the expression indicates $n\lambda^3 = \text{const}$, which implies that the interparticle separation and the de Broglie wavelength are of the same order of magnitude.

CHAPTER 43

1. The Fermi energy is some 7 eV, and that is very much smaller than the energy equivalent of the rest mass of an electron, namely about 0.5 MeV. All energies that arise in this example are much less than this value. That guarantees nonrelativistic behavior throughout.
4. In intrinsic semiconductors the density of n- and p-carriers is the *same*.

CHAPTER 44

2. The de Broglie wavelength is much smaller than the closest distance to the nucleus, so this criterion alone might not suggest that quantum-mechanical effects would play an important role. But remember the atom is a highly quantum-mechanical entity, and the α particle penetrates to a distance well within the innermost Bohr radius. In fact, quantum mechanics does play an important role in the effect of the α on the electrons.
4. Nuclei and neutron stars consist only of nucleons (protons and neutrons), and in both cases they are packed closely—in one instance by the attractive nuclear forces, in the other by the pressure of gravity. This close packing automatically implies that the densities are of the same order of magnitude. The force of gravity caused by the huge mass squeezes the neutrons a bit more than the nuclear forces do. In fact, if the mass of the neutron star is large enough, it will collapse enough to end up as a black hole.
6. $S_n(^{56}\text{Fe}) = [M(n) + M(^{55}\text{Fe}) - M(^{56}\text{Fe})]c^2$. While there is no particular odd-even effect in this case, we might expect ^{55}Mn to be more stable than ^{55}Fe because of its smaller nuclear charge and smaller mutual Coulomb repulsion.

The neutron separation energy is 11.2 MeV rather than the 10.2 MeV for the proton separation energy, so our conjecture is correct; it does take more energy to remove a neutron than a proton.

7. The nuclide ^{209}Pb gives up its neutron rather easily in order to reach the special double magic number of $Z = 82$ and $N = 126$. The liquid-drop model does not take into account the existence of magic numbers, which are, however, a feature of the shell model.

8. With a density of 19,000 kg/m^3, we have a volume of $1.6\ \text{kg}/(1.9 \times 10^4\ \text{kg/m}^3) = 0.8 \times 10^{-4}\ \text{m}^3$. This is about 80 cm^3, or a cube of about 2 in on a side. The mass of a basketball would be almost 1000 kg, far too much to lift without machinery.

9. The answer is (b). The ^{237}Np nucleus is so much more massive than the α particle that it must move very slowly in order to conserve linear momentum. But kinetic energy contains a v^2 factor, so the α particle carries off most of the kinetic energy.

11. The 2 protons can be used again for another proton cycle. The gamma rays contribute to the star's internal energy and can either leave the star as radiant energy or result in another reaction. The neutrinos interact weakly and are likely to leave the star.

12. Very roughly, atomic energies are a million times smaller than nuclear energies. For a given power output, 10^6 times more chemical reactions must take place, and this means that whatever burns in the Sun must be used up in 10^{-6} the time estimated for the Sun's burnout time from nuclear reactions. Since that naïve estimate was 10^{11} yr, we obtain 100,000 years, much shorter than geological times.

CHAPTER 45

1. The electron is about 2000 times lighter than a proton. A collision of the alpha particle with an electron corresponds to a collision between a car and a pigeon. Such a collision could not deflect a car by 90°, and an electron cannot be responsible for the deflection of α particles.

2. In the center-of-mass frame, the proton has equal and opposite momentum to the electron, corresponding to a value of pc that is also 5 GeV. The proton rest mass is 0.94 GeV, and thus its motion, too, is relativistic.

5. The experiment is never carried out exactly at the threshold energy. At any energy above the threshold energy, the protons and antiprotons move away from each other, and the annihilation does not occur.

7. When we go from particle to antiparticle, Q and B change signs. Thus the first two particles cannot be their own antiparticles. The third one has $Q = B = 0$, so it can be its own antiparticle, just like the photon. This is evident from the quark assignment—just change the c-quark to its antiparticle and vice versa.

8. This is not entirely a fair question, because it requires some knowledge about the quantum numbers involved and their conservation. The conservation of quark-type in the strong interactions means that in a reaction that has no t quark in the initial state, the t quark in the final state must be accompanied by a $\bar{t}$. This means that the initial energy of each of the incoming particles is equal to that of the rest energy of the top quark rather than half that value.

9. The mass of a single hydrogen atom is M. Thus the number of atoms per cubic meter is

$$\frac{\text{atoms}}{\text{m}^3} = \frac{\text{kg}}{\text{m}^3} \times \frac{\text{atoms}}{\text{kg}}$$
$$= \frac{\text{kg}}{\text{m}^3} \times \left(\frac{\text{kg}}{\text{atom}}\right)^{-1} = \frac{\rho}{M}.$$

With $M = 1.67 \times 10^{-27}$ kg, this turns out to be about 0.6 atoms per cubic meter.

Answers to Odd-Numbered Understanding the Concepts Questions

CHAPTER 1

1. Yes; a vector $\vec{V}$ will have 4 components: $\vec{V} = (V_x, V_y, V_z, V_w)$.
3. Mine is 180 cm, or 1.80 m.
5. They are dimensionless.
7. Define $c = 299{,}792{,}458$ m/s, then 1 m $= c(1/299{,}792{,}458$ s$)$.
9. Yes.

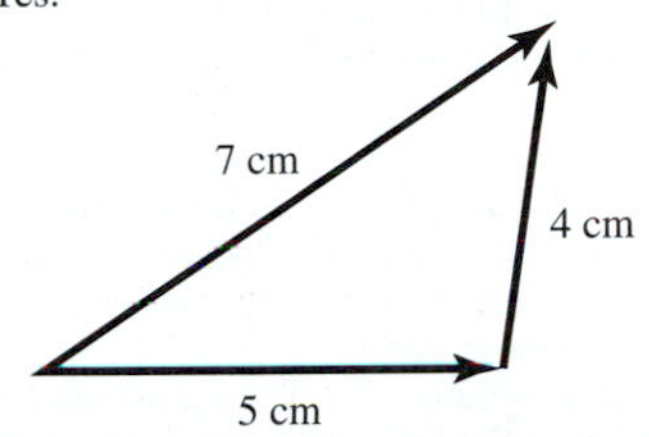

11. No; if the angle between the two unit vectors are 60°.
13. About 99.6% accurate.
15. About 0.3% (assuming a weight of 150 lb and an uncertainty of about 0.5 lb).
17. The surface area of a mouse is large relative to its body size, and the energy radiated through the skin needs to be compensated with sufficient food intake.
19. No.
21. Three; infinite number of sets.

CHAPTER 2

1. In case the car in front of you suddenly stops, your car would travel further before you react, and it would take longer for it to stop after you apply the brake.
3. t is proportional to $g_x^{-1/2}$, and v is proportional to $g_x^{1/2}$.
5. No.
7. $h_{\text{Moon}}/h_{\text{Earth}} = g_{\text{Earth}}/g_{\text{Moon}}$, $h_{\text{Moon}} \approx 5$ m.
9. True.
11. 0; yes.
13. $a =$ constant and $v_0 = 0$.
15. False.
17. 3rd, 2nd, 1st; $v_1 = v_3 > v_2$.
19. Need to measure the diameter d of each wheel; number of rotations per unit time $= v/\pi d$.
21. A box sliding up or down a straight ramp, two unequal masses connected by a string hanging over a fixed pulley, a vehicle accelerating uniformly down a straight road, etc.
23.

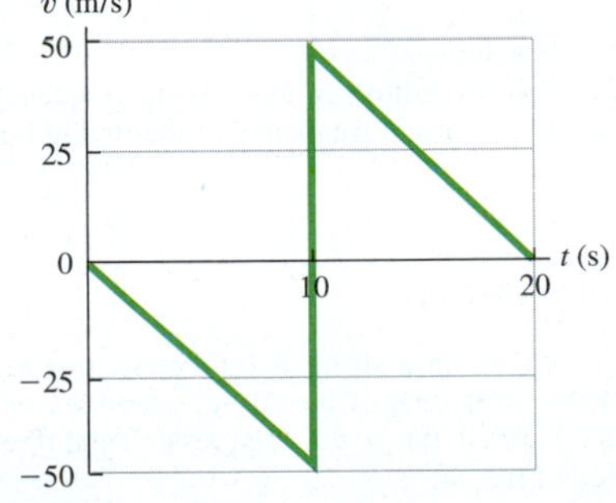

$v = 0$ again at $t = 20$ s, when it returns to the point where it was initially dropped.

CHAPTER 3

1. A little higher.
3. A strong wind can alter the result significantly.
5. No.
7. R is proportional to v_0^2.
9. The ball will land ahead of the car.
11. No.
13. Tilt the umbrella forward at an angle θ, with $\tan\theta =$ running speed/raindrop speed.
15. (a) Outside the edge of the merry-go-round, (b) on yourself or next to you.
17. There is an initial eastward velocity.
19. Only (a) does.
21. No.

CHAPTER 4

1. A projectile motion with horizontal initial velocity.
3. Use a spring scale.
5. The reading is first greater than your weight, then becomes equal to it before getting less than it, and eventually becomes equal to it again.
7. False.
9. The tableware does not accelerate as fast as the tablecloth.
11. The direct cause is the normal force exerted by the person on the table.
13. Everything except gravitational and magnetic forces.
15. Gravity is balanced by the drag force of the air.
17. For example, pull the mass with a spring scale with a constant force F and measure the acceleration a.
19. The greater cross-sectional area of the rubber ball results in greater air resistance.
21. The friction force exerted by the road on the tires accelerates the vehicle; the engine forces the tires to turn, initiating the friction from the road.
23. She will land at a point further than half the radius away from the center of the merry-go-round.
25. Centrifugal force, the upward force that enables an astronaut to "float" inside a space shuttle, the backward force that keeps you at rest on the driver's seat as you accelerate your car forward.
27. Gravity speeds up the marble on its way down and slows it down on its way up, the bowl's contact force keeps it in a circular path, friction slows it down and eventually stops it.
29. It exceeds your weight.
31. Yes; the downward force from the branch.
33. Balance an object and check to see if $\Sigma\vec{F} = 0$.
35. The engine forces the driving tires to turn, which cause the road to exert a forward force of friction on the tires.
37. No.
39. Forces may not necessarily initiate motion.

CHAPTER 5

1. The tension in the wire must have a vertical component to balance the weight of the walker.
3. The mass of the string is not negligible, and/or the observer is in a non-inertial frame of reference.
5. The tension in the rope would vary; friction increases the tension necessary to move the load.
7. To increase the static friction from the road that serves as centripetal force.
9. No; yes.
11. Both help keep the boat from moving sideways; the keel also helps prevent the boat from overturning, while the centerboard is much lighter.
13. Yes in principle; friction reduces efficiency.
15. The weight of additional people causes the scull to sit lower in water, increasing the drag force of the water.
17. All of the statements are true.
19. Gravity serves as the centripetal force; no.
21. Yes.
23. No; the net force is never zero; tension in the string and the weight of the bob.
25. No.
27. Less static friction is available from the ground.
29. The lowest point, where the tension in the vine is the greatest.
31. The die will curve away sideways while descending the bowl's wall.
33. The first term dominates at low speeds ($v \ll b/c$), the second one dominates at high speeds ($v \gg b/c$).

CHAPTER 6

1. Use $(\vec{F}_{\text{drag}} + \vec{F}_g)\cdot dr = dK = d(mv^2/2)$ and integrate to find the speed of the baseball.
3. Yes; the force opposes the motion and reduces the kinetic energy of the object.
5. The centripetal force does zero work and does not change the speed (and the kinetic energy) of the object.
7. No.
9. The energy for the work done on the smokestack comes from the engine, not the man; the work done on the man by the deck equals the work he does on the smokestack.
11. No (if terminal speed has been reached).
13. Yes. For example, the friction of the road that accelerates a car from rest.
15. An amount equal in magnitude to the (negative) work of the drag forces attributed to your weight.
17. Friction from the wall does positive work on the participants while their speeds increase; no net work is done on them once they reach constant speed.
19. Measure the maximum height attained by the first acrobat and hence the kinetic energy imparted to him.
21. The one chopping the bone.
23. To the moving observer $\vec{F}$ and $\vec{v}$ are opposite in direction, so $\Delta K = W < 0$.

25. The mass m of the parachutist and the height h of the jump.
27. The net work done by the drag force is negative, rather than zero, for one complete round trip.

CHAPTER 7

1. No.
3. It means that U of the person at the bottom of the well is lower than that at sea level.
5. Loss of mechanical energy due to air friction as well as during the collision with the floor; a ball thrown down with a large initial speed has a significant amount of initial kinetic energy.
7. Yes; energy released from sugar and oxygen may be converted into the mechanical energy of motion, the heating of muscles and bones, and that in the chemical products that the body produces, etc.
9. No; if yes the spring would be able to oscillate forever (inside a vacuum tube).
11. Measure the final speed v of a falling object from height h (in a vacuum tube) to see if $mgh = mv^2/2$; observe the motion of satellites and planets.
13. Yes; air friction and the collision force from the floor.
15. The normal force is always perpendicular to the direction of motion (which is tangential) so it does no work; this can be verified from $E_i = E_f$.
17. Not necessarily.
19. If they were, then as the ball comes back to where it started its kinetic energy (and speed) would return to the same value as before.
21. None, although the energy of Earth as a whole is fairly constant, at least over time intervals that's much shorter than geological scales.
23. Back to the height reached after the first bounce.

CHAPTER 8

1. Yes (for the Sun-comet system).
3. Yes, the center of mass does stop; no, since $\vec{v}_{cm}$ can be zero before the collision.
5. No. It follows a parabolic path.
7. The body is curved like a horseshoe "draped" over the bar, so only part of the body is above the bar.
9. Yes; $x_{cm} = m_{point} R/(m_{circle} + m_{point})$.
11. Yes.
13. Yes; yes (albeit imperceptibly).
15. Short enough so that our ordinary senses cannot detect a finite time interval.
17. Assuming elastic collisions, the smaller ball can reach 9 times the initial height of the larger one.
19. Conservation of momentum of the gun-ammunition system.
21. It moves slightly forward.
23. As the parachute moves faster it collides with more air molecules (per unit time) moving at higher speeds towards it.
25. 45° (with no energy loss); less than 45° from the table (with energy loss).

CHAPTER 9

1. Down (with right-hand rule); up (with left-hand rule); yes, but one has to be consistent.
3. The angular momentum of the rotation stabilizes the motion.
5. The wheel moves downstream to an observer on the shore and backward relative to the boat.
7. Lower.
9. The one filled with water.
11. Shorten r by a factor of $1/\sqrt{2}$.
13. The force of the water against the side of the canoe can tip it over.
15. The angular speed, which cannot exceed a certain amount or the wheel will disintegrate.
17. To increase the torque applied by increasing the lever arm.
19. The longer wrench allows you to exert a greater torque by providing a longer lever arm.
21. The spring will be further stretched.
23. Two.
25. 1:1.
27. Not enough centripetal force available to support the circular motion.

CHAPTER 10

1. Yes.
3. The wheels have angular momentum so they tend to maintain their orientation.
5. To keep the center of mass as low as possible to increase stability.
7. Considerable torque is required to change the direction of the large spin angular momentum of the fan.
9. Little torque is exerted on the person using the bent leg technique; lifting weight with bent back requires you to counter the torque of the weight.
11. Acquire an initial rotation by using the contact force from the diving board to provide a torque about his/her center of mass.
13. The angular velocity is in the same direction as that of the torque from the cue about the center of mass; the spin angular momentum acquired equals the angular impulse delivered by the cue.
15. No. The friction from the ground exerts a torque on you.
17. Zero.
19. Yes.
21. The pilot must arrange his ailerons and rudder to introduce a force that will push the nose up.
23. The student will also turn, but by only half as much.
25. Slow down.
27. It is absorbed by Earth.
29. (b).

CHAPTER 11

1. Pushing against the slope results in a reactionary force from the slope that tends to push the climber away from it.
3. The force of gravity on the mass would produce a net torque about the point of suspension if the rope is not vertically aligned.
5. Stable equilibrium.
7. No; an object is approximately rigid if its deformation is considerably smaller than its dimensions.
9. True.
11. $\Sigma F_x = ma_x$, $\Sigma F_y = ma_y$, $\Sigma \tau_A = I_A \alpha$ (A is the contact point between the motorcycle and the curb); possible only if the curb extends below the center of the wheels.
13. Yes.
15. The one which makes an angle of 60° with the horizontal.
17. No.
19. No.
21. Unstable equilibrium.
23. It increases the tensile strength of bricks.
25. By heating up the diamond.
27. A structure (e.g., graphite) featuring layers of planes is more vulnerable to shear forces applied along the planes; yes.

CHAPTER 12

1. Drag forces in fluids, the tension in ropes or rods, the normal force between two objects in contact, etc.
3. Yes.
5. Launch the satellite near the equator, in the direction of Earth's rotation; launch it in the direction of the orbital motion of Earth around the Sun.
7. No.
9. A straight line passing through the Sun.
11. Yes.
13. As the Moon completes one revolution around Earth it also completes one rotation about its own axis.
15. Yes.
17. Consider Earth as a uniform sphere plus some local variation of density, and study the deviation of the g data from its uniform value to probe the density variation.
19. Assuming uniform density, g would be uniform on its surface if Earth were a perfect sphere.
21. (b) is correct.
23. The surface of Jupiter is not as rigid as that of Earth,
25. Zero.
27. Yes.
29. $g = 0$ inside a hollow Earth.

CHAPTER 13

1.

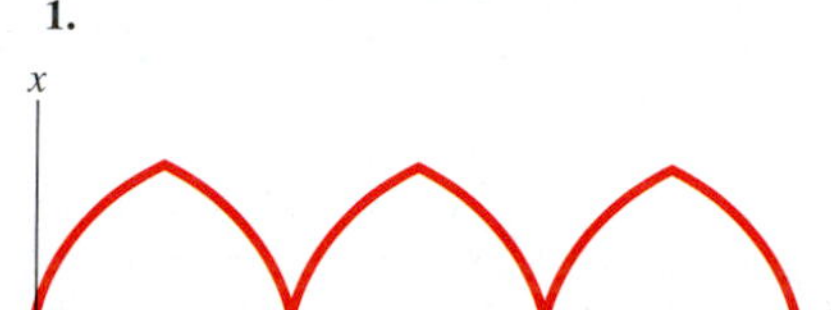

3. K_{max} increases by a factor of 4.
5. The thinner air on the mountain top provides less drag on the pendulum, while the g value there is lower—these two factors affect T in opposite directions.
7. Yes; the positive work done by the external driving force.
9. It should be shortened.
11. Increase.
13. T is decreased by a factor of $(3/2)^{1/2} \approx 1.22$; A is decreased by a factor of $(3/2)^{1/4} \approx 1.11$.
15. (a): Driven harmonic motion while wind is blowing; (b): driven harmonic motion; (c) driven harmonic motion while the child is pumping and damped while he or she is not; (d) and (e): damped harmonic motion.
17. Yes.
19. Yes; longer.
21. You are adjusting the driving frequency to match the natural frequency of the diving board.
23. No.

CHAPTER 14

1. Waves on a string is transverse, due to the elastic stretching of the string; sound waves are longitudinal, due to the compression and rarefaction of the air.

3. For more time, as no energy is transferred to the air to produce the sound wave.
5. (a) is correct.
7. For the same wavelength, the frequency of sound is higher in helium, as the speed of sound in helium is higher due to its lower density.
9. Sound waves cannot propagate in vacuum.
11. $h(x, t) = h_0 \sin(kx - \omega t + \phi_0)$.
13. Yes.
15. The wavelength must be decreasing, causing the frequency to increase.
17. Increase the tension in the wire.
19. No; v is more than doubled.

CHAPTER 15

1. Yes.
3. The pulses approach each other and superimpose as they meet, reinforcing each other if they are on the same side, canceling if they are on opposite sides, and forming a tilted pulse if they are perpendicular to each other. After passing through each other they move on, maintaining their integrity.
5. Stick the gum on one fork and listen to the beat again. If the beat frequency gets lower then this fork oscillates at a higher frequency than the other one.
7. f is unchanged, while λ decreases.
9. A single pulse is the superposition of many harmonic waves of different wavelengths, which travel at different speeds and cause the wave to spread into many ripples.
11. The energy of the wave on the string takes kinetic form rather than elastic potential form at the moment the two pulses cancel each other out, very much like the moment the string is flat in a standing wave.

CHAPTER 16

1. Hot air is less dense than cold air and the buoyant force of the surrounding cold air lifts the balloon.
3. The weight in water is less than that in air by that of the water displaced, from which we know the volume, and hence density, of the body. Fat content is then deduced from the density data.
5. It weighs more as the buoyant force is less than the weight of the compressed air inside.
7. Yes.
9. Its high density allows for a relatively short column to balance air pressure.
11. The head would implode if $p < 0$. The value of $\rho v^2/2$ is always positive; $\rho g h$ can be both positive or negative; p is usually positive but it can be assigned any value as long as $\Delta p = p_1 - p_2$ is unaffected.
13. $h_{max} = P_{air}/\rho_{water} g \approx 10.3$ m.
15. Yes if the sails are curved, which is not absolutely necessary but works better.
17. Styrofoam has greater volume and weighs less in air; estimate the volume and find the density from m/V.
19. The sides of the can are pushing in on the liquid.
21. The oil is closer to the center of the centrifuge—think of the centrifugal force as an effective gravity pointing away from the center of rotation.
23. The water pressure on one's chest and lungs prevents air from entering a long tube deep under the water surface.
25. It remains the same.

CHAPTER 17

1. $\approx 4 \times 10^{23}$ molecules.
3. The pressure increases as depth increases, and more air molecules are present in a given volume at a higher pressure.
5. An equilibrium is reached among all the three phases at the triple point, much like how ice and liquid water can coexist at 0°C.
7. Stand in front of a radiant heater and compare the heat it generates before and after it is turned on.
9. A very cold tire may have exceptionally low pressure, and boosting it to the nominal value while it is so cold could cause the pressure to be dangerously high when the tire is heated up by driving.
11. No.
13. The ball expands upon being heated so it cannot pass through the ring; heating the ring will work.
15. Measure the temperature of each substance to see if it has reached a common value.
17. As vast numbers of tiny molecules collide randomly with the wall of the container, there is no perceptible net force on the wall.
19. Nothing, except for the possible confusion it might cause.
21. Icebergs would sink to the bottom of the ocean. As enough ice builds up below the water surface, it could largely deplete the space occupied by liquid water, and that would be devastating to sea life.

CHAPTER 18

1. Work done on the air as it undergoes adiabatic compression causes its temperature to increase.
3. Yes.
5. (a) and (c) .
7.

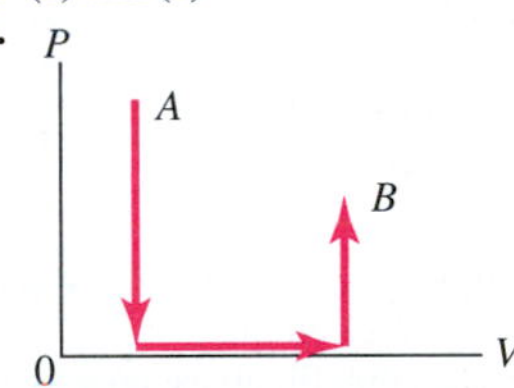

9. To improve thermal insulation with the air trapped in between the layers.
11. An incredible amount of shaking would be required for any significant temperature increase, even if the container is thermally insulated.
13. Fluids with very large specific heat: as thermal reservoir to store and transfer heat and to limit temperature variation; fluids with very small specific heat: as lubricant in machinery.
15. $\Delta Q_{net} = c_1 m_1 \Delta T_1 - c_2 m_2 \Delta T_2 = 0$, so $\Delta T_1/\Delta T_2 = -c_2 m_2/c_1 m_1$. If $c_1 = c_2 = c_{water}$ then $\Delta T_1/\Delta T_2 = -m_2/m_1$; and $\Delta T_{Hg}/\Delta T_{water} = -c_{water}\, m_{water}/c_{Hg}\, m_{Hg} = -m_{water}/[c_{Hg} m_{Hg}/c_{water}] = -m_{water}/m_{eff}$, i.e., 100 g of mercury behaves effectively like water of mass $c_{Hg}(100\text{ g})/c_{water} = 3.3$ g.
17. It takes far more energy to completely separate water molecules from each other than to slightly "loosen them up".
19. $W_{net,\,2} = 2W_{net,\,1}$.
21. Not if $c(T)$ is known.
23. Friction turns 100% of the mechanical energy into heat, and a rapid process prevents thermal equilibrium from being established.
25. Yes. $c_{Hg} \ll c_{water}$ so $\Delta T_{Hg} \gg \Delta T_{water}$.
27. Keeping the engine idling greatly limits the airflow across the radiator surface that can help cool the engine, and winter air is typically drier and has lower specific heat.

CHAPTER 19

1. T increases; the gas molecules on average speed up due to the extra momentum imparted upon them by the inward movement of the piston.
3. It is the probability per a range of one meter.
5. The intermolecular collisions are largely elastic and leave the average molecular kinetic energy unchanged.
7. Thermal equilibrium can still be established between the wall and the gas molecules, yet relatively large fluctuations from the average value of molecular kinetic energy would make T less well defined.
9. Not quite meaningful as an indicator of average molecular kinetic energy, which is a statistical quantity involving a large number of molecules.
11. The intermolecular collisions are largely elastic which conserve momentum, so the same total momentum is carried by the molecules to the wall after the collisions.
13. The miniscule size of the molecules makes it harder for them to hit each other.
15. Air.
17. The inward moving piston collide with gas molecules to increase their average speed (and hence energy).
19. The pane of glass is bombarded on both sides with equal pressure.
21. The airflow helps more energetic water molecules to escape from the water surface
23. Measure the heights of a large group of people and divide the heights into small and equal-sized increments, then check the number of people whose heights fall within each increment. No absolute certainty.
25. The average molecular rotational kinetic energy is a function of temperature.

CHAPTER 20

1. Yes if the ocean temperature is higher than that of the air, as in winter times.
3. Only in the sense that it is associated with the "randomization" in spontaneous physical processes.
5. No.
7. In warm climates.
9. Gas X will permeate the bottle and diffuse throughout the room.
11. Cream poured into a cup of coffee diffuses and mixes up with the coffee, for example. Entropy is increased.
13. They can convert essentially 100% of the electrical energy into heat, but not to work.
15. Only part of the heat released from the chemical energy is converted into mechanical work.
17. No, the room would get even warmer.
19. Extensive.

CHAPTER 21

1. The midpoint of the line joining the two charges.
3. The charges acquired by the balloon through rubbing attracts the opposite charges induced in the wall.

5. The force on each charge and its mass; if $m_1 = 5m_2$ then $d_1 = (1/5)d_2$.
7. The ones farther form the nucleus.
9. The "peanuts" are charged through rubbing and attract the opposite charges induced in your hand.
11. The equilibrium at the center of the ring is unstable.
13. Take the charged cork ball and touch one of the three uncharged ones, then simultaneously touch it with the two remaining balls.
15. The hand becomes charged after rubbing.
17. No.
19. Yes if each contains an equal number of electrons and protons, but that would probably not be the case.
21. The forces $\vec{F}_{21}$ and $\vec{F}_{23}$ both double, so $\vec{F}_1$ and $\vec{F}_2$ will both increase in magnitude and rotate counterclockwise.

CHAPTER 22

1. To prevent a dangerous charge buildup in the gasoline-carrying vehicle by channeling the excess charge to the ground.
3. For the same reasons for introducing the electric field; the gravitational field lines can only end at (not start from) matter.
5. $Q_1 : Q_2 = -5 : 1$.
7. Two locations, one at -2 cm $< x < +4$ cm and the other at $x > 10$ cm.
9. The electric field lines above the surface of Earth point down.
11. The charges on the comb cause the molecules in the paper to polarize, resulting in a net attraction.
13. The charge distribution is that of two dipoles touching at one end, resulting in a net dipole field.
15. The density of the field lines a distance r from the charge is proportional to $1/r^2$, which would be inconsistent with the electric field if it goes like $1/r^{2+\delta}$.
17. A sphere uniformly charged to some negative net charge would do.
19. Yes; the field is not zero since the two dipoles are not at the same location.
21. If $F_E > F_g$ it will accelerate upward, otherwise downward. The initial height is irrelevant.
23. Just find the field due to q_1 and neglect those due to q_1 and q_2.

CHAPTER 23

1. No.
3. The net charge enclosed by the surface is zero, but the electric field on the surface may not be.
5. For a spherical Gaussian surface of radius r centered at the location of a point charge, Φ would be $4\pi cr$, which depends on r of the Gaussian surface, rather than just the charge enclosed.
7. Zero.
9. Zero.
11. The electric field in between the two plates doubles while that elsewhere vanishes. The net flux over a Gaussian surface remains the same.
13. Only that it is invariant under a rotation about the axis of symmetry of the wire.
15. No net charge is present in the region.
17. The electric field is not uniform over the Gaussian cylinder for a charged line of finite length.
19. The charge density is independent of z.
21. No.

CHAPTER 24

1. 1 Joule.
3. Insert a uniformly charged plane within the sphere.
5. The electrostatic potential energy; the work that was done in assembling the charges together.
7. At $\theta = 0$.
9. It is always an equipotential except during the brief time interval when equilibrium is being established as charges redistribute themselves on its surface.
11. The person is charged and the hair (with like charges) repel each other; to prevent a dangerous current from running through the person.
13. No; $E_x = \Delta V/\Delta x$.
15. Yes.
17. The work done by the electrostatic force along any enclosed loop is zero.
19. Yes.
21. Yes.

CHAPTER 25

1. No.
3. If the field drops abruptly to zero outside the plates then the voltage drop around a closed path consisting of one segment leading from one plate to another and another segment that closes the path from outside would be non-zero.
5. The potential difference becomes infinitely large, as it takes an infinite amount of work to concentrate a charge on an infinitely thin wire.
7. The charge on one plate is Q and that on the other is $-Q$.
9. V decreases, C increases, and U decreases.
11. Parallel plates.
13. It reduces C.
15. Yes.
17. Not in classical physics.

CHAPTER 26

1. Yes.
3. The thinner one, as its resistance is greater.
5. The density of free electrons.
7. While water must first fill up the hose before streaming out, a piece of metal is already loaded with free electrons.
9. Yes, but it is a very small effect.
11. No.
13. Eq. (26–25) gives $\rho \approx 10^{-7}\ \Omega \cdot \text{m}$ with $\tau \approx 10^{-14}$ s and $n_e \approx 10^{29}\text{m}^{-3}$.
15. It melts as too much heat is generated to be entirely dissipated in time.
17. $R_{eq} = R_1 + R_2$.
19. Bulb 2 is brighter when the switch is open; bulb 1 is brighter when the switch is closed.
21. Increase.
23. To increase its length (and hence resistance) without taking up too much space.

CHAPTER 27

1. A large current can flow through the body as tap water is a good conductor.
3. The equation representing the loop rule is unchanged.
5. No.
7. $C \approx 1000\ \mu$F, $R \approx 10\ \Omega$, $RC \approx 0.01$ s.
9. If I reverses Q would become $-Q$.
11. Put the two emf's in series to drive the two lightbulbs in parallel.
13. I increases to $3I$.
15. R_2 and R_3 are in series (branch 1), as are R_5 and R_6 (branch 2). Combine these with R_4 (branch 3), with the three branches in parallel, to obtain a single equivalent resistance, R_{eq}, which forms a one-loop circuit along with R_1 and the emf. Impossible for Fig. 27–8(b).
17. The steady-state I_1 decreases.

CHAPTER 28

1. The positive charges (ions) have no collective velocity but the free electrons do.
3. By an electric field; $\vec{v}$ must change direction if the force is magnetic.
5. No.
7. No.
9. Incoming charged particles are concentrated in the polar region where the magnetic field is the strongest.
11. A single circular loop.
13. Repel; attract if one of the current is reversed.
15. Attract.
17. No; yes.
19. No.
21. (d).

CHAPTER 29

1. The needle will maintain an orientation that follows the circular path of the compass itself.
3. The direction of the magnetic force depends on that of the current flow.
5. No.
7. An extended wire is made of many segments, each producing a magnetic field of its own magnitude and direction.
9. It is much easier to measure force accurately.
11. Replace ε_0 with ε.
13. The field doubles for two loops with currents flowing in the same sense, and is zero if they flow in opposite sense.
15. The magnets attract when adjacent poles are opposite and repel when they are the same. The fields add up when unlike poles are adjacent and cancel when like poles are adjacent.

CHAPTER 30

1. No.
3. (e).
5. Yes; the magnetic field may increase in one part of the region and decrease in another part.
7. (a) No, (b) yes, (c) no, (d) no.
9. No.
11. Yes; to compensate for the thermal energy dissipated by the induced current in the metal plate.
13. No change.
15. The falling water gives up its gravitational potential energy and drive the rotary turbine blades in a magnetic field to generate induced current.
17. It falls with $a < g$ as it is above the tube and is entering it, and with $a = g$ when completely inside the tube.
19. An induced current in the ring antiparallel to that in the coil is induced; the ring will not jump.

21. An induced current in the ring antiparallel to that in the coil is induced; the size of the ring must allow the magnetic force on it to cancel with its weight.
23. The residual magnetic field in the iron frames of the moving car induced a current in the wire loop.

CHAPTER 31

1. The nucleus is much more massive than the electrons.
3. Make the core of a solenoid out of the material and check to see if the magnetic field increases or decreases due to the presence of the core.
5. Yes.
7. Yes; very difficult to measure due to ferromagnetism, try search for it at high temperature.
9. Magnetically hard material.
11. The alignment of magnetic moments causing paramagnetism becomes less effective as temperature increases.
13. Pulled into the region.
15. The orbital angular momentum of the electron does not change if $\vec{B}$ is perpendicular to its plane of motion, and changes if $\vec{B}$ is parallel to the plane.
17. 0.
19. The field of a bar magnet is non-uniform.

CHAPTER 32

1. Not necessarily.
3. (b), (a), (c).
5. Yes.
7. (b).
9. Construct an RL circuit and measure its time constant to find L.
11. Due to the current flow.
13. It depends on the total electromagnetic energy U in the circuit, or I_{max}, or V_{max} across the capacitor.
15. Yes.
17. There is no dissipation mechanism to lower the magnetic energy, so I does not drop.
19. $\frac{1}{2}mv^2$ and $\frac{1}{2}kx^2$.
21. The regions in between two wires carrying the same current has lower magnetic pressure than the region outside both wires, resulting in a net attraction between the wires.

CHAPTER 33

1. It must amplify and confine the magnetic field and have sufficient mechanical strength.
3. $X_C = 1/\omega C$ approaches zero as ω is very high.
5. The time average of the current is zero but that of the power is not.
7. The time average of the current is zero so one must use I_{rms}.
9. They can be reversed in principle; step-up becomes step-down, and vice versa.
11. (c).
13. An inductor.
15. No, unless they are of the same kind (say, both are capacitive).
17. The impedances are not infinite for any finite frequency.
19. To prevent signal loss due to abrupt changes in impedance.
21. When $\omega = (1/LC)^{1/2}$.
23. No; transformers can be used.

CHAPTER 34

1. No.
3. The ionosphere has an abundance of free charges, like a metal.
5. Yes.
7. No.
9. They must have very large area, lower density, and high reflectivity; it's difficult to sail into the solar wind.
11. Rotate a polarizer to see if you can get a near-zero transmission.
13. It goes out.
15. All of it.
17. The angular momentum is zero for a linearly polarized electromagnetic wave, and non-zero if it's circularly polarized.
19. Reflective.
21. $F_B \ll F_E$ since $v/c \ll 1$.

CHAPTER 35

1. Through reflection.
3. Extremely difficult; having three mutually perpendicular mirrors ensures that the incident beam is reflected straight back.
5. The bodyguard appears taller.
7. It appears to be shorter and closer to the surface of the water.
9. (d).
11. Not if you tilt your head toward the direction of the coin.
13. A distant house appears as two images, one of which inverted.
15. No air is present in space to scatter the sunlight.
17. As you scan downward from the vertical line you first see his upper body (which appears to be stretched), then you see the reflection of his lower body (upside-down), followed by a direct view of his lower body.
19. No.
21. Light reflected from the Moon has relatively little red component.

CHAPTER 36

1. The sign reads "AMBULANCE" for drivers who see a vehicle arrive from behind in the rear-view mirror.
3. No, but the image is dimmer.
5. $(x, y, z) \rightarrow (-x, -y, z)$.
7. Convex.
9. It is a convex mirror in which images are smaller than the objects, allowing a wider field of view.
11. Yes.
13. Yes.
15. Not very, except the one originated from the focal point.
17. It accentuates any imperfection of the eye by exposing the edge of the lens.
19. Not very plausible.

CHAPTER 37

1. There is no reduction in intensity due to the slits.
3. The fringes are in principle visible without a screen, but they are usually so close together in the field of view since the eye is a converging lens. A screen makes it much easier.
5. Adjacent wavelengths resulting in destructive interferences differ by a factor of 3 but $\lambda_{max} < 3\lambda_{min}$ for visible light.
7. Light rays are parallel when viewed from afar, simplifying the analysis.
9. The total energy over the entire space remains conserved.
11. No.
13. To minimize the reflection $n_{coat} = (n_{air}n_{glass})^{1/2}$.
15. Energy is intensified in other locations.
17. A tiny change in λ can result in a measurable shift in the interference pattern.
19. Total destructive interference requires identical amplitudes of two waves.
21. The slit spacing is much greater than λ.

CHAPTER 38

1. The diffraction pattern spreads out and gets dimmer overall.
3. No image distortion due to the atmosphere.
5. Short wavelength.
7. No.
9. The spreading of the diffraction pattern is proportional to λ, which is longer for red light.
11. When in place it blocks all the light except the two beams at the edges of the single slit.
13. A continuous spectrum (rainbow bands).
15. It interacts with the atoms.
17. One sees "around" the object as one changes the viewing angle of a holographic image.

CHAPTER 39

1. No.
3. [energy/speed^2] = [mass].
5. No.
7. It would indicate that there is no relative motion between Earth and ether during the measurement.
9. No.
11. No.
13. Yes.
15. No; the wire is not charge-neutral in the moving frame and produces both an electric and a magnetic field.
17. An entire pattern of wavelengths signifying a certain element (e.g., hydrogen) has been redshifted.
19. No, light falls only in an accelerated frame or in the presence of gravity.
21. No.

CHAPTER 40

1. Wave-like behaviors are prominent only when a photon interacts with objects comparable in size with its wavelength.
3. rate $\approx (P/hf)(r/2R)^2$, where P = power emitted by the cigarette, f = photon frequency, r = radius of the retina, R = 500 m.
5. The momentum (and hence wavelength) of an electron is different in different reference frames.
7. Yes; W increases with e.
9. T.
11. They don't; τ is longer according to an observer moving relative to the particles.
13. Yes, $L = \frac{1}{2}\lambda$.
15. Yes, but the probability is negligibly small.
17. No.
19. ΔE can be very large during a very short time interval Δt, but not for long.
21. No.

CHAPTER 41

1. $n = 0$ would lead to $\Delta p = 0$, in violation of the uncertainty principle.

3. λ_{min} occurs as the electron jumps from $n = \infty$ to $n = 1$. λ_{max} can be infinity.

5. No.

7. The different isotopes of hydrogen have different reduced mass and hence slightly different spectral lines.

9. The exclusion principle does not apply to two identical atoms so far apart that their electron clouds do not overlap.

11. The effective Z for neutral He is about twice as much as that for He^+, and ΔE is proportional to Z^2.

13. When the potential energy is minimum, the force vanishes.

15. Antiparallel.

17. No.

19. $T = 1.7 \times 10^6$ K; the actual fraction of atoms in the first excited state is much lower than 50% at room temperatures.

CHAPTER 42

1. Most of the electrons would be in the ground state at room temperature, and many chemical reactions would not be possible.

3. The decay of an excited atom with the emission of a photon has a rate that is enhanced by a factor of N^2 in the presence of N other photons of the same momentum and polarization.

5. If $E_F \gg kT$ then only those electrons with energies near E_F can be thermally excited. If $E_F \ll kT$ then all the electrons can participate in thermal excitation, and quantum statistical effects are negligible.

7. Yes; $P_F \sim m_n c$.

9. No.

11. In the absence of the exclusion principle, the atoms in distant galaxies would have their electrons mostly in the ground state, and there would be no resemblance between their spectra and spectra on Earth.

13. The sequence of transitions is much more likely.

15. The pumping mechanism must involve large energy input; the cavity in which the X-rays are to be reflected back and forth is difficult to make; de-excitation may occur for electrons from various states other than the desired excited state; the decay time from a highly excited state is very short.

17. No.

CHAPTER 43

1. No, they are identical.

3. To a good approximation, yes.

5. Holes can be spontaneously filled by stray electrons or by directing a beam of low-energy electrons to the surface.

7. It makes a transition from the top of the valence band to the bottom of the conduction band.

9. Each acceptor atom in a p-type extrinsic semiconductor provides an empty space for an electron from the top of the valence band to jump up to, requiring very little excitation energy. The vacancies left by these electrons form holes.

11. The energy always comes from the voltage source. The variation in the current that flows into the base amplifies the current into the collector by acting as a "gatekeeper" for the current that is generated by the biasing voltage, but does not itself provide energy.

13. The electronic excitation energy in a typical insulator is greater than that of a visible-light photon; the energy gap in a semiconductor is higher than that of an infrared photon but lower than that of a visible-light photon; and the electrons in a partially filled conduction band in a metal can absorb both infrared and visible-light photons.

CHAPTER 44

1. Measure the deflection of the alpha particles in an electric and magnetic field to determine the charge-to-mass ratio (Q/M), then combine with the Rutherford scattering data to determine Q and M.

3. The total A is conserved.

5. Strip an atom of one or more electrons and we get a positively charged ion. To get a negatively charged one we must either give the atom an extra electron, making it very unstable, or take a proton away from its nucleus, requiring a relatively large amount of energy.

7. The study of the inner structure of a nucleus requires high-energy alpha particles, which could disintegrate upon collision; nuclear forces between the alpha particles and the nuclei are not known as precisely as electromagnetic forces.

9. ^{16}O.

11. Radiation adversely affects blood cells. A radioactive material in the bone marrow affects the place where these cells are created, and there the effect is particularly strong.

13. It would take a minimum energy of about 1 MeV for neutrons to cause ^{238}U to fission.

15. In fission a number of divisions of (A, Z) into (A_1, Z_1) and (A_2, Z_2) are possible, as long as $M(A, Z) > M(A_1, Z_1) + M(A_2, Z_2)$. It is very unlikely that both combinations (A_1, Z_1) and (A_2, Z_2) will be stable.

CHAPTER 45

1. Sum of the interior angles in the triangle $= 2\pi + \alpha > 2\pi$, where α is the angle at which the two meridians cross each other at the North Pole.

3. Reactions such as $n \rightarrow p + e^- + \bar{\nu}$ are possible if the total baryon number is conserved and other applicable conservation laws are also observed.

5. (a) Electromagnetic and gravitational force, (b) electromagnetic force, (c) gravitational and electromagnetic force, (d) strong nuclear force and electromagnetic force, (e) electromagnetic force, (f) gravitational and strong nuclear force.

7. No; no.

9. Advantage: energies not accessible in the laboratory can be attained; disadvantages: particles cannot be controlled, and the rate at which they arrive decreases very rapidly with their energy.

11. The conservation laws do not apply to a virtual process.

13. There would be about one proton decay per year in one cubic meter of water.

15. Absorption due to intergalactic dust, the Hubble constant could be smaller for larger u.

17. A boson, (a) no, (b) no.

Answers to Odd-Numbered Problems

CHAPTER 1

1. 3.0×10^3 green jelly beans.
3. 3.5600×10^4 cents
5. 10^7; 10^{14}.
7. 1.8×10^{10} atoms.
9. \$1.18/kg.
11. 32.2 ft/s^2.
13. 3.33 g/cm^3.
15. 34 mi/gal; 10 mi/gal.
17. 0.0402%; 8.49×10^{-6}%.
19. $0.2(1 \pm 30\%)$ m^3.
21. 5%.
23. $[MLT^{-1}]$.
25. $[ML^2T^{-1}]$.
27. No.
29. (a) $[L^{-1}]$, (b) $[L^5T^{-2}]$.
31. (a) $[M^{1/2}L^{3/2}T^{-1}]$, (b) $[ML^2T^{-2}]$.
33. 2 tons.
35. 3.5×10^{-6} cm^2.
37. (a) 10^6 mechanics, (b) 10^6 mechanics, (c) 10^6 mechanics.
39. 1×10^{21} droplets.
41. 1×10^{-19}.
43. 9×10^6 automobiles.
45. 30 cm.
47. 1.2×10^{57} hydrogen atoms; 8.2×10^{56} hydrogen atoms;
49. $\approx$100 m.
51. Her position can be described by the clockwise angle ϕ from the north-south line drawn from the center to the starting point (south end of the lake). Her direction of travel will be tangent to the circle with a constant speed of 3 m/s at an angle ϕ clockwise to the south direction.
53. $(1.2\hat{\imath} + 8.2\hat{\jmath})$ paces (8.3 paces, 82° north of east).
55. Take the origin where the catch was made, with the x-axis to the right and the y-axis upfield. Catch: $0\hat{\imath} + 0\hat{\jmath}$; first turn: $0\hat{\imath} + 0\hat{\jmath} + 15\hat{\jmath} = 15\hat{\jmath}$; second turn: $0\hat{\imath} + 15\hat{\jmath} - 15\hat{\imath} = -15\hat{\imath} + 15\hat{\jmath}$; third turn: $-15\hat{\imath} + 15\hat{\jmath} + 10\hat{\jmath} = -15\hat{\imath} + 25\hat{\jmath}$; fourth turn: $-15\hat{\imath} + 25\hat{\jmath} + 20\hat{\imath} = 5\hat{\imath} + 25\hat{\jmath}$; fifth turn: $5\hat{\imath} + 25\hat{\jmath} - 5\hat{\imath} = 25\hat{\jmath}$; touchdown: $25\hat{\jmath} + 65\hat{\jmath} = 90\hat{\jmath}$.
57. $V_x = +V\cos\alpha$.
59. $\overrightarrow{AB} = (7, 0) = 7$ at an angle of 0°; $\overrightarrow{BC} = (-7, 7) = 7\sqrt{2}$ at an angle of 135°; $\overrightarrow{CA} = (0, -7) = 7$ at an angle of 270°.
61. (a) $\vec{A} = -4\hat{\imath} + 2\hat{\jmath}$, $\vec{B} = -\hat{\imath} + 4\hat{\jmath}$, $\vec{C} = 2\hat{\imath} + 2\hat{\jmath}$, $\vec{D} = 5\hat{\imath} - 3\hat{\jmath}$, (b) $-11\hat{\imath} + 9\hat{\jmath}$, 9.2

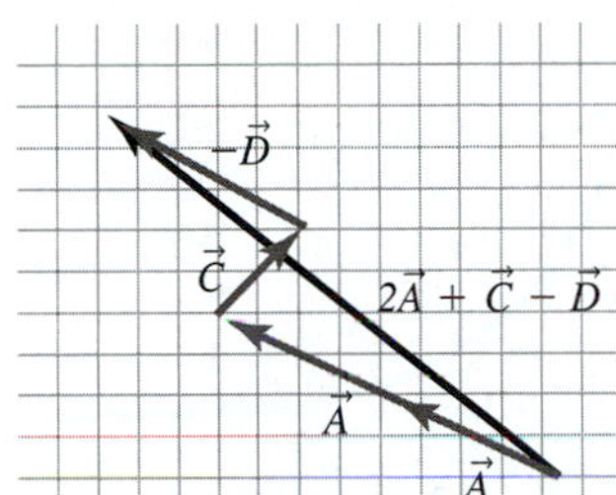

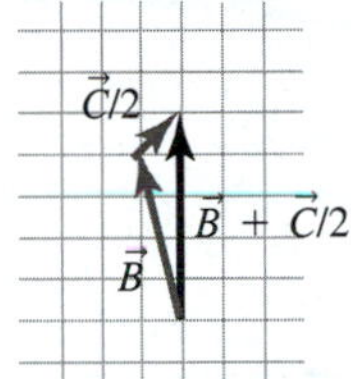

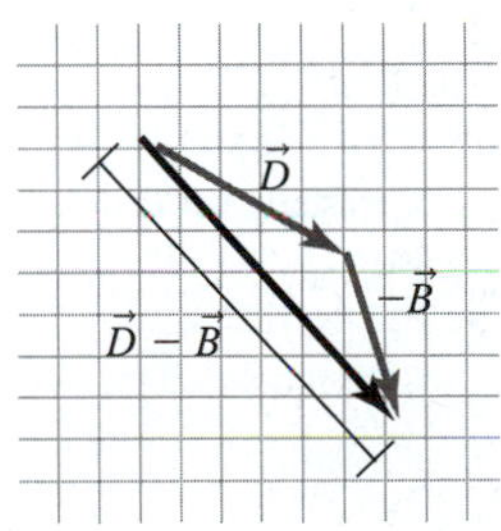

63. $A_x = A\sin(60° - \theta)/\sin 120°$; $A_y = A\sin\theta/\sin 120°$;

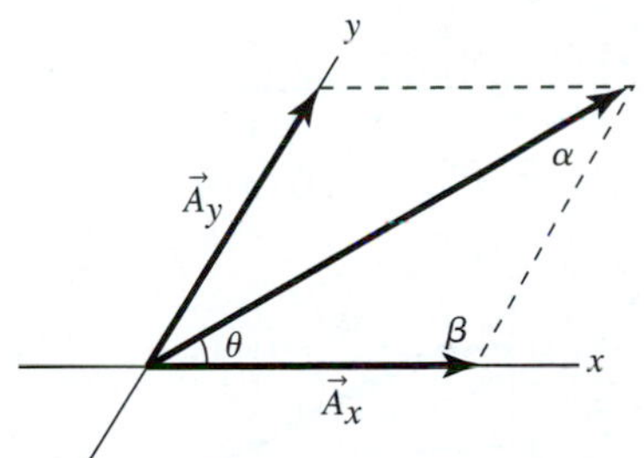

65. (a) 10^{-6} metric tons/g, (b) 10^{-6} m^3/cm^3.
67. 13 min.
69. 2×10^{27} molecules.
71. 3×10^{-8} cm.
73. 2×10^{41} kg/galaxy; 1.2×10^{68} H atoms.
77. 10^3 trucks/day; 2×10^3 trucks/day.
79. 1.1×10^8 km; 29 yr.
81. 10^{44} molecules.
83. (a) $\vec{v} = -v\sin\theta\,\hat{\imath} + v\cos\theta\,\hat{\jmath}$, or $\vec{v} = +v\sin\theta\,\hat{\imath} - v\cos\theta\,\hat{\jmath}$.

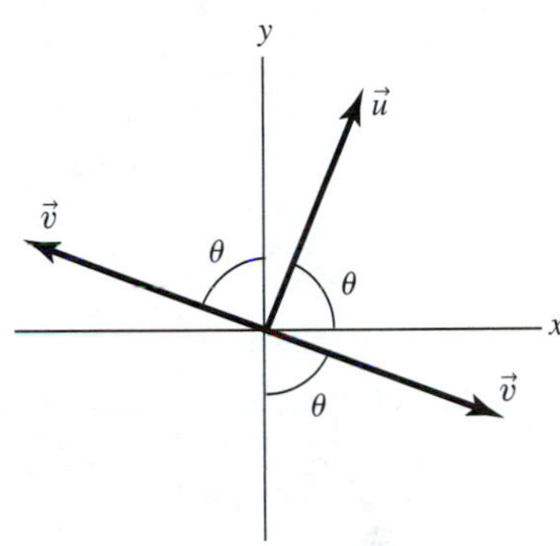

85. (a) $r\sin\theta\cos\phi$, (b) $r\cos\theta$, (c) $r\sin\theta\sin\phi$.
87. $t_0 = (\text{a constant})\ell(\lambda/t)^{1/2}$.

CHAPTER 2

1. +21 cm; 21 cm from the origin in the positive direction.
3. 252 m; 0;

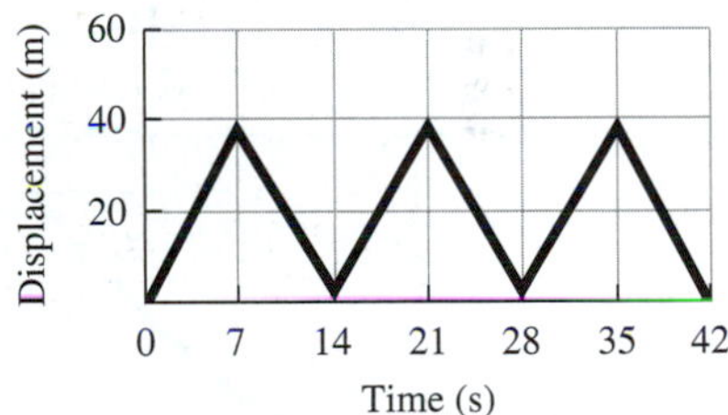

5.
(a)

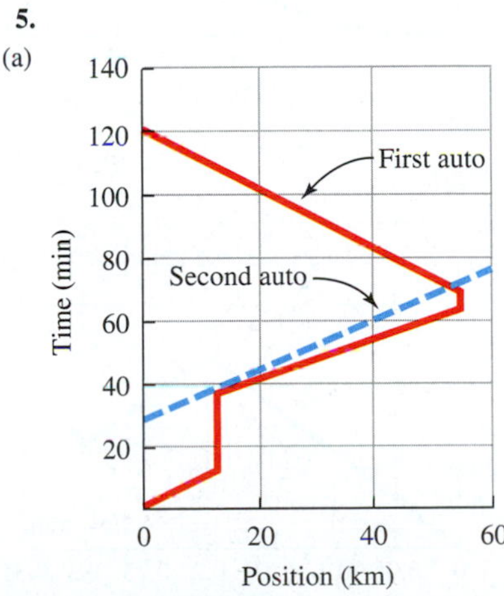

(b) 37 min and 15 km; 68 min and 53 km.

7.

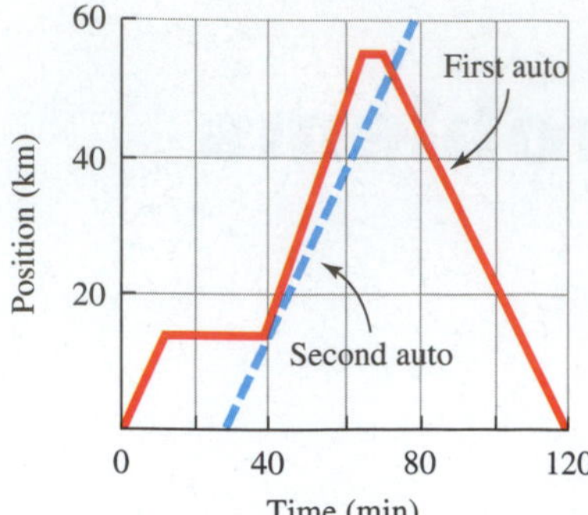

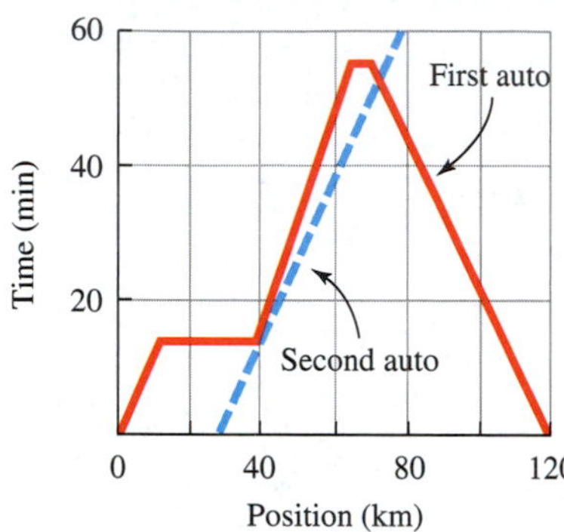

9. (a) $4.3\hat{j}$ mi; (b) $0.56\hat{j}$ mi/min.

11.

t (s)	v (m/s)	x (m)
0.0	0.00	0.00
0.5	0.75	0.19
1.5	1.75	1.44
2.5	8.75	6.69
3.5	21.75	21.94
4.5	39.75	52.69
5.5	62.75	103.94
6.5	90.75	180.69
7.5	122.75	287.44

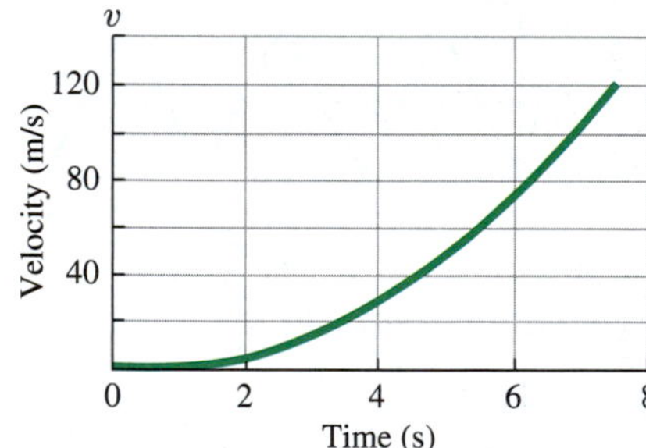

13. 80 km.

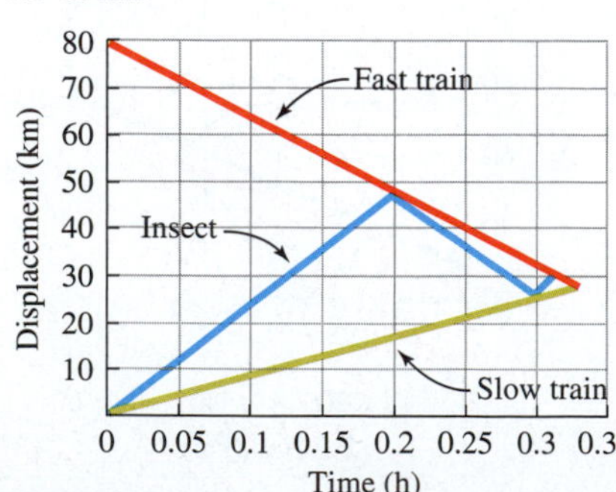

15. $1.4\hat{\imath}$ cm/s; $1.8\hat{\imath}$ cm/s; $3.5\hat{\imath}$ cm/s; $2.0\hat{\imath}$ cm/s.
17. 39 s.
19. $0.30g$.
21. $1.6\ \text{m/s}^2$.
23. $A\omega\cos(\omega t)$; $-A\omega^2\sin(\omega t)$.
25. The particle never gets farther from the origin than A; it oscillates back and forth through the origin. The magnitude of the velocity is maximum at the origin and zero at $x = \pm A$. The magnitude of the acceleration is maximum at $x = \pm A$ and zero at the origin.

27.

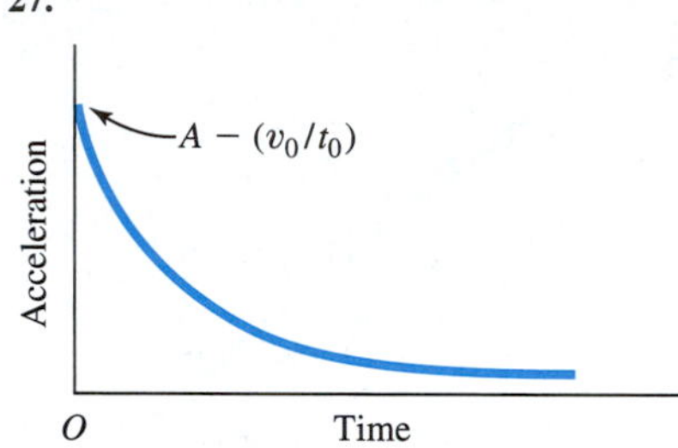

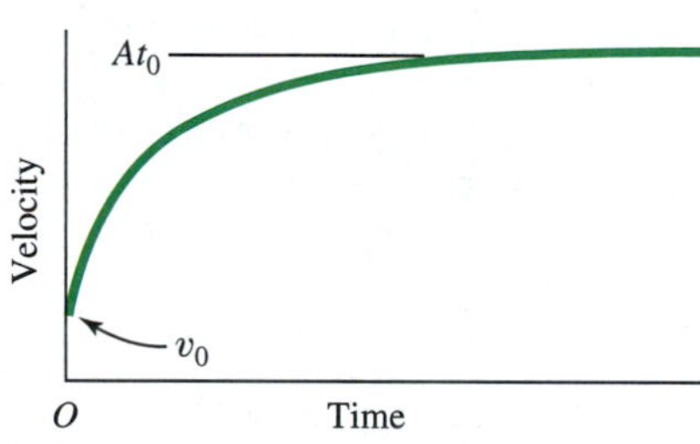

29. $2.02\ \text{ft/s}^2$ ($0.61\ \text{m/s}^2$).
31. 0.75 m/s.
33. (a) $8.0 - 0.5t$, easterly with t in s, v in m/s, (b) 34 m to the east.
35. $1 \times 10^4\ \text{m/s}^2$; 7×10^{-4} s.

37.
(a)

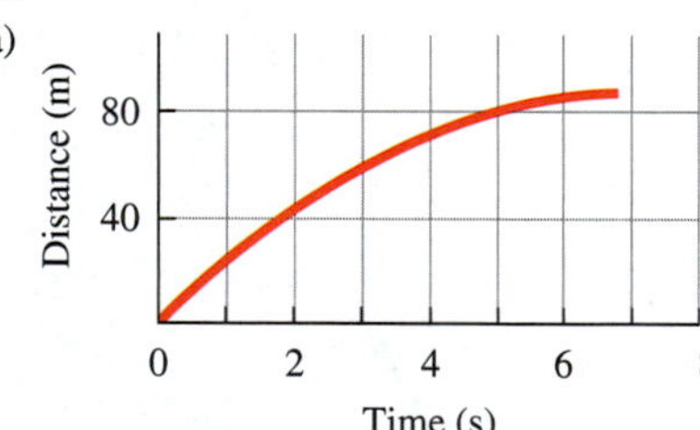

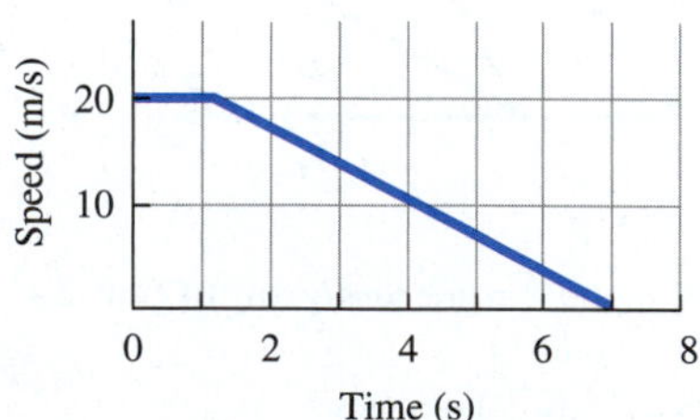

(b) $-3.6\ \text{m/s}^2$.
39. 1.92 s.
41. (a) $2.96\ \text{m/s}^2$, (b) 97.2 s.

43. (a)

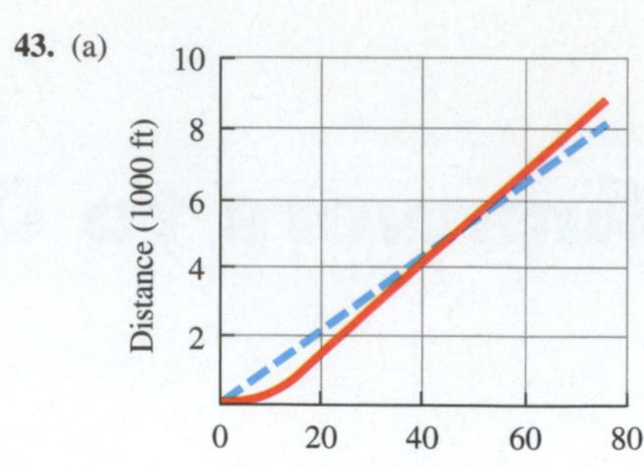

(b) 5940 ft (1.13 mi); (c) 54 s.
45. 3.65 s.
47. (a) 8.0×10^{-9} s; (b) $6.3 \times 10^{14}\ \text{m/s}^2$.
49. $4.0\ \text{m/s}^2$; 18 m.
51. 54 m.
53. $9.0 \times 10^5\ \text{m/s}^2$; 6.7×10^{-4} s.
55. Not much time to say anything!
57. 1.1 s.
59. 3.5 s.
61. -22 m/s; 25 m; 20 m/s.
63. 28.4 m.
65. $-4e^{-0.5t}$; $(2 - v/2)$, with t in s, v in m/s, and a in m/s^2.

67. (a)

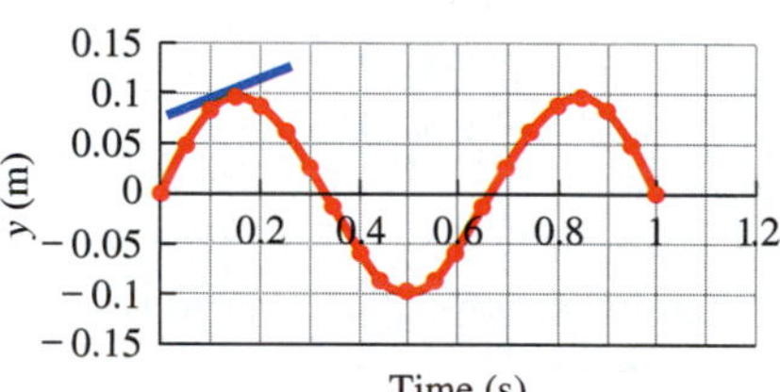

(b) 0.15 m/s, (c) $0.94\ \text{m/s}^2$,

(d)

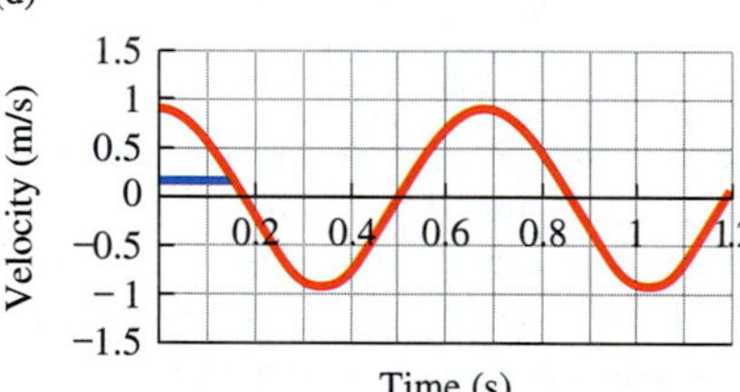

69. $(3v_f/\alpha)^{1/3}$.
71. 0.7 m; 3 m.
73. 12 m.
75. $1.6\ \text{m/s}^2$; 15 m.
77. $x_0 + v_0 t + (1/2)a_0 t^2 + (1/6)Jt^3$.
79. (a) $-(14\ \text{m/s})\hat{j}$, (b) $(13\ \text{m/s})\hat{j}$, (c) $(1.4 \times 10^4\ \text{m/s}^2)\hat{j}$
81. 1.6 m; 0.08 s.

CHAPTER 3

1. $(15\hat{\imath} + 15\hat{j})$ km; $(30\hat{\imath} + 15\hat{j})$ km; $(30\hat{\imath} + 43\hat{j})$ km; 52 km, 55° N of E.
3. $\vec{r}_A = 0$; $\vec{r}_B = 25\hat{\imath}$ m; $\vec{r}_C = (5\hat{\imath} + 35\hat{j})$ m; $\vec{r}_D = 35\hat{j}$ m.

5. $(1.3\hat{i} + 2.7\hat{j})$ km; $(0.1\hat{i} + 4.9\hat{j})$ km.

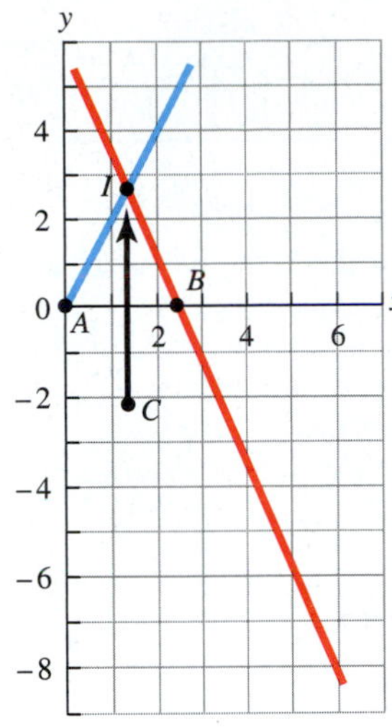

7. $(3.60 \text{ m/s})t\hat{i} + (6.00 \text{ m/s})t\hat{j}$, $0 \le t \le 41.6$ s; $[-(141 \text{ m}) + (7.00 \text{ m/s})t]\hat{i} + (250 \text{ m})\hat{j}$, $41.6 \text{ s} \le t \le 77.3$ s.

9. $(2.0\hat{i} - 3.5\hat{j})$ m; 4.0 m; $-4.0\hat{j}$ m; $d = 4.0$ m; $4.0\hat{i}$ m; $d = 4.0$ m; $\theta(t) = -\pi t/T$.

11. 4.0 m/s^2, opposite to the direction of the velocities.

13. $(3.7 \text{ m/s}^2)\hat{j} - (2.4 \text{ m/s}^2)t\hat{k}$.

15. 30 km/h, 31.6° north of west, $(-4.3\hat{i} + 2.7\hat{j})$ km.

17. (a) $[(0.000225 \text{ m}^2/\text{s}) - (0.0009 \text{ m}^2/\text{s}^2)t]\hat{j}/[(0.0169 \text{ m}^2) + (0.00045 \text{ m}^2/\text{s})t - (0.0009 \text{ m}^2/\text{s}^2)t^2]^{1/2}$, (b) $0.017\hat{j}$ m/s; 0.

19. $\vec{v} = [-(4\pi/T)\sin(\pi t/T)\hat{i} - (4\pi/T)\cos(\pi t/T)\hat{j}]$ m/s; $\phi = \pi/2 - \pi t/T$.

21.

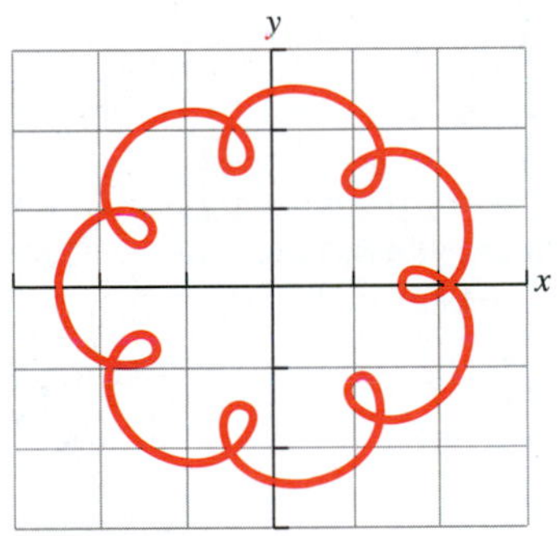

23. $[T^{-1}]$; $v_0 = 0$; $v \to -u$; $a_0 = -Bu$; $a \to 0$.

25. 12 m/s (horizontal); $h = 7.4$ m.

27. (a) 3.64 s, (b) 11.5 m, (c) − 14.7 m/s (down); + 14.7 m/s (up).

29. (a) 0.43 s, (b) 5.4 ft/s², to the left.

31. 5.2 m/s.

33. 187 m/s.

35. 25° below the horizontal.

39. 76°; 14°.

41. (a) 34 m/s, (b) 30 m, (c) 60 m.

43. 329 m.

45. (a) 1.4 s, (b) successful; 8.6 m above the bar.

47. 15° and 75°; 0° and 90°.

49. (a) 7.8×10^3 m/s, (b) 9.1 m/s^2 toward Earth's center.

51. 2.5 m/s^2.

53. 9.5 s.

55. 0.86 km.

57. (a) (0.15 m, −0.32 m), (b) $(-9.3 \times 10^2 \text{ m/s}^2)\hat{i}$, (c) $(-9.3 \times 10^2 \text{ m/s}^2)\hat{j}$.

59. $1.21R$.

61. 15.8 km/h, 18° south of east.

63. 50 km/h.

65. (a) 15 ft/s, (b) 6.7 ft/s.

67. (a) 843 km/h, (b) 781 km/h, 26° W of S, (c) 3900 km, 26° W of S.

69. (a) 29.5 km/s, (b) 30.5 km/s, (c) 30.0 km/s.

71. (a) 45 m/s at 65°, (b) 84 m.

73. $(10\cos\theta + 6)t\hat{i} + (10\sin\theta)t\hat{j}$, with r in km and t in hr; 127°; 68 s.

75. The hammer will hit the deck.

77. (a) 4.4 m/s, (b) 0.88 m going up; 0.88 m coming down; 0, (c) 1.8 m.

79. $\vec{r} = [(5.0 \text{ m/s})t - (0.36 \text{ m})\sin\theta]\hat{i} + [(0.36 \text{ m})(1 - \cos\theta)]\hat{j}$; $\vec{v} = (5.0 \text{ m/s})[(1 - \cos\theta)\hat{i} + \sin\theta\,\hat{j}]$; $\vec{a} = (69 \text{ m/s}^2)(\sin\theta\,\hat{i} + \cos\theta\,\hat{j})$.

CHAPTER 4

1. (a) Force of gravity (toward Earth), (b) force of gravity (down), normal force from the ice (up); and a small friction force from the ice (opposite to the motion), (c) essentially none.

3. 600 N in the $-y$-direction.

5. Yes.

7. (a) 3.0×10^{-4} N, (b) 3.0×10^{-4} N.

9. (a) The acceleration opposite to the motion is due to a retarding force. (b) The observer sees the car (initially at rest) move backward with increasing speed until it reaches $\vec{v}_0$. She would say that this is due to a backward force from the wind, etc.

11.

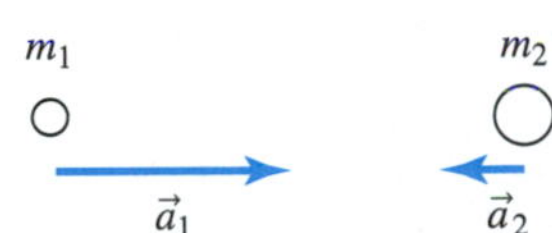

13. $-5.7 \times 10^5 \text{ m/s}^2$; -1.1×10^3 N.

15. $(F/m)\hat{j}$; no change.

17. $(0.91 \text{ N})\hat{i} + (2.3 \text{ N})\hat{j}$.

19. 6.5×10^2 N.

21. $T = F$; $0.433F$.

23. 78 N.

25. $1.6 \times 10^{-21} \text{ m/s}^2$.

27. (a) Earth, (b) the ice and Earth, (c) none.

29. (a) half, (b) twice.

31. (a) 0.14 m/s^2, (c) 1.2×10^4 N backward.

33. $(\hat{i} + \hat{j} + 2\hat{k})$ N.

35. Yes; yes; yes; 0.70 m/s^2; not real.

37. (a) Parallel to the window edge, (b) at an angle given by $\tan\theta = a'/g$, (c) parallel to the window edge.

39.

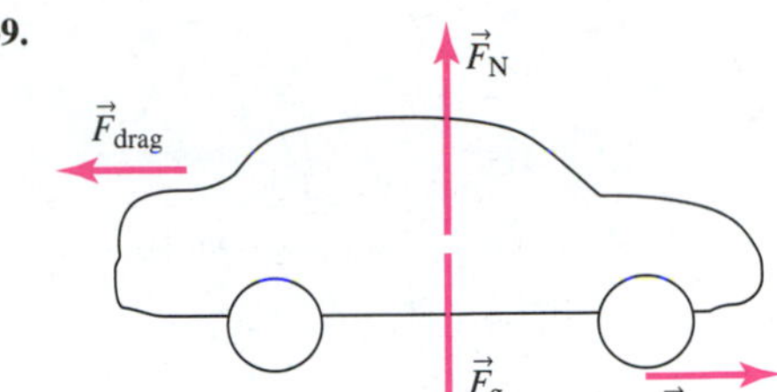

41.

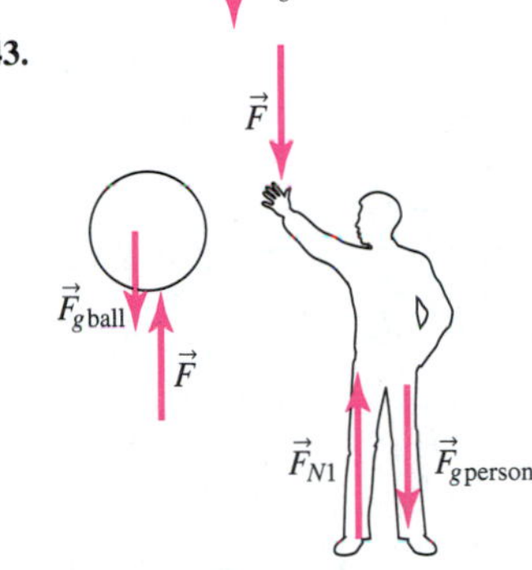

43.

$\vec{F}$

$\vec{F}_{g\,ball}$

$\vec{F}$

$\vec{F}_{N1}$

$\vec{F}_{g\,person}$

(a)

$\vec{F}_{g\,ball}$

$\vec{F}_{N2}$

$\vec{F}_{g\,person}$

(b)

45.

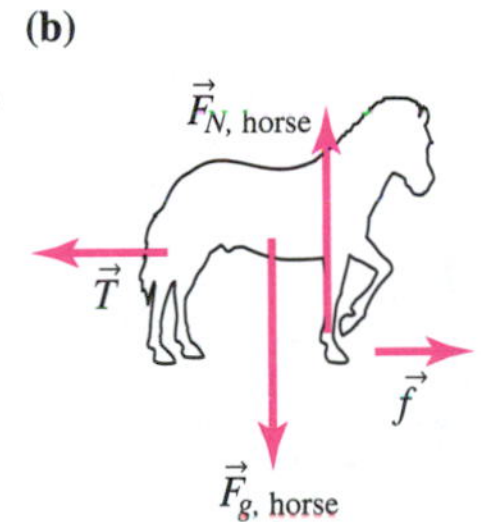

(a)

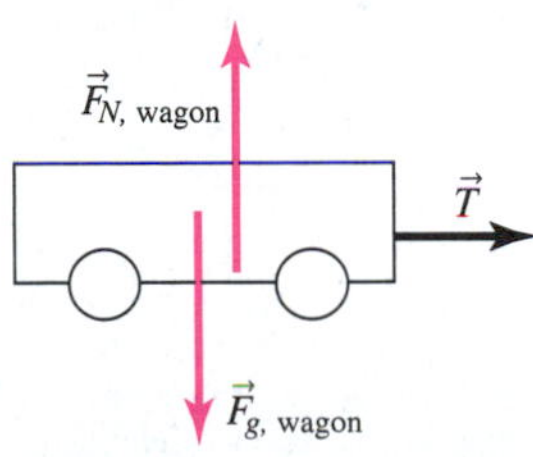

(b)

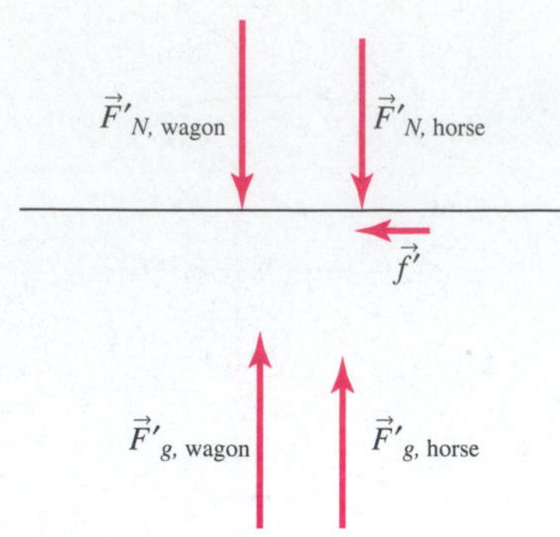

(c)

(d) a forward force from Earth on the horse's hooves.

47.

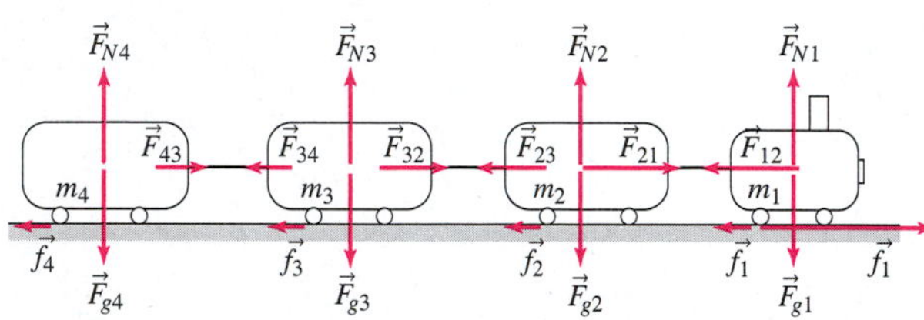

49.

y x $\vec{F}_N$ θ $\vec{f}$ θ $\vec{F}_g$

$\Sigma F_x = F_g \sin 21° - f = ma$ (x-direction);
$\Sigma F_y = F_N - F_g \cos 21° = 0$ (y-direction).

51. There is a net horizontal force from the wall.

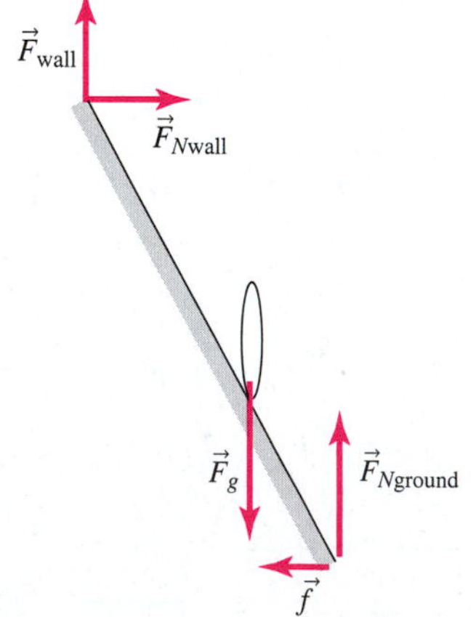

53.

$\vec{F}_e$ $\vec{F}_g$

55. (a) $F_g = mg$ down; $F_N = F_g = mg$ up, (b) mg down and $2mg$ up, (c) mg down.

57. $(1/M)(F_1^2 + F_2^2)^{1/2}$; $\tan^{-1}(F_2/F_1)$.

59. $(-1.1 \times 10^{-2} t^{-5/4})$ N, t in seconds.

61. (a) $c = (k/m)^{1/2}$, (b) $A = 0$; $B = +v_0(m/k)^{1/2}$.

63. (a)

y $\vec{T}_1$ $\vec{T}_2$ 65° 65° x $M\vec{g}$

(b) 82.2 N up, (c) 1.55 m/s² up.

65. $T = (m_W g + m_B g)/4$; $a = [4T - (m_W g + m_B g)]/(m_W + m_B)$.

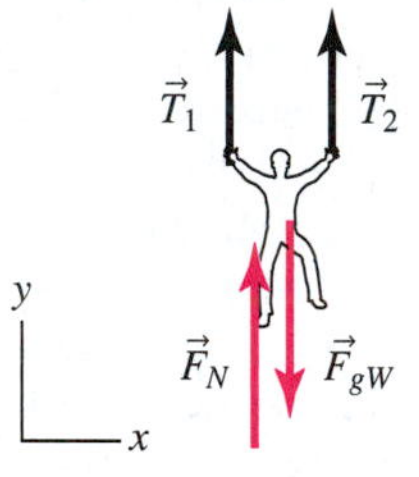

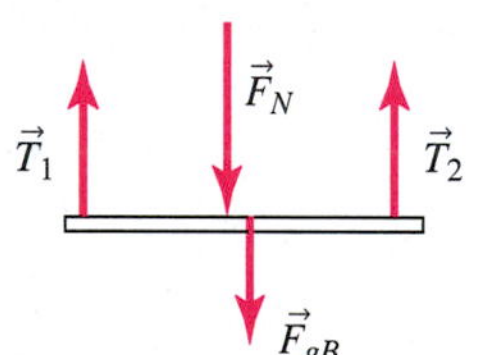

67. (a) Backwards; 3.0×10^4 N, (b) 1.6×10^4 N; 0.18 m/s², (c) 1.1×10^4 N from the first car; 5.2×10^3 N from the third; 0.18 m/s².

69.

(a)

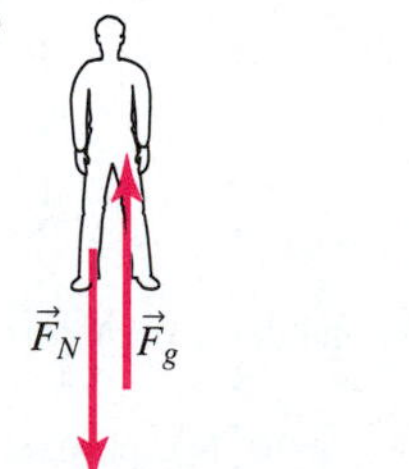

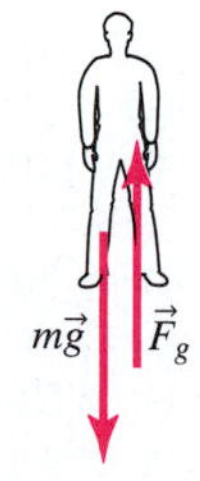

(b) same forces, but the upward force would be fictitious.

71. (a) $[A] = [ML^{-1}]$ with units of kg/m, (b) $(Av^2 - mg)/m = (Av^2/m) - g$, (c) $v_t = (mg/A)^{1/2}$.

73.

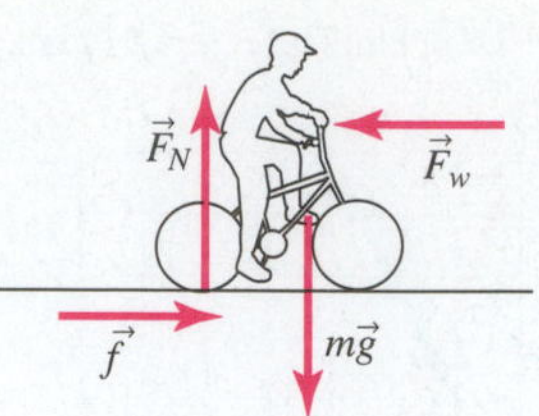

(a) force of gravity: mg (down); normal force of ground: F_N (up); friction force of ground: f (forward); wind resistance: F_w (backward), (b) 0, (c) 0; $v_w + v$; 0, (d) $v + v_w$; 0; 0.

75. (a) 0.40 m/s²; 0.20 m/s², (b) 0.067 m/s²; 0.13 m/s², (c) 8.6 g.

77. $\phi = \theta$.

CHAPTER 5

1. 1.4×10^4 N.
3. 980 N.
5. 0.095 s.
7. (a) 24 N, (b) 7.8 N.
9. (a) 66 N, (b) 5.1 m/s² up.
11. (a) 2.5 m/s² forward, (b) 3.5×10^3 N forward.
13. $M_{\text{max}} = 9.2$ kg; $M_{\text{min}} = 4.5$ kg; 0.
15. 9.8 mm/s²; 2.5 mm/s².
17. $a_1 = [(m_1m_2 + m_1m_3 - 4m_2m_3)/(m_1m_2 + m_1m_3 + 4m_2m_3)]g$; $a_2 = [(m_1m_2 - 3m_1m_3 + 4m_2m_3)/(m_1m_2 + m_1m_3 + 4m_2m_3)]g$; $a_3 = [(-3m_1m_2 + m_1m_3 + 4m_2m_3)/(m_1m_2 + m_1m_3 + 4m_2m_3)]g$; $T_1 = [8m_1m_2m_3/(m_1m_2 + m_1m_3 + 4m_2m_3)]g$; $T_2 = [4m_1m_2m_3/T_2 = [4m_1m_2m_3/(m_1m_2 + m_1m_3 + 4m_2m_3)]g$. If $m_2 = m_3 \neq m_1$, then $a_1 = [(m_1 - 2m_2)/(m_1 + 2m_2)]g$; $a_2 = a_3 = [(2m_2 - m_1)/(m_1 + 2m_2)]g$; $T_1 = [4m_1m_2/(m_1 + 2m_2)]g$; $T_2 = [2m_1m_2/(m_1 + 2m_2)]g$.
19. 48.0 N; 96.0 N; 2.2 m/s² (up); −0.2 m/s² (down); −1.8 m/s² (down).
21. 0.12.
23. 0.61.
25. 3.6×10^2 N.
27. 0.40.
29. (a) 52 m, (b) 3.9 s.
31. (a) 1.8 s, (b) 2.6 N, (c) 0.65.
33. (a) $370/(\cos\theta + 0.75 \sin\theta)$ N, (b) 37°; 2.9×10^2 N.
35. 0.077.
37. 9.5×10^2 N.
39. 0.060 m².
41. 0.57 m/s.
43. 0.67.
45. 5 m/s.
47. 1.3 m/s² toward the center.
49. 1.4×10^2 N toward the center.
51. $v_{\text{max}} = 32$ m/s, $v_{\text{min}} = 5.8$ m/s.
53. 9.5 cm.
55. 0.26.
57. 1.0×10^3 m/s tangent to the orbit.

59. 4.43 m/s.

61. (a)

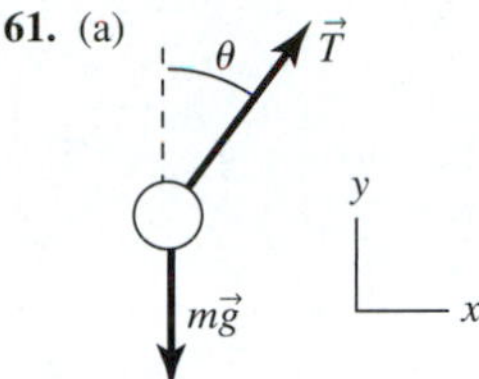

(b) 0.4 m/s.

63. 4.5 m.

65. 0.72 m.

67. 55 m/s.

69. (a) 21 N, (b) 18 N.

71. 1.6×10^{-12} kg/m^3.

73. 1.2×10^2 N.

75. (a) 0.5 N, (b) 1.2 N.

77. (a) $\Delta x_2 = -\Delta x_1/2$, (b) $a_1 = -1.78$ m/s^2 (down); $a_2 = +0.89$ m/s^2 (up), (c) 9.6 N.

79. -0.3 m/s^2, towards mass-1.

81. $\omega = (g/\mu_s R)^{1/2}$.

83. $\omega = [g/(\ell \cos\theta)]^{1/2}$.

85. 26 m.

87. 11 m/s.

89. 2.3 m.

CHAPTER 6

1. (a) 39 J, (b) 1.0 m/s; 89 m/s, (c) 0.40 m/s.

3. (a) The force of gravity = 98 N (down); the upward pull = 98 N; $F_{net} = 0$, (b) 0, (c) 98 J.

5. 1.0×10^4 J.

7. (a) 2.4×10^2 J, (b) friction force, (c) 0.

9. 730 J; -730 J.

11. -3.73 J.

13. 1.8×10^3 J.

15. 2.8×10^{11} J.

17. 2.5 m; $(0.916)^n(3.00$ m$)$.

19. 0.127 J; 0.159 J; 0.175 J.

21. $-mgR(1 - \cos\theta)$.

23. +3.

25. 0.89 m/s.

27. 2.53.

29. 3.9×10^3 J.

31. 6.1×10^3 J.

33.

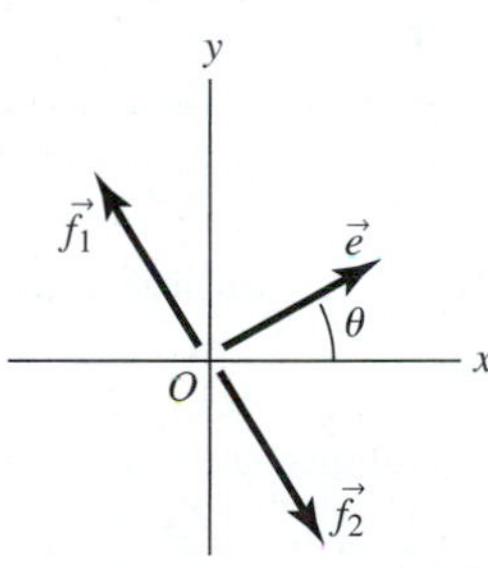

$-\sin\theta\,\hat{i} + \cos\theta\,\hat{j}$; $\sin\theta\,\hat{i} - \cos\theta\,\hat{j}$.

35. 3.4.

37. -24 J.

39. $F_x L$.

41. 0.067 J.

43. $2g_1 - 4g_2$.

45. 8.6 m/s.

47. 8.1×10^{-2} J.

49. 14 J.

51. 2.1×10^3 J.

53. -15 J, -5 J; conservative.

55. (a) 6.3×10^{-2} J, (b) 0, (c) 5.9×10^{-2} J, (d) -5.9×10^{-2} J.

57. (a) 4 J; 6 J; 6 J; 4 J, (b) +5 m, (c) conservative.

59. Constant forces are included; not a function of position only.

61. (a) $0.16C$ J; 0, (b) $0.32C$ J.

63. 22 bulbs.

65. 0.7 kW.

67. 52 kW.

69. 5.0 m/s; 1.3 m/s.

71. 4.7×10^7 W.

73. (a) 16 m/s, (b) 0.5 s, (c) 12 s.

75. 5.8×10^{-12} J; 5.8×10^{-9} J.

77. 0.62; 0.31.

79. -0.2 J.

81. 0.034.

83. (a) 2.2×10^7 J, (b) -2.2×10^{-7} J.

85. (a) $mg = 26$ N down; $F_N = 22$ N perpendicular to plane (up); $f_k = 5.4$ N parallel to plane (down), (b) -18 J; 0; -7.0 J, (c) 4.4 m/s.

87. 5.3 J.

89. (a) 0, (b) -0.20 J, (c) 9.8 J.

91. $mgH(1 + \mu_k \cot\theta)$.

93. $W_g = mgL(\cos\theta_f - \cos\theta_i)$; $v = [2gL(1 - \cos\theta_i)]^{1/2}$; $W_T = 0$.

95. $0.01mK/x$.

CHAPTER 7

1. (a) 0.20 kJ, (b) 29 J; 2.3 kJ.

3. 25 m.

5. 1.2×10^{-2} J.

7. (a) $-8x$ J, with x in m, (b) +18 J, (c) 4.1 m/s.

9.

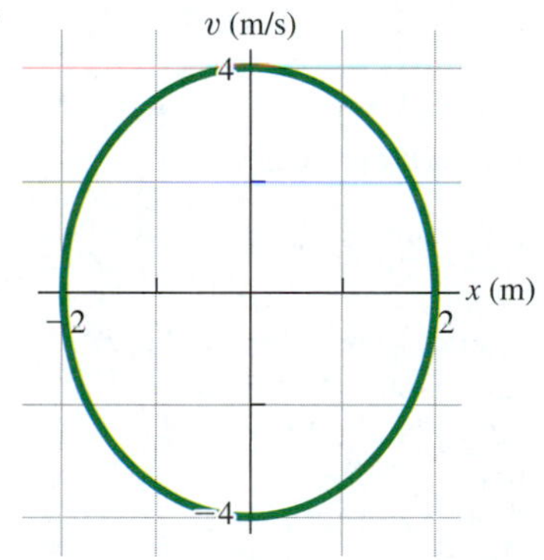

11. 0.541 m; 0.468 m.

13. 0.43 m; 0.48 m.

15. (a) 7.1 m/s, (b) -0.25 J, (c) 0.73.

17. $mg_0 y - (1/2)mg'y^2$.

19.

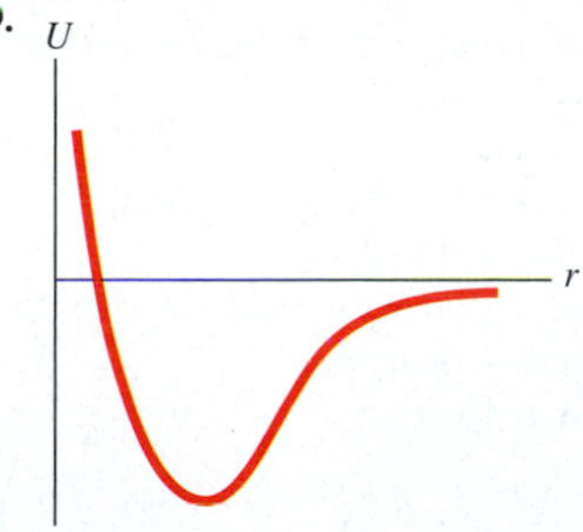

21. $E/(0.8$ N$)$, $-E/(2.5$ N$)$.

23. (a) To the right, (b) $(k/m)^{1/2}x$, (c) to the left; $-(k/m)^{1/2}x$.

25.

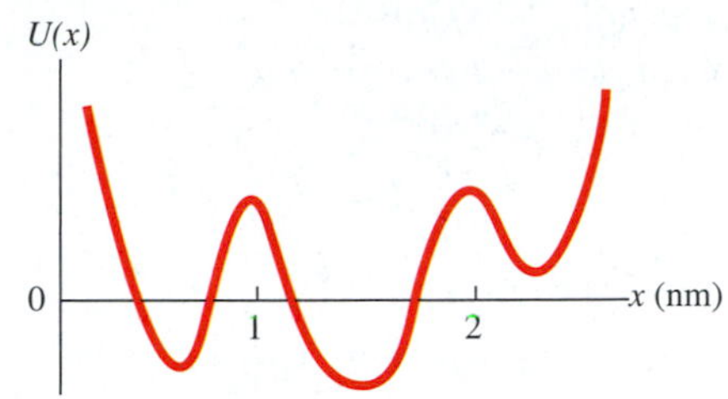

27. $x = 0$ is unstable and $x = \pm 0.24$ m is stable.

29. 503 m/s.

31. 130 m/s; same.

33. (b) $(2gH)^{1/2}$, (c) $2(gH)^{1/2}$.

37. -5.0×10^4 J.

39. (a) 1.4 m/s, (b) 1.1 m/s, (c) 8 cm from point a.

41. (a) $U(x) = \frac{1}{2}k\left(\sqrt{h^2 + x^2} - L\right)^2$.

(b) $U(x) = -\dfrac{k\left(\sqrt{h^2 + x^2} - L\right)x}{\sqrt{h^2 + x^2}}$.

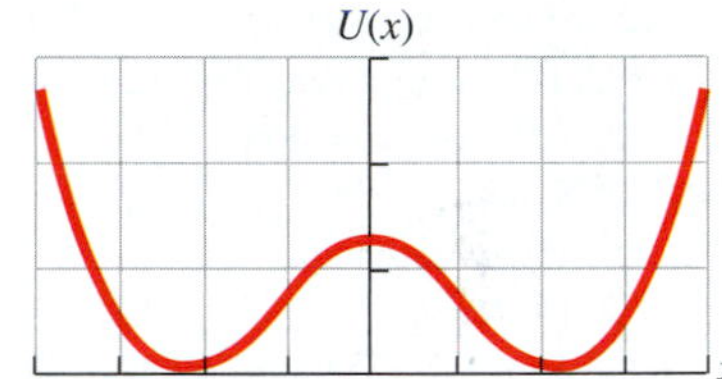

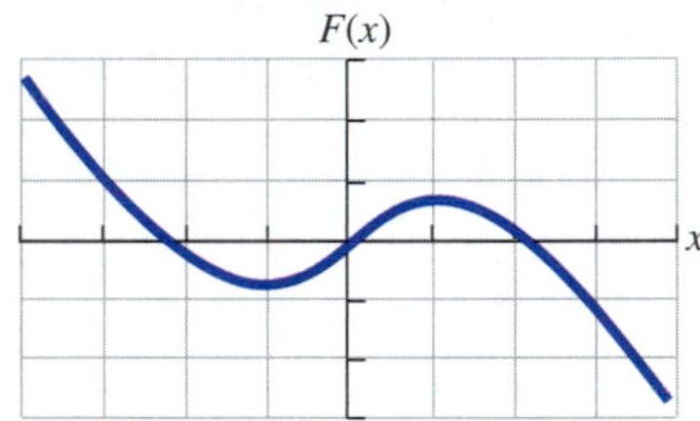

43. (a) GMm/r^2 toward M, (b) $GMm/2r$, (c) $-GMm/2r$.

47. -6.2×10^4 J.

49. ≈ 20 cents.

51. 5.9×10^2 J.

53. (a) $(3 + 3x^2/2 - 0.2x^3/3)$ J, with x in m.

(b) U(J)

25 20 15 10 5

0 1 2 3 4 x (m)

55. (a) Conservative, (b) not conservative.

57. $-A/3$; no.

59. -2 J; -2 J; $+2$ J.

61. (a) 7.7 m/s; 14.7 m/s (b) the skier leaves the surface.

63. (a) 4.4 m/s, (b) 3.7 m/s, (c) 5.9 N; 4.2 N.

65. (a) Conservative force, (b) $-C/r$, (c) 2.1×10^7 m/s.

67. (a) +1.13 J, (b) −0.083 J, (c) −0.30 J, (d) 0.70 m/s, (e) 8.8 cm, (f) 5.0 cm.
69. 1.1 m/s.
73. (a) $(1/2)mv^2 + (1/2)kr^2$, (b) $R(k/m)^{1/2}$, (c) $(2E/mr^2 - k/m)^{1/2}$.
75. $L[\sin\alpha + 2\sin\alpha\cos^2\alpha + 2(\cos^3\alpha - \cos^6\alpha)^{1/2}]$.

CHAPTER 8

1. (a) 1.2 kg · m/s, (b) 5.1 kg · m/s, (c) 7.0×10^2 kg · m/s, (d) 7.6×10^2 kg · m/s.
3. (a) 4.2×10^2 kg · m/s, (b) 6.0×10^6 kg · m/s, (c) 1.2×10^4 kg · m/s, (d) 3.3×10^{-22} kg · m/s, (e) 1.4×10^{-2} kg · m/s.
5. (a) −1.0 m/s, (b) 3.5 J.
7. (a) $(-1.9\hat{\imath} - 10.8\hat{\jmath})$ kg · m/s, (b) $(-4.9\hat{\imath} - 6.8\hat{\jmath})$ m/s, (c) $(-0.90\hat{\imath} - 2.9\hat{\jmath} - 1.2\hat{k})$ m/s, (d) 24 J; 64 J; 19 J.
9. 40 m/s.
11. −12 kg · m/s, opposite to the original motion; 1.7×10^4 N, opposite to the original motion.
13. -1.2×10^3 kg · m/s, down; 1.3×10^4 N, up.
15. 2.2×10^2 kg · m/s, up.
17. 7.5×10^2 N, up.
19. (a) 5.7×10^2 N, (b) 1.4×10^2 J, (c) 0.011 s.
21. $0.60\sqrt{g(0.20 + 0.10N)}$ kg · m/s.
23. −1.3 kg · m/s, down; −0.65 kg · m/s, down.
25. −0.13 m/s, opposite to the direction of the first cart.
27. 0.74 m/s.
29. $m_1/m_2 = 3$.
31. 4×10^{-9}.
33. (a) $M/(m + M)$, (b) $[(m + M)/m]\sqrt{2gh}$.
35. (a) $v[m_2^2 + M(m_1 + m_2)]/M(m_2 + M)$ forward, (b) $v[m_1^2 + M(m_1 + m_2)]/M(m_1 + M)$ forward, (c) $v(m_1 + m_2)/M$ forward.
37. (a) −0.20 m/s, (b) inelastic; 64%.
39. 2.59 m/s.
41. 15°.
43. (a) $\sqrt{2gh}$, (b) $\sqrt{2gh}$, (c) $[(m - 3M)/(m + M)]\sqrt{2gh}$, up, (d) $[(m - 3M)/(m + M)]^2h$, (e) $9h$.
45. 1.2 m/s 60° below the $-x$-axis, not elastic.
47. $(2.00\text{ m/s})\hat{\imath} + (1.00\text{ m/s})\hat{\jmath}$; elastic.
49. 2.6 m/s.
51. 8.7 m/s, 20° above the horizontal.
53. The driver of the car was speeding!
55. 0.78 m from the heavier sphere.
57. $X = 0.38$ m; $Y = -0.66$ m.
59. 0.
61. $X = 0$; $Y = 1.0$ m from the bottom of the handle.
65. $4R/3\pi$ from the center of the arc, along the bisector.
67. $X = Y = 0.6$ m; $X = Y = 0.5$ m.
71. -3.3×10^3 kg/s.
73. 1% of mass must be discarded.
75. 2.5×10^3 m/s.
77. 5.0 m.
79. (a) Spit out seeds one at a time, (b) 6×10^{-3} m/s.
81. (a) 7.9 m/s (to the right), (b) 7.9 m/s (to the right), (c) −1.9 m/s (to the left); +1.1 m/s (to the right), (d) 2.1×10^4 N (to the right).
83. (a)

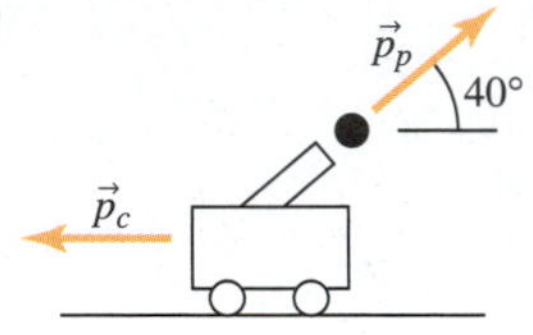

(b) 4.1 m/s recoil, (c) upward impulse provided by the ground.
85. (a) $[8M(M - m)/(M + m)^2]h$, (b) $-[4mM/(M + m)^2]h$.
87. 26 km/h.
89. $R/6$ from the center of the styrofoam sphere.
91. (a) 0.11 m/s, (b) 0, (c) −0.074 m, (d) 0.

CHAPTER 9

1. 0.68 rad/s^2.
3. −17 rad/s^2.
5. 7.3×10^{-5} rad/s from the South Pole to the North Pole; 2.0×10^{-7} rad/s perpendicular to the orbital plane.
7. (a) -2.62×10^2 rad/s^2, (b) 1.67×10^2 m.
9. (a) 0.10 rad/s^2, (b) 11 rev, (c) 2.1×10^2 m.
11. (a) $a = 1.5$ m/s^2; $x = x_0 + (0.75\text{ m/s}^2)t^2$, (b) 100 rad/s^2, (c) $\omega = (100\text{ rad/s}^2)t$.
13. 1.35×10^{-4} m.
15. α is linearly proportional to t.
19. 9.
21. 2.8×10^{-3} kg · m^2.
23. 1.3×10^2 J; 4.0×10^2 J.
25. (a) 0.021 kg · m^2, (b) 0.083 kg · m^2.
27. $I_{\text{Earth}} = 9.8 \times 10^{37}$ kg · m^2; $I_{\text{Neutron star}} = 1.0 \times 10^{39}$ kg · m^2 $\approx 10\, I_{\text{Earth}}$.
29. $M(R_1^2 + R_1^2)/2$.
31. $(3/10)(\tan^2\alpha)MH^2$.
33. $(8/15)\pi[(\rho_1 - \rho_2)R_1^5 + \rho_2 R^5]$.
35. $(16/45)MR^2$.
37. 24 lb.
39. 98 kg.
41. $L(F_{1y} + F_{2y})/2$ perpendicular to the rod.
43. (a) 2.3 N · m, (b) change the 7-N force to 5.4 N.
45. 0.38 N · m.
47. (a) 2.13 kg · m^2, (b) 8.84 kg · m^2/s, (c) 0.886 rev, (d) 18.3 J.
49. 3.8 rad/s up; rotate in the direction of the original rotation of the wheel.
51. -2.46×10^{-5}.
53. 2 s.
55. 0.12.
57. 13.4 m; 17.9 m.
59. 0.95 m.
61. 0.050 m/s^2.
63. 2.3 rev.
65. 5.1×10^6 m.
67. $(3v_0^2/4g)\sin\theta$.
69. m_1R_1/R_2; no.
71. (a) 0.72 kg · m^2/s, (b) 3.8 J, (c) 7.6 J.
73. $3\mu MgR/(R + 2r)$.
75. $MgR^2/(R^2 + 2r^2)$.

CHAPTER 10

1. (a) 1.1×10^{10} kg · m^2/s down, (b) 7.9×10^9 kg · m^2/s down.
3. (a) 5×10^4 kg · m^2/s down, (b) 5×10^4 kg · m^2/s down.
5. 1.5 kg · m^2/s north.
7. $-7.4t^2\hat{k}$ kg · m^2/s (into the page).
9. $(mt^2/2)(bv\hat{\imath} - aw\hat{\jmath} - av\hat{k})$.
11. $\vec{r} \times [m_2m_1/(m_1 + m_2)]\, d\vec{r}/dt = \vec{r} \times \mu\, d\vec{r}/dt$.
13. (a) $6.4 \times 10^{-2}\,\hat{k}$ kg · m^2/s (along ω-direction), (b) same.
15. (a) $m\omega d^2$ along the axis of rotation, (b) $3m\omega d^2/4$ along the axis of rotation, (c) $m\omega d^2/2$ along the axis of rotation.
17. $(17\hat{\imath} - 19\hat{\jmath} + 14\hat{k})$ N · m.
19. $140\hat{k}$ N · m (perpendicular to table).
21. $MgR\sin(\omega t)$.
23. $mvd\sin(2\theta)$ up; 0; impulsive force from the wall.
25. $-7\hat{\imath} + 9\hat{\jmath} + 10\hat{k}$.
27. 1.1 rad/s.
29. 30 m/s.
31. 0.103 rad (0.016 rev).
33. 0.15 rad/s.
35. 5.3×10^2 N · m along the axis.
37. 409 rad/s.
39. 0.14 J.
41. $r_n = (n^2\hbar^2/mk)^{1/4}$, $n = 1, 2, \ldots$; $v_n = (n^2\hbar^2k/m^3)^{1/4}$, $n = 1, 2, \ldots$; $K_n = \sqrt{k/m}\, n\hbar/2$, $n = 1, 2, \ldots$.
43. No energy level 2.0 eV above the lowest state; possible excitation energies are 10.2 eV, 12.1 eV, 12.75 eV, ...
45. $\omega_p = \Delta\theta/\Delta t = Mg\ell/I\omega$ along the z-axis.
47. 2.8 rad/s.
49. 72 rad/s.
51. 0.62 rad/s; −33 J; friction.
53. 3.3 rad/s.
55. (a) 4 m/s, (b) 0.8 s.
57. h.
59. (a) 19.4 m/s, (b) 0.127, (c) 0.132.
61. 26 kg; 52 rad/s; 12 kg.
63. 4.0×10^3 J.
65. (a) K/L, (b) L^2/MK, (c) $2\pi L^3/MK^2$, (d) MK^3/L^4.
67. $K \approx 1 \times 10^{17}$ J; $L \approx 6 \times 10^{20}$ kg · m^2/s.
69. (a) 2.4 m/s, (b) 2.4×10^2 rad/s.

CHAPTER 11

1. 40 kg.
3. 139 N down; 96 N down.
5. (2, 0, 1) m.
7. 1.7 m.
9. 0.8 m.
11. 14.2 N · m in xy-plane 27° from $-x$-axis ($\perp \vec{r}$).
13. (a) $3L/4$; $11L/12$.
15. 0.57 Mg.
17. 1.2×10^2 N; $F_{N_A} = 0$.
19. (a) 2.9 m.
21. (a) 2.9×10^2 N 13.5° above the horizontal, (b) $F_V = 3.5 \times 10^2$ N down; $F_H = 2.8 \times 10^2$ N.
23. (a) $\tan^{-1} \mu_s$, (b) $\tan^{-1}(\Delta/2L)$.
25. 0.58 Mg outward.
27. 1.2 kN.
29. 1.9×10^2 N right; 73 N left; 23 N left.
31. (a) $F_{N_A} = mg\ell_2/2\ell_1$ down; $F_{N_B} = mg(\ell_2 - \ell_1)/2\ell_1$ up, (b) seat will lose contact at B and turn clockwise, (c) 74 N; 25 N.
33. (a) 56 N, (b) $(164\text{ N})\hat{\imath} + (46\text{ N})\hat{\jmath}$.
35. 353 kg.
37. (a) $F_{N_1} = 0$; $F_{N_3} = F_{N_2} = (M + m)g/2 - mgx/L$; $F_{N_4} = 2mgx/L$, (b) $F_{N_4} = 0; F_{N_1} = -2mgx/L$, $F_{N_3} = F_{N_2} = (M + m)g/2 + mgx/L$.
39. $MgR \sin\theta$.
41. 1.33×10^4 N.
43. (a) 0.255 nm.
45. 0.524; 0.740.
47. 2.97 cm.
49. $(\Delta L/L)_c = 0.014$; 14 mm.
51. 2.5×10^2 N/m^2.
53. $\Delta y \approx 10^{-13}$ m.
55. 0.12° below the horizontal.
57. 1.99928 cm.
61. 9.1×10^{-12} m^2/N.
63. 69 N.
65. (a) $(Mg/2)\cot\theta_0$, (b) $(3g\cos\theta_0)/2L$, (c) $\omega = \sqrt{(3g\sin\theta_0)/L}$.
67. 85 N.
69. 1.36×10^3 N.
71. (a) Near legs: $(1 + 2\mu_k)Mg/4$ up; $(1 + 2\mu_k)\mu_k\,Mg/4$ to the left; far legs: $(1 - 2\mu_k)Mg/4$ up; $(1 - 2\mu_k)\mu_k\,Mg/4$ to the left, (b) 0.
75. $\tan(\theta/2)$.
77. 6.88.
79. Top ball: mg, $0.577mg$, $1.155mg$; bottom ball: mg, $1.155mg$, $2.000mg$, $0.577mg$.

CHAPTER 12

1. 2.975×10^{-19} s^2/m^3.
3. (a) $[ML^4T^{-2}]$, (b) $\sqrt{mh}$, (c) $2\pi\sqrt{m/h}$ = a constant.
5. (a) 1.9×10^{-6} N, (b) must be the same.
7. 4.2×10^{-10} N; $\approx 10^{-7}\,W_{\text{fly}}$.
9. 6.6×10^6 m.
11. 1.51 h.
13. 9.56×10^6 m.
15. 4.9 m.
17. 0.25.
19. 6.18×10^5 m.
21. 26 m/s.
23. 1.6×10^{-3} rad/s.
25. 1.09×10^4 m/s.
27. (a) 1.85×10^8 m, (b) 7.99×10^8 m, (c) impossible.
29. 7.79 km/s tangent to the orbit; 1.52×10^{10} J; 2.56×10^{13} kg · m^2/s perpendicular to the orbit.
31. 1.74 h.
33. 1.49×10^{11} m.
35. 13 km/s.
37. 2.99 km/s.
39.
(a)

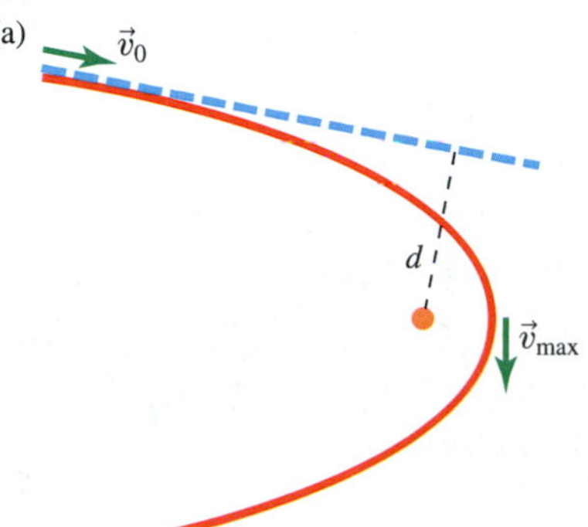

(b) 75.6 km/s perpendicular to the radius, (c) 4.76×10^7 km.
41. (a) 6.91×10^3 m/s, (b) 1.74×10^{13} kg · m^2/s, (c) 8.68×10^{12} kg · m^2/s, (d) the satellite crashes.
43. (a) 7.73×10^3 m/s, (b) 1.03×10^{14} kg · m^2/s, (c) -5.98×10^{10} J, (d) zero, (e) cannot be circular.
45. 0.996860; 0.996868.
47. $(2R^3/GM)^{1/2}$.
49. The tidal force on Jupiter is about 8% that on Earth.
51. 7.9×10^3 m/s.
53. 4.51×10^2 km from the Sun's center.
55. $\theta \approx 3 \times 10^{-10}$ rad; yes.
57. $4\pi(R^3/GM)^{1/2}$.
59. 3.5×10^{-15} m.
61. 0.40 rad/s.
63. (a) 1.66×10^3 m, (b) 2.8 m/s, (c) a speed of 1.76 m/s is needed to orbit the asteroid.
65. (a) Circular orbits are supported, (b) $T^2/r^{n+1} = 4\pi^2(m/k)$ = a constant.
67. $\Delta U \approx -4\pi fr$; $\Delta E = -2\pi fr$; $\Delta K \approx 2\pi fr$.
69. (a) 1.92 m/s relative to himself, which is 1.97 m/s relative to the ship, (b) 1.9 h.
71. (a) 1.68 m/s^2, (b) 1.95 m/s^2.
73. (a) Possible, (b) 1.83×10^7 s, (c) unstable.

CHAPTER 13

1. $\pi/2$ rad.
3.

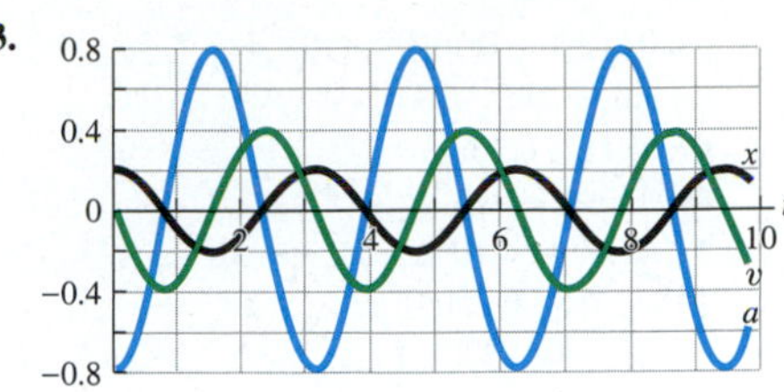

5. 4.09 s.
7. −0.038 m.
9. 7.0137×10^{-6} m
11. $x = A\cos[(2.0\text{ rad/s})t - 1.17\text{ rad}]$, t in s.
13. (a) 1.6 rad/s; 4.0 s, (b) 1.3 m.
15. (a) 4.08 cm/s, (b) 6.41 cm/s; 23.6 cm/s^2.
17. (a) $x = (3.4\text{ cm})\sin[(3.0\text{ rad/s})t + 2.1\text{ rad}]$, (b) 1.8 s, (c) 0; 0.71 s; 1.1 s; 1.8 s,
(d)

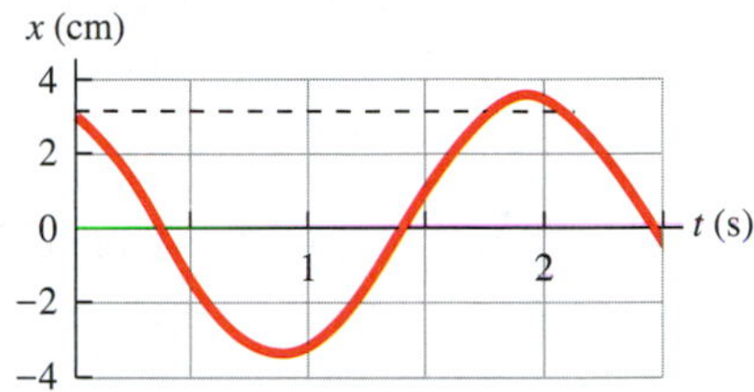

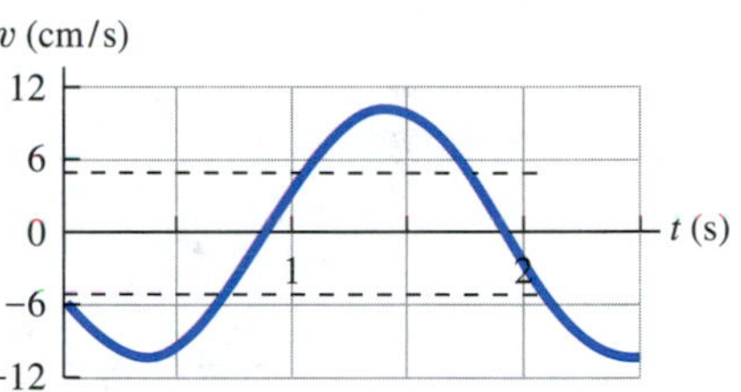

19. R.
21. $x = R\cos(\omega t + \delta)$, with $R = 1.5 \times 10^{11}$ m and $\omega = 1.99 \times 10^{-7}$ rad/s.
25. 2.4 s; 15 cm/s.
27. 5.1 kg.
29. 0.18 N/m.
31. 1.5×10^2 N.
33. 1.2 s.
35. $x = (1.0\text{ cm})\sin[(71\text{ rad/s})t + \pi/2]$.
37. 1.1 s.
39. 0.042 m/s.
41. 6.1 rad/s.
45. 2 m/s; 0.7 m; 0.3 J.
47. (a) $(\ell^3/R_E^3)M_E$, (b) $GM_Em\ell/R_E^3$ toward the center, (c) $-GM_Emx/R_E^3$ toward the center of the tunnel, (d) 1.4 h, (e) 1.4 h.
49. $F_{\text{net}} = -kx^3/2\ell^2$.
51. 9.79 m/s^2.
53. (a) 2.23 m, (b) 430 m.
55. (a) 1.7 s, (b) 0.05 m.
57. $E = mgL(1 - \cos\alpha) \approx mgL\alpha^2/2$ (for $|\alpha| \ll 1$).
59. 0.76 s.
61. $2\pi\sqrt{\dfrac{2(L^2 - 3Ly + 3y^2)}{3(L - 2y)g}}$.
63. (a) $MR^2/2 + M\ell^2$, (b) $-g\ell\theta = (R^2/2 + \ell^2)d^2\theta/dt^2$, (c) $2\pi\sqrt{\dfrac{R^2 + 3\ell^2}{2g\ell}}$, (d) $T \to \infty$; no torque.

65. (a) 2.7 s, (b) 1.0×10^{-4} J, (c) 3.6×10^{-2} m/s.
67. 0.113; 8.4 min.
69. 0.095 kg/s.
71. $k \approx 6 \times 10^4$ kg/s^2; $b \approx 10^3$ kg/s.
73. (2.0 + 0.00051)s.
75. $x_{3.0} = -5.40$ cm; $x_{4.8} = -1.48$ cm; $x_0 = -6.65$ cm.
79. 1.8 Hz.
81. (a) 2.2 N·s/m, (b) 0.25 s, (c) $\Delta\omega = 8.0$ rad/s; $Q = 3.1$.
83. 0.36 m.
85. $\sqrt{2}\,\omega$.
87. 2.5 s.
89. (a) $0.553H$, (b) $0.347H$.
91. (a) 1.2 s, (b) $\pi\sqrt{\frac{2m}{k}}$, (c) $\pi\sqrt{\frac{2m}{k}}$, (d) $2\pi\sqrt{\frac{m}{k}}$.
93. 9.0 s.
95. (a) $2A/e^2$, (b) $U = -e^4/4A + (e^4/4A)x^2$; $e^4/(4A^3m)^{1/2}$.
97. $\frac{1}{2\pi}\sqrt{\frac{g}{2R}}$.
99. $\frac{1}{2\pi}\sqrt{\frac{2k}{3m}}$.

CHAPTER 14

1.

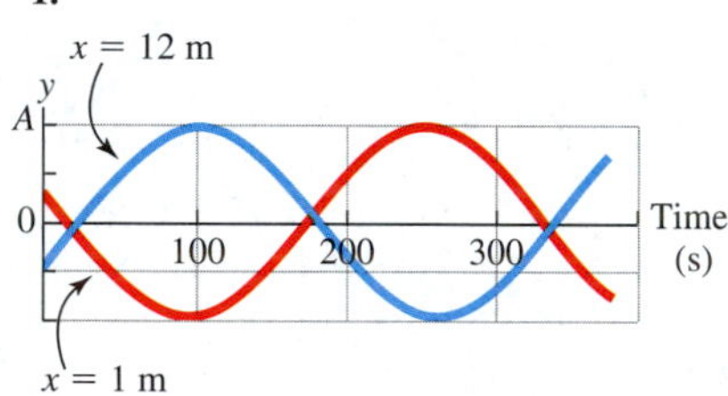

3. 200 m/s.
5. 23 m^{-1}; 48 Hz.
7. 5.1×10^3 m/s; 3.5×10^3 m/s.
9. 7.0×10^4 MN/m^2.
11. 8.03×10^{11} dyne/cm^2.
17. Yes.
19. $Y = 3.2 \times 10^5$ MN/m^2; $G = 8 \times 10^4$ MN/m^2; $L = 1.3 \times 10^4$ km.
21. $\mu\omega^2 R^2$.
23. 25 km/h.
25. $\ell(k/m)^{1/2}$.
27. 2.1×10^3 W; 5.3×10^2 W.
29. 8.9×10^{-10} m.
31. $\mu v \omega^2 z_0^2 \cos^2(kx + \omega t)$.
33. $P = \mu v\left(\frac{\omega}{k}\right)^2\left[\frac{df}{d(x - vt)}\right]^2$.
35. 1.08 m; 0.54 m; 0.36 m;

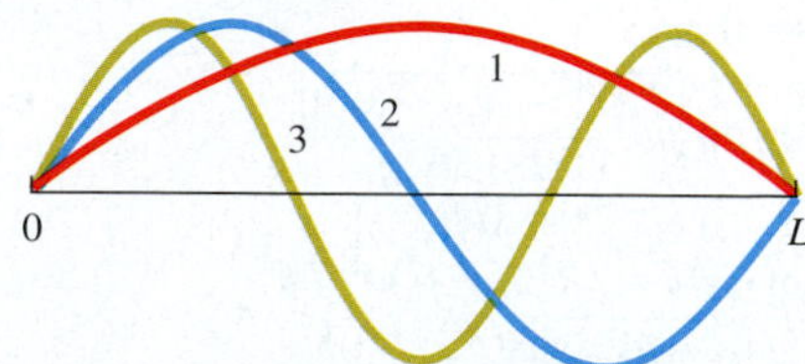

37. 13 N.
39. $m_2/m_1 = 1/9$.
41. 1.
43. 3.5 Hz, 22 rad/s, 0.28 s; 7.1 Hz, 44 rad/s, 0.14 s; 11 Hz, 67 rad/s, 0.094 s.
45. 8.49×10^2 N.
47. 1.28 kg/m^3; 0.085 kg/m^3.
49. 89 dB; 65 dB; 39 dB.
51. 6.3×10^{-2} W.
53. 1.97×10^9 N/m^2.
55. 343 m/s.
57. 147 Hz.
59. (a) 0.874 m; 378 Hz, (b) 372 Hz.
61. 1.5×10^7 m/s.
63. 1.88×10^3 Hz.
65. 440 Hz$/(1 + 0.060t)^{1/2}$.
67. 965 m/s (3.5×10^3 km/h).
69. 1.3 kHz.
71. $\lambda_n = 2.40/(2n - 1)$ m, where $n = 1, 2, 3, \ldots$.
73. 3.2×10^6.
75. 400 Hz; 1.1 m.
77. (a) 484 Hz; 543 Hz, (b) *D*; *B*.
79. 3.5×10^2 m/s.
81.

$$f_1 = \frac{1}{2L}\sqrt{\frac{T}{\mu} + \alpha\frac{\pi^2}{L^2}}; \quad f_2 = \frac{1}{L}\sqrt{\frac{T}{\mu} + 4\alpha\frac{\pi^2}{L^2}};$$

the higher frequency.

CHAPTER 15

1. $A = \sqrt{2}$; $\tan\phi = 1$, or $\phi = \pi/4$.
3. $1.41 z_0 \sin[kx - \omega t + (\pi/4)]$.
5. 3.6 cm.
7. $\delta = 2\pi/3$ and $-2\pi/3$.
9. $2z_0 \sin[kx + (\pi/4)]\cos[\omega t + (\pi/4)]$, which represents a standing wave; $x(\text{nodes}) = [n - (1/4)](\pi/k)$, where $n = 0, \pm1, \pm2, \ldots$.
11. (a) 11 Hz; 7.3 m/s; (b) $z_\ell = 3\sin(kx + \omega t)$; $z_r = 3\sin(kx - \omega t)$.
15. 6 m.
17. 506 Hz.
19. 422 Hz.
21. −1.6%.
23. $z = 2A\cos(\omega t - kL)$.
25. 0.59 cm.
27. (a) 37°; (b) 17.5°; (c) 37°.
31. (a) −1 cm, 0, +1 cm; (b) (6 cm) $\sin(\pi r - \pi v t)$, where $r = (y^2 + 4)^{1/2}$.
33. (d) Total energy is conserved.
35.

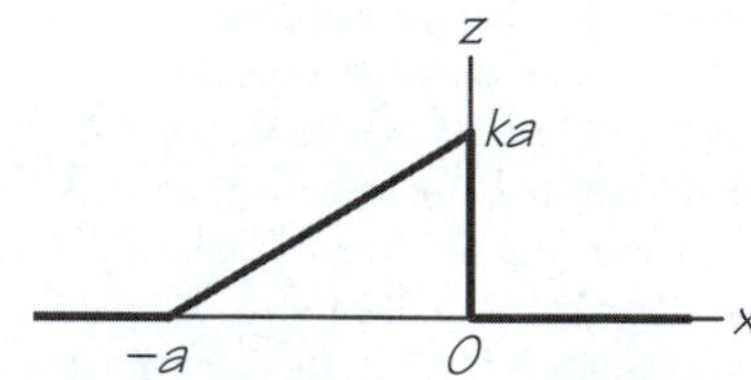

$z(x, t) = 0$ for $x < (vt - a)$ and $x > vt$; $z(x, t) = k(x - vt + a)$ for $(vt - a) < x < vt$.

37. $v_t = z_0\left[2(x - vt)\left(\frac{v}{\alpha^2}\right)\right]e^{-(x-vt)^2/\alpha^2}$;

$a_t = z_0\left[-\frac{2v^2}{\alpha^2} + 4(x - vt)^2\left(\frac{v^2}{\alpha^4}\right)\right]e^{-(x-vt)^2/\alpha^2}$;

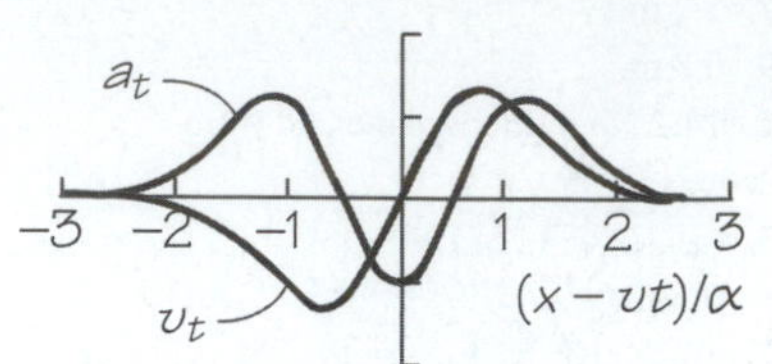

39.

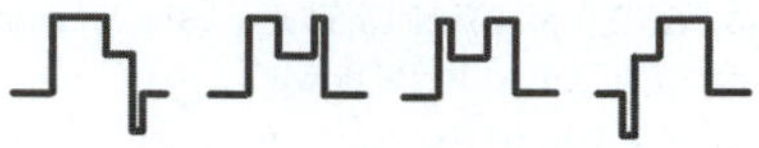

41. $-0.069A$; $1.07A$.
43. 6.6×10^{-3}.
45. $\frac{B}{A} = \frac{k_2 - k_1}{k_2 + k_1} = \frac{\sqrt{\mu_2} - \sqrt{\mu_1}}{\sqrt{\mu_2} + \sqrt{\mu_1}}$; $\frac{C}{A} = \frac{2k_2}{k_2 + k_1} = \frac{2\sqrt{\mu_1}}{\sqrt{\mu_2} + \sqrt{\mu_1}}$; $z_i = A\cos(k_1 x - \omega t)$; $z_r = B\cos(k_1 x + \omega t + \pi)$; $z_t = C\cos(k_2 x - \omega t)$.
47. 1 MHz.
49. (a) 400 Hz; 0.4 m; 160 m/s; (b) 7.7 cm.
51. $\Delta T/T = 9.1 \times 10^{-3}$.
53. $69\cos(4\pi t)$ Hz.
55. 90 Hz; 101 Hz; 111 Hz.
57. 6.0×10^{-7} m; 3.9×10^{-7} m.
59. (a) and (b) are linear.
61. $z(\text{envelope}) \simeq \sin[\Delta k(x/2) - \Delta\omega(t/2) + (\pi/4)]$; $f = (\omega_1 - \omega_2)/2\pi$; $\lambda = 2\pi/(k_1 - k_2)$.

CHAPTER 16

1. 1.15×10^3 kg/m^3.
3. 50 L.
5. $\Delta d_1 = af/\sqrt{2}$; $\Delta d_2 = -af/\sqrt{2}$;
7. 2.1×10^5 N/m^2.
9. (a) 5.1×10^4 N, (b) 1.3×10^2 N (equivalent to lifting 13 kg).
11. 7.8×10^2 N/m^2; p will decrease.
13. 1.11×10^5 Pa; 2.02×10^5 Pa; 1.11×10^6 Pa; 1.01×10^8 Pa.
15. 1.02×10^5 Pa; 1.05×10^5 Pa.
17. $h_{\text{Venus}} = 0.838$ m; $h_{\text{Neptune}} = 0.619$ m.
19. 52 N; 2.2 mm.
21. Sink.
23. 8.4×10^2 kg/m^3.
25. 9.3 m.
27. 0.71 m^3.
29. 13 m^2.
31. 0.96×10^3 kg/m^3.
33. $\rho_1 < \rho_3 < \rho_2$; $D/H = (\rho_3 - \rho_1)/(\rho_2 - \rho_1)$.
35. 8.1×10^2.

37.

(a)

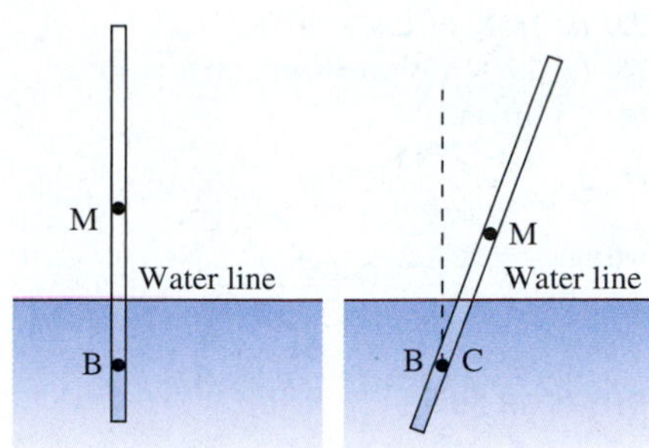

(b)

39. 2.9 m/s; 0.23 kg/s.
41. 57 mi/h.
43. 0.81 cm.
45. 1.6×10^2 W (0.22 hp).
47. -4.7×10^2 Pa (gauge); 4.7×10^2 Pa.
49. 9×10^{-2} N; rise.
51. (a) 7.7 m/s; (b) 4.3 h.
53. (a) 2 m/s; (b) 2.2×10^5 Pa (2.2 atm).
55. $p_0 + (1/2)\rho v_0^2 + \rho g h_0 = p_1 + (1/2)\rho v_1^2 + \rho g h_0$; $p_0 + (1/2)\rho v_0^2 + \rho g h_0 = p_2 + (1/2)\rho v_2^2 + \rho g h_0$; $v_0 D^2 = v_1 d_1^2 + v_2 d_2^2$.
57. $y = y_0 + (\omega^2/2g)r^2$ (which describes a parabola).
59. $a = 1, b = 1$.
61. 5.4 cm.
63. 3.9 atm; less air.
65. (a) 27 m/s; (b) ≈200 times the rate of flow of the hose; (c) 1.9×10^{-3} m²; (d) 3.65×10^5 Pa (gauge).
67. Yes.
69. (a) 6.0×10^2 kg/m³, (b) $-0.39 \times 10^3 \Delta$ N (up), (c) $0.39 \times 10^3 \Delta$ N (down), (d) 1.44 Hz.
71. (a) 1.8×10^5 Pa, (b) 4.8×10^5 Pa, (c) 7.0×10^{-4} m/s, (d) 13 m/s; (e) 0.37 L/s.
73. $[2g(h_1 - h_0)]^{1/2}$.
75. 8.6 min.

CHAPTER 17

1. (b), (e), (f) are in thermal equilibrium.
3. The extensive variable are (a) and (d), and the intensive variable are (b) and (c).
5. (a) $t_F = -40$°F; $t_C = -40$°C, (b) none, (c) $t_F = +575$°F, $T = +575$K.
7. 95.9°F to 108.5°F.
9. $\Delta p = 0.29$ Pa.
11. (a) 457°C; 854°F, (b) −196°C; −321°F, (c) 297K; 24°C, (d) 2.2°C; 276K.
13. $T = 1.25 t_R + 273$; $t_F = 2.25 t_R + 32$.
15. 672°R; 7.6°R; 10,800°R; 492°R; 430°R; 447°R.
17. 52°C (126°F).
19. 0.5 mm.
21. 8.0×10^{-4} K⁻¹.
23. 7.0 cm³.
25. (a) $\Delta T = 104$°C, (b) 57×10^{-6} K⁻¹, (c) 5.9×10^{-3}.
27. 1.3×10^{-4} cm².
29. (a) steel, (b) 0.99971 Hz $\leq f_{aluminum} \leq$ 1.00029 Hz; 0.99979 Hz $\leq f_{copper} \leq$ 1.00021 Hz; 0.99987 Hz $\leq f_{steel} \leq$ 1.00013 Hz.
31. 2.1%.
33. 1.4×10^3 cm³.
35. 7.4×10^4 m³.
37. 0.47 kg/m³.
39. $p_2/p_1 = 0.562$.
41. 1.30×10^5 Pa; 0.28×10^5 Pa; 1.58×10^5 Pa.
43. 3.1×10^{-10} m.
45. (a) 6.1×10^4 Pa; 308K, (b) 3.5×10^{-3} m³, (c) 0.77 atm.
47. (a) $pV = NKT$, (b) $P(V - bN) = N(kT - aN/V)$, (c)

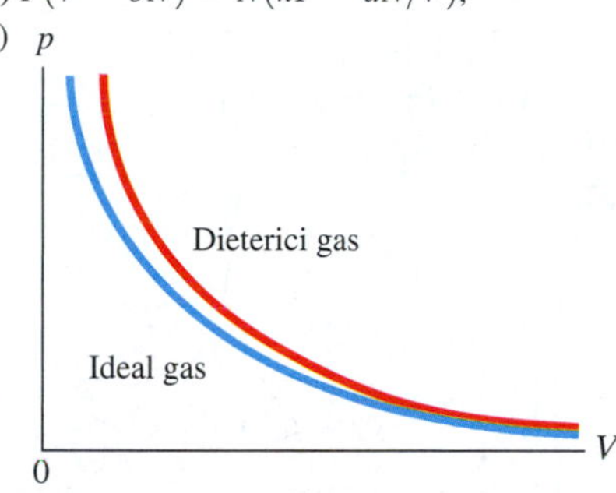

49. 22.4×10^{-3} m³; 1.29 kg/m³.
51. 6 kg/m³.
53. 2.42×10^9 molecules.
55. (a) 71 atm, (b) 71 atm.
57. 40.5 cm³; 142 cm³;

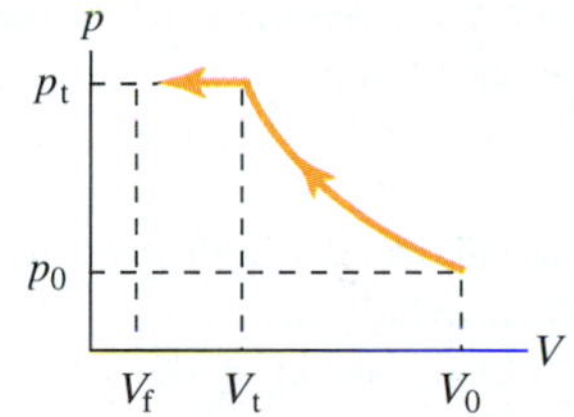

59. 1.4 kg/m³; 0.74 m³/kg.
61. (a) $p_2 = 10^5$ Pa; $V_2 = 600$ cm³; $T_2 = 568$K, (b) $p_3 = 1.30 \times 10^5$ Pa; $V_3 = 600$ cm³; $T_3 = 762$K, (c) $p_4 = 1.30 \times 10^5$ Pa; $V_4 = 300$ cm³; $T_4 = 381$K.
63. 4.0×10^{-2} m³.
65. 2.56×10^3 J; 2.15×10^3 J; the energy of the gas is ≈10^4 the energy of the ball.
67. 373.15K.
69. (b) 421K (148°C).
71. $E_{3200°C} = 1.84 \times 10^4 E_{25°C}$.
73. (a) 6.25×10^{10} Hz, (b) 5.83×10^{12} Hz, (c) 1.67×10^{13} Hz, (d) 6.25×10^{13} Hz.
75. 9.4×10^{-6} m.
77. 4.47×10^{26} W.
79. 2.05×10^{17} W; almost the same.
81. 5.67×10^{-8} W/m²·K⁴.
83. 37 kg.
85. $a = (T_1 Z_2 - T_2 Z_1)/(Z_1 - Z_2)$; $b = (T_1 - T_2)/(Z_1 - Z_2)$; not necessarily.
87. 0.1 m³.
89. 26 m/s.
91. 0.754×10^{-23} J/K′.

CHAPTER 18

1. Reversible transformations: (c), (e) and (h); irreversible transformations: (a), (b), (d), (f), and (g).

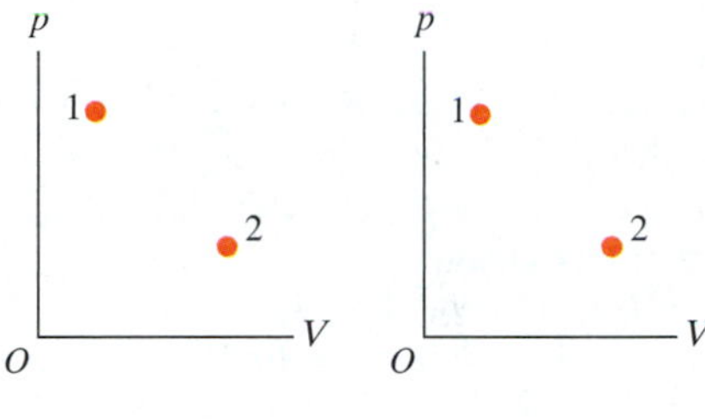

(a) (b)

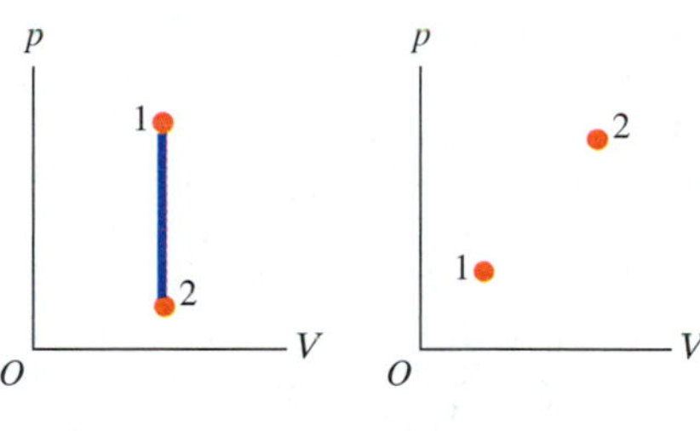

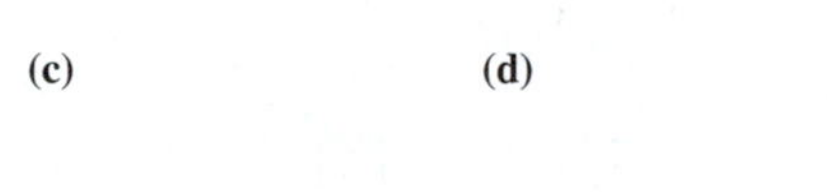

(c) (d)

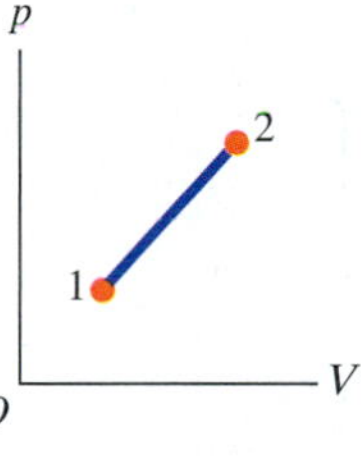

(e)

3. 5×10^2 cal/s.
5. $c_p' pV/R$.
7. (a) 262.8 g of water and 37.2 g of ice at 0°C, (c) 300 g of water at 7.6°C.
9. (a) 0.22 cal/g·K; 5.9 cal/mol·K, (b) 0.103 cal/g·K; 5.8 cal/mol·K, (c) 0.04 cal/g·K; 8 cal/mol·K.
11. 308K.
13. 167K.

15. 7.6%.
17. 287K (14°C).
19. 7.4 kW.
21. Pine: 144 Btu/h; Fiberglas: 62 Btu/h; $R_{eff} = 16\ ft^2 \cdot h \cdot °F/Btu$.
23. (a) $\kappa_{eff} = \kappa_1\kappa_2(L_1 + L_2)/(\kappa_2 L_1 + \kappa_1 L_2)$, (b) $(\kappa_1 L_2 T_1 + \kappa_2 L_1 T_2)/(\kappa_1 L_2 + \kappa_2 L_1)$.
29. $\frac{1}{L}\frac{dQ}{dt} = \frac{2\pi\kappa(T_2 - T_1)}{\ln(1 + \alpha/R_1)}$.
31. 6.7°C.
33. 33 times.
35. 1 cal = 4.14 J.
37. (a) 11.9°C, (b) 10.1°C, (c) ΔT of the water is less than in part (a), (d) net negative temperature change.
39. 1.0×10^6 J.
41. 3.0×10^5 J.
43. 0.012 m³; 0.0034 m³; -1.6×10^3 J.
45. (a) $[ML^{-4}T^{-2}]$, (b) $(V_2 - V_1)[p_0 - (V_2 + V_1)\beta/2]$, (c) same as part (b).
47. 4.1×10^3 J.
49. 633 cal.
51. (a) 2.7×10^4 cal, (b) -3.5×10^4 cal; liberated.
53. (a) $U_A = (c'_V/R)p_A V_A$; $U_B = (c'_V/R)p_B V_B$; $U_C = (c'_V/R)p_B V_A$, (b) $(1/2)(p_A - p_B)(V_B - V_A)$.
55. (a) 2.8×10^2 K; 6.1×10^2 K, (b) 1.2×10^3 J, (c) 2.3×10^2 J, (d) 1.4×10^3 J (into the gas).
57. −0.72K (cooled).
59. 5.0×10^4 J.
63. 20°C.
65. (a) No, (b) no, (c) no.
67. $(p_0V_0 - p_1V_1)/(\gamma - 1)$.
69. $\Delta p_{adiabatic}/\Delta p_{isothermal} = 1.24$.
71. $C_V/C_p = 0.683$; $W_{on} = 1.43p_1V_1$; $\Delta U = 1.43p_1V_1$.
73. 5.4 atm; 210°C.
75. 61.7 cm³; 151 cm³.
79. (a)

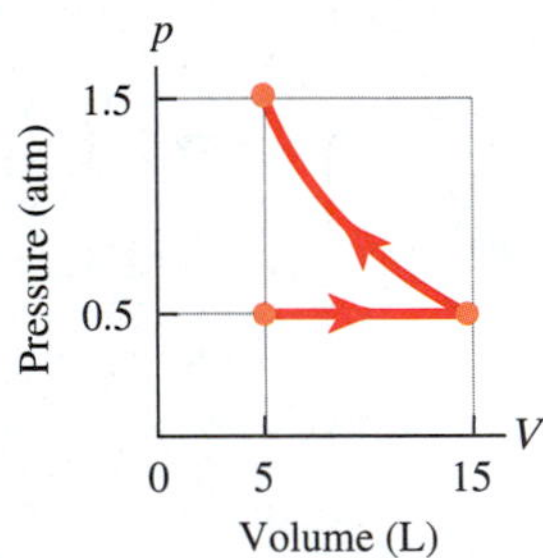

(b) 1.5 atm, (c) -3.2×10^2 J.
81. (a) $+7.1 \times 10^3$ J, (b) $+2.5 \times 10^3$ J.
83. $slope_{adiabatic}/slope_{isothermal} = \gamma$.
85. 255K.
87. 3.85×10^3 J.
89. $dm/dt = 20 \times 10^3$ kg/s.
91. (a) 10 atm, (b) -5.7×10^3 J, (c) -5.7×10^3 J (from the gas).
93. (a) $C_p = 3R/2 + RV(pV^2 - a)/(pV^3 - aV + 2ab)$, (b) $C_V = 3R/2$, (c) $\gamma = 1 + 2V(pV^2 - a)/[3(pV^3 - aV + 2ab)]$.
95. 32°F.
97. $+2.48 \times 10^3$ J.
99. (a) 7.18×10^5 Pa (7.1 atm), (b) 5.9×10^{-2} m³, (c) 2.0×10^4 J, (d) 137K, (e) $+2.0 \times 10^4$ J, (f) $+1.2 \times 10^4$ J.
101. 0.90 W.

CHAPTER 19

1. (a) 1.0×10^{-4}; 5.8×10^{-2}, (b) 6.1×10^{-3}; 0.23, (c) 1.0×10^{-12}; 1.2×10^{-4}, (d) 6.1×10^{-11}; 4.9×10^{-4}.
3. 4×10^{24} components; 4×10^5 m².
5. 6 Pa.
7. (a) 3.8×10^3 J, (b) 304K, (c) 3.8×10^5 m²/s², (d) 6.1×10^2 m/s.
9. $PA = U$.
11. 11 m/s.
13. 4×10^{-17} Pa.
15. (a) 290K, (b) 2.90×10^5 K, (c) 3.6 J.
17. (a) 6.1×10^{-21} J, (b) 3.5×10^{-10} m/s.
19. 7.0×10^5 m/s.
21. 484 m/s; 517 m/s; 412 m/s; 1934 m/s.
23. (a) 1.2×10^4 m/s, (b) 1.9×10^3 m/s.
25. 4.9×10^{-3} m/s.
27. 7.2×10^{73} K.
29. (a) 61.8 mi/h, (b) 62.4 mi/h.
31. (a) 1/1296, (b) 1/324.
33. (a)

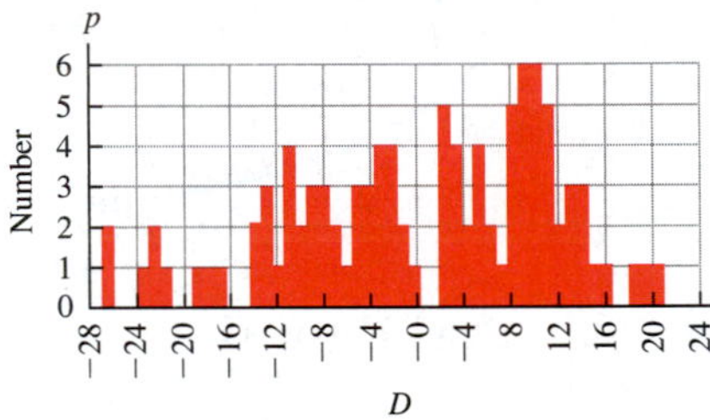

(b) 124.
35. (a) +6.4, (b) 474, (c) +20.
39. 0.984; 0.970; independent of the temperature.
41. $v_{av} = 2(2kT/\pi m)^{1/2}$.
43. $\langle \vec{v} \rangle = \vec{u}$; $\langle \vec{v}^2 \rangle = v^2_{rms,0} + u^2$.
45. $s = 5$.
47. monatomic gas; dumbbell shape; atoms not linear; linear arrangement.
49. (a) 17 kJ, (b) 0.45 kJ/kg · K.
55. 1×10^{-3} atm; 1×10^{-7} atm.
57. ≈100 m.
59. 6×10^{19} atoms; 6×10^{16} atoms; 30 Pa.
61. 0.63; no change.
63. (a) $v_{esc} = 4.1 \times 10^3$ m/s, (b) 1350 K; v_{rms} and T both decrease, (c) similar effect, (d) lighter component will make up a smaller fraction of the atmosphere.
65. (a) Yes, (b) $x/(1 - x)$, (c) $x^{-1/2}$; no.
67. (a) 7.0×10^5 m, (b) The mean free path decreases by a factor of 1000 to 7.0×10^2 m.
69. 2×10^{-3} m/s.
71. $p = nE/3$.
73. (a) 285K, (b) 2.8×10^4 Pa.
75. (a) 3.2×10^{11} atoms/m³, (b) 6×10^4 s, (c) 7×10^7 m.
77. $P = \frac{4}{\pi}\int_{u_0}^{\infty} u^2 e^{-u^2}\,du$, where $u_0 = \sqrt{3/2}\,v_{escape}/v_{rms}$.

CHAPTER 20

1. 50%.
3. (a) The possible outcomes are

1234	1235	1236	1245	1246
1256	1345	1346	1356	1456
2345	2346	2356	2456	3456

(b) 1/15, (c) 3/5.
5. 252 times more likely.
7. 29%.
9. (a) 38%, (b) 0.38 cal.
11. (a) 325K (52°C), (b) 9.1 kWh; 4.1 kWh.
13. 270 kg/s.
15. Violation of the first law of thermodynamics.
17. 60.
19. Engine A is not reversible.
21. 127 W.
23. 13.7; 9.1.
25. $C_V(T_2 - T_3)$.
27. 5.5 W (0.0074 hp).
29. The statement cannot be trusted.
31. ≈1 cent.
33. (a) $T_3 = p_L V_U/nR$; $T_4 = p_L V_L/nR$, (b) $\Delta U_{IV} = -C_V p_L(V_U - V_L)/nR$, (c) $Q_{IV} = +[1 + (C_V/nR)]p_L(V_U - V_L)$, (d) $\eta = nR(p_U - p_L)(V_U - V_L)/[C_V(p_U V_U - p_L V_L) + nRp_U(V_U - V_L)]$.
37. $1 - [(T_3 - T_4)/\gamma(T_2 - T_1)]$.
39. 7.9×10^{-2} J/K · min; irreversible.
41. 31°C; +10 J/K.
43. +0.75 kJ/K.
45. +8 J/K.
47. $\Delta S_{gas} = +44$ J/K; $\Delta S_{universe} = +44$ J/K.
49. 20 J/K.
51. The entropy of the gas decreases.
53. $\Delta T = 0$; $\Delta S = +32.5$ J/K.
55. (a)

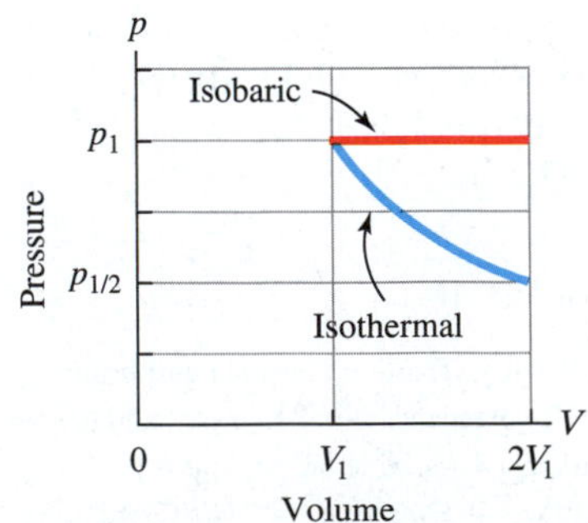

(b) $W_{isobaric} = p_1V_1$; $W_{isothermal} = 0.693p_1V_1$, (c) $\Delta S_{isobaric} = 0.693nc'_p$; $\Delta S_{isothermal} = 0.693nR$.

57. Process $A \rightarrow B$ (isobaric):
$Q = (nR(T_B - T_A), W = nR(T_B - T_A)$,
$\Delta U = (3/2)nR(T_B - T_A), \Delta T = T_B - T_A$,
$\Delta S = (5/2)nR \ln(T_B/T_A)$;
Process $B \rightarrow C$ (adiabatic): $Q = 0$,
$\Delta U = (3/2)nR(T_C - T_B)$,
$W = (3/2)nR(T_B - T_C)$,
$\Delta T = T_C - T_B, \Delta S = 0$;
Process $C \rightarrow D$ (isothermal): $\Delta T = 0$,
$\Delta U = 0, W = (3/2nRT_C \ln(T_A^{2/3}T_C/T_B^{5/3})$,
$Q = (3/2)nRT_C \ln(T_A^{2/3}T_C/T_B^{5/3})$,
$\Delta S = (3/2)nR \ln(T_A^{2/3}T_C/T_B^{5/3})$;
Process $D \rightarrow A$ (constant volume):
$Q = (3/2)nR(T_A - T_C), W = 0$,
$\Delta U = (3/2)nR(T_A - T_C), \Delta T = T_A - T_C$,
$\Delta S = (3/2)nR \ln(T_A/T_C)$.
59. $\Delta S_{gas} = Q/T < 0$ (decreases); $\Delta S_{reservoir} = -Q/T > 0$ (increases); no conflict because this is a reversible process.
61. (a) $T = T_0$, (b) $+nR \ln 3$, (c) $+nR \ln 3$, (d) 0.
63. (a) +87 J, (b) 2.7K, (c) +0.38 J/K.
65.
(a)
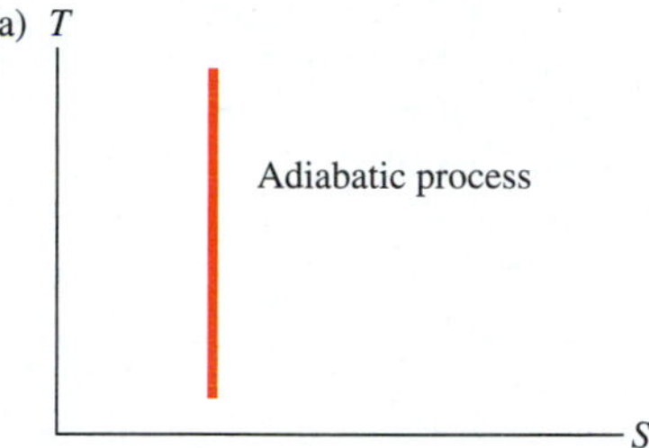

(b)
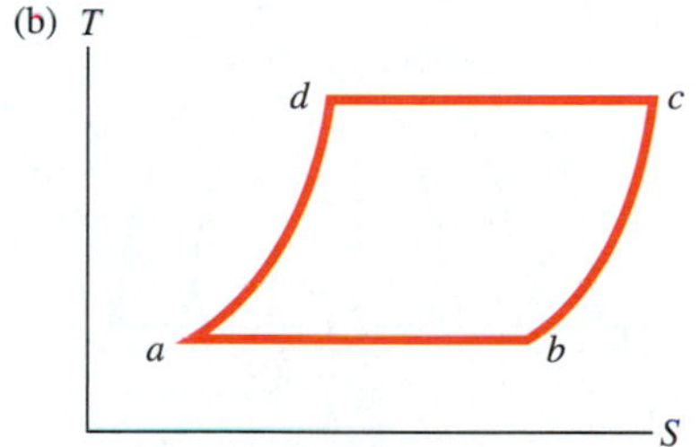

67. (a) 2.25×10^{-2} m^3, (b) 273K, (c) -11.5 J/K.
69. (a) 3.07×10^{-2} m^3, (b) 373K, (c) 0.
71. (a) $Q_{h,\text{ engine}} = 400$ J; $Q_{c,\text{ engine}} = 300$ J; $Q_{c,\text{ refrigerator}} = 300$ J; $Q_{h,\text{ refrigerator}} = 400$ J; $Q_{h,\text{ net}} = 0$; $Q_{c,\text{ net}} = 0$,
(b) $Q_{h,\text{ engine}} = 500$ J; $Q_{c,\text{ engine}} = 400$ J; $Q_{c,\text{ refrigerator}} = 300$ J; $Q_{h,\text{ refrigerator}} = 400$ J; $Q_{h,\text{ net}} = -100$ J (from the reservoir); $Q_{c,\text{ net}} = +100$ J (to the reservoir),
(c) $Q_{h,\text{ engine}} = 400$ J; $Q_{c,\text{ engine}} = 300$ J; $Q_{c,\text{ refrigerator}} = 240$ J; $Q_{h,\text{ refrigerator}} = 340$ J; $Q_{h,\text{ net}} = -60$ J (from the reservoir); $Q_{c,\text{ net}} = +60$ J (to the reservoir).
73. T decreases.
75. $dQ_h/dt = 1070$ MW; $dQ_c/dt = 770$ MW; $dm/dt = 1.5 \times 10^4$ kg/s.
77. 0.30; 60% of the Carnot efficiency.
79. nc_pT.
81. (a) $T_1 = 437$K (164°C); $T_3 = 1020$K (747°C),
(b) $Q_{abs}/m = 5.38 \times 10^6$ J/kg; $Q_{rej}/m = -3.73 \times 10^6$ J/kg; $Q_{net}/m = 1.65 \times 10^6$ J/kg (absorbed);
(c) 1.65×10^6 J/kg.

CHAPTER 21

1. 6.2×10^9 fewer electrons.
3. 4.82×10^4 C.
5. 7.9×10^{-12} per atom.
7. -2×10^{-10} C, 1.25×10^9 electrons; -1×10^{-10} C, 6.2×10^8 electrons; -1×10^{-10} C, 6.2×10^8 electrons.
9. 2.2×10^{-10}.
11. (a) 1.8×10^{51} electrons, (b) 3.5×10^{-39}.
13. (a) conserved, (b) not conserved, (c) conserved, (d) not conserved.
15. -8.5×10^{-24} C.
17. 46 N repulsion, 23 N attraction.
19. $q = 1.6 \times 10^{-19}$ C, 1 electron.
21. 3.5×10^{-10}, mass much larger.
23. (a) 3×10^9 esu in 1 C, (b) 4.8×10^{-10} esu.
25. $q_1/q = q_2/q = \frac{1}{2}$.
27. (a) 2.6×10^{-9} N toward the proton (centripetal), (b) 9.2×10^5 m/s, (c) 4.9×10^{14} Hz, (d) 8.6 N/m.
29. 1.4×10^{-8} C.
31. $q_1 = +1.1 \times 10^{-7}$ C, $q_2 = -3.2 \times 10^{-7}$ C, $q_3 = +5.3 \times 10^{-7}$ C.
33. 0.9 m.
35. $x = 44.4$.
37. zero.
39. $\vec{F}_+ = 39$ N 30° above the line joining the two "up" quarks, $\vec{F}_- = 40$ N toward the center of the line joining the two "up" quarks.
41. (a) $\Sigma\vec{F} = 0$, (b) unstable, (c) stable.
43. (a) $\sqrt{3}kq^2/2L^2$, 9.7° above the $-x$-axis, (b) $3.2kqQ/L^2$, 6.3° below the $-x$-axis.
45. $(2kq\lambda/x_0)\hat{\imath}$.
47. $kqQ/d(L + d)$, away from the rod.
49. Let the radius of each ring be R and the separation between them be $2L$. If $Qq(R^2 - 2L^2) < 0$ then the equilibrium is stable, otherwise unstable.
51. 1.4 N away from the center.
53. 3.0×10^{-7} C.
55. 0.52 N.
57. $3kq^2[3(2.25 + y_0^2)^{-3/2} - (0.25 + y_0^2)^{-3/2}]\hat{\imath}$, y_0 in cm.
59. 5.7 N/m^2; same.
61. 1.5×10^{-3} C.
63. 6.6°.
65. (a) $v = (e^2/4\pi\varepsilon_0 mR)^{1/2}$, (b) $L = (e^2mR/4\pi\varepsilon_0)^{1/2}$, (c) $v = e^2/4\pi\varepsilon_0 L$, (d) $R = 4\pi\varepsilon_0 L^2/me^2$, (e) $\tau = 32\pi^3\varepsilon_0^2L^3/me^4$, (f) $v = 2.2 \times 10^6$ m/s; $R = 5.3 \times 10^{-11}$ m; $\tau = 1.5 \times 10^{-16}$ s.
67. (a) $F_{net} = kq^2[\ell(\ell - 2x)/x^2(\ell - x)^2]$ away from the closer charge; $x = \ell/2$, (b) $F_{net} = -2kq^2\ell\,\Delta x/[(\ell/2)^2 - (\Delta x)^2]^2$, (c) $f = (1/2\pi)(32kq^2/\ell^3m)^{1/2}$.
71. (a) $\vec{F} = kqQR \sum_{n=-\infty}^{\infty} \left\{\frac{1}{[(na)^2 + R^2]^{3/2}}\right\}\hat{\jmath}$,
(b) $\vec{F} = \frac{k\lambda Q}{R}\int_{-\infty}^{\infty}\frac{du}{(u^2 + 1)^{3/2}}\hat{\jmath}$.

CHAPTER 22

1. $1.8 \times 10^7(-0.60\hat{\imath} + 0.80\hat{\jmath})$ N/C.
3. $(1.14 \times 10^{11}$ N/C$)\hat{r}$; $-(1.82 \times 10^{-8}$ N$)\hat{r}$ (toward the nucleus).
5. 3.19×10^6 N/C, 78° above the $+x$-axis.
7. (a) $-(1/4\pi\varepsilon_0)(8q/\ell^2)\hat{\imath}$, (b) 0, (c) $(1/4\pi\varepsilon_0)\{2qd/[d^2 + (\ell/2)^2]^{3/2}\}$ away from the origin.
9.
$$\vec{E} = \frac{2\vec{p}}{4\pi\varepsilon_0 r^3}\left\{\frac{1}{[1 + (L/2r)]^2[1 - (L/2r)]^2}\right\}.$$
$$\vec{E} = \frac{2\vec{p}}{4\pi\varepsilon_0 r^3}.$$
11. Stable; $(1/2\pi)(Qq/\pi\varepsilon_0 a^3m)^{1/2}$.
13. Tripled.
15.
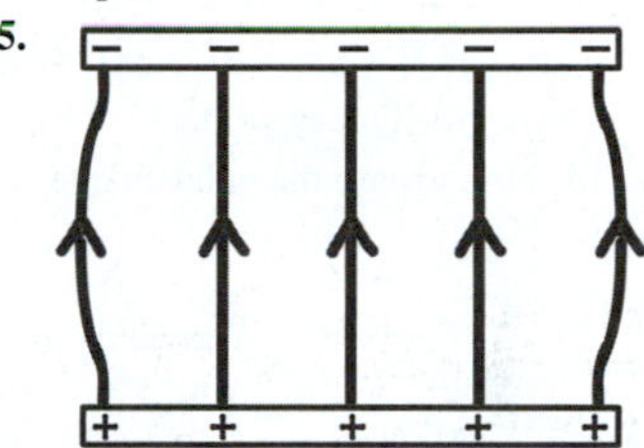

17.

19.
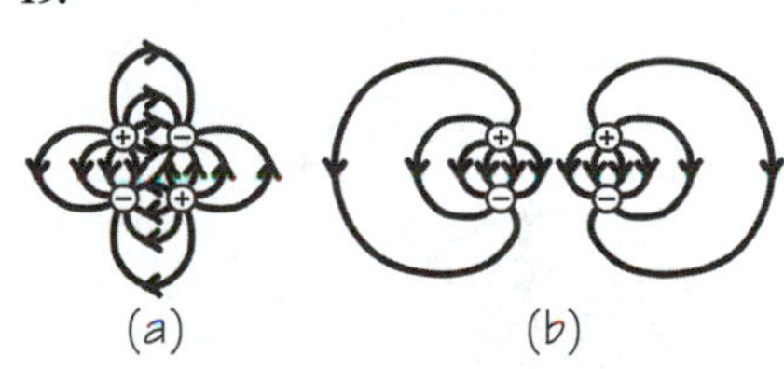

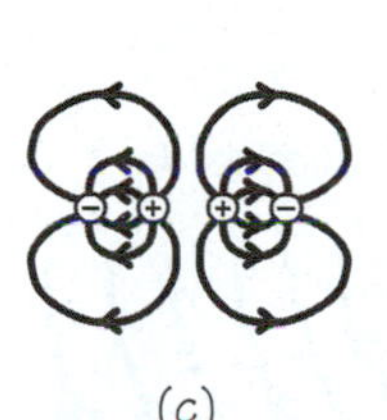

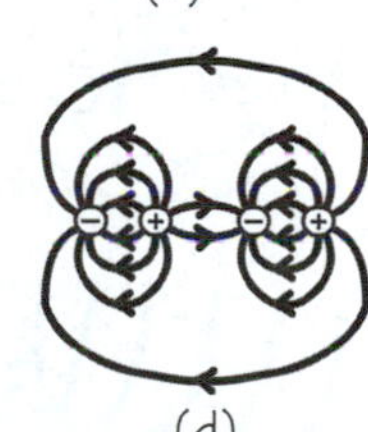

21. 2.7×10^4 N/C perpendicular to and away from the line.
23.
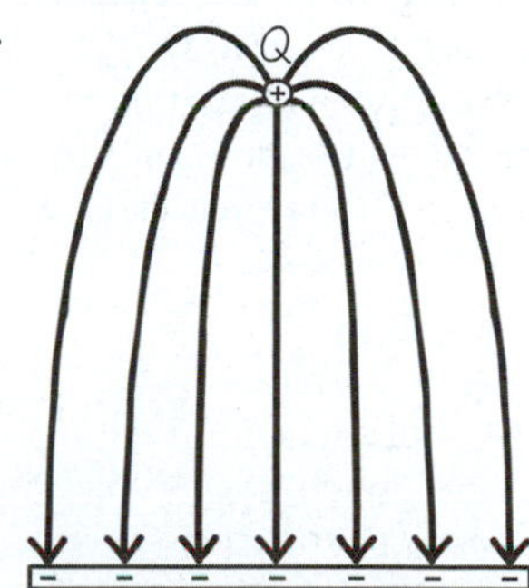

25.

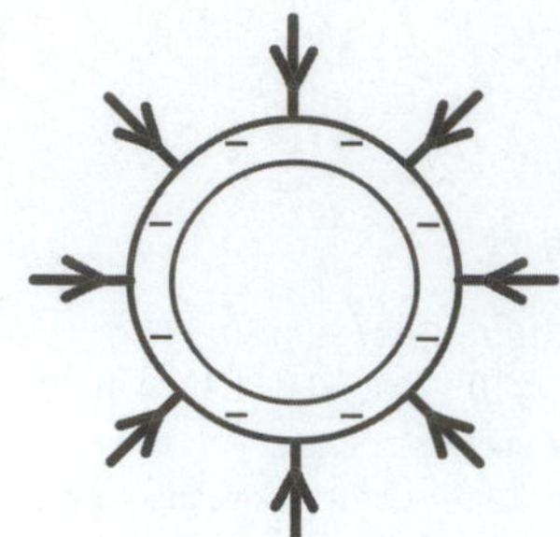

27.

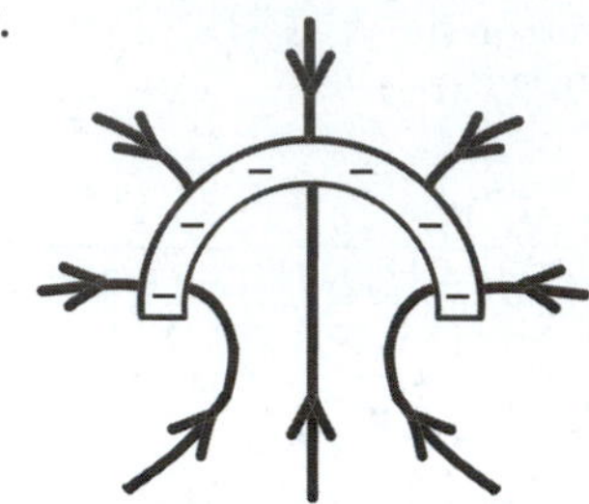

29. σ/ε_0 perpendicular to the plates and away from them; 0.

31. (a) $(z_0 Q/2\pi\varepsilon_0 R^2)[1/z_0 - (R^2 + z_0^2)^{-1/2}]\hat{k}$, (b) $(Q/4\pi\varepsilon_0 R^2)\hat{k}$, (c) $(Q/2\pi\varepsilon_0 R^2)\hat{k}$.

33. 7.3×10^4 N/C toward the point charge.

35. $Q/4\pi r^2 h$.

37. (a) $\dfrac{\sigma L z_0}{2\pi\varepsilon_0}\hat{k}\displaystyle\int_{-L}^{L}\frac{dx}{(x^2 + z_0^2)\sqrt{x^2 + L^2 + z_0^2}}$, (b) $(\sigma/2\varepsilon_0)\hat{k}$, (c) $(\sigma/2\varepsilon_0)\hat{k}$.

39. 0.65 μC.

41. 1.7×10^{-7} C/m^2.

43. $(q\lambda/2\pi\varepsilon_0 m)^{1/2}$.

47. 3.4 cm.

49. 5.7 s^{-1}.

51. $-(1.41 \times 10^{-6}$ N·m$)\hat{k}$.

53. 2×10^6 N/C.

55. $-6p^2/4\pi\varepsilon_0 r^4$ (attraction).

57.

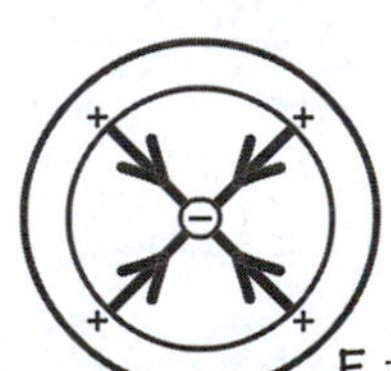

59.

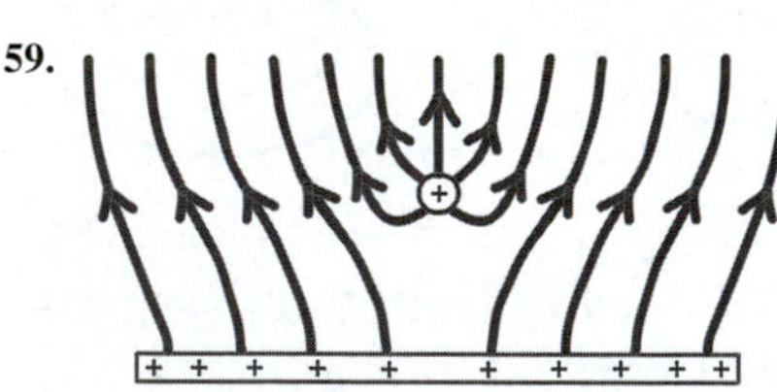

61. (a) $\{(\lambda R/2\pi\varepsilon_0)/[y^2 - (R/2)^2]\}\hat{j}$, (b) $-\{(\lambda R/2\pi\varepsilon_0)/[x^2 + (R/2)^2]\}\hat{j}$.

63. $[(1 + 8.5t^2)\hat{i} + (1 - 14t^2)\hat{j}]$ m, with t in s.

65. (a) 7.2×10^{-14} N away from the plate, (b) -3.2×10^{-13} J, (c) 4.4 m.

67.

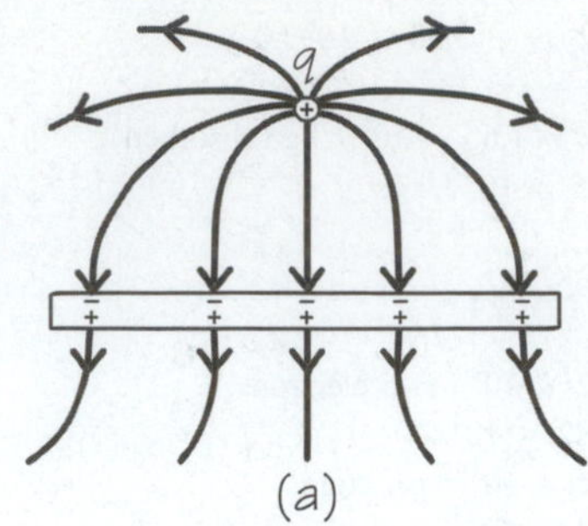

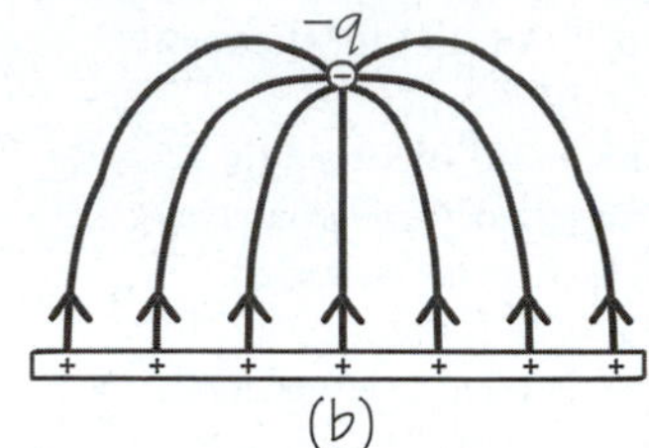

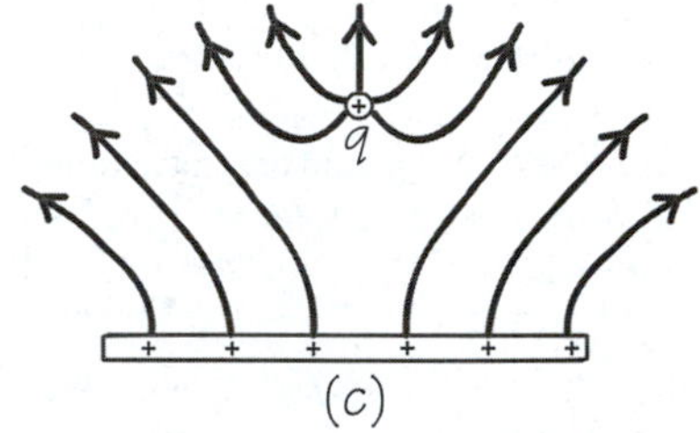

69. 19 kN; 0.

71. $\lambda_0 L^2/6$.

CHAPTER 23

1. (a) $\sigma\pi R^2/2\varepsilon_0$, (b) $0.866\,\sigma\pi R^2/2\varepsilon_0$.

3. $\lambda h/\varepsilon_0$.

5. $+6$ N·m^2/C.

7. $\pi E_0 R^2/3$.

9. $(q/\varepsilon_0)(1 + R^2/h^2)^{1/2}$.

13. (a) Zero, (b) 1.13×10^8 N·m^2/C.

15. (a) Net flux is zero.

17. (a) -3.54×10^{-9} C, (b) -3.54×10^{-9} C, (c) -3.54×10^{-9} C.

19. 9.4×10^2 N·m^2/C out of the sides parallel to the xy- or yz- planes; 10.6×10^2 N·m^2/C out of the side perpendicular to the $-x$-axis; 8.2×10^2 N·m^2/C out of the side perpendicular to the $+x$-axis.

21. $\rho R^2/2\varepsilon_0 r$.

23. $E_{\text{rod}}/E_{\text{pt chge}} = 0.02 = 2\%$.

25. -2.3×10^{-4} C.

27. $(1.8 \times 10^4)\hat{r}$ N/C; $(1.8 \times 10^4)\hat{r}$ N/C.

29. $E = 0$ for $r < r_1$; $[\rho(r^2 - r_1^2)/2\varepsilon_0 r]\hat{r}$ for $r_1 < r < r_2$; $[\rho(r_2^2 - r_1^2)/2\varepsilon_0 r]\hat{r}$ for $r_2 < r$.

31. $E = 0$ for $r < R_1$; $[Q(r^3 - R_1^3)/4\pi\varepsilon_0(R_2^3 - R_1^3)r^2]\hat{r}$ for $R_1 < r < R_2$; $(Q/4\pi\varepsilon_0 r^2)\hat{r}$ for $R_2 < r$.

33. (a) Point the x-axis from plate 1 (where $x = 0$) to plate 2 (where $x = a$), perpendicular to both plates: $\vec{E} = -[(\sigma_1 + \sigma_2)/2\varepsilon_0]\hat{i}$ $(x < 0)$; $\vec{E} = [(\sigma_1 - \sigma_2)/2\varepsilon_0]\hat{i}$ $(0 < x < a)$; $\vec{E} = [(\sigma_1 + \sigma_2)/2\varepsilon_0]\hat{i}$ $(x > a)$, (b) same; must also note that $E = 0$ inside the metallic plate.

35. 1st quadrant: E at $-\theta$; 2nd quadrant: E at $180° + \theta$; 3rd quadrant: E at $180° - \theta$; 4th quadrant: E at θ; $E = 3.3 \times 10^5$ N/C; $\theta = 31°$.

37. $\vec{E} = (-6.7 \times 10^8)\vec{r}$ N/C with r in m, where $r < 3$ cm; $\vec{E} = [(-1.8 \times 10^4)/r^2]\hat{r}$ N/C with r in m, where 3 cm $< r <$ 8 cm; $\vec{E} = [(2.7 \times 10^4)/r^2]\hat{r}$ N/C, with r in m, where 8 cm $< r$.

39.

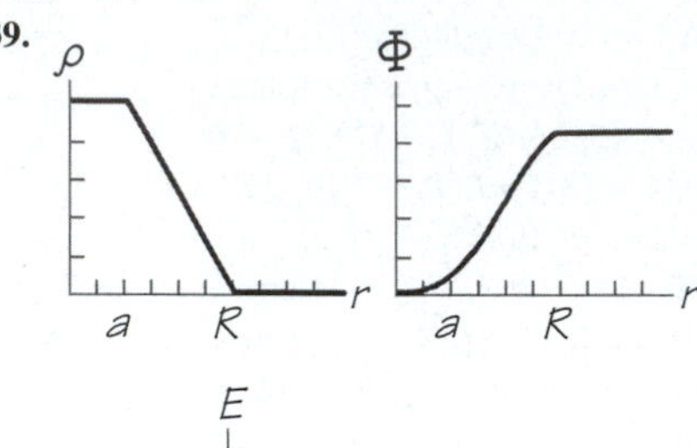

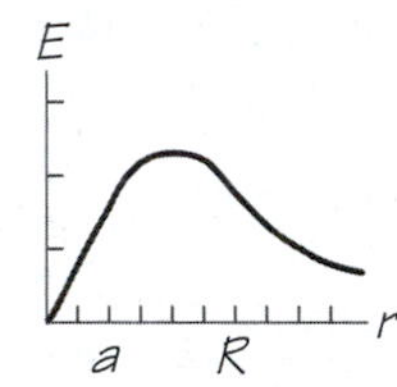

41. $r < R$: $\vec{E} = 0$; $R < r < 2R$: $\vec{E} = (q/4\pi\varepsilon_0 r^2)\hat{r}$; $2R < r$: $\vec{E} = -(q/4\pi\varepsilon_0 r^2)\hat{r}$.

43. 2.7×10^{-5} C/m^2.

45. $\sigma_{\text{sphere}}/\sigma_{\text{shell}} = 1.96$.

47. $\sigma_{\text{inner sphere}} = Q/4\pi a^2$; $\sigma_{\text{shell, inside}} = -Q/4\pi b^2$; $\sigma_{\text{shell, outside}} = +Q/2\pi R^2$.

49.

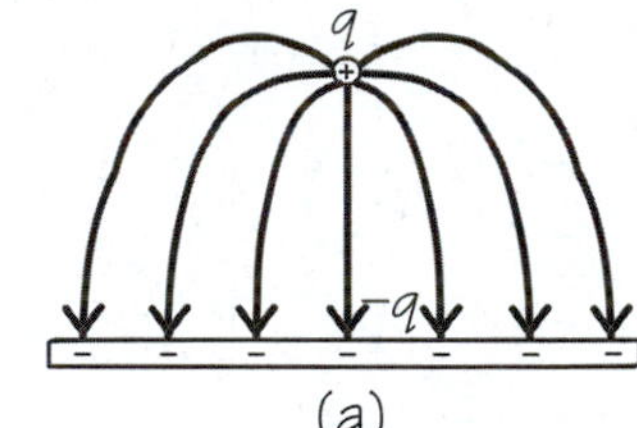

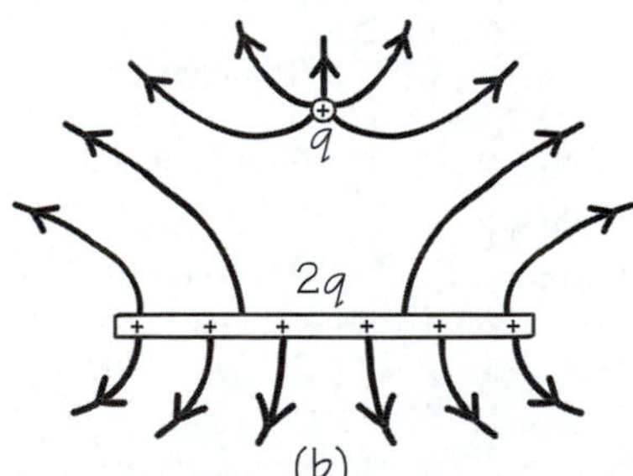

51. $\Phi_{x=0} = 0$; $\Phi_{x=a} = ba^4$; $\Phi_{y=0} = \Phi_{y=a} = \Phi_{z=0} = \Phi_{z=a} = 0$; $q = \varepsilon_0 ba^4$.

53. (a) $0.067Q/\varepsilon_0$, (b) $Q/4\pi\varepsilon_0 R^2$.

57. $0.433EL^2$.

59. $\rho(r) = \varepsilon_0 E/r$.

61. $+6000\varepsilon_0$ C/m^3 (constant).

CHAPTER 24

1. 4.6×10^{-14} J.

3. 9.5×10^{-2} J.

5. (a) 0, (b) -1.35 J, (c) $+1.35$ J.

7. 1.36×10^{-3} J; 0.

9. -4.8×10^{-2} J.

11. $r = \infty$.

13. 3.3×10^{-14} C.

15. $+4.2 \times 10^{-3}$ J.

17. -5.0×10^{4} J.

19. (a) $+5.0 \times 10^{3}$ V, (b) -1.0×10^{-3} J.

21. $+2.4 \times 10^{5}$ V.

23. $+4.3 \times 10^{5}$ V.

25. (a) -3.7×10^{5} V, (b) $V(0, 0) = +1.6 \times 10^{5}$ V; $\vec{E}(0, 0) = -(6.5 \times 10^{7}\ \text{V/m})\hat{j}$.

27. (a) $E_A = 0$; $E_B = (1.13 \times 10^{6}\ \text{N/C} \cdot \text{m})x$, $x < 1$ cm; $E_C = 1.13 \times 10^{4}$ N/C, $x > 1$ cm, (b) $V_B - V_A = -(5.65 \times 10^{5}\ \text{V/m}^2)x^2$, $x < 1$ cm; $V_C = +5.65\ \text{V} - (1.13 \times 10^{4}\ \text{V/m})x$, $x > 1$ cm, (c)

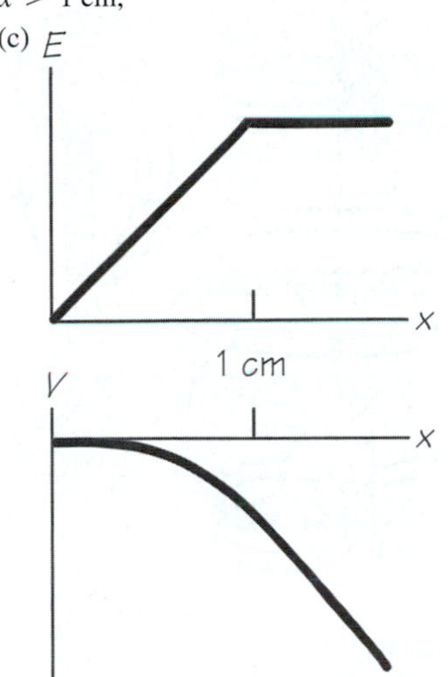

29. $V_{\text{outside}} = Q/4\pi\varepsilon_0 r\ (r > R)$; $(Q/8\pi\varepsilon_0 R)[3 - (r/R)^2]\ (r < R)$.

31.

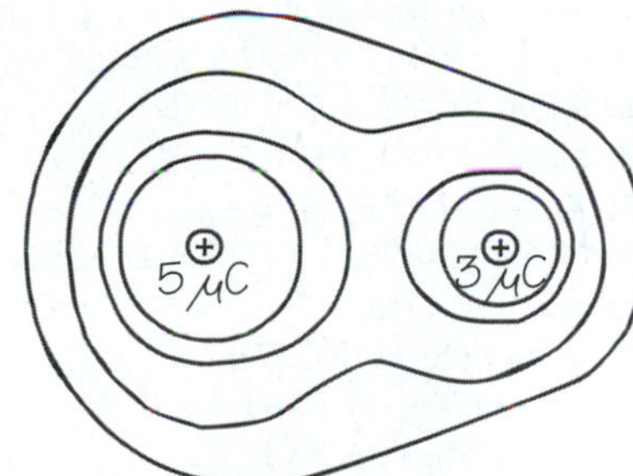

33.

35.

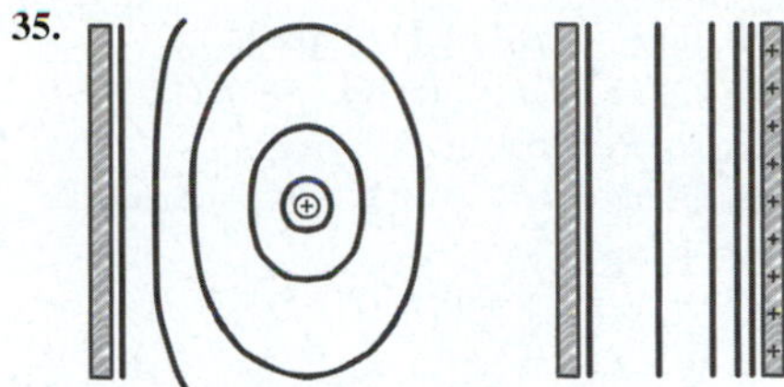

37.

39. $(Q/4\pi\varepsilon_0 x^2)\hat{i}$.

41. $(6.68\ \text{V/m})\hat{i}$.

43. $-(p/4\pi\varepsilon_0 r^3)\hat{i}$.

45. $\vec{E}_{r<R} = (Q/4\pi\varepsilon_0)(6r/R^3)\hat{r}$; $\vec{E}_{r>R} = (Q/4\pi\varepsilon_0 r^2)\hat{r}$.

47. $-(2Qa_0^2/4\pi\varepsilon_0 L r^3)(\sin\theta\hat{i} + 3\cos\theta\hat{j})$.

49. 1.0×10^{-10} C.

51. $-(2\lambda/4\pi\varepsilon_0)\ln(R) + (\text{a constant})$.

53. 9.9×10^{4} V.

55. (a) $\dfrac{q_0}{4\pi\varepsilon_0}\left(\dfrac{3}{x - x_0} - \dfrac{1}{x + x_0/2}\right)$, (b) $\dfrac{q_0}{4\pi\varepsilon_0}\left(\dfrac{2}{x} + \dfrac{7}{2}\dfrac{x_0}{x^2} + \dfrac{11}{4}\dfrac{x_0^2}{x^3} + \dfrac{25}{8}\dfrac{x_0^3}{x^4} + \cdots\right)$, (c) $q_{\text{net}} = 2q_0$, $p = 7q_0x_0/2$, (d) $|x| > 12.1x_0$.

57. $+4.2 \times 10^{-4}$ J.

59. $(q_1r_2 - q_2r_1)/(r_1 + r_2)$.

61. 3.7×10^{-4} C; 5.5 MeV; 8.8×10^{-13} J; 3.2×10^{7} m/s.

63. $Q_1 = 3Q/4$; $Q_2 = Q/4$.

65. 4.4 km.

67. (a) 2.2 μC; 11 μC, (b) 4.95×10^{7} V/m, radial; 9.90×10^{6} V/m, radial.

69. (a) $+5.5 \times 10^{6}$ eV (8.8×10^{-13} J), (b) 3.3×10^{7} m/s.

71. (a) 3.9×10^{6} V, (b) 3.9 MeV, (c) 5.6×10^{-4} C.

73. $(Q/2\pi\varepsilon_0 R^2)\{[R^2 + (a + x)^2]^{1/2} + [R^2 + (a - x)^2]^{1/2} - 2a\}$.

75. $Qa/2\pi\varepsilon_0 x^2$.

77. $U = -(qQ/8\pi\varepsilon_0 R)(3 - r^2/R^2)$; $k = qQ/4\pi\varepsilon_0 R^3$.

79. 3.4×10^{-9} N toward the other Na^+; 1.2×10^{-18} J (7.4 eV).

81. $[Qx/4\pi\varepsilon_0(R^2 + x^2)^{3/2}]\hat{i}$.

83. $\Delta U = -7.68 \times 10^{-19}$ J; $\Delta K = +3.84 \times 10^{-19}$ J; $\Delta E = -3.84 \times 10^{-19}$ J.

85. $(1/4\pi\varepsilon_0)(q_1 + q_2)/r\quad (r_2 < r)$. $(1/4\pi\varepsilon_0)(q_2/r_2 + q_1/r)\quad (r_1 < r < r_2)$. $(1/4\pi\varepsilon_0)(q_1/r_1 + q_2/r_2)\quad (r < r_1)$.

87. (a) 9.9×10^{-3} kg, (b) $(0.023/\sin\theta) + 0.078(1 - \cos\theta)$.

89. $-\rho r^2/4\varepsilon_0 - (\rho R^2/2\varepsilon_0)[\ln(R/a) - 1/2]$ $(r < R)$; $-(\rho R^2/2\varepsilon_0)\ln(r/a)(r > R)$.

91. $4\pi\rho^2 R^5/15\varepsilon_0$.

CHAPTER 25

1. (a) 44 pF, (b) 40 cm.

3. (a) 1 V, (b) 2.5 V, (c) 250 V.

5. 0.33 mm.

7. (a) $C \rightarrow 4\pi\varepsilon_0 r$, (b) $C \rightarrow 4\pi\varepsilon_0 r^2/(R - r) = \varepsilon_0 A/d$.

9. $(4.43 \times 10^{-7}\ \text{m}) + (8.85 \times 10^{-10}\ \text{m/s})t$.

11. 6 kV.

13. 99 J.

15. 600 μF.

17. (a) 5.67×10^{-10} F, (b) 2.83×10^{-4} J; 2.83×10^{-2} J.

19. (a) $(\varepsilon_0 A V_0^2/2d_0^2)(d_1 - d_0)$, (b) $(\varepsilon_0 A V_0^2/2d_0^2)(d_1 - d_0)$, (c) $(\varepsilon_0 A V_0^2/2d_1 d_0)(d_0 - d_1)$, (d) energy has been stored in the battery.

21. $\lambda^2/8\pi^2\varepsilon_0 r^2$.

23. 1.1×10^{-2} J.

25. 1.7×10^{-7} C; $1.0 \times 10^{-5}/r^4$ J/m^3, with r in m; 7.0×10^{-4} J.

27. (a) 1.8 pF, (b) 2.3×10^{-7} J.

29. (a) 3.00×10^{5} V/m, (b) 1.06×10^{-7} C, (c) 1.59×10^{-2} N, (d) 1.59×10^{-5} J.

31. 2.04 μF, 4.46 μF.

33. 1.93 μF.

35. 0.45 pF; 1.33 pF.

37. (a) 1.34 μF, (b) $Q_1 = 165\ \mu$C, $Q_2 = Q_5 = 237\ \mu$C, $Q_3 = Q_7 = 402\ \mu$C.

39. $1.6C$; C; $2C$.

41. (a) 5.0 μF, (b) 25 μC.

43. 370 V.

45. 2.1×10^{12} m^2.

47. (a) 10 nF, (b) 1.3, (c) a hyperbola.

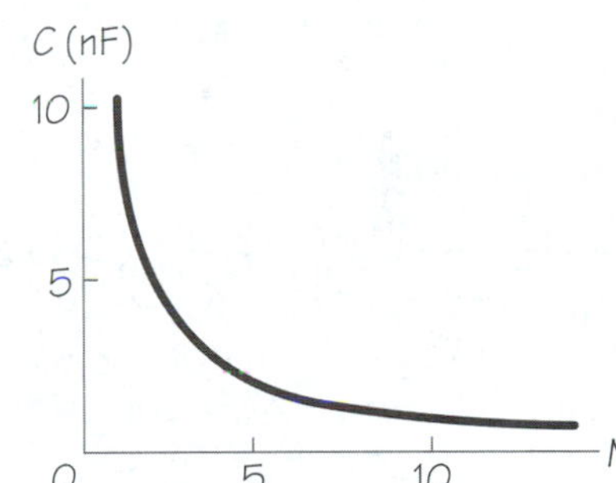

49. $C - C_0 = (\kappa - 1)4\pi\varepsilon_0 R$; $\sigma_{\text{ind}}/\sigma = (\kappa - 1)/\kappa$.

51. $q = 2.3q_0$.

53. $\kappa\varepsilon_0 A/[(d + \kappa(D - d))]$.

55. air: 5.3×10^{-10} C; paper: $Q = 2.0 \times 10^{-9}$ C; neoprene: $Q = 3.6 \times 10^{-9}$ C; Bakelite: $Q = 2.6 \times 10^{-9}$ C; strontium titanate: $Q = 1.8 \times 10^{-7}$ C.

57. $(1 - \kappa)Q^2/2C$ (a decrease).

59. 0.32 nF; 0.90 nF.

63. 2.96 μF.

65. (a) $V_1 = C_2V/(C_1 + C_2)$; $V_2 = C_1V/(C_1 + C_2)$, (b) 17 V.

67. $Q \simeq 10^{-6}$ C.

69. (a) $\frac{1}{2}C_1V^2/N$, (b) $\frac{1}{2}NC_1V^2$, (c) $U_{\text{series}} = \frac{1}{2}Q^2N/C_1$; $U_{\text{parallel}} = \frac{1}{2}Q^2/NC_1$.

71. $\frac{1}{2}(\kappa_0 + \kappa_1)(\varepsilon_0 L^2/d)$.
73. $V_A = 8.5 \times 10^2$ V; $V_B = 7.2 \times 10^2$ V; $V_C = 1.43 \times 10^3$ V.
75. 3.4 J; −2.2 J.
77. $(\kappa_1 - \kappa_0)[\varepsilon_0 L^2/D \ln(\kappa_1/\kappa_0)]$.

CHAPTER 26

1. 1.2×10^5 A/m²; 0.46 C.
3. 1.0×10^4 s (2.8 h).
5. 5.8×10^{17} electrons.
7. 8.7×10^{-6} m/s; 2.0×10^{-5} m/s.
9. 4.5×10^{21} electrons; 4.2×10^9 electrons.
11. $\pi J_0 R^2/2$.
13. $v_+ = 1.7 \times 10^{-6}$ m/s, $v_- = -2.5 \times 10^{-6}$ m/s.
15. 3.1×10^{-15} A.
17. 2.8 A; 5.6×10^4 A/m²; 5.8×10^{-6} m/s.
19. 8.5×10^{-5} m/s; 2.3×10^{-4} m/s.
21. $J_{tube} = I/2\pi R d$ along the tube; $J_{plate} = I/2\pi r d$ radial.
23. $R_2 = 2R_1$.
25. (a) 1.2 Ω, (b) 0.15 cm.
27. 3.27×10^{-2} Ω.
29. 9.6×10^{-4} Ω; 3.0×10^{-4} Ω.
31. 29 m.
33. −0.27.
35. $I_{25} = 19.6$ A; $I_{400} = 17.1$ A.
37. $r_{Cu}/r_{Al} = 1.74$.
39. 188 turns.
41. 2.0 Ω; 0.24 kg; 0.48 kg.
43. 33.4 kg.
45. 1.82×10^{-3} Ω; $r = 0.17$ cm; masses are the same.
47. Same brightness for bulbs 1 and 2; bulb 1 is 4 times as bright as bulbs 2 and 3.
49. 5 Ω; 24 V.
51. 90 Ω; 0.18 A; 2.8 W.
53. $R_{eq} = 100$ Ω.
55. 6 Ω.
57. $I_{2\,\Omega} = 4.7$ A, $V_{2\,\Omega} = 9.3$ V; $I_{4\,\Omega} = 2.3$ A, $V_{4\,\Omega} = 9.3$ V; $I_{6\,\Omega} = 9.7$ A, $V_{6\,\Omega} = 58$ V; $I_{8\,\Omega} = 7.0$ A, $V_{8\,\Omega} = 56$ V; $I_{12\,\Omega} = 10.3$ A, $V_{12\,\Omega} = 124$ V; $I_{24\,\Omega} = 2.7$ A, $V_{24\,\Omega} = 66$ V.
59. (a) 91.3 Ω, (b) 4.38 V, (c) 0.131 A.
61. 6.5×10^{-8} m.
63. $\rho = E/[nev_0(1 - r/R)]$.
65. 2.2 Ω.
67. 1¢.
69. (a) 0.18 A, (b) 35 mA.
71. $(PR)^{1/2}$.
73. 1.9×10^2 J.
75. 84 m.
77. 1.94×10^4 J (5.4 kWh).
79. Change the temperature.
81. 170 V; 43 Ω.
83. The power decreases by 3/4.
85. (a) $\rho L/[\pi r_0(r_0 + \alpha L)]$.
87. $I = [24(4\ \Omega + R_x)/(24\ \Omega + 5R_x)]$ A; $P_4 = [48(4\ \Omega + R_x)/(24\ \Omega + 5R_x)]^2$ W.
89. 0.9 kW; 35 s; 23 min.
91. (a) $R = (\rho_0 L/A)(1 + \alpha k t^2)$, (b) $VI_0/(1 + \alpha k t^2)$, (c) No.
93. $R_2/R_1 = 5.4$.
95. $dT/dt = k/[1 + \alpha(T - T_0)]$, where $k = V^2/mcR_0$; $I(t) = V/R_0\{1 + \alpha[T(t) - T_0]\}$.

CHAPTER 27

1. 0.15 Ω.
3. 0.0075 Ω.
5. 20 A.
7. 0.030 Ω; 300 W.
9. 11.8 V; 0.17 W; 8.7 V; 33 W.
11. $R_2 = R_1$; $R_2 \to \infty$.
13. $V_{gf} = 0$; $V_{ag} = -4.0$ V; $V_{ca} = 5.5$ V.
15. $I_4 = +0.5$ A; $I_5 = +1.0$ A; $I_6 = -1.0$ A.
17. (a) 7.5 Ω, (b) 7.0 Ω.
19. $I_1 = +0.45$ A; $I_2 = +0.38$ A; $I_3 = -0.068$ A.
21. $I_1 = 9.6$ mA; $I_2 = 5.7$ mA; $I_3 = 4.7$ mA.
23. $I_a = \mathscr{E}/[R + (r/N)]$; $I_b = \mathscr{E}/[r + (R/N)]$.
25. 0.38 V; 0.04 V; 0.004 V.
27. One independent junction; $I_1 = 3\mathscr{E}/7R$; $I_2 = I_3 = I_4 = \mathscr{E}/7R$.
29. (a) 60.5 mW; (b) 6.8 mW.
31. (a) 0.075 A, (b) 0.075 A.
33. 2.35 A; 2.52 A; no current.
35. $2V/5R$; $V/5R$.
37. 0.732 V; 0.737 V.
39. 0.010 Ω.
41. (a) 10 Ω, (b) 5×10^4 Ω, (c) 10^5 Ω.
43. 25 V.
47. $R = R_1R_2/R_3$.
49. $R_x = (V/I)/(1 - V/IR_V)(R_V \gg R_x)$.
51. 20 μF.
53. 3.3 Ω.
57. $(\mathscr{E}/R_1)e^{-t/R_1C} + \mathscr{E}/R_2$.
59. 300-Ω resistors in parallel with each other and in series with the 250-Ω resistor and the two capacitors.
61. 6.8×10^3 s (1.9 h).
63. 0.417 A; 0.500 A; 0.167 A.
65. Six 600-Ω resistors in parallel; six 16.7-Ω resistors in series, two series 150-Ω resistors in parallel with two more sets of two series 150-Ω resistors; three series 66.7-Ω resistors in parallel with three series 66.7-Ω resistors.
67. (a) 0.25 kA, (b) 12.7 V, (c) 1.1×10^6 J.
69. $I = \{-(R + \alpha) + [(R + \alpha)^2 + 4\beta\mathscr{E}]^{1/2}\}/2\beta$.

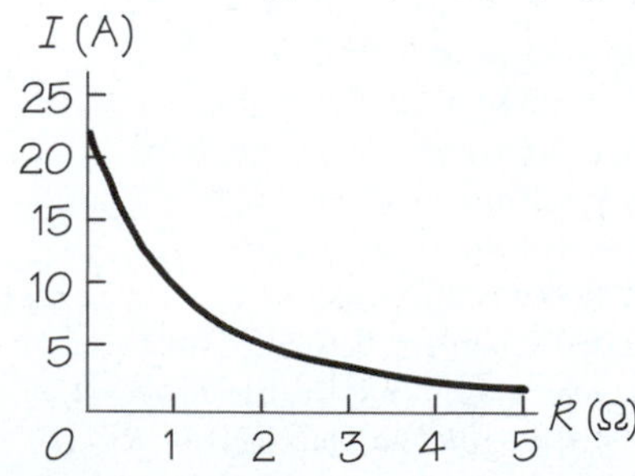

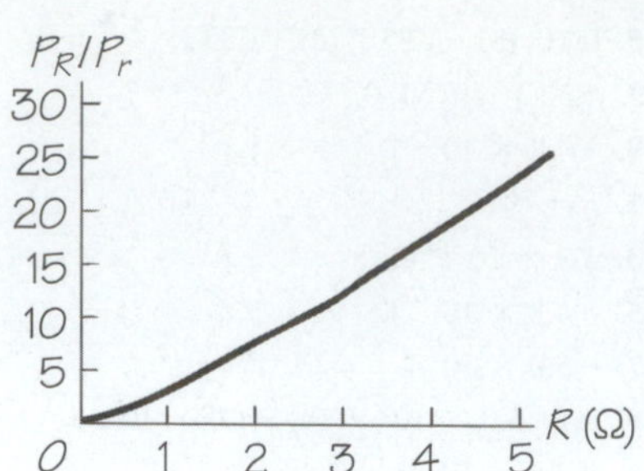

71. 667 W.
73. $I_{mixer} = 6.67$ A; $I_{vacuum} = 5.00$ A; $I_{chandelier} = 5.00$ A; 7 bulbs.
75. 0, $\mathscr{E}/2R$; $\mathscr{E}/R$.
79. (a) 21.4 μs, (b) 2.4 mC.
81. (a) V_0/d, (b) $\sigma\pi r^2 V_0/d$.

CHAPTER 28

1.

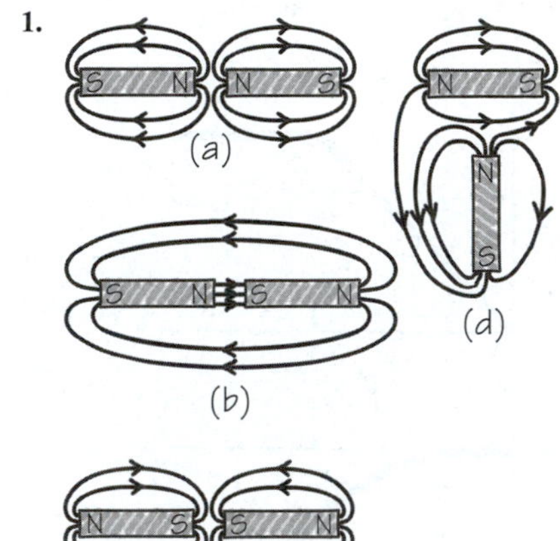

3. −z-direction.
5. $-(7.2 \times 10^{-3}\ \text{T})\hat{k}$.
7. 34 μT.
9. (a) $\theta = (eB\,\Delta t)/m$, (b) 0.15 T.
11. −4.2 mm (horizontal).
13. (a) 2.25×10^{-3} T, (b) 2.5×10^{-2} T.
15. 5.7×10^{-14} m; 1.4×10^{-6} N.
17. Magnetic field; 0.40 T; 1.20 T.
19. $R_e = 0.57$ m; $R_p = 10.4$ m; 1.04 km.
21. (a) 5.9×10^7 m/s, 3.4° from the original direction, (b) 1.2 cm.
23. (a) 24 J (15×10^{19} eV); (b) 16×10^{-8} kg·m/s; 48 J, (c) 8×10^{-8} kg·m/s; 24 J.
25. 6.8×10^{-5} m.
27. 500 eV.
29. The circular motion is parallel to the xz-plane with radius 5.7×10^{-5} m. The motion in the y-direction has a constant acceleration of -2.6×10^{14} m/s².
31. (a) 1.5×10^7 Hz, (b) 4.8×10^7 m/s tangential, (c) 1.9×10^{-12} J (1.2×10^7 eV), (d) 120, (e) 8.0×10^{-6} s.
33. 9.9×10^{-16} kg·m/s < p < 45×10^{-16} kg·m/s; 1.9 TeV < E < 8.4 TeV; 4.2×10^{-3} m/s.

35. 3.54 T.
37. $F_B = 2.4 \times 10^{-16}$ N;
$F_B/F_g = 2.6 \times 10^{13}$;
$K = 5.2 \times 10^{-43}$ J $(3.2 \times 10^{-24}$ eV$)$.
39. (a) $\vec{F}_B$ is to the left and $\vec{F}_E$ is to the right, (b) $F = q|E - vB|$, (c) $E = vB$.
41. $d\omega = qB/2m$.
43. 0.03 T.
45. 8.7×10^{-10} N perpendicular to the wire and to $\vec{B}$.
47. 5.0×10^{-2} N perpendicular to the wire and to $\vec{B}$.
49. $\Delta y = ILB/2k$.
51. 0.25 T.
53. $\mu = 0.11$ A·m² perpendicular to the loop; $\tau = 0.024$ N·m parallel to the loop and perpendicular to $\vec{B}$.
55. $IN\pi R^2 B(1 - \cos\theta)$.
57. -10^{-22} J $\leq U \leq +10^{-22}$ J.
59. (a) 5.0 mA, (b) $6 \times 10^{-5} \cos\theta_i$ J.
61. (a) 4×10^{10} m/s, (b) 10^{-23} N·m.
63. (a) $2IRB\,\hat{j}$, (c) $-2IRB\,\hat{j}$, (c) 0, (d) 0.
65. 0.43 rad/s.
67. 1.5×10^{27} carriers/m³.
69. 0.11 mA.
71. 5.8×10^{11} N/C; 2.6×10^5 T.
73. 0.037 T.
75. $\Delta r/r = 0.05$ (decrease).
77. $e/m = 2V/B^2R^2$.
79. (a) $L = NmvR$ perpendicular to the orbit, (b) $\frac{1}{2}NevR$, (c) $L/\mu = 2m/e$.
81. (b) horizontal direction, (c) $mg\hat{k} + qB(v_y\hat{\imath} + v_z\hat{j}) \times \hat{\imath} = m(dv_y/dt)\hat{j} + m(dv_z/dt)\hat{k}$; $\vec{v} = (mg/qB)\hat{j} + (mg/qB)(\sin\omega t\,\hat{k} - \cos\omega t\,\hat{j})$.

CHAPTER 29

1.

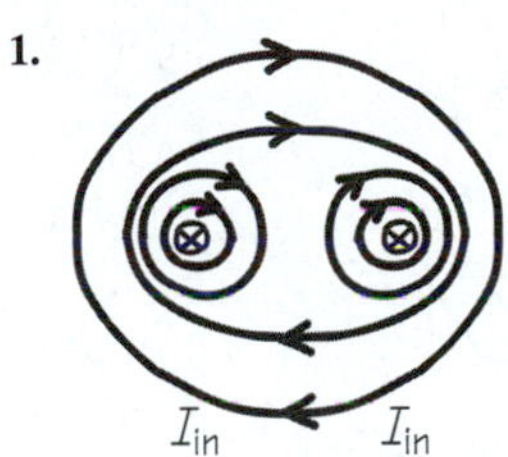

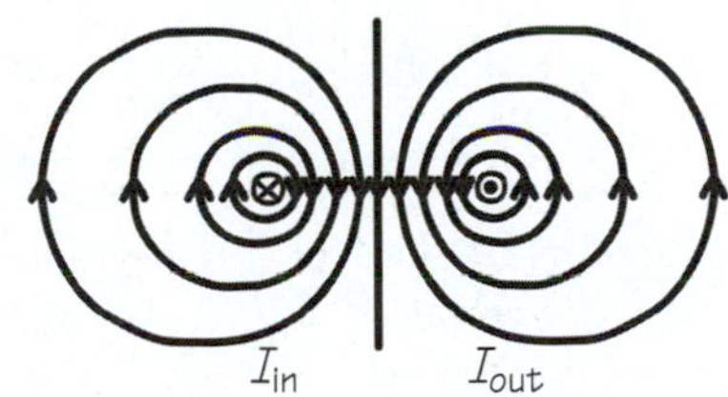

3. 1.7×10^3 N/m.
5. 0.25 μN/m attraction.
7. $[\mu_0] = [MLC^{-2}]$; $[\varepsilon_0] = [C^2M^{-1}L^{-3}T^2]$; 3.00×10^8 m/s.
9. (a) $r = R$, (b) $B_{max} = \mu_0 I/2\pi R$ at $r = R$, (c) $B_{min} = 0$ at $r = 0$ and $r = \infty$,

(d)

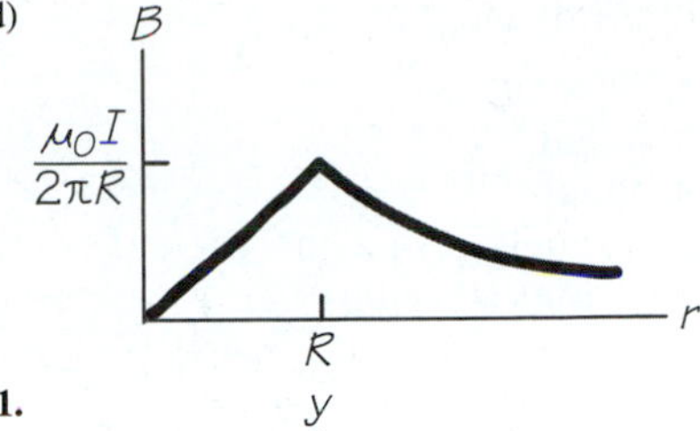

11.

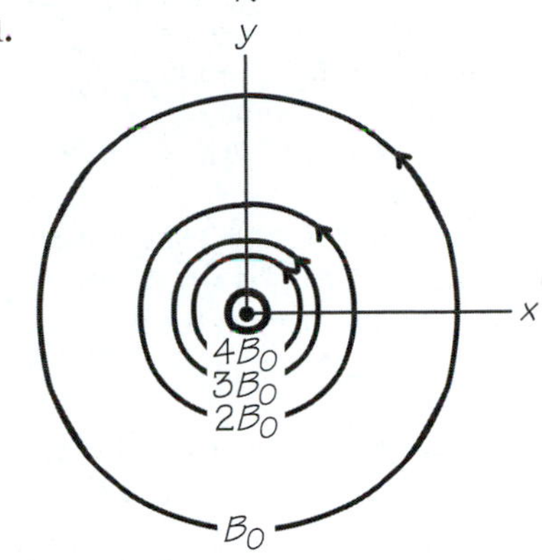

13. 3.0×10^{-3} T.
15. (a) $\dfrac{\mu_0 I}{2\pi(x^2 + y^2)}(-y\hat{\imath} + x\hat{j})$,

(b) $$\frac{\mu_0 I}{2\pi}\left\{-\left[\frac{y}{(x-a)^2 + y^2} + \frac{y}{(x+a)^2 + y^2}\right]\hat{\imath} + \left[\frac{x-a}{(x-a)^2 + y^2} + \frac{x+a}{(x+a)^2 + y^2}\right]\hat{j}\right\},$$

(c) $$\frac{\mu_0 I}{2\pi}\left\{\left[\frac{-y}{(x-a)^2 + y^2} + \frac{y}{(x+a)^2 + y^2}\right]\hat{\imath} + \left[\frac{x-a}{(x-a)^2 + y^2} - \frac{x+a}{(x+a)^2 + y^2}\right]\hat{j}\right\}.$$

17. 6.7×10^{-4} T circular.
19. (a) away from the wire, (b) 5.9×10^{-3} m, (c) same energy.
21. $(\mu_0 Ia/2\pi)\ln[(a + d)/a]$.
25. 4.8×10^{-3} T.
27. 0.35 m.
29. 6.0×10^{-5} Wb.
31. 64 mA.
33. (a) 4.3×10^{-6} T, (b) 2.1×10^{-2} T.
35. $(\mu_0 NIL/2\pi)\ln[(R + L)/R]$.
37. $\sqrt{2}(\mu_0 I/2\pi x)$ in the yz-plane 45° below the y-axis.
39. $\sqrt{2}(\mu_0 I/4\pi H)$ in the xy-plane 45° from the $-x$-axis and the $-y$-axis.
41. (a) $-(2.2 \times 10^{-4}$ T/m$)\,dL\,\hat{j}$, (b) $(0.56 \times 10^{-4}$ T/m$)\,dL\,\hat{k}$, (c) 0, (d) $-(0.20 \times 10^{-4}$ T/m$)\,dL\,\hat{j}$.
43. 2.0×10^{-7} T along the axis.
45. $B = 2\mu_0 I(a^2 + b^2)^{1/2}/\pi ab$;
$B = 2\sqrt{2}\mu_0 I/\pi a\ (a = b)$;
$B \approx 2\mu_0 I/\pi a\ (b \gg a)$.
47. $(5.14 \times 10^{-7}$ T·m/A$)\ I/R$ out of the page.
49. 2.8×10^{-5} T into the page.
51. Circular path; $R = 2\pi mvd^3/q\mu\mu_0$.
53. 2.1×10^{-4} T into the page.
55. $\frac{1}{3}Q\omega R^2\hat{k}$.
57. $(\mathscr{E}/R)e^{-t/RC}$.
59. 5.0×10^{-8} T·m.
61. $-CV_0\omega\sin(\omega t)$.
63. $I = I_d = -4\pi R^2\sigma_0/t_0$.
65. $(\mu_0 I/D)(2/\pi + 1)$ into the paper.
67. (a) $B = (\mu_0 I/2\pi R_1^2)r$ circular CCW, $r < R_1$, (b) $B = \mu_0 I/2\pi r$ circular CCW, $R_1 < r < R_2$, (c) $B = (\mu_0 I/2\pi r)(R_3^2 - r^2)/(R_3^2 - R_2^2)$ circular CCW, $R_2 < r < R_3$, (d) $B = 0$, $R_3 < r$.
69. (a) $-6.0 \times 10^{-5}\,\hat{j}$ N (attraction).
71. (a) $F_B/F_E = 4.3 \times 10^{-26}$, (b) 2.1×10^8 electrons/cm, (c) 2.1×10^{-13}.
73. (a) $$\frac{\mu_0 I}{2R}\left\{\frac{1}{[1 + (x/R)^2]^{3/2}} + \frac{1}{[2 - 2x/R + (x/R)^2]^{3/2}}\right\},$$
$B(0) = 0.677\mu_0 I/R$; $B(R/4) = 0.713\mu_0 I/R$; $B(R/2) = 0.716\mu_0 I/R$.
75. $NI = 336$ A·turns.

CHAPTER 30

1. Rotate the magnet around the coil; increase the speed of rotation.
3. 1.05×10^{-4} V.
5. $d\vec{B}/dt = (0.78$ T/s$)\,\hat{\imath}$.
7. $-BLv$ (clockwise) $(0 < t < L/v)$; $+2BLv$ (counterclockwise) $(L/v < t < 2L/v)$; $-BLv$ (clockwise) $(2L/v < t < 3L/v)$.
9. 18 V.
11. $-2B_0v_0^2t\ (0 < t < L/v_0)$.
13. $-CDv^2t/2\alpha(D + vt)$ clockwise.
15. $-CD^2v/4\alpha(D + vt)$ clockwise.
17. $BA\omega\sin(\omega t)$.
19. 8×10^{-6} V.
21. 2.9×10^{-4} V.
23. 0.20 T.
25. 1.5×10^{-2} V, bottom end at the higher potential.
27. (a) 1.2×10^{-21} N, (b) 2×10^{-5} V.
29. $-2Bv(2Rvt - v^2t^2)^{1/2}\ (0 < t < 2R/v)$.
31.
(a)

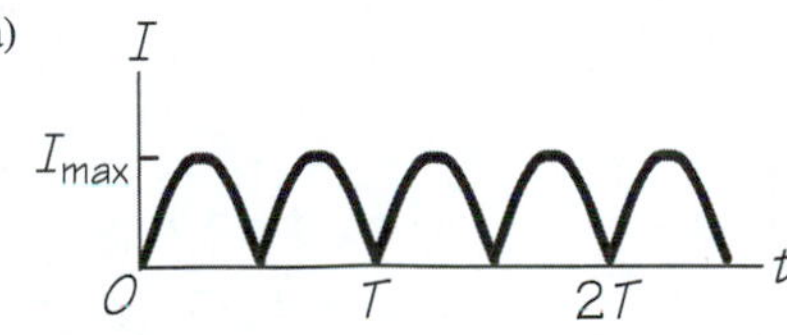

(b) $2NBA\omega/R\pi$, (c) $\pi I_{av}R/2NA\omega$.
33. 1.3×10^{-2} N; 0.46 W.
35. (a) 0.085 N, (b) 0.051 W.
37. $mgR/(BL)^2$.
39. (a) Speed increases until v_t is reached.
41. $(d\Phi_B/dt)^2/R$.
43. aAB_0e^{-at}.
45. $E = +(\mu_0 nI_0\omega r/2)\sin(\omega t)$, circular.
47. (a) 2.7×10^{-5} A, (b) in the same sense as the current in the solenoid.
49. -0.030 V (counterclockwise).

51. (a) 54 mA, (b) 4.4 mW.
53. 5.7 m/s.
55. Constant speed of E/B in the $-y$-direction; $B > 3 \times 10^{-5}$ T.
57. (a) -9.4×10^{-6} V (up), (b) 47 μA, (c) 9.7×10^{-10} J; external force.
59. (a) $(\pi r^2 \mu_0 N I \omega / 2R) \sin(\omega t)$, (b) 90°.
61. Counterclockwise; away from the wire.
63. 0.50 C.
65. $Q = (4.5 \times 10^{-9}\text{ C})[1 - e^{-t/(3.0 \times 10^{-5}\text{ s})}]$, with lower plate positive; $I = (1.5 \times 10^{-4}\text{ A})[1 - e^{-t/(3.0 \times 10^{-5}\text{ s})}]$, clockwise.
67. $\mathcal{E} = (B\pi a^2 \omega/2)\sin(\omega t)$; $P_{av} = B^2\pi^2 a^4 \omega^2/8R$.
69. 2.4×10^{-3} V.
71. (a) 1.8×10^5 m/s, (b) 1.4×10^{-20} J, (d) -10%.

CHAPTER 31

1. 1.0×10^{-2} A·m² in the direction of the magnetic field.
3. 992.
5. 5 turns/m.
7. 4.6×10^2 A/m.
9. 0.014 A.
11. (a) $-(2.4 \times 10^{-5})\vec{B}_0$, (b) $+(2.6 \times 10^{-4})\vec{B}_0$.
13. 0.064 A.
15. (a) 2.2×10^{25} electrons, (b) 2.2 A/m.
17. $\lambda \pi R^3 \omega$, along the axis of the ring.
19. $v = 4.1 \times 10^{10}$ m/s $> c$.
21. $M = 1.4 \times 10^6$ A/m; $\mu/\mu_0 = 3.6 \times 10^2$.
23. $(2.4 \times 10^{-3}\text{ T})\chi_m$.
25. (a) 0.16 T, (b) 2.5×10^{-4} T.
27. 1.45×10^2 A/m; 0.88 T.
29. 289 A/m.
31. 6.3×10^{-4} T.
35. -2.2×10^{-4}.
37. $H = (1.6 \times 10^{-3}\text{ A})/r$ circular; $B = (2.0 \times 10^{-9}\text{ T}\cdot\text{m})/r$ circular; $\Delta B = (1.3 \times 10^{-12}\text{ T}\cdot\text{m})/r$.
39. 0.38 A/m; 4.7×10^{-7} T.
41. (a) With $\vec{B} = B\hat{k}$, $dm_x/dt = g_S m_y B$; $dm_y/dt = -g_S m_x B$; $dm_z/dt = 0$.
43. 4.2×10^6 Hz.
45. 1.96 T; 0.16 T.
49. 1.4 A·m²/cm³.
51. $m = 8 \times 10^{22}$ A·m²; $M = 6 \times 10^2$ A/m; 2×10^9 A.
53. $m_{ring}/m_{disk} = 2$.
55. (a) $C = (NmB/2\pi kT)/(e^{mB/kT} - e^{-mB/kT})$, (b) $\langle \cos\theta \rangle = \dfrac{e^{mB/kT} + e^{-mB/kT}}{e^{mB/kT} - e^{-mB/kT}} - \dfrac{kT}{mB}$, (c)

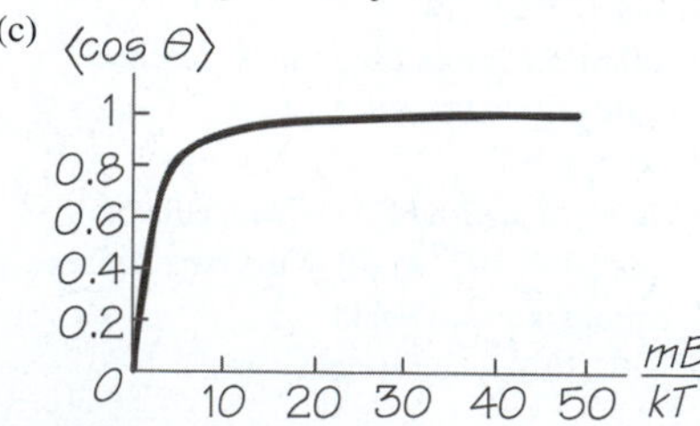

CHAPTER 32

1. 6.0×10^{-5} Wb.
3. 3.1 μH.
5. 30 mH.
7. (a) 1.0 mH, (b) 3×10^{-3} Wb.
9. (a) $\mu_0 A N_1^2/\ell$ or $\mu_0 A N_2^2/\ell$, (b) $\mu_0 A(N_1 + N_2)^2/\ell$, (c) $\frac{1}{4}\mu_0 A(N_1 - N_2)^2/\ell$, (d) $\mu_0 A N_1 N_2/\ell$.
13. $I = 0$ for $t < 0$; $I = -0.40$ μA for $0 < t < 0.30$ s; $I = +0.40$ μA for $0.30\text{ s} < t < 0.60$ s; $I = 0$ for $t > 0.60$ s.
15. $\mathcal{E} = -10$ mV for $0 < t < T/4$, $3T/4 < t < 5T/4, \ldots$; $\mathcal{E} = +10$ mV for $T/4 < t < 3T/4$, $5T/4 < t < 7T/4, \ldots$.
19. (a) 3.9×10^{-3} H, (b) 16 H.
21. $L = \dfrac{\mu_0 \ell}{\pi}\left[\dfrac{1}{2} + \ln\left(\dfrac{a}{r}\right)\right]$. The radius of the wire cannot be neglected.
23. $(\mu_0 L/2\pi)\ln[(a + b)^2/b(2a + b)]$.
25. 1.2 kW.
27. 24 A.
29. $0.64\, e^{-(0.04/\text{s})t}$ mW.
31. (a) Same, (b) same.
33. (a) 6.4×10^3 J, (b) 2.4 L.
35. 2.6×10^2 J.
37. 11 cm.
41. There is an upward force on the cylinder before it completely drops into the solenoid.
43. 1.6×10^{-9} J/m.
49. 25 Ω; 5.6 mH.
51. $V^2 L/2R$.
53. 1.6 mF.
55. 1.5 μF; 0.36 mH.
57. (a) $\omega_1/10$, (b) $\alpha_1/100$, (c) $\omega_2'^2 = (\omega_1^2/100) - (\alpha_1/100)^2$.
59. $\omega' = 6.6 \times 10^4$ rad/s.
61. $2Q_0^2/(LC^3)^{1/2} e^{-2\alpha t}$.
65. $T + \frac{1}{4}\pi R^2 (C^3/L)^{1/2}$.
67. (a) 47 mA, (b) 2.2×10^{-8} J, (c) $U_{L\max}/U_{C\max} = 1$.
69. (a) 4.0×10^{-3} Ω, (b) 1.5×10^{-7} J at $t = 0$; 4.2×10^{-9} J at $t = 100$ oscillations, (c) Joule heating.
73. (a) $I_2 = 0$; $I_1 = I_{\mathcal{E}} = I_L = 2.4$ mA, (b) 0.28 mH, (c) $I_1 = I_{\mathcal{E}} = 0$; $I_L = I_2 = 0.86$ mA.
75. $(\mu_0 I^2/4\pi)[(1/4) + \ln(r_2/r_1)]$.
77. (b) c.
79.

V
Voltage
Time
t

81. 3.2×10^{-2} H; the iron core increases B and M, and concentrates flux.
83. $L = 0.1$ mH; $C_1 = 3.9$ μF; $C_2 = 5.0$ μF.

CHAPTER 33

1. 9000 V.
3. 60 turns.
5. 450 turns; 27 A.
7. 0.40 A.
9. 50 turns.
11. 13 kHz.
13. 1.1×10^4 rad/s.
15. 22.8 μF.
17. $X_{C\max} = 1.7 \times 10^3$ Ω; $X_{L\max} = 300$ Ω.
19. 15 Ω; 95 Hz.
21. (a) $\vec{D}$ is more advanced in phase, (b) 90°; $\vec{D}$ leads, (c) $\tan\delta = A/B$.

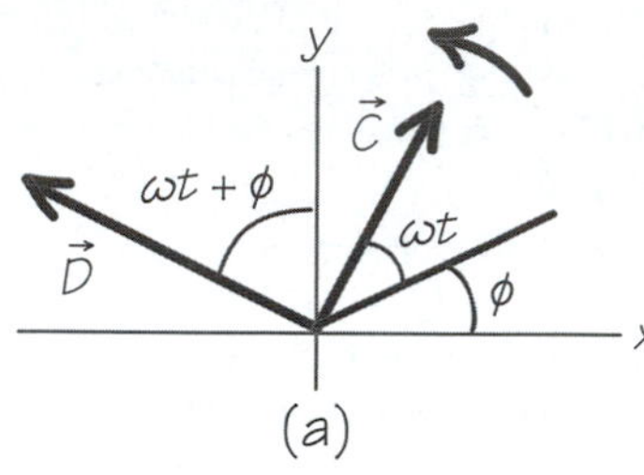

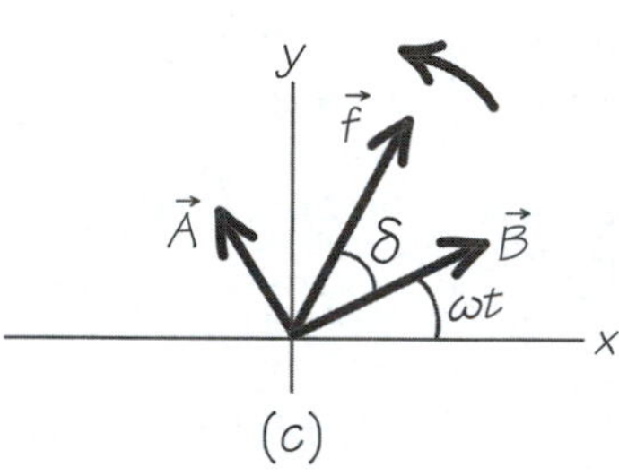

23. 35 pF.
25. $R_2 = \frac{1}{2}R_1$.
27. 2.7×10^{-4} C; 4.0 V.
29. 6.4 H.
31. $X_C = 66.3$ Ω; $X_L = 694$ Ω; $Z = 802$ Ω; $Q_{max} = 13$ μC; $\phi = +51.5°$; $I_{max} = 0.10$ A.
33. $V_C = -4.05$ V; $V_L = +43.1$ V.
35. $I_{max} = 92$ mA; $V_{R\max} = 55.5$ V; $V_{C\max} = 123$ V; $V_{L\max} = 27.9$ V.
37. Double.
39.

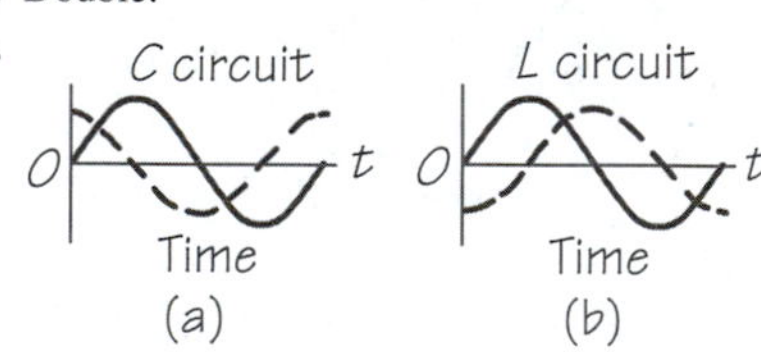

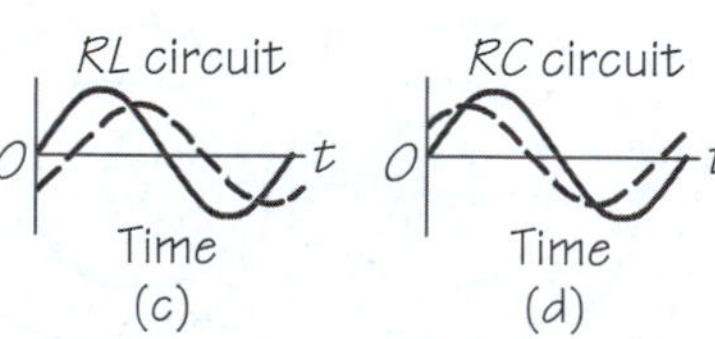

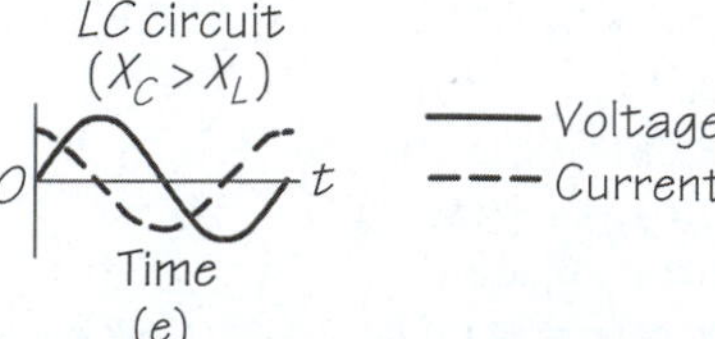

41. $V_{C\max} = 66$ V; $V_{RL\max} = 67$ V.

43. 11 μH.

45.

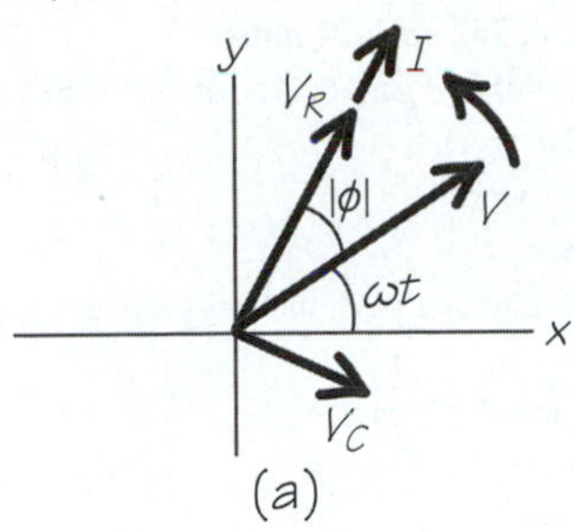

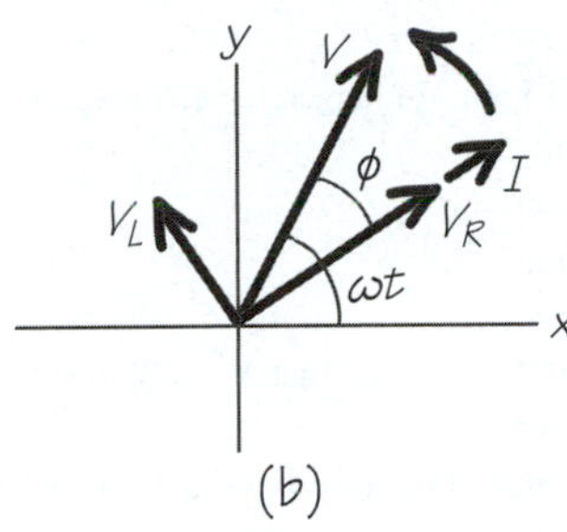

47. 5.0 A.

49. (a) $\langle P \rangle \to \frac{1}{2}V_0^2 R/\omega L \to 0$,
(b) $\langle P \rangle \to \frac{1}{2}V_0^2 R\omega C \to 0$; no current through the capacitor.

51. (a) 0, (b) 0, (c) 1.

53. (a) $R/(X_L^2 + R^2)^{1/2}$, (b) $R/(X_C^2 + R^2)^{1/2}$, (c) 0.

55. (a) 1.6 MHz; (b) 0.034 Ω.

57. 27 μF; 46 μF.

59. 345 V; 9.05 kW.

61. 9.7 kW; 16.7 kW.

63. $R = 22.2\ \Omega$; $X_L = 98\ \Omega$; $X_C = 168\ \Omega$.

65. (a) 31 W, (b) 17 W, (c) 34 W.

67.

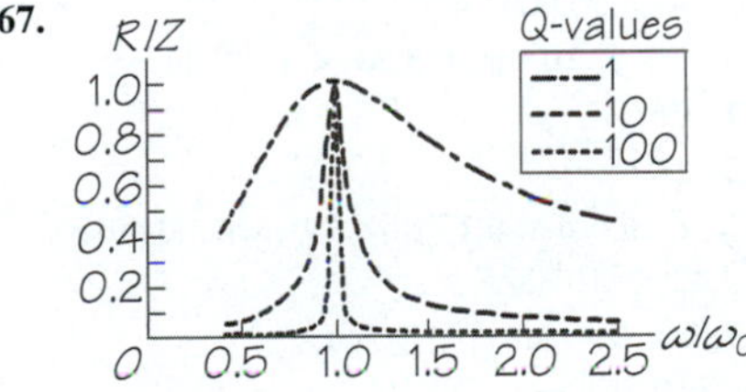

69. (a) 0.20 V, (b) 0.19 V, (c) 5.3×10^{-6} V.

71. $R/L = 5.0 \times 10^4\ \mathrm{s}^{-1}$.

73. $\langle I \rangle = I_0/\pi$; $I_{\mathrm{rms}} = \frac{1}{2}I_0$,

75. At low frequency $Z \to X_C$; at high frequency $Z \to R$.

77. $I = (V_0/Z)\cos(\omega t + \phi)$;
$I_C = (V_0/X_C)[(R/Z)\sin(\omega t + \phi) - \sin(\omega t)]$;
$I_L = (V_0/X_L)$
$[-(R/Z)\sin(\omega t + \phi) + \sin(\omega t)]$; there is a resonant frequency.

79. (a) $V_{C1} = 5.14$ V; $V_{C2} = 2.20$ V, (b) 2.34 V.

83. 29 mW.

85. (a) 5.7 V, (b) 18 V.

87. 2.4 μH; 0.060 Ω.

89. 177 Ω.

91. (a) $Q_c(t) = CV_0e^{i\omega t}$; $I_c(t) = i\omega CV_0e^{i\omega t}$.

93. (a) $Q_{0C} = V_{0C}/\{\omega[i(L\omega - 1/\omega C) + R]\}$;
(b) $Q_c(t) = -i(V_0/\omega Z)e^{i(\omega t - \phi)}$;
$I_c(t) = (V_0/Z)e^{i(\omega t - \phi)}$,
(c) $Q = -(V_0/\omega Z)\cos(\omega t - \phi)$;
$I = (V_0/Z)\sin(\omega t - \phi)$.

95. $\omega = [(1/LC) - (R^2/4L^2)]^{1/2} + i(R/2L)$.

CHAPTER 34

3. $\oint \vec{E}\cdot d\vec{s} = \mu_0 \frac{dM}{dt} - \frac{d}{dt}\iint_s \vec{B}\cdot d\vec{A}$.

5. $\vec{B} = (E_0/c)\cos(kz + \omega t)\hat{i}$, traveling in the $-z$-direction.

7. 94×10^6 Hz; 3.2 m.

13. $+\frac{\partial B_y}{\partial x} = \mu_0\varepsilon_0\frac{\partial E_z}{\partial t}$ and $+\frac{\partial B_y}{\partial t} = \frac{\partial E_z}{\partial x}$.

15. Plane formed by the z-axis and the line $y = -(\tan\theta)x$.

17. $(3\ \mathrm{V/m})\cos[(1.05 \times 10^7\ \mathrm{m}^{-1})z - (3.15 \times 10^{15}\ \mathrm{rad/s})t]\hat{i}$.

19. $\lambda = 2L/n$, $n = 1, 2, 3, \ldots$.

21. 2.2×10^{-4} T.

23. $2.6 \times 10^{-5}\ \mathrm{W/m^2}$.

25. $8.0 \times 10^{-11}\ \mathrm{N/m^2}$.

27. (a) $1.3 \times 10^3\ \mathrm{W/m^2}$, (b) $4.5 \times 10^{-6}\ \mathrm{N/m^2}$.

29. $1.2 \times 10^{-2}\ \mathrm{W/m^2}$; 0.070 J.

31. $E_0 = 134$ V/m; $E_{\mathrm{rms}} = 95$ V/m;
$B_0 = 4.47 \times 10^{-7}$ T; $B_{\mathrm{rms}} = 3.16 \times 10^{-7}$ T.

33. $2.0 \times 10^{-3}\ \mathrm{N/m^2}$.

35. 17 kg · m/s; 3 kg · m/s (about 1/5 of the solar result).

37. 2.5×10^{-5} N · m.

39. $P = 5.9 \times 10^5$ W; vaporize.

41. $\Delta p/A = 2(S/c)\cos^2\theta\,\Delta t$.

43. $\vec{E}(x, t) = (4.8 \times 10^3\ \mathrm{V/m})$
$\sin\{2\pi[(20\ \mathrm{m}^{-1})x + (6 \times 10^9\ \mathrm{Hz})t]\}\hat{k}$;
$\vec{B}(x, t) = (1.6 \times 10^{-5}\ \mathrm{T})$
$\sin\{2\pi[(20\ \mathrm{m}^{-1})x + (6 \times 10^9\ \mathrm{Hz})t]\}\hat{j}$.

45. $\tau_{\mathrm{average}} = 3.1 \times 10^{-16}$ N · m.

47. 4.0×10^{-7} s.

49. (a) $+y$-direction, (b) $-x$-direction, (c) $+z$-direction, (d) electric field in $-y$-direction; magnetic field in $+x$-direction; Poynting vector in $+z$-direction.

51. (a) no solution, (b) 39°, (c) 57°, (d) 72°.

53. 37°.

55. $3.5 \times 10^5\ \mathrm{W/m^2}$.

57. 0; 1/8.

59. (a) $\frac{1}{2}I_0$, (b) 90°, (c) $(1/8)I_0\sin^2(2\theta)$.

61. The superposition is acceptable since the sum of solutions to the Maxwell's equations is itself a solution.

63. $4.7 \times 10^{-5}\ \mathrm{W/m^2}$.

65. (a) $\vec{B} = \mu_0 nI_0\cos(\omega t)\hat{k}$ (along the axis),
(b) $\vec{E} = \frac{1}{2}\mu_0 nI_0\omega r\sin(\omega t)$ (circular),
(c) $0 < t < \frac{1}{4}T$: $\vec{S}$ is in $+\hat{r}$;
$\frac{1}{4}T < t < \frac{1}{2}T$: $\vec{S}$ is in $-\hat{r}$;
$\frac{1}{2}T < t < \frac{3}{4}T$: $\vec{S}$ is in $+\hat{r}$;
$\frac{3}{4}T < t < T$: $\vec{S}$ is in $-\hat{r}$.

69. $\vec{E} = (5.19 \times 10^2\ \mathrm{V/m})$
$\cos[(2.54 \times 10^6\ \mathrm{m}^{-1})x + (1.47 \times 10^6\ \mathrm{m}^{-1})y - (2\pi \times 10^{14}\ \mathrm{s}^{-1})t]\hat{k}$.

71. 5.8×10^5 J; 0.25 kg.

73. $4.5 \times 10^{13}\ \mathrm{photons/m^3}$.

75. $f' = [(c - v)/(c + v)]f$.

77. $a_J = 2.3 \times 10^{-7}\ \mathrm{m/s^2}$;
$a_S = 7.1 \times 10^{-8}\ \mathrm{m/s^2}$; 93 years.

79. $\frac{\partial^2 E_x}{\partial z^2} = \varepsilon_0\mu_0\frac{\partial^2 E_x}{\partial t^2} + \frac{\mu_0}{\rho}\frac{\partial E_x}{\partial t}$.

81. (a) $\vec{B} = (\mu_0 r/2\pi R^2)\,dQ/dt$ circular, for $r < R$, (b) $\vec{S} = -(Qr/2\pi^2R^4\varepsilon_0)(dQ/dt)\hat{r}$.

CHAPTER 35

1. $v_{\mathrm{ice}} = 2.29 \times 10^8$ m/s;
$v_{\text{ethyl alcohol}} = 2.21 \times 10^8$ m/s;
$v_{\mathrm{benzene}} = 2.00 \times 10^8$ m/s;
$v_{\mathrm{diamond}} = 1.24 \times 10^8$ m/s.

3. 492 nm; 4.62×10^{14} Hz.

5. 2.9×10^4 rev/min.

7. 6 m.

9. 1.0 ns.

11. 1.3 cm.

13. 1.62.

15. 590 m.

17. 0.031.

19. 1.4°.

21. $\theta_{\mathrm{water}} = 45.7°$; $\theta_{\mathrm{glass}} = 38.5°$.

23. Cannot be changed.

25. 7.4°; no total internal reflection.

27. Outside the "shadow" of the prism, there is direct illumination. For a distance of 0.57 cm from each edge, there will be no illumination. For the next distance of 0.77 cm − 0.57 cm = 0.20 cm, there will be illumination from one half. For the next distance of 0.73 cm to the center of the pattern, there will be illumination from both halves.

29. 8.3°.

31. 19 cm below the upper surface of the glass.

33. 74° from the normal.

35. 5.3 mm.

37. 0.86 m.

39. Prism will not be totally reflecting; $n_2 \le 1.283$.

41. 1.5.

43. 52%.

51. 0.48°.

53. (a) $\omega \le 12.5 \times 10^{15}$ rad/s, (b) 27.7°.

55. 0.009.

57. 1.31 cm.

59. $\Delta n = [(\cos\theta_4)(\cos\theta_2)/\sin(2\phi)]\,\Delta\theta$, where $\cos\theta_2 = \cos\{\sin^{-1}[(\sin\phi)/n]\}$ and
$\cos\theta_4 =$
$\cos\{\sin^{-1}[n\sin(2\phi - \{\sin^{-1}[(\sin\phi)/n]\})]\}$.

61. $\theta_i = \sin^{-1}(n\sin\phi)$.

63. (a) 4.6×10^6 km, (b) 3.1×10^8 m/s.

CHAPTER 36

1. 3 images.

3. 82.5% of the beam has been dissipated.

5. $0.577a$, a.

7. $i_2 = 26.3$ cm; $R = 30$ cm.

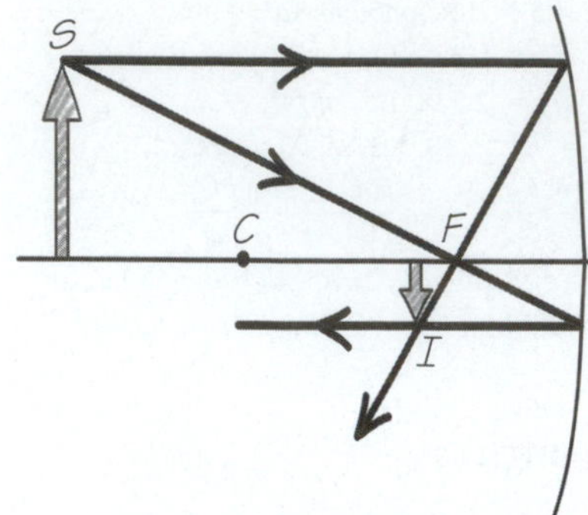

9. 0.80 cm tall, inverted, in front of the mirror.
11. $i = -400$ cm (behind the mirror); +50 cm tall.
13. $-f^2v/(s - f)^2$; $s = 2f$.
17. $i = -90$ cm (in air in front of glass).
19. $i = -30$ cm (behind the glass).
21. $i = +18.4$ cm; 43 cm from the surface.
23.

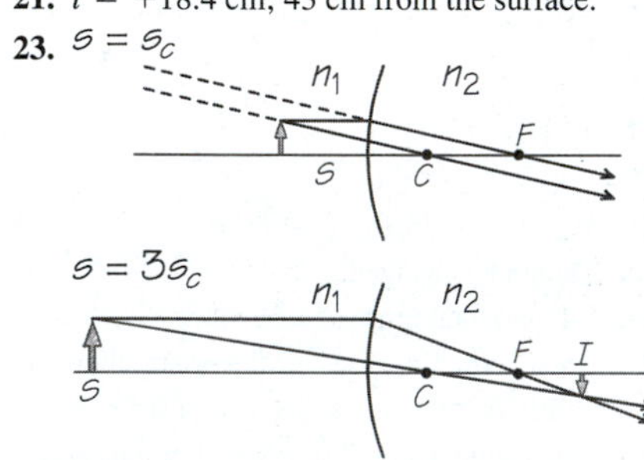

25. (b) $s = n_1R/(n_2 - n_1)$, (c) image is very far in front of the boundary, (d) image approaches the boundary.
31. (a) +16 cm, (b) converging, (c) inverted, (d) −2.1.
33. (a) 8.9 cm in front of the lens, (b) no, (c) yes, (d) +0.60.
35. (a) +38 cm, (b) +13 cm.
37. Virtual image.

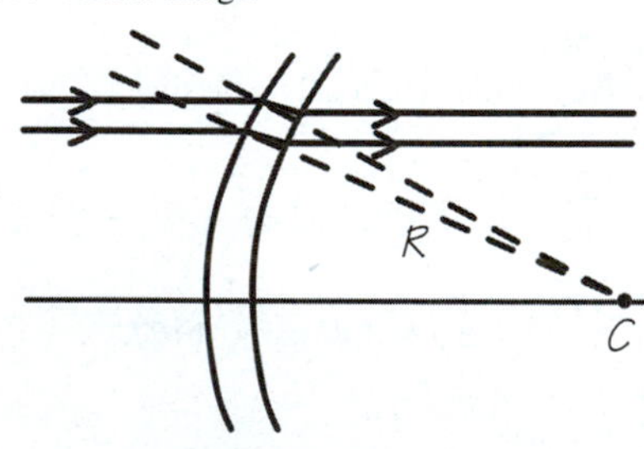

39. Lens *a*: (a) $f_a = 32.1$ cm, 33 cm beyond the lens, (b) inverted and real, (c) $M-$; Lens *b*: (a) $f_b = +77.9$ cm, 84 cm beyond the lens, (b) inverted and real, (c) $M-$; Lens *c*: (a) $f_c = -32.1$ cm, 31 cm in front of the lens, (b) upright and virtual, (c) $M+$; Lens *d*: (a) $f_d = -77.9$ cm, 73 cm in front of the lens, (b) upright and virtual, (c) $M+$.
41. Lens *a*: image is 63.4 cm to the left of the lens, inverted, real, with $M = -0.98$; Lens *b*: image is 393 cm to the right of the lens, upright, virtual, with $M = +6.0$; Lens *c*: image is 21.5 cm to the right of the lens, upright, virtual, with $M = +0.33$; Lens *d*: image is 35.4 cm to the right of the lens, upright, virtual, with $M = +0.55$.
45. +53 cm.
47. +6 cm.
49. 9.6 cm.
51. f_1/f_2.
55. 1.09 cm.
57. Near *positive* lens: image is 15.8 cm in front of the negative lens, or 0.8 cm from the positive lens on the object side; near *negative* lens: image is 65.4 cm in front of the positive lens, or 50.4 cm from the negative lens on the object side; the order of the lenses is important.
59. (a) First image is 12.4 cm behind M_1, (b) second image is 7.9 cm in front of M_2.

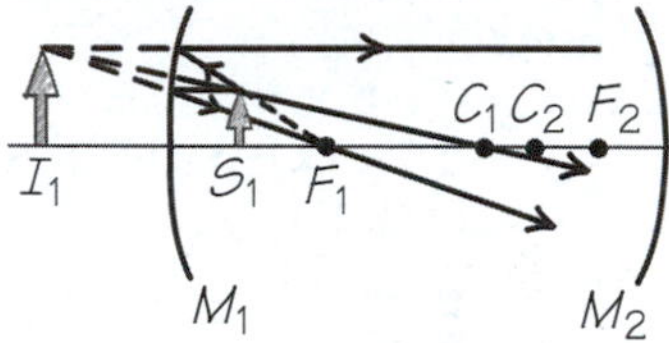

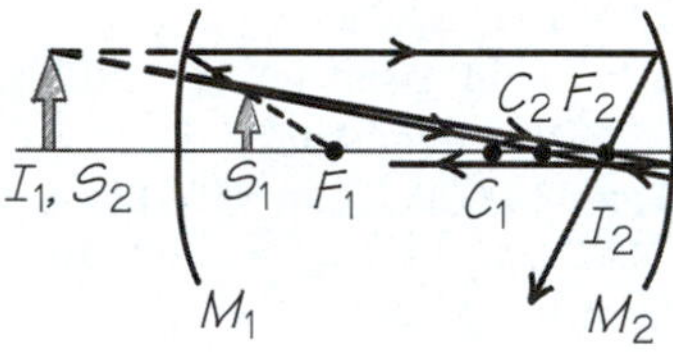

61. $f = [f_1(d^2 - f_1d - 2f_2d + f_1f_2)]/(d^2 - 2f_1d - 2f_2d + 2f_1f_2 + f_1^2)$.
63. $d \rightarrow R/(n - 1)$.
65. The plate may be placed anywhere between the lens and the screen, with thickness 13.1 cm.

CHAPTER 37

1. 4.7×10^2 nm.
3. 0.24 m.
5. 12.2 m.
7. (a) Doubles, (b) reduces by $\frac{1}{2}$, (c) doubles, (d) no change.
9. 6.4×10^2 nm.
11. Inward.
13. $v_{\text{fringe, ripple}} = 3 \times 10^{-6}$ m/s; $v_{\text{fringe, optical}} = 1.2 \times 10^6$ m/s.
15. (b) $\theta = \tan^{-1}\{\lambda[1/(d^2 - \lambda^2) + 1/4R^2]^{1/2}\}$.
17. 3.0×10^3 W/m².
19. 151°; 525 nm.
21. $I_{av} = 2I_0$.
23. $\Delta\theta = 2\sin^{-1}(\lambda/4d)$.
25. (a) along the perpendicular bisector, (b) 2.0×10^{-3} W/m², (c) 9.5°.
27.

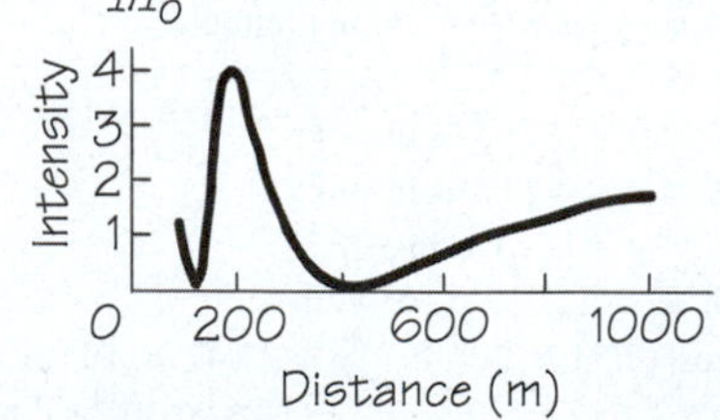

29. 33.6 μm.
31. 81 nm, 242 nm, 404 nm.
33. $t_{min} = 0.208$ μm; 0.416 μm and 0.625 μm would also work.
35. $r = (\lambda R)^{1/2}$.
37. 26 μm.
39. 1000 maxima pass; the rings move in to the center.
41. 6.4 mm; 8.5 mm.
43. $x = (n\lambda R)^{1/2}$.
45. 118 nm.
47. 790 nm.
49. 621 nm.
51. (a) 115 nm, (b) longer, (c) no reflection of green light.
53. 120 nm.
55. 120 nm.
57. 633 nm.
59. 1.00028, assuming that $\lambda = 632.8$ nm (for He-Ne laser).
61. 100 fringes must be counted to 0.1 of a fringe.
63. Uniform.
65. $u \propto \sin^2[2\pi f(x - \frac{1}{2}L)/c]$.
67. $\sin\theta' = \sin\theta - (m\lambda/d)$, $m = 0, \pm1, \pm2, \ldots$
69. 34.0.
73. Two: 0 or $4I_0$; three: I_0; four: 0.
75. $\theta = \sin^{-1}\{n^2 - [\lambda(m - \frac{1}{2})/2h]^2\}^{1/2}$, $m = 1, 2, 3, \ldots$.

CHAPTER 38

1. 450 nm.
3. 39.7 lines/cm.
5. 3.0×10^3 lines/cm.
7. 15,000, 30,000, 45,000; 0.028 nm, 0.014 nm; 0.0093 nm.
9. 3.1×10^4 lines; 8.83×10^{-4} rad/nm.
11. 76°.
13. 18.6°.
15. First order: 6.17° to 9.79°; second order: 12.42° to 19.88°.
17. $\sin\beta = \sin\alpha - m\lambda/d$, $m = 0, \pm1, \pm2, \ldots$.
19. 0.51 m.
21. (a) Maxima: −68°, 0°, 68°; minima: −38°, 38°, (b) 2.0 m.
23. $I_0/I_1 = 22.2$; 1.19°; increase θ.
27. $\Delta\theta_h/\Delta\theta_1 = 0.443$.
29. $\sin\theta = \sin\theta_i - m\lambda/a$, $m = \pm1, \pm2, \ldots$; there is a "central maximum."
31. (b) $\alpha_{max1} = 257.43°$, midway between 1st and 2nd minima = 270°; $\alpha_{max2} = 442.61°$, while midway between 2nd and 3rd minima = 450°.
33. 5.0 cm; 83 m.
35. 70 m.
37. 0.73 m; shorter wavelengths.
39. 2.4 mm.
41. 21 ft.
43. Third order.

45. I/I_0: 3.97, 1.79, 2.43, 0.26.
47. $a = 0.1$ mm; $d = 0.3$ mm.
49. $n = 3m = 3, 6, 9, \ldots$; 0.084°, 0.169°, 0.253°, ...
51. 0.0274°.
53. 0.21 nm.
55. 29.0°.
57. 15×10^3 m.
59. (a) 1.0×10^{-5} rad $= (5.7 \times 10^{-4})°$, (b) 25 μm.
61. (a) $I_0\{\sin[(N\pi \cos\theta)/(N-1)]/\sin[(\pi\cos\theta)/(N-1)]\}^2$, (b) slight decrease in I for θ close to 0° and 180° and then a buildup to the maxima at 90° and 270°.
63. No.
65. 0.45 mm.
67. 22°.

CHAPTER 39

1. 4.04 h; 4.02 h.
3. 3.6×10^8 m/s; no violation.
5. 0.011 s.
7. 3.7×10^9 s (120 yr).
9. (a) 3.3×10^{12} s (10^5 yr), (b) 3.3×10^5 s ($\simeq$ 4 days).
11. 4.2×10^2 m/s.
13. 0.66 h.
15. (a) 24 m, (b) 13×10^{-8} s, (c) 10×10^{-8} s, (d) 5.0×10^{-8} s.
17. Not regularly; the rate is lower on his way out and higher as he returns.

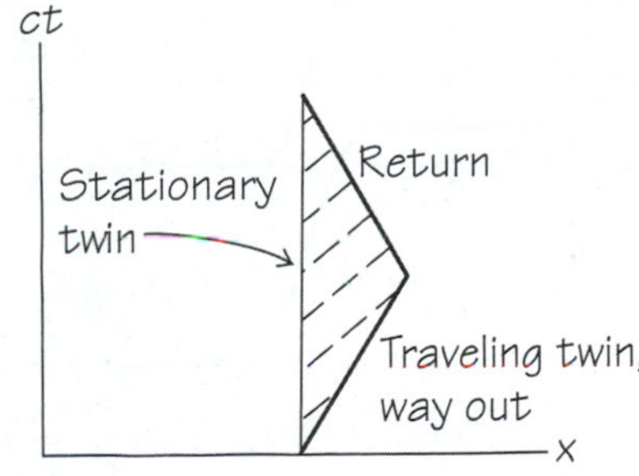

21. 5.4×10^8 s (17 yr).
23. 2.3×10^4 m/s toward Earth.
25. 437 m.
27. (a) 1.2×10^7 m/s, (b) 4.8×10^{24} m (5.1×10^8 ly).
29. $f_2 - f_0 = -1.3 \times 10^{10}$ Hz.
31. 5 yr; (1/5) yr; 20 yr.
33. 0.69c.
35. (a) Distance between the bags is not a measured length, (b) 140 m.
37. (a) $x = 0$, $t = 24$ s. (b) 9.6×10^9 m; 40 s.
39. 0.23c.
43. $(c/n)(1 + un/c)/(1 + u/cn)$.
45. Same form.
47. 2.1×10^{-21} kg·m/s.
49. 1.1×10^{-4} kg/s (3.5×10^3 kg/yr).
51. 4.4×10^9 kg/s.
53. (a) 0.047 MeV, (b) 36 MeV, (c) 110 MeV, (d) not possible.
55. (a) $[1 - (7.9 \times 10^{-7})]c$, (b) 745 GeV.
59. 767 MeV/c^2.
61. (a) 212.19 MeV, (b) 210.42 MeV, (c) 210.50 MeV.
63. $K_M = (M^* - M)^2c^2/2M^*$.
65. $\vec{F} = m\gamma(d\vec{u}/dt) + [m(\gamma^3u/c^2)\, du/dt]\vec{u}$.
67. $\phi = \frac{1}{2}R^2\omega^2$; slow.
69. c.
71. 2×10^{-3} nm.
73. 400 s; 365 s.
75. (a) 0.90c, (b) 2.33 GeV.
77. (a) p; south, (b) $\frac{1}{2}c(M^2 - 4m^2)^{1/2}$.
79. $G(f) \propto e^{-mc^2(f-f_0)^2/2kT}$.
81. (a) 0.933c, (b) 3.6×10^2 MeV/c (1.9×10^{-19} kg·m/s), (c) 12 m.
83. 6.5 GeV/c (3.47×10^{-18} kg·m/s).

CHAPTER 40

1. 0.2 MeV.
3. (a) 8.3×10^{-5} eV, (b) 1.9 eV, (c) 4.0×10^{-7} eV, (d) 7.3 keV.
5. 288 nm; only cesium is useful.
7. 1.8 eV.
9. $\lambda_{max,\,Al} = 290$ nm; $\lambda_{max,\,Cs} = 579$ nm; $\lambda_{max,\,Ni} = 241$ nm; $\lambda_{max,\,Pb} = 292$ nm.
11. (a) 2.4 eV, (b) 12 keV, (c) 1.2×10^{-4} eV, (d) 1.2×10^{-7} eV, (e) 1.2×10^{-8} eV.
13. 1.0×10^{45} photons/s.
17. $I = 6 \times 10^{-14}$ W/m^2; $I/I_0 = 4 \times 10^{-17}$.
19. 5.8×10^3 K.
21. $W = 4.60$ eV (7.37×10^{-19} J); $K_{max} = 5.75$ eV (9.21×10^{-19} J).
23. 88.4°.
25. 2.0×10^{-2} J/m^3; 2.6×10^{-2} J/m^3; 0.77.
27. 5045K.
29. (a) 0.55 nm, (b) 0.17 nm, (c) 5.5×10^{-3} nm, (d) 2.49×10^{-15} m, (e) same size as the wavelengths.
31. 4.0 nm.
33. (a) 1.8×10^{-16} J (1.1×10^3 eV), (b) 9.6×10^{-20} J (0.60 eV).
35. $h/(2mE)^{1/2}$; $hc/(E^2 + 2mc^2E)^{1/2}$.
37. 1.4×10^{-14} m; 4.6×10^{-15} m; 8.5×10^{-16} m.
39. 2.4×10^{-17} J (150 eV); 1.3×10^{-20} J (0.082 eV).
41. 0.36.
43. 0.
47. 10^{-19} J (0.6 eV).
49. 1.5×10^{-14} J (0.092 MeV).
51. 8.8×10^{-21} kg·m/s; 2.3×10^{-14} J (0.14 MeV).
53. 2.7×10^{-4} nm.
55. 3×10^8 m; 6×10^5 m.
57. $\Delta y = a + hD/\pi pa$; $a_{min} = (hD/\pi p)^{1/2}$.
59. −0.14 MeV.
61. 0.33.
63. 1.57×10^4 yr ago; 13,700 B. C.
65. Does support the assertion.
67. 1.6×10^{-7} Ci.
69. 0.20%.
71. $\lambda = 0.027$ nm; $d = 3.5$ nm; $T = 0.02$K.
73. (a) 480 nm, (b) 4500K.
75. 1.2%.
77. $\lambda = h/p = h/\{2m_e[E - U(x)]\}^{1/2}$.

CHAPTER 41

1. 91.2 nm.
3. 12.1 eV, 103 nm; 0.97 eV, 1280 nm.
5. -2.86×10^3 eV.
7. (a) $n = 2$, 3 and 4, (b) 122 nm; 103 nm; 97 nm; ultraviolet region.
9. 7.58 nm.
11. Paschen series: 1876 nm, 1287 nm, 1094 nm, 1005 nm, 955 nm, ..., 820 nm; Bracket series: 4052 nm, 2626 nm, 2165 nm, 1945 nm, ..., 1459 nm; Pfund series: 7460 nm, 4654 nm, 3740 nm, ..., 2280 nm.

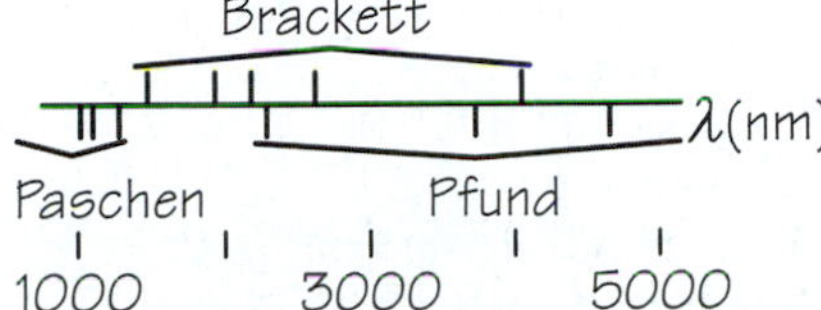

13. (a) $\alpha = 0.00731$ (a dimensionless constant); $1/\alpha = 137$, (b) $-m_ec^2\alpha^2/2n^2$, (c) αc.
15. (a) $E_{Cu} = 10.7$ keV; $E_W = 72.5$ keV.
17. $\frac{2}{3}(e^2/4\pi\varepsilon_0)^7 m_e^2/c^3L^8$.
19. $f = (m_e/4\pi\hbar^3)(e^2/4\pi\varepsilon_0)^2[1/n^2 - 1/(n+k)^2]$; $f \simeq (m_e/2\pi L^3)(e^2/4\pi\varepsilon_0)^2k$ for $n \gg k$; $f(n \gg k) = k(1/T)$.
21. $(3.3 \times 10^{-4})°$.
23. 0.033 nm.
25. $-(6.8\text{ eV})/n^2$; all transitions are in the UV for $n = 1$ series; all transitions in the IR for $n = 2$ series; all other transitions (up to $n \simeq 80$) in the IR.
27. $n = 1$, $\ell = 0$; $n = 2$, $\ell = 0$; $n = 2$, $\ell = 1$; $n = 3$, $\ell = 0$; $n = 3$, $\ell = 1$.
29. Na ($Z = 11$).
31. $Z = 54$ (Xe).
33. (a) 3 spectral lines, (b) 7 spectral lines.
35. $Z = 1$ (H); $Z = 9$ (F); $Z = 27$ (Co, should be Cl, $Z = 17$); $Z = 59$ (Pr, should be Br, $Z = 35$).
37. 7.4×10^{-3} eV.
39. 24 K.
41. 119 kg/s^2, greater than 97 kg/s^2 from Eq. (41–18).
43. $I_1 = 1.43 \times 10^{-46}$ kg·m^2; $I_0 = 1.48 \times 10^{-46}$ kg·m^2; potential has nonlinear terms.
47. 9.48×10^{33}.
49. 1.5×10^{-5} eV; 5.6×10^{-6}.
51. 2.3×10^{-33} J (1.4×10^{-14} eV).
55. $\Delta x \simeq 1 \times 10^{-10}$ m $\simeq 2a_0$.
57. (a) $3U_0(r/a)^4$, (b) $3(\hbar^2/4ma^2)^{2/3}U_0^{1/3}n^{4/3}$.

CHAPTER 42

1. 6.15 eV.
3. Na: 3.25 eV; K: 2.13 eV; Al: 11.7 eV.
5. 1.31×10^{29} electrons/m^3.
7. $n_1 = n_2 = n_3 = 94$; 0.999 eV; $\Delta E = 7.1 \times 10^{-3}$ eV; 1.006 eV.
9. $p_F = \hbar(3\pi^2 n_e)^{1/3}$; $v_F = (\hbar/m_e)(3\pi^2 n_e)^{1/3}$.
11. 5.3×10^{13} N/m^2.
13. (a) $n_e \leq 5.9 \times 10^{32}$ electrons/m^3, (b) $Z \leq 14$.
15. (a) $(\pi^2\hbar^2/8m_e)(N/L)^2$, (b) $\pi^2\hbar^2 N^3/24m_e L^2$, (c) $(\pi^2\hbar^2/2m_e L^2)N$.
17. 36 MeV.
19. (a) $(\hbar^2/5m)(3\pi^2)^{2/3}(N/V)^{5/3}$, (b) 5.5×10^9 N/m^2.
21. $(\hbar^2\pi^2/2m_e)n_e$.
23. (a) $12/N_{sm}^{1/3}$ km, where N_{sm} = number of solar masses, (b) $E_F = (\hbar^2/2m_n)(3\pi^2 n_n)^{2/3}$ $(\hbar^2/2m_n)(3\pi^2)^{2/3}(N/V)^{2/3} = 60N_{sm}^{4/3}$ MeV; $p_F = 335N_{sm}^{2/3}$ MeV/c, (c) $4.7M_{Sun}$.
25. $(\hbar^2\pi/m_e)n_e$; 1.57.
27. 634 nm; 4.73×10^{14} Hz.
29. 3.0 W.
31. (a) $\pi^2\hbar^2 N/2mL^2$, (b) $3\pi^2\hbar^2/2mL^2$.
33. (a) 0.090 W, (b) 6.2×10^{-5} N/m^2.
37. 48.3 μV.
41. 2.5 μV.
43. 3.20 K.
45. 6.96×10^7 m.
47. $E_F = \hbar c n_e^{1/3}$.
53. (a) $0.19N_{sm}^{4/3}$ MeV, where N_{sm} = number of solar masses, (b) $R = (7.2 \times 10^3)/N_{sm}^{1/3}$ km; $R/R_{Sun} = 0.01$.
55. $M < 3 \times 10^{30}$ kg ($\simeq$1.5 solar masses).

CHAPTER 43

1. 5.5×10^{-11} n-carriers/cm^3 at T = 100K; 1.7×10^9 n-carriers/cm^3 ($\simeq 10^{-14}$ that of copper) at T = 300K.
3. (a) 1, (b) 0.99986, (c) 0.76, (d) 0.24, (e) 1.2×10^{-13}.
5. (a) 1.5 eV.
7. 1.03×10^{-12}.
9. (a) 6.9 μm, (b) 1.8 μm, (c) 0.867 μm.
11. 9.3×10^{11}/cm^3.
13. 0.65 eV.
15. $E_c - E_F = 0.080$ eV $\ll E_g$.
17. $n_i = n_n = n_p = 3.0 \times 10^{18}$ m^{-3} (at T = 100K); $n_i = n_n = n_p = 1.6 \times 10^{22}$ m^{-3} (at T = 300K).
19. 0.18 $(\Omega \cdot \text{m})^{-1}$.
21. (b) $n(v)\,dv = \left(\dfrac{8\pi m_e^3}{h^3}\right) e^{-[(m_e v^2/2)-E_F]/kT} v^2\,dv$.

23.

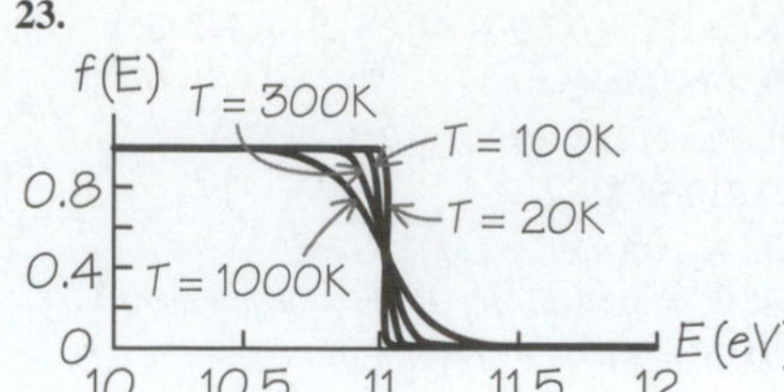

25.

$e^{eV_{ext}/kT} - 1$; eV_{ext} = 0.02 eV; T(K)

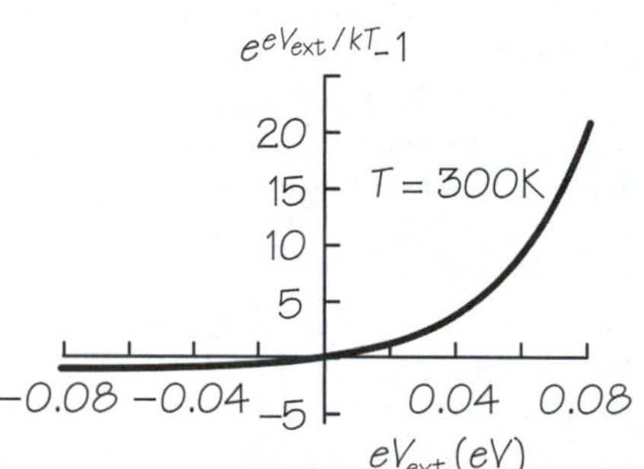

29. (a) 0.64 eV, (b) 0.45 eV.
31. (a) 3.5 mW, (b) 1.6×10^{16} photons/s.
33. 0.61 m_e.
35. B is 0.19 nm closer to the tip than point A.
37. 1.2×10^{-7}.
39. 4.5×10^{10} carriers/m^3.
41. $1.2 \times 10^3 a_0$.
43. 6.8×10^{19} impurities/cm^3; 1.5×10^{-3} (0.15 atomic %).
45. 2.1×10^{13} rad/s.
47. 10^{-100} electrons/m^3 at T = 90K; 6.1×10^{-12} electrons/m^3 at T = 300K.
49. (a) 4×10^{15} photons/s, (b) 4×10^{11} photons.
51. (a) Rapid increase in the resistance, (b) 1.10×10^3 nm, (c) increases.
53. (b) The tunneling current will increase, (c) tunneling current will increase.

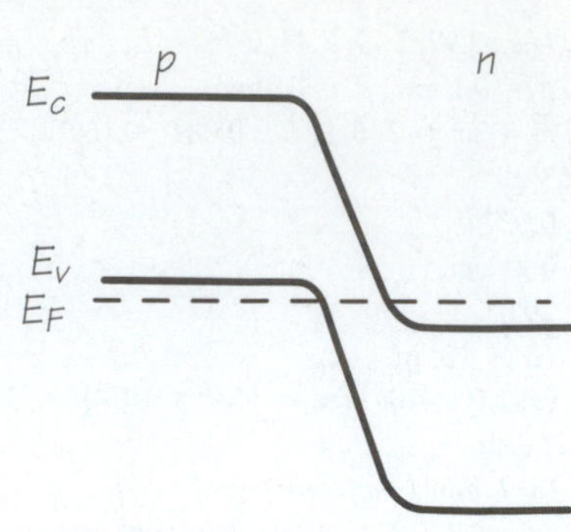

(a) No external voltage

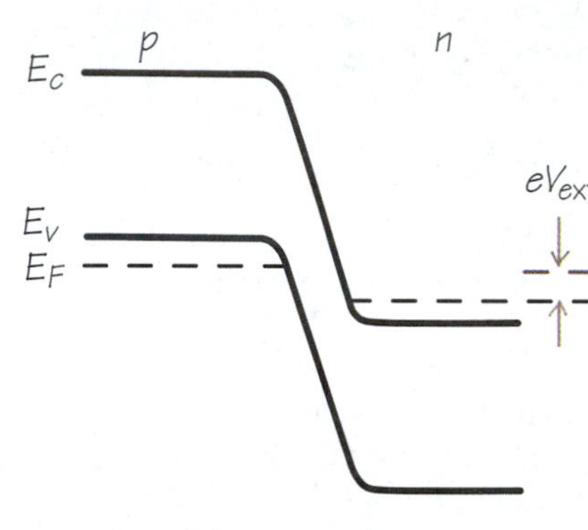

(b) Reverse bias

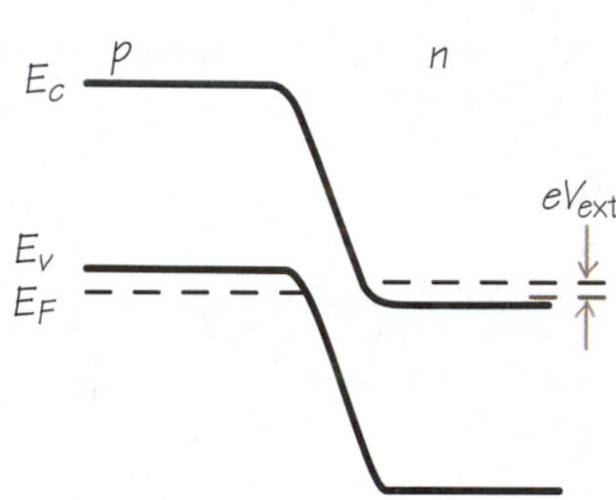

(c) Small forward bias

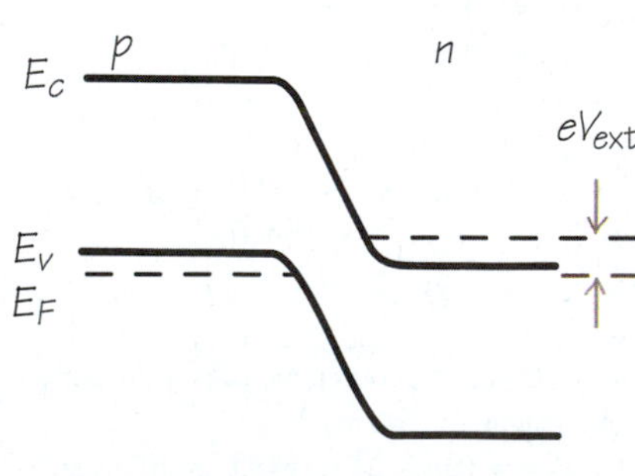

(d) Larger forward bias

CHAPTER 44

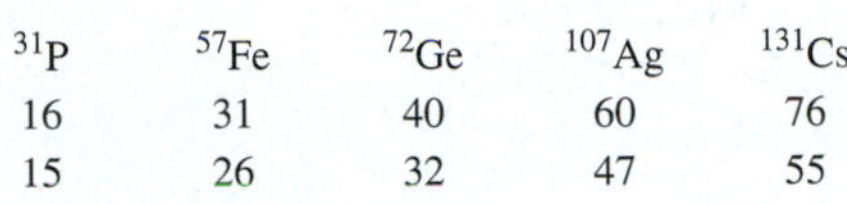

1. nucleus	^{9}Be	^{13}C	^{22}Na	^{31}P	^{57}Fe	^{72}Ge	^{107}Ag	^{131}Cs	^{208}Pb	^{241}Am
neutrons	5	7	11	16	31	40	60	76	126	146
protons	4	6	11	15	26	32	47	55	82	95

3. 8.1 barns.
5. $r_{\min} = 4Z(e^2/4\pi\varepsilon_0)/mv^2$; $v = 2[Z(e^2/4\pi\varepsilon_0)/mR]^{1/2}$.
7. (a) 18 fm, (b) 11 fm.
9. $(Z_1Z_2e^2/4\pi\varepsilon_0)^2[4m^2/(\Delta p)^4]$.
13. ^{7}Li and ^{7}Be: 3.67 fm; ^{9}B and ^{9}Be: 3.50 fm; ^{11}B and ^{11}C: 2.16 fm; ^{21}Ne and ^{21}Na: 1.86 fm; ^{23}Na and ^{23}Mg: 1.75 fm.
15. ^{15}N: $N = 8$; ^{39}K: $N = 20$; ^{40}Ca: $Z = 20$, $N = 20$; ^{56}Fe: neither; ^{140}Ce: $N = 82$; there are other important factors that contribute to the stability of a nucleus, the shell effects just give local small peaks on the curve of binding energy as a function of A.
17. (a) 14 N, (b) 560 N.
19. 1.0×10^{-3}.
21. $A/(2.0 + 0.015A^{2/3})$.
23. 3.80 MeV.
25. 18.3 MeV.
27. 4.7 MeV.
29. 10.49 MeV.
31. 6.4×10^{16} nuclei formed.
37. 0.031 cm^3/yr.
39. 0.784 MeV.
41. 13,700 B.C.
45. $dN_a/dt = -N_a/\tau_1$; $dN_b/dt = -N_b/\tau_2 + N_a/\tau_1$; $dN_c/dt = -N_c/\tau_3 + N_b/\tau_2$; $dN_d/dt = -N_d/\tau_4 + N_c/\tau_3; \ldots$.
49. (a) 1.3 keV, (b) 0.7 MeV.
51. (a) 11.5 metric tons, (b) 4.2×10^{16} J, (c) 5.5×10^{19} fissions/s.
53. (a) 24.8 MeV released, (b) 7.5×10^{47} MeV (1.2×10^{34} J).
55. 2.5×10^2 MeV.
57. 60 fm.
59. (a) Repulsive Coulomb force between the nuclei, (b) $K = 0.7$ MeV for each helium atom.
63. $E_p = (\hbar^2/10m_pr_0^2)(3^7\pi^2Z^5/16A^2)^{1/3}$; $E_n = (\hbar^2/10m_nr_0^2)[3^7\pi^2(A - Z)^5/16A^2]^{1/3}$.

CHAPTER 45

1. 2.5 fm.
3. $q = p_i + \frac{1}{2}Mc$; by a factor of $(4p_i + Mc)/(2p_i + Mc)$.
5. $\theta \simeq 0$, and $q \simeq p_i$.
7. (a) $\sigma(\theta) = Z_1^2Z_2^2(e^2/4\pi\varepsilon_0)^2[4m^2/(\Delta^2)^2]$, (b) σ would be smaller than for a point charge.

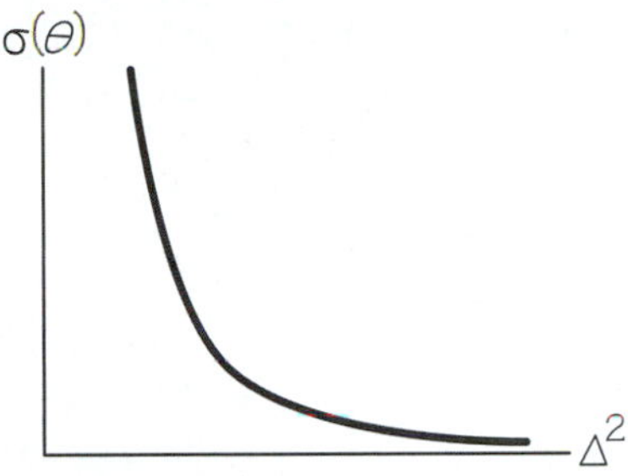

9. 3.44 TeV.
11. Charge: +1; baryon number: $B = 0$; lepton number: $L = 0$.
13. Reaction does not occur.
15. Unstable.
17. $(u\bar{u})$ or $(d\bar{d})$ or a combination of these two.
19. 2.5×10^{-3} fm.
21. (a) 0.26 fm, (b) 0.27 GeV.
23.

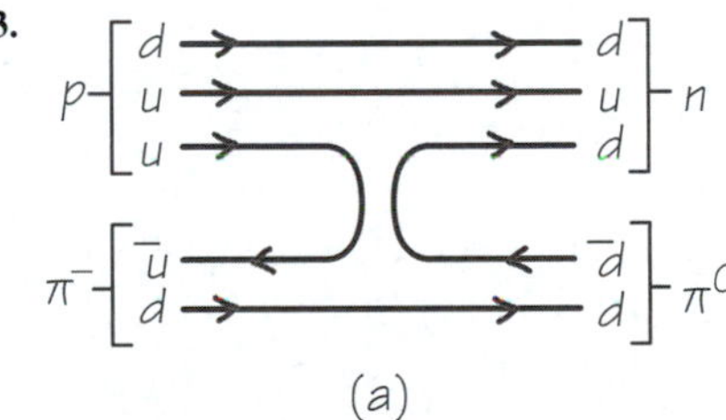

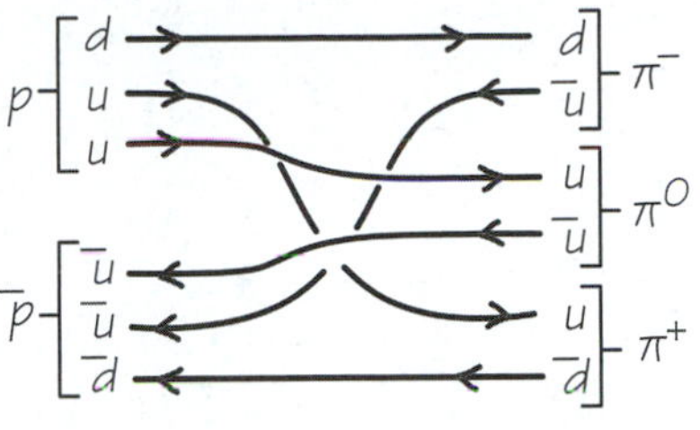

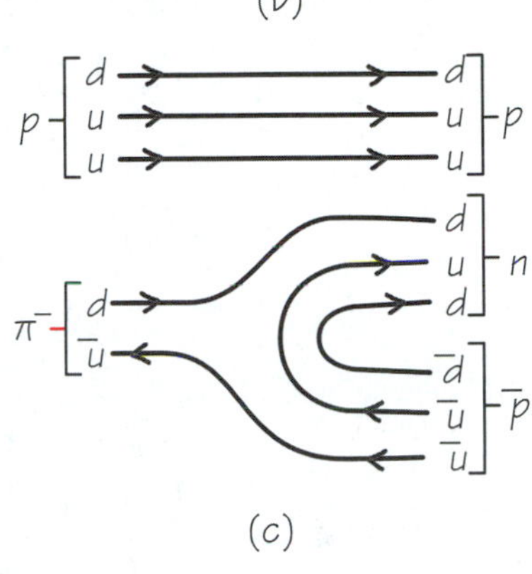

25. 460 MeV.
27. 11.9 GeV/c.
29. 8.1×10^9 MeV.
31. 9.5×10^4 s^{-1}.
33. 3.75 GeV.
35. 1.6 GeV; 180°.
39. $t = (\hbar G/c^5)^{1/2}$.
41. $H = n/t$.
43. 1.2×10^9 yr.
47. 1.3×10^5 K; 6.3×10^5 K.
49. 1.1×10^2 km.
51. 570 mesons (285 pairs).
57. (a) 1.25×10^4 eV, (b) 58 fm.

Photo Credits

FRONTMATTER Page ixL Courtesy: IBM Research, Almaden Research Center. Unauthorized use not permitted. **Page ixR** Diaphor Agency/Index Stock Imagery, Inc. **Page xL** NASA, J.J. Hester, Arizona State University/NASA Headquarters **Page xR** Index Stock Imagery, Inc. **Page xiL** Roger Wood/Corbis/Bettmann **Page xiRT** StockTrek/Getty Images, Inc.-Photodisc. **Page xiRB** Richard Megna/Fundamental Photographs **Page xiiTL** Ben Rose/Getty Images Inc.-Image Bank **Page xiiBL** Dohrn/Science Photo Library/Photo Researchers, Inc. **Page xiiR** J. Kim Vandiver, Massachusetts Institute of Technology **Page xiiiL** Steve Kaufman/ Peter Arnold, Inc. **Page xiiiR** Adam Hart Davis/Photo Researchers, Inc. **Page xivR** Ralph H Wetmore II/Getty Images Inc.-Stone AllStock **Page xivR** AP/Wide World Photos **Page xvRB** Richard Megna/Fundamental Photographs **Page xvRT** GSFC/MITI/ERSDAC/ JAROS, and U.S./Japan ASTER Science Team/NASA Headquarters **Page xviL** Richard Megna/Fundamental Photographs **Page xviR** Michael Freeman **Page xxii** Tom Pantages **Page xxiii** Martin-Baker Aircraft Co Limited **Page xxxi** Paul M Fishbane, Stephen G. Gasiorowicz, Stephen T. Thornton **CHAPTER 1 CO–1** Vince Streano/CORBIS-NY **1–2a** Courtesy: IBM Research, Almaden Research Center. Unauthorized use not permitted. **1–2b** Martyn Goddard/CORBIS-NY **1–2c** NASA/Goddard Space Flight Center **1–3a** David Gray/REUTERS/CORBIS-NY **1–3b–c** AP/Wide World Photos **1–5a** Tom Cogill/Stephen T. Thornton **1–7** Richard Megna/Fundamental Photographs **1–8** Courtesy of the Library of Congress **1–9** EyeWire Collection/ Getty Images-EyeWire, Inc. **CHAPTER 2 CO–2** Randy Faris/CORBIS-NY **2–1** G. Giansanti/ Corbis/Sygma **2–16a** Peter Steehouwer **2–18a,b** Richard Megna/ Fundamental Photographs **2–23** Richard Megna/Fundamental Photographs **CHAPTER 3 CO–3** Diaphor Agency/ Index Stock Imagery, Inc. **3–1a** Reuters/ Gary Hershorn © Reuters New Media Inc./ CORBIS-NY **3–3a** Michael J. Doolittle/ The Image Works **3–11** Richard Megna/ Fundamental Photographs **3–15** Richard Megna/Fundamental Photographs **3–18** NASA Headquarters **3–19** Loren M. Winters/North Carolina School of Science and Mathematics **CHAPTER 4 CO–4** The Bridgeman Art Library International Ltd./Hannah, Robert (1812–1909): "Master Isaac Newton", 1905 (oil on canvas). The Royal Institution, London, UK. The Bridgeman Art Library International Ltd. **4–1** AP/Wide World Photos **4–5** The Image Works **4–6a–c** Richard Megna/Fundamental Photographs **4–12** Martin-Baker Aircraft Co Limited **4–13a** NASA Headquarters **4–14a,b** Tom Cogill/Stephen T. Thornton **4–17a–e** Richard Megna/Fundamental Photographs **4–26** Richard Megna/Fundamental Photographs **CHAPTER 5 CO–5** Ray Ooms/Masterfile Corporation **5–1** John Kelly/The Image Bank/Getty Images Inc. **5–11** Richard Megna/ Fundamental Photographs **5–16a** Tom Cogill/Stephen T. Thornton **5–17** John Schnack/SuperStock, Inc. **5–19a,b** Paul Silverman/Fundamental Photographs **5–21** Jack Novak/ SuperStock, Inc. **5–26a** Photofest **5–29** Loren Winters/Visuals Unlimited **CHAPTER 6 CO–6** James L. Amos/ CORBIS-NY **6–1** Wally McNamee/ CORBIS-NY **6–2a,b** Lawrence Manning/ CORBIS-NY **6–4** Movie Star News **6–7** Tom Cogill/Stephen T. Thornton **6–16** Peter Wilson/Dorling Kindersley Media Library **6–24a** The Granger Collection, New York **6–25** Picture Desk, Inc./Kobal Collection **CHAPTER 7 CO–7** Photograph by Mark Bakalor. Copyright © 2003 International Jugglers' Association, Inc. Dennis MacDonald/ PhotoEdit **7–13** Stuart Westmorland/ CORBIS-NY **7–17** James Leynse/Corbis/ SABA Press Photos, Inc. **CHAPTER 8 CO–8** Bob Collins/The Image Works **8–6** Tony Freeman/PhotoEdit **8–12** NASA, J.J. Hester, Arizona State University/NASA Headquarters **8–14** Photographer's Mate 3rd Class Todd Frantom/U.S. Navy News Photo **8–20** Copyright © Harold & Esther Edgerton Foundation, 2002, courtesy of Palm Press, Inc. **8–24** Richard Megna/ Fundamental Photographs **8–26** Loren M. Winters/North Carolina School of Science and Mathematics **CHAPTER 9 CO–9** Daniel Forster/Duomo Photography Incorporated **9–7a** Tom Pantages **9–8** Photograph courtesy of Norman Heath © Copyright The Museum of Power **9–17** Tom Pantages **9–23** Richard Megna/ Fundamental Photographs **9–28** Claudia Dhimitri/Index Stock Imagery, Inc. **9–33** Richard Megna/Fundamental Photographs **9–35a** Thomas J. Cutitta/ International Imaging/Stephen T. Thornton **9–37a** Richard Megna/Fundamental Photographs **CHAPTER 10 CO–10** NASA and The Bubble Heritage Team (STScI/AURA)/ NASA/Media Services **10–16a** PASCO Scientific/Courtesy of PASCO Scientific **10–17** Gravity Probe B at Stanford University **10–22a** Gerard Lacz/Peter Arnold, Inc. **10–22b(1–2)** Brian Bahr/ Getty Images, Inc.-Allsport Photography **10–23c,d** Tom Cogill/ Stephen T. Thornton **10–24** Jeff Hester and Paul Scowen, Arizona State University, and Palomar Observatory **10–27a** Thomas J. Cutitta/ International Imaging/Stephen T. Thornton **10–28** Thomas J. Cutitta/ International Imaging/Stephen T. Thornton **CHAPTER 11 CO–11** AP/Wide World Photos **11–1** Addison Geary/Stock Boston **11–3** Peter Chadwick/Dorling Kindersley Media Library **11–6** Tom Cogill/Stephen T. Thornton **11–7** Roger Wood/CORBIS-NY **11–8a–c** Paul Silverman/Fundamental Photographs **11–13a,b** Tom Cogill/ Stephen T. Thornton **11–14** E.R. Degginger/Color-Pic, Inc. **11–18a,b** Pearson Education/PH College **11–19a,b** Pearson Education/PH College **CHAPTER 12 CO–12** StockTrek/Getty Images, Inc.-Photodisc. **12–5b** Stephen T. Thornton **12–8a,b** NASA/John F. Kennedy Space Center **12–9** The Burndy Library, Dibner Institute for the History of Science and Technology, Cambridge, Massachusetts, and the Grace K. Babson Collection of the Works of Sir Isaac Newton. **12–11** National Optical Astronomy Observatories **12–12** NASA Headquarters **12–20a,b** John Elk III/Stock Boston **12–28** National Radio Astronomy Observatory **CHAPTER 13 CO–13** Photodisc/Getty Images **13–4** Loren M. Winters, North Carolina School of Science and Mathematics **13–6** Werner H. Muller/ Peter Arnold, Inc. **13–13** Richard Megna/ Fundamental Photographs **CHAPTER 14** **CO–14** AP/Wide World Photos **14–4** Richard Gross/Corbis/Stock Market **14–6** National Film Board of Canada **14–7** Dohrn/Science Photo Library/Photo Researchers **14–18** Mitch Diamond/ PhotoBank, Inc. **14–23** AP/ Wide World Photos **14–28** Ben Rose/ Getty Images Inc.-Image Bank **14–35** J. Kim Vandiver, Massachusetts Institute of Technology **14–36** Roger Ressmeyer/CORBIS BETTMANN **14–37** Jim Kahnweiler/ Positive Images **CHAPTER 15** **CO–15** Visuals Unlimited **15–4a,b** Tom Pantages **15–8** Color-Pic, Inc. **15–12a,b** Thomas J. Cuttita/ International Imaging/ Stephen T. Thornton **15–14** Telex Communications, Inc. **15–15** Reproduced from PSSC Physics, 2nd ed. (1965), D. C. Health & Company with Education Development Center Inc., Newton, Massachussetts **15–23** Howard Sochurek/ Medical Images Inc. **15–27** Tom Cogill/ Stephen T. Thornton **CHAPTER 16** **CO–16** Woods Hole Oceanographic Institution **16–1** Marty Honig/Getty Images, Inc./Photodisc **16–7** Michael Grecco/Stock Boston **16–8** Sargent-Welch/VWR International **16–14a** Johner/ Amana America, Inc. **16–14b** Tom Cogill/ Stephen T. Thornton **16–15a–c** Tom Cogill/Stephen T. Thornton **16–17** Pearson Education/PH College **16–18** Françoise Sauze/Science Photo Library/ Photo Researchers, Inc. **16–22a,b** Tom Cogill/Stephen T. Thornton **16–23a** Sargent-Welch/VWR International **16–25b** Tom Cogill/Stephen T. Thornton **16–30** Steve Kaufman/Peter Arnold, Inc. **CHAPTER 17** **CO–17** Steve Kahn/ Getty Images, Inc.-Taxi **17–2** Glass/Peter Glass Photography **17–3a** Adam Hart-Davis/Science Photo Library/Photo Researchers, Inc. **17–3b** Adam Hart-Davis/Science Photo Library/Photo Researchers, Inc. **17–4** Tom Cogill/ Stephen T. Thornton **17–9** Daemmrich/ The Image Works **17–10a–c** Tom Cogill/ Stephen T. Thornton **17–11** Tom Pantages **17–17** NASA/ Science Source/Photo Researchers, Inc. **CHAPTER 18** **CO–18** Charles E. Rotkin/ CORBIS-NY **18–5** The Granger Collection, New York **18–6a** Grant Heilman/Grant Heilman Photography, Inc. **18–6b** Hans Pfletschinger/Peter Arnold, Inc. **CHAPTER 19** **CO–19** Kevin R. Morris © Bohemian Nomad Picturemakers/ CORBIS-NY **19–13a–c** Paul Silverman/Fundamental Photographs

CHAPTER 20 **CO–20** AP/ Wide World Photos **20–16a** Richard Megna/ Fundamental Photographs **20–16b** AP/ Wide World Photos **CHAPTER 21** **CO–21** The Granger Collection, New York **21–1a** Tom Cogill/Stephen T. Thornton **21–2** Stephen T. Thornton **21–5** The British Library **21–7a** Sargent–Welch/ VWR International/ Courtesy of Central Scientific **21–8** LBNL/Photo Researchers, Inc. **CHAPTER 22** **CO–22** Adam Hart Davis/Photo Researchers, Inc. **22–2a** Harold M. Waage **22–11a** Harold M. Waage/Permission of the Princeton University Physics Department **22–12a** Harold M. Waage/ Permission of the Princeton University Physics Department **CHAPTER 23** **CO–23** © 2003 Peter Menzel/ menzelphoto.com **23–4a** Harold M. Waage/Permission of the Princeton University Physics Department **23–20** © Jim Krider/Arizona State University **CHAPTER 24** **CO–24** Ralph H Wetmore II/Getty Images Inc.-Stone Allstock **24–10** U.S. Geological Suney **24–24** Harold M. Waage/ Princeton University **24–25** Adam Hart–Davis/Science Photo Library/Photo Researchers, Inc. **24–26** John Woodcock/ Dorling Kindersley Media Library **24–27** Lester V. Bergman and Associates, Inc. **24–29a** B. Loberg and H. Norden **24–30** Dr. Erwin Mueller/SPL/Photo Researchers, Inc. **24–34** National Institute of Standards and Technology (NIST). Geometric and Electronic Properties of Cs Structures on III–V (110) Surfaces: From 1–D and 2–D Insulators to 3–D Metals, L.J. Whitman, J.A. Stroscio, R.A. Dragoset, and R.J. Celotta, Phys. Rev. Le tt. 66, 1338 (1991). **24–35** Courtesy of International Business Machines Corporation. Unauthorized use not permitted. **CHAPTER 25** **CO–25** Randy Montoya/Sandia National Laboratories **25–4a** Harold M. Waage/ Princeton University **25–8a-b** Tom Cogill/ Stephen T. Thornton **CHAPTER 26** **CO–26** AP/ Wide World Photos **26–9** Tom Pantages **26–10b** The Image Works **26–21b** Dan McCoy/Rainbow **26–22** Dorling Kindersley Media Library **CHAPTER 27** **CO–27** AP/Wide World Photos **27–1a** NASA Headquarters **27–1b** Still Pictures/Peter Arnold, Inc. **27–1c** Loren Winters/Visuals Unlimited **27–13** Michael Dalton/Fundamental Photographs **27–14a** Tom Cogill/Stephen T. Thornton **27–15a** Tom Cogill/Stephen T. Thornton **27–20c** International Imaging Inc./Stephen T. Thornton **CHAPTER 28** **CO–28** David Miller, National Geophysical Data Center/NOAA Central Library Photo

Collection **28–2a** Phil Degginger/ Color–Pic, Inc. **28–2b** Alfred Pasieka/ SPL/Photo Researchers, Inc. **28–2c** Tom Pantages **28–3a-b** International Imaging Inc./Stephen T. Thornton **28–12** Brookhaven National Laboratory/Science Photo/Photo Researchers, Inc. **28–13b** Lawrence Berkeley National Laboratory **28–16a** The Granger Collection, New York **CHAPTER 29** **CO–29** Princeton University **29–1a-b** Thomas J. Cutitta/ Interantional Imaging Inc./Stephen T. Thornton **29–2a, c** Tom Cogill/Stephen T. Thornton **29–09** Tom Cogill/Stephen T. Thornton **29–15b** Richard Megna/ Fundamental Photographs **29–17a** Richard Megna/Fundamental Photographs **29–25a** Richard Megna/Fundamental Photographs **CHAPTER 30** **CO–30** The Royal Institution **30–19a** IBM Research/Peter Arnold, Inc. **CHAPTER 31** **CO–31** Courtesy of American Superconductor Corporation *(www.amsuper.com)* **31–6** Dr. John Unguris **31–11** John Unguris/National Institute of Standards and Technology (NIST) **31–13** Richard Megna/ Fundamental Photographs **31–17a** Jack Plekan/Fundamental Photographs **31–17b** Berwin, Derek/Getty Images Inc.-Image Bank **CHAPTER 32** **CO–32** Richard Megna/Fundamental Photographs **32–2b** Sargent–Welch/VWR International **32–4** Michael A. Gallitelli/ Metroland Photo, Inc. **32–7** Tom Cogill/ Stephen T. Thornton **CHAPTER 33** **CO–33** Taxi (RM)/Getty Images, Inc. **33–13a-b** Tom Pantages **33–20b-d** Tom Cogill/Stephen T. Thornton **CHAPTER 34** **CO–34** GSFC/MITI/ERSDAC/ JAROS, and U.S./Japan ASTER Science Team/NASA Headquarters **34–7a** Courtesy of the Library of Congress **34–8a.1** Chandra X–Ray Center/NASA/CXC/UMd./ A.Wilson et al. **34–8a.2** Ultraviolet Imaging Telescope/ UIT Science Team/ NASA Headquarters **34–8a.3** Todd Boroson/NOAO/AURA/NSF. Copyright WIYN Consortium, Inc., all rights reserved. **34–8a.4** Infared Space Observatory/ESA/ISO, CAM, M. Sauvage et al./ **34–8a.5** ESA/ISO. CAM. M. Sauvage et al./Chandra X–Ray Center/ NASN/CXC/SAO **34–14** Stephen T. Thornton/Tom Cogill **34–15a-c** Stephen T. Thornton/Tom Cogill **34–20a-b** Leonard Lessin/Peter Arnold, Inc. **CHAPTER 35** **CO–35** Goddard Space Flight Center/ Image courtesy of NASA - National Aeronautics and Space Administration **35–1** The New York Public Library Photographic Services/Art Resource

35–4b Courtesy of the Bancroft Library University of California, Berkeley **35–5b** Runk/Schoenberger/ Grant Heilman Photography, Inc. **35–6** Lawrence Livermore National/Photo Researchers, Inc. **35–11** NASA Headquarters **35–12a** Stephen T. Thornton/Tom Cogill **35–13** Stephen T. Thornton/Tom Cogill **35–17b** Thomas J. Cutitta/International Imaging/ Stephen T. Thornton **35–19a** Sargent-Welch/VWR International **35–24a** Grant Heilman Photography, Inc. **35–27b** Randy O' Rourke/Corbis/Stock Market **CHAPTER 36** **CO–36** Richard Megna/ Fundamental Photographs **36–5** Richard Megna/Fundamental Photographs **36–8c** Stephen T. Thornton/ Tom Cogill **36–9d** Stephen T. Thornton/ Tom Cogill **36–10b** Stephen T. Thornton/ Tom Cogill **36–11b** Stephen T. Thornton/ Tom Cogill **36–12b** Stephen T. Thornton **36–22a-b** E. R. Degginger/Color-Pic, Inc. **36–32b** Jeff J. Daly/Fundamental Photographs **36–33a** Art Resource, N.Y. **36–36** Tom Pantages **36–39** M. C. Escher's, "Hand with Reflecting Sphere" © 2003 Cordon Art B. V.-Baarn-Holland. All rights reserved. **CHAPTER 37** **CO–37** Diane Hirsch/ Fundamental Photographs **37–7** Reproduced by permission from OHG. Copyright © 1962 by Springer-Verlag GmbH & Co KG. **37–13b** Richard Megna/Fundamental Photographs **37–13c** Tom Pantages **37–15** Color-Pic, Inc. **37–16** Richard Megna/Fundamental Photographs **37–20b** Courtesy of PASCO Scientific **37–21a** Sargent-Welch/VWR International **37–22a-b** Sargent-Welch/ VWR International **CHAPTER 38** **CO–38** Georg Fischer/ Bilderberg/Peter Arnold, Inc. **38–1** Pearson Education/PH College **38–3** Ken Kay/Fundamental Photographs **38–5a** Pearson Education/PH College **38–6b** Bausche Lamb **38–9** Don Farrall/Getty Images, Inc.-Photodisc. **38–11** Reproduced by permission. Copyright by Springer-Verlag GmbH & Co KG. **38–14** Tom Pantages **38–15a-c** Pearson Education/PH College **38–17** Richard J. Wainscoat **38–19** Reproduced by permission. Copyright by Springer-Verlag GmbH & Co KG. **38–20** Photo Researchers Inc. **38–22a** Omikron/ Photo Researchers, Inc. **38–23b** The Burndy Library, Dibner Institute for the History of Science and Technology, Cambridge, Massachusetts **38–31** Pascal Goetgheluck/SPL/Photo Researchers, Inc. **38–32** Michael Freeman **CHAPTER 39** **CO–39a-b** A. C. Searle and C. M. Savage, The Australian National University. Animations at *http://www.anu.edu.au/ Physics/Searle.* **39–03** Tom Pantages **39–09** AP/Wide World Photos **39–21** Lawrence Berkeley Laboratory/ Science Photo Library/Photo Researchers, Inc. **39–23b** NASA Headquarters **CHAPTER 40** **CO–40** American Institute of Physics/Emilio Segre Visual Archives **40–1** Maison Albert Schweitzer/ American Institute of Physics/ Emilio Segre Visual Archives **40–4a** American Institute of Physics/Emilio Segre Visual Archives **40–14** American Institute of Physics/Emilio Segre Visual Archives **CHAPTER 41** **CO–41** Lester Lefkowitz/CORBIS-NY **CHAPTER 42** **CO–42** Lawrence Livermore National Laboratory **42–02** The Niels Bohr Archive, Copenhagen **42–03** NASA Headquarters **42–04** NASA Headquarters **42–08b** Dan McCoy/Rainbow **42–09** Laboratory for Laser Energetics/University of Rochester Public Relations **42–10** Carl E. Wieman/The JILA BEC Group **42–16** St. Andrew's University **CHAPTER 43** **CO–43** Hewlett-Packard Laboratories/SPL/Photo Researchers, Inc. **43–12** Jerome Wexler/Photo Researchers, Inc. **43–19** JPL/NASA Headquarters **43–20** Nick Holonyak Jr./University of Illinois at Chicago **43–21** Astrid & Hans Freider/Photo Researchers, Inc. **43–22** Texas Instruments Incorporated, 1989 **43–26** Copyright © Photographer Axel Scherer by Bellcore. Reproduced with permission. All rights reserved. **43–27** 3M **43–32** Delf University of Technology/SPL/Photo Researchers, Inc. **CHAPTER 44** **CO–44** Peter H. Buckley/Pearson Education/PH College **44–11** The Granger Collection, New York **44–18** Shaen Adey; Gallo Images/CORBIS-NY **44–20** Walter Dickerman/Sandia National Laboratories **CHAPTER 45** **CO–45** © CERN Geneva **45–A** Joe Stancampiano/National Geographic Image Collection **45–01** Shaffer/Smith/SuperStock, Inc. **45–02a** Particle Data Group, Lawrence Berkeley National Laboratory **45–10** © CERN Geneva **45–13** Fermilab Visual Media Services **45–14** © CERN Geneva **45–15** C. Powell, P. Fowler & D. Perkins/Science Photo Library/Photo Researchers, Inc. **45–16** Laurent Guiraud/© CERN Geneva **45–19** Harvard-Smithsonian Center for Astrophysics

Index

Note: Italics indicate a definition or primary entry for multiple entries, where applicable.